SIGMA-ALDRICH

Library of Rare Chemicals

STRUCTURE INDEX

published by ALDRICH CHEMICAL COMPANY, INC.

Printed in the United States of America
Library of Congress Catalog Card Number 93-071303
ISBN 0-941-633-36-5

PREFACE

The Sigma-Aldrich Library of Rare Chemicals contains over 37,000 research samples acquired from some of the world's ablest chemists—Professors R. B. Woodward, Tadeus Reichstein, L. F. Fieser, Henry Gilman, Melvin Newman, Oskar Jeger, George Wittig, Karl Dimroth, Ralph A. Raphael, to mention just a few. In keeping with our motto, "Chemists Helping Chemists in Research and Industry," we have supplied the scientific community for years with thousands of difficult-to-obtain compounds. We will continue our commitment in providing a unique service to the researcher by adding many compounds each year, and by offering this exciting Structure Index which will enable the chemist to quickly "window shop" for compounds by chemical structure. This Structure Index is an essential resource for that rare chemical needed to complete a special project, or to locate that 'active' compound in a crucial screening program.

The compounds in our Rare Chemical Library are generally sold in standard unit sizes at $35.00 each. Unit sizes are indicated for each structure in the index. Other unit sizes can be made available and, in special cases, it may be possible to offer different quantities at special prices.

The Library includes research products from many laboratories around the world and intermediates from our own research and development laboratories. Unlike the products in our Catalog/Handbook, they have ***not*** *been analyzed by our Quality Control Department and are offered only on an "as is" basis.*

The Library products are for laboratory use only and are not for drug, household or other uses. Since most of the compounds are research samples which are not commercially available, limited hazard data are known. As with all products of unknown toxicity, appropriate precautions must be observed when handling. These chemicals should be handled only by technically qualified persons.

INTRODUCTION

This book has three sections to help you find a specific rare chemical:

a) Structure Index

b) Molecular Formula Index

c) Product Catalog Number to Page Number Cross Reference

Structure Index

The Sigma-Aldrich Library of Rare Chemicals Structure Index is arranged into forty-eight functional group classifications with subgroups as listed in the Table of Contents. Each chemical is listed only once in the Index.

This Structure Index is based on three criteria:

1) A specially assigned code for each structure.
2) The number of carbon atoms in the molecule.
3) The molecular weight of the chemical.

Due to the many complex combinations possible, there could be instances where large numbers of seemingly similar compounds could vary in the computer sorted functional group classification.

Molecular Formula Index

All Compounds are separately referenced in the Molecular Formula Index starting on page 1145, in ascending order of the molecular formula and, for your convenience, the chemical name and corresponding page number where the structure can be located.

The arrangement used in this index is based on the general molecular formula "$C_xH_yElement_z$" where the order of elements is alphabetical:

* Inorganics precede carbon-containing compounds.
* Carbon-containing inorganics such as Ag_2CO_3 will be found with C listed first, i.e., CAg_2O_3.
* Inorganics containing hydrogen will have hydrogen listed first.
* Organics lacking hydrogen such as C_6N_4 are listed before any C_6H's.
* Water of hydration is not included.
* Salts of amines, except quaternary salts, are listed under the formulas of the free amines.
* Simple salts of organic acids are listed under the formulas of the free acids.

Cross Reference

For your convenience, we have also included a product catalog number to page number cross reference. This cross reference can be found on page 1372.

To Find a Specific Chemical Structure

1) Determine the molecular formula. Look under the correct molecular formula heading and find the specific chemical in the alphabetical listing under that heading. Each name has a page reference next to it where the structure can be found.
2) If you have the product catalog number, simply look in the Product Catalog Number to Page Number cross reference section which starts on page 1372.
3) If you are interested in looking through groups of structures with particular functional groups or ring systems, use the Table of Contents to locate the group or groups you wish to browse.

Sigma-Aldrich Library of Rare Chemicals

Chemists Helping Chemists in Research and Industry

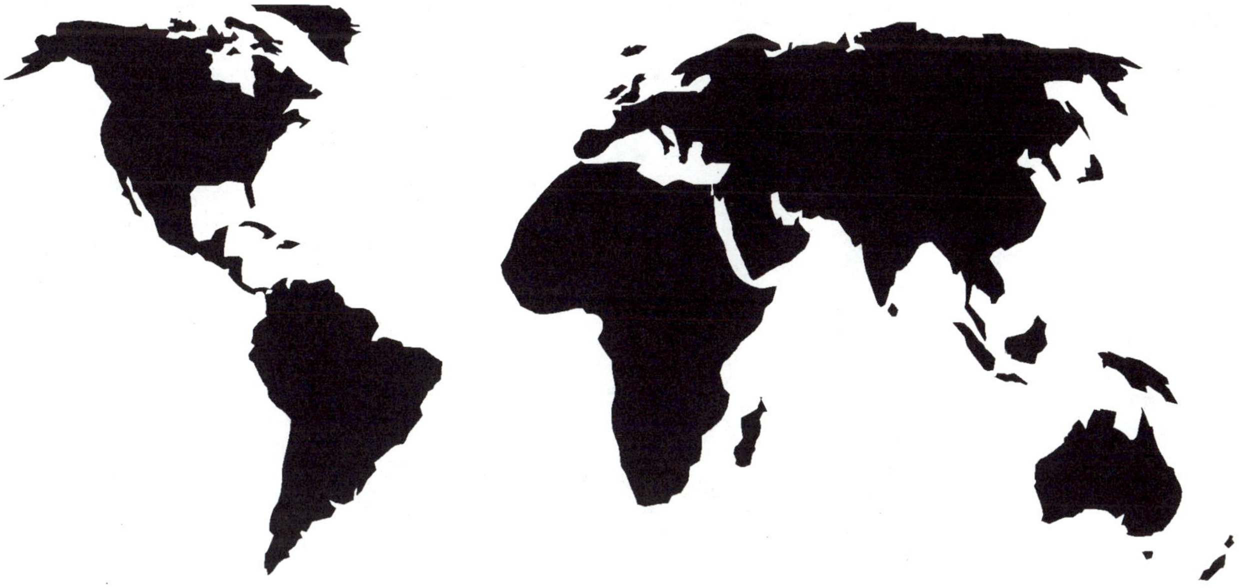

Worldwide Offices

Australia
Aldrich Chemical
Unit 2, 10 Anella Avenue
Castle Hill, NSW 2154
Telephone: 61-2-899-9977
FAX: 61-2-899-9742

Belgium
Aldrich Chemie
K. Cardijnplein 8
B-2880 Bornem
Telephone: 32-3-899-13-01
FAX: 32-3-899-13-11

Brazil
Aldrich-Brazil
Rua Sabara, 566-conj.53
01239-010-Sao Paulo, SP
Telephone: 55-11-231-1866
FAX: 55-11-257-9079

Czech Republic
Aldrich, s.r.o.
Krizikova, 27
180 00 Prague 8
Telephone: 42-2-236-69-73
FAX: 42-2-236-41-41

France
Aldrich Chimie S.a.r.l.
L'Isle D'Abeau Chesnes
B.P. 701
38297 St Quentin Fallavier
Cedex
Telephone: 33-74-82-28-00
FAX: 33-74-95-68-08

Germany
Aldrich-Chemie GmbH & Co. KG
Riedstrasse 2
Post Box 1120
89555 Steinheim
Telephone: 49-73-29-870
FAX: 49-73-29-87-139/239

Holland
Aldrich Chemie
Telephone: 32-3-899-13-01
FAX: 32-3-899-13-11

India
Aldrich-India
B4/158 Safdarjung Enclave
New Delhi 110 029
Telephone: 91-11-688-6872
FAX: 91-11-688-6873

Italy
Aldrich Chimica
Via Gallarate, 154
20151 Milano
Telephone: 39-2-3341-7310
FAX: 39-2-3801-0737

Japan
Aldrich Japan
Kyodo Bldg. Shinkanda
10 Kanda-Mikuracho
Chiyoda-Ku, Tokyo
Telephone: 81-3-3258-0155
FAX: 81-3-3258-0157

Spain
Aldrich Quimica
Apt. Correos, 161
28100 Alcobendas
Madrid
Telephone: 34-1-661-9977
FAX: 34-1-661-9642

Switzerland
Aldrich Chemie
Industriestrasse 25
P.O. Box 260
CH-9470 Buchs
Telephone: 41-81-755-2723
FAX: 41-81-756-7420

United Kingdom
Aldrich Chemical Co. Ltd.
The Old Brickyard, New Road
Gillingham, Dorset SP8 4JL
Telephone: 44-7478-22211
FAX: 44-7478-23779

United States
Aldrich Chemical Co., Inc.
P.O. Box 355
Milwaukee, WI 53201 USA

Telephone:
USA/Canada 800-558-9160
414-273-3850
TWX:
Aldrichem MI 910-262-3052
TELEX:
Aldrich MI 26 843
FAX: 800-962-9591
414-273-4979
Customer Inquiries:
800-558-9160
Technical Information:
800-231-8327

TABLE OF CONTENTS

Preface ... iii

Introduction ... iv

Worldwide Offices ... v

Functional Group Classification

Non-Aromatic Hydrocarbons ... 1
1) Straight Chain ... 1
2) Branched Chain ... 1
3) Unsaturated Straight Chain ... 1
4) Unsaturated Branched Chain ... 1-2
5) Two or more Double bonds ... 2
6) Cyclic Saturated ... 2-3
7) Cyclic Unsaturated ... 3

Non-Aromatic Halogenated Hydrocarbons ... 4
1) Terminal, Straight Chain, Monohalogenated ... 4
2) Nonterminal and Branched, Monohalogenated ... 4
3) Dihalogenated ... 4
4) Polyhalogenated ... 4
5) Unsaturated ... 5
6) Cyclic ... 5-7

Non-Aromatic Alcohols ... 8
1) Primary ... 8
2) Secondary ... 8-9
3) Tertiary ... 9-10
4) Diols ... 10-11
5) Unsaturated ... 11-12
6) Cyclic ... 12-15
7) Halogenated ... 15
8) Polyalcohols and Sugars ... 16

Non-Aromatic Ethers, Acetals and Epoxides ... 17
1) Saturated Open Chain Ethers ... 17
2) Unsaturated Ethers ... 17-18
3) Acetals, Ketals and Ortho Esters ... 18-19
4) Hydroxy ... 19-20
5) Epoxides ... 20-21
6) Cyclic Ethers, Acetals and Ketals ... 21-23

Non-Aromatic Mercaptans and Sulfides ... 24
1) Mercaptans ... 24
2) Sulfides ... 24-27

Non Aromatic Amines ... 28
1) Primary Saturated ... 28
2) Primary Saturated Diamines ... 28
3) Secondary Saturated ... 28-29
4) Tertiary Saturated ... 29
5) Saturated Polyamines ... 29-30
6) Carbocyclic ... 30-31
7) Unsaturated ... 31-32
8) Halogenated ... 32-33
9) Hydroxy or Alkoxy ... 33-35
10) Mercaptan and Sulfide ... 35-36
11) Heterocyclic ... 36-42
12) Quaternary Ammonium Salts ... 42-44

Non-Aromatic Nitro and Nitroso ... 45
1) Nitrite ... 45
2) Nitrate ... 45
3) Nitroso ... 45
4) Nitro ... 45-47

Non-Aromatic Ketones ... 48
1) Saturated ... 48-50
2) Unsaturated ... 50
3) Halogenated ... 50
4) Hydroxy ... 50
5) Acetal ... 50
6) Diketones ... 50-51
7) Cyclic ... 51-57
8) Amino ... 57-58
9) Heterocyclic ... 58-59

Non-Aromatic Aldehydes ... 60

Non-Aromatic Carboxylic Acids ... 61
1) Saturated Open Chain ... 61
2) Saturated Open Chain Diacids ... 61
3) Unsaturated Open Chain ... 61
4) Halogenated Open Chain ... 61-62
5) Hydroxy or Alkoxy ... 62-63
6) Mercaptan or Sulfide ... 63
7) Keto or Aldehyde ... 63-64
8) Cyclic ... 64-67
9) Miscellaneous ... 67-68

Non-Aromatic Acid Salts ... 69
1) Carboxylic Acid Salts ... 69-70
2) Metal Acetylacetonate Chelates ... 70

Non-Aromatic Amino Acids ... 71

Non-Aromatic Esters and Lactones ... 73
1) Saturated Open Chain ... 73-74
2) Carbonates ... 74
3) Saturated Polyesters ... 74-77
4) Cyclic ... 77-78
5) Unsaturated ... 78-81
6) Halogenated ... 81-84
7) Hydroxy or Alkoxy ... 84-86
8) Mercaptan or Sulfide ... 86-87
9) Amino and Quaternary Ammonium ... 87-89
10) Heterocyclic ... 89-91
11) Keto ... 91-94
12) Carboxy ... 94
13) Lactones ... 95-97

Non-Aromatic Anhydrides ... 98
1) Open Chain ... 98
2) Cyclic ... 98-99

Non-Aromatic Acid Halides ... 100
1) Saturated ... 100
2) Unsaturated ... 100
3) Chloroformates and Ester Acid Halides ... 100
4) Diacid Halides ... 100

Non-Aromatic Amides 101
1) Primary Open Chain 101-102
2) Secondary Open Chain 102-106
3) Tertiary Open Chain 106-107
4) Heterocyclic 107-110
5) Oxamides 110
6) Diamides 110-111
7) Hydrazides 111-112
8) Carbamates 112-117
9) Amidic Amino Acids 117-124
10) Lactams 124-125
11) Imides 125-126
12) Ureas 126-130
13) Cyclic Amides (not Lactams) 130-136
14) Imines, Guanidines, Biguanides, Amidines, etc. 136-138
15) Thioamides 138-143

Non Aromatic Nitriles and Cumulated Double Bonds 144
1) Nitriles 144-150
2) Isocyanates 150
3) Thiocyanates 150-151
4) Isothiocyanates 151
5) Carbodiimides 151
6) Azides and Diazonium Salts 151
7) Inorganic Cyanides 151

Non-Aromatic Sulfur-Oxygen Compounds 152
1) Sulfoxides and Sulfones 152-155
2) Sulfites 155
3) Sulfonic Acids and Salts 155-156
4) Sulfonate Esters 156
5) Sulfonyl Halides 156-157
6) Sulfonamides 157
7) Sulfates 157

Non-Aromatic Phosphorus Compounds 158
1) Phosphines 158
2) Phosphites 158
3) Phosphonates 158-159
4) Phosphoramides 159-160
5) Chlorophosphates 160
6) Thiophosphates 160
7) Phosphates 160-161

Aromatic Hydrocarbons 162
1) Alkyl Substituted 162-164
2) Alkenyl Substituted 164-165
3) Non-fused Multiple Ring 165
4) Fused Ring 165-168

Aromatic Halogenated Hydrocarbons 169
1) Mono-substituted 169
2) Ortho-disubstituted 169
3) Meta-disubstituted 169
4) Para-disubstituted 169-170
5) Trichloro or Trifluoromethyl 170
6) Other substituted 170-171
7) Multiple Ring 171-175

Aromatic Ethers, Epoxides and Acetals 176
1) Aryl-Alkyl 176-182
2) Diaryl 182-185
3) Miscellaneous 185-189
4) Epoxides 189
5) Acetals 189

Aromatic Alcohols and Phenols 190
1) Phenols 190-192
2) Catechols 192-193
3) Resorcinols 193
4) Hydroquinones 193
5) Polycyclic Phenols 193-197
6) Aryl-aliphatic Alcohols 197-208

Aromatic Mercaptans and Sulfides 209
1) Mercaptans 209
2) Sulfides 209-214

Aromatic Amines 215
1) Single Ring Anilines 215-220
2) Phenylenediamines 220-222
3) Multiple Ring Anilines 222-234
4) Hydrazines 234-236
5) Aryl-aliphatic Amines 236-244
6) Heterocyclic 244-254
7) Quaternary Ammonium 254-256

Aromatic Nitro and Nitroso Compounds 257
1) Nitroso 257
2) Arylnitroalkenes 257-258
3) Ring Nitro 258-281

Aromatic Ketones 282
1) Unconjugated 282-287
2) Aryl-alkyl 287
3) ortho-Hydroxy Aryl-alkyl 287-299
4) Aryl-alkenyl 299-306
5) Diketones 306-309
6) Diaryl 309-315
7) ortho-Hydroxy Diaryl 309-315
8) Quinones 315-319
9) Chromanones, Chromones, Flavanones, Flavones ... 319-320

Aromatic Aldehydes 321
1) Unconjugated 321
2) Cinnamaldehydes 321
3) Other Conjugated 321-324

Aromatic Carboxylic Acids 325
1) Unconjugated 325-334
2) Cinnamic Acids 334-336
3) Other Conjugated 336-347

Aromatic Amino Acids and Acid Salts 348
1) Amino Acids 348-349
2) N-Protected Amino Acids 349
3) Carboxylic Acid Salts 349-352

Aromatic Esters and Lactones 353
1) Unconjugated 353-363
2) β-Keto 363-364
3) Cinnamates 364
4) Other Conjugated 365-376
5) Phenyl Esters 376-381
6) Carbonates 381-382
7) Lactones 382-387

Aromatic Anhydrides 388

Aromatic Acid Halides **389**

Aromatic Amides **390**
1) N-Phenyl Amides 390-442
2) Conjugated 442-445
3) β-Keto 445-483
4) Polycyclic Amides 483-491
5) Hydrazides 491-504
6) Ureas 504-604
7) Carbamates 604-689
8) Lactams 689-693
9) Other Cyclic Amides 693-714
10) Imides 714-723
11) Amidines 723-727
12) Thioamides, Thioureas, etc. 727-750

Aromatic Nitriles and Cumulated Double Bonds **751**
1) Unconjugated Nitriles 751-758
2) Conjugated Nitriles 758
3) Isocyanide 758-768
4) Thiocyanates 768
5) Isothiocyanates 768
6) Isocyanates 768-769
7) Carbodiimides 769
8) Azides and Diazonium Salts 769-770

Aromatic Sulfur-Oxygen Compounds **771**
1) Sulfoxides 771
2) Sulfones 771-786
3) Sulfinates 786-787
4) Sulfonic Acids 787-790
5) Sulfonate Salts 790-796
6) Sulfonate Esters 796-801
7) Sulfonyl Halides 801-804
8) Sulfonamides 804-829

Aromatic Phosphorus Compounds **830**
1) Phosphines 830
2) Quaternary Phosphorus 830-832
3) Phosphoranes 832-833
4) Phosphites, Phosphonites, etc. 833
5) Pentavalent Phosphorus Compounds 833-837

5-Membered Aromatic Heterocyclics **838**
1) Pyrroles 838-843
2) Furans 843-855
3) Thiophenes 855-859
4) Pyrazoles 859-864
5) Imidazoles 864-869
6) Triazoles 869-870
7) Tetrazoles 870
8) Oxazoles and Oxadiazoles 870-872
9) Thiazoles, Thiadiazoles, etc. 872-880

5-Membered Fused Aromatic Heterocyclics **881**
1) Indoles 881-891
2) Carbazoles 891-892
3) Benzofurans 892-897
4) Benzothiophenes 897-899
5) Indazoles 899-901
6) Benzimidazoles 901-908
7) Benzotriazoles 908-909
8) Benzoxazoles 909-910
9) Benzothiazoles and Benzoselenazoles 910-917
10) Benzothiadiazoles and Benzoselenadiazoles 917-920
11) Purines 920-924

6-Membered Aromatic Heterocyclics **925**
1) Pyridines 925-967
2) Pyridazines 967
3) Pyrimidines 968-977
4) Pyrazines 977-978
5) Triazines 978-980

6-Membered Fused Aromatic Heterocyclics **981**
1) Quinolines 981-995
2) Isoquinolines 995
3) Acridines 996-998
4) Cinnolines, Phthalazines, Quinazolines, Quinoxalines, Phenazines, etc. 998-1002
5) Phenanthrolines 1002
6) Miscellaneous 1002-1004

Heterocyclic N-Oxides **1005**

Oximes **1008**

Alkynes **1015**

Schiff Bases, Azobenzenes, Indicators and Dyes **1020**
1) Schiff Bases 1020-1079
2) Azobenzenes 1079-1093
3) Phthaleins 1093
4) Fluoresceins 1093
5) Sulfonephthaleins 1093-1094
6) Miscellaneous 1094-1095

Silanes **1096**

Boranes **1124**

Organometallic and Arsenic Compounds **1126**
1) Arsenic 1126-1128
2) Antimony 1128-1129
3) Tin 1129-1133
4) Lead 1133-1136
5) Mercury 1136-1139
6) Organometallic Complexes 1139
7) Carbonyl Complexes 1140
8) Miscellaneous 1140-1143

Molecular Formula Index **1145**

Product Catalog Number to Page Number Cross Reference **1372**

The Sigma-Aldrich Library of Rare Chemicals

"Chemists Helping Chemists in Research and Industry"

best describes our mission of providing the scientific community with small quantities of rare and difficult-to-obtain compounds.

The Sigma-Aldrich Library of Rare Chemicals contains 38,000 research samples acquired from all over the world—many from some of the world's best chemists:

Professors R.B. Woodward, Henry Gilman, Tadeus Reichstein, L.F. Fieser, Melvin Newman, Oskar Jeger, George Wittig, Karl Dimroth, Donald Cram, to mention just a few.

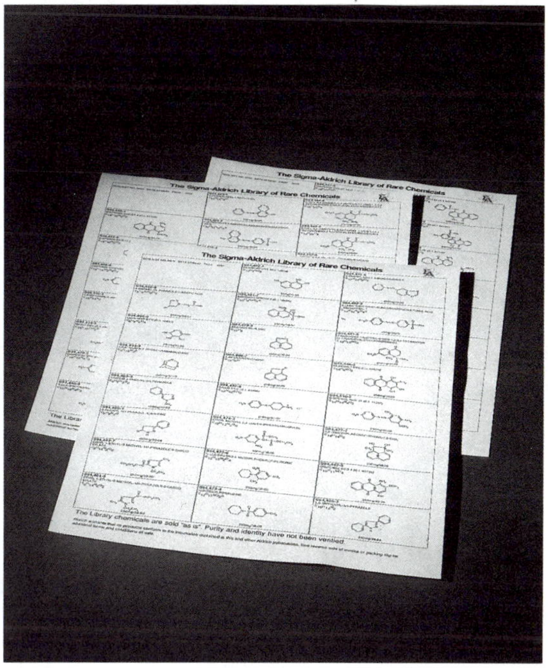

- All of our compounds can be searched by computer providing you a valuable tool for locating compounds with a particular structural fragment or functionality.
- Simply call or write us with any combination of the following parameters:

structural fragments	functional groups
rings systems	elements
molecular weight range	

- A computer-generated list containing names, molecular formulas, prices and structures of all the compounds in the Library that match the criteria will be mailed to you.
- Each sample costs $35.00. Sample amounts differ depending upon the quantities we have in stock. Sample sizes are clearly marked on the computer printouts.

In order to continue our program of ***Chemists Helping Chemists***, we are in constant need of new chemicals for our Library. Please don't let your research samples be lost; allow us to purchase them and make them available to hundreds of scientists around the world.

Please call us at 1-800-231-8327 and we can discuss your special situation.

Tools to aid your search of the entire Aldrich inventory and to find compounds for your research.

Z11,838-9 **The Aldrich Microfiche Library of Chemical Indices**
24 microfiche in a convenient, hardbound holder

Z11,842-7 **The Aldrich Microfiche Library of Chemical Indices**
24 microfiche (without holder)

Z17,000-3 **ALDRICHEM Data Search™ for IBM®PC computers**

Z17,533-1 **ALDRICHEM Data Search™ for Apple® Macintosh® computers**

Z20,446-3 **Aldrich Catalog on CD-ROM, IBM version**
Includes 38,000 listings from the Library of Rare Chemicals

Sigma-Aldrich Library of Rare Chemicals
for your screening and research needs

Modeling made easier

Cochranes molecular models

An extensive model system offering 15 atomic-bonding configurations in 12 colors of plastic atoms. The carbon atom is represented in nine configurations: linear, trigonal (in five angle variations), tetrahedral, octahedral, trigonal-bipyramidal.

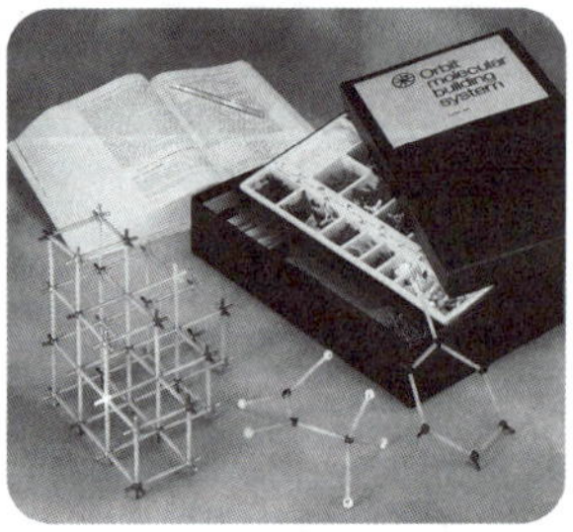

Orbit Sets

Large Set, includes some of every center in the system **Z11,963-6**

Basic organic, for the study of smaller organic compunds **Z11,966-0**

Organic/inorganic, for molecules, complex ions, polymerization, with booklet. (Z12,907-0) **Z18,475-6**

Chem-Tutor student modeling system

Specially designed model set by Professor S.G. Levine of North Carolina State University for use by organic chemistry students during examinations, etc. Contains 80 Orbit-size atom centers, 56 bond tubes, and text in a sturdy plastic tub.

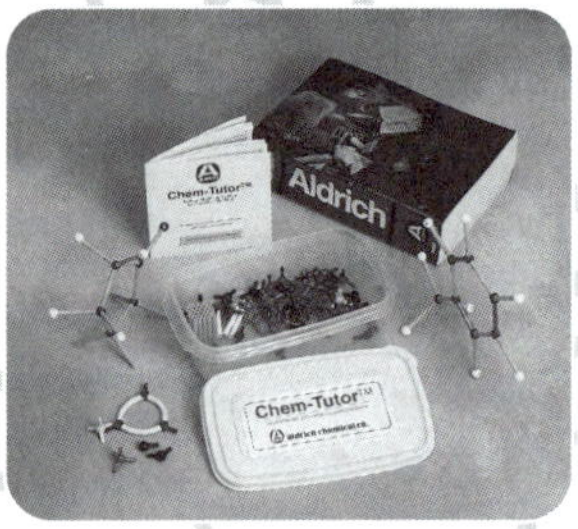

Z22,249-6

Fieser molecular model research kit

Low-cost molecular models designed by Professor Louis F. Fieser. The models consist of sturdy, color-coded plastic and aluminum parts that snap together to form bonds. They are easily assembled and handled, and may be disassembled for repeated use.

Fieser models are highly effective for assessing conformational and steric effects, and geometrical relationships.

Each kit consists of:

- 30 tetrahedral carbon atoms (black)
- 6 pairs of doubled-bonded carbon atoms (black)
- 5 oxygen (or sulfur) atoms (red)
- 2 trivalent nitrogen atoms (blue)
- 2 sulfur atoms (yellow)

Z10,400-0

Giant-size molecular models

Large, lightweight, molecular models for classroom demonstration of organic and inorganic structures. For building models to scale of 20cm/A and 8cm/A. Bond lengths range from 5.5cm (for average-size classroom demonstrations) to 27cm (for larger groups).

Atomic centers are 5cm in diameter, finished in standard atom colors, and have a reinforced vinyl receptor at each bond position for insertion of connecting rods (bonds).

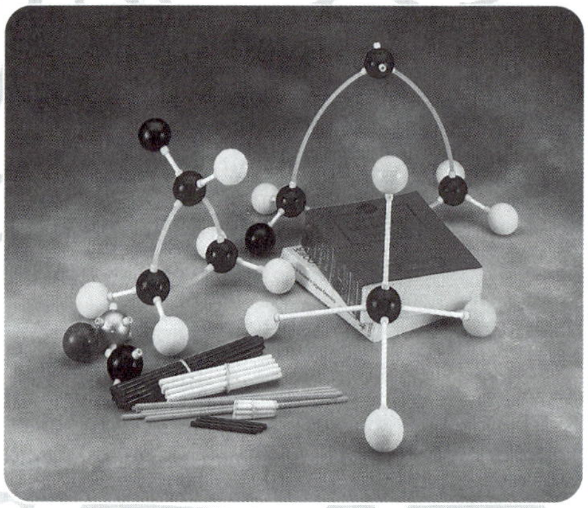

Basic Kit (98 pieces)
Z23,131-2
Standard Kit (162 pieces)
Z12,660-8
Deluxe Kit (196 pieces)
Z23,132-0
All kits include storage box and instructions

Molecular Visions, Darling flexible model kit

By Professor Stephen D. Darling. Flexible pieces show strain due to angle distortion or torsional stress, and whether two atoms in a molecule can approach each other.

103 piece set contains:

- 2 180°
- 4 bond extenders and 8 double bonds
- 4 90° hybrid orbitals
- 36 109° hybrid orbitals (black-carbon), sp3
- 16 120° hybrid orbitals (grey-carbon), sp3
- 4 109° hybrid orbitals (blue-nitrogen), sp3
- 12 109° hybrid orbitals (red-oxygen), sp3
- 5 109° vinyl rubber pieces, sp3
- 12 marker balls (3 each of white, red, green, and blue)

Z10,800-6

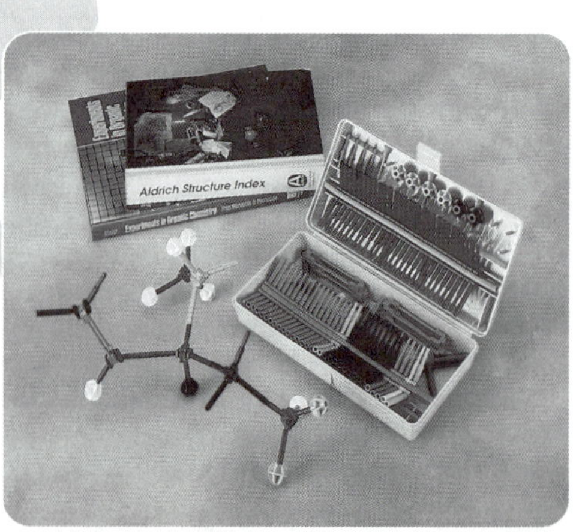

For more modeling kits - see the 1992-1993 Aldrich Catalog Handbook of Fine Chemicals

S45,600-4
3-ETHYLPENTANE

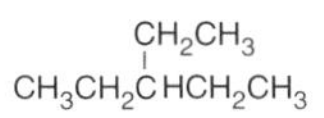

250 mg

S42,780-2
3-METHYLHEPTANE

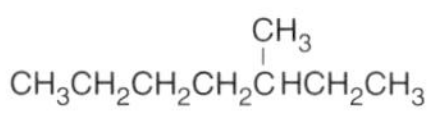

5 gm

S92,221-8
2,4-DIMETHYLHEXANE

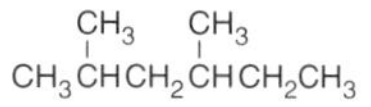

250 mg

S41,032-2
2,2,5-TRIMETHYLHEXANE

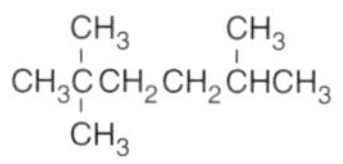

5 gm

S43,093-5
2-METHYLOCTANE

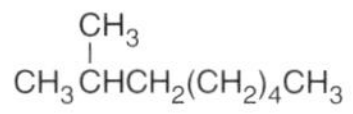

5 gm

S45,658-6
2,2-DIMETHYLHEPTANE

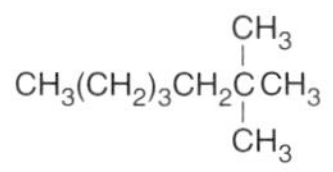

250 mg

S41,205-8
2-METHYLNONANE

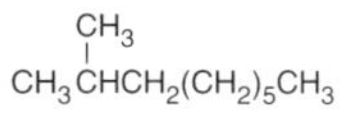

5 gm

S41,206-6
3-METHYLNONANE

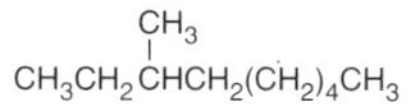

5 gm

S43,164-8
4-METHYLNONANE

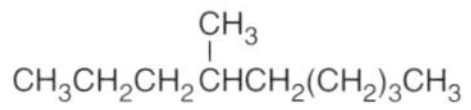

5 gm

S76,540-6
R-(-)-2,6-DIMETHYLOCTANE

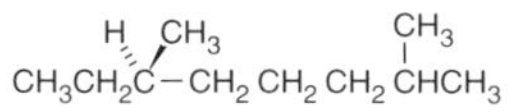

5 gm

S68,045-1
2,2,4,6,6-PENTAMETHYLHEPTANE

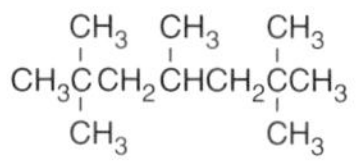

250 mg

S60,322-8
2,6,10-TRIMETHYLDODECANE

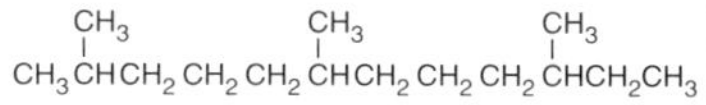

250 mg

S83,900-0
18-HEPTADECYLHEXATRIACONTANE

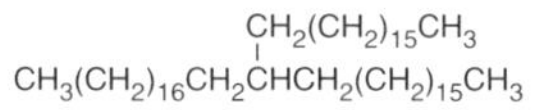

250 mg

S44,259-3
CIS-2-OCTENE

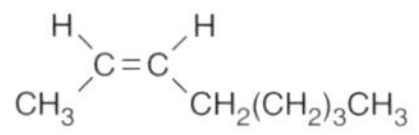

5 gm

S43,674-7
TRANS-2-DECENE

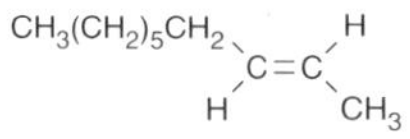

250 mg

S45,347-1
1-DOCOSENE

$CH_3(CH_2)_{18}CH_2CH{=}CH_2$

5 gm

S57,097-4
17-PENTATRIACONTENE

$CH_3(CH_2)_{14}CH_2CH{=}CHCH_2(CH_2)_{15}CH_3$

5 gm

S43,396-9
4-METHYL-2-PENTENE, MIXTURE OF CIS AND TRANS

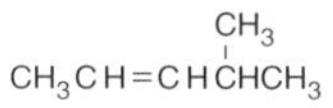

5 gm

S41,192-2
4,4-DIMETHYL-2-PENTENE, MIXTURE OF CIS AND TRANS

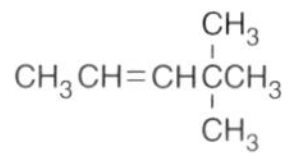

5 gm

S41,202-3
2,3-DIMETHYL-2-PENTENE

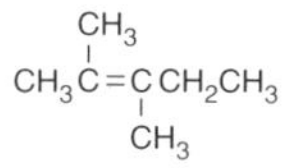

5 gm

S41,911-7
2-METHYL-2-HEXENE

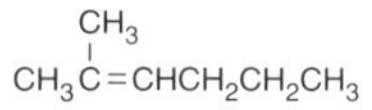

5 gm

S43,744-1
CIS-4,4-DIMETHYL-2-PENTENE

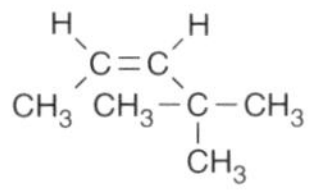

250 mg

S43,746-8
3-ETHYL-2-PENTENE

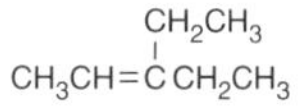

250 mg

S44,091-4
4,4-DIMETHYL-1-PENTENE

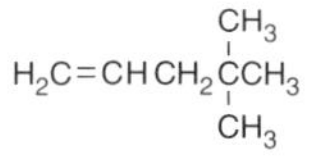

5 gm

S44,870-2
2,4-DIMETHYL-1-PENTENE

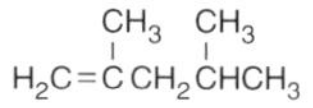

5 gm

S36,971-3
TRANS-2-METHYL-3-HEPTENE

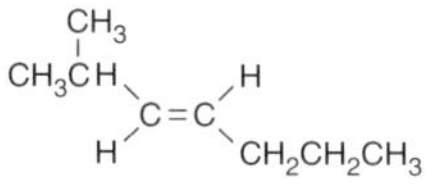

5 gm

S42,954-6
CIS-2,5-DIMETHYL-3-HEXENE

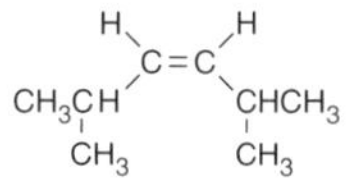

250 mg

S43,154-0
TRANS-2,2-DIMETHYL-3-HEXENE

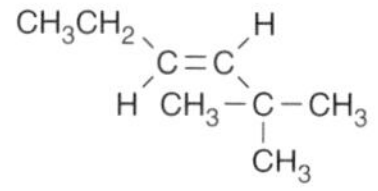

5 gm

S43,217-2
2-METHYL-2-HEPTENE

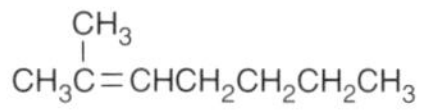

250 mg

S50,471-8
TRANS-4-METHYL-2-HEPTENE

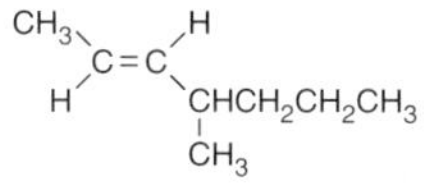

5 gm

S50,474-2
4-METHYL-2-HEPTENE, MIXTURE OF CIS AND TRANS

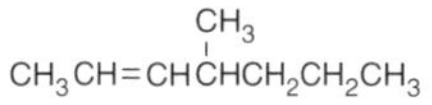

5 gm

S91,807-5
3,4-DIMETHYL-3-HEXENE

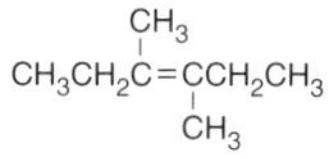

250 mg

S41,198-1
TRANS-2,2-DIMETHYL-3-HEPTENE

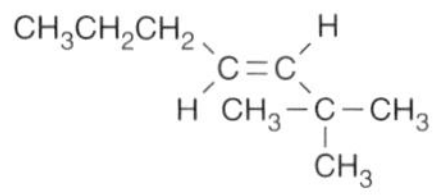

5 gm

S43,159-1
3,5,5-TRIMETHYL-1-HEXENE

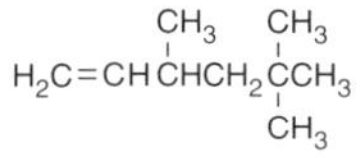

5 gm

S43,215-6
3,5-DIMETHYL-3-HEPTENE, MIXTURE OF CIS AND TRANS

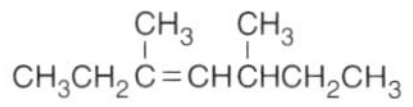

5 gm

S41,780-7
2-METHYL-1-NONENE

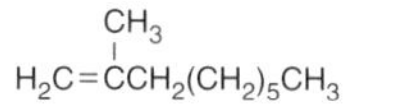

5 gm

S41,542-1
1,1-DINEOPENTYLETHYLENE
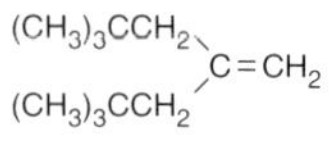
5 gm

S43,001-3
TRIISOBUTYLENE,
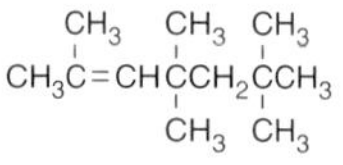
5 gm

S44,677-7
2,2,4,6,6-PENTAMETHYL-3-HEPTENE
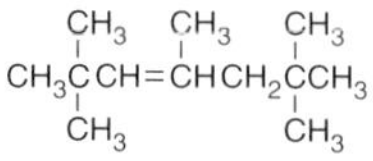
5 gm

S81,262-5
18-HEPTADECYL-17-HEXATRIACONTENE
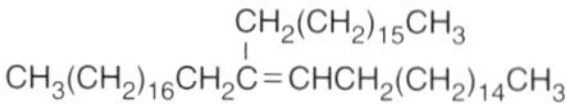
250 mg

S43,916-9
1,5-HEPTADIENE
$H_2C=CHCH_2CH_2CH=CHCH_3$
250 mg

S41,898-6
5-METHYL-1,3,6-HEPTATRIENE
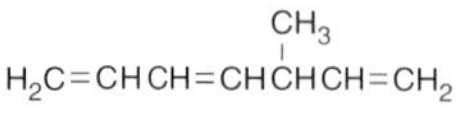
5 gm

S41,043-8
3-METHYL-1,5-HEPTADIENE
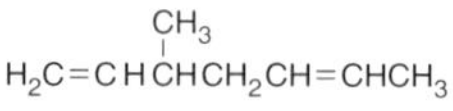
5 gm

S43,218-0
1,4-OCTADIENE, MIXTURE OF CIS AND TRANS
$H_2C=CHCH_2CH=CHCH_2CH_2CH_3$
5 gm

S88,199-6
4-METHYL-1,3-OCTADIENE
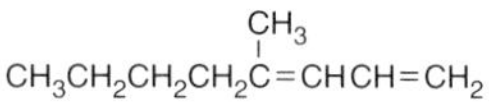
250 mg

S52,843-9
3,7-DIMETHYL-1,3,6-OCTATRIENE
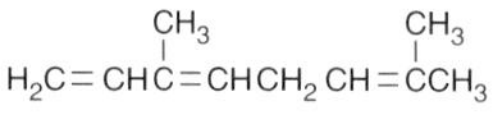
5 gm

S44,145-7
2,6-DIMETHYL-2,CIS-6-OCTADIENE
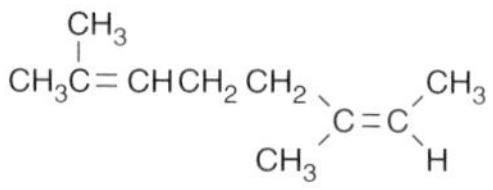
250 mg

S88,425-1
5-ETHYL-6-METHYL-1,10-UNDECADIENE
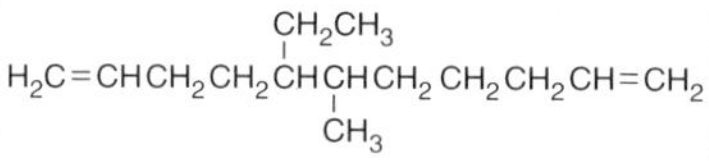
250 mg

S43,464-7
DIMETHYLTRICYCLO[5.2.1.0(2,6)]-DECANE (MIXTURE OF ISOMERS)
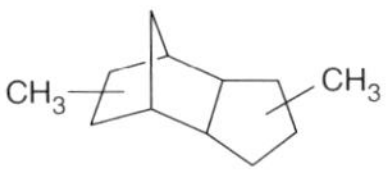
5 gm

S41,877-3
SEC.-BUTYLCYCLOHEXANE
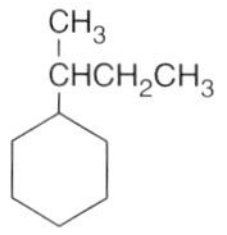
5 gm

S87,988-6
1-TERT-PENTYLCYCLOHEXANE
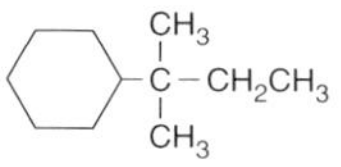
5 gm

S77,499-5
9-(2-METHYLCYCLOHEXYL)-PERHYDROFLUORENE
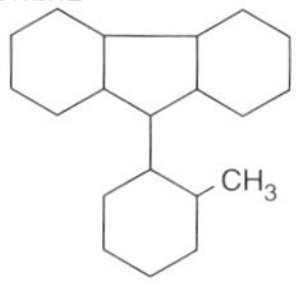
250 mg

S57,330-2
1,3-DIISOPROPYLCYCLOHEXANE
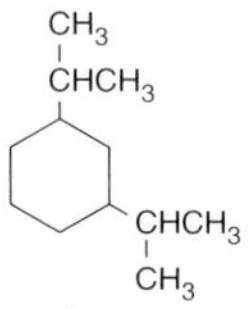
5 gm

S57,134-2
1,4-DIISOPROPYLCYCLOHEXANE

5 gm

S55,347-6
1,2,4-TRIMETHYLCYCLOHEXANE
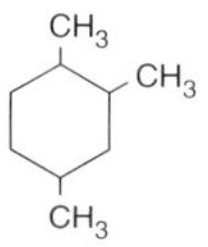
5 gm

S56,486-9
ISOBUTYLCYCLOHEXANE
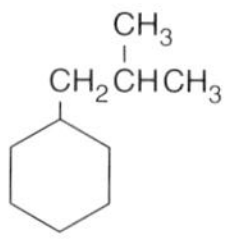
5 gm

S43,166-4
PENTYLCYCLOHEXANE
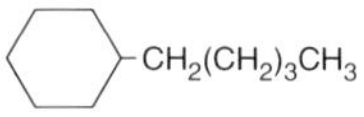
5 gm

S41,576-6
1-CYCLOHEXYLHEXANE
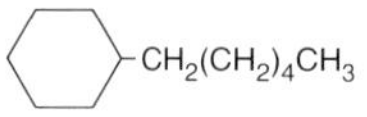
5 gm

S41,242-2
1-CYCLOHEXYLHEPTANE
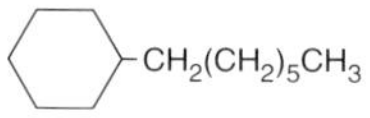
5 gm

S70,600-0
1,2-DICYCLOHEXYLETHANE
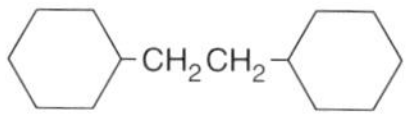
5 gm

S43,640-2
CYCLODODECANE

5 gm

S45,727-2
HEXAHYDROINDAN
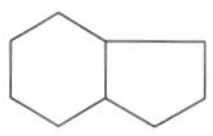
5 gm

S87,570-8
TETRACYCLO[2.2.1.0(2,6).0(3,5)] HEPTANE
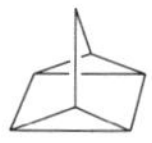
250 mg

S40,581-7
PERHYDROPHENANTHRENE
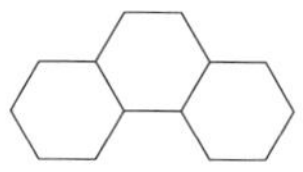
5 gm

S43,648-8
1,3-CYCLOOCTADIENE
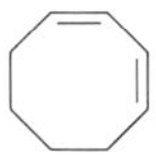
5 gm

S76,551-1
2,5,9-TRIMETHYL-1,5,9-CYCLODODECATRIENE
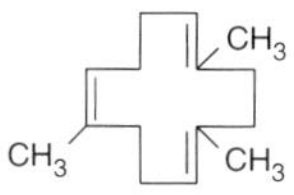
5 gm

S86,643-1
6,6-DIMETHYL-2-METHYLENEBICYCLO [3.1.1]HEPTANE
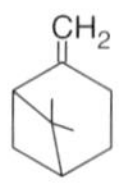
250 mg

S36,731-1
BETA-CARYOPHYLLENE
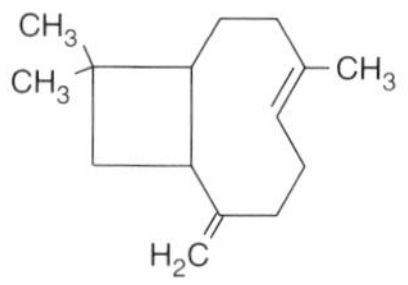
5 gm

S91,806-7
ISOPROPYLIDENECYCLOPENTANE
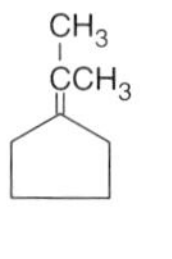
250 mg

S91,815-6
CYCLOPENTYLIDENECYCLOPENTANE
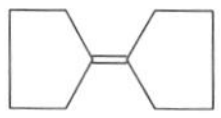
250 mg

S69,785-0
3-METHYL-2-NORBORNENE
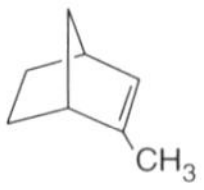
250 mg

S76,270-9
TRICYCLO[5.2.1.0(2,6)]DEC-8-ENE
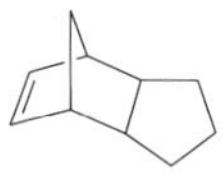
250 mg

S89,099-5
1,2,5B,6,7,10B,10C,10F-OCTAETHYL-OCTAHYDROPENTALENO(2,1,6-CDE:5,4,3-C'D'E')DIAZULENE

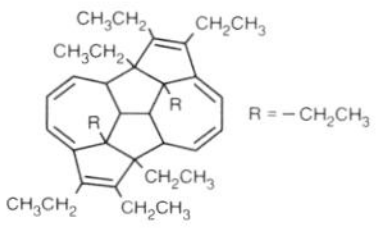

250 mg

S86,659-8
1,2,3,4,4A,5,8,8A-OCTAHYDRO-1,4-METHANONAPHTHALENE

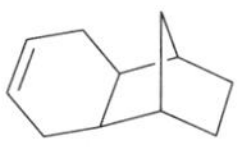

250 mg

S43,155-9
ETHYLIDENECYCLOPENTANE

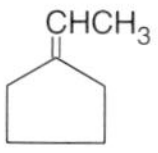

5 gm

S45,630-6
GUAIENE

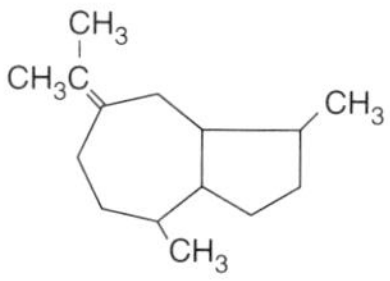

5 gm

S41,622-3
3-ETHYLCYCLOPENTENE

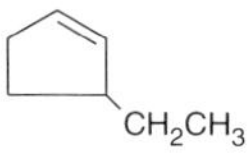

5 gm

S42,852-3
1-ETHYL-1-CYCLOPENTENE

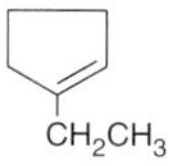

250 mg

S45,891-0
BICYCLO[4.3.0]NONA-3,7-DIENE

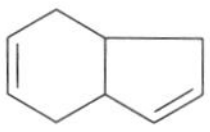

5 gm

S40,475-6
DEHYDRATED DIHYDROCAROTOL

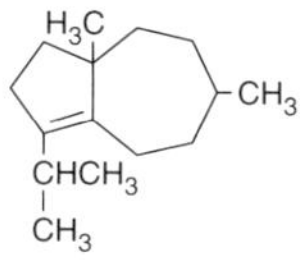

250 mg

S91,817-2
1-(SEC-BUTYLIDENE)CYCLOHEXANE

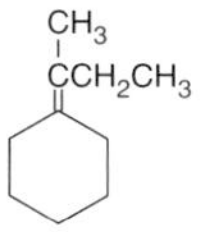

250 mg

S80,568-8
1,2-DICYCLOHEXYLETHYLENE

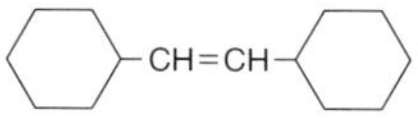

250 mg

S77,097-3
1,4-DICYCLOHEXYL-1,3-CYCLOHEXADIENE

250 mg

S43,152-4
ALLYLCYCLOHEXANE

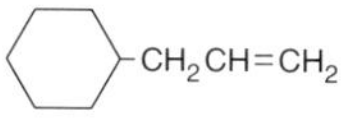

5 gm

S70,150-5
CYCLOHEXYLIDENECYCLOHEXANE

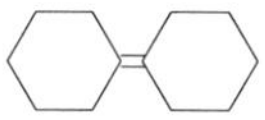

250 mg

S45,551-2
(+)-P-MENTH-1-ENE

250 mg

S50,488-2
4-(TERT-PENTYL)-1-CYCLOHEXENE

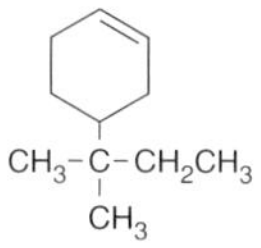

5 gm

S43,420-5
1,2-BIS(3-CYCLOHEXEN-1-YL)-ETHYLENE, MIXTURE OF CIS AND TRANS

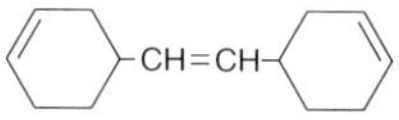

5 gm

S44,181-3
1-METHYL-1-CYCLOHEPTENE

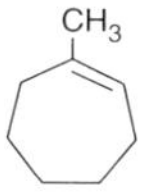

5 gm

S92,880-1
1,5-DIMETHYL-5-ETHYL-1,3-CYCLOHEXADIENE

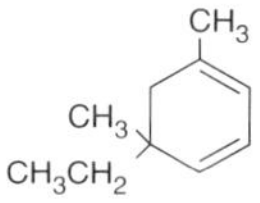

250 mg

S46,297-7
CIS-CYCLODECENE

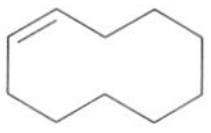

250 mg

S61,544-7
7,7'-BI(1,3,5-CYCLOHEPTATRIENE)

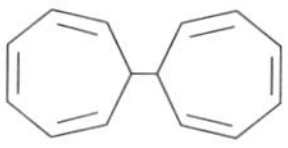

250 mg

S45,266-1
1,5,9-TRIMETHYL-1,5,9-CYCLODODECATRIENE (MIXTURE OF ISOMERS)

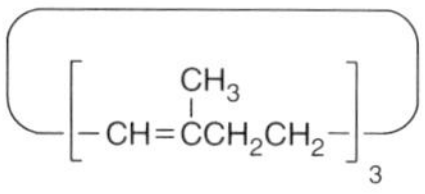

5 gm

S46,201-2
ISOTETRALIN

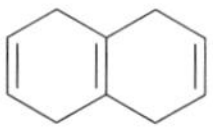

250 mg

S74,814-5
2,3-DIMETHYL-1,4,5,8-TETRAHYDRONAPHTHALENE

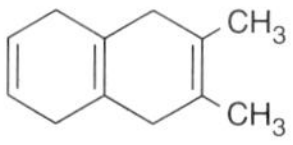

250 mg

S92,876-3
ALPHA-PYRONENE

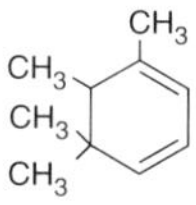

250 mg

S92,877-1
3-PYRONENE

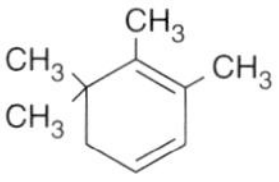

250 mg

S76,290-3 1-BROMOHEPTADECANE $CH_3(CH_2)_{15}CH_2Br$ 250 mg	**S57,035-4** 1-FLUORODECANE $CH_3(CH_2)_8CH_2F$ 5 gm	**S57,030-3** 1-FLUORODODECANE $CH_3(CH_2)_{10}CH_2F$ 5 gm	**S42,937-6** 1-CHLORODODECANE $CH_3(CH_2)_{10}CH_2Cl$ 5 gm	**S92,286-2** 1-FLUOROTETRADECANE $CH_3(CH_2)_{12}CH_2F$ 5 gm
S35,147-4 2-IODO-2-METHYLBUTANE 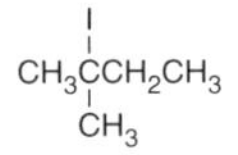 5 gm	**S56,452-4** 1-CHLORO-3-METHYLBUTANE 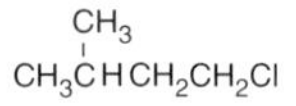 5 gm	**S41,663-0** 1-CHLORO-3,3-DIMETHYLBUTANE 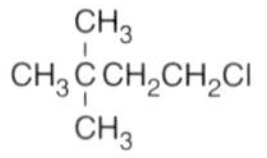 5 gm	**S43,151-6** 3-(CHLOROMETHYL)HEPTANE 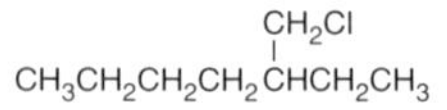 5 gm	**S64,850-7** 2-CHLOROPENTANE 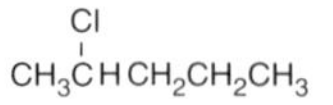 5 gm
S56,332-3 4-CHLOROHEPTANE 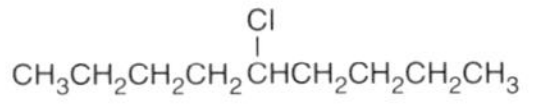 5 gm	**S56,451-6** 2-CHLOROOCTANE 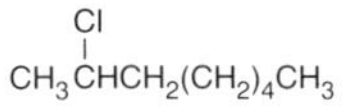 5 gm	**S55,755-2** 4-BROMOOCTANE 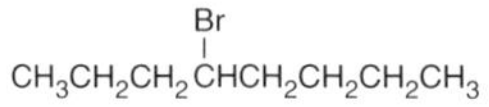 5 gm	**S56,092-8** 2-IODOOCTANE 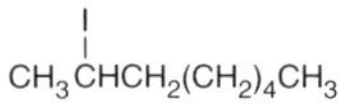 250 mg	**S43,875-8** 2-BROMONONANE 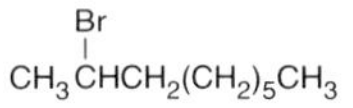 5 gm
S45,257-2 2-BROMOUNDECANE 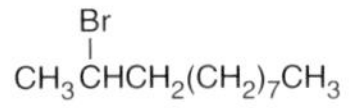 5 gm	**S56,285-8** HEXADECYL BROMIDE (MIXTURE OF ISOMERS) $C_{16}H_{33}Br$ 250 mg	**S55,922-9** TERT-OCTADECYL BROMIDE 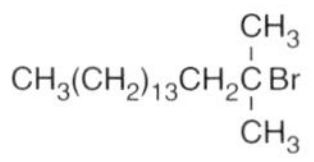 250 mg	**S86,601-6** 2,4-DIBROMO-2-METHYLBUTANE 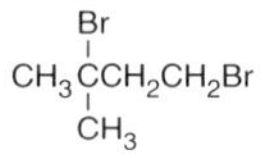 250 mg	**S80,386-3** 1,2-DIBROMODECANE 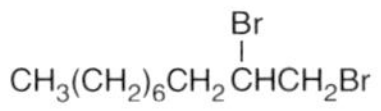 250 mg
S43,723-9 1,2-DICHLOROBUTANE 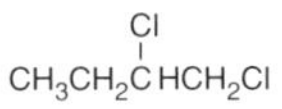 5 gm	**S43,411-6** 1,3-DICHLORO-3-METHYLBUTANE 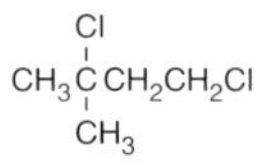 5 gm	**S76,450-7** 1,9-DIIODONONANE $ICH_2(CH_2)_7CH_2I$ 5 gm	**S76,449-3** 1,11-DIIODOUNDECANE $ICH_2(CH_2)_9CH_2I$ 5 gm	**S43,755-7** 1-BROMO-1-CHLOROETHANE 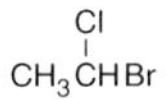 250 mg
S72,208-1 2,2-DICHLOROBUTANE 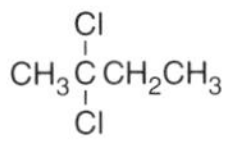 5 gm	**S88,192-9** 2-METHYL-1,2,3-TRIBROMOPROPANE 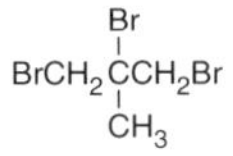 250 mg	**S90,601-8** 2,2-BIS(BROMOMETHYL)-1-BROMOBUTANE 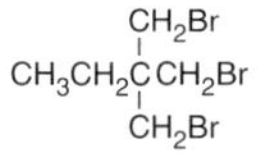 250 mg	**S65,866-9** 2,2,3,3-TETRAFLUOROPROPYL IODIDE $ICH_2CF_2CHF_2$ 250 mg	**S37,330-3** 1,2-DIBROMO-2,3-DICHLOROPROPANE 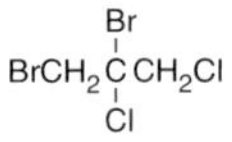 5 gm
S39,205-7 2-METHYL-1,2,3-TRICHLOROPROPANE 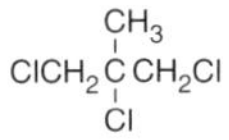 5 gm	**S36,017-1** 2,3-DIBROMO-1,4-DICHLOROBUTANE 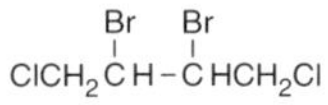 5 gm	**S57,182-2** 2-CHLORO-1,4-DIBROMO-1,1,2-TRIFLUOROBUTANE 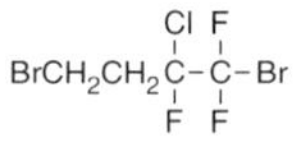 5 gm	**S41,856-0** 1,2,3,4-TETRABROMOBUTANE 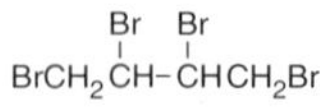 5 gm	**S41,781-5** PENTAERYTHRITYL TETRAIODIDE 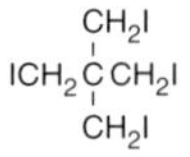 5 gm
S55,640-8 1,2,6,7-TETRABROMOHEPTANE 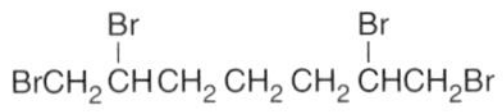 5 gm	**S81,567-5** 1,1,2,2,3,3,4,4,5,5,6,6,7,7,8,8,8-HEPTADECAFLUOROOCTANE $CF_3(CF_2)_5CF_2CHF_2$ 250 mg	**S56,301-3** 1,1,2-TRICHLOROPROPANE 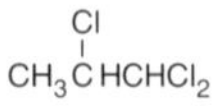 5 gm	**S50,878-0** 1,1,2,3,3-PENTACHLOROPROPANE 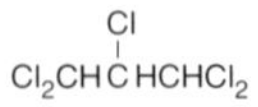 5 gm	**S45,894-5** 1,1,2,3,4,4-HEXACHLOROBUTANE 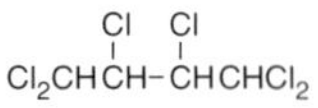 5 gm

S43,072-2
1,2-DIBROMO-1,1,2-TRICHLOROETHANE
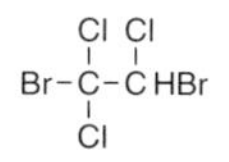
5 gm

S44,925-3
1,2-DICHLOROHEXAFLUOROPROPANE
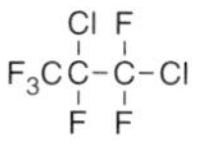
5 gm

S44,019-1
1,1,1,3-TETRACHLOROTETRAFLUOROPROPANE
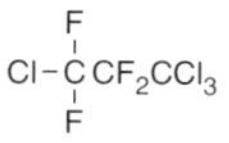
5 gm

S51,977-4
1,1,1,3,3-PENTACHLOROTRIFLUOROPROPANE
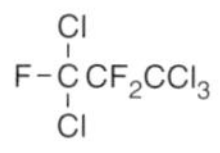
5 gm

S51,976-6
2,2-DIFLUOROHEXACHLOROPROPANE
$Cl_3CCF_2CCl_3$
5 gm

S37,331-1
2,3-DICHLOROOCTAFLUOROBUTANE
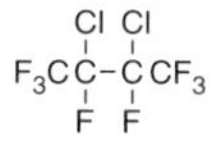
5 gm

S35,037-0
HEPTAFLUORO-2,2,3-TRICHLOROBUTANE
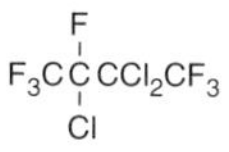
5 gm

S42,699-7
1,3-DICHLOROHEXAFLUOROPROPANE
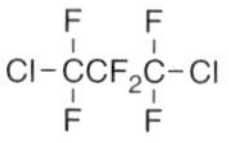
5 gm

S52,097-7
1,1,2,2,3-PENTAFLUOROTRICHLOROPROPANE
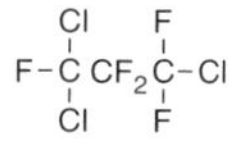
5 gm

S51,942-1
1,1,3,3-TETRACHLOROTETRAFLUOROPROPANE
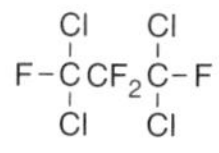
5 gm

S88,120-1
1-BROMO-1-HEXENE
$CH_3CH_2CH_2CH_2CH{=}CHBr$
250 mg

S86,611-3
6-IODO-1-HEPTENE
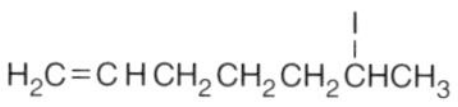
250 mg

S88,194-5
5-BROMO-1,4-OCTADIENE
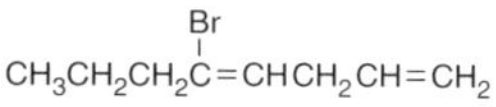
250 mg

S88,031-0
2-BROMO-1,12-TRIDECADIENE
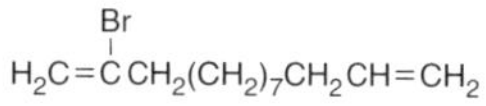
250 mg

S89,400-1
3-BROMO-1,1,2-TRICHLORO-1-PROPENE
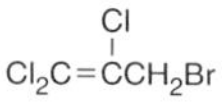
250 mg

S87,755-7
2,6-DIBROMO-2-HEXENE
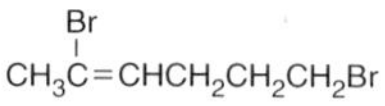
250 mg

S88,125-2
2,6-DIBROMO-1-HEXENE
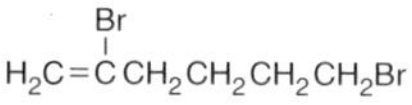
250 mg

S88,505-3
1,4-DIBROMO-2,3-DIMETHYL-2-BUTENE
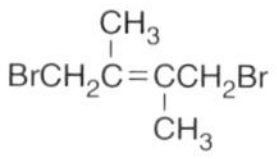
250 mg

S79,305-1
9-BROMO-2-METHYL-2-NONENE
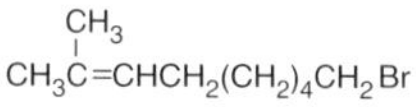
250 mg

S76,193-1
18-BROMO-6,9-OCTADECADIENE
$CH_3(CH_2)_4CH{=}CHCH_2CH{=}CH(CH_2)_7CH_2Br$
250 mg

S37,273-0
1,2,3-TRICHLOROPROPENE
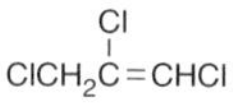
5 gm

S93,165-9
1-BROMO-3-HEXENE
$CH_3CH_2CH{=}CHCH_2CH_2Br$
250 mg

S44,172-4
2,3-BIS(BROMOMETHYL)-1,4-DIBROMO-2-BUTENE
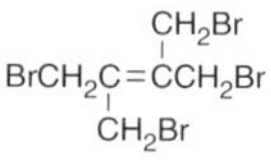
250 mg

S89,393-5
1,2,3,4,5,5,6,6-OCTACHLORO-1-HEXENE
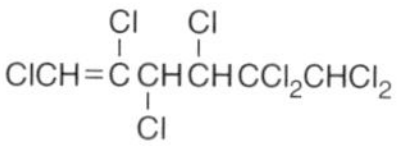
250 mg

S40,751-8
1,2-DICHLORO-3,3-DIFLUORO-1-PROPENE
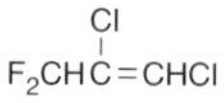
5 gm

S89,367-6
OCTACHLORO-1,3-PENTADIENE
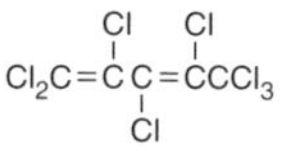
250 mg

S72,214-6
1,1,1-TRICHLORO-2-NONENE
$CH_3(CH_2)_4CH_2CH{=}CHCCl_3$
5 gm

S82,629-4
PERCHLORO-1,3,5-HEXATRIENE
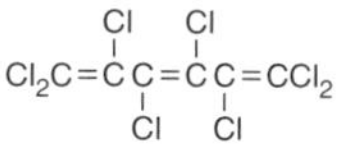
250 mg

S82,630-8
PERCHLORO-1,5-HEXADIENE
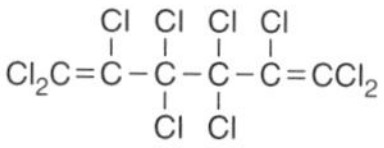
250 mg

S81,566-7
3,4,5,5,6,6,7,7,8,8,9,9,10,10,10-PENTADECAFLUORO-3-DECENE
$CH_3CH_2CF{=}CFCF_2(CF_2)_4CF_3$
250 mg

S81,563-2
1-IODONONADECAFLUORO-1-DECENE
$CF_3(CF_2)_6CF_2CF{=}CFI$
250 mg

S81,569-1
2-METHYL-3,4,4,5,5,6,6,7,7,8,8,9,9,10,10,11,11,11-OCTADECAFLUORO-2-UNDECENE

250 mg

S45,005-7
CYCLOHEPTYL CHLORIDE
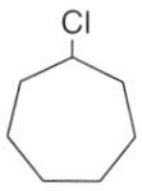
5 gm

S89,417-6
OCTACHLORO-1,3,5-CYCLOHEPTATRIENE
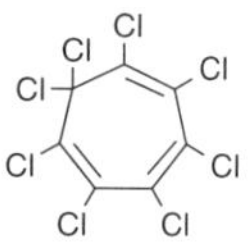
250 mg

S70,612-4
CHLOROCYCLOOCTANE
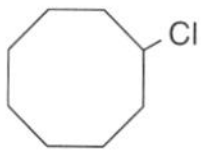
5 gm

S50,900-0
1,2-DIBROMOCYCLOOCTANE
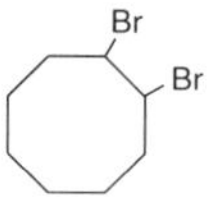
5 gm

S81,564-0
1-BROMOPENTADECAFLUOROCYCLOOCTANE
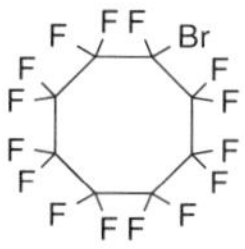
250 mg

S81,565-9
1-IODOPENTADECAFLUOROCYCLOOCTANE
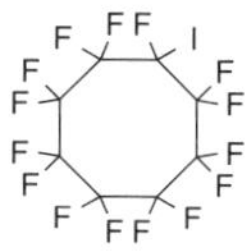
250 mg

S87,592-9
2,5,6,7,8,9,9-HEPTACHLORO-1,2,3,4,4A,5,8,8A-OCTAHYDRO-1,4:5,8-DIMETHANONAPHTHALENE
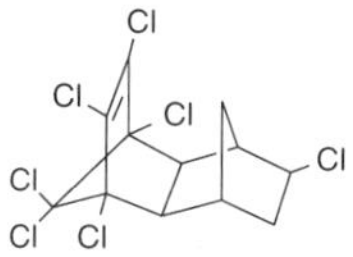
250 mg

S57,150-4
1-(2-BROMOETHYL)-1-ETHYLCYCLOPENTANE
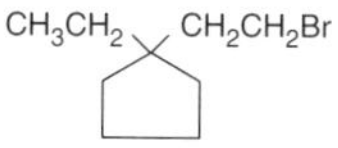
5 gm

S34,548-2
5,6-BIS(CHLOROMETHYL)-1,2,3,4,7,7-HEXACHLOROBICYCLO[2.2.1]HEPT-2-ENE

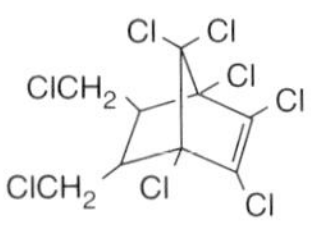

5 gm

S85,242-2
ENDO-2-(BROMOMETHYL)-5-NORBORNENE

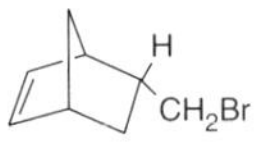

250 mg

S66,931-8
(-)-5-ENDO-BROMOMETHYL-1,2,3,4,7,7-HEXACHLORO-2-NORBORNENE

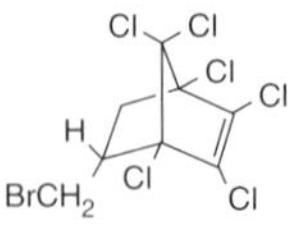

5 gm

S89,331-5
1,4-BIS(BROMOMETHYL)CYCLOHEXANE

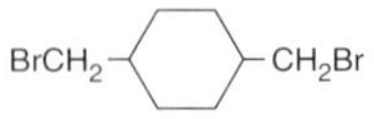

250 mg

S51,969-3
1,2-DIBROMO-4-(1,2-DIBROMOETHYL)-CYCLOHEXANE

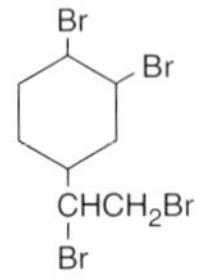

5 gm

S34,501-6
3-CYCLOHEXYLPROPYL BROMIDE

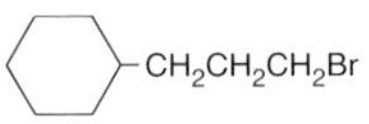

5 gm

S94,279-0
1-BROMO-3-(1-BROMOETHYL)-5,7-DIMETHYLADAMANTANE

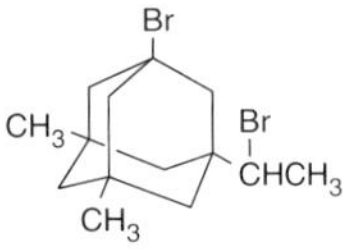

250 mg

S87,608-9
2-(BROMOMETHYL)-6,6-DIMETHYLBICYCLO[3.1.1]HEPT-2-ENE

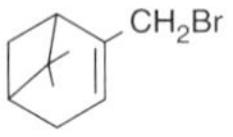

250 mg

S36,027-9
2,2-DICHLOROBICYCLO[4.1.0]HEPTANE

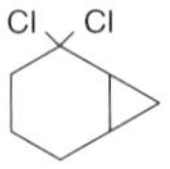

5 gm

S86,478-1
2-BROMO-3-(TRICHLOROMETHYL)-NORBORNANE

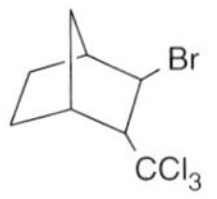

250 mg

S86,428-5
HEXACHLOROCYCLOPROPANE

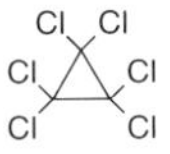

250 mg

S51,083-1
5,5-DICHLOROTETRACYCLO[6.2.1.0(2,7)0(4,6)]UNDEC-9-ENE

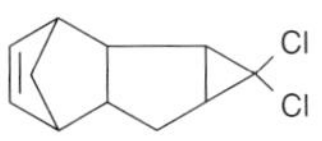

5 gm

S79,220-9
7,7-DIBROMOBICYCLO[4.1.0]HEPTANE

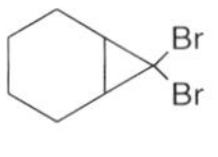

250 mg

S88,529-0
9,9-DICHLORO-1,4,5,8-TETRAHYDRO-4A,8A-METHANONAPHTHALENE

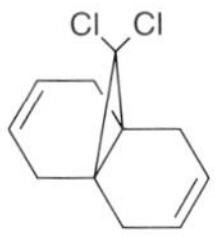

250 mg

S66,875-3
9,9-DIBROMOBICYCLO[6.1.0]NONANE

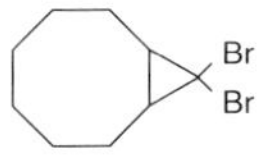

5 gm

S64,345-9
1,2-DICHLORO-1,2,3,3,4,4-HEXAFLUOROCYCLOBUTANE

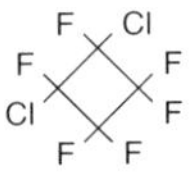

5 gm

S82,628-6
3,4-BIS(DICHLOROMETHYLENE)-1,2-DICHLORO-1-CYCLOBUTENE

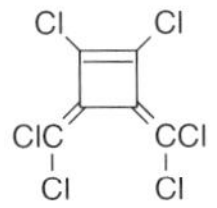

250 mg

S51,509-4
3,4-DICHLOROTETRAMETHYLCYCLOBUTENE

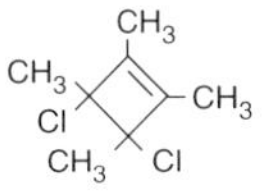

5 gm

S86,668-7
2-BROMO-5,6,7,8,9,9-HEXACHLORO-OCTAHYDRO-1,4:5,8-DIMETHANONAPHTHALENE

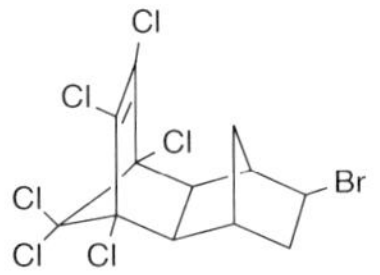

250 mg

S86,667-9
2,3-DIIODO-5,6,7,8,9,9-HEXACHLORO-OCTAHYDRO-1,4:5,8-DIMETHANONAPHTHALENE

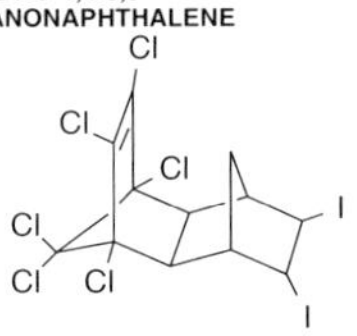

250 mg

S70,604-3
BORNYL BROMIDE

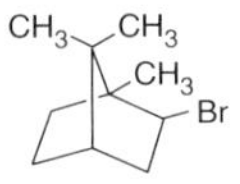

250 mg

S86,502-8
2-CHLORO-3-METHYLENE-4,7,7-TRIMETHYLNORBORNANE

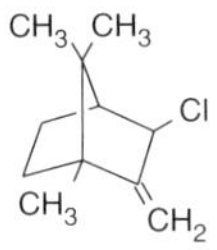

250 mg

S56,662-4
1,2-DICHLOROHEXAFLUORO-CYCLOPENTENE

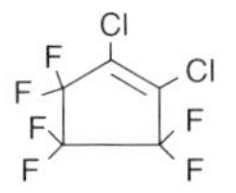

5 gm

S37,014-2
OCTACHLOROCYCLOPENTENE

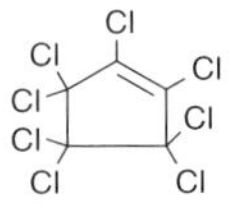

5 gm

S89,453-2
3-(DICHLOROMETHYLENE)-1,2,4,4,5,5-HEXACHLORO-1-CYCLOPENTENE

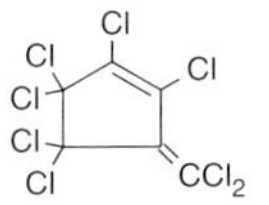

250 mg

S89,395-1
DECACHLORO-1,3A,6,6A-TETRAHYDROPENTALENE

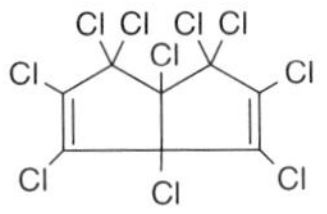

250 mg

S89,392-7
5-METHYL-1,2,3,4,5-PENTACHLORO-1,3-CYCLOPENTADIENE

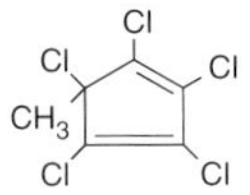

250 mg

S89,391-9
5-(DICHLOROMETHYLENE)-1,2,3,4-TETRACHLORO-1,3-CYCLOPENTADIENE

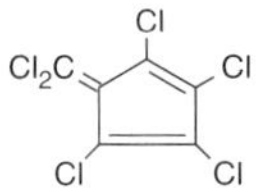

250 mg

S34,237-8
BIS(PENTACHLOROCYCLOPENTADIENE)

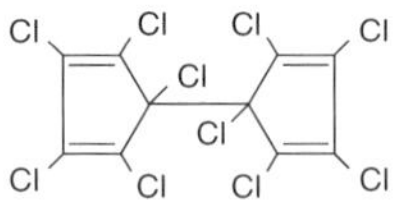

5 gm

S57,169-5
1-CHLORO-1-ETHYLCYCLOHEXANE

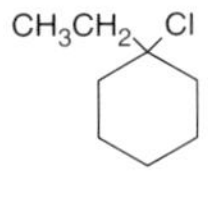

5 gm

S72,202-2
1,4-DICHLOROBICYCLO[2.2.2]OCTANE

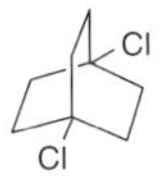

250 mg

S89,618-7
1-BROMO-2-FLUORO-1-CYCLOHEXENE

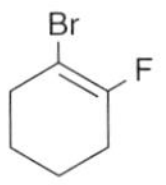

250 mg

S71,434-8
CHLOROCAMPHENE

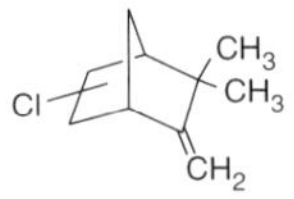

5 gm

S44,639-4
3-BROMOTRICYCLO[2.2.1.0(2,6)]-HEPTANE

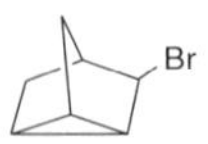

250 mg

S87,569-4
2,3-DIBROMO-5,6,7,8,9,9-HEXACHLORO-1,2,3,4,4A,5,8,8A-OCTAHYDRO-1,4:5,8-DIMETHANONAPHTHALENE

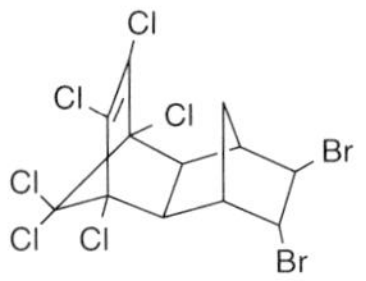

250 mg

S59,698-1
1-BROMO-3-METHYLADAMANTANE

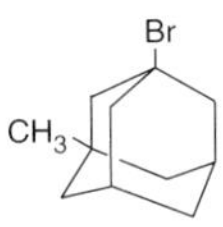

5 gm

S93,808-4
1-(2-BROMOETHYL)ADAMANTANE

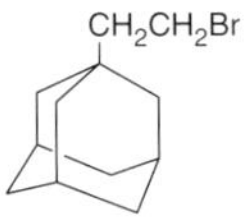

250 mg

S59,701-5
1,3-DIBROMO-5,7-DIMETHYLADAMANTANE

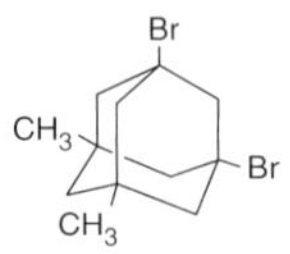

250 mg

S94,085-2
1-BROMO-3-(1-BROMOETHYL)ADAMANTANE

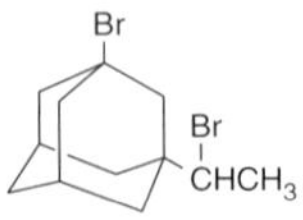

250 mg

S59,595-0
1-BROMO-3,5,7-TRIMETHYLADAMANTANE

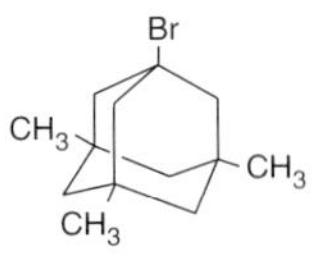

100 mg

S85,253-8
6-CHLORO-1,2,3,5,6,7,8,8A-OCTAHYDRO-1,2,8A-TRIMETHYLNAPHTHALENE

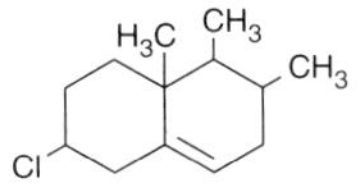

250 mg

S40,970-7
PERCHLOROPENTACYCLODECANE

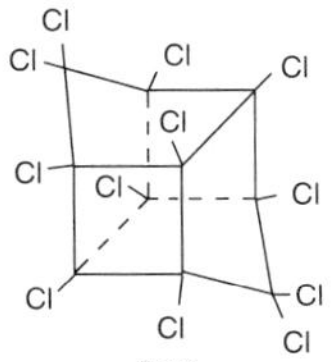

5 gm

S95,783-6
1,9,10,11,12,12-HEXACHLOROTRICYCLO[7.2.1.0(2,8)]DODEC-10-ENE

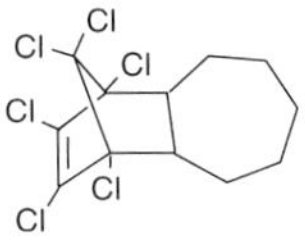

50 mg

S71,312-0 2,3-DIMETHYL-1-BUTANOL 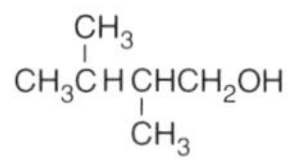 5 gm	**S65,896-0** 2,2-DIMETHYL-1-PENTANOL 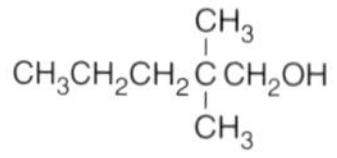 250 mg	**S66,337-9** 2-ETHYL-3-METHYL-1-BUTANOL 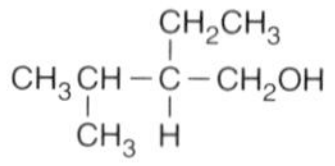 5 gm	**S71,313-9** 2,4-DIMETHYL-1-PENTANOL 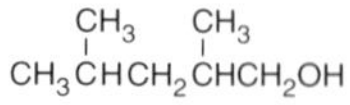 5 gm
S44,404-9 2,2,4-TRIMETHYL-1-PENTANOL 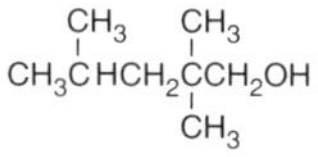 5 gm	**S71,314-7** 2-ETHYL-2-METHYL-1-PENTANOL 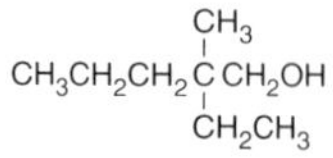 250 mg	**S36,053-8** 2,2-DIETHYL-1-PENTANOL 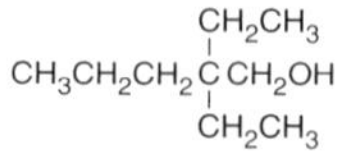 5 gm	**S36,382-0** 2-METHYL-2-PROPYL-1-PENTANOL 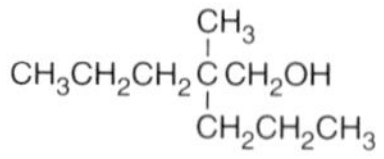 5 gm
S87,658-5 2,2,5-TRIMETHYL-3-HEXANOL 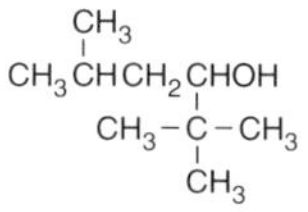 250 mg	**S37,070-3** 2-ISOPROPYL-5-METHYLHEXANOL 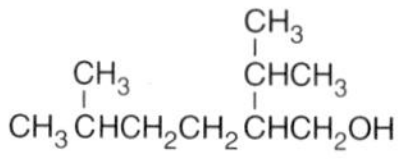 250 mg	**S36,191-7** 2-ETHYL-2-PROPYL-1-HEXANOL 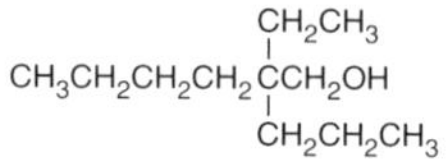 5 gm	**S57,288-8** 2-PENTYL-1-NONANOL 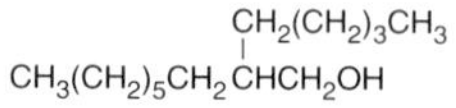 5 gm
S63,226-0 7-ETHYL-2-METHYLUNDECANOL 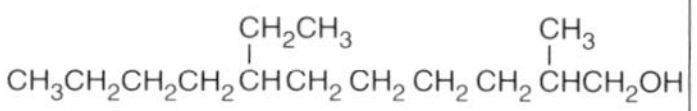 5 gm	**S36,186-0** 3-ETHYL-2-PENTANOL 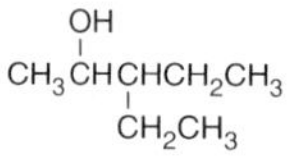 5 gm	**S40,928-6** 4-METHYL-3-HEXANOL 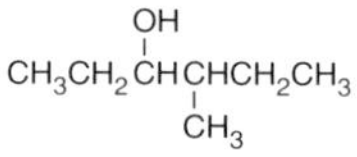 5 gm	**S43,337-3** 3-METHYL-2-HEXANOL 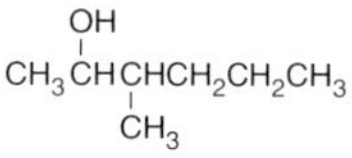 5 gm
S44,745-5 5-METHYL-3-HEXANOL 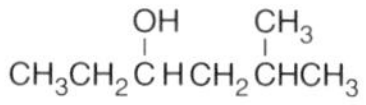 5 gm	**S45,512-1** 4-HEPTANOL 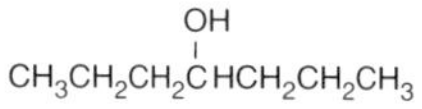 5 gm	**S87,726-3** 4-METHYL-2-HEXANOL 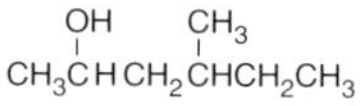 250 mg	**S37,419-9** 3-METHYL-2-HEPTANOL 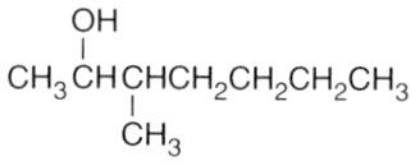 5 gm
S40,805-0 2,4-DIMETHYL-3-HEXANOL 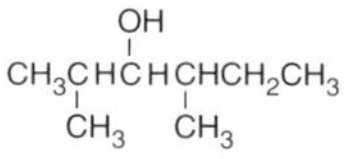 5 gm	**S40,927-8** 2-METHYL-3-HEPTANOL 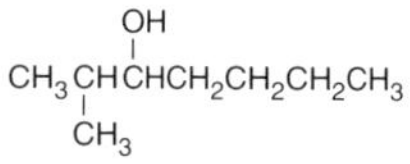 5 gm	**S41,197-3** 2,2,4-TRIMETHYL-3-PENTANOL 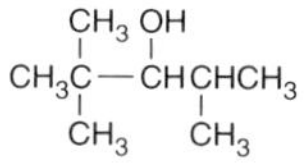 5 gm	**S41,243-0** 3-METHYL-4-HEPTANOL 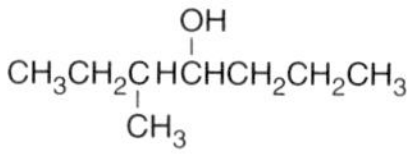 5 gm
S41,698-3 3,4-DIMETHYL-2-HEXANOL (MIXTURE OF ISOMERS) 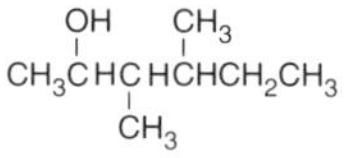 5 gm	**S43,085-4** 5-METHYL-3-HEPTANOL 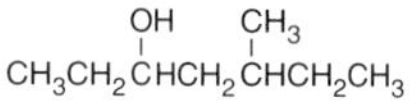 5 gm	**S43,153-2** 2,2-DIMETHYL-3-HEXANOL 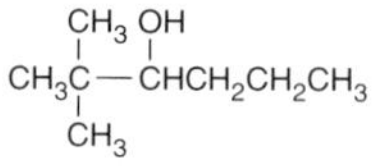 5 gm	**S43,555-4** 4-OCTANOL 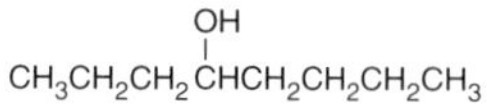 5 gm
S43,742-5 2,5-DIMETHYL-3-HEXANOL 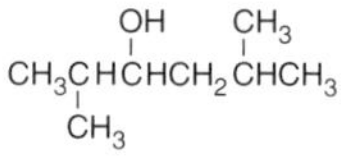 5 gm	**S44,252-6** 4-ETHYL-3-HEXANOL 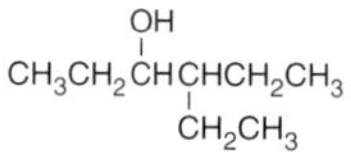 5 gm	**S41,455-7** 5-ETHYL-2-HEPTANOL 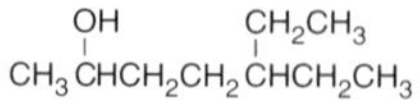 5 gm	**S41,697-5** 2,2-DIMETHYL-3-HEPTANOL 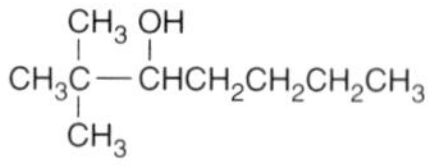 5 gm
S41,746-7 3-METHYL-4-OCTANOL 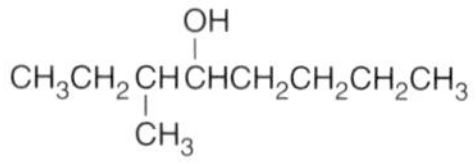 5 gm	**S41,949-4** 2,4-DIMETHYL-3-HEPTANOL 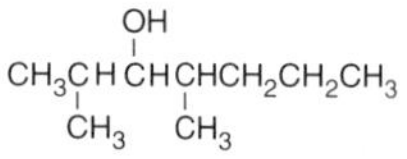 5 gm	**S42,785-3** 5-NONANOL 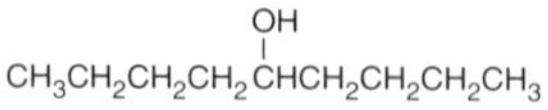 5 gm	**S43,165-6** 2-METHYL-3-OCTANOL 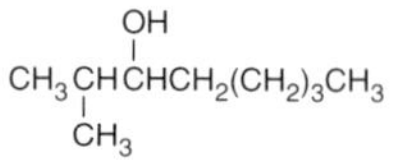 5 gm

S45,653-5 4-NONANOL 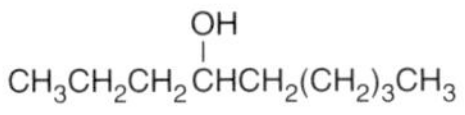 250 mg	**S37,417-2** 6-ETHYL-3-OCTANOL 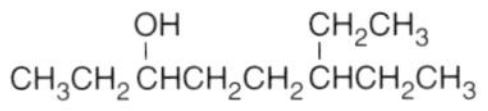 5 gm	**S41,201-5** 2,2-DIMETHYL-3-OCTANOL 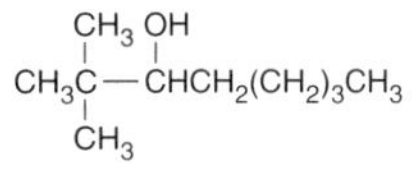 5 gm	**S41,742-4** 2-METHYL-3-NONANOL 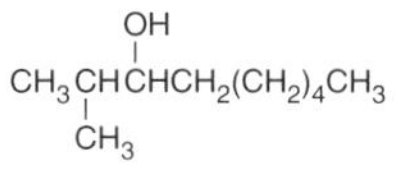 5 gm
S41,743-2 2-METHYL-4-NONANOL 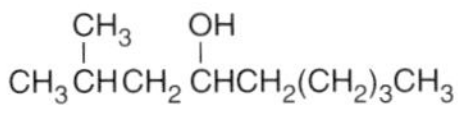 5 gm	**S41,744-0** 3-METHYL-4-NONANOL 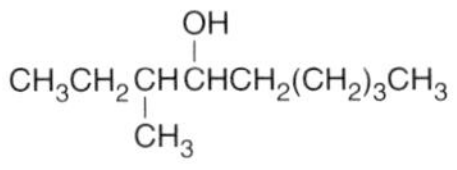 5 gm	**S42,959-7** 3-DECANOL 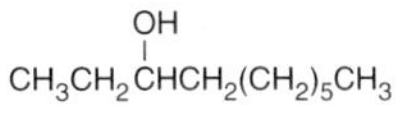 5 gm	**S50,500-5** 3,5-DIMETHYL-4-OCTANOL 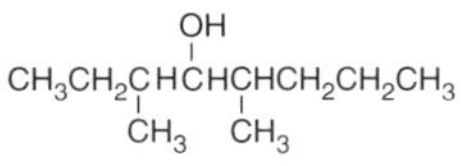 5 gm
S36,895-4 5-ETHYL-2-NONANOL 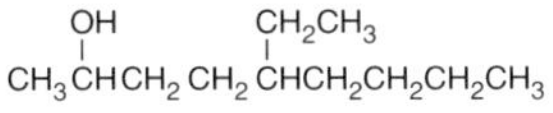 5 gm	**S54,680-1** 6-UNDECANOL 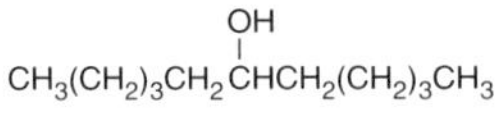 5 gm	**S57,274-8** 3,3,5,5-TETRAMETHYL-4-HEPTANOL 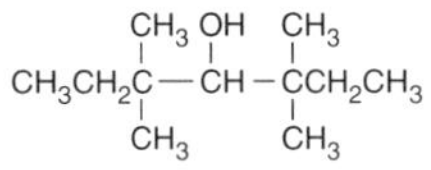 5 gm	**S39,206-5** 6-ETHYL-3-DECANOL 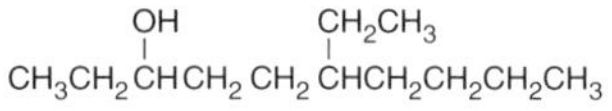 5 gm
S54,421-3 7-TRIDECANOL 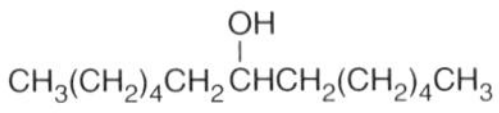 250 mg	**S78,688-8** 2-TRIDECANOL 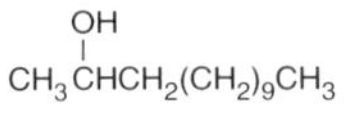 250 mg	**S34,940-2** 7-ETHYL-2-METHYL-4-UNDECANOL 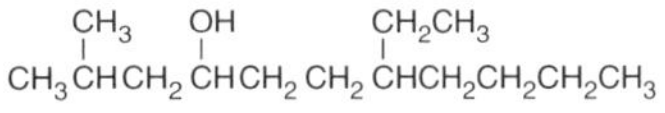 5 gm	**S45,875-9** 2-PENTADECANOL 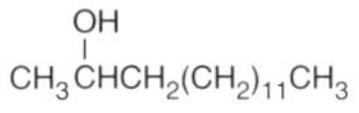 5 gm
S54,420-5 8-PENTADECANOL 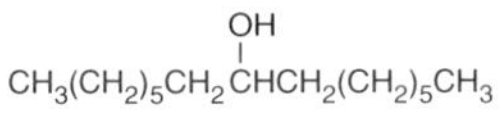 5 gm	**S54,389-6** 5-HEXADECANOL 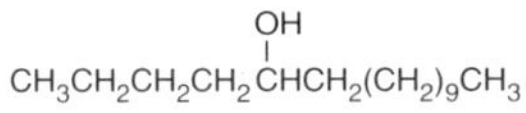 250 mg	**S34,618-7** 3,9-DIETHYL-6-TRIDECANOL 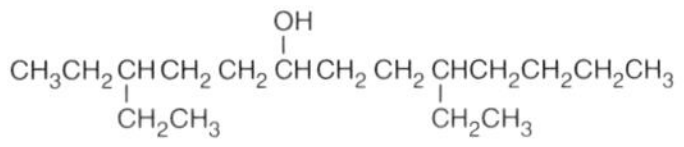 5 gm	**S45,469-9** 2-NONADECANOL 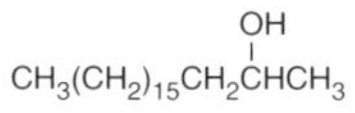 5 gm
S54,679-8 11-HENEICOSANOL 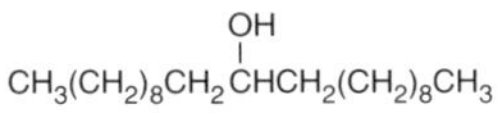 250 mg	**S73,929-4** 12-TRIACONTANOL 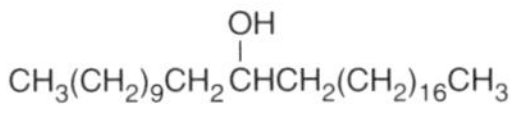 5 gm	**S57,904-1** 18-PENTATRIACONTANOL 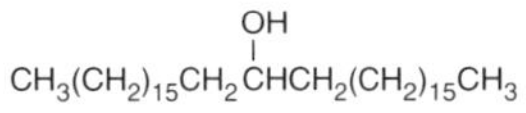 5 gm	**S42,686-5** 3-METHYL-3-HEXANOL 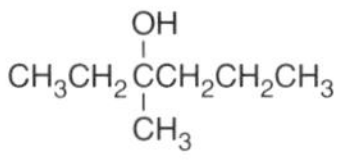 5 gm
S67,917-8 2,4-DIMETHYL-2-PENTANOL 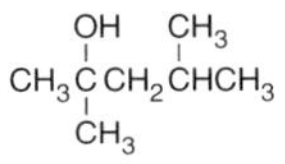 5 gm	**S41,189-2** 3,4-DIMETHYL-3-HEXANOL 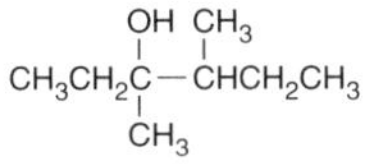 5 gm	**S41,190-6** 3,5-DIMETHYL-3-HEXANOL 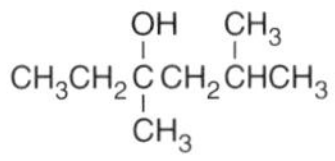 5 gm	**S43,156-7** 2-METHYL-2-HEPTANOL 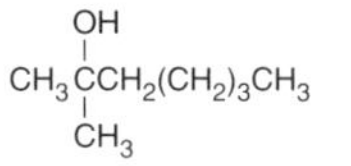 5 gm
S43,162-1 4-METHYL-4-HEPTANOL 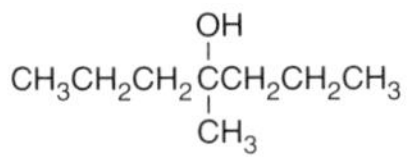 5 gm	**S43,216-4** 2,5-DIMETHYL-2-HEXANOL 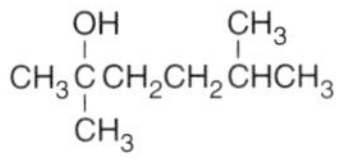 5 gm	**S43,318-7** 2,3-DIMETHYL-2-HEXANOL 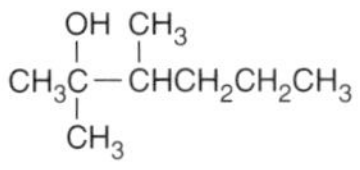 5 gm	**S43,321-7** 3-METHYL-3-HEPTANOL 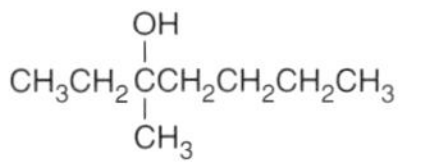 5 gm
S41,203-1 3-ETHYL-3-HEPTANOL 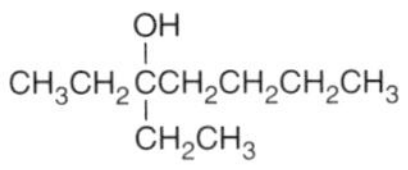 5 gm	**S43,014-5** 4-METHYL-4-OCTANOL 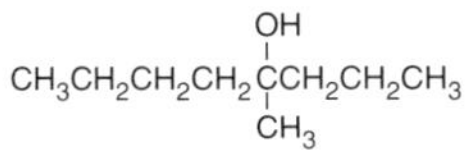 5 gm	**S50,495-5** 3,6-DIMETHYL-3-HEPTANOL 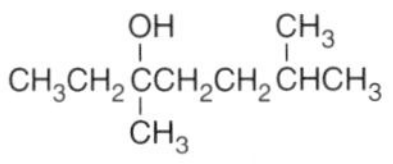 5 gm	**S50,499-8** 2-METHYL-2-OCTANOL 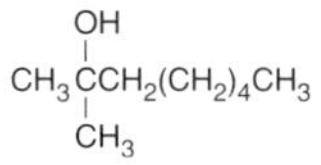 5 gm

S57,735-9
3-METHYL-3-OCTANOL
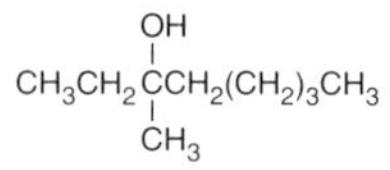
5 gm

S68,052-4
4-METHYL-4-OCTANOL
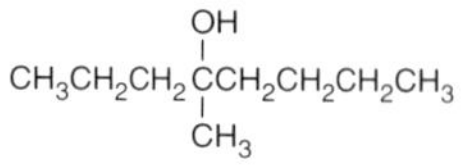
250 mg

S87,763-8
2,6-DIMETHYL-2-HEPTANOL
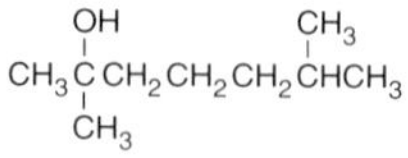
250 mg

S41,745-9
4-METHYL-4-NONANOL
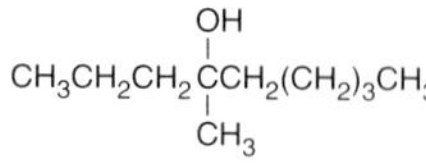
5 gm

S50,478-5
3,4,5-TRIMETHYL-4-HEPTANOL
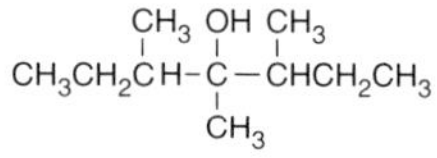
5 gm

S50,502-1
5-METHYL-5-NONANOL
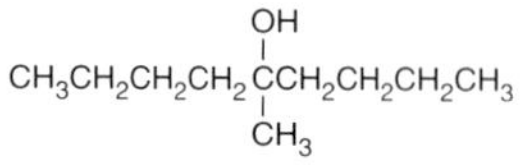
5 gm

S51,932-4
3,4-DIETHYL-3-HEXANOL
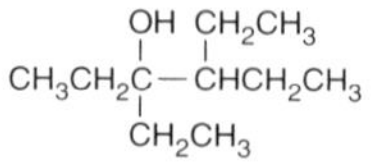
250 mg

S56,304-8
3,6-DIMETHYL-3-OCTANOL
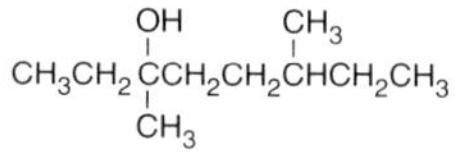
5 gm

S60,312-0
2,3,4,5-TETRAMETHYL-3-HEXANOL
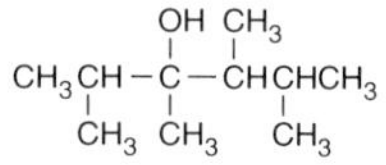
5 gm

S61,244-8
2,4-DIMETHYL-4-NONANOL
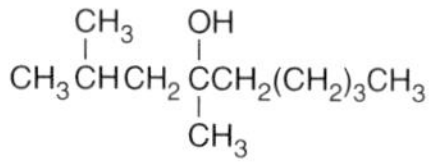
5 gm

S87,712-3
(3RS)(5S)-2,2,3,5-TETRAMETHYL-3-HEPTANOL
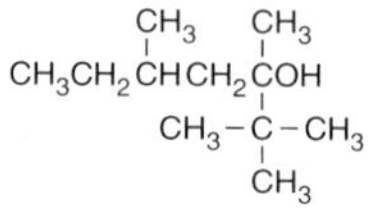
250 mg

S56,813-9
3-ETHYL-3-DODECANOL
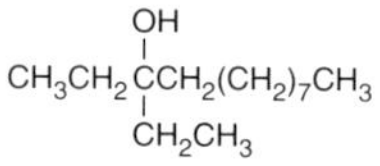
250 mg

S57,817-7
3-METHYL-3-TETRADECANOL
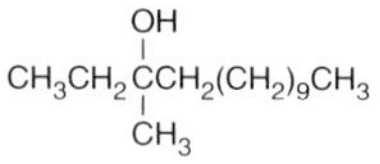
5 gm

S65,639-9
3,7,11-TRIMETHYL-3-DODECANOL
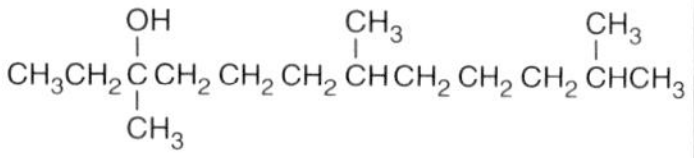
5 gm

S54,612-7
2-METHYL-2-HEPTADECANOL
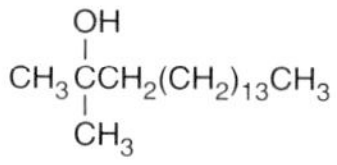
250 mg

S55,635-1
2-METHYL-2-NONADECANOL
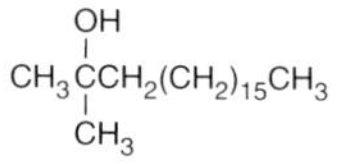
250 mg

S56,249-1
3-ETHYL-3-OCTADECANOL
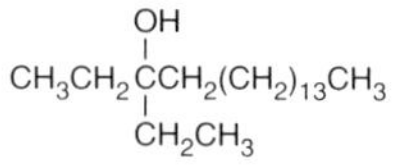
250 mg

S58,849-0
3,7,11,15-TETRAMETHYL-3-HEXADECANOL
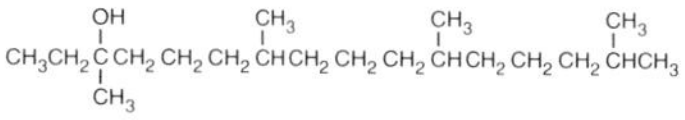
5 gm

S56,250-5
3-ETHYL-3-EICOSANOL
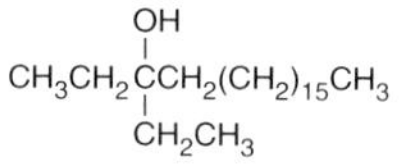
250 mg

S46,087-7
9-OCTYL-9-HEPTADECANOL
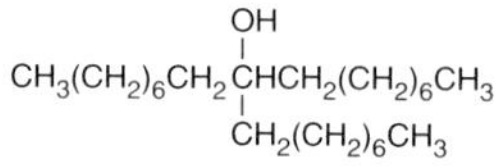
5 gm

S54,670-4
3-ETHYL-3-TETRACOSANOL
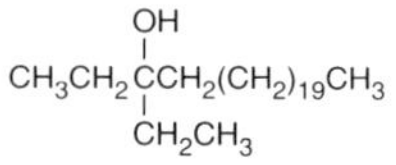
250 mg

S57,803-7
5-BUTYL-5-DOCOSANOL
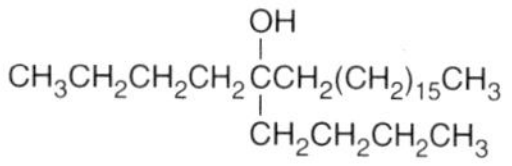
5 gm

S83,901-9
18-HEPTADECYL-18-HEXATRIACONTANOL
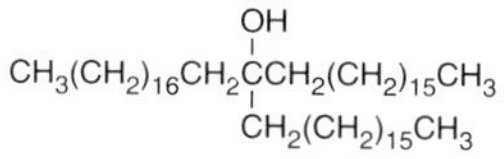
250 mg

S36,847-4
2,2-DIMETHYL-1,3-BUTANEDIOL
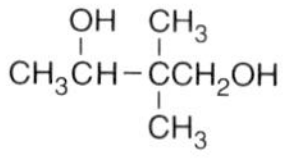
5 gm

S34,308-0
2-SEC-BUTYL-2-METHYL-1,3-PROPANEDIOL
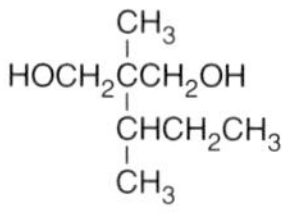
5 gm

S40,505-1
3-METHYL-2,4-HEPTANEDIOL
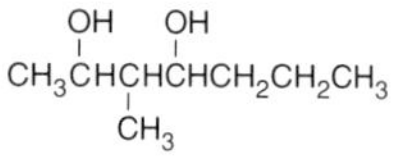
5 gm

S34,932-1
5-ETHYL-3-METHYL-2,4-HEPTANEDIOL
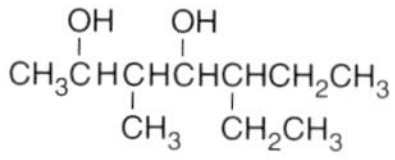
5 gm

S57,331-0
3,6-DIMETHYL-3,6-OCTANEDIOL
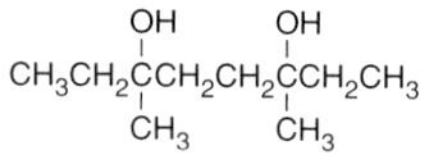
5 gm

S60,374-0
2,7-DIMETHYL-2,7-OCTANEDIOL
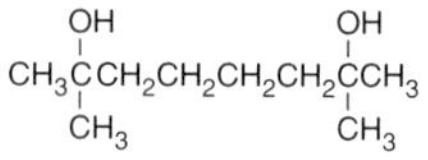
5 gm

S36,890-3
5-ETHYL-3-METHYL-2,4-NONANEDIOL
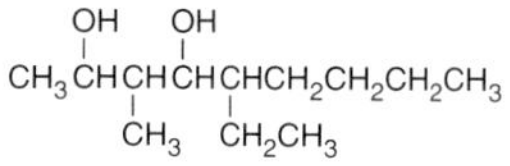
5 gm

S39,343-6
7-ETHYL-2-METHYL-4,6-NONANEDIOL
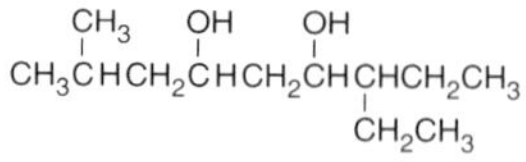
5 gm

S39,346-0
2-BUTYL-1,3-OCTANEDIOL
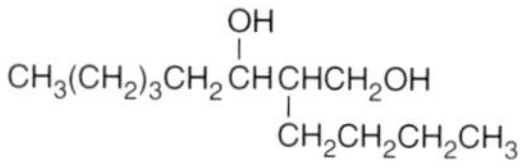
5 gm

S45,654-3
2-METHYL-1,4-BUTANEDIOL
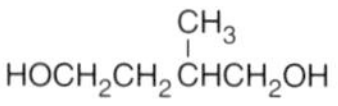
5 gm

S93,737-1
DL-2-METHYL-2,3-BUTANEDIOL
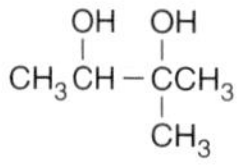
250 mg

S46,039-7
PINACOL HEXAHYDRATE
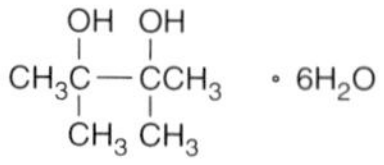
5 gm

S83,832-2
1,4-HEPTANEDIOL
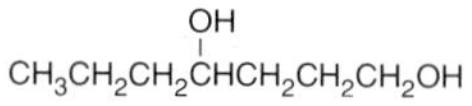
250 mg

S37,852-6 1,1-DIETHYL-1,2-ETHANEDIOL 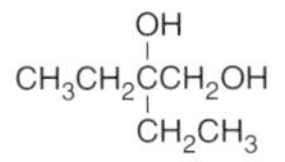 250 mg	**S37,517-9** 2,4,4-TRIMETHYL-2,3-PENTANEDIOL 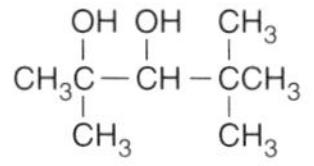 250 mg	**S52,972-9** 1,2-ALKANEDIOLS, C15 TO C18, MIXED 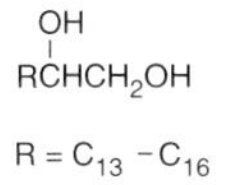 5 gm	**S42,972-4** 3-METHYL-3-BUTEN-2-OL 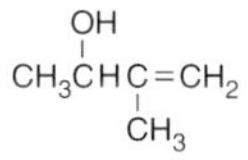 250 mg
S41,833-1 5-HEXEN-3-OL 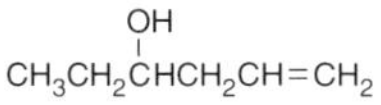 250 mg	**S43,750-6** 2-METHYL-4-PENTEN-2-OL 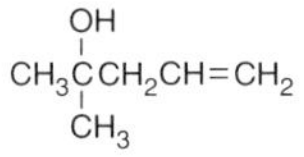 5 gm	**S44,206-2** 4-METHYL-1-PENTEN-3-OL 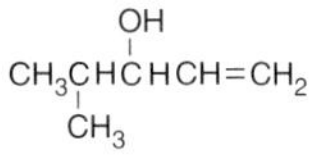 5 gm	**S84,339-3** 5-HEXEN-2-OL 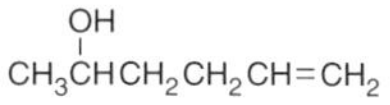 250 mg
S88,870-2 2-METHYL-2-PENTEN-1-OL 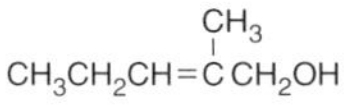 250 mg	**S45,069-3** 3-HEXENE-2,5-DIOL 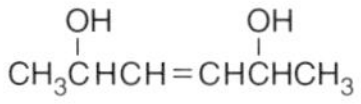 5 gm	**S50,501-3** 1,5-HEPTADIEN-4-OL 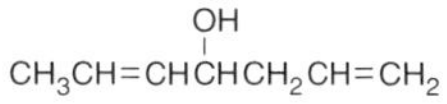 5 gm	**S88,116-3** 2-METHYL-1,5-HEXADIEN-3-OL 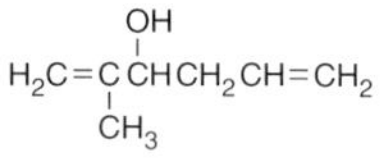 250 mg
S35,038-9 1-HEPTEN-4-OL 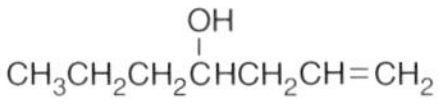 5 gm	**S41,195-7** 3-METHYL-5-HEXEN-3-OL 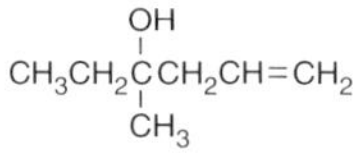 5 gm	**S41,738-6** 2-HEPTEN-4-OL 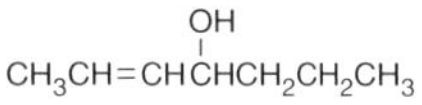 5 gm	**S43,336-5** 2,3-DIMETHYL-4-PENTEN-2-OL 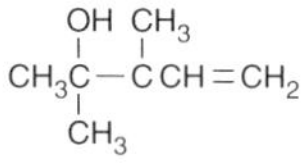 5 gm
S43,917-7 1-HEPTEN-3-OL 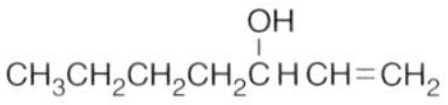 5 gm	**S50,482-3** 3-METHYL-1-HEXEN-3-OL 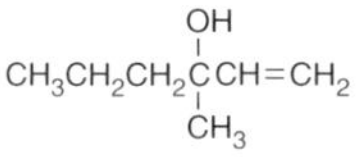 5 gm	**S41,739-4** 4-METHYL-1,6-HEPTADIEN-4-OL 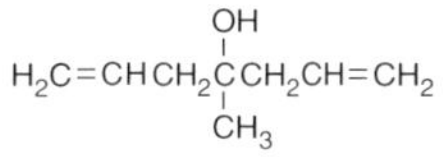 5 gm	**S41,779-3** 2,5-DIMETHYL-1,5-HEXADIEN-3-OL 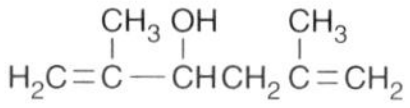 5 gm
S50,494-7 2-METHYL-1,5-HEPTADIEN-4-OL 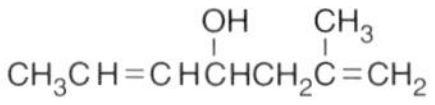 5 gm	**S88,195-3** 2,5-DIMETHYL-3,5-HEXADIEN-2-OL 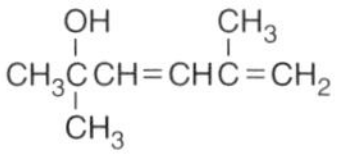 250 mg	**S88,534-7** 2-METHYL-1,6-HEPTADIEN-3-OL 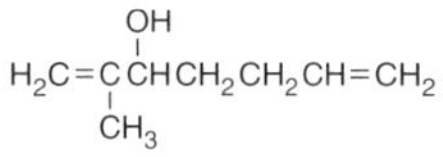 250 mg	**S36,429-0** 2-OCTEN-4-OL 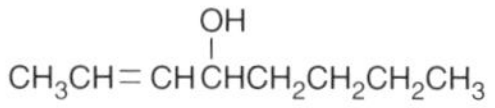 5 gm
S41,737-8 3-ETHYL-5-HEXEN-3-OL 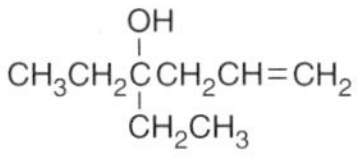 5 gm	**S41,740-8** 2-METHYL-1-HEPTEN-3-OL 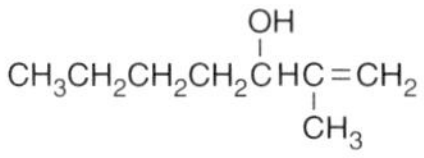 5 gm	**S41,741-6** 4-METHYL-1-HEPTEN-4-OL 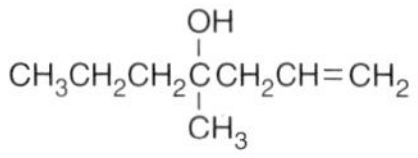 5 gm	**S43,319-5** 3,5-DIMETHYL-5-HEXEN-3-OL 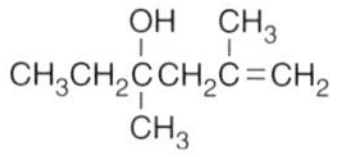 5 gm
S43,410-8 2-ETHYL-2-HEXEN-1-OL, 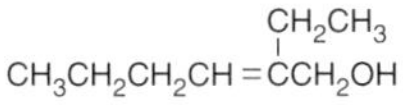 5 gm	**S44,833-8** 4,4-DIMETHYL-2-METHYLENE-1-PENTANOL 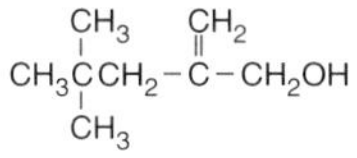 5 gm	**S50,491-2** 2,2-DIMETHYL-5-HEXEN-3-OL 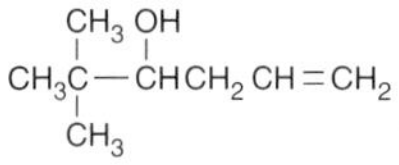 5 gm	**S50,504-8** 5-METHYL-1-HEPTEN-4-OL 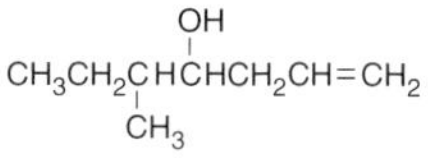 5 gm
S50,479-3 2,6-DIMETHYL-1,6-HEPTADIEN-4-OL 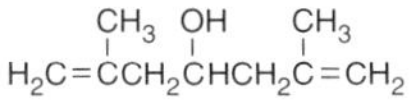 5 gm	**S41,747-5** 1-NONEN-4-OL 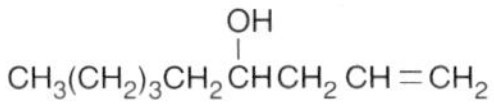 5 gm	**S43,575-9** 1-NONEN-3-OL 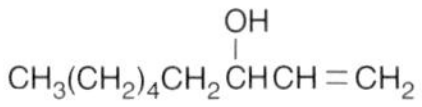 250 mg	**S40,813-1** 3,7-DIMETHYL-2,6-OCTADIEN-1-OL, MIXTURE OF CIS AND TRANS 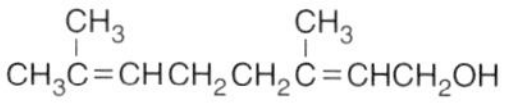 5 gm

S50,496-3
2,4,6-TRIMETHYL-1,6-HEPTADIEN-4-OL
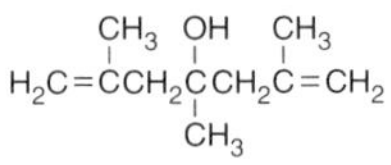
5 gm

S35,159-8
2-ISOPENTYL-3-METHYL-2-BUTEN-1-OL
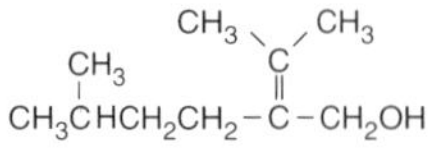
5 gm

S37,208-0
RICINOLEYL ALCOHOL, TECH.
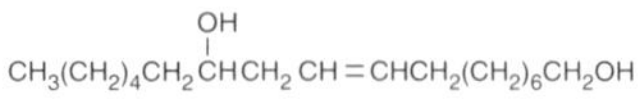
5 gm

S58,703-6
1,1-DIALLYL-1-DOCOSANOL
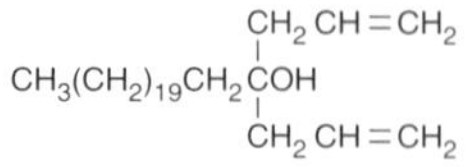
5 gm

S88,453-7
2-(2-METHYL-6-METHYLENECYCLOHEXYL)-1-ETHANOL
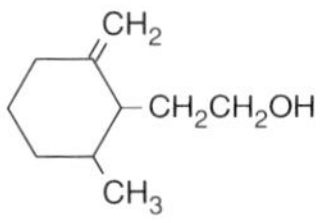
250 mg

S41,194-9
1-METHYLCYCLOHEPTANOL
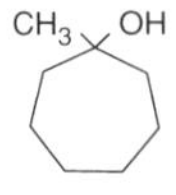
5 gm

S42,982-1
1-METHYLCYCLOOCTANOL
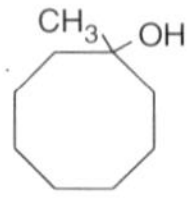
250 mg

S93,198-5
TRANS-5-CYCLODECENOL
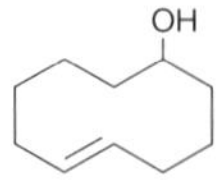
250 mg

S41,379-8
CYCLODECANOL
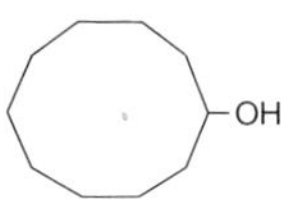
250 mg

S41,656-8
4-CIS-8-TRANS-CYCLODODECADIENE-1,2-TRANS-DIOL
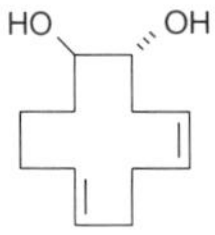
5 gm

S57,812-6
1-ETHYL-1-CYCLODODECANOL
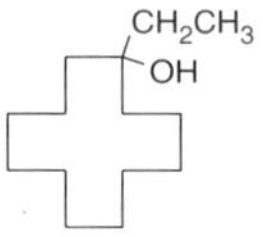
250 mg

S75,746-2
1,1'-BI-1-CYCLOPENTANOL
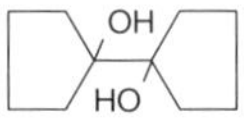
5 gm

S88,683-1
2,3-DECALINDIOL
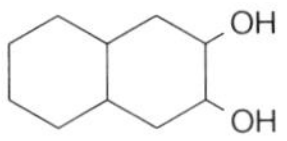
250 mg

S94,203-0
4A,9A-DIHYDROXYPERHYDRODIBENZO(B,E)-(1,4)DIOXIN
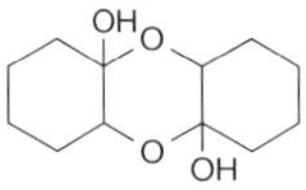
250 mg

S86,374-2
3,5-DIHYDROXYTRICYCLO[2.2.1.0(2,6)]-HEPTANE
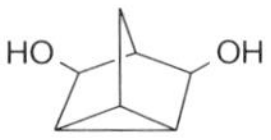
250 mg

S74,822-6
2,7-NORBORNANEDIOL
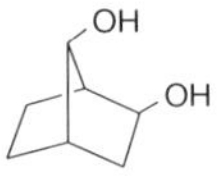
250 mg

S74,812-9
CIS-5,5-DIMETHYL-1,3-CYCLOHEXANEDIOL
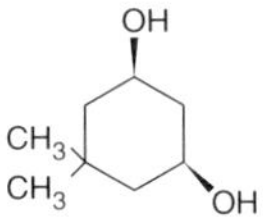
250 mg

S94,686-9
1,3-ADAMANTANEDIOL
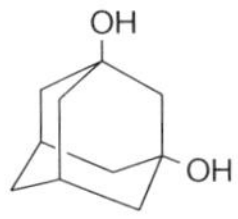
250 mg

S94,932-9
CIS-3,3-DICHLORO-1,2-CYCLOPROPANEDIMETHANOL
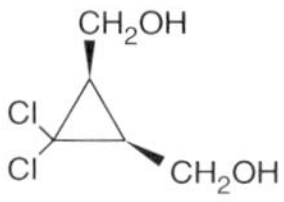
250 mg

S45,481-8
4-METHYL-1,2-CYCLOHEXANEDIMETHANOL
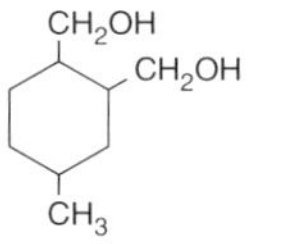
5 gm

S87,656-9
CIS-4-CYCLOHEXENE-1,2-DIMETHANOL
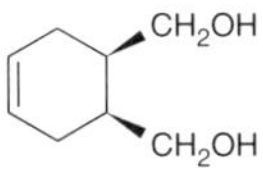
250 mg

S77,114-7
BICYCLO[2.2.2]OCTANE-1,4-DIOL
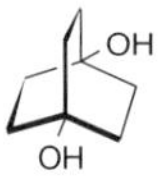
250 mg

S68,031-1
CYCLOPENTYLCYCLOHEXANE-1,1'-DIOL
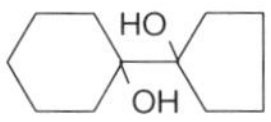
250 mg

S75,779-9
1,1-CYCLOHEXANEDIETHANOL
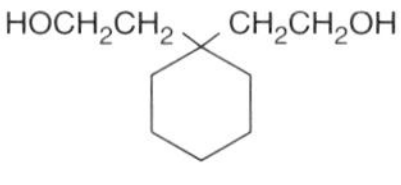
5 gm

S68,035-4
(1,1'-BICYCLOHEXYL)-1,1'-DIOL
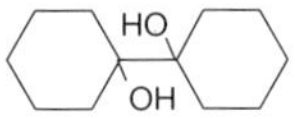
250 mg

S46,030-3
1,3-CYCLOHEXANEDIMETHANOL
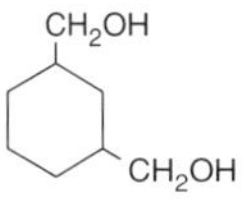
250 mg

S88,371-9
1,2,3,4,6,7,8,8A-OCTAHYDRO-1,6-NAPHTHALENEDIOL
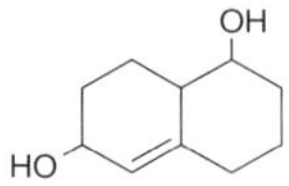
250 mg

S88,527-4
8A-METHYL-1,2,3,4,6,7,8,8A-OCTAHYDRO-1,6-NAPHTHALENEDIOL
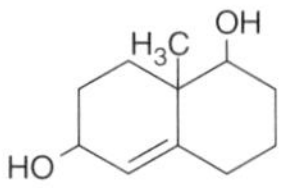
250 mg

S88,556-8
4A,5,5,8A-TETRAMETHYL-1,6-DECALINDIOL
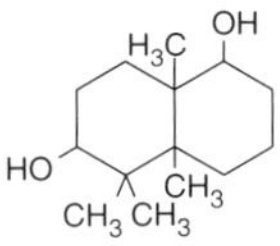
250 mg

S94,819-5
1,3-ADAMANTANEDIMETHANOL
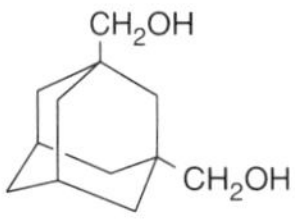
250 mg

S43,095-1
5-CIS-9-TRANS-CYCLODODECADIENE-1,2-CIS-DIOL
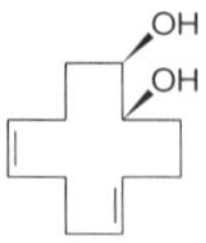
5 gm

S42,710-1
CIS-1,2-CYCLODODECANEDIOL
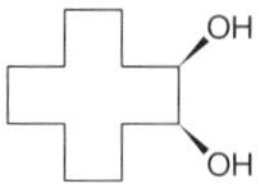
5 gm

S35,281-0
3-METHYL-1,2-CYCLOPENTANEDIOL (MIXTURE OF ISOMERS)
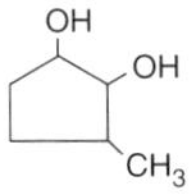
5 gm

S43,405-1
TRICYCLO[5.2.1.0(2,6)]DECANE-3,4-DIOL
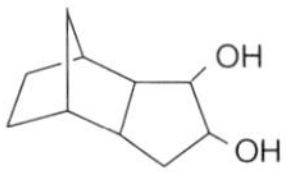
5 gm

S35,279-9
1-METHYL-TRANS,-1,2-CYCLOHEXANEDIOL
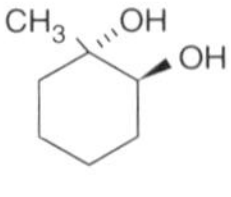
1 gm

S45,556-3
2-NORBORNANEETHANOL
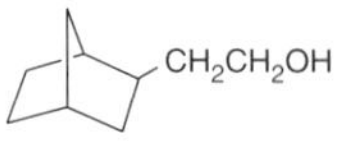
5 gm

S67,117-7
2-CYCLOPROPYL-2-PROPANOL
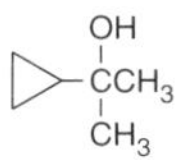

5 gm

S43,528-7
DICYCLOPROPYL CARBINOL
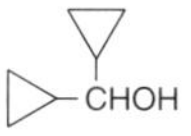

5 gm

S84,376-8
CIS-BICYCLO[3.1.0]HEXAN-2-OL BRUCINE SALT
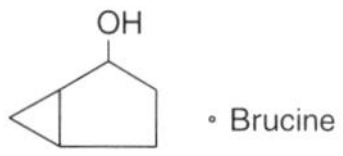

250 mg

S41,713-0
CYCLOHEXYL DICYCLOPROPYL CARBINOL
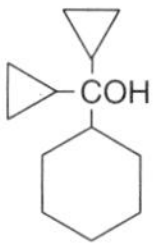

5 gm

S45,550-4
CIS-MYRTANOL
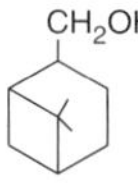

5 gm

S41,735-1
1-ETHYLCYCLOPENTANOL
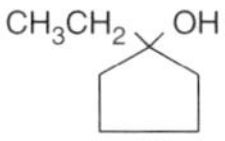

5 gm

S41,922-2
2-CYCLOPENTYLETHANOL
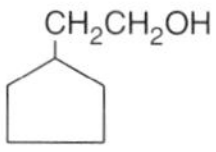

250 mg

S45,660-8
1-CYCLOPENTYLETHANOL
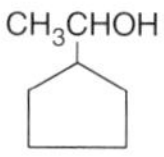

250 mg

S91,489-4
2-ALLYL-1-CYCLOPENTANOL
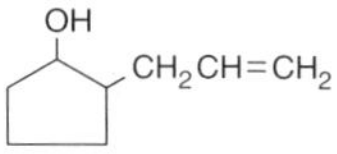

250 mg

S41,699-1
1-PROPYL-1-CYCLOPENTANOL
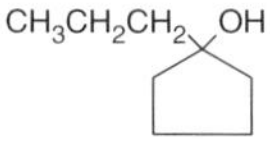

5 gm

S41,187-6
1-BUTYL-1-CYCLOPENTANOL
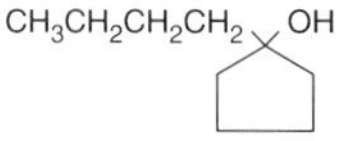

5 gm

S41,245-7
1-AMYLCYCLOPENTANOL
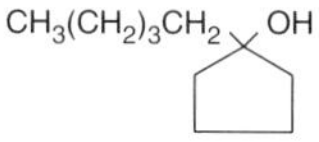

5 gm

S79,307-8
3-TERT-BUTYL-2-METHYLCYCLOPENTANOL
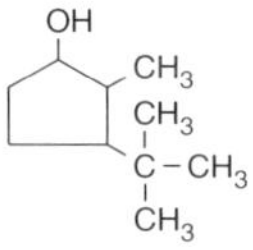

250 mg

S35,132-6
TRICYCLO[5.2.1.0(2,6)]-4-DECENE-8-OL

5 gm

S88,461-8
1,3A,4,5,6,6A-HEXAHYDRO-1-PENTALENOL
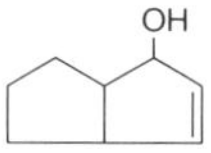

250 mg

S36,260-3
METHYLTRICYCLO[5.2.1.0(2,6)]-DECANE-4-METHANOL
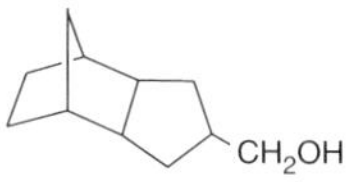

5 gm

S36,249-2
4-HYDROXYMETHYLTRICYCLO-[5.2.1.0(2,6)]DECANE-8-OL
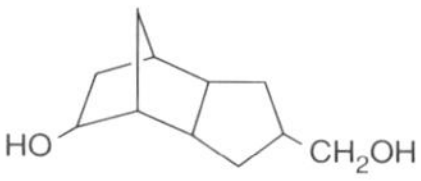

5 gm

S35,050-8
HEXAHYDRO-5-INDANOL,TECH.
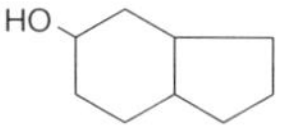

5 gm

S43,009-9
CEDROL
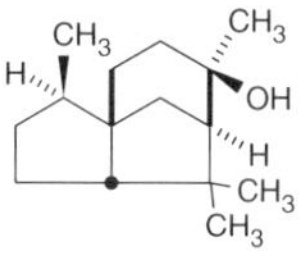

5 gm

S46,280-2
DL-PATCHENOL
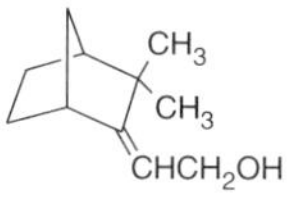

250 mg

S44,861-3
TRI-2-NORBORNYLMETHANOL
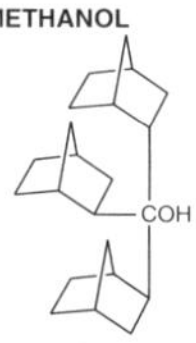

5 gm

S90,639-5
DECAHYDRO-4-AZULENOL
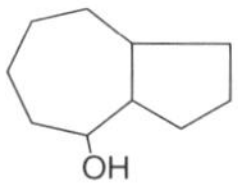

250 mg

S95,776-3
TRICYCLO[5.2.1.0(2,6)]DECA-3,8-DIEN-10-OL
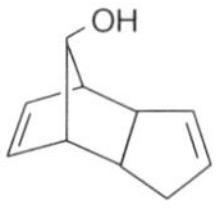

50 mg

S89,619-5
2-FLUORO-1-CYCLOHEXANOL
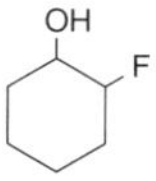

250 mg

S40,846-8
1-ETHYLCYCLOHEXANOL
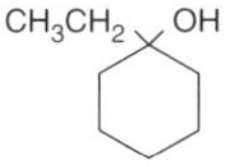

5 gm

S43,682-8
2,4-DIMETHYLCYCLOHEXANOL (MIXTURE OF ISOMERS)
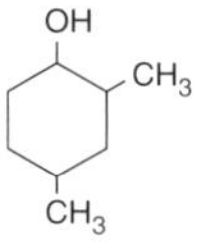

5 gm

S53,486-2
4-VINYL-1,2-CYCLOHEXANEDIOL
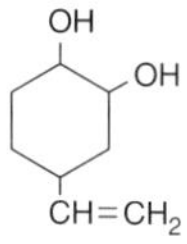

5 gm

S42,910-4
1-ALLYL-1-CYCLOHEXANOL
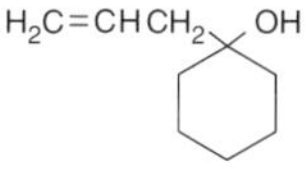

5 gm

S37,411-3
2,6-DIMETHYLCYCLOHEXANEMETHANOL
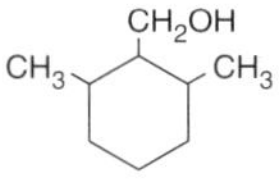

5 gm

S41,900-1
3,3,5-TRIMETHYLCYCLOHEXANOL
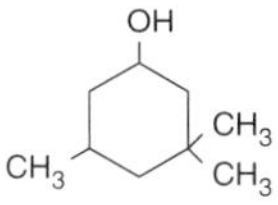

5 gm

S34,601-2
1,2-DIETHYL-1-CYCLOHEXANOL, CIS AND TRANS
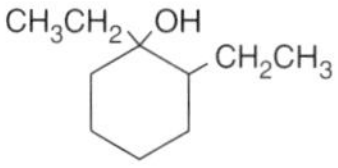

5 gm

S40,695-3
1-BUTYLCYCLOHEXANOL
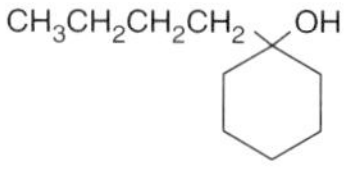

5 gm

S44,700-5
1-CYCLOHEXYL-1-BUTANOL
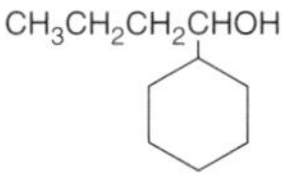

5 gm

S87,659-3
1-CYCLOHEXYL-2-METHYLPROPANOL
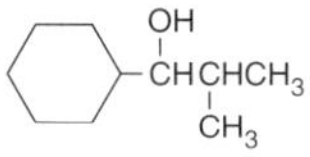

250 mg

S37,370-2
3,3-PENTAMETHYLENE-1,5-PENTANEDIOL
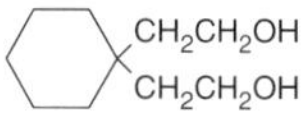

250 mg

S50,055-0
2-HYDROXY-1,1,3,3-CYCLOHEXANETETRAMETHANOL
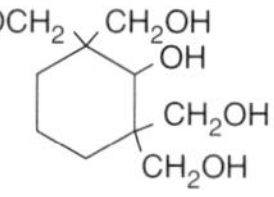

5 gm

S35,925-4
2-TERT-BUTYL-5-METHYLCYCLOHEXANOL

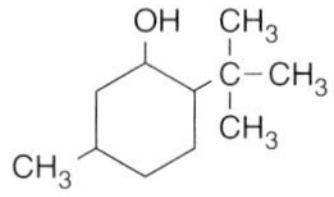

5 gm

S87,663-1
1-CYCLOHEXYL-2,2-DIMETHYL-1-PROPANOL

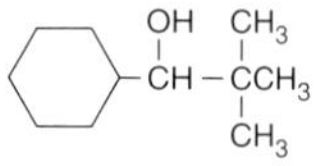

250 mg

S87,666-6
1-CYCLOHEXYL-1-PENTANOL

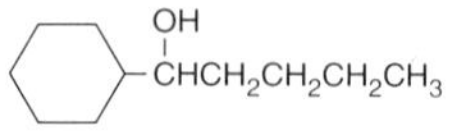

250 mg

S87,667-4
1-CYCLOHEXYL-2,2-DIMETHYL-1-PROPANOL

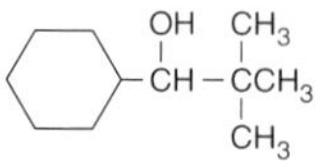

250 mg

S87,713-1
1-CYCLOHEXYL-3-METHYL-1-BUTANOL

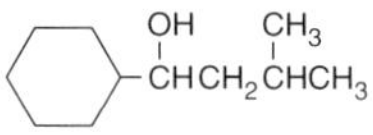

250 mg

S45,783-3
1-CYCLOHEXYLCYCLOHEXANOL

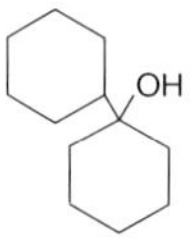

5 gm

S34,917-8
2-(2-ETHYLHEXYL)-1-CYCLOHEXANOL

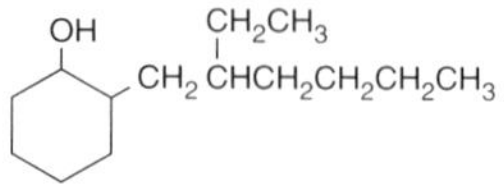

5 gm

S39,345-2
2-HYDROXY-6-PROPYL-1,3,5-TRIETHYLCYCLOHEXANEMETHANOL

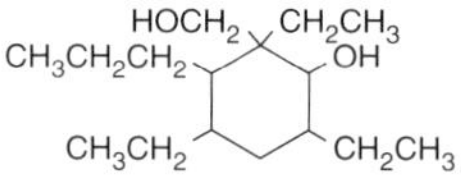

5 gm

S95,634-1
(-)-1,2,3,4,4A,5,8,8A-OCTAHYDRO-1,1,4A-TRIMETHYL-2-NAPHTHOL

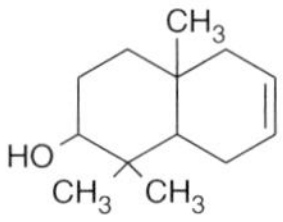

50 mg

S36,241-7
2-(1-HYDROXYETHYL)CYCLOHEXANOL

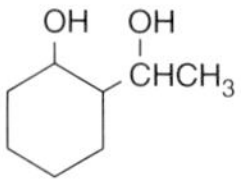

5 gm

S36,576-9
2-PROPYL-1-CYCLOHEXANOL, MIXTURE OF CIS AND TRANS

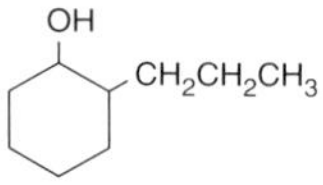

5 gm

S35,912-2
2-BUTYLCYCLOHEXANOL (MIXTURE OF ISOMERS)

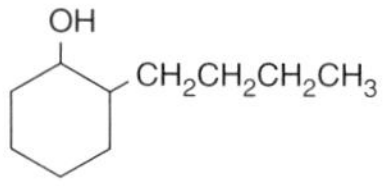

5 gm

S35,068-0
2-(1-HYDROXYBUTYL)CYCLOHEXANOL

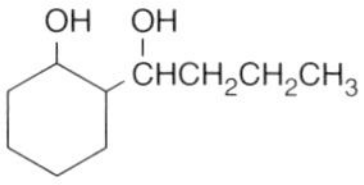

5 gm

S50,774-1
2-CYCLOHEXYLCYCLOHEXANOL

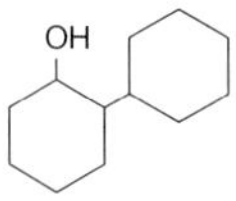

5 gm

S36,153-4
2-(2-ETHYLBUTYL)-1-CYCLOHEXANOL

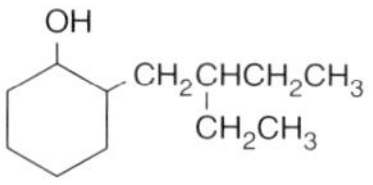

5 gm

S56,072-3
ALPHA-(1-ETHYLPROPYL)-2-HYDROXYCYCLOHEXANEMETHANOL

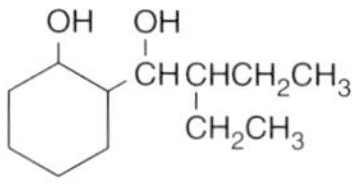

5 gm

S44,131-7
1,3-CYCLOHEXANEDIOL (80% AQU. SOLN.)

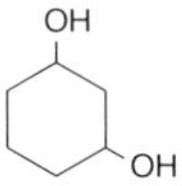

5 gm

S50,493-9
CIS-4-ETHYLCYCLOHEXANOL

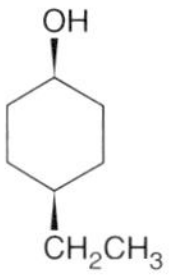

5 gm

S41,409-3
CIS-1,4-CYCLOHEXANEDIMETHANOL,

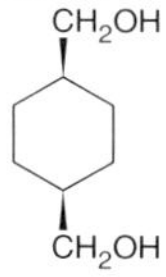

5 gm

S79,971-8
4-PROPYL-1-CYCLOHEXANOL

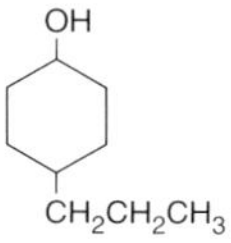

250 mg

S85,920-6
TRANS-4-BUTYLCYCLOHEXANOL

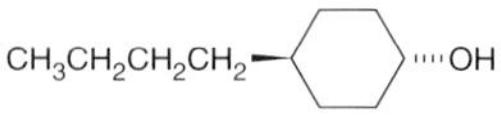

250 mg

S82,970-6
4-CYCLOHEXYLCYCLOHEXANOL

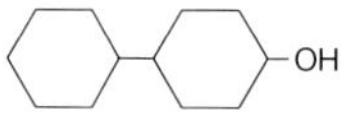

250 mg

S50,266-9
ALPHA,ALPHA-DIMETHYL-4-ISOPROPYLCYCLOHEXANEMETHANOL

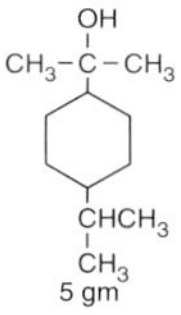

5 gm

S59,108-4
4,4'-ISOPROPYLIDENEDICYCLOHEXANOL

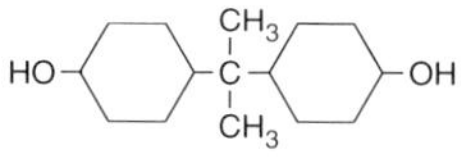

5 gm

S64,348-3
4-TERT-AMYLCYCLOHEXANOL

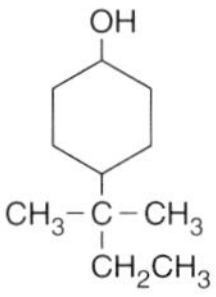

5 gm

S95,363-6
TRANS-4-TERT-BUTYLCYCLOHEXANOL

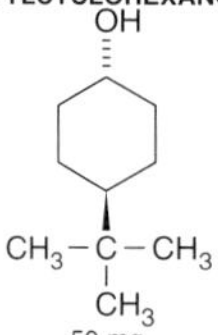

50 mg

S46,311-6
9-CYCLOHEXYLBICYCLO[3.3.1]-NONAN-9-OL

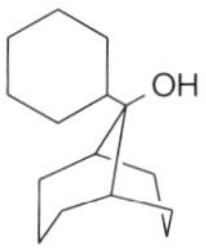

250 mg

S37,532-2
1-METHYL-3-CYCLOHEXENE-1-METHANOL

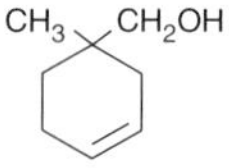

5 gm

S86,487-0
2-(2-CYCLOHEXEN-1-YL)-2-PROPANOL

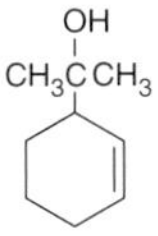

250 mg

S86,488-9
1-(2-CYCLOHEXEN-1-YL)-1-PROPANOL

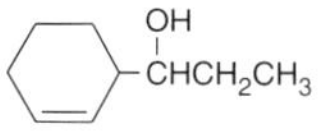

250 mg

S37,520-9
6-METHYL-3-CYCLOHEXENE-1,1-DIMETHANOL

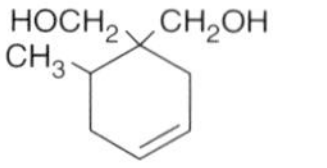

5 gm

S46,095-8
(+)-LIMONEN-10-OL

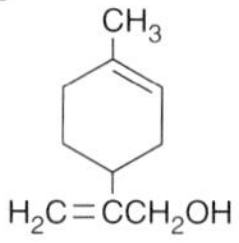

250 mg

S86,349-1
5-ISOPROPYL-2-METHYL-2-CYCLOHEXEN-1-OL

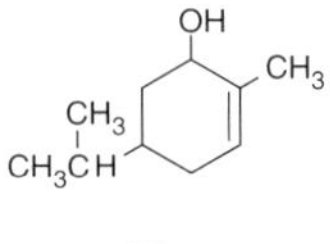

250 mg

S65,487-6
HOMOLIMONENOL

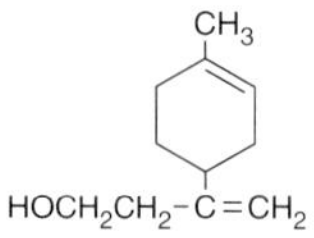

5 gm

S37,580-2
2,2'-DIMETHYL-1-HYDROXYMETHYL-1,1',2,2',3,3',6,6'-OCTAHYDROBENZHYDROL

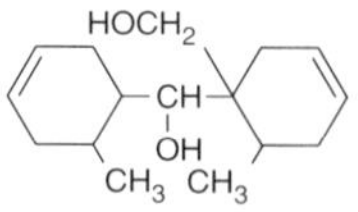

5 gm

S36,932-2
1,1'-BIS(3,5,5-TRIMETHYL-2-CYCLOHEXEN-1-OL)

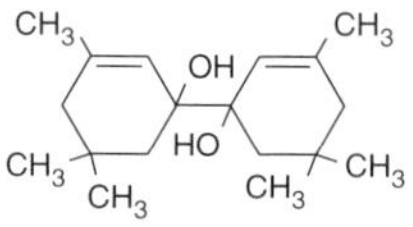

5 gm

S43,838-3
5-NORBORNENE-2,2-DIMETHANOL

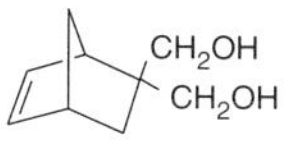

5 gm

S88,363-8
7-ISOPROPYL-2,3,4,4A,5,6,7,8-OCTAHYDRO-2-NAPHTHOL

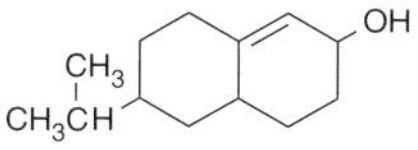

250 mg

S34,580-6
DICYCLOHEXYL CARBINOL

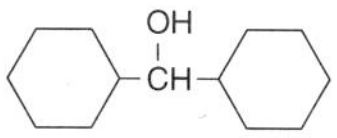

5 gm

S76,451-5
2-DECALINOL

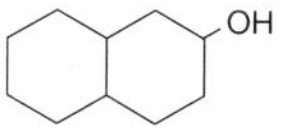

5 gm

S92,677-9
1-DECALINOL

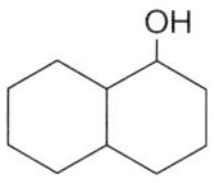

250 mg

S55,604-1
7-ALPHA-HYDROXYMANOOL

250 mg

S59,479-2
PERHYDRO-9B-PHENALENOL (MIXTURE OF ISOMERS)

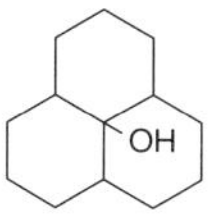

5 gm

S92,642-6
2-METHYL-2-NORBORNANOL

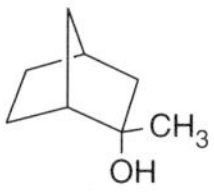

250 mg

S59,610-8
3-METHYL-1-ADAMANTANOL

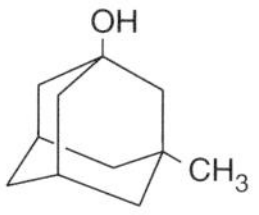

5 gm

S59,607-8
5-METHYL-1,3-ADAMANTANEDIOL

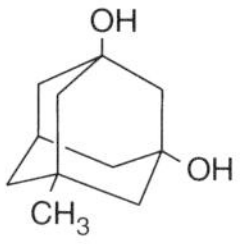

250 mg

S52,744-0
5,7-DIMETHYL-1,3-ADAMANTANEDIOL

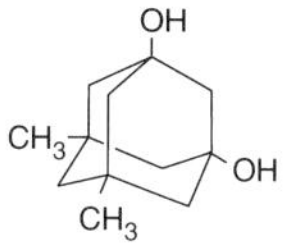

5 gm

S46,021-4
3,5,7-TRIMETHYL-1-ADAMANTANOL

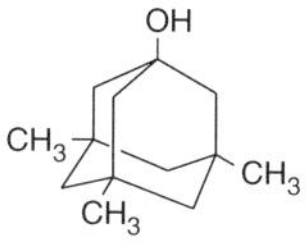

250 mg

S74,044-6
1,3-DIMETHYL-4,5,6,7-TETRAHYDRO-2-INDANOL

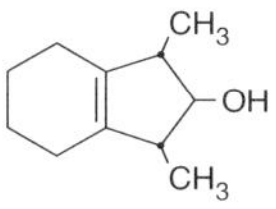

250 mg

S53,125-1
TRICYCLO[5.2.1.0(2,6)]DECAN-8-OL

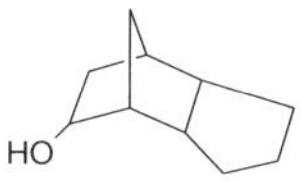

5 gm

S95,784-4
1,2,3,4,4A,5,8,8A-OCTAHYDRO-1,4:5,8-DIMETHANO-2-NAPHTHOL

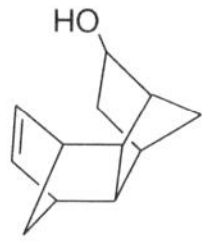

50 mg

S66,124-4
1-CHLORO-2-METHYL-2-PROPANOL

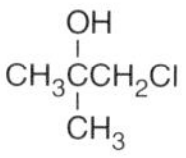

5 gm

S55,119-8
2,2,2-TRIS-(CHLOROMETHYL)-ETHANOL

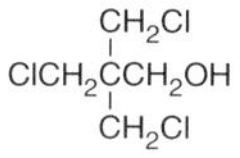

5 gm

S57,831-2
3-BROMO-2,2-BIS(BROMOMETHYL)-1-PROPANOL

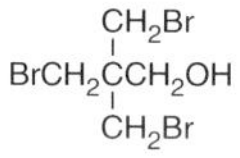

250 mg

S88,155-4
12-BROMO-2-DODECEN-1-OL

$BrCH_2(CH_2)_7CH_2CH{=}CHCH_2OH$

250 mg

S81,880-1
2,2-DIFLUOROETHANOL

F_2CHCH_2OH

250 mg

S36,492-4
3,3,4,4-TETRAFLUORO-2-BUTANOL,

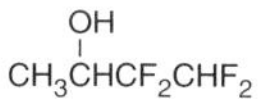

5 gm

S46,285-3
2-METHYL-3,3,4,4-TETRAFLUORO-2-BUTANOL

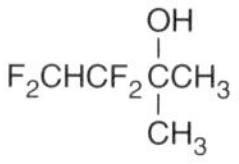

250 mg

S41,282-1
2-METHYL-3,3,4,4,5,5,6,6-OCTAFLUORO-2-HEXANOL,

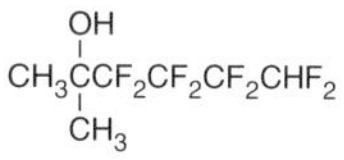

5 gm

S37,389-3
3,3,4,4,5,5,6,6,7,7,8,8-DODECAFLUORO-2-OCTANOL

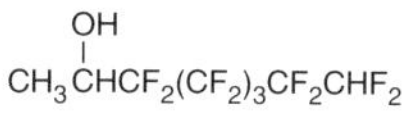

5 gm

S43,971-1
1,1,1-TRIFLUORO-2-PROPANOL

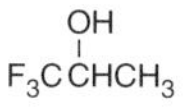

250 mg

S79,366-3
1,1,1-TRICHLORO-3,3,3-TRIFLUORO-2-PROPANOL

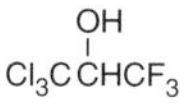

250 mg

S89,403-6
2,2,3,3,3-PENTACHLORO-1-PROPANOL

$Cl_3CCl_2C\,CH_2OH$

250 mg

S91,418-5
1,1,1,3,3,3-HEXACHLORO-2-PROPANOL

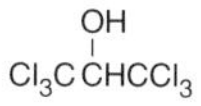

250 mg

S60,023-7
2-METHYL-1,1,1-TRIFLUORO-2-PROPANOL

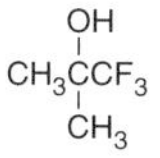

250 mg

S89,404-4
1,1,1,3,4,4-HEXACHLORO-3-BUTEN-2-OL

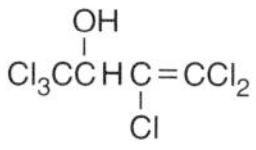

250 mg

S56,607-1
2-METHYL-1,1,1-TRIBROMO-2-PROPANOL

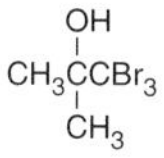

5 gm

S59,316-8
1-CHLORO-1,1-DIFLUORO-2-PENTANOL

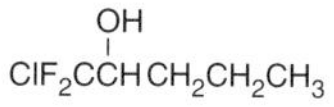

250 mg

S60,293-0
1-CHLORO-1,1-DIFLUORO-2-PENTANOL

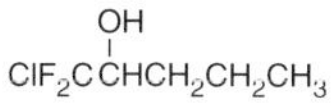

250 mg

S41,786-6
1-(TRICHLOROMETHYL)CYCLOHEXANOL

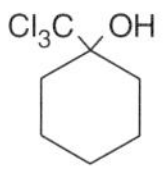

5 gm

S88,097-3
3-BROMO-2-PROPEN-1-OL

$BrCH{=}CHCH_2OH$

250 mg

S63,763-7
2-BROMO-2-CHLORO-1-PROPANOL

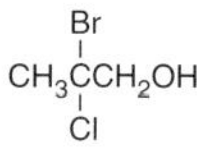

5 gm

S87,145-1
3-CHLORO-2-BUTANOL
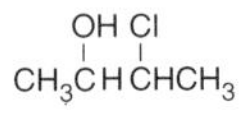
250 mg

S87,753-0
3-BROMO-3-BUTEN-1-OL
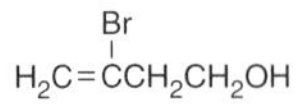
250 mg

S88,129-5
3-BROMO-2-METHYL-2-PROPEN-1-OL
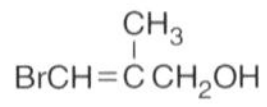
250 mg

S51,857-3
2,2,3-TRICHLORO-1,1-BUTANEDIOL
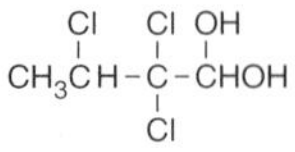
5 gm

S89,414-1
2,3-DIBROMO-5,5-DICHLORO-4-METHYL-2,4-PENTADIEN-1-OL
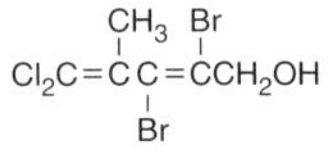
250 mg

S68,571-2
2,3,5,6-TETRABROMO-1,7-HEPTANEDIOL
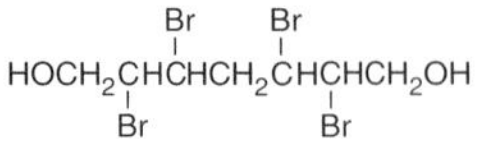
5 gm

S86,108-1
2,2-DICHLORO-1,3-PROPANEDIOL
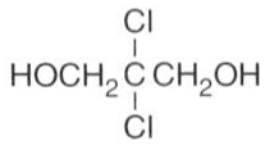
250 mg

S42,989-9
1,4,5,6,7,7-HEXACHLORO-5-NORBORNENE-2,3-DIMETHANOL
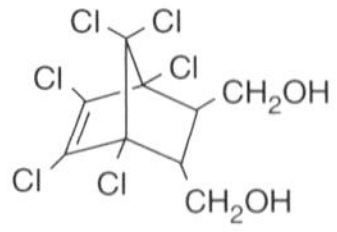
5 gm

S57,392-2
2-CHLORO-7,7-DIMETHYL-1-NORBORNANEETHANOL
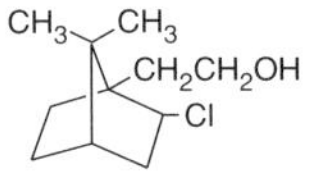
250 mg

S87,591-0
5,6,7,8,9,9-HEXACHLORO-1,2,3,4,4A,5,8,8A-OCTAHYDRO-1,4:5,8-DIMETHANONAPHTHALEN-2-OL
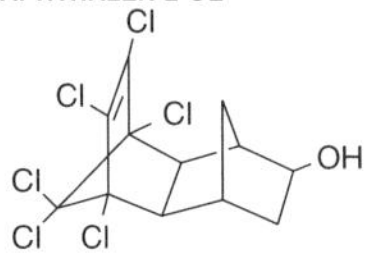
250 mg

S94,832-2
3-BROMO-1-ADAMANTANEETHANOL
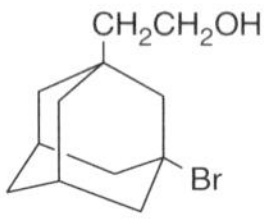
250 mg

S41,024-1
D(+)-TREHOLOSE
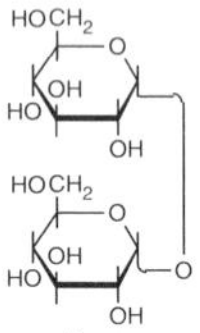
5 gm

S55,100-7
3-(HYDROXYMETHYL)-2,4-PENTANEDIOL
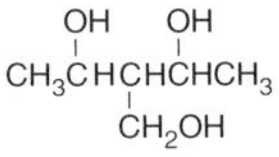
5 gm

S93,930-7
1-O-METHYL-BETA-D-XYLOSIDE
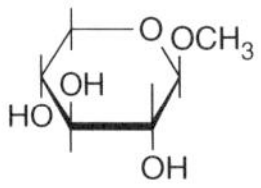
250 mg

S78,311-0
METHYL-ALPHA-D-GALACTOPYRANOSIDE HYDRATE
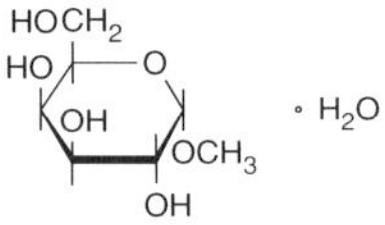
5 gm

S67,488-5
9,10-DIHYDROXYOCTADECANOL
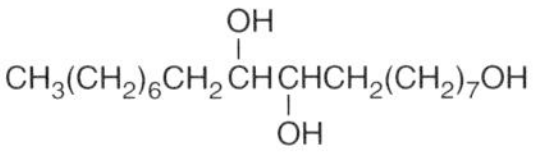
250 mg

S93,902-1
L-MANNITOL
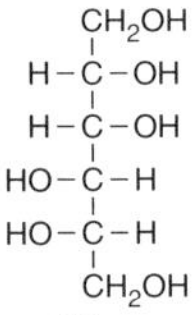
250 mg

S91,745-1
1,2:3,4-DI-O-ISOPROPYLIDENE-D-MANNITOL
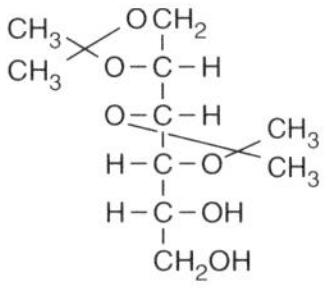
250 mg

S77,916-4
4-CYCLOPENTENE-1,3-DIOL
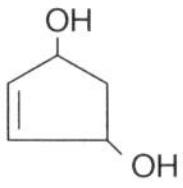
250 mg

S56,691-8
1,2,3,6-HEXANETETROL
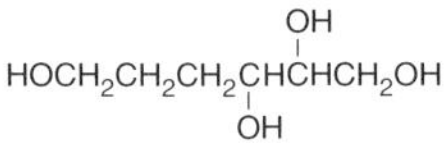
5 gm

S43,296-2
2,5-DIMETHYL-1,2,6-HEXANETRIOL
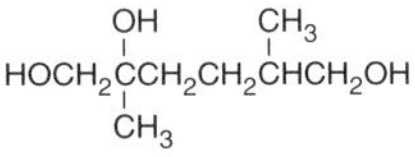
5 gm

S64,122-7
D-ALPHA,ALPHA-GLUCOOCTITOL
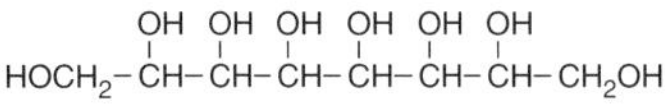
250 mg

S36,261-1
5-HYDROXYMETHYL-1,5,6,7,11-UNDECANEPENTOL
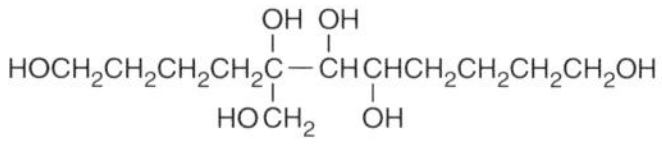
5 gm

S35,804-5
DL-ARABINOSE
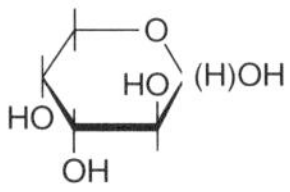
5 gm

S64,175-8
LEVO-RHAMNITOL
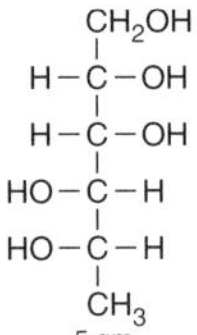
5 gm

S64,146-4
ALPHA-GLUCOHEPTITOL
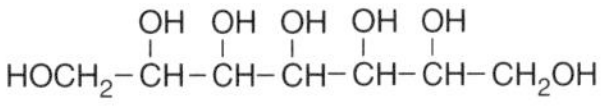
250 mg

S50,930-2
2,3:4,5-DIISOPROPYLIDENE-BETA-D-FRUCTOPYRANOSIDE
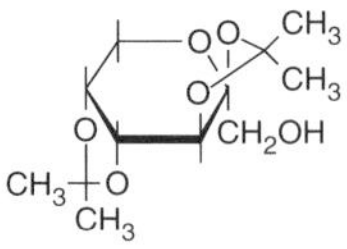
5 gm

S91,742-7
1,2-O-ISOPROPYLIDENE-3,5,6-TRI-O-METHYL-ALPHA-D-GLUCOFURANOSIDE
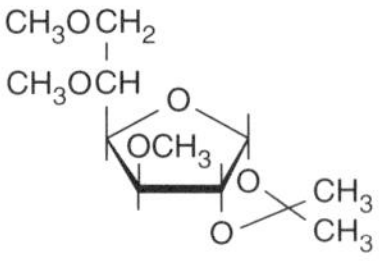
250 mg

S93,912-9
1,3:4,6-DI-O-METHYLENE-D-MANNITOL
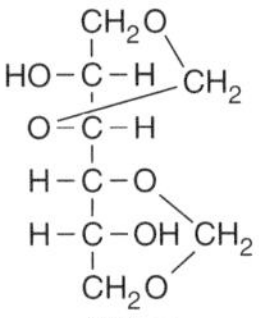
250 mg

S36,527-0
1,3:2,5:4,6-TRI-O-METHYLENE-D-MANNITOL
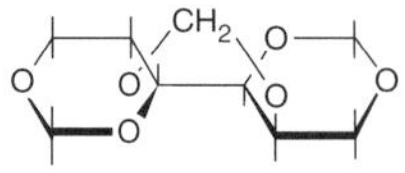
5 gm

S55,214-3
2-CHLORO-1,1,2-TRIFLUOROETHYL ETHYL ETHER

F F
$ClCHCOCH_2CH_3$
F

250 mg

S80,958-6
2-IODOETHYL ETHER

$ICH_2CH_2OCH_2CH_2I$

250 mg

S41,826-9
4-CHLOROBUTYL ETHER

$ClCH_2(CH_2)_2CH_2OCH_2(CH_2)_2CH_2Cl$

5 gm

S58,838-5
CHLOROMETHYL DECYL ETHER

$CH_3(CH_2)_8CH_2OCH_2Cl$

250 mg

S70,641-8
1-BROMO-11-(1-ETHOXYETHOXY)UNDECANE

CH_3
$CH_3CH_2OCHOCH_2(CH_2)_9CH_2Br$

5 gm

S51,178-1
PENTACHLOROETHYL 1,1,2,2-TETRACHLOROETHYL ETHER

$CCl_3CCl_2OCCl_2CHCl_2$

5 gm

S51,982-0
2-BROMOPROPYL ETHYL ETHER

Br
$CH_3CHCH_2OCH_2CH_3$

250 mg

S34,440-0
1,8-DIMETHOXYOCTANE

$CH_3OCH_2(CH_2)_6CH_2OCH_3$

5 gm

S56,571-7
HEPTYL ETHER

$CH_3(CH_2)_5CH_2OCH_2(CH_2)_5CH_3$

5 gm

S34,523-7
DECYL ETHER

$CH_3(CH_2)_8CH_2OCH_2(CH_2)_8CH_3$

5 gm

S34,793-0
DODECYL ETHER

$CH_3(CH_2)_{10}CH_2OCH_2(CH_2)_{10}CH_3$

5 gm

S35,569-0
TETRADECYL ETHER

$CH_3(CH_2)_{12}CH_2OCH_2(CH_2)_{12}CH_3$

5 gm

S35,045-1
HEXADECYL ETHER

$CH_3(CH_2)_{14}CH_2OCH_2(CH_2)_{14}CH_3$

5 gm

S35,407-4
N-OCTADECYL ETHER

$CH_3(CH_2)_{16}CH_2OCH_2(CH_2)_{16}CH_3$

5 gm

S69,792-3
DIPROPOXYMETHANE

$CH_3CH_2CH_2OCH_2OCH_2CH_2CH_3$

250 mg

S80,962-4
4-BROMO-1,7-DIETHOXYHEPTANE

Br
$CH_3CH_2OCH_2CH_2CH_2CHCH_2CH_2CH_2OCH_2CH_3$

250 mg

S68,008-7
1,2-DIDODECYLOXYETHANE

$CH_3(CH_2)_{10}CH_2OCH_2CH_2OCH_2(CH_2)_{10}CH_3$

5 gm

S90,622-0
1,2:7,8-BIS(DICHLOROMETHANO)-1,8-DIMETHOXY-DODECAHYDROBIPHENYLENE

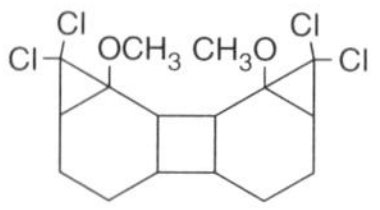

250 mg

S57,808-8
BIS(DECYLOXY)METHANE

$CH_3(CH_2)_8CH_2OCH_2OCH_2(CH_2)_8CH_3$

5 gm

S56,825-2
DIHEXADECYLOXYMETHANE

$CH_3(CH_2)_{14}CH_2OCH_2OCH_2(CH_2)_{14}CH_3$

250 mg

S50,299-5
1,2-PROPANEDIYLDIOXYBIS(TRICYCLO [5.2.1.0(2,6)]DECANE)

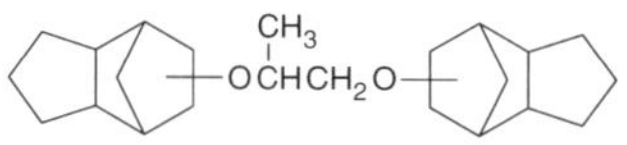

5 gm

S55,650-5
1-METHOXY-4-(TERT-PENTYL)-CYCLOHEXANE

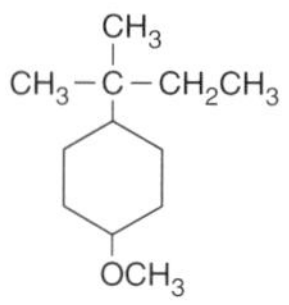

250 mg

74,156-6
7,7-DIMETHOXY-1,4,5,6-TETRACHLORONORBORNANE

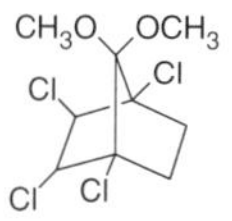

250 mg

S35,955-6
1-CHLORO-5-METHOXY-2-PENTENE

$CH_3OCH_2CH_2CH=CHCH_2Cl$

5 gm

S87,758-1
ALLYL 3-BROMO-2-BUTEN-1-YL ETHER

Br
$H_2C=CHCH_2OCH_2CH=CCH_3$

250 mg

S88,181-3
2-BROMOALLYL METHALLYL ETHER

Br CH_3
$H_2C=CCH_2OCH_2C=CH_2$

250 mg

S86,481-1
1-METHOXY-2-METHYL-1,4-CYCLOHEXADIENE

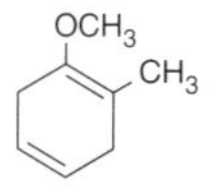

250 mg

S50,292-8
2-BUTENYL ETHER

$CH_3CH=CHCH_2OCH_2CH=CHCH_3$

5 gm

S50,293-6
2-BUTENYL 1-METHYLALLYL ETHER

CH_3
$CH_3CH=CHCH_2OCHCH=CH_2$

5 gm

S87,600-3
3,4-DIMETHYL-1-IODO-4-METHOXY-1-PENTENE

CH_3O CH_3
$CH_3C-CHCH=CHI$
CH_3

250 mg

S88,074-4
2-BROMOALLYL 5-HEXEN-1-YL ETHER

Br
$H_2C=CCH_2OCH_2CH_2CH_2CH_2CH=CH_2$

250 mg

S86,495-1
TERT-BUTYL 1,4-CYCLOHEXADIEN-1-YL ETHER

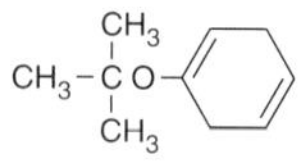

250 mg

S88,160-0
4-BROMO-4-PENTEN-1-YL 2,4-HEXADIEN-1-YL ETHER

Br
$H_2C{=}CCH_2CH_2CH_2OCH_2CH{=}CHCH{=}CHCH_3$

250 mg

S87,772-7
5-BROMO-5-HEXEN-1-YL 2,4-HEXADIEN-1-YL ETHER

Br
$H_2C{=}CCH_2CH_2CH_2CH_2OCH_2CH{=}CHCH{=}CHCH_3$

250 mg

S86,686-5
1-METHOXY-1-CYCLODODECENE

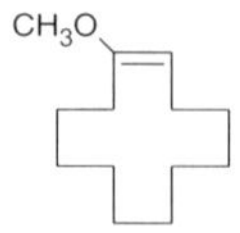

250 mg

S87,780-8
2-BROMOALLYL 10-UNDECEN-1-YL ETHER

Br
$H_2C{=}CCH_2OCH_2(CH_2)_7CH_2CH{=}CH_2$

250 mg

S45,671-3
3,6,9,12,15-PENTAOXA-1-HEPTADECENE

$CH_3CH_2O(CH_2CH_2O)_4CH{=}CH_2$

5 gm

S89,095-2
2-METHYL-4,6,9-TRIOXA-1-DECENE

CH3
$CH_3OCH_2CH_2OCH_2OCH_2C{=}CH_2$

250 mg

S94,984-1
TETRAETHYLENE GLYCOL DIALLYL ETHER

$H_2C{=}CHCH_2O(CH_2CH_2O)_4CH_2CH{=}CH_2$

1 gm

S86,604-0
1,4-DIMETHOXY-2-BUTENE

$CH_3OCH_2CH{=}CHCH_2OCH_3$

250 mg

S85,194-9
5,5-DIMETHOXY-1,2,3,4-TETRACHLORO-1,3-CYCLOPENTADIENE

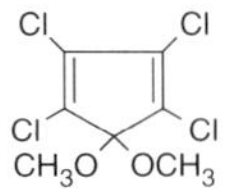

250 mg

S43,071-4
1,4-BIS(VINYLOXY)BUTANE

$H_2C{=}CHOCH_2CH_2CH_2CH_2OCH{=}CH_2$

5 gm

S39,972-8
ISOOCTYL VINYL ETHER

CH3
$CH_3CHCH_2(CH_2)_3CH_2OCH{=}CH_2$

5 gm

S39,973-6
HEXADECYL VINYL ETHER

$CH_3(CH_2)_{14}CH_2OCH{=}CH_2$

5 gm

S39,970-1
OCTADECYL VINYL ETHER

$CH_3(CH_2)_{16}CH_2OCH{=}CH_2$

5 gm

S60,305-8
TETRAETHYLENE GLYCOL METHYL VINYL ETHER

$CH_3O(CH_2CH_2O)_4CH{=}CH_2$

5 gm

S43,847-2
1-(2-ETHOXYETHOXY)-2-(VINYLOXY)-ETHANE

$CH_3CH_2OCH_2CH_2OCH_2CH_2OCH{=}CH_2$

5 gm

S37,523-3
ETHYLENE GLYCOL 1,2-BIS-(DIHYDRODICYCLOPENTADIENYL ETHER)

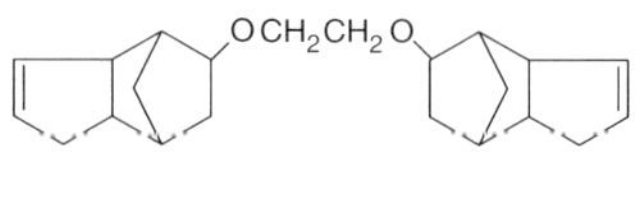

5 gm

S86,673-3
6,6-DIMETHYL-3-(METHOXYMETHYL)-2-METHYLENEBICYCLO[3.1.1]HEPTANE

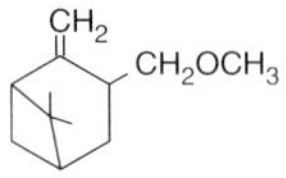

250 mg

S86,662-8
6,6-DIMETHYL-2-(METHOXYMETHYL) BICYCLO[3.1.1]HEPT-2-ENE

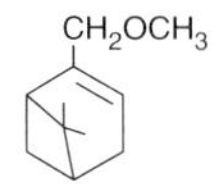

250 mg

S50,269-3
ALLYLOXYTRICYCLO[5.2.1.0(2,6)]-DEC-3-ENE

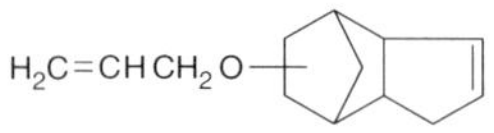

5 gm

S70,250-1
3-METHOXY-1-CYCLOPENTENE

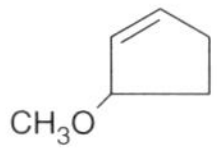

5 gm

S34,504-0
CYCLOHEXYL VINYL ETHER

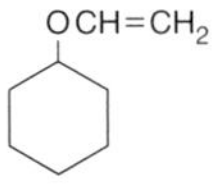

5 gm

S88,463-4
CYCLOHEXYLIDENEMETHYL METHYL ETHER

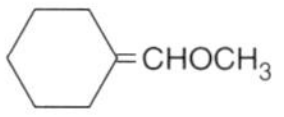

250 mg

S86,483-8
1-CYCLOHEXEN-1-YL ETHYL ETHER

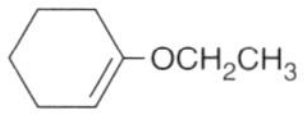

250 mg

S86,621-0
2-METHOXY-3-METHYL-1-CYCLOHEXENE

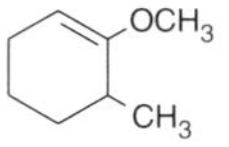

250 mg

S88,169-4
2-BROMOALLYL 2-CYCLOHEXEN-1-YLMETHYL ETHER

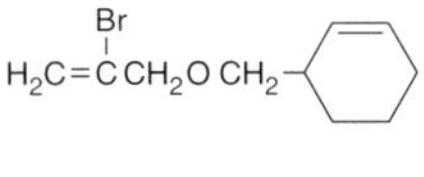

250 mg

S94,597-8
MENTHYL METHYL ETHER

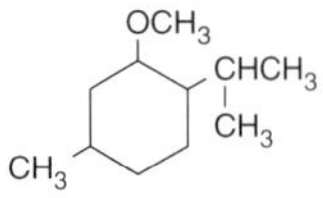

250 mg

S87,576-7
6,6-DIMETHYL-2-[2-(METHALLYLOXY)-ETHYL]BICYCLO[3.1.1]HEPT-2-ENE

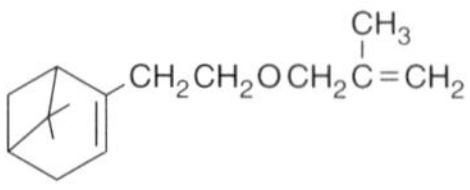

250 mg

S52,809-9
2,6-DIMETHYL-8-(1-METHOXYETHOXY)-2-OCTENE

OCH3 CH3 CH3
$CH_3CHOCH_2CH_2CHCH_2CH_2CH{=}CCH_3$

5 gm

S37,588-8
BIS(6-METHYL-3-CYCLOHEXENYLMETHOXY)-6-METHYL-3-CYCLOHEXENYLMETHANE

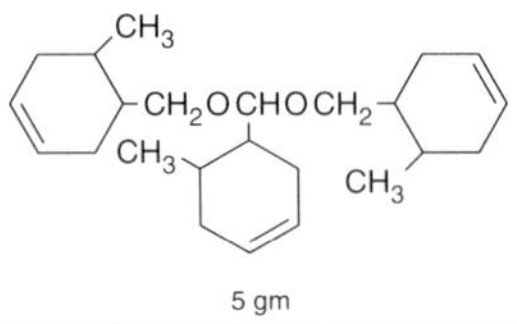

5 gm

S55,128-7
1,1,3,3-TETRAALLYLOXYPROPANE

$OCH_2CH{=}CH_2$
$H_2C{=}CHCH_2OCHCH_2CHOCH_2CH{=}CH_2$
$OCH_2CH{=}CH_2$

5 gm

S55,131-7
TRIOCTADECYL ORTHOFORMATE

$OCH_2(CH_2)_{16}CH_3$
$CH_3(CH_2)_{16}CH_2OCHOCH_2(CH_2)_{16}CH_3$

5 gm

S86,489-7
2,2-DIMETHOXYNORBORNANE

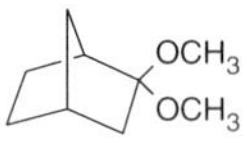

250 mg

S95,017-3
1,1,3,3-TETRAMETHOXYBUTANE

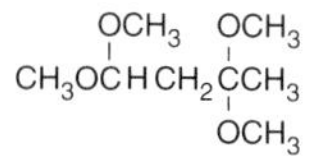

50 mg

S95,020-3
2-BROMO-1,1,3-TRIETHOXYPROPANE

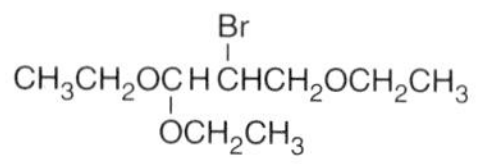

50 mg

S70,543-8
1,1,3-TRIETHOXYBUTANE

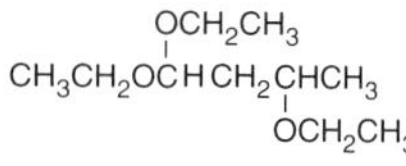

5 gm

S54,070-6
GLYOXAL BIS(DIBUTYL ACETAL)

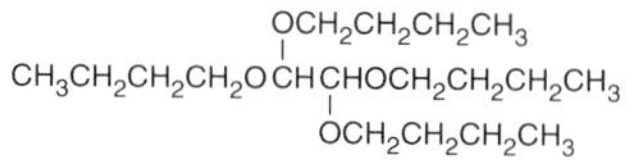

5 gm

S54,071-4
GLYOXAL BIS[BIS(2-ETHYLHEXYL)-ACETAL]

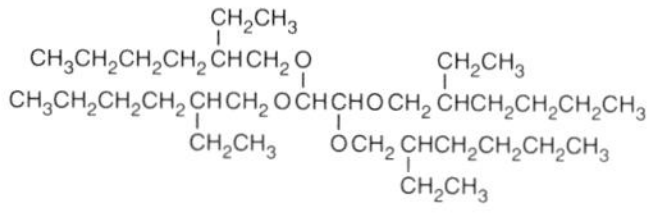

5 gm

S62,644-9
TRIBUTYL ORTHOFORMATE

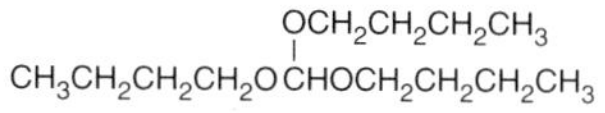

5 gm

S62,645-7
TRIPENTYL ORTHOFORMATE

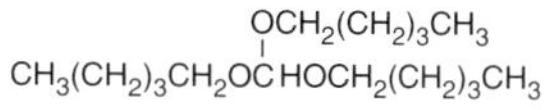

5 gm

S64,867-1
3-ETHOXY-1,1,3-TRIMETHOXYPROPANE

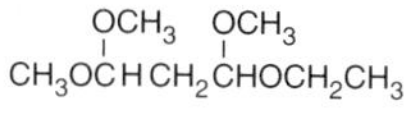

5 gm

S88,161-9
1,3,5-OCTATRIENE-1-CARBOXALDEHYDE DIMETHYL ACETAL

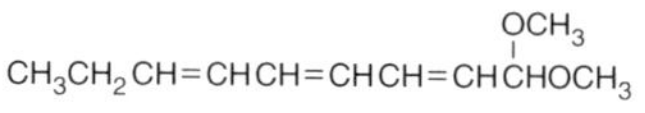

250 mg

S70,544-6
TRIOCTYL ORTHOFORMATE

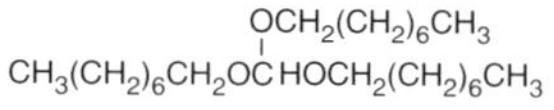

5 gm

S61,545-5
3-CHLORO-2-HYDROXYPROPYL METHYL ETHER

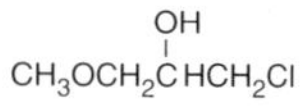

5 gm

S55,317-4
1-ALLYLOXY-3-CHLORO-2-PROPANOL

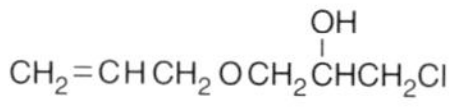

5 gm

S55,319-0
1-CHLORO-3-ISOPROPOXY-2-PROPANOL

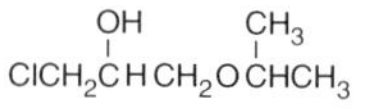

5 gm

S37,922-0
1-CHLORO-3-ETHOXY-2-PROPANOL

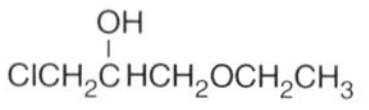

5 gm

S56,257-2
1-CHLORO-3-(4-ETHYL-1-ISOBUTYLOCTYLOXY)-2-PROPANOL

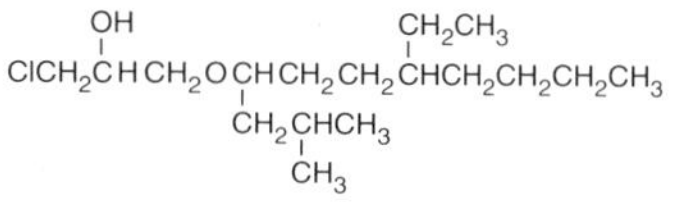

250 mg

S56,812-0
1-CHLORO-3-ISOHEXADECYLOXY-2-PROPANOL

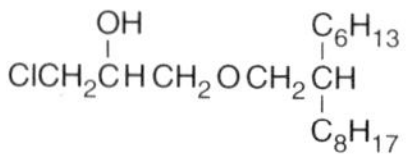

250 mg

S56,819-8
1-CHLORO-3-ISOOCTADECYLOXY-2-PROPANOL

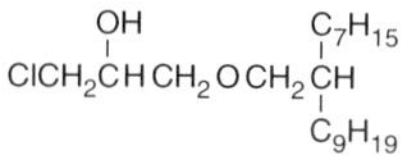

250 mg

S77,342-5
1,1'-(HEXAMETHYLENEDIOXY)BIS(2,2,2-TRICHLOROETHANOL)

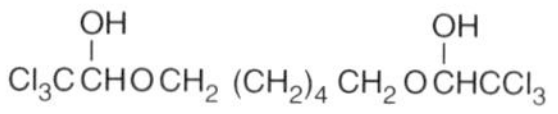

250 mg

S34,798-1
1-ETHOXY-3-PENTANOL

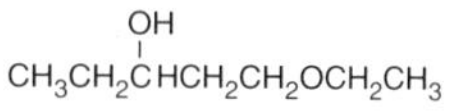

5 gm

S67,114-2
6-METHOXY-2-HEXANOL

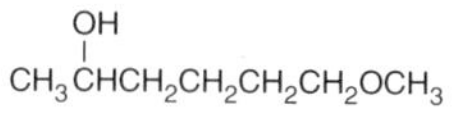

5 gm

S56,700-0
2-ETHOXYMETHYL-2-HYDROXYMETHYL-1,3-PROPANEDIOL

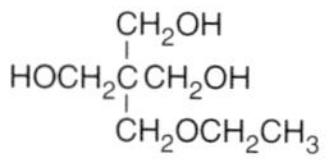

5 gm

S58,803-2
3,5-DIMETHYL-5-METHOXY-3-HEXANOL

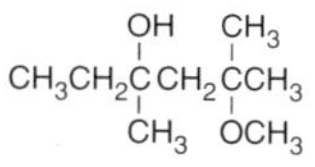

5 gm

S55,322-0
1,3-DIISOPROPOXY-2-PROPANOL

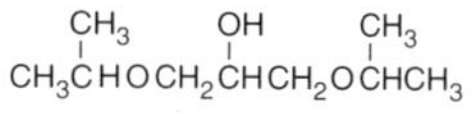

5 gm

S63,107-8
2,4-DIMETHYL-2-METHOXYMETHYL-1,5-PENTANEDIOL

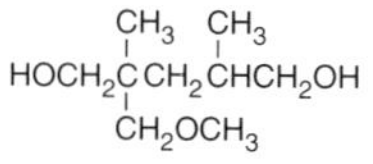

5 gm

S70,527-6
TRIETHYLENE GLYCOL ETHYL ETHER

$HOCH_2CH_2OCH_2CH_2OCH_2CH_2OCH_2CH_3$

5 gm

S34,841-4
ETHYLENE GLYCOL MONOALLYL ETHER

$HOCH_2CH_2OCH_2CH{=}CH_2$

5 gm

S34,314-5
1-BUTOXY-2-PROPANOL

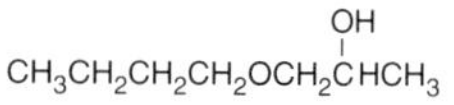

5 gm

S43,329-2
1,3-DIETHOXY-2-PROPANOL

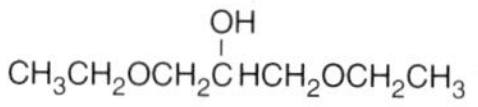

5 gm

S34,603-9
DIETHYLENE GLYCOL MONO-2-HYDROXYPROPYL ETHER

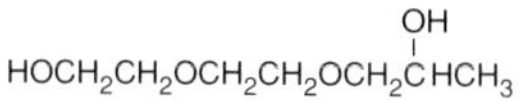

5 gm

S56,657-8
2-(2-ETHYLHEXYLOXY)-ETHANOL

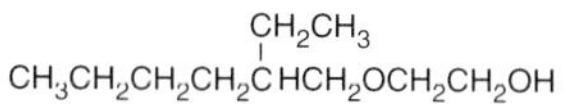

5 gm

S35,652-2
TRIETHYLENE GLYCOL BUTYL ETHER

$HOCH_2CH_2OCH_2CH_2OCH_2CH_2OCH_2CH_2CH_2CH_3$

5 gm

S50,261-8
2-METHOXY-5,9-CYCLODODECADIEN-1-OL

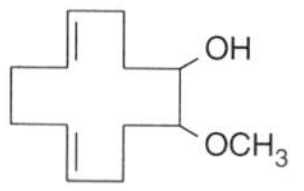

5 gm

S67,733-7
2-METHOXYCYCLOHEXANOL

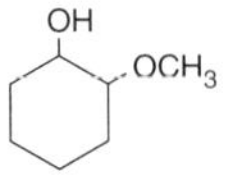

5 gm

S43,409-4
TRANS-4-(2-HYDROXYETHOXY)-1-DECALOL

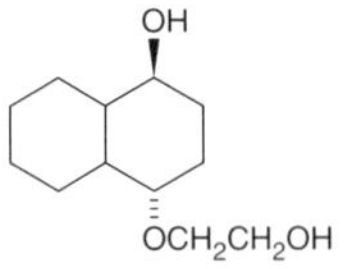

5 gm

S86,687-3
1-(2-METHOXY-2-CYCLOHEXEN-1-YL)-1-HEXANOL

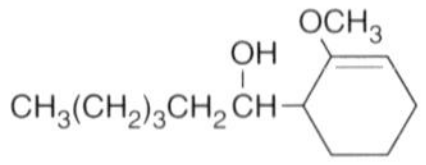

250 mg

S67,116-9
5-METHOXY-2-PENTANOL

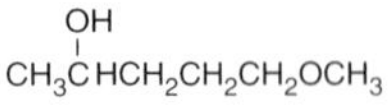

5 gm

S88,167-8
4-(5-BROMO-5-HEXEN-1-YLOXY)-2-BUTEN-1-OL

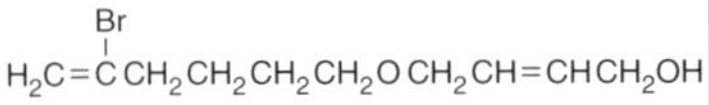

250 mg

S95,782-8
10-BROMO-8-METHOXYTRICYCLO[5.2.1.0(2,6)]DEC-4-EN-3-OL

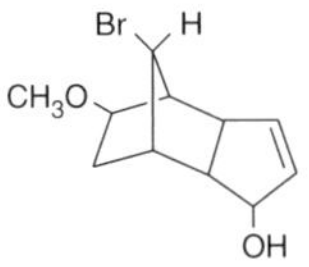

50 mg

S67,273-4
2-ISOPROPOXYCYCLOHEXANOL

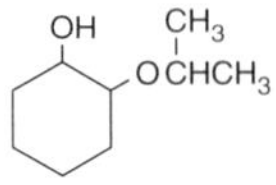

5 gm

S43,408-6
TRANS-4-ETHOXY-1-DECALOL

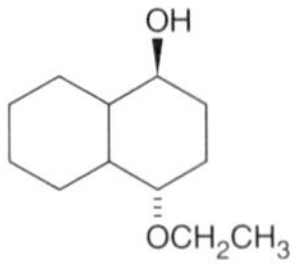

5 gm

S50,281-2
TRANS-4-ALLYLOXYDECAHYDRO-1-NAPHTHOL

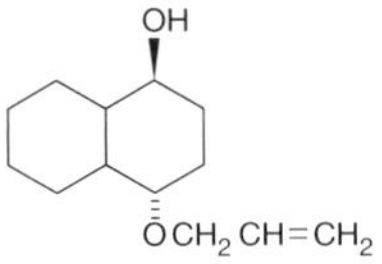

5 gm

S50,298-7
4-BUTOXYDECAHYDRO-1-NAPHTHOL

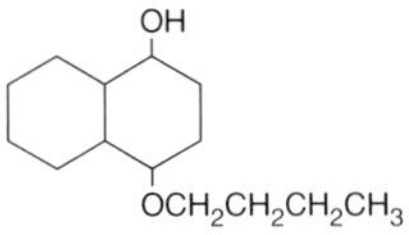

5 gm

S35,070-2
HYDROXYCITRONELLAL DIMETHYL ACETAL

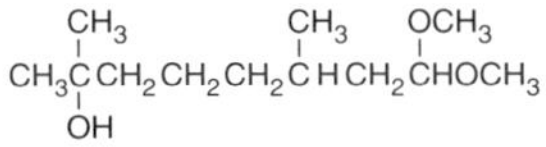

5 gm

S84,266-4
1,9-DIETHOXYNONANE

$CH_3CH_2OCH_2(CH_2)_7CH_2OCH_2CH_3$

250 mg

S86,103-0
1,1-DICHLORO-3,4-EPOXY-4-METHYL-1-PENTENE

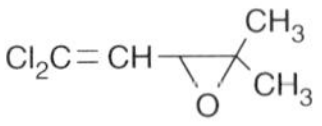

250 mg

S73,953-7
2-METHYLGLYCIDOL

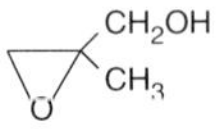

5 gm

S50,268-5
3,4-EPOXYTRICYCLO[5.2.1.0(2,6)]-DECANOL

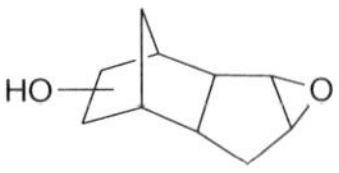

5 gm

S34,283-1
1,3-BUTANEDIOL DIGLYCIDYL ETHER, TECH.

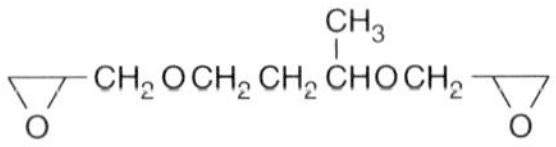

5 gm

S37,574-8
1,4-DIBUTOXY-2,3-EPOXYBUTANE

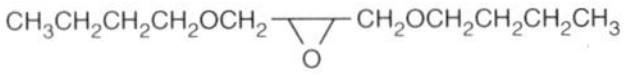

5 gm

S41,365-8
GLYCIDYL LAURYL ETHER

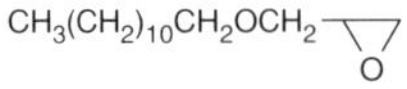

5 gm

S50,279-0
ALLYLOXY-3,4-EPOXY-TRICYCLO[5.2.1.0(2,6)]DECANE

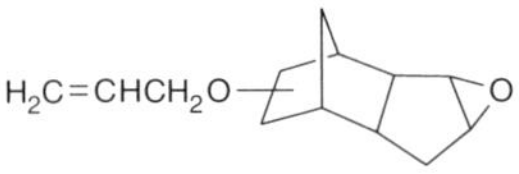

5 gm

S59,641-8
2,6,6-TRICHLORO-1,2-EPOXYCYCLOHEXANE

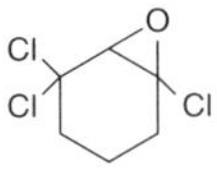

5 gm

S41,427-1
1,2-EPOXY-5,-TRANS-9,-CIS-CYCLODODECADIENE

5 gm

S52,818-8
CEDRENE EPOXIDE

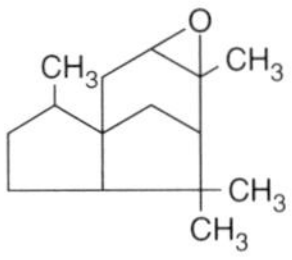

5 gm

S45,300-5
4-VINYLCYCLOHEXENE DIOXIDE

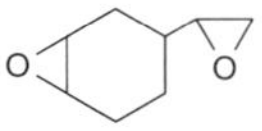

5 gm

S37,439-3
5,6-EPOXY-4,7-METHANO-1-OXASPIRO-[2.5]-OCTANE

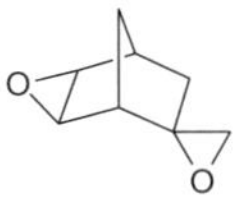

250 mg

S94,933-7
1,4,4A,8A-TETRAHYDRO-1,4-METHANONAPHTHALENE-5,8-DIONE-6,7-EPOXIDE

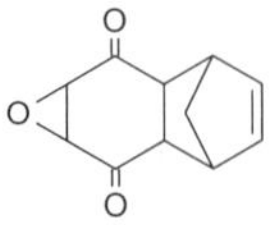

250 mg

S41,863-3
ALLOOCIMENE DIOXIDE, MIXTURE OF CIS AND TRANS

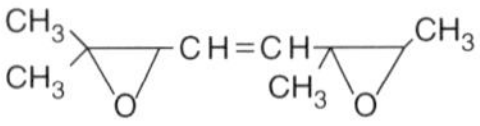

5 gm

S50,270-5
4,5-EPOXY-1,9-DECADIENE

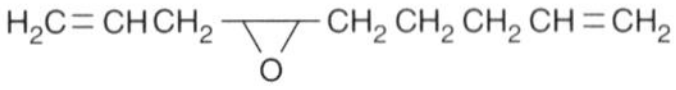

250 mg

S50,300-2
4,5,9,10-DIEPOXY-1-DECENE

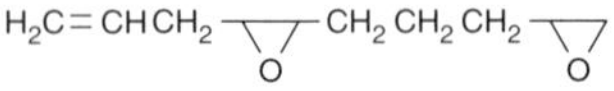

5 gm

S43,407-8
1,2,4,5,9,10-TRIEPOXYDECANE

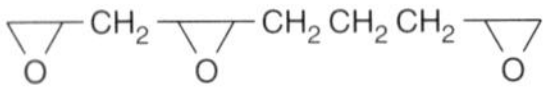

5 gm

S67,263-7
9,10-EPOXYOCTADECANE

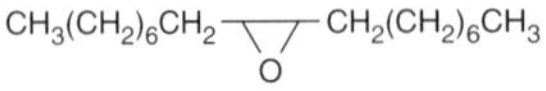

250 mg

S37,524-1
2-EPOXYETHYLBICYCLO[2.2.1]HEPTANE

5 gm

S45,261-0
3-VINYL-7-OXABICYCLO[4.1.0]HEPTANE

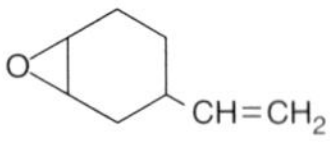

5 gm

S41,967-2
DIPENTENE DIOXIDE
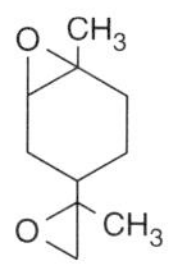
5 gm

S56,696-9
8-OXABICYCLO[5.1.0]OCTANE
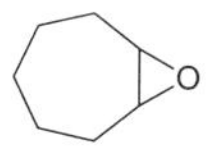
5 gm

S50,256-1
9-OXABICYCLO[6.1.0]NON-4-ENE
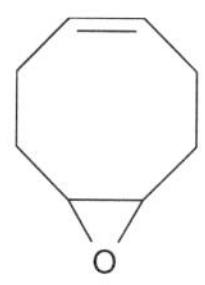
5 gm

S76,506-6
CIS-1,2-EPOXYCYCLODECANE
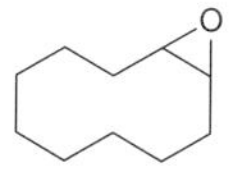
5 gm

S41,998-2
DICYCLOPENTADIENE DIOXIDE, MIXTURE OF ENDO AND EXO
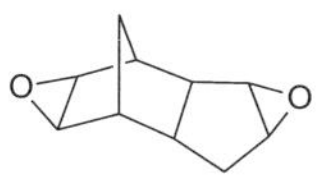
5 gm

S88,452-9
1-(2-CHLOROETHOXY)SPIRO(2.5)OCTANE

250 mg

S62,925-1
3,9-BIS(BROMOMETHYL)-2,4,8,10-TETRAOXA(5.5)SPIROUNDECANE
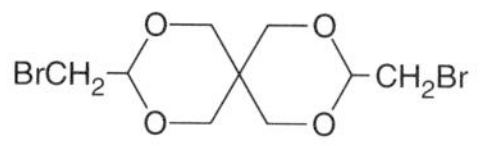
250 mg

S83,831-4
2-(3-CHLOROPROPYL)TETRAHYDROFURAN
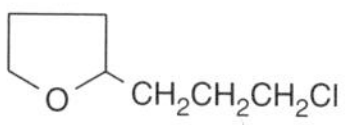
250 mg

S86,467-6
4-(CHLOROMETHYL)-2,2-DIMETHYL-1,3-DIOXOLANE
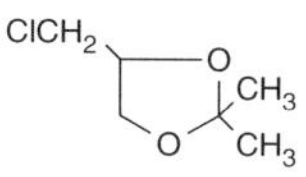
250 mg

S86,600-8
2-(CHLOROMETHYL)-1,3-DIOXANE
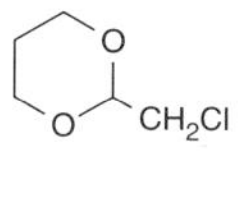
1 gm

S79,191-1
2-(TRICHLOROMETHYL)-1,3,6-TRIOXOCANE
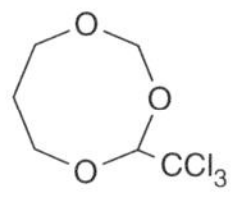
250 mg

S75,326-2
2-(TRICHLOROMETHYL)-1,3-DIOXANE-5,5-DIMETHANOL
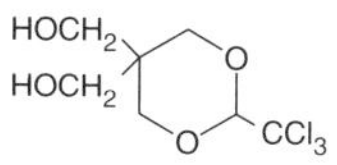
250 mg

S86,603-2
2-OXEPANOL
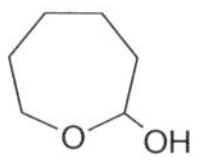
250 mg

S80,618-8
2,5-TETRAHYDROFURANDIOL
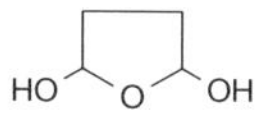
250 mg

S86,040-9
1,2-O-CYCLOHEXYLIDENEINOSITOL
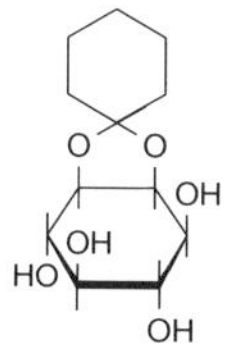
250 mg

S88,308-5
2-HYDROXY-2-(HYDROXYMETHYL)-TETRAHYDROFURAN
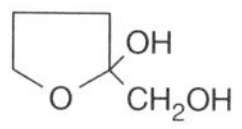
250 mg

S86,605-9
2-METHOXY-5-TETRAHYDROPYRANOL
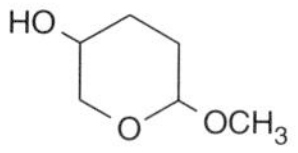
250 mg

S88,528-2
4-METHOXY-7A-METHYL-5,6,7,7A-TETRAHYDRO-1-INDANOL
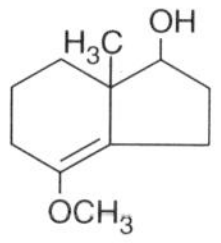
250 mg

S41,751-3
2,5-DIMETHOXYTETRAHYDROFURAN-3,4-CIS-DIOL
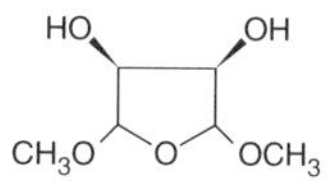
250 mg

S41,497-2
5-HYDROXYMETHYL-5-METHYL-1,3-DIOXANE
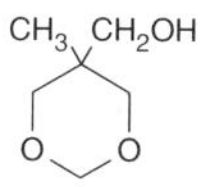
5 gm

S37,609-4
6,8-DIOXA-1-(HYDROXYMETHYL)BICYCLO[3.2.1]OCTANE
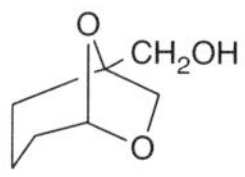
250 mg

S66,427-8
2,2-BIS(HYDROXYMETHYL)-TETRAHYDROFURAN
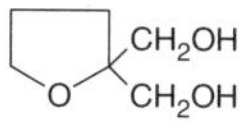
1 gm

S40,449-7
1,5-ANHYDROGLUCITOL
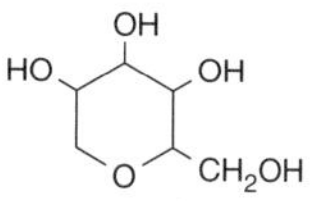
250 mg

S37,536-5
2,2-BIS(HYDROXYMETHYL)-TETRAHYDROPYRAN
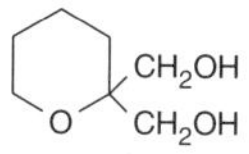
250 mg

S36,094-5
2,5-DIMETHYL-2,3-DIHYDRO-4H-PYRAN-2-METHANOL
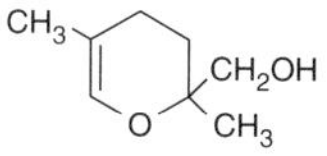
250 mg

S37,425-3
2,5-DIMETHYLTETRAHYDROPYRAN-2-METHANOL
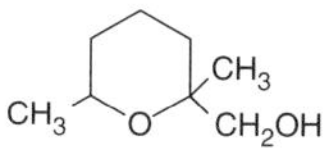
250 mg

S36,902-0
1,3-DIOXAN-5-OL
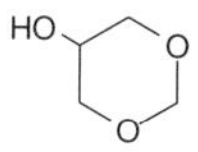
5 gm

S69,784-2
2,4-DIMETHYL-1,3-DIOXANE-2-METHANOL (MIXTURE OF ISOMERS)
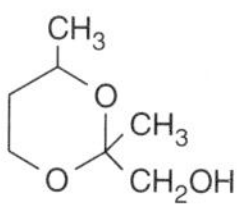
250 mg

S89,901-1
3,3-DIMETHYL-6-(2,2-DIMETHYL-1,3-DIOXOLAN-4-YL)-2,4-DIOXABICYCLO-[3.1.1]HEPTAN-1-OL
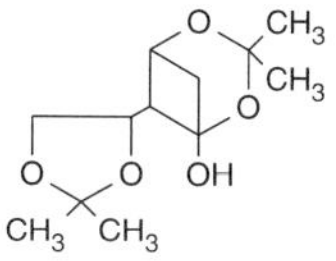
250 mg

S44,373-5
LINALOOL OXIDE
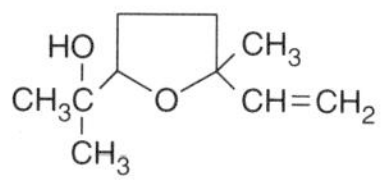
5 gm

S50,034-8
1,5-DIOXASPIRO(5.5)UNDECANE-3,3-DIMETHANOL
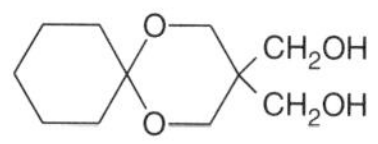
5 gm

S41,484-0
4-HYDROXYTETRAHYDROPYRAN-3,3,5,5-TETRAMETHANOL
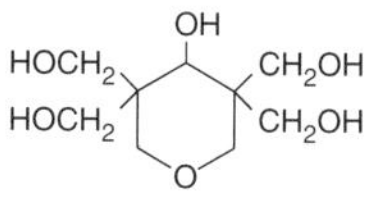
5 gm

S86,435-8
9-OXABICYCLO[3.3.1]NONAN-1-OL
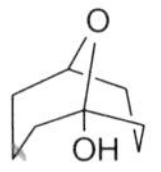

250 mg

S86,039-5
(-)-1,2:5,6-DI-O-CYCLOHEXYLIDENE-L-INOSITOL
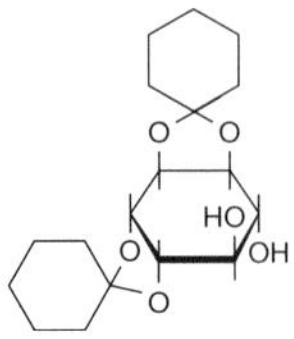

250 mg

S86,665-2
1,3-BIS(1,3-DIOXAN-2-YLOXY)PROPANE
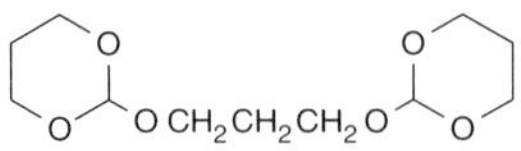

250 mg

S45,974-7
2,5-DIMETHOXYTETRAHYDRO-2-FURALDEHYDE DIMETHYL ACETAL
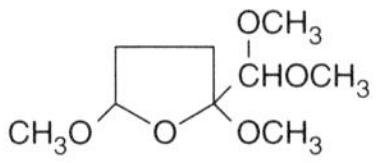

250 mg

S44,768-4
2,5-DIHYDRO-2,5-DIMETHOXY-2-(DIMETHOXY-METHYL)FURAN
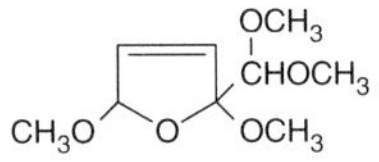

5 gm

S46,279-9
2-METHOXYTETRAHYDROFURAN
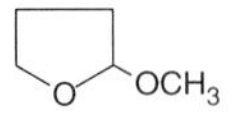

250 mg

S88,080-9
2-(ALLYLOXY)TETRAHYDROPYRAN
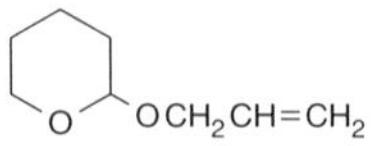

250 mg

S41,138-8
2,6-DIETHOXYTETRAHYDROPYRAN
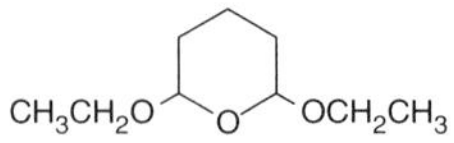

5 gm

S62,738-0
TETRAHYDROPYRAN-2-YL ETHER
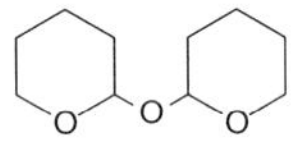

250 mg

S41,177-9
1,5-BIS(5-METHYL-1,3-DIOXAN-5-YL)-2,4-DIOXAPENTANE
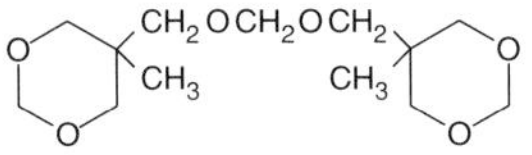

5 gm

S86,636-9
2-(BUTOXYMETHYL)TETRAHYDROFURAN
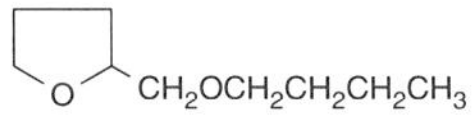

250 mg

S87,605-4
2-ETHOXY-3A,4,5,6,7,7A-HEXAHYDRO-1,3-BENZODIOXOLE
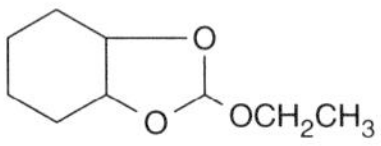

250 mg

S40,857-3
ETHYL TETRAHYDROFURFURYL ETHER
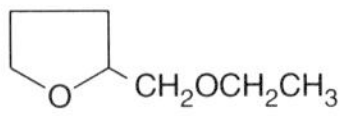

5 gm

S87,729-8
CYCLOHEXYL METHYL ETHER
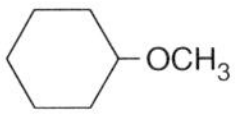

250 mg

S53,487-0
3,6-DIVINYL-2-METHYLTETRAHYDROPYRAN
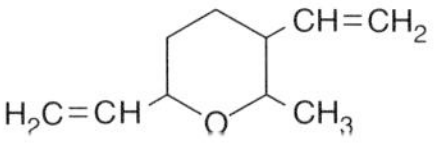

5 gm

S36,124-0
1,3-DIOXEPANE
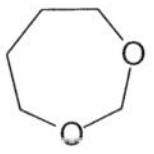

5 gm

S37,447-4
7,12-DIOXASPIRO(5.6)DODEC-9-ENE
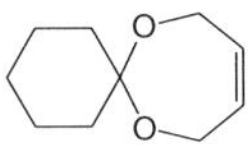

5 gm

S35,903-3
2-ETHOXYTETRAHYDROPYRAN
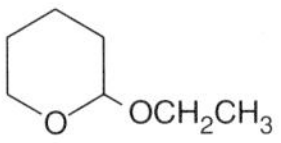

5 gm

S83,477-7
OCTAHYDRO-2,2'-BI-2H-PYRAN
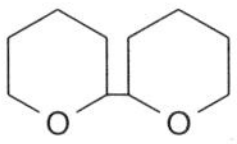

250 mg

S88,520-7
HEXAHYDRODISPIRO[1,3-DIOXOLANE-2,1'(4'H)-NAPHTHALENE-4',2''-(1,3) DIOXOLANE]
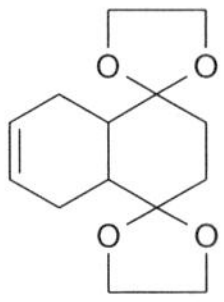

250 mg

S50,248-0
2,3-5,6-DIEPOXY-1,2,3,4,5,6-HEXAMETHYLBICYCLO[2.2.0]HEXANE
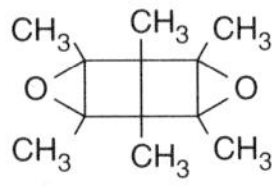

5 gm

S50,258-8
8,9-EPOXY-4-VINYL-3,5-DIOXATRICYCLO[5.2.1.0(2,6)]DECANE
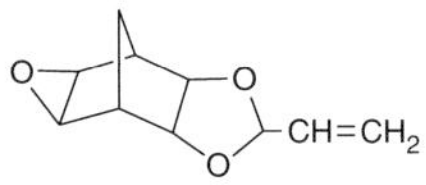

5 gm

S94,752-0
1,4-DIOXASPIRO(4.4)NONANE
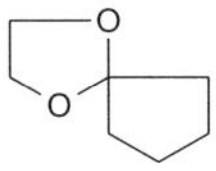

5 gm

S36,037-6
3,4-DICHLOROTETRAHYDROFURAN
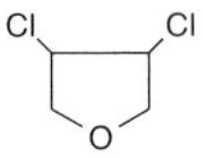

250 mg

S36,466-5
2-PROPYL-1,3-DIOXOLANE
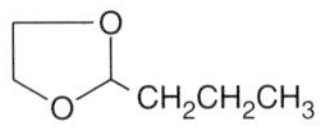

5 gm

S78,704-3
2,2,4-TRIMETHYL-1,3-DIOXOLANE
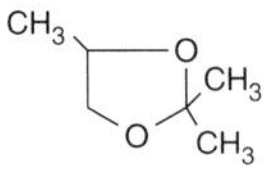

250 mg

S42,777-2
4-METHYL-2-PROPENYL-1,3-DIOXOLANE
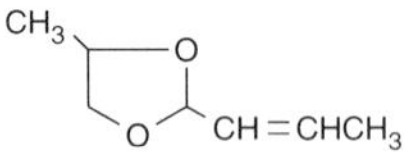

5 gm

S36,380-4
4-METHYL-2-PROPYL-1,3-DIOXOLANE
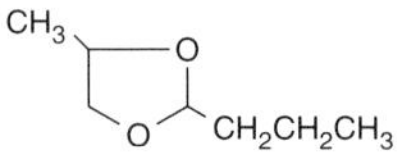

5 gm

S69,822-9
2-TERT-BUTYL-1,3-DIOXOLANE
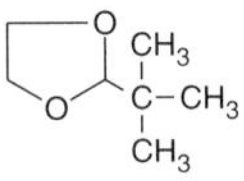

250 mg

S69,755-9
2-TERT-BUTYL-2-METHYL-1,3-DIOXOLANE
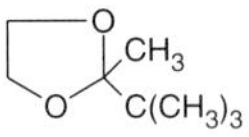

250 mg

S54,081-1
2,2-DIISOPROPYL-1,3-DIOXOLANE
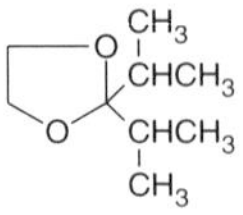

5 gm

S69,826-1
2-TERT-BUTYL-2,4-DIMETHYL-1,3-DIOXOLANE
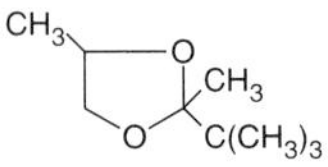

250 mg

S60,904-8
1,6-BIS(1,3-DIOXOLAN-2-YL)-3-HEXENE

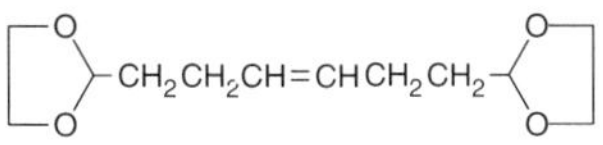

250 mg

S60,902-1
1,10-BIS(1,3-DIOXOLAN-2-YL)-3,7-DECADIENE

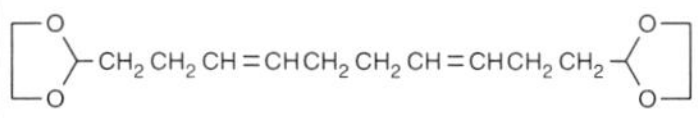

250 mg

S51,438-1
6-BROMO-1,3-DIOXASPIRO(4.5)DECANE

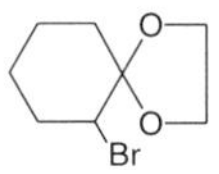

5 gm

S36,368-5
4-METHYL-2,2-PENTAMETHYLENE-1,3-DIOXOLANE

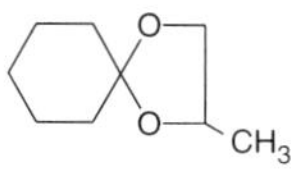

5 gm

S40,759-3
3,9-DI-(3-CYCLOHEXENYL)-2,4,8,10-TETRAOXASPIRO(5.5)UNDECANE

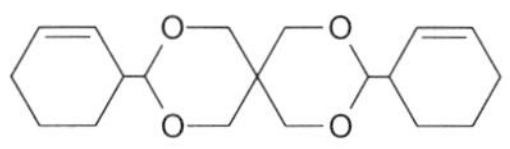

5 gm

S45,885-6
2-METHYLTETRAHYDROPYRAN

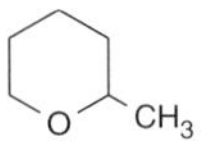

5 gm

S52,864-1
2,5-DIVINYLTETRAHYDROPYRAN

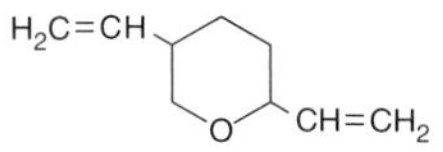

5 gm

S88,766-8
2,2',6,6'-TETRAMETHYL-DELTA4,4'-BI-4H-PYRAN

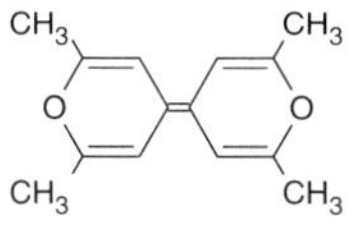

250 mg

S34,691-8
2,4-DIMETHYL-1,3-DIOXANE

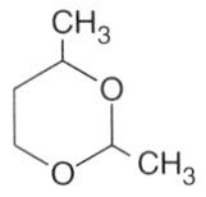

5 gm

S86,471-4
2-ISOPROPYL-1,3-DIOXANE

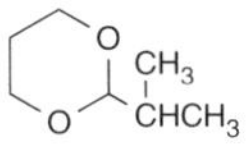

250 mg

S55,324-7
4,4,6-TRIMETHYL-2-VINYL-1,3-DIOXANE

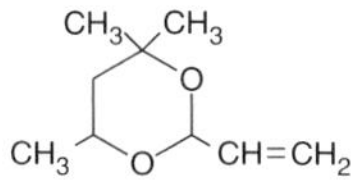

5 gm

S69,790-7
2-ISOPROPYL-4,4,6-TRIMETHYL-1,3-DIOXANE

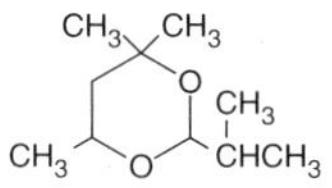

250 mg

S69,927-6
4A,5,6,7,8,8A-HEXAHYDRO-1,3-BENZODIOXANE

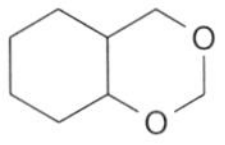

250 mg

S35,034-6
HEPTACHLORO-1,4-DIOXANE

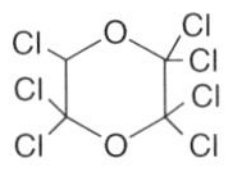

5 gm

S46,276-4
2,2-DICHLORO-1-ETHANETHIOL

Cl_2CHCH_2SH

5 gm

S46,223-3
1,2-PROPANEDITHIOL

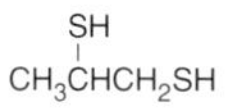

5 gm

S60,005-9
2-CHLOROALLYL MERCAPTAN

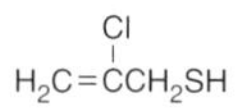

250 mg

S93,967-6
2,3-DIMETHYL-2-BUTANETHIOL

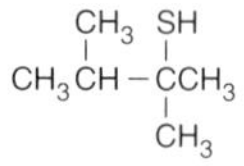

250 mg

S66,343-3
2,4,4-TRIMETHYL-1-PENTANETHIOL

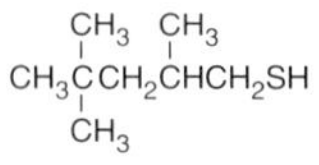

5 gm

S66,969-5
TERT-NONYL MERCAPTAN

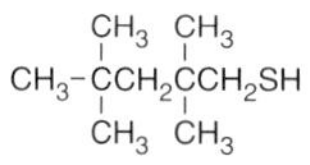

5 gm

S57,059-1
1-TRIDECANETHIOL

$CH_3(CH_2)_{11}CH_2SH$

5 gm

S57,814-2
5,7,7-TRIMETHYL-2-(1,3,3-TRIMETHYLBUTYL)-1-OCTANETHIOL

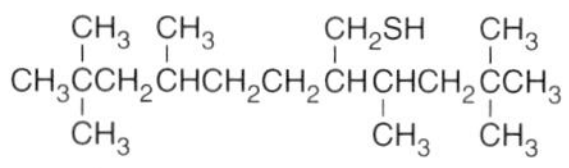

250 mg

S82,604-9
2-TETRAHYDROFURANPROPANETHIOL

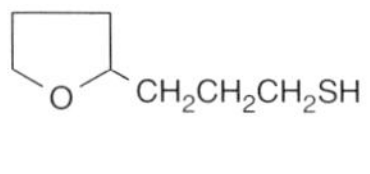

250 mg

S40,913-8
2,9-P-MENTHANEDITHIOL

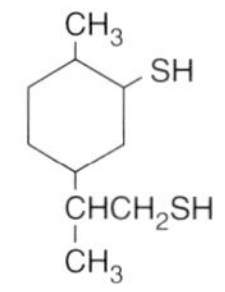

5 gm

S74,152-3
TRITHIOACETALDEHYDE

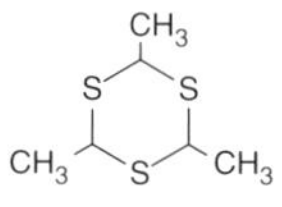

250 mg

S86,614-8
2-ISOPROPYL-1,3-DITHIANE

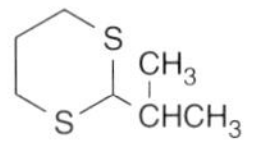

250 mg

S91,819-9
1,1-BIS(METHYLTHIO)-1-OCTENE

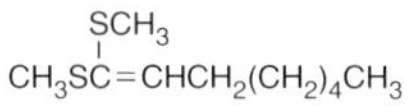

250 mg

S86,680-6
2-OCTYL-1,3-DITHIANE

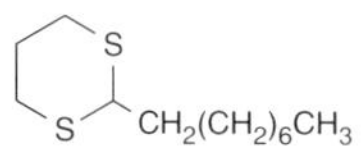

250 mg

S39,160-3
CHLOROMETHYL 2-CHLORO-1,1,2-TRIFLUOROETHYL SULFIDE

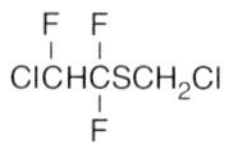

250 mg

S84,093-9
3-CHLOROPROPYL DODECYL SULFIDE

$CH_3(CH_2)_{10}CH_2SCH_2CH_2CH_2Cl$

250 mg

S77,595-9
3-CHLOROPROPYL CYCLOHEXYL SULFIDE

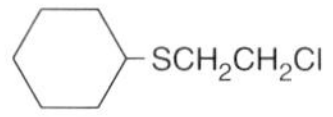

250 mg

S82,068-7
BIS(CHLOROMETHYL)METHYL DISULFIDE

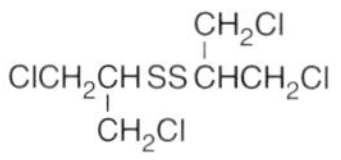

250 mg

S65,773-5
5'-CHLOROMETHYL-1,7,7-TRIMETHYLSPIRO[NORBORNANE-2,2'-(1,3)OXATHIOLANE]

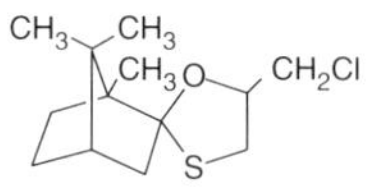

5 gm

S64,898-1
2-CHLOROMETHYL-1-OXA-4-THIASPIRO-(4.5)DECANE

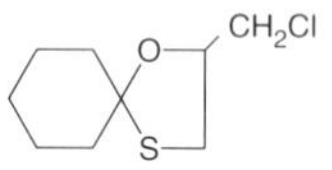

250 mg

S64,906-6
2-CHLOROMETHYL-6-ISOPROPYL-9-METHYL-1-OXA-4-THIASPIRO(4.5)DECANE

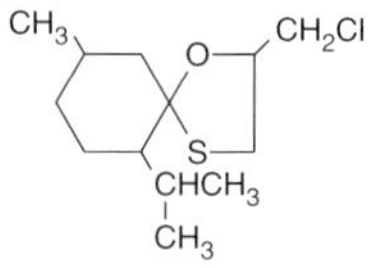

250 mg

S39,154-9
2-CHLORO-1,1,2-TRIFLUOROETHYL HEXADECYL SULFIDE

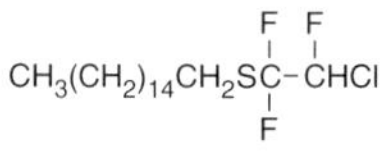

250 mg

S35,566-6
1,2,2,2-TETRACHLOROETHYL DISULFIDE

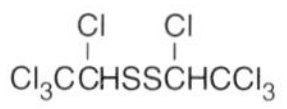

5 gm

S78,556-3
METHYL-BETA-D-THIOGALACTOSIDE

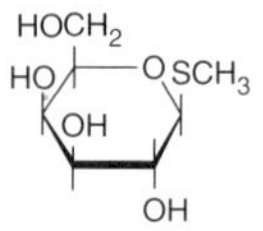

250 mg

S86,083-2
5,7-BIS(METHYLTHIO)-6-HEPTENE-2,4-DIOL

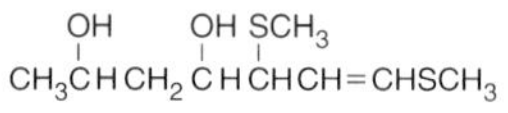

250 mg

S40,885-9
2-HYDROXYETHYL PROPYL SULFIDE

$CH_3CH_2CH_2SCH_2CH_2OH$

5 gm

S45,070-7
2-HYDROXYETHYL ISOPROPYL SULFIDE

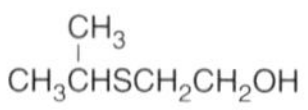

5 gm

S52,740-8
2-(BUTYLTHIO)ETHANOL

$CH_3CH_2CH_2CH_2SCH_2CH_2OH$

5 gm

S35,792-8
AMYL 2-HYDROXYETHYL SULFIDE

$CH_3(CH_2)_3CH_2SCH_2CH_2OH$

5 gm

S93,166-7
6-(METHYLTHIO)-1-HEXANOL

$CH_3SCH_2(CH_2)_4CH_2OH$

250 mg

S36,234-4
HEXYL 2-HYDROXYETHYL SULFIDE

$CH_3(CH_2)_4CH_2SCH_2CH_2OH$

5 gm

S36,222-0
HEPTYL 2-HYDROXYETHYL SULFIDE

$CH_3(CH_2)_5CH_2SCH_2CH_2OH$

5 gm

S36,915-2
2-(OCTYLTHIO)ETHANOL

$CH_3(CH_2)_6CH_2SCH_2CH_2OH$

5 gm

S54,013-7
DECYL 2-HYDROXYETHYL SULFIDE

$CH_3(CH_2)_8CH_2SCH_2CH_2OH$

5 gm

S36,135-6
DODECYL 2-HYDROXYETHYL SULFIDE

$CH_3(CH_2)_{10}CH_2SCH_2CH_2OH$

5 gm

S57,231-4
3-DODECYLTHIO-1-PROPANOL

$CH_3(CH_2)_{10}CH_2SCH_2CH_2CH_2OH$

250 mg

S54,032-3
2-(TETRADECYLTHIO)ETHANOL

$CH_3(CH_2)_{12}CH_2SCH_2CH_2OH$

5 gm

S56,816-3
2-(DODECYLTHIOMETHYL)-2-PROPANOL

OH
$CH_3(CH_2)_{10}CH_2SCH_2CCH_3$
CH_3

250 mg

S50,737-7
2-(HEXADECYLTHIO)ETHANOL

$CH_3(CH_2)_{14}CH_2SCH_2CH_2OH$

250 mg

S54,014-5
2-HYDROXYETHYL OCTADECYL SULFIDE

$CH_3(CH_2)_{16}CH_2SCH_2CH_2OH$

5 gm

S50,196-4
TRIS-(2-HYDROXYETHYL)-
SULFONIUM CHLORIDE

$HOCH_2CH_2$
$HOCH_2CH_2-S^+$ Cl^-
$HOCH_2CH_2$

250 mg

S46,117-2
TETRAHYDROTHIOPYRAN-3-OL

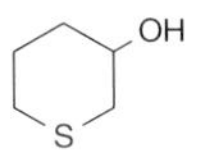

5 gm

S86,612-1
1-METHOXY-4-[(METHYLTHIO)METHOXY]-
2-BUTENE

$CH_3OCH_2CH=CHCH_2OCH_2SCH_3$

250 mg

S69,824-5
1-METHOXY-2-(METHYLTHIO)ETHANE

$CH_3OCH_2CH_2SCH_3$

250 mg

S61,238-3
1,3-BIS(METHYLTHIO)-2-
METHOXYPROPANE

OCH_3
$CH_3SCH_2CHCH_2SCH_3$

5 gm

S87,599-6
6-(METHYLTHIOMETHOXY)-2-HEXENE

$CH_3CH=CHCH_2CH_2CH_2OCH_2SCH_3$

250 mg

S84,096-3
BIS[3-(DODECYLTHIO)PROPYL] ETHER

$[CH_3(CH_2)_{10}CH_2SCH_2CH_2CH_2-]_2O$

250 mg

S37,966-2
2-(METHYLTHIO)ACETALDEHYDE
DIETHYL ACETAL

OCH_2CH_3
$CH_3CH_2OCHCH_2SCH_3$

5 gm

S62,032-7
ALLYL ETHYL SULFIDE

$CH_3CH_2SCH_2CH=CH_2$

5 gm

S86,602-4
METHYL 2-METHYLALLYL SULFIDE

CH_3
$CH_3SCH_2C=CH_2$

250 mg

S36,285-9
ISOBUTYL METHYL SULFIDE

CH_3
$CH_3CHCH_2SCH_3$

5 gm

S44,742-0
BUTYL METHYL SULFIDE

$CH_3CH_2CH_2CH_2SCH_3$

5 gm

S35,739-1
ALLYL PROPYL SULFIDE

$CH_3CH_2CH_2SCH_2CH=CH_2$

5 gm

S35,152-0
ISOAMYL METHYL SULFIDE

CH_3
$CH_3CHCH_2CH_2SCH_3$

5 gm

S35,795-2
AMYL METHYL SULFIDE

$CH_3(CH_2)_3CH_2SCH_3$

5 gm

S57,070-2
ISOPROPYL PROPYL SULFIDE

CH_3
$CH_3CHSCH_2CH_2CH_3$

5 gm

S34,787-6
3,6-DITHIAOCTANE

$CH_3CH_2SCH_2CH_2SCH_2CH_3$

5 gm

S36,630-7
ALLYL SEC-BUTYL SULFIDE

CH_3
$CH_3CH_2CHSCH_2CH=CH_2$

5 gm

S36,719-2
BUTYL PROPYL SULFIDE

$CH_3CH_2CH_2CH_2SCH_2CH_2CH_3$

5 gm

S54,015-3
HEXYL METHYL SULFIDE

$CH_3(CH_2)_4CH_2SCH_3$

5 gm

S54,713-1
ISOBUTYL PROPYL SULFIDE

CH_3
$CH_3CHCH_2SCH_2CH_2CH_3$

5 gm

S54,720-4
ISOBUTYL ISOPROPYL SULFIDE

CH_3 CH_3
$CH_3CHCH_2SCHCH_3$

5 gm

S57,060-5
BUTYL ISOPROPYL SULFIDE

CH_3
$CH_3CHSCH_2CH_2CH_2CH_3$

5 gm

S57,067-2
SEC-BUTYL PROPYL SULFIDE

CH_3
$CH_3CH_2CHSCH_2CH_2CH_3$

5 gm

S57,068-0
TERT-BUTYL ISOPROPYL SULFIDE

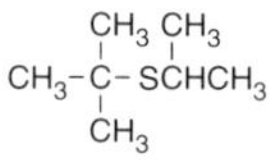

5 gm

S37,589-6
1,5-BIS(METHYLTHIO)-PENTANE

$CH_3SCH_2(CH_2)_3CH_2SCH_3$

5 gm

S83,594-3
TRIS (ETHYLTHIO)METHANE

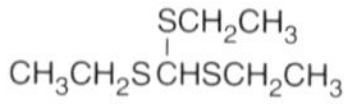

250 mg

S43,268-7
METHALLYL SULFIDE

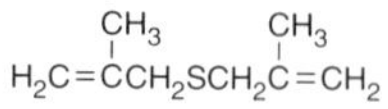

5 gm

S77,107-4
2-BUTENYL BUTYL SULFIDE

$CH_3CH_2CH_2CH_2SCH_2CH=CHCH_3$

250 mg

S57,056-7
BUTYL ISOBUTYL SULFIDE

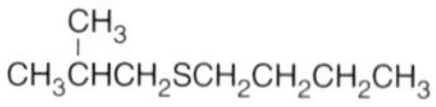

5 gm

S57,057-5
BUTYL TERT-BUTYL SULFIDE

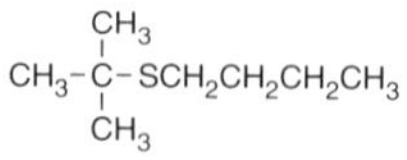

5 gm

S57,061-3
BUTYL SEC-BUTYL SULFIDE

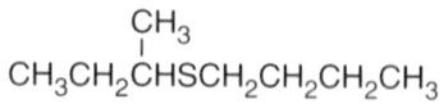

5 gm

S57,072-9
HEPTYL METHYL SULFIDE

$CH_3(CH_2)_5CH_2SCH_3$

5 gm

S34,782-5
2,9-DITHIADECANE

$CH_3SCH_2(CH_2)_4CH_2SCH_3$

5 gm

S34,783-3
3,8-DITHIADECANE

$CH_3CH_2SCH_2CH_2CH_2CH_2SCH_2CH_3$

5 gm

S34,784-1
4,7-DITHIADECANE

$CH_3CH_2CH_2SCH_2CH_2SCH_2CH_2CH_3$

5 gm

S91,930-6
METHYL TERT-OCTYL SULFIDE

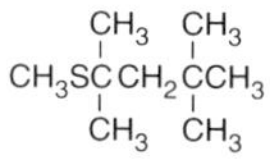

250 mg

S56,733-7
PENTYL SULFIDE

$CH_3(CH_2)_3CH_2SCH_2(CH_2)_3CH_3$

5 gm

S57,047-8
ISOPENTYL SULFIDE

5 gm

S57,065-6
ETHYL OCTYL SULFIDE

$CH_3(CH_2)_6CH_2SCH_2CH_3$

5 gm

S34,786-8
4,9-DITHIADODECANE

$CH_3CH_2CH_2SCH_2CH_2CH_2CH_2SCH_2CH_2CH_3$

5 gm

S54,016-1
DECYL METHYL SULFIDE

$CH_3(CH_2)_8CH_2SCH_3$

5 gm

S54,715-8
ETHYL NONYL SULFIDE

$CH_3(CH_2)_7CH_2SCH_2CH_3$

5 gm

S54,692-5
DECYL ETHYL SULFIDE

$CH_3(CH_2)_8CH_2SCH_2CH_3$

5 gm

S41,313-5
HEPTYL SULFIDE

$CH_3(CH_2)_5CH_2SCH_2(CH_2)_5CH_3$

5 gm

S84,090-4
DODECYL ETHYL SULFIDE

$CH_3(CH_2)_{10}CH_2SCH_2CH_3$

250 mg

S84,091-2
DODECYL PROPYL SULFIDE

$CH_3(CH_2)_{10}CH_2SCH_2CH_2CH_3$

250 mg

S41,292-9
2-ETHYLHEXYL SULFIDE

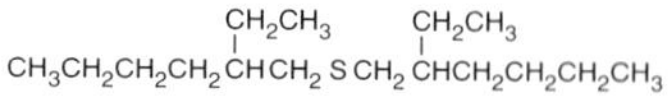

5 gm

S84,094-7
BUTYL DODECYL SULFIDE

$CH_3(CH_2)_{10}CH_2SCH_2CH_2CH_2CH_3$

250 mg

S57,028-1
NONYL SULFIDE

$CH_3(CH_2)_7CH_2SCH_2(CH_2)_7CH_3$

5 gm

S35,997-1
DECYL SULFIDE

$CH_3(CH_2)_8CH_2SCH_2(CH_2)_8CH_3$

5 gm

S54,694-1
UNDECYL SULFIDE

$CH_3(CH_2)_9CH_2SCH_2(CH_2)_9CH_3$

5 gm

S60,262-0
TERT-DODECYL SULFIDE

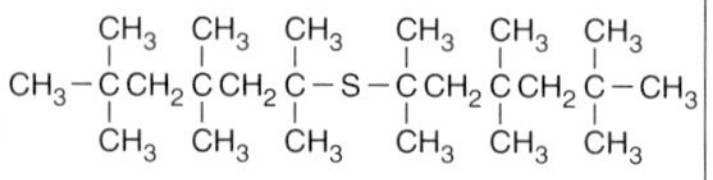

5 gm

S35,570-4
TETRADECYL SULFIDE

$CH_3(CH_2)_{12}CH_2SCH_2(CH_2)_{12}CH_3$

5 gm

S84,099-8
1,6-BIS(DODECYLTHIO)HEXANE

$[CH_3(CH_2)_{10}CH_2SCH_2CH_2CH_2-]_2$

250 mg

S57,050-8
HEXADECYL SULFIDE

$CH_3(CH_2)_{14}CH_2SCH_2(CH_2)_{14}CH_3$

5 gm

S43,362-4
OCTADECYL SULFIDE

$CH_3(CH_2)_{16}CH_2SCH_2(CH_2)_{16}CH_3$

5 gm

S86,674-1
6,6-DIMETHYL-2-METHYLENE-3-(METHYLTHIOMETHYL)BICYCLO[3.1.1]-HEPTANE

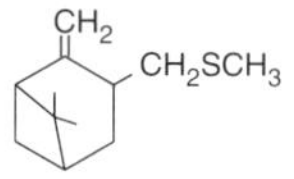

250 mg

S87,610-0
2-(ALLYLTHIOETHYL)-6,6-DIMETHYLBICYCLO[3.1.1]HEPT-2-ENE

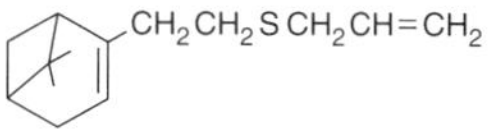

250 mg

S87,609-7
6,6-DIMETHYL-2-(METHALLYLTHIOETHYL)-BICYCLO[3.1.1]HEPT-2-ENE

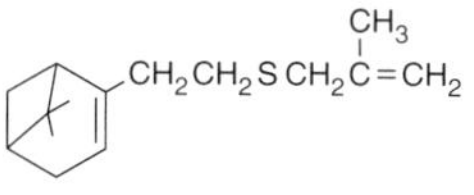

250 mg

S57,055-9
ISOPENTYL DISULFIDE

5 gm

S54,732-8
1,1-DIMETHYLBUTYL DISULFIDE

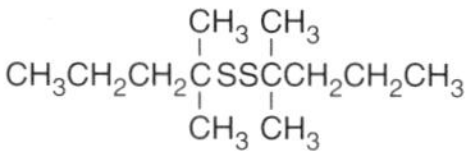

5 gm

S57,074-5
HEXYL DISULFIDE

$CH_3(CH_2)_4CH_2SSCH_2(CH_2)_4CH_3$

5 gm

S60,080-6
TERT-HEPTYL DISULFIDE

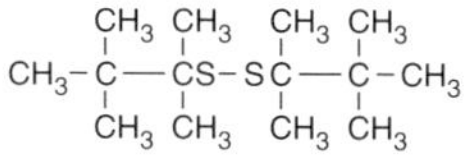

5 gm

S54,719-0
UNDECYL DISULFIDE

$CH_3(CH_2)_9CH_2SSCH_2(CH_2)_9CH_3$

5 gm

S61,885-3
TRIMETHYLSULFONIUM TETRAFLUOROBORATE

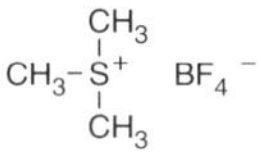

5 gm

S61,879-9
TRIETHYLSULFONIUM TETRAFLUOROBORATE

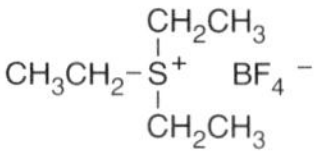

5 gm

S83,749-0
METHYLOCTADECYLPROPYLSULFONIUM IODIDE

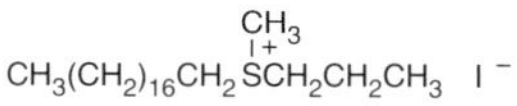

250 mg

S90,604-2
OCTAHYDROCYCLOPENTA(C)THIOPYRAN

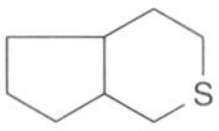

250 mg

S89,617-9
DELTA-2,2'-BI(HEXAHYDRO-4,7-METHANO-1,3-BENZODITHIOLE)

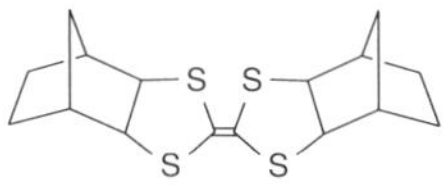

250 mg

S69,787-7
2-ETHYL-2-METHYL-1,3-DITHIOLANE

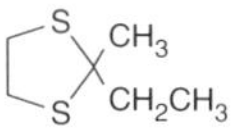

250 mg

S88,519-3
HEXAHYDRODISPIRO[1,3-DITHIOLANE-2,1'(4'H)-NAPHTHALENE-4',2''-(1,3) DITHIOLANE]

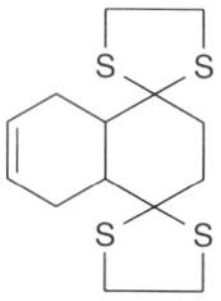

250 mg

S34,739-6
2,6-DIMETHYL-1,4-THIOXANE

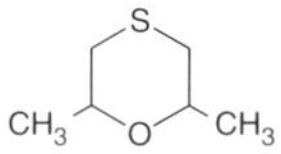

5 gm

S79,369-8
NEOPENTYLAMINE HYDROCHLORIDE
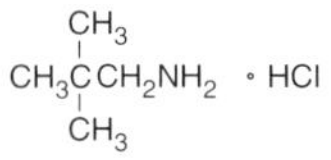
250 mg

S56,329-3
SEC-HEXYLAMINE
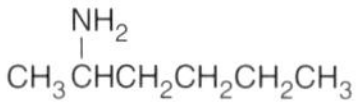
5 gm

S40,815-8
1,4-DIMETHYLPENTYLAMINE

5 gm

S43,298-9
1,3-DIMETHYLPENTYLAMINE
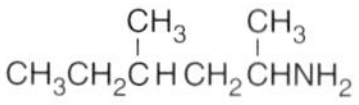
5 gm

S55,129-5
1-ISOPROPYL-2-METHYLPROPYLAMINE
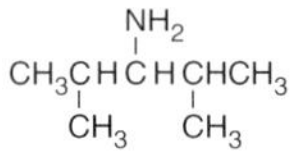
5 gm

S56,417-6
1-METHYLHEXYLAMINE SULFATE
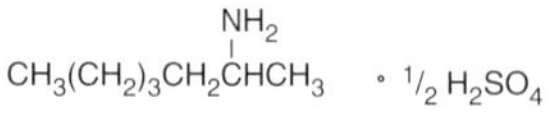
5 gm

S78,693-4
2-AMINOOCTANE
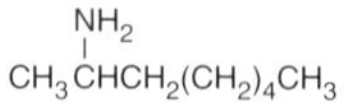
250 mg

S54,029-3
1-METHYLDECYLAMINE
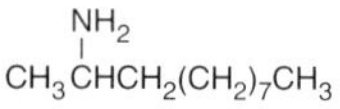
5 gm

S91,825-3
2-METHYL-2-TRIDECYLAMINE HYDROCHLORIDE
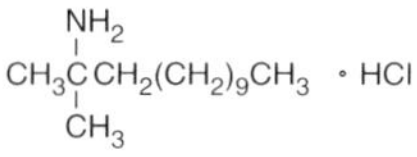
250 mg

S54,023-4
1-HEPTYLOCTYLAMINE
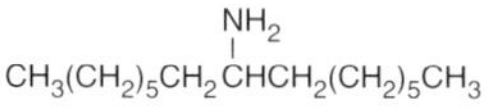
5 gm

S81,456-3
1,2-DIETHYLHYDRAZINE
$CH_3CH_2NHNHCH_2CH_3$
250 mg

S76,488-4
N,N'-DIISOPROPYLHYDRAZINE MONOHYDROCHLORIDE
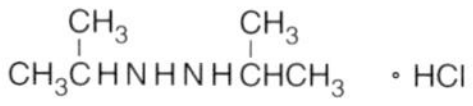
5 gm

S57,301-9
ETHYLENEDIAMINE DIHYDROIODIDE
$H_2NCH_2CH_2NH_2$ • 2HI
5 gm

S64,150-2
DIISOPROPYLAMINE HYDROCHLORIDE
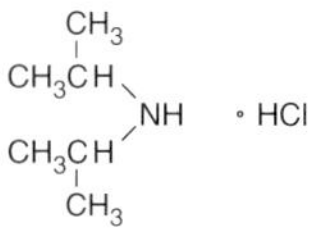
250 mg

S68,071-0
N-ISOPROPYLPROPYLAMINE
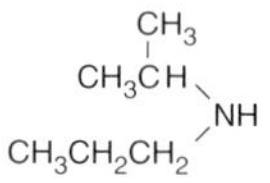
5 gm

S87,151-6
METHYL-TERT-PENTYLAMINE HYDROCHLORIDE
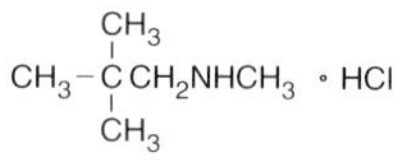
250 mg

S68,070-2
N-ISOPROPYLBUTYLAMINE
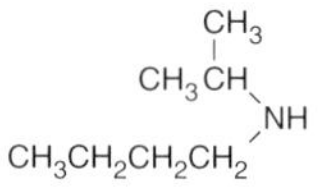
5 gm

S37,997-2
N,1-DIMETHYLHEXYLAMINE
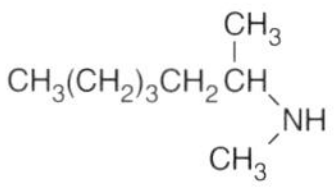
5 gm

S56,615-2
DIBUTYLAMINE HYDROCHLORIDE
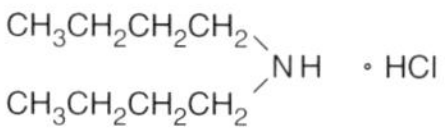
5 gm

S83,567-6
DIISOBUTYLAMINE HYDROCHLORIDE
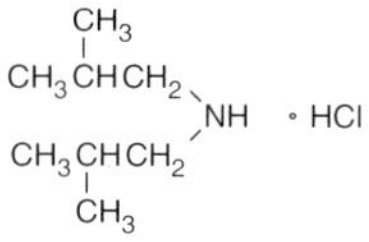
250 mg

S70,589-6
DIISOAMYLAMINE
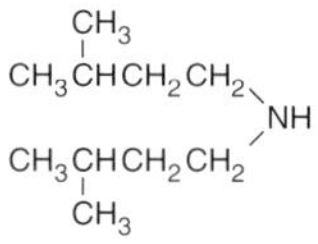
5 gm

S65,495-7
N-ISOPROPYL-1,5-DIMETHYLHEXYLAMINE
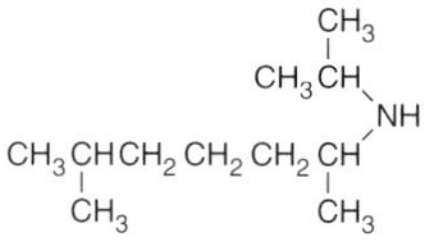
5 gm

S55,330-1
BIS(1,3-DIMETHYLBUTYL)AMINE
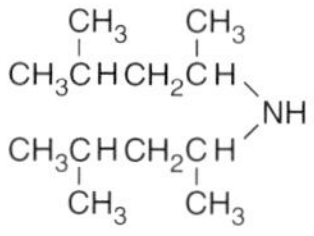
5 gm

S86,681-4
TERT-BUTYL-TERT-OCTYLAMINE
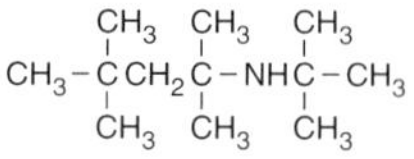
250 mg

S55,941-5
DIHEPTYLAMINE
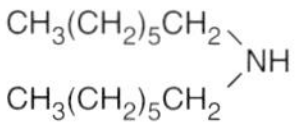
5 gm

S55,578-9
N-METHYLHEXADECYLAMINE
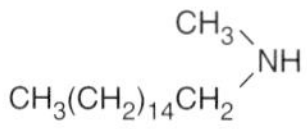
250 mg

S59,579-9
N-METHYLHEXADECYLAMINE HYDROCHLORIDE
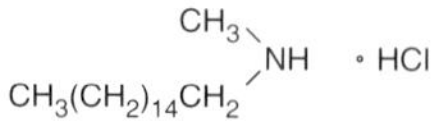
250 mg

S50,589-7
N-ISOPROPYLHEXADECYLAMINE
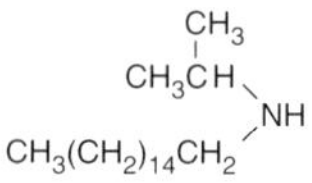
5 gm

S51,651-1
DIUNDECYLAMINE
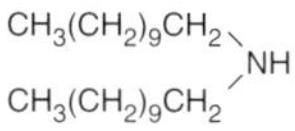
5 gm

S41,830-7
DIDODECYLAMINE
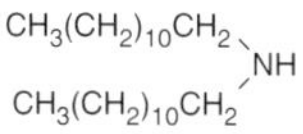
250 mg

S50,958-2
DITRIDECYLAMINE
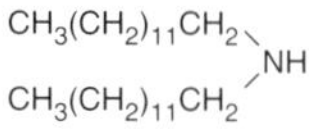
5 gm

S40,830-1
DITETRADECYLAMINE
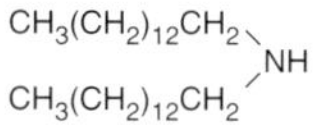
5 gm

S81,180-7
DIOCTADECYLAMINE HYDROCHLORIDE
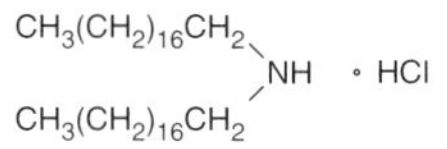
250 mg

S56,322-6
DIHEXYLAMINE HYDROCHLORIDE
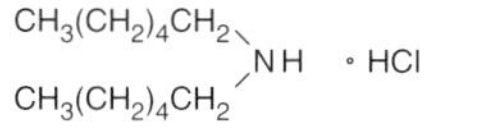
5 gm

S44,283-6
DINONYLAMINE
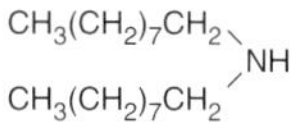
250 mg

S64,782-9
N-METHYLHYDROXYLAMINE OXALATE
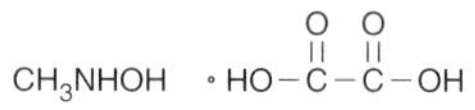
250 mg

S43,463-9
TRIMETHYLAMINE FORMATE
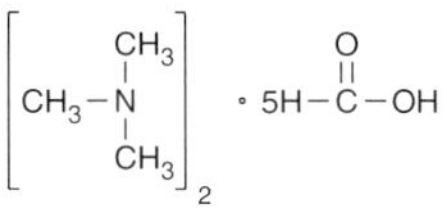
5 gm

S86,314-9
TRIETHYLAMINE METHANESULFONATE
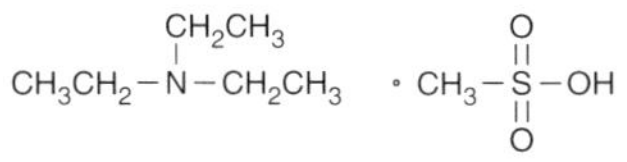
250 mg

S88,214-3
DIISOPROPYL(1-ETHYLPROPYL)AMINE
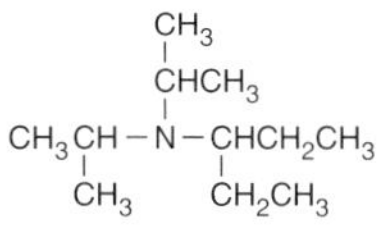
250 mg

S55,428-6
N-METHYLDIHEXYLAMINE
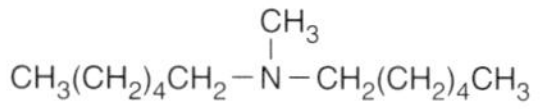
250 mg

S83,094-1
DIBUTYLPENTADECYLAMINE
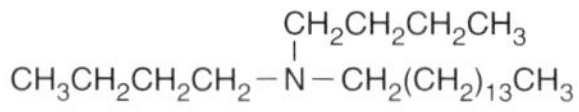
250 mg

S81,062-2
DIBUTYLHEXADECYLAMINE
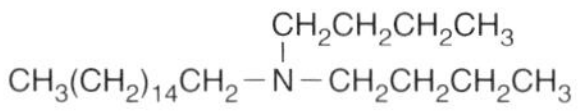
250 mg

S83,097-6
DIBUTYLHEPTADECYLAMINE
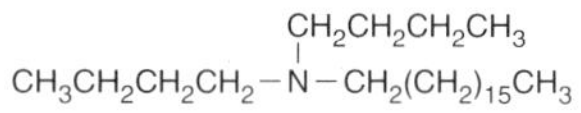
250 mg

S81,199-8
DIBUTYLOCTADECYLAMINE
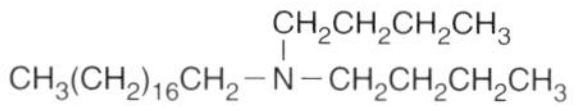
250 mg

S83,899-3
TRIDODECYLAMINE HYDROCHLORIDE
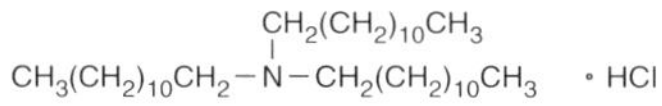
250 mg

S81,181-5
TRIOCTADECYLAMINE
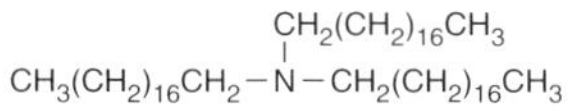
250 mg

S82,824-6
TRIOCTADECYLAMINE HYDROCHLORIDE
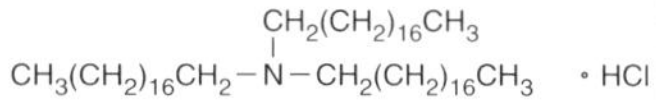
250 mg

S79,269-1
N-CYCLOHEXYLDIMETHYLAMINE N-OXIDE
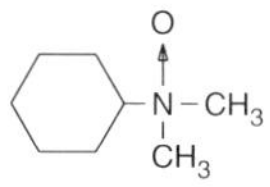
250 mg

S94,830-6
N,N-3,5-TETRAMETHYL-1-ADAMANTANEMETHYLAMINE HYDROCHLORIDE
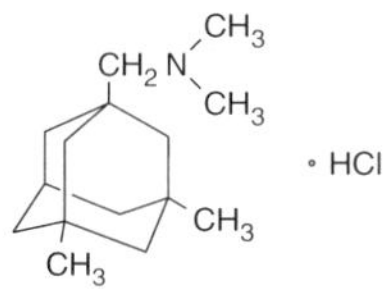
250 mg

S94,831-4
N,N-DIETHYL-3,5-DIMETHYL-1-ADAMANTYLMETHYLAMINE HYDROCHLORIDE
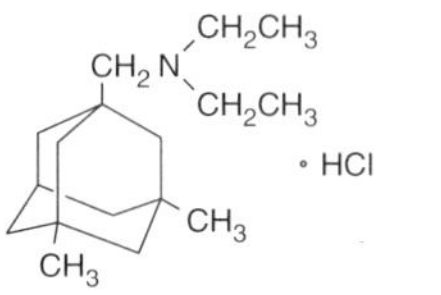
250 mg

S94,833-0
N,N-DIBUTYL-2-(1-ADAMANTYL)-ETHYLAMINE HYDROCHLORIDE
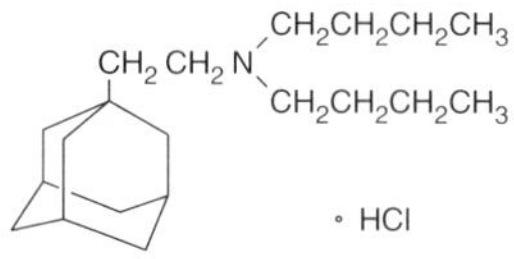
250 mg

S93,161-6
1,4,8,11-TETRAAZAUNDECANE TETRAHYDROCHLORIDE

$H_2NCH_2CH_2NHCH_2CH_2CH_2NHCH_2CH_2NH_2$
• 4HCl

250 mg

S93,160-8
5-METHYL-1,4,8,11-TETRAAZAUNDECANE TETRAHYDROCHLORIDE
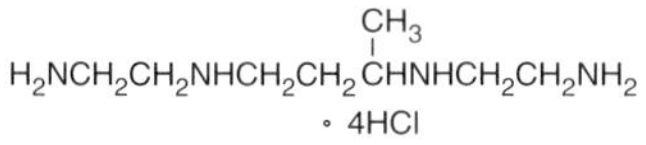
250 mg

S93,162-4
1,5,8,12-TETRAAZADODECANE TETRAHYDROCHLORIDE

$H_2NCH_2CH_2CH_2NHCH_2CH_2NHCH_2CH_2CH_2NH_2$
• 4HCl

250 mg

S93,164-0
1,4,9,12-TETRAAZADODECANE TETRAHYDROCHLORIDE

$H_2NCH_2CH_2NHCH_2CH_2CH_2CH_2NHCH_2CH_2NH_2$
• 4HCl

250 mg

S51,641-4
N-(SEC-BUTYL)-ETHYLENEDIAMINE
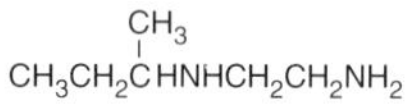
5 gm

S58,835-0
N-BUTYLETHYLENEDIAMINE

$CH_3CH_2CH_2CH_2NHCH_2CH_2NH_2$

250 mg

S37,565-9
N-TETRADECYLETHYLENEDIAMINE

$CH_3(CH_2)_{12}CH_2NHCH_2CH_2NH_2$

5 gm

S39,400-9
1,1-DIMETHYLDIETHYLENETRIAMINE
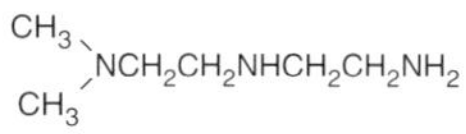
5 gm

S43,947-9
BIS(2-AMINOPROPYL)AMINE
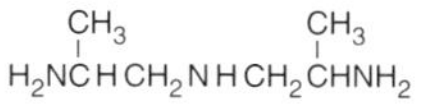
5 gm

S38,850-5
N1,N1-DIMETHYL-1,2-PROPANEDIAMINE
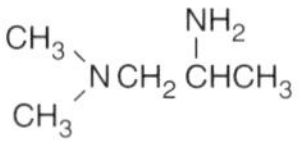
5 gm

S38,864-5
N2,N2-DIMETHYL-1,2-PROPANEDIAMINE
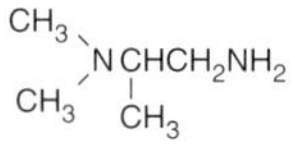
5 gm

S38,872-6
N,N-DIISOPROPYLETHYLENEDIAMINE
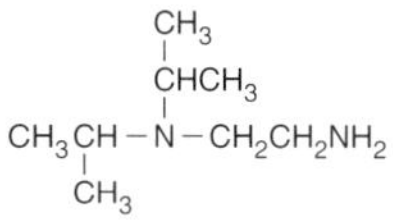
5 gm

S52,274-0
2,2-DIETHYL-1,3-PROPANEDIAMINE
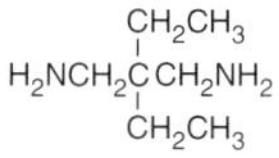
250 mg

S51,187-0
N-(2-ETHYLHEXYL)-1,3-PROPANEDIAMINE

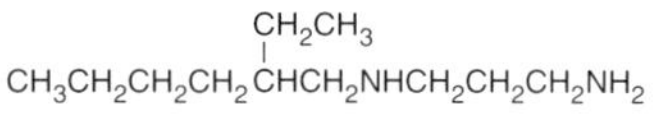

5 gm

S36,604-8
3-(TERT-DODECYLAMINO)-PROPYLAMINE

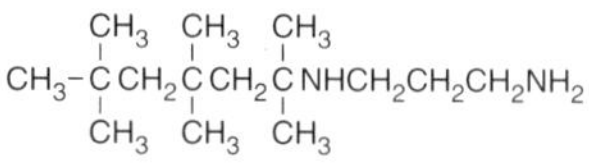

5 gm

S51,192-7
N-OCTADECYL-1,3-PROPANEDIAMINE

$CH_3(CH_2)_{16}CH_2NHCH_2CH_2CH_2NH_2$

5 gm

S55,094-9
N,N-DIISOBUTYL-1,6-HEXANEDIAMINE

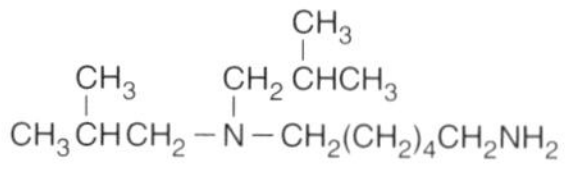

5 gm

S38,847-5
N-ETHYL-N'-METHYLETHYLENEDIAMINE

$CH_3CH_2NHCH_2CH_2NHCH_3$

5 gm

S56,648-9
N,N'-DIPROPYLETHYLENEDIAMINE

$CH_3CH_2CH_2NHCH_2CH_2NHCH_2CH_2CH_3$

5 gm

S51,643-0
N,N'-DI-TERT-BUTYL-ETHYLENEDIAMINE

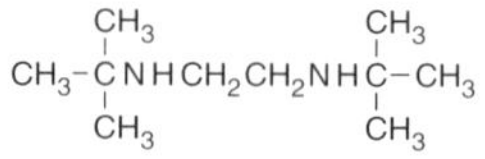

250 mg

S54,050-1
1,7-DI-(SEC-BUTYL)-DIETHYLENETRIAMINE

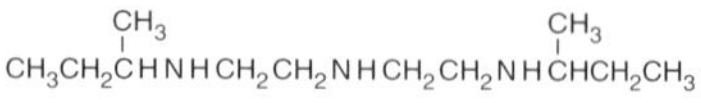

5 gm

S56,442-7
N,N-DIBUTYL-N'-METHYLETHYLENEDIAMINE

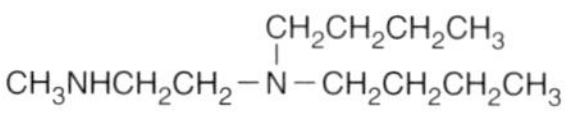

5 gm

S38,848-3
1,4,7-TRIMETHYLDIETHYLENETRIAMINE

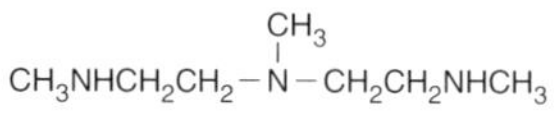

5 gm

S55,691-2
N,N'-BIS(2,4-DIMETHYL-3-PENTYL)-ETHYLENEDIAMINE

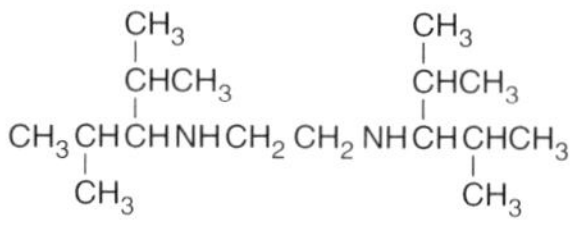

5 gm

S74,575-8
N,N,N',N'-TETRAMETHYLETHYLENEDIAMINE DIHYDROCHLORIDE

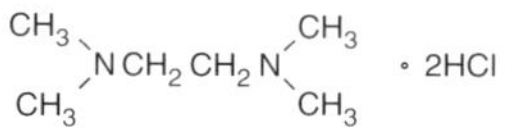

250 mg

S36,498-3
N,N,N',N'-TETRAMETHYL-1,2-PROPANEDIAMINE

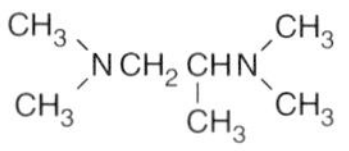

250 mg

S38,844-0
N,N'-DIETHYL-N,N'-DIMETHYLETHYLENEDIAMINE

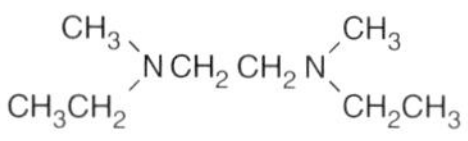

5 gm

S38,852-1
N,N-DIETHYL-N',N'-DIMETHYLETHYLENEDIAMINE

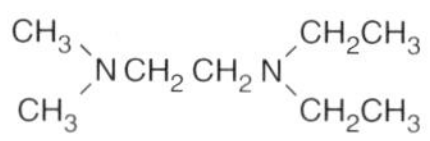

5 gm

S38,834-3
4-METHYL-1,1,7,7-TETRAETHYLDIETHYLENETRIAMINE

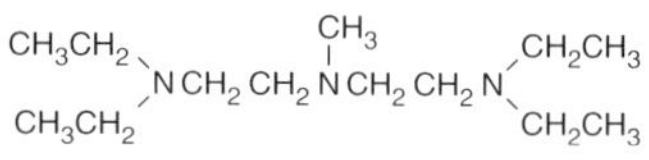

5 gm

S56,350-1
N,N'-DIMETHYL-1,3-PROPANEDIAMINE

$CH_3NHCH_2CH_2CH_2NHCH_3$

5 gm

S38,868-8
N,N,N'-TRIMETHYL-1,3-PROPANEDIAMINE

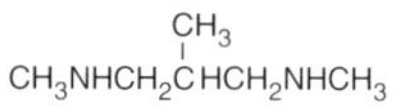

5 gm

S37,925-5
N,N-DIETHYL-N'-METHYL-1,3-PROPANEDIAMINE

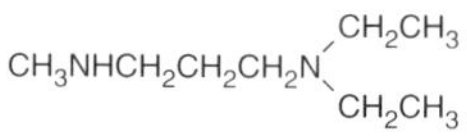

5 gm

S56,328-5
N,N-DIBUTYL-N'-METHYL-1,3-PROPANEDIAMINE

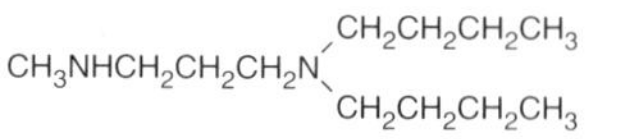

5 gm

S38,855-6
N,N'-DIETHYL-1,4-BUTANEDIAMINE

$CH_3CH_2NHCH_2CH_2CH_2CH_2NHCH_2CH_3$

5 gm

S38,846-7
N,N'-DIETHYL-1,6-HEXANEDIAMINE

$CH_3CH_2NHCH_2(CH_2)_4CH_2NHCH_2CH_3$

5 gm

S39,395-9
N,N'-DIPROPYL-1,6-HEXANEDIAMINE

$CH_3CH_2CH_2NHCH_2(CH_2)_4CH_2NHCH_2CH_2CH_3$

5 gm

S43,302-0
N,N,N',N'-TETRAETHYLMETHANEDIAMINE

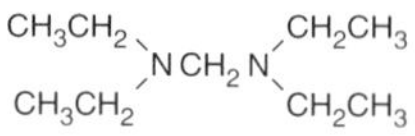

5 gm

S38,860-2
N,N,N',N'-TETRABUTYLMETHANEDIAMINE

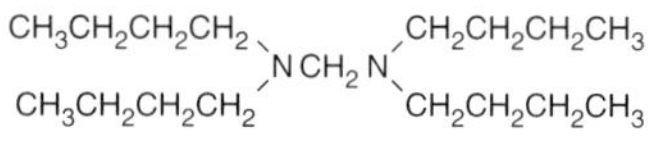

5 gm

S37,159-9
METHYLHYDRAZINE SULFATE

$CH_3NHNH_2 \cdot H_2SO_4$

5 gm

S87,142-7
METHYLHYDRAZINE HYDROCHLORIDE

$CH_3NHNH_2 \cdot HCl$

250 mg

S87,143-5
1-TERT-PENTYLHYDRAZINE HYDROCHLORIDE

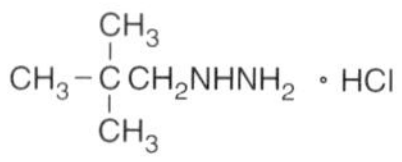

250 mg

S85,958-3
HEPTYLHYDRAZINE OXALATE

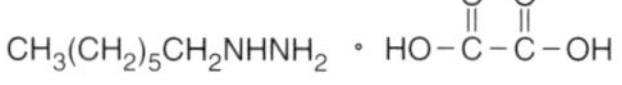

250 mg

S43,211-3
5-AMINO-1,3,3-TRIMETHYLCYCLOHEXANEMETHYLAMINE

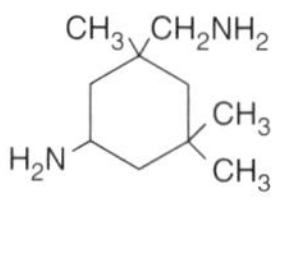

5 gm

S93,883-1
1,4-DIAMINOCYCLOHEXANE, MIXTURE OF CIS AND TRANS

5 gm

S50,936-1
TRANS,TRANS-4,4'-METHYLENEBIS-(CYCLOHEXYLAMINE)

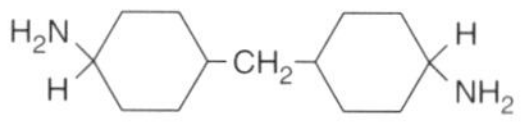

250 mg

S50,938-8
CIS,TRANS-4,4'-METHYLENEBIS-(CYCLOHEXYLAMINE)

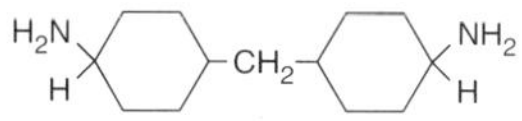

250 mg

S45,710-8
(+)-3-PINANEMETHYLAMINE HYDROCHLORIDE

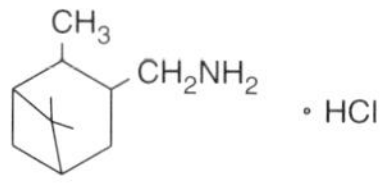

5 gm

S45,788-4
(-)-3-PINANEMETHYLAMINE HYDROCHLORIDE

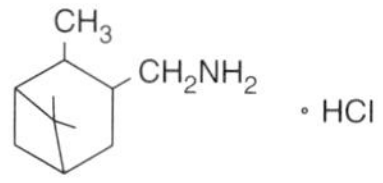

5 gm

S76,285-7
2-AMINOSPRIO(4.5)DECANE HYDROCHLORIDE

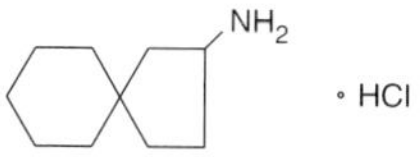

250 mg

S51,286-9
CYCLOHEXYLAMINE DIBORATE MONOHYDRATE

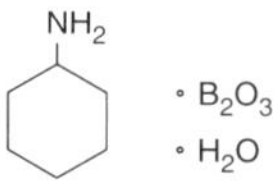

5 gm

S51,287-7
CYCLOHEXYLAMINE TRIBORATE HYDRATE

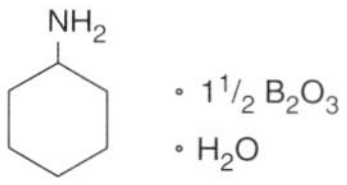

5 gm

S51,288-5
CYCLOHEXYLAMINE TETRABORATE TETRAHYDRATE

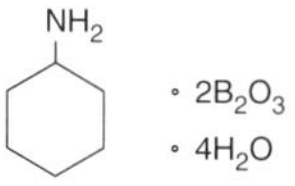

5 gm

S66,982-2
CYCLOHEXYLAMINE HYDROBROMIDE

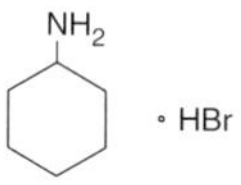

5 gm

S92,650-7
CYCLOHEXYLAMINE HYDROCHLORIDE

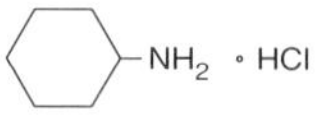

250 mg

S43,031-5
1,4-BIS(AMINOMETHYL)CYCLOHEXANE (MIXTURE OF ISOMERS)

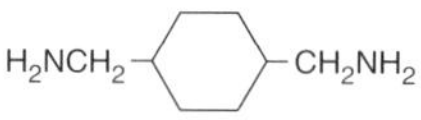

5 gm

S57,089-3
3,3,5-TRIMETHYLCYCLOHEXYLAMINE

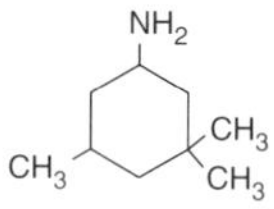

5 gm

S43,782-4
CYCLOHEXANEBUTYLAMINE

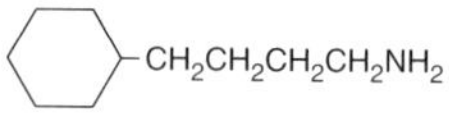

250 mg

S51,997-9
2-BICYCLOHEXYLAMINE

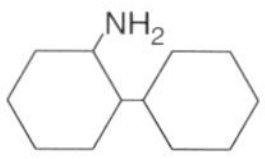

5 gm

S41,823-4
4-TERT-BUTYLCYCLOHEXYLAMINE

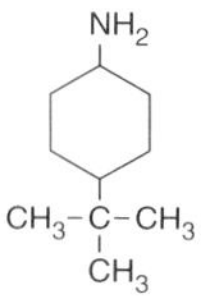

5 gm

S75,794-2
3-AMINOSPIRO(5.5)UNDECANE HYDROCHLORIDE

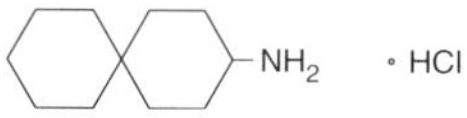

5 gm

S76,239-3
3-AMINOSPIRO(5.5)UNDECANE

250 mg

S92,269-2
1-(AMINOMETHYL)CYCLOOCTANE

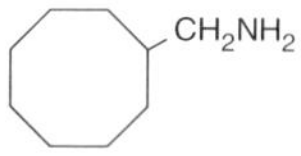

5 gm

S45,546-6
1-ADAMANTANAMINE CITRATE

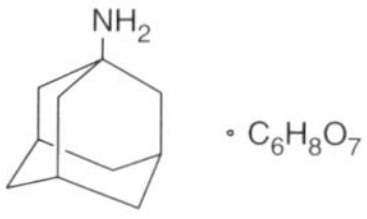

5 gm

S68,341-8
N-CYCLOOCTYLMETHYLAMINE

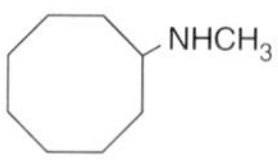

5 gm

S42,848-5
N-METHYLCYCLODODECYLAMINE

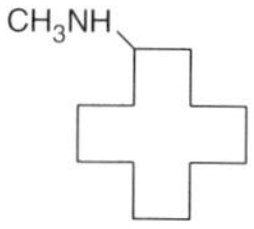

250 mg

S52,211-2
TRANS-N,N'-DIETHYL-1,4-CYCLOHEXYLENEDIAMINE HYDROCHLORIDE

250 mg

S42,970-8
N-METHYLCYCLOPENTYLAMINE

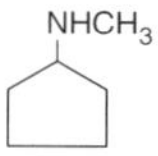

5 gm

S40,591-4
DICYCLOPENTYLAMINE

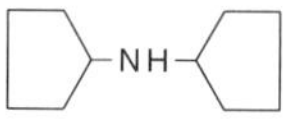

250 mg

S85,972-9
DICYCLOHEXYLAMINE PHOSPHATE

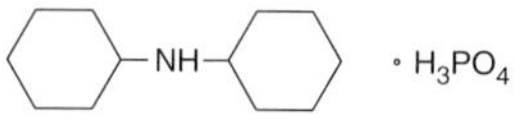

250 mg

S51,836-0
BIS(4-CYCLOHEXYLBUTYL)AMINE

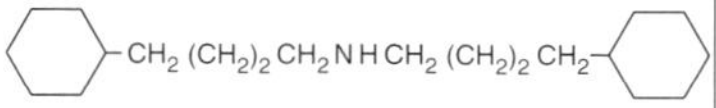

5 gm

S59,883-6
N-CYCLOHEXYLHEXADECYLAMINE

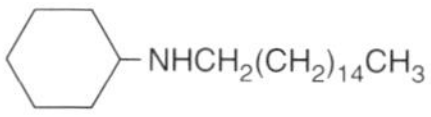

5 gm

S79,250-0
N-CYCLOHEXYLTRICYCLOHEXYLMETHYLAMINE

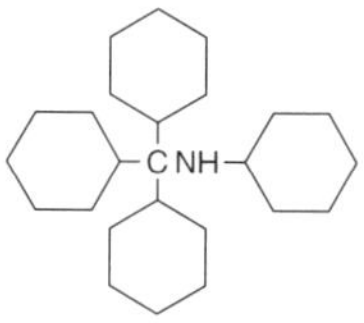

250 mg

S36,967-5
N-METHYLDICYCLOPENTYLAMINE

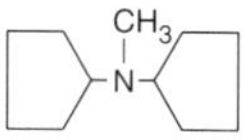

5 gm

S94,656-7
N,N-DIMETHYL-4-CYCLOOCTENE-1-AMINE

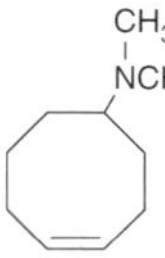

250 mg

S85,943-5
CYCLOHEXYLHYDRAZINE OXALATE

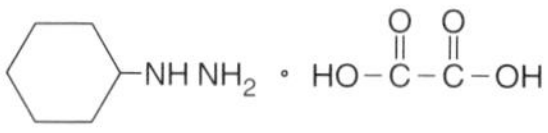

250 mg

S62,243-5
2-AMINO-2-BUTYLDECALIN

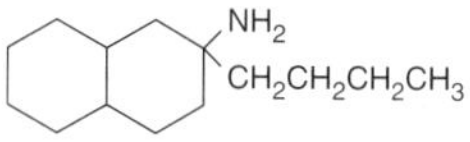

250 mg

S38,164-0
N-(1,5-DIMETHYL-4-HEXENYLIDENE)-ISOBUTYLAMINE

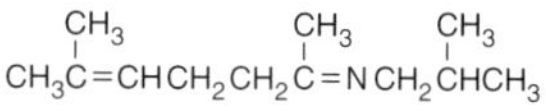

250 mg

S86,676-8
[2-(1,4-CYCLOHEXADIEN-1-YLOXY)-ETHYL]DIETHYLAMINE

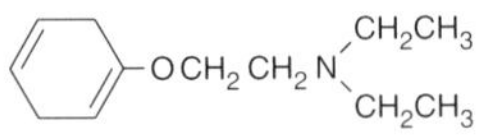

250 mg

S62,139-0
N-PROPYL-1,4-DIAMINO-2-BUTENE

$CH_3CH_2CH_2NHCH_2CH{=}CHCH_2NH_2$

250 mg

S91,495-9
2-ALLYLCYCLOPENTYLAMINE

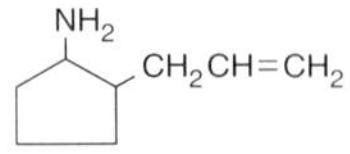

250 mg

S87,364-0
1,2-BIS(DIETHYLAMINO)ETHYLENE

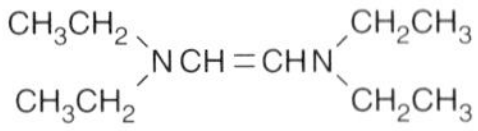

250 mg

S38,858-0
N,N'-DIMETHYL-2-BUTENE-1,4-DIAMINE

$CH_3NHCH_2CH{=}CHCH_2NHCH_3$

5 gm

S43,300-4
N,N,N',N'-TETRAETHYL-2-BUTENE-1,4-DIAMINE

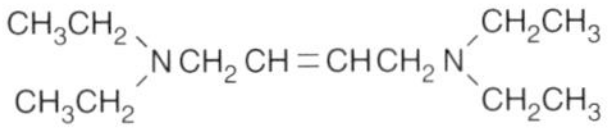

5 gm

S45,678-0
BIS(2-METHYLALLYL)AMINE

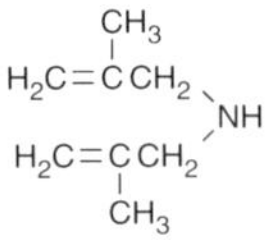

250 mg

S88,472-3
ALLYLHEXYLAMINE

$H_2C{=}CH\ CH_2NHCH_2(CH_2)_4CH_3$

250 mg

S38,106-3
1,5-DIMETHYL-N-ISOPROPYL-4-HEXENYLAMINE

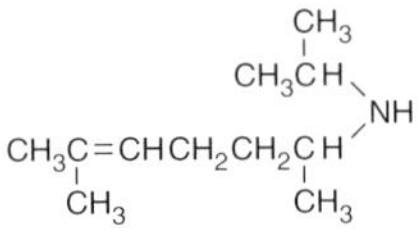

250 mg

S38,547-6
N-BUTYL-1,5-DIMETHYL-4-HEXENYLAMINE

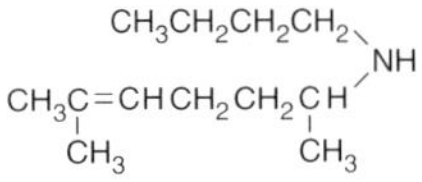

250 mg

S36,529-7
N,N,2-TRIMETHYLPROPENYLAMINE

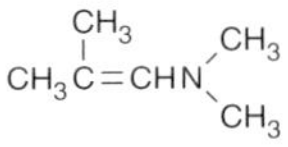

5 gm

S79,986-6
N,N-DIISOPROPYL-2-METHYL-1-PROPENYLAMINE

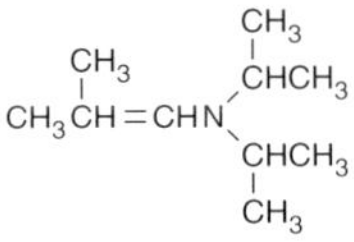

250 mg

S86,685-7
6,6-DIMETHYL-2-[2-(DIMETHYLAMINO)-ETHYL]BICYCLO[3.1.1]HEPT-2-ENE

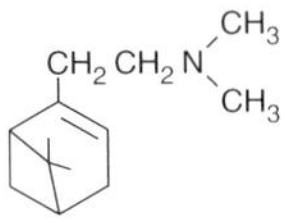

250 mg

S95,041-6
11-AZATRICYCLO[4.4.1.0(1,6)] UNDECA-3,8-DIENE

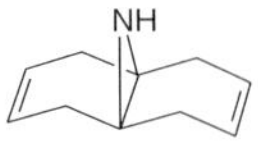

50 mg

S71,101-2
N-(2-CHLOROETHYL)ISOPROPYLAMINE HYDROCHLORIDE

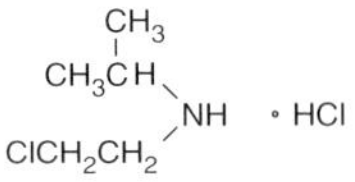

5 gm

S45,439-7
3-DIETHYLAMINOPROPYL CHLORIDE HYDROCHLORIDE

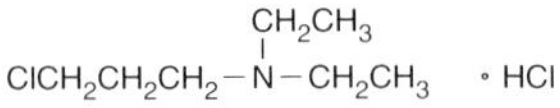

5 gm

S37,626-4
N-(2-CHLOROETHYL)DIBUTYLAMINE HYDROCHLORIDE

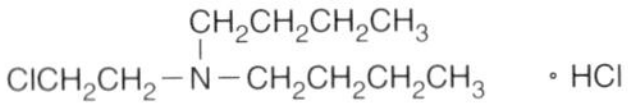

5 gm

S39,156-5
(1-METHYL-2,2,3,3-TETRAFLUOROPROPYL)HYDRAZINE

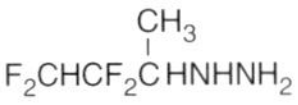

250 mg

S45,345-5
1,1,1,3,3,3-HEXAFLUORO-2,2-PROPANEDIAMINE

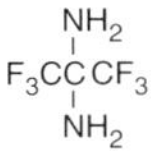

250 mg

S74,201-5
4-(4-IODOCYCLOHEXYL)CYCLOHEXYLAMINE

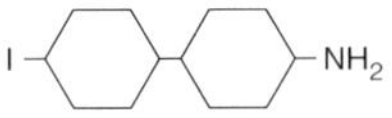

5 gm

S93,809-2
3-CHLORO-1-ADAMANTANEMETHYLAMINE HYDROCHLORIDE

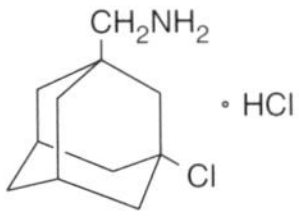

250 mg

S37,387-7
2-CHLORO-N,N,N',N'-TETRAETHYL-1,3-PROPANEDIAMINE

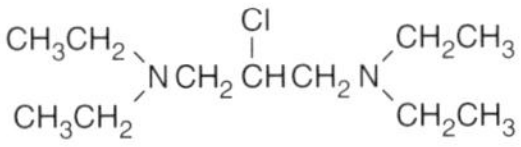

5 gm

S74,200-7
N,N-DIMETHYL-4-(4-IODOCYCLOHEXYL)-CYCLOHEXYLAMINE

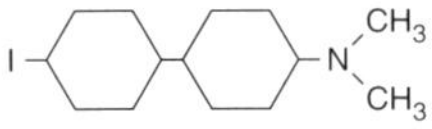

5 gm

S87,926-6
N-DODECYLGLUCAMINE

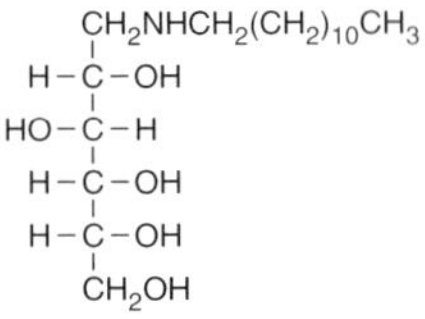

5 gm

S92,205-6
1-AMINO-1-DEOXY-D-DULCITOL

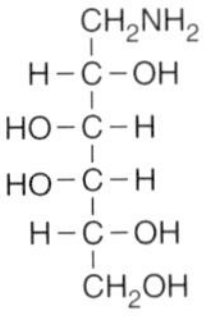

250 mg

S90,680-8
2-(AMINOMETHYL)-2-ETHYL-1,3-PROPANEDIOL

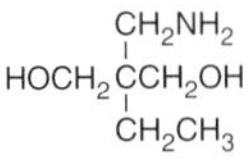

250 mg

S76,481-7
N,N-BIS(2-HYDROXYETHYL)-ETHYLENEDIAMINE DIHYDROCHLORIDE

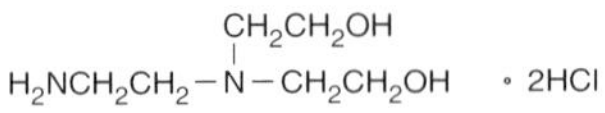

5 gm

S77,048-5
N,N-BIS(2-HYDROXYETHYL)-ETHYLENEDIAMINE HYDROCHLORIDE

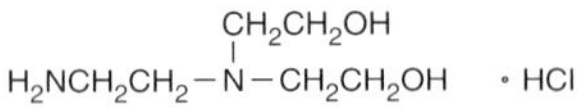

250 mg

S70,540-3
N-(2-HYDROXYETHYL)-N,N',N'-TRIS(3-HYDROXYPROPYL)ETHYLENEDIAMINE

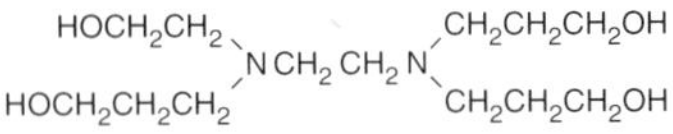

5 gm

S34,543-1
3,3'-DIALLYLOXY-2,2'-DIHYDROXYDIPROPYLAMINE

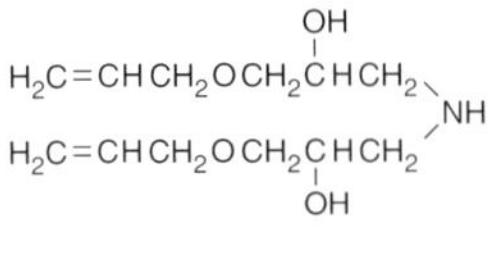

5 gm

S35,002-8
2-(2-DIETHYLAMINOETHOXY)ETHANOL

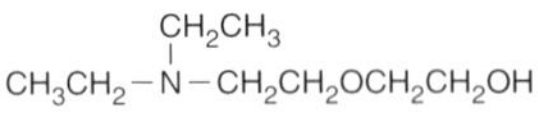

5 gm

S57,389-2
NEAMINE

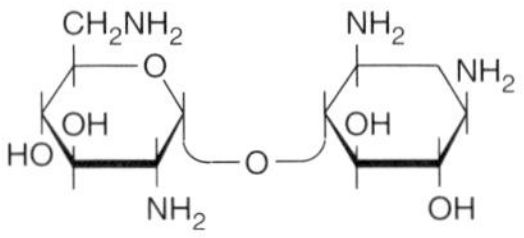

250 mg

S64,138-3
2-(DIMETHYLAMINO)ETHYL-BETA-D-GLUCOSIDE HYDROCHLORIDE
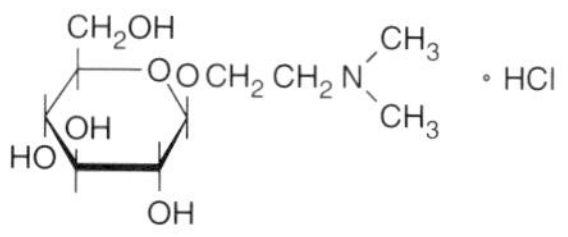
250 mg

S43,832-4
1-(2-AMINOETHYLAMINO)-2-PROPANOL
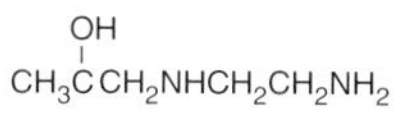
5 gm

S35,075-3
N1-(2-HYDROXYETHYL)-2-METHYL-1,2-PROPANEDIAMINE
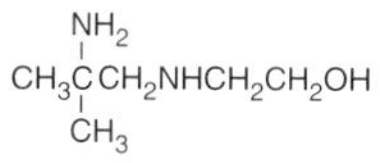
5 gm

S34,702-7
N1-(1,1-DIMETHYL-2-HYDROXYETHYL)-2-METHYL-1,2-PROPANEDIAMINE
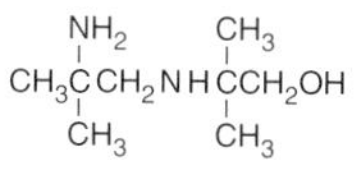
5 gm

S44,683-1
1,3-DIAMINO-N-(2-HYDROXYETHYL)-PROPANE
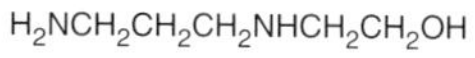
5 gm

S50,894-2
2-AMINO-2-METHYL-3-HEXANOL
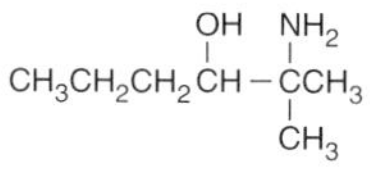
250 mg

S52,244-9
2-[2-(METHYLAMINO)ETHYLAMINO]-ETHANOL
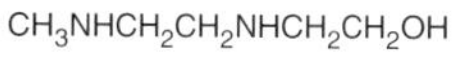
5 gm

S38,862-9
2-(2-DIETHYLAMINOETHYLAMINO)-ETHANOL
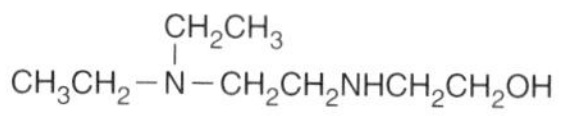
5 gm

S67,756-6
1,3-BIS(TERT-BUTYLAMINO)-2-PROPANOL HYDROCHLORIDE
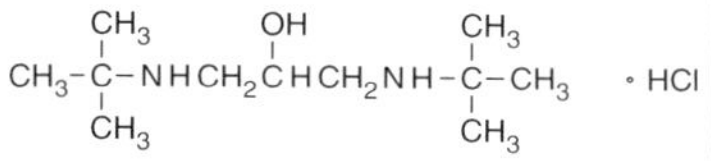
250 mg

S40,840-9
3-(2-HYDROXYETHYLAMINO)-1-PROPANOL
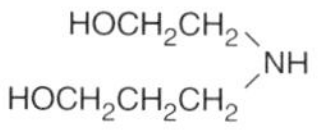
5 gm

S37,340-0
1-ETHYLAMINO-2-METHYL-2-PROPANOL HYDROCHLORIDE
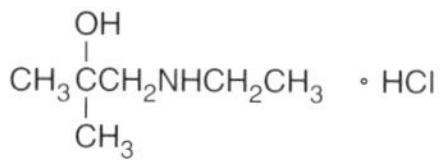
5 gm

S41,837-4
2-(BUTYLAMINO)-ETHANOL
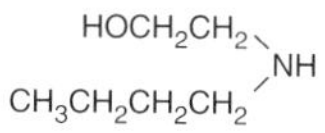
5 gm

S88,414-6
4-(ALLYLAMINO)-4-METHYL-2-PENTANOL
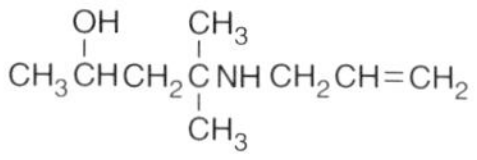
250 mg

S88,410-3
4-(2-BUTEN-1-YLAMINO)-4-METHYL-2-PENTANOL
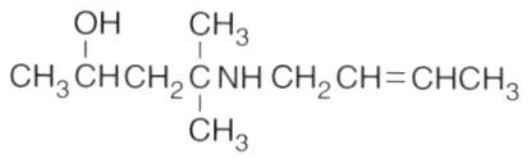
250 mg

S57,122-9
5,5'-IMINODI-1-PENTANOL
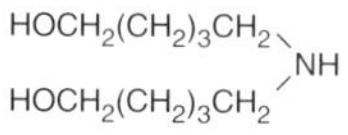
250 mg

S36,137-2
1-(2-HYDROXYETHYLAMINO)-2-PROPANOL

5 gm

S61,994-9
1-PROPYLAMINO-3-BUTEN-2-OL
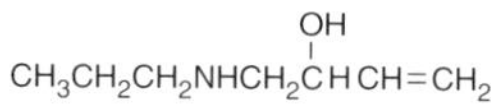
250 mg

S39,396-7
2-(2-METHYLBUTYLAMINO)ETHANOL
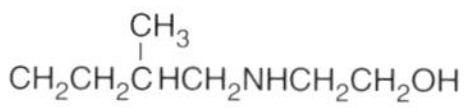
5 gm

S39,398-3
2-(ISOPENTYLAMINO)ETHANOL
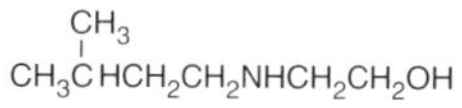
5 gm

S39,399-1
2-(PENTYLAMINO)ETHANOL
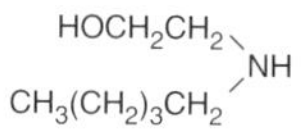
5 gm

S50,907-8
1-ISOPROPYLAMINO-2-METHYL-2-PROPANOL
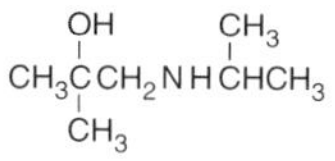
5 gm

S51,189-7
2-BUTYLAMINO-1-PROPANOL
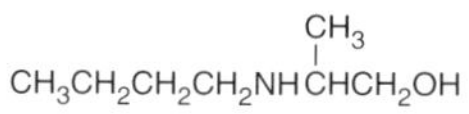
5 gm

S55,812-5
3,3'-IMINODI-2-BUTANOL
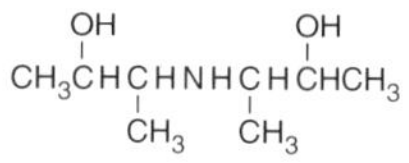
5 gm

S51,166-8
1,1'-DIETHYL-2,2',2''-NITRILOTRIETHANOL
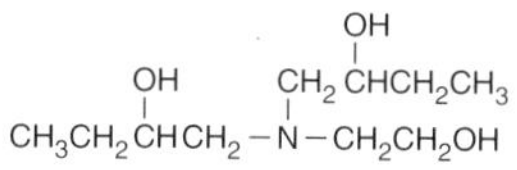
5 gm

S54,028-5
2-(1-METHYLOCTYLAMINO)ETHANOL
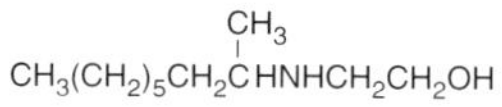
5 gm

S51,173-0
1,1'-DIMETHYL-1,1'-DIPROPYL-2,2'-IMINODIETHANOL
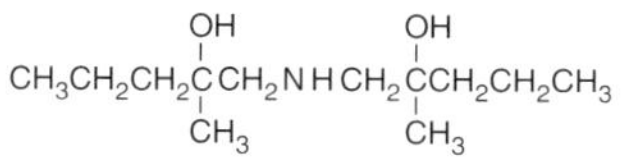
5 gm

S50,909-4
1-CYCLOHEXYLAMINO-2-PROPANOL
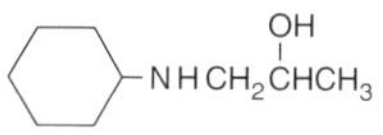
5 gm

S50,910-8
1-CYCLOHEXYLAMINO-2-PROPANOL HYDROCHLORIDE
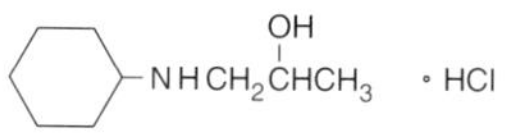
5 gm

S51,193-5
N-CYCLOHEXYL-2,2'-IMINODIETHANOL
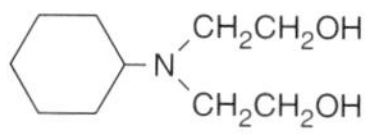
5 gm

S42,869-8
2-CYCLOHEXYLAMINOETHANOL
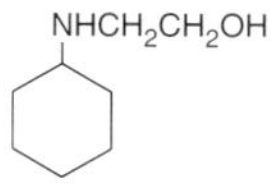
5 gm

S37,842-9
4-DIMETHYLAMINO-1-BUTANOL
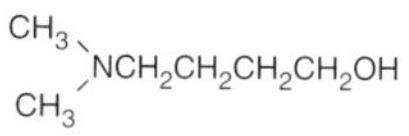
5 gm

S40,797-6
2-DIMETHYLAMINO-2-METHYL-1-PROPANOL
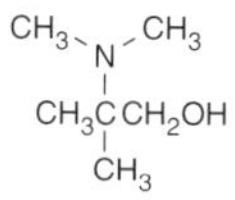
5 gm

S37,949-2
5-DIMETHYLAMINO-1-PENTANOL
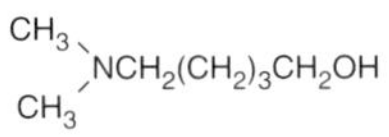
250 mg

S55,318-2
N-ALLYL-2,2'-IMINODIETHANOL
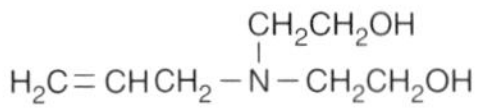
5 gm

S61,996-5
1-DIETHYLAMINO-3-BUTEN-2-OL
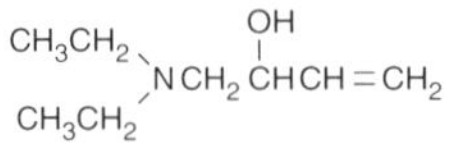
5 gm

S37,792-9
3-DIETHYLAMINO-2,2-DIMETHYL-1-PROPANOL
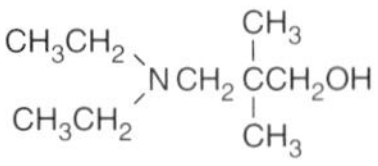
5 gm

S37,666-3
4-DIISOPROPYLAMINO-1-BUTANOL
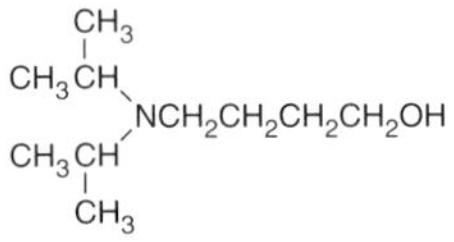
5 gm

S38,045-8
4-DIPROPYLAMINO-1-BUTANOL
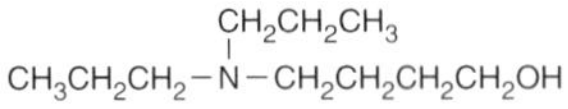
5 gm

S38,107-1
3-DIBUTYLAMINO-1-PROPANOL
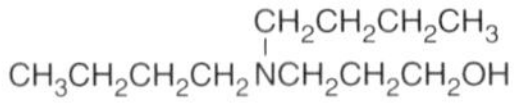
5 gm

S38,130-6
5-DIPROPYLAMINO-1-PENTANOL
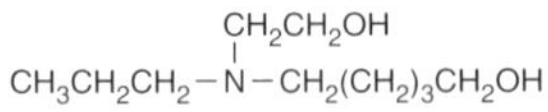
5 gm

S37,723-6
4-(DIISOBUTYLAMINO)-1-BUTANOL
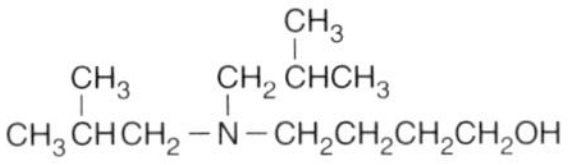
5 gm

S37,758-9
6-DIPROPYLAMINO-1-HEXANOL
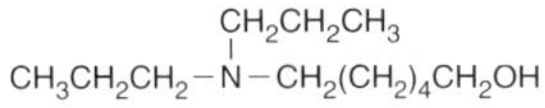
5 gm

S39,189-1
4-DIBUTYLAMINO-1-BUTANOL
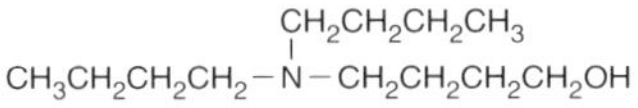
5 gm

S57,312-4
2-(DIMETHYLAMINO)ETHANOL HYDROCHLORIDE
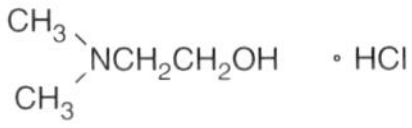
5 gm

S57,326-4
2-(DIMETHYLAMINO)ETHANOL BITARTRATE
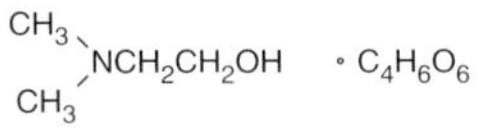
5 gm

S37,930-1
N-ISOPROPYL-2,2'-IMINODIETHANOL
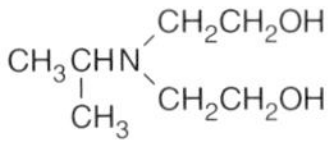
5 gm

S37,985-9
2-(DIPROPYLAMINO)ETHANOL
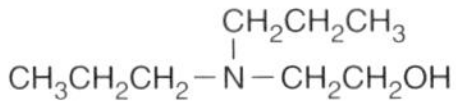
5 gm

S50,428-9
2,2'-(TERT-BUTYLIMINO)DIETHANOL
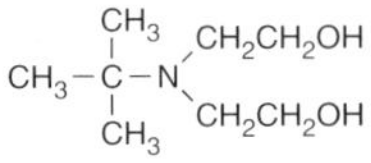
5 gm

S34,229-7
N,N-BIS(2-HYDROXYPROPYL)-ETHANOLAMINE
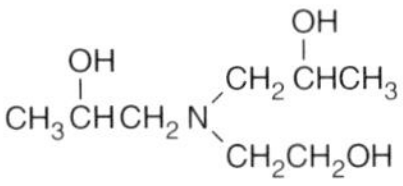
5 gm

S34,777-9
1-DIPROPYLAMINO-2-PROPANOL
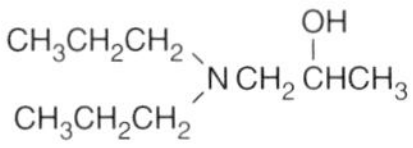
5 gm

S61,995-7
1-DIPROPYLAMINO-2-BUTANOL
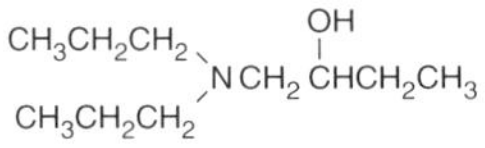
5 gm

S34,885-6
1-DIBUTYLAMINO-2-PROPANOL
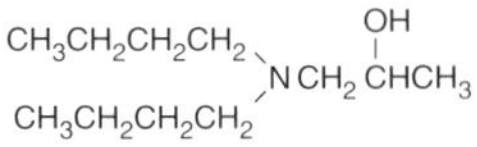
5 gm

S34,221-1
1,3-BIS(DIETHYLAMINO)-2-PROPANOL
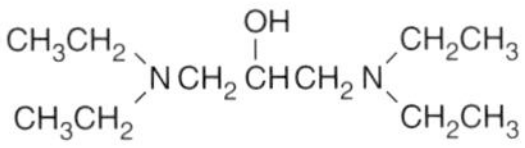
5 gm

S61,998-1
1-DIBUTYLAMINO-3-BUTEN-2-OL
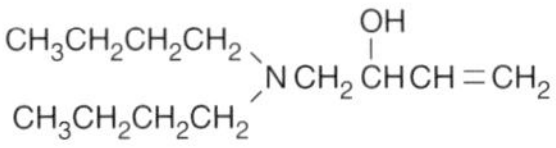
5 gm

S61,997-3
1-DIBUTYLAMINO-2-BUTANOL
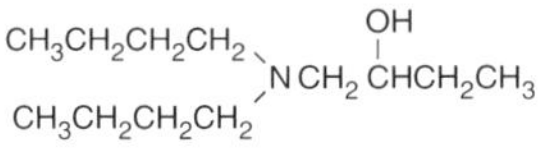
5 gm

S56,610-1
2-(DIOCTYLAMINO)ETHANOL
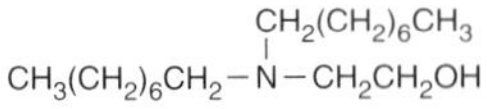
5 gm

S51,191-9
1,1'-CYCLOHEXYLIMINODI-2-PROPANOL
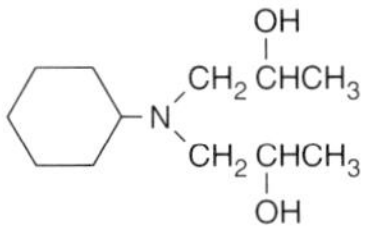
5 gm

S67,094-4
2-HYDROXYBUTYLHYDRAZINE
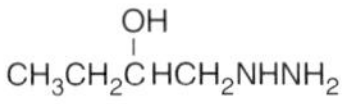
5 gm

S44,106-6
N,N-DIMETHYL-2-(VINYLOXY)-ETHYLAMINE
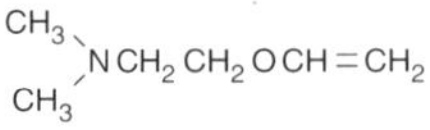
5 gm

S43,763-8
3-(2-METHOXYETHOXY)PROPYLAMINE

$CH_3OCH_2CH_2OCH_2CH_2CH_2NH_2$

5 gm

S50,290-1
BI-[3-(2-METHOXYETHOXY)-PROPYLAMINE]

$H_2NCH_2CH_2CH_2(OCH_2CH_2)_3OCH_2CH_2CH_2NH_2$

250 mg

S37,751-1
3,6,9-TRIOXAPENTADECYLAMINE

$CH_3(CH_2)_4CH_2OCH_2CH_2OCH_2CH_2OCH_2CH_2NH_2$

250 mg

S38,204-3
2-[2-(NONYLOXY)ETHOXY]ETHYLAMINE

$CH_3(CH_2)_7CH_2OCH_2CH_2OCH_2CH_2NH_2$

250 mg

S38,391-0
2-[2-(DODECYLOXY)ETHOXY]-ETHYLAMINE

$CH_3(CH_2)_{10}CH_2OCH_2CH_2OCH_2CH_2NH_2$

250 mg

S39,938-8
4-METHOXYBUTYLAMINE
$CH_3OCH_2CH_2CH_2CH_2NH_2$
250 mg

S52,761-0
3-ISOBUTOXYPROPYLAMINE
$CH_3CH(CH_3)CH_2OCH_2CH_2CH_2NH_2$
5 gm

S39,202-2
3-HEXYLOXYPROPYLAMINE
$CH_3(CH_2)_4CH_2OCH_2CH_2CH_2NH_2$
5 gm

S39,207-3
3-(2-ETHYLBUTOXY)PROPYLAMINE
$CH_3CH_2CH(CH_2CH_3)CH_2OCH_2CH_2CH_2NH_2$
5 gm

S39,210-3
3-(2-ETHYLHEXYLOXY)PROPYLAMINE
$CH_3CH_2CH_2CH_2CH(CH_2CH_3)CH_2OCH_2CH_2CH_2NH_2$
5 gm

S39,208-1
3-DODECYLOXYPROPYLAMINE
$CH_3(CH_2)_{10}CH_2OCH_2CH_2CH_2NH_2$
5 gm

S51,654-6
3,3'-(DECAMETHYLENEDIOXY)BIS-(PROPYLAMINE)
$H_2NCH_2CH_2CH_2OCH_2(CH_2)_8CH_2OCH_2CH_2CH_2NH_2$
5 gm

S45,871-6
2-ETHOXYETHYLAMINE
$CH_3CH_2OCH_2CH_2NH_2$
250 mg

S42,150-2
2-DODECYLOXY-N-METHYLETHYLAMINE
$CH_3(CH_2)_{10}CH_2OCH_2CH_2NHCH_3$
250 mg

S43,355-1
N-(METHOXYMETHYL)DIETHYLAMINE
$CH_3CH_2-N(CH_2CH_3)-CH_2OCH_3$
5 gm

S50,317-7
1,1-DIALLYLOXYTRIMETHYLAMINE
$(CH_3)_2NCH(OCH_2CH=CH_2)OCH_2CH=CH_2$
5 gm

S62,145-5
1-DIETHYLAMINO-5-METHOXY-2-PENTENE
$(CH_3CH_2)_2NCH_2CH=CHCH_2CH_2OCH_3$
250 mg

S62,155-2
1-DIISOPROPYLAMINO-5-METHOXY-2-PENTENE
$[(CH_3)_2CH]_2NCH_2CH=CHCH_2CH_2OCH_3$
250 mg

S68,582-8
2,2'-IMINOBIS(ACETALDEHYDE DIETHYL ACETAL)
$[(CH_3CH_2O)_2CHCH_2]_2NH$
5 gm

S65,865-0
METHYLAMINOACETALDEHYDE DIETHYL ACETAL
$CH_3NHCH_2CH(OCH_2CH_3)OCH_2CH_3$
250 mg

S54,729-8
2,2'-METHYLIMINOBIS(ACETALDEHYDE DIMETHYL ACETAL)
$CH_3CH(OCH_3)CH_2-N(CH_3)-CH_2CH(OCH_3)OCH_3$
5 gm

S34,644-6
DIISOPROPYLAMINOACETALDEHYDE DIETHYL ACETAL
$[(CH_3)_2CH]_2NCH_2CH(OCH_2CH_3)OCH_2CH_3$
5 gm

S79,289-6
3,3-OXETANEBIS(METHYLAMINE)-DIHYDROBROMIDE
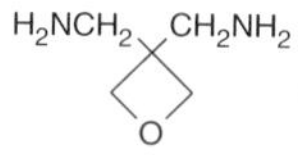
• 2 HBr
250 mg

S60,278-7
3,9,BIS(3-AMINOPROPYL)-2,4,8,10-TETRAOXASPIRO(5,5)UNDECANE
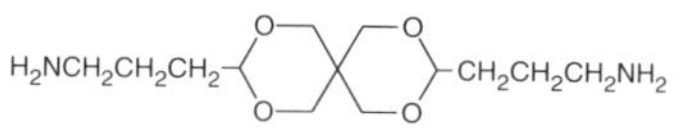
5 gm

S41,397-6
2-AMINOMETHYL-3,4-DIHYDRO-2H-PYRAN
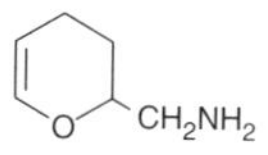
5 gm

S62,737-2
DITETRAHYDROFURFURYLAMINE
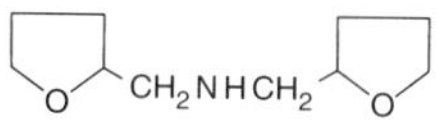
250 mg

S55,431-6
N-METHYLTETRAHYDROFURFURYLAMINE
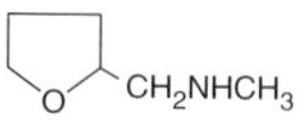
5 gm

S81,593-4
N-GLYCIDYLDIETHYLAMINE
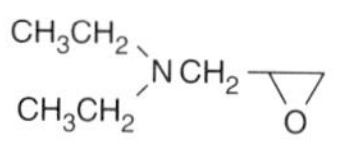
250 mg

S41,635-5
N,N-DIMETHYLFORMAMIDE ETHYLENE ACETAL
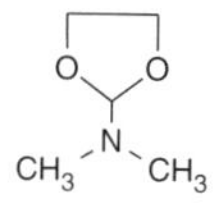
5 gm

S43,388-8
N,N,5,5-TETRAMETHYL-1,3-DIOXAN-2-AMINE
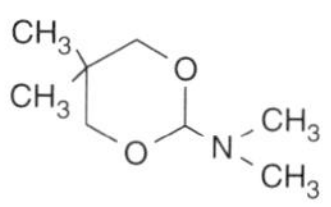
5 gm

S62,306-7
N,N-DIETHYLHYDROXYLAMINE HEMIOXALATE
$CH_3CH_2-N(CH_2CH_3)-OH \cdot {}^1/_2\ HO-C(=O)-C(=O)-OH$
250 mg

S44,953-9
2-[(2-AMINOETHYL)THIO]ETHANOL
$H_2NCH_2CH_2SCH_2CH_2OH$
5 gm

S83,413-0
1-(DIETHYLAMINO)-3-[3-(DIETHYLAMINO)PROPYLTHIO]-2-PROPANOL
$(CH_3CH_2)_2NCH_2CH(OH)CH_2SCH_2CH_2CH_2N(CH_2CH_3)_2$
250 mg

S83,402-5
1-[2-(DIETHYLAMINO)ETHYLTHIO]-3-BUTEN-2-OL
$(CH_3CH_2)_2NCH_2CH_2SCH_2CH(OH)CH=CH_2$
250 mg

S43,426-4
2-AMINOETHANETHIOL BITARTRATE
$H_2NCH_2CH_2SH \cdot C_4H_6O_6$
5 gm

S45,459-1
2-CYCLOHEXYLAMINOETHANETHIOL
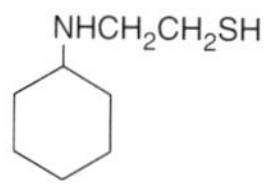
5 gm

S40,356-3
3-(DODECYLTHIO)PROPYLAMINE
$CH_3(CH_2)_{10}CH_2SCH_2CH_2CH_2NH_2$
250 mg

S40,350-4
2-(DODECYLTHIO)ETHYLAMINE

$CH_3(CH_2)_{10}CH_2SCH_2CH_2NH_2$

250 mg

S82,238-8
3-(CYCLOHEXYLTHIO)PROPYLAMINE

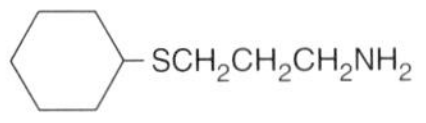

250 mg

S82,202-7
BETA-(3-DIETHYLAMINOPROPYLTHIO)-TRIETHYLAMINE DIHYDROCHLORIDE

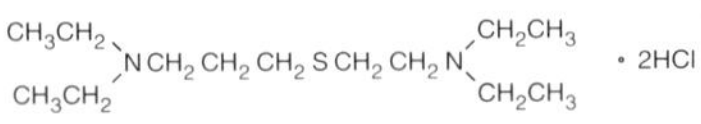

250 mg

S82,199-3
2-(DIETHYLAMINO)ETHYL DISULFIDE DIHYDROCHLORIDE

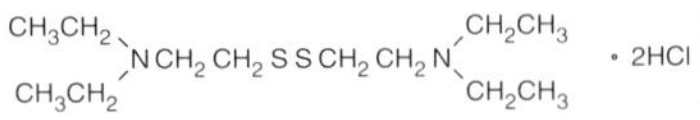

250 mg

S58,068-6
1,3,5,7,2,4,6,8-TETRATHIATETRAZOCINE

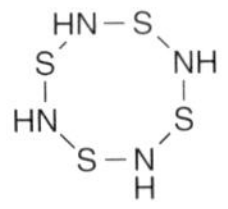

250 mg

S64,989-9
THIALDINE

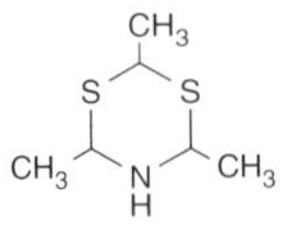

5 gm

S57,629-8
HEXAMETHYLENETETRAMINE, COMPOUND WITH ALLYL IODIDE

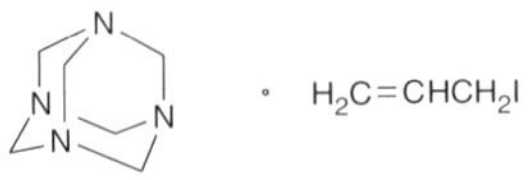

5 gm

S62,195-1
1-(3-CHLOROPROPYL)-4-METHYLPIPERAZINE DIHYDROCHLORIDE

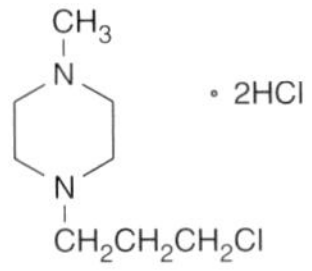

5 gm

S38,885-8
2-(TRICHLOROMETHYL)-OXAZOLIDINE

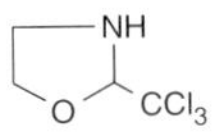

5 gm

S75,456-0
4,4'-DIHYDROXY-1,1'-DIMETHYL-4,4'-ETHYLENEDIPIPERIDINE

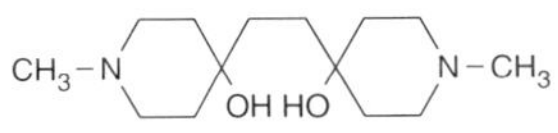

250 mg

S78,926-7
1-CYCLOHEXYL-1,2-DIHYDROTHYMINE

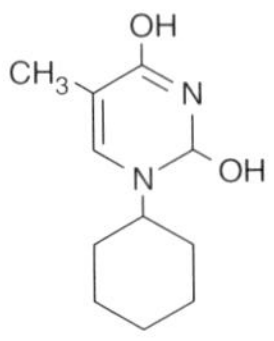

250 mg

S37,656-6
1-(2-METHOXYETHYL)-3-PYRROLIDINEMETHANOL

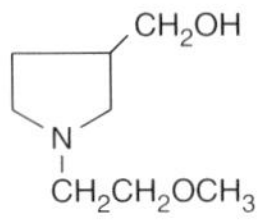

250 mg

S38,524-7
1-(2-METHOXYETHYL)-3-METHYL-3-PYRROLIDINEMETHANOL

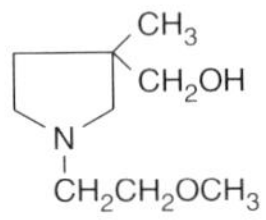

5 gm

S52,748-3
2,6-DIMETHYL-4-MORPHOLINEETHANOL

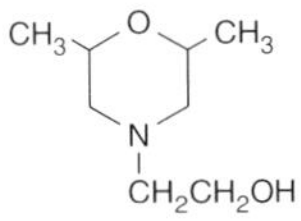

5 gm

S56,582-2
2-MERCAPTO-4,6,6-TRIMETHYLHEXAHYDROPYRIMIDIN-4-OL

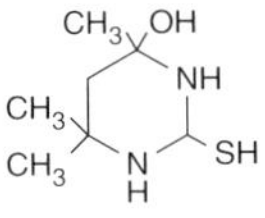

5 gm

S88,306-9
3,4-DIHYDRO-3-(6-HYDROXYHEXYL)-2H-THIAZOL-1-IUM BROMIDE

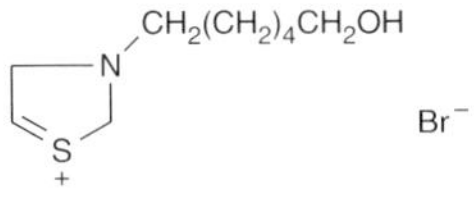

250 mg

S78,677-2
1-(3-AMINOPROPYL)-4-(2-HYDROXYETHYL)PIPERAZINE

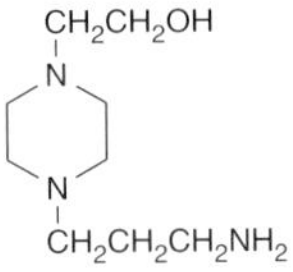

250 mg

S78,882-1
4-AMINO-1,6-DIHYDRO-2-HYDROXY-1-METHYLPYRIMIDINE

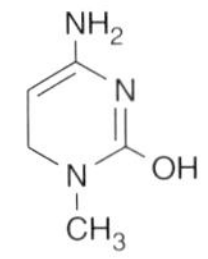

250 mg

S94,806-3
BETA-AMINO-1-ADAMANTANEETHANOL HYDROCHLORIDE

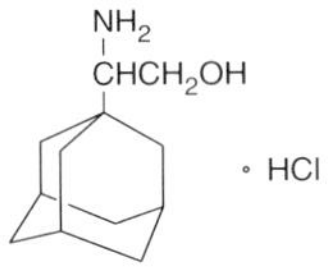

250 mg

S36,244-1
2-[4-(2-HYDROXYETHYL)-PIPERAZINOMETHYL]-3-HYDROXYPIPERIDINE

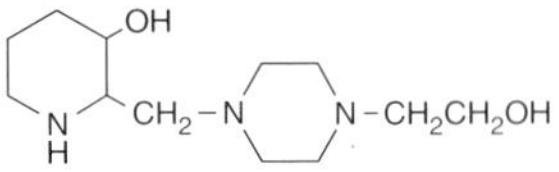

5 gm

S51,611-2
2-(2-MORPHOLINOETHYLAMINO)ETHANOL

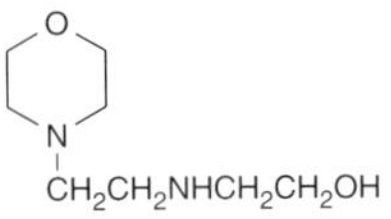

5 gm

S40,793-3
2-(DIMETHYLAMINOMETHYL)-3-HYDROXY-1-(2-HYDROXYETHYL)PIPERIDINE

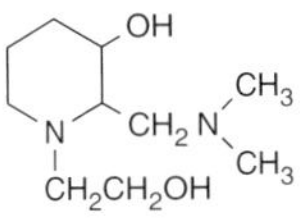

250 mg

S42,822-1
1,4-BIS(2-HYDROXYPROPYL)-2,5-DIMETHYLPIPERAZINE

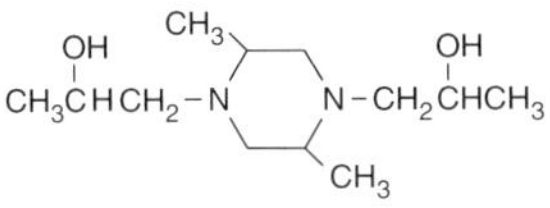

5 gm

S39,218-9
1,3-DIMORPHOLINO-2-PROPANOL

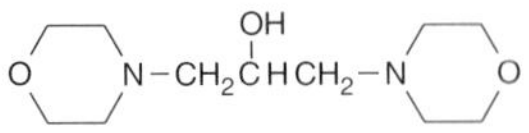

5 gm

S41,990-7
4-[3-(4-PIPERIDYL)PROPYL]-1-PIPERIDINEETHANOL

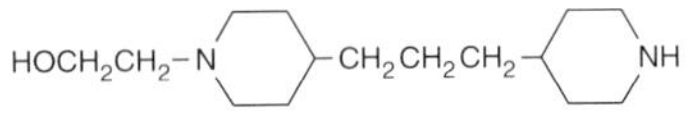

5 gm

S35,875-4
1-(2-[N,N-BIS(2-HYDROXYPROPYL)-AMINO]ETHYL)-4-(2-HYDROXYPROPYL)-PIPERAZINE

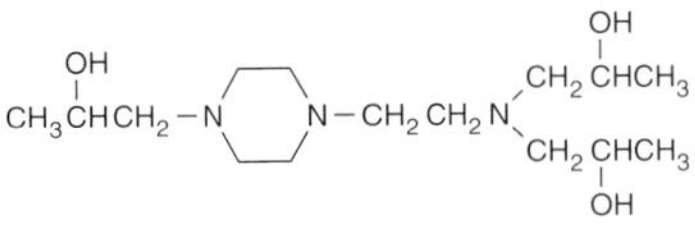

5 gm

S36,232-8
2-(HEXAMETHYLENEIMINO)ETHANOL

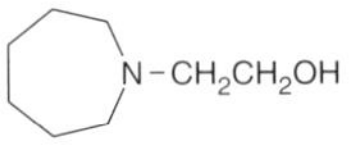

5 gm

S35,052-4
1-(HEXAMETHYLENEIMINO)-2-PROPANOL

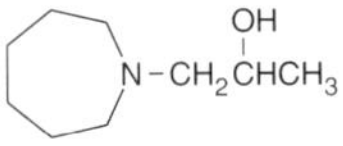

5 gm

S84,143-9
1-ISOPROPYL-3-AZETIDINOL HYDROCHLORIDE

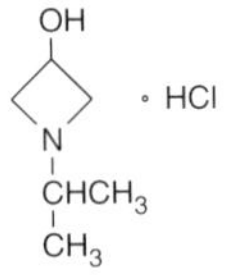

250 mg

S59,318-4
1-PYRROLIDINEETHANOL

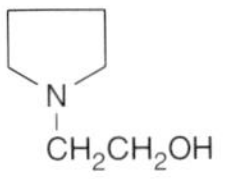

5 gm

S37,057-6
1-PYRROLIDINO-2-PROPANOL

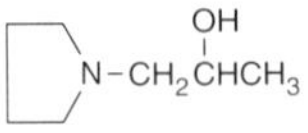

5 gm

S45,465-6
1-ETHYL-2-PYRROLIDINEMETHANOL

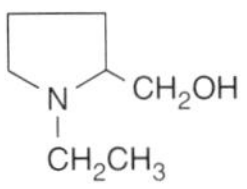

250 mg

S34,230-0
1,4-BIS(2-HYDROXYPROPYL)-2-METHYLPIPERAZINE

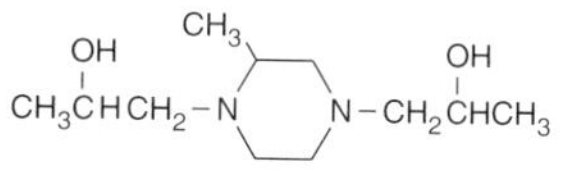

5 gm

S36,272-7
4-(2-HYDROXYPROPYL)MORPHOLINE

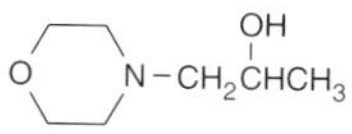

5 gm

S34,712-4
ALPHA,2,6-TRIMETHYL-4-MORPHOLINEETHANOL

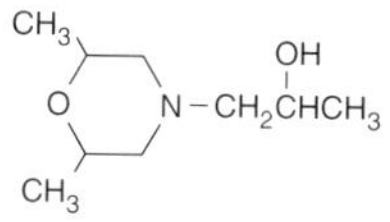

5 gm

S43,331-4
3-(PIPERIDINOMETHYL)-2-NORBORNANOL, MIXTURE OF ENDO AND EXO

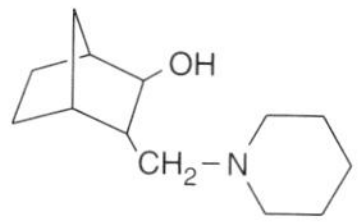

5 gm

S45,076-6
3-(MORPHOLINOMETHYL)-2-NORBORNANOL, MIXTURE OF ENDO AND EXO

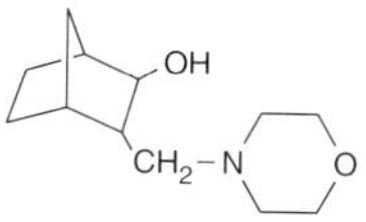

5 gm

S37,951-4
1-ETHYL-3-PYRROLIDINEMETHANOL

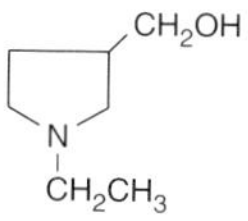

250 mg

S37,786-4
BETA,BETA-DIMETHYL-1-PYRROLIDINEPROPANOL

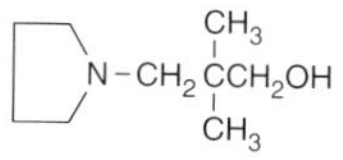

5 gm

S89,696-9
3-PIPERIDINO-1-CYCLOHEXANOL HYDROCHLORIDE

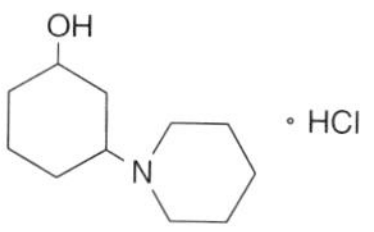

250 mg

S41,885-4
ALPHA-METHYL-1-PIPERIDINEETHANOL

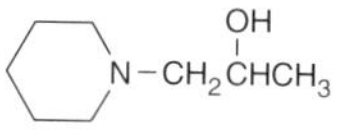

5 gm

S43,481-7
1-PIPERIDINEPROPANOL

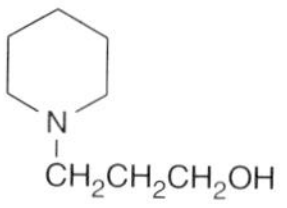

5 gm

S35,342-6
2-METHYL-1-PIPERIDINEPROPANOL

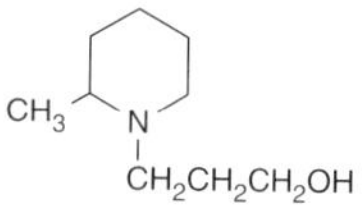

5 gm

S38,033-4
BETA,BETA-DIMETHYL-1-PIPERIDINEPROPANOL

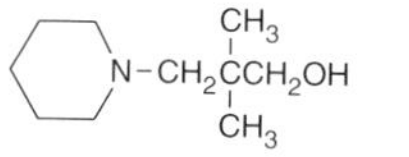

5 gm

S87,773-5
5-METHYL-3-PIPERIDINO-4-HEXEN-1-OL

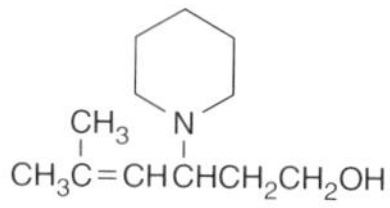

250 mg

S88,034-5
2-METHYL-5-PIPERIDINO-3-HEXEN-1-OL

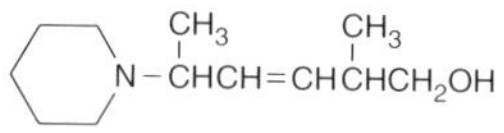

250 mg

S88,029-9
6-METHYL-4-PIPERIDINO-5-HEPTEN-2-OL

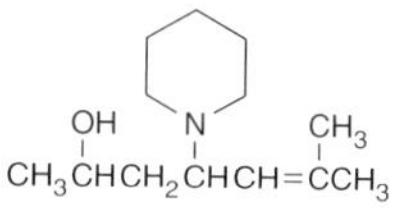

250 mg

S87,795-6
6-(PIPERIDINOMETHYL)-5-OCTEN-3-OL

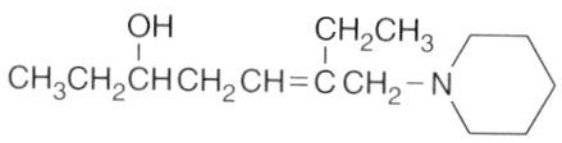

250 mg

S88,021-3
6-METHYL-7-PIPERIDINO-5-OCTEN-3-OL

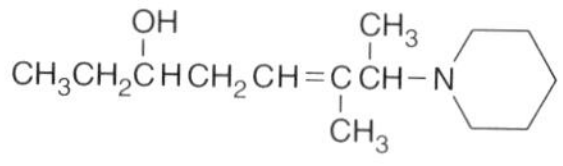

250 mg

S45,517-2
3-(4-METHYLPIPERAZINOMETHYL)-2-NORBORNANOL DIHYDROCHLORIDE

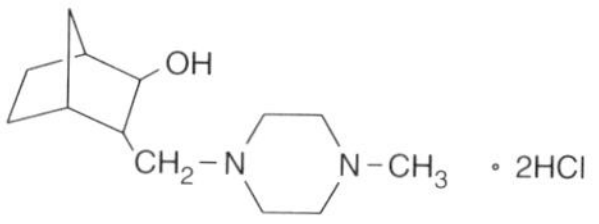

5 gm

S38,553-0
BETA,BETA-DIMETHYL-4-MORPHOLINEPROPANOL

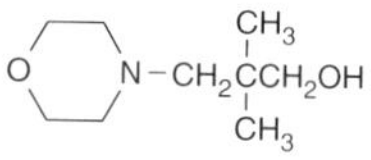

5 gm

S88,050-7
5-METHYL-6-MORPHOLINO-4-HEXEN-2-OL

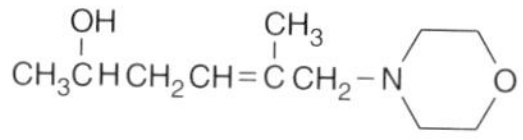

250 mg

S57,137-7
2,4-DIMETHYL-2-OXAZOLINE-4-METHANOL

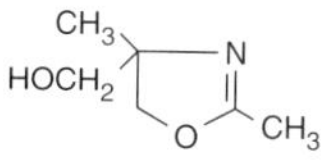

5 gm

S44,612-2
3-HYDROXYPIPERIDINE

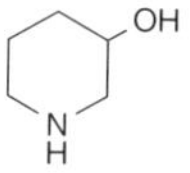

5 gm

S79,978-5
1,3-DIMORPHOLINO-2-PROPANOL DIHYDROCHLORIDE

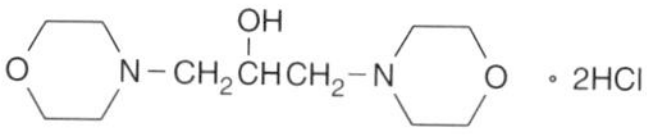

250 mg

S75,748-9
3-AZASPIRO(5.5)UNDECANEPROPANOL

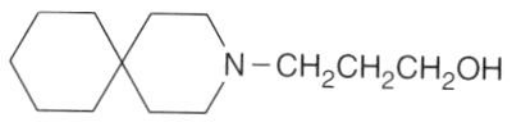

5 gm

S35,343-4
METHYL 3-PIPERIDYL CARBINOL

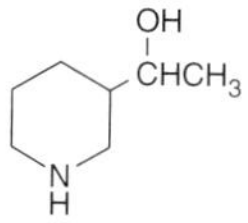

5 gm

S87,779-4
2,5-DIMETHYL-5-PIPERIDINO-3-HEXEN-1-OL

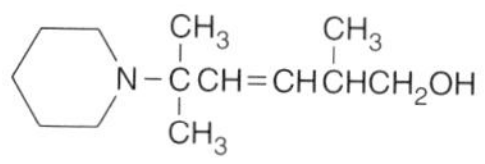

250 mg

S46,124-5
2-METHYL-3-QUINUCLIDINOL

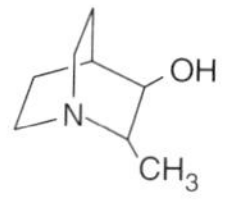

5 gm

S86,647-4
1-(4-METHOXY-2-BUTEN-1-YLOXYMETHYL)-PYRROLIDINE

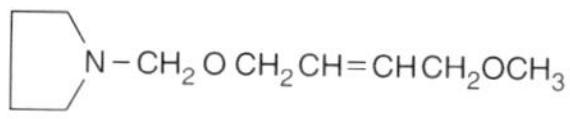

250 mg

S50,296-0
3-ALLYLOXYMETHYL-5-METHYL-1-AZA-4,6-DIOXABICYCLO[3.3.0]OCTANE

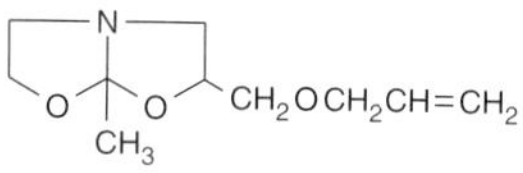

5 gm

S42,531-1
ISOQUINIDINE

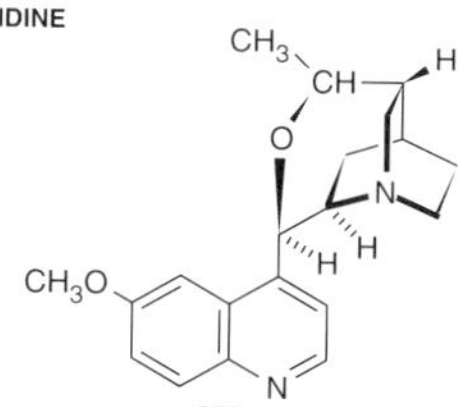

250 mg

S86,615-6
1-(ETHOXYMETHYL)PYRROLIDINE

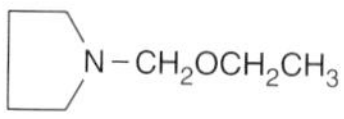

250 mg

S86,635-0
1-(2-BUTEN-1-YLOXYMETHYL)-PYRROLIDINE

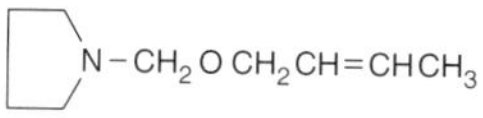

250 mg

S86,666-0
1-(2-HEXEN-1-YLOXYMETHYL)-PYRROLIDINE
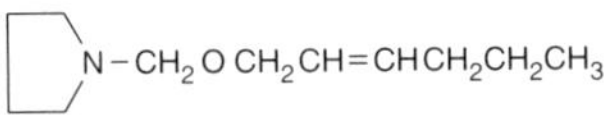
250 mg

S86,691-1
1-(OCTYLOXYMETHYL)PYRROLIDINE
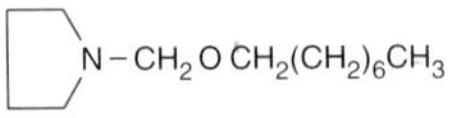
250 mg

S89,100-2
2-METHYL-1-PIPERIDINO-5-(2-TETRAHYDROPYRANYLOXY)-2-PENTENE
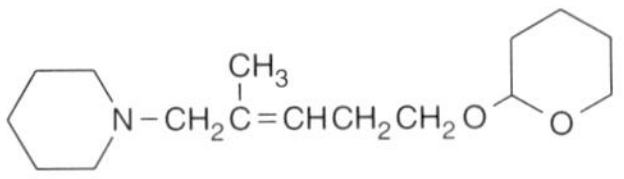
250 mg

S62,138-2
5-METHOXY-1-PIPERIDINO-1-PENTENE
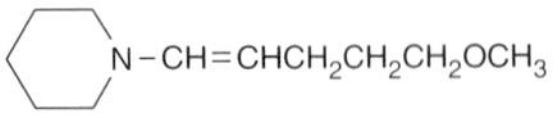
250 mg

S62,142-0
5-METHOXY-1-PIPERIDINO-2-PENTENE
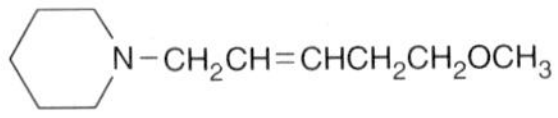
250 mg

S62,144-7
5-METHOXY-1-MORPHOLINO-1-PENTENE
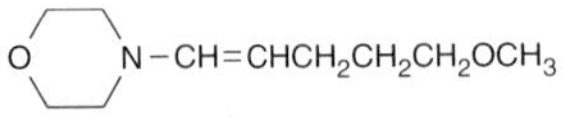
250 mg

S35,535-6
1-PYRROLIDINEACETALDEHYDE DIETHYL ACETAL
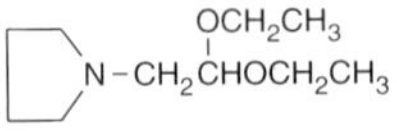
5 gm

S44,550-9
1-AZA-4,6-DIOXABICYCLO[3.3.0]OCTANE
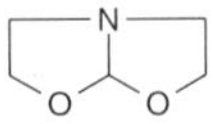
250 mg

S45,526-1
1-AZA-4,6-DIOXA-5-METHYLBICYCLO-[3.3.0]OCTANE
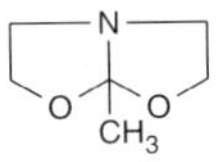
5 gm

S43,462-0
1-PIPERIDINEPROPIONALDEHYDE DIMETHYL ACETAL
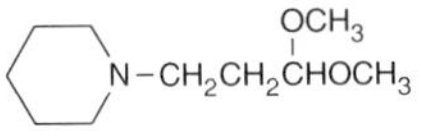
5 gm

S87,796-4
5-METHYL-5-PIPERIDINO-3-HEXENAL DIMETHYL ACETAL
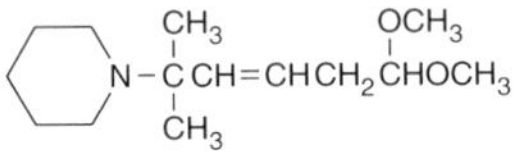
250 mg

S87,783-2
6-METHYL-6-PIPERIDINO-4-HEPTENAL DIMETHYL ACETAL
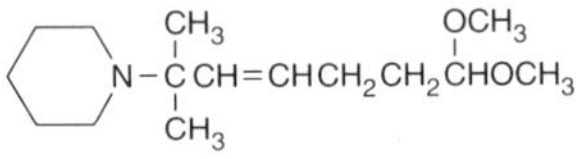
250 mg

S88,008-6
5-PIPERIDINO-3-NONENAL DIMETHYL ACETAL
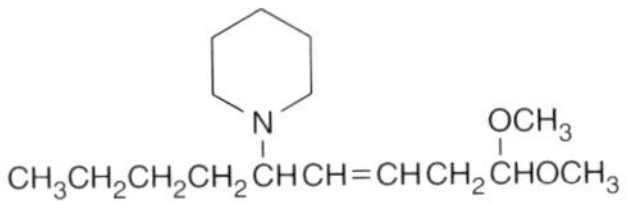
250 mg

S87,774-3
5-MORPHOLINO-3-HEXENAL DIMETHYL ACETAL
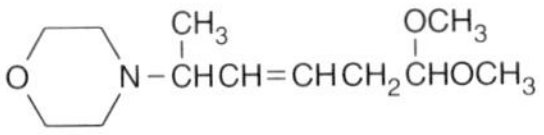
250 mg

S76,715-8
2-MORPHOLINOACETALDEHYDE DIETHYLACETAL
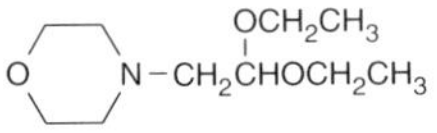
250 mg

S88,597-5
2,3-BIS(4-ISOPROPENYL-2-OXAZOLIN-2-YL)BUTANE
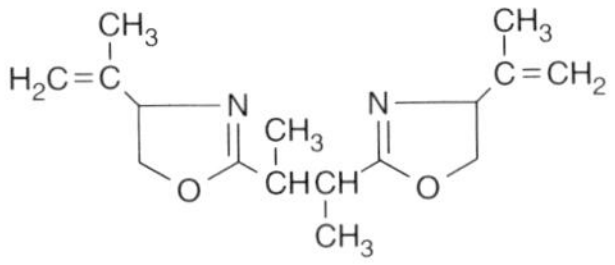
250 mg

S51,077-7
1-AMINO-1,4-DIHYDRO-4,4,6-TRIMETHYL-2-PYRIMIDINETHIOL
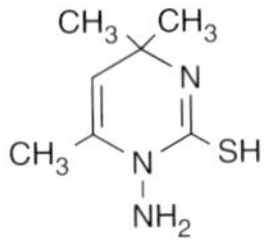
5 gm

S36,948-9
4,5-DIHYDRO-4,4,6-TRIMETHYL-2-PYRIMIDINETHIOL
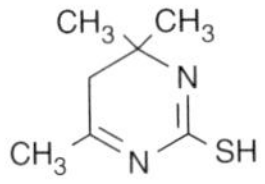
5 gm

S54,704-2
3,4-DIHYDRO-4,4,6-TRIMETHYL-2-PYRIMIDINETHIOL
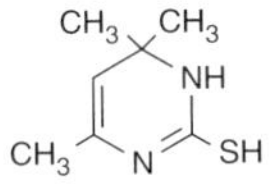
5 gm

S82,201-9
BETA-(MORPHOLINOETHYLTHIO)-TRIETHYLAMINE DIHYDROCHLORIDE
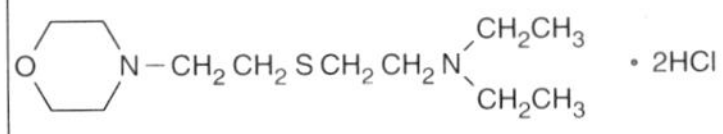
250 mg

S50,650-8
1-(BUTYLTHIOMETHYL)PIPERIDINE
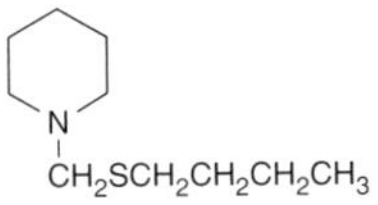
5 gm

S65,838-3
1,4-BIS(2-IMIDAZOLIN-2-THIO)BUTANE DIHYDROBROMIDE
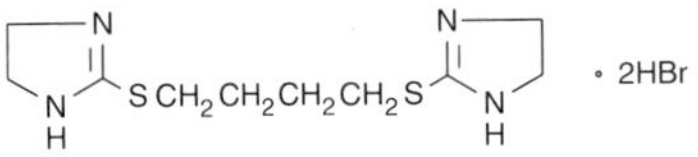
250 mg

S76,211-3
BIS(1-IMIDAZOLIN-2-YLMETHYL)-SULFIDE DIHYDROCHLORIDE
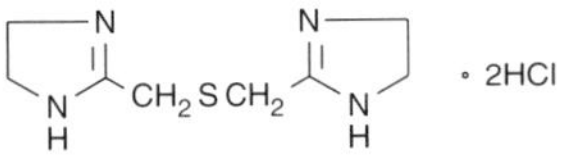
250 mg

S36,463-0
N-(4-PIPERIDYLMETHYL)-ETHYLENEDIAMINE
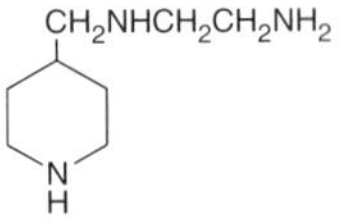
5 gm

S46,275-6
3-AMINO-N-ETHYLPIPERIDINE
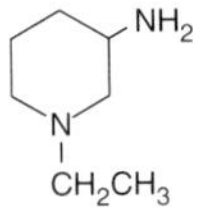
250 mg

S87,147-8
4-AMINO-1,4-DIHYDRO-1-METHYLPYRIDINE DIHYDROCHLORIDE
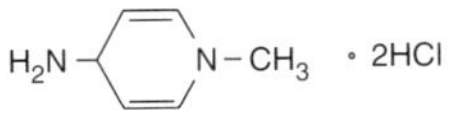
250 mg

S41,985-0
1-PIPERIDINEPROPYLAMINE
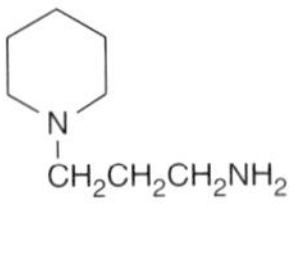
5 gm

S43,544-9
3-HEXAMETHYLENEIMINO-1-PROPYLAMINE
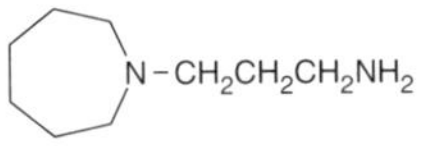
5 gm

S42,724-1
3-AMINO-3-PYRAZOLINE DIHYDROCHLORIDE
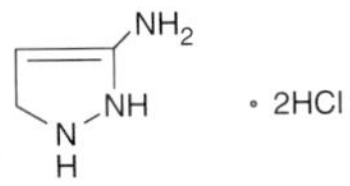
5 gm

S94,633-8
2-AMINO-2-IMIDAZOLINE HYDROBROMIDE
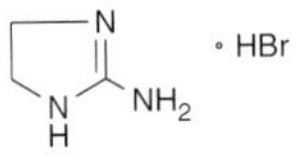
250 mg

S41,464-6
1-(3-AMINOPROPYL)-4-METHYLPIPERAZINE
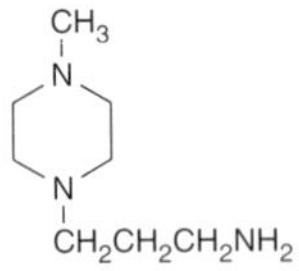
250 mg

S44,504-5
3,4-DIMETHYLFURAZAN
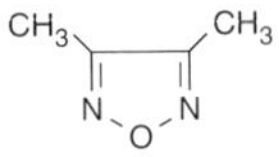
5 gm

S60,327-9
5,7-DIMETHYL-2,3,6,7-TETRAHYDRO-1H-1,4-DIAZEPINE PERCHLORATE

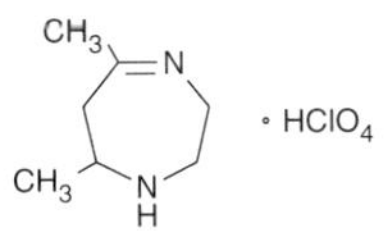

250 mg

S42,542-7
LEPTOCLADINE

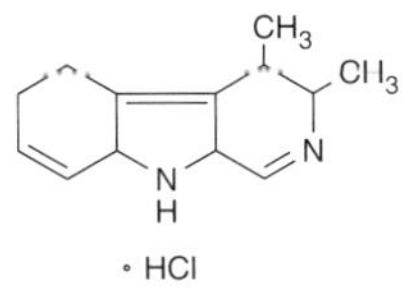

250 mg

S36,462-2
N-(2-PIPERIDYLMETHYL)DIETHYLAMINE

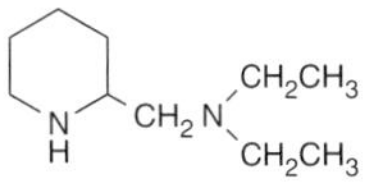

5 gm

S53,553-2
1-BUTYLPIPERAZINE DIHYDROBROMIDE

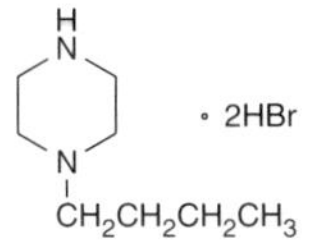

250 mg

S65,126-5
2-(2-TETRAHYDROFURYLAMINOMETHYL)-PIPERIDINE

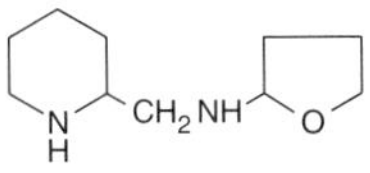

250 mg

S34,700-0
1,4-DIMETHYLHOMOPIPERAZINE

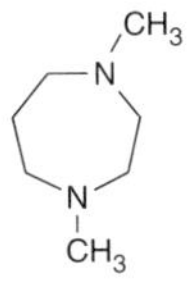

5 gm

S35,496-1
N-(2-PIPERIDYLMETHYL)DIMETHYLAMINE

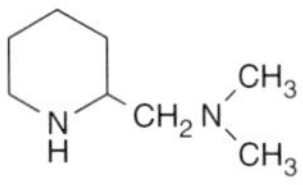

5 gm

S35,490-2
2-(PIPERIDINOMETHYL)PIPERIDINE

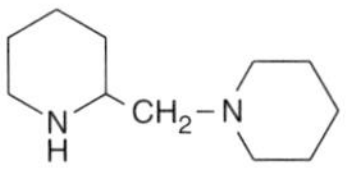

5 gm

S40,993-6
1-PROPYLPIPERAZINE DIHYDROCHLORIDE

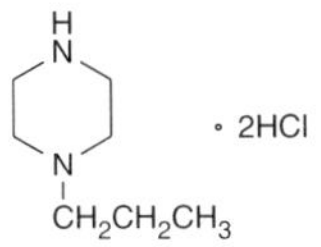

5 gm

S36,081-3
N-(2-DIMETHYLAMINOETHYL)-N'-METHYLPIPERAZINE

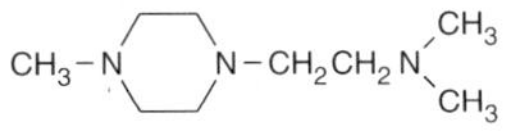

5 gm

S87,388-8
2,2'-BIS(5,6-DIHYDRO-4H-1,3-THIAZINE) DIHYDROCHLORIDE

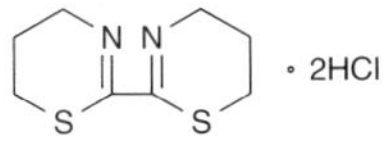

250 mg

S41,371-2
2,3,5,6-TETRAMETHYLPIPERAZINE

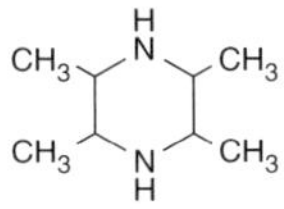

5 gm

S74,556-1
1,1'-ETHYLENEDIPIPERAZINE

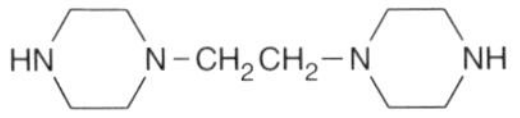

250 mg

S41,259-7
4,4'-ETHYLENEDIMORPHOLINE

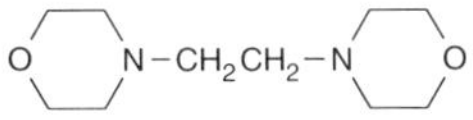

5 gm

S87,358-6
1-(1-AZACYCLOHEPT-2-EN-1-YL)-2-MORPHOLINO-1-CYCLOOCTENE

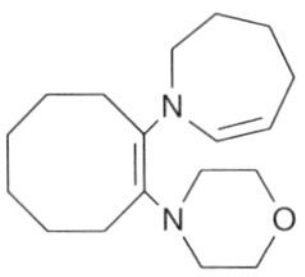

250 mg

S87,353-5
1-(1-AZACYCLOHEPT-2-EN-1-YL)-2-MORPHOLINO-1-CYCLODODECENE

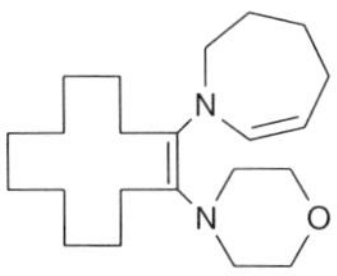

250 mg

S87,352-7
1-(1-AZACYCLOHEPT-2-EN-1-YL)-2-MORPHOLINO-1-CYCLOTETRADECENE

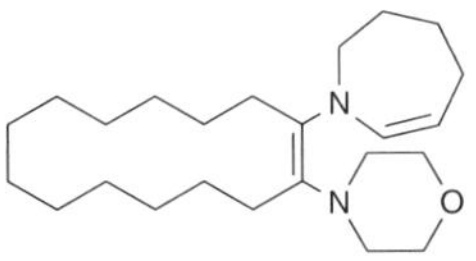

250 mg

S38,794-0
2-(2-PYRROLIDINOETHYL)PIPERIDINE

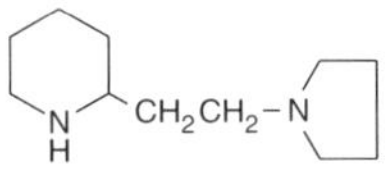

5 gm

S53,562-1
4-(DIMETHYLAMINO)PIPERIDINE DIHYDROCHLORIDE

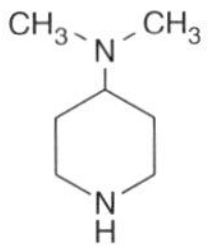

250 mg

S53,562-1
4-(DIMETHYLAMINO)PIPERIDINE

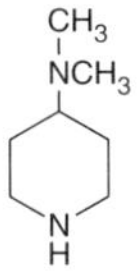

250 mg

S87,290-3
1-METHYL-4(1H)-PYRIDONE HYDRAZONE DIHYDROCHLORIDE

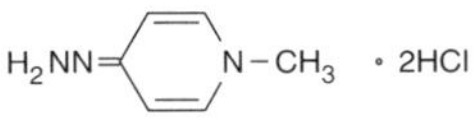

250 mg

S77,447-2
N-[3-(DIMETHYLAMINO)PROPYL]-CAMPHIDINE DIHYDROCHLORIDE

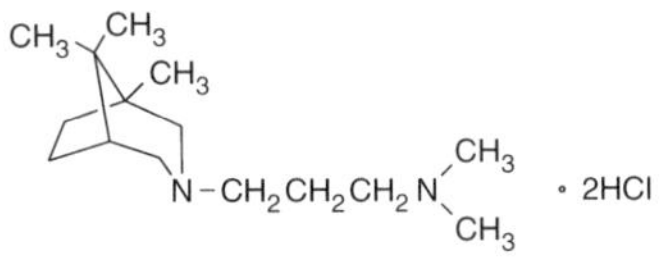

250 mg

S35,538-0
4-PYRROLIDINOPIPERIDINE DIHYDROCHLORIDE

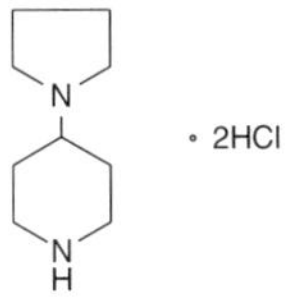

5 gm

S87,217-2
2-(2-PIPERIDINO-1-CYCLOHEXEN-1-YL)-AZACYCLONON-1-ENE

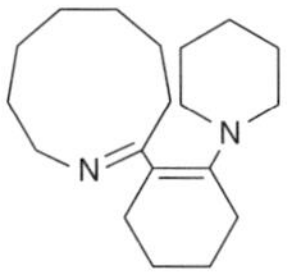

250 mg

S38,792-4
1,2'-ETHYLENEDIPIPERIDINE

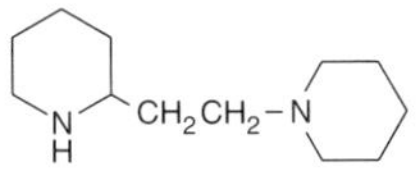

5 gm

S79,397-3
2-CHLORO-1,3-DIPIPERIDINOPROPANE DIHYDROCHLORIDE

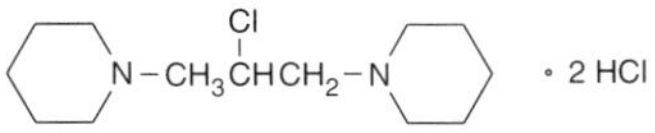

250 mg

S89,309-9
PERHYDRO-1,4,7,9B-TETRAAZAPHENALENE

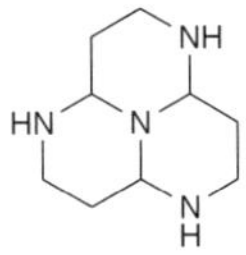

250 mg

S79,398-1
2-CHLORO-1,3-DIMORPHOLINOPROPANE DIHYDROCHLORIDE

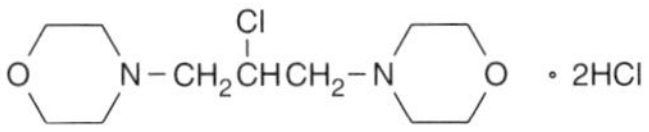

250 mg

S76,701-8
2,2'-ETHYLENEBIS(1,4,5,6-TETRAHYDROPYRIMIDINE) DITOSYLATE

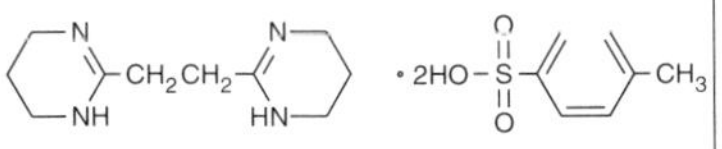

250 mg

S35,588-7
2,2,4,6-TETRAMETHYL-1,2-DIHYDROPYRIDINE

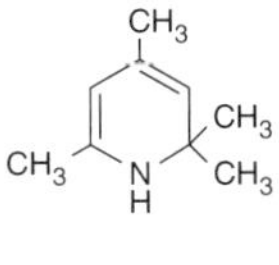

5 gm

S45,758-2
DODECAMETHYLENEIMINE

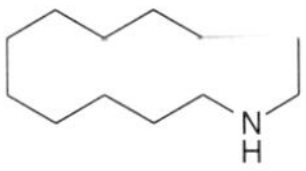

5 gm

S93,558-1
2-METHYL-2-(METHYLAMINOMETHYL)-1,3-DIOXOLANE

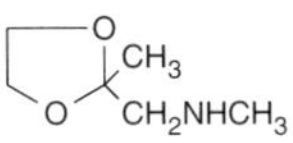

250 mg

S77,130-9
4,7-EPOXY-7-METHYL-2,3,3A,4,5,6,7,7A-OCTAHYDRO-1H-ISOINDOLE HYDROCHLORIDE

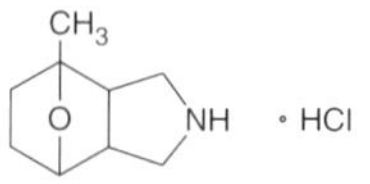

250 mg

S36,134-8
DODECAHYDROCARBAZOLE

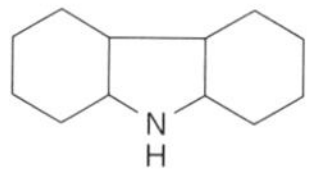

5 gm

S94,230-8
4,4-DIMETHYLOXAZOLIDINE

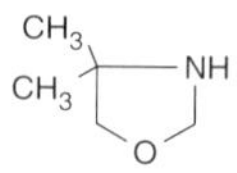

5 gm

S37,942-5
2,6-DIMETHYLPIPERIDINE HYDROCHLORIDE

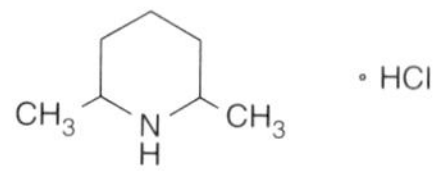

5 gm

S37,972-7
2,6-DIMETHYLPIPERIDINE HYDROBROMIDE

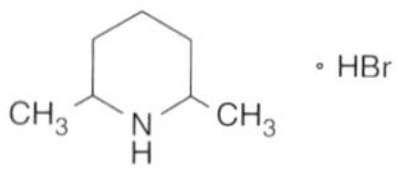

5 gm

S40,994-4
4-PROPYLPIPERIDINE

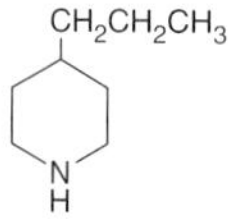

5 gm

S56,800-7
4,4'-METHYLENEDIMORPHOLINE

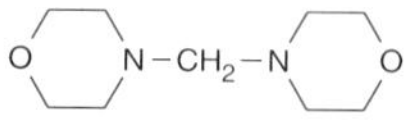

250 mg

S43,253-9
1-(2-ETHYL-1-BUTENYL)PYRROLIDINE

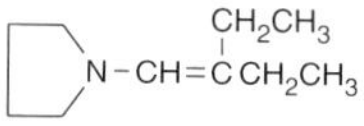

5 gm

S43,254-7
1-(2-ETHYL-1-BUTENYL)PIPERIDINE

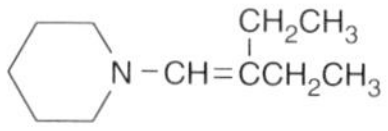

5 gm

S41,569-3
4-(2-ETHYL-1-BUTENYL)MORPHOLINE

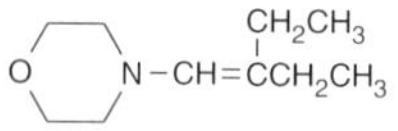

5 gm

S37,585-3
2,5-BIS(MORPHOLINMETHYL)-1,4-DIOXANE

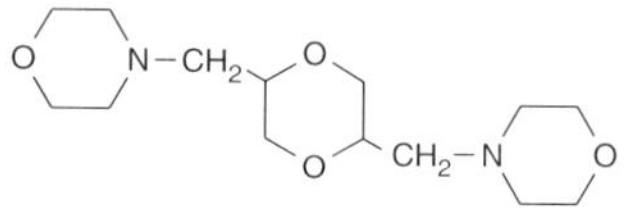

5 gm

S76,183-4
4-GLYCIDYLMORPHOLINE

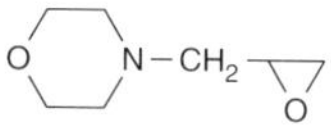

250 mg

S77,133-3
4,7-EPOXY-2-METHYL-2,3,3A,4,5,6,7,7A-OCTAHYDRO-1H-ISOINDOLE

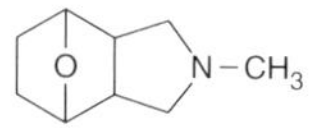

250 mg

S46,268-3
1,2-DIMETHYLPYRROLIDINE

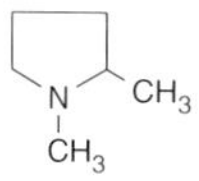

250 mg

S34,285-8
1-(3-BUTENYL)PYRROLIDINE

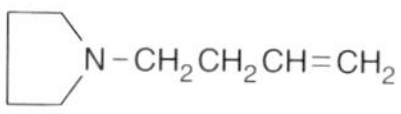

5 gm

S86,504-4
1-(2-METHYL-1-PYRROLIDINYL)-CYCLOHEXENE

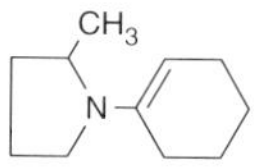

250 mg

S88,760-9
1,3-DIPYRROLIDINODODECAHYDRO-1H-PHENALENIUM TETRAFLUOROBORATE

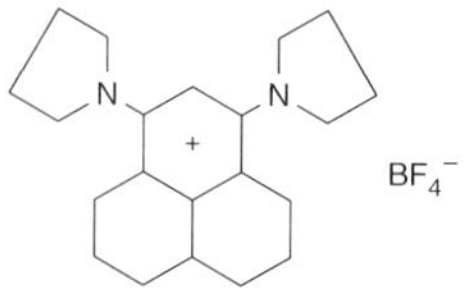

250 mg

S46,310-8
1,3,5-TRICYCLOHEXYLHEXAHYDRO-S-TRIAZINE

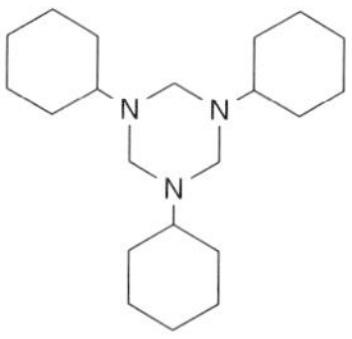

5 gm

S39,800-4
4-CYCLOHEXYLMORPHOLINE

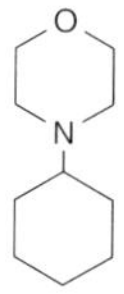

5 gm

S44,365-4
3-CHLORO-1-ETHYLPIPERIDINE HYDROCHLORIDE

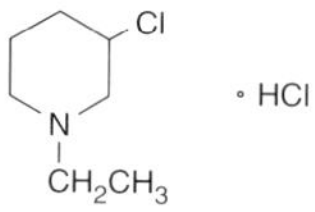

250 mg

S64,771-3
1-ETHYL-4-METHYLPIPERIDINE

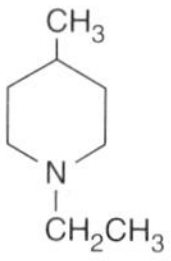

5 gm

S35,488-0
1-PIPERIDINO-1-ISOBUTENE

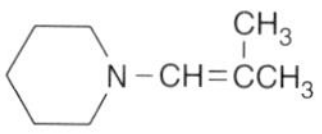

5 gm

S64,774-8
1-PROPYL-4-METHYLPIPERIDINE

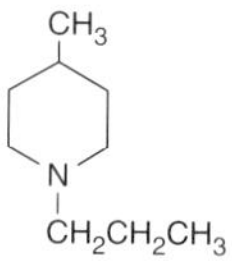

5 gm

S64,775-6
1-BUTYL-4-METHYLPIPERIDINE

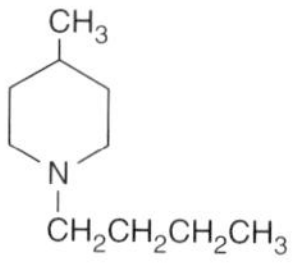

5 gm

S44,475-8
N,N'-METHYLENEBIS(PIPERIDINE)

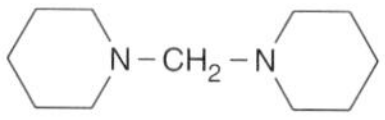

5 gm

S88,043-4
1-(2,4-DIMETHYL-1,4-PENTADIEN-1-YL)-PIPERIDINE

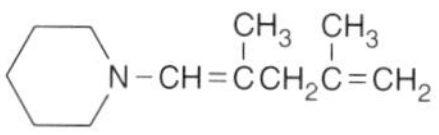

250 mg

S88,045-0
1-(5-METHYL-2,4-HEXADIEN-1-YL)-PIPERIDINE

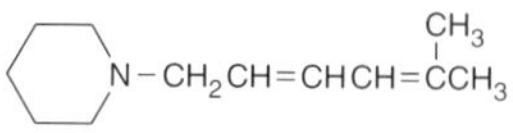

250 mg

S64,776-4
1-HEXYL-4-METHYLPIPERIDINE

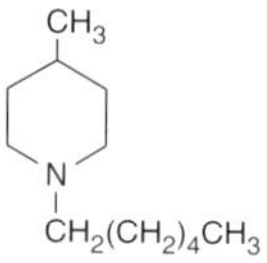

5 gm

S88,027-2
1-(2,5-DIMETHYL-1,4-HEXADIEN-1-YL)-PIPERIDINE

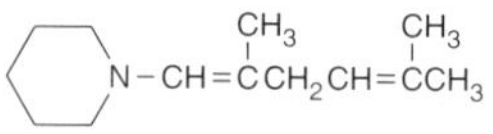

250 mg

S88,105-8
1-(2,5-DIMETHYL-2,5-HEXADIEN-1-YL)-PIPERIDINE

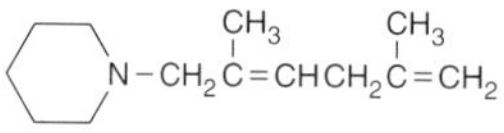

250 mg

S87,794-8
5-METHYL-2-PIPERIDINO-3-OCTENE

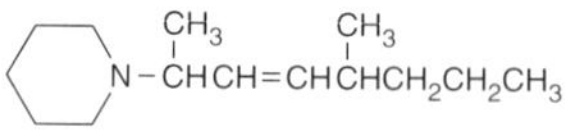

250 mg

S88,005-1
1,5-DIPIPERIDINO-2-METHYL-3-HEXENE

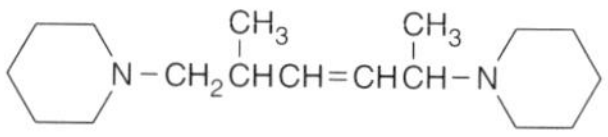

250 mg

S34,961-5
N-ETHYL-1,2,5,6-TETRAHYDROPYRIDINE
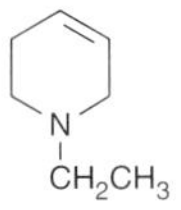
5 gm

S36,528-9
1,2,4-TRIMETHYLPIPERAZINE
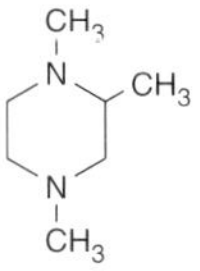
5 gm

S91,957-8
4-ETHYL-OCTAFLUOROMORPHOLINE
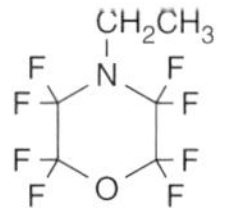
250 mg

S35,369-8
4-(2-METHYL-1-PROPENYL)MORPHOLINE
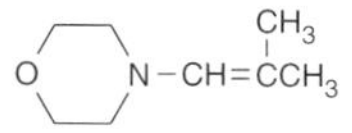
250 mg

S51,840-9
4-ISOBUTYLMORPHOLINE
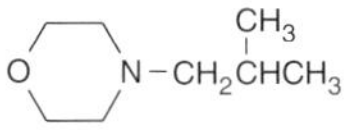
5 gm

S88,042-6
4-(2,5-DIMETHYL-2,5-HEXADIEN-1-YL)-MORPHOLINE
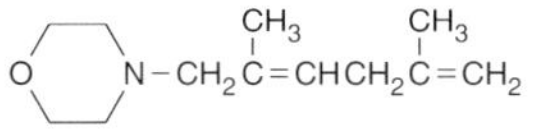
250 mg

S88,156-2
4-(2,7-OCTADIEN-1-YL)MORPHOLINE
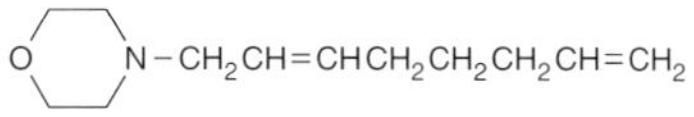
250 mg

S88,028-0
4-(2,6-DIMETHYL-2,5-HEPTADIEN-1-YL)-MORPHOLINE
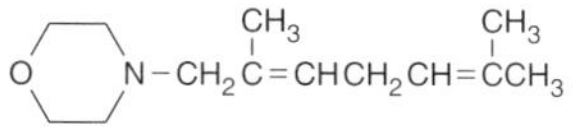
250 mg

S88,022-1
4-(2,9-DECADIEN-1-YL)MORPHOLINE
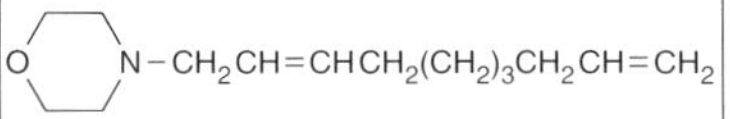
250 mg

S87,993-2
1-PYRROLIDINOL
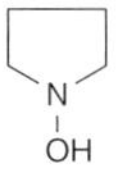
5 gm

S75,744-6
1,8,8-TRIMETHYL-3-AZABICYCLO[3.2.1]-OCTANE (CAMPHIDINE)
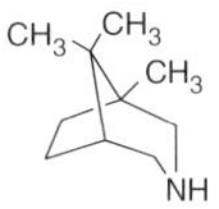
5 gm

S65,203-2
2,5-DIMETHYLPYRROLIDINE HYDROBROMIDE
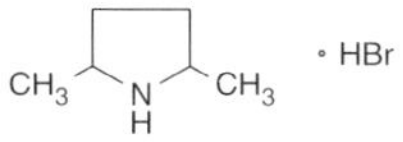
250 mg

S51,150-1
2,4-DIMETHYL-3-ETHYLPYRROLIDINE
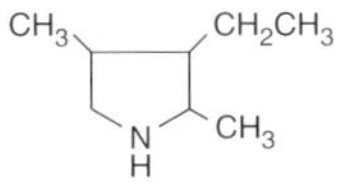
5 gm

S87,580-5
1,1'-METHYLENEBIS(PYRROLIDINE)
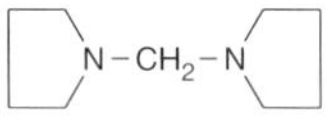
250 mg

S95,825-5
2-PYRROLIDINOBICYCLO[3.3.1]NONAN-9-ONE ETHYLENE KETAL
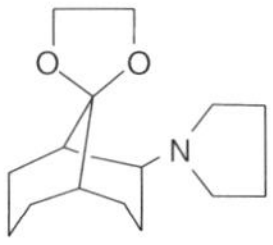
50 mg

S86,431-5
7-AZABICYCLO[2.2.1]HEPTANE HYDROCHLORIDE
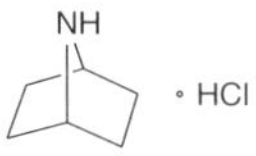
250 mg

S66,858-3
4-PIPERIDONE ETHYL HEMIKETAL
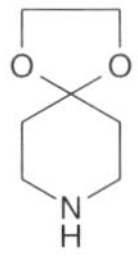
5 gm

S82,233-7
3-PYRAZOLINE HYDROCHLORIDE
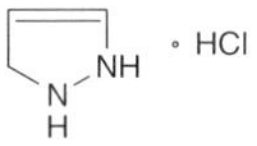
250 mg

S54,981-9
2-METHYL-2-IMIDAZOLINE TARTRATE
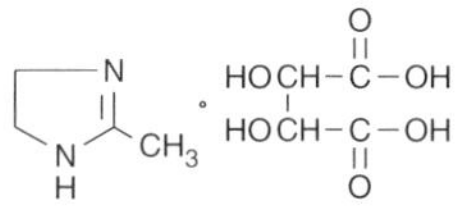
5 gm

S58,069-4
2,4,4-TRIMETHYL-2-OXAZOLINE
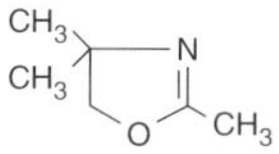
5 gm

S61,691-5
4,4-DIMETHYL-2-ETHYL-2-OXAZOLINE
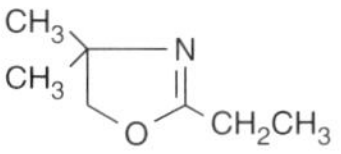
250 mg

S45,955-0
1-AZA-3,5-DIMETHYL-4,6-DIOXABICYCLO[3.3.0]OCTANE
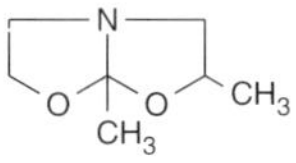
5 gm

S50,271-5
5-ETHYL-1-AZA-4,6-DIOXABICYCLO-[3.3.0]OCTANE
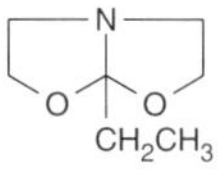
5 gm

S50,272-3
5-ETHYL-3-METHYL-1-AZA-4,6-DIOXABICYCLO[3.3.0]OCTANE
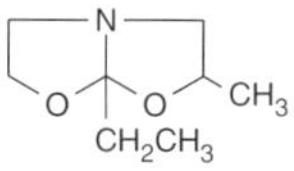
5 gm

S90,620-4
HEXAHYDRO-1,3,5-TRITHIA-6B-AZACYCLOPENTA(CD)PENTALENE
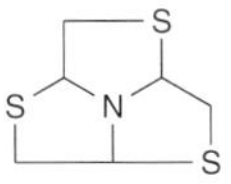
250 mg

S41,411-5
1-PIPERIDINO-1-CYCLOHEXENE
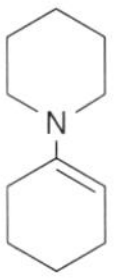
5 gm

S53,128-6
4,4'-TRIMETHYLENEBIS[1-(1-CYCLOHEXEN-1-YL)PIPERIDINE]
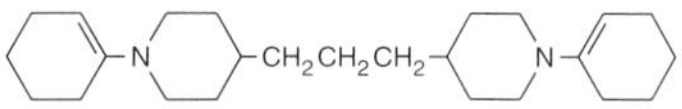
250 mg

S52,565-0
4,4-DIMETHYLPIPERIDINE HYDROCHLORIDE
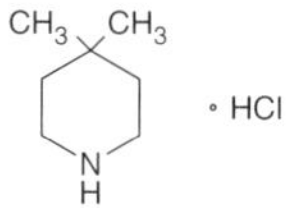
250 mg

S36,184-4
5-ETHYL-2-METHYLPIPERIDINE
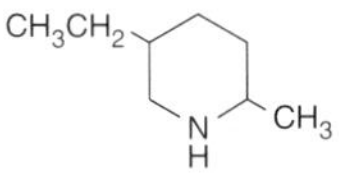
5 gm

S40,726-7
N-CYCLOHEXYLPIPERIDINE
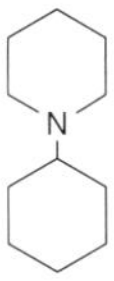
5 gm

S55,320-4
2,3,4,5-TETRAHYDRO-2,2,4,6-TETRAMETHYLPYRIDINE
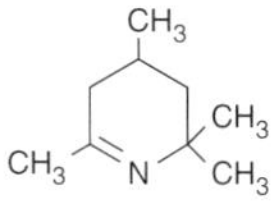
5 gm

S76,684-4
2,2'-METHYLENEBIS-(HEXAHYDROPYRIMIDINE)
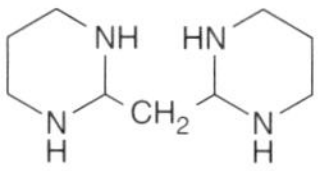
250 mg

S68,447-3
HEXAMETHYLENETETRAMINE MANDELATE
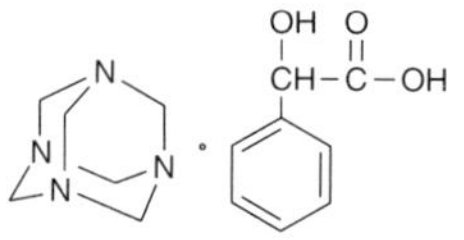
250 mg

S61,717-2
5,6-DIHYDRO-4,4,6-TRIMETHYL-2-VINYL-1,3-OXAZINE
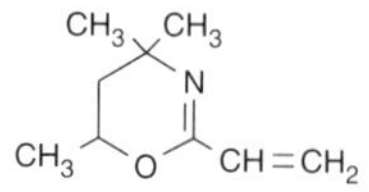
5 gm

S68,339-6
MORPHOLINE HYDROCHLORIDE
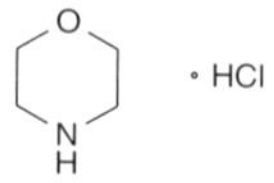
5 gm

S60,082-2
4-(2-CHLOROALLYL)MORPHOLINE
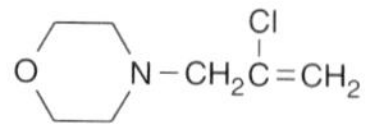
5 gm

S55,196-1
4-BUTYLMORPHOLINE
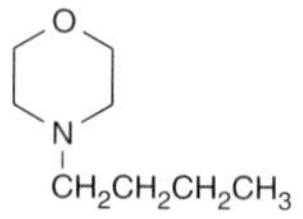
250 mg

S36,131-3
4,4'-DITHIODIMORPHOLINE
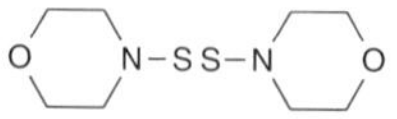
5 gm

S94,688-5
SPIRO(DIAZIRIDINE-3,2'-ADAMANTANE)
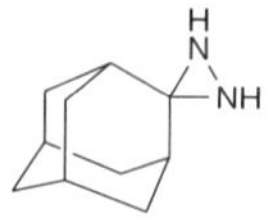
250 mg

S93,802-5
1-METHYL-SPIRO(DIAZIRIDINE-3,2'-ADAMANTANE)
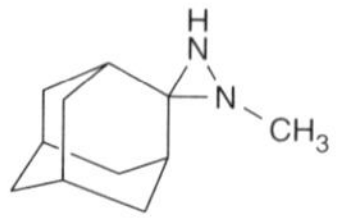
250 mg

S72,242-1
TRANS-DECAHYDROQUINOLINE HYDROBROMIDE
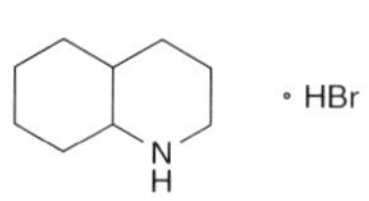
250 mg

S80,807-5
PERHYDROQUINOXALINE
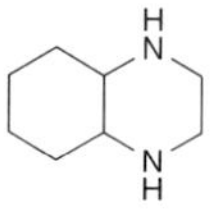
250 mg

S52,057-8
2,3,6,7-TETRAHYDRO-5,7,7-TRIMETHYL-5H-PYRROLO(1,2-A)IMIDAZOLE
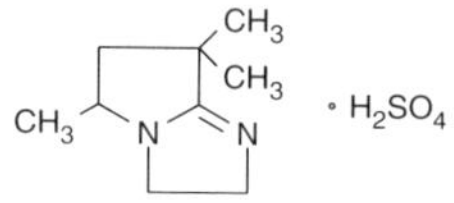
5 gm

S61,889-6
(IODOMETHYL)-TRIMETHYLAMMONIUM IODIDE
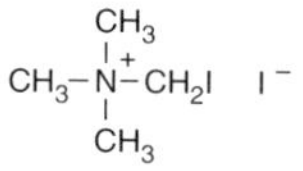
5 gm

S74,231-7
TETRAKIS(2-CHLOROETHYL)-AMMONIUM CHLORIDE
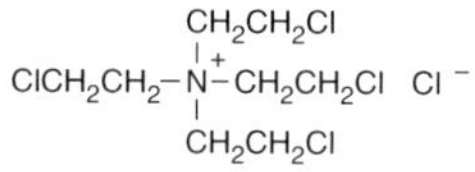
5 gm

S40,023-8
N,N-BIS(2-CHLOROETHYL)-N-HYDROXY-N-METHYLAMMONIUM CHLORIDE
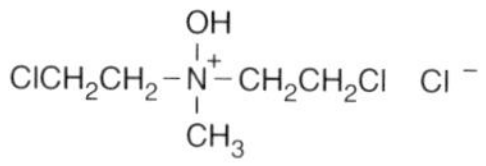
5 gm

S79,296-9
[3-(BROMOMETHYL)OXETAN-3-YLMETHYL]-TRIMETHYLAMMONIUM BROMIDE
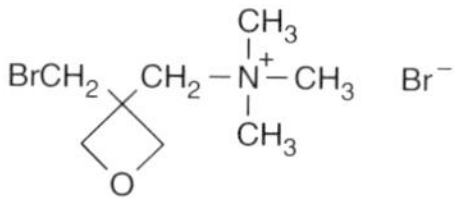
250 mg

S86,606-7
1-(IODOMETHYL)-1-METHYLPYRROLIDINIUM IODIDE
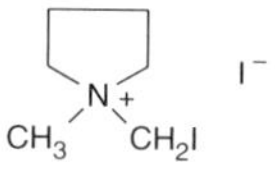
250 mg

S79,310-8
N,N'-[2,2-BIS(HYDROXYMETHYL)-TRIMETHYLENE]BIS-(TRIMETHYLAMMONIUM BROMIDE)
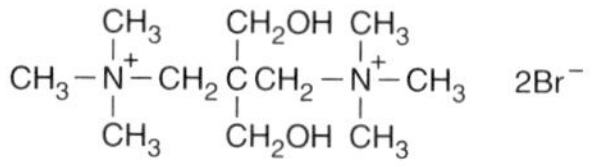
250 mg

S79,318-3
2-HYDROXYETHYLIDYNETRIS (TETRAMETHYLAMMONIUM BROMIDE)
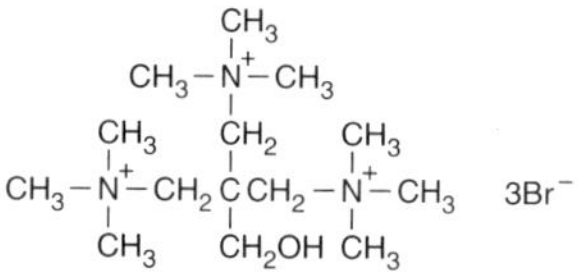
250 mg

S93,739-8
(3-HYDROXYPROPYL)-TRIMETHYLAMMONIUM IODIDE
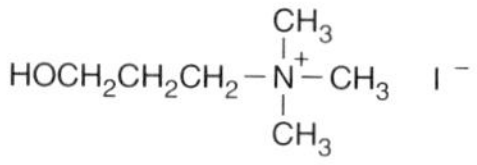
250 mg

S34,969-0
2-HYDROXYETHYLTRIMETHYLAMMONIUM BICARBONATE (50% AQUEOUS SOLUTION)
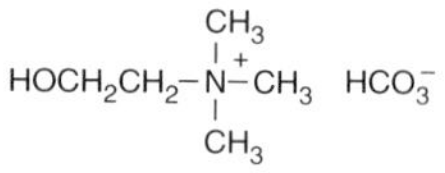
5 gm

S64,853-1
TETRAKIS(2-HYDROXYETHYL)-AMMONIUM HYDROXIDE
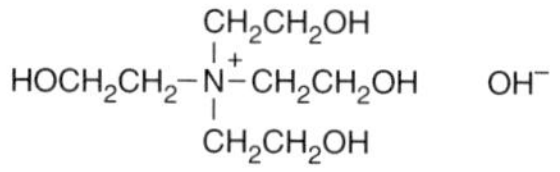
5 gm

S43,918-5
(2-HYDROXYETHYL)-TRIETHYLAMMONIUM IODIDE
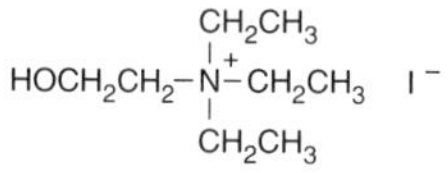
5 gm

S87,382-9
1-(5,5-DIMETHYL-3-ETHOXY-2-CYCLOHEXEN-1-YLIDENE)PYRROLIDINIUM TETRAFLUOROBORATE
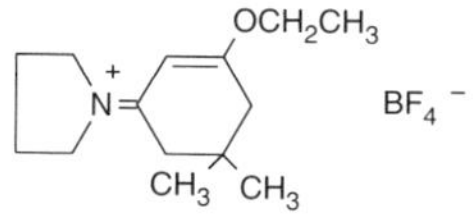
250 mg

S87,169-9
4-(5,5-DIMETHYL-3-METHOXY-2-CYCLOHEXEN-1-YLIDENE)-MORPHOLINIUM IODIDE
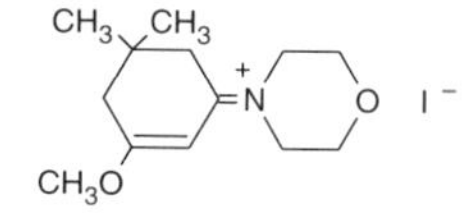
250 mg

S63,033-0
4-(1-METHYLTHIOETHYLIDENE)-MORPHOLINIUM IODIDE
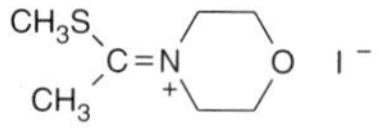
5 gm

S95,508-6
[1,4-DIOXA-8-THIASPIRO(4.5)DEC-7-YLTHIOMETHYL]-TRIMETHYLAMMONIUM IODIDE
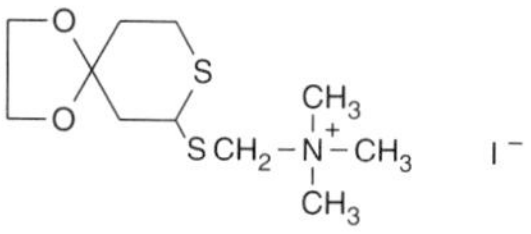
50 mg

S87,195-8
4-(5,5-DIMETHYL-3-MORPHOLINO-2-CYCLOHEXEN-1-YLIDENE)-MORPHOLINIUM IODIDE
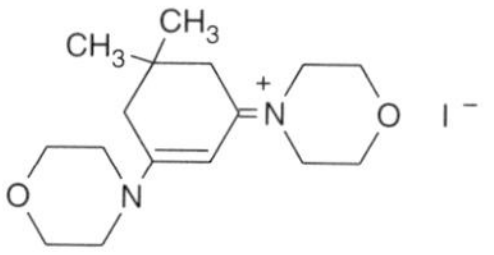
250 mg

S42,642-3
(-)-SPARTEINE METHIODIDE
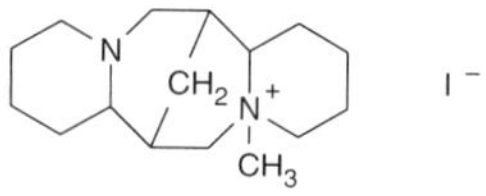
250 mg

S79,327-2
METHANETETRAYLTETRAKIS (TETRAMETHYLAMMONIUM BROMIDE)
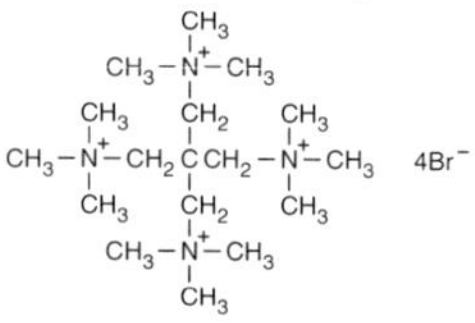
250 mg

S50,389-4
DIBUTYLAMMONIUM HEXAFLUOROPHOSPHATE
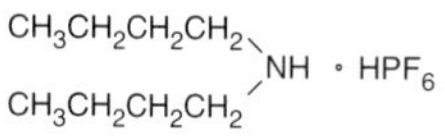
5 gm

S34,887-2
DIBUTYLAMMONIUM HEXAFLUOROARSENATE
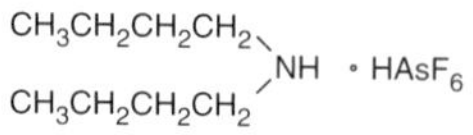
5 gm

S50,377-0
DIPENTYLAMMONIUM TETRAFLUOROBORATE
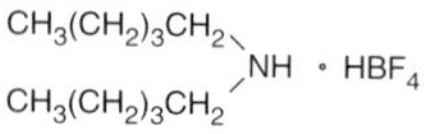
5 gm

S50,381-9 BIS(2-ETHYLHEXYL)AMMONIUM TETRAFLUOROBORATE 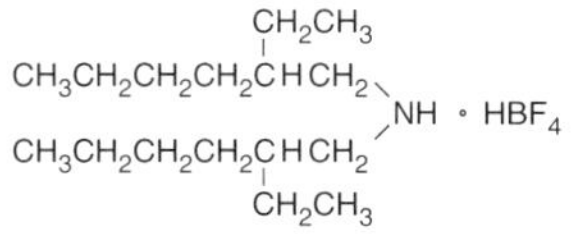 5 gm	**S50,383-5** BIS(2-ETHYLHEXYL)AMMONIUM HEXAFLUOROPHOSPHATE 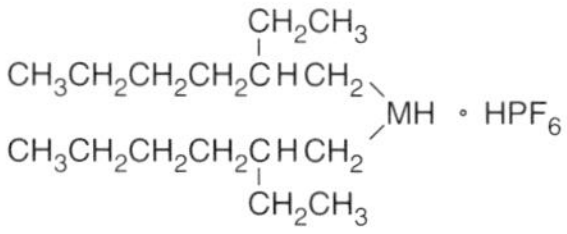 5 gm	**S34,888-0** DIBUTYLAMMONIUM HEXAFLUOROSILICATE 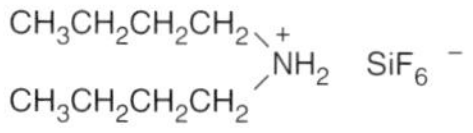 5 gm	**S41,629-0** TETRAMETHYLAMMONIUM CHLORIDE (50% AQUEOUS SOLUTION) 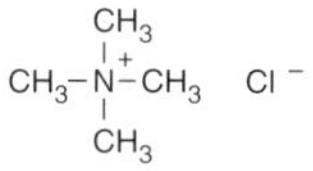 5 gm
S93,338-4 TETRAMETHYLAMMONIUM OCTAHYDROTRIBORATE 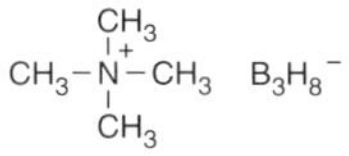 250 mg	**S34,292-0** BUTYLAMMONIUM TETRAFLUOROBORATE 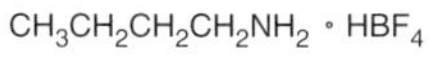 5 gm	**S50,362-2** BUTYLAMMONIUM HEXAFLUOROPHOSPHATE 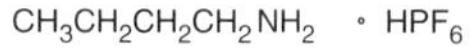 5 gm	**S43,376-4** TETRAMETHYLAMMONIUM TRIBROMIDE 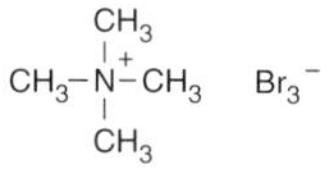 5 gm
S88,739-0 TERT-BUTYL(METHOXYIMINO)-METHYLAMMONIUM TETRAFLUOROBORATE 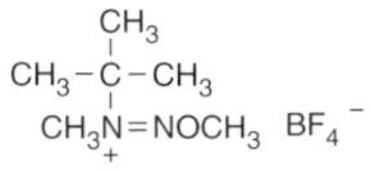 250 mg	**S76,570-8** DIMETHYLETHYLPROPYLAMMONIUM BROMIDE 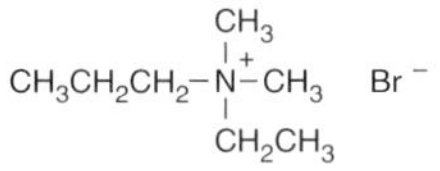 5 gm	**S94,982-5** N-BUTYLTRIMETHYLAMMONIUM IODIDE 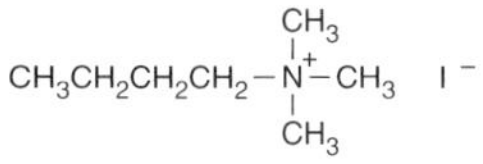 250 mg	**S76,571-6** DIETHYLMETHYLPROPYLAMMONIUM BROMIDE 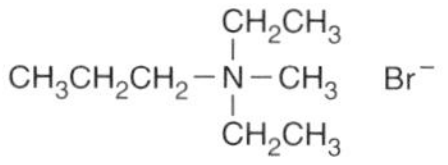 5 gm
S36,786-9 DIBUTYLAMMONIUM TETRAFLUOROBORATE 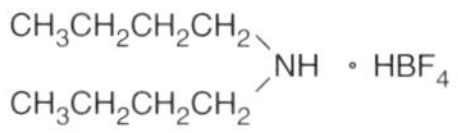 250 mg	**S44,942-3** ALLYLTRIETHYLAMMONIUM BROMIDE 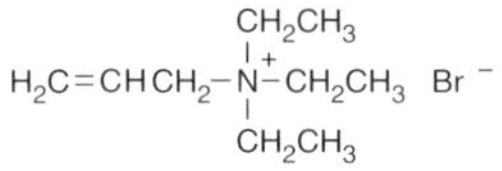 5 gm	**S50,363-0** TRIPROPYLAMMONIUM HEXAFLUOROPHOSPHATE 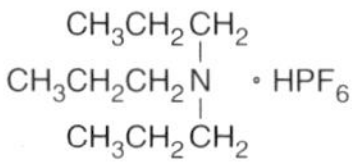 5 gm	**S35,675-1** TRIPROPYLAMMONIUM HEXAFLUOROARSENATE 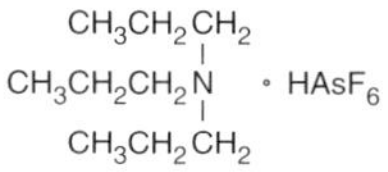 5 gm
S86,973-2 [2-(1,4-CYCLOHEXADIEN-1-YL)ETHYL]-TRIMETHYLAMMONIUM PERCHLORATE 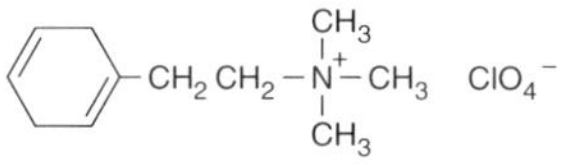 250 mg	**S86,974-0** [2-(2,5-CYCLOHEXADIEN-1-YL)ETHYL]-TRIMETHYLAMMONIUM PERCHLORATE 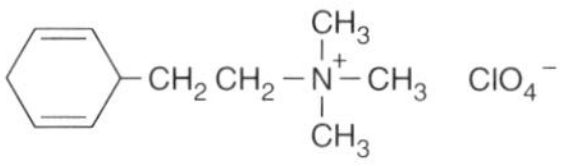 250 mg	**S86,972-4** [2-(2,5-CYCLOHEXADIEN-1-YLIDENE)-ETHYL]TRIMETHYLAMMONIUM IODIDE 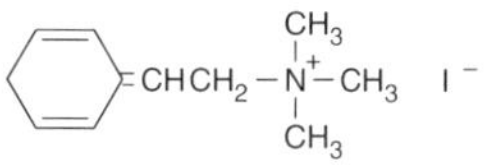 250 mg	**S86,975-9** [2-(1,4-CYCLOHEXADIEN-1-YL)ETHYL]-TRIMETHYLAMMONIUM IODIDE 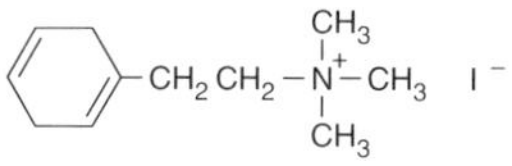 250 mg
S72,709-1 SEC-OCTYLTRIMETHYLAMMONIUM IODIDE 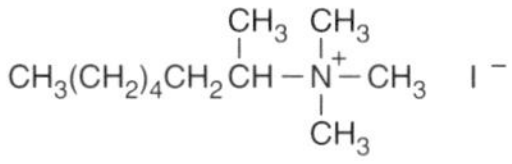 250 mg	**S88,489-8** TETRAPROPYLAMMONIUM CHLORIDE 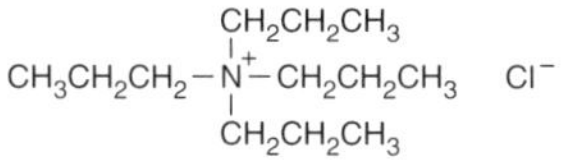 250 mg	**S35,597-6** TETRAPROPYLAMMONIUM TETRAFLUOROBORATE 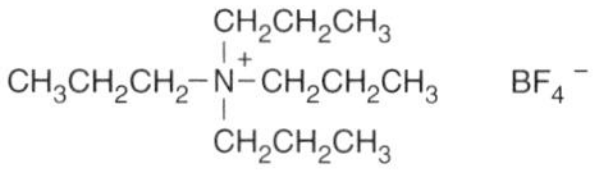 5 gm	**S50,380-0** TRIBUTYLAMMONIUM TETRAFLUOROBORATE 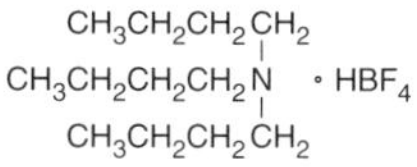 5 gm
S56,660-8 N,N'-HEXAMETHYLENEBIS-(TRIMETHYLAMMONIUM CHLORIDE) 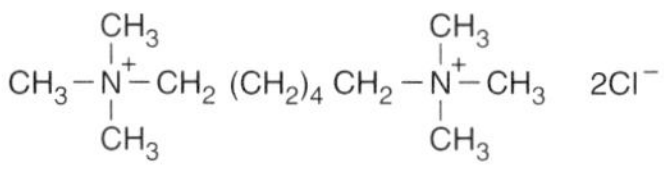 5 gm	**S86,988-0** [4-(2,5-CYCLOHEXADIEN-1-YLIDENE)-2-BUTEN-1-YLIDENE]DIMETHYLAMMONIUM PERCHLORATE 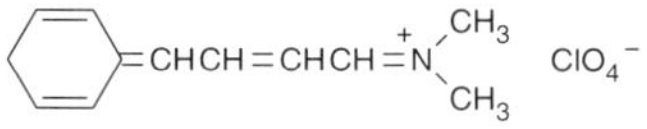 250 mg	**S50,364-9** TRIBUTYLAMMONIUM HEXAFLUOROPHOSPHATE 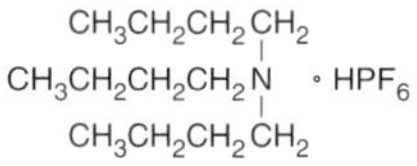 5 gm	**S50,382-7** TETRAPROPYLAMMONIUM HEXAFLUOROARSENATE 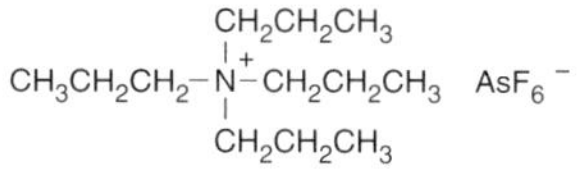 5 gm
S87,008-0 [4-(2,5-CYCLOHEXADIEN-1-YLIDENE)-2-BUTEN-1-YL]TRIMETHYLAMMONIUM IODIDE 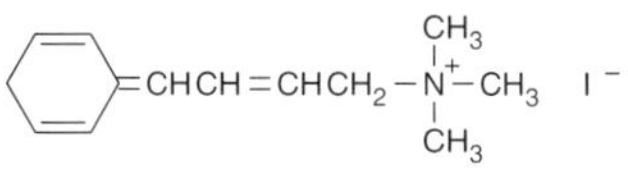 250 mg	**S45,232-7** METHYLTRIBUTYLAMMONIUM IODIDE 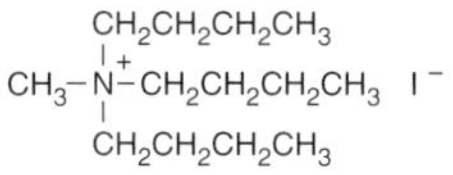 5 gm	**S76,569-4** PENTYLTRIBUTYLAMMONIUM BROMIDE 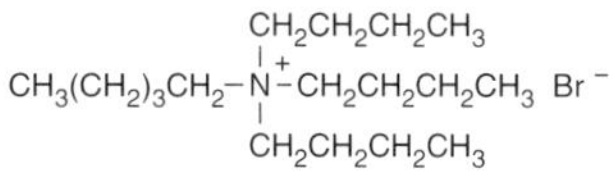 5 gm	**S34,384-6** CETYLTRIMETHYLAMMONIUM BOROHYDRIDE 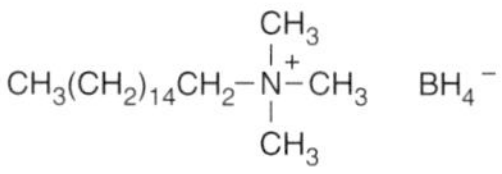 5 gm

S91,527-0
HEXADECYLTRIETHYLAMMONIUM BROMIDE

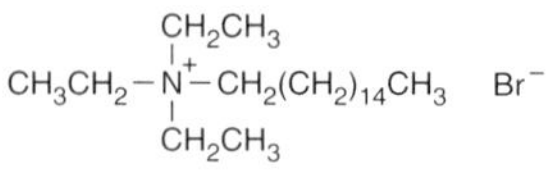

250 mg

S81,064-9
DIBUTYLMETHYLTRIDECYLAMMONIUM IODIDE

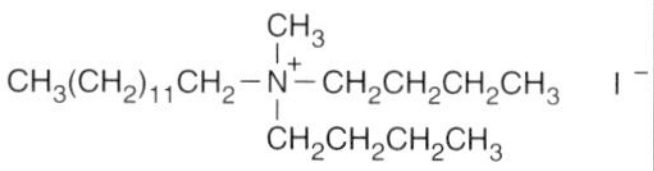

250 mg

S36,394-4
METHYLTRIOCTYLAMMONIUM BOROHYDRIDE

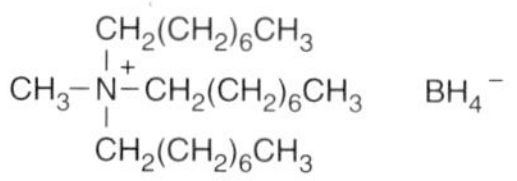

5 gm

S68,887-8
N,N'-METHYLENEBIS-(DIMETHYLTETRADECYLAMMONIUM BROMIDE)

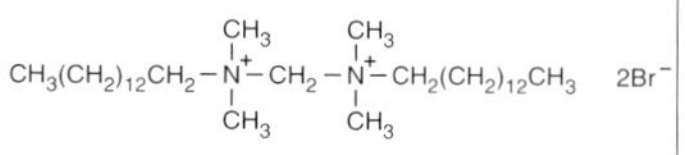

5 gm

S86,118-9
4-CYCLOHEXYLIDENEMORPHOLINIUM PERCHLORATE

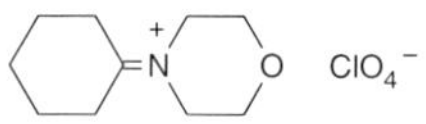

250 mg

S76,574-0
N,N-PENTAMETHYLENEBIS-(TRIMETHYLAMMONIUM) DIIODIDE

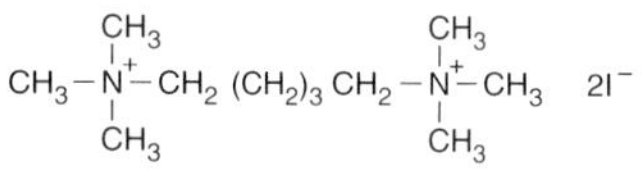

5 gm

S35,991-2
(CYCLOPROPYLMETHYL)-TRIMETHYLAMMONIUM IODIDE

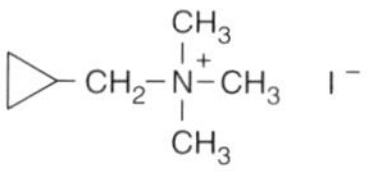

250 mg

S61,881-0
3,4,4-TRIMETHYL-2-OXAZOLINIUM IODIDE

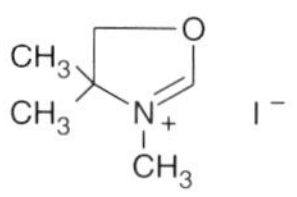

5 gm

S50,360-6
CYCLOHEXYLDIETHYLAMMONIUM HEXAFLUOROARSENATE

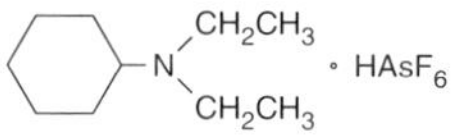

5 gm

S61,704-0
CYCLOHEXYLTRIETHYLAMMONIUM IODIDE

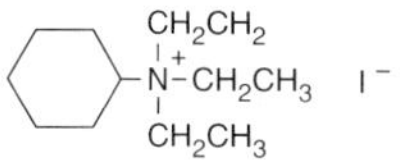

250 mg

S34,731-0
1,1-DIMETHYLPIPERIDINIUM IODIDE

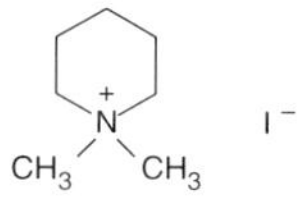

5 gm

S88,217-8
1-ETHYL-1-(2,4-PENTADIEN-1-YL)-PIPERIDINIUM BROMIDE

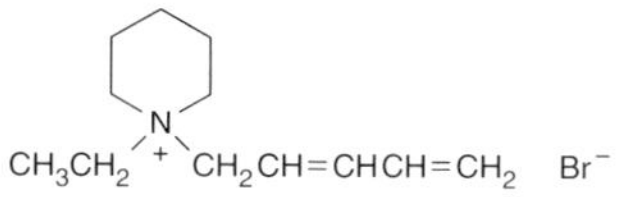

250 mg

S88,014-0
1-METHYL-1-(2-METHYL-2-OCTEN-1-YL)-PIPERIDINIUM IODIDE

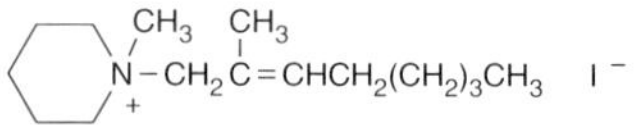

250 mg

S72,711-3
1,5-DIMETHYL-1,5-DIAZONIABICYCLO-[3.2.2]NONANE DIBROMIDE

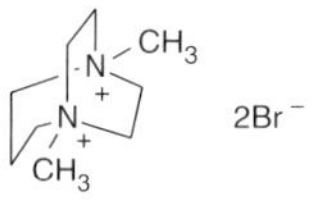

5 gm

S50,357-6
MORPHOLINIUM TETRAFLUOROBORATE

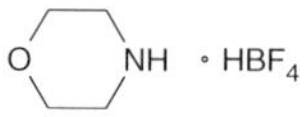

5 gm

S88,151-1
4-METHYL-4-(2-OCTEN-1-YL)-MORPHOLINIUM IODIDE

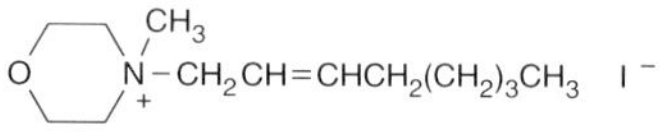

250 mg

S88,754-4
DIMETHYL(ETHOXYIMINO)AMMONIUM TETRAPHENYLBORATE

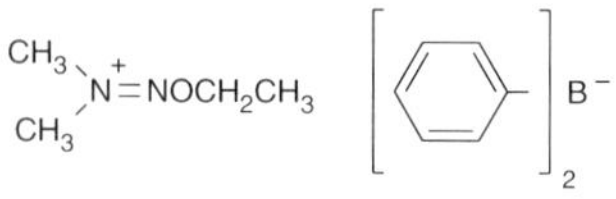

250 mg

S72,897-7
1,4-DIMETHYL-1,4-DIAZONIABICYCLO-[2.2.2]OCTANE DIBROMIDE

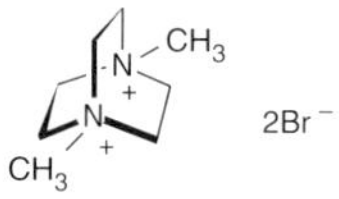

5 gm

S95,826-3
N-(9,9-ETHYLENEDIOXYBICYCLO[3.3.1]-NONAN-2-YL)-N-METHYLMORPHOLINIUM IODIDE

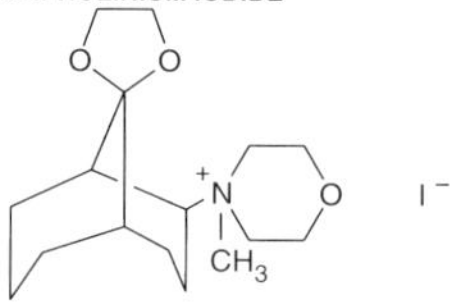

50 mg

S76,443-4
4,7-ENDOMETHYLENE-2-METHYL-2-(3-MORPHOLINOPROPYL)TETRAHYDRO-ISOINDOLINIUM IODIDE METHIODIDE

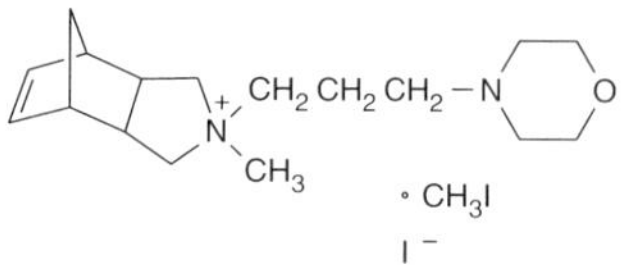

5 gm

S84,941-3
BUTYL NITRATE

$CH_3CH_2CH_2CH_2ONO_2$

250 mg

S91,767-2
OCTYL NITRATE

$CH_3(CH_2)_6CH_2ONO_2$

250 mg

S64,841-8
CYCLOHEXYL NITRATE

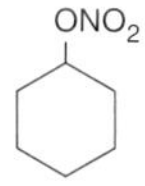

5 gm

S44,451-0
N-NITROSODICYCLOHEXYLAMINE

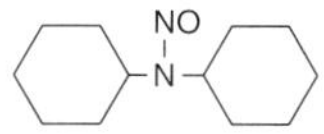

5 gm

S52,636-8
N-NITROSODIETHYLAMINE

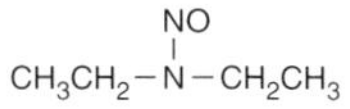

5 gm

S64,258-4
1-NITROSOPIPERIDINE

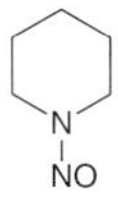

250 mg

S88,509-6
1,2-DICHLORO-3-NITROPROPANE

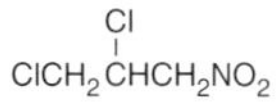

250 mg

S75,633-4
TRIS(CHLOROMETHYL)NITROMETHANE

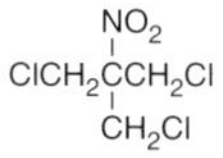

5 gm

S84,887-5
2-NITRO-2-PROPYL-1,3-PROPANEDIOL

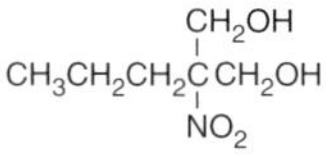

250 mg

S84,888-3
2-ISOPROPYL-2-NITRO-1,3-PROPANEDIOL

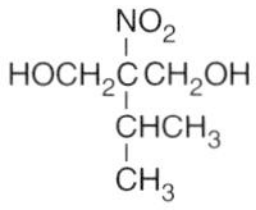

250 mg

S84,902-2
2-ETHYL-2-NITRO-1,3-PROPANEDIOL

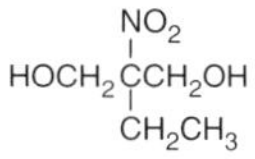

250 mg

S84,945-6
3,3,5,8,10,10-HEXANITRO-1,12-DODECANEDIOL

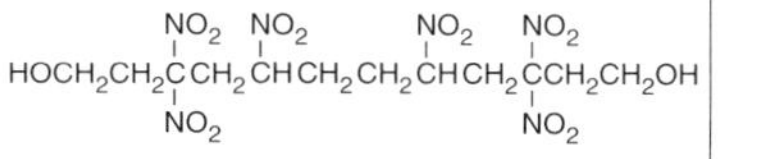

250 mg

S84,884-0
2,2,5,5-TETRANITRO-1,6-HEXANEDIOL

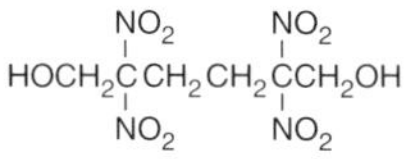

250 mg

S45,634-9
2-METHYL-2-NITRO-1,3-PROPANEDIOL

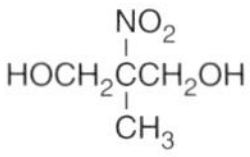

5 gm

S54,786-7
1-DEOXY-1-NITRO-D-MANNITOL

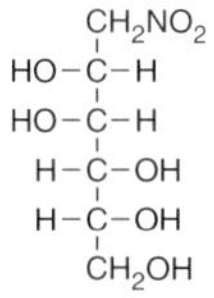

250 mg

S45,212-2
2,2-DINITROPROPANOL (30% SOLUTION IN ETHYLENE CHLORIDE)

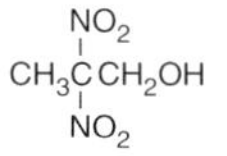

5 gm

S42,933-3
2-NITRO-1-BUTANOL

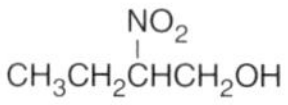

5 gm

S84,886-7
4-NITRO-3-HEXANOL

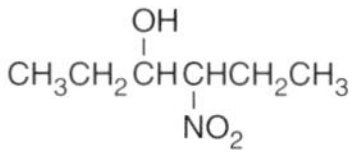

250 mg

S35,391-4
1-(NITROMETHYL)CYCLOHEXANOL

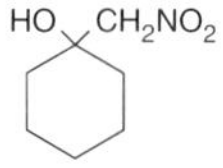

5 gm

S91,912-8
1-(1-NITROCYCLOHEXYL)-1-ETHANOL

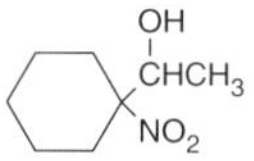

250 mg

S86,071-9
1,3-BIS(NITROMETHYL)-4,4-DIMETHYLCYCLOHEXANOL

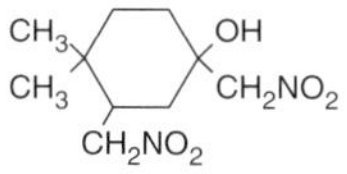

250 mg

S39,263-4
2,2-DIMETHYL-5-NITRO-1,3-DIOXANE-5-METHANOL

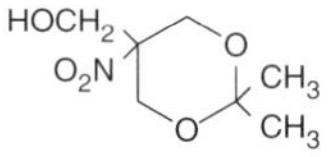

5 gm

S91,818-0
1-METHOXY-2,4,5-TRIMETHYL-2,4,5-TRINITROHEXANE

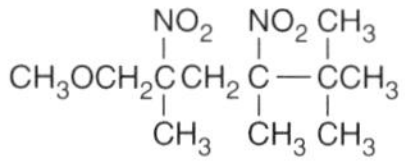

250 mg

S91,804-0
2,3-DIMETHYL-2,3-DINITRO-1-METHOXYBUTANE

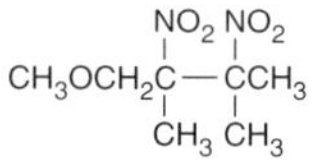

250 mg

S88,602-5
ISOBUTYL(2-NITRO-1-PROPEN-1-YL)AMINE

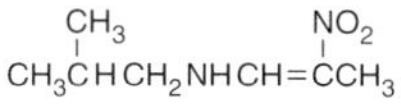

250 mg

S68,173-3
N,N-DIMETHYL-2-NITROETHYLAMINE

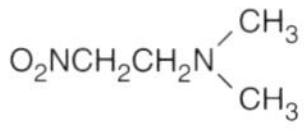

5 gm

S85,991-5
4,4-DIMETHYL-2-(ACI-NITROMETHYL)-2-OXAZOLIN-3-IUM HYDROXIDE, INNER SALT

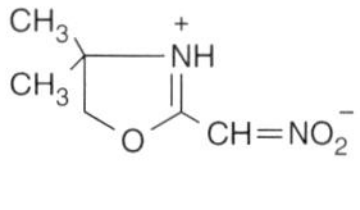

250 mg

S91,769-9
2,3-DIMETHYL-2,3-DINITROPENTANE

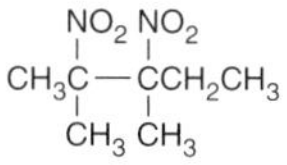

250 mg

S84,906-5
3,4-DIMETHYL-3,4-DINITROHEXANE

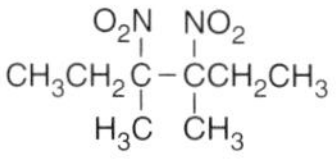

250 mg

S91,851-2
1-(1-METHYL-1-NITROETHYL)-1-NITROCYCLOPENTANE

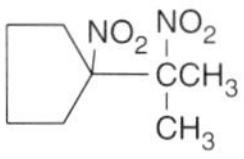

250 mg

S91,810-5
1-(1-METHYL-1-NITROPROPYL)-1-NITROCYCLOPENTANE

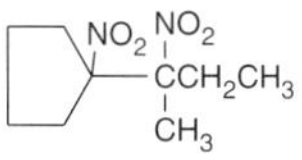

250 mg

S91,816-4
1,1'-DINITRODICYCLOPENTYL

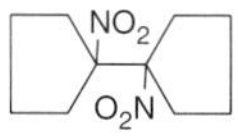

250 mg

S91,891-1
1-(1-METHYL-1-NITROETHYL)-1-NITROCYCLOHEXANE
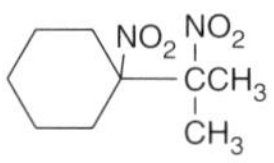
250 mg

S91,857-1
1-(1-METHYL-1-NITROPROPYL)-1-NITROCYCLOHEXANE
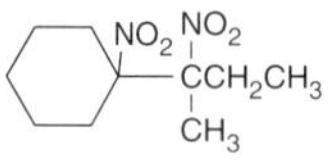
250 mg

S91,823-7
1,1'-DINITRODICYCLOHEXYL
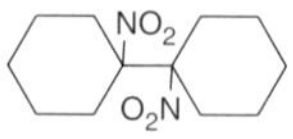
250 mg

S91,865-2
1-NITRO-1-(1-NITROCYCLOHEXYL)-CYCLOHEPTANE
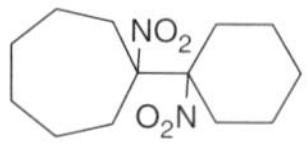
250 mg

S52,124-8
2,2-DIMETHYL-1,3-DINITROPROPANE
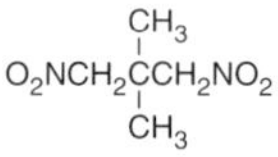
5 gm

S52,118-3
1,3-DINITRO-2-ETHYL-2-METHYLPROPANE
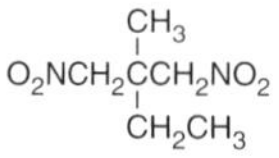
5 gm

S84,940-5
1,4-DINITROBUTANE
$O_2NCH_2CH_2CH_2CH_2NO_2$
250 mg

S84,881-6
1,4-DINITRO-1,1,4,4-TETRACHLOROBUTANE
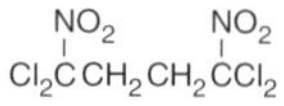
250 mg

S84,901-4
1,4-DINITRO-1,1,4,4-TETRABROMOBUTANE
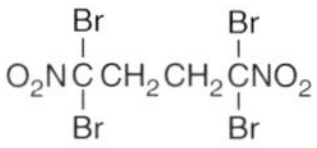
250 mg

S41,945-1
1,1-DINITROETHANE (30% SOLUTION IN ETHYLENE CHLORIDE)
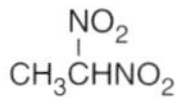
5 gm

S84,925-1
DIISOPROPYLNITRAMINE
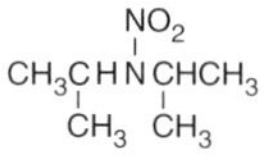
250 mg

S56,596-2
2-CHLORO-2-NITROPROPANE
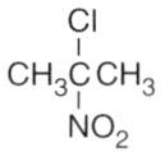
5 gm

S70,581-0
1-CHLORO-1-NITROPROPANE
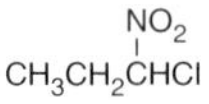
250 mg

S56,409-5
1,1-DICHLORO-1-NITROPROPANE
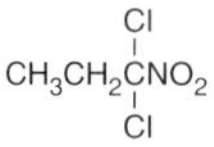
5 gm

S84,882-4
2-METHYL-1-NITROPROPANE
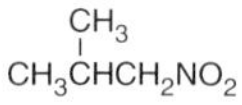
250 mg

S91,917-9
2-NITROBUTANE
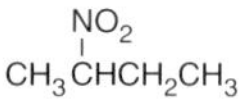
250 mg

S91,801-6
2,2-DINITROBUTANE
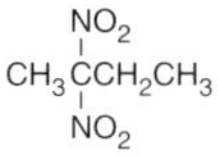
250 mg

S84,942-1
1,1,6,6-TETRANITROHEXANE
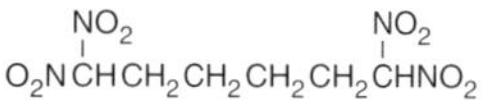
250 mg

S84,883-2
1,6-DINITRO-1,1,6,6-TETRABROMOHEXANE
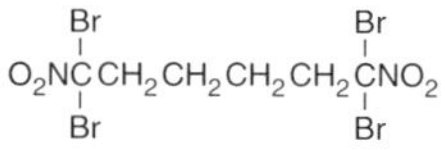
250 mg

S91,909-8
1-NITROCYCLOHEPTANE
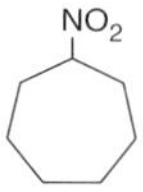
250 mg

S91,910-1
1-NITROHEPTANE
$CH_3(CH_2)_5CH_2NO_2$
250 mg

S52,177-9
2,2-DIETHYL-1,3-DINITROPROPANE
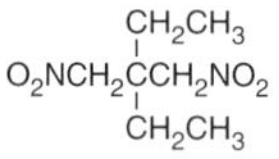
5 gm

S84,889-1
2-NITROOCTANE
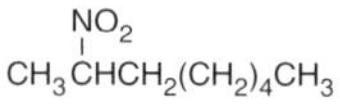
250 mg

S91,768-0
1-NITROOCTANE
$CH_3(CH_2)_6CH_2NO_2$
250 mg

S52,306-2
1,3-DINITRO-2-ISOBUTYL-2-METHYLPROPANE
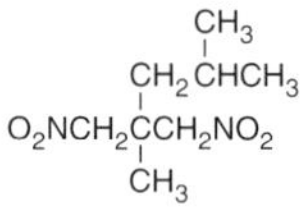
5 gm

S91,808-3
1,1-DINITROOCTANE
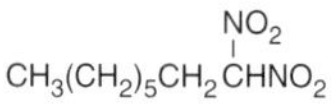
250 mg

S91,849-0
1-BROMO-1-NITROOCTANE
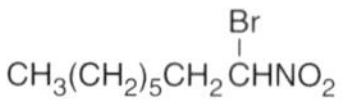
250 mg

S52,665-7
2-BUTYL-2-ETHYL-1,3-DINITROPROPANE
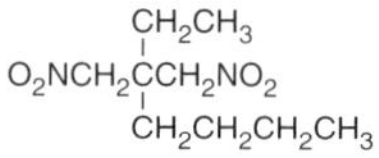
5 gm

S52,666-5
2,2-DIPROPYL-1,3-DINITROPROPANE
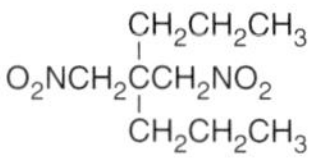
5 gm

S91,894-6
1,1-DINITROCYCLODODECANE
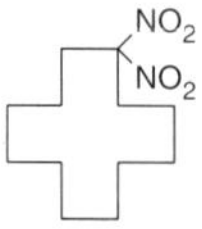
250 mg

S91,751-6
2-METHYL-2-NITROTRIDECANE
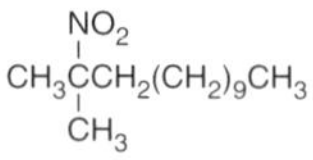
250 mg

S91,802-4
1,1-DINITROCYCLOPENTANE
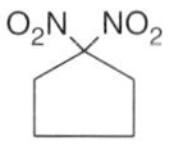
250 mg

S83,648-6
1-CYCLOPENTYL-2-NITROETHENE

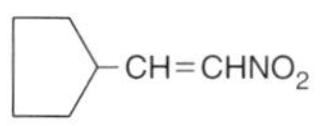

250 mg

S91,803-2
1,1-DINITROCYCLOHEXANE

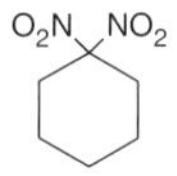

250 mg

S69,789-3
2-ISOPROPYL-5-NITRO-1,3-DIOXANE

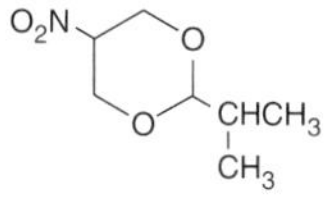

250 mg

S84,911-1
3-(HEPTAFLUOROPROPYL)-2-NITRO-5-NORBORNENE

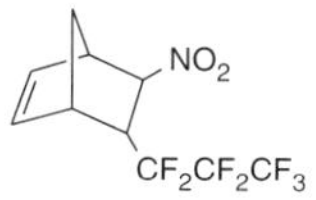

250 mg

S61,426-2 (ETHYLTHIO)ACETONE 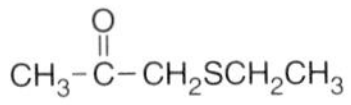 5 gm	**S88,598-3** 5-NITRO-2-PENTANONE 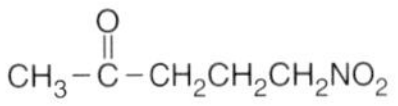 250 mg	**S76,472-8** 4,4-DIMETHYL-5-NITRO-2-PENTANONE 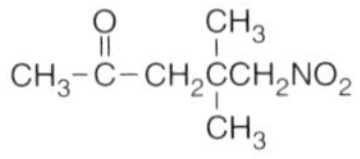 5 gm	**S91,919-5** 2-NITRO-2,4,4-TRIMETHYL-3-PENTANONE 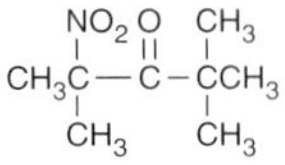 250 mg
S91,928-4 5-NITRO-2-OCTANONE 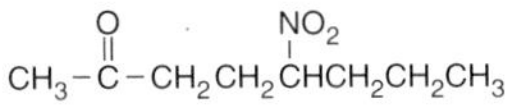 250 mg	**S41,179-5** 4-METHYL-3-HEXANONE 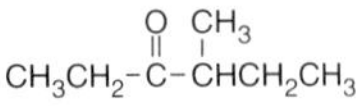 5 gm	**S43,734-4** 3-METHYL-2-HEXANONE 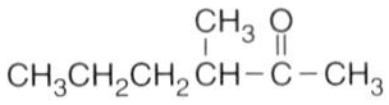 5 gm	**S45,748-5** 2-METHYL-3-HEXANONE 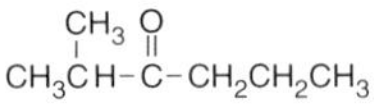 5 gm
S37,355-9 3-ISOPROPYL-2-PENTANONE 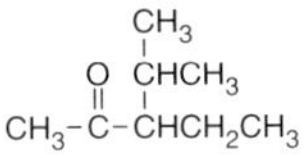 5 gm	**S43,057-9** 3-METHYL-2-HEPTANONE 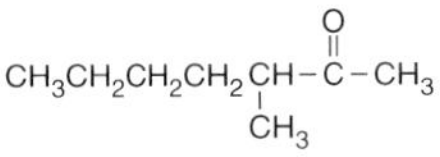 250 mg	**S43,111-7** 2,4-DIMETHYL-3-HEXANONE 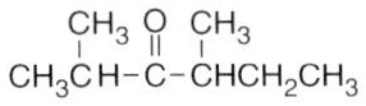 5 gm	**S45,648-9** 4-OCTANONE 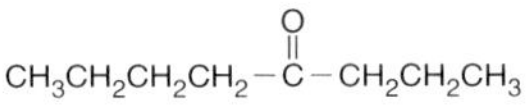 250 mg
S50,475-0 2-METHYL-4-HEPTANONE 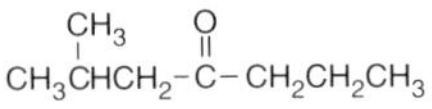 5 gm	**S60,336-8** 6-METHYL-2-HEPTANONE 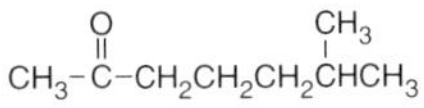 250 mg	**S35,794-4** AMYL ISOPROPYL KETONE 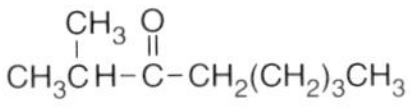 5 gm	**S41,182-5** 4-NONANONE 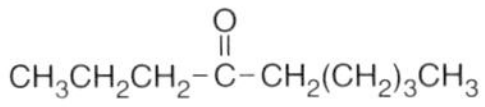 5 gm
S43,148-6 3-METHYL-4-OCTANONE 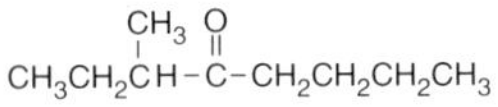 5 gm	**S43,317-9** 2,4-DIMETHYL-3-HEPTANONE 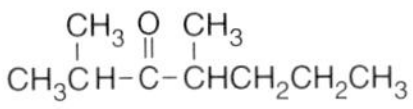 5 gm	**S45,049-9** 2,6-DIMETHYL-3-HEPTANONE 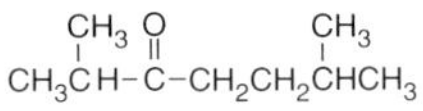 5 gm	**S50,476-9** 3-ETHYL-4-HEPTANONE 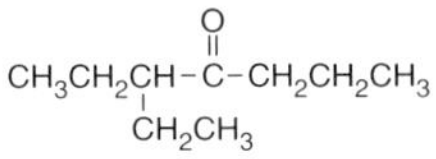 5 gm
S35,057-5 HEXYL ISOPROPYL KETONE 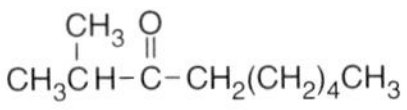 5 gm	**S35,172-5** 3-ISOPROPYL-2-HEPTANONE 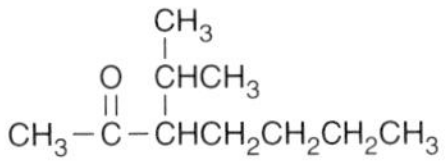 5 gm	**S35,793-6** AMYL ISOBUTYL KETONE 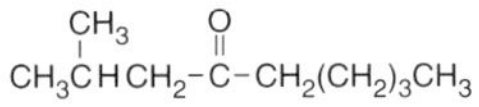 5 gm	**S43,147-8** 3-METHYL-4-NONANONE 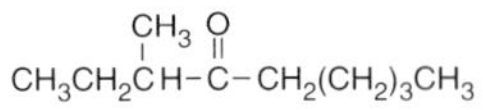 5 gm
S50,469-6 3,5-DIMETHYL-4-OCTANONE 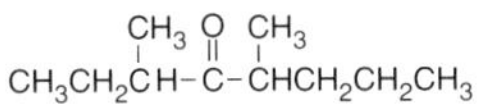 5 gm	**S35,040-0** HEPTYL ISOPROPYL KETONE 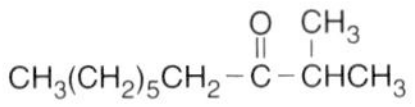 5 gm	**S35,056-7** HEXYL ISOBUTYL KETONE 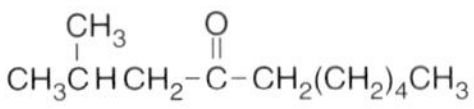 5 gm	**S50,487-4** 4-METHYL-5-DECANONE 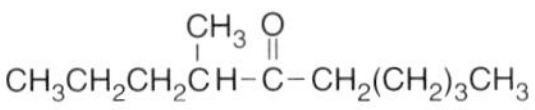 5 gm
S54,718-2 3-BUTYL-2-HEPTANONE 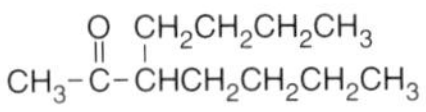 5 gm	**S70,418-0** 2,8-DIMETHYL-5-NONANONE 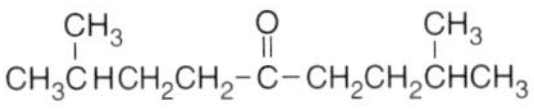 5 gm	**S35,039-7** HEPTYL ISOBUTYL KETONE 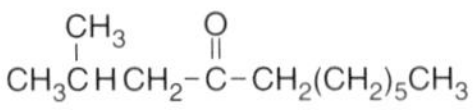 5 gm	**S37,344-3** 7-ETHYL-2-METHYL-4-NONANONE 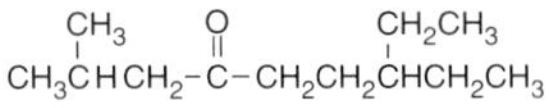 5 gm

S54,404-3
3-DODECANONE
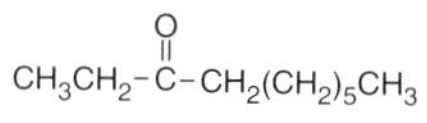
5 gm

S54,407-8
5-DODECANONE
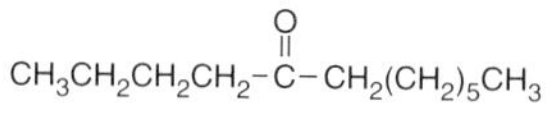
5 gm

S55,938-5
2-DODECANONE
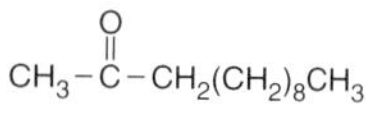
250 mg

S86,678-4
2-METHYL-3-UNDECANONE
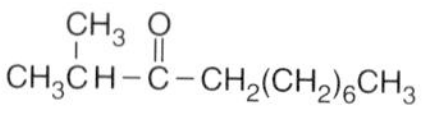
250 mg

S87,929-0
2-METHYL-4-UNDECANONE
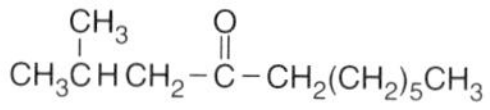
5 gm

S54,676-3
3-TRIDECANONE
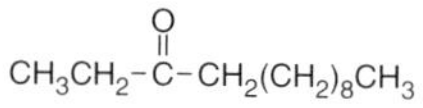
250 mg

S86,690-3
2,2-DIMETHYL-3-UNDECANONE
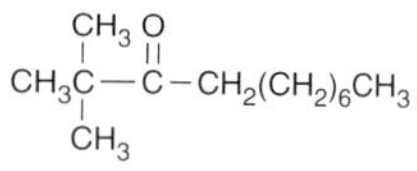
250 mg

S34,941-0
7-ETHYL-2-METHYL-4-UNDECANONE
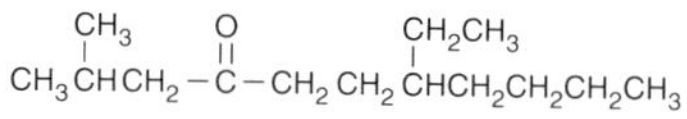
5 gm

S54,398-5
3-TETRADECANONE
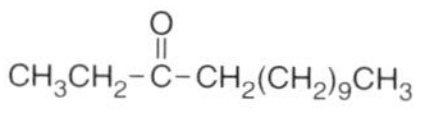
5 gm

S54,378-0
7-PENTADECANONE
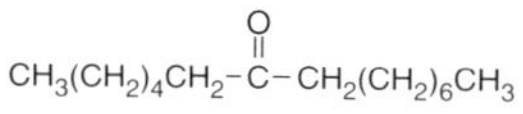
5 gm

S54,385-3
4-PENTADECANONE
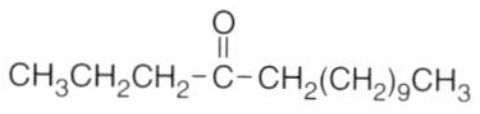
250 mg

S54,409-4
6-PENTADECANONE
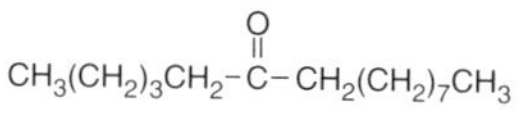
250 mg

S54,673-9
5-PENTADECANONE
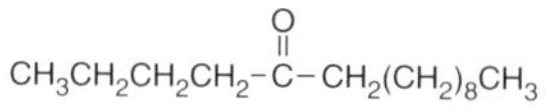
250 mg

S82,840-8
8-PENTADECANONE
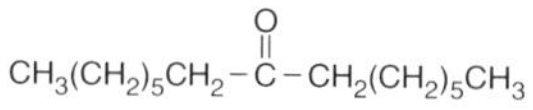
250 mg

S54,393-4
3-HEXADECANONE
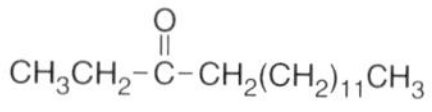
250 mg

S54,408-6
7-HEXADECANONE
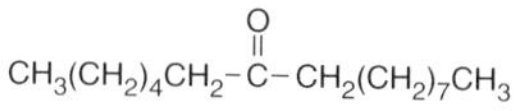
5 gm

S54,392-6
4-HEPTADECANONE
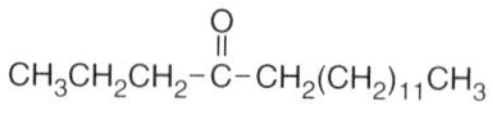
5 gm

S54,410-8
6-HEPTADECANONE
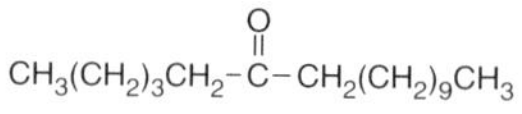
5 gm

S54,411-6
7-HEPTADECANONE
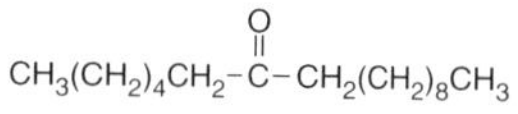
5 gm

S54,412-4
8-HEPTADECANONE
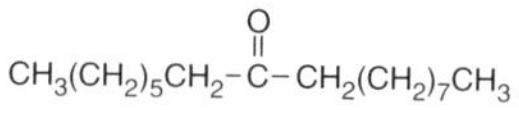
250 mg

S76,291-1
2-HEPTADECANONE
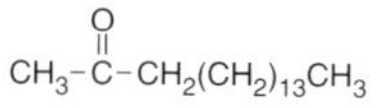
250 mg

S36,098-8
2,6-DIMETHYL-10-HEXADECANONE
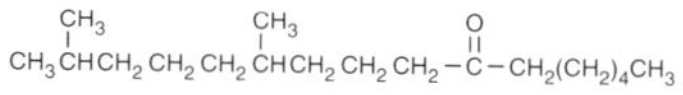
5 gm

S54,018-8
3-OCTADECANONE
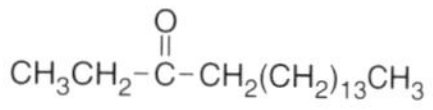
5 gm

S54,021-8
5-OCTADECANONE
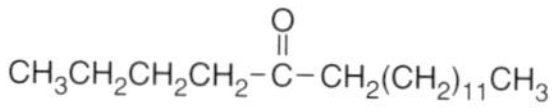
5 gm

S54,025-0
9-OCTADECANONE
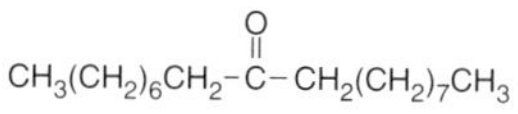
5 gm

S54,011-0
2-NONADECANONE
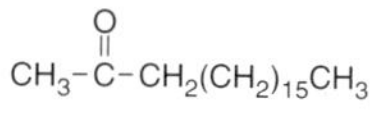
5 gm

S82,843-2
9-NONADECANONE
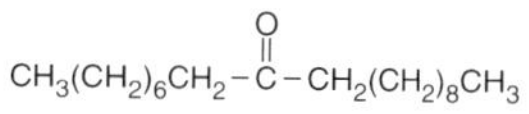
250 mg

S57,300-0
3-EICOSANONE
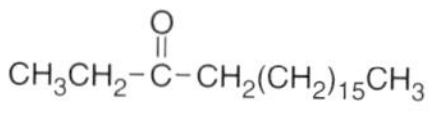
5 gm

S44,971-7
11-HENEICOSANONE
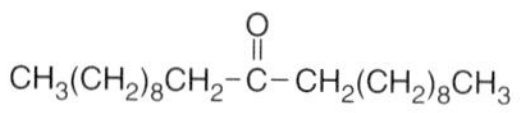
5 gm

S75,750-0
10-HENEICOSANONE
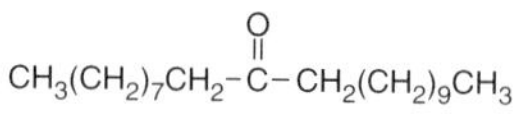
5 gm

S41,128-0
12-TRICOSANONE
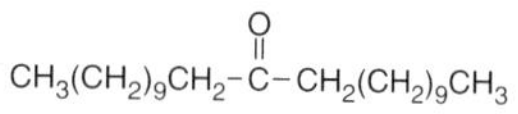
5 gm

S56,290-4
10-TRICOSANONE
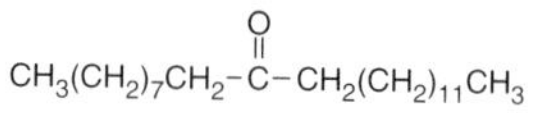
250 mg

S56,291-2
6-TRICOSANONE
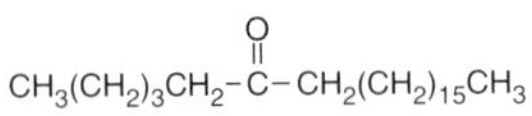
250 mg

S54,418-3
14-HEPTACOSANONE
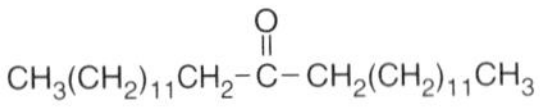
5 gm

S51,852-2
17-TRITRIACONTANONE
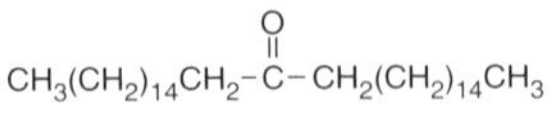
5 gm

S57,284-5
18-PENTATRIACONTANONE
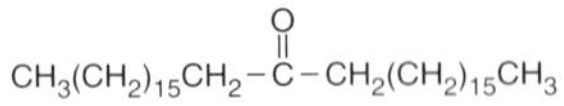
5 gm

S80,894-6
4-HYDROXY-3-METHYL-3-BUTEN-2-ONE, SODIUM SALT
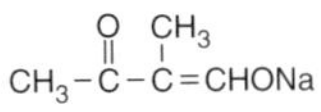
250 mg

S94,667-2
2-METHYL-3-ISOPROPYL-5-HEPTEN-4-ONE
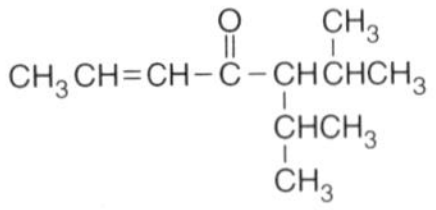
250 mg

S56,695-0
3-METHYL-3-BUTEN-2-ONE
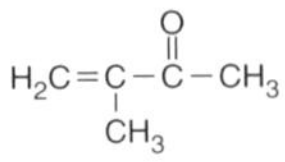
5 gm

S36,170-4
5-ETHYL-3-HEPTEN-2-ONE
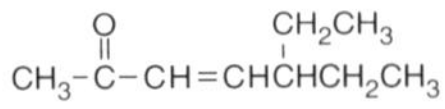
5 gm

S84,347-4
2,7-DIMETHYL-2,5-OCTADIEN-4-ONE
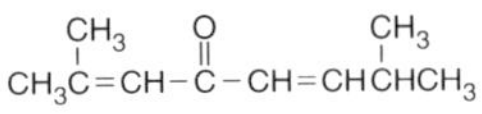
250 mg

S94,986-8
3-ETHYL-4-METHYL-3-HEPTEN-5-ONE
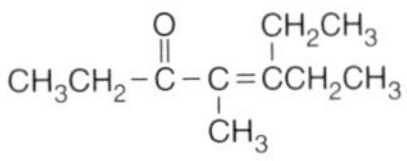
1 gm

S78,667-5
PSEUDOIONONE
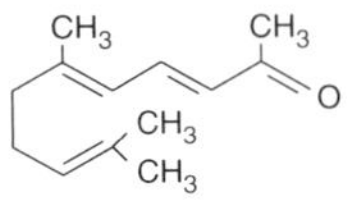
250 mg

S36,031-7
1,7-DICHLORO-4-HEPTANONE,TECH.
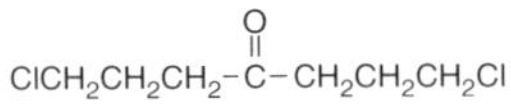
5 gm

S46,265-9
1,1,3,3-TETRACHLOROACETONE
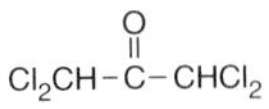
250 mg

S79,968-8
4-METHYL-5,5,5-TRIFLUORO-2-PENTANONE
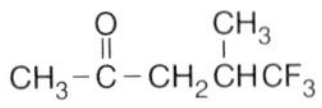
250 mg

S79,962-9
4-METHYL-5,5,5-TRIFLUORO-3-PENTEN-2-ONE
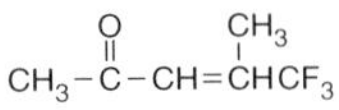
250 mg

S44,414-6
3-BROMO-2-BUTANONE
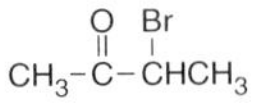
250 mg

S91,918-7
2-BROMO-2,4,4-TRIMETHYL-3-PENTANONE
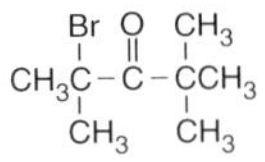
250 mg

S53,598-2
2,6-DIHYDROXY-1,1,1,7,7,7-HEXACHLORO-4-HEPTANONE
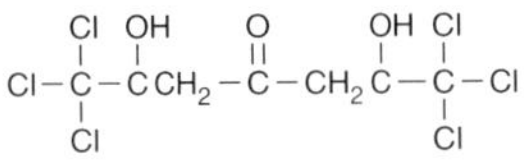
5 gm

S67,903-8
2,6-DIHYDROXY-2,6-DIMETHYL-4-HEPTANONE
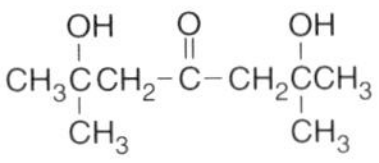
5 gm

S64,117-0
D-GALTOSE
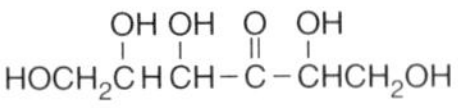
5 gm

S70,550-0
5-HYDROXY-4-OCTANONE
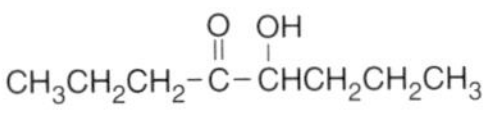
5 gm

S93,340-6
2,2,5,5-TETRAMETHYL-4-HYDROXY-3-HEXANONE
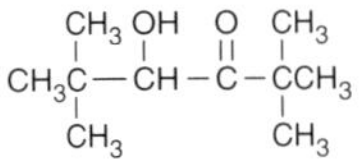
250 mg

S46,135-0
4-HYDROXY-3-METHYL-2-BUTANONE
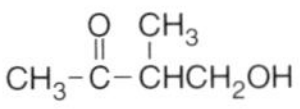
5 gm

S84,352-0
2,7-DIMETHYL-6-HYDROXY-2-OCTEN-4-ONE
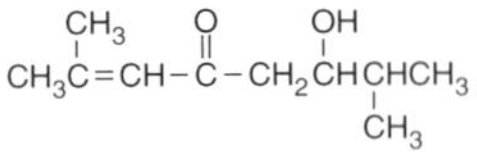
250 mg

S56,189-4
3-CHLORO-5-HYDROXY-2-PENTANONE
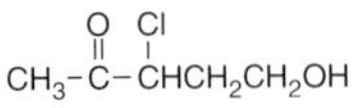
5 gm

S44,755-2
1,5-DIHYDROXY-2,2,4,4-TETRACHLORO-3-PENTANONE
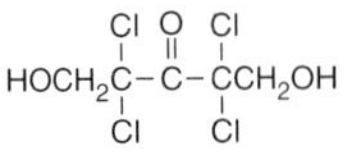
5 gm

S54,103-6
PYRUVIC ALDEHYDE DIBUTYL ACETAL
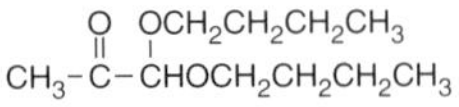
250 mg

S83,955-8
1,1,1-TRIFLUORO-2,4-PENTANEDIONE, DIMETHYLTHALLIUM SALT
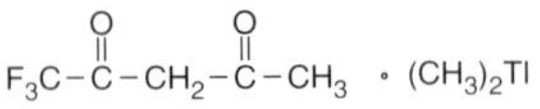
250 mg

S83,957-4
6-METHYL-1,1,1-TRIFLUORO-2,4-HEPTANEDIONE, DIMETHYLTHALLIUM SALT
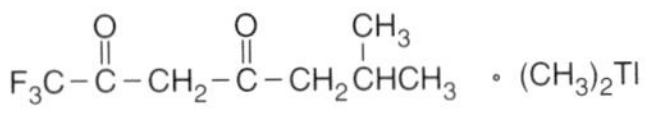
250 mg

S83,958-2
1,1,1-TRIFLUORO-2,4-OCTANEDIONE, DIMETHYLTHALLIUM SALT
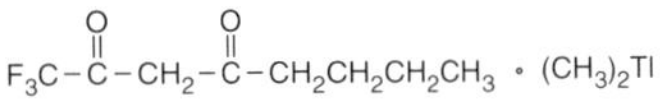
250 mg

S83,961-2
1,1,1-TRIFLUORO-2,4-NONANEDIONE, DIETHYLTHALLIUM SALT
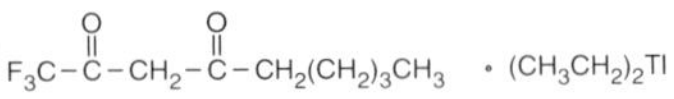
250 mg

S87,930-4
2,2,4,6,6-PENTAMETHYL-3,5-HEPTANEDIONE
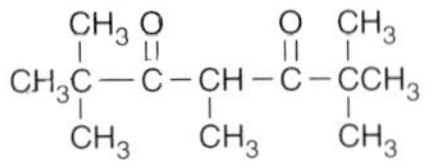
5 gm

S82,655-3
7,16-DOCOSANEDIONE
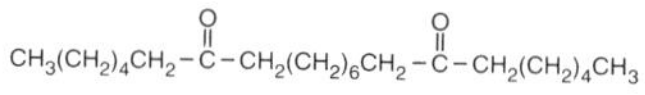
250 mg

S73,241-9
2,2,5,5-TETRAMETHYL-3,4-HEXANEDIONE
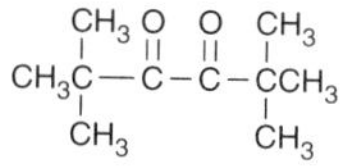
250 mg

S62,245-1
2,4,6-HEPTATRIONE
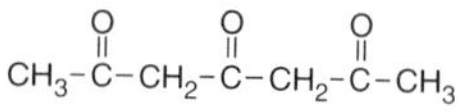
250 mg

S34,049-9
3-ACETYL-5-METHYLHEXAN-2-ONE
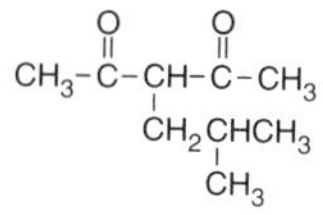
250 mg

S35,725-1
3-ACETYL-2-HEPTANONE
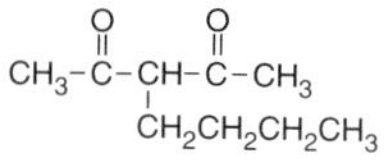
5 gm

S83,956-6
2,4-HEPTANEDIONE, DIMETHYLTHALLIUM SALT
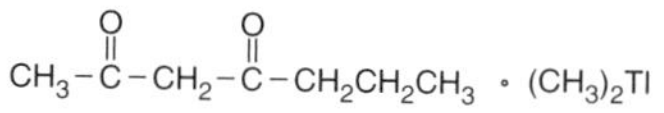
250 mg

S34,052-9
3-ACETYL-2-OCTANONE
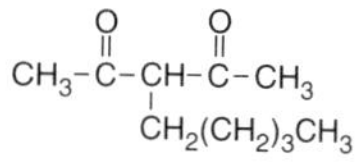
250 mg

S83,959-0
2,4-OCTANEDIONE, DIMETHYLTHALLIUM SALT
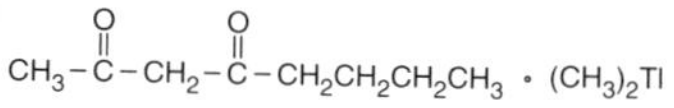
250 mg

S67,892-9
1,1,2,2-TETRAACETYLETHANE
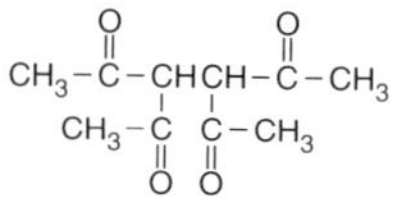
5 gm

S87,248-2
3,9-UNDECANEDIONE
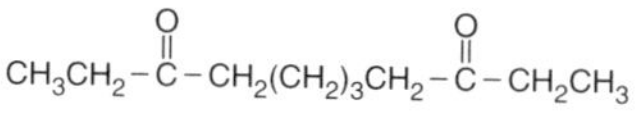
250 mg

S36,836-9
3,4-DIISOPROPYL-2,5-HEXANEDIONE
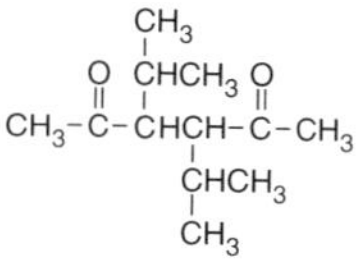
250 mg

S79,234-9
CHLOROMETHYL 2-HYDROXY-5-METHYLCYCLOHEXYL KETONE
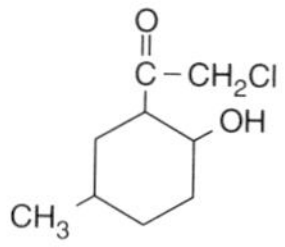
250 mg

S95,603-1
4A-(DICHLOROMETHYL)-2,4A,5,6,7,8-HEXAHYDRO-2-NAPHTHALENONE
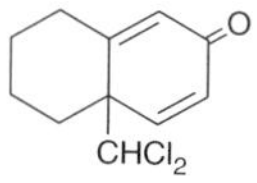
50 mg

S95,779-8
3,8-DIBROMOCAMPHOR
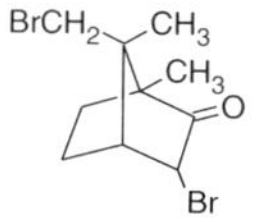
50 mg

S95,777-1
8-BROMO-2,3-BORNANEDIONE
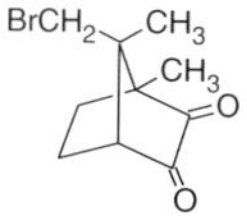
50 mg

S55,606-8
4-DICHLOROMETHYL-4-METHYL-2,5-CYCLOHEXADIEN-1-ONE
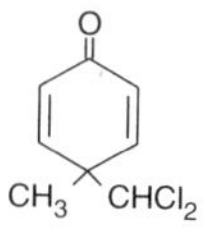
250 mg

S67,140-1
4-(TRICHLOROMETHYL)-2,4,5-TRIMETHYL-2,5-CYCLOHEXADIEN-1-ONE
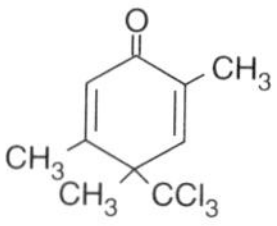
5 gm

S75,785-3
3-(TRIFLUOROMETHYL)CYCLOHEXANONE
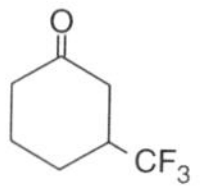
5 gm

S80,498-3
2,5-DIBROMO-3,6-DIMETHYL-1,4-BENZOQUINONE
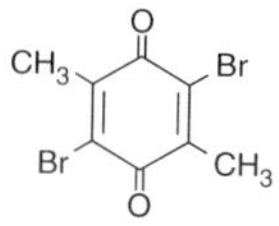
250 mg

S79,293-4
3,4-DIHYDROXY-2-METHYL-2-CYCLOPENTEN-1-ONE
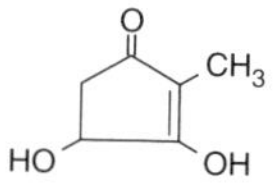
250 mg

S95,705-4
2,3,4,4A,4B,5,6,7,8,8A-DECAHYDRO-6,7-DIHYDROXY-8A-METHYL-2-PHENANTHRENONE
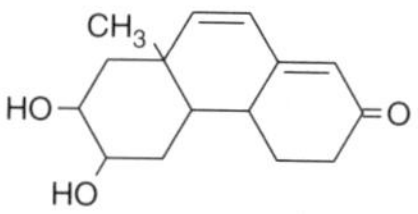
50 mg

S95,706-2
TRANS-2,3,4,4A,4B,5,6,7,8,8A-DECAHYDRO-6,7-DIHYDROXY-1,8A-DIMETHYL-2-PHENANTHRENONE
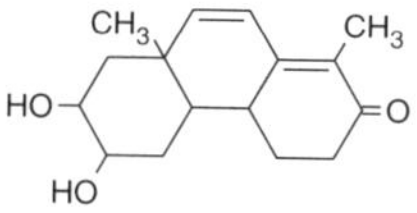
50 mg

S95,708-9
CIS-2,3,4,4A,4B,5,6,7,8,8A-DECAHYDRO-6,7-DIHYDROXY-1,8A-DIMETHYL-2-PHENANTHRENONE
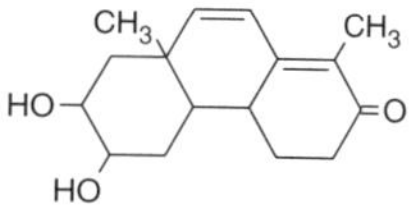
50 mg

S85,040-3
2,3'-DIHYDROXY-2',3,5,6'-TETRAMETHYLDIPHENOQUINONE, SODIUM SALT
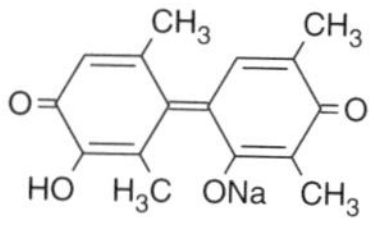
250 mg

S76,269-5
9,10-DIHYDROXY-4,6A-DIMETHYL-DODECAHYDROBENZ(A)ANTHRACEN-3 (2H)-ONE
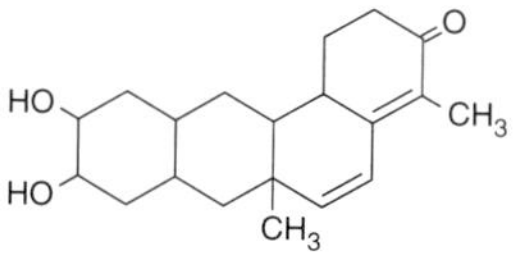
250 mg

S76,278-4
2,3,4,4A,4B,5,6,7,8,8A-DECAHYDRO-6,7-DIHYDROXY-1-METHYL-2-PHENANTHRENONE
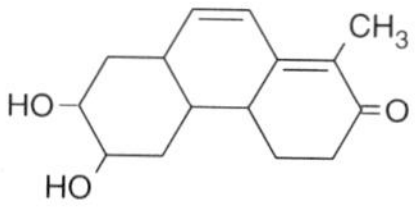
250 mg

S89,657-8
2,2'-METHYLENEBIS(5,5-DIMETHYL-3-HYDROXY-2-CYCLOHEXEN-1-ONE)
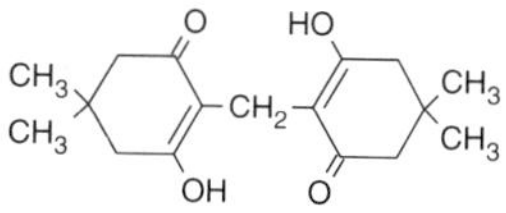
250 mg

S94,998-1
8,8-DIHYDROXY-OCTACHLORO-3A,4,7,7A-TETRAHYDRO-4,7-METHANO-1H-INDEN-1-ONE
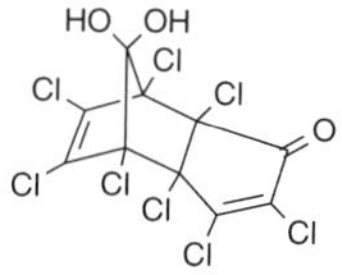
1 gm

S45,586-5
3,3-DICHLORO-2,2-DIHYDROXYCYCLOHEXANONE
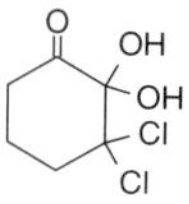
5 gm

S95,056-4
4-HYDROXY-3-METHOXY-2-METHYL-2-CYCLOPENTEN-1-ONE
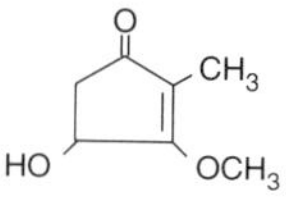
50 mg

S79,802-9
2-HYDROXY-4-METHOXYCYCLOHEXYL METHYL KETONE
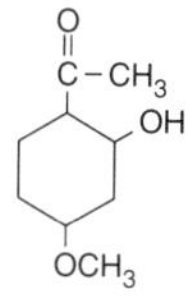
250 mg

S44,119-8
CHLORANILIC ACID, DISODIUM SALT DIHYDRATE
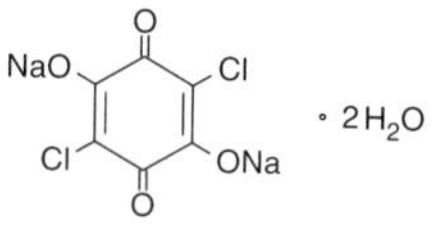
5 gm

S95,319-9
2,5-DIMETHYL-3-HYDROXY-1,4-BENZOQUINONE

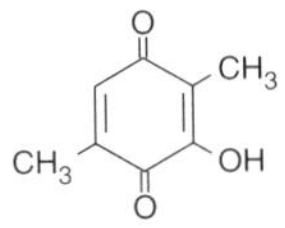

50 mg

S35,981-5
1-CYCLODECANOL-2-ONE

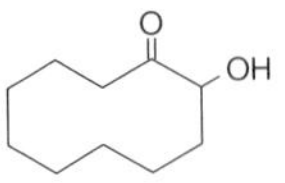

250 mg

S56,567-9
2-HYDROXYCYCLODODECANONE

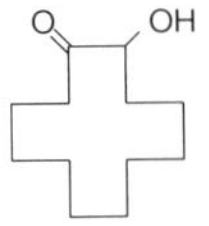

5 gm

S56,603-9
2-HYDROXYCYCLOPENTADECANONE

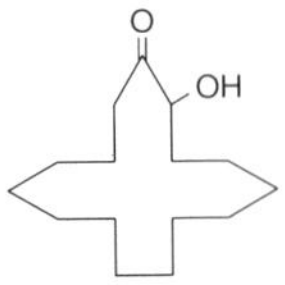

5 gm

S56,633-0
2-HYDROXYCYCLOHEXADECANONE

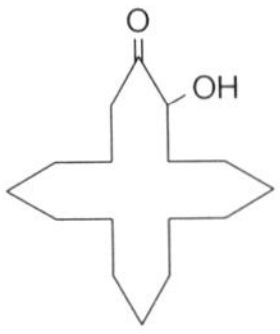

5 gm

S43,050-1
2-HYDROXY-3-METHYL-2-CYCLOPENTEN-1-ONE HYDRATE

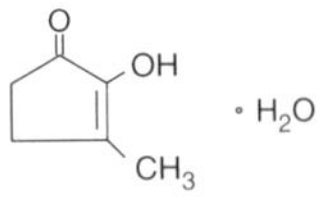

5 gm

S61,393-2
2,3,5,6-TETRAHYDROXY-1,4-BENZOQUINONE

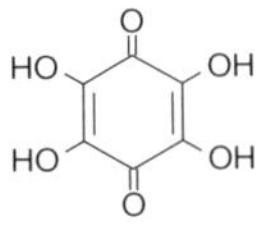

250 mg

S61,536-6
2,7-DIHYDROXY-1,2,4A,8A-TETRAHYDRO-2,4,6,8-TETRAMETHYL-1,4-ETHENONAPHTHALEN-3(4H)-ONE

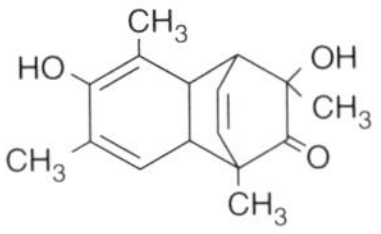

5 gm

S88,260-7
2-ALLYL-4-HYDROXY-3-METHYL-2-CYCLOPENTEN-1-ONE

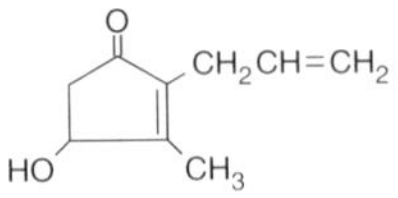

250 mg

S89,695-0
2-BROMO-5,5-DIMETHYL-3-METHOXY-2-CYCLOHEXEN-1-ONE

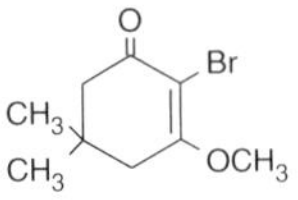

250 mg

S84,350-4
2-ALLYL-2,3-DIMETHYL-4-HYDROXY-1-CYCLOPENTANONE

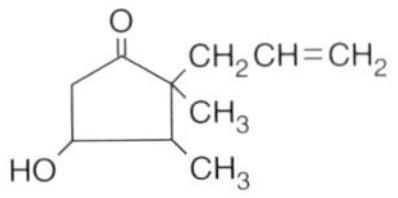

250 mg

S86,698-9
2-(2-HYDROXYETHYL)-2,6,6-TRIETHYL-1-CYCLOHEXANONE

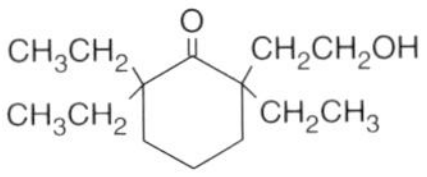

250 mg

S95,627-9
3-ACETYL-4-HYDROXY-8A-METHYL-1,5,8,8A-TETRAHYDRO-1-NAPHTHALENONE

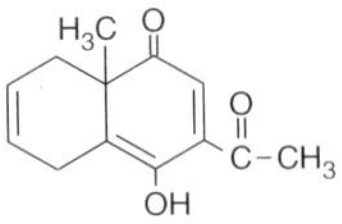

50 mg

S34,044-8
1-ACETYLCYCLOHEXANOL

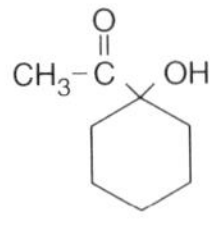

5 gm

S94,204-9
2,3,3A,4,5,7,8,9,9A,9B-DECAHYDRO-3-HYDROXY-3A-METHYL-1H-BENZ(E)-INDEN-7-ONE

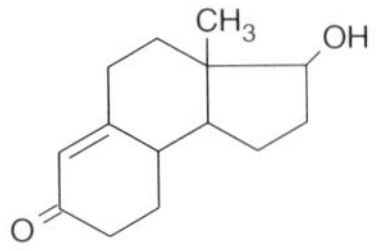

250 mg

S79,311-6
2-METHYL-3,5,5-TRIETHOXY-2-CYCLOPENTEN-1-ONE

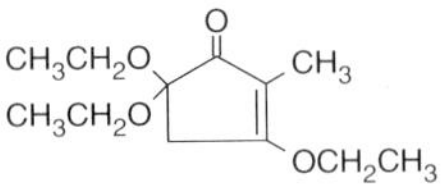

250 mg

S79,291-8
2-(3-METHYL-2-BUTENYL)-3,5,5-TRIETHOXY-2-CYCLOPENTEN-1-ONE

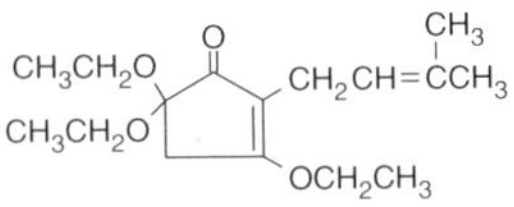

250 mg

S85,984-2
2,6-DICHLORO-3,5-DIMETHOXY-1,4-BENZOQUINONE

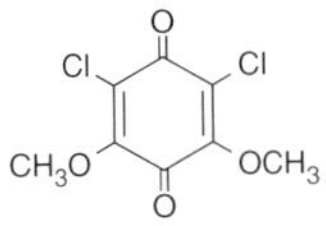

250 mg

S87,153-2
4,5-DIMETHOXY-3-CYCLOHEXENE-1,2-DIONE

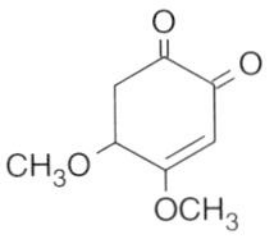

250 mg

S85,041-1
2,3-DIMETHOXY-5,6-DIMETHYL-1,4-BENZOQUINONE

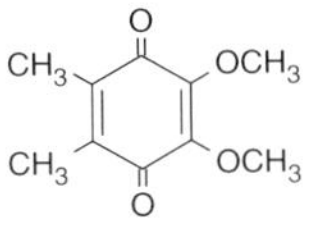

250 mg

S80,554-8
4-METHOXYCYCLOHEXYL 4-METHOXYCYCLOHEXYLMETHYL KETONE

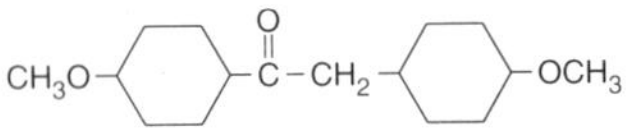

250 mg

S95,630-9
TRANS-4A,7-DIMETHYL-1,4,4A,5,8,8A-HEXAHYDRO-2-METHOXY-1,4-NAPHTHALENEDIONE

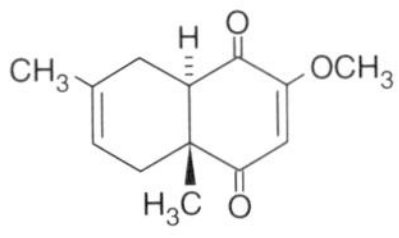

50 mg

S88,525-8
5-METHOXY-8A-METHYL-3,4,8,8A-TETRAHYDRO-1,6(2H,7H)-NAPHTHALENEDIONE

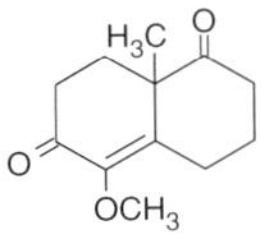

250 mg

S86,609-1
2-METHOXY-2-CYCLOHEXEN-1-ONE

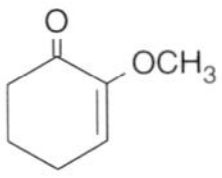

250 mg

S70,343-5
2-ETHOXY-2-CYCLOHEXEN-1-ONE

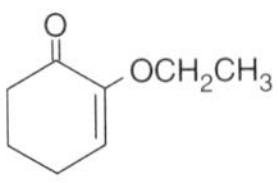

5 gm

S75,685-7
2-METHOXY-5-METHYL-1,4-BENZOQUINONE

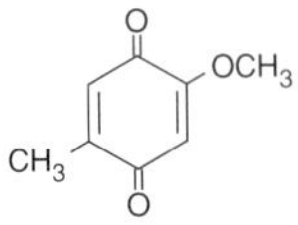

5 gm

S74,021-7
1-(1-CYCLOPENTEN-1-YLOXY)-2-PROPANONE

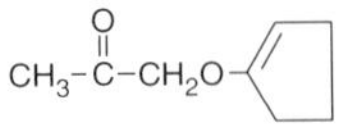

250 mg

S67,113-4
2,4-DIBUTYL-3-ETHOXY-2-CYCLOBUTEN-1-ONE

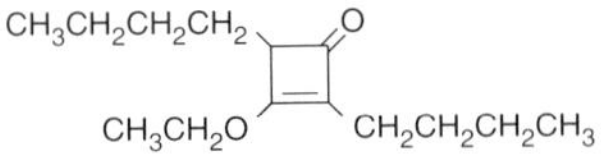

250 mg

S86,497-8
3-ISOBUTOXY-2-CYCLOHEXEN-1-ONE

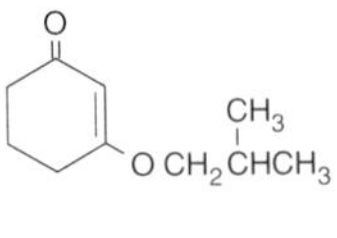

250 mg

S87,385-3
5,5-DIMETHYL-3-ETHOXY-2-CYCLOHEXEN-1-ONE

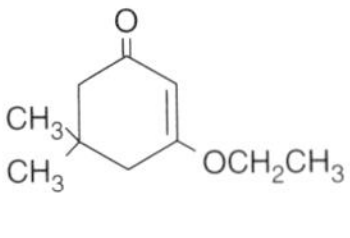

250 mg

S75,686-5
1,4,4A,5,8,8A-HEXAHYDRO-3-METHOXY-8A-METHYL-1-NAPHTHALENONE

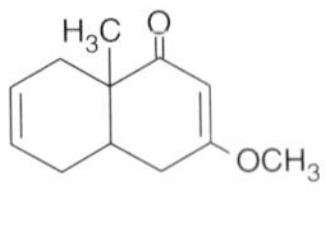

5 gm

S52,504-9
2,6-DIMETHOXY-P-BENZOQUINONE

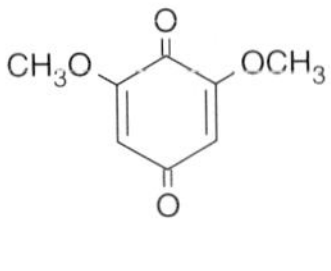

250 mg

S52,405-0
2,6-DIBROMO-3,5-DIMETHOXY-P-BENZOQUINONE

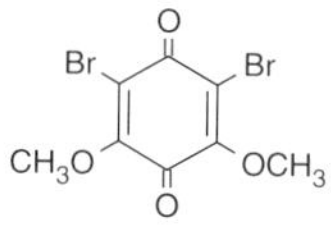

5 gm

S95,720-8
DELTA-4-2,2-PROPYLENEDITHIO-4,6A,9,9-TETRAMETHYLPERHYDRO-PHENANTHRO(2,3-D)(1,3)DIOXOL-3-ONE

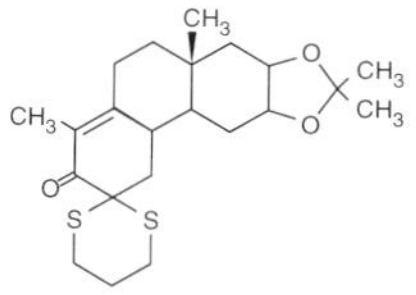

50 mg

S95,385-7
2-(ISOPROPOXYMETHYLENE)-6-METHYLCYCLOHEXANONE

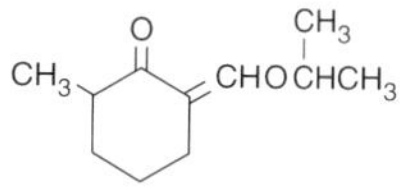

50 mg

S95,518-3
3,5,7-TRIBROMOTROPOLONE SODIUM SALT

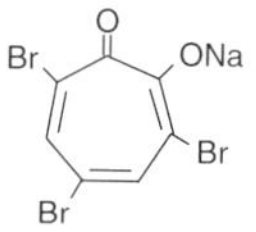

50 mg

S91,490-8
3,3-(ETHYLENEDIOXY)CAMPHOR

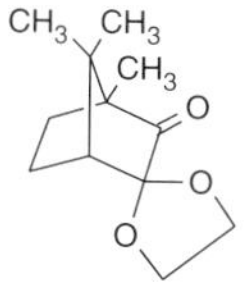

250 mg

S95,055-6
2,2-DIBROMO-4-CYCLOPENTENE-1,3-DIONE

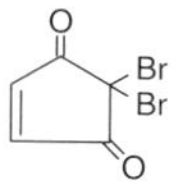

50 mg

S95,612-0
8A-METHYL-1,2,3,4,4A,5,8,8A-OCTAHYDRO-1,3-NAPHTHALENEDIONE

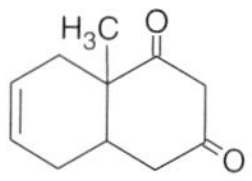

50 mg

S95,623-6
4A,8-DIMETHYL-1,2,3,4,4A,5,6,7-OCTAHYDRO-1,7-NAPHTHALENEDIONE

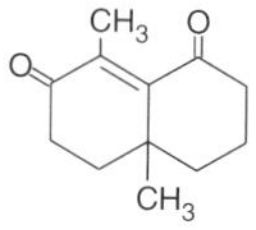

50 mg

S88,557-6
2-(BUTYLTHIOMETHYLENE)-6-METHYL-1-CYCLOHEXANONE

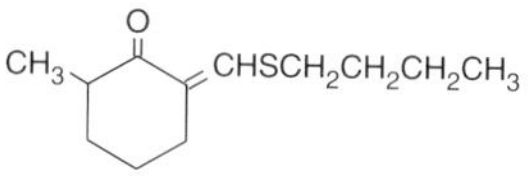

250 mg

S95,780-1
2,4,5,7-TETRACHLOROTRICYCLO-[6.2.1.0(2,7)]UNDECA-4,9-DIENE-3,6-DIONE

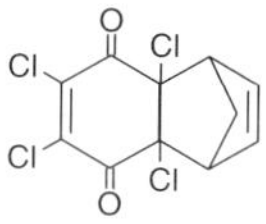

50 mg

S86,096-4
2,5-DIMETHYL-3,6-DINITRO-1,4-BENZOQUINONE

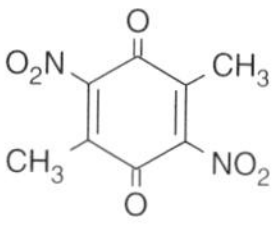

250 mg

S84,639-2
2-METHYL-5-MORPHOLINO-1,4-BENZOQUINONE

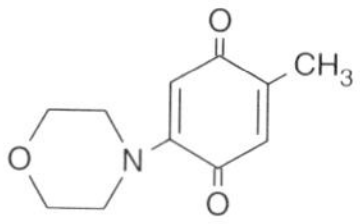

250 mg

S84,647-3
2,5-DIMORPHOLINO-1,4-BENZOQUINONE

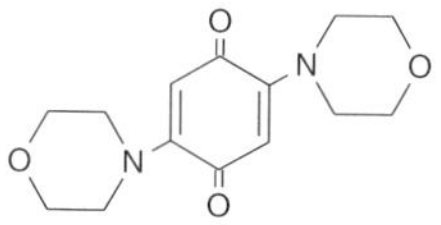

250 mg

S93,548-4
2-(DIMETHYLAMINOMETHYL)-1-CYCLOHEXANONE HYDROCHLORIDE

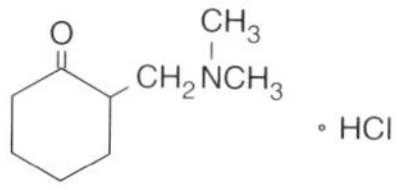

250 mg

S42,847-7
(+)-3-NITROCAMPHOR, SODIUM SALT

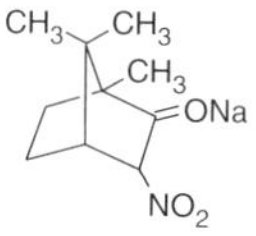

250 mg

S94,358-4
2-BROMO-1-CYCLODODECANONE

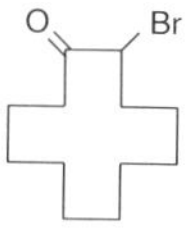

250 mg

S37,107-6
2-BROMOCYCLODODECANONE

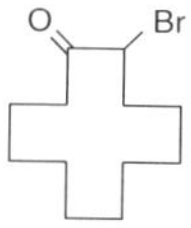

5 gm

S43,107-9
2,12-DIBROMOCYCLODODECANONE

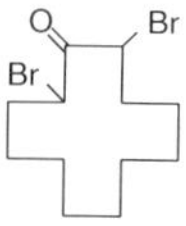

5 gm

S35,565-8
2,2,4,5-TETRACHLOROCYCLOPENTENE-1,3-DIONE

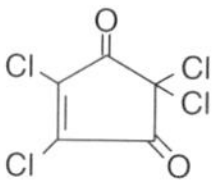

5 gm

S62,663-5
2,3,4,4,5-PENTACHLORO-2-CYCLOPENTEN-1-ONE

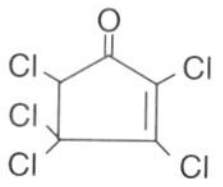

5 gm

S36,904-7
HEXACHLORO-2-CYCLOPENTEN-1-ONE

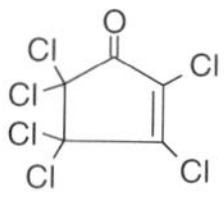

5 gm

S46,116-4
PERCHLORO-2-CYCLOBUTEN-1-ONE

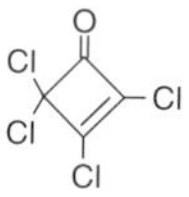

5 gm

S89,406-0
2,5-DIBROMO-2,3,4,5-TETRACHLORO-3-CYCLOPENTEN-1-ONE

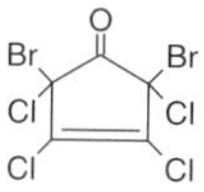

250 mg

S42,865-5
CHLORANILIC ACID, BARIUM SALT

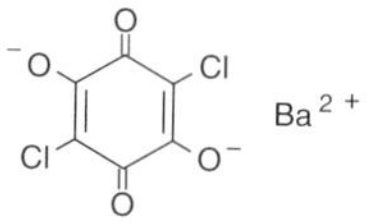

5 gm

S51,981-2
TRICHLORO-1,4-BENZOQUINONE

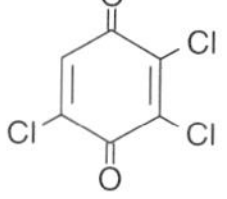

250 mg

S92,903-4
2,2-DIMETHYL-4-CYCLOPENTEN-1,3-DIONE

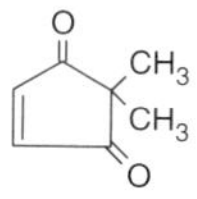

250 mg

S35,728-6
6-ACETYL-3-METHYL-2-CYCLOHEXEN-1-ONE

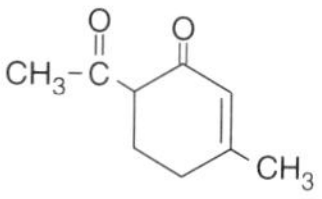

5 gm

S86,056-5
6,7-DIHYDRO-7-METHYL-1,4(5H)-INDANDIONE

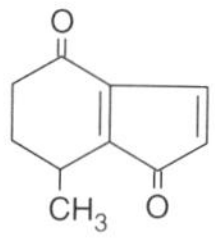

250 mg

S89,418-4
1,3,4,5,6,6,7-HEPTACHLOROBICYCLO-[3.2.0]HEPT-3-EN-2-ONE

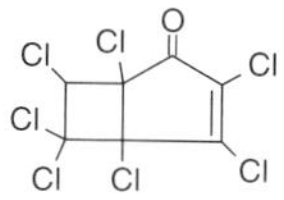

250 mg

S84,349-0
5-ALLYL-4,5-DIMETHYL-2-CYCLOPENTEN-1-ONE

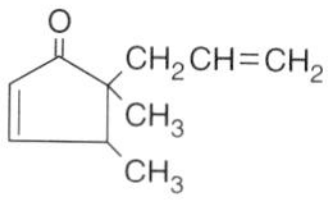

250 mg

S84,356-3
2,5-DIALLYL-4,5-DIMETHYL-2-CYCLOPENTEN-1-ONE

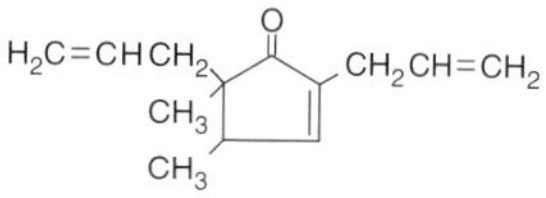

250 mg

S95,775-5
TRICYCLO[5.2.1.0(2.6)]DECA-4,8-DIEN-3-ONE

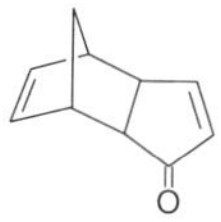

50 mg

S95,715-1
DODECAHYDRO-4,6A,9,9-TETRAMETHYLPHENANTHRO(2,3-D)(1,3)-DIOXOL-3-ONE
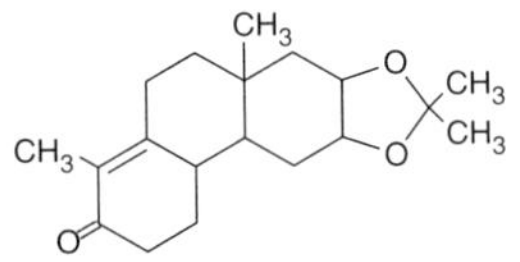
50 mg

S95,704-6
2,3,4,4A,4B,5,8,8A,9,10-DECAHYDRO-8A-METHYL-2-PHENANTHRENONE
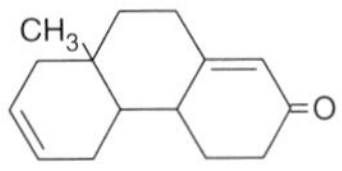
50 mg

S34,918-6
2-(2-ETHYLHEXYLIDENE)-1-CYCLOHEXANONE
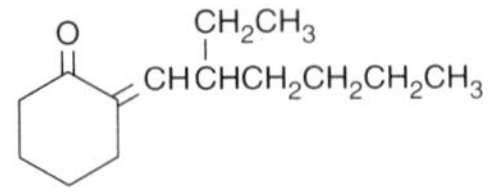
5 gm

S88,477-4
3-CHLORO-2-METHYL-2-CYCLOHEXEN-1-ONE
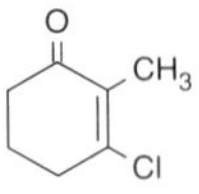
250 mg

S87,154-0
3-CHLORO-5,5-DIMETHYL-2-CYCLOHEXEN-1-ONE
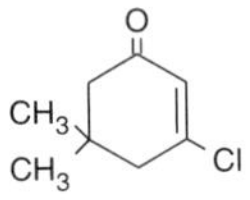
250 mg

S71,443-7
DIETHYLDIMETHYL-2-CYCLOHEXEN-1-ONE
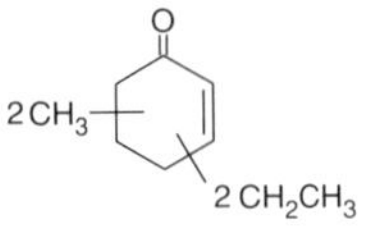
5 gm

S77,052-3
2-BUTYL-3,5,5-TRIMETHYL-2-CYCLOHEXEN-1-ONE
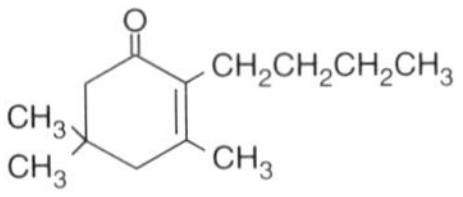
250 mg

S37,205-6
1-METHYL-BETA-IONONE
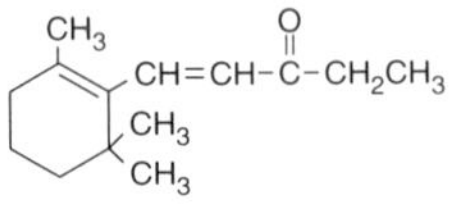
5 gm

S37,206-4
1-METHYL-ALPHA-IONONE
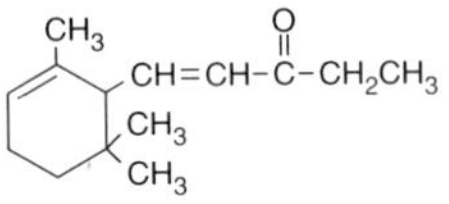
5 gm

S40,929-4
3-METHYL-ALPHA-IONONE
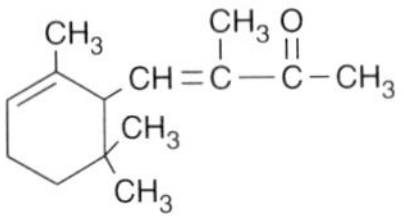
5 gm

S40,930-8
3-METHYL-BETA-IONONE
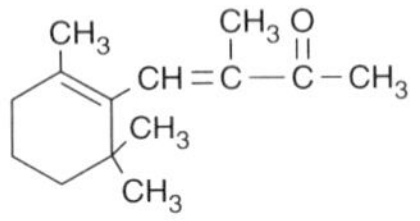
5 gm

S95,609-0
(+)-1,2,4A,5,8,8A-HEXAHYDRO-4A-METHYL-2-NAPHTHALENONE
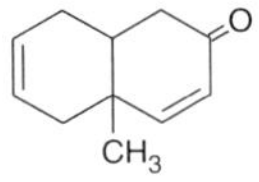
50 mg

S95,610-4
(-)-1,2,4A,5,8,8A-HEXAHYDRO-4A-METHYL-2-NAPHTHALENONE
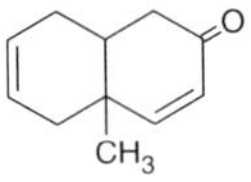
50 mg

S94,205-7
1,4A-DIMETHYL-4,4A,5,6,7,8-HEXAHYDRO-2(3H)-NAPHTHALENONE
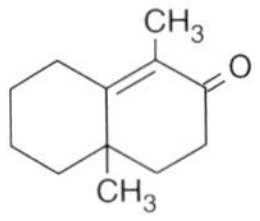
5 gm

S95,622-8
6,8A-DIMETHYL-1,2,4A,5,8,8A-HEXAHYDRO-2-NAPHTHALENONE
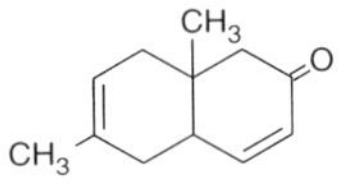
50 mg

S95,632-5
1,2,4A,5,8,8A-HEXAHYDRO-1,1,4A-TRIMETHYL-2-NAPHTHALENONE
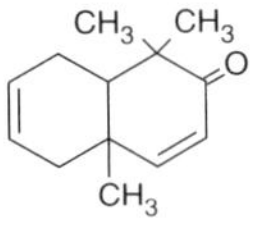
50 mg

S95,703-8
8A-METHYL-2,3,4,4A,4B,5,8,8A-OCTAHYDRO-2-PHENANTHRENONE
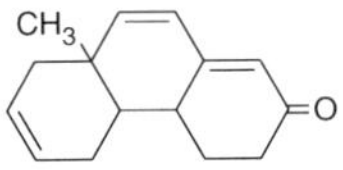
50 mg

S61,731-8
2-BROMO-6-METHYL-1,4-BENZOQUINONE
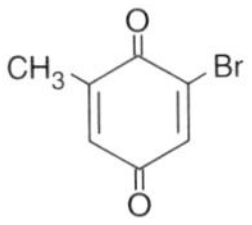
5 gm

S84,398-9
2,5-DIMETHYLBENZOQUINONE
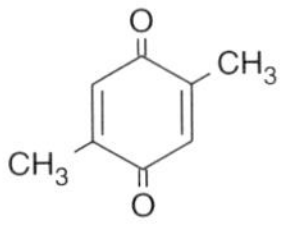
250 mg

S84,387-3
2,6-DIBROMO-3,5-DIMETHYL-1,4-BENZOQUINONE
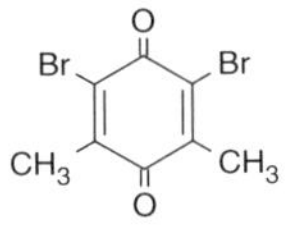
250 mg

S84,388-1
2,3-DIBROMO-5,6-DIMETHYL-1,4-BENZOQUINONE
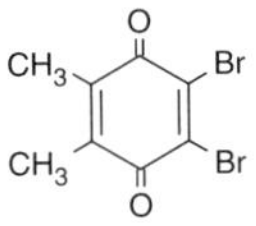
250 mg

S84,684-8
2-TERT-BUTYL-5-METHYL-1,4-BENZOQUINONE
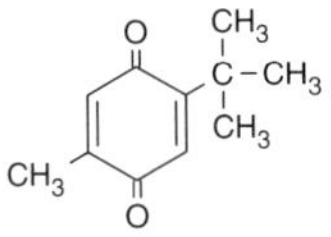
250 mg

S84,650-3
3,3'-DI-TERT-BUTYL-5,5'-DIMETHYLDIPHENOQUINONE
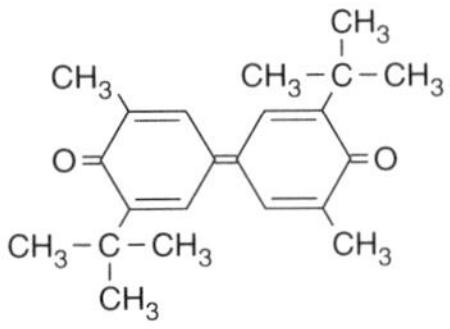
250 mg

S84,651-1
3,3',5,5'-TETRA-TERT-BUTYLDIPHENOQUINONE
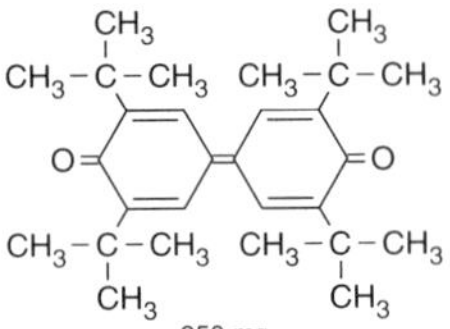
250 mg

S75,690-3
DL-1,8A-DIMETHYL-2,3,4,4A,4B,5,8,8A-OCTAHYDRO-2-PHENANTHRENONE
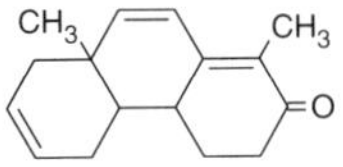
5 gm

S79,301-9
3-BUTYL-1,2,4-CYCLOPENTANETRIONE
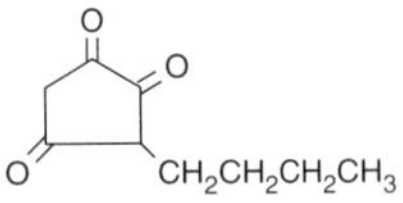
250 mg

S79,303-5
3-(3-METHYL-2-BUTEN-1-YL)-1,2,4-CYCLOPENTANETRIONE
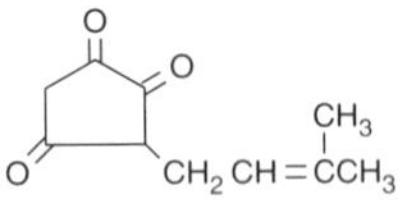
250 mg

S92,912-3
DL-TRANS-2,3-DECALINDIONE
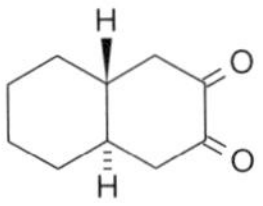
250 mg

S87,400-0
2-DECYL-1,3-CYCLOTETRADECANEDIONE
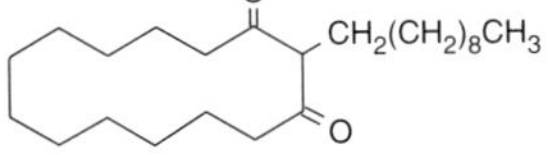
100 mg

S87,397-7
2-HEXADECYL-1,3-CYCLOTETRADECANEDIONE
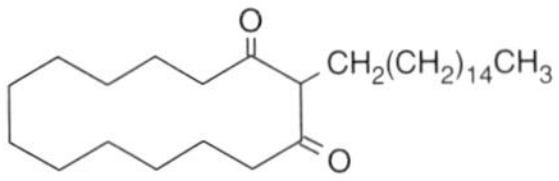
250 mg

S87,396-9
TETRAMETHYLENE-2,2'-BIS(1,3-CYCLOTETRADECANEDIONE)
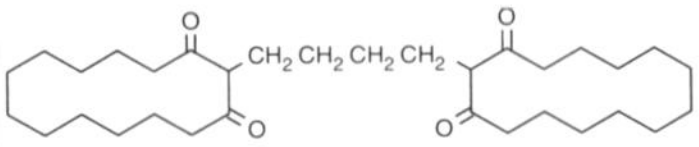
250 mg

S87,473-6
OCTADECAMETHYLENE-2,2'-BIS(1,3-CYCLOTETRADECANEDIONE)
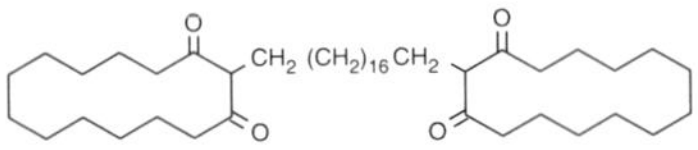
250 mg

S89,674-8
2,2-BIS(3-OXOBUTYL)-5,5-DIMETHYL-1,3-CYCLOHEXANEDIONE
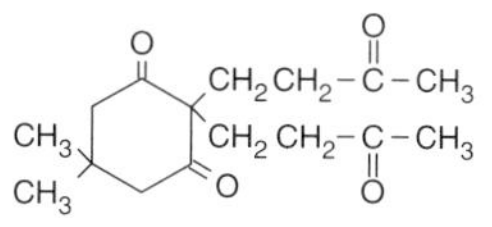
250 mg

S87,287-3
1,8-BIS(2-OXOCYCLOHEXYL)-1,8-OCTANEDIONE
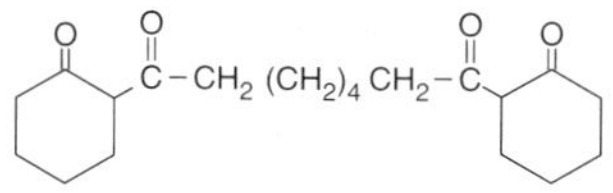
250 mg

S87,284-9
1,10-BIS(2-OXOCYCLOHEXYL)-1,10-DECANEDIONE
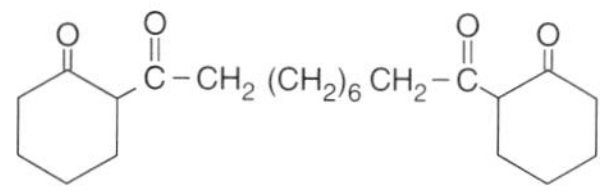
250 mg

S88,496-0
3A,4,7,7A-TETRAHYDRO-1,3-INDANDIONE
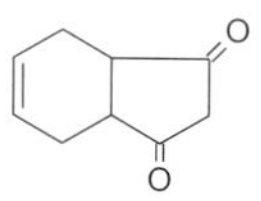
250 mg

S81,211-0
2-BROMO-5,5-DIMETHYL-1,3-CYCLOHEXANEDIONE
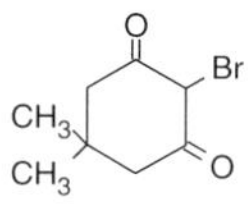
250 mg

S89,749-3
2,2-DIBROMO-5,5-DIMETHYL-1,3-CYCLOHEXANEDIONE
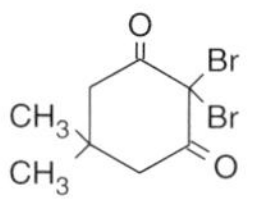
250 mg

S66,260-7
2,2,4,4,6,6-HEXAMETHYL-1,3,5-CYCLOHEXANETRIONE
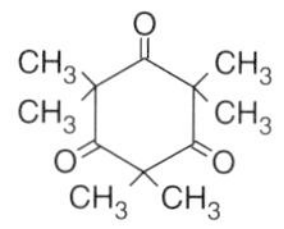
250 mg

S87,242-3
2,6-DIPROPIONYL-1-CYCLOHEXANONE
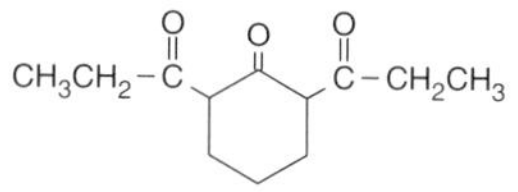
250 mg

S72,195-6
D-ALPHA,ALPHA'-BICAMPHOR
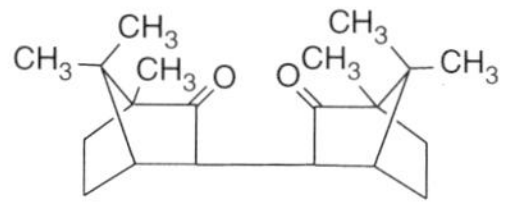
5 gm

S90,668-9
1,5-DECALINDIONE
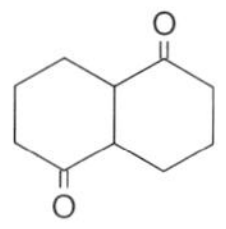
250 mg

S95,792-5
1,4,4A,4B,5,8,8A,8B,9,10-DECAHYDRO-1,4:5,8-DIMETHANOANTHRACENE-9,10-DIONE
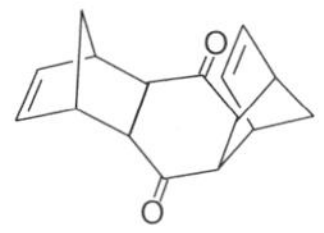
50 mg

S87,932-0
1,4-CYCLOOCTANEDIONE
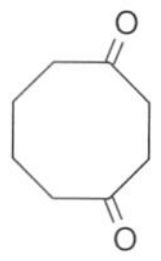
5 gm

S79,328-0
1,4,5,8-TETRAHYDRO-2,3,6,7-TETRAMETHYL-9,10(9H,10H)-ANTHRACENEDIONE
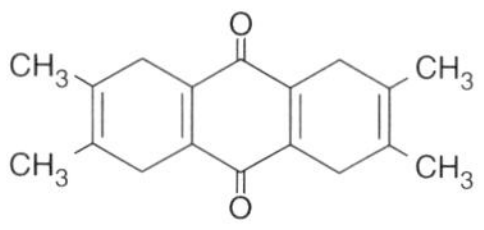
250 mg

S70,547-0
METHYL 2-METHYLCYCLOPENTYL KETONE
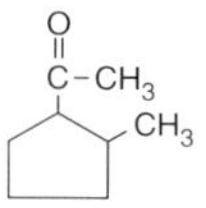
5 gm

S84,348-2
2-ALLYL-2,3-DIMETHYL-1-CYCLOPENTANONE
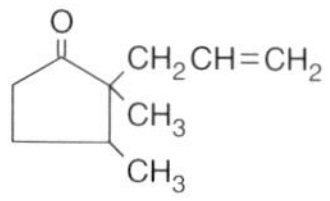
250 mg

S36,221-2
2-HEPTYLCYCLOPENTANONE
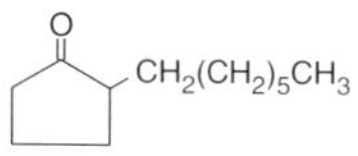
5 gm

S86,654-7
1,4,4A,7,8,8A-HEXAHYDRO-1,4-METHANONAPHTHALEN-5(6H)-ONE
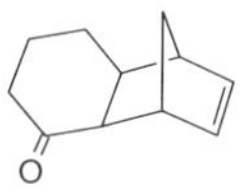
250 mg

S37,319-2
L-3-BROMOCAMPHOR
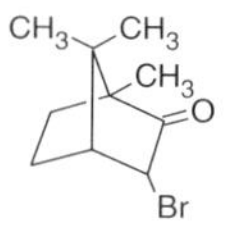
5 gm

S60,307-4
2-ACETYL-3,3-DIMETHYLBICYCLO[2.2.1]-HEPTANE
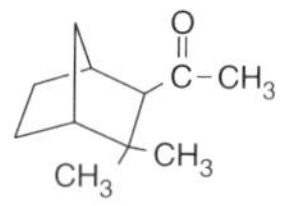
5 gm

S92,916-6
SPIRO(4.5)DECAN-1,6-DIONE
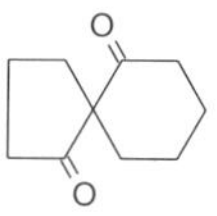
250 mg

S92,917-4
SPIRO(4.5)DECAN-1,7-DIONE
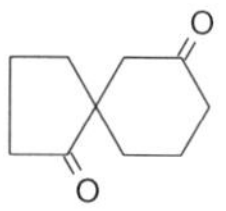
250 mg

S40,458-6
ISOACORONE
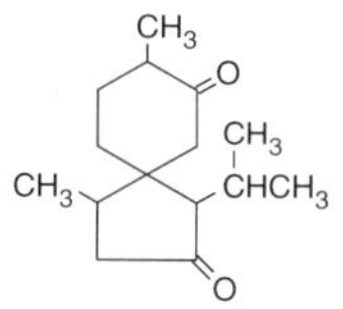
250 mg

S57,022-2
D-FENCHONE
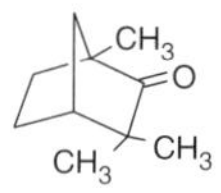
5 gm

S95,788-7
1,4,4A,4B,5,8,8A,9A-OCTAHYDRO-1,4:5,8-DIMETHANOFLUOREN-9-ONE, ISOMER 2
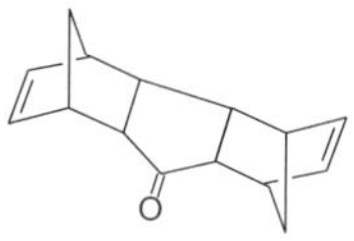
50 mg

S87,588-0
PERHYDRO-2-PENTALENONE
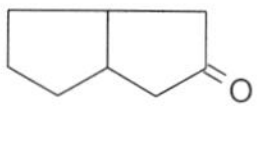
250 mg

S95,787-9
1,4,4A,4B,5,8,8A,9A-OCTAHYDRO-1,4:5,8-DIMETHANOFLUOREN-9-ONE, ISOMER I
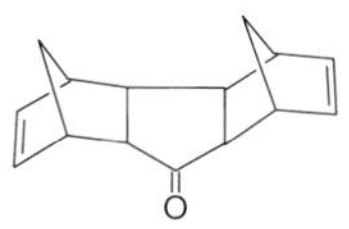
50 mg

S95,786-0
1,2,3,4,5,6,7,8,10,10,11,11-DODECA-CHLORO-1,4,4A,4B,5,8,8A,8B-OCTA-H-1,4:5,8-DIMETHANOFLUOREN-9-ONE
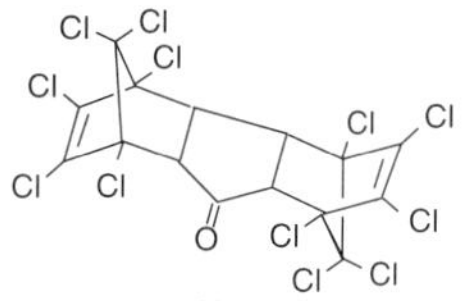
50 mg

S55,680-7
2,2,6,6-TETRACHLOROCYCLOHEXANONE
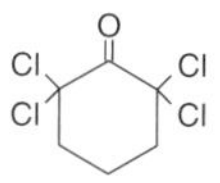
5 gm

S88,536-3
3-CHLORO-2-METHYL-1-CYCLOHEXANONE
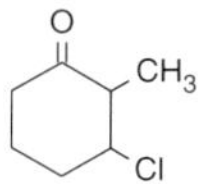
250 mg

S41,905-2
3,5-DIMETHYLCYCLOHEXANONE (MIXTURE OF ISOMERS)
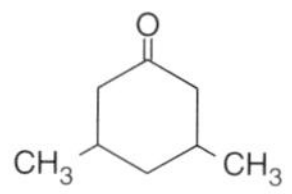
5 gm

S64,974-0
CYCLOHEXYL METHYL KETONE
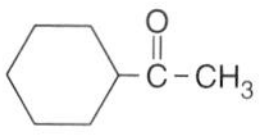
5 gm

S43,041-2
CYCLOHEXYLACETONE
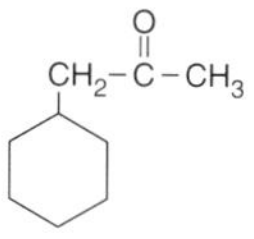
5 gm

S43,567-8
3,3,5-TRIMETHYLCYCLOHEXANONE

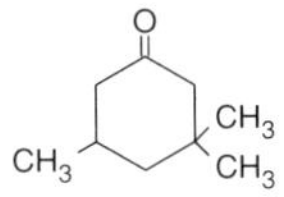

5 gm

S88,378-6
3-ISOPROPYL-1-CYCLOHEXANONE

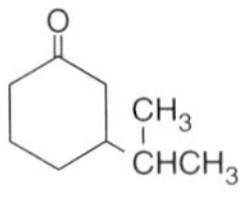

250 mg

S57,669-7
2-TERT-BUTYLCYCLOHEXANONE

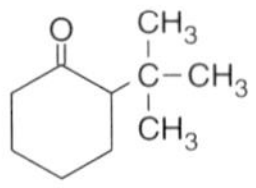

250 mg

S52,844-7
4-(TERT-PENTYL)CYCLOHEXANONE

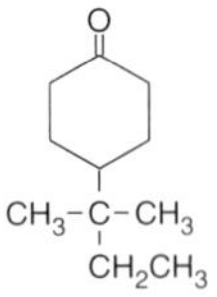

5 gm

S62,653-8
2-TERT-BUTYL-4-METHYLCYCLOHEXANONE

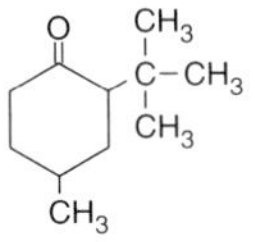

5 gm

S86,663-6
2-PENTYL-1-CYCLOHEXANONE

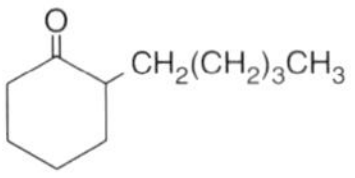

250 mg

S87,665-8
1-CYCLOHEXYL-3-METHYL-1-BUTANONE

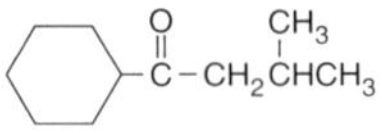

250 mg

S86,677-6
2,2,6-TRIETHYL-1-CYCLOHEXANONE

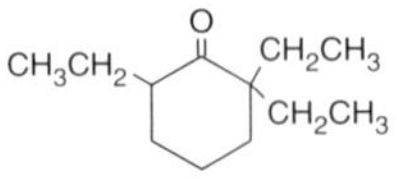

250 mg

S63,353-4
2,6-DI-SEC-BUTYLCYCLOHEXANONE

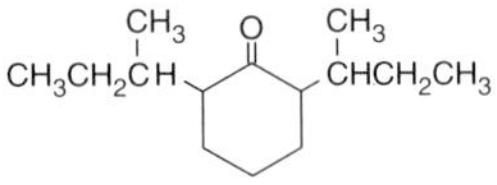

5 gm

S88,812-5
2-OCTYL-1-CYCLOHEXANONE

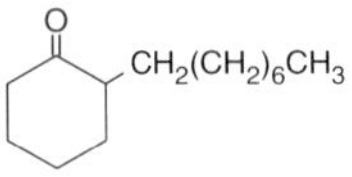

250 mg

S58,681-1
2,4-DI(TERT-PENTYL)CYCLOHEXANONE

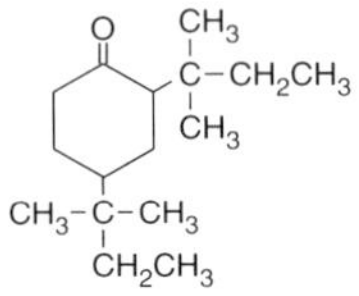

250 mg

S60,029-6
1,2,3,4A,4B,5,6,7,8,10,10A-DECAHYDROPHENANTHREN-4(3H)-ONE

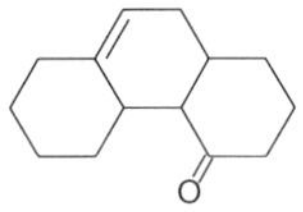

5 gm

S95,614-7
4A-METHYL-1,2,3,4,4A,5,8,8A-OCTAHYDRO-2-NAPHTHALENONE

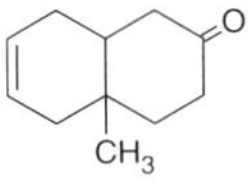

50 mg

S35,913-0
2-BUTYLCYCLOHEXANONE

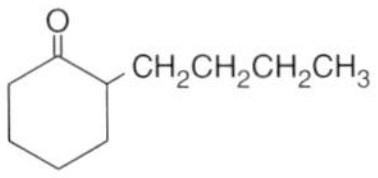

5 gm

S62,651-1
2-SEC-BUTYLCYCLOHEXANONE

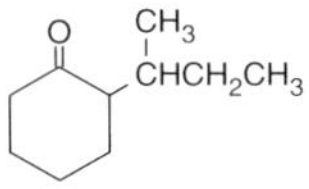

5 gm

S41,274-0
2-(2-ETHYLBUTYLIDENE)-1-CYCLOHEXANONE

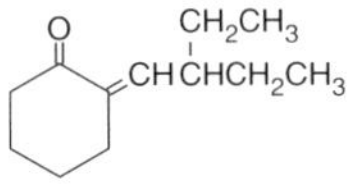

5 gm

S51,113-7
2-(2-METHYLPENTYL)CYCLOHEXANONE

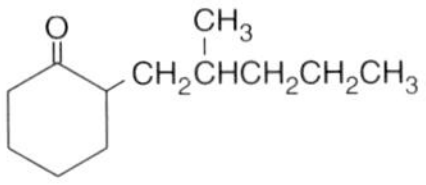

5 gm

S61,732-6
2-(2-ETHYLBUTYL)-1-CYCLOHEXANONE

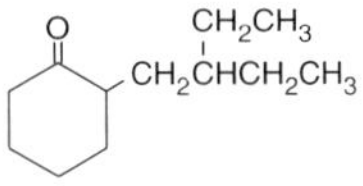

5 gm

S51,116-1
2-(2-ETHYLHEXYL)CYCLOHEXANONE

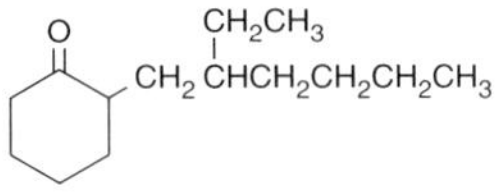

5 gm

S59,794-5
3-PENTADECYLCYCLOHEXANONE

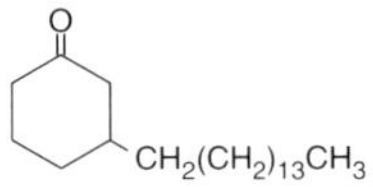

5 gm

S52,862-5
4-CYCLOHEXYLCYCLOHEXANONE

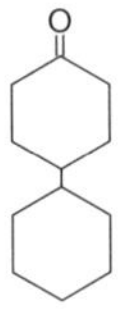

5 gm

S94,216-2
2,2,3,6-TETRAMETHYLCYCLOHEXANONE

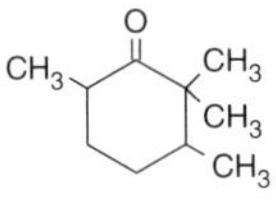

250 mg

S88,523-1
4A,8A-DIMETHYL-2-DECALONE

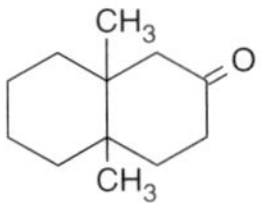

250 mg

S75,747-0
SPIRO(5.5)UNDECAN-2-ONE

5 gm

S76,463-9
SPIRO(5.5)UNDECAN-3-ONE

5 gm

S95,615-5
4A-METHYLDECAHYDRONAPHTHALEN-2-ONE

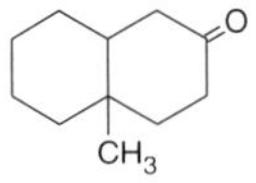

50 mg

S89,372-2
3-CYCLOHEXEN-1-ONE

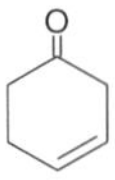

250 mg

S94,206-5
4,4A,5,6,7,8-HEXAHYDRO-4A,8,8-TRIMETHYL-2(3H)-NAPHTHALENONE

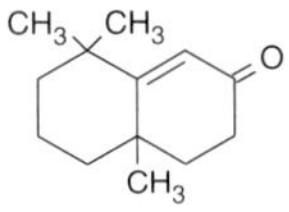

250 mg

S95,693-7
1,4,4A,4B,5,8,8A,9A-OCTAHYDRO-9(10H)-ANTHRACENONE

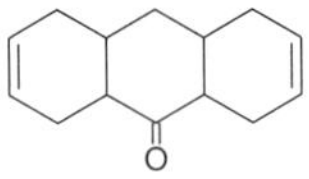

50 mg

S93,978-1
4-CYCLODECEN-1-ONE

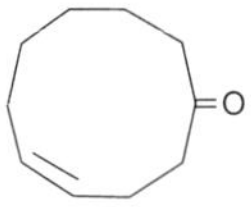

5 gm

S53,497-8
METHYLTRIBROMO-P-BENZOQUINONE

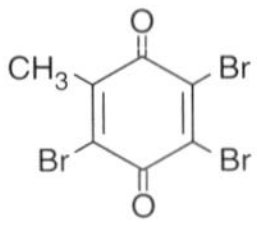

250 mg

S40,531-0
3,3',5,5'-TETRAMETHYLDIPHENOQUINONE

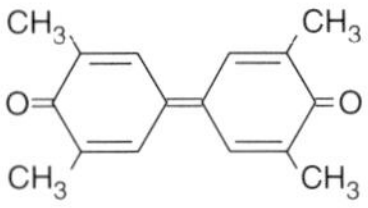

5 gm

S45,072-3
3-METHYL-2-DECALONE (MIXTURE OF ISOMERS)

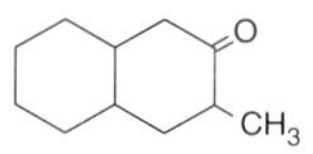

5 gm

S95,624-4
1,4A-DIMETHYLDECAHYDRONAPHTHALEN-2-ONE

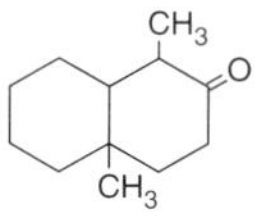

50 mg

S36,436-3
3,5,5,8,8-PENTAMETHYL-2-DECALONE,TECH.

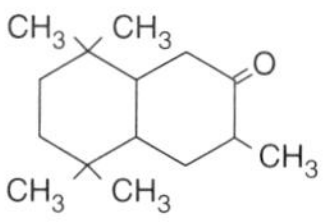

5 gm

S37,488-1
NORTRICYCLENONE-3

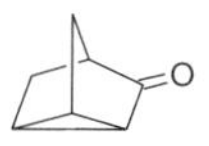

250 mg

S41,941-9
8-KETOTRICYCLO[5.2.1.0(2,6)]-4-DECENE

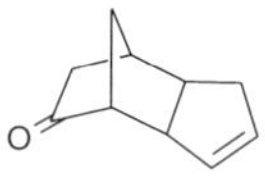

250 mg

S44,603-3
PENTACYCLO[6.2.1.O(2,7).O(4,10).0(5,9)]UNDECANE-3,6-DIONE

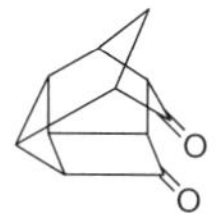

250 mg

S40,879-4
1,2,3,4,9,9-HEXACHLORO-1,4,4A,8A-TETRAHYDRO-1,4-METHANONAPHTHALENE-5,8-DIONE

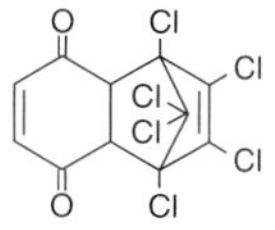

5 gm

S53,119-7
2-HEXYLIDENECYCLOPENTANONE

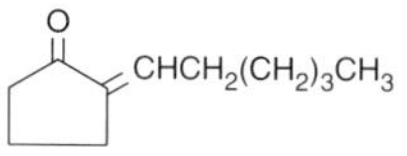

5 gm

S92,267-6
5,8A-DIMETHYL-3,4,8,8A-TETRAHYDRO-1,6(2H,7H)-NAPHTHALENEDIONE

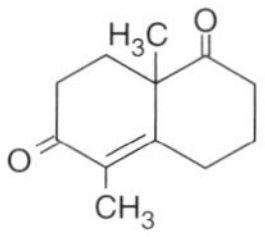

250 mg

S86,930-9
1,2-CYCLOBUTANEDIONE

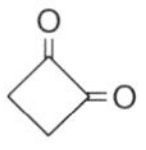

250 mg

S86,111-1
3,6-DIHYDROXY-4-(METHOXYIMINO)-2-METHYL-2,5-CYCLOHEXADIEN-1-ONE N-OXIDE

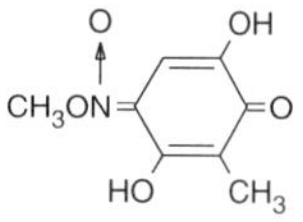

250 mg

S87,429-9
3-HYDROXY-2-(METHYLAMINO)-4-(METHYLIMINO)-2-CYCLOBUTEN-1-ONE

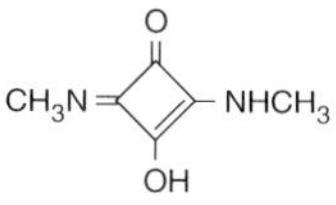

250 mg

S89,700-0
5,5-DIMETHYL-3-(2-HYDROXYETHYLAMINO)-2-CYCLOHEXEN-1-ONE HYDROCHLORIDE

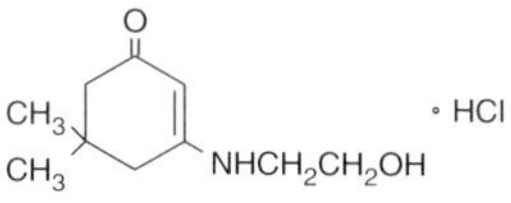

250 mg

S89,666-7
5,5-DIMETHYL-3-[(2-HYDROXYETHYL)-METHYLAMINO]-2-CYCLOHEXEN-1-ONE HYDROCHLORIDE

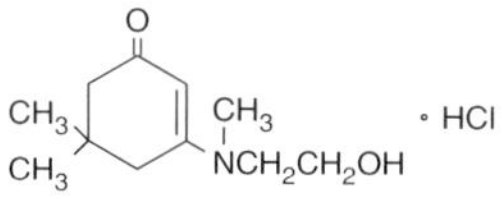

250 mg

S89,541-5
4-AMINO-3-METHYL-3-PENTEN-2-ONE

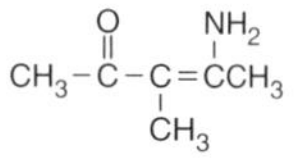

250 mg

S89,667-5
2-(1-AMINOETHYLIDENE)-5,5-DIMETHYL-1,3-CYCLOHEXANEDIONE

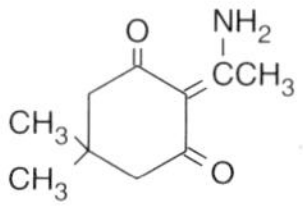

250 mg

S74,269-4
3-AMINOCAMPHOR HYDROCHLORIDE

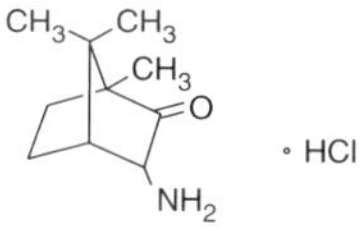

5 gm

S89,712-4
2,2'-METHYLENEBIS(3-AMINO-5,5-DIMETHYL-2-CYCLOHEXEN-1-ONE)

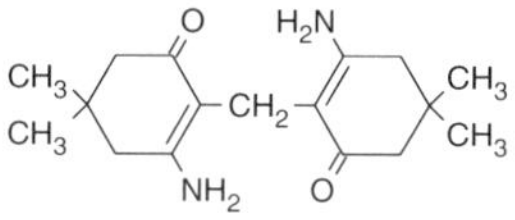

250 mg

S89,763-9
3-AMINO-2-CYCLOHEXEN-1-ONE HYDROBROMIDE

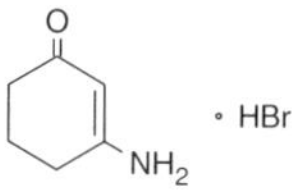

250 mg

S89,735-3
3-AMINO-2,5,5-TRIMETHYL-2-CYCLOHEXEN-1-ONE

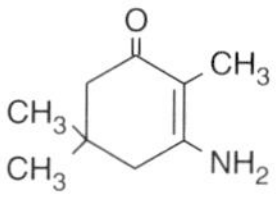

250 mg

S89,745-0
3-AMINO-5,5-DIMETHYL-2-CYCLOHEXENE-1-THIONE

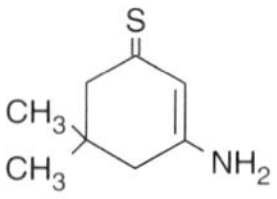

250 mg

S87,430-2
3,4-BIS(METHYLAMINO)-3-CYCLOBUTENE-1,2-DIONE

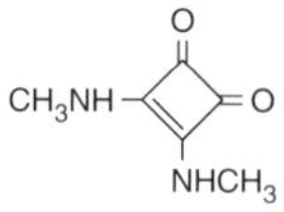

250 mg

S87,438-8
3,4-BIS(BUTYLAMINO)-3-CYCLOBUTENE-1,2-DIONE

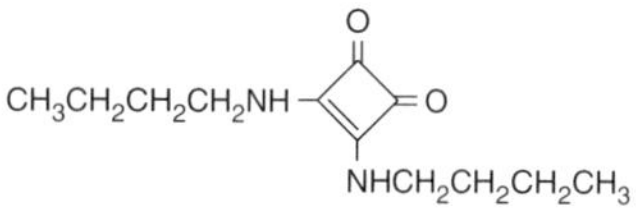

250 mg

S89,776-0
2,2'-METHYLENEBIS[5,5-DIMETHYL-3-(METHYLAMINO)]-2-CYCLOHEXEN-1-ONE

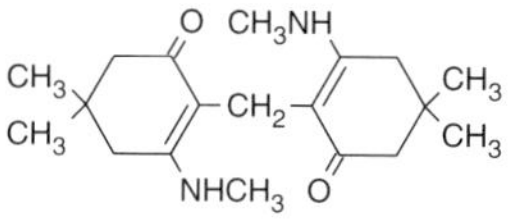

250 mg

S89,665-9
3-(CYCLOHEXYLAMINO)-2-CYCLOHEXEN-1-ONE

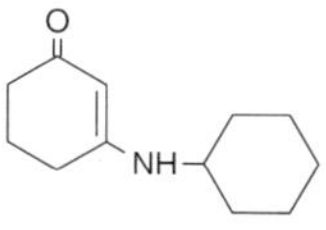

250 mg

S89,738-8
5,5-DIMETHYL-3-(METHYLAMINO)-2-CYCLOHEXEN-1-ONE HYDROCHLORIDE

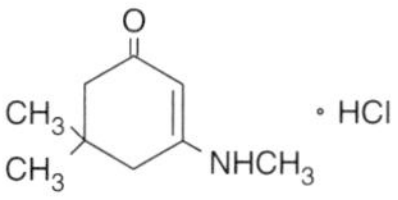

250 mg

S44,813-3
1-DIPROPYLAMINO-2-PROPANONE

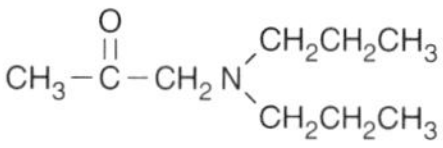

5 gm

S57,378-7
4-(DIETHYLAMINO)-2-BUTANONE HYDROCHLORIDE

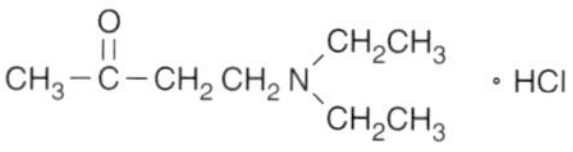

250 mg

S68,100-8
2,2-DIMETHYL-5-DIMETHYLAMINO-3-PENTANONE HYDROCHLORIDE

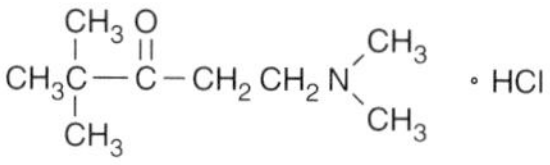

5 gm

S70,345-1
2-(DIMETHYLAMINOMETHYL)-1-CYCLOHEXANONE

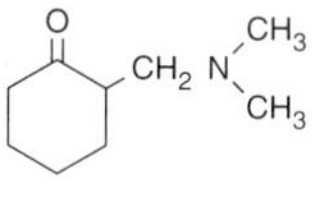

5 gm

S34,223-8
2,6-BIS(DIMETHYLAMINOMETHYL)-CYCLOHEXANONE DIHYDROCHLORIDE

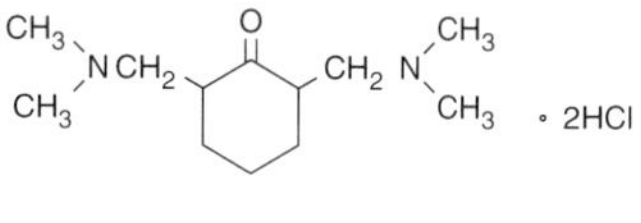

5 gm

S91,501-7
(3-OXOBUTYL)-TRIMETHYLAMMONIUM BROMIDE

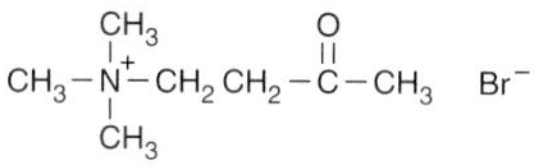

250 mg

S89,699-3
5,5-DIMETHYL-3-(2,2-DIMETHYLHYDRAZINO)-2-CYCLOHEXEN-1-ONE

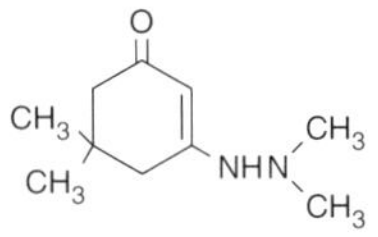

250 mg

S36,323-4
ACETYLACETONE IMIDE

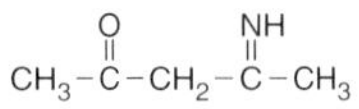

5 gm

S94,299-5
5,5-DIMETHYL-2-(DIMETHYLAMINOMETHYLENE)-1,3-CYCLOHEXANEDIONE

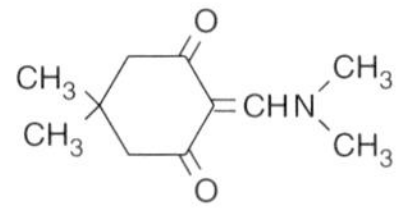

5 gm

S92,996-4
2-AZACYCLODECANONE

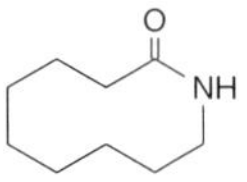

250 mg

S45,495-8
6-(DIETHYLAMINOMETHYL)KOJIC ACID HYDROCHLORIDE

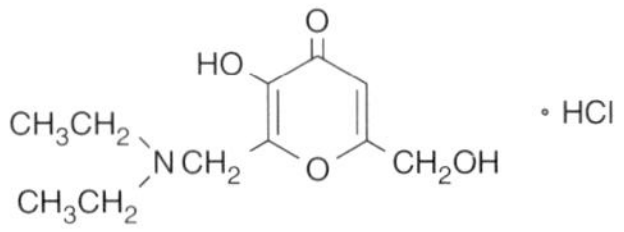

5 gm

S95,504-3
5,6-DIHYDRO-4-HYDROXY-2H-THIOPYRAN-3-YL 2-ETHOXYETHYL KETONE

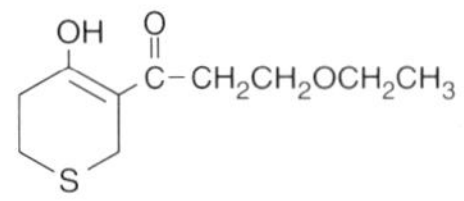

50 mg

S65,769-7
2,4-O-ETHYLIDENE-D-ERYTHROSE

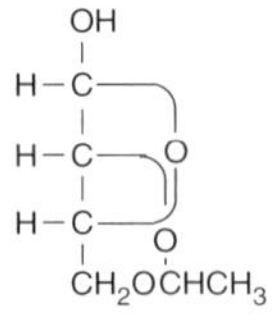

250 mg

S42,571-0
6-(4-METHYLPIPERAZINOMETHYL)-KOJIC ACID

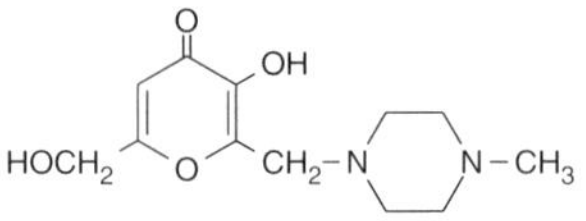

5 gm

S42,583-4
6-(MORPHOLINOMETHYL)KOJIC ACID

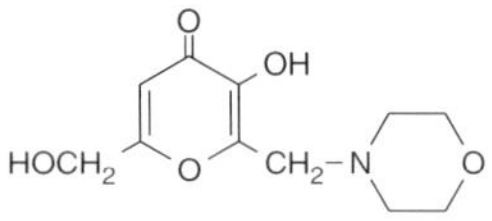

5 gm

S42,637-7
6-(PYRROLIDINOMETHYL)KOJIC ACID HYDROCHLORIDE

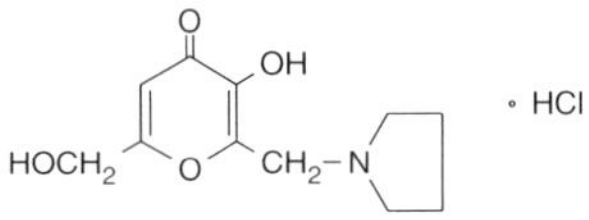

5 gm

S42,623-7
6-(PIPERIDINOMETHYL)KOJIC ACID

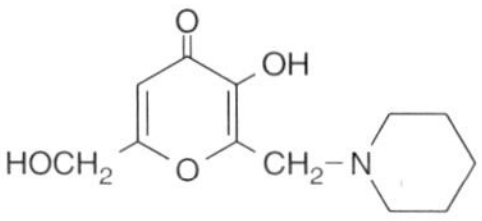

5 gm

S95,507-8
4'A-HYDROXYOCTAHYDROSPIRO[1,3-DIOXANE-2,3'-THIOPYRANO(2,3-B)-THIOPYRAN]-6'-ONE

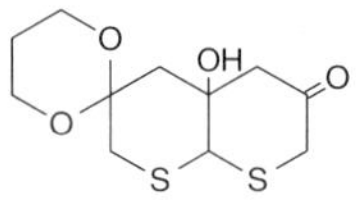

50 mg

S95,501-9
2,3,4,6,7,8A-HEXAHYDRO-5-HYDROXYTHIOPYRANO(2,3-B)-THIOPYRAN-4-ONE

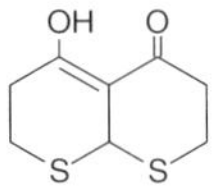

50 mg

S40,444-6
8,10-DIETHYLLOBELIONOL HYDROCHLORIDE

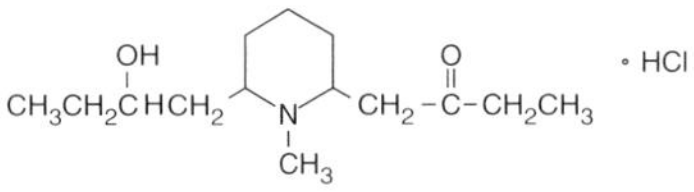

100 mg

S51,508-6
2,5-DIMETHYL-2-HYDROXYMETHYL-3(2H)-FURANONE

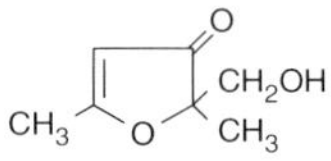

250 mg

S94,166-2
5-(1-HYDROXY-1-METHYLETHYL)-2-METHYL-2-VINYLTETRAHYDROFURAN

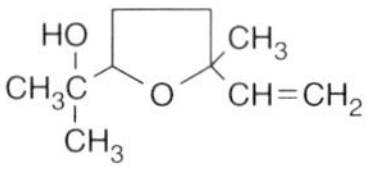

5 gm

S74,468-9
1,4-DIHYDRO-6,7-DIMETHOXY-1-METHYL-4-CINNOLINONE PICRATE

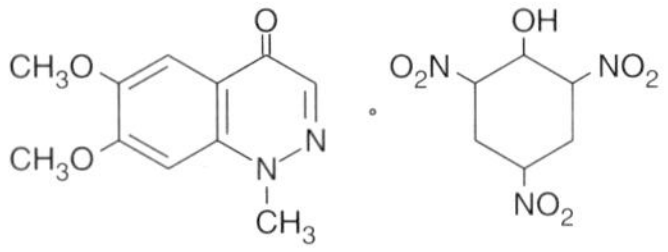

250 mg

S56,108-8
1-(1,3-DIOXOLAN-2-YL)-2-PROPANONE

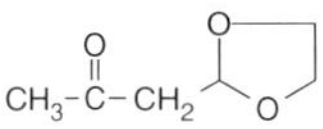

5 gm

S67,245-9
1,4-CYCLOHEXANEDIONE BIS(1,3-PROPYLENE KETAL)

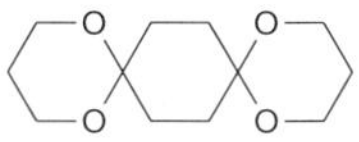

5 gm

S95,043-2
2,3-EPOXY-3A-METHYL-3A,4,7,7A-TETRAHYDROINDAN-1-ONE

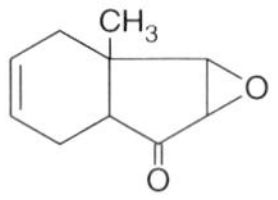

50 mg

S51,507-8
2-CYCLOHEXYL-5-METHYL-3(2H)-FURANONE

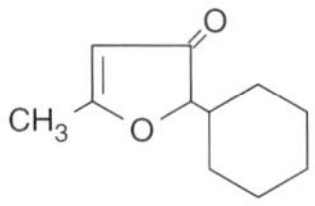

250 mg

S95,505-1
9-(ACETYLMETHYLTHIO)-1,5-DIOXA-8-THIASPIRO(5.5)UNDECAN-10-ONE

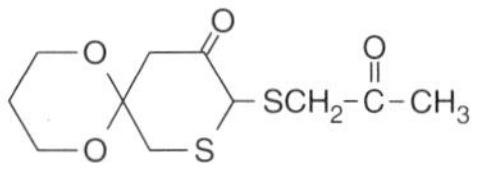

50 mg

S86,946-5
6,7-DIHYDRO-2-METHYLCYCLOPENTA(B)-PYRAN-4(5H)-ONE

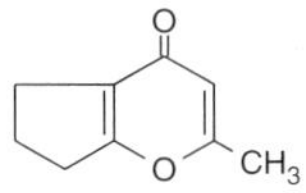

250 mg

S34,539-3
3,5-DIACETYL-2,4,6-TRIMETHYL-1,4-DIHYDROPYRIDINE

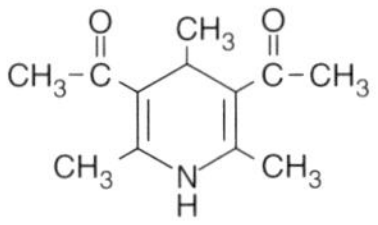

5 gm

S94,952-3
3-ETHYL-6,7-DIHYDRO-2-METHYL-5-METHYLENE-4(5H)INDOLONE

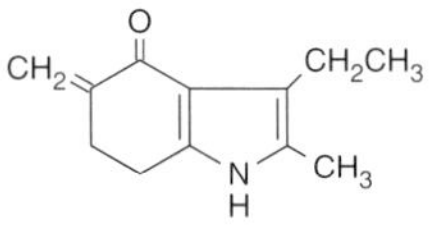

100 mg

S94,966-3
3,5-DIBROMO-2,2,6,6-TETRAMETHYL-4-PIPERIDONE

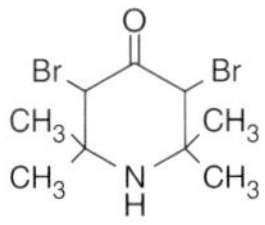

250 mg

S42,468-4
8,10-DIETHYLNORLOBELIDIONE HYDROCHLORIDE

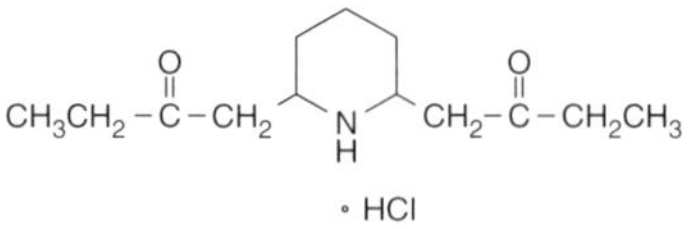

250 mg

S89,694-2
5,5-DIMETHYL-3-PYRROLIDINO-2-CYCLOHEXEN-1-ONE

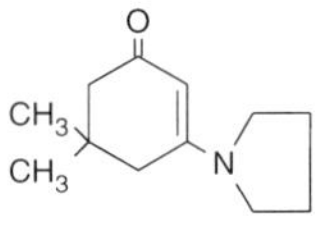

250 mg

S89,709-4
HEXAHYDRO-1',4,4,7',7'-PENTAMETHYLSPIRO[CYCLOHEXANE-1,3'-(2'H)QUINOLINE]-2,5',6-TRIONE

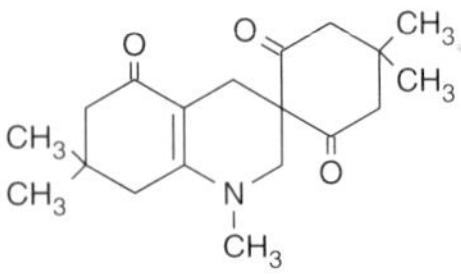

250 mg

S36,459-2
3-(PIPERIDINOMETHYL)-2-NORBORNANONE

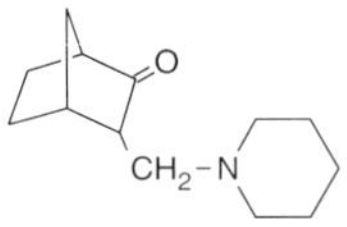

5 gm

S45,075-8
3-(4-METHYLPIPERAZINOMETHYL)-2-NORBORNANONE

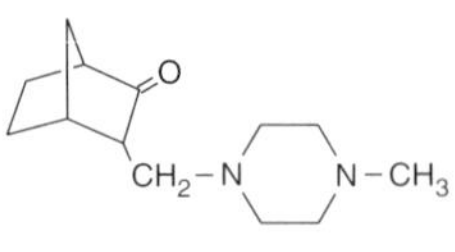

5 gm

S39,590-0
2-(MORPHOLINOMETHYL)-CYCLOPENTANONE HYDROCHLORIDE

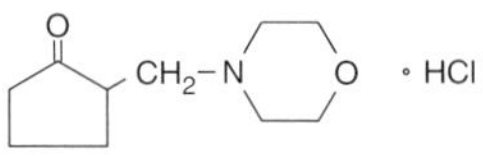

5 gm

S67,730-2
2-PIPERIDINOCYCLOHEXANONE
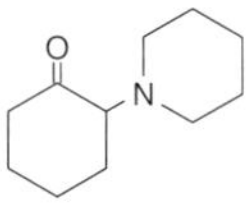
250 mg

S56,361-7
1-ISOPROPYL-4-PIPERIDONE
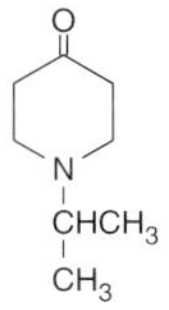
5 gm

S42,996-1
4-PIPERIDINO-2-BUTANONE HYDROCHLORIDE
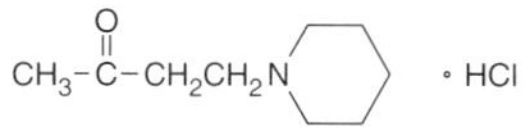
5 gm

S43,473-6
1-BUTYL-4-PIPERIDONE
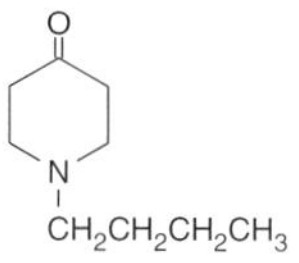
5 gm

S88,437-5
4-METHYL-3-(PIPERIDINOMETHYL)-2-PENTANONE
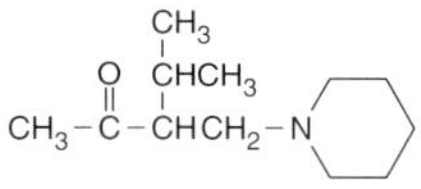
250 mg

S42,464-1
8,10-DIETHYLLOBELIDIONE
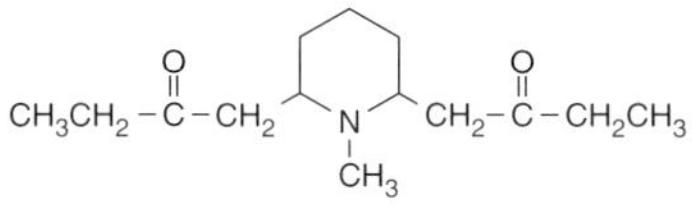
250 mg

S43,943-6
TROPINONE HYDROBROMIDE
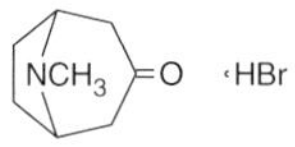
250 mg

S40,949-9
3-(MORPHOLINOMETHYL)-2-NORBORNANONE
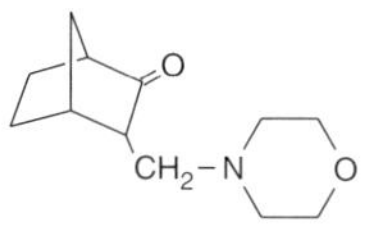
5 gm

S43,431-0
1-METHYL-1-(8-OXO-2-BICYCLO[3.2.1]-OCTYL)PYRROLIDINIUM IODIDE
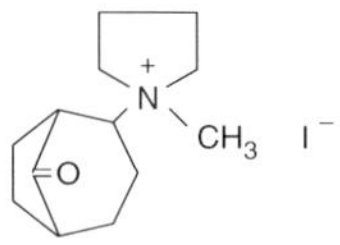
5 gm

S89,711-6
HEXAHYDRO-4,4,7',7'-TETRAMETHYLSPIRO[CYCLOHEXANE-1,3'-(2'H)-QUINOLINE]-2,5'6-TRIONE
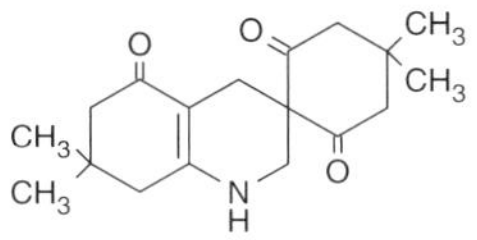
250 mg

S95,714-3
DECAHYDRO-4,6A,9,9-TETRAMETHYLPHENANTHRO(2,3-D)(1,3)-DIOXOL-3-ONE
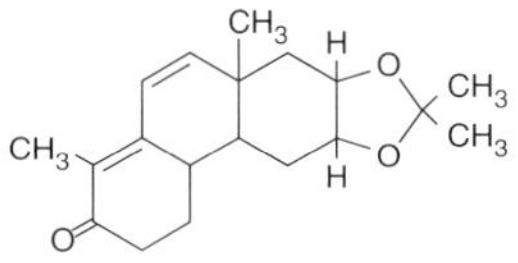
50 mg

S88,522-3
8A'-METHYL-3',4',8',8A'-TETRAHYDROSPIRO[1,3-DIOXOLANE-2,1'-(2'H)NAPHTHALENE]6'(7'H)ONE
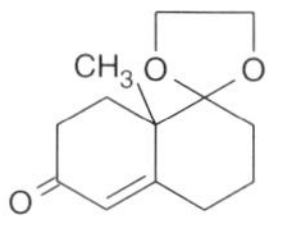
250 mg

S86,961-9
2-METHYL-5,6,7,8-TETRAHYDROCHROMONE
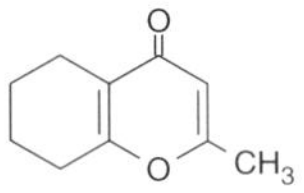
250 mg

S87,160-5
5,5-DIMETHYL-3-MORPHOLINO-2-CYCLOHEXEN-1-ONE
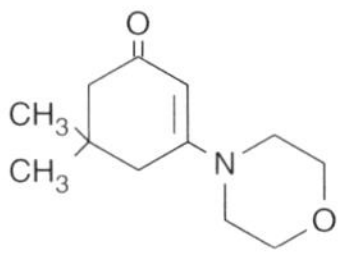
250 mg

S87,432-9
4H-THIOPYRAN-4-ONE
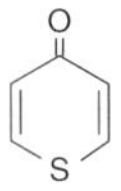
250 mg

S55,325-5
3-ACETYL-2,2-DIMETHYLOXIRANE
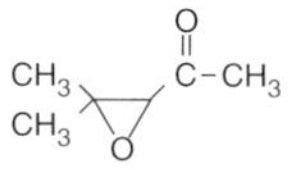
5 gm

S86,043-3
SPIRO(BICYCLO[3.2.0]HEPT-3-ENE-6,2'-(1,3)DITHIOLAN)-7-ONE
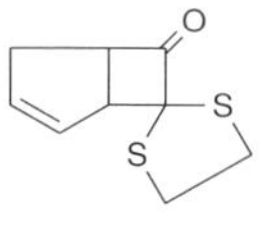
250 mg

S79,309-4
2,4-BIS(1,3-DITHIOLAN-2-YL)-1-CYCLOPENTANONE
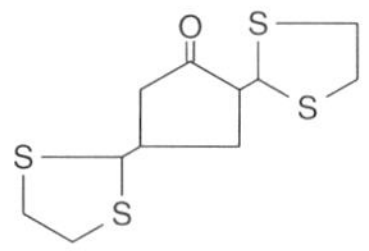
250 mg

S87,243-1
2-(4,5,6,7-TETRAHYDRO-3H-AZEPIN-2-YL)-1-CYCLOHEXANONE
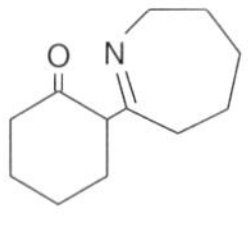
250 mg

S87,183-4
2-(AZACYCLONON-1-EN-2-YL)-1-CYCLOHEXANONE
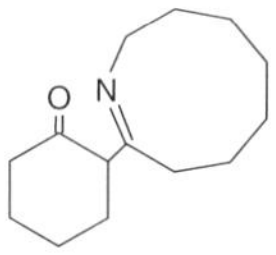
250 mg

S87,206-7
2-(AZACYCLODODEC-1-EN-2-YL)-1-CYCLOHEXANONE
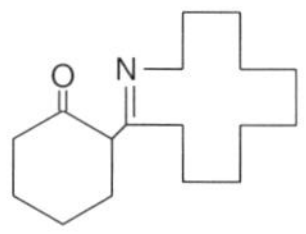
250 mg

S87,200-8
2-(AZACYCLOTRIDEC-1-EN-2-YL)-1-CYCLOHEXANONE
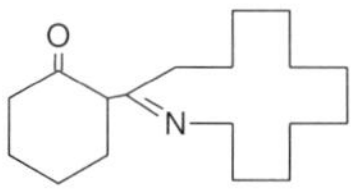
250 mg

S95,824-7
2-MORPHOLINOBICYCLO[3.3.1]-NONAN-9-ONE
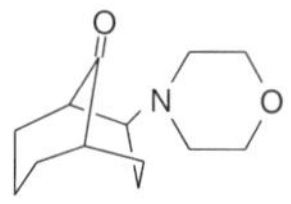
50 mg

S93,700-2
3,3A,4,6,7,8,8A,8B-OCTAHYDRO-1(2H)-AS-INDACENONE
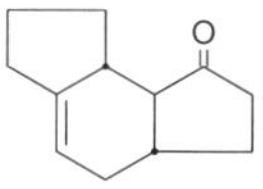
250 mg

S87,433-7
4H-PYRAN-4-THIONE
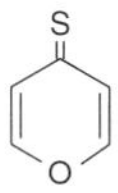
250 mg

S87,427-2
2,6-DIMETHYL-4H-PYRAN-4-THIONE
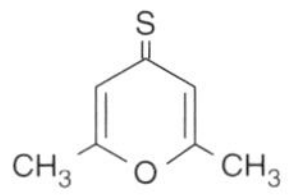
250 mg

S87,431-0
4H-THIOPYRAN-4-THIONE
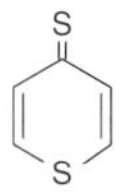
250 mg

S46,013-3
L-(-)-GLYCERALDEHYDE

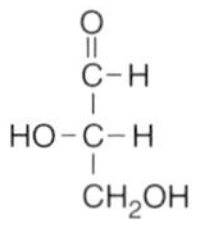

250 mg

S64,119-7
D-GLUCO-L-TALOOCTOSE

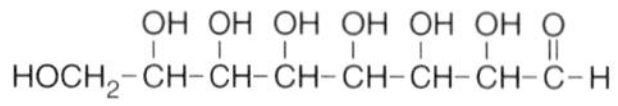

5 gm

S61,262-6
2-THIOGLYCERALDEHYDE

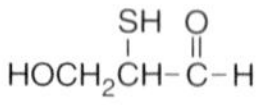

5 gm

S42,682-2
ALDOL

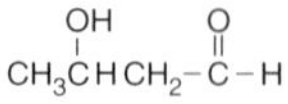

5 gm

S66,132-5
2-HYDROXYMETHYL-2-METHYLBUTYRALDEHYDE

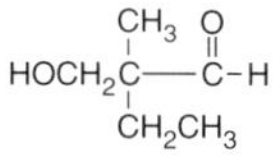

250 mg

S35,069-9
HYDROXYCITRONELLAL

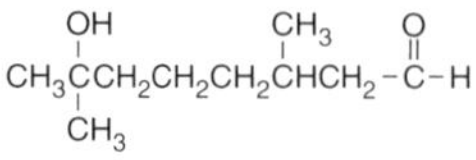

5 gm

S52,810-2
4-(4-HYDROXY-4-METHYLPENTYL)-3-CYCLOHEXENE-1-CARBOXALDEHYDE

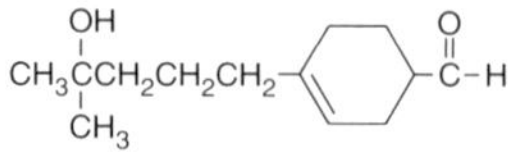

5 gm

S89,748-5
3,3'-IMINOBIS(METHACROLEIN)

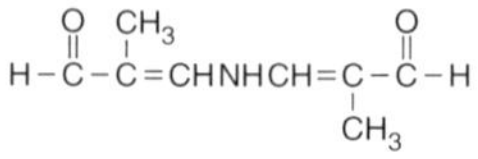

250 mg

S90,647-6
1-FORMYL-6-(DIMETHYLAMINO)FULVENE

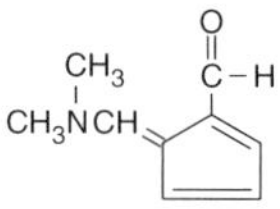

250 mg

S88,126-0
4-PENTENAL

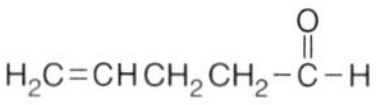

250 mg

S45,515-6
3-METHYLVALERALDEHYDE

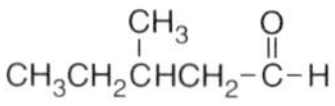

250 mg

S41,627-4
2-ETHYLISOVALERALDEHYDE

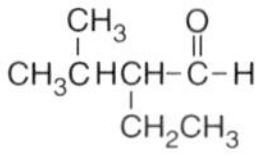

5 gm

S57,621-2
2,2-DIMETHYL-3,4-OCTADIENAL

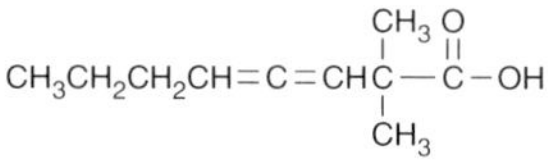

5 gm

S45,792-2
ALPHA-CITRONELLAL

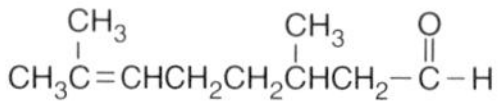

250 mg

S61,993-0
2-ETHYLCAPRYLALDEHYDE

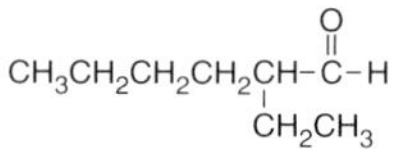

5 gm

S34,301-3
2-BUTYL-2-ETHYL-5-METHYL-3,4-HEXADIENAL

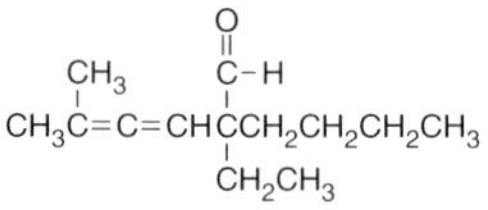

5 gm

S43,190-7
TETRADECYL ALDEHYDE (50% SOLN IN DIETHYL PHTHALATE)

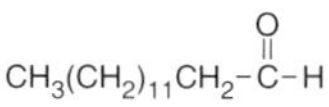

5 gm

S89,398-6
TRICHLOROACROLEIN

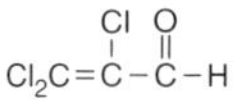

250 mg

S88,095-7
4-METHYL-1,3-PENTADIENE-1-CARBOXALDEHYDE

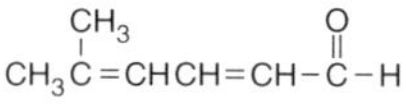

250 mg

S89,419-2
1,2,3,4,5,6,6-HEPTACHLORO-1,3,5-HEXATRIENE-1-CARBOXALDEHYDE

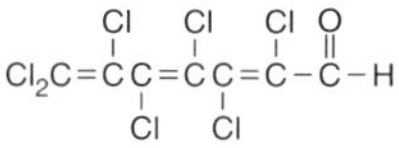

250 mg

S51,855-7
2-ETHYL-2-HEXENAL

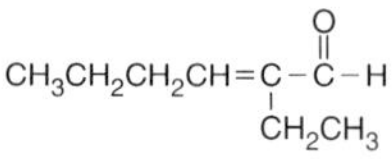

5 gm

S88,170-8
2,4-NONADIENE-2-CARBOXALDEHYDE

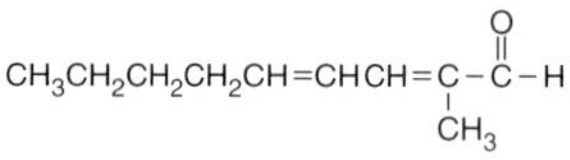

250 mg

S86,041-7
7-METHYL-1,4-DIOXASPIRO(4.5)DEC-7-ENE-8-CARBOXALDEHYDE

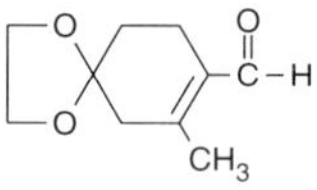

250 mg

S89,565-2
3-METHYL-1-CYCLOHEXENE-1-CARBOXALDEHYDE

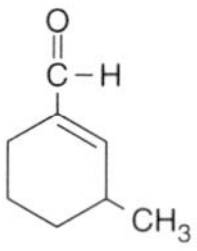

250 mg

S88,454-5
1,4-DIOXASPIRO(4.5)DECANE-7-ACETALDEHYDE

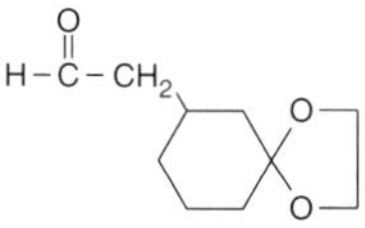

250 mg

S43,015-3
6-METHYL-3-CYCLOHEXENE-1-CARBOXALDEHYDE

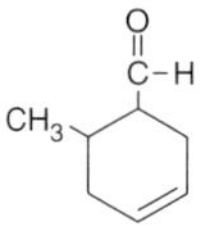

5 gm

S52,836-6
3,5,6-TRIMETHYL-3-CYCLOHEXENE-1-CARBOXALDEHYDE

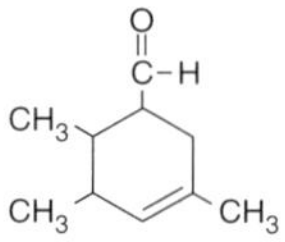

5 gm

S88,885-0
2,4,6-TRIMETHYL-3-CYCLOHEXENE-1-CARBOXALDEHYDE

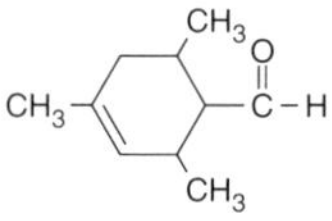

250 mg

S52,838-2
4-(4-METHYL-2-PENTENYL)-3-CYCLOHEXENE-1-CARBOXALDEHYDE

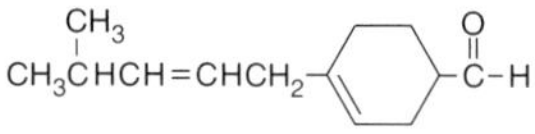

5 gm

S86,624-5
2,2-DIETHYLBUTYRIC ACID
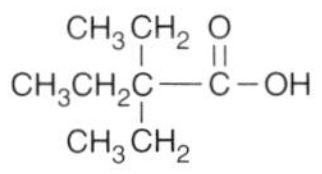
250 mg

S88,535-5
2-METHYLHEPTANOIC ACID
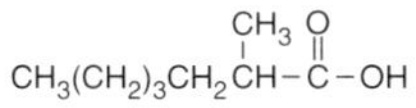
250 mg

S87,657-7
2-ETHYL-2-METHYLHEXANOIC ACID
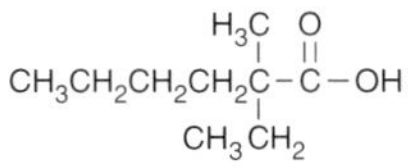
250 mg

S41,500-6
3,5,5-TRIMETHYLHEXANOIC ACID
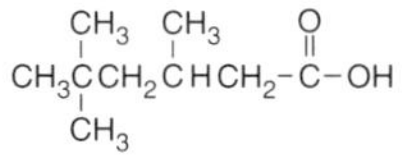
5 gm

S87,462-0
3-METHYLOCTANOIC ACID
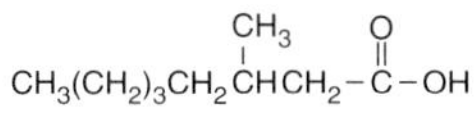
250 mg

S87,458-2
2-METHYLNONANOIC ACID
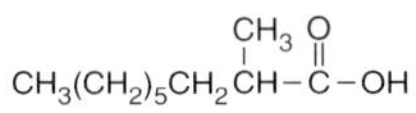
250 mg

S87,459-0
(S)-3-METHYLNONANOIC ACID
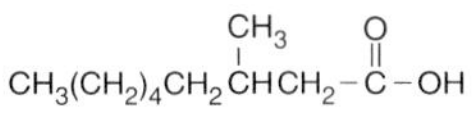
250 mg

S87,457-4
7-METHYLDECANOIC ACID
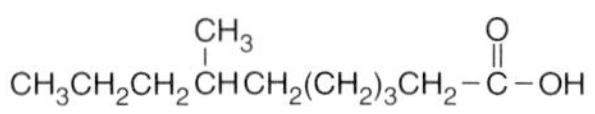
250 mg

S87,455-8
3-METHYLUNDECANOIC ACID
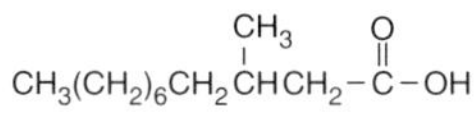
250 mg

S65,895-2
2-METHYLDODECANOIC ACID
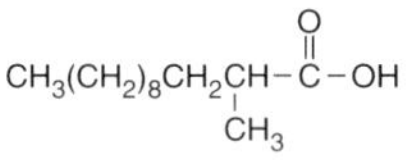
5 gm

S87,453-1
4-METHYLDODECANOIC ACID
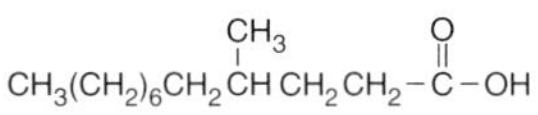
250 mg

S51,619-8
2-METHYLOCTADECANOIC ACID
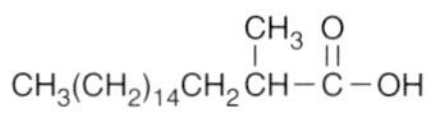
250 mg

S40,969-3
PENTACOSANOIC ACID
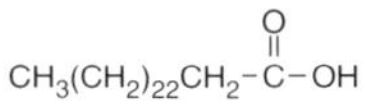
250 mg

S56,744-2
ISOPROPYLMALONIC ACID
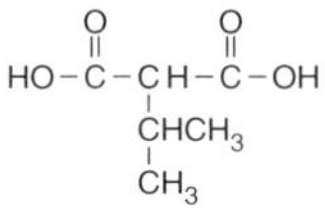
5 gm

S40,804-2
2,3-DIMETHYLGLUTARIC ACID
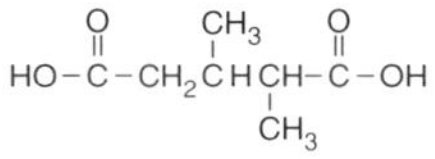
5 gm

S34,931-3
3-ETHYL-3-METHYLGLUTARIC ACID
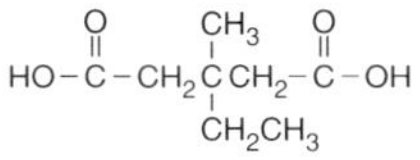
5 gm

S42,802-7
2,5-DIMETHYLADIPIC ACID
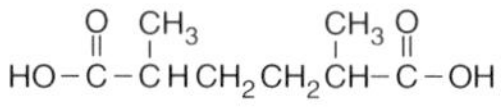
250 mg

S41,966-4
1,3,5-PENTANETRICARBOXYLIC ACID
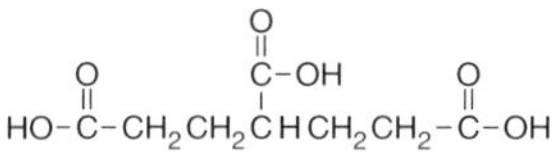
5 gm

S52,762-9
3-(TERT-BUTYL)ADIPIC ACID
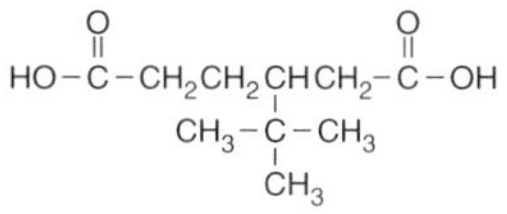
5 gm

S43,981-9
2,4,6-OCTATRIENOIC ACID
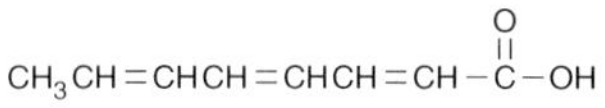
250 mg

S36,171-2
3-ETHYL-2-HEXENOIC ACID
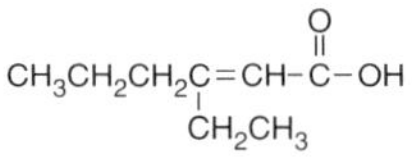
5 gm

S35,401-5
TRANS-2-NONENOIC ACID
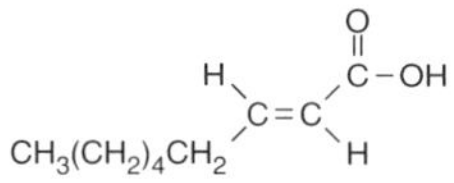
5 gm

S73,476-4
5,9-DIMETHYL-2,8-DECADIENOIC ACID
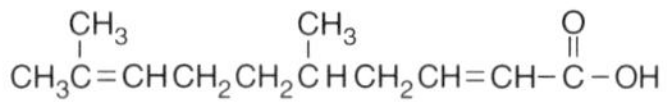
250 mg

S60,369-4
3-BUTENE-1,2,3-TRICARBOXYLIC ACID
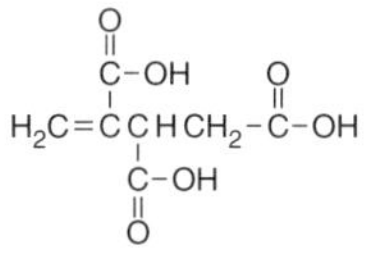
5 gm

S43,831-6
CIS,CIS-2,5-DIMETHYLMUCONIC ACID
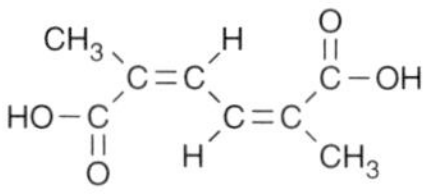
5 gm

S88,458-8
2,3-DIMETHYL-4-PENTENOIC ACID
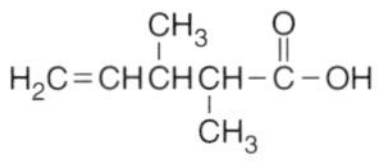
250 mg

S88,086-8
2,4-DIMETHYL-1,3-PENTADIENE-1-CARBOXYLIC ACID
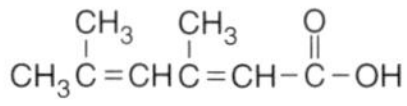
250 mg

S88,464-2
2,2,3-TRIMETHYL-4-PENTENOIC ACID
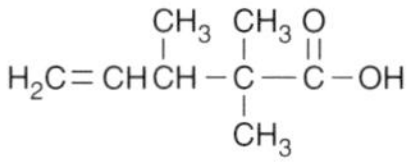
250 mg

S88,413-8
2-(1-METHYLALLYL)-3-HEXENOIC ACID
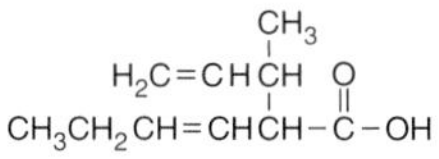
250 mg

S87,343-8
TRANS-9-UNDECENOIC ACID
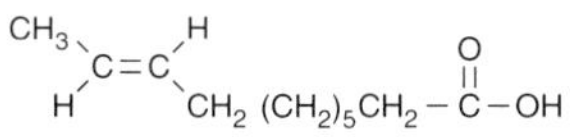
250 mg

S87,344-6
CIS-9-UNDECENOIC ACID
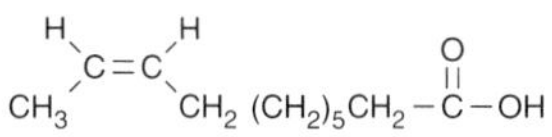
250 mg

S57,302-7
LINOLEIC ACID, UREA ADDUCT
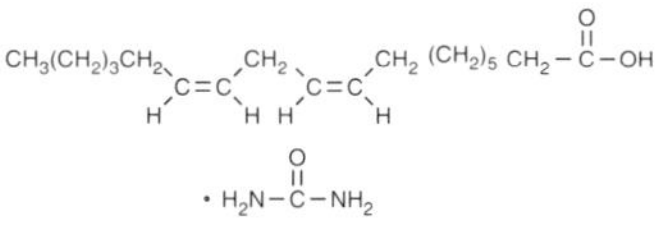
5 gm

S87,104-4
21-DOCOSENOIC ACID
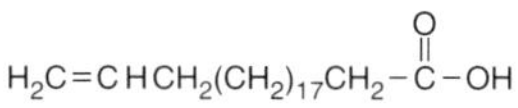
250 mg

S36,226-3
1-HEXADECENYLSUCCINIC ACID
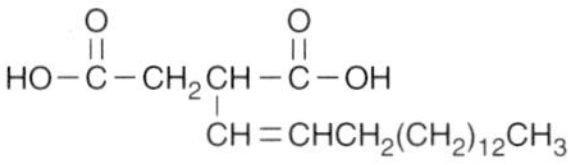
5 gm

S65,878-2
2-(1-OCTADECENYL)SUCCINIC ACID
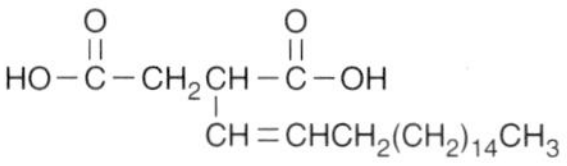
5 gm

S43,949-5
2,3-DICHLOROPROPIONIC ACID
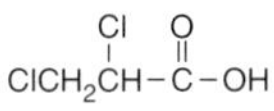
5 gm

S88,131-7
2,3-DIBROMO-2-METHYLPROPIONIC ACID

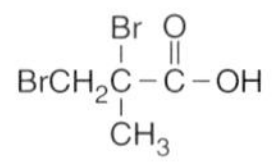

250 mg

S88,133-3
2,3,4-TRIBROMOBUTYRIC ACID

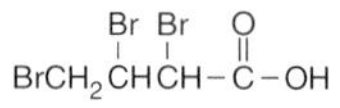

250 mg

S95,004-1
3-(BROMOMETHYL)CROTONIC ACID

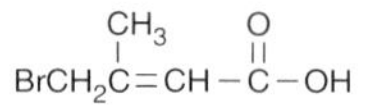

50 mg

S72,189-1
3-BROMO-2,2-DIETHYLPROPIONIC ACID

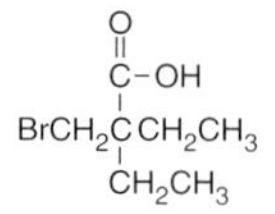

250 mg

S84,602-3
10-BROMODECANOIC ACID

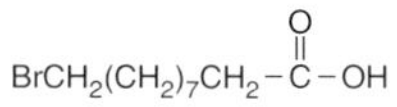

250 mg

S44,834-6
2,2,3,3,4,4,5,5,6,6,7,7,8,8,9,9,-HEXADECAFLUORONONANOIC ACID

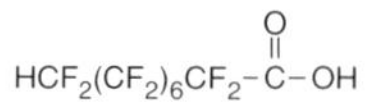

5 gm

S89,445-1
CIS-3,4,4,4-TETRACHLOROCROTONIC ACID

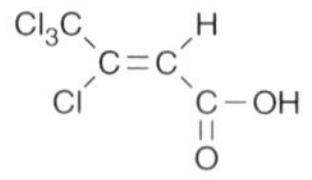

250 mg

S88,346-8
3-(TRIFLUOROMETHYL)-2-BUTENOIC ACID

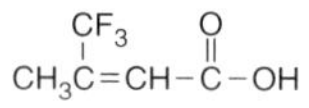

250 mg

S81,893-3
4,4,4-TRIFLUOROBUTYRIC ACID

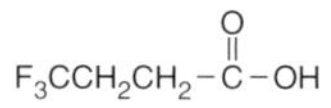

250 mg

S86,462-5
CHLOROFUMARIC ACID

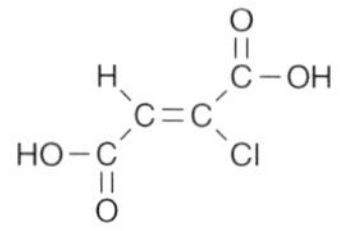

250 mg

S45,576-8
BROMOMALEIC ACID

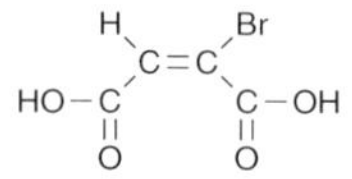

250 mg

S89,408-7
2,3,4-TRICHLOROGLUTACONIC ACID

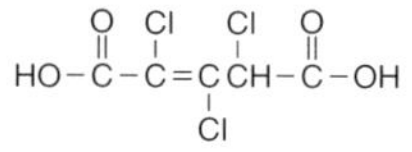

250 mg

S89,399-4
TRICHLOROACRYLIC ACID

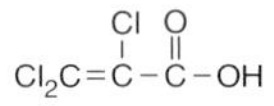

250 mg

S86,463-3
2-BROMO-2-BUTENOIC ACID

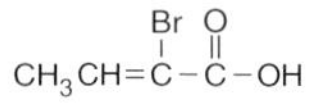

250 mg

S81,996-4
ALPHA,ALPHA'-DIBROMOSUCCINIC ACID

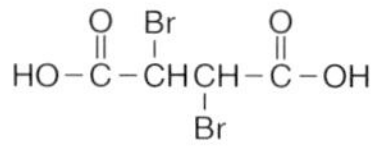

250 mg

S70,690-6
2,3-DIBROMOBUTYRIC ACID

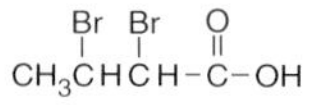

5 gm

S70,555-1
2-BROMO-3-METHYLVALERIC ACID

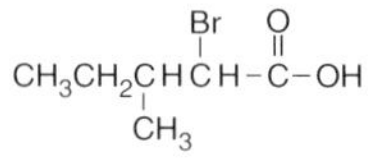

5 gm

S45,904-6
2-BROMOLAURIC ACID

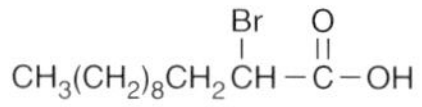

250 mg

S81,271-4
2-CHLOROSTEARIC ACID

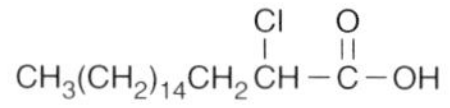

250 mg

S87,752-2
3-BROMO-2-METHYLACRYLIC ACID

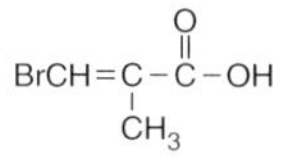

250 mg

S89,368-4
1-BROMO-4,4-DICHLORO-1,3-BUTADIENE-1-CARBOXYLIC ACID

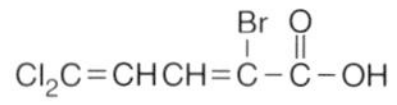

250 mg

S35,423-6
PENTACHLORO-CIS-2,4-PENTADIENOIC ACID

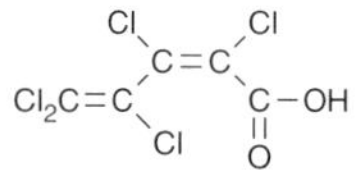

5 gm

S89,452-4
4-BROMO-1,2,3,4-TETRACHLORO-1,3-BUTADIENE-1-CARBOXYLIC ACID

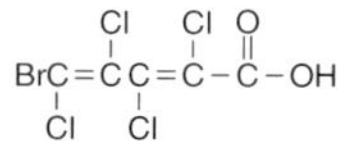

250 mg

S86,002-6
4,5-DIBROMO-2-HEXENOIC ACID

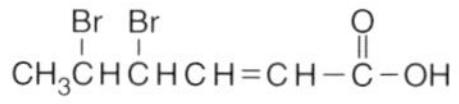

250 mg

S90,603-4
3,3-DICHLOROGLUTARIC ACID

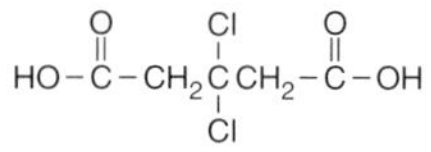

250 mg

S40,749-6
3,4-DICHLOROADIPIC ACID

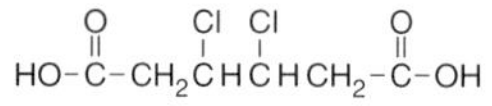

5 gm

S44,344-1
TRANS-3-CHLOROCROTONIC ACID

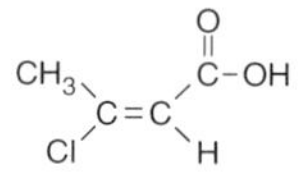

250 mg

S41,962-1
3-CHLOROBUTYRIC ACID

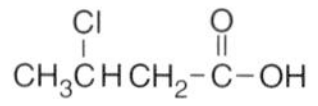

5 gm

S56,132-0
2-CHLOROBUTYRIC ACID

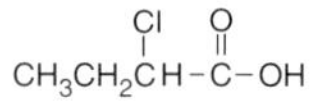

5 gm

S63,122-l
2,5-DIBROMOADIPIC ACID

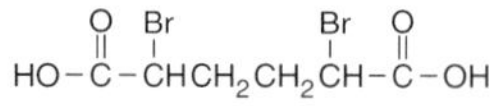

250 mg

S41,586-3
2-BROMOSTEARIC ACID, TECH.

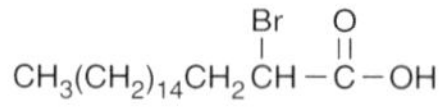

5 gm

S37,158-0
9,10,12,13-TETRABROMOOCTADECANOIC ACID

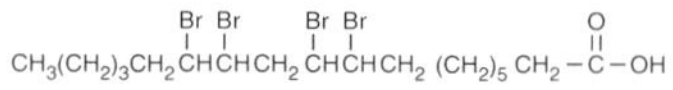

5 gm

S56,745-0
9,10,12,13,15,16-HEXABROMOOCTADECANOIC ACID

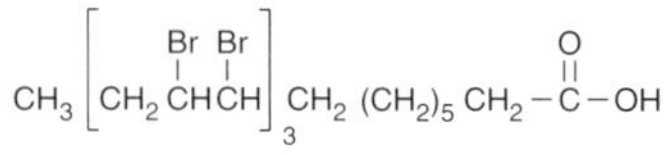

5 gm

S91,841-5
ETHYL 2-CHLORO-2-METHYLPROPIONATE

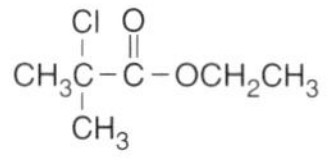

250 mg

S73,941-3
3-HYDROXY-4,4,4-TRICHLOROBUTYRIC ACID

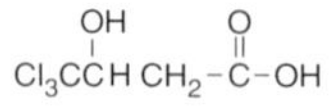

5 gm

S60,028-8
3-HYDROXY-4,4,4-TRIFLUOROBUTYRIC ACID

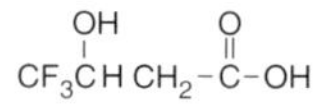

250 mg

S78,904-6
3-BUTYLMALIC ACID MONOSODIUM SALT HYDRATE
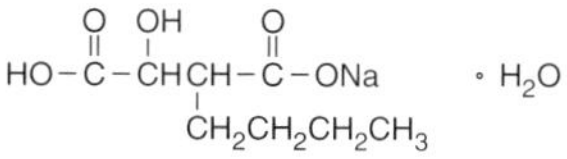
250 mg

S67,630-6
MALIC ACID
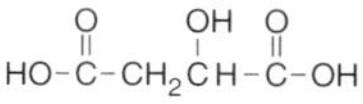
250 mg

S37,354-0
2-HYDROXYVALERIC ACID
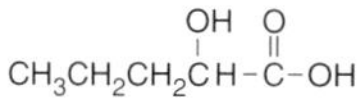
250 mg

S57,684-0
2-HYDROXYOCTANOIC ACID
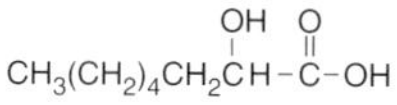
250 mg

S74,280-5
2-HYDROXYDODECANOIC ACID
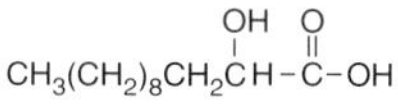
5 gm

S43,919-3
2-HYDROXYMYRISTIC ACID
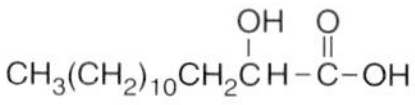
250 mg

S44,224-0
2-HYDROXYPALMITIC ACID
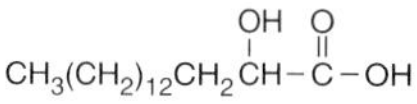
250 mg

S56,164-9
ALEURITIC ACID
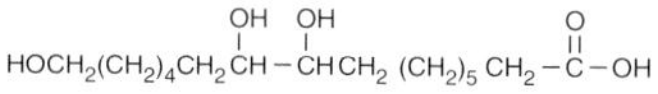
5 gm

S41,537-5
DL-CITRAMALIC ACID
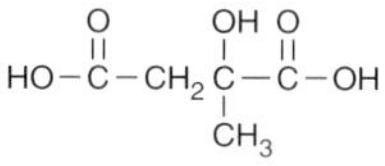
250 mg

S95,035-1
(-)-3-HYDROXYBUTYRIC ACID, QUININE SALT
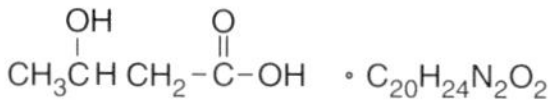
50 mg

S88,188-0
2,2-DIMETHYL-3-HYDROXYPROPIONIC ACID
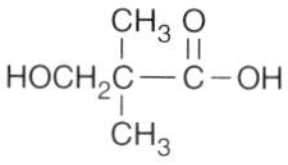
250 mg

S94,983-3
3,3-DIMETHYL-3-(1,1-DIMETHYL-2-HYDROXYETHYLAMINO)PROPIONIC ACID
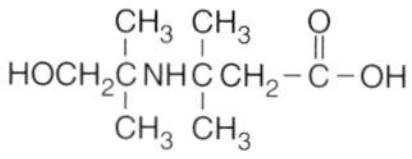
250 mg

S88,808-7
9-HYDROXYDECANOIC ACID
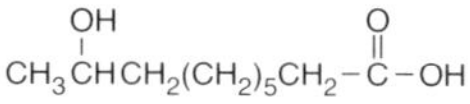
250 mg

S93,543-3
(D)-9-HYDROXYSTEARIC ACID
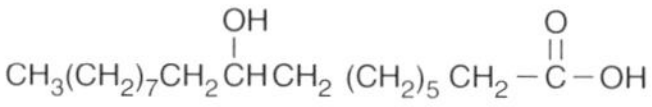
250 mg

S88,473-1
2-METHOXYPROPIONIC ACID
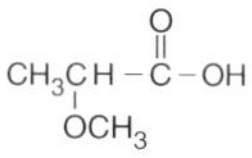
250 mg

S95,009-2
2-(METHOXYMETHYL)SUCCINIC ACID
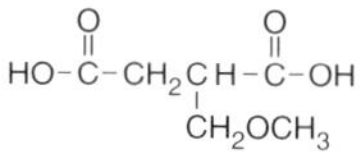
50 mg

S88,506-1
3-ETHOXY-2-BUTENOIC ACID
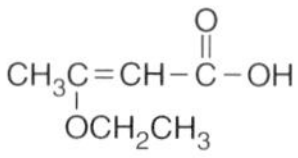
250 mg

S70,557-8
3-METHOXYBUTYRIC ACID
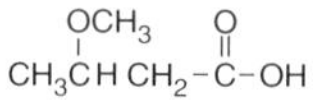
5 gm

S37,697-3
4-METHOXYBUTYRIC ACID
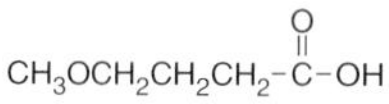
250 mg

S44,874-5
3-ETHOXYPROPIONIC ACID
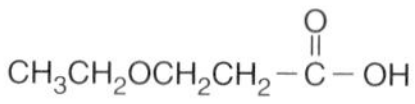
5 gm

S53,071-9
3-(OCTYLOXY)PROPIONIC ACID
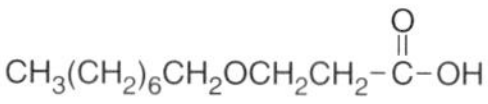
250 mg

S35,934-3
2-(CARBOXYMETHYLTHIO)ETHYL ETHER
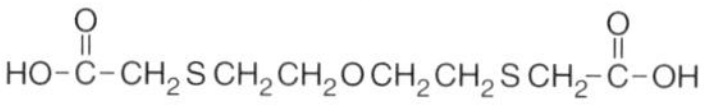
5 gm

S62,023-8
2-(ETHYLTHIO)ACETIC ACID
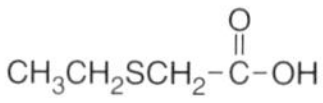
250 mg

S53,730-6
3-(DODECYLTHIO)PROPIONIC ACID
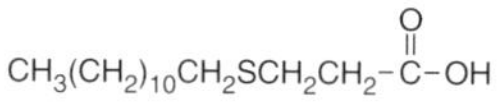
5 gm

S53,765-9
11-(BUTYLTHIO)UNDECANOIC ACID
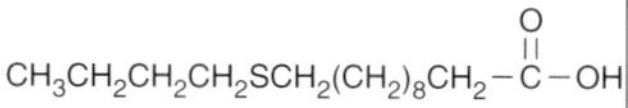
5 gm

S60,537-9
4-(DODECYLTHIO)BUTYRIC ACID

250 mg

S35,289-6
METHYLENEBIS(THIOGLYCOLLIC ACID)
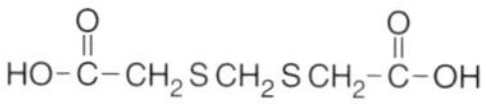
5 gm

S57,087-7
2,2'-(ETHYLENEDITHIO)DIACETIC ACID
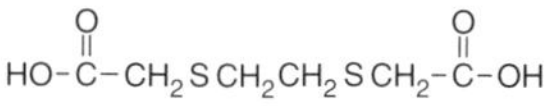
5 gm

S61,451-3
S,S'-METHYLENEBIS(3-MERCAPTOPROPIONIC ACID)
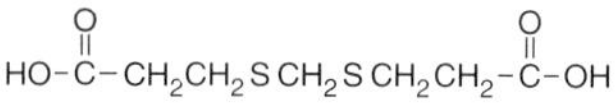
5 gm

S35,906-8
1,4-BUTANEBIS(THIOGLYCOLLIC ACID)
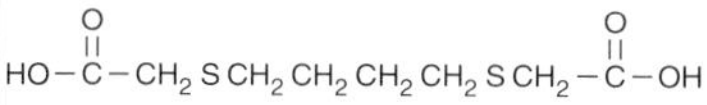
5 gm

S41,019-5
2,2'-THIODISUCCINIC ACID
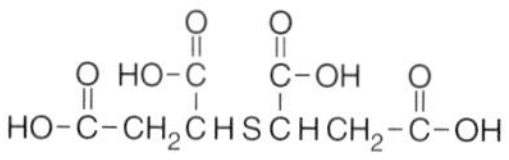
5 gm

S74,203-1
S,S'-ETHYLIDENEBIS-(THIOGLYCOLIC ACID)
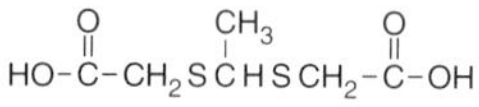
5 gm

S56,324-2
3,5-DIBROMOLEVULINIC ACID
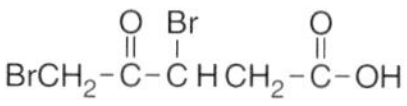
5 gm

S87,255-5
3-OXO-1,8-OCTANEDICARBOXYLIC ACID
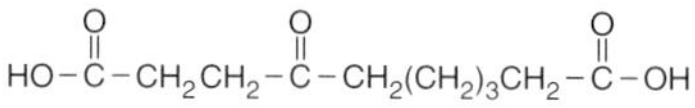
250 mg

S95,003-3
4-OXO-2-PENTENOIC ACID
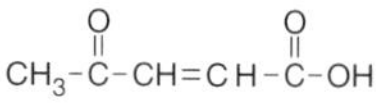
50 mg

S95,010-6
3,3-DIMETHYL-4-OXOVALERIC ACID
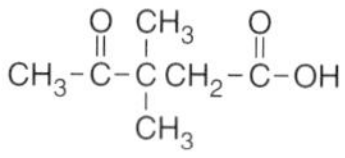
50 mg

S86,950-3
7-OXO-8-NONENOIC ACID
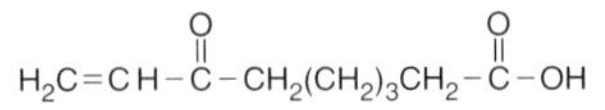
250 mg

S69,803-2
4-METHYL-5-OXOHEXANOIC ACID
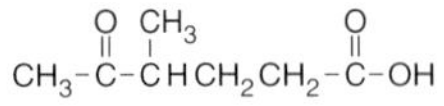
250 mg

S88,600-9
5-METHYL-4-OXOHEXANOIC ACID
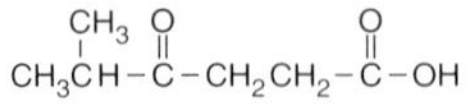
250 mg

S69,797-4
2,4-DIMETHYL-5-OXOHEXANOIC ACID
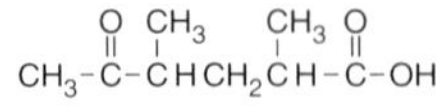
250 mg

S87,460-4
2-METHYL-7-OXONONANOIC ACID
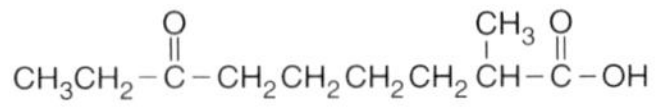
250 mg

S87,461-2
3-METHYL-7-OXONONANOIC ACID
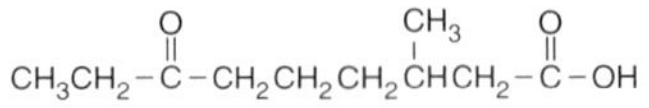
250 mg

S82,649-9
10-OXOUNDECANOIC ACID
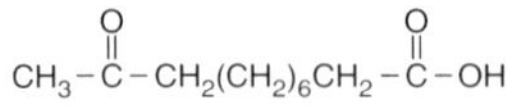
250 mg

S86,991-0
7-OXO-11-DODECENOIC ACID
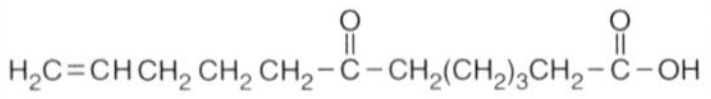
250 mg

S87,512-0
13-OXOPENTADECANOIC ACID
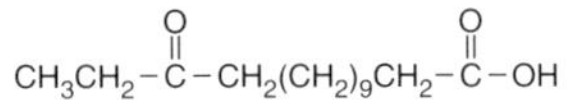
250 mg

S87,047-1
6-OXO-15-HEXADECENOIC ACID
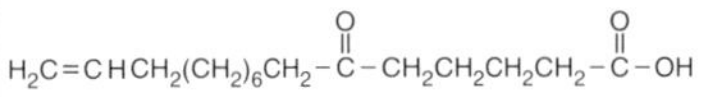
250 mg

S87,070-6
7-OXO-16-HEPTADECENOIC ACID
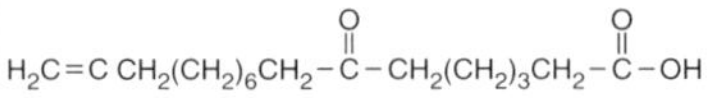
250 mg

S87,492-2
13-OXOOCTADECANOIC ACID
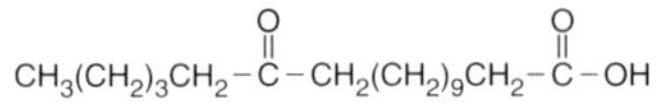
250 mg

S87,288-1
7-OXOTETRACOSANOIC ACID
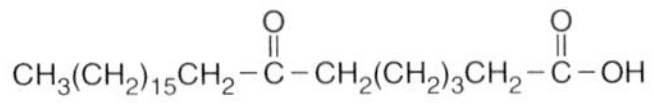
250 mg

S87,487-6
13-OXOTETRACOSANOIC ACID
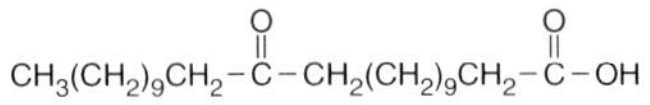
250 mg

S87,484-1
13-OXOTRIACONTANOIC ACID
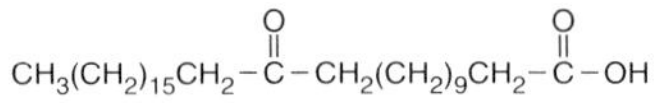
250 mg

S87,251-2
4-OXO-1,9-NONANEDICARBOXYLIC ACID
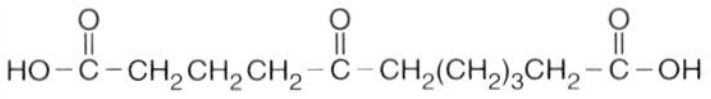
250 mg

S87,257-1
6-OXO-1,11-UNDECANEDICARBOXYLIC ACID
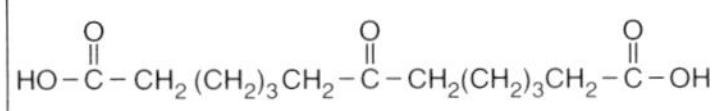
250 mg

S87,271-7
6-OXO-1,12-DODECANEDICARBOXYLIC ACID
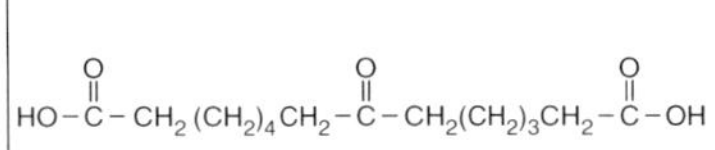
250 mg

S87,273-3
6-OXO-1,14-TETRADECANEDICARBOXYLIC ACID
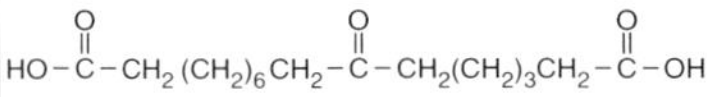
250 mg

S87,279-2
6,12-DIOXO-1,17-HEPTADECANEDICARBOXYLIC ACID
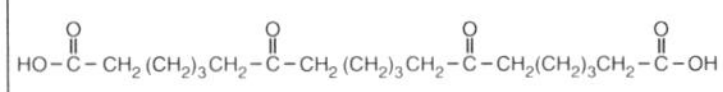
250 mg

S87,280-6
6,15-DIOXO-1,20-EICOSANEDICARBOXYLIC ACID
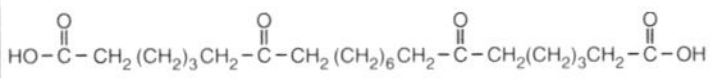
250 mg

S87,481-7
13,20-DIOXODOTRIACONTANEDIOIC ACID
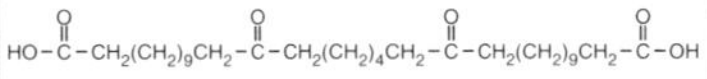
250 mg

S87,472-8
13,34-DIOXOHEXATETRACONTANEDIOIC ACID
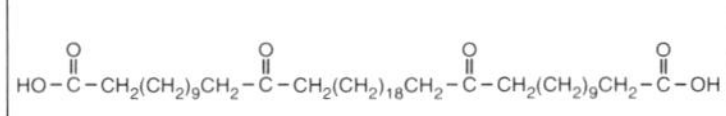
250 mg

S57,409-0
4-OXOHEXANOIC ACID
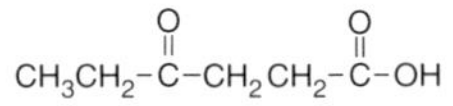
250 mg

S78,430-3
2-OXONONANOIC ACID
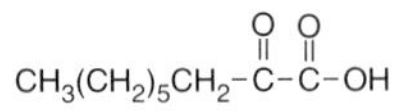
250 mg

S58,051-1
7-OXOTRIDECANOIC ACID
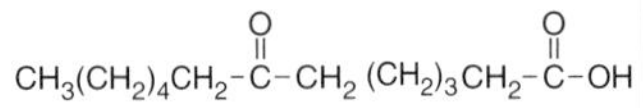
250 mg

S81,784-8
10-OXOPALMITIC ACID
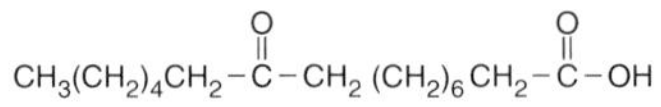
250 mg

S54,896-0
3-METHYL-4-OXO-2-PENTENOIC ACID
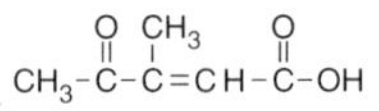
250 mg

S87,619-4
3,3,3-TRIMETHYLPYRUVIC ACID
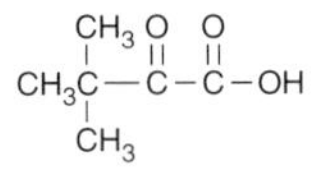
250 mg

S89,499-0
6-OXOTRICOSANOIC ACID
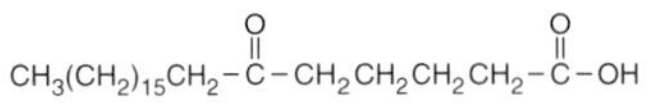
250 mg

S95,357-1
2-(BROMOMETHYL)-2,3-DIMETHYL-4-OXOCYCLOHEXANECARBOXYLIC ACID
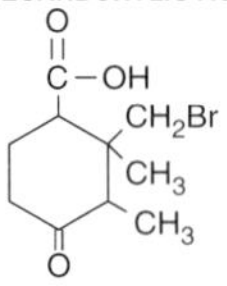
50 mg

S94,817-9
3-BROMO-ALPHA-HYDROXY-1-ADAMANTANEACETIC ACID
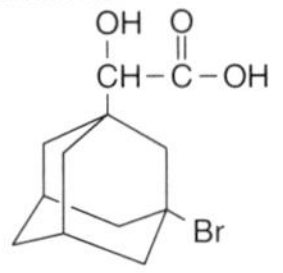
250 mg

S91,447-9
CIS-2-HYDROXY-1-CYCLOHEXANECARBOXYLIC ACID
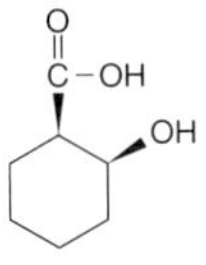
250 mg

S67,153-3
METHYL CIS-2-HYDROXYCYCLOHEXANECARBOXYLATE
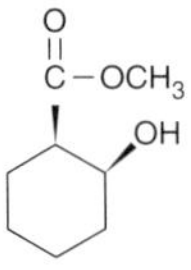
5 gm

S95,606-6
1,4,4A,5,8,8A-HEXAHYDRO-5-HYDROXY-8-OXO-1-NAPHTHALENECARBOXYLIC ACID
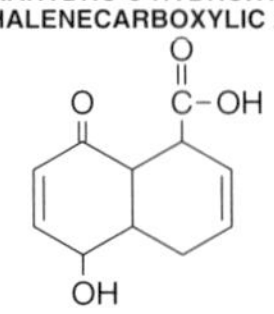
50 mg

S45,917-8
1-HYDROXYCYCLOHEPTANECARBOXYLIC ACID
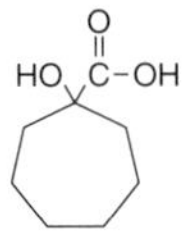
250 mg

S40,476-4
C14 ACID DERIVED FROM APOAROMADENDRONE
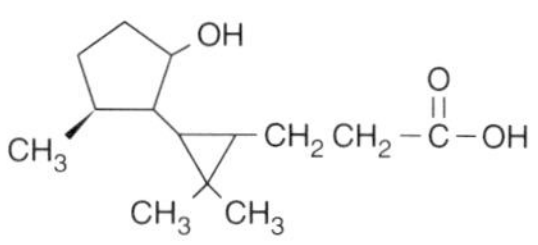
250 mg

S44,123-6
(+/-)-PINOLIC ACID
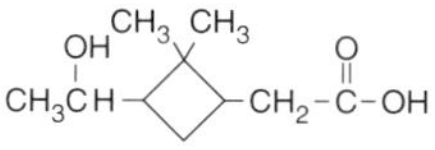
5 gm

S55,831-1
5-(TERT-BUTYL)-2-HYDROXYCYCLOHEXANECARBOXYLIC ACID
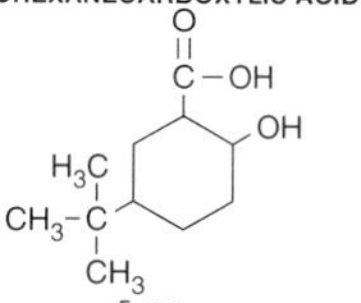
5 gm

S59,608-6
3-HYDROXY-1-ADAMANTANEACETIC ACID
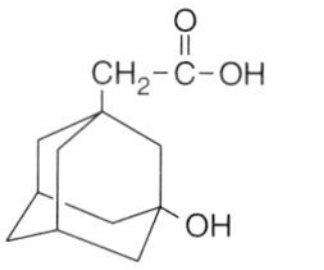
250 mg

S59,598-5
ALPHA-HYDROXY-3,5,7-TRIMETHYL-1-ADAMANTANEACETIC ACID
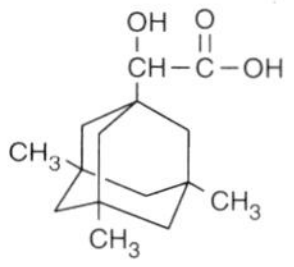
5 gm

S95,350-4
2,3-DIHYDRO-4-ETHOXYPHTHALIC ACID
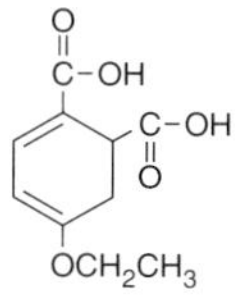
50 mg

S95,384-9
2-METHYL-2-(4-METHYL-3-OXOCYCLOHEXYL)MALONIC ACID
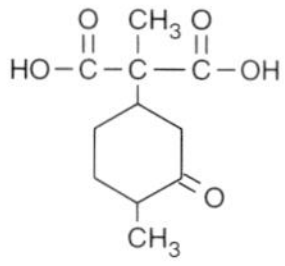
50 mg

S95,356-3
2-(2-CYCLOHEXEN-1-YL)-2-METHYLMALONIC ACID
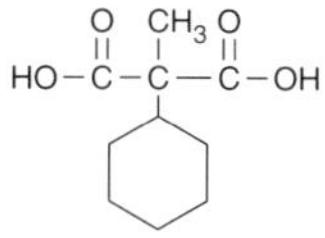
50 mg

S87,231-8
4-OXO-4-(2-OXOCYCLOHEXYL)-BUTYRIC ACID
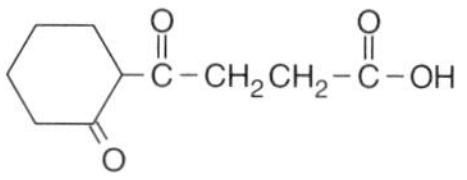
250 mg

S86,485-4
5-OXOTRICYCLO[2.2.1.0(2,6)]HEPTANE-2,3-DICARBOXYLIC ACID
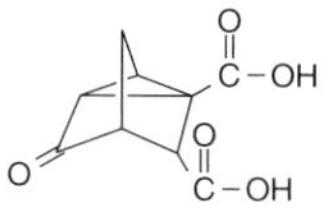
250 mg

S95,602-3
5,8-DIOXO-1,4,4A,5,8,8A-HEXAHYDRO-1-NAPHTHALENECARBOXYLIC ACID
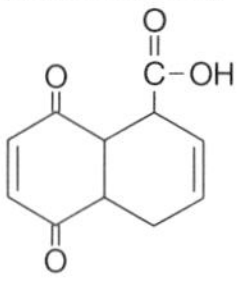
50 mg

S95,848-4
(-)-CIS-ISOKETOPINIC ACID
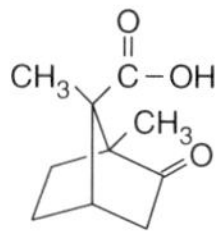
50 mg

S89,509-1
2,3-NORBORNANEDICARBOXYLIC ACID
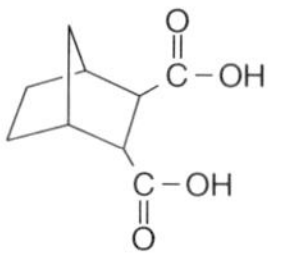
250 mg

S95,815-8
TRICYCLO[4.2.2.0(2,5)]DECA-3,9-DIENE-7,8-DICARBOXYLIC ACID
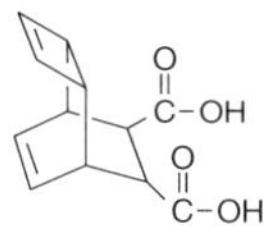
50 mg

S36,761-3
CYCLOHEXANE-1,1-BIS-(MERCAPTOACETIC ACID)
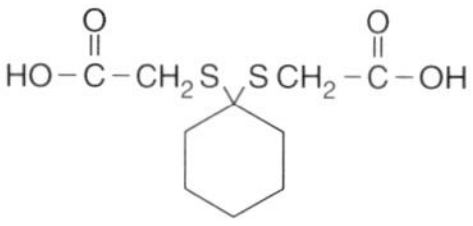
5 gm

S35,983-1
CYCLOHEXANE-1,1-BIS(3-MERCAPTOPROPIONIC ACID)
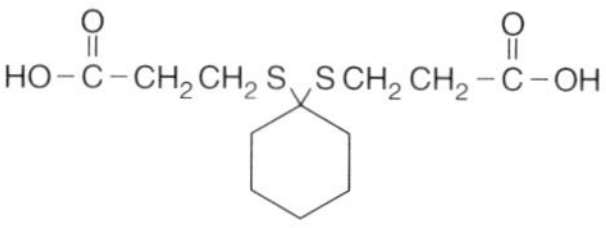
5 gm

S89,540-7
2,2'-IMINOBIS(1-CYCLOPENTANECARBOXYLIC ACID)
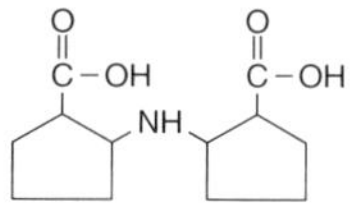
250 mg

S87,546-5
CIS-2-(ETHYLAMINO)-1-CYCLOHEXANECARBOXYLIC ACID
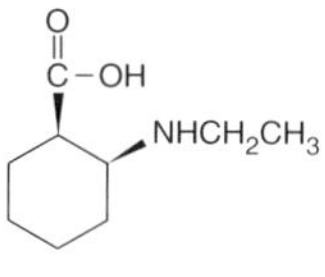
250 mg

S64,581-8
1-FORMYL-3-CYCLOHEXENE-1-PROPIONIC ACID
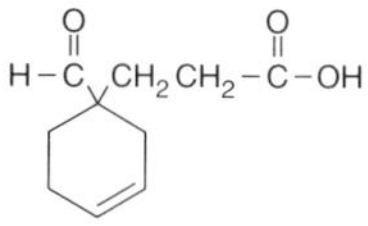
250 mg

S95,713-5
2,3,4,4A,4B,5,8,8A,9,10-DECAHYDRO-8A-METHYL-2-OXO-1-PHENANTHRENEPROPIONIC ACID
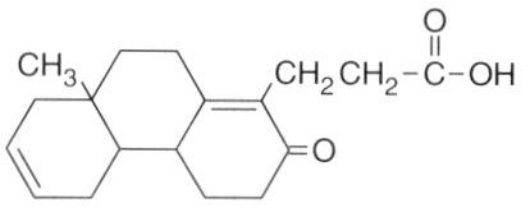
50 mg

S95,712-7
8A-METHYL-2,3,4,4A,4B,5,8,8A-OCTAHYDRO-2-OXO-1-PHENANTHRENEPROPIONIC ACID
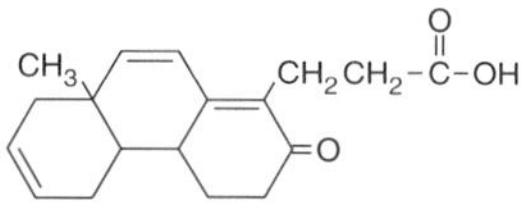
50 mg

S45,299-8
(+/-)-3-CAMPHORCARBOXYLIC ACID
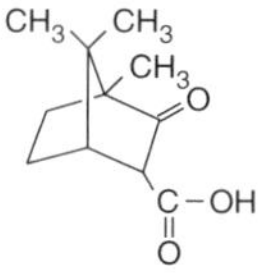
5 gm

S74,542-1
1-CARBOXY-4-OXO-1-CYCLOHEXANEACETIC ACID
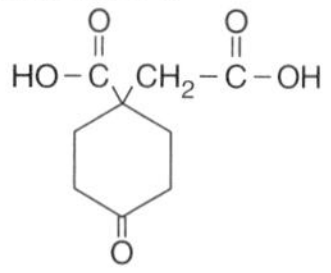
250 mg

S88,479-0
3-OXO-1-CYCLOPENTANEACETIC ACID
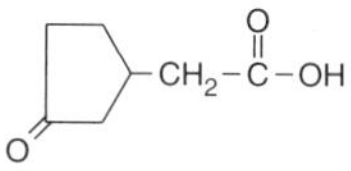
250 mg

S78,240-8
2-OXO-1-CYCLOHEXANEPROPIONIC ACID
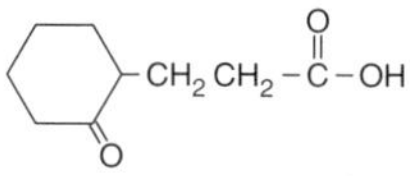
5 gm

S79,833-9
4-(4,4-DIMETHYL-2-OXOCYCLOHEXYL)-BUTYRIC ACID
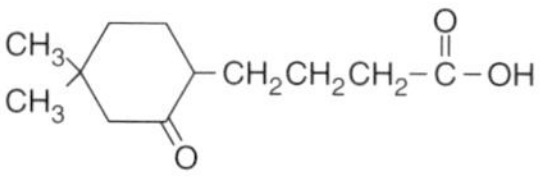
250 mg

S43,477-9
2-OXO-1,1,3,3-CYCLOHEXANETETRAPROPIONIC ACID
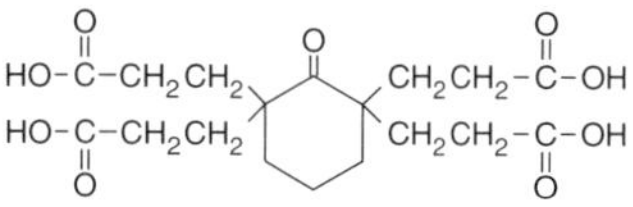
5 gm

S95,359-8
1-METHYL-2-OXOCYCLOHEXANEPROPIONIC ACID
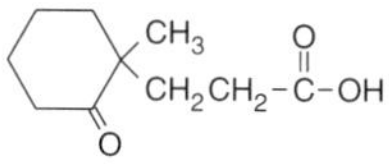
50 mg

S94,218-9
3-(2-OXO-1,3,3-TRIMETHYLCYCLOHEXYL)-PROPIONIC ACID
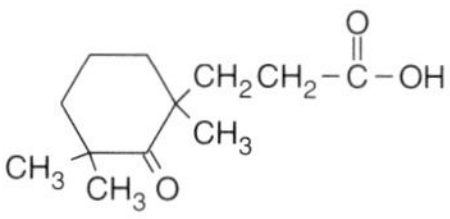
250 mg

S59,738-4
ALPHA-CHLORO-5-METHYL-1,3-ADAMANTANEDIACETIC ACID
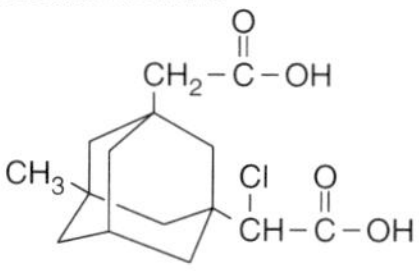
5 gm

S59,700-7
ALPHA-CHLORO-5,7-DIMETHYL-1,3-ADAMANTANEDIACETIC ACID
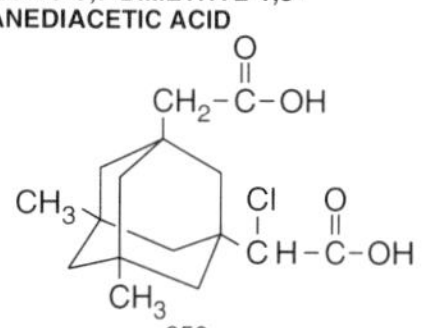
250 mg

S46,052-4
ALPHA-CHLORO-2-NORBORNANEACETIC ACID
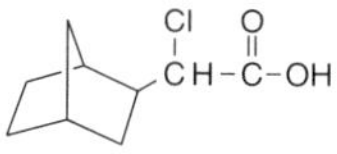
250 mg

S59,709-0
ALPHA-CHLORO-3,5-DIMETHYL-1-ADAMANTANEACETIC ACID
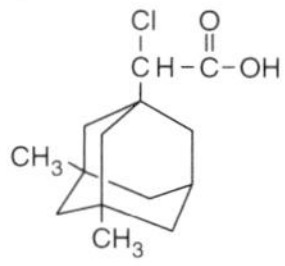
250 mg

S85,225-2
2,5-NORBORNADIENE-2,3-DICARBOXYLIC ACID
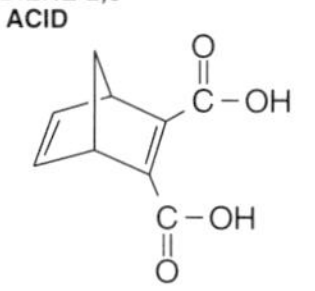
250 mg

S43,442-6
1-CYCLOUNDECENE-1-CARBOXYLIC ACID
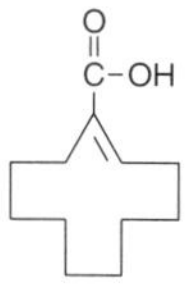
5 gm

S52,009-8
3,5-DIMETHYL-1-CYCLOHEXENE-1,2-DICARBOXYLIC ACID
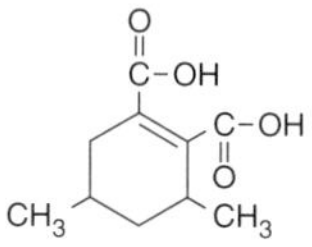
250 mg

S75,793-4
CYCLOPENTYLIDENEACETIC ACID
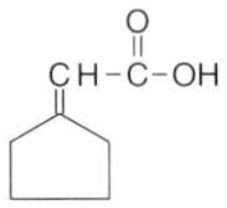
5 gm

S88,280-1
1-CYCLOPENTENE-1-CARBOXYLIC ACID
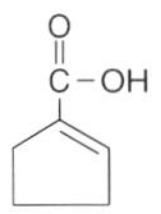
250 mg

S40,547-7
2-CYCLOHEXYLIDENEBUTYRIC ACID
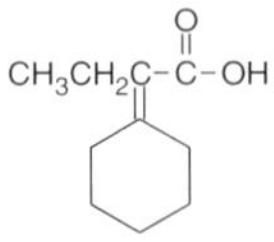
5 gm

S70,157-2
2-(4-METHYL-1-CYCLOHEXYLIDENE)ACETIC ACID
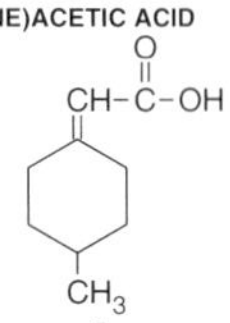
5 gm

S46,120-2
5-NORBORNENE-2-ACRYLIC ACID
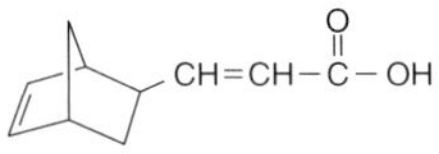
5 gm

S67,873-2
1,1-CYCLOPENTANEDICARBOXYLIC ACID
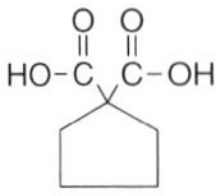
5 gm

S43,350-0
(2-CYCLOPENTEN-1-YL)-MALONIC ACID
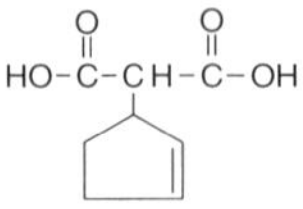
5 gm

S84,600-7
1,1-CYCLOHEXANEDICARBOXYLIC ACID
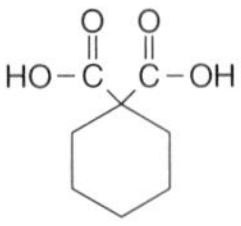
250 mg

S76,280-6
TRANS-NORPINIC ACID
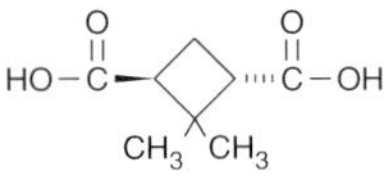
250 mg

S76,279-2
PINIC ACID
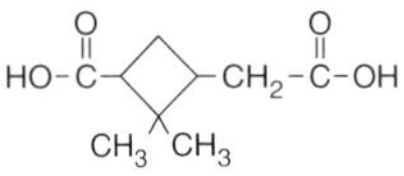
250 mg

S84,628-7
PERHYDROCYCLOBUTA(1,2:3,4)-DICYCLOPENTENE-2,5-DICARBOXYLIC ACID
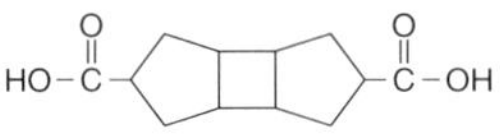
250 mg

S75,780-2
3-METHYL-1,1-CYCLOPENTANEDIACETIC ACID
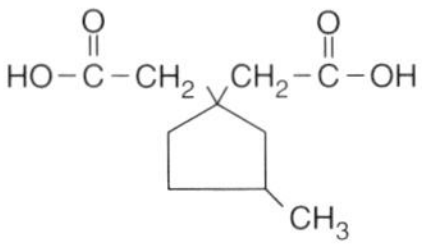
5 gm

S72,230-8
BICYCLO[2.2.1]HEPTANE-3,4-DICARBOXYLIC ACID
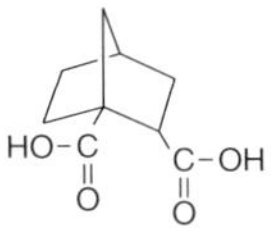
5 gm

S75,795-0
4-METHYL-1,1-CYCLOHEXANEDIACETIC ACID
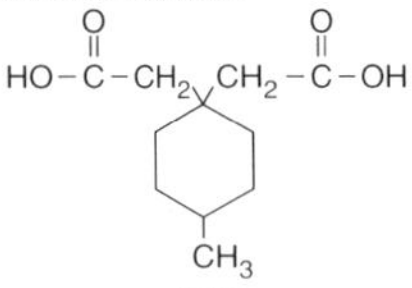
5 gm

S76,460-4
3-METHYL-1,1-CYCLOHEXANEDIACETIC ACID
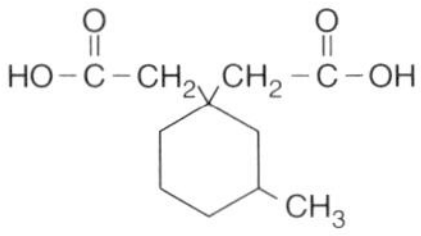
5 gm

S75,789-6
4-CARBOXY-1,1-CYCLOHEXANEDIACETIC ACID
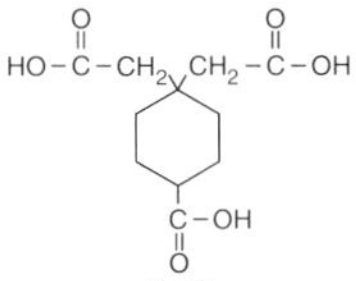
5 gm

S75,743-8
TRANS-1,2-CYCLOHEXANEDIACETIC ACID
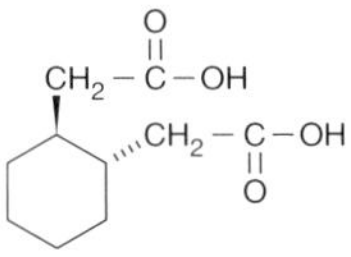
5 gm

S76,281-4
SPIRO(5.5)UNDECANE-2,2-DIACETIC ACID
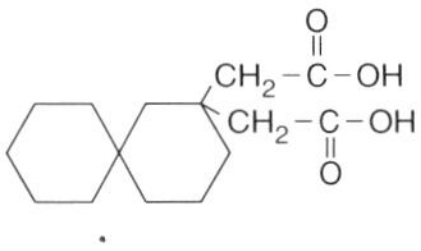
250 mg

S74,163-9
1-CARBOXY-3-CYCLOHEXENE-1-ACETIC ACID
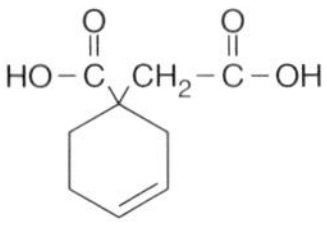
250 mg

S77,059-0
4-METHYL-4-CYCLOHEXENE-1,2-DICARBOXYLIC ACID
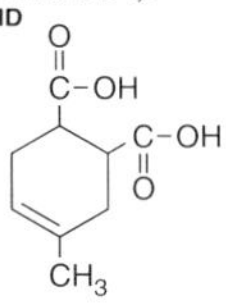
250 mg

S44,441-3
4-CYCLOHEPTENE-1-CARBOXYLIC ACID
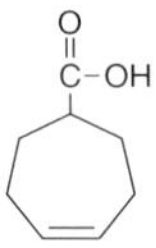
250 mg

S43,524-4
CYCLOUNDECANECARBOXYLIC ACID
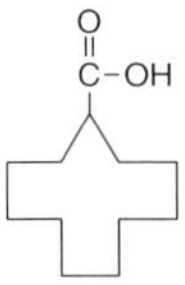
5 gm

S58,096-1
1,1-CYCLOHEPTANEDIACETIC ACID
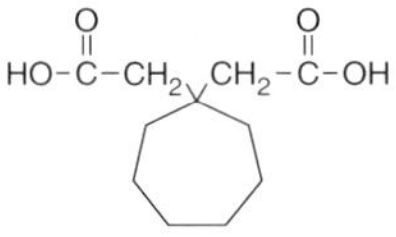
250 mg

S95,063-7
3-CARBOXY-1,2,2-TRIMETHYLCYCLOPENTANEACRYLIC ACID
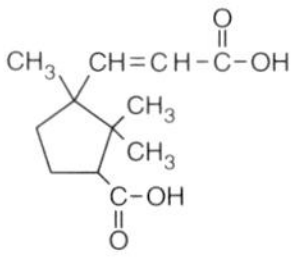
50 mg

S44,788-9
HEXAHYDRO-4-METHYLPHTHALIC ACID
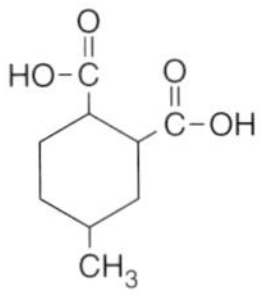
5 gm

S44,382-4
CIS-1,2-CYCLOHEXANEDICARBOXYLIC ACID
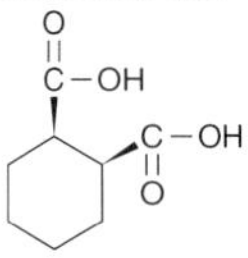
5 gm

S95,355-5
6-CARBOXY-6-METHYL-3-CYCLOHEXENE-1-ACETIC ACID
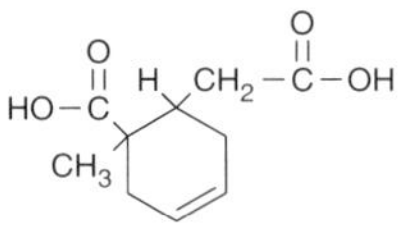
50 mg

S85,243-0
BICYCLO[4.1.0]HEPTANE-7-CARBOXYLIC ACID
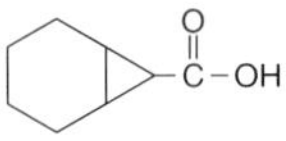
250 mg

S45,490-7
(-)-3-PINANECARBOXYLIC ACID
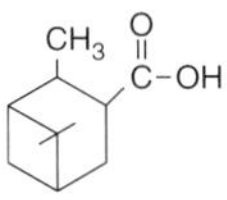
5 gm

S46,037-0
(+)-3-PINANECARBOXYLIC ACID
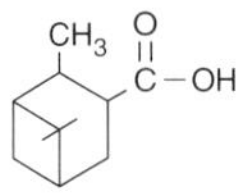
250 mg

S43,523-6
2-CYCLOPENTYLHEXANOIC ACID
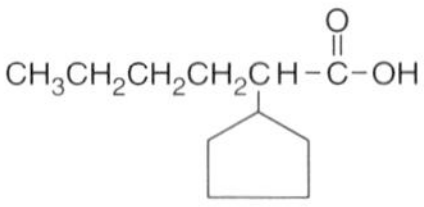
5 gm

S86,682-2
5,6,7,8,9,9-HEXACHLORO-OCTAHYDRO-1,4:5,8-DIMETHANONAPHTHALENE-2-CARBOXYLIC ACID
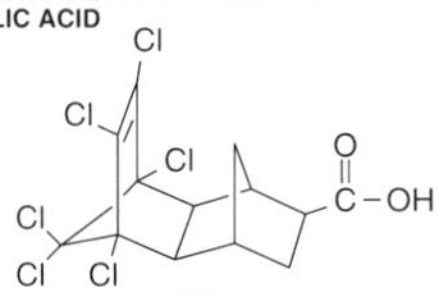
250 mg

S86,477-3
2-CHLORO-5-NORBORNENE-2-CARBOXYLIC ACID
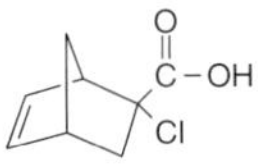
250 mg

S75,799-3
SPIRO(4.5)DECANE-8-CARBOXYLIC ACID
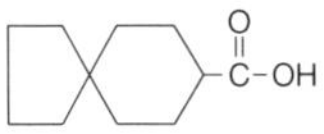
250 mg

S92,867-4
BICYCLO[2.2.2]OCTANE-2-CARBOXYLIC ACID
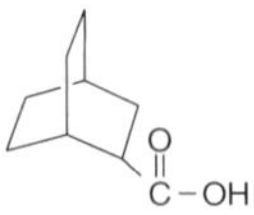
250 mg

S93,687-1
TRANS-3-METHYLBICYCLO[2.2.2]OCTANE-2-CARBOXYLIC ACID
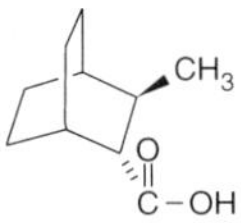
250 mg

S65,887-1
1-CYCLOPENTENYLACETIC ACID
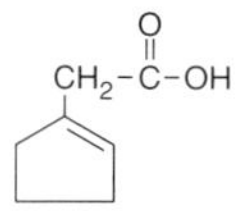
250 mg

S37,162-9
ALPHA-PROPYL-1-CYCLOPENTENE-1-ACETIC ACID
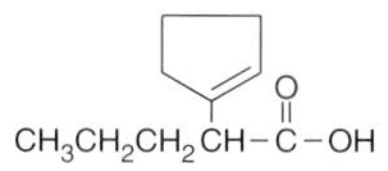
5 gm

S95,061-0
(-)-1,2,3-TRIMETHYL-2-CYCLOPENTENEACRYLIC ACID
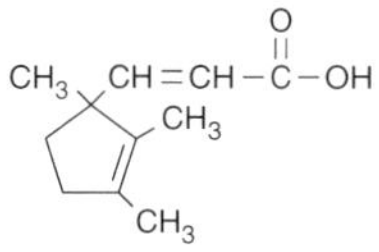
50 mg

S89,913-5
2-CYCLOPENTENE-1-TRIDECANOIC ACID
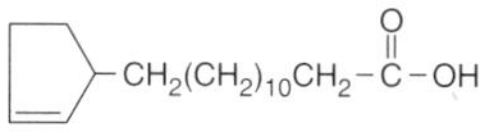
250 mg

S66,232-1
EXO-BICYCLO[2.2.1]HEPT-5-ENE-2-CARBOXYLIC ACID
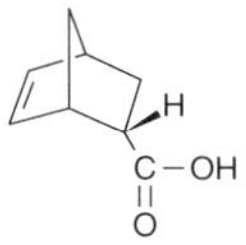
5 gm

S51,865-4
1-BROMOCYCLOHEXANECARBOXYLIC ACID
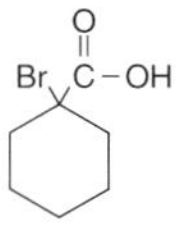
250 mg

S43,522-8
2-CYCLOHEXYLBUTYRIC ACID
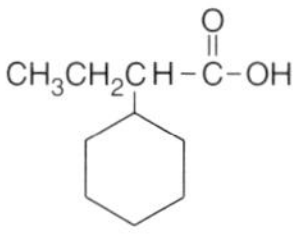
5 gm

S40,724-0
2-(1-CYCLOHEXENYL)-BUTYRIC ACID
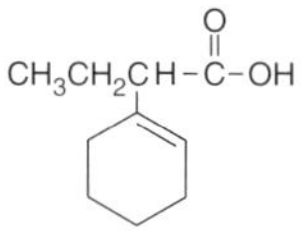
5 gm

S94,766-0
3-BUTYL-5-(2,6,6-TRIMETHYL-1-CYCLOHEXEN-1-YL)-2,4-PENTADIENOIC ACID
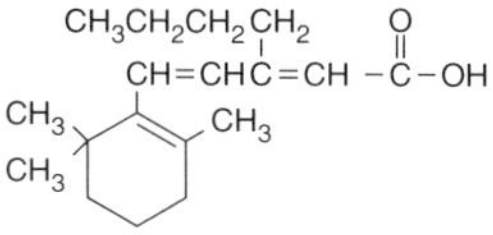
5 gm

S66,492-8
BICYCLO[2.2.1]HEPT-5-ENE-2-CARBOXYLIC ACID
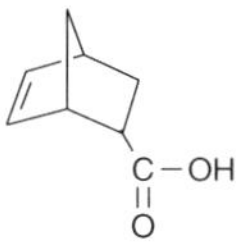
5 gm

S67,034-0
2-METHYL-5-NORBORNENE-2-CARBOXYLIC ACID
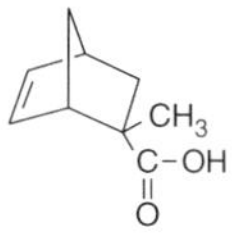
5 gm

S94,800-4
4-TRICYCLO[4.3.1.1(3,8)]-UNDECANECARBOXYLIC ACID
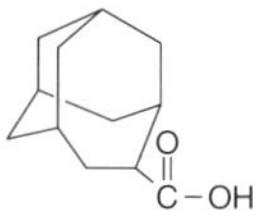
250 mg

S93,807-6
ALPHA-BROMO-1-ADAMANTANEACETIC ACID
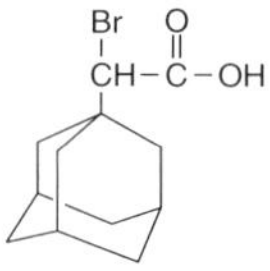
250 mg

S94,282-0
3-BROMO-1-ADAMANTANEACETIC ACID
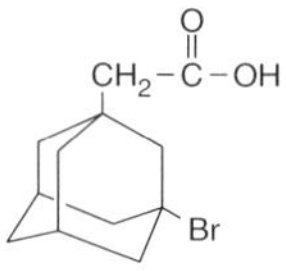
250 mg

S59,708-2
3,5-DIBROMO-7-METHYL-1-ADAMANTANEACETIC ACID
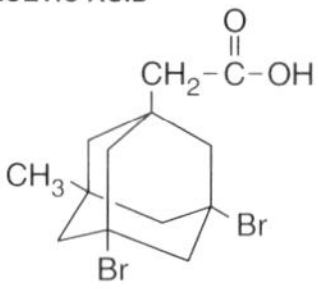
250 mg

S59,713-9
3,7-DIMETHYL-1-ADAMANTANEACETIC ACID
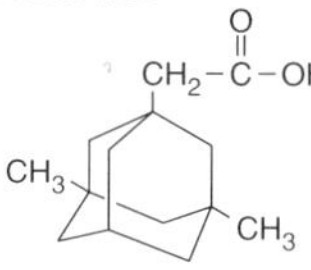
5 gm

S59,737-6
3,5,7-TRIMETHYL-1-ADAMANTANEACETIC ACID
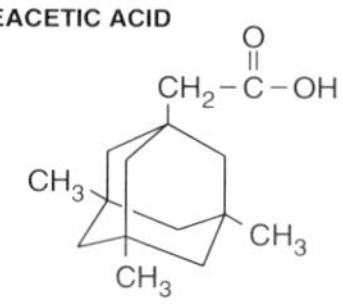
250 mg

S45,022-7
ALPHA-CHLORO-3,5,7-TRIMETHYL-1-ADAMANTANEACETIC ACID
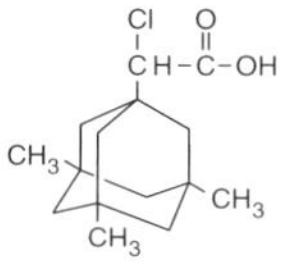
5 gm

S43,649-6
D-(+)-GLUCURONIC ACID
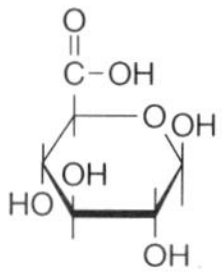
5 gm

S64,152-9
AMMONIUM GLYCYRRHIZINATE
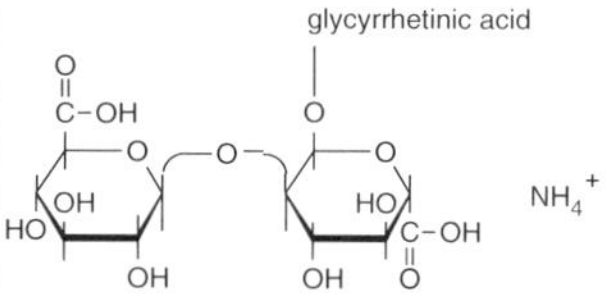

250 mg

S36,903-9
GLYCYRRHIZIC ACID, MONOSODIUM SALT
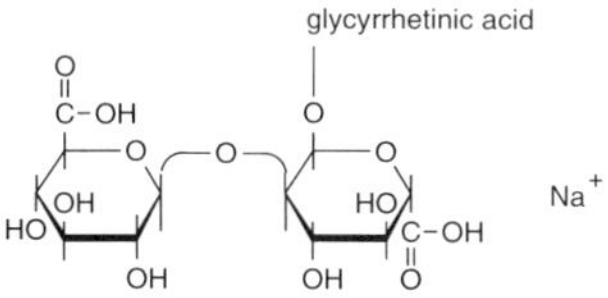

5 gm

S88,790-0
DELTA4,4'-BI-4H-PYRAN-2,2',6,6'-TETRACARBOXYLIC ACID
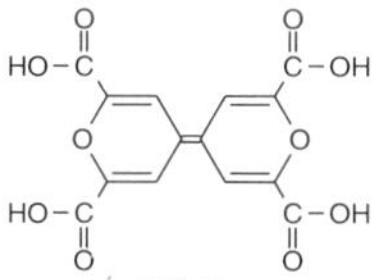
250 mg

S89,504-0
7-OXABICYCLO[2.2.1]HEPT-5-ENE-2,3-DICARBOXYLIC ACID
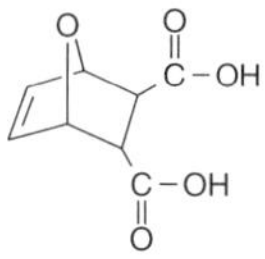
250 mg

S88,906-7
1,2,4-BUTANETRICARBOXYLIC ACID
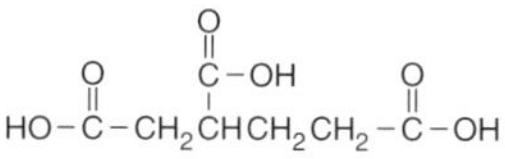
250 mg

S89,330-7
4-AMINO-1-METHYL-4-PIPERIDINECARBOXYLIC ACID
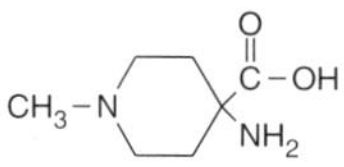
250 mg

S87,198-2
18-AMINOOCTADECANOIC ACID
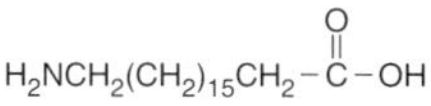
250 mg

S42,955-4
4-(AMINOMETHYL)-CYCLOHEXANECARBOXYLIC ACID
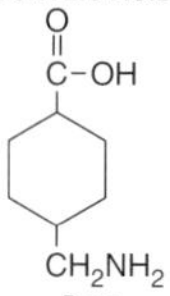
5 gm

S42,228-2
ETHYLENEDIAMINETETRAACETIC ACID, DIMORPHOLINIUM SALT TETRAHYDRATE
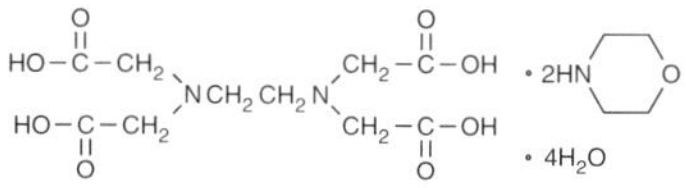
250 mg

S39,821-7
1,4-DIHYDRO-3,5-DIIODO-4-OXO-1-PYRIDINEACETIC ACID, DIETHYLAMINE SALT
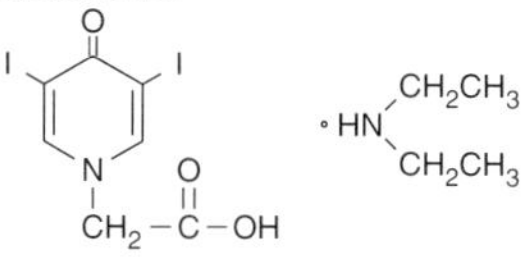
250 mg

S37,954-9
5-(DIMETHYLAMINO)VALERIC ACID HYDROCHLORIDE
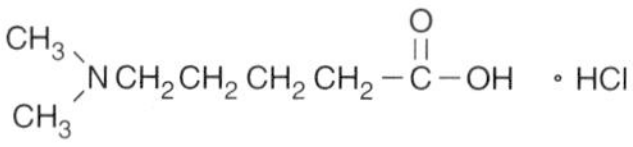
250 mg

S91,431-2
(CARBOXYMETHYL)-TRIMETHYLAMMONIUM CHLORIDE
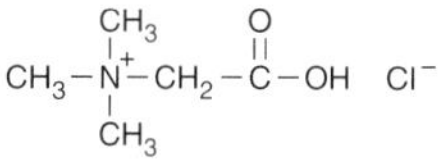
250 mg

S91,463-0
(5-CARBOXYPENTYL)-TRIMETHYLAMMONIUM CHLORIDE
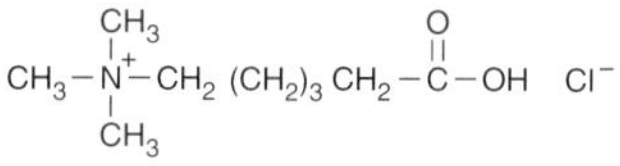
250 mg

S86,110-3
2-NITROPROPIONIC ACID
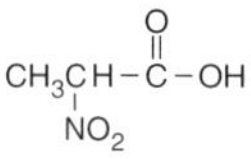
250 mg

S95,050-5
4A,5-EPOXY-3-OXO-14H-4,6A,9,9-4-METHYL-PHENANTHRO(2,3-D)(1,3)DIOXOLE-4-PROPIONIC ACID
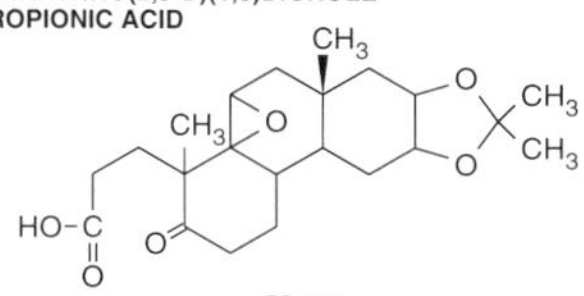
50 mg

S95,721-6
3-OXO-TETRAMETHYL-PERHYDROPHENANTHRO(2,3-D)(1,3)-DIOXOLE-4-PROPIONIC ACID
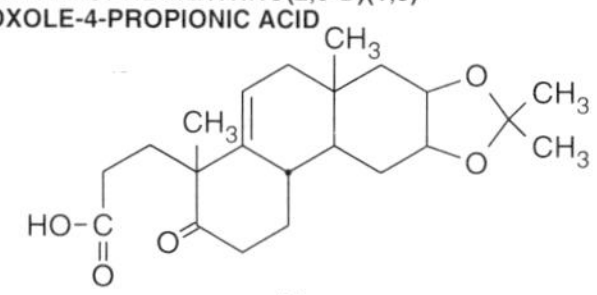
50 mg

S84,368-7
1,1,2-ETHANETRICARBOXYLIC ACID
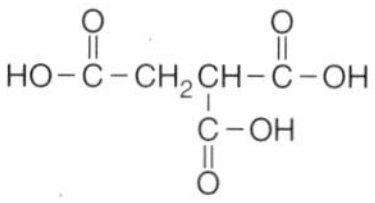
250 mg

S76,560-0
2,2,5,5-TETRAKIS(CARBOXYMETHYLTHIO)-1,4-DITHIANE

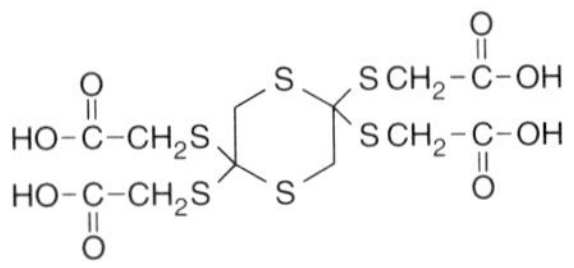

5 gm

S95,275-3
5,5-DIMETHYL-2,4-THIAZOLIDINEDICARBOXYLIC ACID

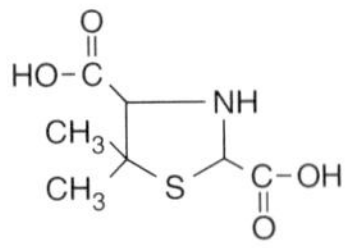

50 mg

S70,451-2
D-PENICILLAMINE ACETONE ADDUCT HYDROCHLORIDE HYDRATE

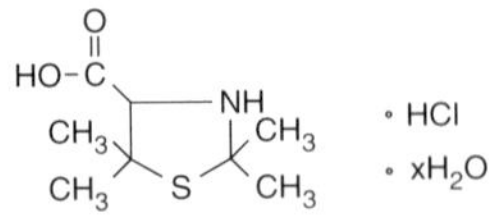

5 gm

S81,246-3
9,10-EPOXYOCTADECANOIC ACID

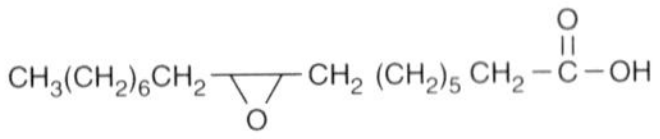

250 mg

S81,249-8
9,10-EPITHIOOCTADECANOIC ACID

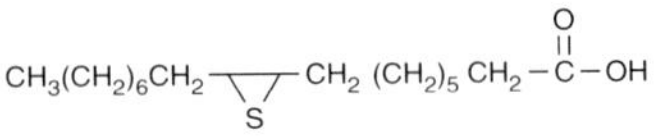

250 mg

S62,741-0
TETRAHYDRO-2-FUROIC ACID

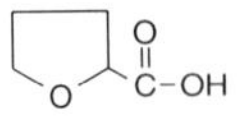

250 mg

S95,282-6
2-METHYL-1,3-DITHIOLANE-2-PROPIONIC ACID

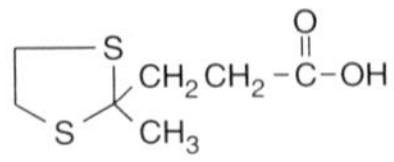

50 mg

S74,026-8
2-(2,4,5-TRIMETHYL-1,3-DITHIOLAN-2-YL)-4-PENTENOIC ACID

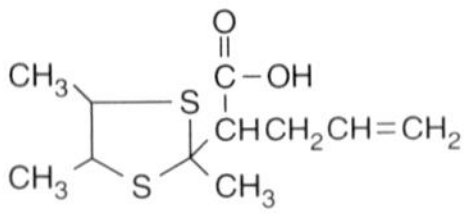

250 mg

S94,231-6
2,2-DIMETHYL-4-THIAZOLIDINECARBOXYLIC ACID

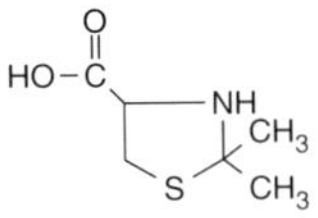

250 mg

S68,050-8
TETRAHYDROPYRAN-4-CARBOXYLIC ACID

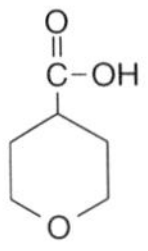

250 mg

S82,698-7
2-CHLOROACETIC ACID, DODECYLAMINE SALT

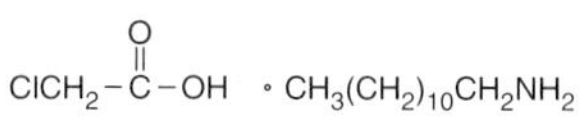

250 mg

S82,831-9
TRIFLUOROACETIC ACID, THALLIUM(I) SALT

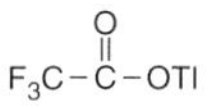

250 mg

S83,950-7
BARIUM TRIFLUOROACETATE

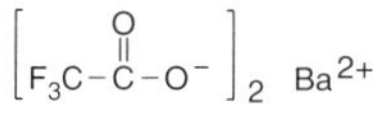

250 mg

S52,010-1
MALIC ACID, DIAMMONIUM SALT

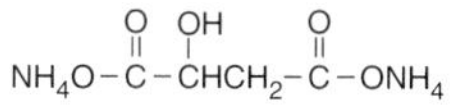

250 mg

S55,613-0
D-GLUCARIC ACID, CALCIUM SALT

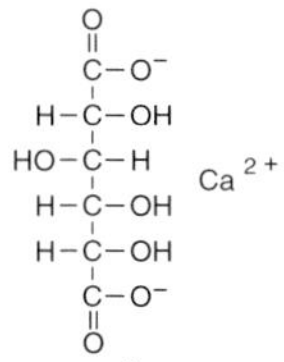

5 gm

S46,045-1
GLUCONIC ACID, STRONTIUM SALT

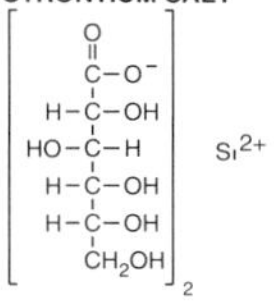

5 gm

S60,325-2
GLUCONIC ACID, MANGANESE(II) SALT

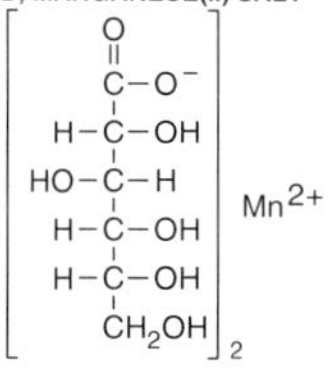

5 gm

S56,570-9
HEPTAGLUCONIC ACID, CALCIUM SALT

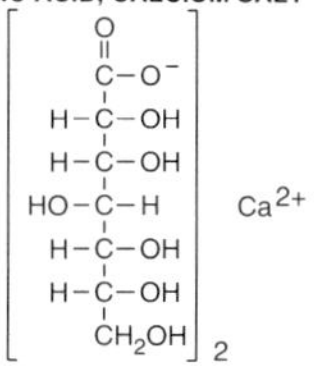

5 gm

S87,997-5
N,N-BIS(CARBOXYMETHYL)GLUCAMINE, DISODIUM SALT

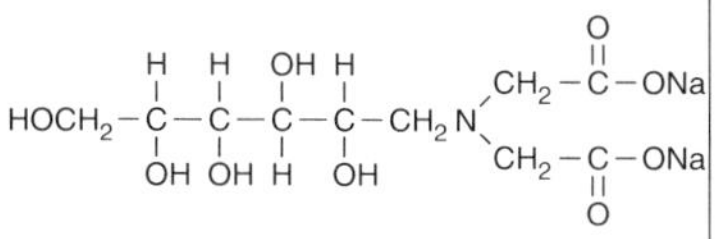

5 gm

S51,919-7
TARTARIC ACID, MONOPOTASSIUM SALT

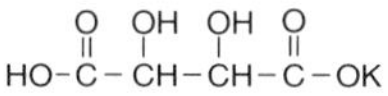

5 gm

S64,140-5
ZINC BORNEOL GLUCURONIC ACID (FROM URINE)

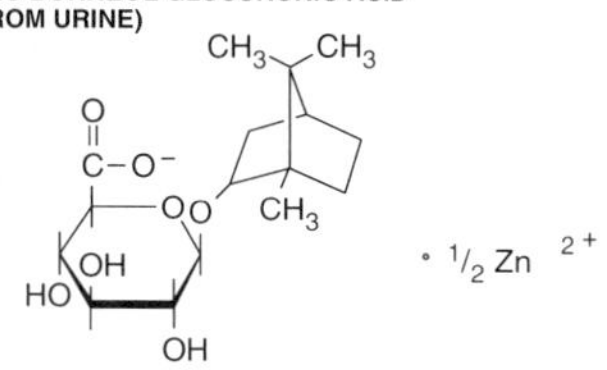

250 mg

S40,008-4
2-PROPYLCITRIC ACID, TRISODIUM SALT

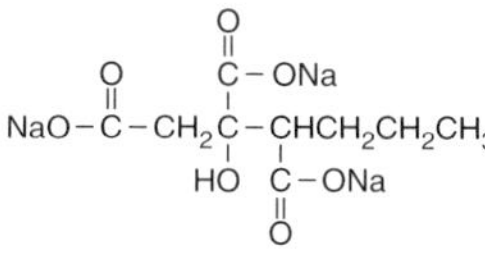

250 mg

S50,350-9
CITRIC ACID, TIN(II) SALT

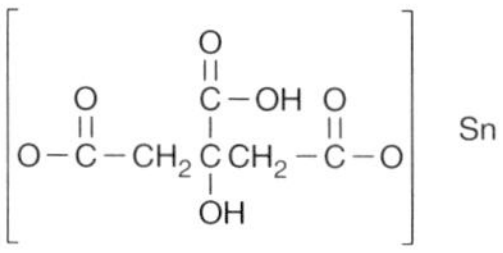

5 gm

S95,323-7
2-HYDROXY-2-(4-OXOTETRAHYDROPYRAN-3-YL)MALONIC ACID, DIPOTASSIUM SALT

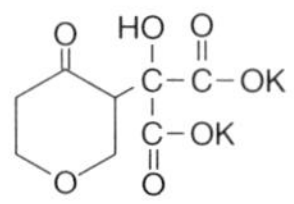

50 mg

S76,501-5
L-2-HYDROXYGLUTARIC ACID, DISODIUM SALT

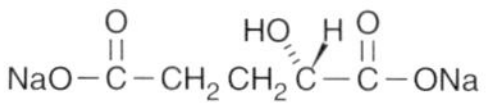

5 gm

S57,952-1
GLUCURONIC ACID, SODIUM SALT

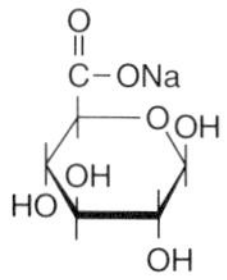

5 gm

S57,305-1
L-(+)-LACTIC ACID, CALCIUM SALT, HYDRATE

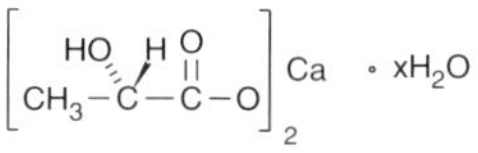

5 gm

S93,326-0
L-(+)-LACTIC ACID, LITHIUM SALT

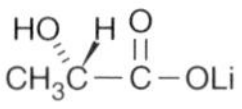

250 mg

S41,987-7
N-(2-HYDROXYETHYL)-ETHYLENEDIAMINETRIACETIC ACID, TRISODIUM SALT, TECH., 40%

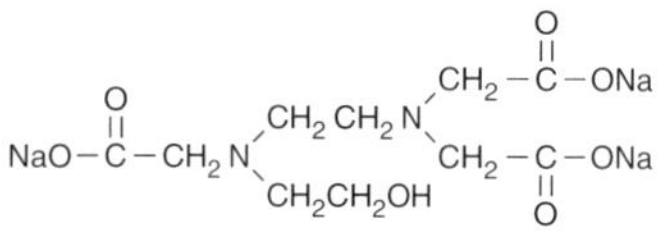

5 gm

S95,048-3
3-(3-BENZYLOXY-4-METHOXYPHENYL)-GLYCIDIC ACID, SODIUM SALT

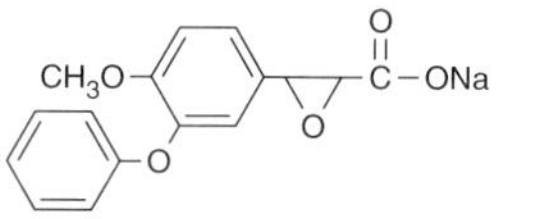

50 mg

S41,866-8
CHELIDONIC ACID, MONOSODIUM SALT, DIHYDRATE

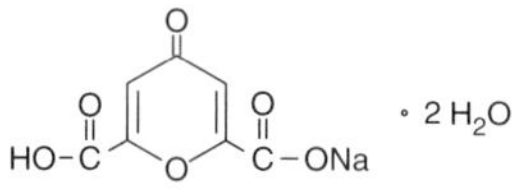

5 gm

S95,042-4
7-OXABICYCLO(4.1.0)HEPTANE-1-CARBOXYLIC ACID, SODIUM SALT

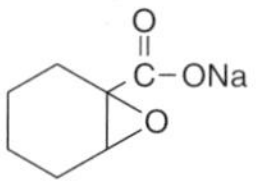

50 mg

S53,488-9
MERCAPTOACETIC ACID, CALCIUM SALT

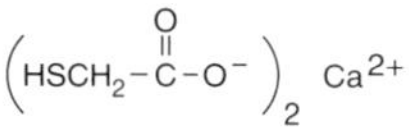

5 gm

S72,252-9
2-(TETRAHYDROPYRAN-2-YLTHIO)ACETIC ACID, POTASSIUM SALT

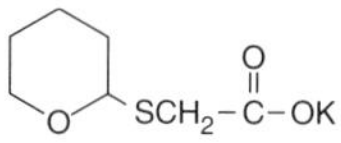

5 gm

S42,024-7
4-CARBOXY-3,3-DIMETHYLTHIAZOLIDINIUM HYDROXIDE, INNER SALT

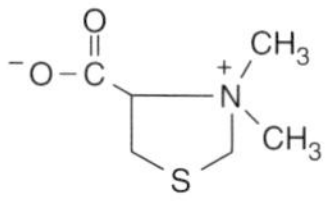

250 mg

S57,314-0
GLYOXYLIC ACID, BARIUM SALT

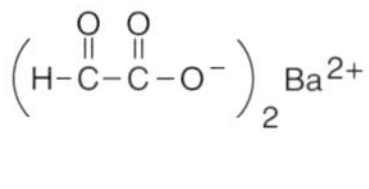

5 gm

S87,399-3
3-OXOTETRACOSANOIC ACID, SODIUM SALT

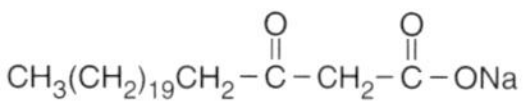

250 mg

S56,601-2
ADIPIC ACID, DISILVER SALT

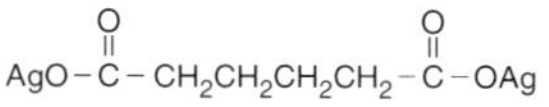

5 gm

S82,697-9
FORMIC ACID, OCTADECYLAMINE SALT

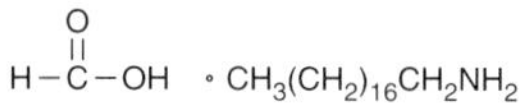

250 mg

S83,321-5
PROPIONIC ACID, OCTADECYLAMINE SALT

250 mg

S82,699-5
BUTYRIC ACID, OCTADECYLAMINE SALT

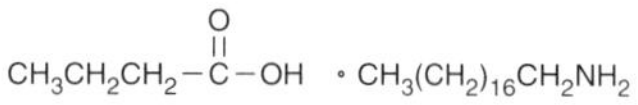

250 mg

S82,701-0
HEXANOIC ACID, OCTADECYLAMINE SALT

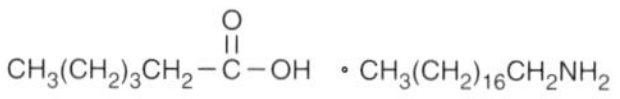

250 mg

S82,702-9
OCTANOIC ACID, OCTADECYLAMINE SALT

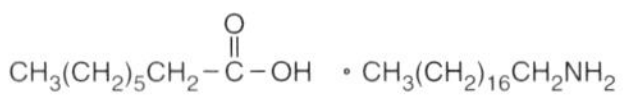

250 mg

S82,703-7
DECANOIC ACID, OCTADECYLAMINE SALT

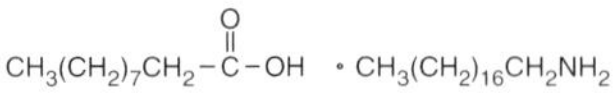

250 mg

S82,599-9
LAURIC ACID, OCTADECYLAMINE SALT

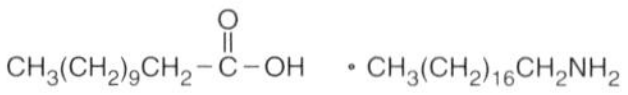

250 mg

S83,325-8
PALMITIC ACID, OCTADECYLAMINE SALT

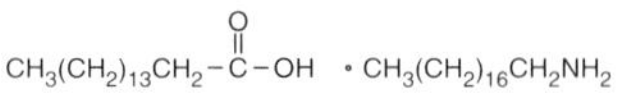

250 mg

S83,326-6
PALMITIC ACID, DODECYLAMINE SALT

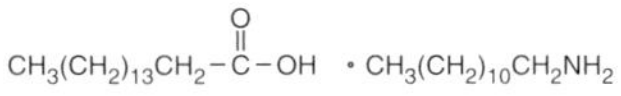

250 mg

S82,597-2
STEARIC ACID, OCTADECYLAMINE SALT

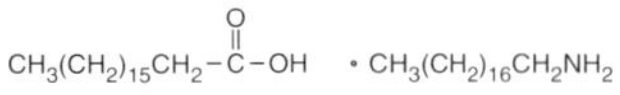

250 mg

S82,598-0
STEARIC ACID, DODECYLAMINE SALT

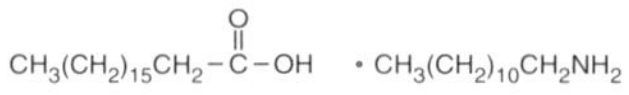

250 mg

S42,759-4
SUCCINIC ACID, DIAMMONIUM SALT

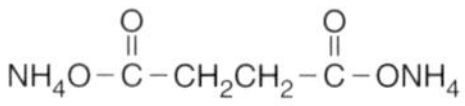

250 mg

S81,395-8
LITHIUM BUTYRATE

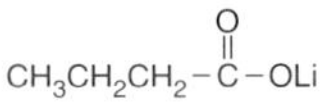

250 mg

S69,315-4
3-METHYLBUTYRIC ACID, STRONTIUM SALT

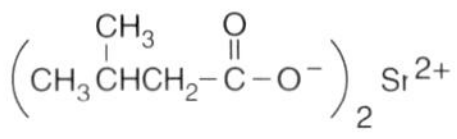

5 gm

S58,143-7
OLEIC ACID, CALCIUM SALT

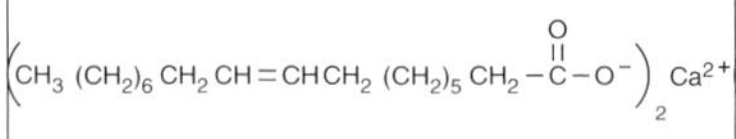

5 gm

S65,478-7
OLEIC ACID, COPPER(II) SALT

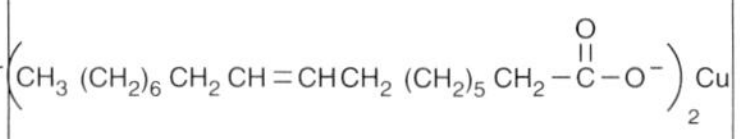

5 gm

S68,343-4
STEARIC ACID, DIHYDROXYALUMINUM SALT

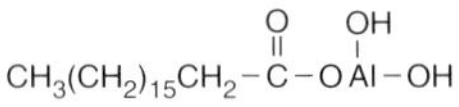

5 gm

S81,484-9
STEARIC ACID, SODIUM SALT

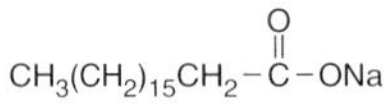

250 mg

S83,682-6
STEARIC ACID, CALCIUM SALT

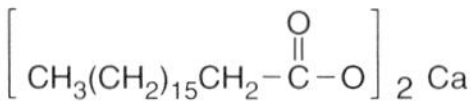

250 mg

S46,220-9
LITHIUM 4-CYCLOHEXANEBUTYRATE

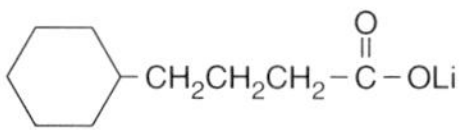

5 gm

S46,266-7
SODIUM 4-CYCLOHEXANEBUTYRATE

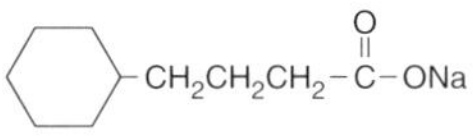

250 mg

S43,985-1
DIETHYLDITHIOCARBAMIC ACID, LEAD(II) SALT, MOIST

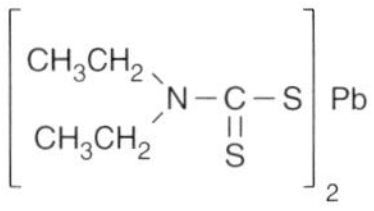

5 gm

S36,566-1
MAGNESIUM TRIFLUOROACETYLACETONATE MONOHYDRATE

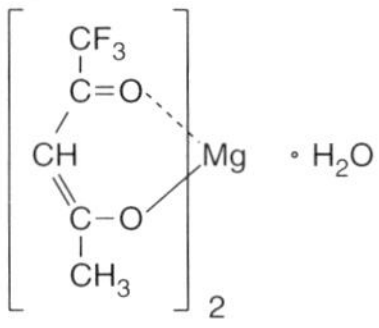

250 mg

S36,563-7
FERRIC TRIFLUOROACETYLACETONATE

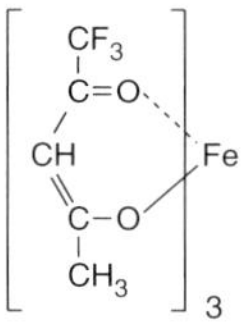

250 mg

S89,539-3
3-(1-CARBOXY-4-HYDROXYCYCLOHEXYL)-2-HYDROXYPROPIONIC ACID, BARIUM SALT

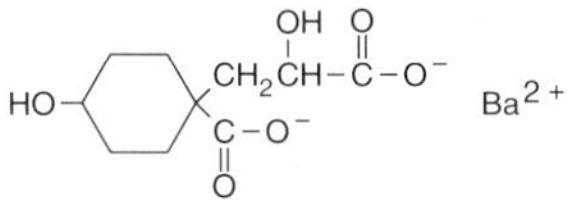

250 mg

S40,674-0
BERYLLIUM ACETYLACETONATE

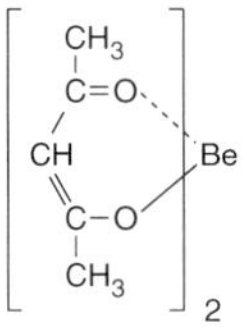

5 gm

S43,345-4
ETHYL ACETOACETATE, COPPER(II) DERIVATIVE

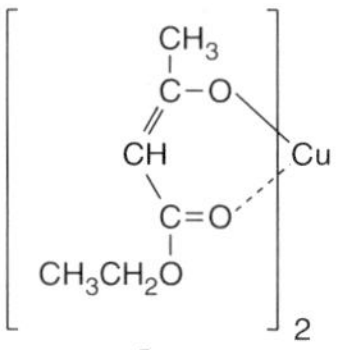

5 gm

S51,390-3
METHYL 3-OXOEICOSANOATE, COPPER(II) DERIVATIVE

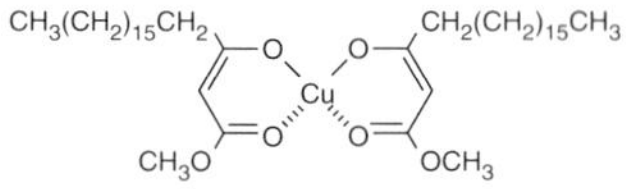

250 mg

S89,556-3
BERYLLIUM ACETATE

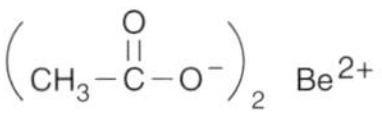

250 mg

S45,375-7
4,4,4-TRIFLUORO-DL-VALINE
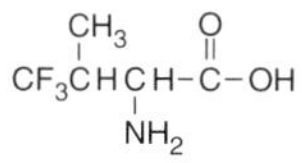
250 mg

S45,468-0
2-(METHYLAMINOMETHYL)TARTRONIC ACID
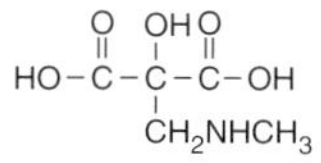
5 gm

S35,353-1
DL-2-METHYLSERINE
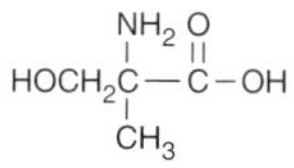
5 gm

S88,869-9
N-(2-HYDROXYETHYL)-BETA-ALANINE
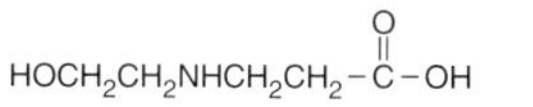
250 mg

S36,331-6
3-METHOXY-DL-VALINE
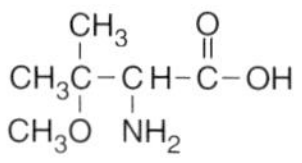
250 mg

S85,923-0
3-(AMINOOXY)PROPIONIC ACID HYDROCHLORIDE
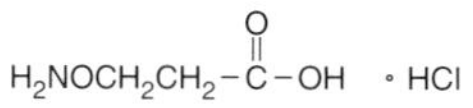
250 mg

S78,701-9
PENICILLAMINE HYDROCHLORIDE
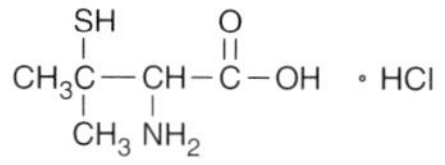
250 mg

S60,088-1
DL-METHIONINE METHYLSULFONIUM CHLORIDE
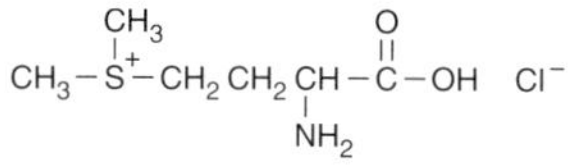
5 gm

S78,898-8
DL-CYSTATHIONINE
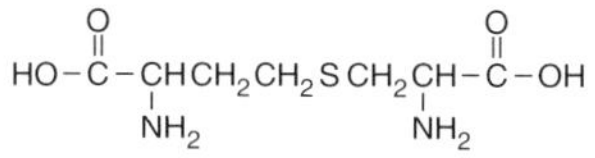
250 mg

S44,922-9
DJENKOLIC ACID DIHYDROCHLORIDE
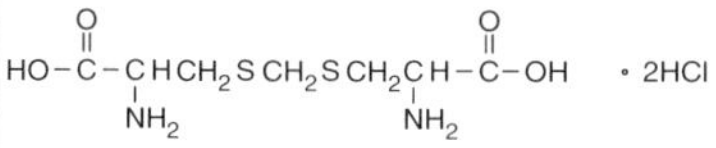
5 gm

S36,161-5
S-ETHYL-L-CYSTEINE
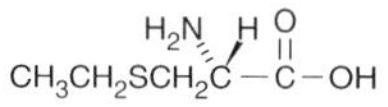
250 mg

S78,792-2
ALPHA-METHYL-DL-METHIONINE
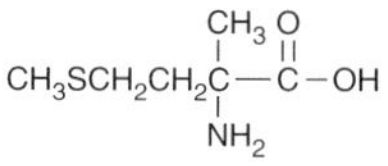
250 mg

S56,418-4
DL-HOMOCYSTINE
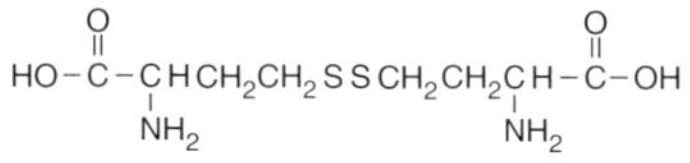
5 gm

S81,087-8
CYSTINE DIETHYL ESTER DIHYDROCHLORIDE
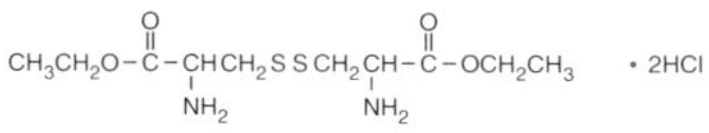
250 mg

S58,202-6
(3-AMINO-3-CARBOXYPROPYL)-DIMETHYLSULFONIUM BROMIDE
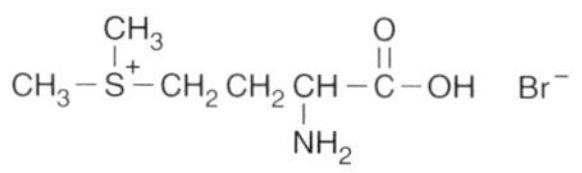
250 mg

S85,942-7
METHIONINE METHIODIDE
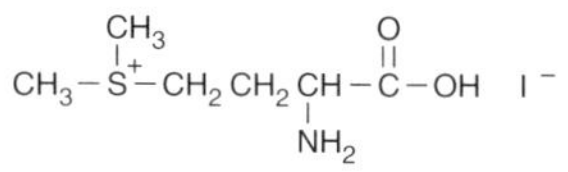
250 mg

S45,208-4
D(-)-GLUTAMINE
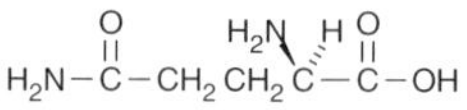
5 gm

S44,658-0
DL-ORNITHINE MONOHYDROCHLORIDE
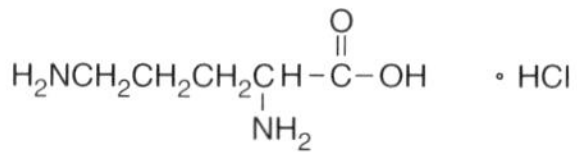
250 mg

S59,488-1
DL-ORNITHINE HYDROBROMIDE
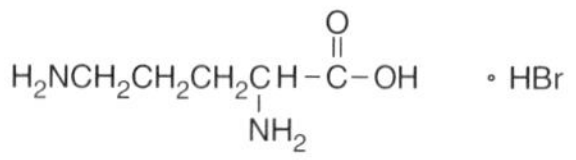
250 mg

S94,816-0
ALPHA,ALPHA'-DIAMINO-1,3-ADAMANTANEDIACETIC ACID
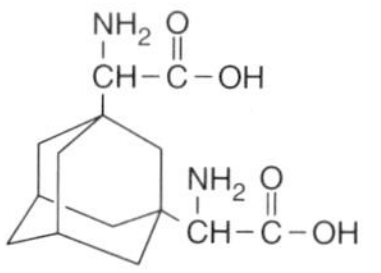
250 mg

S86,992-9
12-AMINO-7-OXODODECANOIC ACID
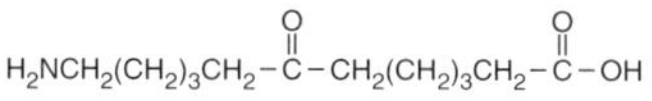
250 mg

S87,239-3
12-AMINO-7-OXODODECANOIC ACID HYDROCHLORIDE
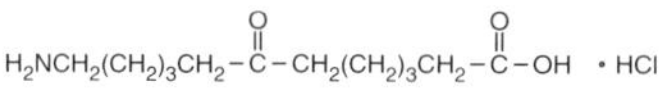
250 mg

S87,168-0
13-AMINO-7-OXOTRIDECANOIC ACID HYDROCHLORIDE
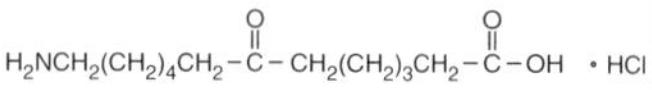
250 mg

S87,182-6
14-AMINO-7-OXOTETRADECANOIC ACID
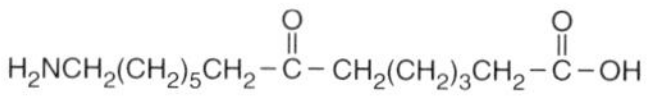
250 mg

S87,207-5
17-AMINO-7-OXOHEPTADECANOIC ACID
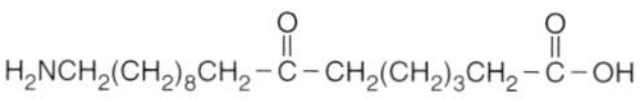
250 mg

S87,197-4
18-AMINO-7-OXOOCTADECANOIC ACID
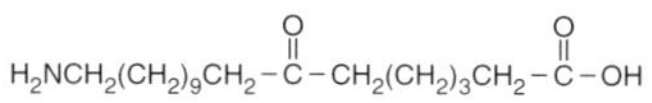
250 mg

S90,678-6
5-AMINOAZELAIC ACID
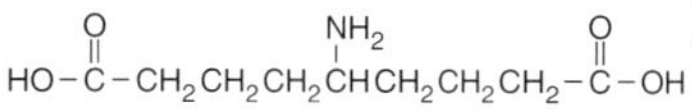
250 mg

S51,808-5
ASPARTIC ACID, DISODIUM SALT
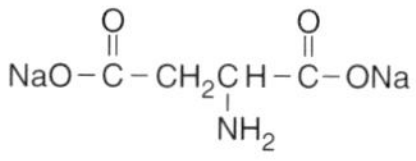
5 gm

S60,076-8
L-ASPARTIC ACID, POTASSIUM MAGNESIUM SALT
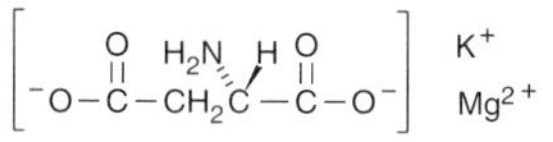
5 gm

S37,222-6
L-GLUTAMIC ACID, MONOPOTASSIUM SALT
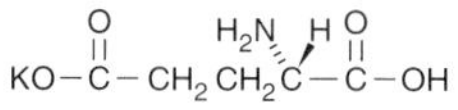
5 gm

S59,517-9
GLUTAMIC ACID, MAGNESIUM SALT DIHYDRATE
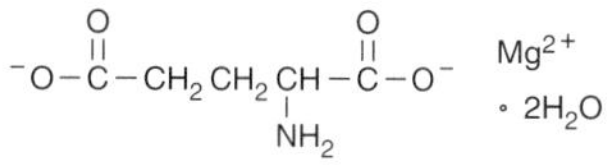
5 gm

S60,368-6
GLUTAMIC ACID, CALCIUM SALT TRIHYDRATE
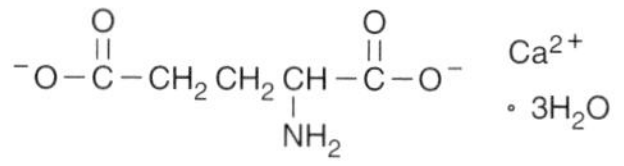
5 gm

S80,580-7
4-ETHYLGLUTAMIC ACID HYDROCHLORIDE

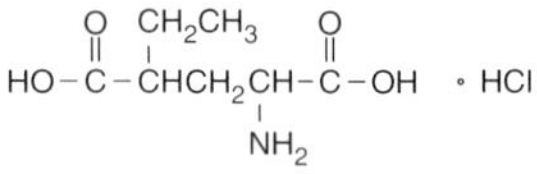

250 mg

S41,577-4
GLYCINE, CUPRIC SALT

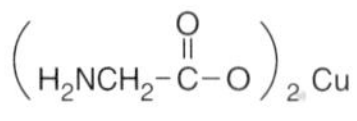

5 gm

S80,289-1
2-ALLYL-L-GLYCINE

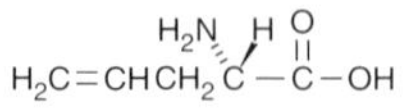

250 mg

S78,899-6
DL-2-AMINOHEPTANOIC ACID

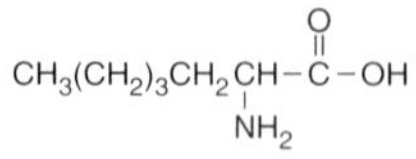

250 mg

S94,834-9
(+/-)-ALPHA-AMINO-1-ADAMANTANEPROPIONIC ACID

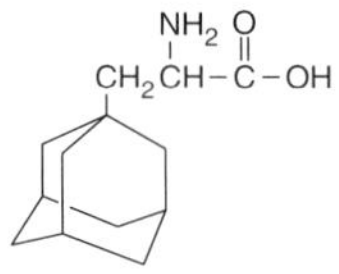

250 mg

S89,505-9
3-AMINO-2-NORBORNANECARBOXYLIC ACID

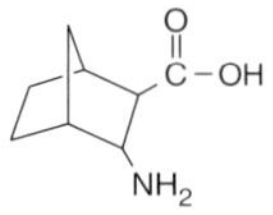

250 mg

S87,554-6
TRANS-2-AMINO-1-CYCLOHEXANECARBOXYLIC ACID

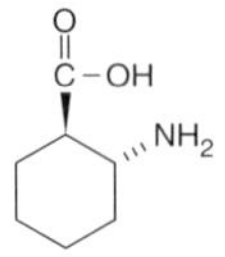

250 mg

S89,508-3
3-AMINOBICYCLO[2.2.2]OCT-5-ENE-2-CARBOXYLIC ACID

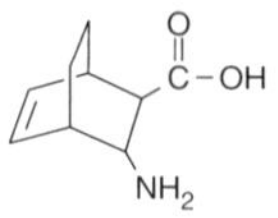

250 mg

S89,507-5
3-AMINOBICYCLO[2.2.2]OCTANE-2-CARBOXYLIC ACID

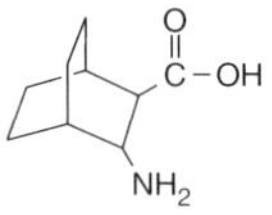

250 mg

S94,280-4
ALPHA-AMINO-3,5-DIMETHYL-1-ADAMANTANEACETIC ACID

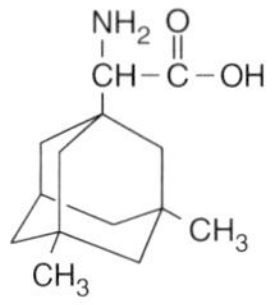

250 mg

S94,818-7
ALPHA-AMINO-3,5,7-TRIMETHYL-1-ADAMANTANEACETIC ACID

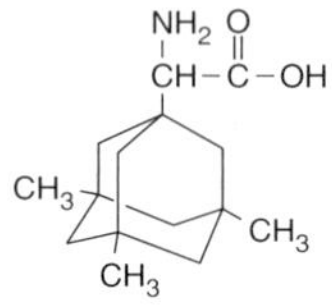

250 mg

S78,505-9
N-METHYL-DL-ASPARTIC ACID

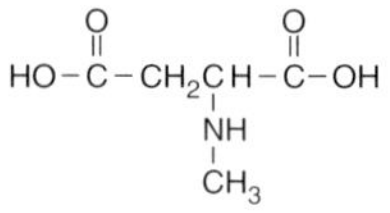

250 mg

S78,323-4
N-METHYL-L-ISOLEUCINE

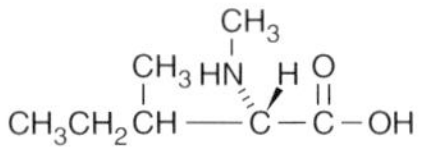

5 gm

S37,383-4
N-CYCLOHEXYL-BETA-ALANINE

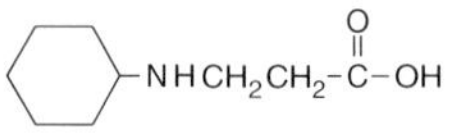

5 gm

S59,740-6
N-BUTYL-2-(3,5,7-TRIMETHYL-1-ADAMANTYL)GLYCINE

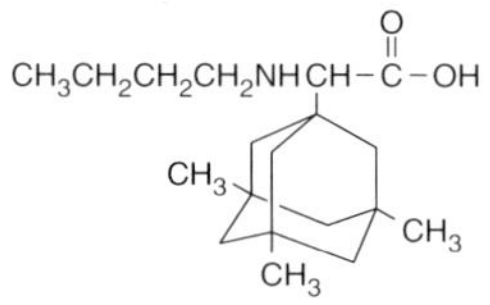

250 mg

S93,641-3
6-DIMETHYLAMINO-4-KETOHEXANOIC ACID HYDROCHLORIDE

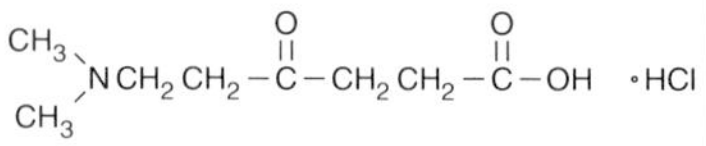

250 mg

S94,981-7
N,N-DIALLYLGLYCINE

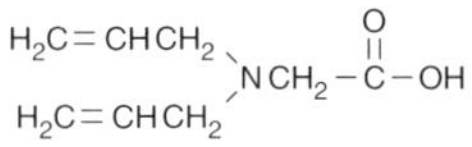

1 gm

S64,540-0
4-DIETHYLAMINOBUTYRIC ACID HYDROCHLORIDE

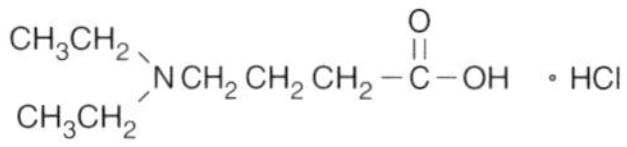

250 mg

S92,871-2
N,N-DIMETHYL-L-ALANINE

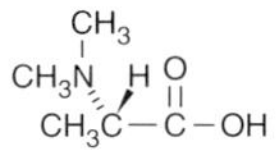

250 mg

S92,870=4
N,N-DIMETHYLVALINE

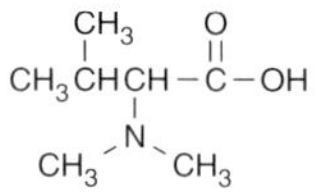

250 mg

S45,853-8
SEC-BUTYL ACETATE
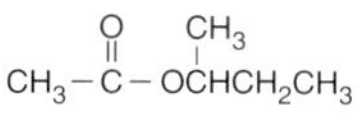
5 gm

S84,885-9
2-NITROBUTYL ACETATE
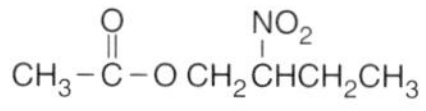
250 mg

S91,792-3
ETHYL 2-METHYL-2-NITROPROPIONATE
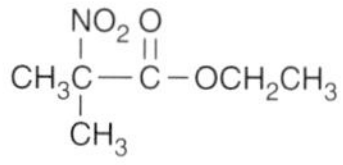
250 mg

S88,601-7
2-NITROPENTYL ACETATE
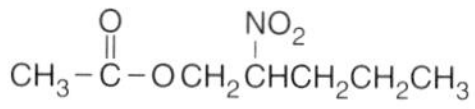
250 mg

S91,845-8
ETHYL 2-NITROVALERATE
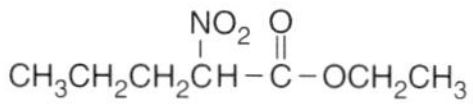
250 mg

S91,886-5
ETHYL 2-NITROHEXANOATE
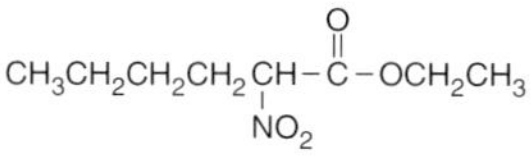
250 mg

S91,913-6
TERT-BUTYL 2-METHYL-2-NITROPROPIONATE
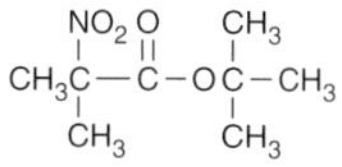
250 mg

S91,926-8
ETHYL 3-NITROPROPIONATE
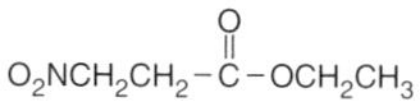
250 mg

S91,927-6
ETHYL 4-NITROBUTYRATE
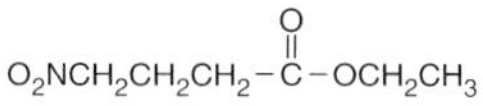
250 mg

S91,843-1
ETHYL 2,2-DIMETHYL-3-NITROPROPIONATE
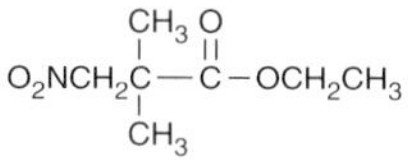
250 mg

S91,848-2
ETHYL 2,2-DIMETHYL-3-NITROBUTYRATE
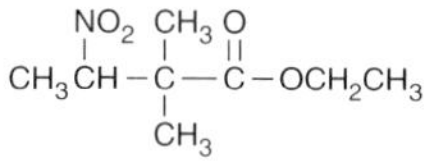
250 mg

S91,853-9
ETHYL 3-NITRO-2,2,3-TRIMETHYLBUTYRATE
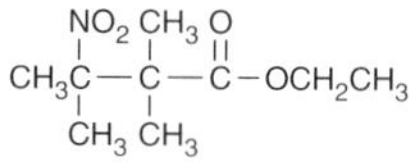
250 mg

S91,854-7
ETHYL 2,4-DIMETHYL-4-NITROVALERATE
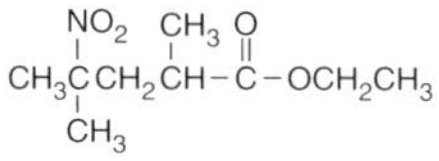
250 mg

S91,883-0
TERT-BUTYL 3-NITRO-2,2,3-TRIMETHYLBUTYRATE
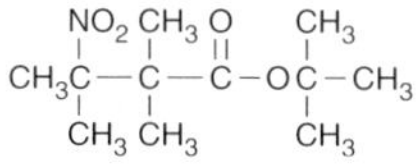
250 mg

S91,880-6
ETHYL 2,2-DIMETHYL-3-NITRODECANOATE
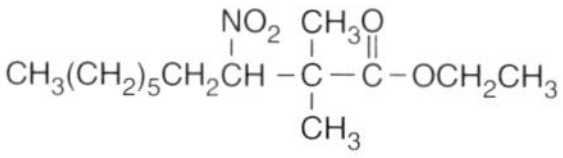
250 mg

S42,677-6
ISOAMYL FORMATE
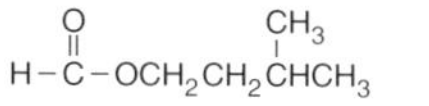
5 gm

S85,198-1
TERT-AMYL ACETATE
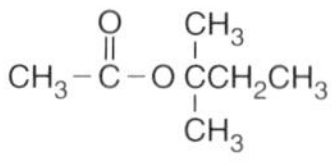
250 mg

S88,459-6
METHYL 3,3-DIMETHYLBUTYRATE
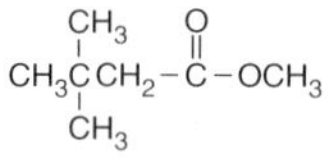
250 mg

S54,709-3
ETHYL 3,3-DIMETHYLBUTYRATE
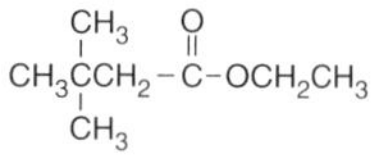
5 gm

S56,130-4
ISOBUTYL BUTYRATE
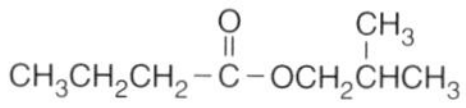
5 gm

S76,539-2
BUTYL ISOBUTYRATE
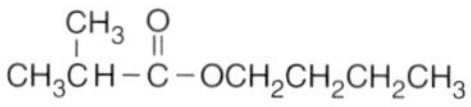
5 gm

S76,541-4
PROPYL ISOVALERATE
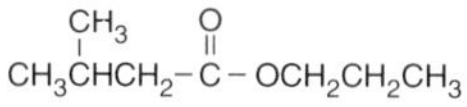
5 gm

S88,375-1
PENTYL PROPIONATE
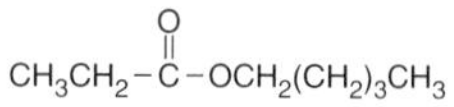
250 mg

S43,171-0
HEPTYL ACETATE
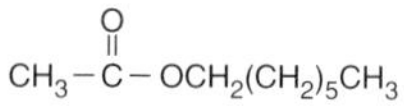
5 gm

S52,855-2
HEXYL ISOBUTYRATE
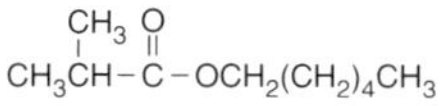
5 gm

S54,083-8
2-ETHYLHEXYL ACETATE
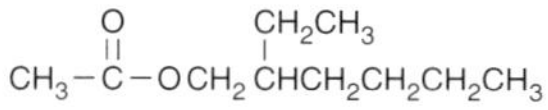
5 gm

S86,650-4
ETHYL 2,2-DIETHYLBUTYRATE
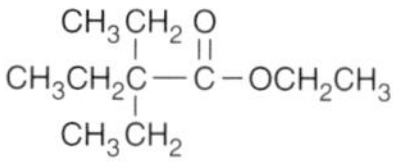
250 mg

S36,425-8
NONYL ACETATE
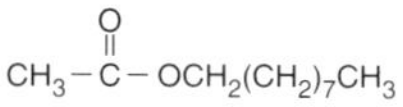
5 gm

S50,921-3
ISOPENTYL HEXANOATE
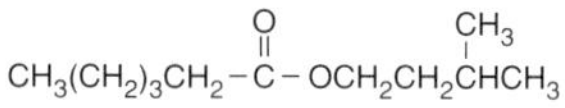
5 gm

S52,853-6
3,5,5-TRIMETHYLHEXYL ACETATE
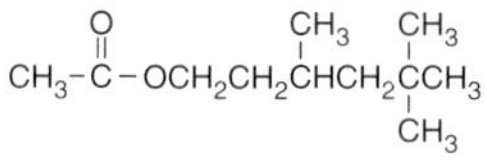
5 gm

S57,058-3
PENTYL HEXANOATE
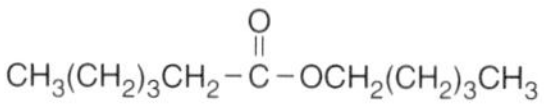
5 gm

S63,038-1
HEPTYL ISOVALERATE
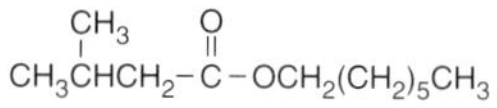
250 mg

S35,994-7
DECYL ACETATE
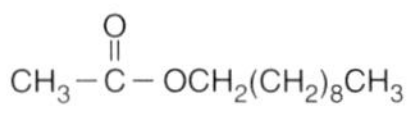
5 gm

S52,807-2
3,7-DIMETHYLOCTYL ACETATE
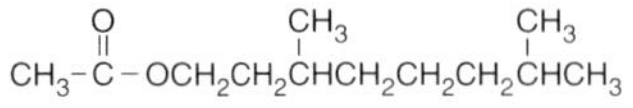
5 gm

S91,923-3
TERT-BUTYL 2,2,3,3-TETRAMETHYLBUTYRATE
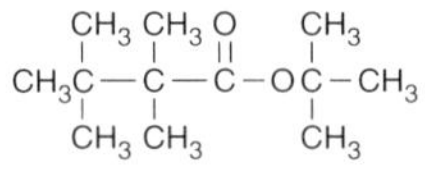
250 mg

S88,405-7
2-METHYL-2-NONEN-1-YL PROPIONATE
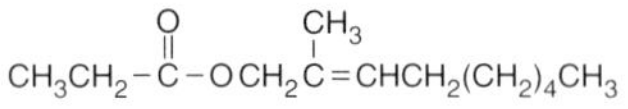
250 mg

S50,942-6
1-METHYLHEPTYL HEXANOATE
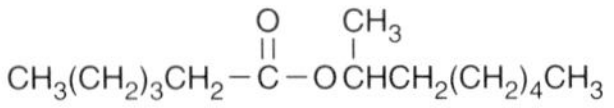
250 mg

S83,888-8
ETHYL LAURATE
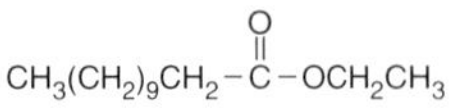
250 mg

S43,358-6
ETHYL HEPTADECANOATE
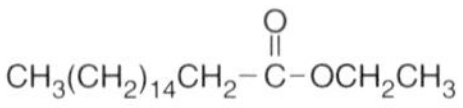
5 gm

S57,286-1
ETHYL NONADECANOATE
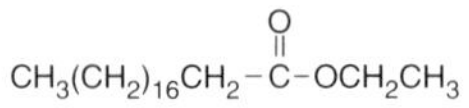
250 mg

S57,317-5
ISOPROPYL STEARATE
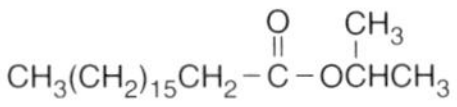
5 gm

S43,514-7
BUTYL STEARATE
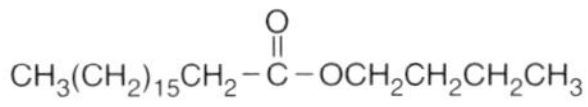
5 gm

S57,091-5
DECYL STEARATE
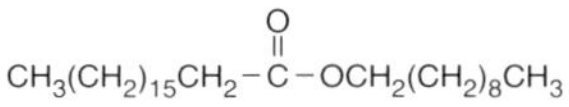
5 gm

S79,803-7
STEARYL STEARATE
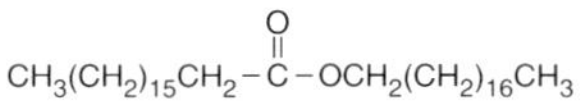
250 mg

S84,088-2
ETHYL 2-(OCTADECYL)OCTADECANOATE
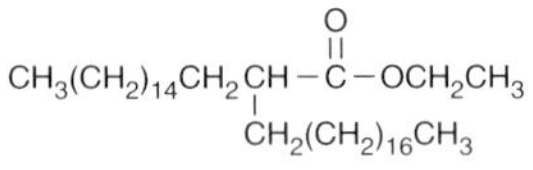
250 mg

S86,694-6
5,6,7,8,9,9-HEXACHLORO-1,2,3,4,4A,5,8,8A-OCTAHYDRO-1,4:5,8-DIMETHANONAPHTH-2-YL ACETATE
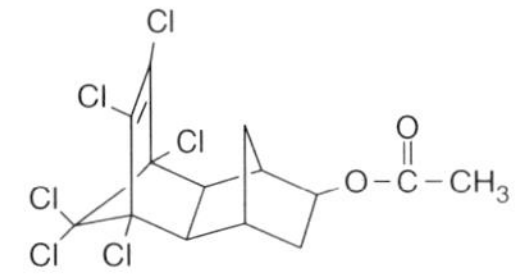
250 mg

S39,412-2
HEXYLXANTHOGENIC ACID, POTASSIUM SALT
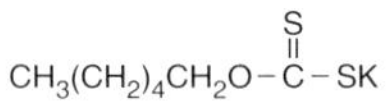
250 mg

S34,840-6
ETHYLENE GLYCOL BIS(METHYL CARBONATE)
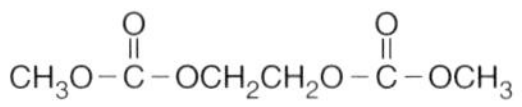
5 gm

S52,007-1
DIISOBUTYL CARBONATE
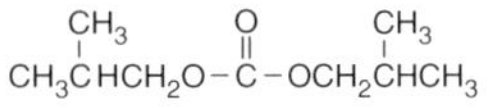
250 mg

S34,584-9
DIDODECYL CARBONATE
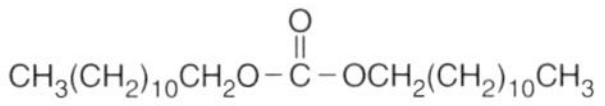
5 gm

S82,832-7
O-ETHYLXANTHIC ACID, THALLIUM SALT
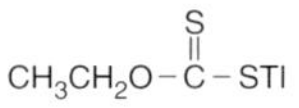
250 mg

S86,019-0
GLYCEROL 1,2-DISTEARATE
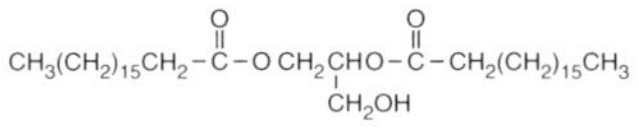
250 mg

S42,932-5
TRIBUTYL CITRATE
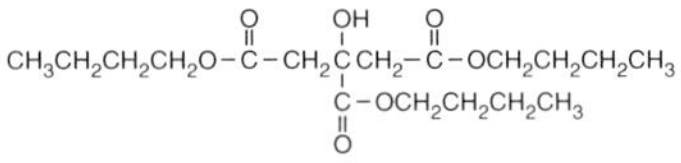
5 gm

S44,292-5
ETHYLENEBIS(2-OXYETHYL ACETATE)
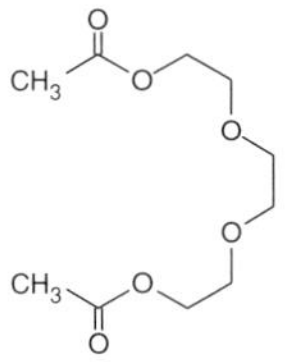
5 gm

S94,666-4
DIMETHYL 2-(DIMETHOXYMETHYL)-2-ETHYLGLUTARATE
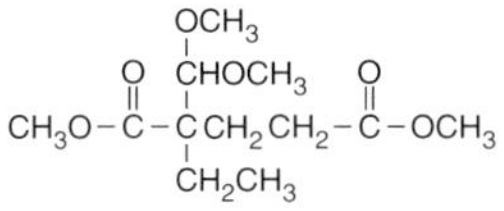
250 mg

S92,233-1
TRICAPRIN
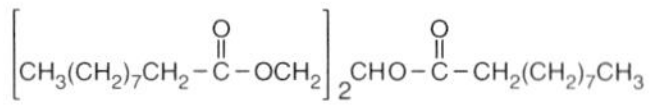
250 mg

S91,741-9
2-DEOXY-1-O-METHYL-3,5,6-TRI-O-ACETYL-ALPHA,BETA-D-GALACTOFURANOSIDE
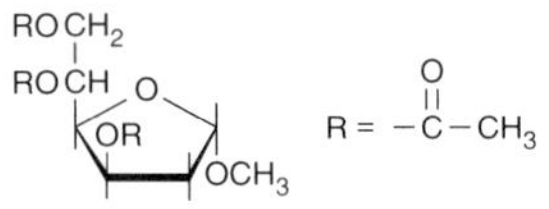
250 mg

S92,890-9
1-(2-ACETOXYETHOXY)ETHYL ACETATE
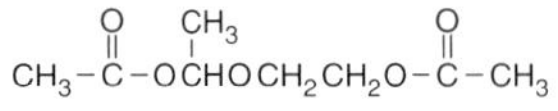
250 mg

S92,891-7
3-(ACETOXYMETHOXY)PROPYL ACETATE
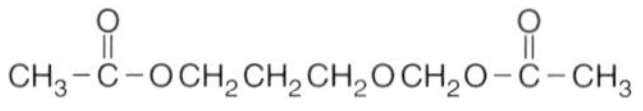
250 mg

S50,445-9
O,O-DIETHYL DITHIOBIS(THIOFORMATE)

5 gm

S93,923-4
1-O-METHYL-2,3,4,6-TETRA-O-ACETYL-ALPHA-D-GLUCOSIDE
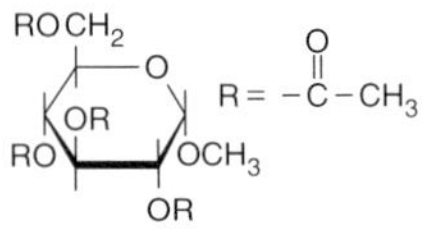
250 mg

S60,324-4
GLYCEROL TRILAURATE
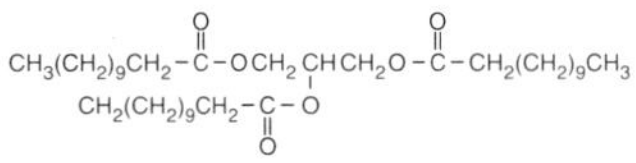
250 mg

S35,620-4
1,2,5-TRIACETOXYPENTANE
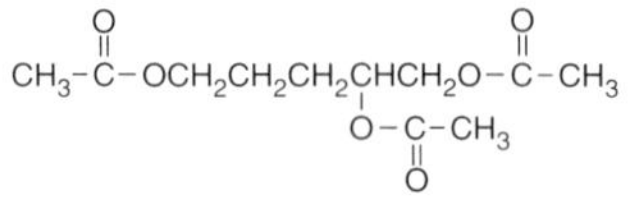
5 gm

S70,602-7
ETHYLENE GLYCOL DIPROPIONATE
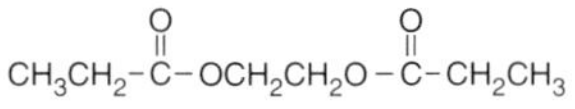
5 gm

S79,375-2
2-METHYL-2,3-BUTANEDIOL DIACETATE
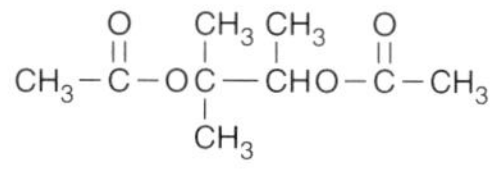
250 mg

S39,678-8
1-(BUTOXYCARBONYL)ETHYL BUTYRATE
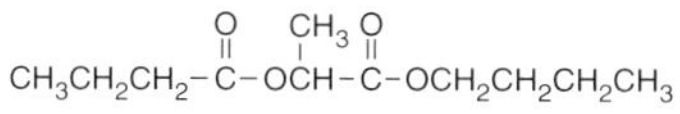
5 gm

S84,896-4
2-ISOPROPYL-2-NITRO-1,3-DIMETHYLENE BIS(BUTYRATE)
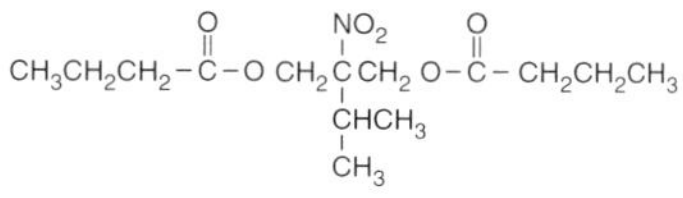
250 mg

S50,726-1
1,3-BUTANEDIOL DIACETATE
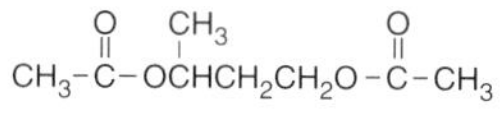
5 gm

S71,341-4
DIETHYL 2-BUTYL-2-METHYLMALONATE
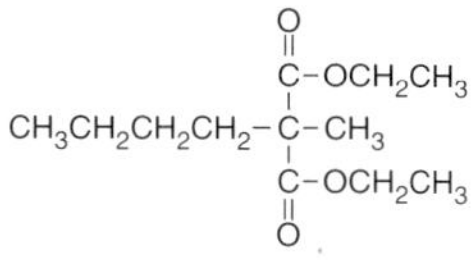
5 gm

S75,331-9
PENTAERYTHRITOL TETRAACETATE
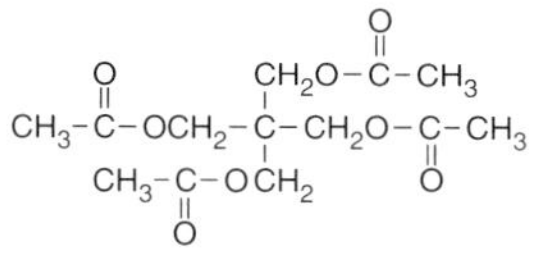
5 gm

S45,772-8
DIETHYL METHYLSUCCINATE
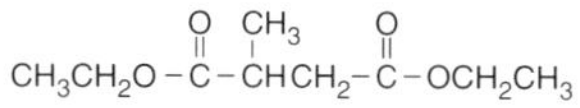
250 mg

S88,218-6
DIETHYL 2,3-DIETHYLSUCCINATE
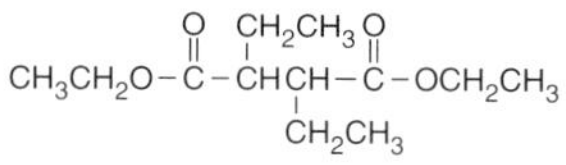
250 mg

S94,664-8
3-(METHOXYCARBONYL)-PROPIONIC ANHYDRIDE
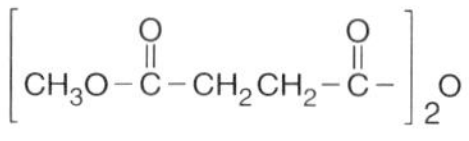
250 mg

S94,158-1
2,5-DIMETHYL-2,5-HEXANEDIOL DIACETATE
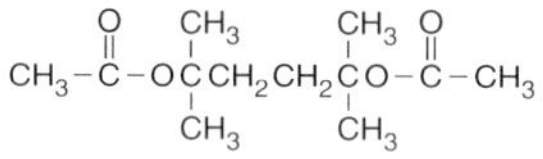
5 gm

S93,983-8
TETRAETHYL 1,2,2,3-PROPANETETRACARBOXYLATE
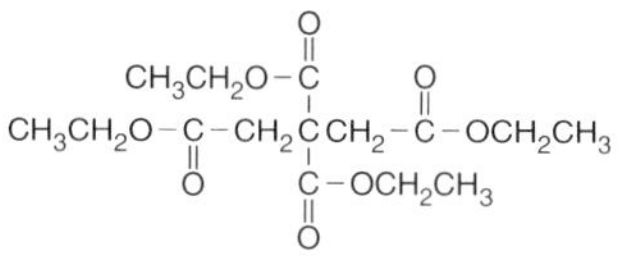
250 mg

S39,148-4
METHYL 9-(ACETYLTHIO)OCTADECANOATE
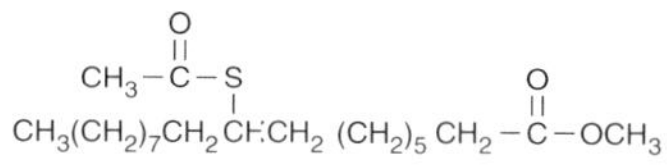
250 mg

S91,847-4
DIETHYL 2-METHYL-2-NITROMALONATE
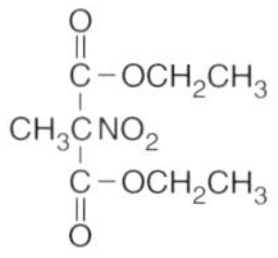
250 mg

S91,882-2
DIETHYL 4-NITROPIMELATE
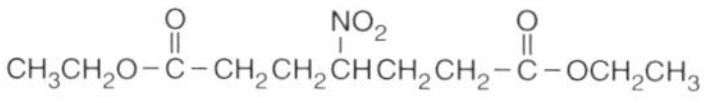
250 mg

S86,098-0
METHYL 5-OXOHEXANOATE
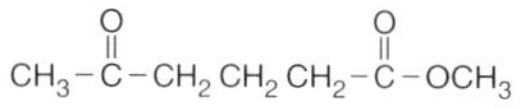
250 mg

S51,810-7
PROPYL OXALATE
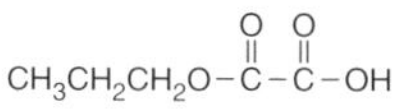
5 gm

S68,002-8
DIDODECYL OXALATE
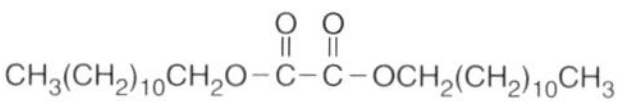
250 mg

S43,131-1
DIETHYL 2-[2-(METHOXYCARBONYL)-PROPYL]MALONATE
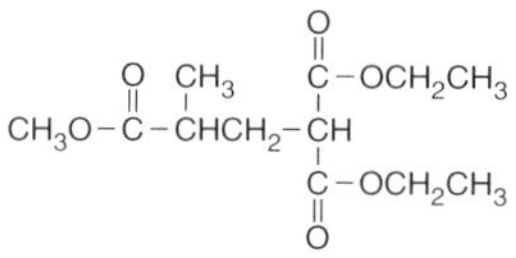
5 gm

S56,134-7
TRIETHYL 1,1,2-PROPANETRICARBOXYLATE
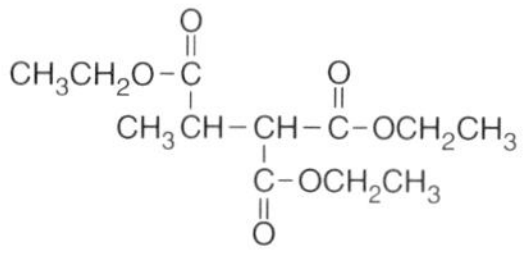
5 gm

S35,653-0
TRIETHYL 2,2,4-PENTANETRICARBOXYLATE
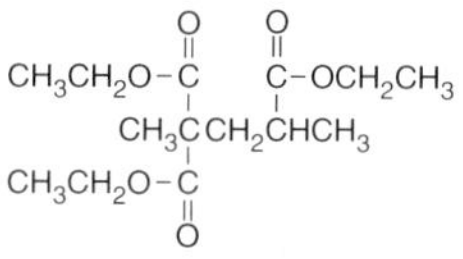
5 gm

S70,558-6
TETRAETHYL 1,1,2,3-PROPANETETRACARBOXYLATE
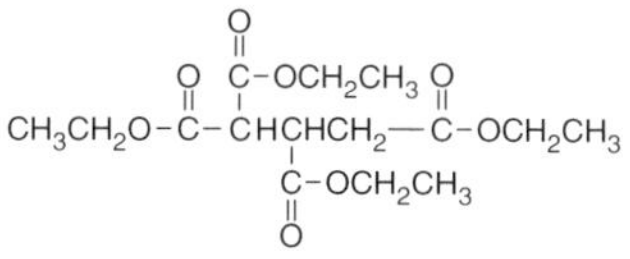
5 gm

S60,257-4
DIETHYL MALONATE, THALLIUM(I) SALT
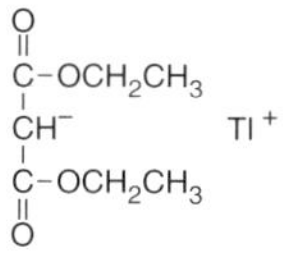
5 gm

S81,088-6
BUTYL ETHYL MALONATE
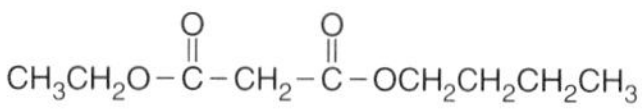
250 mg

S65,070-6
DIETHYL 2-(ETHOXYCARBONYL)MALONATE
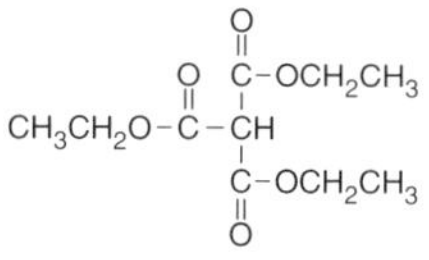
5 gm

S51,939-1
DIETHYL 2-METHYL-2-PROPYLMALONATE
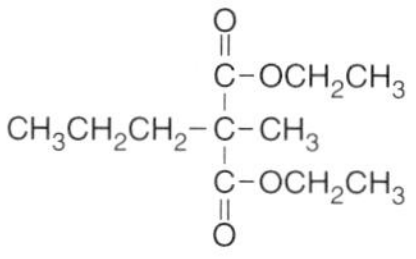
250 mg

S51,947-2
DIETHYL ISOBUTYLMALONATE
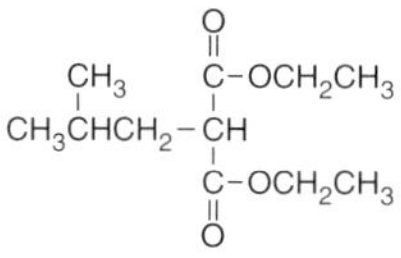
5 gm

S37,688-4
DIETHYL TERTPENTYLMALONATE
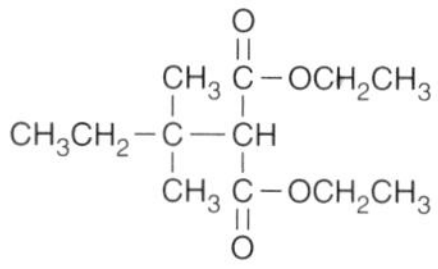
250 mg

S37,689-2
DIETHYL ISOPENTYLMALONATE
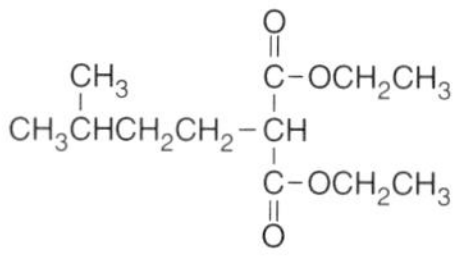
5 gm

S38,168-3
DIETHYL PENTYLMALONATE
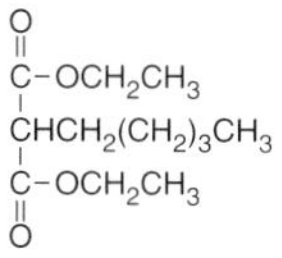
5 gm

S36,050-3
DIETHYL DIPROPYLMALONATE
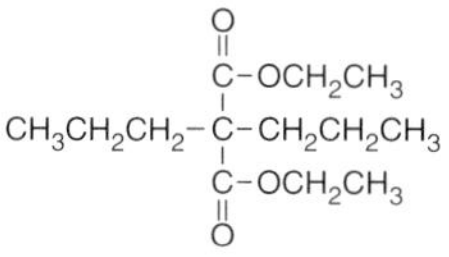
5 gm

S51,928-6
DIETHYL (2-METHYLPENTYL)MALONATE
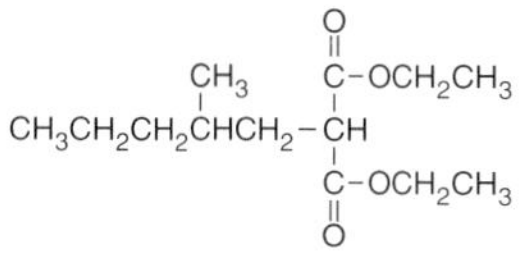
5 gm

S51,948-0
DIETHYL (2-ETHYLBUTYL)MALONATE
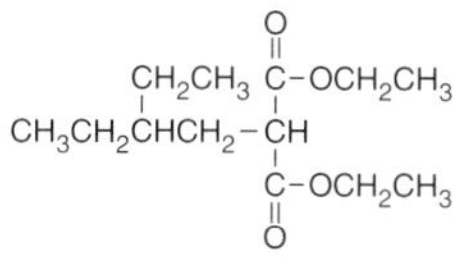
5 gm

S34,604-7
DIETHYL 2-ETHYL-2-ISOAMYLMALONATE

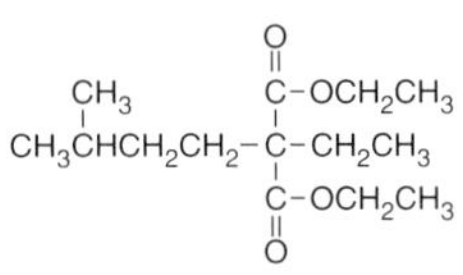

5 gm

S70,592-6
DIETHYL HEPTYLMALONATE

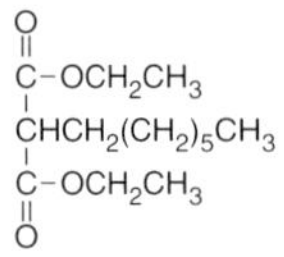

5 gm

S38,350-3
DIETHYL DIBUTYLMALONATE

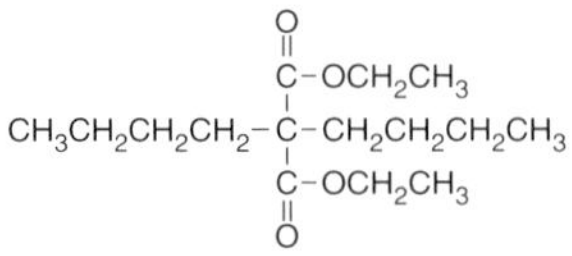

5 gm

S38,572-7
DIBUTYL DIETHYLMALONATE

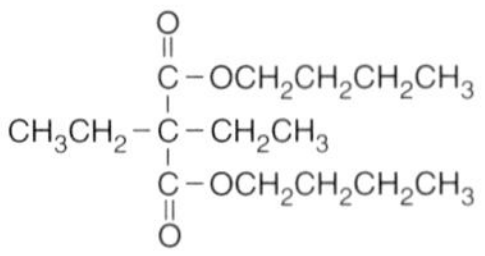

5 gm

S51,885-9
DIETHYL DIISOPENTYLMALONATE

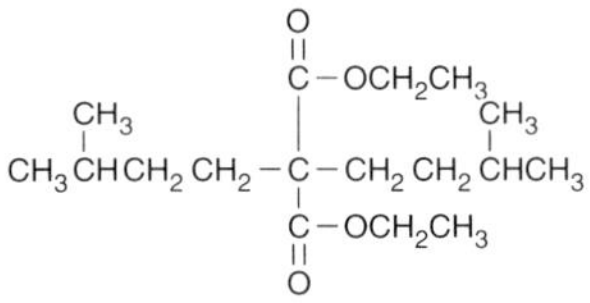

5 gm

S70,563-2
DIHEPTYL MALONATE

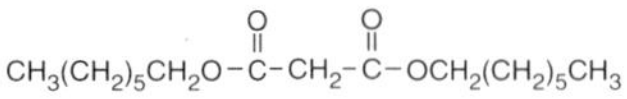

5 gm

S61,768-7
DIETHYL 2-TETRADECYLMALONATE

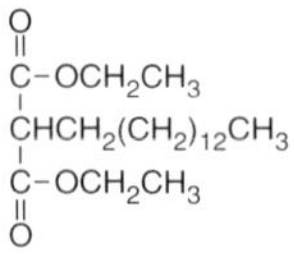

250 mg

S65,603-8
DIETHYL 2-HEXADECYLMALONATE

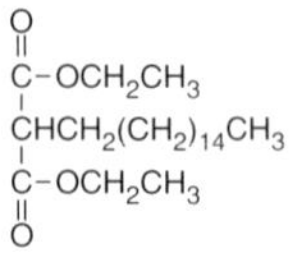

250 mg

S52,144-2
DIETHYL 2-OCTADECYLMALONATE

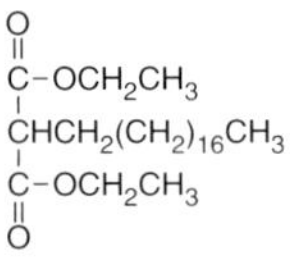

5 gm

S50,950-7
DIETHYL DIDODECYLMALONATE

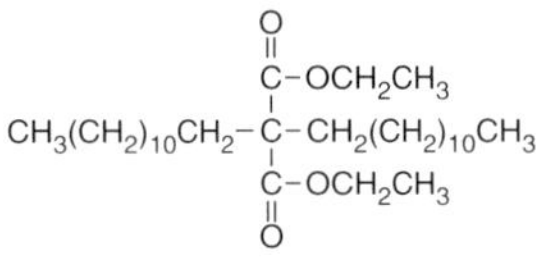

250 mg

S41,200-7
DIMETHYL 2-METHYLADIPATE

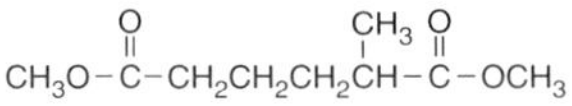

5 gm

S41,700-9
DIMETHYL 2,5-DIMETHYLADIPATE

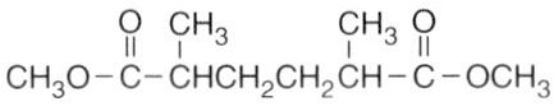

5 gm

S40,765-8
DIETHYL 2,4-DIMETHYLGLUTARATE (MIXTURE OF ISOMERS)

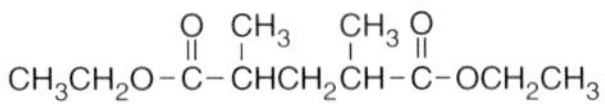

5 gm

S56,388-9
DIISOPROPYL ADIPATE

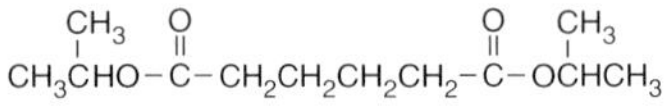

5 gm

S88,255-0
DIMETHYL 2,2,5,5-TETRAMETHYLADIPATE

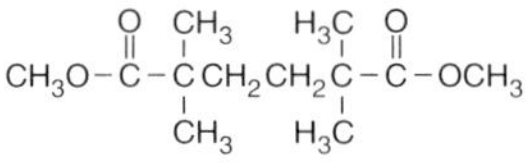

250 mg

S57,135-0
DIMETHYL UNDECANEDIOATE

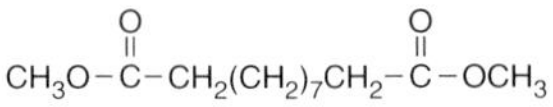

5 gm

S51,978-2
DIPENTYL SUCCINATE

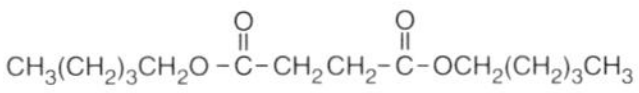

5 gm

S36,523-8
TRIETHYL 1,3,4-PENTANETRICARBOXYLATE

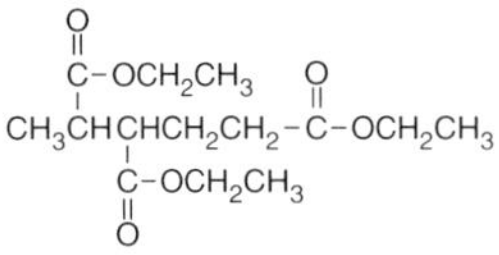

250 mg

S43,565-1
TRIETHYL 1,3,5-PENTANETRICARBOXYLATE

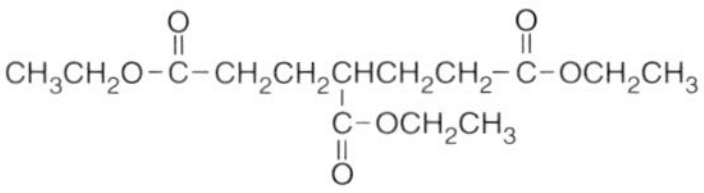

5 gm

S50,945-0
DIPENTYL ADIPATE

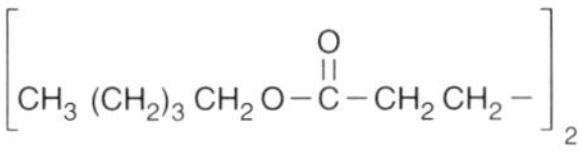

5 gm

S67,728-0
DIETHYL TRIDECANEDIOATE

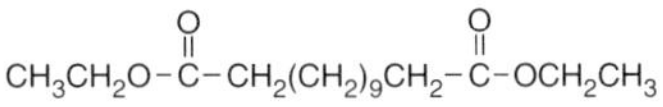

5 gm

S42,783-7
DIBUTYL SEBACATE

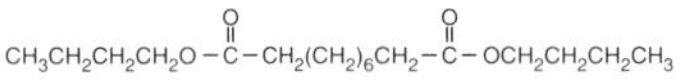

5 gm

S70,598-5
DIBUTYL 3-(BUTOXYCARBONYL)ADIPATE

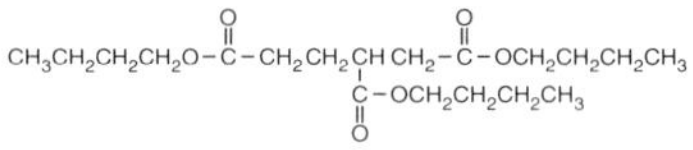

250 mg

S70,706-6
DIETHYL OCTADECANEDIOATE

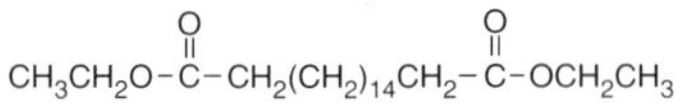

5 gm

S91,480-0
DIMETHYL DOCOSANEDIOATE

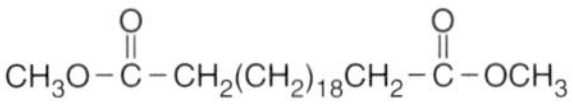

250 mg

S56,580-6
DIOCTYL AZELATE

5 gm

S39,780-6
TRIS(2-ETHYLHEXYL)-1,3,5-PENTANETRICARBOXYLATE

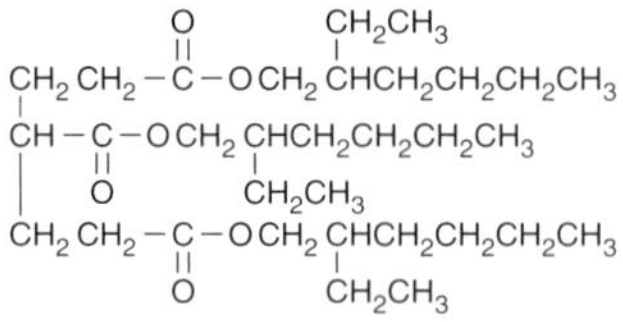

5 gm

S39,242-1
DIDOCOSYL SUCCINATE

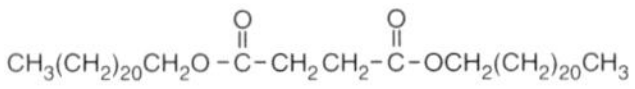

5 gm

S57,966-1
DI-TERT-BUTYL GLUTARATE

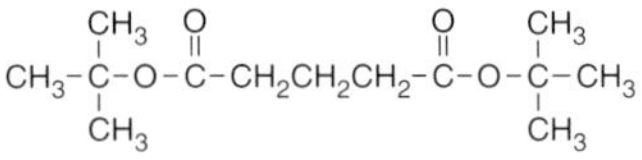

250 mg

S41,175-2
2,2-BIS(ACETOXYMETHYL)-PROPYL ACETATE

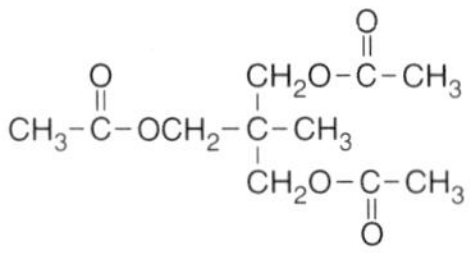

5 gm

S37,531-4
2,5-DIMETHYL-1,6-HEXANEDIOL DIACETATE

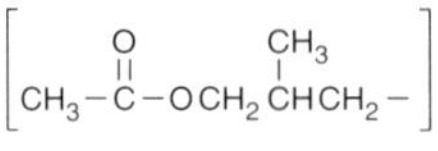

5 gm

S66,493-6
O,O-DIETHYL THIOMALONATE

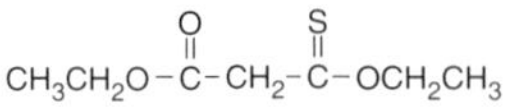

5 gm

S86,697-0
7,7-DIMETHYL-2-(2-METHOXYETHYL)-BICYCLO[3.1.1]HEPTAN-3-YL ACETATE

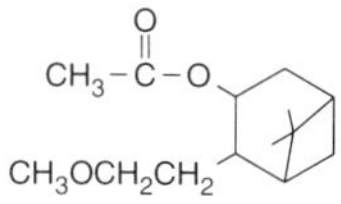

250 mg

S94,722-9
2-CYCLOHEXYL CYCLOHEXYLACETATE

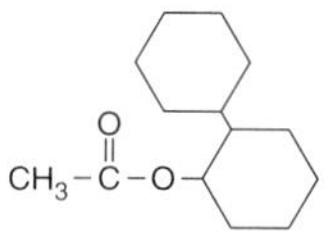

5 gm

S82,607-3
PROPYL TETRAHYDROFURFURYL OXALATE

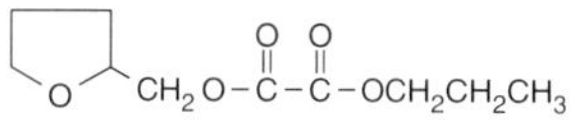

250 mg

S88,205-4
DIMETHYL CIS-1,2-CYCLOHEXANEDICARBOXYLATE

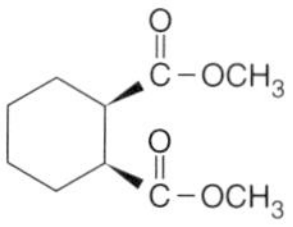

250 mg

S95,051-3
CIS-DIMETHYL 1,2-CYCLOBUTANEDICARBOXYLATE

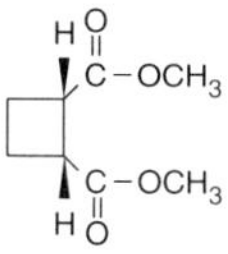

50 mg

S88,442-1
1,4-PERHYDROPENTALENEDIOL DIACETATE

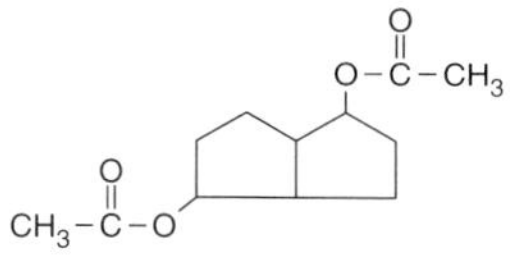

250 mg

S86,024-7
TRANS-1,4-CYCLOHEXANEDIOL DIACETATE

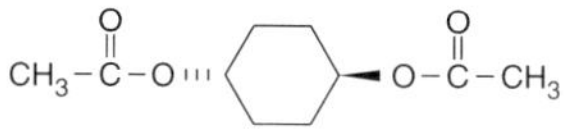

250 mg

S94,992-2
DIMETHYL 1,4,4A,5,8,8A-HEXAHYDRO-4A,8A-NAPHTHALENEDICARBOXYLATE

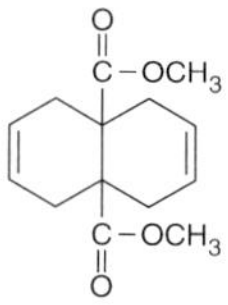

1 gm

S91,852-0
ETHYL 1-NITRO-1-CYCLOHEXANECARBOXYLATE

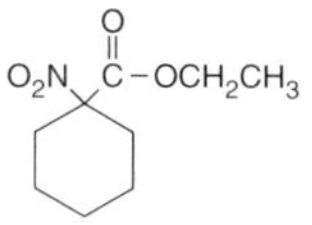

250 mg

S91,862-8
ETHYL 2-METHYL-2-(1-NITROCYCLOHEXYL)PROPIONATE

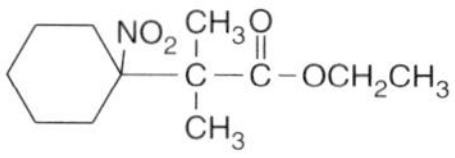

250 mg

S51,920-0
DICYCLOHEXYL OXALATE

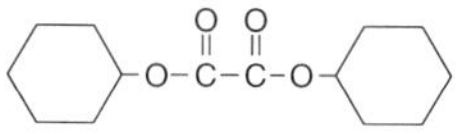

250 mg

S43,617-8
DIETHYL 1,1-CYCLOBUTANEDICARBOXYLATE

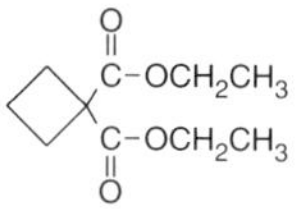

5 gm

S71,324-4
DIETHYL 1,1-CYCLOHEXANEDIACETATE

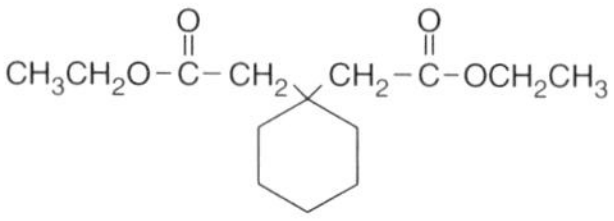

5 gm

S70,618-3
DIETHYL HEXAHYDROPHTHALATE

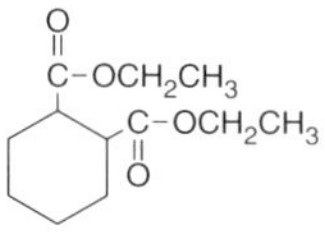

5 gm

S35,982-3
CYCLOHEPTYL FORMATE

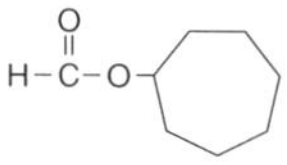

5 gm

S34,506-7
CYCLOOCTYL ACETATE

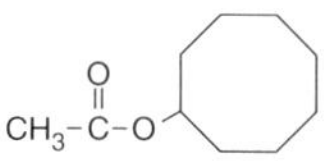

5 gm

S43,446-9
CIS-DIMETHYL PINATE

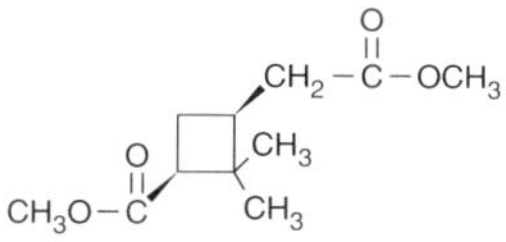

5 gm

S90,605-0
DECAHYDROCYCLOBUTA(1,2:3,4)-DICYCLOPENTENE-1,4-DIOL DIACETATE

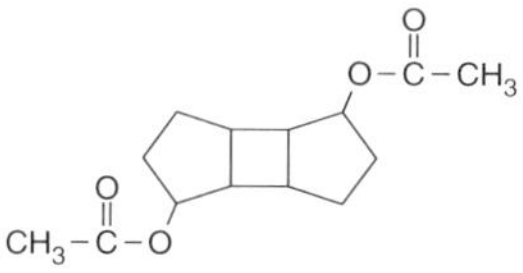

250 mg

S70,574-8
PINAN-10-YLMETHYL 2,2-DIMETHYL-3-(VINYLOXYCARBONYL)-1-CYCLOBUTANEACETATE

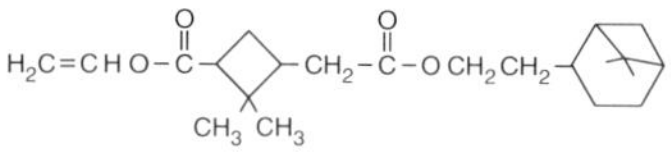

250 mg

S56,180-0
DIMETHYL 1,2,2-TRIMETHYL-1,3-CYCLOPENTANEDICARBOXYLATE

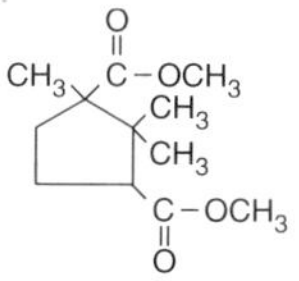

5 gm

S59,712-0
DIMETHYL 1,3-ADAMANTANEDIACETATE

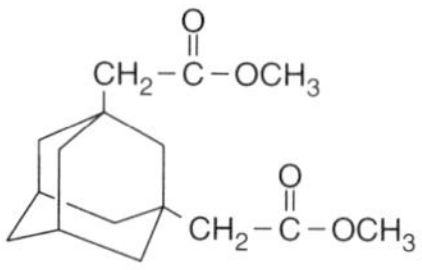

250 mg

S95,521-3
2,6-DIACETOXYBICYCLO[3.3.0]OCTANE

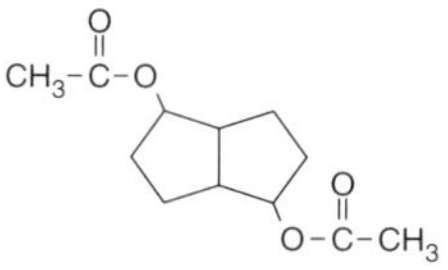

50 mg

S85,195-7
METHYL CYCLOBUTANECARBOXYLATE

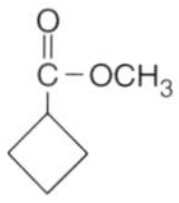

250 mg

S36,350-2
METHYL CYCLOPENTANECARBOXYLATE

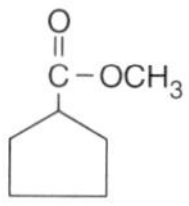

5 gm

S43,310-1
ETHYL CYCLOPENTYLACETATE

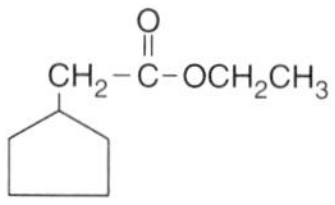

5 gm

S86,646-6
METHYL 2-CYCLOPENTYL-2-METHYLPROPIONATE

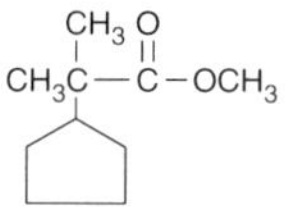

250 mg

S35,154-7
DL-ISOBORNYL ACETATE

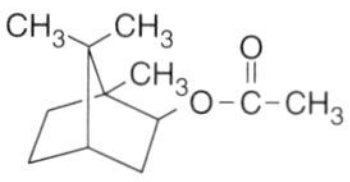

5 gm

S86,376-9
2-METHYL-2-NORBORNYL ACETATE

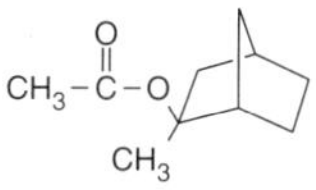

250 mg

S54,534-1
TRICYCLO[5.2.1.0(2,6)]DEC-8-YL FORMATE

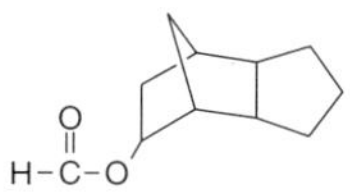

5 gm

S35,280-2
METHYL 2-CYCLOHEXYLBUTYRATE

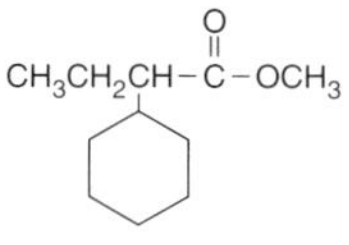

5 gm

S86,664-4
METHYL 2-CYCLOHEXYL-2-METHYLPROPIONATE

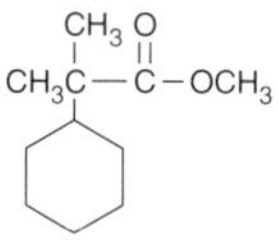

250 mg

S59,716-3
METHYL 4-TERT-BUTYL-2-METHYLCYCLOHEXANECARBOXYLATE

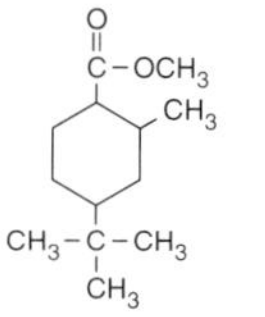

5 gm

S54,533-3
2-METHYLCYCLOHEXYL ACETATE
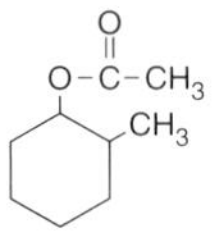
5 gm

S76,476-0
2-TERT-BUTYLCYCLOHEXYL ACETATE
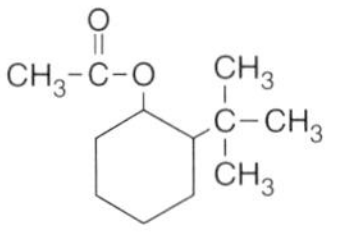
5 gm

S52,839-0
2-(TERT-PENTYL)CYCLOHEXYL ACETATE
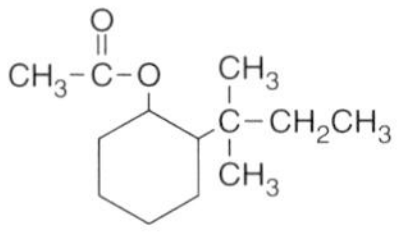
5 gm

S91,866-0
ETHYL 3-METHYL-3-(2-NITROCYCLOHEXYL)BUTYRATE
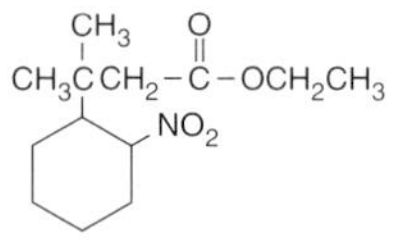
250 mg

S35,993-9
DECAHYDRO-2-NAPHTHYL ACETATE
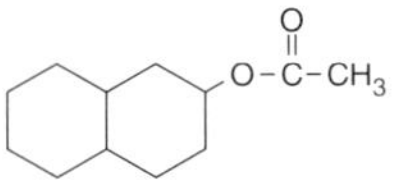
5 gm

S40,932-4
METHYL 2-NORBORNANECARBOXYLATE
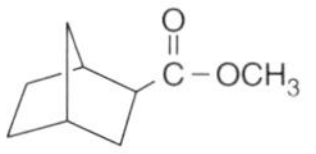
5 gm

S95,774-7
2-NORBORNYL ACETATE
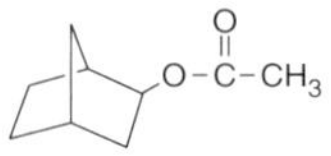
50 mg

S52,825-0
CEDRYL FORMATE
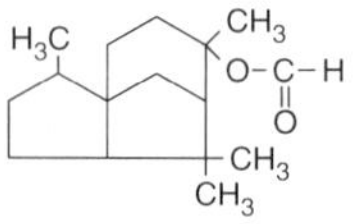
5 gm

S94,836-5
1-ADAMANTYL 1-ADAMANTANEACETATE
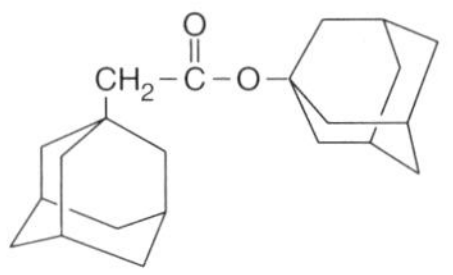
250 mg

S95,451-9
DI-L-MENTHYL CARBONATE
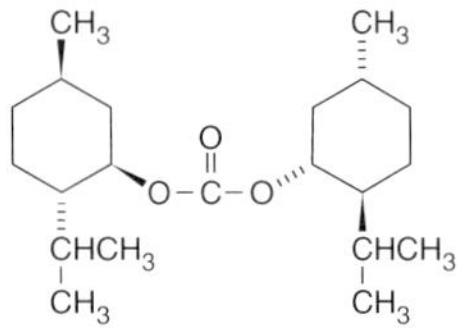
50 mg

S43,124-9
2-NORBORNYL THIOLACETATE
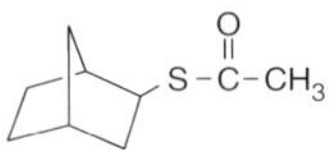
5 gm

S88,401-4
2-METHOXY-2-NONEN-1-YL PENTANOATE
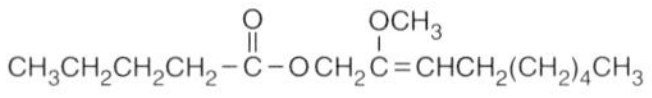
250 mg

S88,106-6
BUTYL 3-BUTOXYMETHACRYLATE
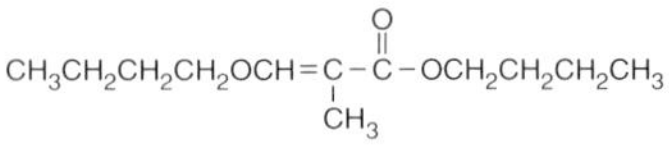
250 mg

S94,150-6
ETHYL 3,3-DIHYDRO-4-ETHOXY-2,2,6-TRIMETHYLBENZOATE
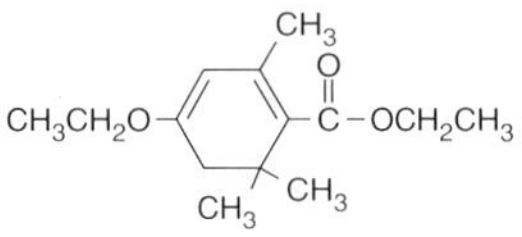
250 mg

S60,071-7
METHYL 3-ACETOXYCROTONATE
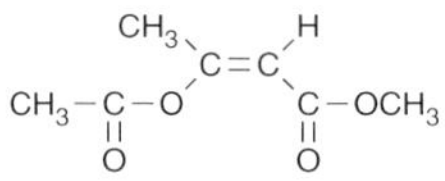
5 gm

S67,710-8
2-METHYL-1-BUTENYL ACETATE, MIXTURE OF CIS AND TRANS
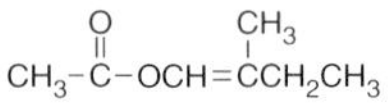
5 gm

S59,532-2
VINYL 3,5,5-TRIMETHYLHEXANOATE
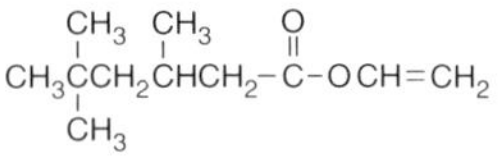
5 gm

S44,923-7
VINYL NEODECANOATE (INHIBITED WITH 5 PPM HYDROQUINONE)
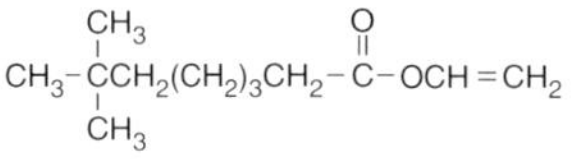
5 gm

S67,735-3
1-CYCLODODECEN-1-YL ACETATE
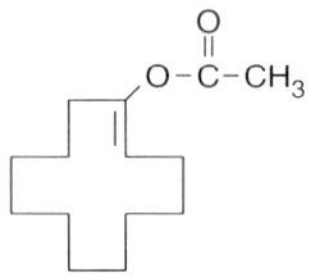
250 mg

S92,132-7
ETHYL 2,3-DIACETOXYCROTONATE
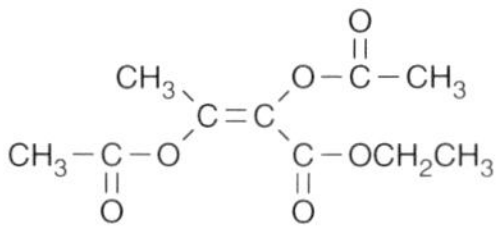
250 mg

S95,719-4
3,4,4A,4B,5,6,7,8,8A,9-DECAHYDRO-1,8A-DIMETHYL-2,6,7-TRIACETOXYPHENANTHRENE
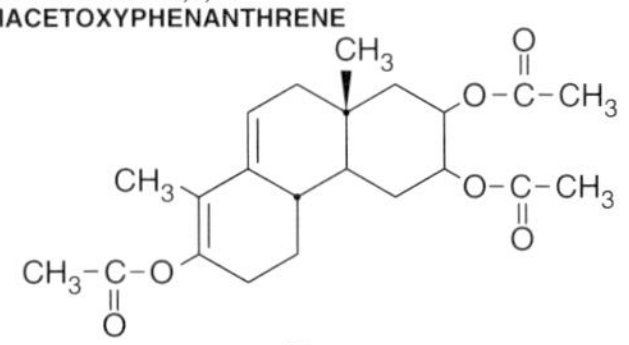
50 mg

S86,149-9
DIMETHYL 2-(4,6-DIACETOXYHEPTYLIDENE)MALONATE
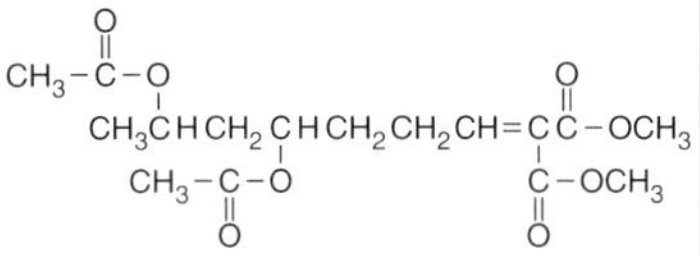
250 mg

S88,213-5
DIETHYL 2-ALLYL-2-METHYLMALONATE
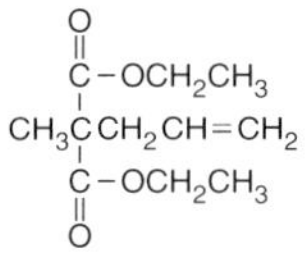
250 mg

S94,931-0
DIMETHYL 3-CYCLOPENTENE-1,1-DICARBOXYLATE
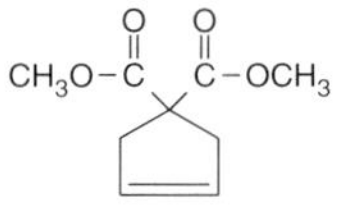
250 mg

S88,053-1
DIMETHYL 2-(3-METHYL-2-BUTEN-1-YLIDENE)SUCCINATE
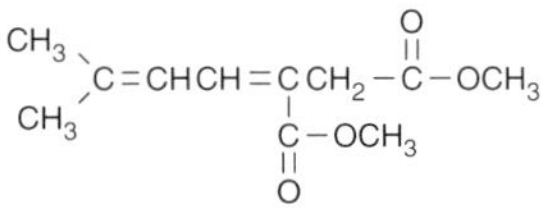
250 mg

S86,351-3
DIMETHYL 5-NORBORNENE-2,3-DICARBOXYLATE
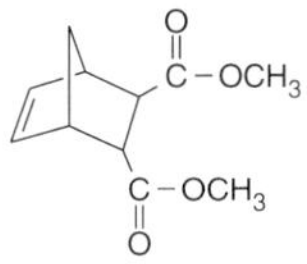
250 mg

S66,138-4
DIETHYL 2-(1-BUTYLIDENE)MALONATE
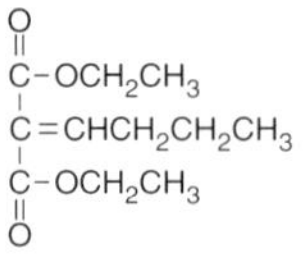
250 mg

S66,136-8
DIETHYL CYCLOPENTYLIDENEMALONATE
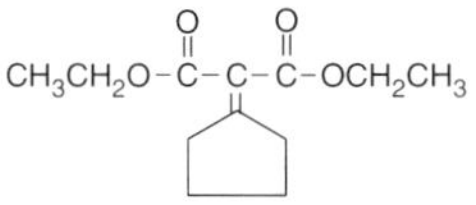
250 mg

S88,172-4
MONOMETHYL 2,4-DIMETHYLMUCONATE

250 mg

S88,177-5
DIMETHYL 2-METHYLMUCONATE

250 mg

S34,542-3
DIALLYL FUMARATE
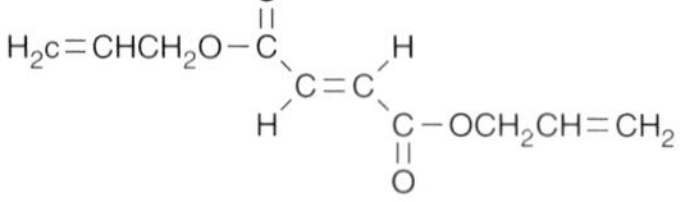
5 gm

S65,312-8
DIPROPYL MALEATE
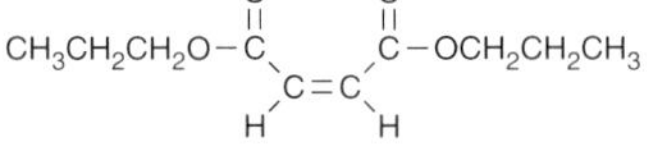
5 gm

S88,211-9
DIMETHYL 1,3,5-HEPTATRIENE-1,6-DICARBOXYLATE

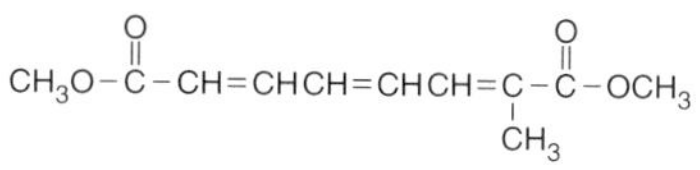

250 mg

S36,020-1
DIBUTYL FUMARATE

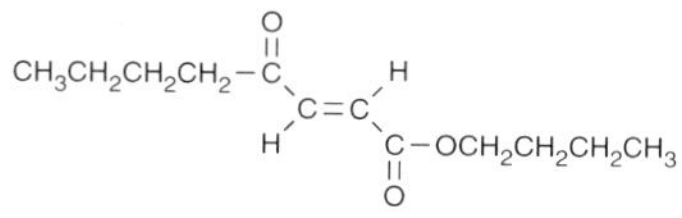

5 gm

S63,153-1
TRIETHYL 1,2,3-PROPENETRICARBOXYLATE

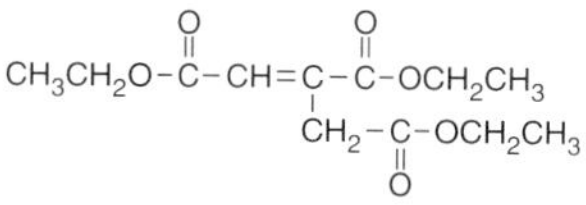

250 mg

S70,572-1
DIBUTYL METHYLENESUCCINATE

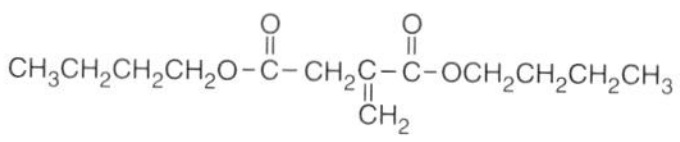

5 gm

S68,026-5
BIS(2-ETHYLHEXYL) MALEATE

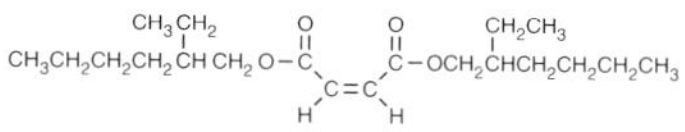

5 gm

S91,514-9
DIMETHYL BICYCLO[3.2.2]NONA-2,6,8-TRIENE-6,7-DICARBOXYLATE

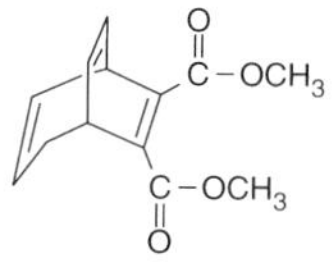

250 mg

S35,737-5
ALLYL CROTONATE

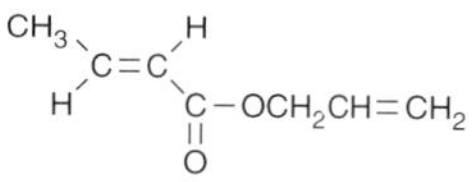

5 gm

S87,757-3
3-BROMO-3-BUTEN-1-YL ACRYLATE

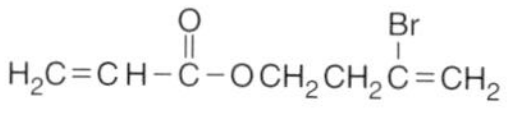

250 mg

S88,168-6
METHYL 4-ETHYL-2-HEPTENOATE

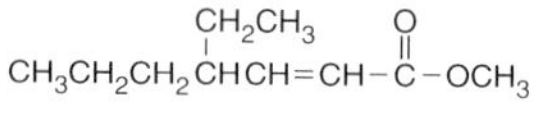

250 mg

S36,367-7
METHYL 2-NONENOATE

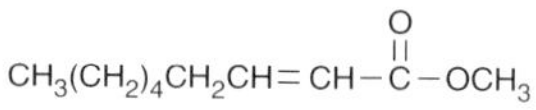

5 gm

S57,318-3
ETHYL 2-OCTENOATE

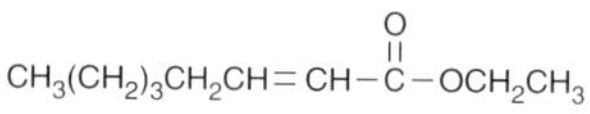

5 gm

S88,162-7
METHYL 4-METHYL-2,4,6-OCTATRIENE-2-CARBOXYLATE

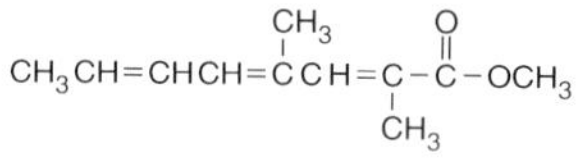

250 mg

S56,063-4
METHYL 2-DECENOATE

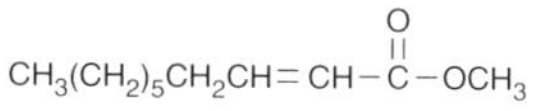

5 gm

S88,440-5
METHYL 2-METHYLENE-4-DECENOATE

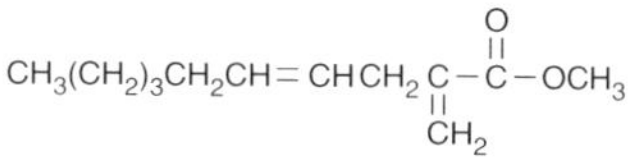

250 mg

S88,441-3
METHYL 2-METHYL-2,4-DECADIENOATE

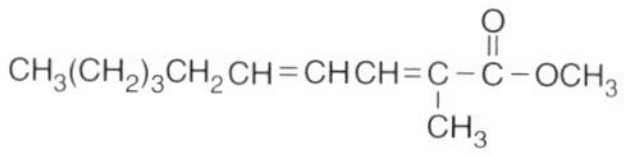

250 mg

S88,438-3
ETHYL 2-METHYL-2-NONENOATE

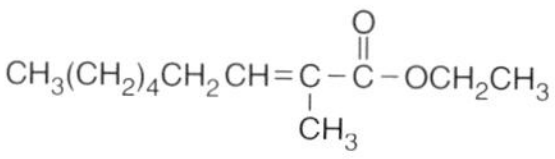

250 mg

S87,790-5
METHYL 5-METHYL-2,4,6-OCTATRIENE-2-CARBOXYLATE

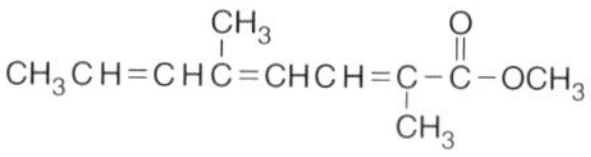

250 mg

S87,792-1
METHYL 2,4,4-UNDECATRIENE-2-CARBOXYLATE

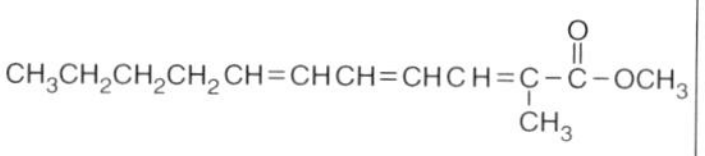

250 mg

S88,430-8
ETHYL 2,3-DIMETHYL-2-NONENOATE

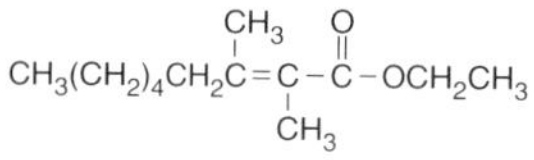

250 mg

S57,815-0
ISO-HEXADECYL ACRYLATE (MIXTURE OF BRANCHED ISOMERS)

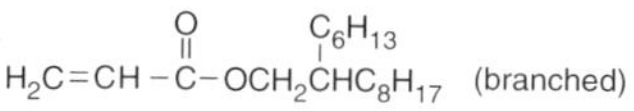

250 mg

S59,118-1
STEARYL METHACRYLATE

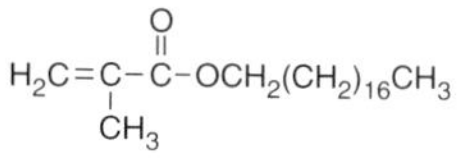

250 mg

S88,147-3
DIMETHYL 2,4,6,8-DECATETRAENE-2,9-DICARBOXYLATE

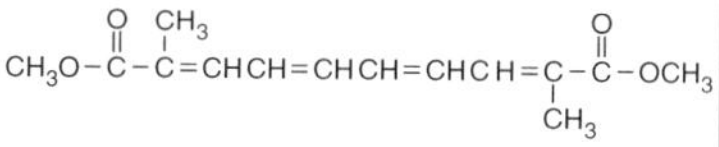

250 mg

S36,172-0
BIS(2-ETHYLHEXYL) FUMARATE

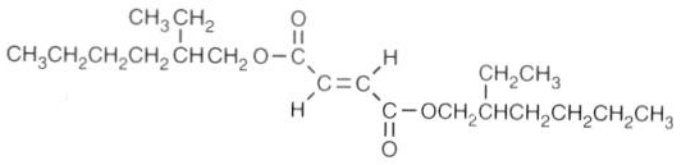

5 gm

S37,456-3
6-METHYL-3-CYCLOHEXENYLMETHYL 2-ETHYL-2-HEXENOATE

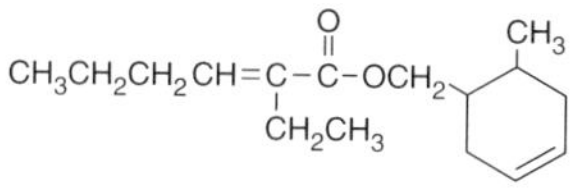

5 gm

S53,485-4
2-NORBORNYL METHACRYLATE

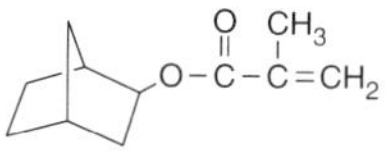

5 gm

S37,457-1
BIS(6-METHYL-3-CYCLOHEXENYLMETHYL)-OXALATE

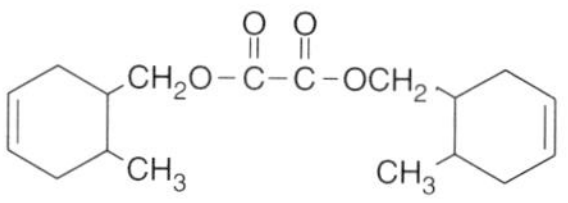

5 gm

S52,158-2
DIETHYL 2-ETHYL-2-PROPENYLMALONATE

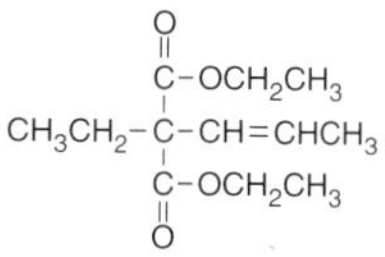

5 gm

S76,447-7
DIETHYL ALLYL-1-(METHYLBUTYL)-MALONATE

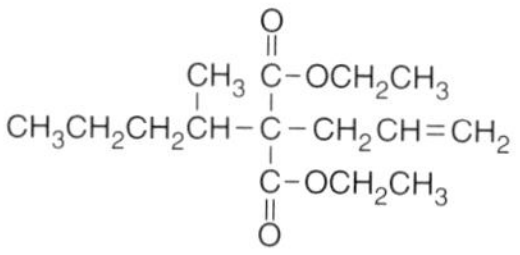

5 gm

S34,602-0
DIETHYL 2-CYCLOPENTENYLMALONATE

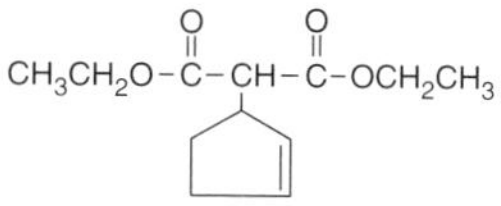

5 gm

S89,554-7
ETHYL 4-ACETOXY-1-(ETHOXYCARBONYL)-2,4-CYCLOPENTADIENE-1-ACETATE

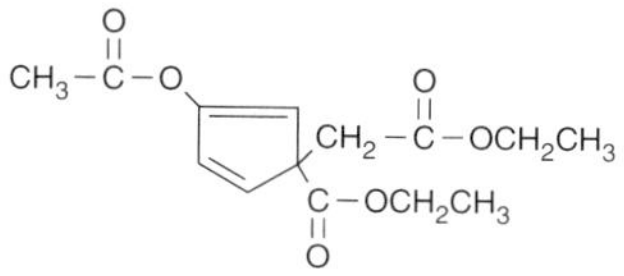

250 mg

S45,983-6
ALLYL ACRYLATE

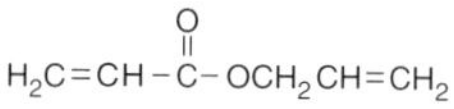

5 gm

S41,375-5
1-METHYLALLYL ACETATE

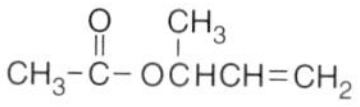

5 gm

S51,034-3
3-BUTENYL ACETATE

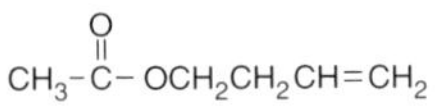

5 gm

S88,379-4
2-BUTEN-1-YL ACETATE

250 mg

S37,526-8
ALLYL 3-BUTENOATE

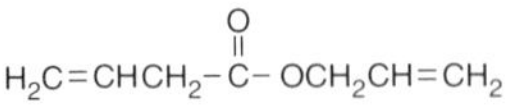

5 gm

S88,093-0
4-PENTEN-1-YL ACETATE

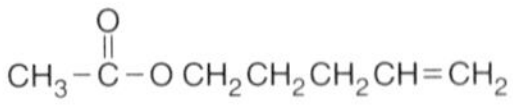

250 mg

S88,117-1
3-METHYL-3-BUTEN-1-YL ACETATE

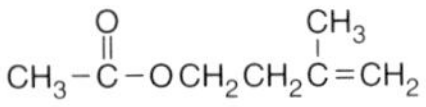

250 mg

S88,374-3
2-BUTEN-1-YL PROPIONATE

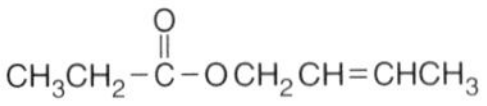

250 mg

S94,665-6
ETHYL 4-PENTENOATE

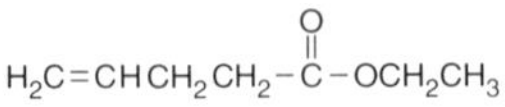

250 mg

S76,592-9
METHYL TRANS,CIS-2,5-HEPTADIENOATE

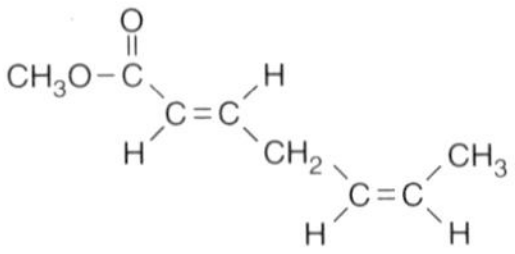

5 gm

S88,087-6
2,4-HEXADIEN-1-YL ACETATE

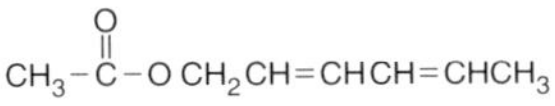

250 mg

S94,316-9
ETHYL 4-METHYL-3-PENTENOATE

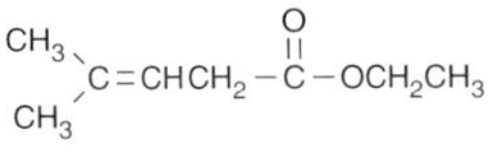

250 mg

S41,430-1
METHYL 5-NORBORNENE-2-CARBOXYLATE

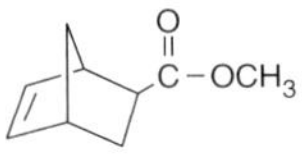

5 gm

S88,411-1
2-BUTEN-1-YL 3-HEXENOATE

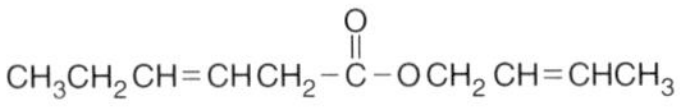

250 mg

S52,830-7
3,7-DIMETHYL-2,6-OCTADIENYL FORMATE

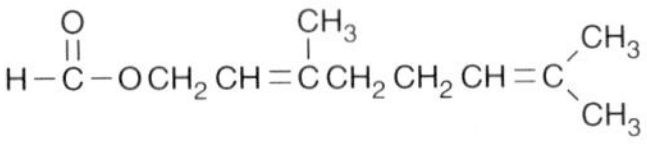

5 gm

S52,815-3
CITRONELLYL FORMATE

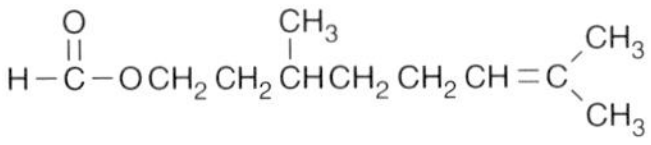

5 gm

S88,049-3
2-METHYL-2-OCTEN-1-YL ACETATE

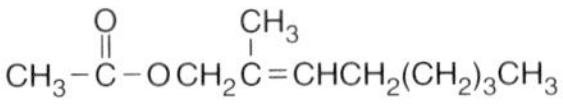

250 mg

S41,664-9
GERANYL ACETATE

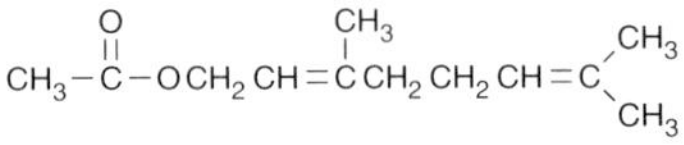

5 gm

S52,840-4
1,1,5-TRIMETHYL-4,6-HEPTADIENYL ACETATE

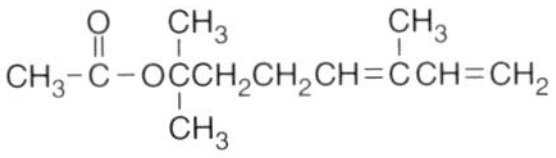

5 gm

S94,668-0
ETHYL 2,5,5-TRIMETHYL-2-CYCLOHEXENE-1-CARBOXYLATE

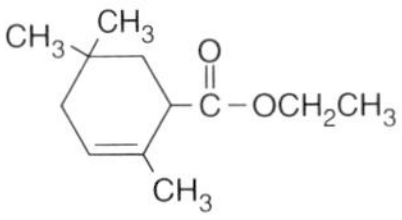

250 mg

S37,453-9
VINYL DECANOATE

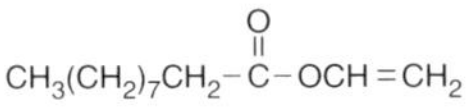

5 gm

S41,869-2
METHYL 10-UNDECENOATE

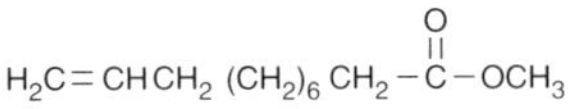

5 gm

S52,828-5
3,7-DIMETHYL-6-OCTENYL ACETATE

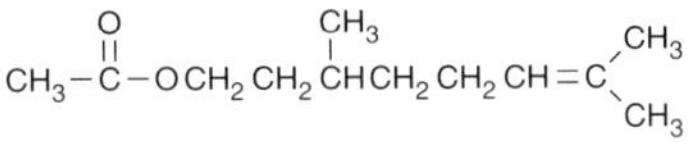

5 gm

S52,829-3
3,7-DIMETHYL-2,6-OCTADIENYL PROPIONATE

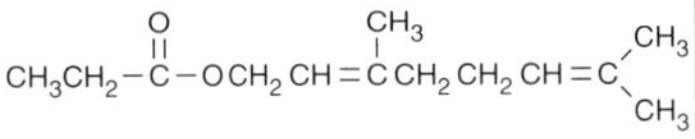

5 gm

S52,806-4
CITRONELLYL PROPIONATE

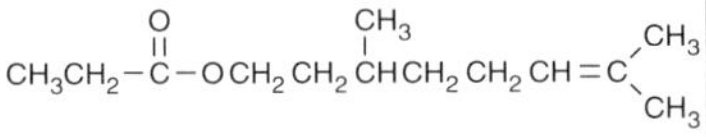

5 gm

S52,805-6
3,7-DIMETHYL-2,6-OCTADIENYL ISOBUTYRATE

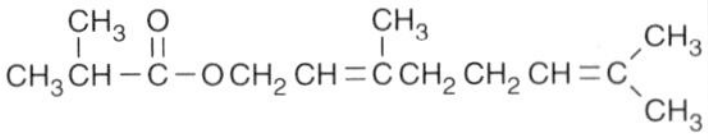

5 gm

S52,822-6
3,7-DIMETHYL-6-OCTENYL ISOBUTYRATE

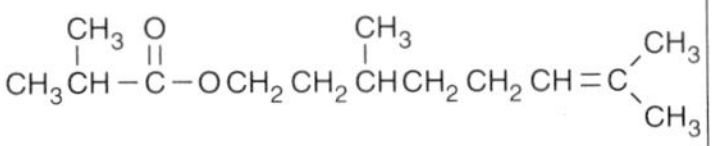

5 gm

S88,486-3
METHYL 2,4-DIMETHYL-3-HEXYL-4-PENTENOATE

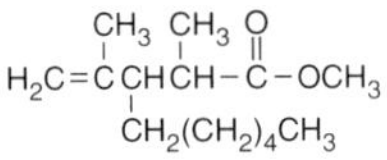

250 mg

S37,104-1
VINYL STEARATE

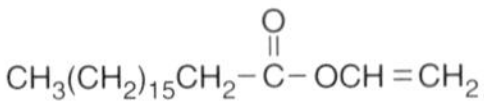

5 gm

S57,071-0
ETHYL ELAIDATE

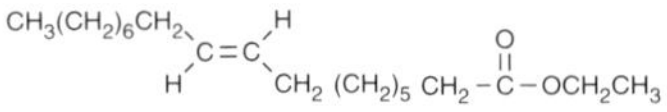

5 gm

S64,878-7
METHYL ABIETATE

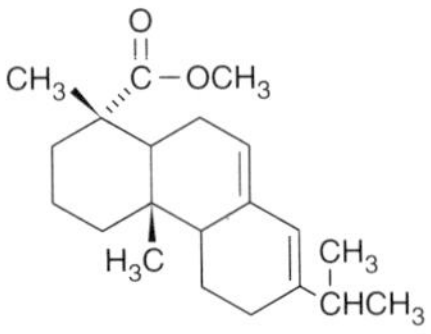

5 gm

S35,260-8
METHYL ACETYLRICINOLEATE

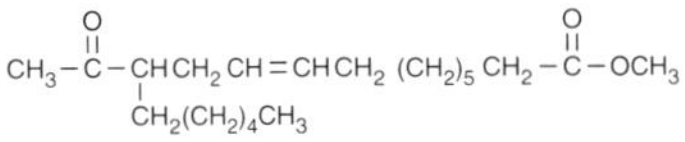

5 gm

S70,562-4
METHYL 13-DOCOSENOATE

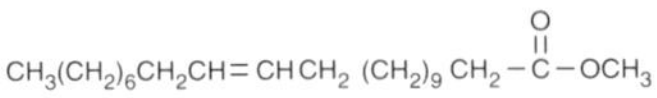

5 gm

S70,594-2
ETHYL 13-DOCOSENOATE

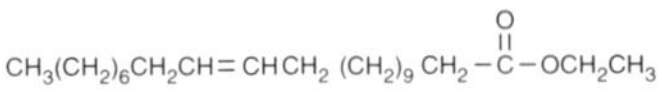

5 gm

S36,099-6
2,5-DIMETHYL-2-HEXENE-1,6-DIOL DIACETATE

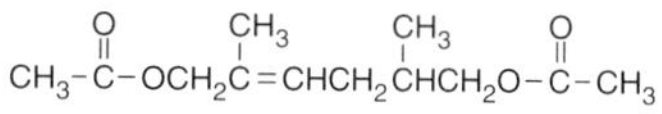

5 gm

S39,986-8
TETRAALLYL CIS,CIS,CIS,CIS-1,2,3,4-CYCLOPENTANETETRACARBOXYLATE

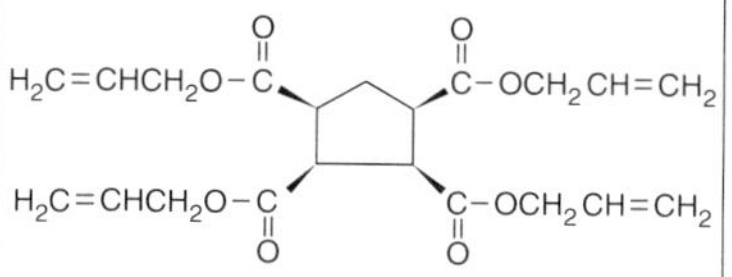

5 gm

S54,461-2
DIMETHYL 4-CYCLOHEXENE-1,2-DICARBOXYLATE

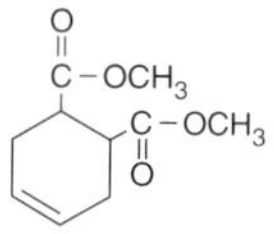

5 gm

S44,548-7
ETHYL 5-NORBORNENE-2-CARBOXYLATE, MIXTURE OF ENDO AND EXO

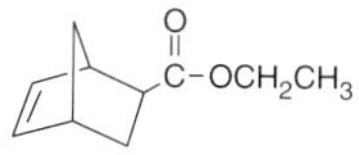

5 gm

S85,226-0
TRICYCLO[5.2.1.0(2,6)]DEC-3-EN-8-YL ACETATE

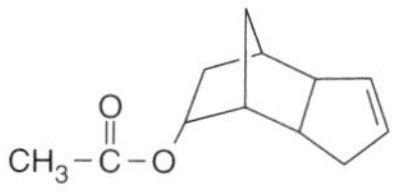

250 mg

S67,731-0
1-CYCLOPENTEN-1-YL ACETATE

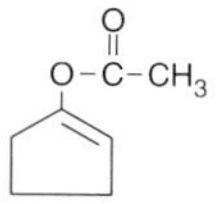

5 gm

S88,084-1
METHYL 2-METHYL-2-CYCLOPENTENE-1-CARBOXYLATE

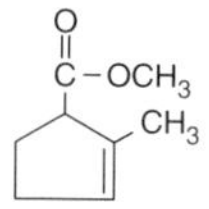

250 mg

S94,287-1
ETHYL 1-CYCLOPENTENE-1-CARBOXYLATE

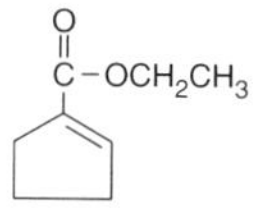

250 mg

S70,619-1
ETHYL 2-CYCLOPENTENE-1-TRIDECANOATE

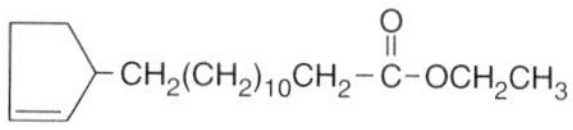

250 mg

S53,032-8
5-NORBORNEN-2-YLMETHYL 5-NORBORNENE-2-CARBOXYLATE

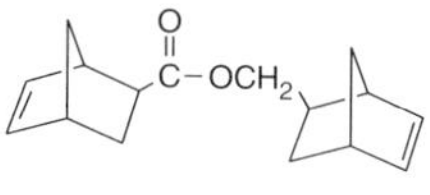

5 gm

S43,059-5
1,2,3,4,4A,5,8,8A-OCTAHYDRO-1,4:5,8-DIMETHANONAPHTH-2-YL ACETATE

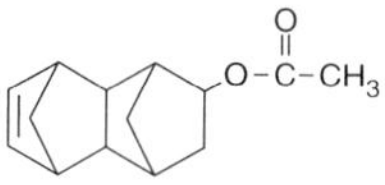

5 gm

S86,619-9
1-CYCLOHEXEN-1-YL ACETATE

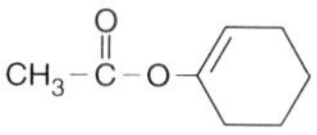

250 mg

S86,620-2
2-CYCLOHEXEN-1-YL ACETATE

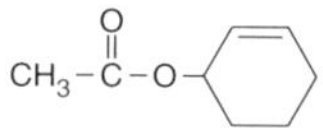

250 mg

S36,883-0
ETHYL 1-CYCLOHEXENE-1-ACETATE

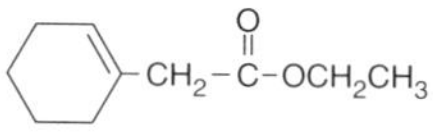

5 gm

S37,528-4
6-METHYL-3-CYCLOHEXENYLMETHYL ACETATE

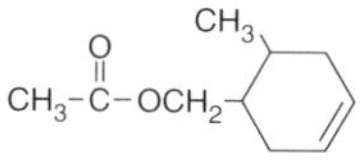

5 gm

S88,446-4
3,5,5-TRIMETHYL-2-CYCLOHEXEN-1-YL ACETATE

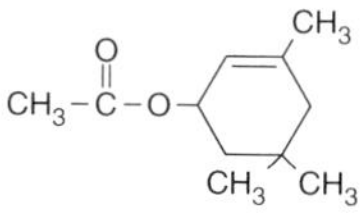

250 mg

S88,447-2
2,3,3-TRIMETHYL-1-CYCLOHEXEN-1-YL ACETATE

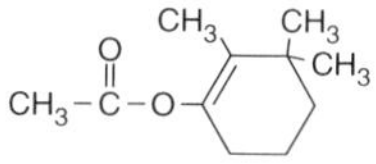

250 mg

S34,825-2
ETHYL 2-(1-CYCLOHEXENYL)BUTYRATE

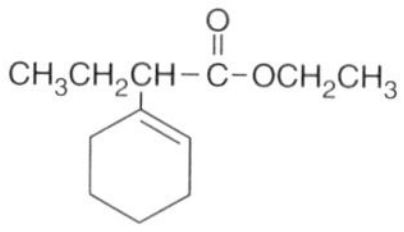

5 gm

S57,990-4
VITAMIN A ACETATE

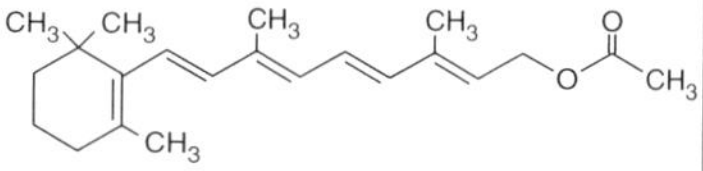

250 mg

S58,073-2
VITAMIN A PALMITATE

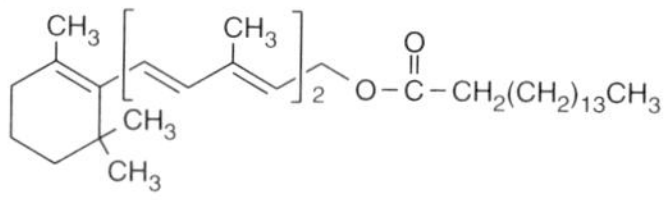

250 mg

S70,609-4
ETHYL ABIETATE

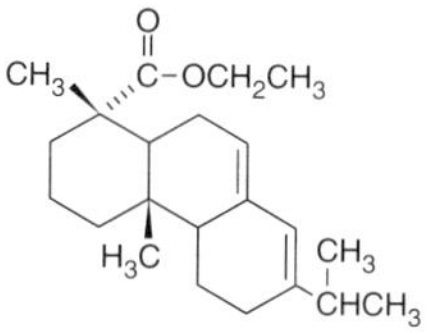

5 gm

S40,877-8
GUAIYL ACETATE

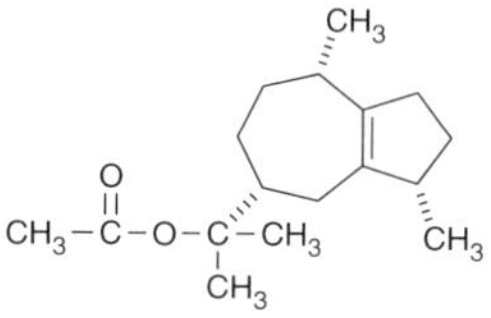

5 gm

S52,700-9
DIETHYL 2-(3-BROMOPROPYL)-2-ISOPENTYLMALONATE

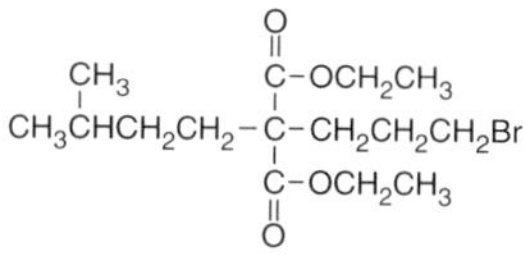

5 gm

S62,257-5
O,O'-ETHYLENE BIS(3-CHLOROPROPIONATE)

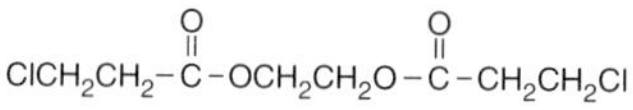

250 mg

S88,113-9
DIETHYL 2-BROMO-2-(BROMOMETHYL)-MALONATE

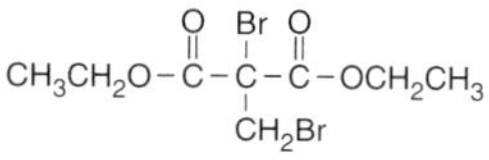

250 mg

S52,053-5
DIETHYL 2-(3-BROMOPROPYL)-2-BUTYLMALONATE

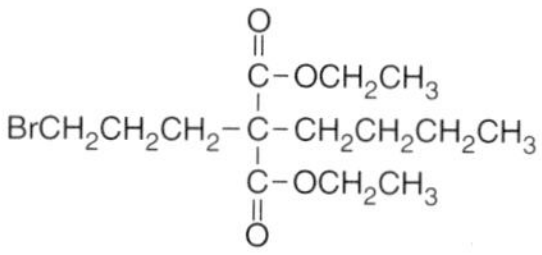

5 gm

S36,738-9
2-CHLOROETHYL FORMATE

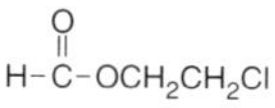

5 gm

S57,898-3
METHYL 2,3-DIBROMOPROPIONATE

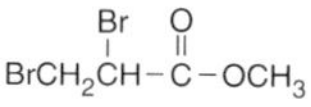

250 mg

S43,144-3
PROPYL CHLOROACETATE

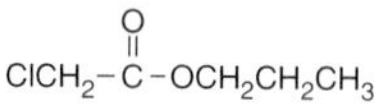

5 gm

S34,830-9
ETHYL 2,3-DICHLOROPROPIONATE

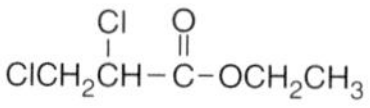

5 gm

S88,128-7
METHYL 2,3-DIBROMO-2-METHYLPROPIONATE

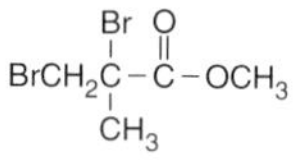

250 mg

S55,639-4
2,3-DIBROMOPROPYL CHLOROACETATE

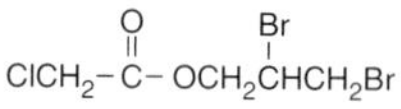

5 gm

S42,877-9
4-CHLOROBUTYL ACETATE
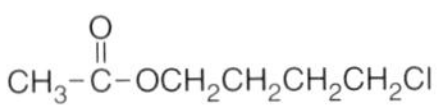
5 gm

S43,145-1
BUTYL CHLOROACETATE

5 gm

S56,384-6
ISOPROPYL 3-CHLOROPROPIONATE
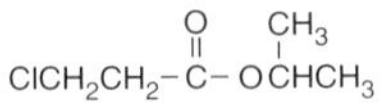
5 gm

S53,599-0
3-CHLORO-1-(CHLOROMETHYL)-PROPYL ACETATE
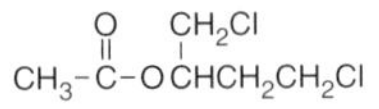
5 gm

S50,249-9
2,3,4-TRICHLOROBUTYL ACETATE
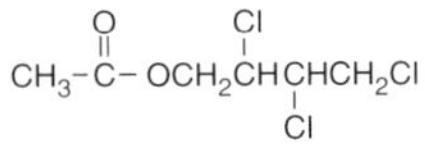
250 mg

S50,277-4
1-CHLOROMETHYL-2,3-DICHLOROPROPYL ACETATE
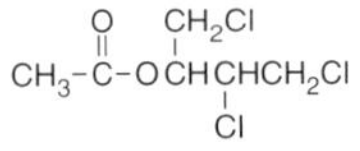
5 gm

S67,721-3
METHYL 5-IODOVALERATE
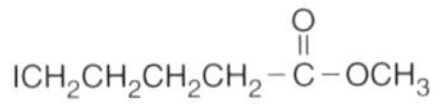
5 gm

S36,163-1
ETHYL 2,3-DIBROMOISOBUTYRATE
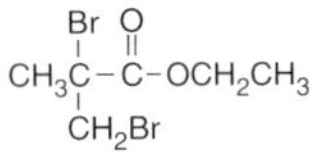
5 gm

S77,019-1
ETHYL 3-BROMO-2-(BROMOMETHYL)-PROPIONATE
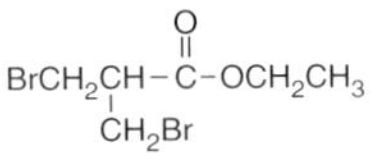
250 mg

S43,149-4
ISOAMYL CHLOROACETATE
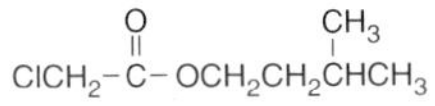
5 gm

S67,717-5
ETHYL 5-IODOVALERATE
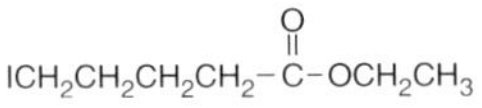
5 gm

S94,318-5
METHYL 2,6-DIBROMOCAPROATE
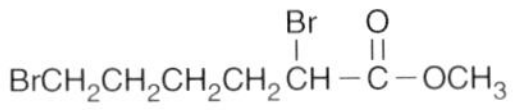
250 mg

S36,289-1
ISOPROPYL 6-CHLOROHEXANOATE
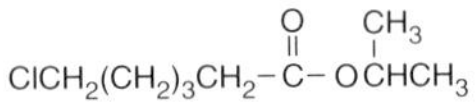
5 gm

S53,072-7
OCTYL CHLOROACETATE
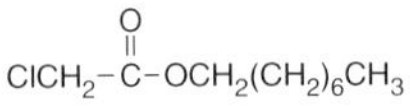
250 mg

S38,159-4
DECYL CHLOROACETATE
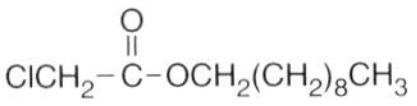
5 gm

S55,749-8
ETHYL 11-BROMOUNDECANOATE
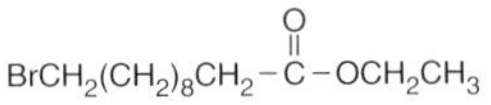
5 gm

S94,813-6
DIMETHYL ALPHA,ALPHA'-DIBROMO-1,3-ADAMANTANEDIACETATE
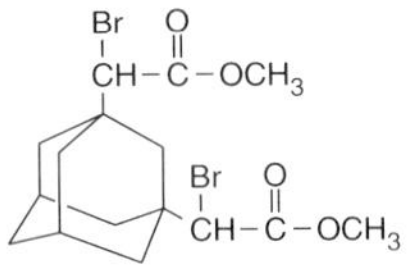
250 mg

S94,812-8
METHYL ALPHA-BROMO-1-ADAMANTANEACETATE
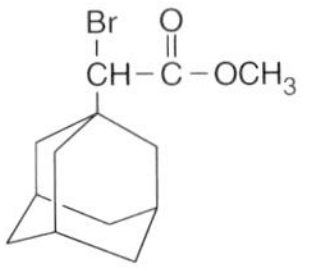
250 mg

S65,934-7
2-CHLOROETHYL CARBONATE
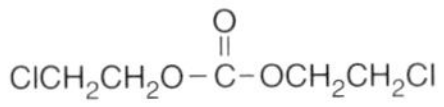
250 mg

S34,439-7
3-CHLORO-2-METHYL-1-PROPANETHIOL ACETATE
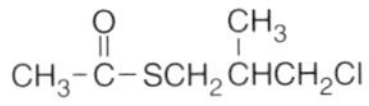
5 gm

S70,601-9
ETHYL DICHLOROACETATE
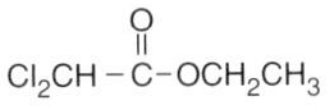
5 gm

S67,282-3
METHYL 3,3-DICHLORO-2,2-DIMETHYLPROPIONATE
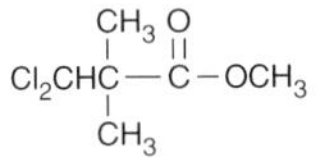
5 gm

S34,047-2
1-(ACETYLMERCAPTO)-2,2-DICHLOROETHANE
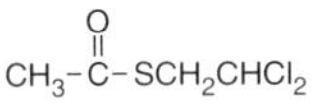
5 gm

S86,106-5
ETHYL 4,4,4-TRICHLORO-2-BUTENOATE
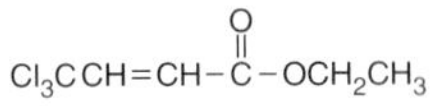
250 mg

S89,405-2
METHYL PENTACHLOROPROPIONATE
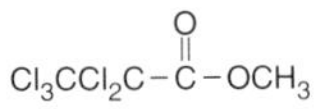
250 mg

S89,409-5
METHYL HEPTACHLOROBUTYRATE
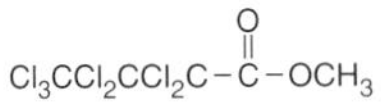
250 mg

S95,040-8
CYCLOPROPYLMETHYL TRIFLUOROACETATE
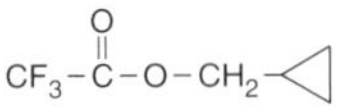
50 mg

S60,017-2
ETHYL 4,4,4-TRIFLUOROBUTYRATE
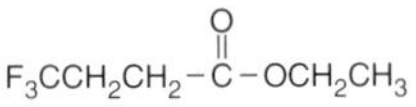
250 mg

S64,832-9
TERT-BUTYL TRICHLOROACETATE
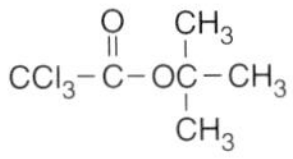
5 gm

S81,896-8
ETHYL 3-BROMO-4,4,4-TRIFLUOROBUTYRATE
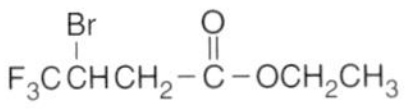
250 mg

S54,140-0
1-(TRICHLOROMETHYL)BUTYL ACETATE
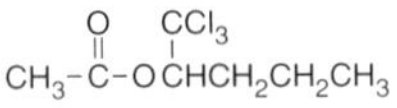
250 mg

S34,507-5
CYCLOOCTYL TRIFLUOROACETATE
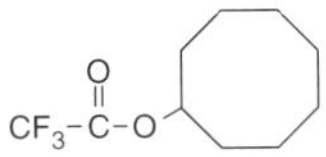
5 gm

S60,382-1
ETHYL PERFLUOROOCTANOATE

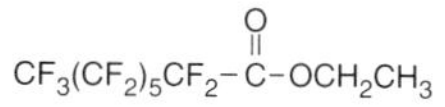

5 gm

S95,400-4
2-ISOPROPYL-5-METHYLCYCLOHEXYL TRIFLUOROACETATE

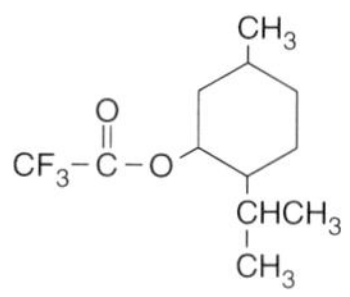

50 mg

S92,889-5
1-(2-ACETOXYETHOXY)-2-CHLOROETHYL ACETATE

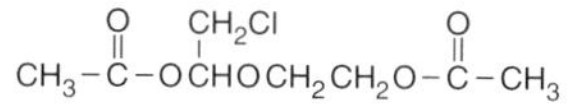

250 mg

S65,840-5
DIMETHYL 2,2-DIBROMOMALONATE

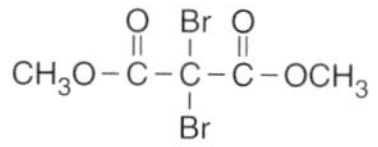

5 gm

S54,721-2
DIETHYL 2,6-DIBROMOHEPTANEDIOATE

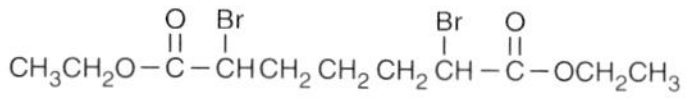

5 gm

S66,342-5
METHYL 2,3-DIBROMOISOBUTYRATE

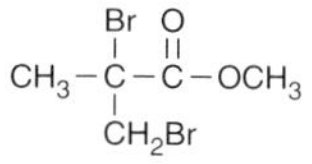

5 gm

S56,193-2
ETHYL 2-BROMO-3-METHYLBUTYRATE

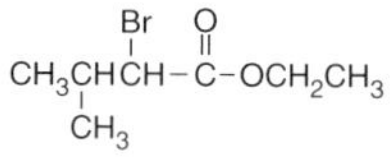

5 gm

S82,489-5
BUTYL 2-BROMO-2-METHYLPROPIONATE

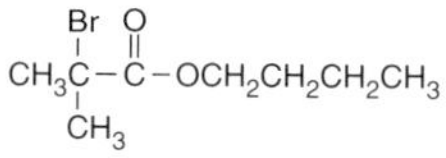

250 mg

S88,417-0
ISOPROPYL 2-BROMOHEXANOATE

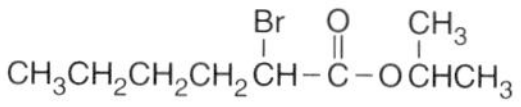

250 mg

S52,687-8
ETHYL 2-BROMO-2-ETHYLHEXANOATE

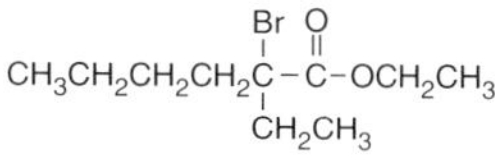

5 gm

S50,305-3
DIMETHYL ALPHA,ALPHA'-DICHLORO-5-METHYL-1,3-ADAMANTANEDIACETATE

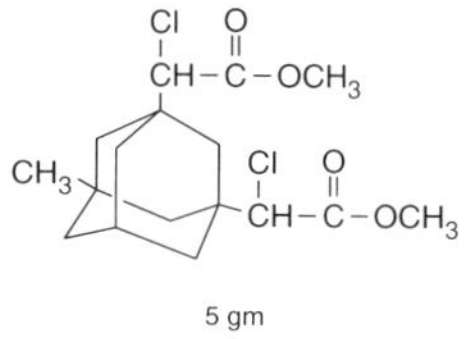

5 gm

S50,306-1
DIMETHYL ALPHA,ALPHA'-DICHLORO-5,7-DIMETHYL-1,3-ADAMANTANEDIACETATE

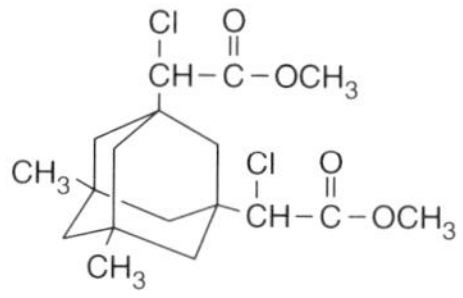

5 gm

S88,475-8
DIMETHYL 2-CHLORO-1-PROPENE-1,3-DICARBOXYLATE

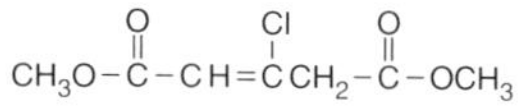

250 mg

S88,132-5
METHYL 3-BROMOACRYLATE

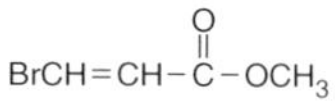

250 mg

S87,754-9
METHYL 3-BROMO-2-METHYLACRYLATE

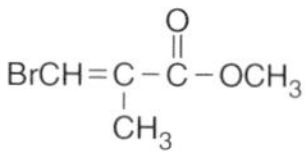

250 mg

S89,371-4
METHYL 1-BROMO-4,4-DICHLORO-1,3-BUTADIENE-1-CARBOXYLATE

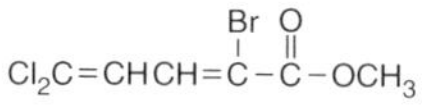

250 mg

S91,770-2
DIETHYL 2-CHLORO-2-METHYLMALONATE

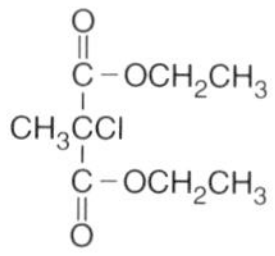

250 mg

S91,509-2
DIMETHYL 2,5-DIBROMOADIPATE

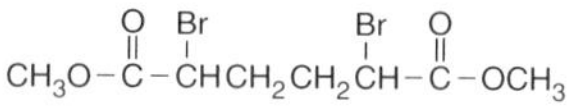

250 mg

S40,855-7
ETHYL PERFLUOROGLUTARATE

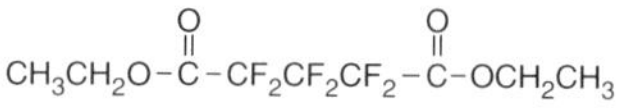

5 gm

S55,945-8
METHYL 3-CHLOROBUTYRATE

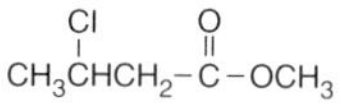

5 gm

S56,394-3
METHYL 2-BROMO-2-METHYLPROPIONATE

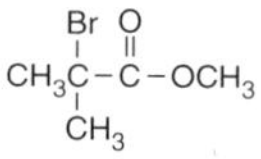

5 gm

S43,137-0
ETHYL 2-FLUOROBUTYRATE

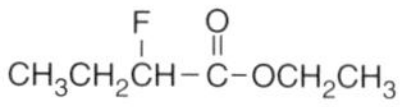

250 mg

S91,840-7
ETHYL 2-CHLOROBUTYRATE

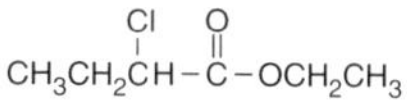

250 mg

S36,162-3
ETHYL 2,3-DIBROMOBUTYRATE

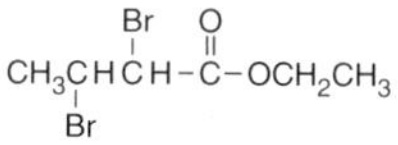

5 gm

S89,412-5
ETHYL PENTACHLORO-3-BUTENOATE

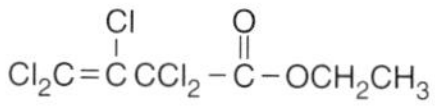

250 mg

S89,411-7
METHYL 2,2,3,4,5,5-HEXACHLORO-4-PENTENOATE

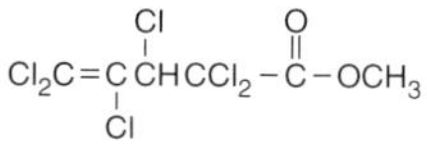

250 mg

S89,422-2
ETHYL 2,2,3,4,5-PENTACHLORO-4-PENTENOATE

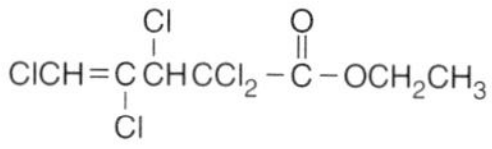

250 mg

S37,649-3
ETHYL 2-BROMO-2-ETHYLBUTYRATE

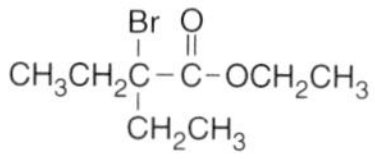

5 gm

S88,407-3
2-BUTEN-1-YL 2-BROMOHEXANOATE

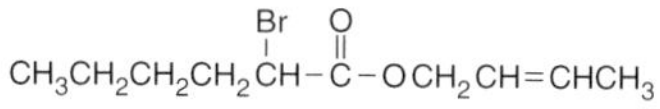

250 mg

S37,740-6
ETHYL 2-BROMODECANOATE

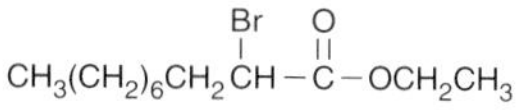

250 mg

S40,842-5
ETHYL 2-BROMOMYRISTATE

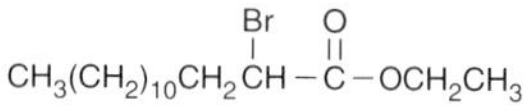

5 gm

S94,319-3
ETHYL 13-ACETOXY-2-CHLOROTRIDECANOATE

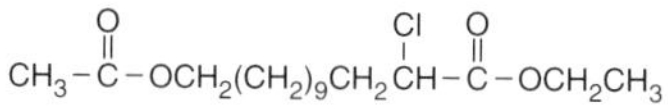

250 mg

S81,058-4
METHYL 3-BROMO-1-CYCLOPENTANECARBOXYLATE

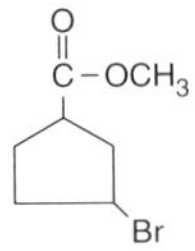

250 mg

S37,819-4
METHYL ALPHA-BROMOCYCLOHEXANEACETATE

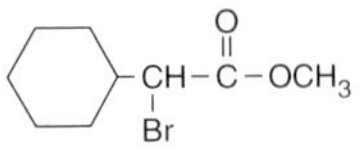

5 gm

S91,850-4
ETHYL 1-BROMO-1-CYCLOHEXANECARBOXYLATE

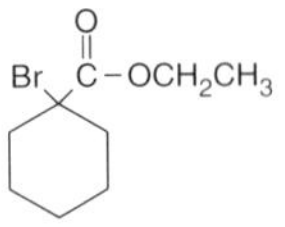

250 mg

S65,116-8
METHYL 1-BROMO-2-DECALINCARBOXYLATE

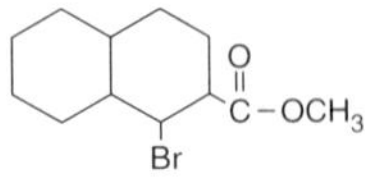

250 mg

S95,640-6
2-ACETOXY-6,7-DIBROMO-1,1,4A-TRIMETHYLDECAHYDRONAPHTHALENE

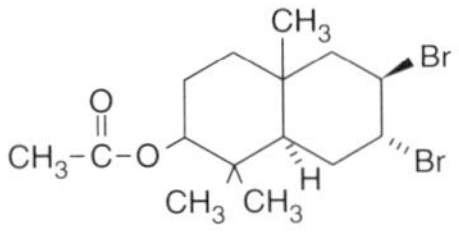

50 mg

S94,815-2
2-BROMO-2-(3-BROMO-1-ADAMANTYL)-ETHYL ACETATE

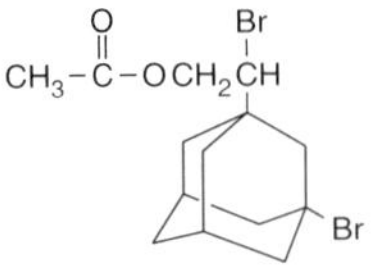

250 mg

S36,579-3
NORMAL-BUTYL 3-FLUOROLACTATE

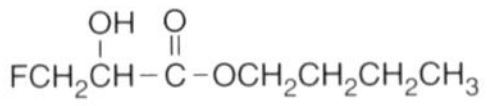

250 mg

S81,897-6
ETHYL 3-HYDROXY-4,4,4-TRIFLUOROBUTYRATE

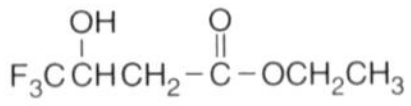

250 mg

S53,939-2
ETHYL 3-HYDROXY-4,4,4-TRICHLOROBUTYRATE

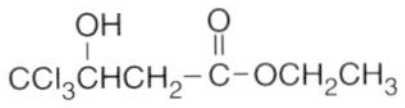

250 mg

S85,667-3
BIS[1-(ETHOXYCARBONYL)-2,2,2-TRICHLOROETHYL] ETHER

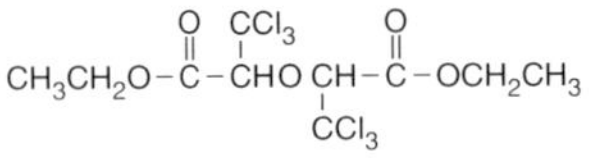

250 mg

S70,624-8
SUCROSE MONOPALMITATE

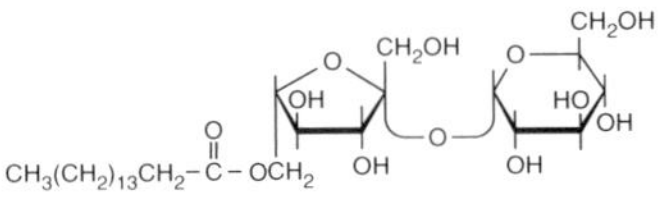

5 gm

S86,037-9
MYO-INOSITOL 3,4,5,6-TETRAACETATE

250 mg

S51,955-3
DIISOAMYL TARTRATE

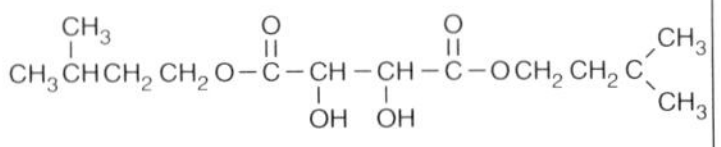

5 gm

S88,804-4
L-MENTHYL 2,2-DIHYDROXYACETATE

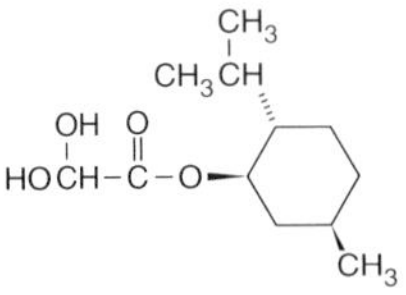

250 mg

S75,323-8
PENTAERYTHRITOL MONOACETATE

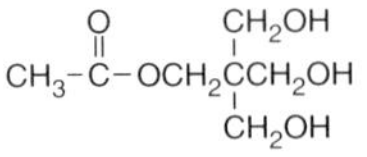

5 gm

S80,693-5
ETHYL 3,5-DIHYDROXY-6-OCTENOATE

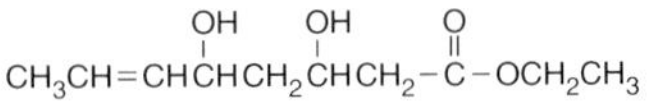

250 mg

S94,383-5
DIMETHYL SHELLOATE

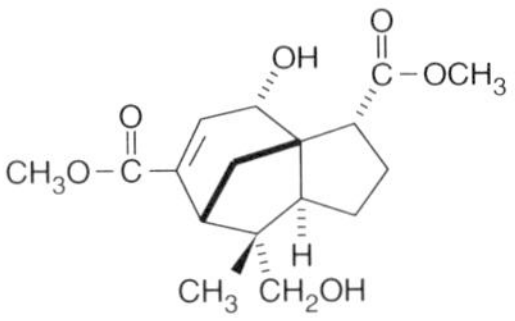

250 mg

S87,727-1
ETHYL 3-HYDROXY-4-METHOXYBUTYRATE

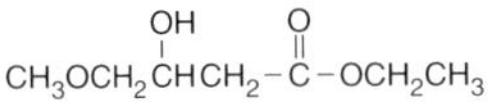

250 mg

S86,030-1
TRANS-1,2-CYCLOHEXANEDIOL MONO[(-)-MENTHOXYACETATE]

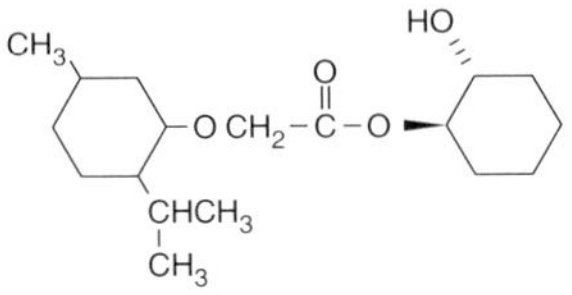

250 mg

S85,913-3
DIETHYL 2-HYDROXYFUMARATE, POTASSIUM SALT

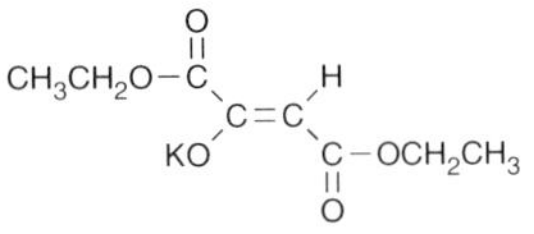

250 mg

S86,938-4
DIETHYL 2-HYDROXYFUMARATE, SODIUM SALT

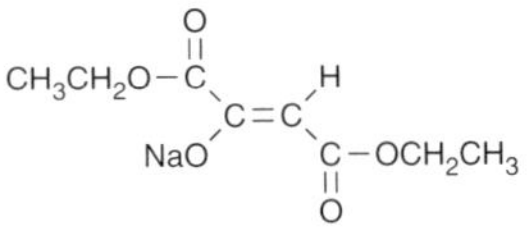

250 mg

S37,923-9
DIMETHYL MALATE

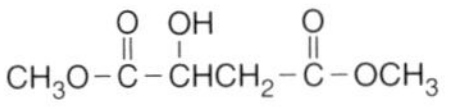

5 gm

S37,791-0
TRIMETHYL CITRATE

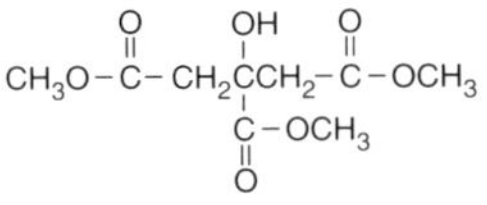

5 gm

S35,263-2
METHYL ARABONATE

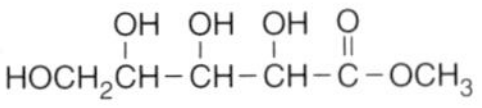

5 gm

S37,436-9
ALLYL 2-HYDROXY-3-BUTENOATE

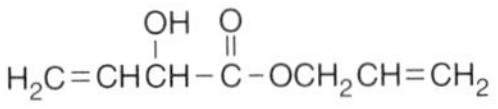

5 gm

S38,548-4
ETHYL 2,2-DIETHYLGLYCOLATE

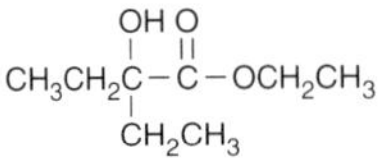

5 gm

S38,099-7
ETHYL 2-ETHYL-2-HYDROXY-5-METHYLHEXANOATE

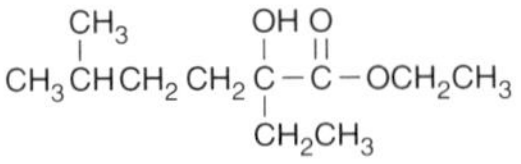

5 gm

S44,921-0
2-HYDROXYPROPYL DODECANOATE

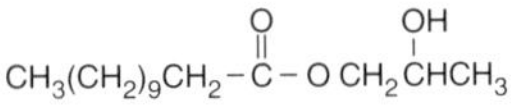

5 gm

S59,117-3
2-HYDROXYPROPYL STEARATE

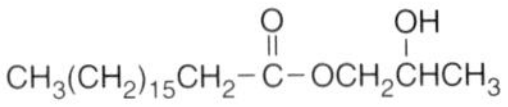

250 mg

S89,413-3
METHYL 3-HYDROXY-2,2,4,5,5-PENTACHLORO-4-PENTENOATE

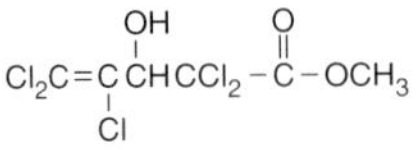

250 mg

S85,304-6
ETHYL 2-(1-HYDROXYCYCLOHEXYL)-ACETATE

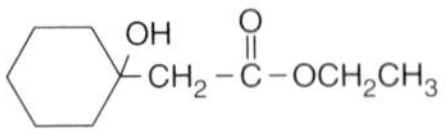

250 mg

S59,736-8
METHYL 4-TERT-BUTYL-2-HYDROXYCYCLOHEXANECARBOXYLATE

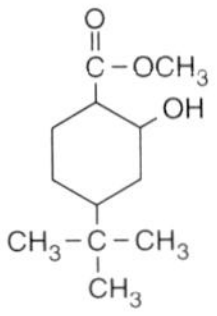

5 gm

S76,461-2
METHYL HEXAHYDROSALICYLATE

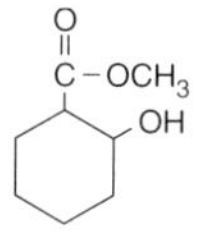

5 gm

S72,219-7
TRANS-4-HYDROXYCYCLOHEXYL ACETATE

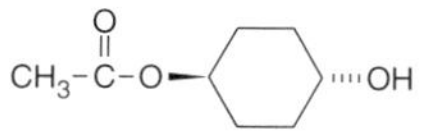

5 gm

S72,220-0
CIS-4-HYDROXYCYCLOHEXYL ACETATE

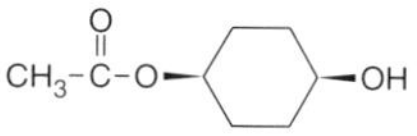

5 gm

S88,073-6
METHYL 6-HYDROXY-2,4-HEPTADIENE-2-CARBOXYLATE

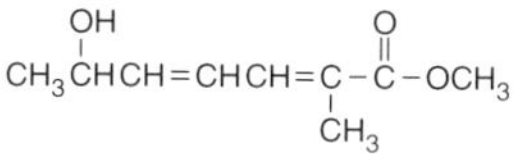

250 mg

S88,064-7
METHYL 6-HYDROXY-6-METHYL-2,4-HEPTADIENE-2-CARBOXYLATE

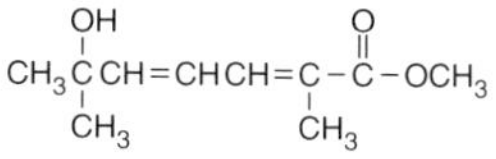

250 mg

S87,722-0
DIETHYL 2-(HYDROXYMETHYLENE)-SUCCINATE

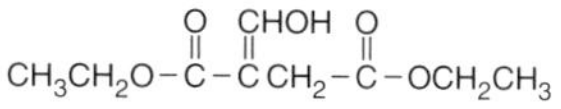

250 mg

S86,072-7
ETHYL 4-HYDROXY-2-METHYL-1-CYCLOHEXENE-1-CARBOXYLATE

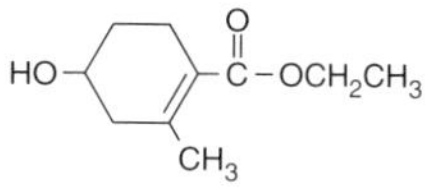

250 mg

S92,868-2
ENT-15ALPHA-ANGELOYLOXYKAUR-16-EN-3BETA-OL, 90% & 16B,17-EPOXYKAURANE-3B-OL, 10%

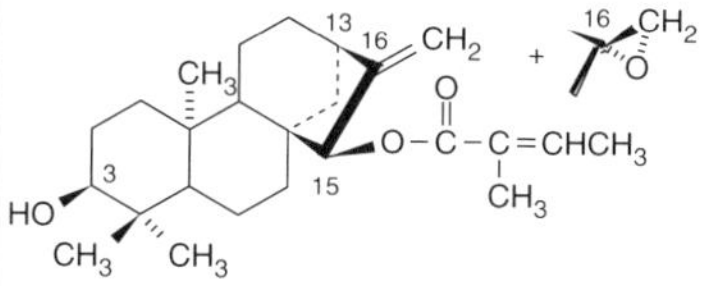

250 mg

S65,617-8
METHYL 2,2-BIS(HYDROXYMETHYL)-PROPIONATE

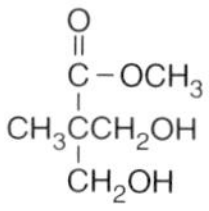

250 mg

S61,861-6
ETHYL 3-ETHYL-3-HYDROXYVALERATE

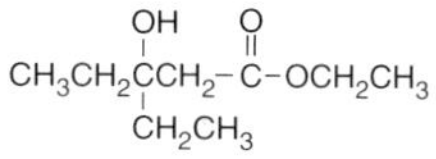

5 gm

S66,318-2
METHYL 3-HYDROXY-2,3,4,4-TETRAMETHYLVALERATE

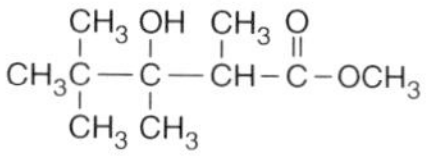

5 gm

S88,406-5
METHYL 5-HYDROXY-2-METHYL-3-DECENOATE

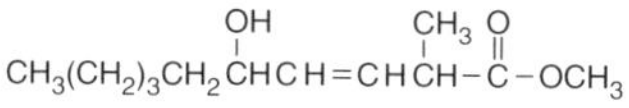

250 mg

S67,726-4
ETHYL 2-HYDROXYCYCLOHEXANEPROPIONATE

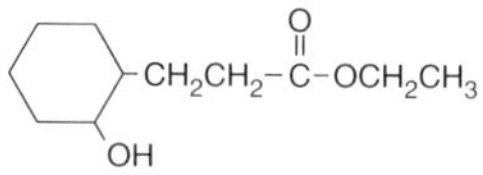

5 gm

S52,085-3
O-BUTYL S-(3-HYDROXYPROPYL)-DITHIOCARBONATE

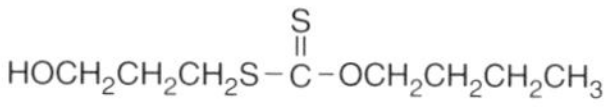

5 gm

S70,596-9
DIETHYL 2,3-DIETHOXYSUCCINATE

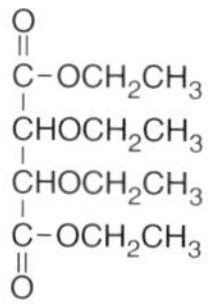

5 gm

S70,565-9
2-(2-METHOXYETHOXY)ETHYL ACETATE

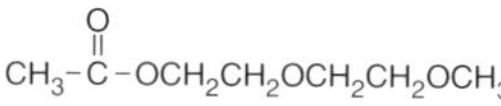

5 gm

S56,506-7
METHYL DIETHOXYACETATE

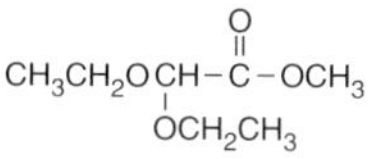

250 mg

S53,556-7
DIETHYL 2-ETHOXYMALONATE

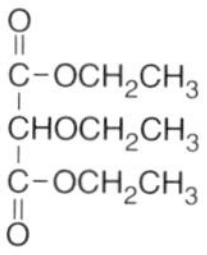

250 mg

S89,563-6
ETHYL 2-CHLORO-2-ETHOXYACETATE

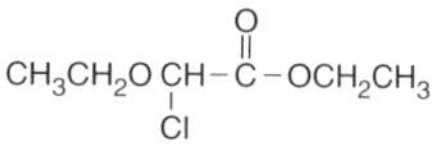

250 mg

S88,470-7
METHYL 2,3-DIMETHYL-2-METHOXY-4-PENTENOATE

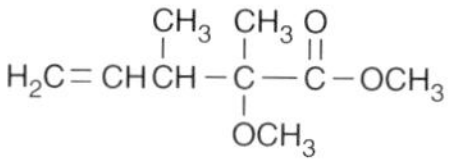

250 mg

S54,067-6
BUTYL BUTOXYACETATE

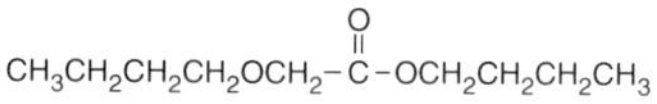

5 gm

S67,736-1
2-METHOXYCYCLOHEXYL ACETATE

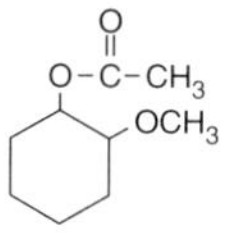

5 gm

S89,569-5
ETHYL 2-CHLORO-2-(CYCLOHEXYLOXY)-ACETATE

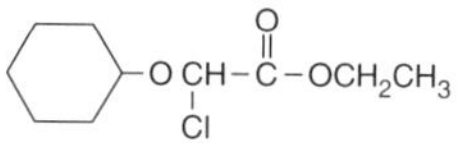

250 mg

S51,936-7
BIS(2-ETHOXYETHYL) CARBONATE

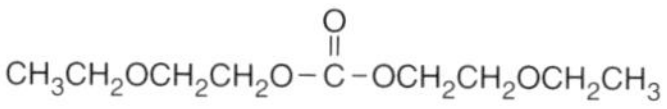

250 mg

S95,015-7
DIMETHYL 2-(METHOXYMETHYL)SUCCINATE

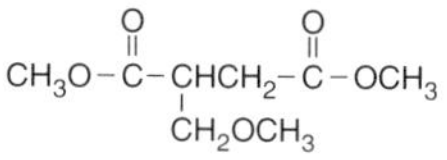

50 mg

S52,152-3
DIETHYL 2-ETHYL-2-(METHOXYMETHYL)-MALONATE

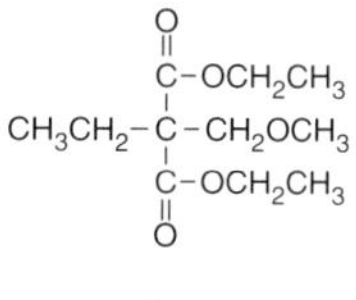

5 gm

S95,027-0
ETHYL 2-(DIETHOXYMETHYL)BUTYRATE

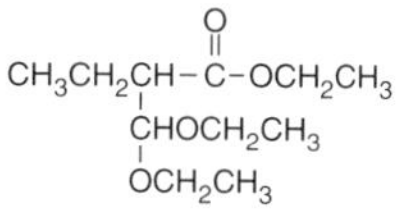

50 mg

S87,730-1
ETHYL 4-METHOXYCROTONATE

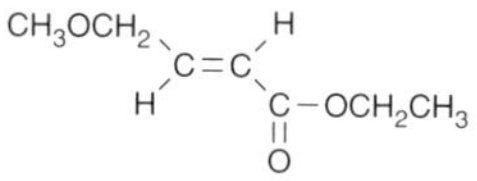

250 mg

S38,073-3
DIETHYL (2-METHOXYETHYL)MALONATE

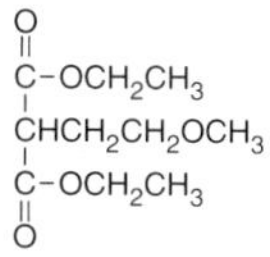

5 gm

S95,064-5
METHYL 4-(1-ETHOXYETHOXY)-2-CYCLOPENTENEACETATE

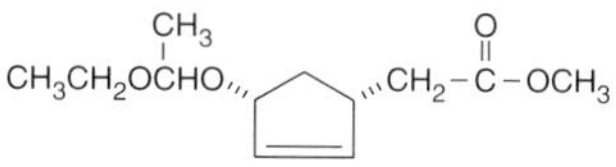

50 mg

S35,055-9
HEXYL 3-HEXYLOXYCROTONATE

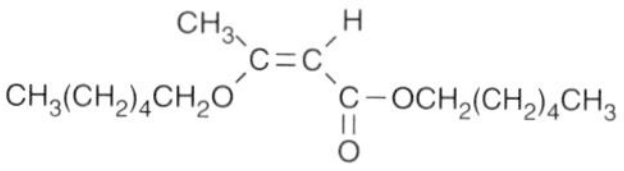

5 gm

S37,454-7
VINYL 3-(2-ETHYLHEXYLOXY)-PROPIONATE

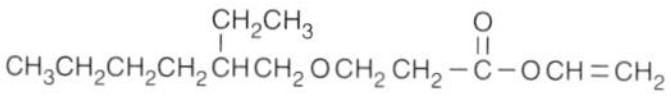

5 gm

S38,242-6
ETHYL 11-METHOXYUNDECANOATE
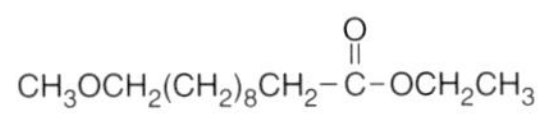
250 mg

S54,069-2
CYCLOHEXYL CYCLOHEXYLOXYACETATE
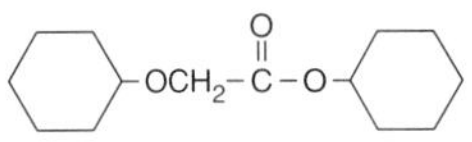
5 gm

S51,317-2
2,2'-OXYDIETHYL BIS(O-BUTYLXANTHATE)
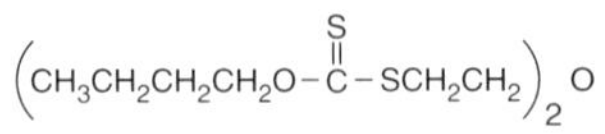
5 gm

S73,932-4
METHYL 4-(DIMETHOXYMETHYL)-4-ETHYL-6-HEPTENOATE
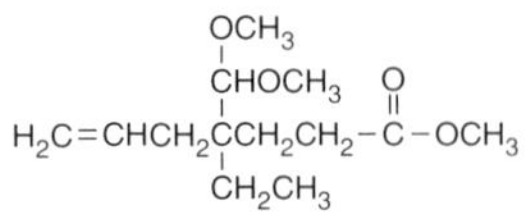
5 gm

S95,031-9
TETRAETHYL 1,4-DIHYDROXY-1,3-BUTADIENE-1,2,3,4-TETRACARBOXYLATE, DISODIUM SALT
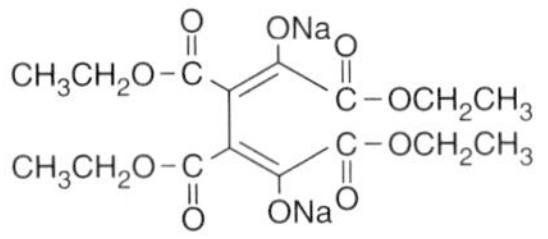
50 mg

S86,038-7
1,2-O-ISOPROPYLIDENE-MYO-INOSITOL 3,4,5,6-TETRAACETATE
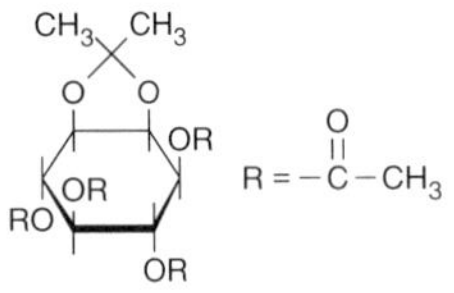
250 mg

S59,401-6
2-HYDROXYETHYL MERCAPTOACETATE
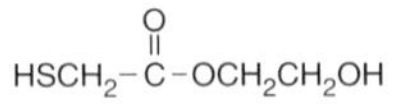
250 mg

S52,101-9
1,1,1-PROPANETRIYL TRIS(MERCAPTOACETATE)
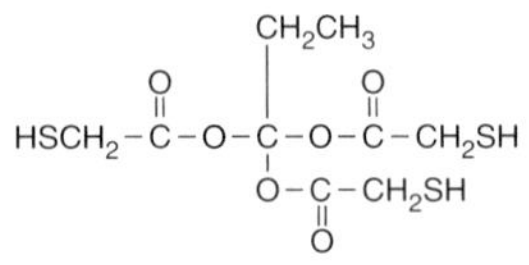
5 gm

S52,102-7
1,1,1-PROPANETRIYL TRIS-(3-MERCAPTOPROPIONATE)
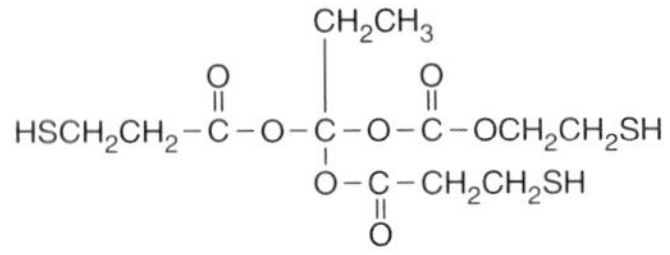
5 gm

S51,171-4
PENTAERYTHRITOL TETRAKIS-(3-MERCAPTOPROPIONATE)
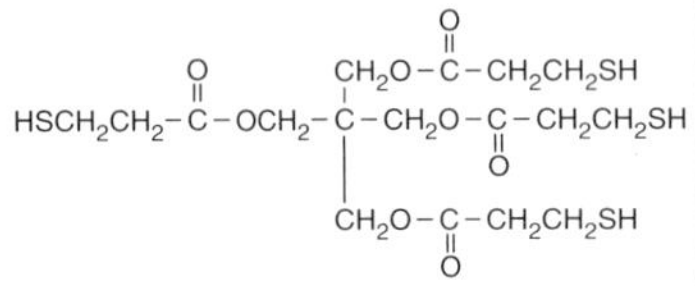
250 mg

S91,743-5
1-THIO-D-GALACTOPYRANOSE TETRAACETATE
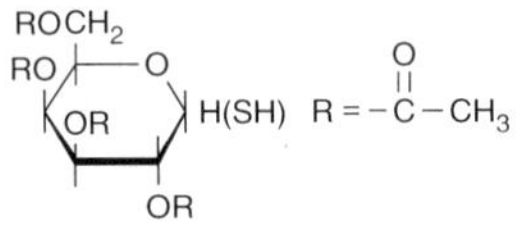
250 mg

S51,170-6
2-HYDROXYMETHYL-2-METHYL-1,3-PROPANEDIOL TRIS(MERCAPTOACETATE)
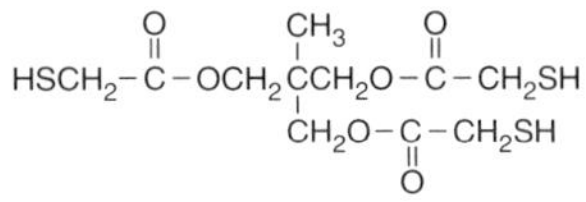
5 gm

S51,114-5
2-HYDROXYMETHYL-2-METHYL-1,3-PROPANEDIOL TRIS(3-MERCAPTOPROPIONATE)
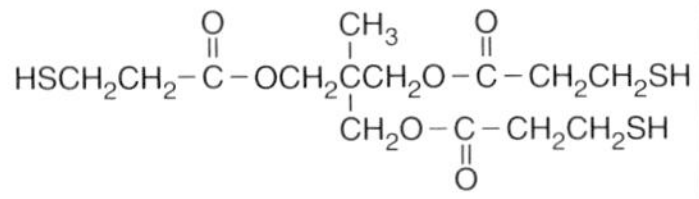
5 gm

S95,026-2
DIETHYL 2-[3-(METHYLTHIO)PROPYL]-MALONATE
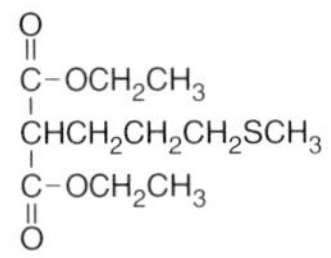
50 mg

S36,304-9
2-MERCAPTOETHYL ACETATE
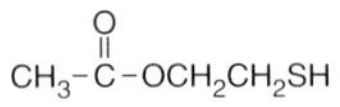
5 gm

S59,393-1
ISOPROPYL MERCAPTOACETATE
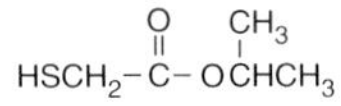
5 gm

S78,668-3
3-MERCAPTOPROPYL ACETATE

250 mg

S57,031-1
ISOPENTYL MERCAPTOACETATE
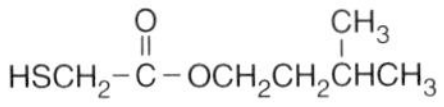
5 gm

S59,397-4
HEXYL MERCAPTOACETATE
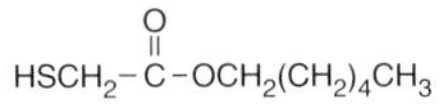
250 mg

S55,904-0
ISOOCTYL MERCAPTOACETATE
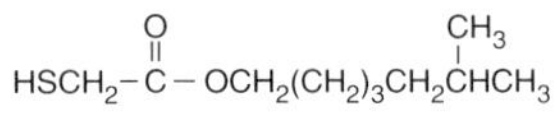
5 gm

S59,486-5
2-ETHYLHEXYL MERCAPTOACETATE
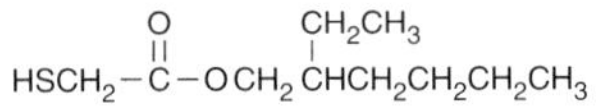
5 gm

S59,396-6
NONYL MERCAPTOACETATE
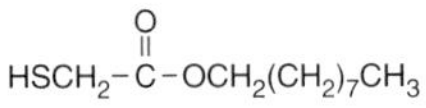
5 gm

S59,485-7
2-ETHYLHEXYL 3-MERCAPTOPROPIONATE
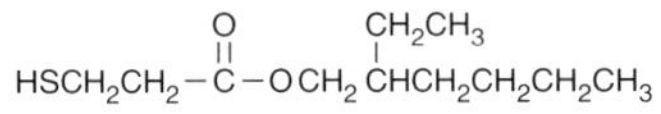
250 mg

S59,481-4
DECYL MERCAPTOACETATE
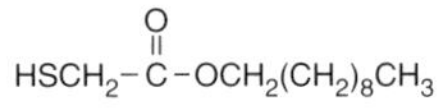
5 gm

S55,928-8
DODECYL MERCAPTOACETATE
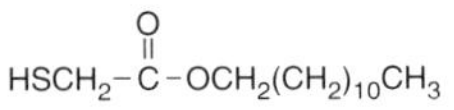
5 gm

S60,042-3
DODECYL 3-MERCAPTOPROPIONATE
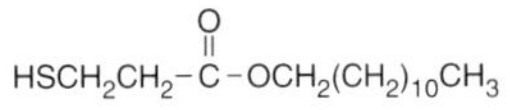
5 gm

S55,753-6
OCTADECYL MERCAPTOACETATE
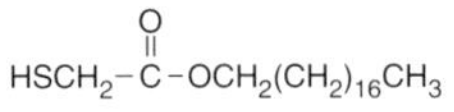
5 gm

S37,572-1
2-METHYLTHIOETHYL ACETATE
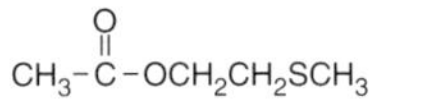
5 gm

S36,568-8
METHYL 3-(ETHYLTHIO)PROPIONATE
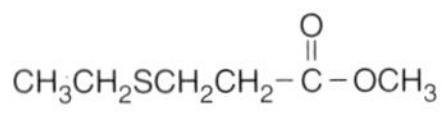
5 gm

S95,016-5
TERT-BUTYL 2-(METHYLTHIO)PROPIONATE
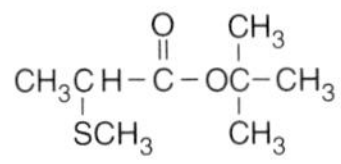
50 mg

S57,802-9
ETHYL 2-(DODECYLTHIO)ACETATE
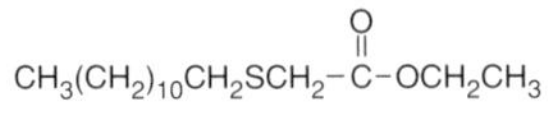
250 mg

S37,576-4
2-METHYLTHIOETHYL PALMITATE
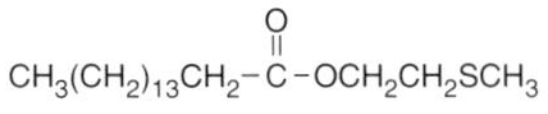
5 gm

S59,480-6
DIETHYL THIODIGLYCOLATE

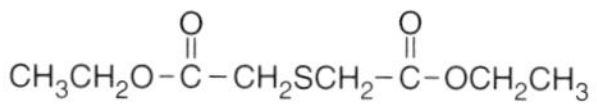

5 gm

S40,766-6
DIETHYL 3,5-DITHIAPIMELATE

5 gm

S45,940-2
DIETHYL 3,3'-THIODIPROPIONATE

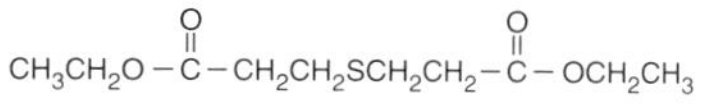

5 gm

S59,400-8
DIBUTYL 2,2'-(METHYLENEDITHIO)-DIACETATE

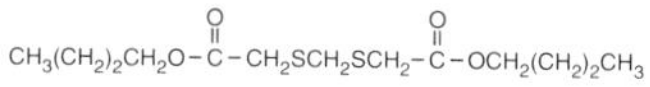

5 gm

S40,786-0
DILAURYL 3,3'-THIODIPROPIONATE

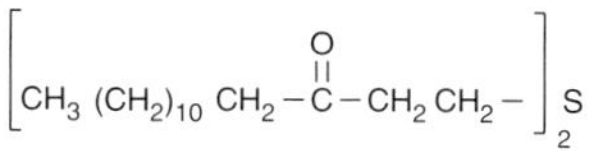

5 gm

S36,122-4
DIOCTADECYL 3,3'-THIODIPROPIONATE

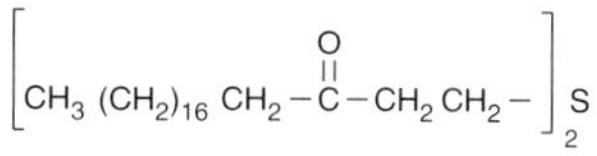

5 gm

S38,720-7
S-HEXADECYL THIOACETATE

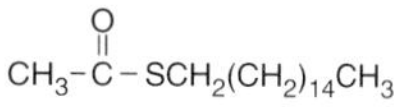

250 mg

S34,889-9
DIBUTYL DITHIODIGLYCOLATE

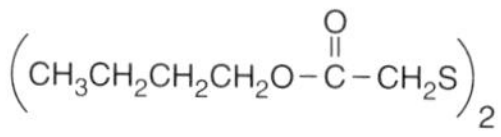

5 gm

S43,835-9
(ETHOXYCARBONYL)METHYL DIMETHYLSULFONIUM BROMIDE

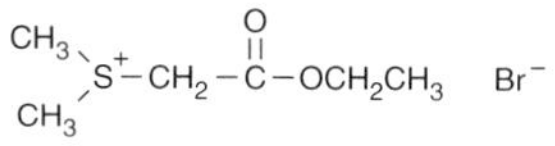

5 gm

S88,122-8
S-ETHYL 2-METHYLTHIOPROPIONATE

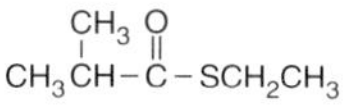

250 mg

S91,811-3
S-METHYL THIOOCTANOATE

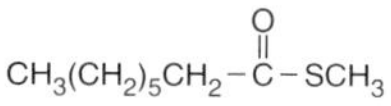

250 mg

S86,498-6
S-METHYL O-(1-METHYLHEPTYL)-DITHIOCARBONATE

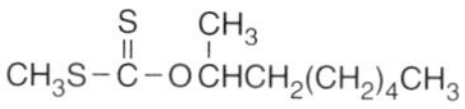

250 mg

S90,626-3
3,3'-[N-(2,3-DIHYDROXYPROPYL)IMINO]-BIS(ETHYL PROPIONATE)

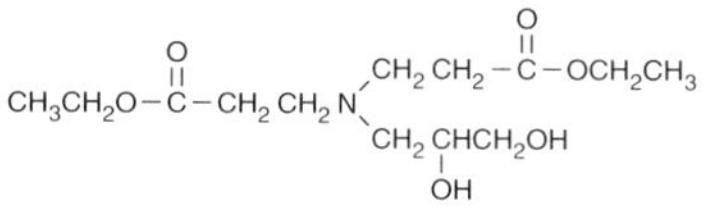

250 mg

S50,189-1
[2-HYDROXY-3-(METHACRYLOYLOXY)-PROPYL]TRIMETHYLAMMONIUM CHLORIDE

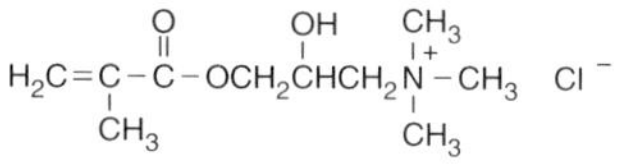

5 gm

S94,987-6
DL-SERINE ISOPROPYL ESTER HYDROCHLORIDE

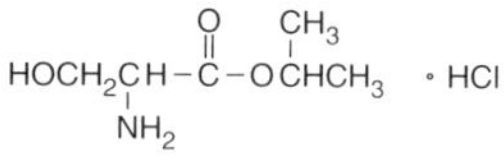

250 mg

S76,543-0
DL-THREONINE METHYL ESTER HYDROCHLORIDE

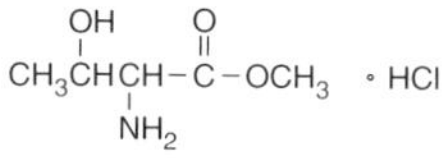

5 gm

S85,239-2
DL-THREONINE METHYL ESTER

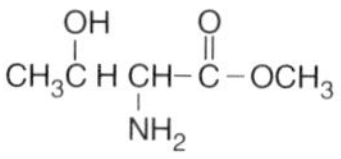

250 mg

S34,893-7
N',N2-DICARBETHOXY-N'-(2-HYDROXYETHYL)-2-METHYL-1,2-PROPANEDIAMINE

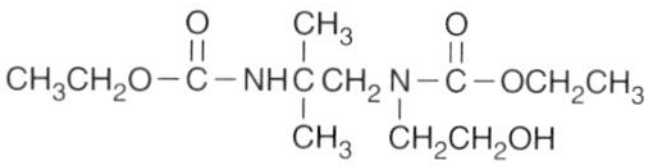

5 gm

S78,790-6
D-METHIONINE METHYL ESTER

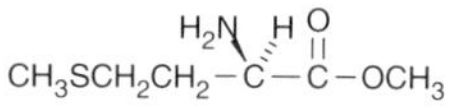

250 mg

S78,515-6
L-METHIONINE TERT-BUTYL ESTER HYDROCHLORIDE

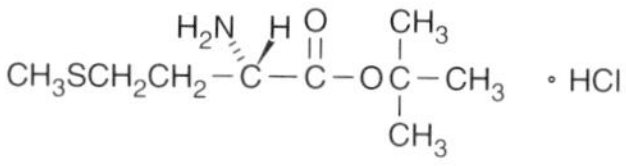

250 mg

S61,388-6
METHYL 3-AMINO-2-BUTENOATE

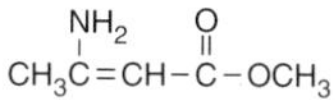

250 mg

S40,792-5
DIMETHYL 3-AMINOGLUTACONATE

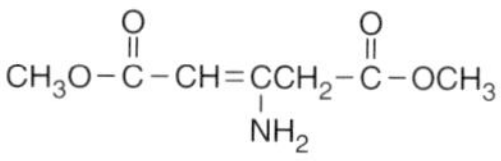

5 gm

S74,529-4
DIMETHYL 1-AMINO-1-CYCLOHEXENE-2,3-DICARBOXYLATE

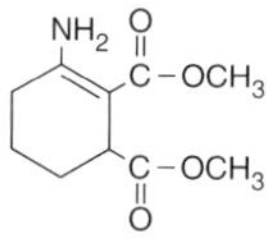

250 mg

S42,372-6
2-AMINOETHYL 2,4-HEXADIENOATE HYDROCHLORIDE

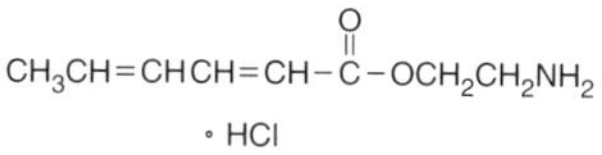

5 gm

S78,721-3
DL-ASPARTIC ACID DIMETHYL ESTER HYDROCHLORIDE

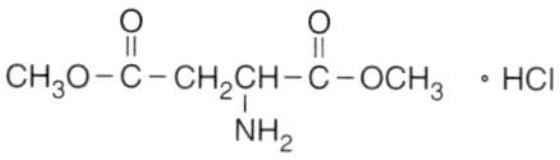

250 mg

S60,114-4
DIETHYL GLUTAMATE

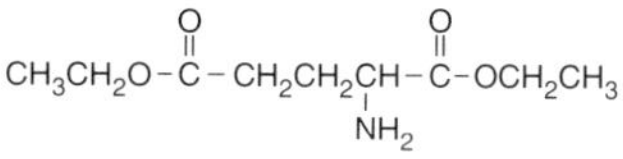

250 mg

S90,657-3
DIMETHYL 3-AMINOGLUTARATE

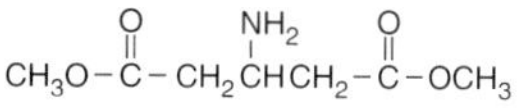

250 mg

S74,229-5
DIMETHYL 2-AMINO-1-CYCLOHEXENE-1,5-DICARBOXYLATE

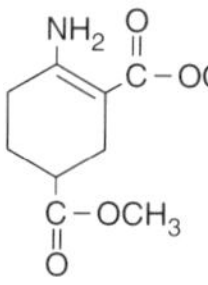

5 gm

S45,734-5
L-ALANINE TERT-BUTYL ESTER HYDROCHLORIDE

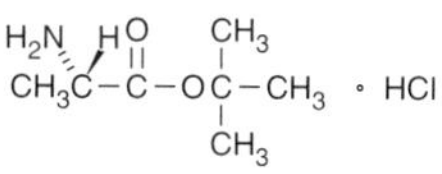

250 mg

S84,601-5
L-LEUCINE ETHYL ESTER

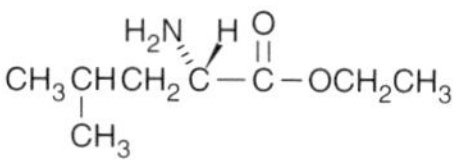

250 mg

S75,673-3
L-VALINE TERT-BUTYL ESTER HYDROCHLORIDE

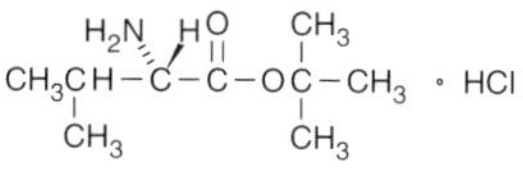

5 gm

S75,684-9
METHYL 6-AMINOHEXANOATE HYDROCHLORIDE

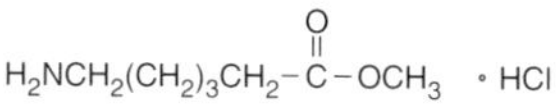

5 gm

S87,545-7
TRANS-ETHYL 2-AMINO-1-CYCLOHEXANECARBOXYLATE

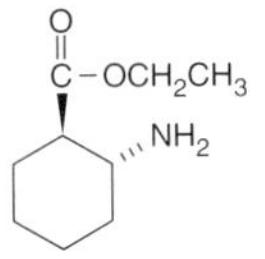

250 mg

S93,806-8
METHYL ALPHA-AMINO-1-ADAMANTANEACETATE HYDROCHLORIDE

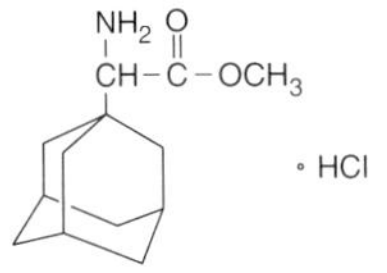

250 mg

S50,429-7
2-(TERT-BUTYLAMINO)ETHYL METHACRYLATE

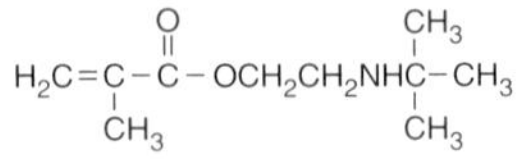

5 gm

S89,689-6
ETHYL 3-(5,5-DIMETHYL-3-OXO-1-CYCLOHEXEN-1-YLAMINO)PROPIONATE

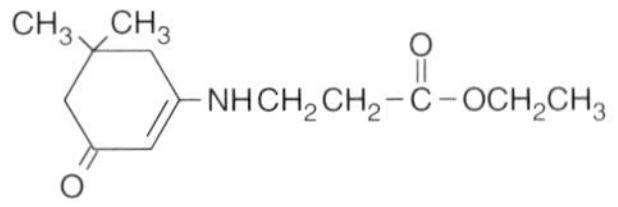

250 mg

S73,726-7
METHYL 3-(METHYLAMINO)CROTONATE

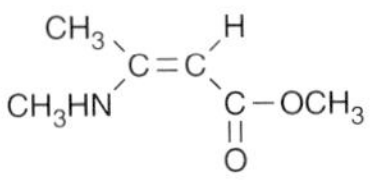

5 gm

S37,343-5
ETHYL 3-METHYLAMINOCROTONATE

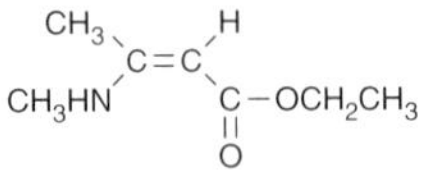

5 gm

S54,121-4
ETHYL 3-(ETHYLAMINO)CROTONATE

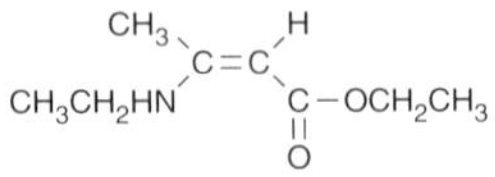

250 mg

S50,908-6
ETHYL 3-(ALLYLAMINO)CROTONATE

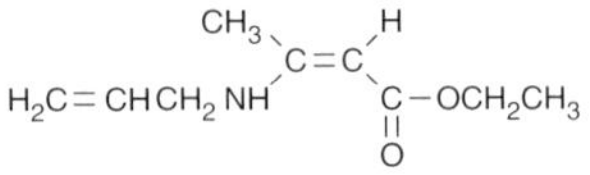

5 gm

S53,560-5
ETHYL 3-(METHYLAMINO)BUTYRATE

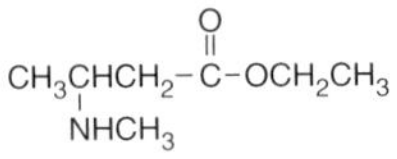

250 mg

S55,363-8
ETHYL 3-(DODECYLAMINO)PROPIONATE

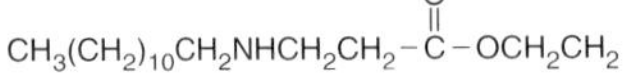

250 mg

S56,307-2
ETHYL 3-(DIETHYLAMINO)CROTONATE

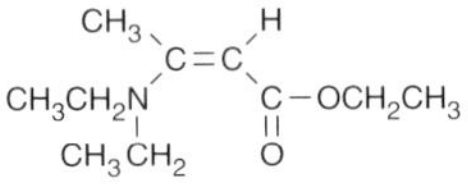

5 gm

S54,100-1
DIETHYL 2-(DIETHYLAMINO)MALONATE

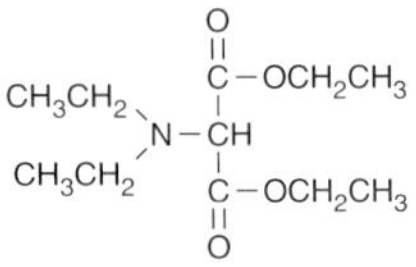

250 mg

S34,665-9
2-(DIMETHYLAMINO)ISOPROPYL ACETATE

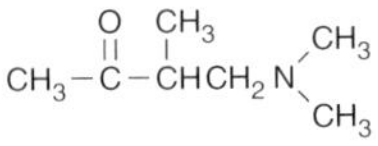

5 gm

S36,042-2
2-DIETHYLAMINOETHYL 2-CYCLOHEXYLBUTYRATE CITRATE

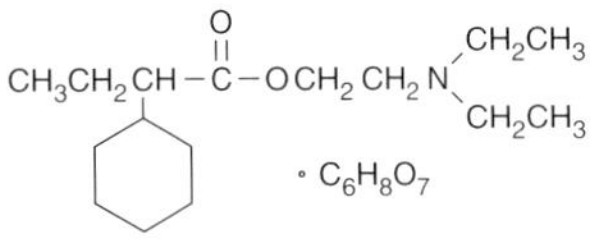

5 gm

S60,274-4
2-(DIETHYLAMINO)ETHYL 2-CYCLOHEXYL-2-ISOBUTYLACETATE

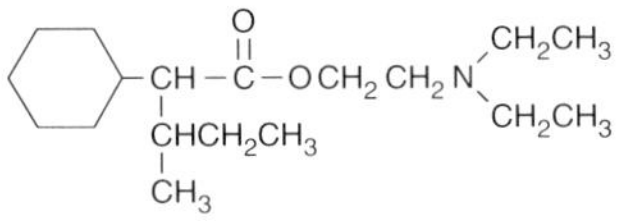

5 gm

S53,290-8
ETHYL 3-(DIMETHYLAMINO)CROTONATE

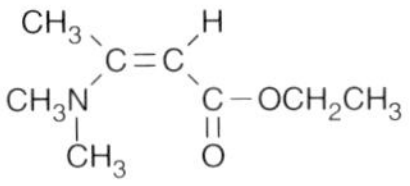

5 gm

S43,241-5
DIETHYL (2-DIMETHYLAMINOETHYL)-MALONATE

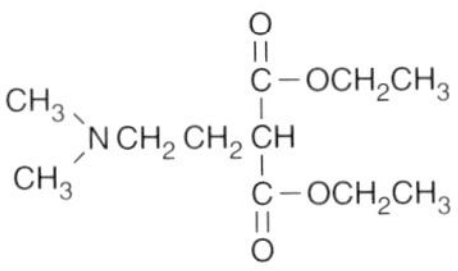

5 gm

S43,242-3
DIETHYL (3-DIMETHYLAMINOPROPYL)-MALONATE

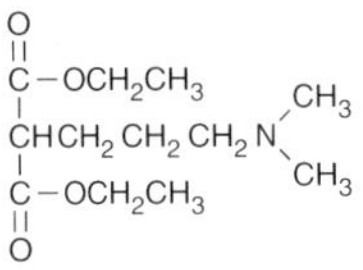

5 gm

S43,240-7
DIETHYL (3-DIETHYLAMINOPROPYL)-MALONATE

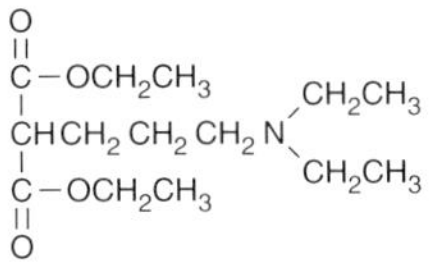

5 gm

S58,839-3
N-HEXADECYL-N-METHYLGLYCINE ETHYL ESTER

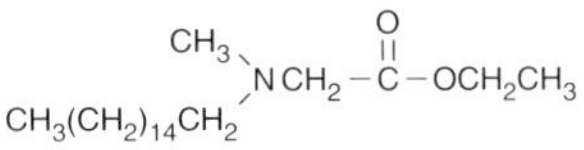

250 mg

S57,332-9
DIETHYL (METHYLIMINO)DIACETATE

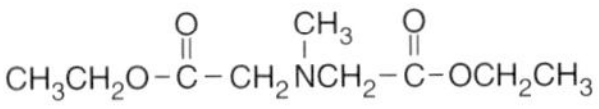

5 gm

S92,608-6
N-(ETHOXYCARBONYLMETHYL)-N-METHYL-BETA-ALANINE ETHYL ESTER

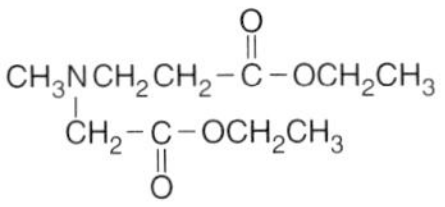

250 mg

S34,837-6
ETHYL 3-(DIPROPYLAMINO)PROPIONATE

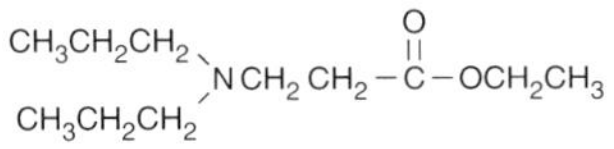

5 gm

S34,827-9
ETHYL 3-(DIBUTYLAMINO)PROPIONATE

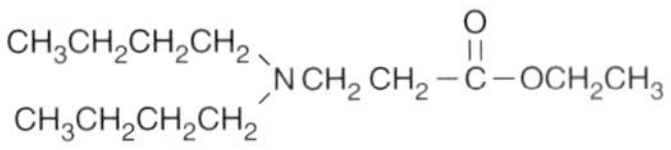

5 gm

S56,796-5
DIETHYL 3,3'-(METHYLIMINO)-DIPROPIONATE

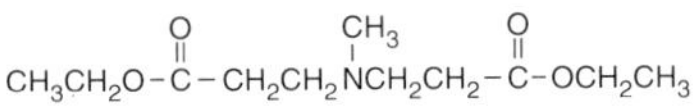

250 mg

S71,337-6
3,3'-ETHYLIMINOBIS(ETHYL PROPIONATE)

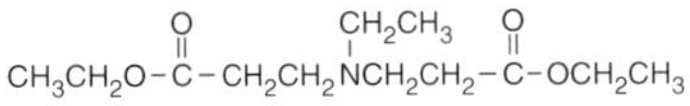

5 gm

S89,693-4
ETHYL 1-(DIMETHYLAMINO)-1-CYCLOHEXANEACETATE HYDROCHLORIDE

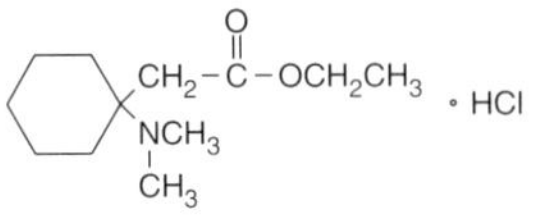

250 mg

S38,952-8
3-DIMETHYLAMINOPROPYL ETHYL CARBONATE

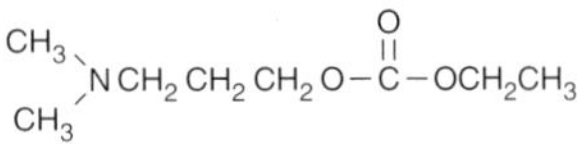

5 gm

S39,851-9
SUCCINYLDICHOLINE DIIODIDE

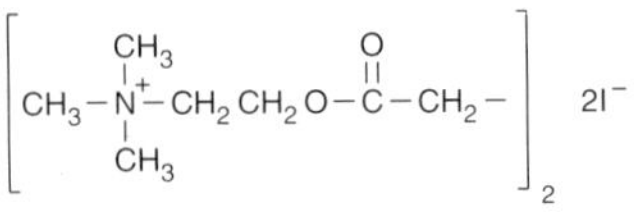

250 mg

S44,263-1
PROPIONYLCHOLINE IODIDE

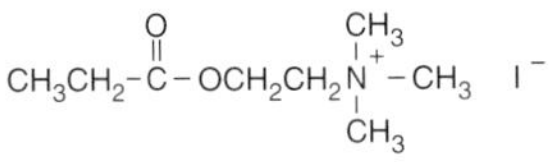

5 gm

S56,033-2
BUTYRYLCHOLINE BROMIDE

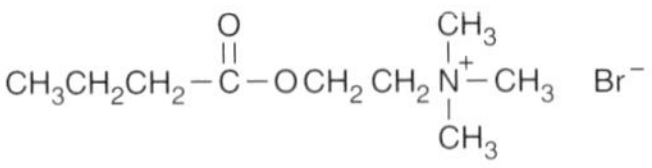

5 gm

S67,899-6
DIETHYL 1,2-DIMETHYL-1,2-HYDRAZINEDICARBOXYLATE

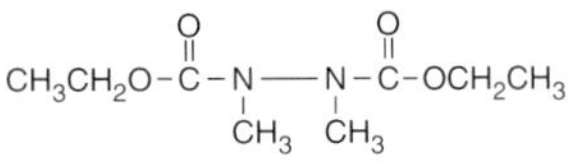

5 gm

S39,272-3
2,2-DINITROPROPYL ACRYLATE
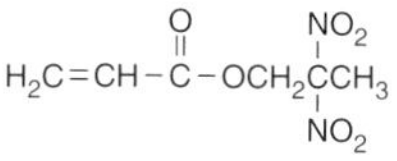
5 gm

S39,270-7
2,2-DINITROPROPYL ACETATE
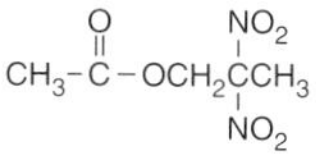
5 gm

S39,271-5
2,2-DINITROPROPYL PROPIONATE
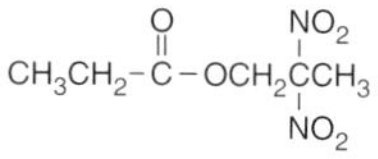
5 gm

S58,847-4
TERT-DECYL 2,4-DIMETHYL-4-NITROVALERATE
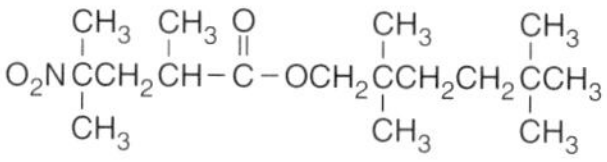
250 mg

S56,811-2
DODECYL 4-METHYL-4-NITROHEXANOATE
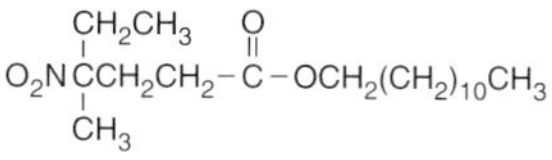
250 mg

S35,310-8
METHYL 1-NITROCYCLOHEXANEPROPIONATE
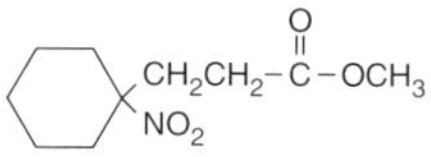
5 gm

S37,890-9
ALANINE METHYL ESTER HYDROCHLORIDE
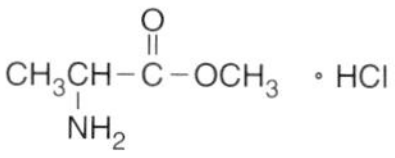
5 gm

S76,579-1
D-VALINE METHYL ESTER
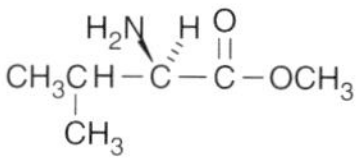
5 gm

S59,889-5
L-LEUCINE ETHYL ESTER HYDROCHLORIDE
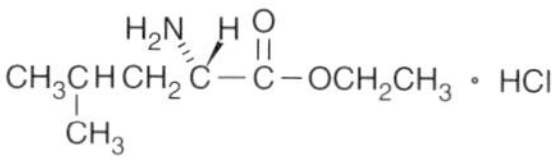
250 mg

S64,145-6
6-IODO-ALPHA-METHYLGLUCOSIDE TRIACETATE
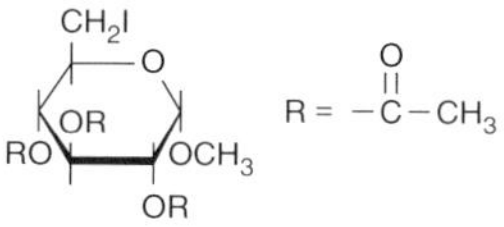
5 gm

S41,665-7
(CHLOROMETHYL)ETHYLENE CARBONATE
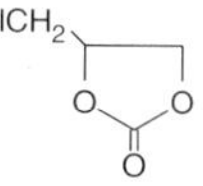
5 gm

S74,955-9
DIETHYL 4-CHLOROMETHYL-1,4-DIHYDRO-2,6-DIMETHYL-3,5-PYRIDINEDICARBOXYLATE
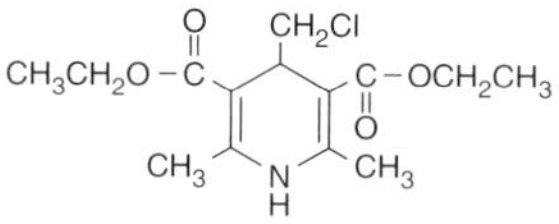
250 mg

S52,799-8
2,2,2-TRICHLOROETHYL 4-HYDROXY-1-PIPERIDINEPROPIONATE
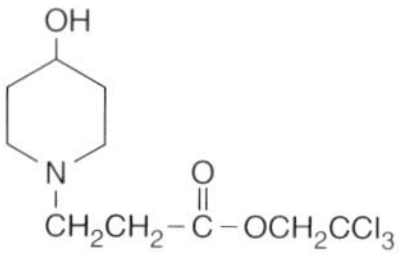
5 gm

S95,241-9
TETRAHYDROFURFURYL 2,2,2-TRIFLUOROACETATE
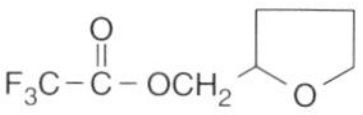
50 mg

S95,837-9
ISOPROPYL 4,6-DIHYDROXY-1,7-EPOXY-1,4,4A,5,6,7,8,8A-OCTAHYDRO-NAPHTHALENE-8-CARBOXYLATE
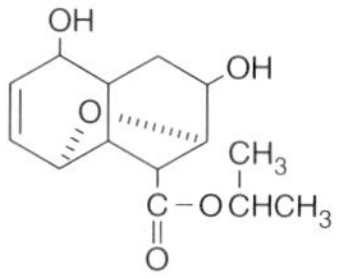
50 mg

S95,133-1
METHYL 4-HYDROXY-2-(METHOXYCARBONYL)-1-PYRROLIDINEPROPIONATE
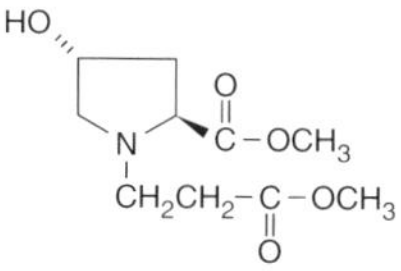
50 mg

S85,244-9
ALPHA-AMINO-GAMMA-METHYL-GAMMA-VALEROLACTONE HYDROCHLORIDE
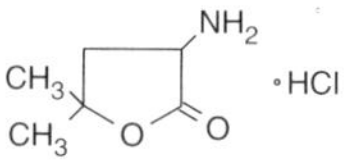
250 mg

S52,044-6
ETHYL 4-HYDROXY-1-PIPERIDINECARBOXYLATE
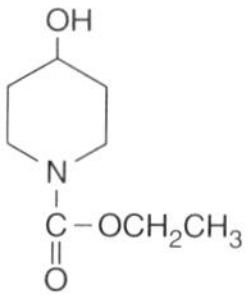
250 mg

S95,247-8
2-(DIACETOXYMETHYL)-2,5-DIHYDRO-2,5-DIMETHOXYFURAN
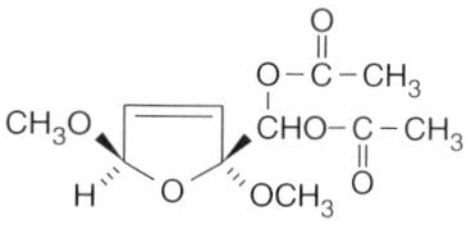
50 mg

S64,116-2
BETA-METHYLISORHAMNOSIDE 2,3,4-TRIACETATE
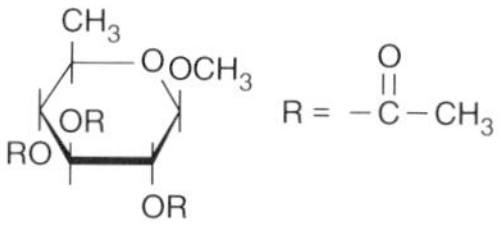
250 mg

S74,538-3
ETHYL 2-ETHOXY-3-TETRAHYDROFURANACETATE
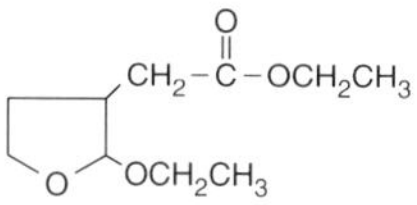
250 mg

S95,512-4
S-(DIMETHOXYTHIOPYRANYL)TRIMETHYL-2-OXABICYCLO[2.2.1]HEPTANE-1-THIOCARBOXYLATE
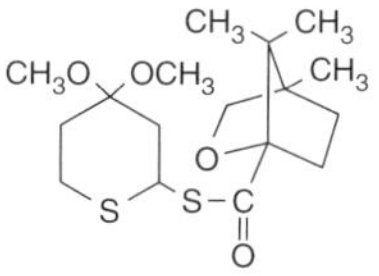
50 mg

S64,172-3
ACETOBROMOCELLOBIOSE
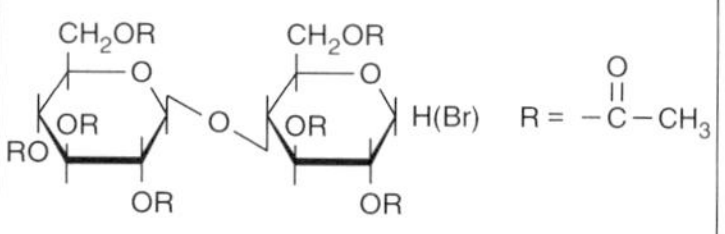
1 gm

S38,773-8
2,5-DIHYDRO-2,5-DIMETHOXYFURFURYL ACETATE
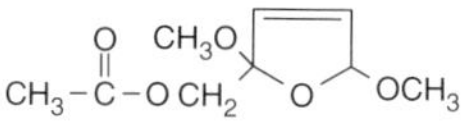
250 mg

S64,120-0
LEVO-GLUCOSANE TRIACETATE
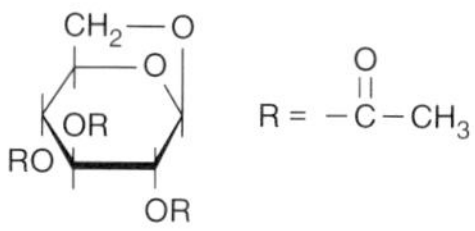
5 gm

S87,574-0
TETRAMETHYL BIS(METHOXYCARBONYL)-CYCLOPROPENYLOXABICYCLO[2.2.1]-HEPTADIENETETRACARBOXYLATE
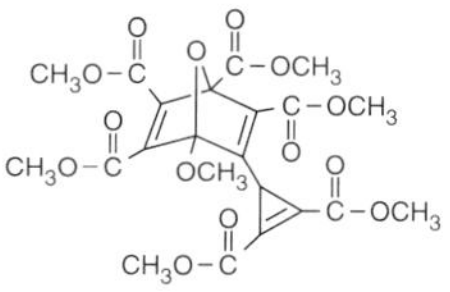
250 mg

S88,789-7
TETRAETHYL DELTA-4,4'-BI-4H-PYRAN-2,2',6,6'-TETRACARBOXYLATE
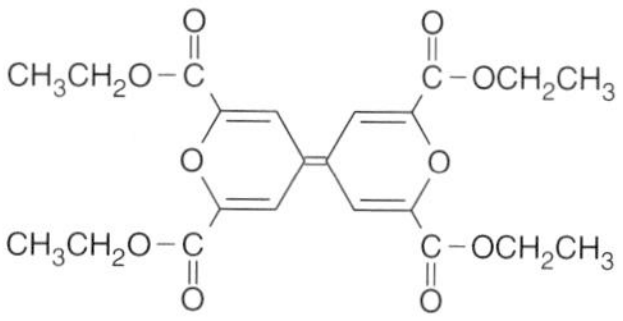
250 mg

S37,516-0
ETHYLENEGLYCOL BIS(2,3-EPOXYBUTYRATE)
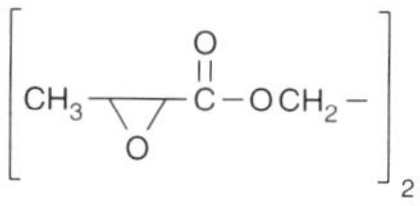
5 gm

S79,856-8
2,3-TETRAHYDROPYRANDIOL DIACETATE
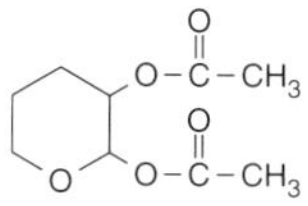
250 mg

S46,234-9
ETHYL 2,3-EPOXYBUTYRATE
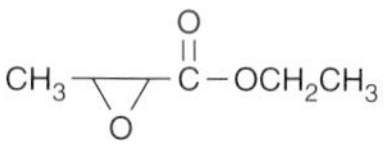
5 gm

S37,441-5
ETHYL 2,3-EPOXY-3-METHYLVALERATE
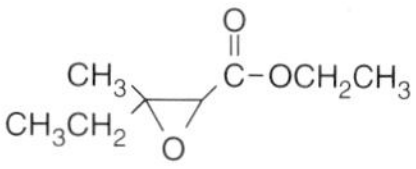
5 gm

S35,269-1
METHYL 3-SEC-BUTYL-3-METHYLGLYCIDATE
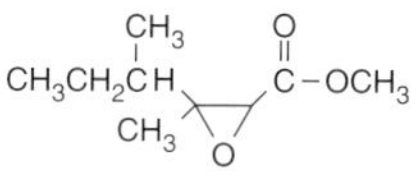
5 gm

S37,450-4
BUTYL 7-OXABICYCLO[4.1.0]HEPTANE-1-CARBOXYLATE
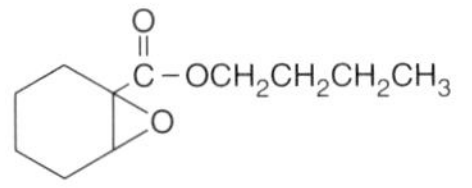
250 mg

S88,476-6
METHYL 2-TETRAHYDROFURANCARBOXYLATE
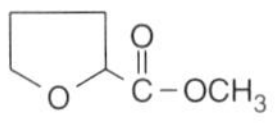
250 mg

S43,091-9
TETRAHYDROFURFURYL ACETATE
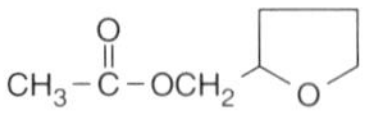
5 gm

S36,495-9
TETRAHYDROPYRAN-2-YLMETHYL ACETATE
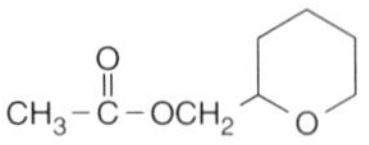
5 gm

S43,801-4
2,5-DIMETHYLTETRAHYDROPYRAN-2-YLMETHYL ACETATE
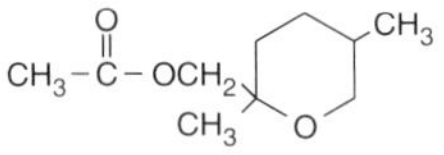
250 mg

S39,490-4
4,4-BIS(ACETOXYMETHYL)-3-ETHYL-2-METHYL-2-OXAZOLINIUM IODIDE
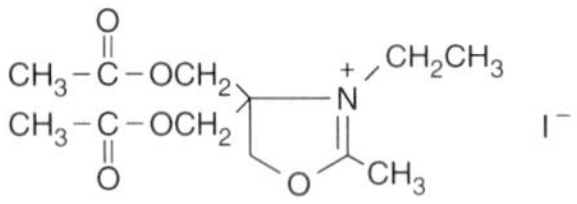
250 mg

S79,246-2
TETRAMETHYL 9AH-QUINOLIZINE-1,2,3,4-TETRACARBOXYLATE
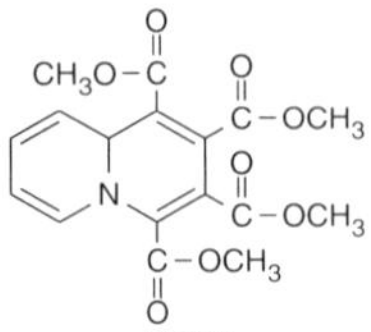
250 mg

S88,469-3
ISOPROPYL 4,5-EPOXY-2-HEXENOATE
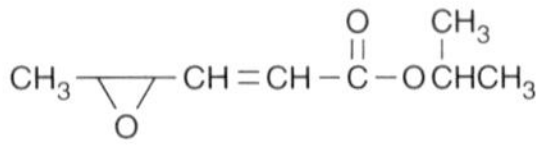
250 mg

S87,581-3
ETHYL 2-METHYL-1,3-DIOXOLANE-2-PROPIONATE
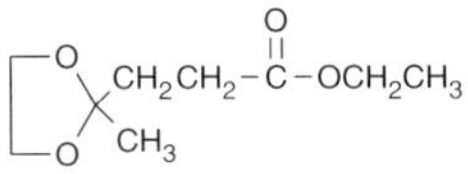
250 mg

S53,325-4
ETHYL 4-(2-PIPERIDYLMETHYL)-1-PIPERAZINECARBOXYLATE
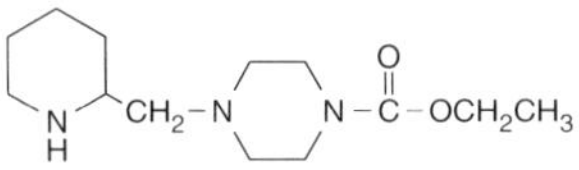
5 gm

S34,943-7
ETHYL 4-MORPHOLINEPROPIONATE
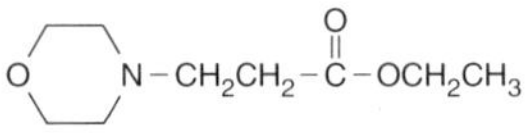
5 gm

S67,207-6
ETHYL 3-PYRROLIDINOCROTONATE
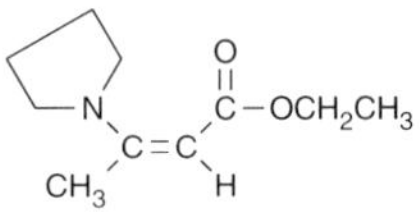
5 gm

S38,109-8
ETHYL BETA-METHYL-1-PIPERIDINEACRYLATE
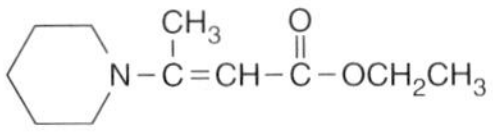
5 gm

S38,136-5
ETHYL 4-METHOXYCARBONYL-1-PIPERIDINEACETATE
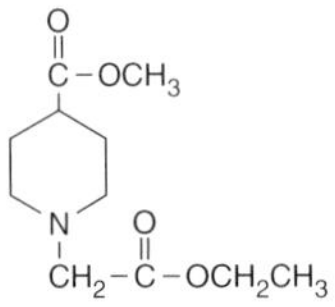
5 gm

S36,346-4
METHYL 4-ETHOXYCARBONYL-1-PIPERIDINEACETATE
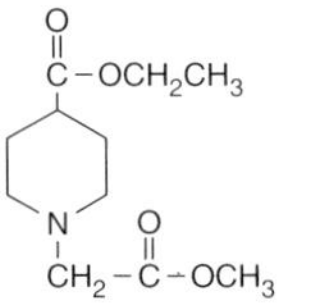
5 gm

S36,359-6
METHYL 3-(HEXAMETHYLENEIMINO)-ISOBUTYRATE
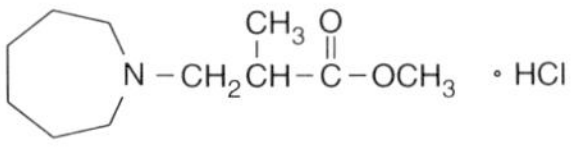
5 gm

S39,446-7
BUTYL 1-METHYL-2-PYRROLIDINEACETATE
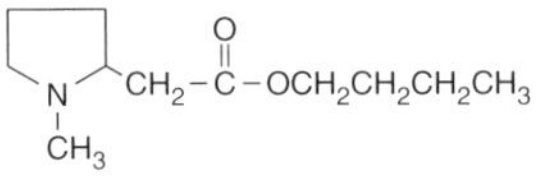
5 gm

S51,839-5
3-(2-ISOPROPYLOXAZOLIDINO)-PROPYL ACETATE
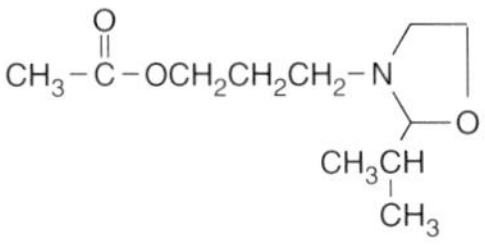
5 gm

S36,378-2
METHYL ALPHA-METHYL-1-PIPERIDINEPROPIONATE HYDROCHLORIDE
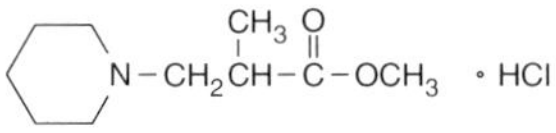
5 gm

S56,184-3
3-QUINUCLIDYL ACETATE HYDROCHLORIDE
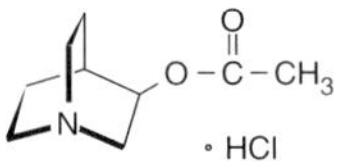
250 mg

S36,364-2
METHYL ALPHA-METHYL-4-MORPHOLINEPROPIONATE HYDROCHLORIDE
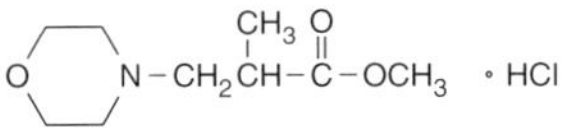
5 gm

S35,286-1
METHYL ALPHA,2,6-TRIMETHYL-4-MORPHOLINEPROPIONATE
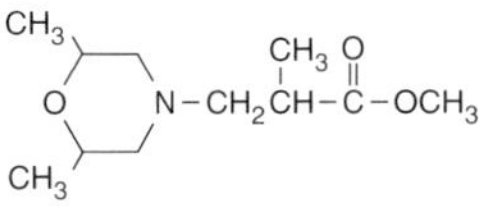
5 gm

S36,167-4
ETHYL 3-(2,6-DIMETHYLMORPHOLINO)-PROPIONATE
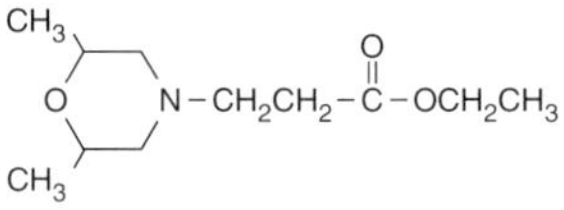
5 gm

S35,304-3
METHYL N-METHYL-3-PIPERIDINECARBOXYLATE METHOBROMIDE
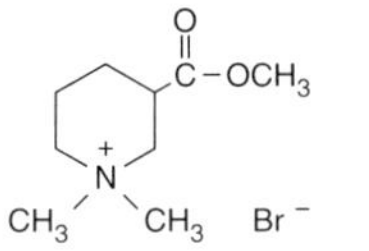
5 gm

S70,577-2
2-TETRAHYDROFURYL ACRYLATE
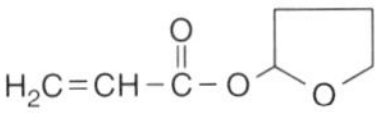
5 gm

S88,492-8
METHYL 5-(2-METHYL-1,3-DIOXOLAN-2-YL)-2-METHYLENEPENTANOATE
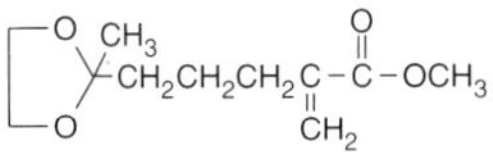
250 mg

S80,878-4
METHYL 2-PYRAZOLINE-3-CARBOXYLATE
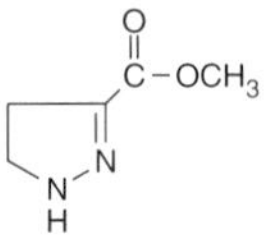
250 mg

S59,948-4
4-(CARBETHOXYMETHYL)MORPHOLINE
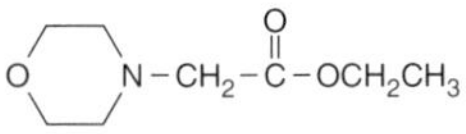
5 gm

S83,830-6
ETHYL 2-(1H-TRIAZIRIN-1-YL)ACETATE
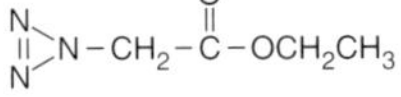
250 mg

S93,662-6
DIMETHYL 1,4-DIHYDRO-1,2,4,5-TETRAZINE-3,6-DICARBOXYLATE
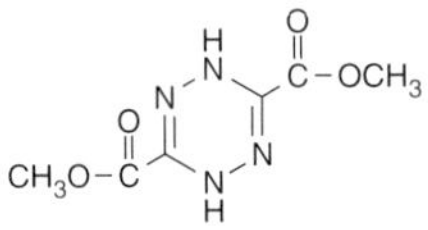
250 mg

S37,584-5
DIETHYL 7-OXATRICYCLO[4.1.0]-HEPTANE-3,4-DICARBOXYLATE
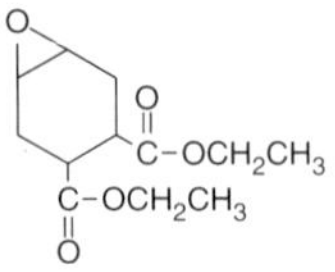
5 gm

S75,574-5
ETHYL 6,7-DIHYDRO-5H-THIAZOLO(3,2-A)PYRIMIDINE-3-ACETATE HYDROCHLORIDE
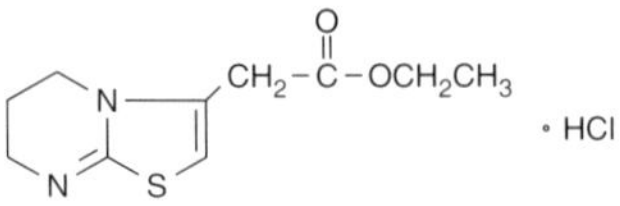
250 mg

S37,438-5
4,5-EPOXYPENTYL ACETATE
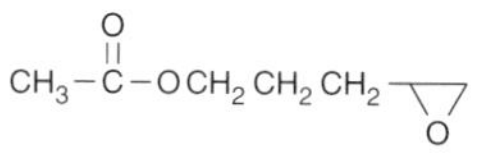
250 mg

S43,006-4
METHYL 10,11-EPOXYUNDECANOATE
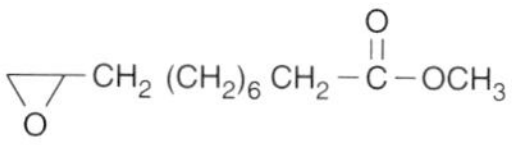
5 gm

S37,446-6
6-ACETOXY-3-OXATRICYCLO-[3.2.1.0(2,4)]OCTANE
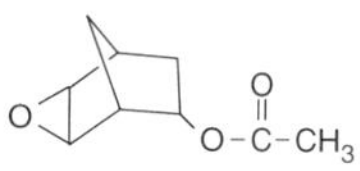
250 mg

S37,527-6
BUTYL 3-OXATRICYCLO[3.2.1.0(2.4)]-OCTANE-7-CARBOXYLATE
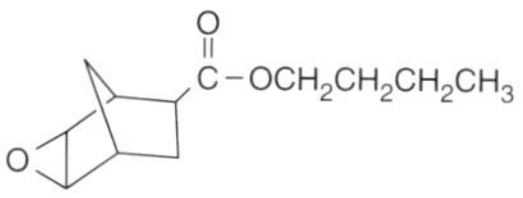
5 gm

S37,571-3
3,4-EPOXY-1-METHYL-6-ACETOXYMETHYLCYCLOHEXANE
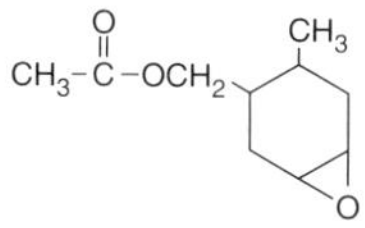
5 gm

S37,586-1
ALLYL 3,4-EPOXY-6-METHYLCYCLOHEXANECARBOXYLATE
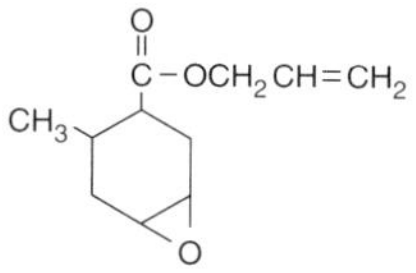
5 gm

S95,066-1
2-ACETOXY-1-METHYL-1-PROPYL-6,10-DITHIASPIRO(4.5)DECANE
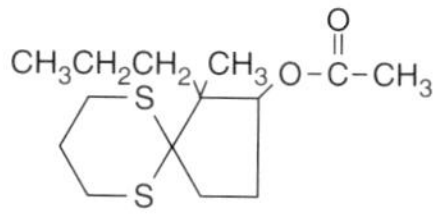
50 mg

S79,302-7
METHYL 2-METHYL-1,3-DIOXOLANE-2-BUTYRATE
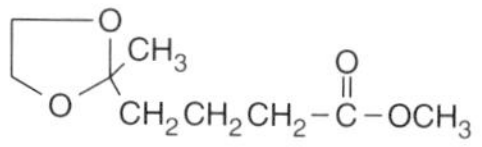
250 mg

S73,931-6
METHYL 4-(1,3-DIOXOLAN-2-YL)-4-ETHYL-6-HEPTENOATE
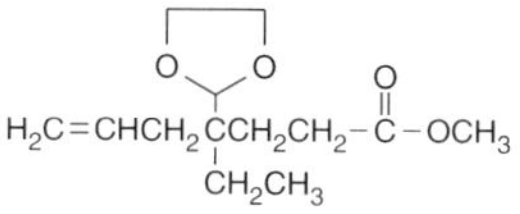
5 gm

S88,450-2
METHYL 1,4-DIOXASPIRO(4.5)-DECANE-7-ACETATE
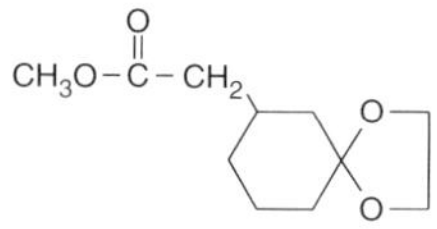
250 mg

S95,409-8
METHYL 3-[10-METHYL-1,4-DIOXASPIRO-(4.5)DEC-7-EN-8-YL]PROPIONATE
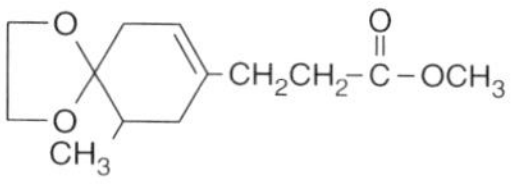
50 mg

S92,031-2
ETHYL 2-THIAZOLIDINEACETATE
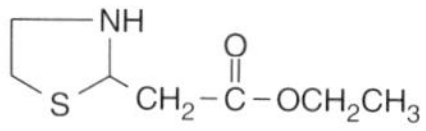
250 mg

S52,837-4
3-PENTYLTETRAHYDROPYRAN-4-YL ACETATE
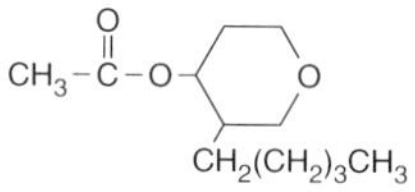
5 gm

S73,448-9
ETHYL 2-(3,4,5,6-TETRAHYDRO-2-PYRIMIDINYL)ACETATE
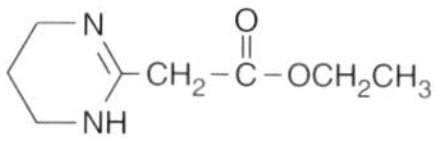
250 mg

S94,192-1
ETHYL BETA-MORPHOLINOCROTONATE
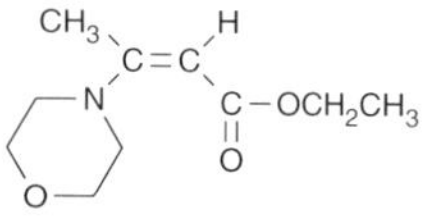
5 gm

S81,901-8
ETHYL 2-METHOXY-3-OXO-4,4,4-TRIFLUOROBUTYRATE
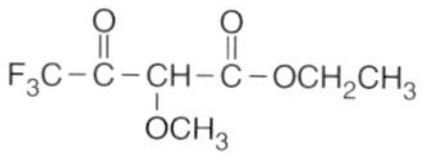
250 mg

S81,308-7
ETHYL 3-OXO-4,4,4-TRIFLUOROBUTYRATE, DIMETHYLTHALLIUM SALT
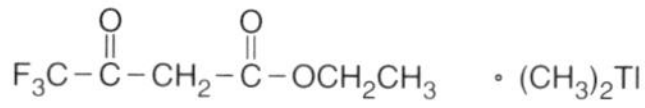
250 mg

S40,461-6
LASERPITIN
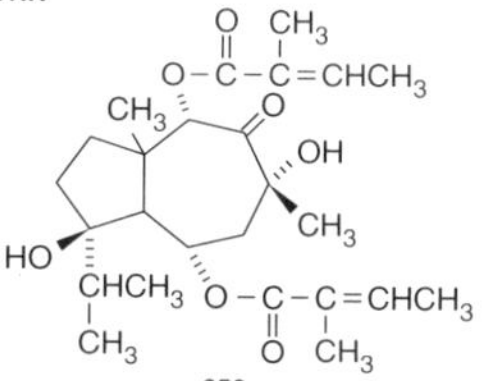
250 mg

S95,499-3
TETRAETHYL ALPHA,ALPHA'-DIHYDROXY-4-OXO-3,5-TETRAHYDROPYRANDIMALONATE
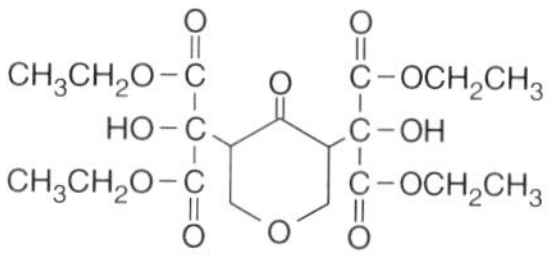
50 mg

S70,705-8
2,3-DIHYDROXYPROPYL 4-OXOSTEARATE
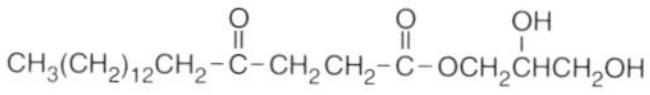
250 mg

S79,816-9
2,6-DIMETHOXY-4-(2-HYDROXYACETYL)-CYCLOHEXYL ACETATE
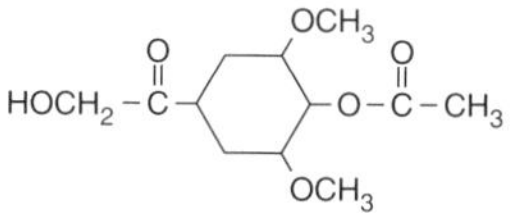
250 mg

S75,979-1
ETHYL 2-HYDROXY-4-OXO-2-PENTENOATE, SODIUM SALT
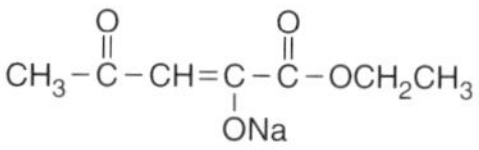
250 mg

S36,361-8
METHYL 2-KETO-L-GULONATE
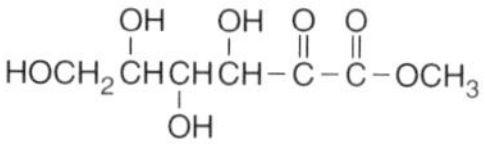
5 gm

S79,308-6
ETHYL 2,5-DIOXO-3-ETHYL-4-HYDROXY-3-CYCLOPENTENE-1-GLYOXYLATE
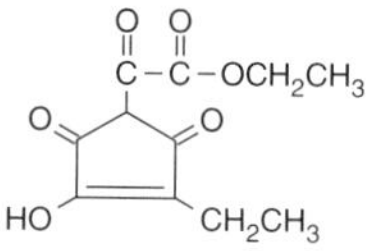
250 mg

S35,187-3
KOJYL PALMITATE
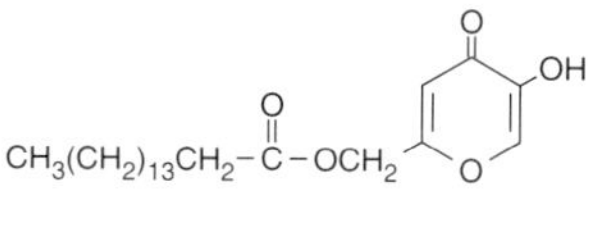
5 gm

S89,731-0
DIMETHYL 2,2-DIMETHYL-4-HYDROXY-6-OXO-4-CYCLOHEXENE-1,3-DICARBOXYLATE
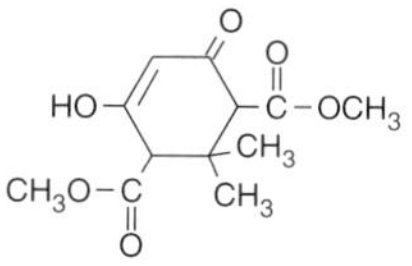
250 mg

S80,694-3
ETHYL 5-HYDROXY-3-OXO-6-OCTENOATE
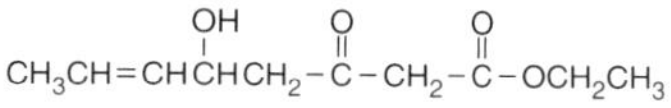
250 mg

S94,783-0
METHYL 3,6-DIMETHYL-4-HYDROXY-2-OXO-3-CYCLOHEXENE-1-CARBOXYLATE
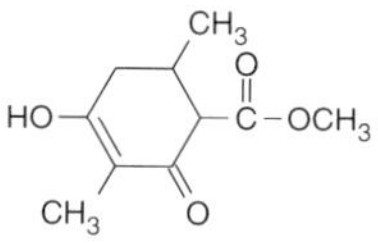
250 mg

S89,543-1
ETHYL 4-HYDROXY-1-METHYL-6-OXO-2,4-CYCLOHEXADIENE-1-CARBOXYLATE
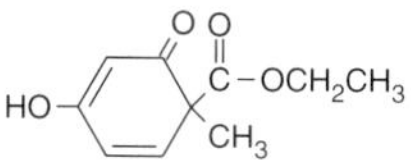
250 mg

S80,809-1
DIMETHYL CIS-DIETHYL-DIHYDROXY-DIOXO-HEXAH-CYCLOBUTA-DICYCLOPENTENEDICARBOXYLATE
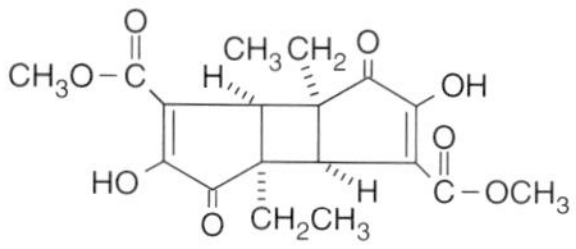
250 mg

S63,895-1
TRIETHYLENE GLYCOL BIS(DECANOATE)
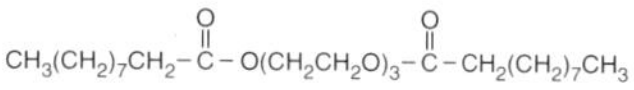
250 mg

S73,458-6
ETHYL 2-(5,5-DIMETHYL-3-OXO-1-CYCLOHEXENYLOXY)ACETATE
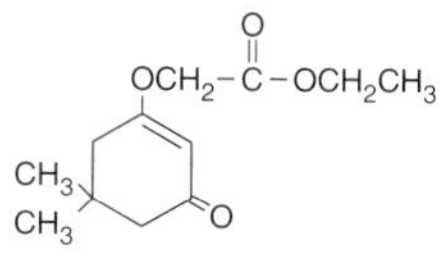
250 mg

S62,254-0
DIETHYLENE GLYCOL DIACETATE
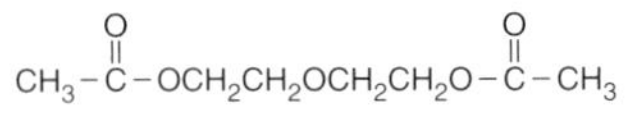
5 gm

S46,267-5
4-HYDROXY-3-PENTEN-2-ONE ACETATE
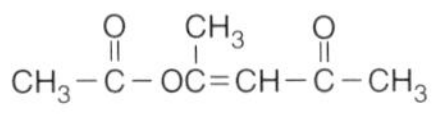
250 mg

S80,669-2
DIETHYL 2-(DIETHOXYMETHYL)-3-FORMYLSUCCINATE
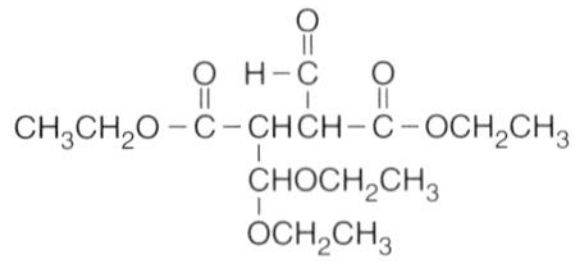
250 mg

S94,217-0
DIMETHYL 4-OXO-4H-PYRAN-2,6-DICARBOXYLATE
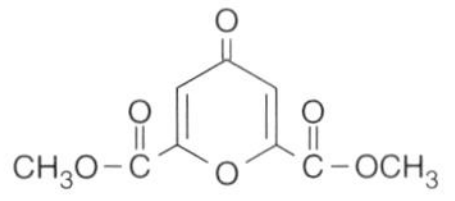
250 mg

S36,240-9
1-HYDROXY-2-BUTANONE ACETATE
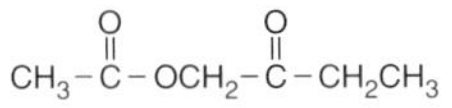
5 gm

S95,049-1
DECAHYDRO-6,7-DIACETOXY-1,8A-DIMETHYL-9,10-EPOXYPHENANTHREN-2-ONE
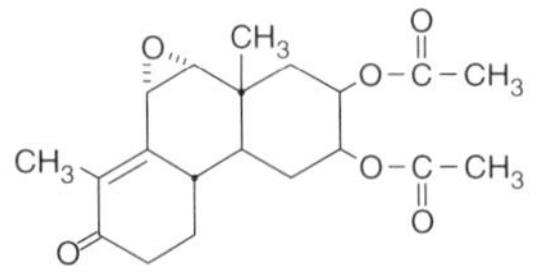
50 mg

S95,629-5
1-ACETOXY-1,2,4A,5,8,8A-HEXAHYDRO-4A-METHYL-2-NAPHTHALENONE
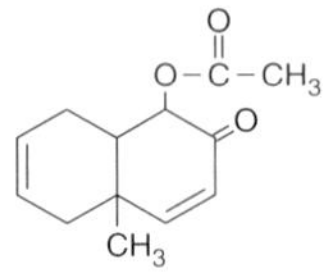
50 mg

S41,617-7
ETHYL 3-QUINUCLIDINONE-2-CARBOXYLATE HYDROCHLORIDE
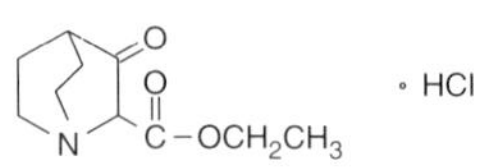
5 gm

S87,520-1
METHYL 2-OXO-1-CYCLODODECANECARBOXYLATE
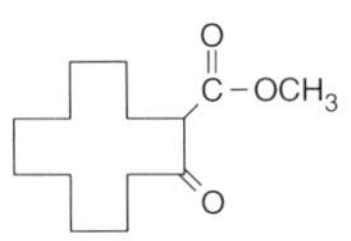
250 mg

S94,296-0
ETHYL 2-OXOCYCLODODECANECARBOXYLATE
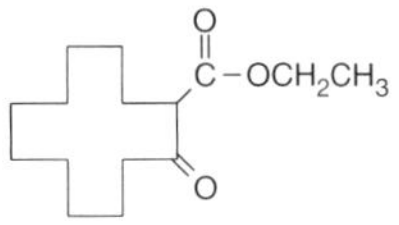
250 mg

S94,157-3
METHYL 6-METHYL-2,4-CYCLOHEXANEDIONECARBOXYLATE
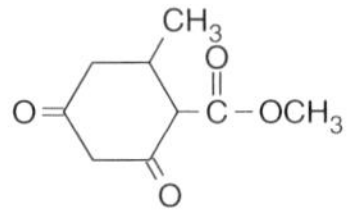
5 gm

S94,098-4
ETHYL 2,4-DIOXO-6-METHYL-1-CYCLOHEXANECARBOXYLATE
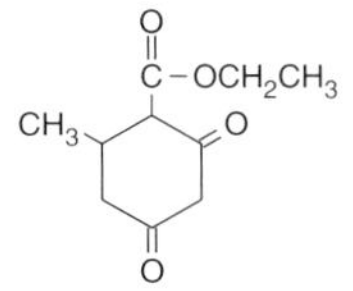
5 gm

S87,498-1
METHYL 2-(1,3-DIOXO-2-CYCLOTETRADECYL)ACETATE
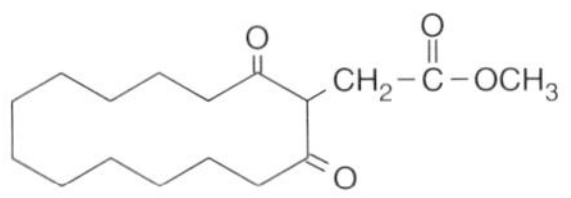
250 mg

S86,084-0
DIMETHYL 4-OXO-1,2-CYCLOPENTANEDICARBOXYLATE
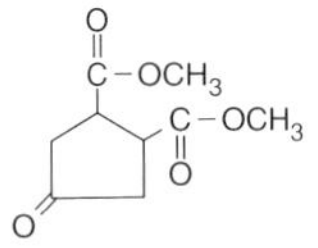
250 mg

S84,630-9
8-OXO-2-DECAHYDRONAPHTHYL ACETATE
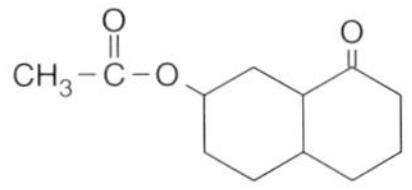
250 mg

S75,466-8
DIETHYL 1,4-DIHYDRO-4-OXO-2,6-PYRIDINEDICARBOXYLATE
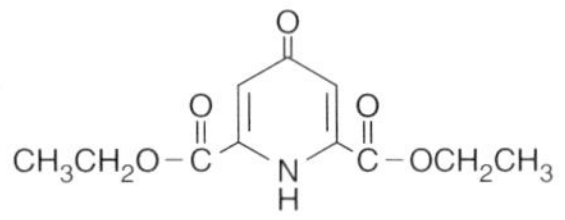
250 mg

S56,443-5
ETHYL 2-ACETYL-3-(DIMETHYLAMINO)-ACRYLATE
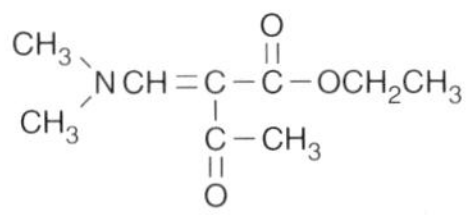
5 gm

S95,619-8
METHYL 5,8-DIOXO-1,4,4A,5,8,8A-HEXAHYDRO-1-NAPHTHALENECARBOXYLATE
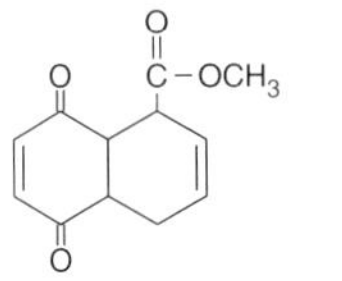
50 mg

S89,410-9
METHYL 3-OXO-2,2,4,5,5-PENTACHLORO-4-PENTENOATE
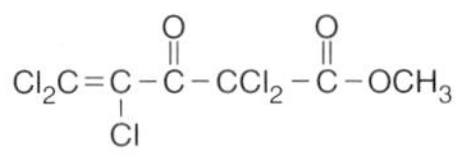
250 mg

S89,420-6
ETHYL 3-OXO-2,2,4,5-TETRACHLORO-4-PENTENOATE
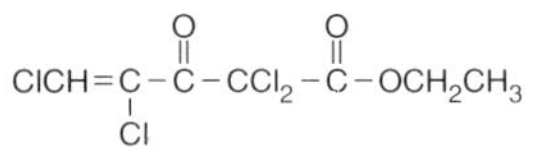
250 mg

S89,425-7
METHYL 2-OXO-3,4,5,6,6-PENTACHLORO-3,5-HEXADIENE-1-CARBOXYLATE
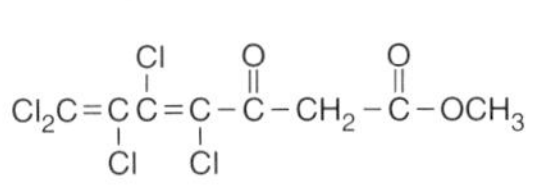
250 mg

S67,158-4
DIMETHYL 9-OXO-2-CYCLONONENE-1,2-DICARBOXYLATE
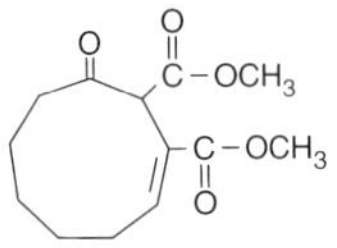
250 mg

S89,416-8
METHYL 5,5-DICHLORO-3-OXO-4-PENTENOATE
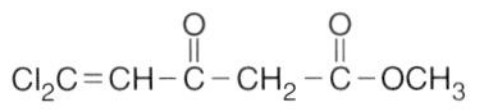
250 mg

S88,468-5
ETHYL 2-ACETYL-2-PENTENOATE
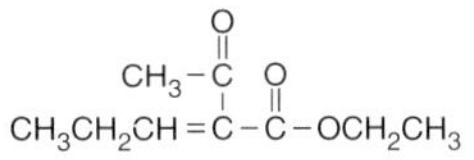
250 mg

S91,484-3
DIMETHYL 3-OXOBICYCLO[3.2.2]NONA-6,8-DIENE-6,7-DICARBOXYLATE
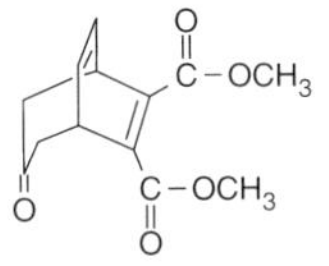
250 mg

S84,723-2
METHYL 5-OXOCYCLOPENTENEHEPTANOATE
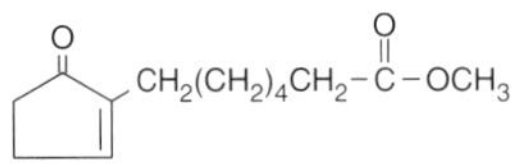
250 mg

S95,633-3
(+)-7-ACETOXY-4A-METHYL-1,2,4A,5,6,7,8,8A-OCTAHYDRO-2-NAPHTHALENONE
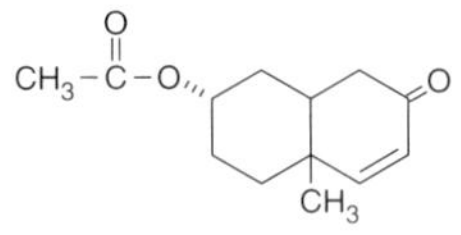
50 mg

S34,339-0
4-CARBETHOXY-3,5-DIMETHYL-2-CYCLOHEXENE-1-ONE
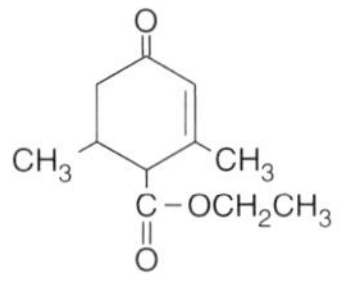
5 gm

S52,066-7
DIETHYL 3-HEPTYL-2-OXOSUCCINATE
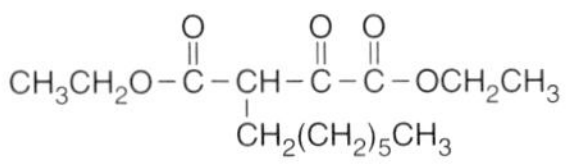
5 gm

S68,304-3
DIETHYL 3-ALLYL-2-OXOSUCCINATE
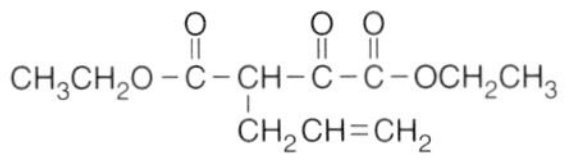
5 gm

S74,263-5
DIETHYL 2-OXO-3-PROPYLSUCCINATE
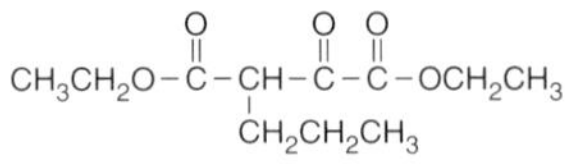
5 gm

S34,826-0
ETHYL 1-CYCLOPENTANONE-2-GLYOXYLATE
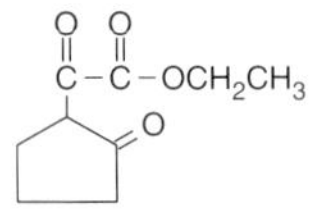
5 gm

S79,297-7
ETHYL 2,3,5-TRIOXO-1-CYCLOPENTANEGLYOXYLATE

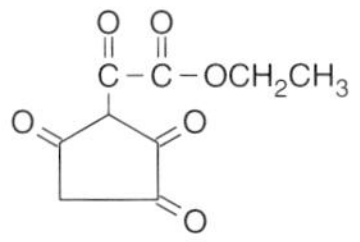

250 mg

S79,316-7
ETHYL 3-BUTYL-2,4,5-TRIOXO-1-CYCLOPENTANEGLYOXYLATE

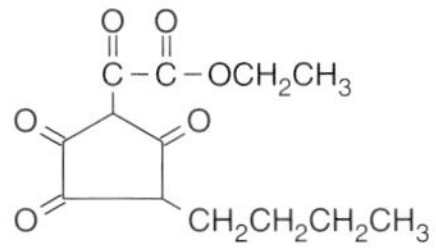

250 mg

S79,292-6
ETHYL 3-OCTYL-2,4,5-TRIOXO-1-CYCLOPENTANEGLYOXYLATE

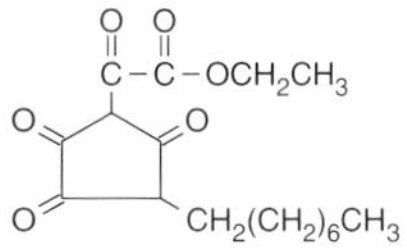

250 mg

S73,468-3
DIETHYL 2-OXOGLUTARATE

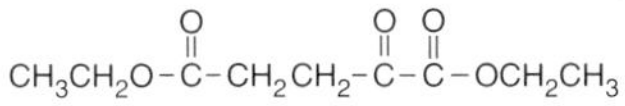

250 mg

S56,421-4
DIMETHYL 2-ACETYLGLUTARATE

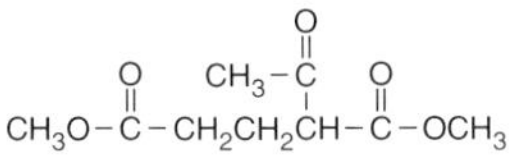

5 gm

S43,255-5
1-ETHYL 5-METHYL 2-ACETYLGLUTARATE

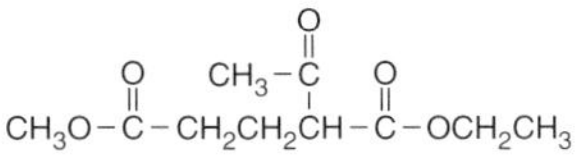

5 gm

S88,451-0
DIETHYL 2-METHYL-3-OXOADIPATE

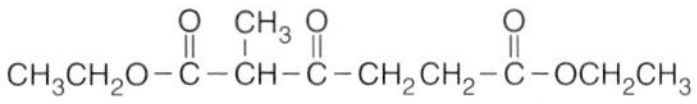

250 mg

S89,584-9
DIETHYL 3-ACETYL-4-OXOPIMELATE

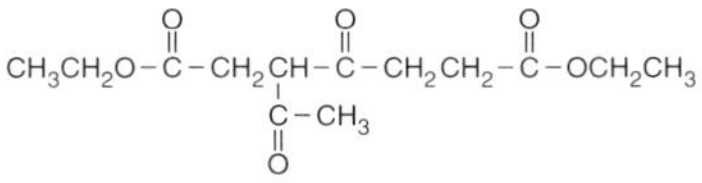

250 mg

S45,535-0
DIETHYL 2-ETHYL-2-(1-METHYL-3-OXOBUTYL)MALONATE

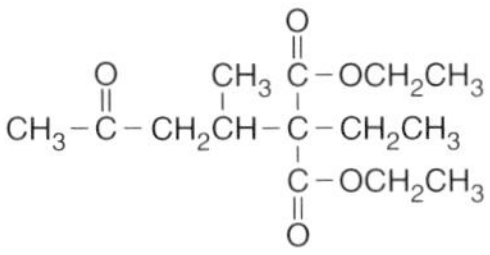

5 gm

S76,061-7
METHYL 4,5-BIS(METHOXYCARBONYL)-3-OXO-2,3,4,5-TETRAHYDRO-2-THIOPHENEACETATE

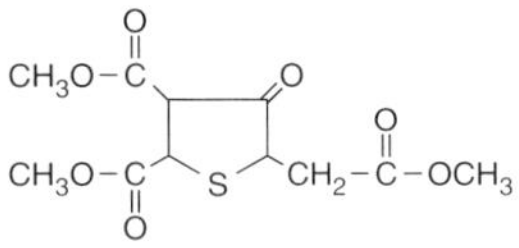

250 mg

S74,202-3
TETRAMETHYL 4-OXO-1,1,3,5-CYCLOHEXANETETRACARBOXYLATE

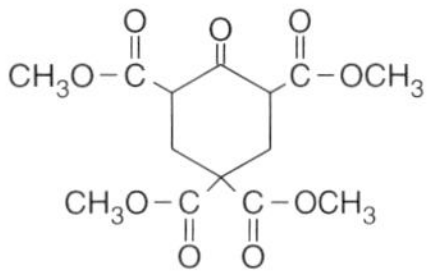

50 mg

S88,433-2
METHYL 2-OXO-1-(3-OXOPENTYL)-1-CYCLOHEXANECARBOXYLATE

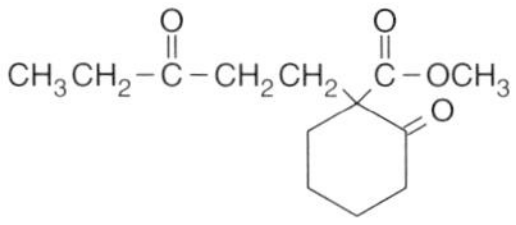

250 mg

S67,386-2
ETHYL 1,2-BIS(METHOXYCARBONYL)-9-OXOCYCLONONANEPROPIONATE

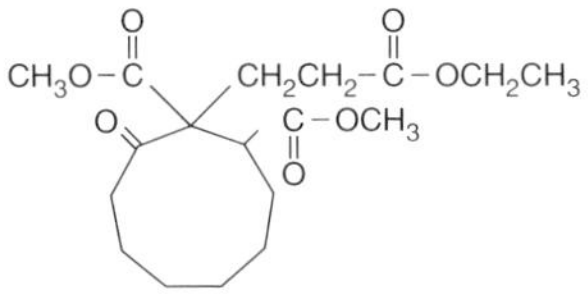

5 gm

S41,950-8
METHYL 2,2-DIMETHYLACETOACETATE

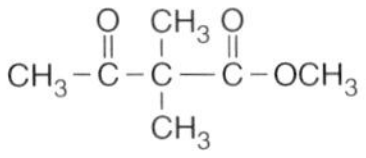

250 mg

S81,307-9
ETHYL 3-OXOBUTYRATE, DIMETHYLTHALLIUM SALT

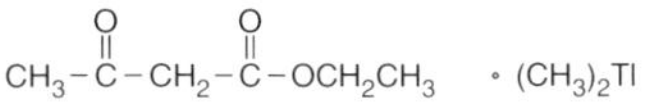

250 mg

S34,805-8
ETHYL 2-ALLYLACETOACETATE

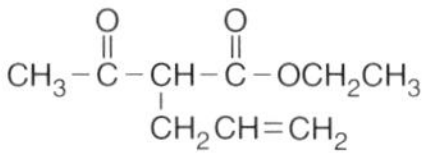

5 gm

S34,925-9
ETHYL 3-OXO-6-HEPTENOATE

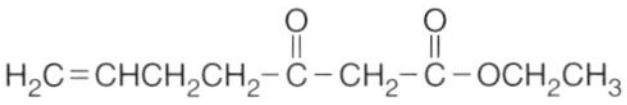

5 gm

S59,524-1
ALLYL 2-ACETYLACETOACETATE

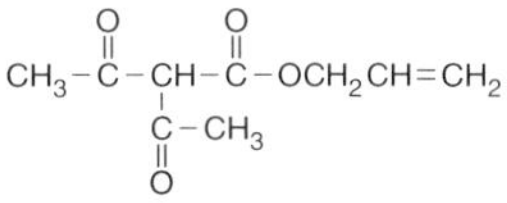

5 gm

S66,117-1
ETHYL 2-BUTYLACETOACETATE

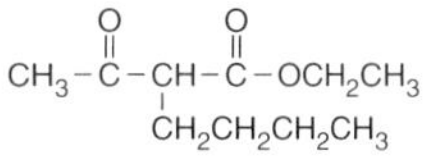

250 mg

S67,109-6
ETHYL 3-OXO-2,2,4-TRIMETHYLVALERATE

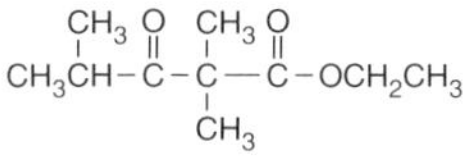

250 mg

S59,529-2
BUTYL 2-ACETYLACETOACETATE

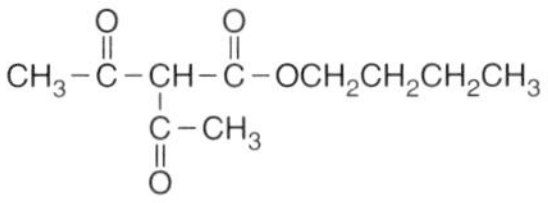

5 gm

S39,935-3
METHYL 3-OXODECANOATE

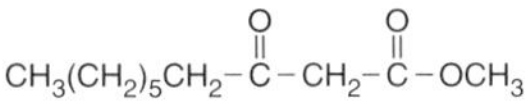

250 mg

S55,690-4
HEPTYL ACETOACETATE

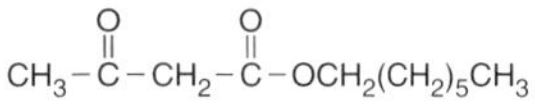

5 gm

S39,944-2
METHYL 3-OXODODECANOATE

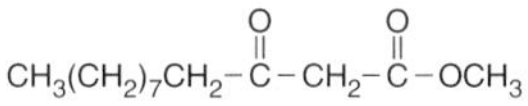

250 mg

S59,533-0
1,4-BUTANEDIOL DIACETOACETATE

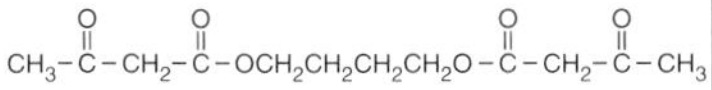

5 gm

S54,114-1
DIETHYL 2-BUTYL-2-HEPTANOYLMALONATE

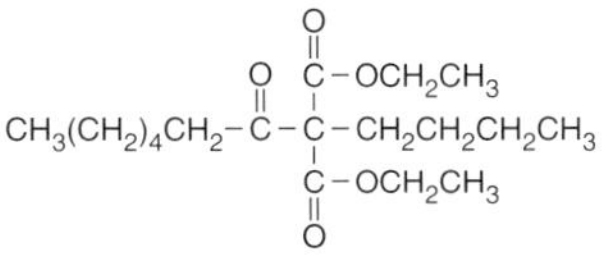

250 mg

S41,012-8
TETRAMETHYL BICYCLO[3.3.1]NONANE-2,6-DIONE-1,3,5,7-TETRACARBOXYLATE

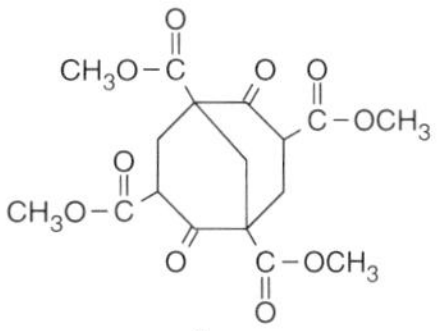

5 gm

S34,294-7
BUTYL 2-CYCLOPENTANONE-1-CARBOXYLATE

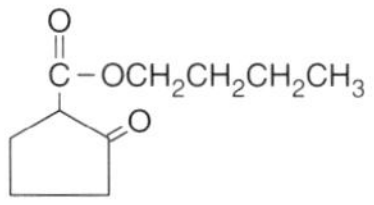

5 gm

S77,057-4
ETHYL 3-OXO-2-(2-PYRROLIDINYLIDENE)-BUTYRATE

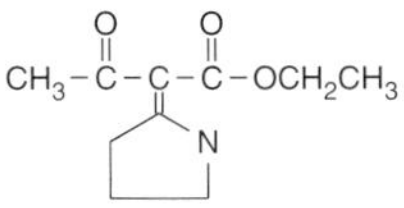

250 mg

S89,698-5
METHYL 2,2-DIMETHYL-4,6-DIOXO-1-CYCLOHEXANECARBOXYLATE

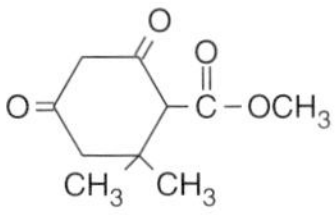

5 gm

S90,681-6
ETHYL 4,6-DIOXO-1,2,2-TRIMETHYL-1-CYCLOHEXANECARBOXYLATE

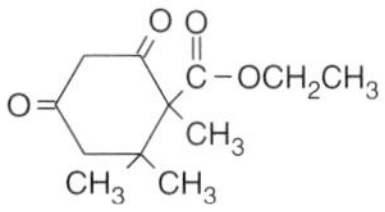

250 mg

S90,602-6
DIETHYL 5-OXOAZELATE

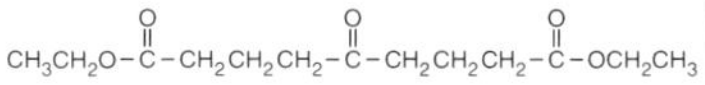

250 mg

S74,528-6
METHYL 1-(METHOXYCARBONYL)-4-OXOCYCLOHEXANEACETATE

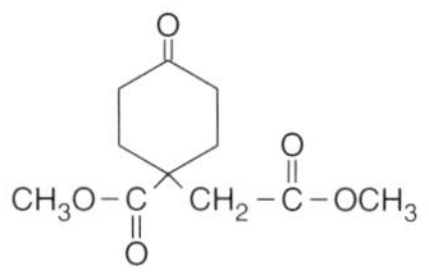

250 mg

S88,462-6
METHYL 3-OXO-1-CYCLOPENTANEACETATE

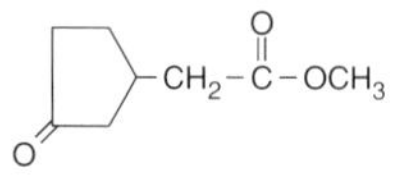

250 mg

S88,467-7
METHYL 3-OXO-1-CYCLOHEXANEACETATE

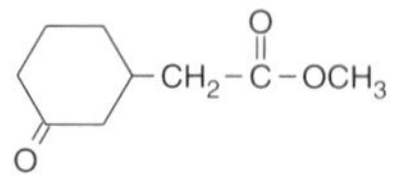

250 mg

S91,605-6
METHYL 3-(3-METHYL-2-OXOCYCLOHEXYL)-PROPIONATE

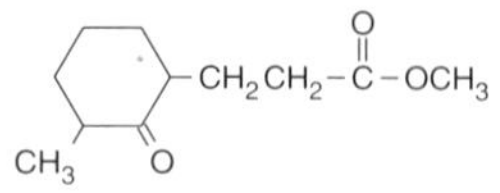

250 mg

S53,592-3
ETHYL 2-ACETYL-5-OXOHEXANOATE

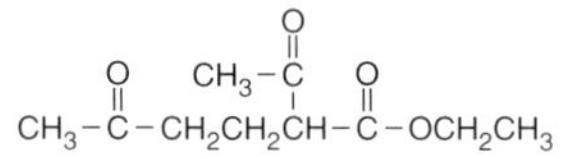

5 gm

S54,053-6
METHYL 14-OXOEICOSANOATE

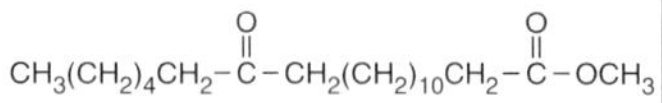

5 gm

S67,389-7
ETHYL 1,2-BIS(METHOXYCARBONYL)-10-OXOCYCLODECANEPROPIONATE

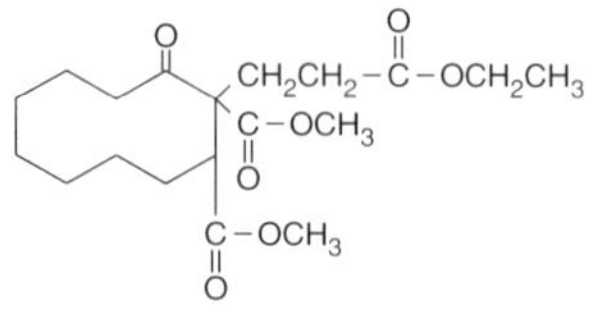

5 gm

S61,913-2
TETRAMETHYL 3,7-DIOXOBICYCLO[3.3.0]-OCTANE-2,4,6,8-TETRACARBOXYLATE

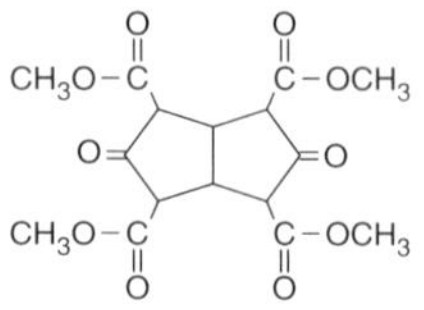

250 mg

S59,510-1
DIETHYL 2-OXO-1,3-CYCLOHEXANEDIPROPIONATE

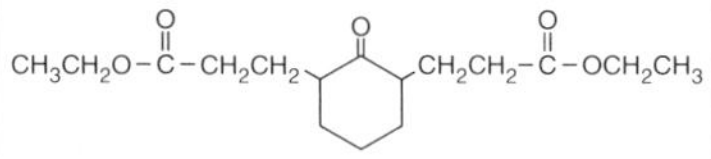

5 gm

S84,722-4
METHYL 2-OXOCYCLOPENTANEHEPTANOATE

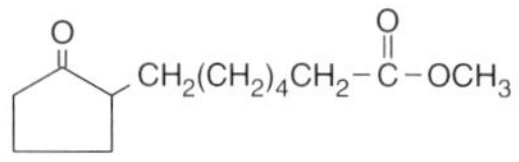

250 mg

S70,973-5
METHYL 2-OXOCYCLOHEXANEPROPIONATE

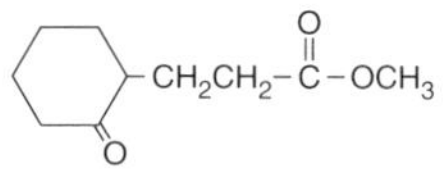

5 gm

S66,914-8
ETHYL 3-(2-OXOCYCLOHEXYL)PROPIONATE

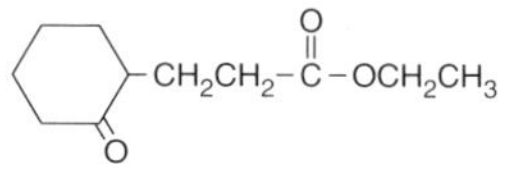

5 gm

S88,387-5
8-OXO-2-PERHYDROPHENANTHRENOL ACETATE

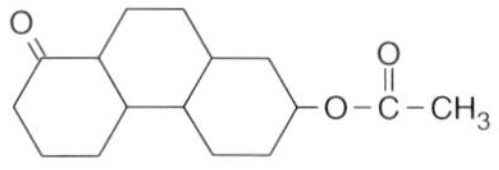

250 mg

S87,644-5
L-MENTHYL 3,3,3-TRIMETHYLPYRUVATE

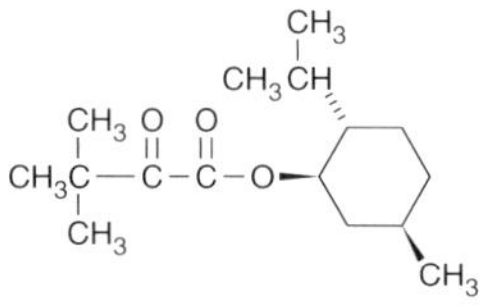

250 mg

S51,985-5
ETHYL 5-METHYL-3-OXOHEXANOATE

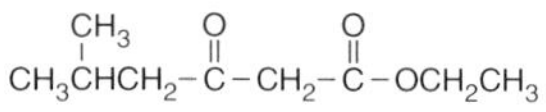

250 mg

S86,645-8
METHYL 7-METHYL-3-OXO-6-OCTENOATE

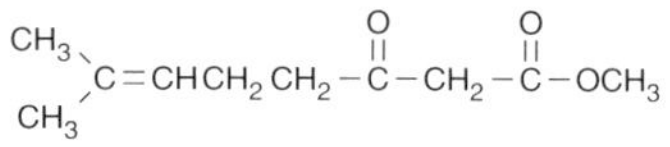

250 mg

S92,610-8
ETHYL 2-(3-BUTEN-1-YL)ACETOACETATE

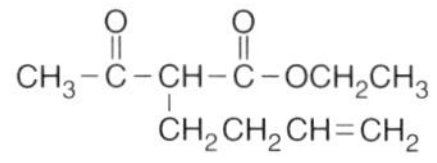

250 mg

S52,827-7
ETHYL 2-HEXYLACETOACETATE

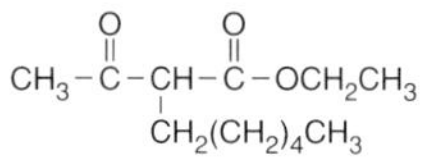

5 gm

S51,230-3
ETHYL 2-HEPTYLACETOACETATE

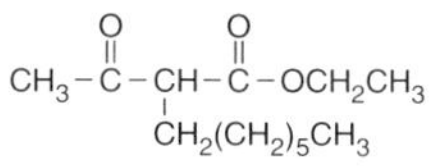

5 gm

S51,075-0
METHYL 3-OXOHEXADECANOATE

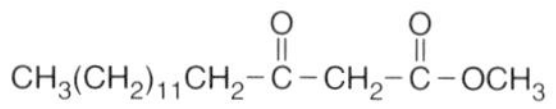

5 gm

S41,996-6
ETHYL STEAROYLACETATE

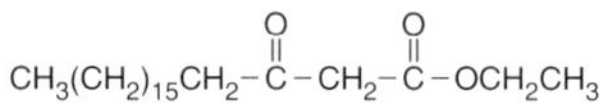

5 gm

S66,864-8
ETHYL 2-METHYL-3-OXOVALERATE

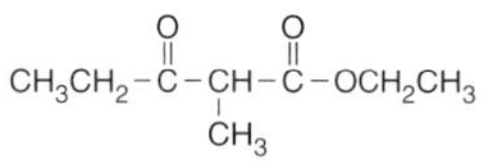

250 mg

S82,815-7
DIETHYL ETHYL(TETRADECYL)MALONATE

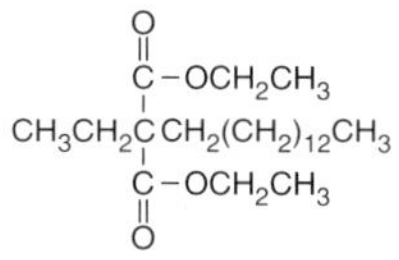

250 mg

S82,820-3
DIETHYL ETHYL(HEXADECYL)MALONATE

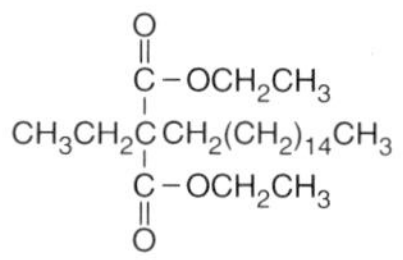

250 mg

S82,821-1
DIETHYL ETHYL(OLEYL)MALONATE

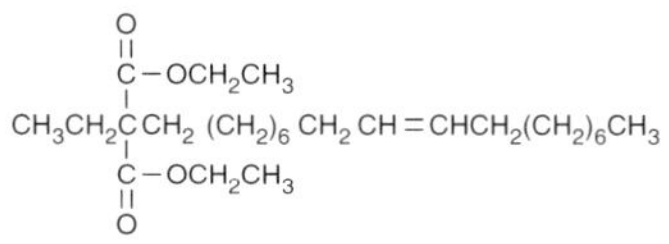

250 mg

S82,823-8
DIETHYL ETHYL(OCTADECYL)MALONATE

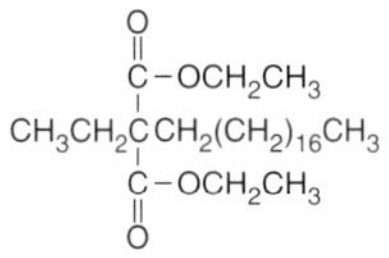

250 mg

S51,431-4
METHYL BETA-OXOCYCLOHEXANEPROPIONATE

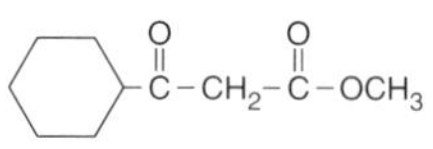

250 mg

S93,327-9
N-ACETYL-L-ALANINE METHYL ESTER

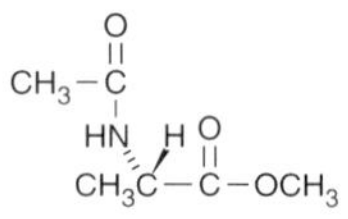

250 mg

S92,579-9
S-(ETHOXYCARBONYLMETHYL)-1,1-DITHIOOXAMATE

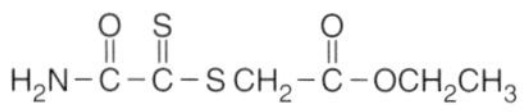

250 mg

S54,446-9
2-BROMOETHYL MALEATE

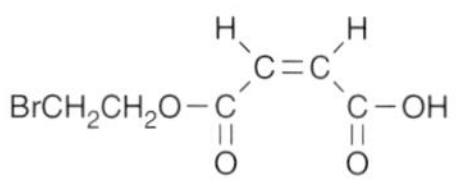

250 mg

S62,237-0
DODECYLOXYETHOXYETHOXYETHOXYETHYL MALEATE

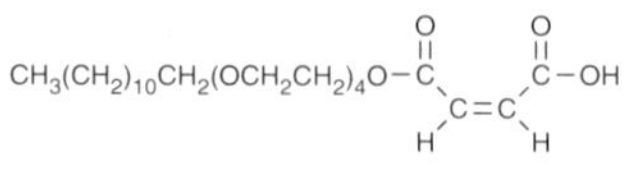

250 mg

S39,032-1
ETHYLXANTHOGENACETIC ACID

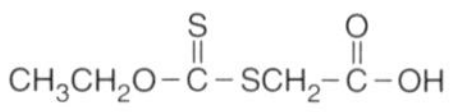

5 gm

S59,986-7
1,2-ETHYLENEGLYCOL BIS(DITHIOGLYCOLATE)

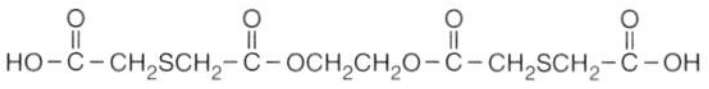

250 mg

S84,340-7
2-OXO-2H-PYRAN-4,6-DICARBOXYLIC ACID

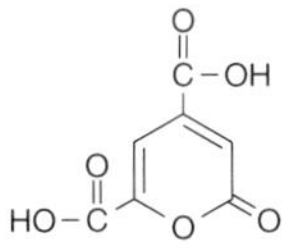

250 mg

S91,591-2
MONOMETHYL HEXAHYDROPHTHALATE

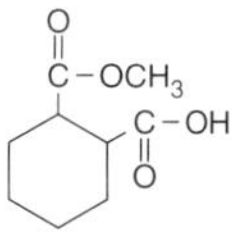

250 mg

S95,007-6
4-METHYL ITACONATE

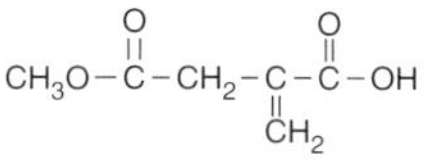

50 mg

S53,723-3
11-(METHOXYCARBONYLMETHYLTHIO)-UNDECANOIC ACID

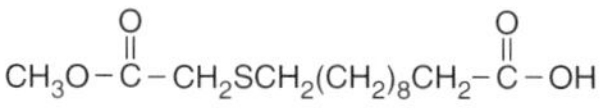

5 gm

S35,727-8
3-(ACETYLMERCAPTO)BUTYRIC ACID

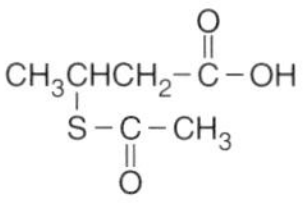

5 gm

S59,696-5
METHYL 3-(CARBOXYMETHYL)-ALPHA-CHLORO-5-METHYL-1-ADAMANTANEACETATE

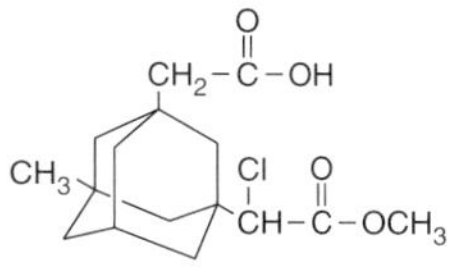

250 mg

S89,094-4
MONO-METHYL 2-METHYLMUCONATE

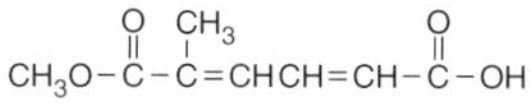

250 mg

S76,562-7
MONOETHYL MALEATE, THALLIUM SALT

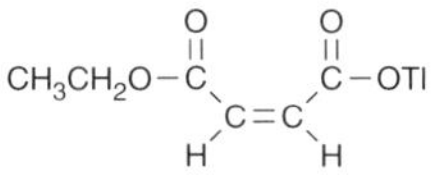

5 gm

S63,000-4
HEXADECYL MALEATE

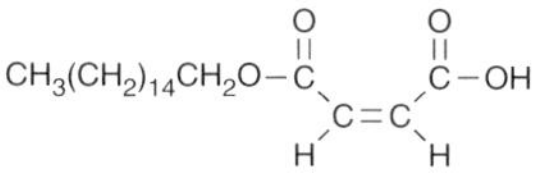

250 mg

S92,609-4
MONOMETHYL METHYLMALONATE

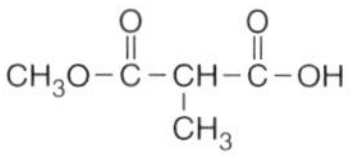

250 mg

S66,976-8
MONOMETHYL SEBACATE

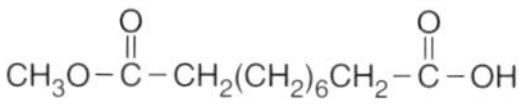

5 gm

S74,226-0
HEXADECYL SUCCINATE

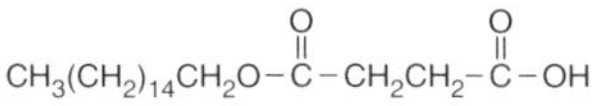

5 gm

S41,401-8
DEXTRO-5-ACETOXY-4-METHYLPENTANOIC ACID

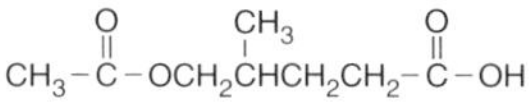

5 gm

S56,850-3
MONODODECYL SUCCINATE

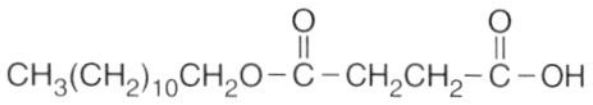

250 mg

S40,704-6
CARBOXYMETHYL PALMITATE

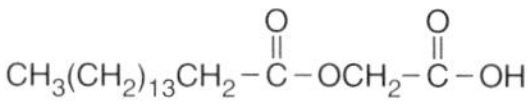

5 gm

S65,588-0
OCTADECYL SUCCINATE

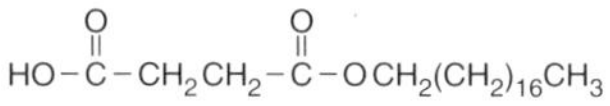

250 mg

S77,069-8
6-(ACETOXYMETHYL)-3-METHYL-3-CYCLOHEXENE-1-CARBOXYLIC ACID

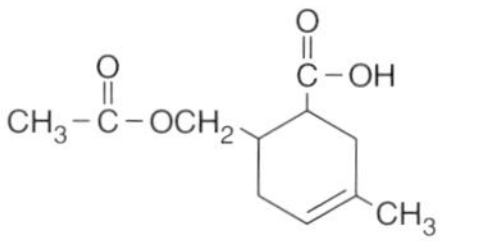

250 mg

S95,398-9
(-)-4-CARBOXY-2,3,3-TRIMETHYL-1-CYCLOHEXEN-1-YL ACETATE

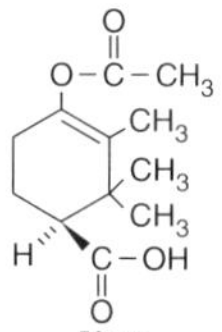

50 mg

S67,555-5
D-MANNONIC GAMMA-LACTONE

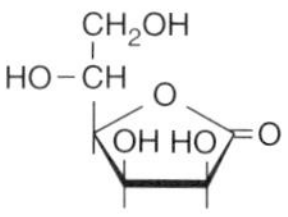

250 mg

S41,968-0
5,5-DIHYDROXY-3,4-EPOXY-3-METHYLHEXANOIC ACID DELTA-LACTONE

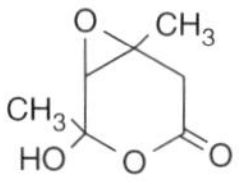

5 gm

S67,287-4
BETA-HYDROXY-GAMMA-BUTYROLACTONE

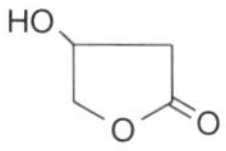

250 mg

S88,340-9
ETHYL 2,5-DIHYDRO-4-HYDROXY-2-OXO-3-FURANATE

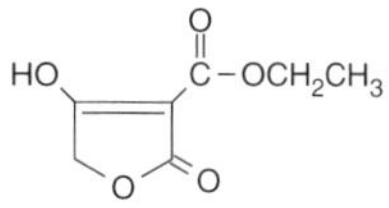

250 mg

S85,938-9
6'-6A-DIHYDRO-G'-(HYDROXYMETHYL)SPIRO-CYCLOHEXANE-1,2'FURO(3,4-D)(1,3)-DIOXOLE-4'-(3A'H)-ONE

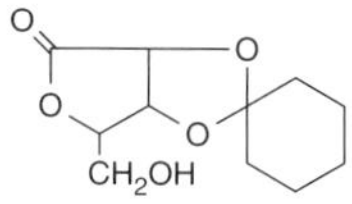

250 mg

S62,042-4
BETA,BETA-DIMETHYL-ALPHA-HYDROXY-GAMMA-VALEROLACTONE

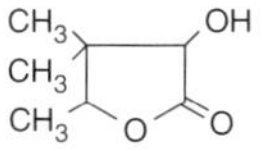

250 mg

S90,686-7
2,2-DIMETHYL-HEXAHYDRO-7-HYDROXY-4,7-METHANO-1,3-DIOXOLO(4,5-C)-OXEPIN-6-ONE

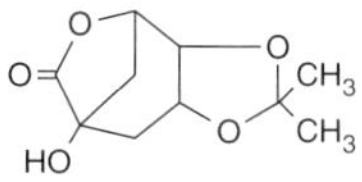

250 mg

S36,215-8
GLYCERINE CARBONATE

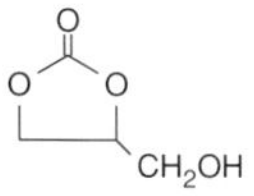

5 gm

S95,236-2
METHYL 2,5-DIHYDRO-4-HYDROXY-2-OXO-3-FURANCARBOXYLATE

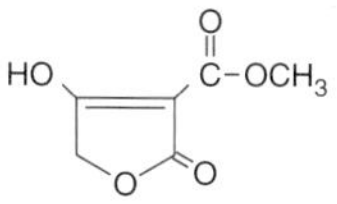

50 mg

S76,345-4
3,6-DIMETHYL-4-HYDROXY-2H-PYRAN-2-ONE

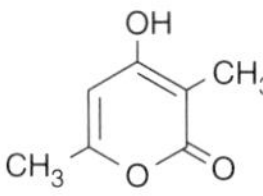

250 mg

S95,605-8
2A,5,5A,6,8A,8B-HEXAHYDRO-6-HYDROXY-2H-NAPHTHO(1,8-BC)-FURAN-2-ONE

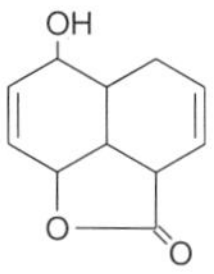

50 mg

S95,242-7
ETHYL ALPHA,2-DIOXO-3-TETRAHYDROFURANACETATE

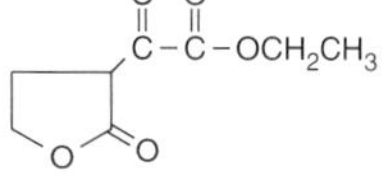

50 mg

S86,104-9
3,4-DICHLORO-5-METHOXY-2(5H)-FURANONE

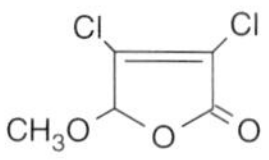

250 mg

S86,001-8
3,4-DIBROMO-5-METHOXY-2(5H)-FURANONE

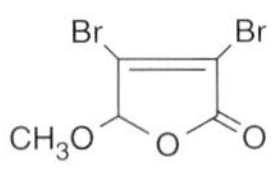

250 mg

S66,315-8
2-METHOXY-4-OXO-2,5,5-TRIMETHYL-1,3-DIOXOLANE

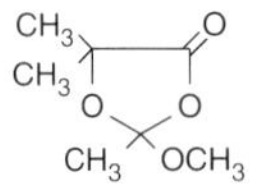

250 mg

S94,901-9
3,3-DIMETHYL-1,4-BENZODIOXAN-2-ONE

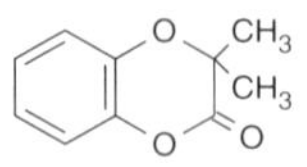

250 mg

S94,165-4
4-ACETYLDIHYDRO-2,3-FURANDIONE

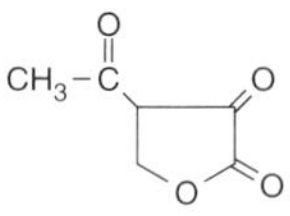

250 mg

S95,811-5
METHYL 4-OXO-1,8,8-TRIMETHYL-3-OXABICYCLO[3.2.1]OCTANE-2-ACETATE

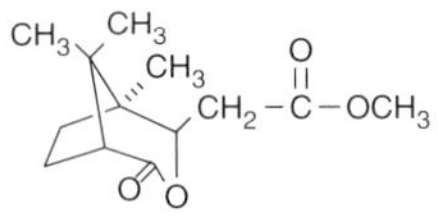

50 mg

S95,239-7
1,4,4A,6,7,7A-HEXAHYDRO-3H-FURO-(3,4-C)THIOPYRAN-1,7-DIONE

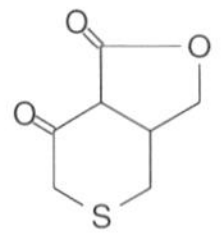

50 mg

S86,055-7
5,6-DIMETHYL-1,3,3A,4,5,7A-HEXAHYDRO-3-OXO-4-ISOBENZOFURANCARBOXYLIC ACID

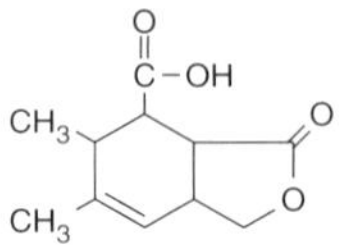

250 mg

S86,140-5
1,3,3A,6,7,7A-HEXAHYDRO-1-OXO-4,5,6-TRIMETHYL-7-ISOBENZOFURANCARBOXYLIC ACID

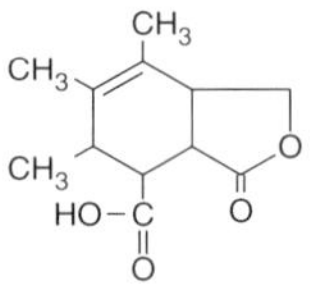

250 mg

S76,502-3
5,10-DIOXA-1,4-CYCLODECANEDIONE

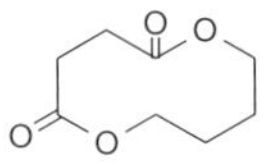

5 gm

S77,442-1
4-BROMO-5-METHYL-7-OXO-6-OXABICYCLO[3.2.1]OCTAN-2-YLMETHYL ACETATE

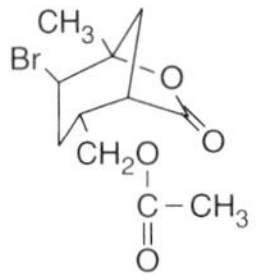

250 mg

S78,403-6
L-HOMOCYSTEINE THIOLACTONE HYDROCHLORIDE

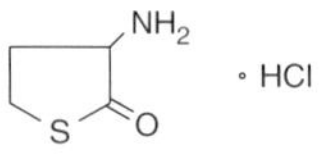

250 mg

S95,248-6
BIS(NITROMETHYL)METHYL-DIHYDRO-DIMETHYLSPIRO[DIOXOLAN-FURO(2,3-D)-DIOXOL]-2-ONE

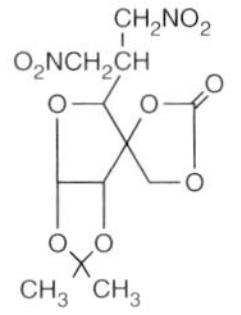

50 mg

S34,610-1
N,N-DIETHYL-2-KETOMORPHOLINIUM CHLORIDE

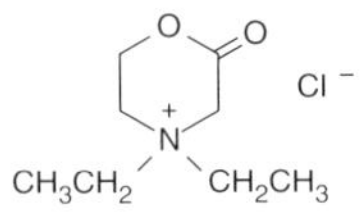

5 gm

S34,363-3
5-CARBOXYDEHYDROACETIC ACID

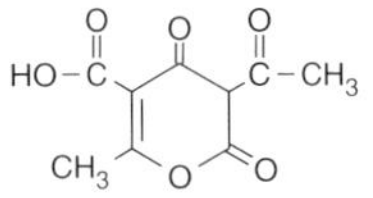

5 gm

S89,426-5
5-ACETYL-3-CHLORO-6-METHYL-2H-PYRAN-2-ONE

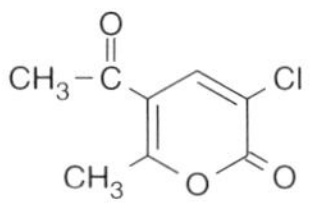

250 mg

S34,629-2
5,6-DIHYDRO-4,6,6-TRIMETHYL-ALPHA-PYRONE

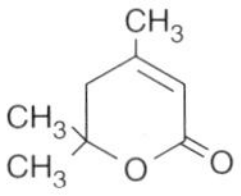

5 gm

S44,120-1
4-HYDROXY-4-METHYL-3-(3-OXOBUTYL)-VALERIC ACID GAMMA-LACTONE

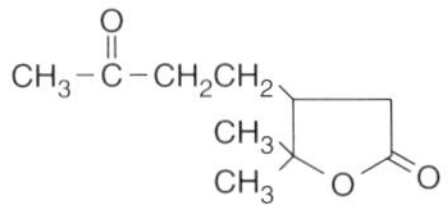

5 gm

S66,490-1
6-EXO-IODO-5-ENDO-HYDROXYNORBORNANE-2-ENDO-CARBOXYLIC ACID LACTONE

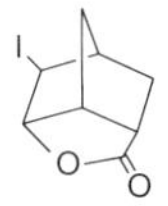

5 gm

S45,750-7
ISODEHYDRACETIC ACID

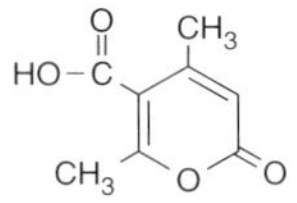

250 mg

S87,245-8
1,2,5A,6,7,8,9,9A-OCTAHYDROCYCLOPENTA(C)(1)-BENZOPYRAN-4(3H)-ONE

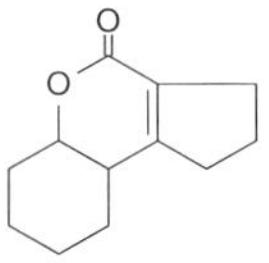

250 mg

S86,107-3
3,4-DICHLORO-5-METHYL-2(5H)-FURANONE

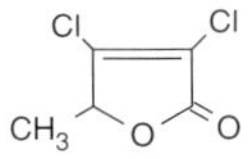

250 mg

S89,447-8
5-(BROMODICHLOROMETHYL)-3,4,5-TRICHLORO-2(5H)-FURANONE

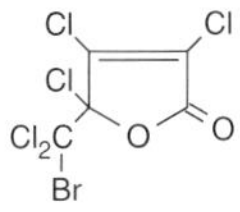

250 mg

S89,448-6
3,4,5,6-TETRACHLORO-2H-PYRAN-2-ONE

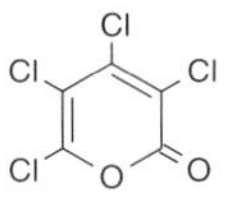

250 mg

S61,529-3
ETHYL 3-BROMO-4,6-DIMETHYL-2H-PYRAN-5-CARBOXYLATE

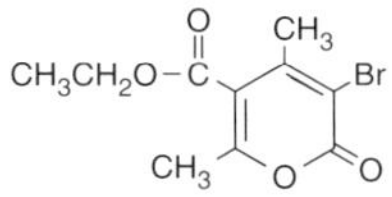

250 mg

S61,546-3
ETHYL 3-BROMO-5,6-DIHYDRO-4,6-DIMETHYL-2H-PYRAN-5-CARBOXYLATE

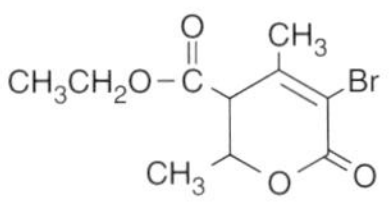

5 gm

S51,878-6
2-ETHOXYCARBONYL-4-HYDROXY-2-METHYLBUTYRIC ACID GAMMA-LACTONE

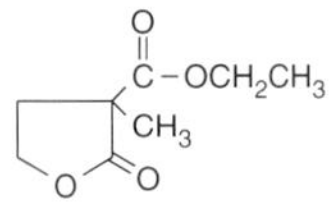

5 gm

S51,925-1
2-ETHOXYCARBONYL-2-(2-HYDROXY-ETHYL)VALERIC ACID GAMMA LACTONE

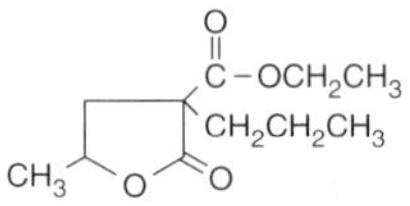

5 gm

S52,117-5
2-ETHOXYCARBONYL-2-(2-HYDROXY-ETHYL)HEXANOIC ACID GAMMA-LACTONE

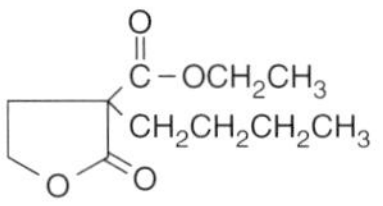

5 gm

S51,864-6
2-ETHOXYCARBONYL-2-(2-HYDROXY-ETHYL)THEPTANOIC ACID GAMMA-LACTONE

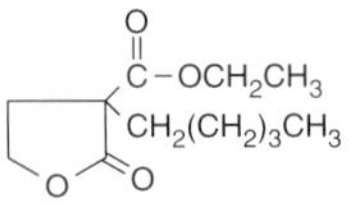

250 mg

S52,129-9
2-ETHOXYCARBONYL-2-(2-HYDROXY-ETHYL)-5-METHYLHEXANOIC ACID GAMMA-LACTONE

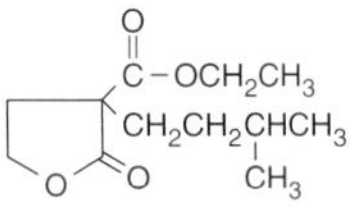

5 gm

S67,658-6
5-HYDROXY-2,3-NORBORNANEDICARBOXYLIC ACID GAMMA-LACTONE

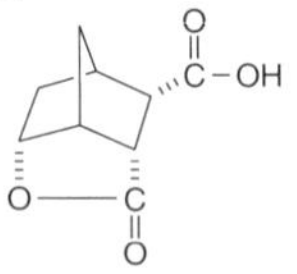

250 mg

S46,253-5
TEREBIC ACID (TETRAHYDRO-2,2-DIMETHYL-5-OXO-3-FURANCARBOXYLIC ACID)

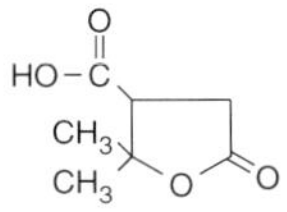

250 mg

S51,870-0
2-ETHYL-2-(2-HYDROXYETHYL)-GLUTARIC ACID GAMMA-LACTONE

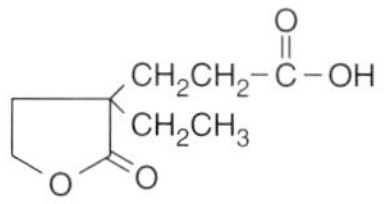

5 gm

S52,167-1
2-(2-HYDROXYETHYL)-2-PROPYLGLUTARIC ACID GAMMA-LACTONE

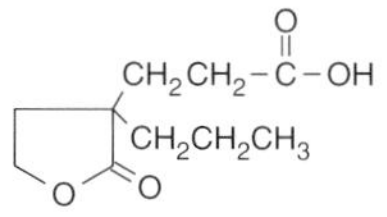

250 mg

S52,127-2
2-(2-HYDROXYETHYL)-2-ISOPENTYLGLUTARIC ACID GAMMA-LACTONE

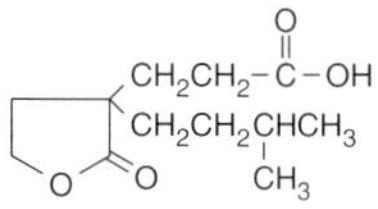

5 gm

S77,053-1
4-IODO-5-METHYL-7-OXO-6-OXABICYCLO-[3.2.1]OCTANE-2-CARBOXYLIC ACID

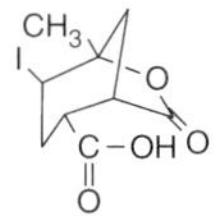

250 mg

S44,318-2
ETHYL 5-OXO-2-ISOXAZOLINE-4-CARBOXYLATE

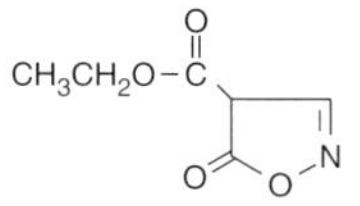

5 gm

S35,742-1
AMBRETTOLIDE

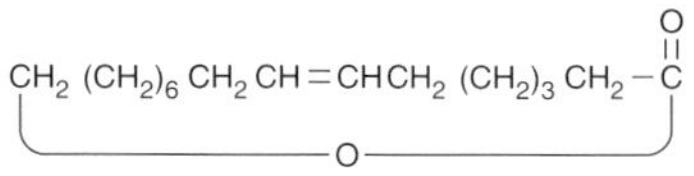

250 mg

S72,213-8
8-HYDROXYOLEIC ACID LACTONE

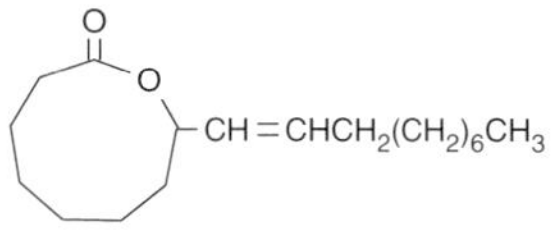

5 gm

S70,344-3
1,4-DIOXACYCLOHEXADECANE-5,16-DIONE

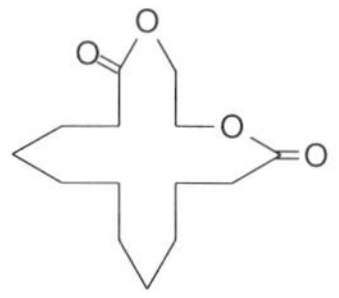

5 gm

S76,624-0
3,4,4A,5,6,8A-HEXAHYDRO-3,6,6,8A-TETRAMETHYL-3,5-ETHANO-1H,3H-PYRANO(3,4-C)PYRAN-1,8-DIONE

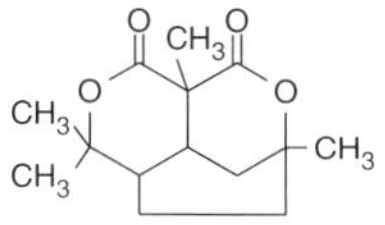

250 mg

S95,238-9
3,3A,6,6A-TETRAHYDRO-2H-CYCLOPENTA-(B)FURAN-2-ONE

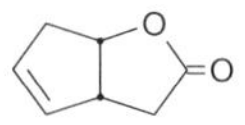

50 mg

S86,010-7
4-(2,2-DICHLOROVINYL)-4,5-DIHYDRO-5,5-DIMETHYL-2(3H)-FURANONE

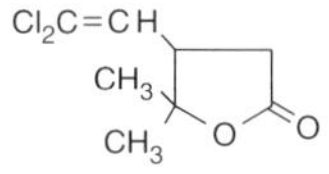

250 mg

S76,798-0
5-HEXYL-2-TETRAHYDROFURANONE

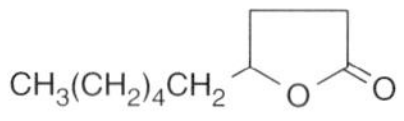

250 mg

S86,375-0
7-OXABICYCLO[4.2.1]NON-4-EN-8-ONE

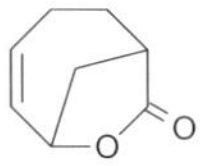

250 mg

S39,787-3
5-METHYL-2-PROPYL-1,3-DIOXOLAN-4-ONE

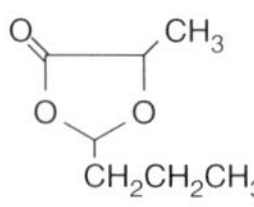

5 gm

S39,779-2
2-(1-ETHYLPROPYL)-5-METHYL-1,3-DIOXOLAN-4-ONE

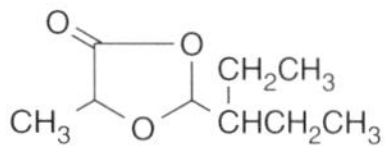

5 gm

S39,786-5
2-(1-ETHYLPENTYL)-5-METHYL-1,3-DIOXOLAN-4-ONE

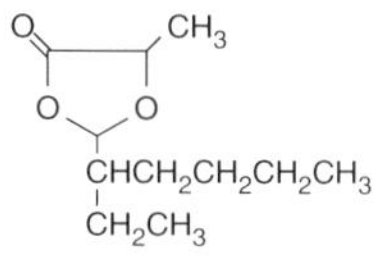

5 gm

S46,264-0
6-HYDROXYTETRAHYDROPYRAN-2-CARBOXYLIC ACID LACTONE

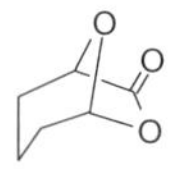

5 gm

S73,949-9
4-ETHYLIDENE-2-METHYL-2-OXAZOLIN-5-ONE

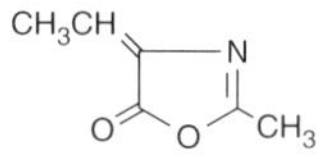

5 gm

S67,832-5
4,4,5,5-TETRACHLORO-1,3-DIOXOLAN-2-ONE

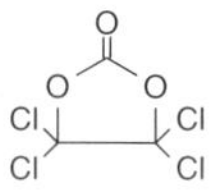

5 gm

S93,966-8
1,3-DIOXAN-2-ONE

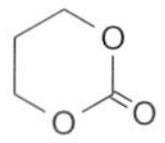

250 mg

S38,942-0
5-METHYL-5-PROPYL-1,3-DIOXAN-2-ONE

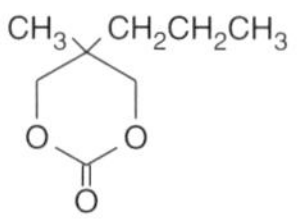

250 mg

S75,628-8
5,5-DIETHYL-1,3-DIOXAN-2-ONE

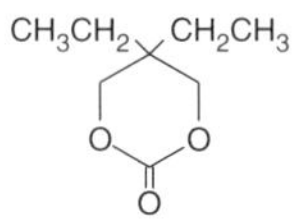

5 gm

S37,583-7
2-ETHYL-3-PROPYLACRYLIC ANHYDRIDE

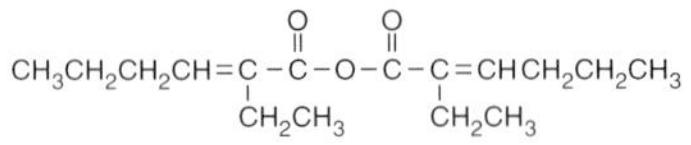

5 gm

S57,098-2
ISOVALERIC ANHYDRIDE

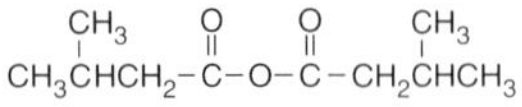

5 gm

S95,822-0
7-ACETYL-4,9-DIOXATRICYCLO-[5.2.2.0.(2,6)]UNDEC-10-ENE-3,5,8-TRIONE

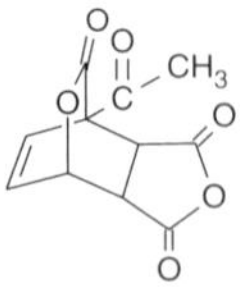

50 mg

S45,447-8
TETRAHYDROFURAN-2,3,4,5-TETRACARBOXYLIC DIANHYDRIDE

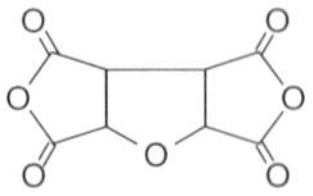

5 gm

S92,717-1
2,3-DIACETYLSUCCINIC ANHYDRIDE

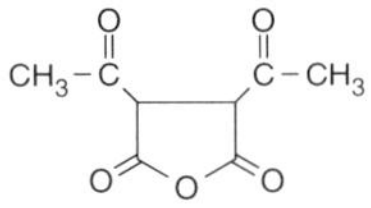

250 mg

S77,449-9
ENDO-3,6-DIMETHYL-3,6-ENDOXOHEXAHYDROPHTHALIC ANHYDRIDE

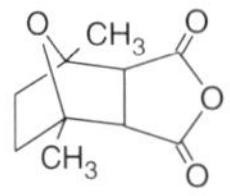

250 mg

S84,008-4
4,7-DIHYDRO-4,7A,7B-TRIMETHYL-4,7-EPOXYISOBENZOFURAN-1,3(7A,7B)-DIONE

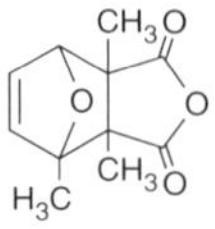

250 mg

S75,796-9
3,6-ENDOXO-3-METHYLHEXAHYDROPHTHALIC ANHYDRIDE

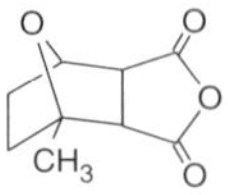

5 gm

S66,265-8
3,6-DIMETHYL-4-CYCLOHEXENE-1,2-DICARBOXYLIC ANHYDRIDE

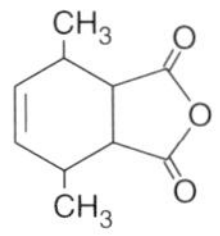

250 mg

S89,526-1
1,5-DIMETHYL-6-(ETHOXYCARBONYL)-BICYCLO[2.2.2]OCT-5-ENE-2,3,7,8-TETRACARBOXYLIC DIANHYDRIDE

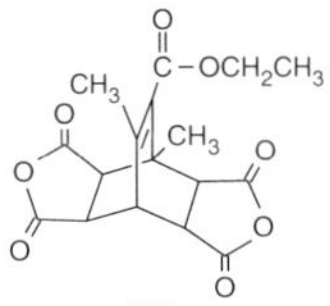

250 mg

S86,491-9
BICYCLO[2.2.2]OCTA-2,5-DIENE-2,3-DICARBOXYLIC ANHYDRIDE

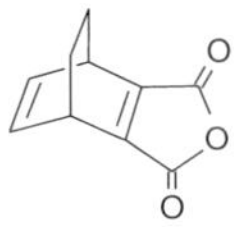

250 mg

S57,914-9
2-(ACETYLTHIO)SUCCINIC ANHYDRIDE

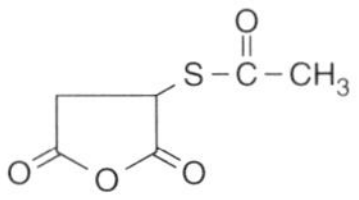

250 mg

S41,930-3
1,2-CYCLOHEXANEDICARBOXYLIC ANHYDRIDE

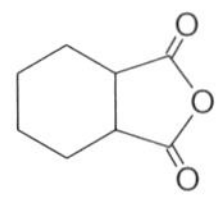

5 gm

S45,739-6
CIS-1,2-CYCLOBUTANEDICARBOXYLIC ANHYDRIDE

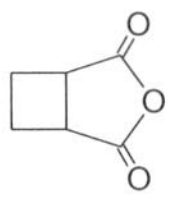

250 mg

S85,223-6
2-BROMO-5-NORBORNENE-2,3-DICARBOXYLIC ANHYDRIDE

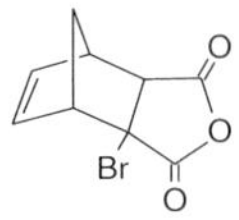

250 mg

S85,224-4
NORBORNANE-2,3-DICARBOXYLIC ANHYDRIDE

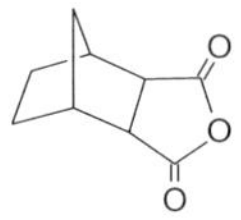

250 mg

S75,733-0
1,3-CYCLOPENTANEDICARBOXYLIC ANHYDRIDE

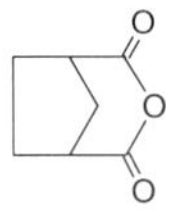

250 mg

S75,740-3
2-METHYL-8-OXASPIRO(4.5)-DECANE-7,9-DIONE

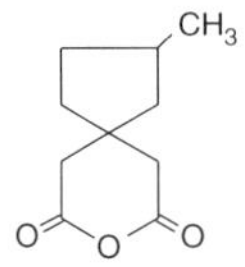

250 mg

S95,234-6
2,3-DIBROMOMALEIC ANHYDRIDE

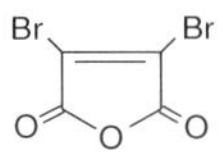

50 mg

S35,412-0
2-(1-OCTENYL)SUCCINIC ANHYDRIDE

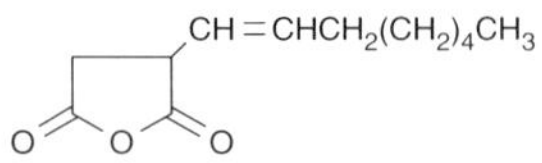

5 gm

S42,881-7
2-DECEN-1-YLSUCCINIC ANHYDRIDE

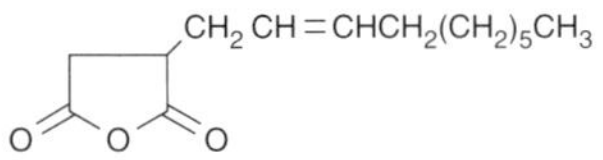

5 gm

S35,996-3
DECYLSUCCINIC ANHYDRIDE

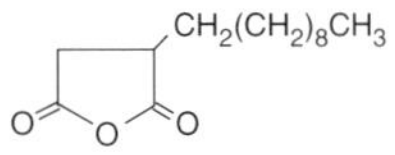

5 gm

S34,795-7
DODECYLSUCCINIC ANHYDRIDE

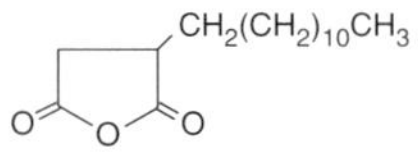

5 gm

S36,491-6
2-(1-TETRADECENYL)-SUCCINIC ANHYDRIDE

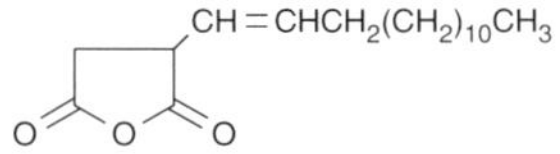

5 gm

S37,067-3
TETRADECYLSUCCINIC ANHYDRIDE

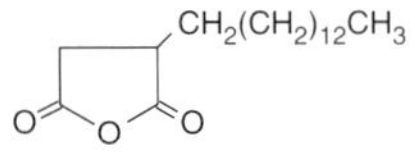

250 mg

S35,044-3
1-HEXADECENYLSUCCINIC ANHYDRIDE

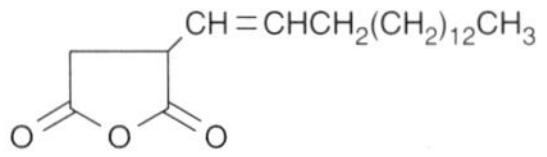

5 gm

S35,048-6
HEXADECYLSUCCINIC ANHYDRIDE

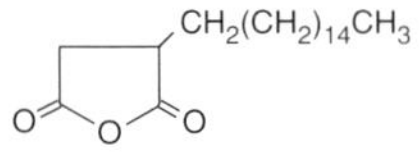

5 gm

S35,406-6
2-(1-OCTADECENYL)SUCCINIC ANHYDRIDE

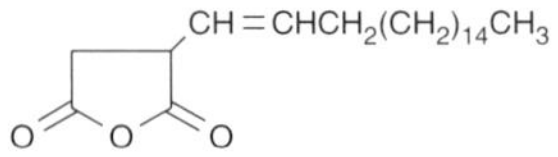

1 gm

S40,967-7
2-OCTADECYLSUCCINIC ANHYDRIDE

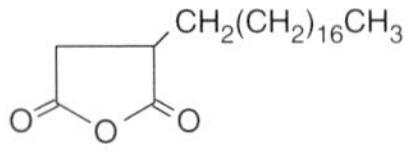

5 gm

S34,792-2
2-(1-DOCOSENYL)SUCCINIC ANHYDRIDE

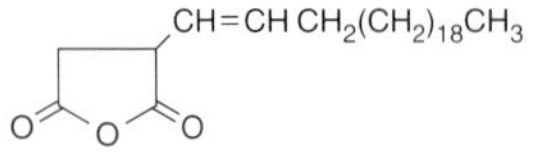

5 gm

S76,282-2
8-METHYL-2-OXASPIRO(4.5)-DECANE-1,3-DIONE

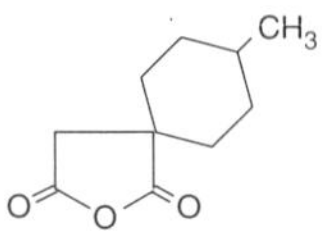

250 mg

S67,842-2
BICYCLO[4.2.2]DEC-7-ENE-9,10-DICARBOXYLIC ANHYDRIDE

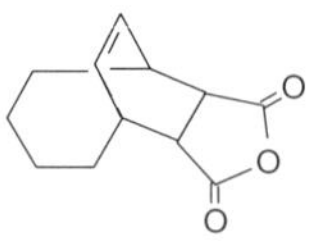

250 mg

S68,570-4
3,4-DICARBOXY-1,2,3,4-TETRAHYDRO-1-NAPHTHALENESUCCINIC ACID DIANHYDRIDE

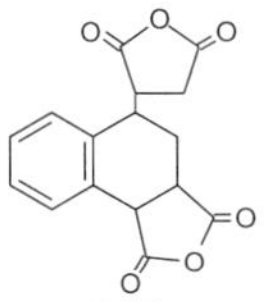

5 gm

S77,055-8
5-METHYL-3A,4,7,7A-TETRAHYDROPHTHALIC ANHYDRIDE

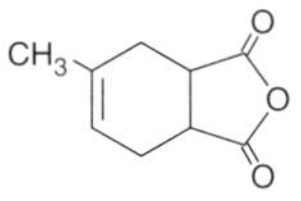

250 mg

S75,770-5
3,6-ENDOXOHEXAHYDROPHTHALIC ANHYDRIDE

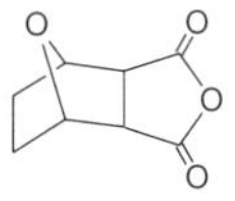

5 gm

S56,159-2
3-CYCLOHEXENE-1,2-DICARBOXYLIC ANHYDRIDE

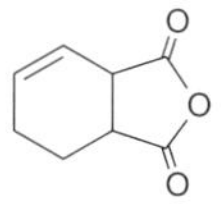

250 mg

S76,452-3
9-TERT-BUTYL-3-OXASPIRO(5.5)-UNDECANE-2,4-DIONE

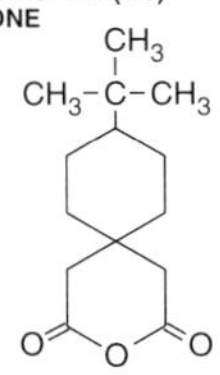

5 gm

S95,813-1
10-OXATETRACYCLO[6.3.0.(2,7).0-(3,6)]TRIDECA-4,12-DIENE-9,11-DIONE, ISOMER II

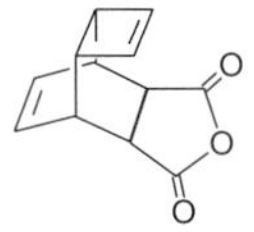

50 mg

S75,784-5
9-ISOPROPYL-3-OXASPIRO(5.5)-UNDECANE-2,4-DIONE

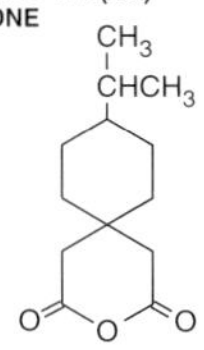

250 mg

S95,812-3
10-OXATETRACYCLO[6.3.0.(2,7).0.-(3,6)]TRIDECA-4,12-DIENE-9,11-DIONE, ISOMER I

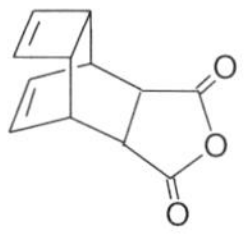

50 mg

S41,219-8
HEXACHLOROHEXAHYDRO-1,4-METHANONAPHTHALENE-6,7-DICARBOXYLIC ANHYDRIDE

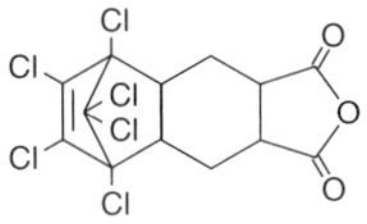

5 gm

S59,290-0
2,2-BIS(CHLOROMETHYL)-PROPIONYL CHLORIDE

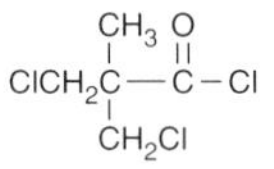

5 gm

S57,835-5
3,3-DICHLOROPIVALOYL CHLORIDE

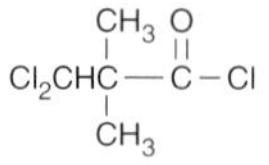

5 gm

S81,890-9
2-BROMO-4,4,4-TRIFLUOROBUTYRYL BROMIDE

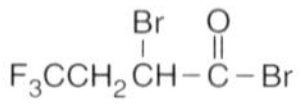

250 mg

S89,397-8
PENTACHLOROPROPIONYL CHLORIDE

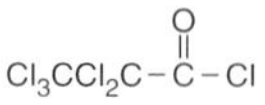

250 mg

S81,892-5
4,4,4-TRIFLUOROBUTYRYL CHLORIDE

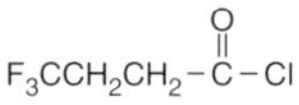

250 mg

S81,891-7
4,4,4-TRIFLUOROBUTYRYL BROMIDE

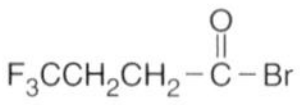

250 mg

S57,809-6
4-METHYL-4-NITROHEXANOYL CHLORIDE

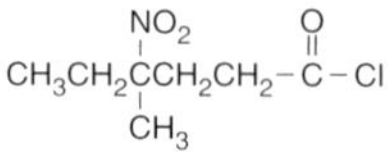

5 gm

S44,960-1
PROPIONYL FLUORIDE

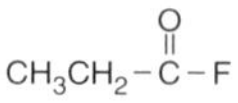

5 gm

S44,339-5
2-ETHYLBUTYRYL CHLORIDE

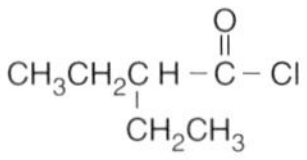

5 gm

S81,889-5
4,4,4-TRIFLUOROCROTONYL CHLORIDE

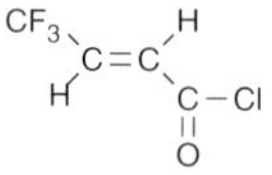

250 mg

S89,390-0
TRICHLOROACRYLOYL CHLORIDE

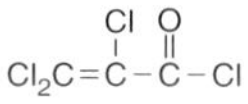

250 mg

S88,408-1
4-DECENOYL CHLORIDE

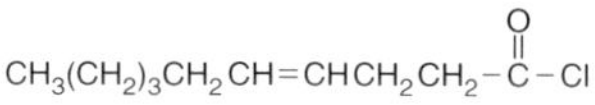

250 mg

S36,465-7
PROPYL CHLOROTHIOLFORMATE

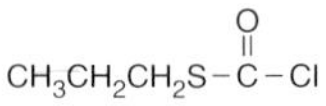

5 gm

S52,062-4
3-CHLOROCYCLOHEXYL CHLOROFORMATE

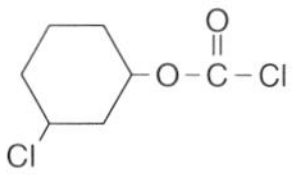

5 gm

S34,626-8
DIGLYCOL CHLOROFORMATE

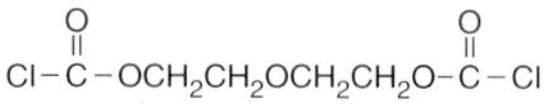

5 gm

S84,914-6
4,4,7,7-TETRANITROSEBACOYL CHLORIDE

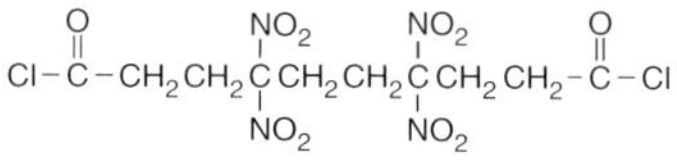

250 mg

S85,294-5
ALPHA,ALPHA'-DIMETHYLSUCCINYL CHLORIDE

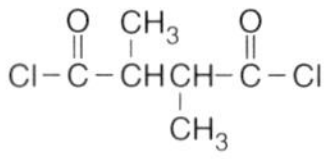

250 mg

S79,381-7
2-MENTHOXYACETYL CHLORIDE

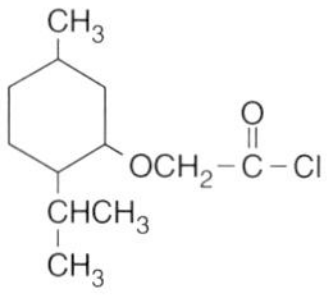

250 mg

S66,238-0
1-METHYL-1-CYCLOHEXANECARBOXYLIC ACID CHLORIDE

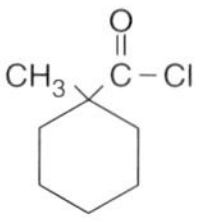

5 gm

S37,375-3
2,2,3-TRICHLOROPROPIONAMIDE
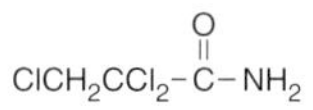
5 gm

S74,996-6
2,3-DIBROMOPROPIONAMIDE
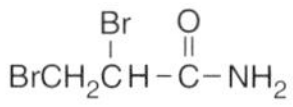
250 mg

S94,810-1
ALPHA-BROMO-1-ADAMANTANEACETAMIDE
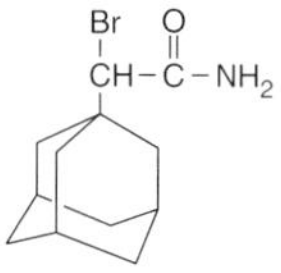
250 mg

S94,809-8
ALPHA-IODO-1-ADAMANTANEACETAMIDE
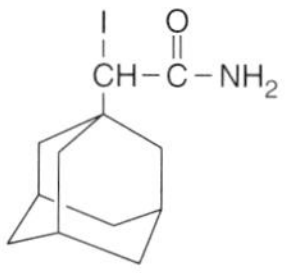
250 mg

S89,402-8
2,2,3,3,3-PENTACHLOROPROPIONAMIDE
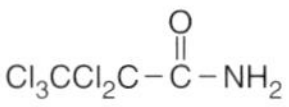
250 mg

S93,905-6
D-GLUCONAMIDE
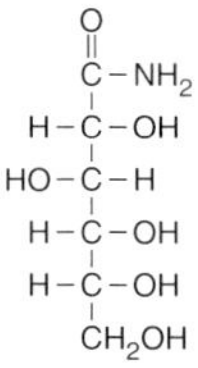
250 mg

S44,979-2
O-(2-HYDROXYETHYL)GLYCOLAMIDE
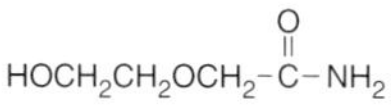
5 gm

S55,686-6
1-HYDROXY-2,2,6,6-TETRACHLOROCYCLOHEXANECARBOXAMIDE
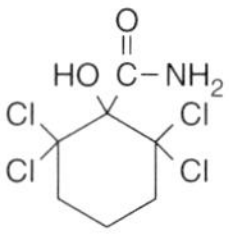
5 gm

S44,148-1
5-HYDROXYVALERAMIDE
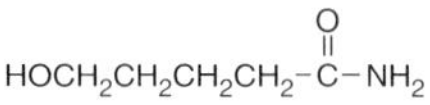
5 gm

S50,168-9
12-HYDROXYOCTADECANAMIDE
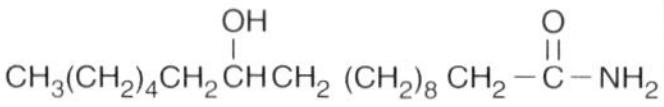
5 gm

S38,412-7
2-(2-[2-(DODECYLOXY)ETHOXY]-ETHOXY)ACETAMIDE

5 gm

S37,187-4
METHOXYACETAMIDE
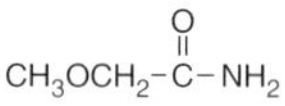
5 gm

S93,743-6
MALONAMIC ACID
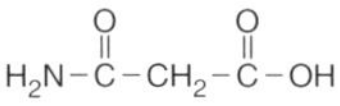
250 mg

S77,774-9
D-VALINAMIDE ACETATE
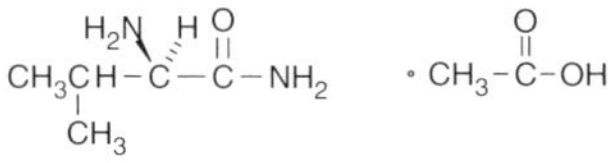
250 mg

S40,725-9
N-CYCLOHEXYLIDENE-2-CARBAMOYLCYCLOHEX-1-ENYLAMINE
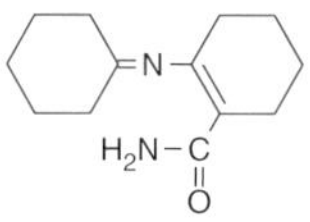
5 gm

S94,820-9
N,N,N',N'-TETRAMETHYL-1,3-ADAMANTANEDICARBOXAMIDE
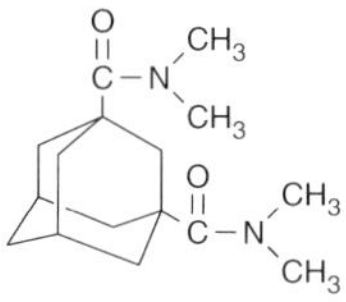
250 mg

S37,854-2
1-PYRROLIDINEACETAMIDE
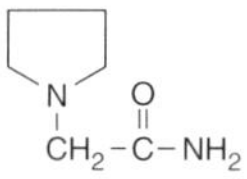
250 mg

S94,803-9
1-ADAMANTANECARBOHYDROXAMIC ACID
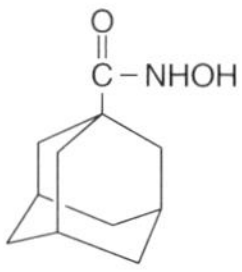
250 mg

S54,024-2
12-OXOOCTADECANAMIDE
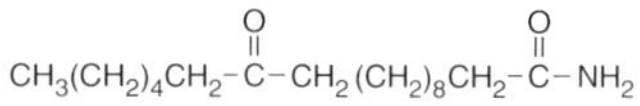
5 gm

S54,030-7
14-OXOEICOSANAMIDE
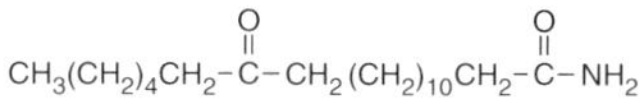
5 gm

S70,416-4
THIOACETYLACETAMIDE HYDROCHLORIDE
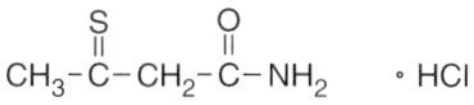
5 gm

S89,407-9
1,2,3,4,4-PENTACHLORO-1,3-BUTADIENE-1-CARBOXAMIDE
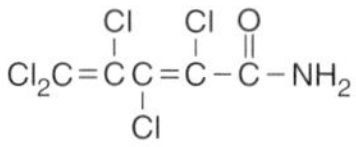
250 mg

S66,263-1
METHYL 2,2-DIISOPROPYLMALAMATE
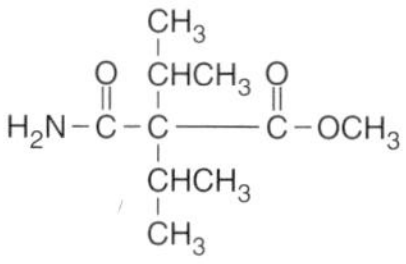
5 gm

S38,618-9
ETHYL 2-BUTYL-2-CARBAMOYLHEXANOATE
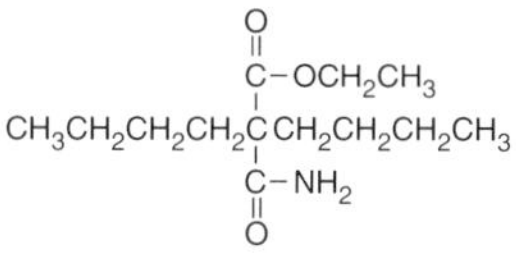
5 gm

S88,094-9
ETHYL 2,N-DIMETHYLMALONAMATE
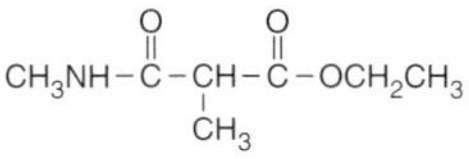
250 mg

S89,510-5
3-CARBAMOYLBICYCLO[2.2.2]OCT-5-ENE-2-CARBOXYLIC ACID, AMMONIUM SALT
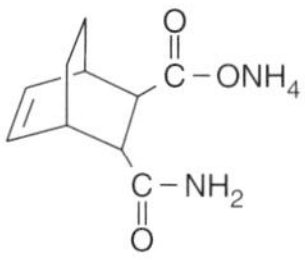
250 mg

S89,511-3
3-CARBAMOYLBICYCLO[2.2.2]OCTANE-2-CARBOXYLIC ACID, AMMONIUM SALT
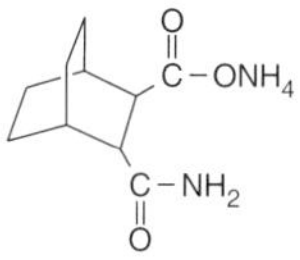
250 mg

S94,821-7
N,N'-DIMETHYL-1,3-ADAMANTANEDICARBOXAMIDE
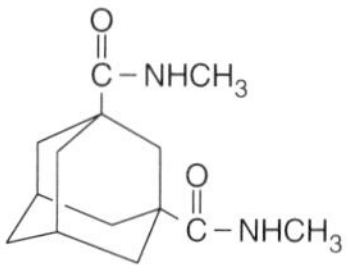
250 mg

S68,074-5
CYCLOBUTANECARBOXAMIDE
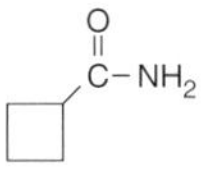
5 gm

S67,129-0
ISOVALERAMIDE
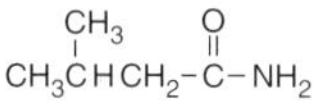
250 mg

S79,963-7
(S)-(+)-2-METHYLBUTYRAMIDE
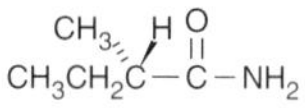
250 mg

S37,848-8
2-ETHYLBUTYRAMIDE
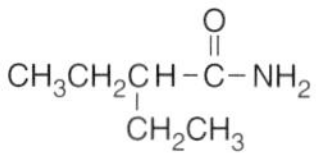
250 mg

S89,370-6
HEPTACHLORO-2,4-PENTADIENE-1-CARBOXAMIDE

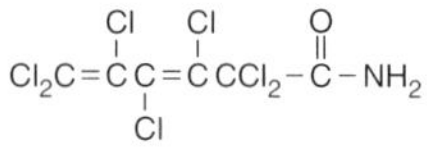

250 mg

S88,088-4
5-METHYL-2,4-HEXADIENE-2-CARBOXAMIDE

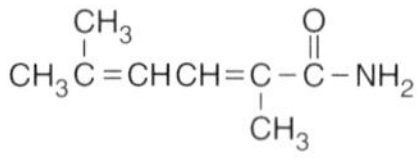

250 mg

S59,526-8
NONANOAMIDE

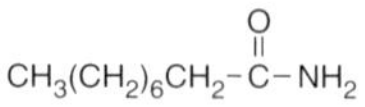

5 gm

S79,835-5
3,7-DIMETHYL-6-OCTENAMIDE

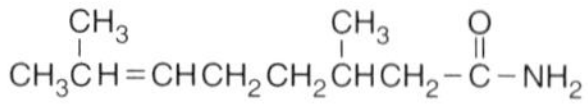

250 mg

S54,987-8
DECANAMIDE

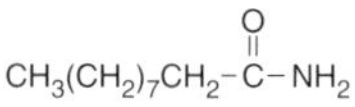

250 mg

S40,834-4
DODECANAMIDE

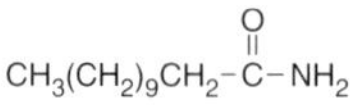

5 gm

S56,300-5
TRIDECANAMIDE

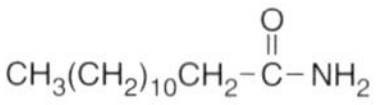

5 gm

S35,567-4
TETRADECANAMIDE

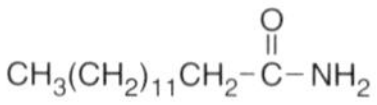

5 gm

S35,043-5
HEXADECANAMIDE

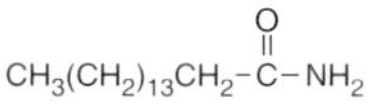

5 gm

S45,598-9
CYCLOPROPANECARBOXAMIDE

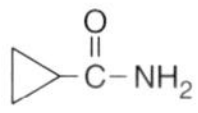

250 mg

S86,433-1
2-CHLORO-5-NORBORNENE-2-CARBOXAMIDE

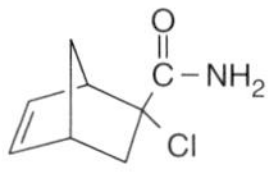

250 mg

S95,023-8
4-METHYL-4-(2-METHYL-1,3-DIOXOLAN-2-YL)VALERAMIDE

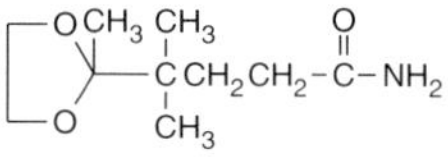

50 mg

S86,436-6
2-METHYLCYCLOHEXANECARBOXAMIDE

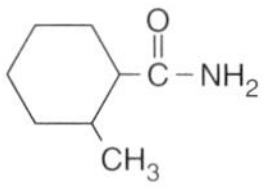

250 mg

S34,491-5
2-(1-CYCLOHEXENYL)BUTYRAMIDE

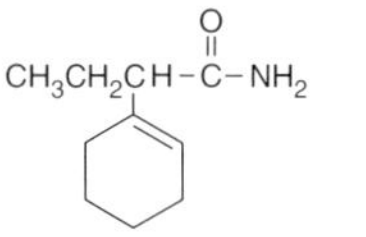

5 gm

S85,953-2
7,7-DIMETHYL-1-NORBORNANECARBOXAMIDE

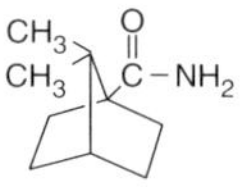

250 mg

S94,690-7
2-METHYL-2-ADAMANTANECARBOXAMIDE

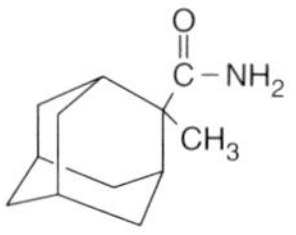

250 mg

S94,283-9
3,5,7-TRIMETHYL-1-ADAMANTANECARBOXAMIDE

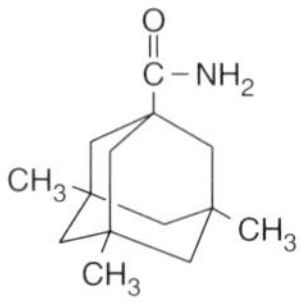

250 mg

S95,519-1
BICYCLO[3.3.0]OCTANE-2-ACETAMIDE

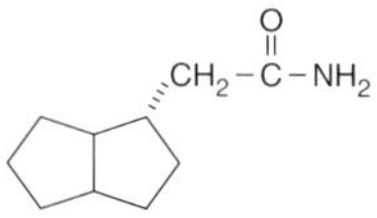

50 mg

S95,520-5
4-METHYLBICYCLO[3.3.0]OCTANE-2-CARBOXAMIDE

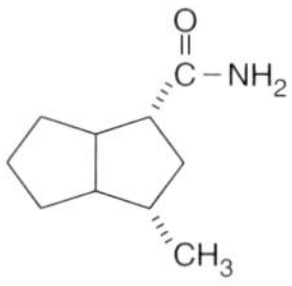

50 mg

S67,266-1
HEXANOHYDROXAMIC ACID

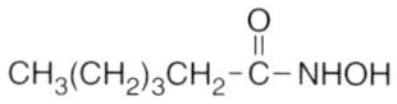

250 mg

S34,399-4
CHLOROACETYL-DL-SERINE

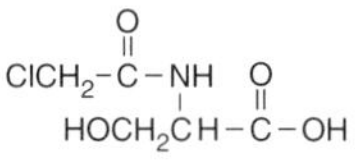

5 gm

S38,381-3
2-CHLORO-N-(2-DODECYLOXYETHYL)-N-METHYLACETAMIDE

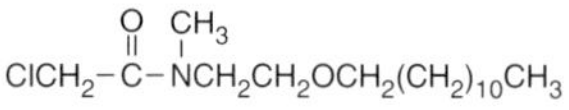

250 mg

S38,396-1
2-CHLORO-N-[2-(DODECYLTHIO)ETHYL]-ACETAMIDE

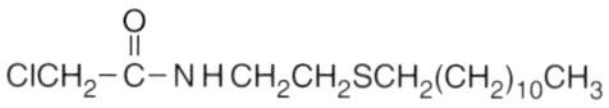

5 gm

S52,947-8
ETHYL 6-(3-CHLOROPROPIONAMIDO)-HEXANOATE

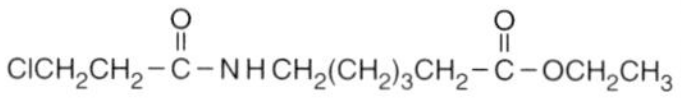

5 gm

S66,847-8
2-CHLORO-N-METHYLACETAMIDE

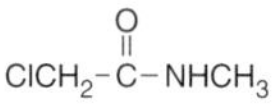

250 mg

S37,908-5
N-ALLYL-2-CHLOROACETAMIDE

5 gm

S77,532-0
N-TERT-BUTYL-2-CHLOROACETAMIDE

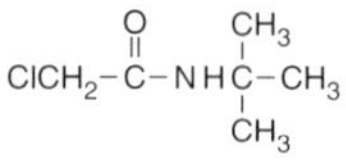

250 mg

S62,769-0
2-CHLORO-N-PENTYLACETAMIDE

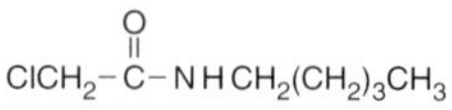

250 mg

S37,994-8
2-CHLORO-N-HEXYLACETAMIDE

5 gm

S38,025-3
2-CHLORO-N-(1-METHYLHEXYL)-ACETAMIDE

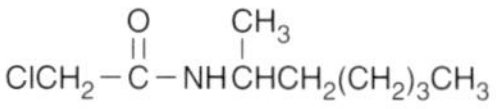

5 gm

S37,683-3
2-CHLORO-N-OCTYLACETAMIDE

5 gm

S38,035-0
2-CHLORO-N-(1,5-DIMETHYLHEXYL)-ACETAMIDE

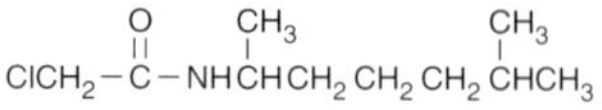

250 mg

S37,717-1
2-CHLORO-N-(1-METHYLOCTYL)-ACETAMIDE

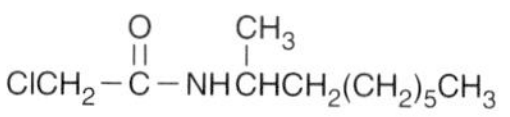

250 mg

S38,137-3
2-CHLORO-N-NONYLACETAMIDE

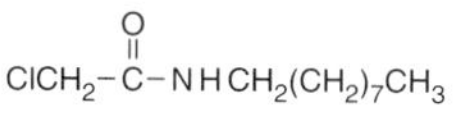

250 mg

S37,734-1
N-DECYL-2-CHLOROACETAMIDE

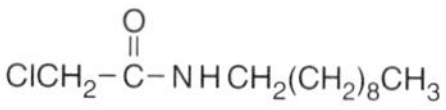

5 gm

S38,198-5
2-CHLORO-N-(10-UNDECENYL)ACETAMIDE

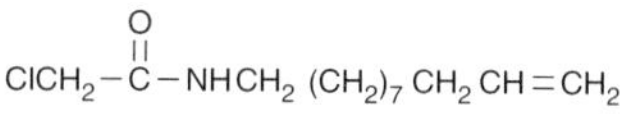

250 mg

S38,183-7
2-CHLORO-N-(1-METHYLDECYL)-ACETAMIDE

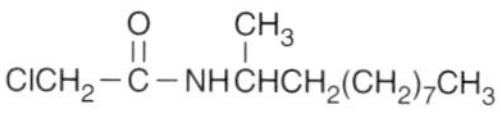

5 gm

S38,214-0
2-CHLORO-N-UNDECYLACETAMIDE

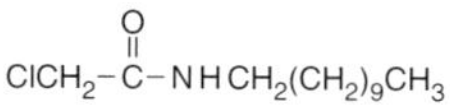

5 gm

S38,328-7
2-CHLORO-N-DODECYLACETAMIDE

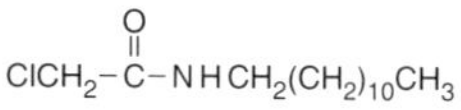

5 gm

S38,345-7
2-CHLORO-N-TRIDECYLACETAMIDE

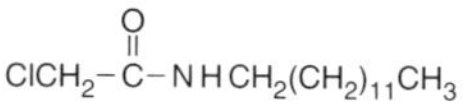

250 mg

S38,370-8
2-CHLORO-N-TETRADECYLACETAMIDE

5 gm

S38,425-9
2-CHLORO-N-OCTADECYLACETAMIDE

5 gm

S37,651-5
2-CHLORO-N-CYCLOHEXYLACETAMIDE

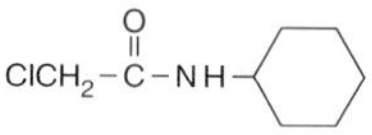

5 gm

S59,707-4
N-(CYCLOHEXYL)-ALPHA-CHLORO-3,5,7-TRIMETHYL-1-ADAMANTANEACETAMIDE

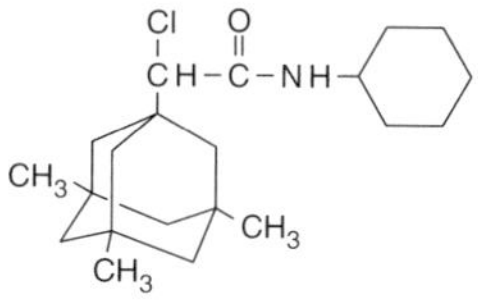

250 mg

S38,068-7
2-CHLORO-N-HEPTYL-N-METHYLACETAMIDE

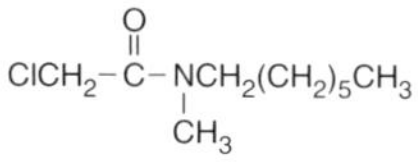

250 mg

S37,686-8
2-CHLORO-N-METHYL-N-(1-METHYLOCTYL)ACETAMIDE

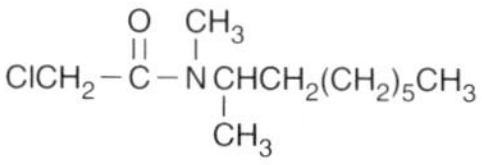

250 mg

S37,707-4
2-CHLORO-N-CYCLOHEXYL-N-METHYLACETAMIDE

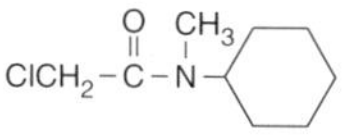

250 mg

S67,237-8
CHLORAL FORMAMIDE

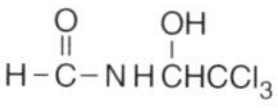

250 mg

S89,242-4
N-(3-THIAZOLIN-2-YL)-2,2,2-TRICHLOROACETAMIDE

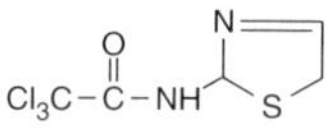

250 mg

S72,028-3
N-CYCLOHEXYL-2,2,2-TRICHLOROACETAMIDE

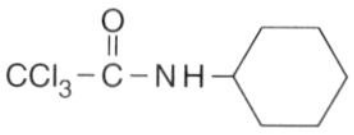

5 gm

S94,679-6
N-(1-ADAMANTYL)TRIFLUOROACETAMIDE

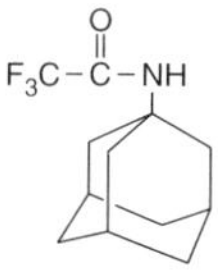

250 mg

S94,691-5
N-(2-ADAMANTYL)-2,2,2-TRIFLUOROACETAMIDE

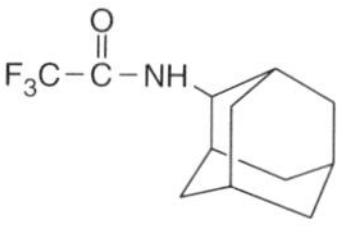

250 mg

S94,683-4
N-(1-ADAMANTYLMETHYL)-TRIFLUOROACETAMIDE

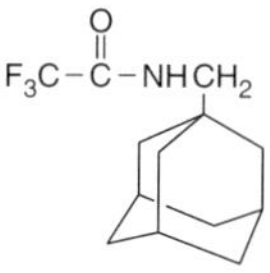

250 mg

S75,758-6
2-ACETAMIDO-1,2-DIDEOXY-1-NITRO-D-MANNITOL

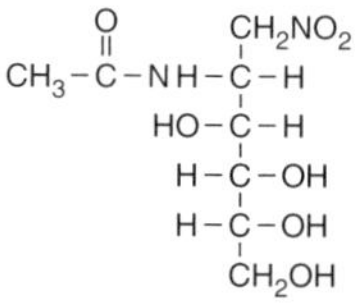

5 gm

S39,487-4
N-[TRIS-(HYDROXYMETHYL)METHYL]-ACETAMIDE

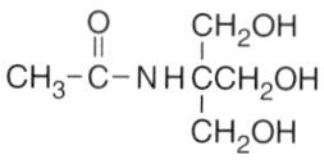

250 mg

S40,521-3
N-[2-(2-HYDROXYETHOXY)ETHYL]-FORMAMIDE

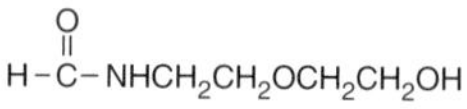

5 gm

S95,410-1
6-(TERT-BUTOXYCARBOXAMIDO)-2,3-EPOXY-2-METHYLCYCLOHEXANEMETHANOL

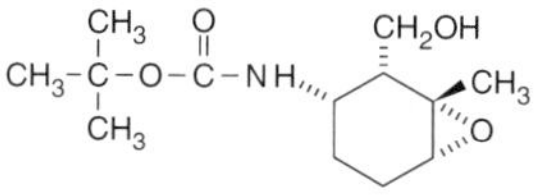

50 mg

S38,518-2
N-HYDROXYMETHYLGLUCONAMIDE

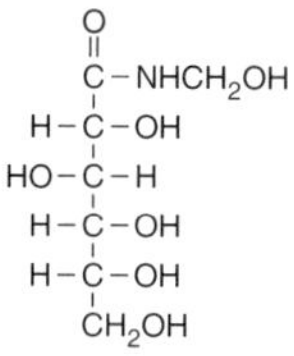

5 gm

S57,962-9
D-PANTOTHENIC ACID

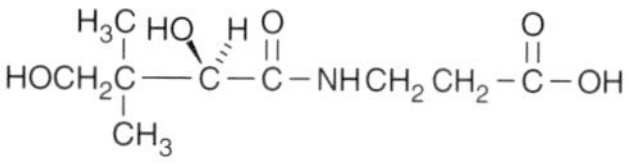

5 gm

S57,913-0
L-2,4-DIHYDROXY-3,3-DIMETHYL-N-(3-HYDROXYPROPYL)BUTYRAMIDE

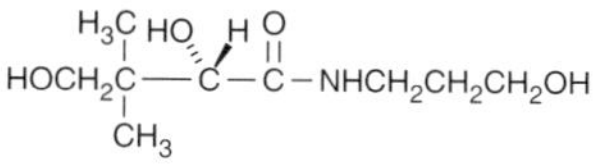

250 mg

S61,643-5
D-2,4-DIHYDROXY-3,3-DIMETHYL-N-(3-HYDROXYPROPYL)BUTYRAMIDE

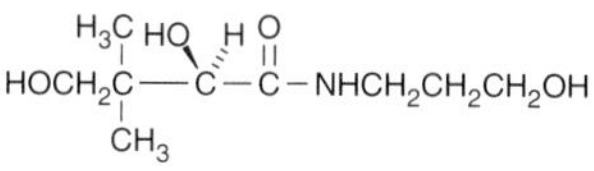

5 gm

S62,230-3
2-HYDROXY-N-OCTYLPROPIONAMIDE

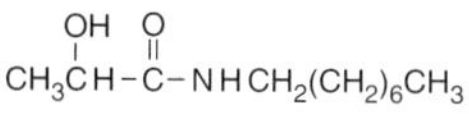

5 gm

S94,807-1
ALPHA-HYDROXY-1-ADAMANTANEACETAMIDE

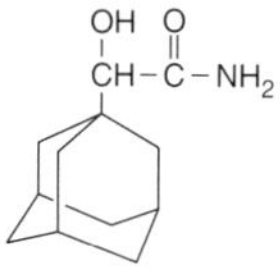

250 mg

S42,221-5
N-(2-HYDROXYETHYL)PALMITAMIDE

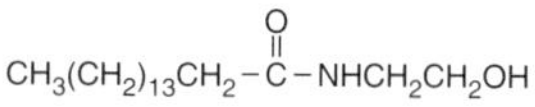

250 mg

S65,898-7
N-(2-HYDROXYETHYL)FORMAMIDE

5 gm

S64,791-8
N-HYDROXYMETHYLMETHACRYLAMIDE

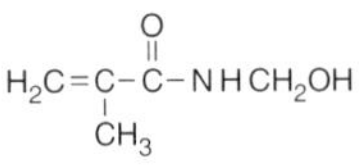

250 mg

S53,697-0
N-ETHYL-2-HYDROXY-2-METHYLPROPIONAMIDE

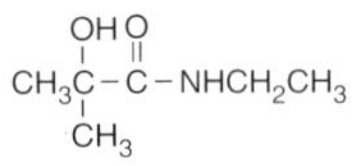

5 gm

S42,091-3
N-(HYDROXYMETHYL)-10-UNDECENOAMIDE

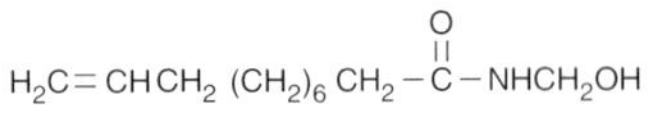

250 mg

S53,579-6
N-(2-HYDROXYETHYL)LAURAMIDE

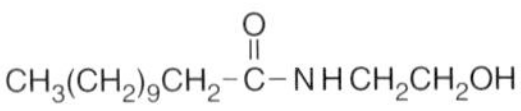

5 gm

S84,961-8
2-METHOXY-6-(2,2-DIMETHYLPROPIONAMIDO)-CYCLOHEXANECARBOXYLIC ACID

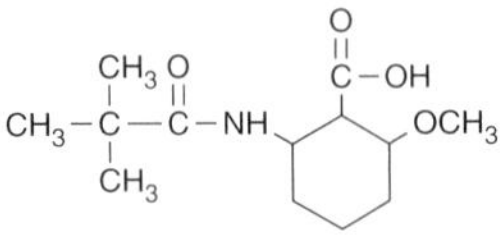

250 mg

S36,328-6
N-(3-METHOXYPROPYL)FORMAMIDE

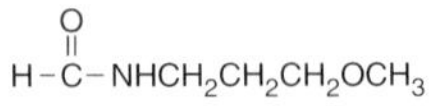

5 gm

S88,823-0
DIETHYL 2-ACETAMIDO-2-(2-FORMYLETHYL)MALONATE

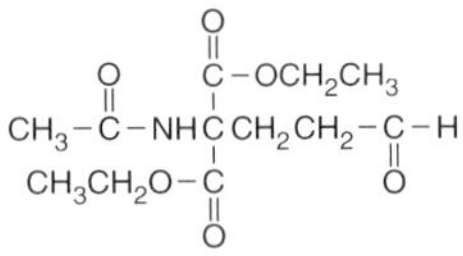

250 mg

S45,936-4
2-ACETAMIDO-2-DEOXY-1,3,4,6-TETRA-O-ACETYL-ALPHA-D-GLUCOPYRANOSE

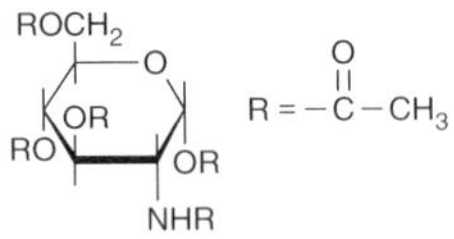

250 mg

S34,997-6
D-N-FORMYLPENICILLAMINE

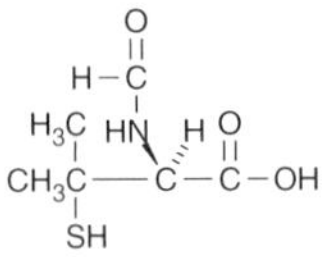

250 mg

S70,513-6
DL-METHIONYL-GLYCINE ETHYL ESTER

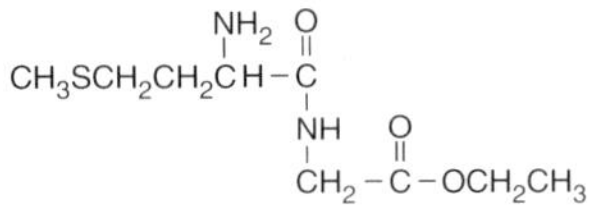

5 gm

S44,180-5
2-METHYLMALONAMIDE

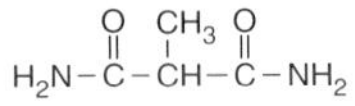

250 mg

S57,053-2
D-LEUCYLGLYCYLGLYCINE

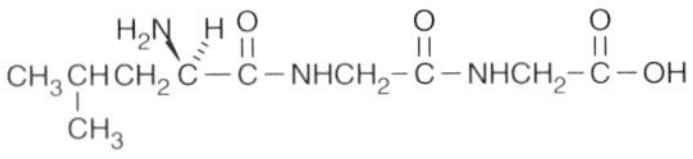

5 gm

S35,033-8
GLYCYL-DL-VALINE

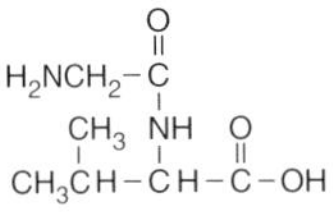

5 gm

S35,019-2
N-GLYCYLANILINE HYDROCHLORIDE

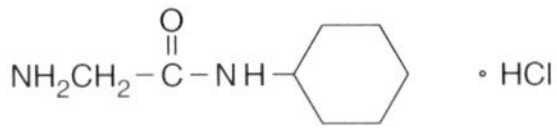

5 gm

S87,105-2
N-[2-(METHYLAMINO)ETHYL]-2-OXONONADECANAMIDE

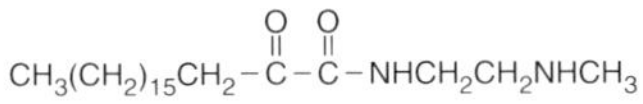

250 mg

S42,303-3
2-DIETHYLAMINO-N-TETRADECYLACETAMIDE CITRATE

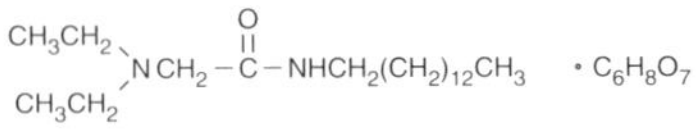

250 mg

S57,624-7
N-(2-DIETHYLAMINOETHYL)OLEAMIDE

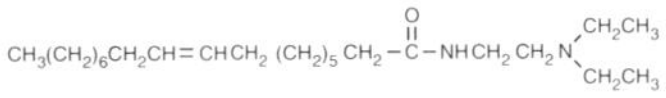

5 gm

S67,891-0
N-[2-(DIETHYLAMINO)ETHYL]STEARAMIDE

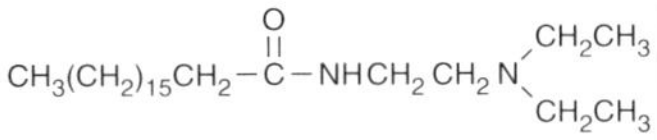

5 gm

S42,183-9
N-[2-(DIMETHYLAMINO)ETHYL]-LAURAMIDE

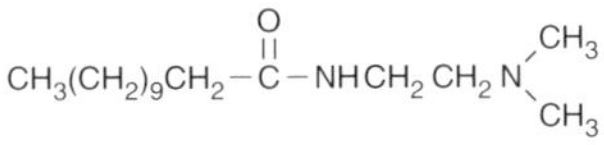

250 mg

S42,212-6
2-DIETHYLAMINO-N-(1-METHYLDECYL)-ACETAMIDE HYDROCHLORIDE

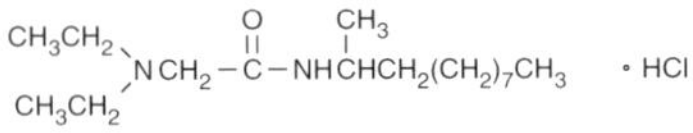

250 mg

S36,233-6
N-[3-(HEXAMETHYLENEIMINO)PROPYL]-ACETOACETAMIDE

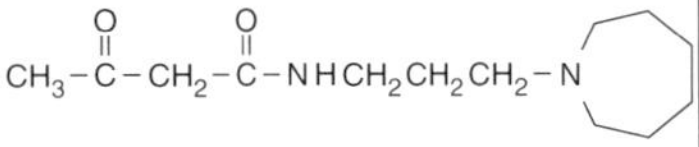

5 gm

S35,494-5
N-(3-PIPERIDINOPROPYL)-ACETOACETAMIDE

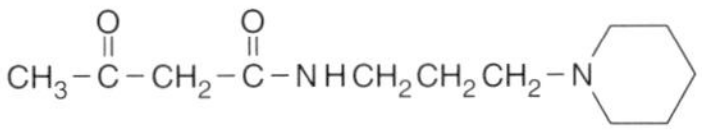

5 gm

S36,085-6
N-(3-DIMETHYLAMINOPROPYL)FORMAMIDE

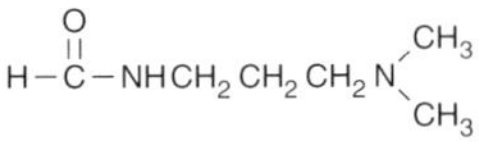

5 gm

S38,019-9
N-(3-DIETHYLAMINOPROPYL)ACETAMIDE

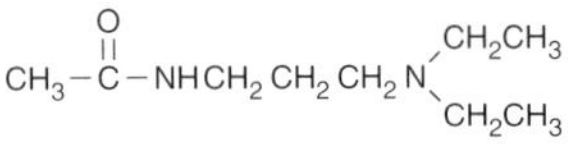

5 gm

S36,084-8
N-(3-DIMETHYLAMINOPROPYL)-DODECANAMIDE

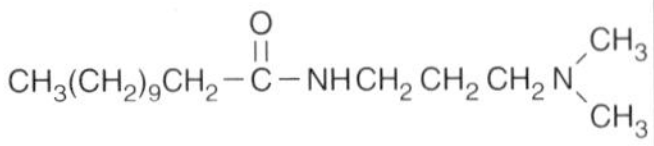

5 gm

S36,086-4
N-(3-DIMETHYLAMINOPROPYL)-OCTADECANAMIDE

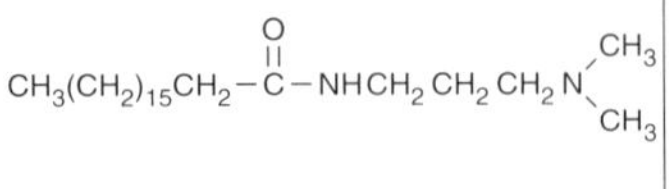

5 gm

S55,186-4
DIETHYL 2-ACETAMIDO-2-(2-CYANOETHYL)MALONATE

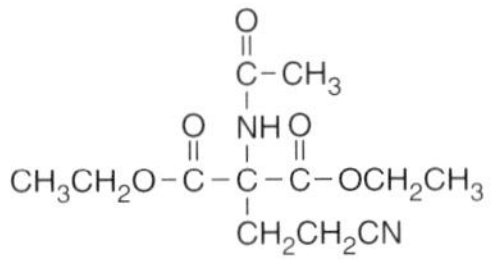

1 gm

S51,119-6
N-METHYLDIACETAMIDE

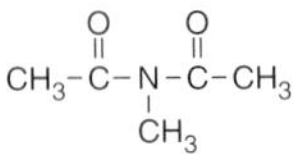

5 gm

S42,326-2
2-DIMETHYLAMINO-N-(1-METHYLHEXYL)-ACETAMIDE HYDROCHLORIDE

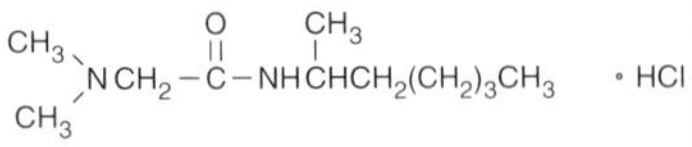

250 mg

S37,991-3
N-(3-DIETHYLAMINOPROPYL)FORMAMIDE

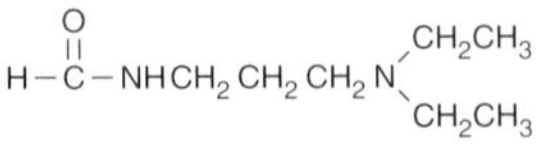

5 gm

S37,872-0
N-ALLYL-2-(DIMETHYLAMINO)ACETAMIDE

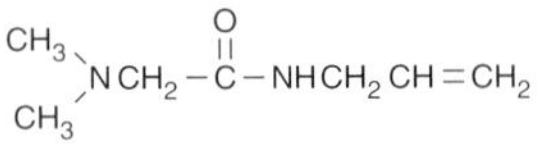

250 mg

S94,979-5
PROPIONOHYDROXAMIC ACID

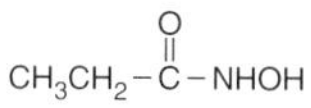

250 mg

S75,619-9
OCTANOHYDROXAMIC ACID, POTASSIUM SALT

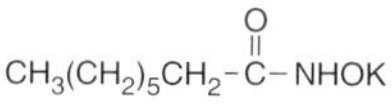

5 gm

S94,835-7
1-ADAMANTYLACETOHYDROXAMIC ACID

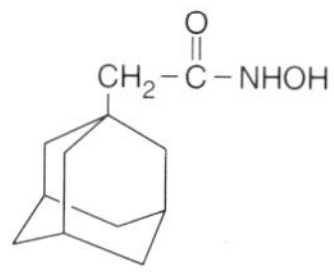

250 mg

S55,098-1
N-BUTYLFORMAMIDE

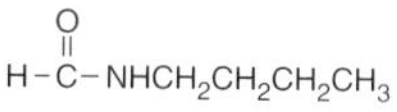

5 gm

S37,262-5
N-CYCLOHEPTYLACETOACETAMIDE

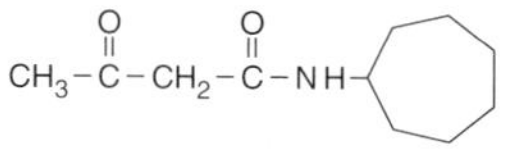

250 mg

S36,109-7
N-(1,4-DIMETHYLPENTYL)-ACETOACETAMIDE

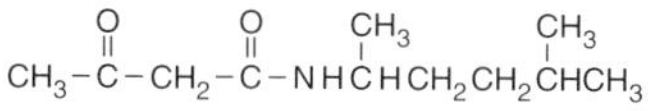

5 gm

S36,430-4
N-(TERT-OCTYL)ACETOACETAMIDE

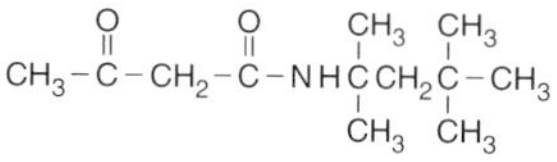

5 gm

S55,692-0
N-(2-ETHYLHEXYL)ACETOACETAMIDE

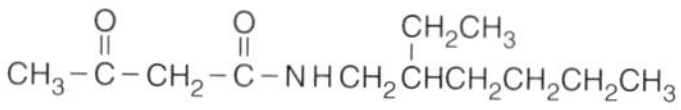

5 gm

S56,126-6
N-OCTYLACETOACETAMIDE

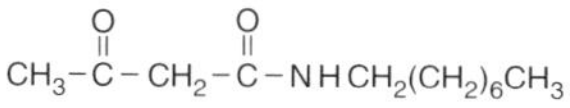

5 gm

S34,511-3
N-CYCLOPENTYLACETOACETAMIDE

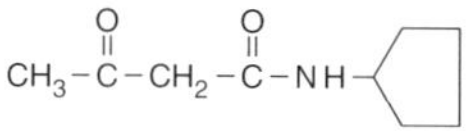

5 gm

S34,029-4
ACETOACETOCYCLOHEXYLAMIDE

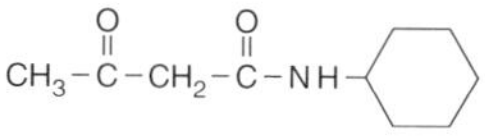

5 gm

S95,025-4
N-TERT-BUTYL-4,4-DIMETHYL-3-OXOPENTANAMIDE

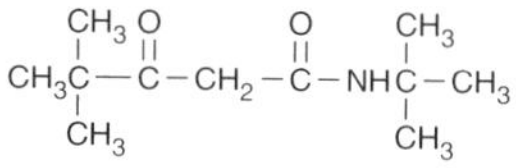

50 mg

S86,965-1
N-BUTYL-2-OXO-1-CYCLOPENTANECARBOXAMIDE

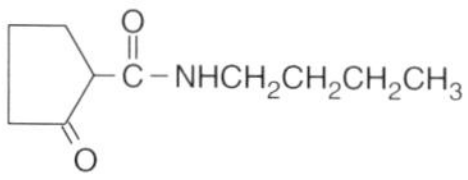

250 mg

S54,833-2
N,N'-(1,4-CYCLOHEXYLENEDIMETHYLENE)-DIMALEAMIC ACID

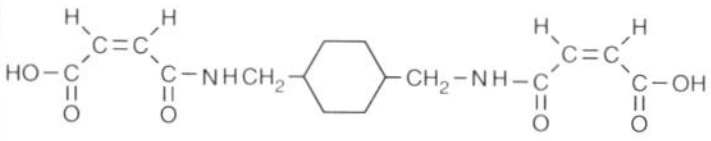

250 mg

S53,631-8
N-CYCLODODECYLMALEAMIC ACID

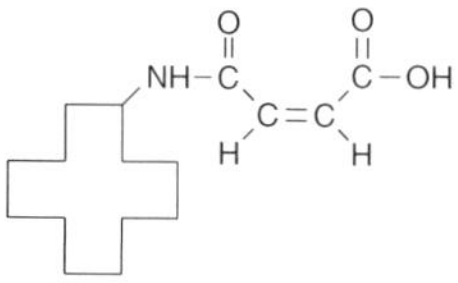

250 mg

S89,455-9
N-METHYL-1,2,3,4,4-PENTACHLORO-1,3-BUTADIENE-1-CARBOXAMIDE

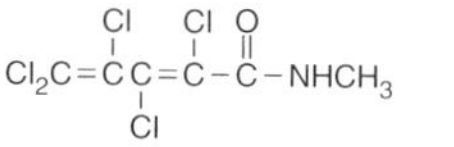

250 mg

S60,149-7
N-(TERT-BUTYL)ACRYLAMIDE

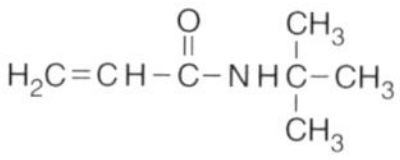

250 mg

S52,868-4
N-(2-NORBORNYL)ACRYLAMIDE

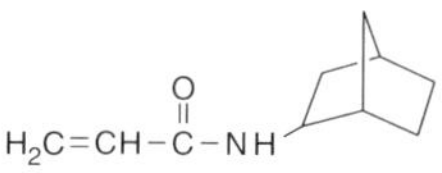

5 gm

S70,097-5
N-TERT-BUTYLSUCCINAMIC ACID

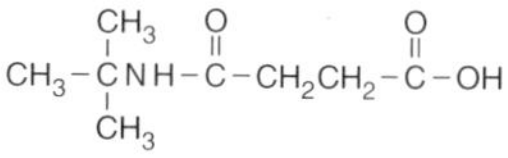

5 gm

S87,820-0
N-OCTADECYLGLUTARAMIC ACID

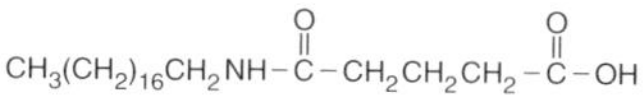

5 gm

S94,837-3
N,N'-DICYCLOHEXYL-1,3-ADAMANTANEDICARBOXAMIDE

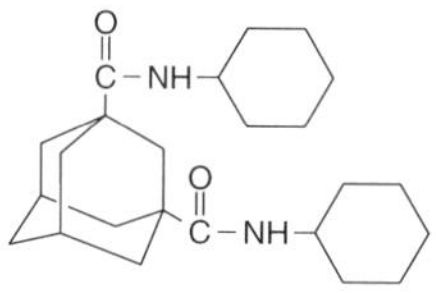

250 mg

S45,131-2
N-ACETYL-DL-LEUCINE

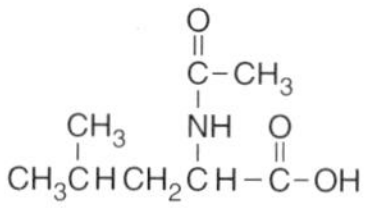

250 mg

S46,018-4
N-ACETYL-L-GLUTAMINE

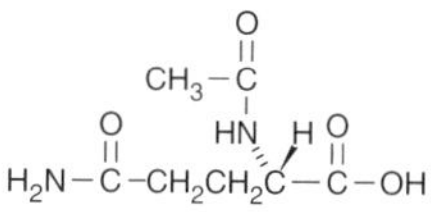

250 mg

S94,989-2
3-ACETAMIDO-3-METHYLBUTYRIC ACID

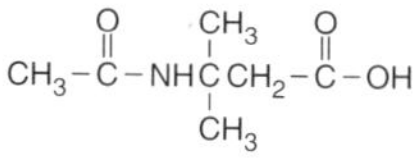

5 gm

S93,813-0
3-ACETAMIDO-1-ADAMANTANECARBOXYLIC ACID

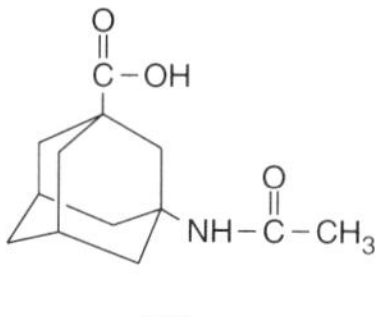

250 mg

S92,109-2
METHYL 3-(N-METHYLCARBAMOYL)-3-DIAZIRINECARBOXYLATE

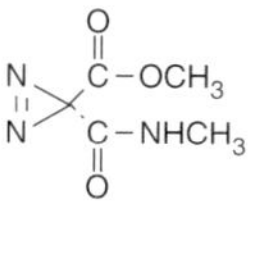

250 mg

S59,523-3
N-(2-ACETOXYETHYL)ACETAMIDE

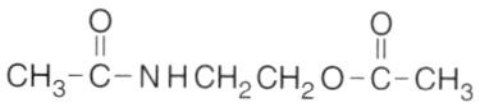

5 gm

S94,822-5
METHYL 3-ACETAMIDO-1-ADAMANTANECARBOXYLATE

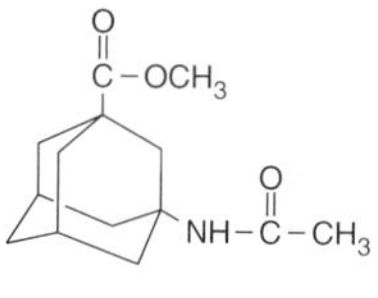

250 mg

S40,758-5
2,4-DICYANO-3-ETHYL-3-METHYLGLUTARAMIDE

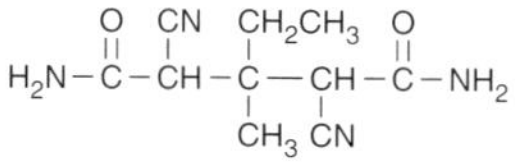

5 gm

S35,730-8
ACETYLURETHAN

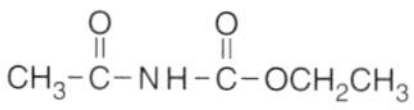

5 gm

S37,838-0
N-ISOPROPYLFORMAMIDE

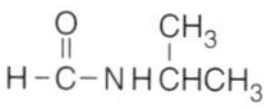

5 gm

S37,899-2
N-ISOPROPYLACETAMIDE

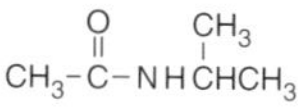

5 gm

S62,039-4
N-SEC-BUTYLFORMAMIDE
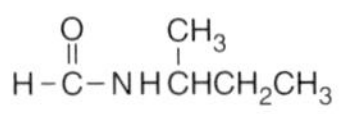
5 gm

S35,736-7
N-ALLYL-2-BROMOISOVALERAMIDE
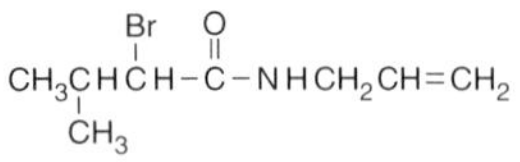
5 gm

S88,471-5
N-ALLYLHEXANAMIDE
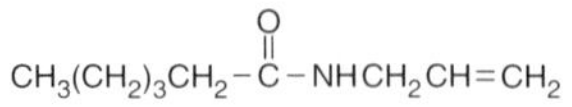
250 mg

S34,071-5
N-ALLYL-2-BROMO-2-ETHYLBUTYRAMIDE
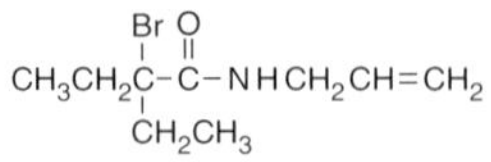
5 gm

S40,341-5
N-DODECYLACETAMIDE
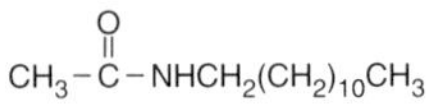
5 gm

S83,327-4
N-OCTADECYLFORMAMIDE
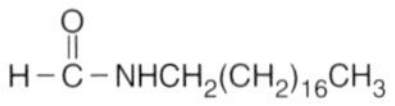
250 mg

S83,328-2
N-OCTADECYLPROPIONAMIDE
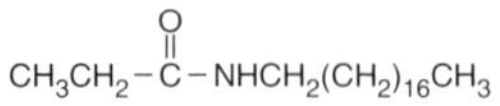
250 mg

S56,841-4
N-(1,1-DIETHYLOCTADECYL)FORMAMIDE
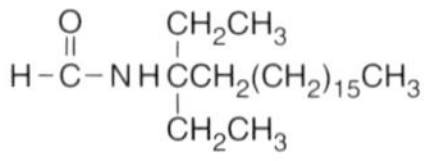
250 mg

S82,706-1
N-DODECYLDODECANAMIDE
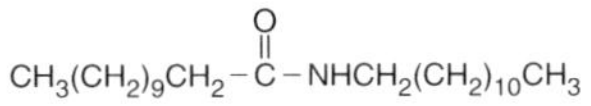
250 mg

S82,709-6
N-OCTADECYLOCTANAMIDE
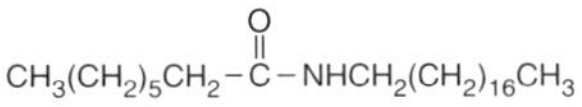
250 mg

S82,711-8
N-DODECYLHEXADECANAMIDE
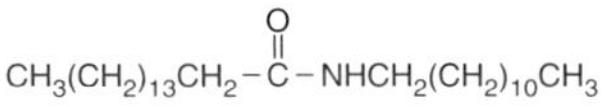
250 mg

S83,331-2
N-OCTADECYLDODECANAMIDE
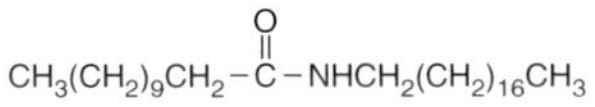
250 mg

S83,332-0
N-DODECYLOCTADECANAMIDE
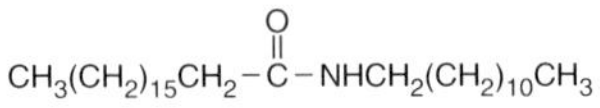
250 mg

S82,713-4
N-OCTADECYLPALMITAMIDE
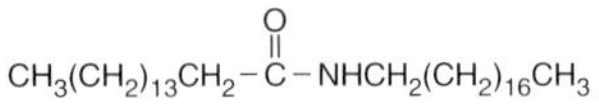
250 mg

S82,714-2
N-OCTADECYLOCTADECANAMIDE
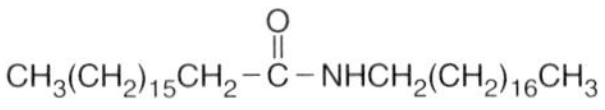
250 mg

S63,762-9
N-ETHYLCYCLOPROPANECARBOXAMIDE
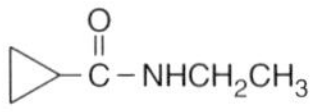
250 mg

S62,135-8
N-PROPYLCYCLOPROPANECARBOXAMIDE
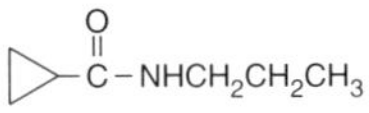
5 gm

S68,257-8
1-BROMO-N-TERT-BUTYL-1-CYCLOPENTANECARBOXAMIDE
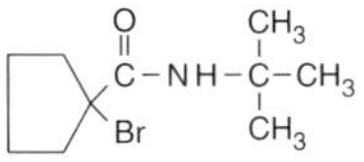
5 gm

S76,563-5
N-(5-CHLORO-2-METHYLCYCLOHEXYL)-ACETAMIDE
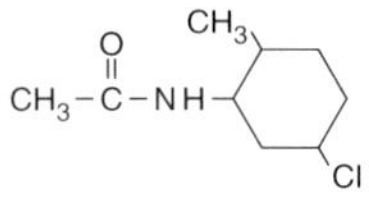
5 gm

S80,513-0
N-(4-BROMO-2,5-DIMETHYLCYCLOHEXYL)-ACETAMIDE
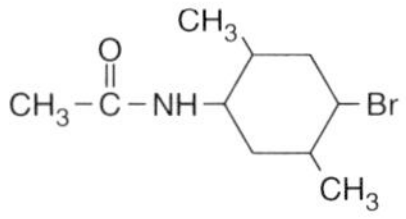
250 mg

S94,083-6
N-METHYL-1-ADAMANTANECARBOXAMIDE
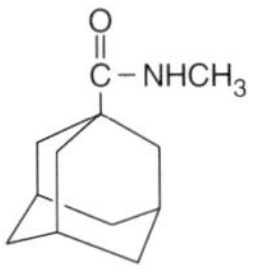
250 mg

S94,685-0
N-(2-ADAMANTYL)-1-ADAMANTANECARBOXAMIDE
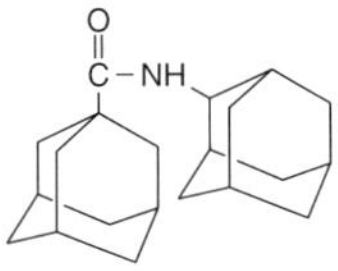
250 mg

S35,268-3
METHYL N-BUTYLCARBAMATE
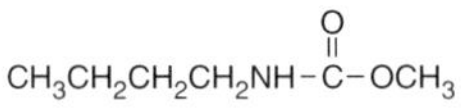
5 gm

S36,159-3
ETHYL CYCLOHEXANECARBAMATE
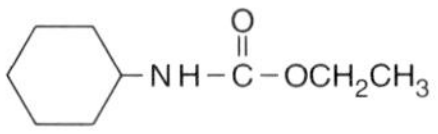
5 gm

S56,025-1
N,N-BIS(HYDROXYMETHYL)-2,3-DIBROMOPROPIONAMIDE
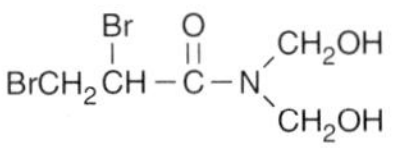
5 gm

S88,092-2
3-BROMO-2,2,N,N-TETRAMETHYLPROPIONAMIDE
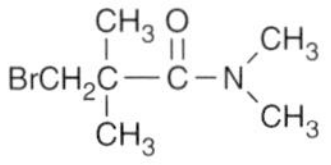
250 mg

S77,199-6
2-CHLORO-N,N-DIALLYLACETAMIDE
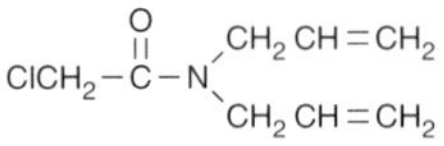
250 mg

S37,965-4
2-CHLORO-N,N-DIISOPROPYLACETAMIDE
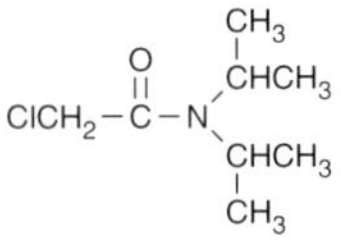
5 gm

S68,464-3
N-CYCLOHEXYL-2,2-DICHLORO-N-METHYLACETAMIDE
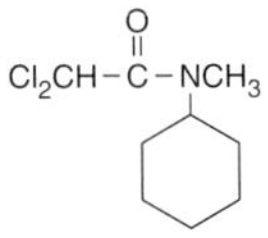
250 mg

S51,433-0
N,N-DIBUTYLLACTAMIDE
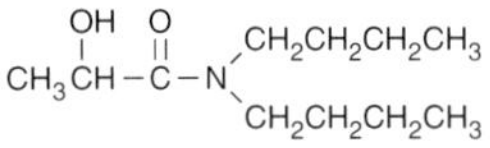
250 mg

S88,404-9
N-ALLYL-N-(1,1-DIMETHYL-3-HYDROXYBUTYL)ISOBUTYRAMIDE
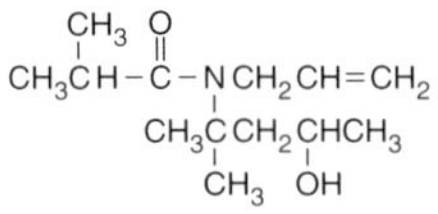
250 mg

S88,418-9
N-(2-BUTEN-1-YL)-N-(1,1-DIMETHYL-3-HYDROXYBUTYL)HEXANAMIDE
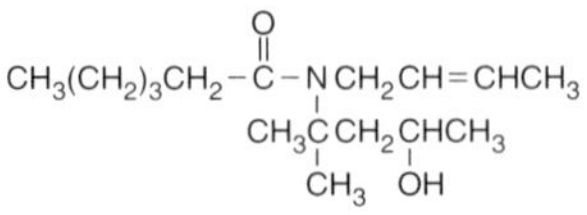
250 mg

S88,421-9
N-ALLYL-N-(1,1-DIMETHYL-3-HYDROXYBUTYL)OCTANAMIDE
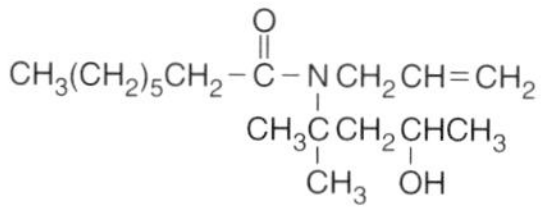
250 mg

S42,184-7
2-DIETHYLAMINO-N-(1-ISOBUTYL-3-METHYLBUTYL)-N-METHYLACETAMIDE HYDROCHLORIDE
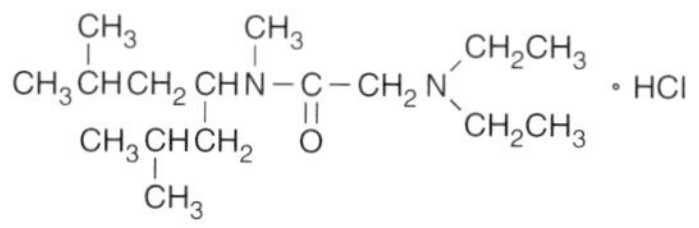
250 mg

S73,387-3
6-AMINO-4-OXO-7,7,N,N-TETRAMETHYL-5-OCTENAMIDE
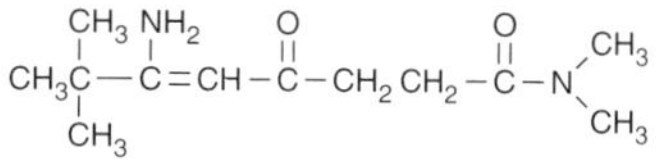
250 mg

S42,185-5
2-DIETHYLAMINO-N-ETHYL-N-OCTYLACETAMIDE HYDROCHLORIDE
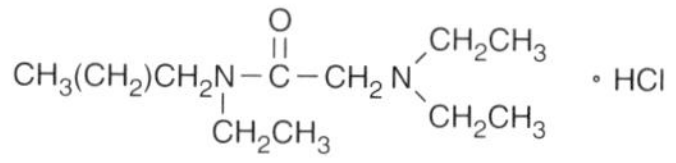
250 mg

S42,186-3
2-DIETHYLAMINO-N-(1,5-DIMETHYLHEXYL)-N-ETHYLACETAMIDE HYDROCHLORIDE
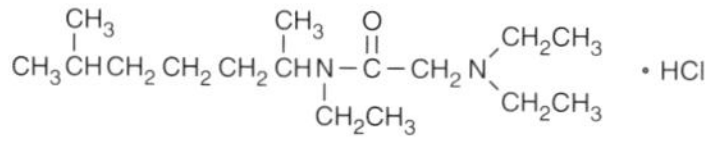
250 mg

S42,187-1
N-BUTYL-2-DIETHYLAMINO-N-(1,3-DIMETHYLBUTYL)ACETAMIDE HYDROCHLORIDE
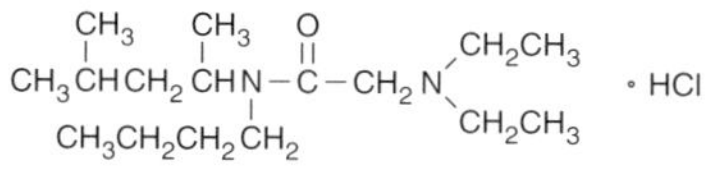
250 mg

S77,503-7
N,N'-ETHYLENEBIS(2,2-DIMETHYL-N-ETHYLPROPIONAMIDE)
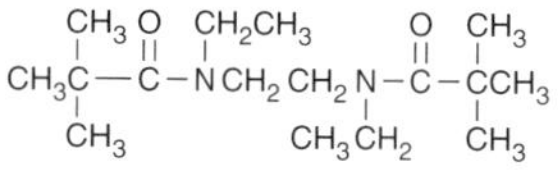
250 mg

S93,981-1
N,N-DIMETHYLVALERAMIDE
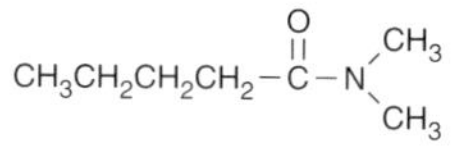
250 mg

S59,229-3
N,N-BIS[2-(DODECANOYL)ETHYL]-DODECANAMIDE
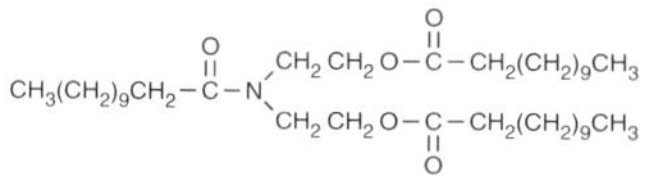
250 mg

S42,117-0
N,N-DIETHYL-2-OXO-4,6,6-TRIMETHYL-3-CYCLOHEXENE-1-CARBOXAMIDE
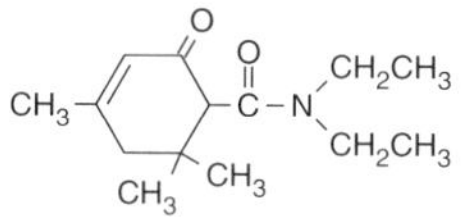
250 mg

S87,996-7
N-(1,1-DIMETHYL-2-OXOPROPYLTHIO)-N-METHYLCARBAMOYL FLUORIDE
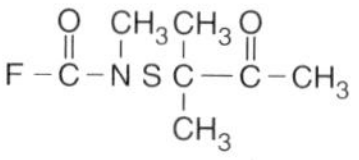
5 gm

S71,340-6
ETHYL N,N-DIETHYLOXAMATE
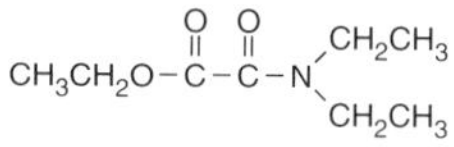
250 mg

S40,762-3
N,N-DIETHYLACETOACETAMIDE
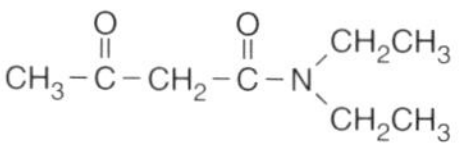
5 gm

S76,959-2
3-(N,N-DIETHYLCARBAMOYL)-5-NORBORNENE-2-CARBOXYLIC ACID
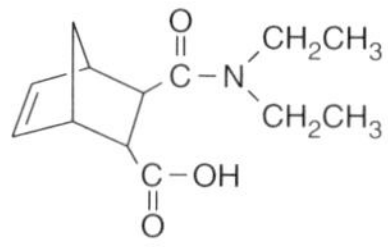
250 mg

S42,934-1
N,N-DIALLYLFORMAMIDE
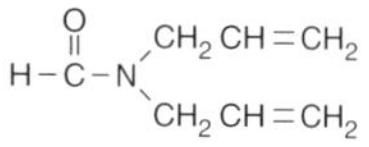
5 gm

S40,370-9
N,N-DIISOPROPYLFORMAMIDE
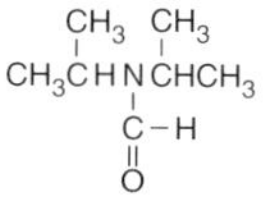
250 mg

S88,089-2
N-ALLYL-N-(2-BROMOALLYL)ACETAMIDE
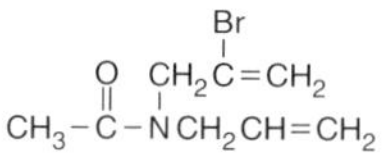
250 mg

S61,390-8
N,N-DIBUTYLACETAMIDE
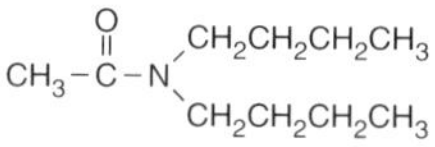
250 mg

S86,651-2
2,2-DIETHYL-N,N-DIMETHYLBUTYRAMIDE
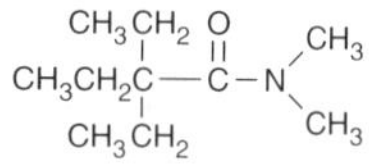
250 mg

S88,429-4
N-ALLYL-N-HEXYLISOBUTYRAMIDE
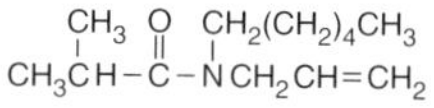
250 mg

S86,184-7
N,N-DIBUTYL-2,2-DIMETHYLPROPIONAMIDE
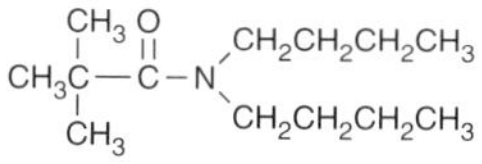
250 mg

S34,698-5
N,N-DIMETHYLDODECANAMIDE
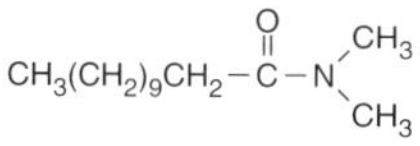
5 gm

S86,186-3
N,N-DIHEXYLACETAMIDE
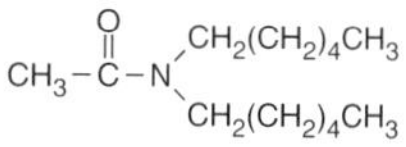
250 mg

S34,738-8
N,N-DIMETHYLTETRADECANAMIDE
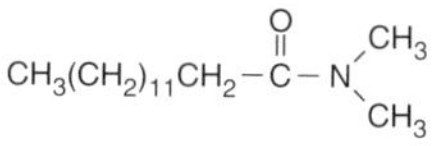
5 gm

S39,007-0
N,N-DIBUTYLLAURAMIDE
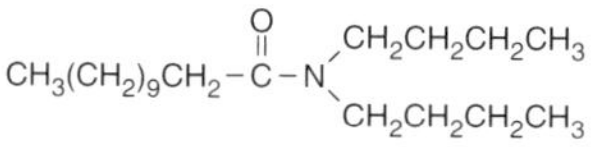
250 mg

S77,462-6
N-CYCLOHEXYL-2,2-DIMETHYL-N-ETHYLPROPIONAMIDE
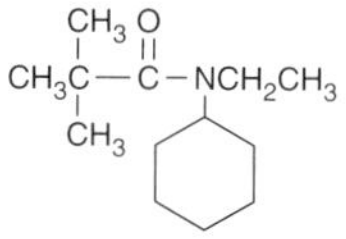
250 mg

S94,082-8
N,N-DIMETHYL-1-ADAMANTANECARBOXAMIDE
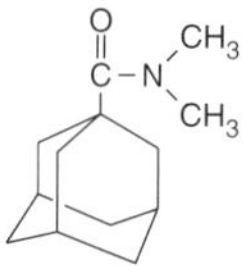
250 mg

S93,811-4
N,N-DIETHYL-1-ADAMANTANECARBOXAMIDE
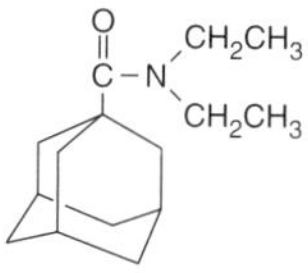
250 mg

S74,936-2
3-(2,2-DICHLOROACETYL)-3-AZABICYCLO[3.2.2]NONANE
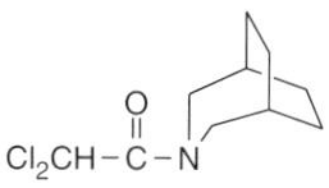
250 mg

S69,542-4
1-(2,2,2-TRICHLOROACETYL)PIPERIDINE
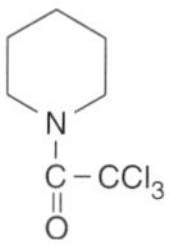
5 gm

S73,017-3
N-(2,2,2-TRICHLOROACETYL)MORPHOLINE
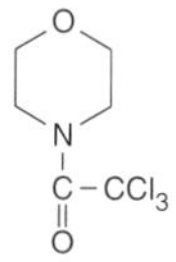
5 gm

S42,412-9
N-ACETYL-3-NORTROPANOL
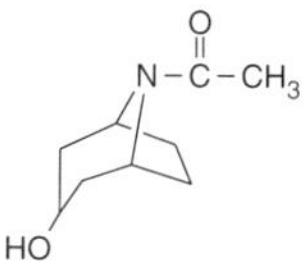
250 mg

S94,630-3
1-ACETYL-3-AMINO-2-PYRAZOLINE
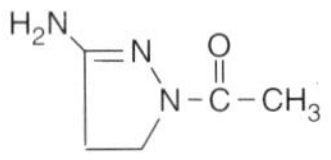
250 mg

S53,585-0
4-(CYCLOHEXYLAMINO)ISONIPECOTAMIDE
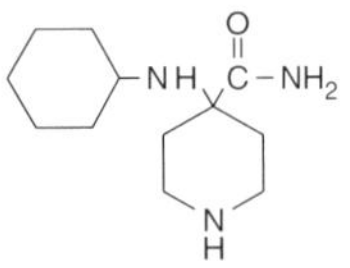
250 mg

S53,586-9
4-(ETHYLAMINO)ISONIPECOTAMIDE
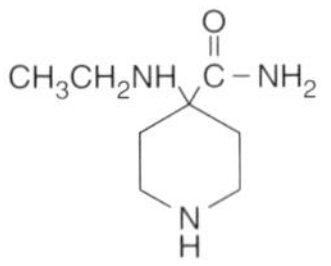
250 mg

S51,123-4
TETRAETHYL 2,2'-(1,4-PIPERAZINEDIYLDIMETHYL)BIS(2-FORMAMIDOMALONATE)
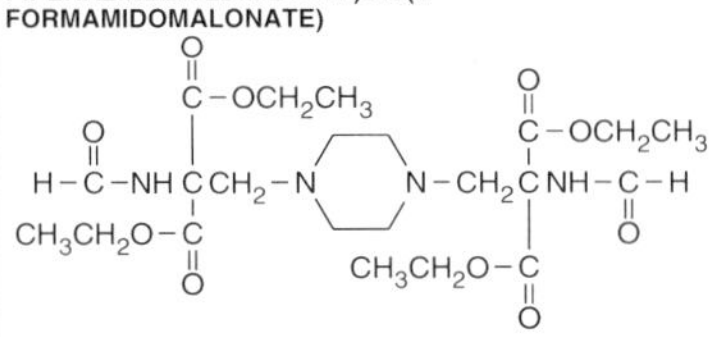
250 mg

S42,154-5
N-ETHYL-N-HEPTYL-1-PYRROLIDINEACETAMIDE HYDROCHLORIDE
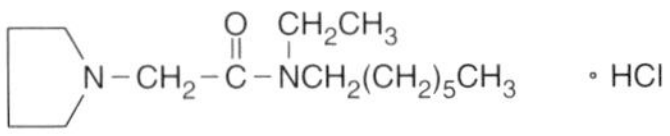
250 mg

S42,192-8
N-(1-ISOBUTYL-3-METHYLBUTYL)-N-METHYL-1-PYRROLIDINEACETAMIDE HYDROCHLORIDE
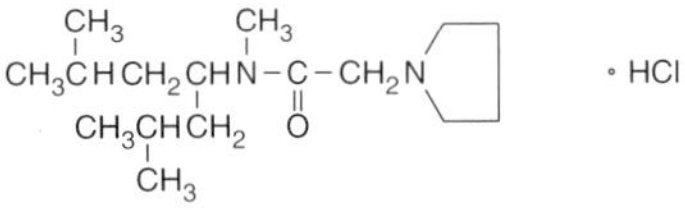
250 mg

S42,222-3
N-METHYL-N-(1-METHYLDECYL)-1-PYRROLIDINEACETAMIDE HYDROCHLORIDE
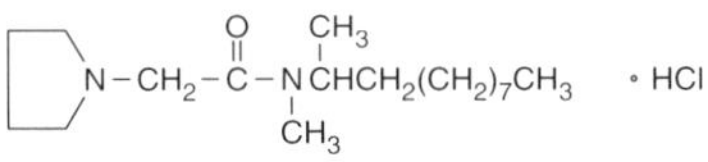
250 mg

S42,156-1
N-METHYL-N-(1-METHYLHEXYL)-1-PIPERIDINEACETAMIDE HYDROCHLORIDE
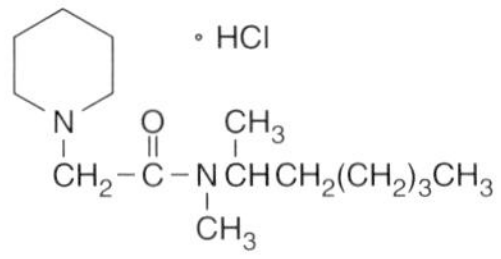
250 mg

S42,258-4
N-ETHYL-N-(1-METHYLDECYL)-1-PYRROLIDINEACETAMIDE HYDROCHLORIDE
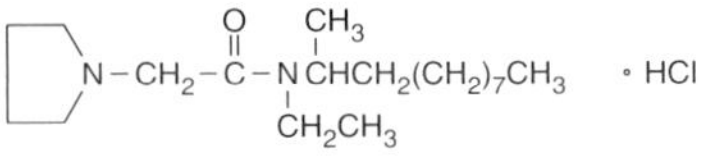
250 mg

S42,215-0
N-BUTYL-N-(1,3-DIMETHYLBUTYL)-1-PIPERIDINEACETAMIDE HYDROCHLORIDE
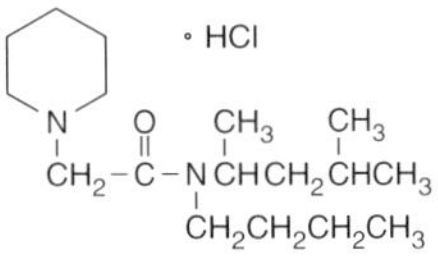
250 mg

S35,931-9
4-CARBAMOYL-4-METHYLAMINOPIPERIDINE
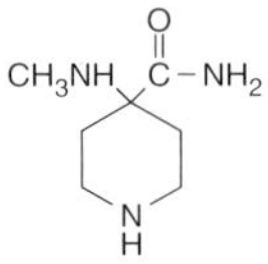
5 gm

S34,332-3
4-CARBAMOYL-4-(ISOPROPYLAMINO)-PIPERIDINE
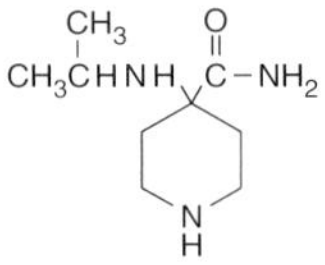
250 mg

S89,512-1
N-ETHYL-2-(PYRROLIDINOMETHYLENE)-BUTYRAMIDE
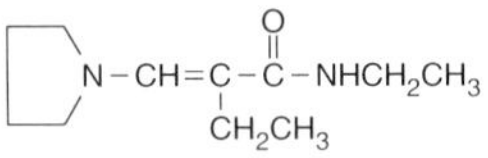
250 mg

S42,276-2
3-CARBAMOYL-N-DODECYL-1-PIPERIDINEACETAMIDE
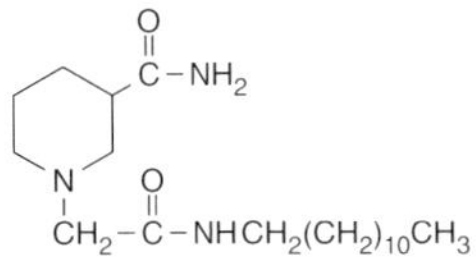
5 gm

S34,325-0
4-CARBAMOYL-4-(DIMETHYLAMINO)-PIPERIDINE DIHYDROCHLORIDE
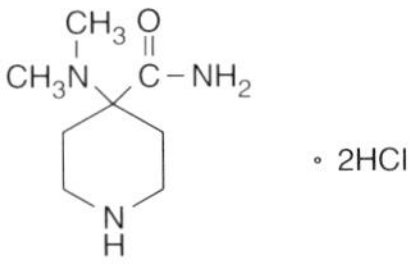
5 gm

S42,051-4
1-PENTYLNIPECOTAMIDE
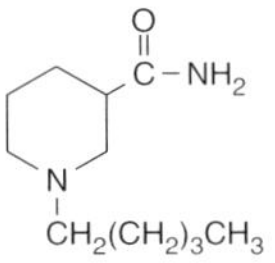
250 mg

S42,052-2
1-ISOPENTYLNIPECOTAMIDE
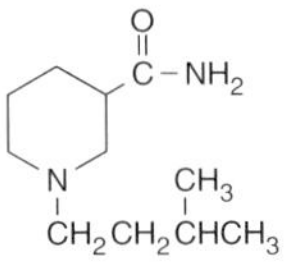
5 gm

S42,111-1
1-HEPTYLNIPECOTAMIDE
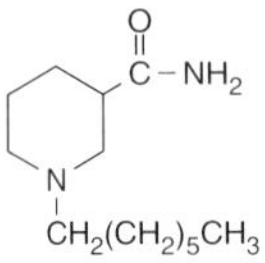
250 mg

S42,226-6
1-DODECYLNIPECOTAMIDE
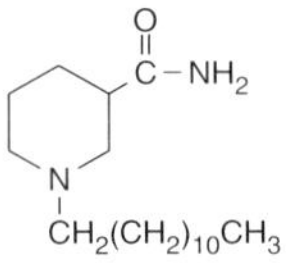
5 gm

S42,230-4
DECYL 3-CARBAMOYL-1-PIPERIDINEACETATE
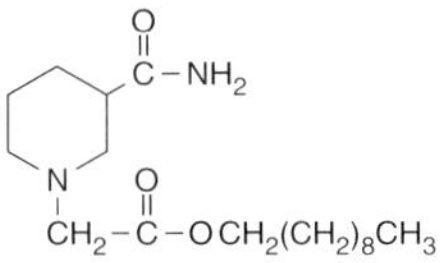
250 mg

S42,344-0
N,N-DIETHYL-1-PROPYLNIPECOTAMIDE HYDROCHLORIDE
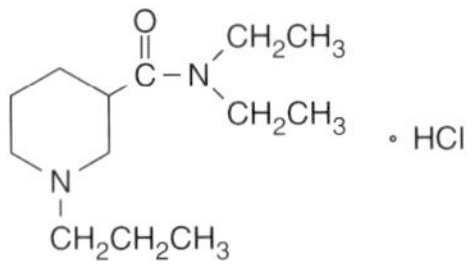
250 mg

S42,275-4
DODECYL 4-CARBAMOYL-1-PIPERIDINEACETATE
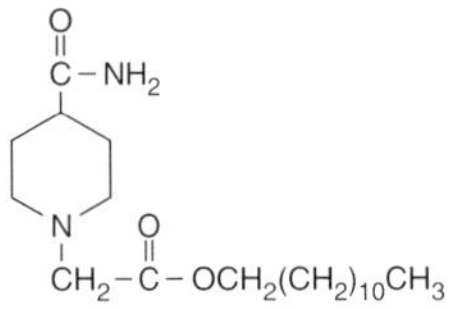
250 mg

S42,229-0
DECYL 4-CARBAMOYL-1-PIPERIDINEACETATE
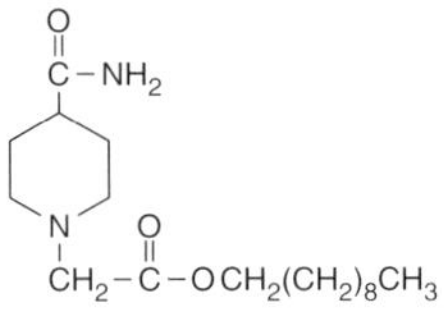
250 mg

S42,049-2
1-ISOPENTYLISONIPECOTAMIDE
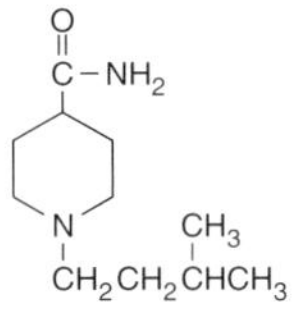
250 mg

S42,050-6
1-PENTYLISONIPECOTAMIDE
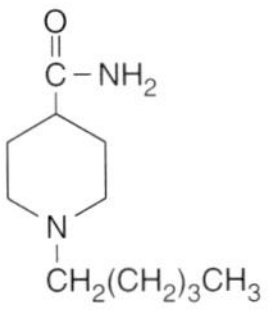
250 mg

S42,090-5
1-HEXYLISONIPECOTAMIDE
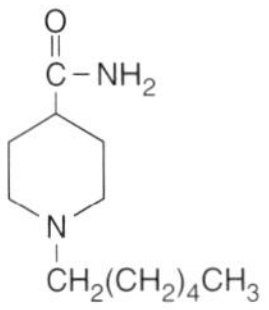
250 mg

S42,159-6
1-NONYLISONIPECOTAMIDE
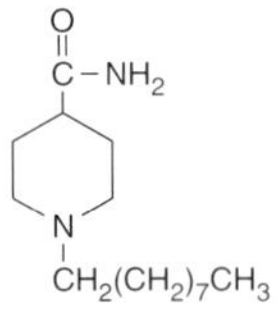
250 mg

S42,227-4
1-DODECYLISONIPECOTAMIDE
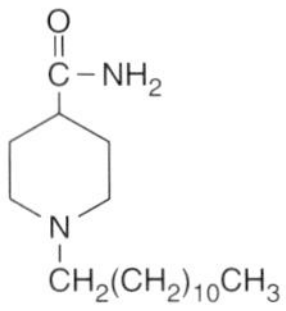
5 gm

S42,290-8
1-HEXADECYLISONIPECOTAMIDE
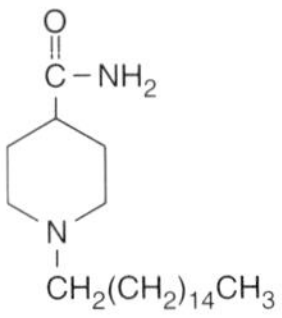
250 mg

S42,300-9
1-OCTADECYLISONIPECOTAMIDE
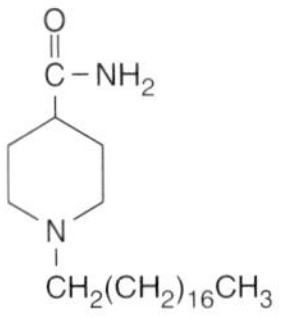
5 gm

S42,301-7
1-OCTADECYLNIPECOTAMIDE
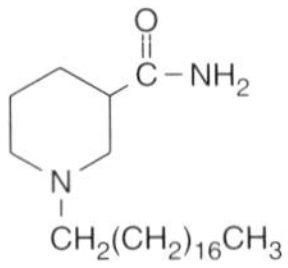
5 gm

S65,942-8
DIMETHYL 2-(PIPERIDINOMETHYLENEACETAMIDO)-MALONATE
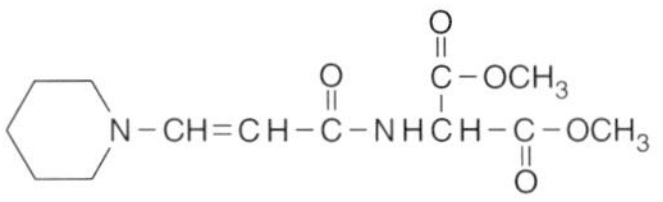
5 gm

S89,375-7
2,3,4,5,5-PENTACHLORO-1-PYRROLIDINO-2,4-PENTADIEN-1-ONE

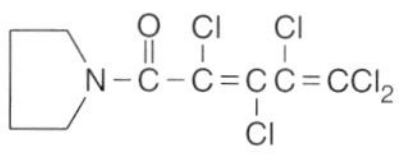

250 mg

S89,431-1
1-MORPHOLINO-2,3,4,5,5-PENTACHLORO-2,4-PENTADIEN-1-ONE

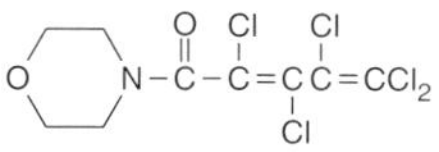

250 mg

S42,273-8
1-TETRADECYLNIPECOTAMIDE

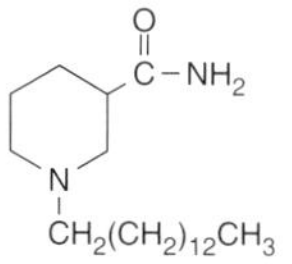

250 mg

S42,110-3
N-(1-METHYLHEXYL)-1-PYRROLIDINEACETAMIDE HYDROCHLORIDE

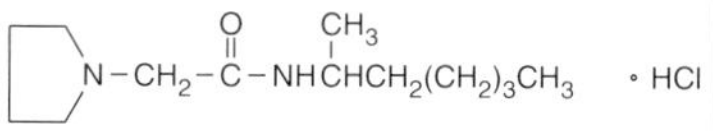

250 mg

S42,195-2
N-(1,3-DIMETHYLBUTYL)-N-PROPYL-1-PIPERIDINEACETAMIDE HYDROCHLORIDE

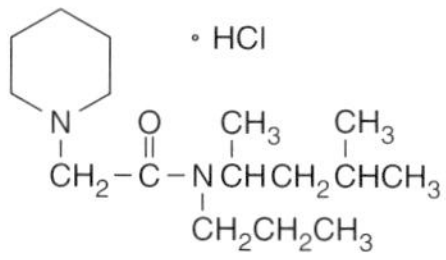

250 mg

S42,225-8
N-ETHYL-N-(1-ISOBUTYL-3-METHYLBUTYL)-1-PIPERIDINEACETAMIDE HYDROCHLORIDE

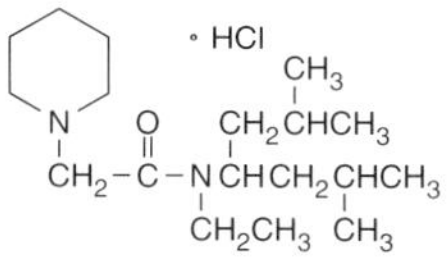

250 mg

S42,259-2
N-METHYL-N-(1-METHYLDECYL)-1-PIPERIDINEACETAMIDE HYDROCHLORIDE

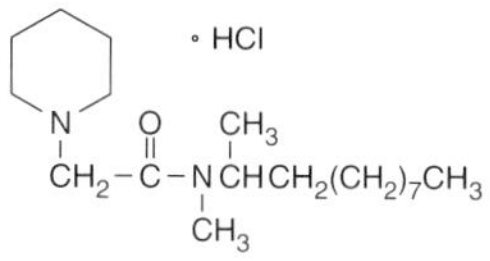

250 mg

S38,937-4
1-FORMYL-4-METHYLPIPERAZINE

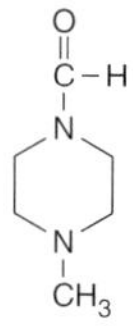

250 mg

S94,281-2
1-(1-ADAMANTYLACETYL)PYRROLIDINE

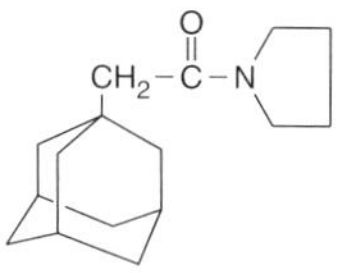

250 mg

S94,829-2
(3,5-DIMETHYL-1-ADAMANTYLMETHYL)-PYRROLIDINE HYDROCHLORIDE

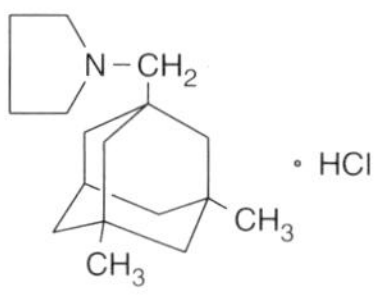

250 mg

S55,687-4
1,2-EPOXY-2,6,6-TRICHLOROCYCLOHEXANECARBOXAMIDE

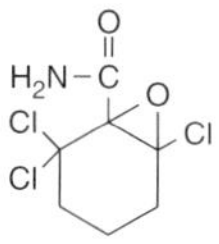

5 gm

S35,051-6
N,N-HEXAMETHYLENEACETOACETAMIDE

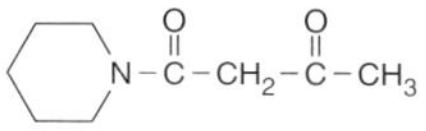

5 gm

S39,083-6
1-ACETOACETYL-2-METHYLPIPERIDINE

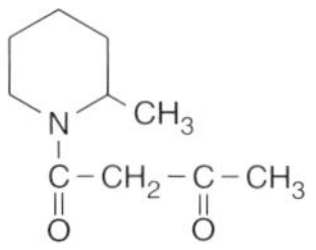

5 gm

S39,084-4
1-ACETOACETYL-3-METHYLPIPERIDINE

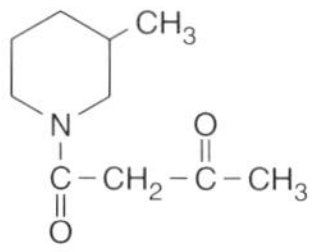

5 gm

S39,085-2
1-ACETOACETYL-4-METHYLPIPERIDINE

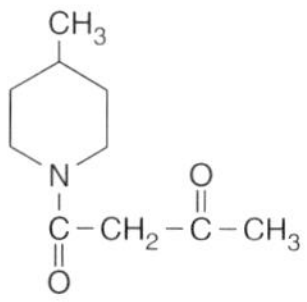

5 gm

S37,599-3
ACETOACETYL-2,6-DIMETHYLPIPERIDIDE

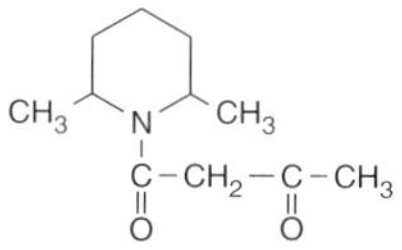

5 gm

S34,209-2
1,4-BIS(ACETOACETYL)PIPERAZINE

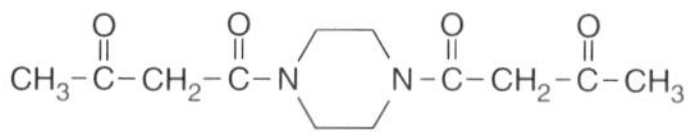

5 gm

S78,787-6
1-ACETOACETYLPIPERIDINE

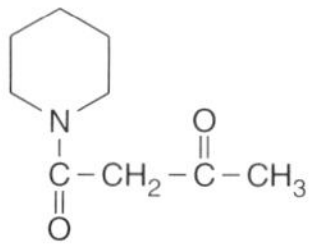

250 mg

S95,670-8
2-ACETYL-8-METHYLDECAHYDROISOQUINOLIN-7-ONE

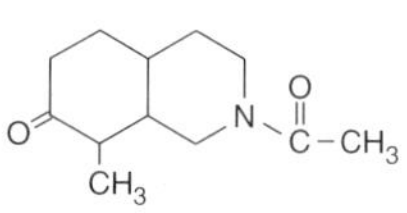

50 mg

S70,132-7
GAMMA-OXO-1-PYRROLIDINEBUTYRIC ACID

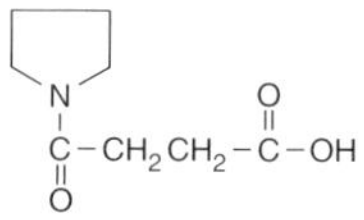

5 gm

S84,947-2
1,6-BIS(1-PYRROLIDINYL)-1,6-HEXANEDIONE

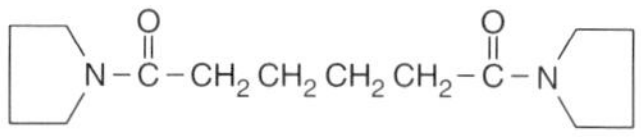

250 mg

S70,450-4
N-FORMYL-L-PENICILLAMINE, ACETONE ADDUCT

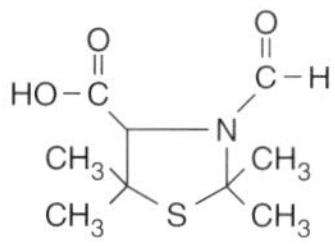

5 gm

S70,452-0
N-FORMYL-DL-PENICILLAMINE, ACETONE ADDUCT

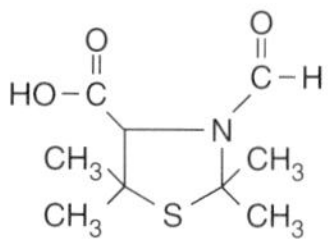

5 gm

S70,271-4
1-ACETYL-4-PIPERIDINECARBOXYLIC ACID

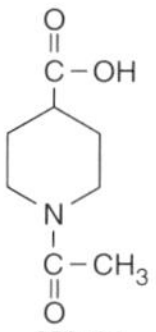

250 mg

S95,046-7
N-(7-OXABICYCLO[4.1.0]HEPT-1-YLCARBONYL)GLYCINE ETHYL ESTER

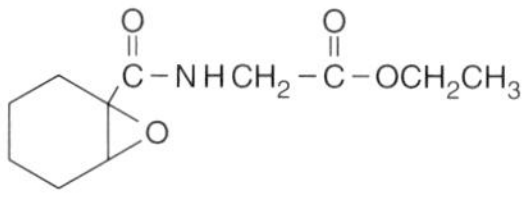

50 mg

S40,699-6
1-BUTYRYLAZIRIDINE

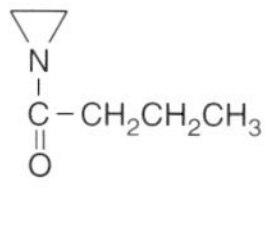

5 gm

S43,045-5
N-FORMYLHEXAMETHYLENEIMINE

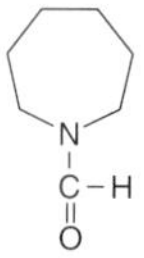

5 gm

S43,282-2
1-ISOBUTYRYLAZIRIDINE

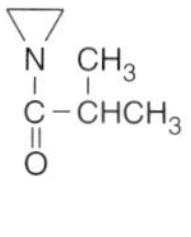

250 mg

S94,827-6
1-(1-ADAMANTANECARBONYL)PYRROLIDINE

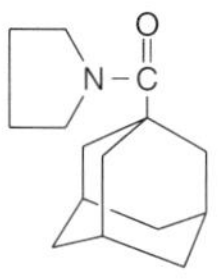

250 mg

S94,814-4
1-[2-(1-ADAMANTYL)-2-BROMOACETYL]-PYRROLIDINE

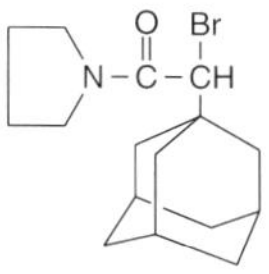

250 mg

S42,194-4
N-(1,5-DIMETHYLHEXYL)-N-METHYL-1-PIPERIDINEACETAMIDE HYDROCHLORIDE

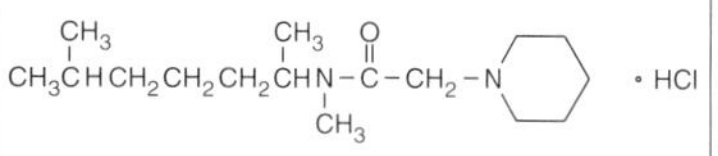

250 mg

S94,826-8
1-(1-ADAMANTANECARBONYL)PIPERIDINE

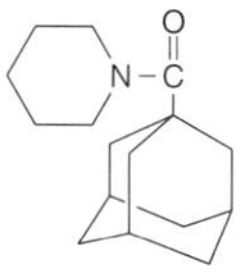

250 mg

S34,261-0
4-(2-BROMOISOVALERYL)MORPHOLINE
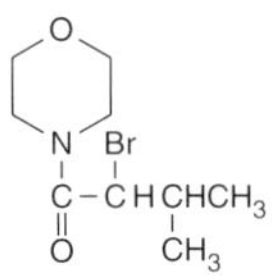
5 gm

S78,682-9
N-(2-BROMO-2,2-DIETHYLACETYL)-MORPHOLINE
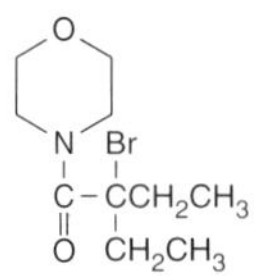
250 mg

S93,812-2
1-(1-ADAMANTANECARBONYL)MORPHOLINE
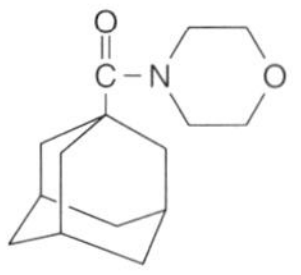
250 mg

S94,828-4
1-(3,5-DIMETHYL-1-ADAMANTANECARBONYL)MORPHOLINE
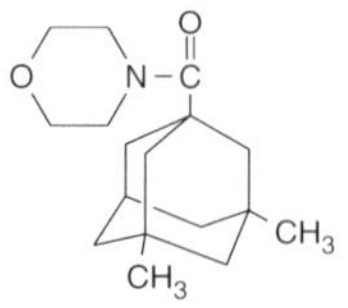
250 mg

S94,823-3
1,1'-(1,3-ADAMANTANEDICARBONYL)-BISMORPHOLINE
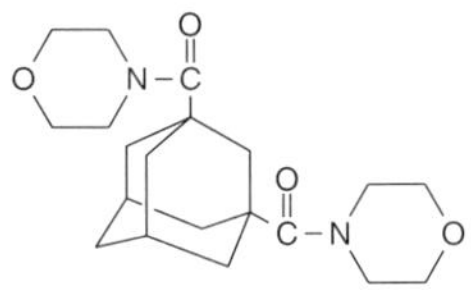
250 mg

S94,824-1
1,1'-(1,3-ADAMANTANEDICARBONYL)-BISPIPERIDINE
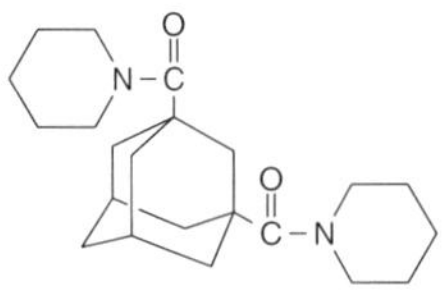
250 mg

S36,189-5
ETHYL N-PIPERIDINOCARBOXYLATE
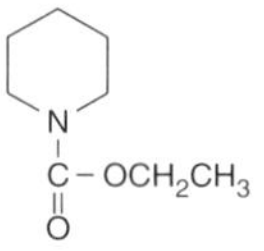
5 gm

S34,942-9
ETHYL N-MORPHOLINOCARBOXYLATE
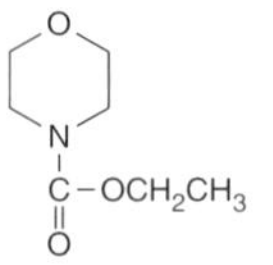
5 gm

S75,624-5
1,4-BIS(1,3-DIHYDROXY-2-METHYL-2-PROPYL)OXAMIDE
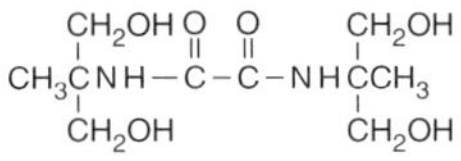
5 gm

S75,625-3
1,4-BIS[TRIS(HYDROXYMETHYL)METHYL]-OXAMIDE
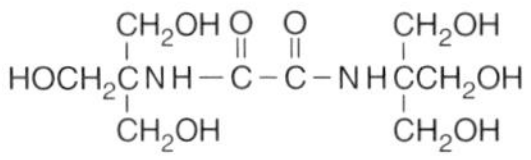
5 gm

S51,018-1
N,N'-BIS(2-HYDROXYETHYL)OXAMIDE
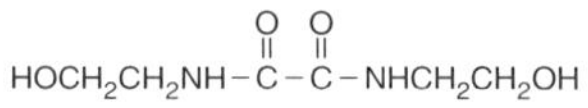
5 gm

S40,968-5
OXAMIC ACID, AMMONIUM SALT
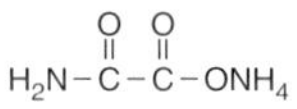
5 gm

S45,464-8
N,N'-DIMETHYLOXAMIDE
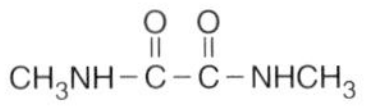
5 gm

S73,051-3
N,N'-DIPROPYLOXAMIDE
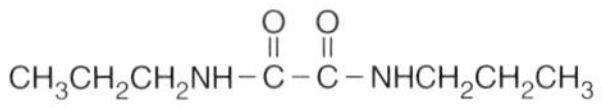
5 gm

S73,049-1
N,N'-DIDECYLOXAMIDE
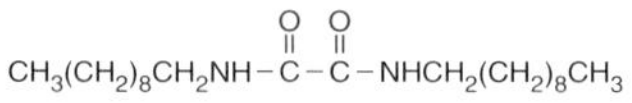
5 gm

S81,183-1
N,N'-DITETRADECYLOXAMIDE
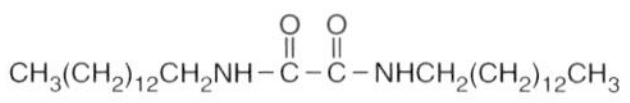
250 mg

S77,472-3
N,N,N',N'-TETRAISOBUTYLOXAMIDE
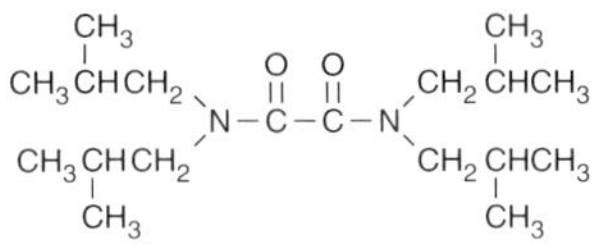
250 mg

S63,291-0
N,N,N',N'-TETRAMETHYLOXALAMIDE
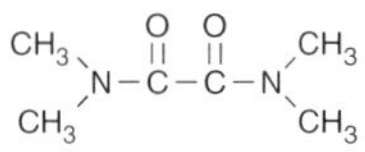
250 mg

S63,336-4
N,N'-BIS(3-CHLOROPROPIONYL)-ETHYLENEDIAMINE
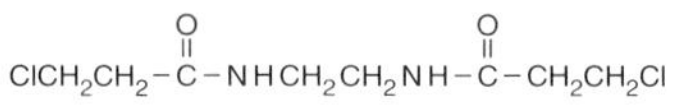
250 mg

S67,550-4
1,4-BIS(TRICHLOROACETYL)PIPERAZINE
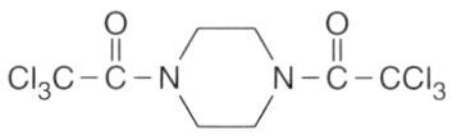
5 gm

S37,223-4
N-(2-HYDROXYETHYL)-N,N'-ETHYLENEBIS(FORMAMIDE)
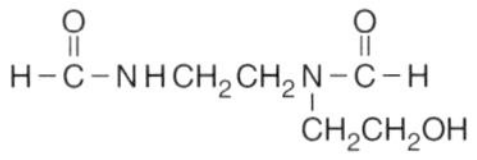
5 gm

S64,056-5
2,3-DIMETHOXY-N,N,N',N'-TETRAMETHYLSUCCINAMIDE
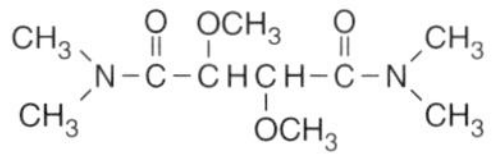
5 gm

S36,503-3
2,2'-THIOBIS(ACETAMIDE)
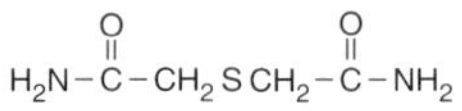
5 gm

S68,010-9
2-AMINOMALONAMIDE
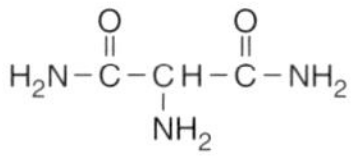
5 gm

S35,687-5
DL-VALYLGLYCYLGLYCINE
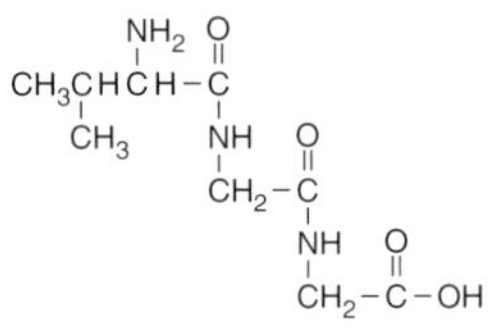
250 mg

S35,858-4
1,2-BIS(ACETOACETAMIDO)ETHANE
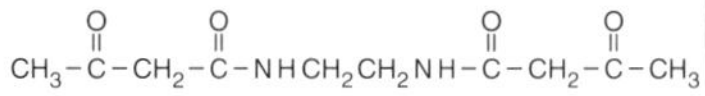
5 gm

S37,603-5
N,N'-BIS(ACETOACETYL)-DIETHYLENETRIAMINE
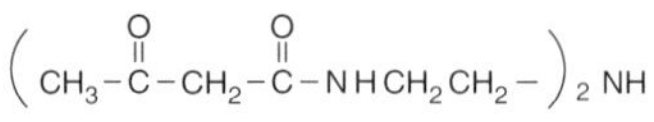
5 gm

S36,168-2
N,N'-ETHYLENEBIS(FORMAMIDE)
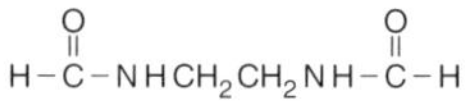
5 gm

S38,938-2
N,N'-ETHYLENEBIS(ACETAMIDE)
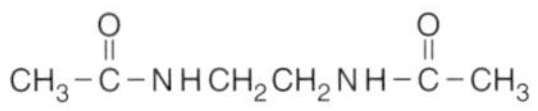
250 mg

S68,392-2
N,N'-ETHYLENEBIS(TETRADECANAMIDE)
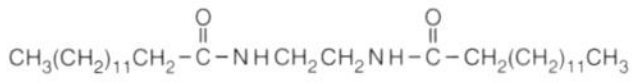
250 mg

S82,715-0
N,N'-ETHYLENEBIS(OCTADECANAMIDE)
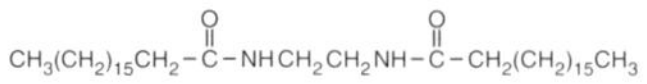
250 mg

S88,082-5
2,N,N,N',N'-PENTAMETHYLMALONAMIDE
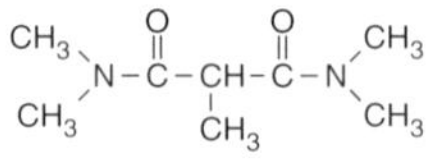
250 mg

S65,195-8
N,N,N',N'-TETRABUTYLMALONAMIDE

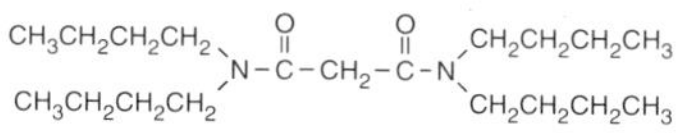

5 gm

S41,141-8
4,4'-MALONYLDIMORPHOLINE

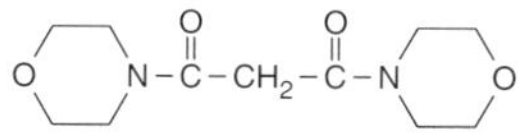

5 gm

S75,325-4
MALEAMIDE

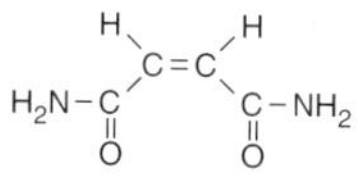

5 gm

S37,485-7
N,N'-DIMETHYLSUCCINAMIDE

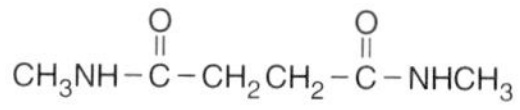

250 mg

S43,330-6
2-METHYL-1,4-PIPERAZINEDICARBOXALDEHYDE

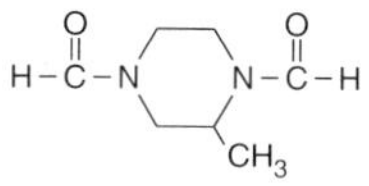

5 gm

S36,356-1
N,N'-METHYLENEBIS(FORMAMIDE)

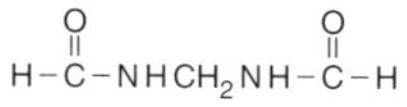

5 gm

S60,874-2
N,N'-METHYLENEDIACETAMIDE

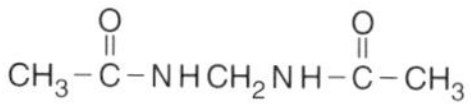

5 gm

S84,140-4
N,N'-ETHYLIDENEBIS(ACETAMIDE)

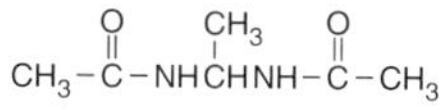

250 mg

S37,592-6
1,4-BIS(ACETOACETYLAMINOMETHYL)-CYCLOHEXANE

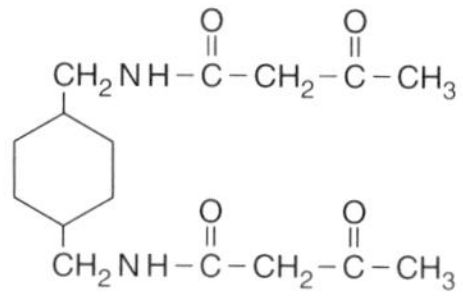

5 gm

S41,172-8
1,4-BIS(ACETOACETYL)-2-METHYLPIPERAZINE, TECH.

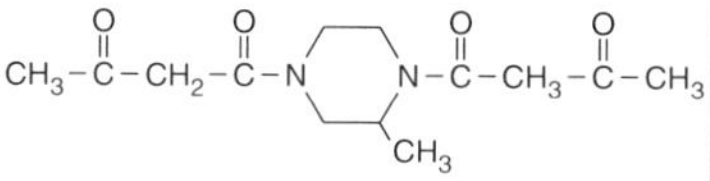

5 gm

S74,268-6
1,3-DIACETAMIDOACETONE

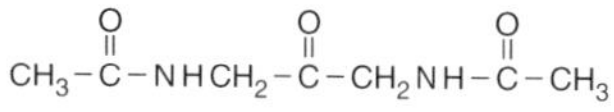

5 gm

S61,660-5
DIBROMOMALONIC DIAMIDE

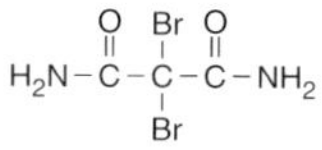

250 mg

S38,936-6
N,N'-DIMETHYLMALONAMIDE

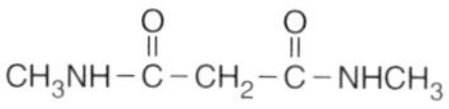

250 mg

S51,544-2
N,N'-HEXAMETHYLENEBIS(METHYL-SUCCINAMATE)

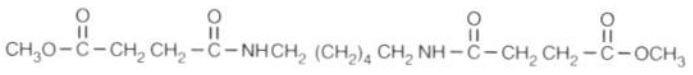

250 mg

S81,537-3
N,N,N',N'-TETRAETHYLSUCCINAMIDE

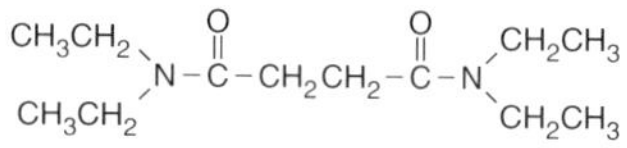

250 mg

S84,932-4
N,N,N',N'-TETRAETHYLADIPAMIDE

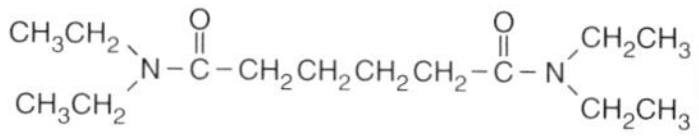

250 mg

S53,270-3
S,S'-[N,N'-METHYLENEBIS(2-CARBAMOYLETHYL)] BIS(THIOACETATE)

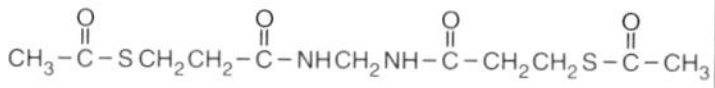

250 mg

S59,530-6
3-METHYLGLUTARAMIDE

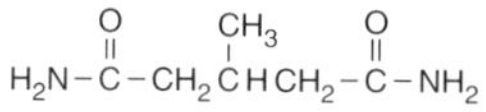

5 gm

S56,896-1
N,N,N',N'-TETRAMETHYLSUCCINAMIDE

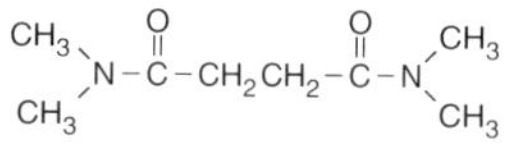

250 mg

S50,931-0
HEXAHYDRO-1,3,5-TRIPROPIONYL-1,3,5-TRIAZINE

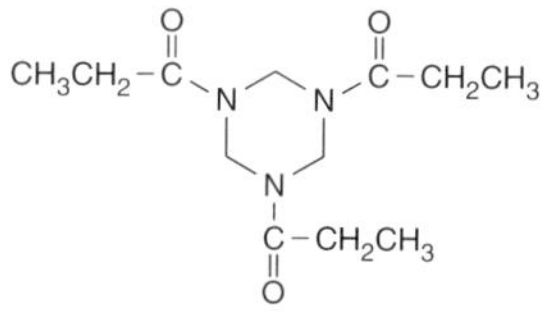

5 gm

S35,866-5
1,2-BIS(DICHLOROACETYL)HYDRAZINE

5 gm

S67,741-8
2-HYDROXY-2-METHYLPROPIONIC HYDRAZIDE

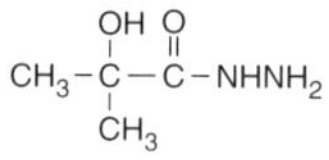

5 gm

S55,031-0
GLUCONIC ACID HYDRAZIDE

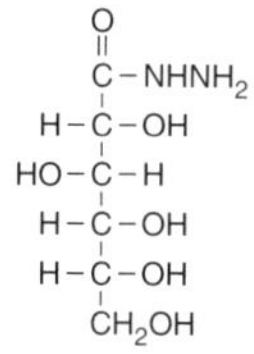

250 mg

S55,027-2
3-HYDROXYBUTYRIC ACID HYDRAZIDE

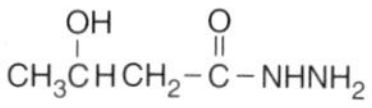

250 mg

S54,788-3
N2-(2-HYDROXYETHYL)STEARIC ACID HYDRAZIDE

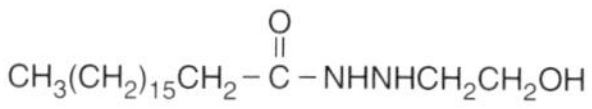

250 mg

S93,689-8
5-ENDO-HYDROXYBICYCLO[2.2.2]OCTANE-2-ENDO-CARBOXYLIC ACID HYDRAZIDE

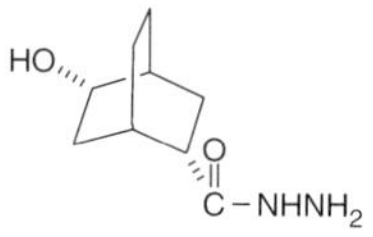

250 mg

S41,785-8
2,5-DIMETHOXYTETRAHYDRO-2-FUROIC ACID HYDRAZIDE, MIXTURE OF CIS AND TRANS

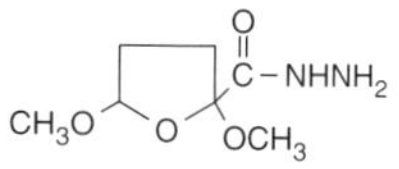

5 gm

S89,975-5
2,2'-OXYBIS(ACETIC ISOPROPYLIDENEHYDRAZIDE)

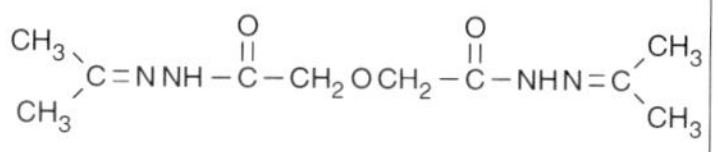

250 mg

S78,705-1
L-GLUTAMIC ACID DELTA-HYDRAZIDE

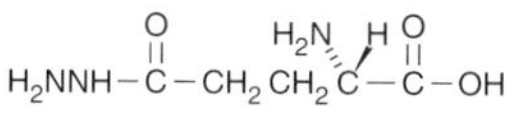

250 mg

S64,281-9
MALEIC ACID DIHYDRAZIDE

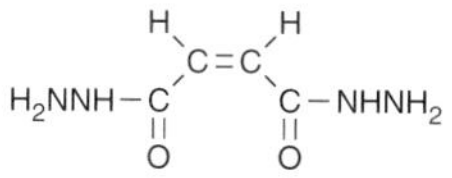

250 mg

S37,591-8
N2,N2-DIMETHYLACETOACETHYDRAZIDE

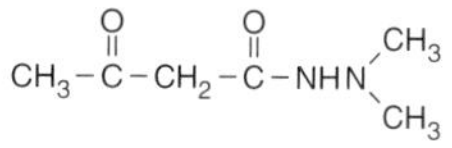

5 gm

S46,170-9
ACETIC 2-(1-METHYL-4-PIPERIDYL)HYDRAZIDE

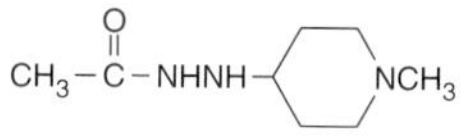

5 gm

S84,920-0
1,2-DIMETHYL-1,2-DIPROPIONYLHYDRAZINE

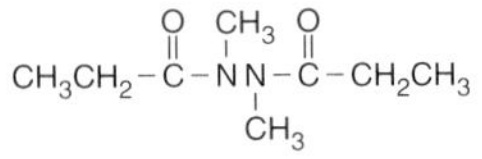

250 mg

S84,922-7
1,2-BIS(2-METHYL-4-PENTENOYL)-HYDRAZINE

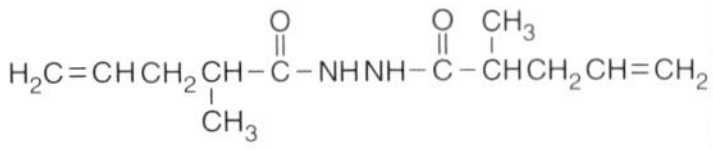

250 mg

S84,921-9
1,2-BIS(2-METHYLPENTANOYL)HYDRAZINE

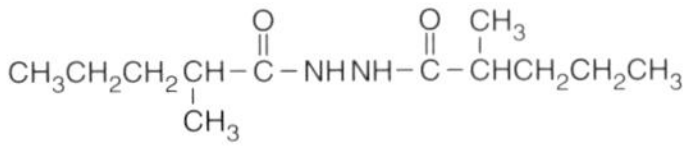

250 mg

S90,148-2
OXALIC BIS[(4-METHYLCYCLOHEXYLIDENE)HYDRAZIDE]

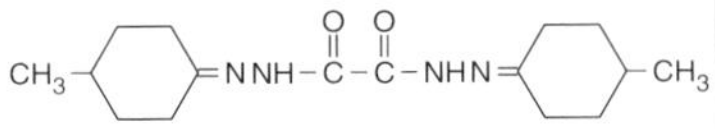

250 mg

S84,401-2
MALONIC BIS(ISOPROPYLIDENEHYDRAZIDE)

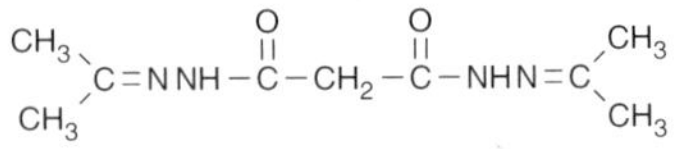

250 mg

S42,022-0
ISOBUTYRIC ISOPROPYLIDENEHYDRAZIDE

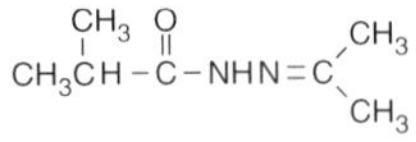

250 mg

S89,542-3
ADIPIC DIALDEHYDE DISEMICARBAZONE

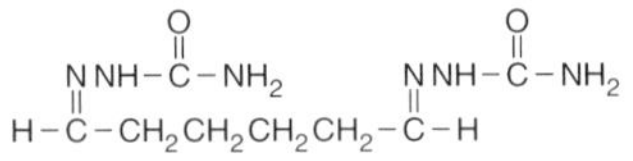

250 mg

S90,327-2
ACETIC DODECYLIDENEHYDRAZIDE

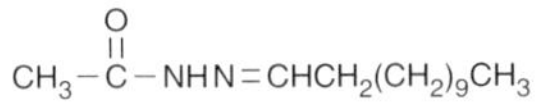

250 mg

S86,898-1
OCTADECANOIC BUTYLIDENEHYDRAZIDE

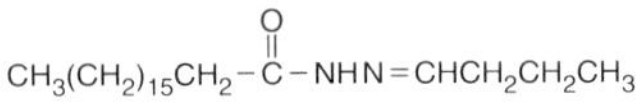

250 mg

S86,907-4
OCTADECANOIC CYCLOPENTYLIDENEHYDRAZIDE

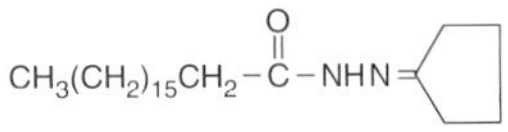

250 mg

S89,958-5
MALONIC DIHYDRAZIDE

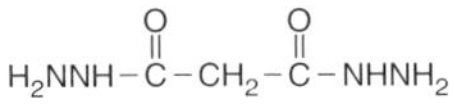

250 mg

S82,807-6
ETHYLMALONIC DIHYDRAZIDE

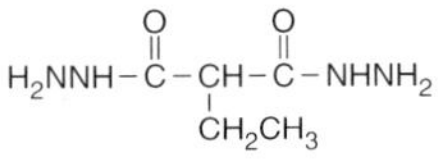

250 mg

S84,273-7
4-CYCLOHEXEN-1,2-DICARBOXYLIC DIHYDRAZIDE

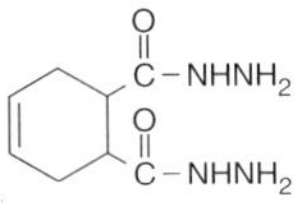

250 mg

S40,342-3
ISOBUTYRIC HYDRAZIDE

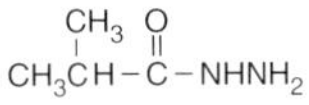

250 mg

S45,321-8
BUTYRIC HYDRAZIDE

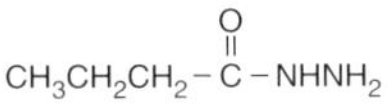

5 gm

S37,989-1
ACETIC ISOPROPYLIDENEHYDRAZIDE

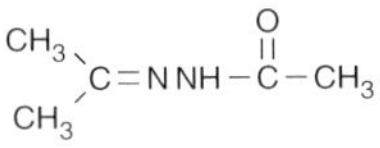

5 gm

S64,537-0
VALERIC HYDRAZIDE

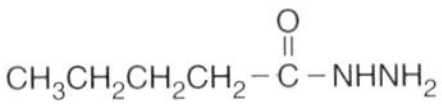

250 mg

S63,766-1
GLUTARIC DIHYDRAZIDE

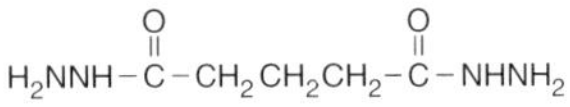

5 gm

S84,903-0
2-METHYLPENTANOIC HYDRAZIDE

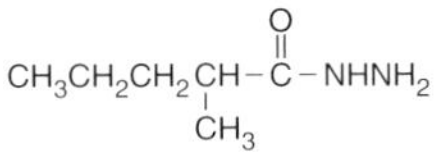

250 mg

S84,904-9
HEXANOIC HYDRAZIDE

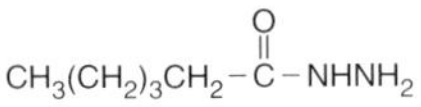

250 mg

S87,150-8
TRIMETHYLACETIC N2-METHYLHYDRAZIDE

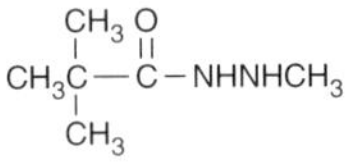

250 mg

S88,959-8
1,2-DIPROPIONYLHYDRAZINE

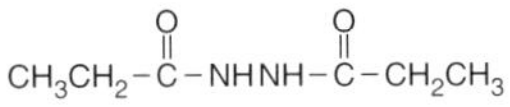

250 mg

S36,457-6
PIMELIC DIHYDRAZIDE

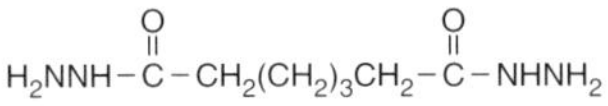

5 gm

S55,028-0
DECANOIC HYDRAZIDE

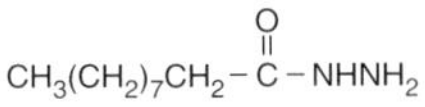

250 mg

S62,847-6
HEXDECANOIC HYDRAZIDE

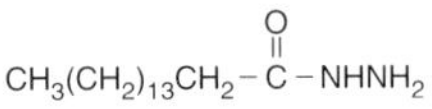

250 mg

S87,615-1
STEARIC (2-BUTYL-2-ETHYL-5-METHYL-3,4-HEXADIEN-1-YLIDENE)HYDRAZIDE

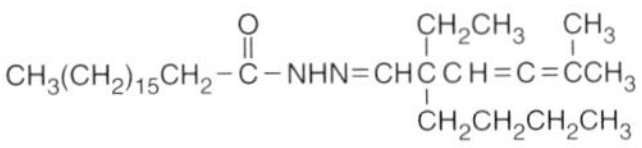

250 mg

S84,916-2
CYCLOHEXANECARBOXYLIC N2-(CYCLOHEXYLMETHYL)HYDRAZIDE

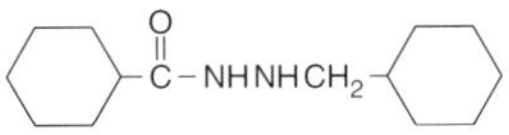

250 mg

S94,802-0
1-ADAMANTANECARBOXYLIC ACID HYDRAZIDE

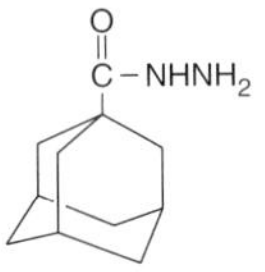

250 mg

S94,804-7
1-ADAMANTANECARBOXYLIC ACID 2,2-DIMETHYLHYDRAZIDE

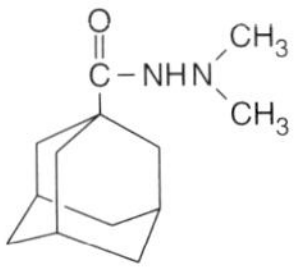

250 mg

S85,701-7
ETHYL N-[1-(2-CHLOROETHOXY)-2,2,2-TRICHLOROETHYL]CARBAMATE

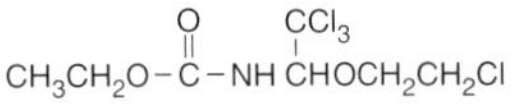

250 mg

S85,690-8
METHYL N-[1-(2-CHLOROETHOXY)-2,2,2-TRICHLOROETHYL]CARBAMATE

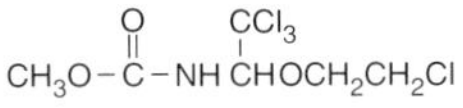

250 mg

S87,959-2
3-(4-BROMOBUTOXY)-2-METHOXYPROPYL N-OCTADECYLCARBAMATE

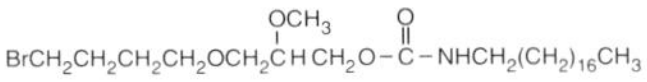

5 gm

S66,697-1
N,N'-HEXAMETHYLENEBIS(2-CHLOROETHYL CARBAMATE)

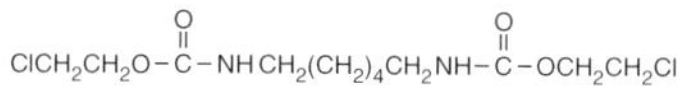

5 gm

S50,278-2
ETHYL N-(2-CHLOROETHYL)-N-PROPIONYLCARBAMATE

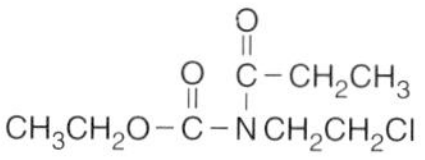

250 mg

S50,262-6
ETHYL N-(2-CHLOROETHYL)-N-ISOBUTYRYLCARBAMATE

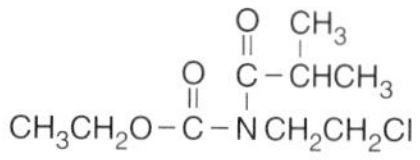

5 gm

S62,493-4
2-CHLOROETHYL N-CYCLOHEXYLCARBAMATE

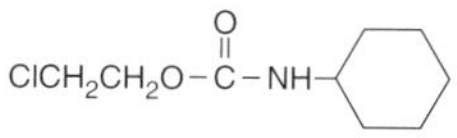

5 gm

S85,744-0
METHYL N-(1-HYDROXY-2,2,2-TRICHLOROETHYL)CARBAMATE

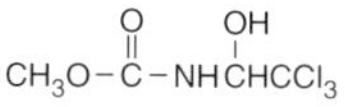

250 mg

S75,763-2
ETHYL N-(1-HYDROXY-2,2,2-TRICHLOROETHYL)CARBAMATE

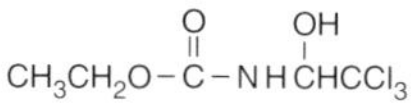

5 gm

S85,745-9
PROPYL N-(1-HYDROXY-2,2,2-TRICHLOROETHYL)CARBAMATE

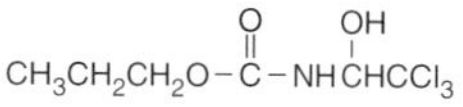

250 mg

S85,746-7
ISOPROPYL N-(1-HYDROXY-2,2,2-TRICHLOROETHYL)CARBAMATE

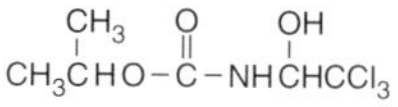

250 mg

S85,747-5
BUTYL N-(1-HYDROXY-2,2,2-TRICHLOROETHYL)CARBAMATE

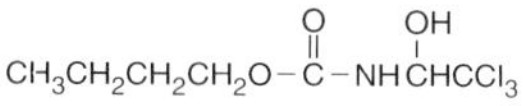

250 mg

S85,681-9
METHYL N-[1-(2-METHOXYETHOXY)-2,2,2-TRICHLOROETHYL]CARBAMATE

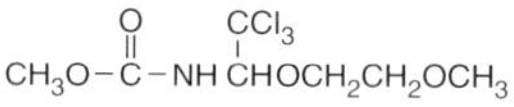

250 mg

S85,680-0
METHYL N-[1-(1-ETHOXYETHOXY)-2,2,2-TRICHLOROETHYL]CARBAMATE

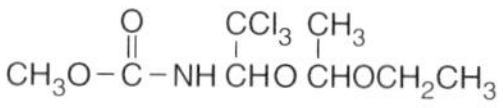

250 mg

S85,663-0
DIMETHYL N,N'-[1,1'-OXYBIS(2,2,2-TRICHLOROETHYL)]BISCARBAMATE

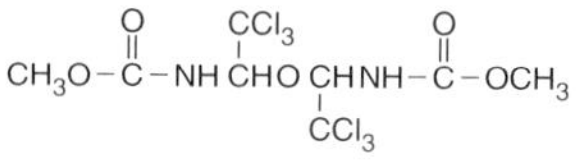

250 mg

S85,664-9
DIPROPYL N,N'-[1,1'-OXYBIS(2,2,2-TRICHLOROETHYL)]BISCARBAMATE

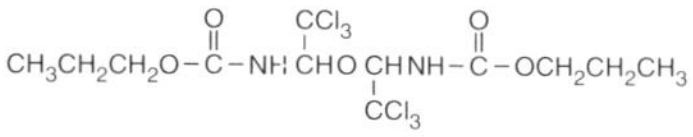

250 mg

S85,677-0
DIISOPROPYL N,N'-[1,1'-OXYBIS-(2,2,2-TRICHLOROETHYL)]BISCARBAMATE

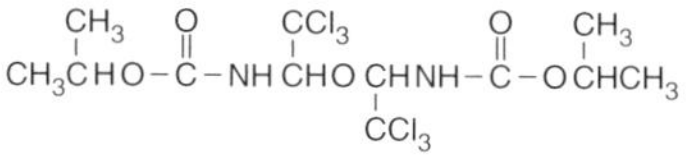

250 mg

S85,670-3
DIBUTYL N,N'-[1,1'-OXYBIS(2,2,2-TRICHLOROETHYL)]BISCARBAMATE

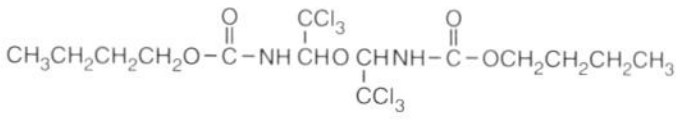

250 mg

S85,671-1
METHYL N-(1-METHOXY-2,2,2-TRICHLOROETHYL)CARBAMATE

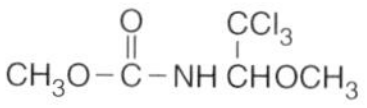

250 mg

S85,661-4
ETHYL N-(1-METHOXY-2,2,2-TRICHLOROETHYL)CARBAMATE

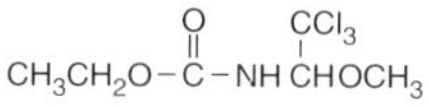

250 mg

S85,673-8
METHYL N-(1-ETHOXY-2,2,2-TRICHLOROETHYL)CARBAMATE

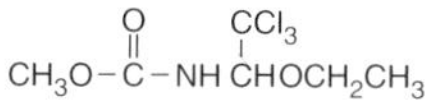

250 mg

S85,674-6
METHYL N-(1-PROPOXY-2,2,2-TRICHLOROETHYL)CARBAMATE

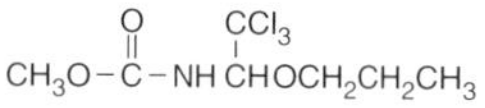

250 mg

S85,695-9
ETHYL N-(1-ETHOXY-2,2,2-TRICHLOROETHYL)CARBAMATE

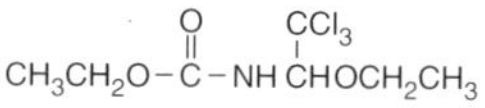

250 mg

S85,710-6
PROPYL N-(1-METHOXY-2,2,2-TRICHLOROETHYL)CARBAMATE

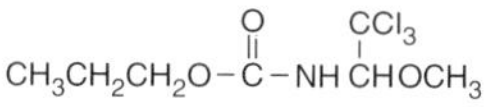

250 mg

S85,682-7
METHYL N-(1-BUTOXY-2,2,2-TRICHLOROETHYL)CARBAMATE

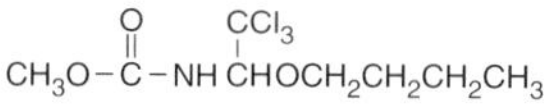

250 mg

S85,685-1
METHYL N-(1-ISOBUTOXY-2,2,2-TRICHLOROETHYL)CARBAMATE

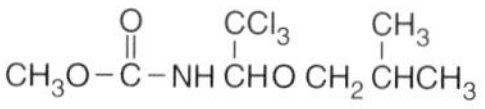

250 mg

S85,704-1
ETHYL N-(1-ISOPROPOXY-2,2,2-TRICHLOROETHYL)CARBAMATE

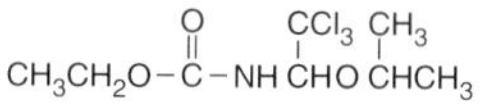

250 mg

S85,706-8
METHYL N-(1-SEC-BUTOXY-2,2,2-TRICHLOROETHYL)CARBAMATE

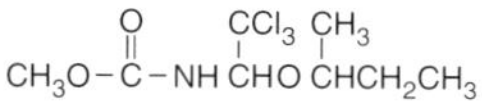

250 mg

S85,739-4
BUTYL N-(1-METHOXY-2,2,2-TRICHLOROETHYL)CARBAMATE

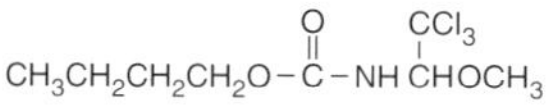

250 mg

S85,689-4
METHYL N-(1-PENTOXY-2,2,2-TRICHLOROETHYL)CARBAMATE

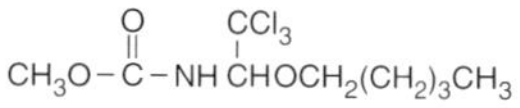

250 mg

S85,697-5
ETHYL N-(1-BUTOXY-2,2,2-TRICHLOROETHYL)CARBAMATE

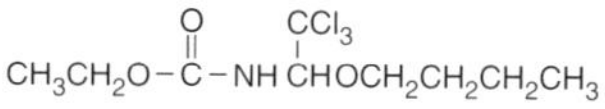

250 mg

S85,698-3
ETHYL N-(1-ISOBUTOXY-2,2,2-TRICHLOROETHYL)CARBAMATE

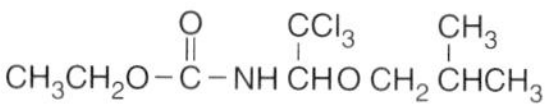

250 mg

S85,707-6
ETHYL N-(1-SEC-BUTOXY-2,2,2-TRICHLOROETHYL)CARBAMATE

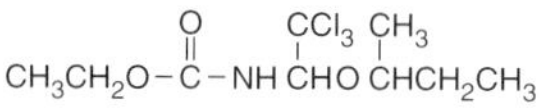

250 mg

S85,702-5
ETHYL N-(1-PENTOXY-2,2,2-TRICHLOROETHYL)CARBAMATE

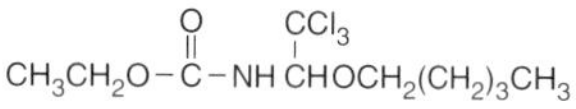

250 mg

S85,729-7
METHYL N-[1-(METHYLTHIO)-2,2,2-TRICHLOROETHYL]CARBAMATE

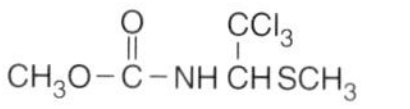

250 mg

S85,723-8
METHYL N-[1-(ETHYLTHIO)-2,2,2-TRICHLOROETHYL]CARBAMATE

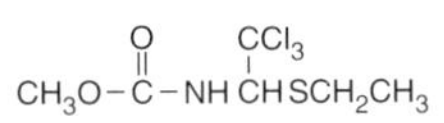

250 mg

S85,730-0
ETHYL N-[1-(METHYLTHIO)-2,2,2-TRICHLOROETHYL]CARBAMATE

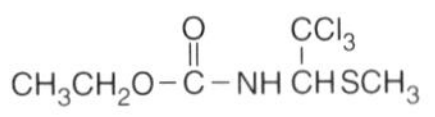

250 mg

S85,728-9
ETHYL N-[1-(PROPYLTHIO)-2,2,2-TRICHLOROETHYL]CARBAMATE

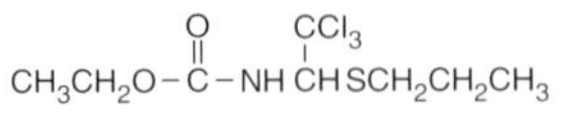

250 mg

S85,734-3
ETHYL N-[1-(BUTYLAMINO)-2,2,2-TRICHLOROETHYL]CARBAMATE

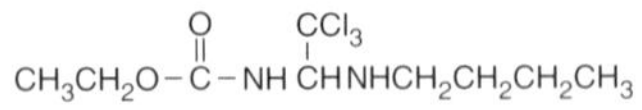

250 mg

S85,748-3
METHYL N-[1-(PENTYLAMINO)-2,2,2-TRICHLOROETHYL]CARBAMATE

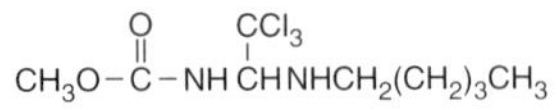

250 mg

S85,749-1
METHYL N-[1-(HEXYLAMINO)-2,2,2-TRICHLOROETHYL]CARBAMATE

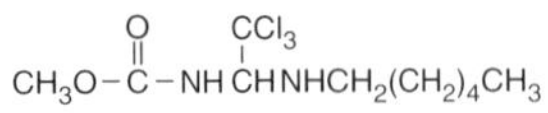

250 mg

S85,716-5
METHYL N-[1-(METHYLAMINO)-2,2,2-TRICHLOROETHYL]CARBAMATE

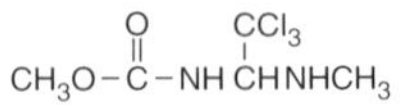

250 mg

S85,717-3
METHYL N-[1-(ETHYLAMINO)-2,2,2-TRICHLOROETHYL]CARBAMATE

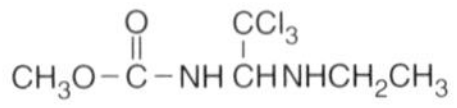

250 mg

S85,718-1
ETHYL N-[1-(ETHYLAMINO)-2,2,2-TRICHLOROETHYL]CARBAMATE

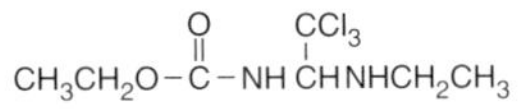

250 mg

S85,726-2
METHYL N-[1-(PROPYLAMINO)-2,2,2-TRICHLOROETHYL]CARBAMATE

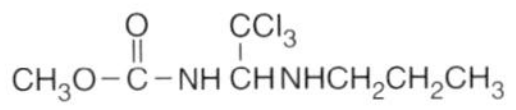

250 mg

S85,727-0
METHYL N-[1-(ISOPROPYLAMINO)-2,2,2-TRICHLOROETHYL]CARBAMATE

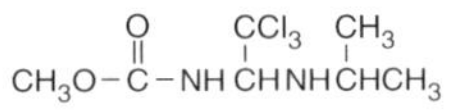

250 mg

S85,721-1
ETHYL N-[1-(PROPYLAMINO)-2,2,2-TRICHLOROETHYL]CARBAMATE

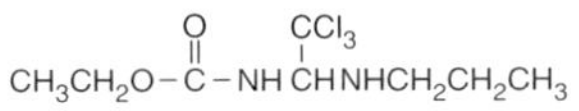

250 mg

S85,733-5
METHYL N-[1-(BUTYLAMINO)-2,2,2-TRICHLOROETHYL]CARBAMATE

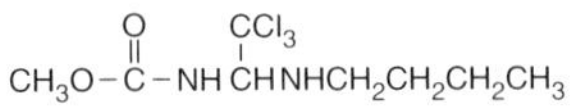

250 mg

S85,743-2
ETHYL N-[1-(ISOPROPYLAMINO)-2,2,2-TRICHLOROETHYL]CARBAMATE

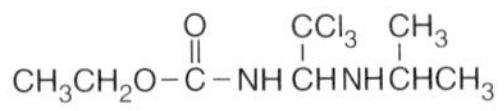

250 mg

S85,737-8
ETHYL N-[1-(DIETHYLAMINO)-2,2,2-TRICHLOROETHYL]CARBAMATE

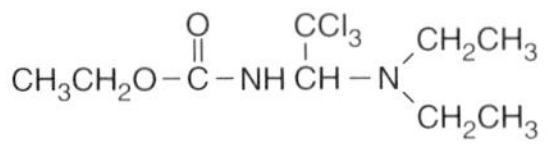

250 mg

S85,738-6
METHYL N-[1-(BUTYLMETHYLAMINO)-2,2,2-TRICHLOROETHYL]CARBAMATE

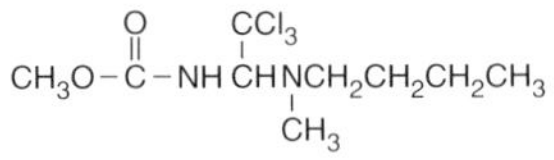

250 mg

S85,720-3
METHYL N-[1-(DIMETHYLAMINO)-2,2,2-TRICHLOROETHYL]CARBAMATE

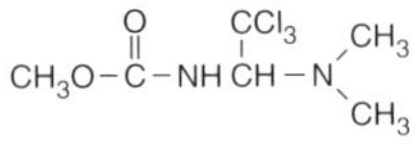

250 mg

S85,735-1
ETHYL N-[1-(DIMETHYLAMINO)-2,2,2-TRICHLOROETHYL]CARBAMATE

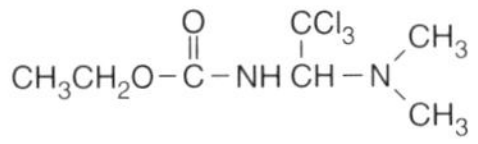

250 mg

S85,740-8
METHYL N-[1-(DIETHYLAMINO)-2,2,2-TRICHLOROETHYL]CARBAMATE

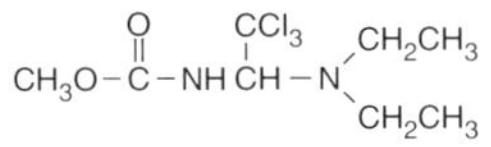

250 mg

S85,659-2
METHYL N-(1-ACETOXY-2,2,2-TRICHLOROETHYL)CARBAMATE

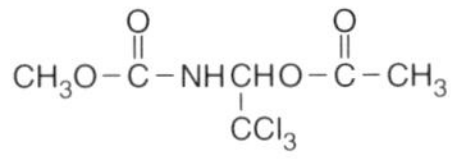

250 mg

S85,683-5
ETHYL N-(1-ACETOXY-2,2,2-TRICHLOROETHYL)CARBAMATE

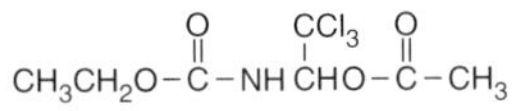

250 mg

S85,666-5
ISOPROPYL N-(1-ACETOXY-2,2,2-TRICHLOROETHYL)CARBAMATE

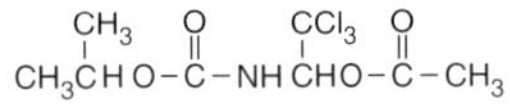

250 mg

S85,684-3
PROPYL N-(1-ACETOXY-2,2,2-TRICHLOROETHYL)CARBAMATE

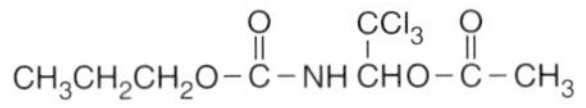

250 mg

S85,700-9
METHYL N-(1-BUTYRYLOXY-2,2,2-TRICHLOROETHYL)CARBAMATE

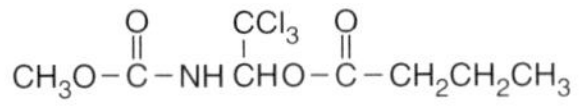

250 mg

S85,703-3
METHYL N-(1-ISOBUTYRYLOXY-2,2,2-TRICHLOROETHYL)CARBAMATE

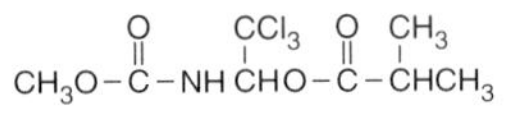

250 mg

S85,679-7
BUTYL N-(1-ACETOXY-2,2,2-TRICHLOROETHYL)CARBAMATE

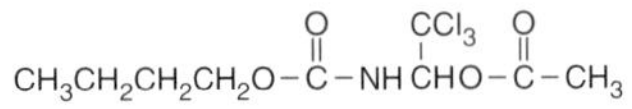

250 mg

S89,893-7
1,2,2,2-TETRACHLOROETHYL-N,N-DIETHYLCARBAMATE

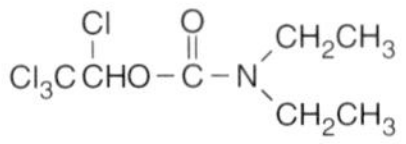

250 mg

S88,926-1
2,2,2-TRICHLOROETHYL-N-METHYLCARBAMATE

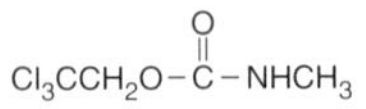

250 mg

S87,992-4
ETHYL N-[1,1-BIS(HYDROXYMETHYL)-PROPYL]CARBAMATE

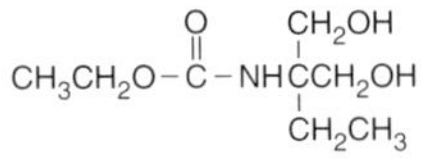

5 gm

S87,958-4
3-(4-HYDROXYBUTOXY)-2-METHOXYPROPYL N-OCTADECYLCARBAMATE

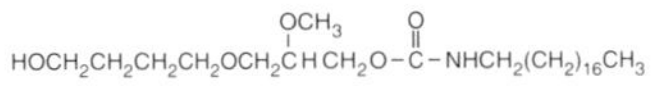

5 gm

S50,014-3
BIS(2-HYDROXYETHYL)-ETHYLENEDICARBAMATE

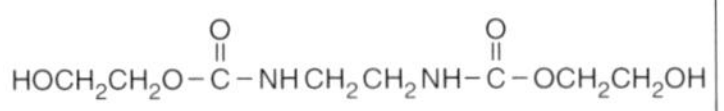

5 gm

S87,963-0
2-(HYDROXYMETHYL)-TETRAHYDROFURFURYL N-OCTADECYLCARBAMATE

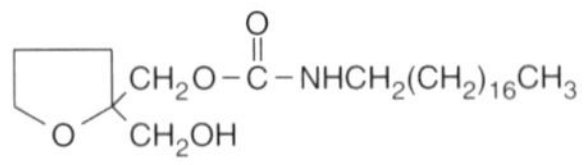

5 gm

S88,292-5
5-(HYDROXYMETHYL)-TETRAHYDROFURFURYL N-OCTADECYLCARBAMATE

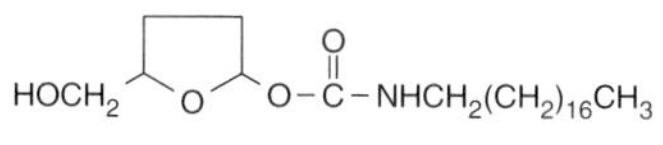

250 mg

S39,265-0
2-HYDROXYETHYL N-METHYLCARBAMATE

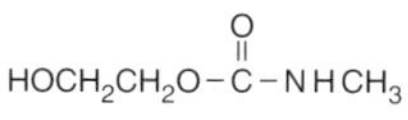

5 gm

S34,894-5
N',N2-DICARBETHOXY-N'-(3-HYDROXYPROPYL)-2-METHYL-1,2-PROPANEDIAMINE

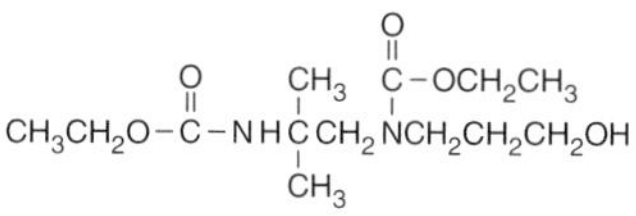

5 gm

S34,892-9
DIETHYL 1,1-DIMETHYL-N2-(1,1-DIMETHYL-2-HYDROXYETHYL)-ETHYLENEDICARBAMATE

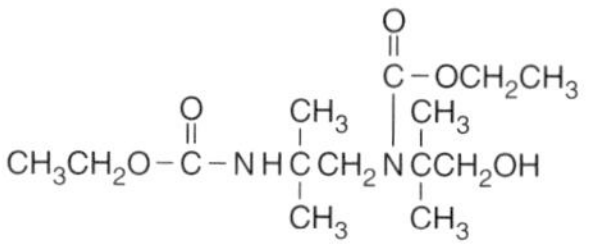

5 gm

S39,198-0
ETHYL N-(2-HYDROXYETHYL)CARBAMATE

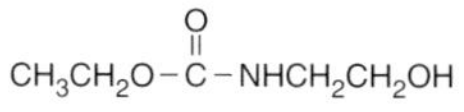

5 gm

S36,173-9
ETHYL N-(HYDROXY-TERT-BUTYL)-CARBAMATE

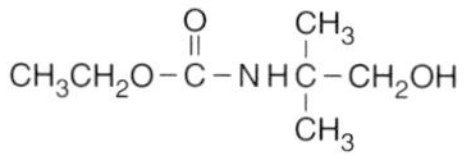

5 gm

S36,174-7
ETHYL 1-(HYDROXYMETHYL)-PROPYLCARBAMATE

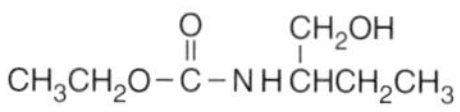

5 gm

S87,995-9
(TETRAHYDROFURAN-2,5-DIYLDIMETHYL)-BIS(N-OCTADECYLCARBAMATE)

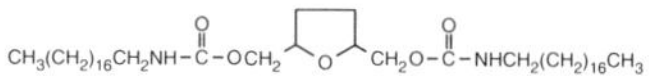

5 gm

S60,032-6
2-METHYL-2-PROPYL-1,3-PROPANEDIOL CARBAMATE CYCLOPROPYLCARBAMATE

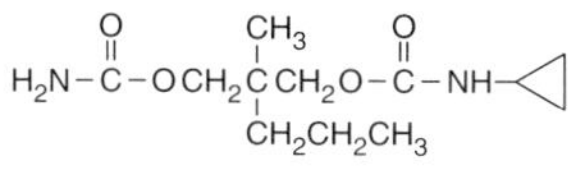

5 gm

S78,742-6
N-(TERT-BUTOXYCARBONYL)-L-METHIONYL-L-ALANINE METHYL ESTER

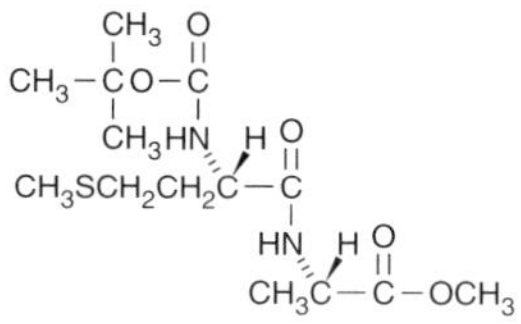

250 mg

S40,899-9
ISOPROPYL CARBAMATE

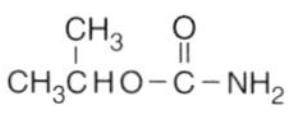

5 gm

S36,631-5
ALLYL CARBAMATE

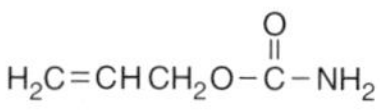

5 gm

S35,205-5
METHALLYL CARBAMATE

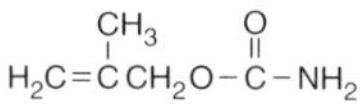

5 gm

S41,814-5
ETHYL 4-PROPYL-1-PIPERAZINECARBOXYLATE

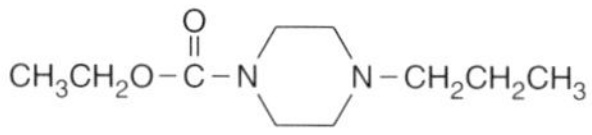

5 gm

S36,051-1
DIETHYL ETHYLENEDICARBAMATE

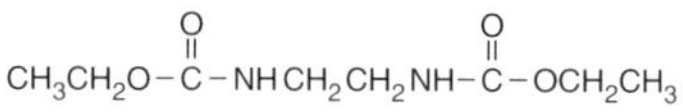

5 gm

S34,895-3
N',N2-DICARBETHOXY-N'-ISOPROPYL-2-METHYL-1,2-PROPANEDIAMINE

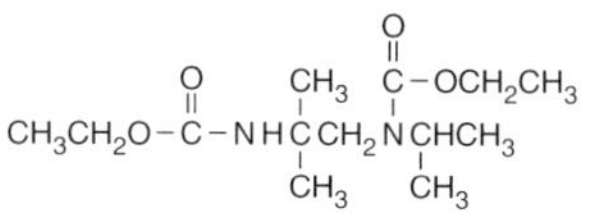

5 gm

S34,828-7
ETHYL N-[3-(DIBUTYLAMINO)PROPYL]-CARBAMATE

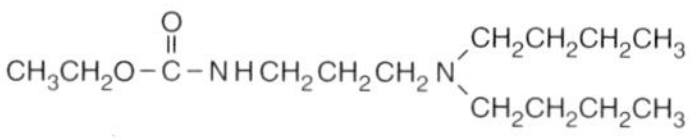

5 gm

S37,563-2
PROPYLENE DIURETHANE

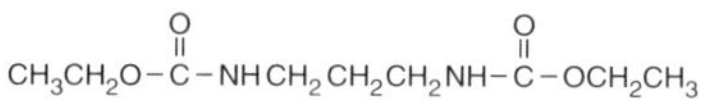

5 gm

S44,715-3
ETHYL 3-NORTRICYCLYLCARBAMATE

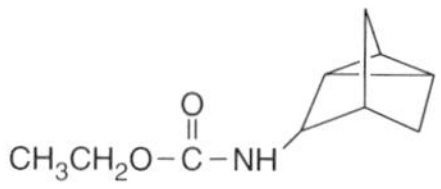

5 gm

S65,021-8
N,N'-BIS(2-DIETHYLAMINOETHOXYCARBONYL)-1,6-HEXAMETHYLENEDIAMINE

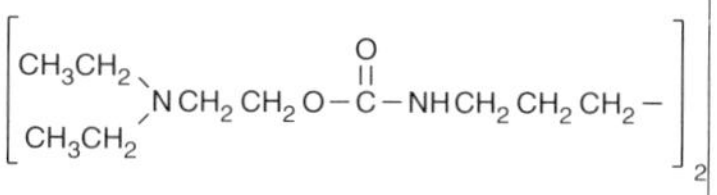

5 gm

S62,497-7
2-(DIMETHYLAMINO)ETHYL N-CYCLOHEXYLCARBAMATE

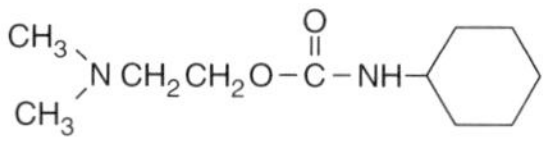

250 mg

S68,072-9
ETHYL N-(ETHOXYCARBONYL)-N-(ETHOXYCARBONYLMETHYL)-3-AMINOPROPIONATE

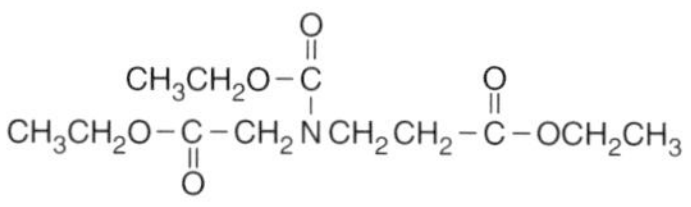

5 gm

S71,728-2
BIS[2-(DIMETHYLAMINO)ETHYL]-N,N'-HEXAMETHYLENEBISCARBAMATE

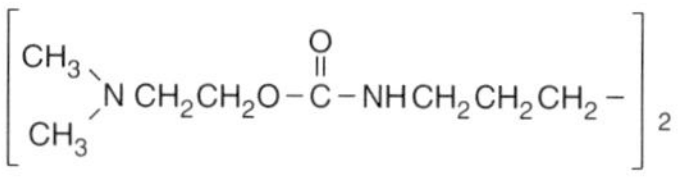

5 gm

S45,760-4
ETHYL 1-PYRROLIDINECARBOXYLATE

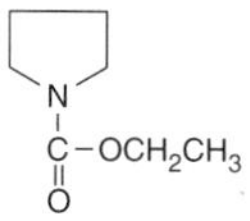

250 mg

S60,827-0
N,N'-METHYLENEBIS(METHYL CARBAMATE)

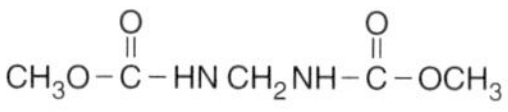

5 gm

S43,855-3
N,N'-METHYLENEBIS(ETHYL CARBAMATE)

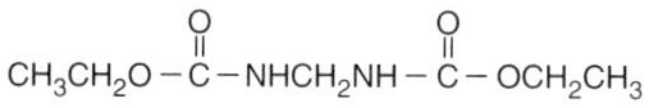

5 gm

S90,250-0
CYCLODODECYL N-(ETHOXYCARBONYLMETHYL)CARBAMATE

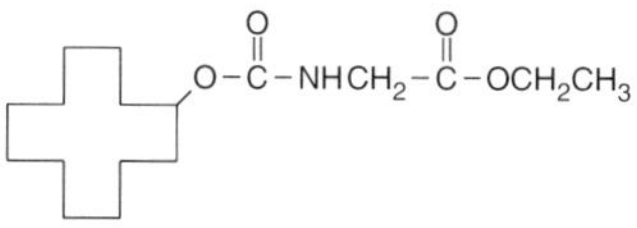

250 mg

S85,927-3
ETHYL N-HYDROXYCARBAMATE, SODIUM SALT

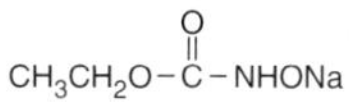

250 mg

S59,175-0
ETHYL ([(2,6-DIMETHYL-4-OXO-4H-PYRAN-3-YL)CARBONYL])CARBAMATE

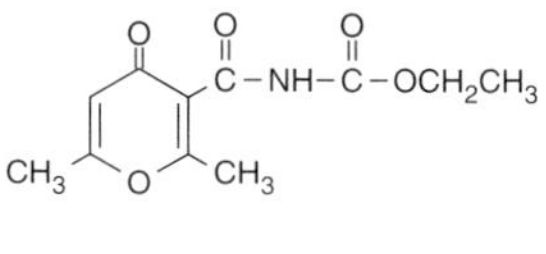

250 mg

S41,170-1
ETHYL 4-ACETOACETYL-1-PIPERAZINECARBOXYLATE

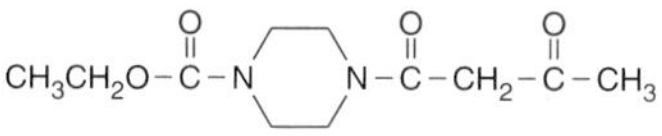

5 gm

S76,492-2
N(ALPHA)-N(EPSILON)-DI-T-BOC-L-LYSINE DICYCLOHEXYLAMINE SALT

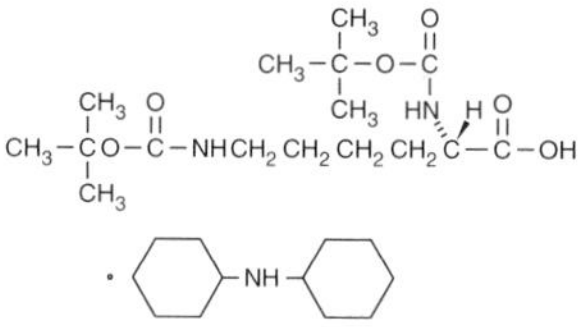

5 gm

S78,728-0
N-(TERT-BUTOXYCARBONYL)-D-VALINE DICYCLOHEXYLAMINE SALT

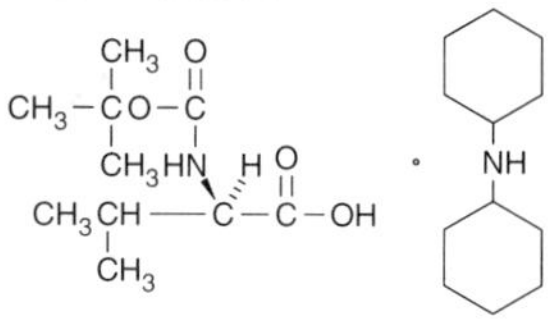

250 mg

S75,593-1
S-(2-CARBOXYETHYL)-N-ALLYLTHIOCARBAMATE
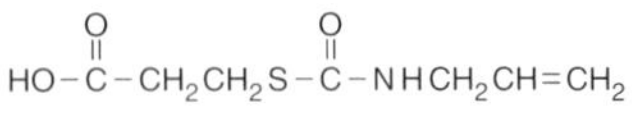
250 mg

S53,589-3
N-(ETHOXYCARBONYL)GLYCINE ETHYL ESTER
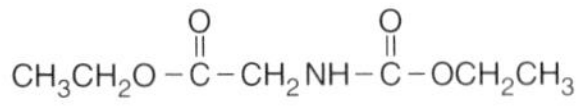
250 mg

S95,426-8
4-(ETHOXYCARBONYLHYDRAZINOCARBONYL)-2,3,3-TRIMETHYL-1-CYCLOHEXEN-1-YL ACETATE
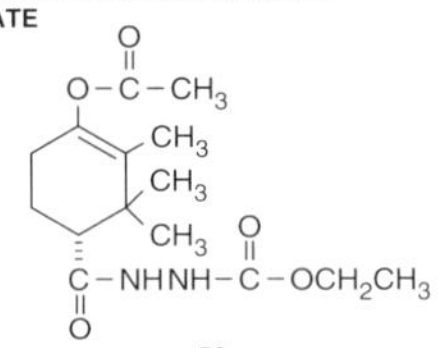
50 mg

S75,759-4
ETHYL N-(ETHOXYCARBONYL)CARBAMATE
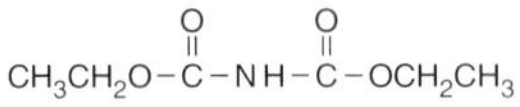
5 gm

S59,043-6
N,N'-HEXAMETHYLENEBIS(METHYLCARBAMATE)
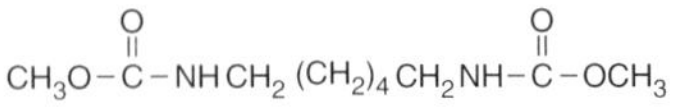
5 gm

S60,864-5
DIETHYL HEPTAMETHYLENEDICARBAMATE
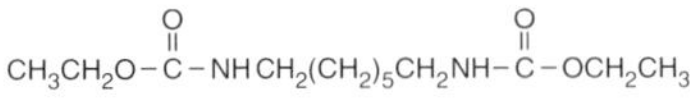
250 mg

S66,695-5
N,N'-HEXAMETHYLENEBIS(ALLYLCARBAMATE)
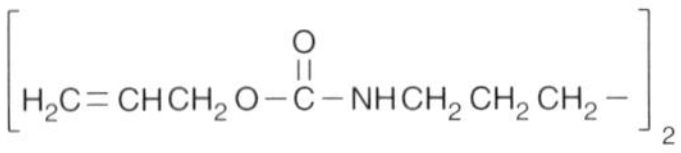
5 gm

S59,046-0
N,N'-HEXAMETHYLENEBIS(PROPYLCARBAMATE)
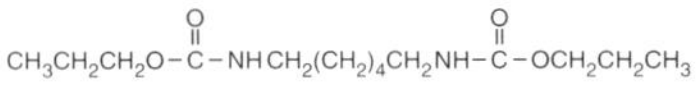
5 gm

S59,054-1
N,N'-HEXAMETHYLENEBIS(ISOPROPYLCARBAMATE)
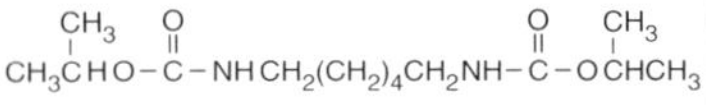
5 gm

S66,722-6
N,N'-HEXAMETHYLENEBIS(BUTYLCARBAMATE)
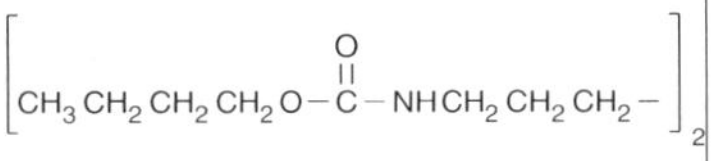
5 gm

S63,215-5
N,N'-BIS(CYCLOHEXYLOXYCARBONYL)-1,6-HEXANEDIAMINE
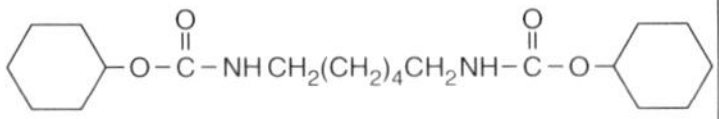
5 gm

S43,212-1
ETHYL 2-NORBORNYLCARBAMATE
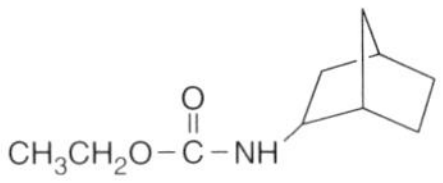
5 gm

S94,684-2
1-ADAMANTYLMETHYL N-ETHYLCARBAMATE
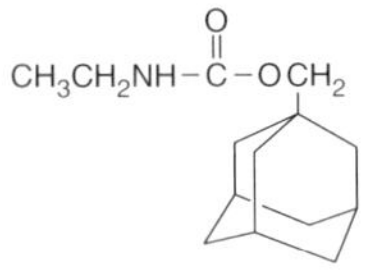
250 mg

S89,881-3
1-CHLOROETHYL N,N-DIPROPYLCARBAMATE
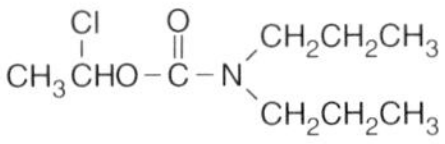
250 mg

S40,698-8
N-BUTYLURETHANE
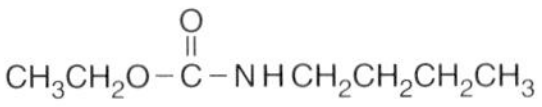
5 gm

S80,614-5
METHYL N-ISOPROPYL-N-(3-METHYL-1,3-BUTADIEN-1-YL)CARBAMATE
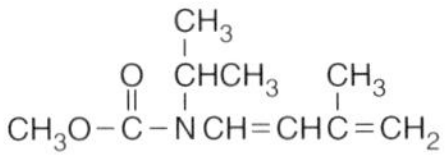
250 mg

S81,592-6
CYCLOHEXYL CARBAMATE
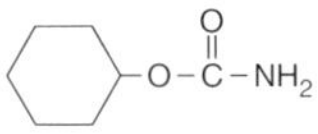
250 mg

S62,479-9
ISOPROPYL N-CYCLOHEXYLCARBAMATE
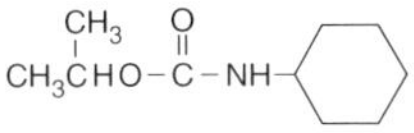
5 gm

S62,480-2
PROPYL N-CYCLOHEXYLCARBAMATE
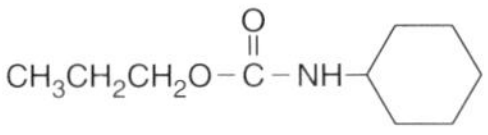
5 gm

S62,495-0
BUTYL N-CYCLOHEXYLCARBAMATE
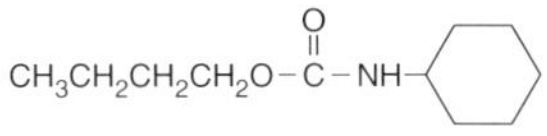
5 gm

S64,579-6
MENTHYL CARBAZATE
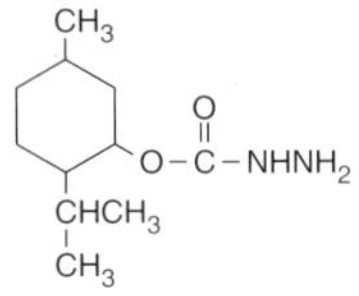
250 mg

S60,702-9
CYCLOHEXYL N-CYCLOHEXYLCARBAMATE
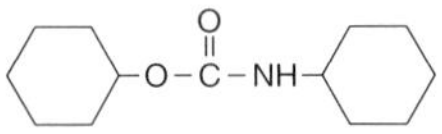
5 gm

S89,963-1
3-METHYLCYCLOHEXYL N-CYCLOHEXYLCARBAMATE
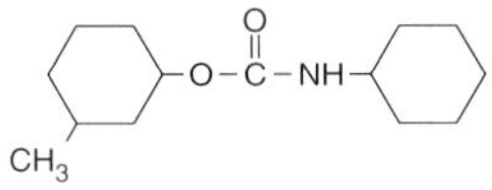
250 mg

S62,488-8
2-METHYLCYCLOHEXYL N-CYCLOHEXYLCARBAMATE
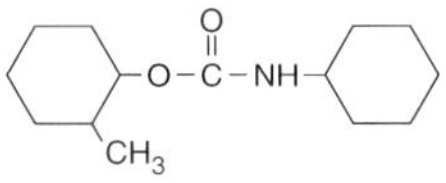
5 gm

S94,848-9
2-(1-ADAMANTYL)ETHYL N-ETHYLCARBAMATE
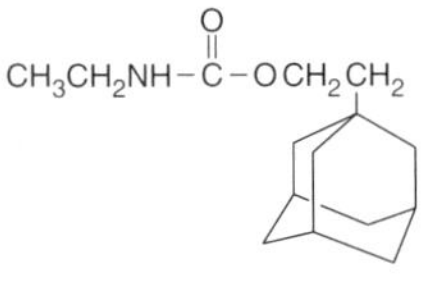
250 mg

S94,847-0
1-(1-ADAMANTYL)PROPYL N-ETHYLCARBAMATE
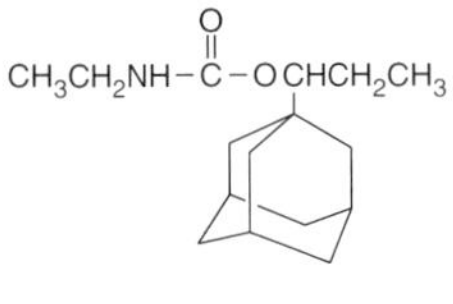
250 mg

S94,849-7
2-(1-ADAMANTYL)ETHYL N-PROPYLCARBAMATE
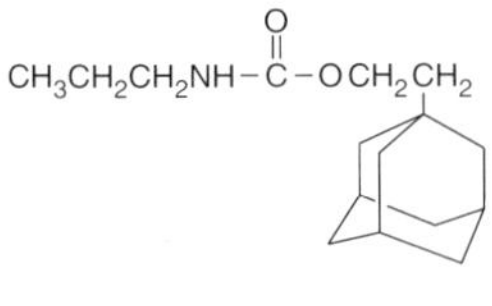
250 mg

S94,845-4
3,5,7-TRIMETHYL-1-ADAMANTYL N-PROPYLCARBAMATE
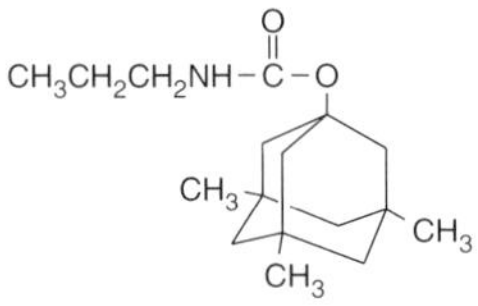
250 mg

S94,846-2
1-(1-ADAMANTYL)PROPYL N-PROPYLCARBAMATE
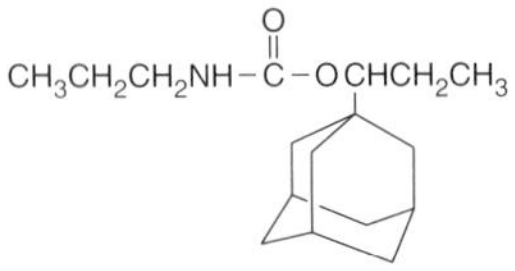
250 mg

S94,844-6
3,5,7-TRIMETHYL-1-ADAMANTYL N-BUTYLCARBAMATE
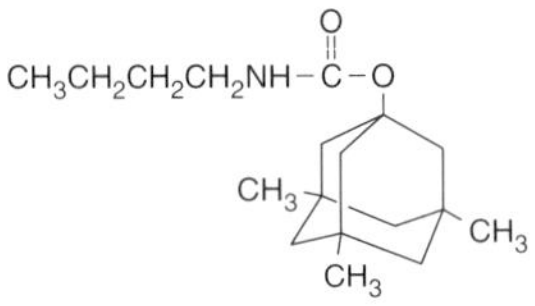
250 mg

S70,009-6
S-DODECYL N-(ETHOXYCARBONYLMETHYL)-THIOCARBAMATE
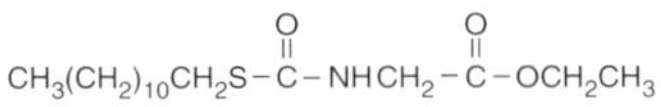
250 mg

S95,785-2
N-(CAMPHANYLIDENE)URETHANE
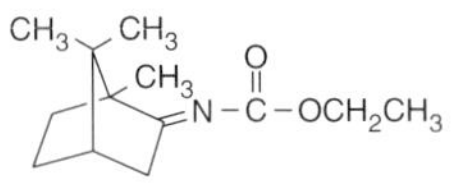
50 mg

S36,284-0
ISOBUTYL CARBAMATE

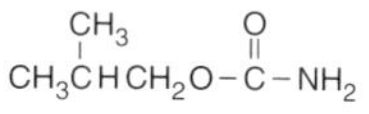

5 gm

S36,144-5
ETHYL N-ALLYLCARBAMATE

5 gm

S36,190-9
ETHYL N-PROPYLCARBAMATE

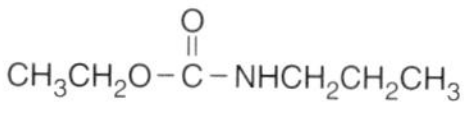

5 gm

S34,949-6
ETHYL TERT-OCTYLCARBAMATE

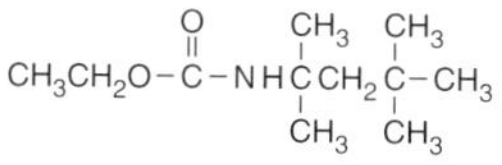

5 gm

S34,607-1
DIETHYL HEXAMETHYLENEDICARBAMATE

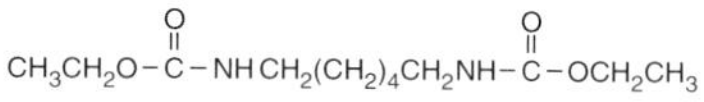

5 gm

S35,417-1
OLEYL CARBAMATE

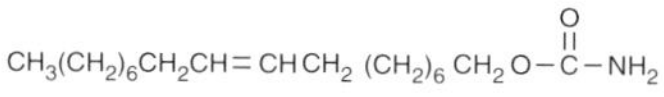

5 gm

S34,614-4
DIETHYL 1,4-PIPERAZINEDICARBOXYLATE

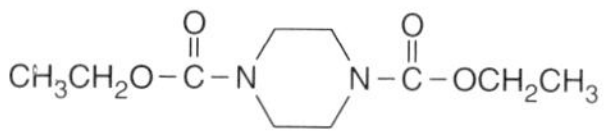

5 gm

S34,391-9
CHLOROACETYL-DL-ASPARAGINE

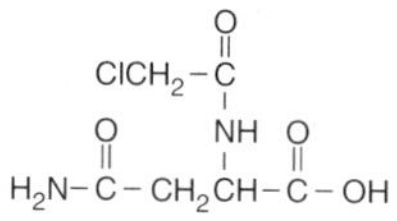

5 gm

S58,286-7
2-(2-CHLOROACETAMIDO)ACRYLIC ACID

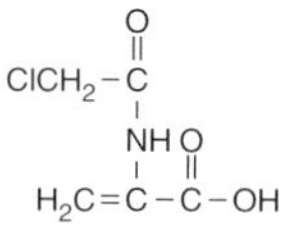

5 gm

S45,177-0
N-(CHLOROACETYL)GLYCINE

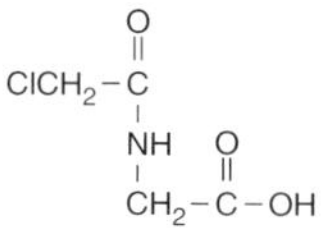

250 mg

S80,397-9
DIETHYL 2-ACETAMIDO-2-(3-BROMO-2-OXOPROPYL)MALONATE

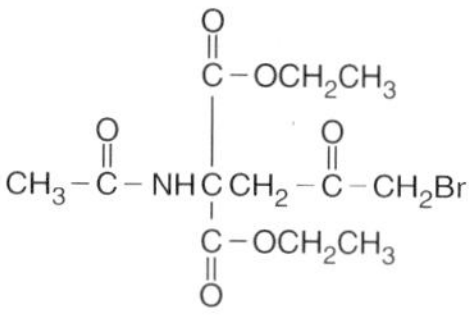

250 mg

S80,460-6
DIETHYL 2-ACETAMIDO-2-(2,3-DIBROMOPROPYL)MALONATE

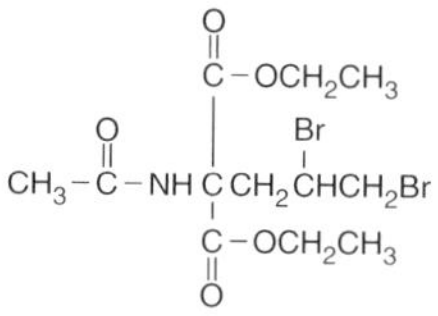

250 mg

S78,788-4
N-(CHLOROACETYL)-L-ASPARTIC ACID

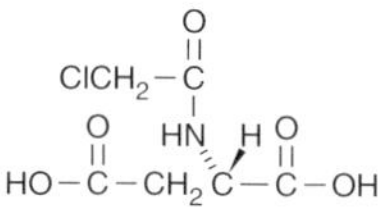

250 mg

S34,388-9
CHLOROACETYL-DL-ALANINE

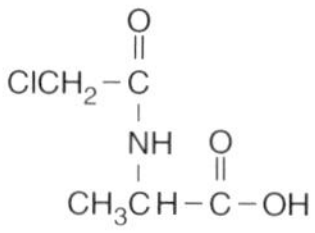

5 gm

S78,979-8
N-CHLOROACETYL-L-ALANINE

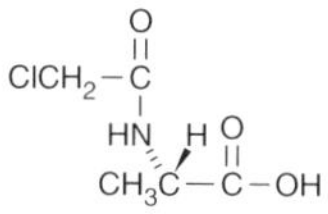

250 mg

S44,945-8
CHLOROACETYL-DL-VALINE

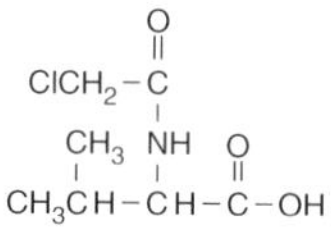

250 mg

S45,083-9
CHLOROACETYL-DL-NORVALINE

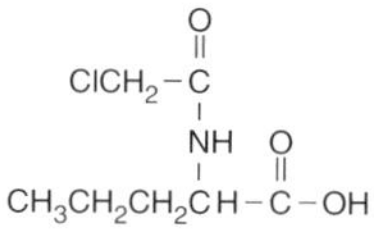

5 gm

S78,285-8
2-CHLOROACETYL-L-VALINE

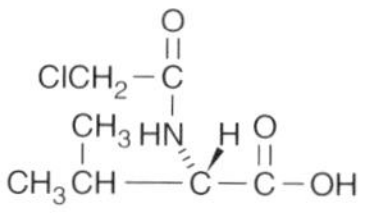

5 gm

S34,394-3
CHLOROACETYL-DL-LEUCINE

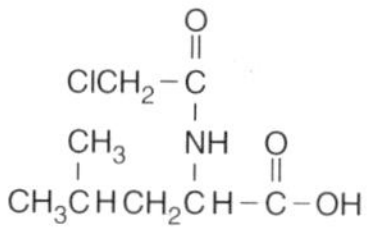

5 gm

S34,395-1
CHLOROACETYL-L-LEUCINE

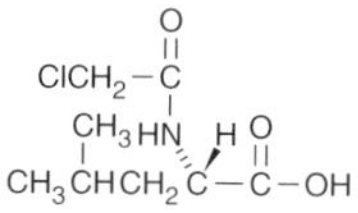

5 gm

S70,519-5
CHLOROACETYL-L-ISOLEUCINE

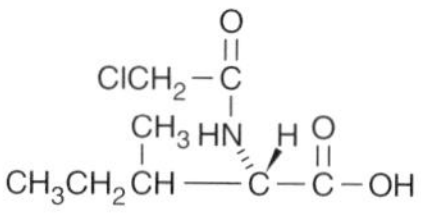

5 gm

S78,397-8
N-(CHLOROACETYL)GLYCYLGLYCINE

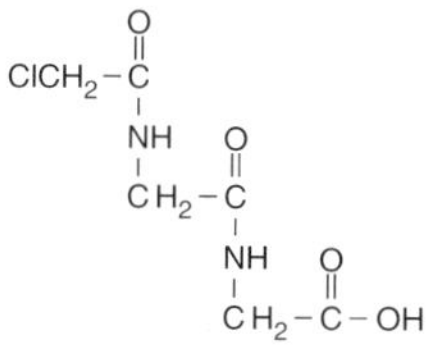

5 gm

S34,393-5
CHLOROACETYLGLYCYL-DL-VALINE

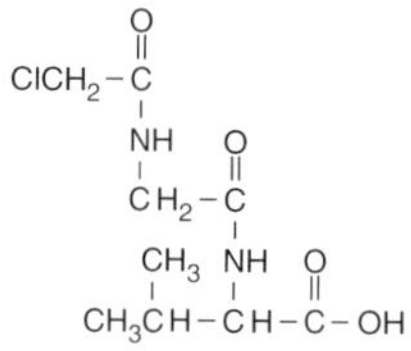

5 gm

S34,397-8
CHLOROACETYL-DL-LEUCYLGLYCINE

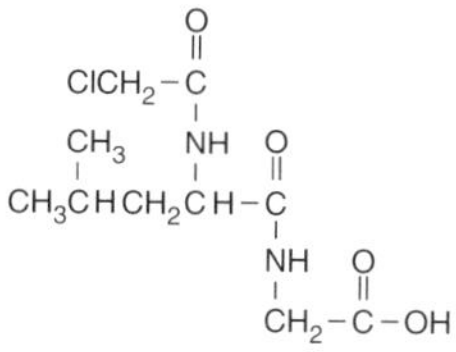

5 gm

S54,688-7
N-DICHLOROACETYL-DL-SERINE, SODIUM SALT

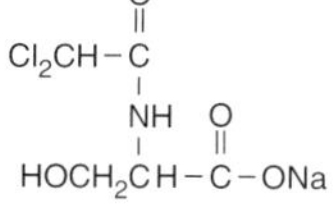

250 mg

S65,264-4
N-(TRIFLUOROACETYL)-DL-ASPARTIC ACID

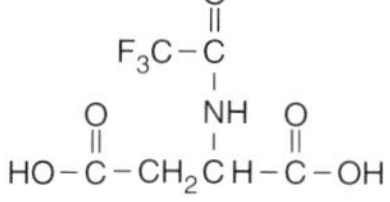

5 gm

S68,102-4
N-(TRIFLUOROACETYL)GLYCINE

5 gm

S51,216-8
2-(N-DODECYLCARBAMOYLMETHYL)-2-HYDROXYSUCCINIC ACID

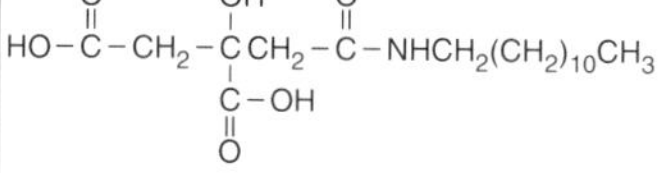

250 mg

S77,749-8
L-LEUCYL-L-SERINAMIDE HYDROCHLORIDE

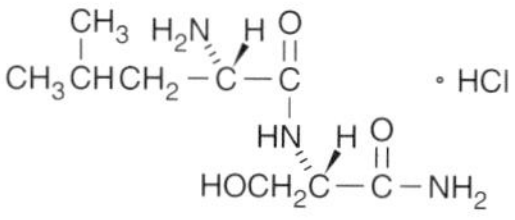

250 mg

S77,649-1
L-2-AMINOBUTYRYL-L-SERINE

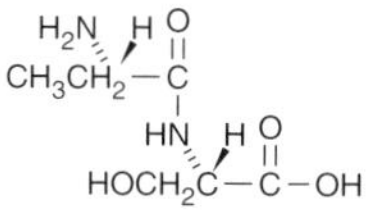

250 mg

S78,416-8
BETA-ALANYL-L-SERINE

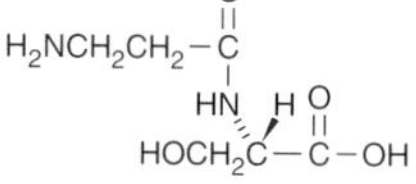

250 mg

S45,156-8
GLYCYL-L-SERINE

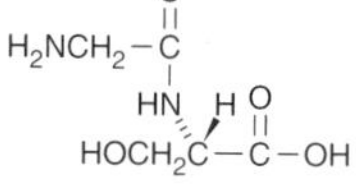

250 mg

S45,122-3
L-ALANYL-L-THREONINE
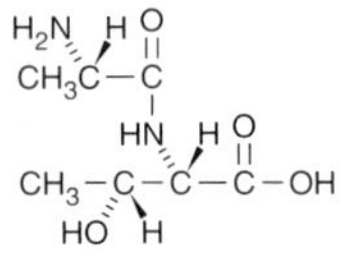
250 mg

S77,730-7
L-THREONYLGLYCYLGLYCINE
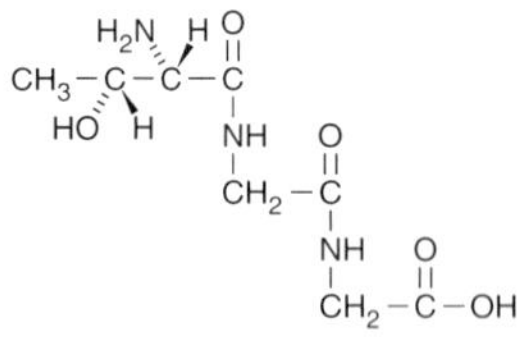
250 mg

S35,026-5
GLYCYL-DL-SERINE
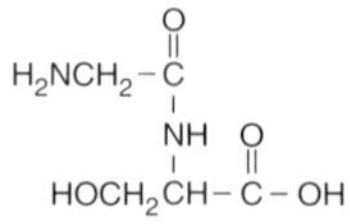
5 gm

S78,133-9
DL-THREONINAMIDE HYDROCHLORIDE
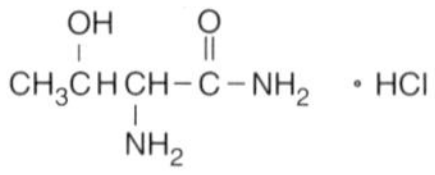
250 mg

S78,156-8
L-2-AMINOBUTYRYL-L-THREONINE
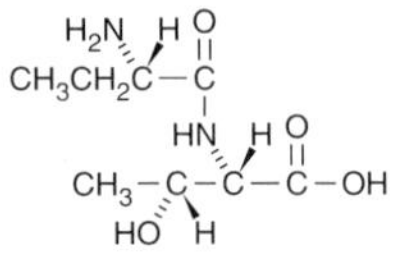
250 mg

S78,507-5
SARCOSYL-L-SERINE
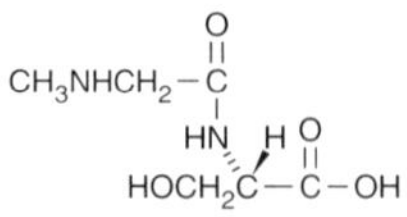
250 mg

S76,599-6
N-ACETYL-DL-THREONINE
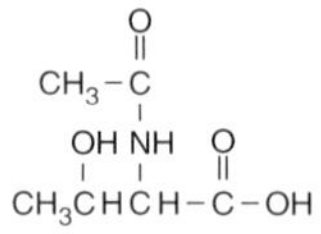
5 gm

S34,278-5
2-BROMOPROPIONYL-DL-SERINE
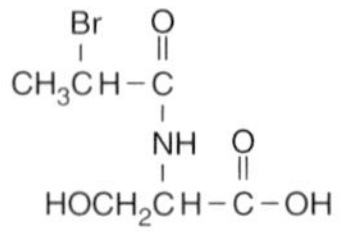
5 gm

S78,405-2
N-ACETYL-DL-SERINAMIDE
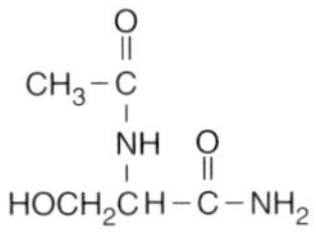
250 mg

S78,717-5
N-ACETYL-L-SERINAMIDE
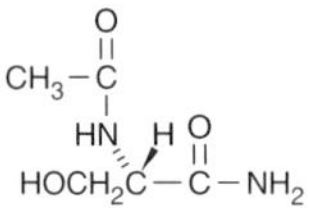
250 mg

S78,912-7
L-ISOLEUCYL-L-SERINAMIDE HYDROCHLORIDE
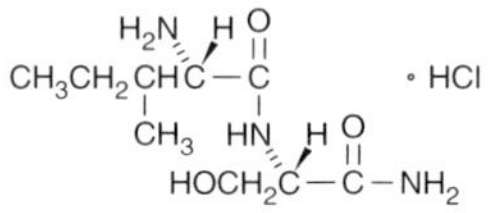
250 mg

S40,630-9
DL-N-ACETYL-BETA-MERCAPTOISOLEUCINE
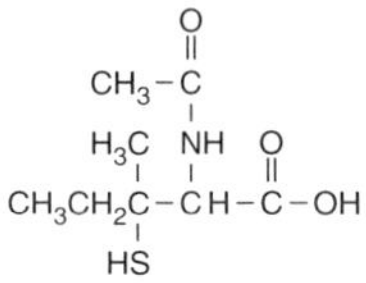
5 gm

S40,915-4
L-METHIONYLGLYCINE
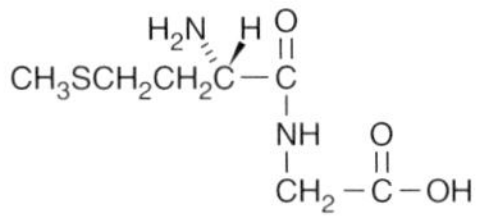
250 mg

S45,154-1
GLYCYL-L-METHIONINE
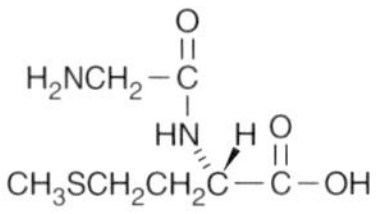
250 mg

S79,059-1
NORVALYLMETHIONINE METHYL ESTER
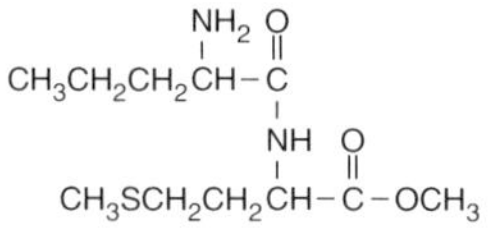
250 mg

S60,104-7
N-FORMYL-DL-METHIONINE
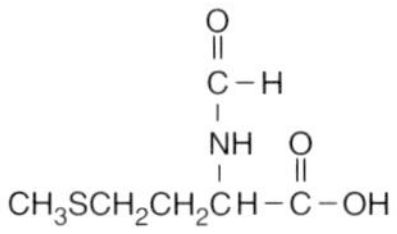
250 mg

S78,536-9
TERT-BUTOXYCARBONYL-L-LEUCYL-L-METHIONINAMIDE
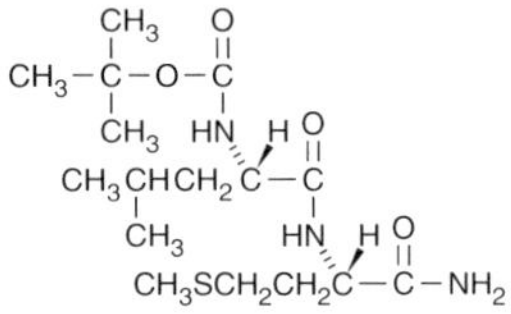
250 mg

S78,413-3
N-ACETYL-S-METHYLCYSTEINE
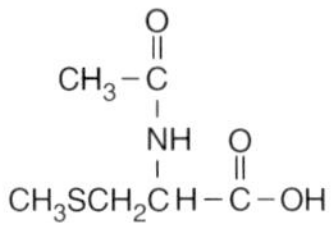
250 mg

S78,375-7
N-ACETYL-DL-ETHIONINE
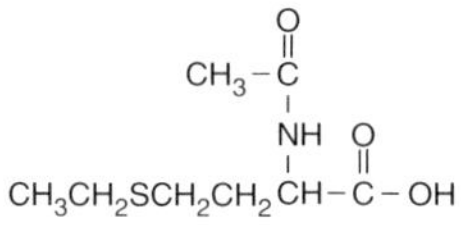
5 gm

S78,424-9
N-ACETYL-L-ETHIONINE
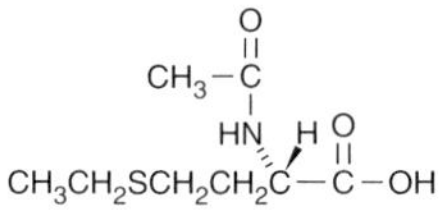
250 mg

S78,664-0
N-ACETYL-D-ETHIONINE
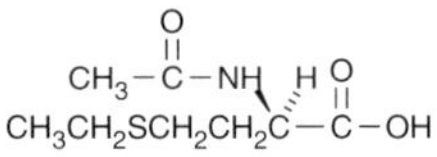
250 mg

S78,470-2
TERT-BUTOXYCARBONYL-L-ISOLEUCYL-L-METHIONINE METHYL ESTER
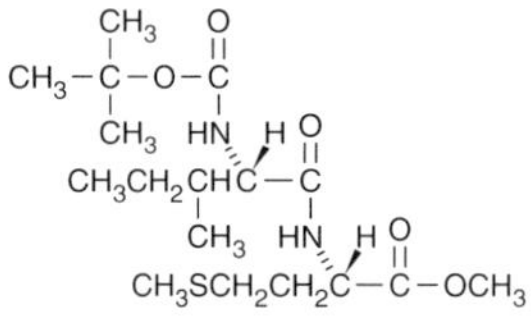
250 mg

S78,750-7
N-(TERT-BUTOXYCARBONYL)-L-LEUCYL-L-METHIONINE METHYL ESTER
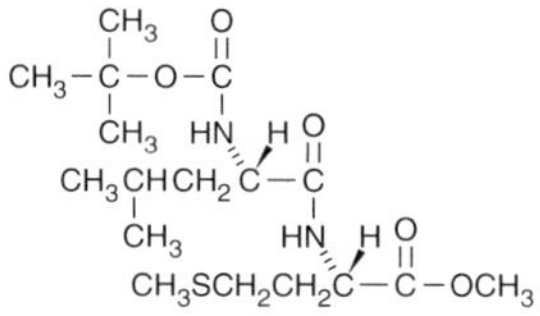
250 mg

S77,784-6
N(ALPHA)-T-BOC-L-ISOLEUCYL-L-METHIONAMIDE
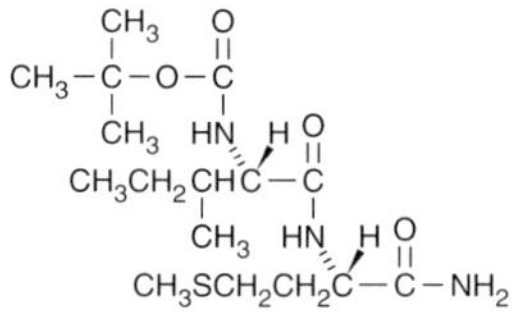
250 mg

S77,688-2
N-ACETYL-DL-METHIONINAMIDE
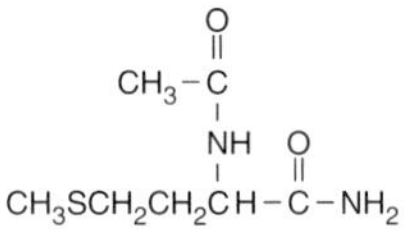
250 mg

S34,623-3
N,N'-DIFORMYL-L-CYSTINE
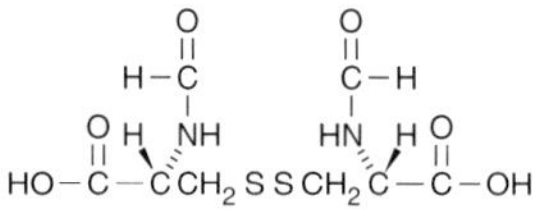
5 gm

S60,112-8
GLYCYL-BETA-ALANINE
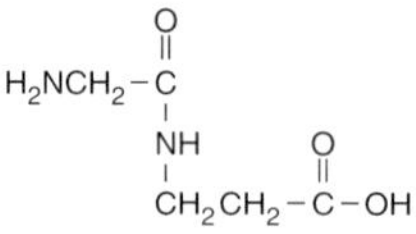
250 mg

S58,366-9
DL-ALANYL-DL-ASPARAGINE MONOHYDRATE
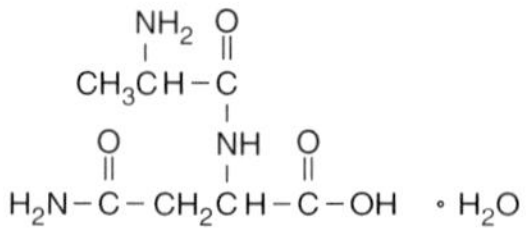
250 mg

S45,116-9
BETA-ALANYLGLYCINE
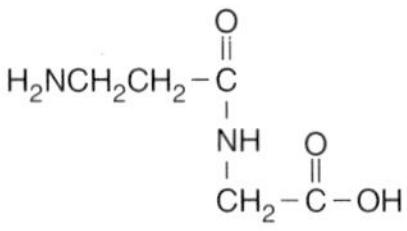
250 mg

S45,113-4
BETA-ALANYL-L-ALANINE
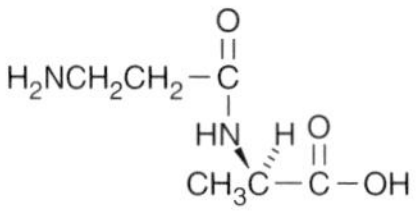
250 mg

S57,911-4
DL-GLUTAMINE
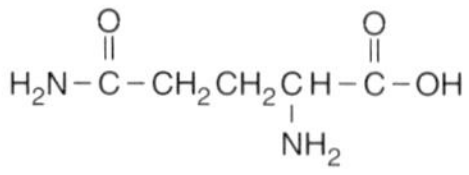
250 mg

S45,111-8
N-ALPHA-ACETYL-L-ORNITHINE
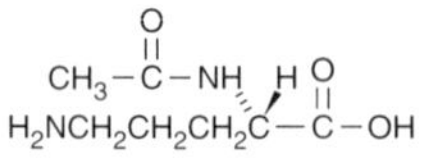
250 mg

S77,687-4
N-(ALPHA)-ACETYL-DL-ORNITHINE
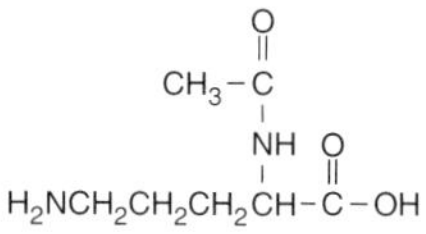
250 mg

S37,867-4
GLYCYLGLYCINE ETHYL ESTER HYDROCHLORIDE
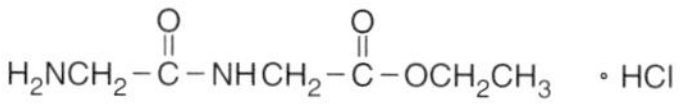
5 gm

S78,340-4
L-2-AMINOBUTYRAMIDE
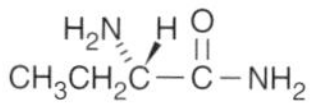
5 gm

S45,192-4
S-(ETHYLCARBAMOYL)-L-CYSTEINE, 98%
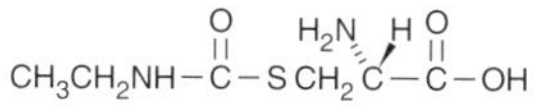
5 gm

S78,927-5
LYSYLVALINE, DICYCLOHEXYLAMINE SALT
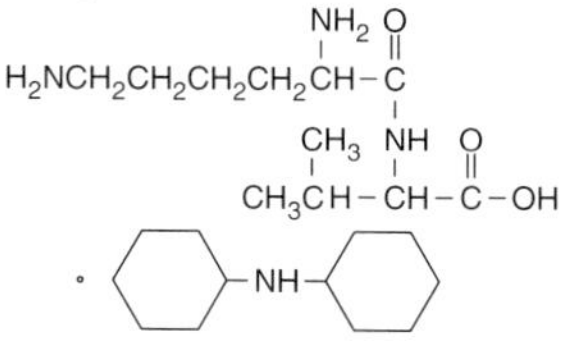
250 mg

S78,793-0
GLYCYL-D-ASPARTIC ACID
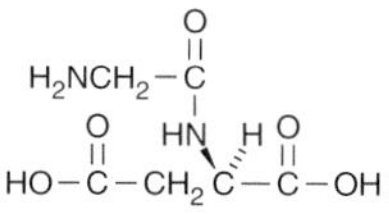
250 mg

S78,572-5
L-2-AMINOBUTYRYL-L-ASPARTIC ACID HYDRATE
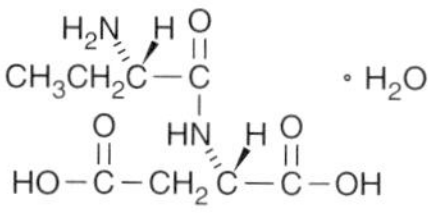
250 mg

S78,800-7
GLYCYL-D-ASPARAGINE
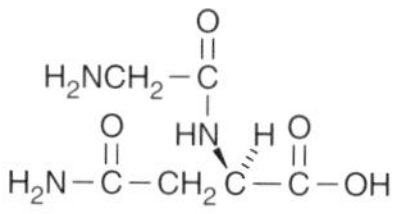
250 mg

S78,896-1
N-ALPHA-GLYCYL-L-GLUTAMINE
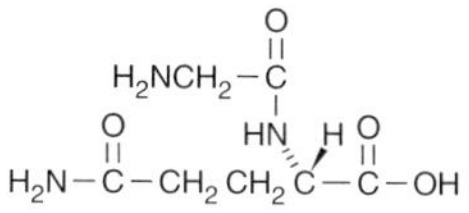
250 mg

S45,125-8
L-ALPHA-ASPARTYLGLYCINE HEMIACETATE
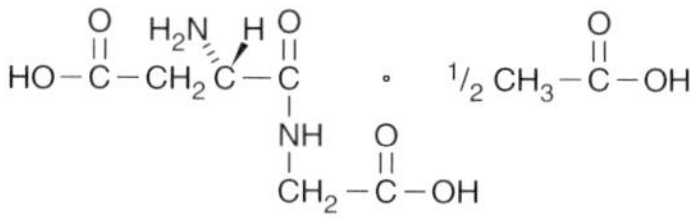
250 mg

S45,507-5
GLYCYL-L-ASPARTIC ACID
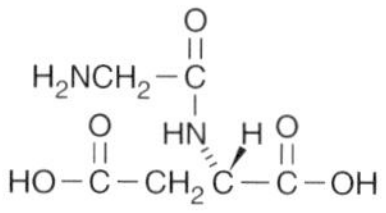
250 mg

S70,526-8
N-EPSILON-CYCLOPENTYLOXYCARBONYL-L-LYSINE
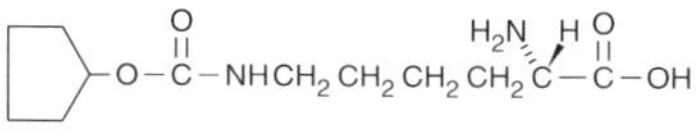
5 gm

S78,180-0
BETA-ALANYL-L-VALINE
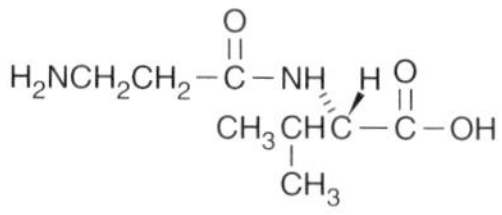
250 mg

S78,357-9
NEPSILON-ACETYL-DL-LYSINE
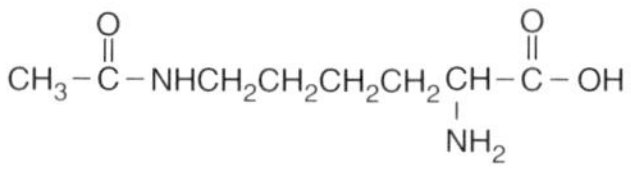
5 gm

S78,425-7
N-ALPHA-ACETYL-DL-LYSINE MONOHYDRATE
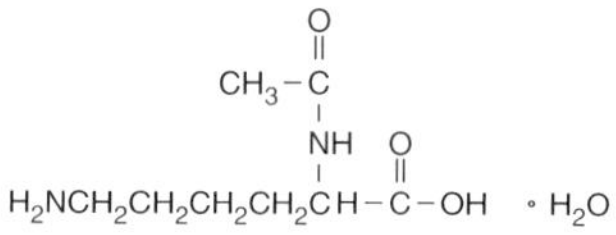
250 mg

S45,135-5
BETA-ALANYL-DL-LEUCINE
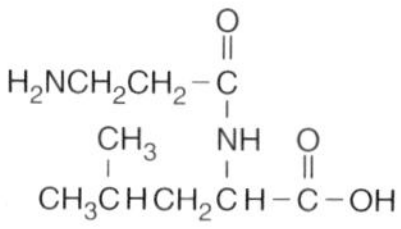
250 mg

S78,918-6
L-ALANYL-L-PROLYLGLYCINE MONOHYDRATE
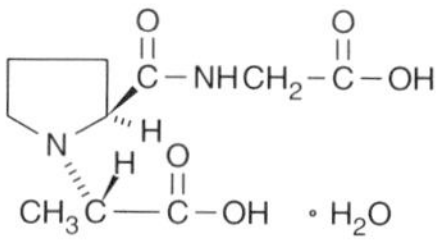
250 mg

S45,682-9
GLYCYLGLYCINE HYDROCHLORIDE MONOHYDRATE
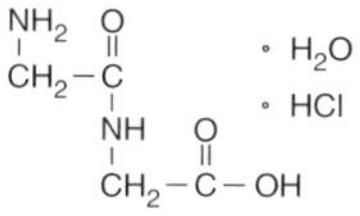
5 gm

S36,216-6
GLYCYL-DL-ALANINE
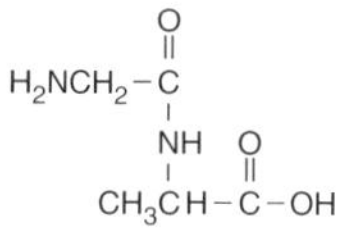
250 mg

S45,115-0
D-ALANYLGLYCINE
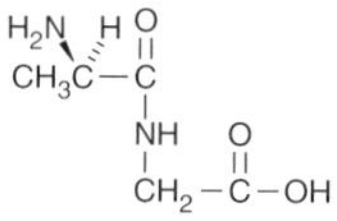
250 mg

S45,185-1
GLYCYL-L-ALANINE
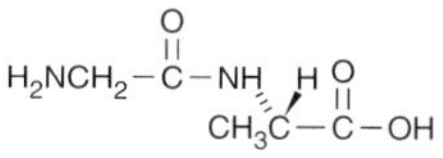
250 mg

S45,504-0
L-ALANYLGLYCINE
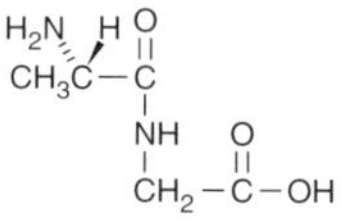
250 mg

S34,069-3
DL-ALANYL-DL-SERINE
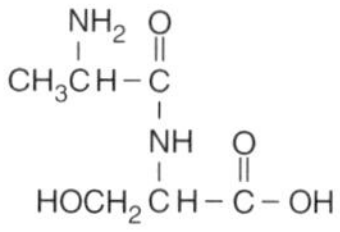
5 gm

S44,057-4
GLYCYLGLYCYLGLYCINE
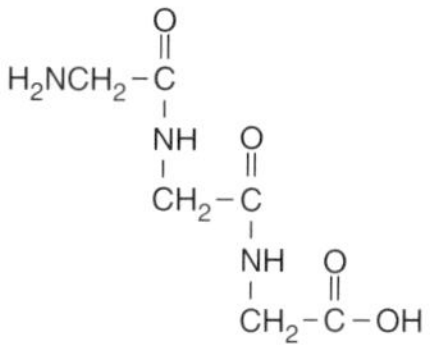
5 gm

S45,355-2
L-VALYLGLYCINE
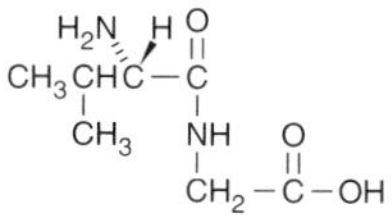
250 mg

S46,252-7
GLYCYL-L-VALINE
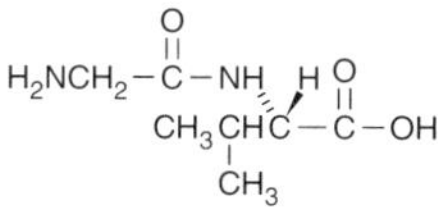
5 gm

S77,648-3
L-2-AMINOBUTYRYL-L-ALANINE
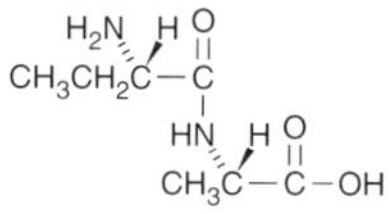
250 mg

S34,066-9
DL-ALANYLGLYCYLGLYCINE
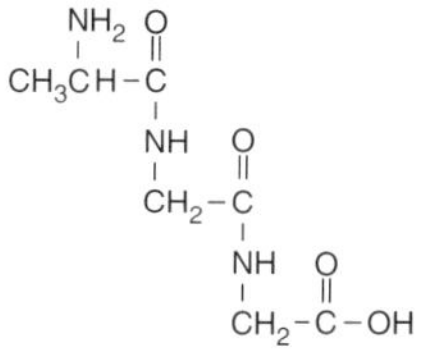
5 gm

S77,610-6
GLYCYLGLYCYL-L-ALANINE
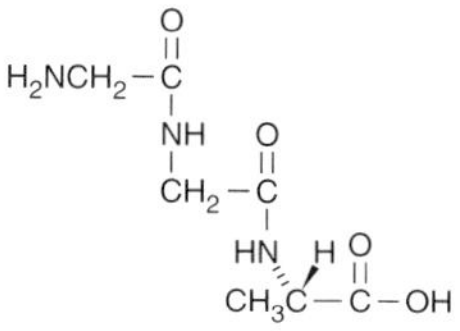
250 mg

S78,420-6
BETA-ALANYLGLYCYLGLYCINE HEMIHYDRATE
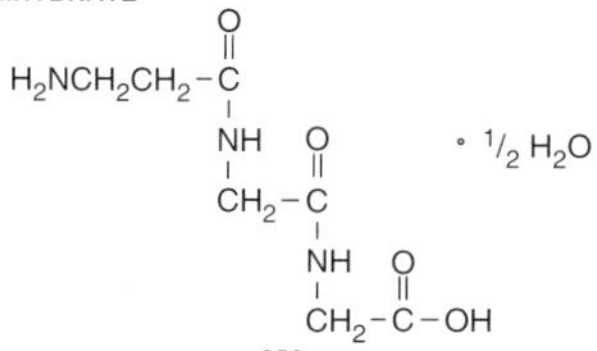
250 mg

S45,279-3
L-VALYL-L-ALANINE
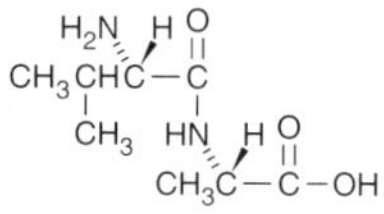
250 mg

S78,511-3
GLYCYL-L-NORLEUCINE
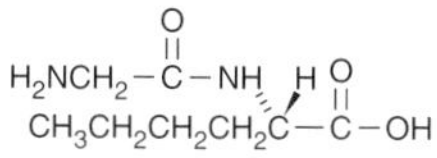
250 mg

S78,514-8
L-2-AMINOBUTYRYLGLYCYLGLYCINE

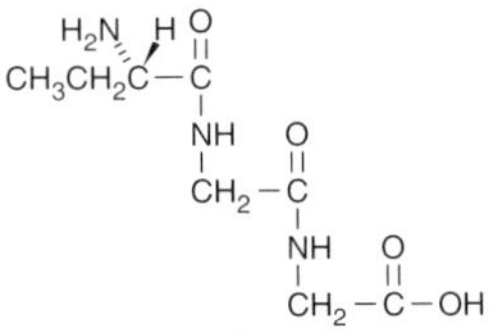

250 mg

S45,119-3
DL-ALANYL-DL-LEUCINE

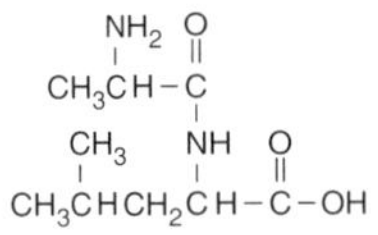

250 mg

S45,120-7
L-ALANYL-L-LEUCINE HYDRATE

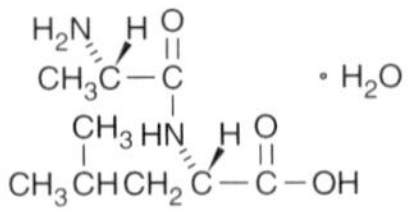

250 mg

S45,503-2
L-ALANYL-L-ISOLEUCINE

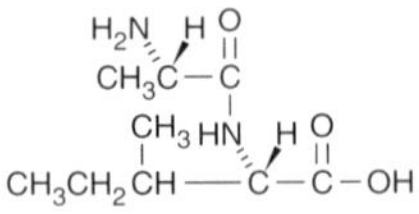

250 mg

S77,777-3
L-NORLEUCYL-L-ALANINE

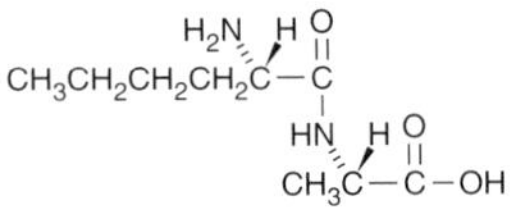

250 mg

S45,502-4
L-ALANYL-L-ALANYL-L-ALANINE

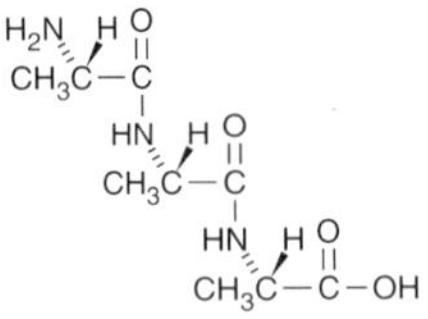

250 mg

S77,776-5
L-ALANYLGLYCYLGLYCYLGLYCINE

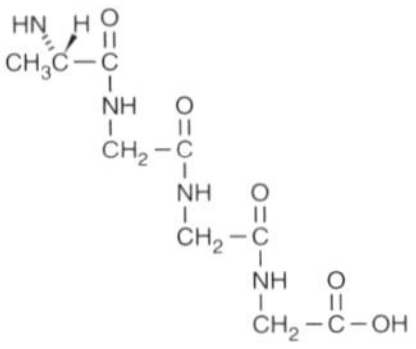

250 mg

S45,280-7
L-VALYL-L-VALINE

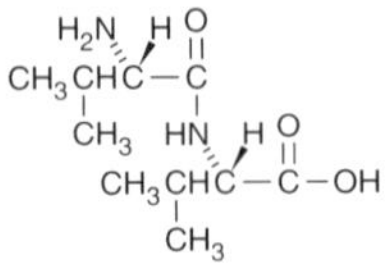

250 mg

S77,628-9
L-NORVALYL-L-NORVALINE

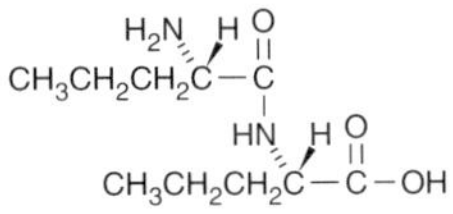

250 mg

S77,778-1
L-VALYL-L-NORVALINE

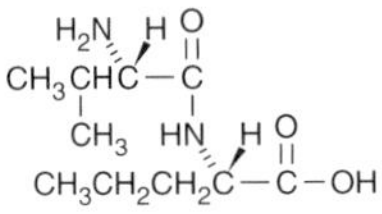

250 mg

S78,431-1
L-2-AMINOBUTYRYL-L-ISOLEUCINE

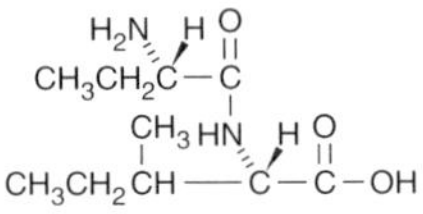

250 mg

S78,433-8
L-2-AMINOBUTYRYL-L-LEUCINE

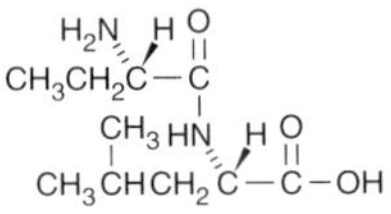

250 mg

S78,437-0
L-NORVALYL-L-VALINE

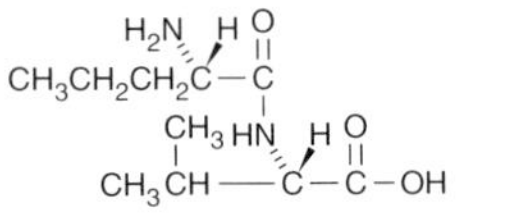

250 mg

S45,158-4
DL-LEUCYLGLYCYLGLYCINE

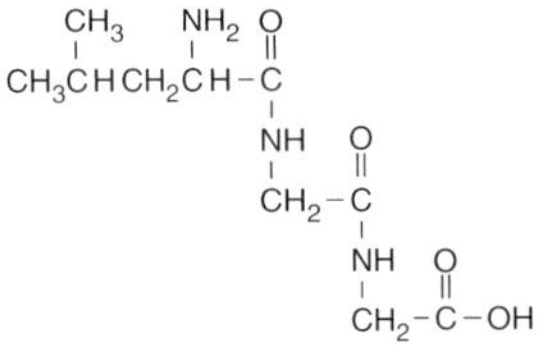

250 mg

S78,196-7
GLYCYLGLYCYL-L-ISOLEUCINE

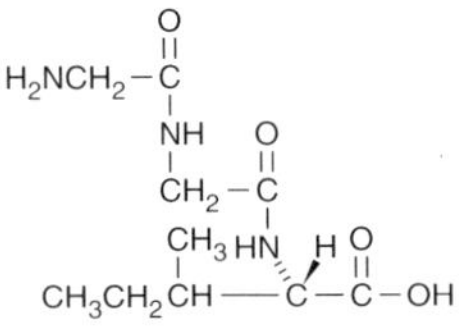

250 mg

S78,171-1
L-NORVALYL-L-ISOLEUCINE

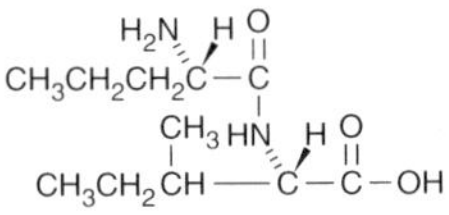

250 mg

S78,517-2
L-NORVALYL-L-LEUCINE ACETATE

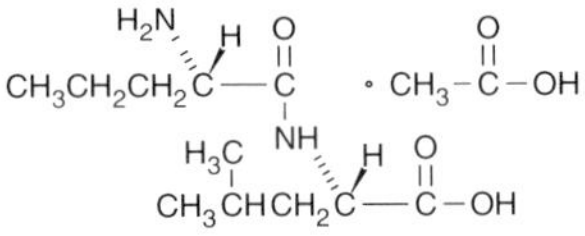

250 mg

S78,518-0
L-NORLEUCYL-L-NORVALINE

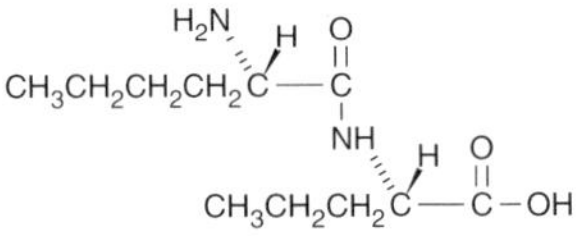

250 mg

S77,614-9
GLYCYL-L-LEUCYLGLYCYLGLYCINE

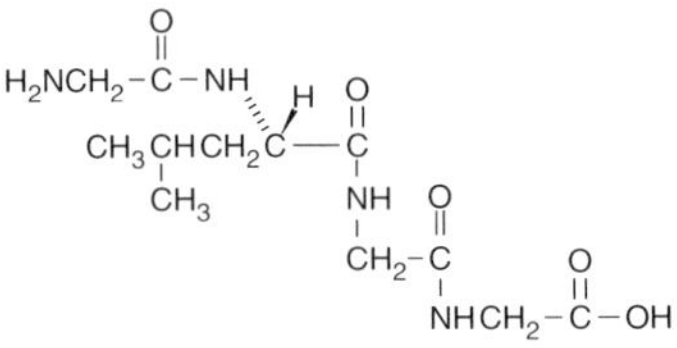

250 mg

S78,803-1
D-VALYL-D-VALYL-D-VALINE

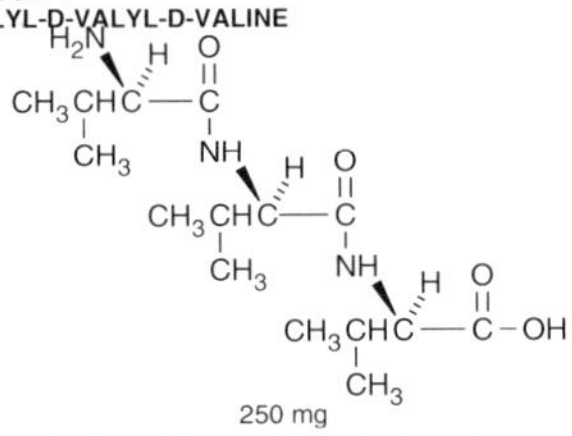

250 mg

S78,971-2
D-ALANYL-D-ALANYL-D-ALANYL-D-ALANYL-D-ALANINE

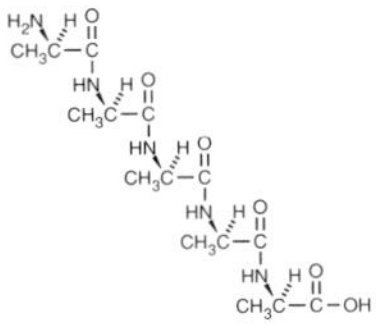

250 mg

S77,650-5
GLYCYL-L-LEUCYLGLYCYL-L-LEUCINE

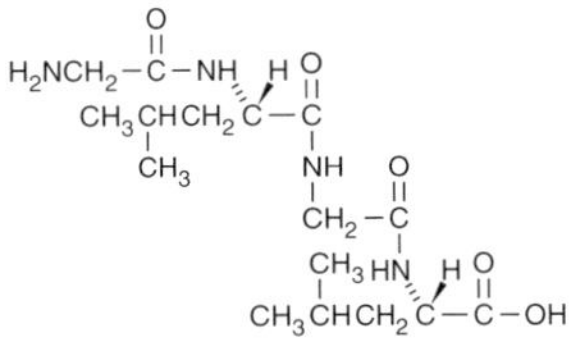

250 mg

S78,540-7
L-ISOLEUCYL-L-ISOLEUCYL-L-ISOLEUCINE HYDRATE

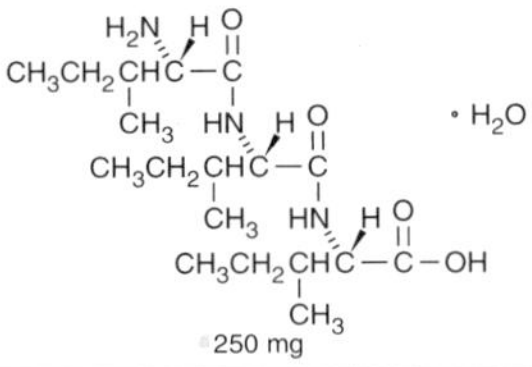

250 mg

S35,191-1
DL-LEUCYLGLYCINE

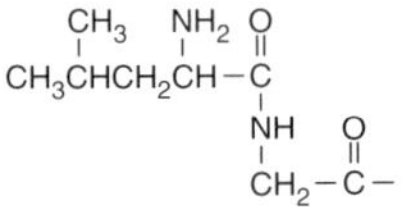

250 mg

S35,021-4
GLYCYLGLYCYL-DL-VALINE

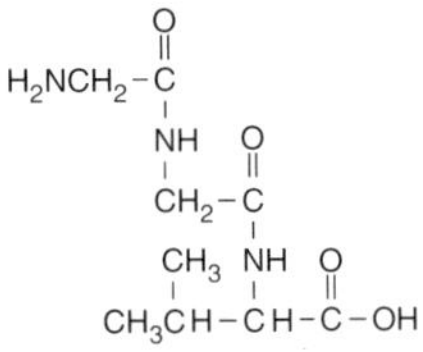

5 gm

S35,195-4
DL-LEUCYL-DL-LEUCINE

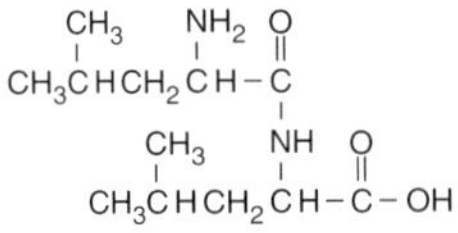

250 mg

S78,414-1
L-ALANYL-BETA-ALANINE

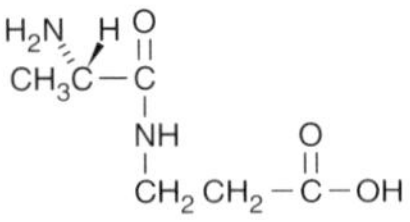

250 mg

S77,627-0
L-ALANYL-L-NORLEUCINE

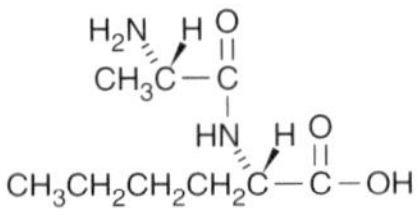

250 mg

S79,043-5
NORVALYLALANINE METHYL ESTER

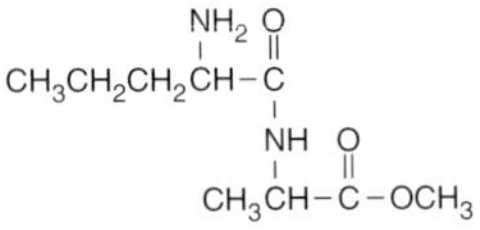

250 mg

S78,264-5
N-ALPHA-ACETYL-D-LYSINE

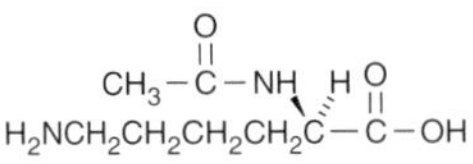

5 gm

S78,890-2
DL-LEUCINAMIDE HYDROCHLORIDE

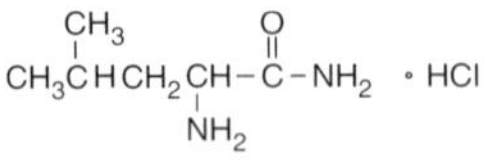

250 mg

S78,417-6
L-ALANYL-BETA-ALANINAMIDE HYDROCHLORIDE

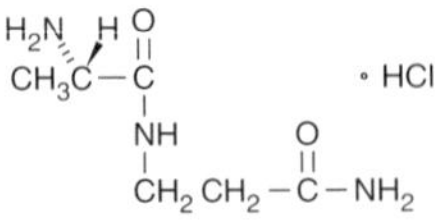

250 mg

S77,615-7
GLYCYLGLYCYL-L-ALANINAMIDE ACETATE HEMIHYDRATE
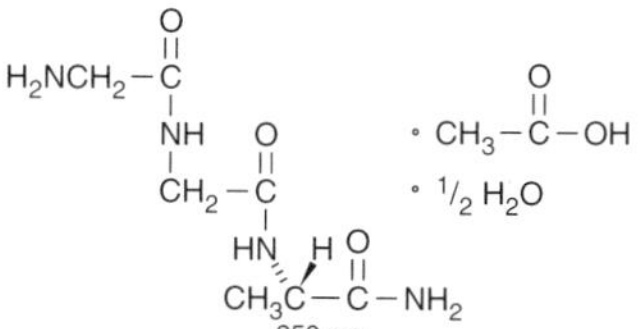
250 mg

S77,689-0
L-VALYL-L-ALANINAMIDE ACETATE
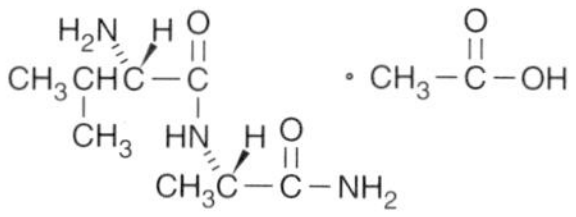
250 mg

S78,199-1
L-ALANYL-L-ISOLEUCINAMIDE ACETATE
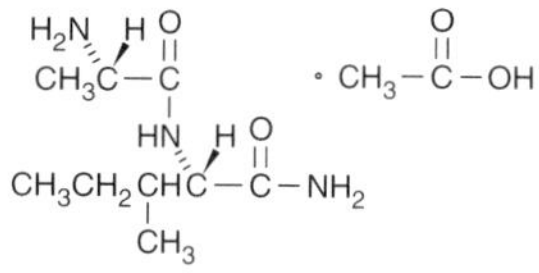
250 mg

S77,725-0
L-VALYLGLYCYLGLYCINAMIDE HYDROCHLORIDE HEMIHYDRATE
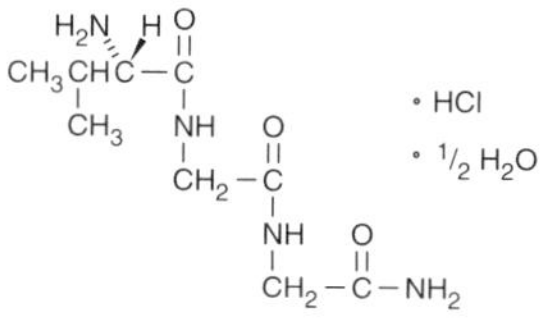
250 mg

S77,780-3
L-ISOLEUCYLGYLCYLGLYCINAMIDE HYDROCHLORIDE
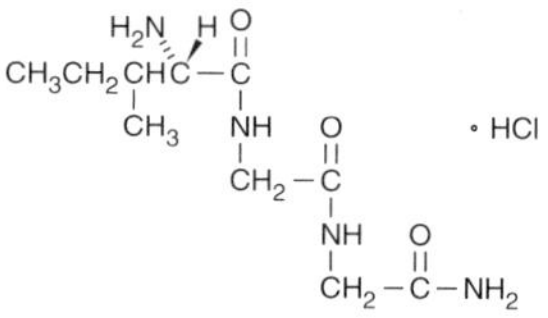
250 mg

S78,168-1
L-LEUCYL-L-ISOLEUCINAMIDE HYDROCHLORIDE
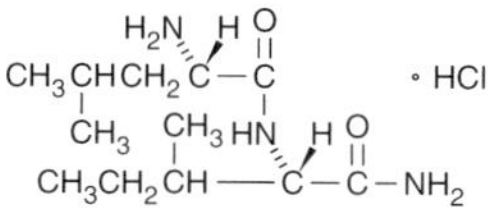
250 mg

S78,170-3
LEUCYLLEUCINAMIDE HYDROCHLORIDE
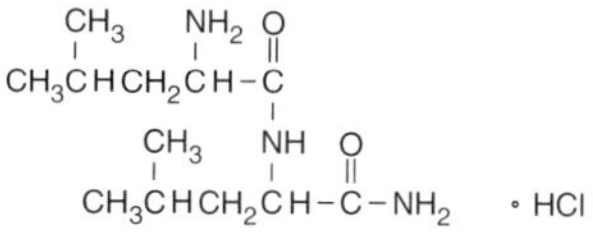
250 mg

S78,404-4
L-2-AMINOBUTYRAMIDE HYDROCHLORIDE
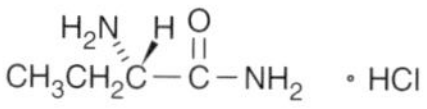
250 mg

S70,381-8
N,N'-ETHYLENEBIS(SUCCINAMIC ACID)
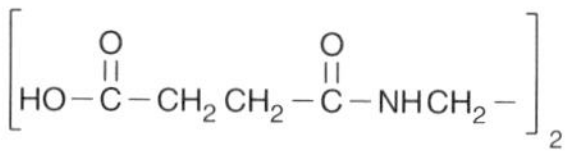
5 gm

S77,647-5
N-ACETYL-DL-PROLINAMIDE
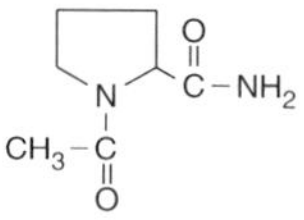
250 mg

S34,273-4
2-BROMOPROPIONYL-DL-ASPARAGINE
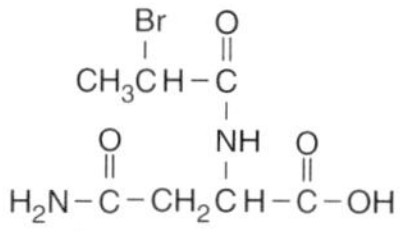
5 gm

S44,999-7
N-(TERT-BUTOXYCARBONYL)-L-LEUCINE MONOHYDRATE
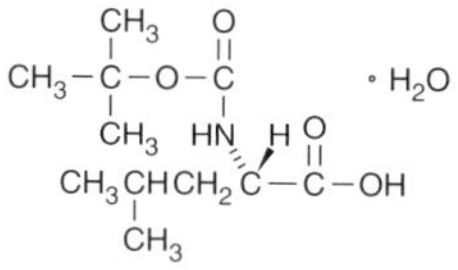
250 mg

S77,743-9
SARCOSYLALANINE
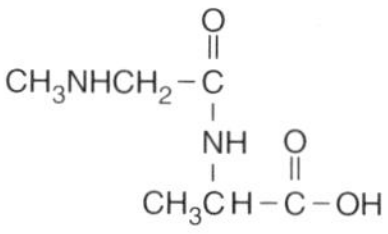
250 mg

S78,510-5
SARCOSYLGLYCYLGLYCINE
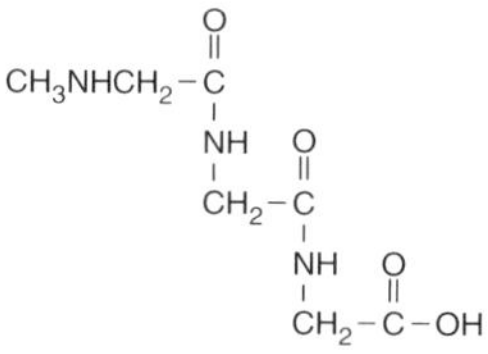
250 mg

S77,656-4
SARCOSYL-L-VALINE
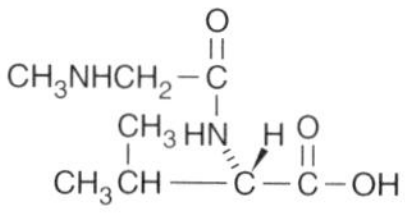
250 mg

S77,724-2
SARCOSYL-L-ISOLEUCINE
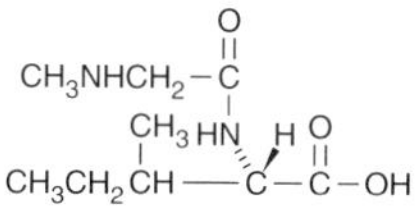
250 mg

S78,154-1
SARCOSYL-L-ISOLEUCINAMIDE HYDROCHLORIDE
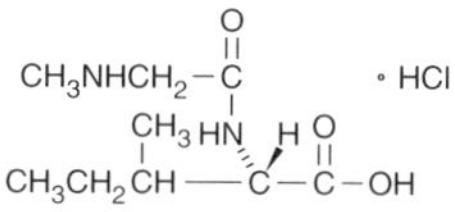
250 mg

S87,737-9
2-(ETHYLAMINO)ACETAMIDE HYDROCHLORIDE
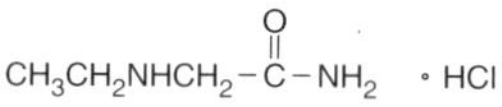
250 mg

S80,583-1
DIETHYL 2-ACETAMIDO-2-(2-OXOPROPYL)-MALONATE
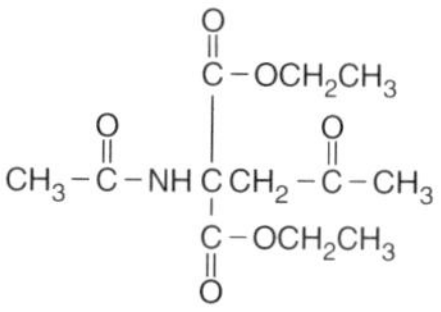
250 mg

S80,242-5
2-ACETAMIDO-2-ALLYLMALONIC ACID
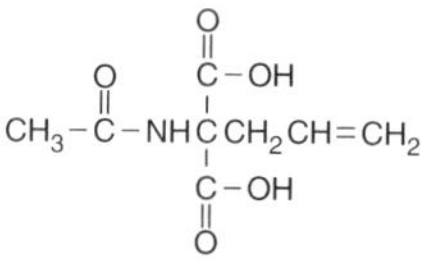
250 mg

S80,328-6
2,2-(DIACETAMIDO)PROPIONIC ACID
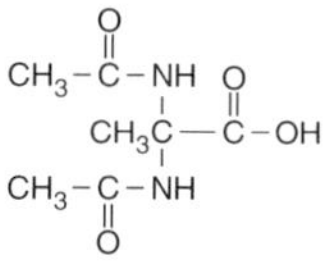
250 mg

S78,288-2
N-ACETYL-DL-LEUCINAMIDE
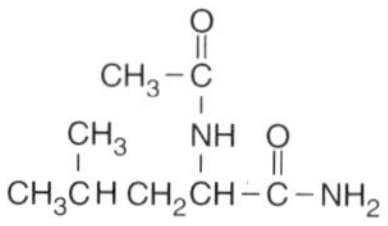
5 gm

S78,521-0
L-NORLEUCYL-L-NORLEUCINAMIDE HYDROCHLORIDE
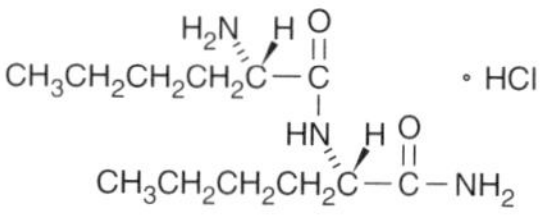
250 mg

S78,280-7
N-ALPHA-ACETYL-N-METHYLGLYCINAMIDE
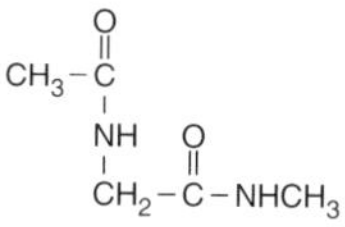
5 gm

S78,503-2
N-HYDROXY-L-ASPARAGINE
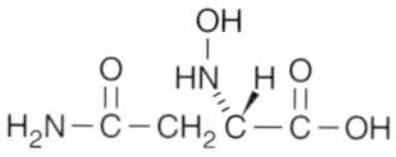
250 mg

S78,564-4
N-ALPHA-HYDROXY-D-ASPARAGINE
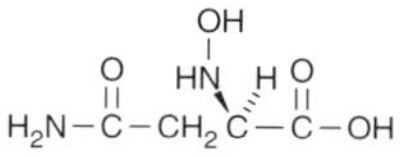
250 mg

S76,409-4
N,N'-DODECAMETHYLENEBIS-(MALEAMIC ACID)
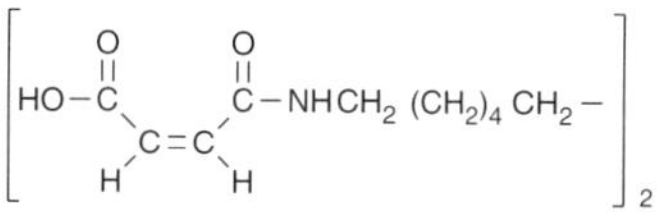
5 gm

S50,593-5
MALEAMIC ACID
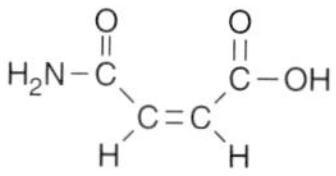
250 mg

S70,351-6
N-DECYLMALEAMIC ACID
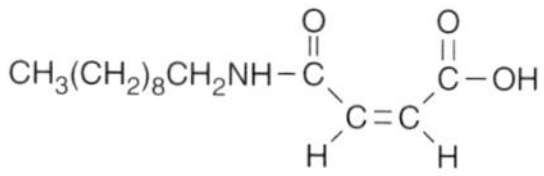
5 gm

S70,354-0
N-DODECYLMALEAMIC ACID
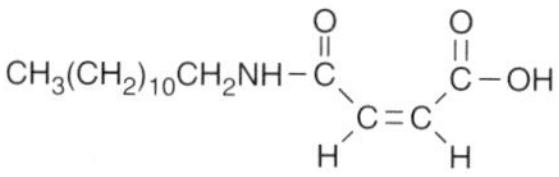
5 gm

S70,349-4
N-HEXADECYLMALEAMIC ACID
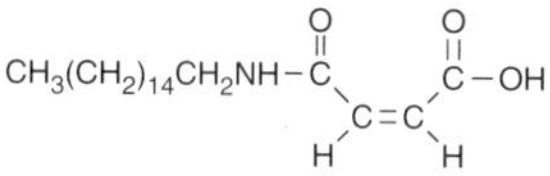
5 gm

S70,350-8
N-OCTADECYLMALEAMIC ACID
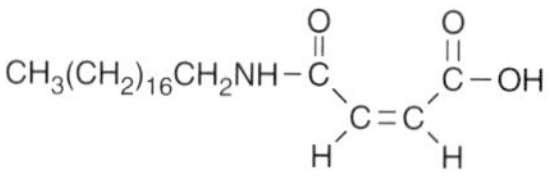
5 gm

S34,333-1
N-CARBAMOYLMALEAMIC ACID
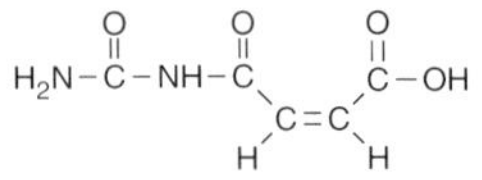
5 gm

S62,681-3
N-METHYLMALEAMIC ACID
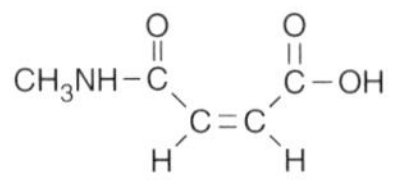
5 gm

S44,139-2
N-ETHYLMALEAMIC ACID
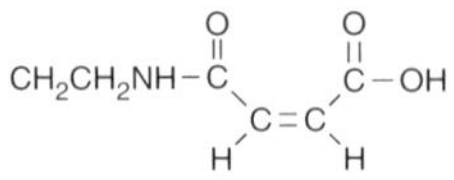
5 gm

S39,388-6
N-ALLYLMALEAMIC ACID
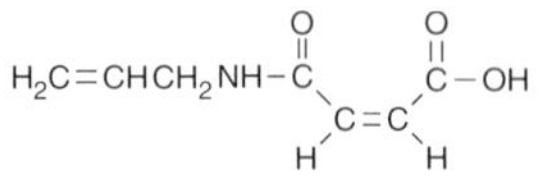
5 gm

S61,884-5
N-PROPYLMALEAMIC ACID
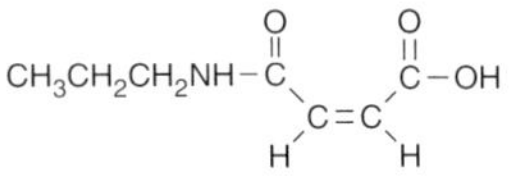
5 gm

S39,392-4
N-BUTYLMALEAMIC ACID
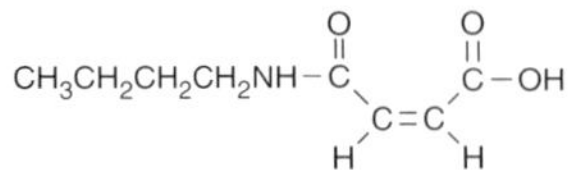
5 gm

S53,058-1
N-OCTYLMALEAMIC ACID
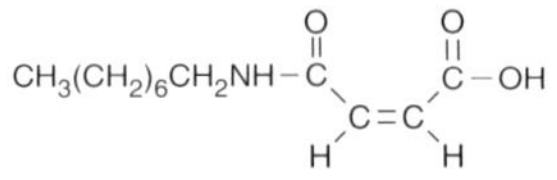
250 mg

S39,389-4
N-CYCLOHEXYLMALEAMIC ACID
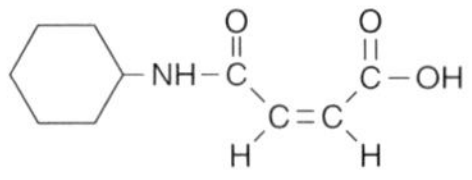
5 gm

S55,336-0
N-(3,3,5-TRIMETHYLCYCLOHEXYL)-MALEAMIC ACID
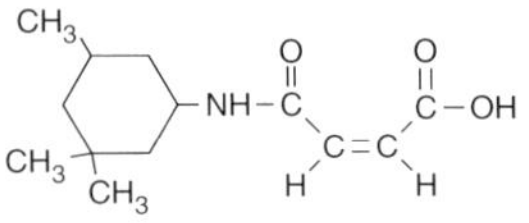
250 mg

S65,297-0
ETHYL 2-ACETAMIDOMALONATE
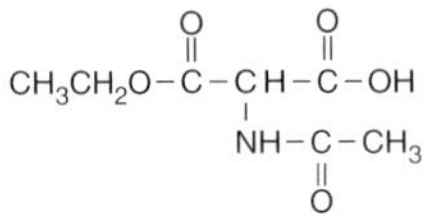
250 mg

S39,969-8
2-CARBAMOYL-2-PROPYLVALERIC ACID
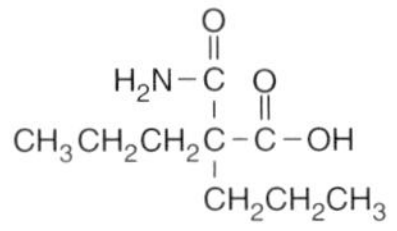
5 gm

S66,294-1
2,2-DIISOPROPYLMALONAMIC ACID
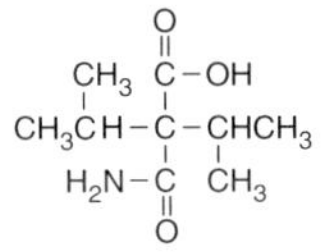
250 mg

S80,466-5
DIETHYL 2-ACETAMIDO-2-[2-(ETHOXYCARBONYL)-1-METHYLETHYL]-MALONATE
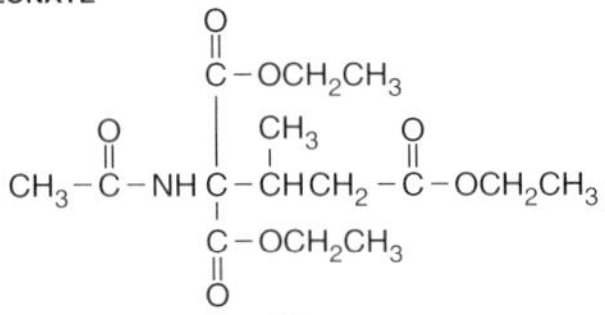
250 mg

S78,383-8
N-ACETYL-DL-ASPARTIC ACID
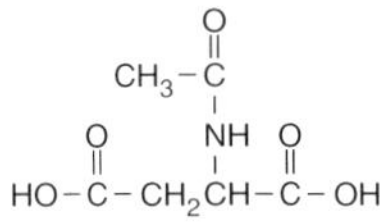
5 gm

S76,418-3
N,N'-DODECAMETHYLENEBIS-(GLUTARAMIC ACID)
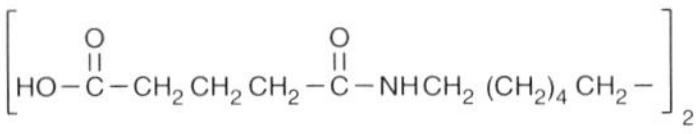
5 gm

S76,428-0
N,N'-DODECAMETHYLENEBIS-(HEXAHYDROPHTHALAMIC ACID)
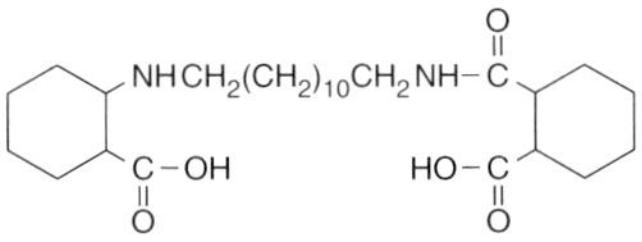
5 gm

S78,991-7
N-TERT-BUTOXYCARBONYL-D-GLUTAMIC ACID 5-METHYL ESTER
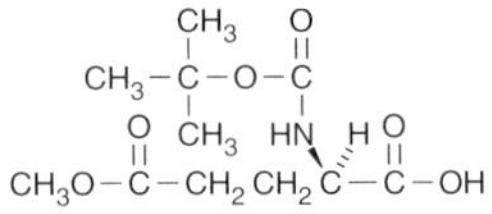
250 mg

S78,290-4
N-ALPHA-ACETYL-D-ASPARAGINE HYDRATE
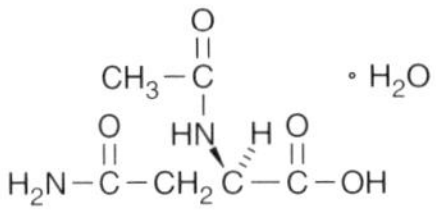
5 gm

S78,666-7
N-ALPHA-ACETYL-DL-ASPARAGINE
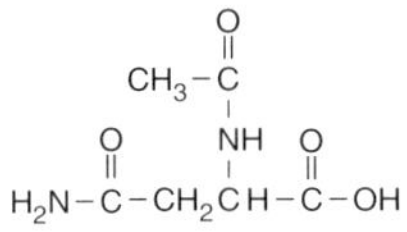
250 mg

S73,063-7
3-(N-CYCLOHEXYLCARBAMOYL)-5-NORBORNENE-2-CARBOXYLIC ACID
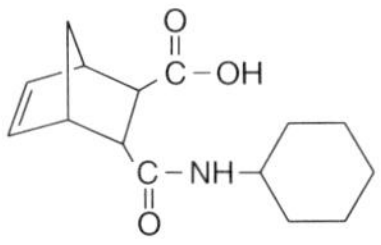
5 gm

S73,065-3
3-(N-PROPYLCARBAMOYL)-5-NORBORNENE-2-CARBOXYLIC ACID
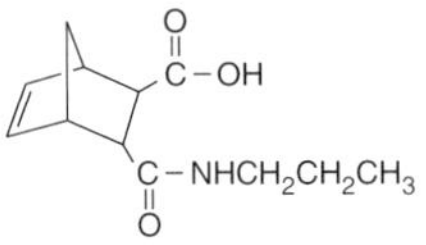
5 gm

S72,063-1
N-CYCLOHEXYLGLUTARAMIC ACID
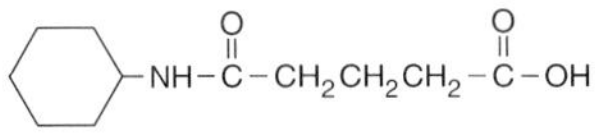
5 gm

S70,277-3
CIS-N-TERT-BUTYLHEXAHYDROPHTHALAMIC ACID
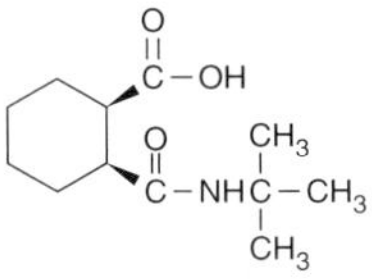
5 gm

S71,911-0
CIS-N-CYCLOHEXYLHEXAHYDROPHTHALAMIC ACID
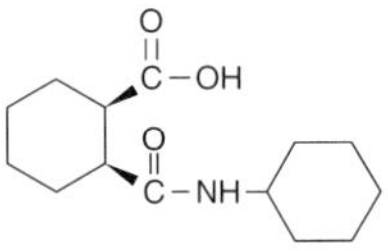
5 gm

S70,353-2
CIS-N-(DODECYL)-HEXAHYDROPHTHALAMIC ACID
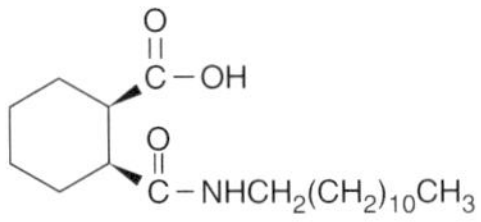
5 gm

S73,072-6
CIS-N-(HEXADECYL)-HEXAHYDROPHTHALAMIC ACID
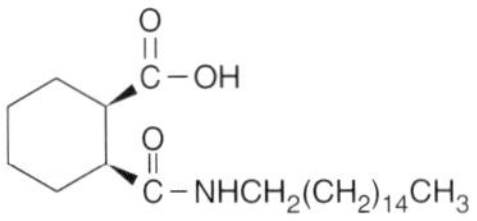
5 gm

S73,073-4
CIS-HEXAHYDRO-N-(OCTADECYL)-PHTHALAMIC ACID
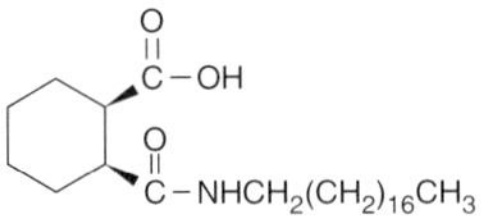
5 gm

S76,430-2
3-(N-HEXADECYLCARBAMOYL)-5-NORBORNENE-2-CARBOXYLIC ACID
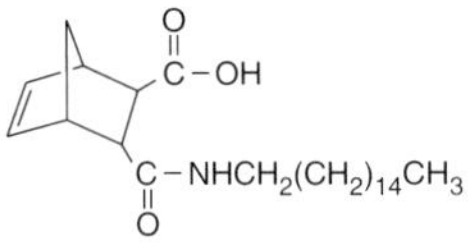
5 gm

S61,530-7
N-ACETYL-DL-GLUTAMIC ACID
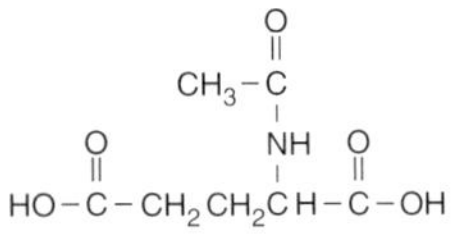
250 mg

S78,729-9
N-(TERT-BUTOXYCARBONYL)-L-NORVALINE DICYCLOHEXYLAMINE SALT
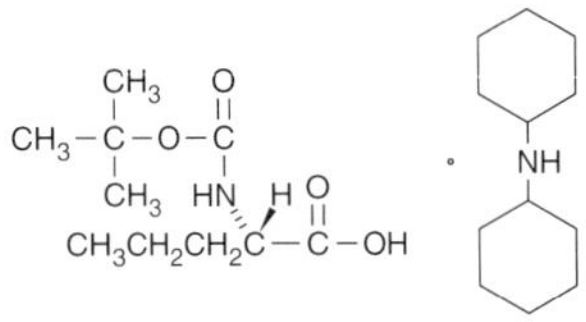
250 mg

S78,309-9
N-ACETYL-DL-ALANINE
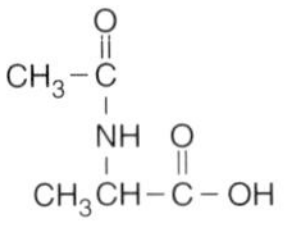
5 gm

S78,888-0
N-FORMYL-L-2-AMINOBUTYRIC ACID
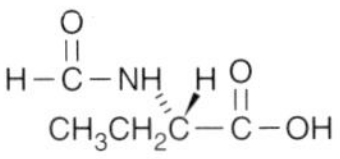
250 mg

S34,274-2
DL-2-BROMOPROPIONYLGLYCINE
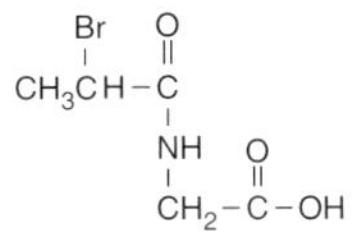
5 gm

S75,672-5
N-FORMYL-DL-VALINE
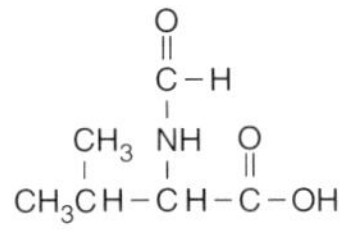
5 gm

S77,768-4
L-2-ACETAMIDOBUTYRIC ACID
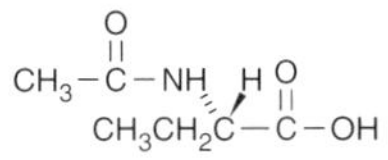
250 mg

S78,798-1
N-FORMYL-L-VALINE
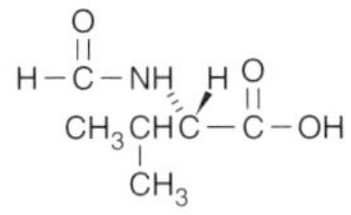
250 mg

S46,261-6
N-ACETYL-L-ASPARTIC ACID
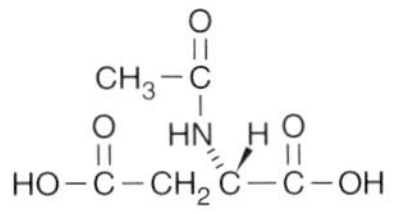
250 mg

S34,272-6
2-BROMOPROPIONYL-DL-ALANINE
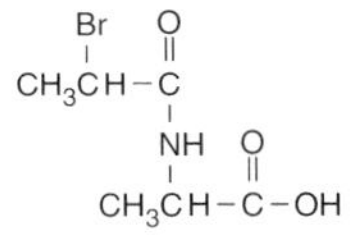
5 gm

S80,291-3
N-ACETYL-2-ALLYL-DL-GLYCINE
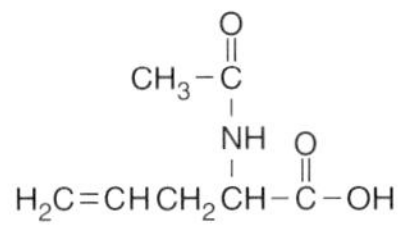
250 mg

S77,727-7
N-ACETYL-D-VALINE
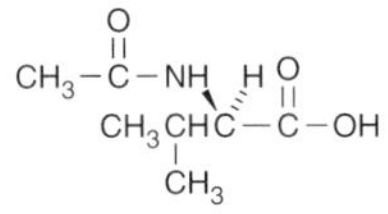
250 mg

S78,570-9
N-ACETYL-DL-NORVALINE
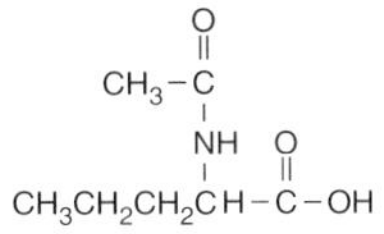
250 mg

S34,434-6
DL-BROMOISOVALERYLGLYCINE
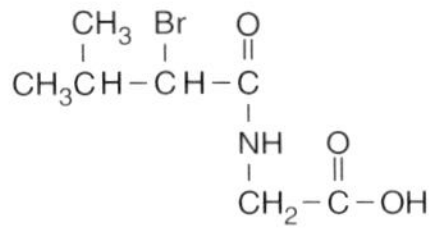
5 gm

S57,943-2
N-ACETYL-L-LEUCINE
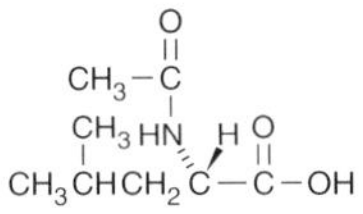
5 gm

S78,308-0
N-ACETYL-D-ALLO-ISOLEUCINE
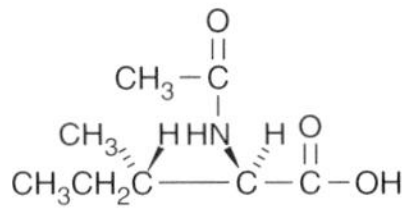
5 gm

S78,350-1
N-ACETYL-L-ISOLEUCINE
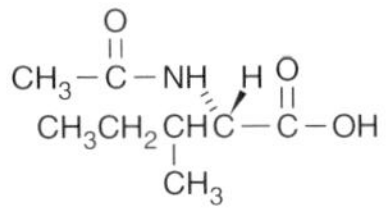
5 gm

S78,707-8
N-ACETYL-DL-NORLEUCINE
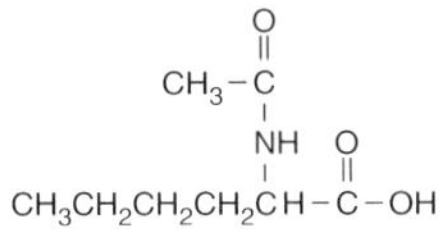
250 mg

S78,726-4
N-ACETYL-D-NORLEUCINE
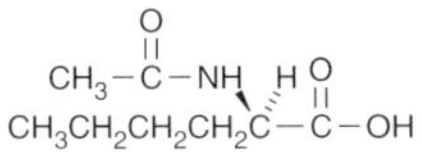
250 mg

S78,905-4
N-ACETYL-D-LEUCINE
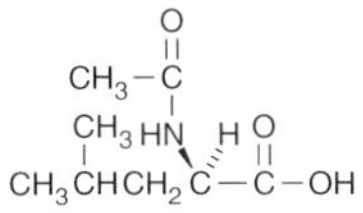
250 mg

S34,256-4
DL-2-BROMOISOCAPROYLGLYCINE
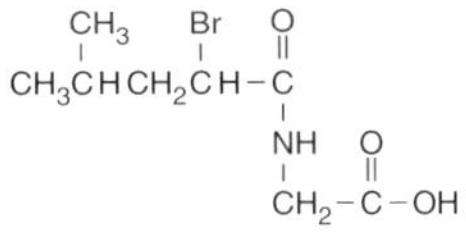
5 gm

S34,276-9
2-BROMOPROPIONYL-DL-NORVALINE
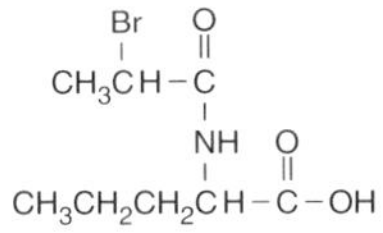
5 gm

S67,430-3
N-OCTANOYLGLYCINE
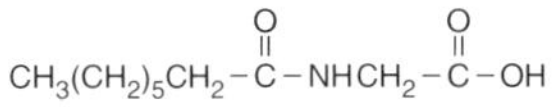
250 mg

S34,259-9
DL-2-BROMOISOCAPROYL-DL-LEUCINE
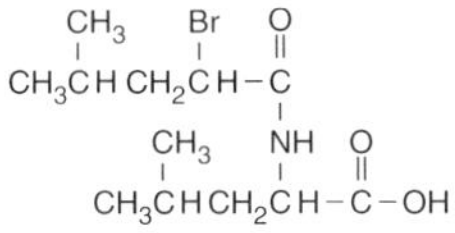
5 gm

S79,041-9
N-ACETYLGLYCYLGLYCINE
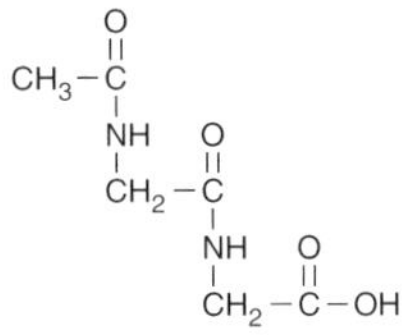
250 mg

S78,509-1
D-2-AMINOBUTYRYL-D-ALANINE
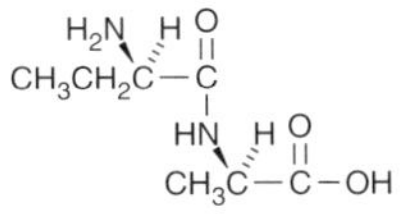
250 mg

S34,275-0
2-BROMOPROPIONYL-DL-GLYCYLGLYCINE
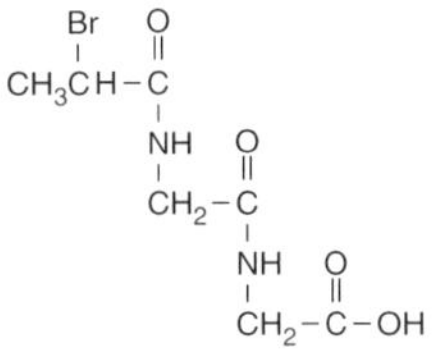
5 gm

S78,906-2
D-ALANYL-D-VALINE
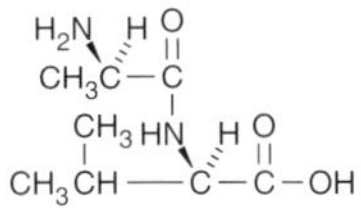
250 mg

S78,336-6
2-AMINOBUTYRYLVALINE
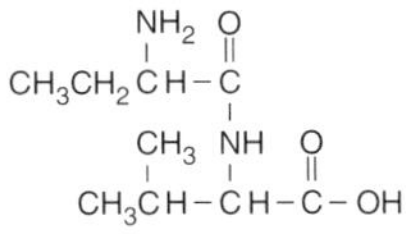
5 gm

S78,910-0
GLYCYLGLYCYL-L-VALINE
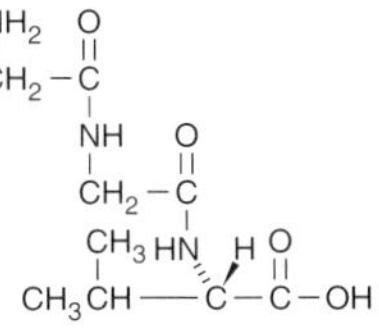
250 mg

S78,919-4
D-VALYL-L-VALINE HYDROCHLORIDE
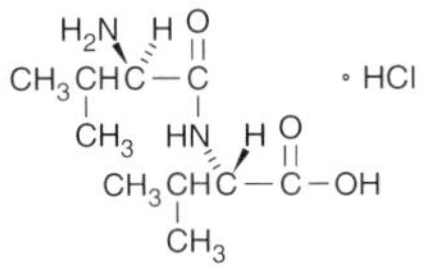
250 mg

S79,007-9
N-ACETYLGLYCYL-L-LEUCINE
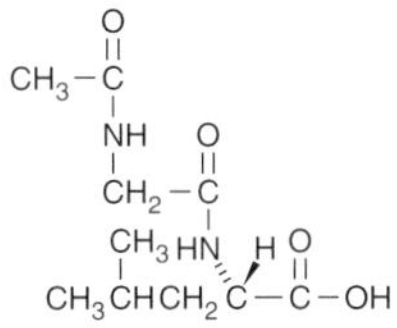
250 mg

S78,436-2
GLYCYLGLYCYLGLYCYLGLYCYLALANINE HEMIHYDRATE
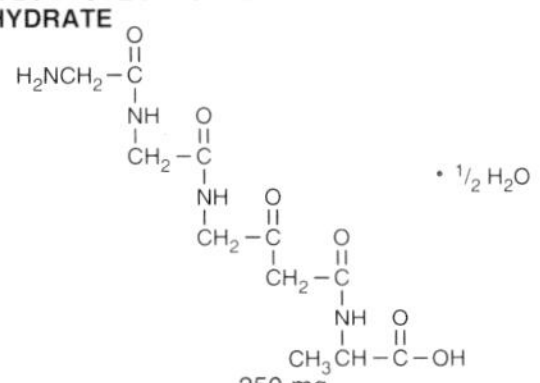
250 mg

S76,530-9
N-L-LEUCYL-L-LEUCINE
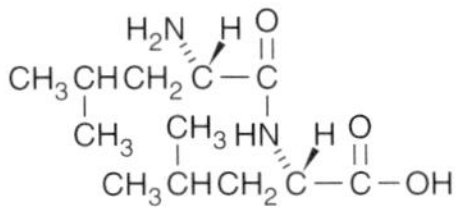
5 gm

S78,935-6
L-LEUCYL-D-LEUCINE
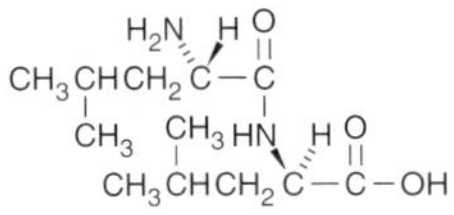
250 mg

S78,936-4
D-LEUCYL-L-LEUCINE
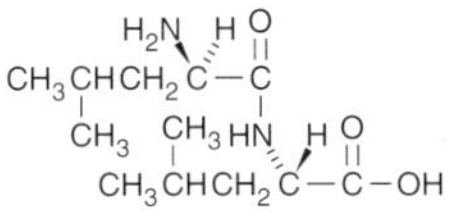
250 mg

S78,903-8
L-PROLYL-L-ALANINE MONOHYDRATE
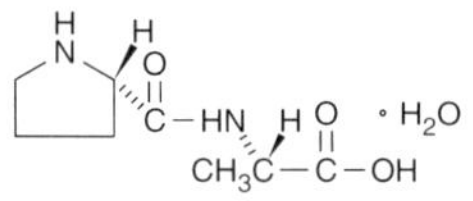
250 mg

S77,726-9
N-ACETYLSARCOSINE
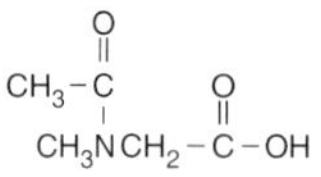
250 mg

S36,578-5
N-LAUROYLSARCOSINE
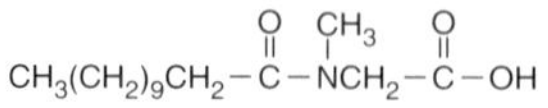
5 gm

S76,559-7
N-GLYCYL-L-PROLINE
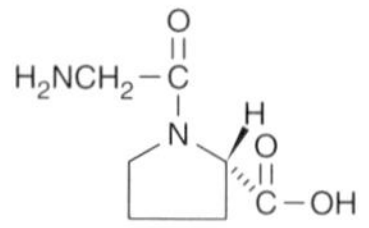
5 gm

S77,770-6
N-ACETYL-DL-PROLINE
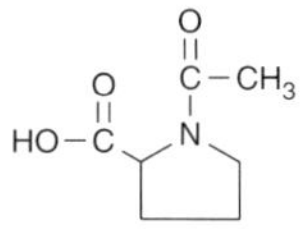
250 mg

S51,044-0
3-(METHYLCARBAMOYL)PROPIONIC ACID
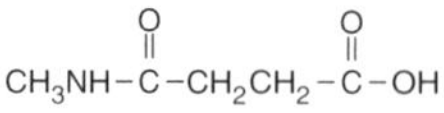
250 mg

S76,595-3
N-ACETYL-BETA-ALANINE
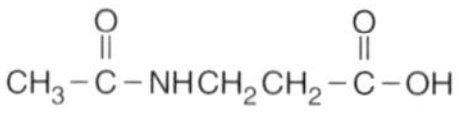
5 gm

S45,836-8
4-ACETAMIDOBUTYRIC ACID
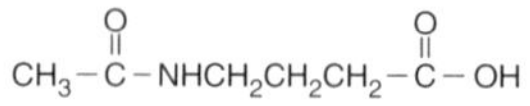
5 gm

S34,260-2
DL-N-(2-BROMO-3-METHYLBUTYRYL)-GLYCYLGLYCINE
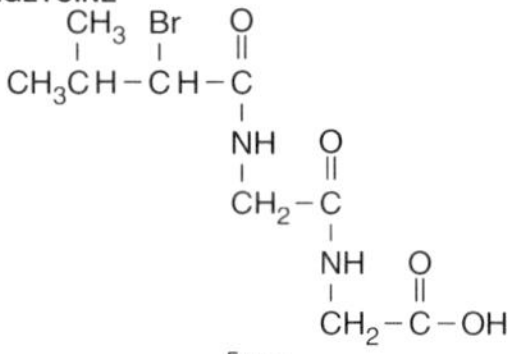
5 gm

S78,911-9
L-LEUCYL-BETA-ALANINE
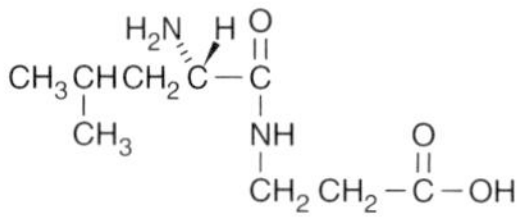
250 mg

S78,568-7
N-ACETYLGLYCINE METHYL ESTER
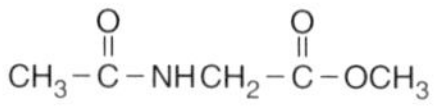
250 mg

S76,581-3
N-ACETYL-L-LEUCINE METHYL ESTER
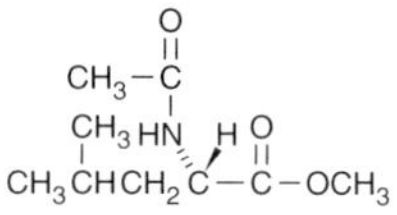
5 gm

S79,042-7
N-ACETYLGLYCYLGLYCINE METHYL ESTER
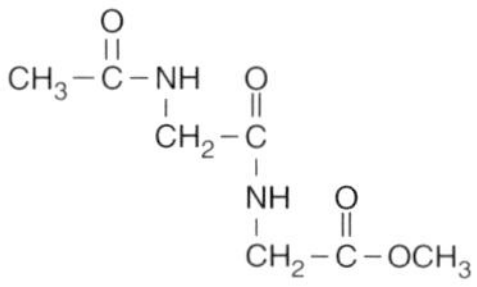
250 mg

S77,783-8
N-ACETYL-L-ALANYL-L-ALANYL-L-ALANYL-L-ALANINE METHYL ESTER
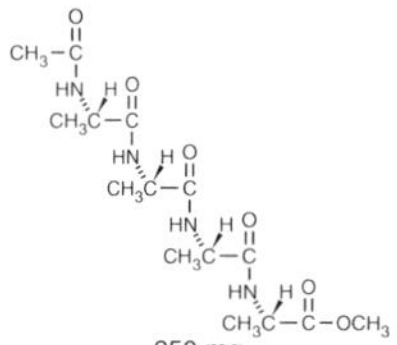
250 mg

S78,406-0
N-ACETYL-DL-ALANINAMIDE
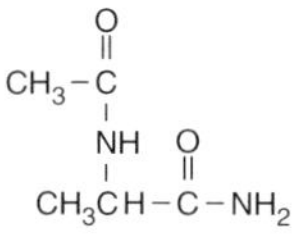
250 mg

S78,426-5
N-ALPHA-ACETYL-D-ALLO-ISOLEUCINAMIDE
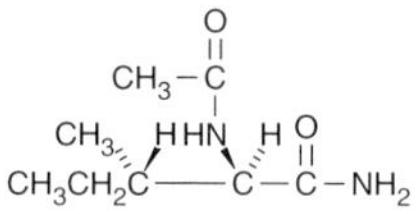
250 mg

S77,625-4
N-ACETYL-L-LEUCINAMIDE
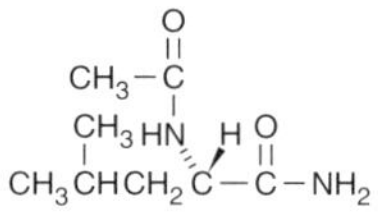
250 mg

S76,544-9
N-GLYCYL-L-LEUCINAMIDE HYDROCHLORIDE
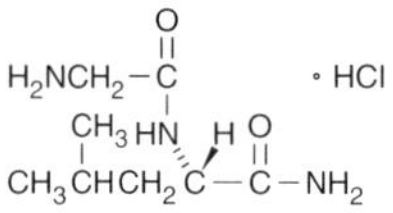
5 gm

S78,423-0
GLYCYLGLYCYLGLYCYLGLYCINAMIDE HYDROCHLORIDE
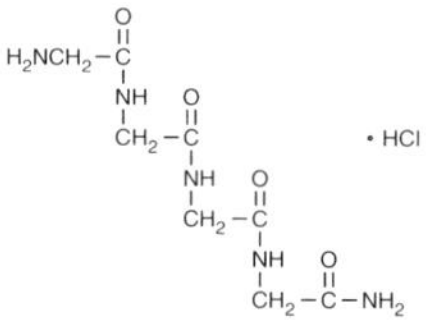
250 mg

S77,753-6
L-VALYL-L-LEUCINAMIDE HYDROCHLORIDE
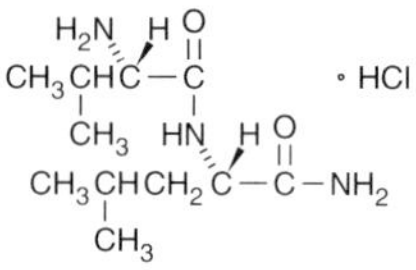
250 mg

S78,440-0
L-ISOLEUCYL-L-ISOLEUCINAMIDE HYDROCHLORIDE
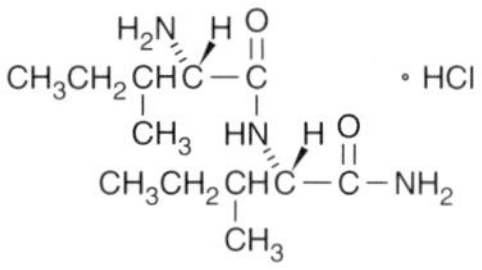
250 mg

S78,178-9
L-PROLYLGLYCYLGLYCINAMIDE
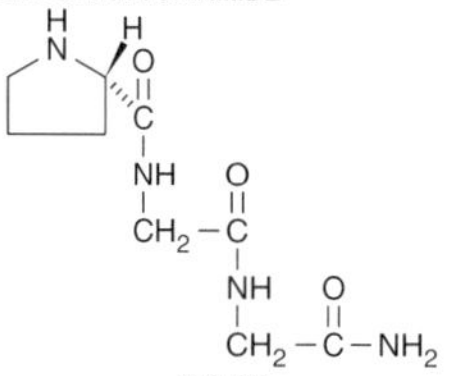
250 mg

S79,058-3
N-ACETYL-L-LEUCINE-N-METHYLAMIDE
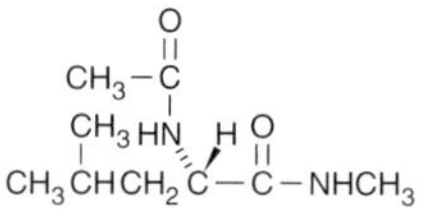
250 mg

S43,229-6
ALLOXANTIN
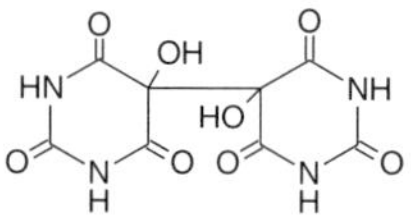
250 mg

S72,742-2
2-(2-HYDROXYETHYL)-6-METHYL-3(2H)-PYRIDAZINONE
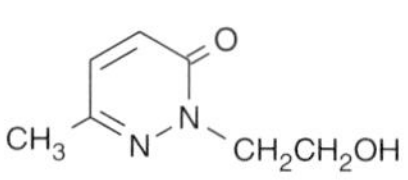
250 mg

S89,564-4
3,4-DIMETHYL-4-HYDROXY-2-PYRROLIDINONE
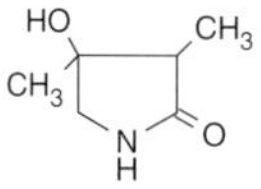
250 mg

S43,760-3
3-HYDROXY-2-PIPERIDONE
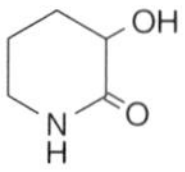
250 mg

S36,255-7
1-(HYDROXYMETHYL)-2-PYRROLIDINONE
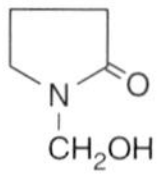
5 gm

S92,745-7
1-METHYL-3-(METHYLTHIO)-1,2,4-TRIAZASPIRO(5,5)UNDEC-2-EN-5-ONE
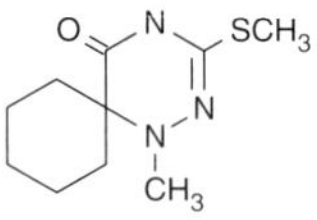
250 mg

S36,432-0
2-OXO-1-PYRROLIDINEBUTYRAMIDE
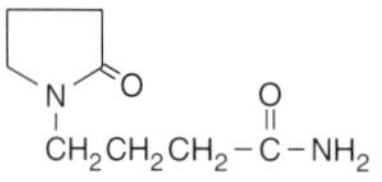
5 gm

S93,494-1
3-METHYL-2-PIPERAZINONE
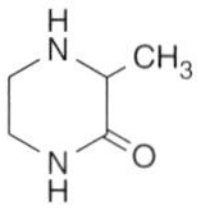
5 gm

S43,422-1
1-BUTYL-2-PYRROLIDINONE
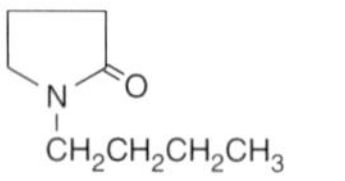
5 gm

S78,250-5
ETHYL 2-OXO-1-PYRROLIDINEACETATE

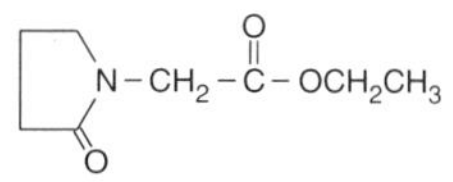

5 gm

S89,503-2
1,4,5-TRIMETHYL-3-PYRROLIN-2-ONE

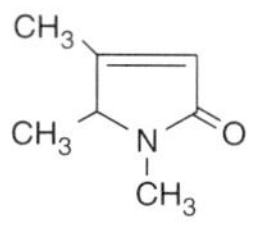

250 mg

S78,434-6
L-PYROGLUTAMYL-L-VALINE

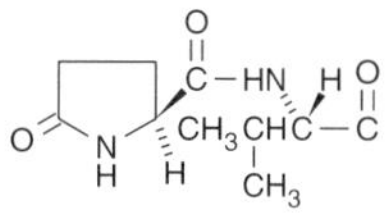

250 mg

S44,744-7
1-METHYL-5-OXO-3-PYRROLIDINECARBOXYLIC ACID

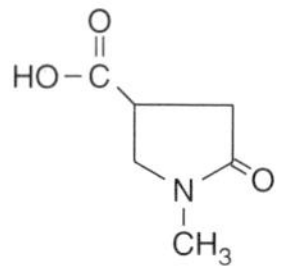

5 gm

S37,706-6
ETHYL 1,3-DIMETHYL-2-OXO-3-PYRROLIDINECARBOXYLATE

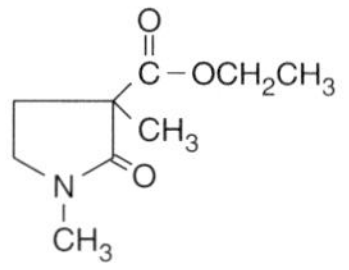

5 gm

S46,262-4
3-AMINO-2-OXAZOLIDINONE SULFATE

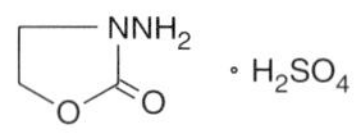

5 gm

S78,876-7
L-PYROGLUTAMIC ACID AMIDE

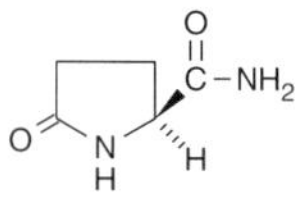

250 mg

S44,882-6
EPSILON-CAPROLACTAM, ZONE REFINED

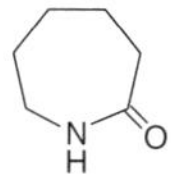

5 gm

S87,148-6
3,3-DIBROMO-4,5,6,7-TETRAHYDRO-1H-AZEPIN-2(3H)-ONE

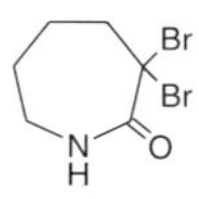

250 mg

S51,511-6
7-BUTYLTETRAHYDRO-2(3H)-AZEPINONE

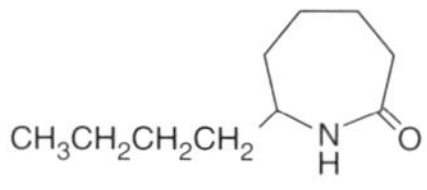

250 mg

S54,075-7
4-(TERT-BUTYL)CAPROLACTAM

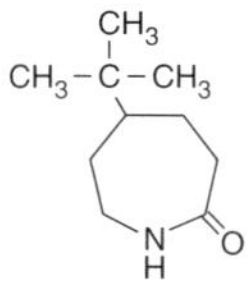

5 gm

S87,166-4
3,3-DIBROMOAZACYCLODODECAN-2-ONE

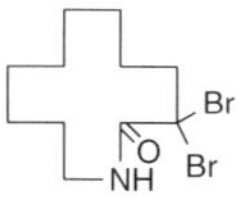

250 mg

S93,496-8
1,7-DIAZASPIRO(4.4)NONAN-2,6-DIONE

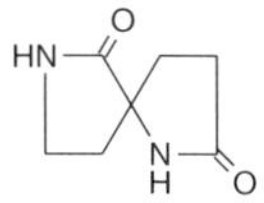

250 mg

S95,091-2
3,3,5-TRIMETHYL-2-PYRROLIDINONE

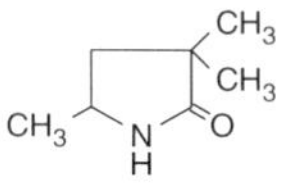

50 mg

S78,254-8
TRANS-DECAHYDROQUINOLIN-2-ONE

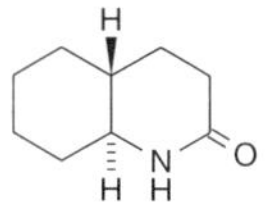

5 gm

S95,095-5
2,3-DIACETOXYSUCCINIMIDE

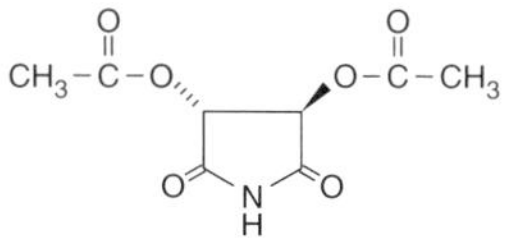

50 mg

S78,451-6
TERT-BUTOXYCARBONYL-L-METHIONINE N-HYDROXYSUCCINIMIDE ESTER

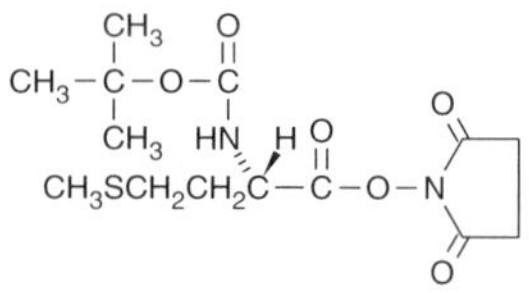

250 mg

S53,484-6
ADIPIMIDE

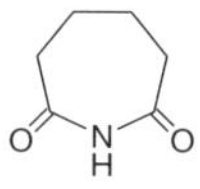

5 gm

S35,880-0
1,1'-BISUCCINIMIDE(N,N'-BISUCCINIMIDYL)

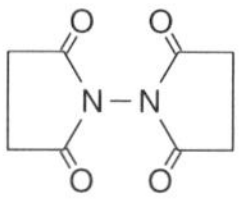

250 mg

S85,996-6
2,6-DIOXO-N-METHOXY-N-METHYL-4-PIPERIDINE CARBOXAMIDE

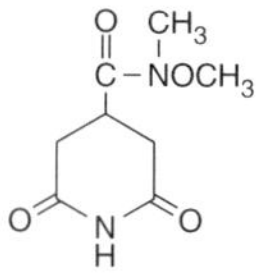

250 mg

S95,085-8
2,2-DIMETHYLSUCCINIMIDE

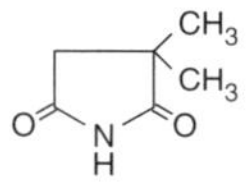

50 mg

S95,476-4
(-)-3-(1,1-DIMETHYL-2-OXOPROPYL)-2,6-PIPERIDINEDIONE

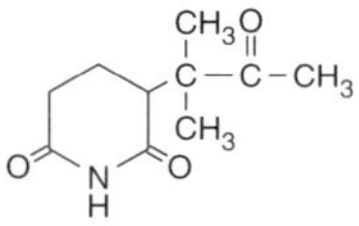

50 mg

S43,243-1
2,5-DIOXO-3-PYRROLINE-1-CARBOXAMIDE

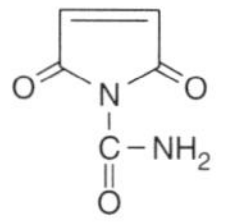

5 gm

S85,997-4
2,6-DIOXO-4-PIPERIDINECARBOXYLIC ACID

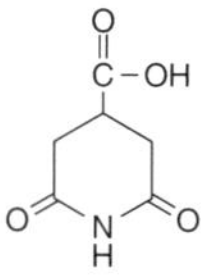

250 mg

S86,073-5
S-BUTYL 2,6-DIOXOPIPERIDINE-4-THIOCARBOXYLATE

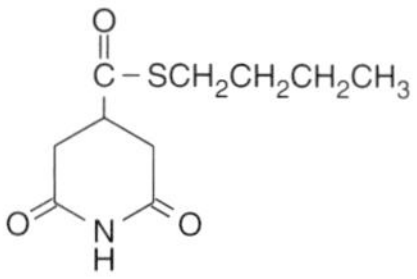

250 mg

S74,933-8
3-SUCCINIMIDOPROPIONIC ACID

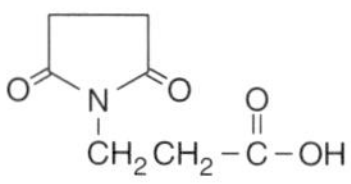

250 mg

S77,608-4
N-TERT-BOC-L-ISOLEUCINE N-SUCCINIMIDE ESTER

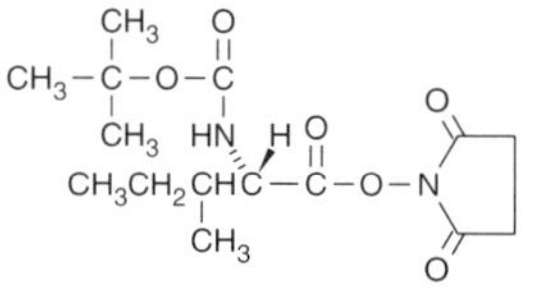

250 mg

S78,271-8
N-(TERT-BUTOXYCARBONYL)-L-PROLINE N-HYDROXYSUCCINIMIDE ESTER

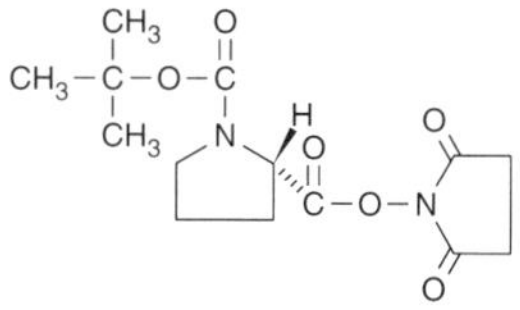

5 gm

S78,450-8
TERT-BUTOXYCARBONYL-L-VALINE N-HYDROXYSUCCINIMIDE ESTER

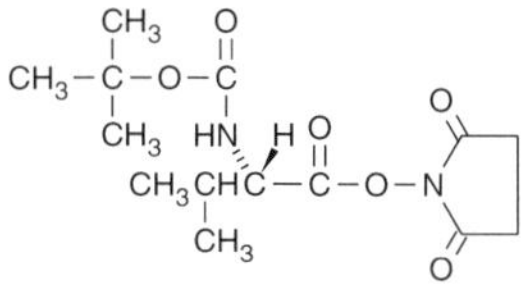

250 mg

S36,177-1
N-(ETHYLMERCURI)-1,4,5,6,7,7-HEXA-CHLOROBICYCLO[2.2.1]HEPT-5-ENE-2,3-DICARBOXIMIDE

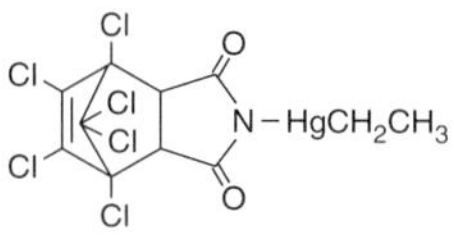

5 gm

S35,589-5
3,3-TETRAMETHYLENEGLUTARIMIDE

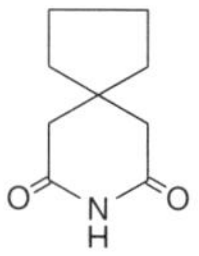

5 gm

S55,161-9
N-CYCLOHEXYLSUCCINIMIDE

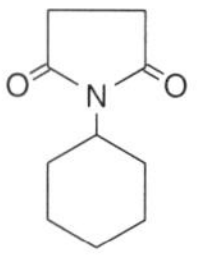

5 gm

S35,427-9
3,3-PENTAMETHYLENEGLUTARIMIDE

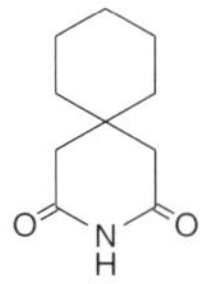

5 gm

S52,550-2
N-BROMOGLUTARIMIDE

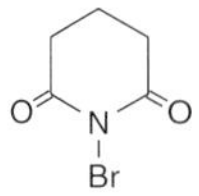

250 mg

S95,802-6
10-OXA-4-AZATRICYCLO[5.2.1.0(2,6)]-DECANE-3,5-DIONE

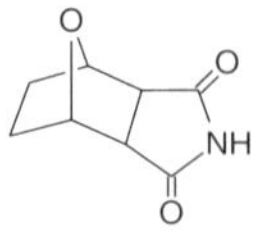

50 mg

S40,579-5
1,4,5,6,7,7-HEXACHLOROBICYCLO-[2.2.1]-5-HEPTENE-2,3-DICARBOXIMIDE

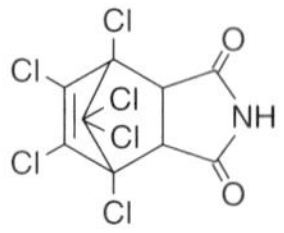

5 gm

S78,119-3
2,2'-[IMINOBIS(CARBOXAMIDO)]-DIETHANOL BIS(2-CHLOROACETATE)

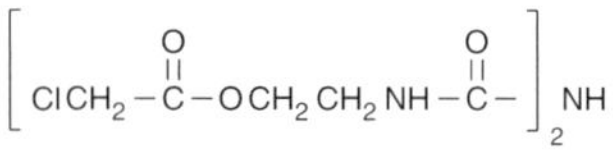

250 mg

S75,321-1
1,5-BIS(2-CHLOROETHYL)BIURET

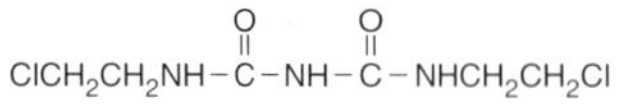

5 gm

S53,931-7
1,3-BIS(1-HYDROXY-2,2,2-TRICHLOROETHYL)UREA

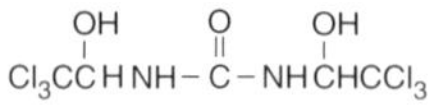

250 mg

S54,792-1
1,1-BIS(2-HYDROXYETHYL)UREA

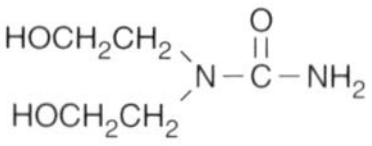

250 mg

S37,348-6
GLUCOSE UREIDE

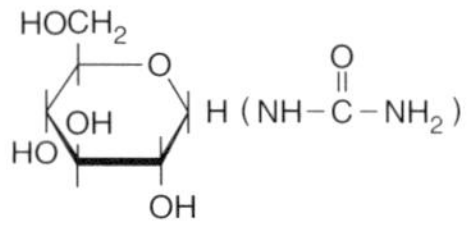

5 gm

S79,294-2
5-HYDROXY-4-METHYL-4-CYCLOPENTENE-1,3-DIONE 1-SEMICARBAZONE

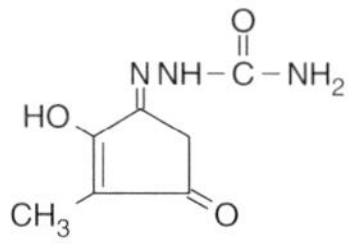

250 mg

S42,011-5
1-GLUCOSYL-3-(HYDROXYMETHYL)UREA

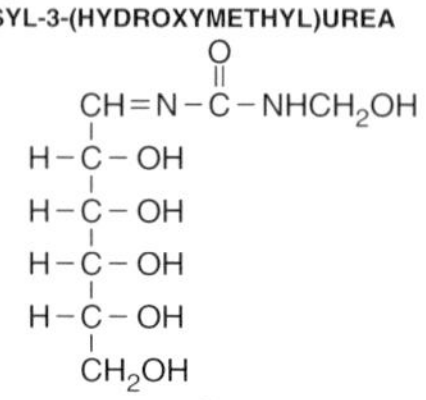

5 gm

S80,473-8
1-ETHYL-3-(3-HYDROXYPROPYL)UREA

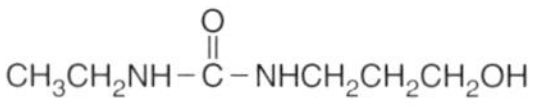

250 mg

S80,474-6
1-ETHYL-3-(2-METHOXYETHYL)UREA

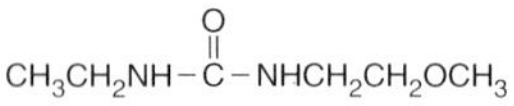

250 mg

S67,908-9
L-CITRULLINE MONOHYDROCHLORIDE

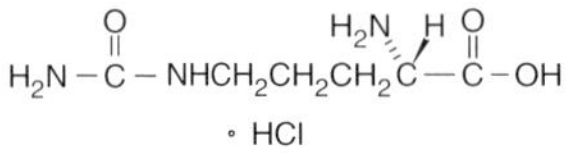

5 gm

S34,809-0
ETHYL 3-AMINO-3-UREIDOBUTYRATE

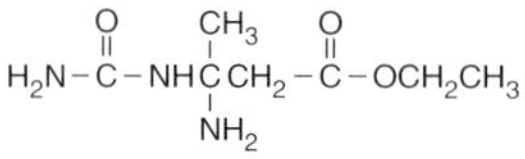

5 gm

S78,665-9
N6-CARBAMOYL-DL-LYSINE

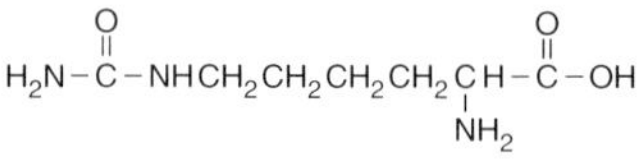

250 mg

S90,041-9
1-CYCLOHEXYL-3-(2-PYRROLIDINOETHYL)UREA

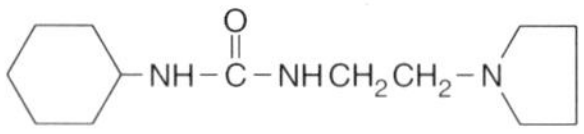

250 mg

S91,342-1
1-CYCLOHEXYL-3-(1-ETHYL-3-PIPERIDINYL)UREA

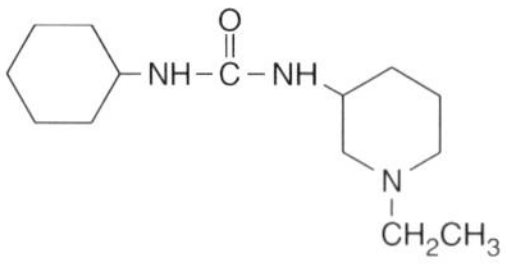

250 mg

S90,011-7
1-ETHYL-3-(2-MORPHOLINOETHYL)UREA

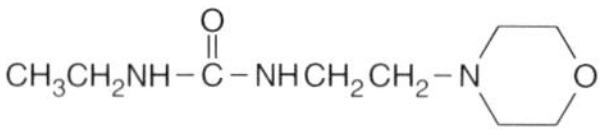

250 mg

S90,038-9
1,1'-HEXAMETHYLENEBIS[3-(2-MORPHOLINOETHYL)UREA]

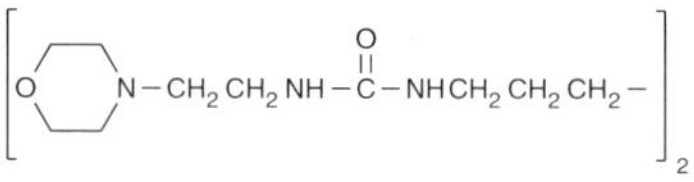

250 mg

S37,569-1
N-METHYL-2-UREIDOACETAMIDE

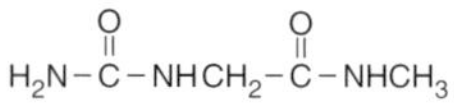

250 mg

S34,790-6
1,2-DIUREIDOETHANE

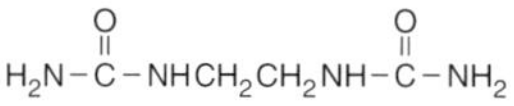

5 gm

S86,469-2
1,2-DIMETHYLENEBIS(3-METHYL-1-UREA)

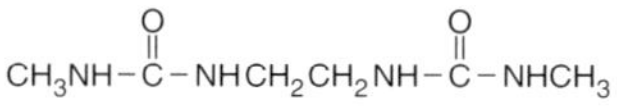

250 mg

S69,457-6
1-ETHYL-3-(3-DIMETHYLAMINOPROPYL)UREA

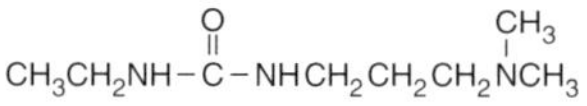

5 gm

S37,472-5
ETHYL HYDANTOATE

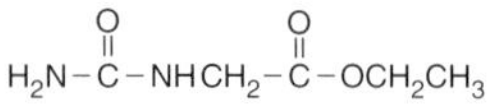

5 gm

S37,504-7
CYCLOHEXYLUREA

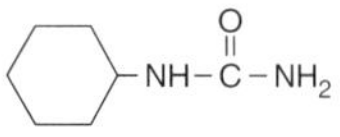

5 gm

S36,048-1
1-(N,N-DIETHYLCARBAMOYL)-4-METHYLPIPERAZINE CITRATE

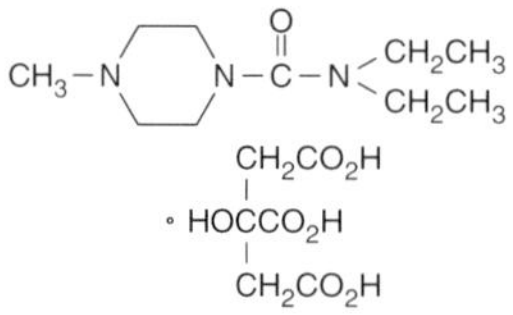

5 gm

S45,423-0
N-ISOPROPYL-4-METHYL-1-PIPERAZINECARBOXAMIDE

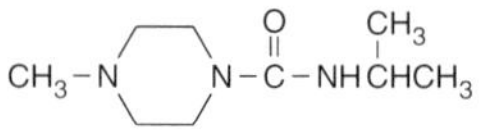

5 gm

S79,559-3
N-TERT-BUTYL-4-METHYL-1-PIPERAZINECARBOXAMIDE

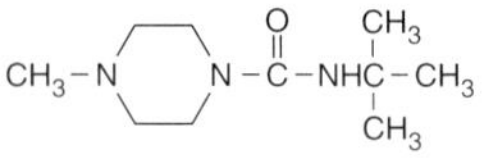

250 mg

S44,978-4
ACETONE SEMICARBAZONE

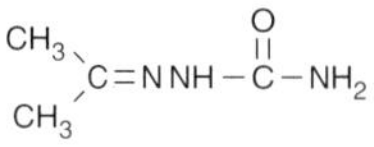

5 gm

S53,474-9
4,4'-BISEMICARBAZIDE

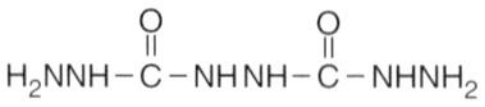

5 gm

S69,851-2
4-PROPYL-1-(10-UNDECENOYL)-SEMICARBAZIDE

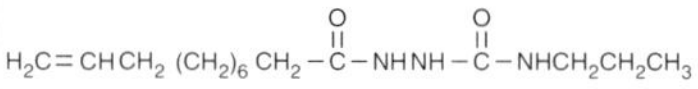

250 mg

S93,815-7
1-(3-METHYL-1-ADAMANTYLCARBONYL)-SEMICARBAZIDE

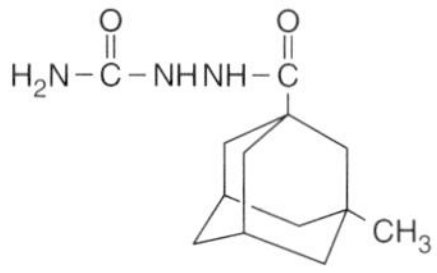

250 mg

S93,817-3
1-(3-METHYL-1-ADAMANTYLACETYL)-SEMICARBAZIDE

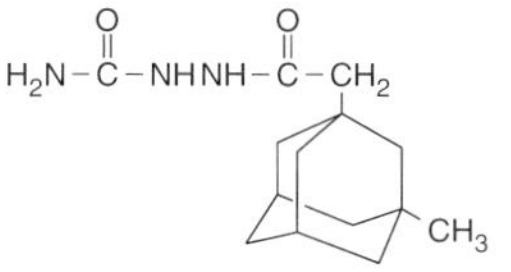

250 mg

S93,820-3
1-(3,5-DIMETHYL-1-ADAMANTYLACETYL)-SEMICARBAZIDE

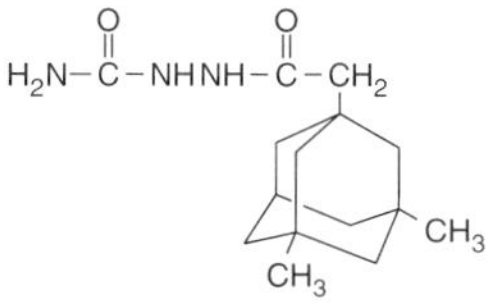

250 mg

S37,146-7
BIUREA

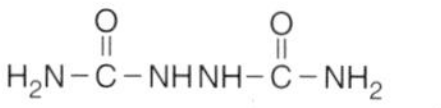

5 gm

S35,196-2
LEVULINIC ACID SEMICARBAZONE

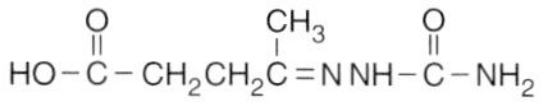

5 gm

S67,601-2
1-(4-METHYL-2-PENTYLIDENE)-SEMICARBAZIDE

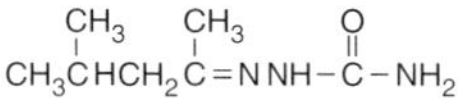

5 gm

S69,247-6
CYCLODODECANONE SEMICARBAZONE

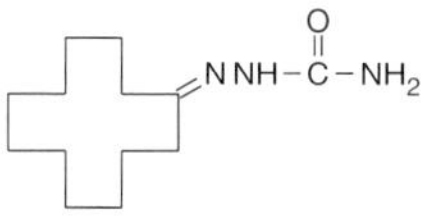

250 mg

S88,559-2
8A-METHYL-2-DECALONE SEMICARBAZONE

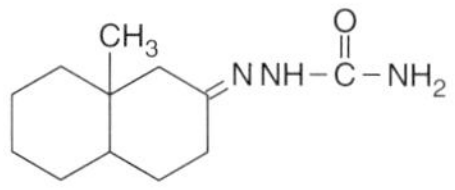

250 mg

S94,756-3
BICYCLO[3.2.0]HEPT-2-EN-6-ONE SEMICARBAZONE

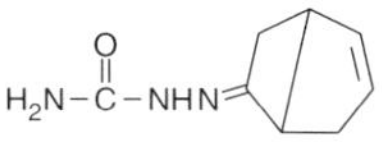

250 mg

S90,229-2
ETHYL 2-(3,3-DIALLYLUREIDO)ACETATE

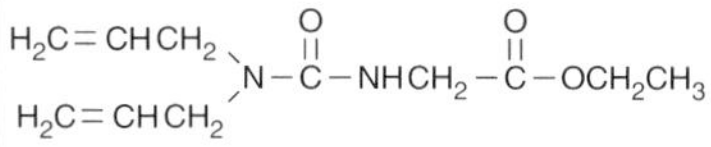

250 mg

S60,446-1
4-[N-(CYCLOHEXYL)CARBAMOYL]-MORPHOLINE

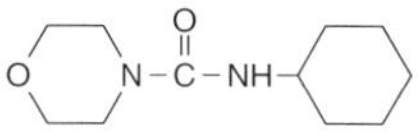

5 gm

S60,788-6
4-(BUTYLCARBAMOYL)MORPHOLINE

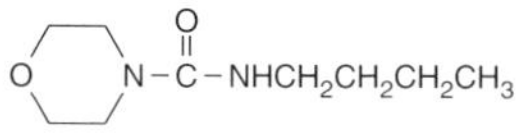

5 gm

S43,624-0
1-NITROBIURET

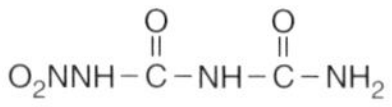

5 gm

S34,967-4
ETHYL 2-(UREIDOMETHYLENE)-ACETOACETATE

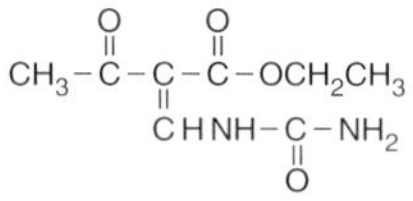

5 gm

S35,685-9
3-UREIDOMETHYLENE-2,4-PENTANEDIONE

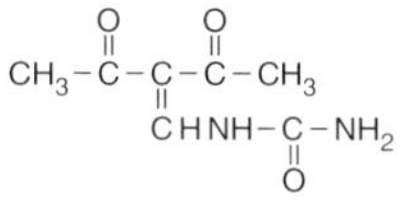

5 gm

S35,299-3
METHYL MALEURATE

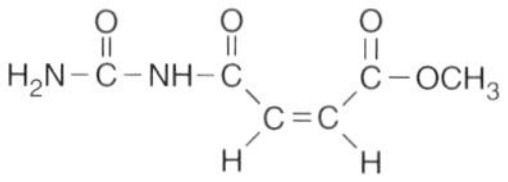

5 gm

S54,779-4
BUTYL N-CARBAMOYLMALEAMATE

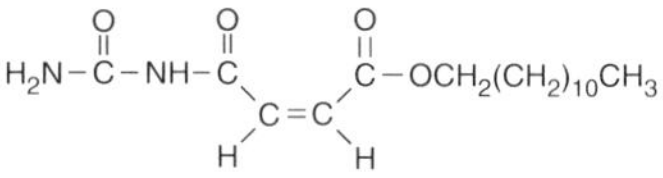

250 mg

S54,779-4
DODECYL N-CARBAMOYLMALEAMATE

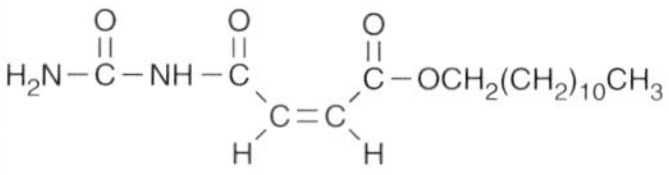

250 mg

S36,158-5
1-(2-ETHYLCROTONOYL)UREA

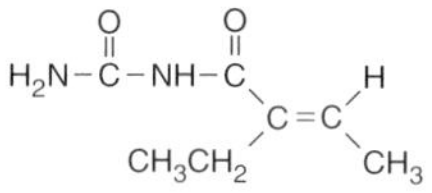

5 gm

S57,689-1
OLEIC ACID UREA ADDUCT

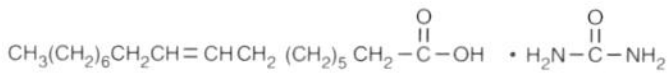

5 gm

S56,759-0
13-DOCOSENOIC ACID UREA ADDUCT

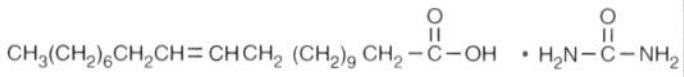

5 gm

S40,677-5
1,3-BIS(5-CARBOXYPENTYL)UREA

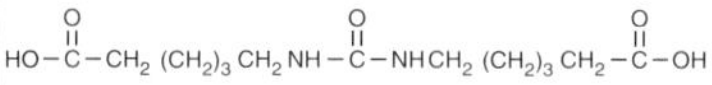

5 gm

S45,873-2
HYDANTOIC ACID

5 gm

S34,214-9
1,3-BIS[5-(CARBOMETHOXY)PENTYL]UREA

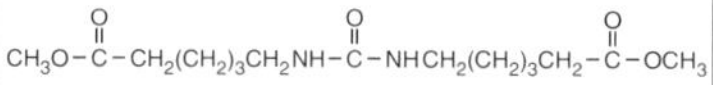

5 gm

S91,257-3
1-(2-OXO-3-OXAZOLIDINYL)-3-PROPYLUREA

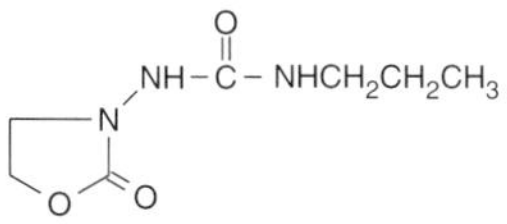

250 mg

S87,617-8
METHYL N-CARBAMOYLCARBAMATE

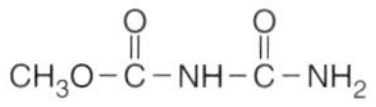

250 mg

S85,957-5
ETHYL 3-UREIDOPROPIONATE

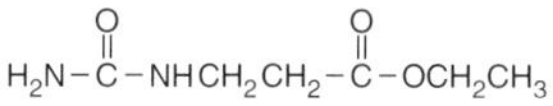

250 mg

S65,554-6
3-CYCLOHEXYLUREIDOMETHYL PROPIONATE

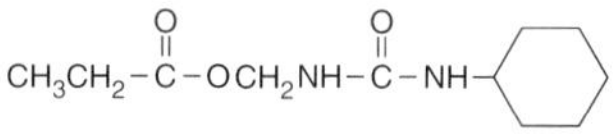

5 gm

S54,789-1
TRIURET

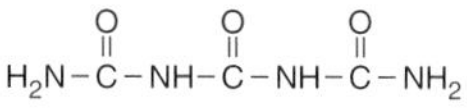

250 mg

S34,490-7
CYCLOHEPTYLUREA

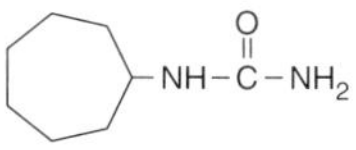

5 gm

S34,508-3
CYCLOOCTYLUREA

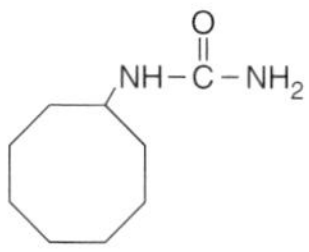

5 gm

S57,778-2
1,6-DIETHYLBIUREA

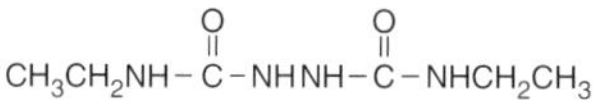

5 gm

S60,754-1
4-BUTYL-1,1-DIMETHYLSEMICARBAZIDE
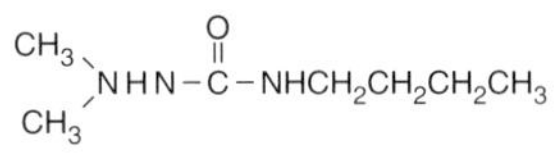
5 gm

S57,441-4
ALLOPHANIC ACID HYDRAZIDE
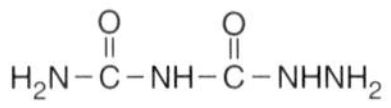
250 mg

S91,339-1
1-ACETYL-4-CYCLOHEXYLSEMICARBAZIDE
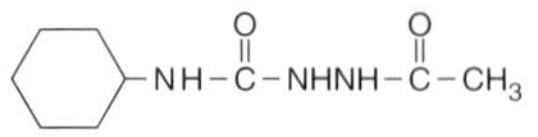
250 mg

S67,937-2
1,3-DIPIPERIDINOUREA
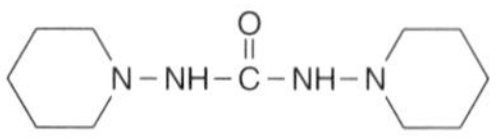
5 gm

S61,387-8
CYCLOHEXANONE SEMICARBAZONE
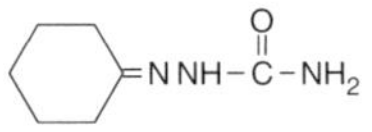
5 gm

S60,746-0
4-CYCLOHEXYL-1,1-DIMETHYLSEMICARBAZIDE
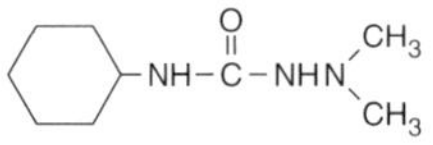
5 gm

S37,257-9
FORMYLUREA
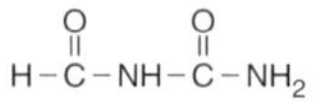
5 gm

S41,186-8
1-ACETYL-3-METHYLUREA
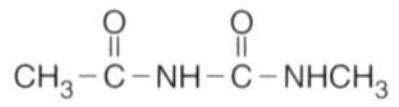
5 gm

S82,106-3
1-DODECANOYLUREA
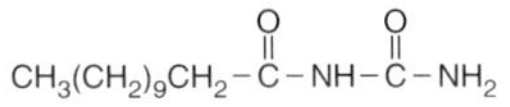
250 mg

S82,107-1
1-TETRADECANOYLUREA
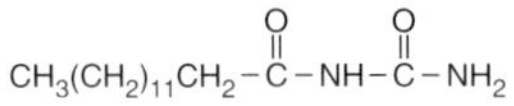
250 mg

S82,112-8
1-HEXADECANOYLUREA
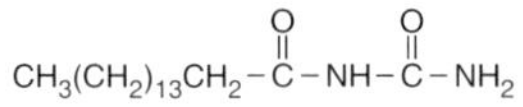
250 mg

S82,110-1
1-OCTADECANOYLUREA
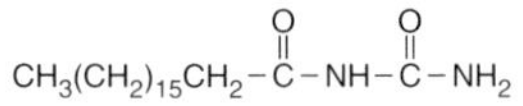
250 mg

S86,168-5
1,1,3,3-TETRABUTYLUREA
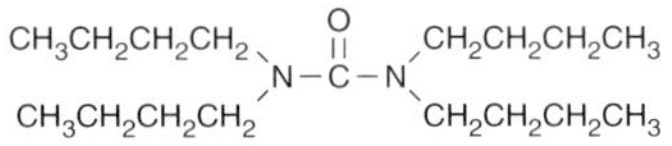
250 mg

S82,529-8
NITROUREA
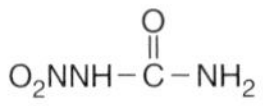
250 mg

S80,299-9
1-METHOXYUREA
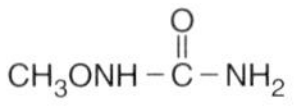
250 mg

S80,306-5
1-ETHOXYUREA
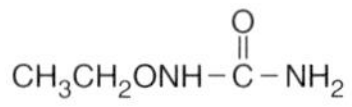
250 mg

S71,375-9
1,1,3-TRIMETHYLUREA
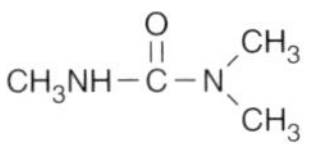
5 gm

S58,618-8
1,1-DIETHYLUREA, SODIUM DERIVATIVE
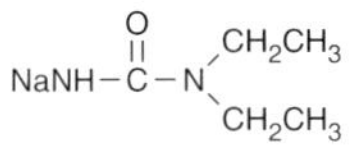
5 gm

S67,235-1
SEC-BUTYLUREA
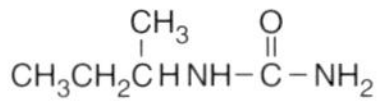
5 gm

S67,238-6
TERT-BUTYLUREA
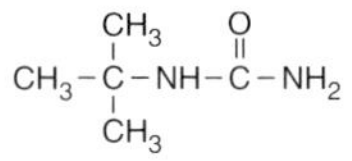
5 gm

S80,310-3
1-ISOBUTYLUREA
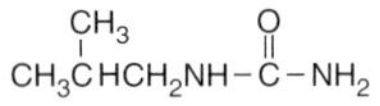
250 mg

S80,467-3
1-ALLYL-3-ETHYLUREA
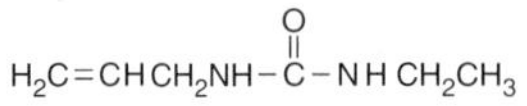
250 mg

S63,176-0
1-(1,1-DIMETHYLPROPYL)UREA
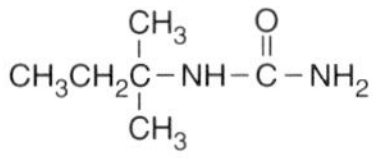
250 mg

S72,475-0
1-(SEC-BUTYL)-3-METHYLUREA
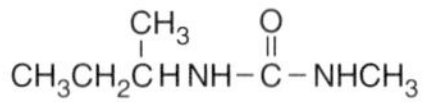
5 gm

S80,471-1
1-ISOBUTYL-3-METHYLUREA
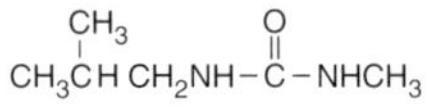
250 mg

S80,585-8
1-ETHYL-3-PROPYLUREA
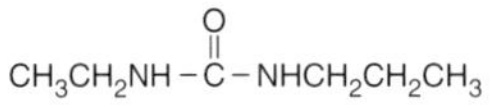
250 mg

S80,579-3
1-ETHYL-3-(2-METHYLALLYL)UREA
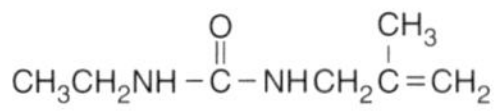
250 mg

S50,255-3
1-(1-ETHYLBUTYL)UREA
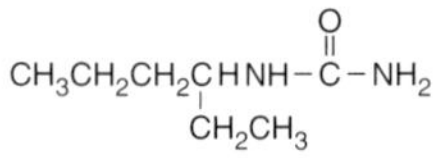
250 mg

S80,207-7
1-BUTYL-3-ETHYLUREA
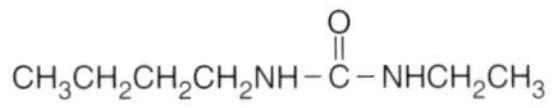
250 mg

S60,736-3
1-ALLYL-3-BUTYLUREA
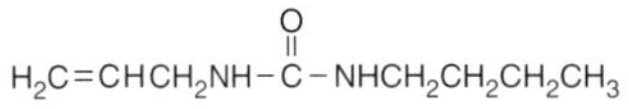
5 gm

S51,664-3
HEPTYLUREA
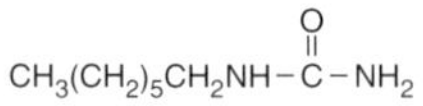
5 gm

S89,279-3
1-HEXYL-3-METHYLUREA
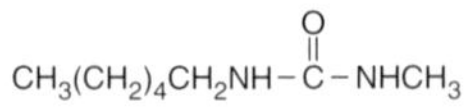
250 mg

S70,886-0
1,1'-HEXAMETHYLENEDIUREA
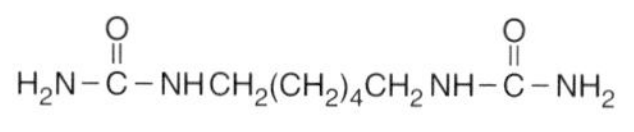
5 gm

S54,657-7
1,3-DIBUTYLUREA
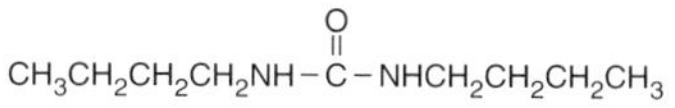
250 mg

S63,633-9
1,1'-HEXAMETHYLENEBIS(3-METHYLUREA)
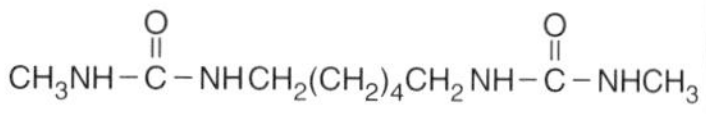
5 gm

S86,976-7
1,3-DIBUTYL-1-ETHYLUREA
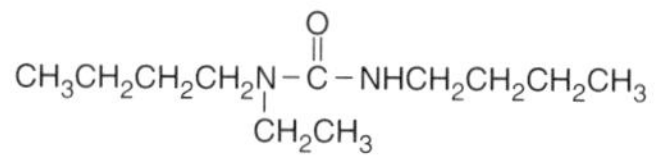
250 mg

S59,035-5
3,3'-HEXAMETHYLENEBIS(1,1-DIMETHYLUREA)
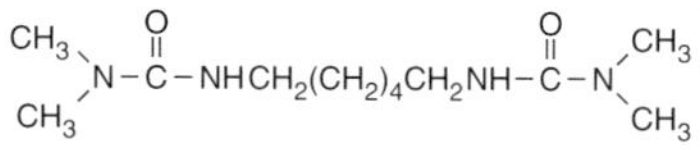
5 gm

S63,214-7
1,1'-HEXAMETHYLENEBIS(3-ETHYLUREA)
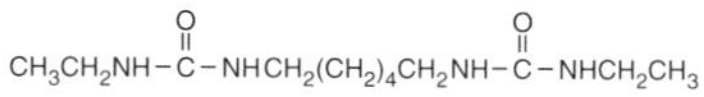
5 gm

S86,185-5
1,1,3-TRIBUTYLUREA
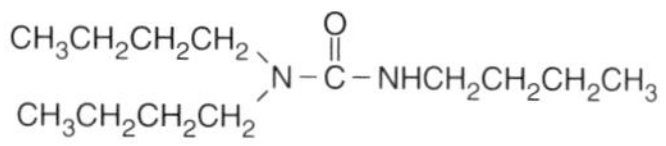
250 mg

S71,132-2
1,1'-HEXAMETHYLENEBIS(3-ALLYLUREA)
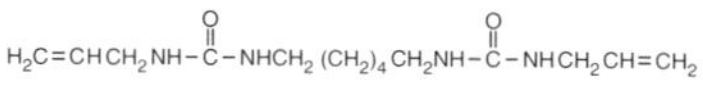
5 gm

S63,200-7
1,1'-HEXAMETHYLENEBIS(3-ISOPROPYLUREA)
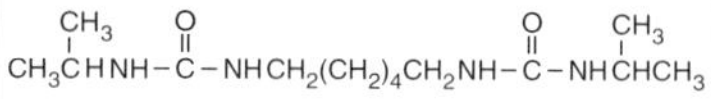
5 gm

S63,223-6
1,1'-HEXAMETHYLENEBIS(3-PROPYLUREA)
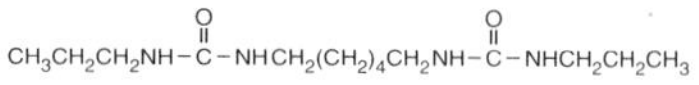
5 gm

S81,186-6
1-TETRADECYLUREA
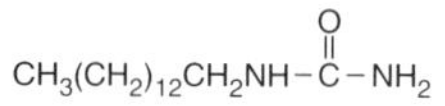
250 mg

S63,212-0
1,1'-HEXAMETHYLENEBIS(3-BUTYLUREA)
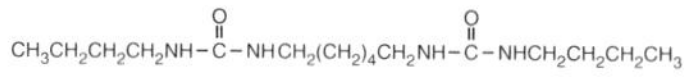
5 gm

S66,740-4
1,1'-HEXAMETHYLENEBIS(3-ISOBUTYLUREA)
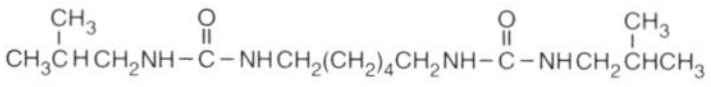
5 gm

S71,100-4
3,3'-HEXAMETHYLENEBIS(1,1-DIETHYLUREA)
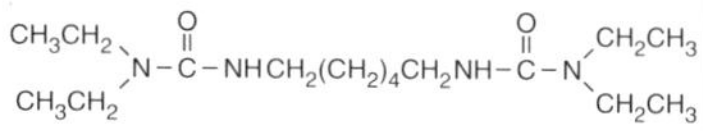
5 gm

S64,722-5
3,3'-HEXAMETHYLENEBIS(1,1-DIALLYLUREA)
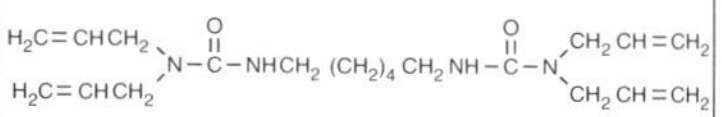
5 gm

S64,713-6
3,3'-HEXAMETHYLENEBIS(1,1-DIISOPROPYLUREA)
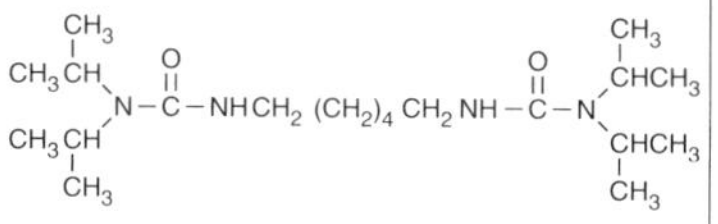
5 gm

S67,483-4
1-(TERT-BUTYL)-3-OCTADECYLUREA
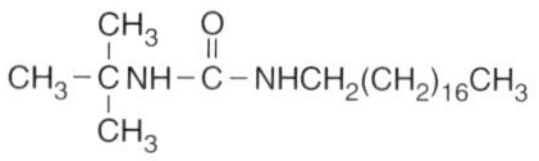
250 mg

S64,723-3
3,3'-HEXAMETHYLENEBIS(1,1-DIBUTYLUREA)
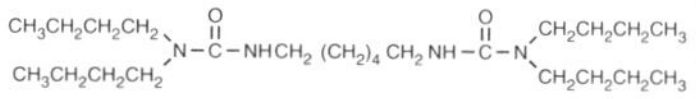
5 gm

S69,464-9
N,N'-HEXAMETHYLENEBIS(2-METHYL-1-AZIRIDINECARBOXAMIDE)
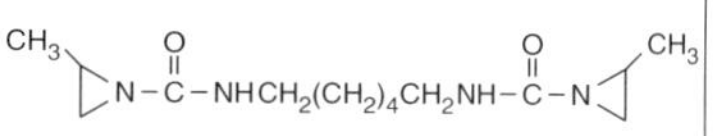
250 mg

S34,514-8
CYCLOPENTYLUREA
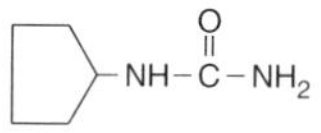
5 gm

S34,512-1
1-CYCLOPENTYL-1-METHYLUREA
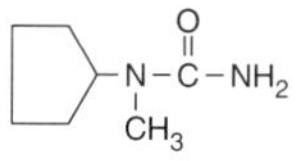
5 gm

S60,456-9
1-CYCLOHEXYL-3-METHYLUREA
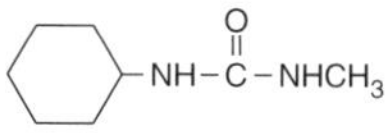
5 gm

S60,356-2
1-CYCLOHEXYL-3-ETHYLUREA
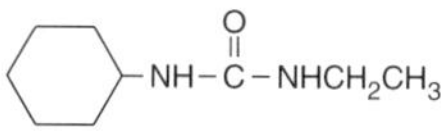
5 gm

S60,454-2
3-CYCLOHEXYL-1,1-DIMETHYLUREA
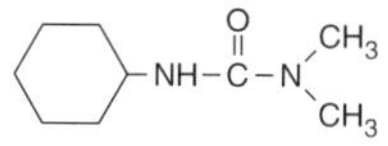
5 gm

S60,345-7
1-ALLYL-3-CYCLOHEXYLUREA
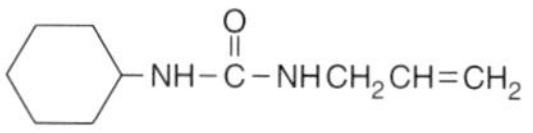
5 gm

S62,428-4
1-CYCLOHEXYL-3-ISOPROPYLUREA
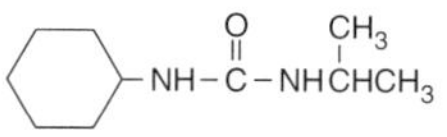
5 gm

S62,429-2
1-CYCLOHEXYL-3-PROPYLUREA
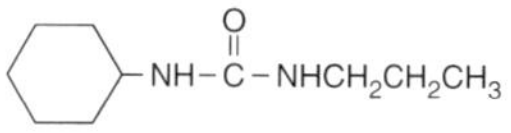
5 gm

S60,340-6
3-CYCLOHEXYL-1,1-DIETHYLUREA
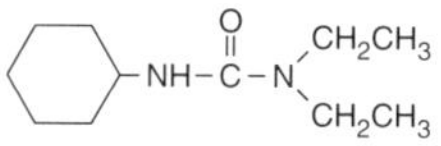
5 gm

S62,431-4
1-BUTYL-3-CYCLOHEXYLUREA
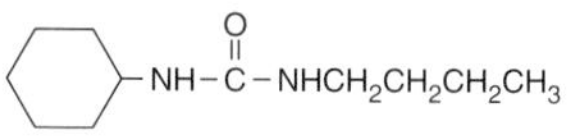
5 gm

S66,069-8
1-CYCLOHEXYL-3-ISOBUTYLUREA
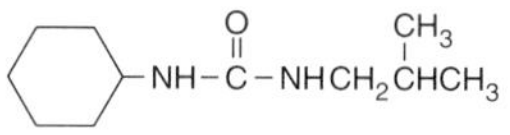
5 gm

S72,163-8
1-TERT-BUTYL-3-CYCLOHEXYLUREA
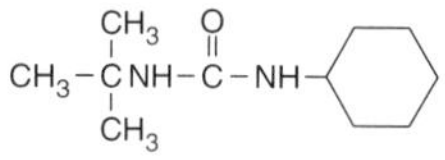
250 mg

S89,945-3
3-TERT-BUTYL-1-CYCLOHEXYL-1-METHYLUREA
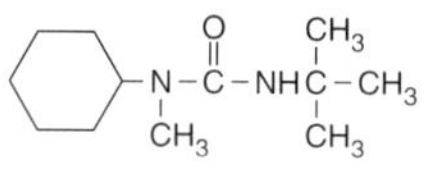
250 mg

S60,461-5
3-CYCLOHEXYL-1,1-DIALLYLUREA

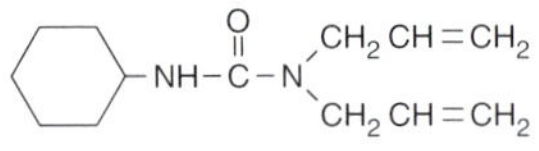

5 gm

S62,432-2
3-CYCLOHEXYL-1,1-DIPROPYLUREA

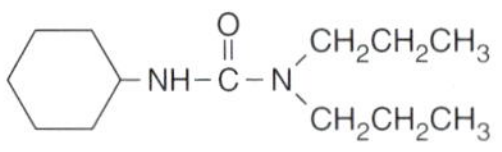

5 gm

S62,433-0
3-CYCLOHEXYL-1,1-DIISOPROPYLUREA

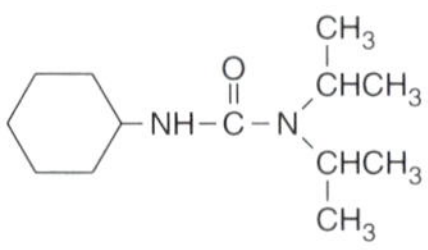

5 gm

S89,949-6
3-TERT-BUTYL-1-CYCLOHEXYL-1-ETHYLUREA

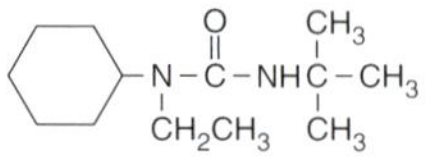

250 mg

S60,645-6
1,3-DICYCLOHEXYL-1-METHYLUREA

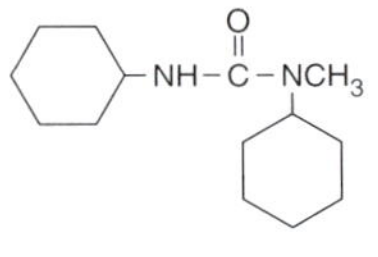

5 gm

S89,278-5
1,1-DICYCLOHEXYL-3-METHYLUREA

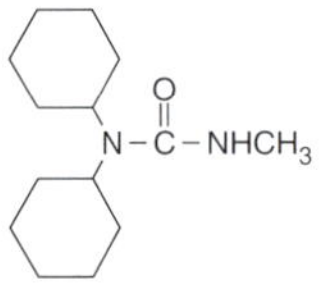

250 mg

S89,942-9
1,3-DICYCLOHEXYL-1-ETHYLUREA

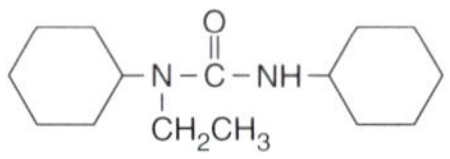

250 mg

S66,060-4
3-CYCLOHEXYL-1,1-DIISOBUTYLUREA

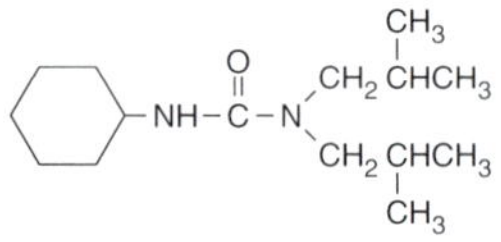

5 gm

S66,547-9
3-CYCLOHEXYL-1,1-DIBUTYLUREA

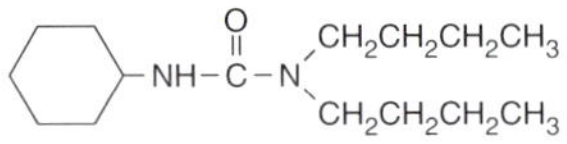

5 gm

S72,112-3
3-BUTYL-1,1-DICYCLOHEXYLUREA

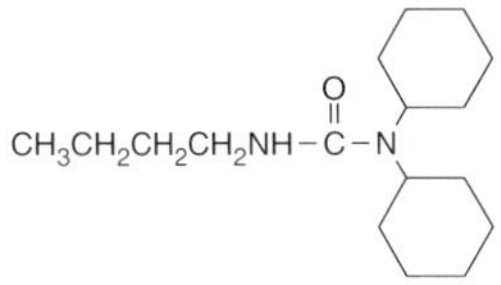

250 mg

S61,079-8
1,1,3-TRICYCLOHEXYLUREA

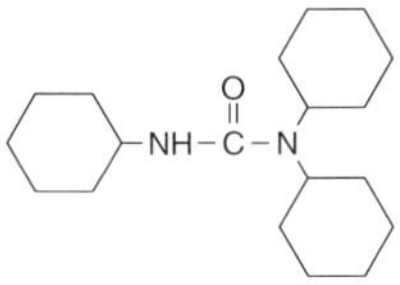

5 gm

S72,511-0
1-CYCLOHEXYL-1-METHYL-3-OCTADECYLUREA

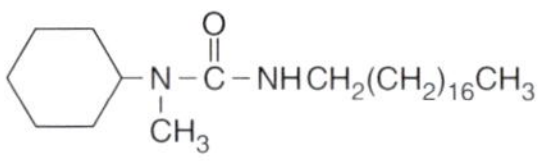

250 mg

S60,359-7
1-(CYCLOHEXYLCARBAMOYL)PIPERIDINE

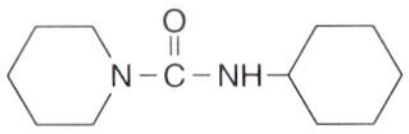

5 gm

S66,049-3
4-[N-(CYCLOHEXYL)CARBAMOYL]-2,6-DIMETHYLMORPHOLINE

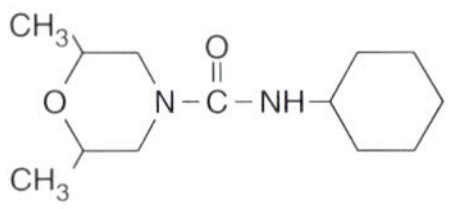

5 gm

S37,955-7
N-METHYL-1-PIPERIDINECARBOXAMIDE

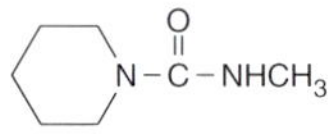

250 mg

S58,622-6
N,N'-HEXAMETHYLENEBIS(1-PIPERIDINECARBOXAMIDE)

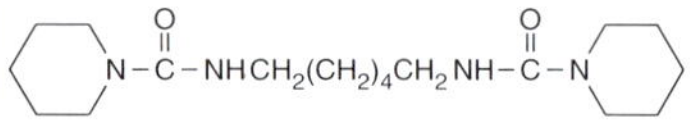

5 gm

S37,897-6
4-MORPHOLINECARBOXAMIDE

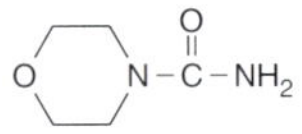

5 gm

S94,681-8
1-(1-ADAMANTYL)-3-ETHYLUREA

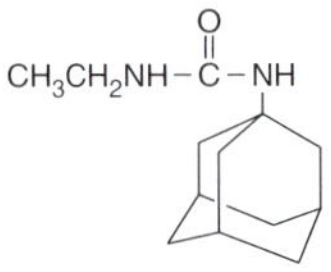

250 mg

S94,850-0
1-(1-ADAMANTYLMETHYL)-3-ETHYLUREA

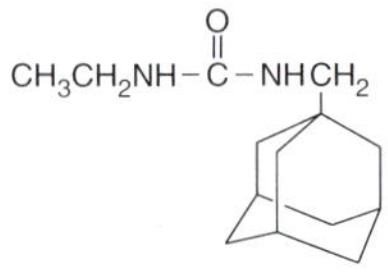

250 mg

S94,853-5
1-(1-ADAMANTYL)-3-PROPYLUREA

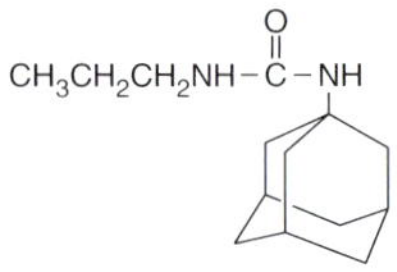

250 mg

S94,854-3
1-(1-ADAMANTYL)-3-ISOPROPYLUREA

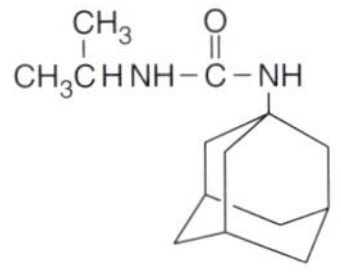

250 mg

S94,682-6
1-(1-ADAMANTYLMETHYL)-3-ISOPROPYLUREA

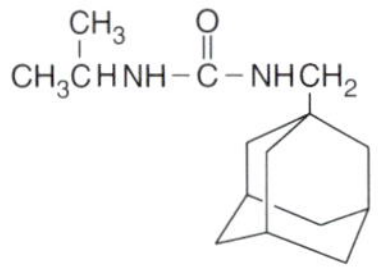

250 mg

S94,852-7
1-(1-ADAMANTYL)-3-BUTYLUREA

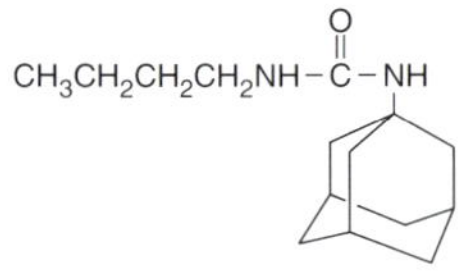

250 mg

S94,851-9
1-(1-ADAMANTYLMETHYL)-3-BUTYLUREA

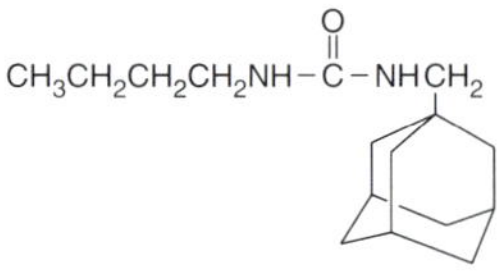

250 mg

S51,704-6
1-(CYCLOHEXYLMETHYL)UREA

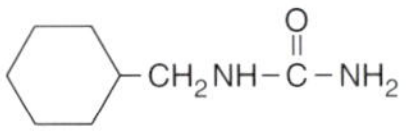

5 gm

S91,385-5
1-ALLYL-6-(CHLOROMETHYLAMINO)URACIL

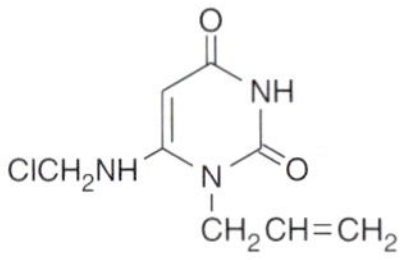

250 mg

S55,441-3
3-AMINO-1-(2,2,3,3-TETRAFLUOROPROPYL)-2-PYRAZOLIN-5-ONE

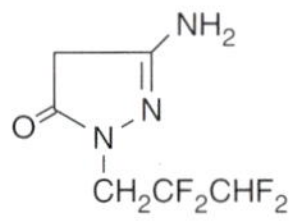

5 gm

S61,255-3
5-(DIBROMOMETHYL)-5-ETHYLHYDANTOIN

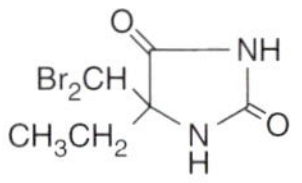

5 gm

S64,240-1
3-(BUTYLAMINO)-1-(2,2,2-TRIFLUOROETHYL)-2-PYRAZOLIN-5-ONE

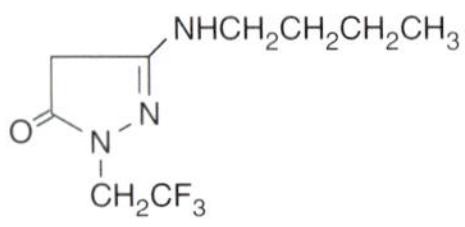

250 mg

S64,243-6
3-(ACRYLAMIDO)-1-(2,2,2-TRIFLUOROETHYL)-2-PYRAZOLIN-5-ONE

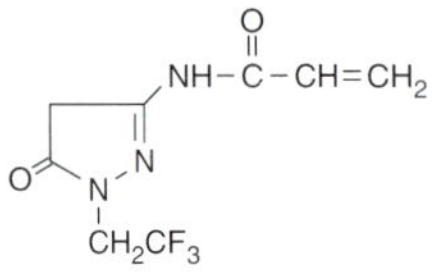

250 mg

S78,881-3
3-METHYLCYTIDINE METHOSULFATE

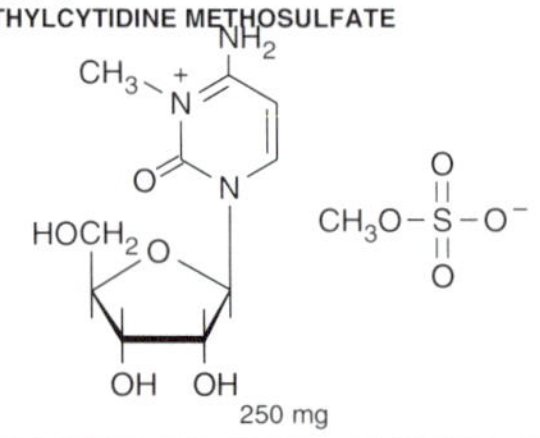

250 mg

S78,175-4
5-(HYDROXYMETHYL)URIDINE

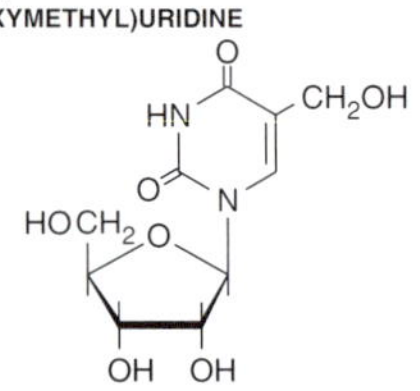

250 mg

S86,051-4
5'-O-ACETYLURIDINE
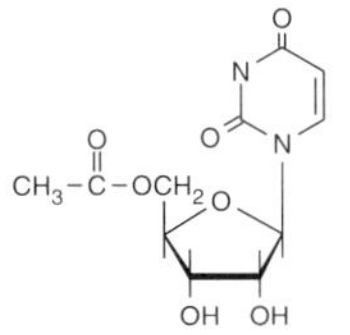
250 mg

S88,849-4
4,4-BIS(HYDROXYMETHYL)-2-OXAZOLIDINONE
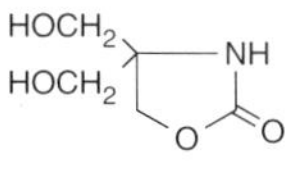
250 mg

S85,936-2
1,3-BIS(2-HYDROXYETHYL)URACIL
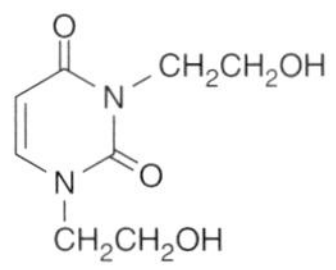
250 mg

S43,221-0
L-4-DIMETHYLAMINO-1,2-DIPHENYL-3-METHYL-2-BUTANOL HYDROCHLORIDE
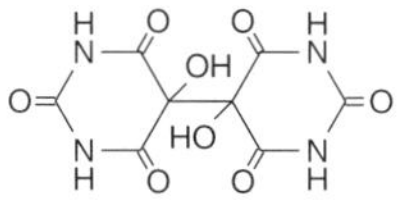
250 mg

S86,058-1
METHYL 4-HYDROXY-2-(PYRROLIDINOCARBONYL)-CYCLOPENTANECARBOXYLATE
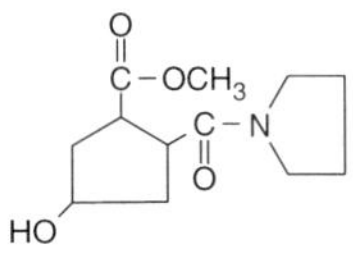
250 mg

S78,869-4
4,4'-DITHIOBIS[1-BETA-D-RIBOFURANOSYL-2(1H)-PYRIMIDINONE]
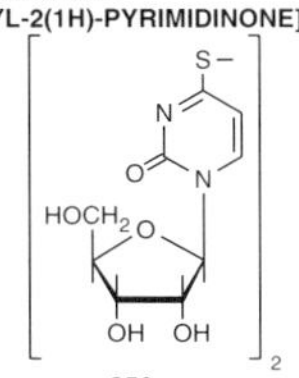
250 mg

S44,924-5
2',3'-ISOPROPYLIDENECYTIDINE HYDROCHLORIDE
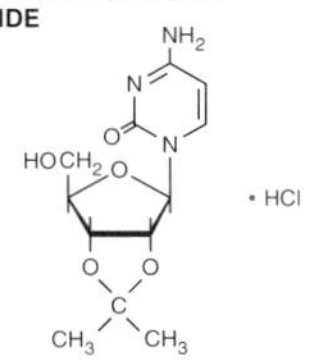
250 mg

S91,863-6
3',5'-DI-O-ACETYLURIDINE
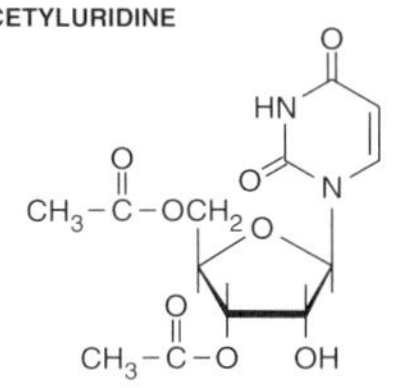
250 mg

S45,499-0
THYMIDINE-5'-MONOPHOSPHATE, DISODIUM SALT DIHYDRATE
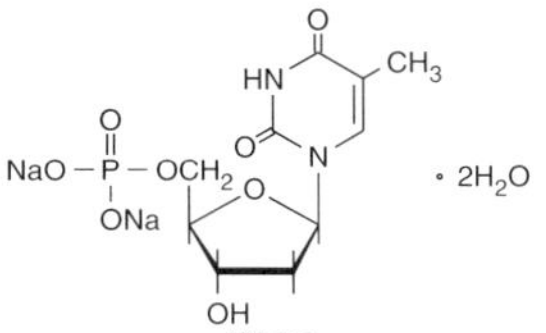
250 mg

S80,578-5
6-AMINO-3-ETHYL-1-(2-HYDROXYETHYL)-5-NITROSOURACIL
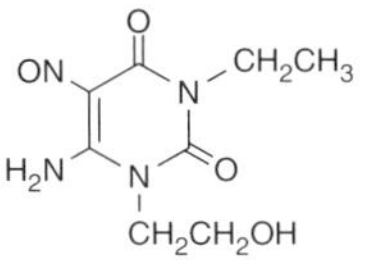
250 mg

S80,361-8
6-AMINO-3-ETHYL-1-(2-HYDROXY-PROPYL)5-NITROSOURACIL
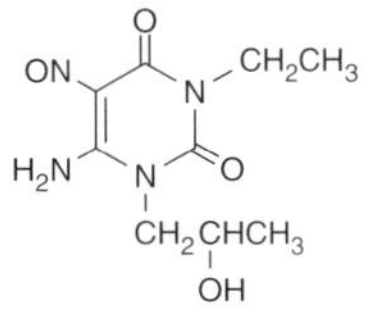
250 mg

S80,362-6
6-AMINO-3-ETHYL-1-(3-HYDROXY-PROPYL)5-NITROSOURACIL
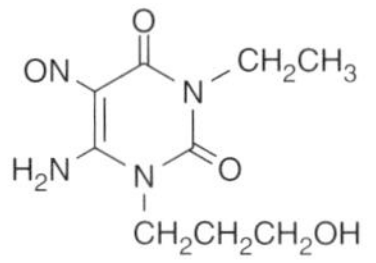
250 mg

S80,459-2
5,6-DIAMINO-3-ETHYL-1-(2-HYDROXYETHYL)URACIL
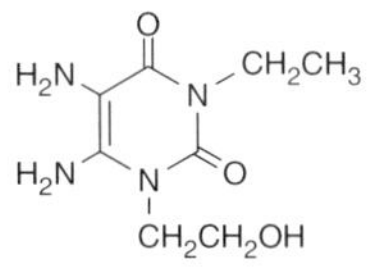
250 mg

S80,380-4
1,3-DIETHYL-6-(2-HYDROXYETHYL-AMINO)URACIL
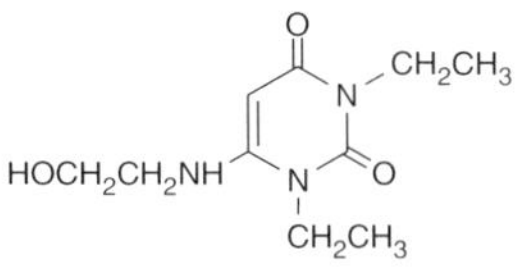
250 mg

S52,104-3
1,2,4-TRIS(2-HYDROXYETHYL)-1,2,4-TRIAZINE-3,5,6(1H,2H,4H)-TRIONE
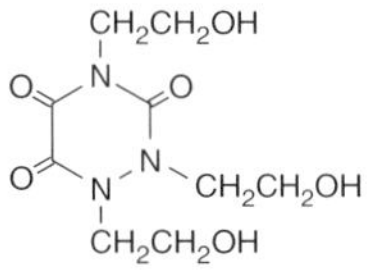
5 gm

S75,622-9
4-HYDROXYMETHYL-4-METHYL-2-OXAZOLIDINONE
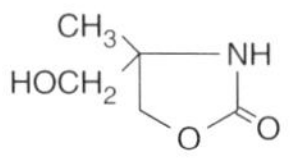
5 gm

S88,096-5
5-HYDROXY-5-METHYLHYDANTOIN
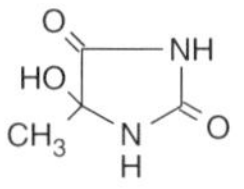
250 mg

S36,105-4
5,5-DIMETHYL-1-HYDROXYMETHYLHYDANTOIN
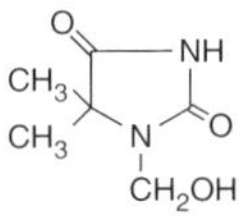
5 gm

S46,210-1
5-HYDROXYMETHYL-6-METHYLURACIL
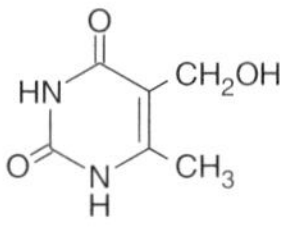
250 mg

S95,470-5
6-HYDROXY-5,5,6-TRIMETHYL-2-PIPERIDONE
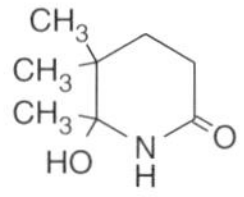
50 mg

S57,138-5
5-BROMO-5,6-DIHYDRO-5-METHYLURACIL
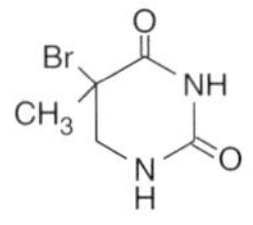
5 gm

S44,783-8
3-METHOXYCARBONYLMETHOXY-1-METHOXY-CARBONYLMETHYL-2(1H)-PYRIDONE
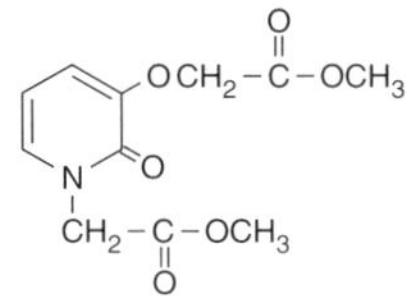
5 gm

S80,219-0
6-AMINO-3-ETHYL-1-(3-METHOXYPROPYL)-5-NITROSOURACIL
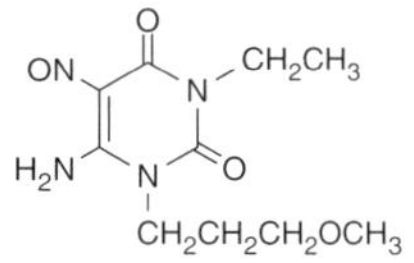
250 mg

S80,454-1
6-AMINO-3-ETHYL-1-(2-METHOXYETHYL)-URACIL
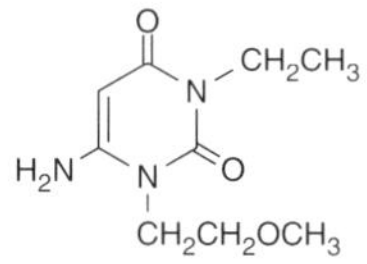
250 mg

S45,173-8
2',3',5'-TRIACETYLURIDINE
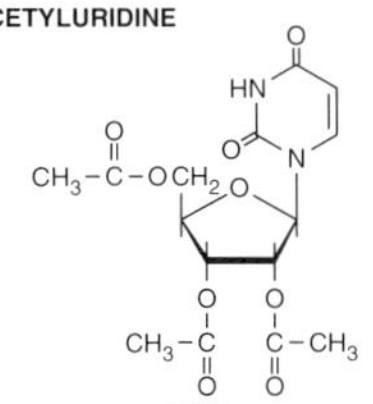
5 gm

S94,882-9
5-ACETYLBARBITURIC ACID
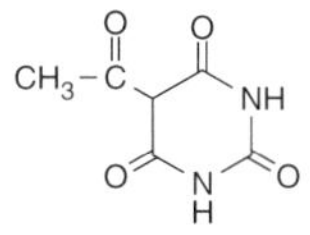
250 mg

S95,475-6
METHYL 1,2-DIMETHYL-5-FORMYL-6-OXO-1,4,5,6-TETRAHYDRO-3-PYRIDINECARBOXYLATE
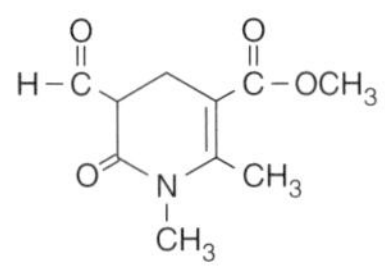
50 mg

S94,164-6
(RS)-HYDROOROTIC ACID
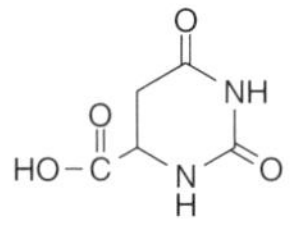
250 mg

S45,170-3
5'-ACETYL-2',3'-ISOPROPYLIDENEURIDINE
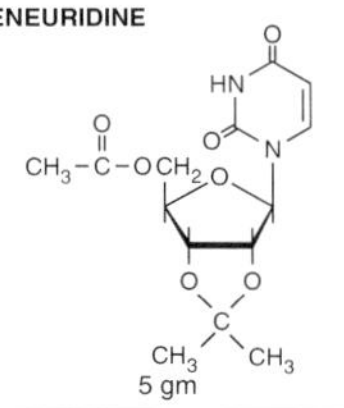
5 gm

S77,714-5
2',3'-ISOPROPYLIDENEURIDINE
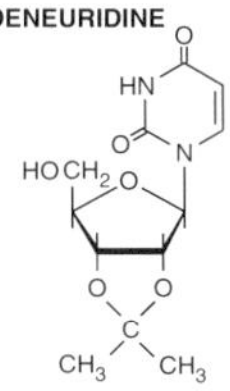
250 mg

S78,892-9
4-AMINO-5-(FORMYLAMINO)-1-METHYL-2-(METHYLTHIO)-6(1H)-PYRIMIDINONE
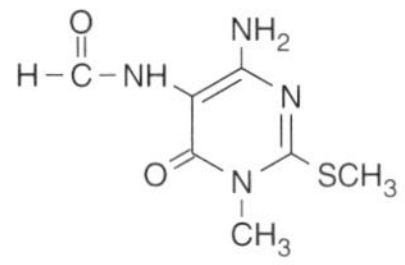
250 mg

S93,172-1
6-AMINO-5-(METHYLAMINO)-1-METHYLURACIL
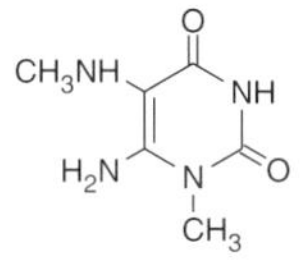
250 mg

S92,838-0
6-AMINO-N1,N3-DIMETHYL-5-(METHYLAMINO)URACIL
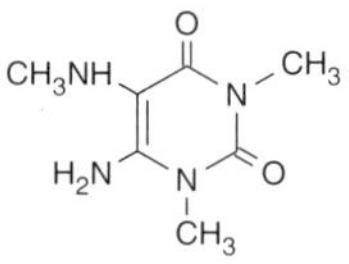
250 mg

S80,417-7
6-AMINO-1-CYCLOPROPYL-3-ETHYL-5-NITROSOURACIL
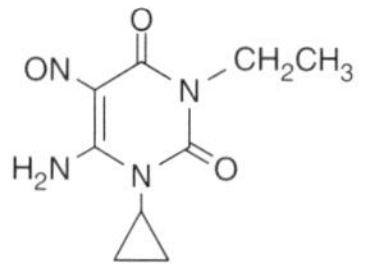
250 mg

S80,414-2
6-AMINO-1-ETHYL-5-NITROSOURACIL
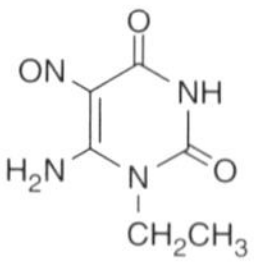
250 mg

S80,355-3
1-ALLYL-6-AMINO-3-ETHYL-5-NITROSOURACIL
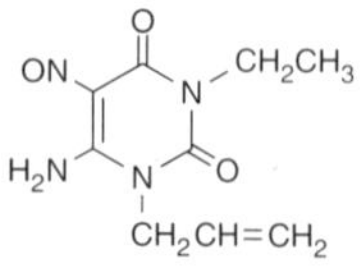
250 mg

S80,419-3
6-AMINO-1-ISOBUTYL-3-METHYL-5-NITROSOURACIL
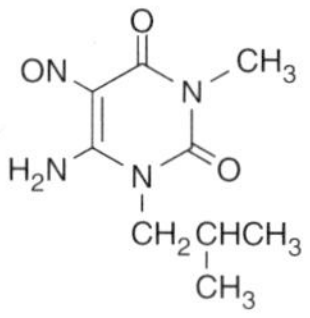
250 mg

S80,367-7
6-AMINO-1,3-DIALLYL-5-NITROSOURACIL
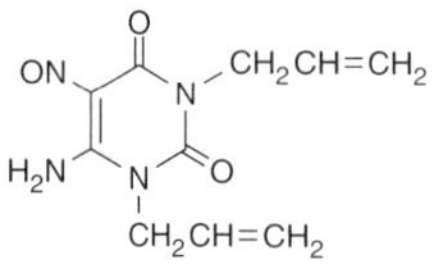
250 mg

S80,420-7
6-AMINO-3-ETHYL-1-ISOBUTYL-5-NITROSOURACIL
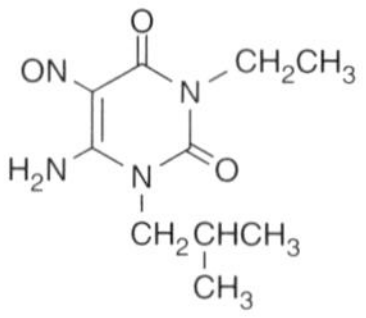
250 mg

S80,396-0
6-AMINO-1,3-DIBUTYL-5-NITROSOURACIL
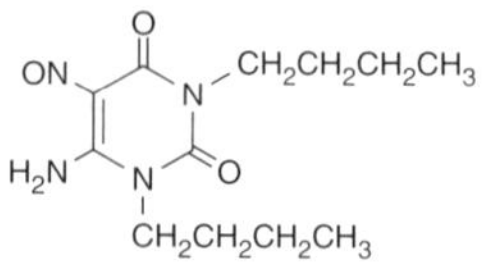
250 mg

S80,339-1
1-ALLYL-6-AMINO-3-METHYL-5-NITROSOURACIL
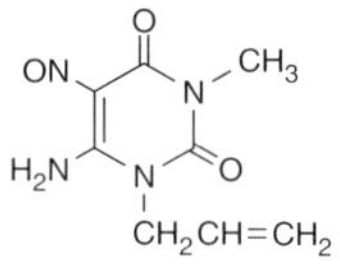
250 mg

S80,284-0
ETHYL 6-AMINO-1-ETHYL-5-NITRO-3-URACILACETATE
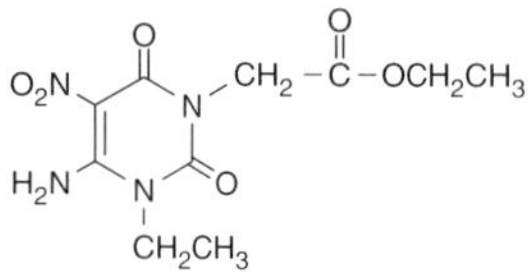
250 mg

S80,203-4
6-AMINO-1-ETHYL-5-NITROURACIL
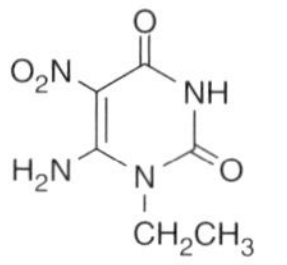
250 mg

S80,243-3
5,6-DIAMINO-1,3-DIETHYLURACIL
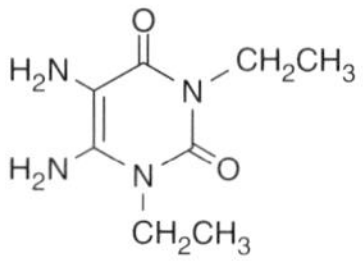
250 mg

S80,325-1
6-AMINO-1-ETHYL-3-METHYL-5-NITROSOURACIL
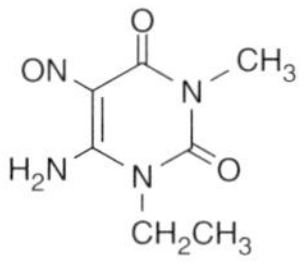
250 mg

S80,394-4
6-AMINO-3-ETHYL-1-(2-METHYLALLYL)-5-NITROSOURACIL
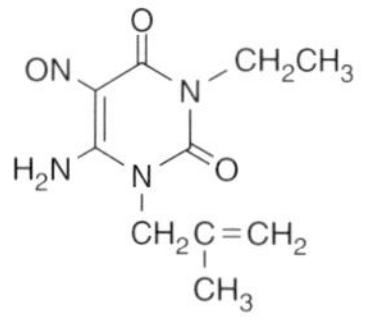
250 mg

S80,476-2
5-AMINO-3-BUTYL-1-ETHYL-6-NITROSOURACIL
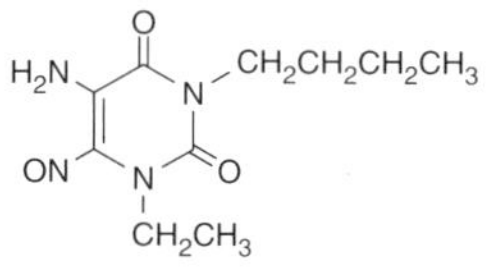
250 mg

S43,870-7
6-AMINO-1,3-DIMETHYL-5-NITROSOURACIL
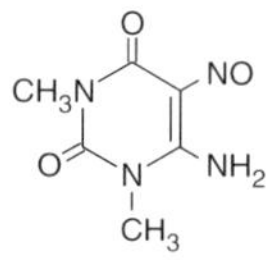
5 gm

S80,461-4
6-AMINO-1,3-DIETHYL-5-NITROSOURACIL
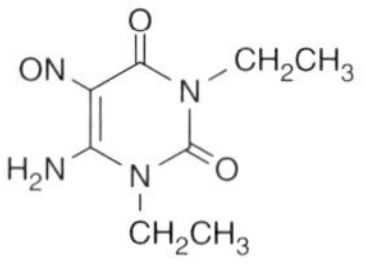
250 mg

S80,311-1
6-AMINO-1-METHYL-5-NITROURACIL
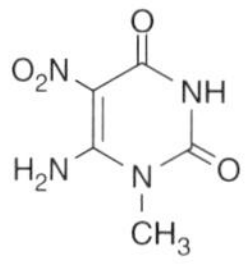
250 mg

S80,400-2
5,6-DIAMINO-1-METHYLURACIL
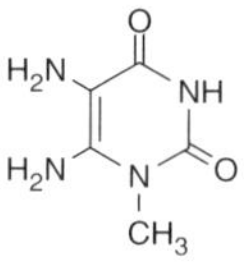
250 mg

S80,418-5
1-ALLYL-5,6-DIAMINO-3-ETHYLURACIL
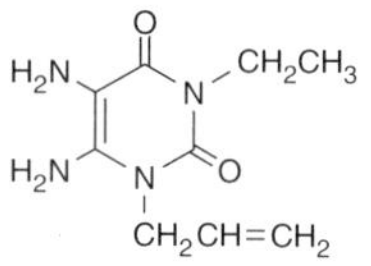
250 mg

S50,438-6
5,6-DIAMINOURACIL
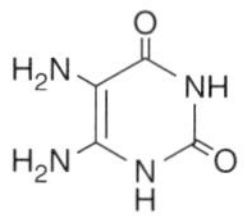
250 mg

S61,773-3
4-AMINO-1,2-DIMETHYL-5-(PENTADECYL)-PYRAZOLIDINONE
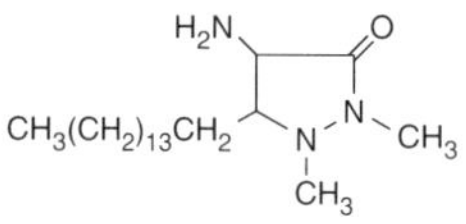
5 gm

S78,724-8
6-AMINO-1,3-DIMETHYL-5-(FORMYLAMINO)URACIL
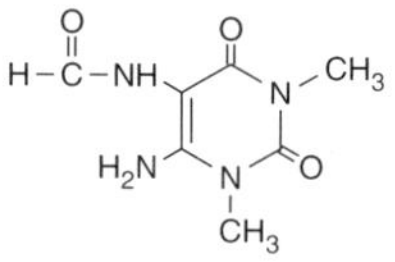
250 mg

S68,577-1
5-AMINOBARBITURIC ACID
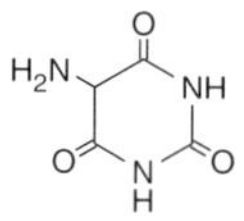
5 gm

S42,003-4
1-ALLYL-6-AMINO-3-ETHYLURACIL
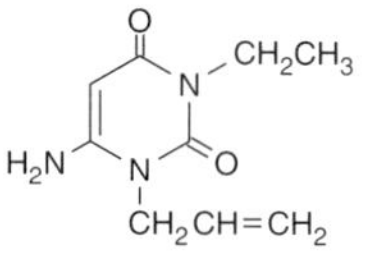
250 mg

S43,571-6
1-ALLYL-6-AMINOURACIL
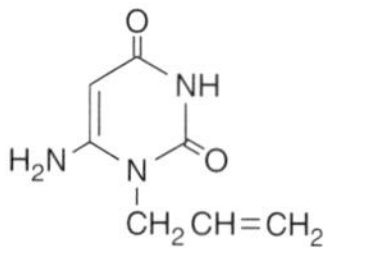
5 gm

S80,401-0
6-AMINO-1-ETHYLURACIL
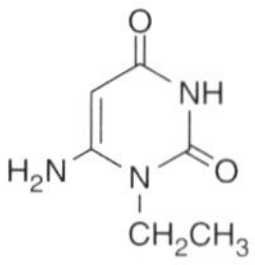
250 mg

S42,371-8
6-AMINO-1,3-DIETHYLURACIL
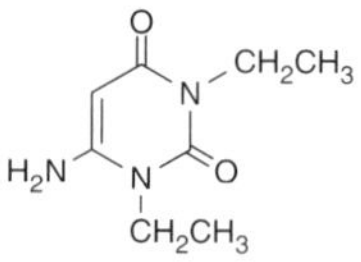
5 gm

S85,956-7
4-AMINO-5,6-DIHYDRO-1-METHYL-2(1H)-PYRIMIDINONE
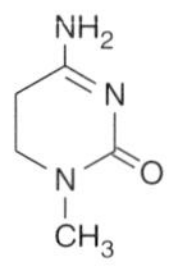
250 mg

S80,440-1
6-AMINO-1,3-DIETHYL-5-FORMYLURACIL
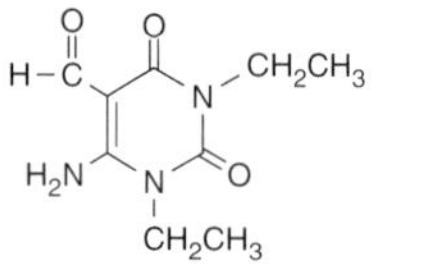
250 mg

S80,239-5
6-AMINO-1,3-DIMETHYL-5-FORMYLURACIL
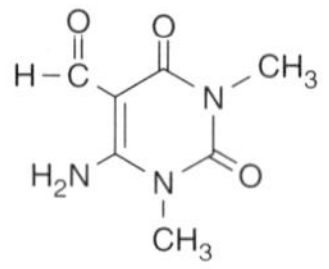
250 mg

S80,327-8
6-AMINO-3-ETHYL-1-METHYLURACIL
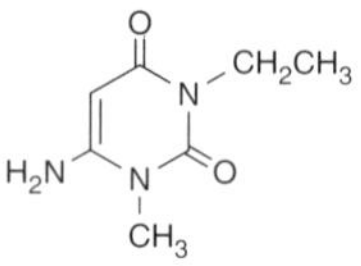
250 mg

S80,548-3
5-AMINO-1,3,6-TRIMETHYLURACIL
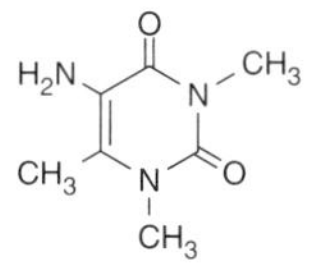
250 mg

S80,342-1
6-AMINO-1-ISOBUTYLURACIL
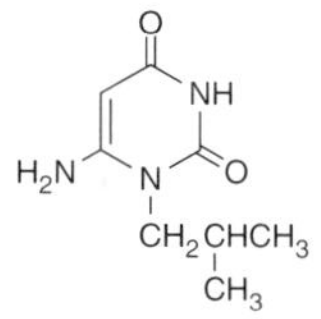
250 mg

S80,458-4
6-AMINO-3-ETHYL-1-(2-METHYLALLYL)-URACIL
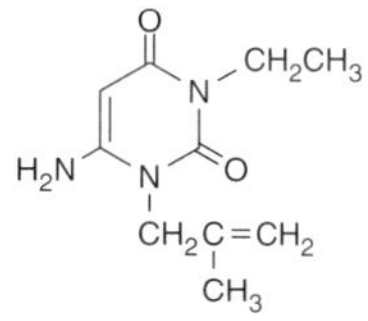
250 mg

S80,379-0
6-AMINO-3-ETHYL-1-ISOBUTYLURACIL
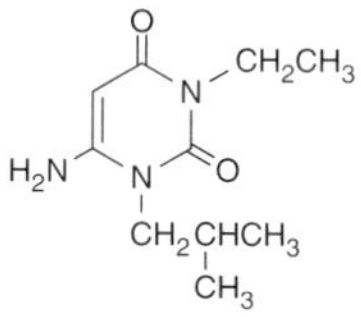
250 mg

S80,478-9
5-AMINO-1,3-DIBUTYLURACIL
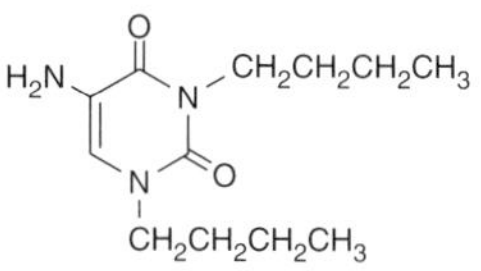
250 mg

S93,267-1
6-AMINO-5-BROMO-3-METHYLURACIL
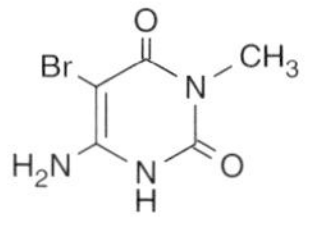
250 mg

S78,554-7
L-ALANYLGLYCYLGLYCINE ANHYDRIDE
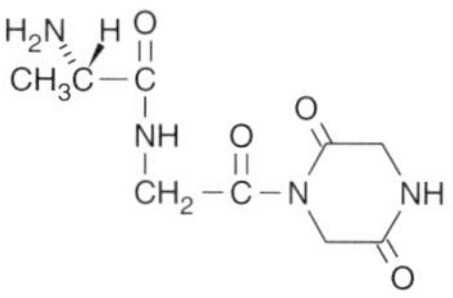
250 mg

S78,909-7
1-(2-TETRAHYDROPYRANYL)CYTOSINE
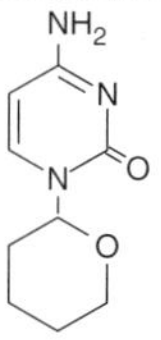
250 mg

S57,009-5
4-AMINO-1,2-DIMETHYL-3-PENTADECYL-3-PYRAZOLIN-5-ONE HYDROCHLORIDE
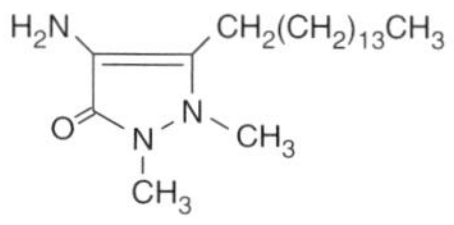
• HCl
5 gm

S45,304-8
1-(2-AMINOETHYL)-2-IMIDAZOLIDONE
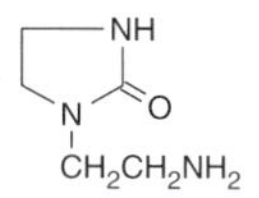
250 mg

S44,914-8
5-METHYLCYTOSINE HYDROCHLORIDE
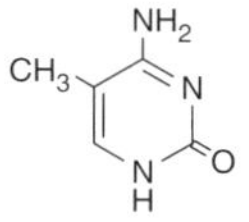
• HCl
250 mg

S94,881-0
1-AMINO-4,6-DIMETHYLPYRIMIDIN-2(1H)-ONE
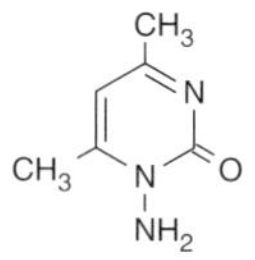
250 mg

S92,202-1
5,6-DIAMINO-1-METHYL-2-THIOURACIL
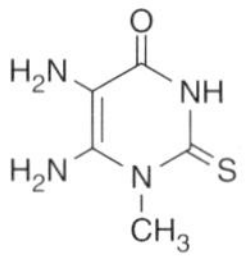
250 mg

S80,278-6
1,3-DIMETHYL-6-(FORMYLHYDRAZINO)-5-NITROURACIL
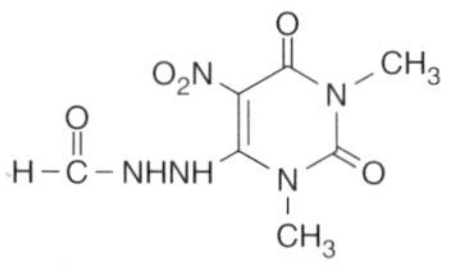
250 mg

S80,279-4
1,3-DIETHYL-6-METHYL-5-NITROURACIL
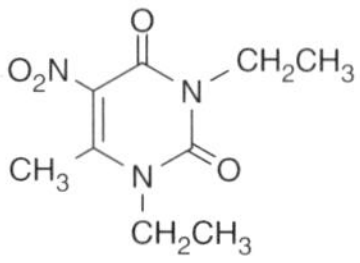
250 mg

S46,078-8
1,3-DIMETHYL-5-NITROURACIL MONOHYDRATE
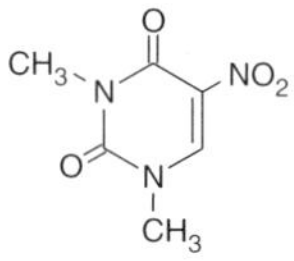
5 gm

S53,356-4
5,6-DIHYDRO-6-METHYLURACIL
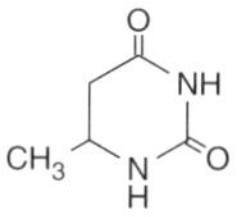
5 gm

S77,448-0
DL-N-[3-(DIMETHYLAMINO)PROPYL]-CAMPHORIMIDE HYDROCHLORIDE
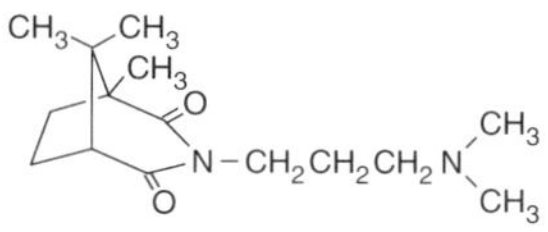
• HCl
250 mg

S41,229-5
5,6-DIHYDRO-5-METHYLURACIL
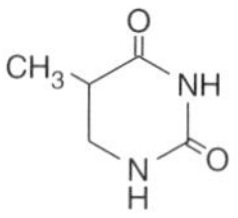
5 gm

S94,229-4
4-IMINO-2-THIAZOLIDINONE
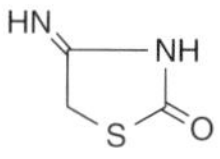
250 mg

S95,154-4
ETHYL 1-(TERT-BUTYL)-4,5-DIOXO-3-PYRROLIDINECARBOXYLATE
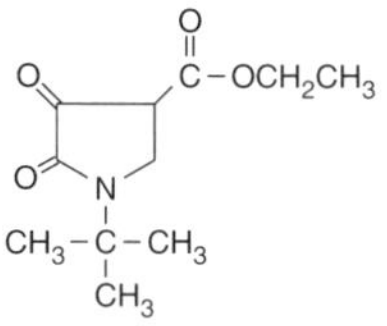
50 mg

S36,930-6
5-ISOBUTYL-5-METHYL-3-VINYLHYDANTOIN
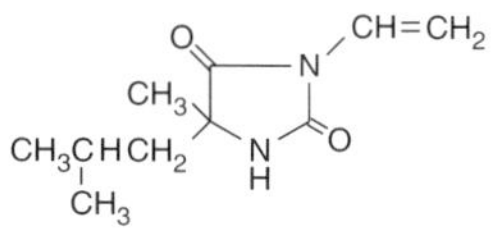
5 gm

S75,737-3
4,7-EPOXY-2-METHYL-2,3,3A,4,5,6,7,7A-OCTAHYDRO-1H-ISOINDOLE-1,3-DIONE
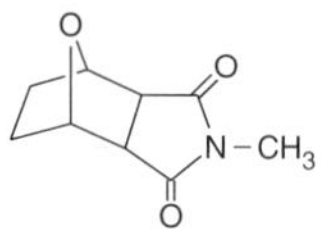
5 gm

S75,738-1
EXO-2,7-DIMETHYL-4,7-EPOXY-2,3,3A,4,5,6,7,7A-OCTAHYDRO-1H-ISOINDOLE-1,3-DIONE
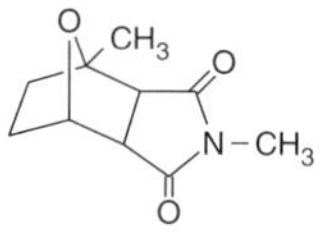
5 gm

S89,751-5
8-METHYL-5,6,7,8-TETRAHYDRO-4H-PYRIMIDO(1,2-A)PYRIMIDIN-2(3H)-ONE
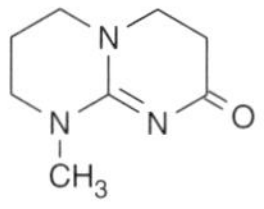
250 mg

S89,703-5
8-METHYL-2,3,7,8-TETRAHYDROIMIDAZO-(1,2-A)PYRIMIDIN-5(6H)-ONE HYDRIODIDE
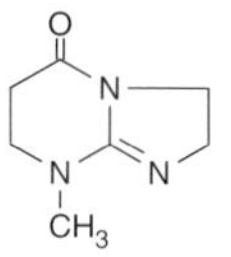
• HI
250 mg

S89,753-1
8-METHYL-2,3,7,8-TETRAHYDROIMIDAZO-(1,2-A)PYRIMIDIN-5(6H)-ONE
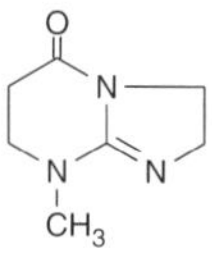
250 mg

S89,759-0
1-METHYL-1,2,3,7-TETRAHYDROIMIDAZO (1,2-A)PYRIMIDIN-5(6H)-ONE HYDROBROMIDE
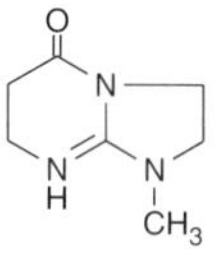
• HBr
250 mg

S50,785-7
1-[2-(DIETHYLAMINO)ETHYL]-3-PENTADECYL-2-PYRAZOLIN-5-ONE
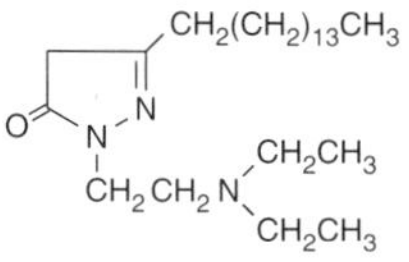
250 mg

S91,606-4
1,4,6-TRIMETHYL-2(1H)-PYRIMIDINONE HYDROCHLORIDE
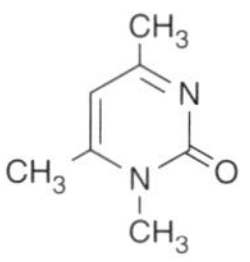
• HCl
250 mg

S91,549-1
3,4-DIHYDRO-1,3,4,6-TETRAMETHYL-2(1H)-PYRIMIDINONE
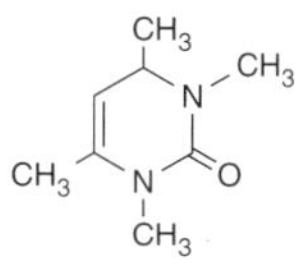
250 mg

S80,301-4
1,3-DIMETHYL-6-HYDRAZINOURACIL
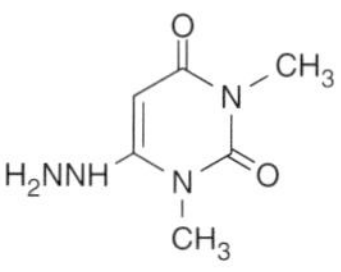
250 mg

S80,345-6
1,3-DIETHYL-6-HYDRAZINOURACIL

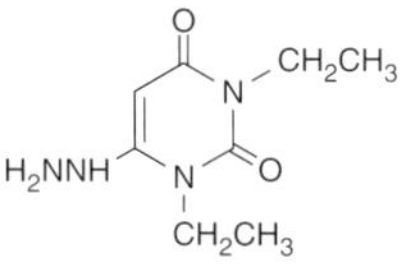

250 mg

S41,849-8
3-AMINO-2-OXAZOLIDINONE

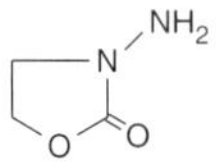

5 gm

S95,256-7
4,4-DIBROMO-3-METHYL-2-PYRAZOLIN-5-ONE

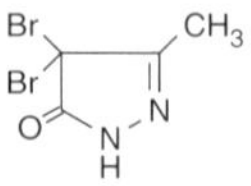

50 mg

S95,801-8
2-NITROSO-4,7,7-TRIMETHYL-2-AZABICYCLO[2.2.1]HEPTAN-3-ONE

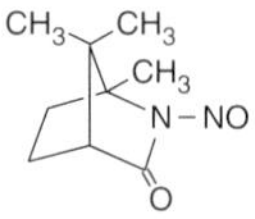

50 mg

S95,269-9
3-NITROSO-1-OXA-3-AZASPIRO(4.5)-DECAN-2-ONE

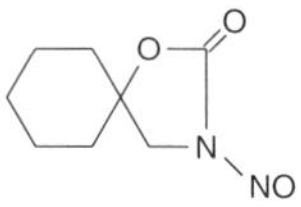

50 mg

S80,277-8
5-NITRO-1,3,6-TRIMETHYLURACIL

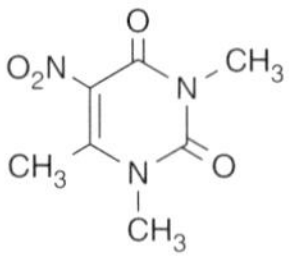

250 mg

S89,692-6
METHYL 4-OXO-4,6,7,8-TETRAHYDRO-1H-PYRIMIDO(1,2-A)PYRIMIDINE-2-BUTYRATE HYDROCHLORIDE

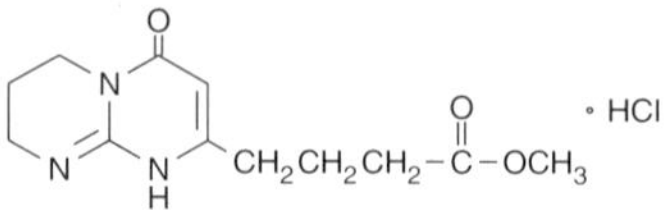

250 mg

S84,898-0
1-(3-OXOBUTYL)-3,6(1H,2H)-PYRIDAZINEDIONE

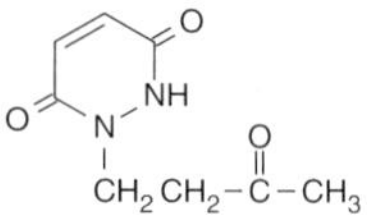

250 mg

S76,210-5
4,5,6,7-TETRAHYDROISATIN

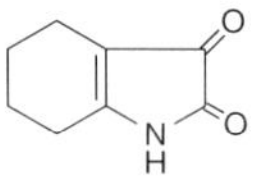

250 mg

S95,809-3
ETHYL N-(2,6-DIOXO-1,8,8-TRIMETHYL-7-AZABICYCLO[3.2.1]OCT-7-YL)-CARBAMATE

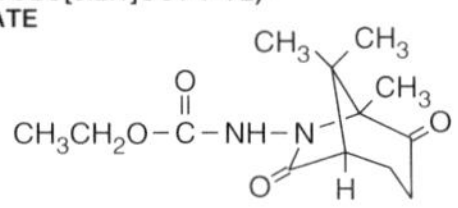

50 mg

S95,807-7
5,8,8-TRIMETHYL-6-AZABICYCLO-[3.2.1]OCTANE-4,7-DIONE

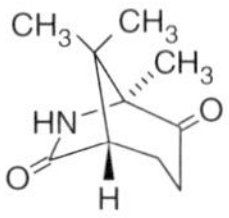

50 mg

S77,056-6
5-(CHLOROCARBONYL)URACIL

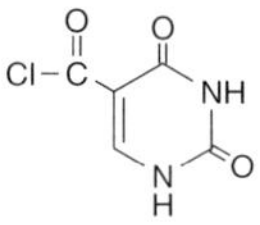

250 mg

S78,257-2
1,3-DIMETHYL-6-URACILCARBOXYLIC ACID

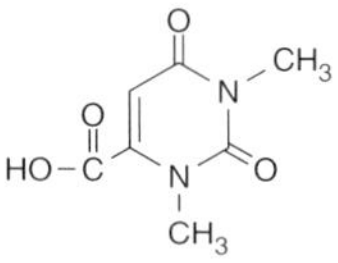

5 gm

S85,934-6
5-BROMO-1-METHYLURACIL

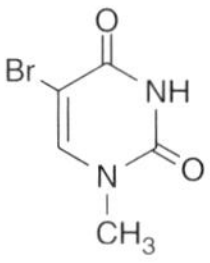

250 mg

S86,342-4
5-CHLORO-1,3-DIMETHYLURACIL

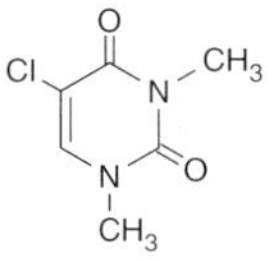

250 mg

S86,470-6
1,3,5-TRIMETHYLURACIL

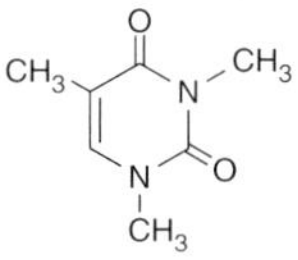

250 mg

S46,303-5
6-CHLORO-1,3-DIMETHYLURACIL

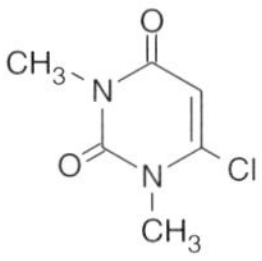

5 gm

S43,309-8
3,5-DIBROMO-2-PYRIDONE, SODIUM SALT

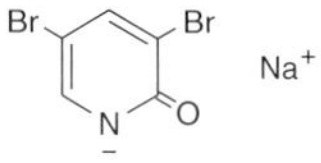

5 gm

S86,466-8
1,3-DIMETHYL-HEXAHYDRO-2,4,5,6-PYRIMIDINETETRAONE

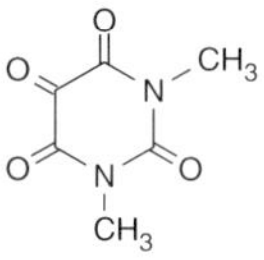

250 mg

S75,621-0
MORPHOLINE-2,3-DIONE

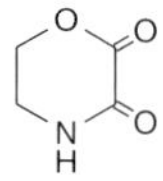

5 gm

S75,626-1
5,6-DIMETHYLMORPHOLINE-2,3-DIONE

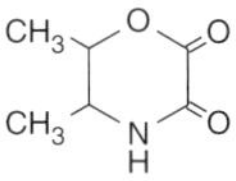

5 gm

S93,262-0
2,3-PIPERAZINEDIONE

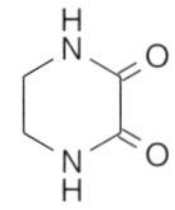

250 mg

S93,263-9
5-METHYL-2,3-PIPERAZINEDIONE

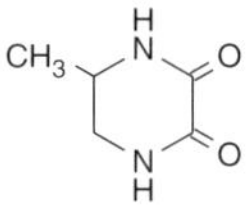

250 mg

S61,765-2
1-BUTYL-3-(ETHOXYCARBONYLMETHYL)-BARBITURIC ACID

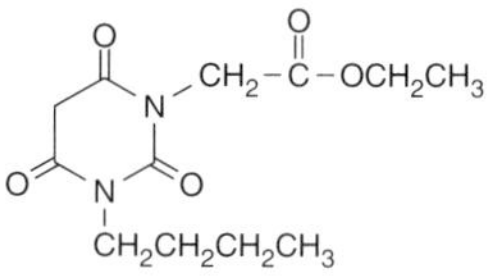

250 mg

S79,854-1
2,3,5A,6,7,8,10,10A-OCTAHYDRO-1H,5H-DIPYRROLO(1,2-A:1',2'D)-PYRAZIN35810TETRALONE

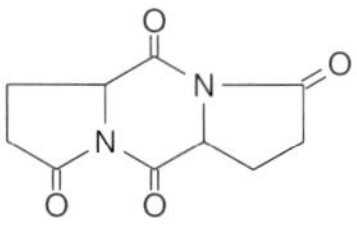

250 mg

S94,883-7
6-BROMO-1,2,4-TRIAZINE-3,5-(2H,4H)-DIONE

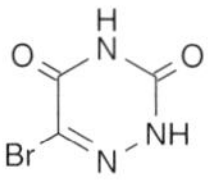

250 mg

S85,992-3
1,2-BUTYL-4,5-DIHYDRO-3,6(1H,2H)-PYRIDAZINEDIONE

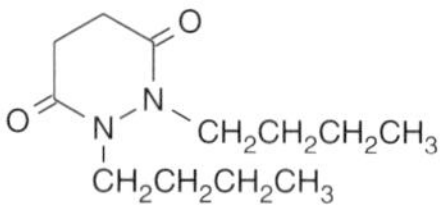

250 mg

S95,472-1
1,6-DIHYDRO-6-OXO-1,2,5-TRIMETHYL-3-PYRIDINECARBOXYLIC ACID

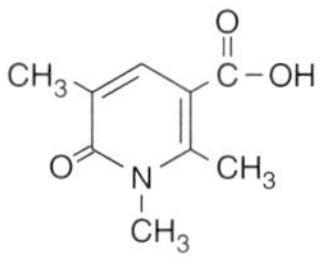

50 mg

S73,202-8
METHYL METHYL-OCTAHYDRO-3,5,8-TRIOXO-2A,4,8A-TRIAZAACENAPHTHYLENE-6-CARBOXYLATE

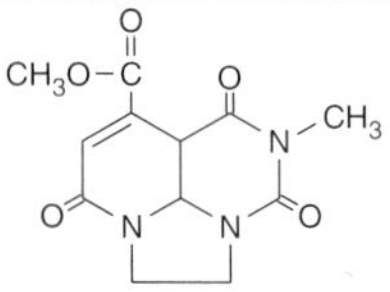

250 mg

S88,875-3
5,5-DIMETHYL-3-ETHYL-2,4-OXAZOLIDINEDIONE

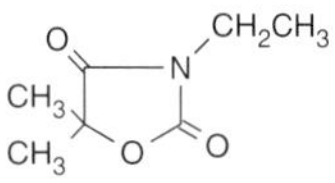

250 mg

S95,474-8
METHYL 2,5-DIMETHYL-6-OXO-1,4,5,6-TETRAHYDRO-3-PYRIDINECARBOXYLATE

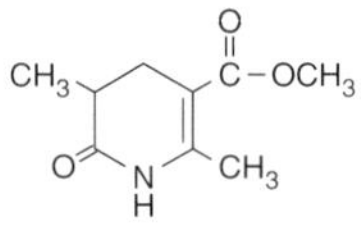

50 mg

S95,468-3
METHYL 1,6-DIHYDRO-1-METHYL-6-OXO-3-PYRIDINECARBOXYLATE

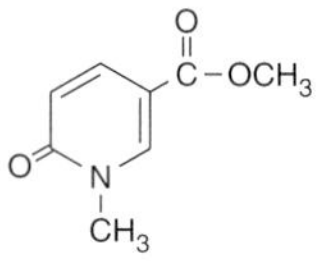

50 mg

S35,590-9
5,5-TETRAMETHYLENEHYDANTOIN

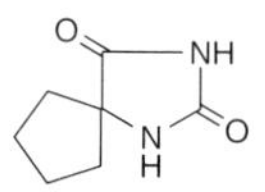

5 gm

S35,173-3
5-ISOPROPYLHYDANTOIN

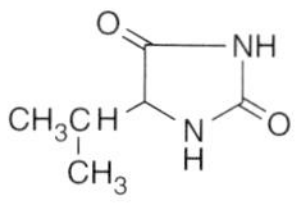

5 gm

S75,223-1
[2-(2,4-DIOXO-5-IMIDAZOLIDINYL)-ACETYL]UREA

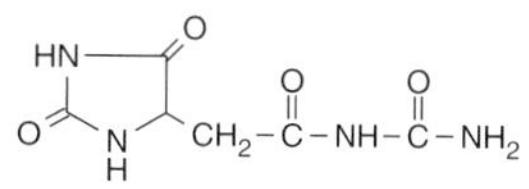

250 mg

S35,428-7
PENTAMETHYLENEHYDANTOIN

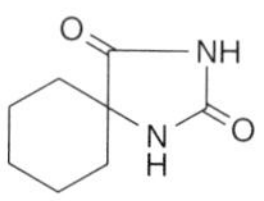

5 gm

S35,429-5
5,5-PENTAMETHYLENE-3-VINYLHYDANTOIN

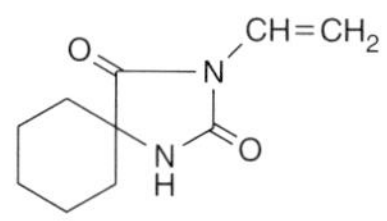

5 gm

S46,128-8
N,N',N,N-TETRAACETYLGLYCOLURIL

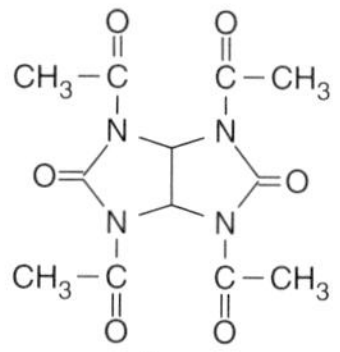

5 gm

S78,915-1
1-CYCLOHEXYL-5-IODOURACIL

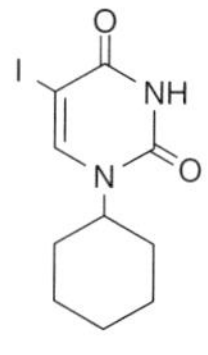

250 mg

S85,955-9
5-BROMO-5,6-DIHYDRO-1-METHYLURACIL

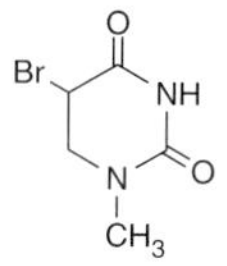

250 mg

S78,797-3
1-ETHYLURACIL

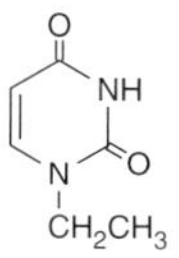

250 mg

S77,598-3
1,3-DIBROMOHYDANTOIN

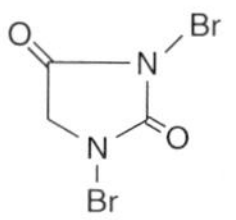

250 mg

S36,972-1
5-METHYLHYDANTOIN

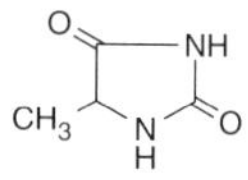

5 gm

S94,972-8
3-BROMO-5,5-DIMETHYLHYDANTOIN

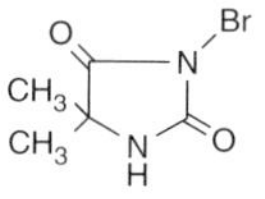

250 mg

S77,810-9
5-ETHYL-5-METHYLHYDANTOIN

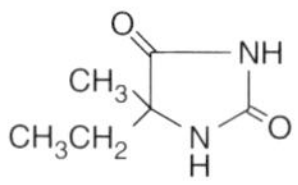

250 mg

S77,599-1
1,3-DIBROMO-5-ETHYL-5-METHYLHYDANTOIN

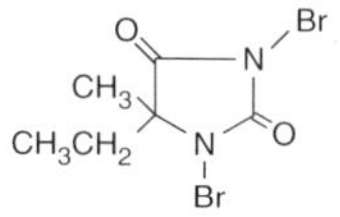

250 mg

S77,812-5
5-ISOPROPYL-5-METHYLHYDANTOIN

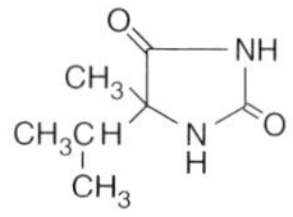

250 mg

S77,813-3
1-BROMO-3,5-DIMETHYL-5-ETHYLHYDANTOIN

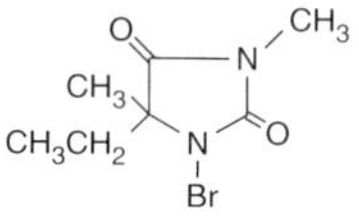

250 mg

S77,814-1
1,3-DIBROMO-5-ISOPROPYL-5-METHYLHYDANTOIN

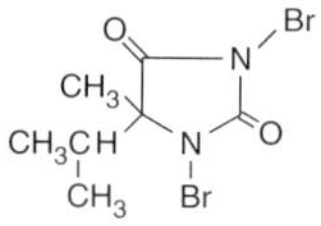

250 mg

S43,732-8
HEXAMETHYLENEHYDANTOIN

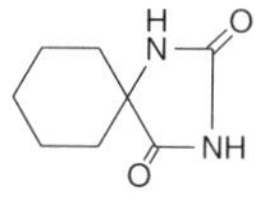

250 mg

S34,791-4
1,3-DIVINYL-5-ISOBUTYL-5-METHYLHYDANTOIN

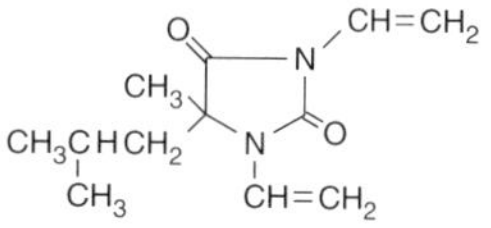

5 gm

S75,668-7
1-CYCLOHEXYL-5,6-DIHYDRO-2,4(1H,3H)-PYRIMIDINEDIONE

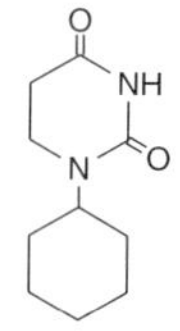

5 gm

S77,685-8
5-BROMO-5,6-DIHYDROURACIL

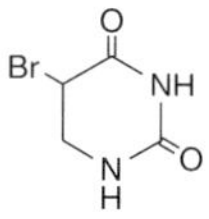

250 mg

S78,256-4
1,3-DIMETHYLURACIL

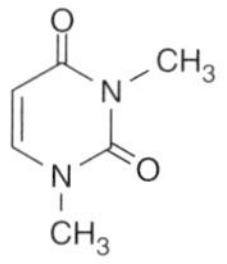

5 gm

S59,085-1
DICHLORO-1,3,5-TRIAZINE-2,4,6 (1H,3H,5H)-TRIONE, POTASSIUM SALT

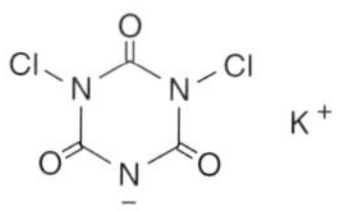

250 mg

S78,563-6
1-METHYL-5-AZAURACIL

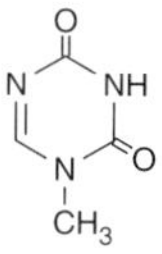

250 mg

S76,293-8
N-METHYL-5-NORBORNENE-2,3-DICARBOXIMIDE

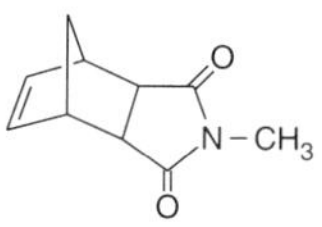

250 mg

S77,128-7
CAMPHORIMIDE(1,8,8-TRIMETHYL-3-AZABICYCLO[3.2.1]OCTANE-2,4-DIONE)

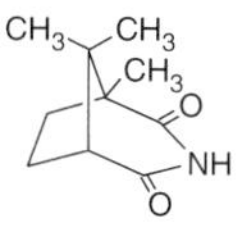

250 mg

S76,283-0
EXO-4,7-EPOXY-2-ETHYL-7-METHYL-2,3,3A,4,5,6,7,7A-OCTAHYDRO-1H-ISOINDOLE-1,3-DIONE

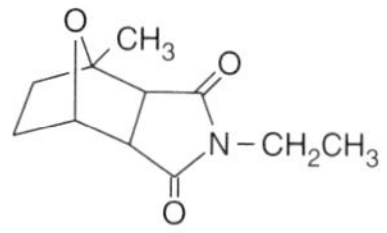

250 mg

S77,129-5
9-TERT-BUTYL-3-AZASPIRO(5.5)-UNDECANE-2,4-DIONE, AMMONIUM SALT

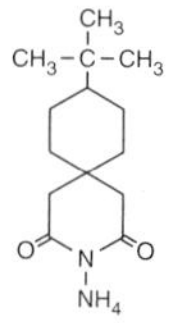

250 mg

S93,269-8
5,6-DIHYDRO-1-ETHYLURACIL

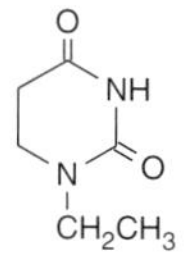

250 mg

S76,774-3
3-AZATRICYCLO[4.2.1.0(2,5)]-NONAN-4-ONE

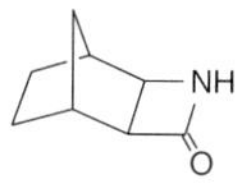

250 mg

S94,801-2
4-AZATRICYCLO[4.3.1.1(3,8)]-UNDECAN-5-ONE

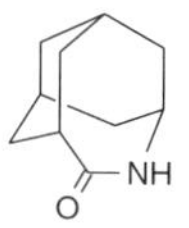

250 mg

S95,660-0
4A-METHYL-1,2,3,4,4A,5,6,7-OCTAHYDRO-2-QUINOLINONE

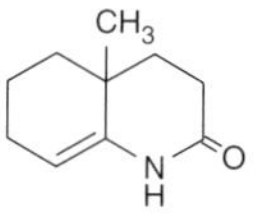

50 mg

S85,970-2
3-METHYL-2-(METHYLTHIO)-4(3H)-PYRIMIDINONE
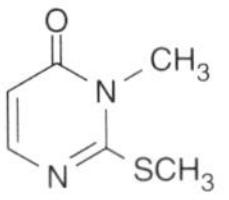
250 mg

S89,758-2
1-(2-IMIDAZOLIN-2-YL)-2-PYRROLIDINONE
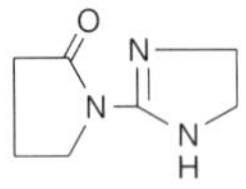
250 mg

S40,014-9
2,6-DIMETHYL-1-PIPERIDINECARBOXALDEHYDE
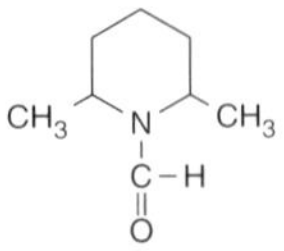
250 mg

S68,171-7
3,4,6,7,8,9-HEXAHYDRO-2H-PYRIDO-(1,2-A)PYRIMIDIN-6-ONE
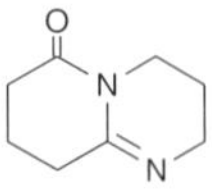
250 mg

S94,950-7
1-(1-ADAMANTYLCARBONYL)-DECAMETHYLPIPERIDINE
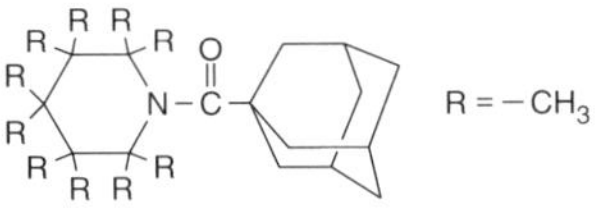
250 mg

S44,908-3
THERMOPSINE
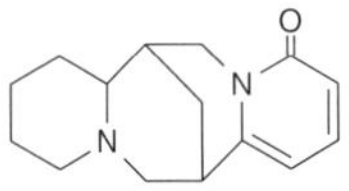
250 mg

S89,752-3
1-METHYL-4,6,7,8-TETRAHYDRO-1H-PYRIMIDO(1,2-A)PYRIMIDIN-2(3H)-ONE
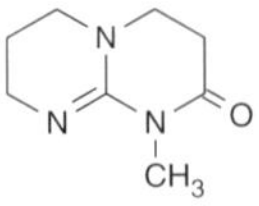
250 mg

S36,867-9
HEXAHYDRO-1,2,4,5-TETRAZINE-3,6-DIONE
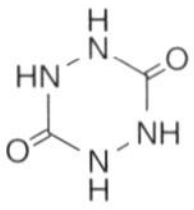
5 gm

S92,200-5
3(2H)-PYRIDAZINONE
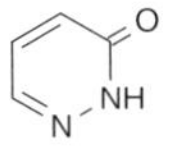
250 mg

S84,918-9
1,2-BIS(1-METHYLPROPYL)-4,5-DIHYDRO-3,6(1H,2H)-PYRIDAZINEDIONE
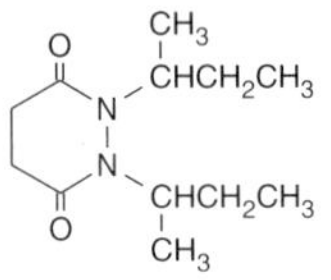
250 mg

S43,572-4
N,N',N'',N'''-TETRACHLOROGLYCOLURIL
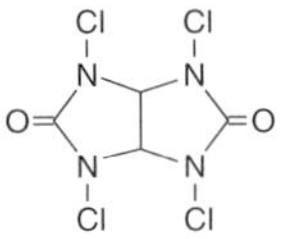
5 gm

S90,037-0
N,N'-HEXAMETHYLENEBIS(4-METHYL-1-PIPERIDINECARBOXAMIDE)
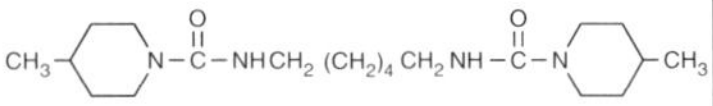
250 mg

S89,223-8
4-METHYL-N-OCTADECYL-1-PIPERIDINECARBOXAMIDE
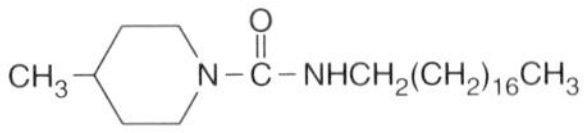
250 mg

S75,629-6
5,6-DIHYDRO-4,6,6-TRIMETHYL-2(1H)-PYRIMIDINONE
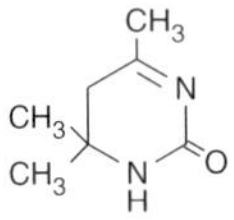
5 gm

S57,596-8
CHLOROFORMAMIDINE HYDROCHLORIDE
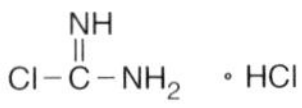
5 gm

S81,547-0
1-METHYLGUANIDINE SULFATE
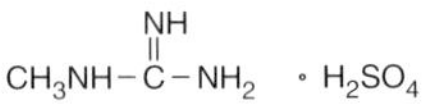
250 mg

S81,549-7
1-ALLYLGUANIDINE SULFATE
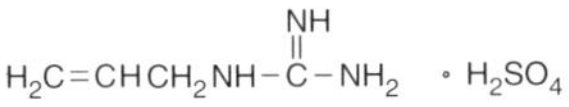
250 mg

S39,091-7
1-ISOPROPYLGUANIDINE SULFATE
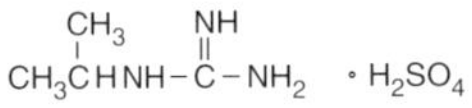
250 mg

S81,548-9
1-PROPYLGUANIDINE SULFATE
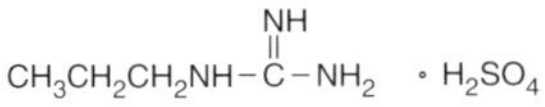
250 mg

S46,049-4
1,1-DIETHYLGUANIDINE HYDROCHLORIDE
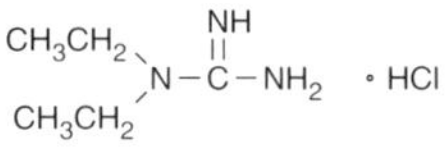
250 mg

S68,274-8
ETHYL 2-METHYLPROPIONIMIDATE-HYDROCHLORIDE
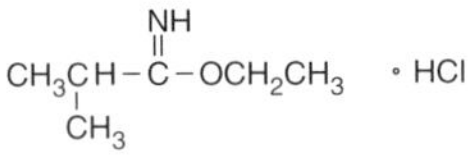
5 gm

S46,011-7
GALEGINE SULFATE
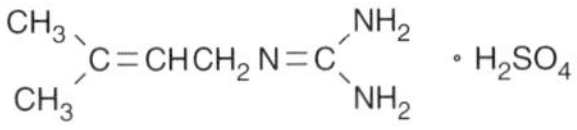
250 mg

S83,734-2
(CHLOROACETYL)GUANIDINE-HYDROCHLORIDE
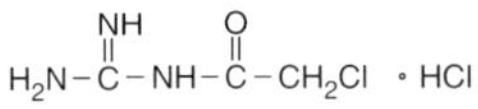
250 mg

S46,051-6
5-CHLOROMETHYL-2-IMINOOXAZOLIDINE
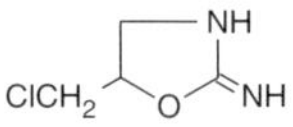
250 mg

S54,780-8
N',2,2,2-TETRACHLOROACETAMIDINE
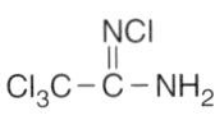
250 mg

S53,940-6
ETHYL 3-HYDROXY-4,4,4-TRICHLOROBUTYRIMIDATE HYDROCHLORIDE
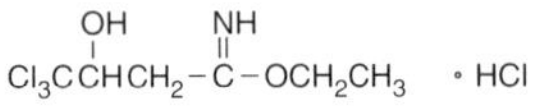
250 mg

S79,367-1
N-(3-HYDROXYPROPYL)-GUANIDINE NITRATE
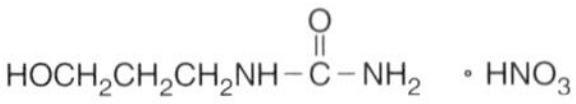
250 mg

S78,711-6
KASUGAMYCIN HYDROCHLORIDE HYDRATE
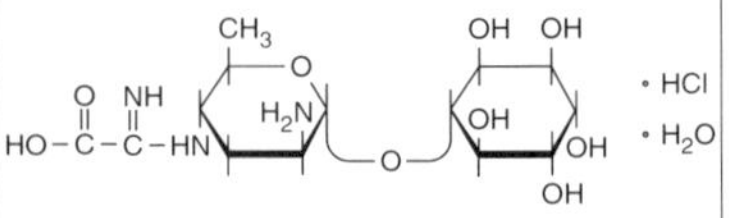
250 mg

S89,767-1
2-(1,4,5,6-TETRAHYDRO-2-PYRIMIDINYLAMINO)-1-ETHANOL
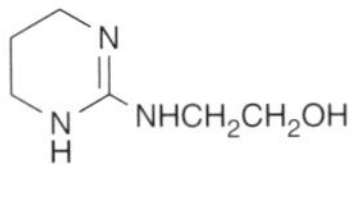
250 mg

S35,087-7
4-HYDROXY-2-IMINO-4-THIAZOLINE-5-ACETIC ACID
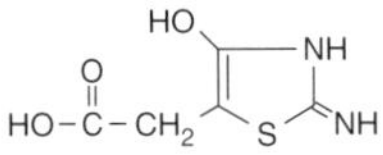
5 gm

S63,324-0
O-METHYLISOUREA HYDROCHLORIDE
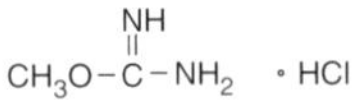
250 mg

S94,926-4
DIETHYL OXALIMIDATE DIHYDROCHLORIDE
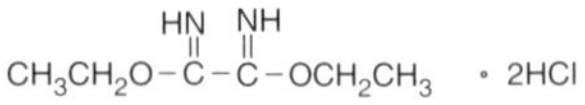
250 mg

S91,537-8
1-(2,2-DIETHOXYETHYL)GUANIDINE
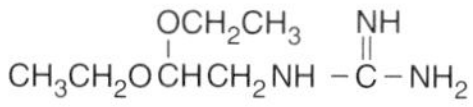
250 mg

S80,294-8
6-AMINO-1,3-DIMETHYL-5-(2-MERCAPTO-4-OXO-2-IMIDAZOLIN-5-YLIDENEAMINO)-URACIL
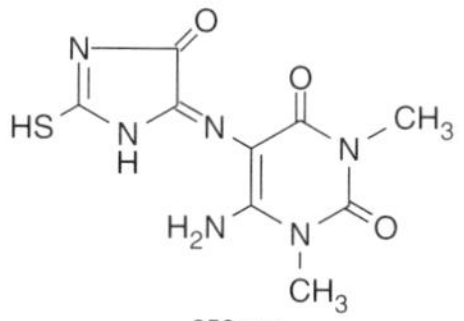
250 mg

S37,940-9
2-(2-MORPHOLINOETHYL)-2-THIOPSEUDOUREA DIHYDROCHLORIDE
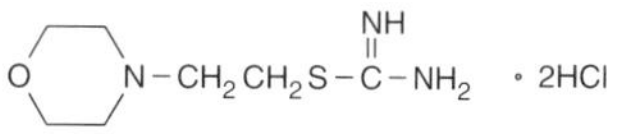
5 gm

S77,127-9
S,S'-PENTAMETHYLENEDIISOTHIURONIUM-DICHLORIDE
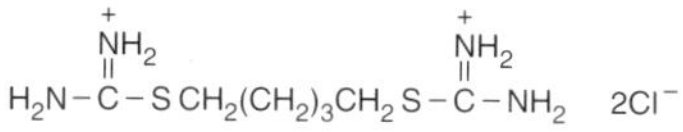
250 mg

S68,388-4
METHYL HYDRAZINOCARBOXIMIDOTHIOATE-TOSYLATE SALT
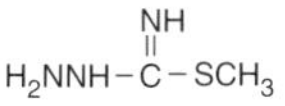
250 mg

S35,181-4
3-ISOTHIOUREIDOPROPIONIC ACID
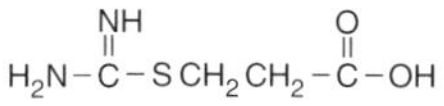
5 gm

S94,928-0
5-(2-THIOPSEUDOUREIDO)PENTANOIC ACID HYDROCHLORIDE
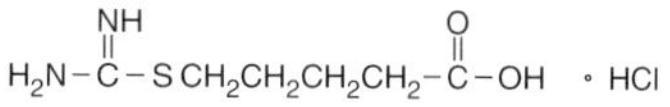
250 mg

S50,053-4
2-BUTYL-2-THIOPSEUDOUREA HYDROBROMIDE
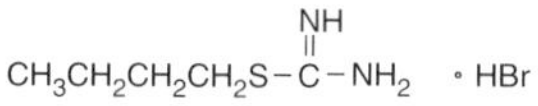
250 mg

S38,523-9
2-OCTYL-2-THIOPSEUDOUREA HYDROBROMIDE
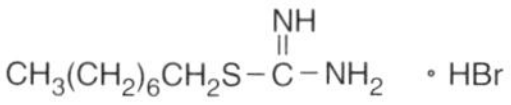
5 gm

S59,271-4
2-[5,7,7-TRIMETHYL-2-(1,3,3-TRIMETHYLBUTYL)OCTYL]-2-THIOPSEUDOUREA HYDROBROMIDE
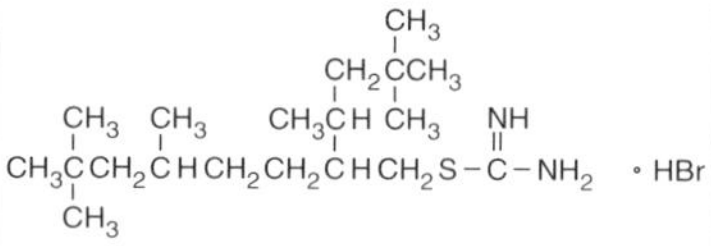
250 mg

S35,805-3
L-ARGININE L-GLUTAMATE
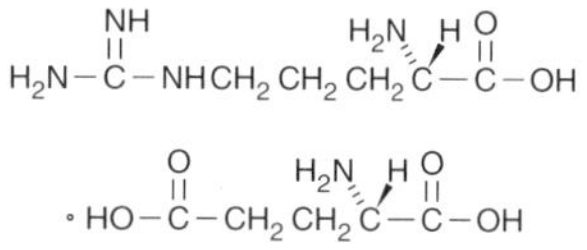
5 gm

S78,289-0
ARGININE ETHYL ESTER HYDROCHLORIDE
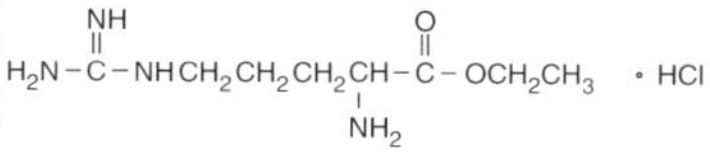
5 gm

S89,766-3
2-AMINO-1-METHYL-2-IMIDAZOLINE HYDROBROMIDE
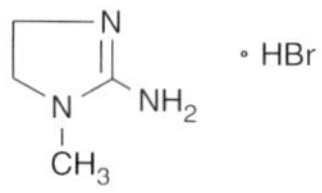
250 mg

S89,775-2
2-AMINO-1-METHYL-1,4,5,6-TETRAHYDROPYRIMIDINE HYDROBROMIDE
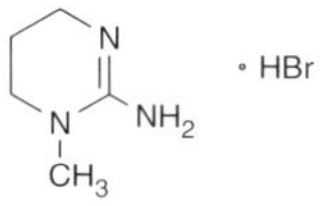
250 mg

S89,697-7
METHYL 2-AMINO-3,4,5,6-TETRAHYDRO-4-PYRIMIDINEBUTYRATE HYDROCHLORIDE
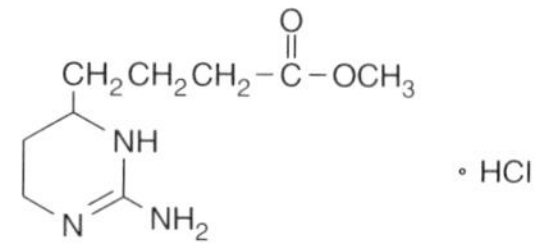
250 mg

S41,498-0
GUANYLUREA PHOSPHATE
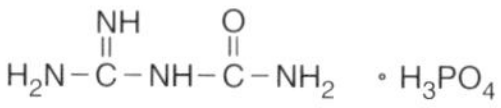
5 gm

S89,770-1
2-AMINO-1,4,5,6-TETRAHYDROPYRIMIDINE HYDROBROMIDE
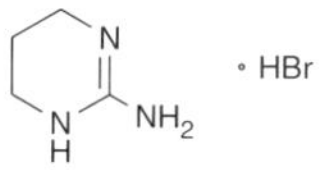
250 mg

S38,238-8
1-[2-(DIETHYLAMINO)ETHYL]-GUANIDINE SULFATE
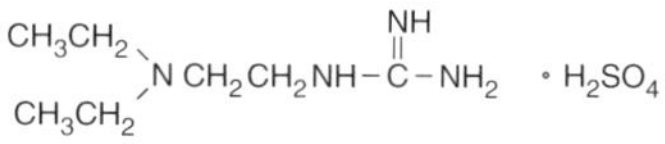
250 mg

S89,756-6
2,3,5,6,7,8-HEXAHYDRO-8-METHYLIMIDAZO(1,2-A)PYRIMIDIN-5-IMINE HYDRIODIDE
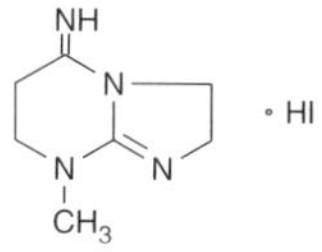
250 mg

S78,571-7
N-ALPHA-FORMYLNITRO-L-ARGININE
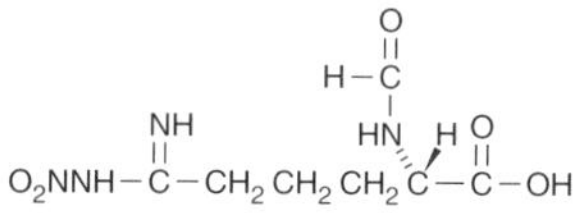
250 mg

S78,352-8
N-ALPHA-ACETYL-DL-ARGININE DIHYDRATE
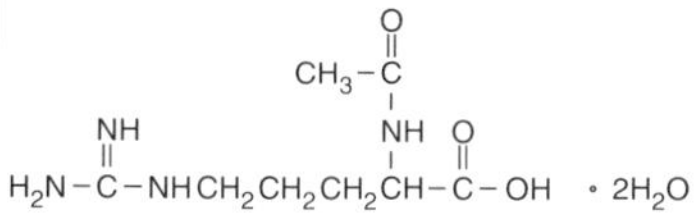
5 gm

S38,566-2
1-BUTYLGUANIDINE SULFATE
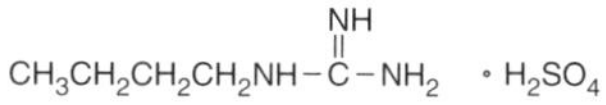
5 gm

S52,514-6
1-CYCLODODECYL-3-NITROGUANIDINE
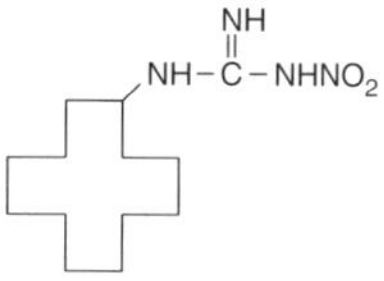
250 mg

S42,063-8
CIS-3A,4,5,6,7,7A-HEXAHYDRO-2-(METHYLAMINO)BENZOXAZOLE MALEATE
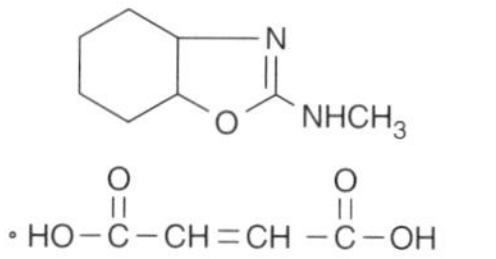
5 gm

S89,760-4
3-(2-IMIDAZOLIN-2-YLAMINO)-PROPIONIC ACID
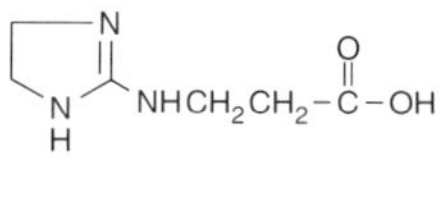
250 mg

S89,743-4
2-(METHYLAMINO)-3,4,5,6-TETRAHYDRO-4-PYRIMIDINEBUTYRIC ACID HYDROCHLORIDE
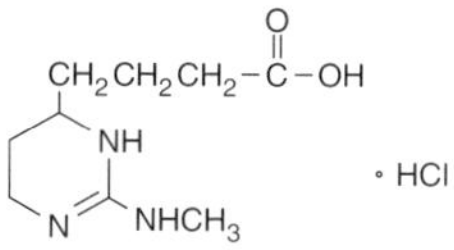
250 mg

S89,708-6
2-(METHYLAMINO)-2-IMIDAZOLINE-HYDRIODIDE
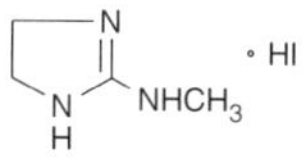
250 mg

S38,542-5
1,1-DIETHYLGUANIDINE SULFATE
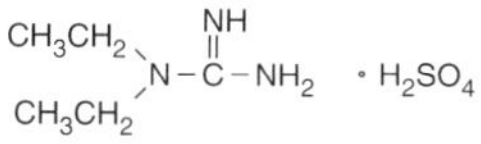
250 mg

S87,291-1
1-METHYL-2-PYRROLIDINONE HYDRAZONE DIHYDROCHLORIDE
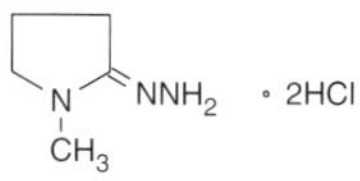
250 mg

S40,867-0
FORMIMIDOYLGLYCINE
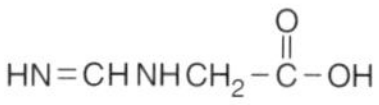
250 mg

S89,765-5
N-(HEXAHYDRO-1-METHYL-2-PYRIMIDINYLIDENE)ACETAMIDE
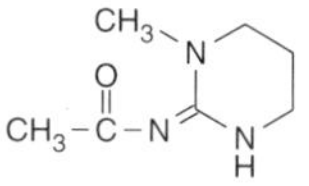
250 mg

S89,734-5
N-(HEXAHYDRO-1-METHYL-2-PYRIMIDINYLIDENE)-2,2,2-TRIMETHYLACETAMIDE HYDROCHLORIDE
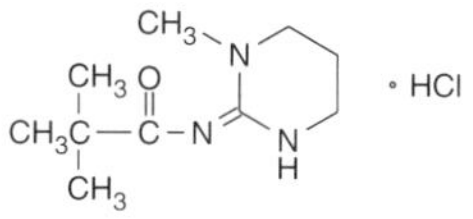
250 mg

S38,519-0
1-PIPERIDINECARBOXAMIDINE SULFATE

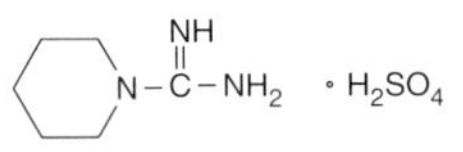

250 mg

S52,768-8
2,2'-ETHYLENEBIS(2-THIOPSEUDOUREA-HYDROBROMIDE)

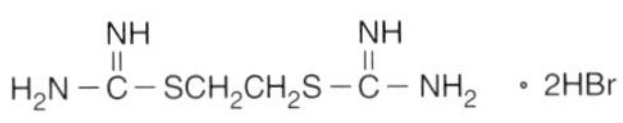

5 gm

S53,802-7
2-(2-METHYLALLYL)-2-THIOPSEUDOUREA HYDROCHLORIDE

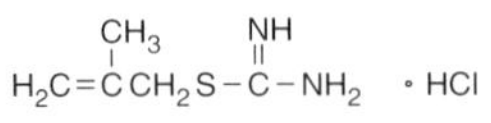

5 gm

S77,444-8
S,S'-HEPTAMETHYLENEDIISOTHIURONIUM-DIBROMIDE

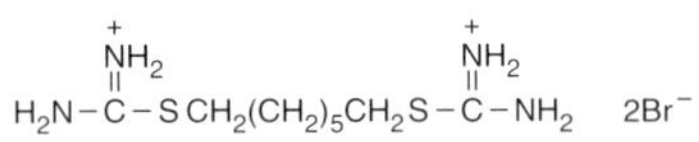

250 mg

S61,740-7
AMINOGUANIDINE SULFATE MONOHYDRATE

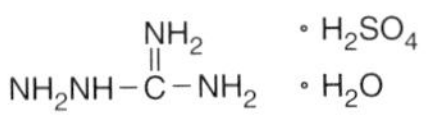

5 gm

S45,680-2
TRIAMINOGUANIDINE HYDROCHLORIDE

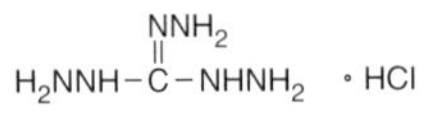

5 gm

S77,172-4
1-(NONYLIDENE)GUANIDINE NITRATE

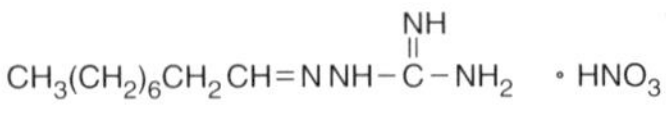

250 mg

S77,181-3
1-(OCTADECYLIDENEAMINO)-GUANIDINE NITRATE

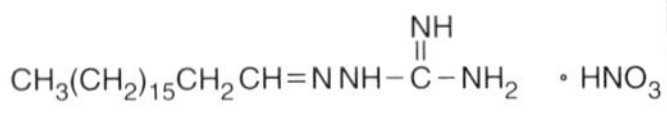

250 mg

S52,730-0
1-NITRO-3-PROPYLGUANIDINE

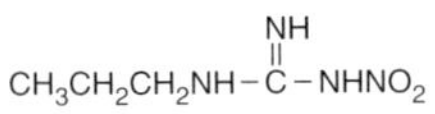

250 mg

S41,327-5
1-ISOAMYL-3-NITRO-1-NITROSOGUANIDINE

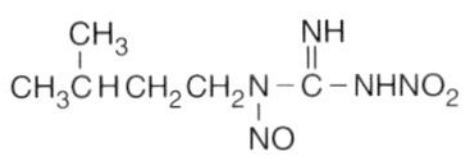

5 gm

S41,366-6
1-HEXYL-3-NITRO-1-NITROSOGUANIDINE

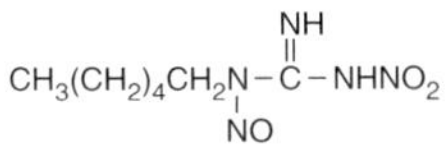

5 gm

S41,329-1
3-NITRO-1-NITROSO-1-OCTYLGUANIDINE

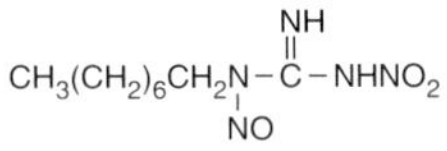

250 mg

S51,884-0
5-ETHYL-2-IMINO-5-ISOBUTYL-4-THIAZOLIDINONE

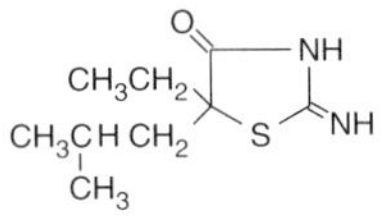

5 gm

S54,495-7
1-(4-OXO-2-THIAZOLIN-2-YL)-GUANIDINE

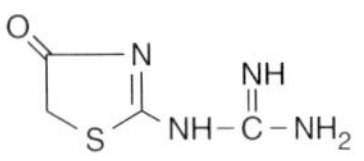

5 gm

S78,722-1
5-GUANIDINOVALERIC ACID HEMIHYDRATE

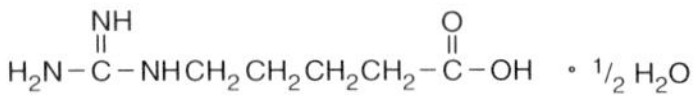

250 mg

S56,614-4
1-DODECYLGUANIDINE ACETATE

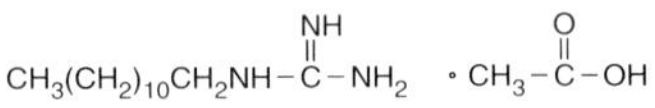

5 gm

S51,055-6
11-(AMIDINOTHIO)UNDECANOIC ACID

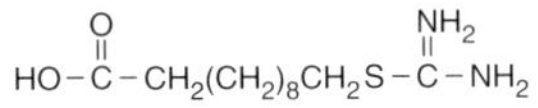

5 gm

S78,418-4
METHYL 4-GUANIDINOBUTYRATE HYDROCHLORIDE

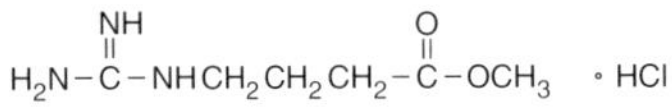

250 mg

S61,457-2
1-FORMYLGUANIDINE

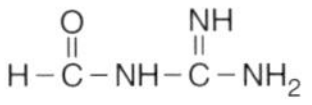

5 gm

S57,591-7
GUANYLUREA

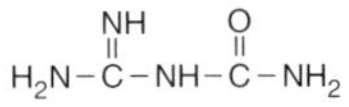

5 gm

S41,478-6
AMIDINOUREA SULFATE

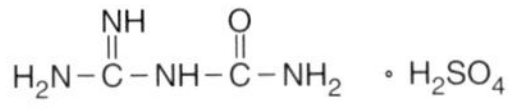

5 gm

S80,435-5
6-CHLORO-1,3-DIETHYL-4-IMINO-1,2,3,4-TETRAHYDRO-2-PYRIMIDINONE

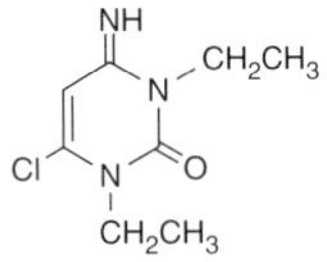

250 mg

S52,975-3
2-METHYL-2-THIOPSEUDOUREA-HYDRIODIDE

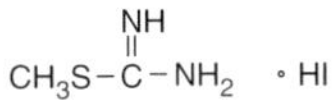

250 mg

S55,715-3
2-BUTYL-2-THIOPSEUDOUREA-HYDROCHLORIDE

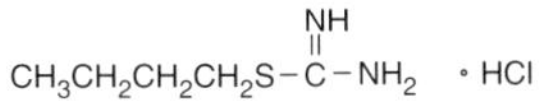

5 gm

S65,121-4
S,S'-HEXAMETHYLENEBIS-(ISOTHIURONIUM BROMIDE)

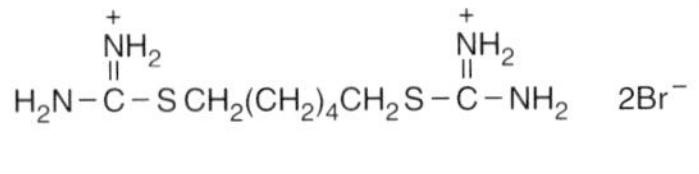

250 mg

S38,193-4
2-DODECYL-2-THIOPSEUDOUREA-HYDROCHLORIDE

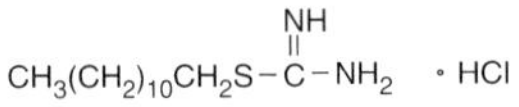

5 gm

S56,993-3
2-(TERT-HEXADECYL)-2-THIOPSEUDOUREA HYDROBROMIDE

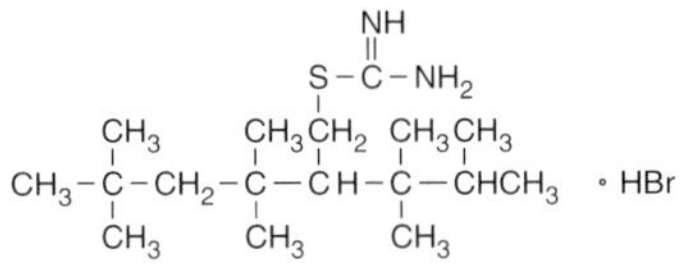

250 mg

S81,546-2
N-(3,4-DIMETHYL-4-THIAZOLIN-2-YLIDENE)ACETAMIDE

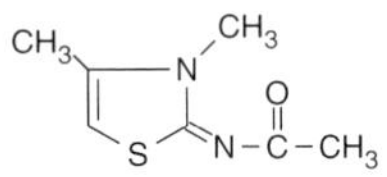

250 mg

S82,265-5
2-CYCLOHEXYL-2-THIOPSEUDOUREA HYDROBROMIDE

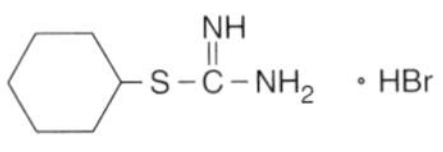

250 mg

S41,979-6
N-CYCLOHEXYLBIGUANIDE HYDROCHLORIDE

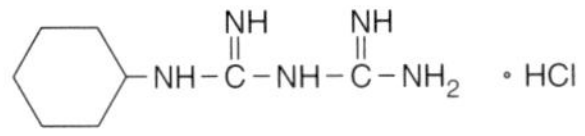

5 gm

S64,930-9
4-ISOPROPYL-3-THIOSEMICARBAZIDE

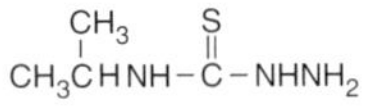

250 mg

S89,873-2
3-(3-CHLOROPROPYL)-1-METHYL-5-THIOXOHYDANTOIN

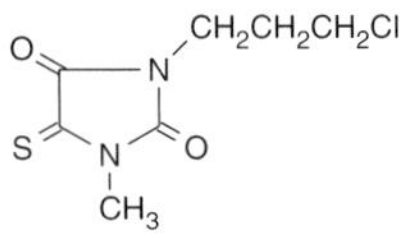

250 mg

S89,875-9
1-(2-CHLOROETHYL)-5-IMINO-3-METHYL-4-THIOHYDANTOIN

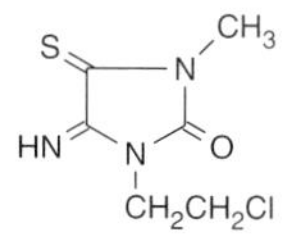

250 mg

S89,874-0
3-(2-CHLOROETHYL)-1-METHYL-5-THIOXOHYDANTOIN

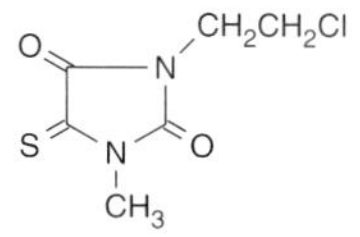

250 mg

S70,499-7
N,N'-BIS(2-HYDROXYETHYL)-DITHIOOXAMIDE

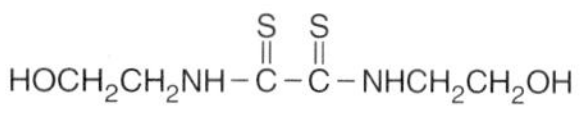

5 gm

S87,304-7
N,N'-BIS(2-HYDROXYPROPYL)-DITHIOOXAMIDE

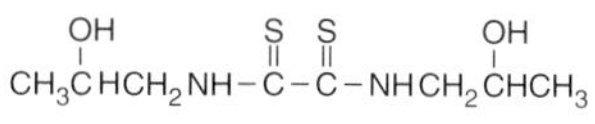

250 mg

S37,210-2
1-(2-HYDROXYETHYL)-2-IMIDAZOLIDINETHIONE

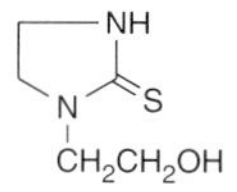

5 gm

S76,196-6
1-(2-HYDROXYETHYL)-3,4,5,6-TETRAHYDROPYRIMIDINE-2(1H)-THIONE

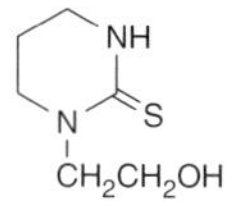

250 mg

S64,746-2
1-(2-HYDROXYPROPYL)-4-THIOXOHEXAHYDRO-1,3,5-TRIAZINE

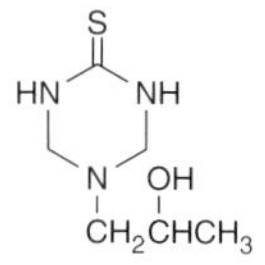

250 mg

S64,749-7
A-ETHYL-A-METHYL-4-THIOXOHEXAHYDRO-1,3,5-TRIAZINE-1-METHANOL

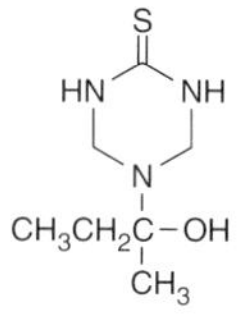

5 gm

S52,146-9
5-(2-HYDROXYETHYL)-5-ISOPENTYL-2-THIOBARBITURIC ACID

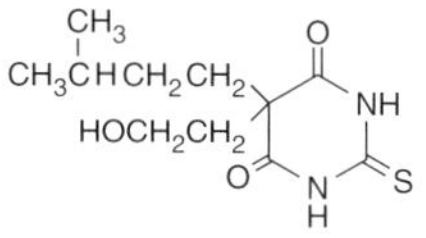

5 gm

S34,797-3
5-(ETHOXYMETHYLENE)-3-ETHYLRHODANINE

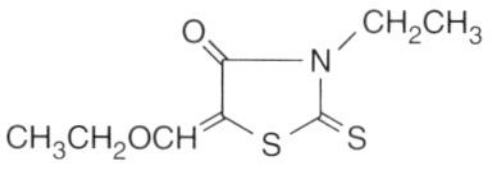

250 mg

S65,816-2
1-(3-METHOXYPROPYL)-4-THIOXOHEXAHYDRO-1,3,5-TRIAZINE

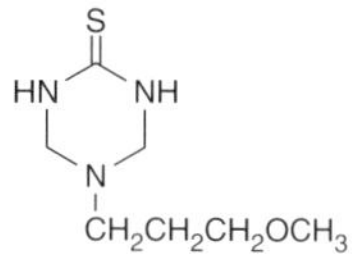

5 gm

S64,913-9
1-(3-METHOXYPROPYL)-2-THIOUREA

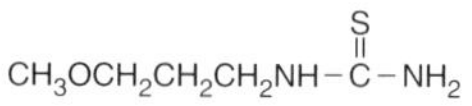

5 gm

S35,763-4
(2-AMINOETHYL)-DITHIOCARBAMIC ACID

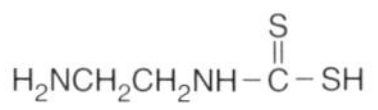

5 gm

S52,987-7
N-[BIS(METHYLTHIO)METHYLENE]-METHYLAMINE HYDRIODIDE

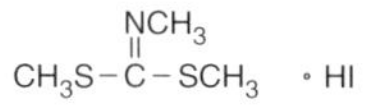

5 gm

S92,001-0
3-METHYL-2-THIAZOLIDINETHIONE

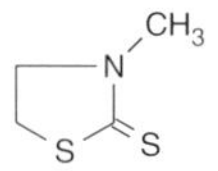

250 mg

S92,000-2
5-METHYL-2-THIAZOLIDINETHIONE

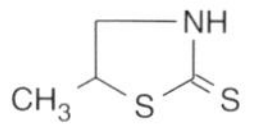

250 mg

S39,122-0
(N,N-DIETHYLTHIOCARBAMOYLTHIO)-ACETIC ACID

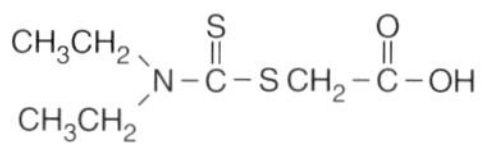

5 gm

S41,312-7
BIS(DIBUTYLTHIOCARBAMOYL)-DISULFIDE

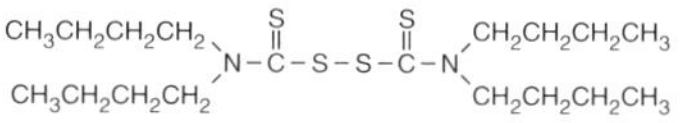

250 mg

S80,346-4
5,6-DIAMINO-1,3-DIETHYL-2-THIOURACIL

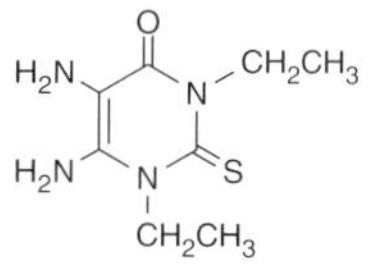

250 mg

S41,013-6
2,2,4,4-TETRAMETHYLIMIDAZOLIDINE-5-THIONE

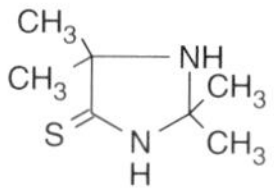

5 gm

S65,818-9
1-(2-DIETHYLAMINOETHYL)-4-THIOXOHEXAHYDRO-1,3,5-TRIAZINE

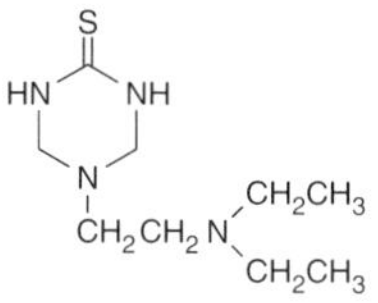

5 gm

S35,878-9
N,N'-BIS(2-PIPERAZINOETHYL)-DITHIOOXAMIDE

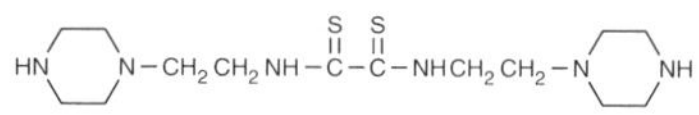

5 gm

S52,115-9
N,N'-BIS(2-MORPHOLINOETHYL)-DITHIOOXAMIDE

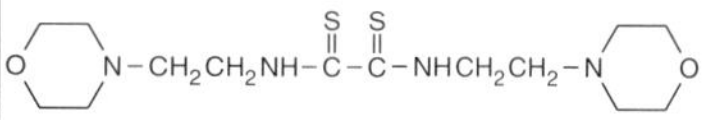

5 gm

S90,488-0
1-(2-MORPHOLINOETHYL)-3-TERT-OCTYL-2-THIOUREA

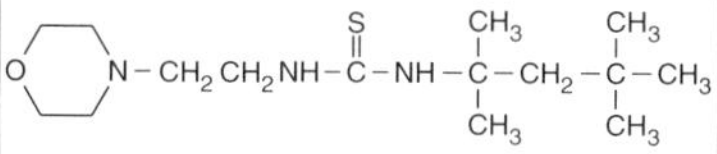

250 mg

S34,697-7
N,N'-DIMETHYLDITHIOOXAMIDE

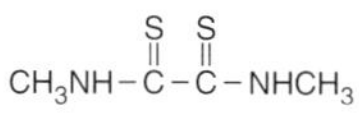

5 gm

S54,752-2
N-[3-(DIMETHYLAMINO)PROPYL]-DITHIOCARBAMIC ACID, AMMONIUM SALT

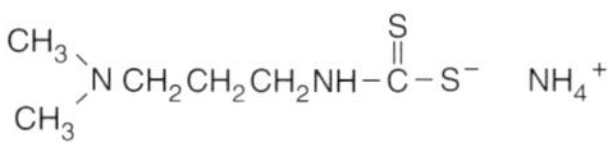

250 mg

S53,633-4
3-CYCLOHEXYL-1,1-DIMETHYL-2-THIOUREA

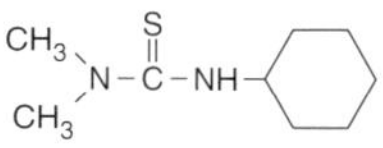

250 mg

S54,949-5
N-(3,3,5-TRIMETHYLCYCLOHEXYL)-DITHIOCARBAMIC ACID

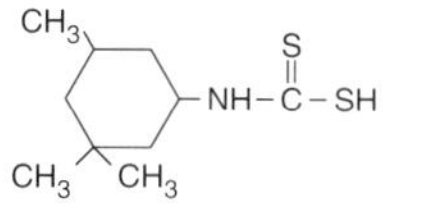

250 mg

S54,609-7
3,5-DIETHYL-3,4,5,6-TETRAHYDRO-2H-1,3,5-THIADIAZINE-2-THIONE

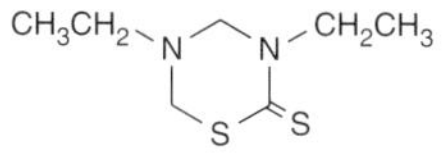

250 mg

S87,360-8
CYCLOHEXYL-2-(PYRROLIDINO-METHYLENE)THIOBUTYRAMIDE

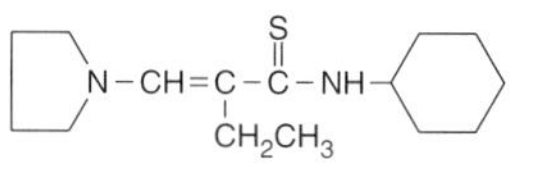

250 mg

S35,740-5
3-ALLYL-2,2,4,4-TETRAMETHYLIMIDAZOLIDINE-5-THIONE-HYDROCHLORIDE

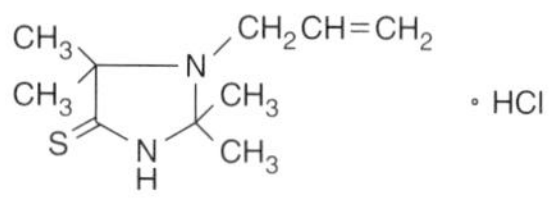

5 gm

S39,197-2
2-(DIMETHYLTHIOCARBAMOYLTHIO)-ACETIC ACID

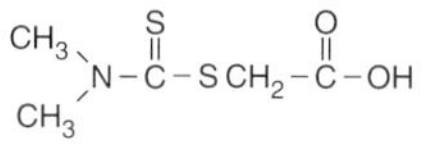

5 gm

S74,411-5
4-THIOXO-2,2,3-TRIMETHYL-5-OXAZOLIDINONE
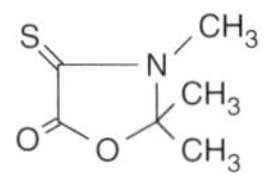
250 mg

S95,011-4
2-(METHOXYCARBONYL)ETHYL-N,N-DIMETHYLDITHIOCARBAMATE
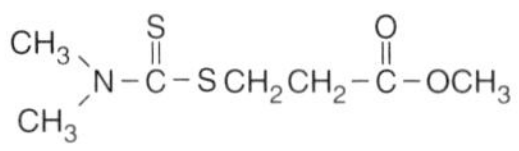
50 mg

S89,772-8
HEXAHYDRO-1-METHYL-2-PYRIMIDINETHIONE
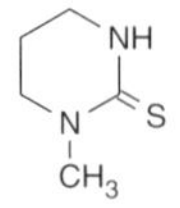
250 mg

S90,487-2
N-DECYL-4-METHYL-1-PIPERAZINETHIOCARBOXAMIDE
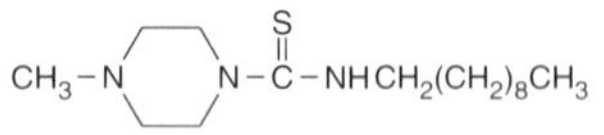
250 mg

S51,388-1
5-METHYLHEXAHYDRO-1,3,5-TRIAZINE-2-THIONE
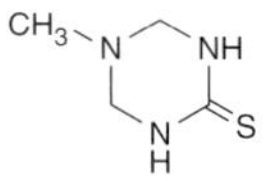
250 mg

S89,560-1
S-METHYL-N,N-DIMETHYLDITHIOCARBAMATE
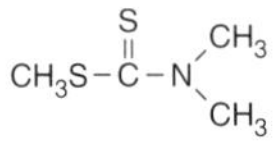
250 mg

S39,638-9
3,5-DIMETHYLTETRAHYDRO-1,3,5-THIADIAZINE-2-THIONE
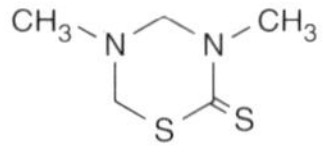
5 gm

S35,988-2
CYCLOPENTAMETHYLENEDITHIOCARBAMIC ACID, SODIUM SALT DIHYDRATE
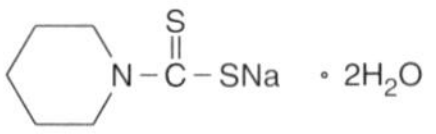
5 gm

S51,017-3
GUANIDINODITHIOCARBAMIC ACID
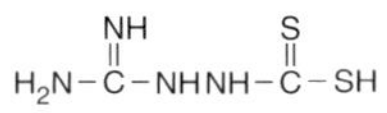
250 mg

S67,686-1
3-(DIMETHYLAMINO)RHODANINE
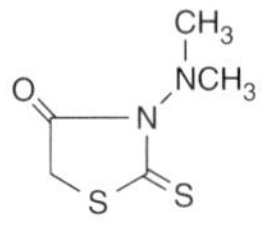
5 gm

S93,818-1
1-(3-METHYL-1-ADAMANTYLACETYL)-3-THIOSEMICARBAZIDE
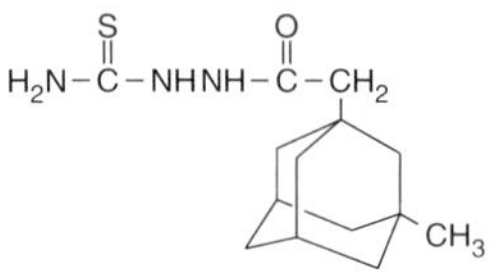
250 mg

S67,762-0
2-THIOBIUREA
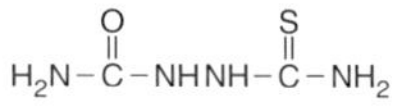
5 gm

S36,770-2
1,6-DIALLYL-2,5-DITHIOBIUREA
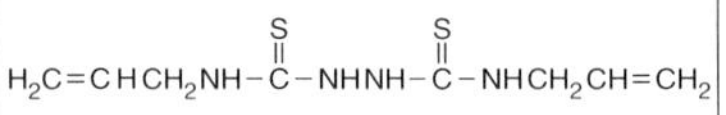
5 gm

S93,816-5
1-(3-METHYL-1-ADAMANTYLCARBONYL)-3-THIOSEMICARBAZIDE
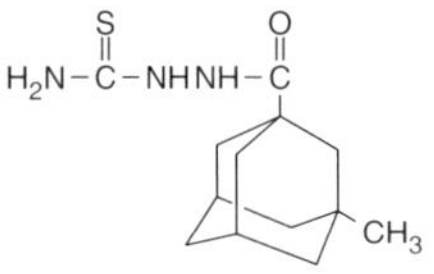
250 mg

S51,214-1
4-(1,1-DIMETHYL-3-OXOBUTYL)-3-THIOSEMICARBAZIDE
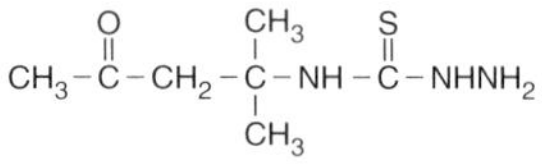
250 mg

S44,230-5
ACETONE THIOSEMICARBAZONE
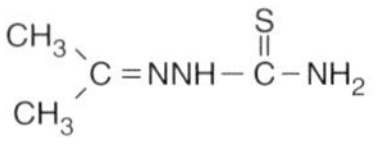
5 gm

S89,256-4
CROTONALDEHYDE THIOSEMICARBAZONE
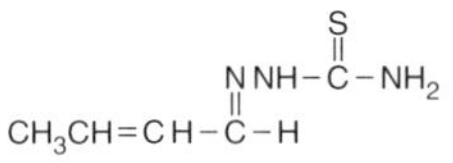
250 mg

S89,257-2
ISOBUTYRALDEHYDE THIOSEMICARBAZONE
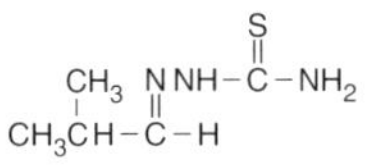
250 mg

S67,668-3
CYCLODODECANONE THIOSEMICARBAZONE
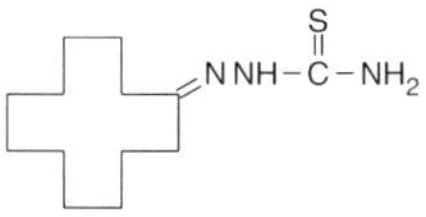
5 gm

S89,261-0
18-PENTATRIACONTANONE THIOSEMICARBAZONE
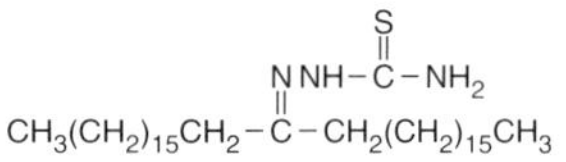
250 mg

S89,258-0
1-CYCLOHEXANONE THIOSEMICARBAZONE
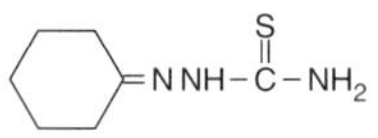
250 mg

S58,810-5
METHYL 3-(3,4,5,6-TETRAHYDRO-2H-AZEPIN-7-YL)DITHIOCARBAZATE
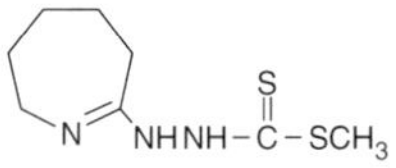
250 mg

S54,569-4
3,5-DICYCLOHEXYL-3,4,5,6-TETRAHYDRO-2H-1,3,5-THIADIAZINE-2-THIONE
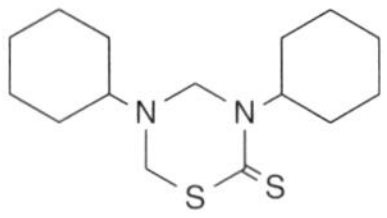
250 mg

S55,337-9
3,5-BIS(3,3,5-TRIMETHYLCYCLOHEXYL)-TETRAHYDRO-1,3,5-THIADIAZINE-2(2H)-THIONE
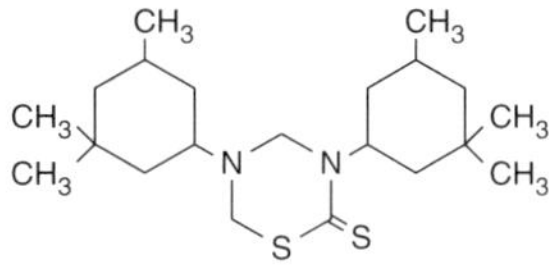
250 mg

S50,049-6
N-(DITHIOETHOXYCARBONYL)GLYCINE
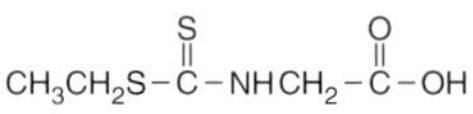
250 mg

S87,555-4
1-FORMYL-4-METHYL-3-THIOSEMICARBAZIDE

250 mg

S35,201-2
4-OXO-2-THIOXOHEXAHYDROPYRIMIDINE-3-PROPIONIC ACID
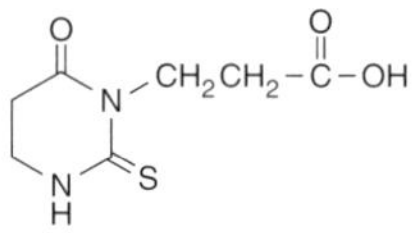
250 mg

S64,275-4
5-ISOBUTYLIDENERHODANINE
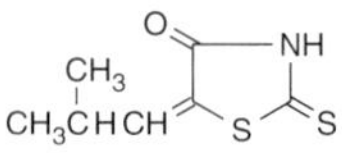
250 mg

S63,037-3
5,5-DIMETHYL-2-THIOXO-4-OXAZOLIDINONE
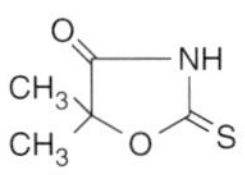
5 gm

S66,181-3
5,5-DIMETHYL-3-ISOPROPYLIMIDAZOLIDINE-2-THIONE
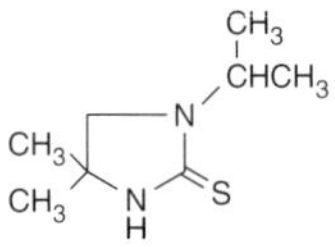
5 gm

S64,748-9
1-(TERT-BUTYL)-4-THIOXOHEXAHYDRO-1,3,5-TRIAZINE
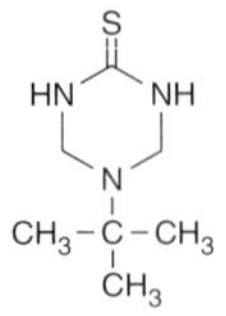
5 gm

S65,827-8
1-(TERT-OCTYL)-4-THIOXOHEXAHYDRO-1,3,5-TRIAZINE
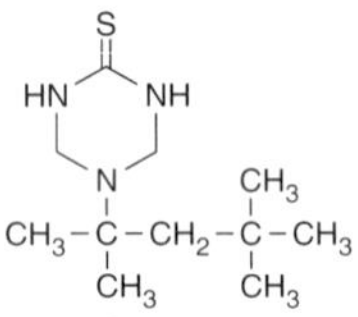
5 gm

S76,156-7
6-METHYL-1,2,3,5-TETRAHYDRO-5-THIOXOIMIDAZO(1,2-C)PYRIMIDIN-7(6H)-ONE
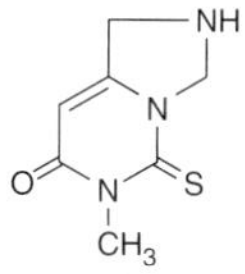
250 mg

S55,615-7
3,6-DIMETHYL-2-THIOURACIL
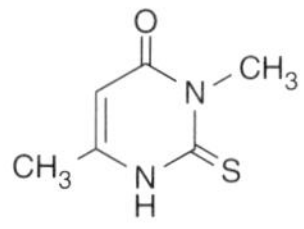
250 mg

S76,362-4
THIOPARABANIC ACID
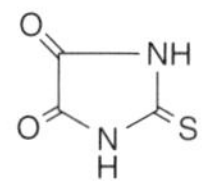
250 mg

S56,450-8
2-THIOBARBITURIC ACID, SODIUM DERIVATIVE
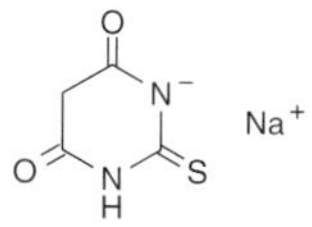
5 gm

S34,930-5
5-ETHYL-5-(2-METHYLALLYL)-2-THIOBARBITURIC ACID
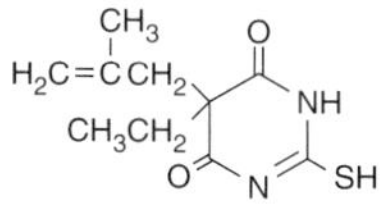
5 gm

S35,862-2
N,N'-BIS(CARBOXYMETHYL)-DITHIOOXAMIDE
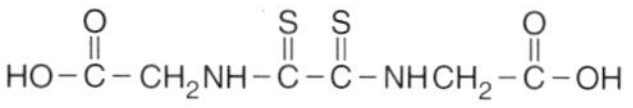
5 gm

S40,612-0
1,3-BIS(5-CARBOXYPENTYL)-2-THIOUREA
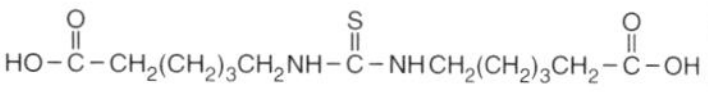
5 gm

S52,119-1
N-(THIOCARBAMOYL)GLYCINE
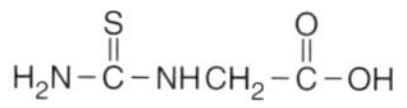
5 gm

S41,005-5
3-RHODANINEPROPIONIC ACID
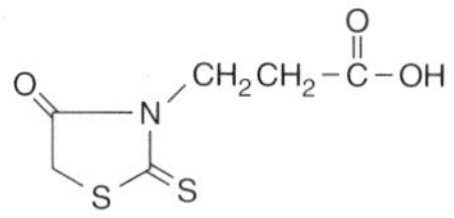
5 gm

S34,788-4
3-(THIOCARBAMYLTHIO)PROPIONIC ACID
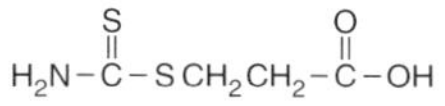
5 gm

S51,694-5
3-(N-ETHYLTHIOCARBAMOYLTHIO)-PROPIONIC ACID
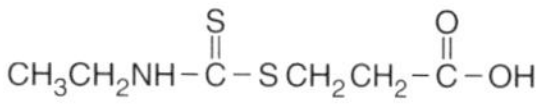
5 gm

S41,290-2
DIETHYL DELTA-5,5'-BIRHODANINE-3,3'-DIACETATE
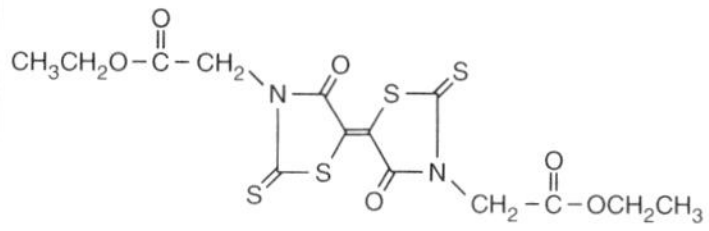
250 mg

S41,289-9
DIETHYL 5,5'-BIRHODANINE-3,3'-DIACETATE
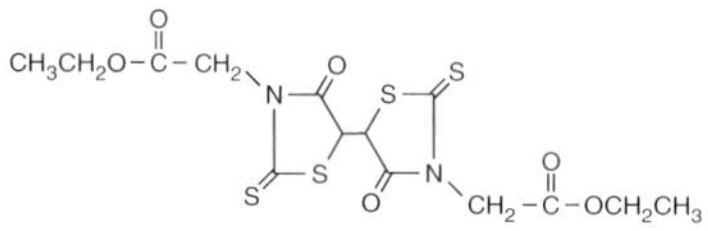
250 mg

S61,841-1
ETHYL 3-ETHYL-2-THIOHYDANTOIN-1-ACETATE
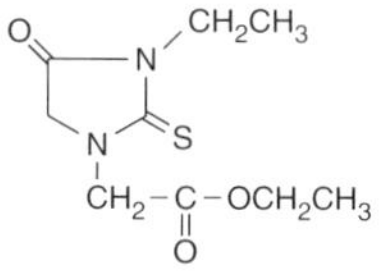
5 gm

S36,193-3
ETHYL 3-RHODANINEACETATE
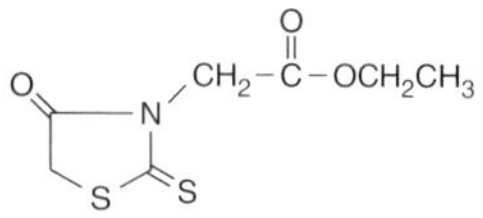
250 mg

S34,962-3
ETHYL 3-THIOCARBAMOYLCARBAZATE
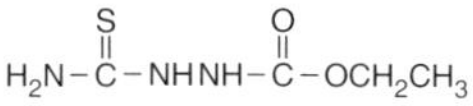
5 gm

S69,157-7
1-(CYCLOHEXYLCARBAMOYL)-3-THIOSEMICARBAZIDE
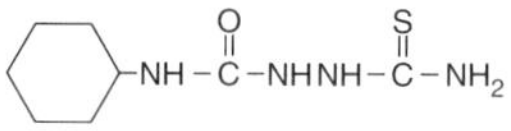
250 mg

S73,757-7
1-CYCLOHEXYL-5-THIOBIUREA
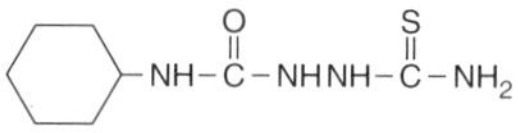
5 gm

S44,841-9
3-DIACETYLAMINORHODANINE
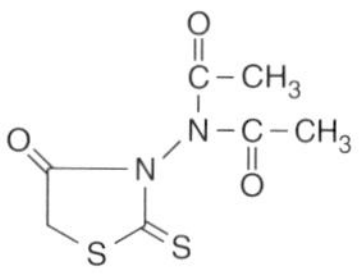
5 gm

S81,345-1
1-HEXADECANOYL-2-THIOUREA
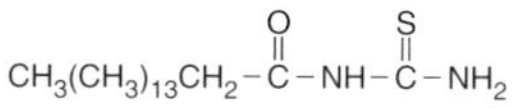
250 mg

S85,917-6
5,6-DIHYDRO-1-METHYL-2-THIOURACIL
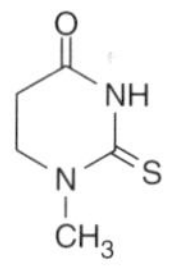
250 mg

S41,288-0
3,3'-DIETHYL-DELTA-5,5'-BIRHODANINE
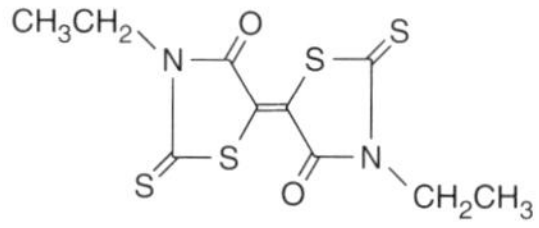
250 mg

S52,371-2
3-CYCLOHEXYLRHODANINE
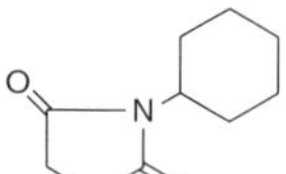
5 gm

S83,101-8
S-TETRADECYL TETRADECANOTHIOIMIDATE HYDROCHLORIDE
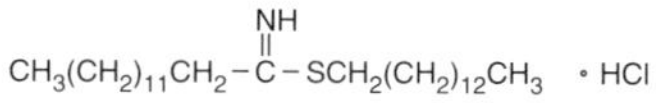
250 mg

S62,287-7
N,N-DIMETHYLTHIOACETAMIDE
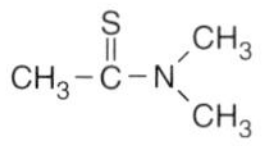
250 mg

S55,367-0
THIOOCTADECANAMIDE
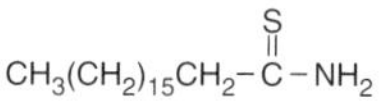
250 mg

S94,117-4
7,14-DIAZADISPIRO(5.1.5.2)-PENTADECA-15-THIONE
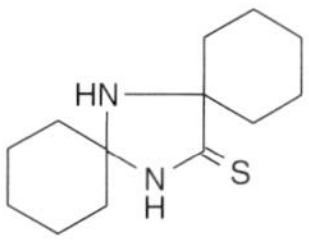
5 gm

S94,805-5
1-ADAMANTANETHIOCARBOXAMIDE
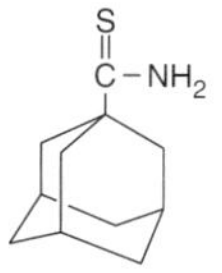
250 mg

S42,880-9
1-FORMYLTHIOSEMICARBAZIDE
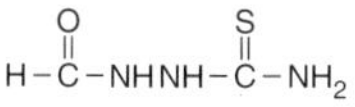
5 gm

S40,625-2
N-ACETAMIDORHODANINE HYDRATE
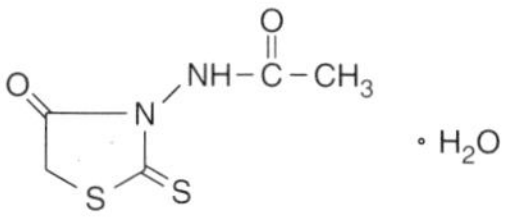
5 gm

S88,851-6
4-BUTYL-1,1-DIMETHYL-3-THIOSEMICARBAZIDE
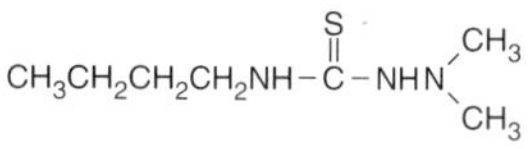
250 mg

S95,314-8
O-CYCLOHEXYL THIOCARBAZATE
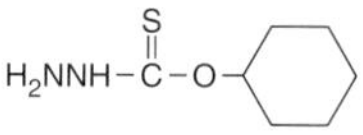
50 mg

S85,941-9
O-METHYL N-(N-CYCLOHEXYLCARBAMOYL)-THIOCARBAMATE

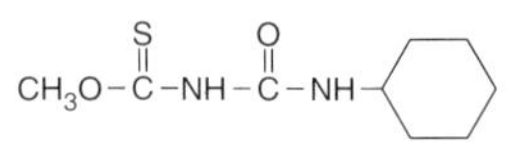

250 mg

S85,940-0
O-ETHYL N-(N-CYCLOHEXYLCARBAMOYL)-THIOCARBAMATE

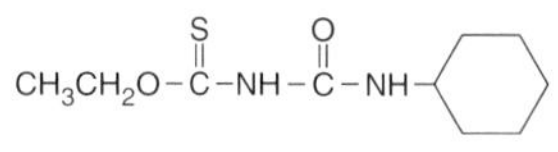

250 mg

S86,252-5
1-AMIDINO-2-THIOUREA OXALATE

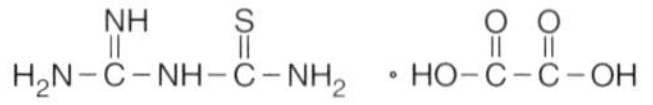

250 mg

S51,217-6
1-METHYL-3-PROPYL-2-THIOUREA

250 mg

S56,597-0
1-BUTYL-2-THIOUREA

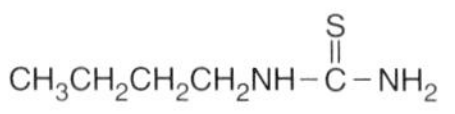

5 gm

S86,257-6
1-AMIDINO-3-PROPYL-2-THIOUREA

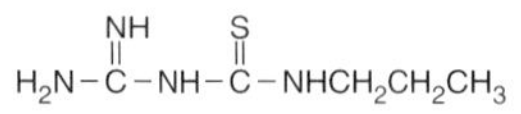

250 mg

S37,112-2
1,3-DIALLYL-2-THIOUREA

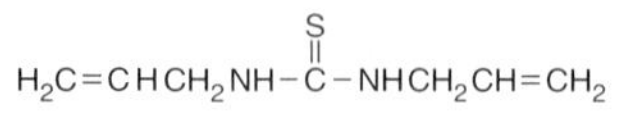

5 gm

S52,126-4
1,3-DIPROPYL-2-THIOUREA

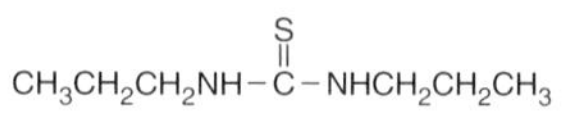

5 gm

S44,682-3
1,3-DI-TERT-BUTYL-2-THIOUREA

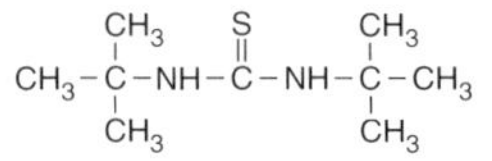

5 gm

S40,354-7
1-DODECYL-2-THIOUREA

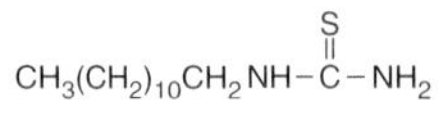

250 mg

S57,038-9
1,3-DIHEXYL-2-THIOUREA

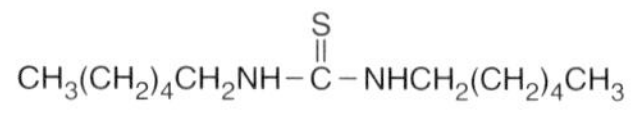

5 gm

S82,113-6
1-DODECANOYL-2-THIOUREA

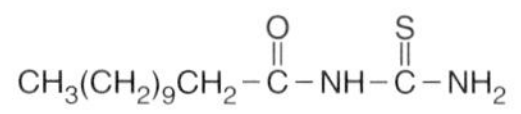

250 mg

S62,236-2
3-ALLYL-1,1-DIPENTYLUREA

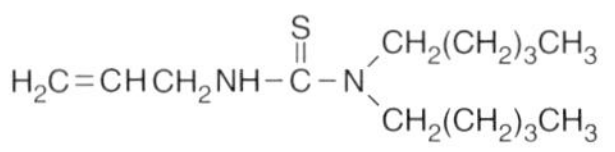

5 gm

S68,565-8
1,3-DIHEPTYL-2-THIOUREA

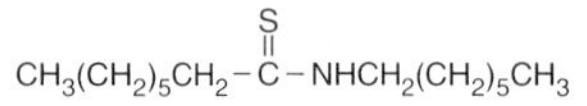

5 gm

S36,123-2
1,3-DIOCTYL-2-THIOUREA

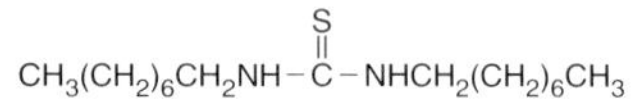

5 gm

S68,043-5
1,3-DIDECYL-2-THIOUREA

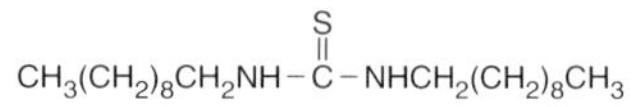

250 mg

S90,010-9
1-ALLYL-3-OCTADECYL-2-THIOUREA

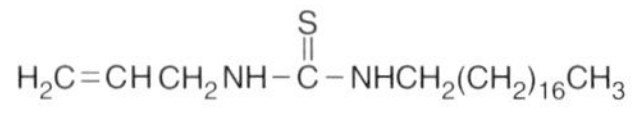

250 mg

S54,031-5
1,3-DIDODECYL-2-THIOUREA

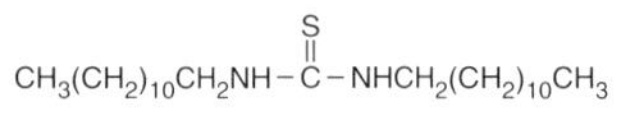

5 gm

S81,195-5
1,3-DITETRADECYL-2-THIOUREA

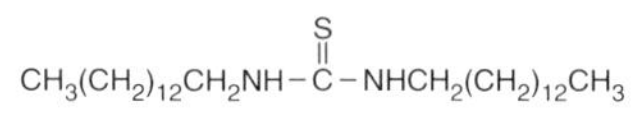

250 mg

S81,182-3
1,3-DIHEXADECYL-2-THIOUREA

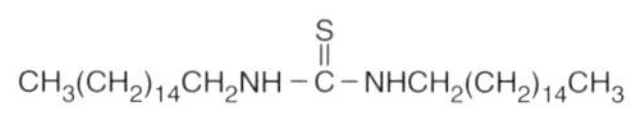

250 mg

S36,766-4
4-CYCLOHEXYL-3-THIOSEMICARBAZIDE

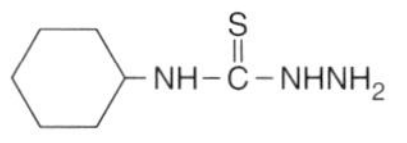

5 gm

S34,581-4
1,3-DICYCLOHEXYL-2-THIOUREA

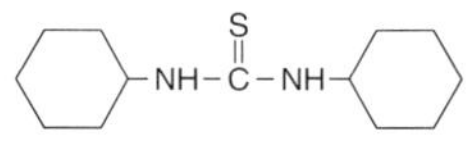

5 gm

S87,144-3
4,6-DIMETHYL-HEXAHYDRO-S-TRIAZINE-2-THIONE

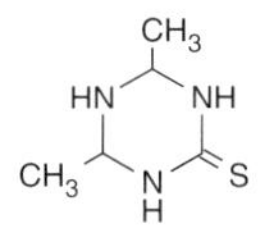

250 mg

S94,855-1
1-(1-ADAMANTYL)-3-METHYL-2-THIOUREA

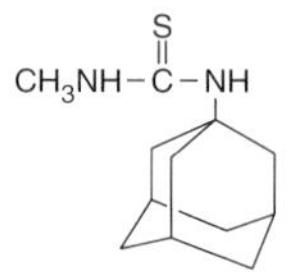

250 mg

S94,838-1
1-ADAMANTYL METHYL KETONE THIOSEMICARBAZONE

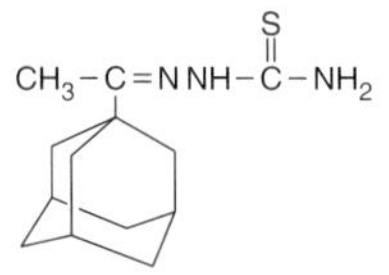

250 mg

S65,140-0
TETRAMETHYLENEBIS-(ISOTHIURONIUM BROMIDE)

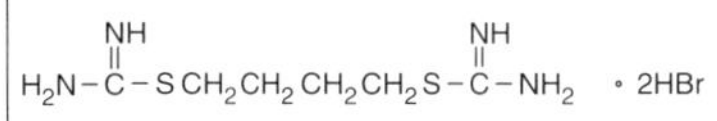

250 mg

S52,732-7
N,N'-(1,4-CYCLOHEXYLENEDIMETHYL)-BIS(DITHIOCARBAMIC ACID, AMMONIUM SALT)

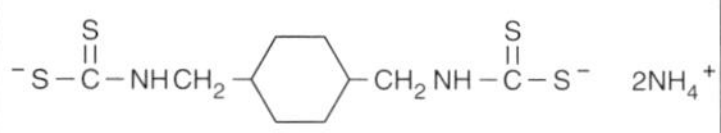

250 mg

S34,509-1
CYCLOPENTAMETHYLENEDITHIOCARBAMIC ACID, CADMIUM SALT

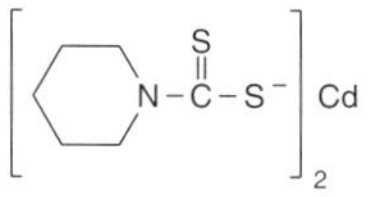

5 gm

S52,421-2
4,4'-METHYLENEBIS(1-PIPERIDINEDITHIOCARBOXYLIC ACID, AMMONIUM SALT)

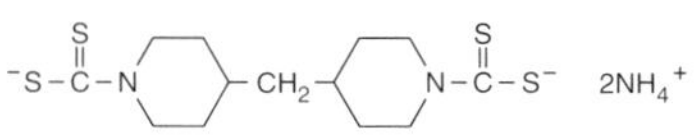

5 gm

S52,372-0
4,4'-TRIMETHYLENEBIS(1-PIPERIDINEDITHIOCARBOXYLIC ACID, AMMONIUM SALT)

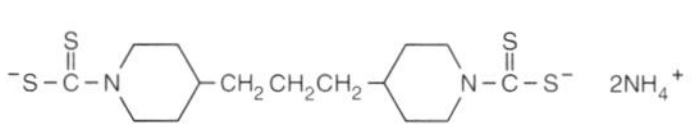

5 gm

S35,484-8
1,4-PIPERAZINEDICARBODITHIOIC-ACID, DISODIUM SALT HEXAHYDRATE

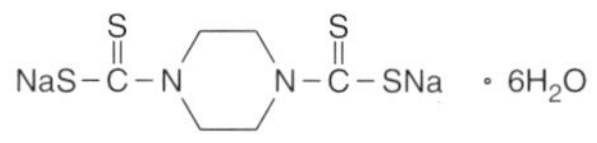

5 gm

S35,362-0
4-MORPHOLINECARBODITHIOIC ACID, SODIUM SALT

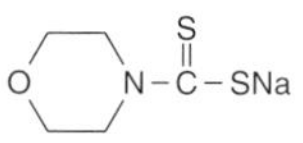

5 gm

S94,572-2 TETRAHYDRO-2H-1,3-THIAZINE-2-THIONE	S51,514-0 4-CYCLOHEXYL-1H-TETRAZOLE-5(4H)-THIONE
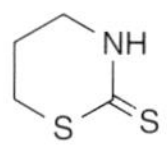	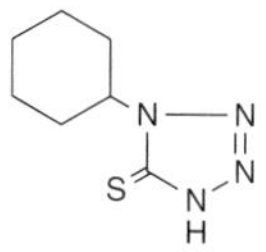
5 gm	250 mg

S88,127-9
3-BROMO-2,2-DIMETHYLPROPIONITRILE

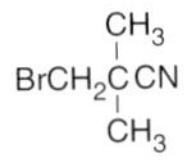

250 mg

S75,762-4
ETHYL N-(1-CYANO-2,2-DICHLOROVINYL)-ARBAMATE

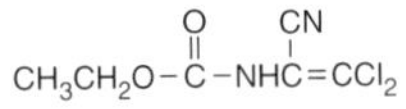

5 gm

S89,444-3
PENTACHLOROPROPIONITRILE

Cl_3CCl_2CCN

250 mg

S95,636-8
ETHYL 3-CYANO-2-HYDROXY-1,4,4A,5,6,7,8,8A-OCTA-H-4-OXO-1-NAPHTHALENECARBOXYLATE

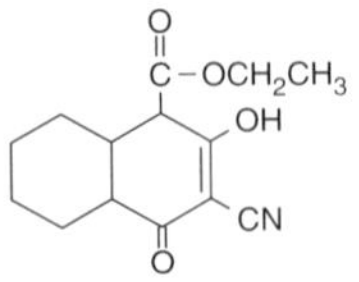

50 mg

S57,818-5
3-(2-HYDROXYETHYLTHIO)PROPIONITRILE

$HOCH_2CH_2SCH_2CH_2CN$

5 gm

S74,230-9
(3-HYDROXY-1-METHYL-2-BUTENYL)-MALONONITRILE

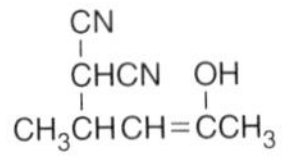

5 gm

S34,378-1
DL-CARNITINENITRILE CHLORIDE

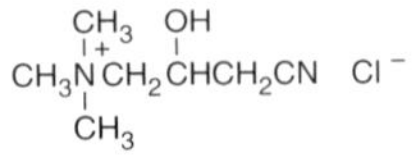

5 gm

S74,164-7
1-CYANO-4-HYDROXY-1-CYCLOHEXANEACETONITRILE

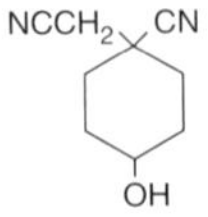

250 mg

S36,180-1
ETHYL METHYL KETONE CYANOHYDRIN

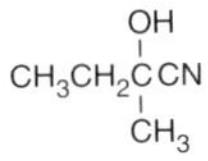

5 gm

S55,683-1
1-HYDROXY-2,2,6,6-TETRACHLOROCYCLOHEXANECARBONITRILE

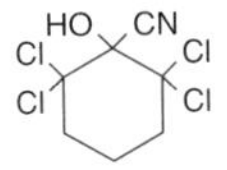

5 gm

S91,475-4
5-CYANO-1-ETHYL-6-HYDROXY-4-METHYL-2(1H)-PYRIDONE

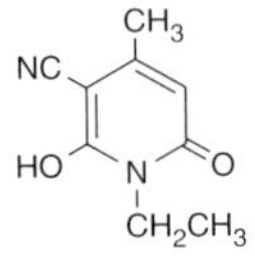

250 mg

S55,070-1
2-CYANO-N-(2-HYDROXYETHYL)-ACETAMIDE

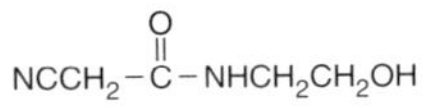

250 mg

S81,538-1
3-CYANO-1-PROPANOL

$NCCH_2CH_2CH_2OH$

250 mg

S88,385-9
2-CYANO-3-HYDROXYACRYLAMIDE, SODIUM SALT

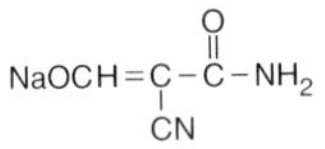

250 mg

S55,582-7
METHYL 2-CYANO-3-ETHOXYACRYLATE

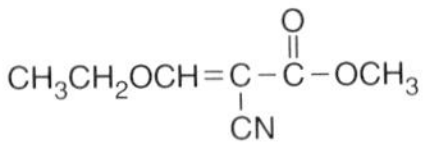

5 gm

S88,439-1
METHYL 9-CYANO-2-METHOXYNONANOATE

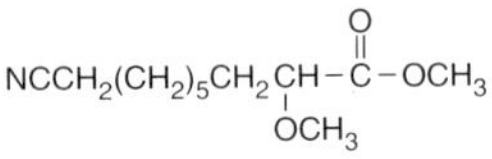

250 mg

S39,950-7
3,3'-(ETHYLENEDIOXY)DIPROPIONITRILE

$NCCH_2CH_2OCH_2CH_2OCH_2CH_2CN$

250 mg

S50,003-8
4,7,10-TRIOXATRIDECANEDINITRILE

$NCCH_2CH_2OCH_2CH_2OCH_2CH_2OCH_2CH_2CN$

5 gm

S36,230-1
1,2,3,4,5,6-HEXAKIS-(2-CYANOETHOXY)-HEXANE

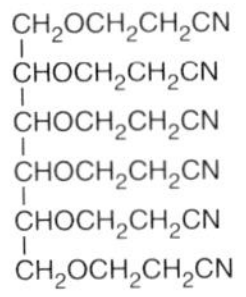

5 gm

S44,463-4
2-CYANO-3-ETHOXYACRYLIC ACID

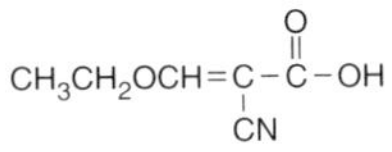

5 gm

S56,318-8
3-ISOPROPOXYPROPIONITRILE

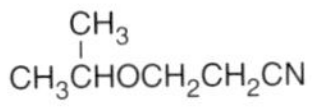

5 gm

S70,546-2
BIS(4-CYANOBUTYL) ETHER

$NCCH_2CH_2CH_2CH_2OCH_2CH_2CH_2CH_2CN$

5 gm

S50,005-4
3-OCTYLOXYPROPIONITRILE

$CH_3(CH_2)_6CH_2OCH_2CH_2CN$

250 mg

S53,030-1
6,6'-OXYDIHEXANENITRILE

$NCCH_2(CH_2)_3CH_2OCH_2(CH_2)_3CH_2CN$

5 gm

S38,347-3
3-(DODECYLOXY)PROPIONITRILE

$CH_3(CH_2)_{10}CH_2OCH_2CH_2CN$

250 mg

S56,259-9
3-(OCTADECYLOXY)PROPIONITRILE

$CH_3(CH_2)_{16}CH_2OCH_2CH_2CN$

250 mg

S58,302-2
3-(TERT-OCTADECYLOXY)PROPIONITRILE

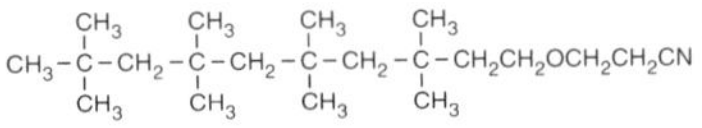

5 gm

S35,562-3
TETRAACETYL-L-RHAMNONITRILE

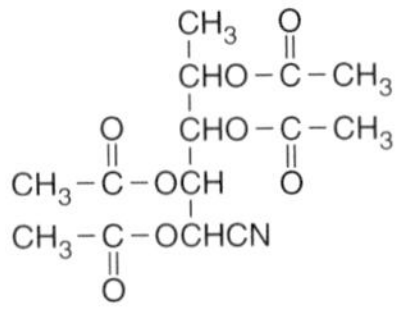

250 mg

S91,740-0
1-S-(2-CYANOETHYL)-2,3,4,6-TETRA-O-ACETYL-1-THIO-BETA-D-GALACTOPYRANOSIDE

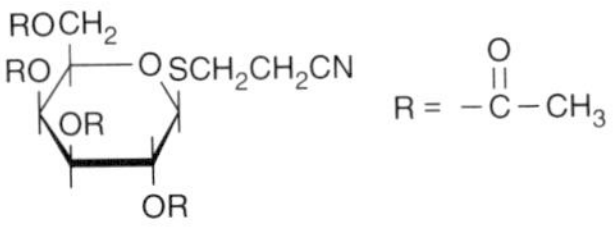

250 mg

S62,867-0
1,4-BUTANEDIOL BIS(CYANOACETATE)

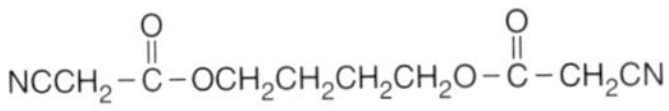

5 gm

S95,638-4
ETHYL 3-CYANO-2-HYDROXY-OCTA-H-4-OXO-1-NAPHTHALENECARBOXYLATE HYDRAZINE SALT

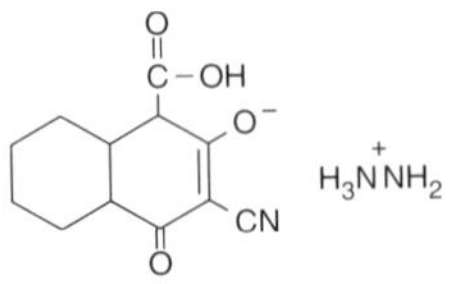

50 mg

S71,343-0
TRIETHYL 3-CYANO-2,3,4-PENTANETRICARBOXYLATE

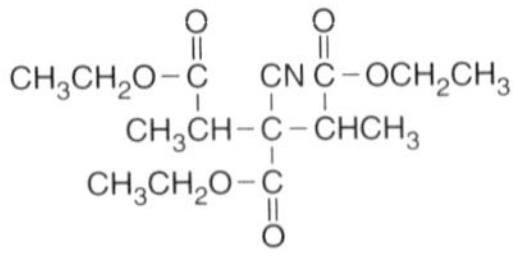

5 gm

S43,517-1
2-CYANOETHYL THIOETHER

$NCCH_2CH_2SCH_2CH_2CN$

5 gm

S91,795-8
3,3'-DITHIOBIS(2,2,3,3-TETRAMETHYLPROPIONITRILE)

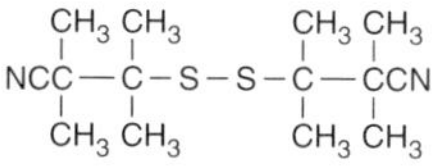

250 mg

S93,830-0
4,7,10-TRIAZADECANENITRILE TRIHYDROCHLORIDE

$H_2NCH_2CH_2NHCH_2CH_2NHCH_2CH_2CN \cdot 3HCl$

250 mg

S94,689-3
2-AMINO-2-ADAMANTANECARBONITRILE HYDROCHLORIDE

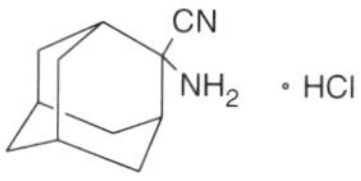

250 mg

S85,968-0
4-AMINO-1-(2-CYANOETHYL)-2(1H)-PYRIMIDINONE

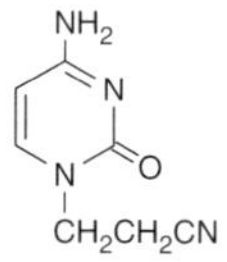

250 mg

S95,637-6
ETHYL 2-AMINO-3-CYANO-1,4,4A,5,6,7,8,8A-OCTAHYDRO-4-OXO-1-NAPHTHALENECARBOXYLATE

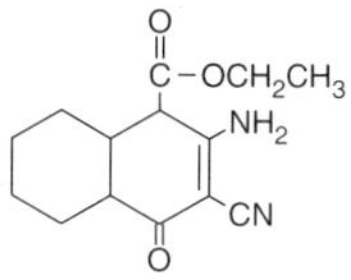

50 mg

S63,911-7
3-AMINO-1-(2-CYANOPROPYL)-2-PYRAZOLIDIN-5-ONE

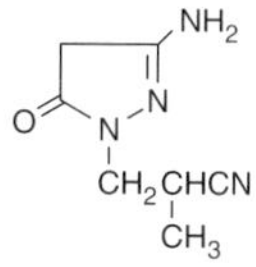

250 mg

S58,275-1
3-AMINO-1-(2-CYANOETHYL)-2-PYRAZOLIN-5-ONE

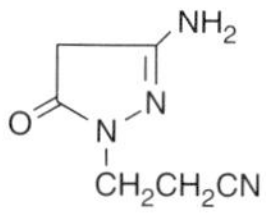

5 gm

S78,716-7
5-AMINO-4-CYANO-1-METHYL-2-PYRAZOLINE

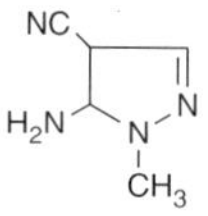

250 mg

S51,659-7
9-AMINONONANENITRILE

$H_2NCH_2(CH_2)_6CH_2CN$

5 gm

S45,428-1
4-(AMINOMETHYL)-CYCLOHEXANECARBONITRILE

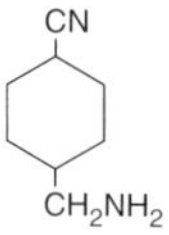

5 gm

S87,420-5
3,4-BIS[(2-CYANOETHYL)AMINO]-3-CYCLOBUTENE-1,2-DIONE

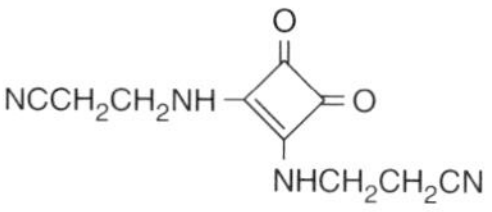

250 mg

S43,016-1
ETHYLENEDIAMINE-N,N,N',N'-TETRAACETONITRILE

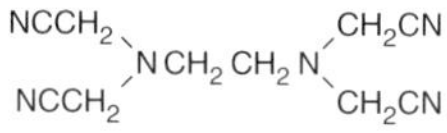

5 gm

S57,088-5
N-METHYLGLYCINONITRILE SULFATE

$CH_3NHCH_2CN \cdot H_2SO_4$

5 gm

S35,155-5
2-(ISOBUTYLAMINO)ACETONITRILE

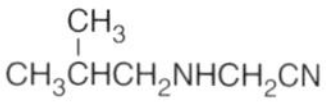

5 gm

S43,294-6
1-ETHYLAMINOCYCLOHEXANECARBONITRILE

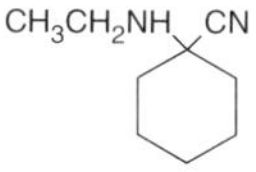

5 gm

S65,795-6
N-(CYCLOHEXYL)-CYCLOHEXANECARBONITRILE

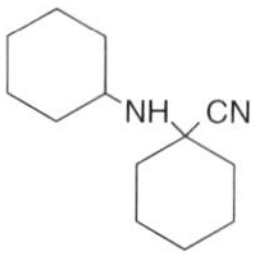

5 gm

S41,838-2
NITRILOTRIACETONITRILE

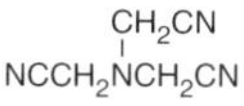

5 gm

S36,907-1
(N-HEXAMETHYLENEIMINO)ACETONITRILE

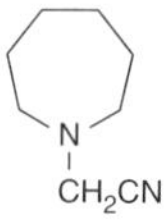

5 gm

S34,300-5
N-BUTYL-N-ETHYLAMINOACETONITRILE

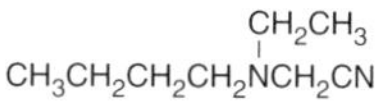

5 gm

S34,645-4
N,N-DIISOPROPYLAMINOACETONITRILE

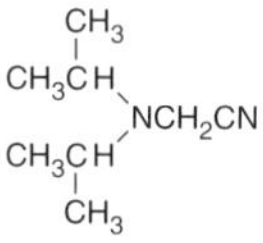

5 gm

S34,774-4
2-(DIPROPYLAMINO)ACETONITRILE

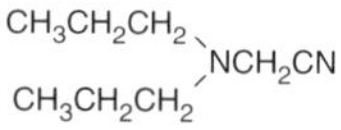

5 gm

S43,380-2
DIISOBUTYLAMINOACETONITRILE

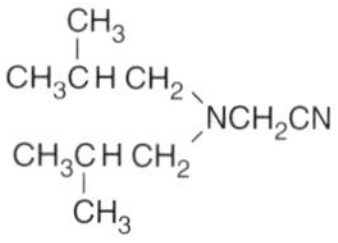

5 gm

S53,557-5
N,N-DIBUTYLGLYCINONITRILE

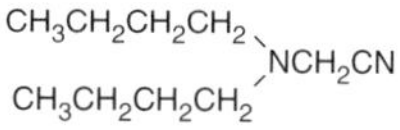

250 mg

S36,477-0
1-PYRROLIDINEACETONITRILE

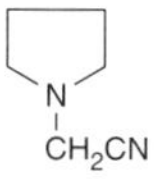

5 gm

S41,150-7
1-DIMETHYLAMINO-CYCLOHEXANECARBONITRILE

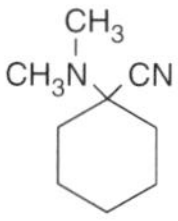

5 gm

S40,987-1
1-PIPERIDINEACETONITRILE

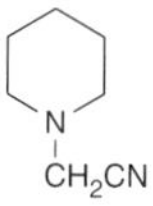

5 gm

S35,364-7
4-MORPHOLINEACETONITRILE

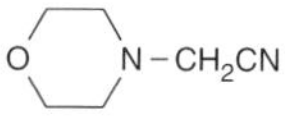

5 gm

S80,307-3
2-CYANO-2-NITROSOACETAMIDE

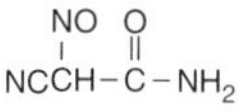

250 mg

S43,425-6
N-(CYANOMETHYL)ACETAMIDE

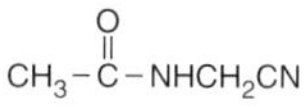

5 gm

S57,005-2
N-(1-CYANO-1-ETHYLPROPYL)ACETAMIDE

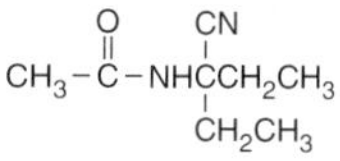

250 mg

S38,330-9
N-(1-CYANOCYCLOHEXYL)-N-METHYLVALERAMIDE

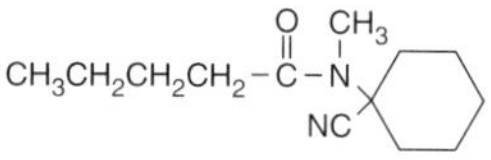

250 mg

S95,516-7
3-CYANO-2-METHYL-3,4,5,6-TETRAHYDRO-2H-1,2-OXAZINE

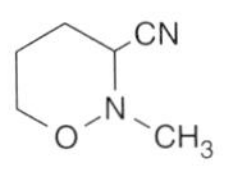

50 mg

S70,500-4
1,3,5-TRIS(CYANOMETHYL)HEXAHYDRO-1,3,5-TRIAZINE

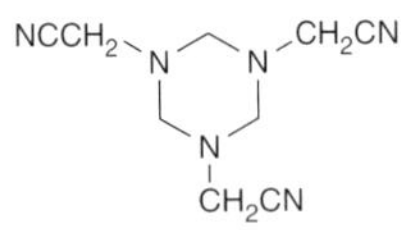

250 mg

S87,428-0
3-[(2-CYANOETHYL)AMINO]-3-CYCLOBUTENE-1,2-DIONE

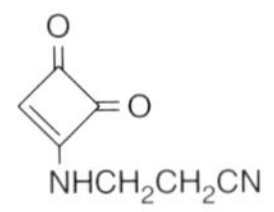

100 mg

S54,139-7
3-(ETHYLAMINO)PROPIONITRILE

$CH_3CH_2NHCH_2CH_2CN$

250 mg

S54,058-7
3,3'-IMINODIPROPIONITRILE HYDROCHLORIDE

$NCCH_2CH_2NHCH_2CH_2CN \cdot HCl$

5 gm

S39,949-3
3-(BUTYLAMINO)PROPIONITRILE

$CH_3CH_2CH_2CH_2NHCH_2CH_2CN$

250 mg

S43,374-8
3-(CYCLOHEXYLAMINO)PROPIONITRILE

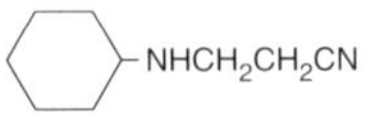

5 gm

S50,645-1
METHYL 2-CYANO-3-(DIMETHYLAMINO)-ACRYLATE

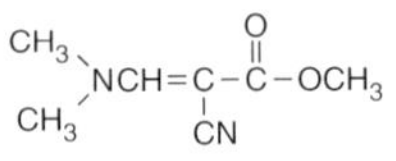

5 gm

S43,400-0
ETHYL 2-CYANO-3-DIMETHYLAMINOACRYLATE

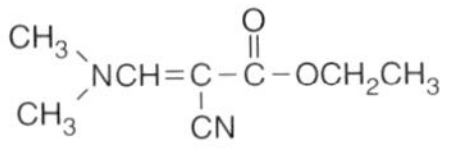

5 gm

S43,893-6
3-DIETHYLAMINOPROPIONITRILE

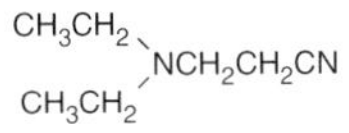

5 gm

S55,177-5
N-(2-CYANOETHYL)SUCCINIMIDE

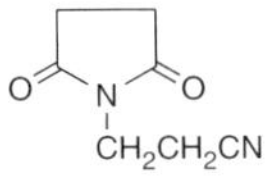

250 mg

S36,909-8
HEXAHYDRO-1H-AZEPINE-1-PROPIONITRILE

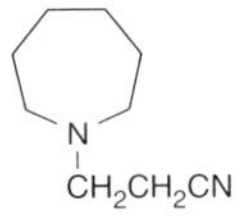

5 gm

S34,778-7
3-(DIPROPYLAMINO)PROPIONITRILE

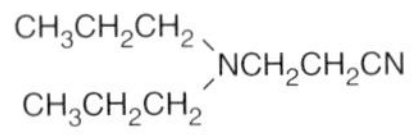

5 gm

S34,886-4
3-(DIBUTYLAMINO)PROPIONITRILE

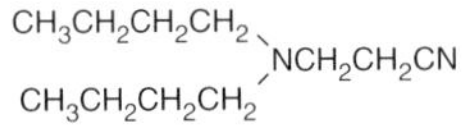

5 gm

S35,539-9
1-PYRROLIDINEPROPIONITRILE

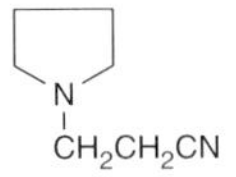

5 gm

S87,541-4
2-(DIETHYLAMINO)-3-CYCLOHEXENE-1-CARBONITRILE PERCHLORATE

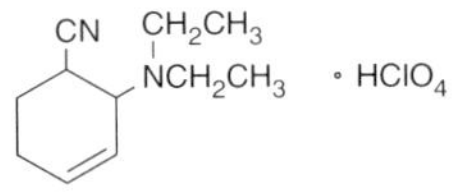

250 mg

S42,015-8
GLYOXAL N-CYANOACETYL-N'-PROPIONYLDIHYDRAZONE

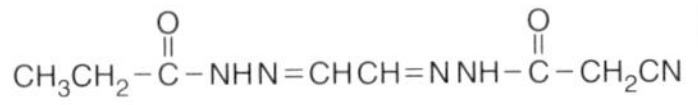

250 mg

S67,957-7
2-CYANO-N-VINYLACETAMIDE

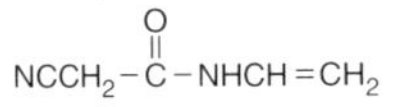

5 gm

S55,338-7
ETHYL 2-CARBAMOYL-2-CYANOACETATE, SODIUM SALT

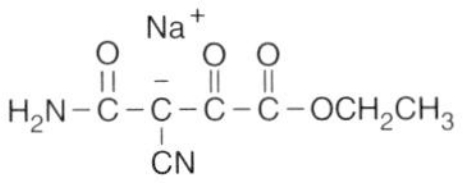

5 gm

S51,612-0
2,6-BIS(2,2-DICYANOETHYL)-CYCLOHEXANONE

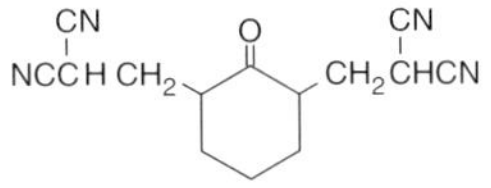

250 mg

S43,183-4
1,2-DIHYDRO-3,6-DIOXO-1-PYRIDAZINEPROPIONITRILE

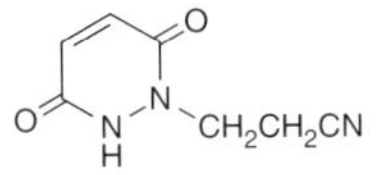

5 gm

S64,582-6
DIMETHYL 3,3-DICYANOPIMELATE

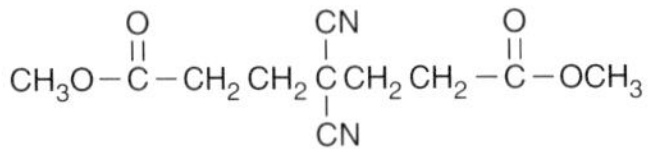

250 mg

S40,015-7
1-BUTYL-3-(CYANOACETYL)UREA

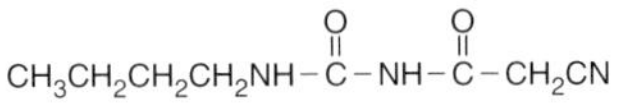

250 mg

S34,578-4
2,4-DICYANO-3,3-PENTAMETHYLENEGLUTARIMIDE

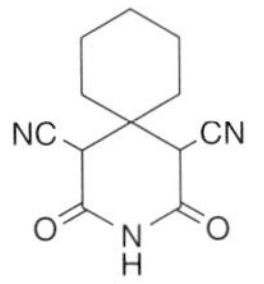

250 mg

S68,178-4
2-CYANO-N-METHYLACETAMIDE

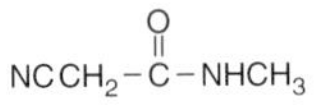

250 mg

S95,006-8
2-CYANO-N-(2-CYANOETHYL)ACETAMIDE

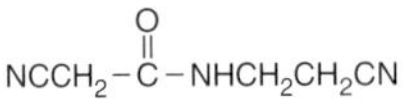

50 mg

S45,842-2
1-CYANOACETYLPYRROLIDINE

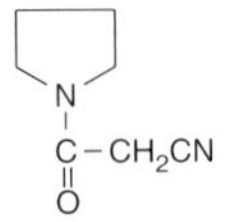

5 gm

S42,988-0
4-CYANOACETYLMORPHOLINE

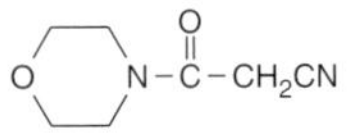

5 gm

S41,164-7
1-AMINO-1,2-DIHYDRO-4,6-DIMETHYL-2-OXONICOTINONITRILE

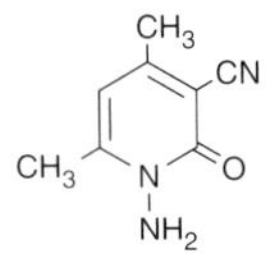

5 gm

S85,965-6
1-(2-CYANOETHYL)UREA

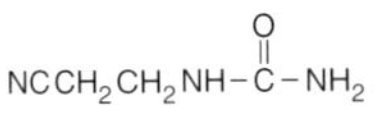

250 mg

S78,313-7
1-(2-CYANOETHYL)-1-METHYLUREA

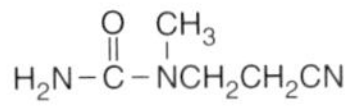

5 gm

S78,376-5
1-(2-CYANOETHYL)-1-CYCLOHEXYLUREA

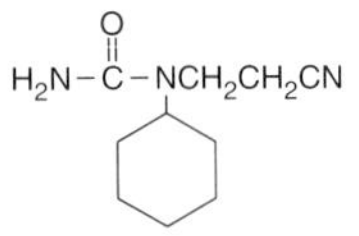

5 gm

S88,493-6
1,1,3,3,5,5-HEXACYANOPENTANE
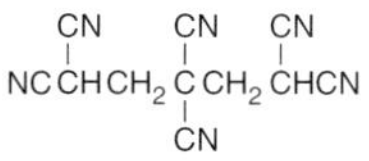
250 mg

S68,334-5
1-(2-CYANOETHYL)AZIRIDINE
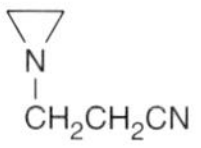
5 gm

S57,316-7
CYCLOHEXYLIDENEMALONONITRILE
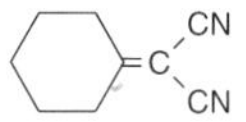
5 gm

S34,713-2
2,6-DIMETHYL-4-MORPHOLINEPROPIONITRILE
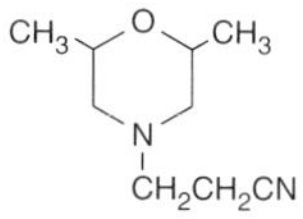
5 gm

S34,662-4
4-DIMETHYLAMINOBUTYRONITRILE
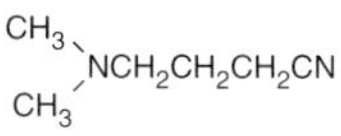
5 gm

S43,892-8
4-(DIETHYLAMINO)BUTYRONITRILE
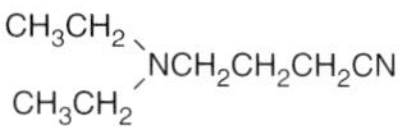
5 gm

S37,350-8
4-(HEXAMETHYLENEIMINO)BUTYRONITRILE
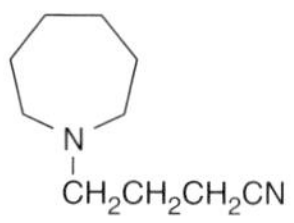
5 gm

S34,646-2
4-(DIISOPROPYLAMINO)BUTYRONITRILE
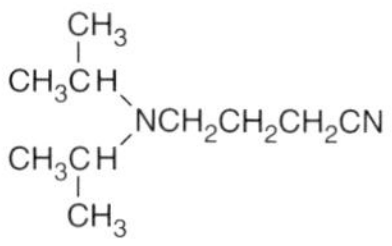
5 gm

S34,775-2
4-(DIPROPYLAMINO)BUTYRONITRILE
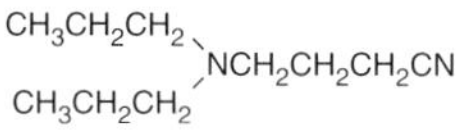
5 gm

S34,913-5
4-(DIBUTYLAMINO)BUTYRONITRILE
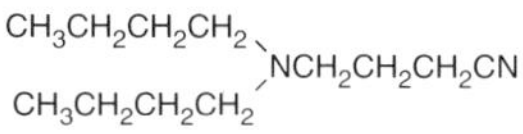
5 gm

S35,485-6
1-PIPERIDINEBUTYRONITRILE
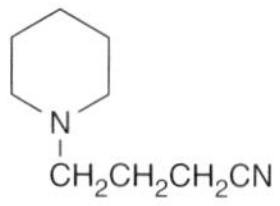
5 gm

S35,366-3
4-MORPHOLINEBUTYRONITRILE
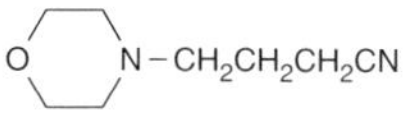
5 gm

S64,770-5
4-METHYL-4-NITROVALERONITRILE
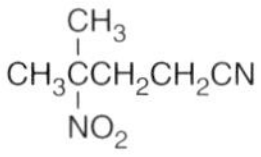
250 mg

S58,846-6
4-METHYL-4-NITROHEXANENITRILE
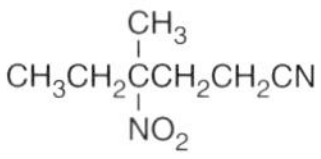
250 mg

S62,044-0
ETHYL 2,3-DICYANO-3-METHYLBUTYRATE
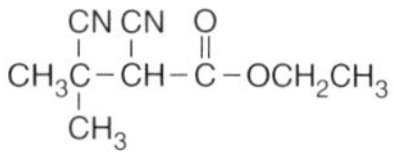
5 gm

S75,969-4
3,6-DIETHYL-4,5-DIHYDRO-2,7-DIMETHYL-1H-AZEPINE-4-CARBONITRILE
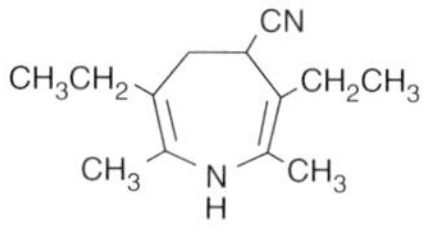
250 mg

S43,513-9
TERT-BUTYLCYANAMIDE
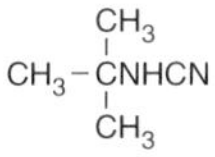
5 gm

S34,823-6
ETHYL N-CYANO-N-METHYLAMINOACETATE
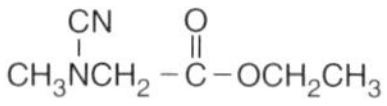
5 gm

S34,541-5
DIALLYLCYANAMIDE
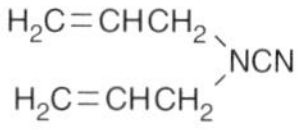
5 gm

S89,664-0
3-[(2-CYANOETHYL)METHYLAMINO]-5,5-DIMETHYL-2-CYCLOHEXEN-1-ONE
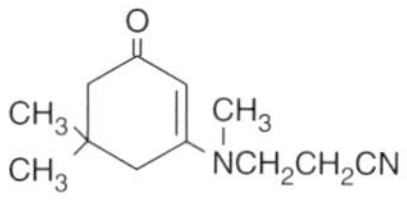
250 mg

S90,873-8
METHYL 2-CYANO-3-PIPERIDINOACRYLATE
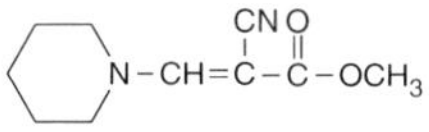
250 mg

S43,127-3
1-PYRROLIDINEVALERONITRILE
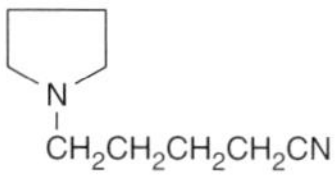
5 gm

S89,557-1
3-CYANOPROPIONALDEHYDE SEMICARBAZONE
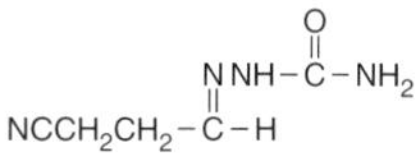
250 mg

S42,026-3
CYANOACETIC ACID ISOPROPYLIDENEHYDRAZIDE
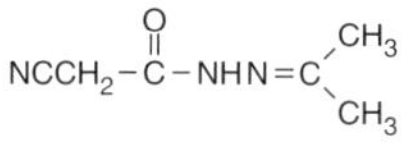
250 mg

S78,115-0
2-CYANOACETIC (3,3,5-TRIMETHYLCYCLOHEXYLIDENE)HYDRAZIDE
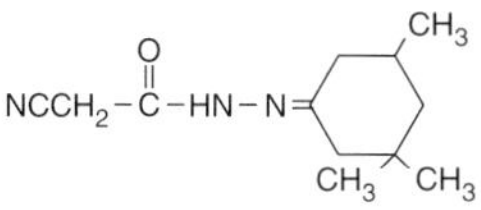
250 mg

S34,323-4
5-CARBAMOYL-3-CYANO-4,4-DIMETHYL-6-IMINO-2-PIPERIDONE
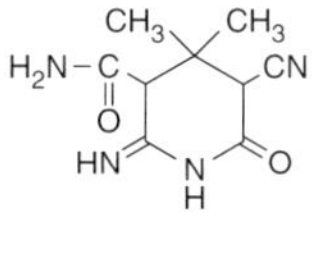
5 gm

S40,331-8
1-CYANO-1-METHYLETHYL ACETATE
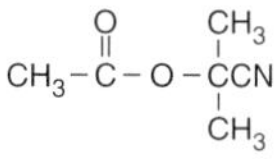
250 mg

S61,513-7
ISOPROPYL CYANOACETATE
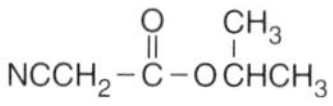
5 gm

S61,511-0
CYCLOHEXYL CYANOACETATE
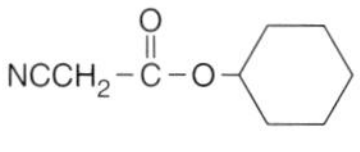
5 gm

S55,685-8
1,2-EPOXY-2,6,6-TRICHLOROCYCLOHEXANECARBONITRILE
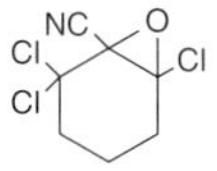
5 gm

S67,713-2
1-HYDROXYAMINO-1-CYCLOHEXANECARBONITRILE
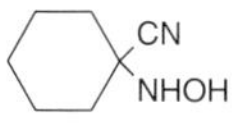
250 mg

S91,772-9
2-METHYL-2-NITROBUTYRONITRILE
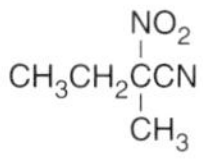
250 mg

S91,759-1
2-METHYL-2-[1-(NITROMETHYL)-CYCLOHEXYL]PROPIONITRILE
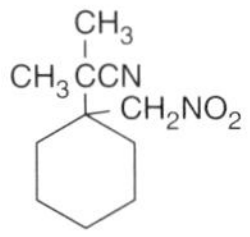
250 mg

S54,595-3
DIETHYL 2-CYANO-3-OXOSUCCINATE
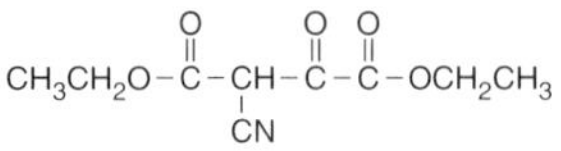
250 mg

S43,114-1
1-ETHYL 5-METHYL 2-ACETYL-2-(2-CYANOETHYL)GLUTARATE
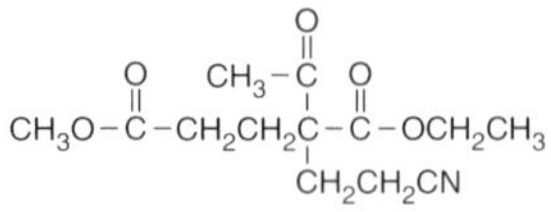
5 gm

S58,087-2
ETHYL 2,2-BIS(2-CYANOETHYL)-ACETOACETATE
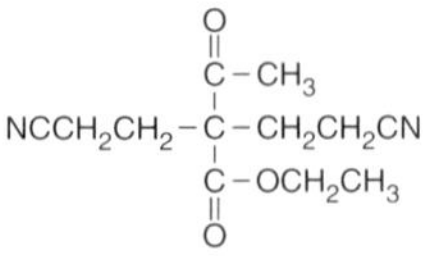
250 mg

S64,940-6
2-(2-CYANOETHYL)-5,5-DIMETHYL-1,3-CYCLOHEXANEDIONE
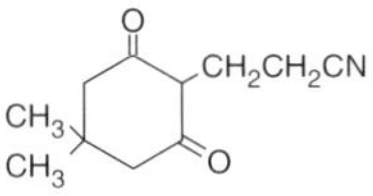
5 gm

S95,024-6
(-)-METHYL 4-CYANO-5,5-DIMETHYL-6-OXOHEPTANOATE
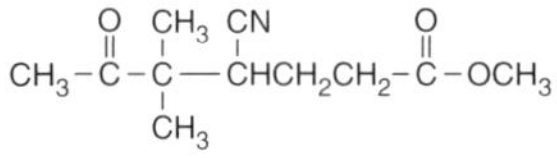
50 mg

S59,354-0
4-ISOPROPYL-7-METHYL-5-OXOOCTANENITRILE
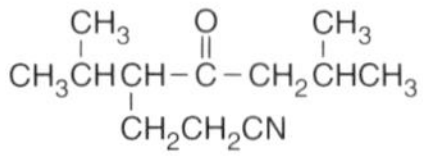
5 gm

S70,638-8
5-CYANO-3-(2-CYANOETHYL)-3-ISOPROPENYL-2-PENTANONE
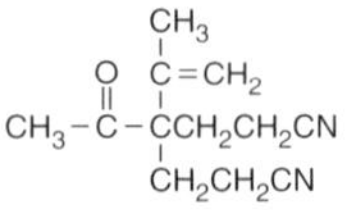
5 gm

S88,630-0
2-OXO-1-CYCLOPENTANECARBONITRILE
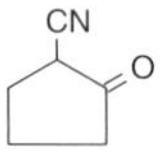
250 mg

S88,478-2
3-(CYANOMETHYL)-1-CYCLOPENTANONE
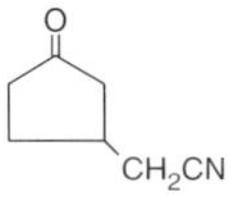
250 mg

S88,631-9
3,3-DIMETHYL-2-OXO-1-CYCLOPENTANECARBONITRILE
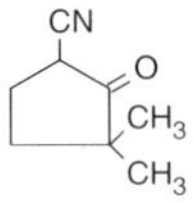
250 mg

S93,794-0
1-CYANO-2,5,6,6-TETRAMETHYLBICYCLO-[3.1.0]HEXAN-4-ONE
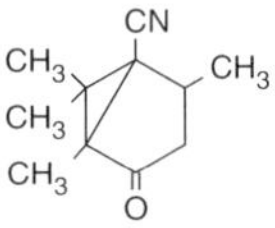
250 mg

S88,324-7
2-OXOCYCLOHEXANECARBONITRILE
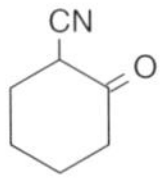
250 mg

S59,888-7
1,2,2-TRICHLORO-6-OXO-1-CYCLOHEXANECARBONITRILE
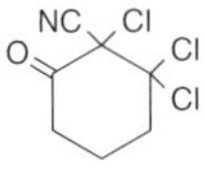
250 mg

S43,303-9
2-OXO-1,3-CYCLOHEXANEDIPROPIONITRILE
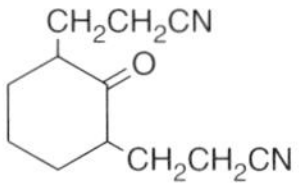
5 gm

S52,671-1
DIETHYL 2-(ALPHA-CYANOCYCLOHEXYLMETHYLENE)MALONATE
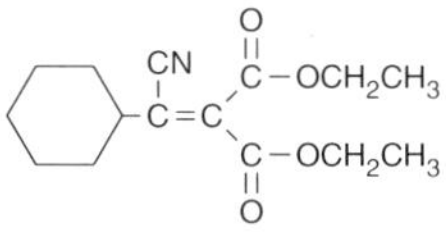
250 mg

S43,519-8
2-CYANO-2-HEXENOIC ACID
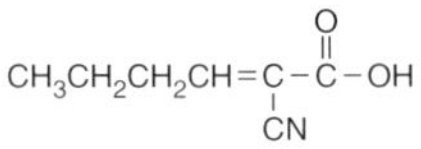
5 gm

S78,242-4
2-CYANO-2-CYCLOHEXYLIDENEACETIC ACID
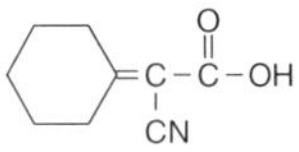
5 gm

S88,322-0
2-CYANO-2-CYCLOHEXYLIDENEACETIC ACID
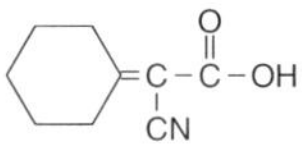
250 mg

S35,278-0
METHYL 2-CYANO-3-METHYLCROTONATE
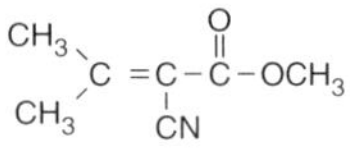
5 gm

S66,120-1
ETHYL 2-CYANO-2-PENTENOATE
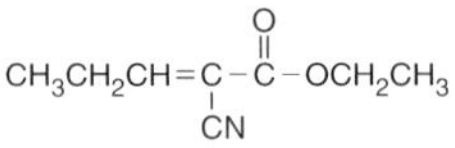
250 mg

S66,415-4
ETHYL 2-CYANO-2-HEXENOATE
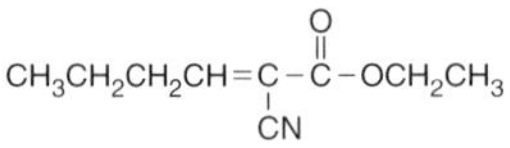
250 mg

S66,145-7
ETHYL 2-CYANO-3-ETHYL-2-PENTENOATE
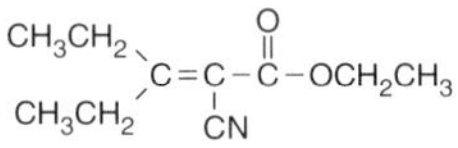
5 gm

S66,116-3
ETHYL 2-CYANO-3-METHYL-2-OCTENOATE
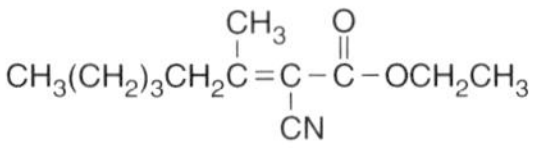
5 gm

S66,126-0
ETHYL 2-CYANO-3-ETHYL-2-HEPTENOATE
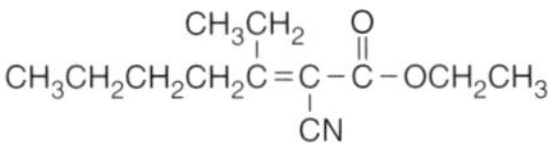
5 gm

S66,142-2
ETHYL 2-CYANO-3-METHYL-2-NONENOATE
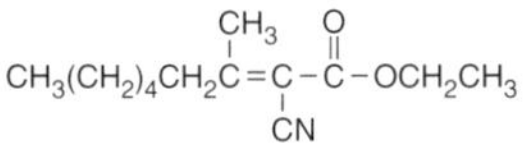
5 gm

S40,922-7
METHYL CYCLOPENTYLIDENECYANOACETATE
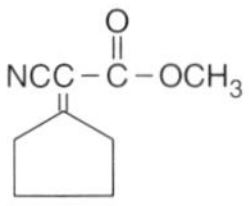
5 gm

S66,098-1
ETHYL 2-CYANO-2-CYCLOPENTYLIDENEACETATE
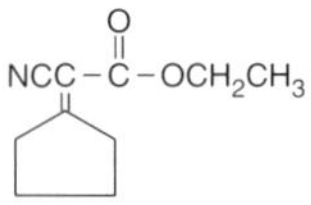
250 mg

S66,128-7
ETHYL 2-CYANO-2-CYCLOHEXYLIDENEACETATE
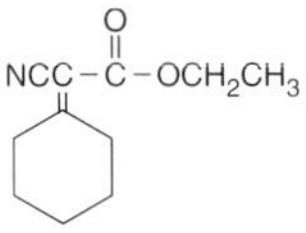
5 gm

S85,285-6
ETHYL 2-CYANO-2-CYCLOHEXYLIDENEACETATE
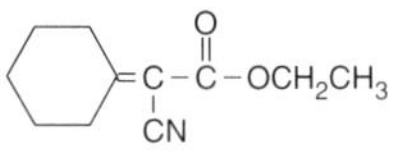
250 mg

S35,650-6
TRIETHYL 1-CYANO-3,3,4-PENTANETRICARBOXYLATE
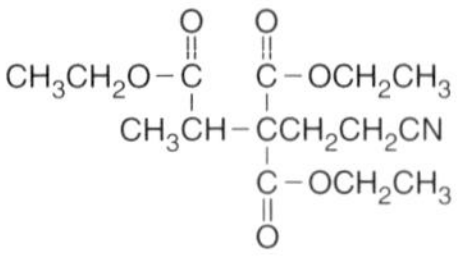
5 gm

S44,807-9
DIETHYL BIS(2-CYANOETHYL)MALONATE
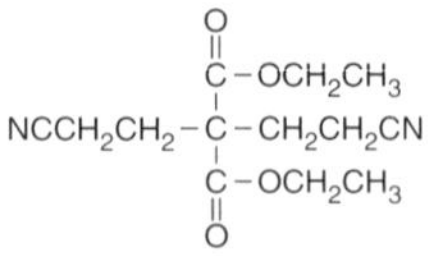
5 gm

S36,092-9
DIMETHYL 3-(CYANOMETHYLENE)-GLUTARATE

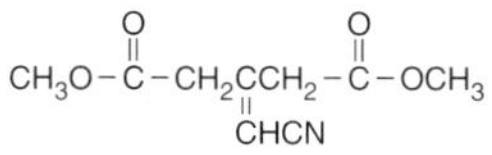

5 gm

S58,252-2
DIETHYL 2-CYANO-2,3-DIMETHYLSUCCINATE

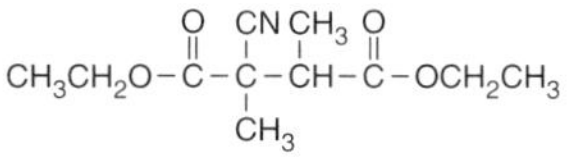

5 gm

S36,759-1
9-CYANOPELARGONIC ACID

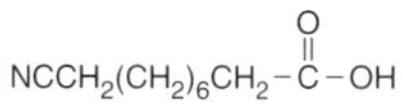

5 gm

S93,498-4
TRANS,TRANS-9-CYANO-2-DECALINCARBOXYLIC ACID

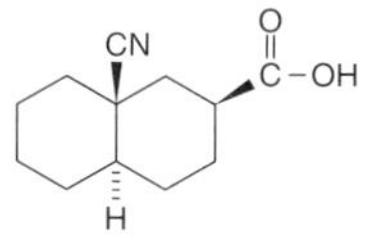

250 mg

S36,964-0
METHYL 3-CYANOPROPIONATE

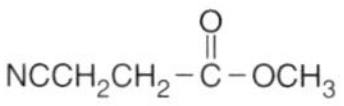

5 gm

S76,208-3
ETHYL 2-CYANO-2-METHYLPROPIONATE

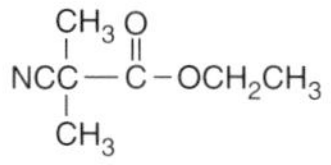

250 mg

S54,970-3
PENTYL CYANOACETATE

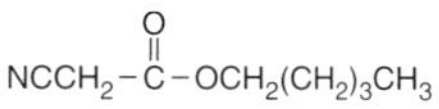

5 gm

S71,334-1
ETHYL 2-CYANO-3-METHYL-2-PENTENOATE

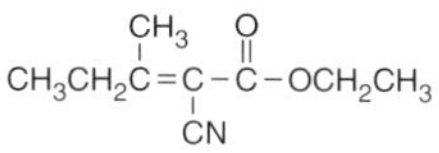

5 gm

S34,816-3
ETHYL ALPHA-SEC-BUTYLCYANOACETATE

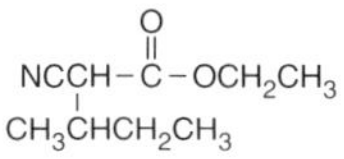

5 gm

S37,426-1
ETHYL 6-CYANOHEXANOATE

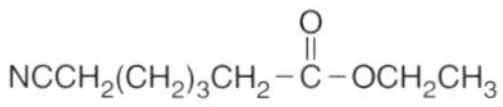

5 gm

S66,144-9
ETHYL ALPHA-BUTYLCYANOACETATE

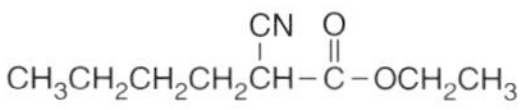

5 gm

S36,154-2
ETHYL SEC-BUTYLETHYLCYANOACETATE

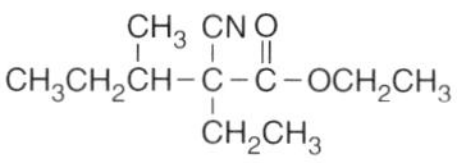

5 gm

S43,143-5
2-ETHYLHEXYL CYANOACETATE

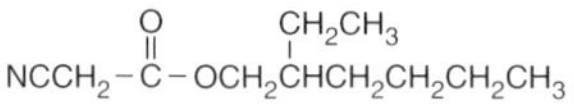

5 gm

S38,525-5
ETHYL 2-BUTYL-2-CYANOHEXANOATE

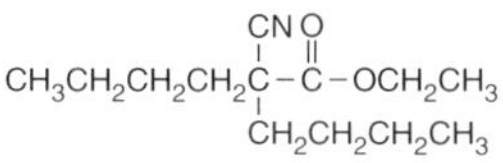

5 gm

S55,127-9
DECYL CYANOACETATE

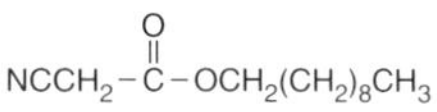

5 gm

S88,562-2
ETHYL 2-(2-CYANOETHYL)-2,3-DICYANO-3,6-DIMETHYL-6-HEPTENOATE

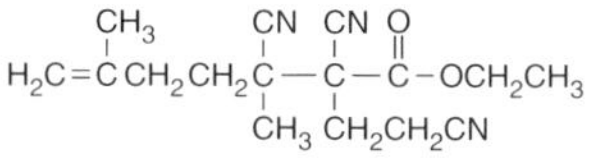

250 mg

S88,566-5
TERT-BUTYL 2-(2-CYANOETHYL)-2,3-DICYANO-3,6-DIMETHYL-6-HEPTENOATE

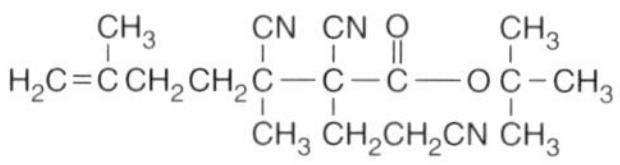

250 mg

S44,792-7
ETHYL 2-CYANO-1-CYCLOPROPANECARBOXYLATE

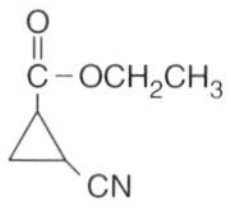

5 gm

S95,397-0
(-)-4-CYANO-1,2-EPOXY-2,3,3-TRIMETHYLCYCLOHEXYL ACETATE

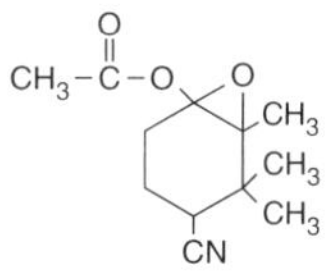

50 mg

S43,103-6
1-CYANOCYCLOPENTYL ACETATE

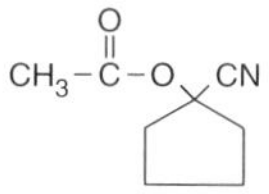

5 gm

S65,150-8
ETHYL 1-CYANO-1-CYCLOPENTANECARBOXYLATE

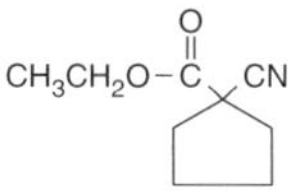

250 mg

S38,956-0
ETHYL ALPHA-CYANO-ALPHA-ETHYLBICYCLO[3.2.1]OCT-2-ENE-2-ACETATE

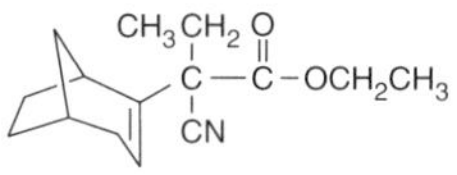

250 mg

S95,149-8
TERT-BUTYL 2-CYANO-2-(2-PYRROLIDINYLIDENE)ACETATE

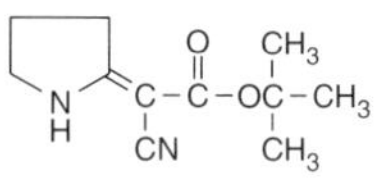

50 mg

S43,104-4
1-CYANOCYCLOHEXYL ACETATE

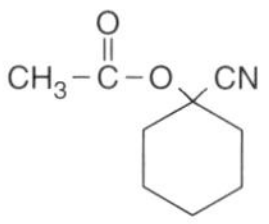

5 gm

S35,228-4
ETHYL ALPHA-(CYCLOHEXYLIDENE)-CYANOACETATE

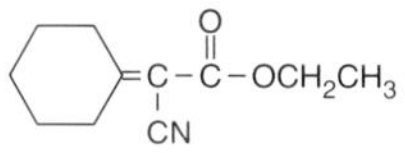

5 gm

S87,155-9
METHYL 2-CYANO-2-(1-CYCLOHEXEN-1-YL)ACETATE

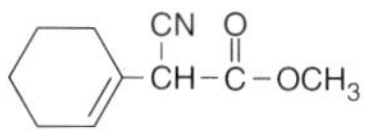

250 mg

S34,822-8
ETHYL ALPHA-CYANO-ALPHA-1-CYCLOHEXENYLBUTYRATE

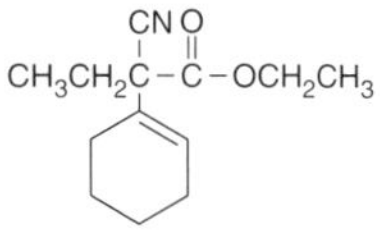

5 gm

S95,396-2
(-)-4-CYANO-2,3,3-TRIMETHYL-1-CYCLOHEXEN-1-YL ACETATE

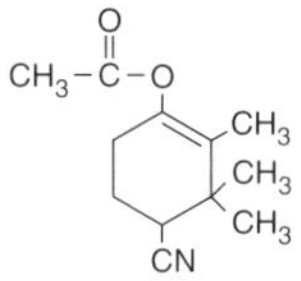

50 mg

S95,496-9
3-CYANO-5,6-DIHYDRO-4,5,5,-TRIMETHYL-2H-PYRAN-2-ONE

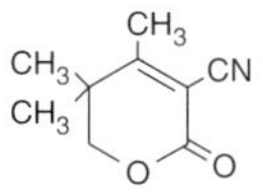

50 mg

S40,844-1
ETHYL ALPHA-CYANO-2-NORBORNANEACETATE

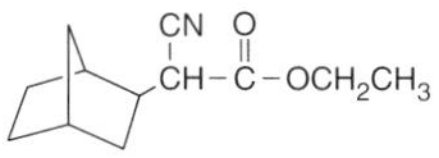

5 gm

S92,873-9
1-(1-CYANO-1-METHYLETHYL)-SEMICARBAZIDE

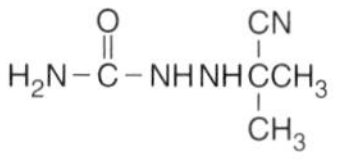

250 mg

S80,488-6
1-(2-CYANOACETYL)-3-CYCLOHEXYL-1-ETHYLUREA

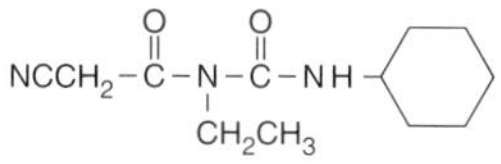

250 mg

S34,573-3
2,4-DICYANO-3-ETHYL-3-METHYLGLUTARIMIDE

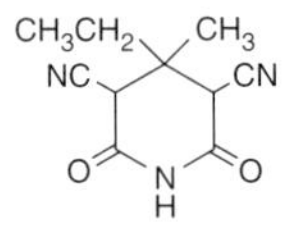

5 gm

S95,100-5
5-CYANO-4,4,5-TRIMETHYL-2-PYRROLIDINONE

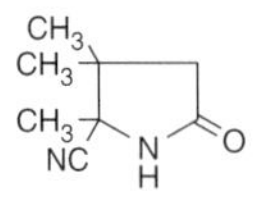

50 mg

S92,386-9
3-(DIMETHYLAMINO)-3-OXOPROPIONITRILE

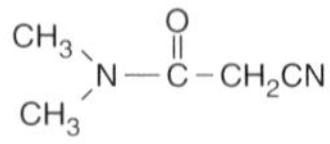

250 mg

S42,092-1
N-(1-CYANOCYCLOHEXYL)-N-METHYLBUTYRAMIDE

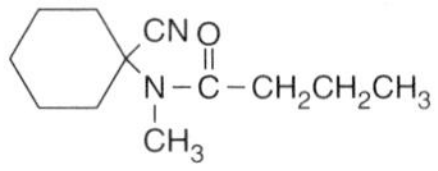

5 gm

S92,387-7
(HYDROXYIMINO)MALONONITRILE, POTASSIUM SALT

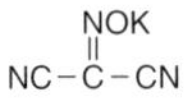

250 mg

S89,449-4
1,2,3,4,4-PENTACHLORO-1,3-BUTADIENE-1-CARBONITRILE

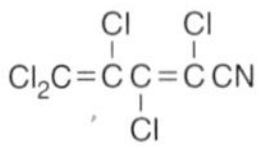

250 mg

S51,142-0
2-HEXENEDINITRILE

$NCCH_2CH_2CH{=}CHCN$

5 gm

S35,649-2
TRIETHYLACETONITRILE

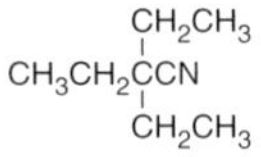

5 gm

S37,428-8
3-ETHYLHEPTANENITRILE

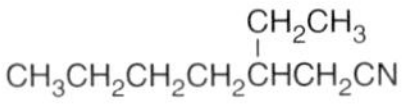

5 gm

S44,821-4
ALPHA-ETHYL-1-CYCLOHEPTENE-1-ACETONITRILE

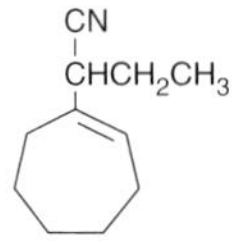

5 gm

S88,634-3
2,6-DIMETHYL-2,6-HEPTANEDICARBONITRILE

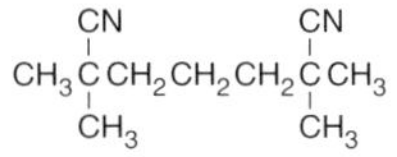

250 mg

S88,637-8
2,7-DIMETHYL-2,7-OCTANEDICARBONITRILE

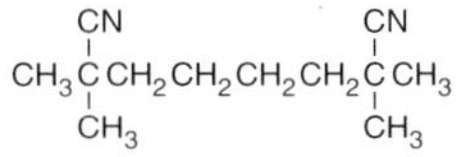

250 mg

S45,006-5
TRIDECANENITRILE

$CH_3(CH_2)_{10}CH_2CN$

5 gm

S56,600-4
TETRADECANENITRILE

$CH_3(CH_2)_{11}CH_2CN$

5 gm

S44,581-9
TETRADECANEDINITRILE

$NCCH_2(CH_2)_{10}CH_2CN$

5 gm

S43,995-9
NONADECANENITRILE

$CH_3(CH_2)_{16}CH_2CN$

5 gm

S88,713-7
2-IMINO-1-CYCLOPENTANECARBONITRILE

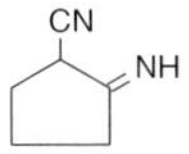

250 mg

S91,888-1
1-CHLORO-1-CYCLOPENTANECARBONITRILE

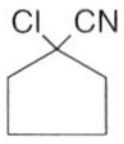

250 mg

S80,859-8
6-(2-CYANOETHYL)-1,4-DIOXASPIRO-(4.5)DECANE

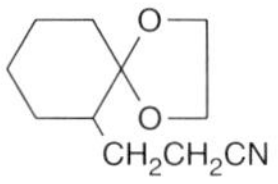

250 mg

S57,666-2
8A-CYANO-2-DECALONE ETHYLENE KETAL

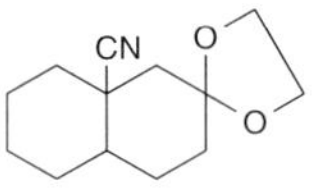

250 mg

S86,607-5
1-CHLOROCYCLOHEXANECARBONITRILE

250 mg

S34,497-4
2-(CYCLOHEXYLIDENE)BUTYRONITRILE

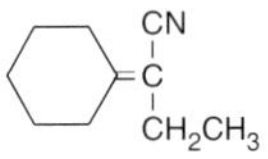

5 gm

S43,263-6
CYCLOHEXANEBUTYRONITRILE

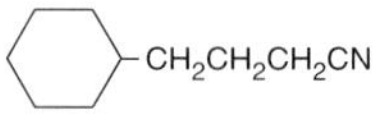

5 gm

S91,764-8
2-CYCLOHEXYL-2-METHYLPROPIONITRILE

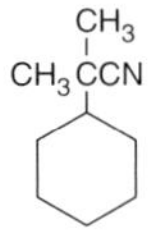

250 mg

S89,568-7
1,2-CYCLOHEXANEDIACETONITRILE

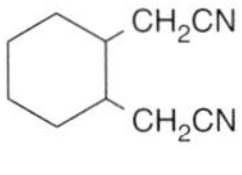

250 mg

S54,090-0
1,4-CYCLOHEXANEDICARBONITRILE

5 gm

S44,977-6
1-PIPERIDINEPENTANENITRILE

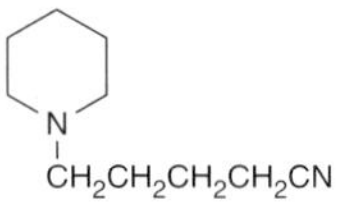

5 gm

S43,128-1
4-MORPHOLINEVALERONITRILE

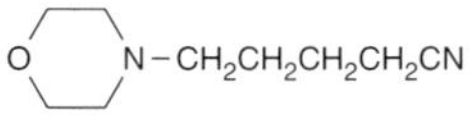

5 gm

S93,702-9
TRANS,TRANS-2,9-DICYANODECALIN

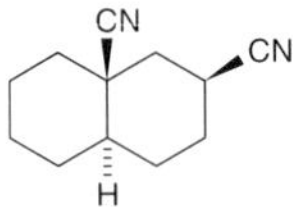

250 mg

S52,867-6
ENDO-5-NORBORNENE-2-CARBONITRILE

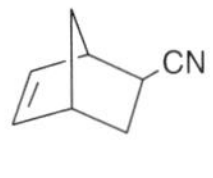

5 gm

S44,825-7
HEXYL ISOCYANIDE

$CH_3(CH_2)_4CH_2NC$

5 gm

S35,907-6
2-(2-BUTOXYETHOXY)ETHYL-THIOCYANATE

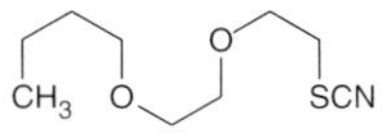

5 gm

S52,174-4
METHYL 2-THIOCYANATOBUTYRATE

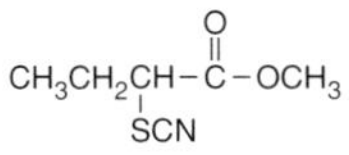

5 gm

S52,166-3
METHYL 3-METHYL-2-THIOCYANATOBUTYRATE

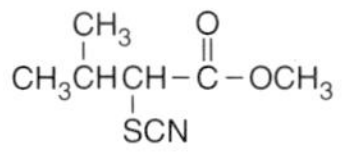

5 gm

S52,183-3
ETHYL 2-THIOCYANATOBUTYRATE

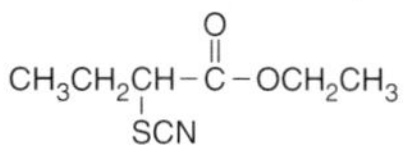

5 gm

S81,373-7
ETHYLENE DITHIOCYANATE

$NCSCH_2CH_2SCN$

250 mg

S44,981-4
NORMAL-BUTYLTHIOCYANATE

$CH_3CH_2CH_2CH_2SCN$

5 gm

S43,766-2
TETRAMETHYLENE DITHIOCYANATE

$NCSCH_2CH_2CH_2CH_2SCN$

5 gm

S34,524-5
DECYL THIOCYANATE

$CH_3(CH_2)_8CH_2SCN$

5 gm

S51,749-6
DODECYL THIOCYANATE

$CH_3(CH_2)_{10}CH_2SCN$

5 gm

S82,605-7
3-(2-TETRAHYDROFURYL)PROPYL THIOCYANATE

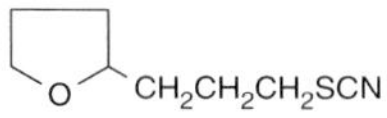

250 mg

S64,896-5
3-METHOXYPROPYL ISOTHIOCYANATE

$CH_3OCH_2CH_2CH_2NCS$

5 gm

S64,897-3
2-(ISOTHIOCYANATOMETHYL)-TETRAHYDROFURAN

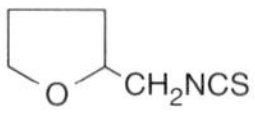

5 gm

S52,141-8
ISOBUTYL ISOTHIOCYANATE

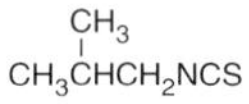

5 gm

S64,905-8
PENTYL ISOTHIOCYANATE

$CH_3CH_2CH_2CH_2CH_2NCS$

5 gm

S64,894-9
3,3,5-TRIMETHYLCYCLOHEXYL ISOTHIOCYANATE

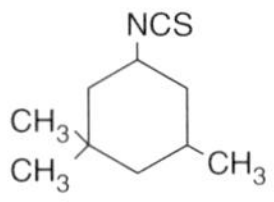

5 gm

S62,671-6
1,3-DICYCLOHEXYLCARBODIIMIDE METHIODIDE

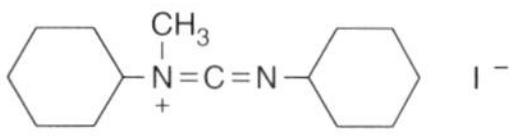

250 mg

S34,505-9
CYCLOOCTANONE AZINE

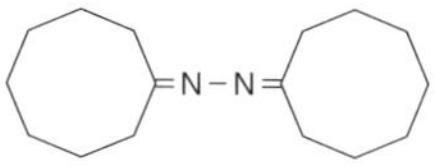

250 mg

S88,747-1
1-METHYL-2-PYRROLIDINONE AZINE

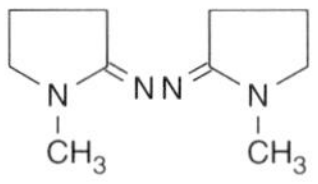

250 mg

S40,761-5
DICYCLOPROPYL KETONE AZINE

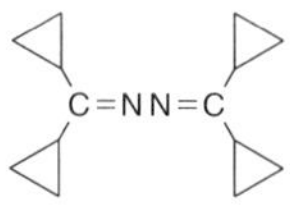

250 mg

S77,806-0
BROMOMETHYL ETHYL SULFONE
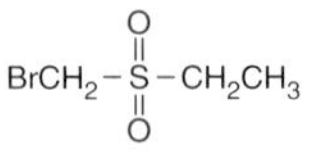
250 mg

S74,193-0
BIS(2-CHLOROETHYL)SULFONE
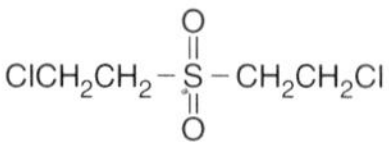
250 mg

S36,704-4
2-BROMOETHYL SULFONE
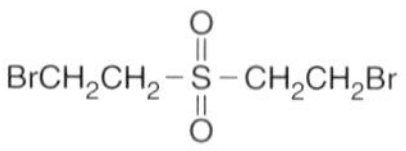
5 gm

S80,895-4
TERT-BUTYL CHLOROMETHYL SULFONE
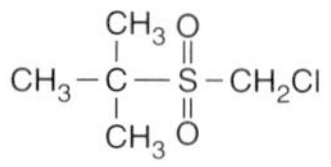
250 mg

S93,878-5
TRANS-N-(4-AMINO-3-SULFOLANYL)-2,2,2-TRICHLOROACETAMIDE HYDROCHLORIDE
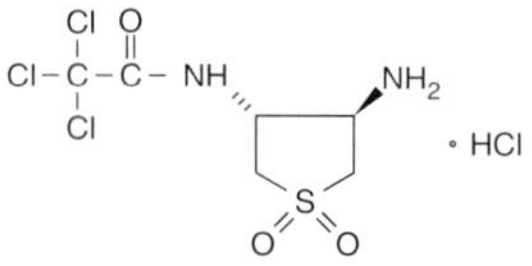
250 mg

S93,873-4
N-(4-CHLORO-3-SULFOLANYL)-2,2,2-TRICHLOROACETAMIDE
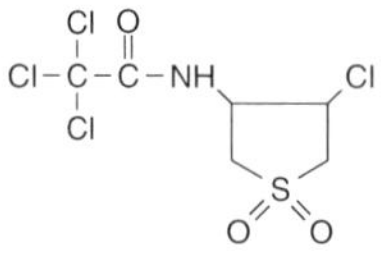
250 mg

S63,361-5
TRICHLOROMETHYL SULFONE
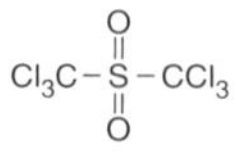
5 gm

S88,347-6
2-BROMOPROPYL TRIFLUOROMETHYL SULFONE
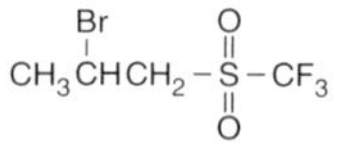
250 mg

S93,233-7
CIS-3,4-DIHYDROXYTETRAHYDROTHIOPHENE-1,1-DIOXIDE
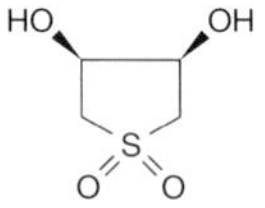
250 mg

S93,241-8
TRANS-3,4-DIHYDROXYTETRAHYDROTHIOPHENE-1,1-DIOXIDE
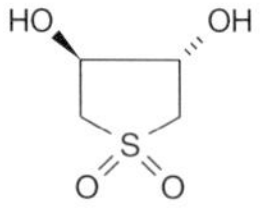
250 mg

S77,546-0
ALPHA,ALPHA'-DI-TERT-BUTYL-2,5-SULFOLANEDIMETHANOL
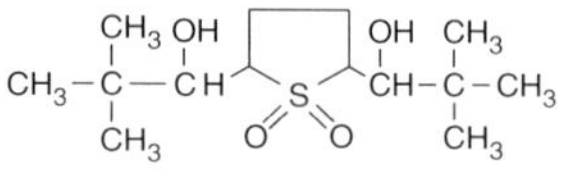
250 mg

S90,073-7
3-(CYCLOHEXYLAMINO)-4-HYDROXYTETRAMETHYLENE SULFONE
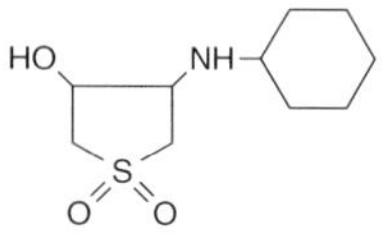
250 mg

S93,880-7
TRANS-3-HYDROXY-4-(ISOPROPYLAMINO)-TETRAHYDROTHIOPHENE-1,1-DIOXIDE
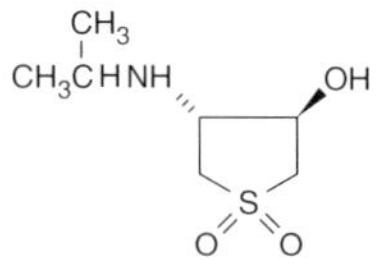
250 mg

S93,881-5
CIS-3-HYDROXY-4-(ISOPROPYLAMINO)-TETRAHYDROTHIOPHENE-1,1-DIOXIDE
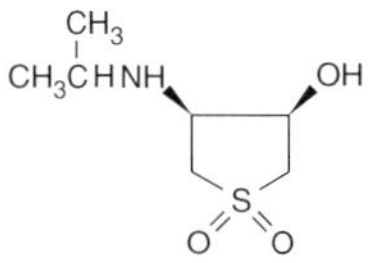
250 mg

S76,893-6
3-HYDROXY-4-(4-METHYLPIPERIDINO)-TETRAMETHYLENE SULFONE
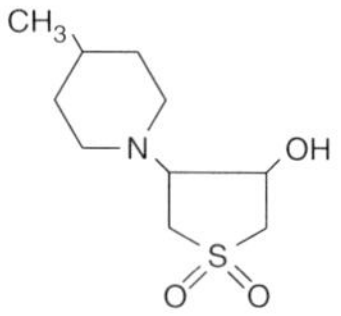
250 mg

S43,063-3
2,2'-SULFINYLDIETHANOL
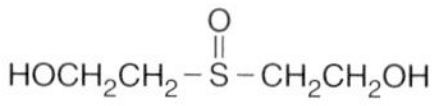
5 gm

S54,036-6
2-(DECYLSULFINYL)ETHANOL
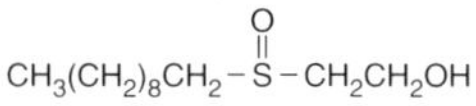
5 gm

S54,012-9
2-HYDROXYETHYL DODECYL SULFOXIDE
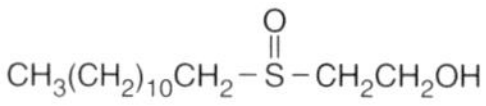
5 gm

S54,026-9
2-(HEXADECYLSULFINYL)ETHANOL
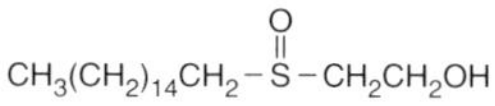
5 gm

S54,033-1
2-(OCTADECYLSULFINYL)ETHANOL
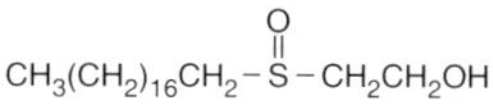
5 gm

S79,910-6
3-(BUTYLSULFONYL)PROPANOL
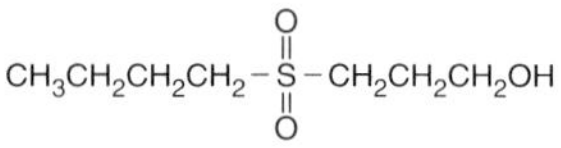
250 mg

S84,092-0
3-(DODECYLSULFONYL)PROPANOL
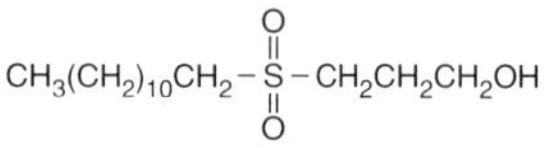
250 mg

S54,054-4
2-(HEXADECYLSULFONYL)ETHANOL
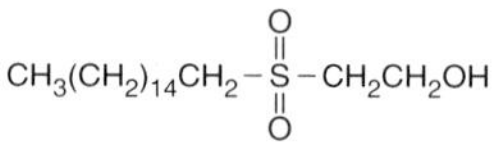
5 gm

S45,248-3
TETRAHYDROTHIOPHENE-3-OL-1,1-DIOXIDE
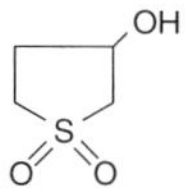
5 gm

S94,859-4
2-BROMO-4-HYDROXY-4,5-DIHYDROTHIOPHENE 1,1-DIOXIDE
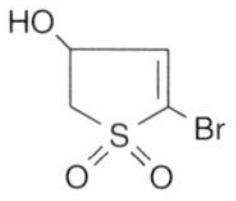
250 mg

S93,874-2
3-BROMO-4-HYDROXYTETRAHYDROTHIOPHENE-1,1-DIOXIDE
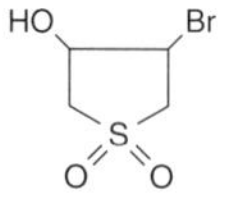
250 mg

S93,245-0
TRANS-3,4-DIMETHOXYTETRAHYDROTHIOPHENE-1,1-DIOXIDE
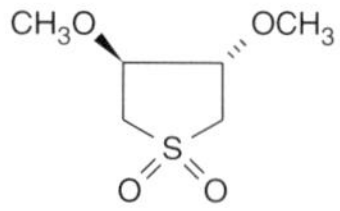
250 mg

S56,707-8
3-[1,3-BIS(ALLYLOXY)-2-PROPOXY]-TETRAHYDROTHIOPHENE 1,1-DIOXIDE
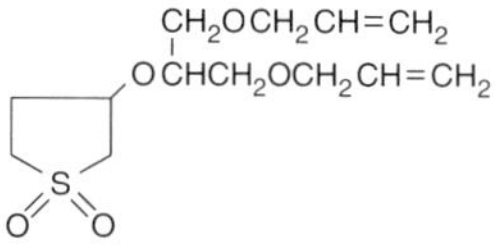
5 gm

S55,673-4
3-ALLYLOXYTETRAHYDROTHIOPHENE 1,1-DIOXIDE
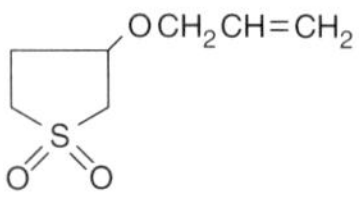
5 gm

S56,703-5
3-(2-CHLOROALLYLOXY)-TETRAHYDROTHIOPHENE 1,1-DIOXIDE
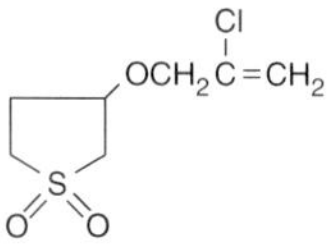
5 gm

S56,699-3
3-(1-METHYLALLYLOXY)-TETRAHYDROTHIOPHENE 1,1-DIOXIDE
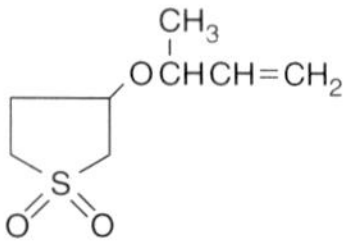
5 gm

S56,701-9
3-(2-BUTENYLOXY)-TETRAHYDROTHIOPHENE 1,1-DIOXIDE
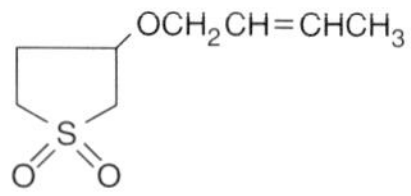
5 gm

S55,671-8
3-BUTOXYTETRAHYDROTHIOPHENE 1,1-DIOXIDE

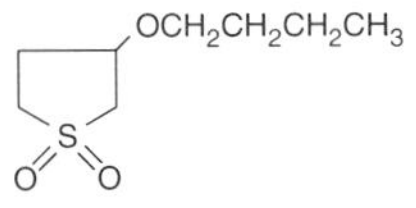

5 gm

S54,979-7
3-SULFOLANYL ETHER

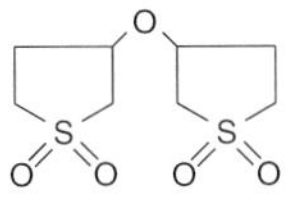

5 gm

S94,675-3
S,S'-METHYLENEBIS-(SULFONYLACETIC ACID)

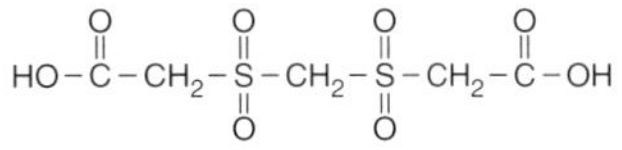

250 mg

S79,304-3
3-METHYL-3-(2-METHYL-1,3-DIOXOLAN-2-YL)-1-(METHYLSULFINYL)-2-BUTANONE

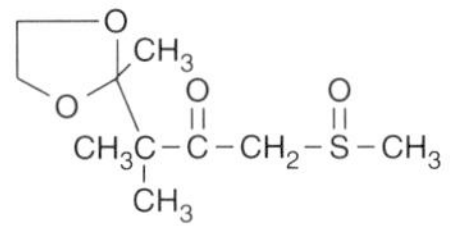

250 mg

S95,503-5
METHYL 1,4-DIOXA-8-THIASPIRO(4.5)-DECANE-6-CARBOXYLATE S-OXIDE

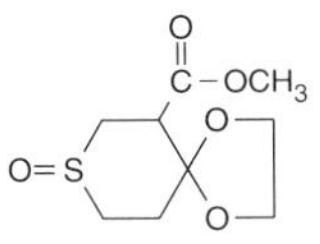

50 mg

S93,865-3
2-(3-SULFOLANYL)MALONIC ACID DIHYDRAZIDE

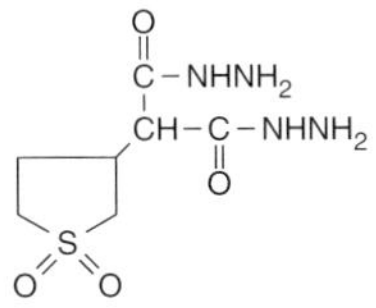

250 mg

S69,783-4
ETHYL (ETHYLSULFINYLMETHYL) SULFIDE

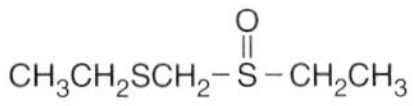

250 mg

S93,879-3
TRANS-3,4-DIAMINOTETRAHYDROTHIOPHENE-1,1-DIOXIDE DIHYDROCHLORIDE

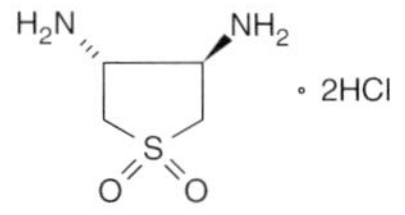

250 mg

S45,541-5
TETRAHYDRO-3-THIOPHENAMINE-1,1-DIOXIDE

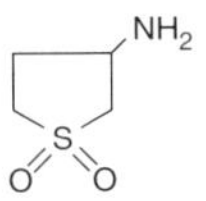

5 gm

S57,934-3
DL-METHIONINE SULFONE

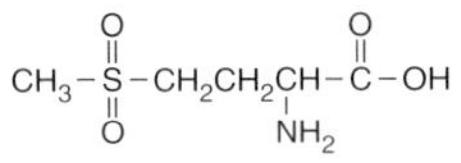

250 mg

S78,355-2
D-ETHIONINE SULFONE

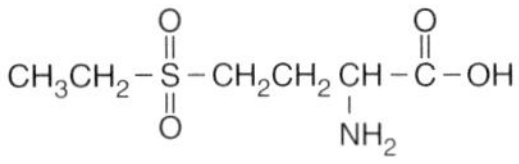

5 gm

S93,246-9
TRANS-3-AMINO-4-CHLOROTETRAHYDROTHIOPHENE-1,1-DIOXIDE HYDROCHLORIDE

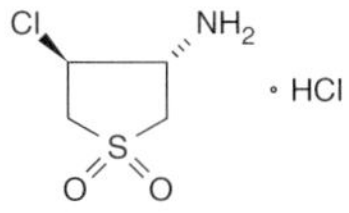

250 mg

S94,863-2
TRANS-4-METHYLTETRAHYDRO-3-THIOPHENAMINE 1,1-DIOXIDE HYDROCHLORIDE

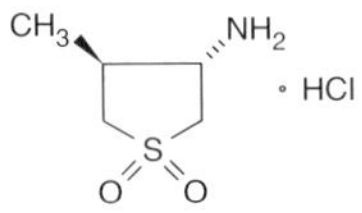

250 mg

S56,704-3
N-ALLYLTETRAHYDRO-3-THIOPHENAMINE 1,1-DIOXIDE

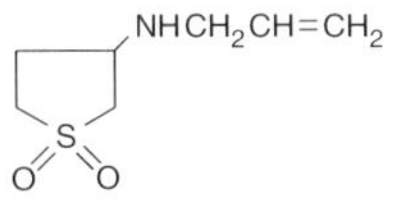

5 gm

S86,329-7
2-CHLORO-3-(DIISOPROPYLAMINO)-2,4,4-TRIMETHYLTHIETANE-1,1-DIOXIDE

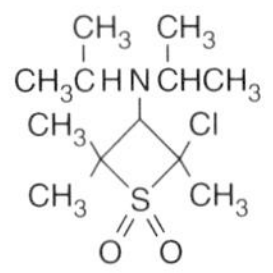

250 mg

S86,320-3
2,2-DICHLORO-4,4-DIMETHYL-3-MORPHOLINOTHIETANE-1,1-DIOXIDE

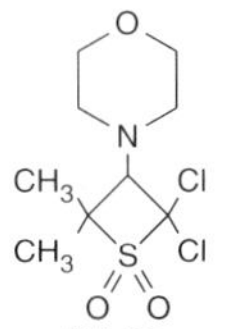

250 mg

S56,686-1
N,N-DIMETHYLTETRAHYDRO-3-THIOPHENAMINE 1,1-DIOXIDE

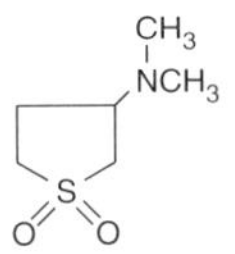

5 gm

S60,264-7
N-(1,1-DIOXOTETRAHYDROTHIOPHEN-3-YL)DODECANAMIDE

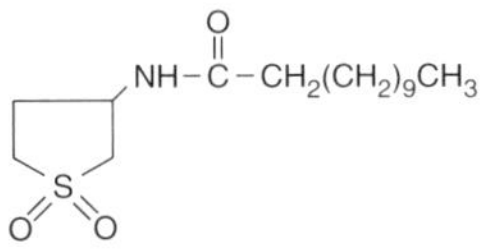

5 gm

S54,985-1
N-METHYL-N-NITROSO-3-TETRAHYDROTHIOPHENAMINE 1,1-DIOXIDE

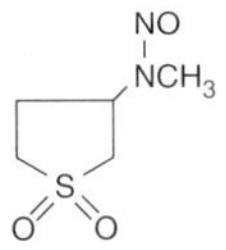

5 gm

S93,863-7
3-(NITROMETHYL)TETRAHYDROTHIOPHENE-1,1-DIOXIDE

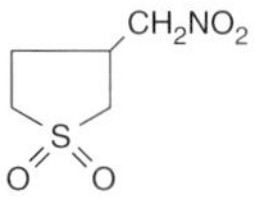

250 mg

S94,094-1
3-OXOSULFOLANE

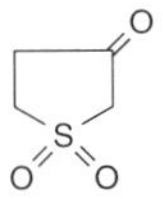

250 mg

S64,615-6
DIMETHYL 3,3'-SULFONYLDIPROPIONATE

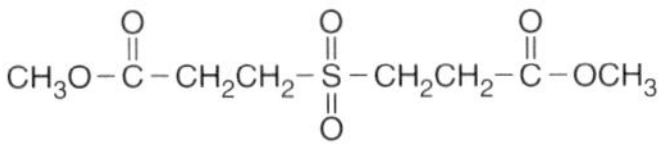

5 gm

S92,441-5
4-CHLOROTETRAHYDROTHIOPHENE-3-OL-1,1-DIOXIDE N-PROPYLCARBAMATE

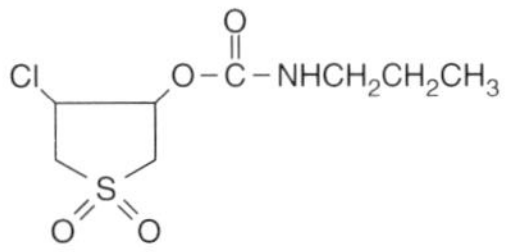

250 mg

S90,072-9
4-BROMO-3-SULFOLANYL N-PROPYLCARBAMATE

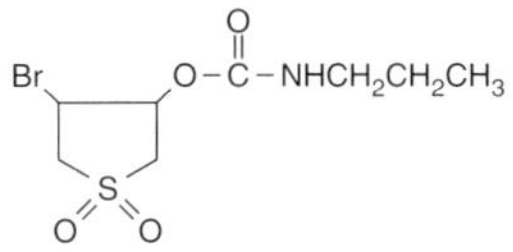

250 mg

S60,270-1
3-ACETOXYTETRAHYDROTHIOPHENE-1,1-DIOXIDE

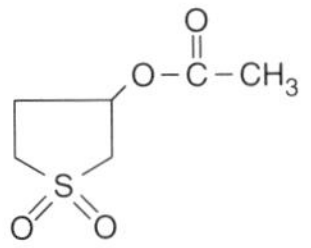

5 gm

S56,694-2
3-ACETOXY-4-CHLOROTETRAHYDROTHIOPHENE-1,1-DIOXIDE

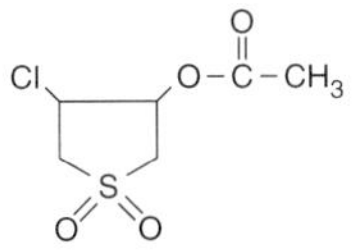

250 mg

S93,872-6
4-METHYL-3-SULFOLANYL ACETATE

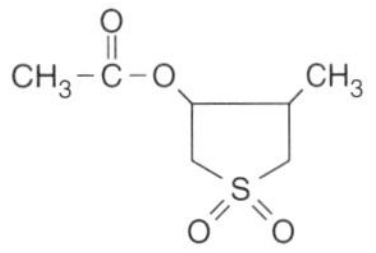

250 mg

S56,685-3
3-HEXANOYLOXYTETRAHYDROTHIOPHENE 1,1-DIOXIDE

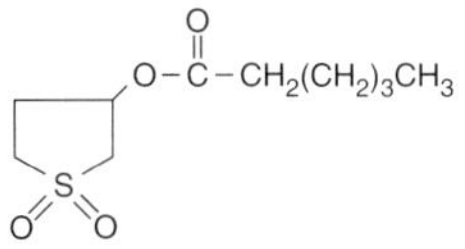

5 gm

S61,447-5
2-(METHYLSULFONYL)ACETAMIDE

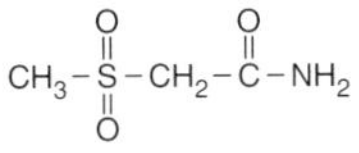

5 gm

S94,861-6
3-ACETAMIDO-2,3-DIHYDROTHIOPHENE-1,1-DIOXIDE

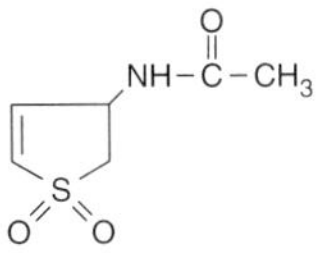

250 mg

S39,408-4
3,3'-SULFONYLDIPROPIONITRILE

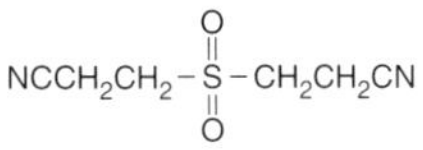

250 mg

S93,855-6
2,3-DIHYDRO-3-THIOCYANATOTHIOPHENE-1,1-DIOXIDE

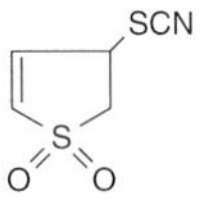

250 mg

S88,500-2
1,3-DITHIANE-1-OXIDE

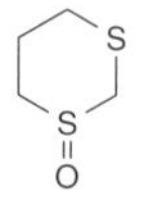

250 mg

S44,097-3
PROPYL SULFOXIDE

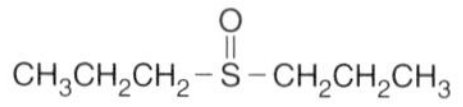

5 gm

S76,790-5
DIISOBUTYL SULFOXIDE

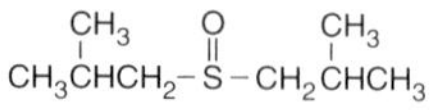

250 mg

S36,928-4
ISOPENTYL SULFOXIDE

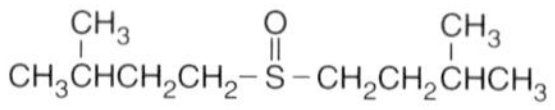

5 gm

S54,035-8
DECYL SULFOXIDE

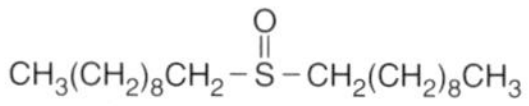

5 gm

S83,329-0
DODECYL SULFOXIDE

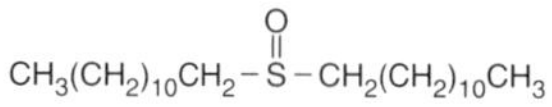

250 mg

S62,073-4
TRIMETHYLENE SULFONE

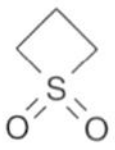

250 mg

S77,890-7
BIS(METHYLSULFONYL)METHANE

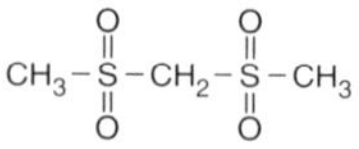

250 mg

S64,616-4
1,3,5-TRITHIANE-
1,1,3,3,5,5-HEXAOXIDE

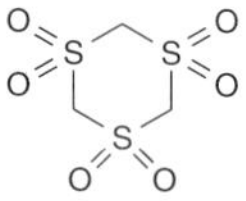

5 gm

S77,905-9
2-IODO-1-PROPENYL METHYL SULFONE

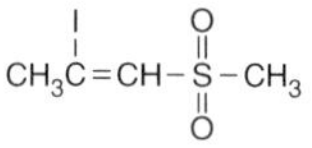

250 mg

S77,573-8
BUTYL ETHYL SULFONE

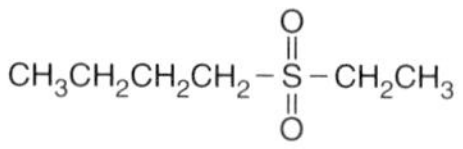

250 mg

S79,909-2
ISOPROPYL SULFONE

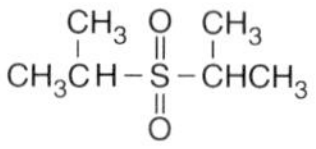

250 mg

S79,951-3
ETHYL PENTYL SULFONE

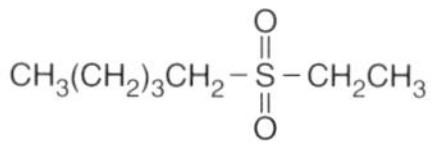

250 mg

S79,948-3
3,3-DIMETHYL-2-IODO-1-BUTENYL
ETHYL SULFONE

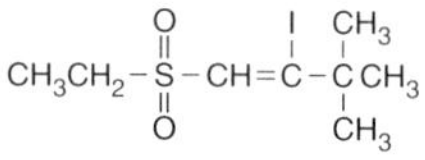

250 mg

S77,547-9
1,2-BIS(ISOPROPYLSULFONYL)-1,2-
DICHLOROETHYLENE

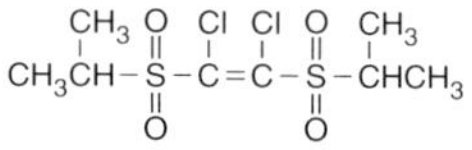

250 mg

S36,666-8
PENTYL SULFONE

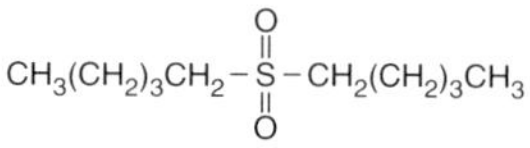

5 gm

S79,903-3
1,4-BIS(TERT-BUTYLSULFONYL)BUTANE

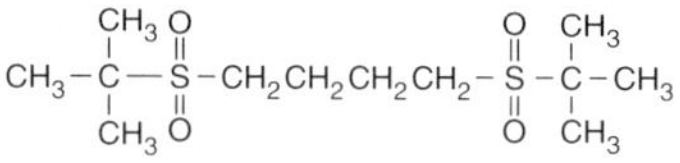

250 mg

S77,845-1
HEXYL OCTYL SULFONE

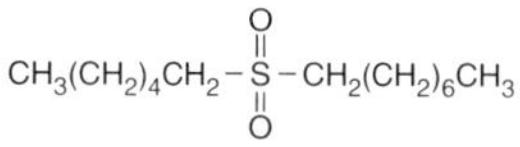

250 mg

S84,097-1
1,6-BIS(DODECYLSULFONYL)HEXANE

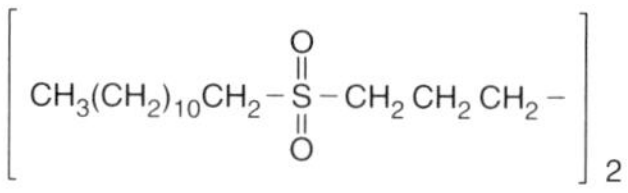

250 mg

S93,868-8
3-(MORPHOLINOSULFONYL)-
TETRAHYDROTHIOPHENE-1,1-DIOXIDE

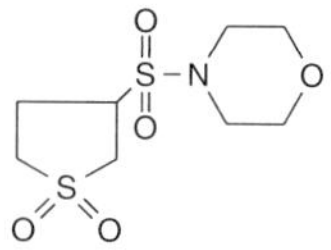

250 mg

S77,542-8
7-THIABICYCLO[4.2.0]OCT-1-ENE-
7,7-DIOXIDE

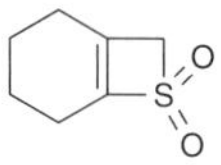

250 mg

S93,248-5
3-CHLORO-2,5-DIHYDROTHIOPHENE-
1,1-DIOXIDE

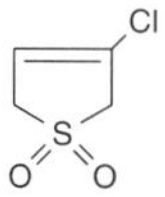

250 mg

S93,862-9
3-CHLOROTETRAHYDROTHIOPHENE-
1,1-DIOXIDE

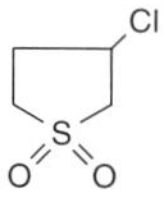

250 mg

S94,858-6
2-BROMO-4,5-DIHYDROTHIOPHENE-
1,1-DIOXIDE

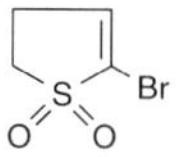

250 mg

S93,231-0
2,5-DIHYDRO-3,3,4,4-
TETRACHLOROTHIOPHENE-1,1-DIOXIDE

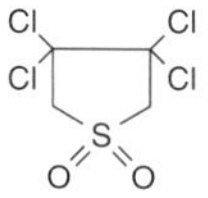

250 mg

S93,864-5
2,3-DIBROMOTETRAHYDROTHIOPHENE-
1,1-DIOXIDE

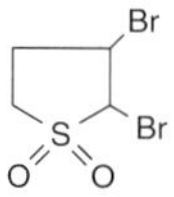

250 mg

S94,860-8
4,5-DIHYDRO-3-METHYLTHIOPHENE-
1,1,-DIOXIDE

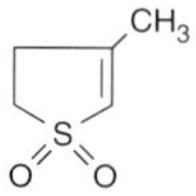

250 mg

S56,665-9
2-METHYLTETRAHYDROTHIOPHENE-
1,1-DIOXIDE

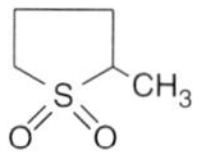

5 gm

S93,230-2
TRANS-3,4-DIBROMO-3-
METHYLTETRAHYDROTHIOPHENE-
1,1-DIOXIDE

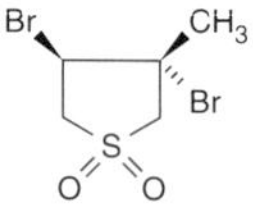

250 mg

S45,970-4
2,4-DIMETHYL-3-SULFOLENE

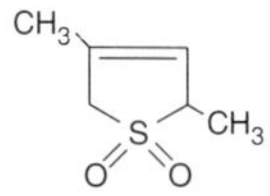

5 gm

S93,249-3
2,5-DIHYDRO-3,4-DIMETHYLTHIOPHENE-
1,1-DIOXIDE

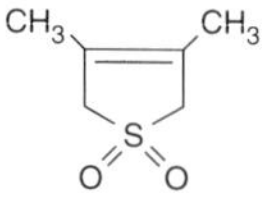

250 mg

S93,871-8
3-(BUTYLSULFONYL)-
TETRAHYDROTHIOPHENE-1,1-DIOXIDE

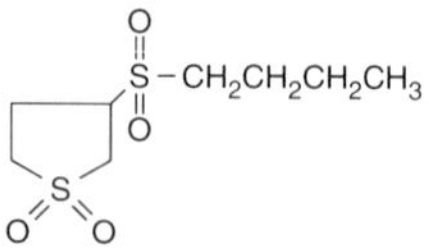

250 mg

S85,192-2
2,2-DIMETHYL-1,3-DITHIOLANE-1,1,3,3-TETROXIDE
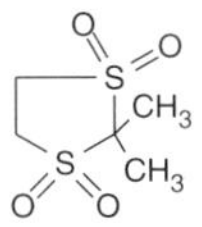
250 mg

S77,565-7
PENTAMETHYLENE SULFONE
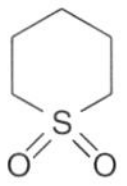
250 mg

S76,674-7
HEXAHYDRO-2-(METHYLSULFONYLMETHYL)-PYRIMIDINE HYDROCHLORIDE
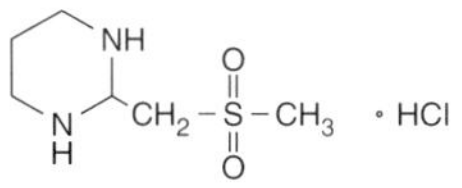
250 mg

S35,917-3
1,3-BUTYLENE GLYCOL SULFITE
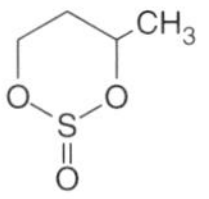
5 gm

S61,847-0
3-SULFOPROPYL 11-BROMOUNDECANOATE, SODIUM SALT
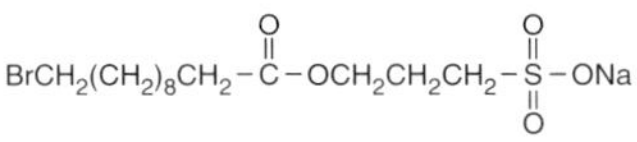
5 gm

S61,394-0
BROMOMETHANESULFONIC ACID SODIUM SALT
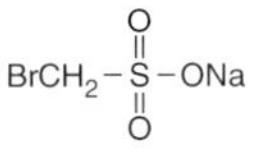
250 mg

S53,408-0
3-BROMO-1-PROPANESULFONIC ACID, SODIUM SALT
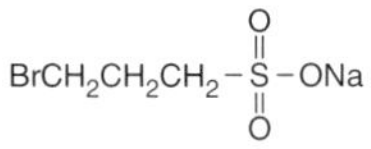
5 gm

S56,834-1
3-IODO-1-PROPANESULFONIC ACID, POTASSIUM SALT
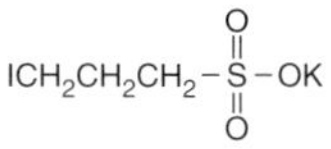
5 gm

S93,831-9
4-CHLORO-1-BUTANESULFONIC ACID, SODIUM SALT
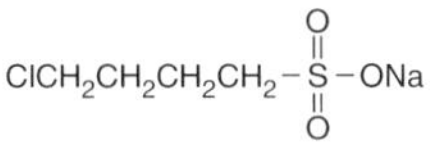
250 mg

S55,448-0
4-BROMO-1-BUTANESULFONIC ACID, SODIUM SALT
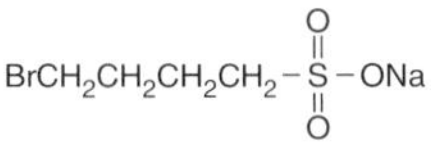
250 mg

S66,890-7
SUCCINALDEHYDE SODIUM BISULFITE-ADDITION COMPOUND
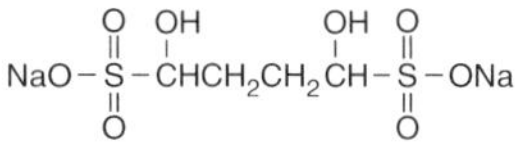
5 gm

S63,320-8
N,N-BIS(2-HYDROXYETHYL)-AMINOMETHANESULFONIC ACID, SODIUM SALT
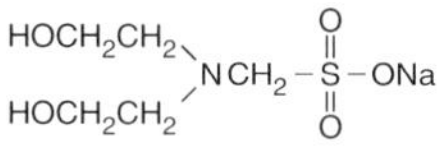
250 mg

S93,236-1
TRANS-4-HYDROXYTETRAHYDROTHIOPHENE-1,1-DIOXIDE-3-SULFONIC ACID, SODIUM SALT
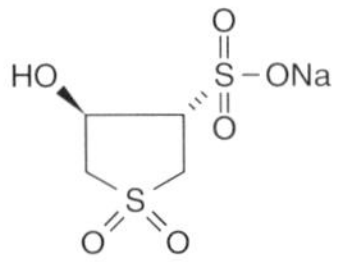
250 mg

S68,928-9
1-HYDROXY-1-METHYL-1-ETHANESULFONIC ACID SODIUM SALT
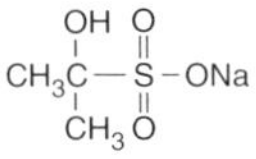
5 gm

S42,931-7
ACETONE SODIUM BISULFITE ADDITION COMPOUND
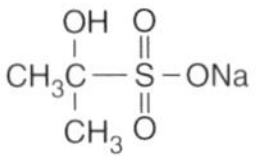
5 gm

S35,067-2
4-HYDROXY-1-BUTANESULFONIC ACID, SODIUM SALT
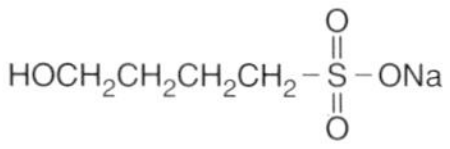
5 gm

S45,694-2
TRIMETHYLACETALDEHYDE SODIUM-BISULFITE ADDITION COMPOUND
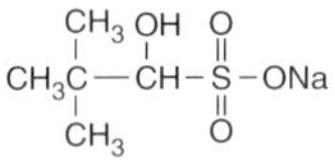
5 gm

S56,282-3
2-HYDROXY-2-NONANESULFONIC ACID, SODIUM SALT
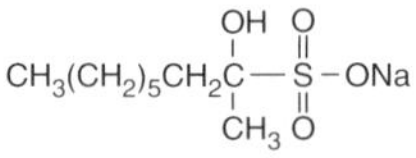
250 mg

S68,037-0
1-HYDROXYHEXADECANE-1-SULFONIC ACID SODIUM SALT
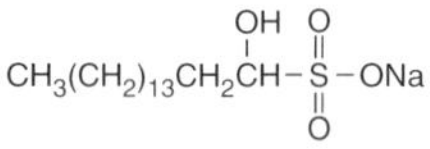
250 mg

S74,278-3
1-HYDROXY-1-OCTADECANESULFONIC ACID, SODIUM SALT
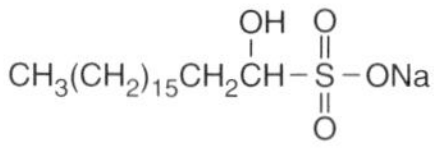
5 gm

S51,024-6
1-HYDROXY-1-CYCLOPENTANESULFONIC ACID, SODIUM SALT
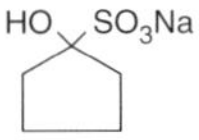
250 mg

S51,030-0
1-HYDROXY-1-CYCLOHEXANESULFONIC ACID, SODIUM SALT
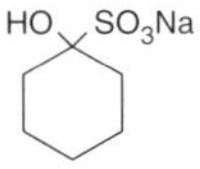
250 mg

S58,709-5
3,3'-THIOBIS(1-PROPANESULFONIC ACID, SODIUM SALT)
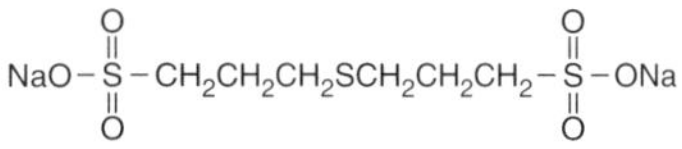
250 mg

S37,877-1
2-AMINOETHYLTHIOSULFONIC ACID
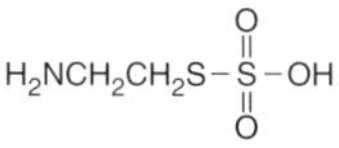
250 mg

S58,807-5
1-AMINO-2-CYANO-1,1-ETHANEDISULFONIC ACID, DISODIUM SALT TRIHYDRATE
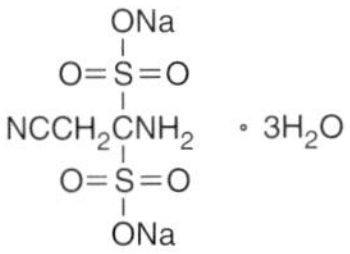
250 mg

S45,207-6
DL-CYSTEIC ACID
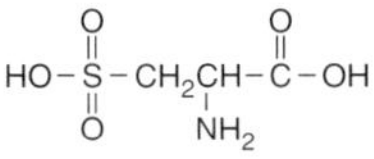
5 gm

S55,173-2
1-PIPERIDINEPROPANESULFONIC ACID
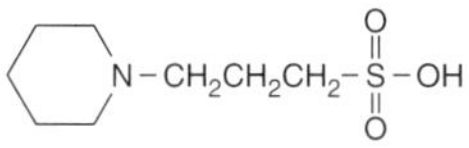
5 gm

S58,818-0
N-METHYL-N-STEAROYLTAURINE
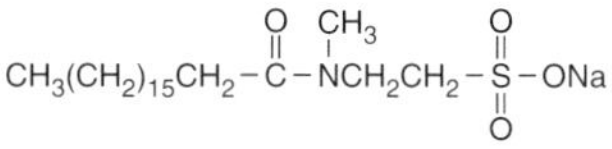
250 mg

S79,950-5
3,3-DIMETHYL-2-OXO-1-BUTANESULFONIC ACID HYDRATE
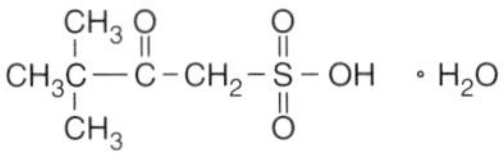
250 mg

S51,194-3
DIETHYL SULFOSUCCINATE, SODIUM SALT
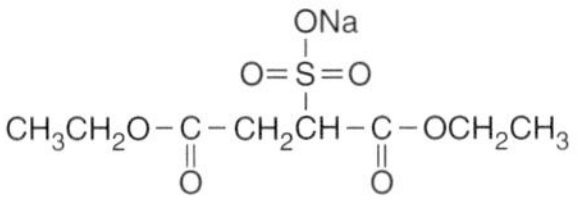
5 gm

S51,181-1
DIBUTYL SULFOSUCCINATE, SODIUM SALT
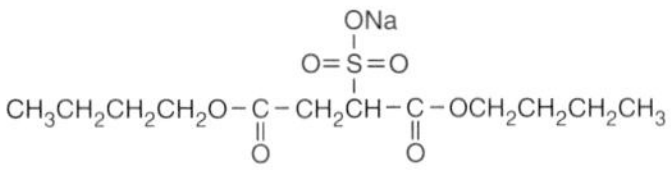
5 gm

S85,280-5
3-SULFOPROPIONIC ACID, DIAMMONIUM SALT
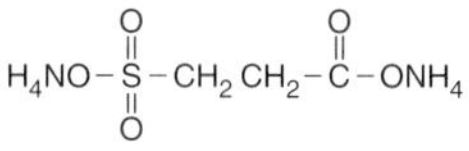
250 mg

S51,645-7
2-METHYL-2-SULFOOCTADECANOIC ACID

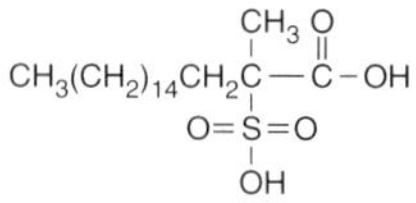

5 gm

S93,544-1
ACETOXYMETHANESULFONIC ACID, POTASSIUM SALT

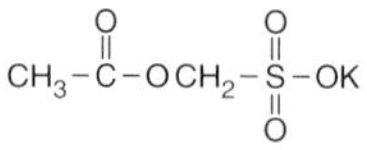

250 mg

S86,313-0
1-(ETHOXYCARBONYL)-1-ETHANESULFONIC ACID, POTASSIUM SALT

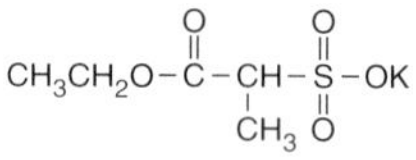

250 mg

S61,796-2
METHYL 2-SULFOOCTADECANOATE, SODIUM SALT

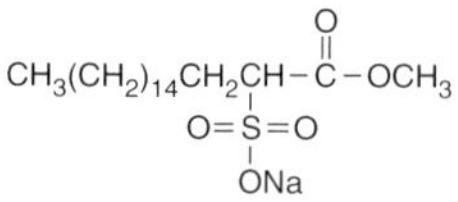

250 mg

S63,313-5
3-SULFOPROPYL ISOOCTADECANOATE, SODIUM SALT

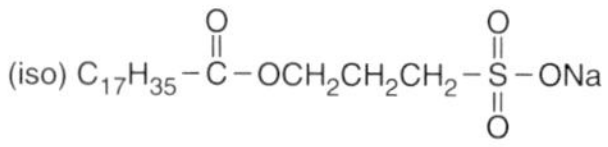

250 mg

S57,758-8
ETHYL 2-SULFODOCOSANOATE, SODIUM SALT

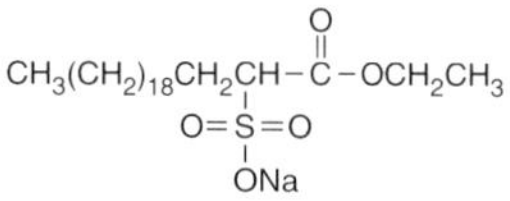

5 gm

S60,869-6
CARBAMOYLSULFONIC ACID, POTASSIUM SALT

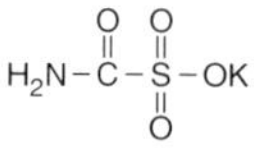

250 mg

S64,221-5
3-CYANOPROPANESULFONIC ACID POTASSIUM SALT

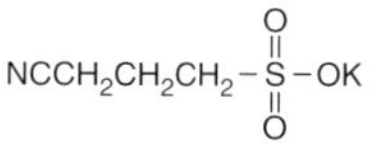

250 mg

S93,252-3
2,3-DIHYDROTHIOPHENE-1,1-DIOXIDE-3-SULFONIC ACID, SODIUM SALT

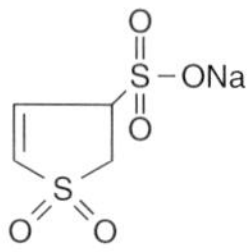

250 mg

S57,932-7
ETHANESULFONIC ACID, MORPHOLINE ADDUCT

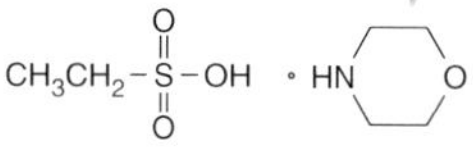

250 mg

S70,665-5
1,2-ETHANEDISULFONIC ACID, DISODIUM SALT

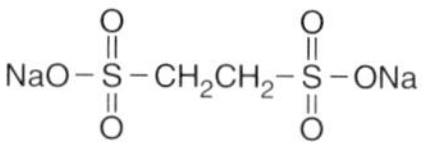

5 gm

S79,992-0
3,3-DIMETHYL-1-BUTANESULFONIC ACID, SODIUM SALT HYDRATE

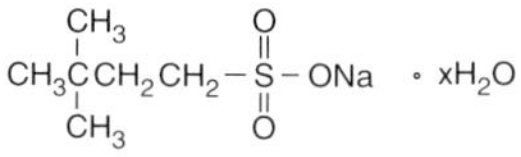

250 mg

S56,793-0
2-HEXYL-1-DECANESULFONIC ACID, SODIUM SALT

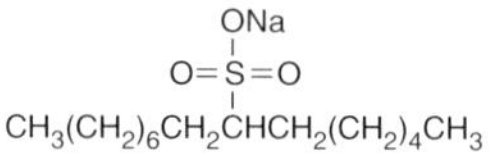

250 mg

S60,501-8
1-HEXADECANESULFONIC ACID, BARIUM SALT

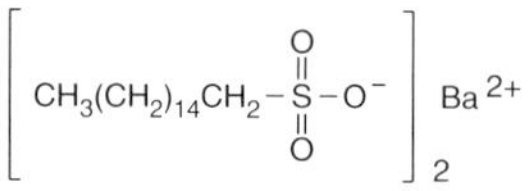

250 mg

S77,808-7
3,3-OXETANEBIS(METHANESULFONIC ACID) DISODIUM SALT

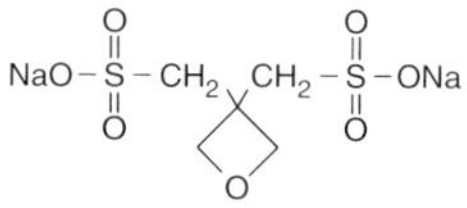

250 mg

S79,961-0
4-(2,5-DIMETHYLCYCLOHEXYL)-1-BUTANESULFONIC ACID, SODIUM SALT

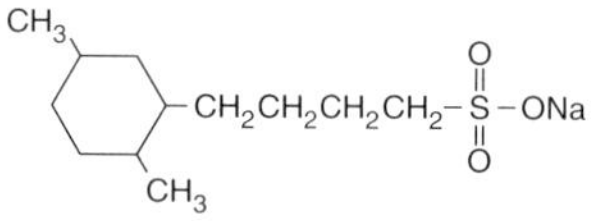

250 mg

S94,662-1
METHANESULFINIC ACID, LITHIUM SALT

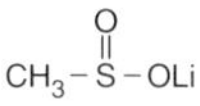

250 mg

S69,463-0
2,2,2-TRIFLUOROETHYL MESYLATE

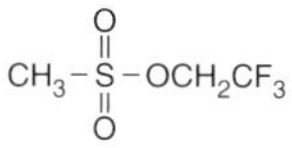

5 gm

S95,656-2
1,4,4A,5,8,8A-HEXAHYDRO-2-METHOXY-4A-METHYL-4-OXO-1-NAPHTHYL-10-CAMPHORSULFONATE

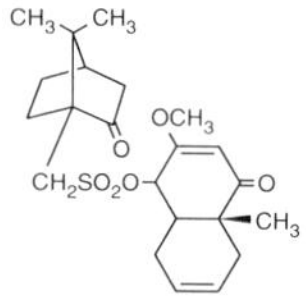

50 mg

S43,952-5
METHYL ALPHA,D-CAMPHORSULFONATE

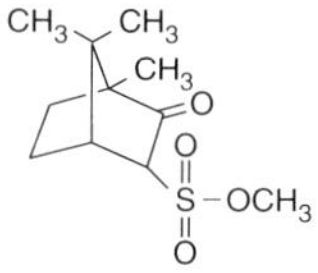

5 gm

S64,165-0
ALPHA-METHYLMANNOSIDE-6-(P-TOSYLATE) 2,3,4-TRIBENZOATE

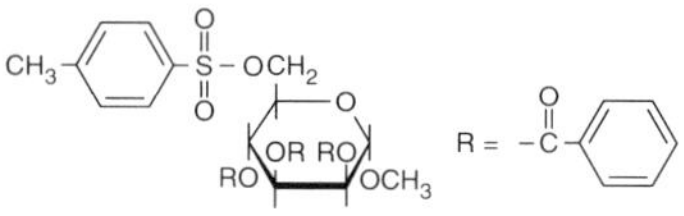

100 mg

S52,778-5
N,N'-METHYLENEBIS(2-CARBAMOYLETHYL)-BIS(METHANESULFONATE)

5 gm

S44,465-0
ACETYL METHANESULFONATE

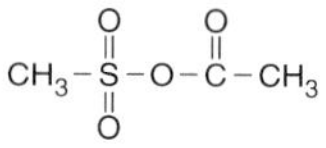

5 gm

S52,777-7
N,N'-(1-METHYLETHYLENE)BIS(2-CARBAMOYLETHYL)-BIS(METHANESULFONATE)

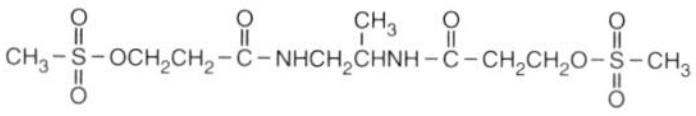

5 gm

S63,236-8
5-TRIDECYL-1,2-OXATHIOLANE-2,2-DIOXIDE

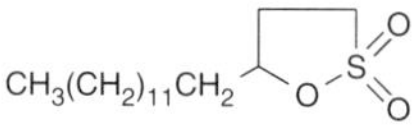

250 mg

S77,879-6
2,2-DIMETHYLPROPYL BROMOSULFONATE

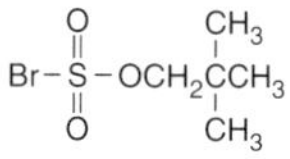

250 mg

S64,772-1
NEOPENTYL 1-BUTANESULFONATE

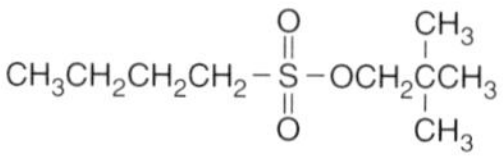

5 gm

S67,256-4
DODECYL METHANESULFONATE

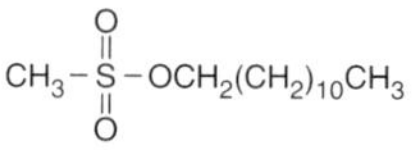

250 mg

S67,258-0
TETRADECYL METHANESULFONATE

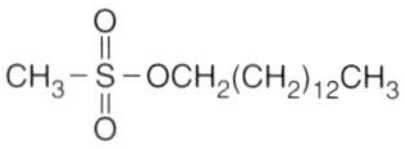

250 mg

S54,137-0
TETRAHYDROFURFURYL METHANESULFONATE

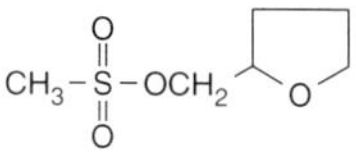

250 mg

S38,569-7
AMMONIUM METHANESULFONATE

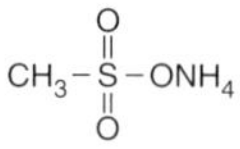

5 gm

S77,882-6
(METHYLSULFONYL)-METHANESULFONYL CHLORIDE

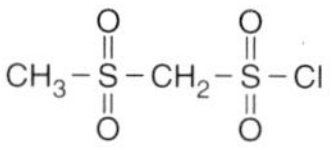

250 mg

S59,278-1
5,7,7-TRIMETHYL-2-(1,3,3-TRIMETHYLBUTYL)-1-OCTANESULFONYL CHLORIDE

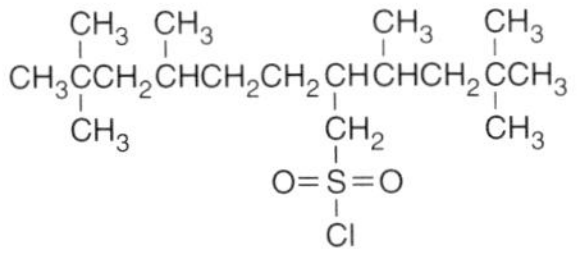

5 gm

S77,807-9
N,N-DIETHYLBROMOMETHANESULFONAMIDE

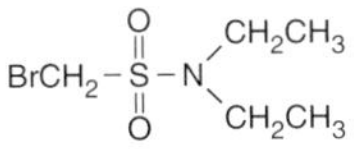

250 mg

S77,579-7
4-(BROMOMETHYLSULFONYL)MORPHOLINE

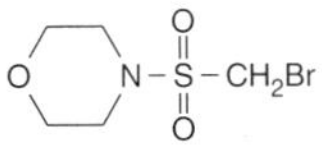

250 mg

S68,112-1
4,5-DIHYDROXY-2,1,3-THIADIAZOLE-2,2-DIOXIDE, DISODIUM SALT DIHYDRATE

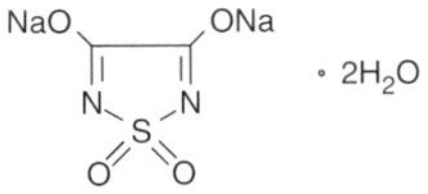

5 gm

S68,027-3
1-(2-SULFAMOYLETHYL)UREA

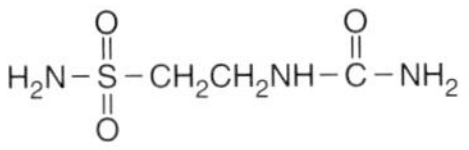

250 mg

S53,750-0
3-(HEXADECYLSULFONAMIDO)-PROPIONIC ACID

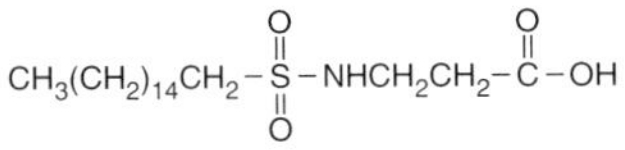

5 gm

S76,780-8
4-METHYL-1-(METHYLSULFONYL)-SEMICARBAZIDE

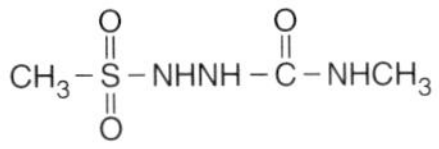

250 mg

S93,869-6
3-SULFOLANESULFONAMIDE

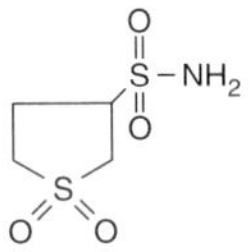

250 mg

S82,830-0
SULFAMIC ACID, CESIUM SALT

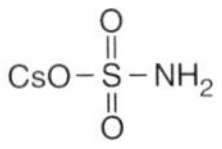

250 mg

S44,032-9
CYCLOHEXYLSULFAMIC ACID

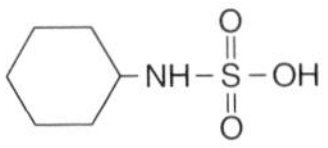

5 gm

S42,825-6
CYCLOHEXYLSULFAMIC ACID, CALCIUM SALT DIHYDRATE

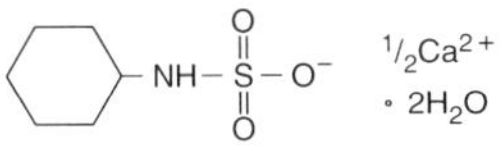

5 gm

S61,667-2
N,N-DIMETHYLMETHANESULFONAMIDE

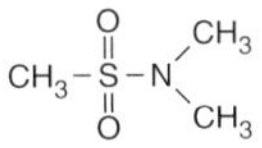

5 gm

S79,949-1
N,N-DIETHYLMETHANESULFONAMIDE

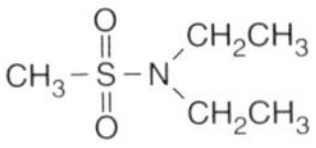

250 mg

S55,373-5
N-DODECYLMETHANESULFONAMIDE

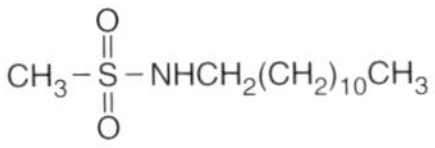

250 mg

S79,997-1
4-(METHYLSULFONYL)MORPHOLINE

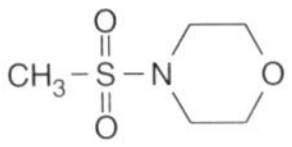

250 mg

S79,999-8
4-(ETHYLSULFONYL)MORPHOLINE

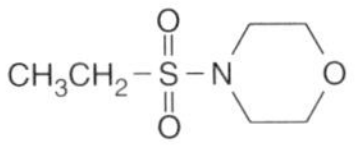

250 mg

S50,454-8
1-DODECANESULFONIC ACID HYDRAZIDE

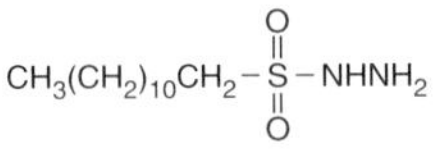

5 gm

S76,361-6
N,N'-DIACETYLSULFAMIDE

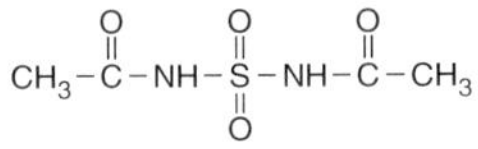

250 mg

S82,611-1
THALLIUM(I) BISULFITE

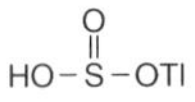

250 mg

S60,407-0
3-IODOPROPYL SULFATE, SODIUM SALT

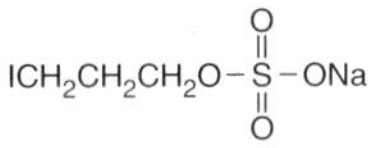

250 mg

S36,542-4
TRIS(ETHYLENEDIAMINE)NICKEL(II) SULFATE

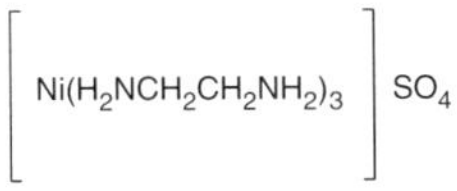

5 gm

S65,154-0
FORMYLTRIMETHYLAMMONIUM METHYL SULFATE

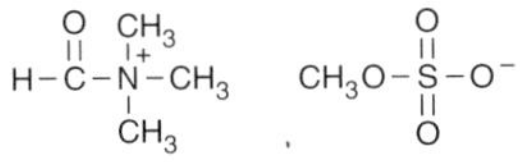

5 gm

S54,781-6
2-AMINO-2-METHYL-1-PROPANOL HYDROGEN SULFATE

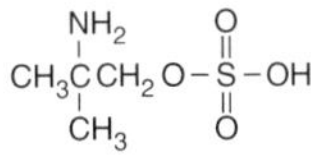

250 mg

S61,714-8
BUTYL SULFATE SODIUM SALT

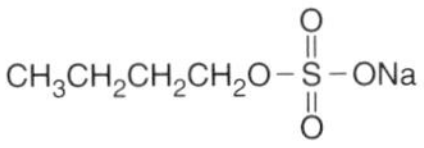

5 gm

S61,716-4
HEXYL SULFATE, SODIUM SALT

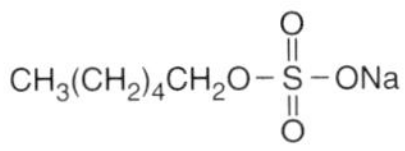

5 gm

S61,894-2
NONYL SULFATE, SODIUM SALT

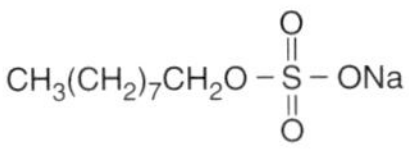

5 gm

S68,335-3
TRIDECYL SULFATE, SODIUM SALT

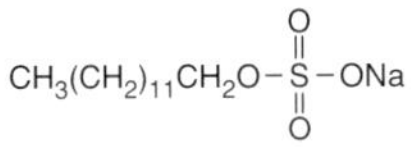

5 gm

S81,250-1
DIDODECYL SULFATE

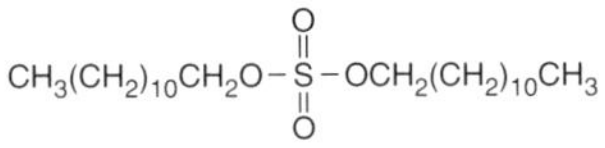

250 mg

S44,338-7
TRIS(HYDROXYMETHYL)PHOSPHINE
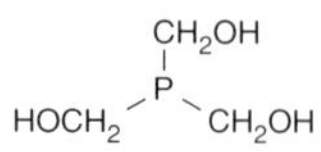
250 mg

S83,568-4
TRIETHYLPHOSPHINE OXIDE
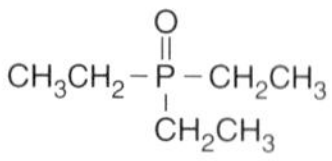
250 mg

S55,001-9
METHYLTRIBUTYLPHOSPHONIUM IODIDE
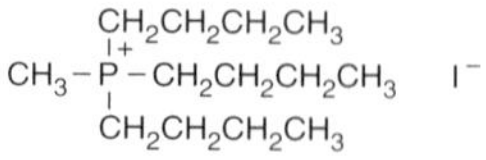
5 gm

S54,472-8
ISOPROPYLTRIBUTYLPHOSPHONIUM IODIDE
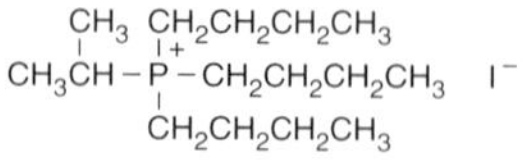
250 mg

S86,195-2
BIS(2-ETHYLHEXYL)PHOSPHINE OXIDE
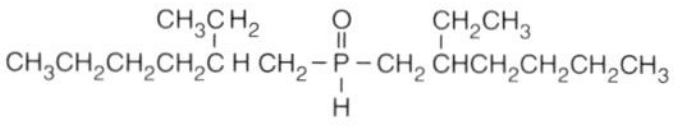
250 mg

S44,065-5
TRIOCTYLPHOSPHINE
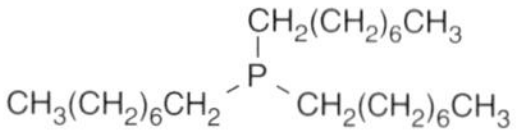
5 gm

S81,266-8
TRI(DODECYL)PHOSPHINE
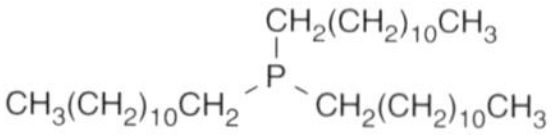
250 mg

S54,334-9
(4-IODOBUTYL)-
TRIETHYLPHOSPHONIUM IODIDE
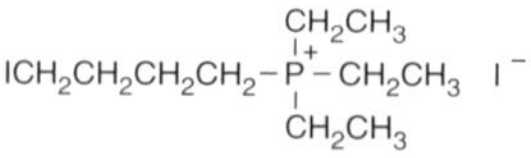
250 mg

S61,902-7
TRIS(DIPROPYLAMINO)PHOSPHINE
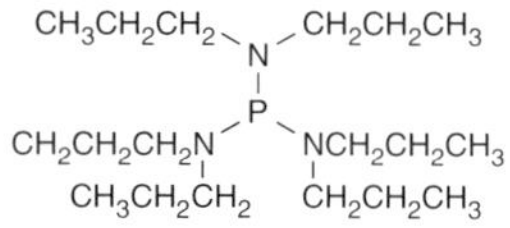
5 gm

S44,822-2
TRIS(2-CARBAMOYLETHYL)-
PHOSPHINE OXIDE
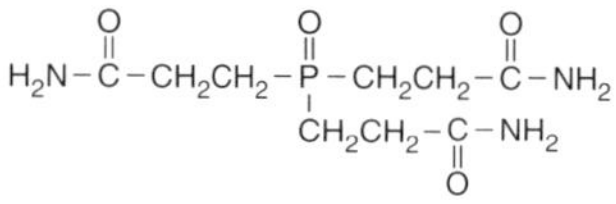
5 gm

S86,190-1
DICYCLOPENTYL(N,N-DIETHYLCARBAMOYL)-
PHOSPHINE OXIDE
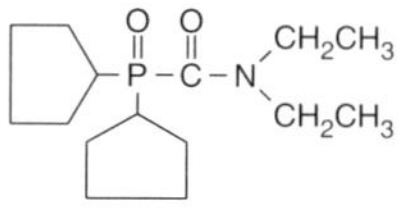
250 mg

S76,464-7
BIS(2-CYANOETHYL)PHOSPHINE
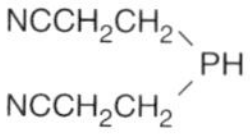
5 gm

S43,603-8
TRIS(2-CYANOETHYL)PHOSPHINE
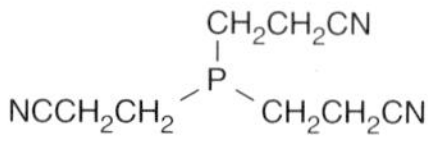
5 gm

S84,072-6
(2-CYANOETHYL)-
TRIETHYLPHOSPHONIUM BROMIDE
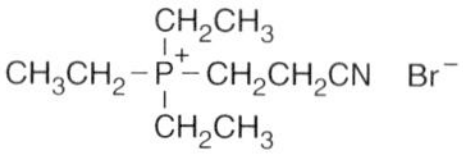
250 mg

S37,401-6
PHOSPHORIC TRIMORPHOLIDE
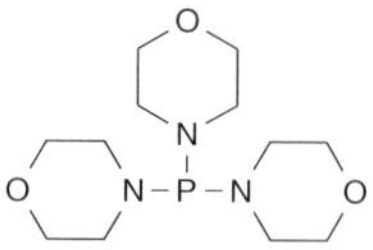
5 gm

S94,493-9
1,3,5-TRIAZA-7-PHOSPHAADAMANTANE
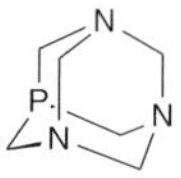
250 mg

S42,895-7
ETHYLENE PHOSPHITE
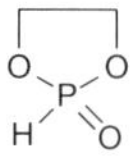
5 gm

S44,317-4
DIALLYL PHOSPHITE
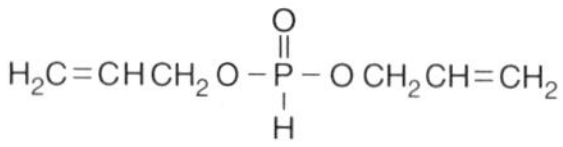
5 gm

S62,642-2
DI-TERT-BUTYL PHOSPHITE
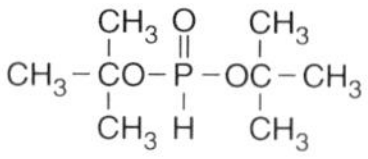
250 mg

S54,591-0
TRIPROPYL PHOSPHITE
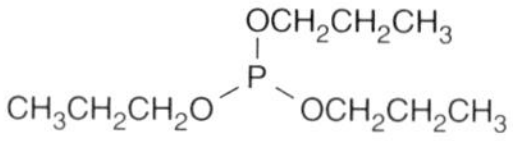
250 mg

S42,939-2
TRILAURYL TRITHIOPHOSPHITE
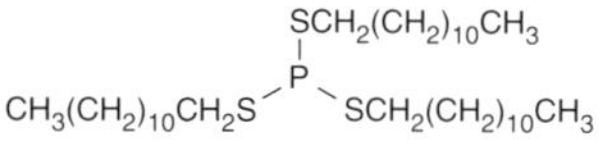
5 gm

S41,296-1
TRIOCTADECYL PHOSPHITE
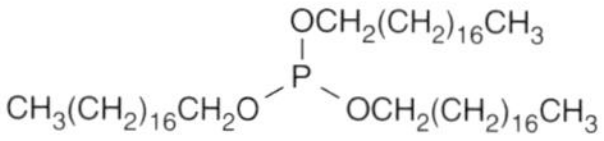
5 gm

S40,682-1
BIS(HYDROXYMETHYL)PHOSPHINIC ACID
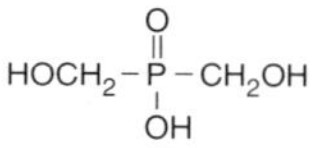
5 gm

S65,894-4
DIMETHYL ETHYLPHOSPHONATE
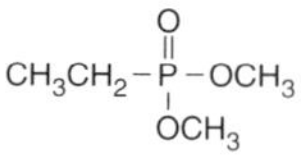
5 gm

S95,012-2
DIETHYL ALLYLPHOSPHONATE
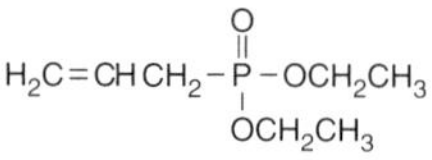
50 mg

S63,289-9
DIETHYL PROPIONYLPHOSPHONATE
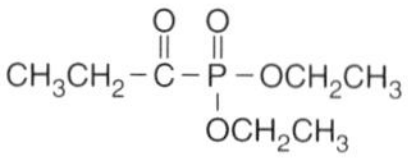
250 mg

S53,429-3
TETRAETHYL ETHYLENEDIPHOSPHONATE
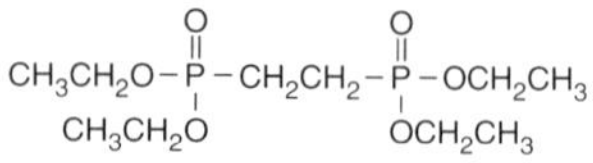
250 mg

S84,946-4
DIBUTYL (1-NITROPENTYL)PHOSPHONATE
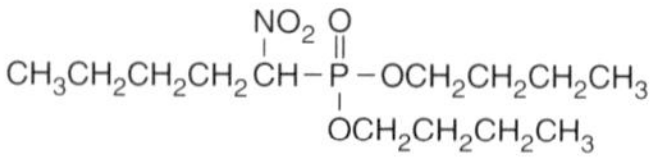
250 mg

S41,209-0
DIAMYL AMYLPHOSPHONATE
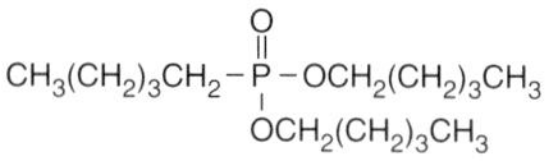
5 gm

S86,167-7
BIS(1,3-DIMETHYLBUTYL)-
SEC-BUTYLPHOSPHONATE
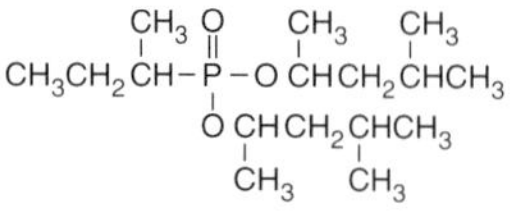
250 mg

S86,176-6
TETRAISOBUTYL
METHYLENEDIPHOSPHONATE
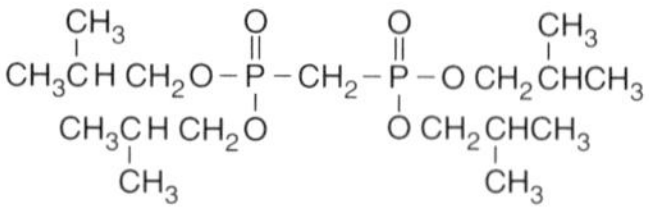
250 mg

S86,178-2
TETRABUTYL
1,3-TRIMETHYLENEDIPHOSPHONATE
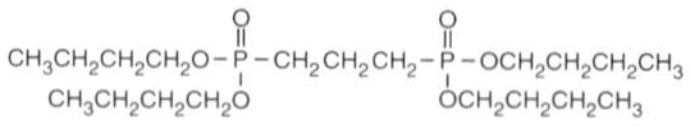
250 mg

S86,177-4
1,1-DIBUTYL 3,3-DIHEXYL METHYLENEDIPHOSPHONATE

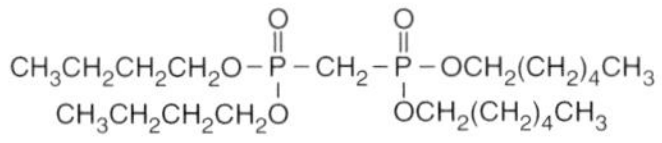

250 mg

S86,175-8
TETRAHEXYL METHYLENEDIPHOSPHONATE

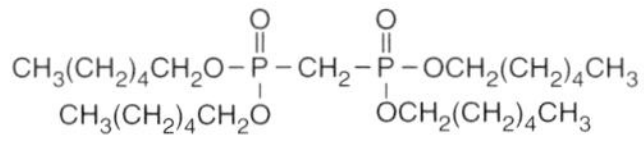

250 mg

S36,685-4
BIS(2-CHLOROETHYL)VINYLPHOSPHONATE

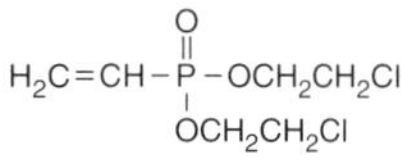

5 gm

S34,779-5
DIPROPYL CHLOROMETHYLPHOSPHONATE

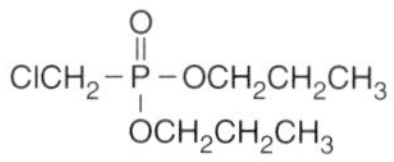

5 gm

S37,274-9
DIBUTYL CHLOROMETHYLPHOSPHONATE

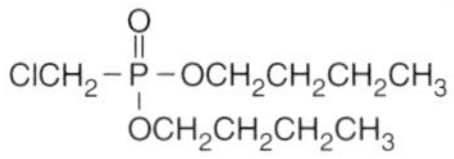

5 gm

S35,865-7
BIS(2-CHLOROISOPROPYL)-ISOPROPENYLPHOSPHONATE

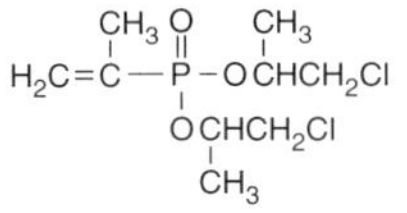

5 gm

S35,864-9
BIS(2-CHLOROISOPROPYL)-2-CHLOROISOPROPYLPHOSPHONATE

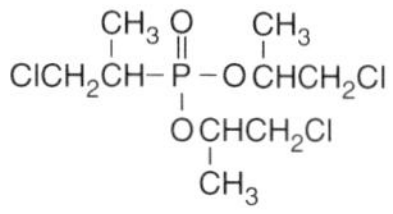

5 gm

S45,877-5
TRICHLOROMETHYLPHOSPHONIC ACID

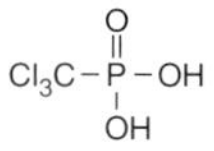

5 gm

S89,907-0
DIETHYL (1,2,2,2-TETRACHLOROETHOXYCARBONYL)-PHOSPHONATE

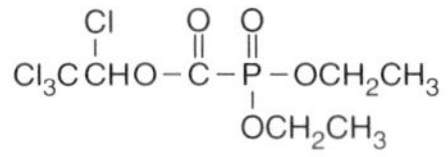

250 mg

S36,091-0
DIMETHYL 1-(BUTYRYLOXY)-2,2,2-TRICHLOROETHYLPHOSPHONATE

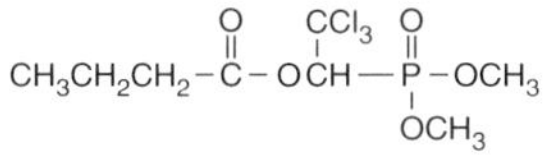

5 gm

S36,046-5
DIETHYL BIS(2-HYDROXYETHYL)-AMINOMETHYLPHOSPHONATE

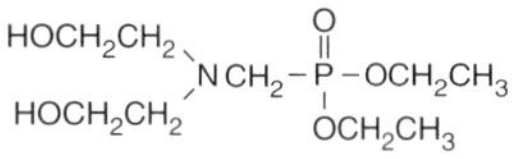

5 gm

S88,377-8
DIETHYL [METHOXY(METHOXYCARBONYL)-METHYL]PHOSPHONATE

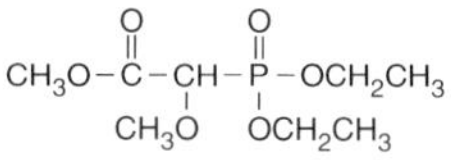

250 mg

S42,035-2
(AMINOMETHYL)PHOSPHONIC ACID, IRON(II) SALT

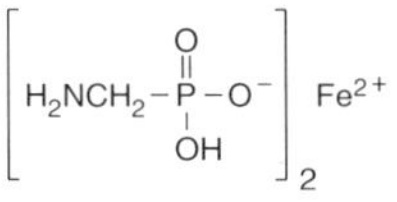

250 mg

S71,066-0
DIETHYL (3-METHOXYCARBONYL-2-PROPENYL)PHOSPHONATE

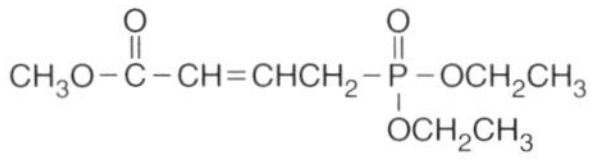

5 gm

S64,257-6
ETHYL 2-(DIETHYLPHOSPHONO)-HEXADECANOATE

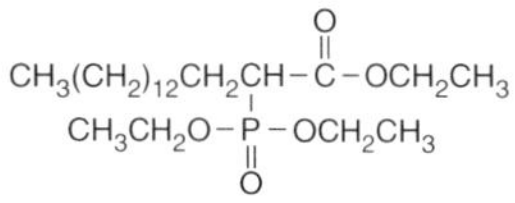

250 mg

S64,578-8
DIETHYL CARBAMOYLMETHYLPHOSPHONATE

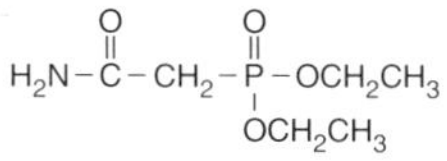

250 mg

S86,125-1
DIISOPROPYL (N,N-DIETHYLCARBAMOYL)-PHOSPHONATE

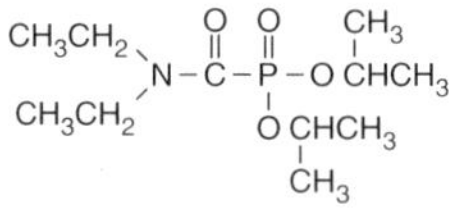

250 mg

S86,192-8
DIISOBUTYL (N,N-DIETHYLCARBAMOYL)-PHOSPHONATE

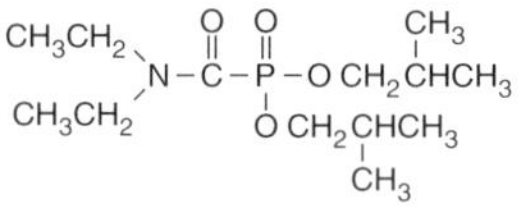

250 mg

S86,187-1
DI-SEC-BUTYL (N,N-DIETHYLCARBAMOYLMETHYL)-PHOSPHONATE

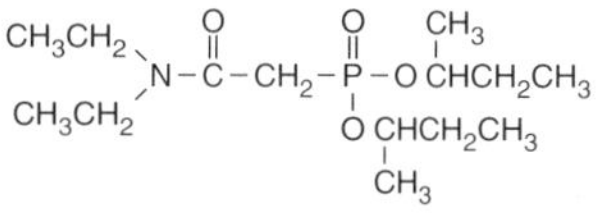

250 mg

S86,189-8
DIISOPROPYL (N,N-DIISOBUTYLCARBAMOYL)-PHOSPHONATE

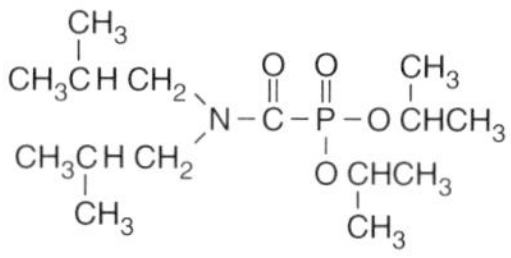

250 mg

S86,193-6
DIBUTYL [1-(N,N-DIETHYLCARBAMOYL)-ETHYL]PHOSPHONATE

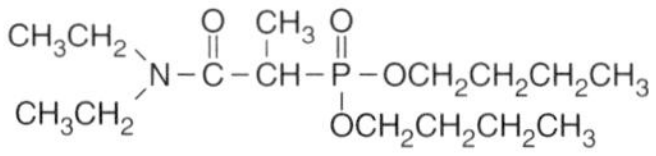

250 mg

S86,169-3
DIBUTYL (N,N-DIBUTYLCARBAMOYLMETHYL)-PHOSPHONATE

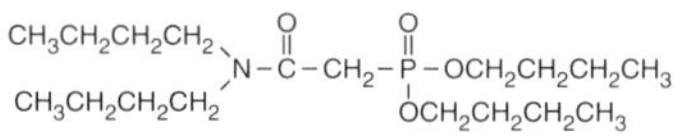

250 mg

S86,170-7
DIOCTYL (N,N-DIMETHYLCARBAMOYLMETHYL)-PHOSPHONATE

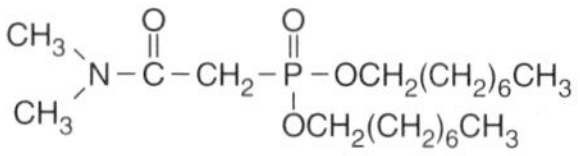

250 mg

S86,171-5
DIOCTYL (N,N-DIETHYLCARBAMOYL)-PHOSPHONATE

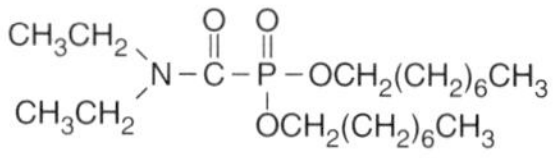

250 mg

S86,172-3
DIHEXYL [1-(N,N-DIPROPYLCARBAMOYL)-ETHYL]PHOSPHONATE

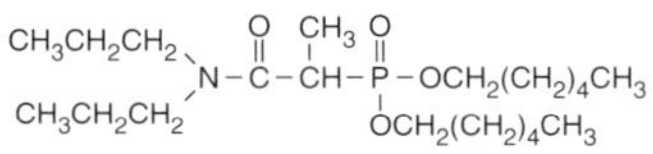

250 mg

S86,180-4
BIS(2,2,4-TRIMETHYLPENTYL)-(N,N-DIETHYLCARBAMOYLMETHYL)-PHOSPHONATE

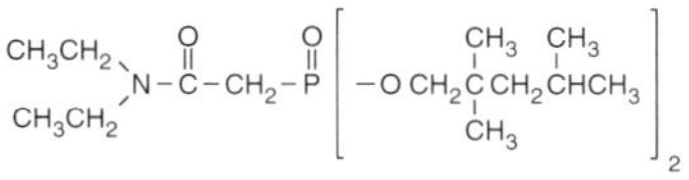

250 mg

S86,181-2
BIS(1,3-DIMETHYLBUTYL)-(N,N-DIISOBUTYLCARBAMOYLMETHYL)-PHOSPHONATE

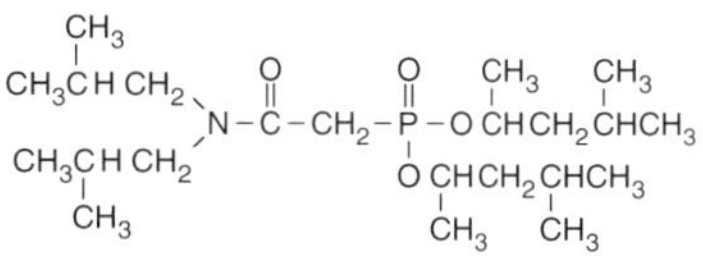

250 mg

S86,182-0
BIS(2-ETHYLHEXYL)-(N,N-DIETHYLCARBAMOYLMETHYL)-PHOSPHONATE

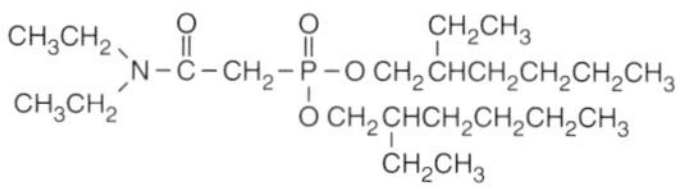

250 mg

S86,183-9
BIS(2-ETHYLHEXYL)-(N,N-DIISOPROPYLCARBAMOYLMETHYL)-PHOSPHONATE

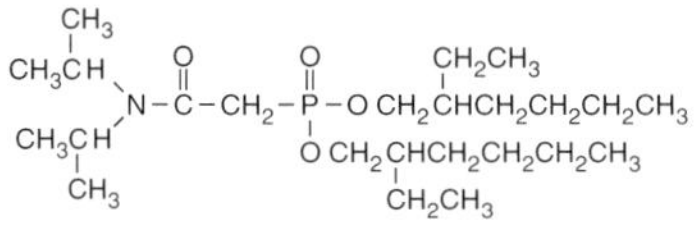

250 mg

S44,272-0
DIETHYL METHYLAMIDOPHOSPHATE

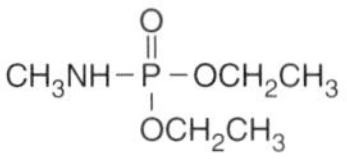

5 gm

S43,054-4
DIETHYL ETHYLAMIDOPHOSPHATE

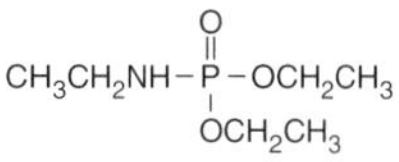

5 gm

S85,912-5
DIISOPROPYL PHOSPHORAMIDATE

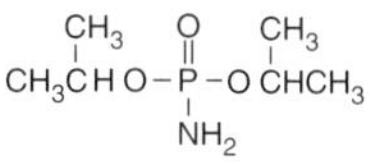

250 mg

S41,981-8
DIETHYL ISOPROPYLAMIDOPHOSPHATE

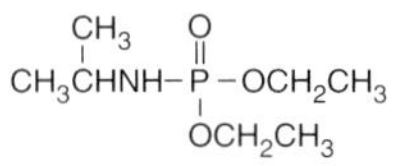

5 gm

S43,195-8
DIETHYL TERT-BUTYLAMIDOPHOSPHATE

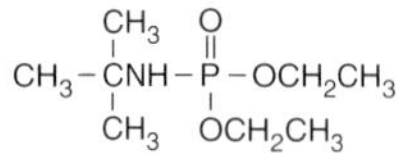

5 gm

S87,607-0
N-METHYL-N',N',N'',N''-TETRAISOPROPYLPHOSPHORIC TRIAMIDE

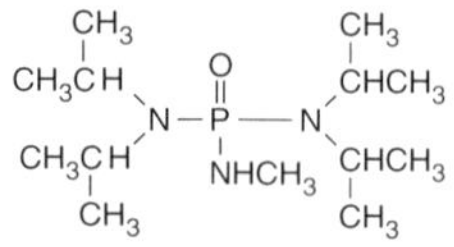

250 mg

S36,840-7
DILAURYL AMIDOPHOSPHATE

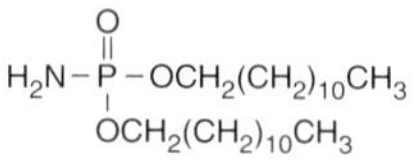

5 gm

S55,638-6
BIS(2,3-DIBROMOPROPYL)-N-(2-HYDROXYETHYL)PHOSPHORAMIDATE

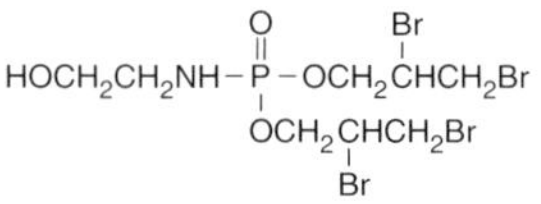

5 gm

S41,228-7
DIETHYL HEXYLAMIDOPHOSPHATE

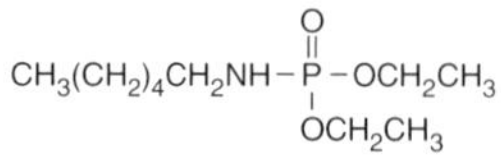

5 gm

S44,214-3
DIETHYL CYCLOHEXYLAMIDOPHOSPHATE

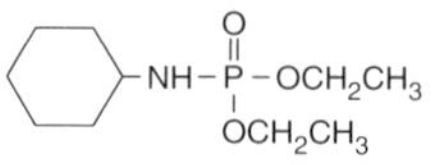

5 gm

S65,256-3
DIPENTYL PHOSPHOROCHLORIDATE

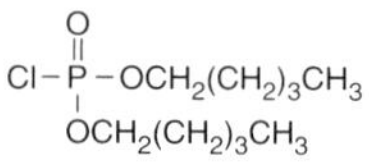

250 mg

S87,606-2
TETRAISOPROPYLPHOSPHORODIAMIDIC CHLORIDE

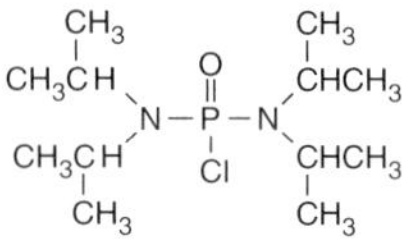

250 mg

S50,157-3
N-ISOPROPYLTHIOPHOSPHORAMIDIC DICHLORIDE

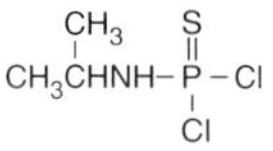

5 gm

S50,164-6
PROPYL PHOSPHORODICHLORIDOTHIONATE

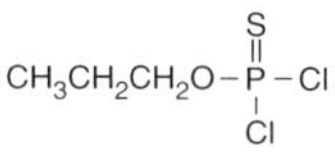

5 gm

S54,331-4
TETRAMETHYLDIPHOSPHINE DISULFIDE

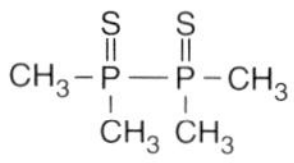

250 mg

S50,160-3
BUTYL PHOSPHORODICHLORIDOTHIONATE

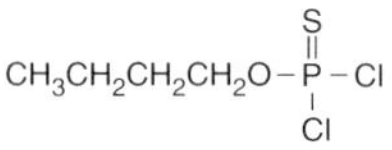

5 gm

S35,419-8
2-CHLORO-2,3-DIHYDRO-3,3,5-TRIMETHYL-1,2-OXAPHOSPHOLE-2-THIONE

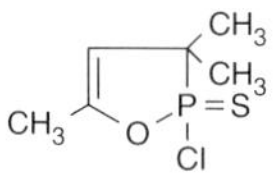

5 gm

S36,130-5
DIPROPYL CHLOROTHIOPHOSPHATE

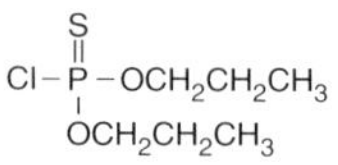

5 gm

S54,330-6
TETRAETHYLDIPHOSPHINE DISULFIDE

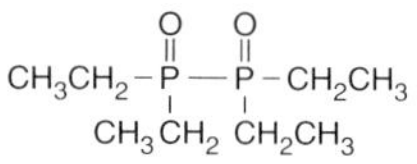

250 mg

S56,469-9
DIBUTYL CHLOROTHIOPHOSPHATE

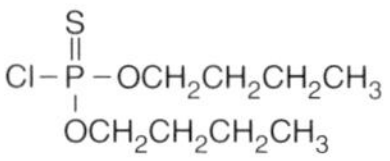

5 gm

S37,068-1
TETRAETHYL DITHIOPYROPHOSPHATE

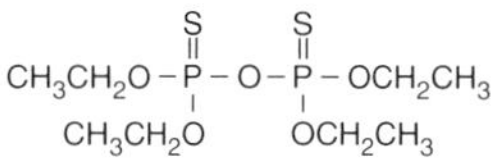

5 gm

S50,158-1
N,N-DIETHYLTHIOPHOSPHORAMIDIC DICHLORIDE

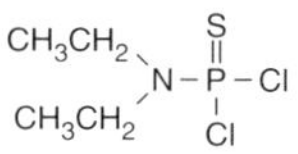

5 gm

S43,382-9
METHYL (DIMETHOXYPHOSPHINOTHIOYLTHIO)-ACETATE

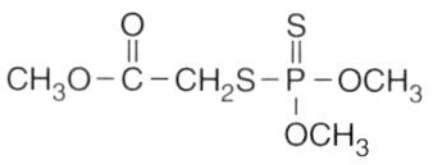

5 gm

S44,432-4
DICYCLOHEXYLDITHIOPHOSPHINIC ACID, AMMONIUM SALT HYDRATE

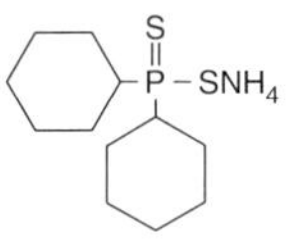

5 gm

S44,691-2
DIMETHYL ACID PYROPHOSPHATE

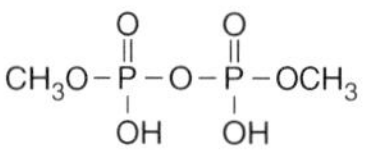

5 gm

S86,033-6
TERT-BUTYL PHOSPHATE, DICYCLOHEXYLAMINE SALT

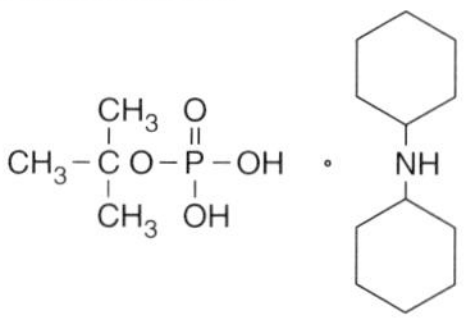

250 mg

S40,580-9
DIETHYL ACID PYROPHOSPHATE

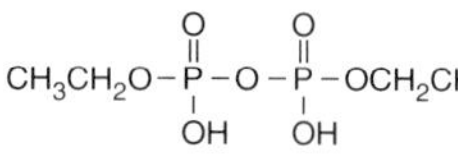

5 gm

S86,129-4
TRIISOBUTYL PHOSPHATE

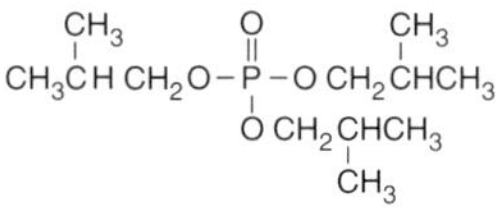

250 mg

S75,522-2
DIDODECYL PHOSPHATE

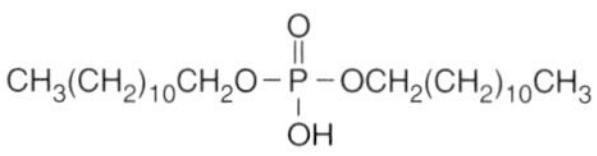

250 mg

S77,453-7
DIMYRISTYL PHOSPHATE

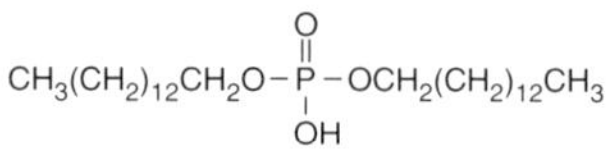

250 mg

S77,093-0
DIPENTADECYL PHOSPHATE

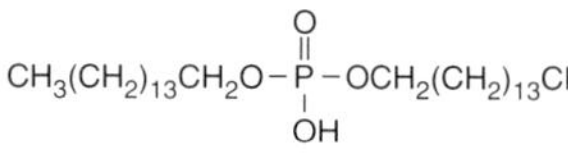

250 mg

S75,568-0
DIHEPTADECYL PHOSPHATE

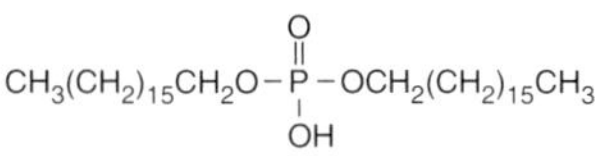

250 mg

S43,905-3
DIOCTADECYL PHOSPHATE

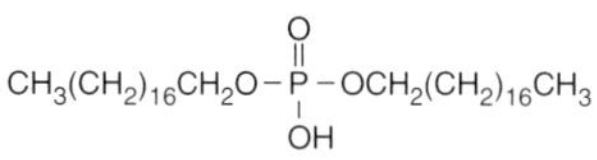

250 mg

S52,084-5
TRIS(1-BROMOMETHYL-2-CHLOROETHYL)-PHOSPHATE

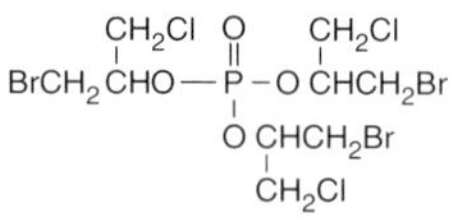

5 gm

S56,075-8
BIS(2,3-DIBROMOPROPYL)-2-(METHACRYLOYLOXY)ETHYL PHOSPHATE

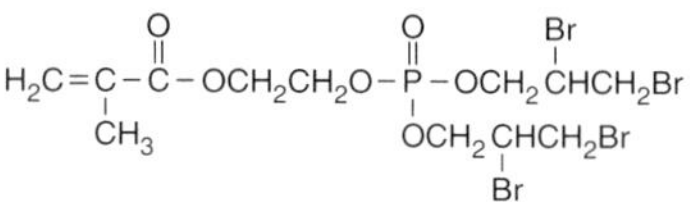

5 gm

S86,036-0
MYO-INOSITOL 1-(DIHYDROGEN PHOSPHATE), BARIUM SALT

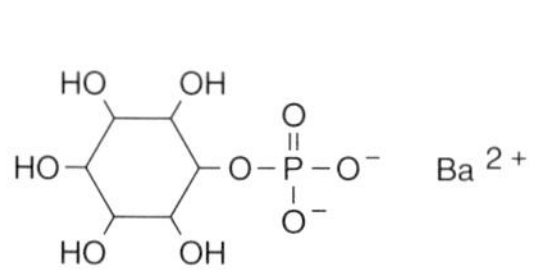

250 mg

S78,560-1
GLUCONIC ACID 6-(DIHYDROGEN PHOSPHATE), BARIUM SALT

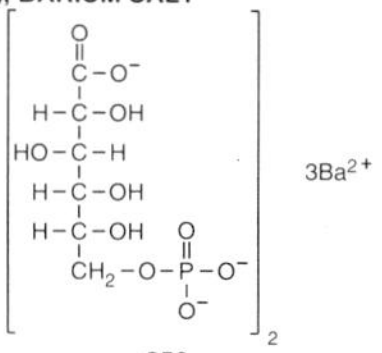

250 mg

S40,868-9
FRUCTOSE-1-PHOSPHATE, BARIUM SALT HEPTAHYDRATE

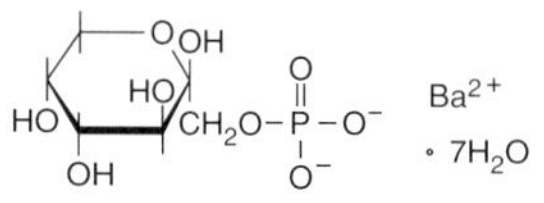

5 gm

S40,869-7
FRUCTOSE-1-PHOSPHATE, BARIUM SALT, TRIHYDRATE, CRYST

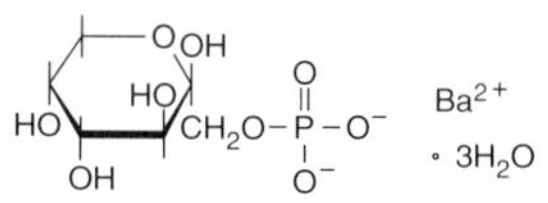

5 gm

S40,870-0
FRUCTOSE-1-PHOSPHATE, DICYCLOHEXYLAMMONIUM SALT

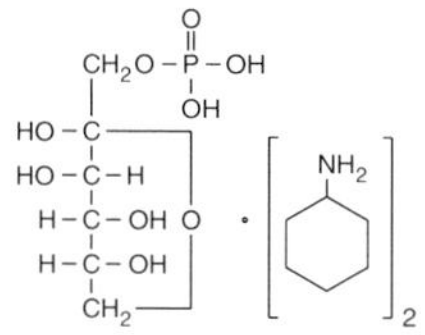

250 mg

S57,929-7
GLYCEROL-1-PHOSPHATE, DISODIUM SALT

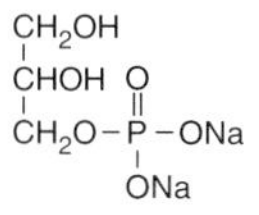

250 mg

S45,167-3
5-METHYL-2'-DEOXYCYTIDINE-5'-MONOPHOSPHORIC ACID

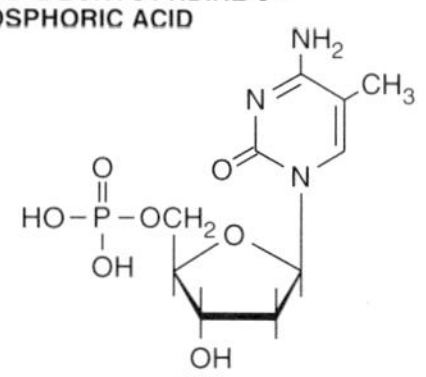

250 mg

S45,166-5
2'-DEOXYURIDINE 5'-MONOPHOSPHATE, DISODIUM SALT DIHYDRATE

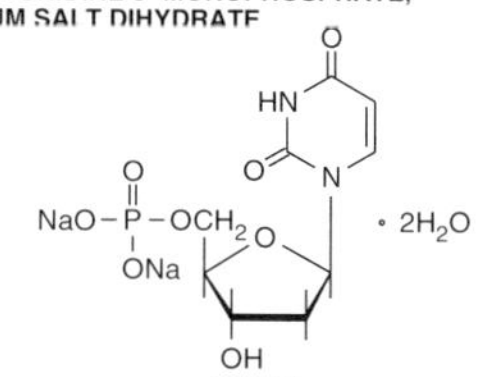

250 mg

S86,027-1
TRANS-1,2-CYCLOHEXANEDIOL MONOPHOSPHATE

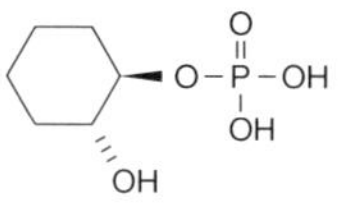

250 mg

S78,559-8
CYTIDYL (3'-5')CYTIDINE AMMONIUM SALT

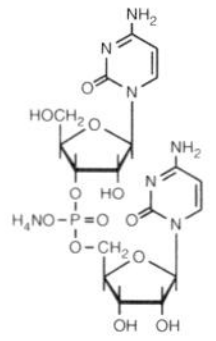

250 mg

S85,918-4
2-METHOXYETHYL PHOSPHATE, BARIUM SALT

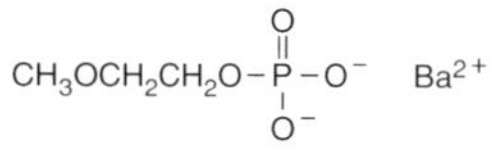

250 mg

S56,817-1
TRIS(2-METHOXYETHYL)PHOSPHATE

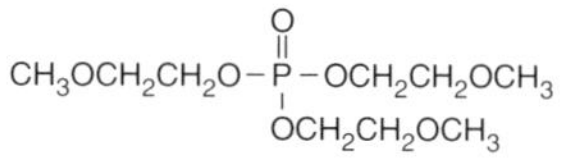

250 mg

S45,164-9
URIDINE-5'-DIPHOSPHOGLUCOSE, DISODIUM SALT

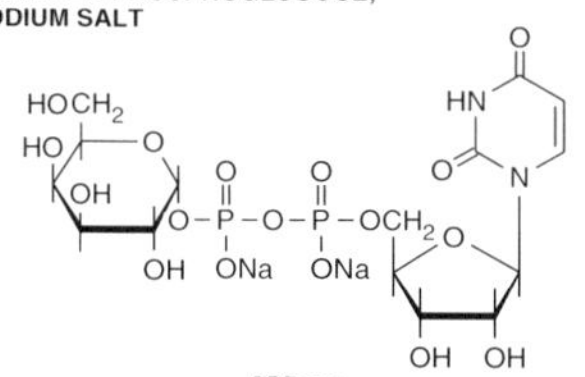

250 mg

S36,544-0
TRIS(TETRAHYDROFURFURYL)PHOSPHATE

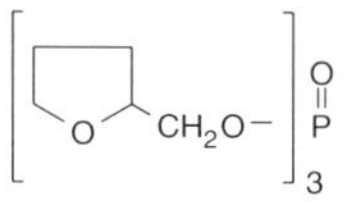

5 gm

S36,482-7
DL-SERINE PHOSPHATE HYDRATE

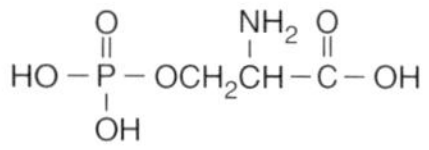

250 mg

S36,506-8
DL-THREONINE PHOSPHATE HYDRATE

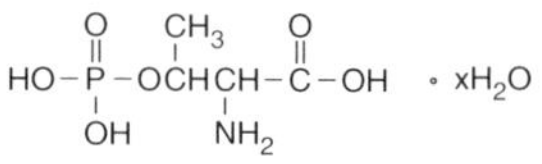

250 mg

S42,938-4
TRIPROPYLAMINE PHOSPHATE

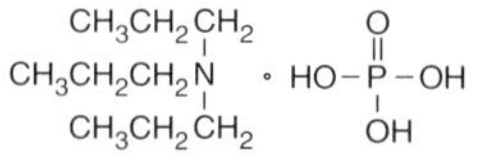

5 gm

S78,201-7
TRIS[2-(DIETHYLAMINO)ETHYL]-PHOSPHATE TRIHYDROCHLORIDE

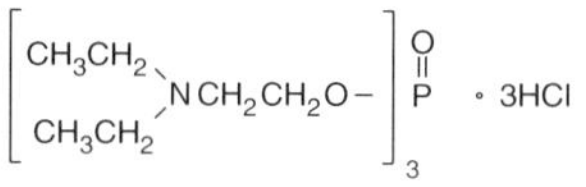

5 gm

S86,648-2
1-CYCLOHEXEN-1-YL DIETHYL PHOSPHATE

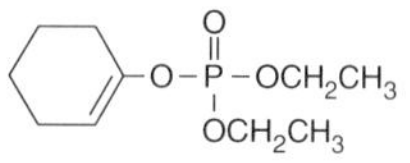

250 mg

S45,036-7
1,4-DIOXANE DIPHOSPHATE

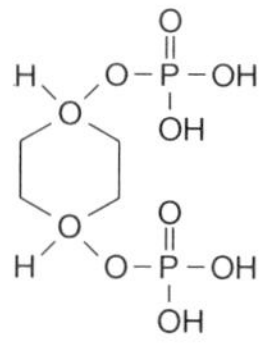

5 gm

S84,648-1
1-TERT-BUTYL-2,3,4,5-TETRAMETHYLBENZENE
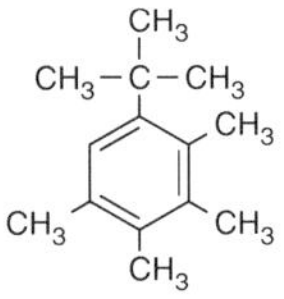

250 mg

S85,268-6
2,3-DIMETHYL-4B,8B,8C,8D-TETRAHYDRODIBENZO(A,F)CYCLOPROPA(CD)PENTALENE
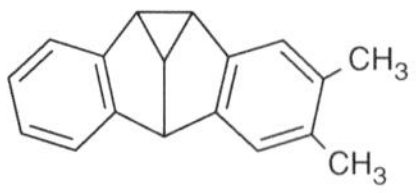

250 mg

S86,413-7
9,10,12,13-TETRAHYDRO-11H-9,10-ENDO-CYCLOPROPANTHRACENE
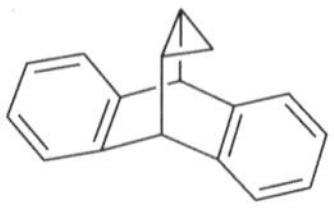
250 mg

S45,564-4
CYCLOPROPYLPHENYLMETHANE
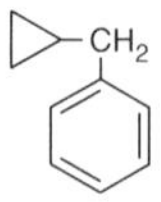

250 mg

S85,269-4
9,10,11,12,13,14-HEXAHYDRO-9,10-(1',2')-ENDO-CYCLOBUTANTHRACENE
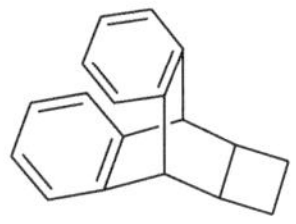
250 mg

S54,931-2
CYCLOPENTYLBENZENE
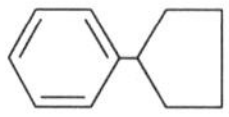
250 mg

S90,650-6
1-BENZYLIDENEINDAN
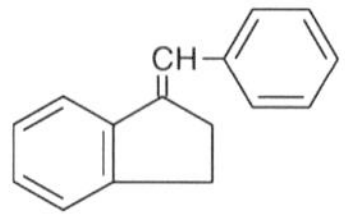

250 mg

S79,222-5
1,4-DIHYDRO-1,4-METHANONAPHTHALENE
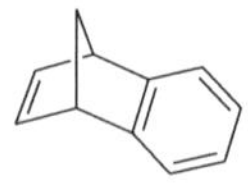
250 mg

S79,230-6
2,3,4,5,6,7,8,9-OCTAHYDRO-1H-TRINDENE
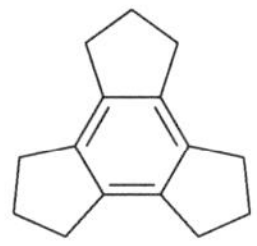
250 mg

S51,174-9
ALPHA-CYCLOHEXYLTOLUENE
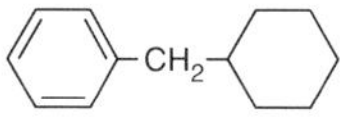

250 mg

S55,270-4
1,1'-(1,4-PHENYLENE)BIS(1-METHYLCYCLOHEXANE)
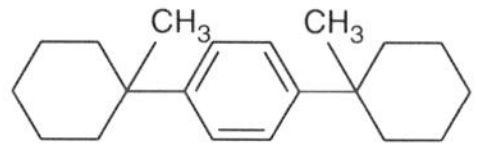

5 gm

S91,462-2
1-(4-CYCLOHEXYLPHENYL)ADAMANTANE
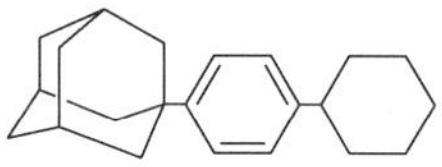
250 mg

S85,209-0
9,10-DIHYDRO-11-METHYL-9,10-ETHANOANTHRACENE
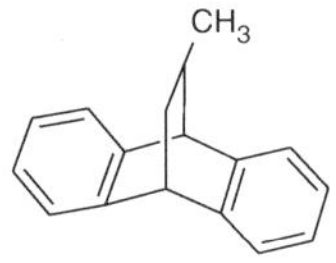

250 mg

S85,289-9
9,10-DIHYDRO-11,12-DIMETHYL-9,10-ETHANOANTHRACENE
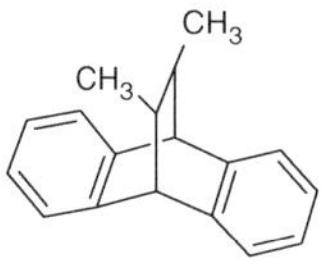

250 mg

S87,571-6
9,9',10,10'-TETRAHYDRO-11,11'-SPIROBI(9,10-ETHANOANTHRACENE)
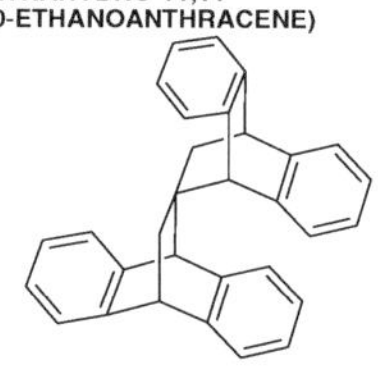
250 mg

S86,410-2
9,10-DIHYDRO-9,10-ETHANOANTHRACENE
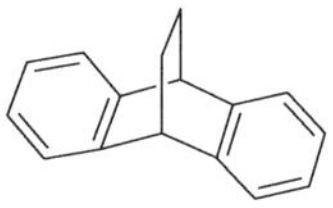
250 mg

S83,382-7
2-PHENYLHEXANE
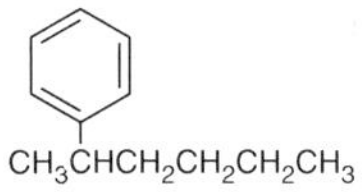

250 mg

S83,362-2
1,3-DIPHENYLPROPANE
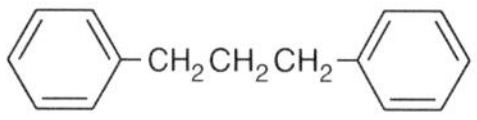

250 mg

S87,696-8
MESO-2,3-DIPHENYLBUTANE
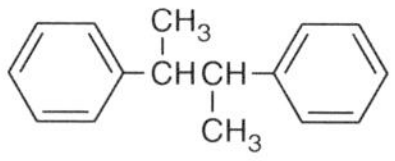

250 mg

S83,187-5
1,4-DIPHENYLHEXANE
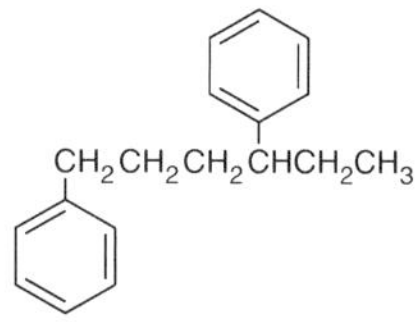

250 mg

S83,384-3
3-PHENYLDODECANE
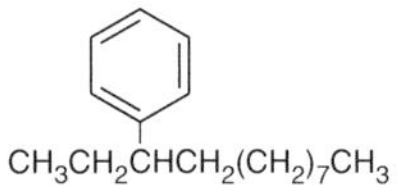

250 mg

S83,385-1
4-PHENYLDODECANE
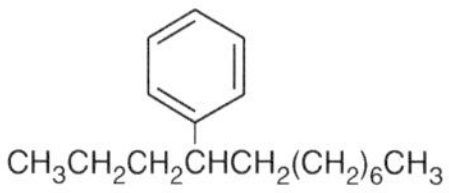

250 mg

S83,387-8
5-PHENYLDODECANE
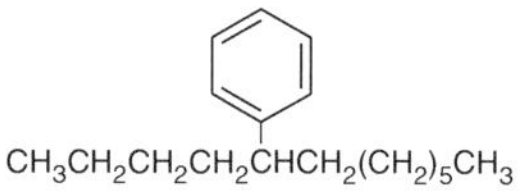

250 mg

S83,388-6
6-PHENYLDODECANE
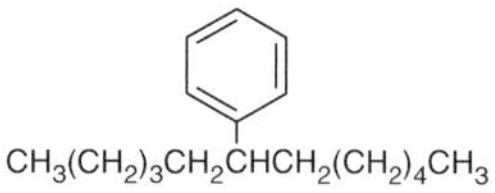

250 mg

S83,368-1
TRIBENZYLMETHANE
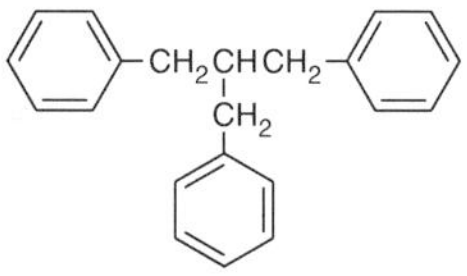

250 mg

S82,812-2
HEXADECYLBENZENE
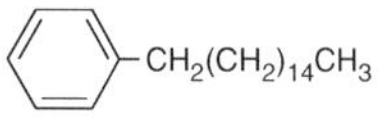

250 mg

S83,678-8
PHENYL(P-TOLYL)METHANE
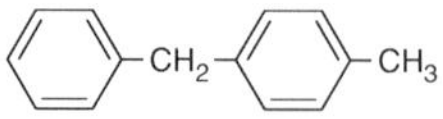

250 mg

S84,241-9
1,4-DIBENZYLBENZENE
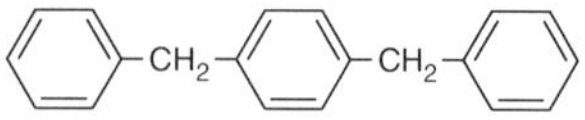

250 mg

S94,556-0
4-PROPYLBIPHENYL
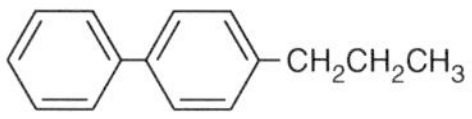

250 mg

S46,289-6
2-METHYLBUTYLBENZENE
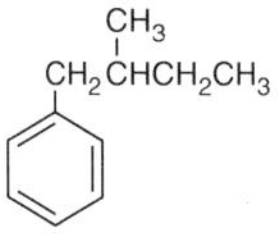

250 mg

S57,095-8
(1-METHYLBUTYL)BENZENE
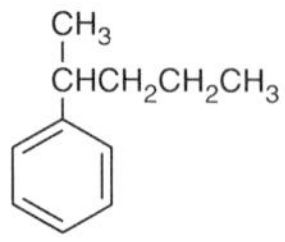

5 gm

S35,475-9
PHENYL-N-TETRADECANE (MIXTURE OF ISOMERS)
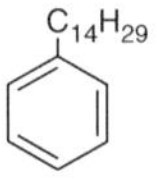

250 mg

S58,088-0
1-PHENYLPENTADECANE
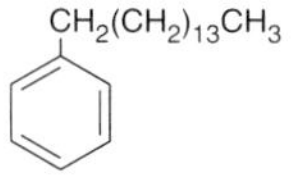
5 gm

S58,079-1
1-PHENYLNONADECANE
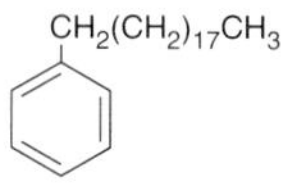
5 gm

S68,575-5
(+)-1-PHENYL-1-XYLYLETHANE
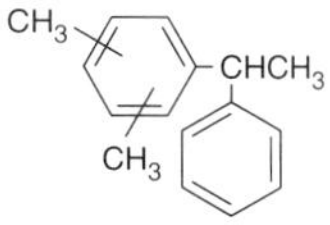
5 gm

S45,715-9
TETRAPHENYLMETHANE
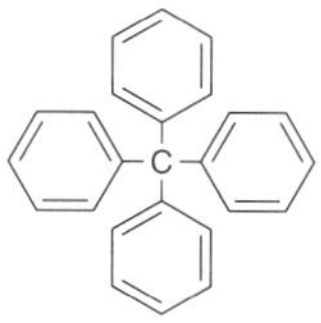
250 mg

S35,476-7
1-PHENYL-1,3,3-TRIMETHYLINDAN
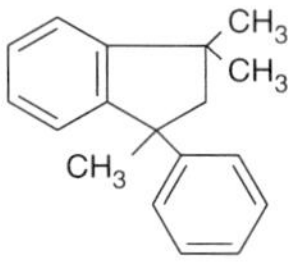
5 gm

S83,196-4
1,1-DIPHENYLPENTANE
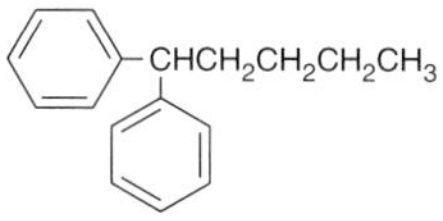
250 mg

S81,958-1
1,1,2-TRIPHENYLETHANE
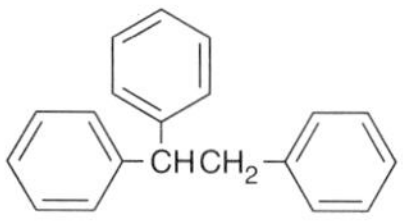
250 mg

S82,653-7
1,1,1-TRIPHENYLETHANE
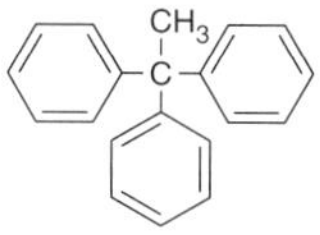
250 mg

S88,671-8
1,1,2-TRIPHENYLPROPANE
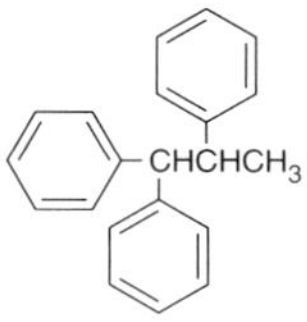
250 mg

S83,198-0
1,1,1-TRIPHENYLPENTANE
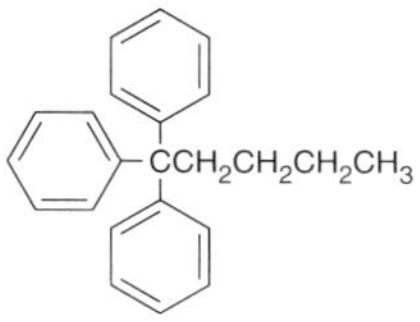
250 mg

S82,323-6
1,1,1,2-TETRAPHENYLETHANE
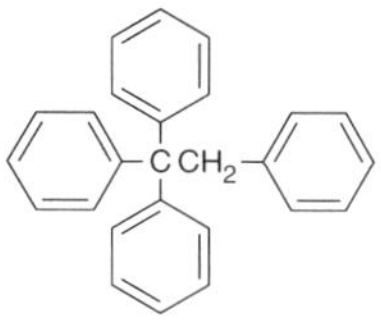
250 mg

S82,322-8
1,1,4,4-TETRAPHENYLBUTANE
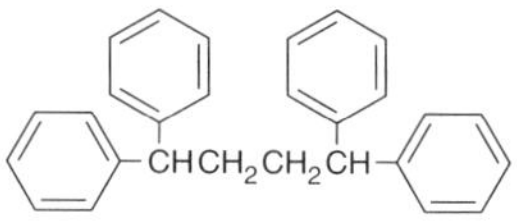
250 mg

S93,801-7
1-PHENYLADAMANTANE
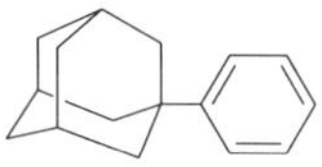
250 mg

S84,237-0
1,2-DIHYDRO-2,3-DIPHENYLNAPHTHALENE
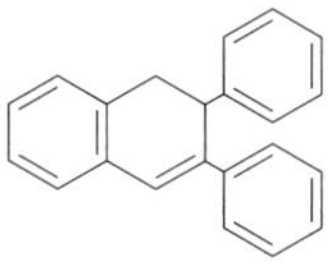
250 mg

S75,884-1
5-PHENYLDIBENZOSUBERANE
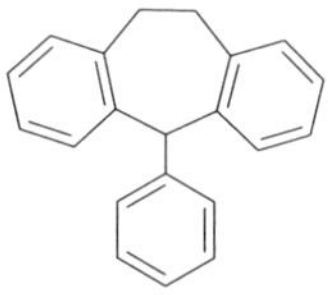
250 mg

S79,356-6
DI-TERT-BUTYLBIBENZYL
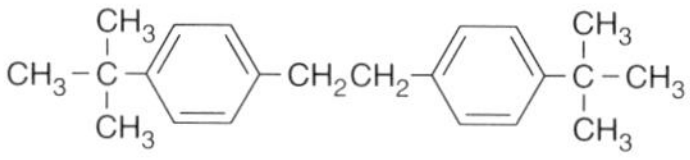
250 mg

S54,708-5
1-(TERT-BUTYL)-4-ETHYLBENZENE
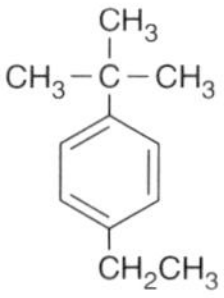
5 gm

S43,260-1
1,2-DI-P-TOLYLETHANE
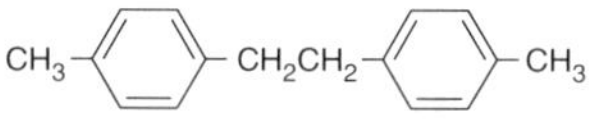
250 mg

S69,846-6
DITOLYLMETHANE (MIXTURE OF ISOMERS)
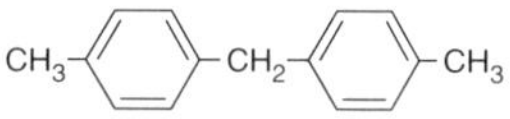
250 mg

S51,631-7
4,4'-ISOPROPYLIDENEDITOLUENE
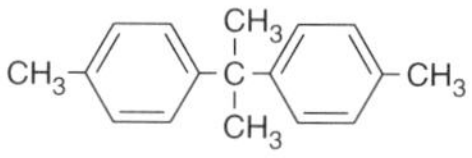
5 gm

S83,182-4
TRI(P-TOLYL)METHANE
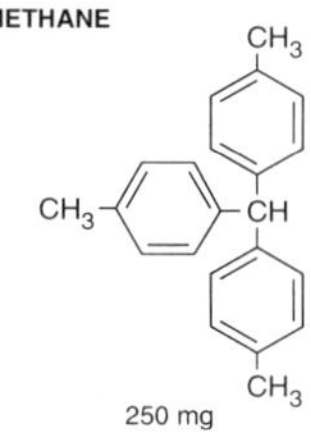
250 mg

S77,516-9
9-(P-TOLYL)PERHYDROFLUORENE
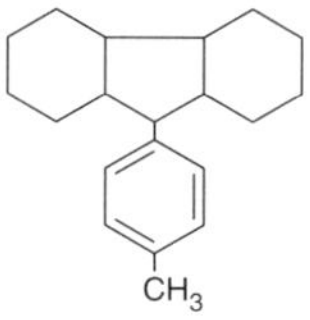
250 mg

S41,268-6
3,4-DIMETHYLCUMENE
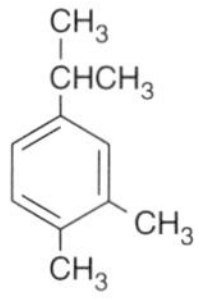
5 gm

S36,839-3
3,5-DIISOPROPYLTOLUENE
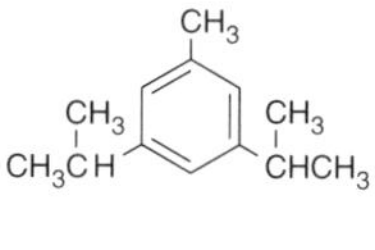
5 gm

S79,828-2
3,3',5,5'-TETRAMETHYLBIBENZYL
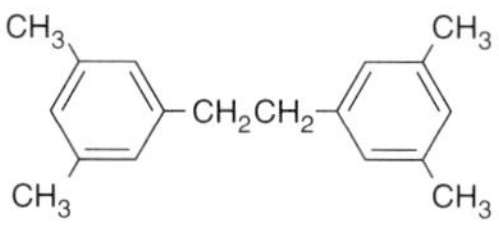
250 mg

S36,937-3
5-ISOPROPYL-M-XYLENE
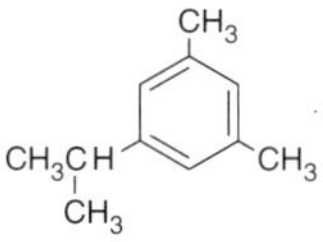
5 gm

S37,321-4
3-TERT-BUTYL-5-ETHYLTOLUENE
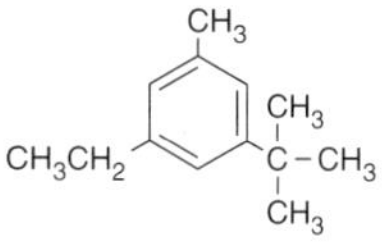
5 gm

S42,811-6
1,2,4,5-TETRAISOPROPYLBENZENE
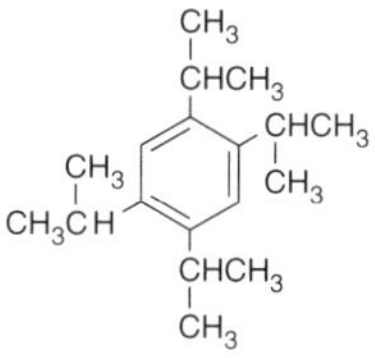
5 gm

S36,074-0
2,5-DIISOPROPYL-P-XYLENE
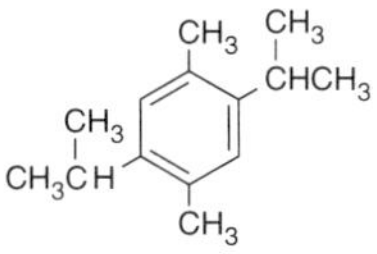
5 gm

S43,346-2
4,6-DIISOPROPYL-1,3-DIMETHYLBENZENE
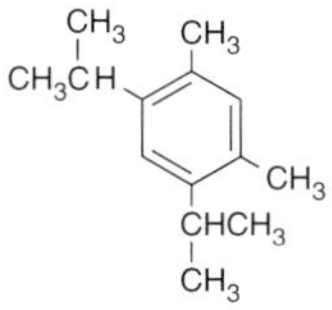
5 gm

S66,330-1
1-(2,2-DIMETHYLPROPYL)-2,4,5-TRIMETHYLBENZENE
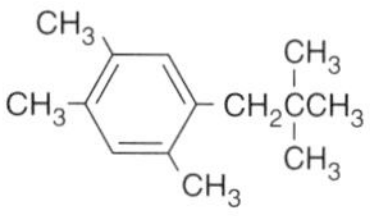
5 gm

S67,280-7
HEXAETHYLBENZENE
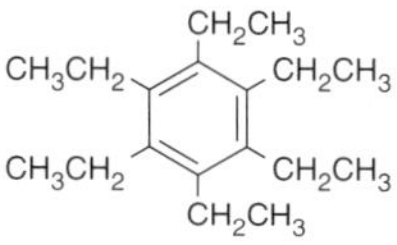
250 mg

S86,578-8
9,10-DIHYDRO-9,10-DIMETHYL-9,10-ETHENOANTHRACENE

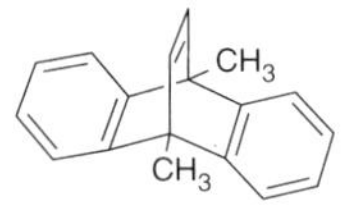

250 mg

S52,463-8
1-(3-METHYLBENZYL)ADAMANTANE

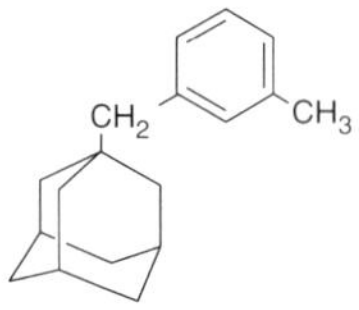

250 mg

S41,988-5
1,2,3,4,5,6,7,8-OCTAHYDROANTHRACENE

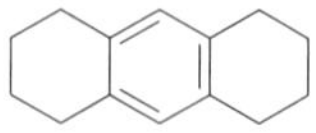

5 gm

S58,063-5
1,1,2,2-TETRAPHENYLETHANE

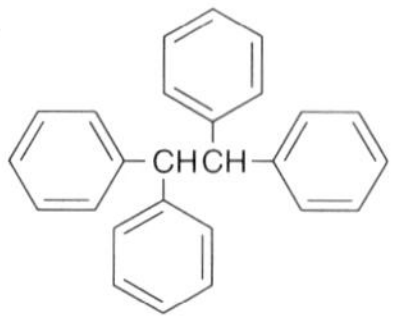

250 mg

S94,798-9
1,1,4,4,5,8-HEXAMETHYL-1,2,3,4-TETRAHYDRONAPHTHALENE

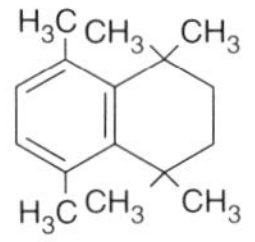

5 gm

S44,932-6
2-PHENYL-2-NORBORNENE

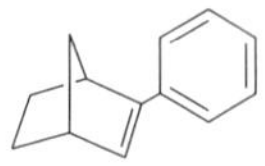

5 gm

S43,576-7
1-PHENYL-1-CYCLOPENTENE

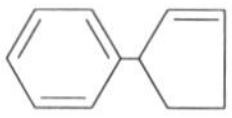

250 mg

S82,971-4
ALPHA-CYCLOHEXYLIDENETOLUENE

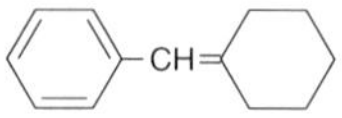

250 mg

S86,510-9
1-METHYLENE-2-PHENYLCYCLOHEXANE

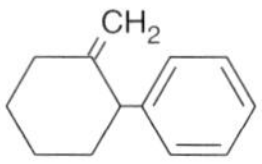

250 mg

S86,354-8
3-PHENYL-1-CYCLOHEXENE

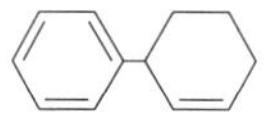

250 mg

S35,609-3
1-(O-TOLYL)-1-CYCLOHEXENE

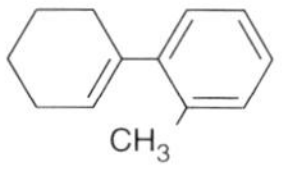

5 gm

S35,616-6
1-(P-TOLYL)-1-CYCLOHEXENE

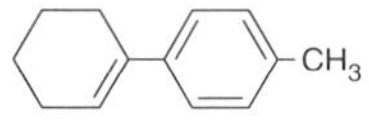

5 gm

S35,615-8
1-(M-TOLYL)-1-CYCLOHEXENE

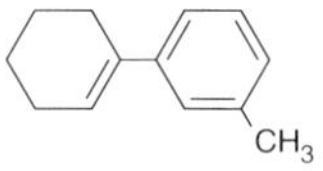

5 gm

S85,314-3
9,10-DIHYDRO-11-METHYLENE-9,10-ETHANOANTHRACENE

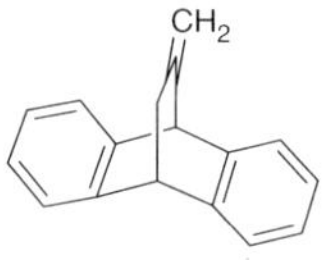

250 mg

S86,511-7
3-PHENYL-1-CYCLOHEPTENE

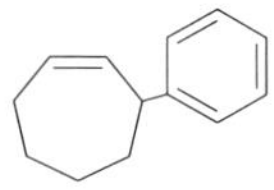

250 mg

S86,530-3
3-PHENYL-1-CYCLOOCTENE

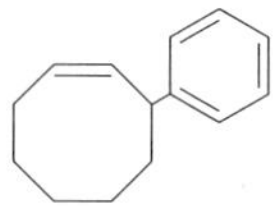

250 mg

S94,961-2
2-METHYL-1-PHENYL-1-BUTENE

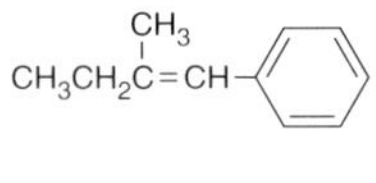

250 mg

S79,362-0
1,2-DIPHENYL-2-BUTENE

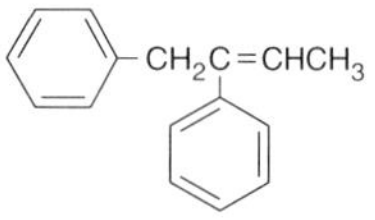

250 mg

S88,139-2
2,3-DIPHENYL-1-BUTENE

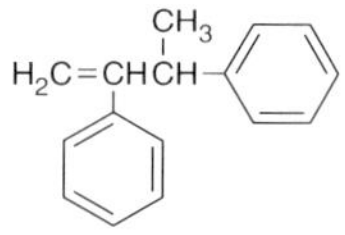

250 mg

S81,476-8
1,1,1,6,6,6-HEXAPHENYL-3-METHYL-3-HEXENE

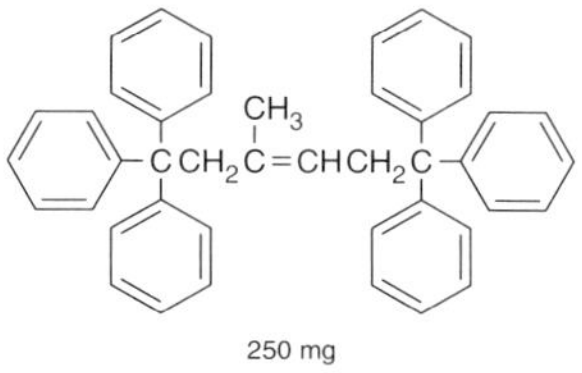

250 mg

S35,157-1
ISOCROTYLBENZENE

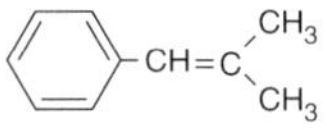

250 mg

S43,966-5
1-PHENYL-3-BUTENE

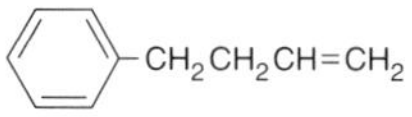

5 gm

S44,177-5
1-PHENYL-2-BUTENE

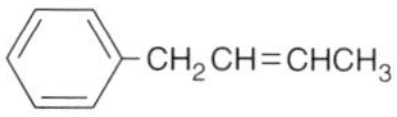

5 gm

S87,769-7
3-METHYL-1-PHENYL-1,3-BUTADIENE

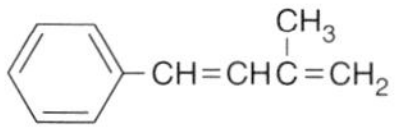

250 mg

S57,964-5
1-PHENYL-1-PENTENE

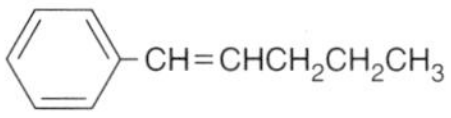

5 gm

S88,158-9
4-METHYL-1-PHENYL-1,3-PENTADIENE

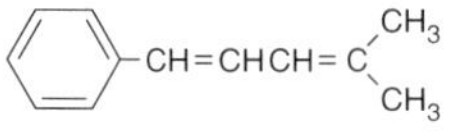

250 mg

S67,403-6
4-METHYL-2-PHENYL-2-PENTENE

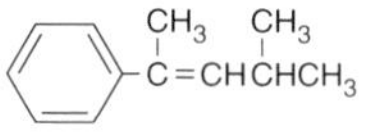

5 gm

S56,256-4
1,1-DIPHENYL-1-DODECENE

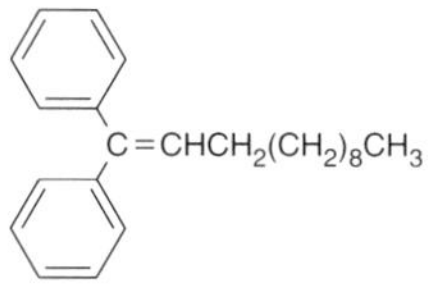

250 mg

S84,211-7
1,1,3,3-TETRAPHENYL-1-PROPENE

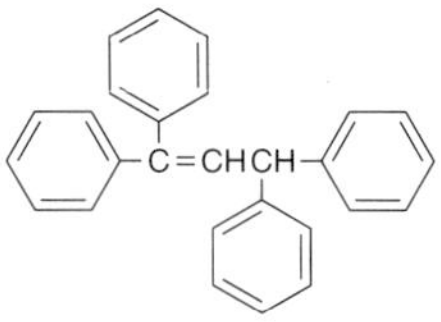

250 mg

S87,768-9
1-(O-TOLYL)-1,3-BUTADIENE

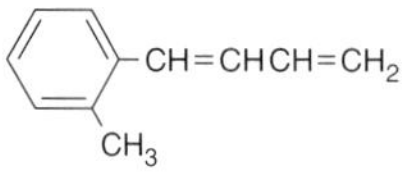

250 mg

S41,752-1
1-(O-TOLYL)CYCLOHEPTENE

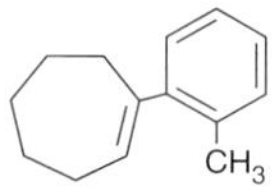

250 mg

S45,698-5
1-ISOPROPENYL-4-ISOPROPYLBENZENE

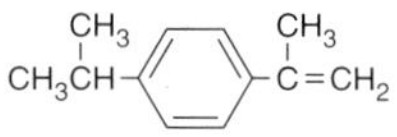

5 gm

S71,316-3
4-TERT-BUTYL-ALPHA-METHYLSTYRENE
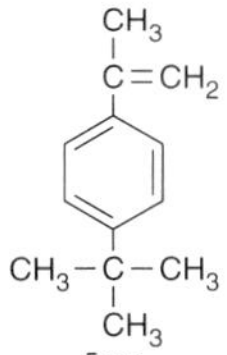

5 gm

S41,223-6
1-(P-TOLYL)CYCLOHEPTENE
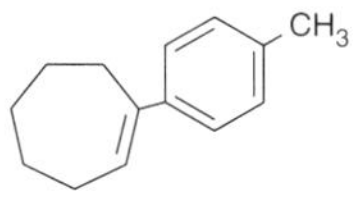

5 gm

S95,651-1
1',2',3,4-TETRAHYDRO-1,1'-BINAPHTHYL
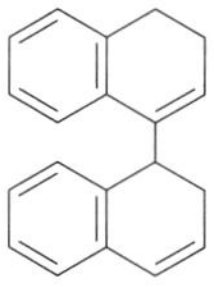
50 mg

S90,899-1
DELTA5,5'-BIDIBENZOSUBERANE
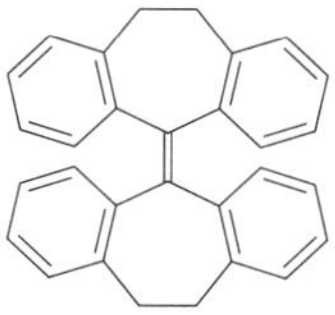
250 mg

S91,496-7
1,2,3-TRIPHENYL-1-CYCLOPROPENE
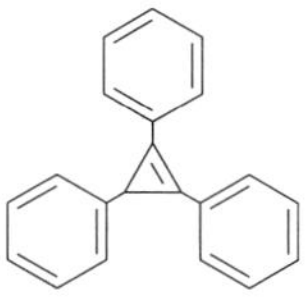
250 mg

S94,478-5
1,1,2,2-TETRAPHENYLCYCLOPROPANE
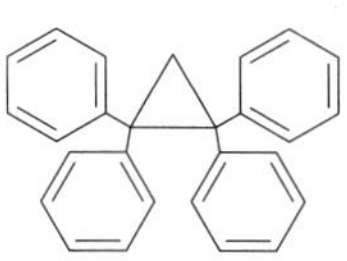
250 mg

S82,960-9
OCTAPHENYLCYCLOBUTANE
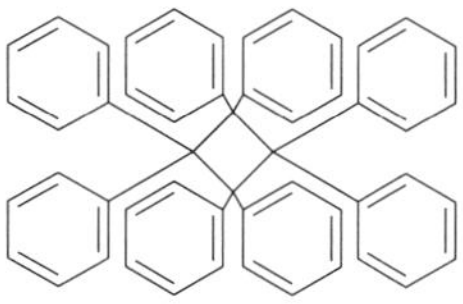
250 mg

S95,069-6
1,2,3-TRIPHENYL-1,3-CYCLOPENTADIENE
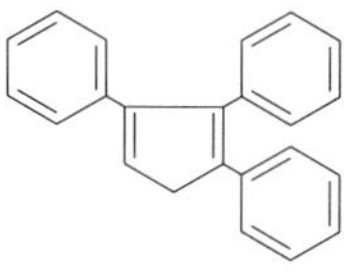
50 mg

S95,074-2
5-BENZYLIDENE-1,2,3-TRIPHENYL-1,3-CYCLOPENTADIENE
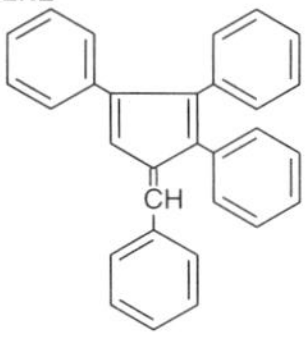

50 mg

S84,613-9
ALPHA-(4-METHYLCYCLOHEXYLIDENE)-TOLUENE
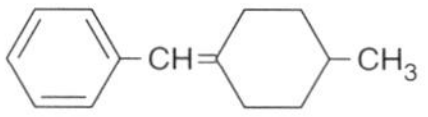

250 mg

S84,677-5
ALPHA-CYCLOHEXYLSTILBENE
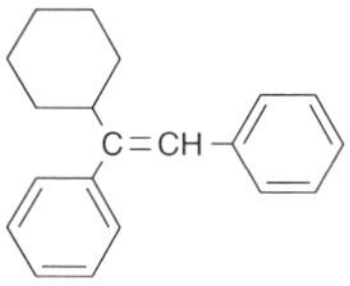

250 mg

S83,080-1
4,4'-BIS(1-CYCLOHEXEN-1-YL)BIPHENYL
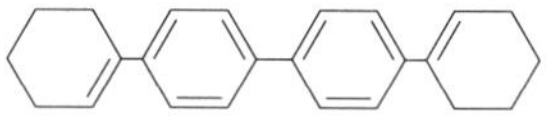
250 mg

S94,561-7
4,4'-BIS(4-PENTYL-1-CYCLOHEXEN-1-YL)BIPHENYL
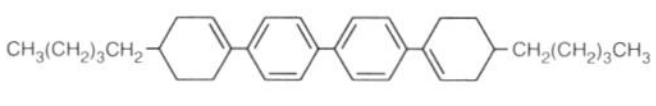

250 mg

S88,010-8
1,4-DIPHENYL-2-METHYL-1,3-BUTADIENE
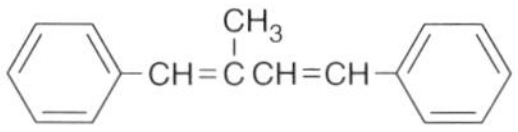

250 mg

S84,214-1
1,1,3,3-TETRAPHENYLALLENE
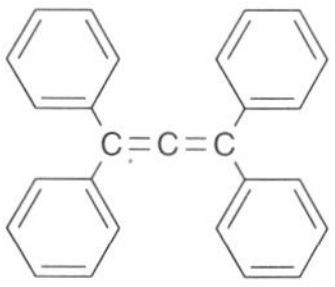

250 mg

S83,107-7
1,2,3,4-TETRAPHENYL-1,3-BUTADIENE
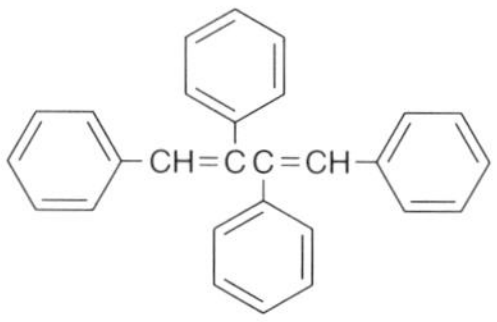

250 mg

S82,792-4
PENTAPHENYLBENZENE
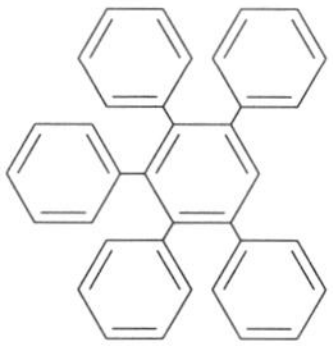
250 mg

S84,677-5
1,1-BIS(O-TOLYL)-2-PHENYLETHYLENE
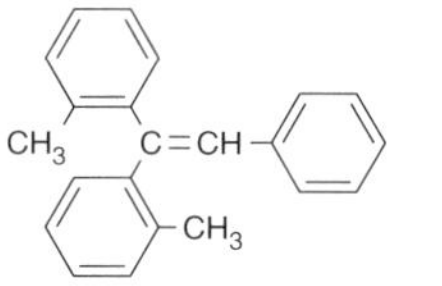

250 mg

S88,268-2
1,3-DIPHENYL-2-(O-TOLYL)-1-PROPENE
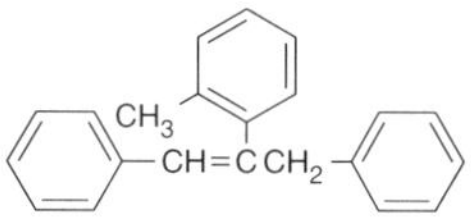

250 mg

S83,197-2
2-METHYL-P-TERPHENYL
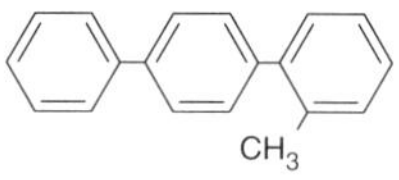

250 mg

S54,702-6
P-DISTYRYLBENZENE

250 mg

S83,685-0
4-OCTADECYLBIPHENYL
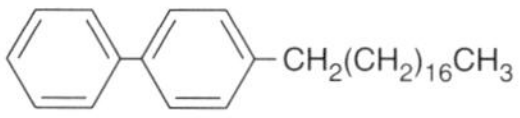

250 mg

S92,680-9
1-TERT-BUTYL-3,5-DIPHENYLBENZENE
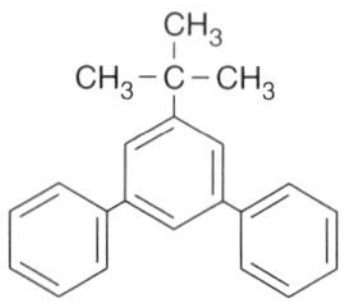

250 mg

S88,652-1
1,1-DIPHENYL-2-METHYLPROPANE
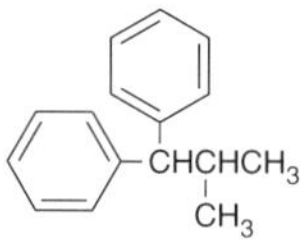

250 mg

S79,217-9
6,7,8,9-TETRAHYDRO-1,2,3,4-TETRAPHENYL-5H-BENZOCYCLOHEPTENE

250 mg

S79,218-7
5,6,7,8,9,10-HEXAHYDRO-1,2,3,4-TETRAPHENYLBENZOCYCLOOCTENE

250 mg

S80,683-8
4,4'-DIMETHYLSTILBENE
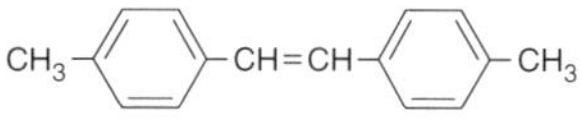

250 mg

S88,226-7
1,4-BIS(P-TOLYL)-1,3-BUTADIENE
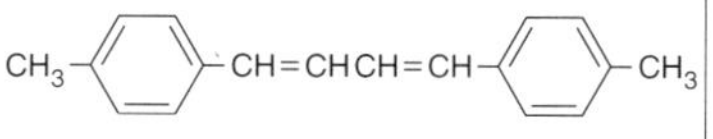

250 mg

S35,877-0
1,4-BIS(4-PHENYL-1,3-BUTADIENYL)-BENZENE
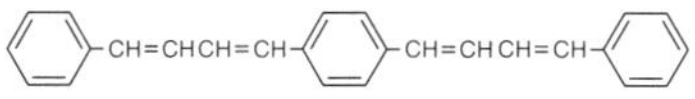

250 mg

S93,581-6
1,2-DIHYDRO-3,5-DIMETHYLBENZ(J)-ACEANTHRYLENE
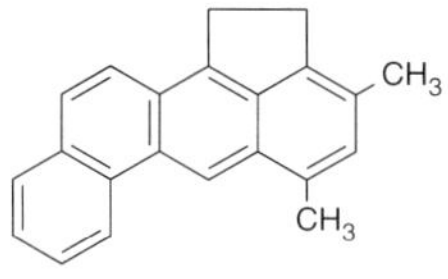

10 mg

S94,651-6
9,9'-BIFLUORENYLIDENE
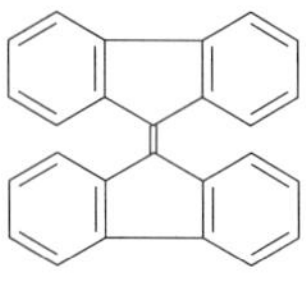
100 mg

S86,395-5
1,3-DIPHENYLINDENE
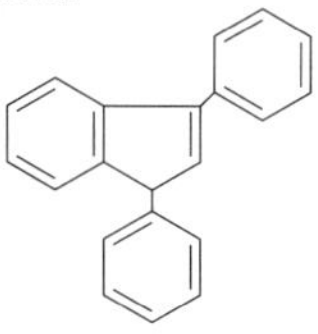
250 mg

S91,749-4
9-CYCLOHEXYL-9-METHYLFLUORENE
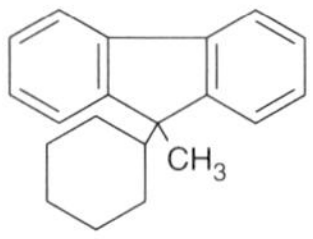

250 mg

S40,953-7
TRANS-1-(1-NAPHTHYL)-2-PHENYLETHYLENE
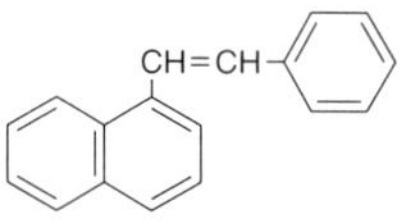

250 mg

S84,238-9
2,3-DIPHENYLNAPHTHALENE
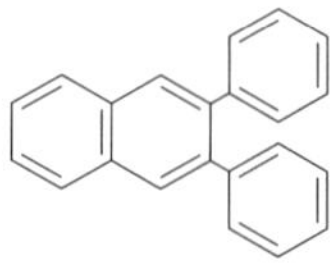
250 mg

S75,543-5
9-CINNAMYLIDENEFLUORENE
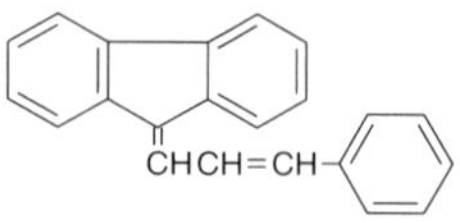

250 mg

S41,983-4
1,3,6,8-TETRAPHENYLPYRENE
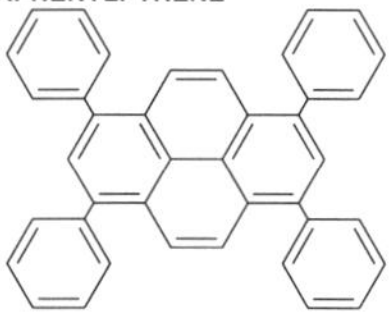
5 gm

S84,213-3
1,1,3-TRIPHENYLINDENE
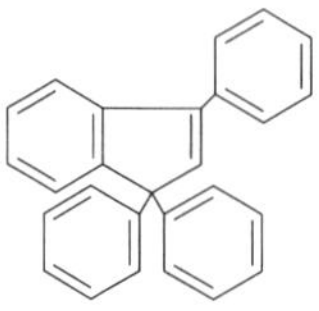
250 mg

S83,500-5
9,9-DIPHENYLFLUORENE
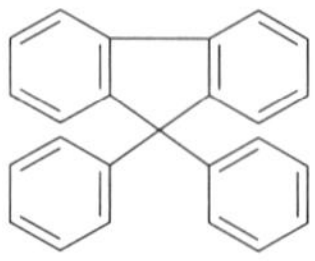
250 mg

S84,209-5
1,2,3-TRIPHENYLACENAPHTHYLENE
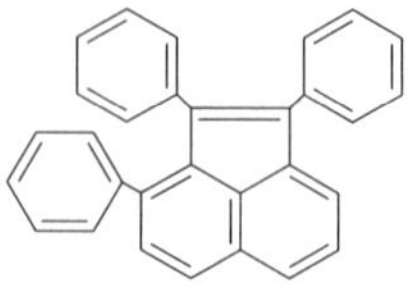
250 mg

S76,802-2
1-BENZYLINDENE
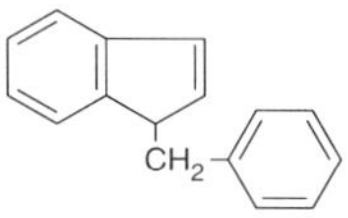

250 mg

S92,373-7
1,2-DIPHENYLINDENE
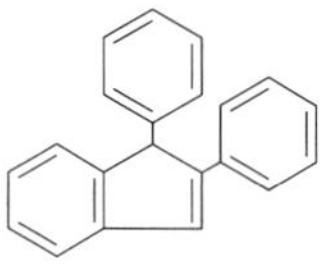
250 mg

S77,035-3
1,2,3-TRIPHENYLNAPHTHALENE

250 mg

S83,691-5
1-BENZYLNAPHTHALENE
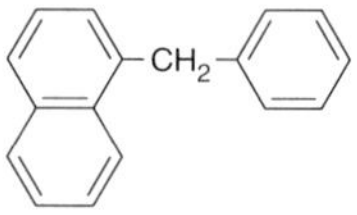

250 mg

S37,242-0
9-BENZYLIDENEFLUORENE
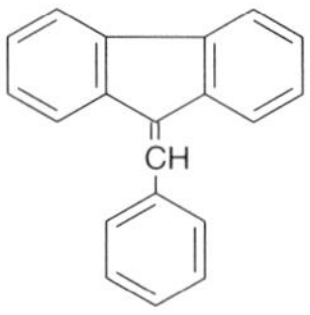

250 mg

S83,200-6
9-METHYL-9-PHENYLFLUORENE
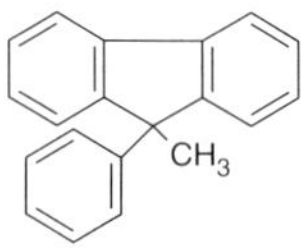

250 mg

S83,199-9
9-BUTYL-9-PHENYLFLUORENE
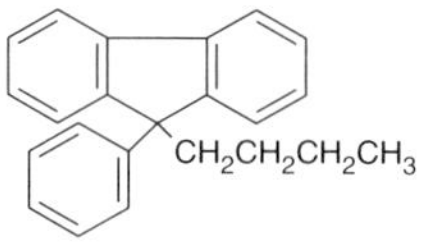

250 mg

S90,663-8
1,2,3,-TRIPHENYLAZULENE
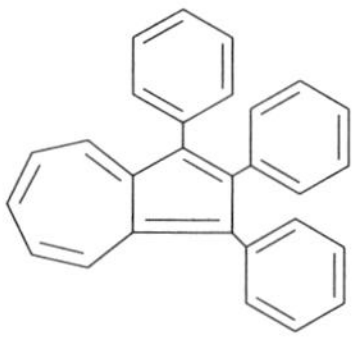
250 mg

S97,304-1
AR-C 1 RUBICENE
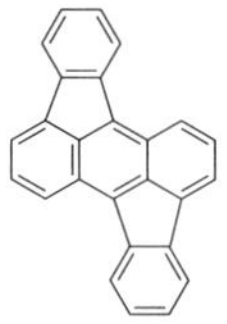
10 mg

S73,129-3
1-ALLYLNAPHTHALENE
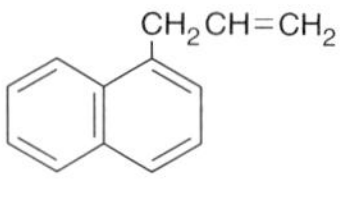

5 gm

S39,332-0
2-ISOPROPYLNAPHTHALENE
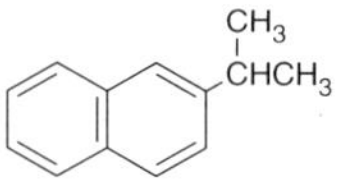

5 gm

S41,340-2
2,2'-BINAPHTHYL
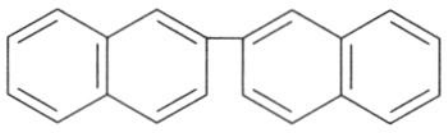
250 mg

S39,333-9
2,6-DIISOPROPYLNAPHTHALENE
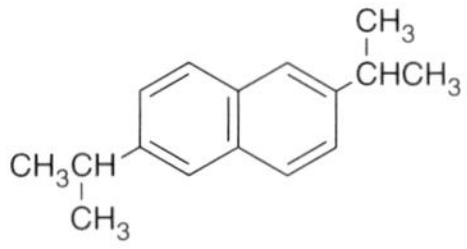

5 gm

S36,565-3
ISOPROPYLNAPHTHALENE (MIXTURE OF ISOMERS)
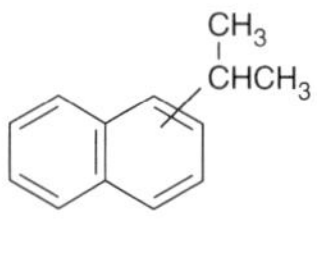

5 gm

S93,614-6
8-METHYLBENZO(A)PYRENE
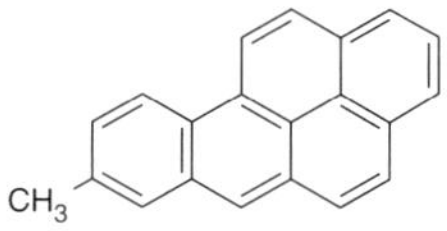

10 mg

S93,615-4
9-METHYLBENZO(A)PYRENE
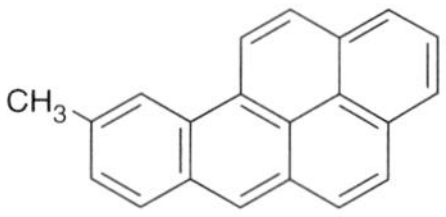

10 mg

S97,536-2
AR-L13 10-METHYLBENZO(A)PYRENE
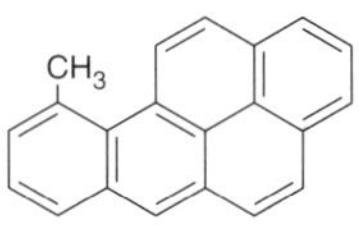

10 mg

S97,548-6
AR-L30 7,10-DIMETHYLBENZO(A)PYRENE
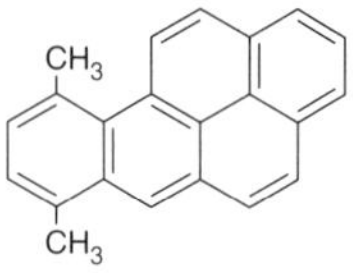

10 mg

S97,452-8
AR-G 3 BENZO(C)PHENANTHRENE
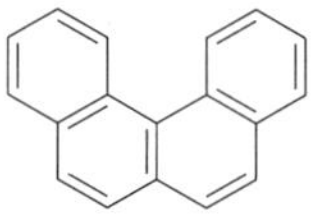
10 mg

S93,601-4
1-METHYLBENZO(C)PHENANTHRENE
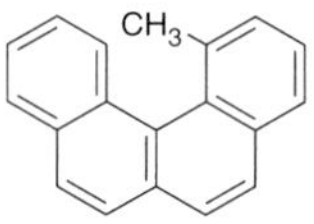

10 mg

S93,602-2
2-METHYLBENZO(C)PHENANTHRENE
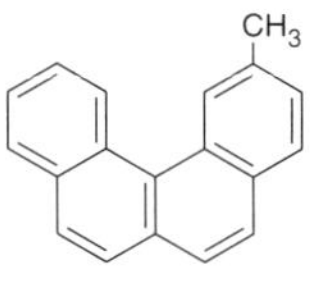

10 mg

S93,603-0
4-METHYLBENZO(C)PHENANTHRENE
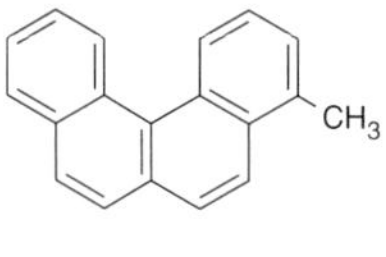

10 mg

S93,604-9
5-METHYLBENZO(C)PHENANTHRENE
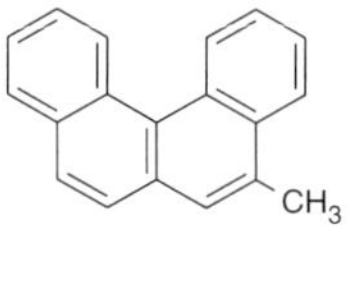

10 mg

S97,474-9
AR-G32 3-METHYLBENZO(C)PHENANTHRENE
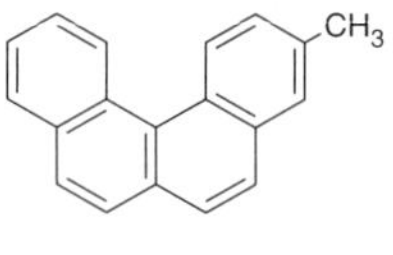

10 mg

S97,494-3
AR-G51 5,8-DIMETHYLBENZO(C)-PHENANTHRENE
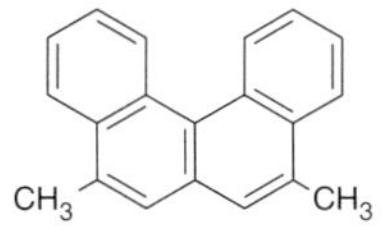

10 mg

S97,495-1
AR-G52 1,12-DIMETHYLBENZO(C)-PHENANTHRENE
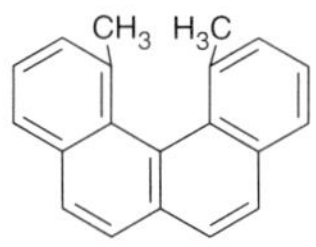

10 mg

S44,270-4
TRUXENE
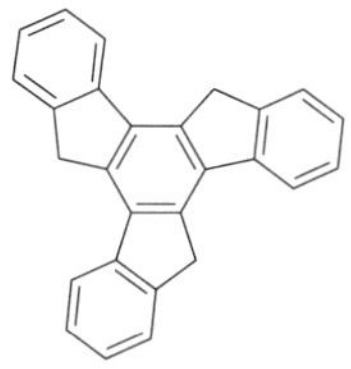
5 gm

S93,605-7
1-METHYLBENZ(A)ANTHRACENE
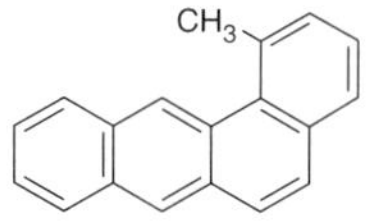

10 mg

S93,606-5
2-METHYLBENZ(A)ANTHRACENE
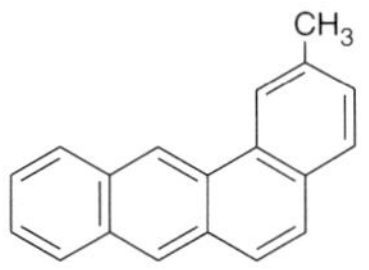

10 mg

S93,607-3
3-METHYLBENZ(A)ANTHRACENE
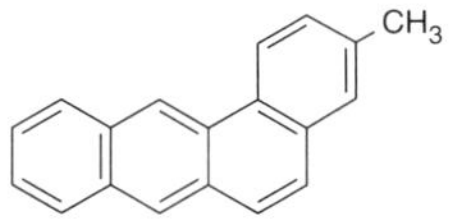

10 mg

S93,608-1
4-METHYLBENZ(A)ANTHRACENE
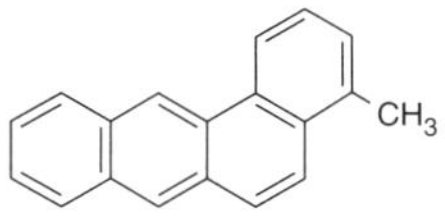

10 mg

S93,610-3
9-METHYLBENZ(A)ANTHRACENE
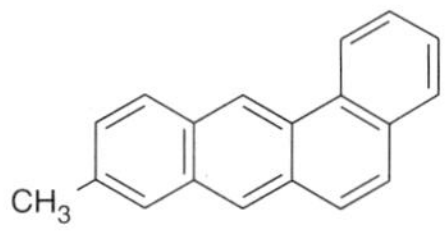

10 mg

S93,611-1
7-METHYLBENZ(A)ANTHRACENE
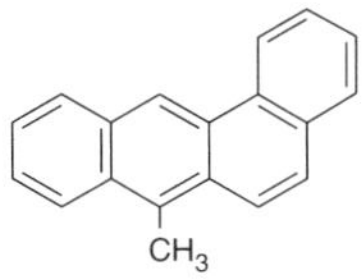

10 mg

S97,377-7
AR-E32 5-METHYLBENZ(A)ANTHRACENE
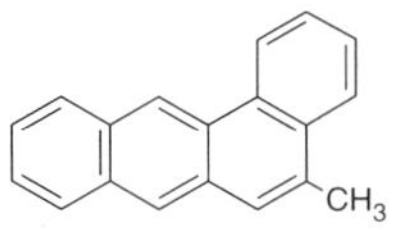

10 mg

S97,378-5
AR-E33 6-METHYLBENZ(A)ANTHRACENE
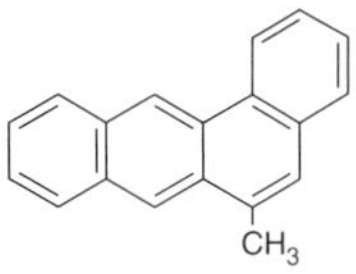

10 mg

S97,381-5
AR-E40 10-METHYLBENZ(A)ANTHRACENE
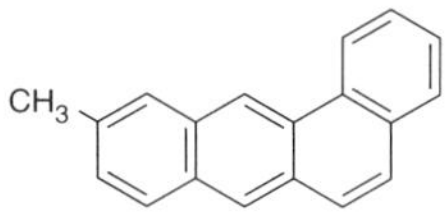

10 mg

S97,558-3
AR-E107 5-METHYLBENZ(A)-ANTHRACENE PICRATE
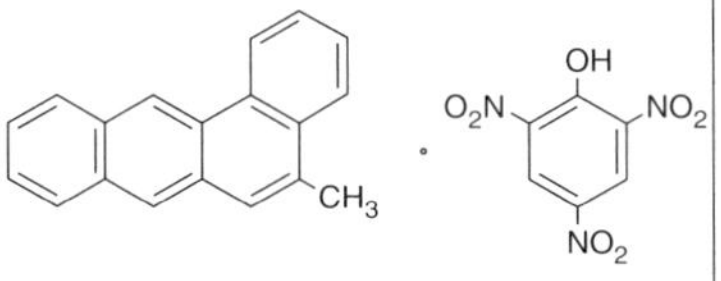

10 mg

S97,405-6
AR-E68 3,9-DIMETHYLBENZ(A)-ANTHRACENE
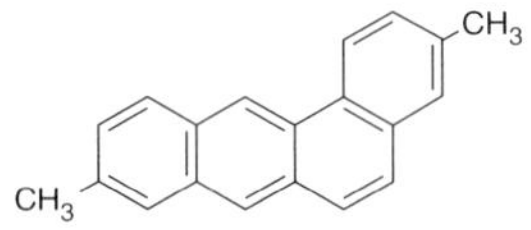

10 mg

S97,413-7
AR-E76 12,12A-DIHYDRO-3,9-DIMETHYLBENZ(A)ANTHRACENE
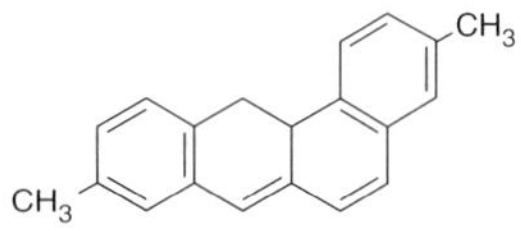

10 mg

S97,564-8
AR-L24 1-VINYLBENZO(A)PYRENE
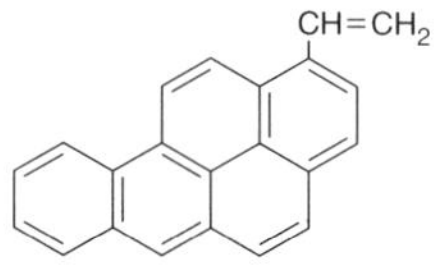

10 mg

S36,059-7
5,12-DIHYDROTETRACENE
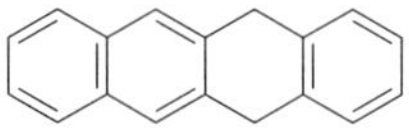
250 mg

S44,354-9
PYRENE, ZONE REFINED
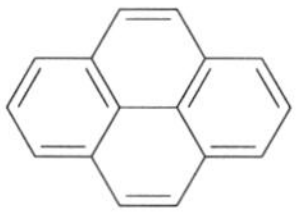
250 mg

S97,514-1
AR-H20 1-ETHENYLPYRENE
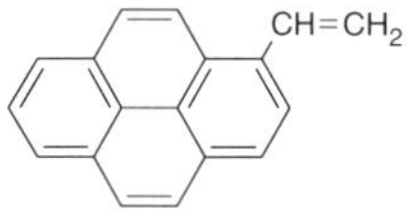

10 mg

S97,551-6
AR-M 2 PHENANTHRO(3,4-C)-PHENANTHRENE
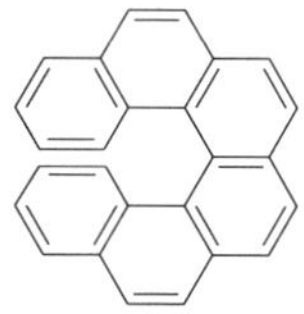
10 mg

S97,555-9
AR-M 7 3-METHYLPHENANTHRO(3,4-C)-PHENANTHRENE
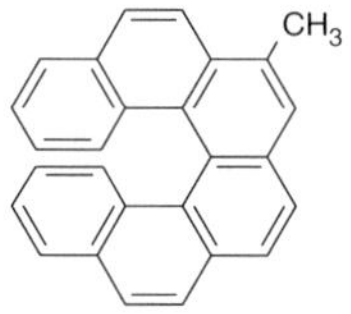

10 mg

S97,556-7
AR-M 8 4A,5,6,12C-TETRAHYDRO-3-METHYLPHENANTHRO(3,4-C)PHENANTHRENE
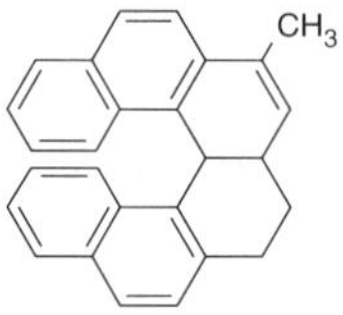

10 mg

S93,582-4
1,2-DIHYDROACEANTHRYLENE
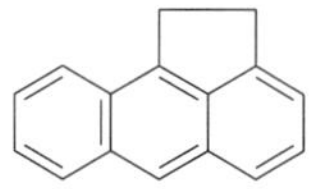
10 mg

S84,292-3
6B,7,7A,13B,14,14A-HEXAHYDROACENAPHTHO(1,2-K)-FLUORANTHENE
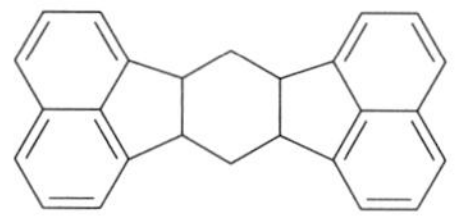
250 mg

S71,338-4
2-ISOBUTYLNAPHTHALENE
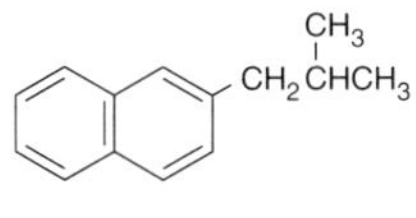

5 gm

S94,380-0
TRANS-1,2-DI(2-NAPHTHYL)ETHYLENE
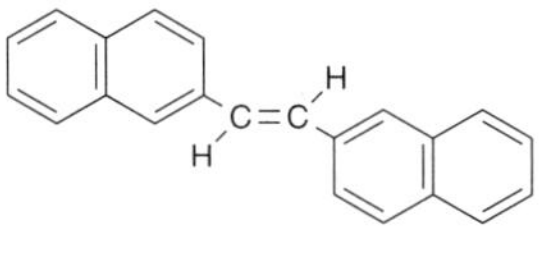

250 mg

S94,381-9
TRANS-1,2-DI(7-METHYL-2-NAPHTHYL)-ETHYLENE
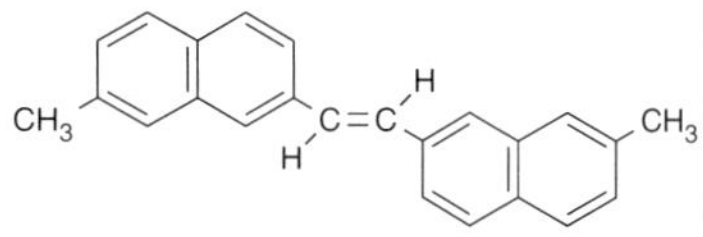

250 mg

S94,382-7
TRANS-1-(6-METHYL-2-NAPHTHYL)-2-(7-METHYL-2-NAPHTHYL)ETHYLENE
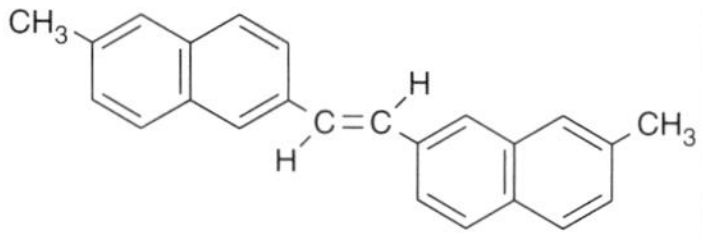

250 mg

S84,242-7
2,7-DI-TERT-BUTYLNAPHTHALENE
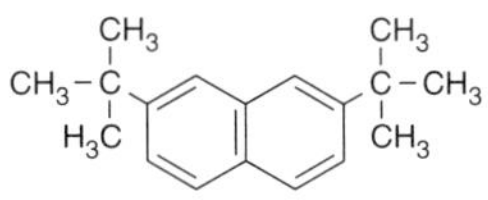

250 mg

S89,918-6
1,2-BENZOFLUORENE
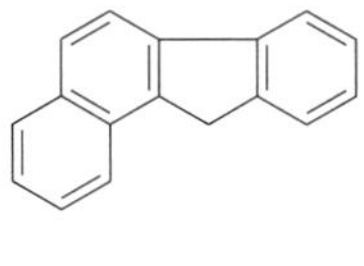
250 mg

S40,675-9
1,1'-BI-(ACENAPHTHENYLIDENE)
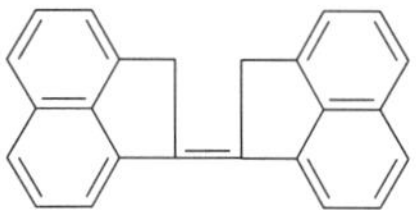
5 gm

S97,402-1
AR-E63 6,8-DIMETHYLBENZ(A)-ANTHRACENE
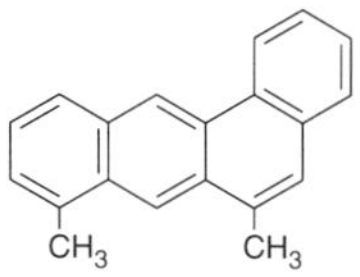

10 mg

S93,580-8
8,9,11-TRIMETHYLBENZ(A)ANTHRACENE

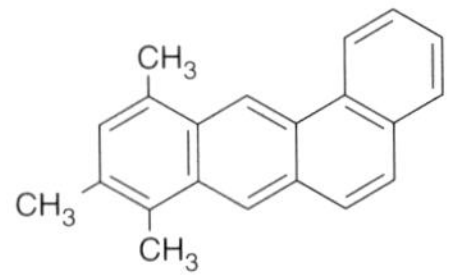

10 mg

S86,363-7
2,3-DIMETHYLANTHRACENE

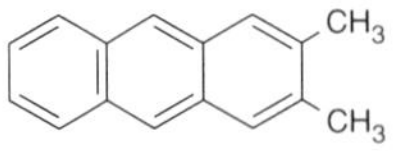

250 mg

S90,698-0
5,6-DIMETHYLCHRYSENE

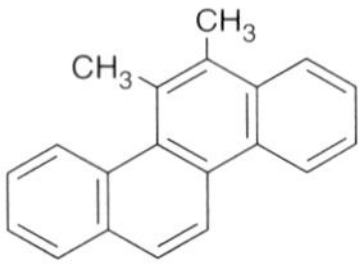

250 mg

S84,256-7
9,9'-BIPHENANTHRENE

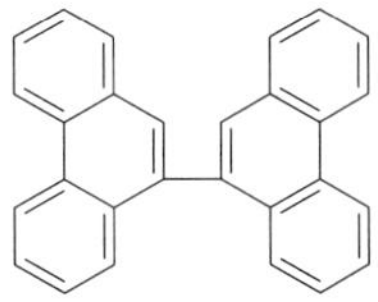

250 mg

S90,661-1
4,6,8-TRIMETHYLAZULENE

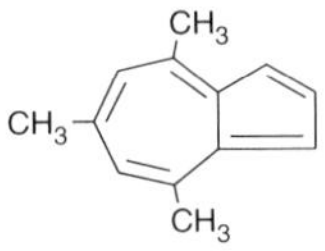

250 mg

S84,678-3
2,3,6,7-TETRAETHYLBIPHENYLENE

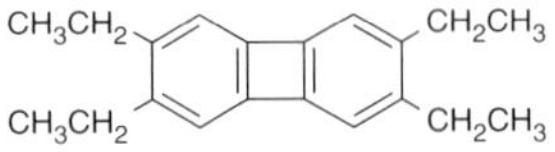

250 mg

S46,121-0
1-BROMOISOPROPYLBENZENE
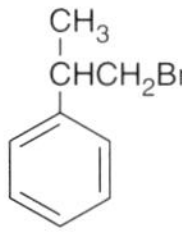

5 gm

S38,742-8
1-CHLORO-3-PHENYLBUTANE
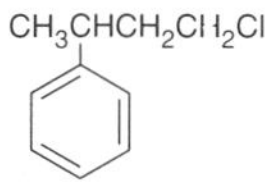

250 mg

S86,640-7
1-BROMO-4-PHENYLBUTANE
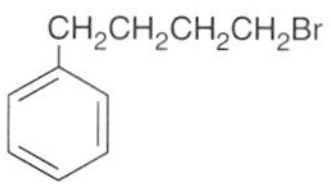

250 mg

S87,624-0
(+)-1-CHLORO-3-METHYL-2-PHENYLBUTANE
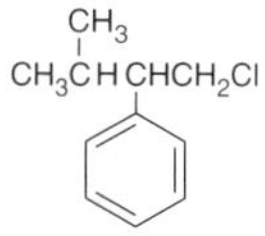

250 mg

S87,628-3
(+)-1-CHLORO-3,3-DIMETHYL-2-PHENYLBUTANE
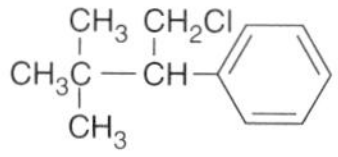

250 mg

S55,416-2
2-IODO-M-XYLENE
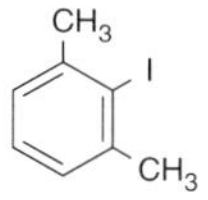

250 mg

S86,657-1
2-BROMO-4-TERT-BUTYL-1-METHYLBENZENE
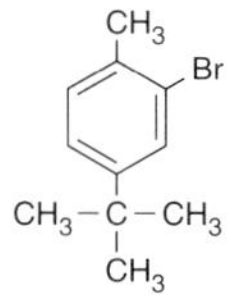

250 mg

S41,972-9
DIPHENYLIODONIUM NITRATE
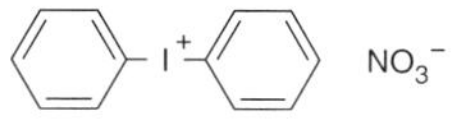

250 mg

S93,389-9
DIPHENYLIODONIUM TETRAFLUOROBORATE
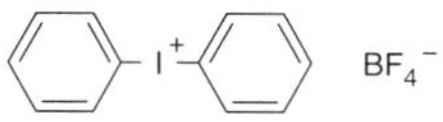

250 mg

S55,206-2
(4-BROMOPHENYL)CYCLOHEXANE
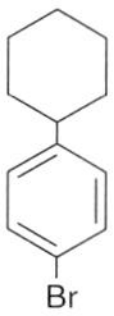

5 gm

S46,235-7
(+/-)-(1-CHLOROETHYL)BENZENE
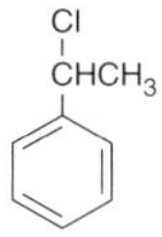

5 gm

S88,179-1
BETA-BROMO-ALPHA-METHYLSTYRENE
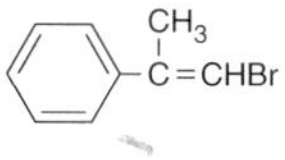

250 mg

S87,788-3
2-BROMO-3-PHENYL-2-BUTENE
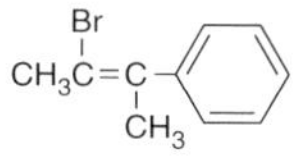

250 mg

S88,157-0
1-BROMO-4-METHYL-1-PHENYL-1,4-PENTADIENE
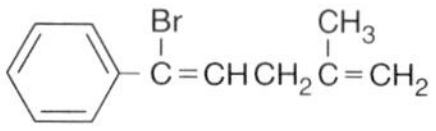

250 mg

S51,249-4
DIPHENYLIODONIUM BROMIDE
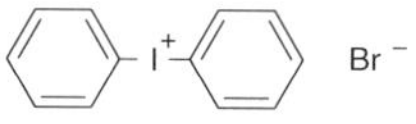

250 mg

S84,979-0
2-BROMO-1-PHENYLNAPHTHALENE
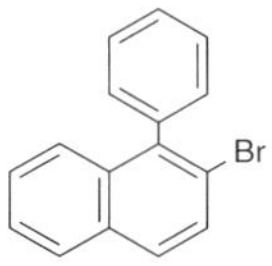

250 mg

S72,223-5
ALPHA-ETHYL-4-METHYLBENZYL CHLORIDE
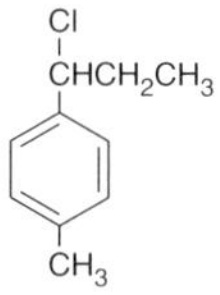

5 gm

S41,369-0
2-CHLOROBENZYL IODIDE
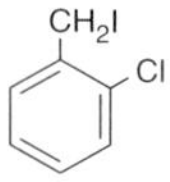

5 gm

S69,770-2
1-BROMO-2-(2-BROMOETHYL)BENZENE
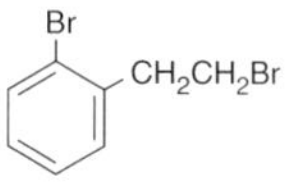

250 mg

S80,980-2
ALPHA-BROMO-2-(2-BROMOETHYL)TOLUENE
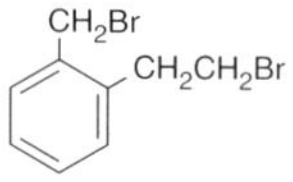

250 mg

S92,128-9
ALPHA,ALPHA,ALPHA',ALPHA'-TETRABROMO-O-XYLENE
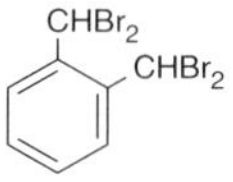

250 mg

S86,960-0
1,2-DIBROMO-3,4,5,6-TETRAMETHYLBENZENE
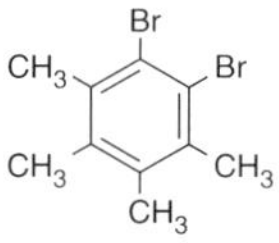

250 mg

S94,389-4
3,4-DIIODOTOLUENE
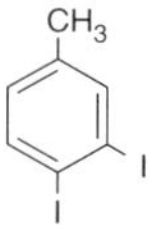

250 mg

S84,984-7
4,5-DIBROMO-O-XYLENE
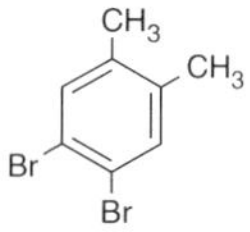

250 mg

S41,707-6
1,3-BIS(1,2-DIBROMOETHYL)BENZENE
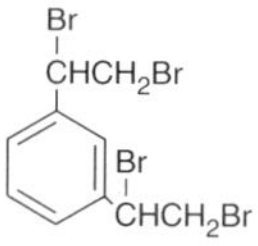

5 gm

S35,941-6
3-CHLORO-ALPHA,ALPHA-DIBROMOTOLUENE
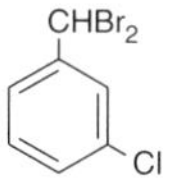

5 gm

S60,070-9
ALPHA,ALPHA,3-TRIBROMOTOLUENE
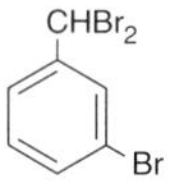

5 gm

S55,306-9
3-(1-BROMOETHYL)-A,A,A-TRIFLUOROTOLUENE
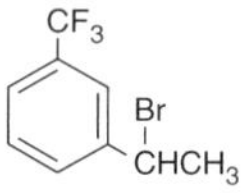

5 gm

S86,959-7
1,3-DIBROMO-2,4,5,6-TETRAMETHYLBENZENE
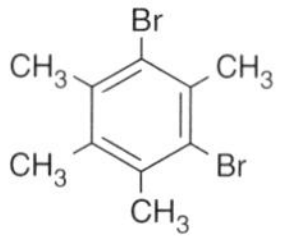

250 mg

S77,802-8
4-BROMO-1-TERT-BUTYL-2-CHLOROBENZENE
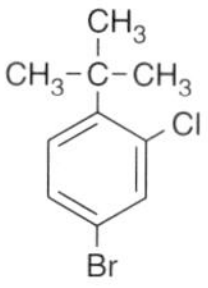

50 mg

S94,593-5
4,6-DIBROMO-M-XYLENE
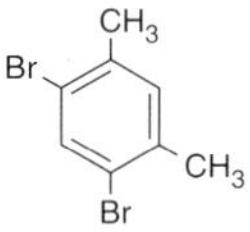

250 mg

S56,842-2
1-BROMO-4-(1,2-DIBROMOETHYL)BENZENE
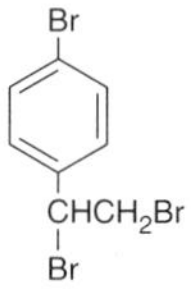

250 mg

S55,923-7
7-CHLORO-P-CYMENE

250 mg

S44,976-8
1,4-BIS(1,2-DIBROMOETHYL)BENZENE
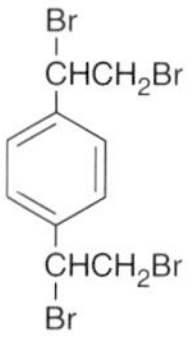

5 gm

S44,126-0
ALPHA,ALPHA,ALPHA',ALPHA'-TETRACHLORO-P-XYLENE

5 gm

S56,414-1
ALPHA,ALPHA,ALPHA',ALPHA'-TETRABROMO-P-XYLENE
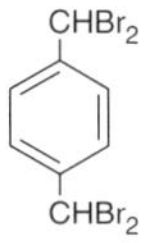

5 gm

S41,301-1
P-CHLOROFLUOROBENZENE

5 gm

S44,567-3
1,4-DIBROMOBENZENE, ZONE REFINED

250 mg

S36,203-4
4-FLUOROCUMENE
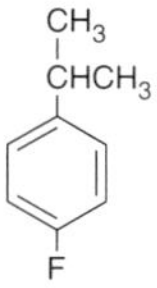

5 gm

S76,787-5
ALPHA-4-DICHLOROPROPYLBENZENE
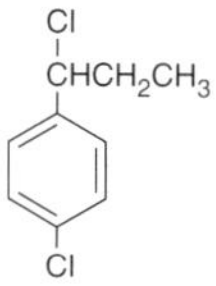

250 mg

S42,808-6
4-CHLORO-BETA,BETA-DIMETHYLSTYRENE
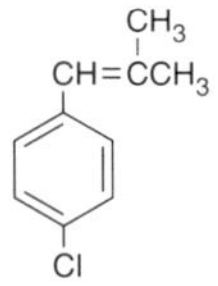
5 gm

S45,637-3
1,3-BIS(TRICHLOROMETHYL)BENZENE
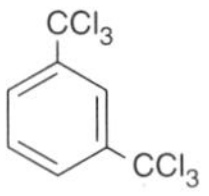
5 gm

S43,959-2
3-CHLOROBENZOTRICHLORIDE
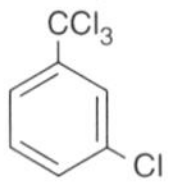
250 mg

S78,215-7
2-(1-BROMOETHYL)-ALPHA,ALPHA,ALPHA-TRIFLUOROTOLUENE
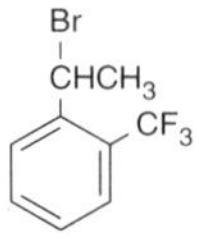
5 gm

S43,000-5
4-FLUORO-ALPHA,ALPHA,ALPHA-TRICHLOROTOLUENE

5 gm

S91,896-2
1,3-BIS(TRIFLUOROMETHYL)-5-(1,1,2-TRIMETHYLPROPYL)BENZENE
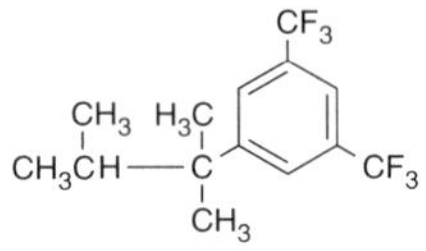
250 mg

S62,686-4
2,3,5,6-TETRAMETHYLBENZYL CHLORIDE
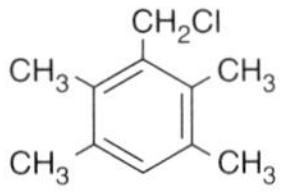
5 gm

S40,753-4
BIS(CHLOROMETHYL)MESITYLENE
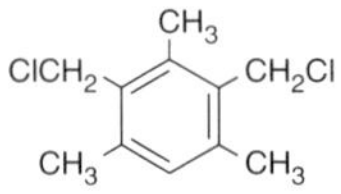
250 mg

S94,760-1
1,3-DIMETHYL-2,4,6-TRIS-(CHLOROMETHYL)BENZENE
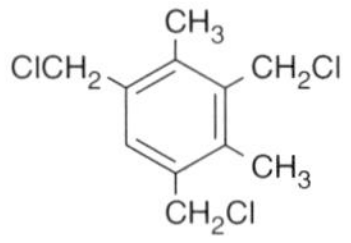
5 gm

S66,289-5
ALPHA1,ALPHA3-DICHLOROHEXAMETHYLBENZENE
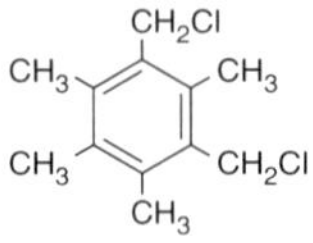
250 mg

S86,989-9
1,2-BIS(CHLOROMETHYL)-3,4,5,6-TETRAMETHYLBENZENE
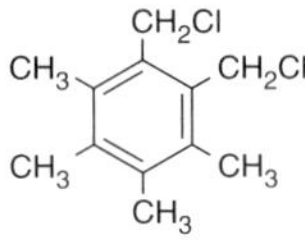
250 mg

S44,701-3
ALPHA,ALPHA',2,4,5,6-HEXACHLORO-M-XYLENE
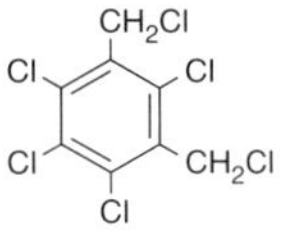
5 gm

S94,309-6
ALPHA,ALPHA'-DIBROMO-2,5-DICHLORO-P-XYLENE
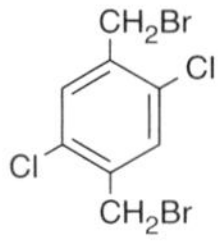
250 mg

S45,235-1
1-(2-CHLOROETHYL)-4-FLUOROBENZENE
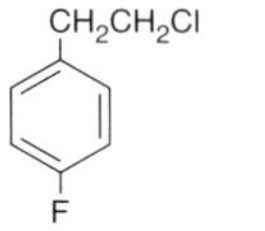
5 gm

S66,335-2
2,4,5-TRIMETHYLBENZYL CHLORIDE
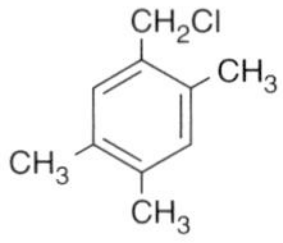
5 gm

S40,754-2
2,5-BIS(CHLOROMETHYL)-P-XYLENE
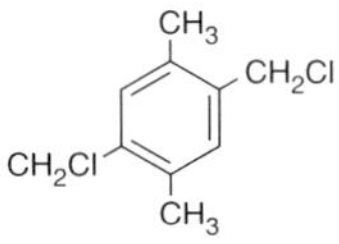
5 gm

S40,755-0
4,6-BIS(CHLOROMETHYL)-M-XYLENE
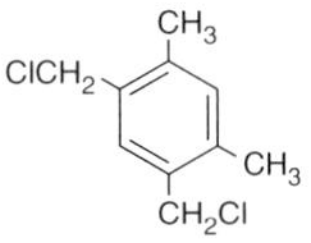
5 gm

S77,417-0
ALPHA,ALPHA,2,3,4,5,6-HEPTAFLUOROTOLUENE
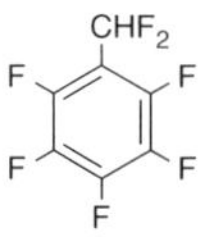
250 mg

S44,983-0
ALPHA,ALPHA,ALPHA',ALPHA'-2,3,5,6-OCTACHLORO-P-XYLENE
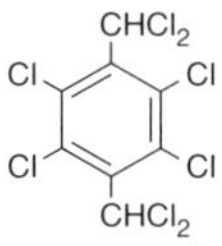
5 gm

S36,489-4
ALPHA,ALPHA,2,6-TETRACHLOROTOLUENE
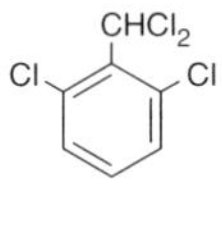
5 gm

S36,025-2
2,4-DICHLOROBENZOTRICHLORIDE
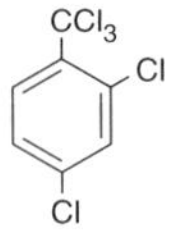
5 gm

S36,026-0
3,4-DICHLOROBENZOTRICHLORIDE
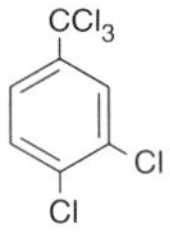
5 gm

S69,771-0
4-CHLORO-1,3-BIS(TRICHLOROMETHYL)-BENZENE
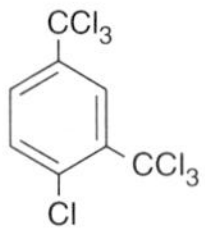
250 mg

S77,582-7
1,4-BIS(TRICHLOROMETHYL)-2-CHLOROBENZENE
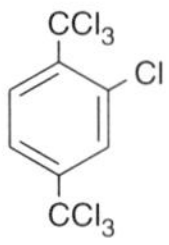
250 mg

S75,817-5
ALPHA,ALPHA,ALPHA,ALPHA',ALPHA',ALPHA',5-HEPTAFLUORO-M-XYLENE
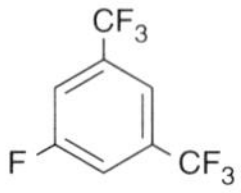
250 mg

S64,347-5
2,3,4,6-TETRAFLUOROBROMOBENZENE
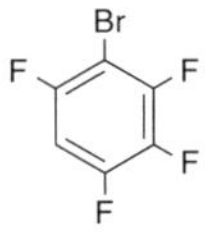
5 gm

S41,075-6
1,3-DIBROMOTETRAFLUOROBENZENE
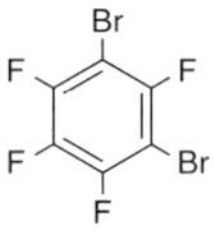
5 gm

S82,496-8
1-IODO-2,3,5,6-TETRACHLOROBENZENE
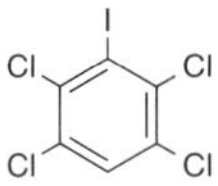
250 mg

S94,313-4
2,4-DICHLORO-1,3,5-TRIBROMOBENZENE
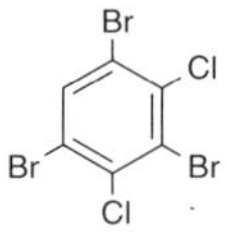
250 mg

S43,561-9
2,3,5,6-TETRACHLORO-P-XYLENE
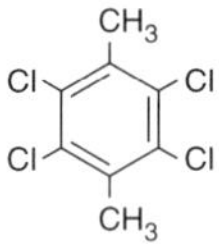
5 gm

S45,510-5
ALPHA-METHYL-2,3,4,5,6-PENTAFLUOROSTYRENE
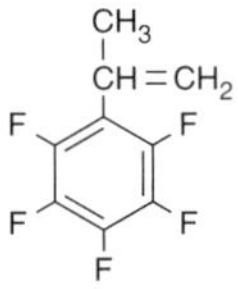
250 mg

S82,827-0
1-ALLYL-2,3,4,5,6-PENTACHLOROBENZENE
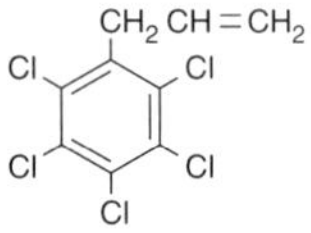
250 mg

S86,941-4
1,2,3-TRIBROMO-4,5,6-TRIMETHYLBENZENE
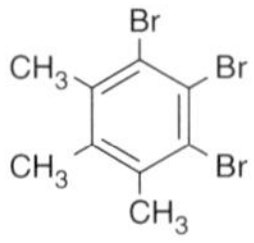
250 mg

S86,942-2
1,2,4-TRIBROMO-3,5,6-TRIMETHYLBENZENE
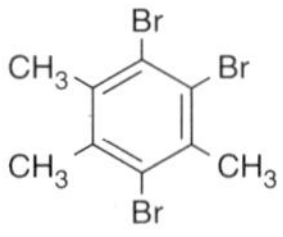
250 mg

S86,943-0
1,3,5-TRIBROMO-2,4,6-TRIMETHYLBENZENE
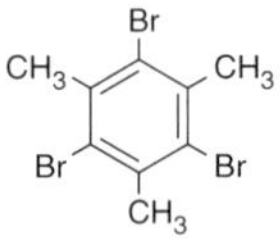
250 mg

S81,429-6
1-BUTYL-2,3,4,5,6-PENTAFLUOROBENZENE
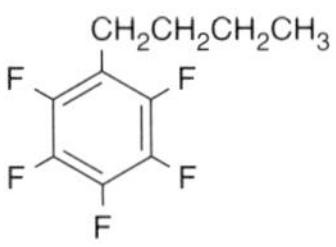
250 mg

S70,676-0
1,4-DIBROMO-2,3,5,6-TETRAMETHYLBENZENE
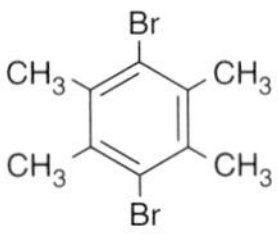
5 gm

S55,764-1
1-BROMOPENTAMETHYLBENZENE
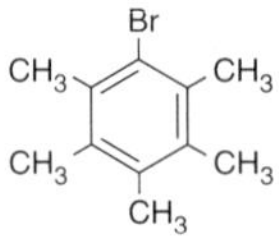
5 gm

S70,616-7
TRIETHYLBROMOBENZENE
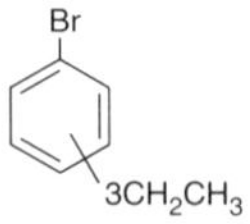
5 gm

S83,540-4
1,4-DI-TERT-BUTYL-2,3,5,6-TETRAFLUOROBENZENE
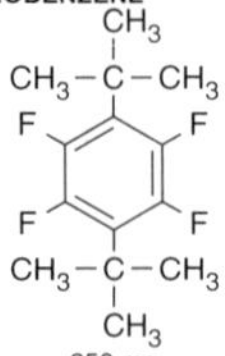
250 mg

S94,308-8
1-BROMO-2,3-DIIODO-5-NITROBENZENE
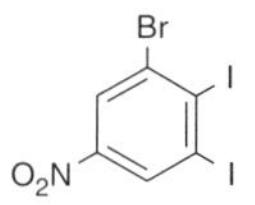
100 mg

S89,427-3
3-(4-BROMOPHENYL)-1,1,2,3-TETRACHLORO-1-PROPENE
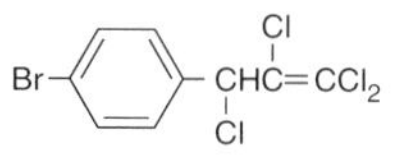
250 mg

S94,301-0
1,3-DICHLORO-2-IODOBENZENE
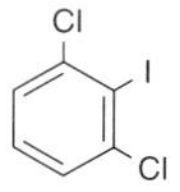
250 mg

S65,402-7
2-CHLORO-6-IODOTOLUENE
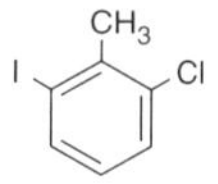
250 mg

S40,687-2
1-BROMO-2,5-DICHLOROBENZENE
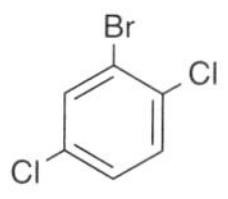
5 gm

S66,269-0
1-CHLORO-2,5-DIBROMOBENZENE
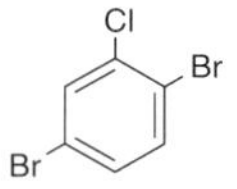
250 mg

S66,861-3
1,2-DICHLORO-4-IODOBENZENE
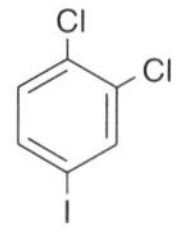
5 gm

S56,741-8
5-BROMO-2-CHLOROTOLUENE
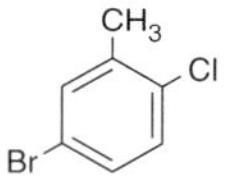
5 gm

S36,899-7
4-FLUORO-2-IODOTOLUENE
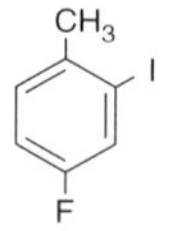
250 mg

S39,139-5
2,3,6-TRICHLOROTOLUENE
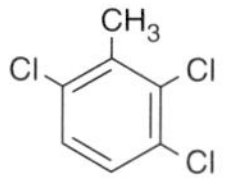
5 gm

S73,205-2
2,3,4-TRICHLOROTOLUENE
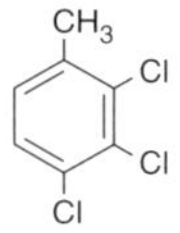
250 mg

S94,303-7
1-BROMO-2,3,5-TRICHLOROBENZENE
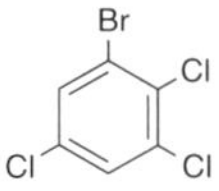
250 mg

S94,304-5
5-BROMO-1,2,3-TRICHLOROBENZENE
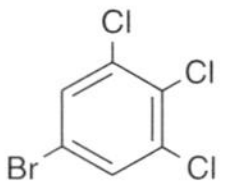
250 mg

S94,306-1
2-BROMO-1,3,5-TRICHLOROBENZENE
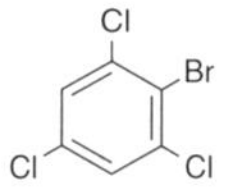
250 mg

S94,305-3
5-BROMO-1,3-DICHLORO-2-IODOBENZENE
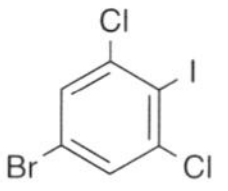
250 mg

S54,726-3
1-IODO-2,4,6-TRIBROMOBENZENE
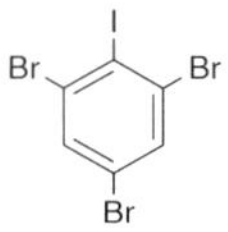
5 gm

S94,300-2
4-CHLORO-2,6-DIBROMOTOLUENE
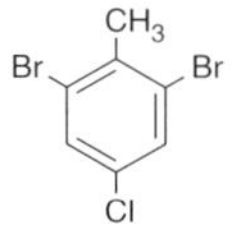
250 mg

S88,850-8
1,3-DIBROMO-2-IODO-5-METHYLBENZENE
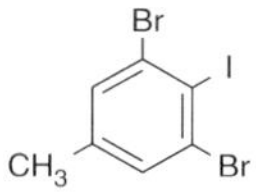
250 mg

S94,311-8
1-BROMO-2,4,5-TRICHLOROBENZENE
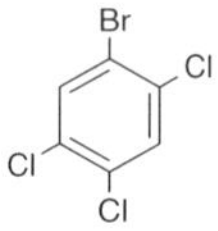
250 mg

S94,312-6
2,4,5-TRIBROMOTOLUENE
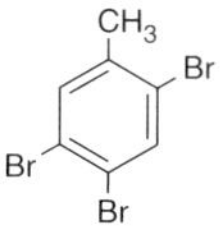
250 mg

S88,184-8
1-BROMO-2,5-DIIODO-4-METHYLBENZENE
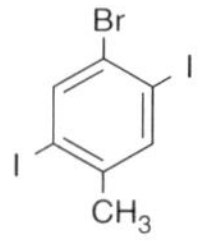
250 mg

S94,359-2
1-CHLORO-2,4,5-TRIMETHYLBENZENE
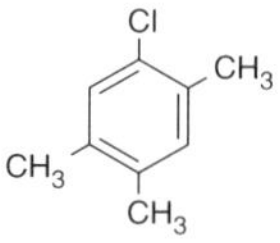
250 mg

S94,360-6
1-IODO-2,4,5-TRIMETHYLBENZENE
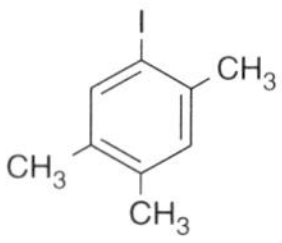
250 mg

S89,428-1
1-PHENYL-1,3,3-TRICHLOROALLENE
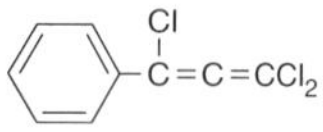
250 mg

S89,458-3
3-PHENYL-1,1,2,3-TETRACHLORO-1-PROPENE
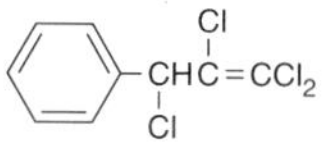
250 mg

S82,958-7
ALPHA,ALPHA'-DICHLOROSTILBENE
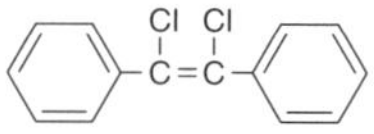
250 mg

S80,699-4
ALPHA,BETA-DIBROMOSTILBENE
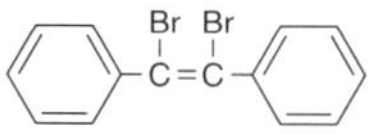
250 mg

S83,084-4
1-CHLORO-1-PHENYL-1-DODECENE
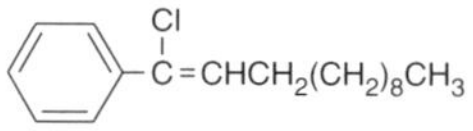
250 mg

S56,729-9
1,3-DIETHYL-2-FLUOROBENZENE
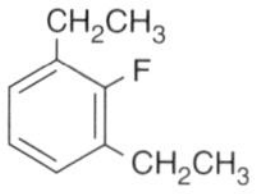
250 mg

S63,898-6
2,6-DIETHYLBROMOBENZENE
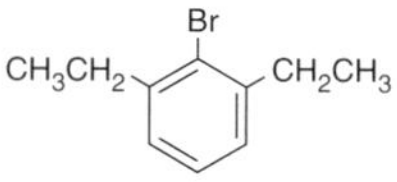
250 mg

S60,386-4
5-CHLORO-2-FLUOROTOLUENE
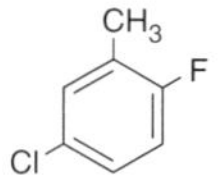
5 gm

S40,686-4
2-BROMO-P-CYMENE
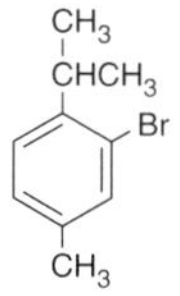
5 gm

S84,698-8
1,2,3,4-TETRAKIS(BROMOMETHYL)-BENZENE
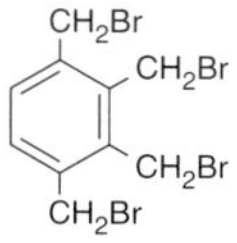
250 mg

S94,302-9
1,2,3,5-TETRABROMOBENZENE
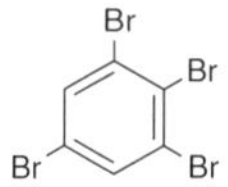
250 mg

S85,286-4
1,1-BIS(4-BROMOPHENYL)-2-CHLOROETHANE
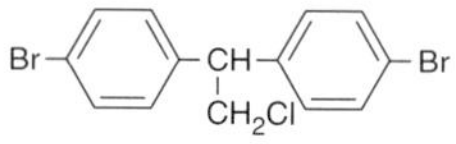
250 mg

S84,264-8
2,2'-BIS(BROMOMETHYL)-4,4'-DIBROMOBIPHENYL
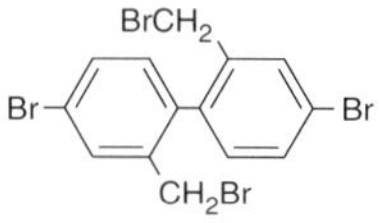
250 mg

S77,003-5
9-(2-BROMOACETYL)PHENANTHRENE
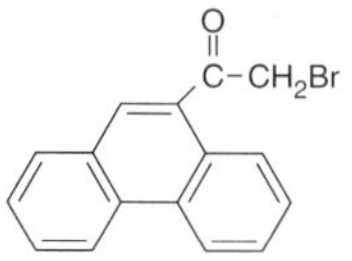
250 mg

S86,450-1
11-BROMO-10-BROMOMETHYL-10,11-DIHYDRO-5,10-METHANO-5H-DIBENZO (A,D)CYCLOHEPTENE
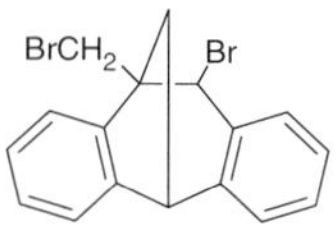
250 mg

S85,210-4
11-(CHLOROMETHYL)-9,10-DIHYDRO-9,10-ETHANOANTHRACENE
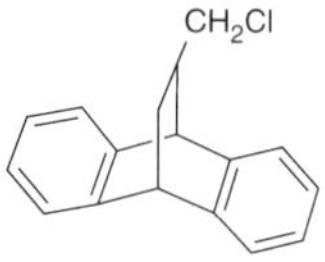
250 mg

S86,368-8
11-(BROMOMETHYL)-9,10-DIHYDRO-9,10-ETHANOANTHRACENE
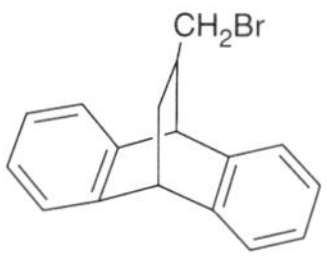
250 mg

S85,318-6
11-(CHLOROMETHYL)-9,10-DIHYDRO-12-METHYL-9,10-ETHANOANTHRACENE
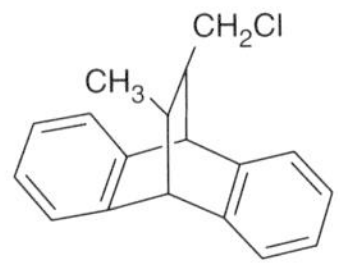
250 mg

S85,265-1
11,12-BIS(CHLOROMETHYL)-9,10-DIHYDRO-9,10-ETHANOANTHRACENE
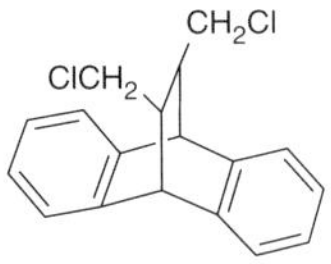
250 mg

S86,585-0
11-(BROMOMETHYL)-9,10-DIHYDRO-12-METHYL-9,10-ETHANOANTHRACENE
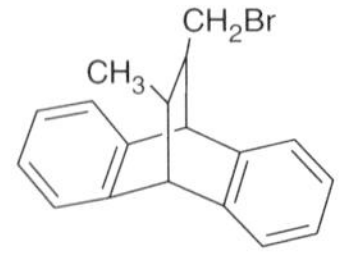
250 mg

S36,618-8
4-BROMOMETHYLBIPHENYL
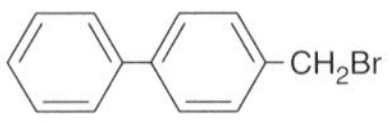
5 gm

S79,238-1
1,2-BIS(BROMOMETHYL)NAPHTHALENE
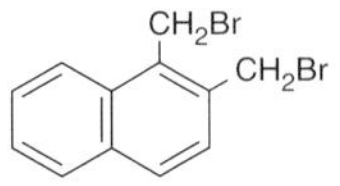
250 mg

S79,240-3
2,3-BIS(BROMOMETHYL)NAPHTHALENE
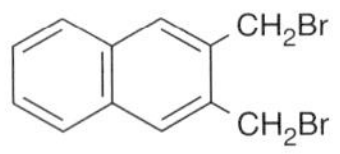
250 mg

S94,376-2
2-BROMOMETHYL-6-METHYLNAPHTHALENE
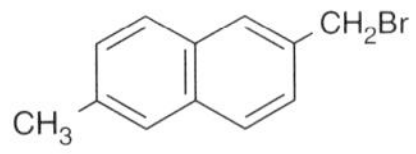
250 mg

S77,421-9
1-BROMOMETHYLFLUORENE
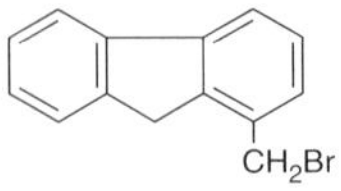
250 mg

S97,368-8
AR-E22 7-(CHLOROMETHYL)BENZ(A)-ANTHRACENE
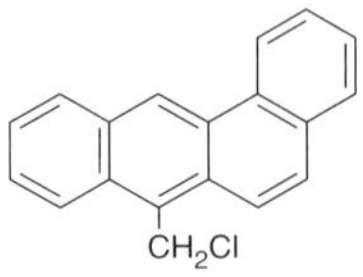
10 mg

S97,395-5
AR-E57 7-CHLOROMETHYL-12-METHYLBENZ(A)ANTHRACENE
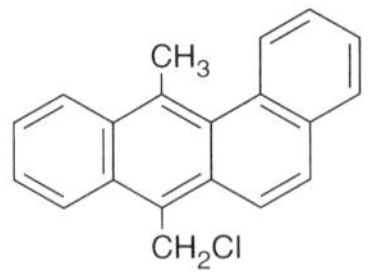
10 mg

S91,828-8
9-(BROMOMETHYL)ANTHRACENE
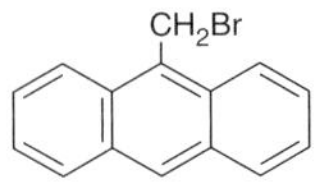
250 mg

S86,362-9
9-(CHLOROMETHYL)-10-METHYLANTHRACENE
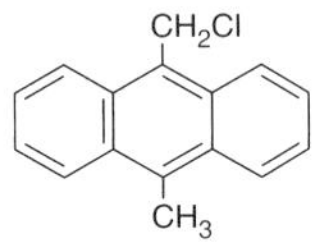
250 mg

S80,965-9
9,10-BIS(BROMOMETHYL)ANTHRACENE
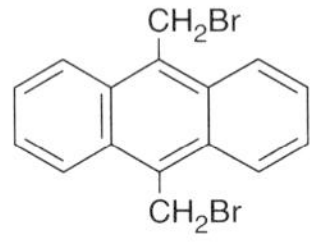
250 mg

S94,767-9
9-BROMOMETHYL-10-METHYLPHENANTHRENE
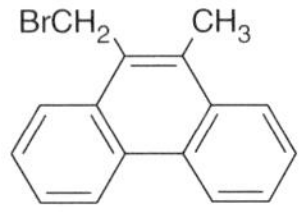
5 gm

S97,332-7
AR-D30 4,5-BIS(BROMOMETHYL)-PHENANTHRENE
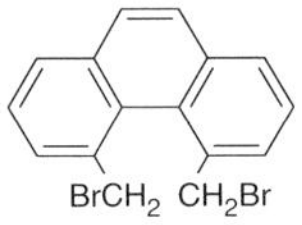
10 mg

S45,056-1
1,1-BIS(4-FLUOROPHENYL)-2,2-DICHLOROETHANE
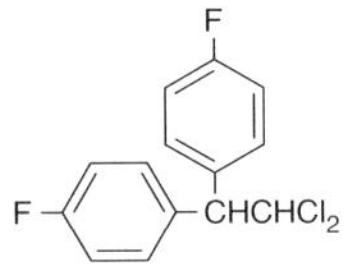
5 gm

S94,207-3
6,7-BIS(DIBROMOMETHYL)-1,1,4,4-TETRAMETHYL-1,2,3,4-TETRAHYDRONAPHTHALENE
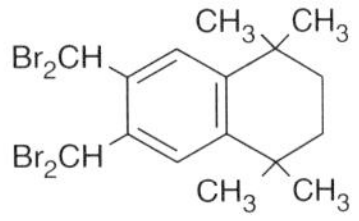
250 mg

S44,931-8
2,2-DIPHENYL-1,1,1-TRICHLOROETHANE
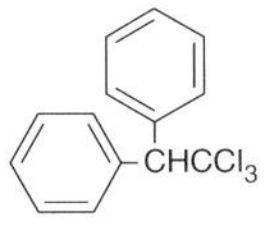
5 gm

S85,200-7
1-(O-TOLYL)-1-(P-TOLYL)-2,2,2-TRICHLOROETHANE
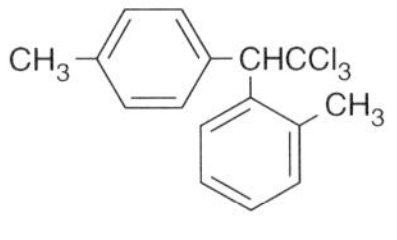
250 mg

S64,162-6
DI-(P-BROMOPHENYL)TRICHLOROETHANE
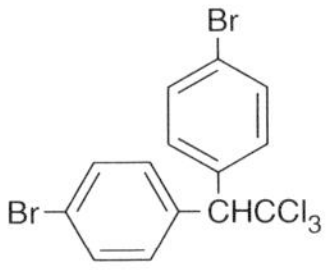
250 mg

S41,052-7
2,2-BIS(4-FLUOROPHENYL)-1,1,1-TRIFLUOROETHANE
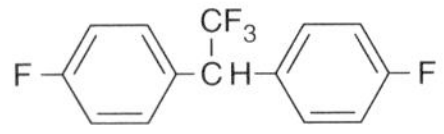
5 gm

S35,902-5
1-(4-BROMOPHENYL)-1-PHENYL-2,2,2-TRICHLOROETHANE
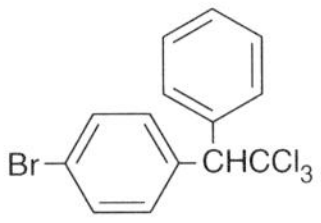
5 gm

S41,444-1
1,1-BIS(4-CHLOROPHENYL)-1,2,2,2-TETRACHLOROETHANE
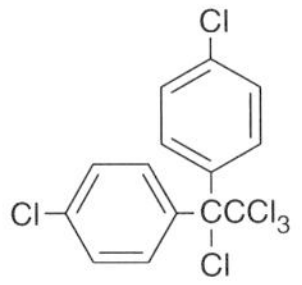
5 gm

S69,759-1
3,3'-BIS(TRIFLUOROMETHYL)BIPHENYL
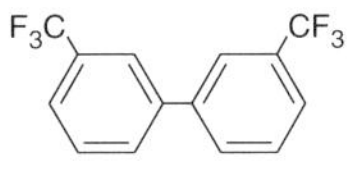
250 mg

S83,014-3
1,4-DIPHENYL-2,3,5,6-TETRACHLOROBENZENE
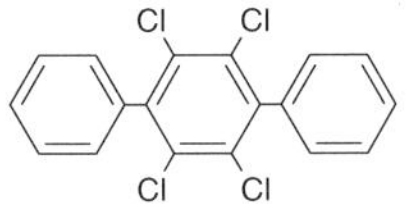
250 mg

S83,077-1
2-BROMO-P-TERPHENYL
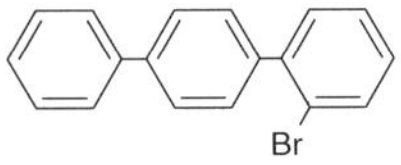
250 mg

S88,026-4
4-BROMOSTILBENE
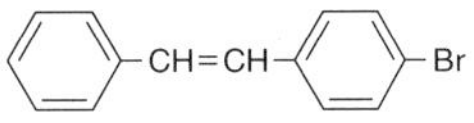
250 mg

S77,454-5
4-FLUORO-ALPHA'-METHYLSTILBENE
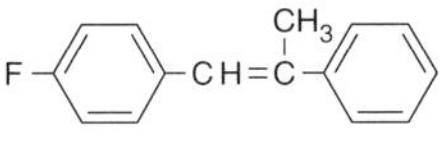
250 mg

S83,686-9
(4-CHLOROPHENYL)PHENYLMETHANE
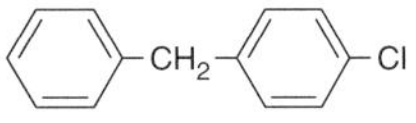
250 mg

S83,880-2
4,4'-DIBROMOTRIPHENYLMETHANE
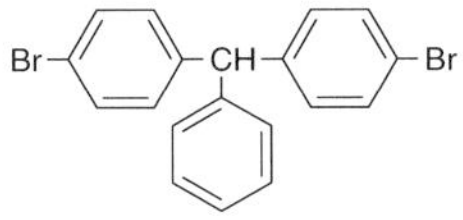
250 mg

S83,079-8
4-IODO-P-TERPHENYL
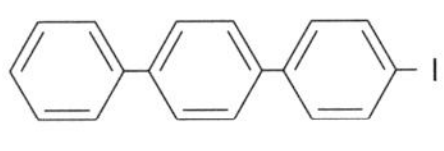
250 mg

S79,956-4
5-BROMO-2-IODOBIPHENYL
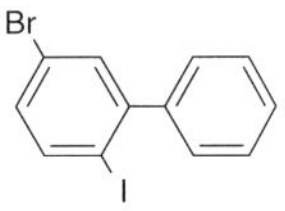
250 mg

S92,681-7
2,4-DICHLORO-1,3,5-TRIPHENYLBENZENE
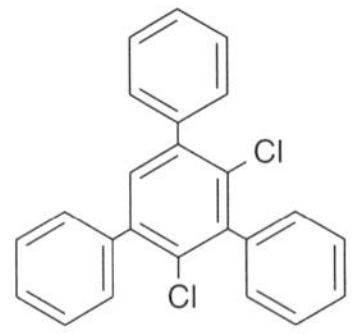
250 mg

S81,063-0
1-BROMO-2,4,6-TRIPHENYLBENZENE
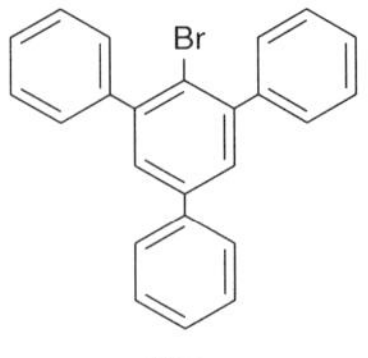
250 mg

S79,957-2
3,5-DIBROMO-2-IODOBIPHENYL
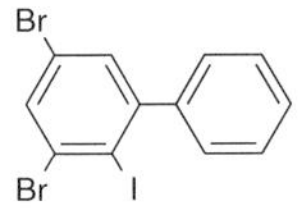
250 mg

S82,382-1
2,3,4,5,6-PENTAFLUOROBIPHENYL
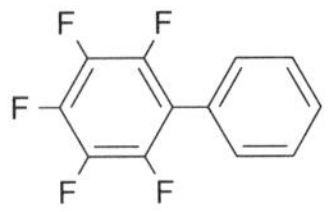
250 mg

S68,879-7
1-(2-BROMOPHENYL)-2-(4-BROMOPHENYL)-1,2-DIBROMOETHANE
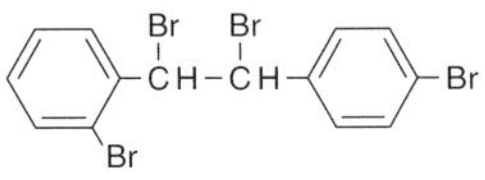
5 gm

S83,195-6
2,2'-DIBROMOBIPHENYL
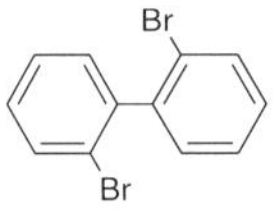
250 mg

S85,981-8
2,2'-DIIODOBIPHENYL
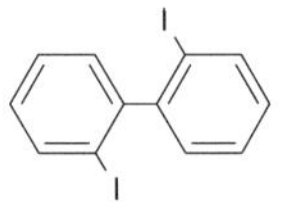
250 mg

S86,523-0
4,4'-DICHLOROBIBENZYL
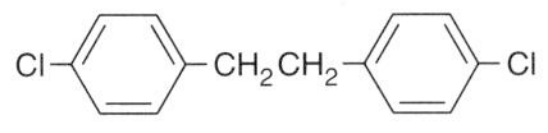
250 mg

S89,439-7
1,2-BIS(4-CHLOROPHENYL)-1,2-DICHLOROETHANE
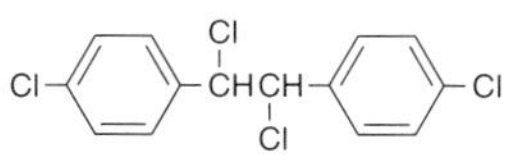
250 mg

S89,438-9
1,2-BIS(4-BROMOPHENYL)-1,2-DICHLOROETHANE
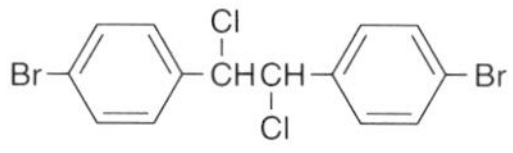
250 mg

S91,445-2
1,1-BIS(4-CHLOROPHENYL)ETHYLENE
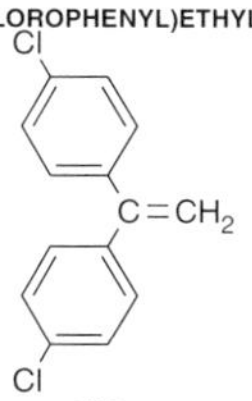
250 mg

S86,359-9
1,1-BIS(4-CHLOROPHENYL)-2-CHLOROETHENE
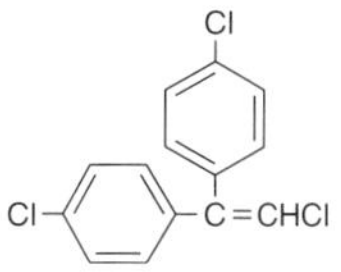
250 mg

S69,926-8
4-TERT-BUTYL-4'-CHLORODIPHENYLMETHANE
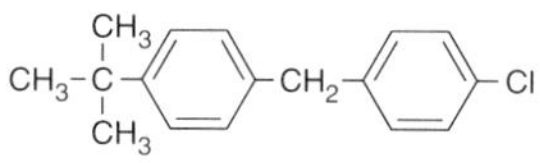
250 mg

S92,456-3
1-(2,4-DICHLOROBENZYLIDENE)-1,4-DIHYDRONAPHTHALENE
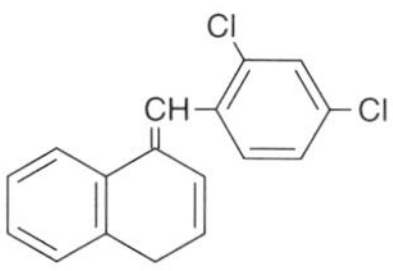
250 mg

S41,072-1
DECAFLUOROBENZHYDRYL BROMIDE
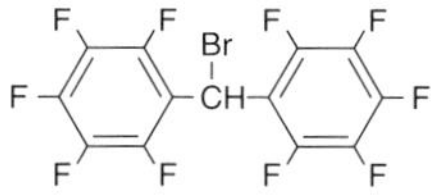
250 mg

S83,169-7
2-BROMO-2',3,3',4,4',5,5',6,6'-NONAFLUOROBIPHENYL
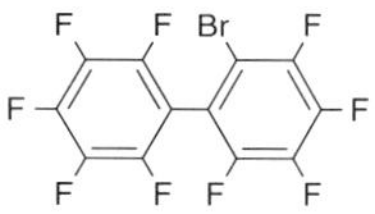
250 mg

S94,558-7
4-BUTOXY-4'-IODOBIPHENYL
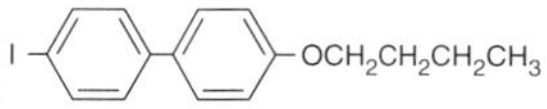
250 mg

S85,267-8
2,3-DICHLORO-4B,8B,8C,8D-TETRAHYDRODIBENZO(A,F)CYCLOPROPA(CD)PENTALENE
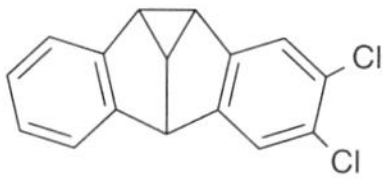
250 mg

S86,412-9
13,13-DICHLORO-9,10,12,13-TETRAHYDRO-11H-9,10-ENDO-CYCLOPROPANTHRACENE
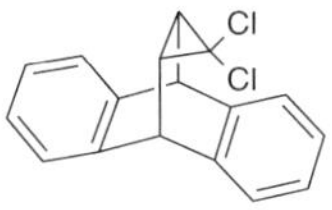
250 mg

S86,562-1
13-BROMO-13-CHLORO-9,10,12,13-TETRAHYDRO-11H-9,10-ENDO-CYCLOPROPANTHRACENE
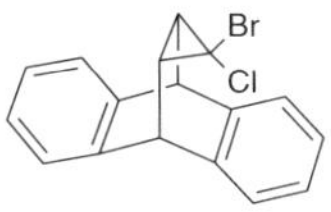
250 mg

S88,373-5
(2,2-DIBROMO-1-METHYLCYCLOPROPYL)-BENZENE
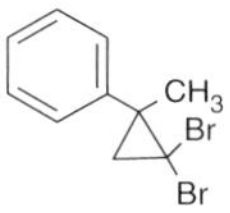
250 mg

S89,630-6
7-BROMO-7H-CYCLOPROPA(A)-ACENAPHTHYLENE
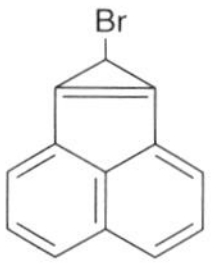
250 mg

S79,219-5
1,2-DIBROMOBENZOCYCLOBUTENE
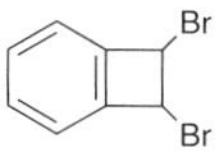
250 mg

S89,394-3
PERCHLOROBENZOCYCLOBUTENE
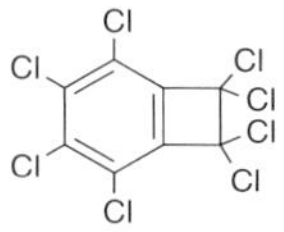
250 mg

S58,859-8
2-IODONAPHTHALENE-BIS-(HEXACHLOROCYCLOPENTADIENE) ADDUCT
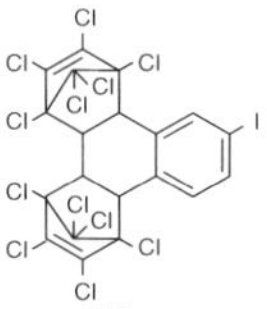
250 mg

S58,864-4
2,3-DIBROMONAPHTHALENE-BIS-(HEXACHLOROCYCLOPENTADIENE) ADDUCT
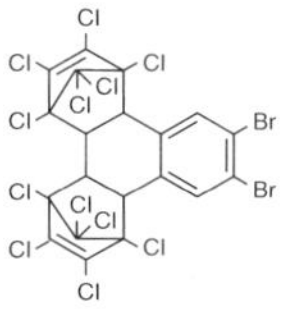
5 gm

S86,350-5
2,3-DICHLORO-1,2,3,4-TETRAHYDRO-1,4-METHANONAPHTHALENE
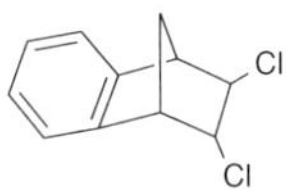
250 mg

S86,378-5
6,7-DIBROMO-5,6-DIHYDRO-5,9-METHANO-9H-BENZOCYCLOHEPTENE
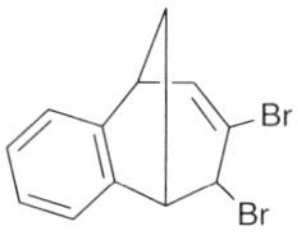
250 mg

S86,449-8
12-CHLORO-10,11-DIHYDRO-5,10-METHANO-5H-DIBENZO(A,D)CYCLOHEPTENE
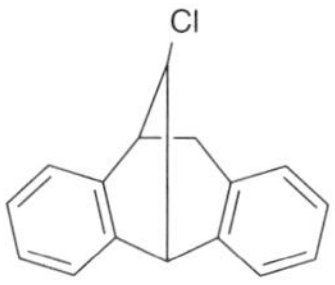
250 mg

S86,407-2
11,12-DICHLORO-10,11-DIHYDRO-5,10-METHANO-5H-DIBENZO(A,D)CYCLOHEPTENE
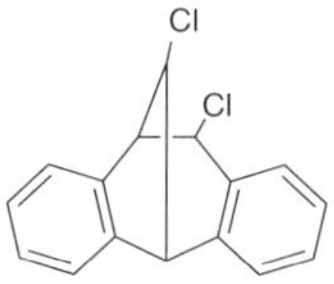
250 mg

S86,385-8
10-CHLORO-11,12-DIBROMO-10,11-DIHYDRO-5,10-METHANO-5H-DIBENZO(A,D)CYCLOHEPTENE
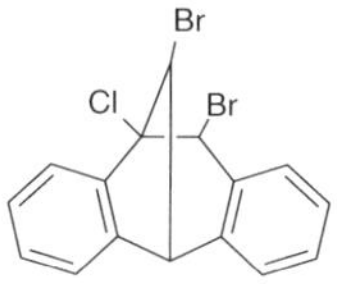
250 mg

S94,583-8
1,1,3-TRIBROMOINDENE
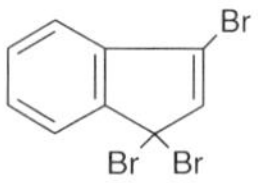
250 mg

S64,848-5
PERCHLOROFULVALENE
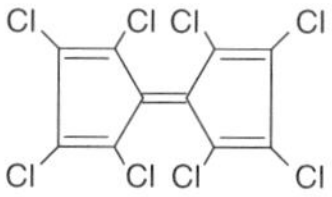
5 gm

S36,204-2
3-FLUOROFLUORANTHENE
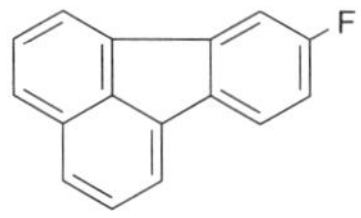
250 mg

S86,327-0
1-CHLORO-3-PHENYLCYCLOHEXANE
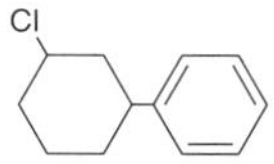
250 mg

S86,552-4
9-CHLORO-9,10-DIHYDRO-9,10-ETHANOANTHRACENE
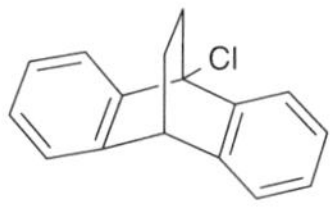
50 mg

S86,361-0
2,12-DICHLORO-9,10-DIHYDRO-9,10-ETHANOANTHRACENE
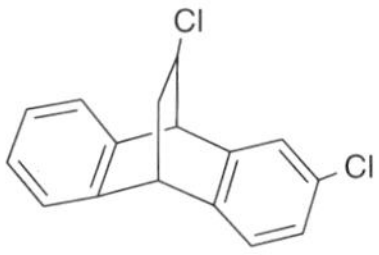
50 mg

S86,406-4
11,12-DICHLORO-9,10-DIHYDRO-9,10-ETHANOANTHRACENE
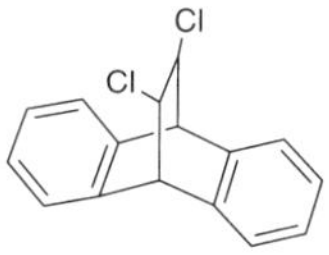
250 mg

S86,551-6
9-BROMO-9,10-DIHYDRO-9,10-ETHANOANTHRACENE
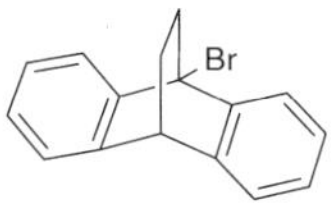
50 mg

S86,542-7
9,10-DIHYDRO-2,3,11,12-TETRACHLORO-9,10-ETHANOANTHRACENE
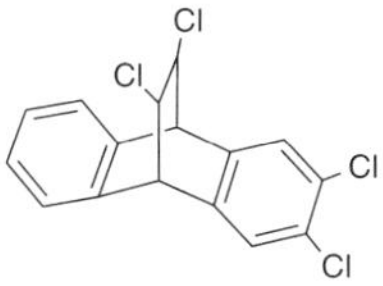
250 mg

S86,566-4
11-BROMO-9,10-DIHYDRO-12-METHYLENE-9,10-ETHANOANTHRACENE
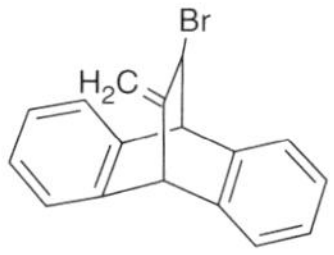
250 mg

S86,579-6
11,12-DICHLORO-9,10-DIHYDRO-9,10-DIMETHYL-9,10-ETHANOANTHRACENE
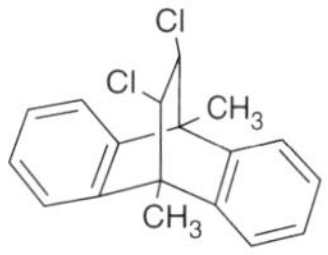
250 mg

S86,567-2
11,13-DICHLORO-10,11-DIHYDRO-5,10-ETHANO-5H-DIBENZO(A,D)CYCLOHEPTENE
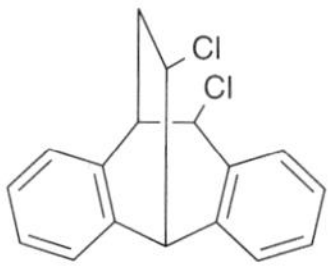
250 mg

S83,733-4
1,2-DICHLORO-1,2-DIPHENYLETHANE
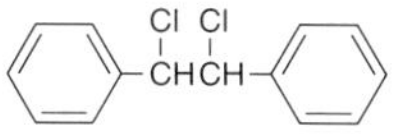
250 mg

S82,959-5
1,2-DIPHENYL-1,1,2,2-TETRACHLOROETHANE
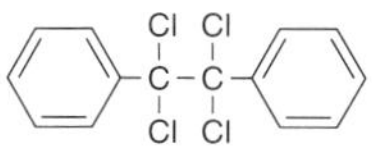
250 mg

S83,938-8
CHLOROTRIBENZYLMETHANE
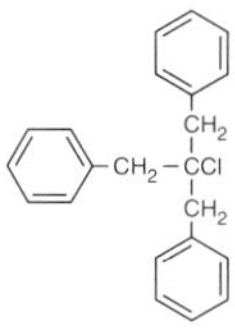
250 mg

S87,100-1
1,2-BIS(ALPHA,BETA-DIBROMOPHENETHYL)BENZENE
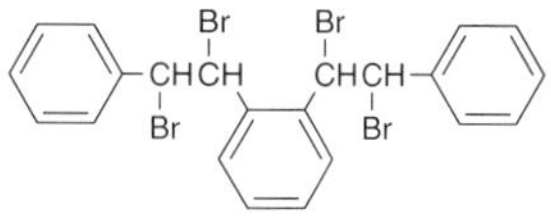
250 mg

S90,633-6
BROMOPHENYL(O-TOLYL)METHANE
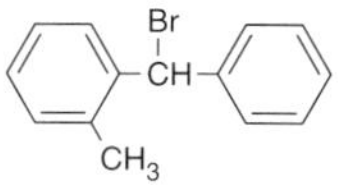
250 mg

S87,012-9
1,1-DICHLORO-2,2-DIPHENYLETHYLENE
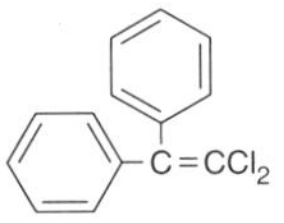
250 mg

S44,200-3
2-BROMO-1,1-DIPHENYLETHYLENE
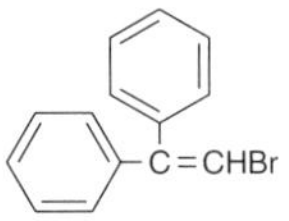
250 mg

S87,011-0
1,1-DIBROMO-2,2-DIPHENYLETHYLENE
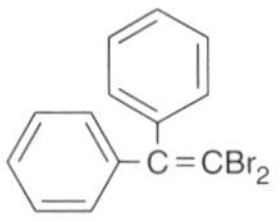
250 mg

S68,881-9
1-CHLOROTRIPHENYLETHYLENE
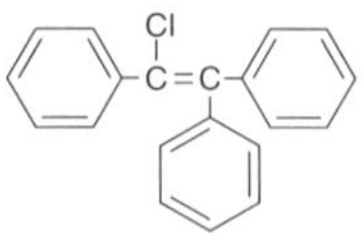
5 gm

S84,212-5
2-BROMO-1,1,3,3-TETRAPHENYL-1-PROPENE
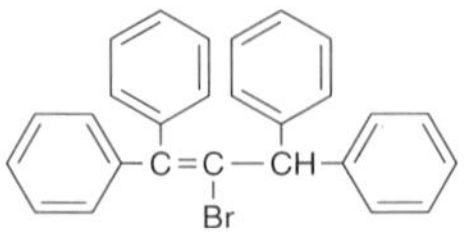
250 mg

S82,633-2
2,3-DIPHENYL-1,5,6,7,7-PENTACHLOROBICYCLO[2.2.1]-HEPTA-2,5-DIENE
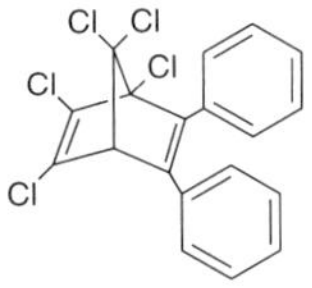
250 mg

S43,529-5
2,2'-DIFLUOROBIPHENYL
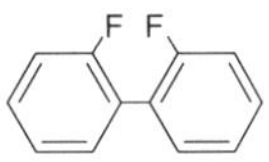
250 mg

S89,442-7
1,2-BIS(P-TOLYL)-1,2-DICHLOROETHANE
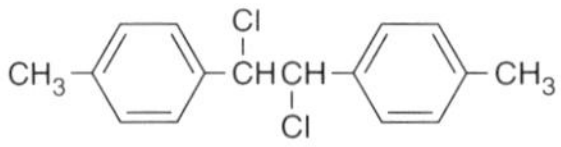
250 mg

S38,180-2
4-BROMO-ALPHA-CHLORODIPHENYL-METHANE
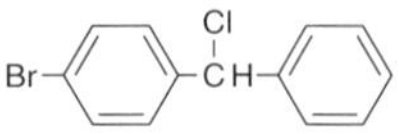
5 gm

S40,692-9
(4-BROMOPHENYL)(4-CHLOROPHENYL)-METHANE
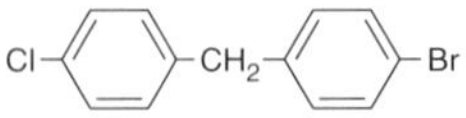
5 gm

S43,193-1
1,1-DICHLORO-2,2-DI-(P-TOLYL)-ETHYLENE
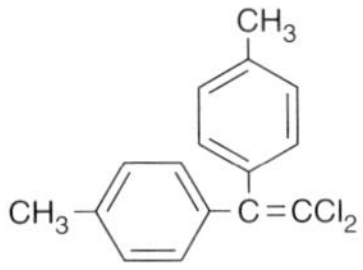
5 gm

S89,384-6
3,3-BIS(P-TOLYL)-1,1,2-TRICHLORO-1-PROPENE
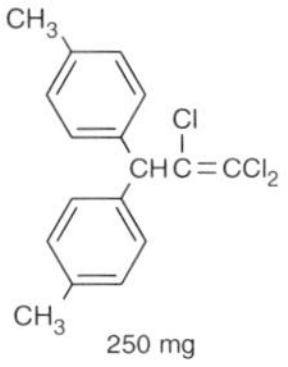
250 mg

S64,626-1
4-BROMO-4'-(TERT-OCTYL)-DIPHENYLMETHANE
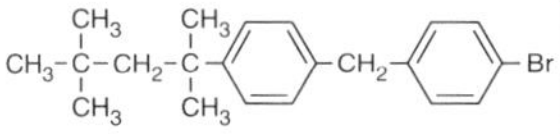
250 mg

S43,510-4
4-BROMO-4'-FLUOROBIPHENYL
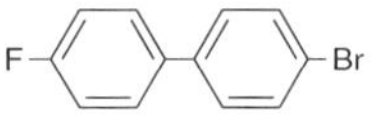
250 mg

S34,405-2
9-(4-CHLOROBENZYLIDENE)FLUORENE
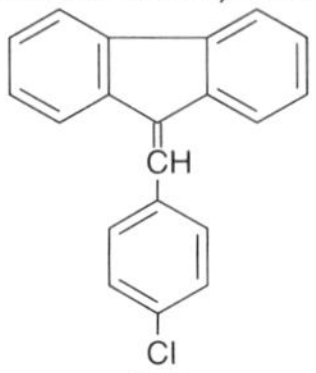
5 gm

S65,497-3
4-BROMO-3-METHYLBIPHENYL
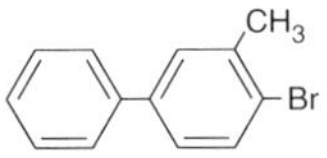
5 gm

S62,734-8
9-(3,4-DICHLOROBENZYLIDENE)FLUORENE
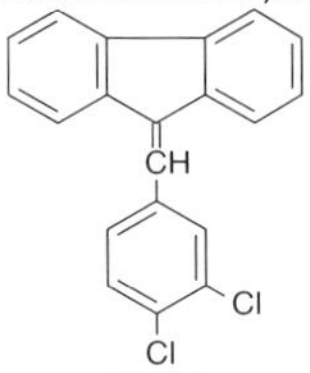
250 mg

S89,443-5
1,2-DICHLORO-1,2-DIMESITYLETHANE
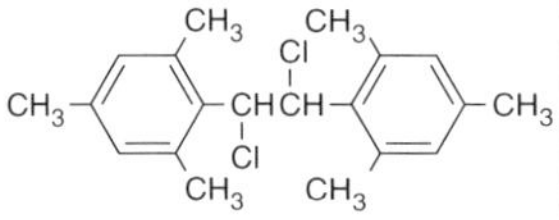
250 mg

S93,963-3
6-CHLOROMETHYL-7-ETHYL-1,2,3,4-TETRAHYDRO-1,1,4,4-TETRAMETHYLNAPHTHALENE
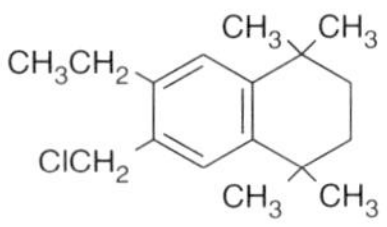
5 gm

S85,256-2
2-CHLORO-9,10-DIHYDRO-9,10-ETHENOANTHRACENE
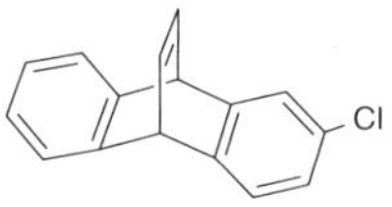
250 mg

S86,543-5
11-CHLORO-9,10-DIHYDRO-9,10-ETHENOANTHRACENE
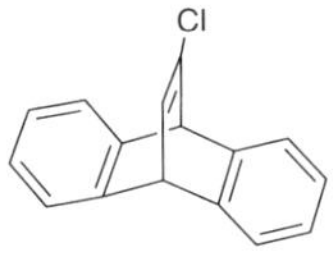
250 mg

S85,309-7
11,12-DICHLORO-9,10-DIHYDRO-9,10-ETHENOANTHRACENE
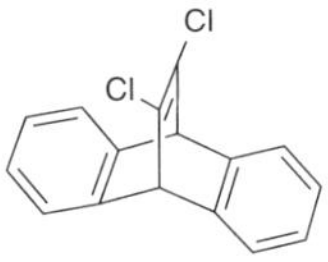
250 mg

S84,295-8
2-BROMOTRIPTYCENE
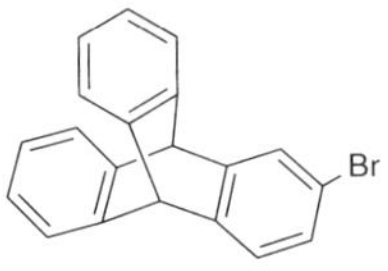
250 mg

S87,572-4
12,13-DICHLORO-9,10-DIHYDRO-9,10-PROPENOANTHRACENE
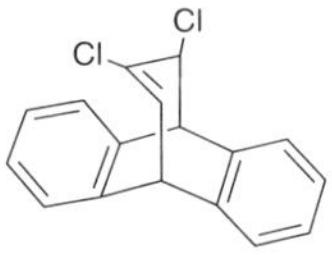
250 mg

S58,000-7
TRANS-1,2-DIBROMO-1,2,3,4-TETRAHYDRONAPHTHALENE
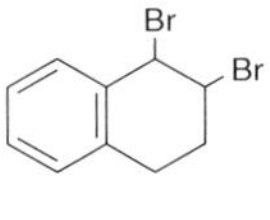
5 gm

S45,384-6
2,3-DIBROMONAPHTHALENE
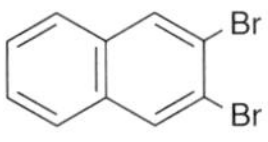
250 mg

S45,555-5
1,2,3,4-TETRACHLORONAPHTHALENE
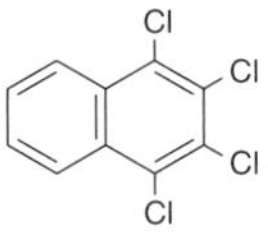
5 gm

S67,157-6
1,5-DIBROMONAPHTHALENE
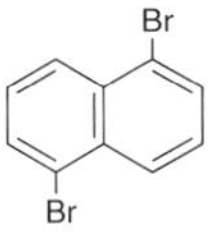
250 mg

S82,461-5
1,4-DIBROMONAPHTHALENE
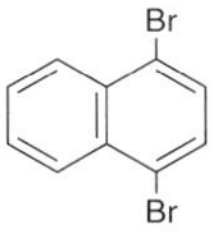
250 mg

S53,443-9
4,4'-DIBROMO-1,1'-BINAPHTHYL
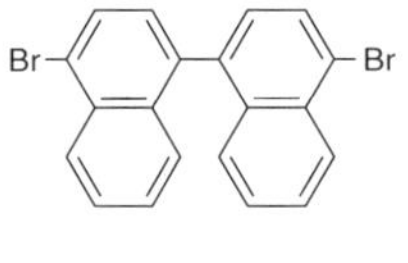
250 mg

S73,146-3
2-IODONAPHTHALENE
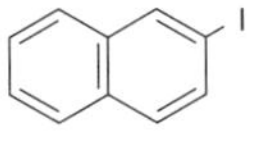
5 gm

S97,499-4
AR-H 1 4-BROMOPYRENE
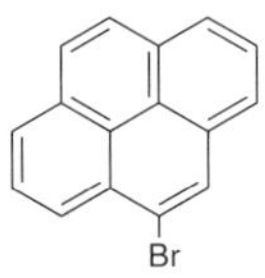
10 mg

S97,472-2
AR-G28 1-FLUORO-12-METHYLBENZO(C)-PHENANTHRENE
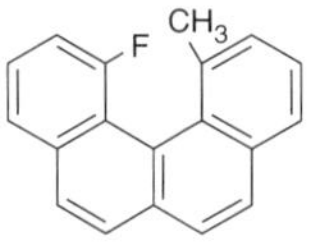
10 mg

S86,506-0
TRANS-1,2-DIBROMOACENAPHTHENE
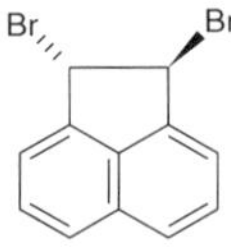
250 mg

S75,431-5
2-FLUOROFLUORENE
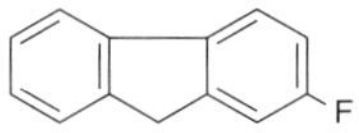
250 mg

S44,198-8
9-CHLOROFLUORENE
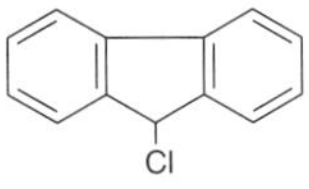
5 gm

S93,320-1
1,3-DICHLOROFLUORENE
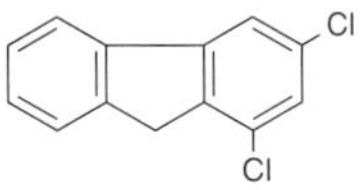
250 mg

S75,542-7
2-IODOFLUORENE
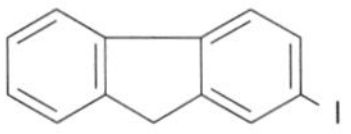
250 mg

S75,155-3
9-BROMO-2,4,7-TRICHLOROFLUORENE
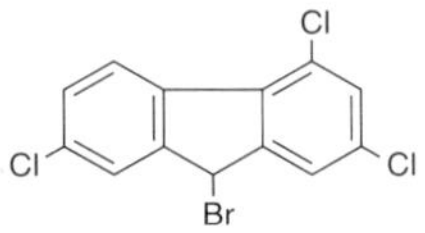
5 gm

S88,892-3
2,7-DIBROMO-9,9-DICHLOROFLUORENE
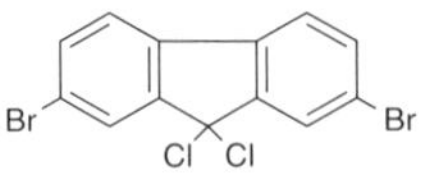
250 mg

S75,424-2
2,7,9-TRIBROMOFLUORENE

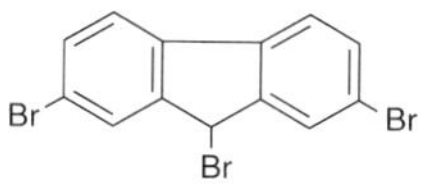

250 mg

S97,355-6
AR-E 7 1-FLUOROBENZ(A)ANTHRACENE

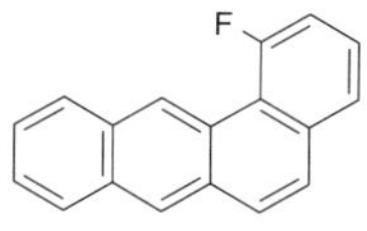

10 mg

S97,354-8
AR-E 4 7-BROMOBENZ(A)ANTHRACENE

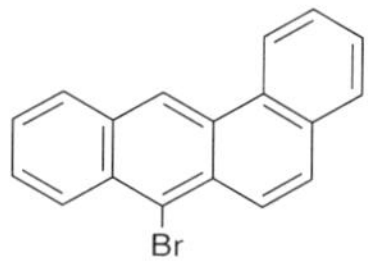

10 mg

S97,374-2
AR-E29 3-FLUORO-7-METHYLBENZ(A)-ANTHRACENE

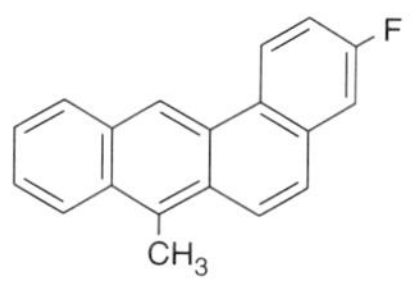

10 mg

S97,376-9
AR-E31 10-FLUORO-7-METHYLBENZ(A)-ANTHRACENE

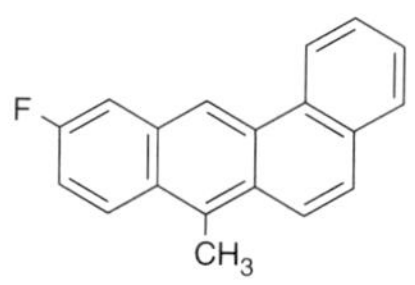

10 mg

S97,369-6
AR-E23 4-CHLORO-7-METHYLBENZ(A)-ANTHRACENE

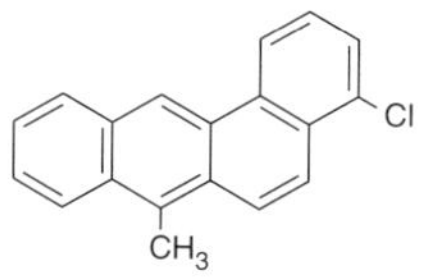

10 mg

S97,371-8
AR-E24 9-CHLORO-7-METHYLBENZ(A)-ANTHRACENE

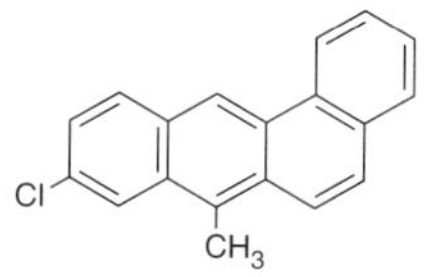

10 mg

S97,365-3
AR-E20 4-BROMO-7-METHYLBENZ(A)-ANTHRACENE

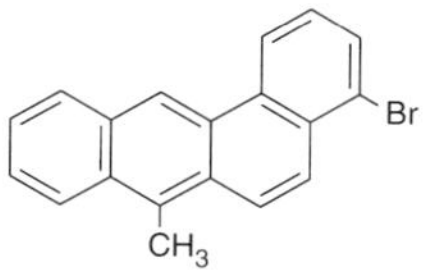

10 mg

S97,366-1
AR-E21 5-BROMO-7-METHYLBENZ(A)-ANTHRACENE

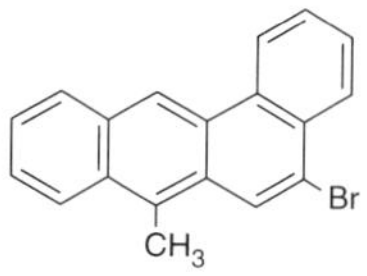

10 mg

S97,396-3
AR-E59 7,12-DIMETHYL-5-FLUOROBENZ(A)ANTHRACENE

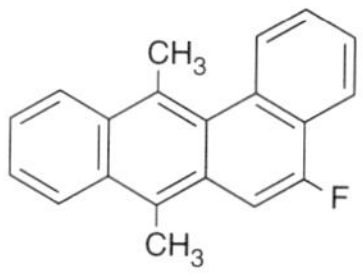

10 mg

S97,397-1
AR-E60 6,8-DIMETHYL-5-FLUOROBENZ(A)-ANTHRACENE

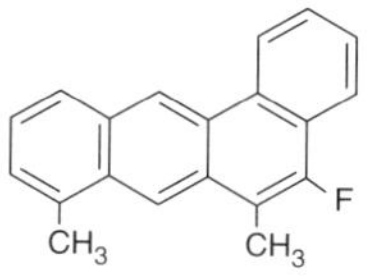

10 mg

S97,400-5
AR-E62 7,12-DIMETHYL-9-FLUOROBENZ(A)ANTHRACENE

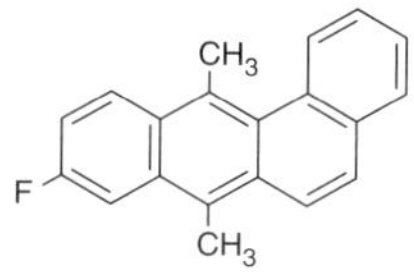

10 mg

S97,401-3
AR-E62A 7,12-DIMETHYL-10-FLUOROBENZ(A)ANTHRACENE

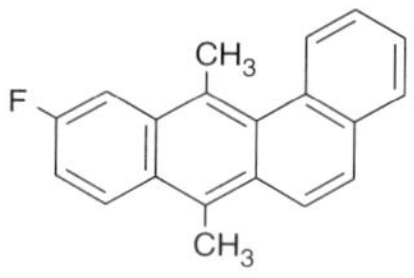

10 mg

S97,394-7
AR-E56 5-BROMO-7,12-DIMETHYLBENZ(A)-ANTHRACENE

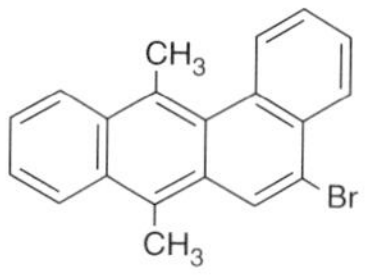

10 mg

S86,381-5
2,6-DICHLOROANTHRACENE

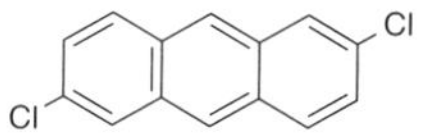

250 mg

S86,441-2
2,3-DICHLOROANTHRACENE

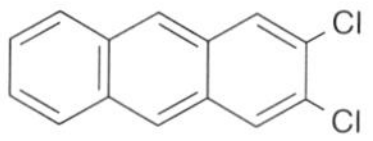

250 mg

S85,254-6
2,3-DICHLORO-9,10-DIHYDROANTHRACENE

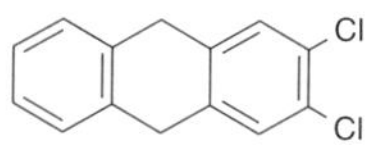

250 mg

S90,733-2
9-BROMO-10-METHYLANTHRACENE

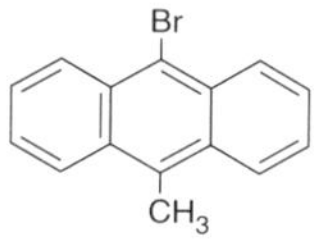

250 mg

S88,256-9
2-TERT-BUTYL-9,10-DIBROMOANTHRACENE

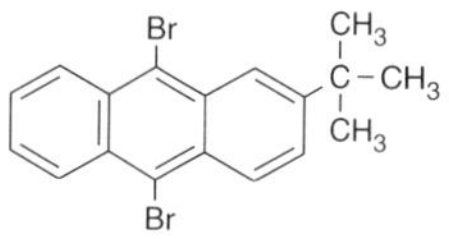

250 mg

S80,964-0
9,10-DIBROMOPHENANTHRENE

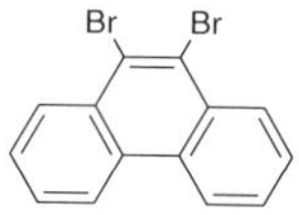

250 mg

S97,309-2
AR-D 5 4,5-DIBROMO-9,10-DIHYDROPHENANTHRENE

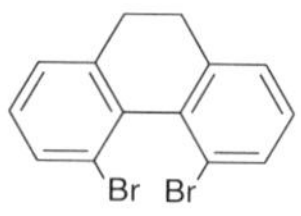

10 mg

S77,030-2
3-CHLORO-7-ISOPROPYL-1-METHYLPHENANTHRENE

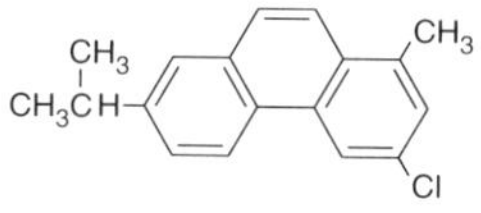

250 mg

S75,641-5
6-FLUOROCHRYSENE

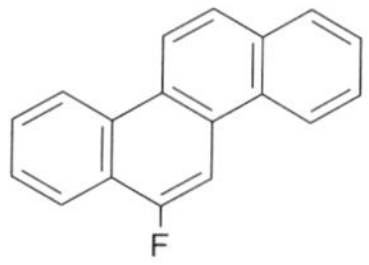

5 gm

S80,997-7
6,12-DIBROMOCHRYSENE

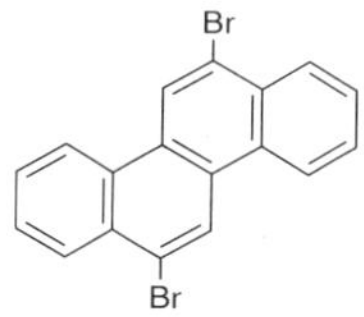

250 mg

S97,550-8
AR-M 1 3-BROMOPHENANTHRO(3,4-C)-PHENANTHRENE

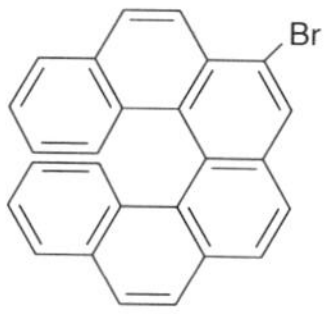

10 mg

S43,623-2
1,3,6,8-TETRACHLOROPYRENE

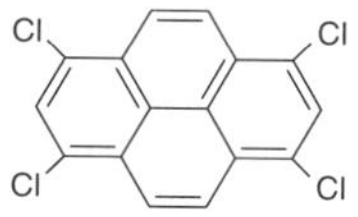

5 gm

S50,865-9
2,3-DIBROMOPROPYL-PENTABROMOPHENYL ETHER

5 gm

S69,704-4
(2-BROMOETHYL) (2-BROMO-4-METHOXYPHENYL) ETHER
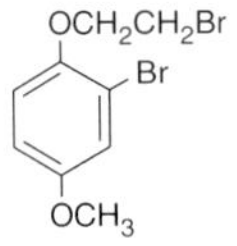
250 mg

S95,373-3
1-BROMO-2-(2,3-DIBROMOPROPYL)-4,5-DIMETHOXYBENZENE
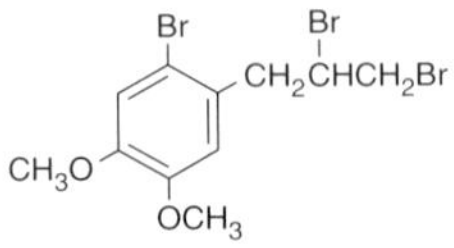
50 mg

S46,300-0
2-BROMOETHYL 4-BROMPHENYL ETHER
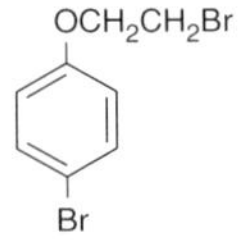
250 mg

S83,340-1
3-CHLOROPROPYL 2,4-DICHLOROPHENYL ETHER
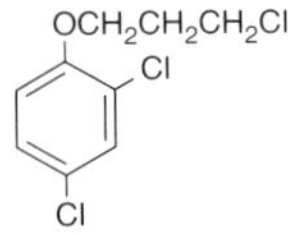
250 mg

S69,769-9
(2-BROMO-4-METHYLPHENYL)-(3-BROMOPROPYL) ETHER
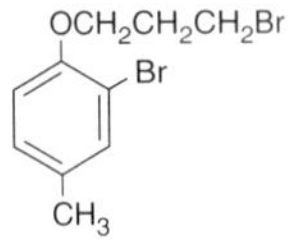
250 mg

S69,781-8
(3-BROMOPROPYL) (2,6-DIBROMO-4-METHYLPHENYL) ETHER
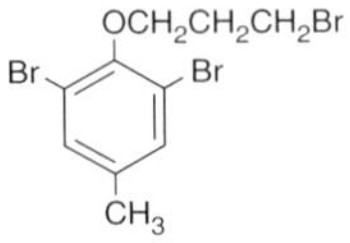
250 mg

S36,370-7
4-(3-METHYLPHENOXY)BUTYL CHLORIDE
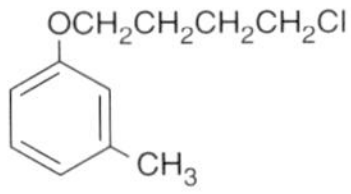
5 gm

S71,440-2
2,5-BIS(CHLOROMETHYL)-1,4-DIMETHOXYBENZENE
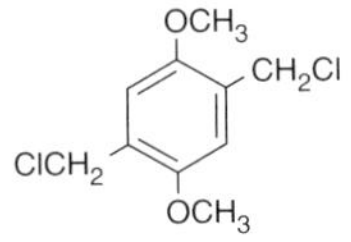
250 mg

S80,872-5
2,5-DIMETHOXYBENZYL CHLORIDE
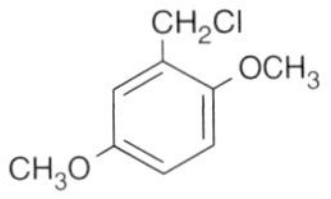
250 mg

S69,712-5
ALPHA,ALPHA'-DIBROMO-2,5-DIMETHOXY-P-XYLENE
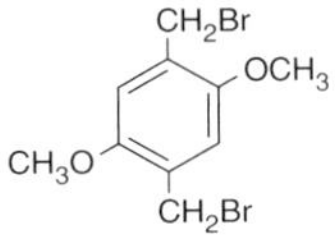
250 mg

S86,009-3
1,4-BIS(CHLOROMETHYL)-2,5-DIETHOXY-BENZENE
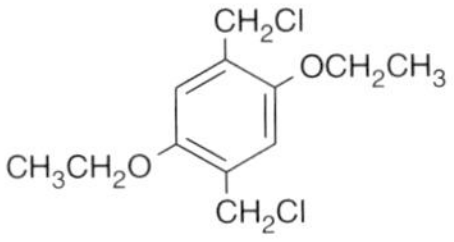
250 mg

S57,367-1
BETA-BROMO-2-METHOXYPHENETOLE
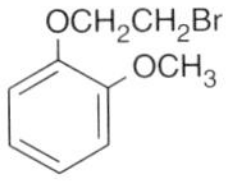
5 gm

S43,061-7
BETA-CHLOROPHENETOLE
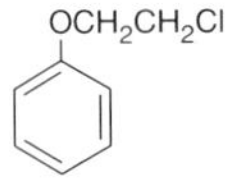
5 gm

S79,298-5
3-CHLOROPROPYL PHENYL ETHER
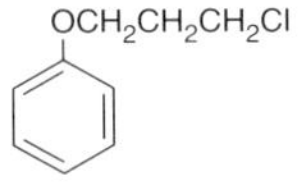
250 mg

S35,320-5
4-(2-METHYLPHENOXY)BUTYL CHLORIDE
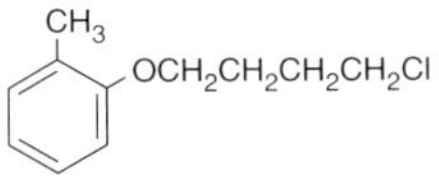
5 gm

S60,515-8
2-BROMOETHYL 4-CHLOROPHENYL ETHER
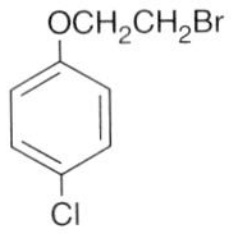
250 mg

S53,571-0
3-CHLOROPROPYL 4-FLUOROPHENYL ETHER
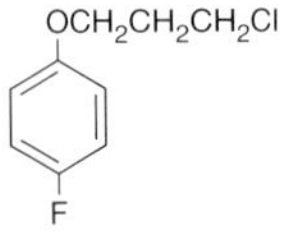
250 mg

S94,393-2
1-BROMO-3-(4-METHOXYPHENYL)PROPANE
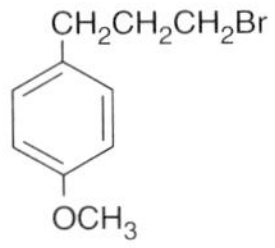
250 mg

S36,371-5
4-(4-METHYLPHENOXY)BUTYL CHLORIDE
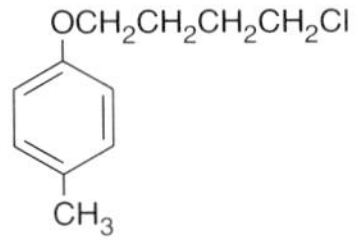
5 gm

S62,889-1
1-(2-BROMOETHOXY)-4-BUTYLBENZENE
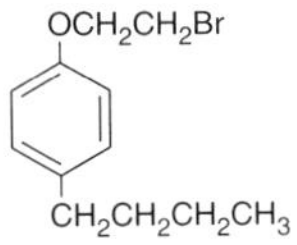
250 mg

S34,234-3
2,2-BIS(4-METHOXYPHENYL)-1,3-DICHLOROPROPANE
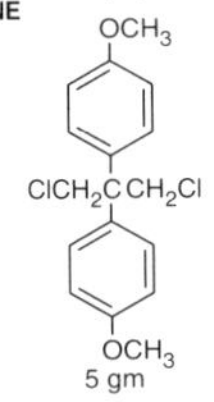
5 gm

S34,232-7
2,2-BIS(4-METHOXY-3-METHYLPHENYL)-1,3-DICHLOROPROPANE
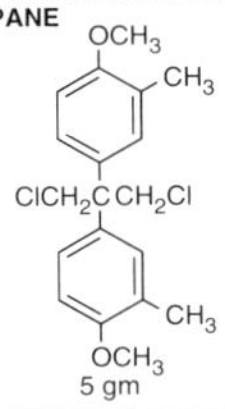
5 gm

S43,235-0
BETA-BROMO-2,4,6-TRIIODOPHENETOLE
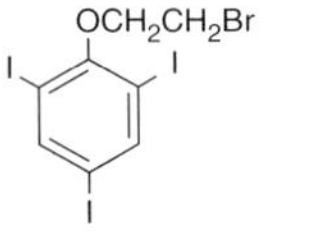
5 gm

S50,957-4
4,4'-ISOPROPYLIDENEBIS(2,6-DIBROMOPHENYL-2,3-DIBROMOPROPYL ETHER)
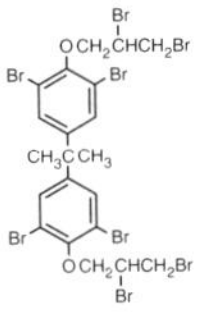
5 gm

S38,895-5
BETA-CHLORO-3-METHYL-ALPHA,ALPHA,BETA-TRIFLUOROPHENETOLE
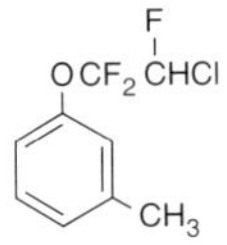
5 gm

S38,894-7
BETA-CHLORO-2-METHYL-ALPHA,ALPHA,BETA-TRIFLUOROPHENETOLE
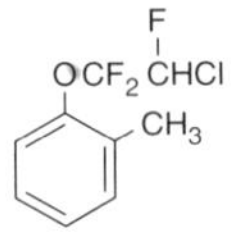
5 gm

S38,717-7
4-BENZYLOXY-ALPHA,ALPHA-DIFLUOROANISOLE
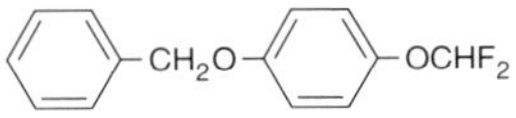
250 mg

S50,863-2
ALLYL PENTABROMOPHENYL ETHER
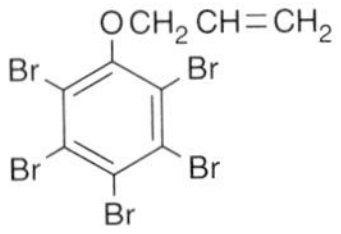
5 gm

S88,364-6
BENZYL 3-BROMOPHENYL ETHER
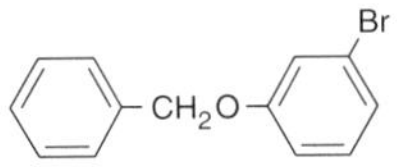
250 mg

S68,145-8
BENZYL 4-BROMOPHENYL ETHER
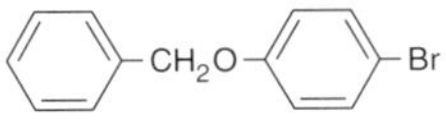
250 mg

S84,681-3
1,4-DIMETHOXY-2,3-DIMETHYL-5-IODOBENZENE
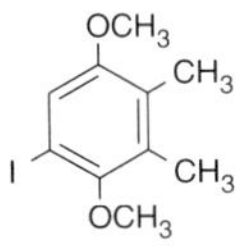
250 mg

S84,680-5
1,2-DIBROMO-3,6-DIMETHOXY-4,5-DIMETHYLBENZENE

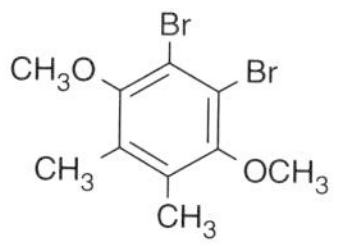

250 mg

S85,029-2
1,2-DIMETHOXY-3,4-DIMETHYL-5-IODOBENZENE

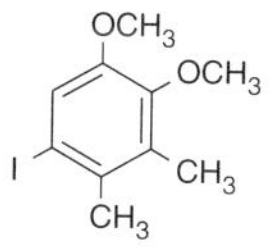

250 mg

S84,638-4
1,4-DIBROMO-2,3,5,6-TETRAMETHOXYBENZENE

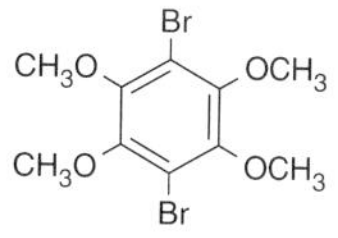

250 mg

S84,637-6
1-BROMO-2,4,5-TRIMETHOXYBENZENE

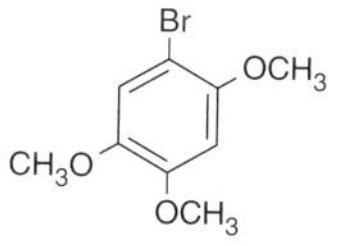

250 mg

S84,272-9
1,2-DIMETHOXY-4-IODOBENZENE

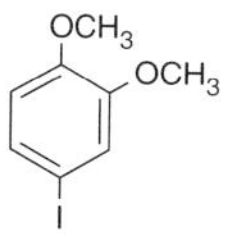

250 mg

S81,382-6
1,2-DIBROMO-4,5-DIMETHOXYBENZENE

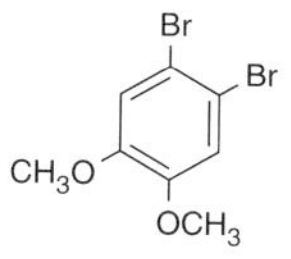

250 mg

S84,397-0
1,2-DIIODO-4,5-DIMETHOXYBENZENE

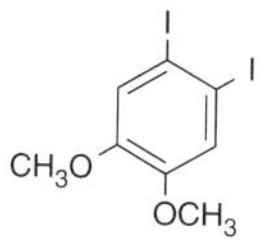

250 mg

S74,494-8
1,2-DIMETHOXY-4-IODO-5-METHYLBENZENE

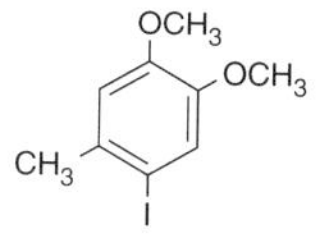

250 mg

S85,986-9
2,2'-DIIODO-4,4',5,5'-TETRAMETHOXYBIPHENYL

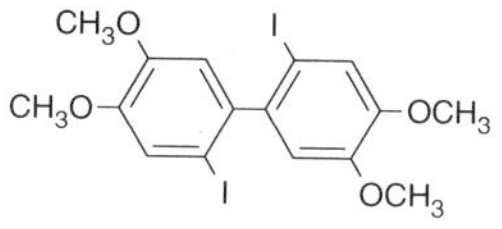

250 mg

S84,719-4
2,2'-DIIODO-3,3'-DIMETHYL-4,4',5,5'-TETRAMETHOXYBIPHENYL

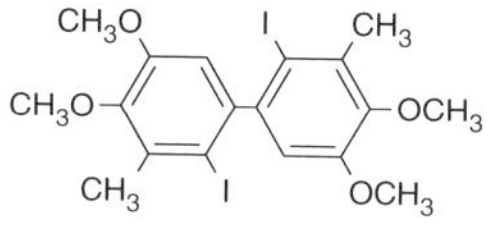

250 mg

S84,674-0
2,2'-DIIODO-3,3',4,4',5,5'-HEXAMETHOXYBIPHENYL

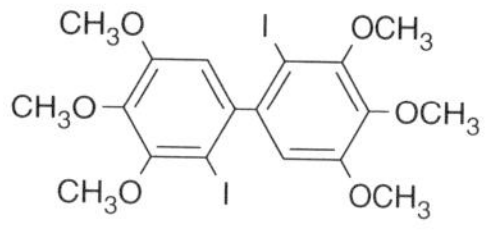

250 mg

S79,320-5
1-(4-FLUOROPHENOXY)-3-PHENOXYPROPANE

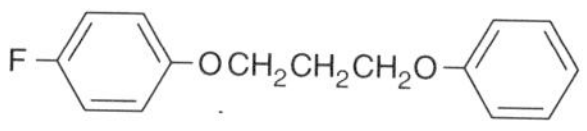

250 mg

S54,150-8
1-BROMO-2,4,6-TRIMETHOXYBENZENE

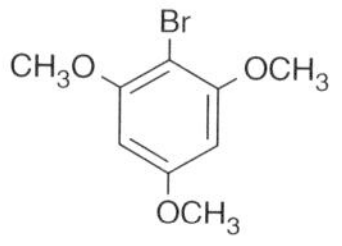

250 mg

S83,780-6
5,5'-DIBROMO-2,2'-DIMETHOXYBIPHENYL

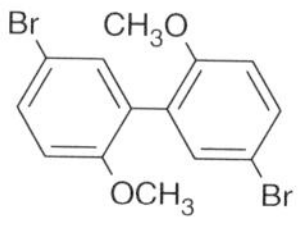

250 mg

S69,722-2
1-(2-BROMO-4-METHYLPHENOXY)-2-PHENOXYETHANE

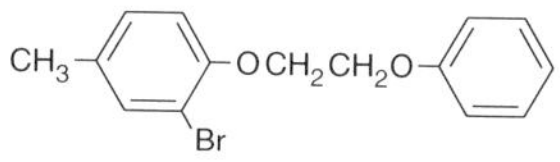

250 mg

S79,201-2
1-(2,6-DIMETHOXY-3-IODOPHENYL)-2-METHYL-1-PHENYLPROPANE

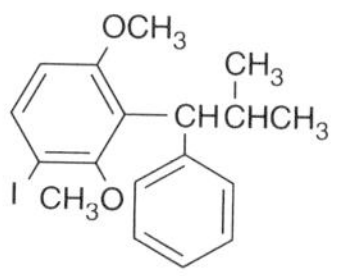

250 mg

S93,704-5
ETHYLENE GLYCOL (4-IODOPHENYL)METHYL ETHER METHYL ETHER

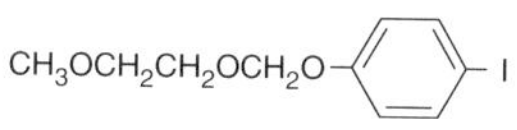

5 gm

S81,809-7
1,4-DIMETHOXY-2-IODO-5-METHYLBENZENE

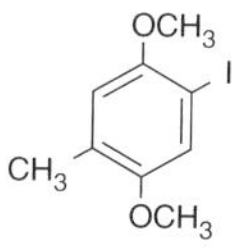

250 mg

S84,706-2
2,2'-DIIODO-4,4'-DIMETHOXY-3,3',5,5'-TETRAMETHYLBIPHENYL

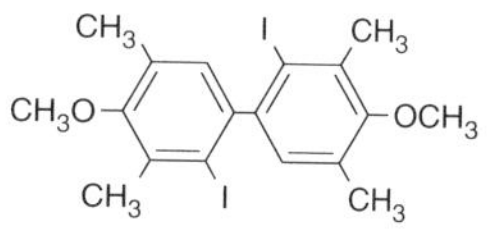

250 mg

S88,038-8
2-BROMOBENZYL PHENYL ETHER

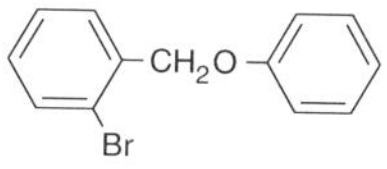

250 mg

S69,766-4
4-IODO-3-PHENYLANISOLE

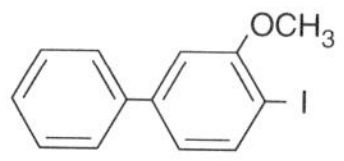

250 mg

S83,085-2
2-CHLOROPHENYL DODECYL ETHER

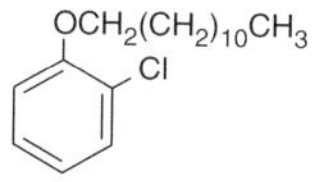

250 mg

S67,111-8
1-BROMO-4-ISOPROPOXYBENZENE

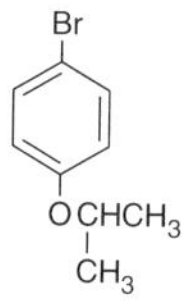

5 gm

S82,586-7
2,6-DIBROMOANISOLE

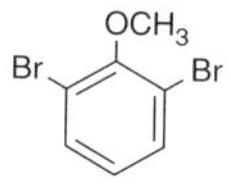

250 mg

S66,038-8
2-BROMO-4-METHYLANISOLE

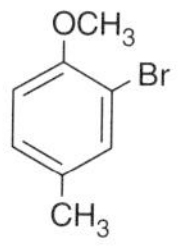

5 gm

S68,321-3
ALLYL 4-BROMO-2-CHLOROPHENYL ETHER

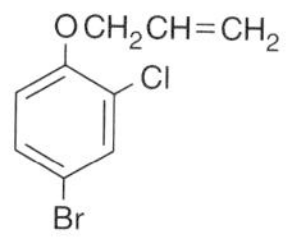

5 gm

S92,615-9
2-CHLORO-5-METHYLPHENYL METHALLYL ETHER

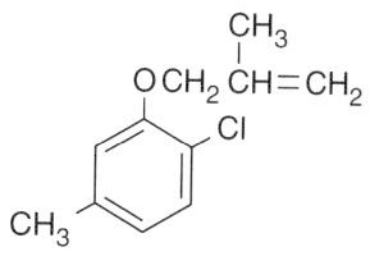

250 mg

S69,760-5
1-(2-BROMO-4-METHYLPHENOXY)-2-BUTENE (MIXTURE OF ISOMERS)

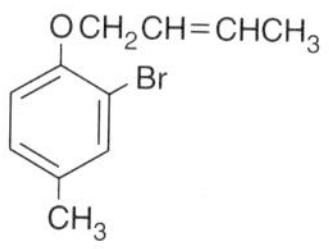

250 mg

S84,018-1
DODECYL 2,4,6-TRIBROMOPHENYL ETHER

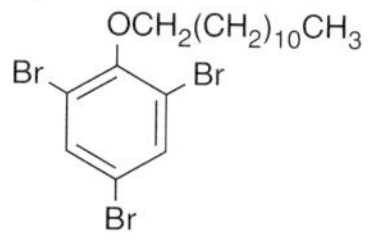

250 mg

S72,041-0
4-CHLORO-2-ISOPROPYL-5-METHYLANISOLE

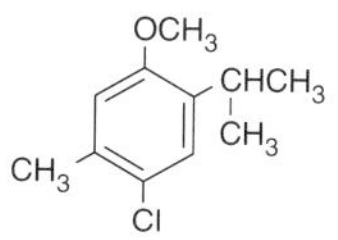

5 gm

S75,755-1
PHENYL-4,6-O-BENZYLIDENE-BETA-D-GLUCOPYRANOSIDE

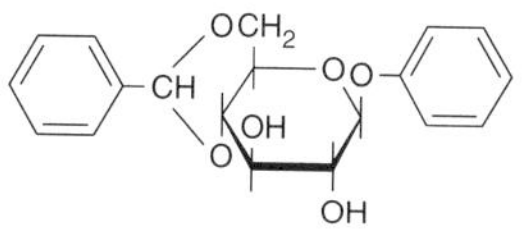

5 gm

S94,960-4
1-(4-METHOXYPHENYL)-2-PHENYLETHANOL

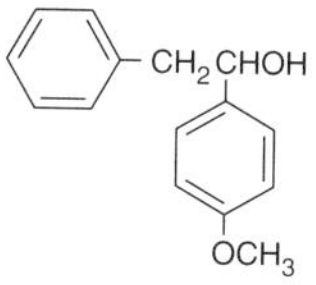

250 mg

S56,090-1
M-METHYLPHENETOLE

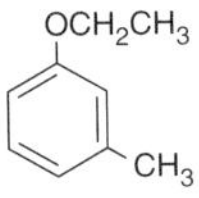

250 mg

S83,624-9
3-ISOPROPYLANISOLE

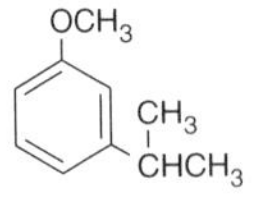

250 mg

S54,106-0
3-(1-BUTENYL)ANISOLE

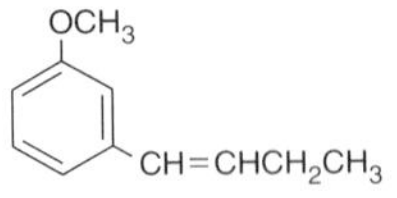

5 gm

S91,739-7
2-DEOXY-1-O-METHYL-3,5,6-TRI-O-BENZYL-ALPHA,BETA-D-GALACTOFURANOSIDE

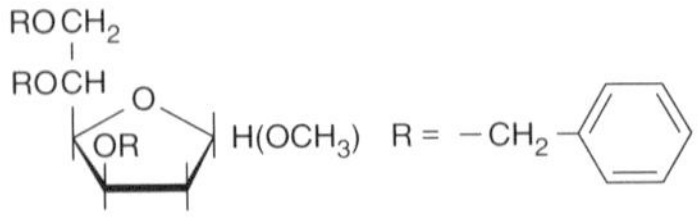

250 mg

S35,847-9
BENZYL 2-METHOXY-4-PROPENYLPHENYL ETHER

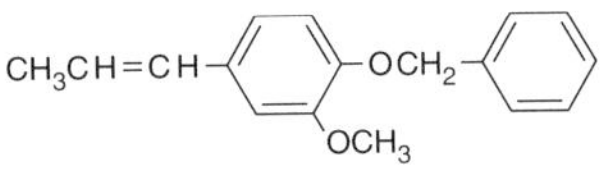

5 gm

S95,075-0
5-METHOXY-1,2,3,4,5-PENTAPHENYL-1,3-CYCLOPENTADIENE

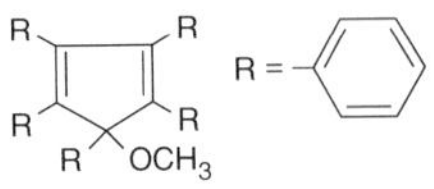

50 mg

S71,310-4
ALPHA,ALPHA-DIMETHYLBENZYL-METHYL ETHER

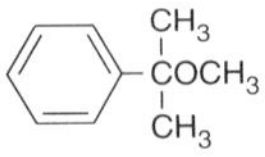

5 gm

S68,030-3
P-CRESYL BENZYL ETHER

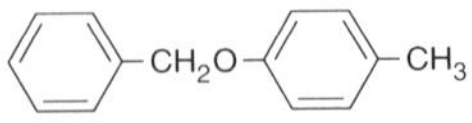

250 mg

S88,221-6
4-BENZYLOXY-BETA,BETA-DIBROMOSTYRENE

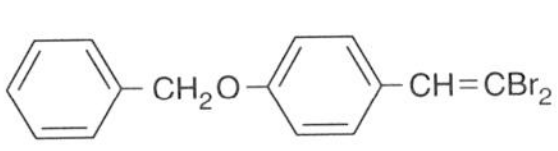

250 mg

S64,638-5
4-BROMOBENZYL PHENYL ETHER

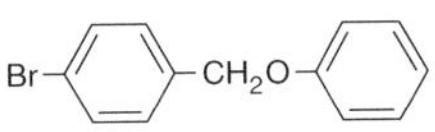

250 mg

S64,803-5
4-METHYLBENZYL PHENYL ETHER

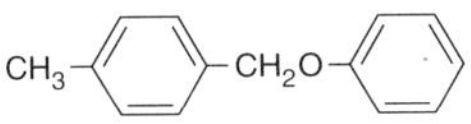

250 mg

S64,816-7
BENZYL 4-TERT-OCTYLPHENYL ETHER

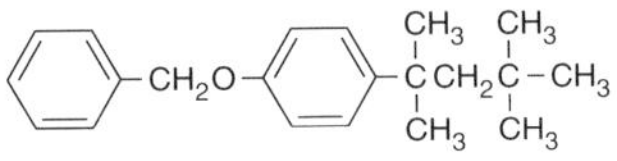

5 gm

S64,637-7
4-METHYLBENZYL 4-(TERT-OCTYL)PHENYL ETHER

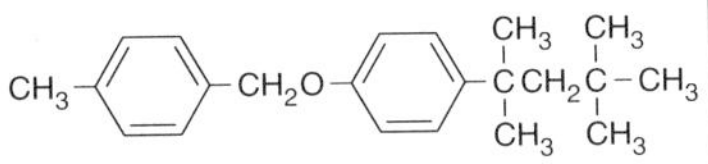

250 mg

S58,684-6
1,3-DIETHOXYBENZENE

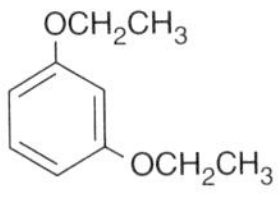

5 gm

S79,209-8
3,3'-ETHYLENE BIS(ANISOLE)

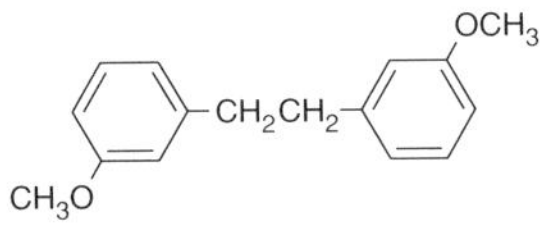

250 mg

S88,883-4
1,2,3,5-TETRAMETHOXYBENZENE

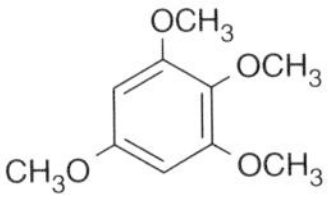

250 mg

S92,034-7
4-(3,4,5-TRIMETHOXYPHENYL)-MORPHOLINE

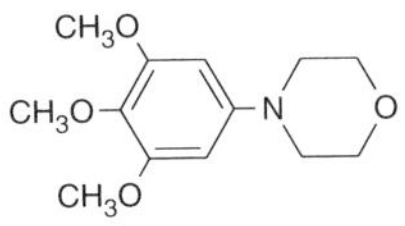

250 mg

S86,421-8
11,12-DICHLORO-9,10-DIHYDRO-2,3-DIMETHOXY-9,10-ETHANOANTHRACENE

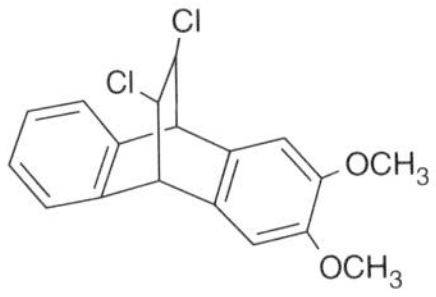

250 mg

S34,078-2
2-ALLYLOXYANISOLE

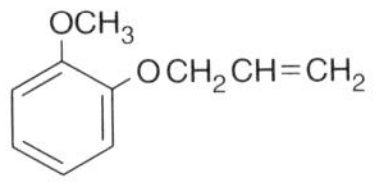

5 gm

S43,021-8
1,2-DIETHOXYBENZENE

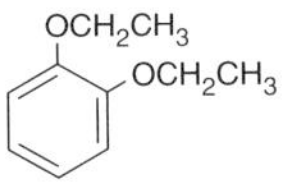

5 gm

S67,976-3
EUGENYL ETHYL ETHER

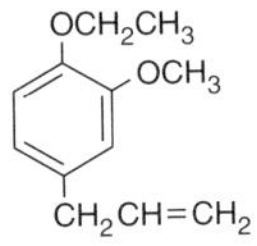

5 gm

S94,185-9
(2,5-DIMETHYLPHENOXY)ACETALDEHYDE DIMETHYL ACETAL

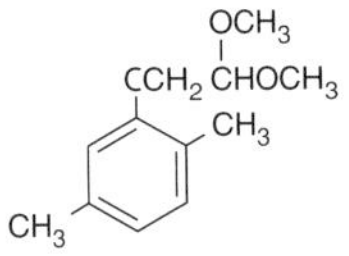

5 gm

S89,386-2
3,3-BIS(3,4-DIMETHOXYPHENYL)-1,1,2-TRICHLORO-1-PROPENE

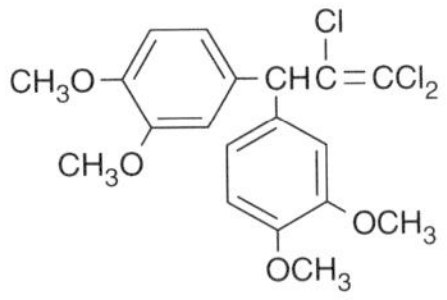

250 mg

S84,653-8
3,3',4,4'-TETRAMETHOXYBIPHENYL

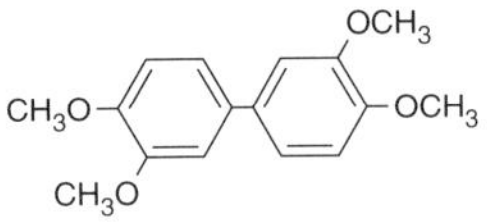

250 mg

S84,700-3
3,3'-DIMETHYL-4,4',5,5'-TETRAMETHOXYBIPHENYL

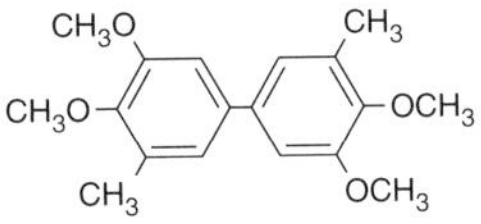

250 mg

S85,207-4
11,12-DICHLORO-9,10-DIHYDRO-2,3-DIMETHOXY-9,10-ETHENOANTHRACENE

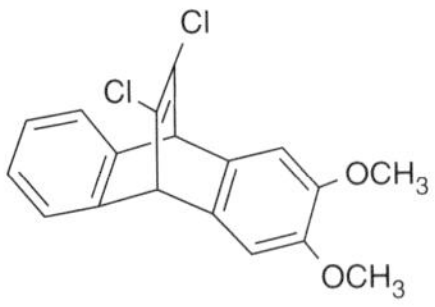

250 mg

S84,660-0
6,7-DIMETHOXYTETRALIN

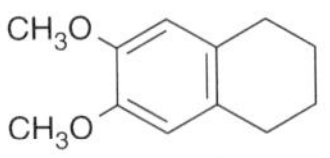

250 mg

S85,245-7
2,3-DIMETHOXYANTHRACENE

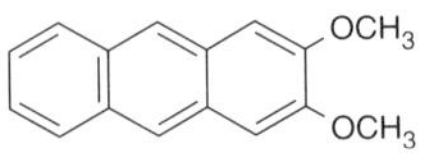

250 mg

S84,656-2
9,10-DIMETHYL-2,3,6,7-TETRAMETHOXYPHENANTHRENE

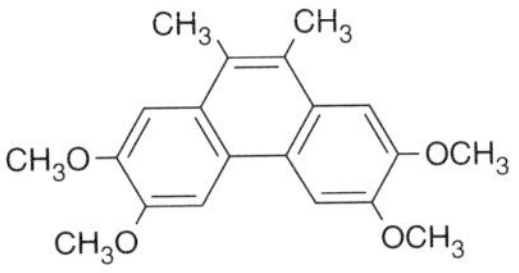

250 mg

S79,321-3
1,3-DIPHENOXYPROPANE

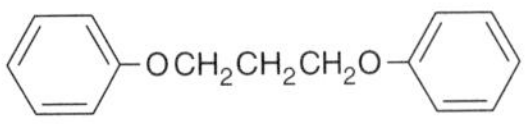

250 mg

S80,883-0
1,3-DIALLYLOXY-4-ETHYLBENZENE

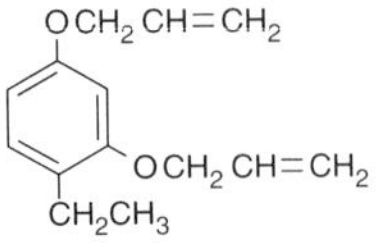

250 mg

S84,676-7
1,2-BIS(3,5-DIMETHOXYPHENYL)ETHANE

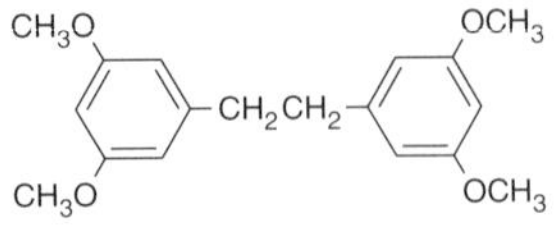

250 mg

S71,445-3
1,4-DIPHENOXYBUTANE

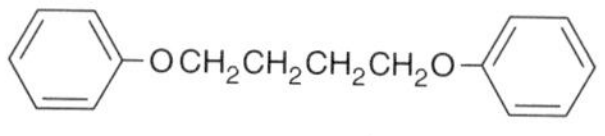

250 mg

S83,756-3
2,2'-DIMETHOXYBIPHENYL

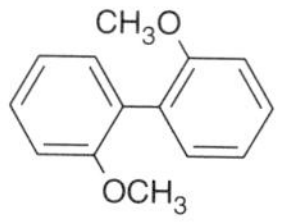

250 mg

S76,662-3
1,4-BIS(OCTYLOXY)BENZENE

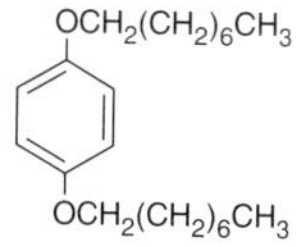

250 mg

S90,708-1
HEXAKIS(4-METHOXYPHENOXYMETHYL)-BENZENE

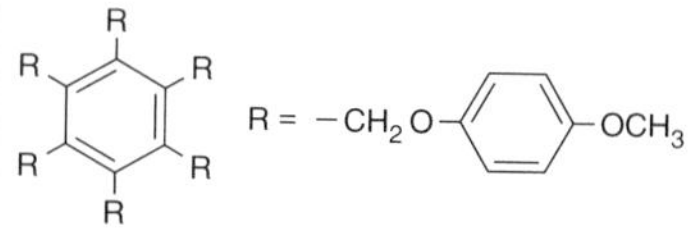

250 mg

S88,711-0
HEXAKIS(4-ISOPROPYLPHENOXYMETHYL)-BENZENE

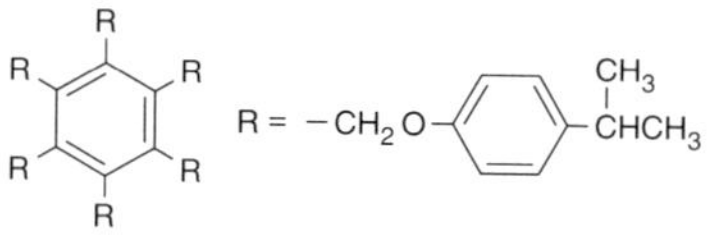

250 mg

S90,717-0
HEXAKIS(4-TERT-BUTYLPHENOXYMETHYL)-BENZENE

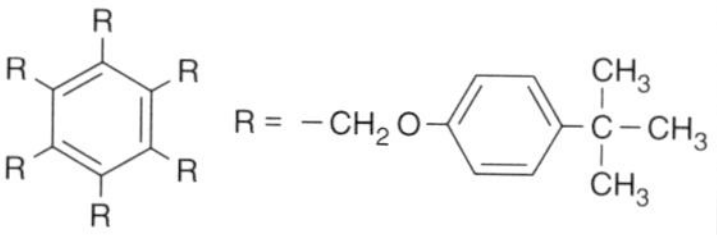

250 mg

S85,028-4
2,5-DIMETHOXYTOLUENE

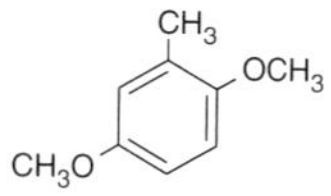

250 mg

S84,682-1
1,4-DIMETHOXY-2,3-DIMETHYLBENZENE

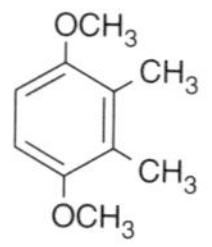

250 mg

S85,046-2
ALPHA-CYCLOPENTYL-4,4'-DIMETHOXYSTILBENE

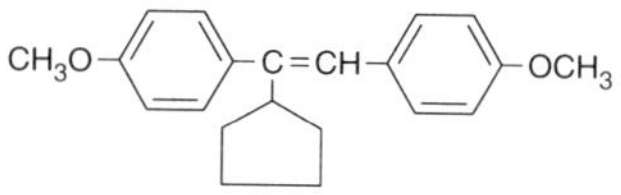

250 mg

S88,515-0
1,4-BIS(4-METHOXYPHENYL)-4-METHYL-1-CYCLOHEXENE

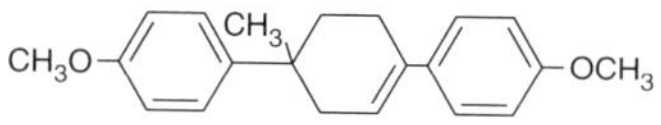

250 mg

S79,329-9
1,6-DIPHENOXYHEXANE

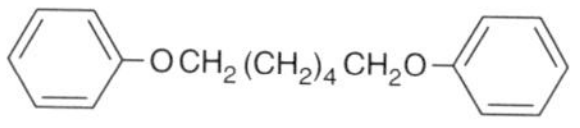

250 mg

S56,368-4
1,4-DIETHOXYBENZENE

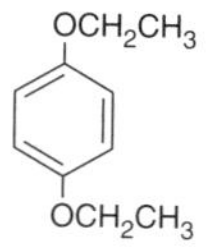

5 gm

S84,325-3
BIS(P-TOLYLOXY)METHANE

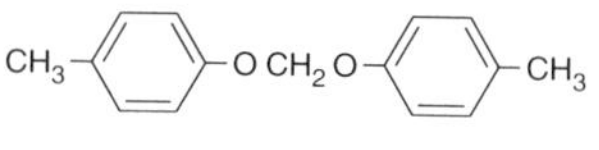

250 mg

S94,222-7
ALPHA,ALPHA'-DIMETHYL-4,4'-DIMETHOXYSTILBENE

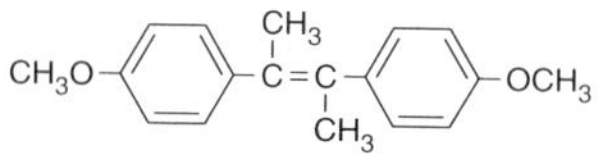

5 gm

S88,228-3
1,4-BIS(4-METHOXYPHENYL)-2-METHYL-1,3-BUTADIENE

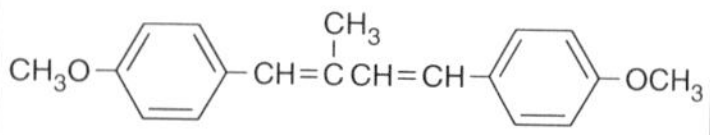

250 mg

S94,323-1
1,6-BIS(4-METHOXYPHENYL)HEXANE

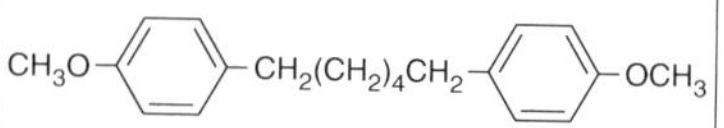

250 mg

S51,237-0
1-HEXADECYLOXY-4-METHOXYBENZENE

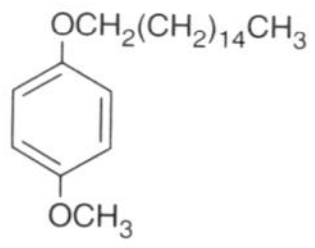

250 mg

S68,018-4
TRI-P-ANISYLMETHANE

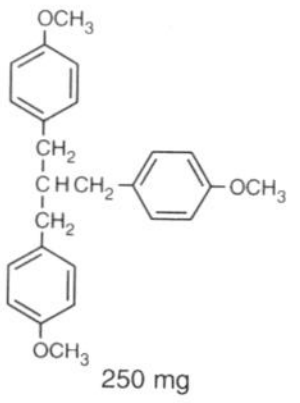

250 mg

S91,991-8
1,3,5-TRIS(4-METHOXYPHENYL)BENZENE

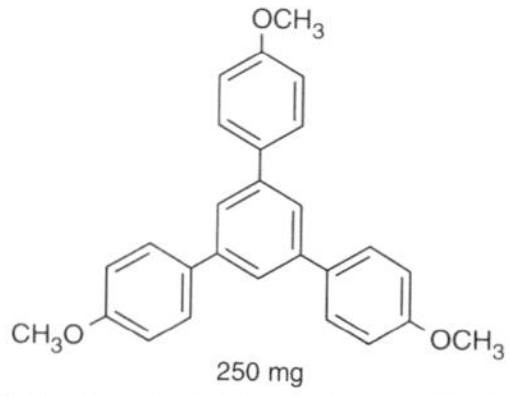

250 mg

S89,385-4
3,3-BIS(4-METHOXYPHENYL)-1,1,2-TRICHLORO-1-PROPENE

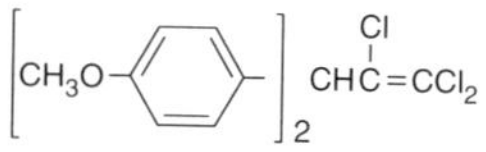

250 mg

S92,669-8
ISOEUGENYL ETHYL ETHER

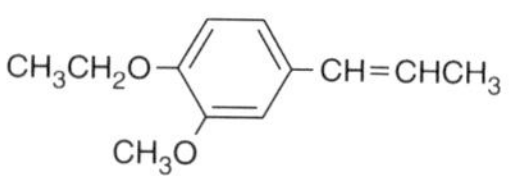

250 mg

S84,696-1
2,2',5,5'-TETRAMETHOXY-3,3',4,4'-TETRAMETHYLBIPHENYL

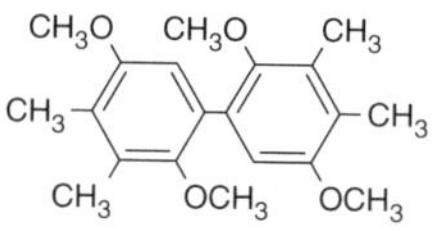

250 mg

S57,552-6
1,2-BIS(M-TOLYLOXY)ETHANE

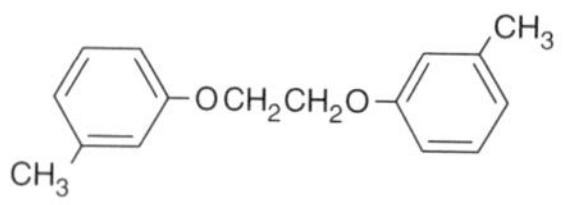

5 gm

S57,459-7
1,2-BIS(2-CHLOROPHENOXY)ETHANE

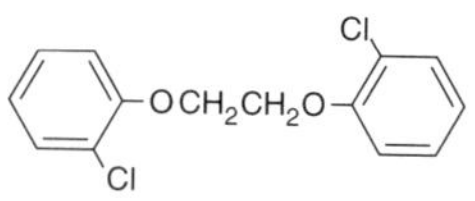

5 gm

S57,586-0
1,2-BIS(O-TOLYLOXY)ETHANE

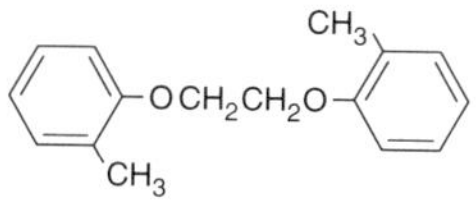

5 gm

S57,537-2
2,2'-(ETHYLENEDIOXY)DIANISOLE

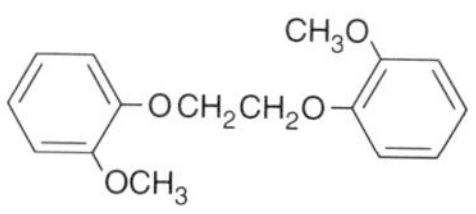

5 gm

S57,469-4
1,2-BIS(4-CHLOROPHENOXY)ETHANE

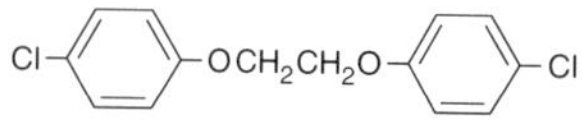

5 gm

S57,489-9
1,2-DI-P-TOLYLOXYETHANE

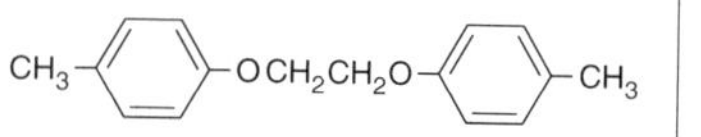

5 gm

S57,462-7
4,4'-(ETHYLENEDIOXY)DIANISOLE

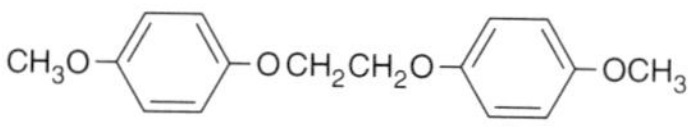

5 gm

S43,369-1
1,2,4-TRIETHOXYBENZENE

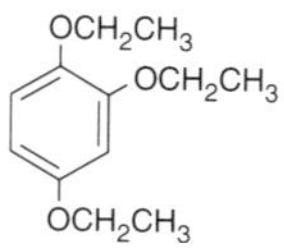

5 gm

S53,442-0
5,5'-DIBROMO-4,4'-BIVERATROLE

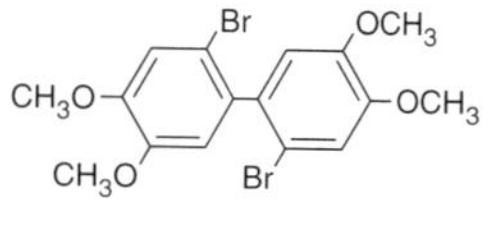

250 mg

S85,044-6
ALPHA-CYCLOPENTYL-4-METHOXYSTILBENE

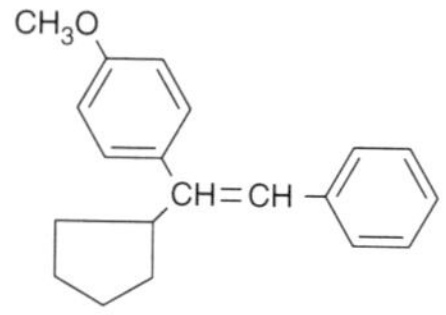

250 mg

S88,000-0
5-(4-METHOXYBENZYLIDENE)-1,2,3,4-TETRAPHENYL-1,3-CYCLOPENTADIENE
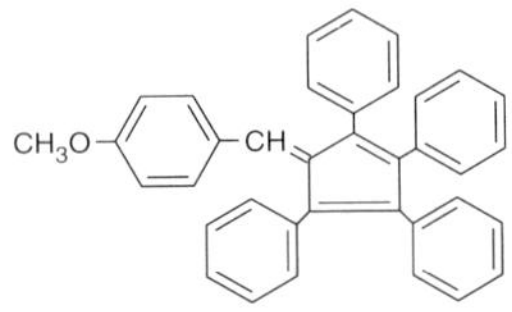

250 mg

S88,543-6
7-METHOXY-1,2,3,4,4A,9,10,10A-OCTAHYDRO-1,1,4A-TRIMETHYLPHENANTHRENE
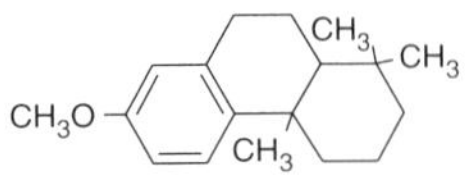

250 mg

S85,045-4
ALPHA-CYCLOHEXYL-4-ETHOXYSTILBENE
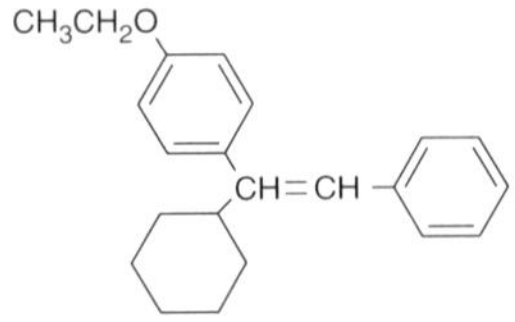

250 mg

S51,853-0
4-CYCLOHEXYLANISOLE
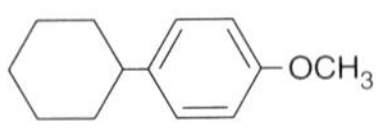

250 mg

S88,563-0
8-METHOXY-1,2,3,4,4A,9,10,10A-OCTAHYDRO-1,1,4A-TRIMETHYLPHENANTHRENE
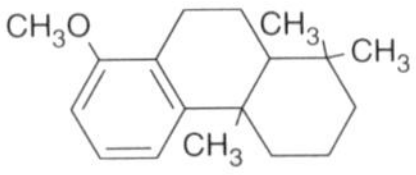

250 mg

S88,514-2
4-(4-METHOXYPHENYL)-4-METHYL-1-PHENYL-1-CYCLOHEXENE
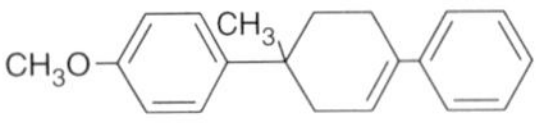

250 mg

S85,307-0
3-(2-METHOXYPHENYL)-1-CYCLOHEXENE
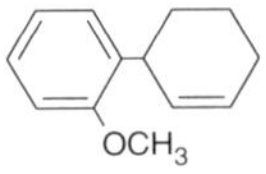

250 mg

S35,232-2
1-(4-METHOXYPHENYL)-1-CYCLOHEXENE
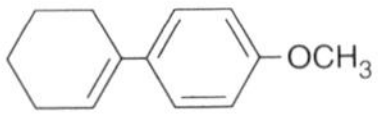

5 gm

S86,415-3
11,12-DICHLORO-9,10-DIHYDRO-2-METHOXY-9,10-ETHANOANTHRACENE
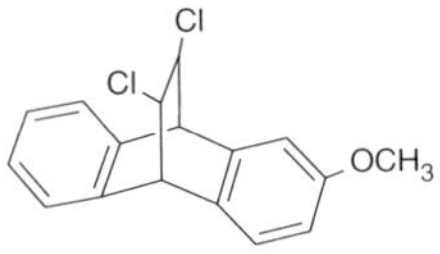

250 mg

S76,783-2
4-PHENOXY-2-BUTENE
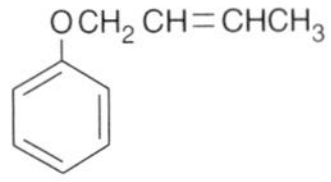

250 mg

S56,337-4
PENTYL PHENYL ETHER
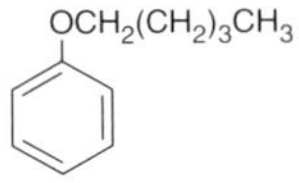

5 gm

S34,794-9
DODECYL PHENYL ETHER
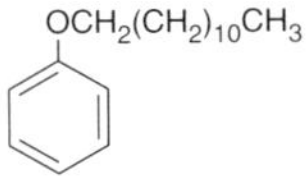

5 gm

S38,686-3
HEXADECYL PHENYL ETHER
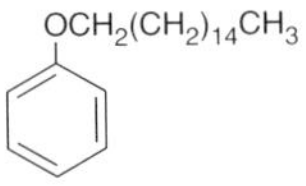

250 mg

S54,294-6
OCTADECYL PHENYL ETHER
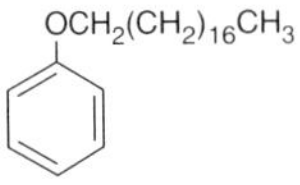

250 mg

S94,391-6
2-METHOXYSTILBENE
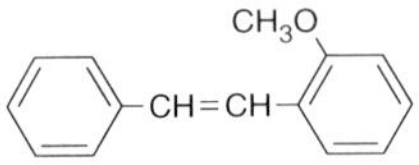

250 mg

S69,007-4
CIS-4-METHOXYSTILBENE
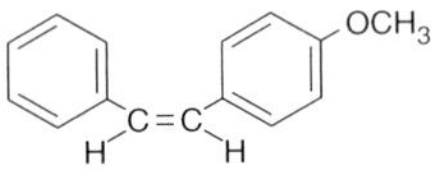

250 mg

S72,236-7
1-(4-METHOXYPHENYL)-1-PHENYLETHYLENE
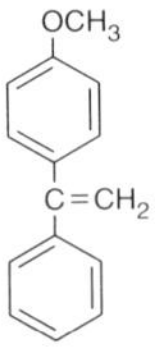

5 gm

S88,015-9
4-METHOXYSTILBENE
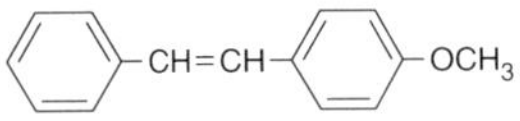

250 mg

S83,688-5
(4-BENZYLPHENYL) METHYL ETHER
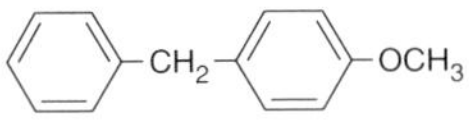

250 mg

S83,687-7
CHLORO(4-METHOXYPHENYL)-PHENYLMETHANE
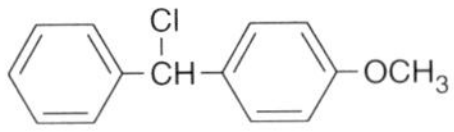

250 mg

S86,536-2
4-METHOXY-ALPHA-PHENYLSTYRENE
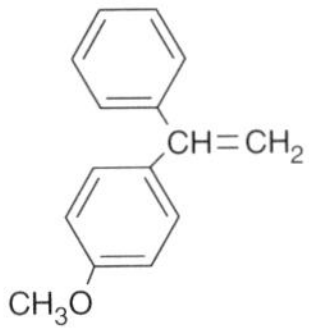

250 mg

S85,015-2
4-METHOXYTRIPHENYLMETHANE
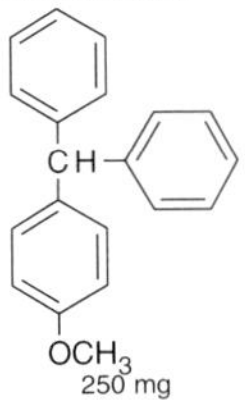

250 mg

S38,984-6
4,4'-BENZYLIDENEDIANISOLE
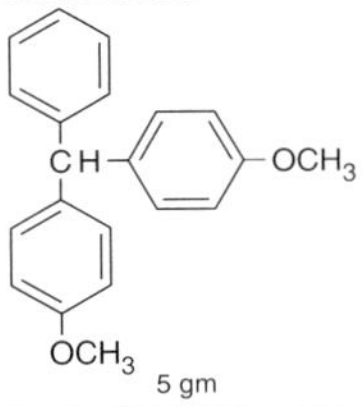

5 gm

S34,239-4
2,4-BIS(1-PHENYLETHYL)ANISOLE
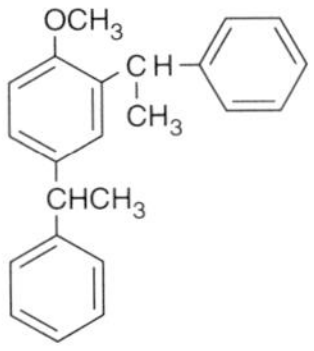

5 gm

S91,989-6
METHYL TRIPHENYLPHENYL ETHER
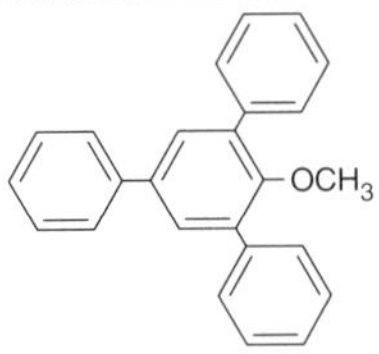

250 mg

S69,686-2
3-METHOXYBIPHENYL
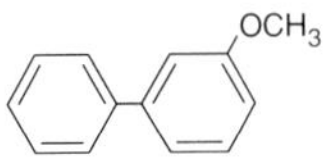

250 mg

S88,190-2
6-METHOXY-1,2,3,4-TETRAPHENYLNAPHTHALENE
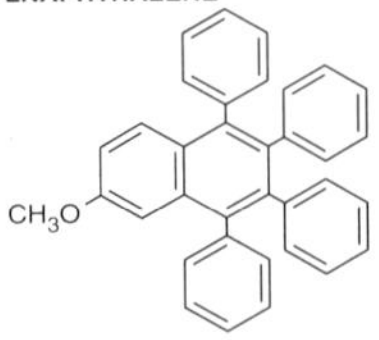

250 mg

S92,062-2
4-METHOXY-9-PHENYLPHENANTHRENE
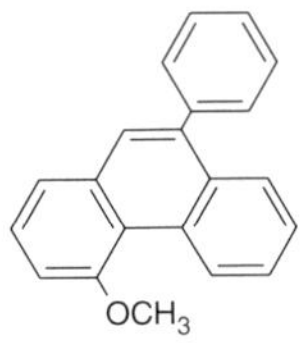

250 mg

S36,208-5
2-FLUOROPHENETOLE
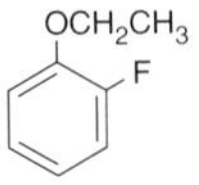

5 gm

S54,474-4
ALLYL 2-CHLOROPHENYL ETHER
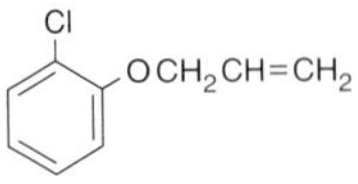

5 gm

S43,291-1
ALLYL O-TOLYL ETHER
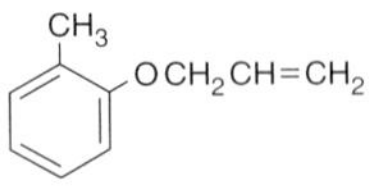

5 gm

S72,216-2
2-METHYLALLYL O-TOLYL ETHER
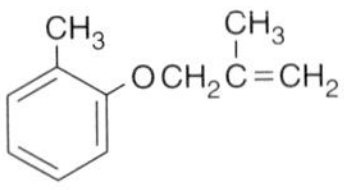

5 gm

S51,847-6
BUTYL O-TOLYL ETHER
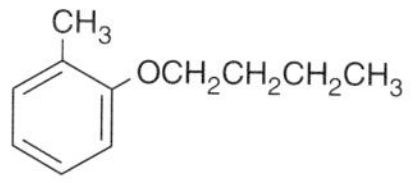

5 gm

S60,845-9
1,4-BIS(2-CHLOROPHENOXY)-2-BUTENE
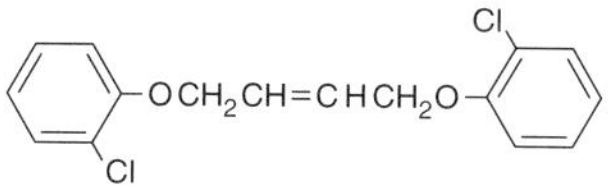

5 gm

S60,848-3
1,4-BIS(O-TOLYLOXY)-TRANS-2-BUTENE
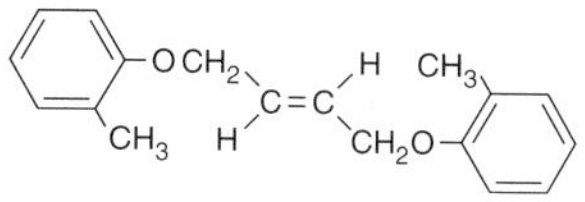

5 gm

S36,209-3
4-FLUOROPHENETOLE
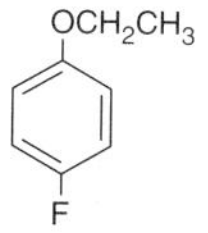

5 gm

S43,304-7
ALLYL P-TOLYL ETHER
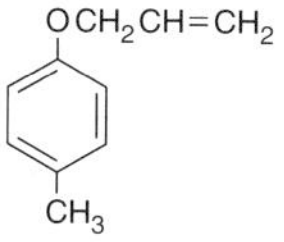

5 gm

S35,500-3
4-PROPYLANISOLE
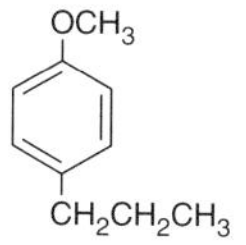

5 gm

S35,910-6
4-TERT-BUTYLANISOLE
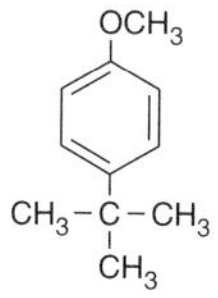

5 gm

S52,738-6
4-(TERT-BUTYL)PHENETOLE
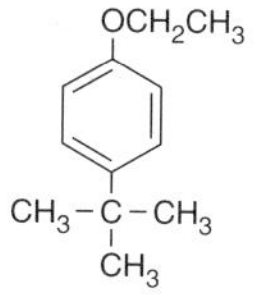

5 gm

S56,339-0
4-(TERT-PENTYL)ANISOLE
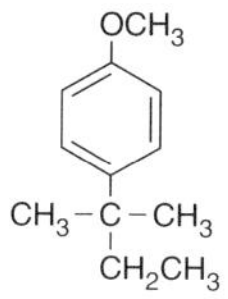

5 gm

S35,850-9
4-BENZYLOXYANISOLE
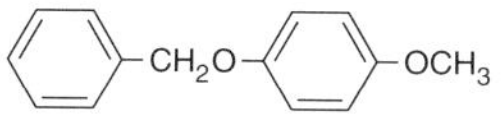

5 gm

S56,439-7
1,4-DIBUTOXYBENZENE
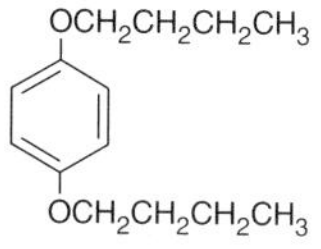

5 gm

S66,185-6
BIS(4-METHOXYPHENYL)-
IODONIUM BROMIDE
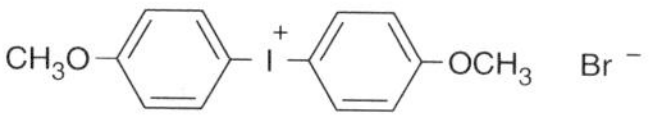

5 gm

S75,694-6
2-(4-METHOXYPHENYL)-6-METHYL-2,5-
HEPTADIENE
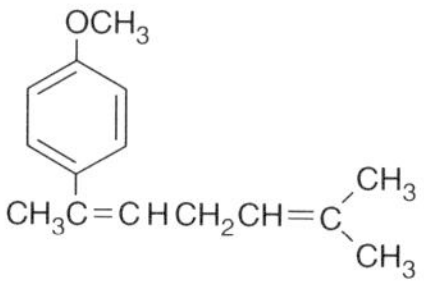

5 gm

S56,342-0
BUTYL 4-(TERT-PENTYL)PHENYL ETHER
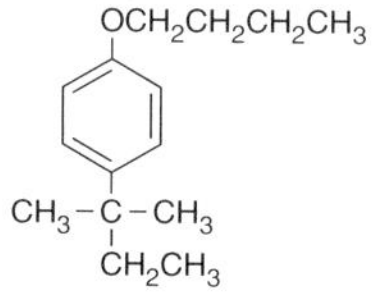

5 gm

S64,828-0
4-(TERT-OCTYL)ANISOLE
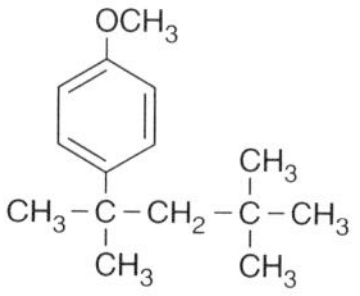

5 gm

S60,826-2
1,4-BIS(4-CHLOROPHENOXY)-2-BUTENE
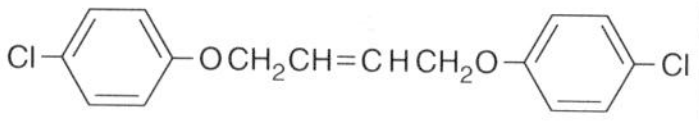

250 mg

S64,827-2
ALLYL 4-TERT-OCTYLPHENYL ETHER
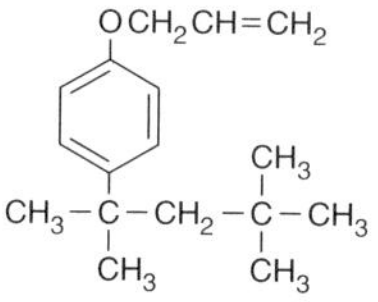

5 gm

S87,786-7
1-(4-METHOXYPHENYL)-4-(P-TOLYL)-
1,3-BUTADIENE
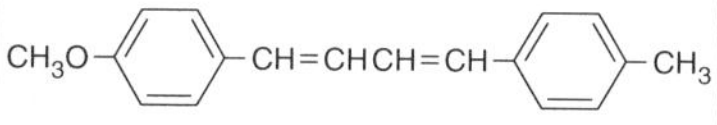

250 mg

S39,627-3
1,4-BIS(4-METHOXYPHENYL)-
1,3-BUTADIENE
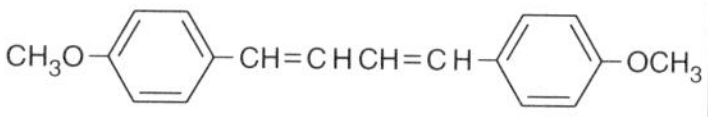

250 mg

S60,843-2
1,4-BIS(P-TOLYLOXY)-2-BUTENE
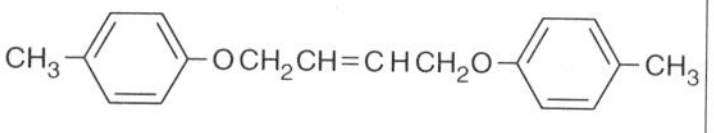

250 mg

S34,796-5
DODECYL P-TOLYL ETHER
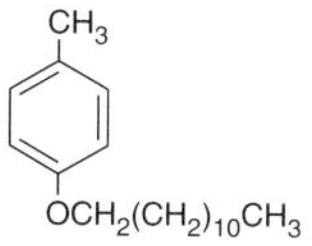

5 gm

S34,615-2
DIETHYLSTILBESTROL DIMETHYL ETHER
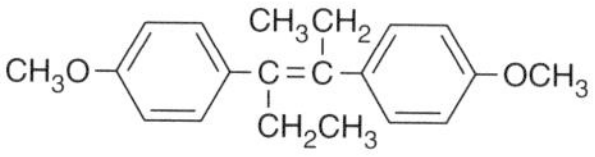

5 gm

S94,220-0
1,1-BIS(4-METHOXYPHENYL)-2-
ETHYL-1-BUTENE
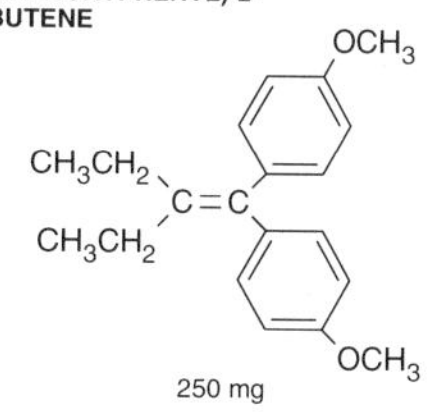

250 mg

S34,311-0
4-TERT-BUTYLPHENYL DODECYL ETHER
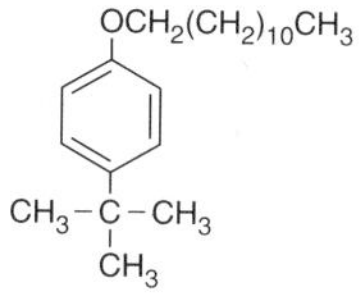

5 gm

S39,315-0
4-CHLOROPHENYL HEXADECYL ETHER
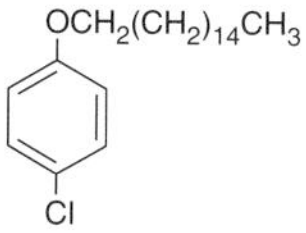

250 mg

S50,203-0
HEXADECYL P-TOLYL ETHER
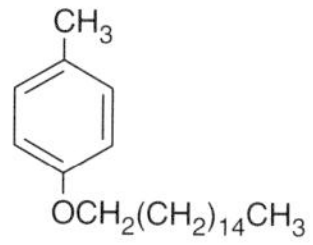

250 mg

S56,228-9
2,4-BIS[4-(HEXADECYLOXY)PHENYL]-
4-METHYL-1-PENTENE
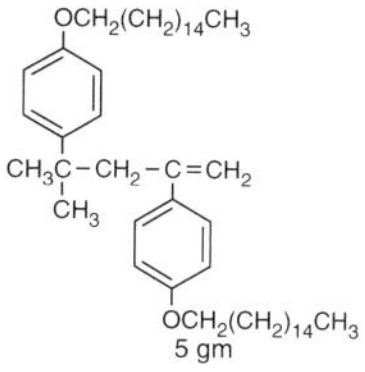

5 gm

S67,171-1
1,1-BIS(4-METHOXYPHENYL)ETHYLENE
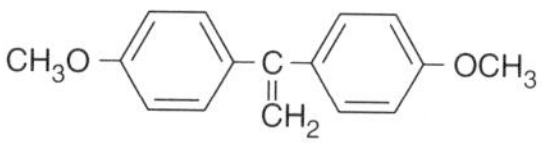

250 mg

S64,623-7
4,4'-DIETHOXYBIPHENYL
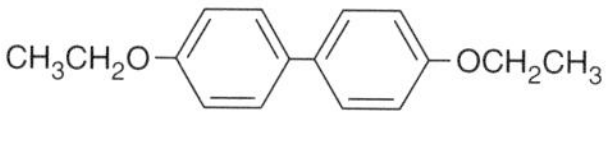

250 mg

S80,532-7
1-BROMO-2-(4-METHOXYPHENYL)INDAN
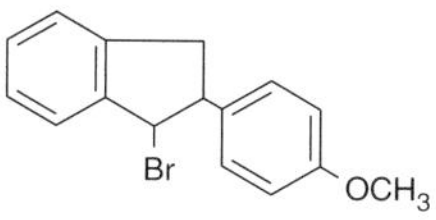

250 mg

S34,588-1
2,6-DIETHOXYTOLUENE
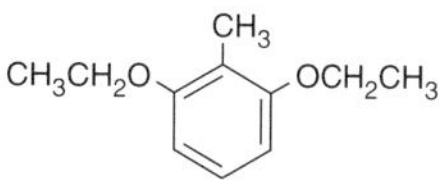

5 gm

S34,545-8
2,6-DIALLYLOXYTOLUENE
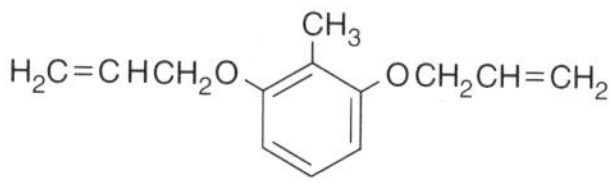

5 gm

S34,780-9
2,6-DIPROPOXYTOLUENE

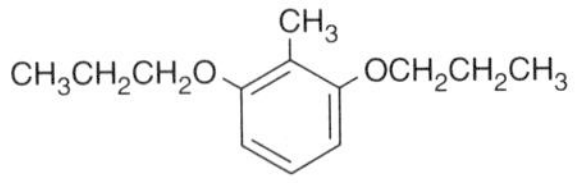

5 gm

S51,428-4
4-CHLORO-3-METHYLANISOLE

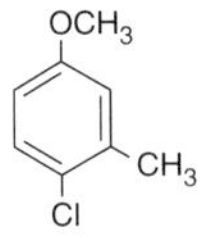

250 mg

S55,075-2
3-CHLORO-4-METHYLANISOLE

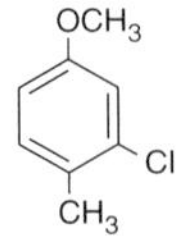

250 mg

S35,738-3
ALLYL 2,4-DICHLOROPHENYL ETHER

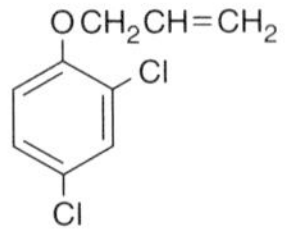

5 gm

S35,942-4
2-CHLORO-1,4-DIETHOXYBENZENE

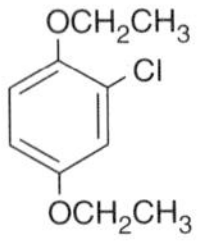

5 gm

S51,342-3
1-CHLORO-2,5-DIPENTYLOXYBENZENE

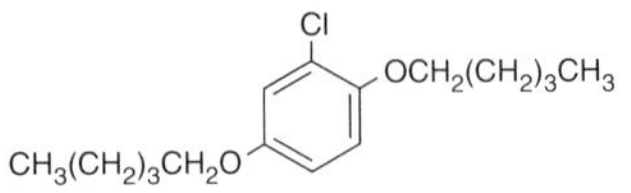

250 mg

S60,844-0
1,4-BIS(2,4-DICHLOROPHENOXY)-2-BUTENE

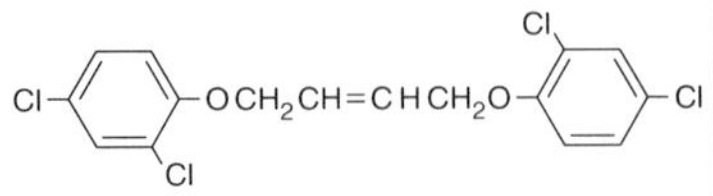

5 gm

S60,841-6
1,4-BIS(4-CHLORO-3-METHYLPHENOXY)-TRANS-2-BUTENE

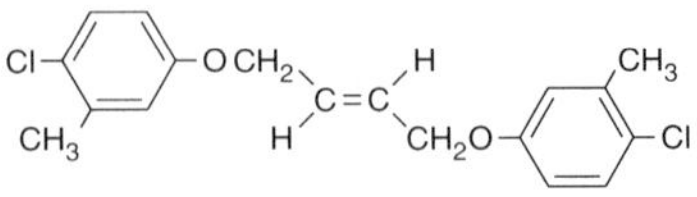

5 gm

S60,840-8
1,4-BIS(2,5-DIMETHYLPHENOXY)-TRANS-2-BUTENE

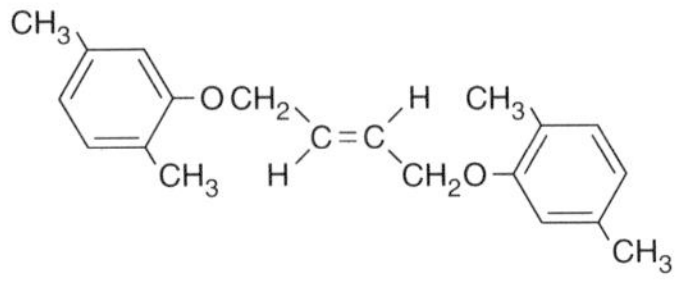

250 mg

S79,826-6
ETHYL 3,5-XYLYL ETHER

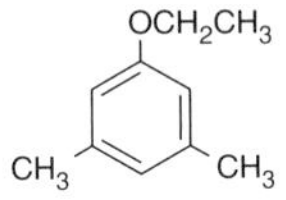

250 mg

S54,045-5
1-CHLORO-2,4,6-TRIMETHOXYBENZENE

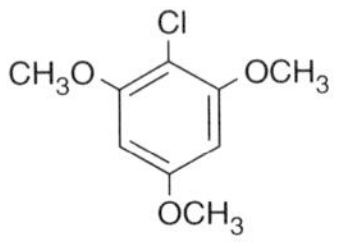

5 gm

S51,964-2
ALLYL 2,4,6-TRIBROMOPHENYL ETHER

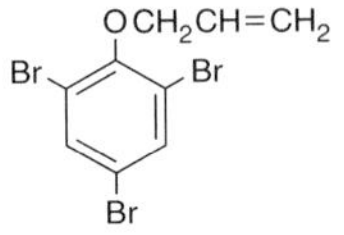

5 gm

S43,368-3
1-ETHYL-2,4,6-TRIMETHOXYBENZENE

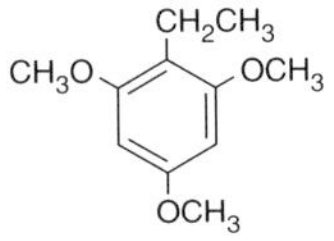

5 gm

S66,320-4
1-ALLYLOXY-2,4-DIMETHYL-6-PROPENYLBENZENE

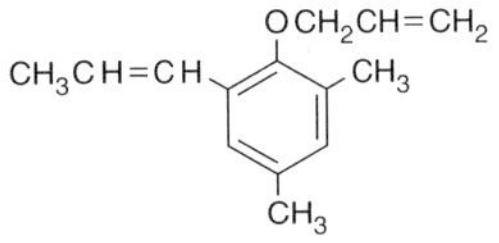

5 gm

S79,347-7
2,6-DI-TERT-BUTYL-4-METHYLANISOLE

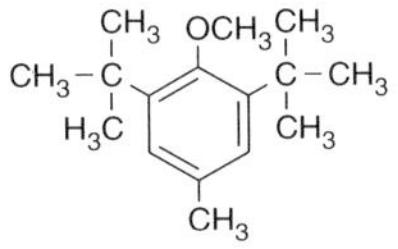

250 mg

S43,367-5
2,4,6-TRIMETHOXYDIPHENYLMETHANE

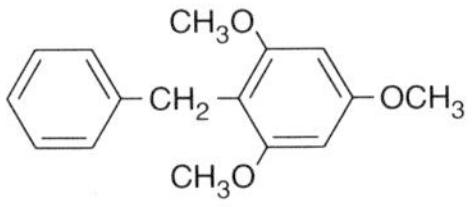

5 gm

S51,967-7
2,2-BIS(4-ALLYLOXY-3,5-DIBROMOPHENYL)PROPANE

5 gm

S36,019-8
2,4-DIBROMO-5-METHYLANISOLE

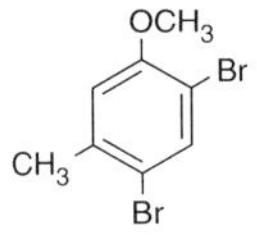

5 gm

S54,835-9
1,4-DIBROMO-2,5-DIMETHOXYBENZENE

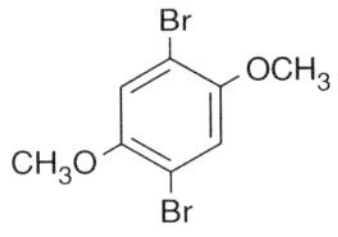

250 mg

S35,946-7
5-CHLORO-2,4-DIMETHOXYIODOBENZENE

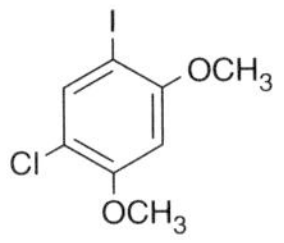

250 mg

S66,106-6
2,5-DI-TERT-BUTYL-1,4-DIMETHOXYBENZENE

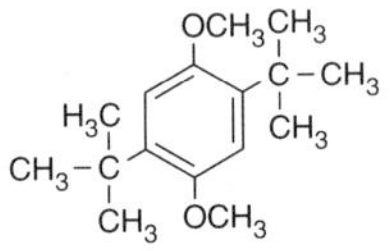

5 gm

S94,799-7
6-METHOXY-1,2,3,4-TETRAHYDRO-1,1,4,4-TETRAMETHYLNAPHTHALENE

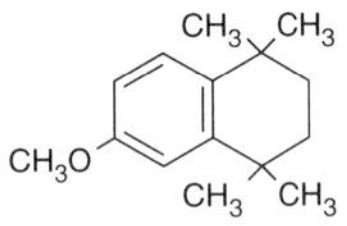

250 mg

S92,313-3
1-BROMO-5-METHOXYNAPHTHALENE

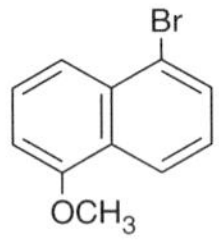

250 mg

S59,222-6
DICHLORODIPHENOXYMETHANE

250 mg

S52,821-8
ALPHA-PENTYLCINNAMALDEHYDE DIETHYL ACETAL

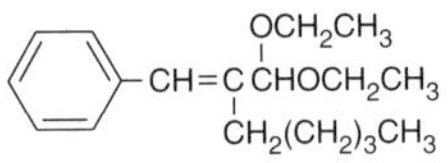

5 gm

S82,513-1
2-BROMOPHENYL ETHER

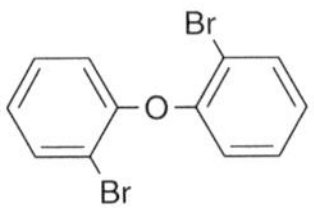

250 mg

S68,317-5
4-BROMO-4'-FLUORODIPHENYL ETHER

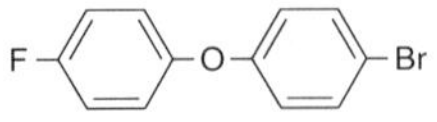

5 gm

S68,213-6
4-BROMOPHENYL 2,4-DI-TERT-BUTYLPHENYL ETHER

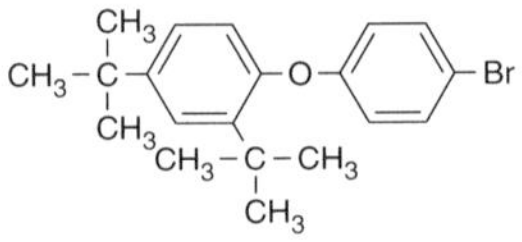

250 mg

S90,700-6
HEXAKIS(PHENOXY)BENZENE

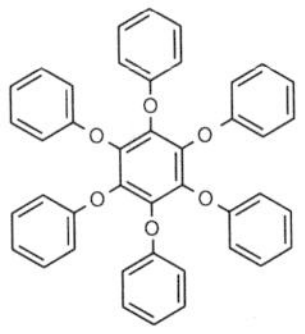

250 mg

S94,784-9
5,7-DIMETHOXY-1,2,3,4-TETRAHYDRO-1,1,4,4-TETRAMETHYLNAPHTHALENE

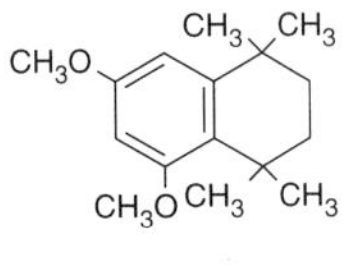

5 gm

S36,325-1
2-METHOXYPHENYL PHENYL ETHER

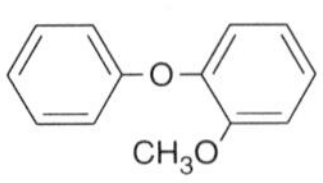

5 gm

S83,233-2
PHENYL O-TOLYL ETHER

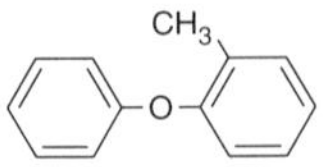

250 mg

S82,973-0
2-BIPHENYLYL 4-BIPHENYLYL ETHER
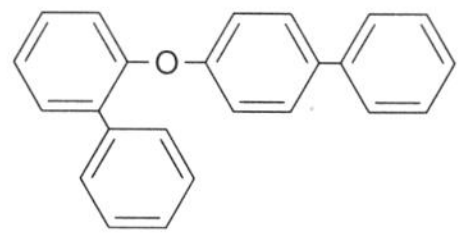
250 mg

S82,972-2
2-BIPHENYLYL PHENYL ETHER
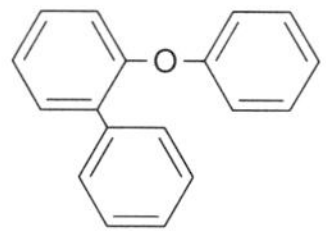
250 mg

S55,124-4
4-FLUOROPHENYL ETHER
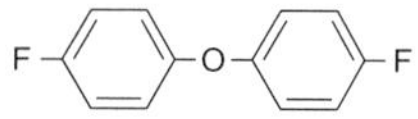
5 gm

S54,689-5
PHENYL 4-VINYLPHENYL ETHER
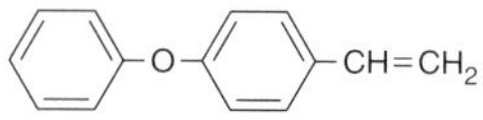
5 gm

S52,737-8
4-(TERT-BUTYL)PHENYL PHENYL ETHER
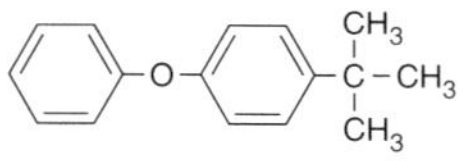
5 gm

S36,446-0
4-PHENOXYBIPHENYL
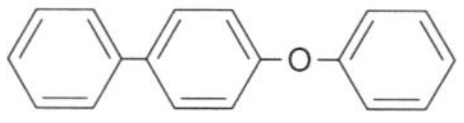
5 gm

S94,736-9
9,10-BIS(4-METHOXYPHENYL)-1-CHLOROANTHRACENE
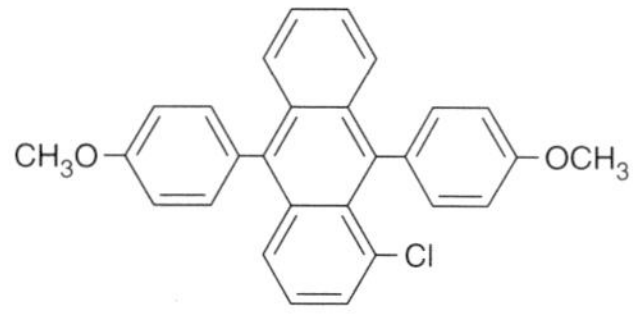
250 mg

S53,552-4
2-BROMOMETHYL-1,4-BENZODIOXANE
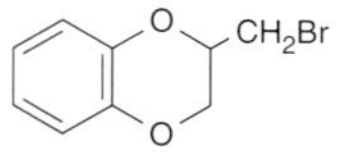
250 mg

S41,807-2
2-BROMOMETHYL-5-METHOXYCOUMARAN
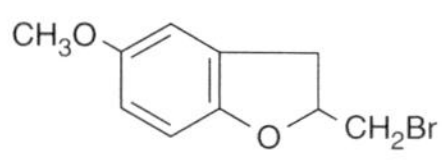
250 mg

S52,623-1
2-(3-CHLOROPROPYL)-2-(4-FLUOROPHENYL)-1,3-DIOXOLANE
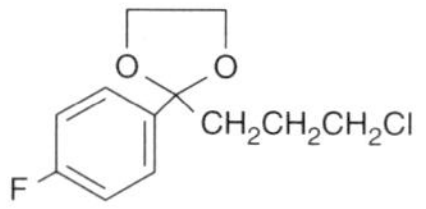
250 mg

S41,806-4
2-BROMOMETHYL-5-FLUOROCOUMARAN
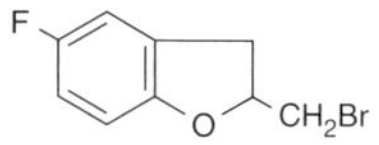
250 mg

S41,805-6
2-BROMOMETHYL-5-METHYLCOUMARAN
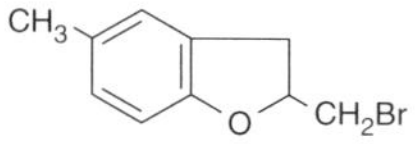
250 mg

S89,820-1
3,9-BIS(TRICHLOROMETHYL)-2,4,8,10-TETRAOXASPIRO(5.5)UNDECANE
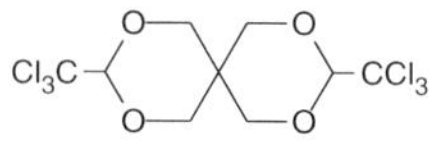
1 gm

S80,627-7
2-(4-BIPHENYLYL)-4-(4-CHLOROPHENOXYMETHYL)-1,3-DIOXOLANE
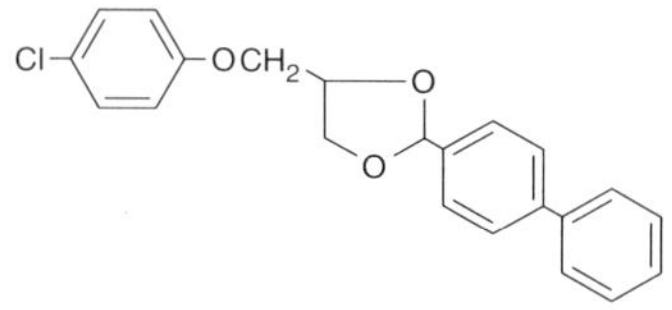
250 mg

S83,605-2
2-(3-BROMOPHENOXY)TETRAHYDROPYRAN
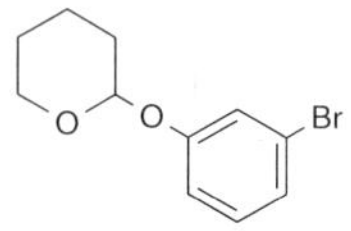
250 mg

S83,533-1
2-(2-BROMOPHENOXY)TETRAHYDROPYRAN
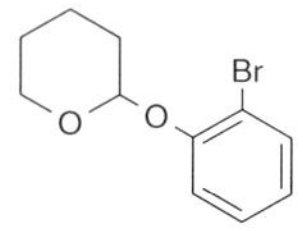
250 mg

S79,867-3
2-(4-BROMOPHENOXY)TETRAHYDROPYRAN
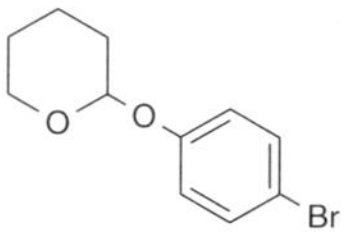
250 mg

S88,299-2
2-(2,4-DICHLOROPHENYL)-5,5-DIMETHYL-2-PHENYL-1,3-DIOXANE
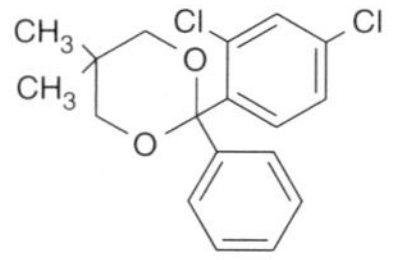
1 gm

S43,219-9
2-(4-FLUOROPHENYL)-1,3-DIOXOLANE
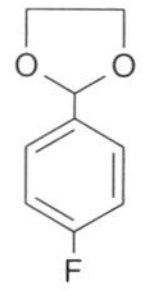
250 mg

S82,767-3
2-(4-BROMOPHENYL)-2-METHYL-1,3-DIOXOLANE
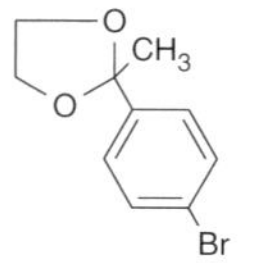
250 mg

S54,770-0
9,10-DIHYDRO-9-HYDROXY-9,10-ETHANOANTHRACENE-11,12-DICARBOXYLIC ANHYDRIDE
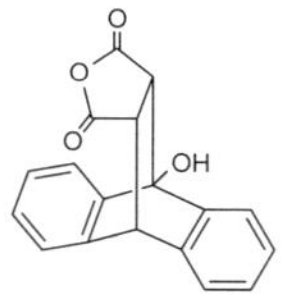
250 mg

S77,320-4
3,4-DIHYDRO-4-(4-HYDROXYPHENYL)-2,2,4-TRIMETHYL-2H-1-BENZOPYRAN
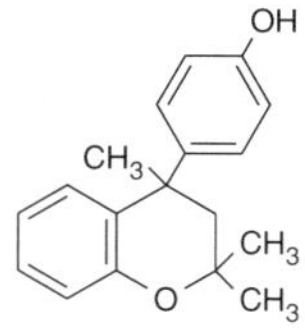
250 mg

S86,035-2
6,7-DIHYDROCYCLOPENTA(1,2-B:1,5-B')-BIS(1)BENZOPYRYLIUM TETRACHLOROFERRATE
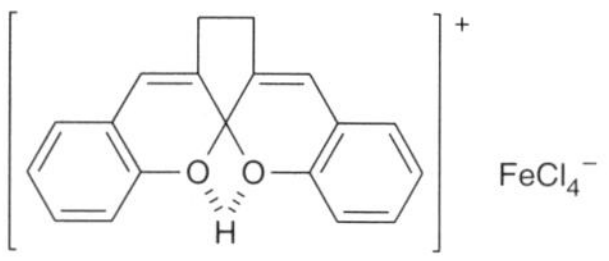
250 mg

S91,965-9
2-ETHOXY-2-PHENYL-1,3-BENZODIOXOLE
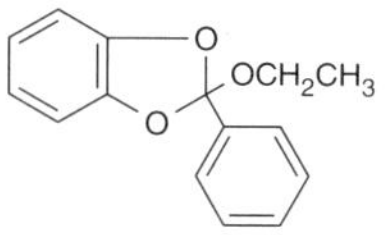
250 mg

S84,216-8
1,1-DIPHENYL-3-ETHOXY-1H,3H-NAPHTHO(1,8-CD)PYRAN
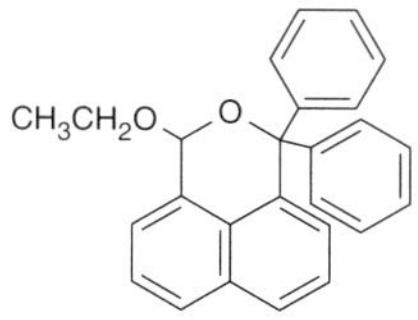
250 mg

S93,914-5
1,6-DI-O-BENZYL-2,4:3,5-DI-O-METHYLENE-L-IDITOL
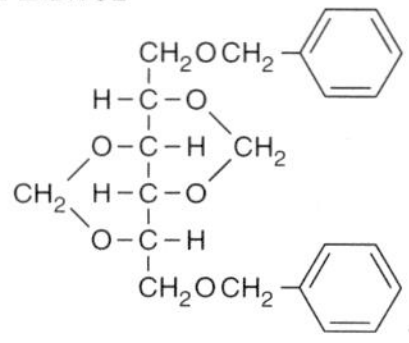
250 mg

S91,951-9
2-ETHOXY-2-METHYL-1,3-BENZODIOXOLE
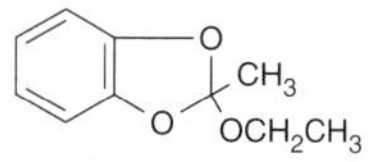
250 mg

S51,782-8
2-BENZYLOXY-2H-PYRAN
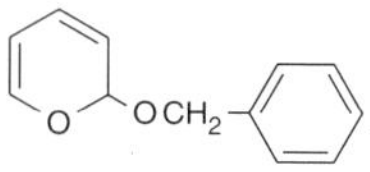
250 mg

S79,312-4
4-(PHENOXYMETHYL)TETRAHYDROPYRAN
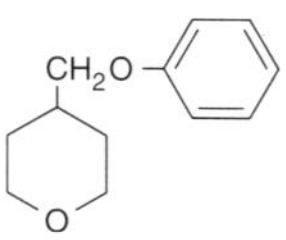
250 mg

S92,798-8
2-METHOXY-2-PHENYLNAPHTHO(2,3-D)-1,3-DIOXOLE
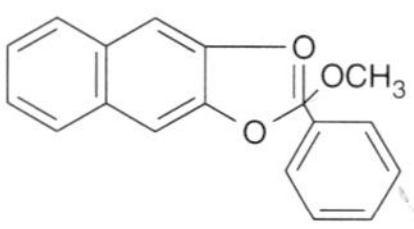
250 mg

S93,911-0
2,5-O-METHYLENE-1,3,4,6-TETRA-O-BENZYL-D-MANNITOL
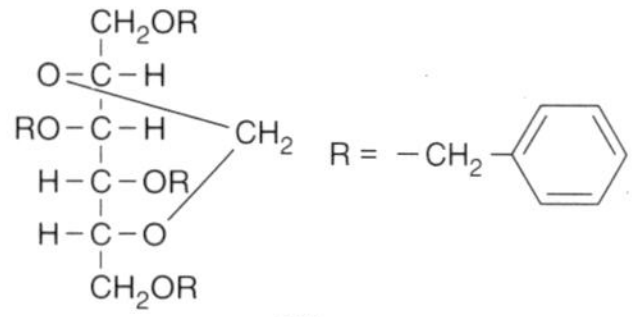
250 mg

S86,555-9
5,12-DIPHENYL-5,5A,6,11,11A,12-HEXAHYDRO-5,12-EPOXY-6,11-METHANONAPHTHACENE
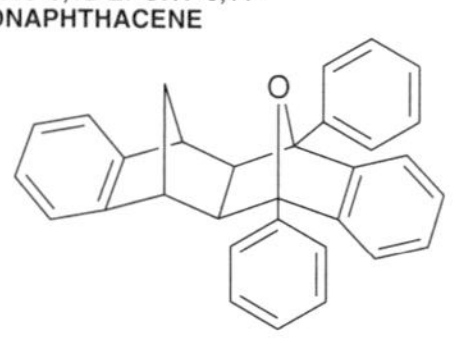
50 mg

S94,130-1
4-PHENYL-2,2-TETRAMETHYLENE-1,3-DIOXOLANE

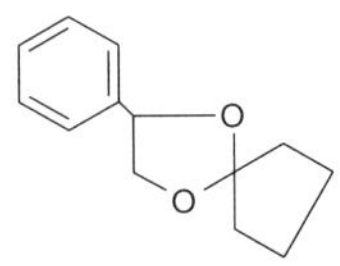

250 mg

S94,132-8
2,2-PENTAMETHYLENE-4-PHENYL-1,3-DIXOLANE

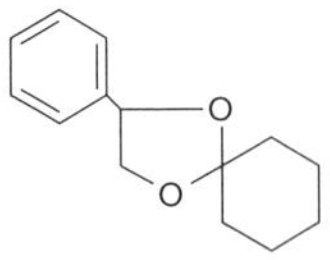

5 gm

S94,123-9
2-ETHYL-4-PHENYL-1,3-DIOXOLANE

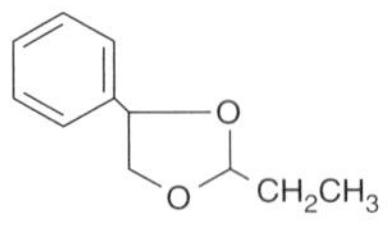

250 mg

S94,121-2
2-ISOPROPYL-4-PHENYL-1,3-DIOXOLANE

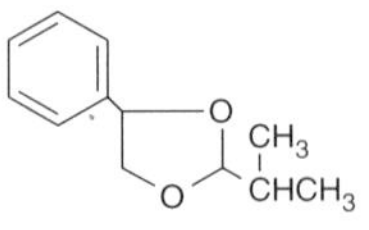

250 mg

S94,124-7
4-PHENYL-2-PROPYL-1,3-DIOXOLANE

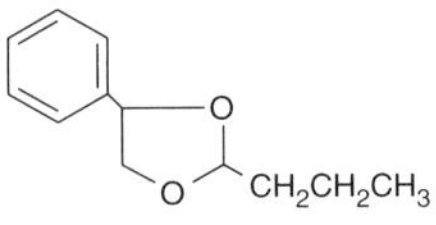

250 mg

S94,125-5
2-BUTYL-4-PHENYL-1,3-DIOXOLANE

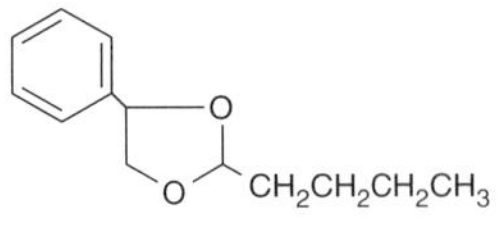

250 mg

S94,126-3
2-HEXYL-4-PHENYL-1,3-DIOXOLANE

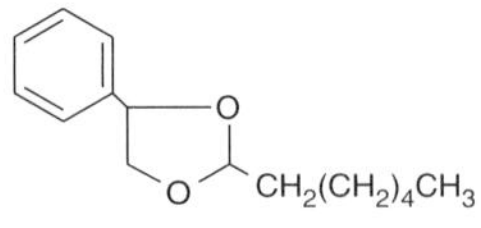

250 mg

S94,129-8
2-HEXYL-2-METHYL-4-PHENYL-1,3-DIOXOLANE

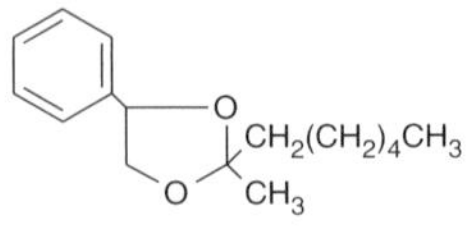

250 mg

S94,127-1
2-OCTYL-4-PHENYL-1,3-DIOXOLANE

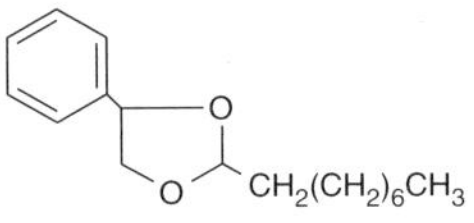

250 mg

S88,302-6
2-PHENYL-2-(O-TOLYL)-1,3-DIOXOLANE

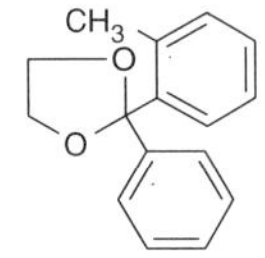

250 mg

S89,653-5
4,4,5,5-TETRAPHENYL-1,3-DIOXOLANE

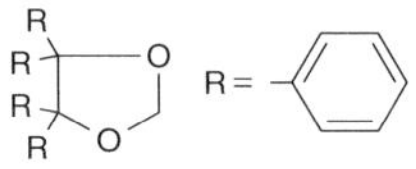

250 mg

S92,820-8
2-(2-BROMOPHENYL)-1,3-DIOXOLANE

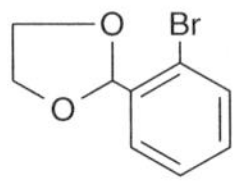

250 mg

S93,922-6
3,4-O-ISOPROPYLIDENE-2-O-MESITYL-1-O-METHYL-BETA-D-ARABINOSIDE

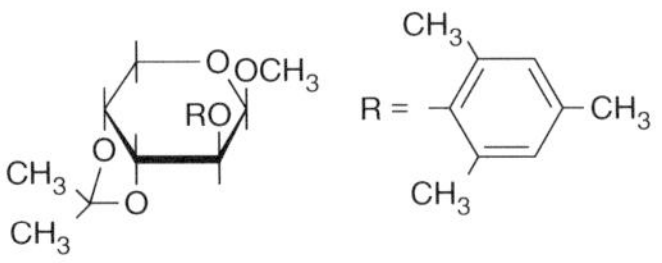

250 mg

S66,338-7
2-CHLOROBENZALDEHYDE ETHYLENE KETAL

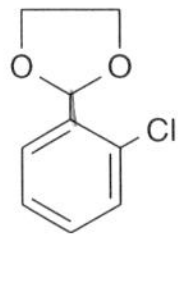

5 gm

S35,355-8
METHYL 2,3,4,6-TETRABENZOYL-ALPHA-D-MANNOPYRANOSIDE

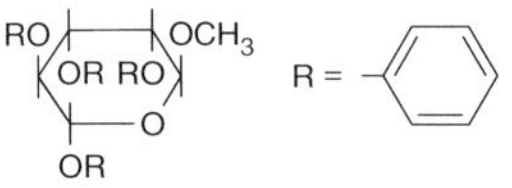

1 gm

S85,024-1
6,7-DIHYDROCYCLOPENTA(1,2-B:1,5-B')-BIS(1)BENZOPYRAN

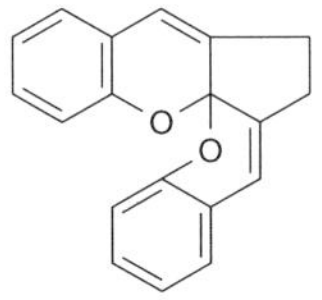

250 mg

S85,038-1
3,4-DIPHENYLTETRAHYDROFURAN

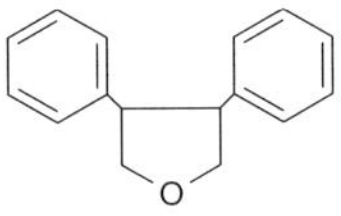

250 mg

S40,827-1
2,2-DIPHENYLTETRAHYDROFURAN

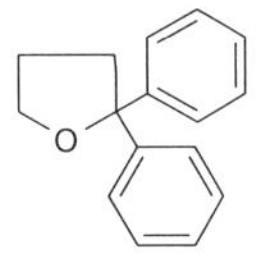

5 gm

S53,581-8
DL-2-(4-FLUOROPHENYL)-2-PHENYLTETRAHYDROFURAN

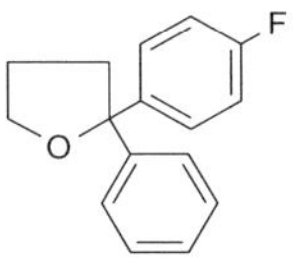

250 mg

S84,048-3
2,3-DIHYDROBENZOFURAN PICRATE

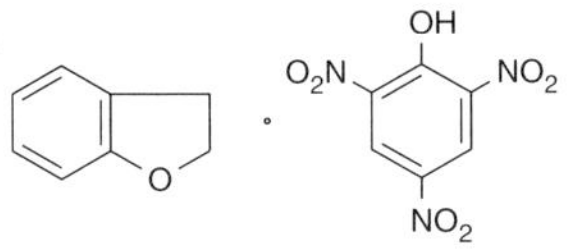

250 mg

S45,583-0
2,3-DIHYDRO-2-METHYLBENZOFURAN

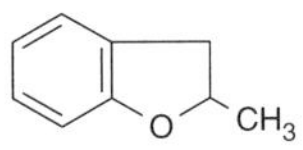

5 gm

S94,575-7
2-METHYL-2-(4-METHYL-3-PENTENYL)-1,3-DIOXOLANE

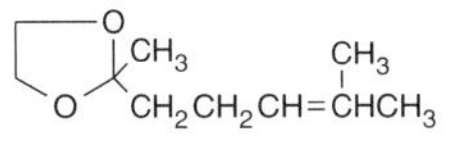

5 gm

S76,692-5
2-PHENYL-2-[4-(TETRAHYDROPYRAN-4-YLMETHYL)PHENYL]-1,3-DIOXOLANE HYDROCHLORIDE

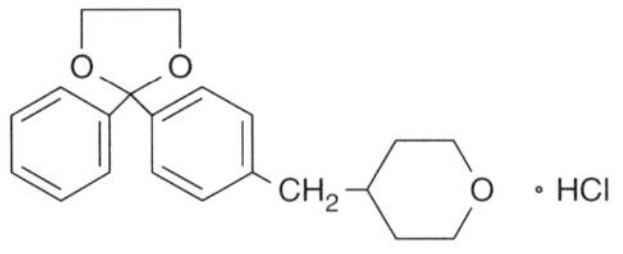

250 mg

S94,753-9
TRANS-4-METHYL-2-(2-PHENYLVINYL)-1,3-DIOXOLANE

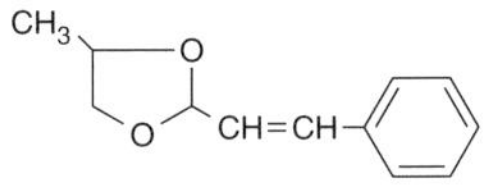

5 gm

S66,481-2
2-(4-BROMOPHENYL)-2-TERT-BUTYL-1,3-DIOXOLANE

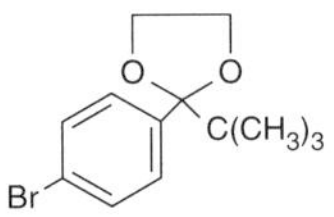

5 gm

S52,534-0
2,3-BENZO-1,4-DIOXASPIRO(4.6)-UNDEC-2-ENE

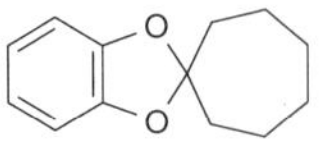

250 mg

S94,785-7
4-METHYL-2-(3,5,5,8,8-PENTAMETHYL-5,6,7,8-TETRAHYDRO-2-NAPHTHYL)-1,3-DIOXOLANE

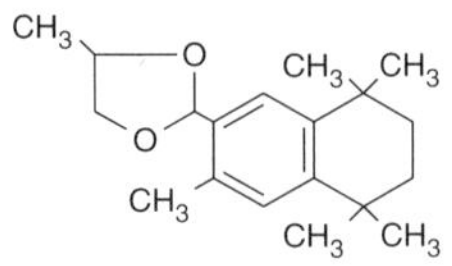

5 gm

S58,901-2
2,3-(METHYLENEDIOXY)NAPHTHALENE

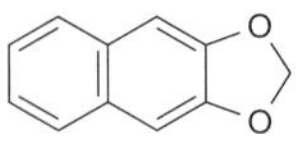

5 gm

S84,377-6
5,6-DIBROMO-1,3-BENZODIOXOLE

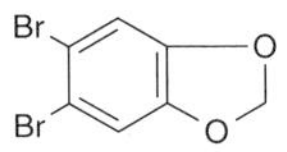

250 mg

S84,304-0
5,6-DIBROMO-2,2-DIMETHYL-1,3-BENZODIOXOLE

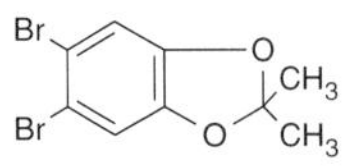

250 mg

S84,293-1
2,2,2',2'-TETRAMETHYL-[5,5'-BI(1,3-BENZODIOXOLE)]

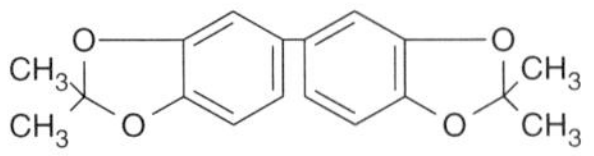

250 mg

S85,023-3
6H,7,8-DIHYDRO(1)BENZOPYRANO(3,2-D)-XANTHENE COMPOUND WITH HYDROCHLORIC ACID

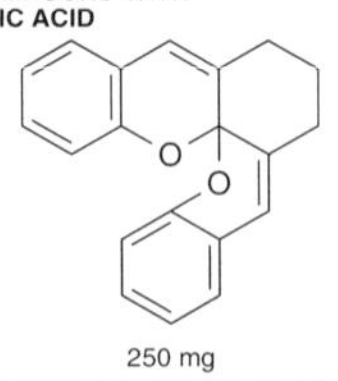

250 mg

S88,788-9
2,2',6,6'-TETRAPHENYL-DELTA4,4'-BI-4H-PYRAN

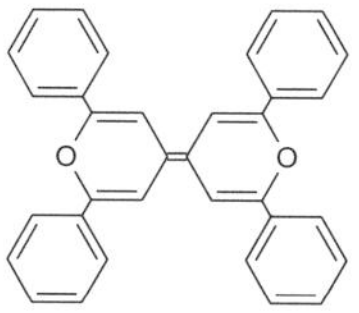

250 mg

S83,237-5
9,9-DIPHENYLXANTHENE

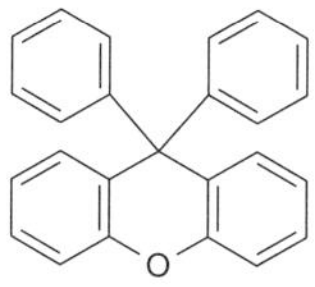

250 mg

S85,031-4
2,2,4-TRIMETHYL-2H-1-BENZOPYRAN

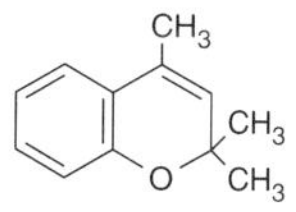

250 mg

S52,549-9
4-(4-CHLOROPHENYL)-4-METHYL-1,3-DIOXANE

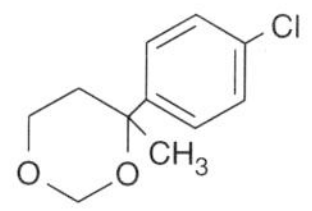

250 mg

S37,577-2
5,5-DIETHYL-2-PHENYL-1,3-DIOXANE

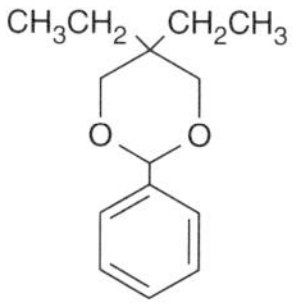

5 gm

S66,962-8
4-METHYL-4-PHENYL-1,3-DIOXANE

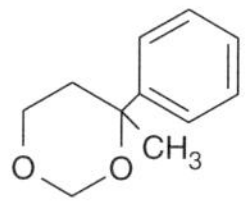

5 gm

S76,736-0
5,5-DIMETHYL-2-(1,1-DIMETHYL-3-BUTENYL)-2-PHENYL-1,3-DIOXANE

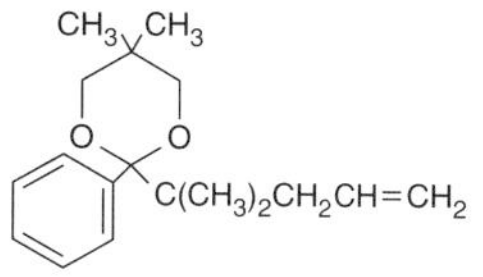

250 mg

S45,886-4
DISALICYLALDEHYDE

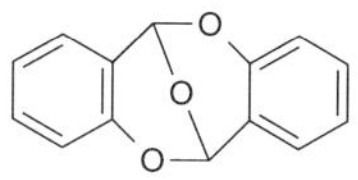

5 gm

S79,814-2
2,3-DIPHENYL-1,4-DIOXANE

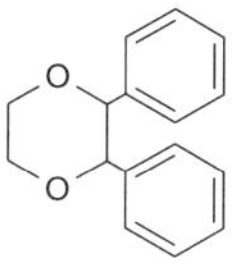

250 mg

S82,052-0
2-IODODIBENZO-P-DIOXIN

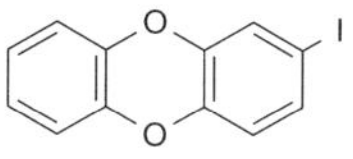

250 mg

S82,062-8
2,7-DIBROMODIBENZO-P-DIOXIN

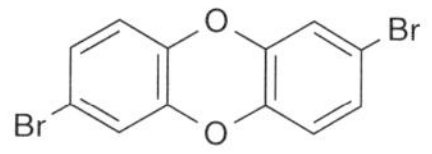

250 mg

S82,060-1
1,6-DIIODODIBENZO-P-DIOXIN

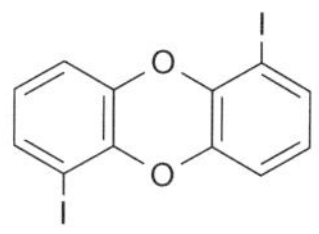

250 mg

S82,065-2
2,3,7,8-TETRAISOPROPYLDIBENZO-P-DIOXIN

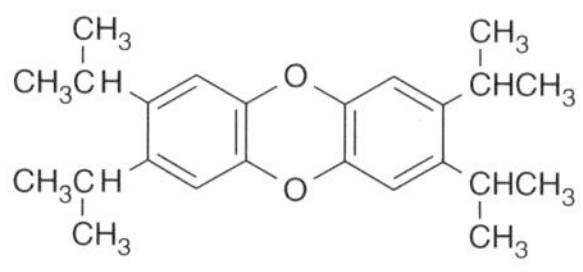

250 mg

S84,124-2
2,2,4,4,6,6-HEXAPHENYL-1,3,5-TRIOXANE

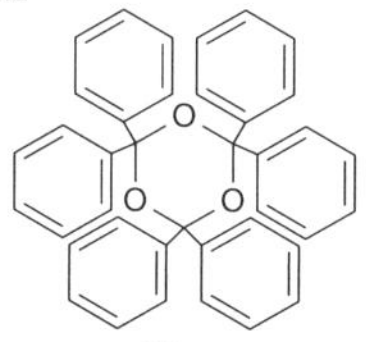

250 mg

S84,125-0
2,2,4,4,6,6,8,8-OCTAPHENYL-1,3,5,7-TETROXOCANE

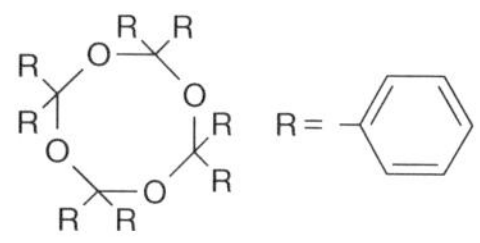

250 mg

S40,924-3
1,2-METHYLENEDIOXY-4-PROPYLBENZENE

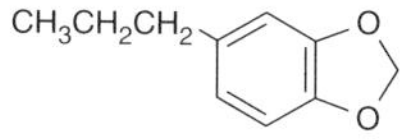

5 gm

S53,081-6
5-(2-BROMOPROPYL)-1,3-BENZODIOXOLANE

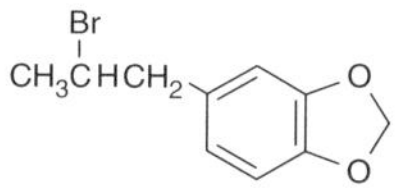

250 mg

S80,966-7
6,7,8,9-TETRAHYDRODIBENZO(B,D)(1,6)-DIOXECIN

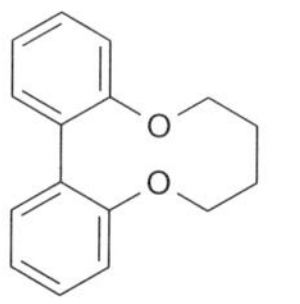

250 mg

S35,248-9
3-METHOXY-3-PHENYLPROPYL CHLORIDE

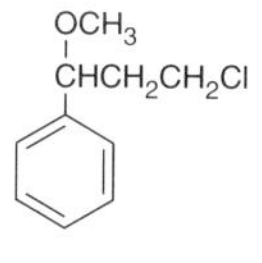

5 gm

S87,973-8
BENZYL 4-BROMOBUTYL ETHER

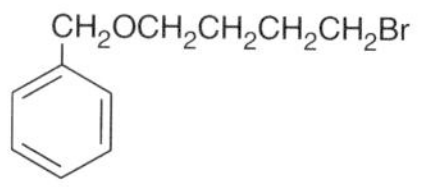

5 gm

S35,168-7
3-ISOPROPOXY-3-PHENYLPROPYL CHLORIDE

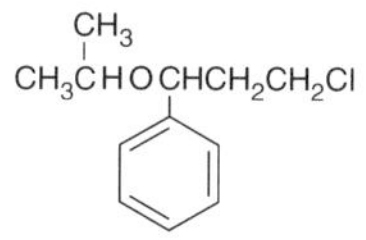

5 gm

S34,448-6
3-(4-CHLOROPHENYL)-3-METHOXYPROPYL CHLORIDE

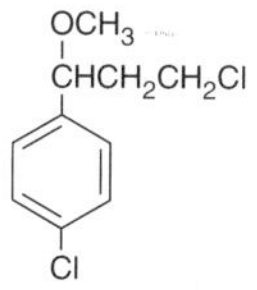

5 gm

S36,329-4
3-METHOXY-3-(4-TOLYL)-PROPYL CHLORIDE

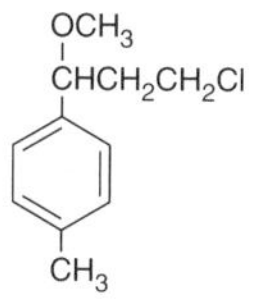

5 gm

S36,287-5
3-ISOPROPOXY-3-(P-TOLYL)-PROPYL CHLORIDE

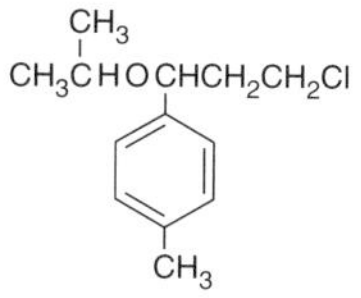

5 gm

S76,190-7
TRIETHYL ORTHO[3-(TRIFLUOROMETHYL)-BENZOATE]

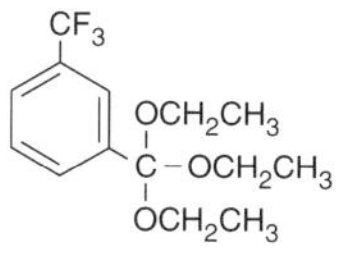

250 mg

S81,383-4
1-(BENZYLOXY)-4-BROMO-2-METHOXYBENZENE

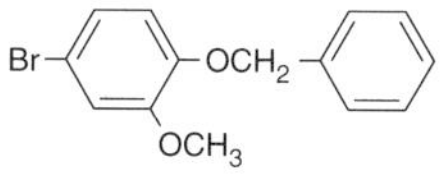

250 mg

S82,374-0
BENZYL 2,4,6-TRIBROMOPHENYL ETHER

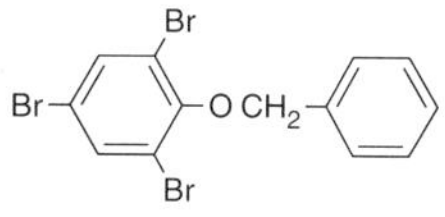

250 mg

S83,344-4
1,2-BIS(2,4-DICHLOROPHENOXY)ETHANE

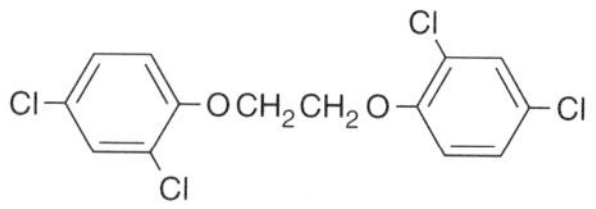

250 mg

S81,954-9
2-(2,4-DICHLOROPHENOXY)ETHYL ETHER

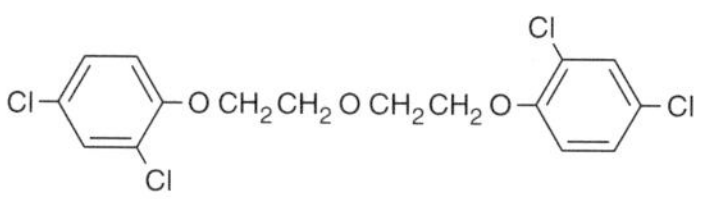

250 mg

S88,580-0
2-IODO-3',4,4',5-TETRAMETHOXYBIPHENYL

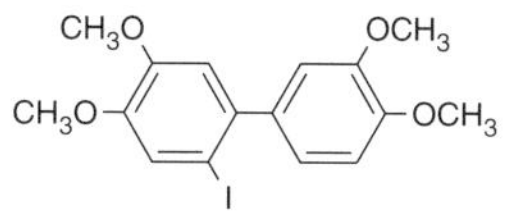

250 mg

S84,285-0
2,2'-DIBROMO-4,4',5,5'-TETRAMETHOXYBIPHENYL

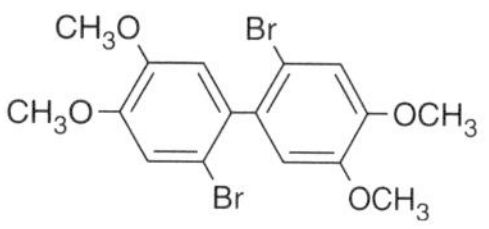

250 mg

S88,582-7
3,3',4,4',5,5'-HEXAMETHOXY-2,2',6-TRIIODOBIPHENYL

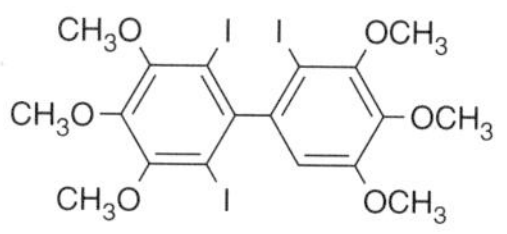

250 mg

S84,279-6
4,4'-DIIODO-3,3'-DIMETHOXYBIPHENYL

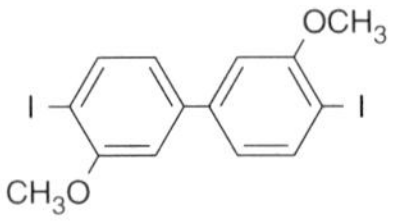

250 mg

S95,799-2
10,11-DIMETHOXY-DODECACHLORO-OCTAHYDRO-1,4:5,8-DIMETHANOTRIPHENYLENE

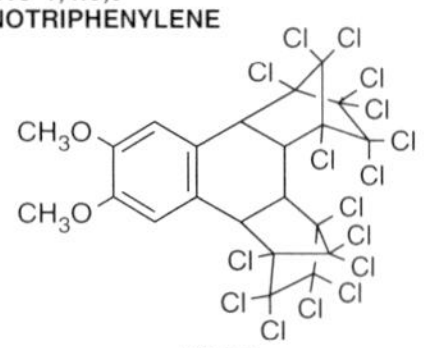

50 mg

S80,960-8
2-BROMOPHENETHYL METHYL ETHER

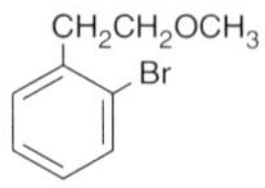

250 mg

S44,985-7
3,6-BIS(METHOXYMETHYL)DURENE

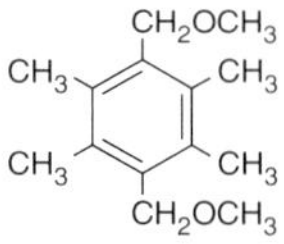

5 gm

S67,959-3
ETHYLENE GLYCOL DIBENZYL ETHER

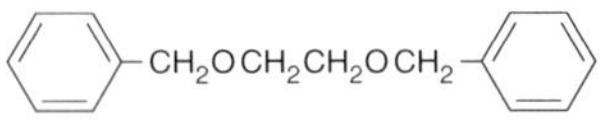

250 mg

S68,025-7
DIVERATRYL ETHER

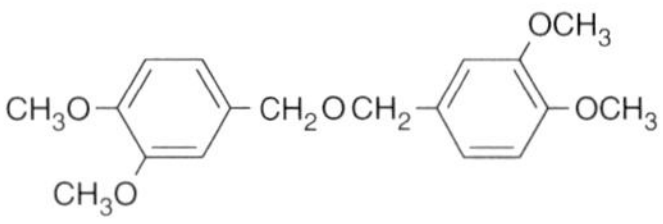

250 mg

S86,705-5
1,4-BIS(BENZYLOXY)-2-BUTENE

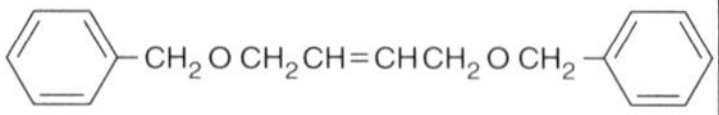

250 mg

S90,703-0
HEXAKIS(BENZOXYMETHYL)BENZENE

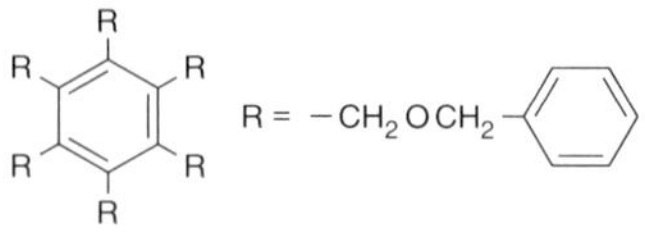

250 mg

S86,683-0
1-METHOXY-4-(4-METHOXYBENZYLOXY)-2-BUTENE

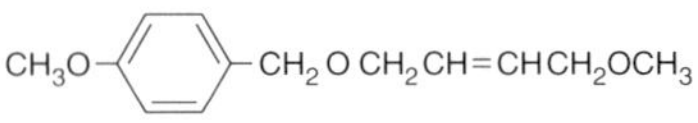

250 mg

S87,586-4
1,1-DIMETHOXY-2,2-DIMETHYL-1-PHENYLPROPANE

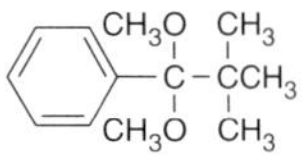

250 mg

S95,455-1
4-METHOXY-ALPHA-PHENYLBENZYL ETHER

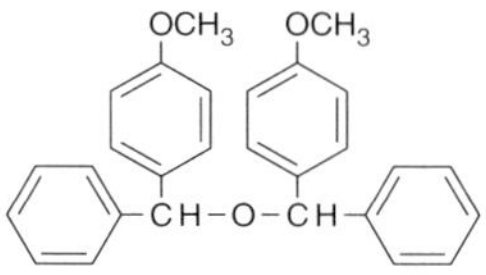

50 mg

S86,628-8
4-METHOXYBENZYL METHYL ETHER

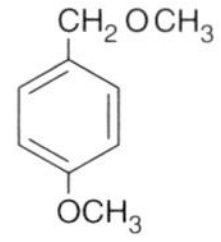

250 mg

S86,493-5
ALLYL BENZYL ETHER

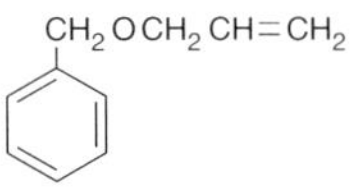

250 mg

S63,026-8
BENZYL ISOPROPYL ETHER

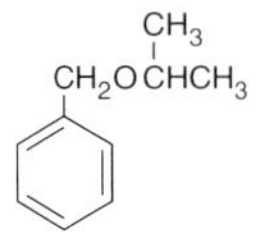

250 mg

S86,653-9
BENZYL METHALLYL ETHER

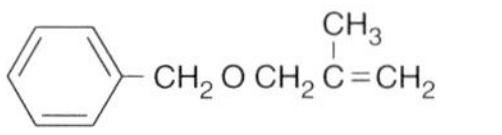

250 mg

S65,385-3
BENZYL TERT-BUTYL ETHER

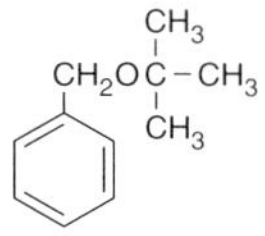

250 mg

S67,986-0
BENZYL ISOBUTYL ETHER

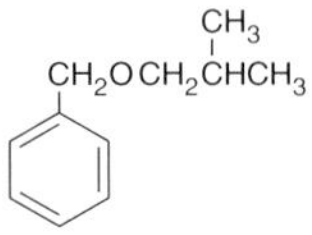

5 gm

S57,733-2
ALPHA-ISOPROPYL-ALPHA-METHYLBENZYL METHYL ETHER

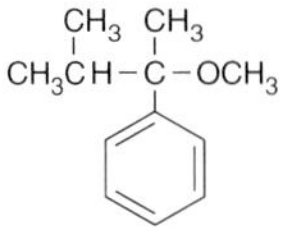

5 gm

S83,383-5
ALPHA,ALPHA-DIETHYLBENZYL METHYL ETHER

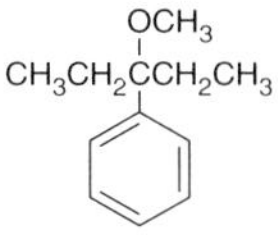

250 mg

S63,016-0
BENZYL HEXYL ETHER

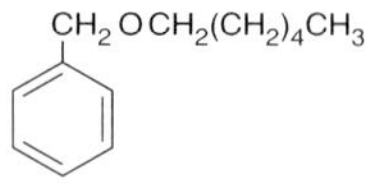

250 mg

S35,266-7
ALPHA-METHYLBENZYL ETHER

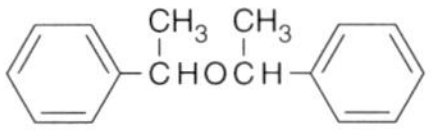

5 gm

S81,960-3
DIPHENYLMETHYL METHYL ETHER

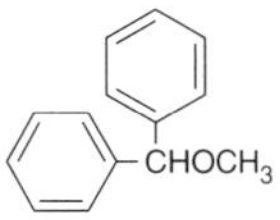

250 mg

S70,608-6
DIPHENYLMETHYL PROPYL ETHER

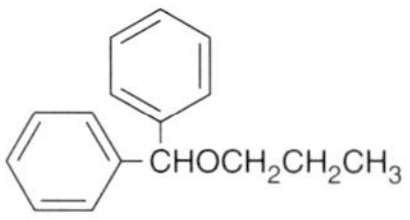

5 gm

S65,358-6
TERT-BUTYL DIPHENYLMETHYL ETHER

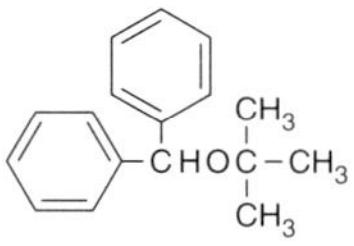

5 gm

S58,110-0
METHYL TRITYL ETHER

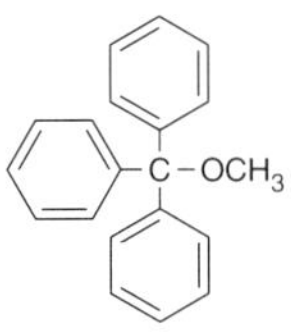

250 mg

S81,197-1
PHENYL TRITYL ETHER

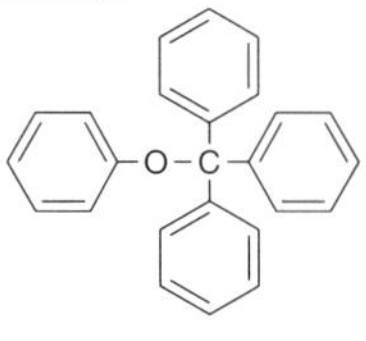

250 mg

S71,444-5
DIBENZHYDRYL ETHER

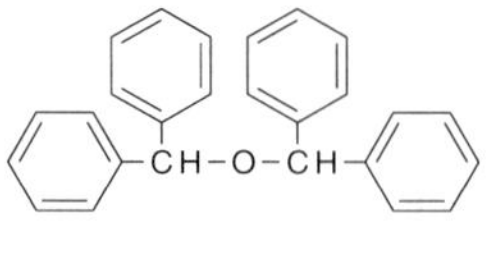

5 gm

S70,977-8
BENZYL 1-METHYL-2-NAPHTHYL ETHER

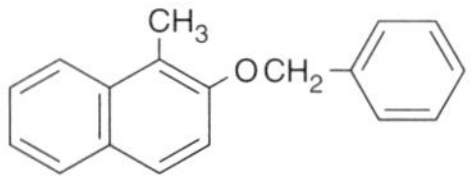

250 mg

S64,809-4
BENZYL 1-NAPHTHYL ETHER

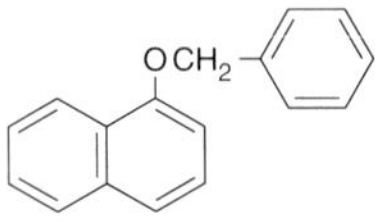

250 mg

S64,621-0
BENZYL 2-NAPHTHYL ETHER

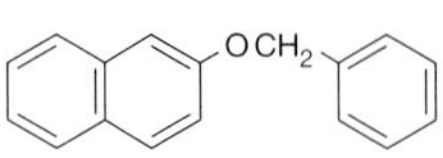

5 gm

S85,306-2
4-TERT-BUTYLBENZYL METHYL ETHER

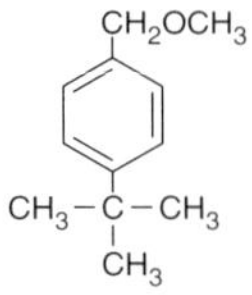

250 mg

S70,605-1
ETHYL 2,4,6-TRIISOPROPYLBENZYL ETHER

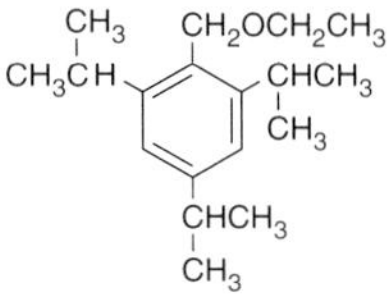

5 gm

S43,196-6
2,5-BIS(METHOXYMETHYL)-P-XYLENE

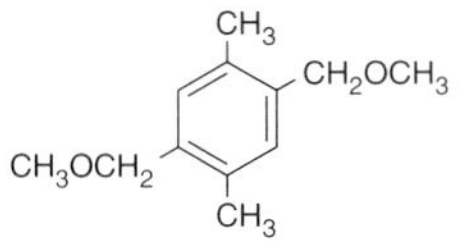

5 gm

S8 8,325-5
1,3'-DIMETHOXY-7-(3,4-DIMETHOXYPHENYL)NAPHTHALENE

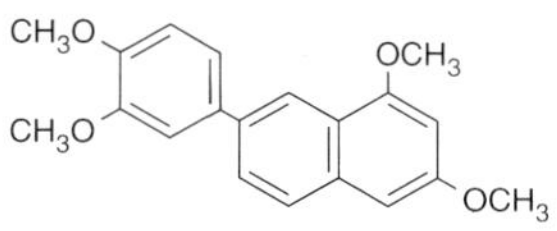

250 mg

S92,992-1
1,4-DIHYDRO-6,7-DIMETHOXY-1,4-METHANONAPHTHALENE

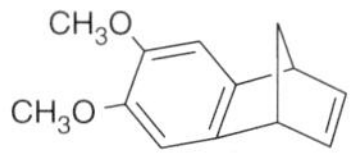

250 mg

S79,300-0
1-METHOXY-2-PHENOXYETHANE

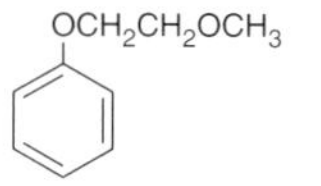

250 mg

S84,287-7
3,3',4,4',5,5'-HEXAMETHOXYBIPHENYL

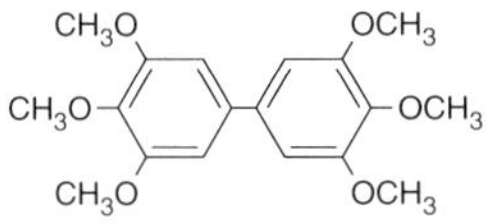

250 mg

S75,970-8
6,7-DIMETHOXY-1-METHYLNAPHTHALENE

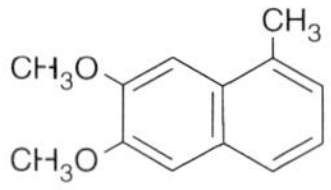

250 mg

S80,971-3
2,2'-DIETHOXYBIPHENYL

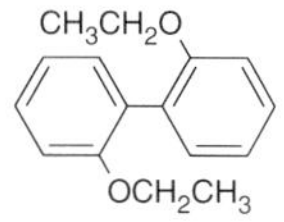

250 mg

S79,386-8
2,2'-DIMETHOXY-1,1'-BINAPHTHYL

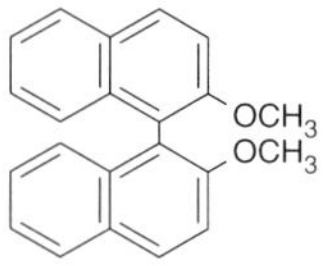

250 mg

S86,456-0
9,10-DIHYDRO-9,10-DIMETHOXY-9,10-DIMETHYLANTHRACENE

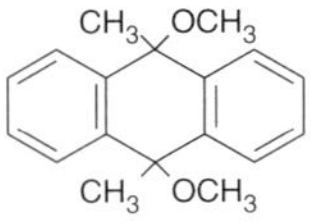

250 mg

S88,395-6
5,6,6A,6B,7,8,12B,13,14,14A-DECAHYDRO-3-ETHOXY-10-METHOXYPICENE

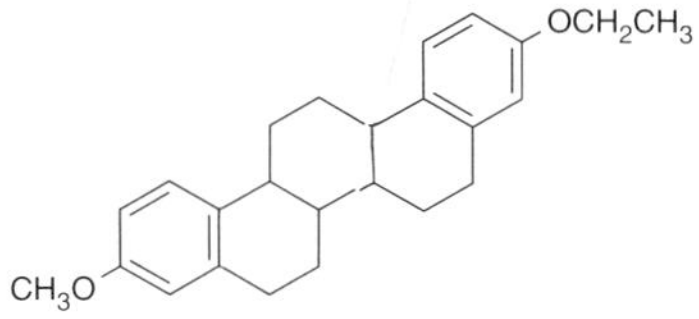

250 mg

S88,570-3
3,10-DIMETHOXY-6A,12B-DIMETHYL-5,6,6A,6B,7,8,12B,13-OCTAHYDROPICENE

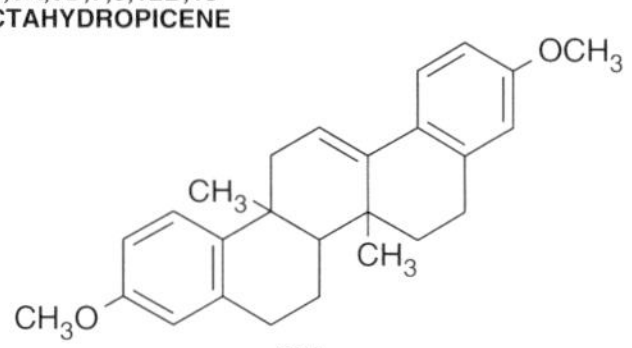

250 mg

S80,985-3
PHENOXYMETHYL ETHER

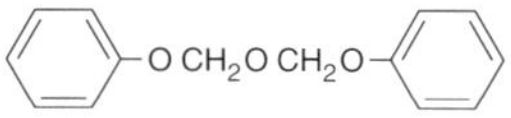

250 mg

S79,207-1
1,1-BIS(2-METHOXYPHENYL)ETHYLENE

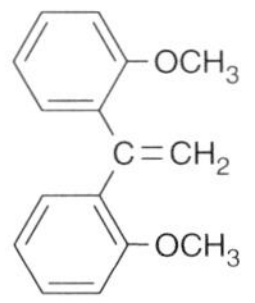

250 mg

S44,512-6
1,4-BENZENEDIMETHANOL DIMETHYL ETHER

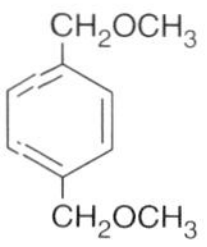

5 gm

S93,552-2
4,4'-BIS(METHOXYMETHOXY)STILBENE

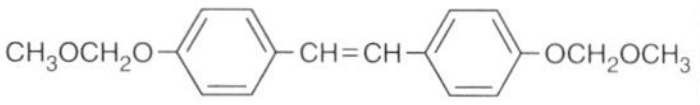

250 mg

S84,290-7
3',6'-DIMETHOXY-2,4,4',6''-TETRAMETHYL-P-TERPHENYL

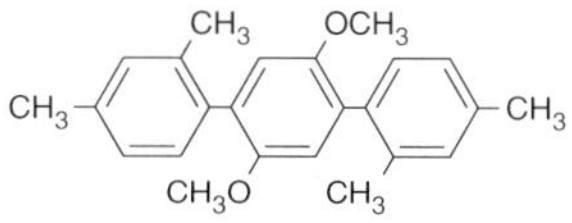

250 mg

S84,296-6
4,4'-DIMETHOXY-3,3',5,5'-TETRAMETHYLBIPHENYL

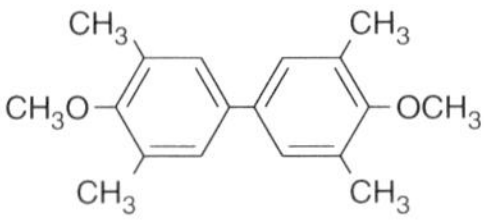

250 mg

S95,655-4
6,6'-DIMETHOXY-1,2'-BINAPHTHYL

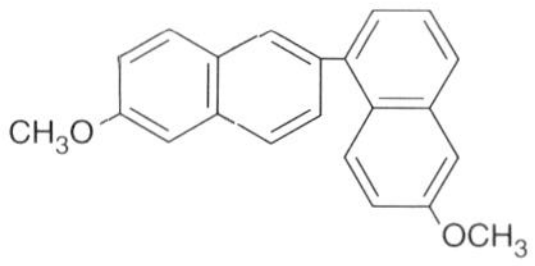

50 mg

S97,339-4
AR-D38 2,7-DIMETHOXYPHENANTHRENE

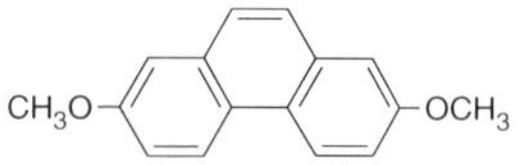

10 mg

S92,665-5
2-METHOXY-2-(2-PHENOXYETHOXY)-PROPANE

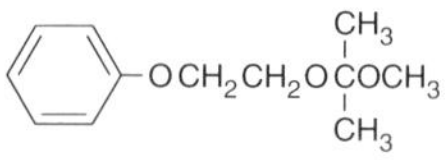

250 mg

S51,844-1
BIS(2-CHLORO-2-PHENOXYETHYL) ETHER

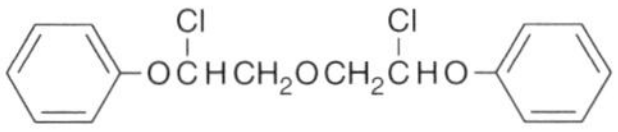

5 gm

S65,209-1
9,10-DIHYDRO-1,2,3,5,6,7-HEXAMETHOXYANTHRACENE

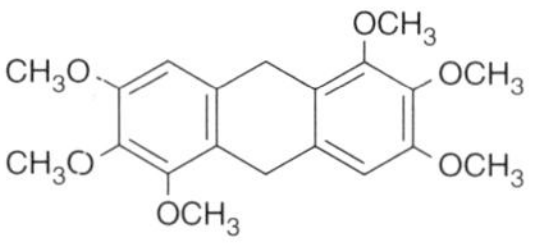

250 mg

S85,211-2
11,12-DICHLORO-9,10-DIHYDRO-9-METHOXY-9,10-ETHANOANTHRACENE

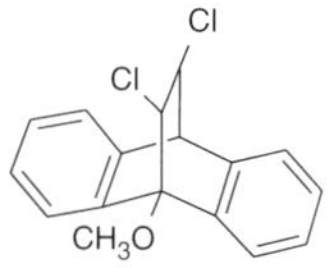

250 mg

S83,712-1
2-(6-BROMO-2-NAPHTHOXY)-TETRAHYDROPYRAN

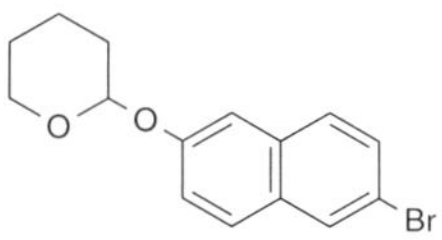

250 mg

S72,207-3
BENZYL METHYL ETHER

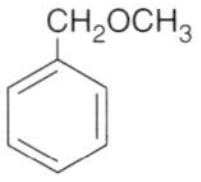

5 gm

S36,369-3
METHYL PHENETHYL ETHER

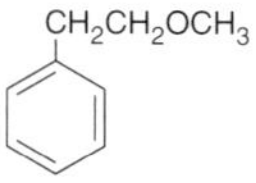

5 gm

S72,210-3
ETHYL STYRYL ETHER

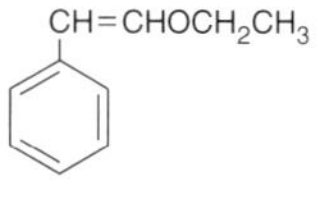

5 gm

S72,217-0
BETA-METHOXY-BETA-METHYLSTYRENE

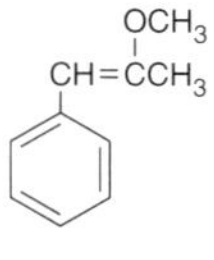

5 gm

S83,918-3
ETHYL PHENETHYL ETHER

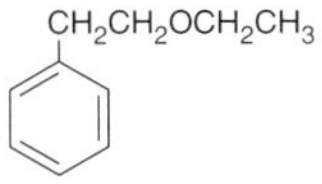

250 mg

S70,149-1
9-ANTHRACENYLMETHYL 4-BENZYLPHENYL ETHER

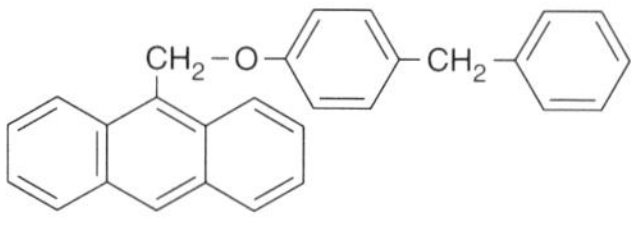

250 mg

S34,966-6
ETHYL TRIPHENYLMETHYL ETHER

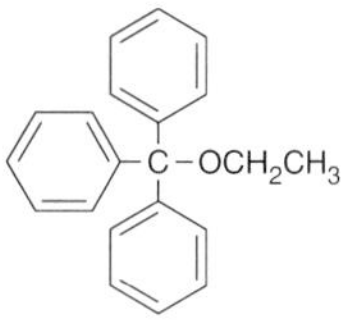

5 gm

S43,335-7
1,4-BENZENEDIMETHANOL DIETHYL ETHER

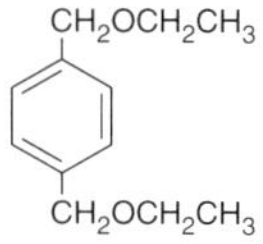

5 gm

S57,148-2
BENZYL 4-CHLOROBENZYL ETHER

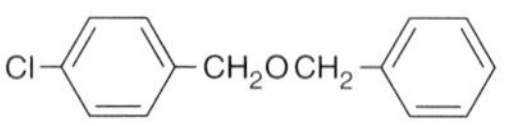

5 gm

S57,142-3
BENZYL 4-METHYLBENZYL ETHER

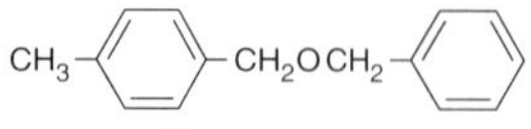

5 gm

S35,210-1
9-(4-METHOXYBENZYLIDENE)FLUORENE

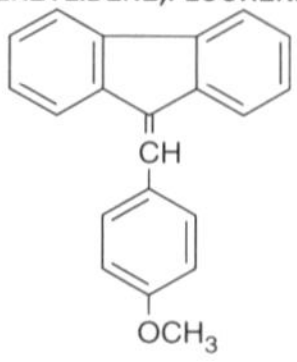

250 mg

S41,214-7
2,6-DICHLOROBENZYL ETHER

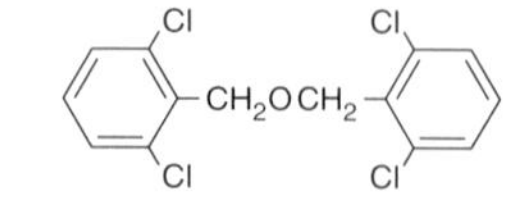

5 gm

S86,366-1
9,10-DIHYDRO-9-METHOXY-9,10-ETHENOANTHRACENE

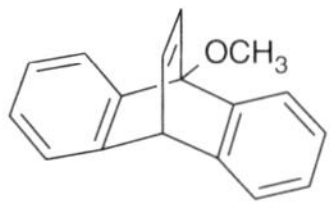

250 mg

S94,797-0
6-ETHYL-7-METHOXYMETHYL-1,2,3,4-TETRAHYDRO-1,1,4,4-TETRAMETHYLNAPHTHALENE

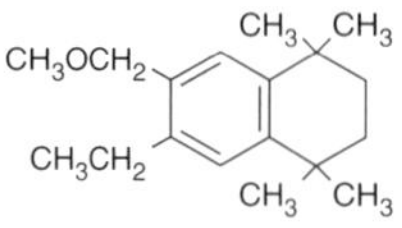

250 mg

S66,297-6
1-CHLORO-3-METHOXYNAPHTHALENE

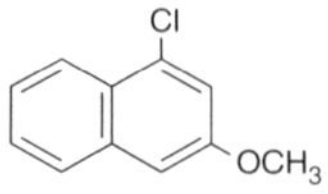

250 mg

S43,531-7
1,7-DIMETHOXYNAPHTHALENE

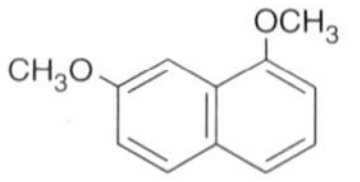

5 gm

S59,591-8
2,4-DIBROMO-1-METHOXYNAPHTHALENE

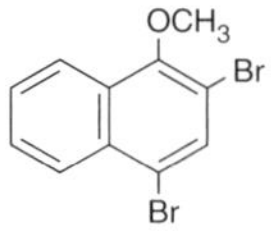

5 gm

S34,944-5
ETHYL 1-NAPHTHYLMETHYL ETHER

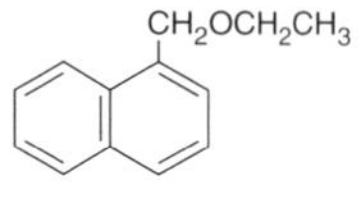

5 gm

S66,333-6
1-(2-BUTOXY)NAPHTHALENE

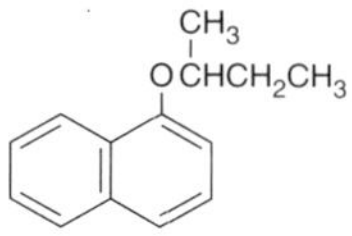

5 gm

S67,703-5
4-METHOXY-1-METHYLNAPHTHALENE

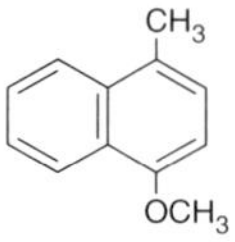

250 mg

S51,676-7
1,5-DIMETHOXYNAPHTHALENE

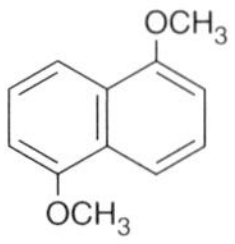

250 mg

S59,240-4
1,4-DIMETHOXYNAPHTHALENE

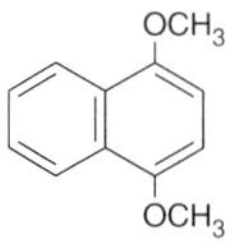

250 mg

S95,626-0
1-ETHYL-6-METHOXYNAPHTHALENE

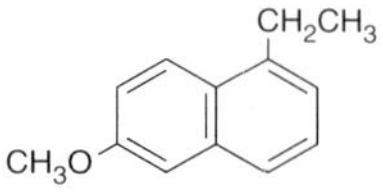

50 mg

S56,197-5
2-ETHOXYNAPHTHALENE

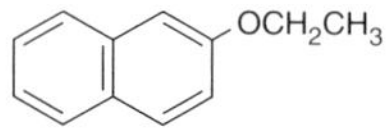

5 gm

S64,805-1
2-NAPHTHYL PROPYL ETHER

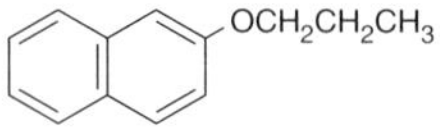

250 mg

S66,161-9
6-METHOXY-2-(1-METHYLVINYL)-NAPHTHALENE

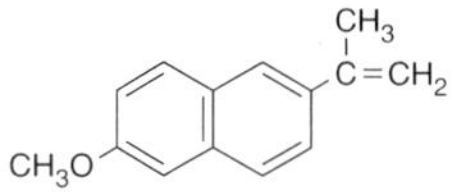

250 mg

S88,358-1
2-ISOPROPYL-6-METHOXYNAPHTHALENE

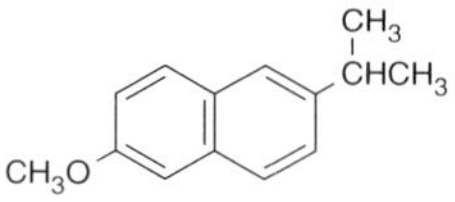

250 mg

S66,240-2
2-METHOXYBENZO(C)PHENANTHRENE

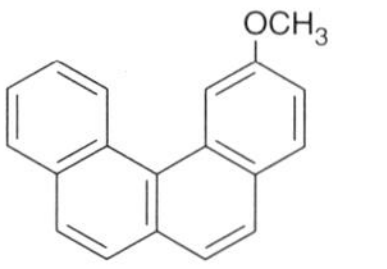

250 mg

S97,519-2
AR-H26 4-(METHOXYMETHYL)PYRENE

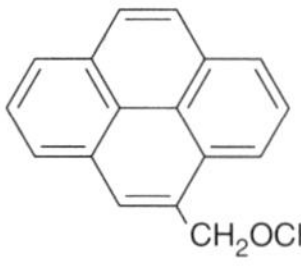

10 mg

S97,446-3
AR-F10 6-METHOXYCHRYSENE

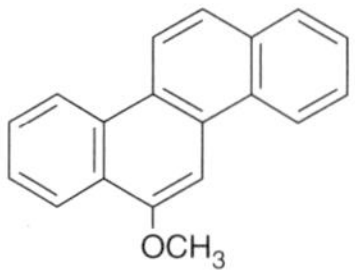

10 mg

S97,475-7
AR-G33 1-METHOXYBENZO(C)-PHENANTHRENE

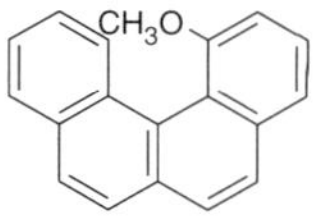

10 mg

S97,477-3
AR-G36 3-METHOXYBENZO(C)-PHENANTHRENE

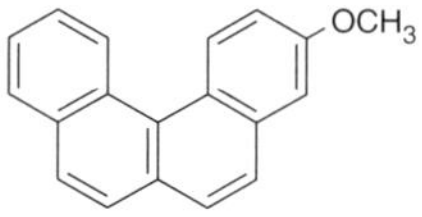

10 mg

S44,444-8
1-BROMO-4-METHOXYNAPHTHALENE

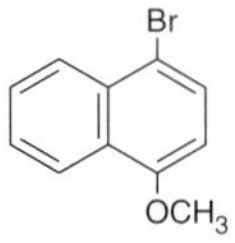

250 mg

S93,964-1
5-METHOXY-1,1,4,4,7-PENTAMETHYL-1,2,3,4-TETRAHYDRONAPHTHALENE

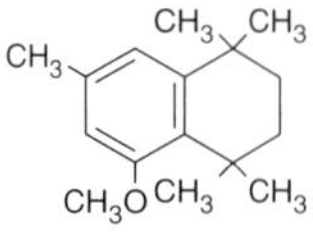

250 mg

S60,842-4
1,4-BIS(2-NAPHTHOXY)-TRANS-2-BUTENE

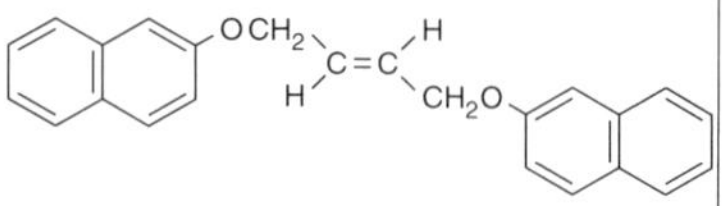

5 gm

S58,865-2
2,2'-DIETHOXY-1,1'-METHYLENEDINAPHTHALENE

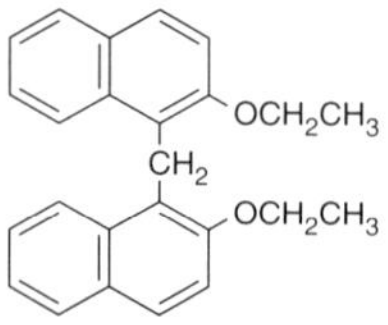

5 gm

S51,766-6
6,6'-DIMETHOXY-2,2'-BINAPHTHYL

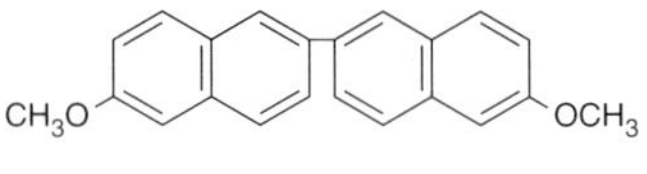

250 mg

S45,410-9
1-METHOXYFLUORENE

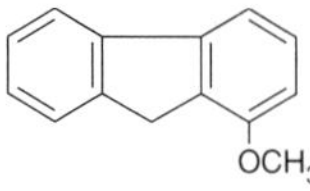

250 mg

S97,382-3
AR-E42 5-METHOXYBENZ(A)ANTHRACENE

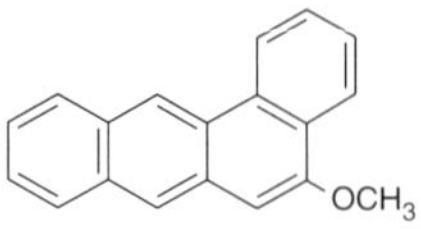

10 mg

S97,386-6
AR-E47 5-METHOXY-7-METHYLBENZ(A)-ANTHRACENE

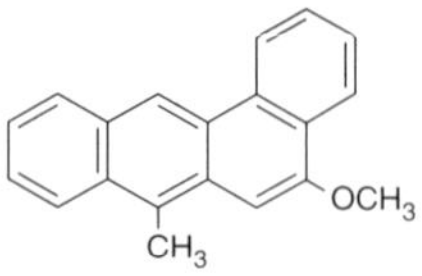

10 mg

S97,408-0
AR-E71 6-METHOXY-12-METHYLBENZ(A)-ANTHRACENE

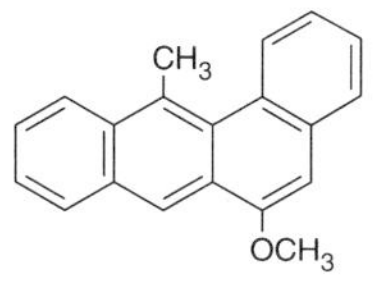

10 mg

S97,410-2
AR-E73 8-METHOXY-7-METHYLBENZ(A)-ANTHRACENE

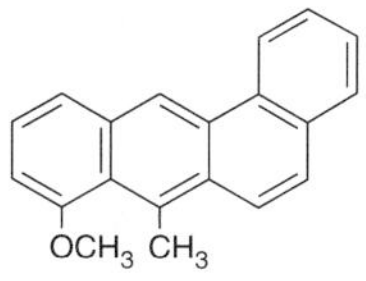

10 mg

S97,436-6
AR-E98 7,12-DIMETHYL-9-METHOXYBENZ (A)ANTHRACENE

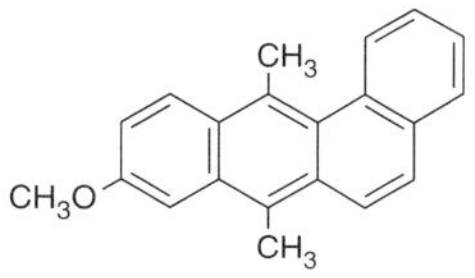

10 mg

S97,435-8
AR-E97 7,12-DIMETHYL-5-ETHOXYBENZ (A)ANTHRACENE

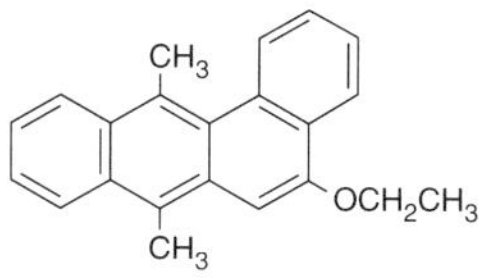

10 mg

S81,127-0
2-METHOXYPHENANTHRENE

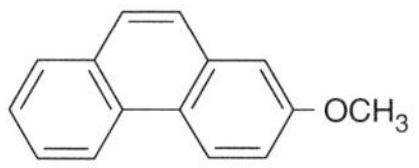

250 mg

S80,967-5
6,7,9,10-TETRAHYDRODIBENZO(H,J)-(1,4,7)TRIOXACYCLOUNDECIN

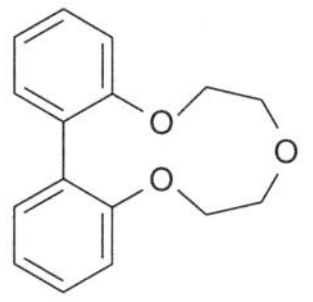

250 mg

S86,655-5
2-(BENZYLOXYMETHYL)-2-METHYLOXIRANE

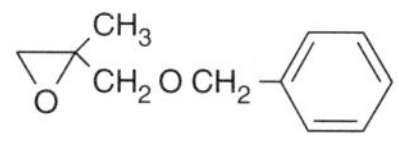

250 mg

S36,747-8
1-(2-CHLOROPHENOXY)-2,3-EPOXYPROPANE

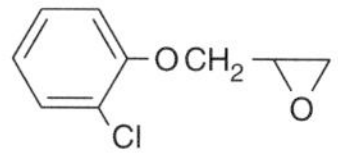

5 gm

S43,439-6
4-BROMOPHENYL 2,3-EPOXYPROPYL ETHER

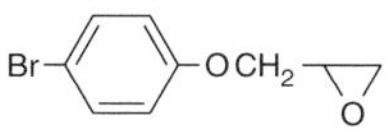

5 gm

S42,942-2
2,3-EPOXYPROPYL 4-ETHYLPHENYL ETHER

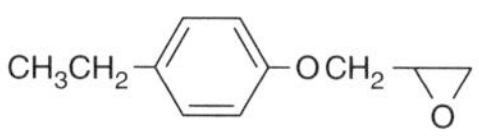

5 gm

S38,166-7
GLYCIDYL MESITYL ETHER

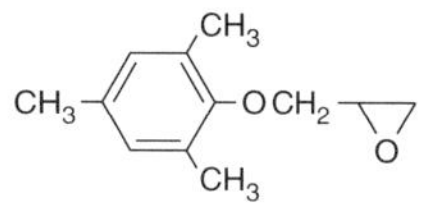

5 gm

S92,599-3
4,4'-BIS(2,3-EPOXYPROPOXY)BIPHENYL

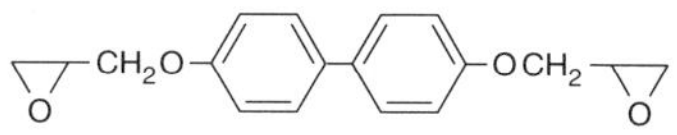

250 mg

S59,971-9
2,4-DICHLOROSTYRENE OXIDE

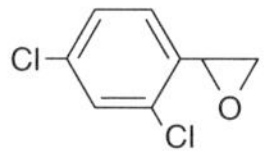

250 mg

S83,531-5
2-(2,4-DICHLOROPHENOXY)-ACETALDEHYDE DIETHYLACETAL

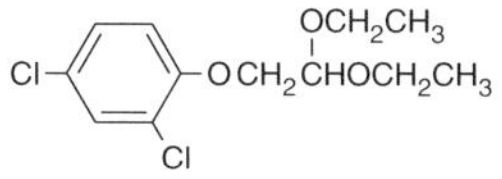

250 mg

S94,772-5
4-ISOPROPYLPHENOXYACETALDEHYDE DIMETHYL ACETAL

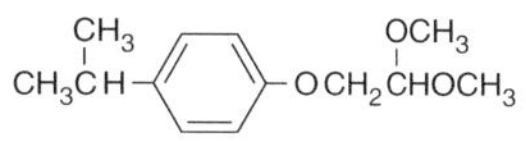

5 gm

S94,993-0
2,4-DIMETHYLPHENOXYACETALDEHYDE DIMETHYL ACETAL

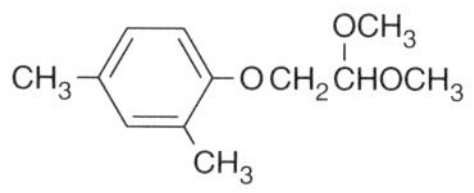

5 gm

S70,273-0
BENZOPHENONE DIMETHYL KETAL

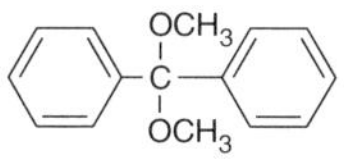

250 mg

S79,315-9
3-PHENOXYPROPIONALDEHYDE DIETHYL ACETAL

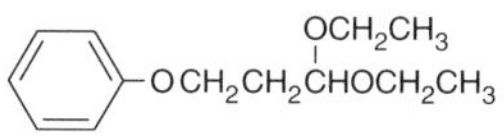

250 mg

S79,353-1
2,2-DIMETHYL-4-PHENYL-1,3-DIOXOLANE

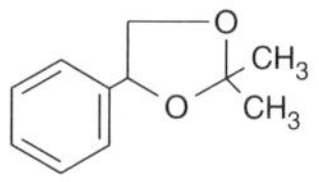

5 gm

S89,321-8
2-PHENYL-1-CYCLOPROPANECARBOXALDEHYDE DIETHYL ACETAL

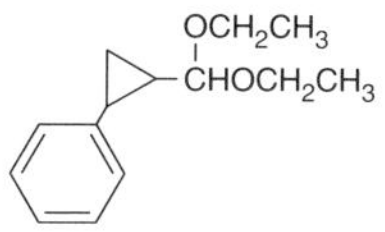

250 mg

S74,018-7
4-[2-(2-CHLOROMETHYL-1,3-DIOXOLAN-2-YL)ETHYL]PHENOL

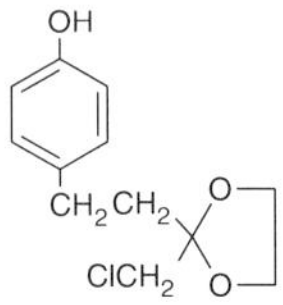

250 mg

S94,762-8
3-CHLOROMETHYL-2,4,6-TRIMETHYLPHENOL

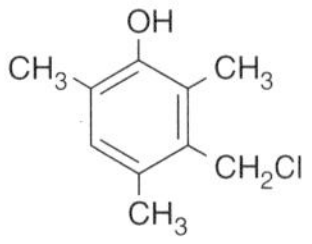

250 mg

S41,867-6
2,3,4,6-TETRACHLOROPHENOL

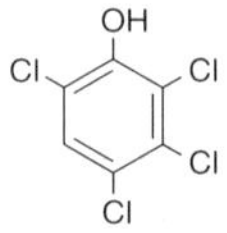

5 gm

S61,401-7
2,3,4,6-TETRACHLOROPHENOL, SODIUM SALT MONOHYDRATE

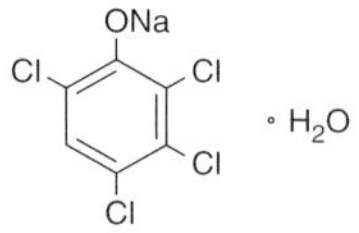

5 gm

S54,861-8
2,4,6-TRIBROMO-M-CRESOL

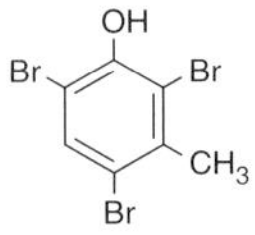

250 mg

S41,264-3
2,4-DICHLORO-3,5-DIMETHYLPHENOL

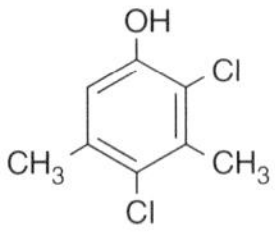

5 gm

S34,741-8
3,5-DIMETHYL-2,4,6-TRICHLOROPHENOL

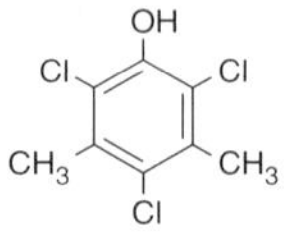

5 gm

S34,462-1
4-CHLORO-2,3,5-TRIMETHYLPHENOL

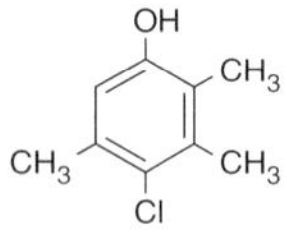

5 gm

S43,106-0
4-CHLORO-2,3,6-TRIMETHYLPHENOL

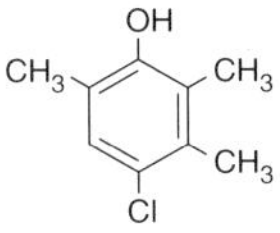

5 gm

S34,571-7
2,4-DICHLORO-3,5,6-TRIMETHYLPHENOL

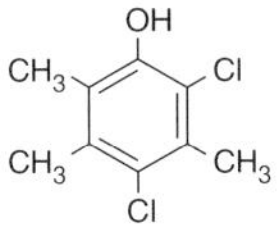

5 gm

S34,911-9
2,4-DICHLORO-3-ETHYL-5-METHYLPHENOL

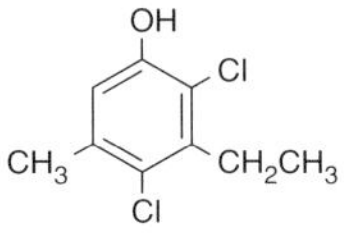

5 gm

S34,912-7
2,4-DICHLORO-5-ETHYL-3-METHYLPHENOL

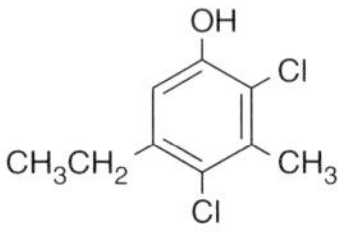

5 gm

S34,939-9
3-ETHYL-5-METHYL-2,4,6-TRICHLOROPHENOL

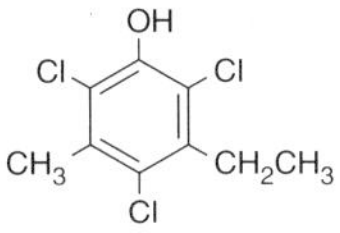

5 gm

S67,626-8
5-METHYL-2,4,6-TRIBROMORESORCINOL

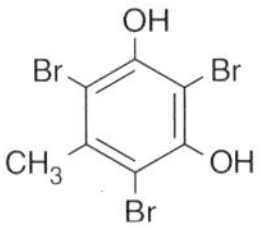

5 gm

S88,575-4
5-BROMO-2-METHOXY-4-METHYLPHENOL

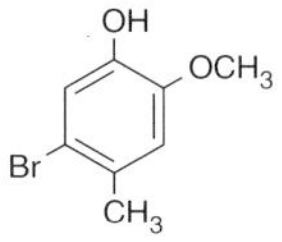

250 mg

S75,587-7
2-CHLORO-5-(2,4,5-TRIMETHYL-1,3-DIOXOLAN-2-YL)PHENOL

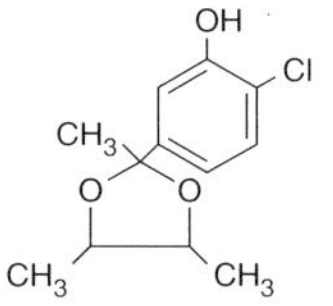

250 mg

S69,740-0
3,4-DICHLOROPHENOL SODIUM SALT

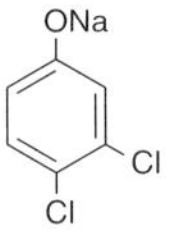

250 mg

S70,661-2
4-BROMO-2-METHYLPHENOL

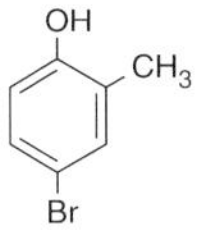

250 mg

S45,380-3
4-TERT-BUTYL-2-CHLOROPHENOL

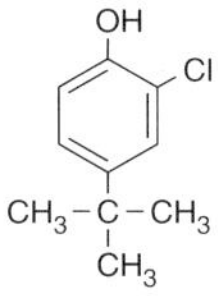

5 gm

S67,697-7
2-CHLORO-4,6-DIBROMOPHENOL

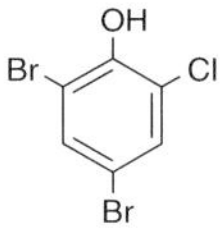

5 gm

S67,747-7
6-BROMO-4-CHLORO-O-CRESOL

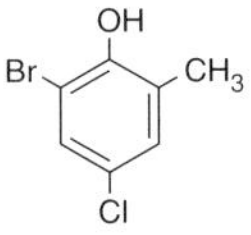

5 gm

S81,057-6
3,5-DIBROMO-2-METHYLPHENOL

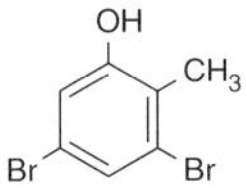

250 mg

S63,311-9
2,6-DIIODO-4-METHYLPHENOL

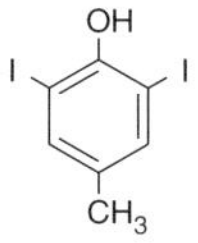

250 mg

S84,270-2
2,6-DIMETHYL-4-IODOPHENOL

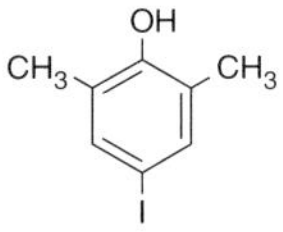

250 mg

S67,705-1
2-TERT-BUTYL-4-BROMO-6-METHYLPHENOL

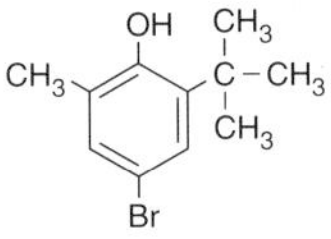

250 mg

S67,729-9
6-BROMO-4-CHLORO-M-CRESOL

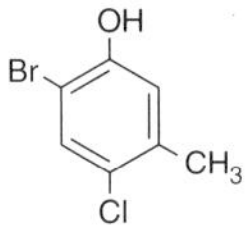

5 gm

S67,752-3
4-BROMO-2,5-DIMETHYLPHENOL

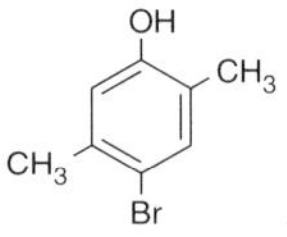

5 gm

S82,587-5
2,4,6-TRIBROMOPHENOL, SODIUM SALT

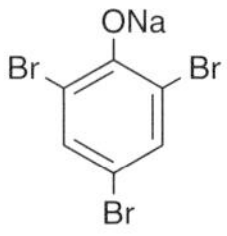

250 mg

S94,370-3
4,4'-PROPYLIDENEDIPHENOL

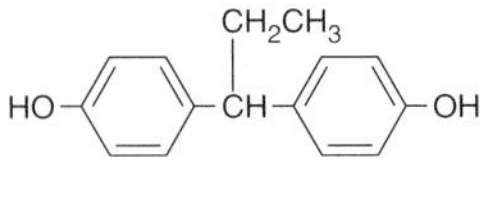

250 mg

S80,879-2
4,4'-(1-ETHYL-2-METHYLTRIMETHYLENE)-BISPHENOL

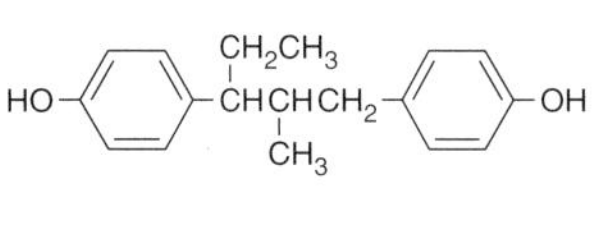

250 mg

S84,016-5
4,4'-METHYLENEBIS(2,6-DIMETHYLPHENOL)

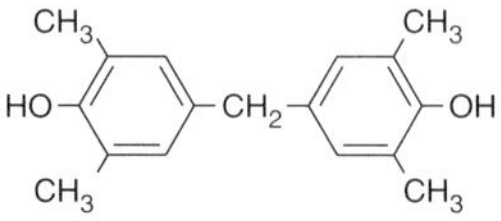

250 mg

S39,259-6
4-(METHOXYMETHYL)PHENOL

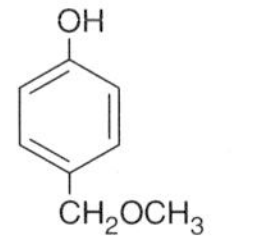

250 mg

S36,788-5
2,6-DI-TERT-BUTYL-ALPHA-METHOXY-P-CRESOL
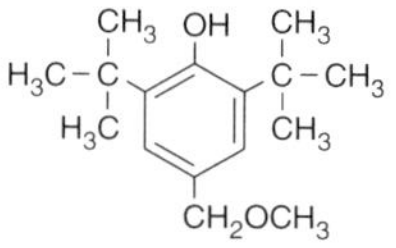
5 gm

S83,924-8
2,5-DIMETHOXYPHENOL
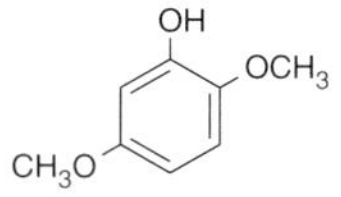
250 mg

S86,094-8
3,5-DIMETHOXYPHENOL, SODIUM SALT
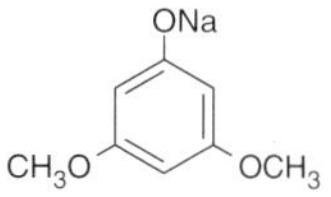
250 mg

S83,460-2
GUAIACOL PICRATE
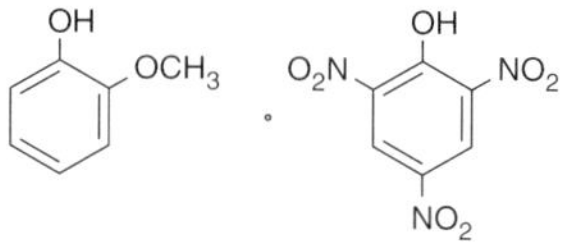
250 mg

S86,164-2
2-ALLYL-4-[2-(1,3-BENZODIOXOL-5-YL)1-METHYLETHYL]-5-METHOXYPHENOL
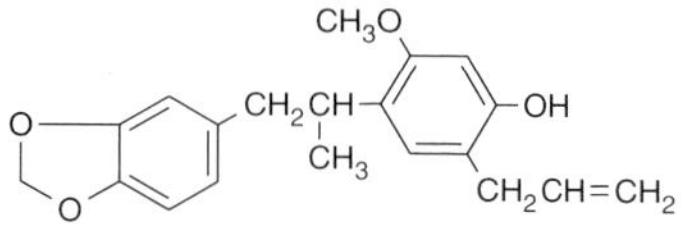
250 mg

S93,553-0
4-(METHOXYMETHOXY)-4'-HYDROXYSTILBENE
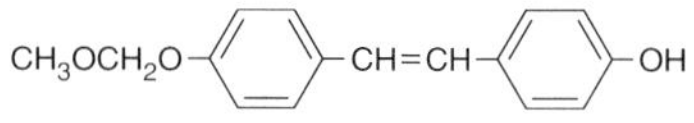
250 mg

S53,744-6
4-(HEXADECYLOXY)PHENOL
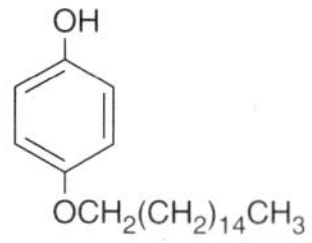
5 gm

S45,211-4
2-CHLORO-5-METHOXYPHENOL
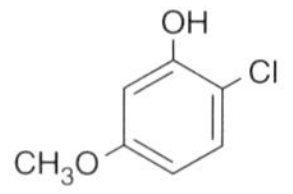
250 mg

S59,979-4
3,5-DIETHOXYPHENOL
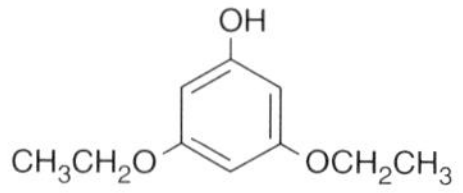
250 mg

S53,481-1
2,3-DICHLORO-4-METHOXYPHENOL
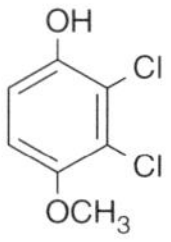
250 mg

S58,808-3
6-BROMO-4-METHOXY-M-CRESOL
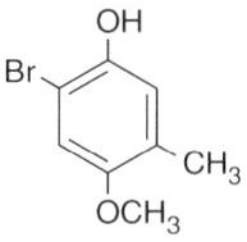
250 mg

S84,720-8
2-(2,3-DIHYDRO-7-ETHYL-2,4,4-TRIMETHYL-4H-1-BENZOPYRAN-2-YL)-5-ETHYLPHENOL
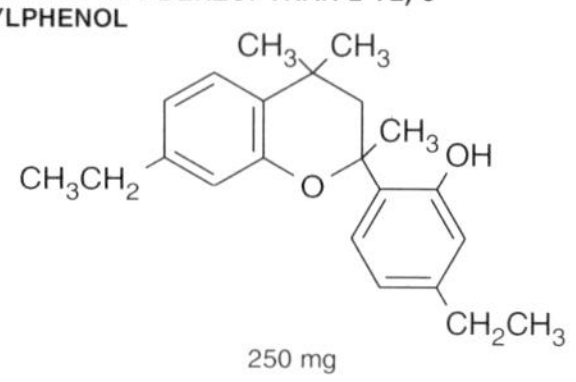
250 mg

S88,348-4
4,5-DIMETHYL-2-(3,4-DIHYDRO-2,4,4,6,7-PENTAMETHYL-2H-1-BENZOPYRAN-2-YL)PHENOL
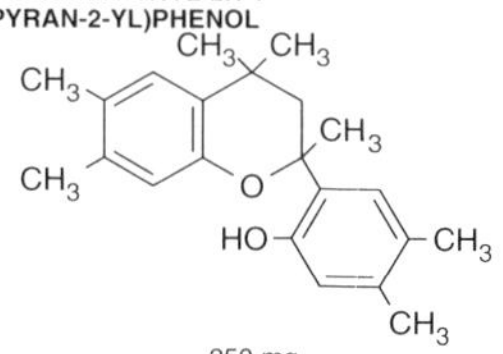
250 mg

S76,255-5
3-(2,4,5-TRIMETHYL-1,3-DIOXOLAN-2-YL)PHENOL
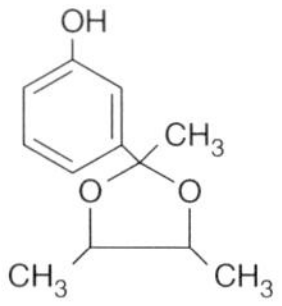
250 mg

S34,510-5
2-(2-CYCLOPENTENYL)PHENOL
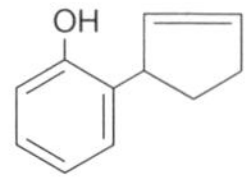
250 mg

S34,582-2
2,4-DI-(2-CYCLOPENTENYL)PHENOL
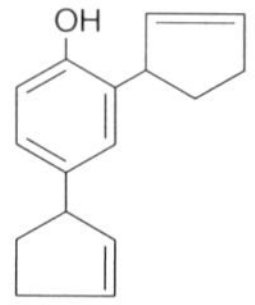
5 gm

S35,090-7
6-HYDROXY-5-ISOPROPENYL-3-METHYL-2,3-DIHYDROBENZOFURAN
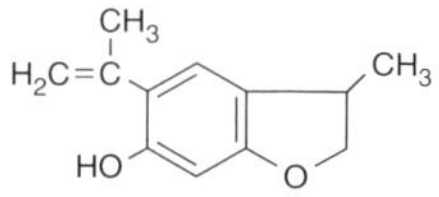
5 gm

S45,862-7
2-CYCLOHEXYLPHENOL
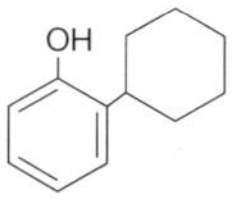
5 gm

S43,379-9
4-CHLORO-2-CYCLOHEXYLPHENOL
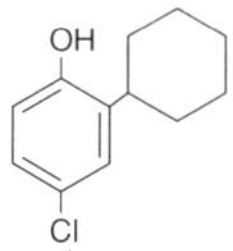
5 gm

S95,608-2
5-METHYL-5,6,7,8-TETRAHYDRO-2-NAPHTHOL
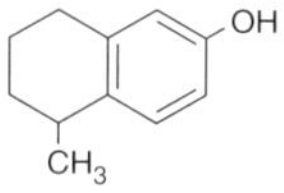
50 mg

S34,757-4
DIPENTENEMONOPHENOL
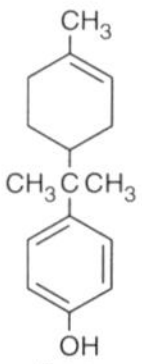
5 gm

S53,029-8
9-(4-HYDROXYPHENYL)-P-MENTH-1-ENE
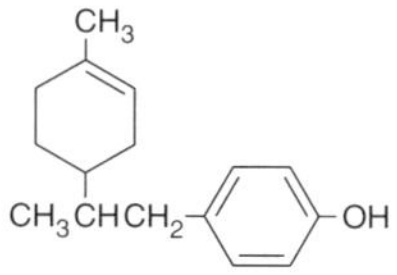
5 gm

S88,103-1
2-STILBENOL
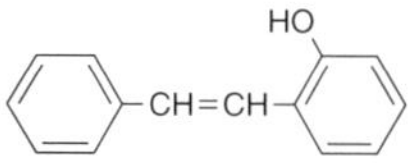
250 mg

S88,024-8
4-STILBENOL
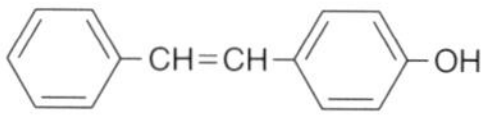
250 mg

S94,371-1
4-(1,1-DIPHENYLPROPYL) PHENOL
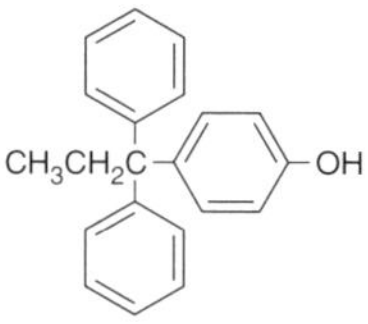
250 mg

S65,892-8
2-(2-BUTENYL)PHENOL
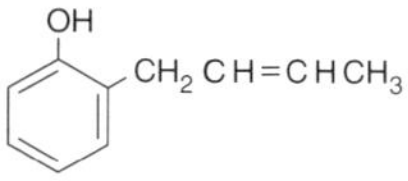
5 gm

S70,534-9
2-(2-METHYLALLYL)PHENOL
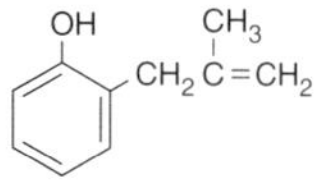
5 gm

S37,134-3
2-(1-METHYLBUTYL)PHENOL
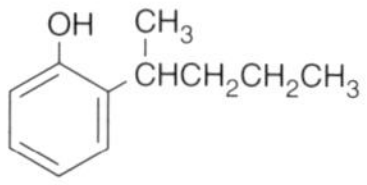
5 gm

S51,479-9
2-DODECYLPHENOL
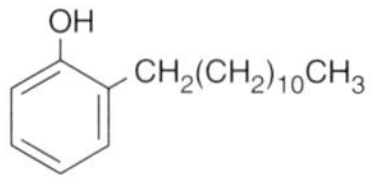
250 mg

S36,963-2
4-(3-METHYLCROTYL)PHENOL
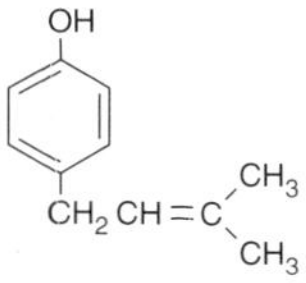
5 gm

S51,180-3
4-ISOPENTYLPHENOL
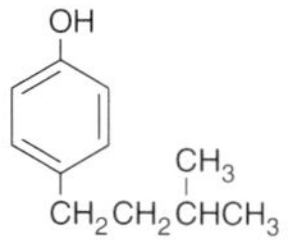
5 gm

S66,230-5
4-AMYLPHENOL
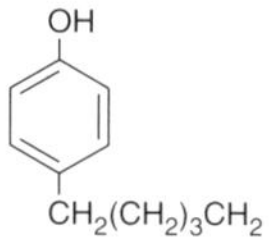
5 gm

S41,965-6
2-CHLORO-6-METHYLPHENOL
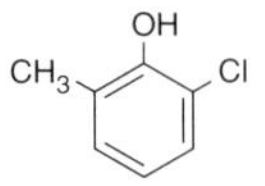
5 gm

S40,550-7
6-PROPYL-O-CRESOL
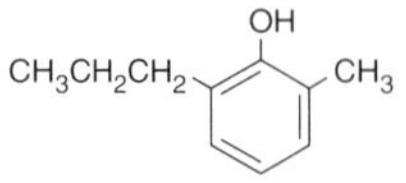
5 gm

S53,302-5
4-FLUORO-2-PROPYLPHENOL
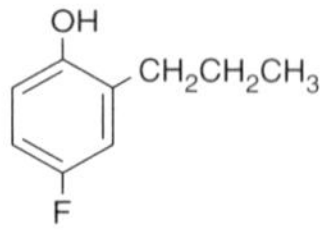
5 gm

S45,854-6
4-TERT-BUTYL-2-METHYLPHENOL
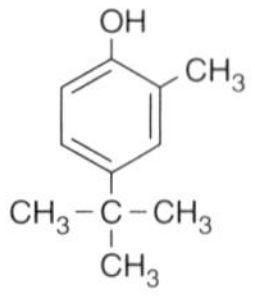
5 gm

S56,598-9
2-CHLORO-4-(TERT-PENTYL)PHENOL
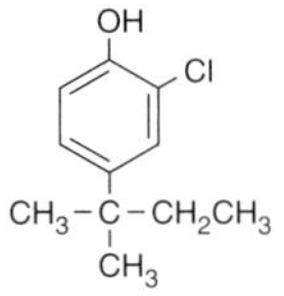
5 gm

S41,482-4
2,5-DIISOPROPYLPHENOL
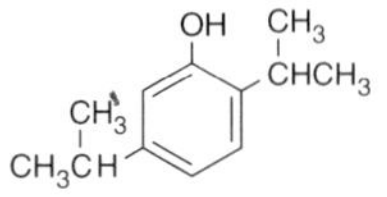
5 gm

S39,215-4
4-(1-ETHYL-1-METHYLPROPYL)-O-CRESOL
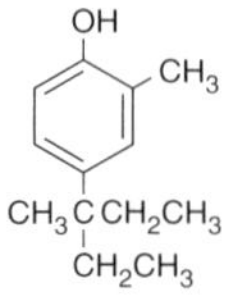
5 gm

S36,790-7
2,4-DI-SEC-BUTYLPHENOL
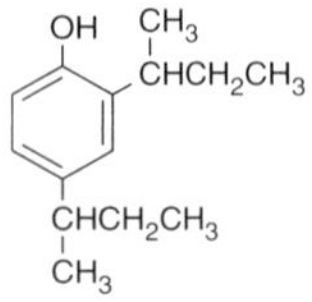
5 gm

S35,317-5
2-METHYL-4-TERT-OCTYLPHENOL
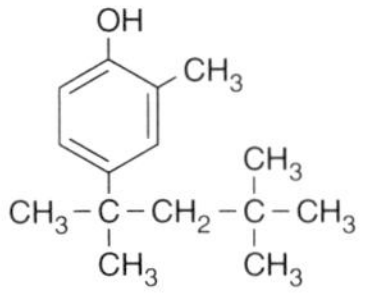
5 gm

S44,941-5
3,5-DIMETHYLPHENOL, ZONE REFINED
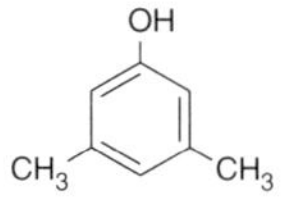
5 gm

S65,909-6
3-TERT-BUTYL-5-METHYLPHENOL
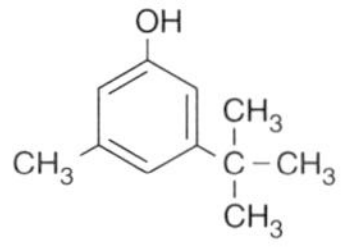
250 mg

S51,130-7
2,6-DIALLYL-M-CRESOL
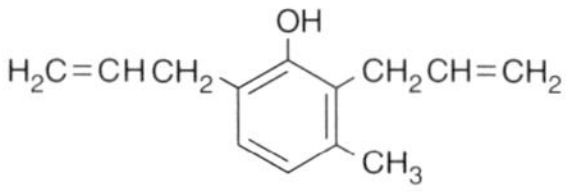
5 gm

S59,629-9
4-BROMO-2,6-DICHLOROPHENOL
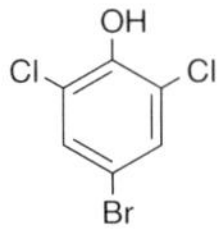
250 mg

S40,518-3
2,4-DICHLORO-6-IODOPHENOL
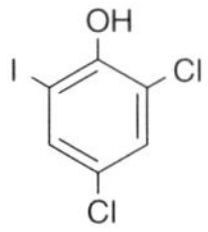
5 gm

S44,378-6
2,4-DICHLORO-6-METHYLPHENOL
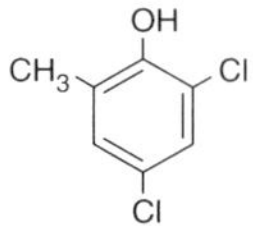
5 gm

S34,424-9
4-CHLORO-3-ETHYL-5-METHYLPHENOL
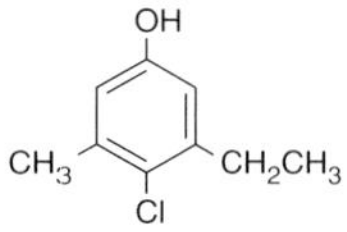
5 gm

S56,343-9
2,6-DICHLORO-4-(TERT-PENTYL)-PHENOL
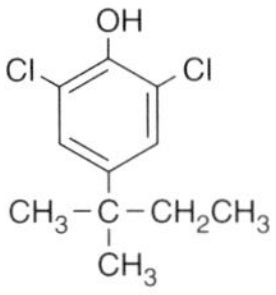
5 gm

S36,837-7
2,6-DIISOPROPYL-4-METHYLPHENOL
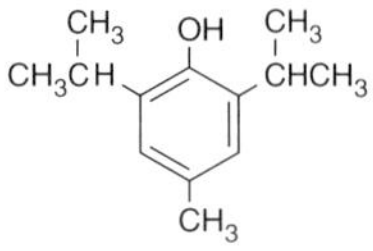
250 mg

S94,373-8
2,6-DIMETHYL-4-(1,1-DIMETHYL-PROPYL)PHENOL
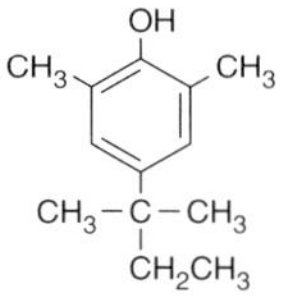
250 mg

S58,339-1
4-CHLORO-2,6-DI-(TERT-BUTYL)PHENOL
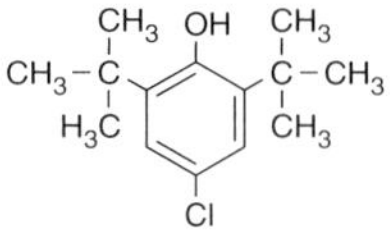
250 mg

S87,991-6
4,6-DI-TERT-BUTYL-2-METHYLPHENOL
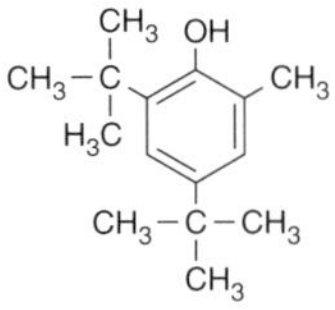
5 gm

S59,496-2
2,4-DIMETHYL-6-ISOOCTYLPHENOL
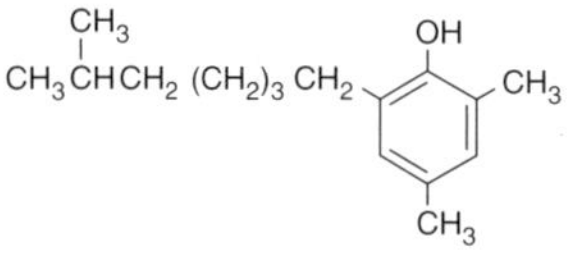
5 gm

S59,497-0
2,4-DIMETHYL-6-ISONONYLPHENOL
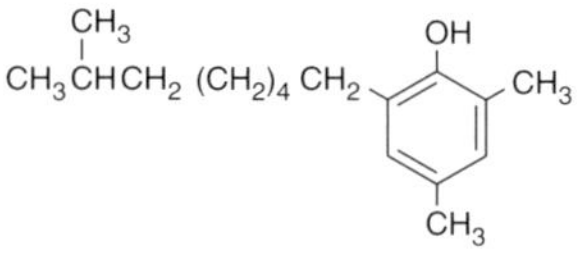
5 gm

S45,878-3
2,4,5-TRIMETHYLPHENOL
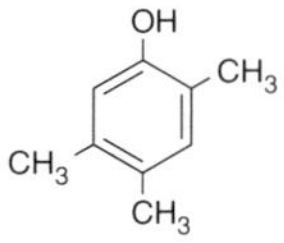
5 gm

S36,096-1
3,4-DIMETHYL-6-ETHYLPHENOL
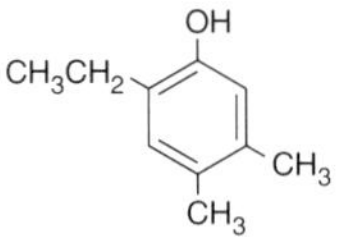
250 mg

S56,491-5
4-CHLORO-5-ISOPROPYL-O-CRESOL
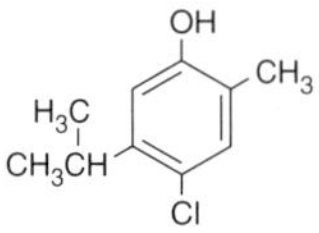
5 gm

S54,475-2
7-CHLORO-4-INDANOL
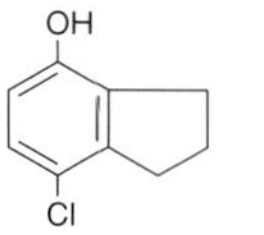
5 gm

S70,651-5
6-TERT-BUTYL-2,3-DIHYDRO-5-HYDROXYINDENE
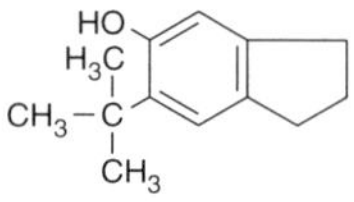
250 mg

S75,973-2
1,2,3,4,4A,9,10,10A-OCTAHYDRO-1,1,4A-TRIMETHYL-6-PHENANTHRENOL
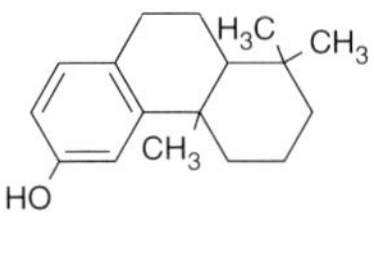
250 mg

S83,525-0
4-CHLOROCATECHOL
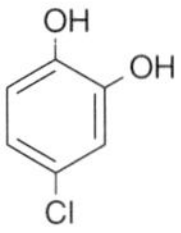
250 mg

S44,887-7
3-METHYLCATECHOL
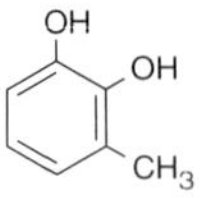
5 gm

S37,237-4
4-TERT-OCTYLCATECHOL
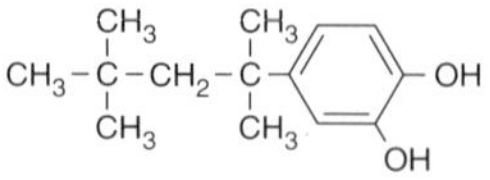
5 gm

S35,178-4
3-ISOPROPYL-6-METHYLCATECHOL

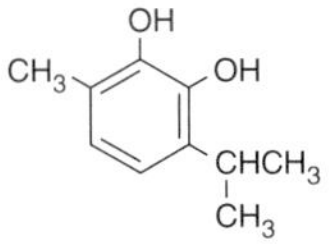

5 gm

S36,716-8
6-TERT-BUTYL-3-METHYLCATECHOL

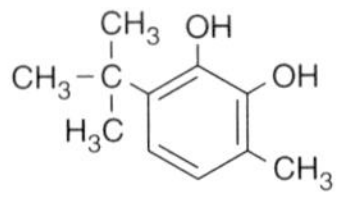

5 gm

S35,924-6
5-TERT-BUTYL-3-METHYLCATECHOL

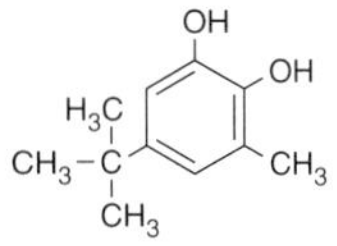

5 gm

S35,315-9
3-METHYL-5-TERT-OCTYLCATECHOL

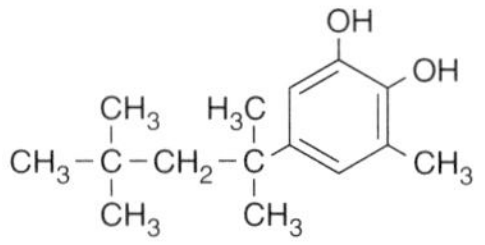

5 gm

S35,316-7
5-METHYL-3-TERT-OCTYLCATECHOL

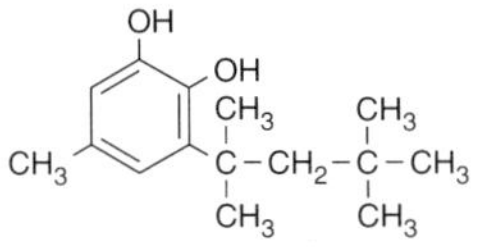

5 gm

S72,239-1
3-PHENYLCATECHOL

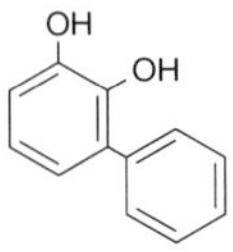

250 mg

S91,998-5
2,5-DIMETHOXYRESORCINOL

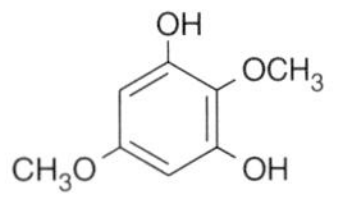

250 mg

S34,502-4
4-CYCLOHEXYLRESORCINOL

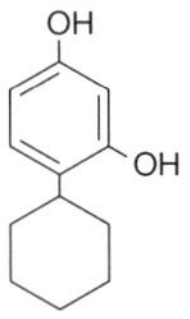

5 gm

S88,813-3
5-PENTADECYLRESORCINOL

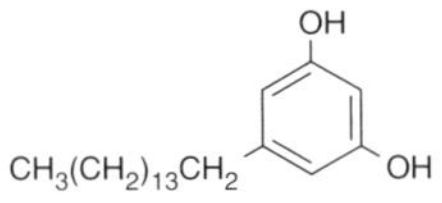

5 gm

S41,491-3
2,4,6-TRIHYDROXYTOLUENE

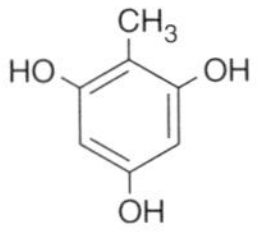

5 gm

S93,394-5
2,5-DIMETHYLRESORCINOL

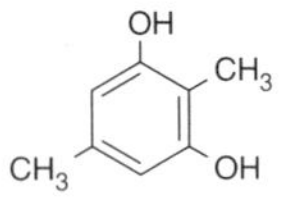

250 mg

S45,876-7
4-PROPYLRESORCINOL

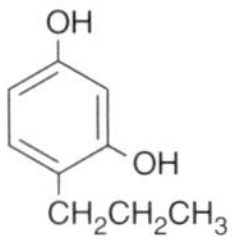

5 gm

S51,965-0
TETRABROMOHYDROQUINONE

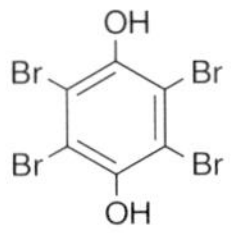

5 gm

S91,945-4
2,5-DICHLOROHYDROQUINONE

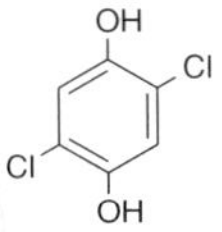

250 mg

S69,640-4
2-BROMO-5-METHYLHYDROQUINONE

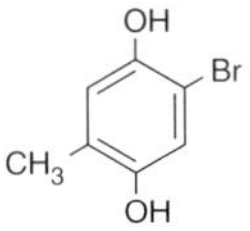

250 mg

S84,288-5
2-PHENYL-5-(2,4-XYLYL)-1,4-HYDROQUINONE

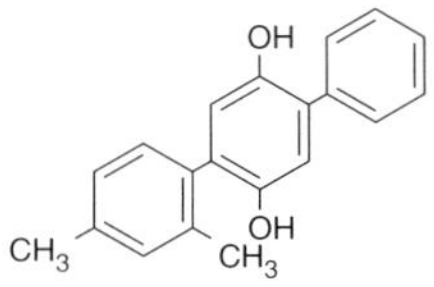

250 mg

S84,126-9
2-ISOPROPYLHYDROQUINONE

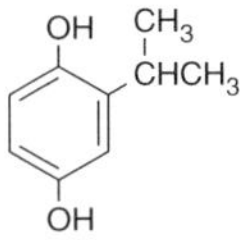

250 mg

S58,272-7
2-DODECYLHYDROQUINONE

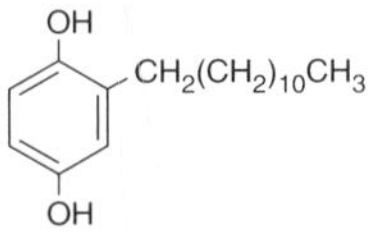

250 mg

S64,291-6
BROMOHYDROQUINONE

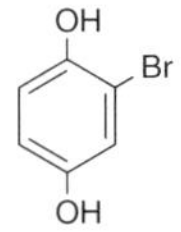

250 mg

S40,736-4
2,5-DI-TERT-AMYLHYDROQUINONE

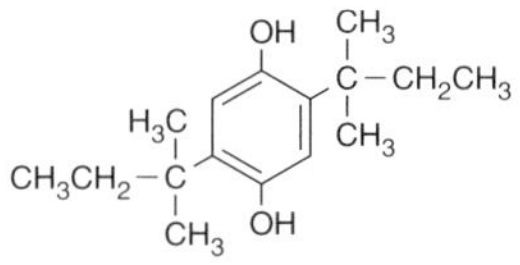

5 gm

S84,275-3
4,4',6,6'-TETRABROMO-2,2'-BIPHENOL

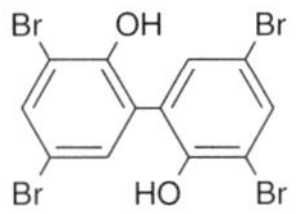

250 mg

S69,839-3
1,1'-(4-CHLOROBENZYLIDENE)DI-(2-NAPHTHOL)

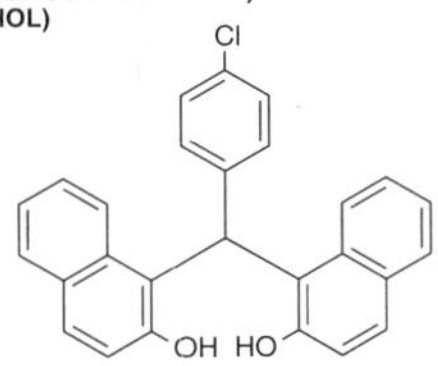

5 gm

S36,354-5
2,2'-METHYLENEBIS(4-CHLORO-6-METHYLPHENOL)

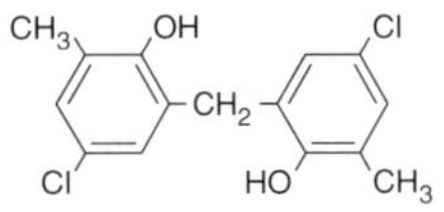

5 gm

S88,578-9
3,3'-DIIODO-2,2',6,6'-TETRAMETHYL-4,4'-BIPHENOL

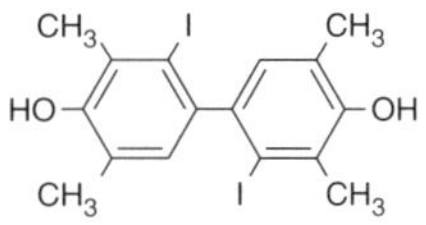

250 mg

S84,646-5
2-BROMO-6-METHOXY-4-PHENYLPHENOL

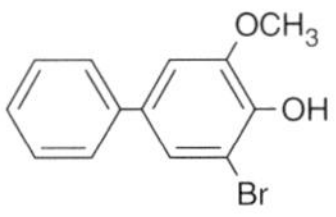

250 mg

S77,526-6
3-CHLORO-2,4,6-TRIPHENYLPHENOL

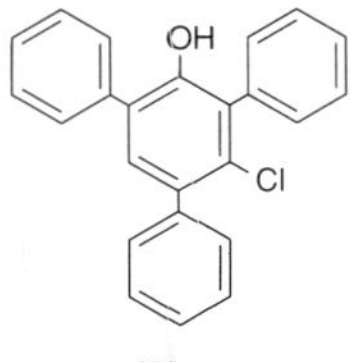

250 mg

S92,374-5
4-(4-BROMOPHENYL)-2,6-DIPHENYLPHENOL

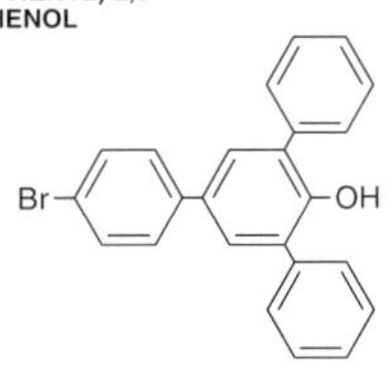

250 mg

S84,641-4
2-BROMO-4-PHENYLPHENOL

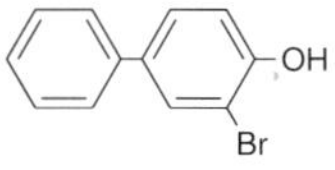

250 mg

S84,686-4
2-IODO-4-PHENYLPHENOL

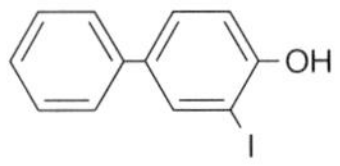

250 mg

S70,533-0
2-(2-CHLOROPHENYL)PHENOL

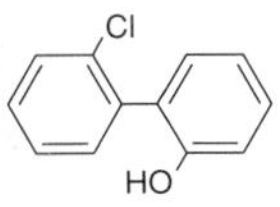

250 mg

S84,640-6
2-(4-BROMOPHENYL)-4,6-DIBROMOPHENOL

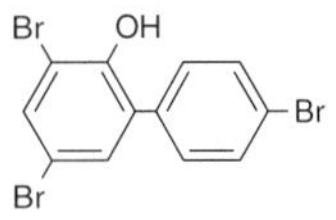

250 mg

S88,581-9
6,6'-DIMETHOXY-4,4'-DIMETHYL-2,2'-BIPHENOL

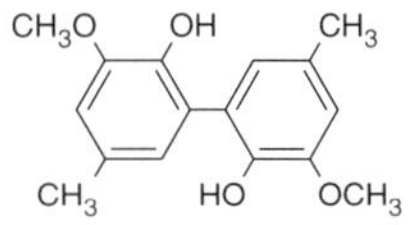

250 mg

S43,028-5
HYDROCOERULIGNONE
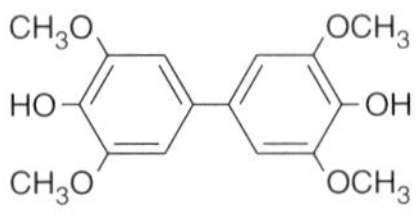
250 mg

S70,140-8
1,1'-(2-METHOXYBENZYLIDENE)DI-2-NAPHTHOL
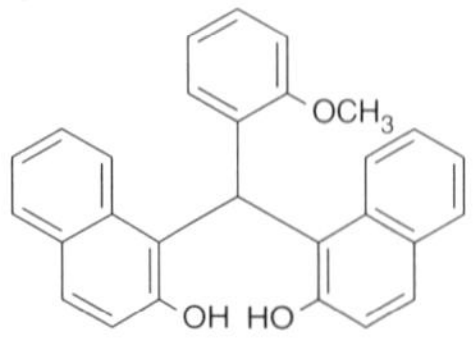
5 gm

S44,022-1
4,4'-DIHYDROXYDIPHENYL ETHER
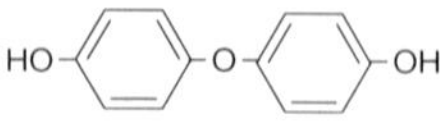
5 gm

S69,625-0
1,1'-(4-METHOXYBENZYLIDENE)DI-2-NAPHTHOL
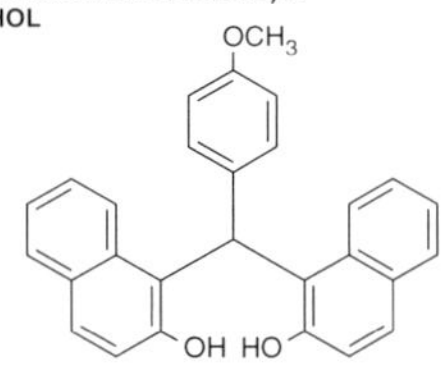
5 gm

S88,574-6
9,10-DIHYDRO-9,10-DIMETHYL-9,10-ETHANOANTHRACENE-2,3,6,7-TETROL
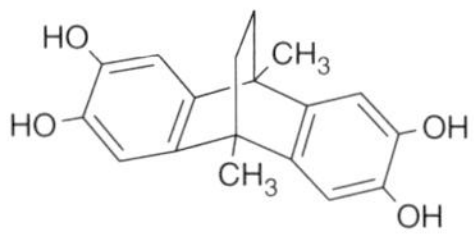
250 mg

S84,675-9
3-(3,4-DIHYDROXYPHENYL)-1-ETHYL-2-METHYL-5,6-INDANDIOL
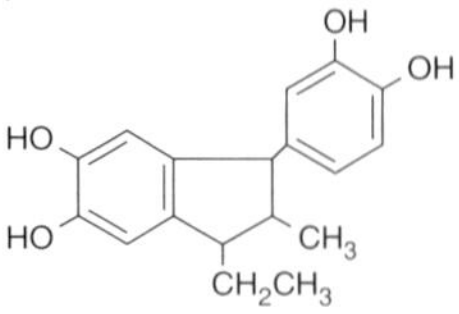
250 mg

S84,335-0
5,6-INDANDIOL
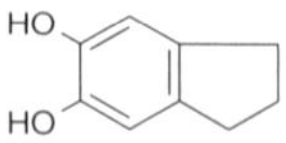
250 mg

S70,693-0
4-PHENYLRESORCINOL
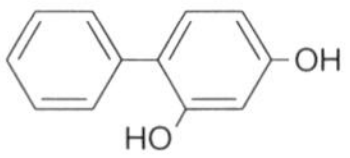
250 mg

S85,037-3
5,6,11,12-TETRAHYDROXYDIBENZO(B,H)-BIPHENYLENE
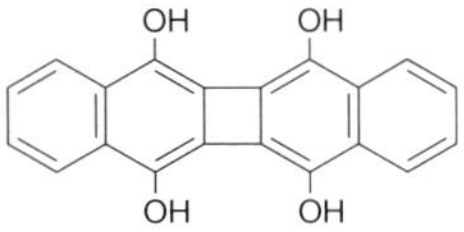
250 mg

S79,396-5
3-(1,2-PROPADIEN-1-YL)-2,2'-BIPHENOL
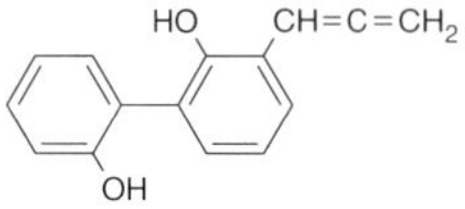
250 mg

S79,395-7
3-ALLYL-2,2'-BIPHENOL
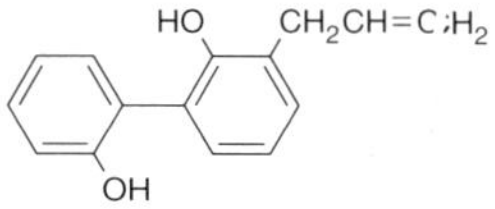
250 mg

S80,377-4
2-BROMO-3-METHYL-1,4-NAPHTHALENEDIOL
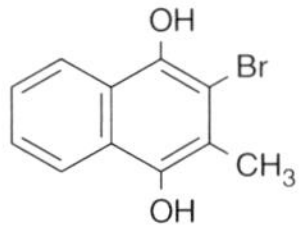
250 mg

S97,312-2
AR-D 8 1,9-DIHYDROXYPHENANTHRENE
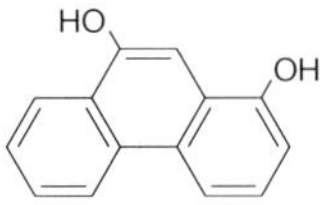
10 mg

S36,691-9
4,4'-CYCLOPENTYLIDENEDIPHENOL
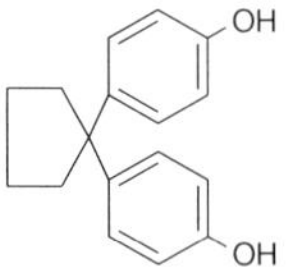
250 mg

S94,770-9
4-[2-HYDROXYOCTAHYDRO-1,3,4-METHENO-2H-CYCLOBUTA(CD)PENTA LEN-2-YL]PHENOL
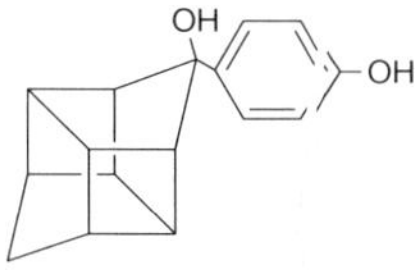
5 gm

S94,927-2
1,1,6,6-TETRAPHENYL-1,6-HEXANEDIOL
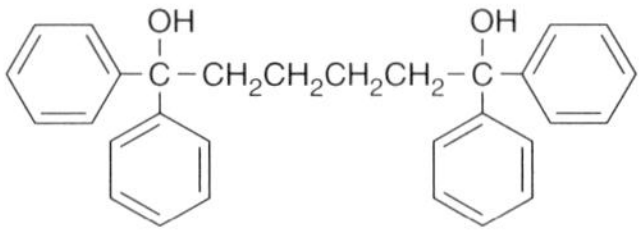
250 mg

S92,117-3
4,4''-DIHYDROXY-3,3'',5,5''-TETRAPHENYL-P-TERPHENYL
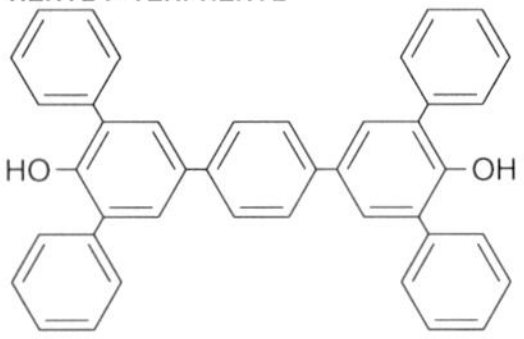
250 mg

S70,621-3
2,4-BIS(4-HYDROXYPHENYL)-4-METHYL-1-PENTENE
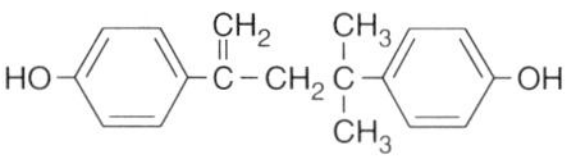
5 gm

S91,978-0
2,4',4''-TRIHYDROXY-P-TERPHENYL
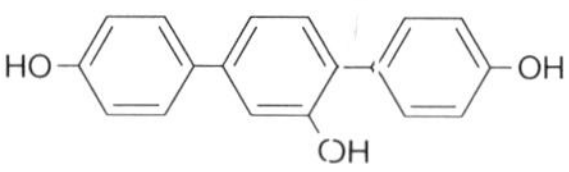
250 mg

S36,692-7
4,4'-(1,3-DIMETHYLBUTYLIDENE)-DIPHENOL
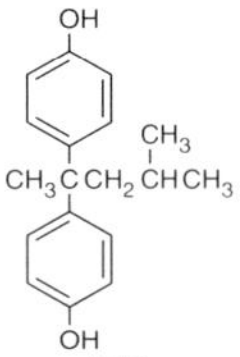
5 gm

S69,841-5
1,1'-(4-METHYLBENZYLIDENE)DI (2-NAPHTHOL)
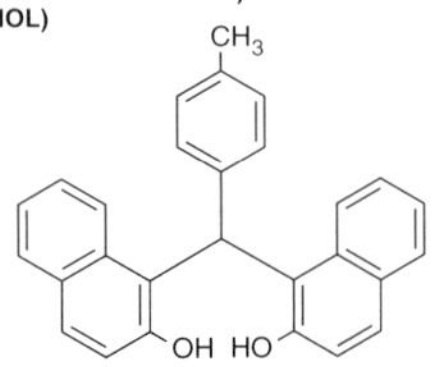
5 gm

S70,327-3
1,1'-(4-ISOPROPYLBENZYLIDENE)DI-2-NAPHTHOL
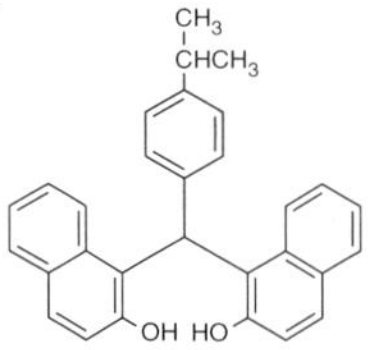
5 gm

S92,673-6
2,6-BIS(2-HYDROXY-5-METHYLBENZYL)-4-METHYLPHENOL
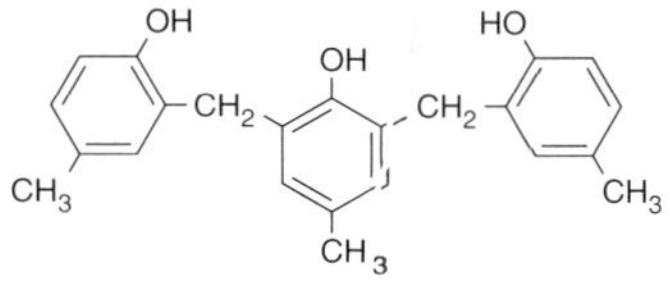
250 mg

S92,674-4
2,2'-METHYLENEBIS(4-METHYLPHENOL)
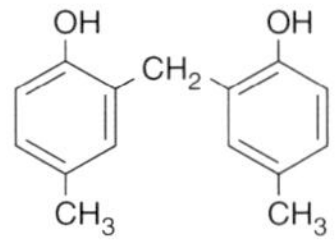
250 mg

S88,313-1
3,3'-DIMETHYL-4,4'-BIPHENOL
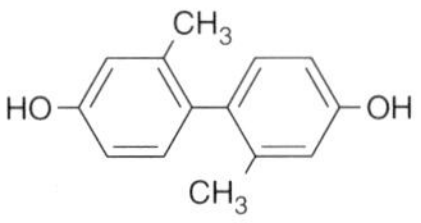
250 mg

S37,309-5
2,2-BIS(3,5-DIMETHYL-4-HYDROXYPHENYL)PROPANE
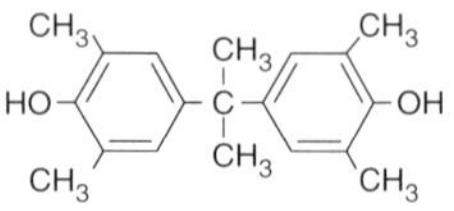
5 gm

S59,651-5
1,1-BIS(2-HYDROXY-3,5-DIMETHYLPHENYL)ISONONANE
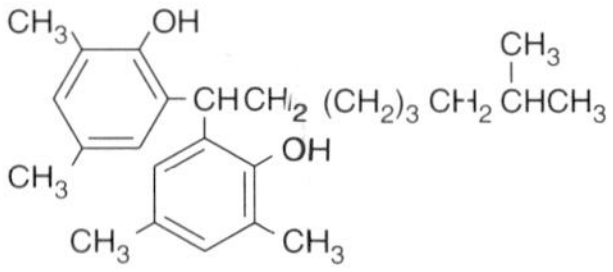
250 mg

S59,498-9
2,2'-BUTYLIDENEBIS(6-TERT-BUTYL-P-CRESOL)
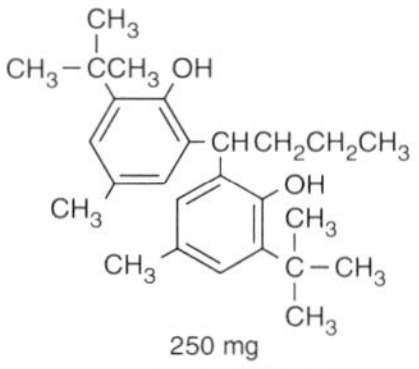
250 mg

S50,284-7
ALPHA2,ALPHA'2-(2,3,5,6-TETRAMETHYL-1,4-PHENYLENE)BIS(6-TERT-BUTYL-2,4-XYLENOL)
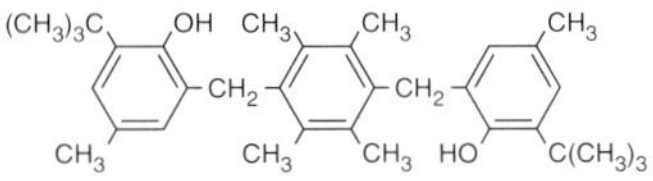
5 gm

S84,705-4
3,3',5,5'-TETRAMETHYL-4,4'-BISPHENOL
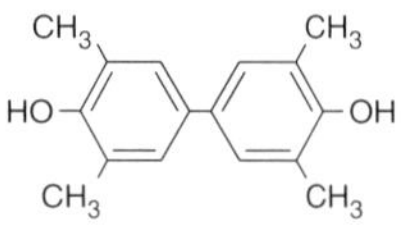
250 mg

S41,256-2
6,6'-DI-TERT-BUTYL-4,4'-BI-O-CRESOL
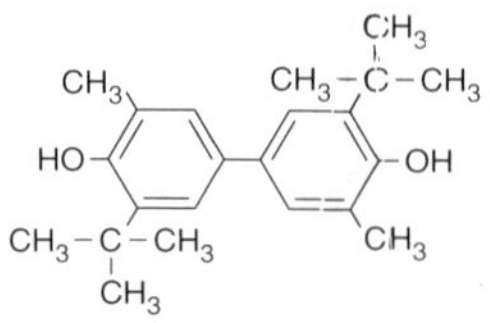
5 gm

S34,640-3
4,4'-DIHYDROXY-3,3',5,5'-TETRA-TERT-BUTYLBIPHENYL
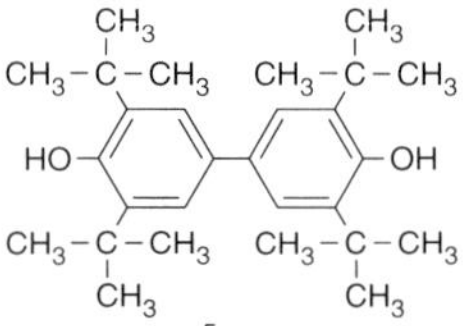
5 gm

S84,670-8
3,3'-BIPHENOL

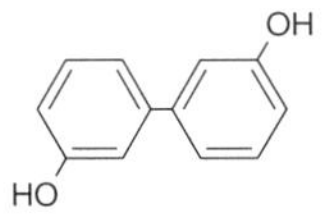

250 mg

S80,940-3
6,6'-DIHYDROXY-5,5'DIMETHYL-2,2'-3,3'-TETRAHYDRO-3,3,3',3'-TETRAMETHYL-1,1'-SPIROBI(1H)INDENE

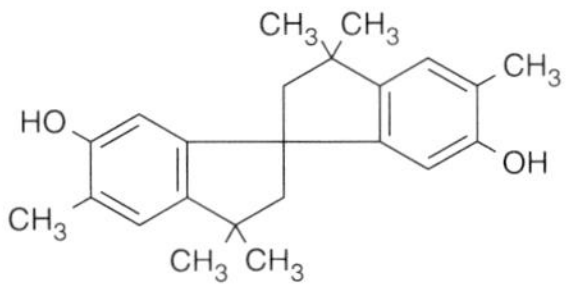

250 mg

S94,477-7
1,2-DIMETHYLACENAPHTHENE-1,2-DIOL

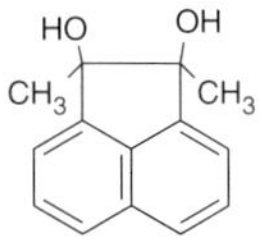

250 mg

S97,505-2
AR-H 8 1,6-DIHYDROXYPYRENE

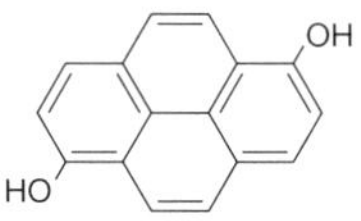

10 mg

S76,237-7
ALPHA,ALPHA-BIS(3-ETHOXY-2-NAPHTHYL)BENZYL ALCOHOL

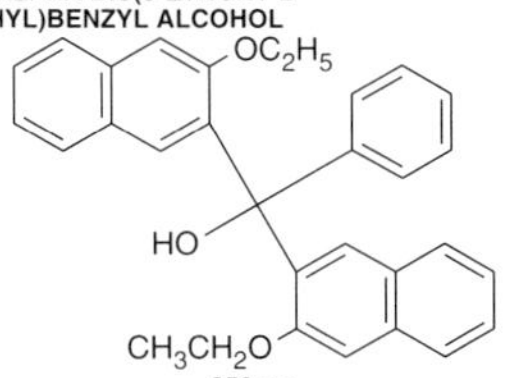

250 mg

S75,971-6
6,7-DIMETHOXY-1-METHYL-2-NAPHTHOL

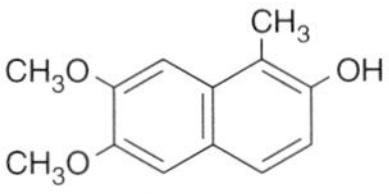

250 mg

S70,095-9
1,1'-(3,4-DIMETHOXYBENZYLIDENE)DI-2-NAPHTHOL

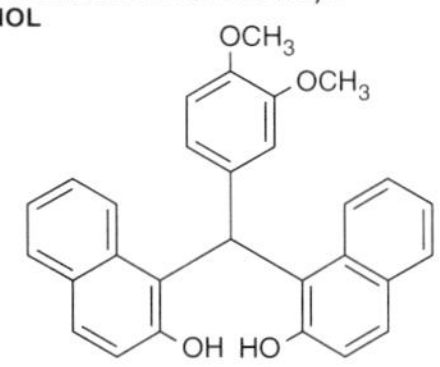

250 mg

S94,776-8
5,5-DIMETHYL-2-(4-HYDROXY-3-METHOXYPHENYL)-1,3-DIOXANE

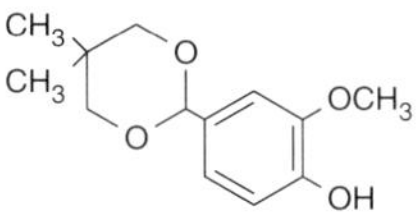

5 gm

S95,600-7
2-HYDROXY-3-METHOXYNAPHTHALENE

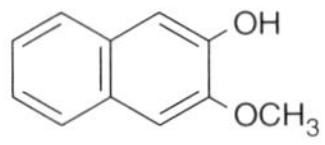

50 mg

S85,308-9
3-METHOXY-2-ANTHRACENOL

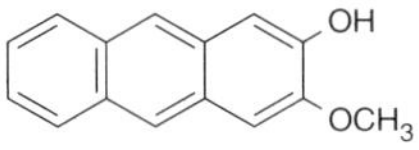

250 mg

S79,363-9
2-(2-ALLYLOXYPHENYL)PHENOL

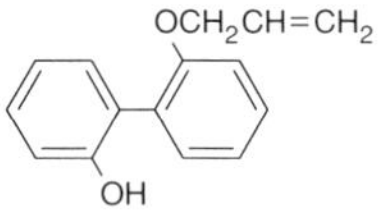

250 mg

S93,726-6
5-METHOXY-1,2,3,4-TETRAHYDRO-1-NAPHTHOL

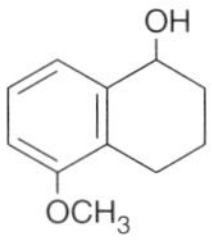

250 mg

S35,095-8
2-BENZYL-5-METHOXYPHENOL

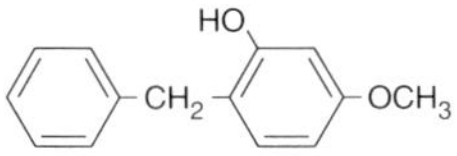

5 gm

S91,408-8
TRIS(P-TOLYL)METHANOL

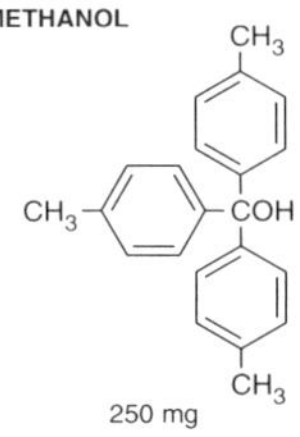

250 mg

S80,929-2
4-HYDROXY-1,3-BENZODIOXOLE

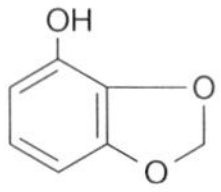

250 mg

S58,897-0
1,4,6-TRIBROMO-2,3-NAPHTHALENEDIOL

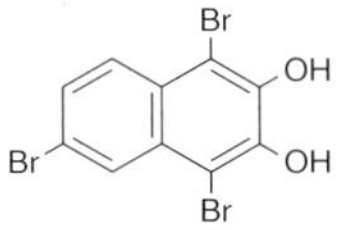

5 gm

S58,895-4
1,4-DIBROMO-2,3-NAPHTHALENEDIOL

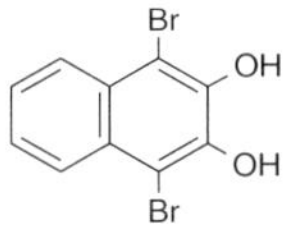

250 mg

S80,935-7
2,3-DIHYDRO-2-(2-HYDROXY-4-METHYLPHENYL)-2,4,4,7-TETRAMETHYL-4H-1-BENZOPYRAN

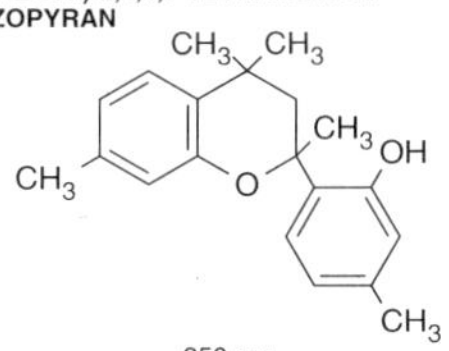

250 mg

S97,310-6
AR-D 6 4,5-DIHYDROXYPHENANTHRENE

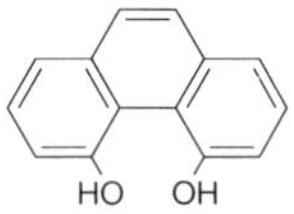

10 mg

S36,458-4
1-PIPERIDINOMETHYL-2-NAPHTHOL

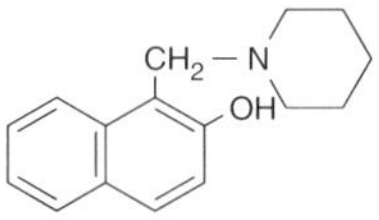

5 gm

S95,645-7
6-HYDROXY-1-PHENYL-1,2,3,4-TETRAHYDRONAPHTHALENE

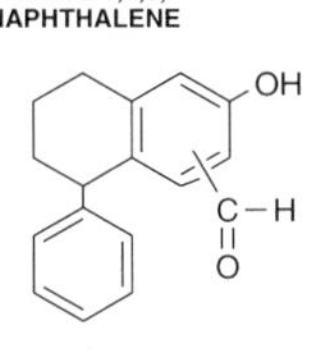

50 mg

S51,924-3
3-METHYL-1,8,9-ANTHRACENETRIOL

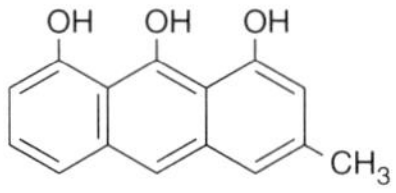

250 mg

S92,840-2
2,3-DIHYDRO-3,3-DIMETHYL-5-HYDROXY-2-(4-MORPHOLINO)BENZOFURAN

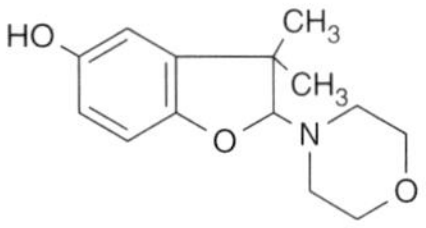

250 mg

S74,206-6
4-(3,5-DIPHENYLCYCLOHEXYL)PHENOL

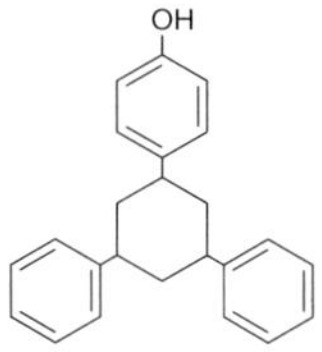

5 gm

S94,737-7
1-(9-PHENANTHRYL)-1-CYCLOHEXANOL

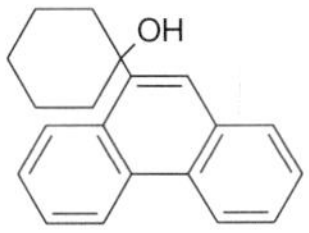

250 mg

S34,756-6
DIPENTENEBISPHENOL

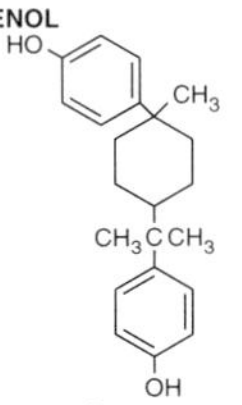

5 gm

S53,025-5
2,9-BIS(4-HYDROXYPHENYL)-P-MENTHANE

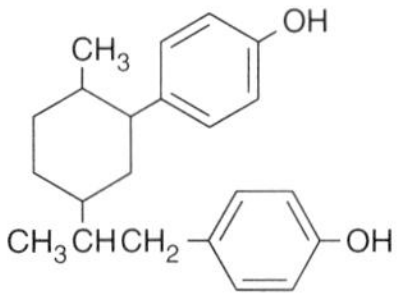

5 gm

S40,681-3
1,1-BIS(4-HYDROXY-3-METHYLPHENYL)-CYCLOHEXANE

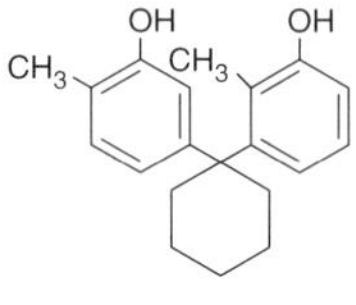

5 gm

S34,755-8
2,9-BIS(4-HYDROXY-3-METHYLPHENYL)-HEXAHYDRO-P-CYMENE

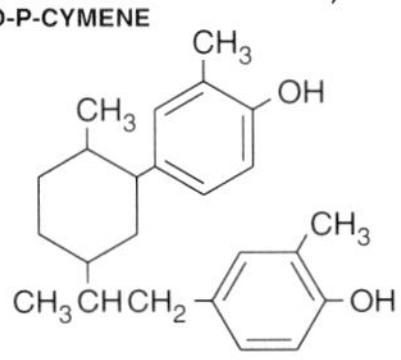

5 gm

S66,154-6
2-HYDROXY-5,6,7,8-TETRAHYDROPHENANTHRENE

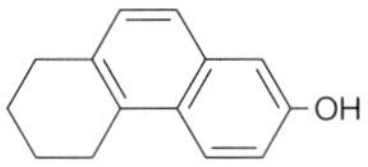

250 mg

S36,266-2
4-(ALPHA-METHYLBENZYL)PHENOL

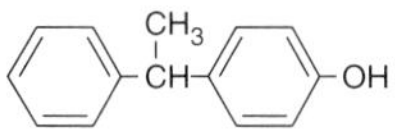

5 gm

S40,669-4
2-BENZYL-4-METHYLPHENOL

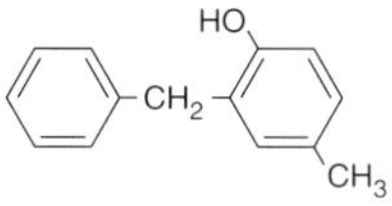

5 gm

S40,663-5
2-BENZYL-4-CHLOROPHENOL

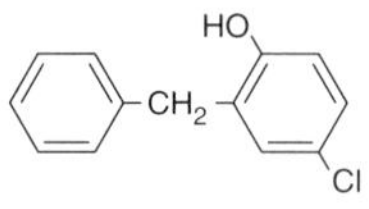

5 gm

S44,756-0
2-CHLORO-4-PHENYLPHENOL

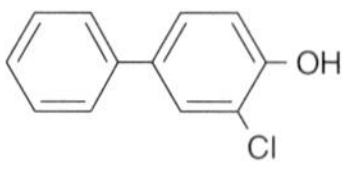

5 gm

S59,499-7
6-(ALPHA-METHYLBENZYL)-2,4-XYLENOL

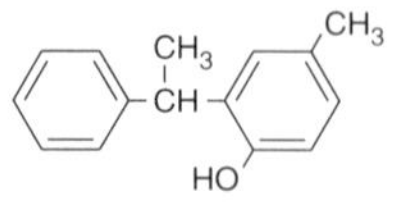

5 gm

S36,015-5
2,6-DIBENZYL-4-METHYLPHENOL

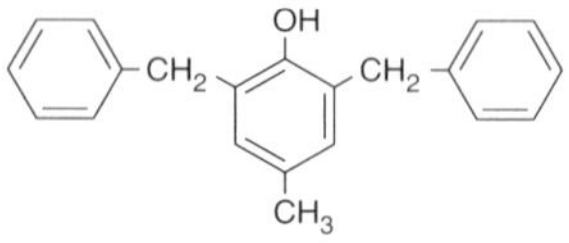

5 gm

S92,082-7
4-(4-BIPHENYLYL)-2,6-DIPHENYLPHENOL

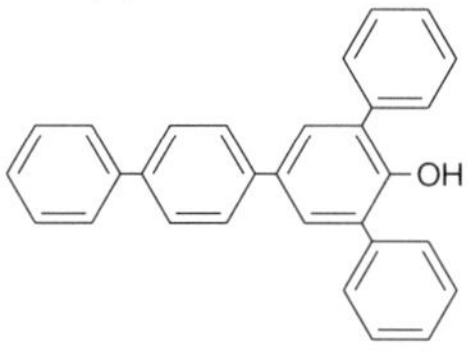

250 mg

S95,641-4
5-PHENYL-5,6,7,8-TETRAHYDRO-2-NAPHTHOL

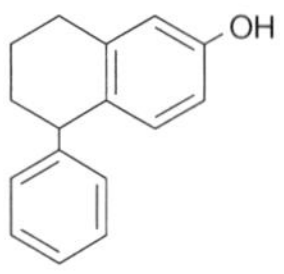

50 mg

S92,134-3
2,4-DIPHENYL-1-NAPHTHOL

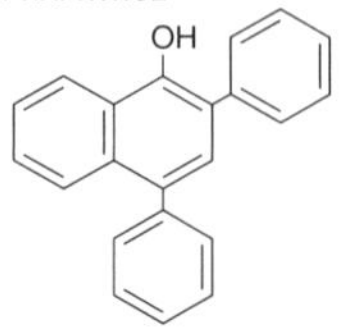

250 mg

S64,635-0
2-(4-METHYLBENZYL)PHENOL

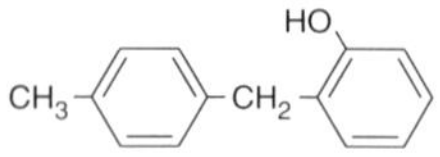

250 mg

S92,691-4
2-(9-FLUORENYLIDENEMETHYL)PHENOL

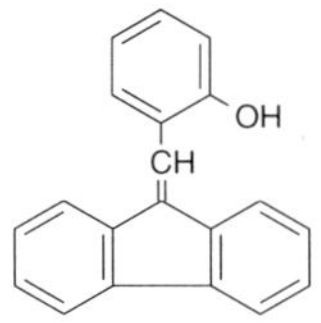

250 mg

S58,009-0
TRANS-ALPHA,ALPHA'-DIETHYL-P,P'-STILBENEDIOL

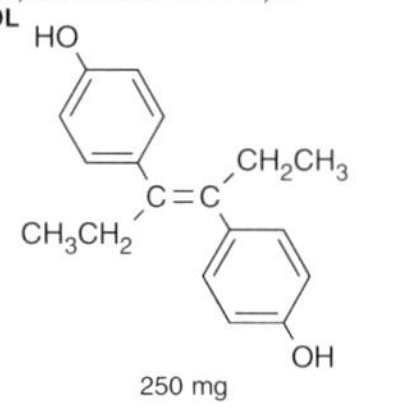

250 mg

S45,436-2
2,2-BIS(4-HYXROXYPHENYL)BUTANE

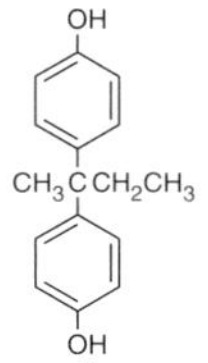

5 gm

S35,137-7
4-(1-INDANYL)PHENOL

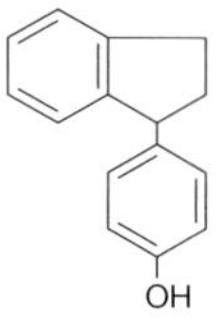

5 gm

S92,692-2
4-(9-FLUOROENYLIDENEMETHYL)PHENOL

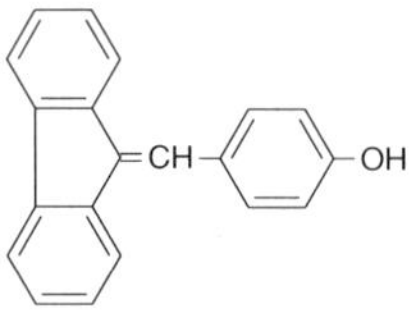

250 mg

S34,226-2
2,2-BIS(4-HYDROXY-3-METHYLPHENYL)-4-METHYLPENTANE

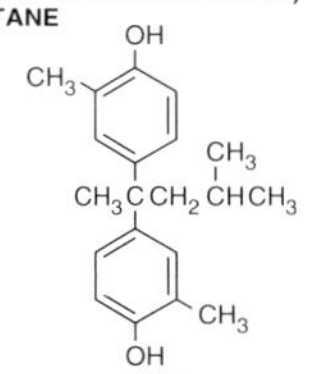

5 gm

S34,212-2
2,6-BIS(3-TERT-BUTYL-2-HYDROXY-5-METHYLBENZYL)-4-METHYLPHENOL

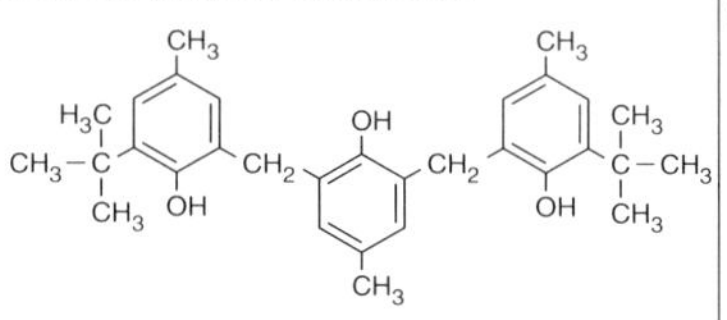

5 gm

S36,530-0
1,3,5-TRIMETHYL-2,4,6-TRIS(3,5-DI-TERT-BUTYL-4-HYDROXYBENZYL)-BENZENE

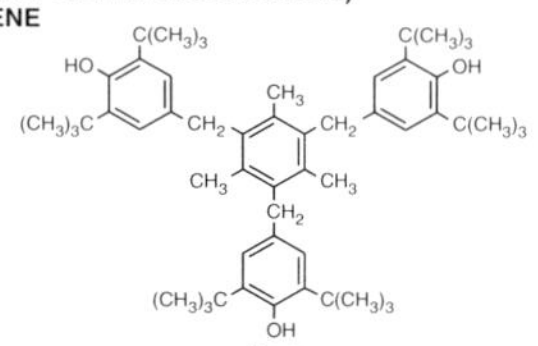

5 gm

S37,322-2
4,4'-BUTYLIDENEBIS(6-TERT-BUTYL-M-CRESOL)

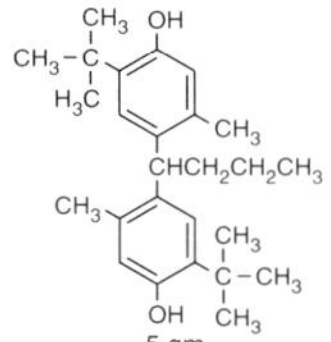

5 gm

S36,532-7
1,2,3-TRIPHENYL-1-INDENOL

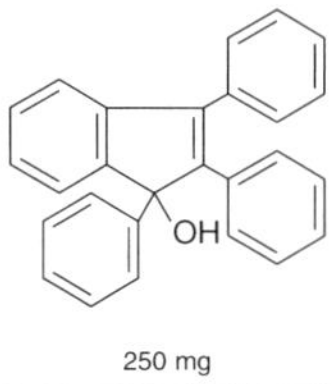

250 mg

S94,793-8
4,5,5,8,8-PENTAMETHYL-5,6,7,8-TETRAHYDRO-2-NAPHTHOL

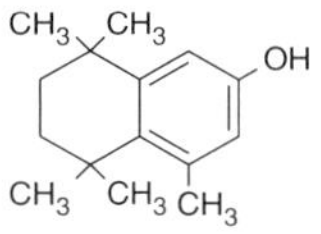

250 mg

S46,286-1
1,7-DIHYDROXYNAPHTHALENE

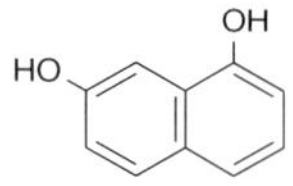

5 gm

S94,655-9
2-BROMO-1,2,3,4-TETRAHYDRO-1-NAPHTHOL

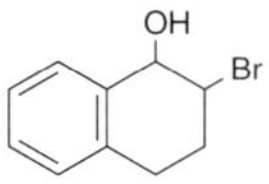

250 mg

S45,001-4
1-IODO-2-NAPHTHOL

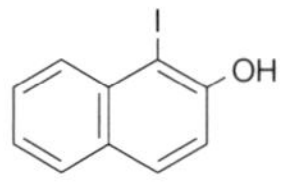

5 gm

S70,978-6
1-METHYL-2-NAPHTHOL

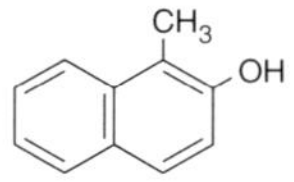

250 mg

S34,877-5
2,6-DIBROMO-1,5-DIHYDROXYNAPHTHALENE

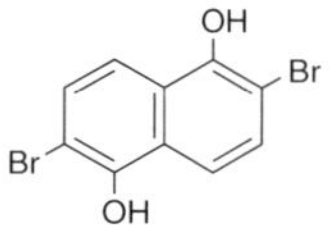

5 gm

S56,631-4
2-NAPHTHOL, BISMUTH (III) SALT

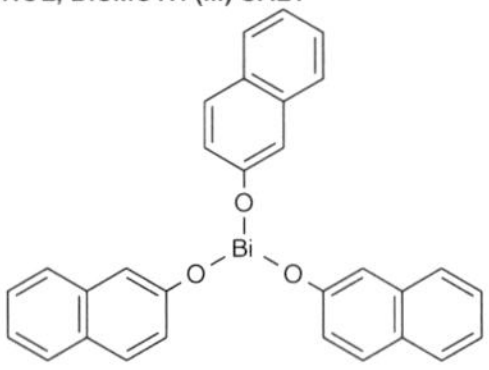

5 gm

S81,948-4
3-CHLORO-2-NAPHTHOL

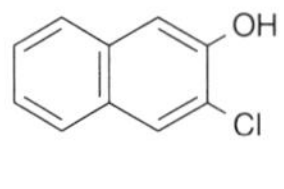

250 mg

S97,533-8
AR-L 7 6-HYDROXYBENZO(A)PYRENE

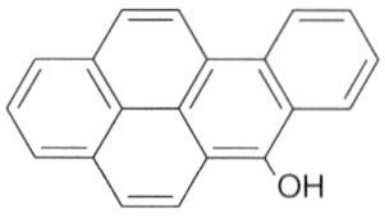

10 mg

S97,502-8
AR-H 5 4-HYDROXYPYRENE

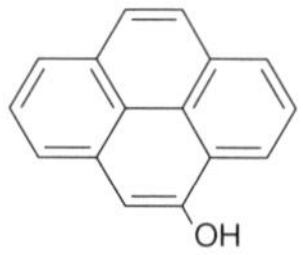

10 mg

S97,503-6
AR-H 6 2-HYDROXYPYRENE

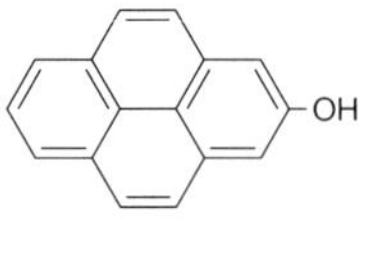

10 mg

S97,453-6
AR-G 4 3-HYDROXYBENZO(C)-PHENANTHRENE

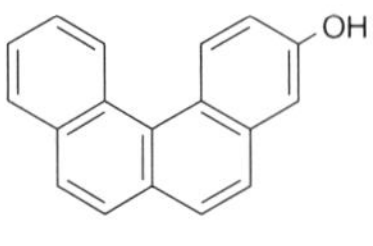

10 mg

S97,454-4
AR-G 5 1-HYDROXYBENZO(C)-PHENANTHRENE

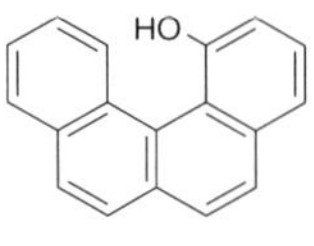

10 mg

S97,455-2
AR-G 6 2-HYDROXYBENZO(C)-PHENANTHRENE

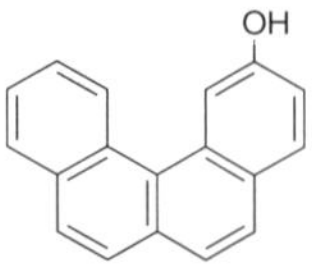

10 mg

S74,914-1
4-HYDROXYFLUORENE

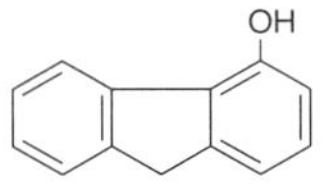

250 mg

S97,411-0
AR-E74 7,12-DIMETHYL-9-HYDROXYBENZ (A)ANTHRACENE

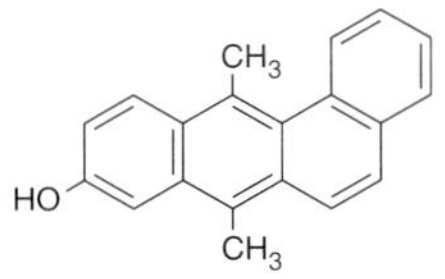

10 mg

S44,162-7
1,1PR-METHYLENEDI-2-NAPHTHOL

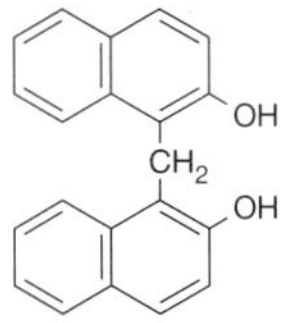

250 mg

S92,672-8
1-CYCLOPENTENE-1,2-DICARBOXYLIC ACID

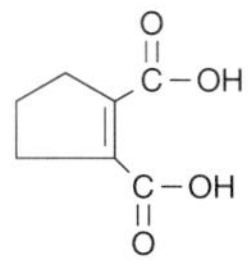

250 mg

S56,430-3
THYMOL IODIDE

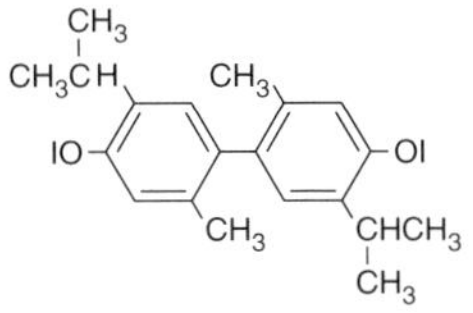

5 gm

S37,824-0
1-CHLORO-3-PHENOXY-2-PROPANOL

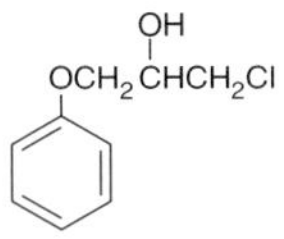

5 gm

S88,342-5
1-CHLORO-3-(5,6,7,8-TETRAFLUORO-1-NAPHTHOXY)-2-PROPANOL

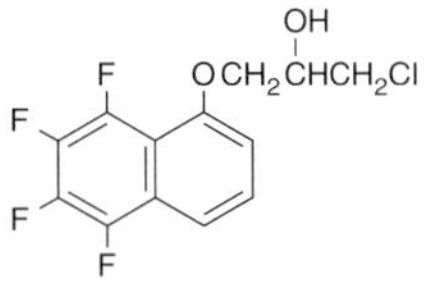

250 mg

S83,986-8
2-CHLORO-1,1-DIPHENYLETHANOL

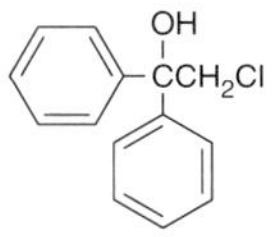

250 mg

S86,526-5
2-BROMO-1,1-DIPHENYL-1-ETHANOL

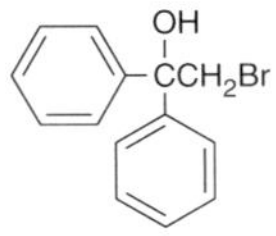

250 mg

S37,754-6
1-CHLORO-3-MESITYLOXY-2-PROPANOL

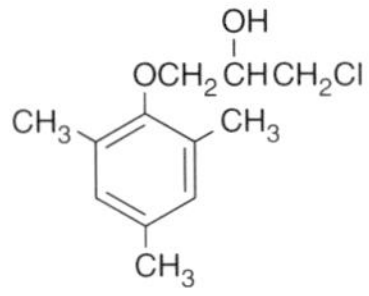

250 mg

S82,574-3
1-CHLORO-3-PHENYL-2-PROPANOL

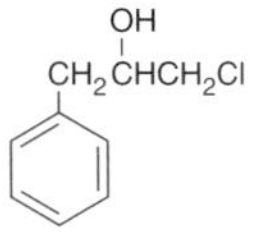

250 mg

S83,843-8
1-CHLORO-3-(P-TOLYL)-2-PROPANOL

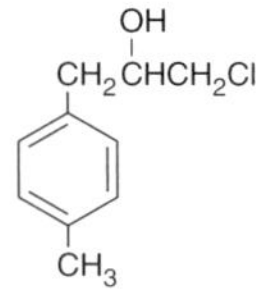

250 mg

S40,752-6
4,4'-DICHLORO-ALPHA-(FLUOROMETHYL)-BENZHYDROL

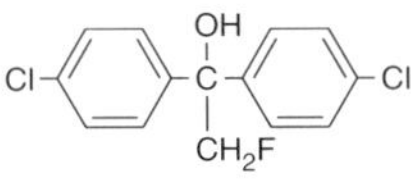

250 mg

S40,714-3
1-(3-CHLOROPHENYL)-2,2-DICHLOROETHANOL

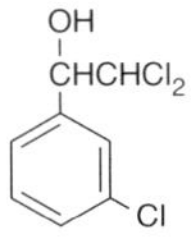

5 gm

S59,328-1
ALPHA,ALPHA-BIS-(DIFLUOROMETHYL)BENZYL ALCOHOL

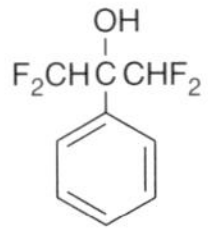

5 gm

S78,672-1
4,4'-DIFLUORO-ALPHA-(TRIFLUOROMETHYL)BENZHYDROL

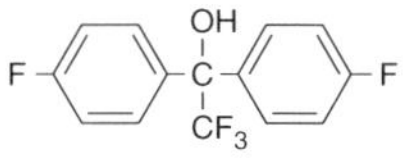

250 mg

S80,685-4
4-CHLORO-(2',3' OR 2',4' OR 2',5' OR 2',6')-BIS(TRIFLUOROMETHYL)-BENZHYDROL

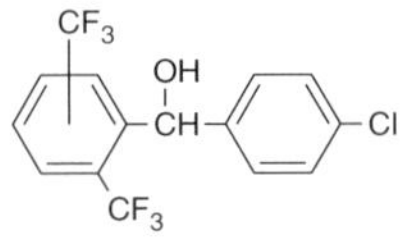

250 mg

S77,017-5
4-FLUORO-4'-(TRIFLUOROMETHYL)-BENZHYDROL

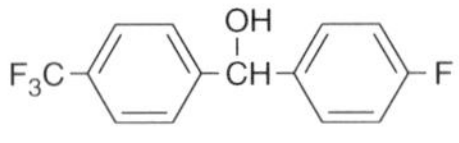

250 mg

S76,478-7
3,4-DICHLORO-4'-(TRIFLUOROMETHYL)-BENZHYDROL

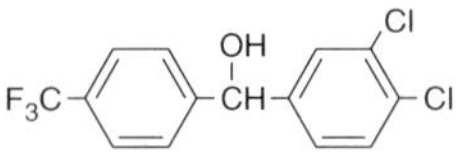

5 gm

S78,204-1
4-(TRIFLUOROMETHYL)BENZHYDROL

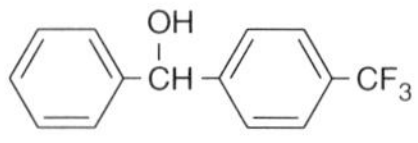

5 gm

S73,528-0
ALPHA-[2,6-BIS(TRIFLUOROMETHYL)-PHENYL]CINNAMYL ALCOHOL

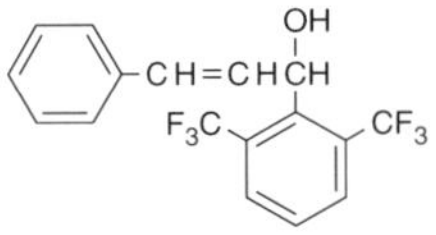

250 mg

S36,524-6
ALPHA-(TRIFLUOROMETHYL)BENZHYDROL

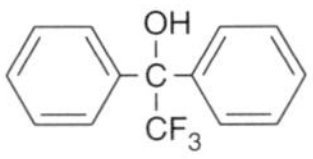

250 mg

S78,206-8
3-(TRIFLUOROMETHYL)BENZHYDROL

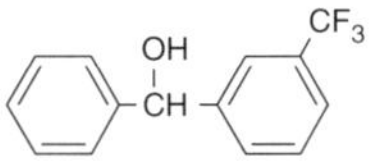

5 gm

S59,358-3
ALPHA-(CHLORODIFLUOROMETHYL)-BENZHYDROL

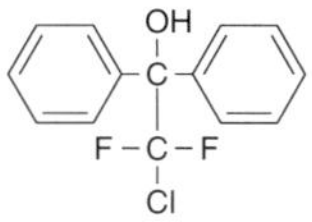

5 gm

S36,557-2
1-(4-BROMOPHENYL)-2,2,2-TRICHLOROETHANOL

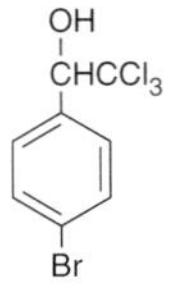

5 gm

S91,812-1
2-[4-(TRIFLUOROMETHYL)PHENYL]-2-PROPANOL

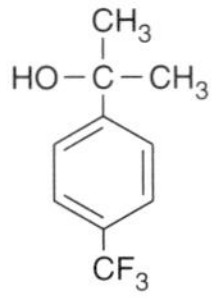

250 mg

S73,247-8
2,5-BIS(TRIFLUOROMETHYL)-ALPHA-ETHYLBENZYL ALCOHOL

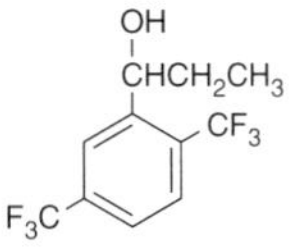

250 mg

S40,717-8
1-(4-CHLOROPHENYL)-2,2,2-TRIFLUOROETHANOL

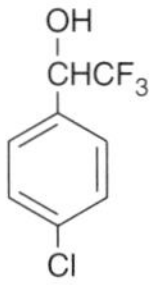

250 mg

S44,716-1
2-(PENTAFLUOROPHENYL)ETHANOL

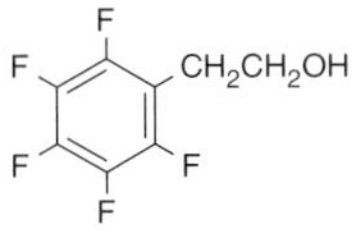

250 mg

S41,115-9
1-(PENTAFLUOROPHENYL)-2-PROPANOL

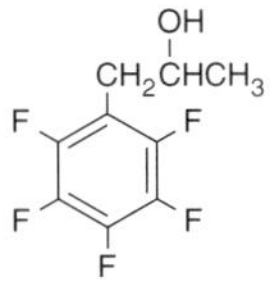

5 gm

S84,661-9
3-BROMO-4-HYDROXY-5-METHOXYBENZYL ALCOHOL

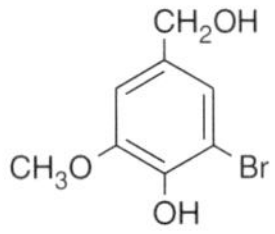

250 mg

S95,381-4
3-(2-BROMO-4,5-DIMETHOXYPHENYL)-1,2-PROPANEDIOL

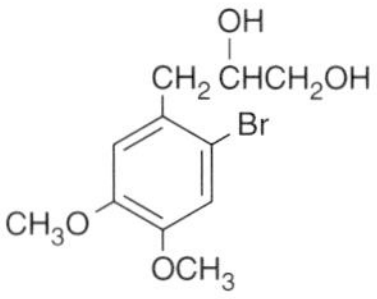

50 mg

S75,187-1
3-(2-FLUOROPHENOXY)-1,2-PROPANEDIOL
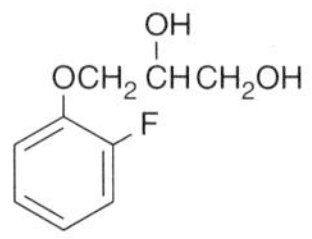
5 gm

S75,218-5
3-(2-CHLOROPHENOXY)-1,2-PROPANEDIOL
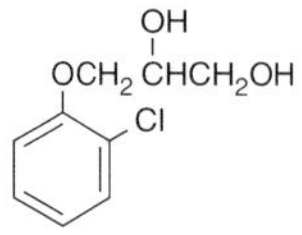
5 gm

S75,205-3
3-(2-CHLORO-6-METHYLPHENOXY)-1,2-PROPANEDIOL
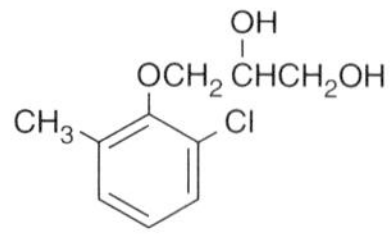
5 gm

S45,529-6
2,3,5,6-TETRACHLORO-P-XYLENE-ALPHA,ALPHA-DIOL
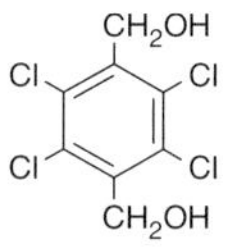
5 gm

S89,440-0
1,2-BIS(4-CHLOROPHENYL)-1,2-ETHANEDIOL
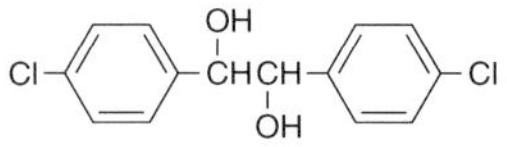
250 mg

S94,995-7
3-(P-CHLOROPHENYL)-2,2-DIETHYL-1,3-PROPANEDIOL
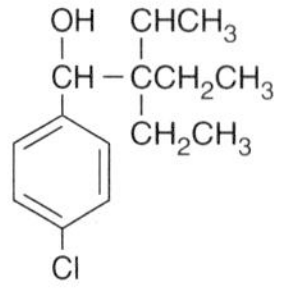
1 gm

S74,209-0
2-(3-CHLORO-4-HYDROXYPHENYL)-ETHYL ALCOHOL
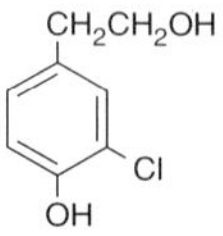
5 gm

S73,922-7
3-BROMO-4-HYDROXYPHENETHYL ALCOHOL
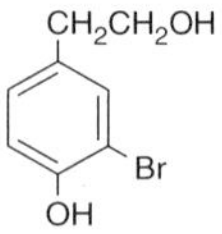
5 gm

S88,594-0
3-BROMO-4,5-DIMETHOXYBENZYL ALCOHOL
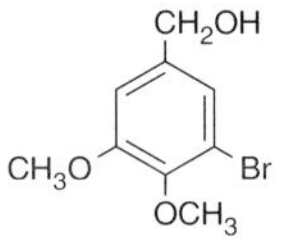
250 mg

S79,214-4
5-BROMO-2,4-DIMETHOXYTRIPHENYLMETHANOL
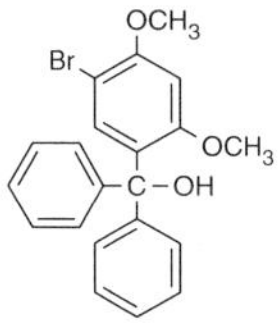
250 mg

S79,213-6
5-BROMO-2-METHOXYTRIPHENYLMETHANOL
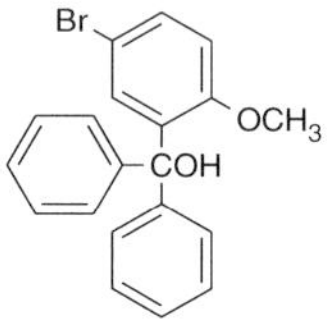
250 mg

S76,753-0
1-(3-CHLOROPHENYL)-1-BUTANOL
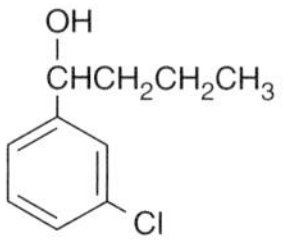
250 mg

S83,317-7
2-FLUOROTRIPHENYLMETHANOL
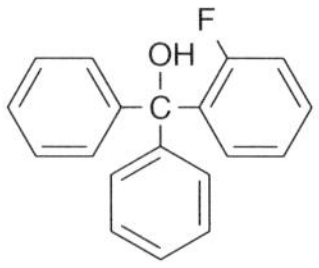
250 mg

S83,316-9
2-CHLOROTRIPHENYLMETHANOL
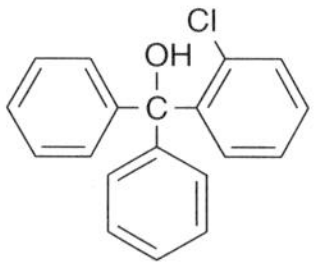
250 mg

S81,477-6
4-CHLOROTRIPHENYLMETHANOL
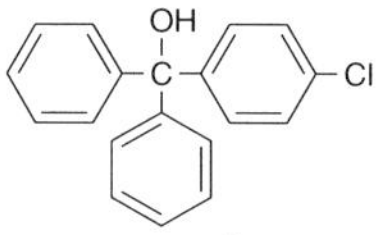
250 mg

S83,881-0
4,4'-DIBROMOTRIPHENYLMETHANOL
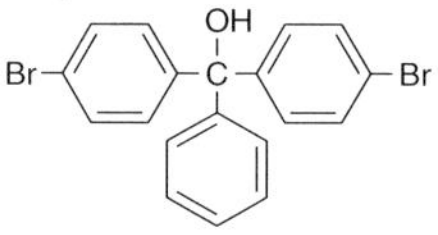
250 mg

S73,449-7
ALPHA-(2,6-DICHLORO-3-METHYLPHENYL)-BENZHYDROL
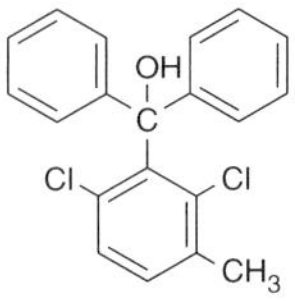
250 mg

S82,368-6
2,3,4,5,6-PENTAFLUOROTRIPHENYLMETHANOL
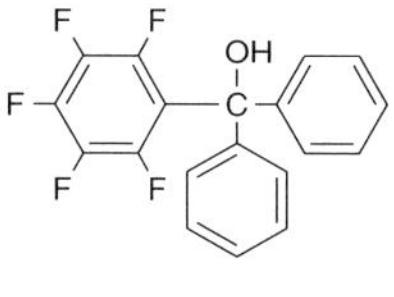
250 mg

S82,369-4
2,3,4,5,6-PENTACHLOROTRIPHENYLMETHANOL
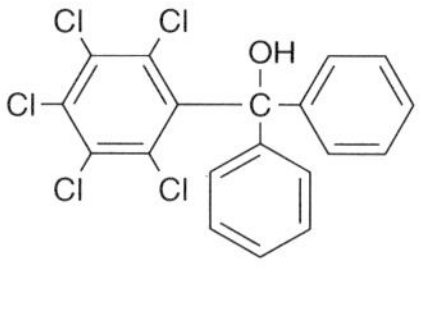
250 mg

S82,580-8
4-(4-BROMOPHENYL)-4-HEPTANOL
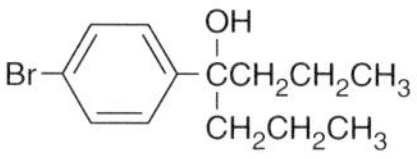
250 mg

S76,735-2
ALPHA-TERT-BUTYL-4-CHLORO-4'-METHYLBENZHYDROL
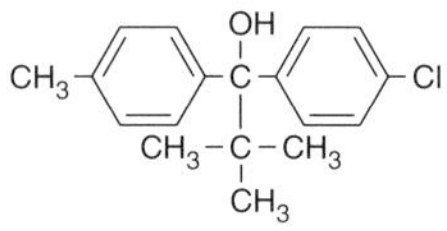
250 mg

S73,345-8
ALPHA-TERT-BUTYL-3,3'-DICHLORO-4,4'-DIMETHYLBENZHYDROL
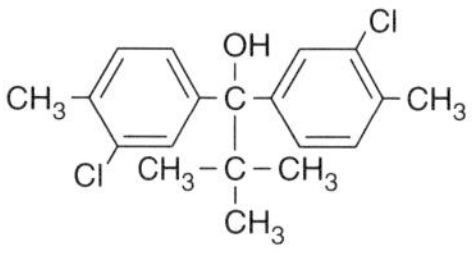
250 mg

S82,983-8
2,3,4,5,6-PENTACHLOROBENZHYDROL
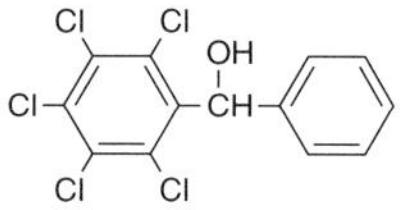
250 mg

S82,636-7
1,1-BIS(PENTAFLUOROPHENYL)-1-ETHANOL
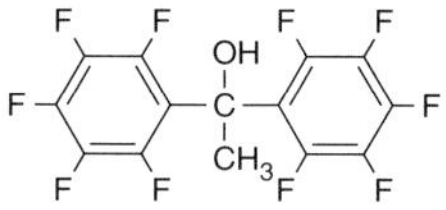
250 mg

S45,483-4
1-(4-CHLOROPHENYL)-1-CYCLOHEXANEMETHANOL
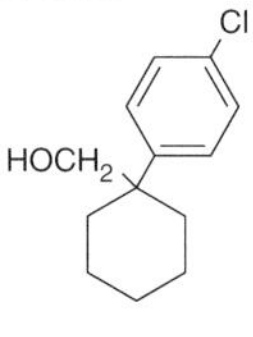
5 gm

S69,924-1
2-CHLOROBENZHYDROL
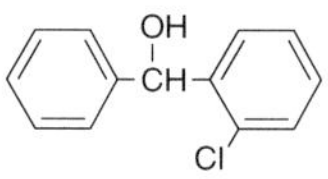
250 mg

S83,015-1
2-BROMOPHENETHYL ALCOHOL

250 mg

S83,018-6
4-(2-CHLOROPHENYL)-1-BUTANOL
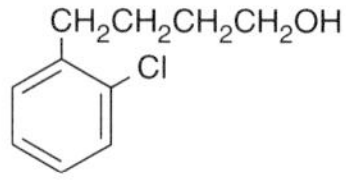
250 mg

S94,669-9
4-CHLORO-ALPHA-TERT-BUTYLBENZYL ALCOHOL
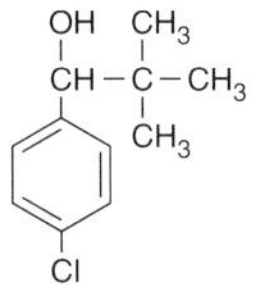
250 mg

S34,977-1
3-FLUOROPHENYL METHYL CARBINOL
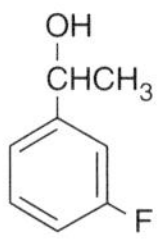
250 mg

S43,516-3
3-CHLOROPHENYL METHYL CARBINOL
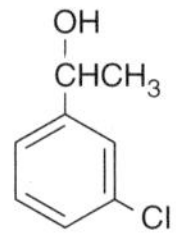
5 gm

S34,553-9
3,4-DICHLOROPHENYL METHYL CARBINOL
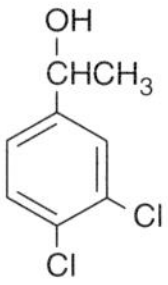
5 gm

S54,699-2
2,5-DICHLORO-ALPHA-METHYLBENZYL ALCOHOL

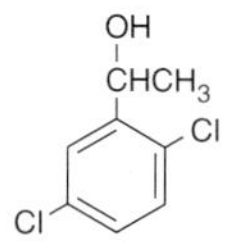

5 gm

S35,897-5
3-BROMOPHENYL METHYL CARBINOL

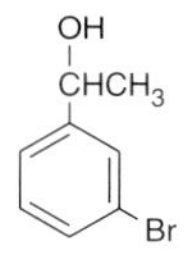

5 gm

S53,024-7
ALPHA,ALPHA,2,2-TETRAMETHYL-6-CHROMANPROPANOL

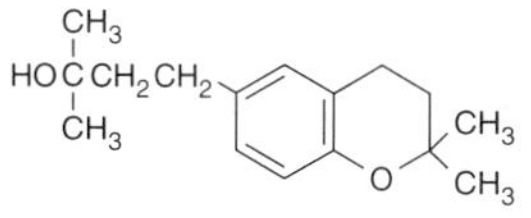

5 gm

S93,730-4
1,2-BIS(HYDROXYMETHYL)-O-CARBORANE

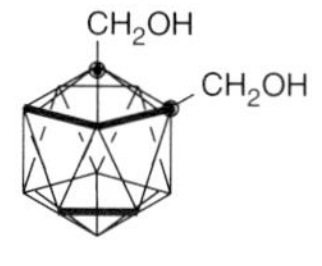

250 mg

S93,731-2
1,2-BIS(2-HYDROXYETHYL)-O-CARBORANE

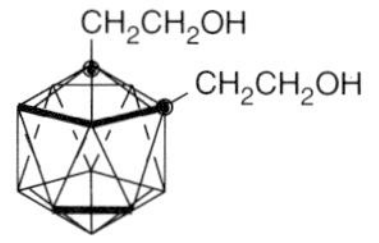

250 mg

S91,433-9
1,3:4,6-DI-O-BENZYLIDENEDULCITOL

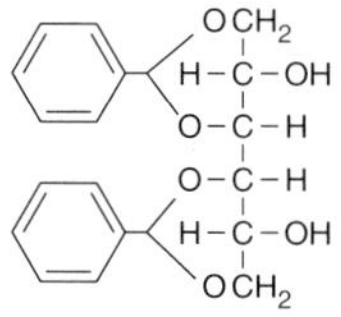

250 mg

S84,133-1
4,6:4',6'-DI-O-BENZYLIDENETREHALOSE

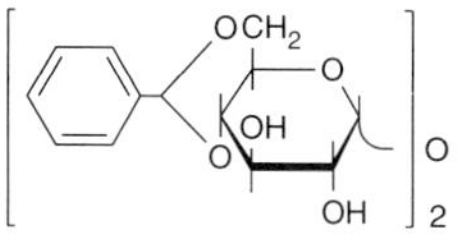

250 mg

S84,135-8
2,2',3,3'-TETRA-O-BENZYLTREHALOSE

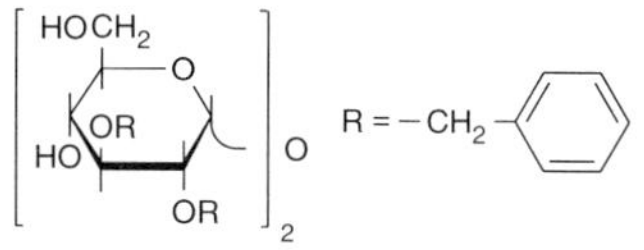

250 mg

S55,185-6
3-CHLORO-4-HYDROXY-5-METHOXYBENZYL ALCOHOL

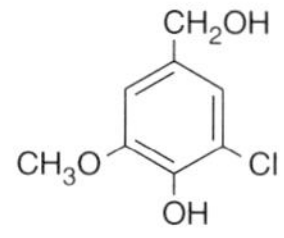

250 mg

S60,834-3
1,4-BIS(2,5-DIMETHYLPHENOXY)-2,3-BUTANEDIOL

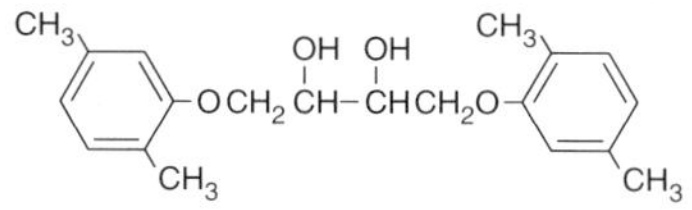

250 mg

S75,936-8
4,5-DIMETHOXY-1,2-BENZENEDIMETHANOL

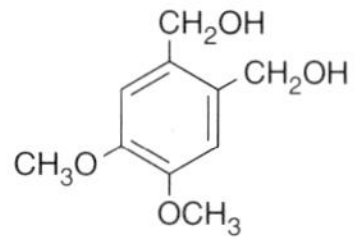

250 mg

S95,441-1
3-(3-BENZYLOXY-4-METHOXYPHENYL)-1,2-PROPANEDIOL

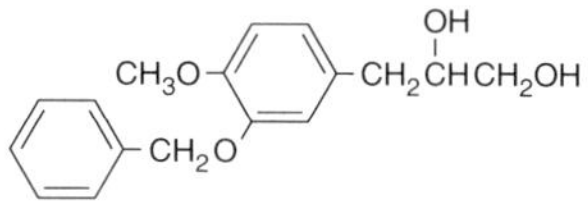

50 mg

S78,520-2
PHENYL ALPHA-D-GLUCOPYRANOSIDE

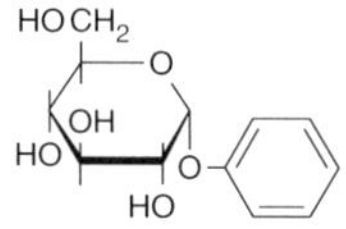

250 mg

S86,021-2
3-TRITYLOXY-1,2-PROPANEDIOL

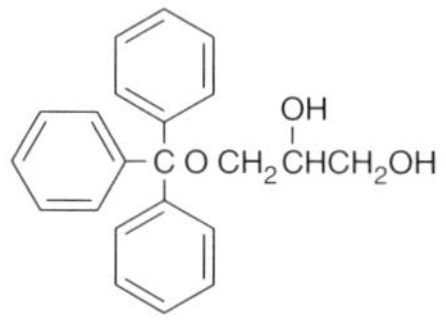

250 mg

S74,947-8
3-(2-ETHYLPHENOXY)-1,2-PROPANEDIOL

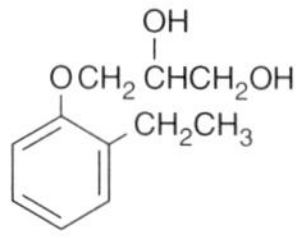

250 mg

S77,092-2
3,3-DIMETHYL-1-PHENYL-1,2-BUTANEDIOL

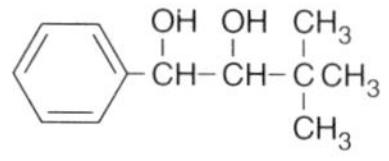

250 mg

S81,457-1
1,1,2,2-TETRAPHENYL-1,2-ETHYLENEGLYCOL

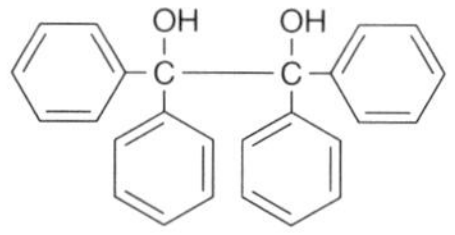

250 mg

S58,145-3
1,1-DIPHENYL-1,2-ETHANEDIOL

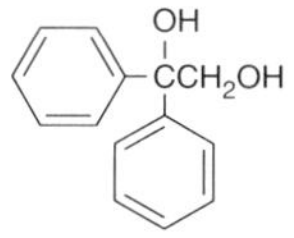

250 mg

S79,364-7
1,1-DIPHENYL-2-METHYL-1,2-PROPANEDIOL

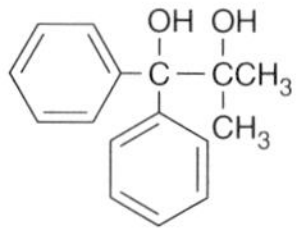

250 mg

S45,472-9
3,4-BIS(4-HYDROXYPHENYL)-3,4-HEXANEDIOL

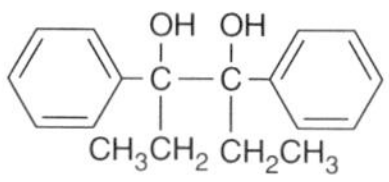

5 gm

S74,819-6
1-PHENYL-1,3-PROPANEDIOL

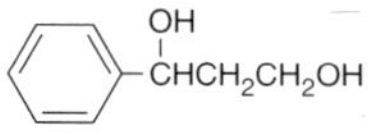

250 mg

S76,660-7
2-(1-HYDROXYCYCLOHEXYL)BENZYL ALCOHOL

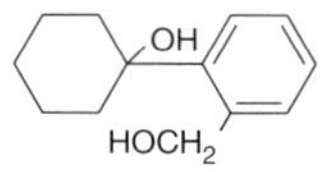

250 mg

S72,184-0
1-PHENYL-1,4-BUTANEDIOL

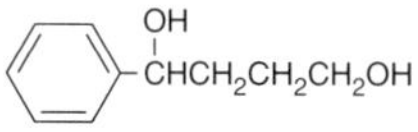

5 gm

S88,320-4
2-(HYDROXYMETHYL)BENZHYDROL

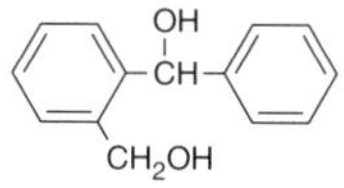

250 mg

S84,230-3
1,2-DIHYDRO-1,2-DIPHENYL-1,4-NAPHTHALENEDIOL

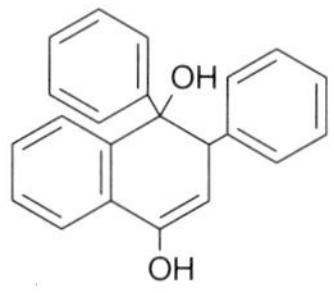

250 mg

S84,221-4
1,2-DIHYDRO-2,3-DIMETHYL-1,2-DIPHENYL-1,4-NAPHTHALENEDIOL

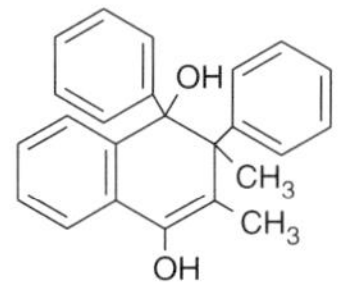

250 mg

S76,174-5
9-[2-(HYDROXYMETHYL)PHENYL]-9-FLUORENOL

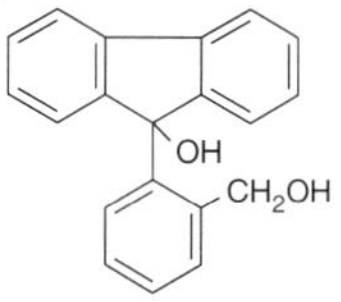

250 mg

S74,827-7
1-PHENYL-1,5-PENTANEDIOL

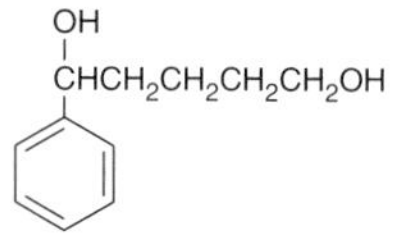

250 mg

S76,368-3
4-(ALPHA-HYDROXYBENZYL)-4-PHENYL-1,7-HEPTANEDIOL

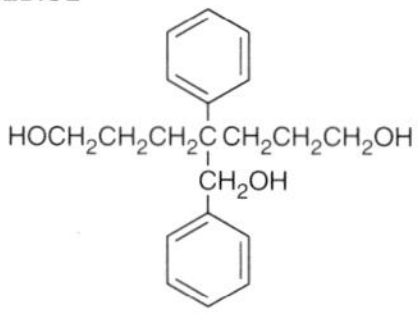

250 mg

S64,640-7
2-HYDROXYTRIPHENYLCARBINOL

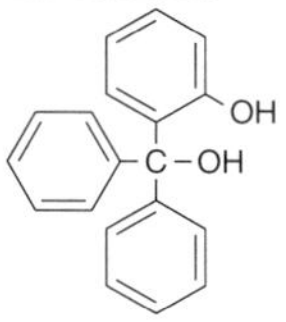

250 mg

S84,223-0
1,2-DIHYDRO-1,2-DIPHENYL-3-METHYL-1,4-NAPHTHALENEDIOL

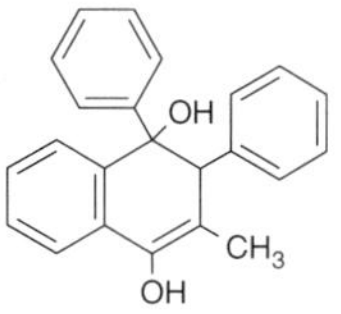

250 mg

S44,134-1
PHENOLPHTHALOL

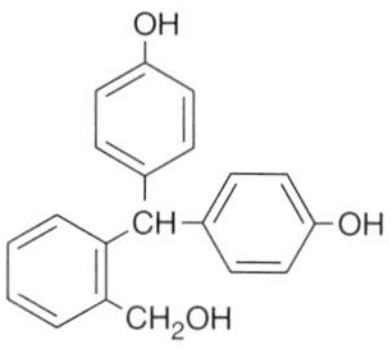

5 gm

S83,320-7
ALPHA,ALPHA,ALPHA',ALPHA'-TETRAPHENYL-(1,1'-BIPHENYL)-2,2'-DIMETHANOL
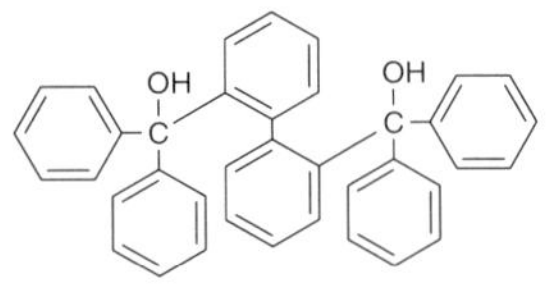
250 mg

S54,848-0
5-CHLOROSALICYL ALCOHOL
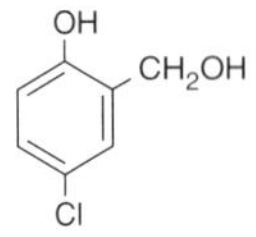
250 mg

S54,850-2
3,5-DICHLOROSALICYL ALCOHOL
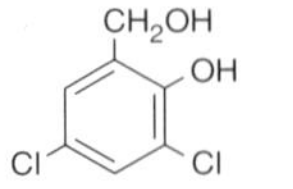
250 mg

S93,344-9
4-HYDROXY-3,5-DI-TERT-BUTYLBENZYL ALCOHOL
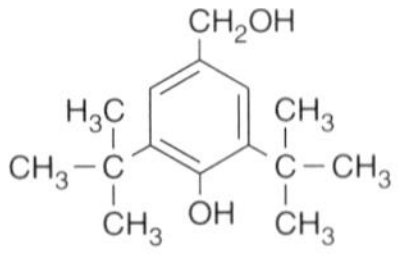
250 mg

S93,909-9
1,3-O-BENZYLIDENE-D-ARABITOL
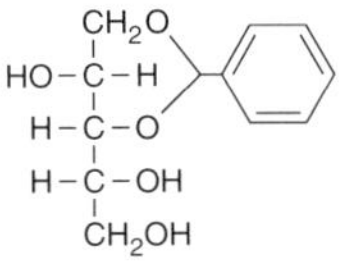
250 mg

S78,464-8
2-NAPHTHYL-BETA-D-GLUCOPYRANOSIDE
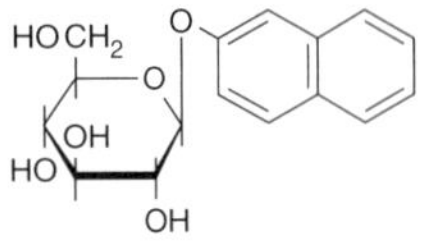
250 mg

S93,916-1
4,6-O-BENZYLIDENE-D-GLUCOPYRANOSIDE
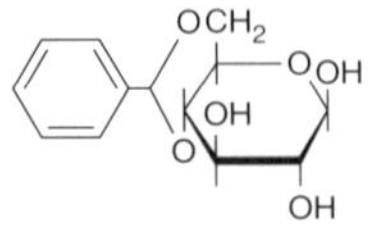
250 mg

S94,325-8
3-(3,4-METHYLENEDIOXYPHENYL)-1,2-PROPANEDIOL
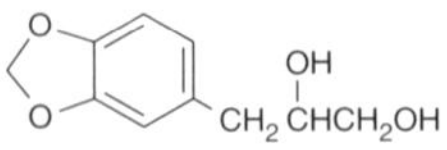
250 mg

S91,974-8
1,4-DIPHENYL-2,3-BUTANEDIOL
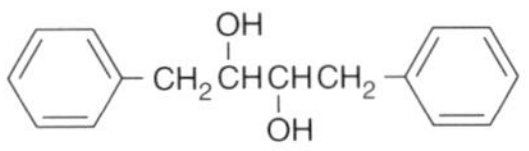
250 mg

S74,505-7
1,2-DIBUTYL-1,2-DIPHENYL-1,2-ETHANEDIOL
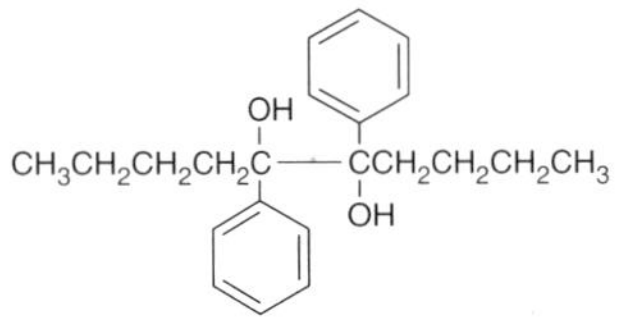
250 mg

S83,991-4
1,1-DIPHENYL-2,2-DI(P-TOLYL)-1,2-ETHANEDIOL
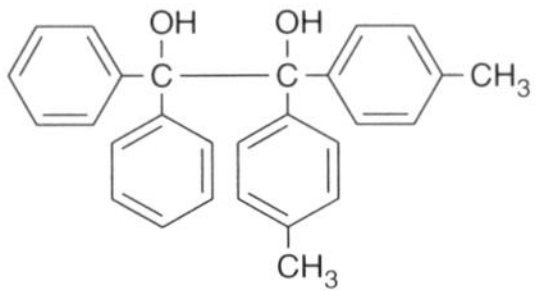
250 mg

S86,411-0
10,11-DIHYDRO-11,12-DIHYDROXY-5,10-METHANO-5H-DIBENZO(A,D)CYCLOHEPTENE
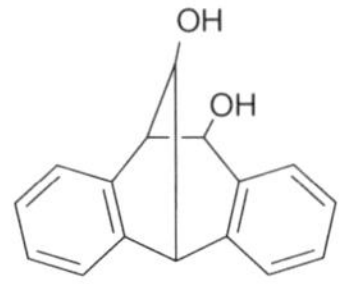
50 mg

S34,144-4
MONO-BENZALPENTAERYTHRITOL
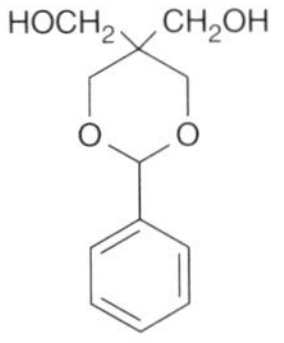
5 gm

S87,683-6
DL-2-(2-HYDROXYPHENYL)-2-PHENYL-1-ETHANOL
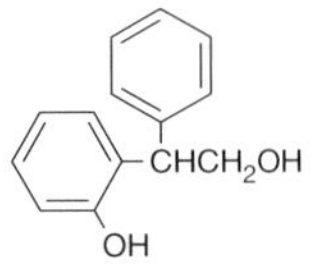
250 mg

S97,415-3
AR-E78 7,12-DIHYDRO-7,12-DIHYDROXY-7,12-DIMETHYLBENZ(A)ANTHRACENE
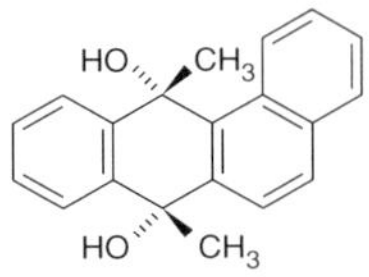
10 mg

S80,995-0
9,10-DIHYDRO-9,10-DIMETHYL-9,10-ANTHRACENEDIOL
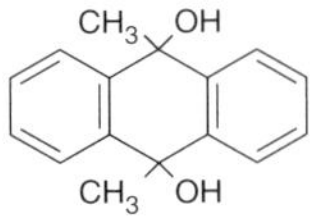
250 mg

S80,977-2
9,10-DIHYDRO-9,10-DIPROPYL-9,10-ANTHRACENEDIOL
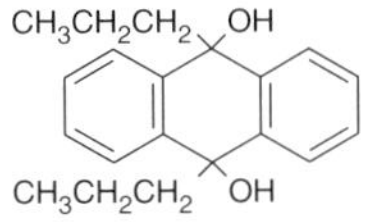
250 mg

S76,686-0
4-(1-HYDROXYETHYL)-4-PHENYL-1,7-HEPTANEDIOL
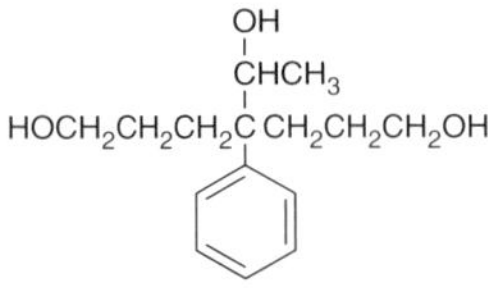
250 mg

S94,728-8
MESO-3,4-DIPHENYL-3,4-HEXANEDIOL
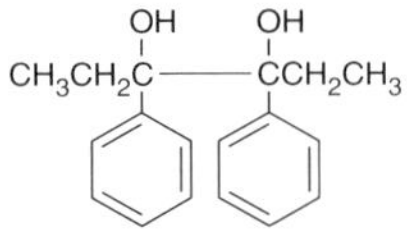
250 mg

S38,348-1
3-(4-HYDROXYPHENYL)-2-PHENYL-1-PROPANOL
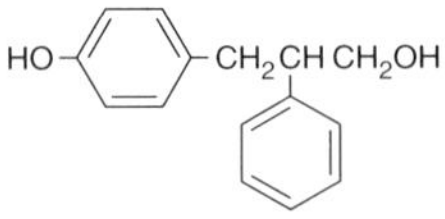
250 mg

S79,216-0
ALPHA,ALPHA'-HEXAMETHYLENEBIS-(BENZHYDROL)
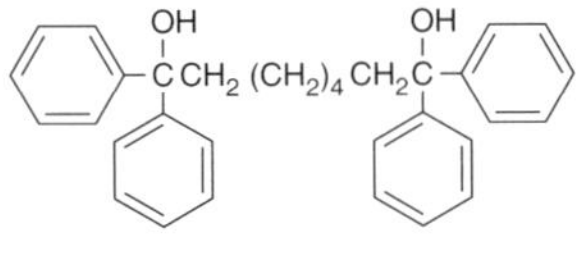
250 mg

S80,975-6
9,10-DIETHYL-9,10-DIHYDRO-9,10-ANTHRACENEDIOL
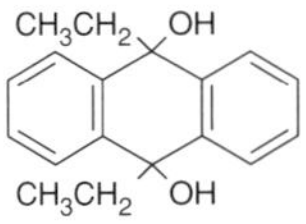
250 mg

S97,343-2
9,10-DIHYDRO-4,5-PHENANTHRENEDIMETHANOL
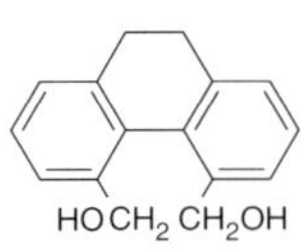
10 mg

S89,655-1
9,10,19,20-TETRAHYDROTETRABENZO-(A,C,G,I)CYCLODODECENE-9,19-DIOL
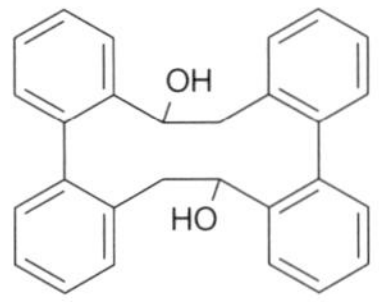
250 mg

S53,855-8
2-(M-TOLYLOXY)ETHANOL
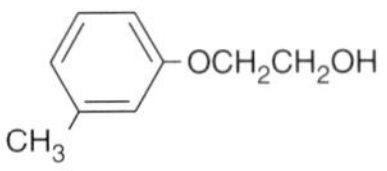
5 gm

S57,811-8
3-(DODECYLOXY)-ALPHA-ETHYL-ALPHA-METHYLBENZYL ALCOHOL
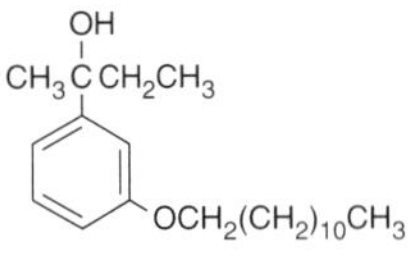
5 gm

S64,124-3
2,3,4,6-TETRA-O-BENZYL-ALPHA-D-GLUCOPYRANOSE
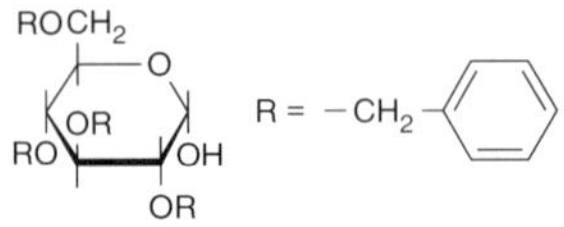
5 gm

S38,105-5
2-[(2-BENZYLOXY)ETHOXY]ETHANOL
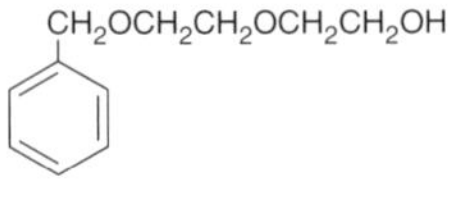
5 gm

S86,020-4
1,3-BIS(TRITYLOXY)-2-PROPANOL
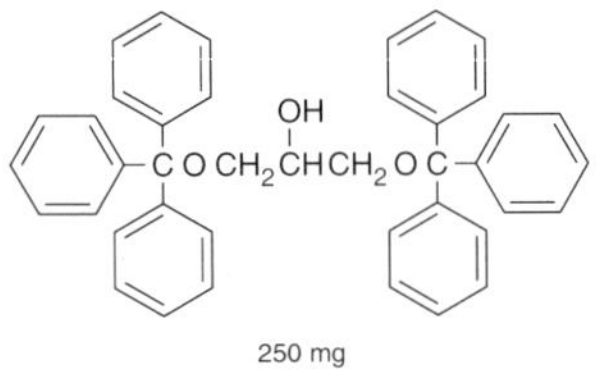
250 mg

S79,212-8
2,6-DIMETHOXYBENZHYDROL
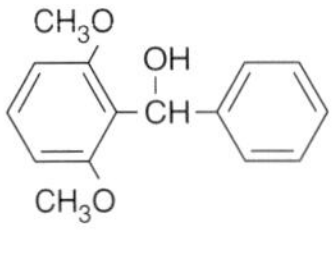
250 mg

S79,203-9
1-(2,6-DIMETHOXYPHENYL)-2-METHYL-1-PHENYL-1-PROPANOL
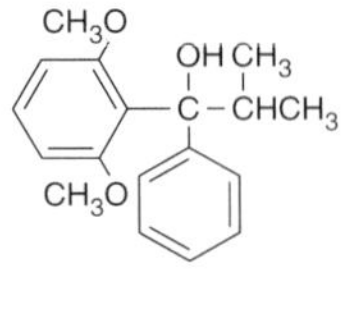
250 mg

S93,703-7
4-(METHOXYMETHOXY)-ALPHA-METHYLBENZYL ALCOHOL
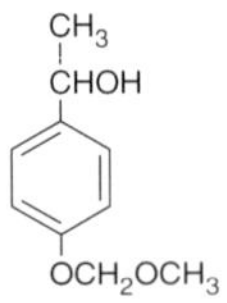
5 gm

S37,662-0
1-BENZYLOXY-2-PROPANOL

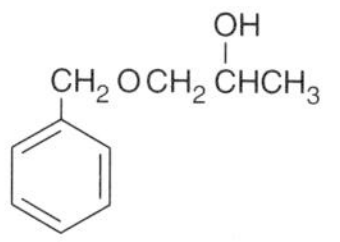

5 gm

S87,684-4
2-PHENOXY-2-PHENYL-1-ETHANOL

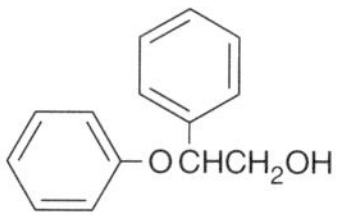

250 mg

S38,401-1
1-[4-(BENZYLOXY)PHENOXY]-2-PROPANOL

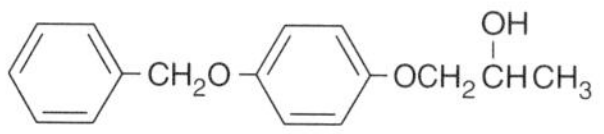

5 gm

S42,057-3
3-(2-METHYLBENZYLOXY)-1,2-PROPANEDIOL

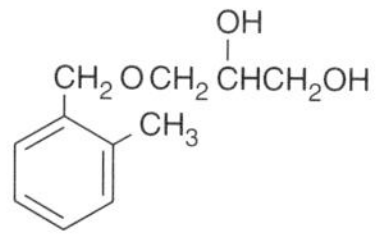

5 gm

S68,009-5
PROPYLENE GLYCOL CUMINYL ETHER

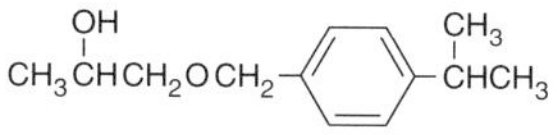

250 mg

S86,696-2
ALPHA-(2-METHOXY-2-CYCLOHEXEN-1-YL)BENZYL ALCOHOL

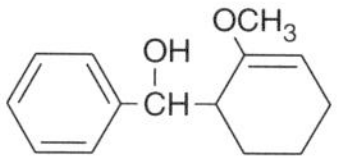

250 mg

S82,210-8
DIPHENYL(2-METHOXYPHENYL)METHANOL

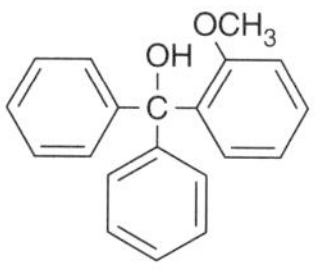

250 mg

S82,851-3
2-PHENOXYTRIPHENYLMETHANOL

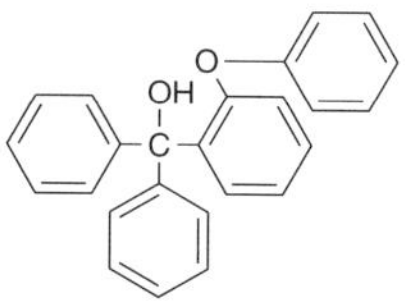

250 mg

S95,354-7
3-(BENZYLOXY)PROPANOL

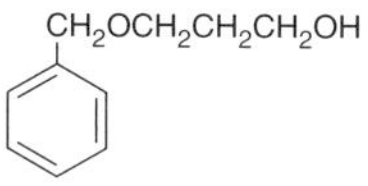

50 mg

S46,076-1
3-METHOXY-ALPHA-METHYLBENZYL ALCOHOL

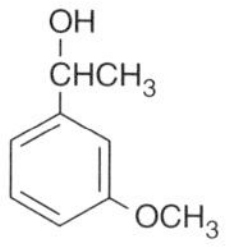

5 gm

S87,957-6
4-BENZYLOXY-1-BUTANOL

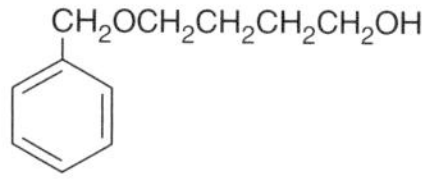

5 gm

S52,483-2
ALPHA-CYCLOPROPYL-4-METHOXYDIPHENYLMETHANOL

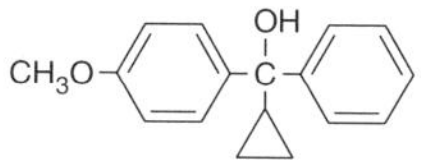

250 mg

S43,073-0
ALPHA-CYCLOPROPYL-4-METHOXYBENZYL ALCOHOL

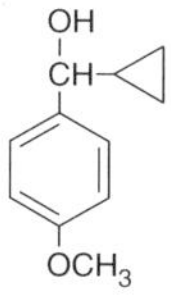

5 gm

S87,720-4
(+)-1-METHOXY-1-PHENYL-2-PROPANOL

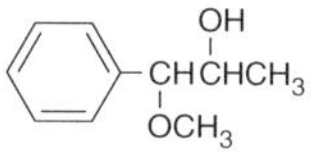

250 mg

S67,794-9
3-(BENZYLOXYMETHYL)-ALPHA-(TERT-BUTYL)BENZYL ALCOHOL

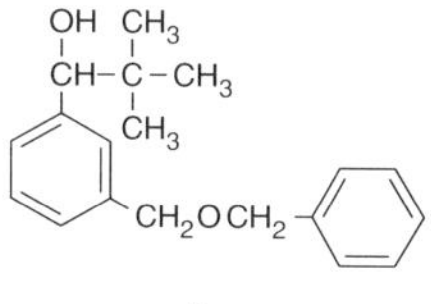

5 gm

S84,969-3
1-[2-(BENZYLOXY)PHENYL]-1-BUTANOL

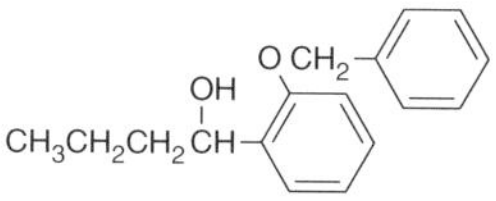

250 mg

S73,527-2
2,2-DIMETHYL-1-(4-PHENOXYPHENYL)-3-(P-TOLYL)-1-PROPANOL

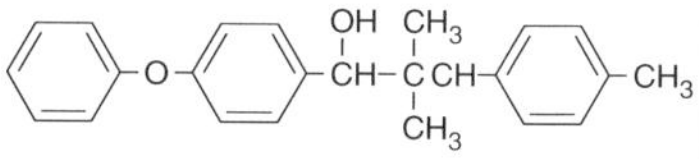

250 mg

S87,701-8
DL-4-METHOXYBENZHYDROL

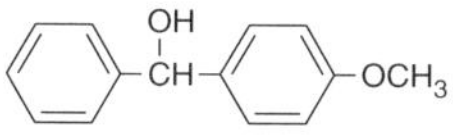

250 mg

S60,147-0
P-ANISYLDIPHENYLCARBINOL

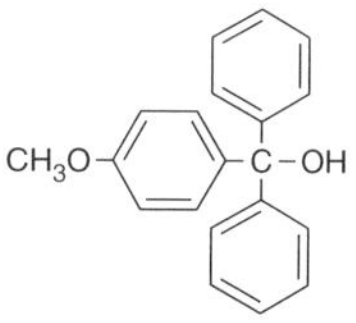

5 gm

S93,194-2
1,1-DIPHENYL-2-(3-METHOXYPHENYL)-ETHANOL

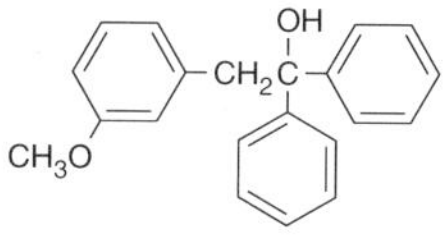

250 mg

S44,115-5
4-METHOXY-ALPHA-METHYLBENZYL ALCOHOL

5 gm

S43,490-6
ALPHA-ETHYL-4-METHOXYBENZYL ALCOHOL

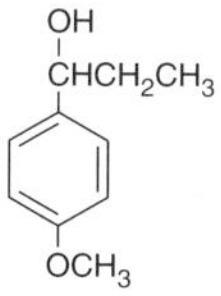

5 gm

S40,713-5
4-CHLOROPHENYL CYCLOPROPYL CARBINOL

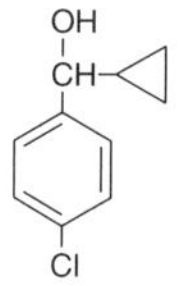

5 gm

S45,444-3
5-CHLORO-2-METHOXYBENZYL ALCOHOL

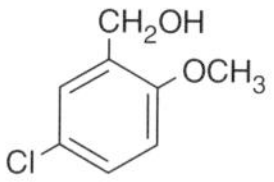

5 gm

S86,423-4
9,10-DIHYDRO-2,3-DIMETHOXY-11-HYDROXY-9,10-ETHANOANTHRACENE

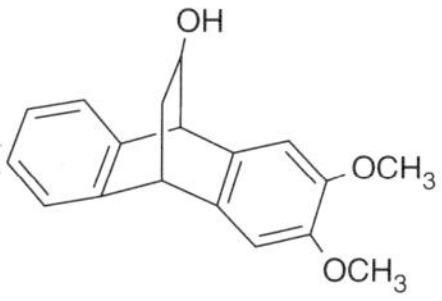

250 mg

S75,972-4
6,7-DIMETHOXY-1,1-DIMETHYL-1,2,3,4-TETRAHYDRO-2-NAPHTHOL

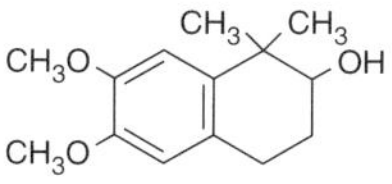

250 mg

S92,995-6
6,7-DIMETHOXY-1,2-DIMETHYL-1,2,3,4-TETRAHYDRO-1,4-METHANO-2-NAPHTHOL

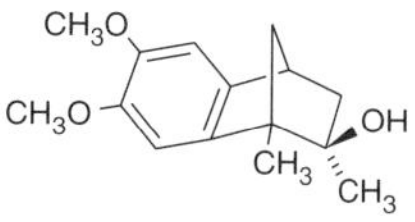

250 mg

S92,994-8
6,7-DIMETHOXY-1,2,3,4-TETRAHYDRO-1,4-METHANO-2-NAPHTHOL

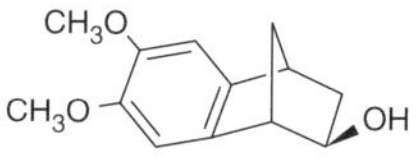

250 mg

S79,210-1
1-(2-ETHOXY-6-METHOXYPHENYL)-2-METHYL-1-PHENYL-1-PROPANOL

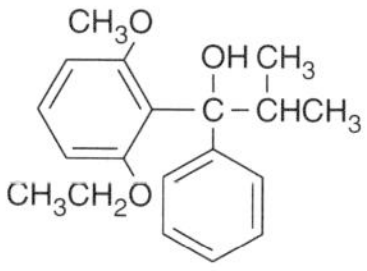

250 mg

S74,976-1
2,6-DIMETHOXYPHENETHYL ALCOHOL

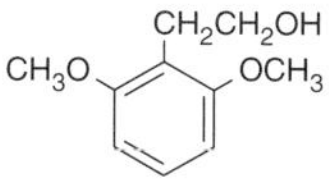

250 mg

S35,961-0
2-(P-CHLOROPHENOXY)ETHANOL

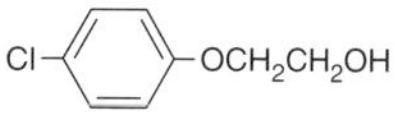

5 gm

S37,060-6
RESORCINOL BIS(2-HYDROXYETHYL)-ETHER

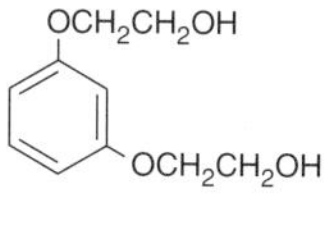

5 gm

S42,047-6
3-BENZYLOXY-1,2-PROPANEDIOL

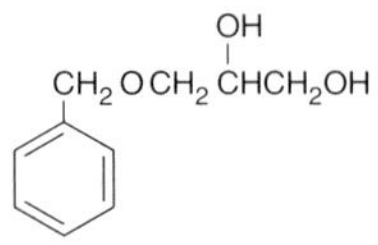

5 gm

S40,348-2
3-PHENOXY-1,2-PROPANEDIOL

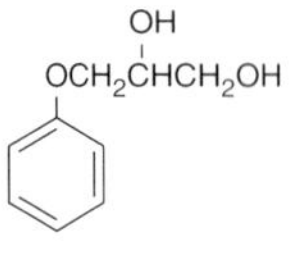

5 gm

S40,385-7
3-(2-METHOXYPHENOXY)-1,2-PROPANEDIOL

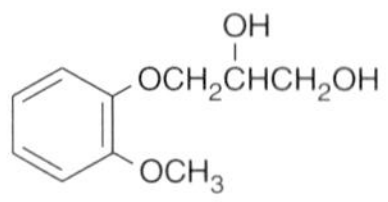

250 mg

S93,969-2
2-(2-ISOPROPYLPHENOXY)ETHANOL

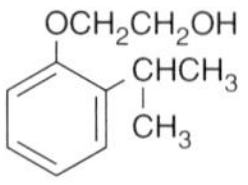

250 mg

S42,058-1
3-(2-METHOXYBENZYLOXY)-1,2-PROPANEDIOL

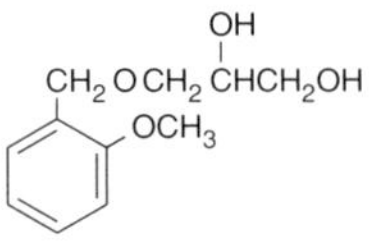

5 gm

S60,882-3
1,4-BIS(2-CHLOROPHENOXY)-2,3-BUTANEDIOL

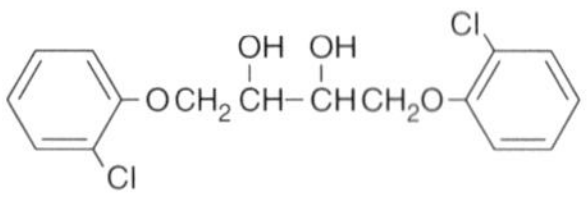

5 gm

S43,386-1
1-(2-BENZYLPHENOXY)-2-PROPANOL

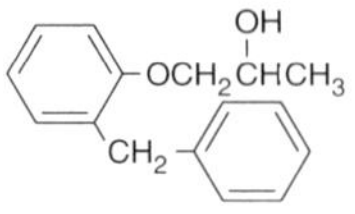

5 gm

S56,252-1
2-(P-TOLYLOXY)ETHANOL

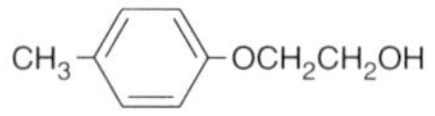

250 mg

S39,397-5
2-[4-(TERT-PENTYL)PHENOXY]ETHANOL

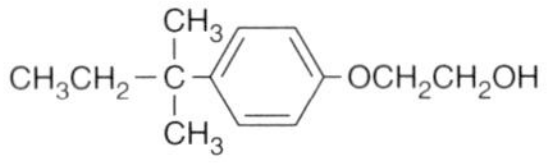

5 gm

S57,997-1
1-[4-(TERT-BUTYL)PHENOXY]-2-PROPANOL

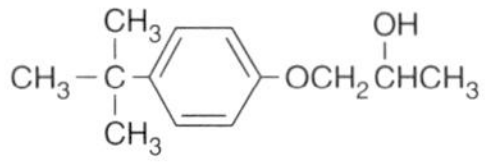

5 gm

S60,832-7
1,4-BIS(4-CHLOROPHENOXY)-2,3-BUTANEDIOL

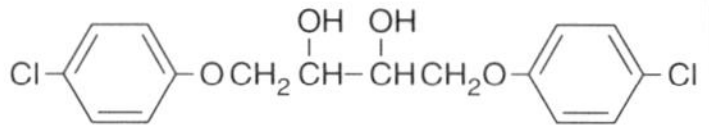

250 mg

S60,828-9
1,4-BIS(4-METHOXYPHENOXY)-2,3-BUTANEDIOL

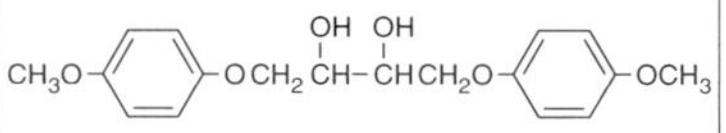

5 gm

S59,732-5
2-(2,6-DICHLOROPHENOXY)ETHANOL

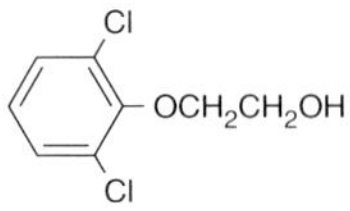

5 gm

S53,859-0
2-(2-METHOXY-4-PROPENYLPHENOXY)-ETHANOL

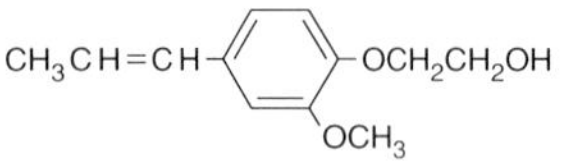

5 gm

S60,833-5
1,4-BIS(2,4-DICHLOROPHENOXY)-2,3-BUTANEDIOL

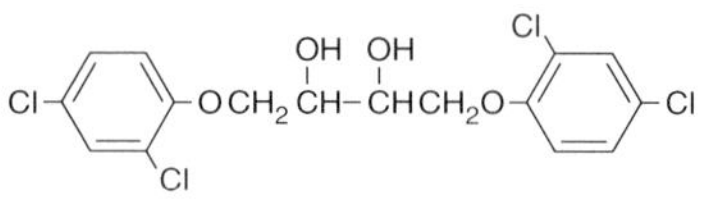

5 gm

S60,839-4
1,4-BIS(3,4-DIMETHYLPHENOXY)-2,3-BUTANEDIOL

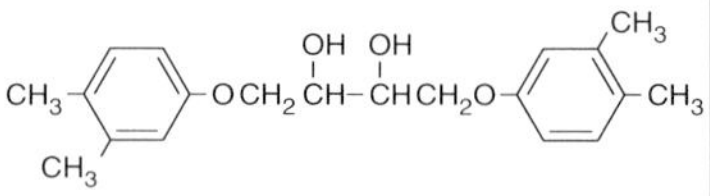

5 gm

S37,963-8
2-(2,4,6-TRIIODOPHENOXY)ETHANOL

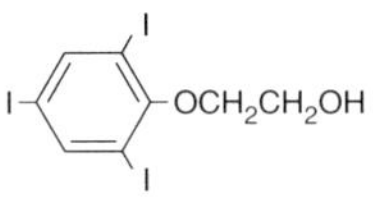

250 mg

S37,733-3
3-MESITYLOXY-1,2-PROPANEDIOL

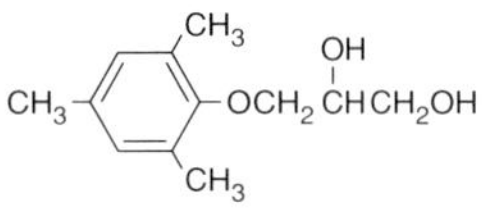

5 gm

S93,971-4
2-(2-NAPHTHYLOXY)ETHANOL

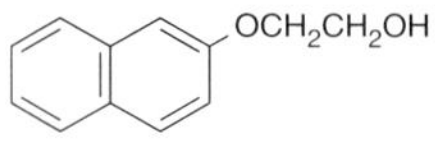

5 gm

S60,815-7
1,4-BIS(2-NAPHTHOXY)-2,3-BUTANEDIOL

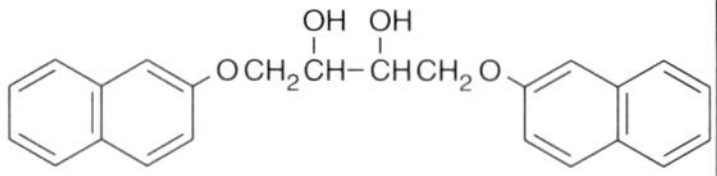

5 gm

S74,594-4
ALPHA-(3-METHOXY-2-NAPHTHYL)-CINNAMYL ALCOHOL

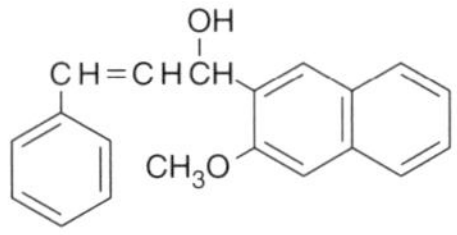

250 mg

S35,231-4
2-(4-METHOXYPHENYL)CYCLOHEXANOL, MIXTURE OF CIS AND TRANS

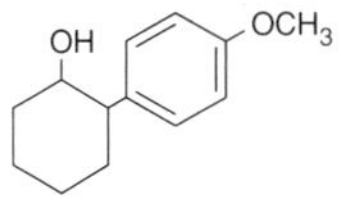

5 gm

S88,436-7
4-(4-METHOXYPHENYL)-2-CYCLOHEXEN-1-OL

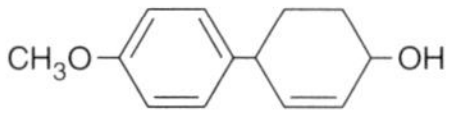

250 mg

S53,853-1
2-(4-METHOXYPHENOXY)ETHANOL

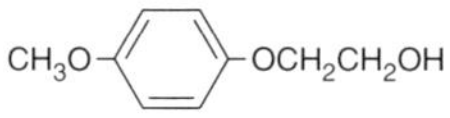

5 gm

S42,764-0
3-(4-CHLOROPHENOXY)-1,2-PROPANEDIOL

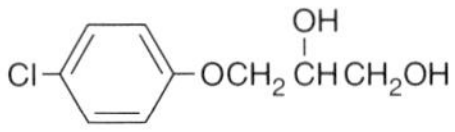

5 gm

S46,073-7
4-ETHOXYPHENETHYL ALCOHOL

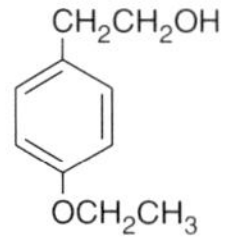

250 mg

S53,862-0
2-[4-(TERT-BUTYL)PHENOXY]ETHANOL

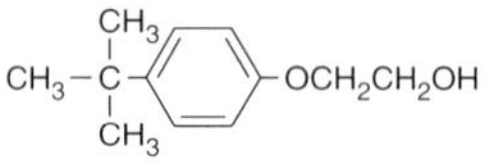

5 gm

S52,764-5
4,4'-OXYDIBENZYL ALCOHOL

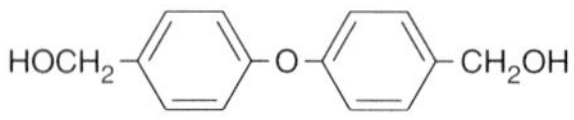

5 gm

S56,242-4
ALPHA,ALPHA-DIMETHYL-4-HEXADECYLOXYBENZYL ALCOHOL

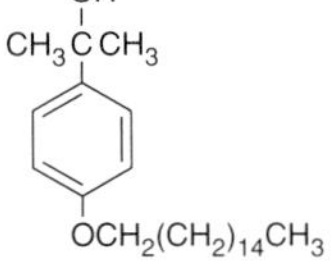

250 mg

S53,864-7
2-(4-BIPHENYLYLOXY)ETHANOL

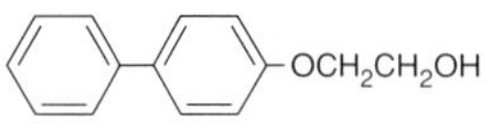

5 gm

S53,852-3
2-[6-(TERT-BUTYL)-O-TOLYLOXY]-ETHANOL

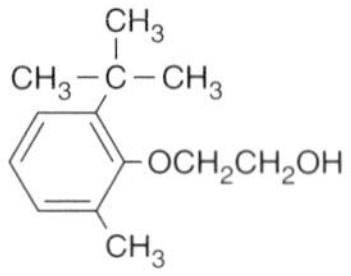

5 gm

S53,863-9
2-(2,4-DICHLOROPHENOXY)ETHANOL

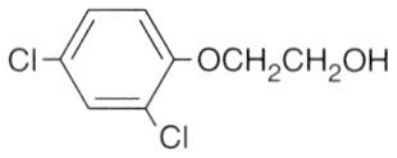

250 mg

S53,865-5
2-(4-CHLORO-M-TOLYLOXY)ETHANOL

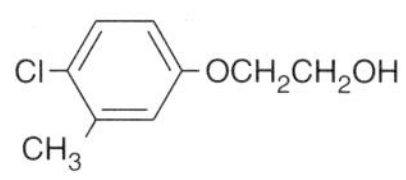

5 gm

S53,861-2
2-(4-CHLORO-3,5-XYLYLOXY)ETHANOL

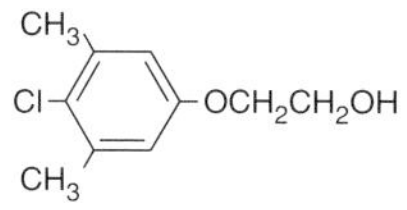

5 gm

S88,357-3
ALPHA,ALPHA-DIMETHYL-6-METHOXY-2-NAPHTHALENEMETHANOL

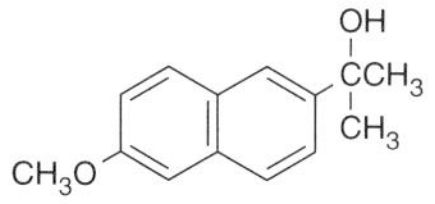

250 mg

S41,127-2
1-(M-TOLYL)CYCLOHEPTANOL

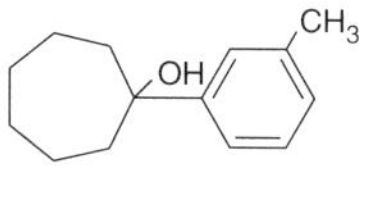

5 gm

S58,841-5
3-(3-BROMOPHENYL)-3-PENTADECANOL

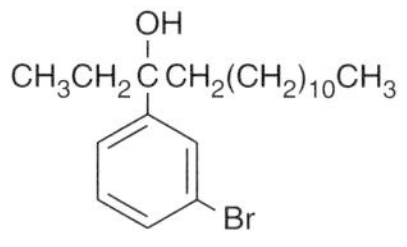

5 gm

S64,134-0
2,3,4,5-DIBENZYLIDENE-DL-XYLITOL

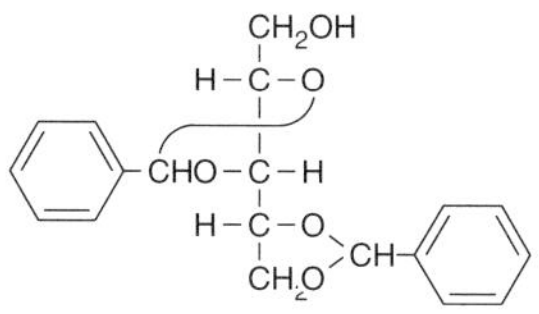

250 mg

S62,706-2
2,3-DIPHENYL-2,3-BUTANEDIOL

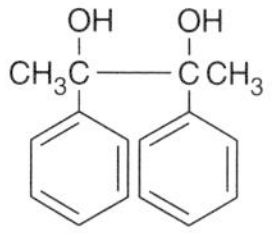

250 mg

S67,750-7
BENZOIN ETHYLENE KETAL

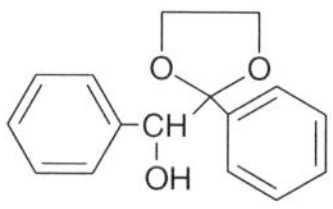

250 mg

S61,537-4
1,1-DIPHENYL-1,2-PROPANEDIOL

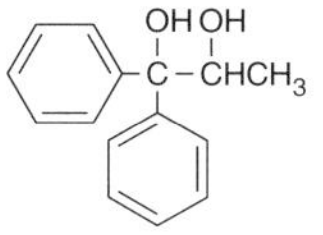

5 gm

S57,147-4
1,2-DIPHENYL-1,2-DI-P-TOLYL-1,2-ETHANEDIOL

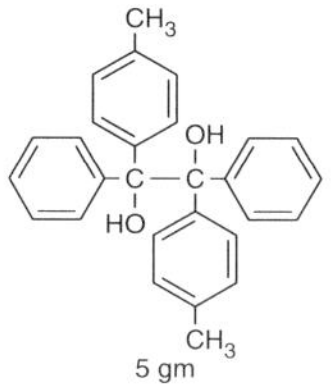

5 gm

S84,977-4
2,3-DIHYDRO-4-PHENYL-4H-1-BENZOPYRAN-4-OL

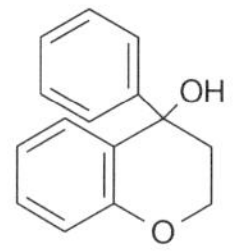

250 mg

S89,651-9
2-ANILINO-1,1,2-TRIPHENYL-1-ETHANOL

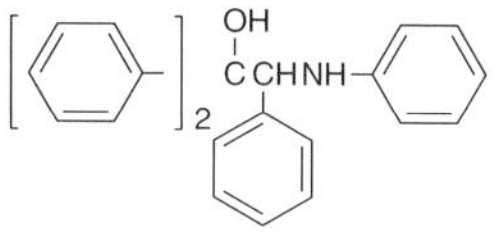

250 mg

S89,654-3
2-(BENZYLIDENEAMINO)-1,1,2-TRIPHENYL-1-ETHANOL

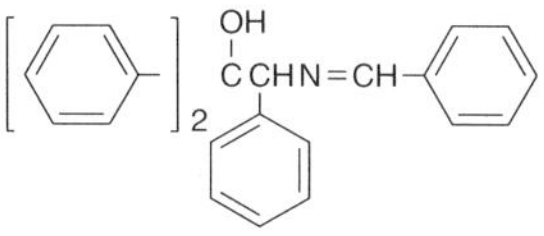

250 mg

S43,266-0
ALPHA-CYCLOPROPYL-ALPHA-METHYLBENZYL ALCOHOL

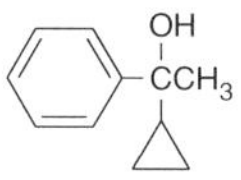

5 gm

S41,720-3
DL-ALPHA-METHYL-ALPHA-(1-METHYLCYCLOPROPYL)BENZYL ALCOHOL

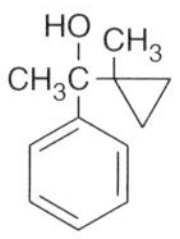

5 gm

S43,199-0
ALPHA,ALPHA-DICYCLOPROPYLBENZYL ALCOHOL

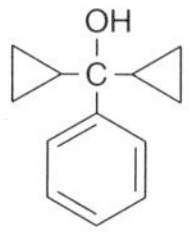

5 gm

S44,991-1
ALPHA-CYCLOPROPYL-4-FLUOROBENZYL ALCOHOL

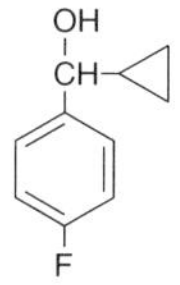

5 gm

S40,693-7
4-BROMOPHENYL CYCLOPROPYL CARBINOL

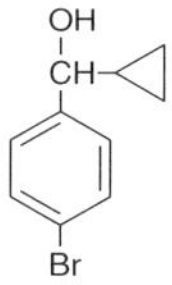

5 gm

S43,167-2
ALPHA-CYCLOPROPYL-4-METHYLBENZYL ALCOHOL

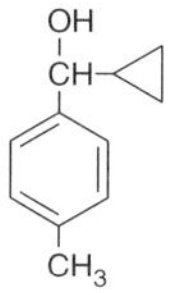

5 gm

S44,729-3
ALPHA-CYCLOPROPYL-4-FLUORO-ALPHA-METHYL-BENZYL ALCOHOL

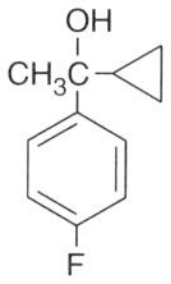

5 gm

S41,718-1
DL-4-CHLORO-ALPHA-CYCLOPROPYL-ALPHA-METHYLBENZYL ALCOHOL

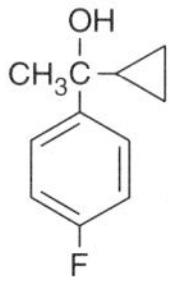

5 gm

S44,719-6
DL-ALPHA-CYCLOPROPYL-ALPHA,4-DIMETHYLBENZYL ALCOHOL

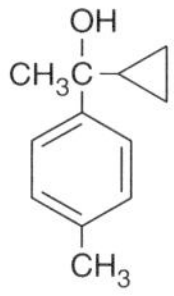

5 gm

S34,520-2
CYCLOPROPYL
4-ISOPROPYLPHENYL CARBINOL

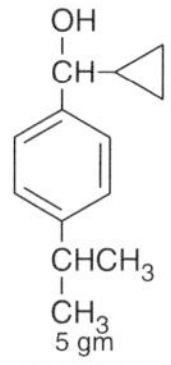

5 gm

S41,809-9
DL-ALPHA,4-DIMETHYL-ALPHA-(1-METHYLCYCLOPROPYL)BENZYL ALCOHOL

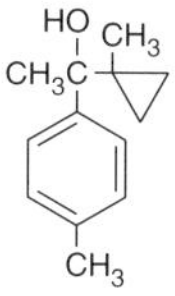

250 mg

S41,715-7
ALPHA,ALPHA-DICYCLOPROPYL-4-FLUOROBENZYL ALCOHOL

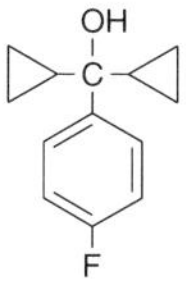

5 gm

S41,714-9
4-CHLORO-ALPHA,ALPHA-DICYCLOPROPYLBENZYL ALCOHOL

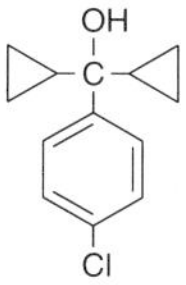

5 gm

S44,712-9
ALPHA,ALPHA-DICYCLOPROPYL-4-METHYLBENZYL ALCOHOL

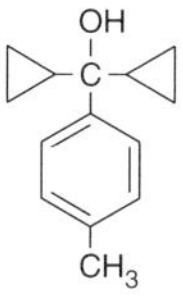

5 gm

S34,515-6
CYCLOPROPYL
2,5-DIMETHYLPHENYL CARBINOL

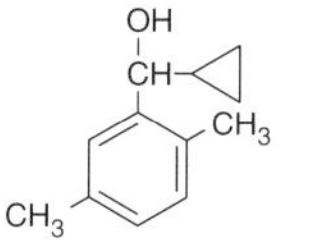

5 gm

S50,466-1
1-PHENYL-1-CYCLOPENTANOL

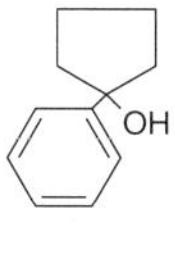

250 mg

S91,461-4
8-PHENYLTRICYCLO[5.2.1.0(2,6)]-DECAN-8-OL

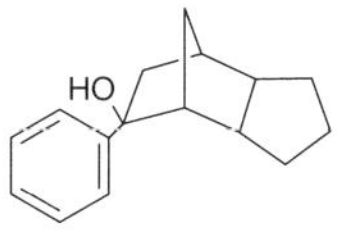

250 mg

S84,206-0
1-[(5-TERT-BUTYL)-1,3-DIHYDRO-1,3,3-TRIPHENYL-1-ISOBENZOFURYL]-3,3-DIMETHYL-1,2-DIPHENYL-2-BUTANOL

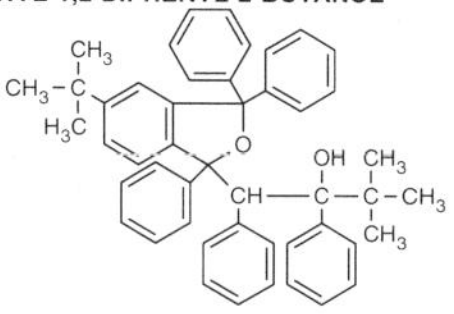

250 mg

S79,342-6
1,2-DIPHENYL-3-(PYRROLIDINOMETHYL)-2-BUTANOL HYDROCHLORIDE

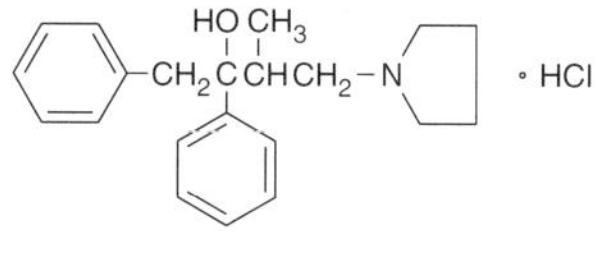

250 mg

S71,311-2
ALPHA-CYCLOHEXYL-ALPHA-METHYLBENZYL ALCOHOL

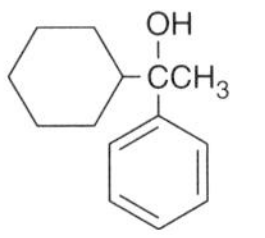

250 mg

S85,982-6
9,10-DIHYDRO-11-HYDROXY-11-PHENYL-9,10-ETHANOANTHRACENE

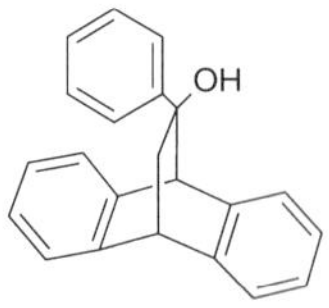

250 mg

S83,570-6
ALPHA-PHENYLALLYL ALCOHOL

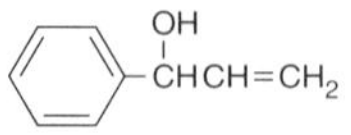

250 mg

S83,731-8
1-PHENYL-1-PROPANOL

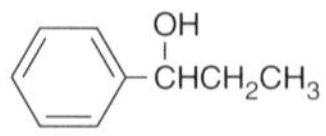

250 mg

S41,208-2
1-PHENYL-1-BUTANOL

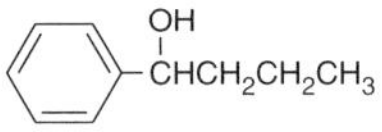

5 gm

S70,532-2
ALPHA-ISOPROPYLBENZYL ALCOHOL

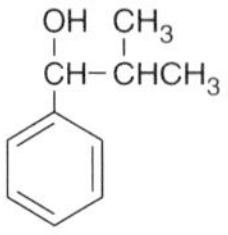

5 gm

S71,009-1
ALPHA-ETHYL-ALPHA-METHYLBENZYL ALCOHOL

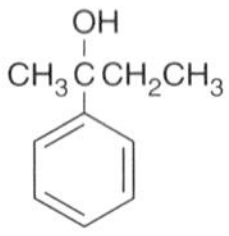

5 gm

S83,732-6
2-PHENYL-4-PENTEN-2-OL

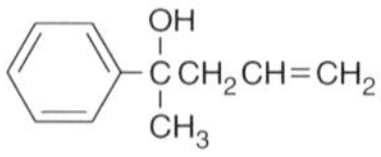

250 mg

S57,157-1
ALPHA-(TERT-BUTYL)BENZYL ALCOHOL

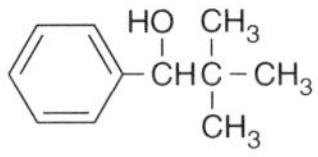

5 gm

S38,558-1
BENZYL PHENYL CARBINOL

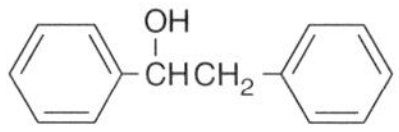

5 gm

S36,822-9
1,3-DIPHENYL-1-BUTANOL

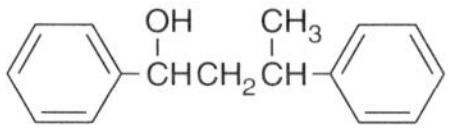

5 gm

S87,646-1
(+)-1,3-DIPHENYL-1-PENTANOL

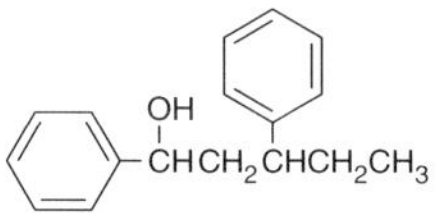

250 mg

S83,550-1
1,1,2-TRIPHENYLETHANOL

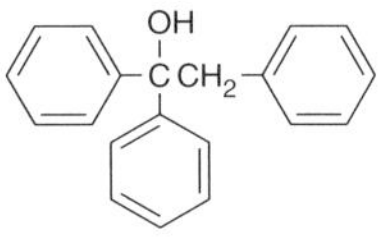

250 mg

S38,761-4
1-PHENYL-1-HEXADECANOL

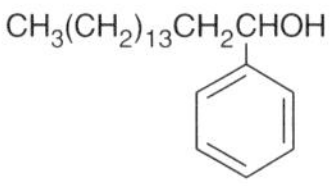

250 mg

S83,367-3
2,4'-DIMETHYLTRIPHENYLMETHANOL

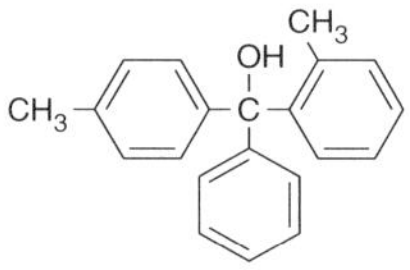

250 mg

S76,301-2
2,2-DIMETHYL-3-PHENYL-1-(P-TOLYL)-PROPANOL

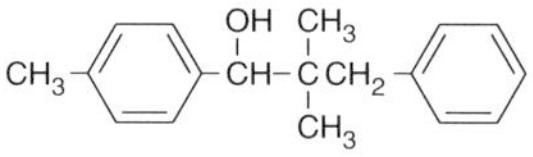

250 mg

S50,949-3
ALPHA,ALPHA-DIPHENYL-4-BIPHENYLMETHANOL

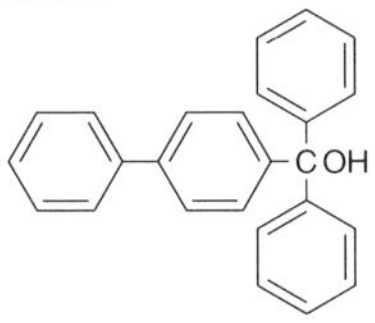

250 mg

S50,943-4
ALPHA-ETHYLDIPHENYLMETHANOL

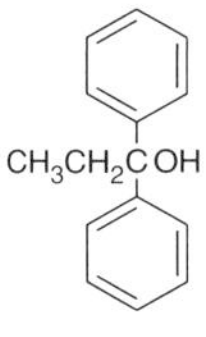

5 gm

S81,838-0
1,1-DIPHENYL-1-PROPANOL

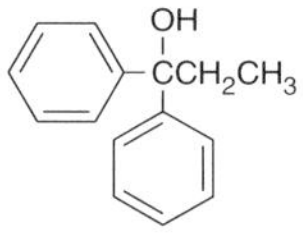

250 mg

S43,262-8
ALPHA-BENZYLBENZHYDROL

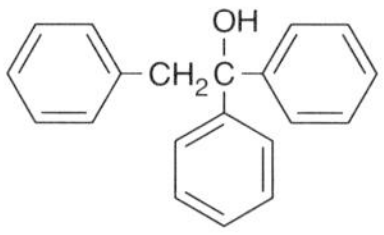

5 gm

S81,047-9
1,1,3-TRIPHENYLALLYL ALCOHOL

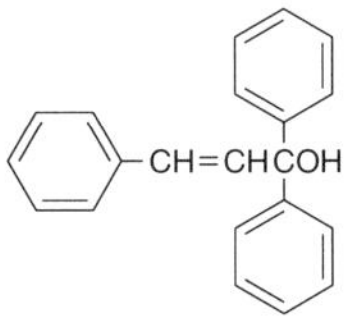

250 mg

S83,571-4
ALPHA,ALPHA-DIPHENYL-1-NAPHTHALENEMETHANOL

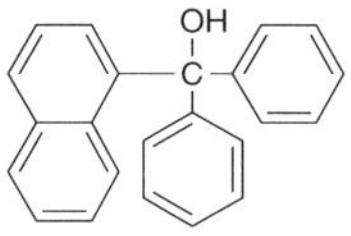

250 mg

S65,936-3
2-PHENYL-2-NORBORNANOL, MIXTURE OF ENDO AND EXO

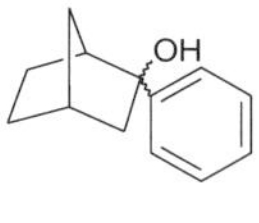

250 mg

S88,349-2
1,2,3,4-TETRAHYDRO-1,2,3-TRIPHENYL-1-NAPHTHOL

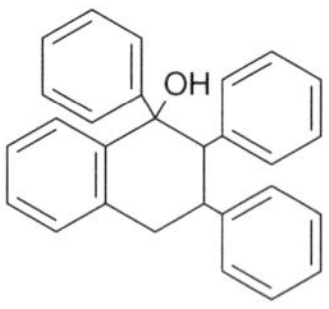

250 mg

S95,644-9
1-(ALPHA-HYDROXYBENZYL)NAPHTHALENE

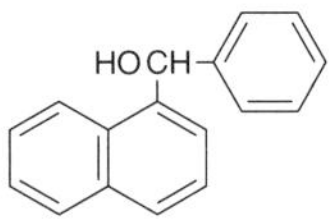

50 mg

S88,351-4
1,2-DIHYDRO-1,2-DIPHENYL-1-NAPHTHOL

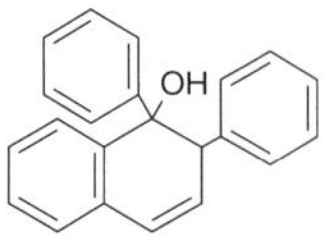

250 mg

S68,396-5
2-(1-CYCLOHEPTEN-1-YL)-BENZYL ALCOHOL

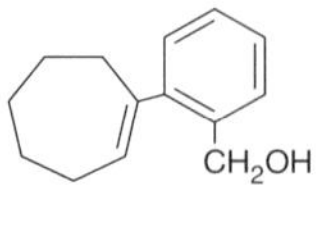

5 gm

S83,377-0
2,3',4''-TRIMETHYLTRIPHENYLMETHANOL

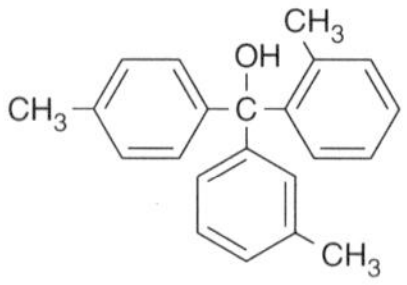

250 mg

S85,951-6
TRIS(O-TOLYL)METHANOL

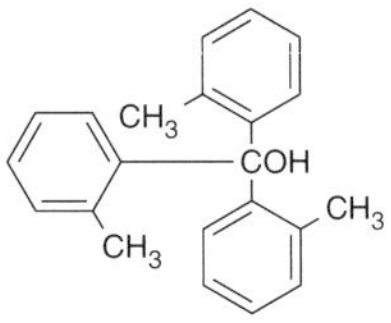

250 mg

S77,498-7
9-HYDROXY-9-(O-TOLYL)-PERHYDROFLUORENE

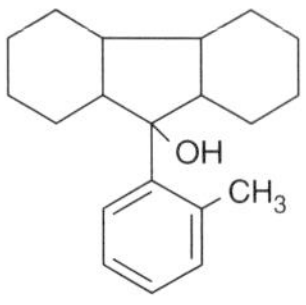

250 mg

S43,488-4
4-CHLORO-ALPHA-ETHYLBENZYL ALCOHOL

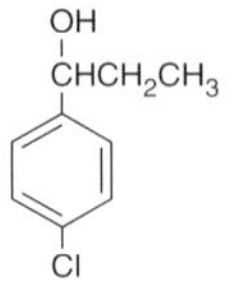

5 gm

S43,489-2
4-BROMO-ALPHA-ETHYLBENZYL ALCOHOL

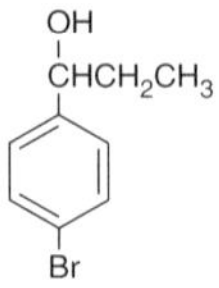

5 gm

S74,818-8
ALPHA-ETHYL-4-METHYLBENZYL ALCOHOL
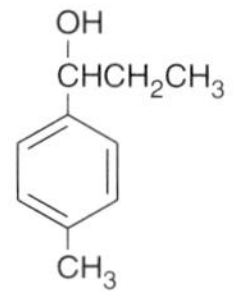
250 mg

S44,722-6
DL-4-FLUORO-ALPHA-PROPYLBENZYL ALCOHOL
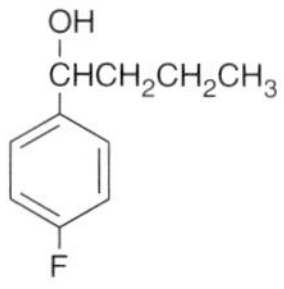
5 gm

S41,808-0
DL-4-CHLORO-ALPHA-PROPYLBENZYL ALCOHOL
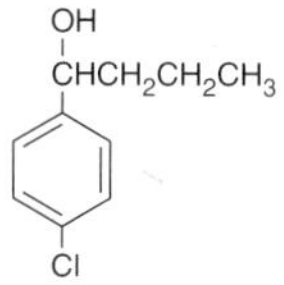
5 gm

S43,187-7
DL-4-BROMO-ALPHA-PROPYLBENZYL ALCOHOL
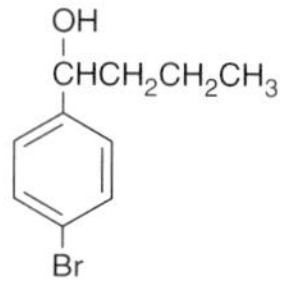
5 gm

S44,695-5
DL-4-METHYL-ALPHA-(PROPYL)BENZYL ALCOHOL
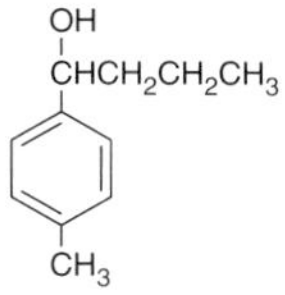
5 gm

S72,190-5
4-TERT-BUTYL-ALPHA,ALPHA-DIMETHYLBENZYL ALCOHOL
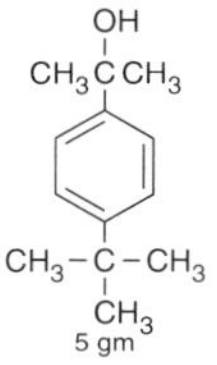
5 gm

S41,091-8
1-(4-FLUOROPHENYL)CYCLOHEPTANOL
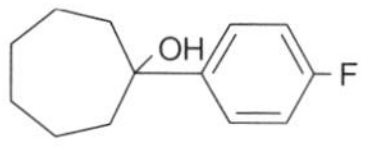
5 gm

S73,348-2
1,3-DI(4-TOLYL)-2,2-DIMETHYLPROPANOL
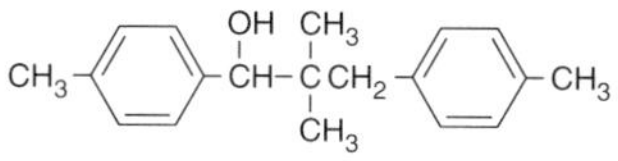
250 mg

S68,068-0
ALPHA,ALPHA,4-TRIMETHYLBENZYL ALCOHOL
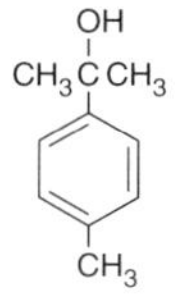
250 mg

S91,460-6
1,1-BIS(P-TOLYL)-3-BUTEN-1-OL
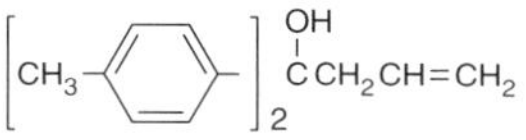
250 mg

S77,504-5
9-HYDROXY-9-(2,6-XYLYL)-PERHYDROFLUORENE
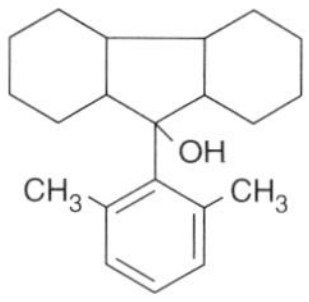
250 mg

S70,530-6
ALPHA,2,5-TRIMETHYLBENZYL ALCOHOL
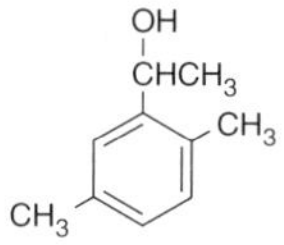
5 gm

S77,497-9
9-HYDROXY-9-(2,4,6-TRIMETHYLPHENYL)-PERHYDROFLUORENE
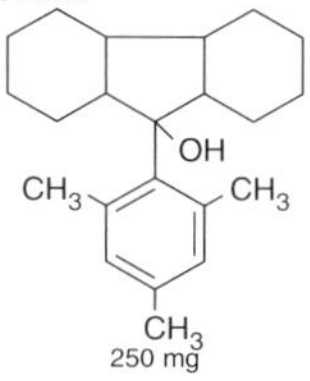
250 mg

S67,227-0
2,4,5-TRIMETHYLBENZYL ALCOHOL
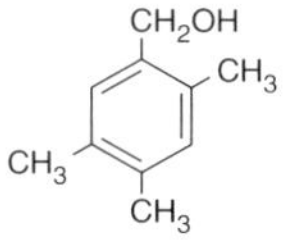
5 gm

S84,139-0
1-(4-BIPHENYLYL)-1-ETHANOL

250 mg

S93,918-8
1,2:4,6-DI-O-BENZYLIDENE-ALPHA-D-GLUCOPYRANOSIDE
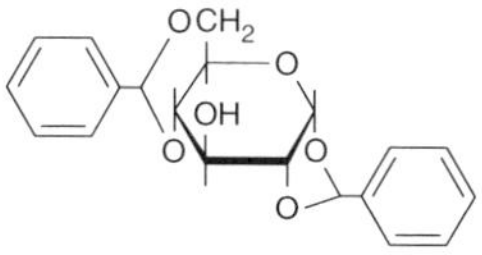
250 mg

S57,928-9
ARBUTIN
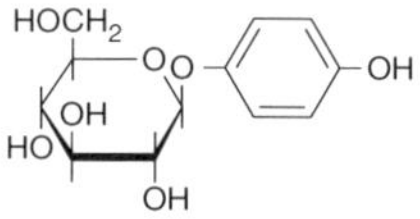
250 mg

S66,155-4
METHYL 4,6-O-BENZYLIDENEGALACTOSIDE
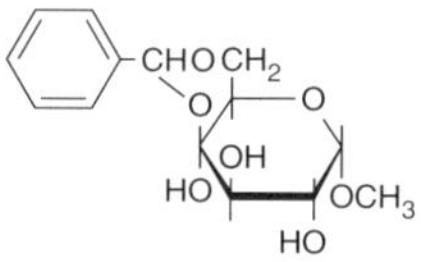
250 mg

S93,921-8
1,2:4,6-DI-O-ISOPROPYLIDENE-ALPHA-D-GLUCOPYRANOSIDE
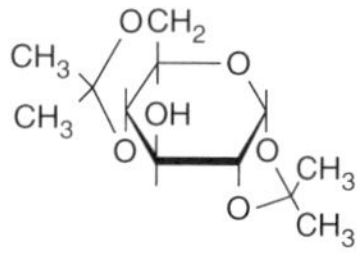
250 mg

S35,060-5
HYDROCINNAMOIN
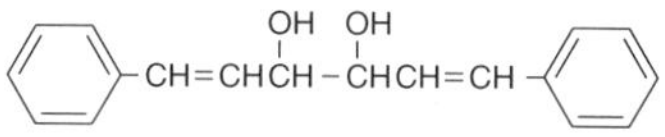
250 mg

S36,070-8
TRANS-2,3-DIHYDROXY-1,2,3,4-TETRAHYDRONAPHTHALENE
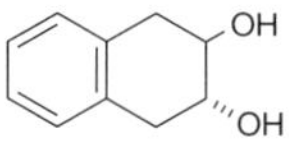
250 mg

S90,891-6
2-(1,2,3,4-TETRAHYDRO-2,2,4,7-TETRAMETHYL-1-QUINOLYL)-1-ETHANOL
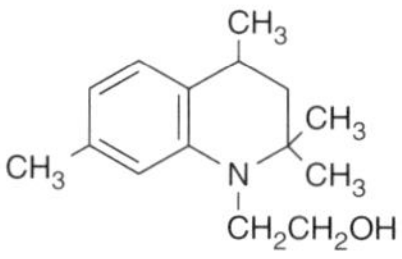
250 mg

S73,274-5
ALPHA-PHENYL-2-PIPERIDINEMETHANOL
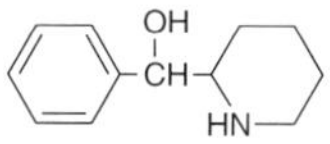
250 mg

S43,126-5
4-FLUORO-ALPHA-METHYL-ALPHA-(1-METHYLCYCLOPROPYL)BENZYL ALCOHOL
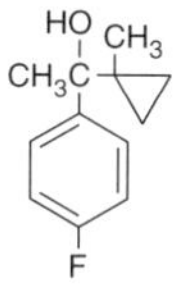
250 mg

S45,482-6
1-PHENYL-1-CYCLOPENTANEMETHANOL
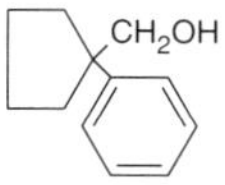
5 gm

S45,491-5
BETA-PHENYLCYCLOPENTANEETHANOL
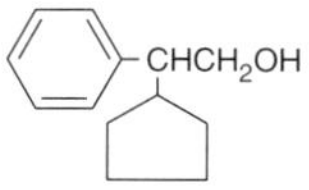
5 gm

S94,500-5
2-(BENZYLOXY)DECACHLORO-OCTAHYDRO-1,3,4-METHENO-2H-CYCLOBUTA(CD)-PENTALEN-2-OL
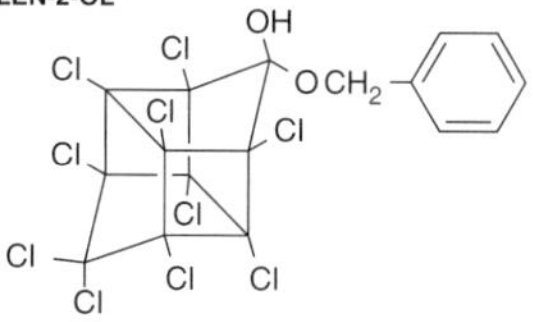
5 gm

S97,300-9
AR-A 2 2,7-DIMETHYL-9-HYDROXYMETHYLFLUORENE
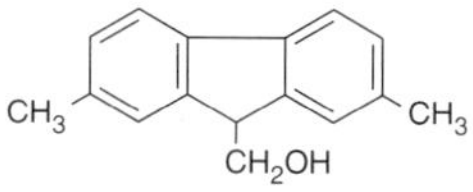
10 mg

S86,557-5
10,11-DIHYDRO-5,10-METHANO-5H-DIBENZO(A,D)CYCLOHEPTEN-11-OL
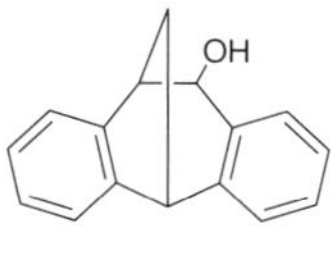
250 mg

S86,553-2
12-CHLORO-10,11-DIHYDRO-5,10-METHANO-5H-DIBENZO(A,D)-CYCLOHEXEN-11-OL
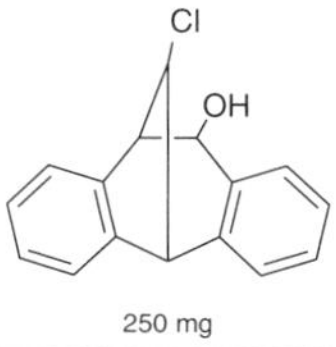
250 mg

S86,587-7
10,11-DIHYDRO-5,11-DIMETHYL-11-HYDROXY-5,10-METHANO-5H-DIBENZO (A,D)CYCLOHEPTENE
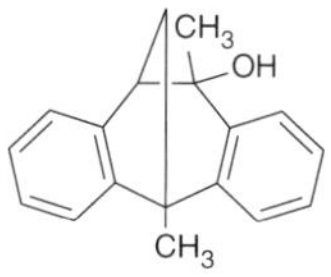
250 mg

S95,073-4
1,2-DIMETHYL-3,4,5-TRIPHENYL-2,4-CYCLOPENTADIENOL
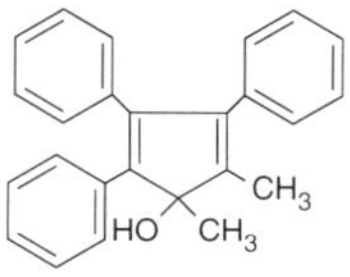
50 mg

S79,229-2
2,3,4,5-TETRAPHENYL-2,4-CYCLOPENTADIENOL
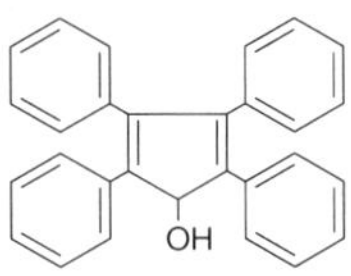
250 mg

S44,897-4
CUBEBIN
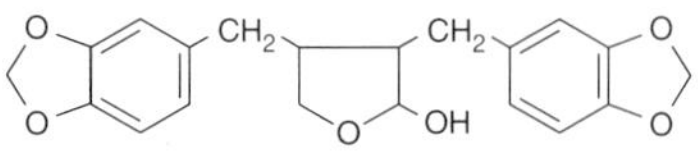
250 mg

S79,966-1
1-PHTHALANOL
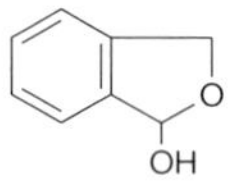
250 mg

S52,750-5
2-METHYL-2-PHENYL-1,3-DIOXOLANE-4-METHANOL
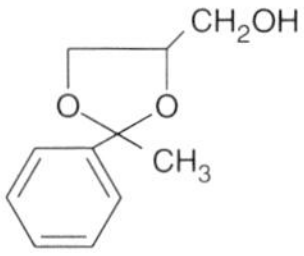
5 gm

S45,329-3
1-PHENYL-1-CYCLOHEXANOL
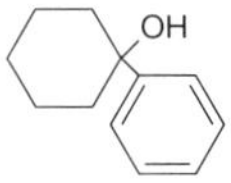
5 gm

S35,607-7
1-(M-TOLYL)CYCLOHEXANOL
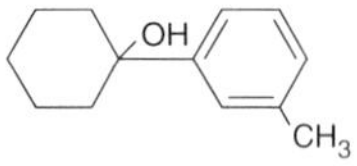
5 gm

S45,676-4
2-PHENYL-1-CYCLOHEXANOL
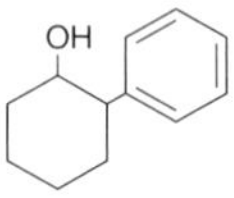
5 gm

S45,977-1
CIS-2-PHENYL-1-CYCLOHEXANOL
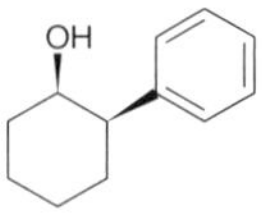
5 gm

S86,514-1
2-PHENYL-1-CYCLOHEXANEMETHANOL
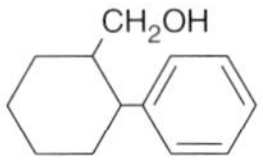
250 mg

S34,482-6
2-(P-CUMENYL)CYCLOHEXANOL
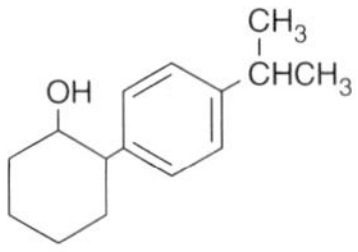
5 gm

S46,003-6
1-PHENYL-1-CYCLOHEXANEMETHANOL
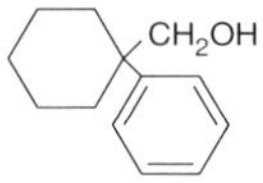
250 mg

S92,657-4
1-BENZYLCYCLOHEXANOL
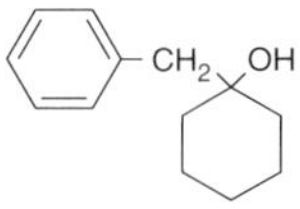
250 mg

S37,082-7
1-(O-TOLYL)CYCLOHEXANOL
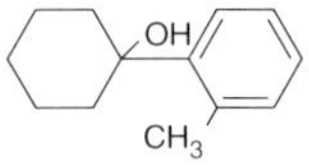
5 gm

S36,211-5
1-(4-FLUOROPHENYL)CYCLOHEXANOL
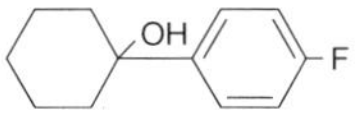
250 mg

S35,963-7
1-(4-CHLOROPHENYL)CYCLOHEXANOL
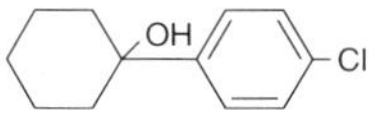
250 mg

S55,267-4
4,4-DIPHENYLCYCLOHEXANOL
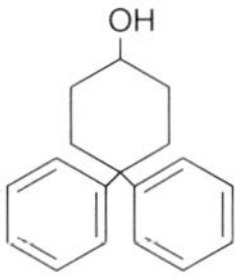
5 gm

S86,717-9
12-HYDROXY-9,9',10,10'-TETRAHYDRO-11,11'-SPIROBI(9,10-ETHANOANTHRACENE)
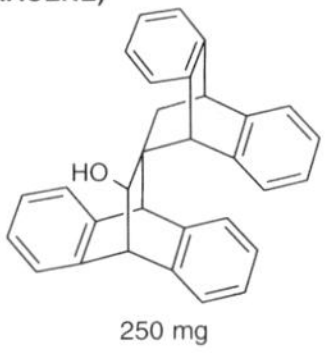
250 mg

S86,409-9
9,10-DIHYDRO-11-HYDROXY-9,10-ETHANOANTHRACENE
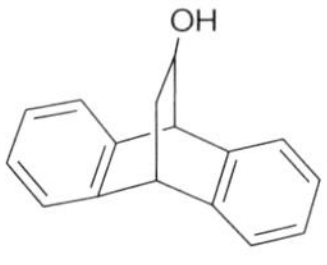
250 mg

S85,312-7
9,10-DIHYDRO-11-HYDROXY-12-METHYL-9,10-ETHANOANTHRACENE
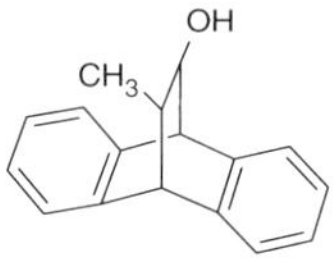
250 mg

S85,313-5
9,10-DIHYDRO-11-(HYDROXYMETHYL)-9,10-ETHANOANTHRACENE
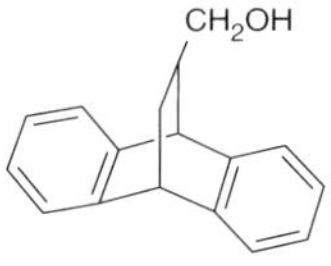
250 mg

S85,217-1
9,10-DIHYDRO-9,10-DIMETHYL-11-HYDROXY-9,10-ETHANOANTHRACENE
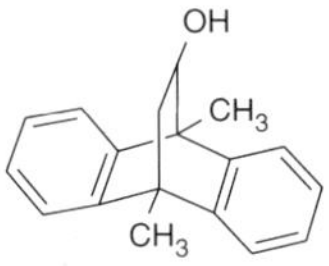
250 mg

S85,262-7
9,10-DIHYDRO-11-(HYDROXYMETHYL)-11-METHYL-9,10-ETHANOANTHRACENE
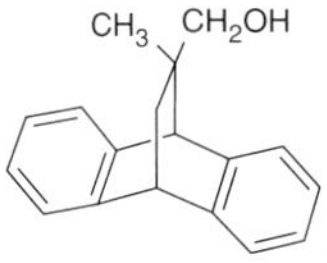
250 mg

S85,263-5
9,10-DIHYDRO-11-(1-HYDROXYETHYL)-9,10-ETHANOANTHRACENE
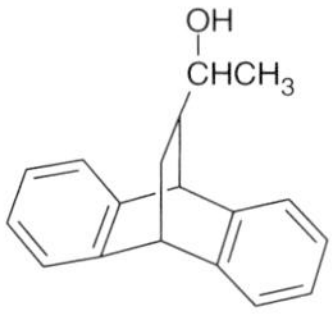
250 mg

S85,317-8
9,10-DIHYDRO-11-(HYDROXYMETHYL)-12-METHYL-9,10-ETHANOANTHRACENE
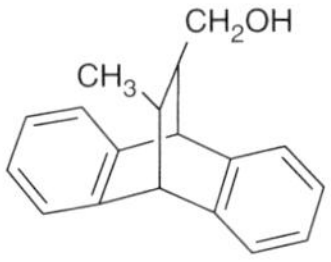
250 mg

S86,455-2
9,10-DIHYDRO-11-(2-HYDROXYETHYL)-9,10-ETHANOANTHRACENE
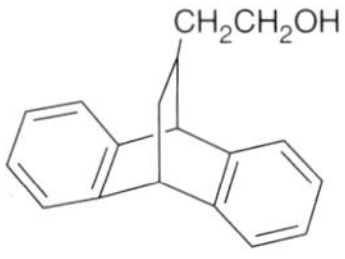
250 mg

S88,201-1
4-PHENYL-3-BUTEN-2-OL
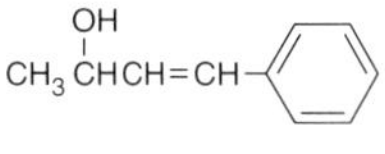
250 mg

S37,238-2
ALPHA,ALPHA-DIMETHYLPHENETHYL ALCOHOL
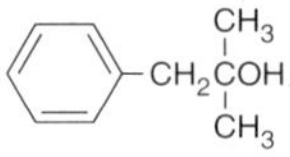
5 gm

S78,203-3
1-PHENYL-2-BUTANOL
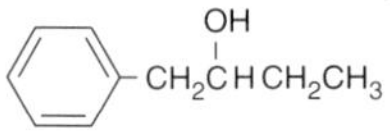
5 gm

S88,107-4
2-METHYL-4-PHENYL-3-BUTEN-2-OL
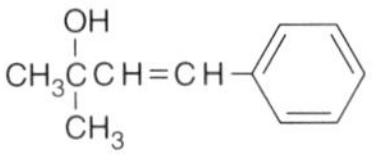
250 mg

S45,887-2
1-PHENYL-1-PENTANOL
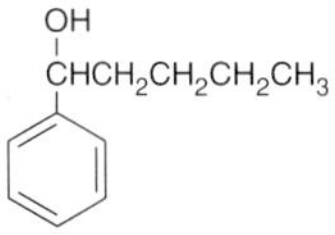
5 gm

S57,111-3
3-PHENYL-3-PENTANOL
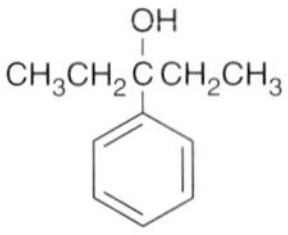
250 mg

S87,625-9
(-)-3-METHYL-2-PHENYL-1-BUTANOL
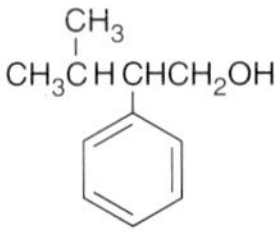
250 mg

S87,626-7
(+)-3-METHYL-2-PHENYL-1-BUTANOL
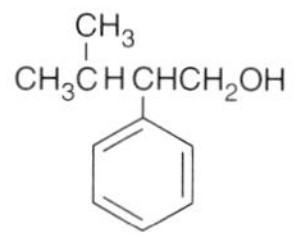
250 mg

S34,935-6
ETHYL METHYL BETA-PHENYLETHYL CARBINOL
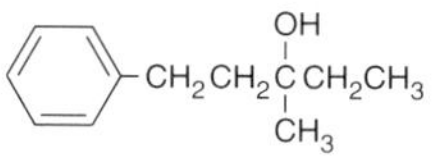
5 gm

S37,764-3
BETA-(SEC-BUTYL)PHENETHYL ALCOHOL
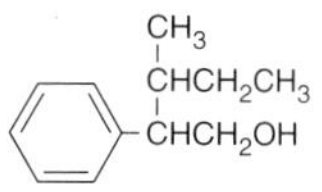
5 gm

S57,745-6
4-METHYL-4-PHENYL-2-PENTANOL
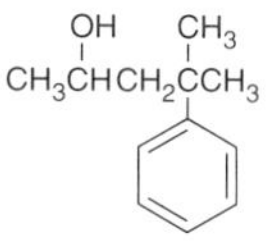
5 gm

S61,762-8
3-BENZYL-3-PENTANOL
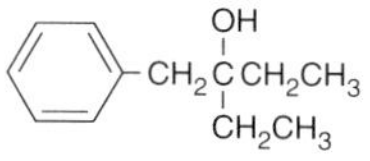
5 gm

S71,327-9
BETA-ISOPROPYL-BETA-METHYLPHENETHYL ALCOHOL
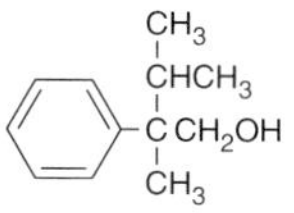
5 gm

S72,215-4
ALPHA-TERT-BUTYLPHENETHYL ALCOHOL
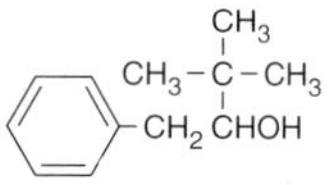
5 gm

S73,453-5
3-BENZYL-3-METHYL-2-BUTANOL
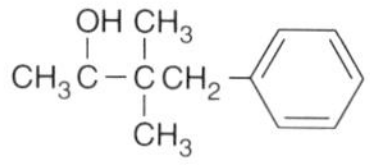
250 mg

S78,202-5
1-PHENYL-2-HEXANOL
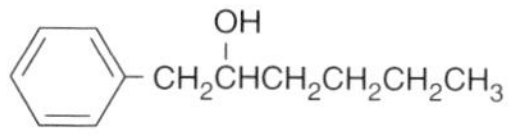
5 gm

S94,670-2
2-METHYL-1-PHENYL-2-PENTANOL
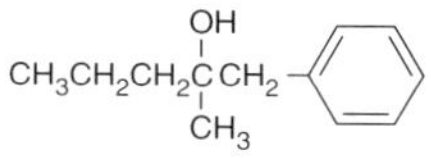
250 mg

S85,983-4
TRANS-2-PHENYLCYCLOHEPTANOL
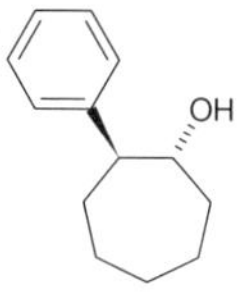
250 mg

S52,812-9
2-BENZYLIDENE-1-HEPTANOL
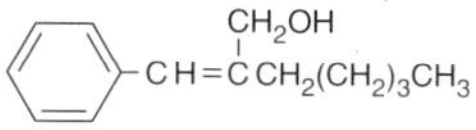
5 gm

S45,485-0
1,3-DIPHENYL-2-PROPANOL
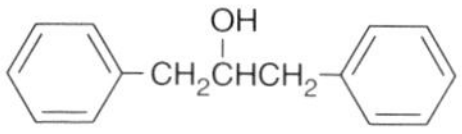
5 gm

S87,640-2
2,2-DIMETHYL-5-PHENYL-3-HEPTANOL
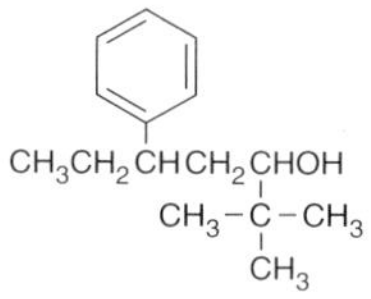
250 mg

S83,920-5
1,1-DIBENZYLETHANOL
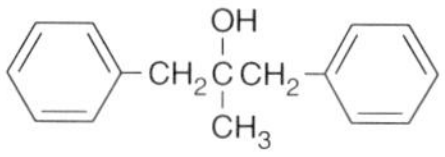
250 mg

S61,771-7
1,1-DIBENZYL-1-PROPANOL
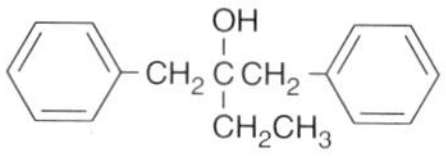
5 gm

S81,386-9
TRIBENZYLMETHANOL
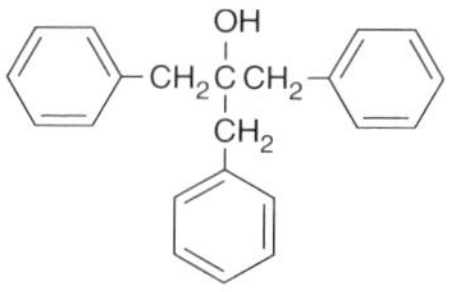
250 mg

S89,622-5
4-TERT-BUTYL-2,6-DIPHENYL-1-CYCLOHEXANOL
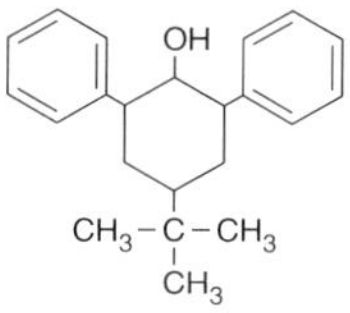
250 mg

S57,806-1
3-PHENYL-3-OCTADECANOL
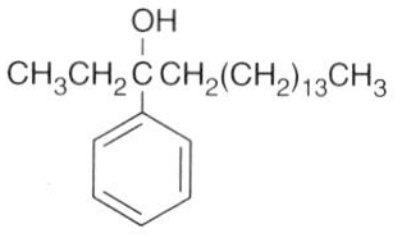
5 gm

S93,565-4
ALPHA-METHYL-(4-BIPHENYLYL)METHANOL
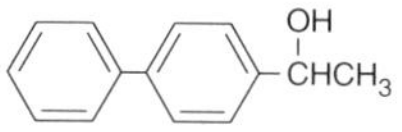
250 mg

S45,683-7
2-(4-BIPHENYLYL)-2-PROPANOL
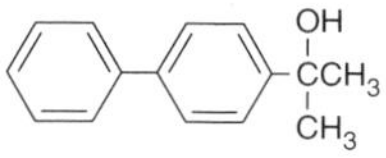
5 gm

S82,720-7
5,5-DIPHENYL-1-PENTANOL
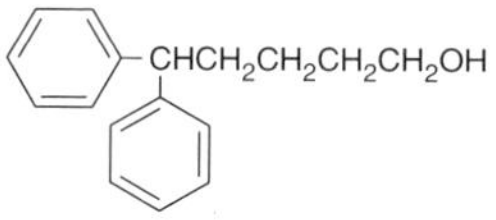
250 mg

S87,741-7
2,2,2-TRIPHENYL-1-ETHANOL
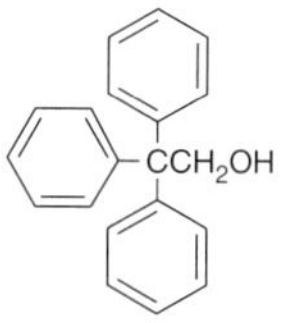
250 mg

S84,659-7
9-HYDROXY-10-PHENYL-4B,9,9A,10-TETRAHYDROINDENO(1,2-A)INDENE
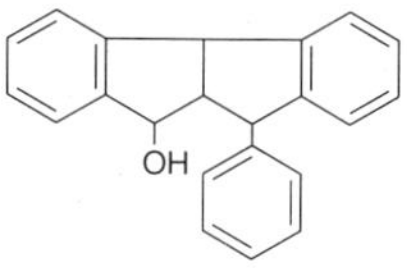
250 mg

S84,231-1
2,3-DIPHENYL-1,2,3,4-TETRAHYDRO-1-NAPHTHOL
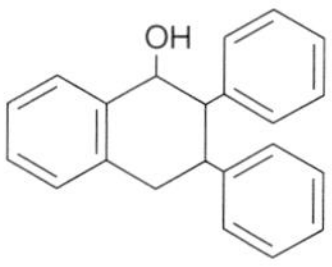
250 mg

S66,273-9
2-HYDROXYMETHYL-3-(1-NAPHTHYL)-3-PHENYL-1-PROPANOL
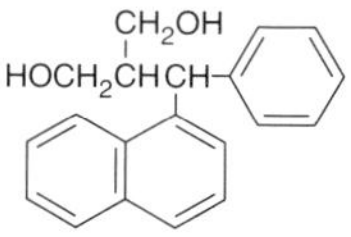
250 mg

S41,262-7
2-BROMOPHENYL METHYL CARBINOL
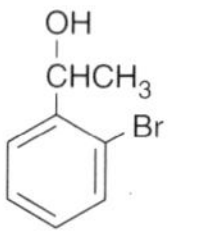
5 gm

S88,058-2
4-(O-TOLYL)-3-BUTEN-2-OL
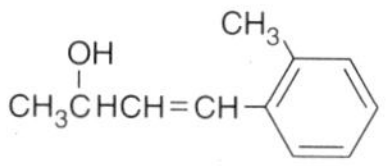
250 mg

S80,972-1
2-[2-(O-TOLYL)ETHYL]-PHENETHYL ALCOHOL
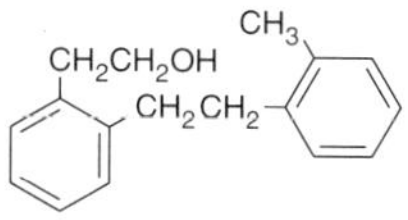
250 mg

S41,928-1
ALPHA,4-DIMETHYLBENZYL ALCOHOL
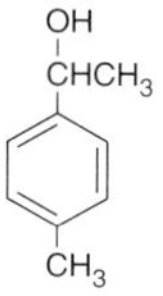
5 gm

S43,487-6
ALPHA-ETHYL-4-FLUOROBENZYL ALCOHOL
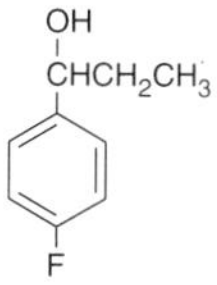
5 gm

S88,056-6
4-(P-TOLYL)-3-BUTEN-2-OL
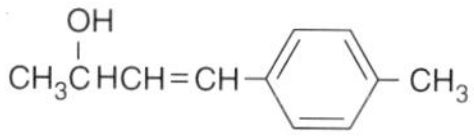
250 mg

S52,858-7
ALPHA,ALPHA,4-TRIMETHYLPHENETHYL ALCOHOL

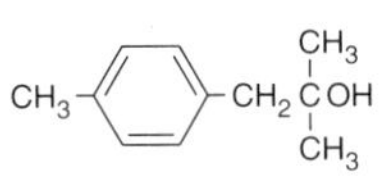

5 gm

S35,088-5
4-(3-HYDROXYISOAMYL)PHENOL

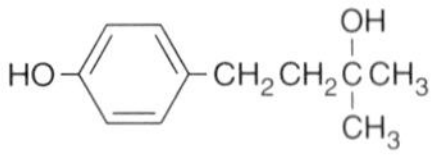

250 mg

S50,252-9
ALPHA,ALPHA-DIMETHYL-4-ISOPROPYLBENZYL ALCOHOL

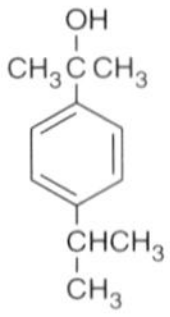

250 mg

S57,622-0
4-(TERT-BUTYL)-ALPHA-METHYLBENZYL ALCOHOL

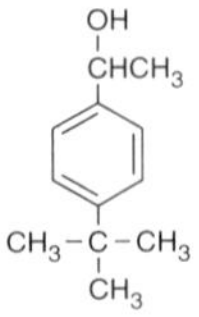

5 gm

S73,278-8
2,2-DIMETHYL-3-(4-TOLYL)-1-PROPANOL

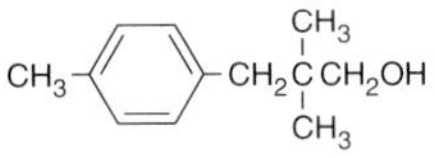

250 mg

S73,452-7
3-METHYL-3-(4-TOLYL)-2-BUTANOL

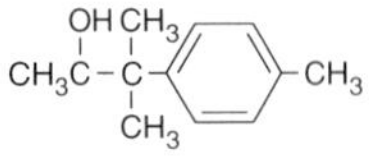

250 mg

S43,234-2
4-BROMOBENZHYDROL

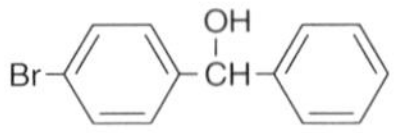

5 gm

S34,724-8
2,5-DIMETHYLPHENETHYL ALCOHOL

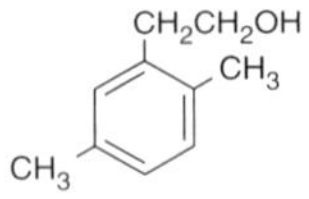

5 gm

S61,617-6
4-ISOPROPYL-2-METHYLPHENETHYL ALCOHOL

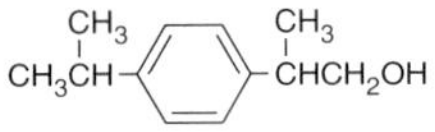

5 gm

S35,203-9
2-MESITYLENEETHANOL

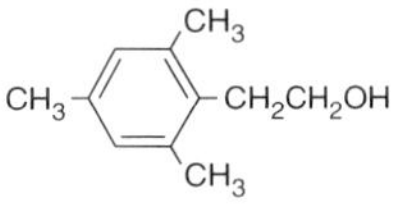

5 gm

S88,589-4
9,10-DIHYDRO-9,10(1',2')-BENZENOANTHRACENE-9-METHANOL

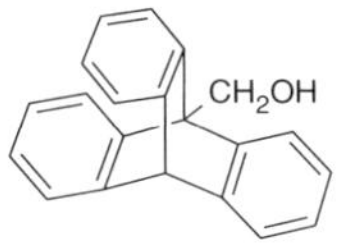

250 mg

S85,232-5
12-CHLORO-10,11-DIHYDRO-5,10-ETHENO-5H-DIBENZO(A,D)-CYCLOHEPTEN-11-OL

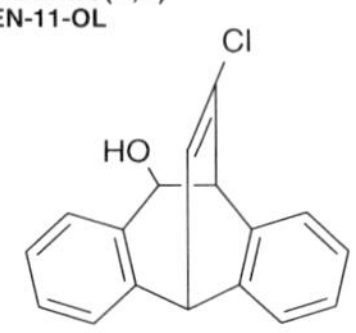

250 mg

S52,564-2
1,1-INDANDIETHANOL

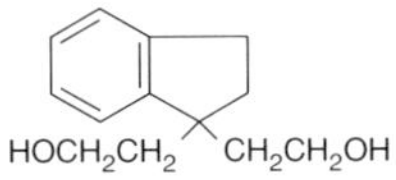

250 mg

S77,496-0
9-HYDROXY-9-(1-NAPHTHYL)-PERHYDROFLUORENE

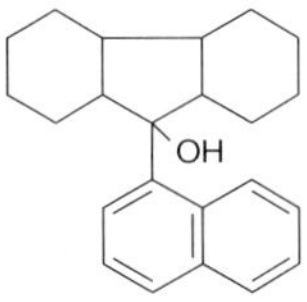

250 mg

S37,402-4
ALPHA,ALPHA-DIMETHYL-1-NAPHTHALENEMETHANOL

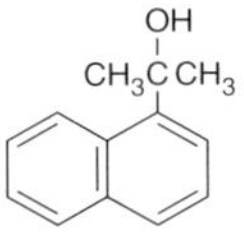

250 mg

S92,911-5
4-METHYL-1-NAPHTHALENEMETHANOL

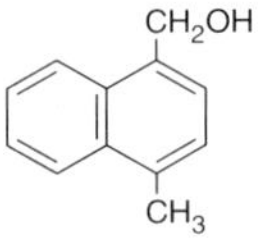

250 mg

S97,539-7
AR-L19 1-(HYDROXYMETHYL)BENZO(A)-PYRENE

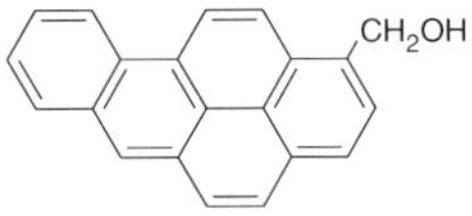

10 mg

S97,549-4
AR-L33 1-(1-HYDROXYETHYL)BENZO(A)-PYRENE

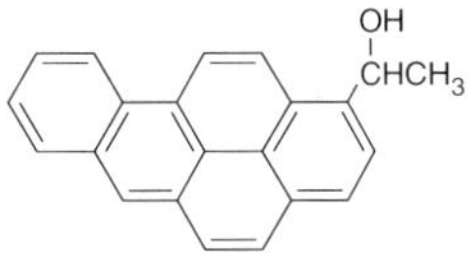

10 mg

S97,512-5
AR-H18 4-HYDROXYMETHYLPYRENE

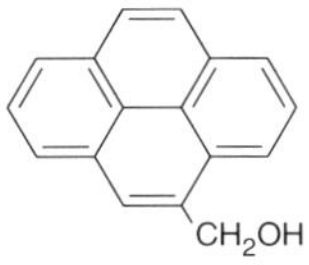

10 mg

S97,518-4
AR-H25 1-(1-HYDROXYETHYL)PYRENE

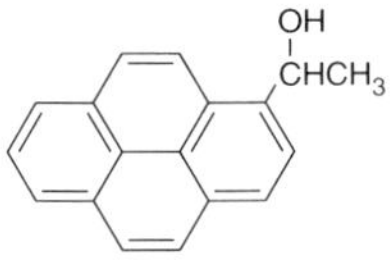

10 mg

S77,429-4
4-FLUORO-9-HYDROXYFLUORENE

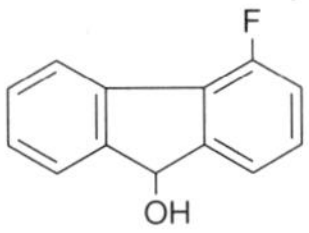

250 mg

S75,422-6
9-HYDROXY-2,4,7-TRICHLOROFLUORENE

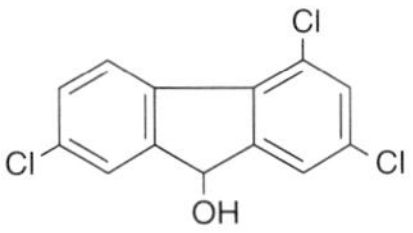

250 mg

S41,974-5
1-FLUORENEMETHANOL

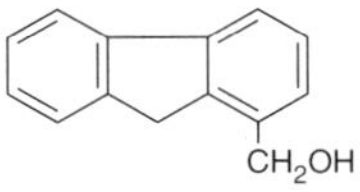

250 mg

S74,888-9
2-FLUORENEMETHANOL

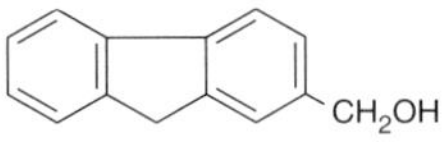

250 mg

S75,615-6
ALPHA-METHYL-2-FLUORENEMETHANOL

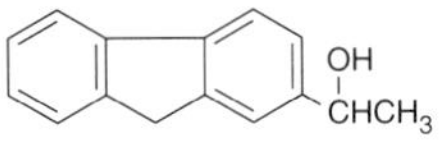

5 gm

S75,564-8
1-(2-FLUORENYL)ALLYL ALCOHOL

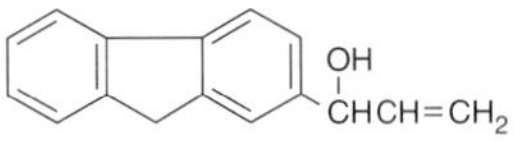

250 mg

S97,432-3
AR-E95 7,12-DIMETHYL-5-HYDROXYMETHYLBENZ(A)ANTHRACENE

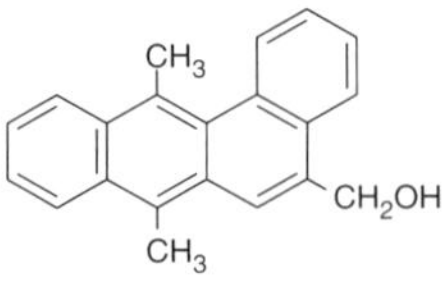

10 mg

S97,317-3
AR-D16 1,2,3,4-TETRAHYDRO-4-HYDROXYPHENANTHRENE

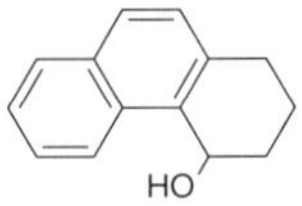

10 mg

S45,782-5
4-PHENANTHRENEMETHANOL

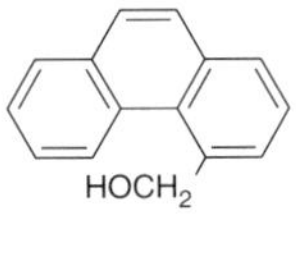

250 mg

S97,330-0
AR-D28 1,2,3,4-TETRAHYDRO-4-HYDROXY-4-METHYLPHENANTHRENE

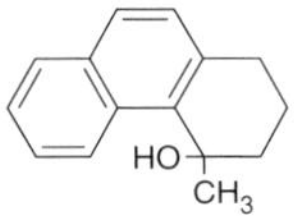

10 mg

S75,923-6
2,3,4,5-TETRAHYDRO-1-BENZOTHIEPIN-5-OL

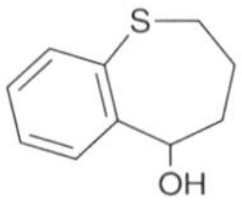

250 mg

S67,163-0
2-BENZYL-2-METHYL-1,3-PROPANEDIOL

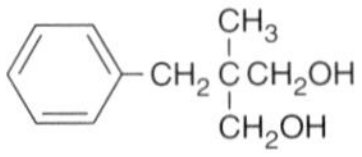

250 mg

S91,931-4
3-(TRIFLUOROMETHYL)BENZALDEHYDE DIMETHYL MERCAPTAL
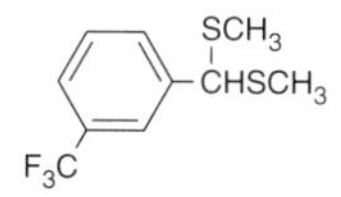
250 mg

S46,145-8
2,4,5-TRICHLOROTHIOPHENOL
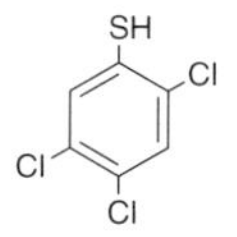
5 gm

S75,909-0
2-MERCAPTOBENZYL ALCOHOL
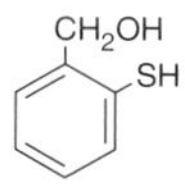
250 mg

S83,115-8
2-PHENOXYTHIOPHENOL
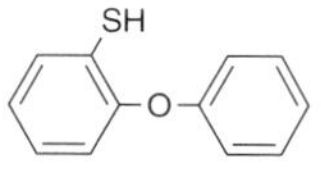
250 mg

S93,732-0
O-CARBORANE-1,2-DITHIOL
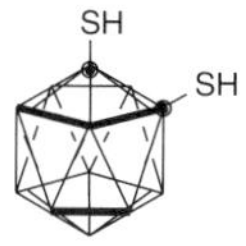
250 mg

S93,733-9
M-CARBORANE-1,7-DITHIOL
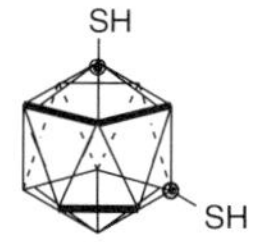
250 mg

S45,362-5
PENTACHLOROTHIOPHENOL
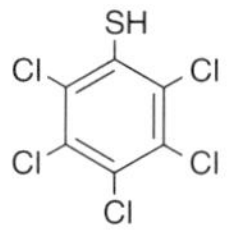
5 gm

S60,009-1
3-METHYLBENZYL MERCAPTAN
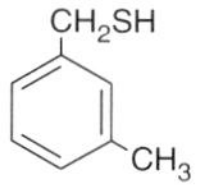
250 mg

S82,834-3
THIOPHENOL, THALLIUM SALT
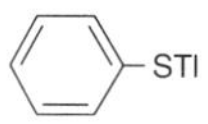
250 mg

S84,657-0
TRIBENZYLMETHANETHIOL
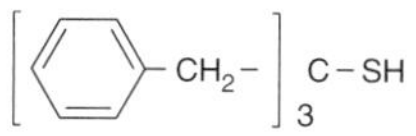
250 mg

S60,008-3
2-METHYLBENZYL MERCAPTAN
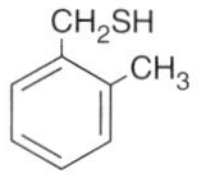
250 mg

S84,267-2
2,2'-BIPHENYLDITHIOL
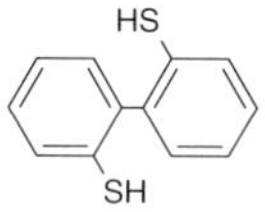
250 mg

S82,837-8
P-THIOCRESOL, THALLIUM SALT
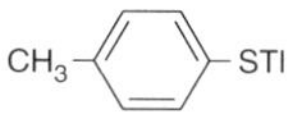
250 mg

S36,949-7
P-XYLENE-ALPHA-THIOL
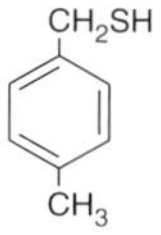
5 gm

S42,786-1
4-TERT-BUTYLTHIOPHENOL
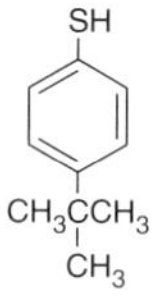
5 gm

S90,696-4
2-(MERCAPTOMETHYL)NAPHTHALENE
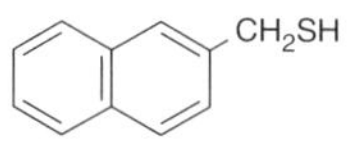
250 mg

S97,313-0
AR-D10 3-MERCAPTOPHENANTHRENE
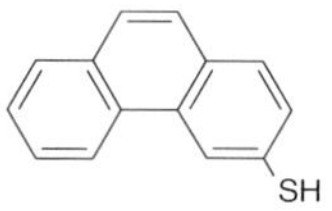
10 mg

S88,709-9
1,1,1-TRIS(2-NAPHTHYLTHIOMETHYL)-ETHANE
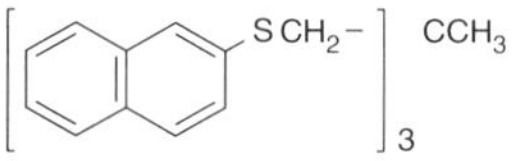
250 mg

S88,791-9
2,2',6,6'-TETRAPHENYL-DELTA 4,4'-BI-4H-THIOPYRAN
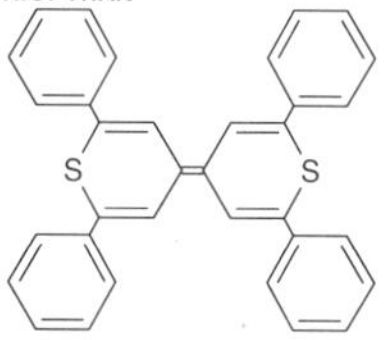
250 mg

S65,173-7
2,2-DIPHENYL-1,3-DITHIANE
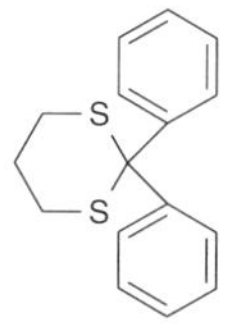
250 mg

S70,710-4
CHLOROMETHYL-PENTACHLOROPHENYL SULFIDE
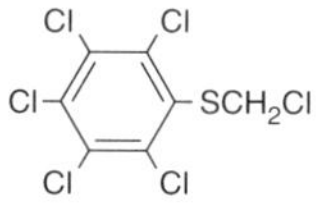
250 mg

S80,372-3
4-CHLOROPHENYL-3-CHLOROPROPYL SULFIDE
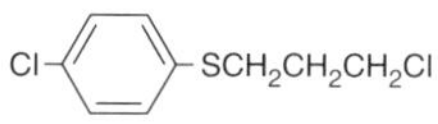
250 mg

S79,920-3
4-BROMOPHENYL 3-CHLOROPROPYL SULFIDE
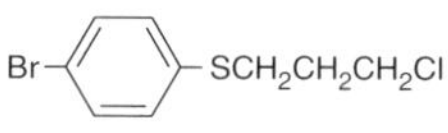
250 mg

S64,895-7
5-CHLOROMETHYL-2-(3,4-METHYLENEDIOXYPHENYL)-1,3-OXATHIOLANE
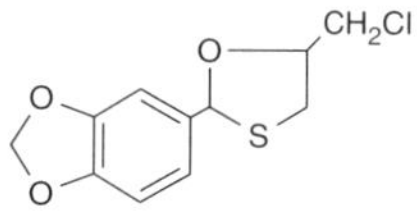
250 mg

S64,903-1
5-CHLOROMETHYL-2-METHYL-2-(3,4-METHYLENEDIOXYPHENYL)-1,3-OXATHIOLANE
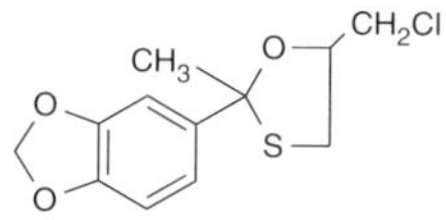
250 mg

S79,928-9
3-CHLOROPROPYL M-TOLYL SULFIDE
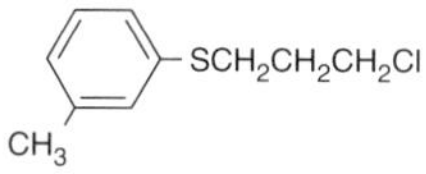
250 mg

S57,725-1
2-CHLOROETHYL PHENYL SULFIDE
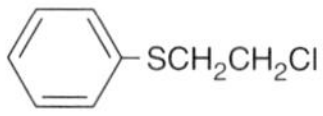
250 mg

S79,921-1
3-CHLOROPROPYL PHENYL SULFIDE
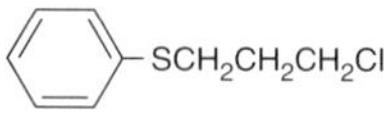
250 mg

S79,927-0
3-CHLORO-1-METHYLPROPYL PHENYL SULFIDE
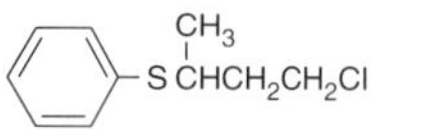
250 mg

S79,929-7
4-BROMOBUTYL PHENYL SULFIDE
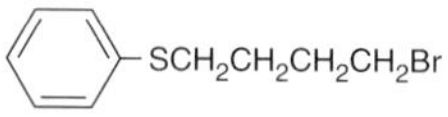
250 mg

S79,934-3
5-CHLOROPENTYL PHENYL SULFIDE
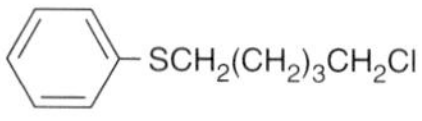
250 mg

S61,404-1
2-CHLOROETHYL-4-CHLOROPHENYL SULFIDE
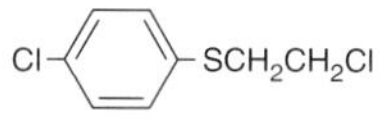
5 gm

S83,280-4
2-CHLOROETHYL P-TOLYL SULFIDE
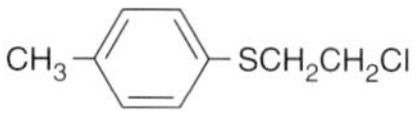
250 mg

S63,793-9
4-CHLOROBENZYL 2-CHLOROETHYL SULFIDE
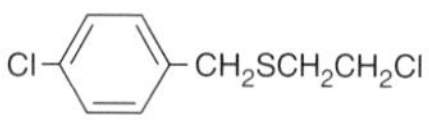
5 gm

S39,129-8
4-CHLOROPHENYL 2-CHLORO-1,1,2-TRIFLUOROETHYL SULFIDE
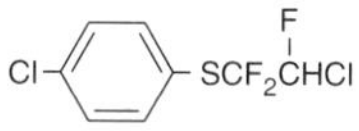
5 gm

S63,887-0
3,4-DICHLOROPHENYL-DIFLUOROMETHYLSULFIDE
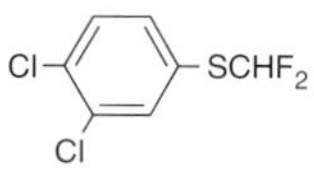
250 mg

S55,357-3
2-CHLORO-1,1,2-TRIFLUOROETHYL-3,4-DICHLOROPHENYL SULFIDE
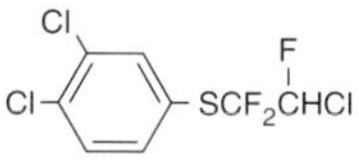
250 mg

S77,297-6
1-[(4-CHLOROPHENYL)THIO]-2,2,2-TRICHLOROETHANOL
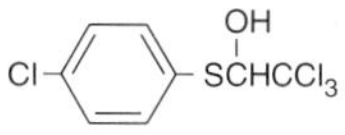
250 mg

S78,105-3
1-(4-CHLOROBENZYLTHIO)-2,2,2-TRICHLOROETHANOL
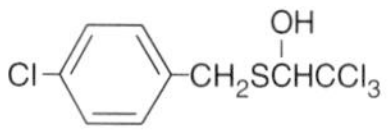
250 mg

S64,901-5
5-CHLOROMETHYL-2-METHYL-2-(2,4-XYLYL)-1,3-OXATHIOLANE
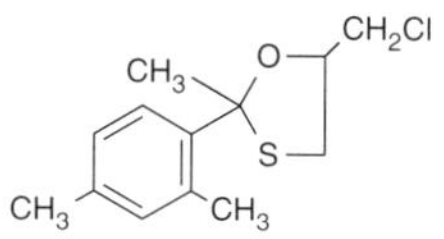
5 gm

S75,184-7
2,2'-THIOBIS(4-CHLOROPHENOL)
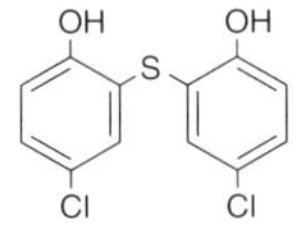
5 gm

S75,181-2
2,2'-THIOBIS(5-CHLORO-3-METHYLPHENOL)
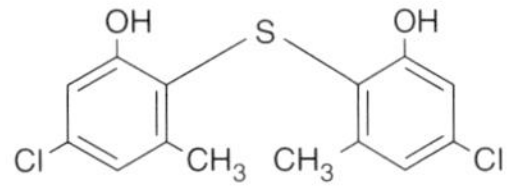
5 gm

S75,183-9
2,2'-THIOBIS(4-CHLORO-6-METHYLPHENOL)
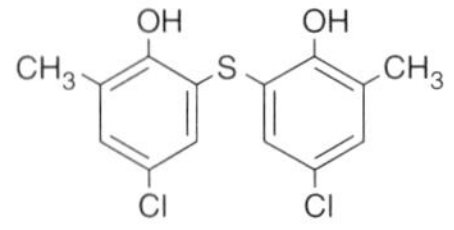
5 gm

S55,677-7
METHYL 2,3,5,6-TETRACHLOROPHENYL SULFIDE
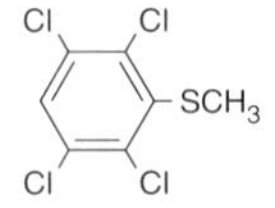
5 gm

S81,160-2
4-BROMOPHENYL PHENYL SULFIDE
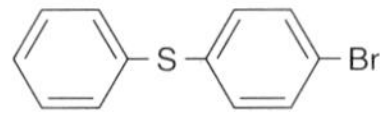
250 mg

S72,194-8
BENZYL 4-CHLOROPHENYL SULFIDE
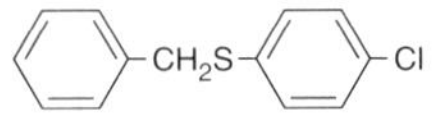
5 gm

S72,203-0
4-CHLOROBENZYL PHENYL SULFIDE
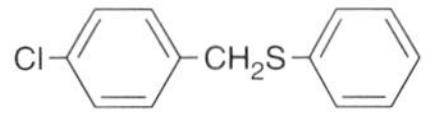
5 gm

S68,131-8
BENZYL 4-BROMOPHENYL SULFIDE
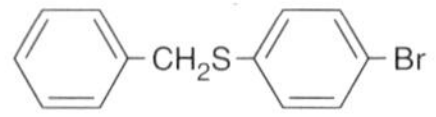
5 gm

S74,599-5
4-CHLORO-ALPHA-ETHYLBENZYL-PHENYL SULFIDE
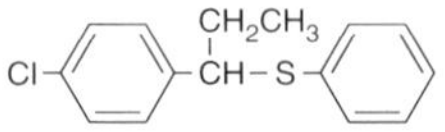
250 mg

S74,810-2
4-CHLORO-ALPHA-METHYLPHENETHYL PHENYL SULFIDE
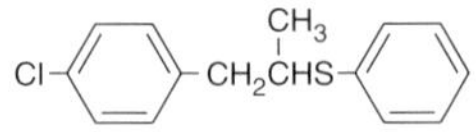
250 mg

S66,339-5
2-BROMOBENZYL METHYL SULFIDE
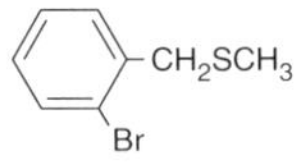
5 gm

S76,789-1
ALLYL 4-BROMOPHENYL SULFIDE

250 mg

S69,632-3
3,4-DICHLOROTHIOANISOLE
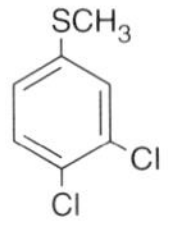
250 mg

S56,794-9
3,4-DICHLOROPHENYL 5,7,7-TRIMETHYL-2-(1,3,3-TRIMETHYLBUTYL)-OCTYLSULFIDE
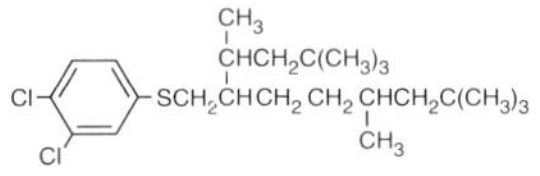
250 mg

S94,398-3
METHYL 2,4,5-TRICHLOROPHENYL SULFIDE
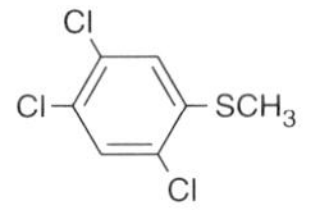
250 mg

S50,435-1
2,2'-DITHIOBIS[4-(1,1,3,3-TETRAMETHYLBUTYL)PHENOL]
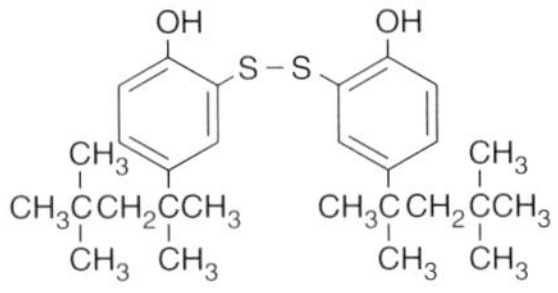
5 gm

S83,087-9
2-(DODECYLTHIO)-1-PHENYLETHANOL
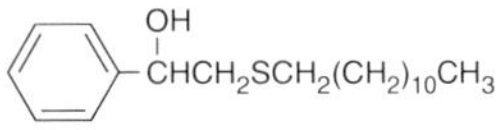
250 mg

S83,096-8
2-(HEXADECYLTHIO)-1-PHENYLETHANOL
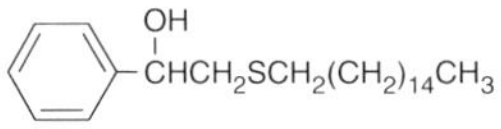
250 mg

S82,258-2
4-(METHYLTHIO)BENZHYDROL
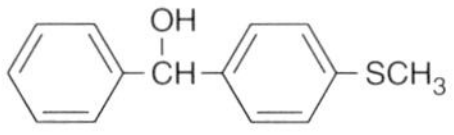
250 mg

S61,463-7
3-PHENYLTHIO-1,2-PROPANEDIOL
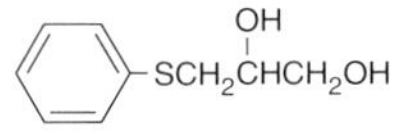
5 gm

S39,155-7
2-(BENZYLTHIO)ETHANOL
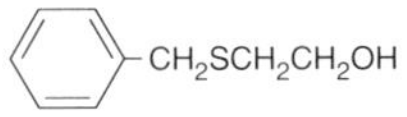
250 mg

S79,923-8
3-(PHENYLTHIO)PROPANOL
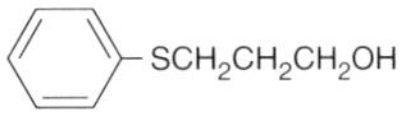
250 mg

S55,026-4
2-(P-TOLYLTHIO)ETHANOL
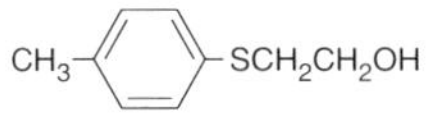
5 gm

S61,409-2
4-CHLOROBENZYL-2-HYDROXYETHYL SULFIDE
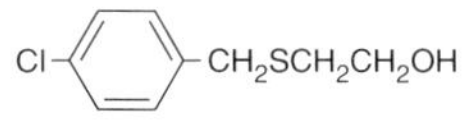
5 gm

S65,580-5
ALPHA,ALPHA-BIS(2-HYDROXYETHYL-THIO)-P-XYLENE
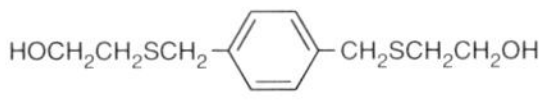
250 mg

S45,896-1
4-(METHYLTHIO)-M-CRESOL
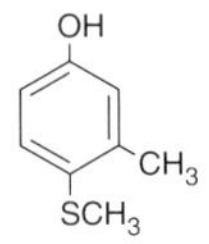
5 gm

S53,377-7
4-METHYLTHIO-O-CRESOL
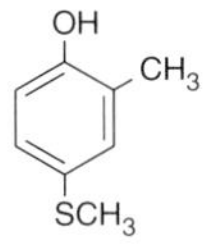
5 gm

S53,860-4
2-[4-(TERT-BUTYL)-O-TOLYTHIO]-ETHANOL
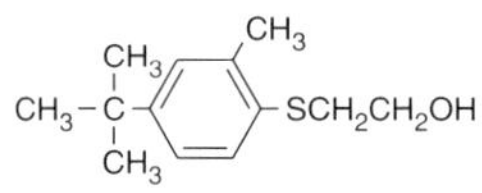
5 gm

S53,379-3
2-HYDROXY-3-METHYL-5-(METHYLTHIO)-BENZYL ALCOHOL
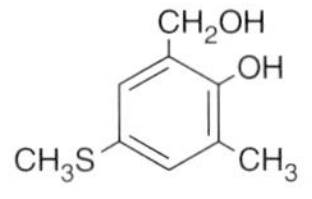
5 gm

S35,920-3
3-TERT-BUTYL-2-HYDROXY-5-METHYLPHENYL SULFIDE
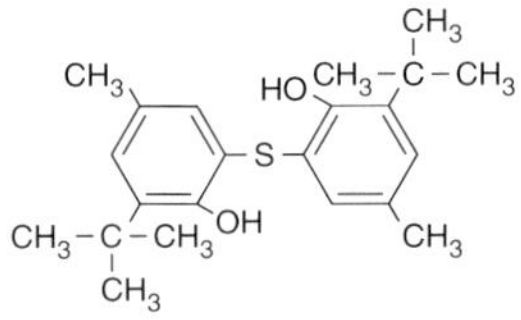
5 gm

S84,955-3
2,3-DIHYDRO-4-HYDROXY-8-METHYL-4H-1-BENZOTHIOPYRAN
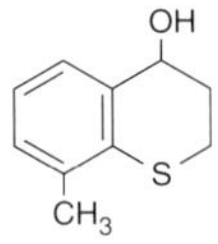
250 mg

S84,959-6
2,3-DIHYDRO-3,4-DIMETHYL-4-HYDROXY-4H-1-BENZOTHIOPYRAN
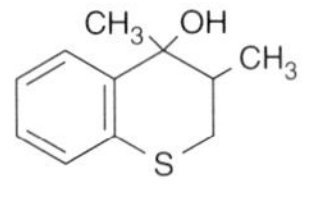
250 mg

S87,582-1
ALPHA,ALPHA-DIMETHOXYBENZYL-PHENYL SULFIDE
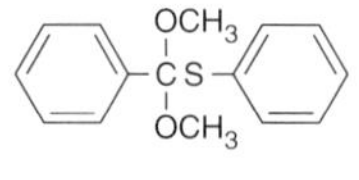
250 mg

S86,629-6
DIMETHOXYMETHYL PHENYL SULFIDE
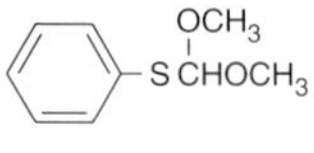
250 mg

S71,744-4
2,3-DIMETHOXYPHENYL DISULFIDE
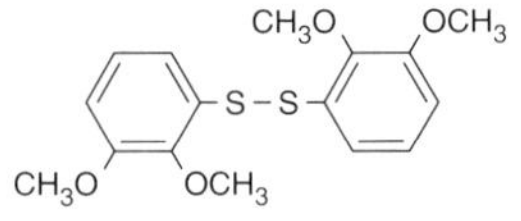
250 mg

S69,795-8
4-[2-(METHYLTHIO)ETHOXY]ANISOLE
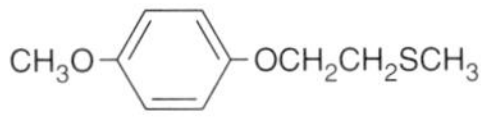
250 mg

S87,598-8
1,1-DIETHOXY-1-(PHENYLTHIO)ETHANE
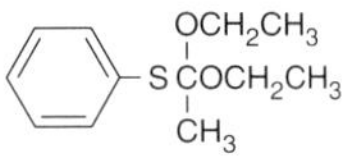
250 mg

S87,596-1
DIBUTOXY(PHENYLTHIO)METHANE
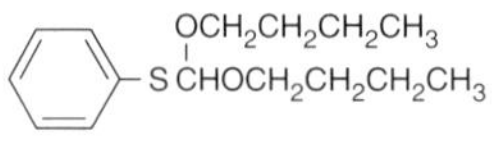
250 mg

S34,019-7
2-ACETAMIDO-4-METHOXYPHENYL-DISELENIDE
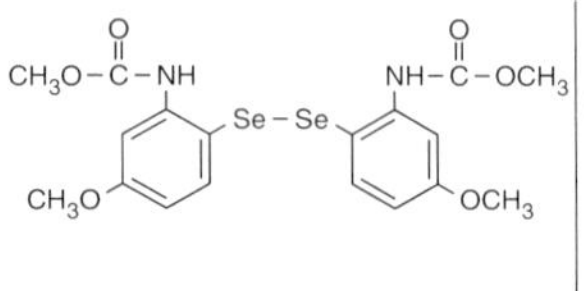
5 gm

S94,070-4
BIS(4-METHOXYPHENYL) DITELLURIDE

250 mg

S94,069-0
BIS(4-ETHOXYPHENYL) DITELLURIDE

250 mg

S69,799-0
2-(METHYLTHIO)ETHYL PHENYL ETHER
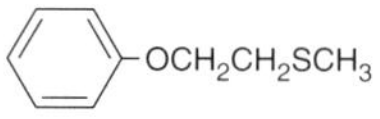

250 mg

S79,317-5
2-(PHENYLTHIO)ETHYL PHENYL ETHER
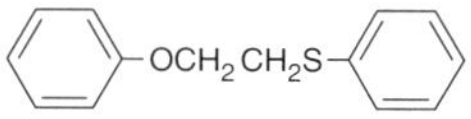

250 mg

S79,324-8
4-(PHENYLTHIO)BUTYL PHENYL ETHER
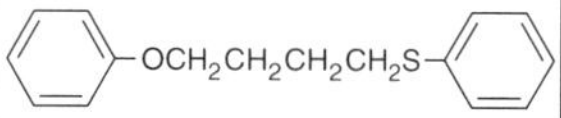

250 mg

S80,730-3
4-METHOXYPHENYL PHENYL SULFIDE
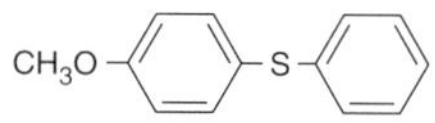

250 mg

S77,120-1
2-(4-METHOXYPHENYL)-1-METHYLETHYL PHENYL SULFIDE
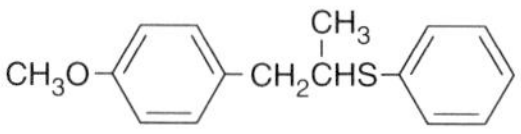

250 mg

S69,793-1
2-(METHYLTHIOMETHYL)ANISOLE
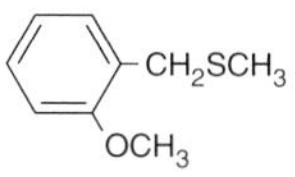

250 mg

S69,800-8
4-(METHYLTHIOMETHYL)ANISOLE
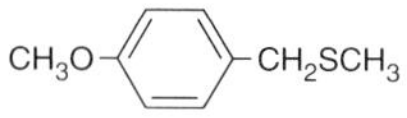

250 mg

S76,799-9
ALLYL 4-METHOXYPHENYL SULFIDE
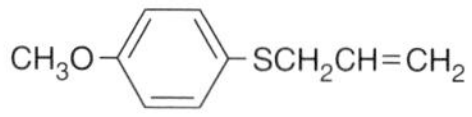

250 mg

S86,064-6
2-(2-METHOXYPHENYL)-1,3-DITHIANE
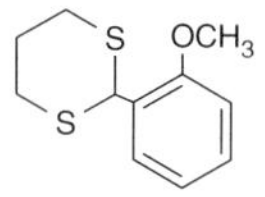

250 mg

S94,068-2
(4-METHOXYPHENYL)TRICHLOROTELLURIUM
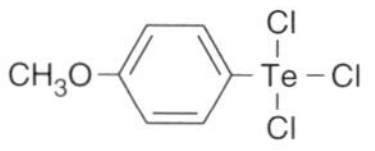

250 mg

S93,834-3
(4-ETHOXYPHENYL)TRICHLOROTELLURIUM
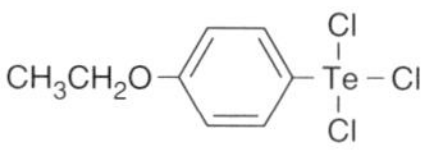

250 mg

S83,483-1
1-METHOXYTHIANTHRENE
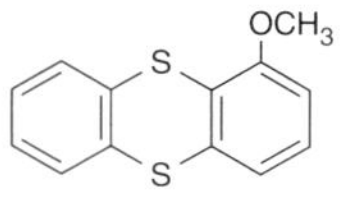

250 mg

S69,788-5
2-(PHENYLTHIO)ACETALDEHYDE-DIETHYL ACETAL
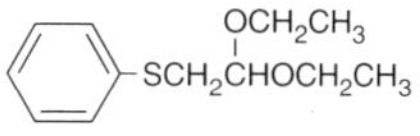

5 gm

S75,760-8
D-GLUCOSE DIBENZYL MERCAPTAL
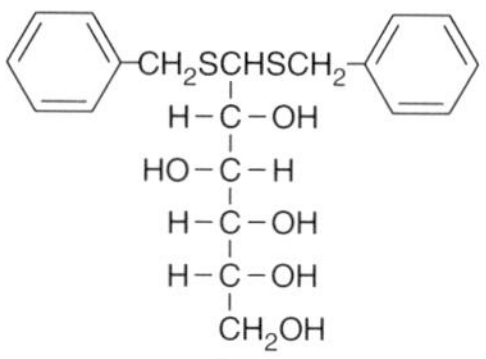

5 gm

S76,786-7
ALLYL M-TOLYL SULFIDE
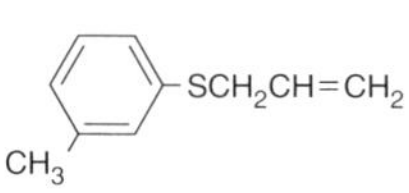

250 mg

S87,425-6
2,6-DIMETHYL-4-(METHYLTHIO)-THIOPYRYLIUM IODIDE
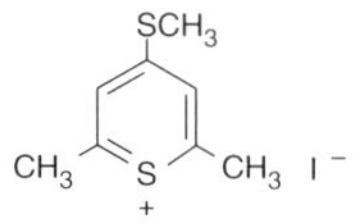

250 mg

S64,619-9
1-(BENZYLTHIO)ACETONE
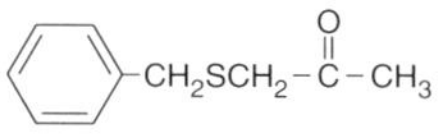

5 gm

S94,918-3
2,4,6-TRIS(METHYLTHIO)TOLUENE
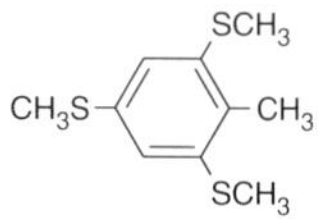

250 mg

S62,031-9
3-(PHENYLTHIO)-1,2-EPOXYPROPANE
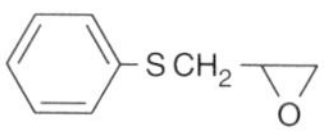

250 mg

S77,567-3
1,2-BIS(P-TOLYLTHIO)-1-CYCLOPENTENE
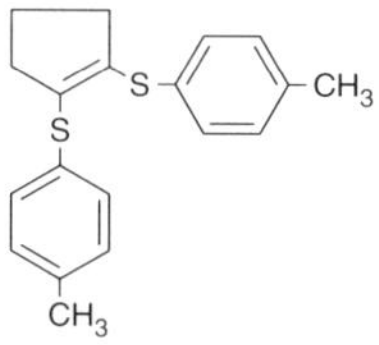

250 mg

S86,486-2
2-(PHENYLTHIO)-1,3-DIOXOLANE
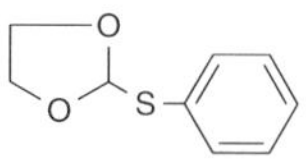

250 mg

S80,707-9
BENZYL CYCLOHEXYL SULFIDE
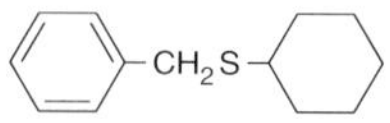

250 mg

S86,356-4
METHYL 2-PHENYLCYCLOHEXYL SULFIDE
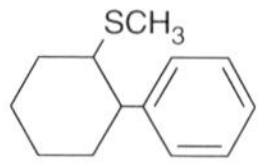

250 mg

S80,711-7
1-PHENYLCYCLOHEXYL P-TOLYL SULFIDE
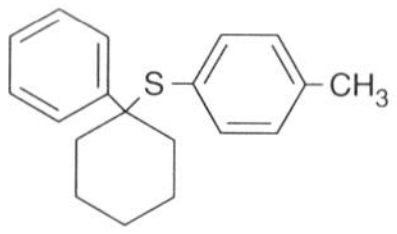

250 mg

S86,670-9
2-CYCLOHEXEN-1-YL PHENYL SULFIDE
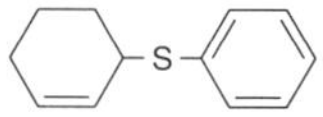

250 mg

S85,259-7
11-CHLORO-9,10-DIHYDRO-12-(PHENYLTHIO)-9,10-ETHANOANTHRACENE
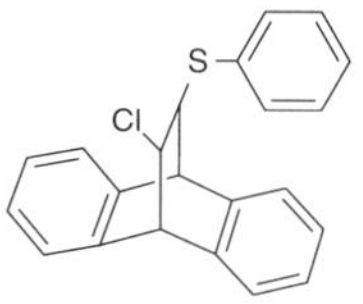

250 mg

S86,500-1
2-(PHENYLTHIO)TETRAHYDROPYRAN
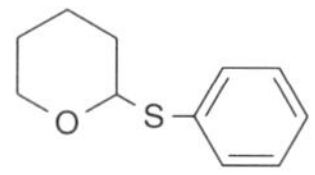

250 mg

S61,424-6
ALLYL PHENYL SULFIDE
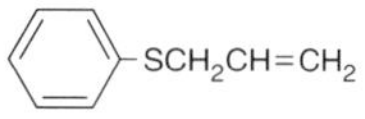

5 gm

S41,050-0
BENZYL ETHYL SULFIDE
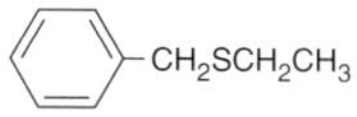

5 gm

S80,988-8
ISOPROPYL PHENYL SULFIDE
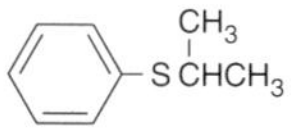

250 mg

S83,595-1
PHENYL PROPYL SULFIDE
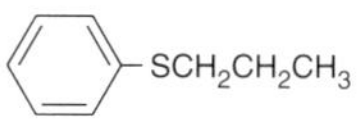

250 mg

S72,224-3
3-CHLOROALLYL PHENYL SULFIDE, MIXTURE OF CIS AND TRANS
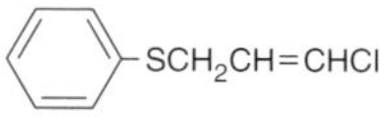

5 gm

S76,788-3
2-CHLOROALLYL PHENYL SULFIDE
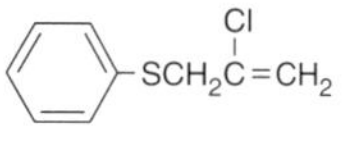

250 mg

S76,784-0
1,3-BUTADIEN-2-YL PHENYL SULFIDE
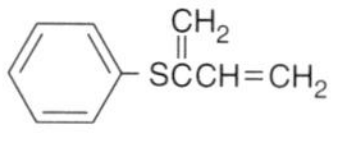

250 mg

S61,420-3
2-METHYLALLYL PHENYL SULFIDE
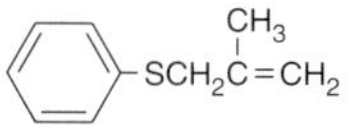

5 gm

S74,620-7
CROTYL PHENYL SULFIDE
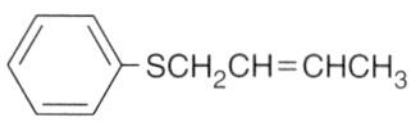

5 gm

S72,211-1
ISOBUTYL PHENYL SULFIDE
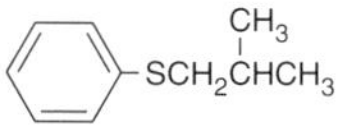

5 gm

S80,989-6
TERT-BUTYL PHENYL SULFIDE
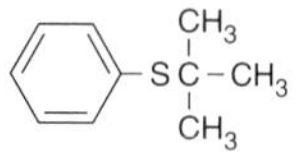

250 mg

S83,598-6
BUTYL PHENYL SULFIDE
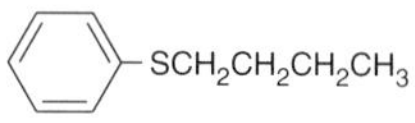

250 mg

S79,938-6
4-PENTENYL PHENYL SULFIDE
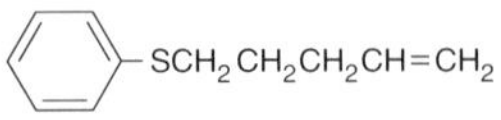
250 mg

S80,700-1
PHENYL STYRYL SULFIDE
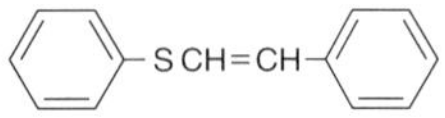
250 mg

S35,462-7
PHENETHYL PHENYL SULFIDE
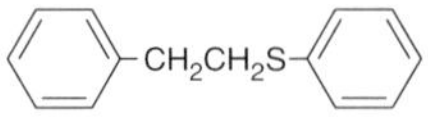
250 mg

S86,525-7
1,2-BIS(PHENYLTHIO)ETHENE
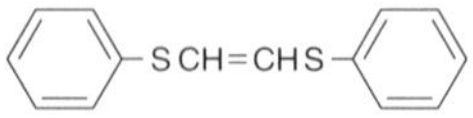
250 mg

S57,707-3
1,2-BIS(PHENYLTHIO)ETHANE
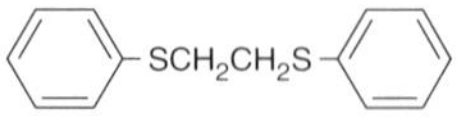
5 gm

S80,590-4
STYRYL M-TOLYL SULFIDE
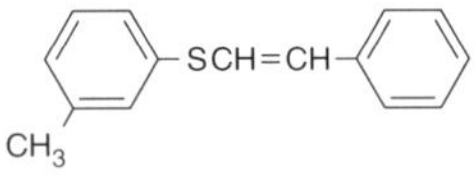
250 mg

S35,463-5
PHENYL 3-PHENYLPROPYL SULFIDE
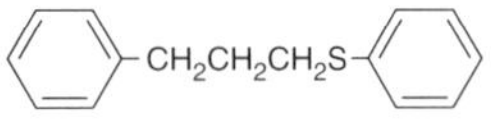
5 gm

S77,119-8
PHENYL 1-PHENYLPROPYL SULFIDE
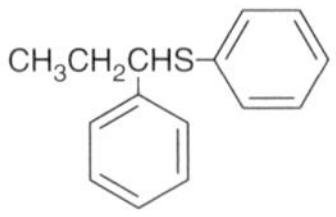
250 mg

S91,906-3
ALPHA,ALPHA-DIMETHYLBENZYL-PHENYL SULFIDE
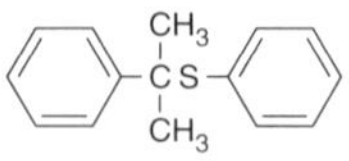
250 mg

S35,455-4
1-PHENYLETHYL SULFIDE
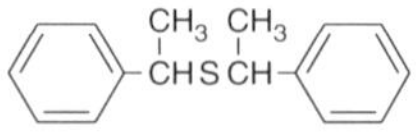
5 gm

S80,745-1
ALPHA,ALPHA-BIS(PHENYLTHIO)TOLUENE
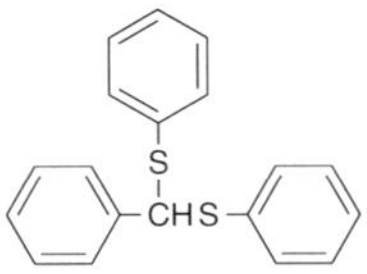
250 mg

S82,810-6
PHENYL TETRADECYL SULFIDE
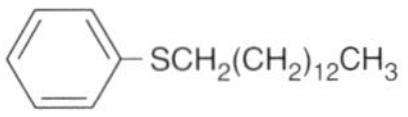
250 mg

S80,983-7
ALPHA,ALPHA'-BIS(PHENYLTHIO)-M-XYLENE
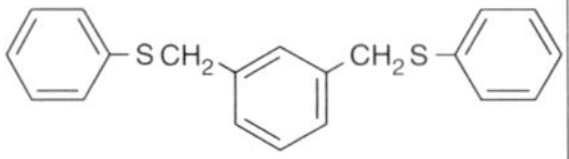
250 mg

S86,458-7
2-CHLORO-1,2-DIPHENYLETHYL-PHENYL SULFIDE
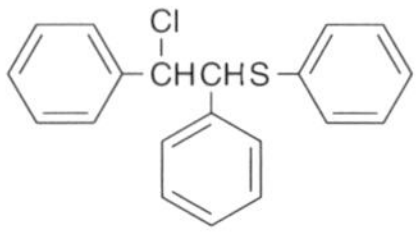
250 mg

S58,239-5
HEXADECYL PHENYL SULFIDE
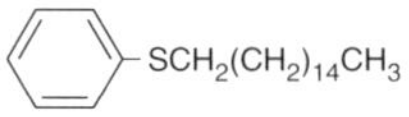
5 gm

S90,702-2
HEXAKIS(PHENYLTHIOMETHYL)BENZENE
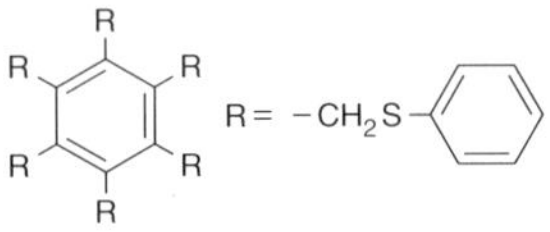
250 mg

S80,589-0
STYRYL O-TOLYL SULFIDE
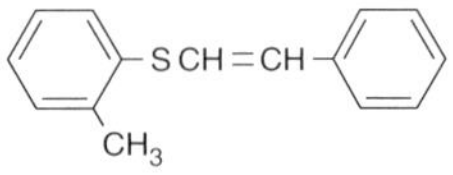
250 mg

S80,986-1
ALPHA,ALPHA'-BIS(BENZYLTHIO)-O-XYLENE
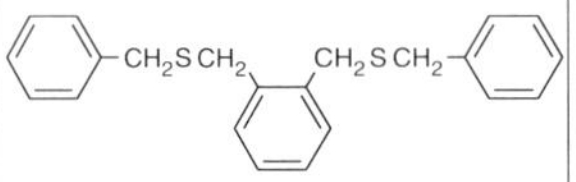
250 mg

S80,982-9
2,2'-BIS(BENZYLTHIO)BIPHENYL
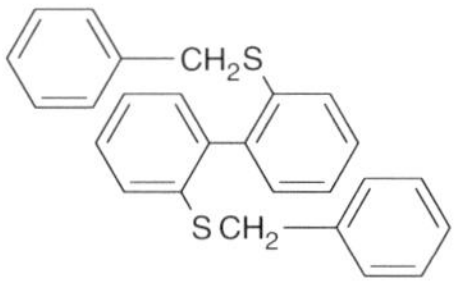
250 mg

S81,159-9
PHENYL P-TOLYL SULFIDE
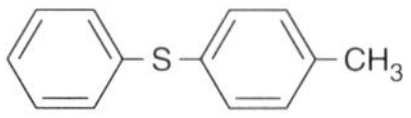
250 mg

S72,185-9
BENZYL P-TOLYL SULFIDE
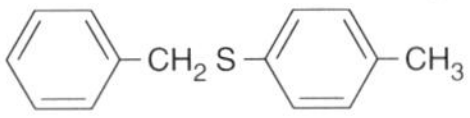
250 mg

S74,809-9
4-METHYLBENZYL PHENYL SULFIDE
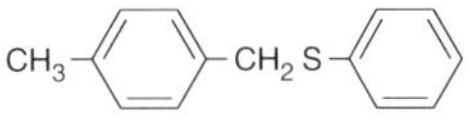
250 mg

S86,541-9
1-(PHENYLTHIO)-2-(P-TOLYLTHIO)-ETHYLENE
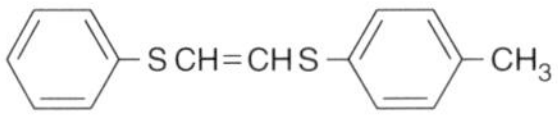
250 mg

S80,984-5
ALPHA,ALPHA'-BIS(BENZYLTHIO)-P-XYLENE
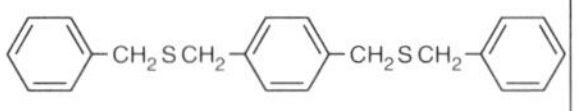
250 mg

S77,858-3
S-[1,1-(DIPHENYL)ETHYL]-P-THIOCRESOL
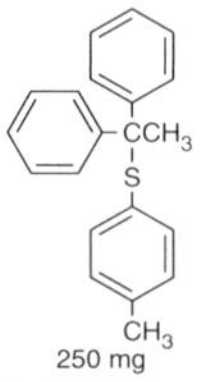
250 mg

S80,704-4
P-TOLYL TRIPHENYLMETHYL SULFIDE
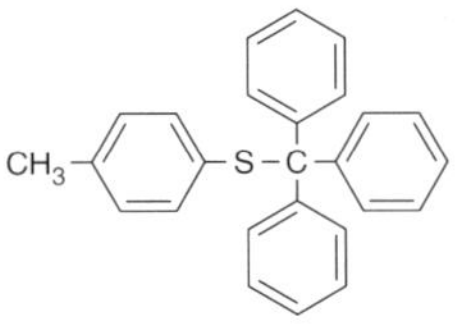
250 mg

S80,719-2
4-BIPHENYLYL PHENYL SULFIDE
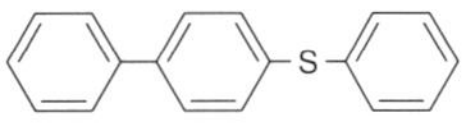
250 mg

S91,898-9
9-(PHENYLTHIO)FLUORENE
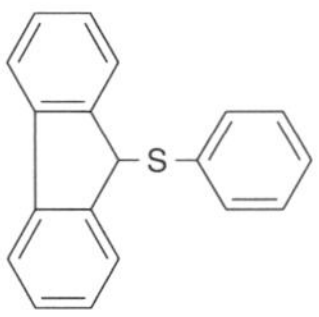
250 mg

S91,836-9
9-ANTHRACENYLMETHYL PHENYL SULFIDE
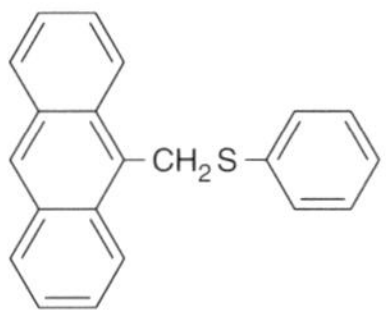
250 mg

S72,221-9
2-METHYLALLYL O-TOLYL SULFIDE
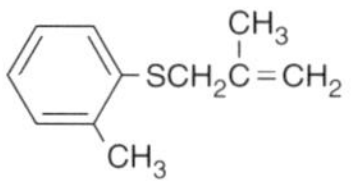
5 gm

S74,623-1
CROTYL O-TOLYL SULFIDE
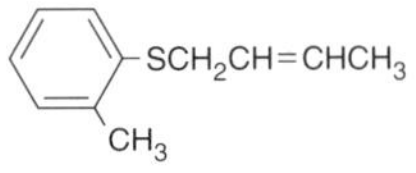
5 gm

S63,365-8
4-CHLOROPHENYL METHYL SULFIDE
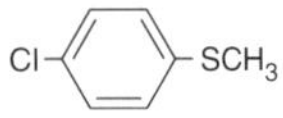
5 gm

S63,794-7
4-CHLOROBENZYL METHYL SULFIDE
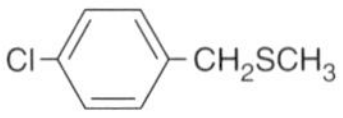
5 gm

S39,168-9
1-CHLORO-4-(METHYLTHIOMETHYLTHIO)-BENZENE
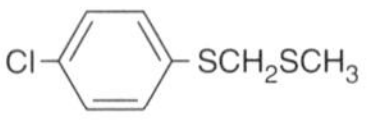
250 mg

S83,281-2
ETHYL P-TOLYL SULFIDE
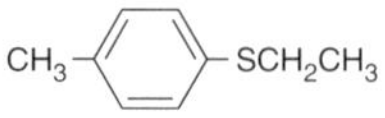
250 mg

S59,274-9
(METHYLTHIO)(P-TOLYLTHIO)METHANE
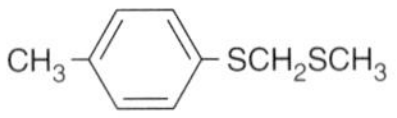
250 mg

S64,838-8
ALLYL P-TOLYL SULFIDE
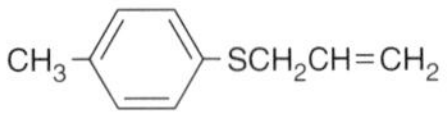
250 mg

S83,283-9
PROPYL P-TOLYL SULFIDE
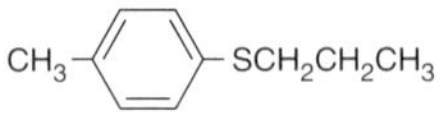
250 mg

S36,573-4
ALPHA,ALPHA-BIS(METHYLTHIO)-P-XYLENE

5 gm

S79,914-9
BUTYL P-TOLYL SULFIDE
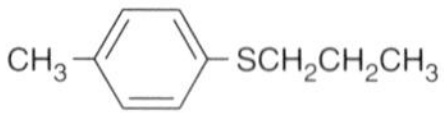
250 mg

S91,820-2
4-TERT-BUTYLPHENYL METHYL SULFIDE
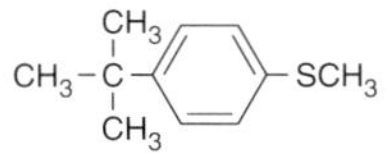
250 mg

S83,285-5
ISOAMYL P-TOLYL SULFIDE
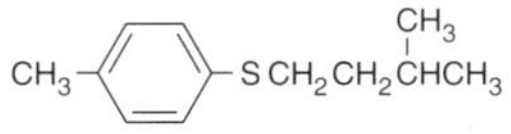
250 mg

S59,639-6
2,4,4',5-TETRACHLORODIPHENYL SULFIDE
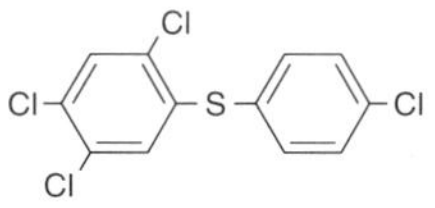
5 gm

S85,296-1
1,2-BIS(P-TOLYLTHIO)ETHENE
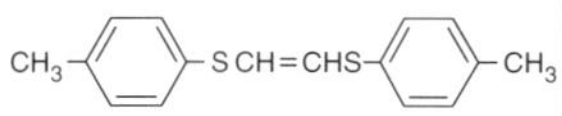
250 mg

S80,697-8
TETRAKIS(P-TOLYLTHIO)ETHYLENE
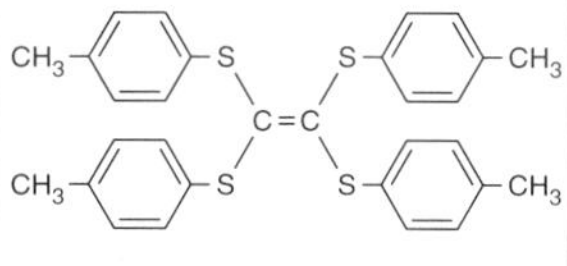
250 mg

S90,710-3
HEXAKIS(4-TERT-BUTYLPHENYLTHIOMETHYL)BENZENE
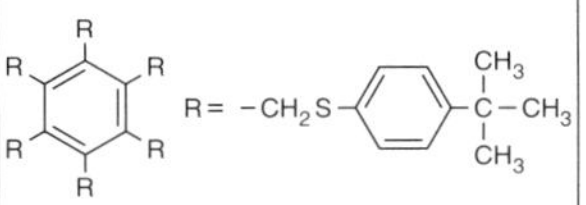
250 mg

S80,981-0
4,4'-BIS(ETHYLTHIO)BIPHENYL
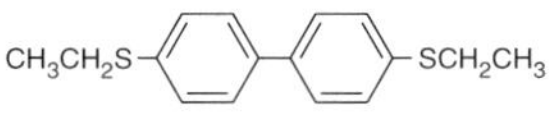
250 mg

S85,322-4
2-NORBORNYL P-TOLYL SULFIDE
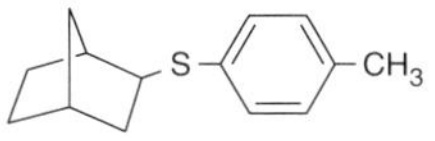
250 mg

S91,839-3
9-ANTHRACENYLMETHYL P-TOLYL SULFIDE
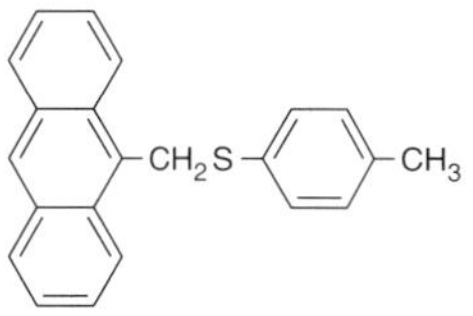
250 mg

S39,314-2
2,5-DICHLOROPHENYL METHYL SULFIDE
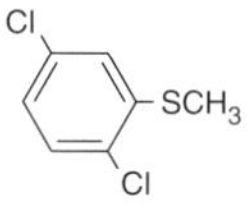
250 mg

S43,401-9
2,4-BIS(METHYLTHIO)-1-CHLOROBENZENE
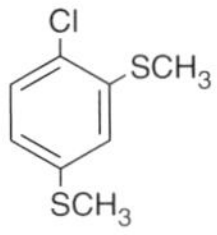
5 gm

S80,998-5
MESITYL METHYL SULFIDE
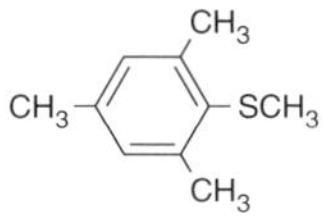
250 mg

S79,917-3
METHYL 2-NAPHTHYLMETHYL SULFIDE
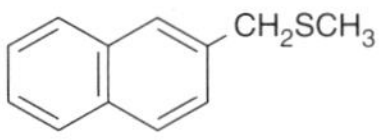
250 mg

S70,147-5
9-ANTHRACENYLMETHYL METHYL SULFIDE
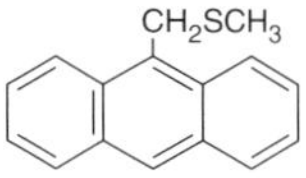
5 gm

S91,837-7
9-ANTHRACENYLMETHYL HEXYL SULFIDE
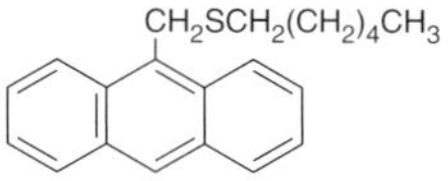
250 mg

S40,936-7
METHYL PHENYL DISULFIDE
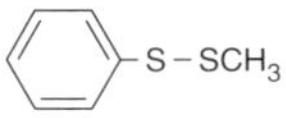
5 gm

S62,011-4
4-CHLOROPHENYL DISULFIDE
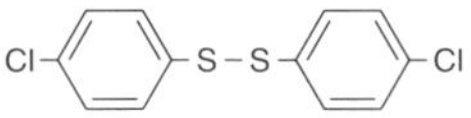
5 gm

S40,512-4
4-BROMOPHENYL DISULFIDE
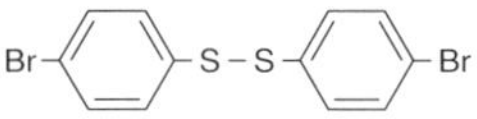
5 gm

S91,778-8
4-TERT-BUTYLPHENYL DISULFIDE
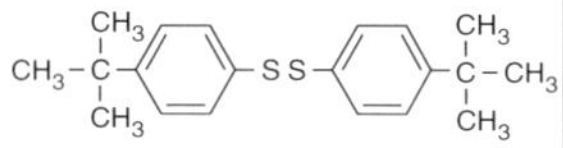
250 mg

S77,856-7
MESITYL DISULFIDE
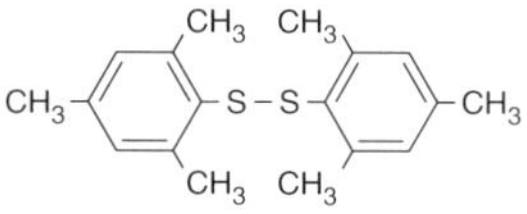
250 mg

S83,859-4
2-NAPHTHYL DISULFIDE
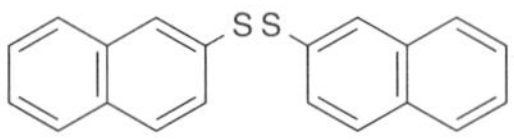
250 mg

S87,479-5
4,4'-BIS(2,6-DIPHENYLTHIOPYRYLIUM TETRAFLUOROBORATE)
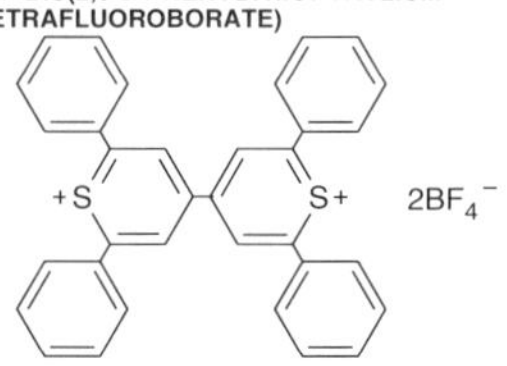
250 mg

S62,690-2
DIPHENYLETHYLSULFONIUM FLUOBORATE
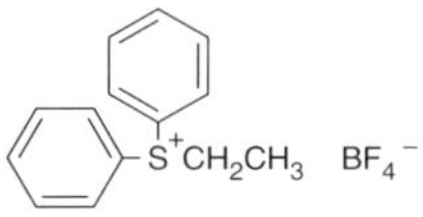
5 gm

S44,203-8
2,1,3-BENZOSELENADIAZOLE
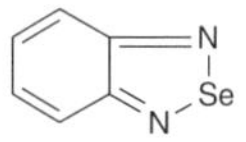
5 gm

S43,532-5
5,6-DIMETHYL-2,1,3-BENZOSELENADIAZOLE
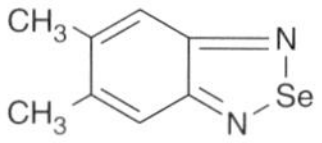
5 gm

S88,919-9
1-PHENYL-1,4-AZASELENETANE
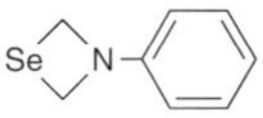
250 mg

S88,884-2
2-METHYL-3-PHENYLSELENAZOLIDINE
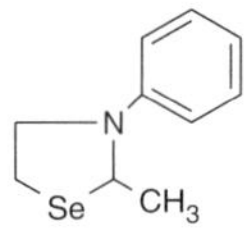
250 mg

S88,895-8
2-ACETAMIDO-3-(BENZYLSELENO)-3-METHYLBUTYRIC ACID
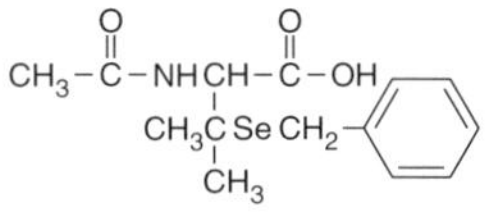
250 mg

S59,191-2
6-NITRO-3,4-XYLYL DISELENIDE
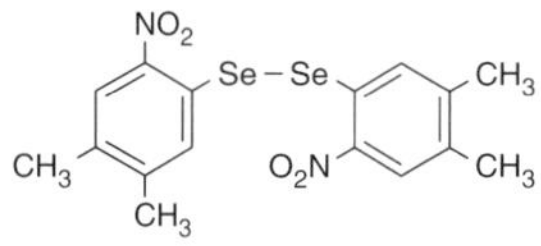
5 gm

S87,948-7
9H-SELENOXANTHEN-9-ONE
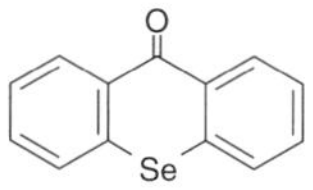
5 gm

S84,985-5
6-BROMO-2H-1-BENZOSELENIN-4(3H)-ONE
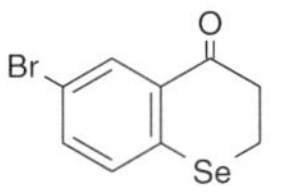
250 mg

S88,833-8
3-METHYL-9H-SELENOXANTHEN-9-ONE
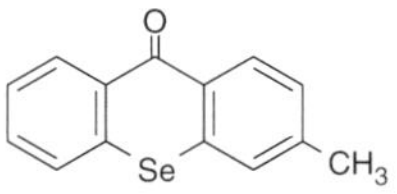
250 mg

S87,416-7
2,6-DIPHENYL-4H-THIOPYRAN-4-THIONE
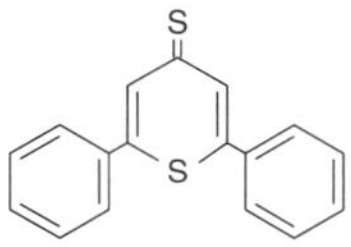
250 mg

S88,915-6
2-(METHYLSELENO)-ALPHA-PHENYLCINNAMIC ACID
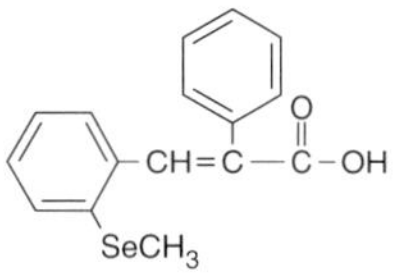
250 mg

S89,310-2
2-(METHYLSELENO)CINNAMIC ACID
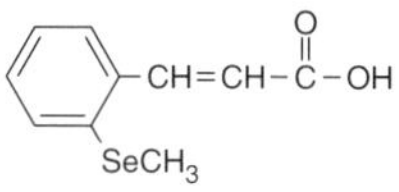
250 mg

S88,921-0
ETHYL 2-(ETHOXYCARBONYLMETHYLSELENO)-BENZOATE
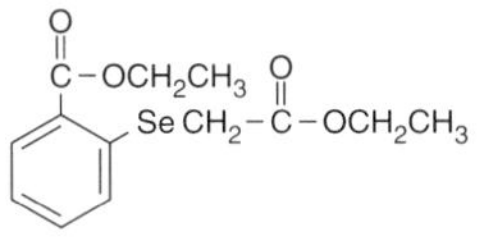
250 mg

S89,098-7
3,3'-DISELENODIPROPIONIC ACID

250 mg

S88,882-6
2-(METHYLSELENO)CINNAMONITRILE
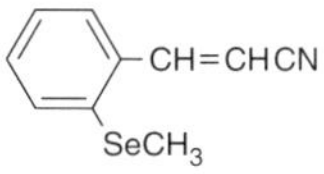
250 mg

S62,295-8
N-(2-SELENOCYANATOETHYL)PHTHALIMIDE
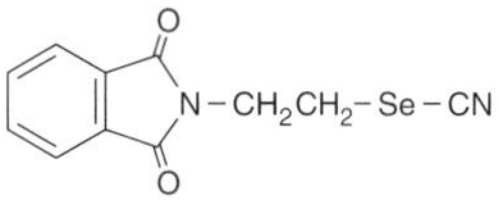
250 mg

S86,007-7
5-BROMO-6-(1,3-DITHIAN-2-YL)-1,3-BENZODIOXOLE
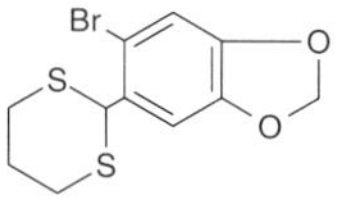
250 mg

S88,923-7
4,4-DIMETHYL-2-[2-(METHYLSELENO)-PHENYL]-2-OXAZOLINE

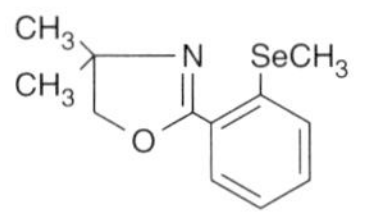

250 mg

S80,873-3
4-PHENYL-1,2-DITHIOLANE SULFATE

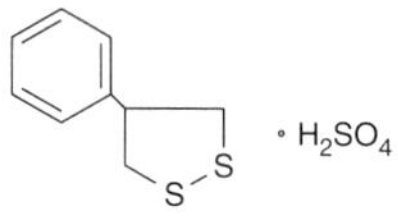

250 mg

S65,384-5
2-PHENYLTHIOXOLANE

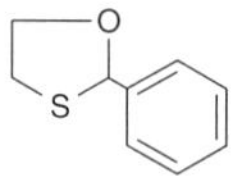

5 gm

S61,479-3
PHENOXATHIIN

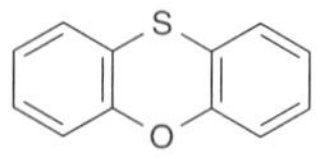

250 mg

S94,067-4
PHENYLTRICHLOROTELLURIUM

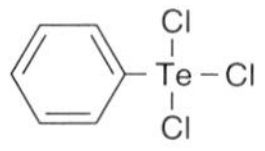

250 mg

S64,256-8
METHYL PHENYL SELENIDE

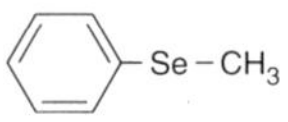

5 gm

S64,972-4
BENZALDEHYDE DIETHYL MERCAPTAL

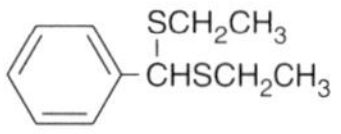

5 gm

S90,697-2
BENZYL SELENIDE

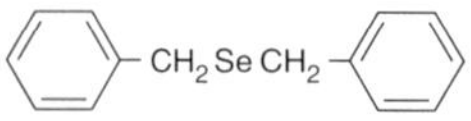

250 mg

S94,071-2
P-TOLYL DITELLURIDE

250 mg

S39,362-2
2,5-DICHLOROPHENYL-METHYLTHIOMETHYL SULFIDE

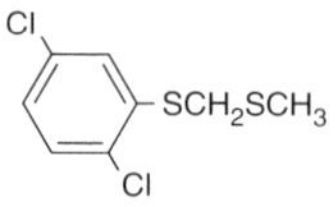

250 mg

S39,373-8
3,4-DICHLOROPHENYL-METHYLTHIOMETHYL SULFIDE

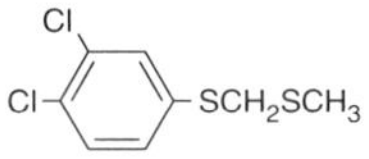

5 gm

S90,716-2
HEXAKIS(2-NAPHTHYLTHIOMETHYL)-BENZENE

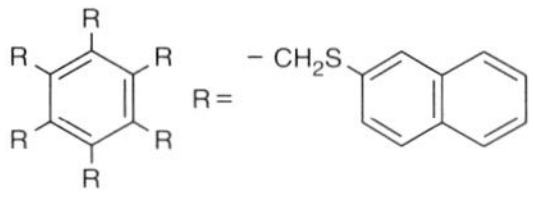

250 mg

S80,963-2
DIBENZO(C,E)(1,2)DITHIIN

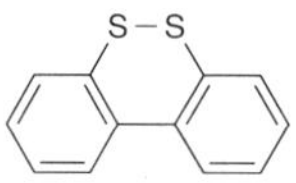

250 mg

S82,279-5
2-BROMOTHIANTHRENE

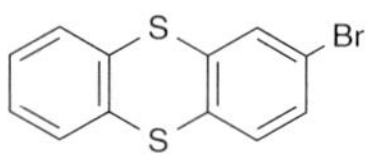

250 mg

S80,961-6
1,4,5,7-TETRAHYDRO-3H-2,6-BENZODITHIONIN

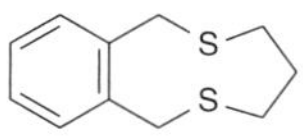

250 mg

S80,969-1
6,7,8,9-TETRAHYDRODIBENZO(B,D)(1,6)-DITHIECIN

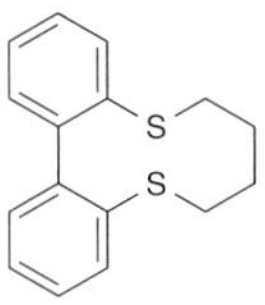

250 mg

S80,976-4
5,16-DIHYDROTRIBENZO(B,D,H)(1,6)-DITHIECIN

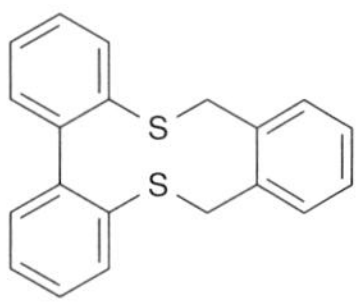

250 mg

S83,844-6
1-CHLORO-3-[4-(DIMETHYLAMINO)-PHENYL]-2-PROPANOL

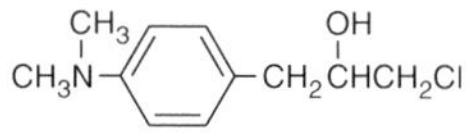

250 mg

S50,035-6
2-[N-(2-BROMOETHYL)ANILINO]ETHANOL

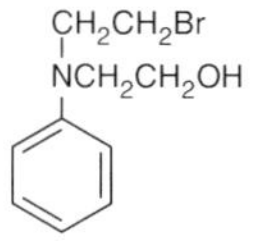

250 mg

S37,477-6
N-(2-BROMOETHYL)ANILINE HYDROBROMIDE

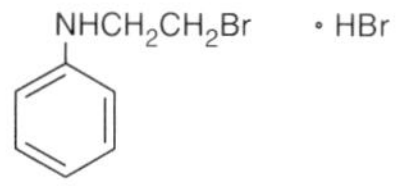

250 mg

S92,312-5
N-(2-BROMOETHYL)ANILINE

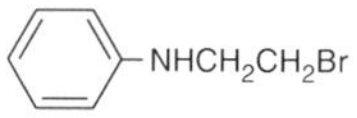

250 mg

S66,008-6
N-(2-CHLOROETHYL)-N-ETHYL-M-TOLUIDINE

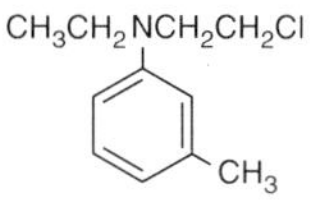

5 gm

S92,902-6
N-(2-BROMOETHYL)-N-ETHYL-M-TOLUIDINE HYDROBROMIDE

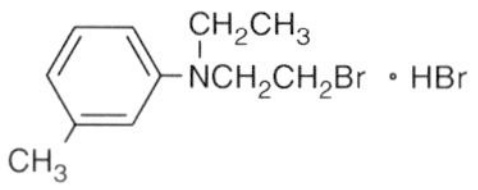

250 mg

S79,338-8
N,N-BIS(2-BROMOETHYL)ANILINE

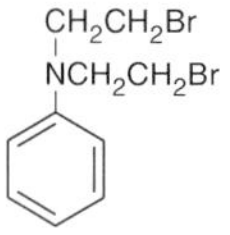

250 mg

S73,966-9
ALPHA-[2-CHLORO-5-(TRIFLUOROMETHYL)-ANILINO]-6-METHOXY-O-CRESOL

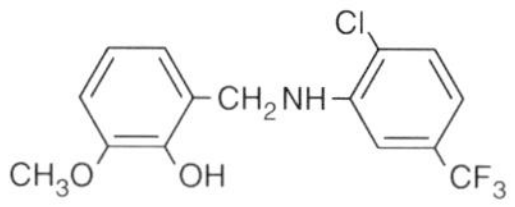

5 gm

S64,910-4
2,6-DIBROMO-3,5-BIS-(TRIFLUOROMETHYL)ANILINE

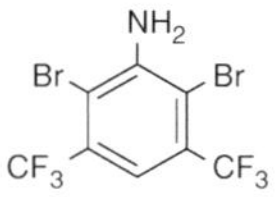

250 mg

S93,799-1
3-AMINO-ALPHA,ALPHA-BIS-(TRIFLUOROMETHYL)-4-METHYLBENZYL ALCOHOL

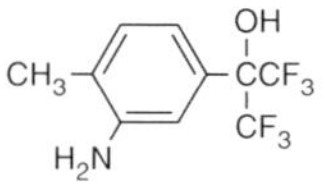

250 mg

S67,797-3
ALPHA,ALPHA,ALPHA-TRIFLUORO-M-TOLUIDINE HYDROCHLORIDE

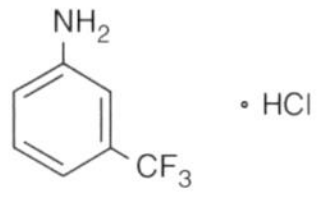

250 mg

S67,796-5
ALPHA,ALPHA,ALPHA-TRIFLUORO-O-TOLUIDINE HYDROCHLORIDE

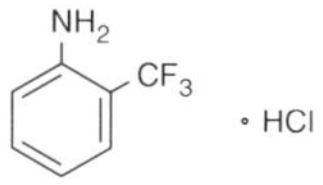

250 mg

S73,426-8
N-PHENYL-ALPHA,ALPHA,ALPHA-TRIFLUORO-M-TOLUIDINE

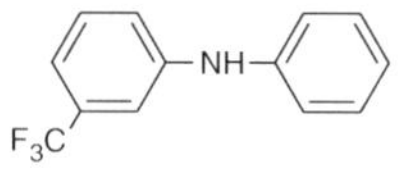

250 mg

S65,627-5
4-AMINO-2,6-DICHLORO-3-METHYLPHENOL

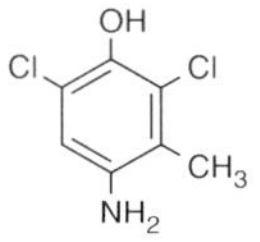

250 mg

S45,856-2
5-CHLORO-2,4-DIMETHOXYANILINE

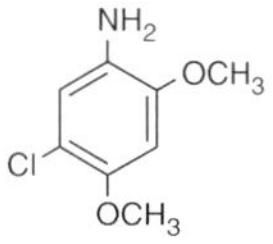

5 gm

S39,230-8
5-CHLORO-O-ANISIDINE HYDROCHLORIDE

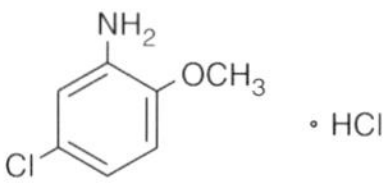

5 gm

S38,726-6
5-CHLORO-2-(HEXADECYLOXY)ANILINE

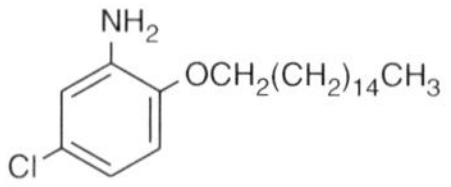

250 mg

S84,983-9
3-BROMO-4-METHOXYANILINE HYDROCHLORIDE

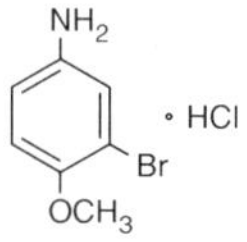

250 mg

S65,883-9
5-CHLORO-2-MERCAPTOANILINE HYDROCHLORIDE

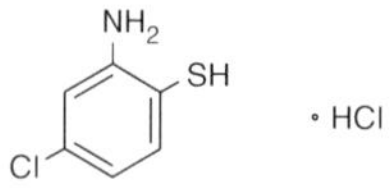

5 gm

S60,575-1
5-CHLORO-2-HEXADECYLTHIOANILINE HYDROCHLORIDE

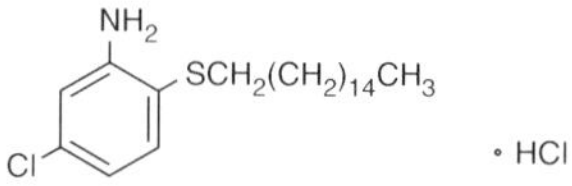

250 mg

S58,831-8
3-CHLORO-4-(METHYLTHIO)ANILINE

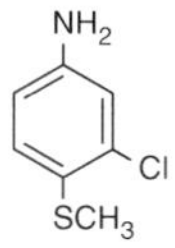

250 mg

S64,266-5
3-CHLOROANILINE HYDROCHLORIDE

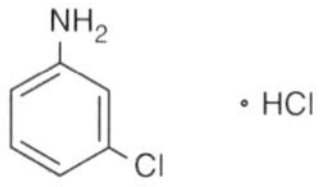

5 gm

S41,152-3
2,4-DIBROMO-3,6-DICHLOROANILINE

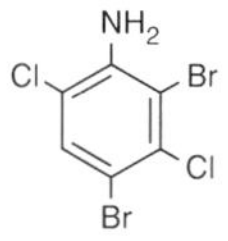

5 gm

S81,396-6
2,3,4,5,6-PENTABROMOANILINE

250 mg

S41,419-0
4-FLUOROANILINE HYDROCHLORIDE

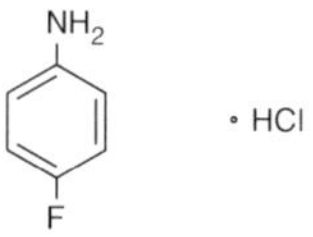

250 mg

S45,796-5
4-BROMOANILINE HYDROCHLORIDE

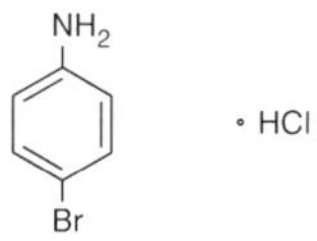

5 gm

S85,193-0
2,4-DICHLOROANILINE HYDROCHLORIDE

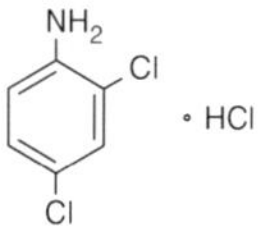

250 mg

S54,159-1
2-CHLORO-4-IODOANILINE

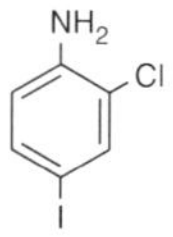

250 mg

S44,458-8
4-FLUORO-2-METHYLANILINE

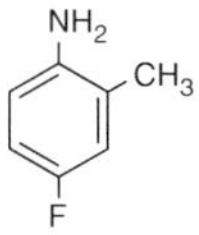

5 gm

S44,542-8
4-FLUORO-3-METHYLANILINE

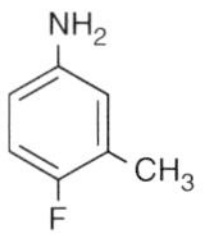

5 gm

S50,798-9
5-CHLORO-O-TOLUIDINE HYDROCHLORIDE

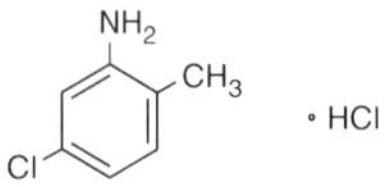

5 gm

S77,036-1
4-CHLORO-M-TOLUIDINE

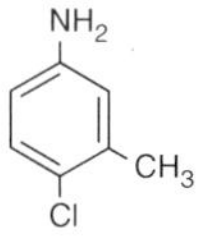

250 mg

S74,194-9
2-BROMO-4-METHYLANILINE HYDROBROMIDE
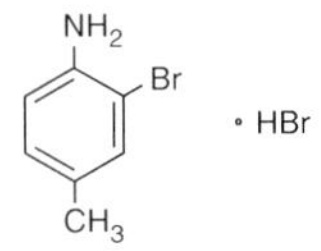
250 mg

S90,863-0
2-BROMO-4-METHYLANILINE HYDROCHLORIDE
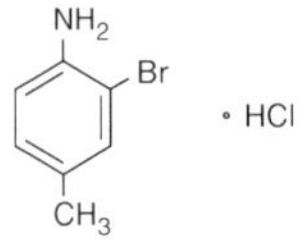
250 mg

S53,253-3
4-IODO-O-TOLUIDINE
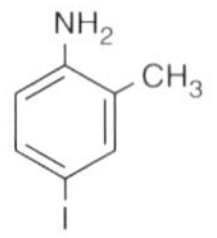
5 gm

S81,584-5
3,5-DIBROMOANILINE
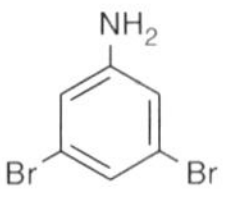
250 mg

S64,632-6
4-BROMO-2,6-DICHLOROANILINE
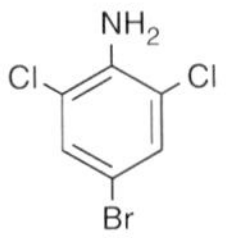
250 mg

S67,692-6
2-CHLORO-4,6-DIBROMOANILINE
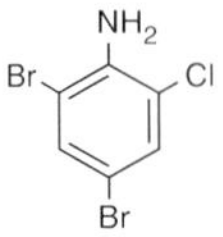
250 mg

S67,748-5
2,6-DIBROMO-4-ETHYLANILINE
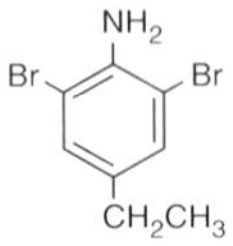
5 gm

S67,693-4
2-BROMO-4,5-DIMETHYLANILINE
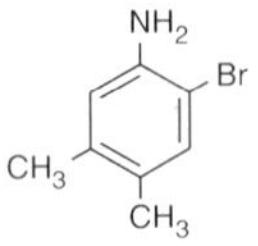
5 gm

S94,698-2
2-BROMO-4,6-DIMETHYLANILINE
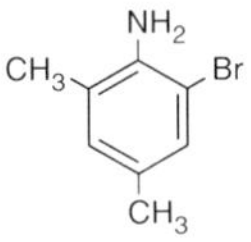
250 mg

S67,715-9
3-CHLORO-N-ETHYLANILINE
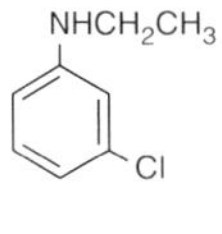
5 gm

S67,727-2
4-CHLORO-N-ETHYLANILINE
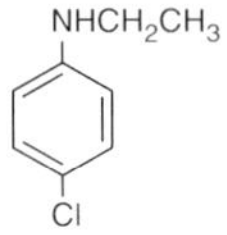
5 gm

S67,708-6
2,3-DICHLORO-N-METHYLANILINE
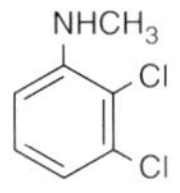
5 gm

S67,709-4
3-CHLORO-N-ETHYL-4-METHYLANILINE
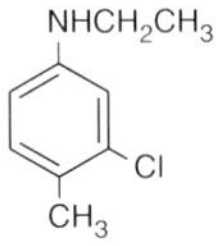
5 gm

S88,078-7
2-BROMO-N-ETHYL-4-METHYLANILINE
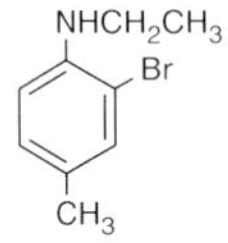
250 mg

S51,362-8
N-(5-CHLORO-M-TOLYL)HEXADECYLAMINE
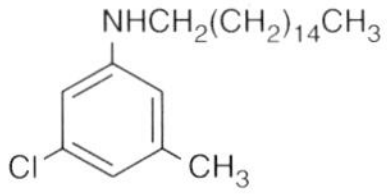
250 mg

S82,616-2
2-BROMO-N,N-DIMETHYLANILINE PICRATE
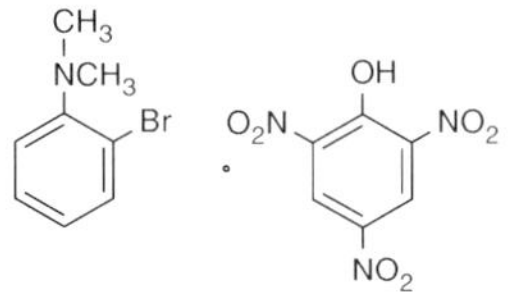
250 mg

S82,072-5
2-BROMO-N,N,5-TRIMETHYLANILINE
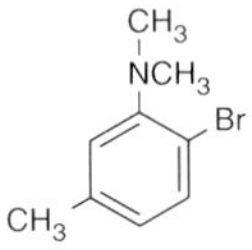
250 mg

S88,680-7
3,5-DICHLORO-N,N-DIMETHYLANILINE
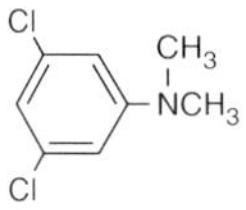
250 mg

S87,756-5
5-IODO-2-METHYLANILINE
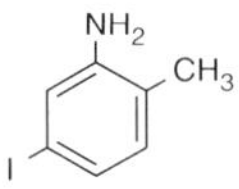
250 mg

S78,442-7
2-AMINOPHENYL BETA-D-GALACTOPYRANOSIDE MONOHYDRATE
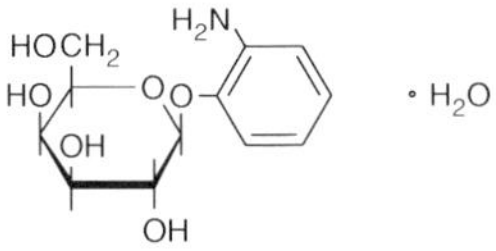
250 mg

S78,194-0
4-AMINOPHENYL ALPHA-D-MANNOPYRANOSIDE
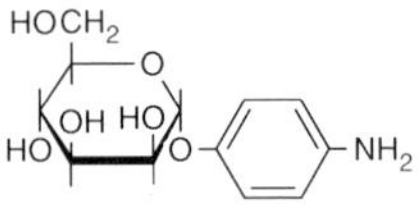
250 mg

S94,025-9
4-(HEXADECYLAMINO)BENZYLAMINE DITARTRATE
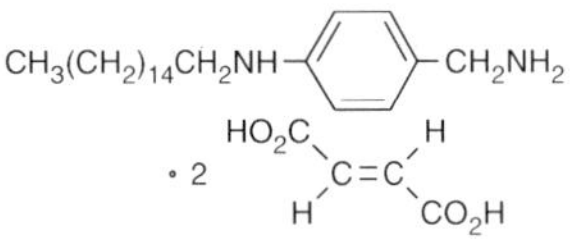
5 gm

S78,302-1
2-AMINOPHENYL-BETA-D-THIOGALACTOPYRANOSIDE
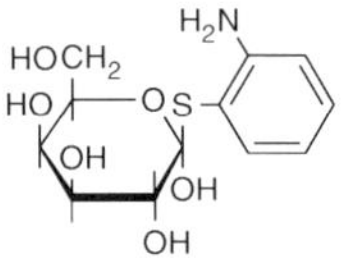
5 gm

S64,242-8
3,5-DIMETHOXY-4-HYDROXYANILINE HYDROCHLORIDE
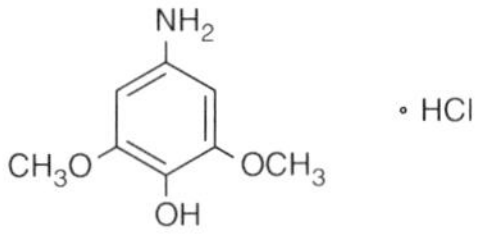
250 mg

S59,863-1
4-AMINOPHENYL 2-HYDROXYETHYL ETHER
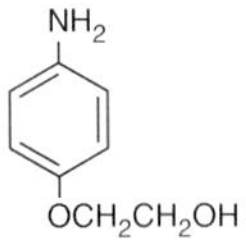
250 mg

S52,782-3
P-PHENETIDINE D-FRUCTOSIDE
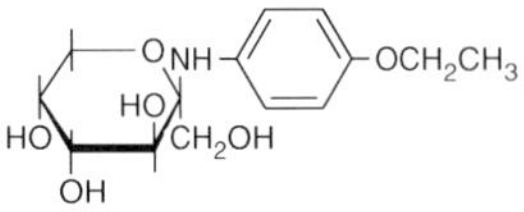
5 gm

S52,783-1
P-PHENETIDINE D-GALACTOSIDE
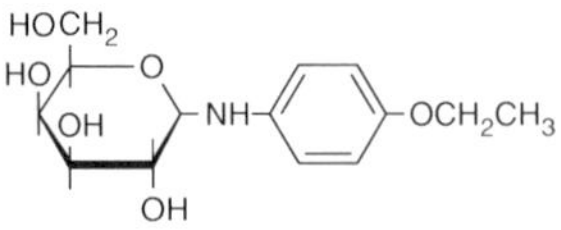
5 gm

S42,620-2
P-PHENETIDINO-L-SORBOSIDE
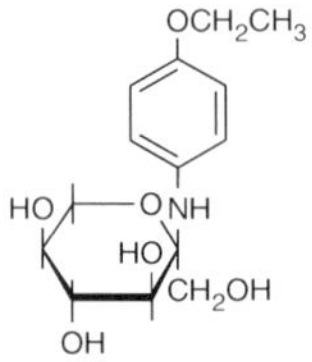
250 mg

S51,135-8
5-AMINO-2-METHYLBENZYL ALCOHOL
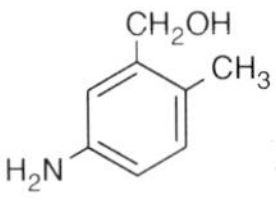
250 mg

S64,249-5
1-(2-AMINOPHENOXY)-3-(TERT-NONYLOXY)-2-PROPANOL
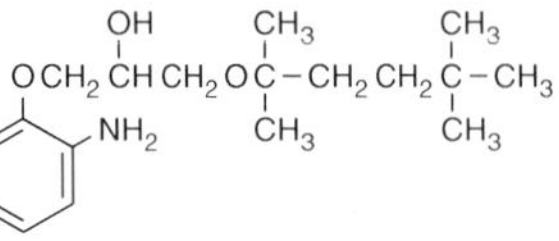
5 gm

S88,323-9
2-AMINO-5-(METHYLTHIO)PHENOL
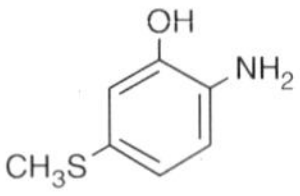
250 mg

S37,856-9
2-AMINOPHENOL HYDROCHLORIDE
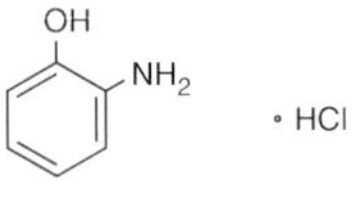
5 gm

S81,888-7
2-AMINO-4-ISOPROPYLPHENOL

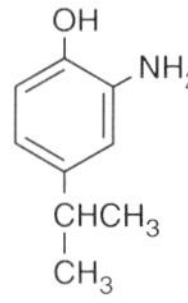

250 mg

S62,679-1
2-AMINO-4-SEC-BUTYLPHENOL

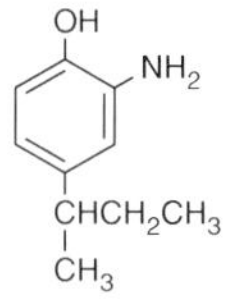

5 gm

S62,680-5
2-AMINO-4-TERT-AMYLPHENOL

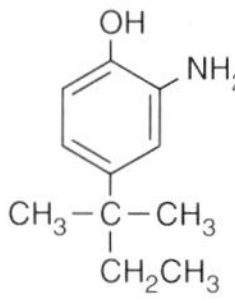

5 gm

S43,004-8
4-AMINO-M-CRESOL HYDROCHLORIDE

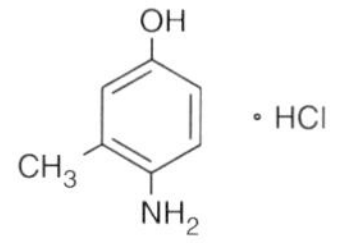

5 gm

S84,137-4
3-AMINO-4-METHYLPHENOL

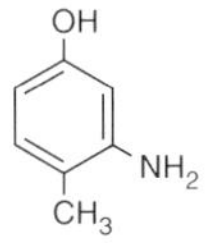

250 mg

S61,972-8
2-ALLYL-4-AMINOPHENOL

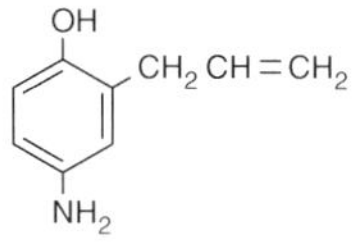

250 mg

S81,887-9
3-AMINO-4-ISOPROPYLPHENOL

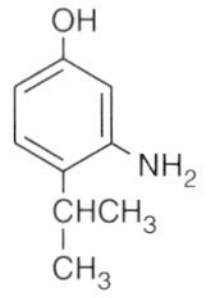

250 mg

S56,563-6
4-AMINOTHYMOL HYDROCHLORIDE

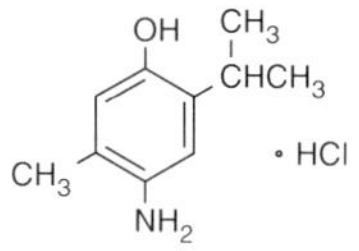

5 gm

S60,558-1
4-AMINO-2-CYCLOHEXYLPHENOL

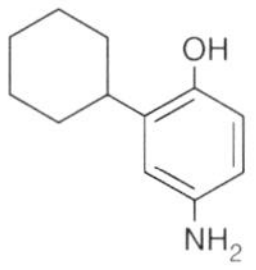

5 gm

S63,734-3
3-CYCLOHEXYL-4-HYDROXYANILINE HYDROCHLORIDE

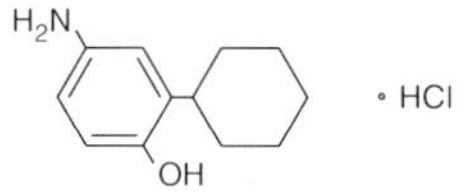

250 mg

S44,082-5
5-AMINO-2-METHYLPHENOL

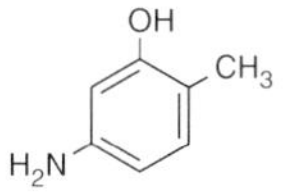

250 mg

S36,149-6
2-(2-ETHYLANILINO)ETHANOL

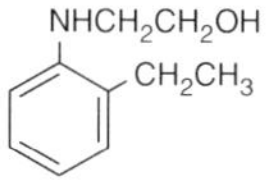

5 gm

S70,537-3
2-(DIMETHYLAMINO)CRESOL

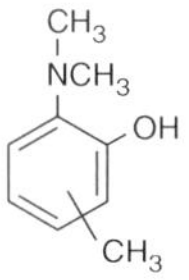

5 gm

S35,617-4
N-(M-TOLYL)DIETHANOLAMINE

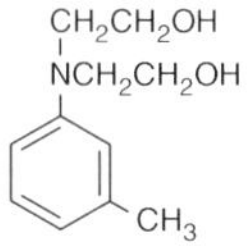

5 gm

S81,886-0
3-(DIMETHYLAMINO)-4-ISOPROPYLPHENOL

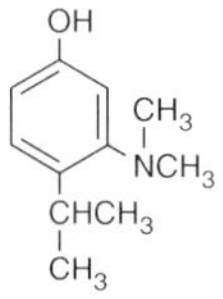

250 mg

S83,745-8
4-(DIMETHYLAMINO)-3-ISOPROPYLPHENOL

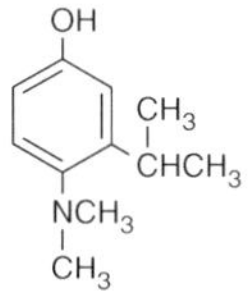

250 mg

S59,634-5
2,6-DICHLORO-4-DIMETHYLAMINOPHENOL HYDROCHLORIDE

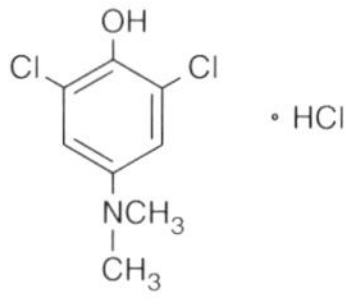

5 gm

S88,102-3
2,4-DIMETHOXY-N-HEXYLANILINE

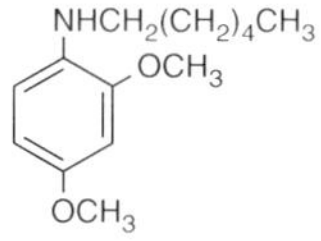

250 mg

S50,783-0
N-(3,5-DIMETHOXYPHENYL)-HEXADECYLAMINE

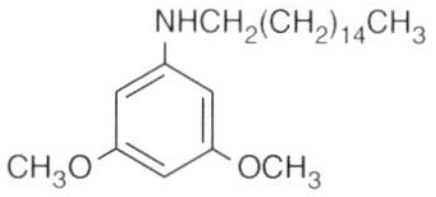

250 mg

S38,740-1
3-(HEXADECYLOXY)ANILINE

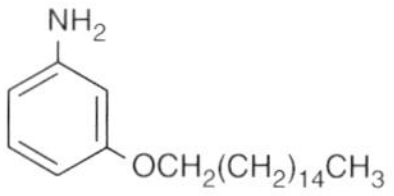

250 mg

S65,310-1
3-PENTYLOXYANILINE

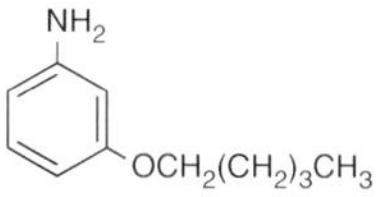

5 gm

S57,790-1
M-(OCTYLOXY)ANILINE

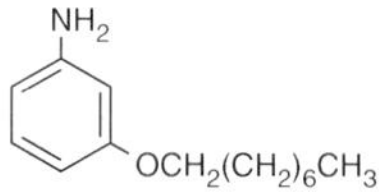

5 gm

S57,490-2
3-(DECYLOXY)ANILINE

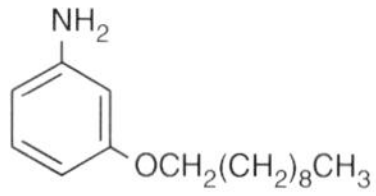

5 gm

S57,565-8
3-(DODECYLOXY)ANILINE

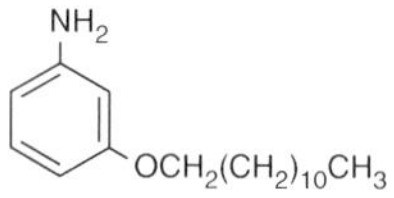

5 gm

S57,531-3
3-(TETRADECYLOXY)ANILINE

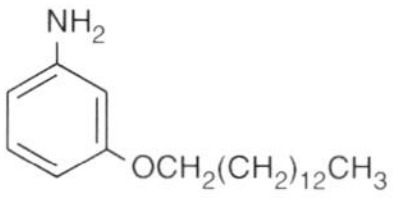

5 gm

S75,175-8
3-(OCTADECYLOXY)ANILINE

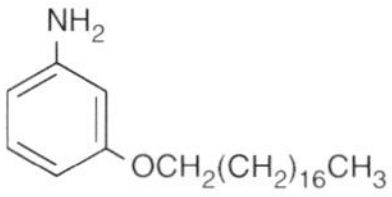

5 gm

S57,816-9
O-(OCTYLOXY)ANILINE

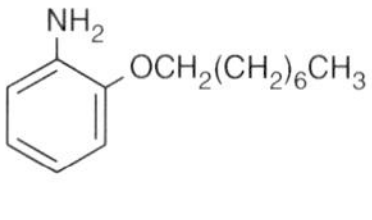

5 gm

S57,335-3
2-(DECYLOXY)ANILINE

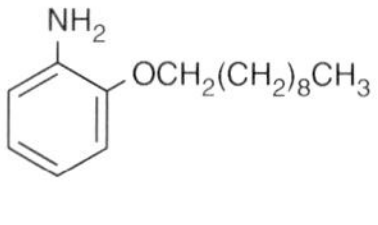

5 gm

S57,563-1
2-(DODECYLOXY)ANILINE

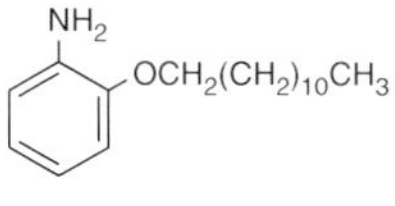

5 gm

S57,567-4
2-(TETRADECYLOXY)ANILINE

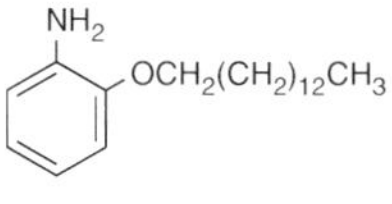

5 gm

S56,286-6
2-[2-(1,3,3-TRIMETHYLBUTYL)-5,7,7-TRIMETHYLOCTYLOXY]ANILINE

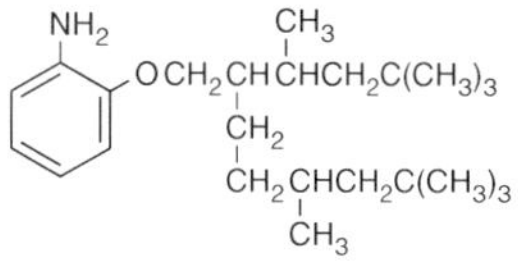

250 mg

S63,505-7
2-AMINO-4-TERT-BUTYLANISOLE

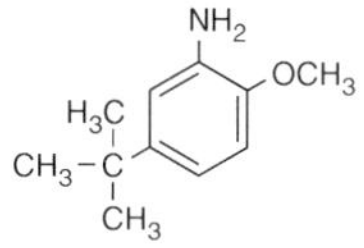

5 gm

S62,656-2
2-AMINO-4-TERT-AMYLANISOLE
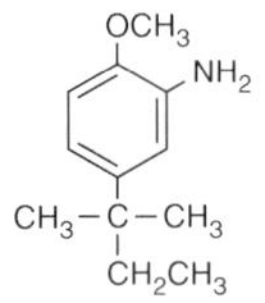
5 gm

S62,646-5
2-AMINO-4-SEC-BUTYLANISOLE
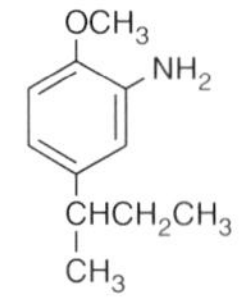
5 gm

S65,542-2
4-CYCLOPENTYLOXYANILINE
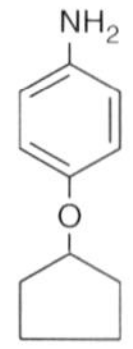
250 mg

S80,549-1
P-ANISIDINE HYDROCHLORIDE
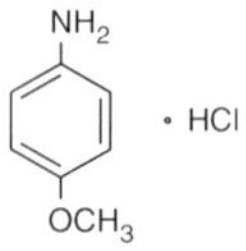
250 mg

S56,629-2
P-PHENETIDINE HYDROCHLORIDE
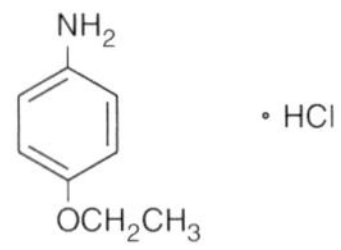
5 gm

S57,525-9
4-(OCTYLOXY)ANILINE
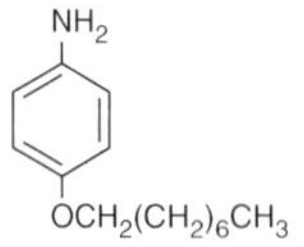
5 gm

S57,338-8
4-(DECYLOXY)ANILINE
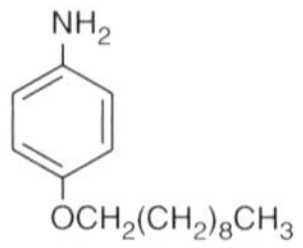
5 gm

S57,532-1
4-(DODECYLOXY)ANILINE
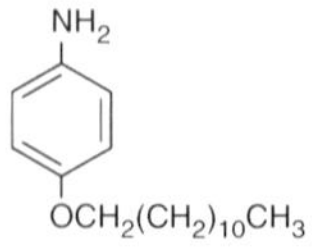
5 gm

S60,510-7
4-(2-BUTYLOCTYLOXY)ANILINE
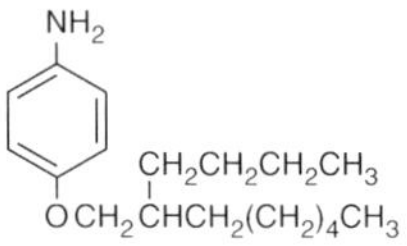
250 mg

S57,478-3
4-(TETRADECYLOXY)ANILINE
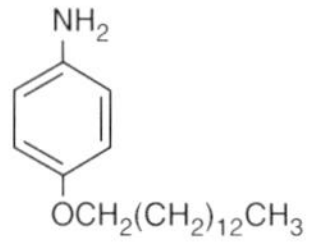
5 gm

S57,339-6
4-(HEXADECYLOXY)ANILINE
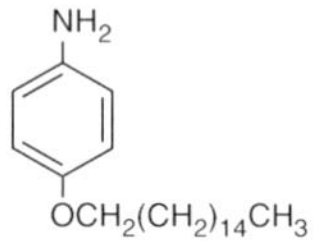
5 gm

S82,813-0
4-(HEXADECYLOXY)ANILINE HYDROCHLORIDE
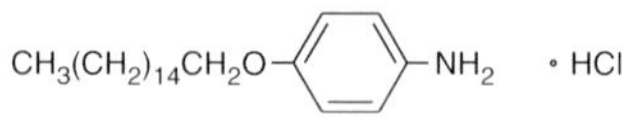
250 mg

S57,477-5
4-(OCTADECYLOXY)ANILINE
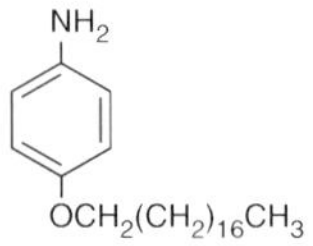
5 gm

S46,247-0
3-METHOXY-4-METHYLANILINE
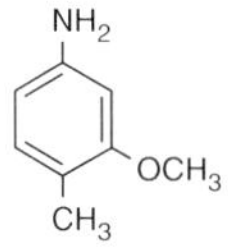
5 gm

S62,647-3
4-AMINO-2-ISOPROPYLANISOLE
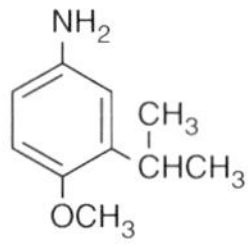
5 gm

S38,275-2
2-(HEXADECYLOXY)ANILINE
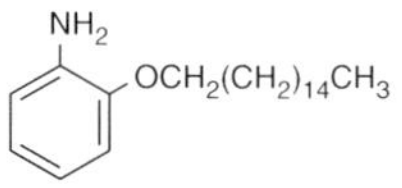
5 gm

S34,070-7
N-ALLYL-P-ANISIDINE,TECH.
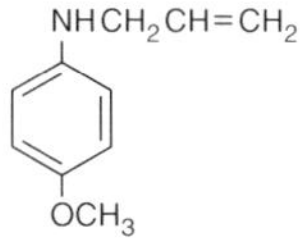
5 gm

S93,774-6
N-(2-METHOXYETHYL)-4-BIPHENYLAMINE
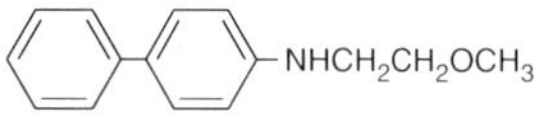
250 mg

S73,366-0
4-HEXYLOXY-N-(1-METHYLHEPTYL)-ANILINE
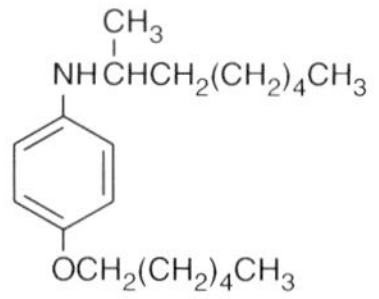
250 mg

S83,737-7
N,N-DIETHYL-M-ANISIDINE
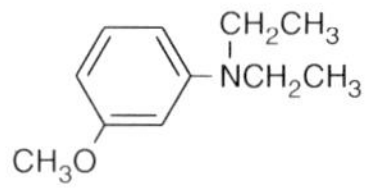
250 mg

S83,741-5
N,N-DIBUTYL-3-METHOXYANILINE
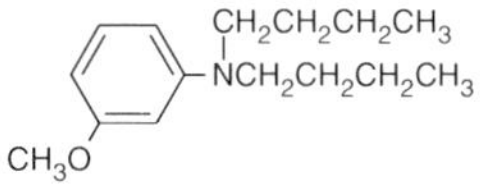
250 mg

S92,131-9
N,N-DIMETHYL-4-METHOXYANILINE
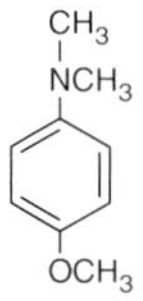
250 mg

S92,660-4
N-TERT-BUTYL-3-METHOXYANILINE
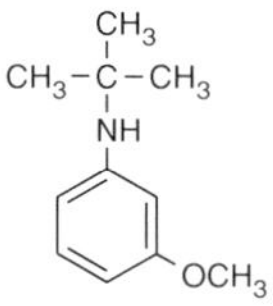
250 mg

S56,255-6
M-(PENTYLTHIO)ANILINE
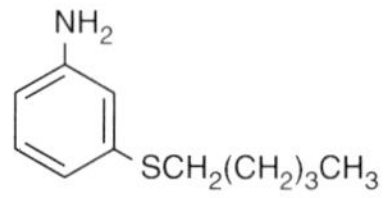
250 mg

S45,018-9
3-HEXADECYLTHIOANILINE
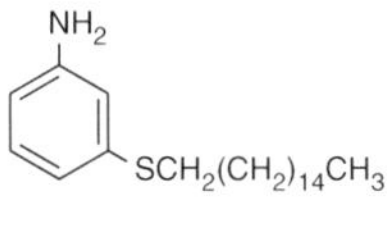
250 mg

S55,889-3
2-(OCTYLTHIO)ANILINE HYDROCHLORIDE
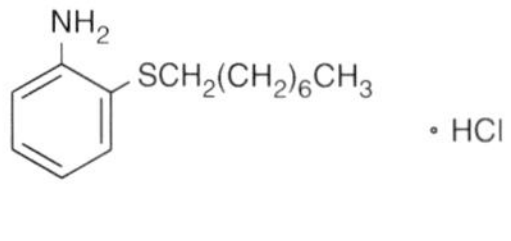
250 mg

S55,876-1
2-(DODECYLTHIO)ANILINE
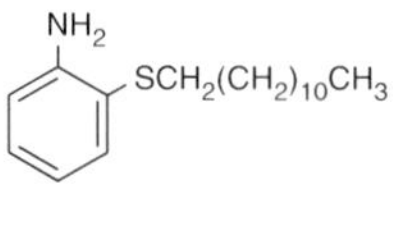
5 gm

S50,536-6
2-(OCTADECYLTHIO)ANILINE HYDROCHLORIDE
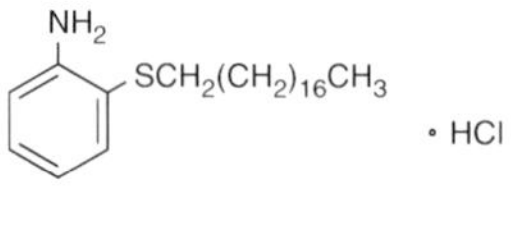
5 gm

S41,859-5
4-METHYLMERCAPTOANILINE HYDROCHLORIDE
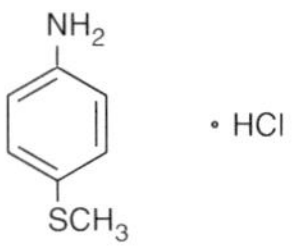
5 gm

S73,491-8
N-OCTYL-4-(OCTYLTHIO)ANILINE HYDROCHLORIDE
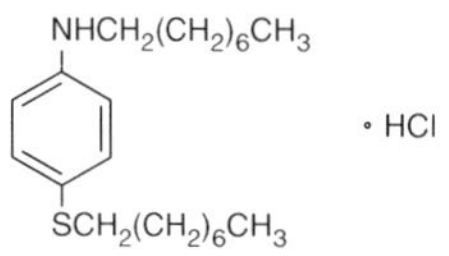
250 mg

S83,597-8
N,N-DIMETHYL-4-(METHYLTHIO)ANILINE
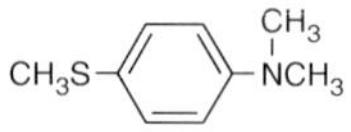
250 mg

S53,279-7
N,N-BIS(2-ETHYLTHIO)ETHYLANILINE
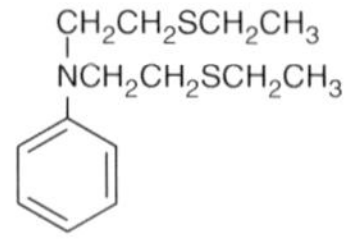
5 gm

S34,101-0
4-AMINO-N,N-DIETHYLANILINE SULFUR DIOXIDE COMPLEX

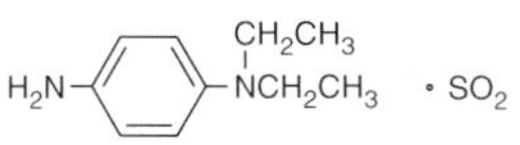

5 gm

S82,071-7
3-ISOPROPYLANILINE

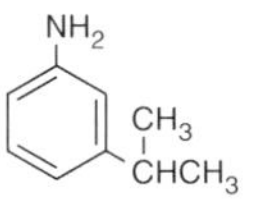

250 mg

S94,361-4
2-BROMO-3,4,6-TRIMETHYLANILINE

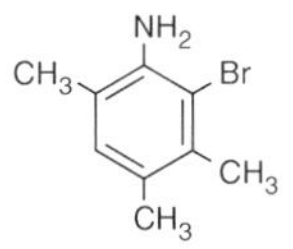

250 mg

S59,616-7
TETRAANILINIUM PYROPHOSPHATE

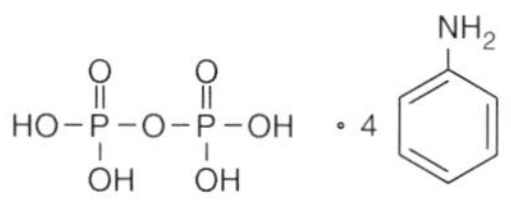

250 mg

S69,832-6
ANILINE HYDROBROMIDE

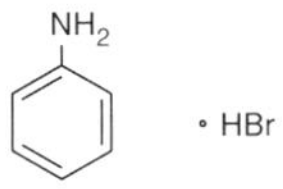

5 gm

S44,520-7
P-TOLUIDINE, ZONE REFINED

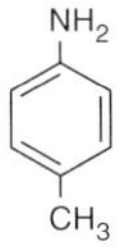

5 gm

S82,797-5
4-ISOPROPYLANILINE HYDROCHLORIDE

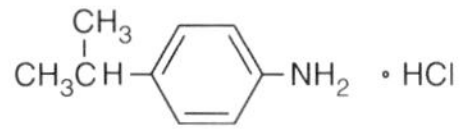

250 mg

S90,881-9
4-DECYLANILINE HYDROCHLORIDE

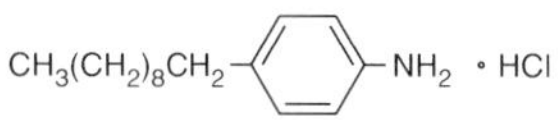

250 mg

S66,928-8
2-TERT-BUTYL-6-METHYLANILINE

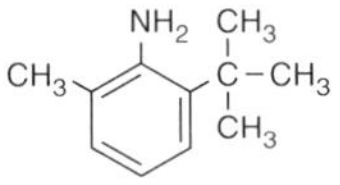

5 gm

S66,929-6
2-ETHYL-6-ISOPROPYLANILINE

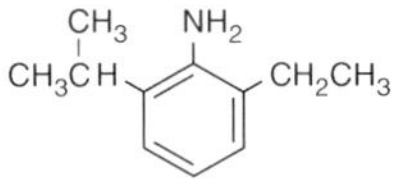

5 gm

S66,937-7
2-SEC-BUTYL-6-ETHYLANILINE

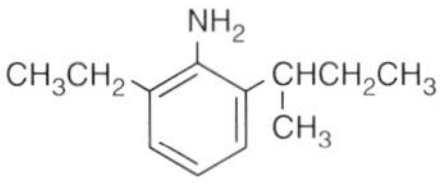

5 gm

S92,211-0
4-METHOXY-2-METHYLANILINE HYDROCHLORIDE

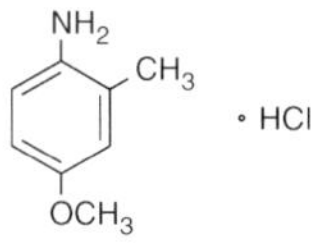

250 mg

S44,239-9
2-AMINO-P-CYMENE

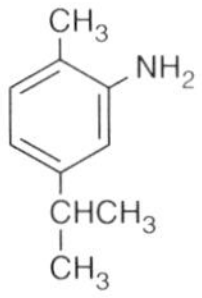

5 gm

S94,592-7
4-ISOPROPYL-3-METHYLANILINE HYDROCHLORIDE

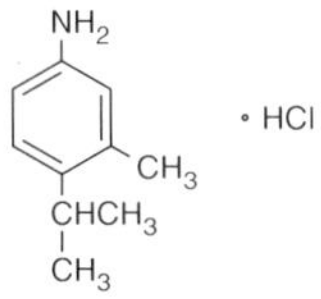

250 mg

S77,537-1
2,6-DIMETHYL-3-ISOPROPYLANILINE HYDROCHLORIDE

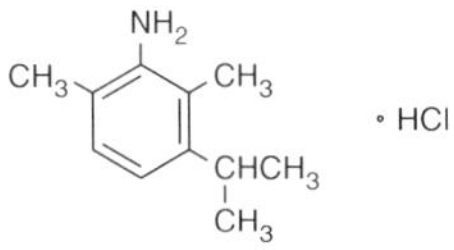

250 mg

S43,996-7
2,4,5-TRIMETHYLANILINE

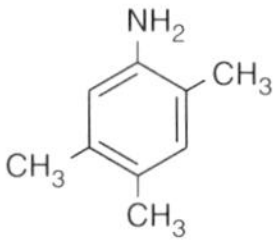

5 gm

S93,334-1
3,4'-DIAMINOBIPHENYL

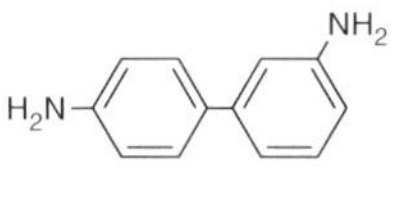

250 mg

S54,894-4
3-(2-METHYL-1,3-DIOXOLAN-2-YL)-ANILINE

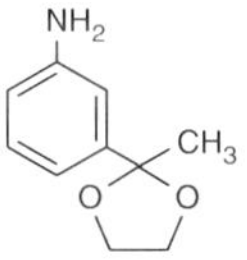

5 gm

S94,687-7
4-(1-ADAMANTYL)ANILINE HYDROCHLORIDE

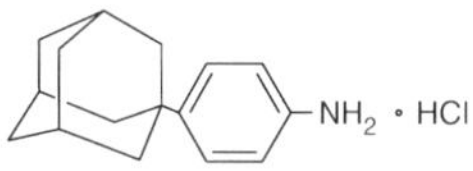

250 mg

S62,202-8
N-DODECYL-M-TOLUIDINE

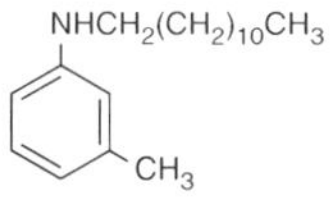

250 mg

S65,747-6
N-ISOPROPYL-M-TOLUIDINE

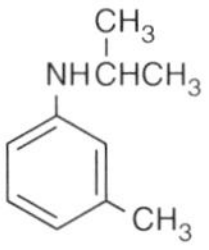

5 gm

S41,864-1
N-ISOPROPYLANILINE

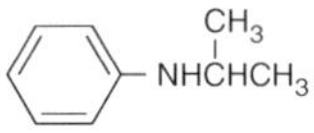

5 gm

S34,478-8
N-CROTYLANILINE,TECH.

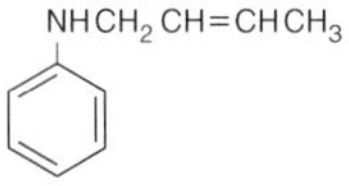

5 gm

S35,204-7
N-METHALLYLANILINE

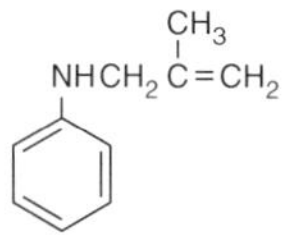

5 gm

S37,229-3
N-SEC-BUTYLANILINE

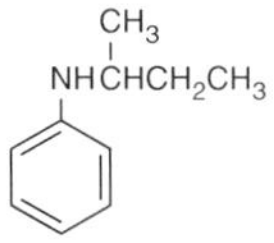

5 gm

S70,578-0
N-ISOPENTYLANILINE

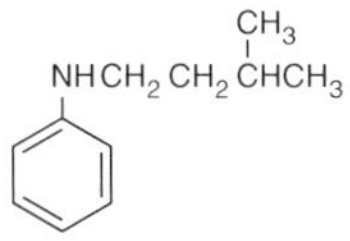

5 gm

S43,276-8
N-ALLYL-O-TOLUIDINE

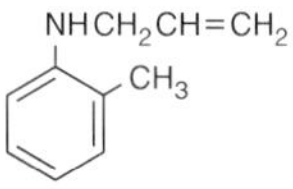

5 gm

S70,575-6
N-METHYL-P-TOLUIDINE

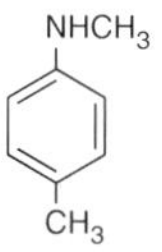

5 gm

S35,271-3
N-METHYLCARVACRYLAMINE

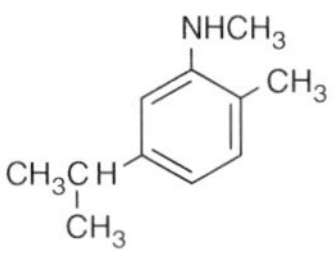

5 gm

S89,304-8
N-METHYLCARVACRYLAMINE OXALATE

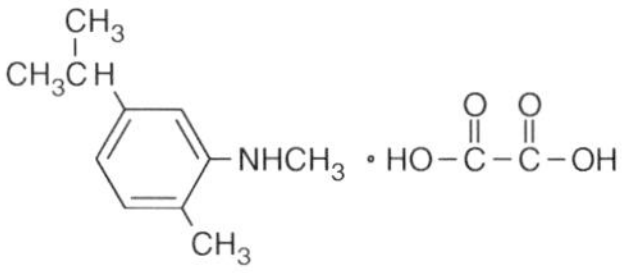

250 mg

S39,805-5
N-CYCLOHEXYLANILINE

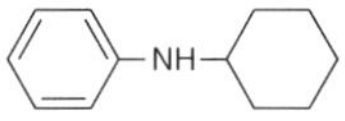

5 gm

S53,277-0
N-PHENYLOCTYLAMINE

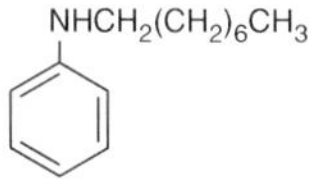

5 gm

S56,905-4
N-PHENYLDODECYLAMINE

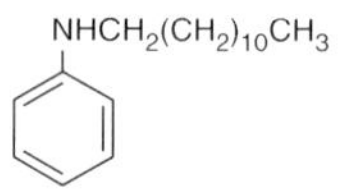

250 mg

S64,859-0
N-METHYL-O-TOLUIDINE

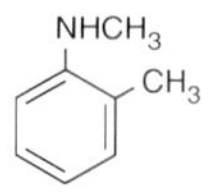

5 gm

S57,101-6
N-ETHYL-O-TOLUIDINE HYDROCHLORIDE

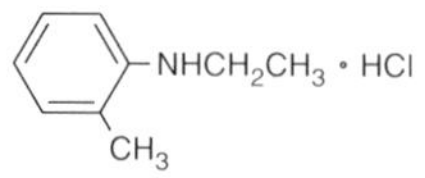

5 gm

S38,118-7
N-BUTYL-O-TOLUIDINE

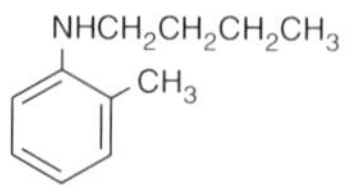

250 mg

S34,481-8
N-CROTYL-P-TOLUIDINE

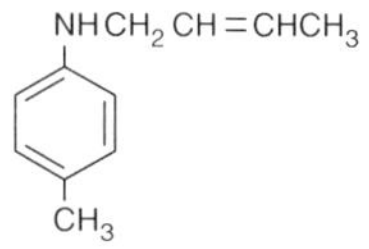

5 gm

S59,502-0
N,3,5-TRIMETHYLANILINE

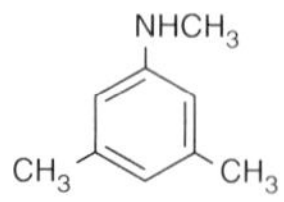

5 gm

S65,309-8
3-BROMO-N-HEXADECYL-5-METHYLANILINE

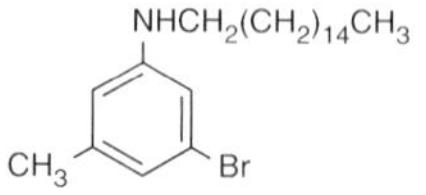

5 gm

S63,115-9
N-HEXADECYL-3,5-XYLIDINE HYDROCHLORIDE

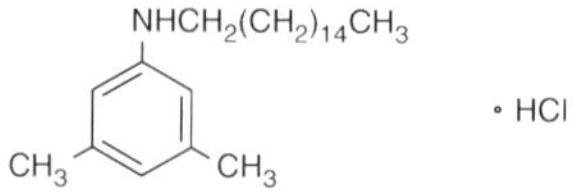

250 mg

S42,861-2
3-CHLORO-N,N-DIMETHYLANILINE

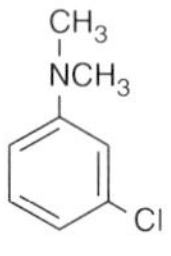

5 gm

S44,481-2
3-CHLORO-N,N-DIETHYLANILINE

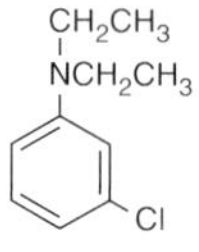

5 gm

S42,813-2
N,N-DIETHYL-M-TOLUIDINE

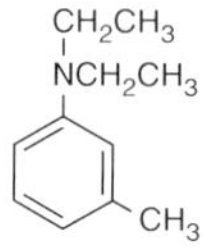

5 gm

S43,867-7
N-ALLYL-N-METHYLANILINE

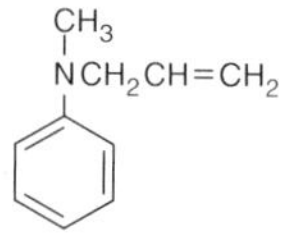

5 gm

S43,950-9
N,N-DI-N-PROPYLANILINE

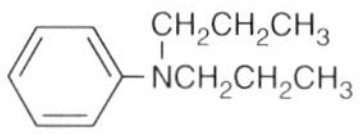

5 gm

S87,103-6
N,N-DIOCTYLANILINE

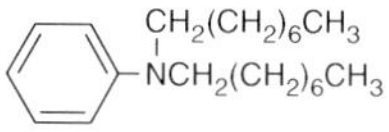

250 mg

S87,122-2
N,N-DIDODECYLANILINE

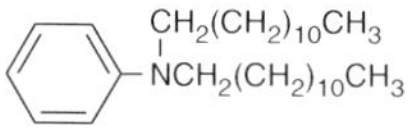

250 mg

S87,124-9
N,N-DIOCTADECYLANILINE

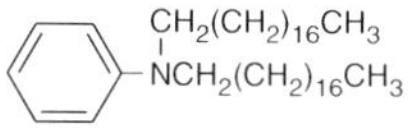

250 mg

S83,750-4
N,N-DIMETHYL-4-(1-HEPTADECEN-1-YL)-ANILINE

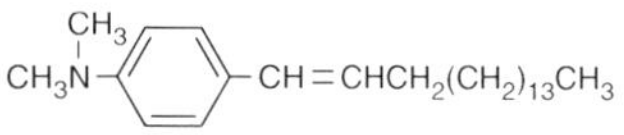

250 mg

S90,832-0
N,N-DIETHYL-3,5-DIMETHYLANILINE

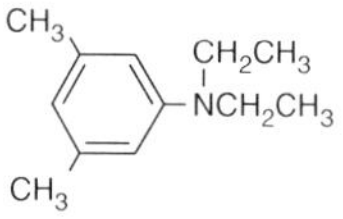

250 mg

S92,897-6
BIS(DIMETHYLAMINO)DIPHENYLMETHANE

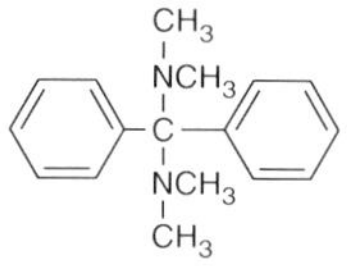

250 mg

S66,325-5
N-ALLYLOXY-N-METHYLANILINE

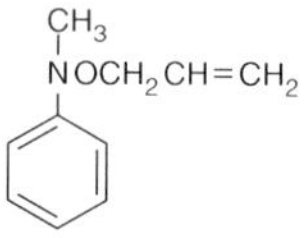

5 gm

S59,291-9
N-ETHYL-N-PHENYLDODECYLAMINE

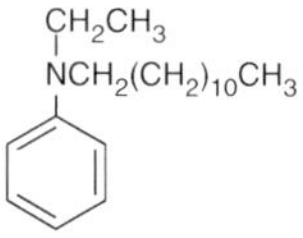

250 mg

S53,034-4
N-PHENYLHEXADECYLAMINE

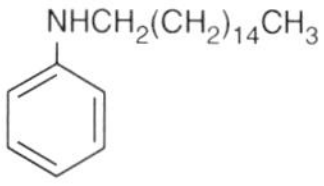

5 gm

S34,540-7
N,N-DIALLYLANILINE

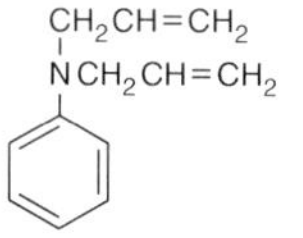

250 mg

S50,414-9
2-CHLORO-N,N-DIMETHYL-P-TOLUIDINE

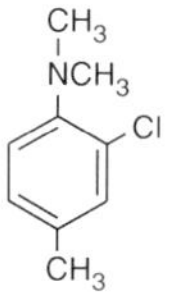

5 gm

S50,424-6
5-CHLORO-N,N-DIMETHYL-O-TOLUIDINE

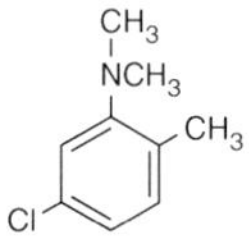

5 gm

S50,282-0
2-CHLORO-N,N-DIETHYL-P-TOLUIDINE

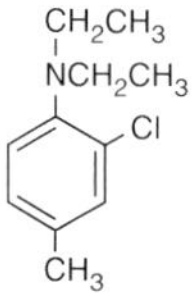

5 gm

S50,283-9
5-CHLORO-N,N-DIETHYL-O-TOLUIDINE

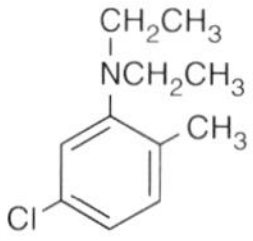

5 gm

S50,642-7
N,N-DIHEXADECYL-3,5-XYLIDINE

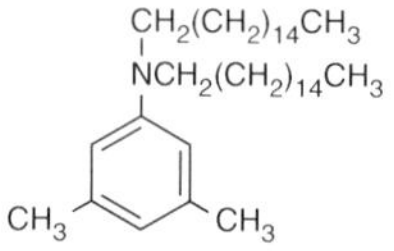

250 mg

S67,802-3
5-CHLORO-N1,N1,-DIBENZYL-2-METHOXY-1,4-PHENYLENEDIAMINE

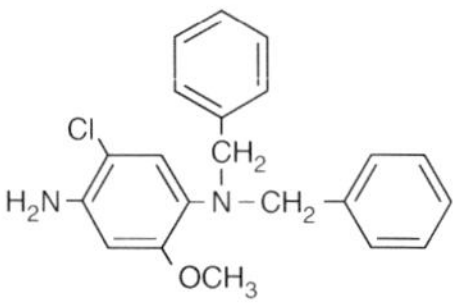

250 mg

S45,764-7
4-CHLORO-1,3-PHENYLENEDIAMINE

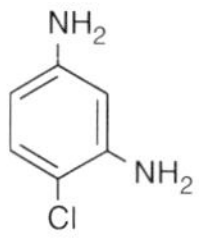

5 gm

S75,177-4
5-CHLORO-1,3-PHENYLENEDIAMINE

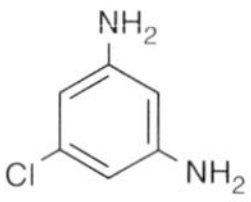

5 gm

S63,730-0
2-CHLORO-4-DIMETHYLAMINOANILINE HYDROCHLORIDE

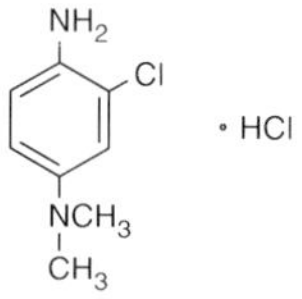

250 mg

S45,531-8
2-CHLORO-N1,N1-DIETHYL-1,4-PHENYLENEDIAMINE MONOHYDROCHLORIDE

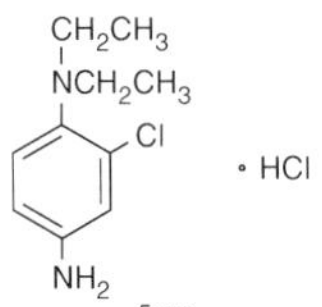

5 gm

S58,298-0
6-(D-RIBITYLAMINO)-3,4-XYLIDINE MONOHYDROCHLORIDE

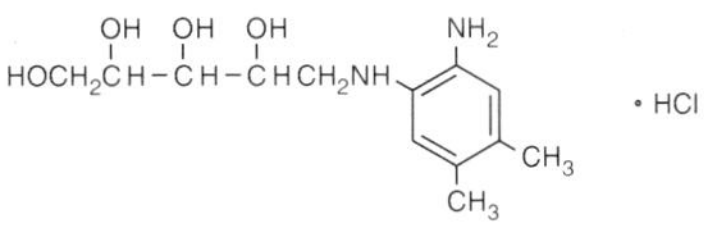

250 mg

S75,767-5
4-(2-HYDROXYETHOXY)-1,3-PHENYLENEDIAMINE DIHYDROCHLORIDE

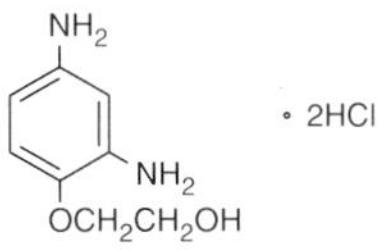

5 gm

S45,781-7
4-(2-HYDROXYETHOXY)-1,3-PHENYLENEDIAMINE SULFATE

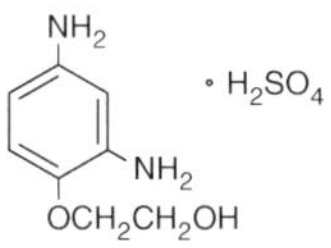

5 gm

S63,758-0
2,4-DIAMINOPHENOL

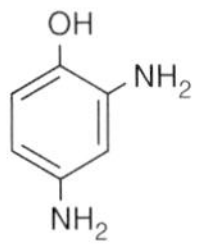

250 mg

S94,397-5
2,4-DIAMINO-6-METHYLPHENOL DIHYDROCHLORIDE

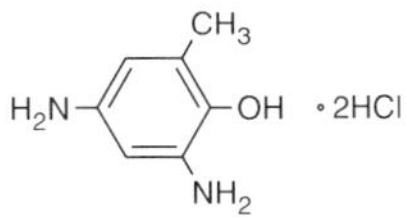

250 mg

S40,735-6
4-(4-AMINO-M-TOLUIDINO)PHENOL

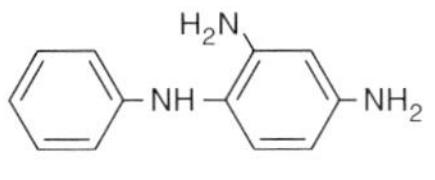

5 gm

S56,438-9
2-(4-AMINO-N1-ETHYL-M-TOLUIDINO)-ETHANOL SULFATE HYDRATE

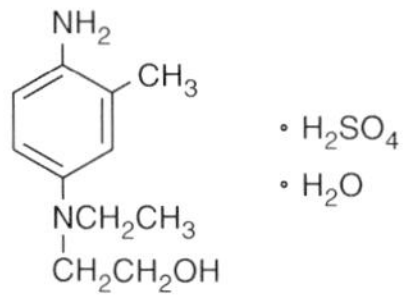

5 gm

S51,176-5
2-(4-AMINO-N-ETHYL-O-TOLUIDINO)-ETHANOL HYDROCHLORIDE

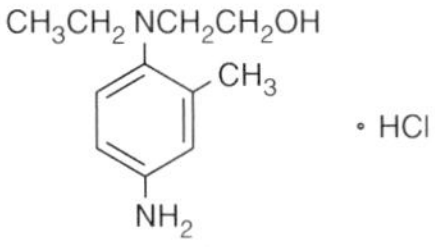

250 mg

S57,900-9
2,5-DIAMINOPHENOL DIHYDROCHLORIDE

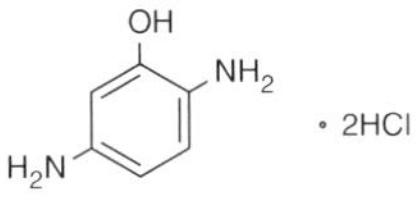

250 mg

S52,063-2
2-[4-(ETHYLAMINO)ANILINO]ETHANOL DIHYDROCHLORIDE

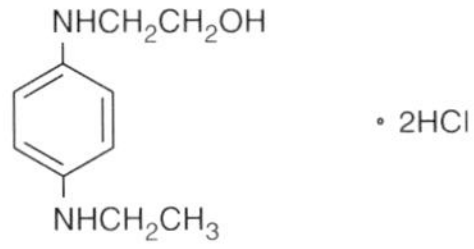

5 gm

S45,827-9
4-ETHOXY-1,3-PHENYLENEDIAMINE SULFATE MONOHYDRATE

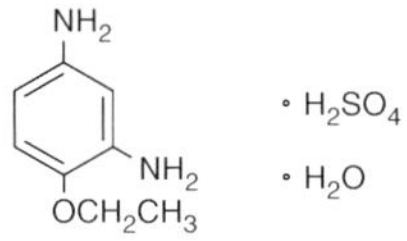

5 gm

S45,493-1
4-ETHOXY-1,3-PHENYLENEDIAMINE DIHYDROCHLORIDE

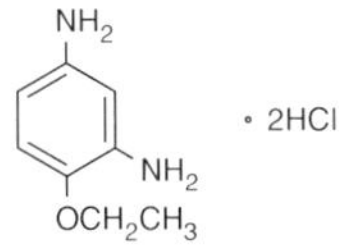

5 gm

S88,305-0
N-(3-METHOXYPROPYL)-1,4-PHENYLENEDIAMINE DIHYDROCHLORIDE

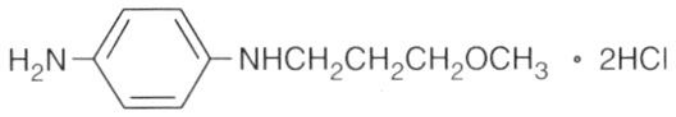

250 mg

S76,741-7
2-ETHOXY-1,3-PHENYLENEDIAMINE DIHYDROCHLORIDE

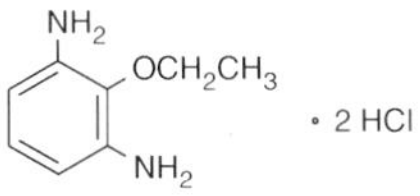

250 mg

S74,806-4
N-(3-ISOPROPOXYPROPYL)-N'-(1-METHYLDECYL)-1,4-PHENYLENEDIAMINE

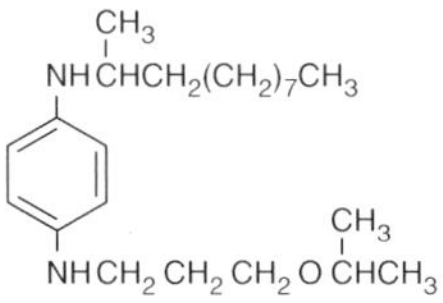

250 mg

S76,645-3
N-(7-ETHOXYHEPTYL)-N'-(1-METHYLHEPTYL)-1,4-PHENYLENEDIAMINE

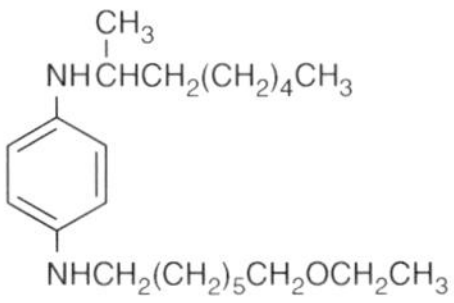

250 mg

S54,506-6
N4,N4-DIMETHYL-1,2,4-BENZENETRIAMINE DIHYDROCHLORIDE

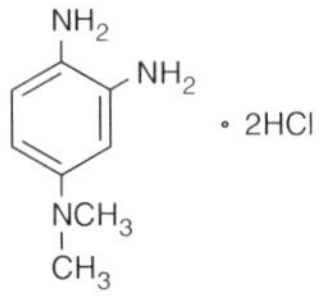

250 mg

S83,616-8
1,2,4-TRIAMINOBENZENE DIHYDROCHLORIDE

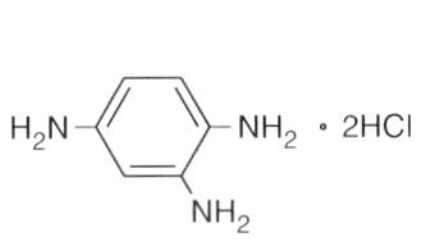

250 mg

S44,614-9
3,4-DIAMINOTOLUENE DIHYDROCHLORIDE

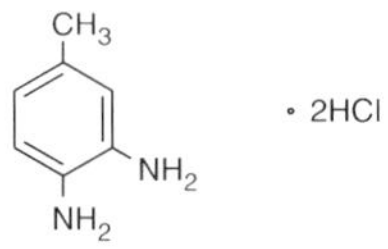

5 gm

S60,409-7
3-(METHYLAMINO)ANILINE SULFATE

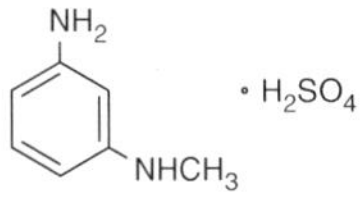

250 mg

S60,608-1
N1,N1,2,3,5,6-HEXAMETHYL-1,4-PHENYLENEDIAMINE DIHYDROCHLORIDE

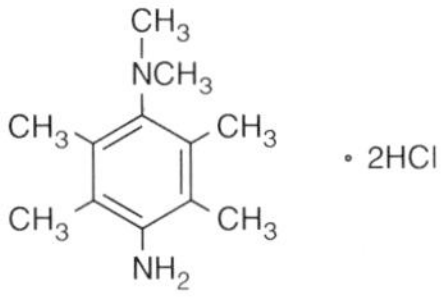

250 mg

S63,743-2
4-AMINO-N,N-DIETHYLANILINE MONOHYDROCHLORIDE

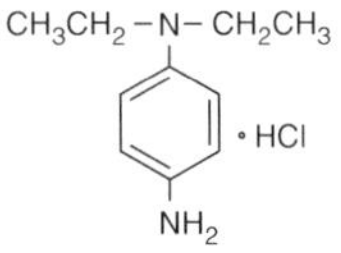

5 gm

S51,367-9
N,N-DIHEXYL-1,4-PHENYLENEDIAMINE HYDROCHLORIDE

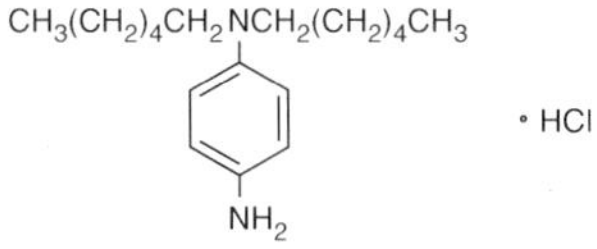

250 mg

S58,718-4
N-DODECYL-N-METHYL-1,4-PHENYLENEDIAMINE DIHYDROCHLORIDE

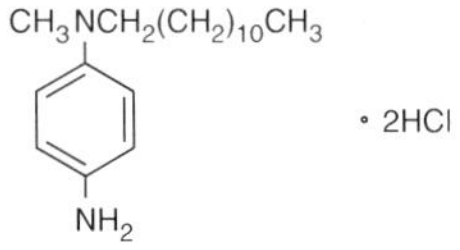

250 mg

S51,351-2
2-CHLORO-N1-CYCLOHEXYL-N1-METHYL-1,4-PHENYLENEDIAMINE HYDROCHLORIDE

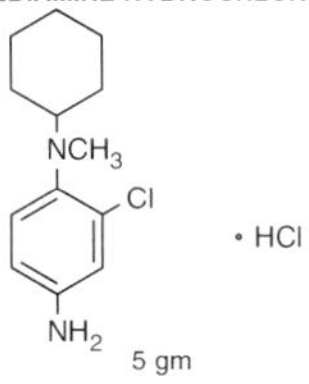

5 gm

S57,327-2
1,4-PHENYLENEDIAMINE SULFATE

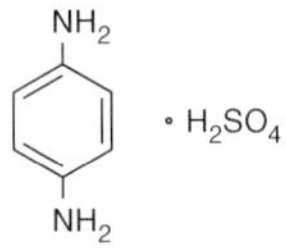

5 gm

S70,669-8
1,4-PHENYLENEDIAMINE OXALATE

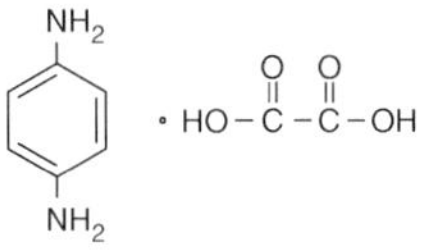

5 gm

S44,227-5
2,5-DIAMINOTOLUENE (2-METHYL-O-PHENYLENEDIAMINE) DIHYDROCHLORIDE

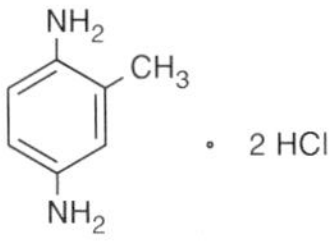

5 gm

S83,739-3
N,N-DIETHYL-N',N'-DIMETHYL-1,3-PHENYLENEDIAMINE

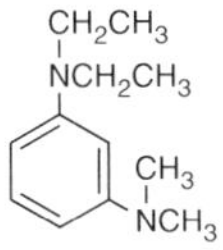

250 mg

S75,894-9
N-[4-(DIETHYLAMINO)-1-METHYLBUTYL]-N'-(1-METHYLHEPTYL)-1,4-PHENYLENEDIAMINE TRIHYDROCHLORIDE

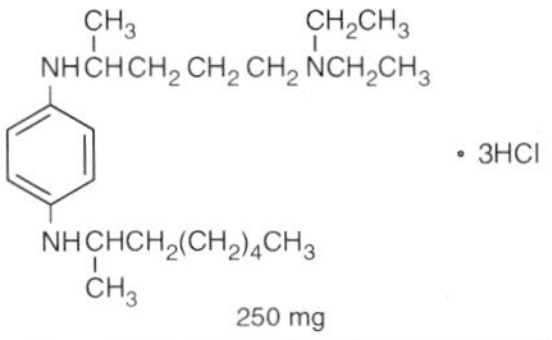

250 mg

S35,876-2
N,N'-BIS(1-METHYLHEPTYL)-1,4-PHENYLENEDIAMINE

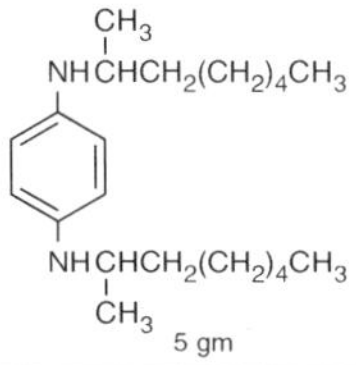

5 gm

S68,140-7
N,N'-BIS(1-METHYLHEPTYL)-1,4-PHENYLENEDIAMINE DIHYDROCHLORIDE

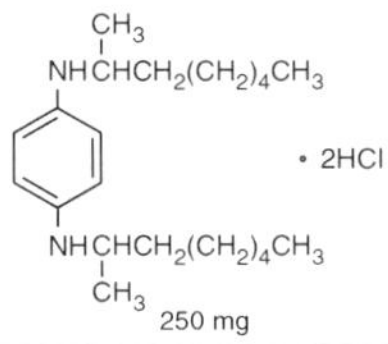

250 mg

S68,174-1
N,N'-DIOCTYL-1,4-PHENYLENEDIAMINE DIHYDROCHLORIDE

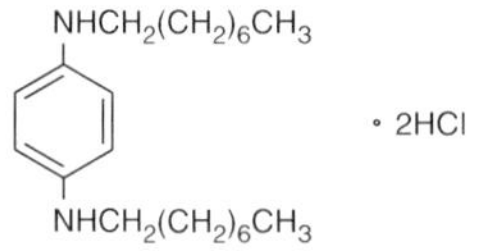

5 gm

S73,481-0
N,N'-BIS(1-METHYLOCTYL)-1,4-PHENYLENEDIAMINE DIHYDROCHLORIDE

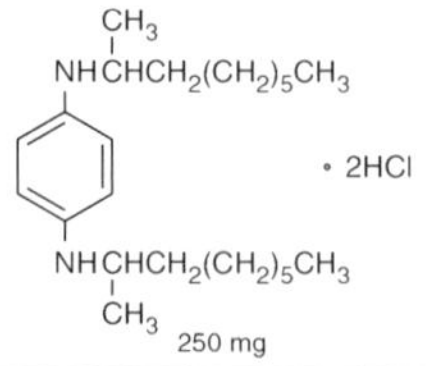

250 mg

S38,184-5
N-(ALPHA-CHLORO-O-TOLYL)-BENZYLAMINE HYDROCHLORIDE

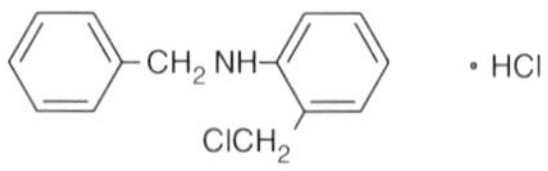

250 mg

S75,523-0
2-[N,N-BIS(CHLOROMETHYL)AMINO]-7-CHLOROFLUORENE

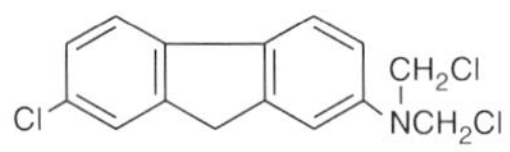

250 mg

S75,555-9
2-[N,N-BIS(2-CHLOROETHYL)AMINO]-FLUORENE

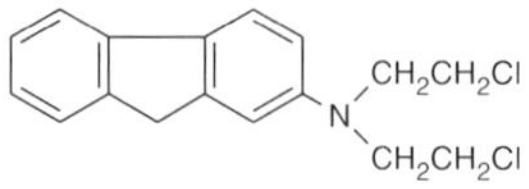

250 mg

S75,409-9
2-[N,N-BIS(2-BROMOETHYL)AMINO]-FLUORENE

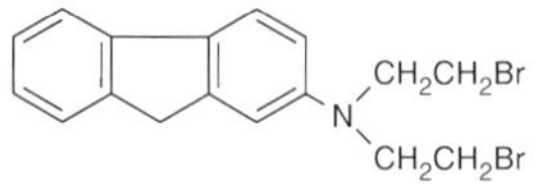

250 mg

S75,497-8
3-[N,N-BIS(BROMOMETHYL)AMINO]-FLUORANTHENE

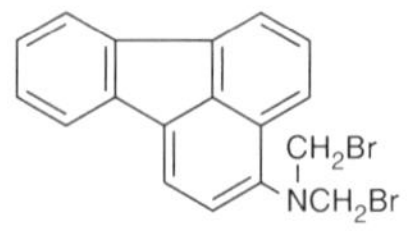

250 mg

S75,565-6
2-[N,N-BIS(CHLOROMETHYL)AMINO]-ANTHRACENE

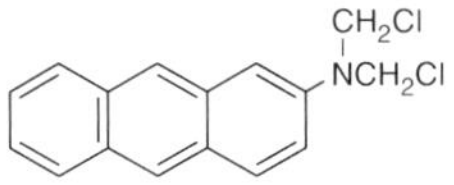

250 mg

S50,861-6
6-(4-CHLOROPHENOXY)-ALPHA,ALPHA,ALPHA-TRIFLUORO-M-TOLUIDINE

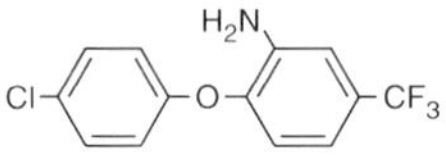

5 gm

S50,864-0
6-(4-CHLOROPHENYLTHIO)-ALPHA,ALPHA,ALPHA-TRIFLUORO-M-TOLUIDINE

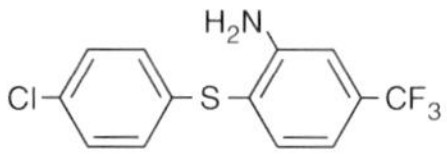

5 gm

S63,105-1
2-AMINO-3',4'-DICHLORO-4-TRIFLUOROMETHYLPHENYL SULFIDE

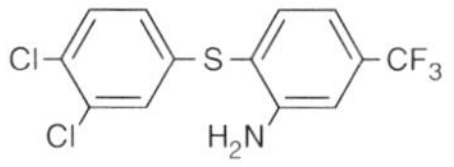

250 mg

S89,260-2
1,1-BIS(4-CHLOROANILINO)-2,2,2-TRICHLOROETHANE

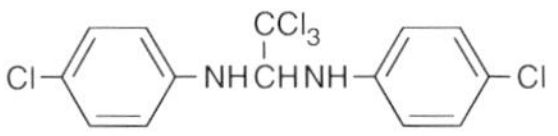

250 mg

S89,266-1
1,1-BIS(3-CHLORO-4-METHYLANILINO)-2,2,2-TRICHLOROETHANE

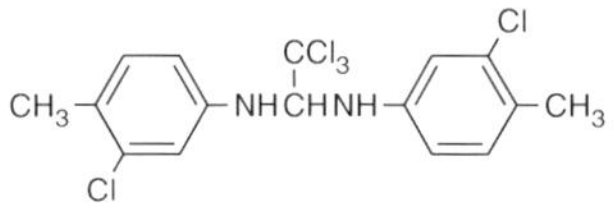

250 mg

S69,624-2
N-[3-(TRIFLUOROMETHYL)PHENYL]-2-HYDROXY-3-METHOXYBENZYLAMINE

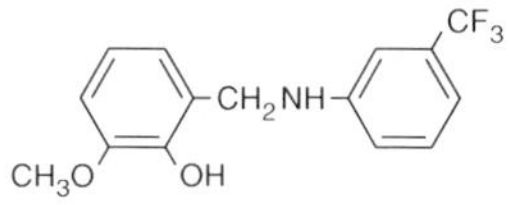

5 gm

S58,412-6
N-(ALPHA,ALPHA,ALPHA-TRIFLUORO-M-TOLYL)VERATRYLAMINE

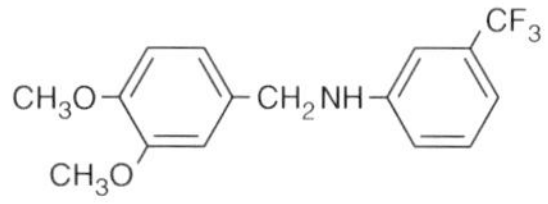

5 gm

S50,867-5
6-(P-TOLYLOXY)-ALPHA,ALPHA,ALPHA-TRIFLUORO-M-TOLUIDINE

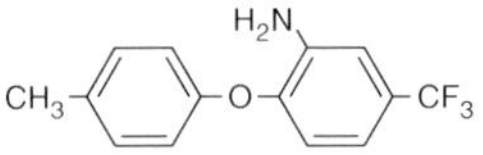

5 gm

S50,795-4
6-[4-(1-ETHYLPROPYL)PHENOXY]-ALPHA,ALPHA,ALPHA-TRIFLUORO-M-TOLUIDINE

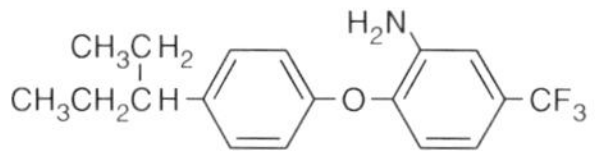

5 gm

S56,289-0
N-BENZYL-ALPHA,ALPHA,ALPHA-TRIFLUORO-M-TOLUIDINE

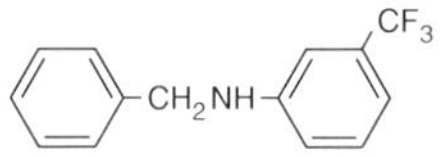

250 mg

S54,443-4
6-CHLORO-N-[A-(HEXADECYLOXY)-BENZYL]-A,A,ATRIFLUORO-M-TOLUIDINE

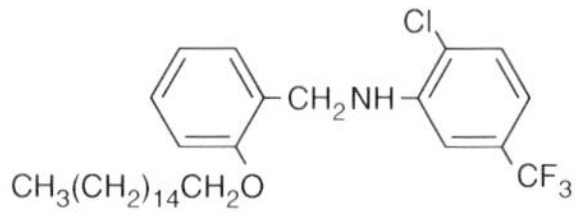

250 mg

S61,947-7
N,N'-(1,4-PHENYLENEDIMETHYLENE)BIS-[3-(TRIFLUOROMETHYL)ANILINE]

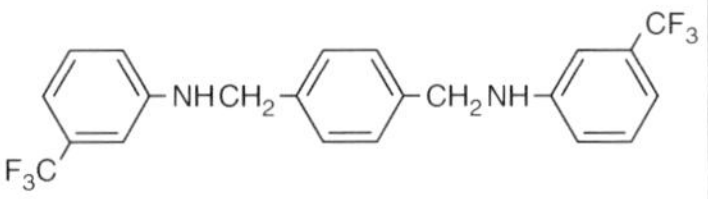

250 mg

S65,741-7
3,5-BIS(TRIFLUOROMETHYL)-DIPHENYLAMINE

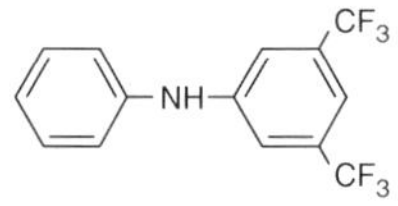

5 gm

S69,627-7
N-(4-FLUOROPHENYL)-2-HYDROXY-3-METHOXYBENZYLAMINE

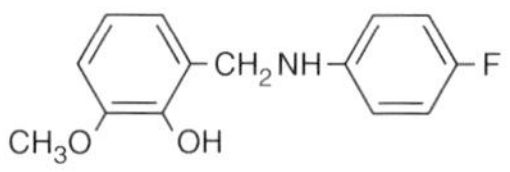

5 gm

S69,535-1
4-BROMO N-(2-HYDROXY-3-METHOXYBENZYL)ANILINE

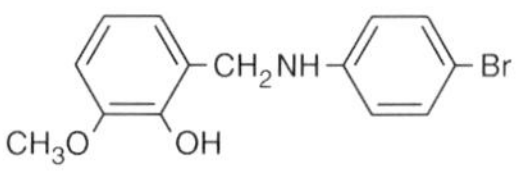

5 gm

S69,626-9
ALPHA-(3-CHLOROANILINO)-6-METHOXY-O-CRESOL

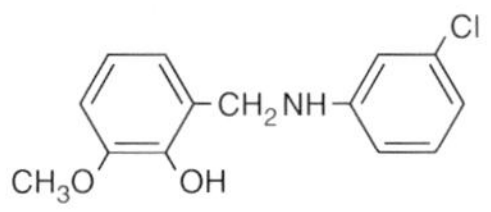

5 gm

S69,279-4
ALPHA-(3,4-DICHLOROANILINO)-6-METHOXY-O-CRESOL

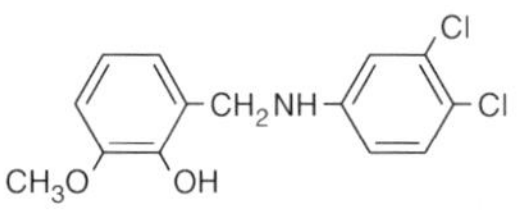

5 gm

S69,837-7
ALPHA-(2,4-DICHLOROANILINO)-6-METHOXY-O-CRESOL

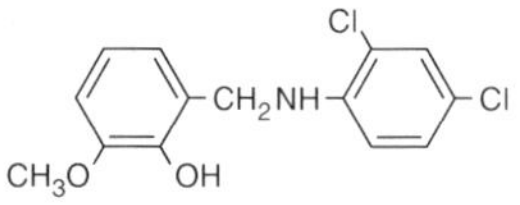

5 gm

S70,115-7
ALPHA-(4-CHLORO-2-METHYLANILINO)-6-METHOXY-O-CRESOL

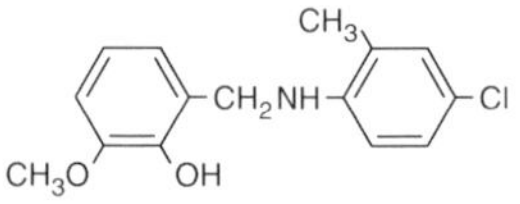

5 gm

S70,133-5
ALPHA-(3-CHLORO-2-METHYLANILINO)-6-METHOXY-O-CRESOL

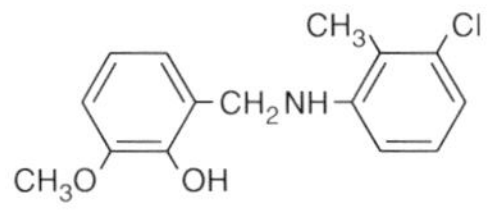

5 gm

S73,964-2
ALPHA-(2-CHLOROANILINO)-6-METHOXY-O-CRESOL

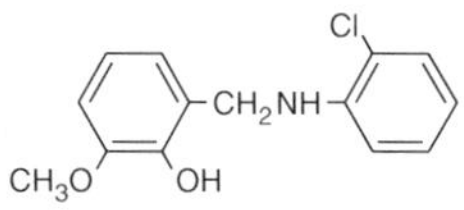

5 gm

S73,965-0
ALPHA-(4-IODOANILINO)-6-METHOXY-O-CRESOL

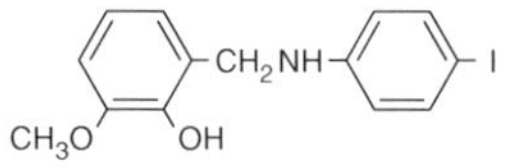

5 gm

S70,138-6
6-METHOXY-ALPHA-(2,4,5-TRICHLOROANILINO)-O-CRESOL

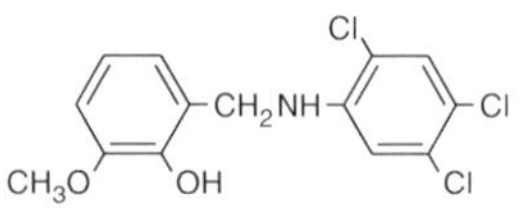

5 gm

S73,968-5
ALPHA-(2,5-DICHLOROANILINO)-6-METHOXY-O-CRESOL
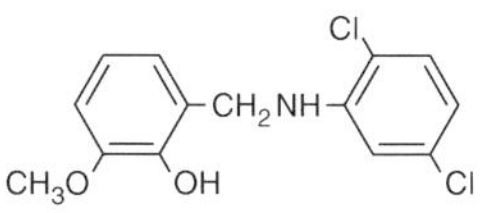
5 gm

S70,094-0
ALPHA-(5-CHLORO-2-METHYLANILINO)-6-METHOXY-O-CRESOL
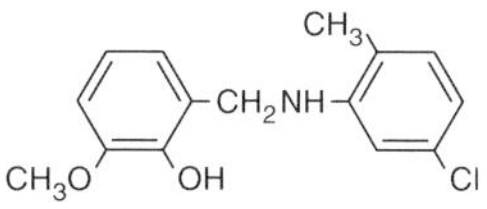
5 gm

S70,319-2
ALPHA-(3-CHLORO-4-METHYLANILINO)-6-METHOXY-O-CRESOL
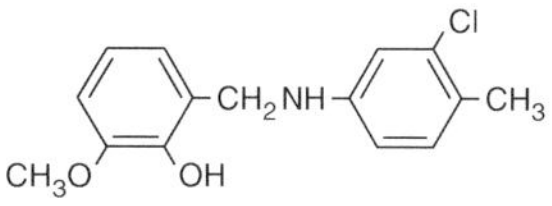
5 gm

S51,603-1
ALPHA-(5-CHLORO-2,4-DIMETHOXYANILINO)-O-CRESOL
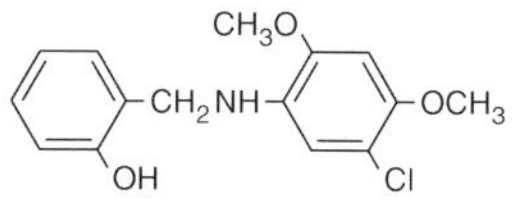
250 mg

S70,142-4
ALPHA-(4-CHLOROANILINO)-6-METHOXY-O-CRESOL
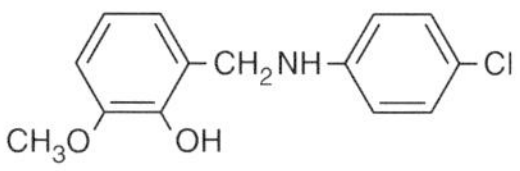
5 gm

S93,098-9
2-BROMO-4-(BIS[4-(DIMETHYLAMINO)-PHENYL]METHYL)-6-METHOXYPHENOL
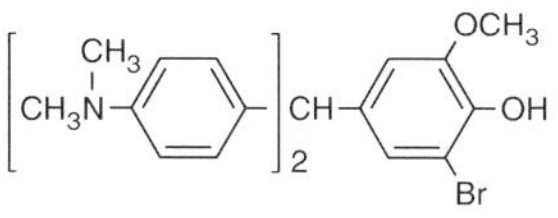
250 mg

S69,270-0
ALPHA-(4-CHLORO-3-NITROANILINO)-P-CRESOL
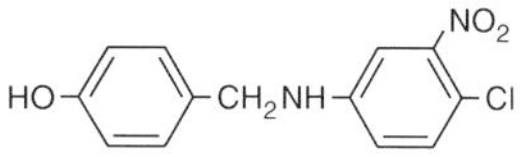
250 mg

S51,407-1
ALPHA-(3-CHLOROANILINO)-O-CRESOL
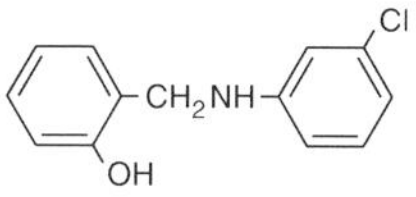
5 gm

S51,409-8
ALPHA-(4-FLUOROANILINO)-O-CRESOL
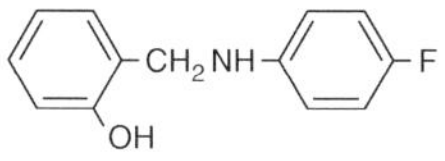
250 mg

S51,601-5
ALPHA-(4-CHLOROANILINO)-O-CRESOL
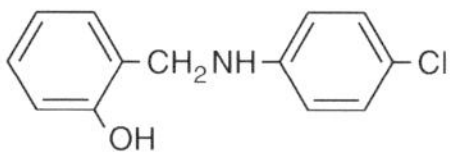
5 gm

S51,244-3
ALPHA-(4-BROMOANILINO)-O-CRESOL
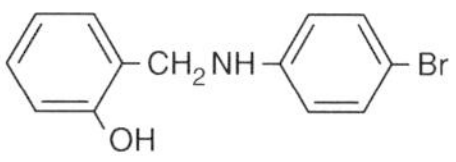
5 gm

S51,410-1
ALPHA-(4-IODOANILINO)-O-CRESOL
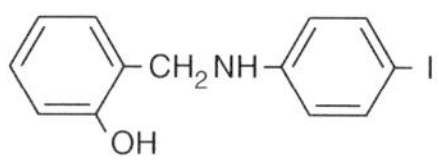
250 mg

S51,600-7
ALPHA-(3-CHLORO-P-TOLUIDINO)-O-CRESOL
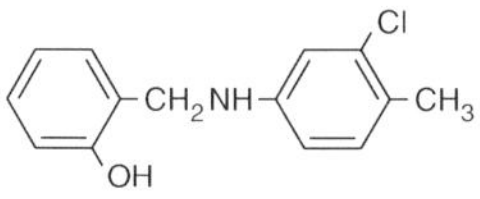
5 gm

S51,734-8
ALPHA-(5-CHLORO-O-TOLUIDINO)-O-CRESOL
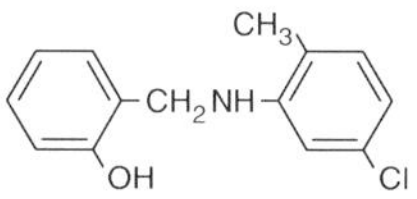
250 mg

S69,105-4
ALPHA-(4-FLUOROANILINO)-P-CRESOL
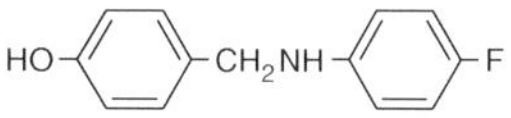
5 gm

S69,654-4
ALPHA-(4-CHLOROANILINO)-P-CRESOL
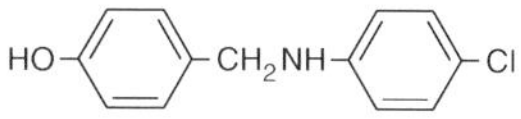
5 gm

S60,121-7
ALPHA-(3-BROMOANILINO)-O-CRESOL
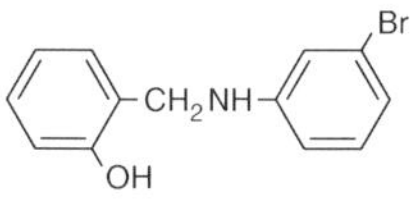
5 gm

S87,377-2
4-(BIS[4-(DIMETHYLAMINO)PHENYL]-METHYL)-2,6-DICHLOROPHENOL
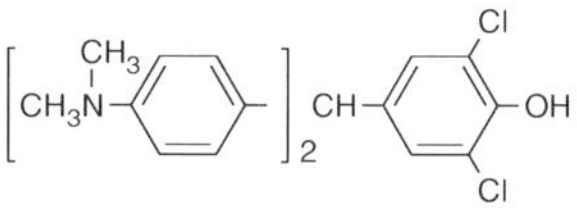
250 mg

S92,905-0
2,4-DIBROMO-6-BIS[4-(DIMETHYLAMINO)-PHENYL]METHYLPHENOL
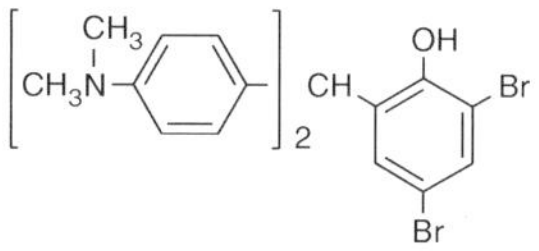
250 mg

S69,111-9
ALPHA-(4-BROMOANILINO)-P-CRESOL
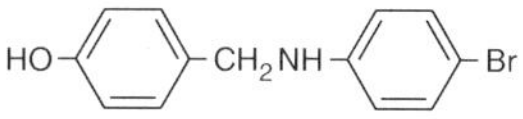
5 gm

S44,204-6
2-AMINO-4-CHLOROPHENYL PHENYL ETHER
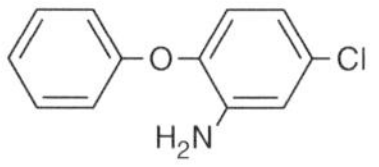
5 gm

S50,827-6
2-(4-CHLOROPHENOXY)ANILINE
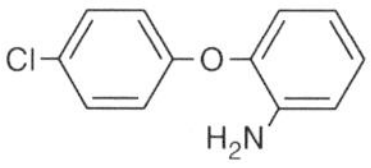
5 gm

S40,756-9
2-(2,4-DICHLOROPHENOXY)ANILINE
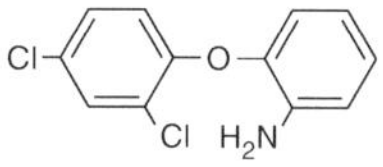
5 gm

S50,857-8
5-CHLORO-2-(2-CHLOROPHENOXY)-ANILINE
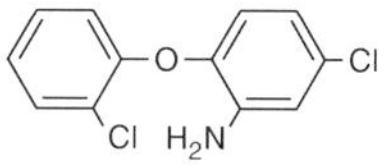
5 gm

S50,868-3
2-(2,4,5-TRICHLOROPHENOXY)ANILINE
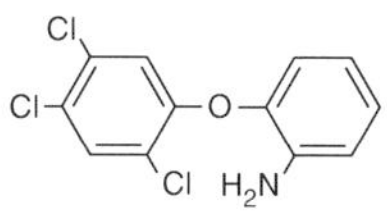
5 gm

S75,182-0
5-CHLORO-2-(4-CHLOROPHENOXY)ANILINE
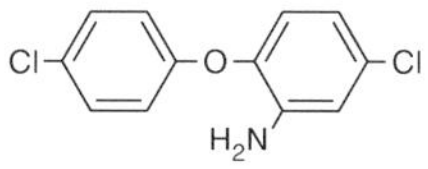
5 gm

S50,790-3
2-(4-CHLORO-M-TOLYLOXY)ANILINE
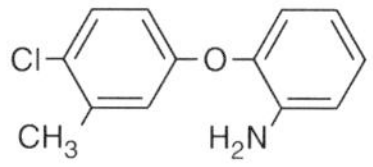
5 gm

S50,809-8
5-CHLORO-2-(P-TOLYLOXY)ANILINE
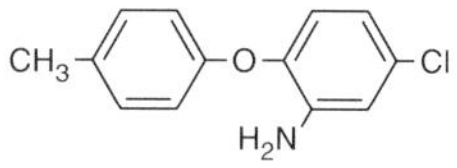
5 gm

S50,871-3
5-CHLORO-2-(4-PENTYLPHENOXY)-ANILINE
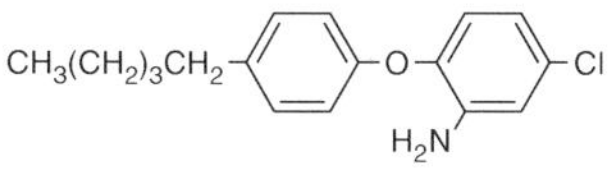
5 gm

S50,822-5
5-CHLORO-2-(4-CHLORO-M-TOLYLOXY)-ANILINE
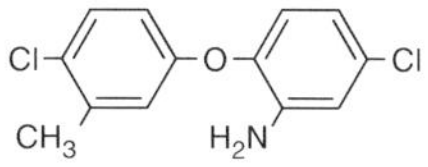
5 gm

S50,797-0
5-CHLORO-2-(2,4,5-TRICHLOROPHENOXY)-ANILINE
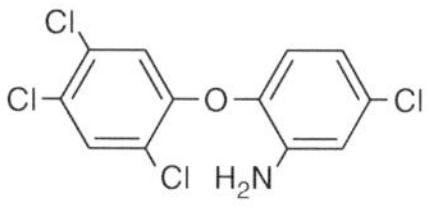
5 gm

S50,800-4
4-(4-CHLORO-M-TOLYLOXY)ANILINE
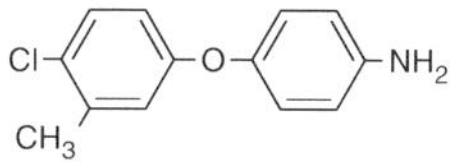
5 gm

S50,830-6
2-CHLORO-5-(4-CHLORO-3,5-XYLYLOXY)-ANILINE
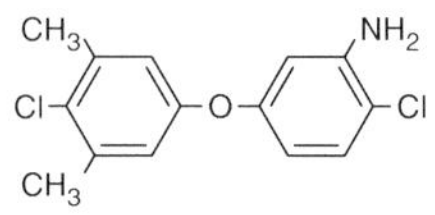
5 gm

S51,470-5
N-ALPHA-(5-CHLORO-2,4-DIMETHOXYPHENYL)-N4,N4-DIMETHYLTOLUENE-ALPHA,4-DIAMINE
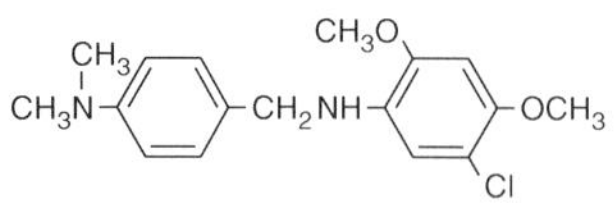
250 mg

S50,399-1
4-CHLORO-N-[2-(P-TOLYLOXY)ETHYL]-ANILINE
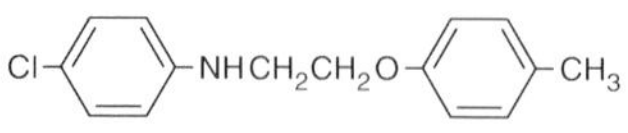
5 gm

S68,460-0
4-CHLORO-N-(4-ETHOXYPHENYL)-BENZYLAMINE
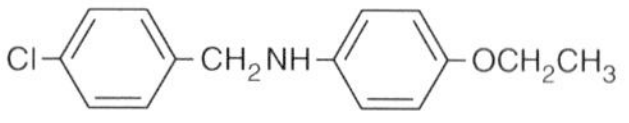
5 gm

S67,810-4
5-CHLORO-N,N-DIBENZYL-2-METHOXYANILINE
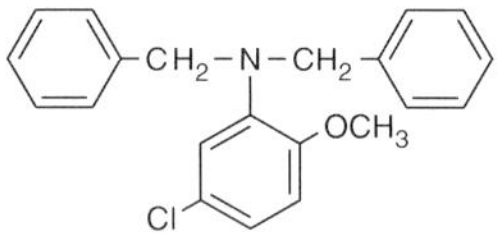
250 mg

S50,823-3
5-CHLORO-2-(4-CHLOROPHENYLTHIO)-ANILINE
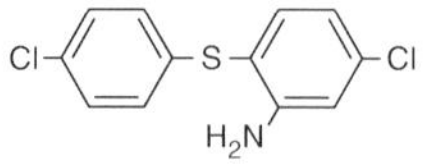
5 gm

S83,627-3
4-(2-CHLOROPHENYLTHIO)ANILINE
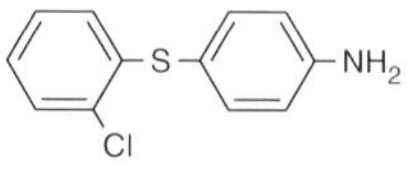
250 mg

S82,889-0
4-(4-CHLOROPHENYLTHIO)ANILINE
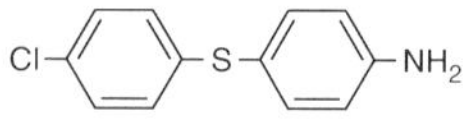
250 mg

S94,196-4
3-CHLORO-2-(P-TOLYLTHIO)ANILINE
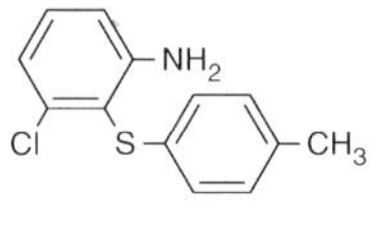
250 mg

S45,696-9
2-AMINO-4-CHLOROPHENYL DISULFIDE
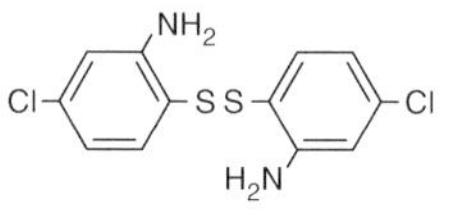
5 gm

S89,643-8
N,N-BIS(2-BROMOPHENYL)-1,2-PHENYLENEDIAMINE
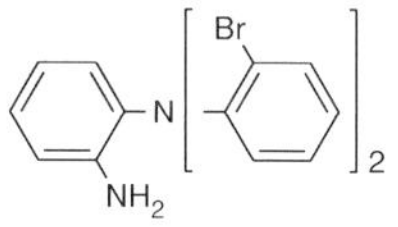
250 mg

S40,732-1
2,2'-DIAMINO-4,4'-DIFLUOROBIBENZYL
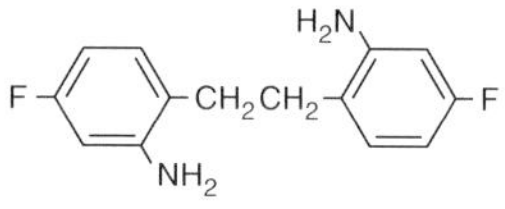
250 mg

S83,378-9
2-CHLORO-4-TRITYLANILINE
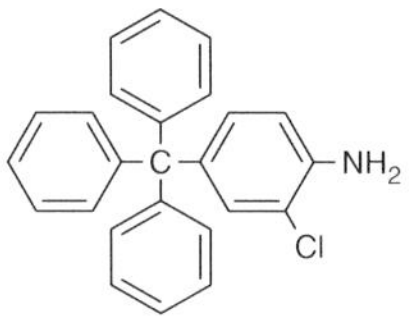
250 mg

S84,669-4
4,4'-DIBROMO-2-BIPHENYLAMINE
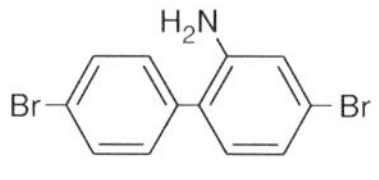
250 mg

S34,113-4
4-AMINO-4'-FLUOROBIPHENYL
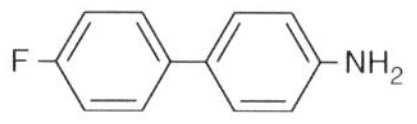
5 gm

S82,645-6
N,N'-DIMETHYL-N,N'-DIPHENYL-2,3,5,6-TETRAFLUORO-P-PHENYLENEDIAMINE
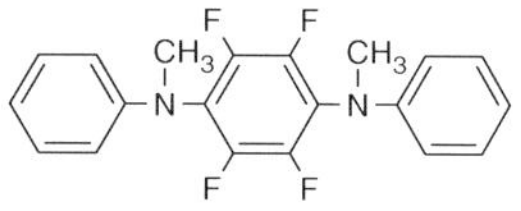
250 mg

S51,415-2
N4,N4-DIMETHYL-N-ALPHA-(4-FLUOROPHENYL)TOLUENE-ALPHA,4-DIAMINE
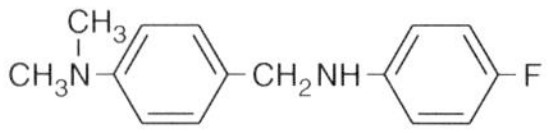
250 mg

S51,312-1
N-ALPHA-(4-CHLOROPHENYL)-N4,N4-DIMETHYLTOLUENE-ALPHA,4-DIAMINE
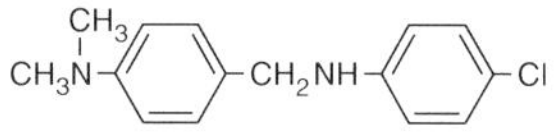
5 gm

S51,416-0
N4,N4-DIMETHYL-N-ALPHA-(4-IODOPHENYL)TOLUENE-ALPHA,4-DIAMINE
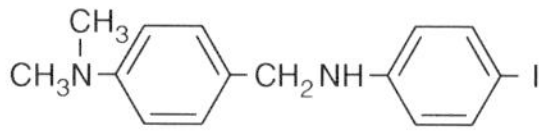
5 gm

S51,322-9
N-ALPHA-(2,3-DICHLOROPHENYL)-N4,N4-DIMETHYLTOLUENE-ALPHA,4-DIAMINE
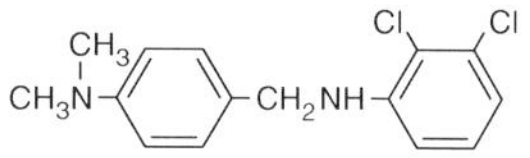
250 mg

S51,423-3
N-ALPHA-(3-CHLORO-O-TOLYL)-N4,N4-DIMETHYLTOLUENE-ALPHA,4-DIAMINE
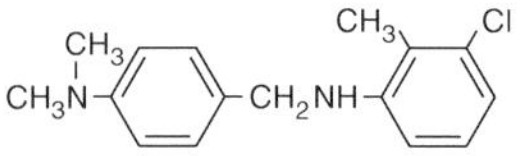
250 mg

S51,732-1
N-ALPHA-(3-CHLORO-P-TOLYL)-N4,N4-DIMETHYLTOLUENE-ALPHA,4-DIAMINE
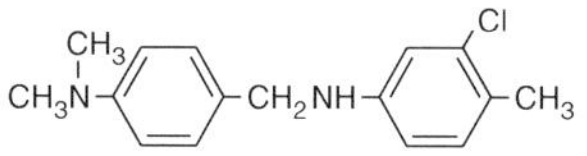
250 mg

S51,736-4
N-ALPHA-(4-CHLORO-O-TOLYL)-N4,N4-DIMETHYLTOLUENE-ALPHA,4-DIAMINE
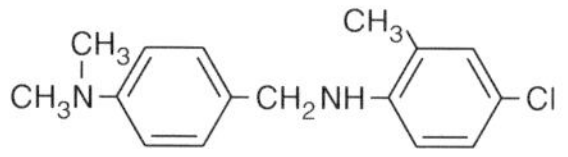
250 mg

S87,865-0
N-(4-BROMO-3-METHYLPHENYL)-4-(DIMETHYLAMINO)BENZYLAMINE
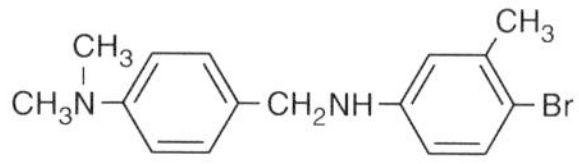
5 gm

S74,675-4
N(ALPHA)-(3,4-DICHLOROPHENYL)-N4,N4-DIMETHYLTOLUENE-ALPHA,4-DIAMINE
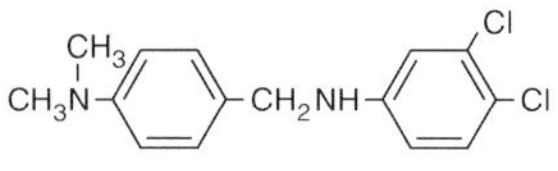
5 gm

S83,492-0
N-[1-(4-BROMOPHENYL)PENTYL]ANILINE HYDROCHLORIDE
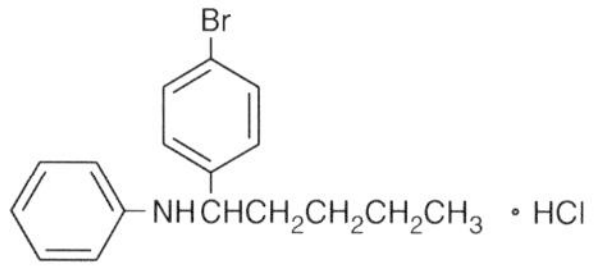
250 mg

S82,283-3
2,2'-DIBROMO-4-PHENYLDIPHENYLAMINE
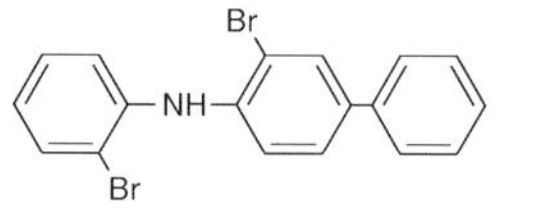
250 mg

S82,282-5
2,2',4,4'-TETRABROMODIPHENYLAMINE
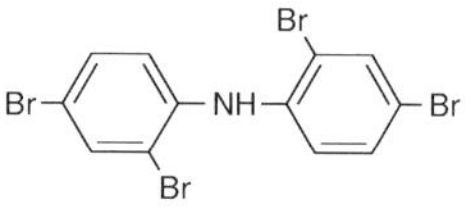
250 mg

S74,569-3
BIS(2-BROMOPHENYL)AMINE
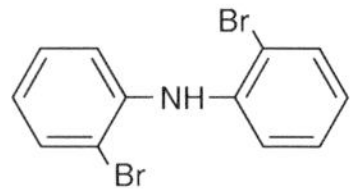
250 mg

S91,414-2
BIS(2,4-DICHLOROPHENYL)AMINE
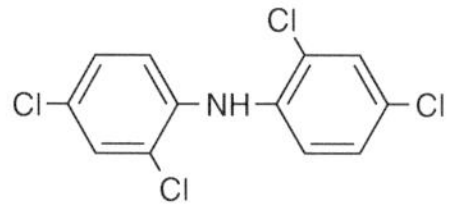
250 mg

S90,772-3
4,4'-[(2-CHLOROPHENYL)METHYLENE]-BIS(N,N-DIMETHYLANILINE)
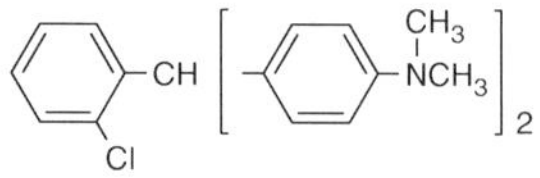
250 mg

S90,778-2
4,4'-[(2-BROMOPHENYL)METHYLENE]BIS-(N,N-DIMETHYLANILINE)
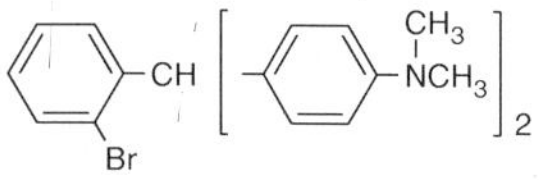
250 mg

S82,286-8
4-CHLORO-2,2'-DIBROMO-N-ETHYLDIPHENYLAMINE

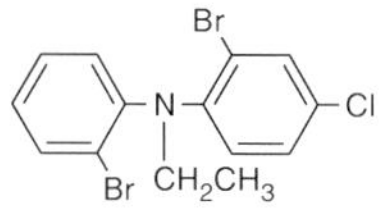

250 mg

S90,875-4
4,4'-[(3-FLUOROPHENYL)METHYLENE]-BIS(N,N-DIMETHYLANILINE)

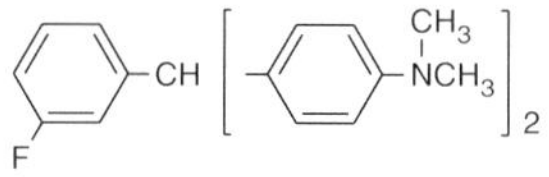

250 mg

S92,906-9
4,4'-(2,3-DICHLOROPHENYLMETHYLENE)-BIS(N,N-DIMETHYLANILINE)

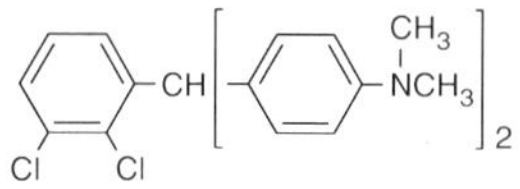

250 mg

S93,101-2
4,4'-[(2,4-DICHLOROPHENYL)-METHYLENE]BIS(N,N-DIMETHYLANILINE)

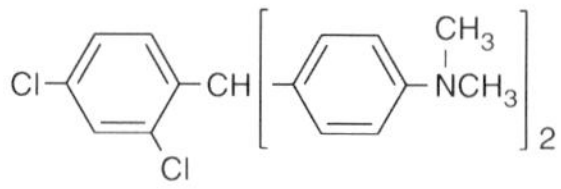

250 mg

S93,102-0
4,4'-[(3,4-DICHLOROPHENYL)-METHYLENE]BIS(N,N-DIMETHYLANILINE)

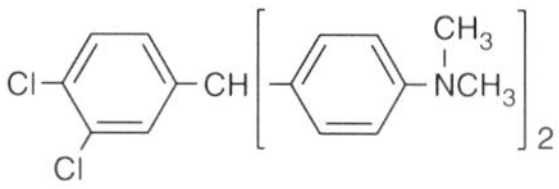

250 mg

S90,779-0
4,4'-[(4-BROMOPHENYL)METHYLENE]BIS-(N,N-DIMETHYLANILINE)

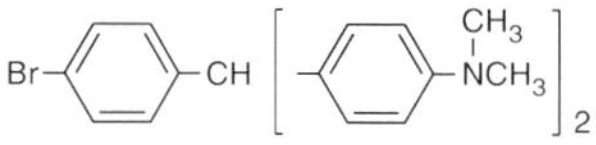

250 mg

S82,284-1
2,2'-DIBROMO-4,4',N-TRIMETHYLDIPHENYLAMINE

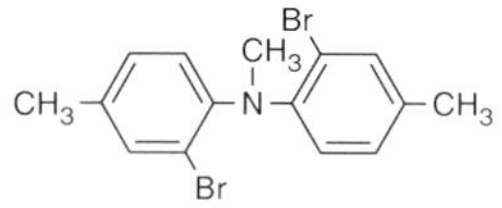

250 mg

S81,837-2
ALPHA-BUTYL-N-PHENYL-2-CHLOROBENZYLAMINE

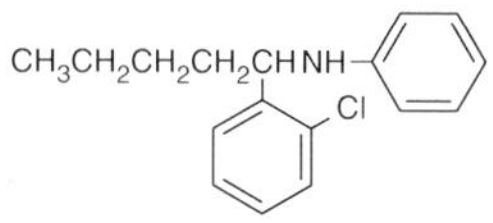

250 mg

S68,592-5
4-CHLORO-N-(3,4-DIMETHYLPHENYL)-BENZYLAMINE

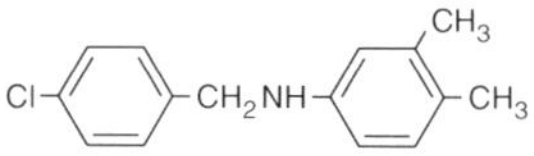

5 gm

S70,367-2
ALPHA-(2-HYDROXY-5-METHYLANILINO)-6-METHOXY-O-CRESOL

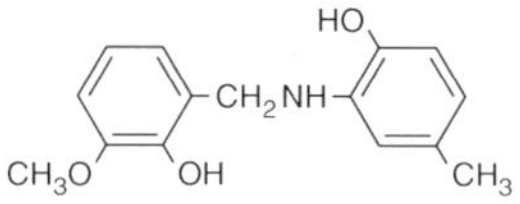

5 gm

S91,747-8
N-PHENYL-D-RIBOSAMINE

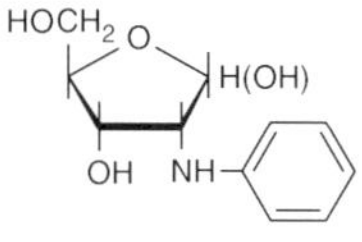

250 mg

S79,883-5
4,4'-IMINODIPHENOL

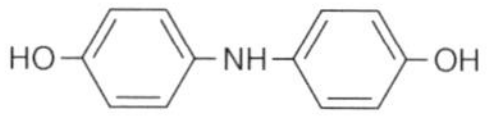

250 mg

S87,381-0
4-(DIMETHYLAMINO)-4'-HYDROXYBENZHYDROL

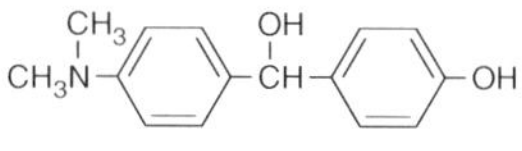

250 mg

S77,403-0
2-[N,N-BIS(2-HYDROXYETHYL)AMINO]-FLUORENE

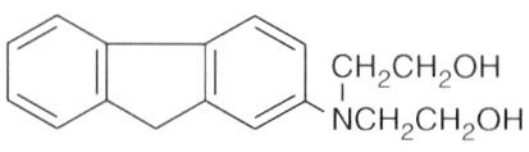

250 mg

S87,324-1
4-(DIMETHYLAMINO)-ALPHA-(4-HYDROXYPHENYL)-1-NAPHTHALENEMETHANOL

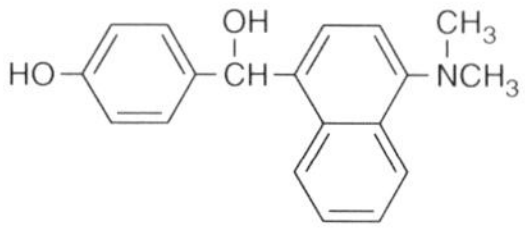

250 mg

S88,758-7
[4-(DIMETHYLAMINO)-1-NAPHTHYL](4-HYDROXY-1-NAPHTHYL)METHANOL

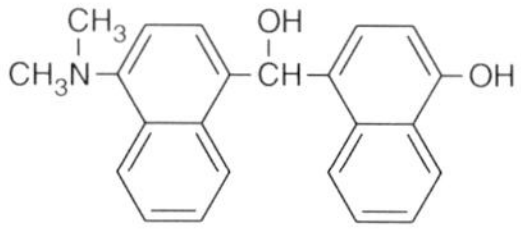

250 mg

S51,110-2
4,4'-BIS(DIMETHYLAMINO)-ALPHA-(1-ETHOXYETHYL)DIPHENYLMETHANOL

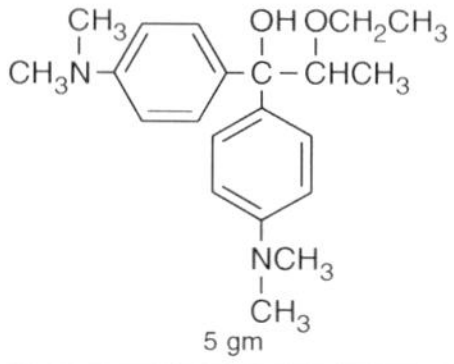

5 gm

S69,644-7
ALPHA-(5-CHLORO-2,4-DIMETHOXYANILINO)-6-METHOXY-O-CRESOL

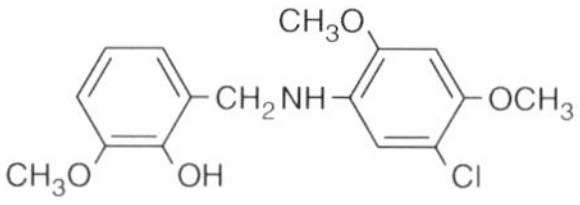

5 gm

S73,969-3
ALPHA-(4-ETHOXYANILINO)-6-METHOXY-O-CRESOL

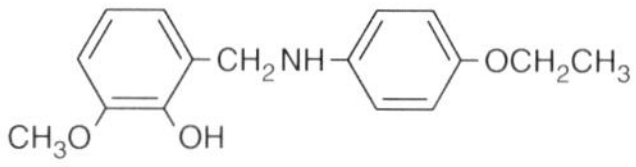

5 gm

S69,754-0
ALPHA-(2-ETHOXYANILINO)-6-METHOXY-O-CRESOL

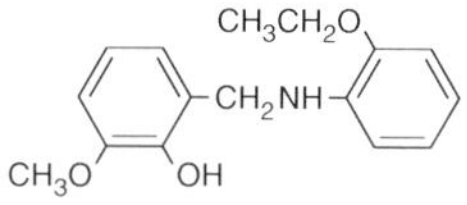

5 gm

S63,913-3
4-[2-(4-AMINOPHENOXY)ETHOXY]-2-TERT-BUTYLPHENOL

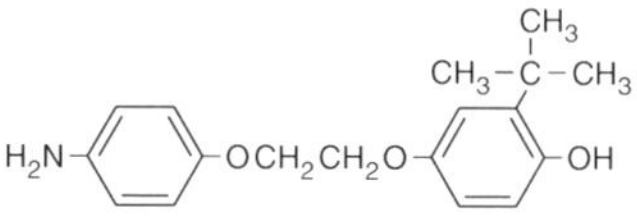

250 mg

S69,641-2
6-METHOXY-ALPHA-(4-METHYLANILINO)-O-CRESOL

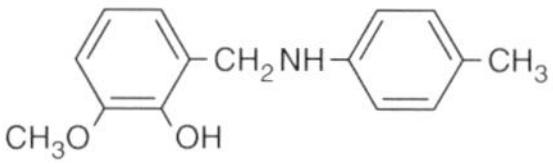

5 gm

S69,643-9
ALPHA-(5-CHLORO-2-METHOXYANILINO)-6-METHOXY-O-CRESOL

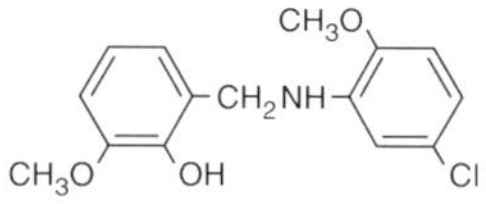

5 gm

S73,972-3
ALPHA-(3,4-DIMETHYLANILINO)-6-METHOXY-O-CRESOL

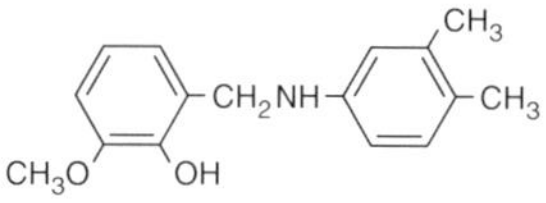

5 gm

S69,796-6
ALPHA-(4-ETHYLANILINO)-6-METHOXY-O-CRESOL

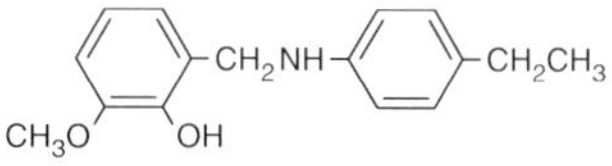

5 gm

S44,464-2
4-(4-AMINOPHENOXY)PHENOL

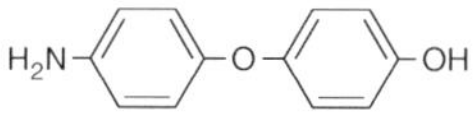

250 mg

S60,122-5
ALPHA-(M-ANISIDINO)-O-CRESOL

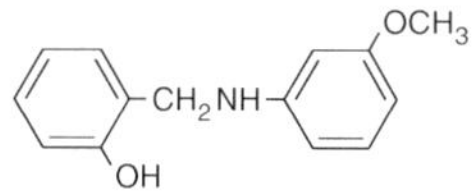

5 gm

S51,413-6
ALPHA-(P-ANISIDINO)-O-CRESOL

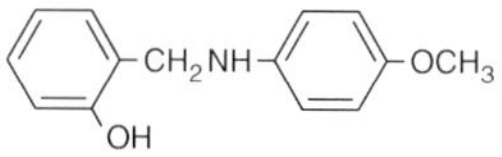

250 mg

S60,116-0
ALPHA-(P-PHENETIDINO)-O-CRESOL

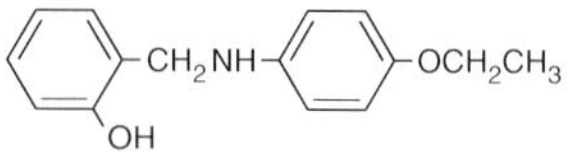

5 gm

S69,656-0
ALPHA-(4-METHOXYANILINO)-P-CRESOL

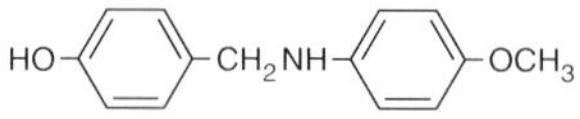

5 gm

S90,770-7
4,4'-BIS(DIMETHYLAMINO)-4''-METHOXY-2''-METHYLTRITYLALCOHOL

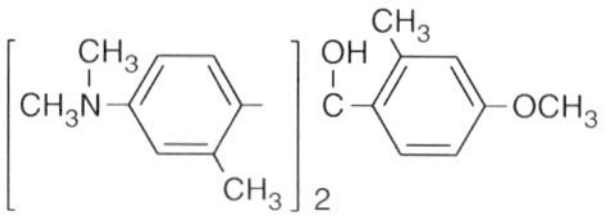

250 mg

S51,807-7
1-(N-METHYL-P-ANISIDINO)-2-NAPHTHOL

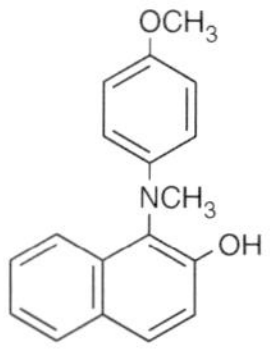

250 mg

S56,135-5
4-CHLORO-ALPHA-[2-(4-DIETHYLAMINO-PHENYL)ETHOXY]-ALPHA-(P-TOLYL)-PHENETHYL ALCOHOL

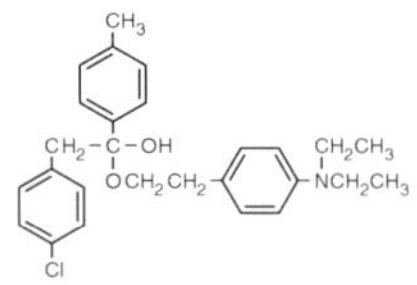

5 gm

S69,274-3
ALPHA-(2-METHOXY-5-METHYLANILINO)-P-CRESOL

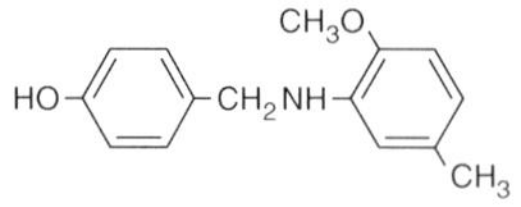

5 gm

S69,112-7
ALPHA-(4-ETHOXYANILINO)-P-CRESOL

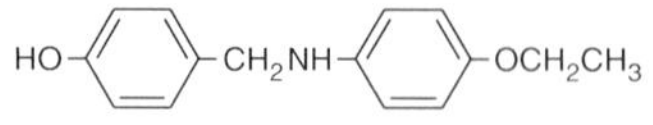

5 gm

S83,409-2
2-(4-AMINOPHENYLTHIO)-1-PHENYLETHANOL

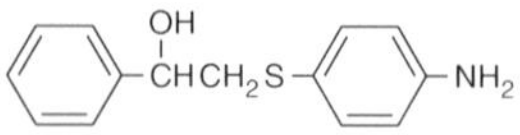

250 mg

S74,474-3
2-AMINO-ALPHA-METHYL-ALPHA-PHENYLBENZYL ALCOHOL

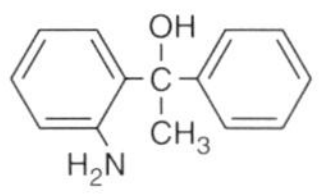

250 mg

S38,243-4
2-(BENZYLAMINO)BENZYL ALCOHOL

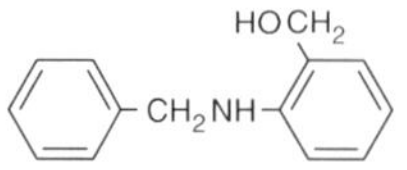

5 gm

S83,921-3
ALPHA-[4-(DIMETHYLAMINO)PHENYL]-PHENETHYL ALCOHOL

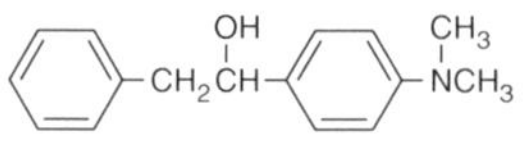

250 mg

S83,604-4
3-[4-(DIMETHYLAMINO)PHENYL]-1,1-DIPHENYLALLYL ALCOHOL

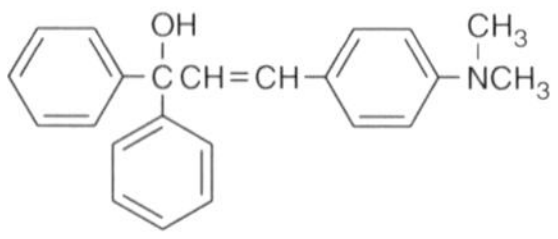

250 mg

S90,769-3
4,4'-BIS(DIMETHYLAMINO)-2,2',4''-TRIMETHYLTRITYL ALCOHOL

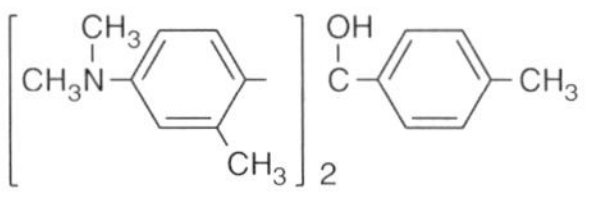

250 mg

S81,048-7
1-[4-(DIMETHYLAMINO)PHENYL]-1,3-DIPHENYLALLYL ALCOHOL

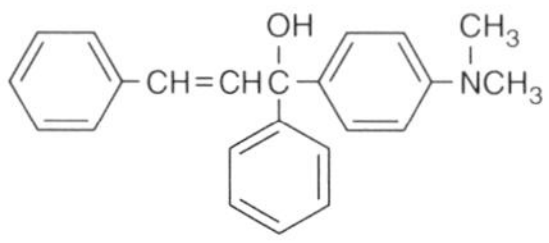

250 mg

S51,602-3
ALPHA-[4-(METHYLTHIO)ANILINO]-O-CRESOL

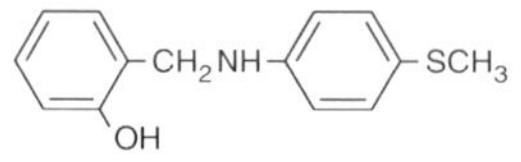

5 gm

S51,729-1
1-[4-(METHYLTHIO)ANILINOMETHYL]-2-NAPHTHOL

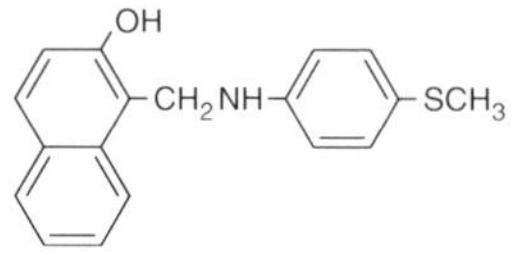

250 mg

S41,048-9
3-AMINO-4-HYDROXYDIPHENYLMETHANE OXALATE

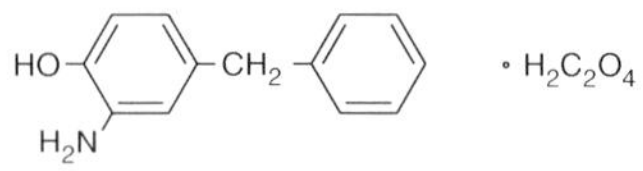

5 gm

S36,642-0
3-AMINO-2-HYDROXYFLUORENE

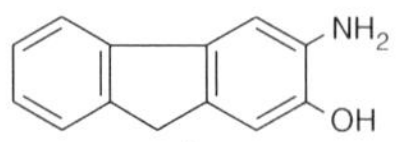

250 mg

S75,518-4
2-AMINO-3-HYDROXYFLUORENE

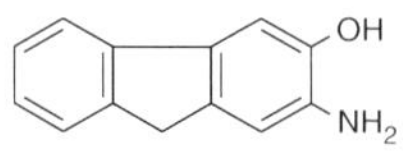

250 mg

S63,597-9
1-(2-AMINO-1-METHYLNAPHTHYL)-2-NAPHTHOL

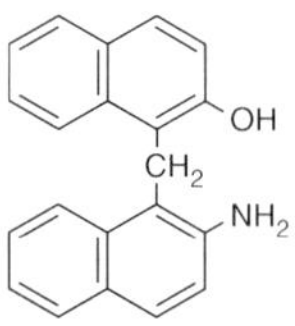

250 mg

S36,652-8
4-AMINO-2-METHYL-1-NAPHTHOL HYDROCHLORIDE

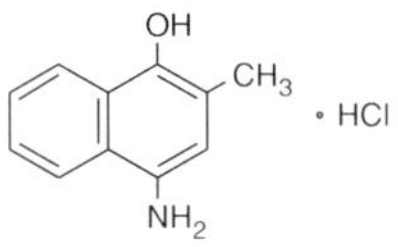

250 mg

S88,820-6
3-AMINO-2-NAPHTHALENEMETHANOL

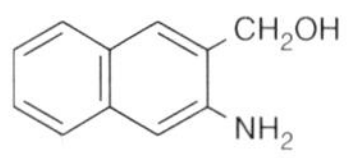

250 mg

S70,679-5
5-AMINO-1-NAPHTHOL HYDROCHLORIDE

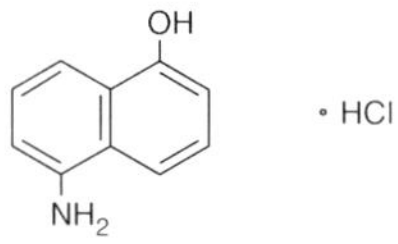

5 gm

S75,312-2
1-AMINO-9-HYDROXYFLUORENE

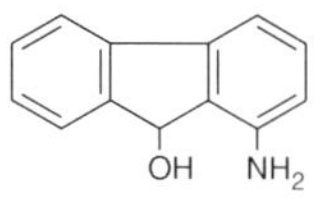

5 gm

S75,442-0
2-AMINO-3,7-DICHLORO-9-HYDROXYFLUORENE

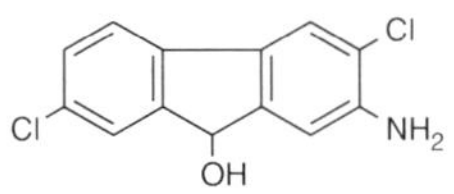

250 mg

S75,601-6
4-AMINO-2,7-DICHLORO-9-HYDROXYFLUORENE

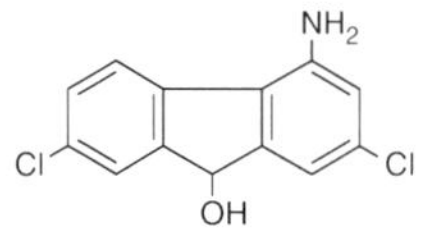

250 mg

S97,315-7
AR-D13 9-AMINO-3-HYDROXYPHENANTHRENE HYDROCHLORIDE

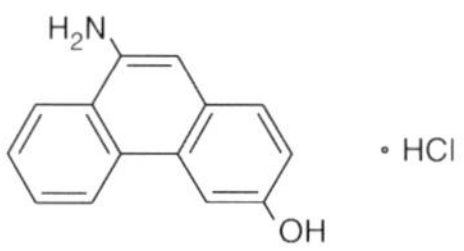

10 mg

S51,418-7
2-[4-(DIMETHYLAMINO)BENZYLAMINO]-PHENOL

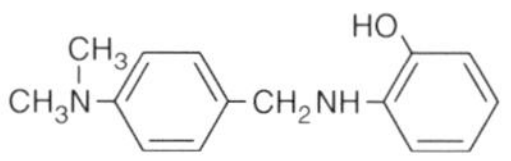

5 gm

S61,644-3
ALPHA-(2-BROMOANILINO)-O-CRESOL

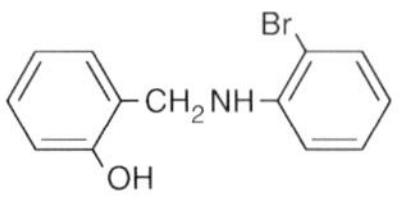

250 mg

S51,607-4
ALPHA-(M-TOLUIDINO)-O-CRESOL

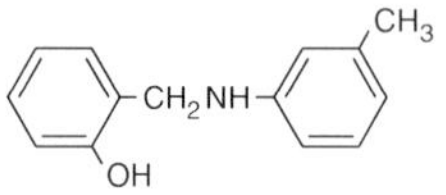

5 gm

S51,596-5
ALPHA-(P-TOLUIDINO)-O-CRESOL

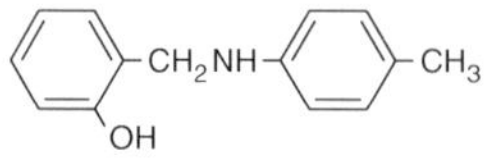

250 mg

S60,118-7
ALPHA-(4-ETHYLANILINO)-O-CRESOL

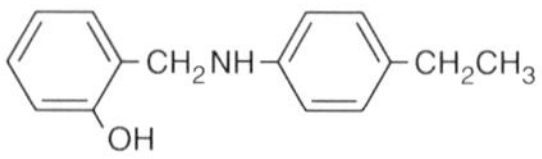

5 gm

S61,377-0
N-(2-HYDROXYBENZYL)-3,5-DICHLOROANILINE

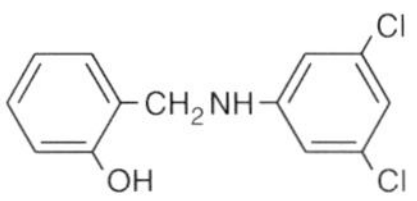

250 mg

S60,131-4
ALPHA-(2,4,6-TRIMETHYLANILINO)-O-CRESOL

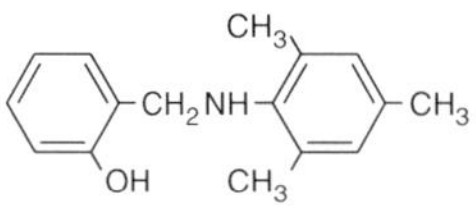

250 mg

S68,894-0
ALPHA-(P-TOLUIDINO)-P-CRESOL

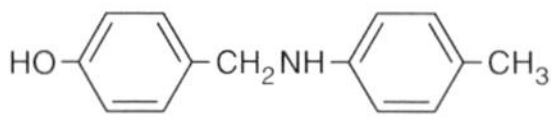

5 gm

S82,016-4
2-ANILINOTRIPHENYLMETHANOL

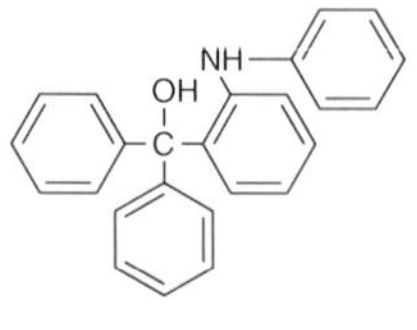

250 mg

S44,865-6
3-(O-TOLUIDINO)PHENOL

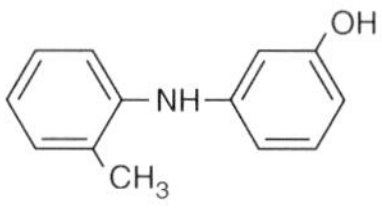

5 gm

S63,748-3
4-HYDROXY-4'-METHYLDIPHENYLAMINE

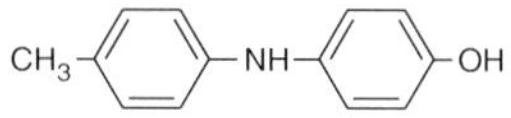

250 mg

S85,961-3
2-(ANILINOMETHYL)PHENOL

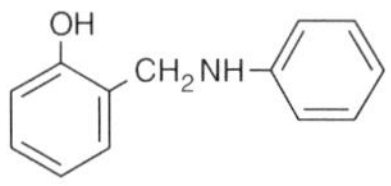

250 mg

S57,902-5
4-HYDROXYDIPHENYLAMINE

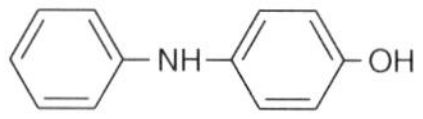

5 gm

S51,825-5
1-(4-CHLOROANILINOMETHYL)-2-NAPHTHOL

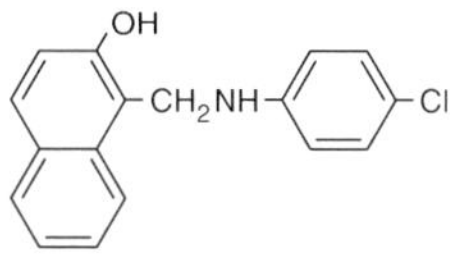

5 gm

S51,730-5
1-(3,4-XYLIDINOMETHYL)-2-NAPHTHOL

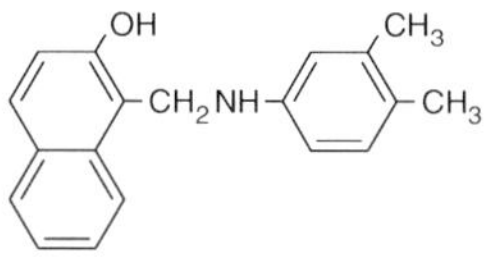

250 mg

S53,875-2
1-(2,5-XYLIDINOMETHYL)-2-NAPHTHOL

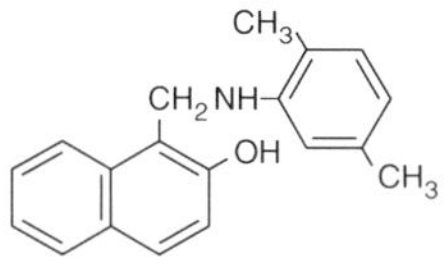

250 mg

S51,822-0
1-(M-TOLUIDINOMETHYL)-2-NAPHTHOL

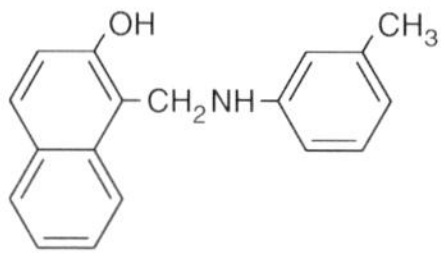

5 gm

S41,254-6
2-(1-NAPHTHYLAMINO)ETHANOL

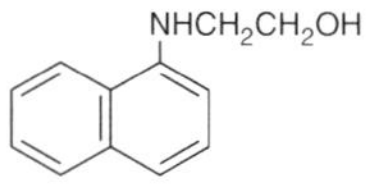

5 gm

S35,126-1
N-(3-HYDROXYPROPYL)-1-NAPHTHYLAMINE

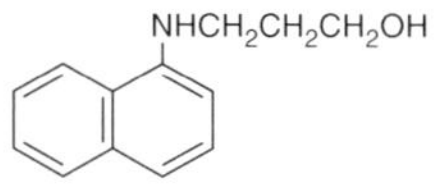

5 gm

S90,744-8
3-(BIS[4-(DIMETHYLAMINO)PHENYL]-METHYL)PHENOL

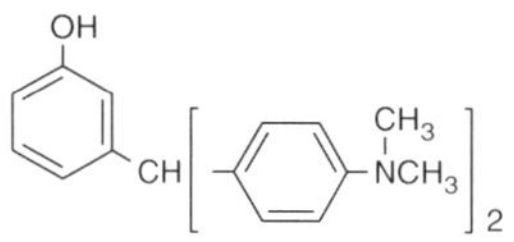

250 mg

S90,761-8
2,2'-METHYLENEBIS[5-(DIMETHYLAMINO)-PHENOL]

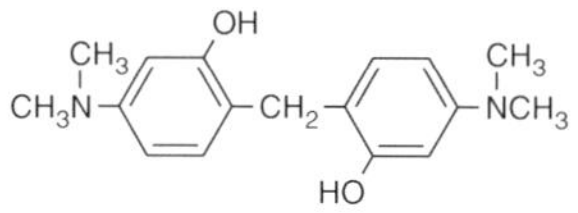

250 mg

S90,735-9
2,2'-ETHYLIDENEBIS[5-(DIMETHYLAMINO)PHENOL]

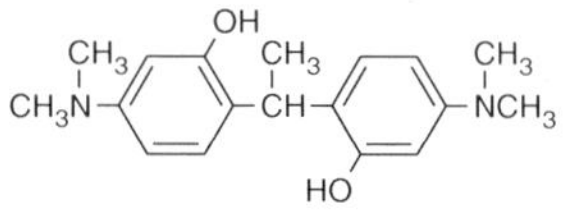

250 mg

S51,809-3
1-(2,3-DICHLORO-N-METHYLANILINO)-2-NAPHTHOL

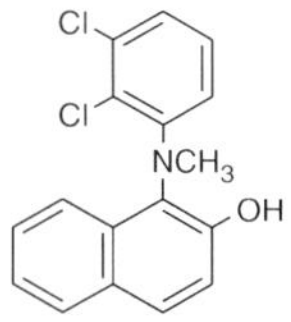

5 gm

S51,806-9
1-(2,5-DICHLORO-N-METHYLANILINO)-2-NAPHTHOL

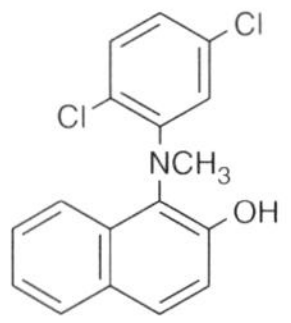

250 mg

S63,268-6
7-DIETHYLAMINO-1-NAPHTHOL

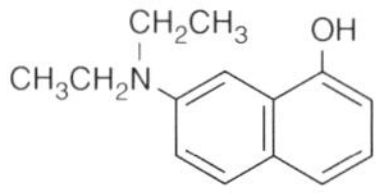

250 mg

S75,816-7
2-(DIMETHYLAMINO)-9-HYDROXYFLUORENE

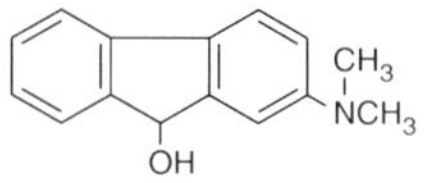

250 mg

S41,040-3
2-DIETHYLAMINO-9-FLUORENOL

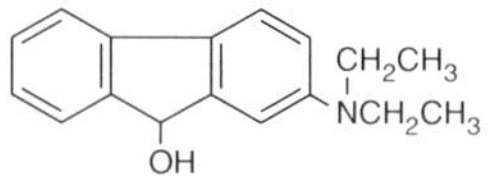

5 gm

S68,903-3
ALPHA-(4-ETHYLANILINO)-P-CRESOL

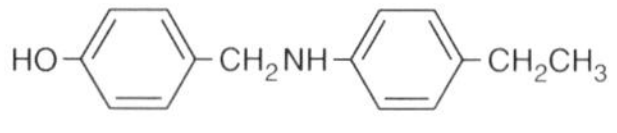

5 gm

S69,271-9
ALPHA-(3,4-DIMETHYLANILINO)-P-CRESOL

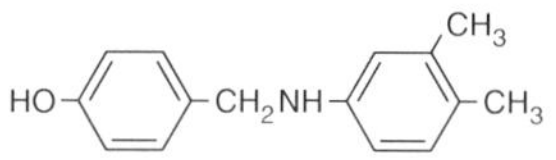

5 gm

S65,512-0
4-BENZYLOXYANILINE

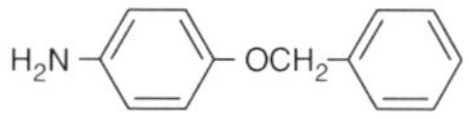

5 gm

S38,709-6
3-BENZYLOXYDIPHENYLAMINE

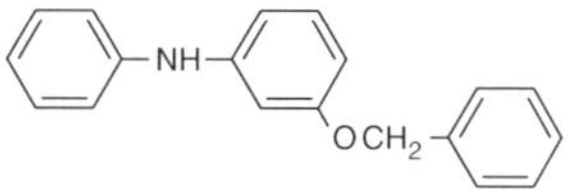

5 gm

S38,415-1
4-BENZYLOXYDIPHENYLAMINE

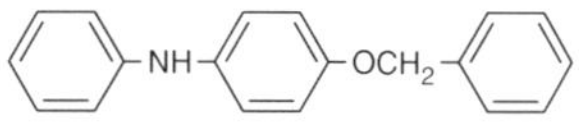

5 gm

S57,573-9
2,2'-(ETHYLENEDIOXY)DIANILINE

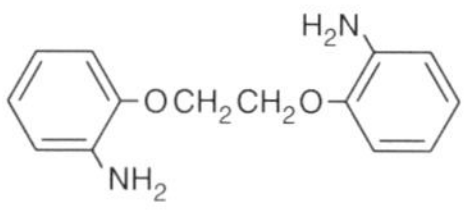

5 gm

S57,481-3
3,3'-(ETHYLENEDIOXY)DIANILINE

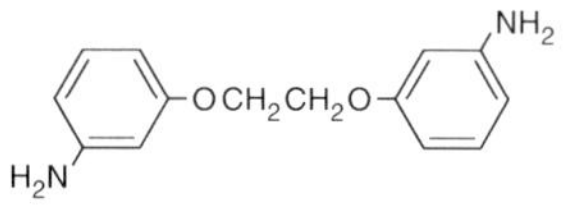

250 mg

S61,803-9
4-(2-PHENOXYETHOXY)ANILINE

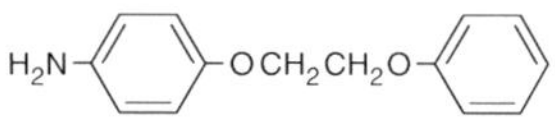

250 mg

S57,512-7
2,2'-(TRIMETHYLENEDIOXY)DIANILINE

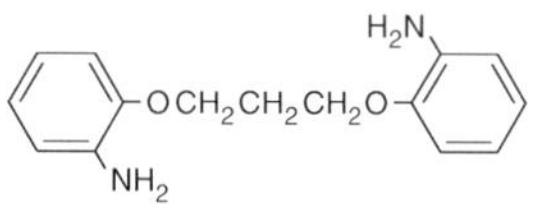

5 gm

S57,360-4
4,4'-(TRIMETHYLENEDIOXY)DIANILINE

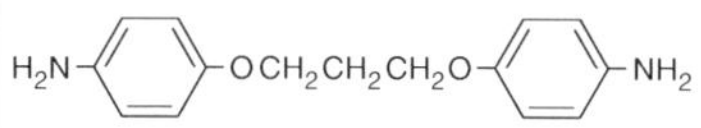

250 mg

S57,559-3
3,3'-(TETRAMETHYLENEDIOXY)DIANILINE

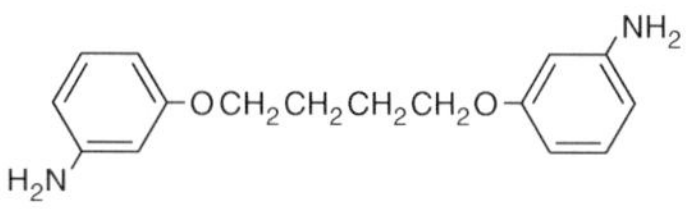

5 gm

S57,529-1
2,2'-(PENTAMETHYLENEDIOXY)DIANILINE

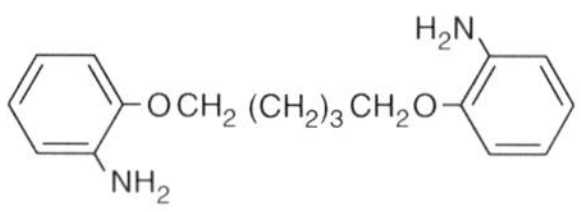

5 gm

S57,476-7
2,2'-(HEXAMETHYLENEDIOXY)DIANILINE

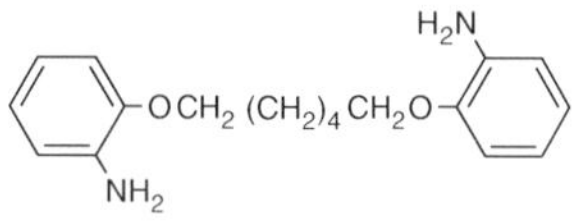

5 gm

S57,535-6
3,3'-(PENTAMETHYLENEDIOXY)DIANILINE

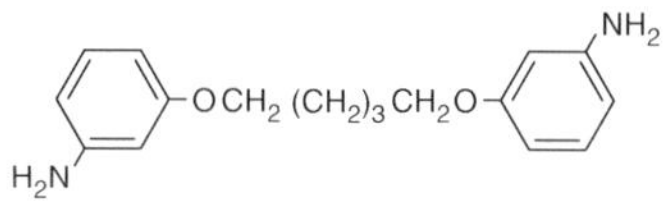

5 gm

S57,527-5
3,3'-(DECAMETHYLENEDIOXY)DIANILINE
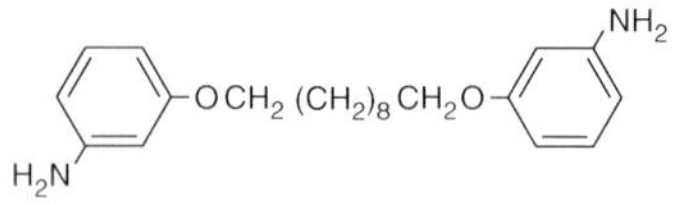
5 gm

S63,359-3
4,4'-(TETRAMETHYLENEDIOXY)DIANILINE
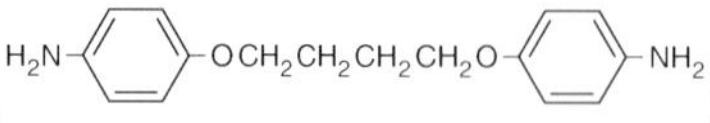
250 mg

S65,510-4
4-(5-PHENOXYPENTYLOXY)ANILINE
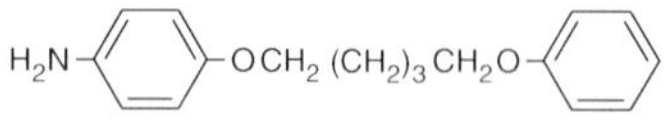
5 gm

S91,963-2
4,4'-DIAMINO-3,3'-DIMETHOXYBIPHENYL
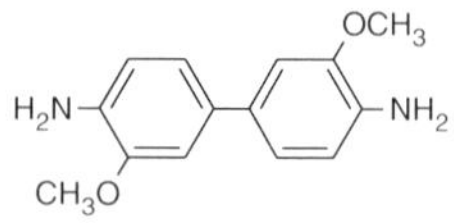
250 mg

S35,658-1
3,4,5-TRIMETHOXY-N-(3',4',5'-TRIMETHOXYBENZYL)ANILINE
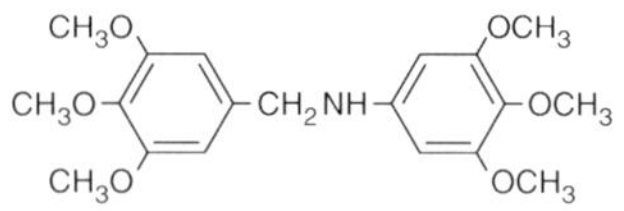
250 mg

S69,561-0
N,N'-METHYLENEBIS(4-ETHOXYANILINE)
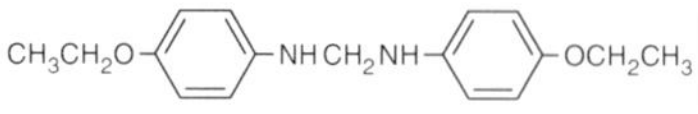
5 gm

S54,130-3
2,3-DIMETHOXY-N-[4-(METHYLTHIO)-PHENYL]BENZYLAMINE
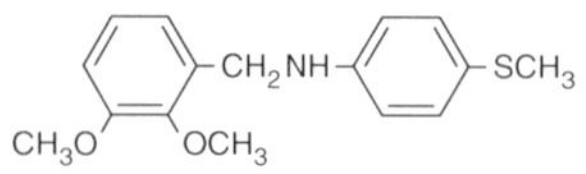
250 mg

S59,195-5
3-(2-PHENOXYETHOXY)ANILINE
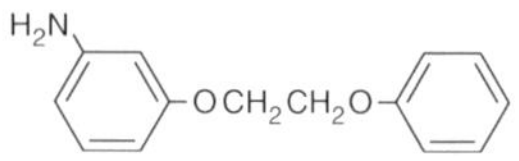
250 mg

S62,907-3
2-[2-(4-TERT-OCTYLPHENOXY)ETHOXY]-ANILINE
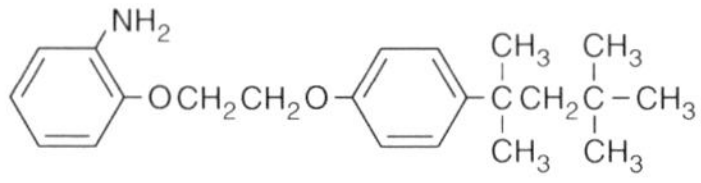
5 gm

S58,005-8
N-(5-CHLORO-2,4-DIMETHOXYPHENYL)-3,4-DIMETHOXYBENZYLAMINE
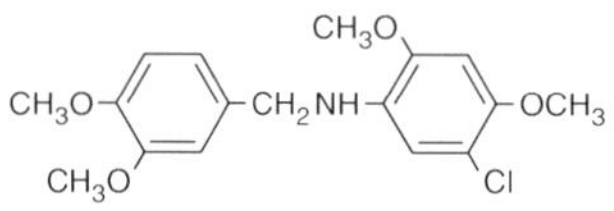
250 mg

S53,159-6
2,3-DIMETHOXY-N-(4-METHOXYPHENYL)-BENZYLAMINE
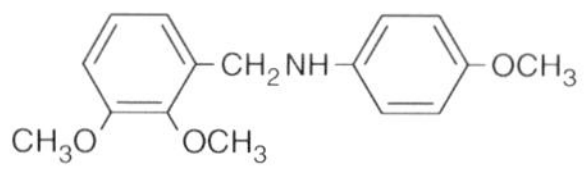
250 mg

S57,723-5
N-(4-FLUOROPHENYL)-3,4-DIMETHOXYBENZYLAMINE
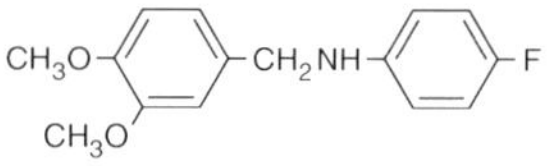
5 gm

S57,641-7
N-(4-CHLOROPHENYL)-3,4-DIMETHOXYBENZYLAMINE
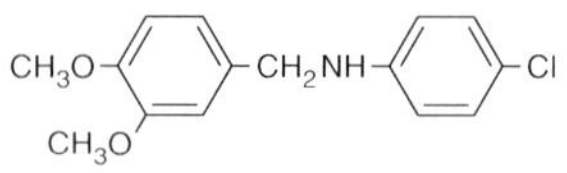
250 mg

S53,144-8
N-(4-BROMOPHENYL)VERATRYLAMINE
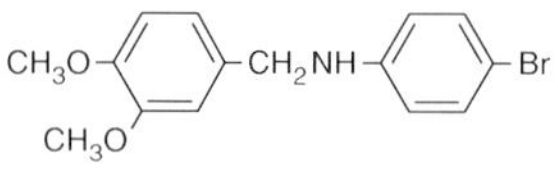
5 gm

S58,408-8
N-(4-IODOPHENYL)VERATRYLAMINE
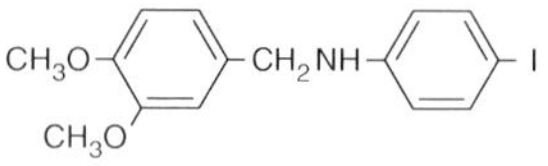
250 mg

S57,644-1
3,4-DIMETHOXY-N-(P-TOLYL)-BENZYLAMINE
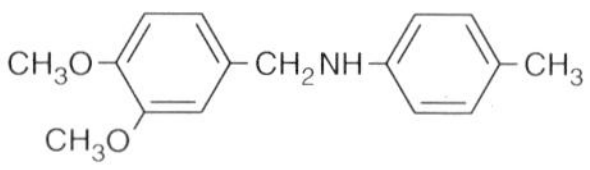
250 mg

S57,650-6
3,4-DIMETHOXY-N-(4-METHOXYPHENYL)-BENZYLAMINE
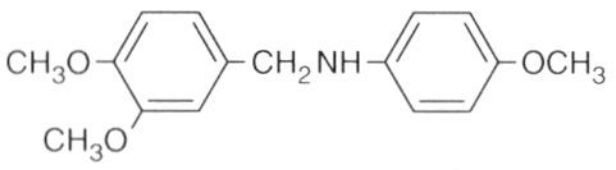
250 mg

S62,317-2
N-(4-ETHYLPHENYL)VERATRYLAMINE
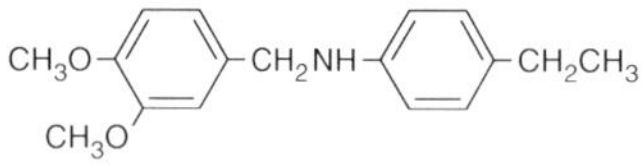
5 gm

S58,949-7
N-(4-ETHOXYPHENYL)VERATRYLAMINE
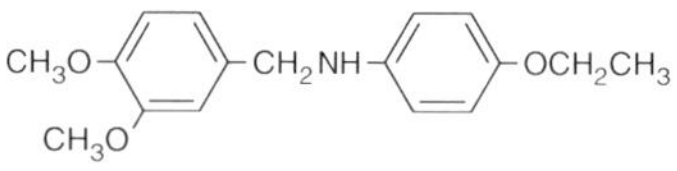
5 gm

S54,377-2
N-(2,3-DICHLOROPHENYL)-2,3-DIMETHOXYBENZYLAMINE
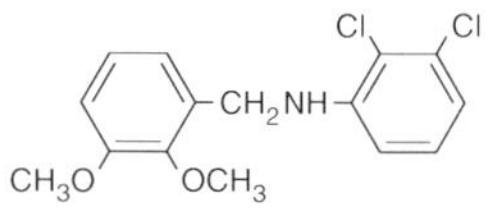
250 mg

S54,302-0
N-(3-CHLORO-P-TOLYL)-2,3-DIMETHOXYBENZYLAMINE
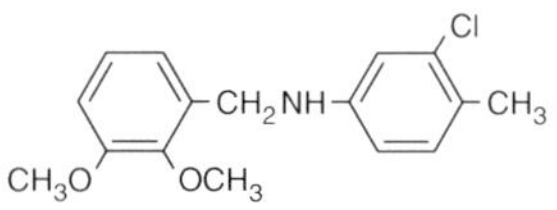
5 gm

S54,133-8
N-(2,3-DIMETHOXYBENZYL)-2,4-XYLIDINE
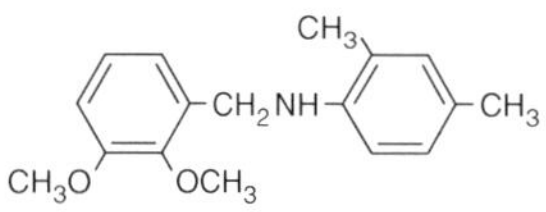
250 mg

S58,406-1
N-(3-CHLOROPHENYL)VERATRYLAMINE
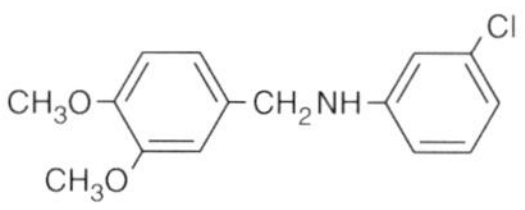
5 gm

S58,969-1
N-(M-TOLYL)VERATRYLAMINE
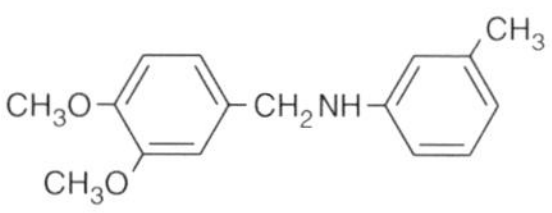
5 gm

S58,902-0
N-VERATRYL-3,4-XYLIDINE
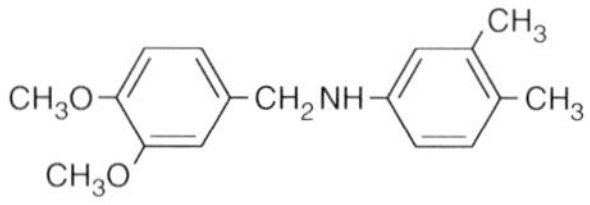
5 gm

S34,083-9
2-AMINO-4-TERT-AMYLPHENYL-PHENYL ETHER
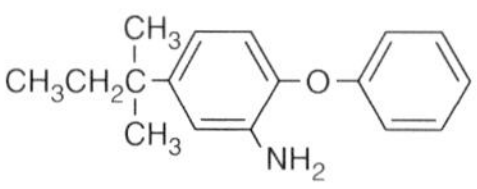
5 gm

S36,512-2
2-(O-TOLYLOXY)ANILINE
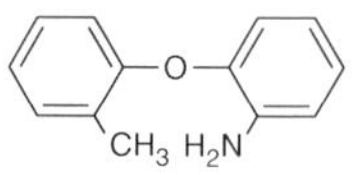
5 gm

S91,994-2
4'-METHOXY-2,3,5-TRIPHENYL-4-BIPHENYLAMINE
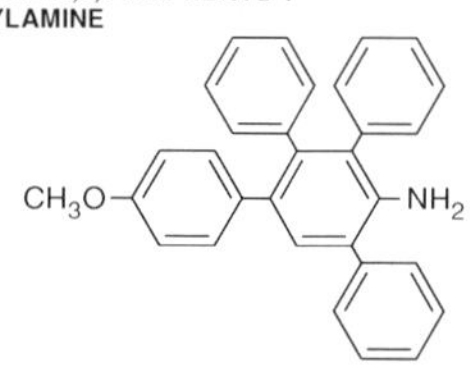
250 mg

S75,826-4
3-AMINO-2-METHOXYFLUORENE
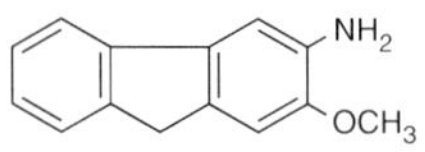
250 mg

S64,289-4
N,N'-BIS(4-METHOXYPHENYL)-ETHYLENEDIAMINE
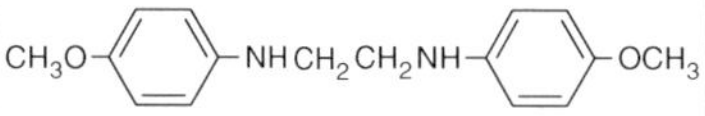
250 mg

S51,826-3
N4,N4-DIMETHYL-N-ALPHA-(P-METHOXYPHENYL)-ALPHA,4-TOLUENEDIAMINE
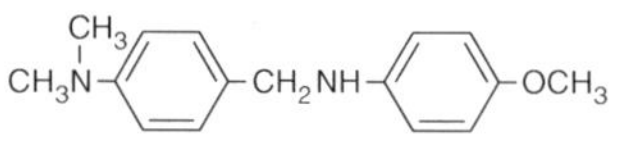
5 gm

S74,677-0
N4,N4-DIMETHYL-N(ALPHA)-(4-ETHOXYPHENYL)TOLUENE-ALPHA,4-DIAMINE
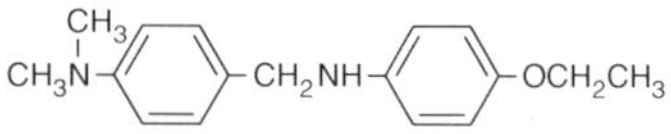
5 gm

S50,333-9
N-[2-(P-TOLYLOXY)ETHYL]-P-TOLUIDINE
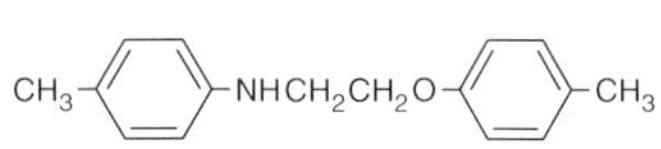
5 gm

S50,402-5
N-[2-(P-TOLYLOXY)ETHYL]-3,4-XYLIDINE
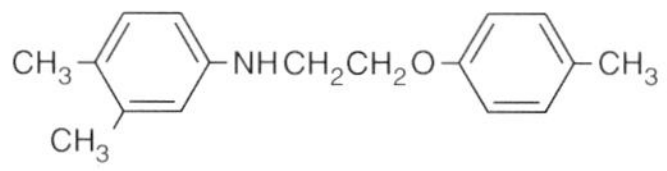
5 gm

S50,448-3
N-[2-(P-TOLYLOXY)ETHYL]-2,4-XYLIDINE
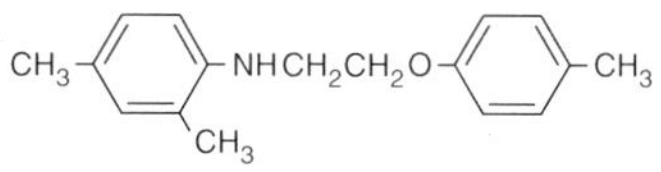
5 gm

S51,376-8
4-OCTYLOXYDIPHENYLAMINE
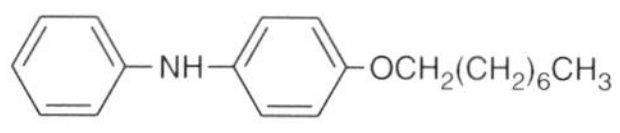
250 mg

S55,631-9
2,4-DICHLORO-N-(2-DODECYLOXYPHENYL)-BENZYLAMINE
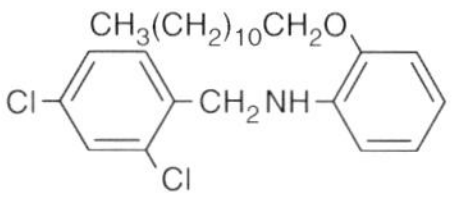
250 mg

S55,608-4
N-(5-CHLORO-2,4-DIMETHOXYPHENYL)-4-METHYLBENZYLAMINE
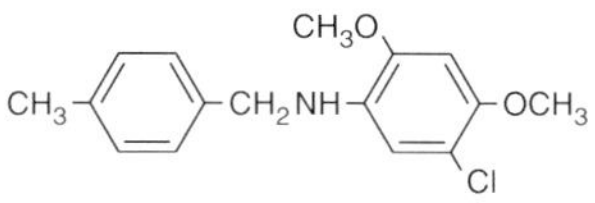
250 mg

S54,646-1
N-(2,5-DICHLOROPHENYL)-2-(HEXADECYLOXY)BENZYLAMINE
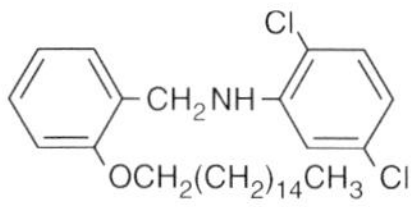
250 mg

S57,240-3
N-(4-FLUOROPHENYL)-4'-METHOXYBENZYLAMINE
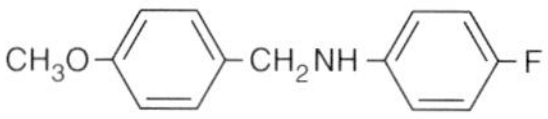
250 mg

S56,473-7
N-(4-BROMOPHENYL)-4'-METHOXYBENZYLAMINE
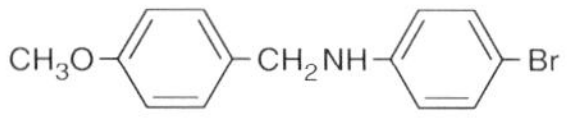
250 mg

S54,619-4
N-(4-METHOXYPHENYL)-4'-METHYLBENZYLAMINE
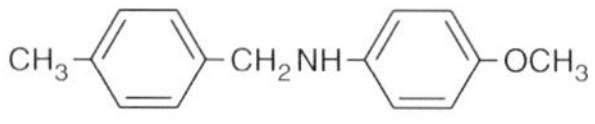
250 mg

S57,243-8
4-METHOXY-N-(4-METHOXYPHENYL)-BENZYLAMINE
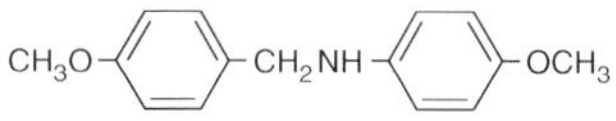
250 mg

S57,236-5
N-(P-METHOXYBENZYL)-P-PHENETIDINE
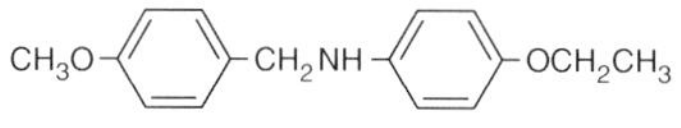
5 gm

S62,545-0
N,N'-[BIS(4-METHOXYPHENYL)]-P-XYLYLENEDIAMINE
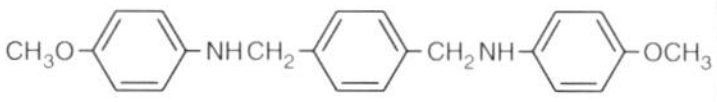
250 mg

S62,611-2
N,N'-BIS(4-ETHOXYPHENYL)-P-XYLYLENEDIAMINE
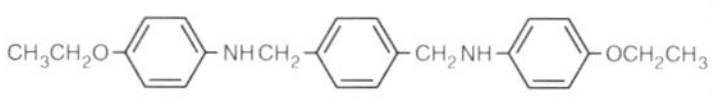
5 gm

S52,366-6
[N-(4-FLUOROPHENYL)]-2,3-DIMETHOXYBENZYLAMINE
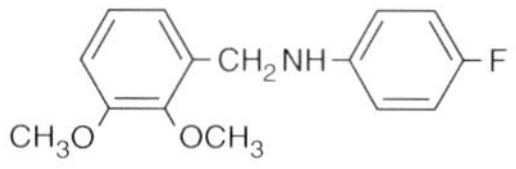
5 gm

S52,363-1
N-(4-CHLOROPHENYL)-2,3-DIMETHOXYBENZYLAMINE
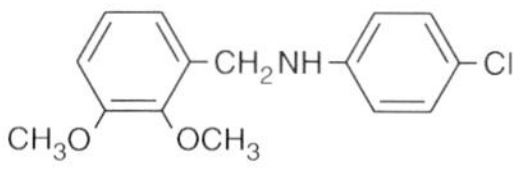
5 gm

S52,383-6
2,3-DIMETHOXY-N-(4-IODOPHENYL)-BENZYLAMINE
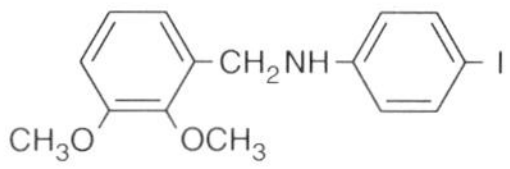
250 mg

S52,361-5
2,3-DIMETHOXY-N-(P-TOLYL)-BENZYLAMINE
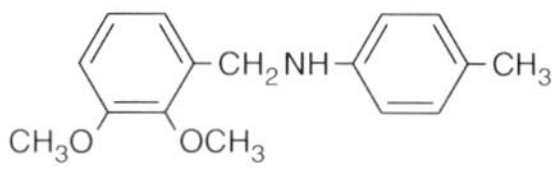
5 gm

S57,386-8
N-(2,4-XYLYL)-4-METHOXYBENZYLAMINE
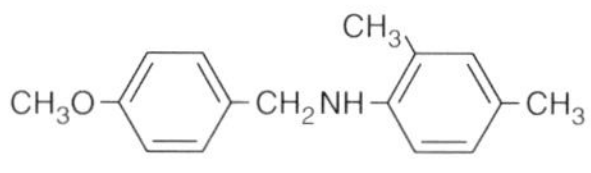
5 gm

S55,871-0
N-(3,4-DICHLOROPHENYL)-4-(HEXADECYLOXY)BENZYLAMINE
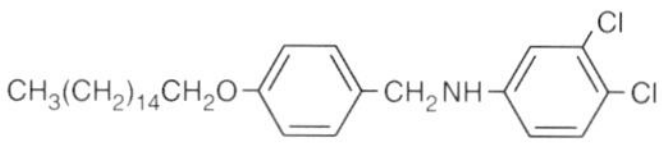
250 mg

S53,289-4
N-(5-CHLORO-2,4-DIMETHOXYPHENYL)-2,3-DIMETHOXYBENZYLAMINE
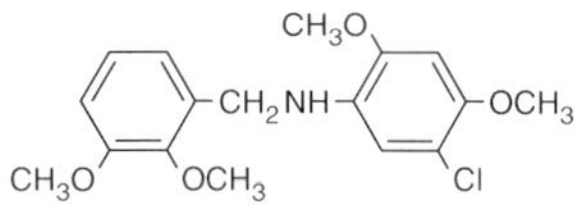
250 mg

S56,925-9
4-METHOXY-N-PHENYLBENZYLAMINE
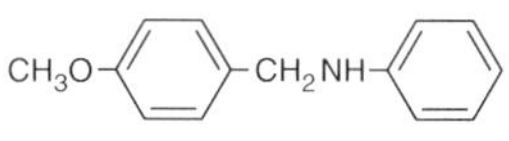
250 mg

S82,076-8
4-METHOXY-4'-METHYLDIPHENYLAMINE
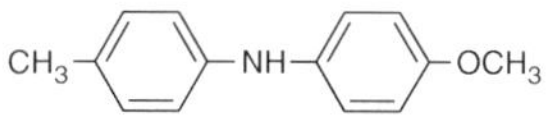
250 mg

S56,927-5
N-(4-CHLOROPHENYL)-4-METHOXYBENZYLAMINE
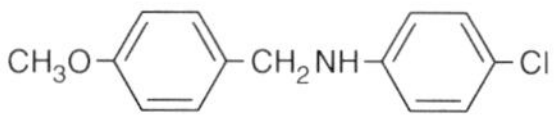
250 mg

S54,730-1
4-ISOPROPOXYDIPHENYLAMINE
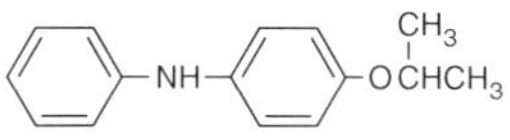
5 gm

S56,928-3
4-METHOXY-N-(P-TOLYL)BENZYLAMINE
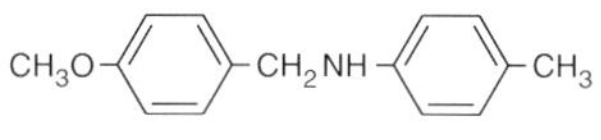
250 mg

S56,857-0
1,4-BIS(4-SEC-BUTYLAMINOPHENOXY)-BENZENE
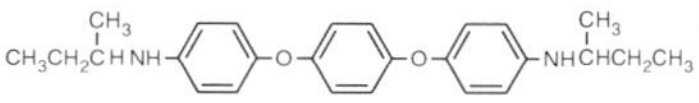
250 mg

S90,765-0
4,4'-OXYBIS(N,N-DIMETHYLANILINE)
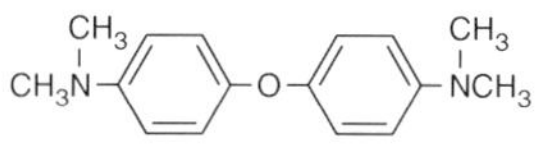
250 mg

S90,874-6
ALPHA,ALPHA-DI(2-METHYL-4-DIMETHYL-AMINOPHENYL)-4'-METHOXYTOLUENE
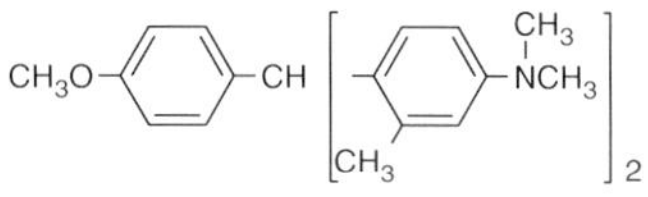
250 mg

S66,020-5
N-ETHYL-N-(2-PHENOXYETHYL)ANILINE HYDROCHLORIDE
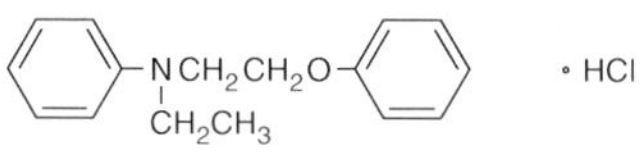
5 gm

S93,557-3
[BIS(4-METHOXYPHENYL)METHYL]-DIMETHYLAMINE
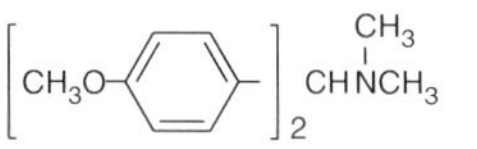
250 mg

S63,314-3
BIS(4-DIMETHYLAMINOPHENYL)METHYL-METHYL ETHER
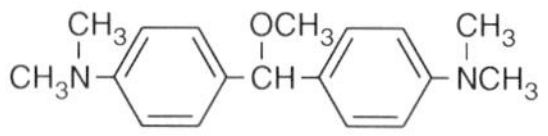
250 mg

S82,042-3
2-ANILINOTHIOPHENOL

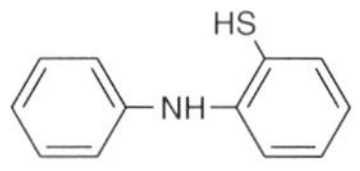

250 mg

S82,909-9
4-(4-ISOPROPYLPHENYLTHIO)-1,3-PHENYLENEDIAMINE

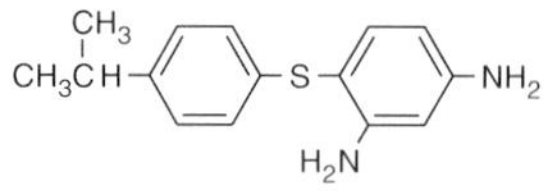

250 mg

S82,899-8
4-(M-TOLYLTHIO)-1,3-PHENYLENEDIAMINE

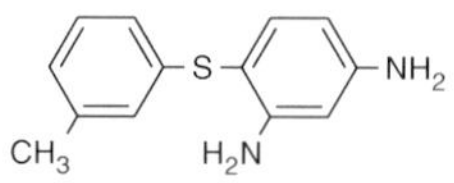

250 mg

S70,412-1
2-(BENZYLTHIO)ANILINE

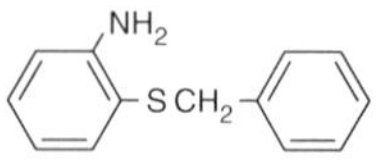

5 gm

S45,032-4
2-(P-TOLYLTHIO)ANILINE, 97%

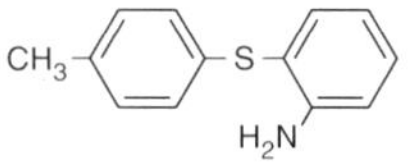

5 gm

S36,661-7
4,4'-THIODIANILINE

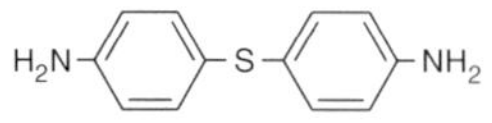

250 mg

S88,710-2
HEXAKIS(4-AMINOPHENYLTHIOMETHYL)-BENZENE

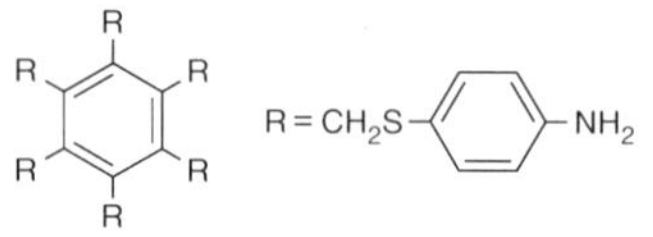

250 mg

S83,628-1
4-(PHENYLTHIO)ANILINE

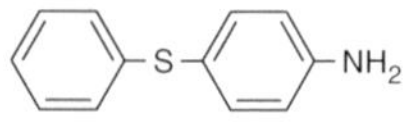

250 mg

S82,896-3
4-(M-TOLYLTHIO)ANILINE

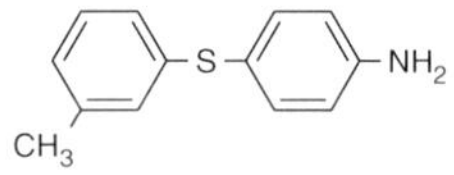

250 mg

S75,557-5
2-AMINO-7-FLUORO-3-(METHYLTHIO)-FLUORENE

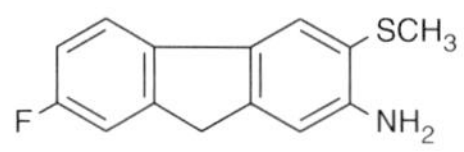

250 mg

S51,737-2
N-(4'-METHYLTHIOPHENYL)-4-DIMETHYLAMINOBENZYLAMINE

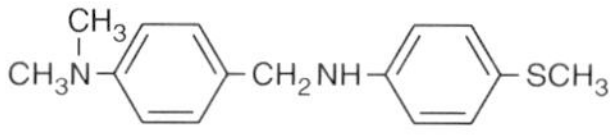

250 mg

S83,599-4
DI[4-(DIMETHYLAMINO)PHENYL]-DISULFIDE

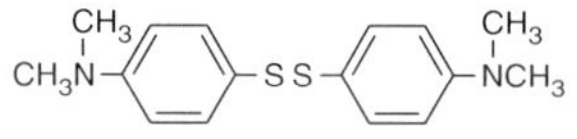

250 mg

S40,735-6
2,4-DIAMINODIPHENYLAMINE

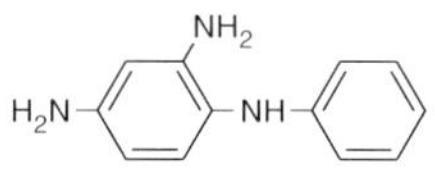

5 gm

S34,102-9
3-AMINO-2-DIMETHYLAMINOFLUORENE

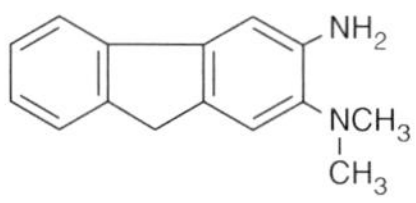

1 gm

S44,076-0
1,2-DIAMINONAPHTHALENE

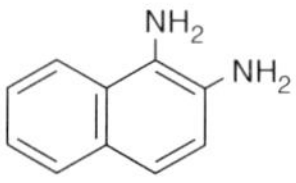

5 gm

S75,519-2
4,5-DIAMINOACENAPHTHENE

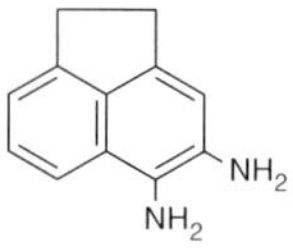

250 mg

S52,003-9
N-PHENYL-1,4-PHENYLENEDIAMINE SULFATE

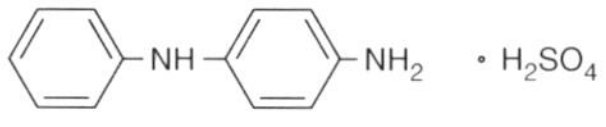

5 gm

S93,571-9
N-BENZYL-1,4-PHENYLENEDIAMINE DIHYDROCHLORIDE

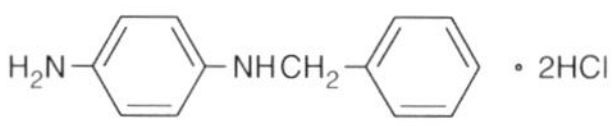

250 mg

S58,798-2
N,N,N',N'-TETRAKIS(4-AMINOPHENYL)-BENZIDINE

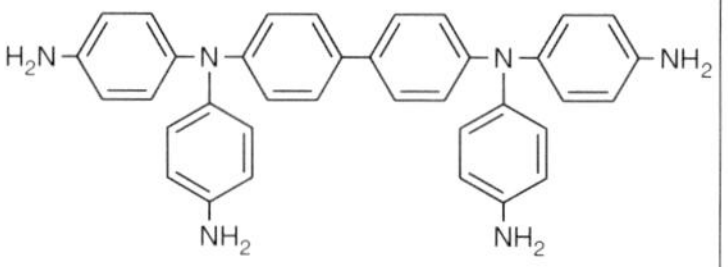

250 mg

S77,445-6
2,2'-DIAMINOBIPHENYL

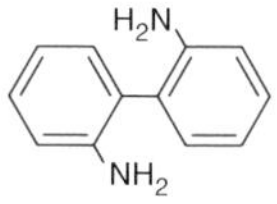

250 mg

S94,695-8
2,2'-DIAMINO-6,6'-DIMETHYLBIPHENYL

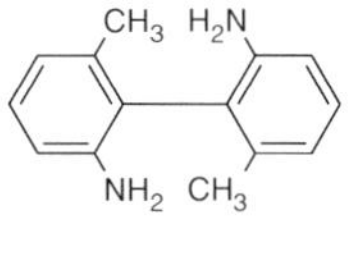

250 mg

S92,371-0
2,2'-DI(4-METHYLANILINO)-PHENYLMETHANE

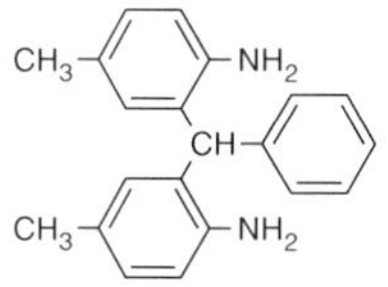

250 mg

S91,997-7
2,3',4,5',6-PENTAPHENYL-3,4'-BIPHENYLDIAMINE

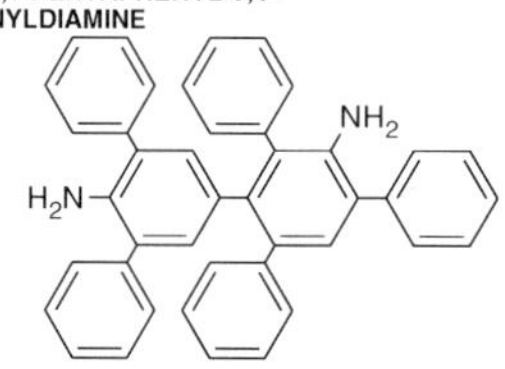

250 mg

S88,004-3
1,6-DI(4-AMINOPHENYL)HEXANE

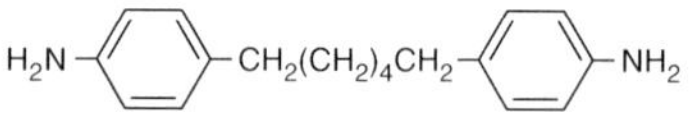

250 mg

S88,312-3
2,2'-DIMETHYLBENZIDINE HYDROCHLORIDE

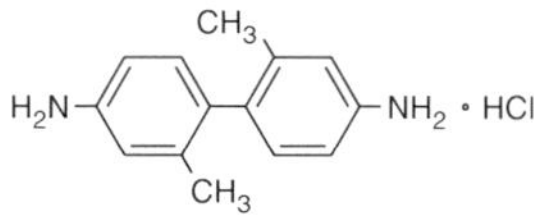

250 mg

S92,195-5
2,3,4,6-TETRAPHENYLANILINE

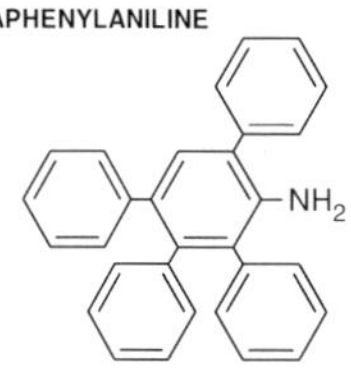

250 mg

S88,025-6
2-AMINOSTILBENE

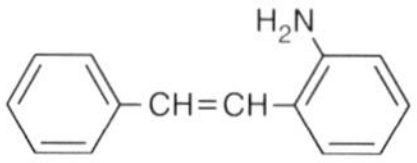

250 mg

S92,110-6
3,5-DIPHENYL-4'-METHYL-2-BIPHENYLAMINE

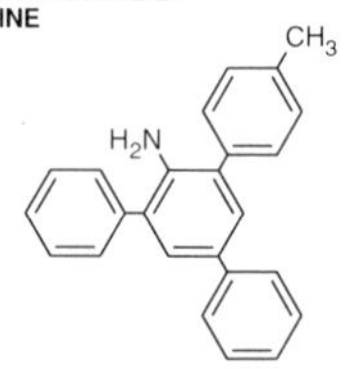

250 mg

S92,675-2
4-(P-TOLYL)-2,3,6-TRIPHENYLANILINE

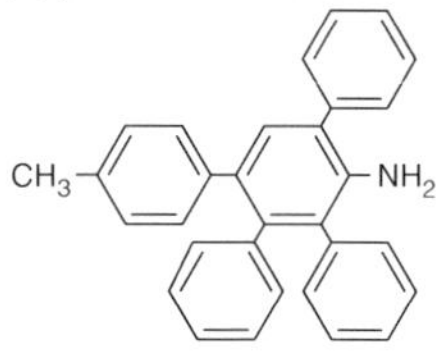

250 mg

S91,983-7
2,4-DIPHENYL-6-METHYLANILINE

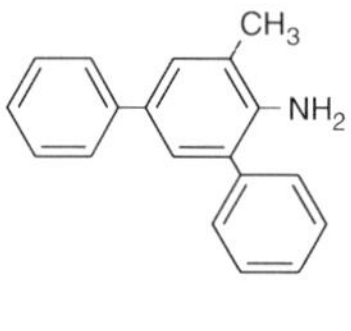

250 mg

S92,186-6
2,4,6-TRIS(4-BIPHENYLYL)ANILINE

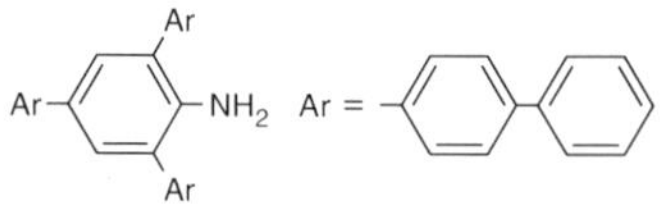

250 mg

S76,304-7
5-AMINO-10,11-DIHYDRO-10-PHENYLDIBENZO(A,D)CYCLOHEPTENE

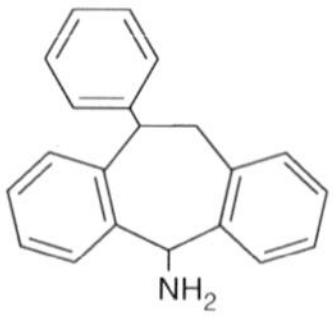

250 mg

S43,970-3
O-TOLIDINE

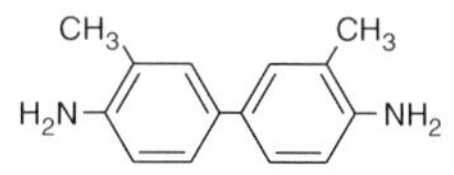

5 gm

S93,600-6
1-AMINOBENZO(C)PHENANTHRENE

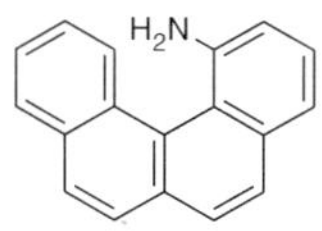

10 mg

S57,125-3
3-AMINOACENAPHTHENE HYDROCHLORIDE

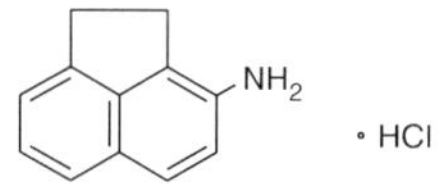

5 gm

S64,002-6
AMINOCHLOROPHENANTHRENE (MIXED ISOMERS)

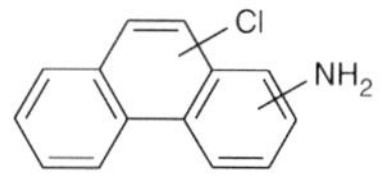

5 gm

S86,004-2
2,4-DIBROMO-1-NAPHTHYLAMINE

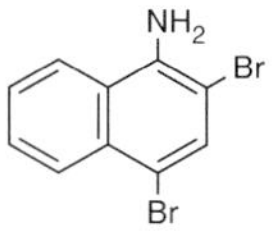

250 mg

S77,915-6
1-AMINO-2-METHYLNAPHTHALENE HYDROCHLORIDE

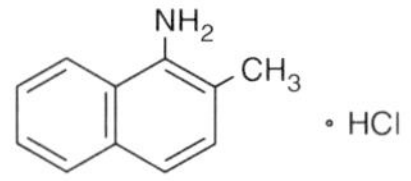

250 mg

S81,060-6
2-AMINONAPHTHALENE HYDROBROMIDE

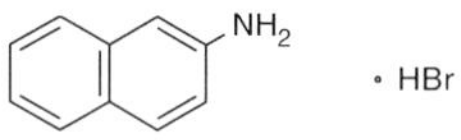

250 mg

S97,506-0
AR-H11 4-AMINOPYRENE

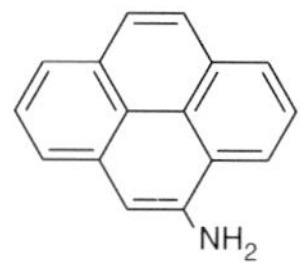

10 mg

S97,508-7
AR-H10 2-AMINOPYRENE

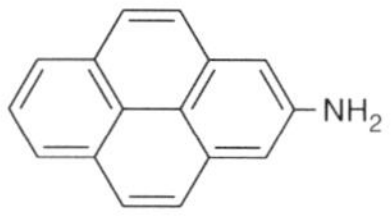

10 mg

S97,457-9
AR-G12 3-AMINOBENZO(C)PHENANTHRENE

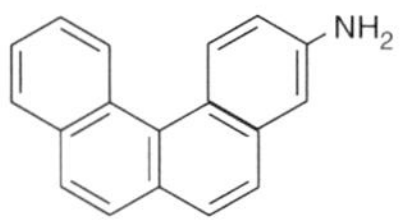

10 mg

S97,459-5
AR-G14 2-AMINOBENZO(C)PHENANTHRENE HYDROCHLORIDE

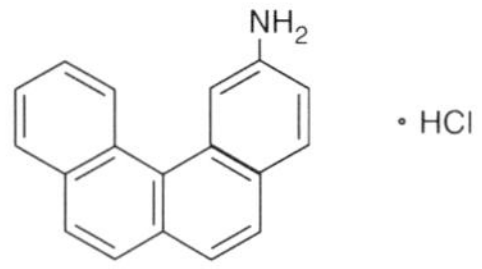

10 mg

S97,460-9
AR-G14A 2-AMINOBENZO(C)PHENANTHRENE

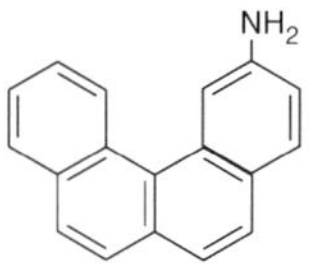

10 mg

S97,461-7
AR-G15 3-AMINOBENZO(C)PHENANTHRENE HYDROCHLORIDE

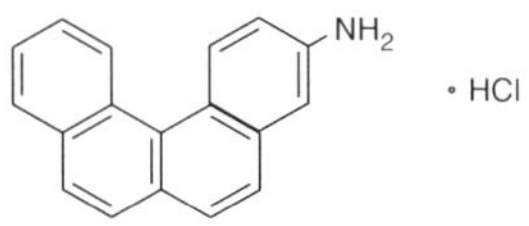

10 mg

S97,463-3
AR-G17 6-AMINOBENZO(C)PHENANTHRENE HYDROCHLORIDE

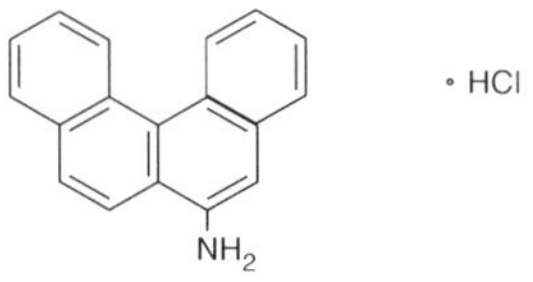

10 mg

S52,080-2
1-BROMO-2-NAPHTHYLAMINE

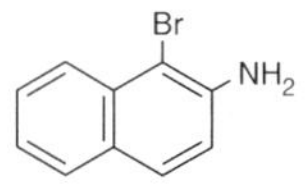

250 mg

S77,084-1
7-AMINO-1-FLUOROFLUORENE

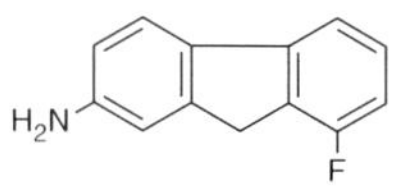

250 mg

S75,553-2
2-AMINO-3-CHLOROFLUORENE

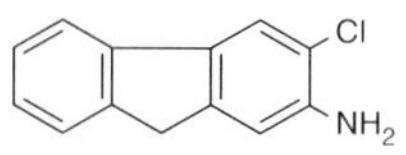

250 mg

S74,952-4
2-AMINO-3-BROMOFLUORENE

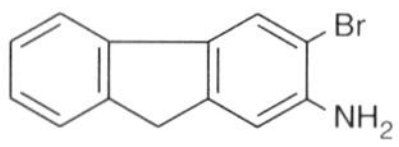

250 mg

S74,928-1
4-AMINO-2,5,7-TRICHLOROFLUORENE

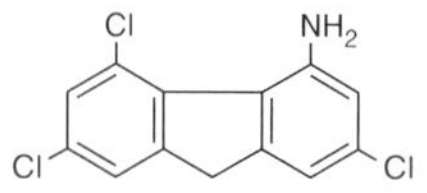

250 mg

S75,491-9
2-AMINO-1-BROMO-3-CHLOROFLUORENE

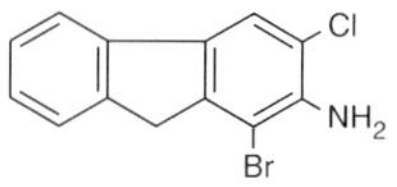

250 mg

S75,488-9
2-AMINO-1-BROMO-3-CHLORO-7-FLUOROFLUORENE

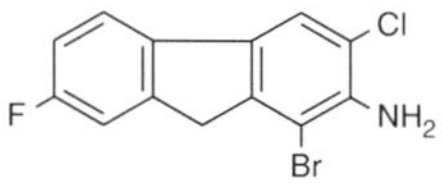

250 mg

S77,423-5
2-AMINO-1,3,4,7-TETRACHLOROFLUORENE

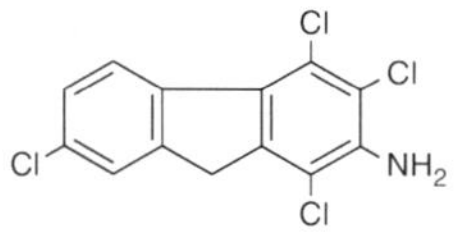

250 mg

S74,905-2
2-AMINO-7-CHLORO-1,3-DIBROMOFLUORENE

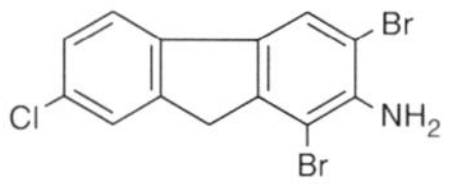

250 mg

S74,916-8
2-AMINO-1,3,7-TRIBROMOFLUORENE

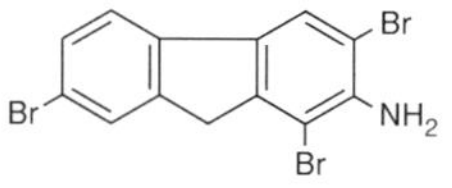

250 mg

S80,444-4
1-AMINO-9,10-DIHYDROANTHRACENE

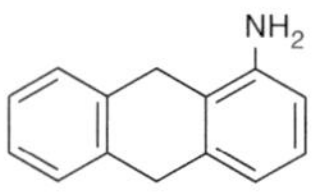

250 mg

S37,507-1
3-AMINOPHENANTHRENE HYDROCHLORIDE

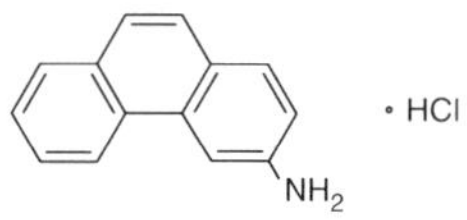

250 mg

S77,420-0
2-AMINOPHENANTHRENE

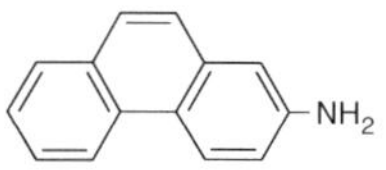

250 mg

S72,253-7
5,6,7,8-TETRAHYDRO-2-PHENANTHRENEAMINE HYDROCHLORIDE

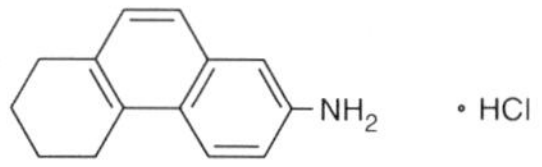

5 gm

S64,015-8
AMINOBROMOPHENANTHRENE (MIXED ISOMERS)

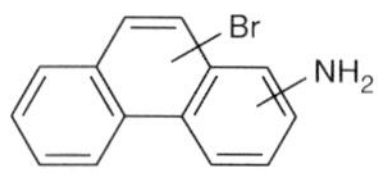

5 gm

S84,265-6
9-AMINO-10-BROMOPHENANTHRENE

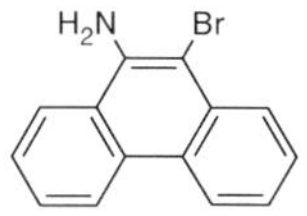

250 mg

S75,432-3
12-AMINO-6-FLUOROCHRYSENE

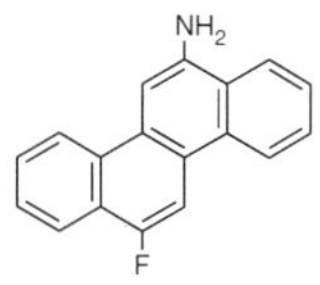

250 mg

S46,025-7
5-AMINOACENAPHTHENE

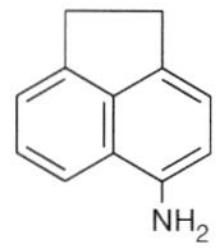

250 mg

S80,555-6
2-AMINO-4'-(DIMETHYLAMINO)-5-ETHYLDIPHENYLAMINE
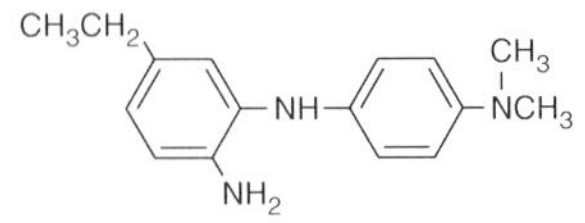
250 mg

S81,609-4
N,N'-DIMETHYL-1,2,N,N'-TETRAPHENYL-1,2-ETHYLENEDIAMINE
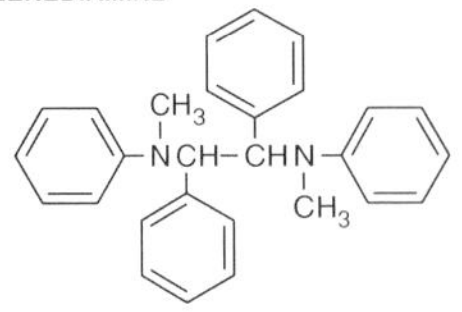
250 mg

S50,557-9
N,N'-DIETHYL-N,N'-DIPHENYLETHYLENEDIAMINE
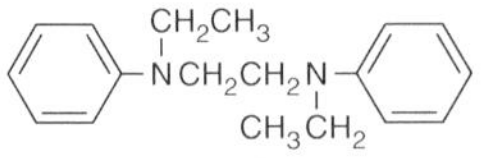
250 mg

S36,295-6
N-ISOPROPYL-N'-PHENYL-1,4-PHENYLENEDIAMINE
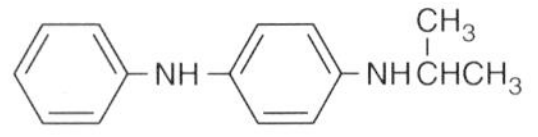
5 gm

S76,099-4
N,N'-BIS(1-METHYLHEPTYL)-1,5-NAPHTHALENEDIAMINE
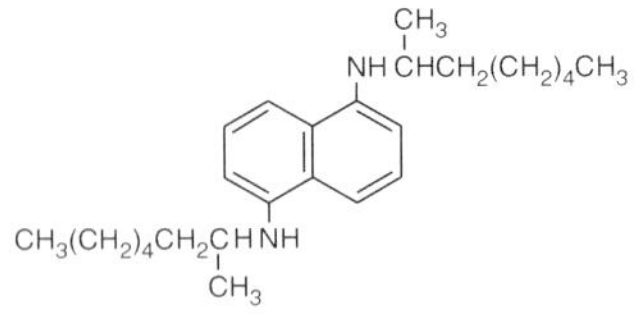
250 mg

S34,179-7
N-BENZYL-4-DIAZO-N-ETHYLANILINE, TIN CHLORIDE DOUBLE SALT
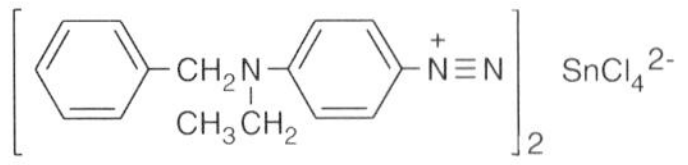
5 gm

S51,417-9
N-PHENYL-4-DIMETHYLAMINOBENZYLAMINE
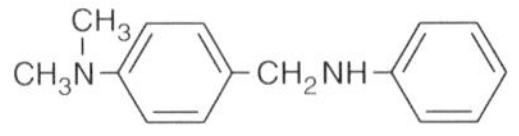
5 gm

S51,424-1
N-(O-TOLYL)-4-DIMETHYLAMINOBENZYLAMINE
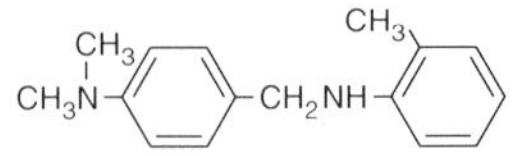
250 mg

S51,293-1
N-(P-TOLYL)-4-DIMETHYLAMINOBENZYLAMINE
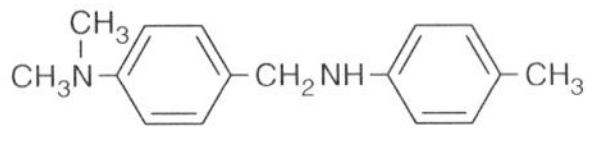
5 gm

S87,866-9
N-(4-ETHYLPHENYL)-4'-DIMETHYLAMINOBENZYLAMINE
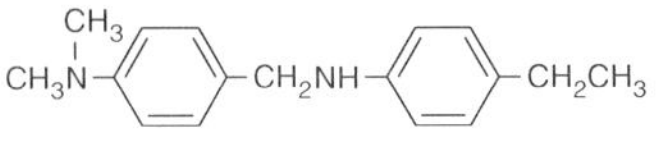
5 gm

S51,731-3
N-(2,4-XYLYL)-4'-DIMETHYLAMINOBENZYLAMINE
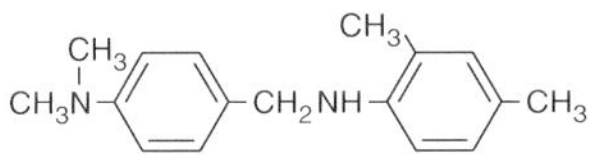
250 mg

S51,735-6
N-(3,4-XYLYL)-4'-DIMETHYLAMINOBENZYLAMINE
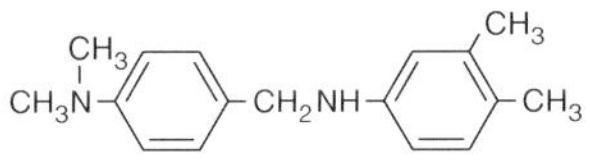
250 mg

S91,491-6
N-(1-PHENYLETHYL)ANILINE
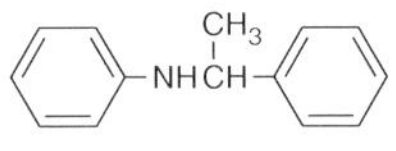
250 mg

S82,522-0
N-BENZHYDRYLANILINE
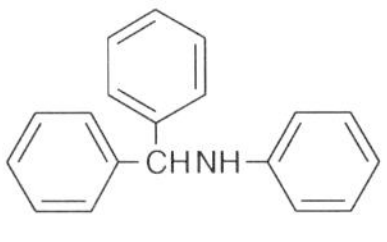
250 mg

S81,989-1
N-TRITYLANILINE
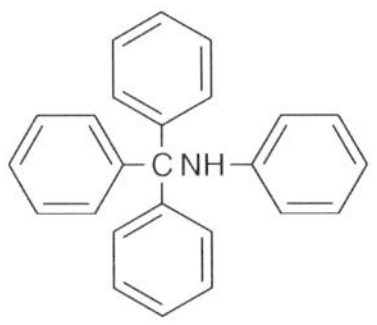
250 mg

S84,107-2
N-(ALPHA-ALLYLBENZYL)ANILINE
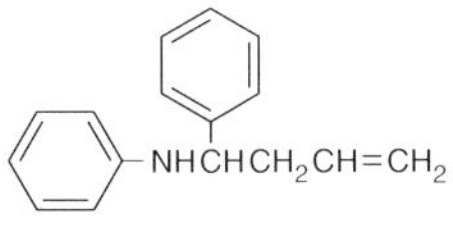
250 mg

S81,061-4
N-(2-NAPHTHYL)TRITYLAMINE
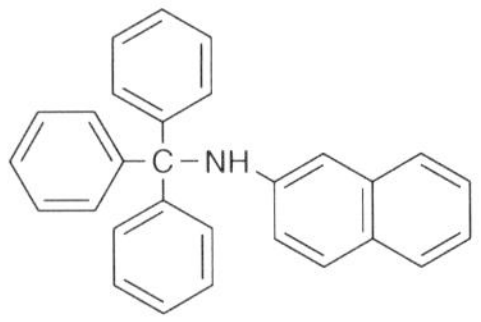
250 mg

S54,842-1
4-METHYL-N-PHENYLBENZYLAMINE
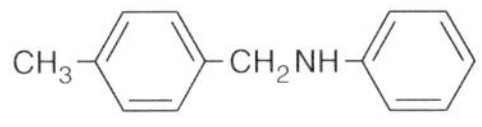
250 mg

S54,802-2
N-(4-FLUOROPHENYL)-4'-METHYLBENZYLAMINE
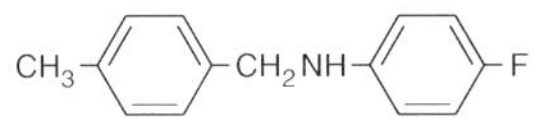
250 mg

S54,614-3
N-(4-CHLOROPHENYL)-4'-METHYLBENZYLAMINE
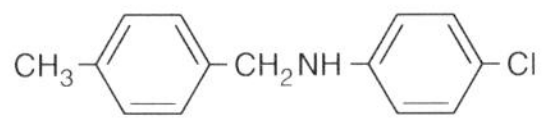
250 mg

S54,615-1
N-(4-BROMOPHENYL)-4-METHYLBENZYLAMINE
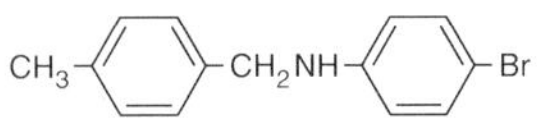
5 gm

S54,839-1
N-(4-IODOPHENYL)-4-METHYLBENZYLAMINE
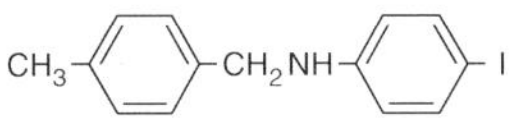
250 mg

S54,623-2
N-(P-TOLYL)-4-METHYLBENZYLAMINE
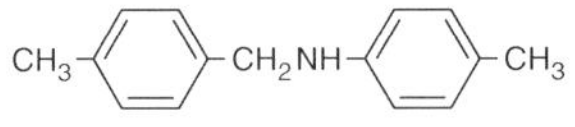
5 gm

S62,118-8
ALPHA,ALPHA'-BIS(4-FLUOROANILINO)-P-XYLENE
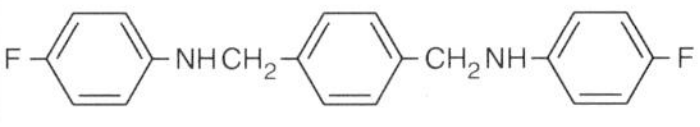
5 gm

S62,108-0
ALPHA,ALPHA'-BIS(4-CHLOROANILINO)-P-XYLENE
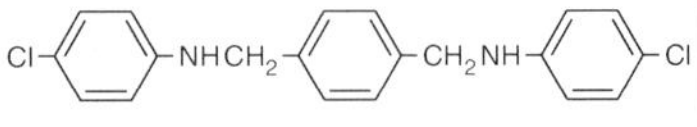
5 gm

S62,625-2
ALPHA,ALPHA'-BIS(4-METHYLANILINO)-P-XYLENE
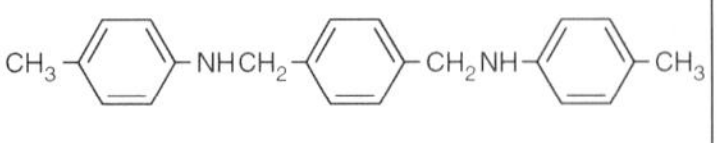
250 mg

S54,878-2
N-(3-CHLORO-P-TOLYL)-4-METHYLBENZYLAMINE
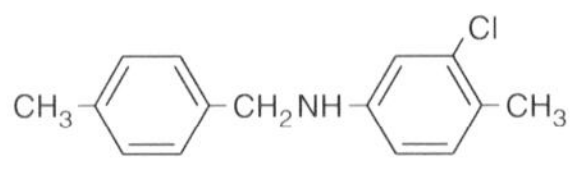
5 gm

S43,116-8
N-METHYL-1-PHENYLCYCLOHEXYLAMINE
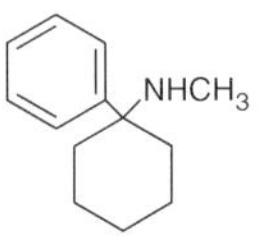
250 mg

S81,185-8
N-(2-NAPHTHYL)-2-PHENYLBENZHYDRYLAMINE
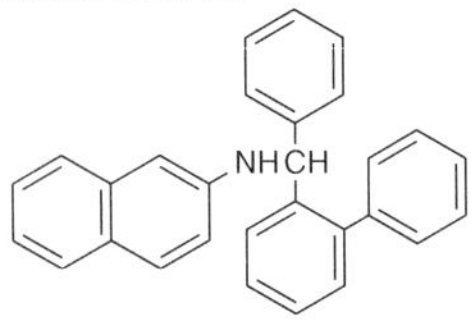
250 mg

S61,754-7
ALPHA,ALPHA'-BIS(4-BROMOANILINO)-P-XYLENE
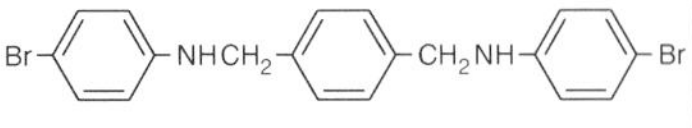
5 gm

S61,949-3
ALPHA,ALPHA'-BIS(4-IODOANILINO)-P-XYLENE
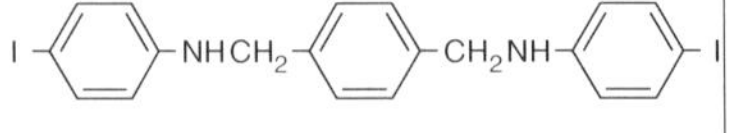
5 gm

S61,956-6
ALPHA,ALPHA-BIS(4-ETHYLANILINO)-P-XYLENE
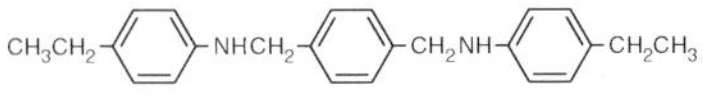
250 mg

S50,544-7
N-(1-NAPHTHYL)HEXADECYLAMINE

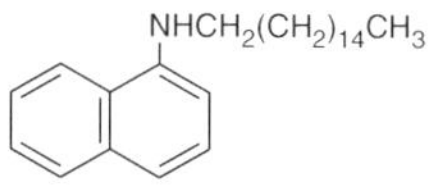

250 mg

S64,873-6
N-ETHYL-2-NAPHTHYLAMINE

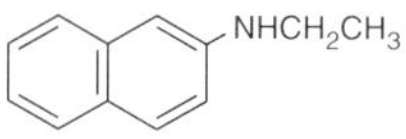

5 gm

S57,202-0
N-(3-CHLOROPHENYL)-2-NAPHTHYLAMINE

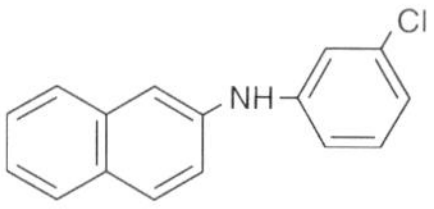

5 gm

S44,723-4
2-METHYLAMINOFLUORENE HYDROCHLORIDE

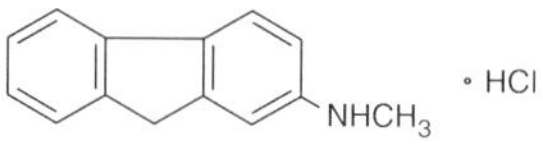

250 mg

S82,808-4
N-ETHYLDIPHENYLAMINE

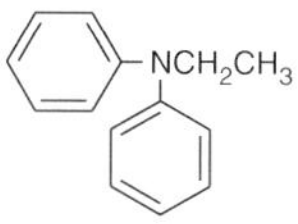

250 mg

S82,816-5
N-DODECYLDIPHENYLAMINE

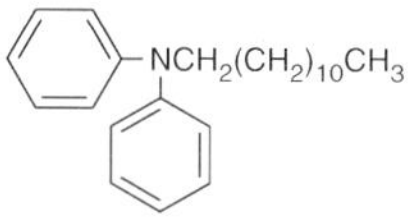

250 mg

S90,776-6
4,4'-(PHENYLMETHYLENE)BIS(N,N,3-TRIMETHYLANILINE)

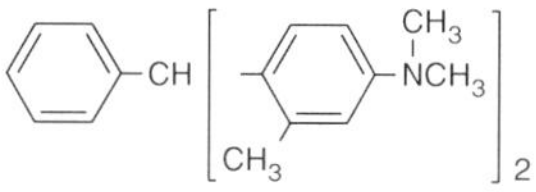

250 mg

S90,775-8
4,4'-[(O-TOLYL)METHYLENE]BIS(N,N-DIMETHYLANILINE)

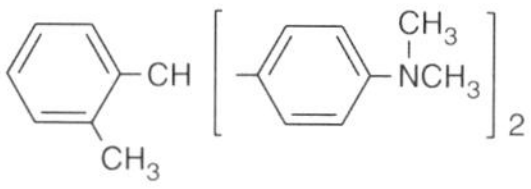

250 mg

S91,976-4
4,4'-(CHLOROMETHYLENE)BIS(N,N-DIMETHYLANILINE)PERCHLORATE

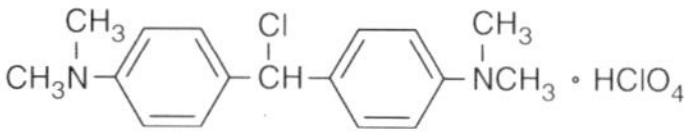

250 mg

S90,777-4
4,4'-[(4-TERT-BUTYLPHENYL)-METHYLENE]BIS(N,N-DIMETHYLANILINE)

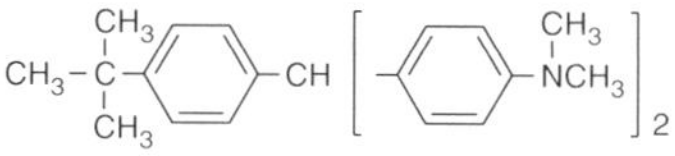

250 mg

S90,740-5
2,6-DIMETHYL-4,4',4''-METHYLIDYNETRIS(N,N-DIMETHYLANILINE)

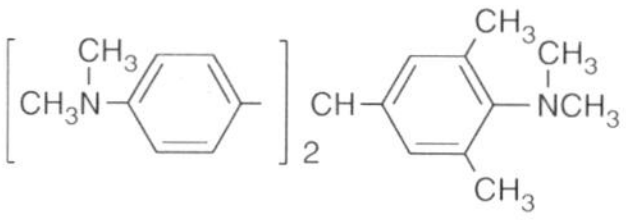

250 mg

S90,741-3
TRIS[4-(DIETHYLAMINO)-2-METHYLPHENYL]CARBENIUM CHLORIDE

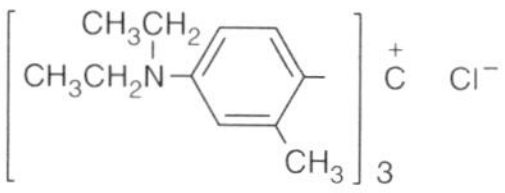

250 mg

S90,746-4
4,4'-METHYLENEBIS(N,N,3,5-TETRAMETHYLANILINE)

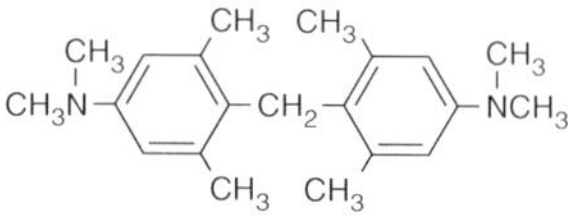

250 mg

S90,747-2
4,4'-METHYLENEBIS(N,N,2,6-TETRAMETHYLANILINE)

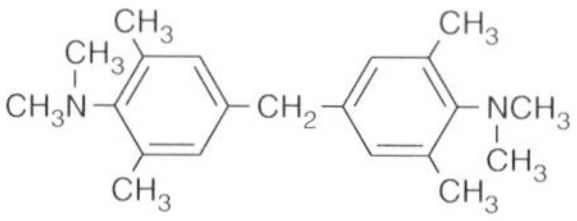

250 mg

S41,670-3
N-BENZYL-N-ETHYL-M-TOLUIDINE

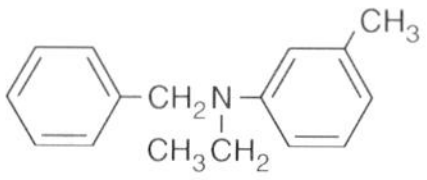

5 gm

S81,779-1
N,N-DIPHENYLBENZYLAMINE

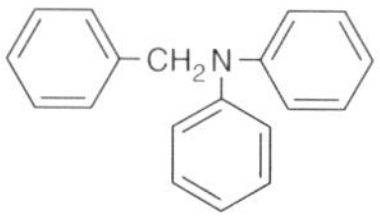

250 mg

S42,707-1
N,N-DIBENZYLANILINE

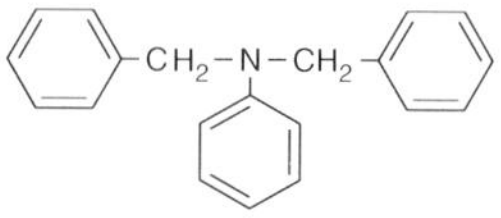

5 gm

S44,696-3
N,N-DIMETHYL-1-PHENYLCYCLOHEXYLAMINE

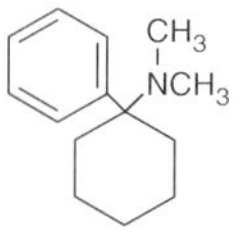

5 gm

S55,857-5
4,4'-CYCLOHEXYLIDENEBIS(N,N-DIMETHYLANILINE)

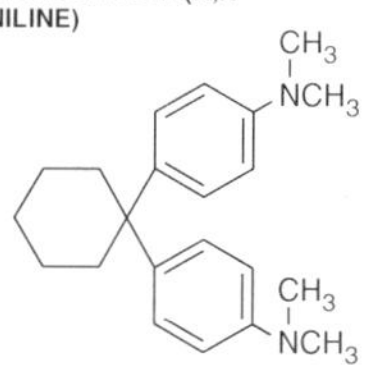

250 mg

S89,387-0
3,3-BIS[4-(DIMETHYLAMINO)PHENYL]-1,1,2-TRICHLORO-1-PROPENE

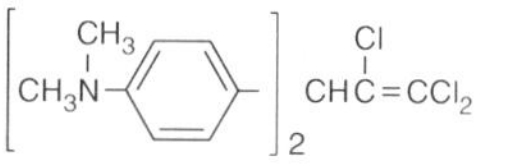

250 mg

S52,175-2
1-(N-METHYL-P-TOLUIDINO)-2-NAPHTHOL

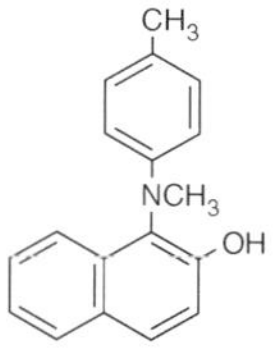

5 gm

S34,660-8
9-[4-(DIMETHYLAMINO)BENZYLIDENE]-FLUORENE

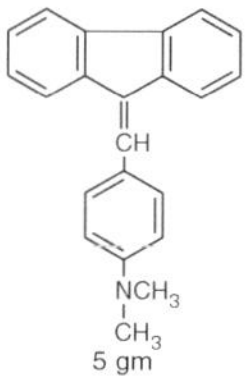

5 gm

S83,391-6
DIETHYL(2-NAPHTHYL)AMINE

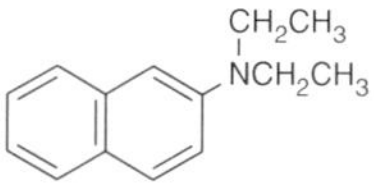

250 mg

S75,617-2
9-(DIMETHYLAMINO)FLUORENE

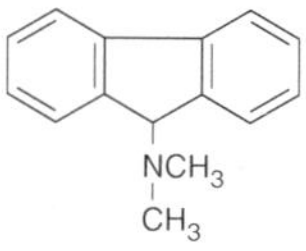

5 gm

S34,252-1
9-BROMO-2-DIMETHYLAMINOFLUORENE HYDROBROMIDE

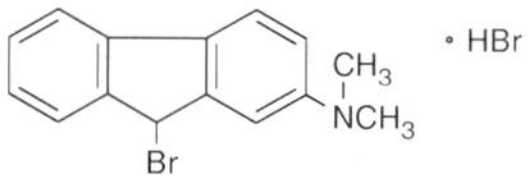

5 gm

S62,729-1
2-DIETHYLAMINOFLUORENE

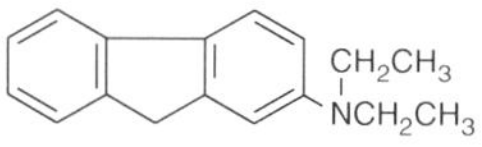

250 mg

S74,592-8
2-DIPROPYLAMINOFLUORENE

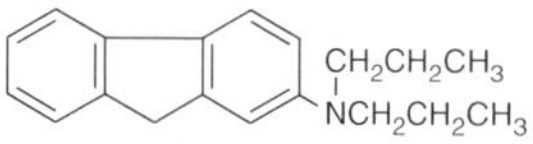

250 mg

S75,560-5
2-[N,N-BIS(2-CHLOROPROPYL)AMINO]-FLUORENE

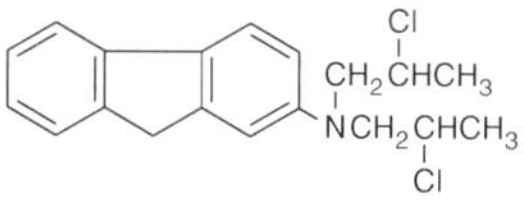

250 mg

S75,531-1
2-[N,N-BIS(2-CHLOROPROPYL)AMINO]-7-FLUOROFLUORENE

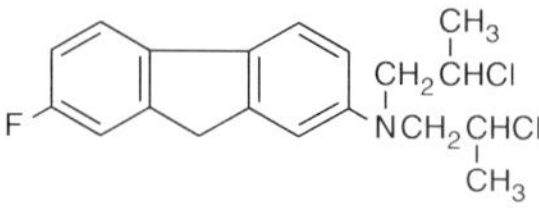

250 mg

S74,940-0
2-[N,N-BIS(2-BROMOPROPYL)AMINO]-FLUORENE

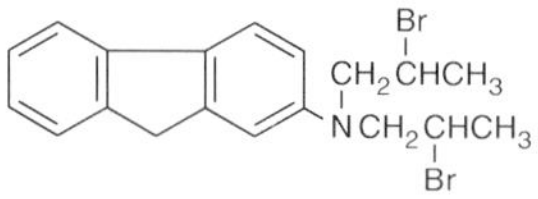

250 mg

S90,748-0
2,7-BIS(DIMETHYLAMINO)-9,10-DIHYDRO-9,9-DIMETHYLANTHRACENE

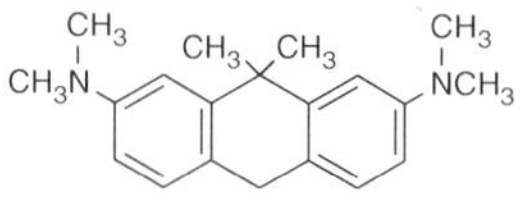

250 mg

S90,897-5
5,5-DIMETHYL-3,7,11-TRIS-(DIMETHYLAMINO)-5H-NAPHTH(3,2,1-DE)-ANTHRACENE

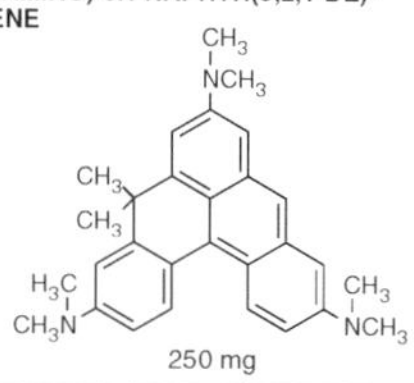

250 mg

S90,731-6
9-BROMO-3-(DIMETHYLAMINO)-PHENANTHRENE
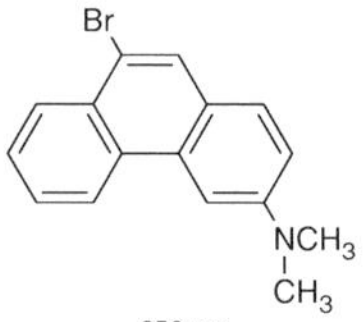
250 mg

S90,738-3
9,13B-DIHYDRO-TETRAMETHYL-TRIS-(DIMETHYLAMINO)-5H-NAPHTH(3,2,1-DE)-ANTHRACEN-13B-YLIUM PERCHLORATE
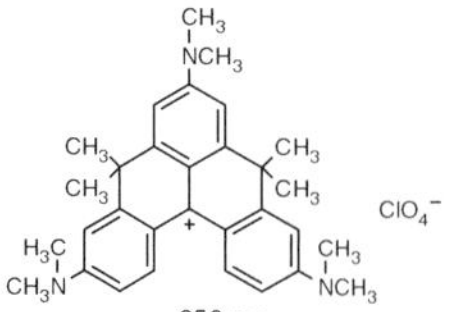
250 mg

S93,551-4
TRIBENZYLAMINE N-OXIDE
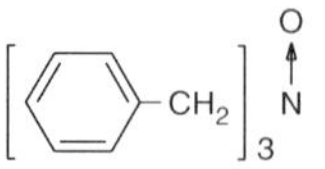
250 mg

S77,077-9
4-FLUORO-4'-(TRIFLUOROMETHYL)-BENZOPHENONE 2-METHOXYPHENYL-HYDRAZONE HYDROCHLORIDE
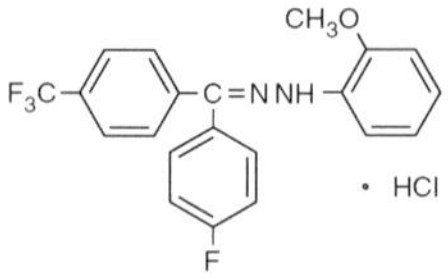
250 mg

S64,909-0
1,2-BIS(3,5-DITRIFLUOROMETHYLPHENYL)HYDRAZINE
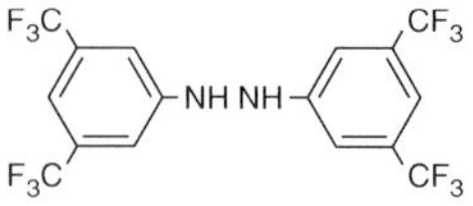
5 gm

S83,532-3
GLYOXAL BIS(2,4,6-TRICHLOROPHENYLHYDRAZONE)
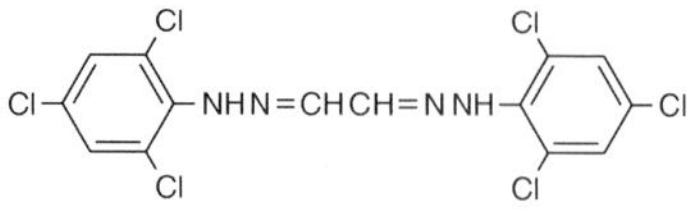
250 mg

S45,509-1
4,4'-DIHYDRAZINOOCTAFLUOROBIPHENYL
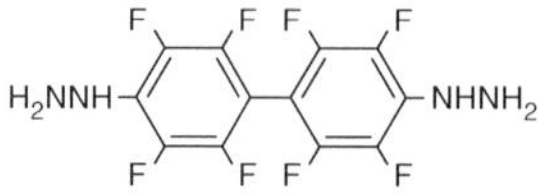
250 mg

S76,154-0
1-(2,6-DICHLOROBENZYL)HYDRAZINE
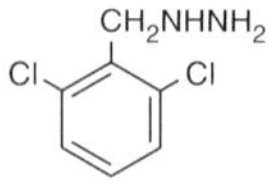
250 mg

S88,615-7
2-CHLOROBENZALDEHYDE PHENYLHYDRAZONE
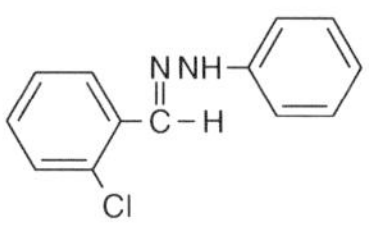
250 mg

S88,614-9
BENZALDEHYDE (4-BROMOPHENYL)-HYDRAZONE
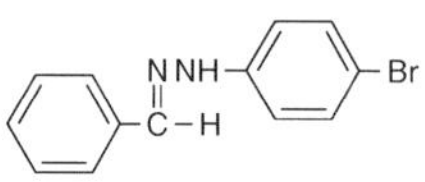
250 mg

S88,612-2
2,4-DICHLOROBENZALDEHYDE PHENYLHYDRAZONE
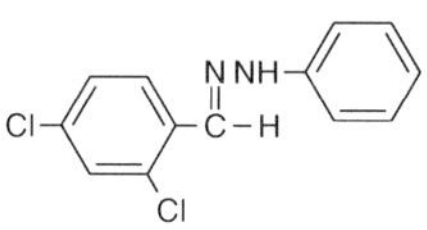
250 mg

S45,536-9
2,3-DICHLOROPHENYLHYDRAZINE HYDROCHLORIDE
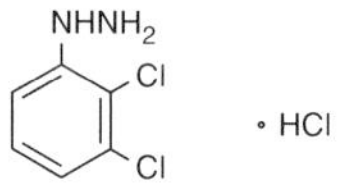
5 gm

S84,894-8
4-BROMOBENZALDEHYDE PHENYLHYDRAZONE
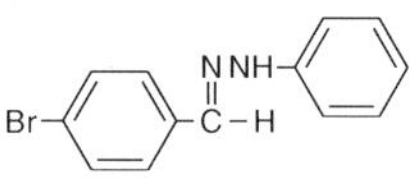
250 mg

S88,873-7
4-CHLOROBENZALDEHYDE HYDRAZONE
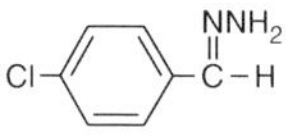
250 mg

S77,165-1
2,6-DICHLOROBENZALDEHYDE (4-BROMOPHENYL)HYDRAZONE
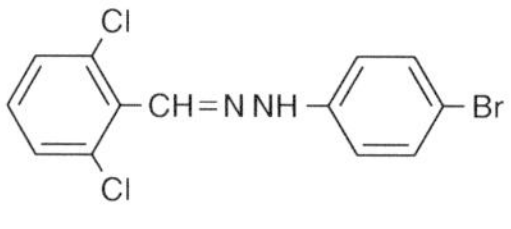
250 mg

S89,250-5
2',4',5'-TRIHYDROXYBUTYROPHENONE PHENYLHYDRAZONE
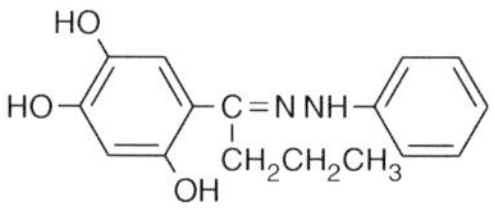
250 mg

S67,744-2
D-MANNO-L-MANNO-OCTOSE PHENYLOSAZONE
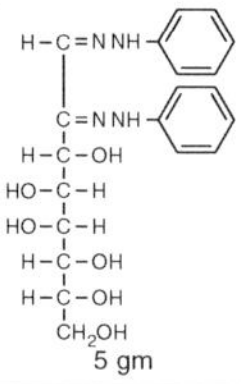
5 gm

S89,441-9
1,2-BIS(PHENYLAZO)-1,2-ETHANEDIOL
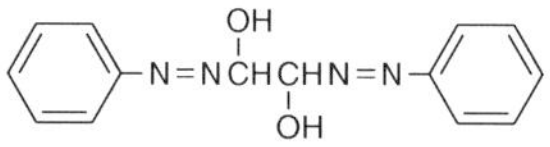
250 mg

S67,288-2
D-MANNO-L-MANNOOCTOSE PHENYLHYDRAZONE
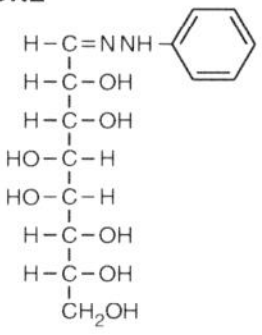
250 mg

S51,015-7
SALICYLALDEHYDE PHENYLHYDRAZONE
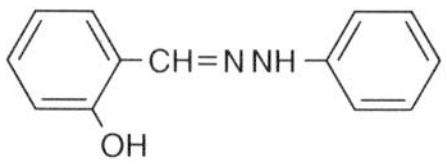
250 mg

S95,647-3
2-HYDROXY-5-METHYL-5,6,7,8,-TETRAHYDRO-1-NAPHTHALENECARBOXALDEHYDE PHENYLHYDRAZONE
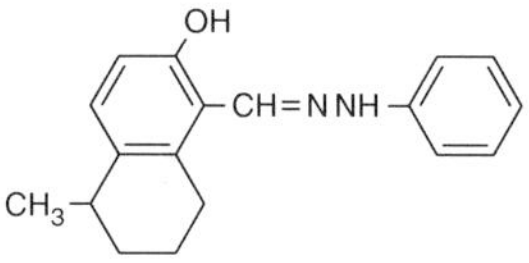
50 mg

S84,724-0
3,4-DIMETHOXYBENZALDEHYDE BENZYLPHENYLHYDRAZONE
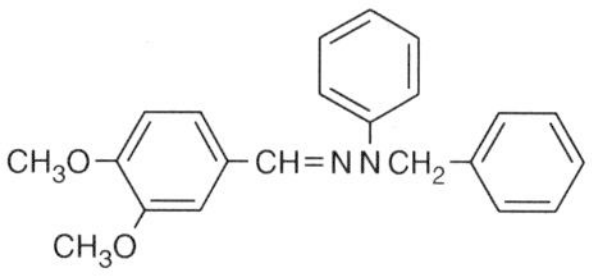
250 mg

S88,625-4
3,4-DIMETHOXYBENZALDEHYDE PHENYLHYDRAZONE
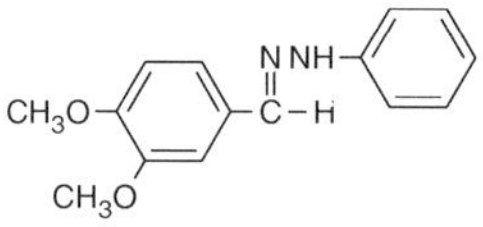
250 mg

S90,085-0
4-ETHOXY-3-METHOXYBENZALDEHYDE AZINE
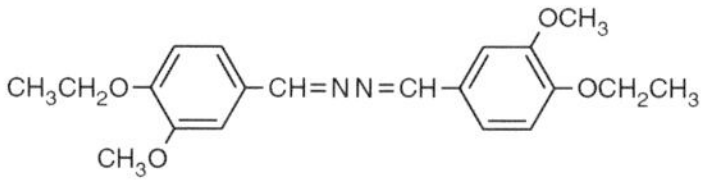
250 mg

S86,220-7
2,4-DIMETHOXYBENZALDEHYDE AZINE
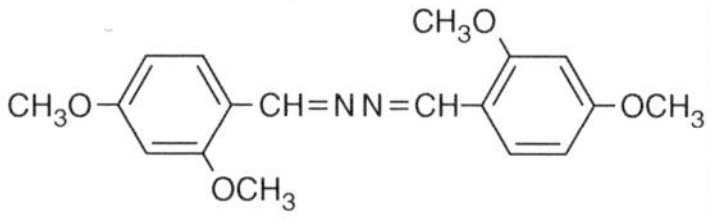
250 mg

S86,221-5
2,4,6-TRIMETHOXYBENZALDEHYDE AZINE
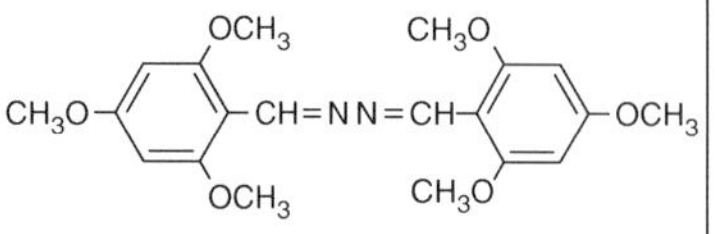
250 mg

S95,431-4
4,4'-DIMETHOXYBENZIL DIHYDRAZONE
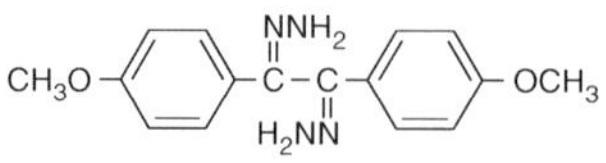
50 mg

S86,218-5
2-METHOXYBENZALDEHYDE AZINE
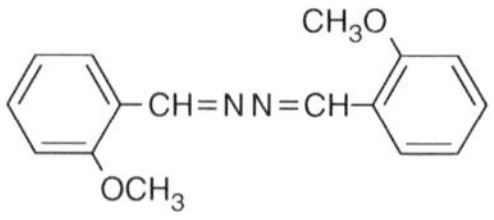
250 mg

S61,838-1
4-HEXADECYLOXY-3-METHOXYBENZALDEHYDE PHENYLHYDRAZONE
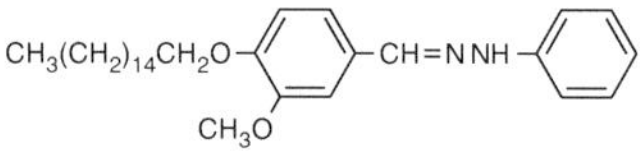
250 mg

S84,725-9
4-BUTOXYBENZALDEHYDE BENZYLPHENYLHYDRAZONE
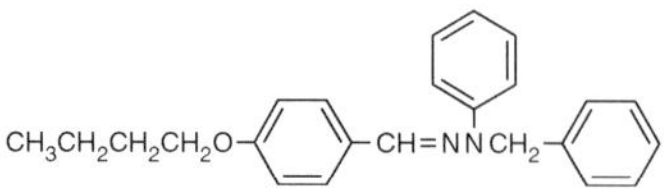
250 mg

S60,615-4
4-METHOXYBENZALDEHYDE PHENYLHYDRAZONE
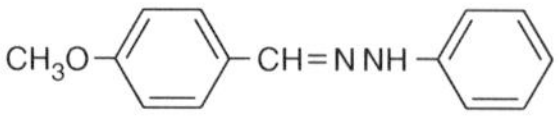
5 gm

S39,055-0
4-NITROBENZALDEHYDE 2-[5-CHLORO-2-(HEXADECYLOXY)PHENYL]HYDRAZONE
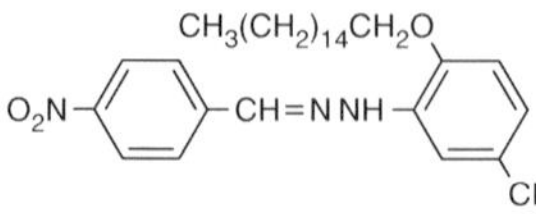
250 mg

S86,013-1
6-METHOXY-1-TETRALONE (P-TOLYL)-HYDRAZONE

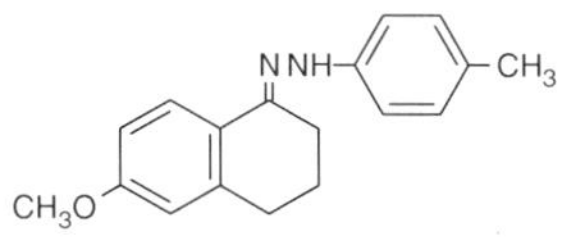

250 mg

S65,824-3
3-BROMO-4-METHOXYBENZALDEHYDE PHENYLHYDRAZONE

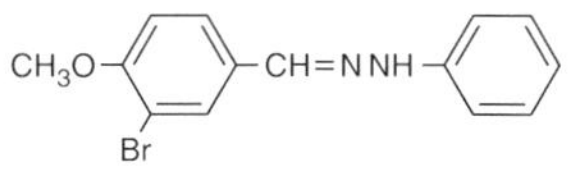

5 gm

S45,017-0
3-(HEXADECYLTHIO)PHENYLHYDRAZINE HYDROCHLORIDE

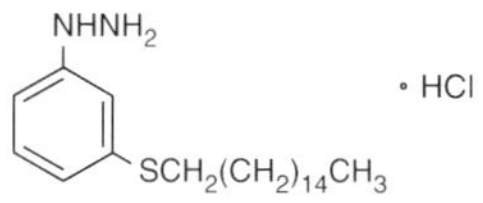

250 mg

S69,633-1
[2-(OCTADECYLTHIO)PHENYL]-HYDRAZINE MONOHYDROCHLORIDE

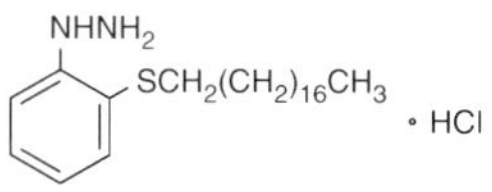

250 mg

S41,314-3
NITRIN

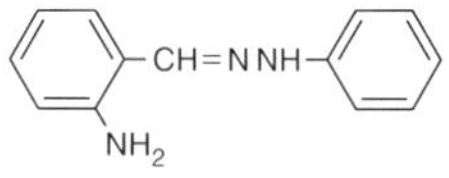

5 gm

S89,624-1
2,3-BIS[2-(TRIPHENYLMETHYL)-HYDRAZINO]NAPHTHALENE

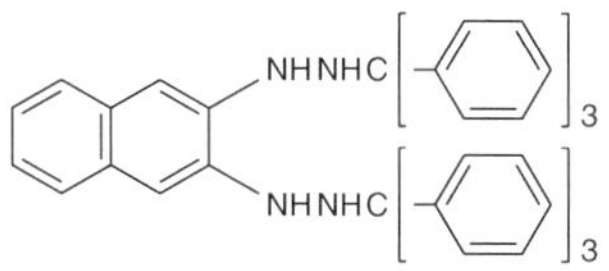

250 mg

S88,704-8
3,5-CYCLOHEXADIENE-1,2-DIONE-BIS(PHENYLHYDRAZONE)

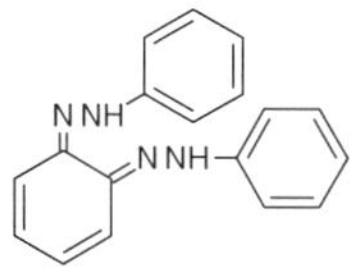

250 mg

S43,506-6
BENZIL DIHYDRAZONE

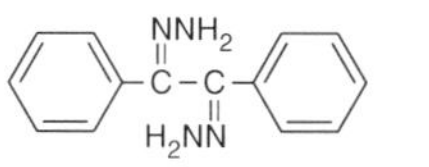

5 gm

S45,596-2
PHENETHYLHYDRAZINE SULFATE

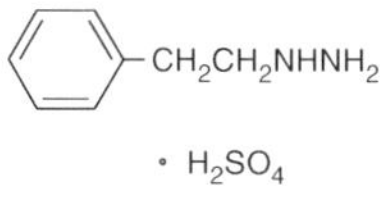

5 gm

S87,445-0
4,4'-BIS(DIMETHYLAMINO)-BENZOPHENONE HYDRAZONE

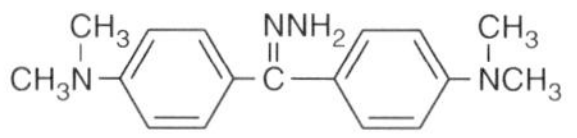

250 mg

S85,935-4
BENZYLHYDRAZINE OXALATE

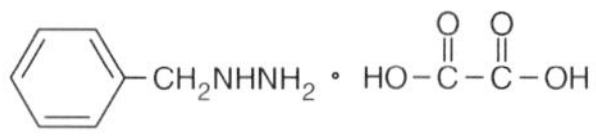

250 mg

S85,974-5
2-BUTANONE (1-METHYL-1-PHENYLPROPYL)HYDRAZONE

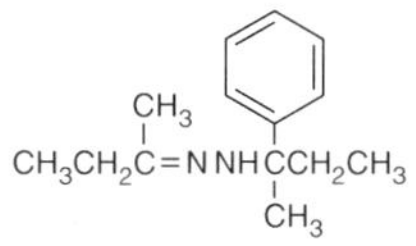

250 mg

S88,626-2
CINNAMALDEHYDE METHYLPHENYLHYDRAZONE

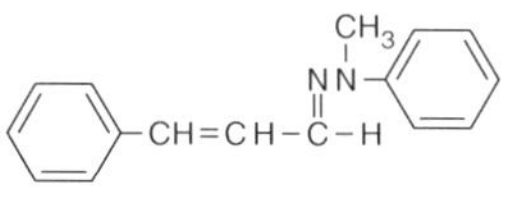

250 mg

S88,620-3
2-METHYLBENZALDEHYDE PHENYLHYDRAZONE

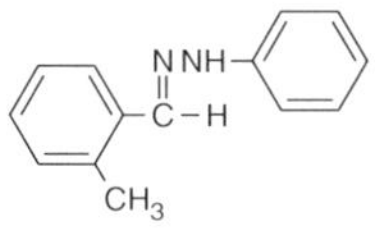

250 mg

S88,627-0
4-ISOPROPYLBENZALDEHYDE PHENYLHYDRAZONE

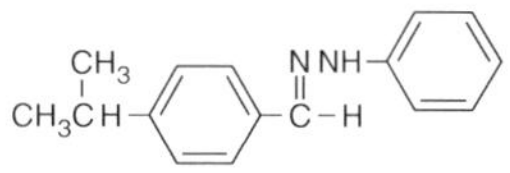

250 mg

S88,629-7
1-NAPHTHALDEHYDE PHENYLHYDRAZONE

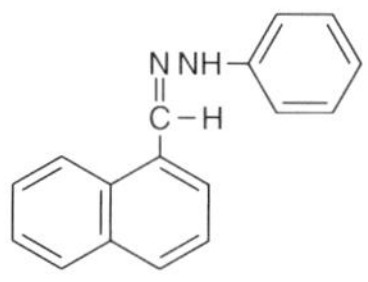

250 mg

S88,628-9
2-NAPHTHALDEHYDE PHENYLHYDRAZONE

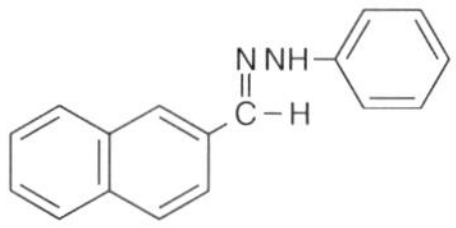

250 mg

S43,781-6
BENZOPHENONE PHENYLHYDRAZONE

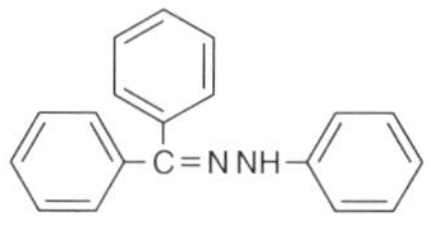

250 mg

S83,267-7
1,2-DIETHYL-1,2-DIPHENYLHYDRAZINE

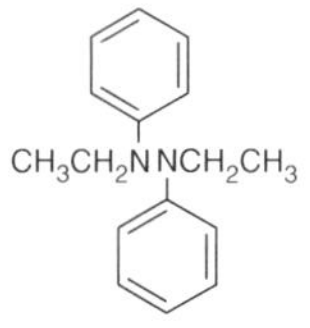

250 mg

S84,915-4
1,2-DIBENZYL-1,2-DIMETHYLHYDRAZINE

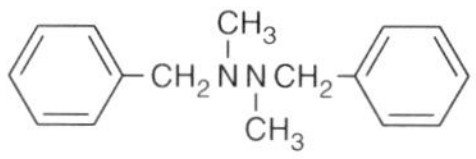

250 mg

S74,035-7
1,2-BIS(TRITYL)HYDRAZINE

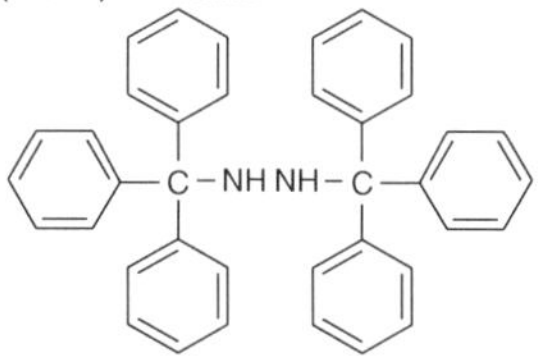

250 mg

S39,006-2
1,2-DI-O-TOLYLHYDRAZINE

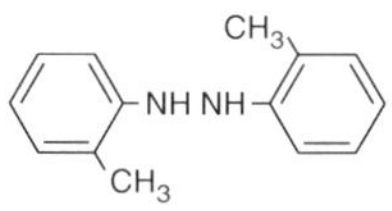

250 mg

S54,310-1
2,2',5,5'-TETRACHLOROHYDRAZOBENZENE

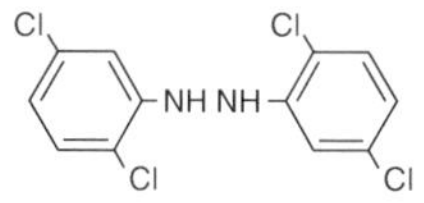

5 gm

S39,936-1
CYCLOHEXANONE PHENYLHYDRAZONE

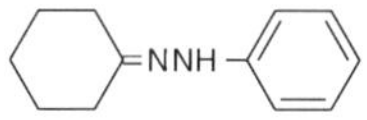

5 gm

S41,632-0
1-METHYL-1-PHENYLHYDRAZINE SULFATE DIHYDRATE

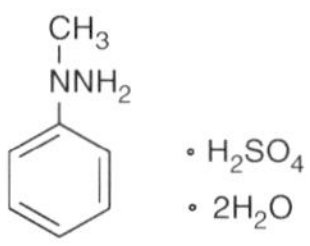

5 gm

S85,931-1
(1-METHYL-1-PHENYLPROPYL)-HYDRAZINE OXALATE

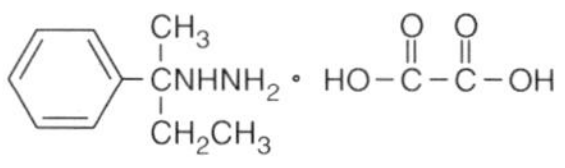

250 mg

S82,319-8
1,1,2-TRIPHENYLHYDRAZINE

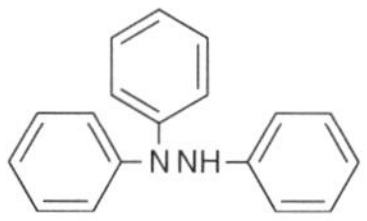

250 mg

S44,199-6
1-(2-CHLOROBENZYL)HYDRAZINE DIHYDROCHLORIDE

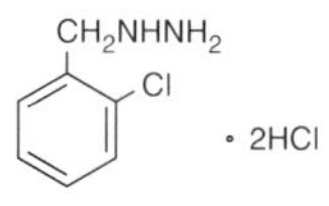

5 gm

S51,324-5
5-CHLORO-O-TOLYLHYDRAZINE

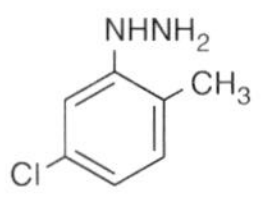

5 gm

S88,874-5
BENZALDEHYDE HYDRAZONE

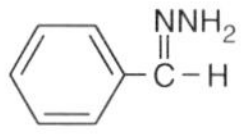

250 mg

S84,890-5
BUTYRALDEHYDE PHENYLHYDRAZONE

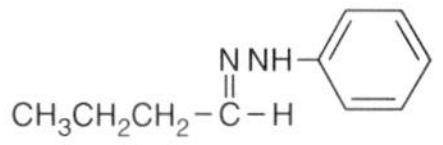

250 mg

S88,109-0
ACETOPHENONE DIMETHYLHYDRAZONE

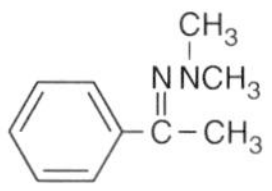

250 mg

S51,072-6
3-METHYL-2-BUTANONE PHENYLHYDRAZONE
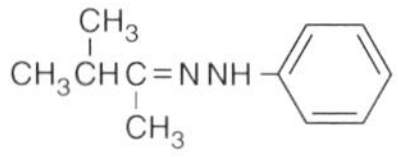
5 gm

S94,576-5
MESITYL OXIDE P-TOSYLHYDRAZONE
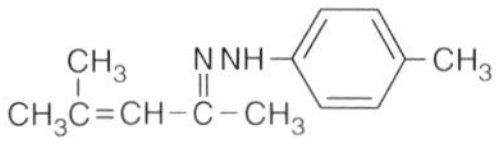
5 gm

S87,371-3
PERINAPHTHENONE (9-FLUORENYLIDENE)-HYDRAZONE
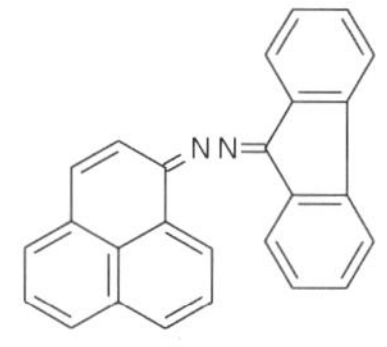
250 mg

S44,144-9
PERINAPHTHENONE HYDRAZONE
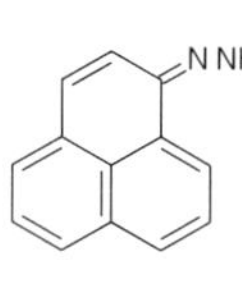
250 mg

S43,237-7
N-(2-CHLOROETHYL)-N-(2-METHOXYETHYL)BENZYLAMINE HYDROCHLORIDE
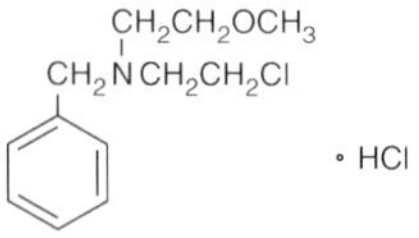
5 gm

S38,160-8
2-CHLORO-N,1'-DIMETHYL-2'-PHENOXYDIETHYLAMINE HYDROCHLORIDE
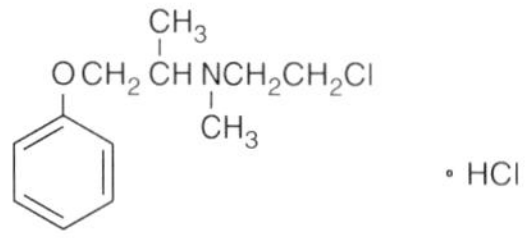
250 mg

S37,787-2
N-(2-CHLOROETHYL)BENZYLAMINE HYDROCHLORIDE
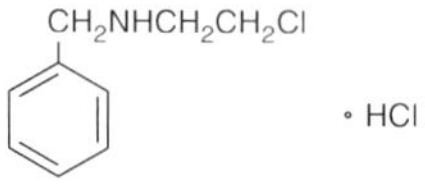
250 mg

S37,790-2
4-CHLORO-N-(2-CHLOROETHYL)-BENZYLAMINE HYDROCHLORIDE
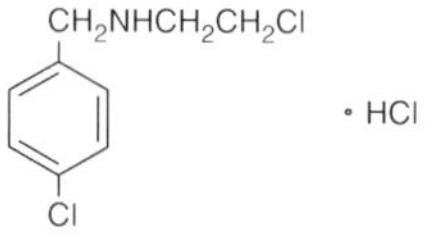
250 mg

S42,717-9
N-(2-CHLOROETHYL)-N-METHYLBENZYLAMINE HYDROCHLORIDE
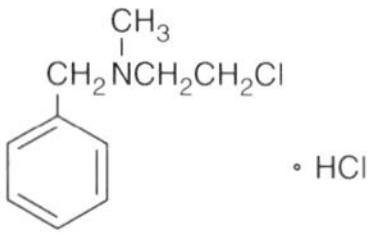
5 gm

S37,684-1
N-(2-CHLOROETHYL)-N-ETHYLBENZYLAMINE HYDROCHLORIDE
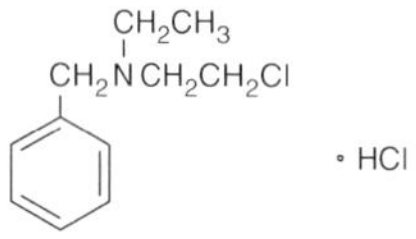
5 gm

S42,735-7
N-(3-CHLOROPROPYL)-N-METHYLBENZYLAMINE HYDROCHLORIDE
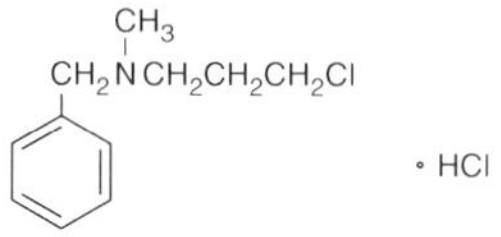
250 mg

S75,930-9
2-(4-CHLOROPHENYL)-5-(HYDROXYAMINO)-1,3-DIOXANE-5-METHANOL
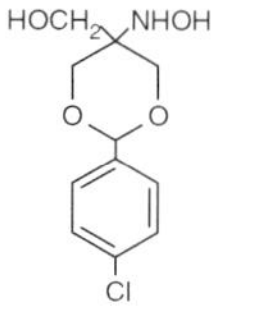
250 mg

S88,204-6
2-BROMO-4,5-DIMETHOXY-N-METHYLBENZYLAMINE
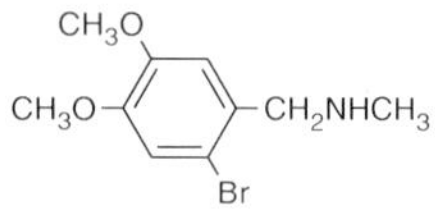
250 mg

S66,234-8
2-(4-BROMOPHENOXY)-N,N,2-TRIMETHYLPROPYLAMINE
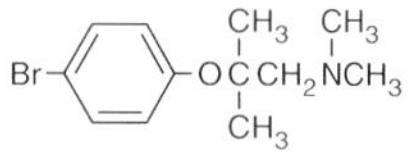
5 gm

S67,918-6
N-(4-FLUOROPHENYL)ETHYLENEDIAMINE
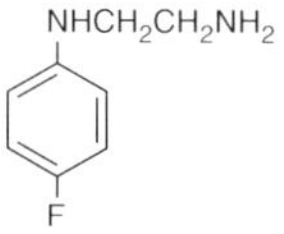
5 gm

S45,577-6
1-(P-CHLOROPHENYL)-CYCLOPROPANEMETHYLAMINE HYDROCHLORIDE
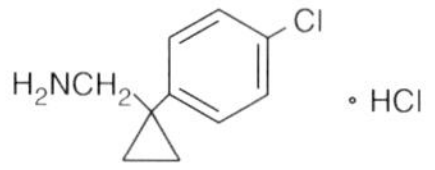
5 gm

S43,629-1
2,3,5,6-TETRACHLORO-P-XYLYLENEDIAMINE
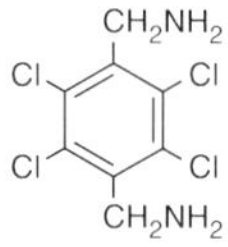
5 gm

S69,774-5
2-BROMOBENZYLAMINE
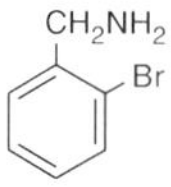
250 mg

S50,927-2
N'-(3-CHLOROPHENYL)-N,N-DIETHYLETHYLENEDIAMINE
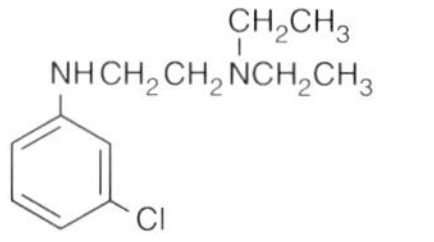
5 gm

S50,928-0
N'-(4-CHLOROPHENYL)-N,N-DIETHYLETHYLENEDIAMINE
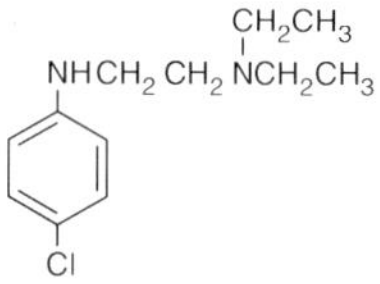
5 gm

S69,316-2
1,4-BIS(2-CHLOROBENZYLAMINOMETHYL)-CYCLOHEXANE DIHYDROCHLORIDE
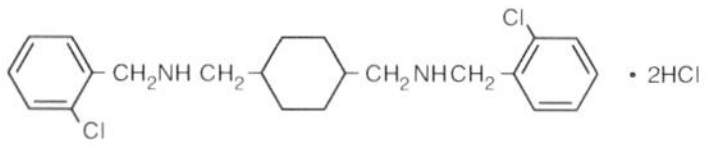
5 gm

S82,623-5
N-[1-(3-BROMOPHENYL)PENTYL]ANILINE HYDROCHLORIDE
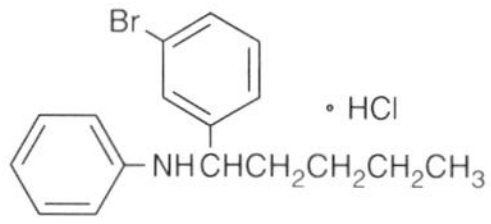
250 mg

S74,264-3
4-BROMO-N,N-DIMETHYLBENZYLAMINE
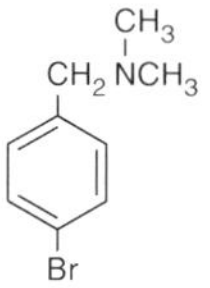
5 gm

S81,379-6
(4-BROMOBENZYL)DIETHYLAMINE HYDROCHLORIDE
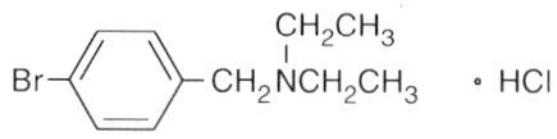
250 mg

S73,338-5
N-(2,6-DICHLOROBENZYL)-N-METHYL-1-NAPHTHALENEMETHYLAMINE
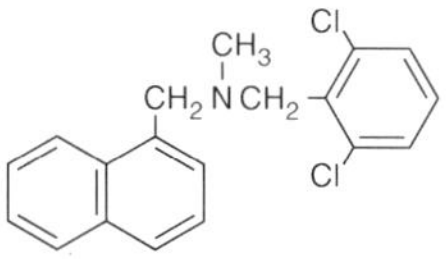
250 mg

S88,057-4
N-ALLYL-2-BROMO-N-METHYLBENZYLAMINE
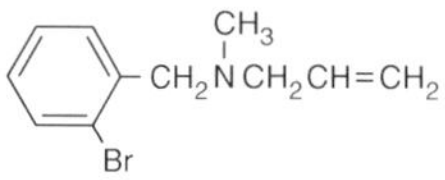
250 mg

S71,838-6
N,N'-BIS(4-FLUOROPHENYL)-DECANEDIAMINE
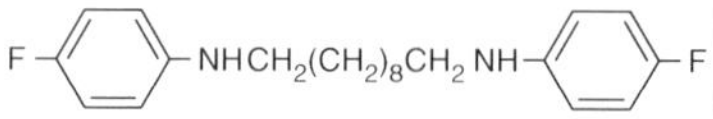
250 mg

S88,638-6
2,6-DICHLOROBENZOPHENONE IMINE
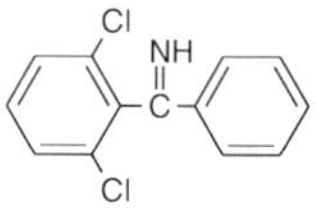
250 mg

S88,639-4
2,6-DICHLOROBENZOPHENONE IMINE HYDROCHLORIDE
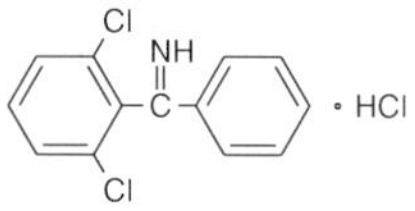
250 mg

S76,425-6
N,N'-DODECAMETHYLENEBIS(2-HYDROXY-3-METHOXYBENZYLAMINE)
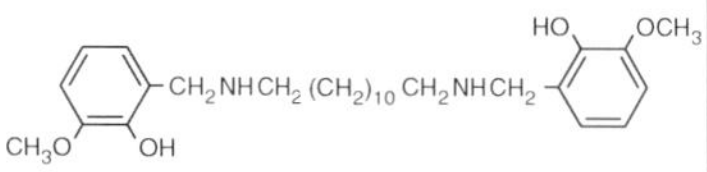
5 gm

S64,892-2
L-NORADRENALINE
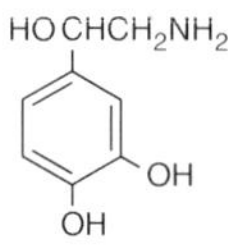
250 mg

S39,321-5
(-)-EPINEPHRINE BITARTRATE
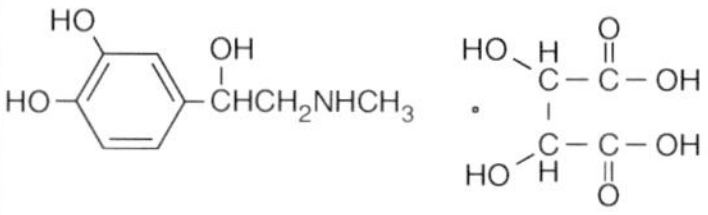
250 mg

S36,935-7
ISOPROTERENOL

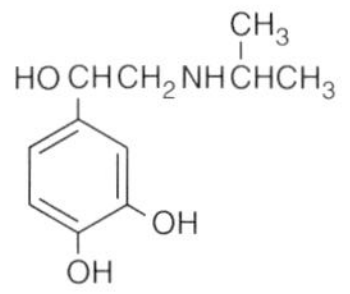

5 gm

S87,970-3
3,5-DIHYDROXY-ALPHA-(ISOPROPYLAMINOMETHYL)BENZYL ALCOHOL HEMISULFATE

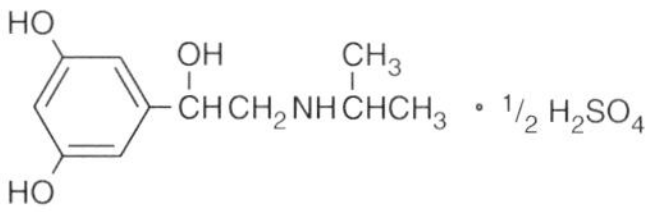

5 gm

S92,225-0
(1R,2S)-(+)-2-AMINO-1-PHENYL-1,3-PROPANEDIOL

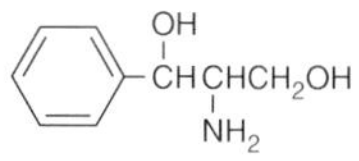

250 mg

S44,726-9
DL-N-ETHYLNORPHENYLEPHRINE HYDROCHLORIDE

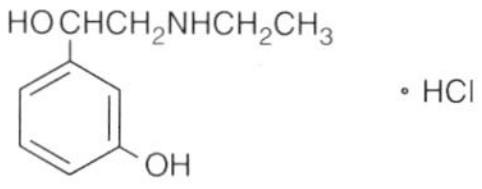

5 gm

S39,254-5
4-HYDROXY-ALPHA-[1-(1-PHENETHYLETHYLAMINO)ETHYL]BENZYL ALCOHOL HYDROCHLORIDE

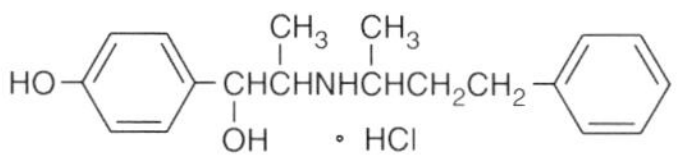

250 mg

S43,323-3
DL-4-HYDROXY-ALPHA-(METHYLAMINOMETHYL)BENZYL ALCOHOL D-TARTRATE

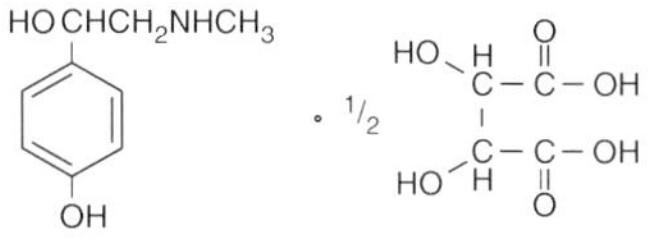

5 gm

S87,661-5
(1S,2S)-2-(DIMETHYLAMINO)-1-PHENYL-1,3-PROPANEDIOL

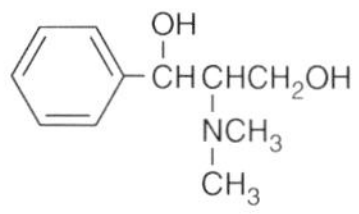

250 mg

S83,573-0
N-METHYL-1-PHENYL-2,2'-IMINODIETHANOL

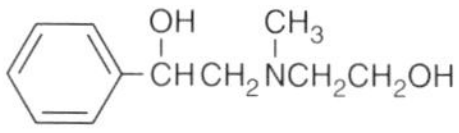

250 mg

S57,977-7
L-NORADRENALIN L-BITARTRATE

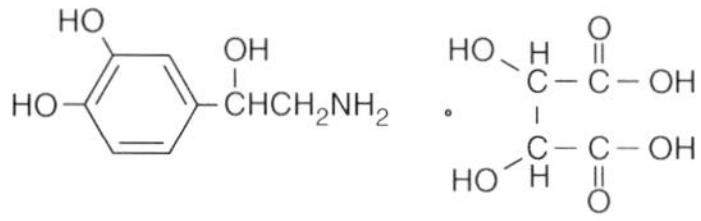

250 mg

S76,426-4
N,N'-DODECAMETHYLENEBIS(2-HYDROXYBENZYLAMINE)

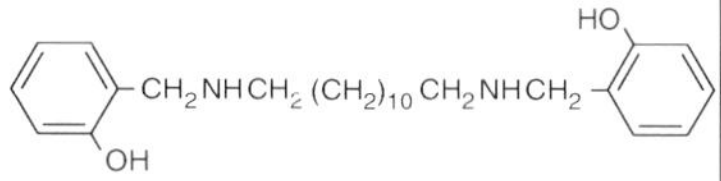

5 gm

S94,994-9
N,N-BIS(2-HYDROXYETHYL)TRITYLAMINE

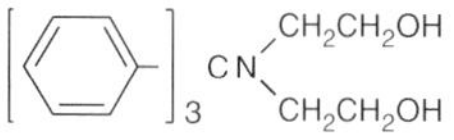

250 mg

S79,263-2
1-(3,5-DIMETHOXYPHENYL)-3-(DIMETHYLAMINOMETHYL)-2-PHENYL-2-BUTANOL HYDROCHLORIDE

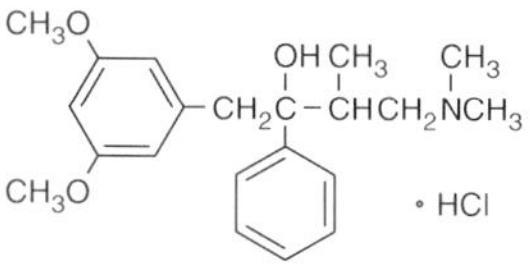

250 mg

S83,410-6
2-[2-(DIETHYLAMINO)ETHOXY]-1-PHENYLETHANOL

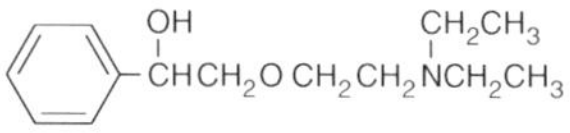

250 mg

S63,989-3
1-(4-BENZYLOXY-3-HYDROXYMETHYL-PHENYL)-2-(TERT-BUTYLAMINO)ETHANOL

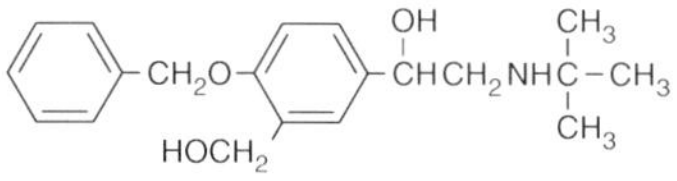

5 gm

S79,261-6
(-)-3-(DIMETHYLAMINO)-1-(2-ETHOXYPHENYL)-2-METHYL-1-PHENYL-1-PROPANOL

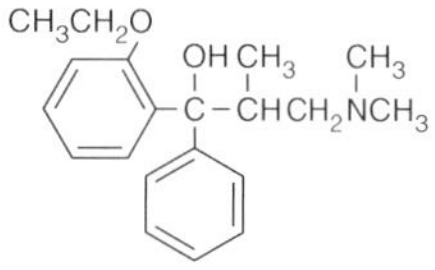

250 mg

S94,959-0
3,5-DIMETHOXY-4-HYDROXYPHENETHYLAMINE

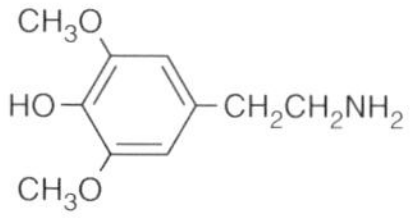

5 mg

S86,068-9
2,6-DIMETHOXY-4-(DIMETHYLAMINOMETHYL)PHENOL

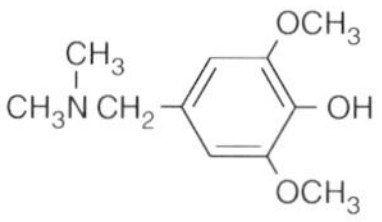

250 mg

S73,967-7
ALPHA-(BENZYLAMINO)-6-METHOXY-O-CRESOL

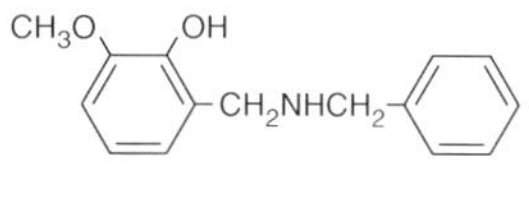

5 gm

S46,022-2
1-(ISOPROPYLAMINO)-3-(1-NAPHTHYLOXY)-2-PROPANOL HYDROCHLORIDE

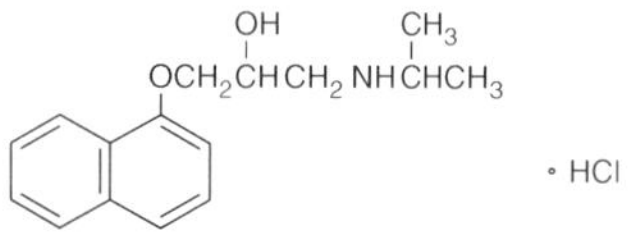

250 mg

S64,975-9
1-(DIETHYLAMINO)-3-(O-TOLYLOXY)-2-PROPANOL

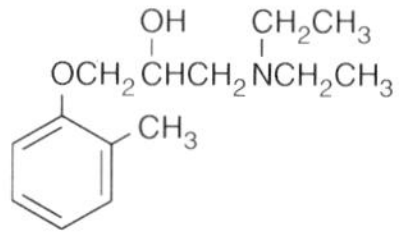

5 gm

S39,798-9
4-[2-(METHYLAMINO)PROPOXY]PHENOL HYDROCHLORIDE

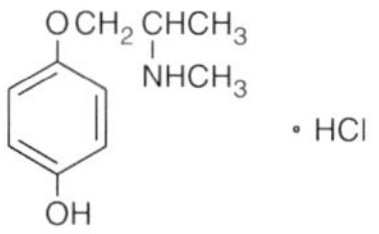

250 mg

S83,742-3
1-([2-(DIETHYLAMINOETHYLTHIO)]-METHYL)-1-[4-(DIMETHYLAMINO)PHENYL]-ETHANOL

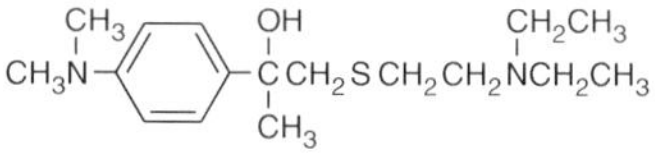

250 mg

S83,412-2
2-[2-(DIETHYLAMINO)ETHYLTHIO]-1-PHENYLETHANOL

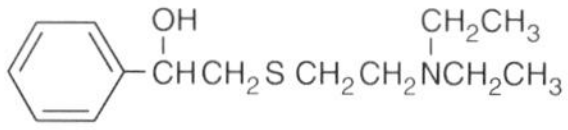

250 mg

S83,414-9
2-[3-(DIETHYLAMINO)PROPYLTHIO]-1-PHENYLETHANOL

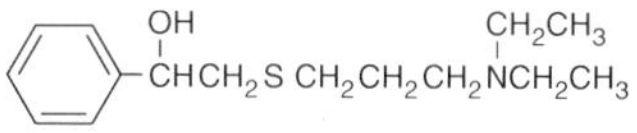

250 mg

S37,260-9
N-2-HYDROXYETHYLNOREPHEDRINE HYDROCHLORIDE

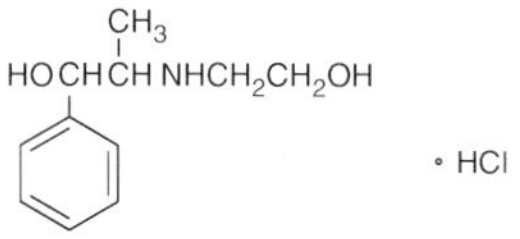

5 gm

S89,710-8
3-(PHENETHYLAMINO)-1-PHENYL-1-CYCLOHEXANOL HYDROCHLORIDE

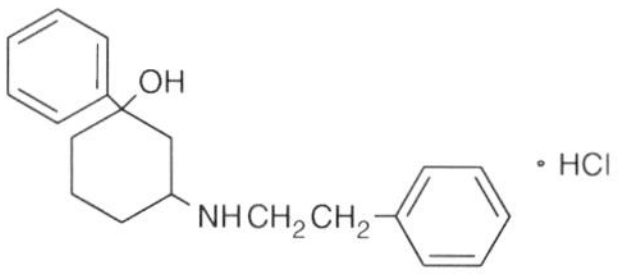

250 mg

S89,364-1
1,2-DIPHENYL-3-METHYL-4-(METHYLPHENETHYLAMINO)-2-BUTANOL HYDROCHLORIDE

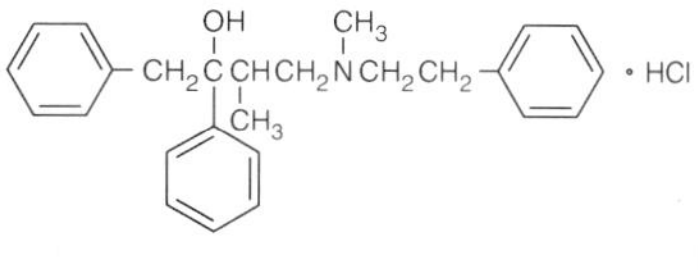

250 mg

S73,385-7
1-[ALPHA-(BENZYLAMINO)BENZYL]-2,2-DIMETHYL-1-PHENYL-1-PROPANOL HYDROCHLORIDE

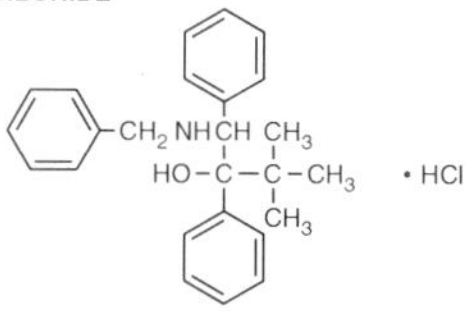

250 mg

S89,343-9
1,2-DIPHENYL-3-METHYL-4-(METHYLAMINO)-2-BUTANOL HYDROCHLORIDE

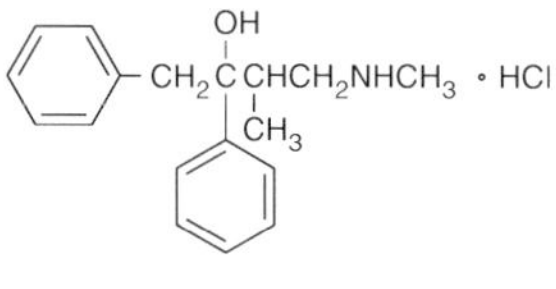

250 mg

S87,750-6
1,1-DIPHENYL-3-(METHYLAMINO)-1-PROPANOL HYDROCHLORIDE DIHYDRATE

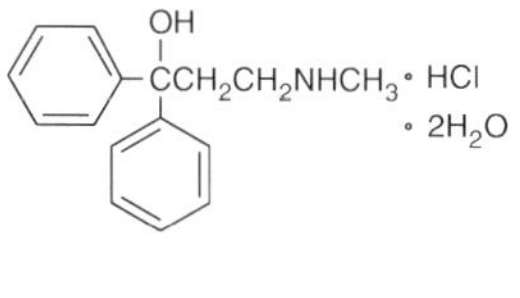

250 mg

S79,394-9
ALPHA-(N-METHYLANILINOMETHYL)BENZYL ALCOHOL HYDROCHLORIDE

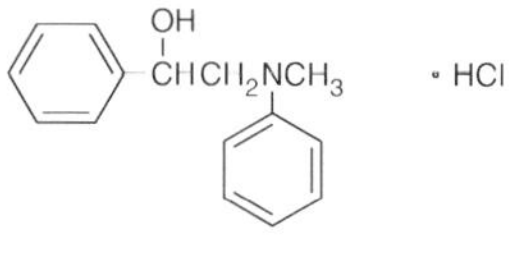

250 mg

S52,925-7
DL-1,2-DIPHENYL-3-METHYL-4-(N-METHYLBENZYLAMINO)-2-BUTANOL HYDROCHLORIDE

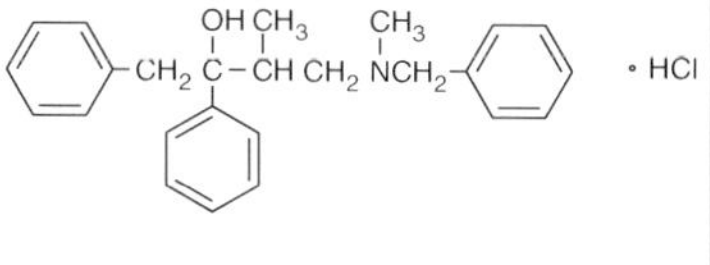

250 mg

S81,597-7
2-(DIETHYLAMINO)-1-PHENYLETHANOL
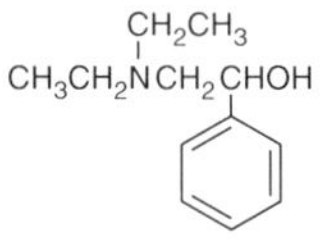
250 mg

S89,669-1
2-[1-(DIMETHYLAMINO)CYCLOHEXYL]-1,1-DIPHENYL-1-ETHANOL
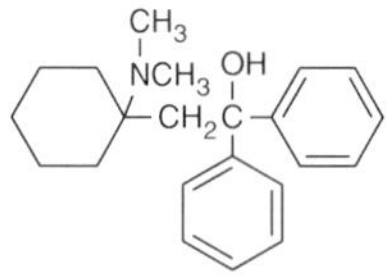
250 mg

S89,777-9
2-[1-(DIMETHYLAMINO)CYCLOHEXYL]-1,1-DIPHENYL-1-ETHANOL HYDROCHLORIDE
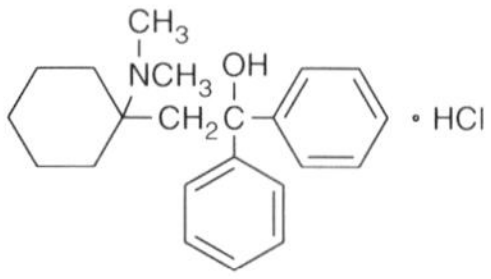
250 mg

S79,359-0
(+)-3-(DIMETHYLAMINO)-1-PHENYLPROPANOL
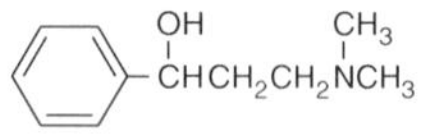
250 mg

S43,220-2
DL-4-DIMETHYLAMINO-1,2-DIPHENYL-3-METHYL-2-BUTANOL HYDROCHLORIDE
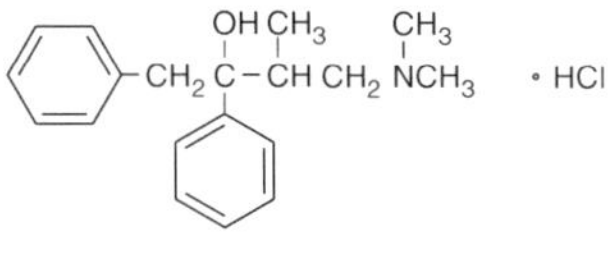
5 gm

S79,264-0
1-(4-TERT-BUTYLPHENYL)-3-(DIMETHYLAMINOMETHYL)-2-PHENYL-2-BUTANOL HYDROCHLORIDE
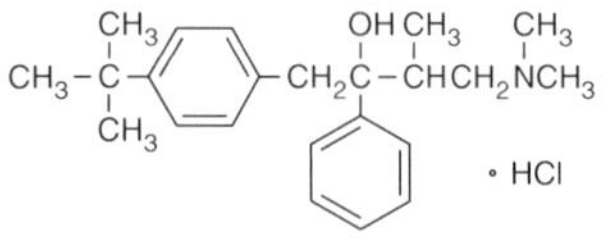
250 mg

S79,265-9
1-(4-BIPHENYLYL)-3-(DIMETHYLAMINOMETHYL)-2-PHENYL-2-BUTANOL HYDROCHLORIDE
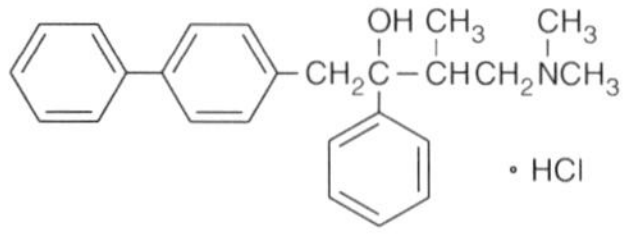
250 mg

S79,352-3
3-(DIMETHYLAMINOMETHYL)-2-PHENYL-1-(3,5-XYLYL)-2-BUTANOL HYDROCHLORIDE
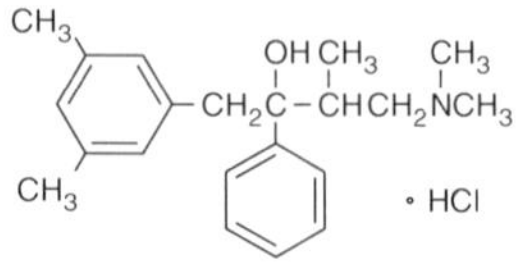
250 mg

S41,222-8
2-(BENZYLMETHYLAMINO)ETHANOL N-OXIDE (50% AQUEOUS SOLUTION)
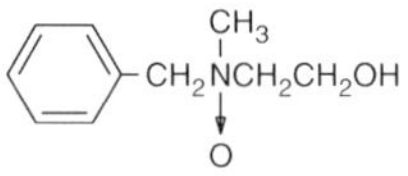
5 gm

S80,623-4
1-TERT-BUTYL-1,3-DIHYDRO-1-[(4-DIMETHYLAMINOMETHYL)PHENYL]-3-METHYL-3-ISOBENZOFURANOL
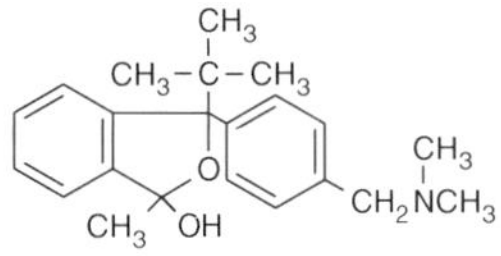
250 mg

S83,407-6
1-(4-AMINOPHENYLTHIO)-3-(DIETHYLAMINO)-2-PROPANOL
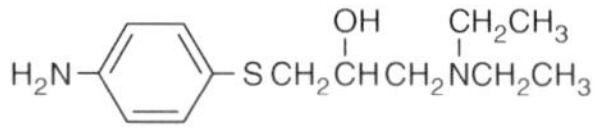
250 mg

S83,406-8
1-(DIETHYLAMINO)-3-(PHENYLTHIO)-2-PROPANOL
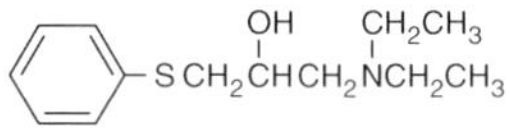
250 mg

S83,411-4
1-(DIETHYLAMINO)-3-(P-TOLYLTHIO)-2-PROPANOL
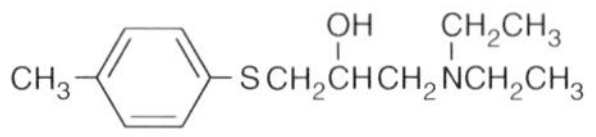
250 mg

S37,759-7
BETA-[2-(DIMETHYLAMINO)ETHYLTHIO]-PHENETHYL ALCOHOL
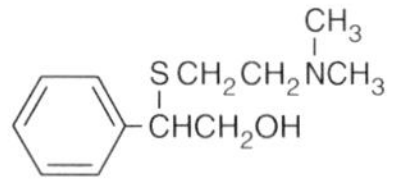
250 mg

S38,225-6
BETA-[2-(DIETHYLAMINO)ETHYLTHIO]-PHENETHYL ALCOHOL
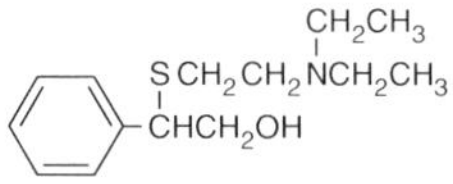
5 gm

S38,236-1
ALPHA-[2-(DIETHYLAMINO)-ETHYLTHIOMETHYL]BENZYL ALCOHOL
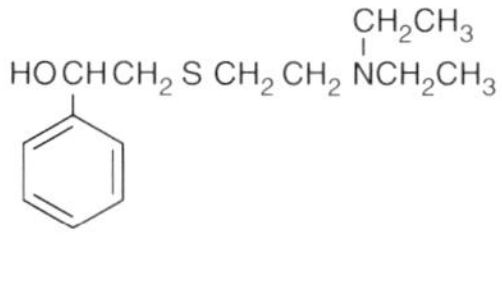
250 mg

S56,127-4
4-AMINO-ALPHA-DIETHYLAMINO-O-CRESOL
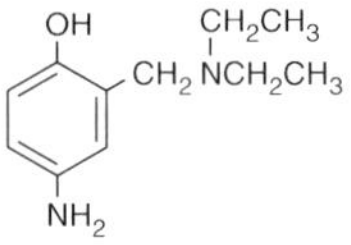
5 gm

S43,466-3
L-(+)-THREO-2-AMINO-1-PHENYL-1,3-PROPANEDIOL
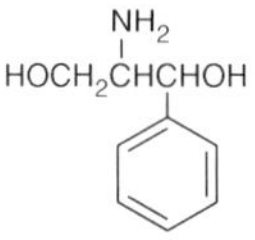
5 gm

S90,652-2
2,2-DIMETHYL-3-(1,2-DIPHENYLETHYLAMINO)-1-PROPANOL
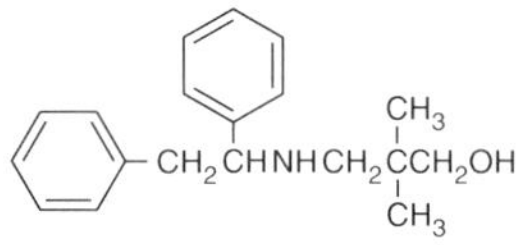
250 mg

S73,977-4
ALPHA-(PHENETHYLAMINO)-O-CRESOL
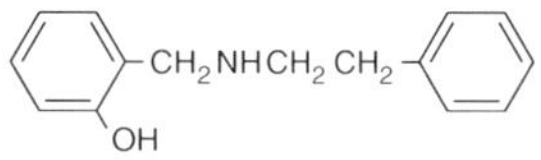
5 gm

S35,985-8
2-CYCLOHEXYLAMINO-1-PHENYLETHANOL
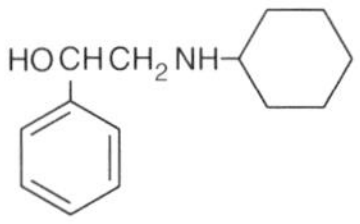
5 gm

S79,358-2
(+)-2-METHYLAMINO-3-PHENYL-1-PROPANOL
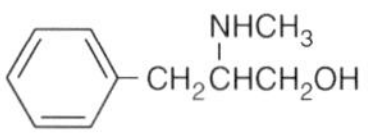
250 mg

S89,719-1
3-(PHENETHYLAMINO)-1-CYCLOHEXANOL HYDROBROMIDE
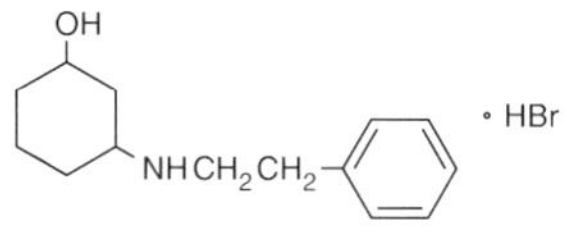
250 mg

S45,225-4
4-[2-(DIMETHYLAMINO)ETHYL]PHENOL SULFATE DIHYDRATE
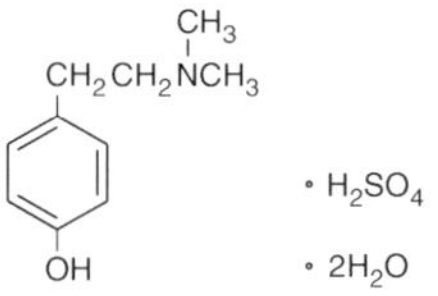
250 mg

S44,168-6
DL-1-BENZYLAMINO-2-PROPANOL
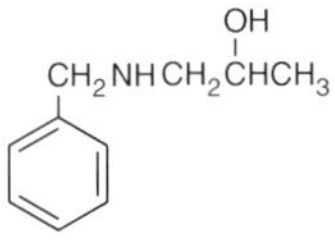
5 gm

S45,451-6
DL-2-BENZYLAMINO-1-PROPANOL
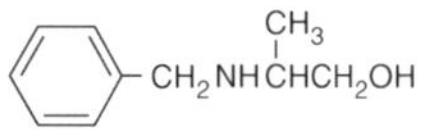
5 gm

S65,905-3
(+)-N-BENZYLALANINOL
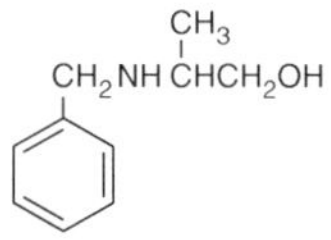
250 mg

S65,906-1
(-)-N-BENZYLALANINOL
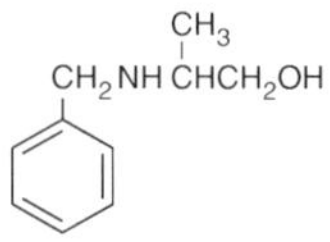
250 mg

S37,829-1
2-(4-CHLOROBENZYLAMINO)ETHANOL HYDROCHLORIDE
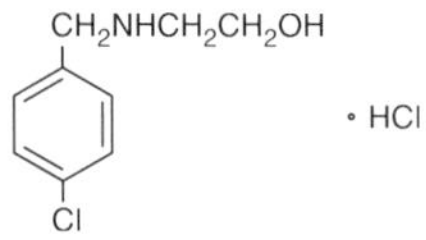
5 gm

S89,728-0
3-(BENZYLAMINO)-1-CYCLOHEXANOL
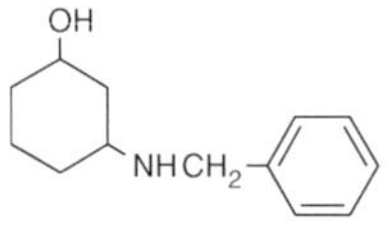
250 mg

S37,631-0
3-BENZYLAMINO-1-PROPANOL HYDROCHLORIDE
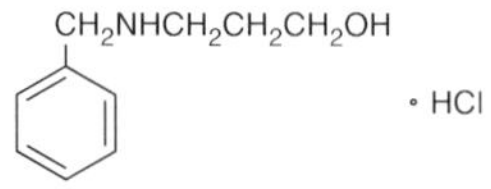
5 gm

S43,188-5
3-BENZYLAMINO-1-PROPANOL
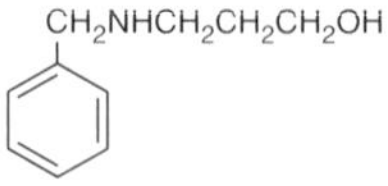
5 gm

S88,434-0
4-(BENZYLAMINO)-4-METHYL-2-PENTANOL

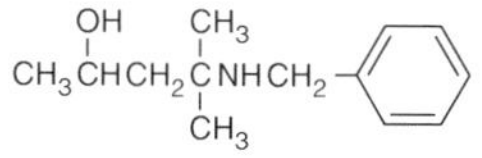

250 mg

S34,158-4
6-BENZYLAMINO-2-METHYL-2-HEPTANOL

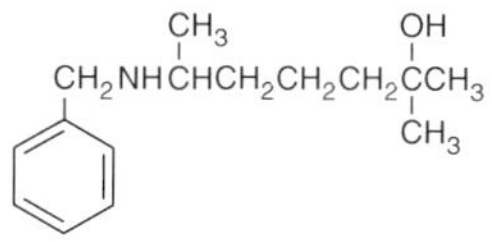

5 gm

S87,044-7
2-(9,10-DIHYDRO-9-ANTHRACENYLAMINO)-1-ETHANOL

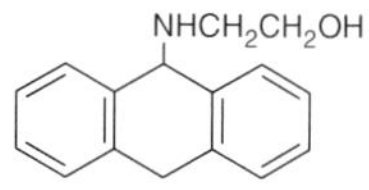

250 mg

S41,155-8
4-[2-(METHYLAMINO)PROPYL]PHENOL HEMISULFATE

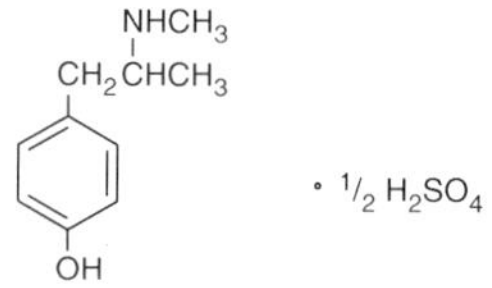

5 gm

S53,881-7
1-(2,4-XYLIDINOMETHYL)-2-NAPHTHOL

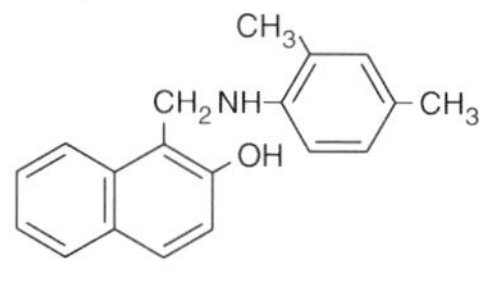

5 gm

S43,185-0
2,2'-(BENZYLIMINO)DIETHANOL N-OXIDE (40% AQUEOUS SOLUTION)

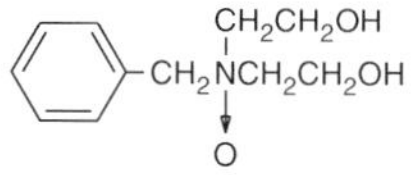

5 gm

S43,948-7
N,N-DIBENZYLETHANOLAMINE

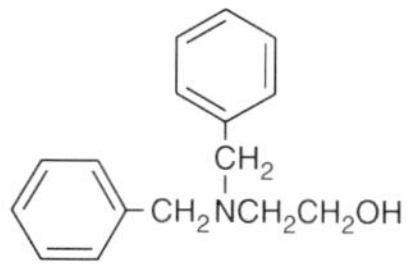

5 gm

S42,681-4
2,2'-(BENZYLIMINO)DIETHANOL

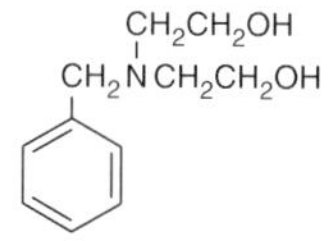

5 gm

S34,688-8
3,6-DIMETHYL-2-DIMETHYLAMINOMETHYLPHENOL

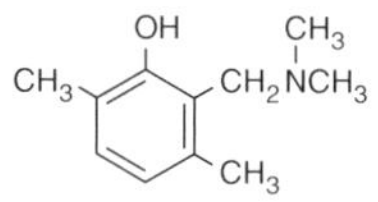

5 gm

S94,505-6
2-(DIETHYLAMINOMETHYL)-3,6-DIMETHYLPHENOL

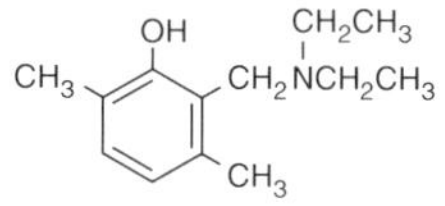

5 gm

S40,746-1
2,6-DI-TERT-BUTYL-4-(DIMETHYLAMINOMETHYL)PHENOL

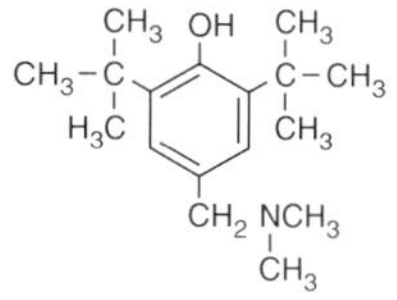

5 gm

S81,656-6
ALPHA-[(DIETHYLAMINO)METHYL]-PHENETHYL ALCOHOL

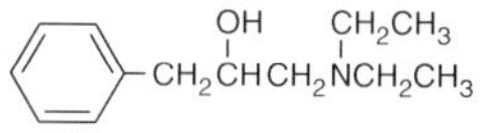

250 mg

S77,132-5
9-[2-(DIAMYLAMINO)-1-HYDROXYETHYL]-1,2,3,4-TETRAHYDROPHENANTHRENE HYDROCHLORIDE

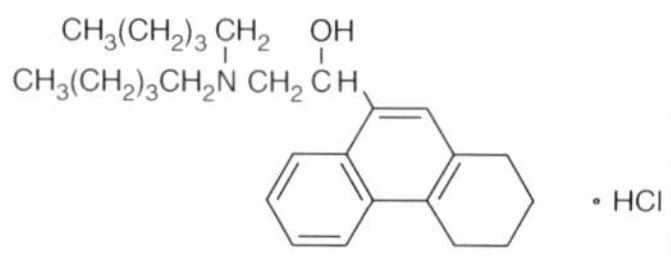

250 mg

S61,616-8
3-BROMO-10-(2-DIHEPTYLAMINO-1-HYDROXYETHYL)PHENANTHRENE HYDROCHLORIDE

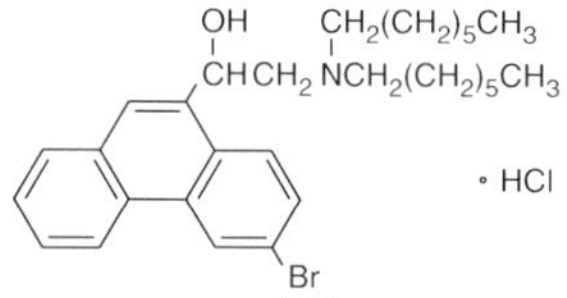

5 gm

S79,383-3
1-BENZYL-2-(DIMETHYLAMINOMETHYL)-1,2,3,4-TETRAHYDRO-1-NAPHTHOL

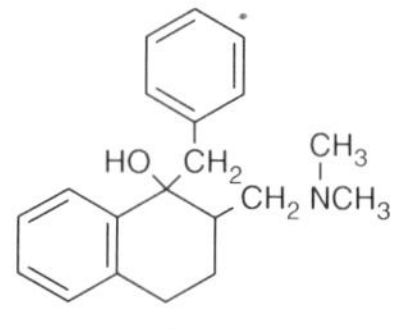

250 mg

S64,149-9
ALPHA-DIAMYLAMINO-1,2,3,4-TETRAHYDRO-9-PHENANTHRENEMETHANOL HYDROCHLORIDE

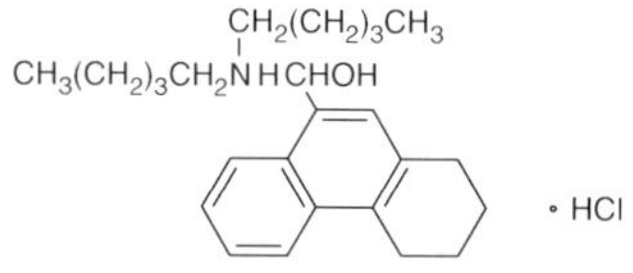

250 mg

S40,794-1
1-DIMETHYLAMINOMETHYL-2-NAPHTHOL

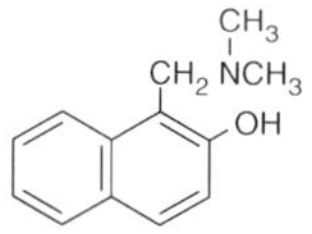

5 gm

S38,040-7
2-BENZYLOXYPROPYLAMINE

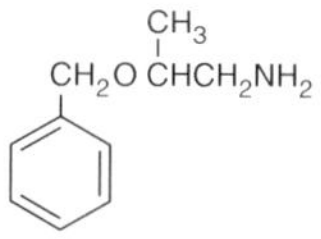

5 gm

S38,377-5
N-(2-BENZYLOXY-1-METHYLETHYL)-BENZYLAMINE

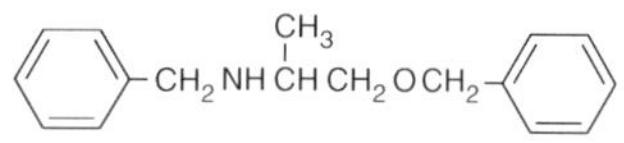

5 gm

S81,647-7
2-(BENZYLOXY)-N,N-DIETHYL-3-PHENYLPROPYLAMINE

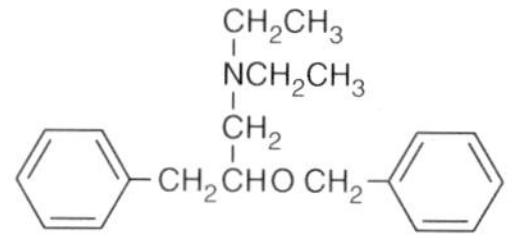

250 mg

S74,488-3
2-(2,4,5-TRIMETHOXYPHENYL)-1,4-BUTANEDIAMINE DIHYDROCHLORIDE

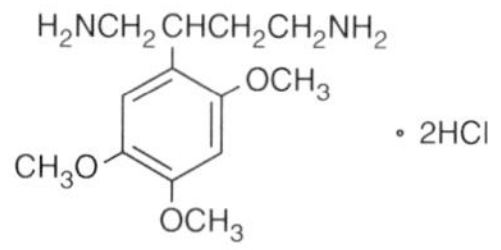

250 mg

S93,556-5
3-BENZYLOXY-4,5-DIMETHOXYPHENETHYLAMINE OXALATE

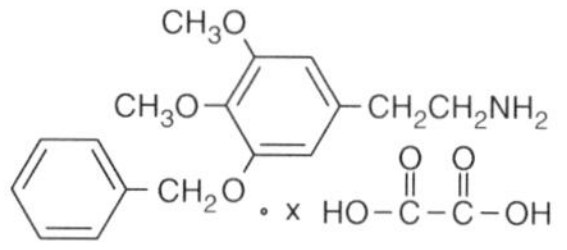

250 mg

S34,073-1
3-ALLYL-2-(2-DIETHYLAMINOETHOXY)-ANISOLE

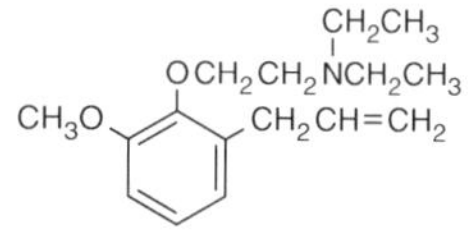

5 gm

S74,485-9
2-(2,5-DIMETHOXYPHENYL)-2-METHYL-1,4-BUTANEDIAMINE DIHYDROCHLORIDE

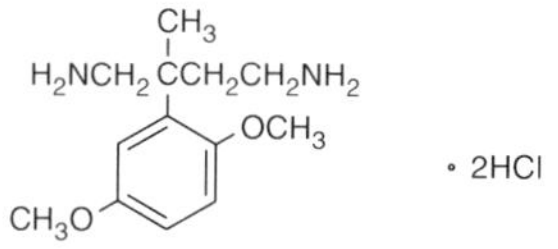

250 mg

S80,860-1
3-(2,5-DIMETHOXYPHENYL)PROPYLAMINE

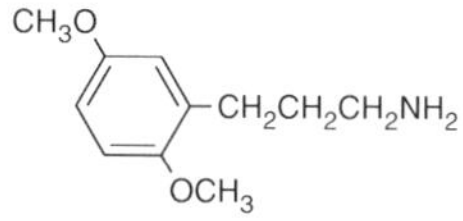

250 mg

S68,332-9
3,3'-DIMETHOXY-ALPHA-METHYLDIPHENETHYLAMINE HYDROCHLORIDE

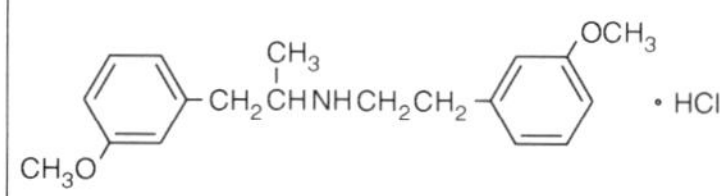

5 gm

S38,451-8
N,N-DIETHYL-N'-[ALPHA-(3,4-DIMETHOXYPHENYL)PHENETHYL]-ETHYLENEDIAMINE

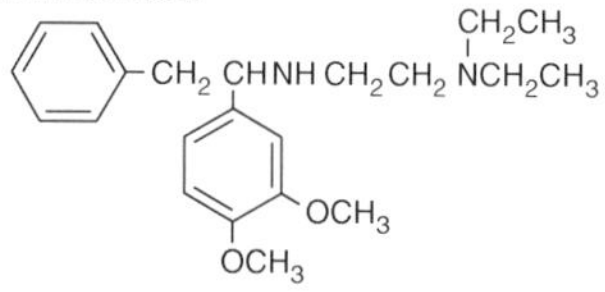

250 mg

S51,752-6
3,4-DIMETHOXY-N-METHYLPHENETHYLAMINE HYDRIODIDE

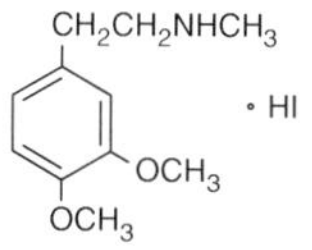

250 mg

S38,375-9
ALPHA-(3,4-DIMETHOXYPHENYL)-N-METHYLPHENETHYLAMINE

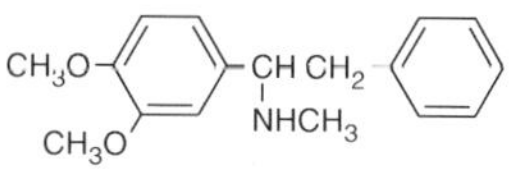

5 gm

S39,828-4
ALPHA-(3,4-DIMETHOXYPHENYL)-N-METHYLPHENETHYLAMINE HYDROCHLORIDE

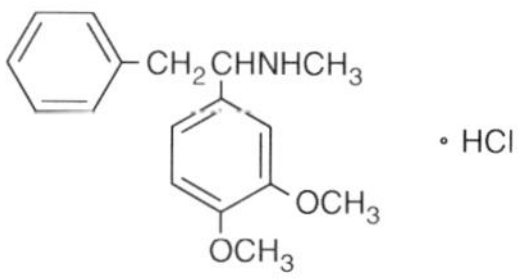

250 mg

S52,559-6
N-CYCLOPROPYLVERATRYLAMINE HYDROCHLORIDE

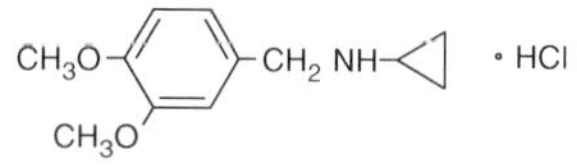

250 mg

S62,784-4
2-AMINO-1-PHENOXYPROPANE

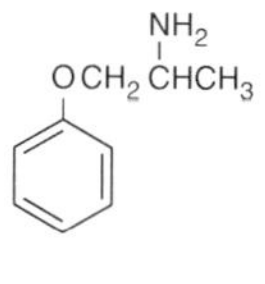

250 mg

S38,734-7
2-HEXADECYLOXY-ALPHA-METHYLBENZYLAMINE
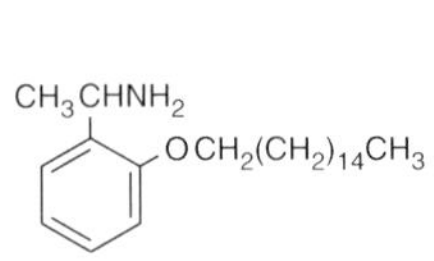
250 mg

S38,711-8
4-HEXADECYLOXY-ALPHA-METHYLBENZYLAMINE
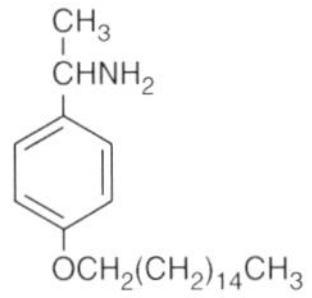
250 mg

S37,708-2
1-(4-BROMOPHENOXYMETHYL)ETHYLAMINE
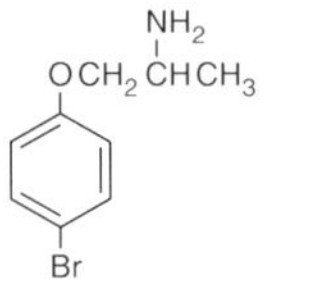
250 mg

S79,314-0
7-PHENOXYHEPTYLAMINE
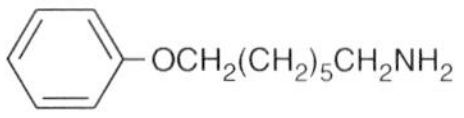
250 mg

S38,585-9
1-HEXADECYLOXY-2-NAPHTHALENEMETHYLAMINE
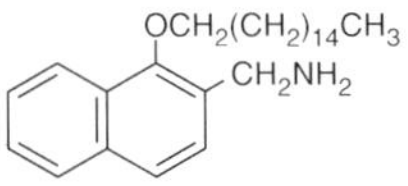
250 mg

S62,796-8
3-(2-HEXADECYLOXY-1-NAPHTHYL)-PROPYLAMINE
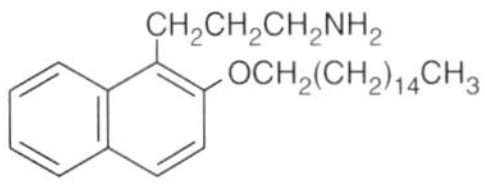
5 gm

S38,439-9
N,N-DIETHYL-N'-[ALPHA-(3-METHOXYPHENYL)PHENETHYL]-ETHYLENEDIAMINE
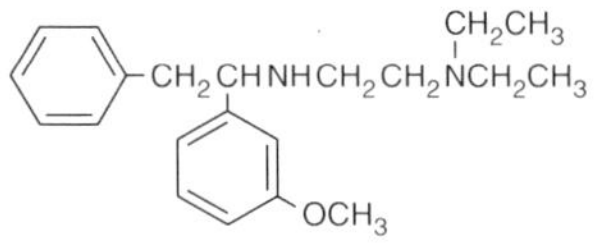
5 gm

S38,452-6
N'-[ALPHA-(4-BENZYLOXYPHENYL)-PHENETHYL]-N,N-DIETHYLETHYLENE-DIAMINE DIHYDROCHLORIDE
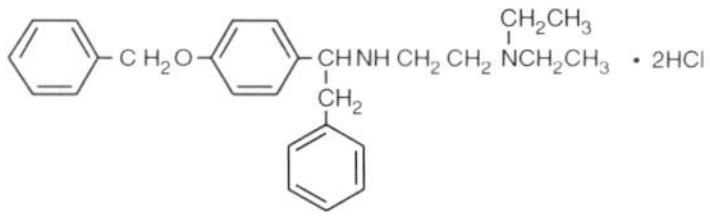
250 mg

S38,453-4
N-BENZYL-ALPHA-(3-METHOXYPHENYL)-PHENETHYLAMINE
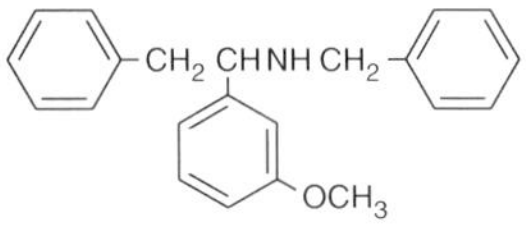
5 gm

S63,987-7
2-BENZYLAMINO-1-(4-METHOXYPHENYL)-PROPANE
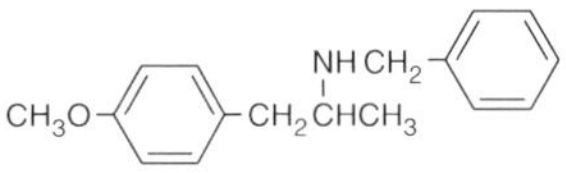
5 gm

S89,713-2
(3-ETHOXYCYCLOHEXYL)PHENETHYLAMINE HYDROCHLORIDE
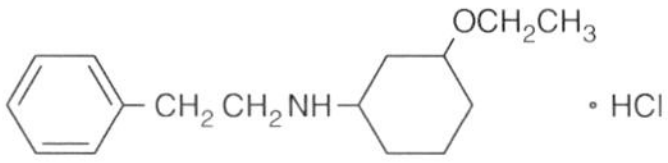
250 mg

S38,371-6
ALPHA-(4-METHOXYPHENYL)-N-METHYLPHENETHYLAMINE
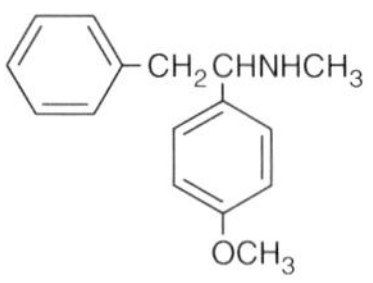
5 gm

S38,376-7
N-ETHYL-ALPHA-(4-METHOXYPHENYL)-PHENETHYLAMINE
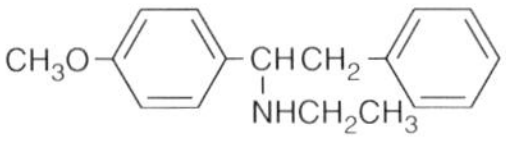
5 gm

S63,575-8
N-(4-METHOXYBENZYL)PHENETHYLAMINE HYDROCHLORIDE
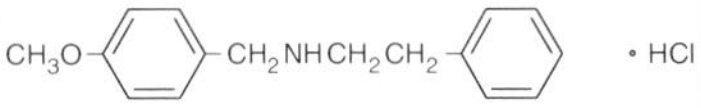
250 mg

S52,556-1
N-CYCLOPROPYL-4-METHOXYBENZYLAMINE HYDROCHLORIDE
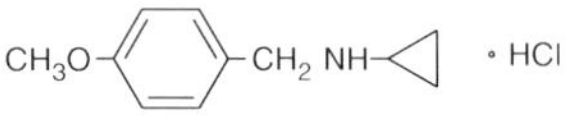
250 mg

S38,047-4
N-ETHYL-4-METHOXYBENZYLAMINE
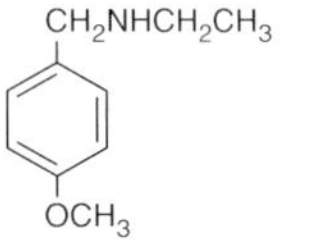
5 gm

S52,420-4
N,N'-BIS(4-METHOXYBENZYL)-ETHYLENEDIAMINE DIHYDROCHLORIDE
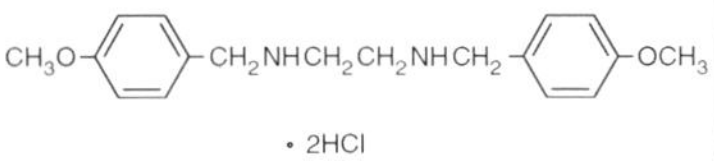
5 gm

S39,190-5
N-METHYL-2-(M-TOLYLOXY)-ETHYLAMINE
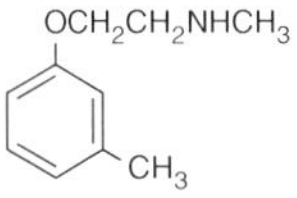
250 mg

S37,711-2
N-METHYL-2-PHENOXYETHYLAMINE
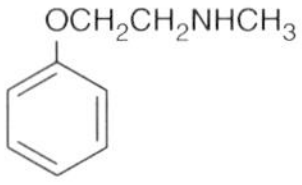
5 gm

S38,219-1
N-(1-METHYL-2-PHENOXYETHYL)-BUTYLAMINE
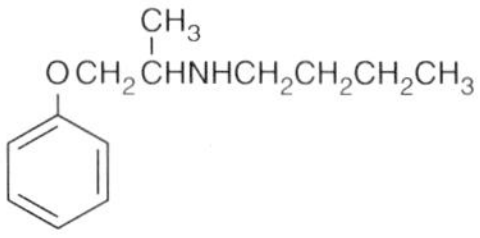
250 mg

S38,400-3
BIS(2-PHENOXYETHYL)AMINE
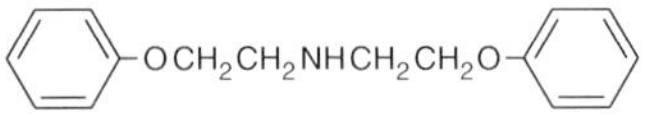
250 mg

S37,710-4
2-(2-CHLOROPHENOXY)-N-METHYLETHYLAMINE
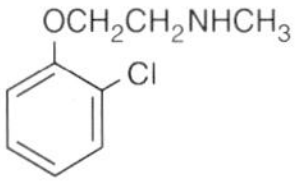
250 mg

S38,573-5
N-METHYL-2-(O-TOLYLOXY)-ETHYLAMINE
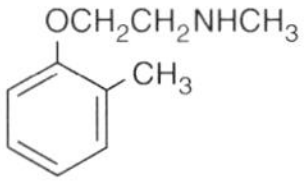
250 mg

S37,801-1
2-(4-CHLOROPHENOXY)-N-METHYLETHYLAMINE
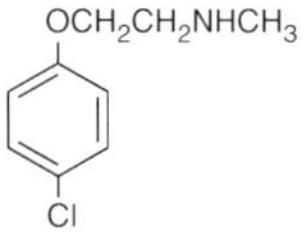
5 gm

S38,634-0
2-(4-CHLOROPHENOXY)-N,1-DIMETHYLETHYLAMINE
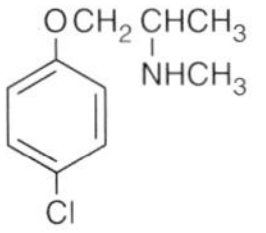
250 mg

S38,052-0
2-(4-CHLORO-M-TOLYLOXY)-N-METHYLETHYLAMINE
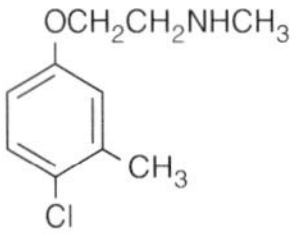
250 mg

S35,236-5
2-METHOXY-ALPHA,N-DIMETHYLPHENETHYLAMINE HYDROCHLORIDE
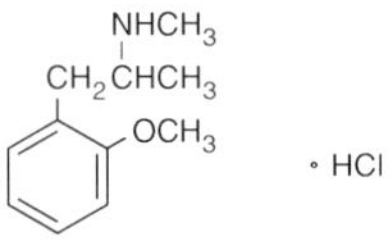
5 gm

S38,405-4
N-METHYL-N-(2-PHENOXYETHYL)-BENZYLAMINE
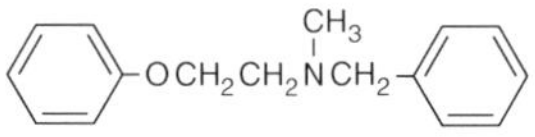
250 mg

S90,667-0
DIBENZYL(3-PHENOXYPROPYL)AMINE HYDROCHLORIDE
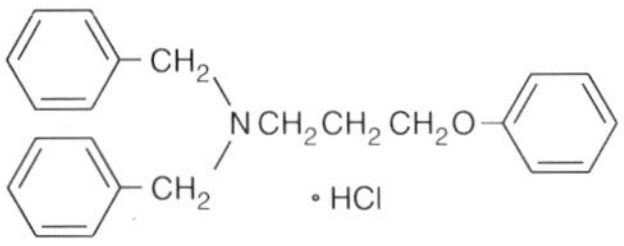
250 mg

S82,530-1
BETA-(PHENOXY)TRIETHYLAMINE
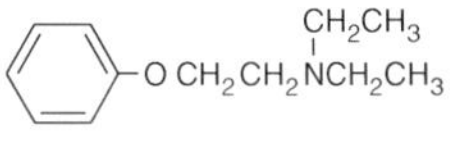
250 mg

S61,236-7
N,N-DIMETHYLFORMAMIDE DIBENZYL ACETAL
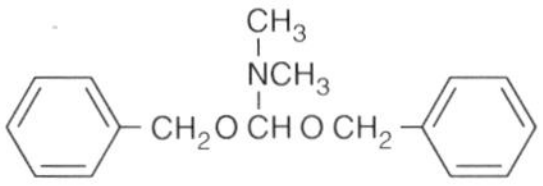
5 gm

S38,390-2
BIS[2-(BENZYLOXY)ETHYL]AMINE
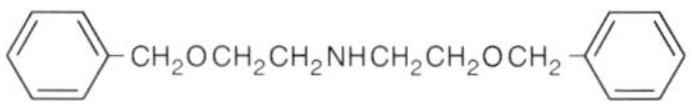
250 mg

S38,084-9
2-(N-BENZYL-N-ETHYLAMINO)-ETHANETHIOL

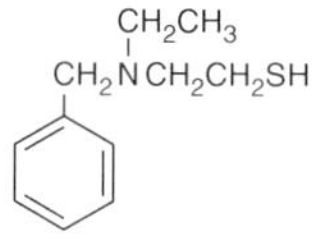

5 gm

S82,532-8
BETA-(P-TOLYLTHIO)TRIETHYLAMINE

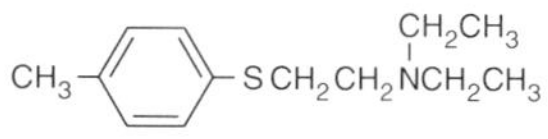

250 mg

S80,574-2
N,N'-BIS(2-AMINOPHENYL)-1,3-PROPANEDIAMINE

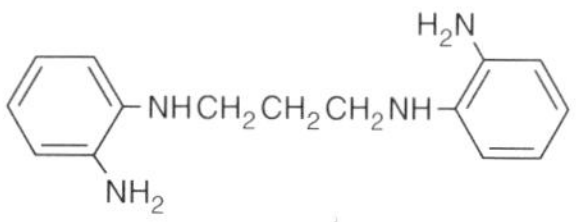

250 mg

S36,985-3
2-METHYL-N-1-PHENYL-1,2-PROPANEDIAMINE

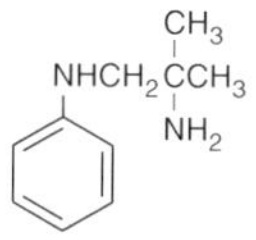

5 gm

S86,948-1
3-(DIMETHYLAMINO)BENZYLAMINE

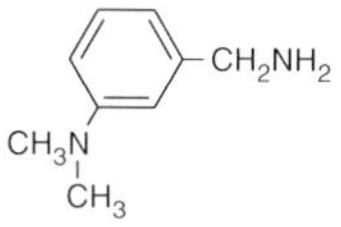

250 mg

S74,479-4
2-PHENYL-1,4-BUTANEDIAMINE DIHYDROCHLORIDE

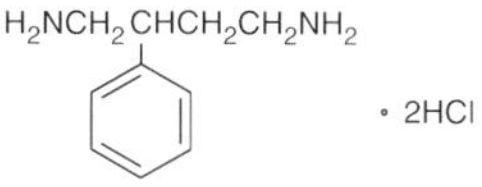

250 mg

S87,966-5
4-METHYL-ALPHA-PHENYLPHENETHYLAMINE HYDROCHLORIDE

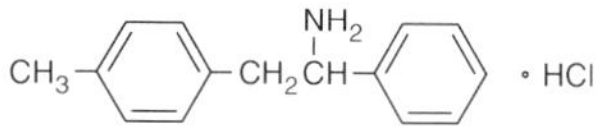

5 gm

S87,983-5
4-METHYL-ALPHA-PHENYLPHENETHYLAMINE

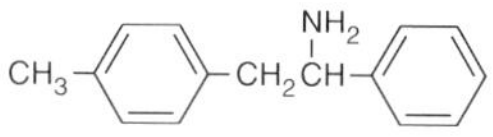

5 gm

S69,313-8
1-PHENYL-1-CYCLOPROPANEMETHYLAMINE HYDROCHLORIDE

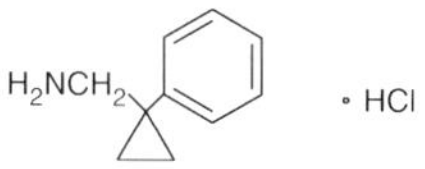

250 mg

S35,593-3
2,3,5,6-TETRAMETHYL-1,4-XYLYLENEDIAMINE DIHYDROCHLORIDE

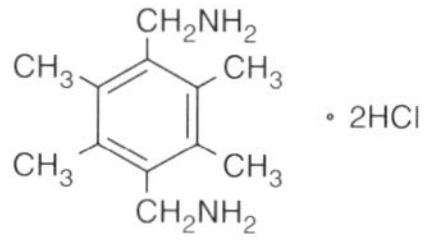

5 gm

S36,863-6
2,5-DIMETHYL-1,4-XYLYLENEDIAMINE DIHYDROCHLORIDE

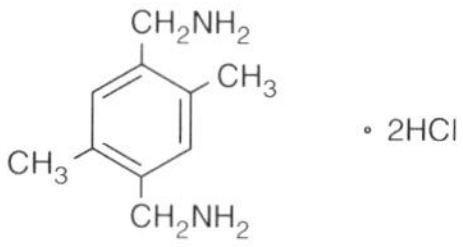

5 gm

S69,668-4
BENZYLAMINE HYDROBROMIDE

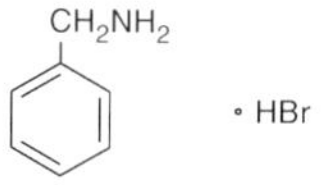

5 gm

S38,112-8
ALPHA-(SEC-BUTYL)BENZYLAMINE

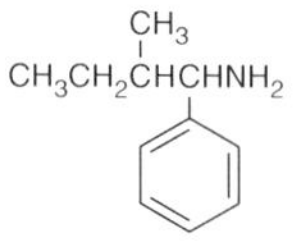

250 mg

S67,757-4
AMINODIPHENYLMETHANE HYDROBROMIDE

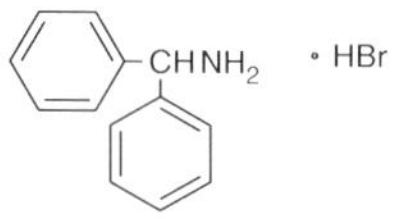

5 gm

S44,808-7
4-FLUORO-ALPHA-METHYLBENZYLAMINE

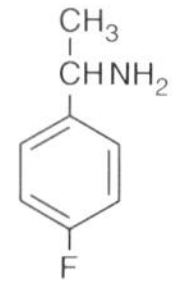

5 gm

S36,288-3
4-ISOPROPYLBENZYLAMINE

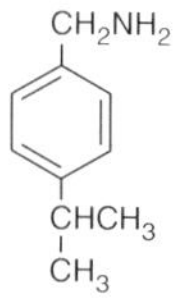

5 gm

S38,736-3
4-HEXADECYL-ALPHA-METHYLBENZYLAMINE

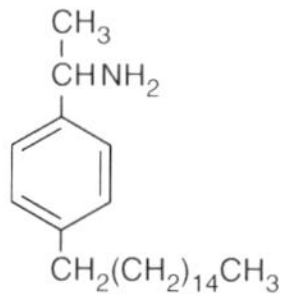

250 mg

S44,243-7
2,4-DIMETHYLBENZYLAMINE

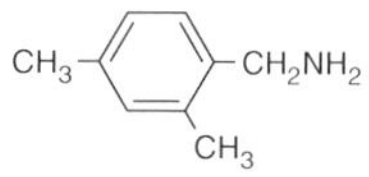

250 mg

S44,519-3
2,5-DIMETHYLBENZYLAMINE

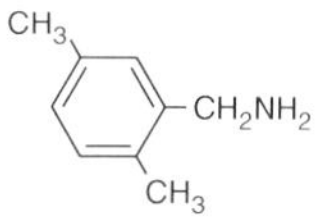

250 mg

S86,559-1
11-AMINO-9,10-DIHYDRO-9,10-ETHANOANTHRACENE

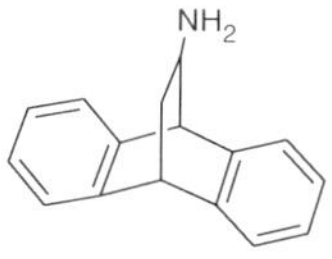

250 mg

S85,311-9
11-(AMINOMETHYL)-9,10-DIHYDRO-9,10-ETHANOANTHRACENE HYDROCHLORIDE HYDRATE

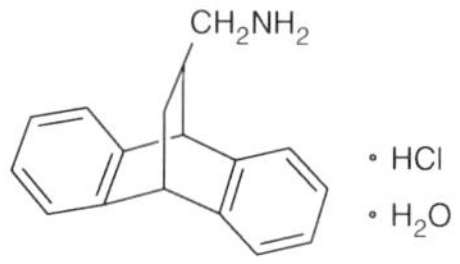

250 mg

S43,783-2
THIOCHROMAN-4-AMINE HYDROCHLORIDE

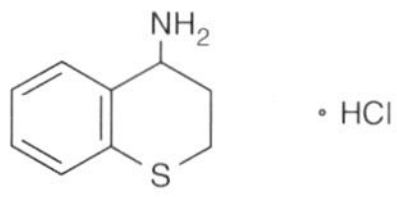

250 mg

S36,376-6
1-METHYL-3-PHENYLPROPYLAMINE HYDROCHLORIDE

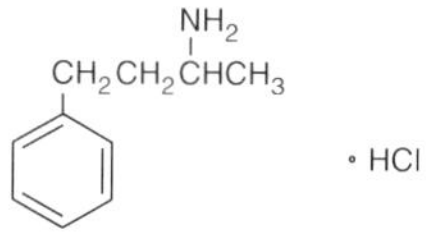

5 gm

S80,661-7
1-AMINO-2-BENZYLINDAN

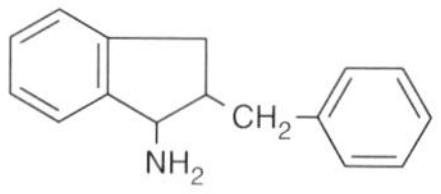

250 mg

S43,486-8
4-BROMO-ALPHA-METHYLBENZYLAMINE

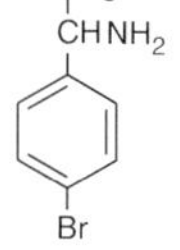

5 gm

S84,990-1
(2-NAPHTHYLMETHYL)AMINE HYDROCHLORIDE

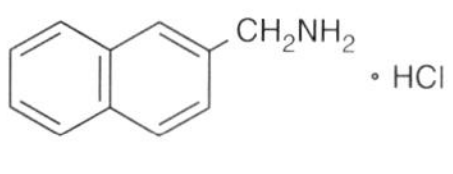

250 mg

S38,730-4
ALPHA-PENTADECYL-2-NAPHTHALENEMETHYLAMINE

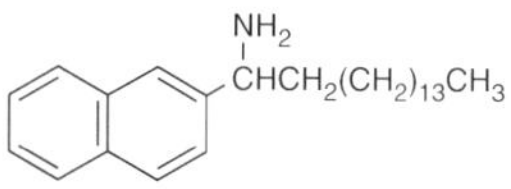

250 mg

S76,744-1
10-(2-AMINOETHYL)-10,11-DIHYDRO-5-METHYLENE-5H-DIBENZO(A,D)-CYCLOHEPTENE

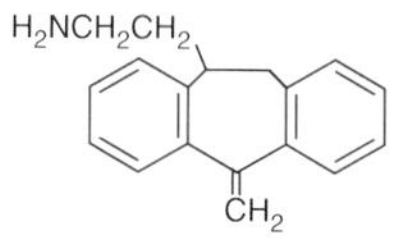

250 mg

S37,750-3
N,N-DIETHYL-N'-PHENYLETHYLENEDIAMINE

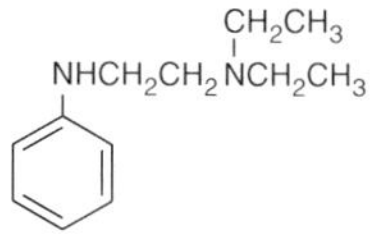

250 mg

S38,839-4
N-BENZYL-N,N'-DIMETHYLETHYLENEDIAMINE

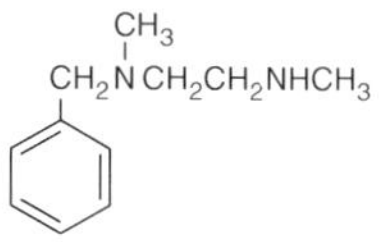

5 gm

S38,186-1
N'-BENZYL-N,N-DIETHYLETHYLENEDIAMINE

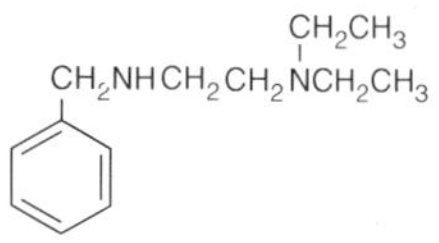

5 gm

S38,209-4
N'-(4-CHLOROBENZYL)-N,N-DIETHYLETHYLENEDIAMINE

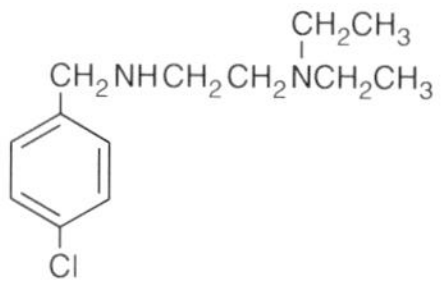

5 gm

S38,191-8
N'-(6-CHLORO-O-TOLYL)-N,N-DIETHYLETHYLENEDIAMINE

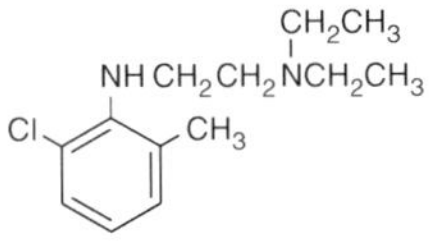

250 mg

S38,399-6
N,N'-DIBENZYLETHYLENEDIAMINE DIHYDROCHLORIDE

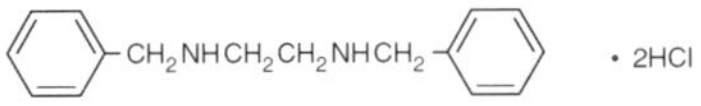

5 gm

S87,744-1
N,N'-BIS(1-PHENYLETHYL)-ETHYLENEDIAMINE

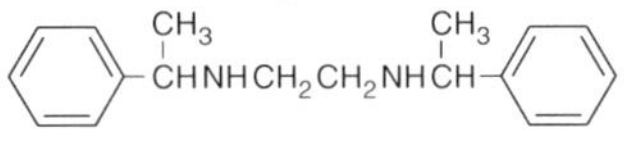

250 mg

S64,268-1
N,N'-BIS(4-CHLOROBENZYL)-ETHYLENEDIAMINE

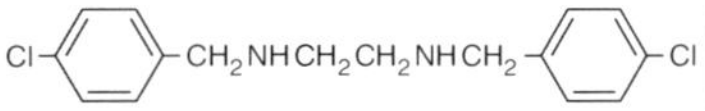

5 gm

S82,533-6
N,N-DIETHYL-N'-METHYL-N'-PHENYLETHYLENEDIAMINE

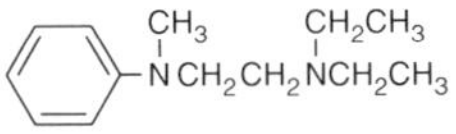

250 mg

S34,181-9
N-BENZYL-N',N'-DIMETHYL-N-PHENYLETHYLENEDIAMINE HYDROCHLORIDE

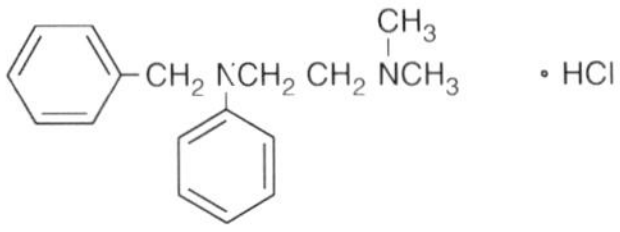

5 gm

S38,840-8
N,N'-DIBENZYL-N,N'-DIMETHYLETHYLENEDIAMINE

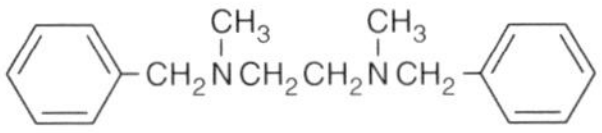

5 gm

S88,262-3
N,N'-BIS(DIPHENYLMETHYLENE)-ETHYLENEDIAMINE

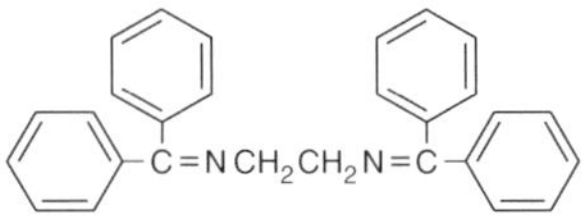

250 mg

S38,444-5
N,N-DIETHYL-N'-(1,2-DIPHENYLETHYL)-1,3-PROPANEDIAMINE

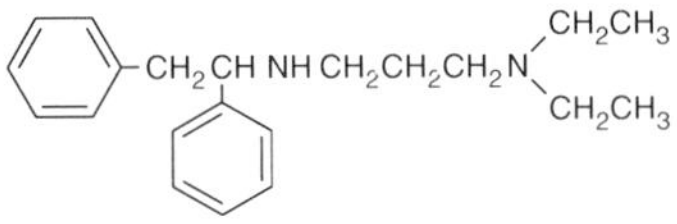

250 mg

S38,339-2
N'-BENZYL-N,N-DIETHYL-1,3-PROPANEDIAMINE

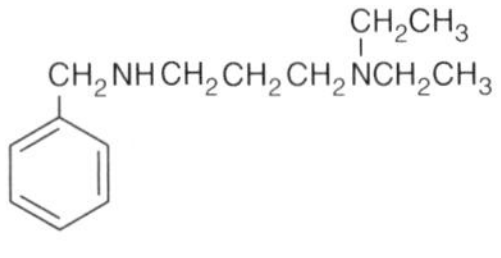

5 gm

S90,711-1
HEXAKIS(ANILINOMETHYL)BENZENE

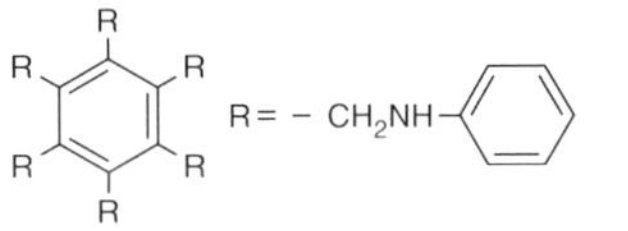

250 mg

S90,699-9
HEXAKIS(N-METHYLANILINOMETHYL)-BENZENE

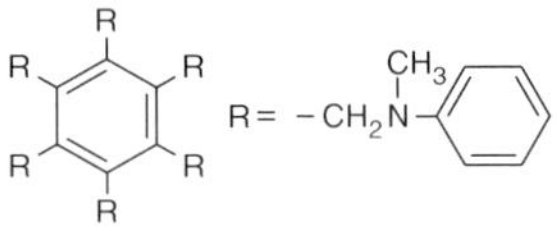

250 mg

S61,981-7
N,N-DIAMYL-N',N'-DIBENZYL-1,4-DIAMINO-2-BUTENE DIHYDROCHLORIDE

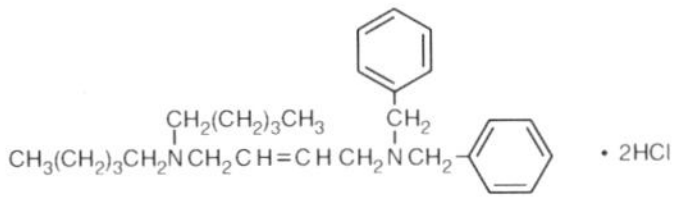

250 mg

S61,979-5
N,N-DIBENZYL-N',N'-DIHEXYL-1,4-DIAMINO-2-BUTENE DIHYDROCHLORIDE

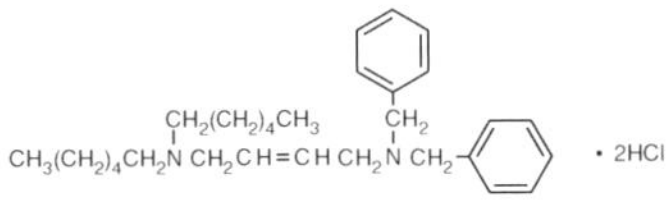

250 mg

S34,762-0
1,2-DIPHENYLETHYL-N-METHYLAMINE

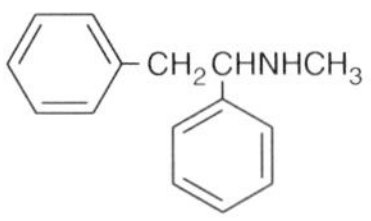

5 gm

S38,358-9
1,2-DIPHENYL-N-METHYLETHYLAMINE HYDROCHLORIDE

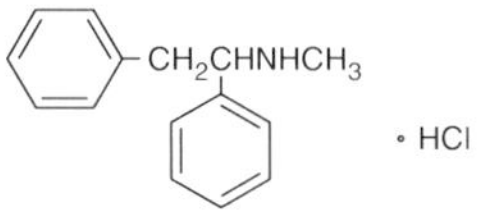

5 gm

S38,413-5
N-BUTYL-ALPHA-PHENYLPHENETHYLAMINE

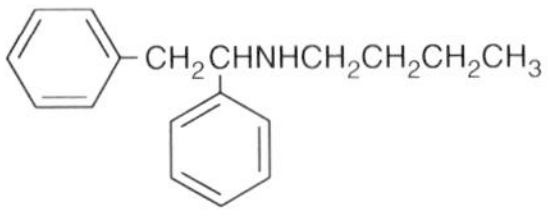

250 mg

S38,432-1
N-BENZYL-ALPHA-PHENYLPHENETHYLAMINE

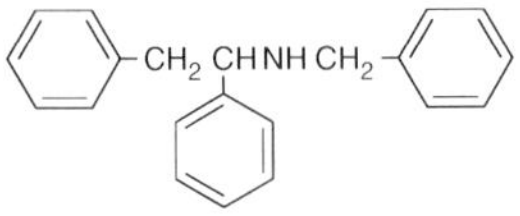

250 mg

S60,052-0
N-(CYCLOPROPYLMETHYL)-4-CHLORO-ALPHA-METHYLPHENETHYLAMINE

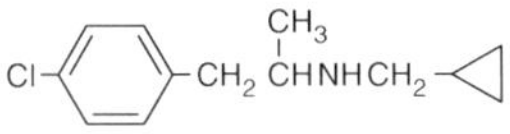

5 gm

S39,766-0
N-(2-NORBORNYL)-(+)-AMPHETAMINE HYDROCHLORIDE

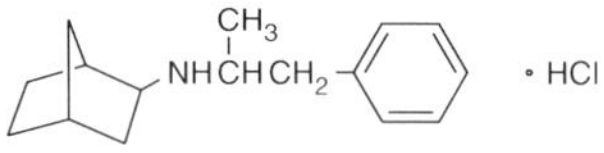

5 gm

S38,364-3
ALPHA-CYCLOHEXYL-N-METHYLPHENETHYLAMINE

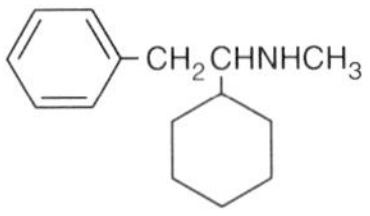

5 gm

S89,721-3
CYCLOHEXYLPHENETHYLAMINE HYDROCHLORIDE

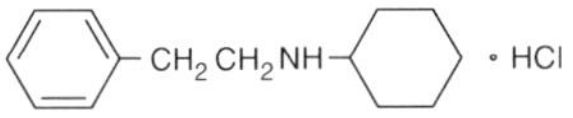

250 mg

S90,615-8
(BETA-METHYLPHENETHYL)PENTYLAMINE HYDROCHLORIDE

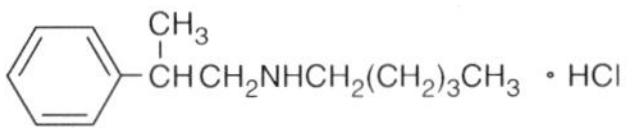

250 mg

S53,976-7
N-STYRYLDIBENZYLAMINE

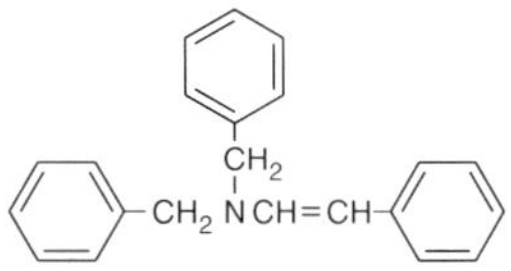

250 mg

S39,765-2
3-[N-METHYL-N-(ALPHA-METHYLPHENETHYL)AMINOMETHYL]-2-NORBORNANONE

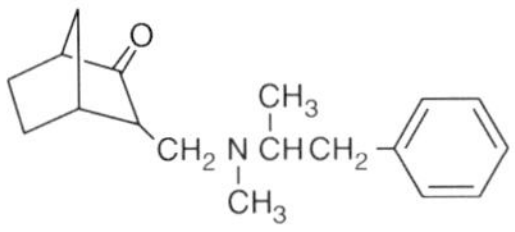

5 gm

S40,972-3
2-(PHENETHYLAMINO)NORBORNANE

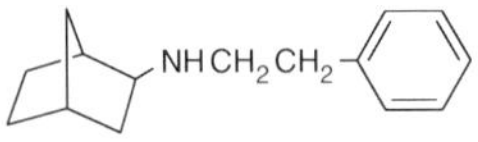

5 gm

S68,330-2
N,N-DIMETHYL-N'-(FLUOREN-9-YL)-ETHYLENEDIAMINE DIHYDROCHLORIDE

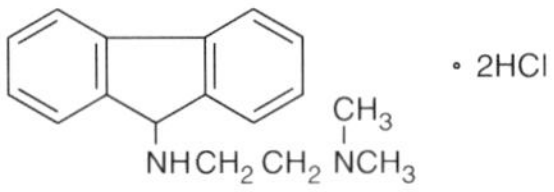

5 gm

S83,657-5
N-BENZYLANILINE HYDROCHLORIDE

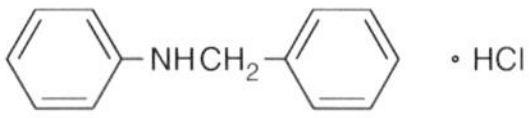

250 mg

S82,974-9
N-(2-PHENYLBENZHYDRYL)ANILINE

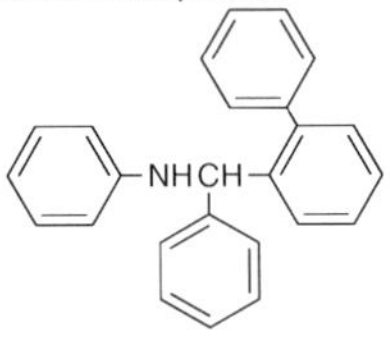

250 mg

S89,640-3
N-(2,4,6-TRIMETHYLBENZYL)ANILINE

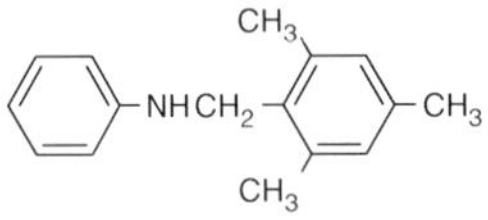

250 mg

S66,878-8
N-BENZYLCYCLOPROPYLAMINE

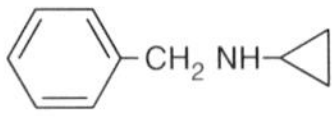

250 mg

S46,244-6
2-(BENZYLAMINO)NORBORNANE

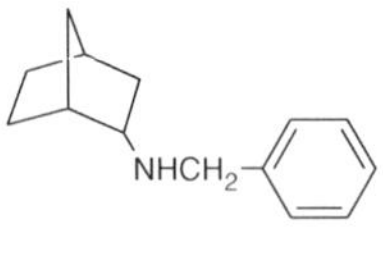

5 gm

S65,152-4
N-BENZYLISOPROPYLAMINE

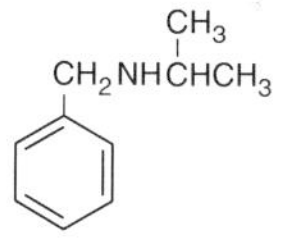

5 gm

S63,315-1
BENZYLISOPROPYLAMINE SALICYCLATE

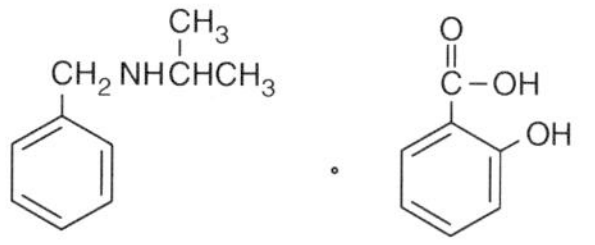

250 mg

S69,666-8
BENZYLISOPROPYLAMINE HYDROBROMIDE

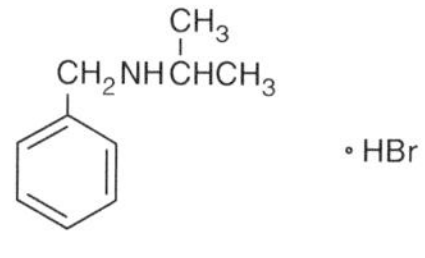

5 gm

S69,750-8
N-ISOPROPYLBENZYLAMINE HYDROCHLORIDE

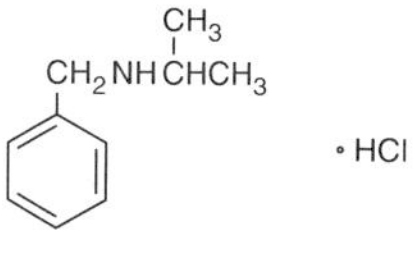

250 mg

S42,828-0
N-BUTYLBENZYLAMINE

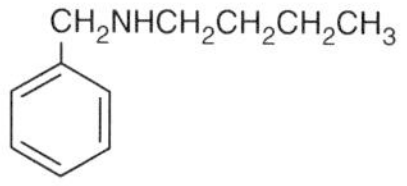

5 gm

S38,152-7
ALPHA-(SEC-BUTYL)-N-METHYLBENZYLAMINE

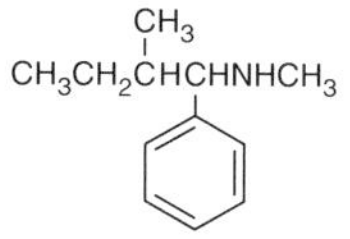

5 gm

S56,196-7
DIBENZYLAMINE HYDROCHLORIDE

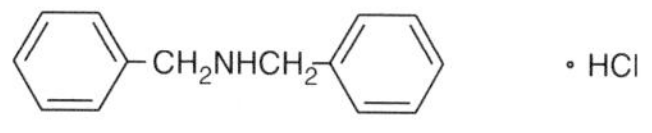

5 gm

S69,845-8
DIBENZYLAMINE HYDROBROMIDE

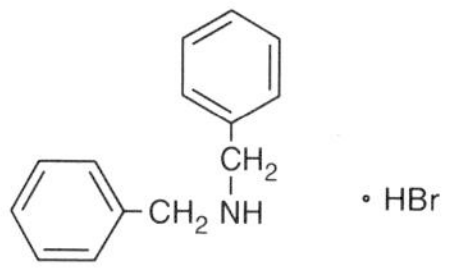

5 gm

S34,683-7
BIS(ALPHA-METHYLBENZYL)AMINE

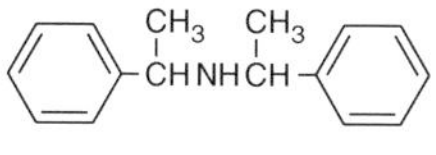

5 gm

S63,986-9
2-BENZYLAMINO-4-PHENYLBUTANE

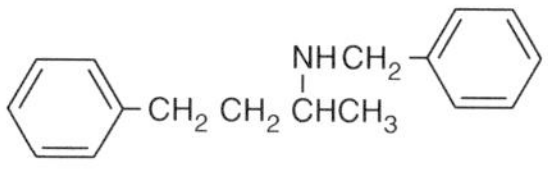

5 gm

S83,086-0
BENZYLDODECYLAMINE

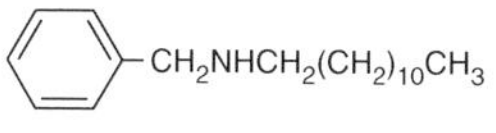

250 mg

S52,561-8
4-CHLORO-ALPHA-ETHYL-N-METHYLBENZYLAMINE HYDROCHLORIDE

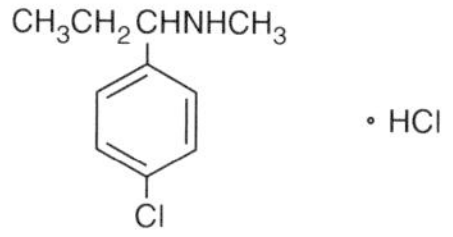

5 gm

S90,013-3
(2,5-DIMETHYLBENZYL)ETHYLAMINE

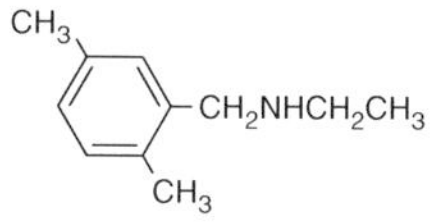

250 mg

S37,675-2
2-PHENOXYDIETHYLAMINE

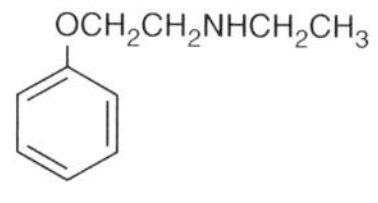

250 mg

S39,767-9
N-(2-NORBORNYL)BENZHYDRYLAMINE HYDROCHLORIDE

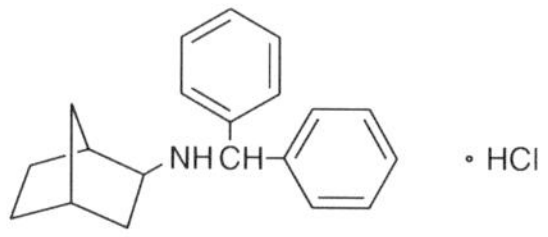

5 gm

S40,707-0
N-(4-CHLOROBENZHYDRYL)-NORBORNANAMINE HYDROCHLORIDE

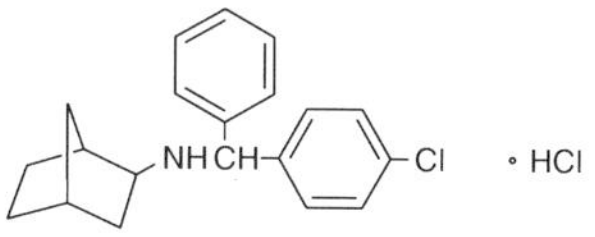

5 gm

S94,298-7
N,N-DIMETHYLBENZYLAMINE HYDROCHLORIDE

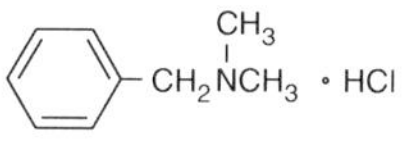

250 mg

S42,807-8
N,N-DIMETHYLBENZYLAMINE-N-OXIDE (50% AQUEOUS SOLUTION)

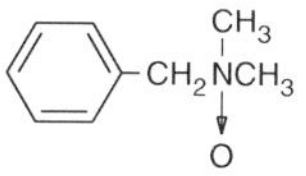

5 gm

S87,974-6
N-ALLYL-N-METHYLBENZYLAMINE

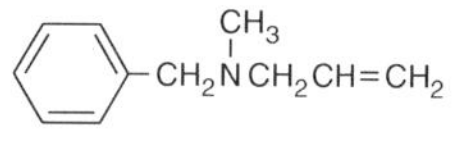

5 gm

S36,014-7
N,N-DIBENZYLMETHYLAMINE

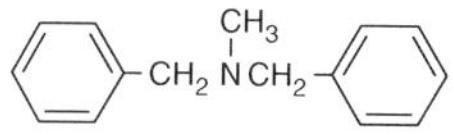

5 gm

S83,098-4
DIBENZYLDODECYLAMINE

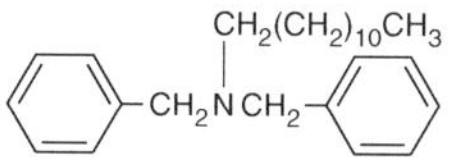

250 mg

S88,865-6
DIMETHYL(DIPHENYLMETHYL)AMINE HYDROCHLORIDE

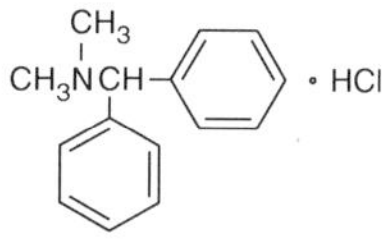

250 mg

S77,489-8
N-BENZYL-N-ISOPROPYLBENZHYDRYLAMINE

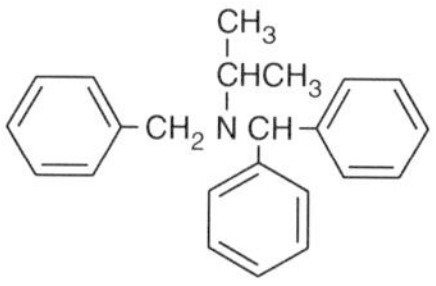

250 mg

S80,954-3
ALPHA,ALPHA'-(DIMETHYLAMINO)-P-XYLENE DIHYDROBROMIDE

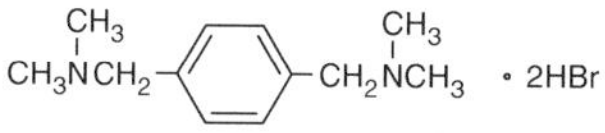

250 mg

S85,315-1
9,10-DIHYDRO-11-(DIMETHYLAMINO)-9,10-ETHANOANTHRACENE

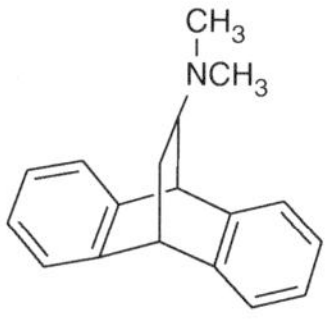

250 mg

S93,339-2
N,N-DIMETHYL-3-PHENYL-2-PROPENAMINE HYDROCHLORIDE

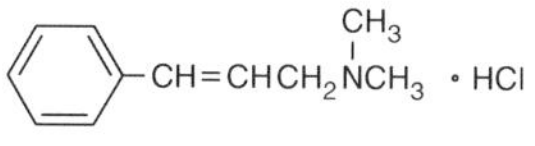

250 mg

S40,319-9
N-ETHYLBIS(3-PHENYLPROPYL)-AMINE CITRATE

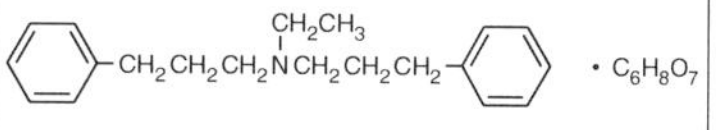

250 mg

S75,984-8
1-[2-(DIMETHYLAMINO)ETHYL]-NAPHTHALENE HYDROCHLORIDE

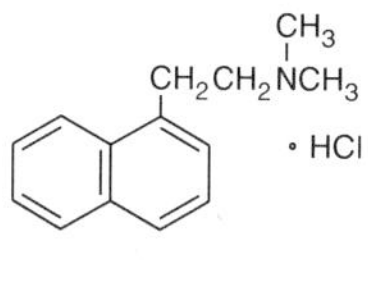

250 mg

S81,995-6
DIETHYL(1-NAPHTHYLMETHYL)AMINE

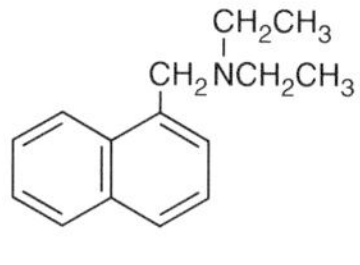

250 mg

S74,841-2
10,11-DIHYDRO-N,N-DIMETHYL-5H-DIBENZO(A,D)CYCLOHEPTENE-10-ETHANAMINE

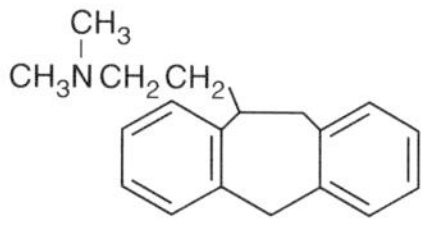

250 mg

S67,843-0
DIHYDRO-10-[2-(DIMETHYLAMINO)ETHYL]-5-METHYL-5H-DIBENZO(A,D)-CYCLOHEPTENE MALEATE

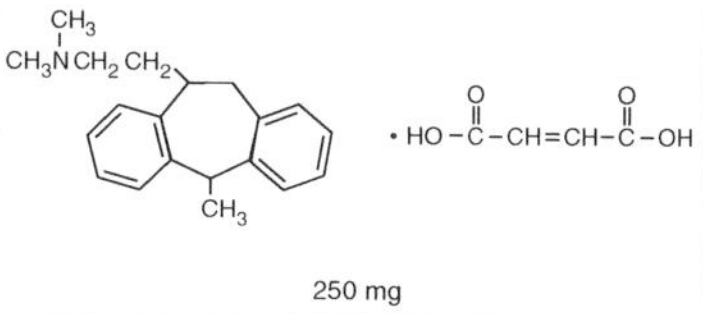

250 mg

S76,014-5
10,11-DIHYDRO-10-(2-DIMETHYLAMINO-ETHYL)-3-FLUORO-5-METHYL-5H-DI-BENZO(A,D)CYCLOHEPTENE MALEATE

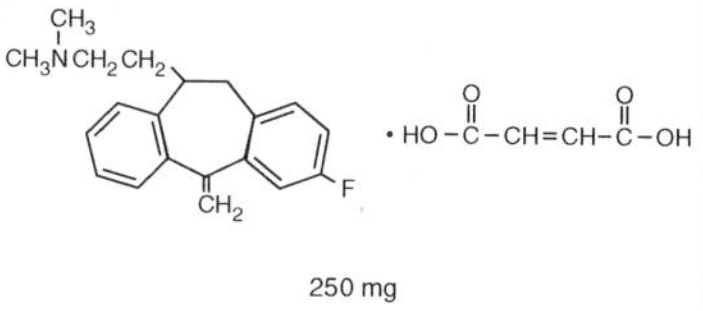

250 mg

S50,391-6
BENZYLDIMETHYLAMMONIUM HEXAFLUOROPHOSPHATE

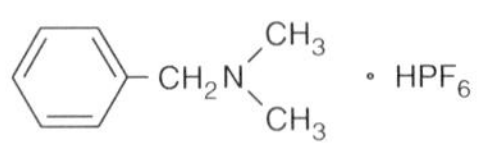

5 gm

S65,830-8
N,N-DIETHYLBENZYLAMINE

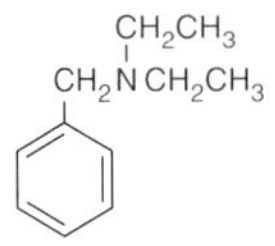

250 mg

S83,204-9
N-(2-NAPHTHYL)BENZOPHENONE IMINE

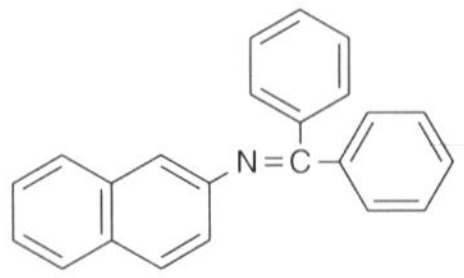

250 mg

S76,119-2
2-(2-CHLOROETHYL)-3-HYDROXY-2,3,6,7-TETRA-HYDROXY-3(3,4-XYLYL)-H-THIAZOLO(3,2-A)PYRIMIDINE HYDROBROMIDE

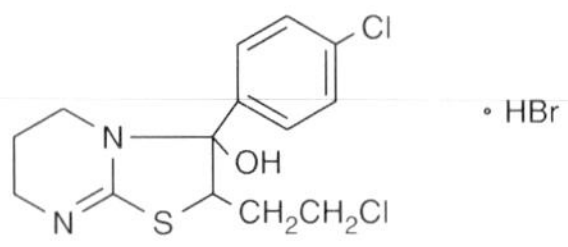

250 mg

S34,438-9
5-CHLOROMETHYL-2-PHENYLOXAZOLIDINE

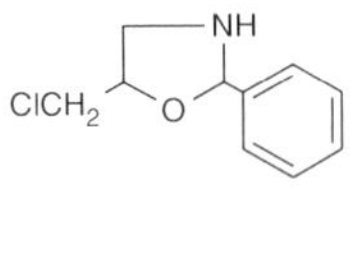

5 gm

S83,408-4
10-(2-CHLOROETHYL)PHENOTHIAZINE

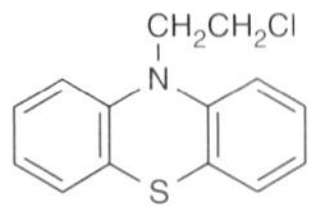

250 mg

S53,574-5
4-(ALPHA,ALPHA,ALPHA-TRIFLUORO-M-TOLYL)-4-PIPERIDINOL

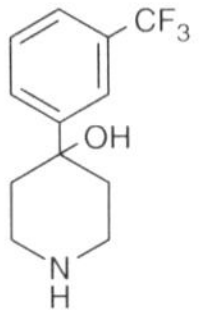

250 mg

S89,352-8
4-[3-(TRIFLUOROMETHYL)PHENYL]-3-PIPERIDINOL HYDROCHLORIDE

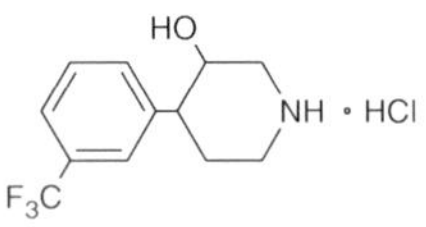

250 mg

S65,106-0
2-(2-[3-(TRIFLUOROMETHYL)ANILINO]-ETHYL)PIPERIDINE

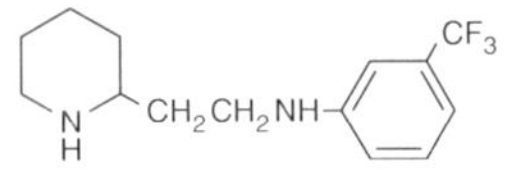

5 gm

S65,103-6
2-(2-[3,5-BIS(TRIFLUOROMETHYL)-ANILINO]ETHYL)PIPERIDINE

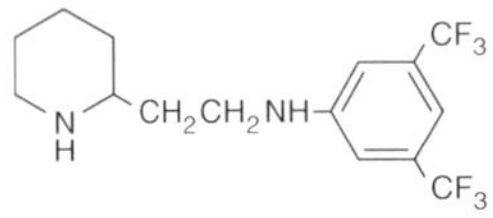

250 mg

S78,211-4
9-[3-(DIETHYLAMINO)PROPYLIDENE]-2-(TRIFLUOROMETHYL)THIOXANTHENE

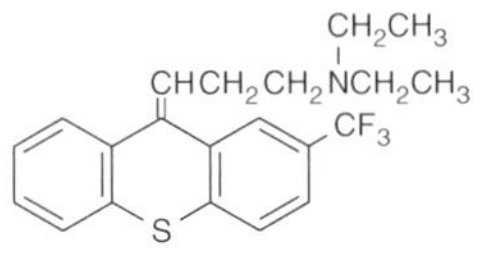

5 gm

S73,242-7
4,5-DICHLORO-2,3-DIHYDRO-2-[4-(TRIFLUOROMETHYL)PHENYL]PYRIDAZINE

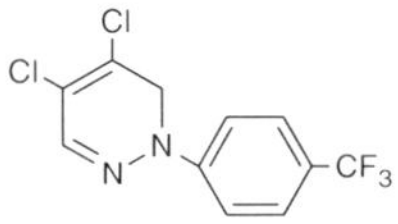

250 mg

S76,445-0
4,4'-OXYBIS[2-CHLORO-6-(MORPHOLMETHYL)PHENOL]

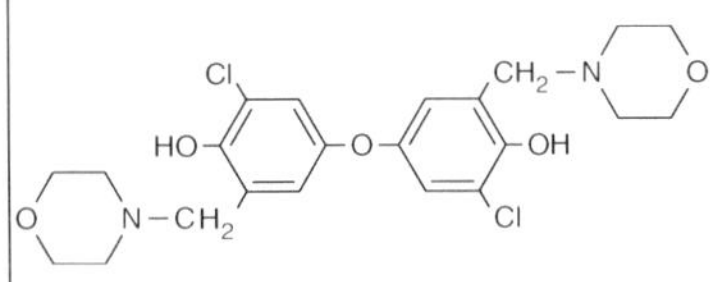

5 gm

S76,398-5
4-CHLORO-ALPHA-(2-IMIDAZOLIN-2-YL)BENZYL ALCOHOL HYDROCHLORIDE

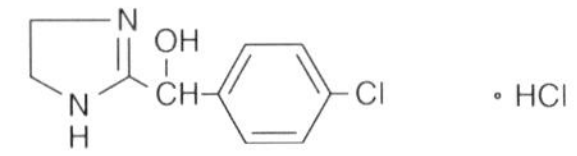

250 mg

S75,832-9
11-(4-CHLOROPHENYL)1,2,3,4,6,11-HEXAHYDRO-11-HYDROXYPYRIDAZINO(1,2-B)PHTHALAZINE

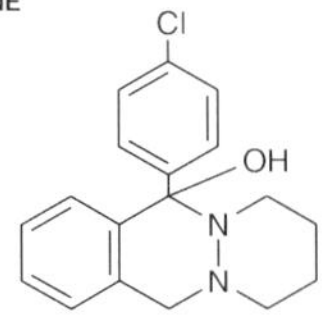

250 mg

S76,630-5
4-CHLORO-ALPHA-(1,4,5,6-TETRAHYDRO-2-PYRIMIDINYL)BENZYL ALCOHOL HYDROCHLORIDE

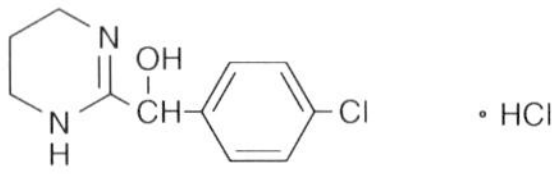

250 mg

S76,135-4
4-CHLORO-ALPHA-(2-QUINOLINYLMETHYL)BENZYL ALCOHOL

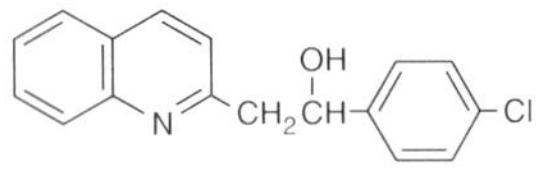

250 mg

S89,326-9
2-[BIS(4-FLUOROPHENYL)METHYL]-3-QUINUCLIDINOL

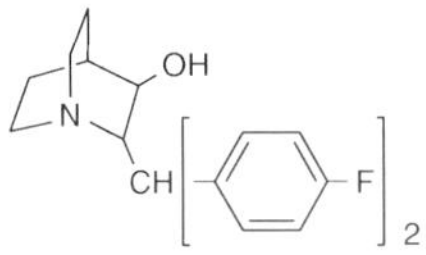

250 mg

S89,344-7
2-[BIS(4-FLUOROPHENYL)METHYL]-3-QUINUCLIDINOL FUMARATE

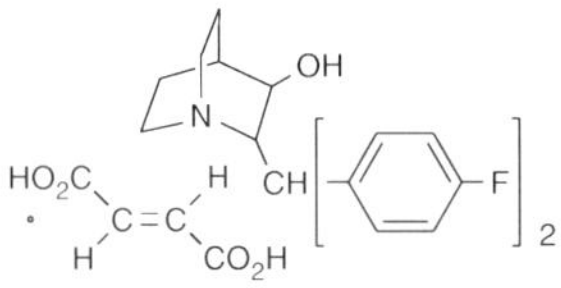

250 mg

S76,171-0
3-(4-CHLOROPHENYL)-2-ISOINDOLINEETHANOL

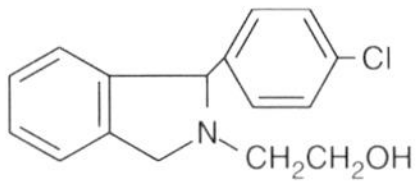

250 mg

S80,656-0
4-CHLORO-ALPHA-[4-(4-TETRAHYDROPYRANYL)-1-PIPERIDINO]-O-CRESOL

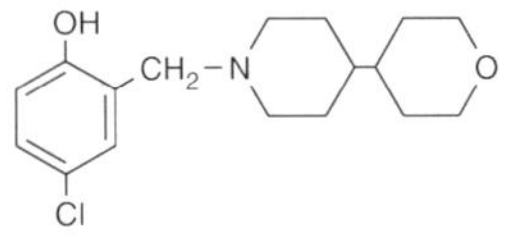

250 mg

S75,919-8
12B-(4-CHLOROBENZYL)2,3-DIMETHOXY-5,6,8,12B-TETRA-H-ISOINDOLO(1,2-A)-ISOQUINOLINE

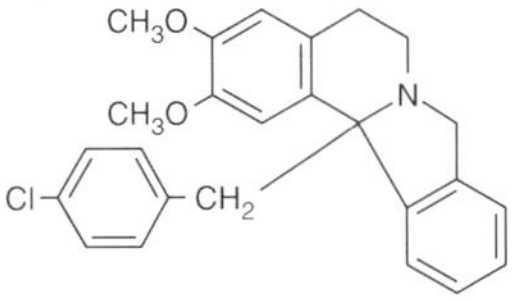

250 mg

S75,452-8
1-(ALPHA-AMINO-4-CHLOROBENZYL)-1,2,3,4-TETRAHYDROISOQUINOLINE DIHYDROCHLORIDE

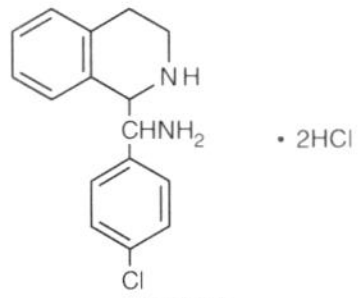

250 mg

S73,526-4
ALPHA-(4-CHLOROPHENYL)-2-(PYRROLIDINOMETHYL)BENZYLAMINE DIHYDROCHLORIDE

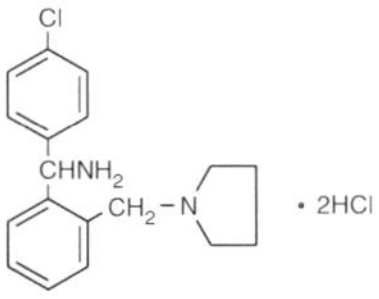

250 mg

S76,376-4
2-(4-BROMOBENZYL)-4,4-DIMETHYL-2-OXAZOLINE

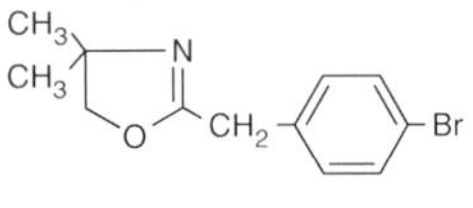

250 mg

S45,296-3
1-(4-CHLOROPHENYL)-2-METHYLPIPERAZINE

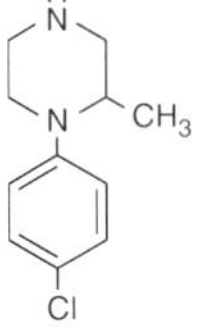

250 mg

S76,621-6
4-(3-FLUOROBENZYL)MORPHOLINE

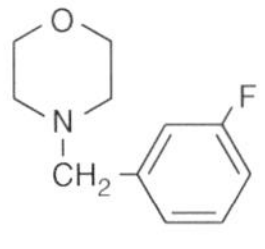

250 mg

S76,613-5
4-(3-CHLOROBENZYL)MORPHOLINE

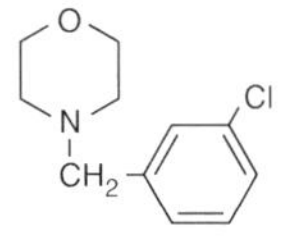

250 mg

S75,838-8
2-TERT-BUTYL-3-(4-CHLOROPHENYL)-AZIRIDINE

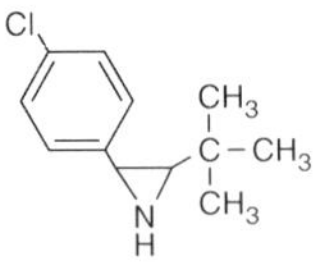

250 mg

S73,376-8
5-(4-CHLOROPHENYL)-2,3-DIHYDRO-5-PHENYL-5H-IMIDAZO(2,1-A)ISOINDOLE

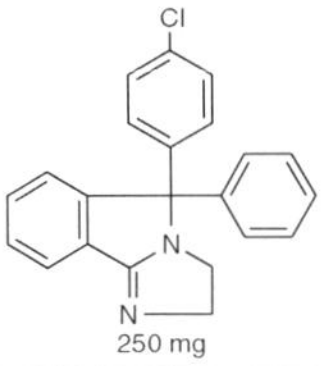

250 mg

S75,903-1
4B-(4-CHLOROPHENYL)-4B,5-DIHYDRO-11H-ISOINDOLO(2,1-A)BENZIMIDAZOLE

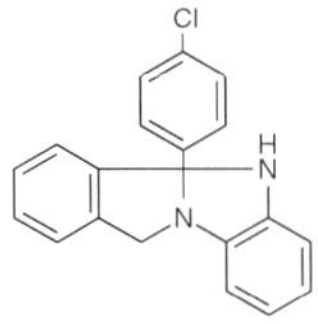

250 mg

S76,625-9
7-CHLORO-2,3-DIHYDRO-5-(4-FLUOROPHENYL)-5-METHYL-5H-IMIDAZO (2,1A)ISOINDOLE

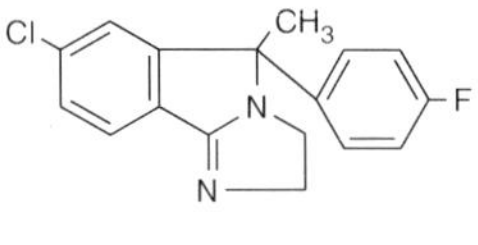

250 mg

S76,120-6
4-(4-CHLOROPHENYL)-1,2-DIHYDRO-2-METHYLPHTHALAZINE
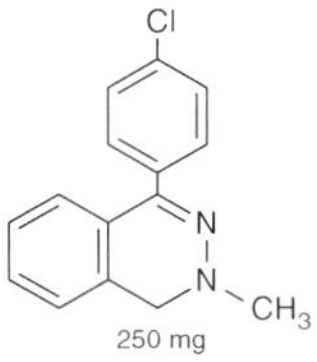
250 mg

S76,010-2
5-BENZYL-2,3,5,6-TETRAHYDROIMIDAZO (2,1-A)ISOQUINOLINE-ALPHA-5-DIOL
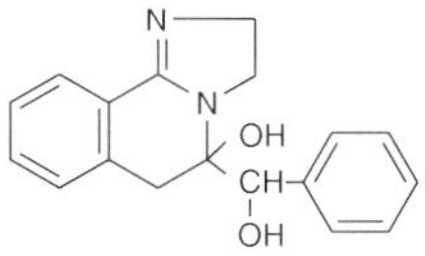
250 mg

S87,318-7
ALPHA-(4-HYDROXYPHENYL)-1-METHYL-1,2,3,4-TETRAHYDRO-6-QUINOLINEMETHANOL
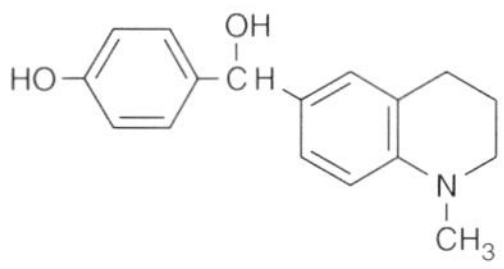
250 mg

S93,616-2
2,3-DIHYDROXYPROMAZINE HYDROCHLORIDE
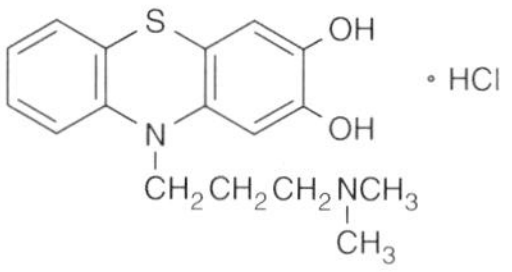
10 mg

S94,745-8
2-CHLORO-7,8-DIHYDROXY-10-(3-DIMETHYLAMINOPROPYL)-10H-PHENOTHIAZINE HYDROCHLORIDE
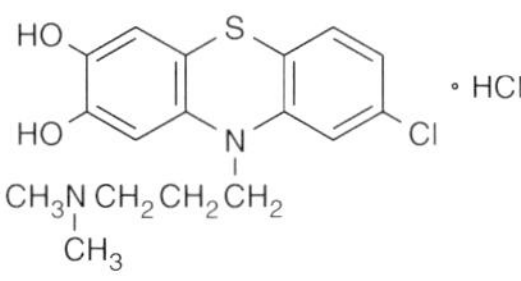
5 mg

S89,353-6
1-BENZYL-4-HYDROXY-3-PIPERIDINEMETHANOL FUMARATE
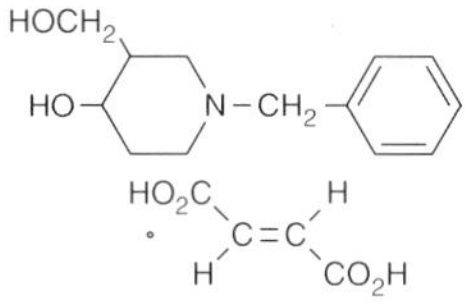
250 mg

S67,436-2
3-(3,4-DIHYDROXYPHENYL)-5,6-DIHYDROIMIDAZO(2,1-B)THIAZOLE HYDROCHLORIDE
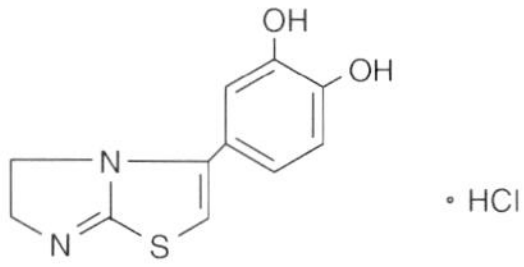
250 mg

S45,892-9
BIS(1,3,3-TRIMETHYL-2-INDOLINYLIDENEMETHYL)-1,3-CYCLOBUTADIENEDIYLIUM-1,3-DIOLATE
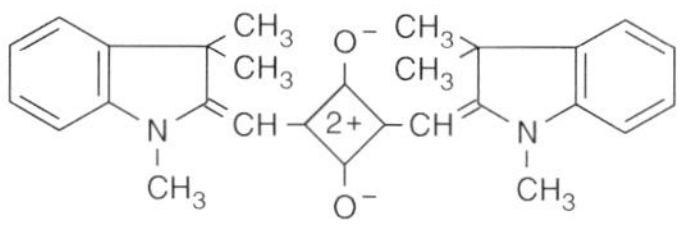
5 gm

S52,058-6
2-(3-[4-(2-METHOXYPHENYL)-PIPERAZINO]PROPOXY)ETHANOL
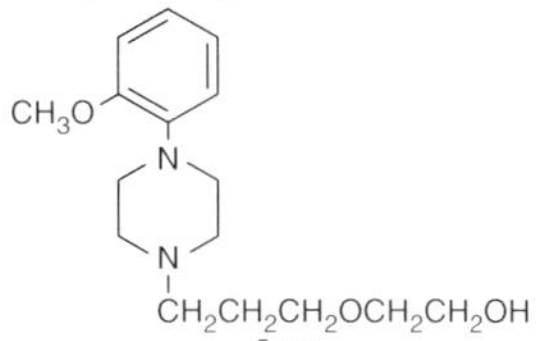
5 gm

S95,673-2
6-(DIMETHYLAMINO)-1-(3-HYDROXY-4-METHOXYBENZYL)-1,2,3,4-TETRAHYDROISOQUINOLINE
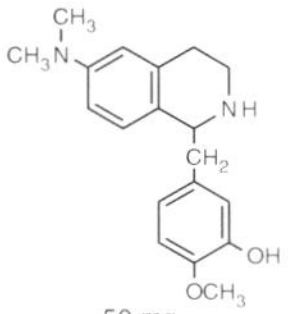
50 mg

S42,608-3
OXYACANTHINE HYDROCHLORIDE TETRAHYDRATE
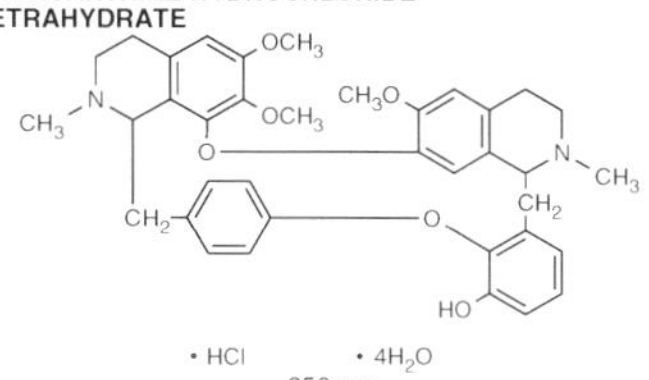
250 mg

S38,226-4
1-(2-BENZYLOXYETHYL)-3-PYRROLIDINEMETHANOL
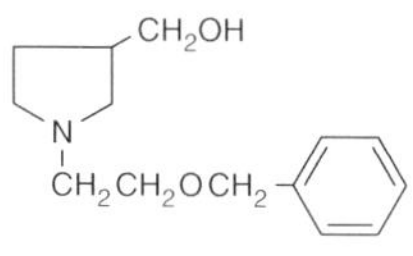
250 mg

S89,363-3
6-METHOXY-2-(4-PHENYL-1,2,5,6-TETRAHYDRO-1-PYRIDYLMETHYL)-1,2,3,4-TETRAHYDRO1NAPHTHOL
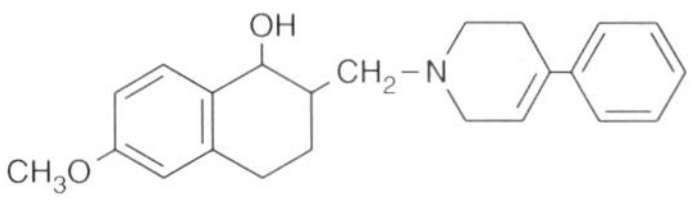
250 mg

S95,517-5
5-AMINO-2-PHENYL-1,3-DIOXANE-5-METHANOL
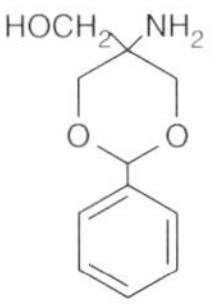
50 mg

S76,389-6
4-METHYL-ALPHA,ALPHA,GAMMA-TRIPHENYL-1-PIPERAZINEPROPANOL
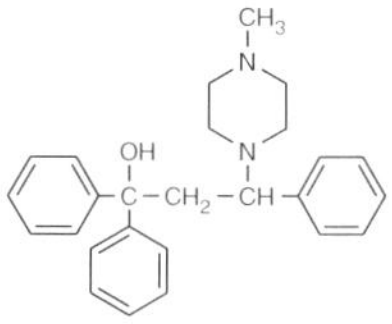
250 mg

S89,658-6
1-PHENYL-3-PIPERIDINO-1-CYCLOHEXANOL
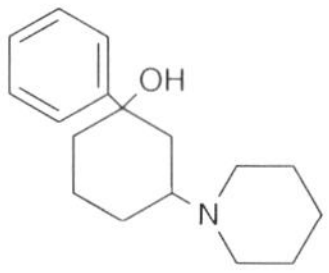
250 mg

S89,346-3
1,2-DIPHENYL-3-METHYL-4-(1,2,5,6-TETRAHYDRO-1-PYRIDYL)-2-BUTANOL HYDROCHLORIDE
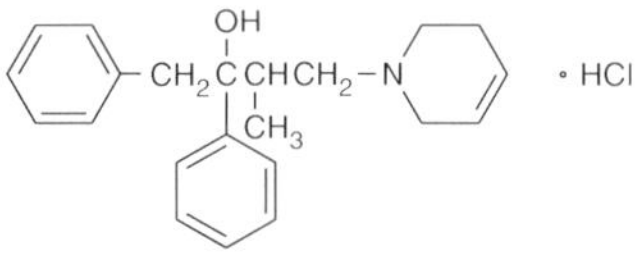
100 mg

S74,883-8
1-BENZYL-2-HYDROXY-2-PHENYL-2,3,5,6-TETRAHYDRO-1H-IMIDAZO(1,2-A)IMIDAZOLE
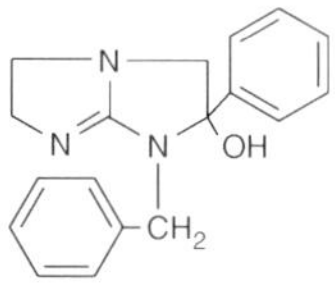
250 mg

S76,360-8
2,3,5,6,7,8-HEXAHYDRO-3-PHENYLTHIAZOLO(3,2-A)(1,3)DIAZEPIN-3-OL HYDROBROMIDE
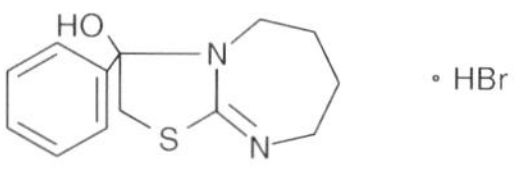
250 mg

S64,588-5
1-PHENYL-3-PYRROLIDINO-1-(P-TOLYL)-1-PROPANOL
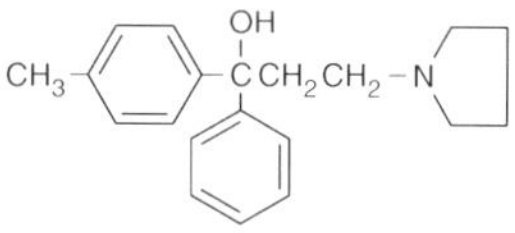
250 mg

S65,719-0
3-(4-CHLOROPHENYL)-2-ETHYL-3-HYDROXY-TETRAHYDRO-5H-THIAZOLO(3,2-A)PYRIMIDINE HYDROBROMIDE
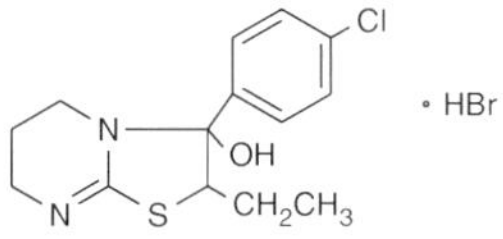
5 gm

S65,723-9
2-BUTYL-3-(4-CHLOROPHENYL)-3-HYDROXY-2,3,6,7-TETRAHYDRO-5H-THIAZOLO(32-A)PYRIMIDINE
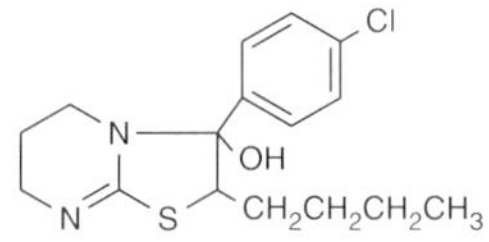
5 gm

S89,705-1
1-BENZYL-4-PHENYL-4-PIPERIDINOL HYDROCHLORIDE
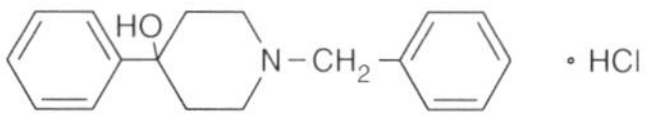
250 mg

S81,067-3
ALPHA-(P-TOLYL)BENZHYDROL
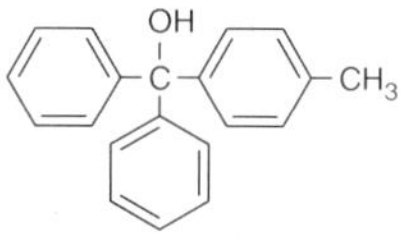
250 mg

S80,622-6
1,3-DIHYDRO-3-[4-(DIMETHYLAMINOMETHYL)PHENYL]-1-METHYL-3-PHENYL-1ISOBENZOFURANOL
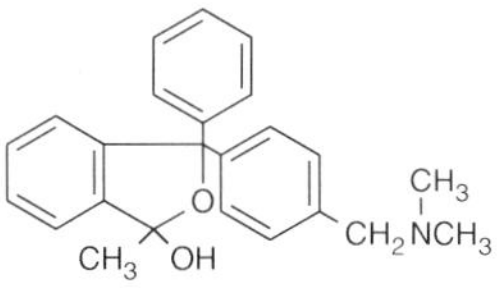
250 mg

S42,632-6
DL-PSEUDOLAUDANINE
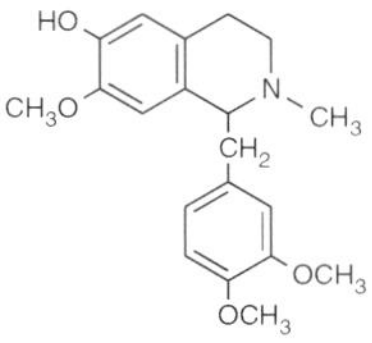
250 mg

S75,931-7
5-(HYDROXYAMINO)-2-PHENYL-1,3-DIOXANE-5-METHANOL
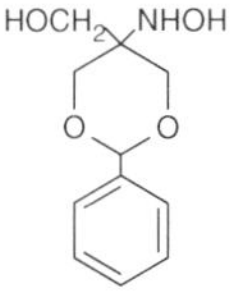
250 mg

S46,069-9
(4S,5S)-(-)-2-ETHYL-5-PHENYL-2-OXAZOLINE-4-METHANOL
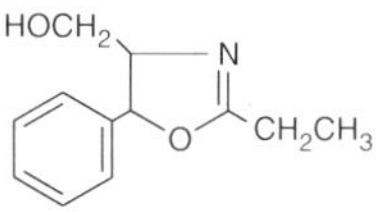
5 gm

S40,450-0
1-PHENETHYLNORLOBELANIDINE
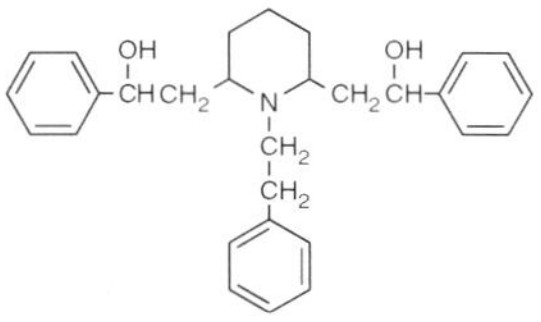
100 mg

S53,572-9
4-(4-FLUOROPHENYL)-4-PIPERIDINOL
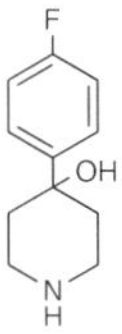
250 mg

S53,573-7
4-(P-TOLYL)-4-PIPERIDINOL
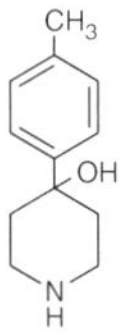
250 mg

S65,987-8
1-BENZHYDRYL-3-HYDROXYAZETIDINE
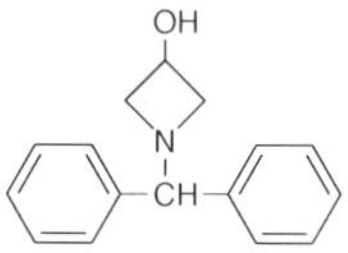
250 mg

S68,066-4
2,6-DI-TERT-BUTYL-4-MORPHOLINOMETHYLPHENOL

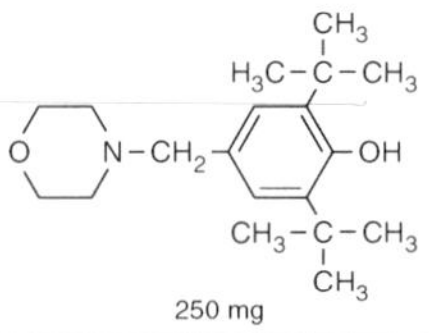

250 mg

S54,893-6
2,5-BIS(MORPHOLINOMETHYL)-HYDROQUINONE

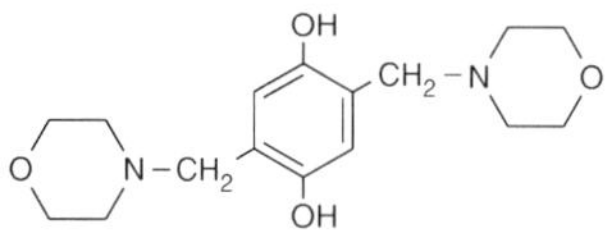

5 gm

S45,223-8
1-BENZYL-3-HYDROXYPIPERIDINE

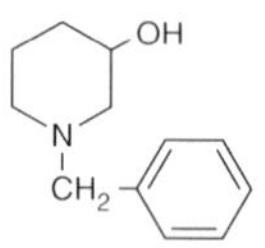

5 gm

S36,997-7
ALPHA-PHENYL-4-MORPHOLINEETHANOL

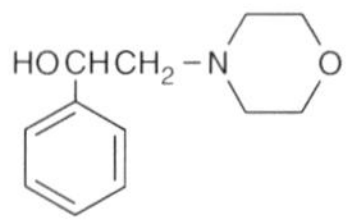

5 gm

S35,843-6
1-BENZYL-4-HYDROXY-4-(P-TOLYL)-PIPERIDINE HYDROCHLORIDE

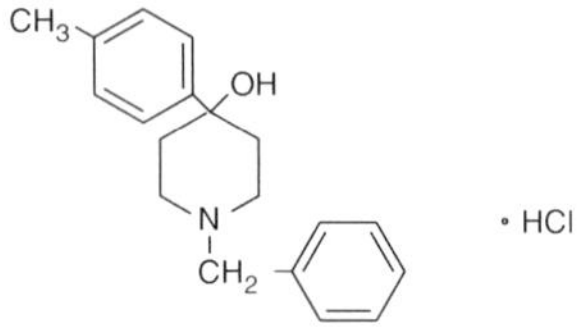

5 gm

S36,395-2
1-MORPHOLINOMETHYL-2-NAPHTHOL

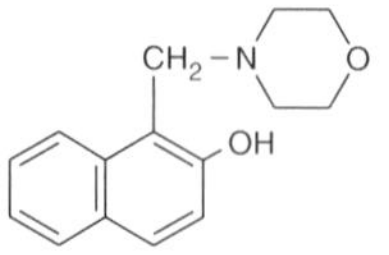

5 gm

S80,654-4
ALPHA-[4-(4-TETRAHYDROPYRANYL)-1-PIPERIDINO]-O-CRESOL

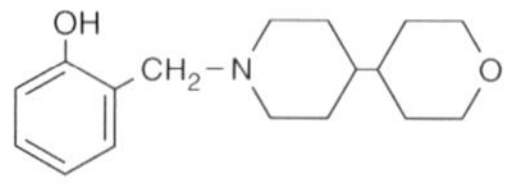

250 mg

S73,463-2
1,3-DIHYDRAZINO-1-HYDROXY-4-PHENYL-1H-PYRIDO(1,2-C)PYRIMIDINE

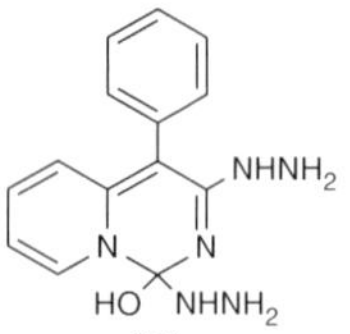

250 mg

S76,649-6
1-(2-PIPERIDINYL)-1-INDANOL HYDROCHLORIDE

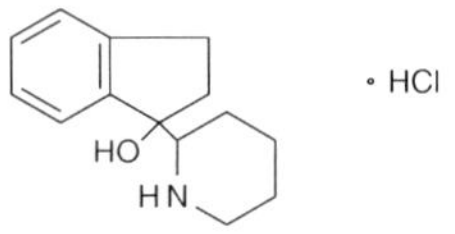

250 mg

S52,922-2
ALPHA-BENZYL-BETA-METHYL-ALPHA-PHENYLTETRAHYDRO-1-PYRIDINE-2-PROPANOL NAPHTHALENESULFONIC ACID

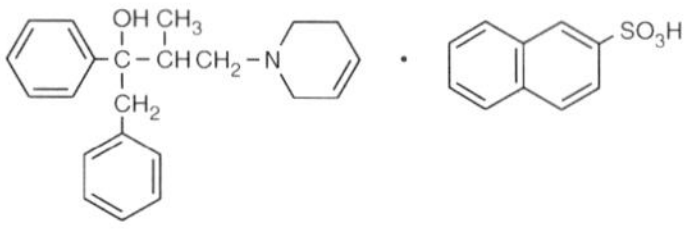

5 gm

S76,320-9
5-(1-ADAMANTYL)-2,3-DIHYDRO-5-HYDROXYIMIDAZO(2,1-A)ISOINDOLE

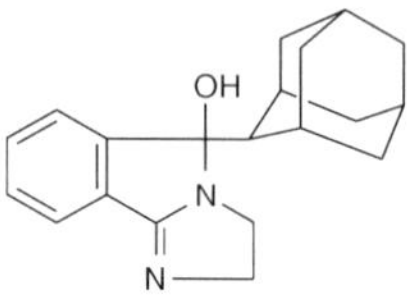

250 mg

S76,054-4
4-(3-METHYL-2-PYRAZOLIN-5-YL)PHENOL

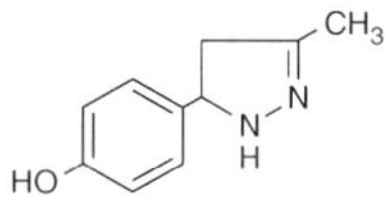

250 mg

S35,121-0
4-HYDROXY-4-(3-PHENYLPROPYL)-PIPERIDINE HYDROCHLORIDE

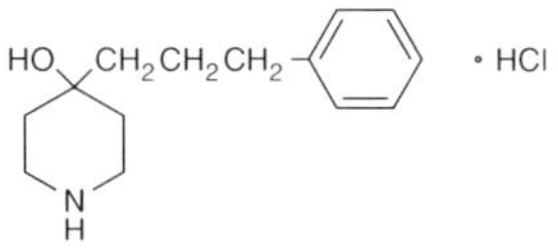

5 gm

S41,800-5
LOBELANIDINE HYDROCHLORIDE

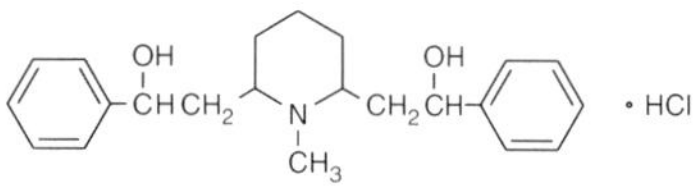

250 mg

S52,030-6
4-PHENYL-1,2,3,6-TETRAHYDRO-1-PYRIDINEETHANOL HYDROCHLORIDE HYDRATE

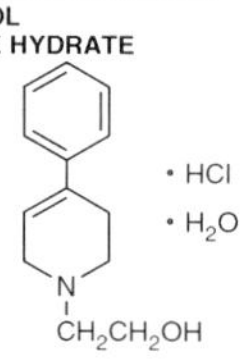

250 mg

S94,741-5
3-HYDROXY-10H-PHENOTHIAZINE

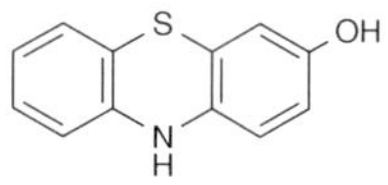

5 mg

S94,004-6
3,3-DIMETHYL-2-(2-HYDROXY-5-METHYLPHENYL)-3H-INDOLE

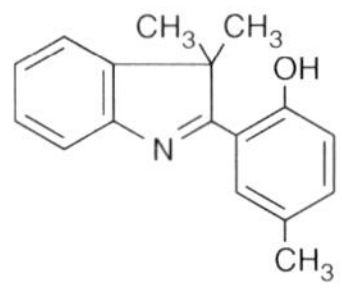

250 mg

S67,831-7
2-(4-BENZYLOXYPHENYL)-3-ETHYLOXAZOLIDINE

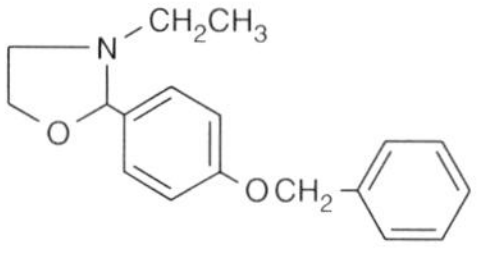

5 gm

S76,315-2
12B-BENZYL-2,3-DIMETHOXY-5,6,8,12B-TETRAHYDROISOINDOLO(1,2-A)-ISOQUINOLINE

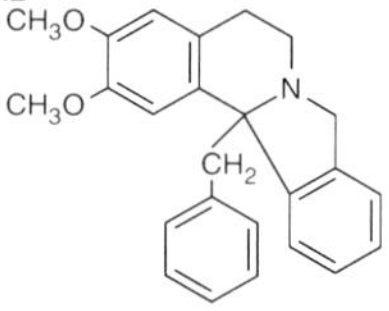

250 mg

S75,907-4
10-CHLORO-2,3-DIMETHOXY-1,3-METHYL-5,6,13,13A-TETRAHYDRO-8H-ISOQUINO[1,2-B]QUINOZOLINE

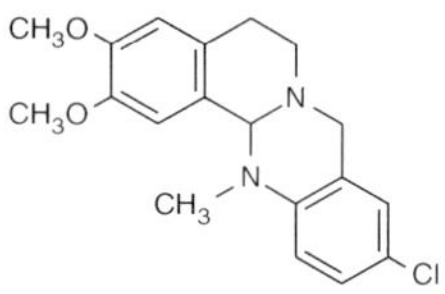

250 mg

S75,917-1
13-BENZYL-10,11-DIMETHOXY-6-METHYL-5,7,8,13-TETRAHYDRO-6H-DIBENZ(C,F)-AZONINE

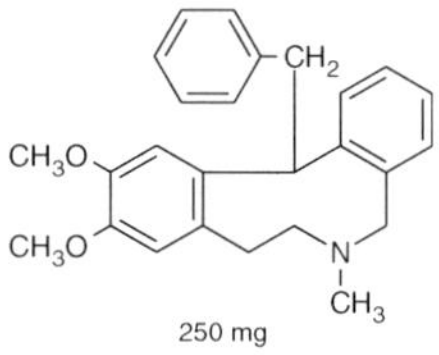

250 mg

S73,427-6
1,2,3,5,6,11,12,13,14,14A-DECAHYDRO-8,9-DIMETHOXYDIBENZO(A,I)-QUINOLIZINE HYDROCHLORIDE

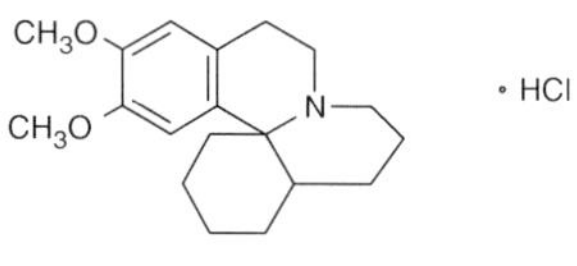

250 mg

S76,083-8
2,3-DIMETHOXY-5,6,8,13A-TETRAHYDRO-13H-ISOQUINO(1,2-B)QUINAZOLINE

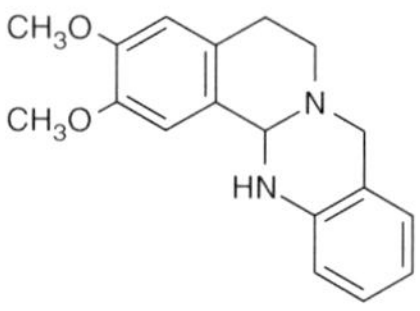

250 mg

S54,209-1
1,3-DIPHENYL-2-(3,4,5-TRIMETHOXYPHENYL)IMIDAZOLIDINE

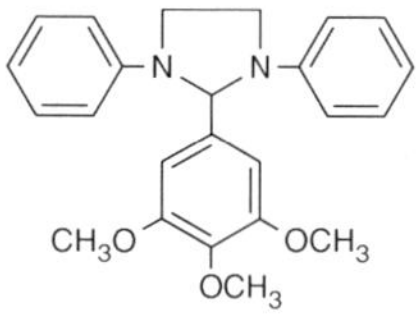

250 mg

S86,148-0
5-BROMO-6-(3,5-DIMETHOXY-4-METHYL-BETA-PIPERIDINOSTYRYL)-1,3-BENZODIOXOLE

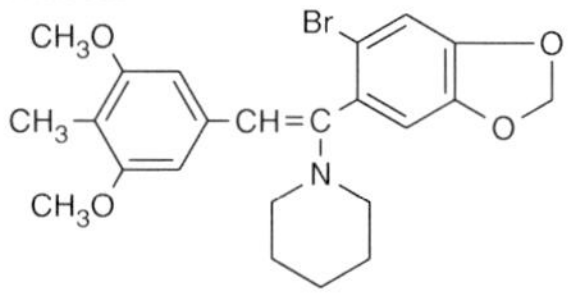

250 mg

S42,432-3
DL-6'-BROMOLAUDANOSINE

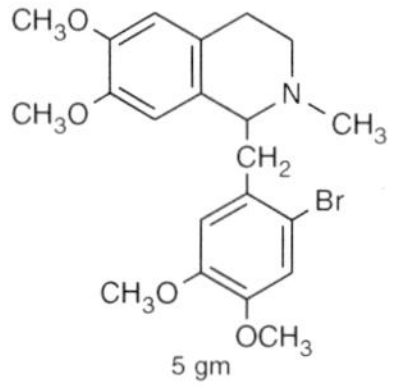

5 gm

S42,471-4
3,4-DIHYDROPAPAVERINE

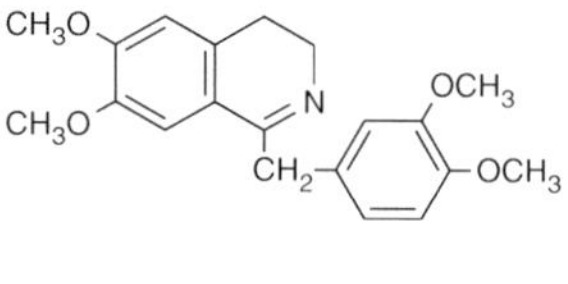

250 mg

S42,616-4
DL-PAVINE HYDROCHLORIDE

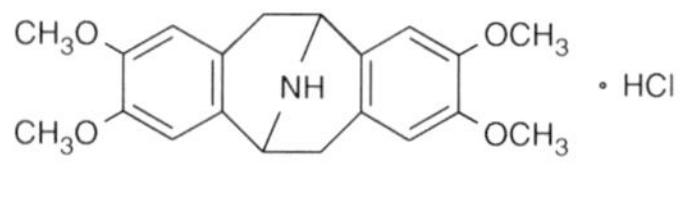

250 mg

S44,906-7
L-TETRAHYDROPALMATINE HYDROCHLORIDE

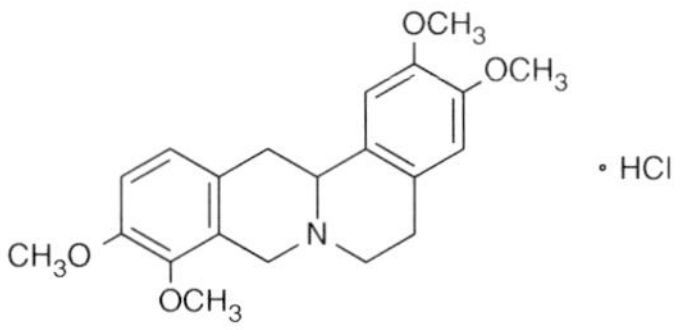

250 mg

S44,917-2
D-TETRAHYDROPALMATINE HYDROCHLORIDE

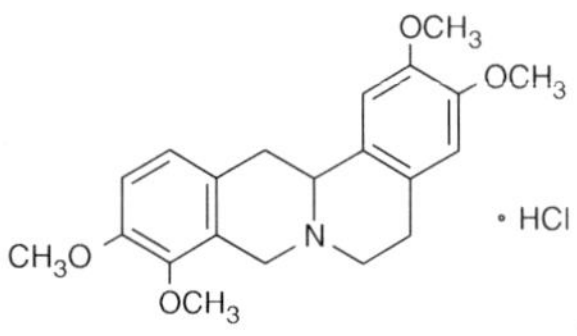

250 mg

S42,449-8
ALPHA-CORALYDINE

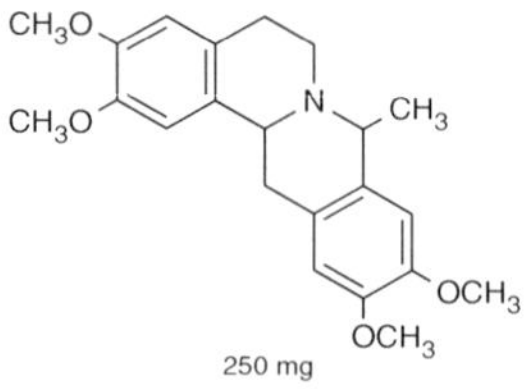

250 mg

S42,450-1
ALPHA-CORALYDINE HYDROCHLORIDE

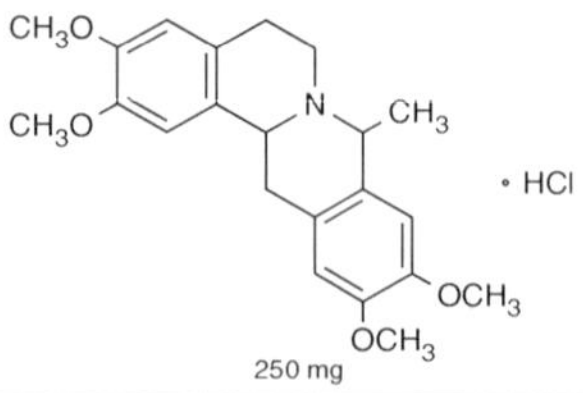

250 mg

S44,912-1
CORYDALINE
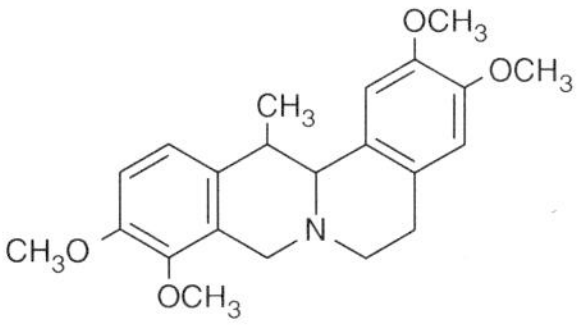
250 mg

S61,507-2
N-PIPERONYL-4-FLUOROANILINE
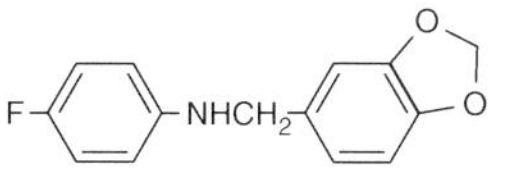
5 gm

S41,755-6
9-DESOXYBERBERINE
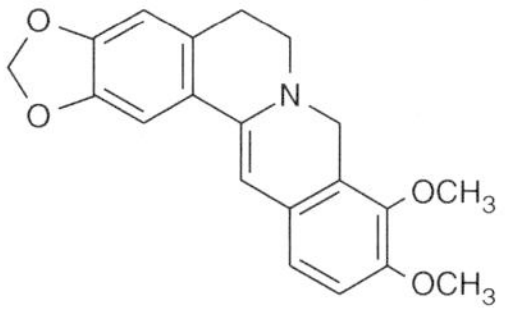
250 mg

S35,220-9
2-METHOXY-4-MORPHOLINOANILINE HYDROCHLORIDE
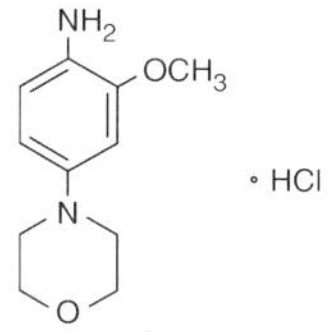
5 gm

S41,263-5
8-(4-ETHOXY-3-METHOXYPHENYL)-1,5-DIAZABICYCLO[3.2.1]OCTANE
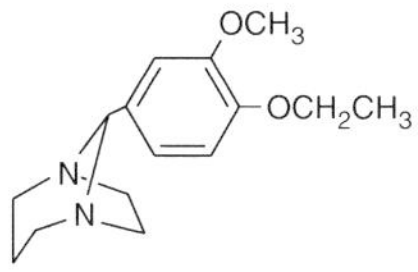
5 gm

S52,779-3
1-(3-CHLORO-P-TOLYL)-4-[3-(4-TERT-OCTYLPHENOXY)PROPYL]PIPERAZINE-HYDROCHLORIDE
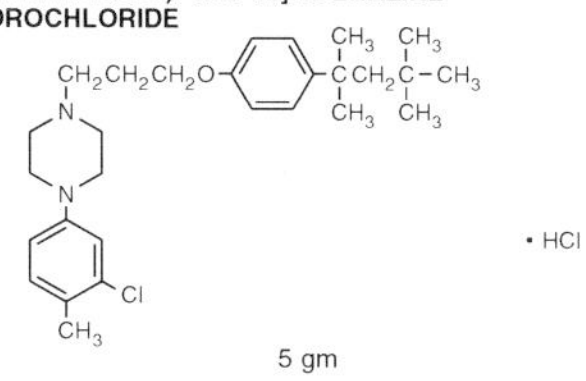
5 gm

S40,227-3
2-CHLORO-10-[3-(2-METHOXYETHYL)-1-METHYL-3-PYRROLIDINYLMETHYL]-PHENOTHIAZINE HYDROCHLORIDE
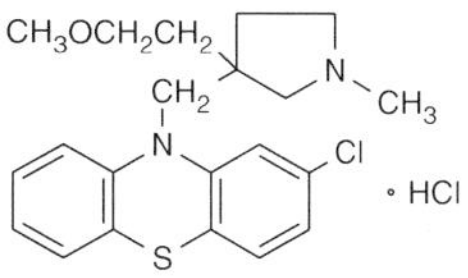
250 mg

S79,870-3
1,2-DIHYDRO-1-[1-(4-METHOXYPHENYL)-4-PIPERIDYL]-2-METHYLENEQUINOLINE
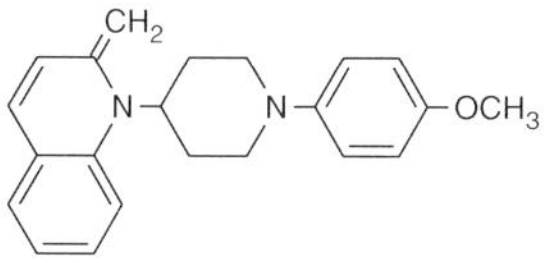
250 mg

S45,409-5
1-(3-METHOXYPHENYL)PIPERAZINE DIHYDROCHLORIDE
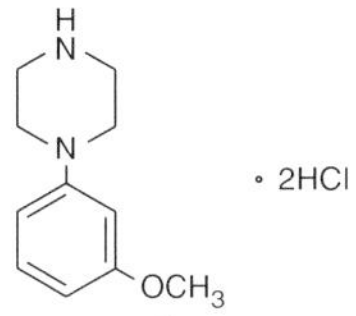
5 gm

S44,213-5
1-(4-METHOXYPHENYL)-2-METHYLPIPERAZINE
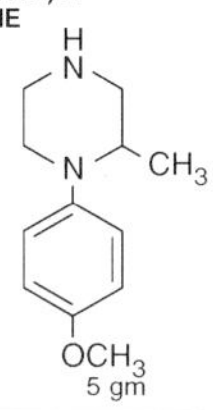
5 gm

S95,739-9
3A,8B-DIMETHYL-2,2-ETHYLENEDITHIO-HEXAHYDRO-6-METHOXYCYCLOPENT(B)-INDOLE
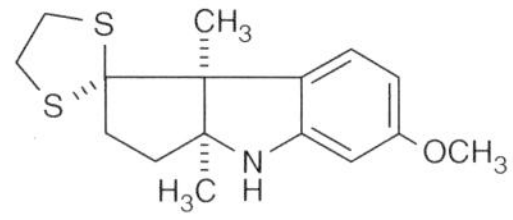
50 mg

S93,620-0
2-METHOXYPHENOTHIAZINE
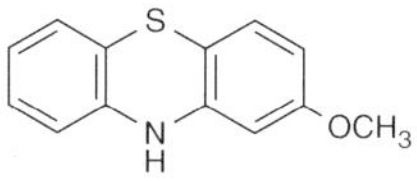
250 mg

S93,512-3
2-CHLORO-7-METHOXY-10H-PHENOTHIAZINE
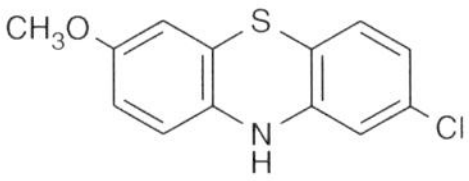
250 mg

S93,789-4
4-CHLORO-1-METHOXY-10H-PHENOTHIAZINE
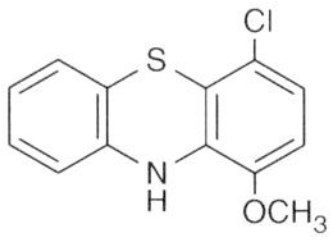
250 mg

S36,058-9
1,2-DIHYDRO-6-ETHOXY-2,2,4-TRIMETHYLQUINOLINE
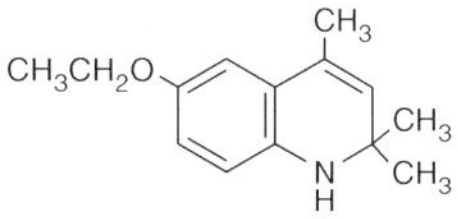
5 gm

S36,958-6
8-(3-METHOXYPHENYL)-1,5-DIAZABICYCLO[3.2.1]OCTANE
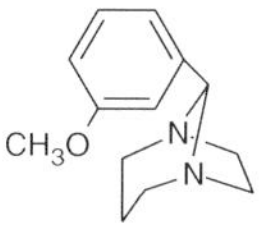
250 mg

S84,994-4
1,3-DIMETHYL-2-(2-METHOXYPHENYL)-IMIDAZOLIDINE
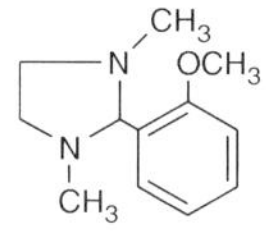
250 mg

S90,651-4
1-(DIPHENYLMETHYL)-3-METHOXYAZETIDINE HYDROCHLORIDE
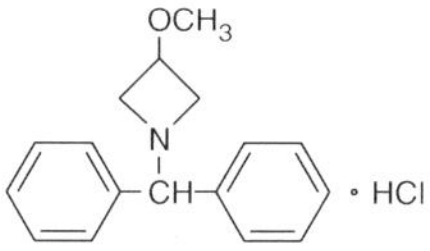
250 mg

S83,738-5
1-(3-METHOXYPHENYL)PIPERIDINE
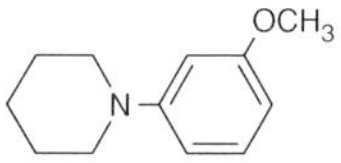
250 mg

S42,440-4
CARNEGINE HYDROCHLORIDE
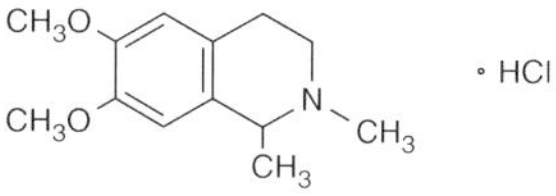
250 mg

S66,182-1
1,4-DIHYDRO-3-ETHOXY-1-PHENYLISOQUINOLINE
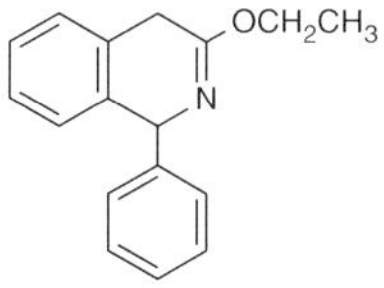
5 gm

S80,738-9
2-ANILINO-1-METHOXY-1,2,3,4-TETRAHYDROISOQUINOLINE
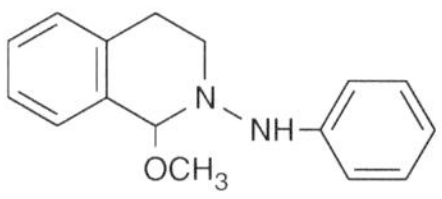
250 mg

S76,609-7
4-TERT-BUTYL-5-(4-METHOXYPHENYL)-2-(METHYLAMINO)-2-OXAZOLINE
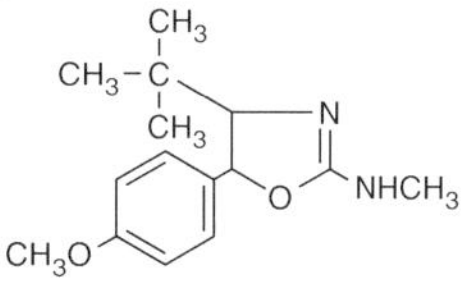
250 mg

S54,268-7
2-(P-ANISIDINO)-2-THIAZOLINE
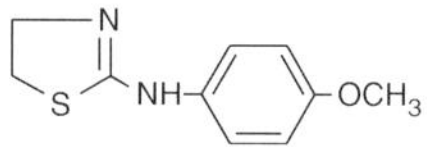
250 mg

S87,232-6
1-ETHOXYISOINDOLINE
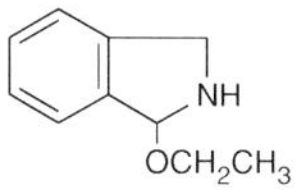
250 mg

S35,469-4
4-PHENYL-4-PROPOXYPIPERIDINE HYDROCHLORIDE
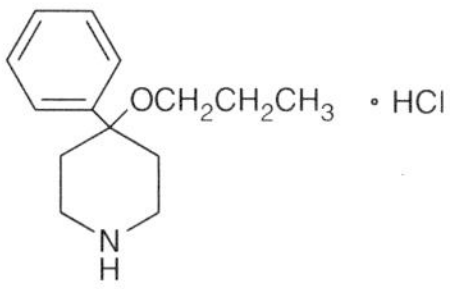
250 mg

S35,235-7
1-(2-METHOXYPHENYL)-4-(3-METHOXYPROPYL)PIPERAZINE DIHYDROCHLORIDE
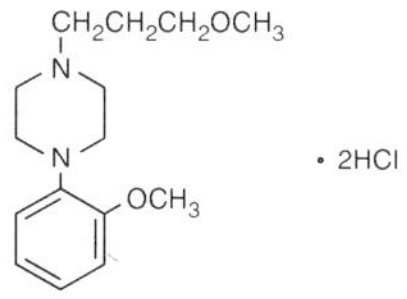
5 gm

S52,781-5
1-(3-CHLORO-P-TOLYL)-4-[6-(4-TERT-PENTYLPHENOXY)HEXYL]-PIPERAZINE HYDROCHLORIDE
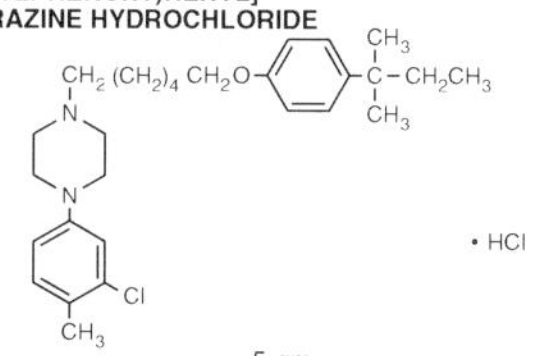
5 gm

S76,760-3
3-TERT-BUTYL-6-METHOXY-4-PHENYL-1,2,3,4-TETRAHYDROISOQUINOLINE
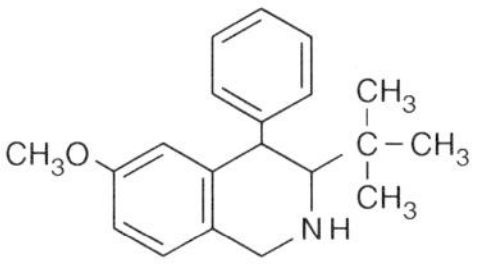
250 mg

S54,202-4
6-(4-CHLOROPHENOXY)-3-[4-(4-CHLOROPHENOXY)PHENYL]-3,4-DIHYDROQUINAZOLINE
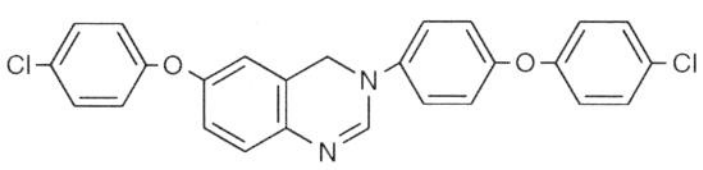
5 gm

S76,626-7
6-METHOXY-1,2,3,4-TETRAHYDROQUINOLINE
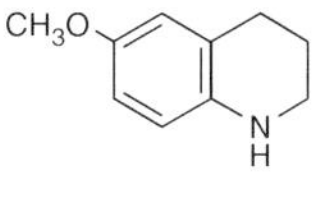
250 mg

S38,823-8
11-METHYLDIBENZ(B,F)OXAZEPINE
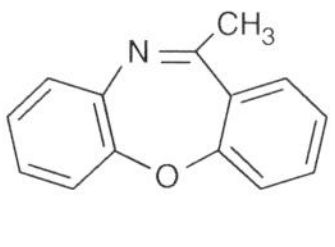
250 mg

S89,328-5
2-(2,2-DIPHENYL-1,3-DIOXOLAN-4-YL)-PIPERIDINE HYDROCHLORIDE
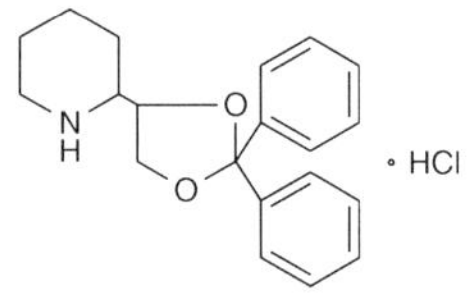
250 mg

S89,325-0
2,2-DIPHENYL-4A,5,6,7,8,8A-HEXAHYDRO-4H-1,3-DIOXINO(5,4-B)-PYRIDINE HYDROCHLORIDE
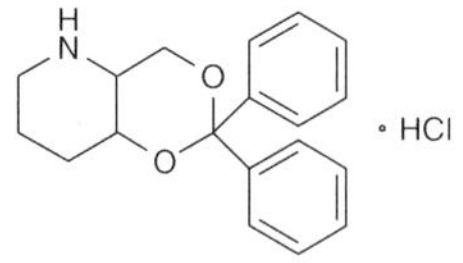
250 mg

S50,276-6
5-ETHYL-3-PHENYL-1-AZA-4,6-DIOXABICYCLO[3.3.0]OCTANE
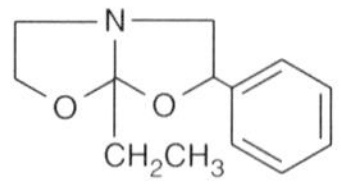
5 gm

S50,442-4
1,4-DIHYDRO-1-PHENYL-4,4,6-TRIMETHYL-2-PYRIMIDINETHIOL
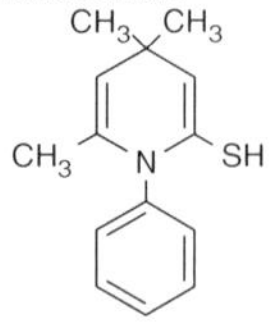
250 mg

S78,209-2
10-[3-(DIMETHYLAMINO)-2-METHYLPROPYL]-2-(METHYLTHIO)-PHENOTHIAZINE
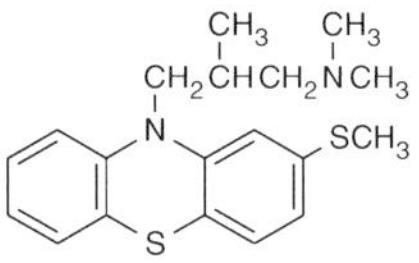
5 gm

S36,447-9
N-PHENYLPHENOTHIAZINE
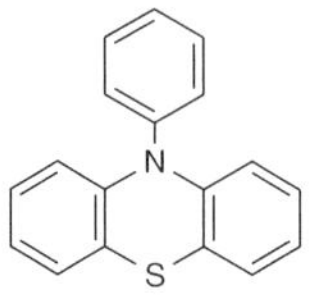
250 mg

S75,932-5
1-BENZYL-4,5-DIHYDRO-2-(METHYLTHIO)-IMIDAZOLE HYDROCHLORIDE
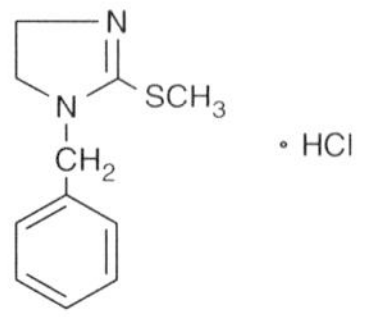
250 mg

S75,888-4
ALPHA,ALPHA'-BIS[(4,5-DIHYDRO-2-IMIDAZOLYL)THIO]-O-XYLENE DIHYDROBROMIDE
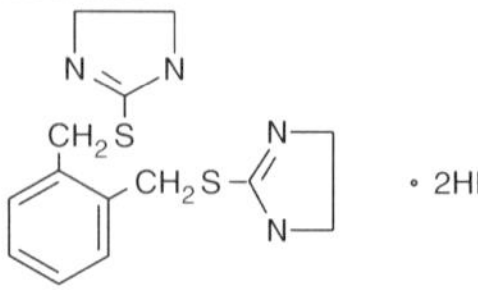
250 mg

S36,638-2
3-AMINO-5,8-DIPHENYL-1,2,4-TRIAZOCINE
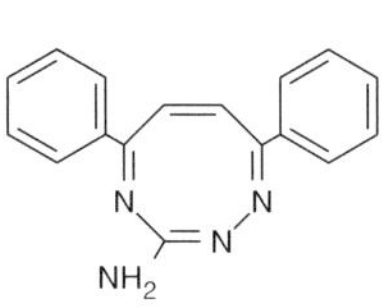
5 gm

S67,789-2
5-AMINO-2,3-DIHYDRO-7-PHENYL-1H-1,4-DIAZEPINE
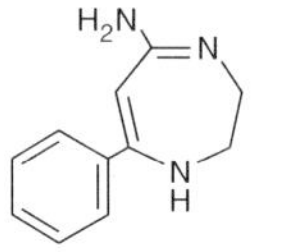
250 mg

S51,126-9
5-AMINO-1,3-DIBENZYL-5-METHYLHEXAHYDROPYRIMIDINE
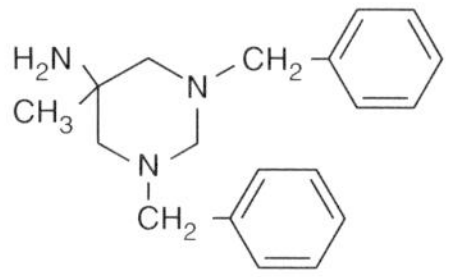
250 mg

S92,835-6
4-AMINO-3-METHYL-2(3H)-PYRIMIDINETHIONE
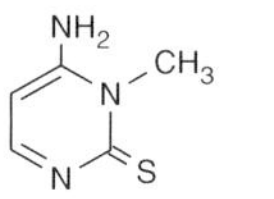
250 mg

S46,194-6
6-AMINOINDOLINE DIHYDROCHLORIDE
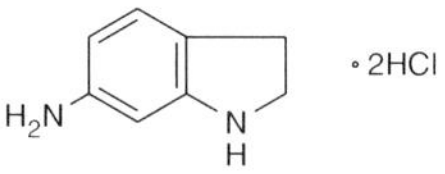
250 mg

S92,391-5
1-(4-AMINOPHENYL)-4-METHYLPIPERAZINE HYDROCHLORIDE
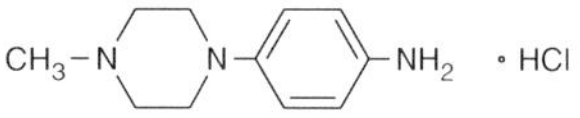
250 mg

S75,873-6
5-(4-AMINOBUTYLIMINO)-7-PHENYL-1,2,3,5-TETRAHYDROIMIDAZO(1,2-C)-THIAZOLE DIHYDROCHLORIDE
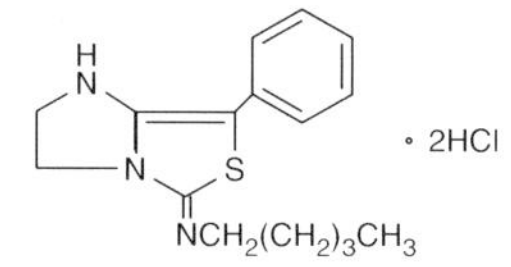
250 mg

S73,559-0
4-(2-IMIDAZOLIN-2-YL)ANILINE
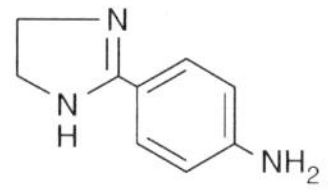
250 mg

S92,648-5
4-MORPHOLINOANILINE HYDROCHLORIDE

250 mg

S83,481-5
1-AMINOTHIANTHRENE
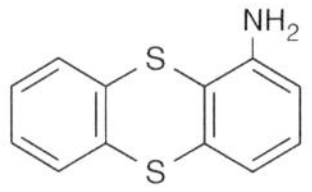
250 mg

S94,884-5
2-AMINO-5-PHENYL-6H-1,3,4-THIADIAZINE HYDROBROMIDE
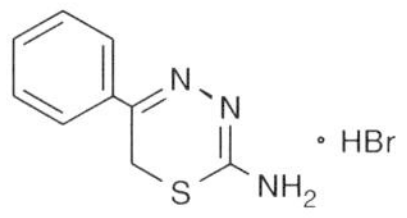
250 mg

S92,899-2
4-(4-METHYLPIPERIDINO)ANILINE HYDROCHLORIDE
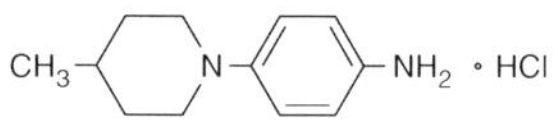
250 mg

S82,054-7
2-AMINODIBENZO-P-DIOXIN
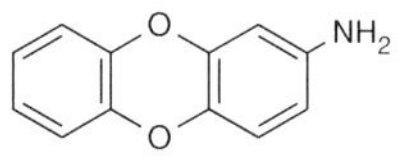
250 mg

S67,790-6
5-TERT-BUTYLAMINO-2,3-DIHYDRO-7-PHENYL-1H,-1,4-DIAZEPINE METHANESULFONATE
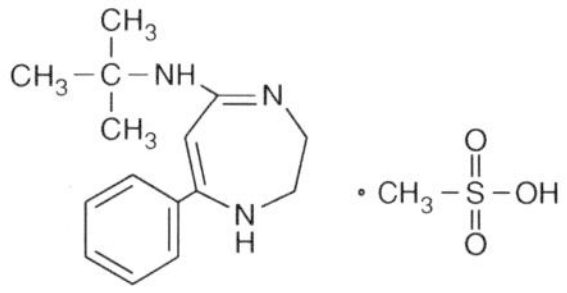
5 gm

S94,904-3
2,4-DIMETHYL-1H-1,5-BENZODIAZEPINE HYDROCHLORIDE
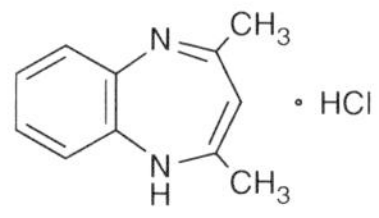
250 mg

S60,035-0
1-(4-CHLORO-ALPHA-PHENYLBENZYL)-HEXAHYDRO-4-METHYL-1H-1,4-DIAZEPINE OCTADECYL SULFATE
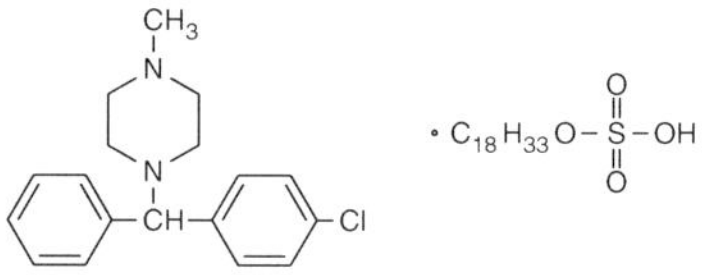
5 gm

S60,308-2
1-(4-CHLORO-ALPHA-PHENYLBENZYL)-HEXAHYDRO-4-METHYL-1H-1,4-DIAZEPINE
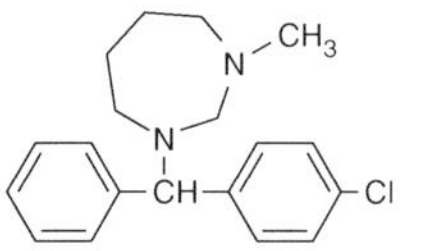
5 gm

S38,216-7
6-CHLORO-N-(2-PYRROLIDINOETHYL)-O-TOLUIDINE
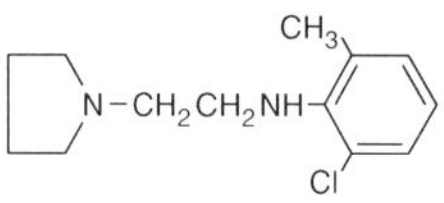
250 mg

S53,559-1
1-(3,4-DICHLOROPHENYL)PIPERAZINE
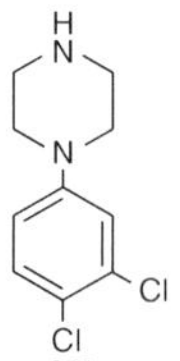
250 mg

S53,576-1
1-(5-CHLORO-O-TOLYL)PIPERAZINE
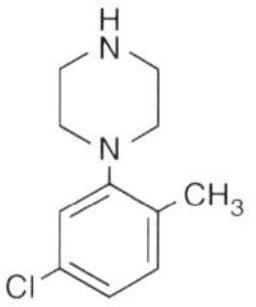
250 mg

S53,566-4
1-(2,4-XYLYL)PIPERAZINE
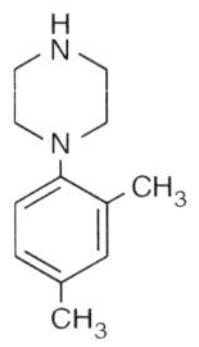
250 mg

S53,568-0
1-(3,4-XYLYL)PIPERAZINE
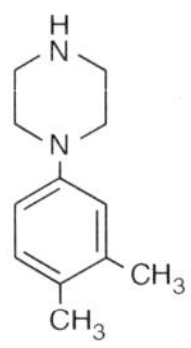
250 mg

S41,682-7
2-METHYL-1-P-TOLYLPIPERAZINE
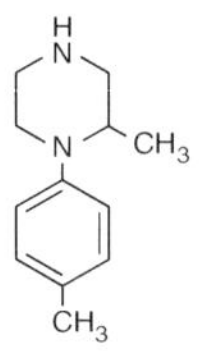
250 mg

S43,507-4
3-BENZYLAMINOQUINUCLIDINE
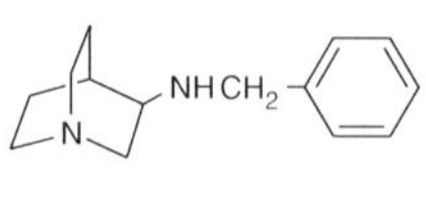
5 gm

S40,166-8
10-(1-METHYL-2-PYRROLIDINYLMETHYL)-PHENOTHIAZINE HYDROCHLORIDE
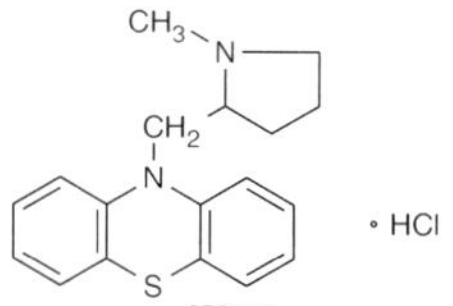
250 mg

S37,398-2
10-(1-BENZYL-3-PIPERIDYL)-2-CHLOROPHENOTHIAZINE FUMARATE
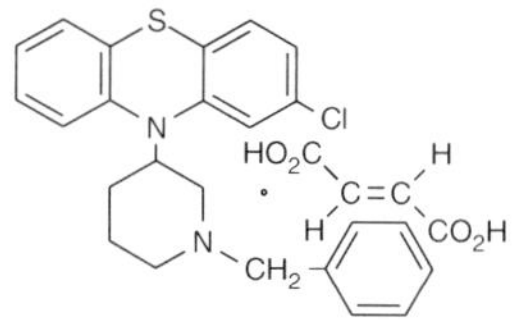
250 mg

S94,743-1
2-CHLORO-10-(2-DIMETHYLAMINOETHYL)-10H-PHENOTHIAZINE HYDROCHLORIDE
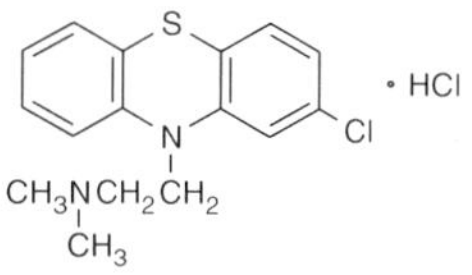
5 mg

S37,298-6
3,4,7,8-TETRAPHENYL-1,2,5,6-TETRAZOCINE
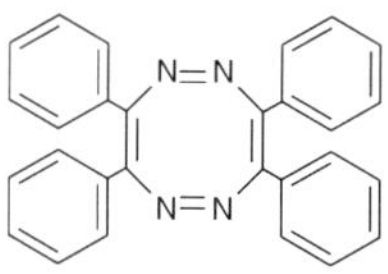
250 mg

S36,412-6
3-NITRO-3H-1,5-BENZODIAZEPINE
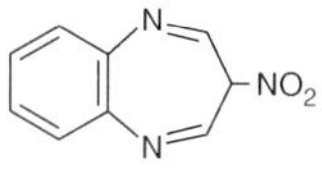
250 mg

S81,189-0
10-(2-PYRROLIDINOETHYL)-PHENOTHIAZINE
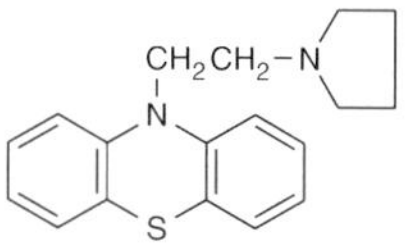
250 mg

S41,027-6
1,2,3,4,6,7-TRIBENZOPHENAZINE
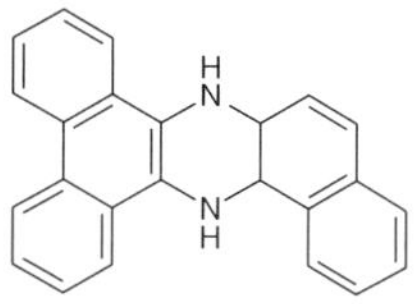
250 mg

S35,586-0
2,4,7,8-TETRAMETHYL-1,5-BENZODIAZEPINE SULFATE
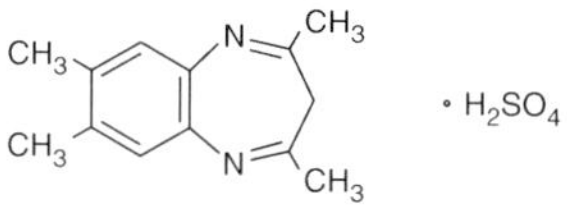
5 gm

S87,451-5
1,4-DIHYDRO-4-[4-(DIMETHYLAMINO)-PHENYLIMINO]-1-ETHYLPYRIDINE HYDROCHLORIDE
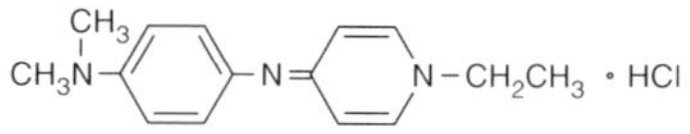
250 mg

S65,127-3
2-[2-(2-PHENETHYLAMINO)ETHYL]-PIPERIDINE
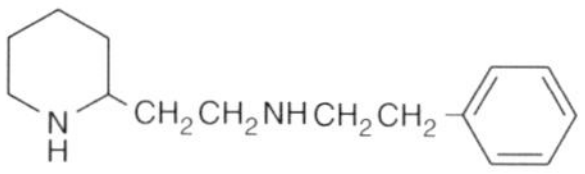
5 gm

S35,799-5
4-ANILINO-1-BENZYLPIPERIDINE
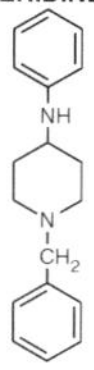
5 gm

S65,109-5
2-(2-ANILINOETHYL)PIPERIDINE
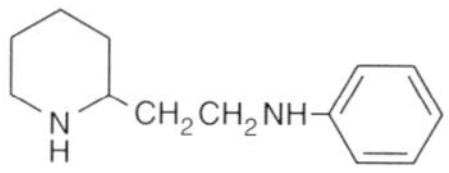
250 mg

S65,102-8
2-[2-(2-CHLOROANILINO)ETHYL]-PIPERIDINE
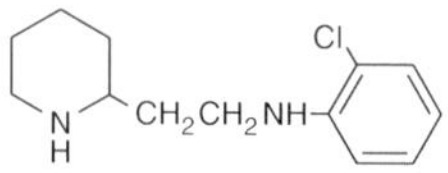
5 gm

S65,107-9
2-[2-(4-METHYLANILINO)ETHYL]-PIPERIDINE
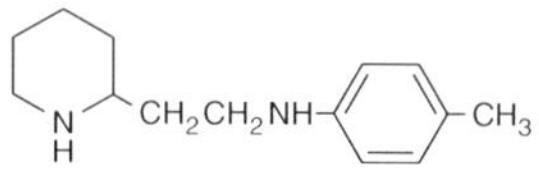
5 gm

S65,133-8
2-[2-(4-ISOPROPYLANILINO)ETHYL]-PIPERIDINE
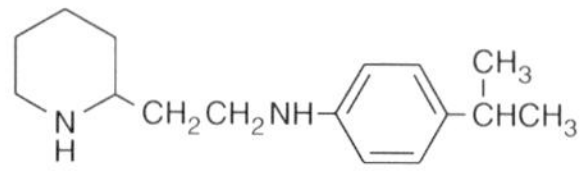
5 gm

S58,065-1
1-BENZYL-4-HYDRAZINOPIPERIDINE
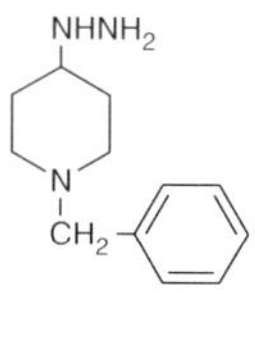
250 mg

S81,188-2
10-[3-(DIETHYLAMINO)PROPYL]-PHENOTHIAZINE
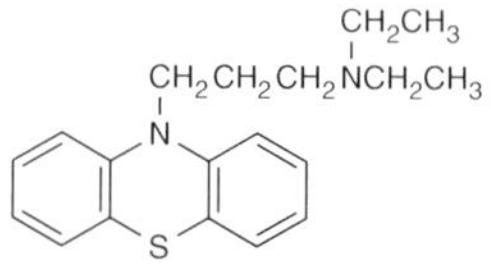
250 mg

S94,746-6
2-CHLORO-10-(3-DIMETHYLAMINOPROPYL)-10H-PHENOTHIAZINE-N-OXIDE MALEATE
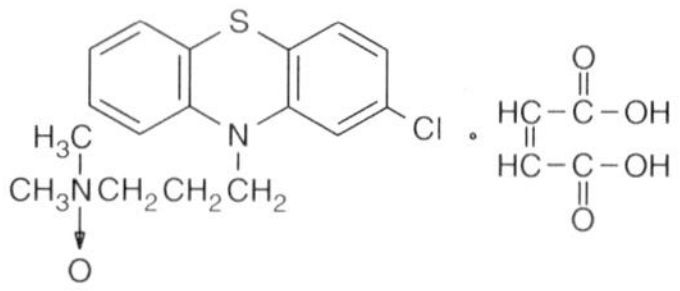
250 mg

S38,431-3
1,3-DIMETHYL-3-(DIPHENYLAMINOMETHYL)-PYRROLIDINE HYDROCHLORIDE
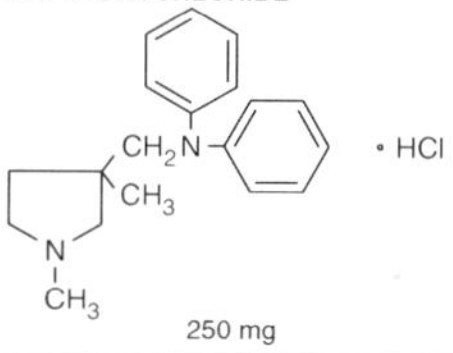
250 mg

S34,200-9
1-BENZYL-4-PYRROLIDINOPIPERIDINE DIHYDROCHLORIDE
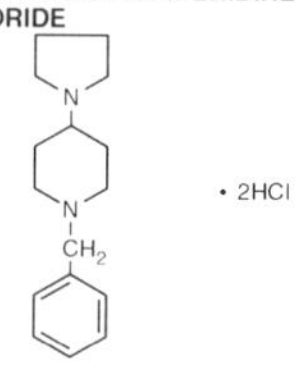
5 gm

S40,185-4
10-(1-ETHYL-3-PYRROLIDINYLMETHYL)-PHENOTHIAZINE HYDROCHLORIDE
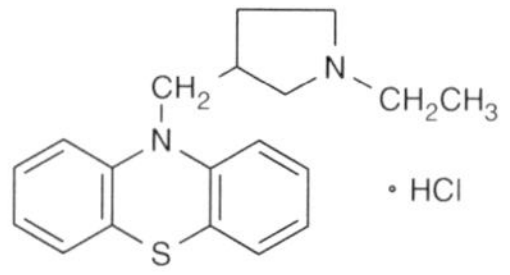
250 mg

S40,210-9
10-(1,2,5-TRIMETHYL-3-PYRROLIDINYLMETHYL)PHENOTHIAZINE HYDROCHLORIDE
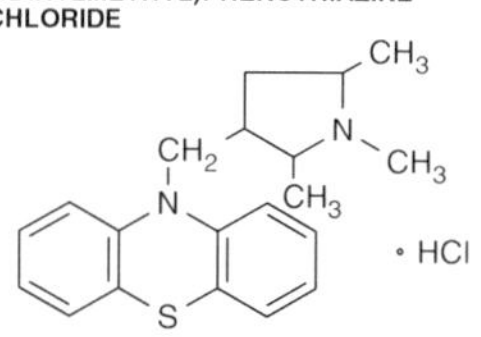
250 mg

S40,233-8
10-(1,3-DIETHYL-3-PYRROLIDINYLMETHYL)PHENOTHIAZINE HYDROCHLORIDE
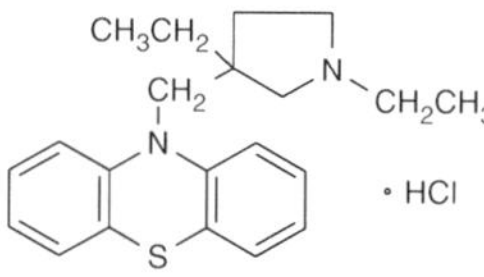
250 mg

S40,260-5
10-(1-ETHYL-3-PROPYL-3-PYRROLIDINYLMETHYL)PHENOTHIAZINE HYDROCHLORIDE
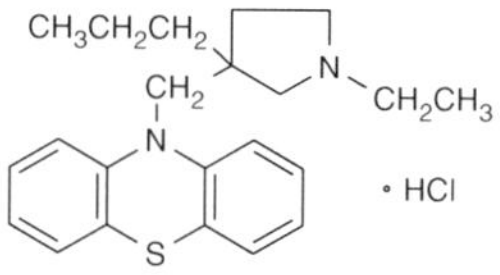
250 mg

S34,087-1
4-AMINO-1-BENZYLPIPERIDINE DIHYDROCHLORIDE HYDRATE
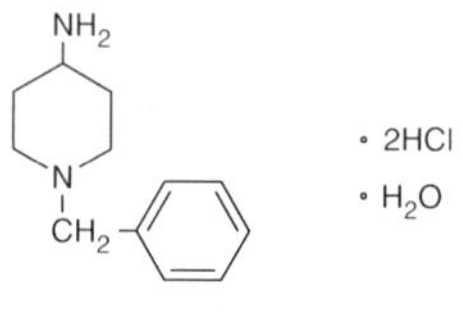
1 gm

S34,177-0
1-BENZYL-4-DIMETHYLAMINOPIPERIDINE DIHYDROCHLORIDE
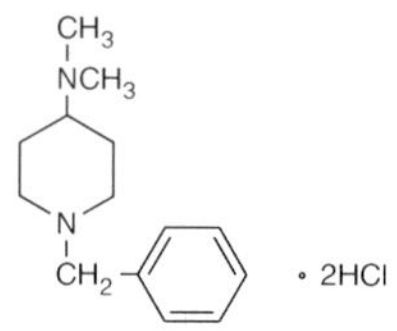
5 gm

S92,061-4
10-[3-(4-METHYL-1-PIPERAZINYL)-PROPYL]-10H-PHENOTHIAZINE
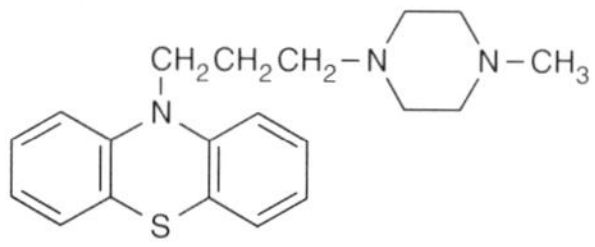
250 mg

S87,380-2
5,12-DIHYDRO-5,12-DIMETHYLQUINOXALINO(2,3-B)-PHENAZINE DIPERCHLORATE
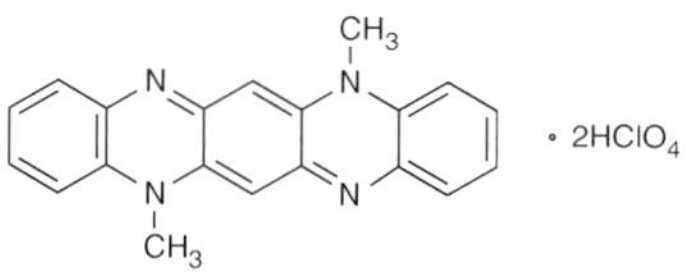
250 mg

S37,311-7
1-BENZYL-4-HYDRAZINOPIPERIDINE DIHYDROCHLORIDE
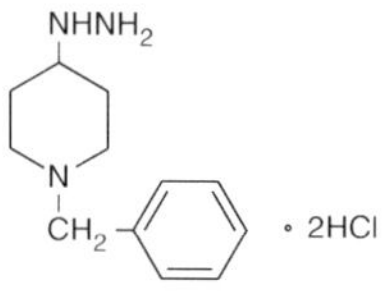
250 mg

S68,193-8
1-(1-METHYLNONYL)-5-(1-METHYLNONYLAMINO)INDOLINE DIHYDROCHLORIDE
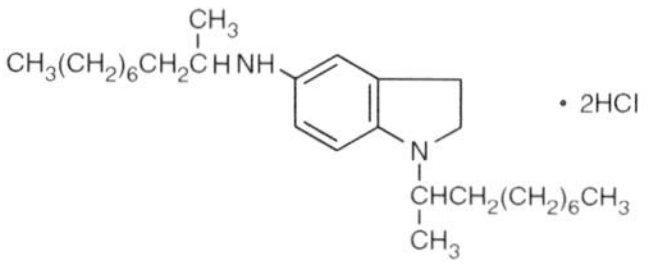
250 mg

S68,224-1
1-(1-METHYLHEPTYL)-5-(1-METHYLHEPTYLAMINO)INDOLINE DIHYDROCHLORIDE
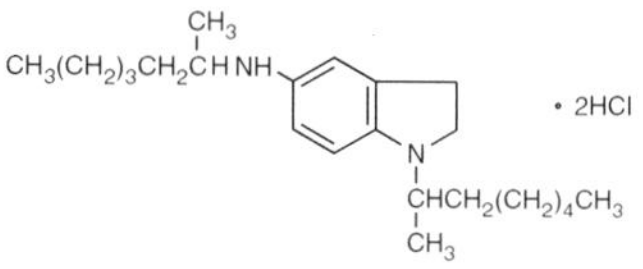
250 mg

S76,040-4
5-(1-METHYLOCTYLAMINO)INDOLINE DIHYDROCHLORIDE
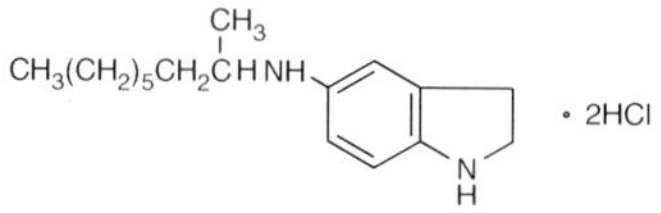
250 mg

S38,953-6
4,4'-(1,4-PHENYLENE)DIMORPHOLINE
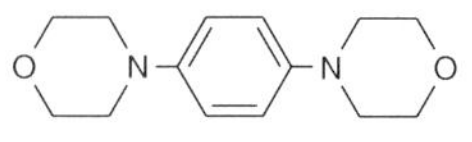
250 mg

S34,202-5
1-BENZYL-1,2,3,4-TETRAHYDROISOQUINOLINE
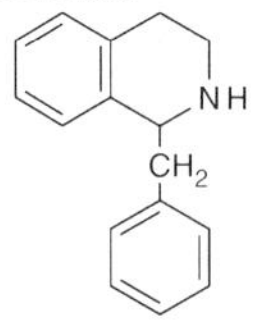
5 gm

S45,528-8
DL-ISOAMARINE
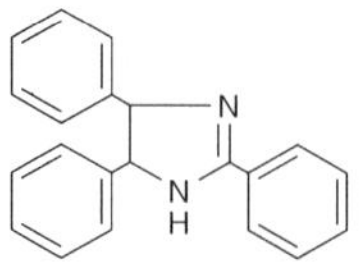
5 gm

S38,385-6
2,3-DIPHENYLPIPERIDINE
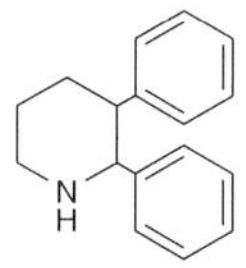
250 mg

S61,696-6
2-BENZYL-5,6-DIHYDRO-4,4,6-TRIMETHYL-1,3-OXAZINE
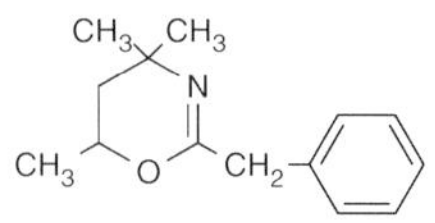
5 gm

S35,781-2
1-BENZYL-3,4-DIHYDROISOQUINOLINE HYDROCHLORIDE
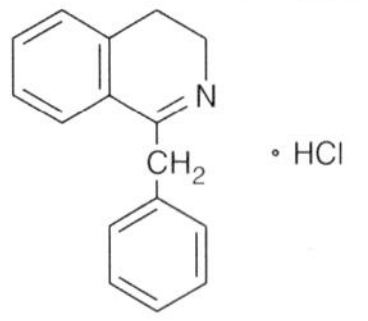
5 gm

S80,662-5
6A-METHYL-5,6,6A,11-TETRAHYDROISOINDOLO-(2,1-A)-QUINAZOLINE HYDROCHLORIDE
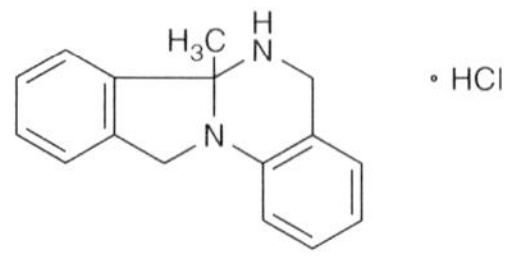
250 mg

S36,450-9
1-PHENYLPIPERAZINE DIHYDROCHLORIDE
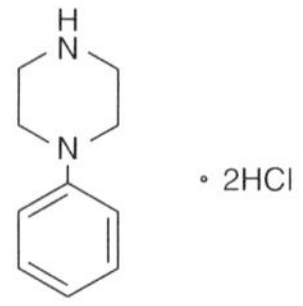
5 gm

S36,375-8
2-METHYL-1-PHENYLPIPERAZINE
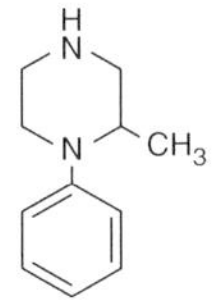
250 mg

S37,083-5
1-(P-TOLYL)PIPERAZINE DIHYDROCHLORIDE
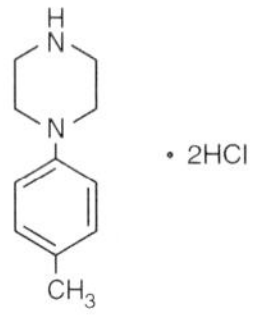
250 mg

S37,186-6
1-BENZYLPIPERAZINE DIHYDROCHLORIDE
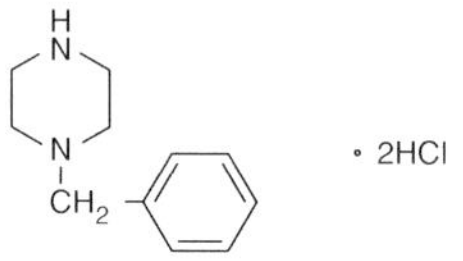
5 gm

S53,924-4
2-ANILINO-2-THIAZOLINE
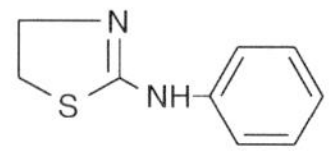
250 mg

S94,958-2
7-CHLORO-2,3-DIHYDRO-5-PHENYL-1H-1,4-BENZODIAZEPINE
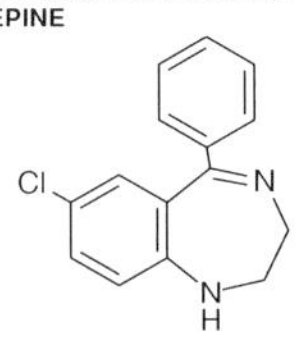
250 mg

S35,425-2
2,2-PENTAMETHYLENEBENZOTHIAZOLINE
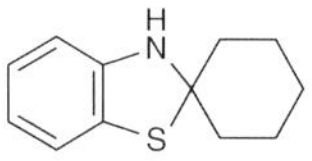
5 gm

S65,914-2
2,3-DIHYDROBENZOTHIAZOLE
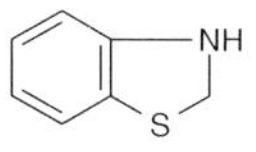
250 mg

S43,533-3
2,2-DIMETHYLBENZOTHIAZOLINE
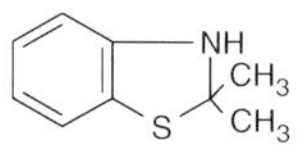
5 gm

S56,111-8
3-METHYL-3-PHENYLPIPERIDINE
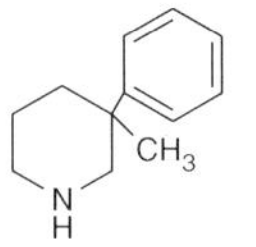
5 gm

S41,860-9
4-(3-PHENYLPROPYL)PIPERIDINE
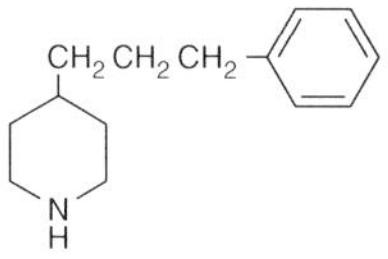
5 gm

S41,797-1
4-(4-FLUOROPHENYL)-1,2,3,6-TETRAHYDROPYRIDINE HYDROCHLORIDE
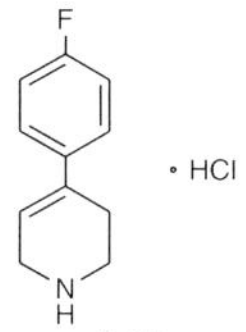
5 gm

S36,513-0
4-(P-TOLYL)-1,2,3,6-TETRAHYDROPYRIDINE HYDROCHLORIDE
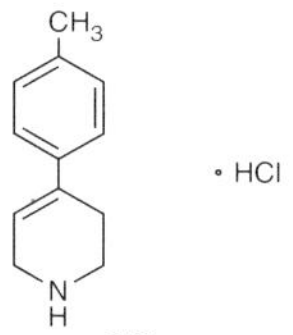
250 mg

S93,750-9
7H-BENZO(C)PHENOTHIAZINE
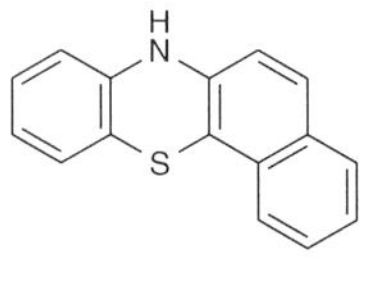
250 mg

S45,067-7
6-FLUOROINDOLINE
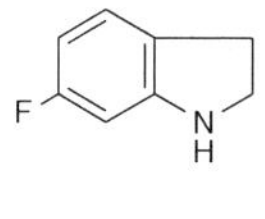
250 mg

S41,269-4
2,3-DIMETHYLINDOLINE (MIXTURE OF ISOMERS)
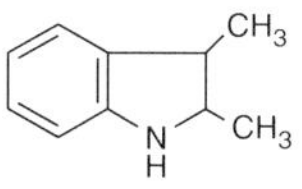
5 gm

S37,364-8
6-METHYL-1,2,3,4-TETRAHYDROQUINOLINE
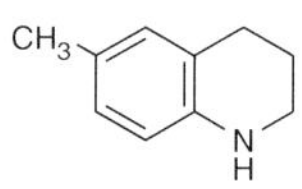
5 gm

S41,010-1
1,2,3,4-TETRAHYDROQUINALDINE
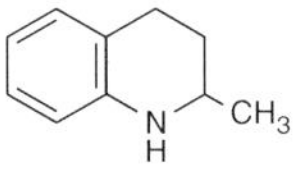
5 gm

S68,184-9
9,10-DIHYDRO-9,9-DIMETHYLACRIDINE
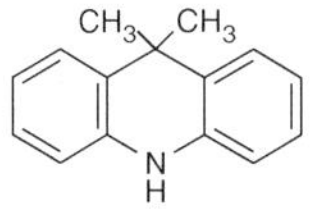
250 mg

S41,133-7
2,3-DIHYDRO-2,2,4-TRIMETHYL-1H-1,5-BENZODIAZEPINE
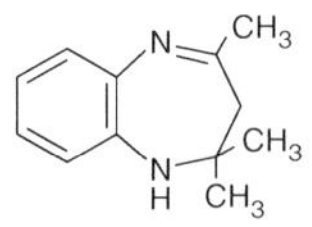
5 gm

S92,567-5
4-PYRROLIDINOANILINE
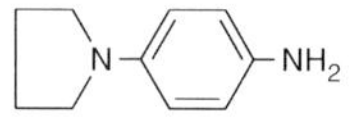
250 mg

S84,119-6
9,10-DIHYDRO-9-[4-(DIMETHYLAMINO)-PHENYL]ACRIDINE
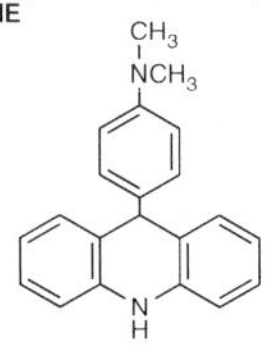
250 mg

S36,007-4
4-(1,5-DIAZABICYCLO[3.2.1]OCT-8-YL)-TOLUENE
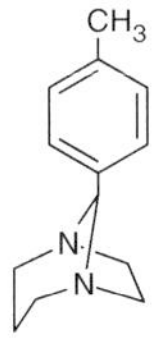
250 mg

S39,614-1
1-[ALPHA-(1-NAPHTHYL)BENZYL]-PIPERIDINE
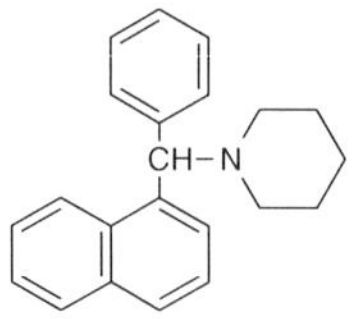
5 gm

S75,752-7
1'-(4-METHYLBENZYL)-3,4,4A,5,6,7,8,8A-OCTAHYDROSPIRO[NAPHTHALENE-2(1H),4'-PIPERIDINE]
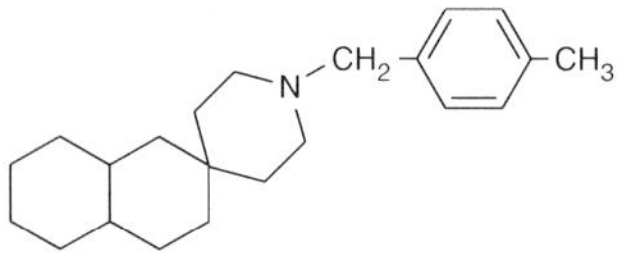
5 gm

S83,293-6
1,2-DIHYDRO-1-METHYL-2-PHENYLQUINOLINE

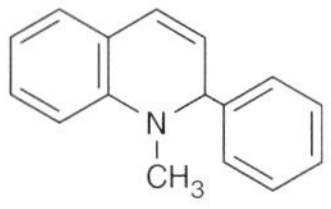

250 mg

S90,762-6
3,6-BIS(DIMETHYLAMINO)-9-XANTHENETHIONE

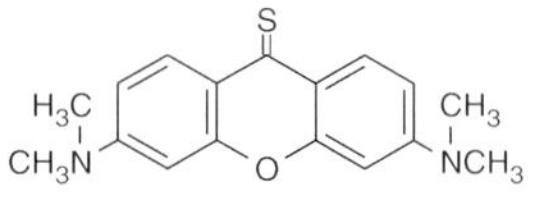

250 mg

S73,324-5
6-CHLORO-2,3,4,4A,9,9A-HEXAHYDRO-1-METHYL-4A-PHENYL-1H-PYRIDO(2,3-B)-INDOLE

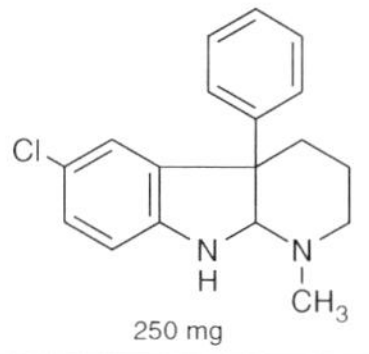

250 mg

S79,235-7
1-PHENYL-3-PYRROLINE

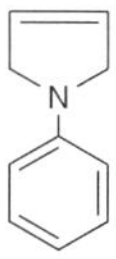

250 mg

S90,640-9
2-PHENYL-1,4,4-TRIMETHYLPYRROLIDINE

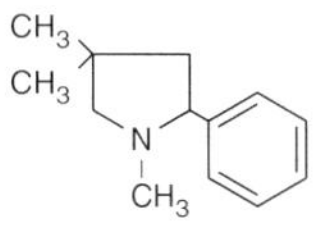

250 mg

S75,859-0
1,2,3,3A,8,8A-HEXAHYDRO-8-METHYL-3A-PHENYLPYRROLO(2,3-B)INDOLE HYDROCHLORIDE

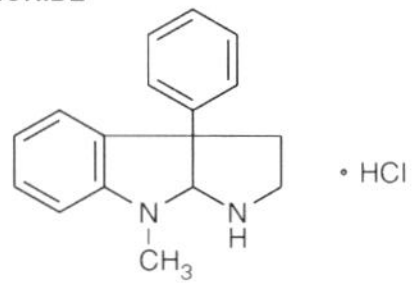

250 mg

S79,236-5
1-(P-TOLYL)-3-PYRROLINE

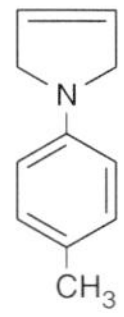

250 mg

S89,474-5
N-METHYLISOINDOLINE

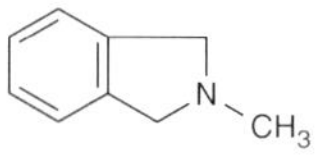

250 mg

S76,097-8
1-[10,11-DIHYDRO-5H-DIBENZO(A,D)-CYCLOHEPTEN-5-YL]-2-METHYLISOINDOLINE

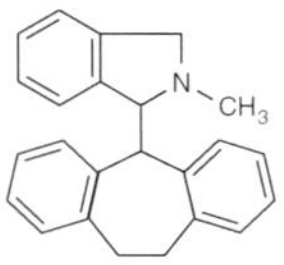

250 mg

S38,818-1
7-CHLORO-1,2,3,4-TETRAHYDROPYRIDO(1,2-A)BENZIMIDAZOLE

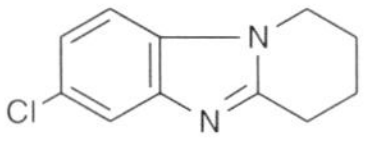

5 gm

S84,997-9
1,2,3-TRIPHENYLIMIDAZOLIDINE

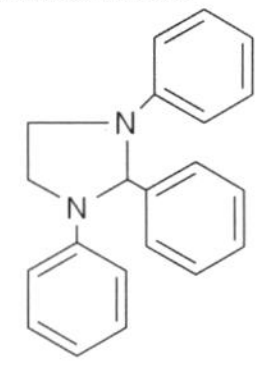

250 mg

S36,823-7
2,3-DIHYDRO-3-METHYL-2-METHYLENEBENZOTHIAZOLE

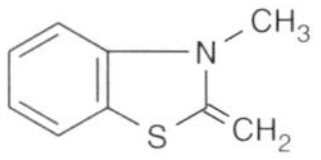

250 mg

S87,083-8
2,7-DIETHYL-1,2,3,3A,4,6,7,8-OCTAHYDROBENZO(L,M,N)(3,8)-PHENANTHROLINE

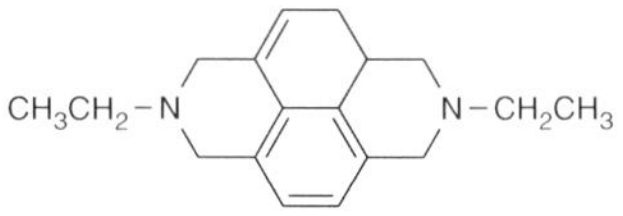

250 mg

S64,778-0
4-BENZYL-1-BUTYLPIPERIDINE

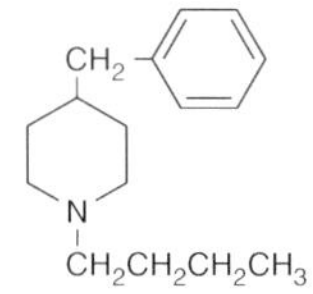

5 gm

S64,777-2
4-BENZYL-1-HEXYLPIPERIDINE

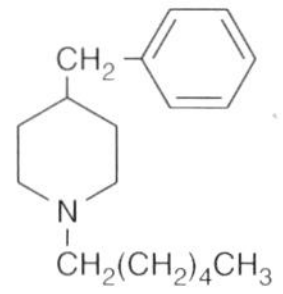

5 gm

S75,858-2
2,3,4,4A,9,9A-HEXAHYDRO-1-METHYL-4A-PHENYL-1H-PYRIDO(2,3-B)INDOLE HYDROCHLORIDE

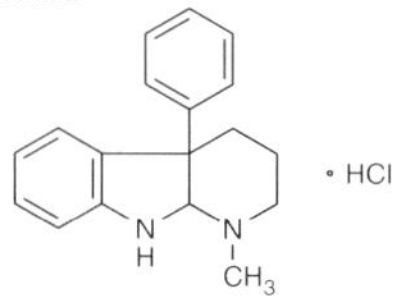

250 mg

S87,367-5
1,1'-BIS(DIPHENYLMETHYL)-4,4'-DIMETHYL-1,1',4,4'-TETRAHYDRO-4,4'-DIPYRIDYL

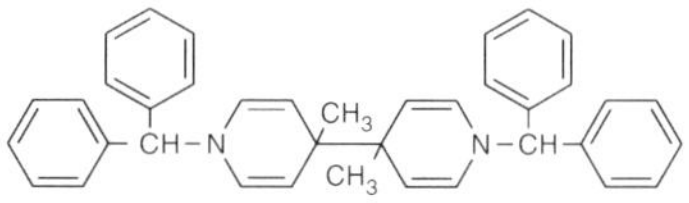

250 mg

S35,356-6
6-METHYL-3-(P-TOLYL)-3,4-DIHYDROQUINAZOLINE

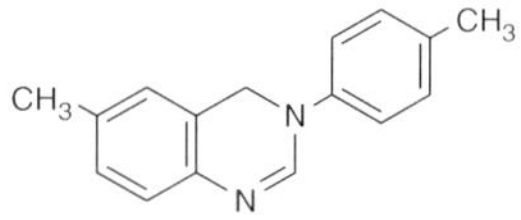

5 gm

S38,393-7
1,4-DIBENZYLPIPERAZINE DIHYDROCHLORIDE

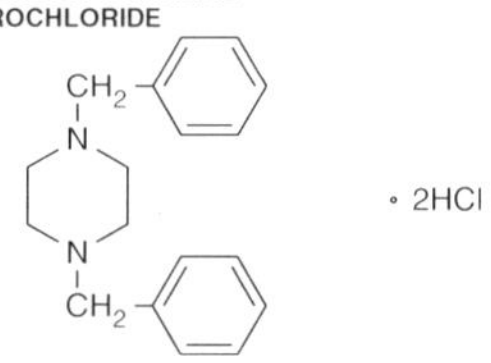

5 gm

S36,126-7
1-(DIPHENYLMETHYL)-4-METHYLPIPERAZINE

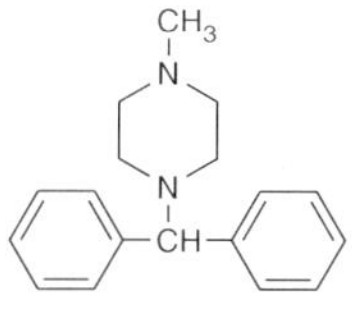

5 gm

S51,149-8
1-(4-CHLORODIPHENYLMETHYL)-4-METHYLPIPERAZINE HYDROCHLORIDE

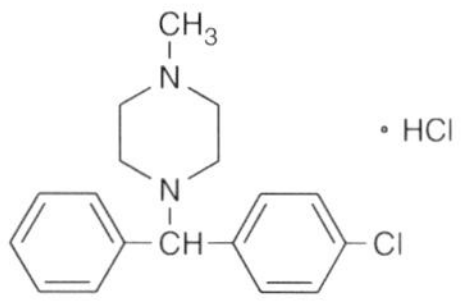

5 gm

S87,784-0
4-(2,3-DIMETHYL-4-PHENYL-2-BUTEN-1-YL)MORPHOLINE

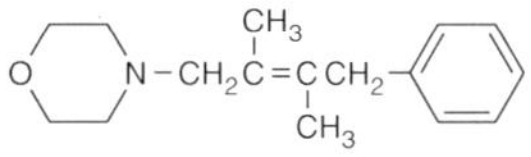

250 mg

S87,075-7
9,10-DIHYDRO-9-MORPHOLINOANTHRACENE

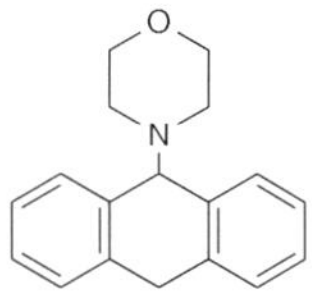

250 mg

S83,298-7
10-ETHYLPHENOXAZINE

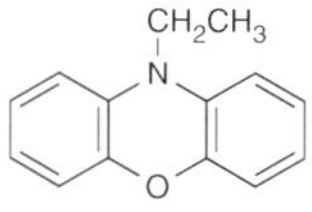

250 mg

S83,610-9
10-(4-BIPHENYLYL)PHENOTHIAZINE

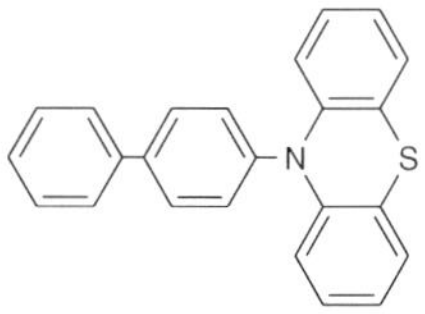

250 mg

S83,606-0
10-METHYLPHENOTHIAZINE

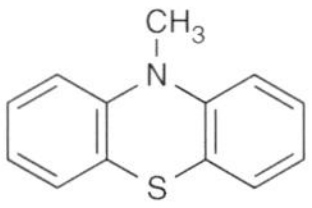

250 mg

S38,954-4
N-ETHYLPHENOTHIAZINE

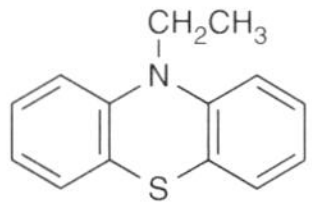

250 mg

S82,407-0
3-CHLORO-10-ETHYLPHENOTHIAZINE

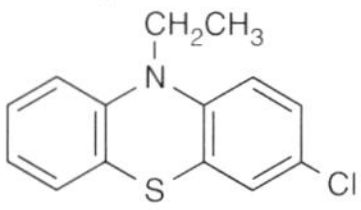

250 mg

S83,608-7
10-ALLYLPHENOTHIAZINE

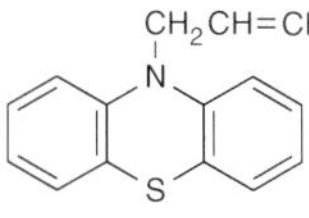

250 mg

S81,922-0
10-ISOPROPYLPHENOTHIAZINE

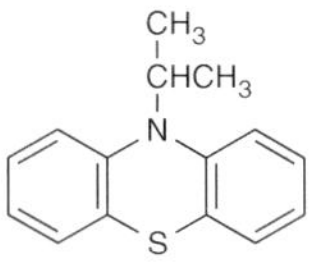

250 mg

S81,923-9
10-PROPYLPHENOTHIAZINE

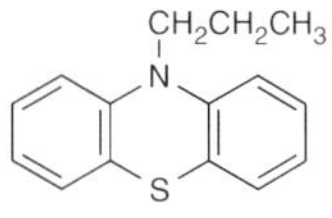

250 mg

S83,609-5
10-DECYLPHENOTHIAZINE

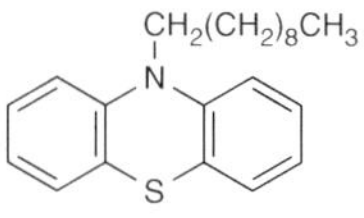

250 mg

S83,613-3
10-OCTADECYLPHENOTHIAZINE

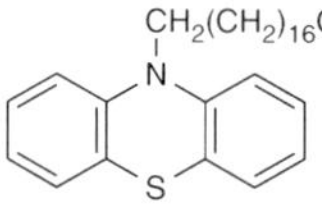

250 mg

S82,408-9
3-CHLORO-10-OCTADECYLPHENOTHIAZINE

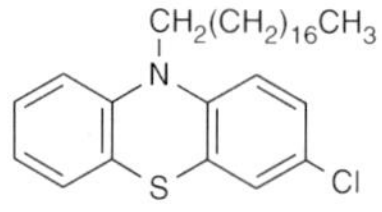

250 mg

S92,651-5
1-TRITYLAZIRIDINE

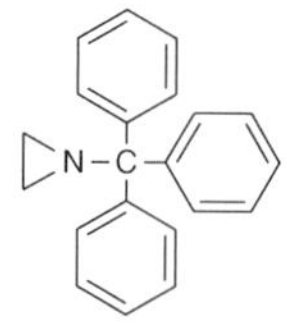

250 mg

S57,179-2
1,2-DIHYDRO-6-PHENYL-1,2,2-TRIMETHYLQUINOLINE

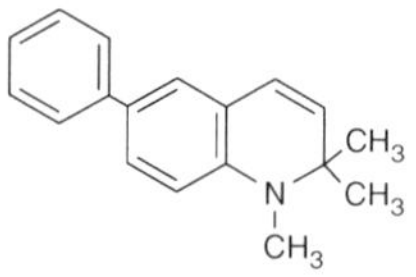

5 gm

S75,874-4
5,6-DIPHENYL-1,2,3,4,5,6-HEXAHYDRO-2-METHYL-2-BENZAZOCINE

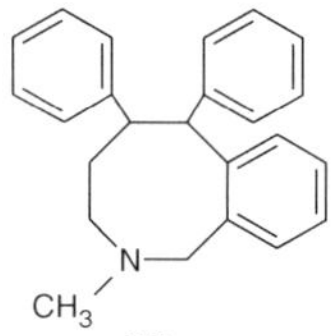

250 mg

S78,678-0
1,2-DIMETHYLINDOLINE

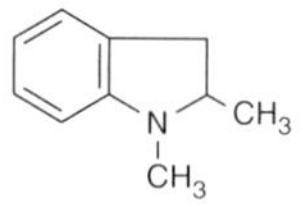

250 mg

S86,990-2
1-BUTYLINDOLINE

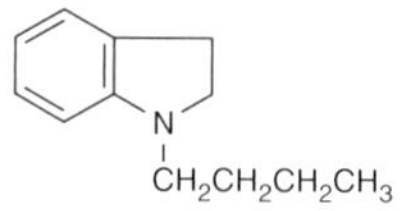

250 mg

S86,971-6
1-PROPYLINDOLINE

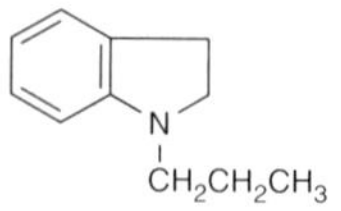

250 mg

S76,009-9
2-CHLORO-5,6,7,9,10,14B-HEXAHYDRO-5-METHYLISOQUINO(2,1-D)(1,4)-BENZODIAZEPINE

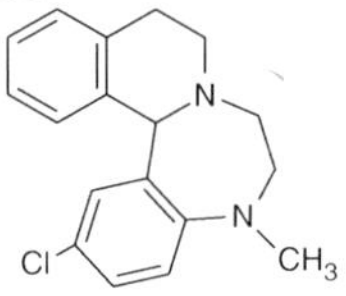

250 mg

S76,118-4
5-METHYL-5,6,6A,7,8,9,10,11-OCTAHYDROPYRIDO(2,1-C)(1,4)-BENZODIAZEPINE DIHYDROCHLORIDE

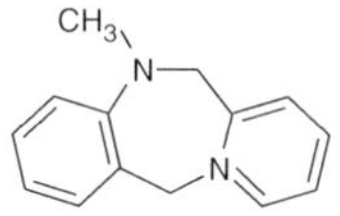

• 2HCl

250 mg

S93,997-8
2-[1-(3,5-DIMETHYLPHENYL)-2-NAPHTHYL]-4,5-DIHYDRO-3,5,5-TRIMETHYLOXAZOLIUM IODIDE

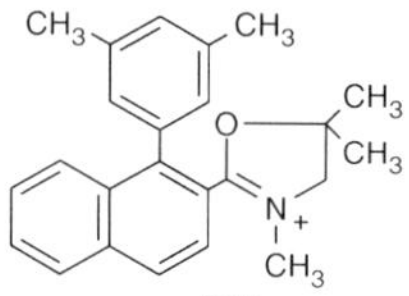

I⁻

5 gm

S85,925-7
1,3-DIPHENYLIMIDAZOLIDINE

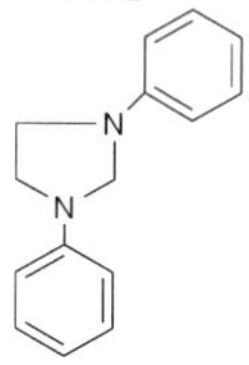

250 mg

S62,860-3
1,3-DIPHENYL-5-(4-IODOPHENYL)-2-PYRAZOLINE

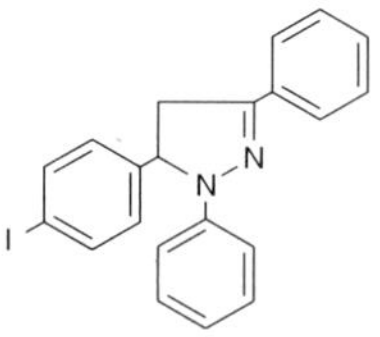

250 mg

S77,528-2
1,3-DIBENZYLIMIDAZOLIDINE

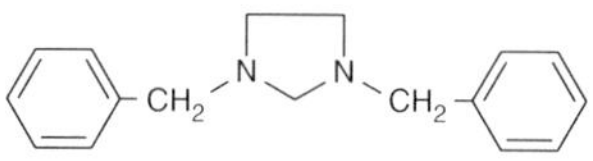

250 mg

S44,015-9
1-BENZYL-4-CHLOROPIPERIDINE HYDROCHLORIDE

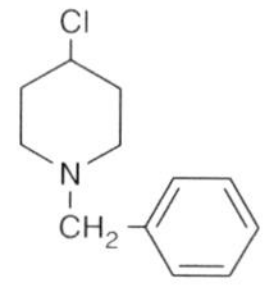

• HCl

5 gm

S75,852-3
DIMETHYL(TETRAHYDRO-ISOQUINOLINYLMETHYL)PROPIOPHENONE-DIMETHYLTRIMETHYLENEKETAL

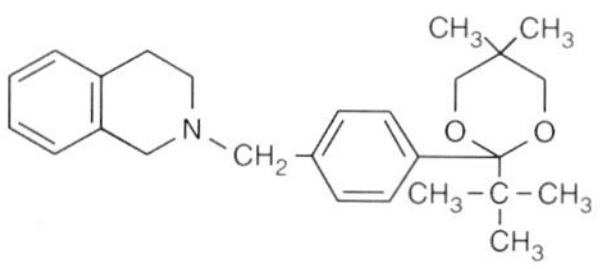

• HCl

250 mg

S76,642-9
4-BENZYLMORPHOLINE

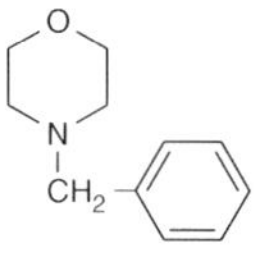

250 mg

S81,151-3
10-BENZYLPHENOXAZINE

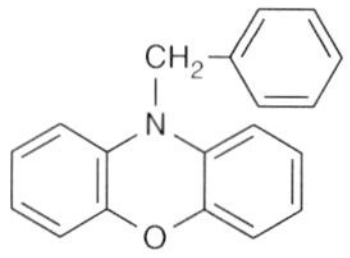

250 mg

S82,330-9
10-BENZYLPHENOTHIAZINE

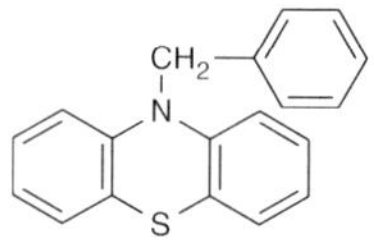

250 mg

S83,136-0
1,2-DIHYDRO-2-PHENYL-2-(P-TOLYL)-QUINOLINE

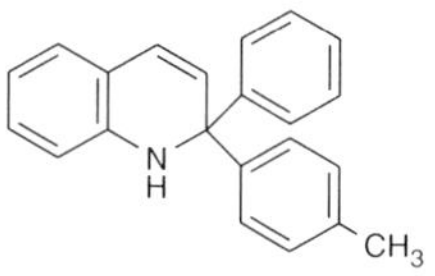

250 mg

S83,135-2
1,2-DIHYDRO-2,2-DIPHENYLQUINOLINE

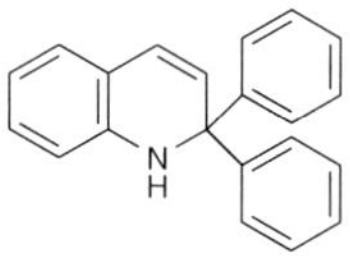

250 mg

S84,063-7
5,6-DIHYDRO-6-PHENYLPHENANTHRIDINE

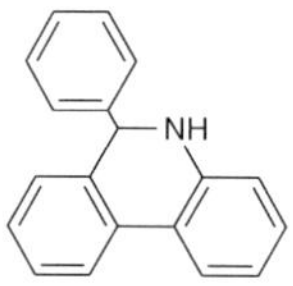

250 mg

S77,855-9
5,6,8,8A,13,14,16,16A-OCTAHYDRO-(1,2,4,5)TETRAZINO(6,1-A:3,4A')-DIISOQUINOLINE

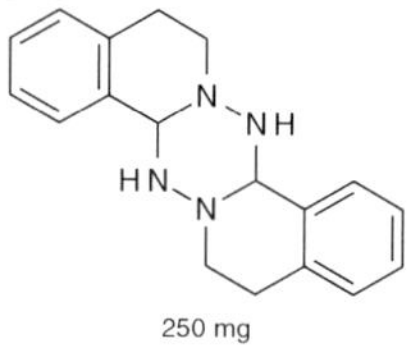

250 mg

S87,990-8
N-AMINO-1,2,3,4-TETRAHYDROISOQUINOLINE

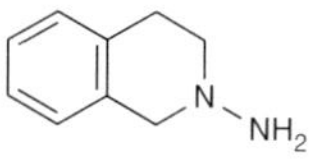

5 gm

S75,887-6
4-(4,5-DIHYDRO-1-IMIDAZOLYL)-BENZOPHENONE

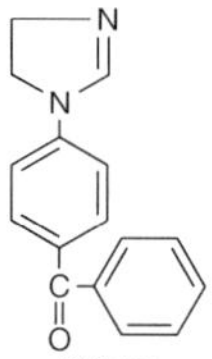

250 mg

S54,158-3
2-(3-CHLOROANILINO)-2-THIAZOLINE

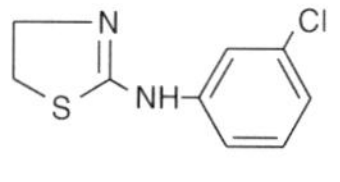

5 gm

S64,939-2
2-(2,4-DICHLOROANILINO)-2-THIAZOLINE

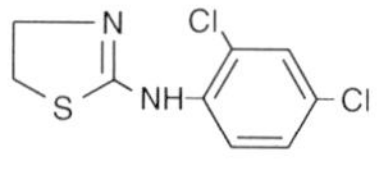

5 gm

S38,991-9
4,5-DIPHENYL-2-METHYL-2-THIAZOLINE

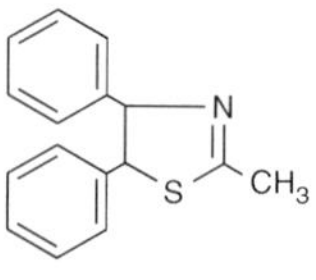

250 mg

S88,307-7
2-PHENYLAZIRIDINE

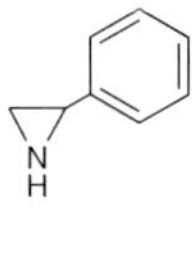

250 mg

S87,158-3
3,3-DIMETHYL-2-PHENYL-3H-AZIRINE

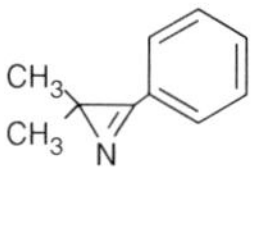

250 mg

S94,643-5
2-PHENYLISOINDOLINE

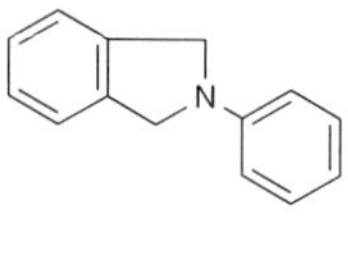

250 mg

S34,763-9
3,3-DIPHENYL-2-ETHYL-1-PYRROLINE

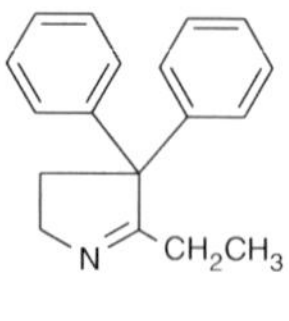

1 gm

S75,877-9
1,2,3,5,6,10B-HEXAHYDRO-10B-PROPYLPYRROLO(2,1-A)-ISOQUINOLINE MALEATE
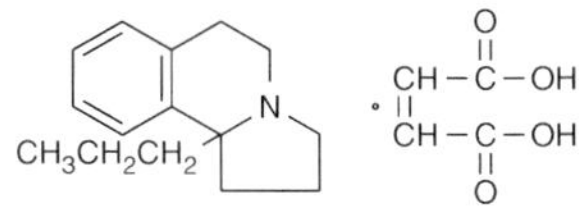
250 mg

S76,228-8
6,7A-DIPHENYL-2,3,5,7A-TETRAHYDROPYRAZOLO(5,1-B)OXAZOLE
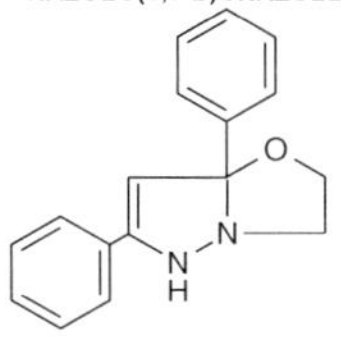
250 mg

S89,637-3
1,2-DIPHENYLPYRAZOLIDINE
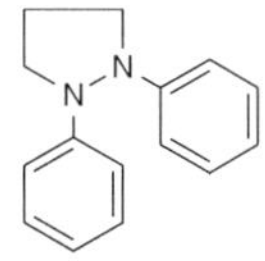
250 mg

S73,249-4
2,3,5,10-TETRAHYDRO-1H-PYRAZOLO(1,2-B)PHTHALAZINE
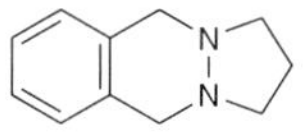
250 mg

S64,610-5
SPIRO(CYCLOHEXANE-1,2'-BENZIMIDAZOLINE)
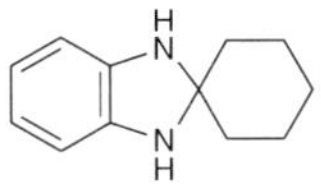
5 gm

S76,060-9
4,5-DIHYDRO-2-STYRYLIMIDAZOLE
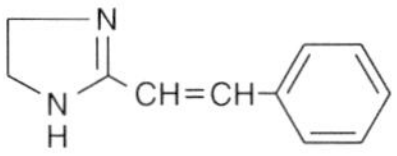
250 mg

S72,250-2
1,1',3,3'-TETRAPHENYL-DELTA2,2'-BIIMIDAZOLIDINE
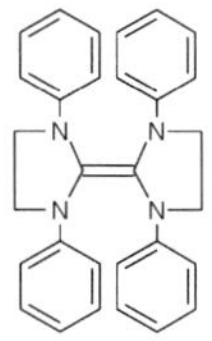
250 mg

S78,669-1
DELTA2,2'-BI-2H-BENZIMIDAZOLE
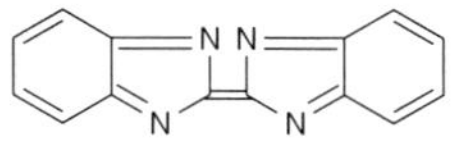
250 mg

S93,998-6
2-[1-(3,5-DIMETHYLPHENYL)-2-NAPHTHYL]-4,5-DIHYDRO-5,5-DIMETHYLOXAZOLE
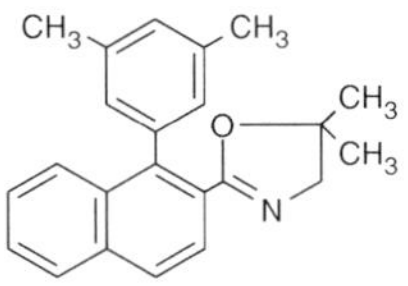
250 mg

S88,818-4
2-(P-TOLYL)OXAZOLIDINE
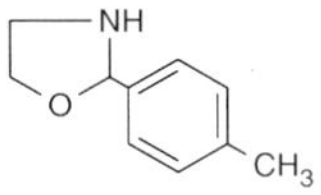
250 mg

S44,632-7
4-BENZYL-4-HYDROXYPIPERIDINE
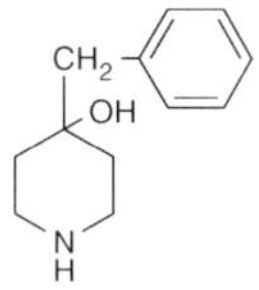
5 gm

S79,980-7
4-(2-PHENYLETHYL)PIPERIDINE HYDROCHLORIDE
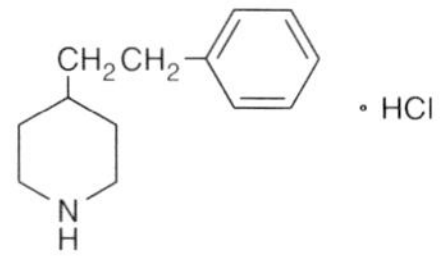
250 mg

S66,872-9
4-PHENYL-1,2,3,6-TETRAHYDROPYRIDINE
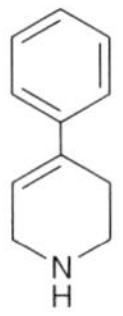
5 gm

S80,675-7
2,4-DIPHENYL-7-METHYL-5,6,7,8-TETRAHYDROQUINAZOLINE
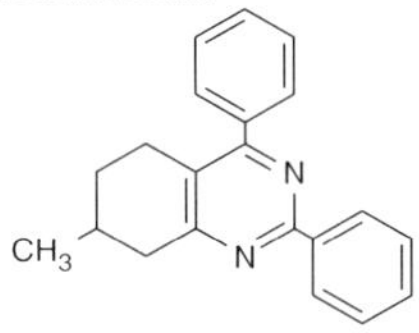
250 mg

S52,544-8
1-METHYL-4-PHENYLPIPERAZINE
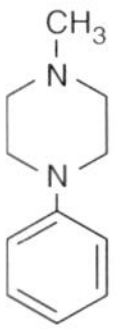
250 mg

S54,113-3
1-(P-CHLOROPHENYL)-4-METHYLPIPERAZINE
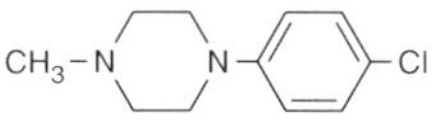
250 mg

S44,405-7
2-METHYL-1-M-TOLYLPIPERAZINE
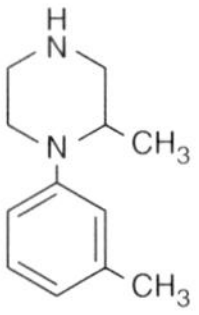
250 mg

S44,980-6
2,3-DIHYDRO-5,6-DIPHENYLPYRAZINE
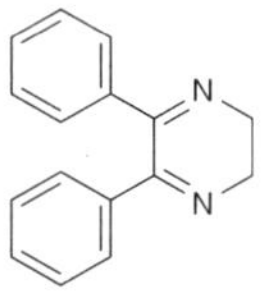
5 gm

S63,108-6
1,4-DIPHENYLPIPERAZINE
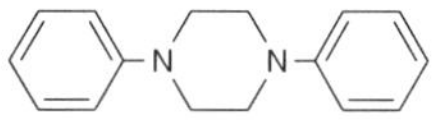
250 mg

S87,159-1
5,6-DIHYDRO-2-PHENYL-4H-1,3-OXAZINE
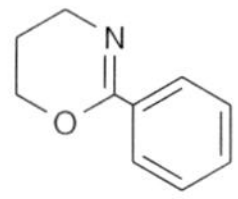
250 mg

S81,152-1
10-PHENYLPHENOXAZINE
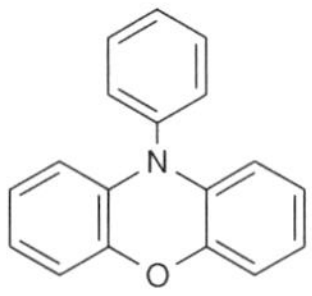
250 mg

S83,952-3
PHENOXAZINE NITRATE
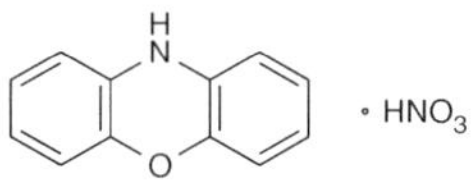
250 mg

S83,614-1
10,10'-(1,1'-BIPHENYL-4,4'-DIYL)-BIS(PHENOTHIAZINE)
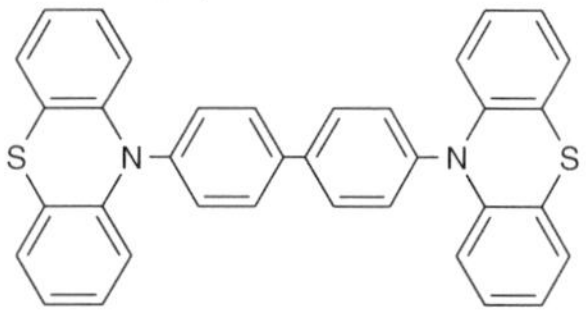
250 mg

S94,742-3
2-CHLORO-10-METHYL-10H-PHENOTHIAZINE
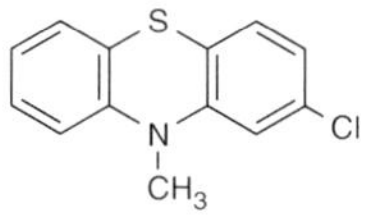
5 mg

S93,618-9
10-ALLYL-2-CHLOROPHENOTHIAZINE
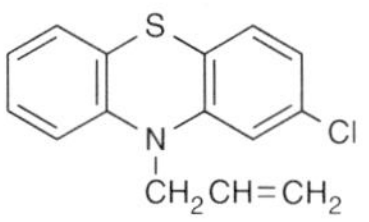
250 mg

S77,861-3
8,16-DIPHENYL-OCTAHYDRO-[1,2,4,5]-TETRAZINO[6,1-A:3,4-A']DIISOQUINOLINE
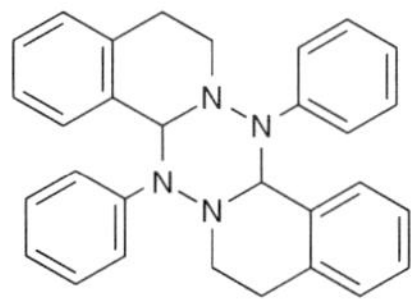
250 mg

S79,879-7
1,2,3,4,4A,10B-HEXAHYDROPHENANTHRIDINE
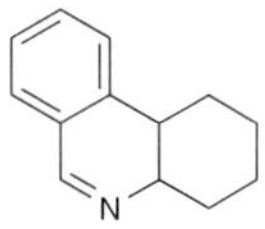
250 mg

S35,667-0
2,3,7-TRIMETHYLINDOLINE
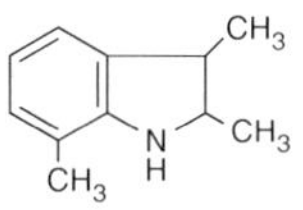
5 gm

S83,482-3
1,2-DIHYDRO-2-(1-PROPENYL)QUINOLINE
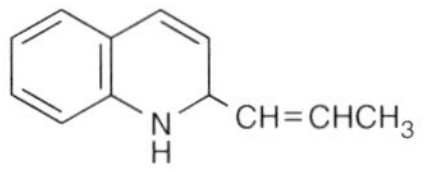
250 mg

S67,102-9
6-CHLORO-1,2-DIHYDRO-2,2,4-TRIMETHYLQUINOLINE
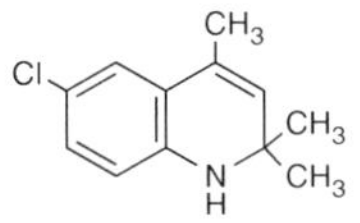
5 gm

S55,842-7
9,9PR-BIACRIDAN
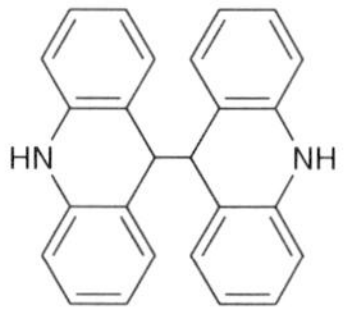
5 gm

S90,840-1
2,3,6,7-TETRAHYDRO-1H,5H-BENZO(I,J)-QUINOLIZINE
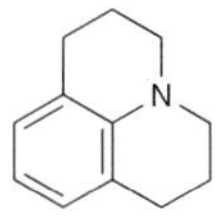
250 mg

S87,622-4
1,2,3,4-TETRAHYDROQUINOXALINE

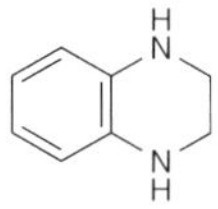

250 mg

S88,271-2
5,6,11,12-TETRAHYDRODIBENZO(B,F)-(1,5)DIAZOCINE

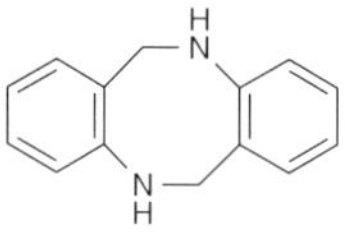

250 mg

S88,266-6
TETRABENZO(B,D,H,J)(1,6)-DIAZACYCLODODECINE

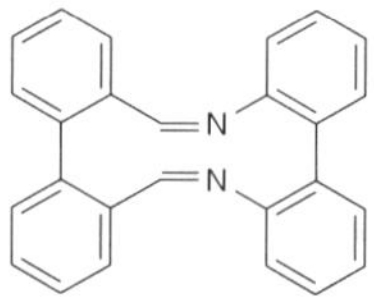

250 mg

S75,937-6
3-(4-CHLOROPHENYL)-3-HYDROXY-7-(2-HYDROXYETHYL)-2,3,5,6-TETRAHYDRO-IMIDAZO(2,1-B)THIAZOL-4-IUM BROMIDE

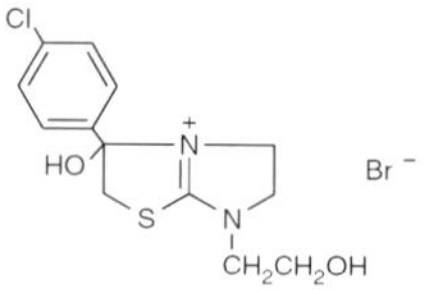

250 mg

S86,206-1
BIS(4-CHLOROBENZYLIDENE)-ETHYLHYDRAZINIUM TETRAFLUOROBORATE

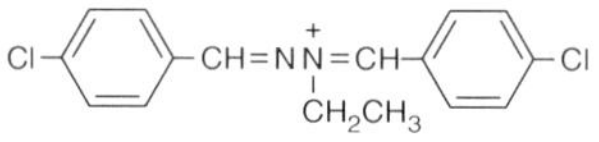

250 mg

S62,901-4
BENZYLBIS(2-HYDROXYETHYL)(2-DODECYLOXYETHYL)AMMONIUM CHLORIDE

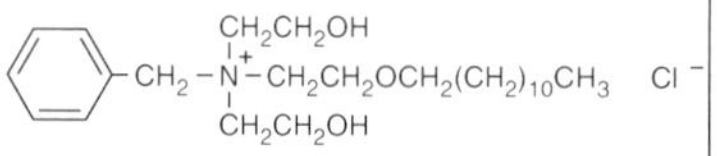

5 gm

S40,292-3
BENZYLBIS(2-HYDROXYETHYL)(2-DODECYLOXYETHYL)AMMONIUM BROMIDE

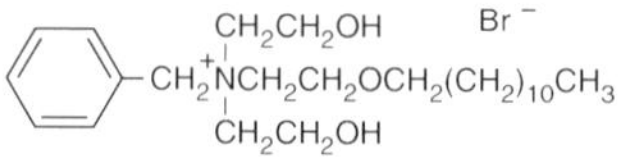

250 mg

S40,305-9
BIS(2-HYDROXYETHYL)CINNAMYL(2-DODECYLOXYETHYL)AMMONIUM CHLORIDE

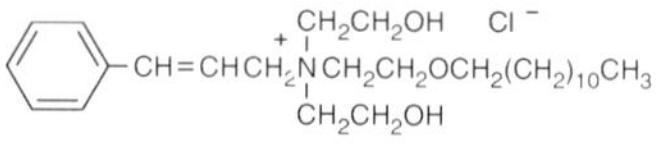

250 mg

S40,881-6
HEXOCYCLIUM

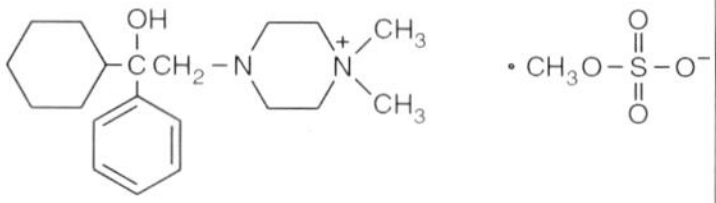

5 gm

S44,842-7
[2-(3,4-DICHLOROPHENYL)-2-HYDROXYETHYL]-DIMETHYLISOPROPYLAMMONIUM IODIDE

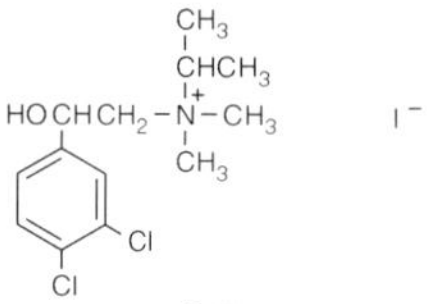

5 gm

S42,439-0
CANDICINE IODIDE

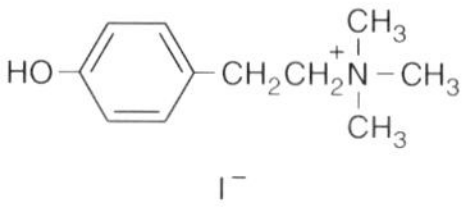

250 mg

S51,593-0
1,1-DIMETHYL-8-HYDROXY-1,2,3,4-TETRAHYDROQUINOLINIUM IODIDE

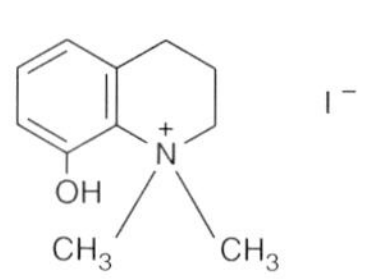

250 mg

S73,292-3
8,9-DIMETHOXY-2,3,5,6-TETRAHYDRO-1H-PYRROLO(2,1-A)-ISOQUINOLINIUM BROMIDE

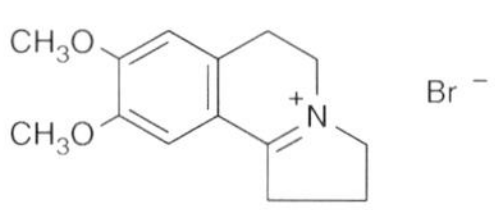

250 mg

S73,349-0
2,3-DIMETHOXY-7-METHYL-OCTAHYDRO-ISOINDOLO(1,2-A)-ISOQUINOLINIUM IODIDE

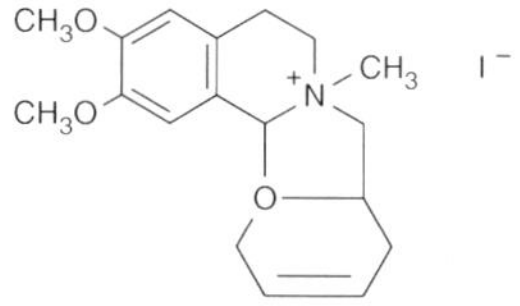

250 mg

S76,167-2
8,9-DIMETHOXY-4-METHYL-2,3,4,5,6,10B-HEXA-H-OXAZOLO(2,3-A)-ISOQUINOLINIUM IODIDE

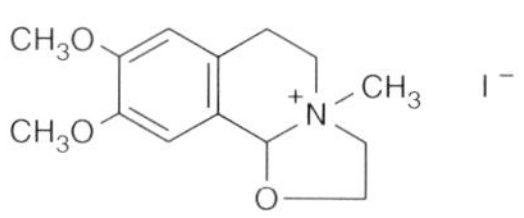

250 mg

S86,203-7
BIS(2,4-DIMETHOXYBENZYLIDENE)-ETHYLHYDRAZINIUM TETRAFLUOROBORATE

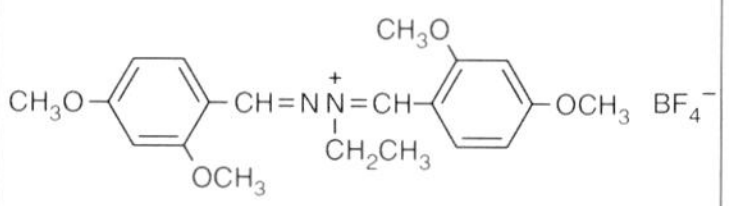

250 mg

S92,901-8
3,6-DIMETHOXY-1,2-PHENYLENEDIAMINE DIHYDROCHLORIDE

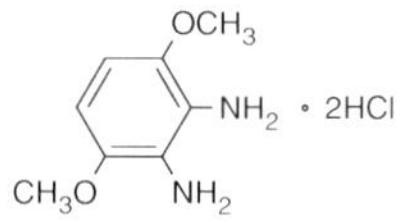

250 mg

S86,204-5
BIS(4-METHOXYBENZYLIDENE)-ETHYLHYDRAZINIUM TETRAFLUOROBORATE

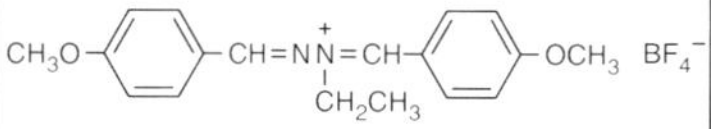

250 mg

S50,232-4
1-[4-HEXADECYLOXY-3-METHOXY-ALPHA-(METHYLTHIO)BENZYLIDENE]-PIPERIDINIUM IODIDE

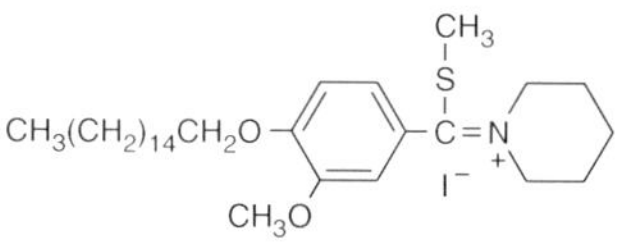

5 gm

S42,477-3
O,O'-DIMETHYLCORYNEINE CHLORIDE

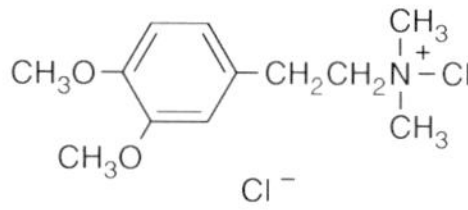

250 mg

S42,478-1
O,O'-DIMETHYLCORYNEINE IODIDE

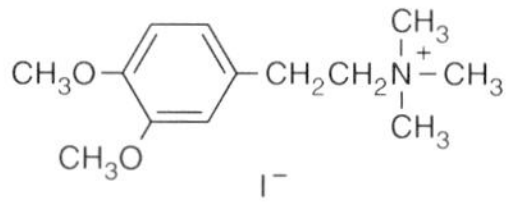

250 mg

S42,567-2
LAUDANOSINE METHIODIDE

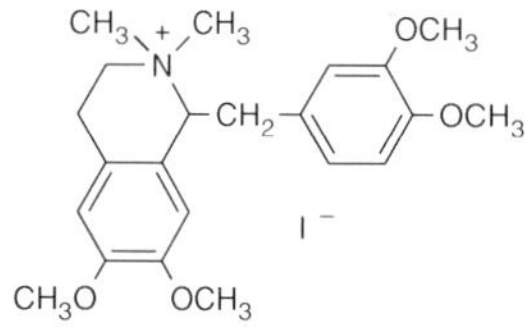

250 mg

S42,455-2
DL-COREXIMINE DIMETHYL ETHER METHIODIDE

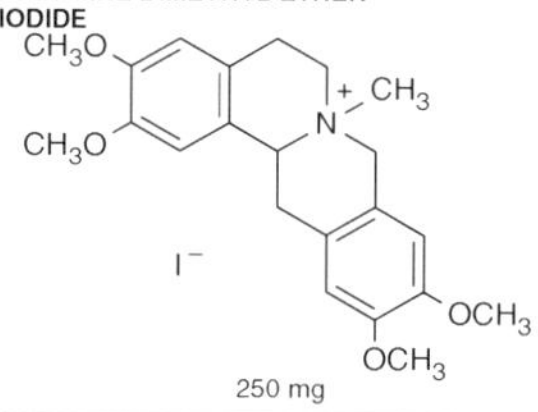

250 mg

S50,537-4
1-(ALPHA-METHYLTHIO-4-PHENOXYBENZYLIDENE)-PIPERIDINIUM IODIDE

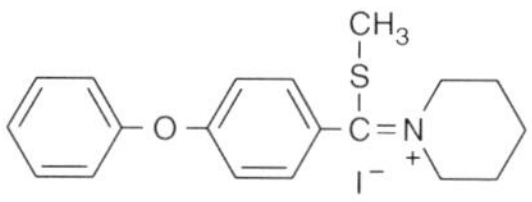

250 mg

S42,562-1
O-METHYLCANDICINE IODIDE

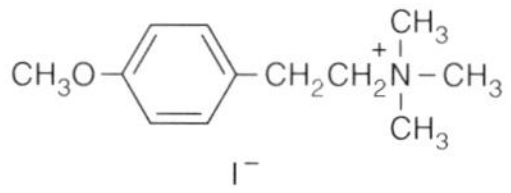

250 mg

S40,275-3
BENZYLDIMETHYL(2-DODECYLOXYETHYL)-AMMONIUM CHLORIDE

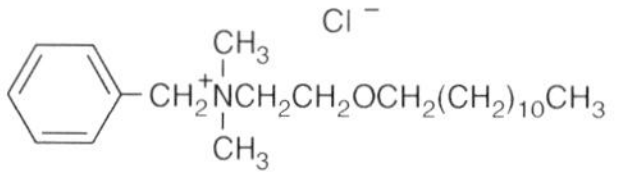

250 mg

S62,214-1
4-(3,5-DIMETHOXYPHENYL)-4-METHYLMORPHOLINIUM METHOSULFATE

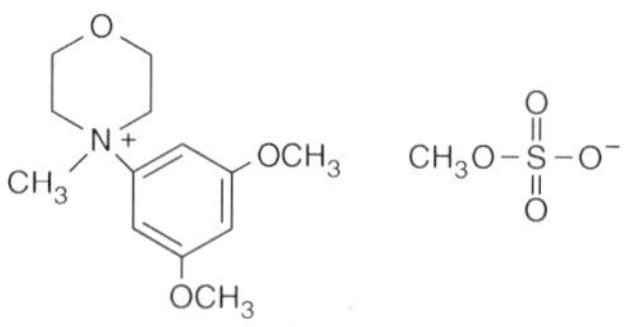

5 gm

S40,303-2
BENZYLDIETHYL(2-[4-(1,1,3,3-TETRAMETHYLBUTYL)PHENOXY]ETHYL)-AMMONIUM CHLORIDE

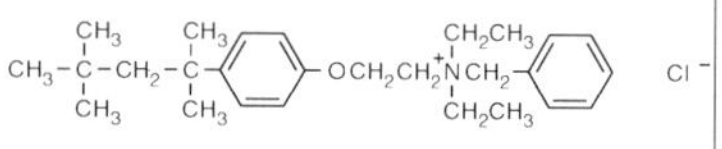

5 gm

S73,474-8
(4-METHOXYPHENYL)METHYL-TETRAHYDRO-1H-PYRAZOLO(1,2-B)-PHTHALAZIN-4-IUM IODIDE

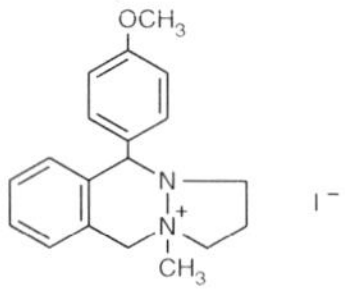

250 mg

S69,645-5
5,6-BENZO-2,3-DIMETHYL-8-ETHOXY-2-AZABICYCLO[2.2.2]-OCT-2-ENIUM IODIDE

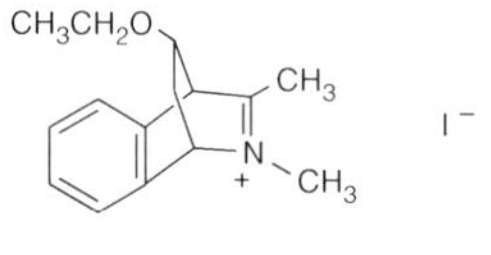

250 mg

S86,152-9
1,1'-(2,5-DIMETHOXY-1,4-PHENYLENEDIMETHYLENE)BIS-(HEXAMETHYLENE)TETRAMMINIUM DICHLORIDE

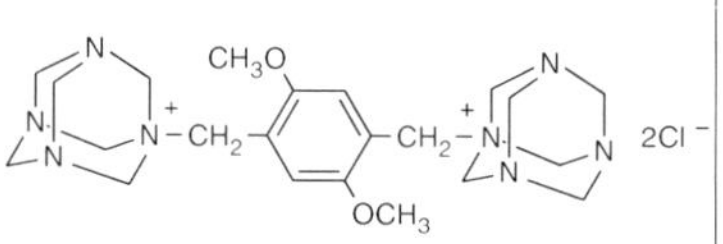

250 mg

S87,346-2
4,5-DIHYDRO-1-(4-METHOXYPHENYL)-3-METHYL-1H-1,2,3-TRIAZOLIUM PERCHLORATE

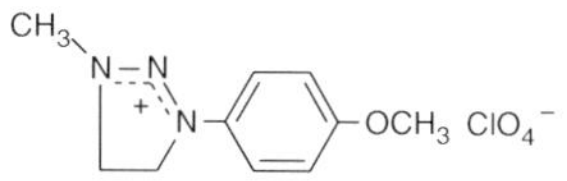

250 mg

S87,333-0
4,5-DIHYDRO-1-(4-METHOXYPHENYL)-3-(P-TOLYL)-1H-1,2,3-TRIAZOLIUM PERCHLORATE

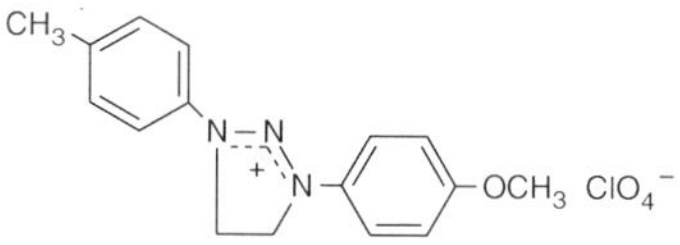

250 mg

S54,698-4
O-(2-BENZYLPHENYL)CHOLINE IODIDE

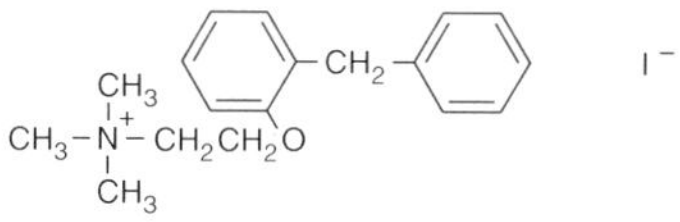

5 gm

S91,504-1
1,2,3-TRIAMINOBENZENE DIHYDROCHLORIDE

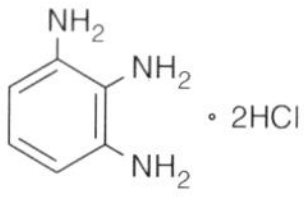

250 mg

S87,283-0
(4-[BIS(P-TOLYL)HYDRAZONO]-2,5-CYCLOHEXADIEN-1-YLIDENE)-DIMETHYLAMMONIUM CHLORIDE

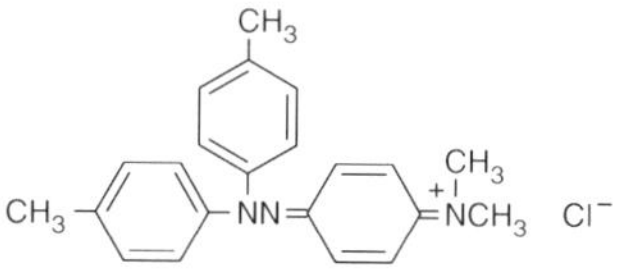

250 mg

S88,780-3
N-CINNAMYLIDENE-N,N'-DIETHYL-N',N'-DIMETHYL-1,4-PHENYLENEDIAMMONIUM DITETRAFLUOROBORATE

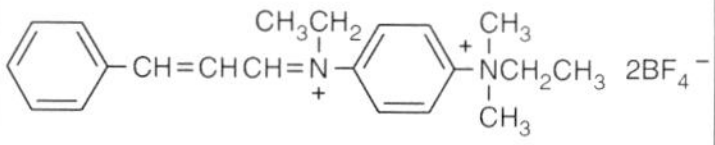

250 mg

S86,962-7
3-METHYL-1-PHENYL-2-IMIDAZOLINIUM IODIDE

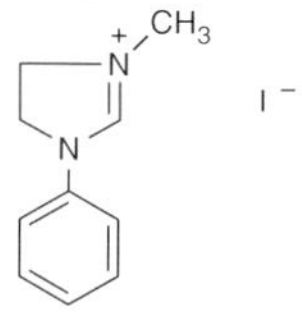

250 mg

S54,572-4
6-CHLORO-3-(4-CHLOROPHENYL)-3,4-DIHYDRO-1-METHYLQUINAZOLINIUM IODIDE

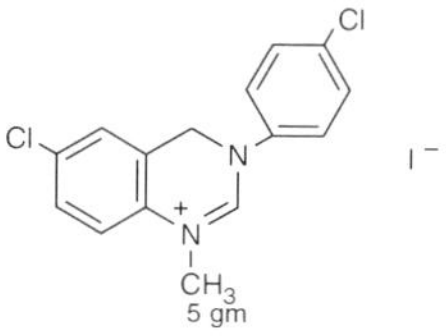

5 gm

S87,007-2
[4-(2,5-CYCLOHEXADIEN-1-YLIDENE)-2-BUTEN-1-YL]TRIMETHYLAMMONIUM PERCHLORATE

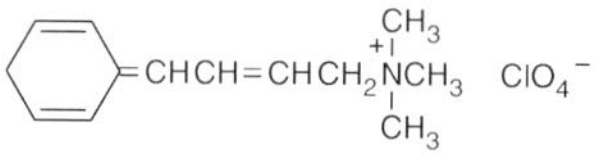

250 mg

S77,591-6
2-BENZYLAMINO-3,4-DIHYDROISOQUINOLINIUM BROMIDE

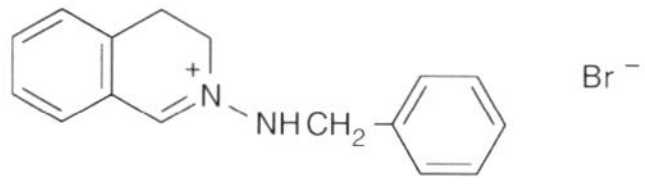

250 mg

S73,346-6
[4-(1,3-DIHYDRO-1-METHYL-1-ISOBENZOFURANYL)BENZYL]-TRIMETHYLAMMONIUM IODIDE

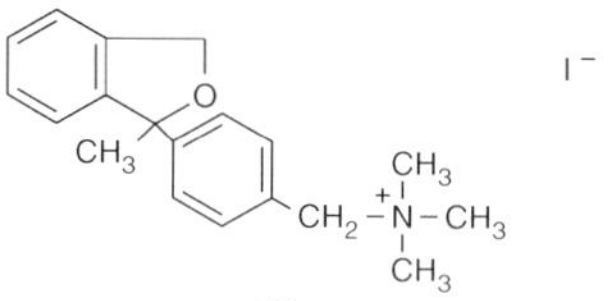

250 mg

S50,352-5
BENZYLDIMETHYLAMMONIUM HEXAFLUOROARSENATE

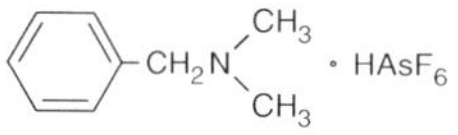

5 gm

S34,206-8
BENZYLTRIMETHYLAMMONIUM HEXAFLUOROPHOSPHATE

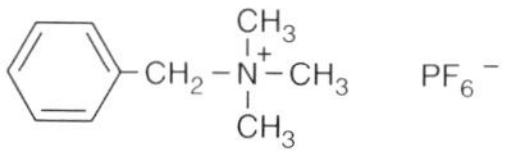

5 gm

S34,204-1
BENZYLTRIMETHYLAMMONIUM HEXAFLUOROANTIMONATE

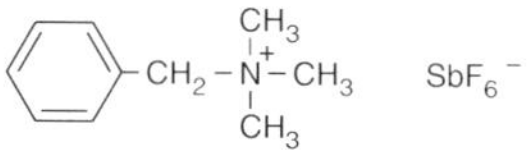

5 gm

S63,869-2
BENZYLDIMETHYLHEXADECYLAMMONIUM BROMIDE

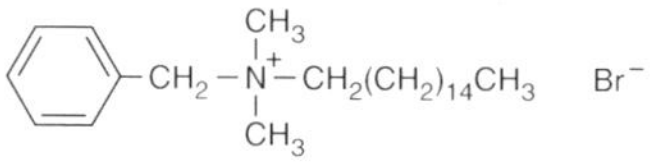

250 mg

S63,892-7
BENZYLDIMETHYLHEXADECYLAMMONIUM DICHROMATE

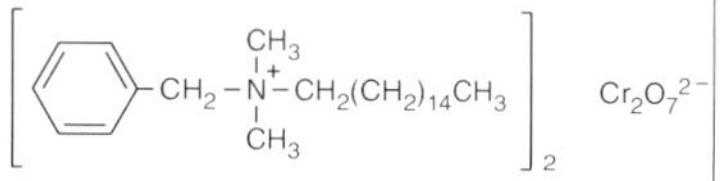

250 mg

S76,137-0
7-CHLORO-5,10-DIHYDRO-3,10-DIMETHYL-5-PHENYL-THIAZOLO(2,3-B)-QUINAZOLIN-10-IUM IODIDE

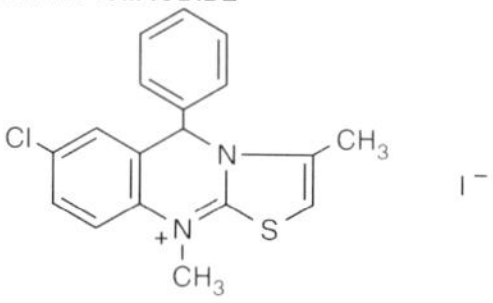

250 mg

S90,616-6
1,1-DIBENZYL-3,3-DIMETHYLAZETIDINIUM BROMIDE

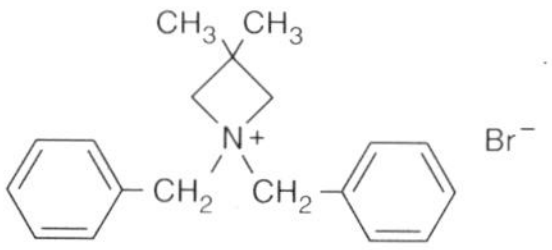

250 mg

S79,247-0
1,3-DIHYDROSPIRO[2H-BENZ(F)-ISOINDOLE-2,1'-PIPERIDINIUM]-BROMIDE

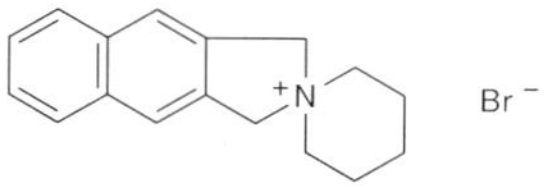

250 mg

S89,628-4
2-METHYL-2-PHENYLISOINDOLINIUM BROMIDE

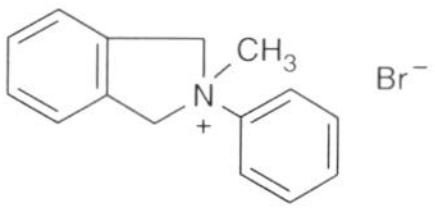

250 mg

S40,938-3
2-METHYL-2-PHENYLISOINDOLINIUM IODIDE

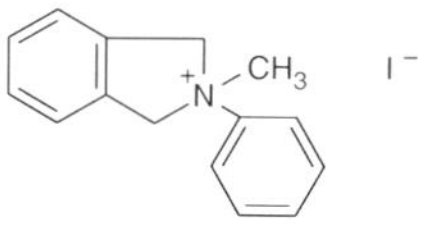

5 gm

S87,368-3
4-PHENYL-PERHYDROPYRROLIZINIUM PERCHLORATE

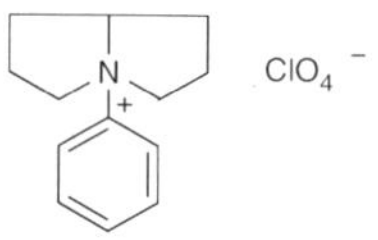

250 mg

S84,623-6
1,1',3,3'-TETRAHYDRO-2,2'-SPIROBI-(2H-ISOINDOLIUM)BROMIDE

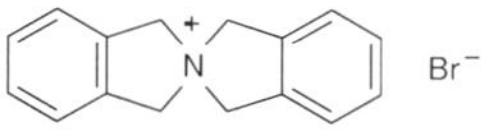

250 mg

S79,241-1
2,3-DIHYDRO-2,2-DIMETHYL-1H-BENZ(E)-ISOINDOLIUM BROMIDE

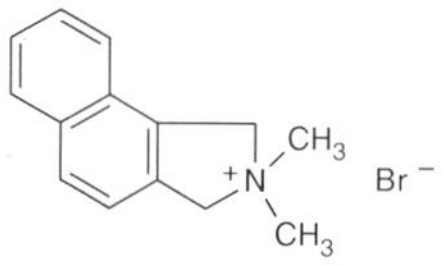

250 mg

S79,243-8
2,3-DIHYDRO-2,2-DIMETHYL-1H-BENZ(F)-ISOINDOLIUM BROMIDE

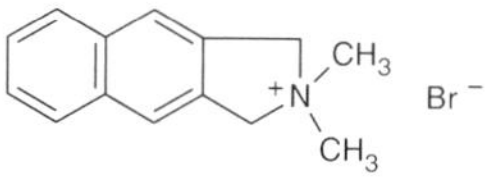

250 mg

S43,830-8
HYDRASTININE CHLORIDE MONOHYDRATE

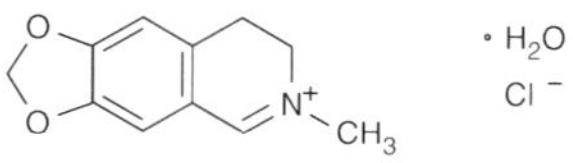

250 mg

S41,120-5
(1-PHENYLCYCLOHEXYL)-TRIMETHYLAMMONIUM IODIDE

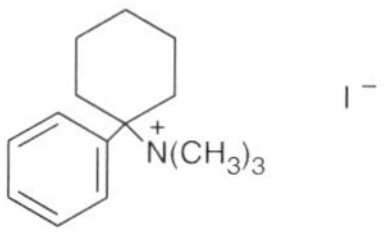

5 gm

S88,136-8
4-METHYL-4-(4-PHENYL-2-CYCLOHEXEN-1-YL)MORPHOLINIUM IODIDE

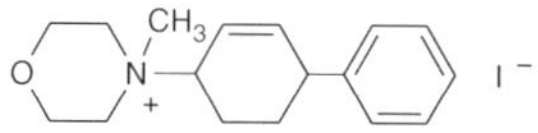

250 mg

S87,332-2
1,3-BIS(P-TOLYL)TETRAHYDRO-1,2,3-TRIAZINIUM PERCHLORATE

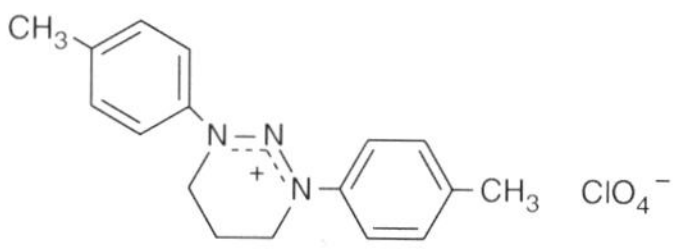

250 mg

S87,348-9
4,5-DIHYDRO-1-METHYL-3-PHENYL-1H-1,2,3-TRIAZOLIUM TETRAFLUOROBORATE

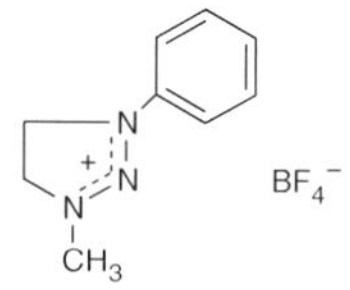

250 mg

S37,036-3
PHENYLTRIMETHYLAMMONIUM HEXAFLUOROPHOSPHATE

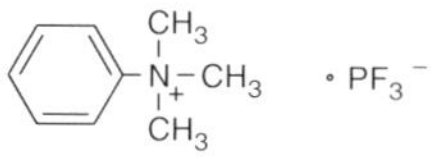

5 gm

S87,341-1
4,5-DIHYDRO-1,3-DIPHENYL-1H-1,2,3-TRIAZOLIUM PERCHLORATE

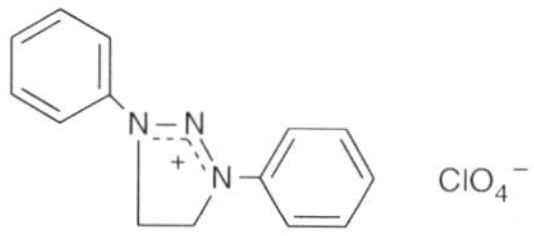

250 mg

S70,662-0
BENZYLDIMETHYLPHENYLAMMONIUM CHLORIDE

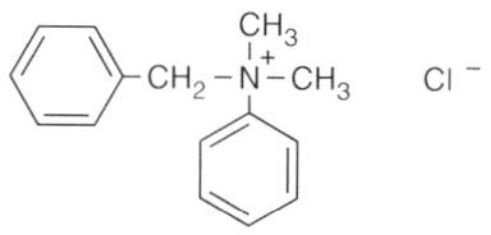

5 gm

S79,858-4
2-(2-BIPHENYLYL)-3,4-DIHYDROISOQUINOLINIUM BROMIDE

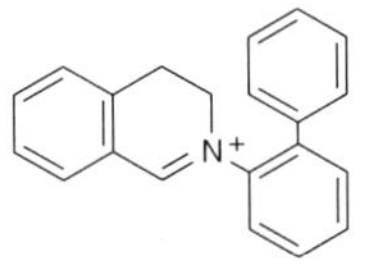

250 mg

S74,826-9
2-ETHYL-11-METHYL-2-PHENYL-2,3,5,10-TETRAHYDRO-1H-PYRAZOLO(12B)-PHTHALAZIN-11-IUM IODIDE

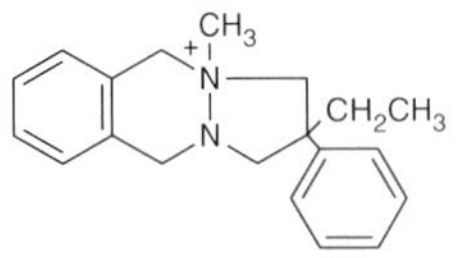

250 mg

S87,347-0
4,5-DIHYDRO-1-METHYL-3-(P-TOLYL)-1H-1,2,3-TRIAZOLIUM PERCHLORATE

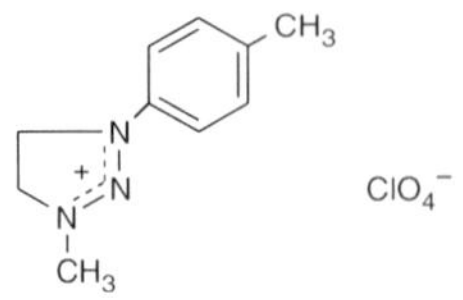

250 mg

S37,388-5
(4-METHYLBENZYL)-TRIMETHYLAMMONIUM CHLORIDE

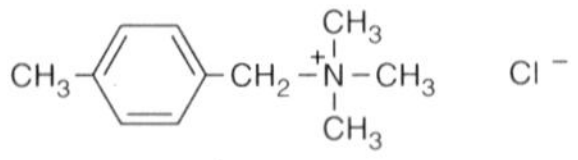

5 gm

S86,205-3
BIS(4-METHYLBENZYLIDENE)-ETHYLHYDRAZINIUM TETRAFLUOROBORATE

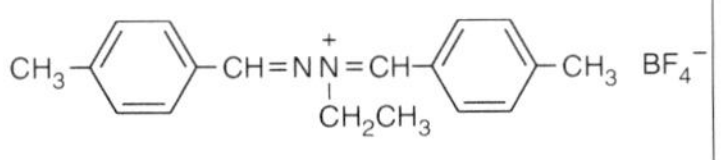

250 mg

S83,190-5
(4-HYDROXY-2-ISOPROPYLPHENYL)-TRIMETHYLAMMONIUM IODIDE

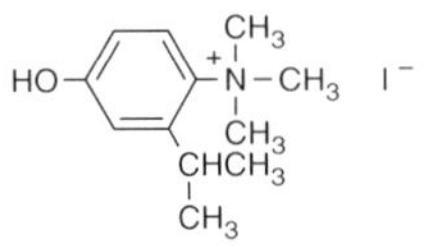

250 mg

S80,905-5
(1,1,2,2,3,3,4,4-OCTAMETHYL-1,2,3,4-TETRAHYDRO-5-NAPHTHYL)-TRIMETHYLAMMONIUM IODIDE

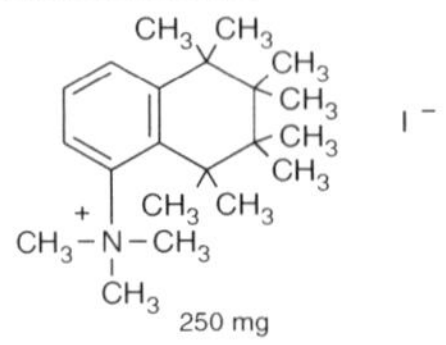

250 mg

S40,291-5
DIMETHYLDODECYL(5,6,7,8-TETRAHYDRO-2-NAPHTHYLMETHYL)AMMONIUM CHLORIDE

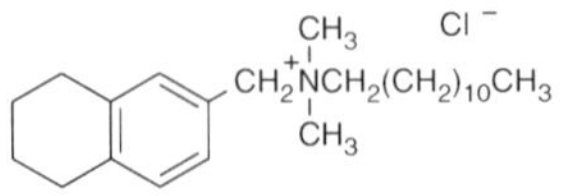

250 mg

S80,750-8
1-NAPHTHYLTRIMETHYLAMMONIUM IODIDE

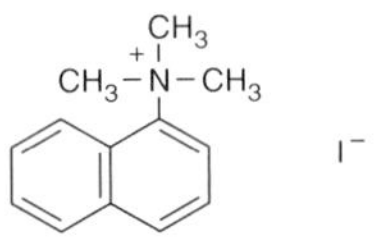

250 mg

S92,908-5
(4-METHYL-1-NAPHTHYLMETHYL)-TRIMETHYLAMMONIUM BROMIDE

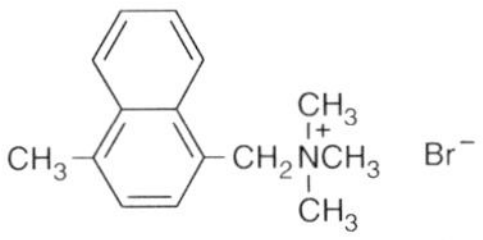

250 mg

S90,653-0
(2-NAPHTHYLMETHYL)-TRIMETHYLAMMONIUM IODIDE

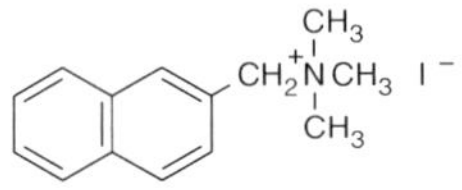

250 mg

S91,834-2
(9-ANTHRACENYLMETHYL)-TRIMETHYLAMMONIUM CHLORIDE

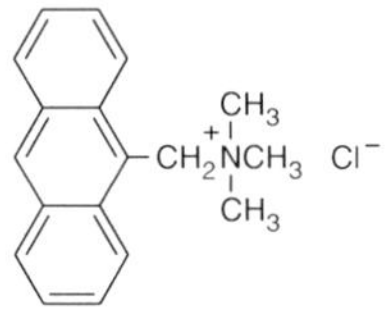

250 mg

S88,276-3
2,2-DIMETHYL-1,2,3,4-TETRAHYDROISOQUINOLINIUM IODIDE

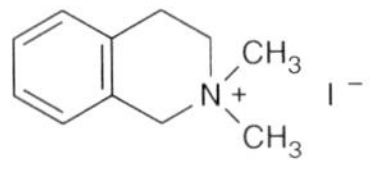

250 mg

S90,666-2
1,3-BIS(TRIMETHYLAMMONIOMETHYL)-AZULENE DIIODIDE

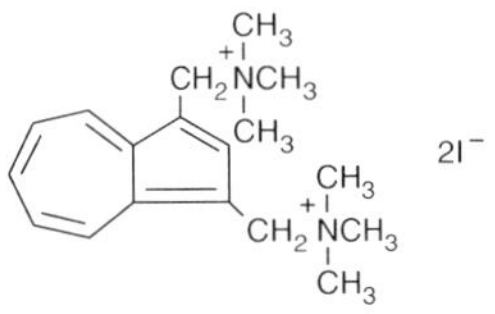

250 mg

S76,186-9
1-BENZYL-1,1-DIMETHYLHYDRAZINIUM CHLORIDE

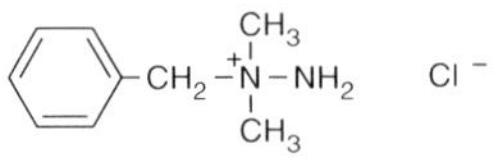

250 mg

S66,451-0
4-BROMO-3-METHYL-1-NAPHTHALENEDIAZONIUM CHLORIDE, MERCURIC BROMIDE COMPLEX

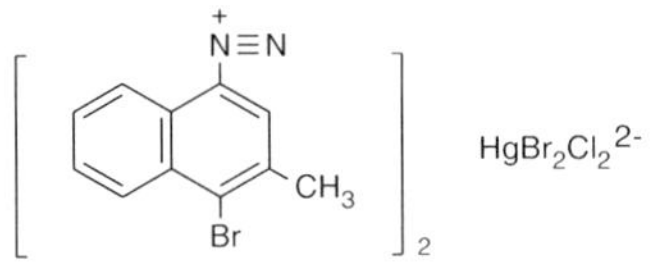

250 mg

S94,763-6
TRIMETHYL(2,4,5-TRIMETHYLBENZYL)-AMMONIUM BROMIDE

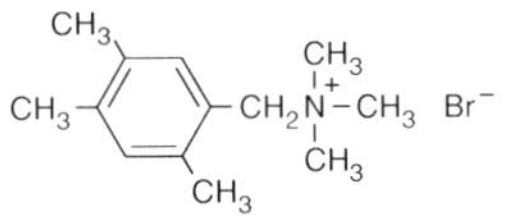

250 mg

S64,262-2
N,N-BIS(2-CHLOROETHYL)-3-METHYL-4-NITROSOANILINE
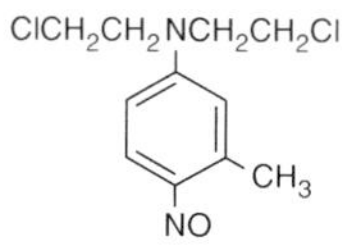
250 mg

S44,926-1
2,4-DINITROSO-1,3-NAPHTHALENEDIOL TETRAHYDRATE
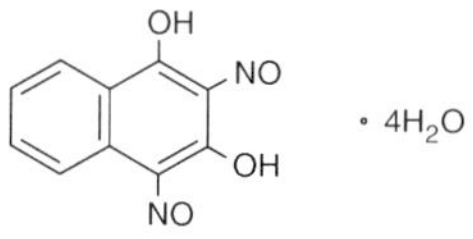
250 mg

S95,352-0
2-ISOPROPYL-5-METHYL-4-NITROSOPHENOL
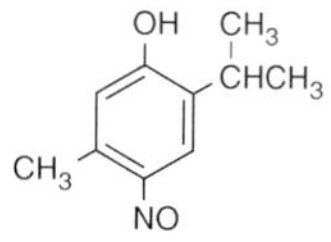
50 mg

S43,271-7
2-CHLORO-4-NITROSOPHENOL, MOIST
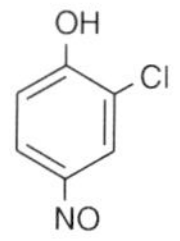
5 gm

S41,892-7
3-METHYL-4-NITROSOPHENOL
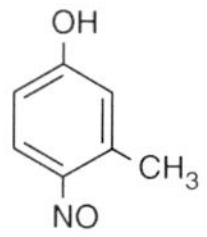
5 gm

S42,978-3
2-METHYL-4-NITROSOPHENOL
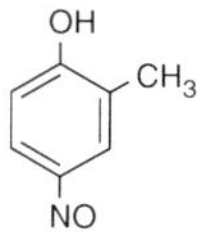
5 gm

S43,272-5
6-CHLORO-4-NITROSO-O-CRESOL, MOIST
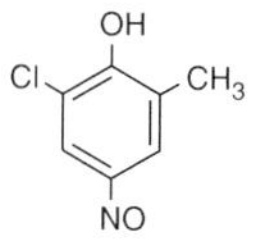
5 gm

S44,998-9
2,6-DIMETHYL-4-NITROSOPHENOL
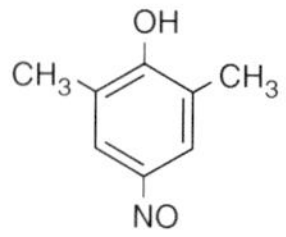
5 gm

S44,431-6
2,6-DI-TERT-BUTYL-4-NITROSOPHENOL
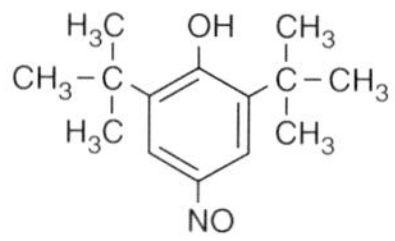
250 mg

S97,306-8
4,5-DINITROSO-1,3,6,8-TETRABROMOPHENANTHRENE
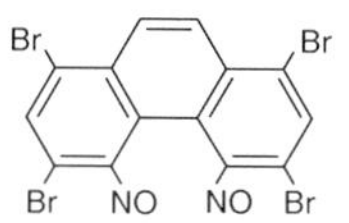
10 mg

S39,636-2
N-METHYL-4-NITROSOANILINE
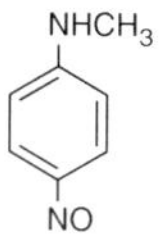
5 gm

S86,964-3
N-TERT-BUTYL-N-NITROSOANILINE
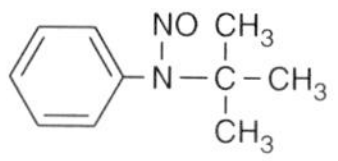
250 mg

S65,138-9
N-NITROSODIBENZYLAMINE
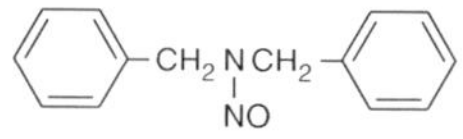
5 gm

S85,228-7
3-CHLORO-2-NITROSO-1,2,3,4-TETRAHYDRO-1,4-METHANONAPHTHALANE DIMER
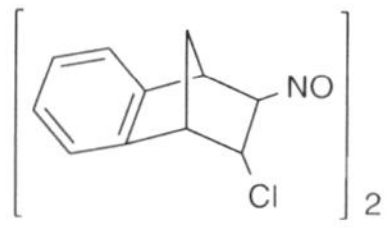
250 mg

S57,719-7
N-NITROSODIPHENYLAMINE
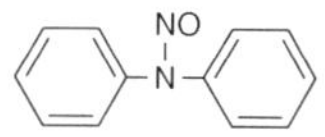
250 mg

S64,821-3
2-HYDROXY-4'-PHENYLACETOPHENONE NITRATE ESTER
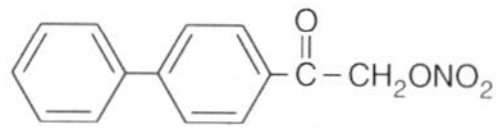
250 mg

S86,124-3
N-BENZYLIDENEANILINE N-OXIDE
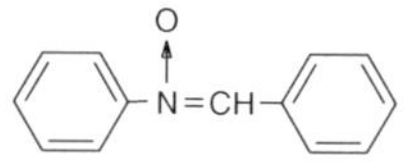
250 mg

S91,897-0
1,3-BIS(TRIFLUOROMETHYL)-5-(2-NITRO-1,1,2-TRIMETHYLBUTYL)BENZENE
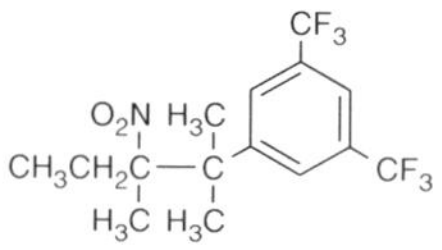
250 mg

S91,824-5
1,3-BIS(TRIFLUOROMETHYL)-5-(1-METHYL-1-NITROETHYL)BENZENE
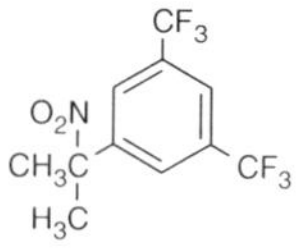
250 mg

S53,656-3
1-CHLORO-4-(2-NITROPROPENYL)-BENZENE
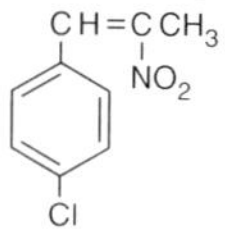
5 gm

S91,900-4
4-BROMO-ALPHA-NITROTOLUENE
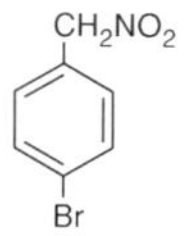
250 mg

S59,463-6
4-HYDROXY-3-METHOXY-BETA-NITROSTYRENE
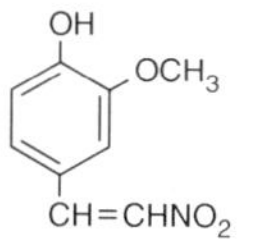
250 mg

S91,914-4
2-NITRO-1-PHENYL-1-PROPANOL
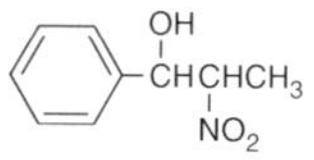
250 mg

S91,459-2
4-(2-NITROVINYL)PHENOL
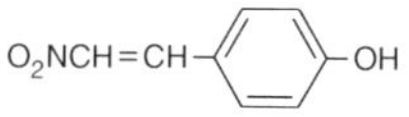
250 mg

S86,075-1
2,3-DIMETHOXY-5-(2-NITROETHYL)-ANILINE
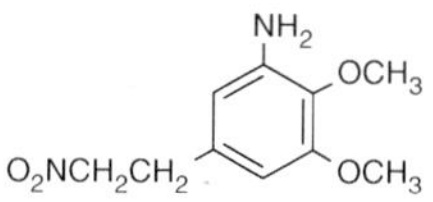
250 mg

S38,064-4
3,4-DIMETHOXY-BETA-NITROSTYRENE
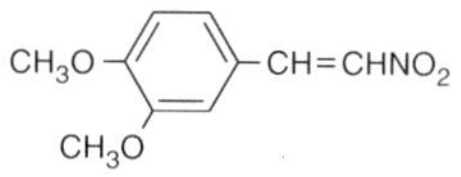
5 gm

S75,221-5
3,4-DIMETHOXY-BETA-NITROSTYRENE
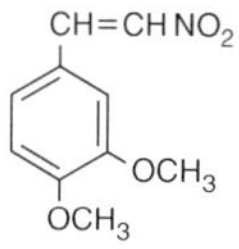
5 gm

S80,863-6
BETA-NITRO-3,4,5-TRIMETHOXYSTYRENE
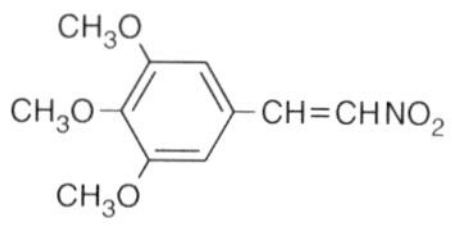
250 mg

S75,219-3
3,5-DIMETHOXY-BETA-NITROSTYRENE
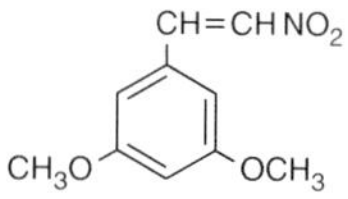
5 gm

S75,216-9
2,5-DIMETHOXY-BETA-NITROSTYRENE
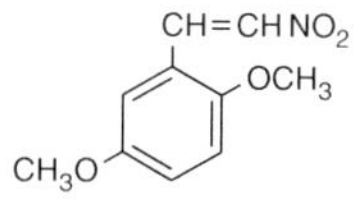
5 gm

S34,653-5
1-(3,4-DIMETHOXYPHENYL)-2-NITRO-1-PROPENE
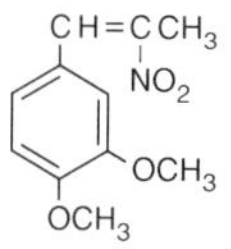
5 gm

S59,462-8
4-METHOXY-BETA-NITROSTYRENE
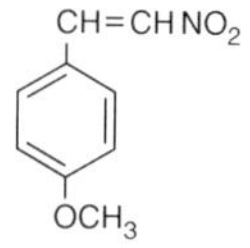
250 mg

S42,984-8
2,5-DIMETHOXY-BETA-METHYL-BETA-NITROSTYRENE
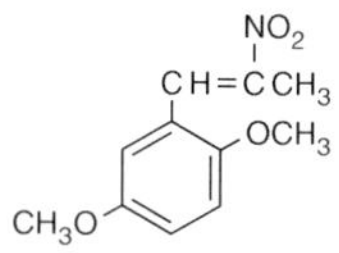
5 gm

S75,695-4
3-METHOXY-BETA-NITROSTYRENE
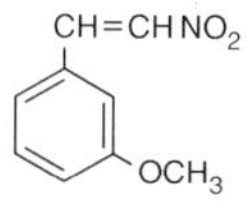
5 gm

S89,920-8
4-METHOXY-BETA-NITROSTILBENE
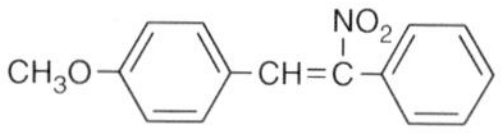
250 mg

S35,240-3
1-(4-METHOXYPHENYL)-2-NITRO-1-PROPENE
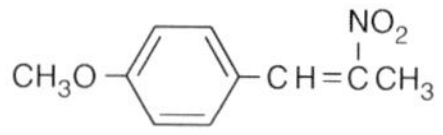
5 gm

S91,822-9
4-(2-NITRO-1,1,2-TRIMETHYLPROPYL)-ANILINE
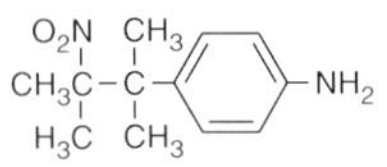
250 mg

S91,877-6
2,3-DIMETHYL-2,3-DINITRO-1,4-DIPHENYLBUTANE
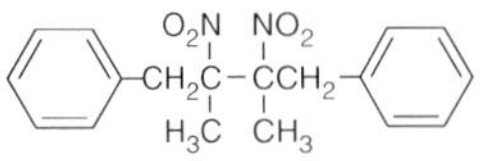
250 mg

S84,934-0
4,4-DIMETHYL-2-(DIPHENYLNITROMETHYL)-2-OXAZOLINE
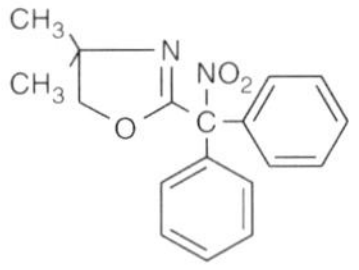
250 mg

S79,881-9
2-NITRO-1-NITROSO-1-PHENYLCYCLOHEXANE
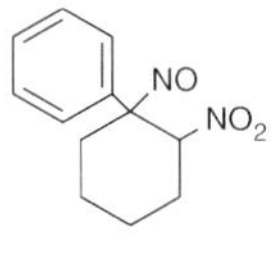
250 mg

S62,080-7
4-DIMETHYLAMINO-BETA-METHYL-BETA-NITROSTYRENE
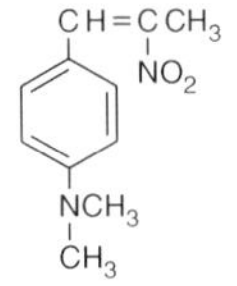
250 mg

S59,466-0
4-DIMETHYLAMINO-BETA-NITROSTYRENE
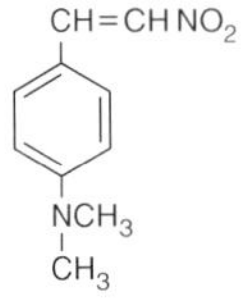
250 mg

S44,709-9
4-FLUORO-BETA-NITROPROPENYLBENZENE
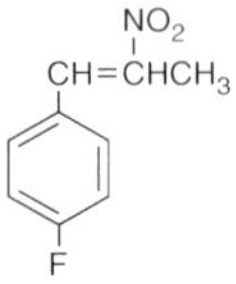
5 gm

S62,078-5
BETA-METHYL-BETA-NITROSTYRENE
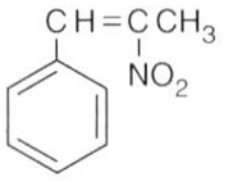
250 mg

S62,079-3
BETA,4-DIMETHYL-BETA-NITROSTYRENE
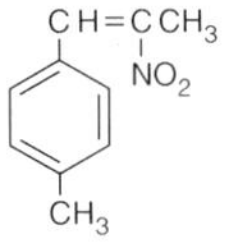
250 mg

S68,266-7
4-METHYL-BETA-NITROSTYRENE
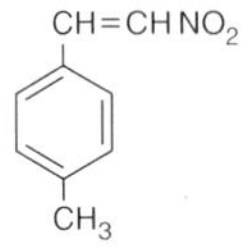
5 gm

S88,618-1
1-(4-METHYL-ALPHA-NITROBENZYLIDENE)-2-PHENYLHYDRAZINE
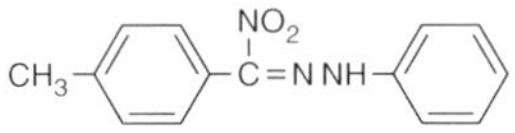
250 mg

S85,995-8
ALPHA-NITRO-2-NAPHTHALDEHYDE PHENYL HYDRAZONE
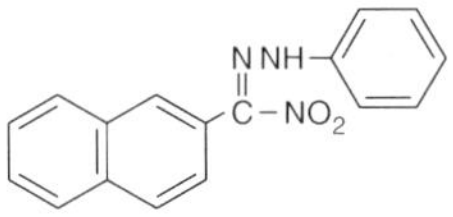
250 mg

S53,395-5
2,6-DICHLORO-BETA-NITROSTYRENE
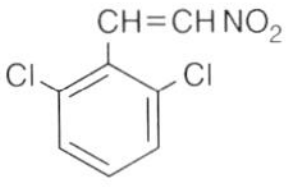
5 gm

S35,291-8
3,4-METHYLENEDIOXY-BETA-NITROSTYRENE
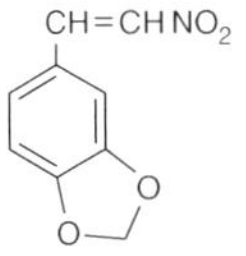
5 gm

S84,935-9
4,4-DIMETHYL-2-[1-(1-NAPHTHYL)-1-NITROPROPYL]-2-OXAZOLINE
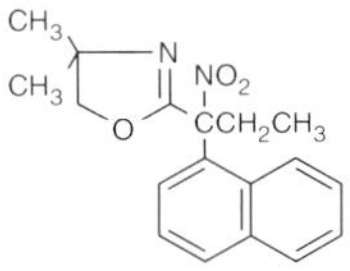
250 mg

S91,901-2
(1-NITROETHYL)BENZENE
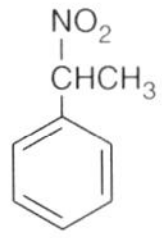
250 mg

S84,907-3
2-NITRO-2-PHENYLPROPANE
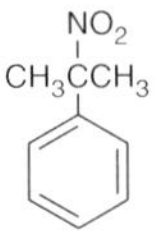
250 mg

S91,890-3
1-(3-NITROPROPYL)BENZENE
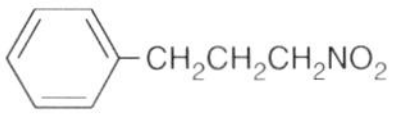
250 mg

S79,204-7
1,1-DIPHENYL-2-NITROETHYLENE
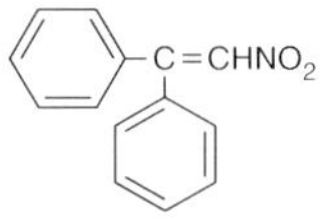
250 mg

S91,832-6
9,10-DIHYDRO-9-(1-METHYL-1-NITROETHYL)ANTHRACENE
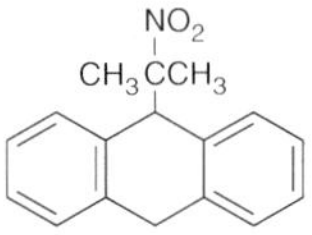
250 mg

S94,512-9
ALPHA-NITROMETHYL-N-PHENYLBENZYLAMINE
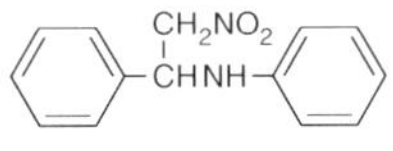
250 mg

S67,805-8
2-(2-BROMOETHOXY)-5-CHLORO-1-NITROBENZENE
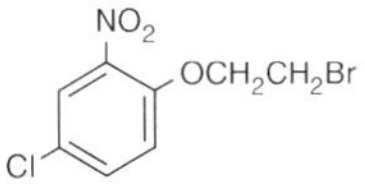
5 gm

S56,937-2
1-[2-(2-CHLOROETHOXY)ETHOXY]-4-NITROBENZENE
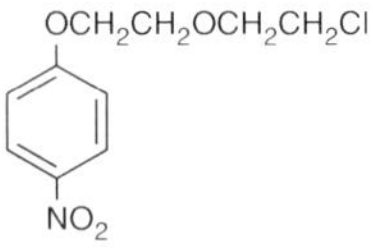
5 gm

S56,939-9
3-CHLOROPROPYL 4-NITROPHENYL ETHER
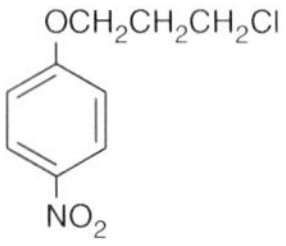
5 gm

S52,500-6
2-CHLOROMETHYL-4-NITROPHENYL HEPTYL ETHER
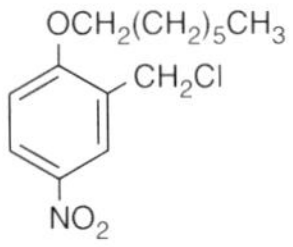
250 mg

S52,640-1
ALPHA-BROMO-5-NITRO-O-TOLYL NONYL ETHER
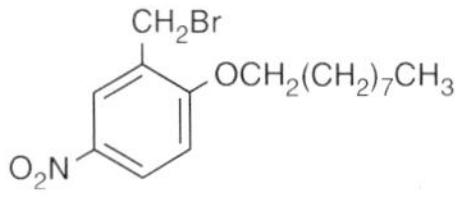
250 mg

S52,644-4
ALPHA-BROMO-5-NITRO-O-TOLYL DECYL ETHER
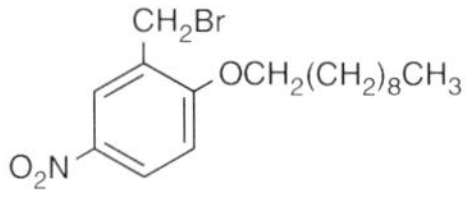
250 mg

S52,637-1
ALPHA-BROMO-5-NITRO-O-TOLYL UNDECYL ETHER
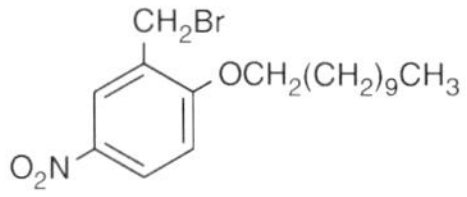
250 mg

S63,367-4
2-CHLOROETHYL 4-NITROPHENYL SULFIDE
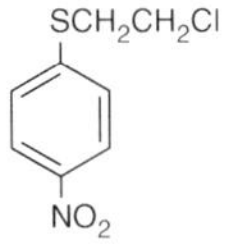
250 mg

S86,316-5
3-CHLOROPROPYL 4-NITROPHENYL SULFIDE
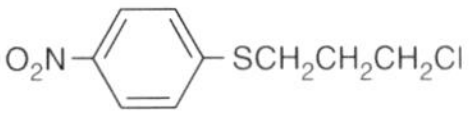
250 mg

S77,589-4
2-(2-BROMOETHYL)BENZALDEHYDE (4-NITROPHENYL)HYDRAZONE
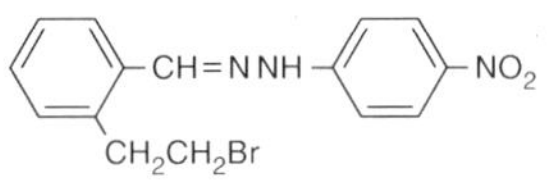
250 mg

S53,476-5
ALPHA-IODO-4-NITROTOLUENE

5 gm

S92,214-5
3-METHYL-4-NITROBENZYL CHLORIDE
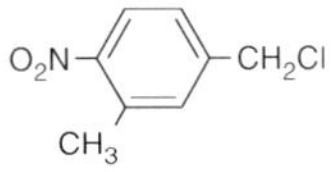
250 mg

S52,026-8
4-TERT-BUTYL-1-CHLOROMETHYL-2-NITROBENZENE
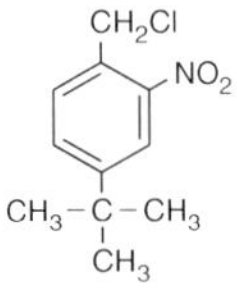
250 mg

S58,254-9
2-CHLORO-1,1,2-TRIFLUOROETHYL 4-NITROPHENYL ETHER
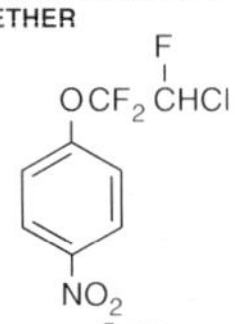
5 gm

S80,387-1
ALPHA,ALPHA-DICHLORO-3-NITROTOLUENE
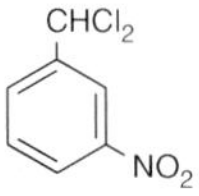
250 mg

S85,293-7
1-(2,2-DIBROMOETHYL)-3-NITROBENZENE
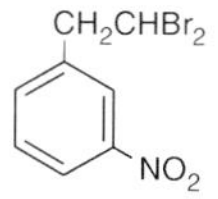
250 mg

S78,210-6
2-BROMOPHENYL 2-NITRO-4-(TRIFLUOROMETHYL)PHENYL ETHER
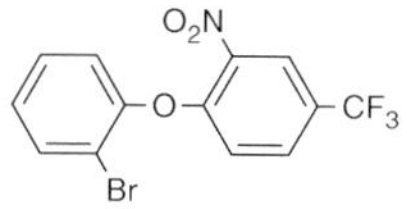
5 gm

S50,875-6
2,4-DICHLOROPHENYL 2-NITRO-ALPHA,ALPHA,ALPHA-TRIFLUORO-P-TOLYL ETHER
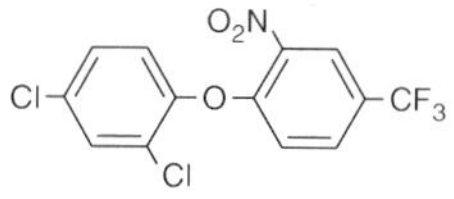
5 gm

S43,333-0
3-TRIFLUOROMETHYL-2,4,6-TRINITROPHENOL
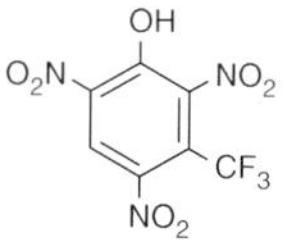
250 mg

S85,902-8
2,6-DINITRO-4-(TRIFLUOROMETHYL)PHENYL METHYL ETHER
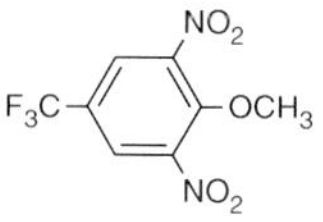
250 mg

S85,903-6
2,6-DINITRO-4-(TRIFLUOROMETHYL)PHENYL PROPYL ETHER
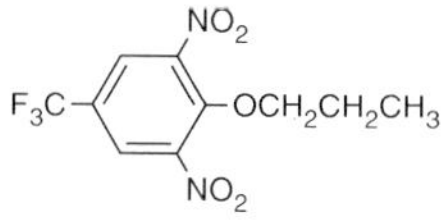
250 mg

S85,904-4
2,6-DINITRO-4-(TRIFLUOROMETHYL)PHENYL ISOPROPYL ETHER
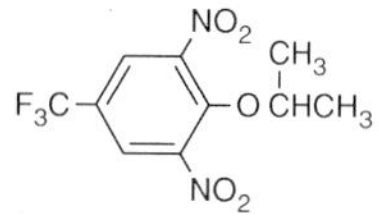
250 mg

S85,905-2
BUTYL 2,6-DINITRO-4-(TRIFLUOROMETHYL)PHENYL ETHER
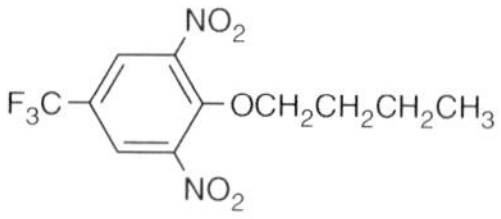
250 mg

S38,595-6
HEXADECYL 4-NITRO-3-(TRIFLUOROMETHYL)PHENYL ETHER
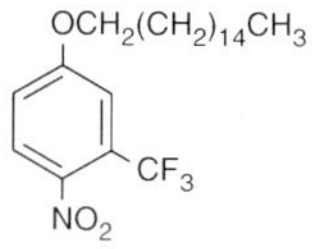
250 mg

S64,934-1
4-METHYLTHIO-3-NITROBENZOTRIFLUORIDE
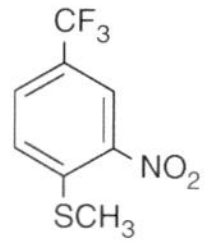
5 gm

S93,828-9
3-NITRO-4-(TRIFLUOROMETHYL)PHENYL SULFIDE
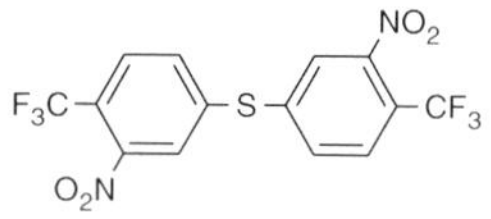
250 mg

S38,315-5
DODECYL 2-NITRO-4-TRIFLUOROMETHYLPHENYL SULFIDE
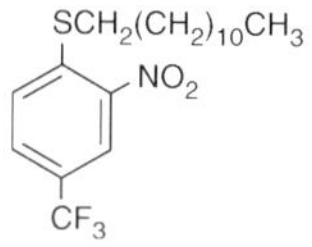
250 mg

S63,061-6
4-NITROPHENYL TRICHLOROMETHYL SULFIDE
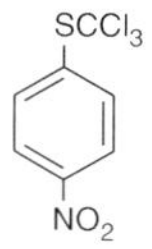
5 gm

S70,341-9
2-NITRO-4-(TRIFLUOROMETHYL)PHENYL DISULFIDE
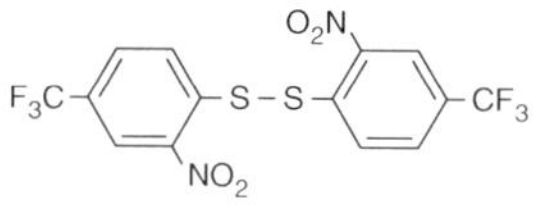
5 gm

S64,928-7
2,6-DINITRO-4-TRIFLUOROMETHYLANILINE
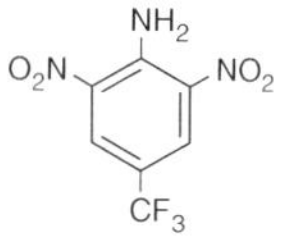
250 mg

S92,647-7
N-METHYL-4-NITRO-2-(TRIFLUOROMETHYL)ANILINE
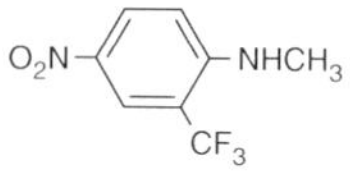
250 mg

S74,666-5
2,4-DINITRO-3'-(TRIFLUOROMETHYL)-DIPHENYLAMINE
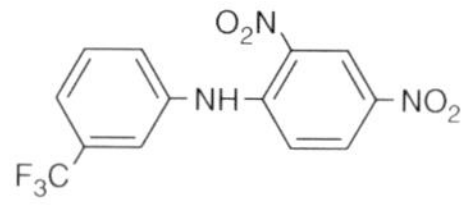
5 gm

S56,909-7
N-PICRYL-ALPHA,ALPHA,ALPHA-TRIFLUORO-O-TOLUIDINE
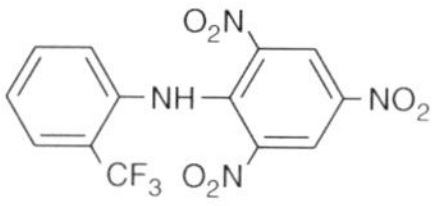
250 mg

S56,912-7
6-CHLORO-N-PICRYL-ALPHA,ALPHA,ALPHA-TRIFLUORO-M-TOLUIDINE
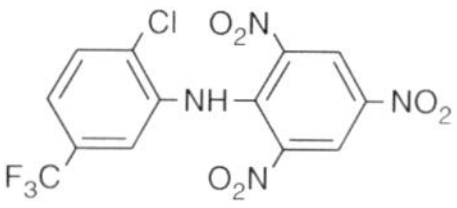
250 mg

S56,910-0
N-PICRYL-ALPHA,ALPHA,ALPHA-TRIFLUORO-M-TOLUIDINE
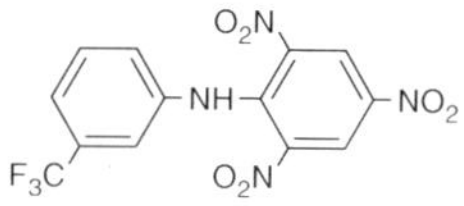
250 mg

S85,837-4
2,6-DINITRO-N-METHYL-4-(TRIFLUOROMETHYL)ANILINE
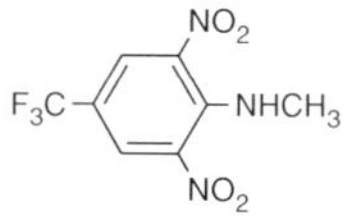
250 mg

S85,838-2
2,6-DINITRO-N-ETHYL-4-(TRIFLUOROMETHYL)ANILINE
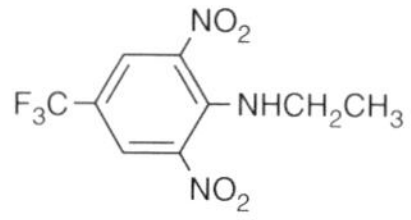
250 mg

S85,839-0
2,6-DINITRO-N-PROPYL-4-(TRIFLUOROMETHYL)ANILINE
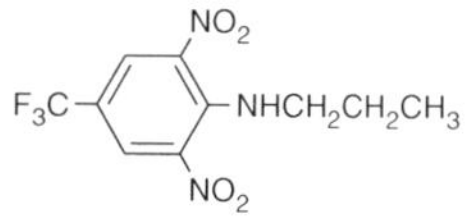
250 mg

S85,840-4
N-BUTYL-2,6-DINITRO-4-(TRIFLUOROMETHYL)ANILINE
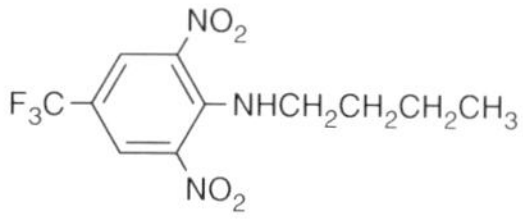
250 mg

S85,841-2
2,6-DINITRO-N-PENTYL-4-(TRIFLUOROMETHYL)ANILINE
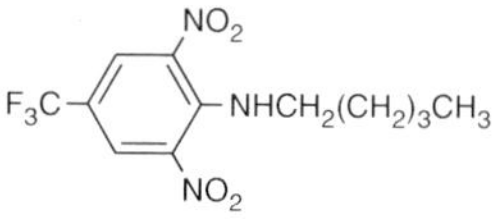
250 mg

S55,311-5
7-NITRO-2-(TRIFLUOROMETHYL)-PHENOTHIAZINE
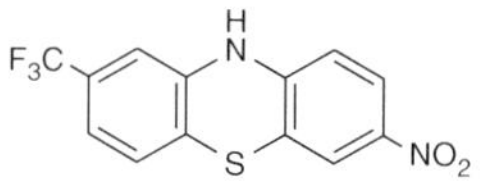
250 mg

S92,223-4
N,N-DIMETHYL-4-NITRO-2-(TRIFLUOROMETHYL)ANILINE
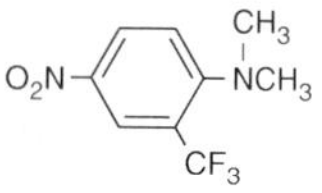
250 mg

S77,050-7
5-[(4-CHLORO-5-ETHYL-2-NITRO)-ANILINO]-1,2,3,4-PENTANETETRAOL
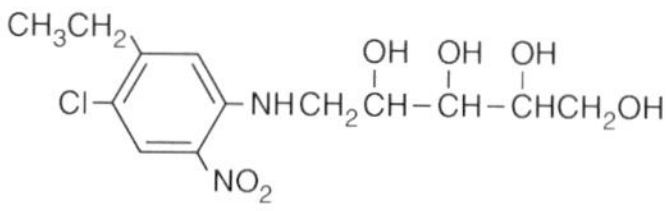
250 mg

S77,045-0
N,N-BIS(2-HYDROXYETHYL)-N'-(4,5-DICHLORO-2-NITROPHENYL)-ETHYLENEDIAMINE
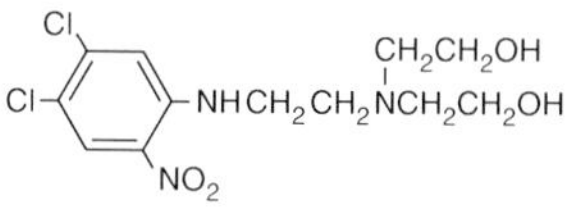
250 mg

S77,049-3
5-[(5-BROMO-4-ETHYL-2-NITRO)-ANILINO]-1,2,3,4-PENTANETETRAOL
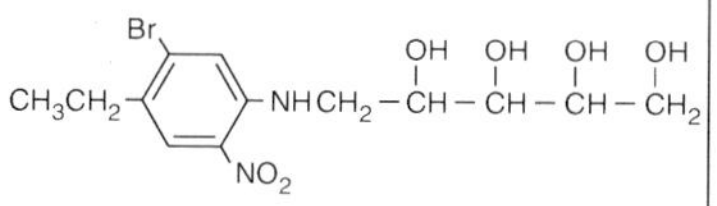
250 mg

S69,534-3
4-CHLORO-N-(2-HYDROXY-3-METHOXYBENZYL)-3-NITROANILINE
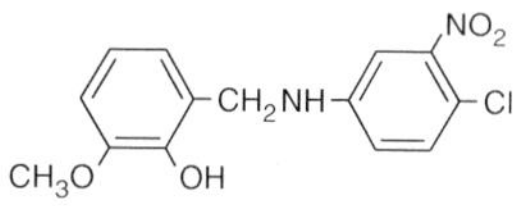
5 gm

S51,833-6
2,2'-(4-CHLORO-6-NITRO-1,3-PHENYLENEDIAMINO)DIETHANOL
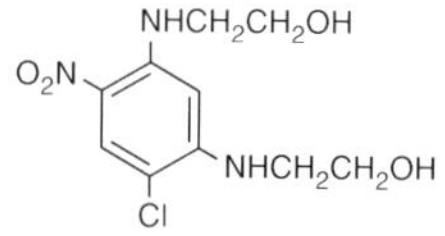
5 gm

S38,272-8
2-(4,5-DICHLORO-2-NITROANILINO)-ETHANOL
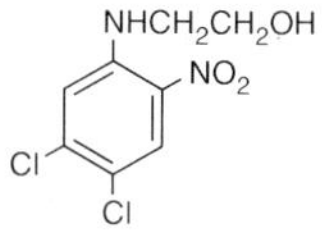
250 mg

S51,567-1
3-(4-CHLORO-2-NITROANILINO)-1-PROPANOL
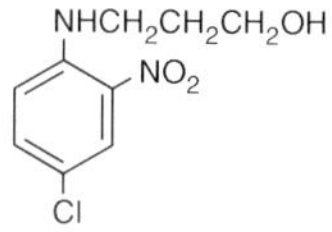
250 mg

S94,699-0
4-CHLORO-2,6-DINITROPHENOL, MOIST
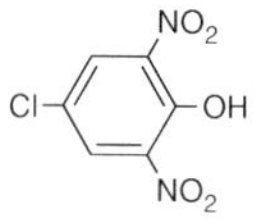
250 mg

S93,097-0
2-BROMO-6-(BIS[4-(DIMETHYLAMINO)-PHENYL]METHYL)-4-NITROPHENOL
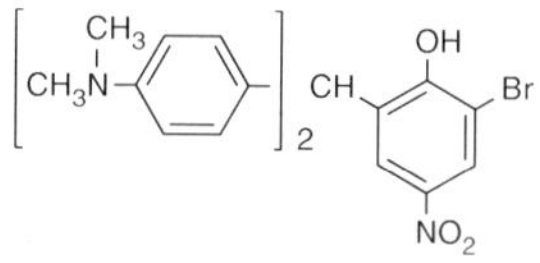
250 mg

S81,393-1
3,6-DIBROMO-2-NITROPHENOL

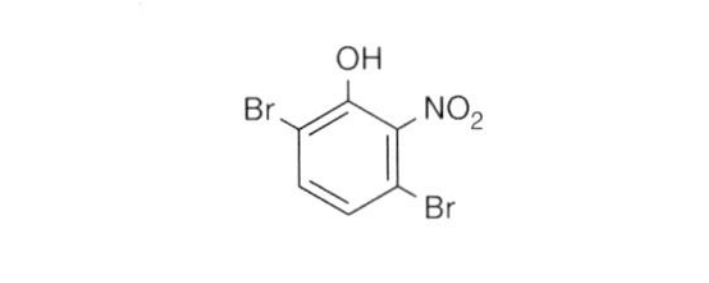

250 mg

S94,197-2
4-(BENZYLOXY)-5-BROMO-2-NITROANILINE

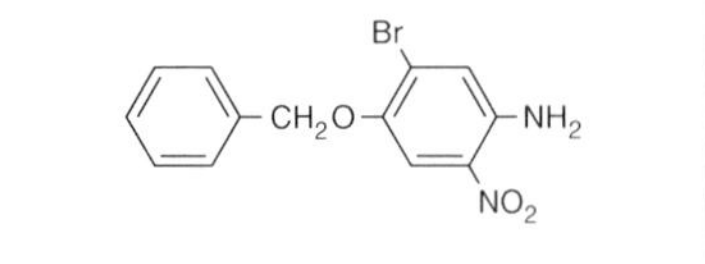

250 mg

S84,396-2
1,2-DIMETHOXY-4-IODO-5-NITROBENZENE

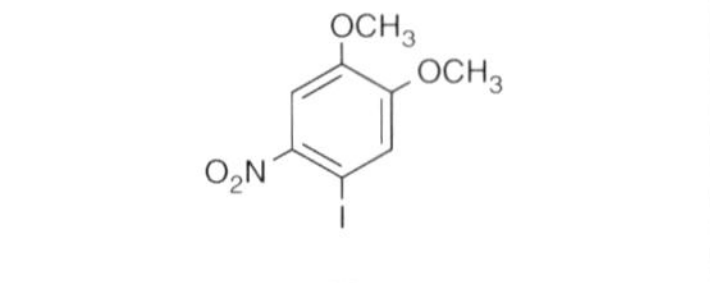

250 mg

S74,674-6
5-CHLORO-2,4-DIMETHOXY-2',4'-DINITRODIPHENYLAMINE

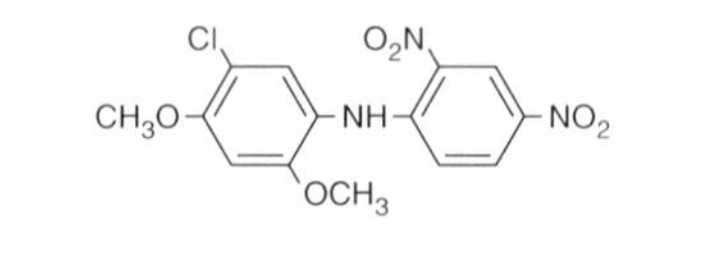

5 gm

S94,187-5
1-BUTOXY-5-CHLORO-4-ETHOXY-2-NITROBENZENE

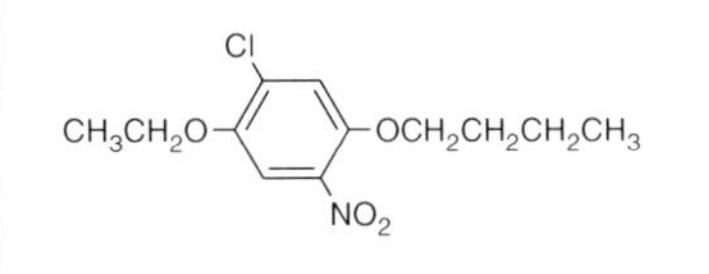

250 mg

S34,406-0
4-CHLORO-2,5-DIETHOXYNITROBENZENE

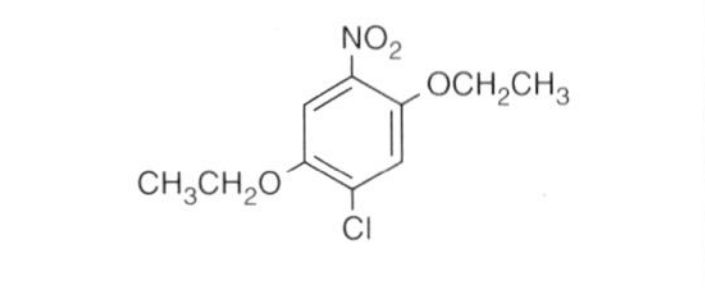

5 gm

S94,186-7
1-CHLORO-5-ETHOXY-4-NITRO-2-PROPOXYBENZENE

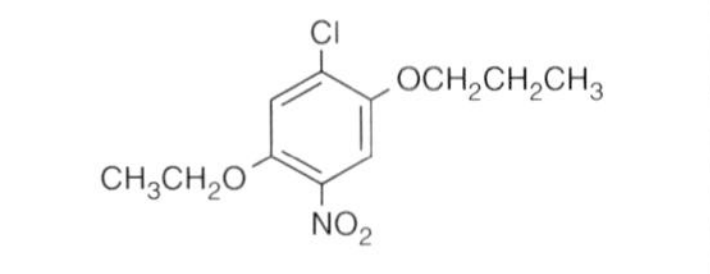

250 mg

S70,723-6
3-BROMO-4-METHOXY-6-NITROANILINE

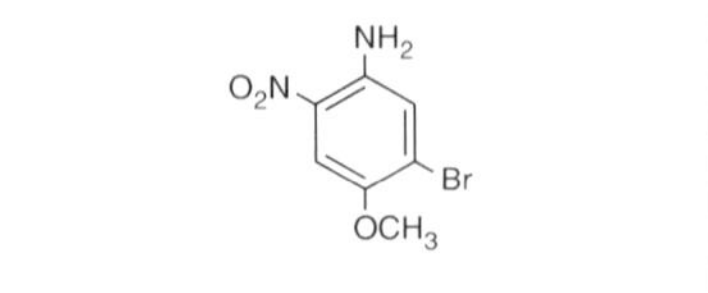

5 gm

S74,971-0
5-ETHOXY-4-FLUORO-2-NITROANILINE

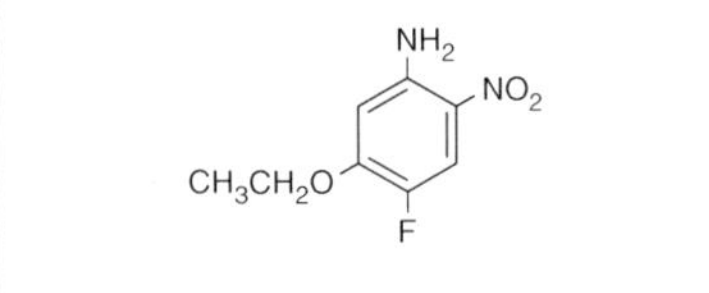

250 mg

S39,966-3
N-(5-CHLORO-2,4-DIMETHOXYPHENYL)-PICRYLAMINE

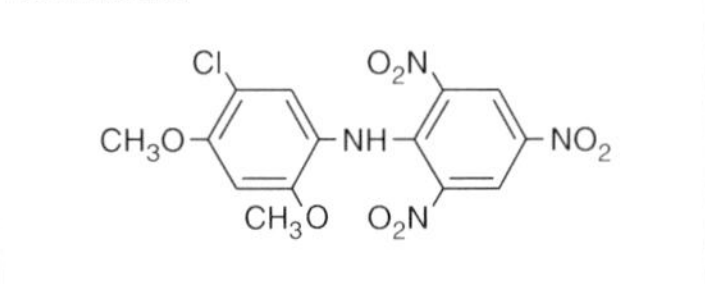

250 mg

S67,811-2
5-CHLORO-N,N-DIBENZYL-2-METHOXY-4-NITROANILINE

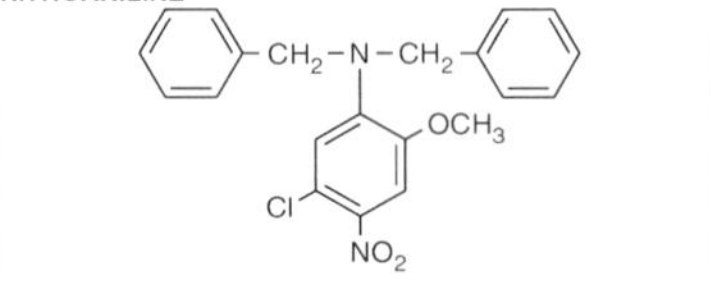

5 gm

S67,809-0
5-CHLORO-2-(4-CHLOROPHENOXY)-N,N-DIETHYL-4-NITROANILINE

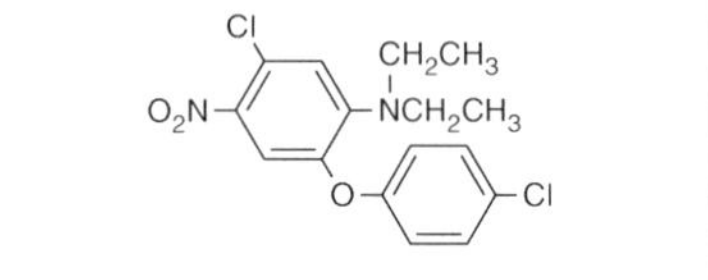

250 mg

S46,136-9
2-CHLORO-6-NITROANISOLE

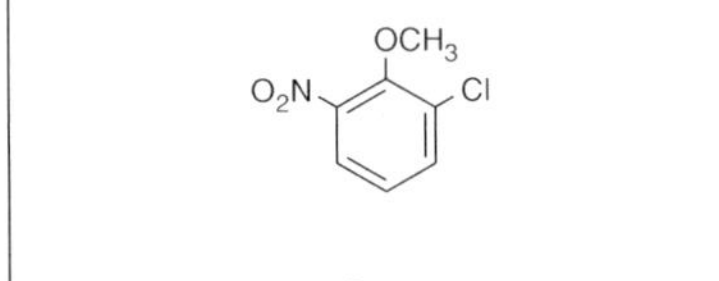

5 gm

S83,922-1
1,4-DIBROMO-2-METHOXY-3-NITROBENZENE

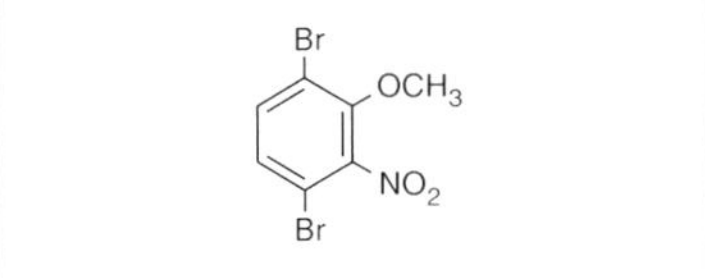

250 mg

S34,442-7
4-CHLORO-4'-NITRODIPHENYL ETHER

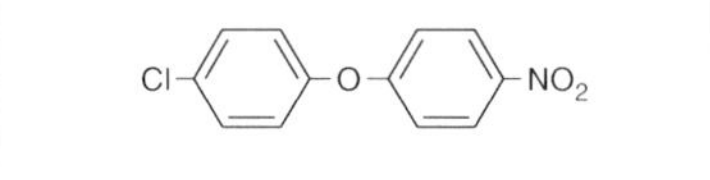

5 gm

S88,904-0
2-BROMO-1-METHOXY-4-NITROBENZENE

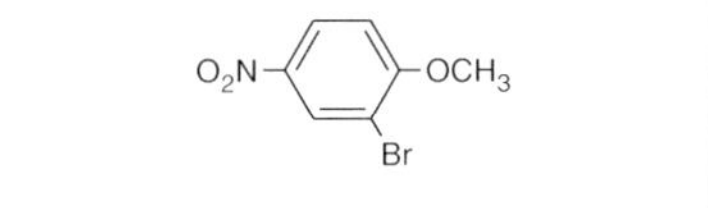

250 mg

S84,380-6
1,2-DIBROMO-5-METHOXY-3-NITROBENZENE

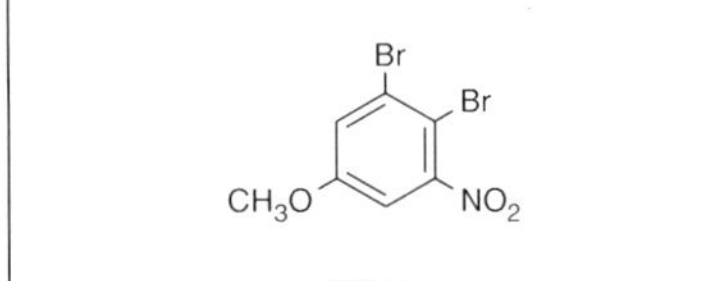

250 mg

S94,184-0
3-CHLORO-4-NITROANISOLE

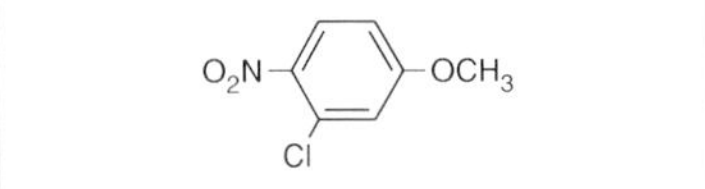

250 mg

S83,625-7
4-CHLOROPHENYL
2,4-DINITROPHENYL SULFIDE

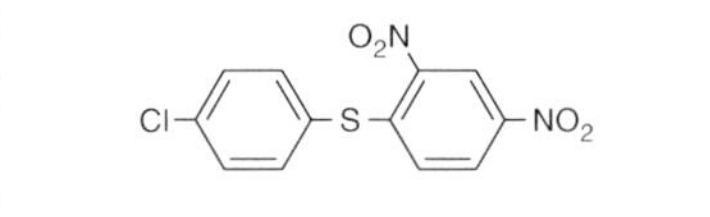

250 mg

S69,638-2
4-(HEXADECYLTHIO)-3-NITROCHLOROBENZENE

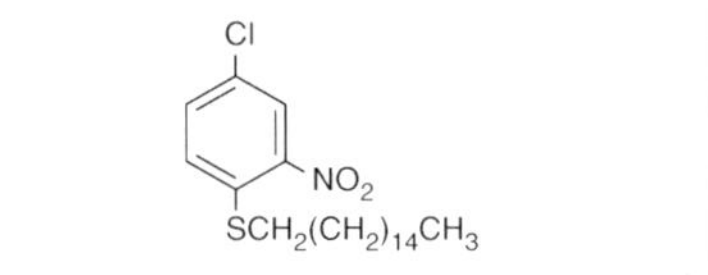

250 mg

S82,796-7
2-CHLOROPHENYL
4-NITROPHENYL SULFIDE

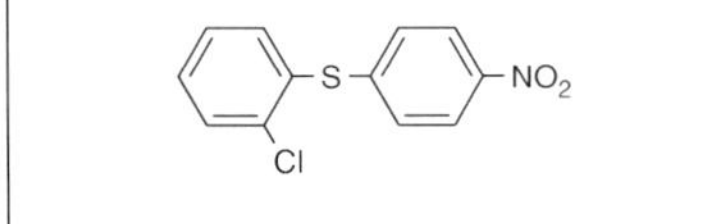

250 mg

S92,111-4
3-BROMO-2,4-DIMETHYL-6-NITROANILINE

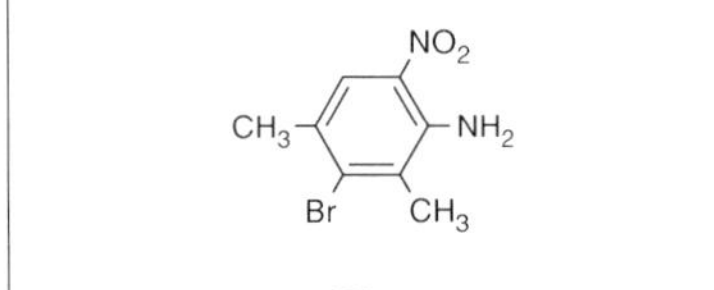

250 mg

S75,536-2
5-FLUORO-2-NITROANILINE

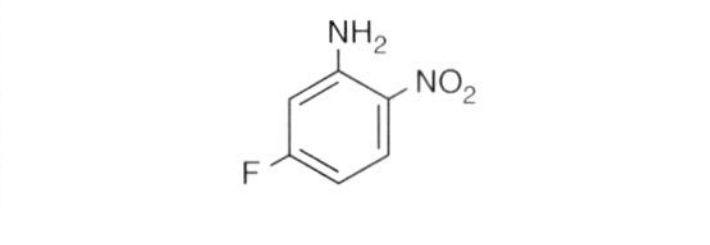

250 mg

S77,038-8
4-CHLORO-6-NITRO-M-TOLUIDINE

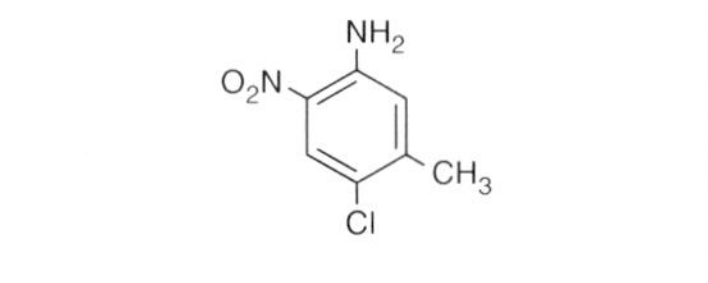

250 mg

S75,199-5
5-CHLORO-4-ETHYL-2-NITROANILINE

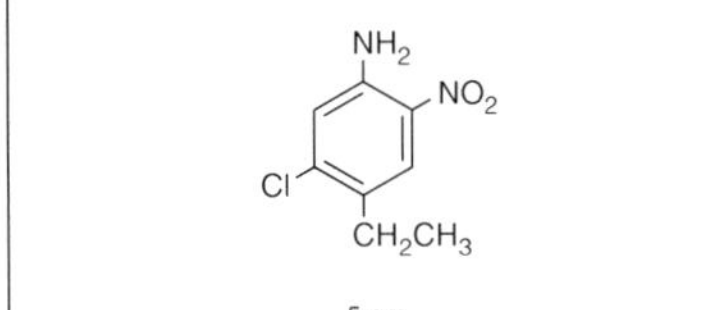

5 gm

S65,528-7
3-NITRO-2,4,6-TRICHLOROANILINE

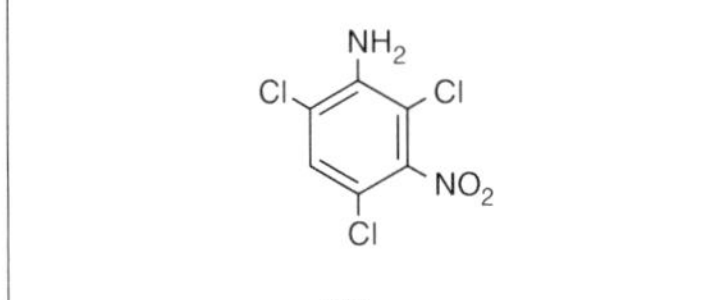

250 mg

S44,526-6
3-IODO-5-NITROANILINE

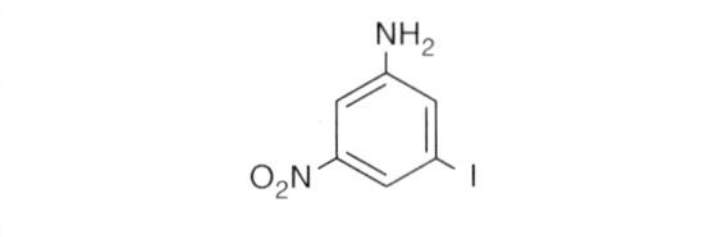

250 mg

S90,858-4
2,4-DICHLORO-5-NITROANILINE

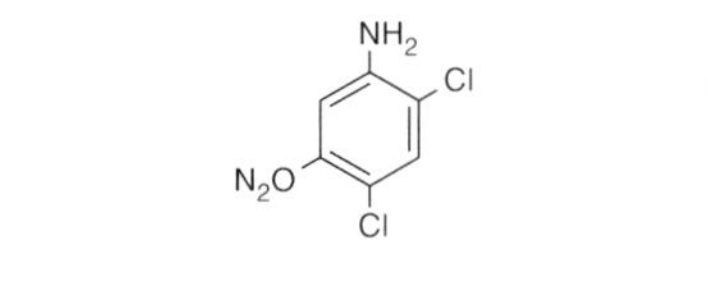

250 mg

S73,950-2
2,6-DIIODO-3-METHYL-4-NITROANILINE

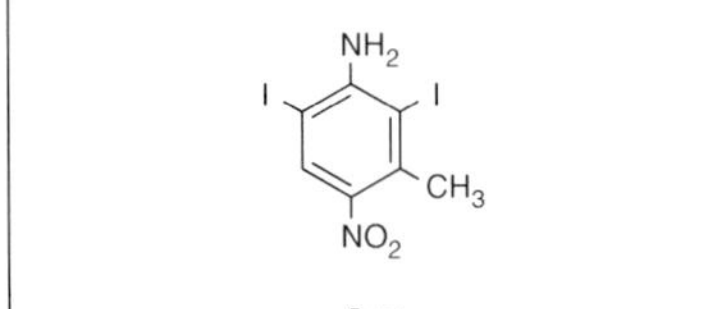

5 gm

S38,912-9
3-CHLORO-4-NITROANILINE

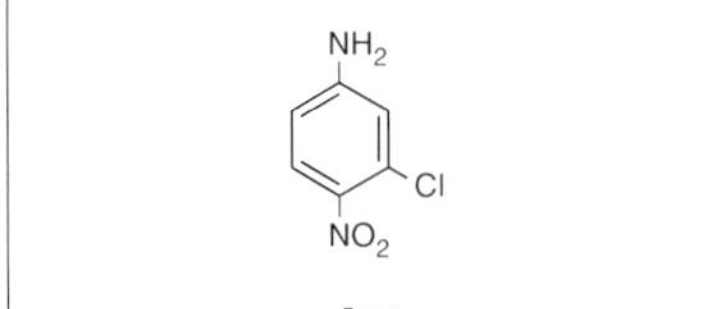

5 gm

S45,453-2
2-BROMO-4-NITROANILINE

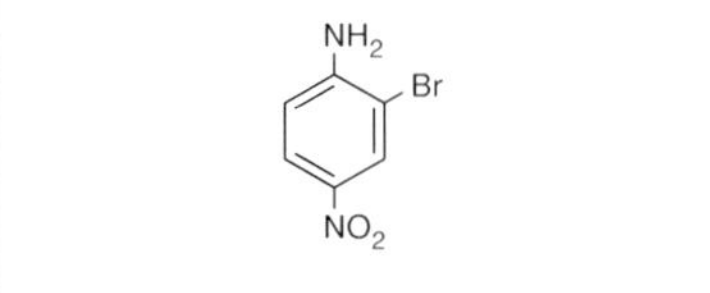

5 gm

S44,228-3
2,5-DICHLORO-4-NITROANILINE

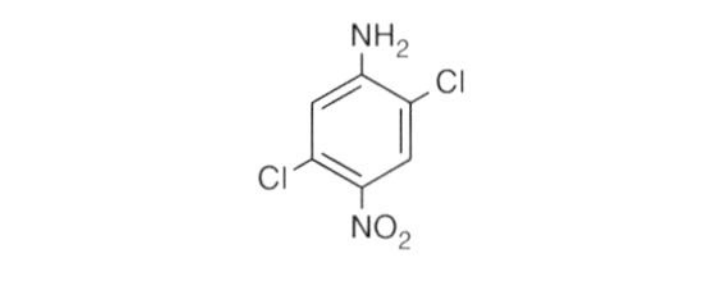

5 gm

S76,651-8
2-BROMO-6-CHLORO-4-NITROANILINE

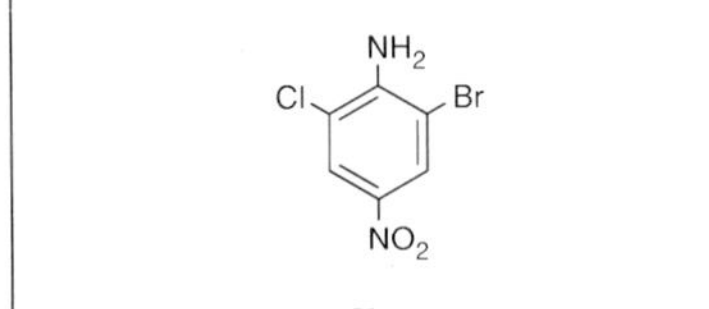

250 mg

S92,207-2
5-BROMO-4-METHYL-2-NITROANILINE

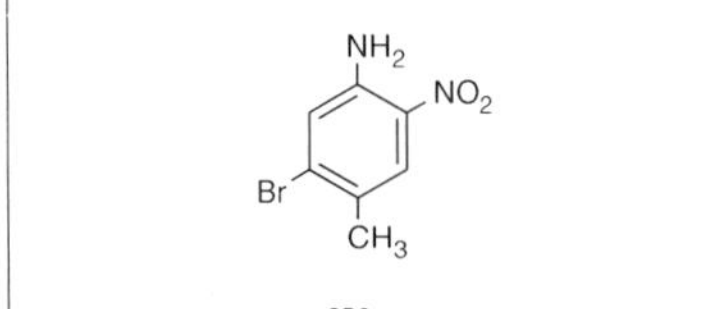

250 mg

S51,832-8
N-(2-CHLOROPHENYL)PICRYLAMINE

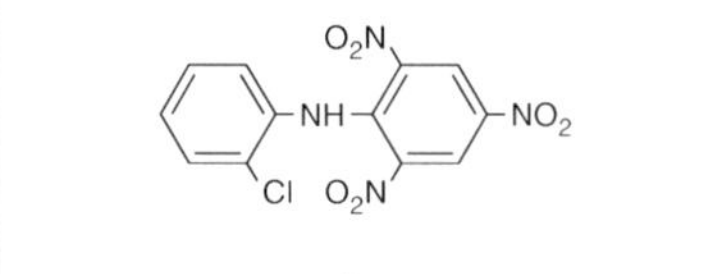

5 gm

S43,207-5
4'-CHLORO-2,4-DINITRODIPHENYLAMINE

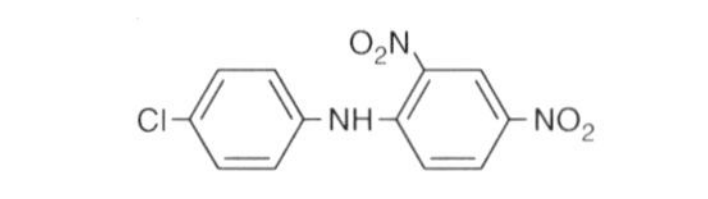

5 gm

S50,329-0
4'-BROMO-2,4-DINITRODIPHENYLAMINE
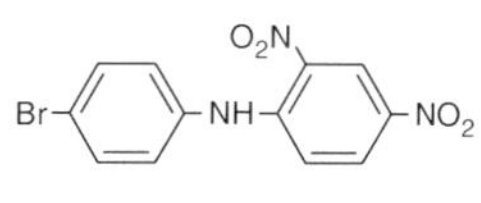
5 gm

S50,323-1
N-(4-CHLOROPHENYL)PICRYLAMINE
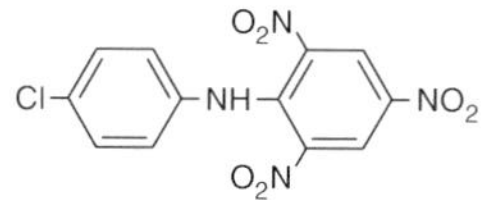
5 gm

S51,823-9
N-(2,3-DICHLOROPHENYL)PICRYLAMINE
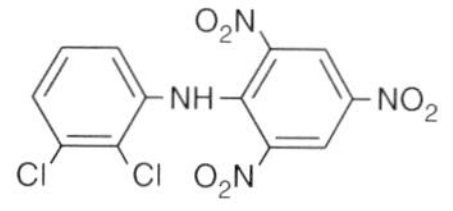
5 gm

S51,830-1
N-(3-CHLORO-O-TOLYL)-PICRYLAMINE
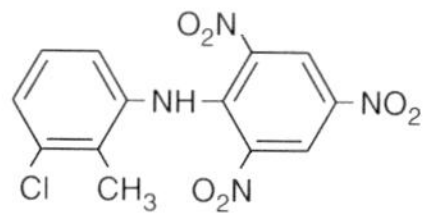
5 gm

S50,334-7
3'-CHLORO-2,4-DINITRODIPHENYLAMINE
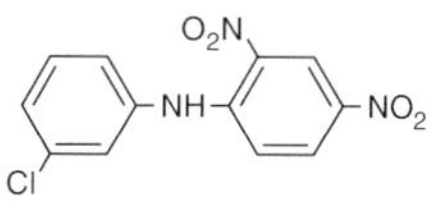
5 gm

S50,455-6
N-(2,4-DICHLOROPHENYL)PICRYLAMINE
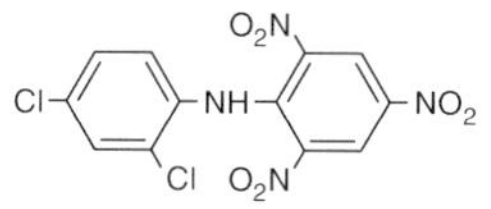
5 gm

S51,824-7
N-(2,5-DICHLOROPHENYL)PICRYLAMINE
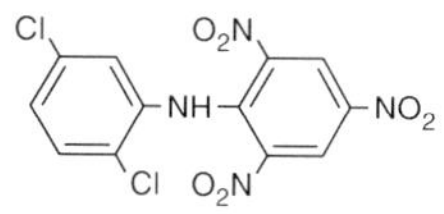
5 gm

S39,925-6
N-(4-CHLORO-O-TOLYL)-PICRYLAMINE
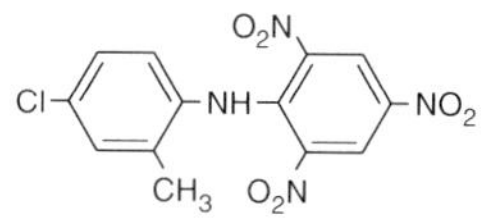
5 gm

S50,447-5
5-CHLORO-N-PICRYL-O-TOLUIDINE
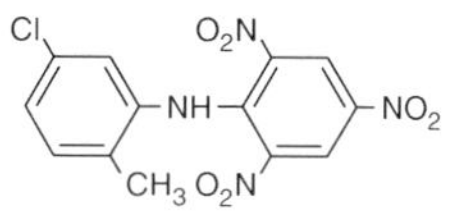
5 gm

S51,674-0
N-(3-CHLOROPHENYL)PICRYLAMINE
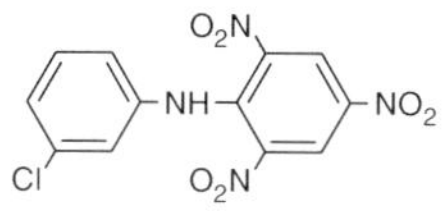
5 gm

S74,664-9
4-CHLORO-2',4'-DINITRO-2-METHYLDIPHENYLAMINE
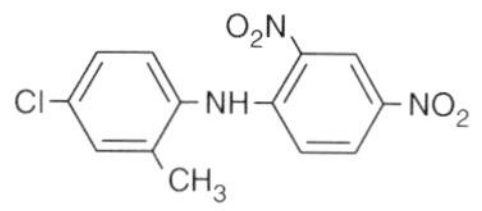
5 gm

S74,673-8
3-CHLORO-2',4'-DINITRO-4-METHYLDIPHENYLAMINE
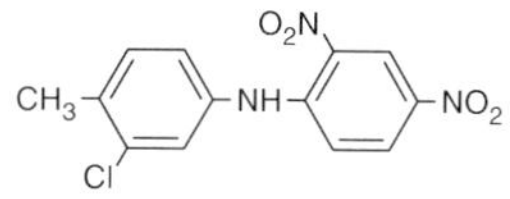
5 gm

S74,671-1
4-BROMO-2',4'-DINITRO-3-METHYLDIPHENYLAMINE
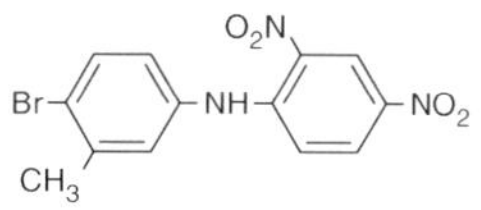
5 gm

S62,534-5
N-(2,4,6-TRICHLOROPHENYL)-PICRYLAMINE
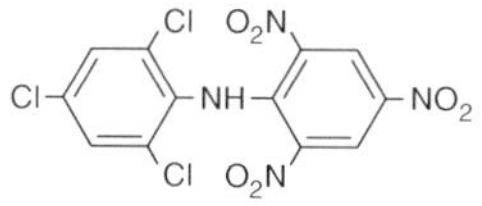
250 mg

S74,637-1
N-(4-BROMO-2-CHLOROPHENYL)-PICRYLAMINE
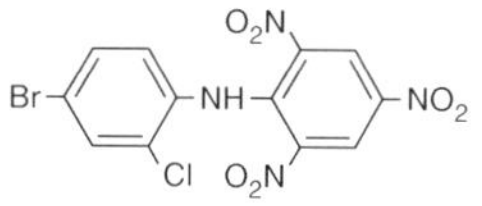
5 gm

S94,368-1
4'-IODO-2,4-DINITRODIPHENYLAMINE
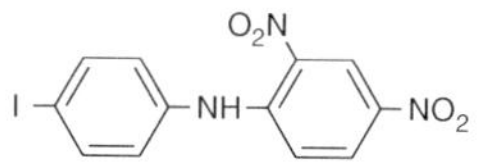
250 mg

S51,834-4
4-CHLORO-N,N'-DIBENZYL-6-NITRO-1,3-PHENYLENEDIAMINE
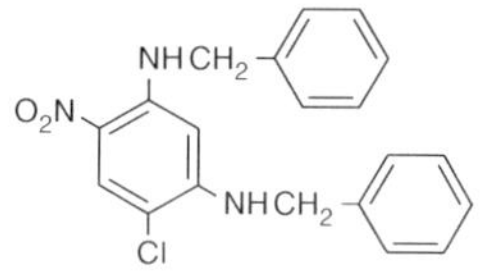
5 gm

S41,069-1
4-CHLORO-N-ETHYL-2-NITROANILINE
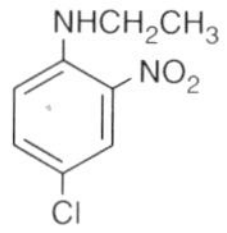
5 gm

S89,641-1
2,2'-DIBROMO-2''-NITROTRIPHENYLAMINE
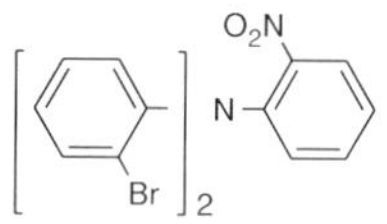
250 mg

S90,120-2
2-CHLOROBENZALDEHYDE (2,6-DINITRO-4-METHYLPHENYL)HYDRAZONE
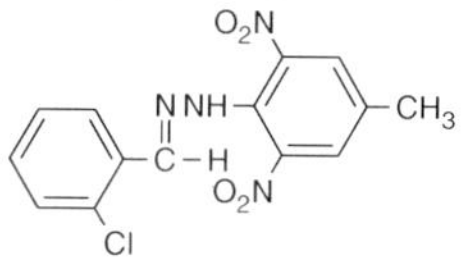
250 mg

S74,158-2
6-CHLORO-2,3-DINITROTOLUENE
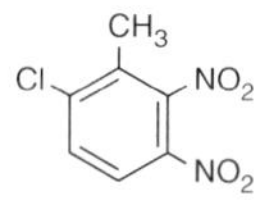
250 mg

S68,484-8
4-CHLORO-N-(4-CHLORO-3-NITROPHENYL)BENZYLAMINE
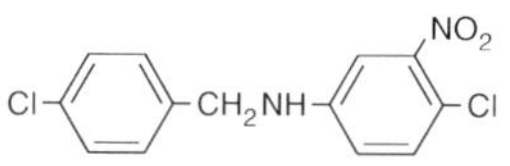
5 gm

S36,866-0
1,3-DINITRO-2,4,5-TRICHLOROBENZENE
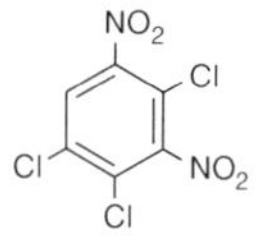
5 gm

S81,583-7
1,3-DINITRO-2,4,5,6-TETRACHLOROBENZENE
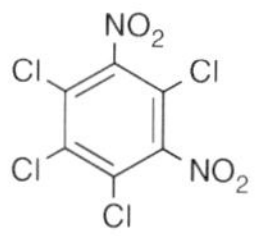
250 mg

S81,595-0
1-BROMO-3,5-DINITRO-2,4,6-TRIMETHYLBENZENE
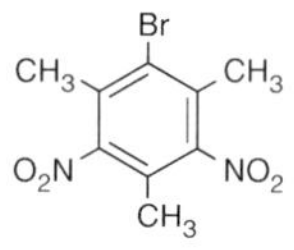
250 mg

S94,315-0
4-CHLORO-3,5-DINITROBIPHENYL
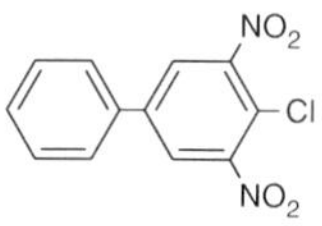
250 mg

S94,394-0
1-BROMO-3,5-DINITROBENZENE
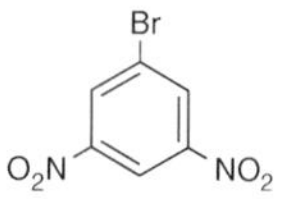
100 mg

S88,679-3
2-CHLORO-N,N-DIMETHYL-4-NITROANILINE
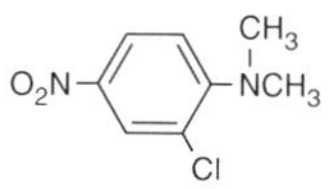
250 mg

S85,465-4
2-CHLOROBENZALDEHYDE (4-NITROPHENYL)HYDRAZONE
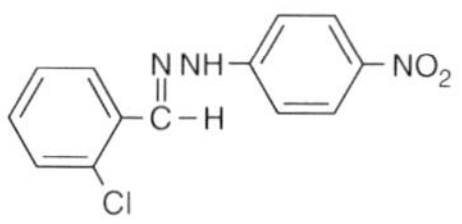
250 mg

S54,459-0
5-BROMO-3,6-DINITROPSEUDOCUMENE
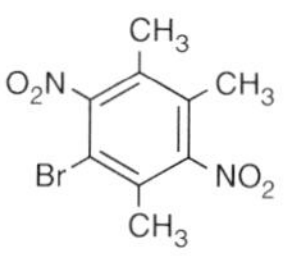
5 gm

S84,274-5
2,2'-DIBROMO-4,4'-DIMETHYL-6,6'-DINITROBIPHENYL
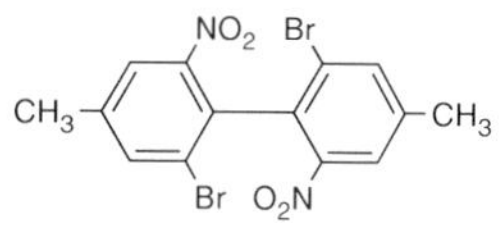
250 mg

S52,048-9
N-(4-BROMOPHENYL)PICRYLAMINE
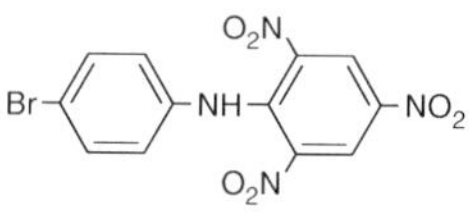
5 gm

S40,712-7
4-CHLORO-2-NITRODIPHENYLAMINE
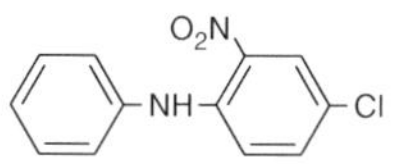
5 gm

S93,128-4
4-FLUORO-N-(4-NITROBENZYL)ANILINE
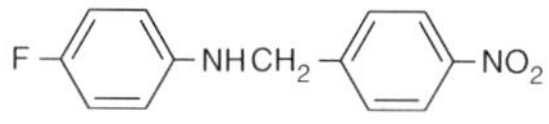
250 mg

S92,904-2
4,4'-(4-BROMO-3-NITROPHENYLMETHYLENE)BIS(N,N-DIMETHYLANILINE)
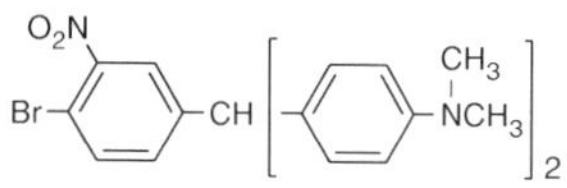
250 mg

S67,807-4
1-(2-CHLORO-4-NITROPHENYL)-4-METHYLPIPERAZINE
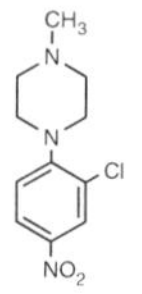
250 mg

S76,572-4
(4-FLUORO-3-NITROPHENYL)-TRIMETHYLAMMONIUM IODIDE
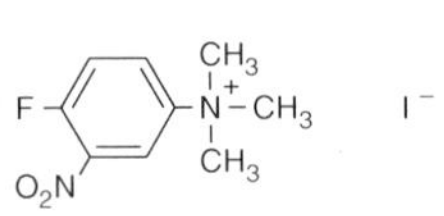
5 gm

S92,575-6
4-CHLORO-2-NITRO-1-PHENYLHYDRAZINE
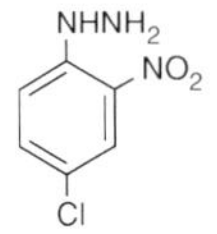
250 mg

S81,582-9
1-NITRO-2,3,4,5,6-PENTABROMOBENZENE

250 mg

S81,596-9
1-BROMO-3-NITRO-2,4,6-TRIMETHYLBENZENE
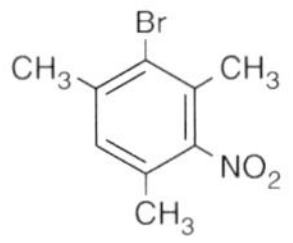
250 mg

S46,031-1
4-IODO-4'-NITROBIPHENYL
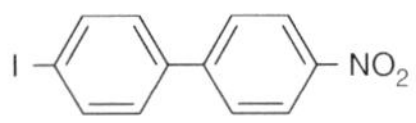
250 mg

S79,202-0
2-IODO-3-NITROTOLUENE
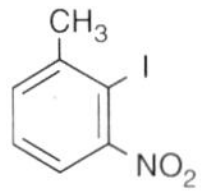
250 mg

S45,941-0
5-CHLORO-2-NITROTOLUENE
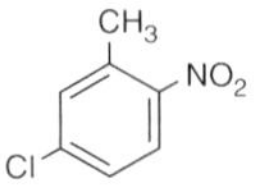
5 gm

S76,665-8
2-CHLORO-5-NITROTOLUENE
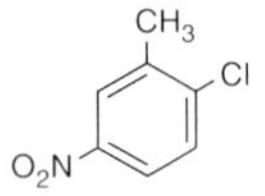
250 mg

S77,039-6
1-CHLORO-2-ETHYL-5-NITROBENZENE
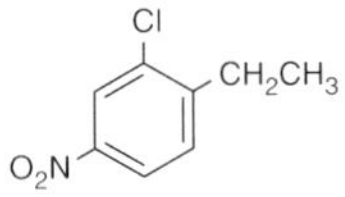
250 mg

S84,289-3
4,4''-DINITRO-2,2',5',6''-TETRAIODO-P-TERPHENYL
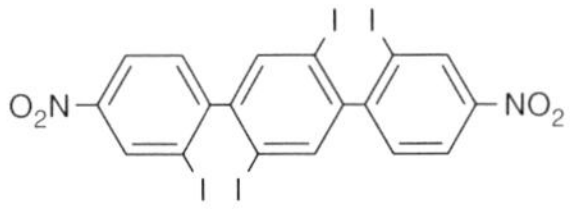
250 mg

S81,397-4
1,3-DIBROMO-5-NITROBENZENE
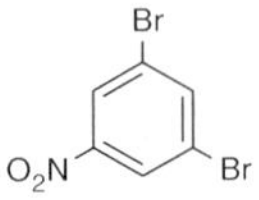
250 mg

S45,757-4
2,6-DICHLORO-3-NITROTOLUENE
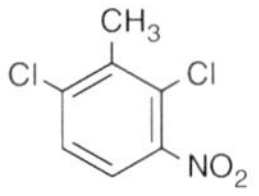
5 gm

S81,594-2
1-BROMO-2,4-DIMETHYL-3-NITROBENZENE
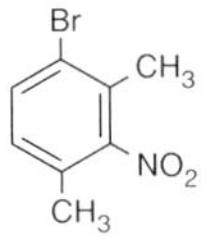
250 mg

S91,944-6
1,3,5-TRIBROMO-2-NITROBENZENE
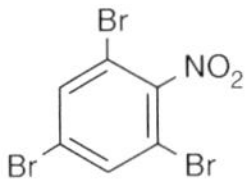
250 mg

S88,186-4
2-BROMO-1,3-DIIODO-5-NITROBENZENE
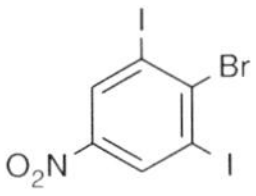
250 mg

S75,209-6
2,4-DICHLORO-5-NITROTOLUENE
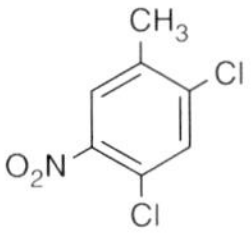
5 gm

S75,303-3
2,5-DICHLORO-4-NITROTOLUENE
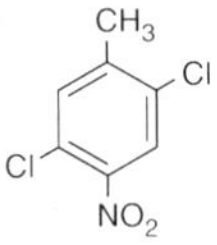
5 gm

S75,196-0
1,3-DICHLORO-6-ETHYL-4-NITROBENZENE
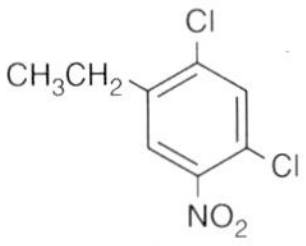
5 gm

S91,999-3
3,3'-DINITRO-OCTAFLUOROBIPHENYL
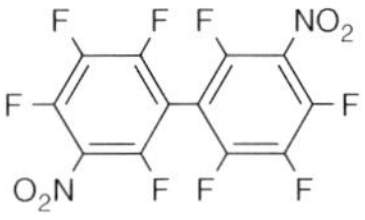
250 mg

S94,696-6
3,5-DIIODO-1-NITROBENZENE
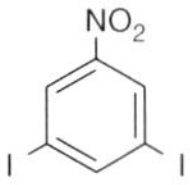
250 mg

S92,583-7
5-BROMO-6-NITRO-1,3-BENZODIOXOLE
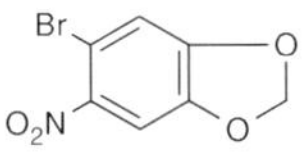
250 mg

S75,153-7
1-FLUORO-2-IODO-5-NITROBENZENE
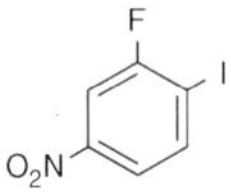
5 gm

S75,530-3
1-CHLORO-2-IODO-5-NITROBENZENE
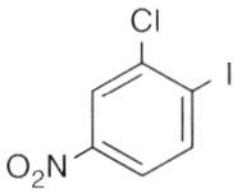
250 mg

S78,130-4
4-NITROPHENYL-BETA-D-FUCOPYRANOSIDE
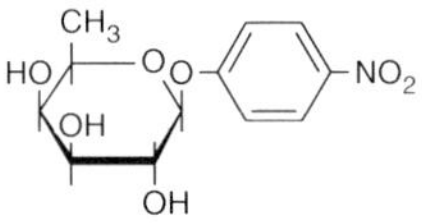
250 mg

S90,720-0
3-(2,4-DINITROPHENOXY)-1,2-PROPANEDIOL
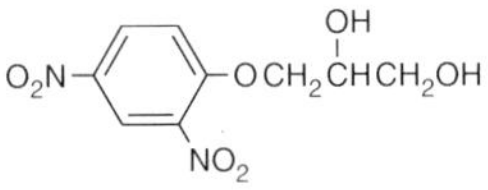
250 mg

S90,682-4
1-O-(3,4-DINITROPHENYL)-BETA-D-CELLOBIOSIDE
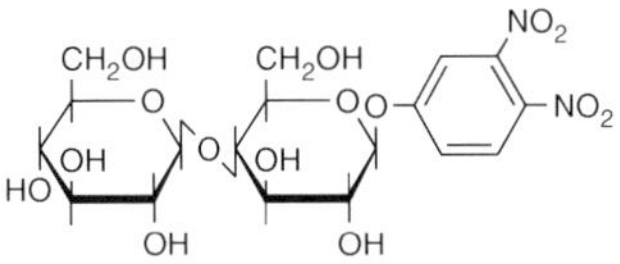
250 mg

S78,878-3
4-NITROPHENYL-BETA-D-THIOGALACTOPYRANOSIDE
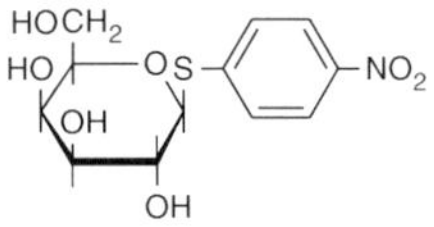
250 mg

S80,354-5
3-(2-NITROANILINO)-1,2-PROPANEDIOL
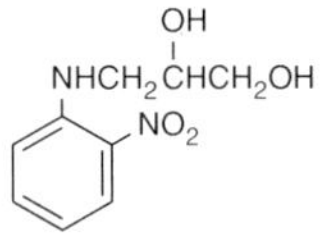
250 mg

S93,567-0
3-(4-NITROANILINO)-1,2-PROPANEDIOL
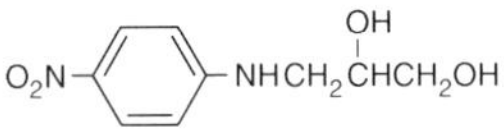
250 mg

S44,000-0
2,4,6-TRINITRORESORCINOL, WET
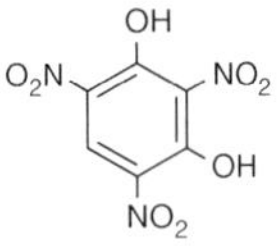
5 gm

S86,101-4
4,6-DINITRO-2-METHYLRESORCINOL
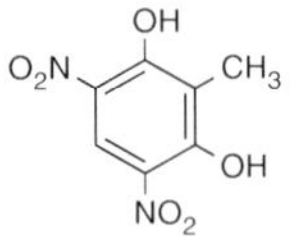
250 mg

S86,100-6
2-METHYL-4-NITRORESORCINOL
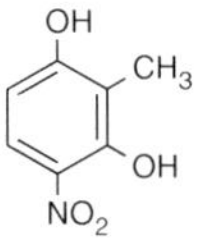
250 mg

S86,105-7
2,6-DINITROHYDROQUINONE
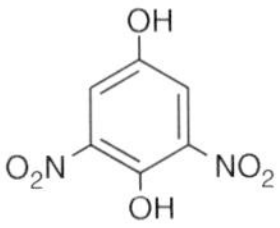
250 mg

S92,125-4
2-[N-(2,4,6-TRINITROPHENOXY)-ANILINO]HYDROQUINONE
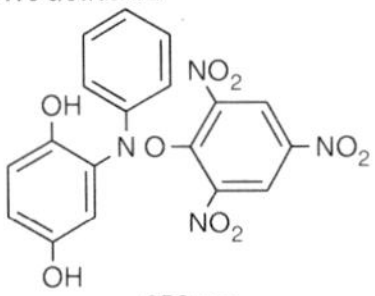
250 mg

S78,579-2
2-NITROPHENYL-BETA-D-THIOFUCOPYRANOSIDE
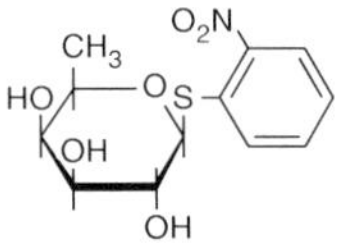
250 mg

S63,247-3
2-HYDROXY-N-(3-HYDROXYPHENYL)-5-NITROBENZYLAMINE
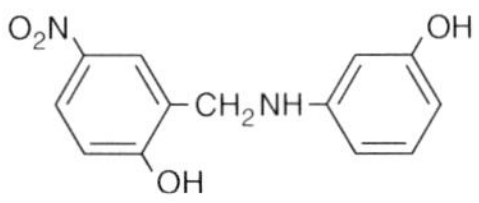
250 mg

S94,765-2
N,N'-BIS(2-HYDROXYETHYL)-N,N'-BIS-(4-NITROPHENYL)ETHYLENEDIAMINE
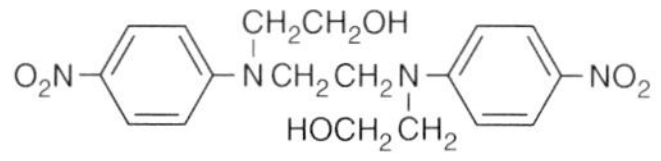
250 mg

S45,992-5
2-NITRO-P-XYLENE-ALPHA,ALPHA'-DIOL
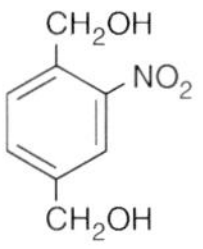
250 mg

S69,840-7
1,1'-(3-NITROBENZYLIDENE)DI-(2-NAPHTHOL)
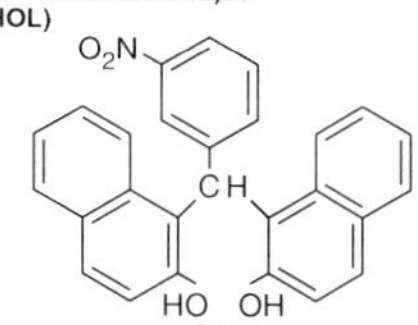
5 gm

S77,043-4
3,4-DIMETHOXY-ALPHA-(NITROMETHYL)BENZYL ALCOHOL
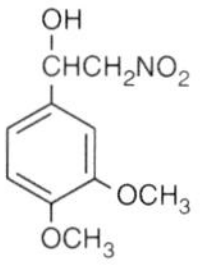
250 mg

S67,857-0
4-(4-NITROBENZYLOXY)BENZYL ALCOHOL
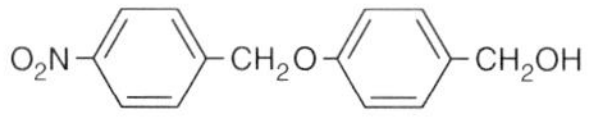
5 gm

S69,528-9
N-(2-HYDROXY-3-METHOXYBENZYL)-2-METHOXY-4-NITROANILINE
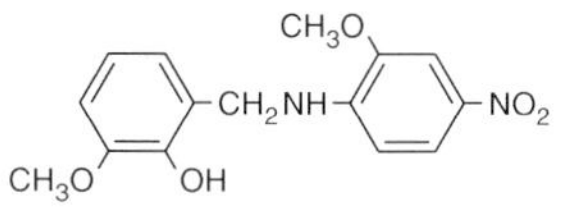
5 gm

S77,771-4
2-NITROPHENYL-BETA-D-GLUCOPYRANOSIDE
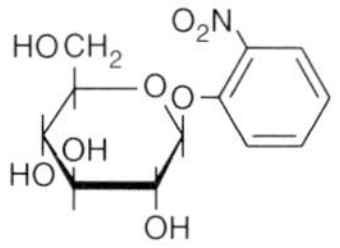
250 mg

S90,804-5
2-(4-AMINO-3-NITROPHENOXY)-1-ETHANOL
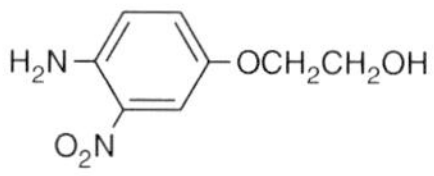
250 mg

S92,216-1
2-(4-AMINO-3-NITROPHENOXY)-1-ETHANOL
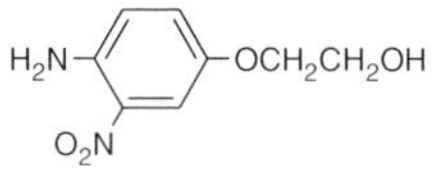
250 mg

S70,134-3
6-METHOXY-ALPHA-(3-NITROANILINO)-O-CRESOL
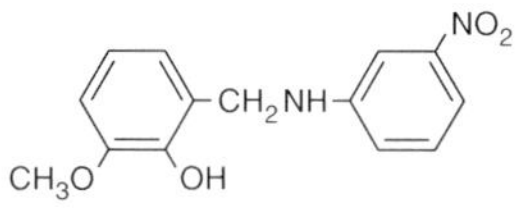
5 gm

S70,088-6
6-METHOXY-ALPHA-(4-NITROANILINO)-O-CRESOL
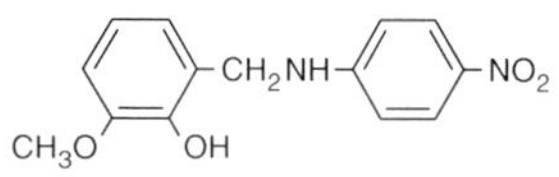
5 gm

S69,642-0
6-METHOXY-ALPHA-(4-METHYL-3-NITROANILINO)-O-CRESOL
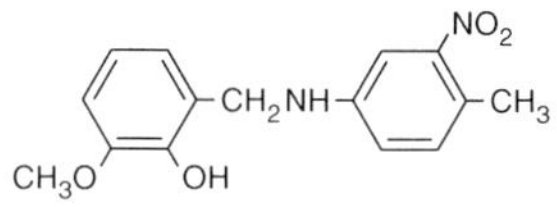
5 gm

S88,576-2
2-METHOXY-4-METHYL-6-NITROPHENOL
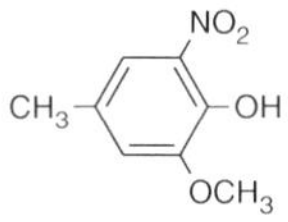
250 mg

S67,826-0
4-METHOXY-3-NITROPHENOL
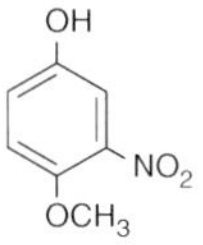
250 mg

S60,480-1
2-HYDROXY-N-(2-METHOXY-5-NITROPHENYL)BENZYLAMINE
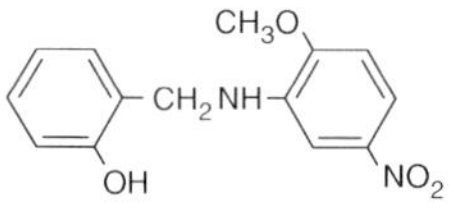
5 gm

S79,097-4
2-ETHOXYMETHYL-4,6-DINITROPHENOL
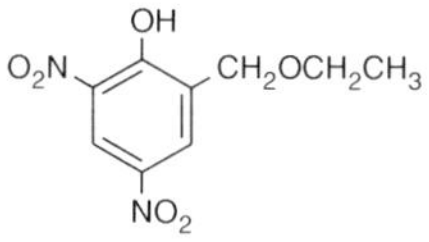
250 mg

S62,116-1
2-HYDROXY-N-(2-METHOXY-4-NITROPHENYL)BENZYLAMINE
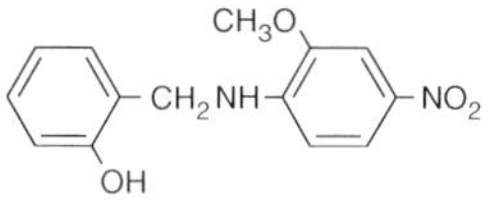
5 gm

S70,314-1
6-METHOXY-ALPHA-(2-METHOXY-5-NITROANILINO)-O-CRESOL
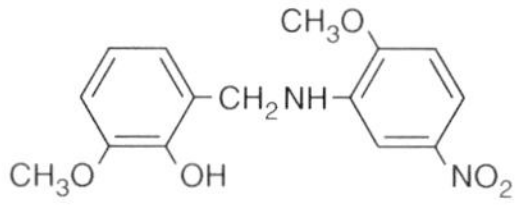
5 gm

S38,128-4
2-BUTOXY-4-NITROBENZYL ALCOHOL
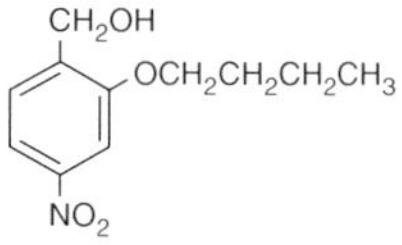
5 gm

S92,032-0
1-(ALPHA-2-NITROBENZYL)-2-(4-NITROPHENYL)HYDRAZINE
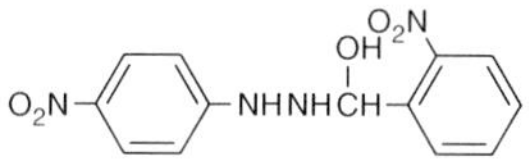
250 mg

S91,904-7
2-METHYL-2-NITRO-1-(3-NITROPHENYL)-1-PROPANOL
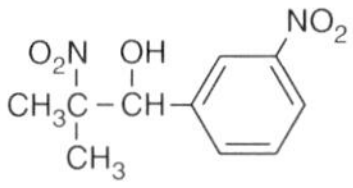
250 mg

S87,703-4
DL-4-NITROBENZHYDROL
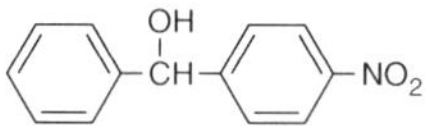
250 mg

S91,903-9
2-METHYL-2-(4-NITROPHENYL)-2-PROPANOL
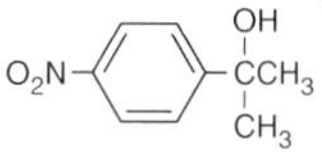
250 mg

S64,800-0
DIPHENYL(4-NITROPHENYL)CARBINOL
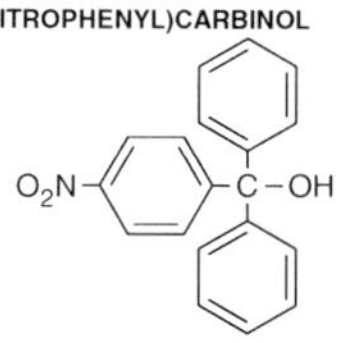
250 mg

S94,195-6
2-AMINO-5-NITRO-4-(P-TOLYLTHIO)-PHENOL
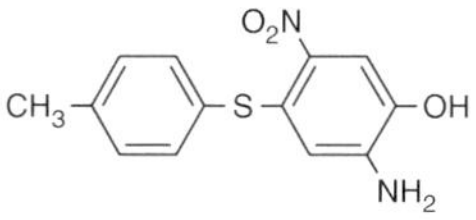
250 mg

S44,899-0
DIMETHYL(2-HYDROXY-5-NITROBENZYL)-SULFONIUM BROMIDE
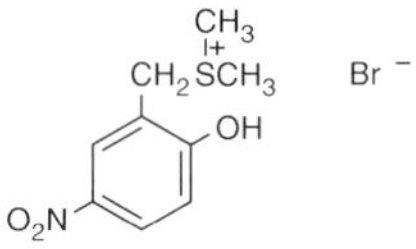
5 gm

S91,479-7
2-(2-AMINO-4-NITROANILINO)-1-ETHANOL
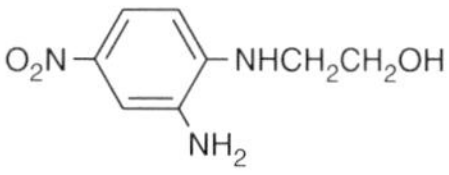
250 mg

S37,498-9
2-AMINO-4-NITRO-4'-HYDROXYDIPHENYLAMINE
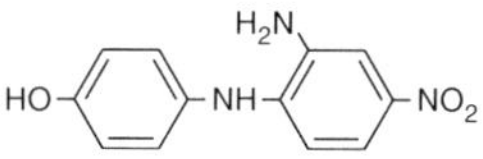
250 mg

S40,648-1
DL-THREO-2-AMINO-1-(P-NITROPHENYL)-1,3-PROPANEDIOL
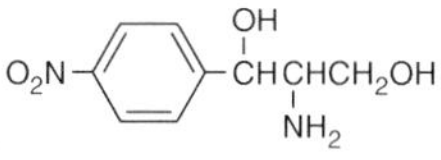
250 mg

S62,299-0
2,4-DINITRO-N-(2-HYDROXYETHYL)-ANILINE
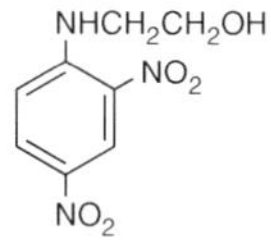
250 mg

S58,967-5
3-(PICRYLAMINO)PHENOL
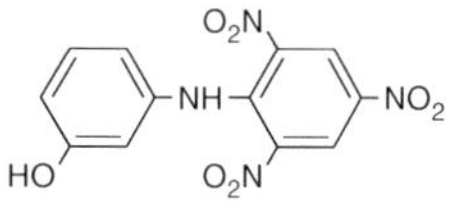
250 mg

S56,067-7
2-(PICRYLAMINO)PHENOL
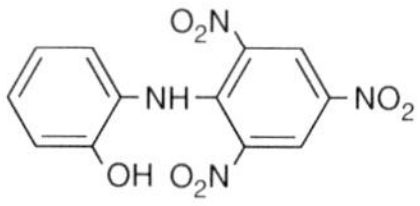
250 mg

S62,535-3
N-(2-HYDROXY-4-METHYLPHENYL)-PICRYLAMINE
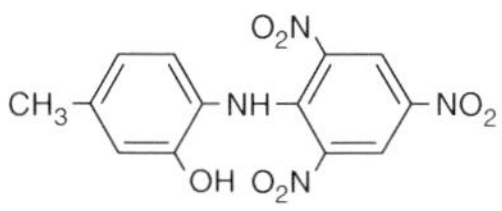
5 gm

S63,785-8
N-(2-HYDROXY-5-METHYLPHENYL)-PICRYLAMINE
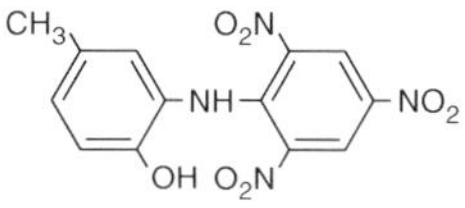
5 gm

S56,906-2
4-(PICRYLAMINO)PHENOL
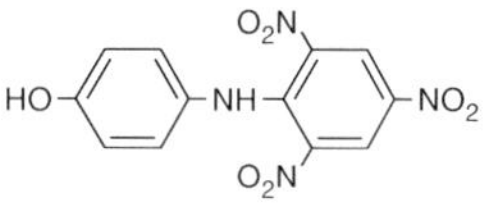
5 gm

S74,670-3
3-(2,4-DINITROANILINO)PHENOL
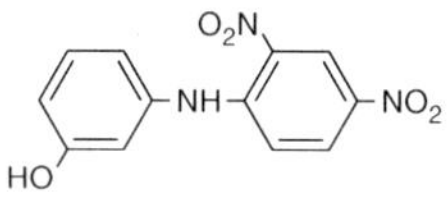
5 gm

S74,667-3
2,4-DINITRO-5'-HYDROXY-2'-METHYLDIPHENYLAMINE
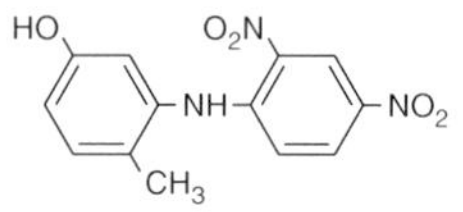
5 gm

S46,293-4
DISPERSE YELLOW 1
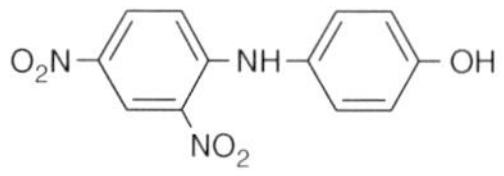
5 gm

S93,574-3
N'-(2-HYDROXYETHYL)-N,N'-BIS(4-NITROPHENYL)ETHYLENEDIAMINE
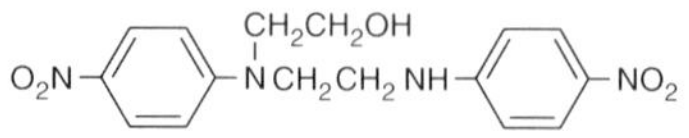
250 mg

S50,411-4
O-(2,4-DINITROANILINO)PHENOL
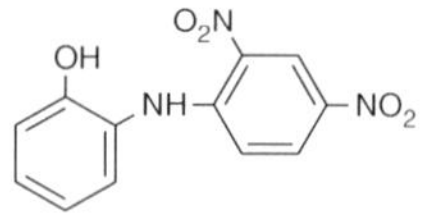
5 gm

S74,390-9
2,6-DINITRO-3'-HYDROXY-4-METHYLDIPHENYLAMINE
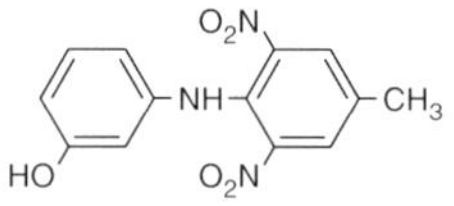
5 gm

S43,859-6
2,3-DINITROPHENOL
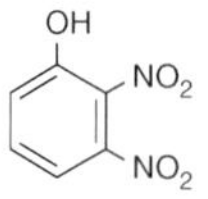
250 mg

S87,832-4
2,4-DINITRO-4'-HYDROXY-2'-METHYLDIPHENYLAMINE
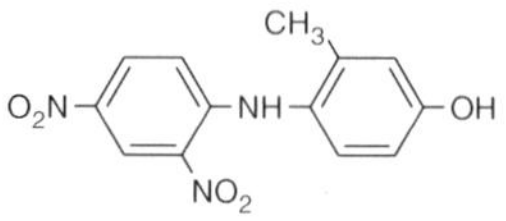
5 gm

S60,483-6
2-HYDROXY-N-(4-METHYL-3-NITROPHENYL)BENZYLAMINE
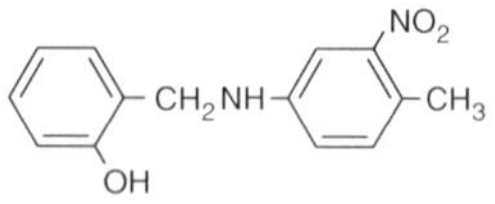
5 gm

S60,484-4
2-HYDROXY-N-(2-METHYL-5-NITROPHENYL)BENZYLAMINE
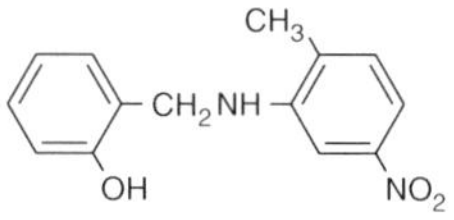
5 gm

S94,750-4
N,N-DIMETHYL-4-NITROBENZYLAMINE-N-OXIDE PICRATE
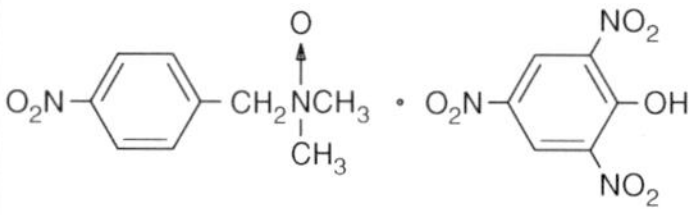
250 mg

S94,751-2
N,N-DIETHYLALLYLAMINE-N-OXIDE PICRATE
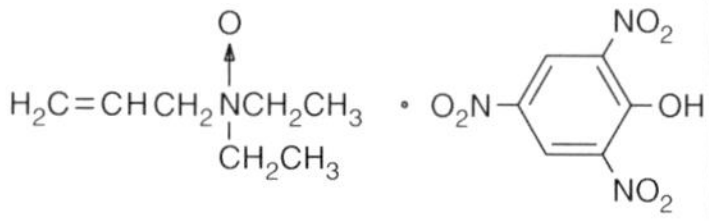
250 mg

S68,908-4
ALPHA-(3-NITROANILINO)-P-CRESOL
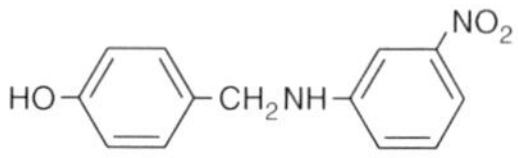
5 gm

S39,642-7
2-(TERT-BUTYL)-4,6-DINITROPHENOL
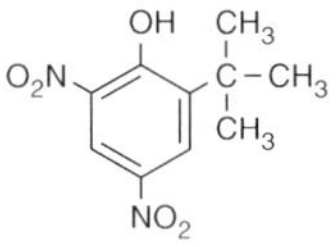
5 gm

S87,895-2
2-[4,5-(METHYLENEDIOXY)-2-NITROBENZYLIDENEAMINO]PHENOL
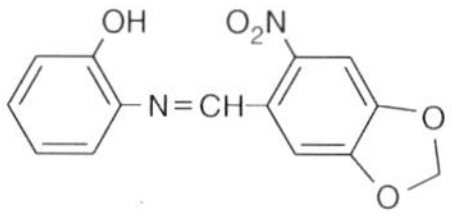
5 gm

S43,284-9
2-CHLORO-4,6-DINITROPHENOL, MOIST
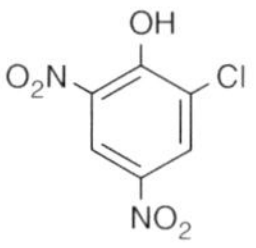
5 gm

S55,485-5
2-(2,4-DINITROSTYRYL)PHENOL
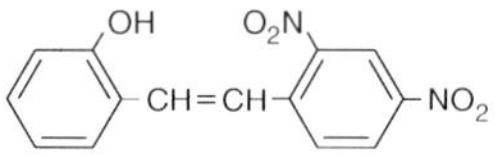
5 gm

S54,095-1
2,6-DINITRO-M-CRESOL
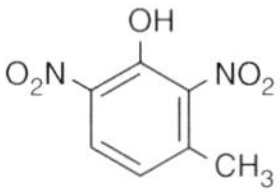
5 gm

S54,099-4
3,5-DINITRO-O-CRESOL
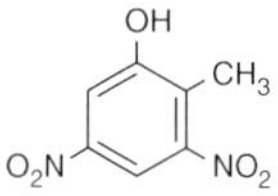
5 gm

S54,899-5
2,4-DINITRO-6-ISOPROPYLPHENOL
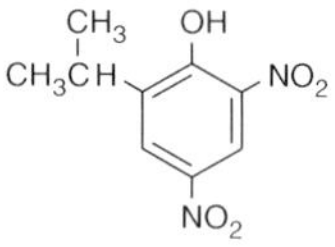
250 mg

S51,108-0
ALPHA-(4-NITROANILINO)-O-CRESOL
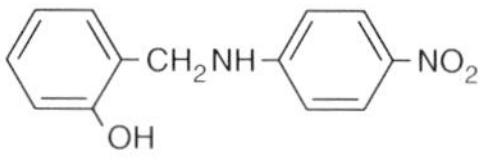
5 gm

S90,805-3
2-(4-NITROANILINO)-1-ETHANOL
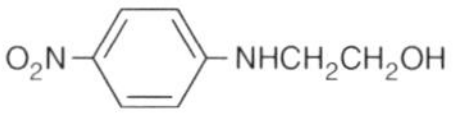
250 mg

S86,123-5
1-(4-NITROPHENYL)-3-PIPERIDINOL
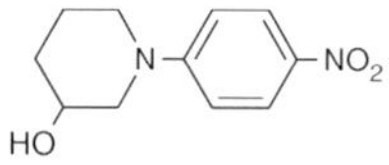
250 mg

S93,572-7
N-(2-HYDROXYETHYL)-N-METHYL-4-NITROANILINE
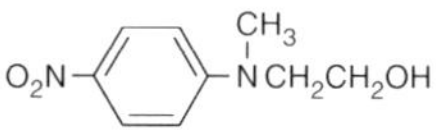
250 mg

S63,145-0
N-(4-NITROPHENYL)-2-HYDROXY-1-NAPHTHYLMETHYLAMINE
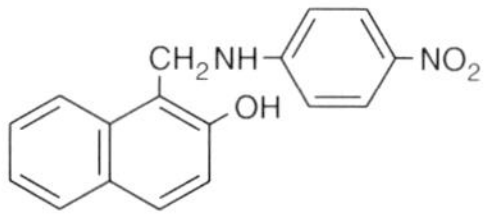
250 mg

S92,189-0
BENZALDEHYDE(2,4-DINITROPHENYL-ALPHA-HYDROXYMETHYL)HYDRAZONE
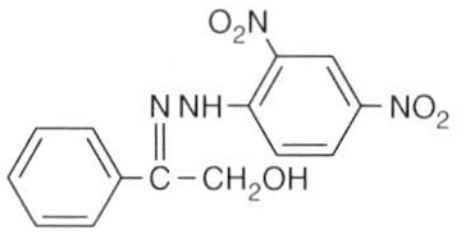
250 mg

S68,281-0
2-NITROPHENOL, SODIUM SALT
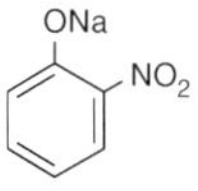
5 gm

S83,810-1
4-ISOPROPYL-2-NITROPHENOL
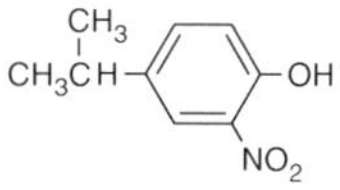
250 mg

S80,917-9
3,5-DIMETHYL-2-NITROPHENOL
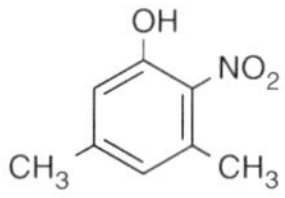
250 mg

S74,284-8
4-SEC-BUTYL-2,6-DINITROPHENOL
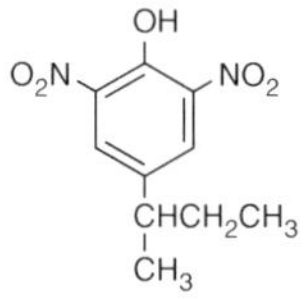
5 gm

S41,103-5
2-NITRO-ALPHA-PHENYL-P-CRESOL
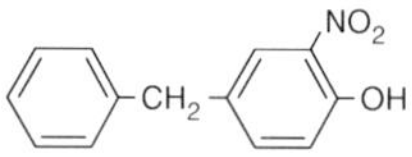
5 gm

S38,969-2
2-NITRO-4-PHENYLPHENOL
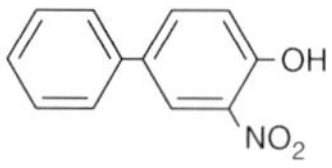
250 mg

S57,721-9
2-CHLORO-6-NITROPHENOL
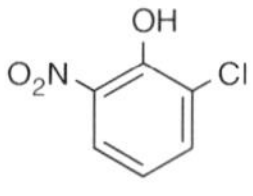
250 mg

S63,352-6
4-SEC-BUTYL-2-NITROPHENOL
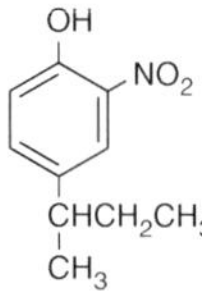
5 gm

S64,013-1
2-NITRO-5-PENTADECYLPHENOL
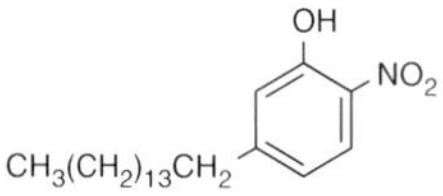
5 gm

S41,575-8
2,4-DIBROMO-6-NITROPHENOL
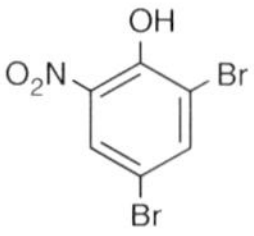
5 gm

S62,657-0
2,4-DI-TERT-BUTYL-6-NITROPHENOL
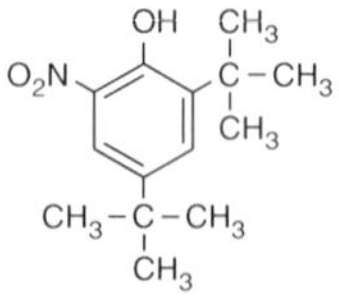
250 mg

S62,660-0
2,4-DI-TERT-AMYL-6-NITROPHENOL

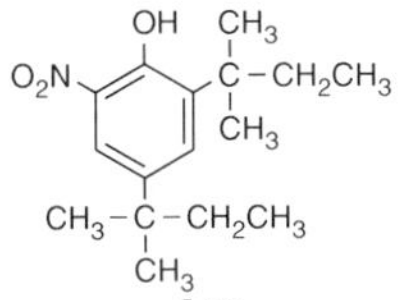

5 gm

S74,951-6
1-HYDROXY-2-NITROFLUORENE

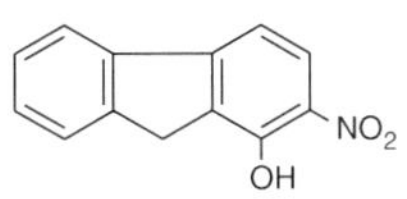

250 mg

S59,232-3
4-NITROPHENOL, POTASSIUM SALT

5 gm

S84,341-5
2-(4-NITROPHENYL)ETHENOL

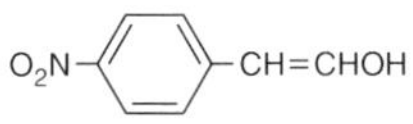

250 mg

S46,292-6
2-CHLORO-4-NITROPHENOL

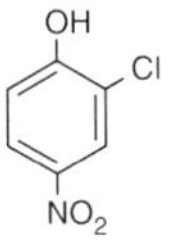

250 mg

S39,991-4
4-NITRO-M-CRESOL, SODIUM SALT

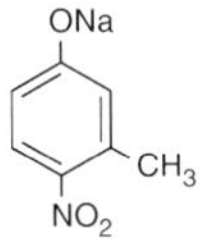

5 gm

S52,125-6
1,3-DIETHYLGUANIDINE PICRATE

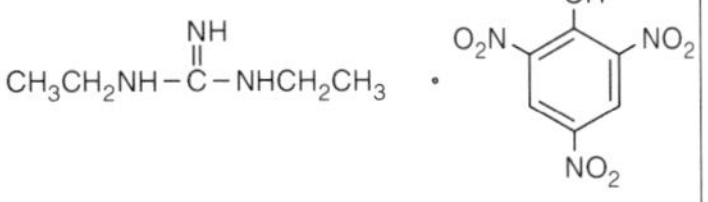

5 gm

S53,102-2
4-CYCLOHEXYL-2-NITROPHENOL

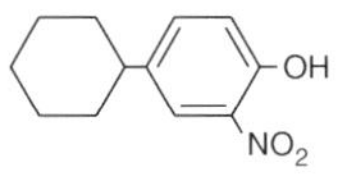

250 mg

S40,965-0
4-NITROPHENYL-ALPHA-D-GALACTOPYRANOSIDE HYDRATE

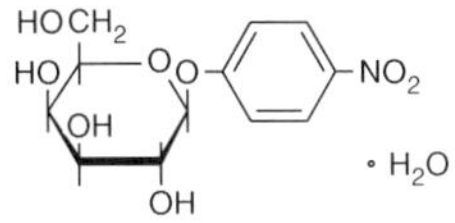

250 mg

S50,892-6
2-CHLORO-6-NITROBENZYL ALCOHOL

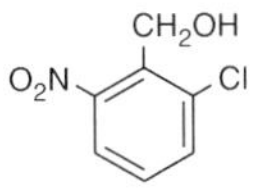

5 gm

S52,222-8
4-NITRO-3-PENTADECYLPHENOL

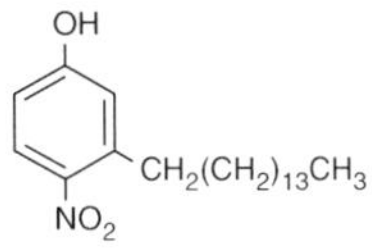

5 gm

S36,264-6
2-HYDROXY-7-NITROFLUORENE

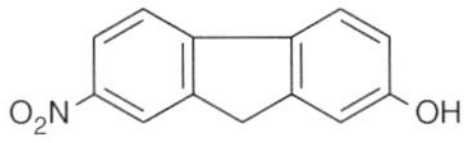

250 mg

S83,553-6
2-METHOXYPHENYL 4-NITROBENZYL ETHER

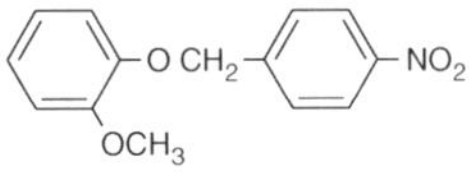

250 mg

S40,673-2
4-BENZYLOXY-2-NITROANILINE

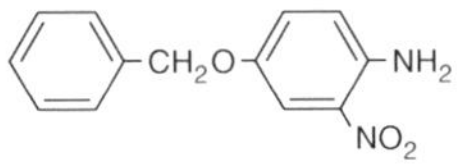

5 gm

S63,983-4
4'-BENZYLOXY-2,4-DINITROSTILBENE

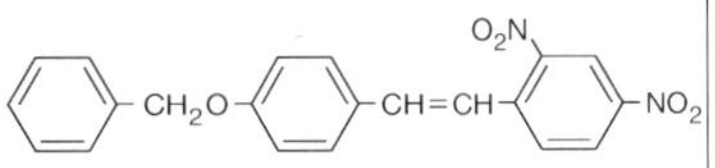

5 gm

S81,651-5
N,N-DIETHYL-BETA-(4-NITROPHENOXY)-PHENETHYLAMINE

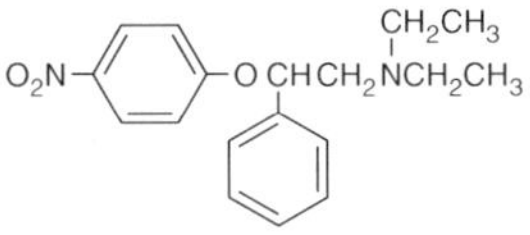

250 mg

S67,261-0
1-METHYL-2-(4-NITROBENZYLOXY)-NAPHTHALENE

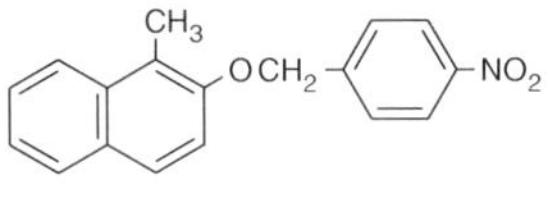

250 mg

S91,907-1
1-METHYL-2-(3-NITROBENZYLOXY)-NAPHTHALENE

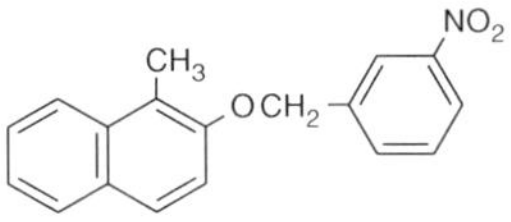

250 mg

S64,831-0
METHYL 4-NITROBENZYL ETHER

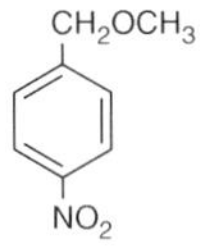

250 mg

S64,624-5
4-NITROBENZYL PHENYL ETHER

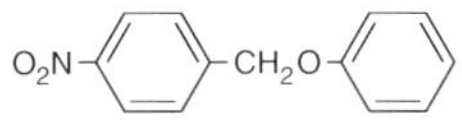

250 mg

S93,126-8
4-TERT-BUTYLPHENYL 4-NITROBENZYL ETHER

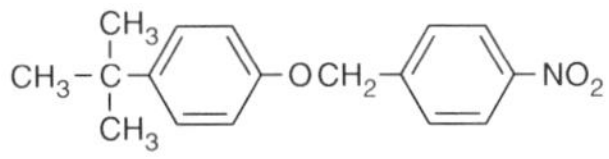

250 mg

S64,634-2
4-NITROBENZYL 4-(TERT-OCTYL)PHENYL ETHER

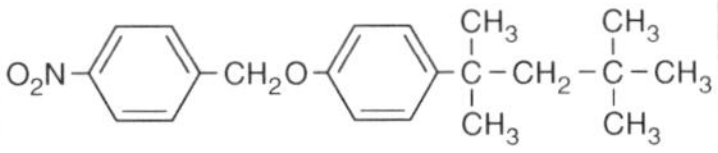

250 mg

S52,484-0
2-METHOXY-5-NITROBENZYL 4-METHOXYPHENYL ETHER

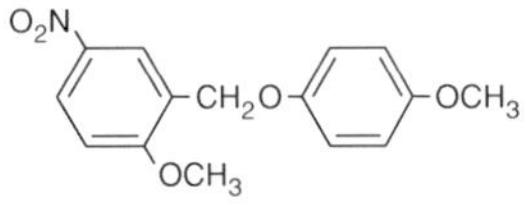

250 mg

S81,485-7
3,3'-DINITRO-2,2',5,5',6,6'-HEXAMETHOXYBIPHENYL

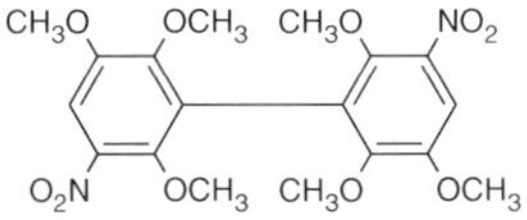

250 mg

S56,269-6
4,5-DIMETHOXY-2-NITROANILINE

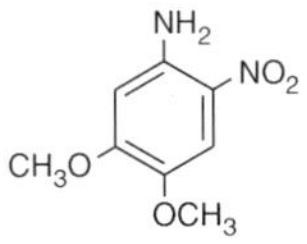

250 mg

S75,220-7
2,3-DIMETHOXYPHENETHYLAMINE PICRATE

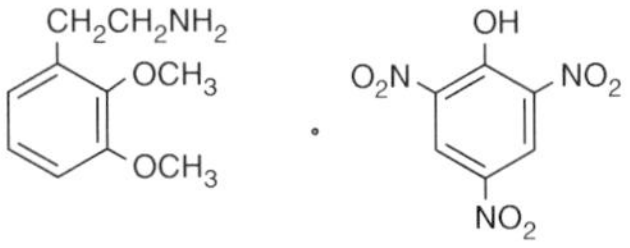

5 gm

S78,216-5
6,7-DIMETHOXY-2-METHYL-1-(2-NITROBENZYL)-1,2,3,4-TETRAHYDROISOQUINOLINE

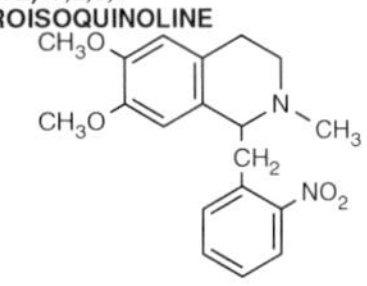

5 gm

S84,664-3
5-NITRO-1,2,3-TRIMETHOXYBENZENE

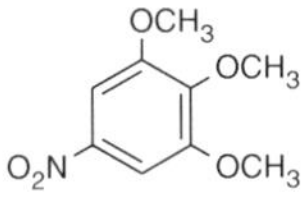

250 mg

S93,424-0
1,5-DICHLORO-2,4-DIMETHOXY-3-NITROBENZENE

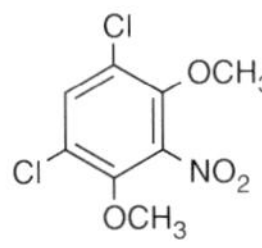

250 mg

S67,803-1
4-(2,5-DIMETHOXY-4-NITROPHENYL)-2,6-DIMETHYLMORPHOLINE

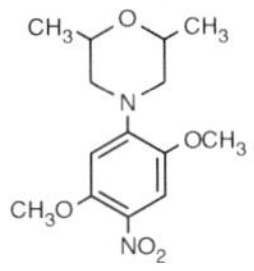

5 gm

S94,189-1
1-CHLORO-2,5-DIBUTOXY-4-NITROBENZENE

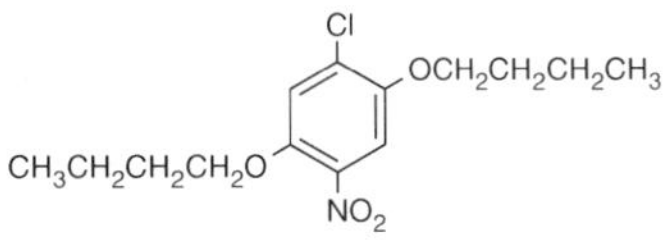

250 mg

S89,919-4
4-AMINO-3,3'-DIMETHOXY-4'-NITROBIPHENYL

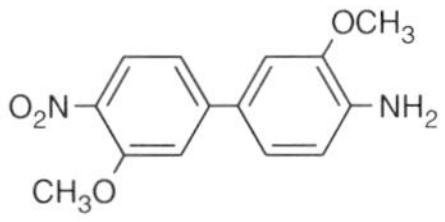

250 mg

S69,524-6
N,N'-METHYLENEBIS(2-METHOXY-4-NITROANILINE)

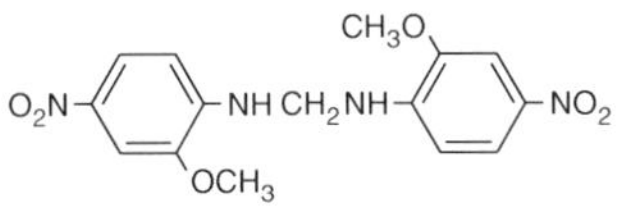

5 gm

S67,815-5
2,5-DIBUTOXY-N-METHYL-4-NITRO-N-(TETRAHYDROFURFURYL)ANILINE

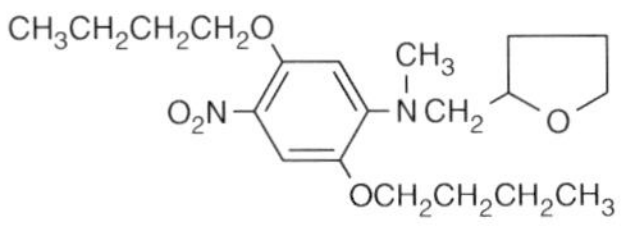

250 mg

S74,290-2
BIS(4-NITROPHENOXY)METHANE

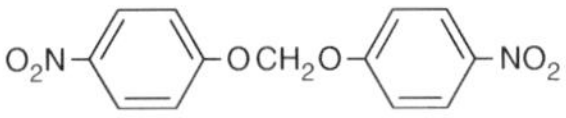

5 gm

S62,931-6
1-(2-AMINO-4-NITROPHENOXY)-2-(4-TERT-OCTYLPHENOXY)ETHANE

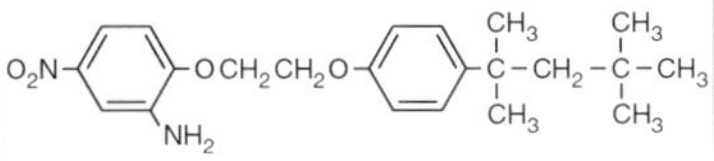

5 gm

S57,767-7
4,5-DINITROVERATROLE

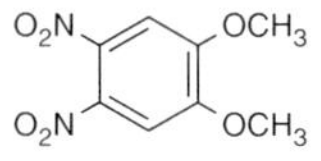

5 gm

S54,131-1
2,3-DIMETHOXY-N-(3-NITROPHENYL)-BENZYLAMINE

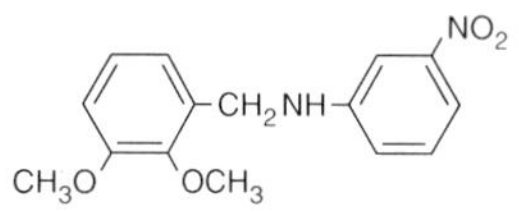

250 mg

S58,928-4
N-(5-NITRO-O-TOLYL)VERATRYLAMINE

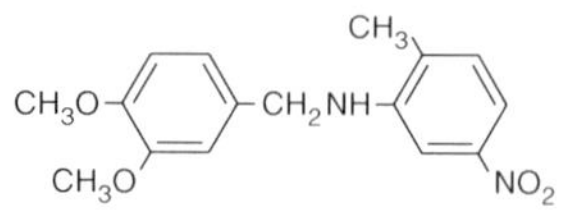

5 gm

S62,918-9
1-(2,4-DINITROPHENOXY)-2-(4-TERT-OCTYLPHENOXY)ETHANE

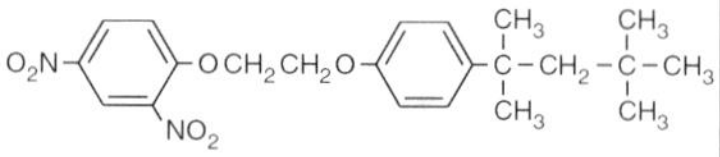

5 gm

S55,574-6
2,3-DIMETHOXY-2',4'-DINITROSTILBENE

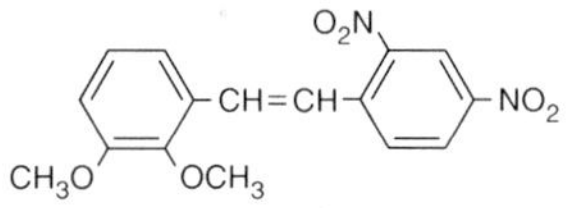

250 mg

S55,407-3
3,4-DIMETHOXY-2',4'-DINITROSTILBENE

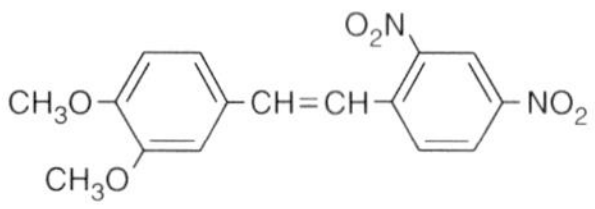

5 gm

S55,487-1
2,4-DINITRO-3',4'-(METHYLENEDIOXY)-STILBENE

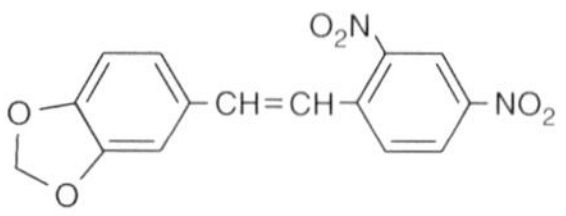

5 gm

S53,315-7
2,3-DIMETHOXY-N-(4-NITROPHENYL)-BENZYLAMINE

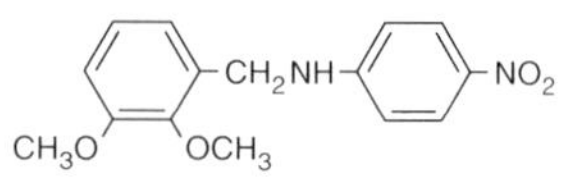

250 mg

S57,523-2
1,2-BIS(2-NITROPHENOXY)ETHANE

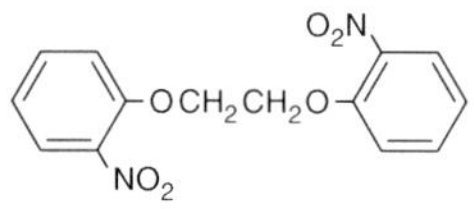

5 gm

S57,451-1
1,2-BIS(3-NITROPHENOXY)ETHANE

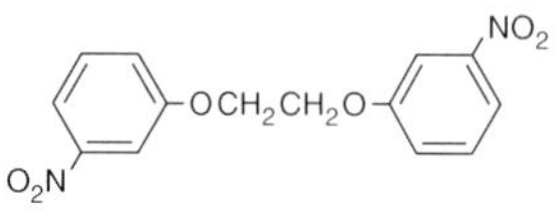

5 gm

S52,527-8
7-NITRO-2-NONYLOXYMETHYL-1,4-BENZODIOXAN

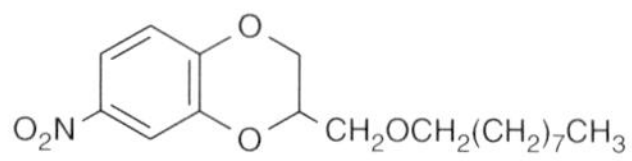

250 mg

S53,327-0
2-(4-NITROPHENOXY)ANISOLE

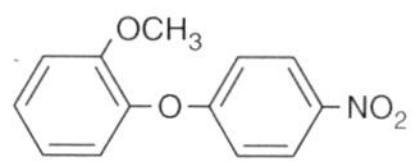

5 gm

S61,799-7
1-(4-NITROPHENOXY)-2-PHENOXYETHANE

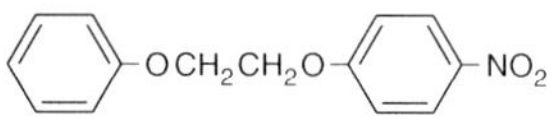

5 gm

S57,470-8
1,2-BIS(4-NITROPHENOXY)ETHANE

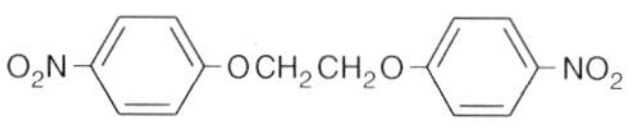

5 gm

S61,810-1
2-(4-NITROPHENOXY)ETHYL PHENYL SULFIDE

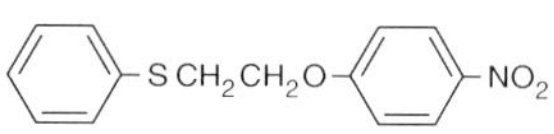

250 mg

S94,194-8
5-METHYL-2-METHOXY-4-NITROPHENYL P-TOLYL SULFIDE

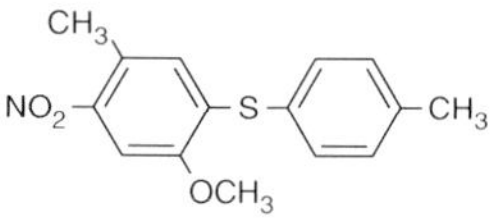

250 mg

S90,862-2
4-METHOXY-2,3-DINITROANILINE

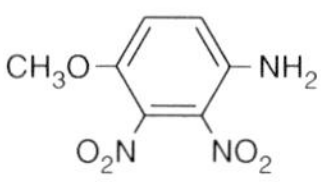

250 mg

S59,951-4
4-AMINO-3-NITROPHENETOLE

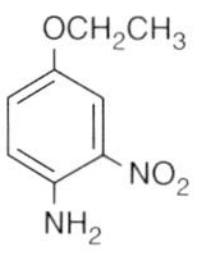

5 gm

S70,711-2
4-METHOXY-3-METHYL-6-NITROANILINE

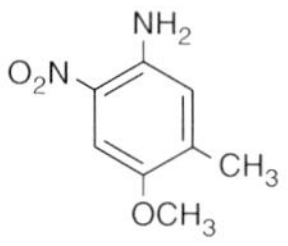

5 gm

S64,589-3
2-ETHOXY-4-NITROANILINE

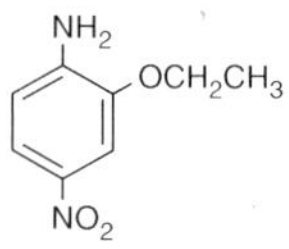

5 gm

S94,188-3
5-CHLORO-2-METHOXY-4-NITROANILINE

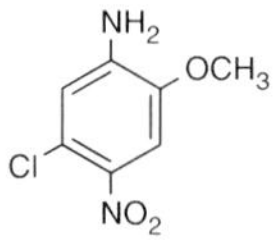

250 mg

S50,332-0
2,4-DINITRO-4'-METHOXYDIPHENYLAMINE

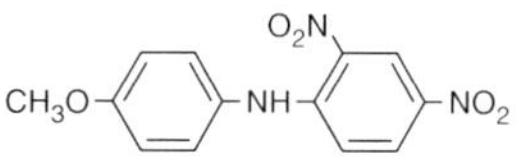

5 gm

S51,464-0
2,4-DINITRO-4'-ETHOXYDIPHENYLAMINE

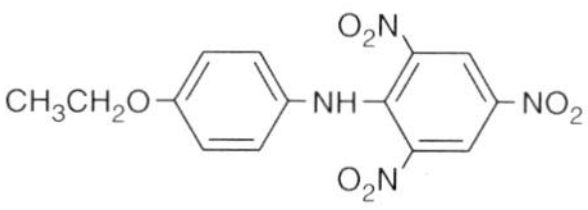

5 gm

S51,462-4
2,4-DINITRO-3'-METHOXYDIPHENYLAMINE

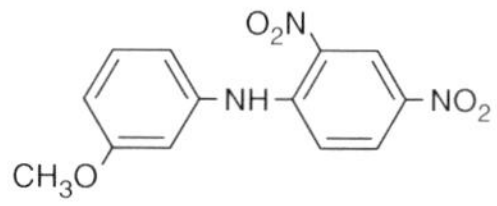

5 gm

S58,382-0
N-(3-CHLORO-2-METHOXYPHENYL)-PICRYLAMINE

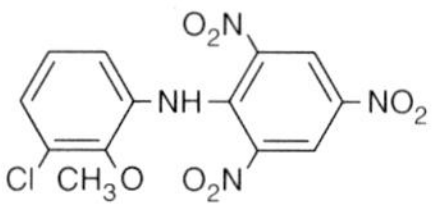

5 gm

S58,383-9
N-(2-METHOXY-5-NITROPHENYL)-PICRYLAMINE

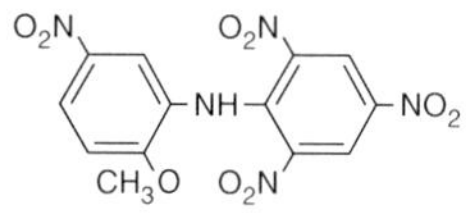

5 gm

S58,931-4
N-(6-METHOXY-M-TOLYL)PICRYLAMINE

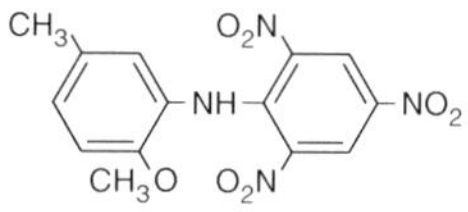

250 mg

S56,918-6
N-(2,5-DIMETHOXYPHENYL)PICRYLAMINE

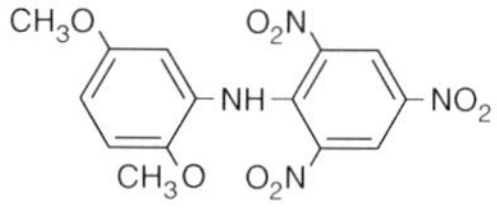

250 mg

S58,946-2
N-(2,4-DIMETHOXYPHENYL)PICRYLAMINE

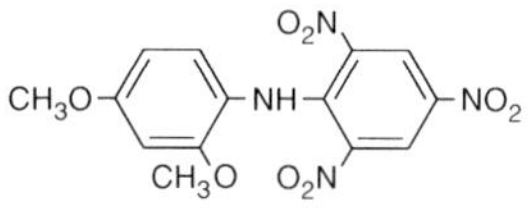

250 mg

S74,668-1
2,4-DINITRO-4'-PHENOXYDIPHENYLAMINE

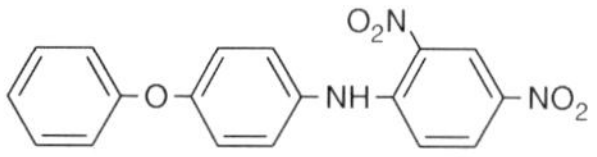

5 gm

S59,647-7
N-(2,4,6-TRINITROPHENYL)-O-PHENETIDINE

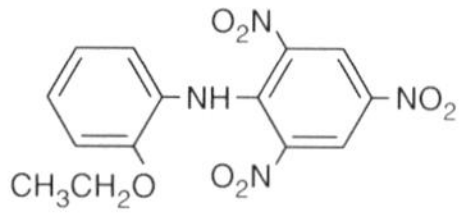

250 mg

S58,947-0
N-(4-ETHOXYPHENYL)PICRYLAMINE

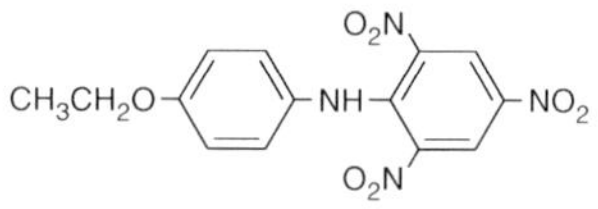

5 gm

S62,544-2
4-(PICRYLAMINO)PHENYL ETHER

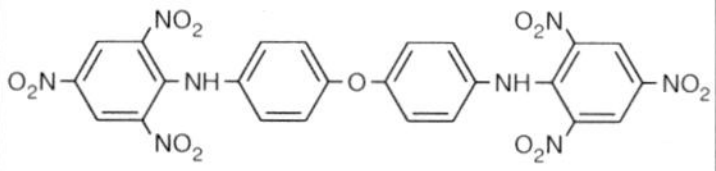

250 mg

S57,023-0
N-(3-METHOXYPHENYL)PICRYLAMINE

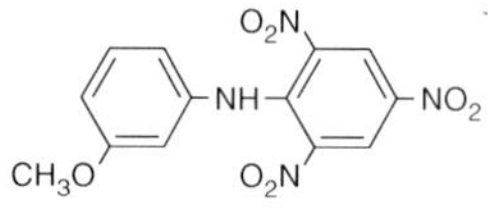

5 gm

S68,490-2
4'-PHENOXY-2,4,6-TRINITRODIPHENYLAMINE

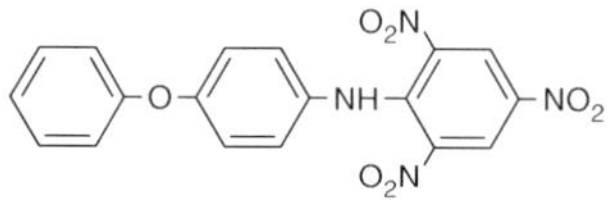

5 gm

S43,209-1
2,4-DINITRO-2'-METHOXYDIPHENYLAMINE

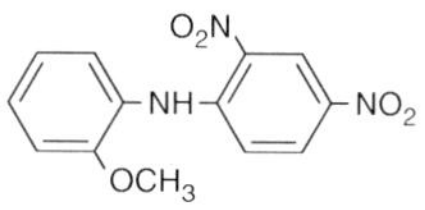

5 gm

S51,463-2
2,4-DINITRO-2'-ETHOXYDIPHENYLAMINE
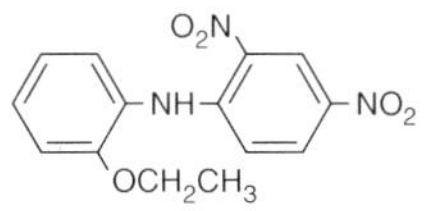
5 gm

S36,117-8
2,4-DINITRO-4'-(2,4-DINITRO-PHENOXY)DIPHENYLAMINE
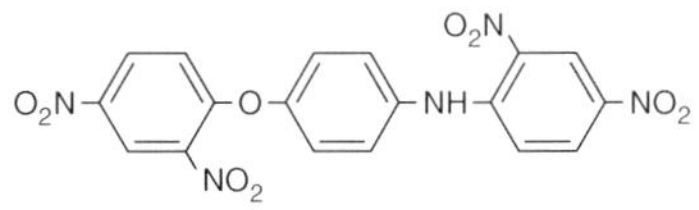
5 gm

S80,468-1
1-METHOXY-3-NITRO-2-NITROSOBENZENE
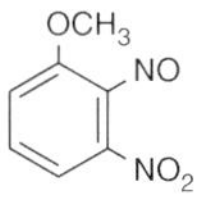
250 mg

S38,760-6
1,4-DIMETHOXY-2,3-DINITROBENZENE
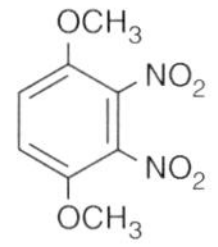
250 mg

S94,191-3
1-(4-METHOXY-2-NITROPHENYL)-PIPERIDINE
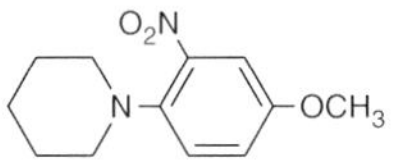
250 mg

S53,611-3
N-(4-METHOXYPHENYL)PICRYLAMINE
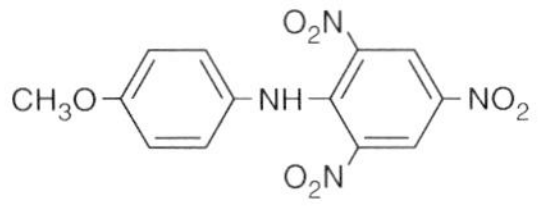
250 mg

S56,929-1
4-METHOXY-N-(3-NITROPHENYL)-BENZYLAMINE
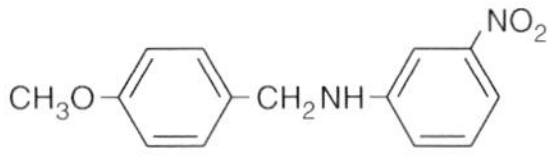
250 mg

S53,498-6
N-(4-METHOXY-O-TOLYL)-PICRYLAMINE
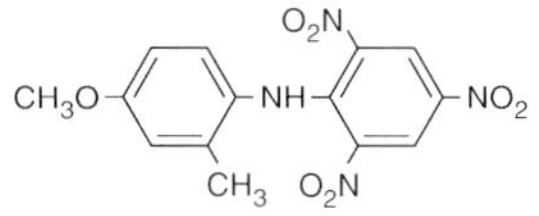
250 mg

S34,299-8
2-TERT-BUTYL-4,6-DINITRO-5-METHYLANISOLE
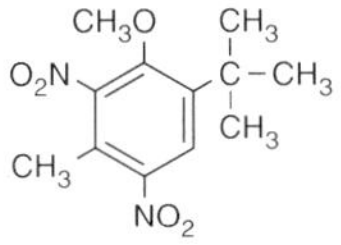
5 gm

S86,402-1
ALLYL 2,4-DINITROPHENYL ETHER
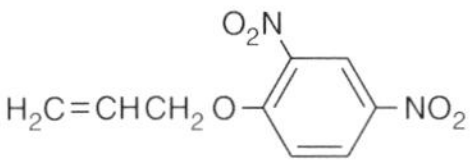
250 mg

S60,445-3
2,4-DINITRO-2'-METHYLDIPHENYL ETHER
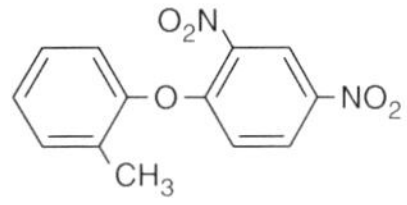
250 mg

S59,474-1
4'-CHLORO-2,4-DINITRODIPHENYL ETHER
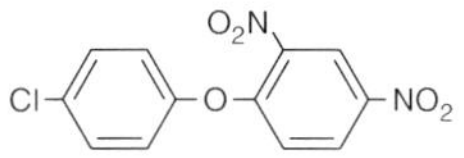
250 mg

S62,763-1
2,4-DINITROANISOLE
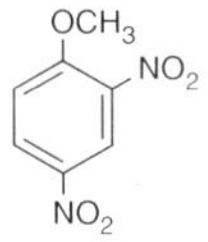
250 mg

S43,328-4
2,4-DINITROPHENETOLE
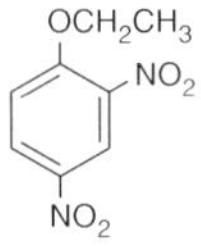
5 gm

S55,406-5
2,4-DINITRO-2'-METHOXYSTILBENE
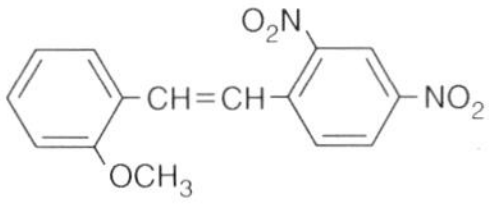
250 mg

S58,411-8
2,4-DINITRO-2'-ETHOXYSTILBENE
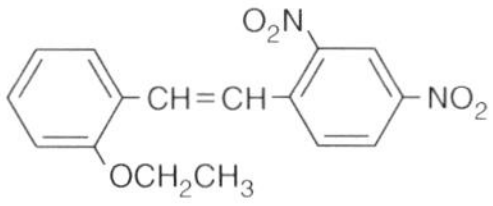
250 mg

S55,566-5
2,4-DINITRO-4'-METHOXYSTILBENE
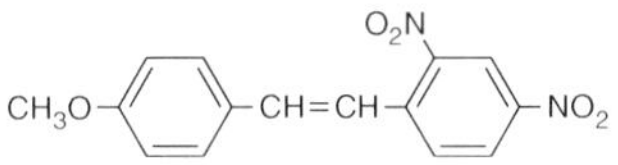
5 gm

S55,410-3
2,4-DINITRO-3'-METHOXYSTILBENE
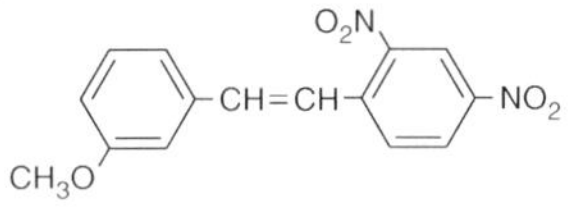
250 mg

S55,404-9
2,5-DIMETHOXY-2',4'-DINITROSTILBENE
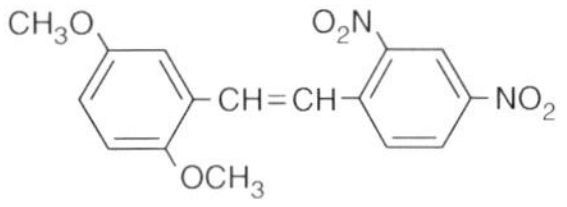
250 mg

S95,326-1
BETA-2-DINITRO-4-METHOXYSTYRENE
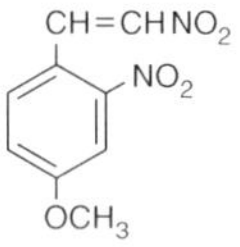
50 mg

S76,365-9
N-(3-METHOXYPROPYL)-4-NITROANILINE
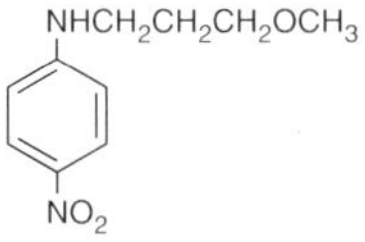
250 mg

S52,632-0
2-DECYLOXY-N-(2,6-DIETHYLPHENYL)-5-NITROBENZYLAMINE
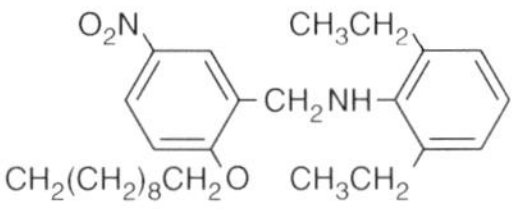
250 mg

S93,570-0
2-METHOXY-N-(4-NITROPHENYL)-BENZYLAMINE
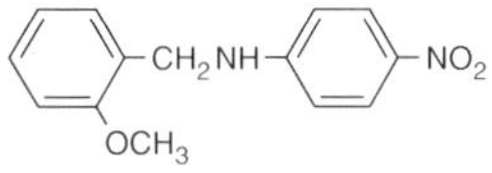
250 mg

S56,926-7
4-METHOXY-N-(4-NITROPHENYL)-BENZYLAMINE
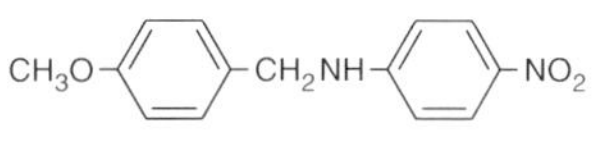
250 mg

S67,822-8
N,N-DIBENZYL-2-ETHOXY-4-NITROANILINE
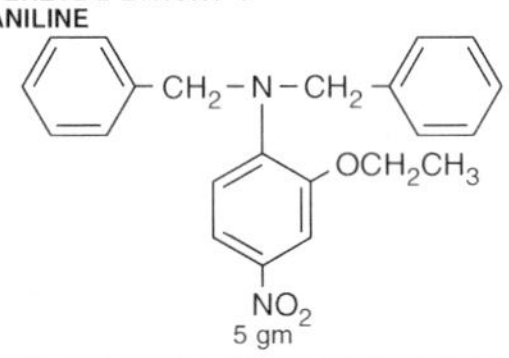
5 gm

S51,811-5
4-(2,5-DIBUTOXY-4-NITROPHENYL)-MORPHOLINE
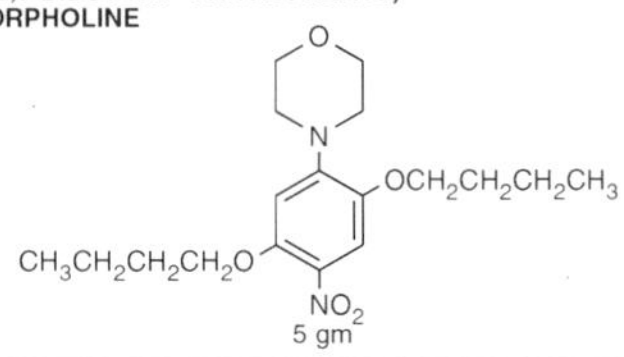
5 gm

S35,370-1
5-MORPHOLINO-2-NITROANISOLE
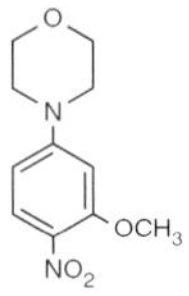
5 gm

S52,532-4
N-(3-CHLORO-O-TOLYL)-2-DECYLOXY-5-NITROBENZYLAMINE
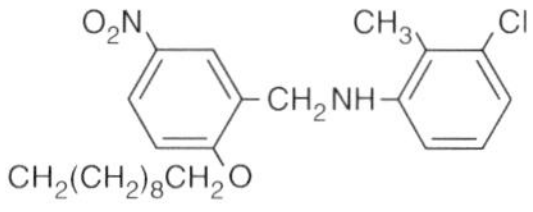
250 mg

S52,627-4
2-BUTOXY-5-NITRO-N-(M-TOLYL)-BENZYLAMINE
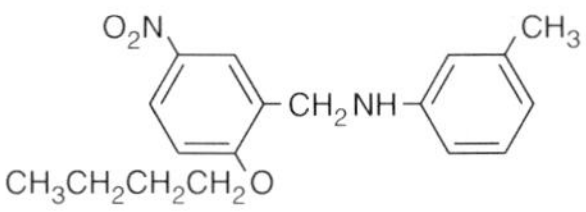
250 mg

S52,531-6
2-DECYLOXY-5-NITRO-N-(2,5-XYLYL)-BENZYLAMINE
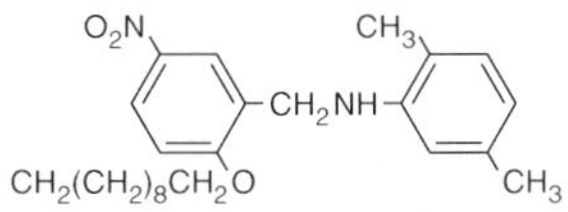
250 mg

S51,838-7
4-(2,5-DIETHOXY-4-NITROPHENYL)-MORPHOLINE
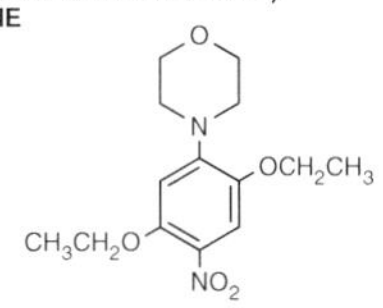
250 mg

S38,811-4
HEXADECYL 2-NITROPHENYL ETHER
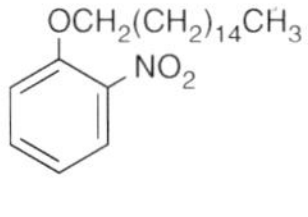
5 gm

S87,636-4
2-NITROPHENYL P-TOLYL ETHER
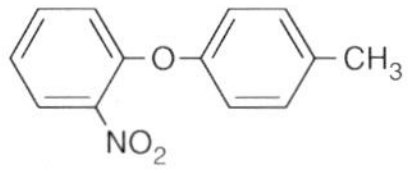
250 mg

S35,908-4
1-BUTOXY-2-NITROBENZENE
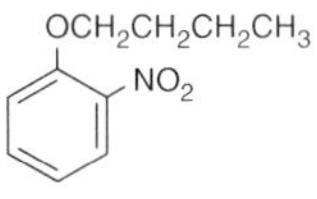
5 gm

S57,472-4
1,3-BIS(2-NITROPHENOXY)PROPANE
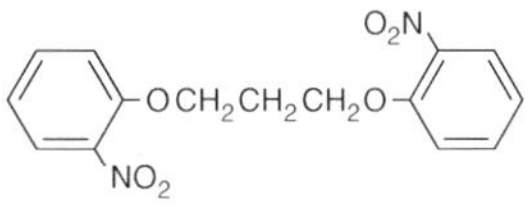
5 gm

S57,546-1
1,4-BIS(2-NITROPHENOXY)BUTANE
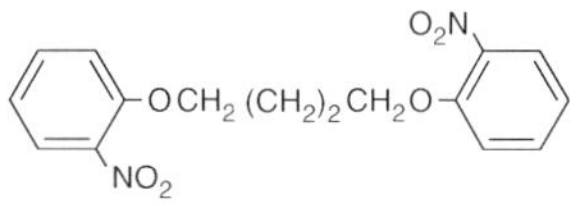
5 gm

S57,496-1
1,5-BIS(2-NITROPHENOXY)PENTANE
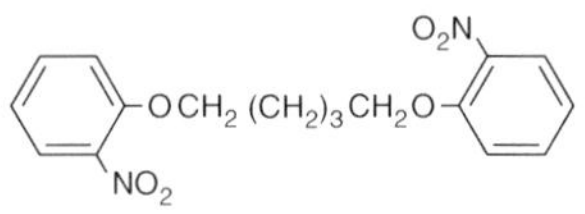
5 gm

S39,150-6
DODECYL 2-NITROPHENYL ETHER
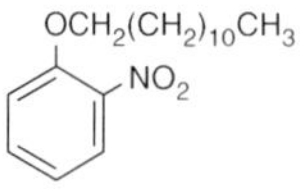
5 gm

S57,458-9
1,6-BIS(2-NITROPHENOXY)HEXANE
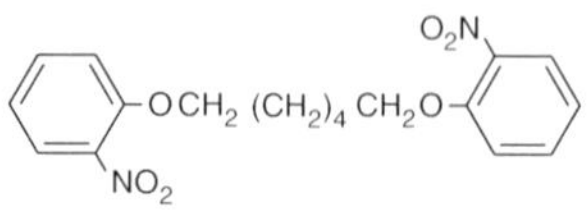
5 gm

S57,577-1
2-NITROPHENYL TETRADECYL ETHER
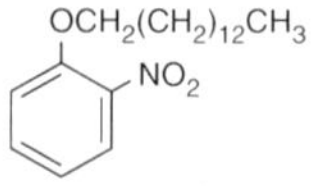
5 gm

S57,555-0
1,10-BIS(2-NITROPHENOXY)DECANE
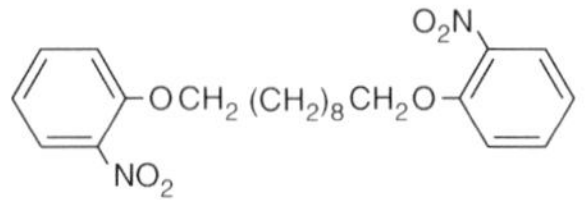
5 gm

S38,705-3
2-NITROPHENYL OCTADECYL ETHER
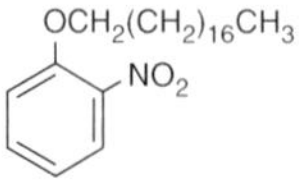
5 gm

S56,254-8
2-NITROPHENYL 2-(1,3,3-TRIMETHYLBUTYL)-5,7,7-TRIMETHYLOCTYL ETHER
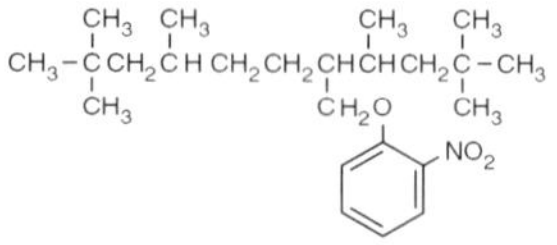
250 mg

S43,483-3
4-CHLOROPHENYL 2-NITROPHENYL ETHER
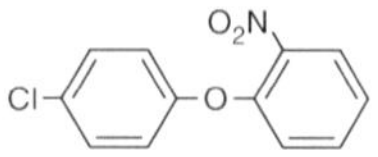
5 gm

S50,815-2
4-CHLOROPHENYL 4-CHLORO-2-NITROPHENYL ETHER
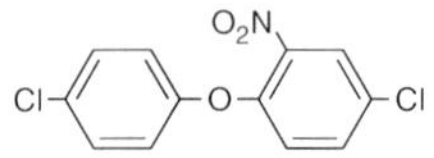
5 gm

S50,836-5
4-CHLORO-2-NITROPHENYL P-TOLYL ETHER
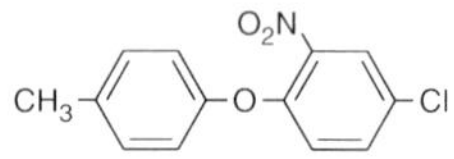
250 mg

S36,078-3
1,4-DIMETHOXY-3-NITROBENZENE
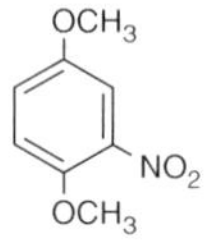
5 gm

S62,659-7
4-TERT-BUTYL-2-NITROANISOLE
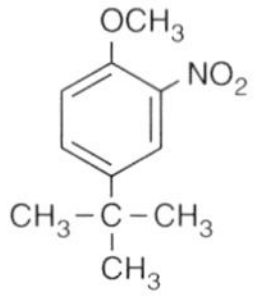
5 gm

S35,947-5
4-CHLORO-2,5-DIMETHOXYNITROBENZENE
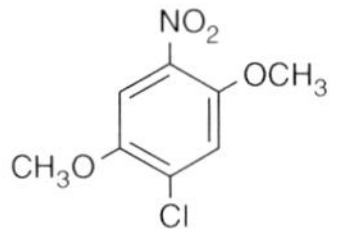
5 gm

S35,948-3
5-CHLORO-2,4-DIMETHOXYNITROBENZENE
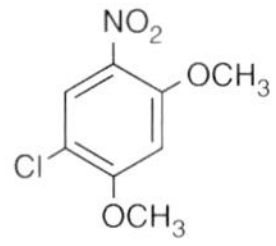
5 gm

S51,345-8
1-CHLORO-2,5-DIPENTYLOXY-4-NITROBENZENE
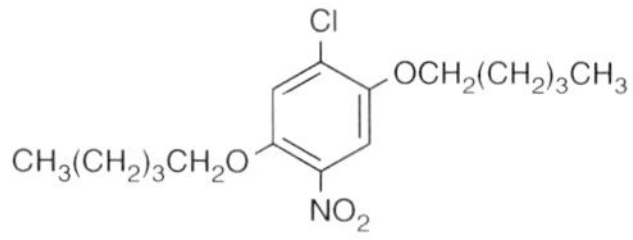
250 mg

S75,521-4
2-METHOXY-3-NITROFLUORENE
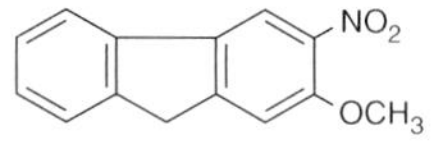
250 mg

S95,424-1
1-[ALPHA,BETA-BIS(1,3-DITHIOLAN-2-YL)BUTYL]-4-METHOXY-3-NITROBENZENE
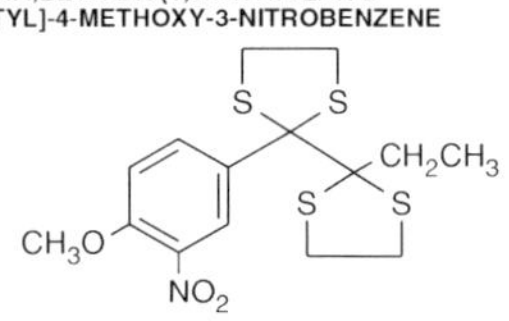
50 mg

S91,993-4
4'-METHOXY-4-NITRO-2,3,5-TRIPHENYLBIPHENYL
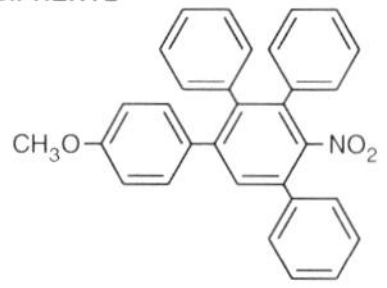
250 mg

S91,899-7
ALPHA,ALPHA-DIMETHYL-4-NITROBENZYL 1-METHYL-2-NAPHTHYL ETHER
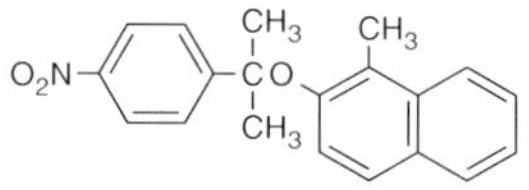
250 mg

S92,012-6
2,4-DINITROPHENYL ETHYL ETHER
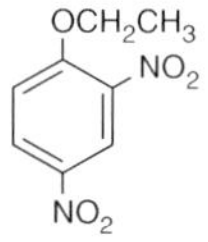
250 mg

S51,937-5
3-NITROPHENETOLE
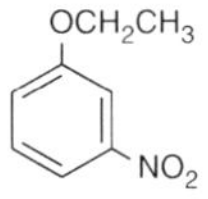
250 mg

S51,686-4
3-NITROPHENYL PENTYL ETHER
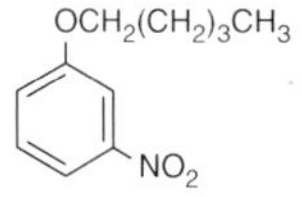
250 mg

S52,981-8
P-NITROPHENYL ETHER
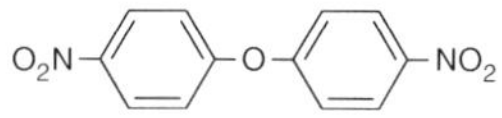
250 mg

S57,345-0
BENZYL 3-NITROPHENYL ETHER
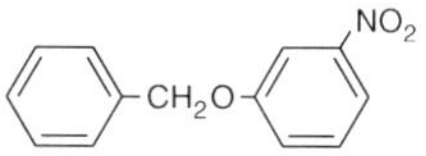
5 gm

S51,687-2
3-NITROPHENYL OCTYL ETHER
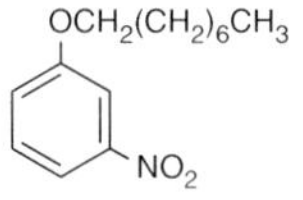
5 gm

S57,508-9
1,3-BIS(3-NITROPHENOXY)PROPANE
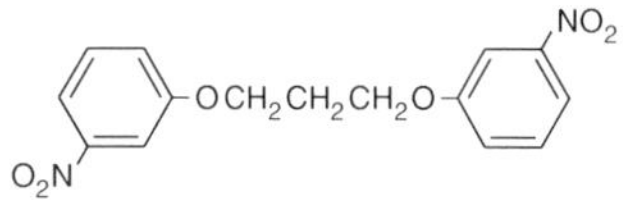
5 gm

S57,511-9
1,5-BIS(3-NITROPHENOXY)PENTANE
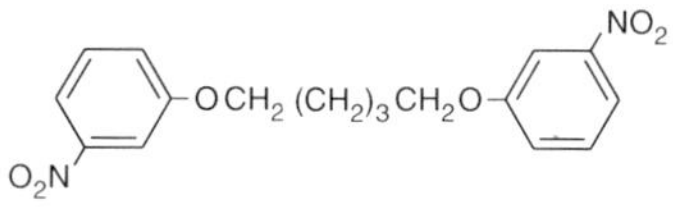
5 gm

S57,333-7
DODECYL 3-NITROPHENYL ETHER
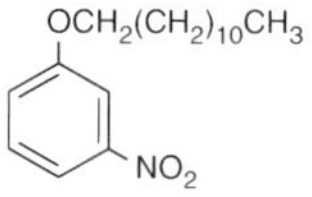
5 gm

S57,500-3
1,6-BIS(3-NITROPHENOXY)HEXANE
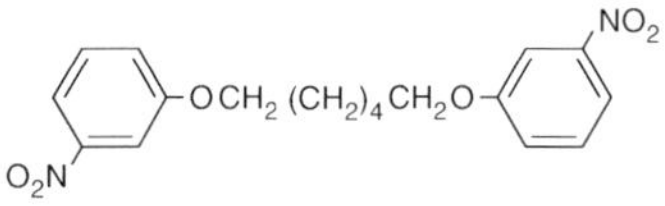
5 gm

S57,542-9
3-NITROPHENYL TETRADECYL ETHER
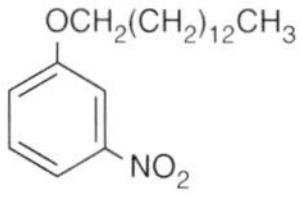
5 gm

S38,829-7
HEXADECYL 3-NITROPHENYL ETHER
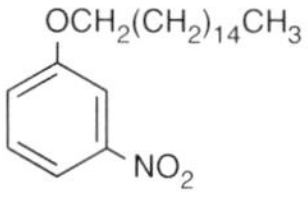
5 gm

S57,455-4
1,10-BIS(3-NITROPHENOXY)DECANE
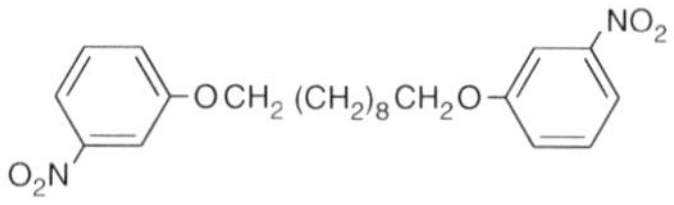
5 gm

S57,454-6
3-NITROPHENYL OCTADECYL ETHER
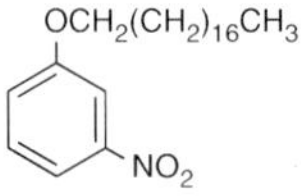
5 gm

S94,190-5
1-(2-METHOXY-4-NITROPHENYL)-PIPERIDINE
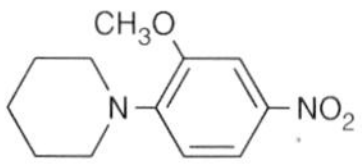
250 mg

S63,357-7
6-NITRO-1,4-BENZODIOXAN
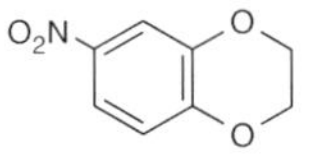
250 mg

S52,501-4
4-[2-(2-BUTOXY-5-NITROBENZYLOXY)-ETHYL]MORPHOLINE OXALATE
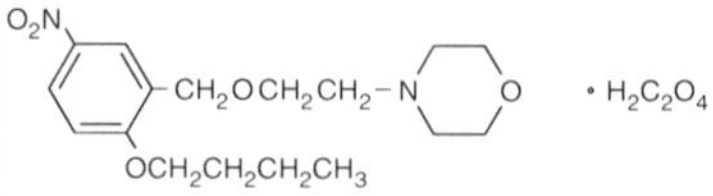
250 mg

S57,431-7
BUTYL 4-NITROPHENYL ETHER
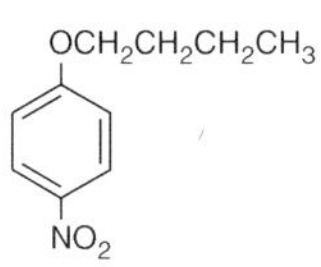
5 gm

S57,580-1
BENZYL 4-NITROPHENYL ETHER
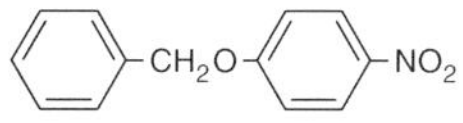
5 gm

S52,642-8
HEPTYL 4-NITROPHENYL ETHER
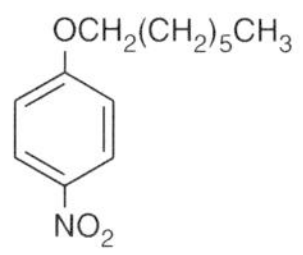
250 mg

S57,509-7
1,3-BIS(4-NITROPHENOXY)PROPANE
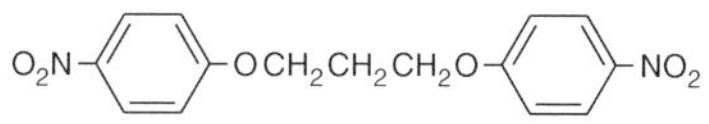
5 gm

S57,344-2
DECYL 4-NITROPHENYL ETHER
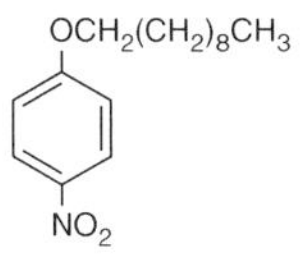
5 gm

S57,461-9
1,4-BIS(4-NITROPHENOXY)BUTANE
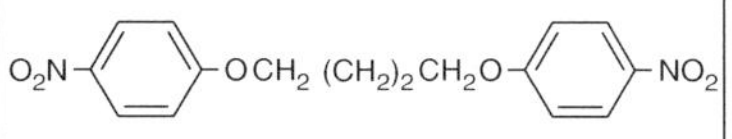
5 gm

S57,364-7
4,4'-(PENTAMETHYLENEDIOXY)-BIS-(NITROBENZENE)
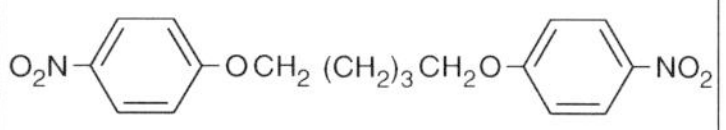
5 gm

S57,549-6
DODECYL 4-NITROPHENYL ETHER
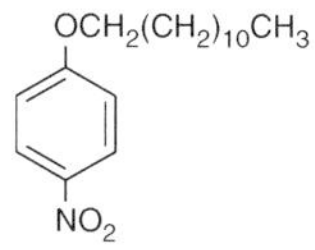
5 gm

S53,132-4
1,3-BIS(4-NITROPHENOXY)BENZENE
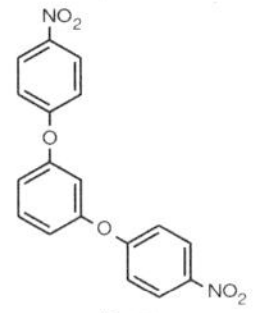
5 gm

S57,499-6
1,6-BIS(4-NITROPHENOXY)HEXANE
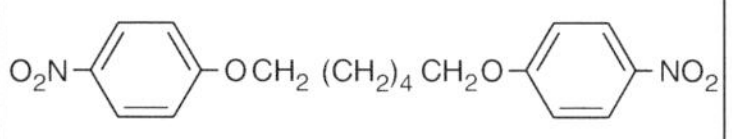
5 gm

S57,539-9
4-NITROPHENYL TETRADECYL ETHER
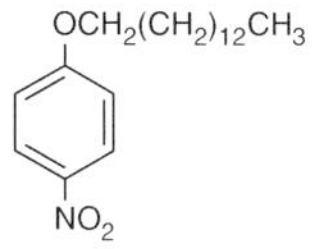
5 gm

S38,284-1
HEXADECYL 4-NITROPHENYL ETHER
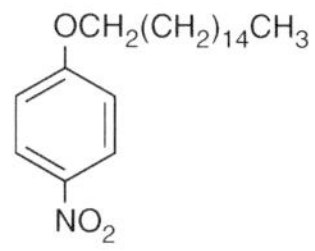
5 gm

S57,505-4
1,10-BIS(4-NITROPHENOXY)DECANE
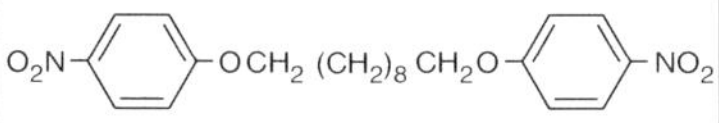
5 gm

S57,445-7
4-NITROPHENYL OCTADECYL ETHER
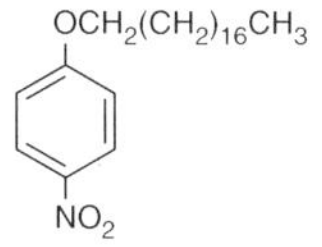
5 gm

S50,799-7
4-CHLORO-M-TOLYL 4-NITROPHENYL ETHER
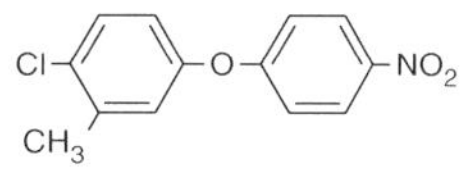
5 gm

S52,499-9
4-CHLOROPHENYL 2-DECYLOXY-5-NITROBENZYL ETHER
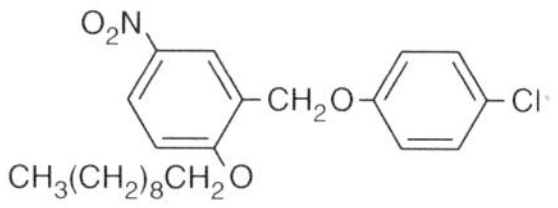
250 mg

S52,496-4
4-BIPHENYLYL 2-METHOXY-5-NITROBENZYL ETHER
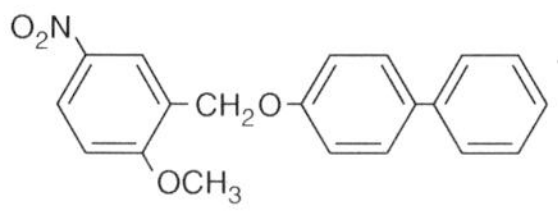
250 mg

S52,498-0
2-HEPTYLOXY-5-NITROBENZYL 2,6-XYLYL ETHER
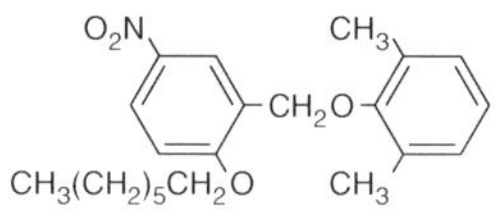
250 mg

S42,683-0
4-CHLORO-2-NITROANISOLE
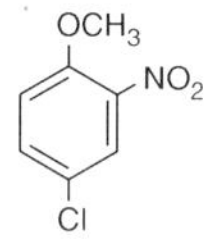
5 gm

S36,281-6
2-IODO-4-NITROANISOLE
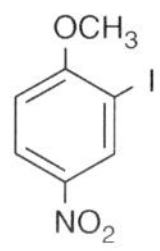
5 gm

S36,282-4
2-IODO-5-NITROANISOLE
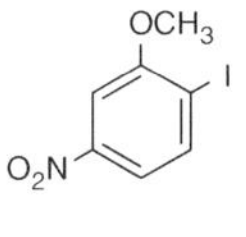
5 gm

S54,498-1
4-IODO-3-NITROANISOLE
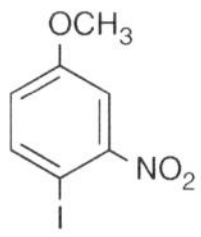
5 gm

S40,787-9
1,3-DIMETHOXY-4-NITROBENZENE
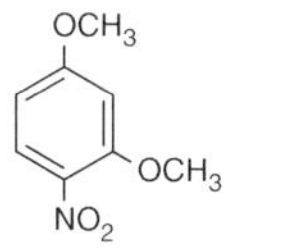
5 gm

S35,849-5
BENZYL 4-NITRO-M-TOLYL ETHER
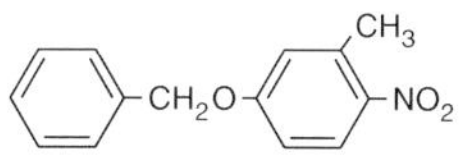
5 gm

S55,408-1
2,4-DIMETHOXY-2',4'-DINITROSTILBENE
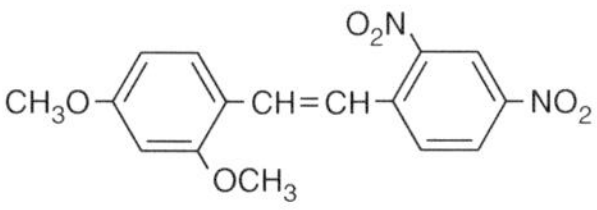
250 mg

S52,503-0
4-CHLORO-O-TOLYL 2-HEPTYLOXY-5-NITROBENZYL ETHER
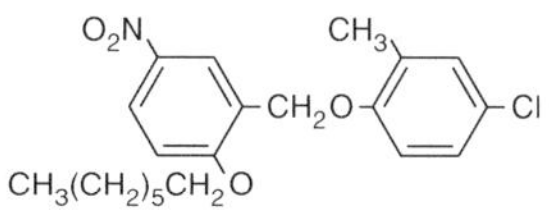
250 mg

S52,502-2
3-CHLOROPHENYL 2-DECYLOXY-5-NITROBENZYL ETHER
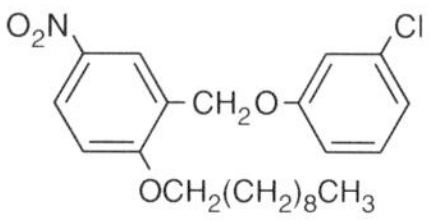
250 mg

S52,639-8
2-DECYLOXY-5-NITROBENZYL 2,5-XYLYL ETHER
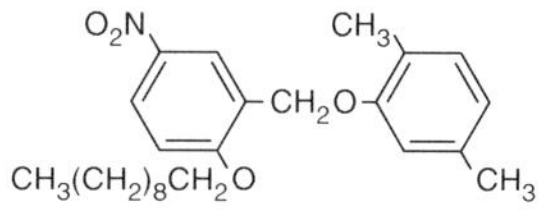
250 mg

S55,352-2
4-HEXADECYLOXY-3-NITROBIPHENYL
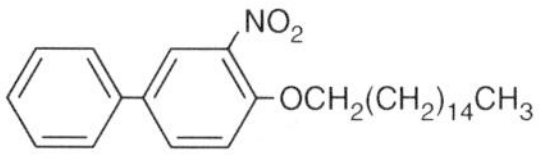
250 mg

S74,391-7
ALPHA-(4-ETHYLANILINO)-3-NITROBENZYL ALCOHOL
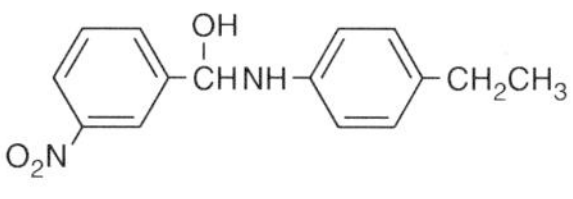
5 gm

S91,499-1
2-(4-NITROPHENYL)-3-PHENYLOXIRANE
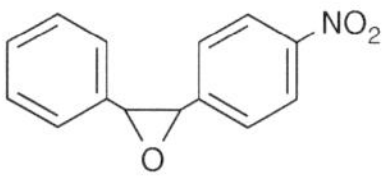
250 mg

S62,900-6
3,9-BIS(3-NITROPHENYL)-2,4,8,10-TETRAOXASPIRO(5.5)UNDECANE
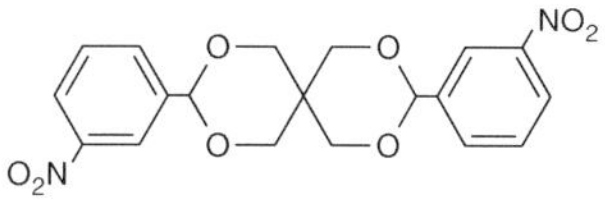
5 gm

S37,004-5
4-NITROPHENYL-BETA-D-GALACTOPYRANOSIDE TETRAACETATE
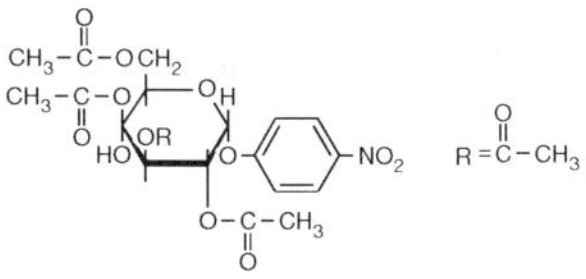
250 mg

S54,787-5
6,8-DINITRO-1,3-BENZODIOXANE
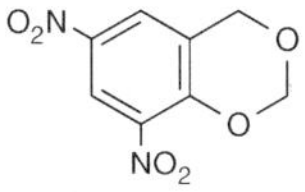
5 gm

S95,335-0
3-NITROBENZALDEHYDE DIMETHYL ACETAL
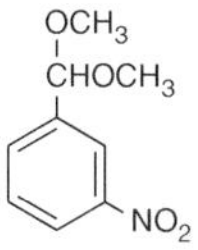
50 mg

S61,746-6
4-NITROBENZYL MERCAPTAN
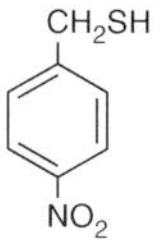
250 mg

S61,748-2
3-NITROBENZYL MERCAPTAN

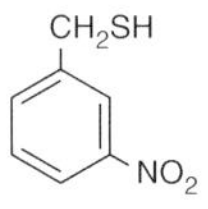

250 mg

S50,328-2
4-(METHYLTHIO)-N-PICRYLANILINE

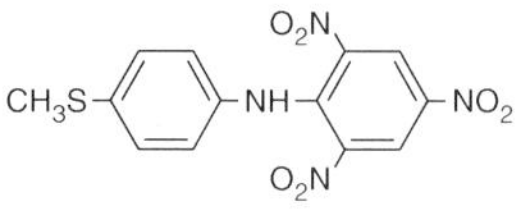

5 gm

S59,391-5
N-(3-METHYLTHIOPHENYL)-2,4,6-TRINITROANILINE

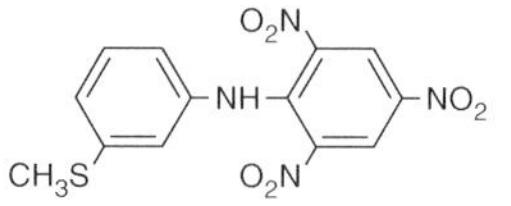

5 gm

S82,892-0
2,4-DINITROPHENYL M-TOLYL SULFIDE

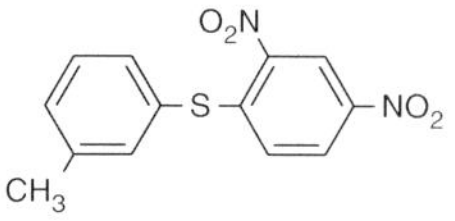

250 mg

S77,096-5
1,3-DINITRO-4-(PHENYLTHIO)-NAPHTHALENE

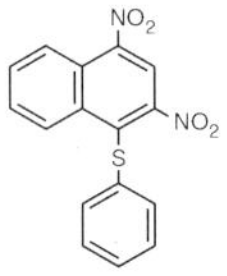

250 mg

S43,573-2
2,4-DINITROPHENYL SULFIDE

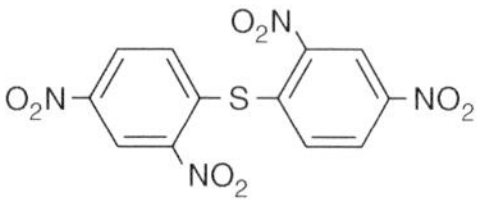

5 gm

S63,281-3
2,4,4',7-TETRANITRODI-1-NAPHTHYL SULFIDE

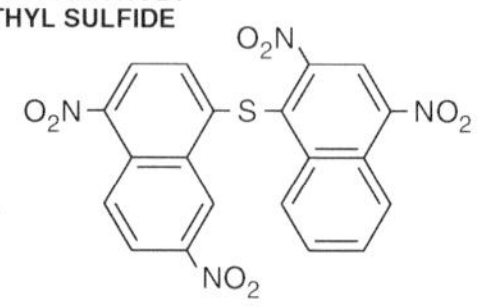

250 mg

S72,057-7
N-(4-NITROPHENYL)-ALPHA-(P-TOLYLTHIO)BENZYLAMINE

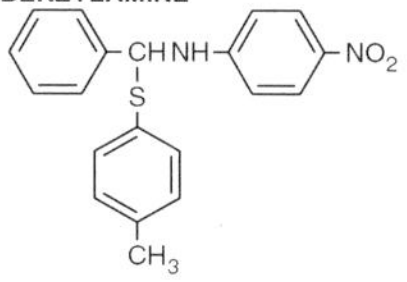

5 gm

S72,182-4
BENZYL 2-NITROPHENYL SULFIDE

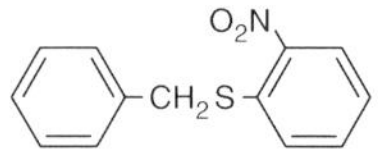

5 gm

S83,631-1
2-NITROPHENYL O-TOLYL SULFIDE

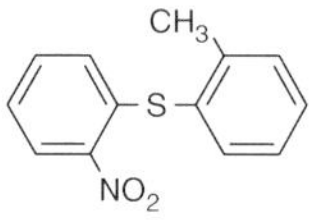

250 mg

S59,253-6
HEXADECYL 2-NITROPHENYL SULFIDE

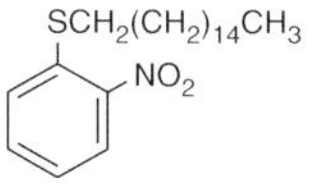

5 gm

S91,935-7
3-NITROPHENYL PHENYL SULFIDE

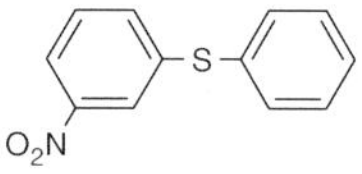

250 mg

S46,100-8
4-NITROPHENYL PHENYL SULFIDE

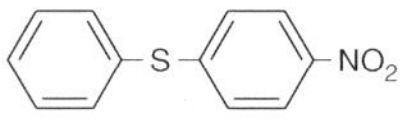

5 gm

S76,797-2
BENZYL 4-NITROPHENYL SULFIDE

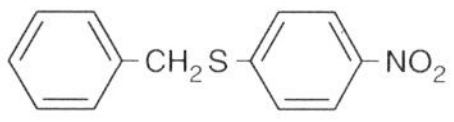

250 mg

S77,122-8
1-METHYL-2-(4-NITROPHENYL)ETHYL PHENYL SULFIDE

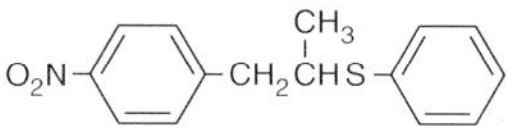

250 mg

S91,790-7
ALPHA,ALPHA-DIMETHYL-4-NITROBENZYL PHENYL SULFIDE

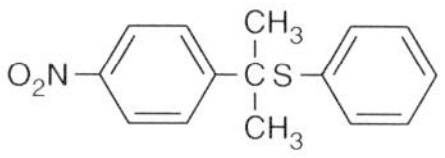

250 mg

S83,630-3
4-NITROPHENYL O-TOLYL SULFIDE

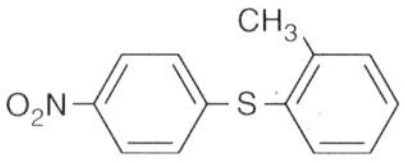

250 mg

S91,933-0
ALPHA,ALPHA-DIMETHYL-4-NITROBENZYL METHYL SULFIDE

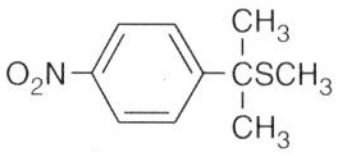

250 mg

S83,634-6
4-NITROPHENYL M-TOLYL SULFIDE

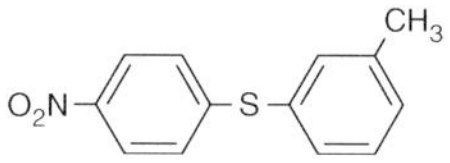

250 mg

S82,906-4
4-ISOPROPYLPHENYL 4-NITROPHENYL SULFIDE

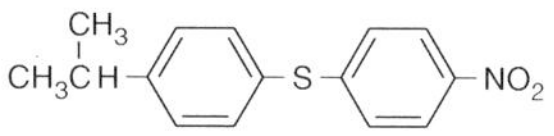

250 mg

S54,983-5
METHYL 3-NITROPHENYL SULFIDE

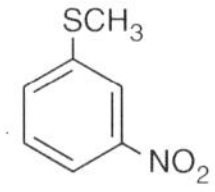

5 gm

S76,254-7
2,4-DINITROPHENYL P-TOLYL DISULFIDE

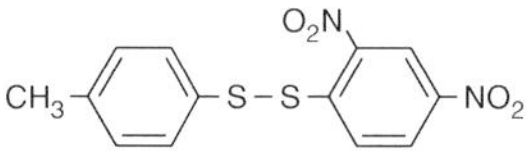

250 mg

S43,798-0
4-CHLORO-2-NITROPHENYL DISULFIDE

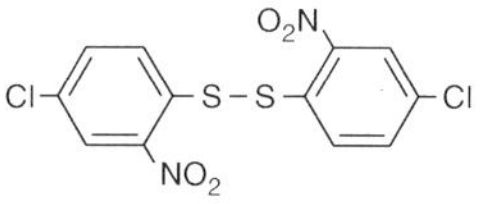

5 gm

S75,597-4
4-NITROPHENYL P-TOLYL DISULFIDE

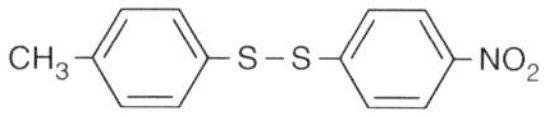

250 mg

S53,099-9
2-CHLORO-5-NITROPHENYL DISULFIDE

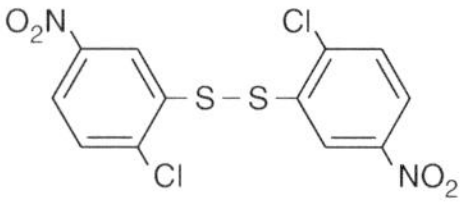

250 mg

S53,100-6
4-CHLORO-3-NITROPHENYL DISULFIDE

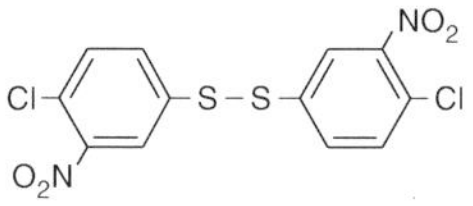

250 mg

S35,672-7
2,4,6-TRINITRO-M-TOLUIDINE

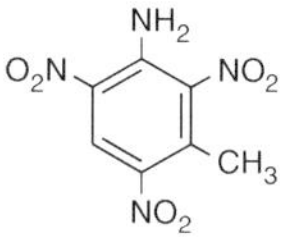

250 mg

S67,824-4
2,4-DIMETHYL-3,6-DINITROANILINE

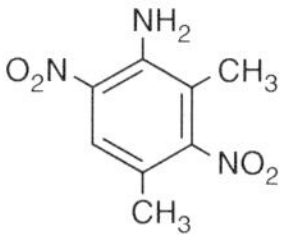

5 gm

S59,509-8
4-NITRO-1,3-PHENYLENEDIAMINE

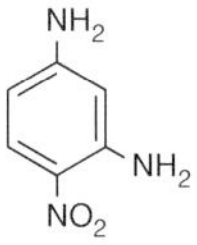

5 gm

S80,561-0
4-NITRO-1,2-PHENYLENEDIAMINE SULFATE

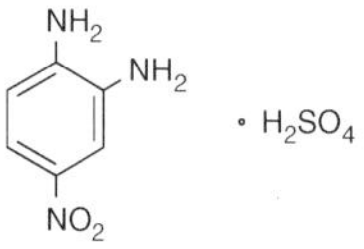

250 mg

S69,710-9
2,4-DINITRO-6-PHENYLANILINE

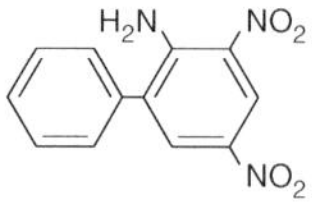

250 mg

S91,468-1
N1-ETHYL-4-NITRO-1,2-PHENYLENEDIAMINE

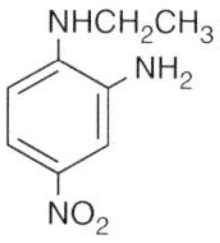

250 mg

S80,320-0
4,5-DINITRO-2-METHYLANILINE

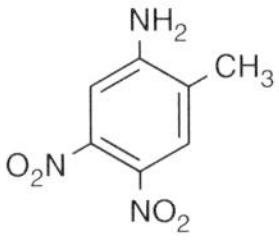

250 mg

S85,003-9
2,3-DINITRO-P-PHENYLENEDIAMINE

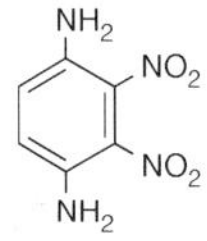

250 mg

S64,260-6
3,3'-DINITROBENZIDINE

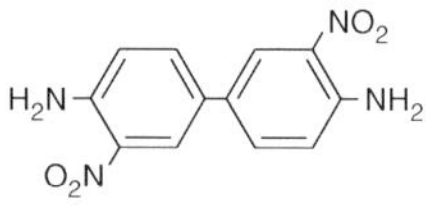

5 gm

S75,302-5
4-ETHYL-2-NITROANILINE

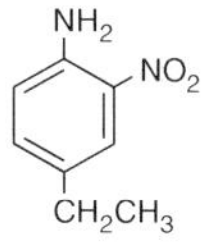

5 gm

S82,073-3
4-ISOPROPYL-2-NITROANILINE
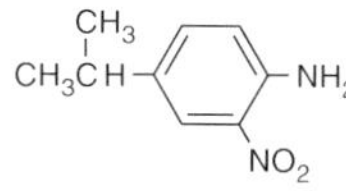
250 mg

S80,237-9
2,3-DINITRO-6-METHYLANILINE
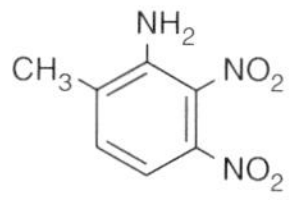
250 mg

S89,906-2
6-ACI-NITRO-1-CYCLOHEXENE-1-AMINE
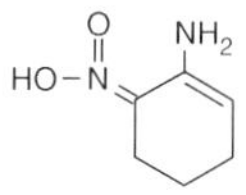
250 mg

S35,779-0
5-AMINO-4-NITROACENAPHTHENE
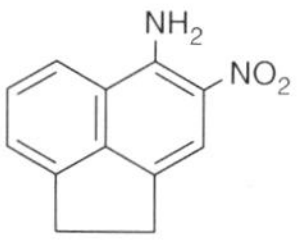
250 mg

S34,747-7
3,3'-DINITROBENZIDINE
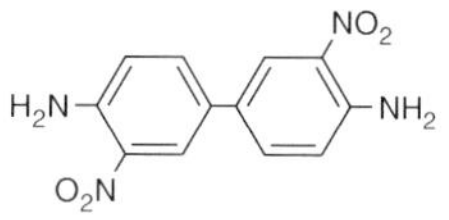
250 mg

S45,352-8
2-AMINO-1-NITRONAPHTHALENE
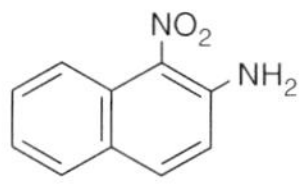
250 mg

S74,954-0
3-AMINO-2-NITROFLUORENE
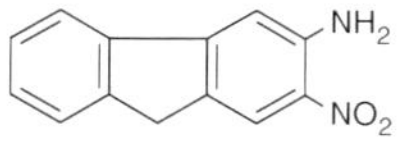
250 mg

S80,569-6
2-AMINO-3-NITROFLUORENE
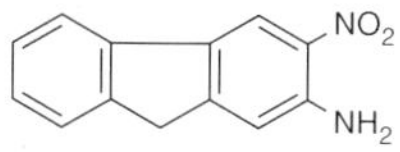
250 mg

S64,817-5
4,4'-DIAMINO-3,3'-DIMETHYL-2,2'-DINITROBIPHENYL
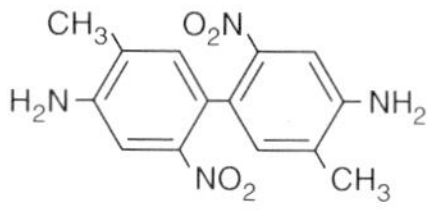
250 mg

S84,281-8
5,5'-DIMETHYL-2,2'-DINITROBENZIDINE
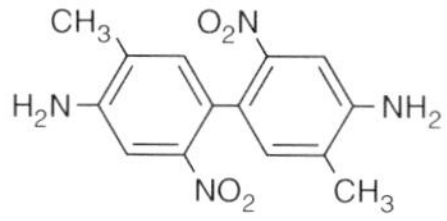
250 mg

S94,090-9
N1-METHYL-4-NITRO-1,2-PHENYLENEDIAMINE
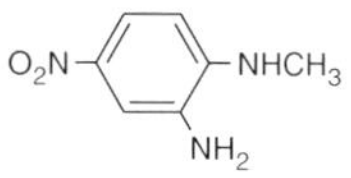
250 mg

S59,267-6
3-NITRO-4,4'-METHYLENEDIANILINE
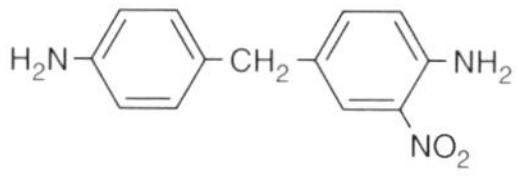
5 gm

S65,324-1
2-NITROBENZIDINE
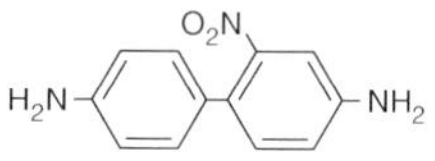
250 mg

S75,211-8
2-ETHYL-5-NITROANILINE
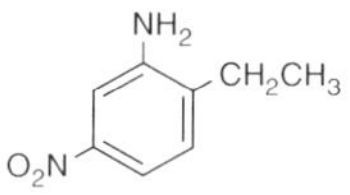
5 gm

S81,474-1
4-ISOPROPYL-3-NITROANILINE
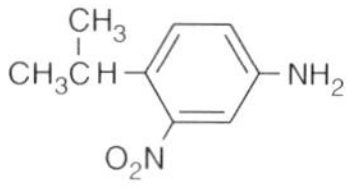
250 mg

S59,825-9
4-(TERT-BUTYL)-3-NITROANILINE
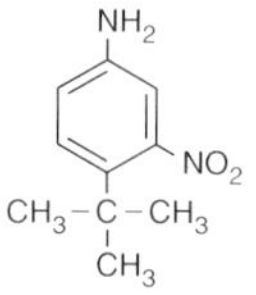
5 gm

S77,014-0
2-BUTYL-5-NITROANILINE
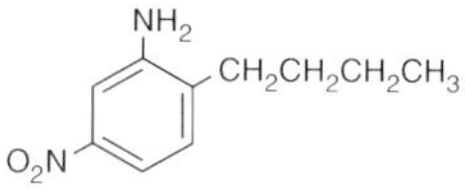
250 mg

S74,036-5
2,4-DIMETHYL-5-NITROANILINE
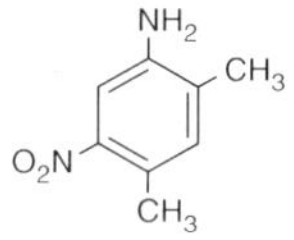
250 mg

S63,125-6
O-NITROBENZYLAMINE HYDROCHLORIDE
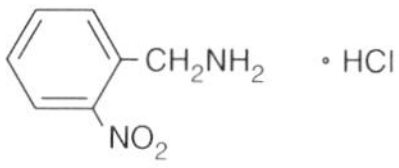
250 mg

S51,560-4
4-NITRO-2,3,5,6-TETRAMETHYLANILINE
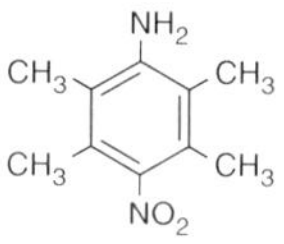
250 mg

S74,491-3
5-NITRO-2-BIPHENYLAMINE
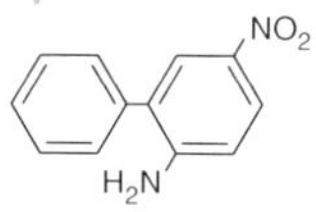
250 mg

S80,238-7
3-METHYL-4-NITROANILINE
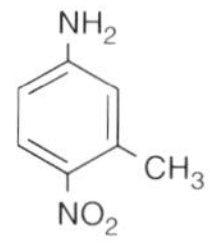
250 mg

S94,697-4
3,5-DIMETHYL-4-NITROANILINE
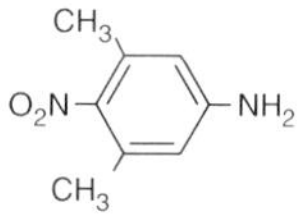
250 mg

S38,510-7
2-BROMO-4-NITRO-1-NAPHTHYLAMINE
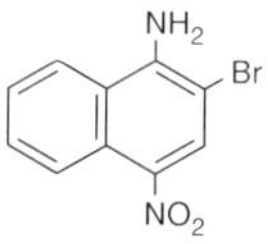
250 mg

S87,013-7
9-AMINO-10-NITROANTHRACENE
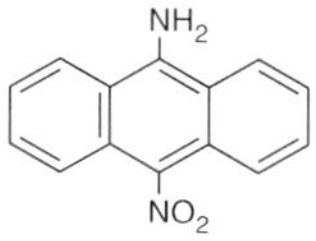
250 mg

S92,192-0
2,6-DIMETHYL-3-NITROANILINE
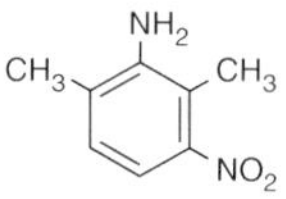
250 mg

S41,468-9
2'-NITRO-4-BIPHENYLAMINE
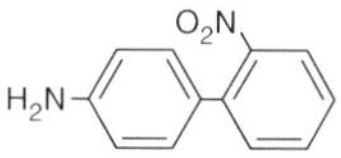
250 mg

S92,649-3
4-NITRO-N-PROPYLANILINE
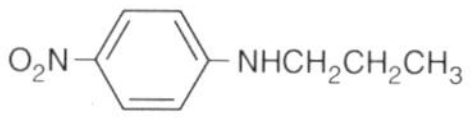
250 mg

S88,141-4
1-(4-AMINOPHENYL)-4-(4-NITROPHENYL)-1,3-BUTADIENE
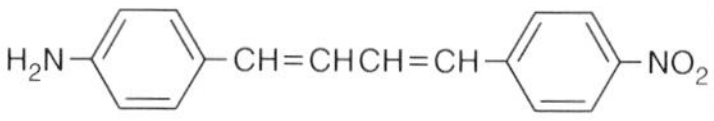
250 mg

S75,425-0
2,5-DIMETHYL-4-NITROANILINE
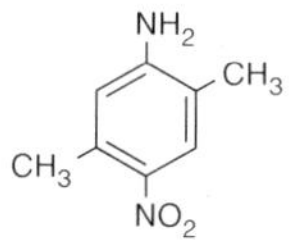
250 mg

S81,464-4
ANILINE PICRATE
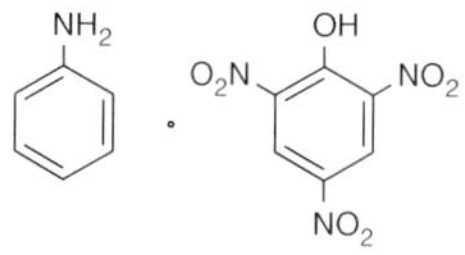
250 mg

S92,849-6
3-AMINO-2,4-DIPHENYL-7-NITRO-1(2H)-ISOQUINOLONE
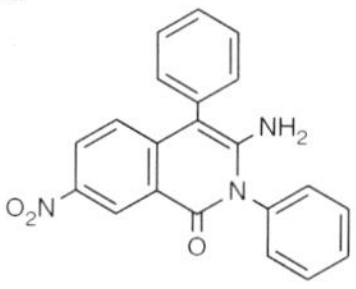
250 mg

S42,833-7
2-AMINO-7-NITROFLUORENE
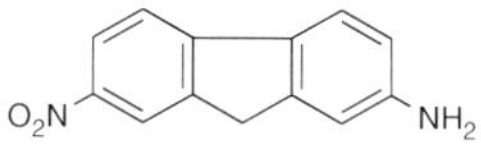
250 mg

S75,436-6
4-AMINO-7-NITROFLUORENE
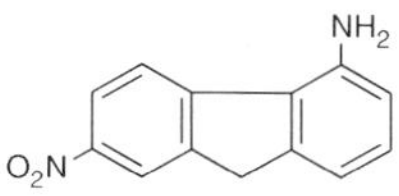
250 mg

S75,506-0
2-AMINO-6-FLUORO-7-NITROFLUORENE
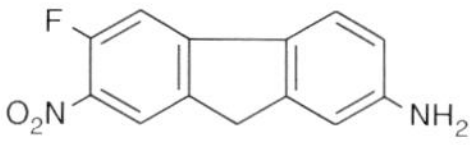
250 mg

S50,770-9
N,N-DIMETHYL-N'-(2,4-DINITROPHENYL)1,4-PHENYLENEDIAMINE
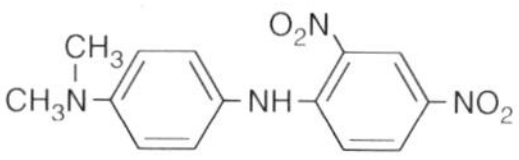
5 gm

S52,631-2
N,N-DIETHYL-N'-(2,4-DINITROPHENYL)-ETHYLENEDIAMINE
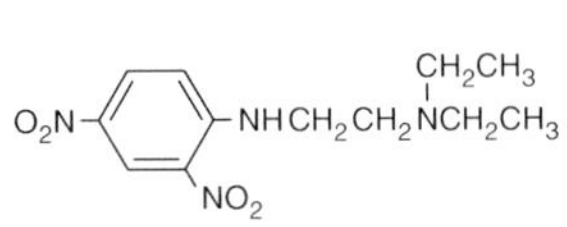
250 mg

S43,347-0
2,4-DINITRO-2'-METHYLDIPHENYLAMINE
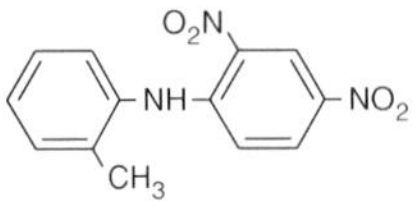
5 gm

S50,335-5
2,4-DINITRO-4'-METHYLDIPHENYLAMINE
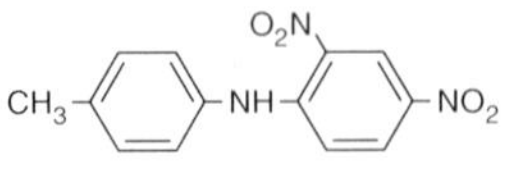
5 gm

S50,325-8
N-PICRYL-P-TOLUIDINE
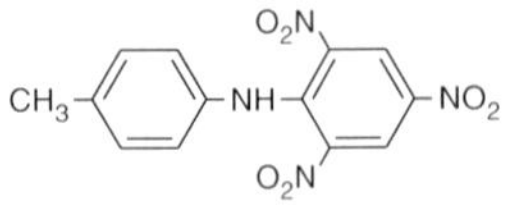
5 gm

S39,975-2
3,4-DIMETHYL-2',4'-DINITRODIPHENYLAMINE
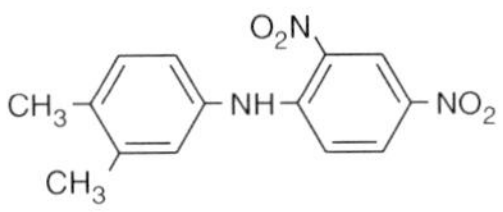
5 gm

S39,977-9
2,4-DIMETHYL-2',4'-DINITRODIPHENYLAMINE
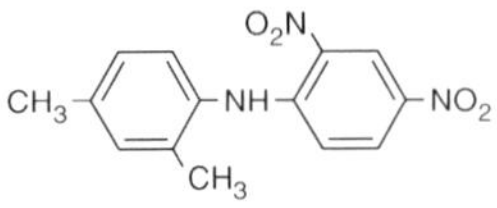
5 gm

S51,412-8
2,5-DIMETHYL-2',4'-DINITRODIPHENYLAMINE
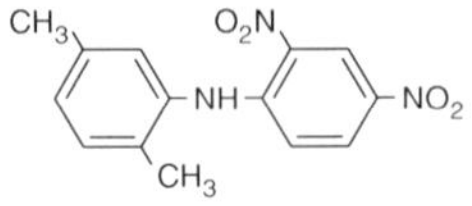
5 gm

S43,363-2
2,3',4,6-TETRANITRODIPHENYLAMINE
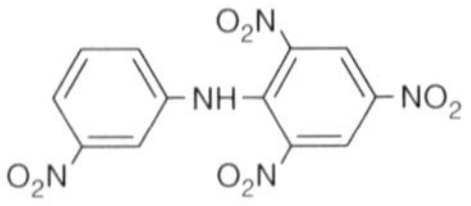
5 gm

S50,321-5
N-PICRYL-M-TOLUIDINE
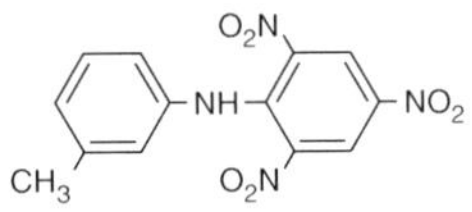
5 gm

S58,957-8
N-(2-FLUOROPHENYL)PICRYLAMINE
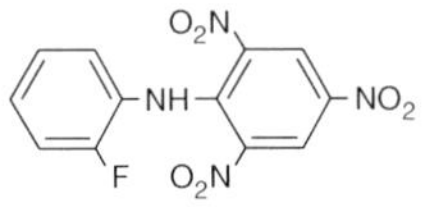
250 mg

S58,920-9
N-(2-BROMOPHENYL)PICRYLAMINE
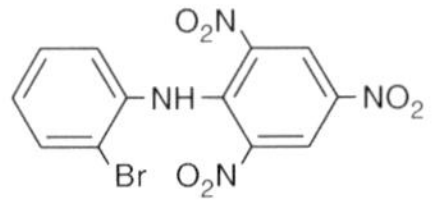
5 gm

S58,963-2
N-(2-ETHYLPHENYL)PICRYLAMINE
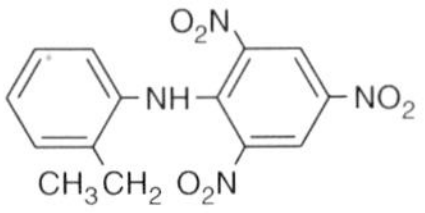
5 gm

S53,621-0
N-(4-FLUOROPHENYL)PICRYLAMINE
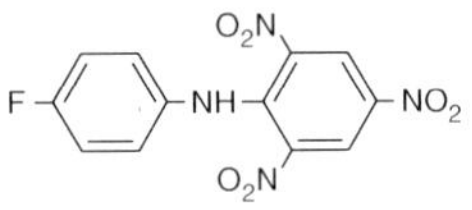
250 mg

S56,077-4
4'-IODO-2,4,6-TRINITRODIPHENYLAMINE
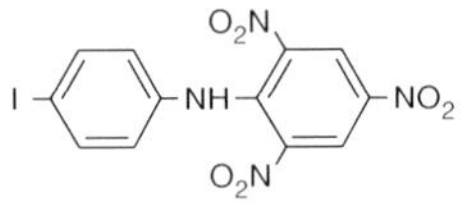
250 mg

S58,954-3
N-(4-ETHYLPHENYL)PICRYLAMINE
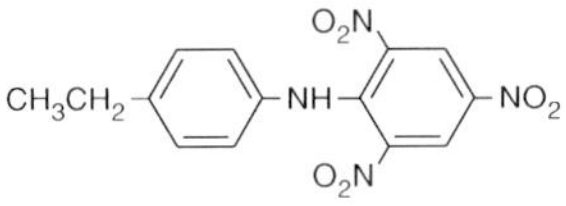
250 mg

S62,539-6
N-(4-BUTYLPHENYL)PICRYLAMINE
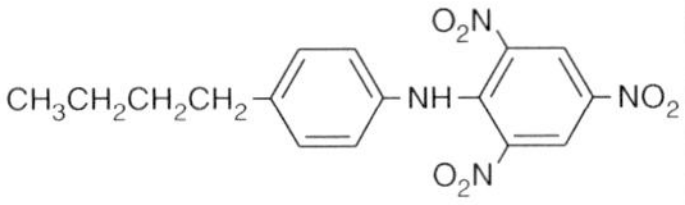
5 gm

S56,908-9
N-PICRYL-2,6-XYLIDINE
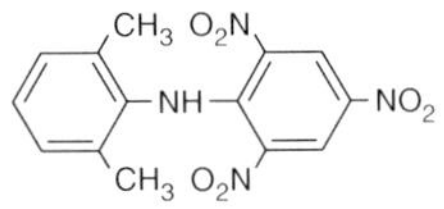
250 mg

S62,933-2
N-PICRYL-2,3-XYLIDINE
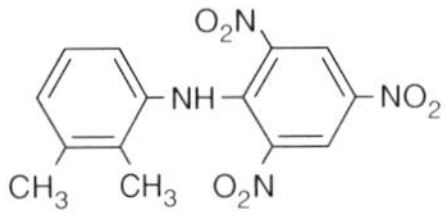
5 gm

S55,828-1
N-(3,4-DICHLOROPHENYL)PICRYLAMINE
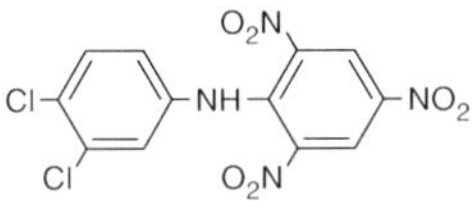
250 mg

S62,449-7
N-(4-CHLORO-3-NITROPHENYL)-PICRYLAMINE
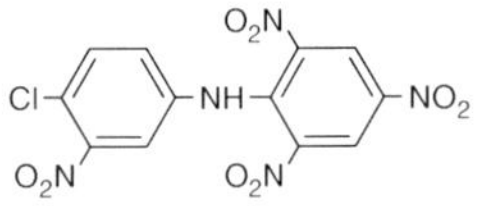
5 gm

S63,786-6
N-(2-CHLORO-5-NITROPHENYL)-PICRYLAMINE
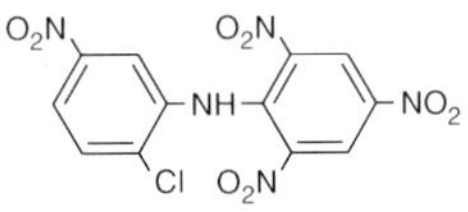
5 gm

S56,064-2
2',5'-DIBROMO-2,4,6-TRINITRODIPHENYLAMINE
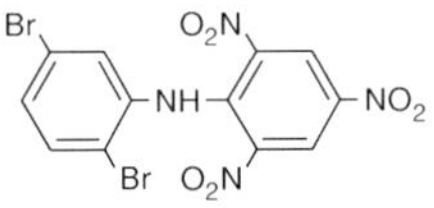
250 mg

S56,917-8
N-(2,4-DIBROMOPHENYL)PICRYLAMINE
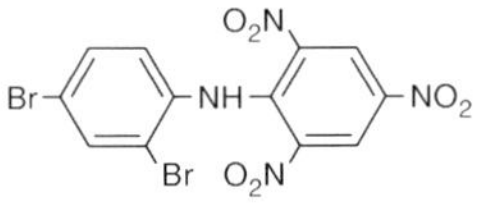
250 mg

S58,965-9
N-(3-CHLORO-P-TOLYL)PICRYLAMINE
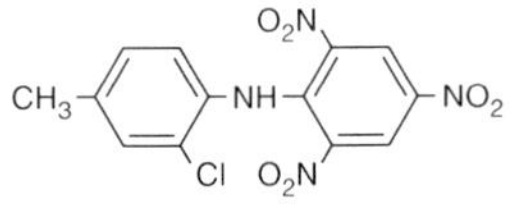
250 mg

S56,913-5
5-NITRO-N-PICRYL-O-TOLUIDINE
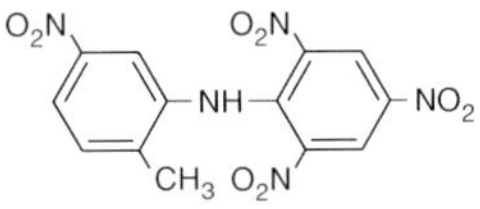
250 mg

S56,914-3
3-NITRO-N-PICRYL-P-TOLUIDINE
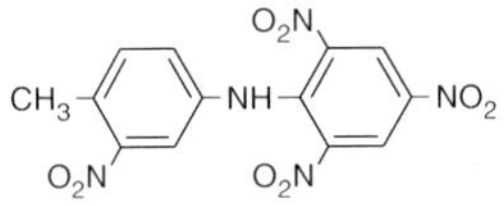
250 mg

S56,073-1
N-(4-BROMO-O-TOLYL)PICRYLAMINE
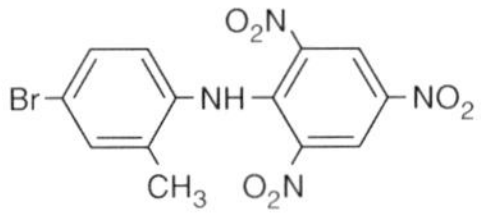
250 mg

S56,911-9
4-BROMO-N-PICRYL-M-TOLUIDINE
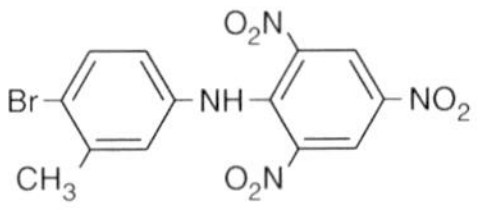
250 mg

S58,948-9
N-(2,5-XYLYL)PICRYLAMINE
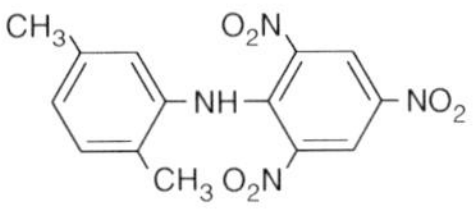
5 gm

S62,121-8
N-(3,5-DICHLOROPHENYL)PICRYLAMINE
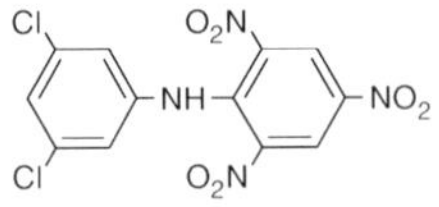
250 mg

S58,964-0
N-(3,5-XYLYL)PICRYLAMINE
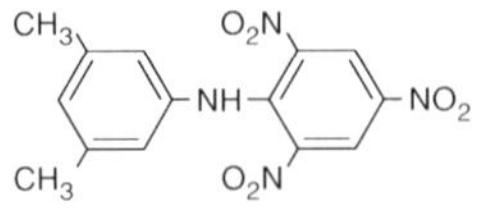
250 mg

S58,938-1
N-(3-FLUOROPHENYL)PICRYLAMINE
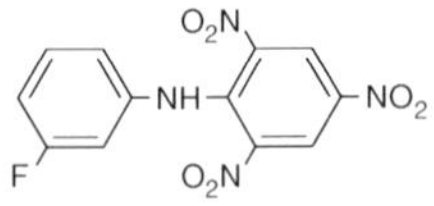
5 gm

S58,945-4
N-(2,4,6-TRIMETHYLPHENYL)-PICRYLAMINE
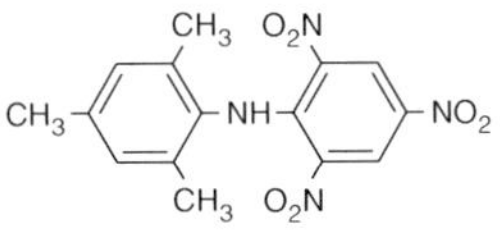
250 mg

S58,980-2
N-(2,4,5-TRIMETHYLPHENYL)-PICRYLAMINE
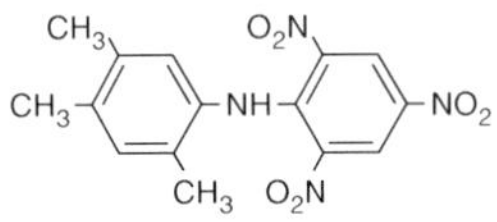
5 gm

S39,902-7
N-PICRYL-1-NAPHTHYLAMINE
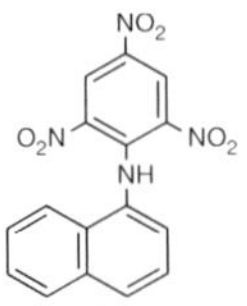
5 gm

S54,193-1
1-(2,4-DINITROPHENYL)PYRROLIDINE
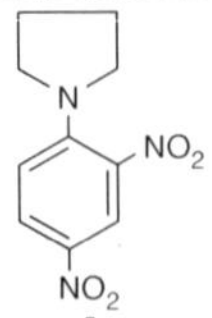
5 gm

S36,120-8
1-(2,4-DINITROPHENYL)PIPERIDINE

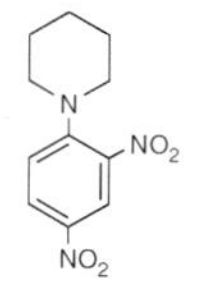

5 gm

S54,195-8
1-(2,4-DINITROPHENYL)-3-METHYLPIPERIDINE

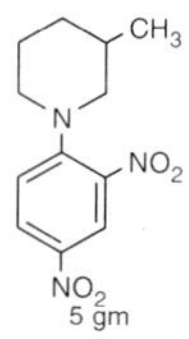

5 gm

s54,197-4
1-(2,4-DINITROPHENYL)-4-METHYLPIPERIDINE

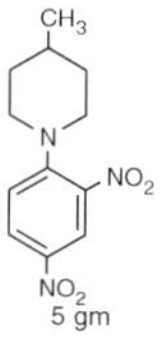

5 gm

S51,835-2
1-CHLORO-2,4-DIMORPHOLINO-5-NITROBENZENE

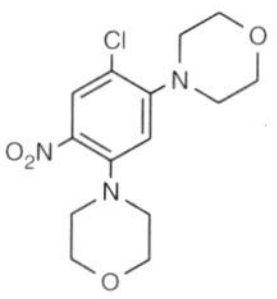

5 gm

S40,527-2
N,N-DIMETHYL-3,4-DINITROANILINE

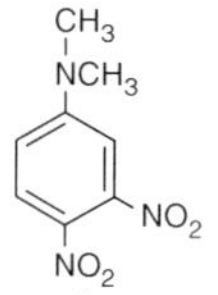

5 gm

S51,701-1
4-CHLORO-6-NITRO-N,N,N',N'-TETRAETHYL-1,3-PHENYLENEDIAMINE

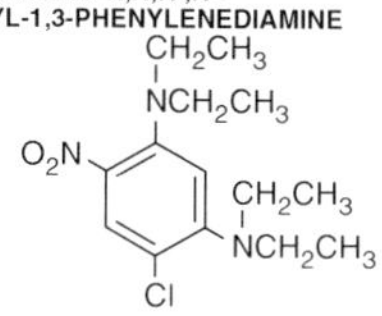

5 gm

S40,822-0
1-(2,4-DINITROPHENYL)-1-METHYLHYDRAZINE

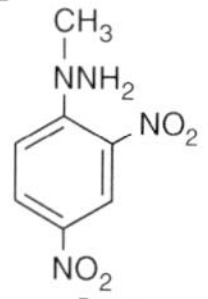

5 gm

S86,440-4
2-METHYL-1-CYCLOHEXANONE (2,4-DINITROPHENYL)HYDRAZONE

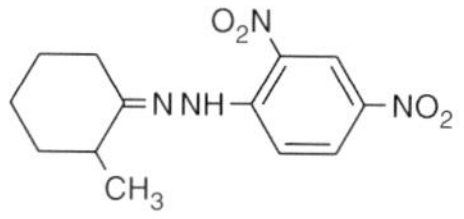

250 mg

S86,512-5
3-METHYLCYCLOHEXANONE (2,4-DINITROPHENYL)HYDRAZONE

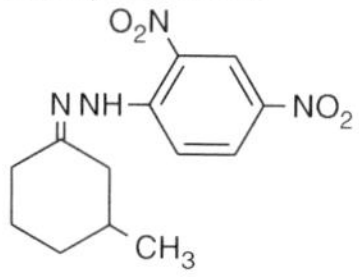

250 mg

S86,439-0
BENZALDEHYDE (2,4-DINITROPHENYL)-HYDRAZONE

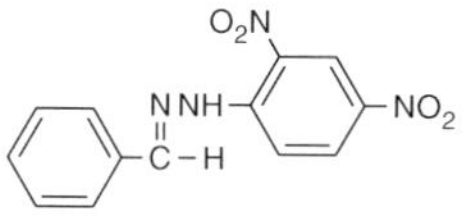

250 mg

S84,250-8
ACETOPHENONE (2,4-DINITROPHENYL)-HYDRAZONE

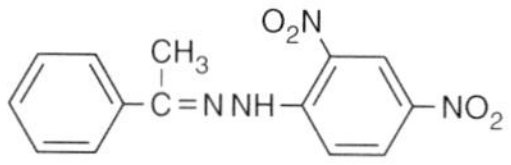

250 mg

S81,930-1
BENZOPHENONE (2,4-DINITROPHENYL)-HYDRAZONE

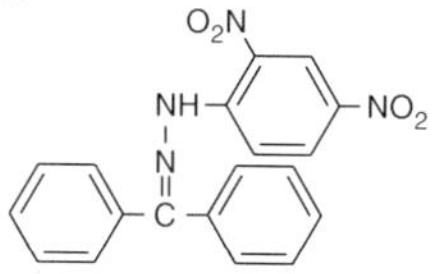

250 mg

S67,234-3
N-(2-BIPHENYLYL)PICRYLAMINE

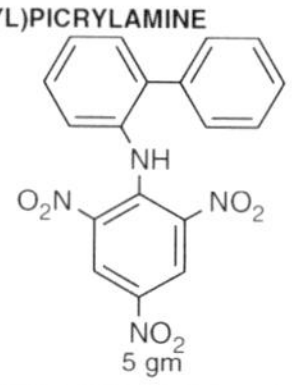

5 gm

S62,188-9
1-(2,4,6-TRINITROPHENYL)PIPERIDINE

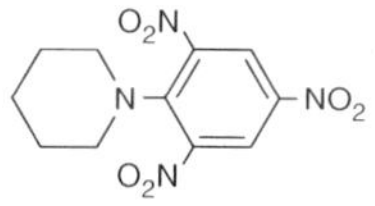

250 mg

S53,430-7
1-CHLORO-2,4-DIPIPERIDINO-5-NITROBENZENE

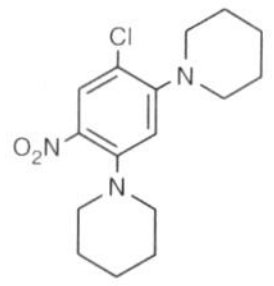

250 mg

S60,572-7
1-CHLORO-3,5-DIPIPERIDINO-2-NITROBENZENE

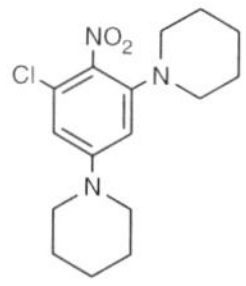

5 gm

S54,198-2
4,4'-TRIMETHYLENEBIS[1-(2,4-DINITROPHENYL)PIPERIDINE]

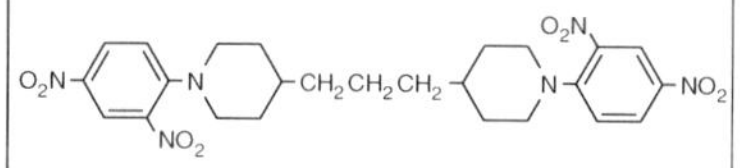

5 gm

S62,187-0
4-(2,4,6-TRINITROPHENYL)MORPHOLINE

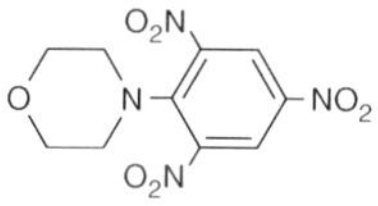

5 gm

S65,380-2
N-PICRYL-O-TOLUIDINE

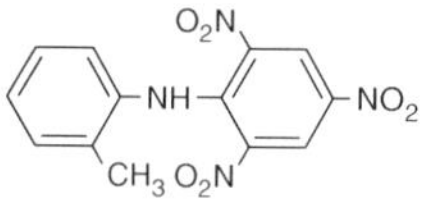

250 mg

S73,011-4
N-PHENETHYLPICRYLAMINE

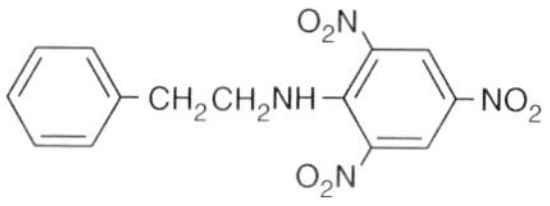

5 gm

S35,848-7
1-BENZYL-4-(2-NITROANILINO)-PIPERIDINE

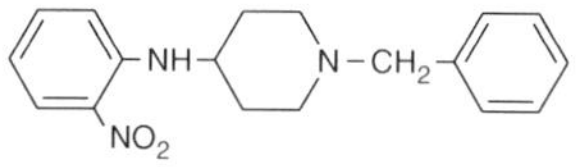

5 gm

S55,816-8
N-(2-METHOXYPHENYL)PICRYLAMINE

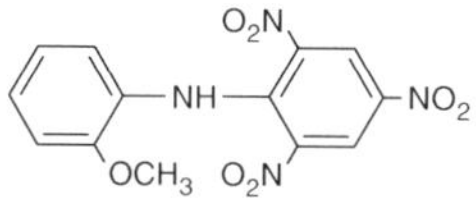

5 gm

S50,059-3
2,4-DINITRO-N-METHYLANILINE

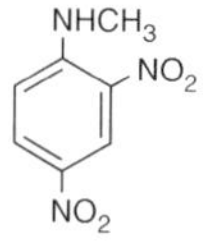

250 mg

S37,128-9
N-BUTYL-2,4-DINITROANILINE

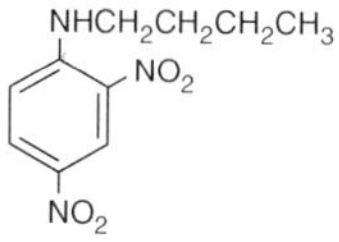

5 gm

S41,730-0
2,4-DINITRO-3'-METHYLDIPHENYLAMINE

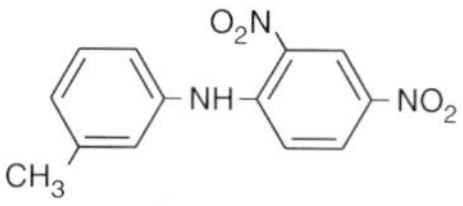

5 gm

S50,318-5
N-PICRYL-3,4-XYLIDINE

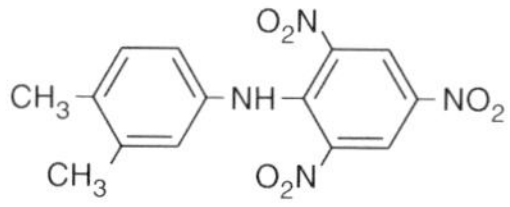

5 gm

S50,319-3
N-PICRYL-2,4-XYLIDINE

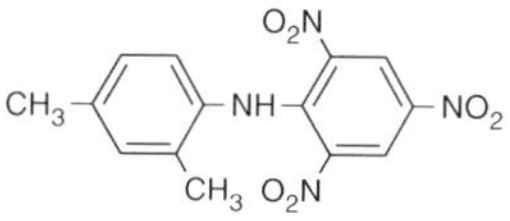

5 gm

S71,906-4
N-PICRYLBENZYLAMINE

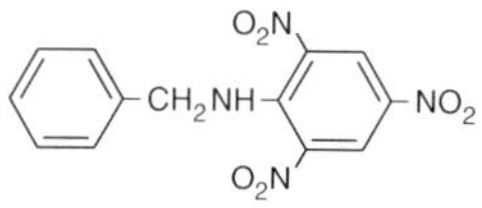

5 gm

S76,424-8
N,N'-DODECAMETHYLENEBIS(2,4,6-TRINTROANILINE)

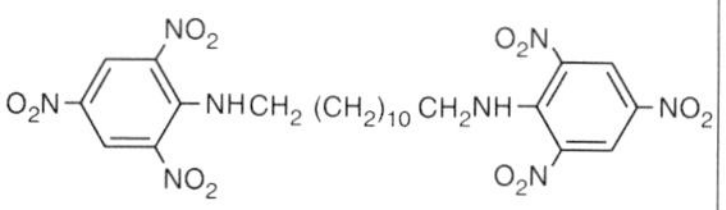

5 gm

S73,012-2
N-CYCLOHEXYLPICRYLAMINE

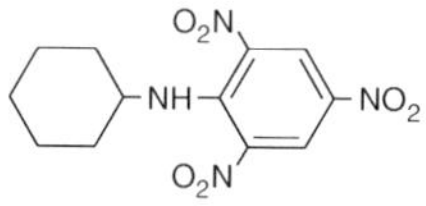

5 gm

S73,995-2
N-(3-MORPHOLINOPROPYL)-2,4,6-TRINITROANILINE

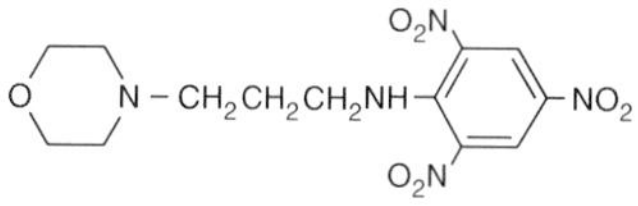

5 gm

S68,522-4
4,4'-METHYLENEBIS[N-(2,4,6-TRINITROPHENYL)ANILINE]

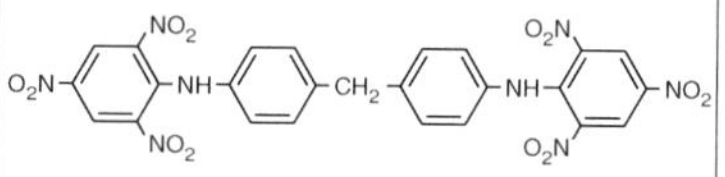

5 gm

S74,660-6
3,5-DIMETHYL-2',4'-DINITRODIPHENYLAMINE

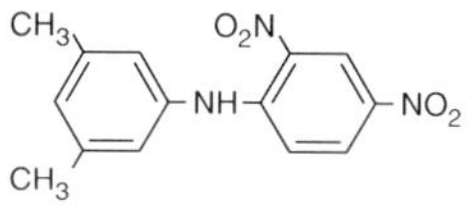

5 gm

S72,062-3
N-DECYL-2,4,6-TRINITROANILINE

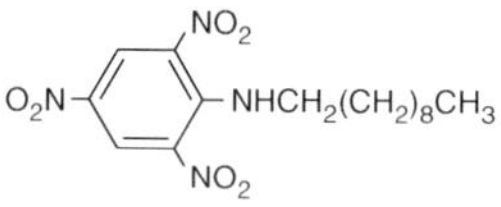

5 gm

S94,734-2
N,N'-BIS(2,4-DINITROPHENYL)-1,2-PHENYLENEDIAMINE

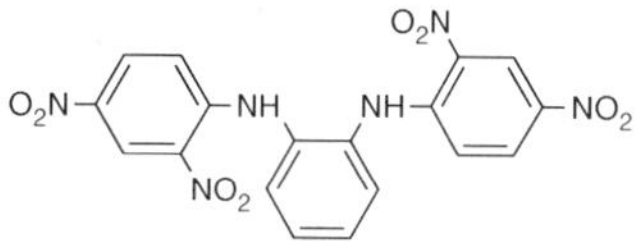

250 mg

S93,575-1
N,N'-BIS(2,4-DINITROPHENYL)-ETHYLENEDIAMINE

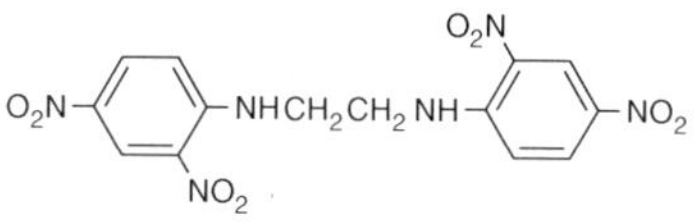

250 mg

S94,733-4
N,N'-BIS(2,4-DINITROPHENYL)-1,3-PHENYLENEDIAMINE
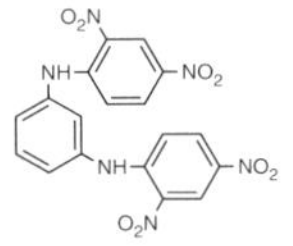
250 mg

S62,321-0
2-CHLORO-4-NITRO-N-(PICRYL)ANILINE
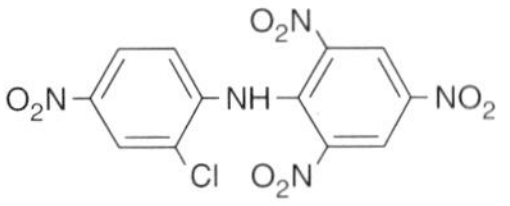
250 mg

S62,322-9
N-PICRYL-2,4,5-TRICHLOROANILINE
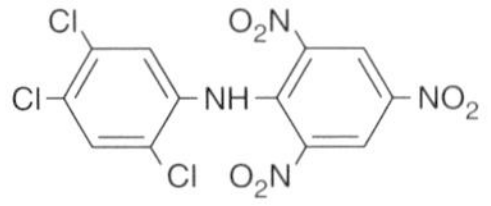
5 gm

S90,784-7
N,N-DIETHYL-2,4-DINITROANILINE
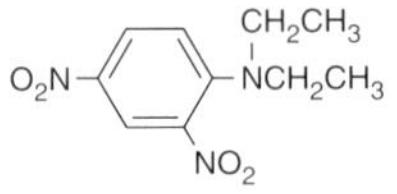
250 mg

S72,060-7
1-(2,4,6-TRINITROPHENYL)PYRROLIDINE
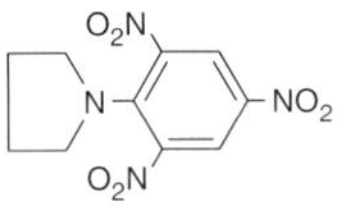
5 gm

S90,836-3
1-(2,4-DINITROPHENYL)-2-METHYLPIPERIDINE
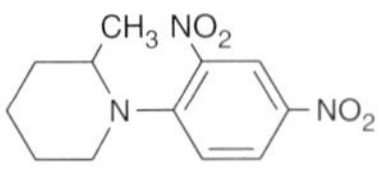
250 mg

S94,379-7
N-BUTYL-2,4-DINITRO-N-METHYL-1-NAPHTHYLAMINE
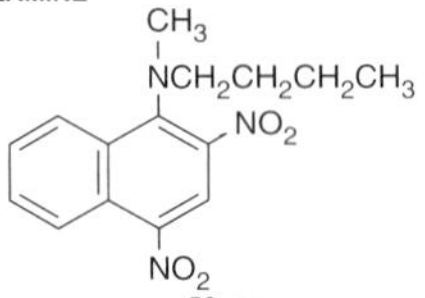
50 mg

S85,251-1
PHENYLACETALDEHYDE (2,4-DINITROPHENYL)HYDRAZONE
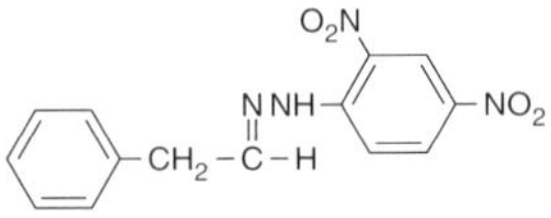
250 mg

S84,251-6
2-PROPANONE N2-(2,4-DINITROPHENYL)-HYDRAZONE
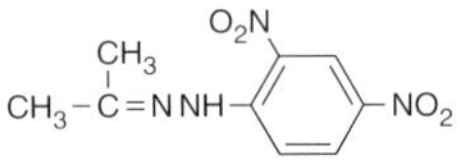
250 mg

S94,843-8
2-ADAMANTANONE (2,4-DINITROPHENYL)-HYDRAZONE
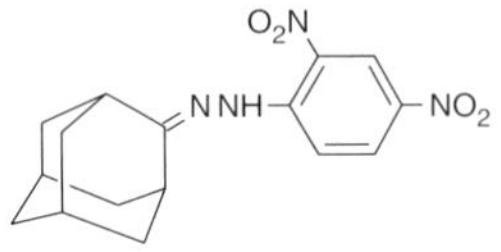
250 mg

S94,840-3
1-ADAMANTYL METHYL KETONE (2,4-DINITROPHENYL)HYDRAZONE
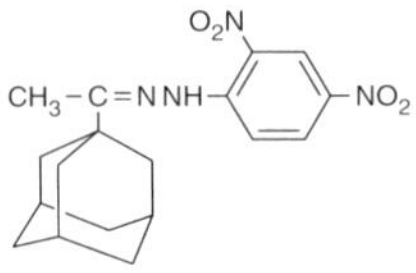
250 mg

S94,841-1
1-ADAMANTYL ETHYL KETONE (2,4-DINITROPHENYL)HYDRAZONE
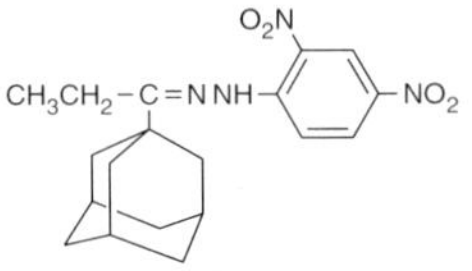
250 mg

S54,982-7
1-(2,4-DINITROPHENYL)-4-PROPYLPIPERIDINE
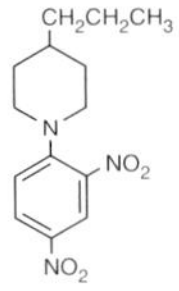
5 gm

S94,472-6
N-(2,4-DINITRO-6-METHYLPHENYL)-PIPERIDINE
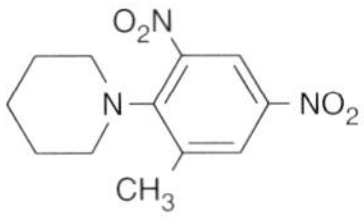
250 mg

S78,221-1
5,7-DINITROINDOLINE
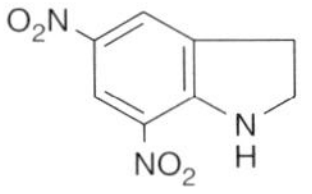
5 gm

S93,573-5
N,N'-DIMETHYL-N,N'-BIS(4-NITROPHENYL)ETHYLENEDIAMINE
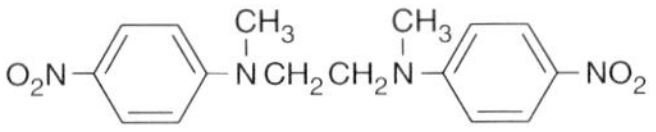
250 mg

S38,680-4
2,6-DINITRO-N-ETHYLANILINE
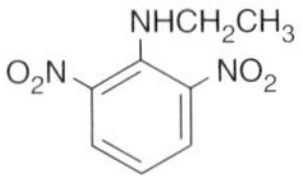
250 mg

S80,733-8
2-[2-(2,4-DINITROANILINO)VINYL]-BENZALDEHYDE PHENYLHYDRAZONE
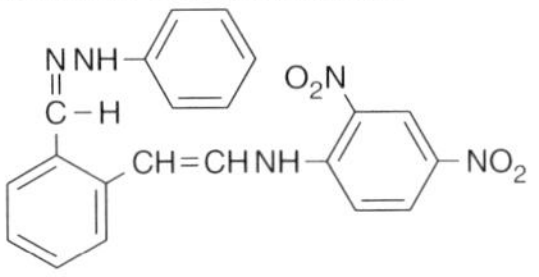
250 mg

S89,976-3
N-CYCLOHEXYL-2,6-DINITRO-4-METHYLANILINE
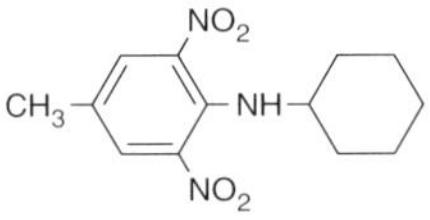
250 mg

S69,545-9
N,N'-METHYLENEBIS(2-NITROANILINE)
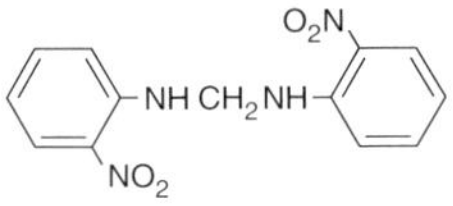
5 gm

S89,638-1
(2-TERT-BUTYLPHENYL)(2-NITROPHENYL)AMINE
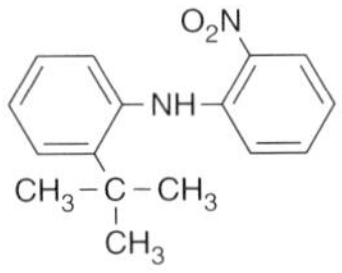
250 mg

S93,578-6
1,3-BIS[(2-NITROPHENYL)-AMINOMETHYL]BENZENE
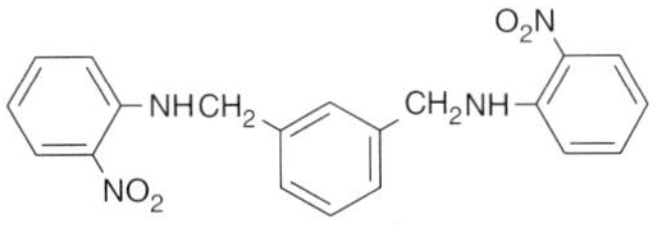
250 mg

S53,391-2
N,N'-BIS(2-NITROPHENYL)-ETHYLENEDIAMINE
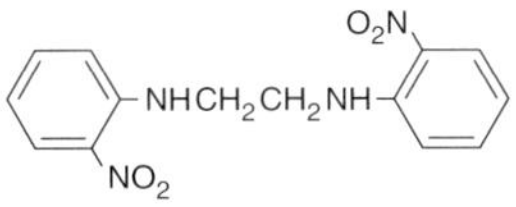
250 mg

S53,620-2
4,5-DICHLORO-N-ETHYL-2-NITROANILINE
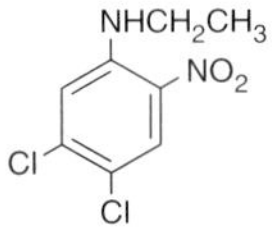
5 gm

S90,797-9
1-METHYL-4-(2-NITROPHENYL)-PIPERAZINE
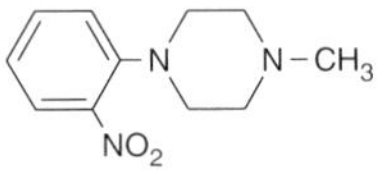
250 mg

S50,065-8
1-(4-BROMO-2-NITROPHENYL)-PYRROLIDINE
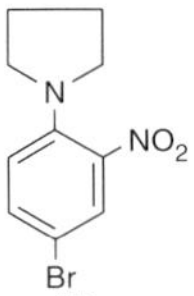
250 mg

S57,833-9
1-NITRO-2,4-DIPIPERIDINOBENZENE
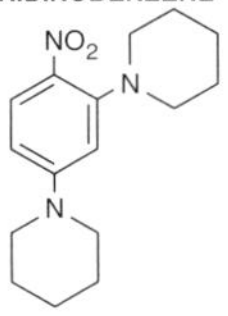
5 gm

S93,577-8
N,N'-DIMETHYL-N,N'-BIS(2-NITROPHENYL)ETHYLENEDIAMINE
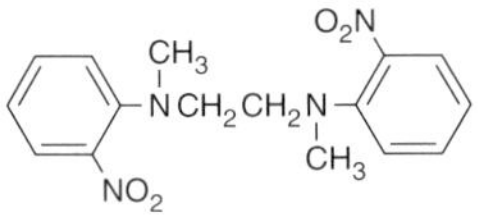
250 mg

S41,167-1
2-DIMETHYLAMINO-1-NITROFLUORENE
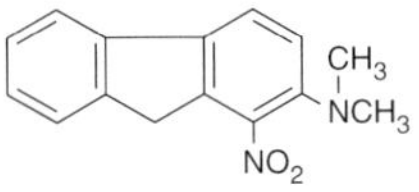
250 mg

S82,069-5
N,N,N',N'-TETRABUTYLETHYLENEDIAMINE PICRATE
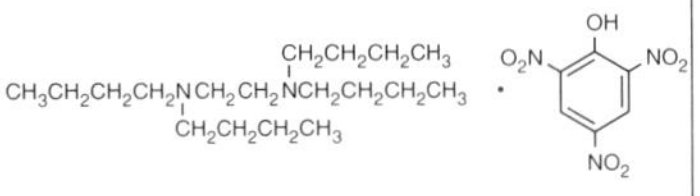
250 mg

S81,534-9
10-(2-NITROPHENYL)PHENOTHIAZINE
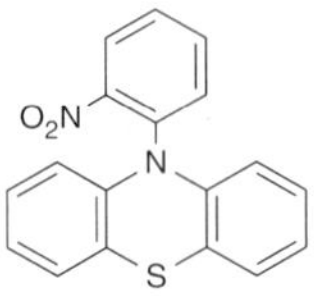
250 mg

S54,201-6
1-(4-BROMO-2-NITROPHENYL)-PIPERIDINE
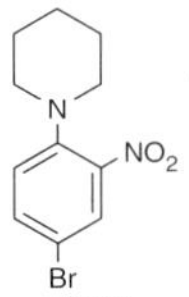
5 gm

S53,958-9
4-METHYL-1-(2-NITROPHENYL)-PIPERIDINE
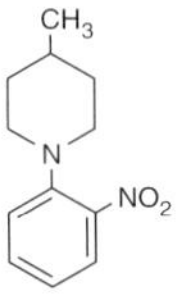
5 gm

S53,914-7
4,4'-TRIMETHYLENEBIS[1-(2-NITROPHENYL)PIPERIDINE]
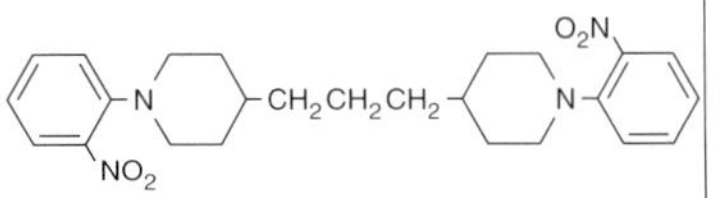
250 mg

S82,066-0
2,3,7-TRINITRODIBENZO-P-DIOXIN
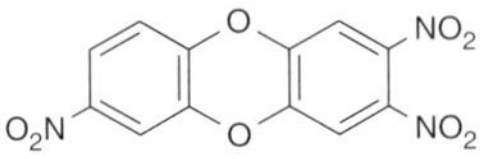
250 mg

S55,946-6
2,4,6-TRINITRODIPHENYLAMINE
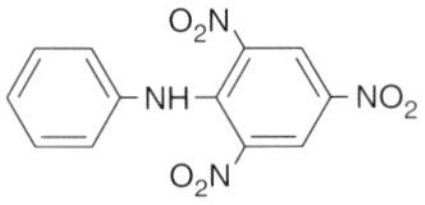
5 gm

S43,402-7
N,N-DIMETHYL-2,4-DINITROSTYRYLAMINE
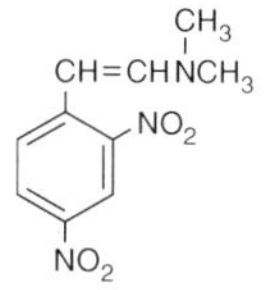
5 gm

S51,314-8
N4,N4-DIMETHYL-N-ALPHA-(3-NITROPHENYL)TOLUENE-ALPHA,4-DIAMINE
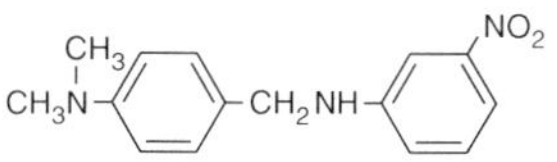
5 gm

S87,839-1
4-(DIMETHYLAMINO)-N-(4-METHYL-3-NITROPHENYL)BENZYLAMINE
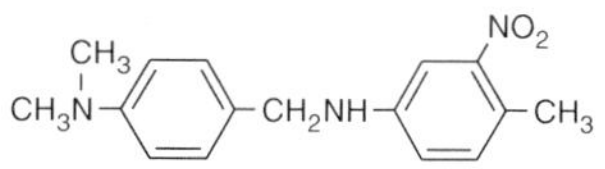
5 gm

S69,294-8
N,N'-METHYLENEBIS(3-NITROANILINE)
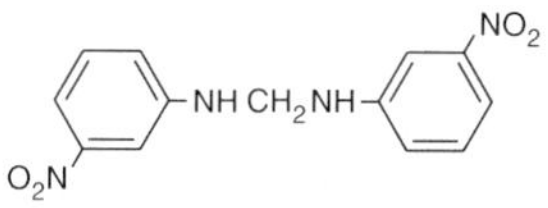
5 gm

S87,864-2
4-BUTYL-2',4'-DINITRODIPHENYLAMINE
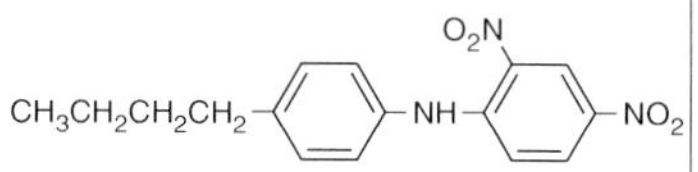
5 gm

S75,809-4
N-(4-METHYL-3-NITROPHENYL)-BENZYLAMINE
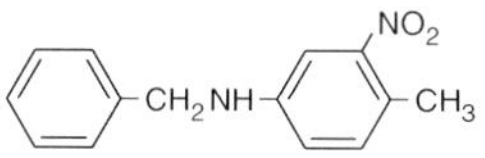
250 mg

S54,618-6
N-(3-NITROPHENYL)-4-METHYLBENZYLAMINE
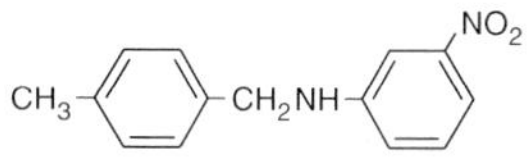
250 mg

S94,735-0
N-SEC-BUTYL-2,4-DINITRO-1-NAPHTHYLAMINE
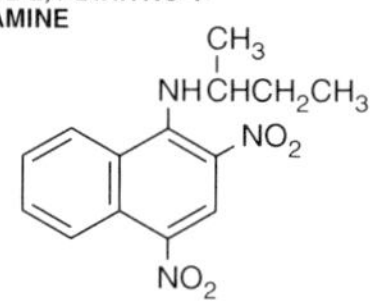
250 mg

S67,187-8
N,N-DIETHYL-3-NITROANILINE
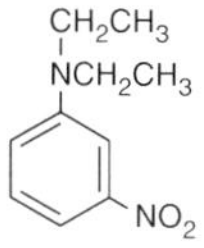
5 gm

S55,333-6
N,N-DIMETHYL-2',4'-DINITRO-4-STILBENAMINE
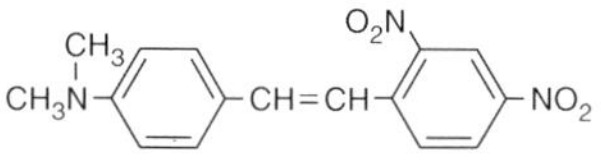
5 gm

S55,577-0
N,N-DIETHYL-4-(2,4-DINITROSTYRYL)-ANILINE
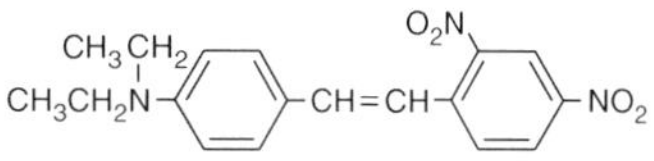
5 gm

S59,233-1
4,4'-METHYLENEBIS(N,N-DIMETHYL-3-NITROANILINE)
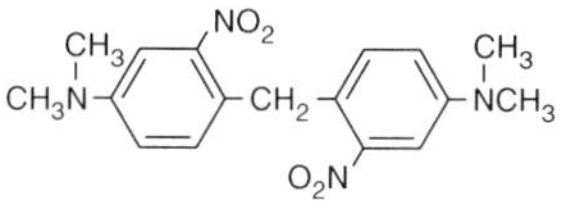
5 gm

S93,752-5
2-CHLORO-10-(3-DIMETHYLAMINO)-PROPYL-10H-PHENOTHIAZINE-5-OXIDE PICRATE
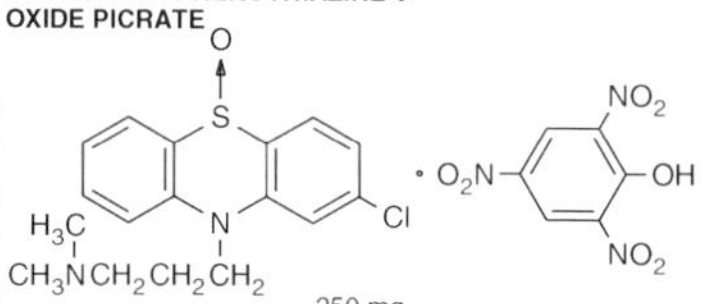
250 mg

S50,526-9
1,1'-TRIMETHYLENEBIS[1-(4-NITROBENZYL)PYRROLIDINIUM CHLORIDE]
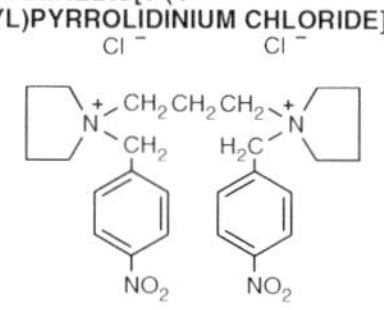
250 mg

S64,730-6
1,2-BIS(4-NITRO-1,1,3,3,5-PENTAMETHYL-6-INDANYL)HYDRAZINE
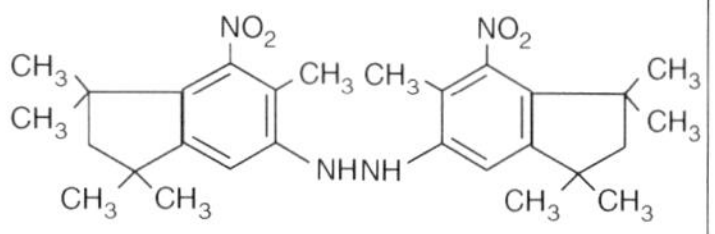
250 mg

S54,898-7
2,4-DINITROMESITYLENE
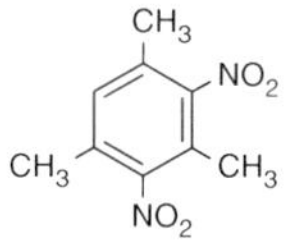
250 mg

S34,297-1
1-TERT-BUTYL-3,5-DIMETHYL-2,4,6-TRINITROBENZENE
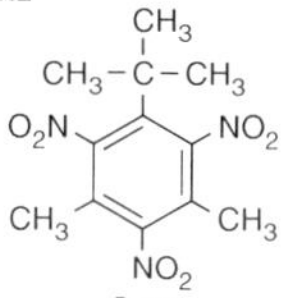
5 gm

S35,916-5
5-TERT-BUTYL-4,6-DINITRO-1,2,3-TRIMETHYLBENZENE
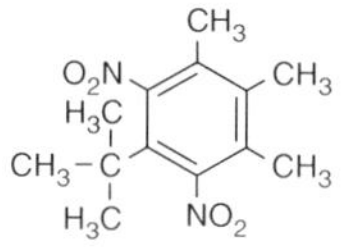
5 gm

S76,719-0
3,4-DIHYDRO-4-(2-NITROPHENYL)-QUINAZOLINE
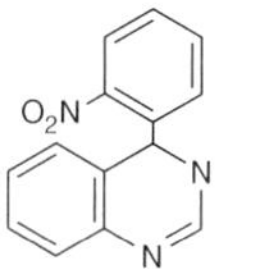
250 mg

S91,439-8
2,4,6-TRINITROSTILBENE
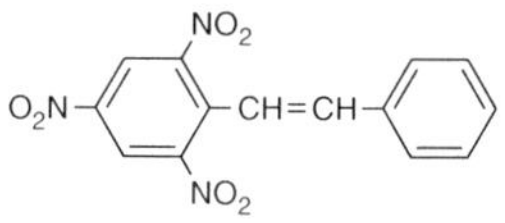
250 mg

S87,631-3
BIS(2,4-DINITROPHENYL)METHANE
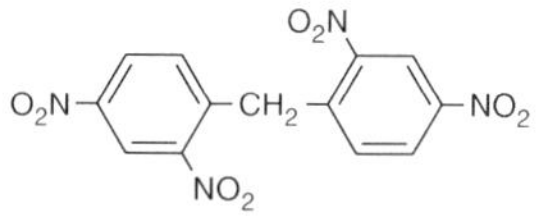
250 mg

S54,588-0
1,3-DISTYRYL-2,4,6-TRINITROBENZENE
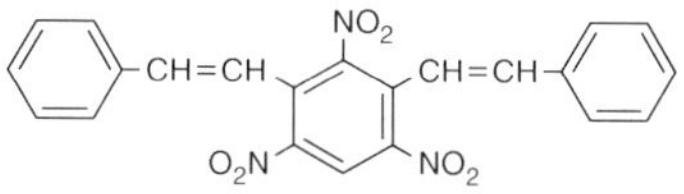
250 mg

S55,401-4
2,4-DINITRO-2'-FLUOROSTILBENE
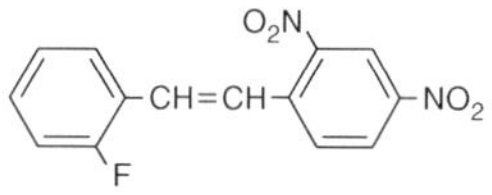
250 mg

S55,565-7
2'-CHLORO-2,4-DINITROSTILBENE
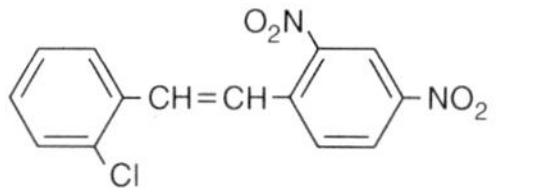
250 mg

S55,403-0
2,2',4-TRINITROSTILBENE
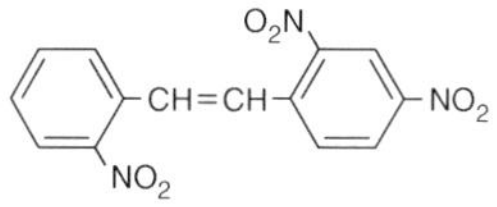
5 gm

S55,413-8
2'-BROMO-2,4-DINITROSTILBENE
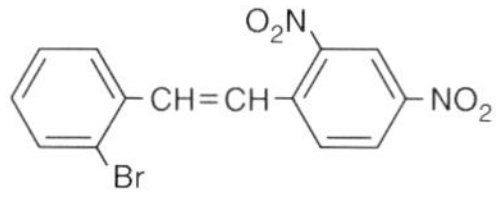
250 mg

S55,335-2
2,4-DINITRO-4'-FLUOROSTILBENE
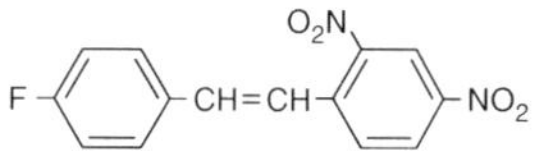
5 gm

S55,405-7
4'-BROMO-2,4-DINITROSTILBENE
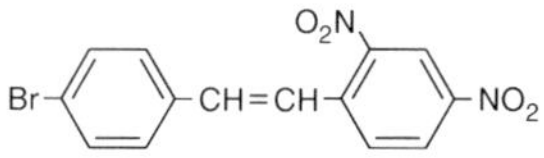
250 mg

S55,334-4
2,6-DICHLORO-2',4'-DINITROSTILBENE
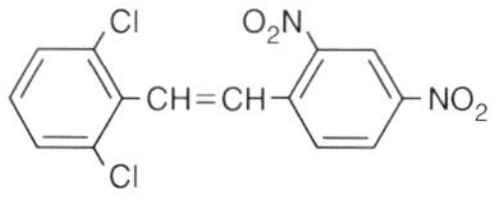
5 gm

S56,191-6
1-BROMO-2,4-DINITROBENZENE
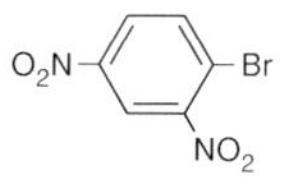
250 mg

S55,398-0
2,4-DINITRO-3'-FLUOROSTILBENE
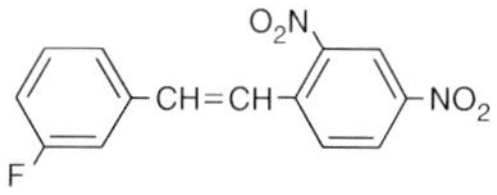
250 mg

S55,402-2
3'-CHLORO-2,4-DINITROSTILBENE
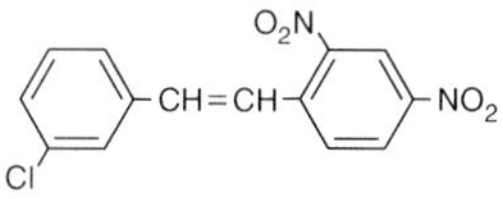
250 mg

S55,576-2
2,3',4-TRINITROSTILBENE
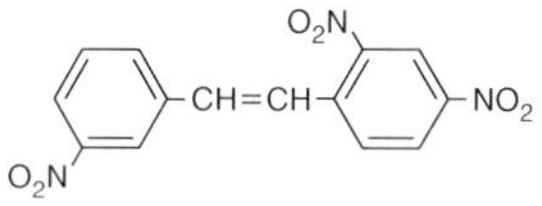
250 mg

S55,484-7
2,4-DICHLORO-2',4'-DINITROSTILBENE
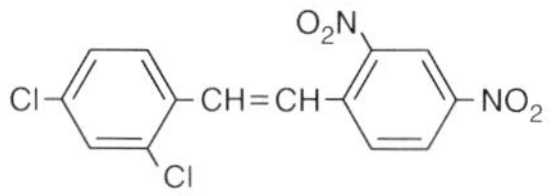
5 gm

S55,400-6
3'-BROMO-2,4-DINITROSTILBENE
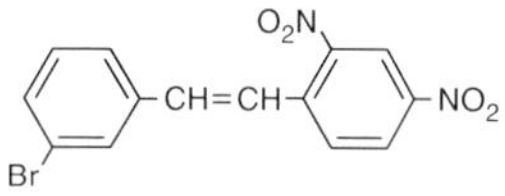
250 mg

S55,570-3
1-(2,4-DINITROPHENYL)-4-PHENYL-1,3-BUTADIENE
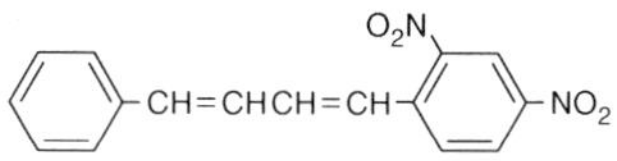
250 mg

S65,620-8
2,4-DINITRO-6-(2,4-DINITROBENZYL)-CHLOROBENZENE

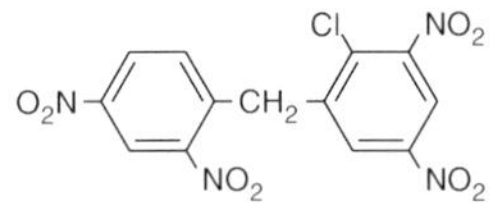

5 gm

S61,374-6
1-(2,4-DINITROSTYRYL)NAPHTHALENE

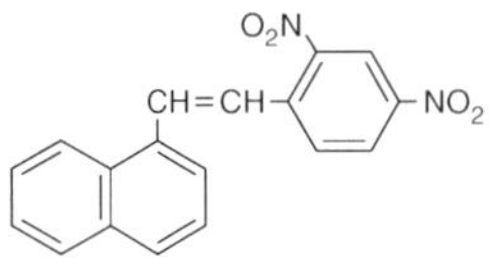

5 gm

S61,951-5
BETA-(9-ANTHRACENYL)-2,4-DINITROSTYRENE

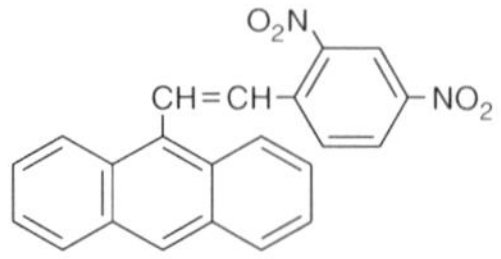

5 gm

S36,118-6
3,5-DINITROIODOBENZENE

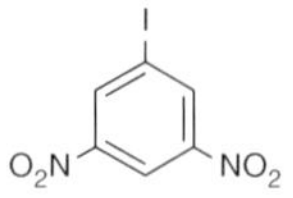

250 mg

S55,102-3
1,3-DICHLORO-4,6-DINITROBENZENE

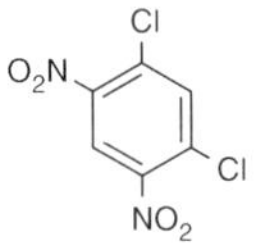

5 gm

S35,891-6
3-BROMO-4,6-DINITROFLUOROBENZENE

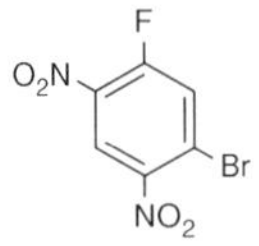

5 gm

S44,011-6
4,6-DINITRO-M-XYLENE

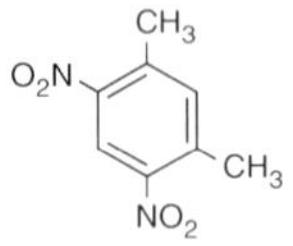

250 mg

S34,750-7
4,6-DINITRO-1,1,3,3,5-PENTAMETHYLINDANE

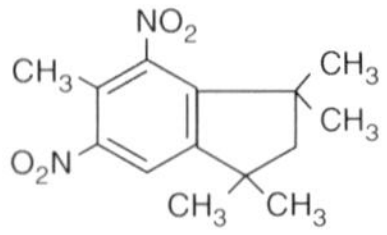

5 gm

S95,598-1
4-METHYL-1,3,8-TRINITRONAPHTHALENE

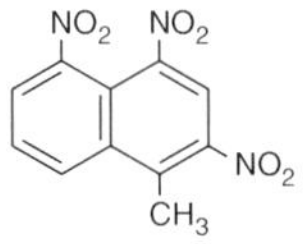

50 mg

S73,443-8
N-(4-NITROPHENYL)PHENETHYLAMINE

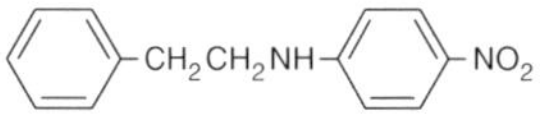

250 mg

S65,782-4
N-BUTYL-4-NITROANILINE

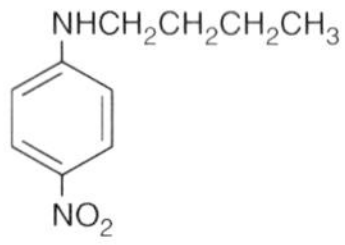

5 gm

S69,514-9
N,N'-METHYLENEBIS(4-NITROANILINE)

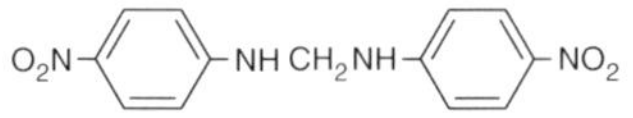

5 gm

S93,569-7
N-(4-NITROPHENYL)BENZYLAMINE

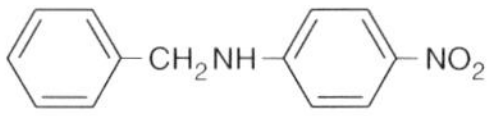

250 mg

S51,399-7
N,N,2,3,5,6-HEXAMETHYL-4-NITROANILINE

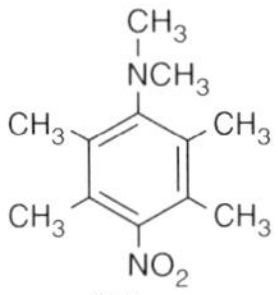

250 mg

S57,128-8
N,N-DIMETHYL-P-NITROANILINE

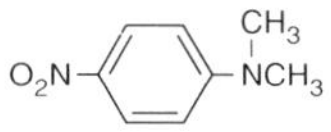

250 mg

S90,783-9
N,N-DIETHYL-4-NITROANILINE

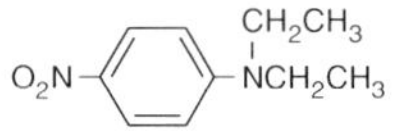

250 mg

S83,936-1
N,N'-TETRAMETHYLENEBIS(N-METHYLANILINE) PICRATE

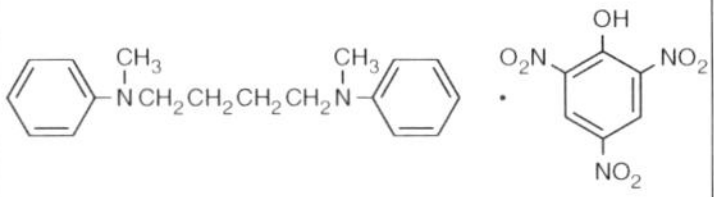

250 mg

S90,835-5
2-METHYL-1-(4-NITROPHENYL)-PIPERIDINE

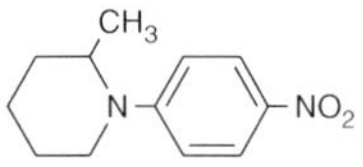

250 mg

S90,826-6
3,5-DIMETHYL-1-(4-NITROPHENYL)-PIPERIDINE

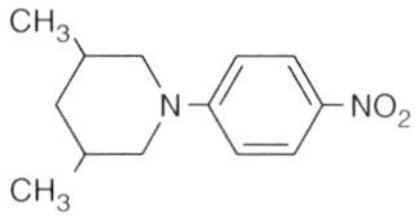

250 mg

S90,767-7
1-(4-NITROPHENYL)-4-PHENYLPIPERAZINE

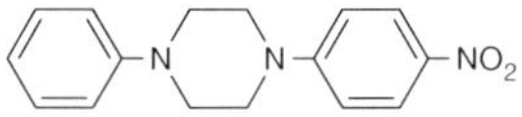

250 mg

S90,798-7
1-METHYL-4-(4-NITROPHENYL)-PIPERAZINE

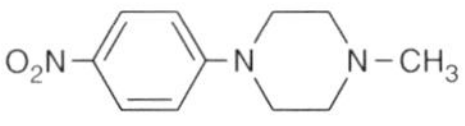

250 mg

S90,833-9
2,6-DIMETHYL-4-(4-NITROPHENYL)-MORPHOLINE

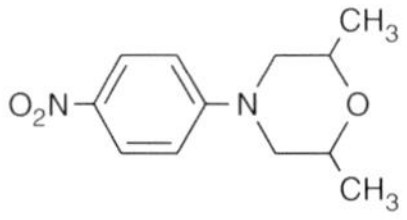

250 mg

S90,834-7
HEXAHYDRO-1-(4-NITROPHENYL)-1H-AZEPINE

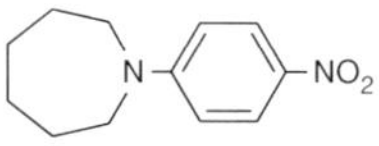

250 mg

S90,827-4
1-(4-NITROPHENYL)OCTAHYDRO-AZOCINE

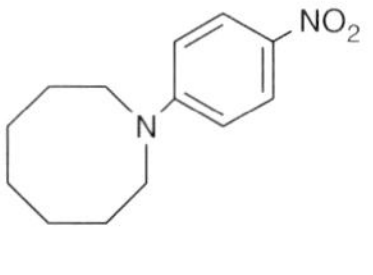

250 mg

S63,277-5
1-(4-NITROPHENYL)PYRROLIDINE

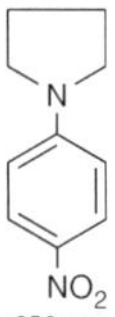

250 mg

S93,555-7
1,4-BIS(4-NITROPHENYL)PIPERAZINE

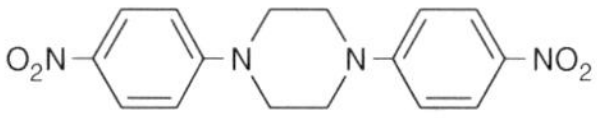

250 mg

S44,762-5
4-(4-NITROPHENYL)MORPHOLINE

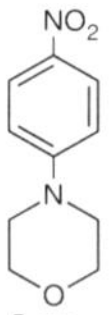

5 gm

S63,274-0
N-DODECYL-N-METHYL-4-NITROANILINE

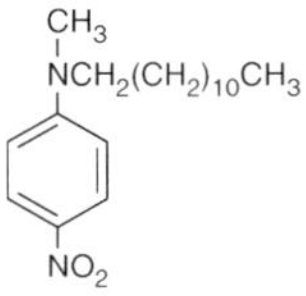

250 mg

S50,032-1
N-HEXADECYL-N-METHYL-4-NITROANILINE

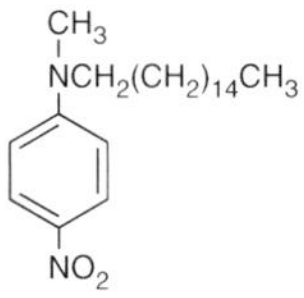

250 mg

S42,838-8
2-CHLORO-N,N-DIETHYL-4-NITROANILINE

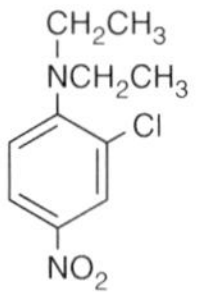

5 gm

S91,905-5
(3-NITROBENZYL)-TRIETHYLAMMONIUM CHLORIDE

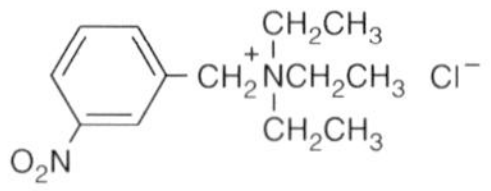

250 mg

S64,797-7
1,2-BIS(4-NITROPHENYL)HYDRAZINE

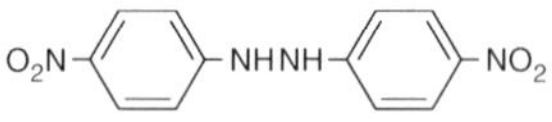

250 mg

S77,028-0
2-CHLOROCYCLOHEXANONE 4-NITROPHENYLHYDRAZONE

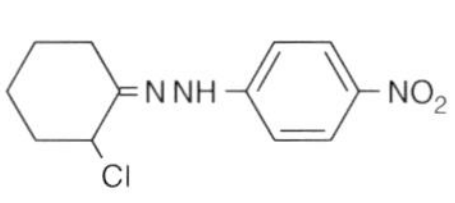

250 mg

S89,725-6
3,5-DIMETHYL-2-CYCLOHEXEN-1-ONE (4-NITROPHENYL)HYDRAZONE

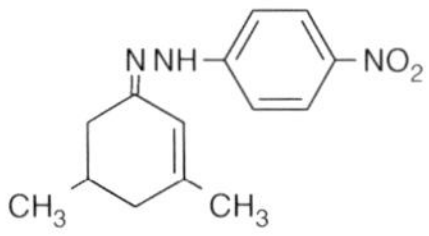

250 mg

S74,487-5
PROPIOPHENONE 4-NITROPHENYLHYDRAZONE

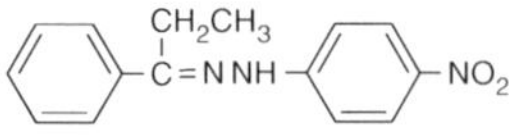

250 mg

S74,510-3
2-BUTANONE 4-NITROPHENYLHYDRAZONE

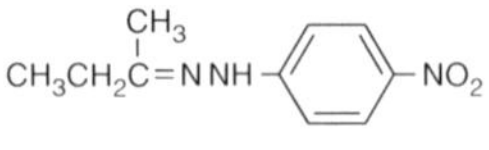

250 mg

S85,466-2
3-NITROBENZALDEHYDE (4-NITROPHENYL)-HYDRAZONE

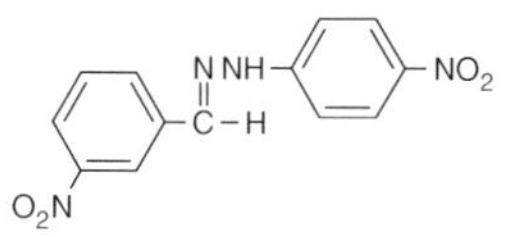

250 mg

S67,954-2
4'-NITROACETOPHENONE 4-NITROPHENYLHYDRAZONE

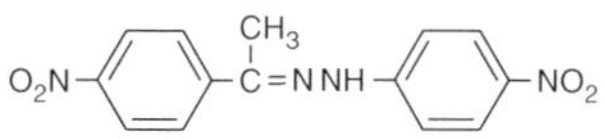

250 mg

S91,902-0
1-(1-METHYL-1-NITROETHYL)-3-NITROBENZENE

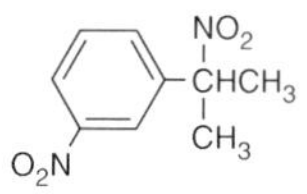

250 mg

S77,274-7
2,3-DIHYDRO-2-(3-NITROPHENYL)-BENZIMIDAZOLE

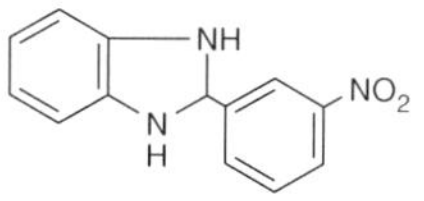

250 mg

S53,919-8
1-(4-NITROPHENYL)PIPERIDINE

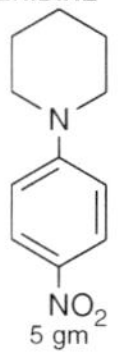

5 gm

S54,199-0
4-METHYL-1-(4-NITROPHENYL)-PIPERIDINE

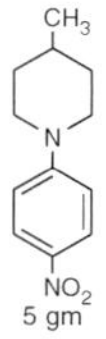

5 gm

S34,417-6
4-CHLORO-1,5-DINITRONAPHTHALENE

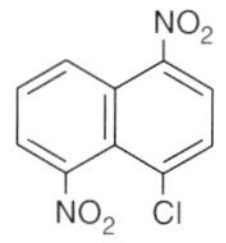

250 mg

S91,821-0
1-NITRO-4-(2-NITRO-1,1,2-TRIMETHYLPROPYL)BENZENE

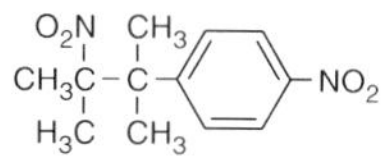

250 mg

S91,755-9
1-(1,2-DIMETHYL-1-ETHYL-2-NITROPROPYL)-4-NITROBENZENE

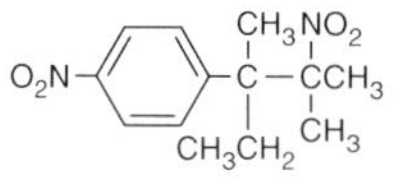

250 mg

S76,346-2
1,3-DIPHENYL-2-NITROPROPANE

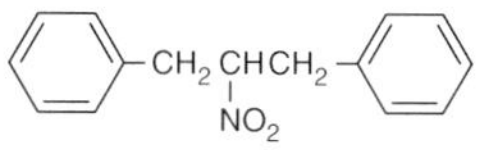

250 mg

S36,415-0
2-NITRO-2-(4-NITROBENZYL)PROPANE

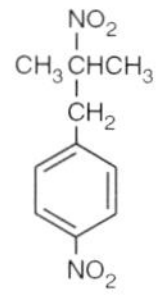

5 gm

S89,559-8
DIETHYLAMINE PICRATE

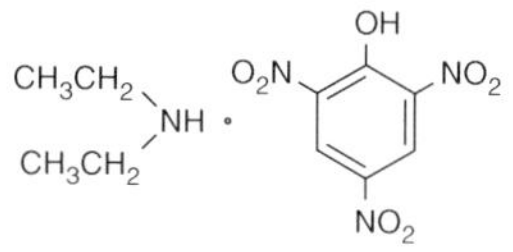

250 mg

S93,130-6
CYCLOHEXYL (4-NITROBENZYL)AMINE

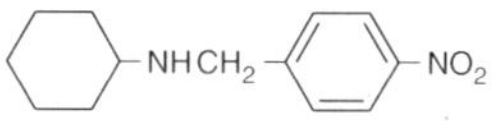

250 mg

S68,005-2
N-ETHYL-4-NITROANILINE

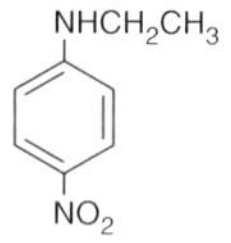

250 mg

S93,129-2
3-METHYL-N-(4-NITROBENZYL)ANILINE

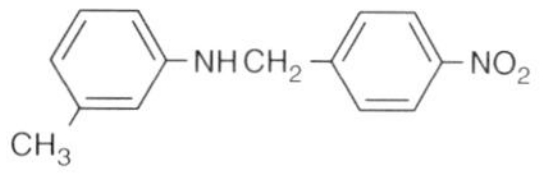

250 mg

S93,127-6
4-NITRODIBENZYLAMINE

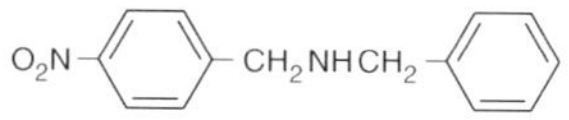

250 mg

S54,884-7
4-METHYL-N-(4-NITROPHENYL)-BENZYLAMINE

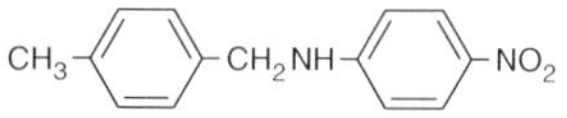

250 mg

S58,691-9
N-ETHYL-5-NITRO-O-TOLUIDINE

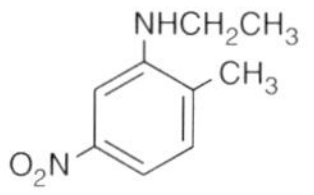

250 mg

S57,260-8
TRIETHYLAMINE N-OXIDE PICRATE

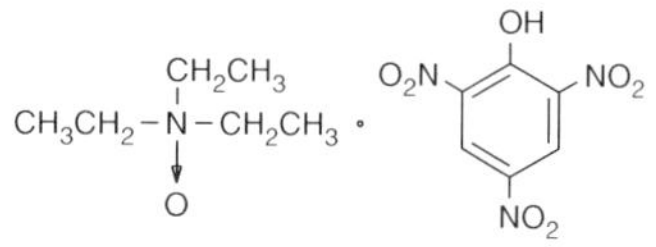

250 mg

S69,724-9
1,4-DIHYDRO-2-METHYL-5-NITROISOQUINOLINIUM TETRAFLUOROBORATE

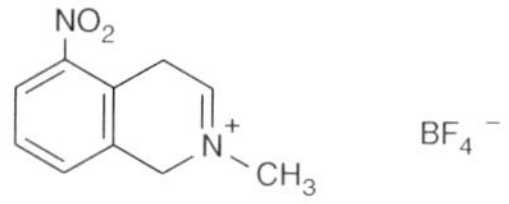

250 mg

S92,348-6
2-[4-(DIMETHYLAMINO)-ALPHA-(4-NITROPHENYLAZO)BENZYLIDENE] 4,5,6-TETRAHYDRO-1,3-BENZODITHIOLE

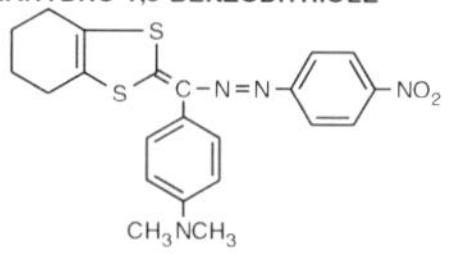

50 mg

S90,773-1
4,4'-[(3-NITROPHENYL)METHYLENE]BIS-(N,N-DIMETHYLANILINE)

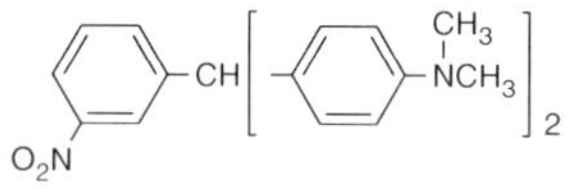

250 mg

S88,075-2
N,N-DIMETHYL-4-NITROBENZYLAMINE

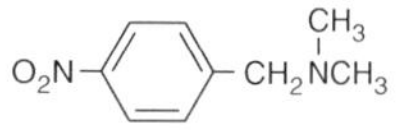

250 mg

S81,110-6
DIETHYL(4-NITROBENZYL)AMINE

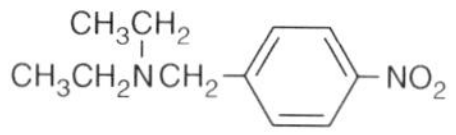

250 mg

S93,554-9
N,N-DIMETHYLBENZYLAMINE-N-OXIDE PICRATE

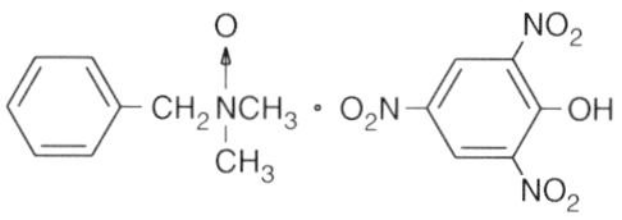

250 mg

S86,404-8
DIETHYL (1-PHENYLETHYL)-AMINE PICRATE

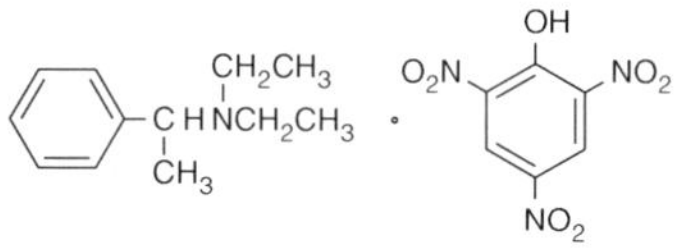

250 mg

S91,826-1
1-[1-METHYL-1-(4-NITROPHENYL)-ETHYL]PIPERIDINE

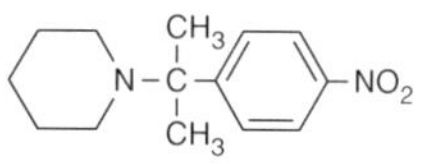

250 mg

S61,375-4
1-METHYL-4-(4-NITROBENZYLIDENE)-1,4-DIHYDROPYRIDINE

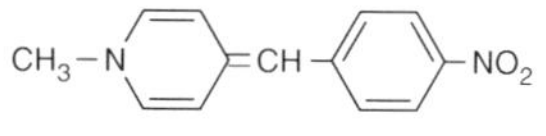

5 gm

S60,863-7
4-DIMETHYLAMINO-4'-NITROSTILBENE

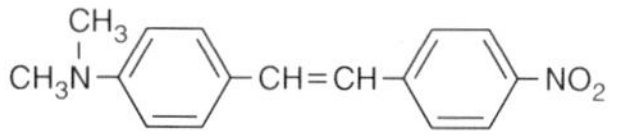

250 mg

S65,591-0
BIS(4-DIMETHYLAMINOPHENYL)(4-NITROPHENYL)METHANE

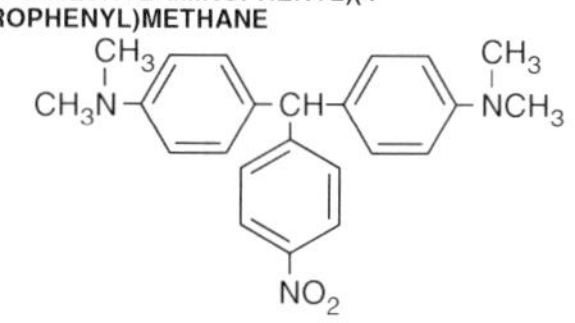

250 mg

S75,318-1
2-(DIMETHYLAMINO)-7-NITROFLUORENE

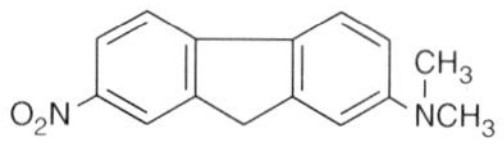

250 mg

S50,529-3
1-METHYL-1-(4-NITROBENZYL)-PIPERIDINIUM CHLORIDE

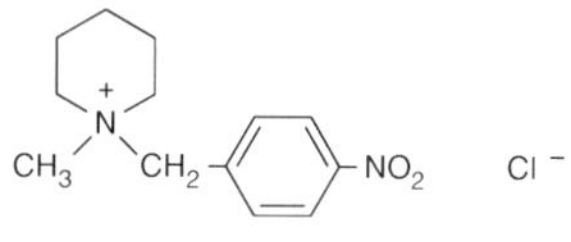

250 mg

S88,617-3
3-NITROBENZALDEHYDE METHYLPHENYLHYDRAZONE

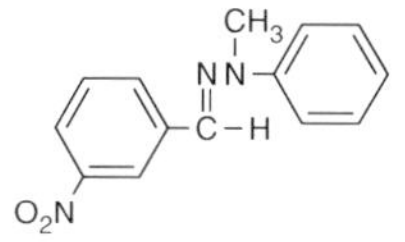

250 mg

S80,109-7
3-NITROBENZALDEHYDE DIMETHYLHYDRAZONE

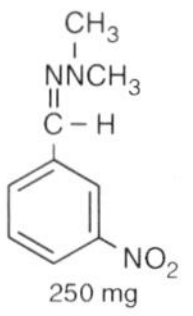

250 mg

S76,027-7
5-(5-METHYL-5-NITRO-1,3-DIOXAN-2-YL)-1,3-BENZODIOXOLE

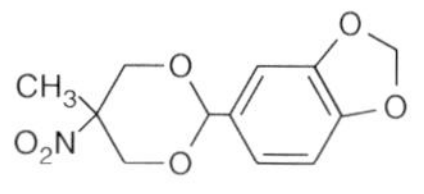

250 mg

S83,633-8
2-NITROPHENYL M-TOLYL SULFIDE

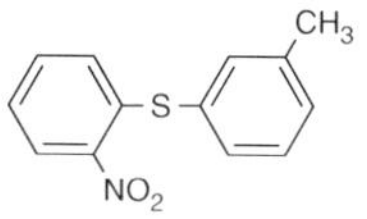

250 mg

S82,070-9
1-ISOPROPYL-3-NITROBENZENE
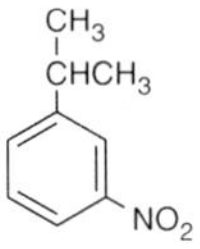
250 mg

S50,916-7
1-(TERT-BUTYL)-3-NITROBENZENE
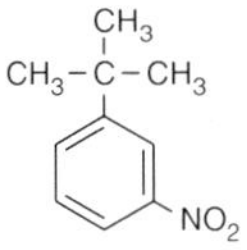
250 mg

S67,525-3
3-(4-NITROPHENYL)-5,6-DIHYDRO-7H-THIAZOLO(3,2-A)PYRIMIDINE
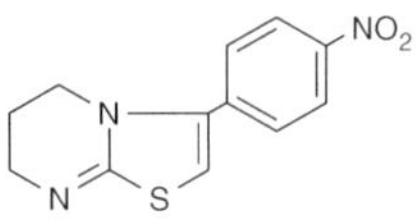
250 mg

S69,744-3
3-(2-NITROPHENYL)PROPYLENE OXIDE
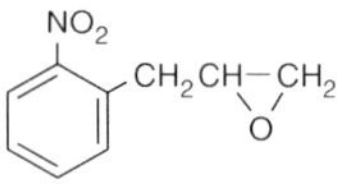
250 mg

S92,652-3
1-(4-NITROPHENYL)AZETIDINE
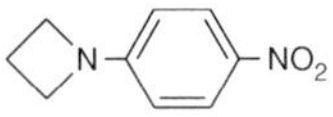
250 mg

S79,223-3
4-CYCLOHEXYL-2-NITROBIPHENYL
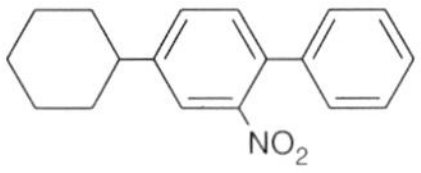
250 mg

S70,646-9
ALPHA-PHENYL-2-NITROTOLUENE
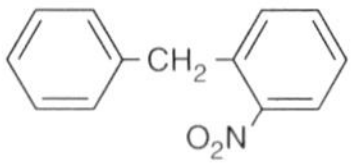
5 gm

S91,420-7
4-NITROSTILBENE
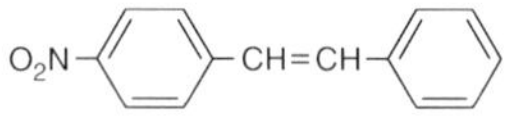
250 mg

S88,222-4
1-(4-NITROPHENYL)-4-PHENYL-1,3-BUTADIENE
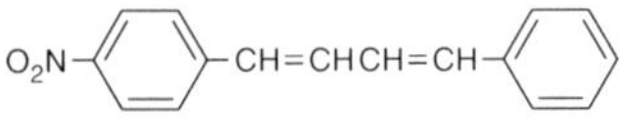
250 mg

S91,779-6
4-NITROBENZALDEHYDE DIPHENYL MERCAPTAL
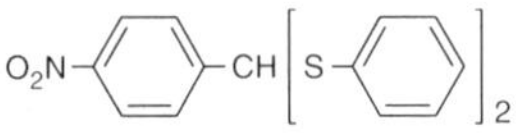
250 mg

S92,108-4
3,5-DIPHENYL-4'-METHYL-2-NITROBIPHENYL
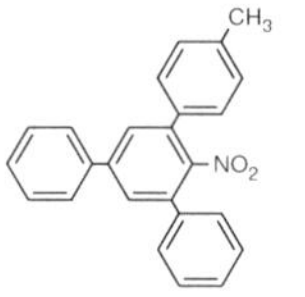
250 mg

S92,078-9
3,5-BIS(P-TOLYL)-4-NITROBIPHENYL
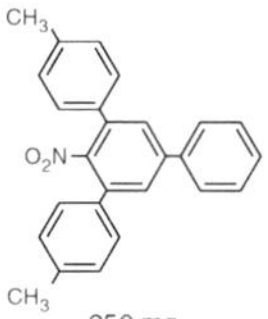
250 mg

S91,750-8
1-(ALPHA,ALPHA-DIMETHYLBENZYL)-4-NITROBENZENE
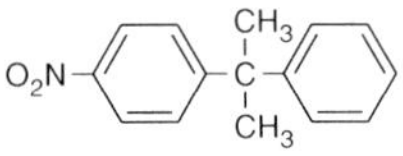
250 mg

S84,672-4
4-NITRO-P-TERPHENYL
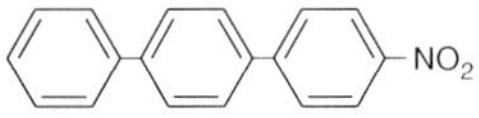
250 mg

S69,757-5
4'-NITRO-M-TERPHENYL
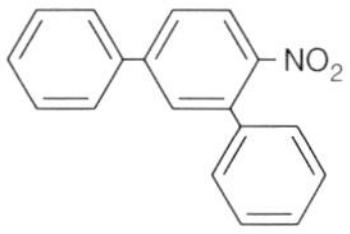
250 mg

S71,160-8
2,6-DIMETHYL-4-PHENYLNITROBENZENE
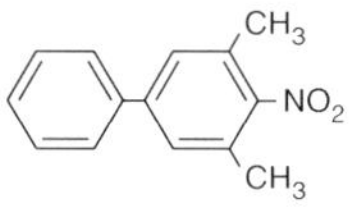
250 mg

S92,126-2
N-(3,5-DIMETHYL-4-NITROBENZYLIDENE)-ANILINE
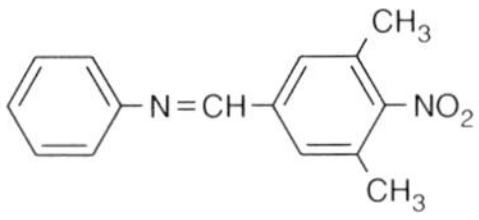
250 mg

S91,996-9
3,4'-DINITRO-2,3',4,5',6-PENTAPHENYLBIPHENYL
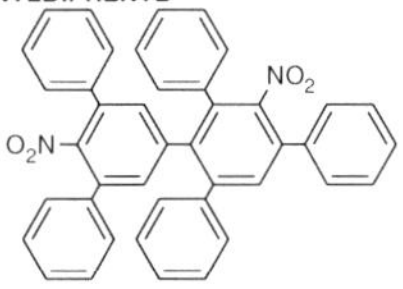
250 mg

S87,799-9
1,6-BIS(2-NITROPHENYL)-1,3,5-HEXATRIENE
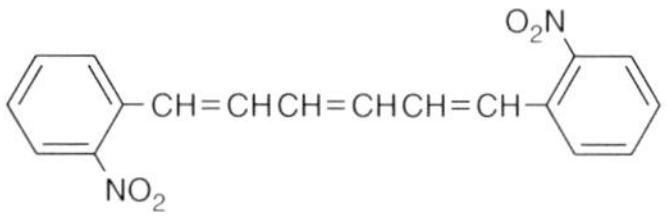
250 mg

S82,968-4
2,4'-DINITROBIPHENYL
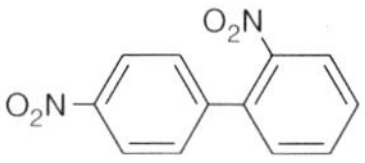
250 mg

S75,400-5
9-(2-NITROCINNAMYLIDENE)FLUORENE
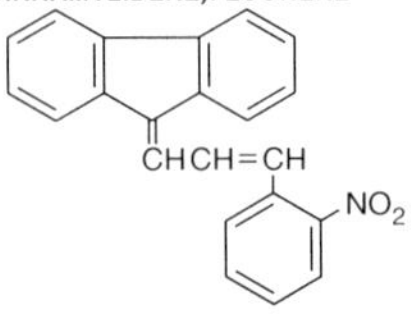
250 mg

S79,983-1
1-(1-CHLOROETHYL)-4-NITROBENZENE
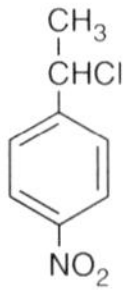
250 mg

S77,451-0
1-(1,1-DIFLUOROETHYL)-4-NITROBENZENE
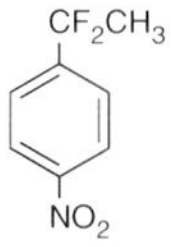
250 mg

S86,472-2
TRANS-BETA-BROMO-4-NITROSTYRENE
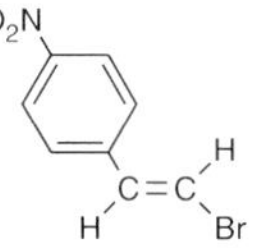
250 mg

S91,814-8
1-TERT-BUTYL-4-NITROBENZENE
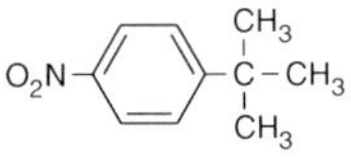
250 mg

S91,765-6
1-(1-METHYL-1-NITROPROPYL)-4-NITROBENZENE
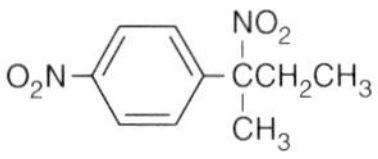
250 mg

S73,393-8
2-TERT-BUTYL-3-(4-NITROPHENYL)-OXAZIRIDINE
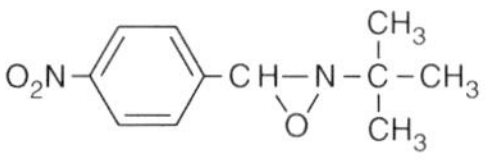
250 mg

S91,761-3
1-(2,2-DIMETHYL-1-NITROPROPYL)-4-NITROBENZENE
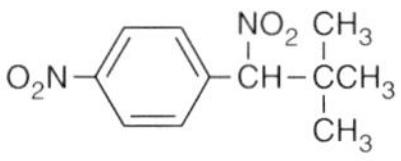
250 mg

S88,013-2
1,4-BIS(4-NITROPHENYL)-1,3-BUTADIENE
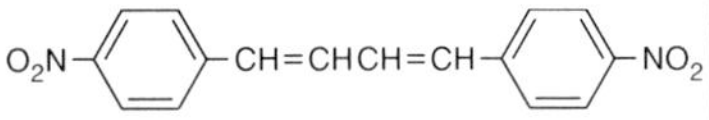
250 mg

S87,785-9
1,6-BIS(4-NITROPHENYL)-1,3,5-HEXATRIENE
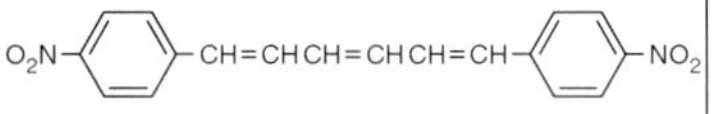
250 mg

S88,227-5
1,6-BIS(4-NITROPHENYL)HEXANE
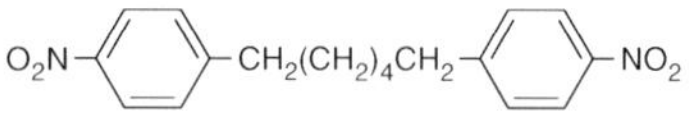
250 mg

S92,121-1
N-(4-NITROBENZYLIDENE)-2,4,6-TRIMETHYLANILINE
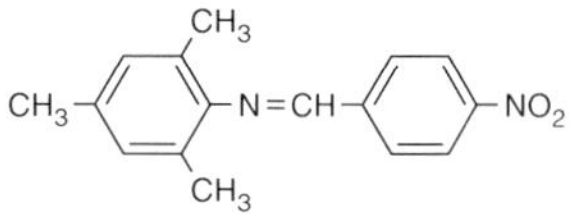
250 mg

S87,634-8
BIS(4-NITROPHENYL)METHANE
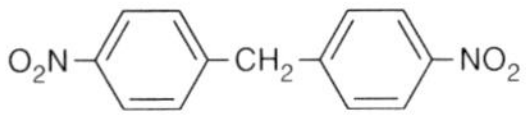
250 mg

S89,383-8
1,1-BIS(4-NITROPHENYL)-2,2-DICHLOROETHENE
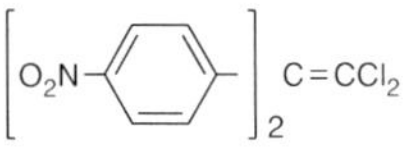
250 mg

S91,776-1
9-(ALPHA,ALPHA-DIMETHYL-4-NITROBENZYL)FLUORENE
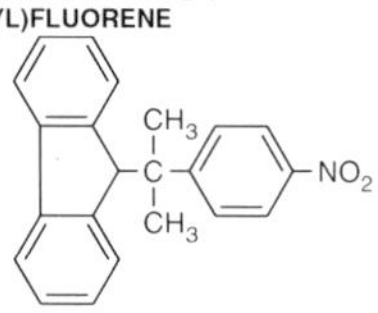
250 mg

S77,862-1
8,16-BIS(4-NITROPHENYL)-OCTAHYDRO-[1,2,4,5]TETRAZINO[6,1-A:3,4-A']-DIISOQUINOLINE
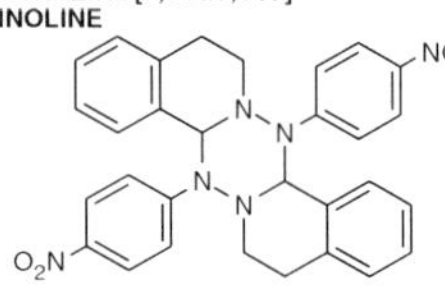
250 mg

S38,005-9
2-NITRO-P-XYLENE
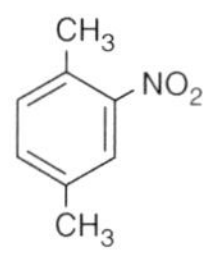
5 gm

S87,931-2
5-TERT-BUTYL-2-NITRO-M-XYLENE
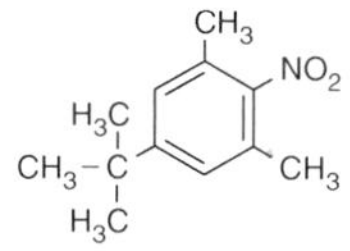
5 gm

S91,458-4
BIS(3-NITROPHENYL)METHANE
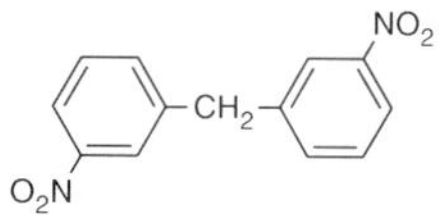
250 mg

S85,010-1
3,3'-DINITROBIPHENYL
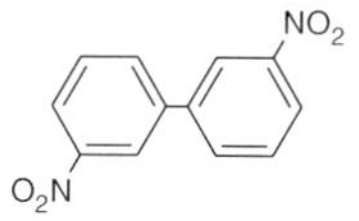
250 mg

S40,958-8
(3-NITROBENZYL)-TRIMETHYLAMMONIUM CHLORIDE
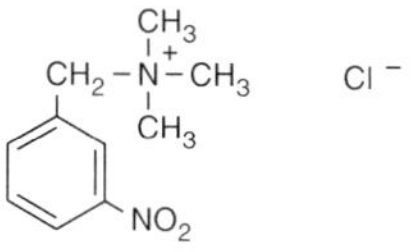
5 gm

S58,860-1
2-CHLORO-3-NITRONAPHTHALENE-BIS-(HEXACHLOROCYCLOPENTADIENE) ADDUCT
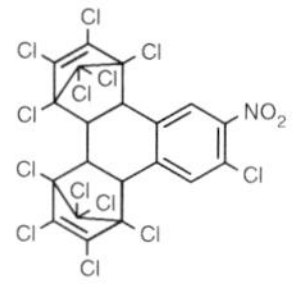
5 gm

S58,862-8
2-IODO-3-NITRONAPHTHALENE-BIS-(HEXACHLOROCYCLOPENTADIENE) ADDUCT
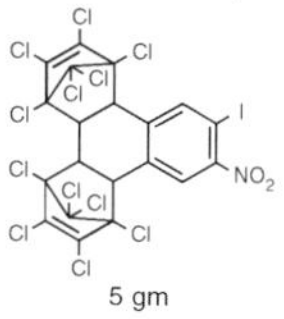
5 gm

S94,584-6
2,3-DIHYDRO-7-NITRO-3,3,4-TRIMETHYLBENZOFURAN
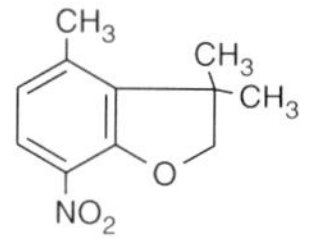
250 mg

S50,112-3
2-(3-NITROPHENYL)-1,3-DIOXOLANE
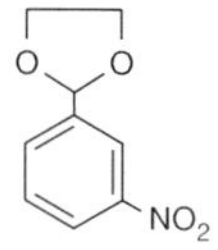
5 gm

S54,903-7
2-METHYL-2-(3-NITROPHENYL)-1,3-DIOXOLANE
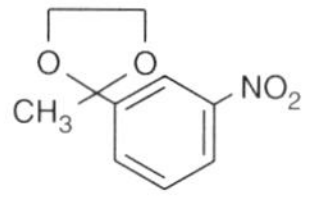
250 mg

S52,535-9
2-ISOBUTYL-2-METHYL-5-NITRO-1,3-BENZODIOXOLE
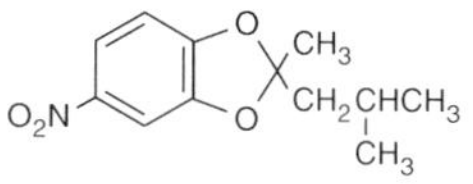
250 mg

S78,243-2
1-NITRO-2-PHENYLCYCLOHEXANE
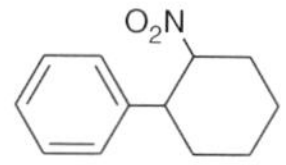
5 gm

S52,944-3
1-(2-NITROPHENYL)PIPERIDINE
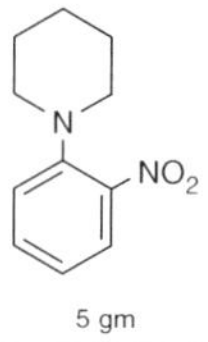
5 gm

S45,430-3
6-NITRO-1,3-BENZODIOXAN
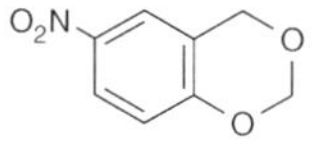
5 gm

S92,678-7
5,7-DIPHENYL-8-NITRO-1,2,3,4-TETRAHYDRONAPHTHALENE
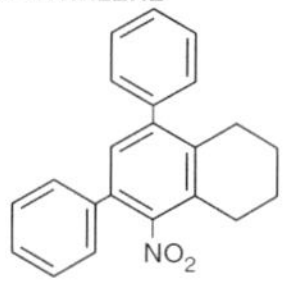
250 mg

S92,244-7
4-NITRO-1,3,9-TRIPHENYLFLUORENE
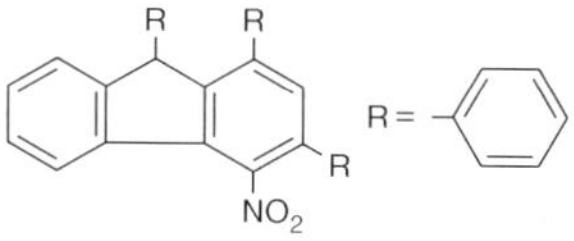
250 mg

S43,359-4
2-ETHYLNITROBENZENE
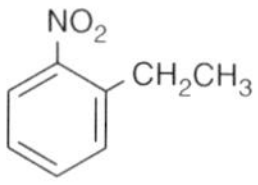
5 gm

S40,534-5
2,2'-DINITROBIBENZYL
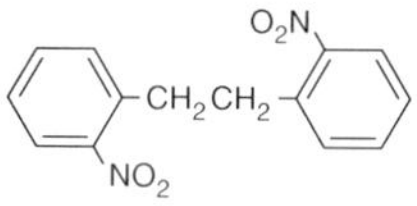
5 gm

S50,641-9
4-BROMO-2'-NITROBIPHENYL
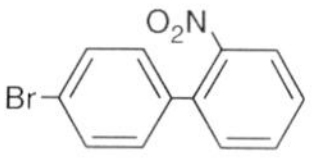
250 mg

S50,914-0
1-(4-NITROPHENYL)-1,3-BUTADIENE
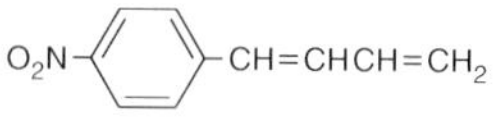
5 gm

S64,620-2
4,4'-DINITROSTILBENE
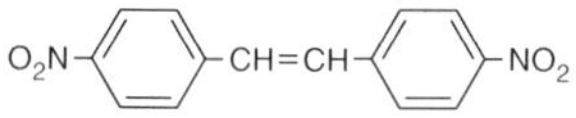
250 mg

S57,985-8
1-DODECYL-4-NITROBENZENE
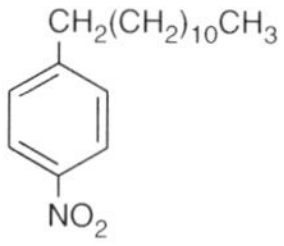
5 gm

S64,631-8
TRIS(4-NITROPHENYL)METHANE
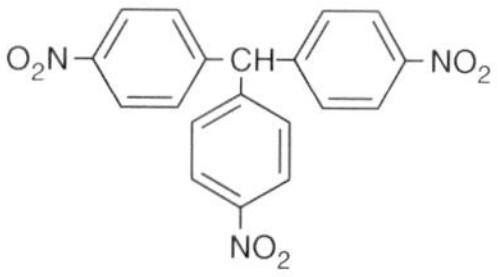
250 mg

S61,391-6
ALPHA-PHENYL-4-NITROTOLUENE
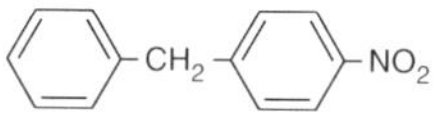
250 mg

S64,812-4
1-(DIPHENYLMETHYL)-4-NITROBENZENE
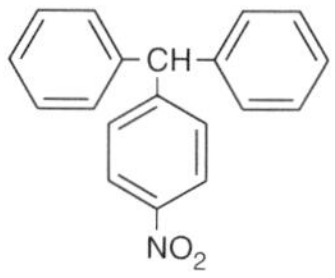
250 mg

S84,703-8
3,4'-DINITROBIPHENYL
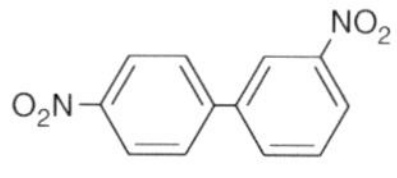
250 mg

S35,384-1
9-(4-NITROBENZYLIDENE)FLUORENE
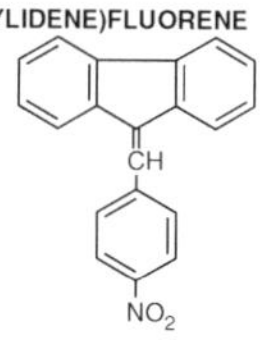
1 gm

S37,541-1
9-(4-CHLOROBENZYLIDENE)-2-NITROFLUORENE
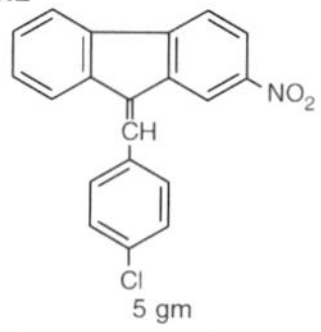
5 gm

S44,593-2
1-BROMO-4-CHLORO-2-NITROBENZENE
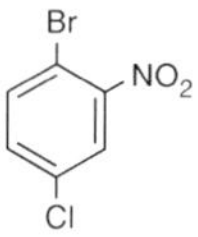
5 gm

S62,231-1
4-BROMO-3-NITROTOLUENE
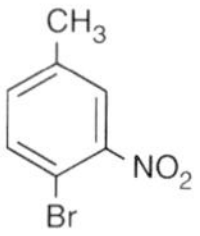
250 mg

S43,293-8
2-IODO-4-NITROTOLUENE
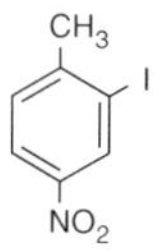
250 mg

S53,673-3
4-IODO-3-NITROTOLUENE
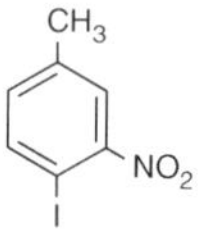
5 gm

S63,354-2
2,2'-DIMETHYL-4,4'-DINITROBIBENZYL
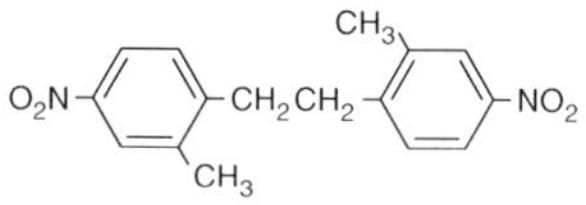
250 mg

S59,521-7
1,3,5-TRICHLORO-2-NITROBENZENE
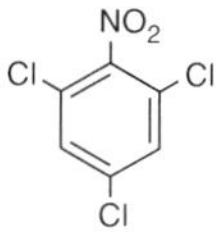
250 mg

S52,002-0
3,5-DINITRO-O-CRESOL, SODIUM SALT
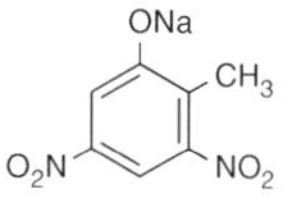
5 gm

S51,429-2
1-NITRO-2,3,5-TRIMETHYLBENZENE
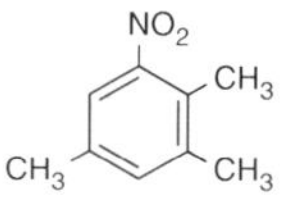
250 mg

S54,895-2
1-NITRO-3,4,5-TRIMETHYLBENZENE
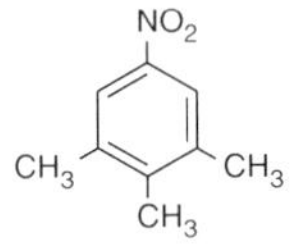
250 mg

S94,764-4
1-NITRO-2,4,6-TRIISOPROPYLBENZENE

250 mg

S53,105-7
2-CHLORO-5-NITRO-P-XYLENE
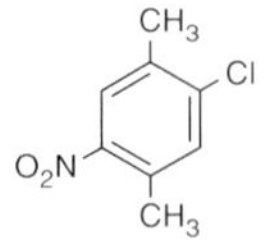
250 mg

S43,360-8
5-NITROPSEUDOCUMENE
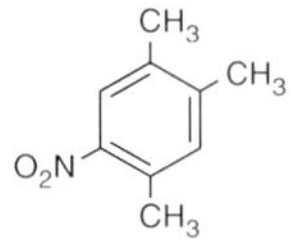
5 gm

S83,463-7
6-METHYL-1-NITRONAPHTHALENE
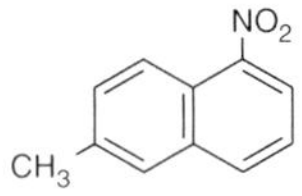
250 mg

S45,051-0
1,2,3,4,5,6-HEXACHLORO-7-NITRONAPHTHALENE
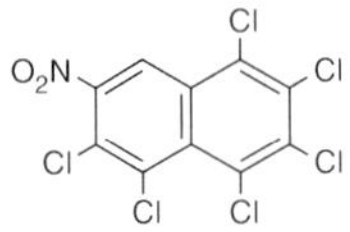
5 gm

S44,694-7
1-FLUORO-4-NITRONAPHTHALENE
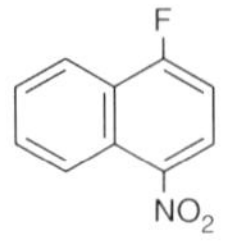
250 mg

S66,302-6
1-BROMO-8-NITRONAPHTHALENE
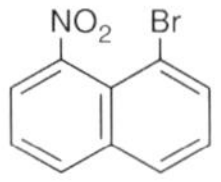
250 mg

S97,444-7
DINITROCHRYSENE
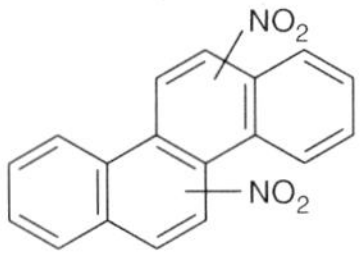
10 mg

S44,236-4
1-BROMO-5-NITRONAPHTHALENE
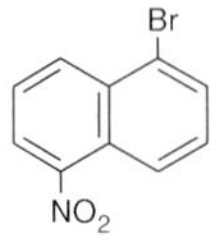
5 gm

S75,509-5
3-NITROFLUORENE
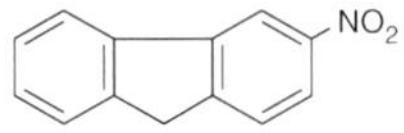
250 mg

S75,545-1
1-FLUORO-2-NITROFLUORENE
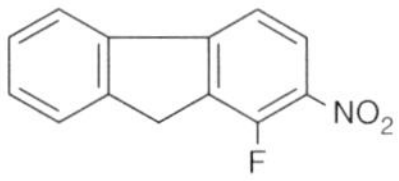
250 mg

S77,401-4
1-FLUORO-7-NITROFLUORENE
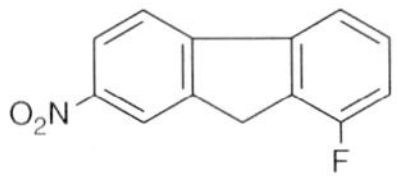
250 mg

S77,405-7
3-FLUORO-2-NITROFLUORENE
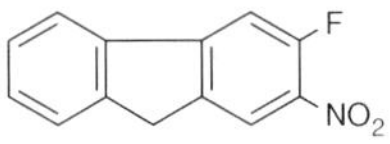
250 mg

S41,270-8
2,5-DINITROFLUORENE
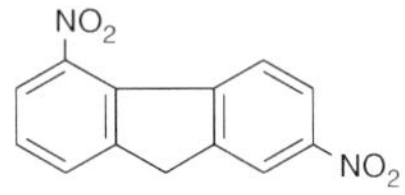
5 gm

S75,825-6
2,6-DINITROFLUORENE
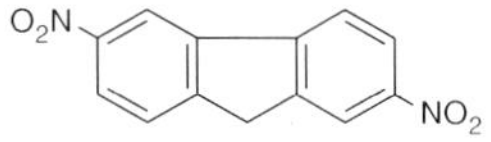
250 mg

S94,133-6
9,9-DICHLORO-2-NITROFLUORENE
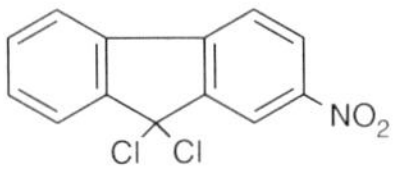
250 mg

S34,265-3
9-BROMO-2-NITROFLUORENE
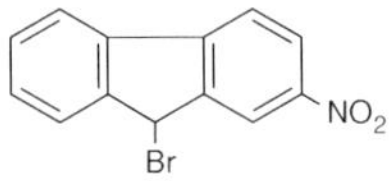
5 gm

S75,484-6
9-BROMO-3-NITROFLUORENE
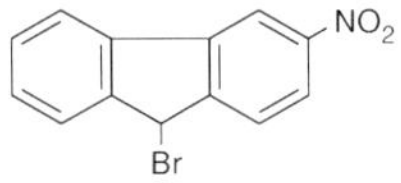
250 mg

S75,612-1
3-BROMO-2-NITROFLUORENE
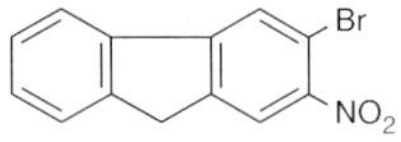
5 gm

S74,917-6
5-NITRO-2,4,7-TRICHLOROFLUORENE
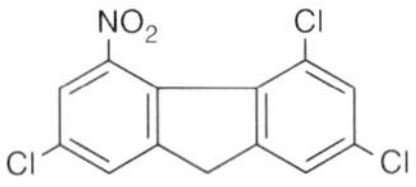
250 mg

S74,895-1
2-IODO-7-NITROFLUORENE
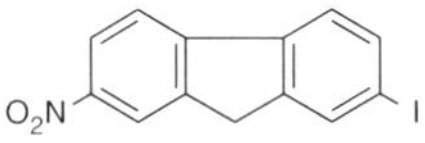
250 mg

S75,215-0
2,7-DIBROMO-4-NITROFLUORENE
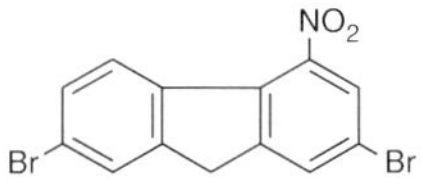
5 gm

S93,107-1
9,9-DIMETHYL-2,7-DINITROFLUORENE
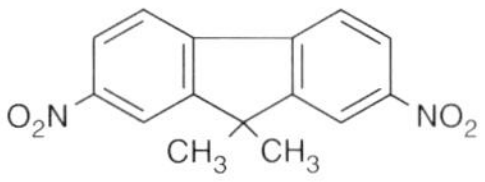
250 mg

S35,383-3
9-(3-NITROBENZYLIDENE)FLUORENE
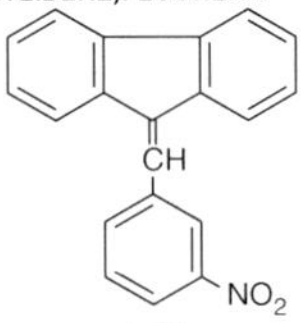
1 gm

S75,830-2
3-NITROPHENANTHRENE
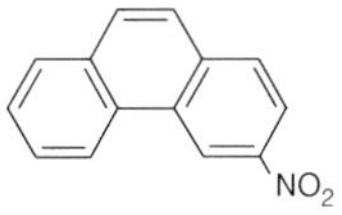
250 mg

S92,399-0
9,10-DIHYDRO-2-NITROPHENANTHRENE
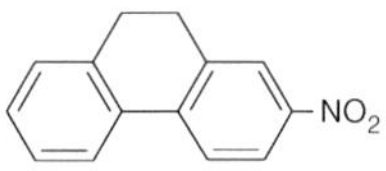
250 mg

S80,992-6
9-CHLORO-10-NITROPHENANTHRENE
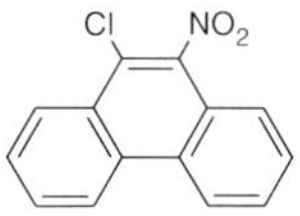
250 mg

S80,991-8
9-BROMO-10-NITROPHENANTHRENE
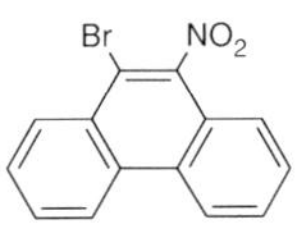
250 mg

S84,254-0
10-BROMO-3-NITROPHENANTHRENE
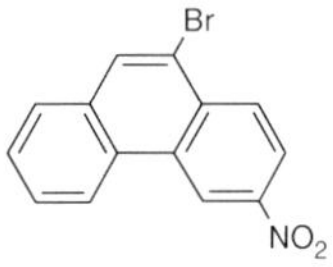
250 mg

S90,825-8
9-BROMO-3-NITROPHENANTHRENE
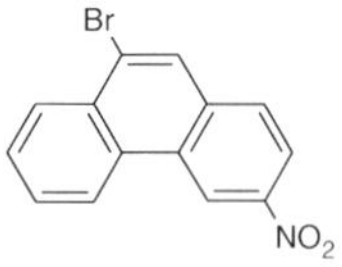
250 mg

S84,257-5
10-NITRO-9,9'-BIPHENANTHRENE
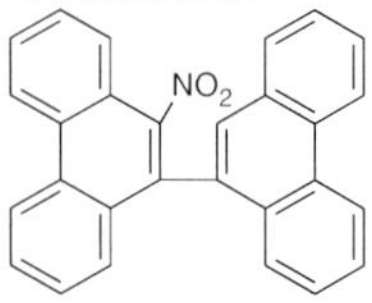
250 mg

S35,387-6
6-NITROCHRYSENE
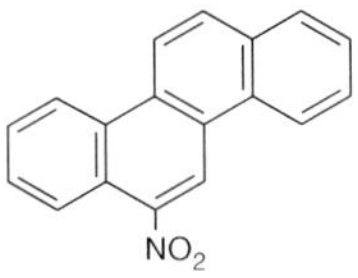
1 gm

S80,974-8
6,12-DINITROCHRYSENE
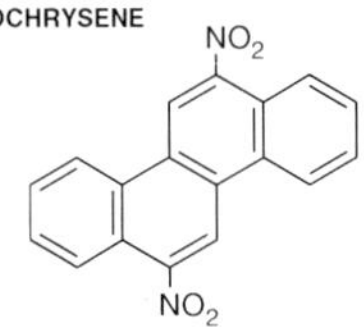
250 mg

S82,401-1
2-NITROTHIANTHRENE
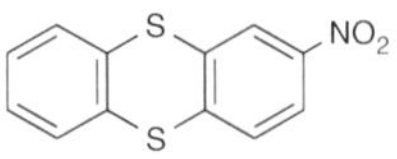
250 mg

S84,713-5
3-NITRODIBENZIODOLIUM IODIDE
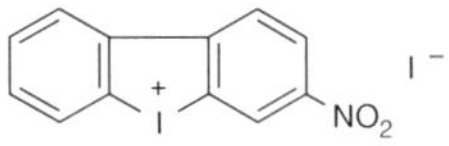
250 mg

S84,704-6 3,7-DINITRODIBENZIODOLIUM HYDROGEN SULFATE 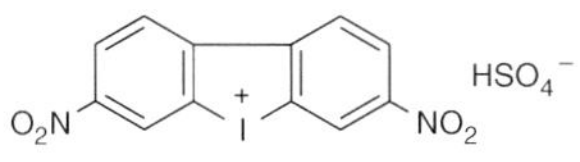 250 mg	**S89,704-3** 1,3,4,6,7,8-HEXAHYDRO-2H-PYRIMIDO (1,2-A)PYRIMIDINE PICRATE 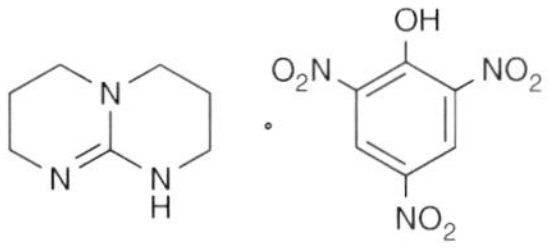 250 mg	**S94,314-2** 4-CHLORO-3-NITROBIPHENYL 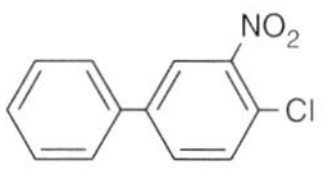 250 mg

S41,132-9
1,2-DIBROMO-5-METHYL-1-PHENYL-3-HEXANONE
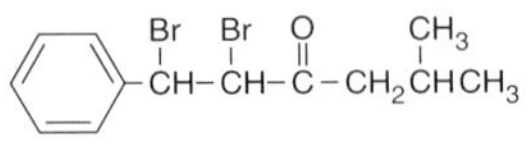
5 gm

S87,649-6
3-BROMO-1,1,1-TRIPHENYLACETONE
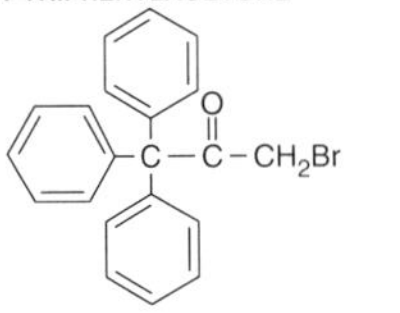
250 mg

S64,927-9
3-(2-NITRO-4-TRIFLUOROMETHYLPHENYLHYDRAZONO)-2,4-PENTANEDIONE
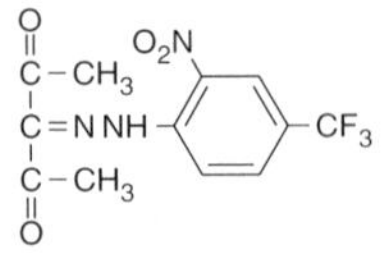
250 mg

S56,717-5
3-CHLORO-4-(ALPHA,ALPHA,ALPHA-TRIFLUORO-M-TOLYL)-2-BUTANONE
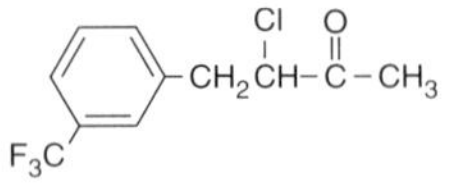
5 gm

S56,721-3
3-CHLORO-3-METHYL-4-(ALPHA,ALPHA,ALPHA-TRIFLUORO-M-TOLYL)-2-BUTANONE
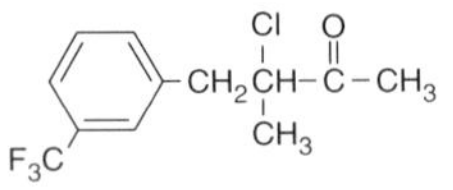
5 gm

S87,766-2
4-[4-(TRIFLUOROMETHYL)PHENYL]-2-BUTANONE
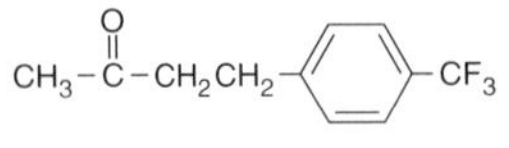
250 mg

S94,293-6
2,2',6,6'-TETRACHLOROBENZOIN
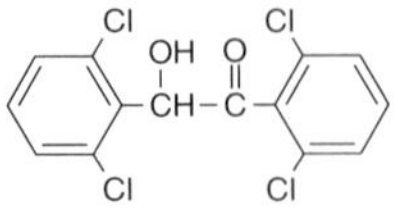
250 mg

S69,717-6
1-(2,3,4,5,6-PENTACHLOROPHENOXY)-2-PROPANONE
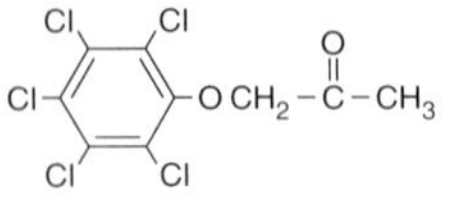
250 mg

S89,683-7
3-(3-CHLOROANILINO)-5,5-DIMETHYL-2-CYCLOHEXEN-1-ONE
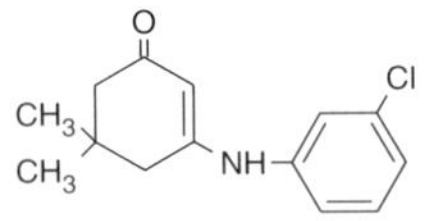
250 mg

S87,869-3
2,6-BIS(4-CHLOROBENZYLIDENE)-1-CYCLOHEXANONE
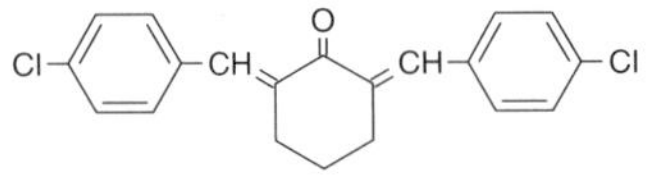
5 gm

S67,862-7
1-(4-CHLOROPHENYL)-5-PHENYL-1,4-PENTADIEN-3-ONE
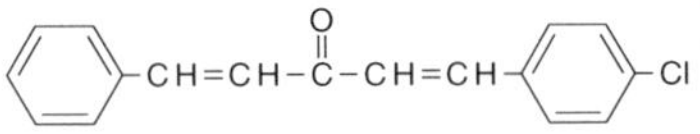
5 gm

S80,826-1
4-(4-FLUOROPHENYL)-5-METHYL-1H-1,2-DIAZEPIN-6(7H)-ONE
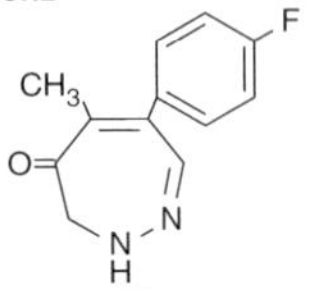
250 mg

S73,993-6
1-(2,4-DICHLOROPHENYL)-4,4-DIMETHYL-1-PENTEN-3-ONE
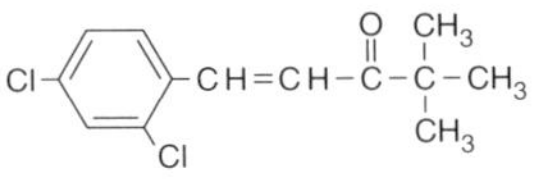
5 gm

S73,486-1
2-(2-CHLOROBENZYLIDENE)-CYCLOHEXANONE
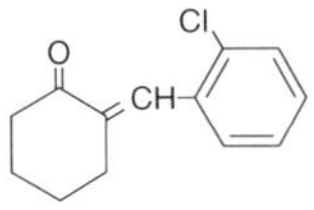
250 mg

S70,092-4
1-(2,4-DICHLOROPHENYL)-5-PHENYL-1,4-PENTADIEN-3-ONE
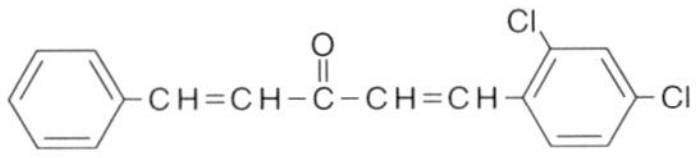
5 gm

S94,729-6
1,5-BIS(4-CHLOROPHENYL)-1,4-PENTADIEN-3-ONE
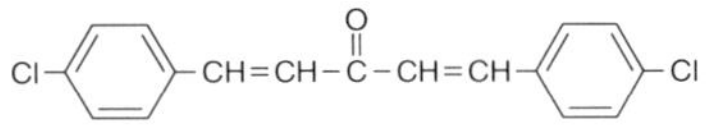
250 mg

S43,247-4
2,6-DIVANILLYLIDENECYCLOHEXANONE
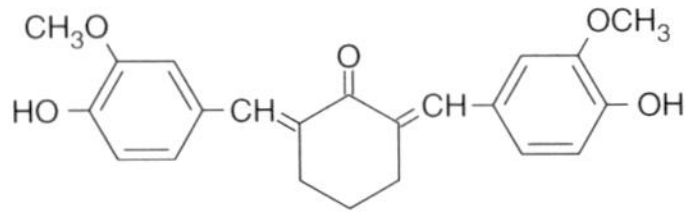
5 gm

S76,733-6
3-HYDROXY-1-(4-METHOXYPHENYL)-3-METHYL-2-BUTANONE
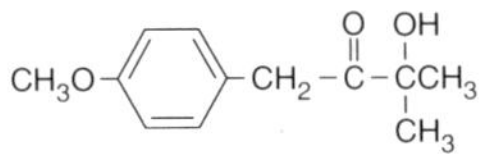
250 mg

S87,043-9
4-HYDROXY-4-PHENYL-3-(PHENYLAZO)-3-BUTEN-2-ONE
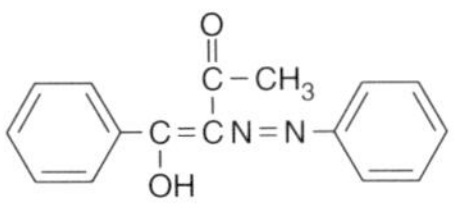
250 mg

S85,201-5
4-HYDROXY-2,3,4-TRIPHENYL-1-CYCLOPENTANONE
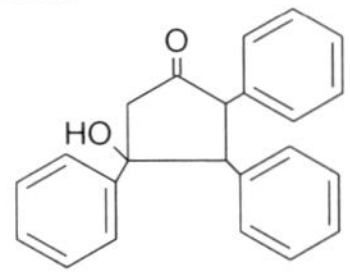
250 mg

S89,715-9
4-(4-HYDROXY-4-PHENYLPIPERIDINO)-3-PENTEN-2-ONE
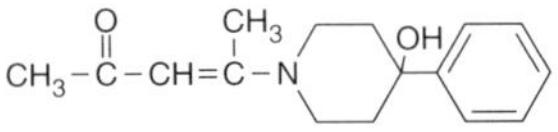
250 mg

S67,609-8
3,4-DIPHENYL-4-HYDROXY-2-CYCLOPENTEN-1-ONE
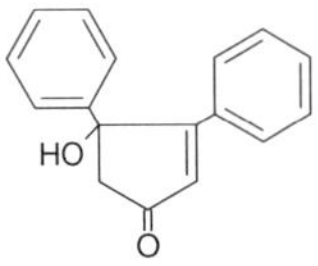
250 mg

S95,072-6
4-HYDROXY-5-METHYL-2,3,4-TRIPHENYL-2-CYCLOPENTENONE
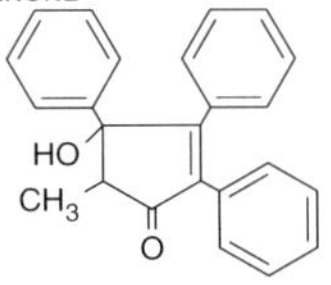
50 mg

S95,601-5
1-HYDROXY-1-METHYLNAPHTHALEN-2(1H)-ONE
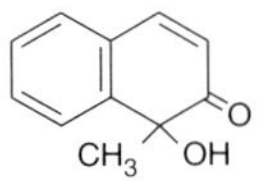
50 mg

S88,690-4
2-(ALPHA-HYDROXY-2-NITROBENZYL)-1-CYCLOHEXANONE
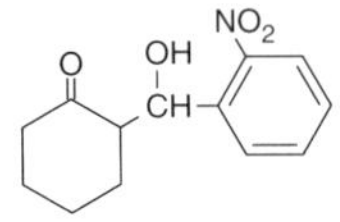
250 mg

S89,681-0
5,5-DIMETHYL-3-(2-HYDROXYANILINO)-2-CYCLOHEXEN-1-ONE
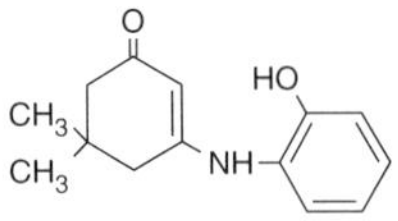
250 mg

S89,724-8
5,5-DIMETHYL-3-(3-HYDROXYANILINO)-2-CYCLOHEXEN-1-ONE
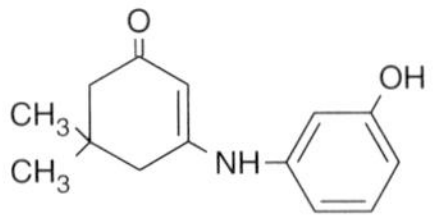
250 mg

S89,680-2
5,5-DIMETHYL-3-(4-HYDROXYANILINO)-2-CYCLOHEXEN-1-ONE
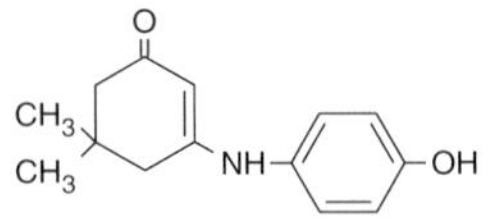
250 mg

S85,019-5
2,5-BIS(2-HYDROXYBENZYLIDENE)-1-CYCLOPENTANONE
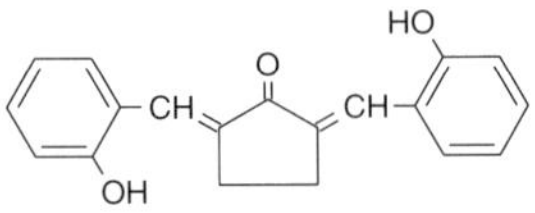
250 mg

S85,020-9
2,5-BIS(2-HYDROXYBENZYLIDENE)-1-CYCLOPENTANONE, DISODIUM SALT
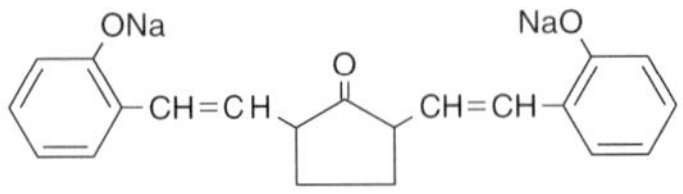
250 mg

S61,886-1
2,5-DIMETHYL-3,4-DIPHENYL-4-HYDROXY-2-CYCLOPENTENONE
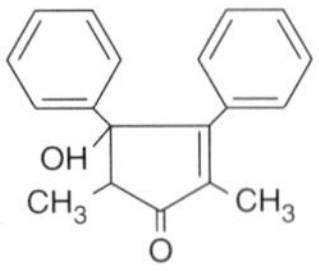
1 gm

S89,621-7
5,5-DIMETHYL-3-(4-HYDROXYSTYRYL)-2-CYCLOHEXEN-1-ONE
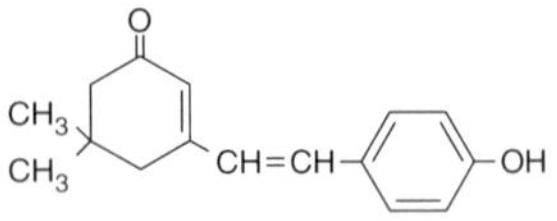
250 mg

S36,071-6
3,5-DIHYDROXY-6-[5-(2,4,6-TRIHYDROXYPHENYL)-2,4-PENTADIENYLIDENE]-2,4-CYCLOHEXADIENONE
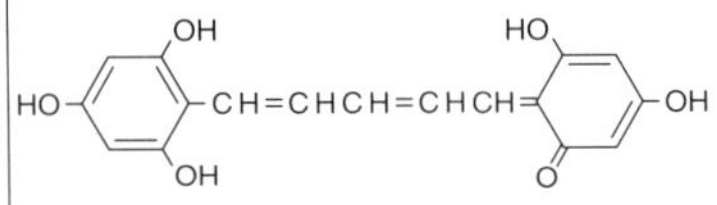
5 gm

S88,553-3
8-HYDROXY-4A-METHYL-4,4A,9,10-TETRAHYDRO-2(3H)-PHENANTHRENONE
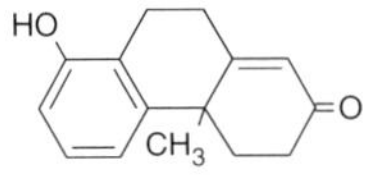
250 mg

S35,065-6
1-(4-HYDROXYBENZYLIDENE)ACETONE
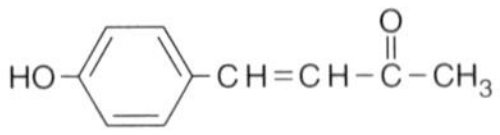
5 gm

S35,066-4
4-HYDROXYCHALCONE
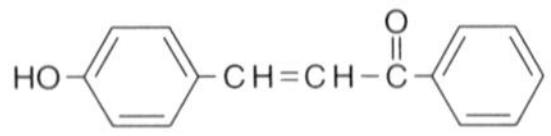
5 gm

S75,879-5
3-HYDROXY-2-(1,2,3,4-TETRAHYDRO-1-ISOQUINOLYL)-2-CYCLOHEXEN-1-ONE
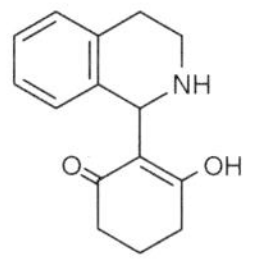
250 mg

S38,427-5
1-(4-HYDROXYPHENYL)-2-PHENYL-3-HEPTANONE
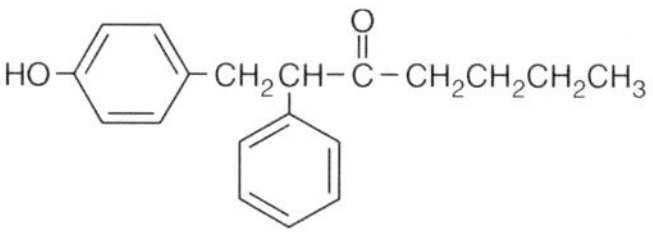
250 mg

S45,324-2
7-SYN-BENZYLOXYMETHYL-5-NORBORNEN-2-ONE
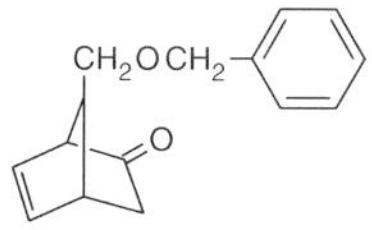
250 mg

S79,326-4
2-PHENYL-3,5,5-TRIETHOXY-2-CYCLOPENTEN-1-ONE
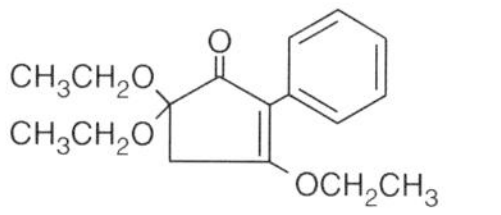
250 mg

S72,225-1
4-(3,4-DIMETHOXY-6-NITROANILINO)-2-BUTANONE
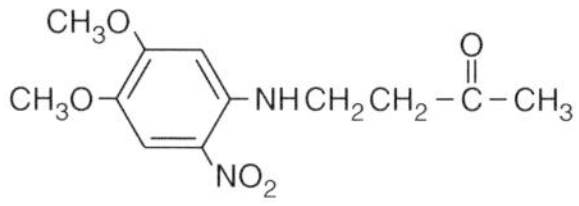
5 gm

S87,886-3
2,6-BIS(3,4-DIMETHOXYBENZYLIDENE)-1-CYCLOHEXANONE
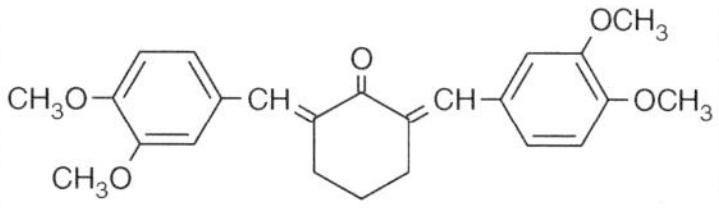
5 gm

S78,244-0
1,5-BIS(3,4-DIMETHOXYPHENYL)-1,4-PENTADIEN-3-ONE
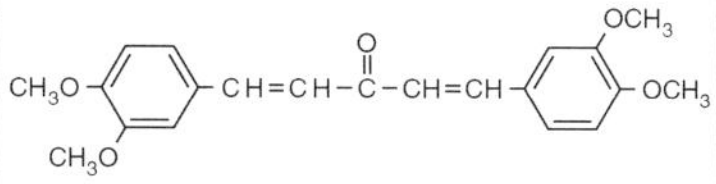
5 gm

S73,962-6
1-(3,4-DIMETHOXYPHENYL)-4,4-DIMETHYL-1-PENTEN-3-ONE
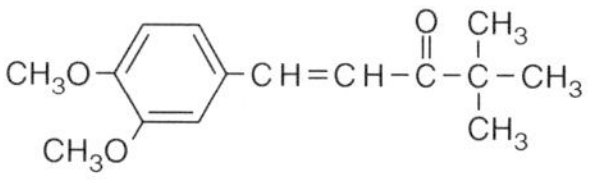
5 gm

S76,139-7
8,9-DIMETHOXY-1,2,3,3A,5,6,10B,12A-OCTAHYDROBENZO(A)CYCLOPENTA(F)-QUINOLIZIN-12(11H)ONE
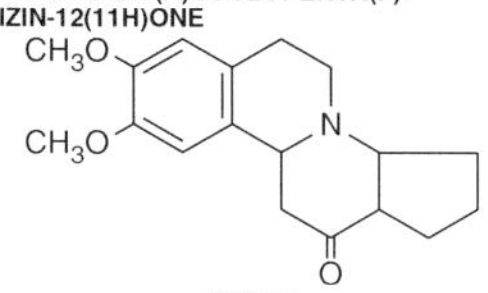
250 mg

S86,145-6
3-CHLORO-2-(2,5-DIMETHOXYPHENYL)-2-CYCLOPENTEN-1-ONE
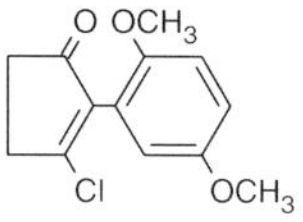
250 mg

S87,885-5
2,6-BIS(2-METHOXYBENZYLIDENE)-1-CYCLOHEXANONE
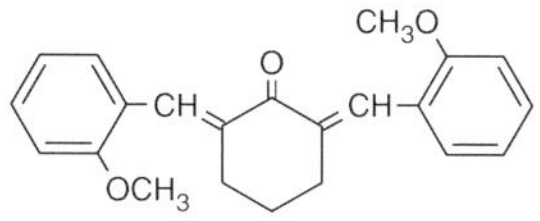
5 gm

S87,883-9
2,6-BIS(4-METHOXYBENZYLIDENE)-CYCLOHEXANONE
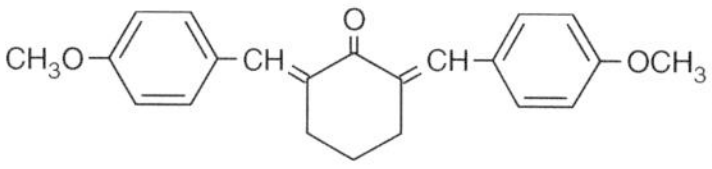
5 gm

S88,571-1
1,4A-DIMETHYL-HEXAHYDRO-7-METHOXY-1-(3-METHOXYPHENETHYL)-2(3H)-PHENAN THRENONE
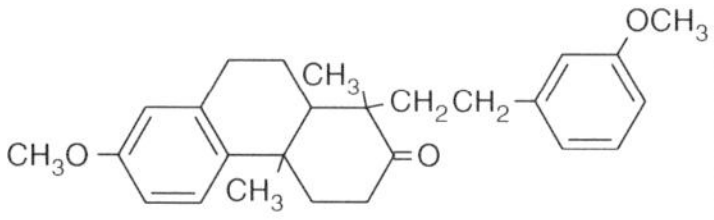
250 mg

S93,736-3
4'-(METHOXYMETHOXY)ACETOPHENONE
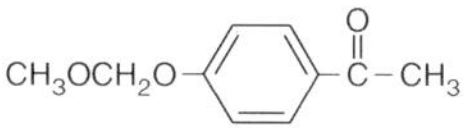
250 mg

S61,960-4
1-CYCLOPROPYL-3-(3,4-METHYLENEDIOXYPHENYL)-2-PROPEN-1-ONE
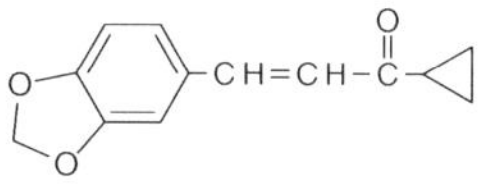
250 mg

S61,987-6
1-CYCLOPROPYL-3-(3,4-DIMETHOXYPHENYL)-2-PROPEN-1-ONE
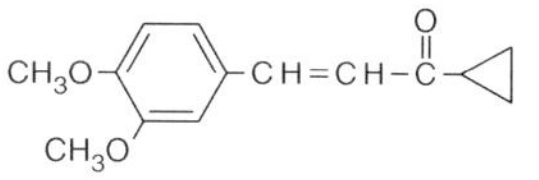
250 mg

S59,451-2
4-(3,4-DIMETHOXYPHENYL)-3-BUTEN-2-ONE
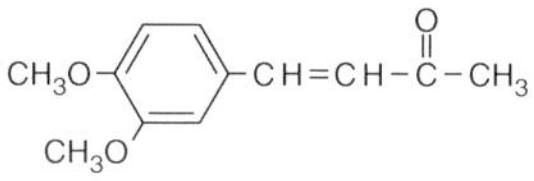
250 mg

S59,571-3
1-(3,4-DIMETHOXYPHENYL)-1-PENTEN-3-ONE
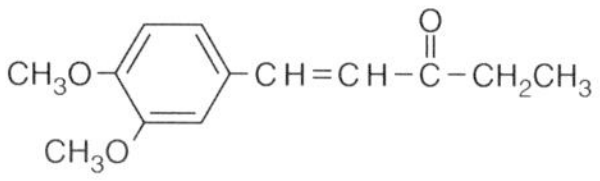
250 mg

S43,233-4
[4-(BENZYLOXY)PHENOXY]-2-PROPANONE
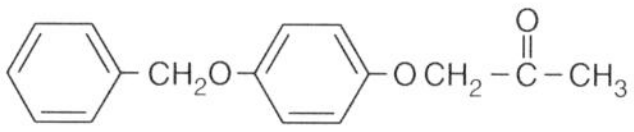
5 gm

S36,548-3
3-VERATRYL-4-HEXANONE
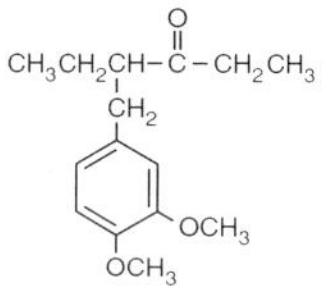
250 mg

S89,676-4
5,5-DIMETHYL-3-(2-METHOXYANILINO)-2-CYCLOHEXEN-1-ONE
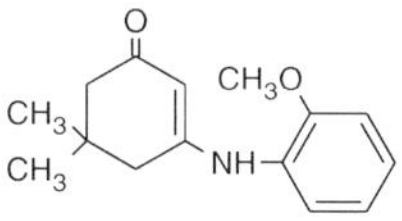
250 mg

S89,677-2
5,5-DIMETHYL-3-(3-METHOXYANILINO)-2-CYCLOHEXEN-1-ONE
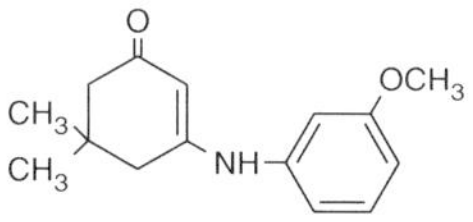
250 mg

S89,675-6
5,5-DIMETHYL-3-(4-METHOXYANILINO)-2-CYCLOHEXEN-1-ONE
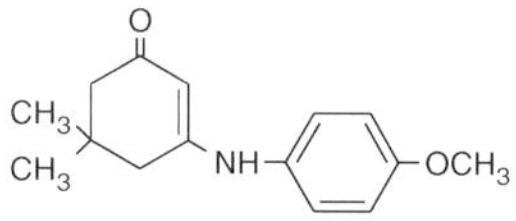
250 mg

S75,589-3
2-(4-METHOXYANILINO)CYCLOHEXANONE
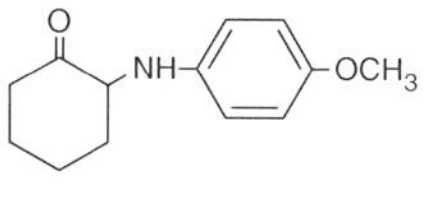
250 mg

S37,821-6
1-(3-NITROPHENOXY)ACETONE
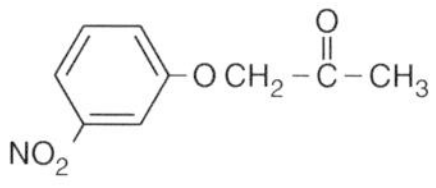
250 mg

S35,234-9
4-(3-METHOXYPHENYL)-3-HEXANONE
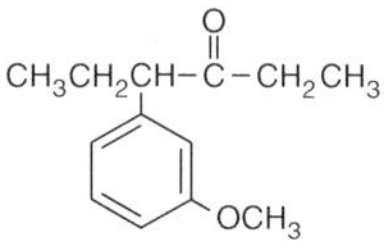
5 gm

S59,459-8
4-(3-METHOXYPHENYL)-3-BUTEN-2-ONE
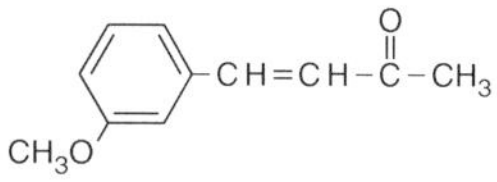
250 mg

S59,454-7
1-(3-METHOXYPHENYL)-4-METHYL-1-PENTEN-3-ONE
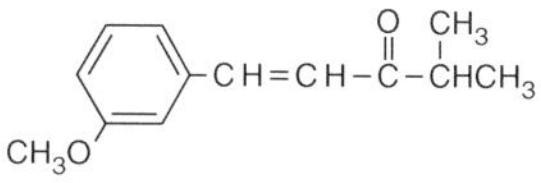
250 mg

S88,554-1
8-METHOXY-3,4,9,10-TETRAHYDRO-1(2H)-PHENANTHRENONE
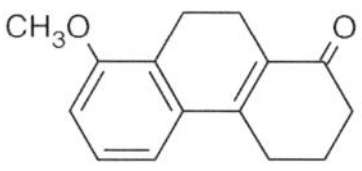
250 mg

S67,092-8
3-(4-METHOXYPHENYL)-5-PHENYL-2-CYCLOHEXEN-1-ONE
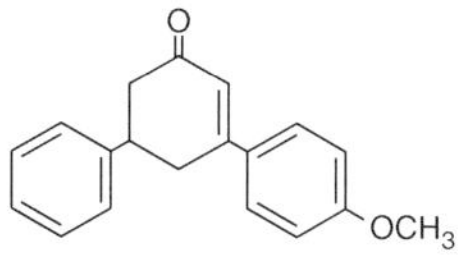
5 gm

S88,545-2
3-(4-METHOXYPHENYL)-6-METHYL-6-PHENYL-2-CYCLOHEXEN-1-ONE
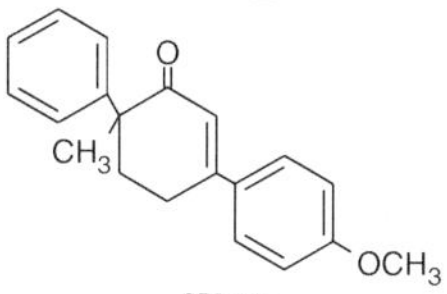
250 mg

S88,518-5
8-METHOXY-4,4A,9,10-TETRAHYDRO-2(3H)-PHENANTHRENONE
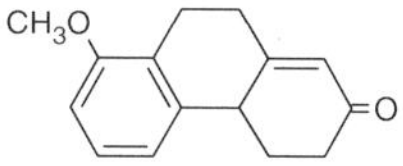
250 mg

S88,540-1
1,4A-DIMETHYL-7-METHOXY-4,4A,9,10-TETRAHYDRO-2(3H)-PHENANTHRENONE
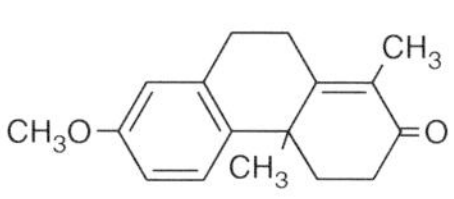
250 mg

S81,728-7
1,1-DIPHENYL-4-(4-METHOXYPHENYL)-3-BUTEN-2-ONE
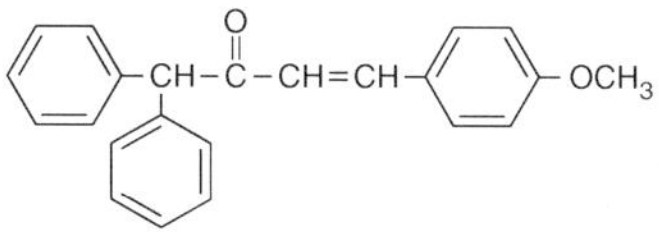
250 mg

S59,572-1
1-(4-METHOXYPHENYL)-1-PENTEN-3-ONE
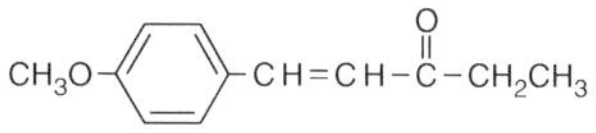
250 mg

S59,457-1
1-(4-METHOXYPHENYL)-4-METHYL-1-PENTEN-3-ONE
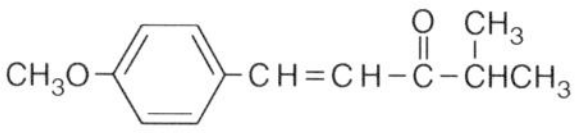
250 mg

S92,676-0
2-[4,4-DIMETHYL-1-(4-METHOXYPHENYL)-3-OXOPENTYL]-1-TETRALONE
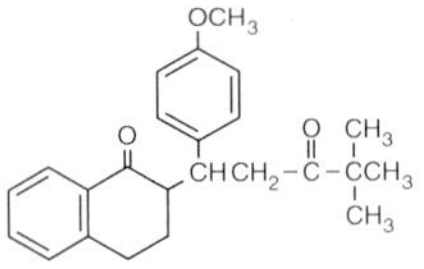
250 mg

S44,940-7
3-(4-METHOXYPHENYL)-2,4,5-TRIPHENYLCYCLOPENTADIENONE
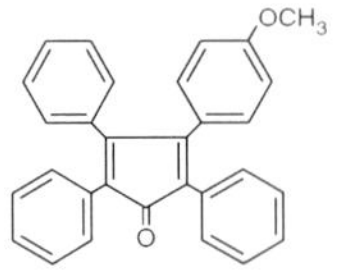
5 gm

S42,416-1
ANHYDROCOTARNINE-ACETONE
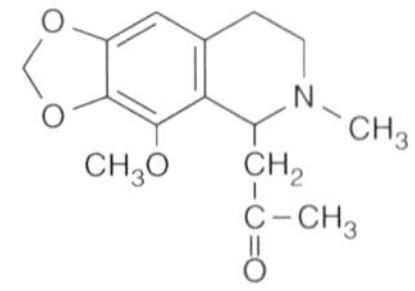
250 mg

S88,542-8
1,4,4A,9,10,10A-HEXAHYDRO-7-METHOXY-1,1,4A-TRIMETHYL-2(3H)-PHENANTHRENONE
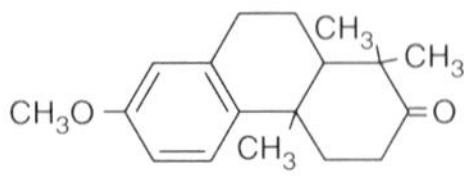
250 mg

S78,702-7
2-(4-METHOXYPHENYL)-1-CYCLOHEXANONE
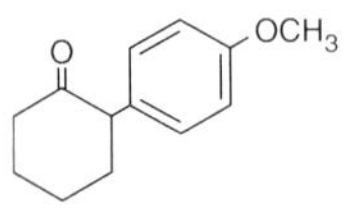
250 mg

S88,513-4
2,5-DIMETHYL-2-(4-METHOXYPHENYL)-5-PHENYL-1-CYCLOHEXANONE
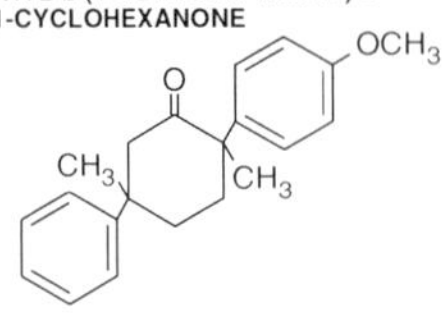
250 mg

S65,718-2
2,4-DIMETHOXY-4B,5,6,7,8,8A-HEXAHYDRO-9(10H)PHENANTHRONE
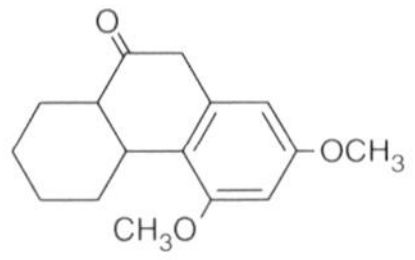
250 mg

S35,233-0
4-(2-METHOXYPHENYL)-3-HEXANONE
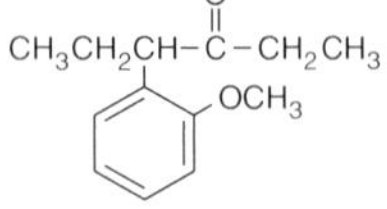
5 gm

S88,210-0
4-(4-METHOXYPHENYL)-2-BUTANONE
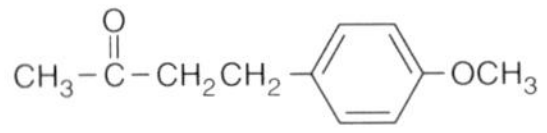
250 mg

S34,652-7
4-(2,4-DIMETHOXYPHENYL)-3-HEXANONE
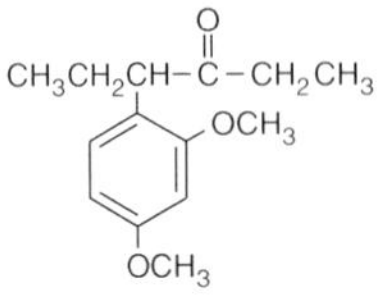
5 gm

S88,526-6
5-METHOXY-1-METHYL-2-TETRALONE
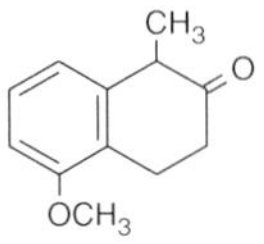
250 mg

S52,004-7
1-(4-METHOXYPHENYL)-7-PHENYL-1,4,6-HEPTATRIEN-3-ONE
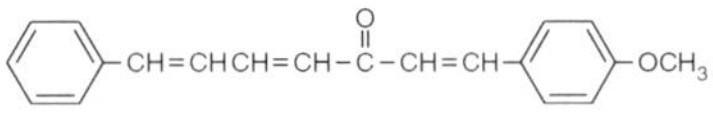
5 gm

S54,587-2
4-(5-BROMO-2-ETHOXYPHENYL)-3-BUTEN-2-ONE
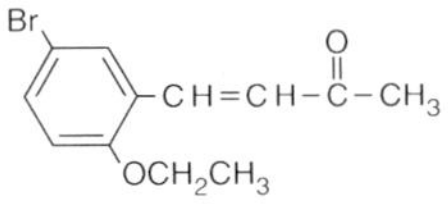
250 mg

S76,018-8
7,7-DIMETHYL-3,4,5,6,7,8-HEXAHYDRO-2-PHENYL-4-PIPERAZINO-2H-1-BENZOPYRAN-5-ONE
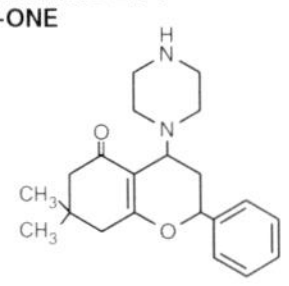
250 mg

S86,135-9
2,6-DISTYRYL-4H-PYRAN-4-ONE
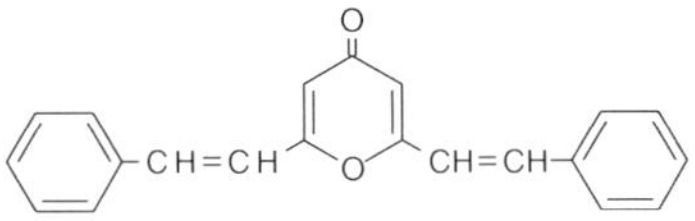
250 mg

S68,259-4
TERT-BUTYL 9-XANTHENYL KETONE
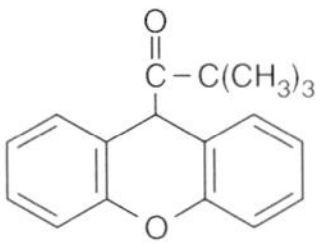
250 mg

S88,565-7
2-(BENZYLTHIOMETHYLENE)-1-DECALONE
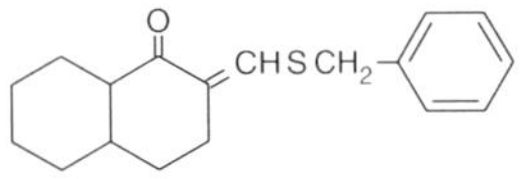
250 mg

S88,567-3
3-(BENZYLTHIOMETHYLENE)-8A-METHYL-2-DECALONE
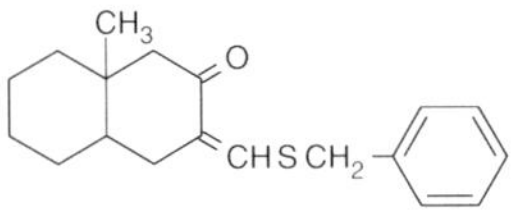
250 mg

S80,370-7
2-(P-TOLYLTHIO)CYCLOPENTANONE
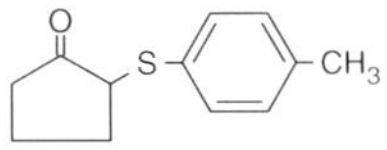
250 mg

S86,669-5
3-(PHENYLTHIO)-1-CYCLOHEXANONE
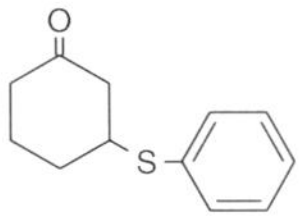
250 mg

S57,697-2
1-(PHENYLTHIO)-2-PROPANONE
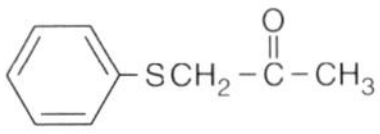
5 gm

S63,503-0
4-PHENYLTHIO-2-BUTANONE
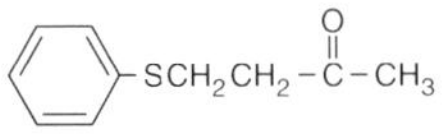
5 gm

S61,410-6
(4-CHLOROPHENYLTHIO)ACETONE
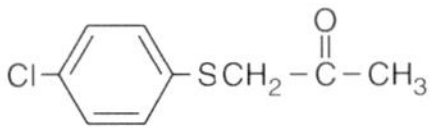
5 gm

S89,679-9
3-(2-AMINOANILINO)-5,5-DIMETHYL-2-CYCLOHEXEN-1-ONE
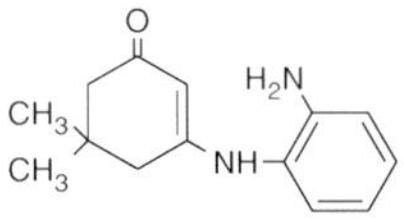
250 mg

S89,678-0
3-(4-AMINOANILINO)-5,5-DIMETHYL-2-CYCLOHEXEN-1-ONE
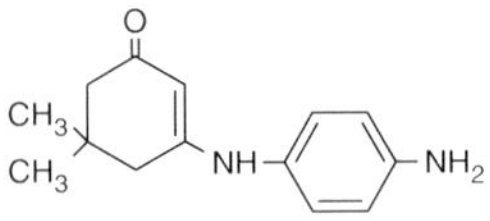
250 mg

S89,685-3
5,5-DIMETHYL-2-NITRO-3-(4-NITROANILINO)-2-CYCLOHEXEN-1-ONE
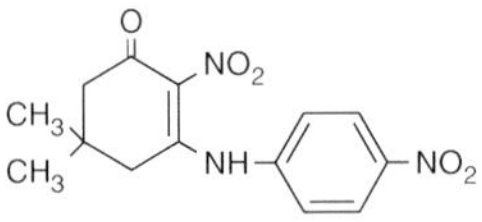
250 mg

S93,337-6
3-MORPHOLINO-4-PHENYL-4-(1,2,3,4-TETRAHYDRO-1-QUINOLINYL)-2-BUTANONE
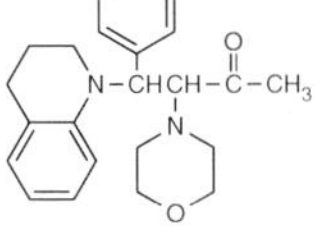
250 mg

S89,663-2
5,5-DIMETHYL-3-(3-NITROANILINO)-2-CYCLOHEXEN-1-ONE
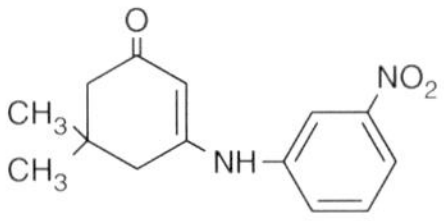
250 mg

S36,192-5
ETHYL PYRUVATE 3-NITROPHENYLHYDRAZONE
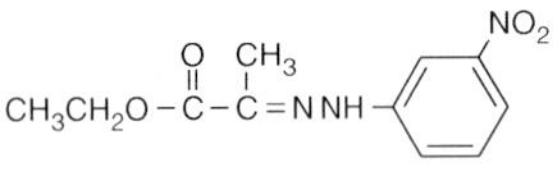
5 gm

S89,691-8
5,5-DIMETHYL-3-(4-NITROANILINO)-2-CYCLOHEXEN-1-ONE
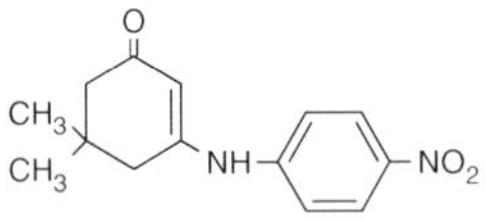
250 mg

S89,717-5
5,5-DIMETHYL-3-(PHENETHYLAMINO)-2-CYCLOHEXEN-1-ONE
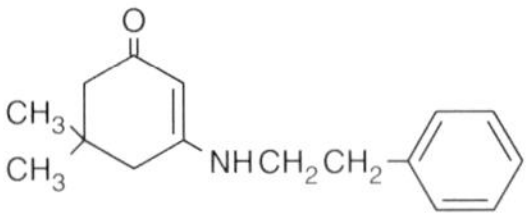
250 mg

S43,508-2
2-BENZYLIDENE-3-QUINUCLIDINONE
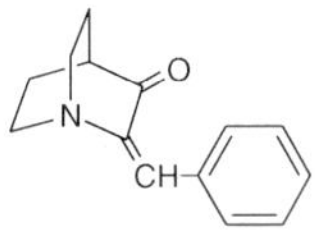
5 gm

S89,661-6
3-(METHYLPHENETHYLAMINO)-2-CYCLOHEXEN-1-ONE HYDROBROMIDE
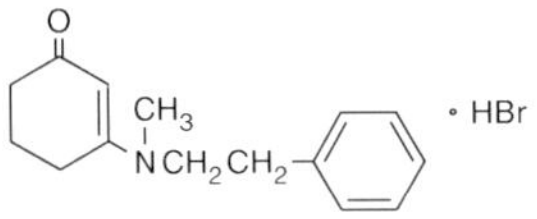
250 mg

S89,714-0
5,5-DIMETHYL-3-(METHYLPHENETHYLAMINO)-2-CYCLOHEXEN-1-ONE
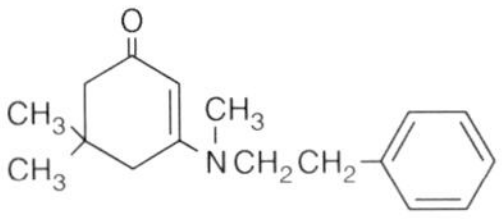
250 mg

S89,718-3
5,5-DIMETHYL-3-(3-METHYLANILINO)-2-CYCLOHEXEN-1-ONE
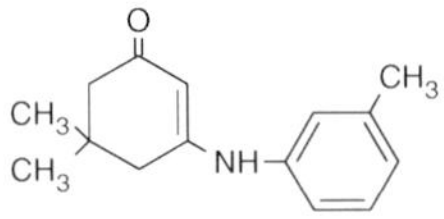
250 mg

S89,716-7
5,5-DIMETHYL-3-(2,4-DIMETHYLANILINO)-2-CYCLOHEXEN-1-ONE
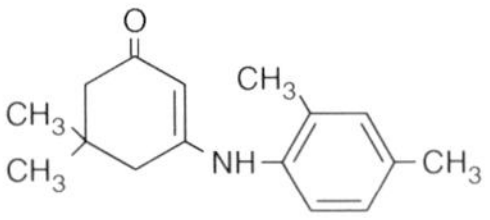
250 mg

S87,887-1
2,6-BIS[4-(DIMETHYLAMINO)-BENZYLIDENE]-1-CYCLOHEXANONE

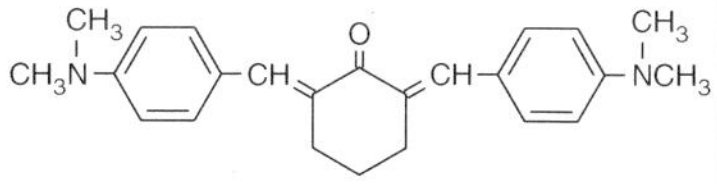

5 gm

S87,378-0
4-[4-(DIMETHYLAMINO)-ALPHA-PHENYLBENZYLIDENE]-2,5-CYCLOHEXADIEN-1-ONE

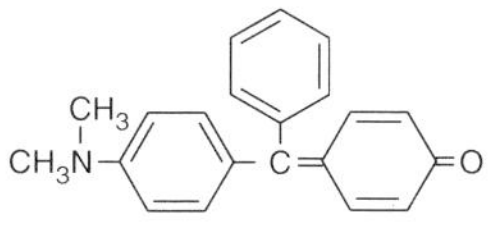

250 mg

S86,446-3
9-BENZYL-9-AZABICYCLO[3.3.1]-NONAN-3-ONE

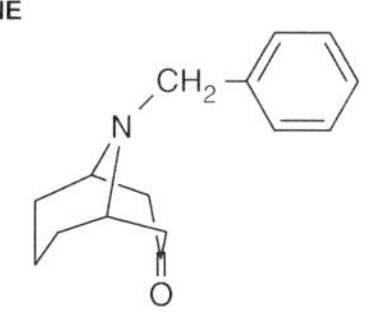

250 mg

S40,668-6
3-(N-BENZYL-N-METHYLAMINOMETHYL)-2-NORBORNANONE

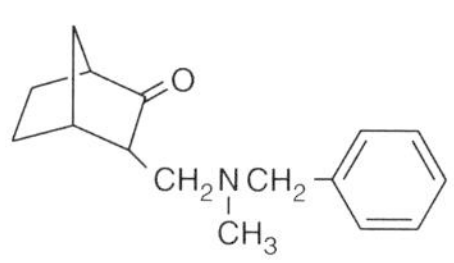

5 gm

S41,549-9
1-BENZYL-3-PIPERIDONE

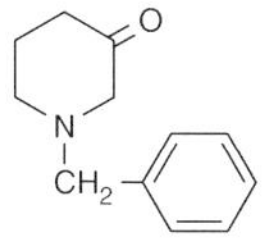

5 gm

S76,746-8
2,6-DIPHENYL-1-METHYL-4-PIPERIDINONE

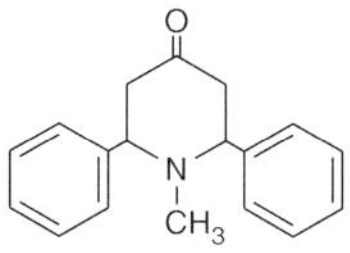

250 mg

S68,269-1
10,11-DIHYDRO-10[2-(4-METHYL-1-PIPERAZINYL)ETHYL]-DIBENZOSUBERONE DIHYDROCHLORIDE

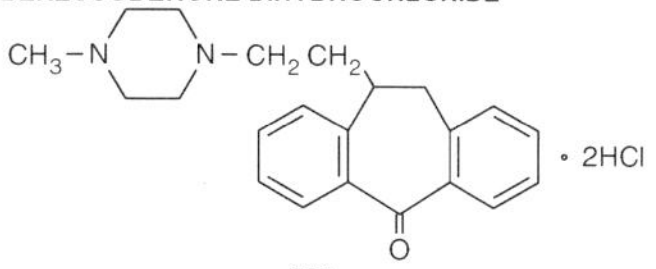

250 mg

S79,365-5
4-(DIMETHYLAMINO)-1,1-DIPHENYL-2-BUTANONE HYDROCHLORIDE

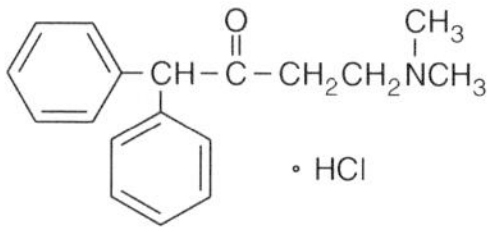

250 mg

S79,344-2
2,2-DIPHENYL-5-(DIMETHYLAMINO)-3-PENTANONE HYDROCHLORIDE

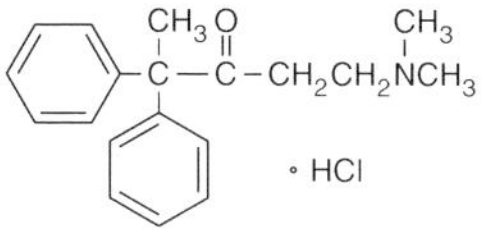

250 mg

S88,623-8
1-PHENYL-1,2-PROPANEDIONE 1-(PHENYLHYDRAZONE)

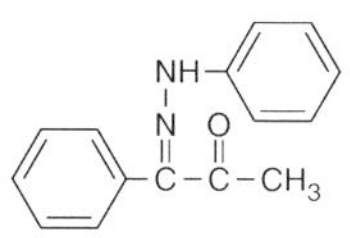

250 mg

S76,414-0
TERT-BUTYL 3-NITROSTYRYL KETONE

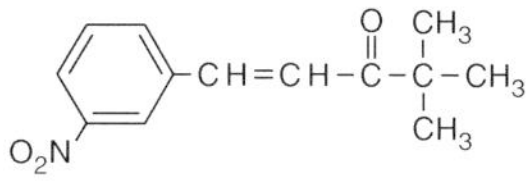

5 gm

S87,870-7
2,6-BIS(3-NITROBENZYLIDENE)-CYCLOHEXANONE

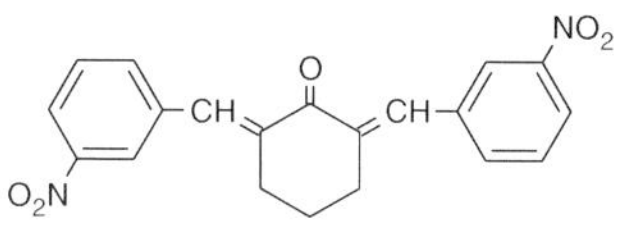

5 gm

S67,869-4
1-(3-NITROPHENYL)-5-PHENYL-1,4-PENTADIEN-3-ONE

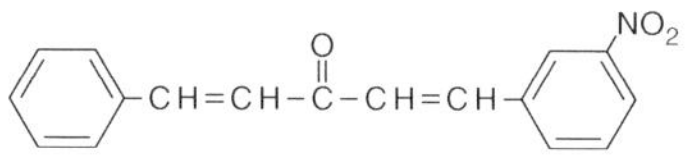

5 gm

S88,689-0
2-METHYL-1-(2-NITROPHENYL)-1-HEXEN-3-ONE

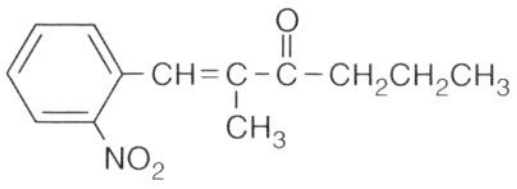

250 mg

S61,988-4
1-CYCLOPROPYL-3-(3-NITROPHENYL)-2-PROPEN-1-ONE

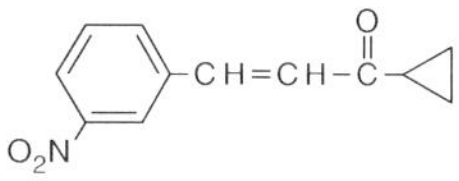

250 mg

S56,718-3
3-CHLORO-4-(3-CHLOROPHENYL)-2-BUTANONE

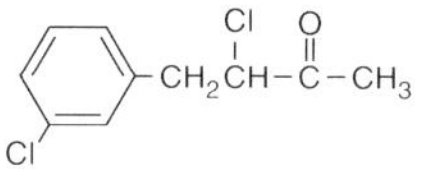

5 gm

S36,670-6
3,4-DIBROMO-4-PHENYL-2-BUTANONE

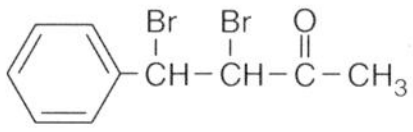

5 gm

S56,724-8
3-CHLORO-3-METHYL-4-PHENYL-2-BUTANONE

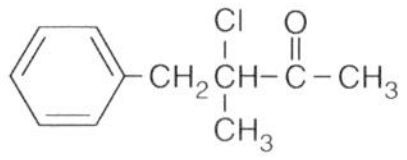

5 gm

S62,014-9
1,3-DIBROMO-1,3-DIPHENYL-2-PROPANONE

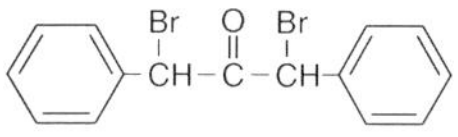

5 gm

S93,342-2
1,5-DIPHENYL-1,2,4,5-TETRABROMO-3-PENTANONE

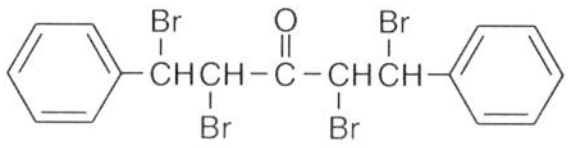

250 mg

S56,719-1
3-CHLORO-4-(2-CHLOROPHENYL)-2-BUTANONE

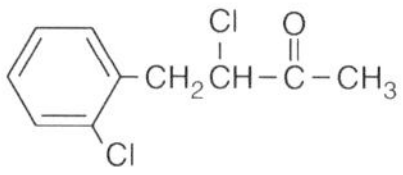

5 gm

S56,716-7
3-CHLORO-4-(4-FLUOROPHENYL)-2-BUTANONE

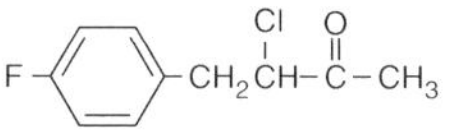

5 gm

S56,714-0
3-CHLORO-4-(4-CHLOROPHENYL)-2-BUTANONE

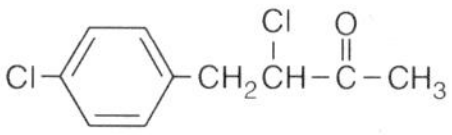

5 gm

S56,712-4
3-CHLORO-4-(4-CHLOROPHENYL)-3-METHYL-2-BUTANONE

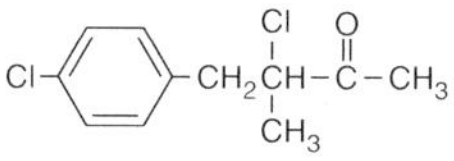

5 gm

S61,950-7
3-(2,4-DICHLOROPHENYL)-1-CYCLOPROPYL-2-PROPEN-1-ONE

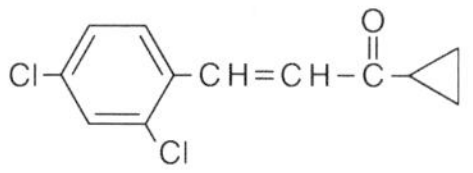

250 mg

S87,890-1
2,6-BIS(3,4-METHYLENEDIOXYBENZYLIDENE)-1-CYCLOHEXANONE

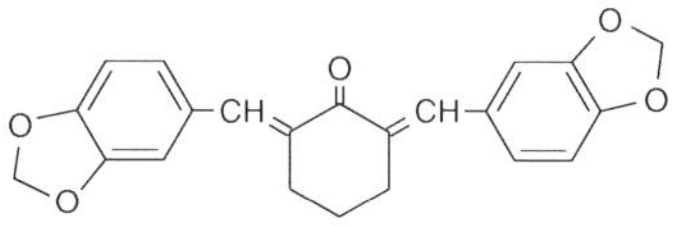

5 gm

S95,071-8
2-METHYL-3,4,5-TRIPHENYL-2,4-CYCLOPENTADIENONE

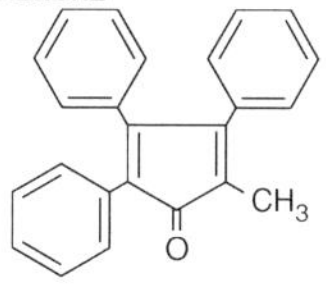

50 mg

S56,042-1
2,6-DIBENZYLIDENECYCLOHEXANONE

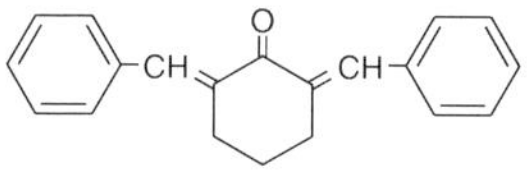

5 gm

S87,871-5
2,6-BIS(4-METHYLBENZYLIDENE)-1-CYCLOHEXANONE

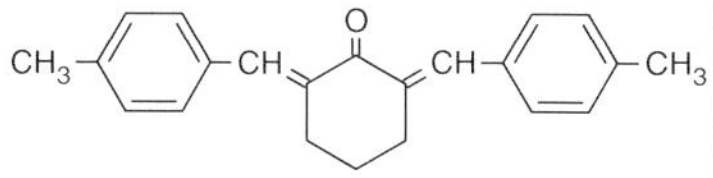

5 gm

S55,117-1
4-METHYL-4-PHENYL-2-CYCLOHEXEN-1-ONE

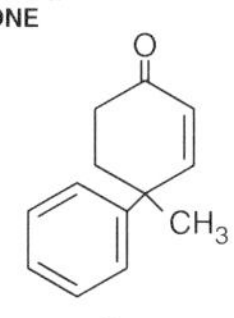

5 gm

S88,355-7
1,4A-DIMETHYL-4,4A,9,10-TETRAHYDRO-2(3H)-PHENANTHRENONE

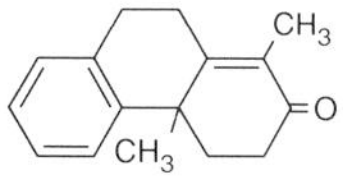

250 mg

S84,235-4
4,6-DIPHENYL-2,3,5,6-TETRAMETHYL-2,4-CYCLOHEXADIEN-1-ONE

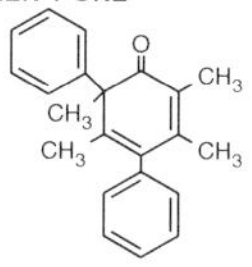

250 mg

S35,321-3
2-METHYL-1-PHENYL-1-BUTEN-3-ONE

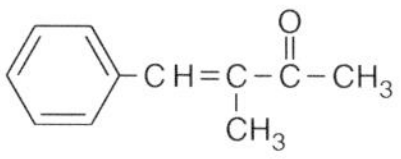

5 gm

S59,460-1
6-PHENYL-3,5-HEXADIEN-2-ONE

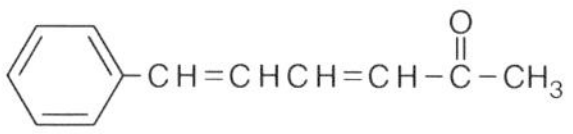

250 mg

S59,456-3
4-METHYL-1-PHENYL-1-PENTEN-3-ONE

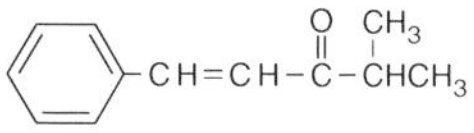

250 mg

S80,828-8
5-METHYL-4-PHENYL-1H-1,2-DIAZEPIN-6(7H)-ONE

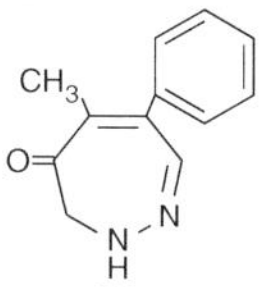

250 mg

S59,584-5
7-PHENYL-4,6-HEPTADIEN-3-ONE
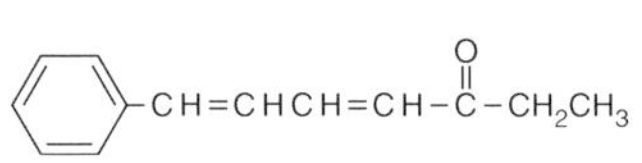
5 gm

S41,101-9
5-METHYL-1-PHENYL-1-HEXEN-3-ONE
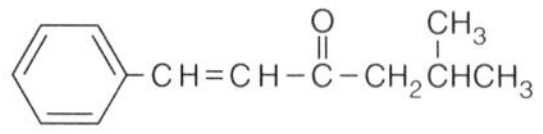
5 gm

S67,698-5
1,5-DIPHENYL-1,4-PENTADIEN-3-ONE
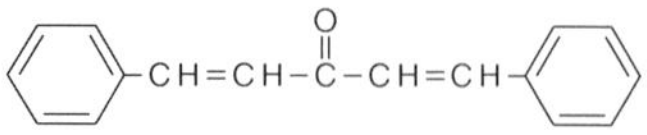
5 gm

S64,846-9
1,9-DIPHENYL-1,3,6,8-NONATETRAEN-5-ONE
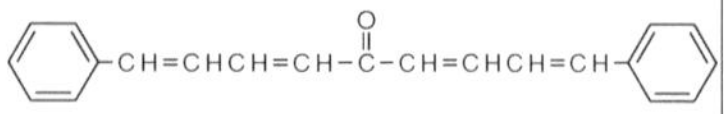
5 gm

S67,868-6
1-(4-METHYLPHENYL)-5-PHENYL-1,4-PENTADIEN-3-ONE
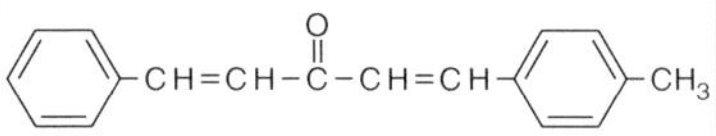
5 gm

S59,455-5
4-(2-CHLOROPHENYL)-3-BUTEN-2-ONE
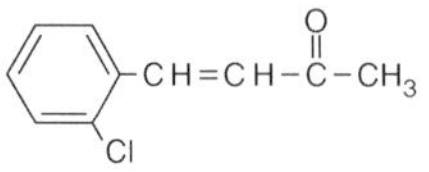
250 mg

S59,575-6
1-(2-CHLOROPHENYL)-1-PENTEN-3-ONE
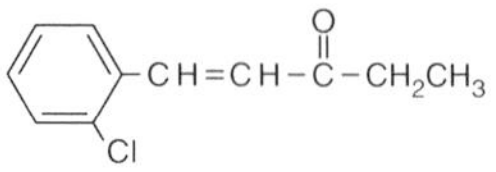
250 mg

S59,452-0
4-(4-CHLOROPHENYL)-3-BUTEN-2-ONE
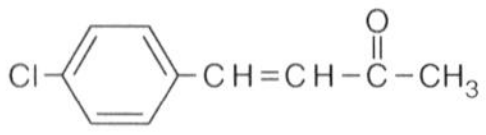
250 mg

S45,521-0
4-(P-TOLYL)-3-BUTEN-2-ONE
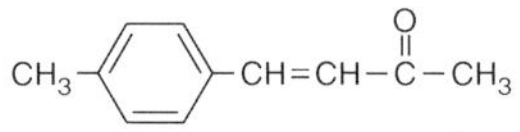
5 gm

S45,004-9
1-(P-TOLYL)-1-PENTEN-3-ONE
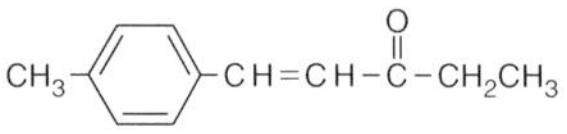
5 gm

S45,014-6
BIS(4-METHYLSTYRYL) KETONE
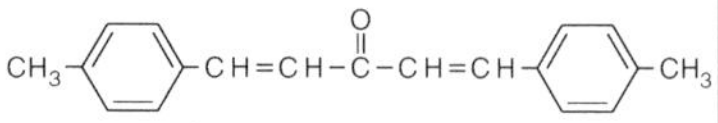
5 gm

S68,262-4
2-(2-NAPHTHYLMETHYLENE)-ALPHA-TETRALONE
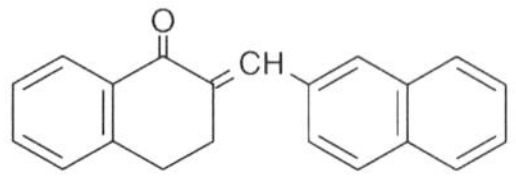
250 mg

S76,136-2
6,7-DIHYDROBENZO(B)NAPHTHO(1,2-D)-THIEPIN-8(9H)-ONE
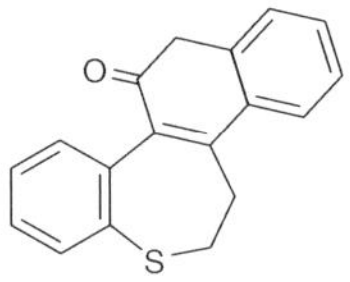
250 mg

S86,565-6
9,10-DIHYDRO-11H-9,10-ENDO-CYCLOPROPANTHRACEN-13(12H)-ONE
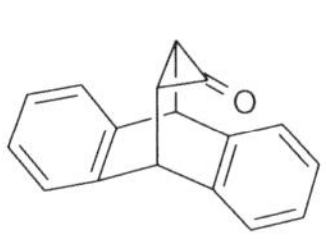
250 mg

S69,930-6
2,2-DICHLORO-2,2A,7,7A-TETRAHYDRO-1H-CYCLOBUT(A)INDEN-1-ONE
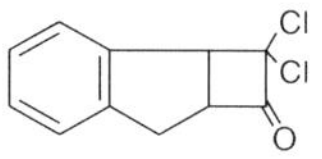
250 mg

S85,030-6
3-PHENYL-1-CYCLOPENTANONE
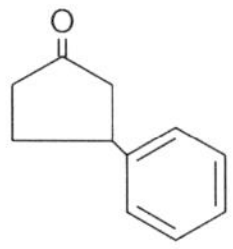
250 mg

S95,842-5
6,6-DIPHENYL-2-DIPHENYLMETHYLENEBICYCLO[3.2.0]-HEPT-3-EN-7-ONE
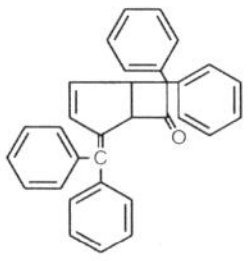
50 mg

S69,013-9
1-BENZYLIDENE-3-PIPERONYLIDENEACETONE
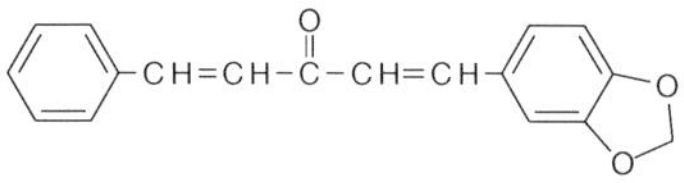
5 gm

S36,294-8
2-(4-ISOPROPYLPHENYL)CYCLOHEXANONE
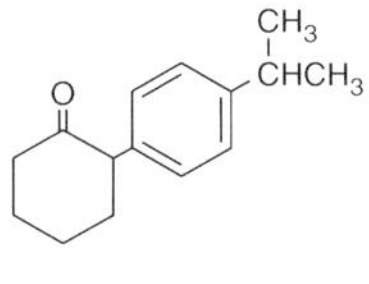
250 mg

S95,483-7
2-(ALPHA-PIPERIDINOBENZYL)-CYCLOHEXANONE
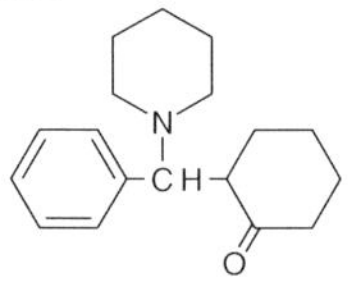
50 mg

S73,371-7
DIPHENYLMETHYL 1-METHYLCYCLOHEXYL KETONE
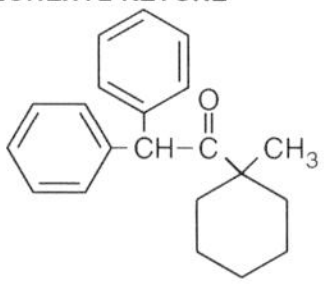
100 mg

S86,549-4
9,10-DIHYDRO-9,10-ETHANOANTRACEN-11-ONE
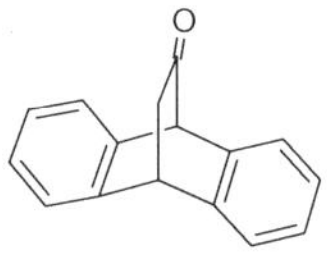
250 mg

S85,258-9
9,10-DIHYDRO-12-METHYL-9,10-ETHANOANTHRACEN-11-ONE
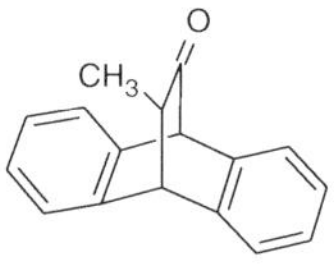
250 mg

S86,581-8
11-ACETYL-9,10-DIHYDRO-9,10-ETHANOANTHRACENE
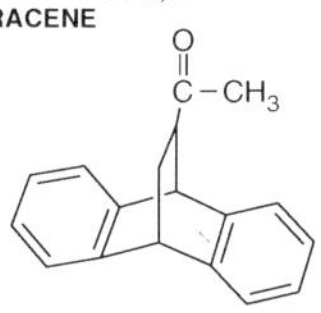
250 mg

S35,468-6
4-PHENYL-4-PROPIONYLPIPERIDINE HYDROCHLORIDE
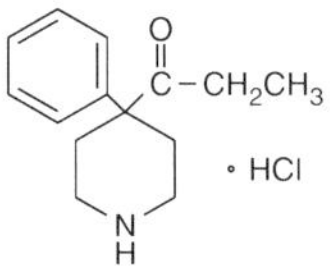
5 gm

S34,315-3
4-BUTYRYL-4-PHENYLPIPERIDINE HYDROCHLORIDE
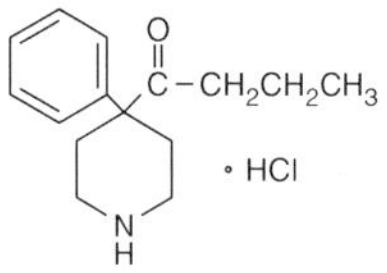
5 gm

S88,111-2
3-PHENYL-2-BUTANONE
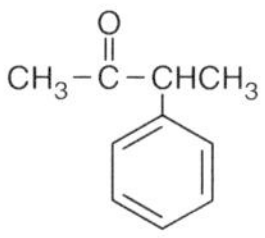
250 mg

S55,747-1
3-METHYL-1-PHENYL-2-BUTANONE
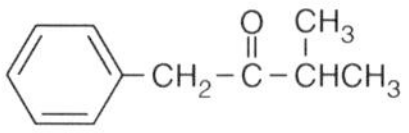
5 gm

S56,198-3
1-PHENYL-2-PENTANONE
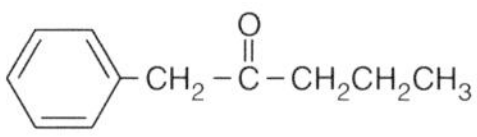
5 gm

S57,033-8
1-PHENYL-3-PENTANONE
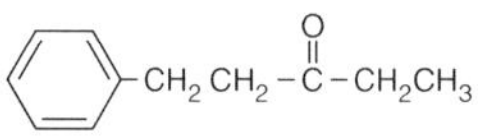
5 gm

S59,351-6
2-PHENYL-3-PENTANONE
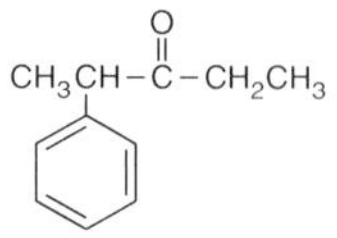
5 gm

S35,457-0
3-PHENYL-4-HEXANONE
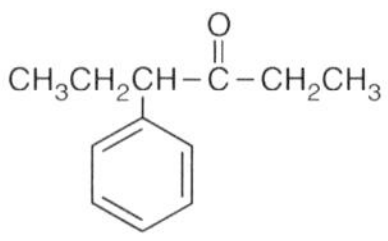
5 gm

S57,309-4
1-PHENYL-2-HEXANONE
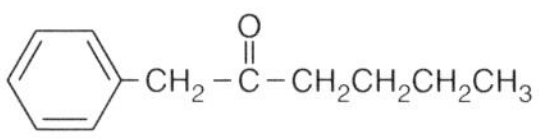
5 gm

S65,740-9
4-METHYL-4-PHENYL-2-PENTANONE
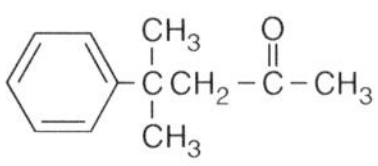
5 gm

S70,411-3
1-PHENYL-3-HEXANONE
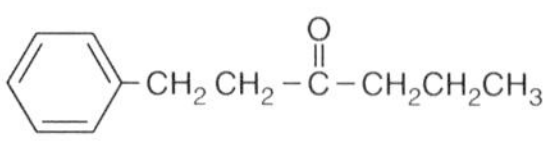
5 gm

S80,825-3
5-METHYL-6-PHENYL-1H-1,2-DIAZEPIN-4(5H)-ONE
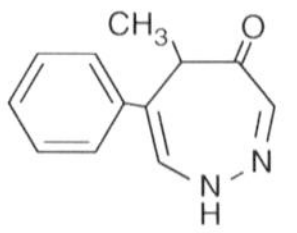
250 mg

S41,100-0
5-METHYL-1-PHENYL-3-HEXANONE

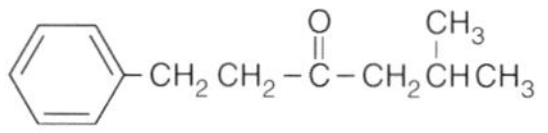

5 gm

S82,658-8
1,1,5,5-TETRAPHENYL-3-PENTANONE

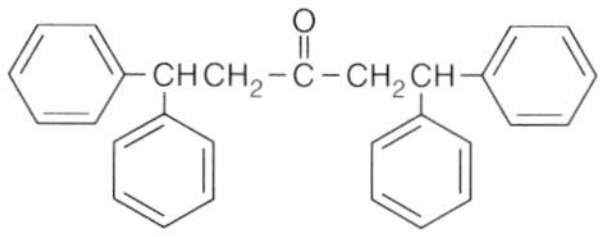

250 mg

S69,513-0
1,1'-TEREPHTHALYLIDENEBIS(4-PHENYL-3-BUTEN-2-ONE)

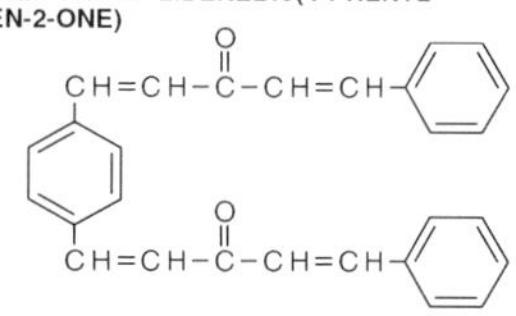

5 gm

S71,441-0
1,1-DIPHENYL-3-PENTANONE

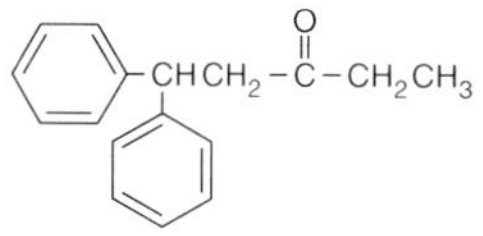

5 gm

S68,248-9
3,3-DIMETHYL-1,1-DIPHENYL-2-BUTANONE

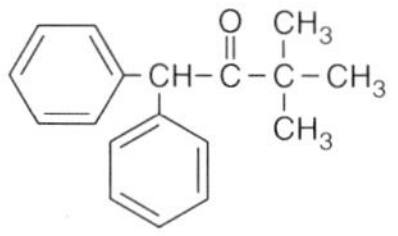

5 gm

S84,240-0
1,1,1-TRIPHENYLACETONE

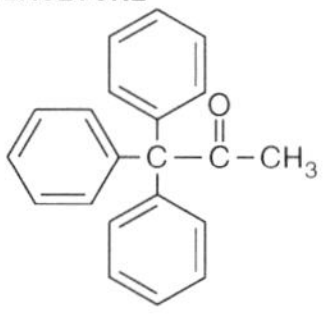

250 mg

S88,108-2
4-(O-TOLYL)-2-BUTANONE

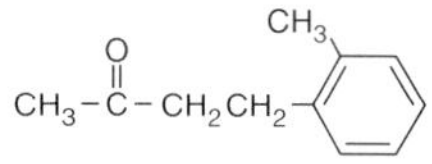

250 mg

S88,209-7
4-(P-TOLYL)-2-BUTANONE

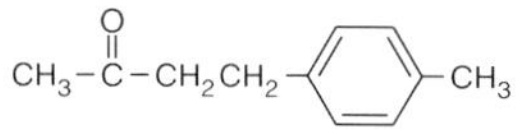

250 mg

S34,976-3
3-(4-FLUOROPHENYL)-4-HEXANONE

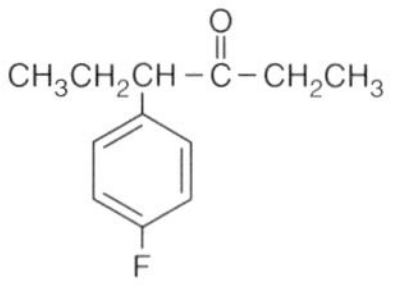

250 mg

S35,610-7
4-(P-TOLYL)-3-HEXANONE

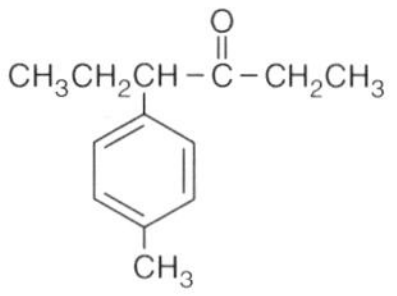

5 gm

S88,648-3
3-MESITYL-3-METHYL-2-BUTANONE

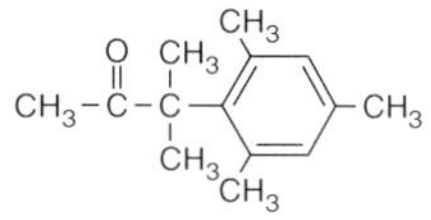

250 mg

S95,817-4
2-ACETYLDIBENZO(E,H)BICYCLO[2.2.2]-OCTA-5,7-DIENE

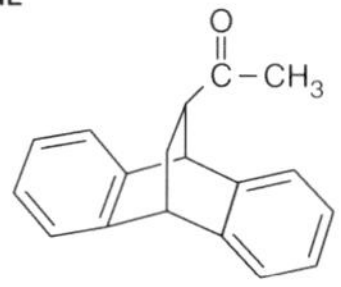

50 mg

S97,301-7
2,3,3A,4,5,9B-HEXAHYDRO-9B-METHYL-1H-BENZ(E)INDEN-5-ONE

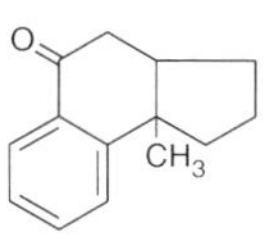

10 mg

S94,480-7
1-ACENAPHTHENONE

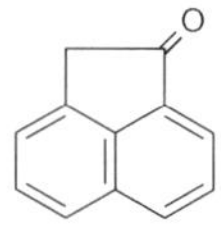

250 mg

S73,316-4
1-(2-INDANYLIDENE)-2-INDANONE

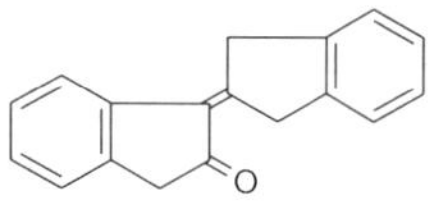

250 mg

S94,213-8
2-BROMO-4,4-DIMETHYL-6-METHOXY-1-TETRALONE

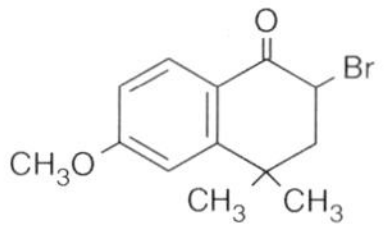

250 mg

S68,272-1
TERT-BUTYL 9-FLUORENYL KETONE

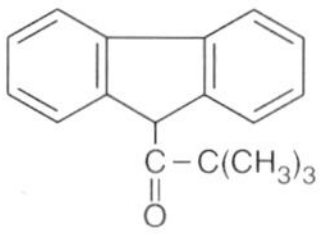

5 gm

S97,329-7
1,2,3,9,10,10A-HEXAHYDRO-6-METHYLPHENANTHREN-3-ONE

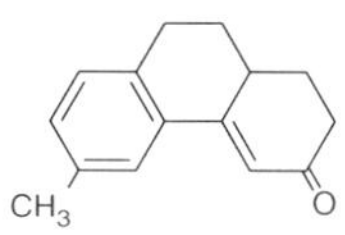

10 mg

S76,375-6
2,3,4,6,7,11B-HEXAHYDRO-1H-DIBENZO(A,F)QUINOLIZIN-13(12H)-ONE

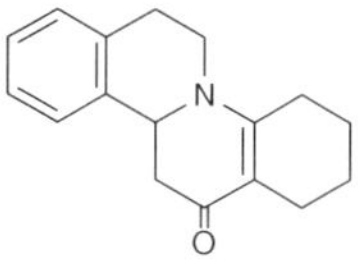

250 mg

S54,716-6
2,5-DIBENZYLIDENECYCLOPENTANONE

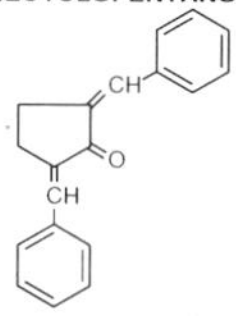

5 gm

S53,805-1
1,5-BIS(4-ISOPROPYLPHENYL)-1,4-PENTADIEN-3-ONE

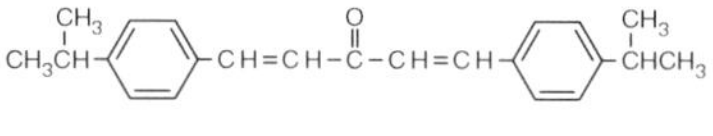

250 mg

S51,419-5
3,4-DICHLORO-4'-METHYLCHALCONE

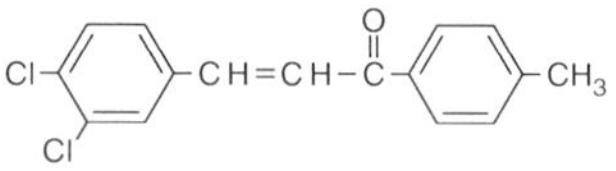

5 gm

S94,774-1
CINNAMYL DODECANOATE

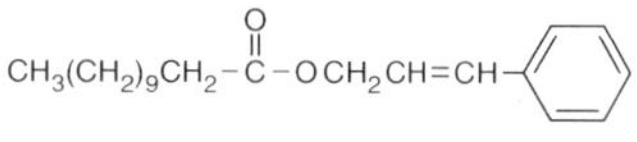

5 gm

S67,239-4
2,3'-DICHLORO-4',6'-DIMETHOXY-2'-HYDROXYACETOPHENONE

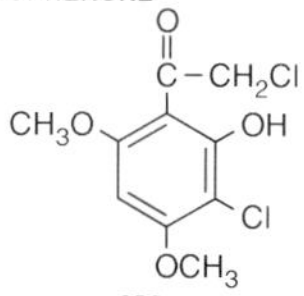

250 mg

S85,240-6
4'-BROMO-2-CHLOROACETOPHENONE

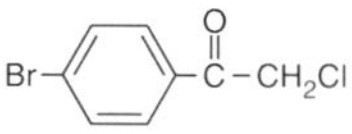

250 mg

S73,126-9
2-CHLORO-2'-HYDROXY-5'-METHOXYACETOPHENONE

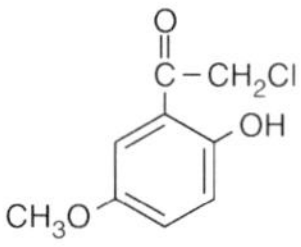

5 gm

S34,430-3
4-CHLORO-4'-METHOXYBUTYROPHENONE

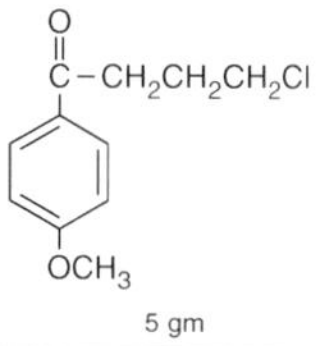

5 gm

S34,409-5
4-CHLORO-2',4'-DIMETHOXYBUTYROPHENONE

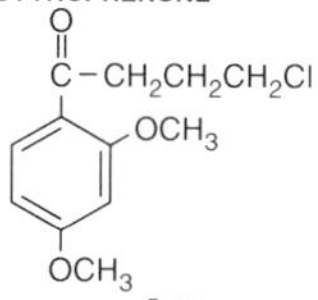

5 gm

S38,725-8
2-BROMO-5'-CHLORO-2'-(HEXADECYLOXY)-ACETOPHENONE

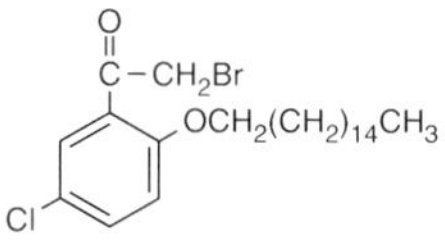

250 mg

S52,927-3
4-CHLORO-4'-PHENOXYBUTYROPHENONE

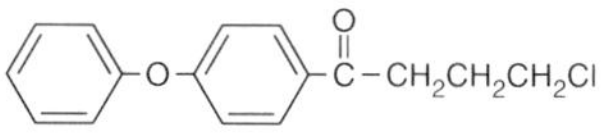

5 gm

S91,921-7
3-CHLORO-2-METHYLPROPIOPHENONE

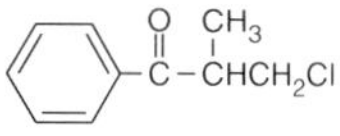

250 mg

S44,579-7
4'-BROMO-3-CHLOROPROPIOPHENONE

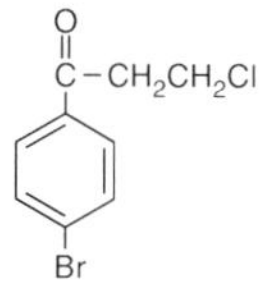

5 gm

S68,110-5
4'-METHYL-3-CHLOROPROPIOPHENONE

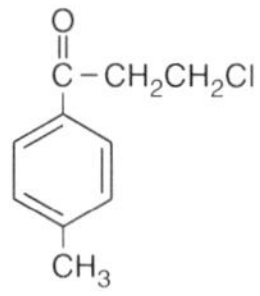

5 gm

S34,433-8
4-CHLORO-4'-METHYLBUTYROPHENONE

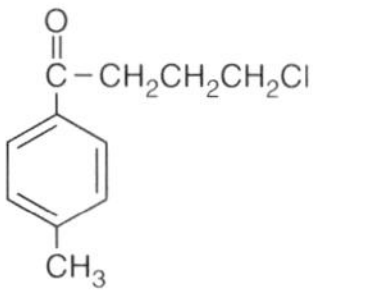

5 gm

S34,420-6
4-CHLORO-4'-ETHYLBUTYROPHENONE

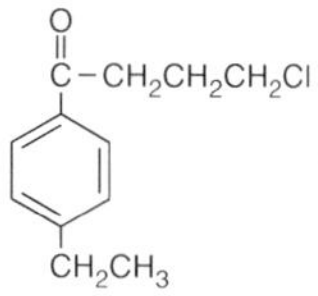

5 gm

S68,205-5
4'-(BROMOMETHYL)-2,2-DIMETHYLPROPIOPHENONE

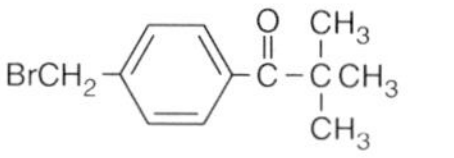

250 mg

S34,428-1
4-CHLORO-4'-ISOPROPYLBUTYROPHENONE
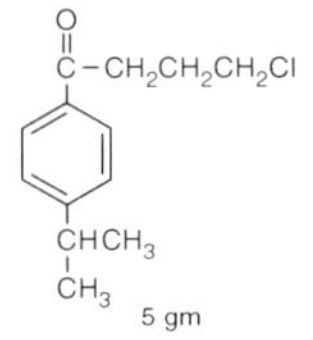
5 gm

S34,414-1
4-CHLORO-3',4'-DIMETHYLBUTYROPHENONE
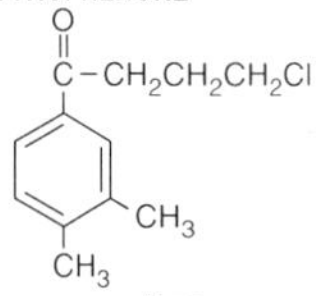
5 gm

S95,530-2
5-(3-CHLOROPROPIONYL)INDAN
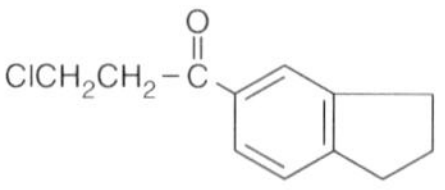
50 mg

S53,558-3
4,4'-DICHLOROBUTYROPHENONE
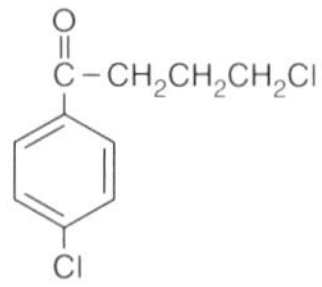
250 mg

S53,967-8
2-BROMO-2',4'-DIMETHYLACETOPHENONE
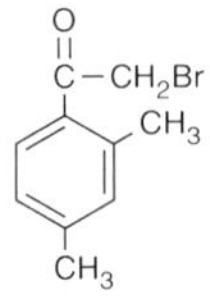
5 gm

S38,199-3
4-BROMO-2',4',6'-TRIMETHYLBUTYROPHENONE
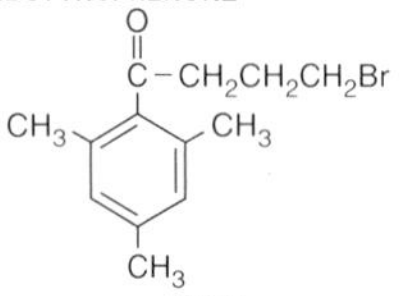
5 gm

S94,618-4
2,2-DICHLORO-3'-DICHLOROMETHYL-4'-TRICHLOROMETHYL-2-PHENYLACETOPHENONE
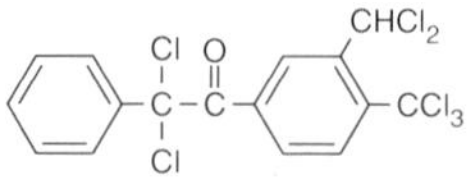
250 mg

S94,619-2
2,2-DICHLORO-2-PHENYL-4'-(TRICHLOROMETHYL)ACETOPHENONE
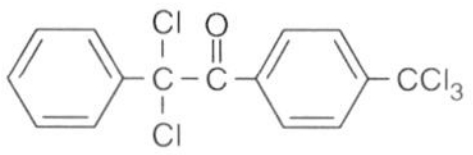
250 mg

S76,667-4
2,2,4'-TRICHLOROACETOPHENONE
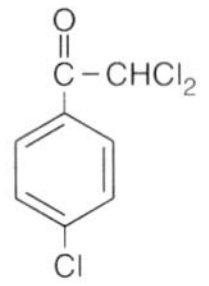
250 mg

S66,889-3
2,2-DIBROMO-2',4'-DIMETHOXYACETOPHENONE
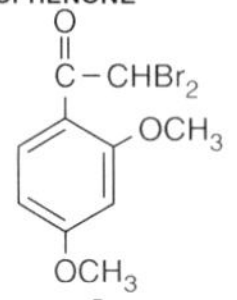
5 gm

S39,770-9
2'-(2-CHLORO-1,1,2-TRIFLUOROETHOXY)-ACETOPHENONE
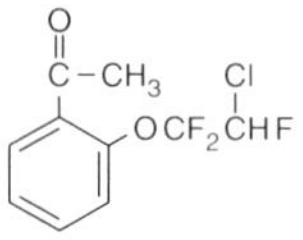
5 gm

S55,875-3
4'-METHOXY-2'-(2-CHLORO-1,1,2-TRIFLUOROETHOXY)ACETOPHENONE
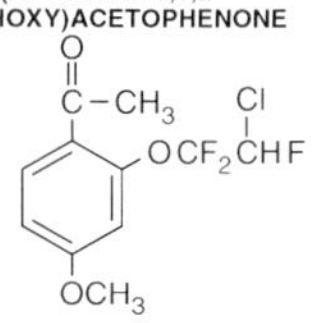
5 gm

S35,629-8
2,2,4'-TRIBROMOACETOPHENONE
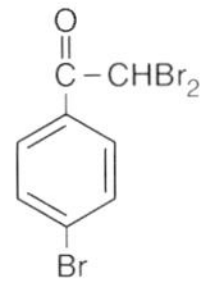
5 gm

S36,488-6
2,2,2',4'-TETRACHLOROACETOPHENONE
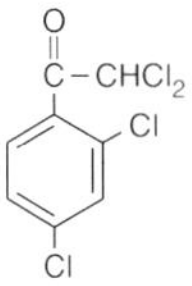
5 gm

S36,330-8
4'-METHOXY-2,2,2-TRIFLUOROACETOPHENONE
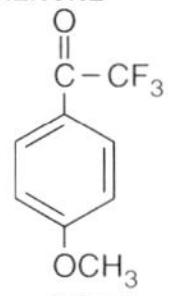
250 mg

S90,794-4
4'-(DIMETHYLAMINO)-2,2,2-TRIFLUOROACETOPHENONE
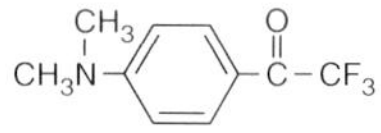
250 mg

S78,217-3
M-(TRIFLUOROMETHYL)PROPIOPHENONE
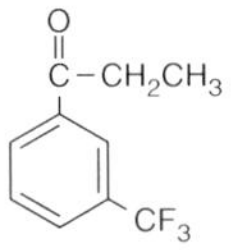
5 gm

S77,026-4
3,6-BIS(TRIFLUOROMETHYL)-9-PHENANTHRYL 2-PIPERIDINYL KETONE
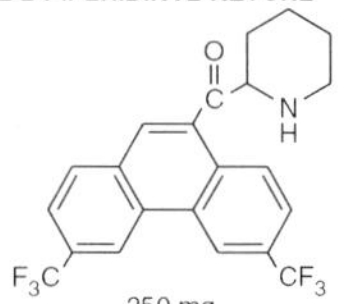
250 mg

S77,061-2
2,2,2-TRIFLUOROACETOPHENONE
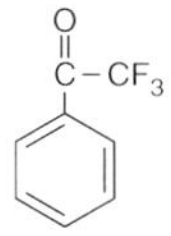
250 mg

S59,325-7
2,2,3,3,3-PENTAFLUOROPROPIOPHENONE
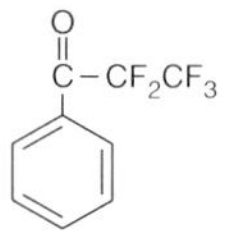
5 gm

S79,975-0
3-METHYL-4,4,4-TRIFLUOROBUTYROPHENONE
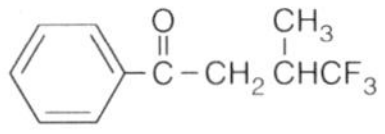
250 mg

S78,218-1
O-(TRIFLUOROMETHYL)PROPIOPHENONE
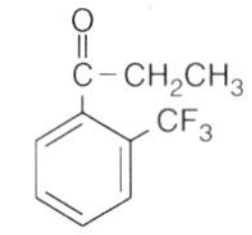
5 gm

S64,633-4
4-BROMOPHENYLGLYOXAL HYDRATE
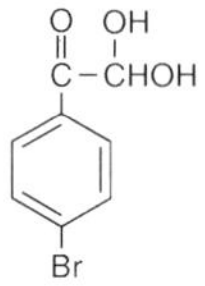
5 gm

S75,579-6
4'-CHLOROBENZOIN
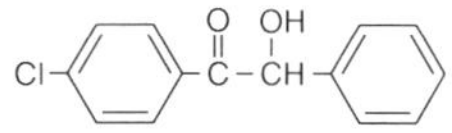
250 mg

S75,592-3
4'-CHLORO-3'-HYDROXYACETOPHENONE
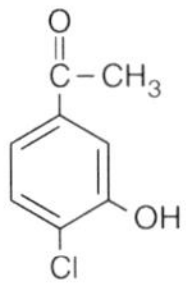
250 mg

S74,480-8
5'-AMINO-4'-CHLORO-2'-NITROACETOPHENONE
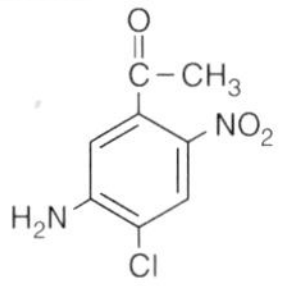
250 mg

S45,994-1
3',5'-DICHLORO-4'-HYDROXYACETOPHENONE
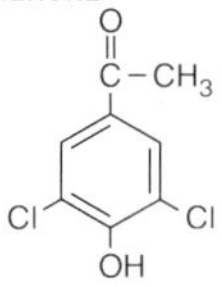
250 mg

S94,912-4
4'-BROMO-2-(2-METHOXYPHENOXY)-ACETOPHENONE
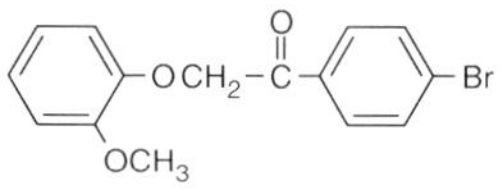
250 mg

S69,678-1
2-(2-BROMO-4-METHYLPHENOXY)-ACETOPHENONE
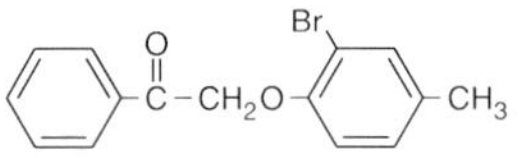
250 mg

S69,674-9
4'-BROMO-3',5'-DIMETHYL-2-PHENOXYPROPIOPHENONE
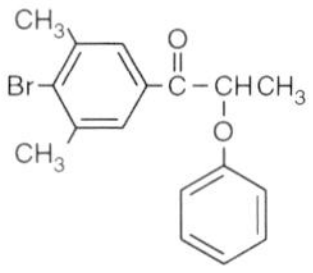
5 gm

S73,482-9
3'-CHLORO-2,2-DIMETHYL-4'-PROPOXYPROPIOPHENONE
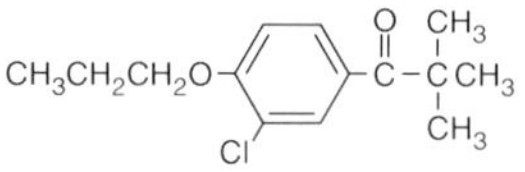
250 mg

S93,378-3
4'-CHLORO-2-(4-CHLOROPHENYLTHIO)-ACETOPHENONE
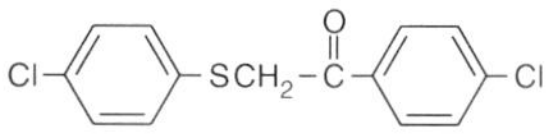
250 mg

S93,391-0
4'-CHLORO-2-(4-TOLYLTHIO)-ACETOPHENONE
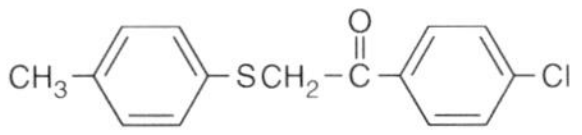
250 mg

S93,386-4
4'-BROMO-2-(4-TOLYLTHIO)-ACETOPHENONE
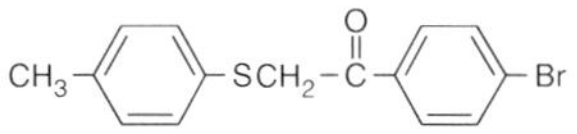
250 mg

S88,696-3
4'-BROMOPHENACYL SULFIDE
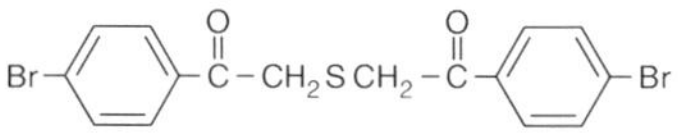
250 mg

S84,948-0
2'-CHLORO-5'-NITROACETOPHENONE
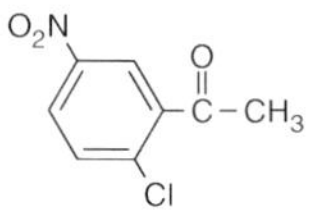
250 mg

S88,678-5
2'-BROMO-5'-NITROACETOPHENONE
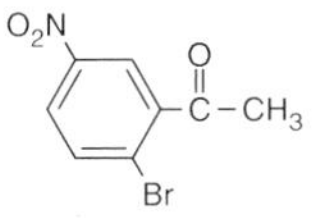
250 mg

S74,830-7
5'-CHLORO-2'-NITROPROPIOPHENONE
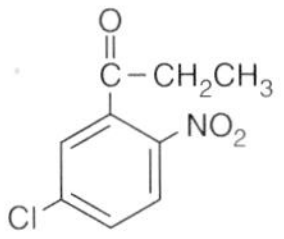
250 mg

S74,471-9
3'-BROMO-4'-METHYL-2'-NITROACETOPHENONE
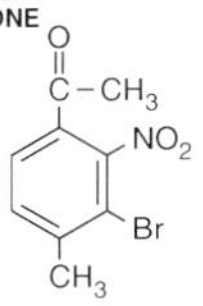
250 mg

S66,853-2
2-BROMO-4'-CHLOROPROPIOPHENONE
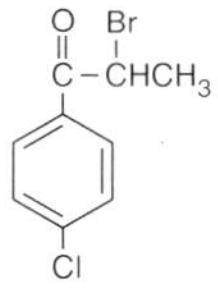
250 mg

S84,002-5
2',3',5',6'-TETRAFLUOROACETOPHENONE
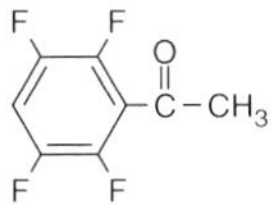
250 mg

S82,865-3
2',3',5',6'-TETRACHLOROACETOPHENONE
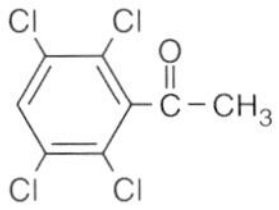
250 mg

S81,786-4
3'-BROMOVALEROPHENONE
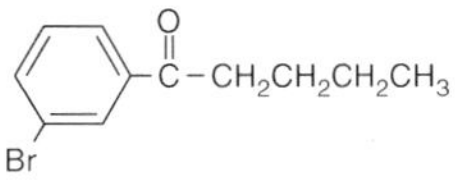
250 mg

S46,163-6
CYCLOBUTYL-4-FLUOROPHENYL KETONE
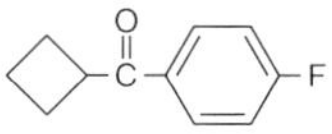
5 gm

S79,232-2
5-CHLORO-3(2H)-BENZOFURANONE
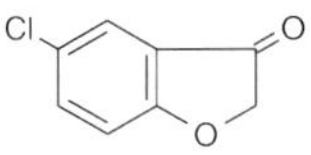
250 mg

S94,755-5
1-BENZOYL-1-BROMOCYCLOHEXANE
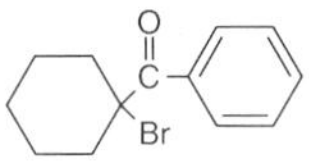
5 gm

S68,303-5
3'-CHLORO-4'-CYCLOHEXYLACETOPHENONE
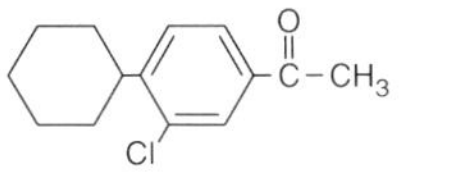
5 gm

S69,681-1
2-(4,4'-DIBROMOBIPHENYL-2-YL)-PROPIOPHENONE
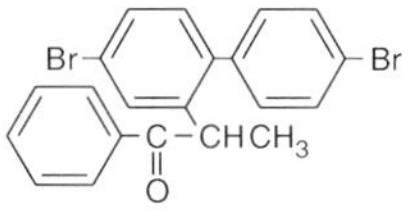
250 mg

S85,027-6
2'-IODOACETOPHENONE
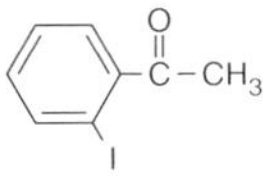
250 mg

S88,855-9
4'-IODOBUTYROPHENONE
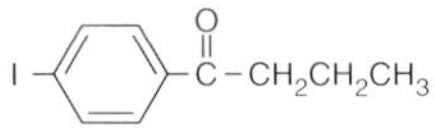
250 mg

S34,465-6
4'-CHLOROVALEROPHENONE
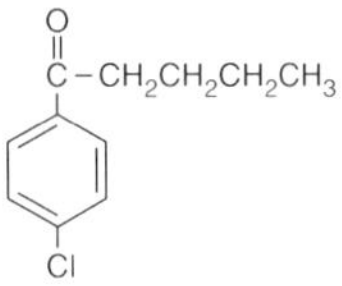
5 gm

S75,929-5
4'-CHLOROHEXANOPHENONE
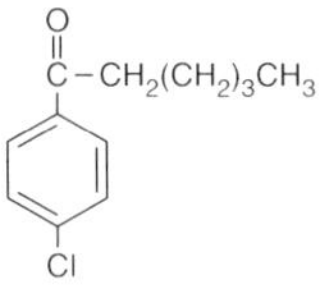
250 mg

S94,620-6
4'-CHLORO-2-(4-CHLOROPHENYL)-ACETOPHENONE
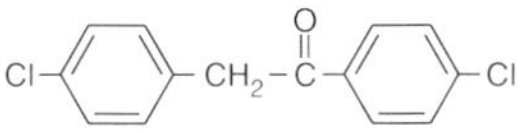
250 mg

S84,082-3
4'-CHLOROOCTADECANOPHENONE
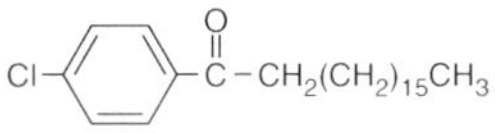
250 mg

S83,894-2
4'-(4-CHLOROPHENYL)-OCTADECANOPHENONE
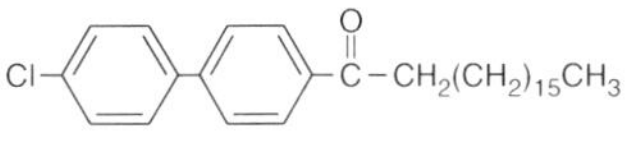
250 mg

S68,313-2
4'-FLUORO-2'-METHYLACETOPHENONE
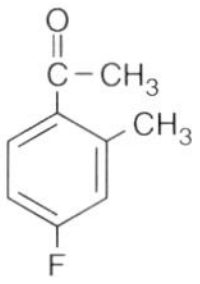
5 gm

S75,721-7
2',4'-DICHLOROPROPIOPHENONE
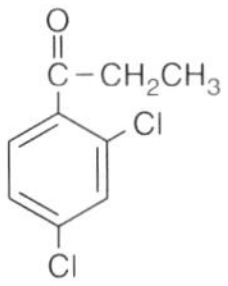
5 gm

S86,348-3
2',4'-DIHYDROXY-2-METHOXYACETOPHENONE
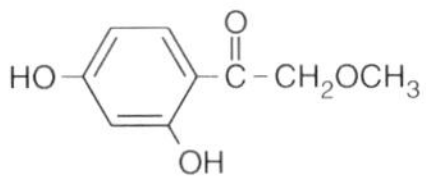
250 mg

S40,775-5
2,4'-DIHYDROXY-3'-METHOXYACETOPHENONE
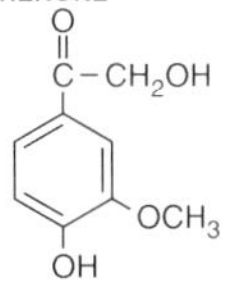
250 mg

S86,445-5
2-(2-METHOXYPHENYL)-2',4',6'-TRIHYDROXYACETOPHENONE
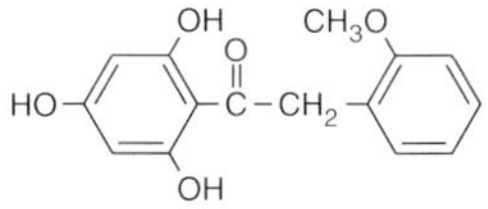
250 mg

S37,357-5
4-METHOXYBENZYL 2,4,6-TRIHYDROXYPHENYL KETONE
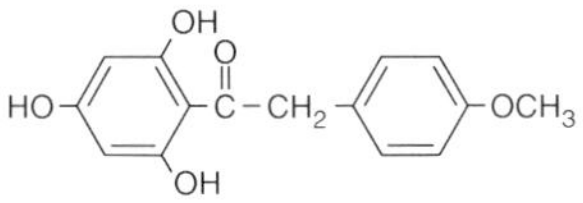
5 gm

S86,540-0
2',4'-DIHYDROXY-2-(3-METHOXYPHENYL)-ACETOPHENONE
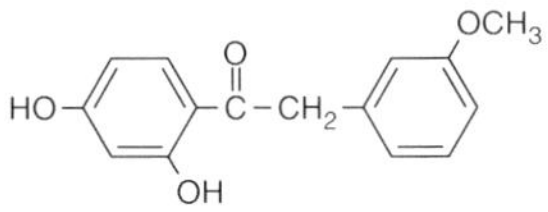
250 mg

S37,093-2
2',4',5'-TRIHYDROXYBUTYROPHENONE
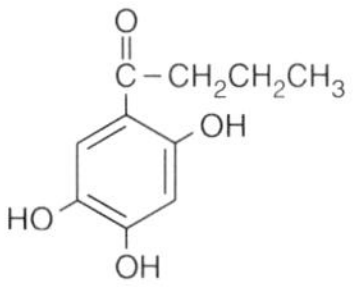
5 gm

S36,426-6
NONYL 2,3,4-TRIHYDROXYPHENYL KETONE
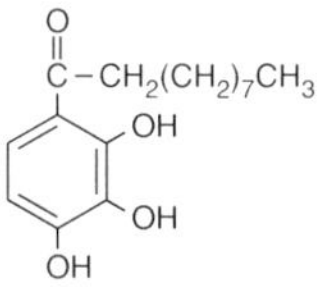
5 gm

S87,971-1
3',5'-DIHYDROXY-2-(ISOPROPYLAMINO)-ACETOPHENONE HEMISULFATE
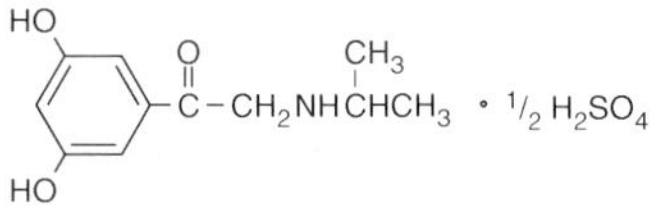
5 gm

S86,003-4
2',4'-DIHYDROXY-3'-METHYL-5'-NITROACETOPHENONE
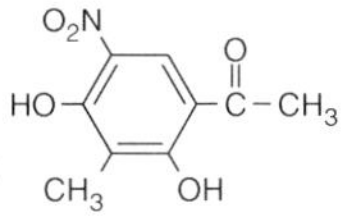
250 mg

S56,386-2
3,4'-DIHYDROXYBUTYROPHENONE
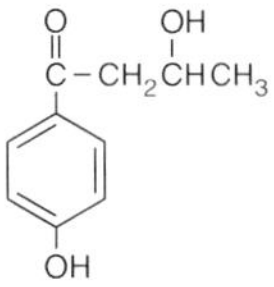
5 gm

S76,322-5
2',4'-DIHYDROXY-6'-METHYLACETOPHENONE
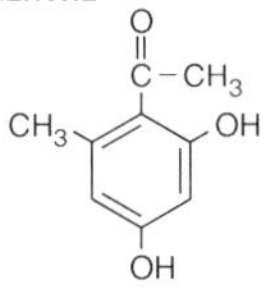
250 mg

S44,663-7
2',4',6'-TRIHYDROXYPROPIOPHENONE MONOHYDRATE
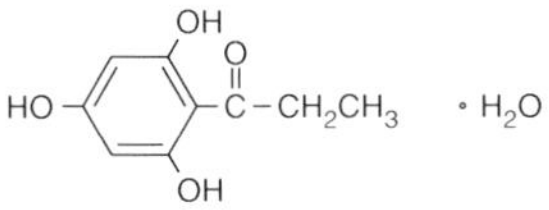
5 gm

S37,252-8
2-HYDROXY-4'-METHOXY-2-PHENYLACETOPHENONE
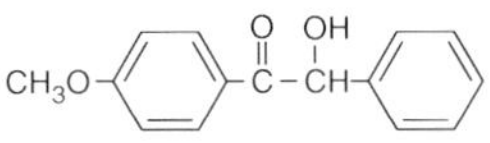
250 mg

S86,360-2
4'-BENZYLOXY-2'-HYDROXYACETOPHENONE
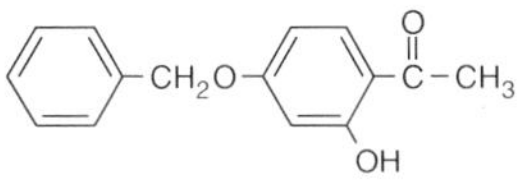
250 mg

S84,668-6
2'-HYDROXY-3',4',6'-TRIMETHOXYACETOPHENONE
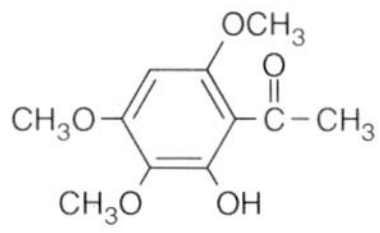
250 mg

S73,970-7
4'-(2-HYDROXY-3-METHOXYBENZYLAMINO)-ACETOPHENONE
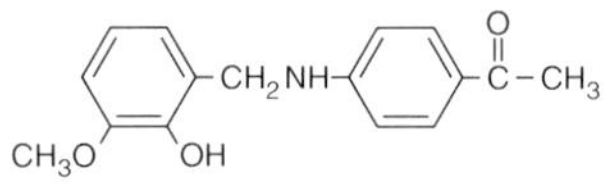
5 gm

S73,973-1
3'-(2-HYDROXY-3-METHOXYBENZYLAMINO)-ACETOPHENONE
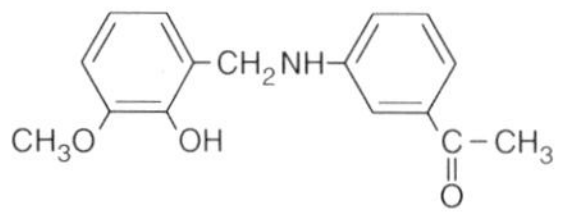
5 gm

S50,459-9
2',4',6'-TRIHYDROXYACETOPHENONE 4'-BETA-NEOHESPERIDOSIDE TRIHYDRATE
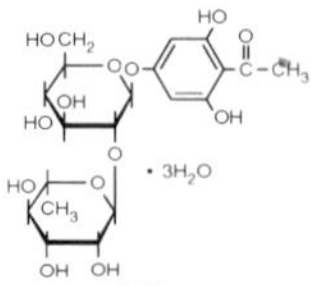
250 mg

S50,462-9
4-METHOXY-2',3,4',6'-TETRAHYDROXYHYDROCINNAMOPHENONE 4'-BETA-NEOHESPERIDOSIDE
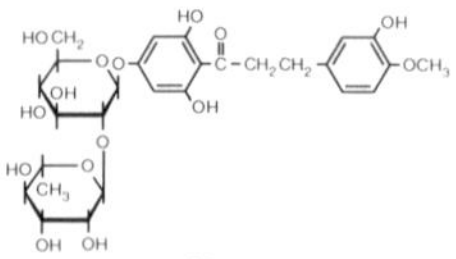
250 mg

S56,224-6
4'-(2-HYDROXYETHOXY)ACETOPHENONE
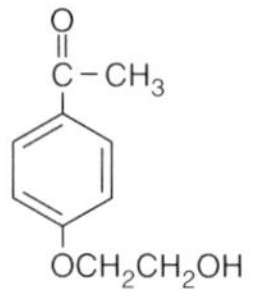
250 mg

S35,504-6
4-HYDROXY-2,3,6-TRIMETHOXY-5H-BENZOCYCLOHEPTEN-5-ONE
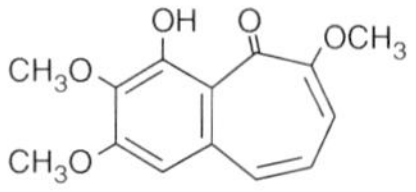
5 gm

S80,888-1
5'-ALLYLOXY-2'-HYDROXYACETOPHENONE
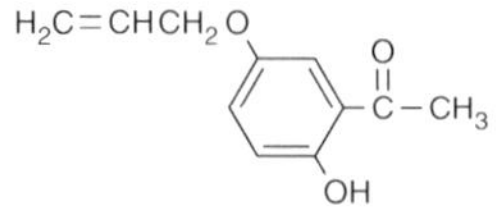
250 mg

S73,235-4
3-HYDROXY-4'-METHOXYACRYLOPHENONE, SODIUM SALT
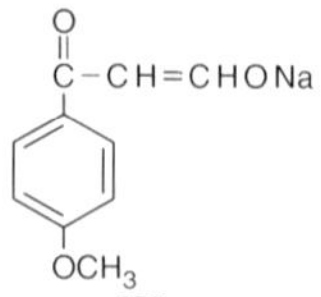
250 mg

S80,946-2
2,6-DIACETYL-3-METHOXYPHENOL
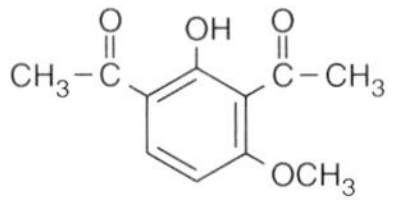
250 mg

S80,885-7
4'-HYDROXY-2'-METHOXY-5'-PROPYLPROPIOPHENONE
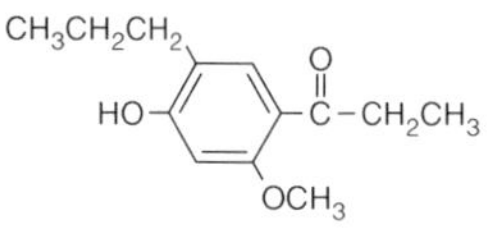
250 mg

S84,233-8
2,6-DI-TERT-BUTYL-3,4-DIPHENYL-4-HYDROXY-ALPHA-TETRALONE
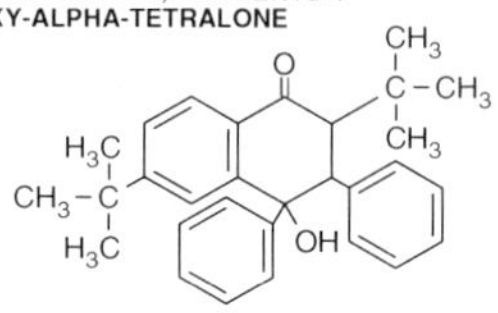
250 mg

S67,798-1
4-(TRIMETHYLACETYL)BENZYL ALCOHOL
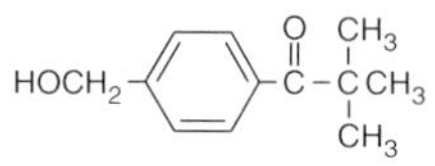
250 mg

S88,878-8
6-HYDROXY-3(2H)-BENZOFURANONE
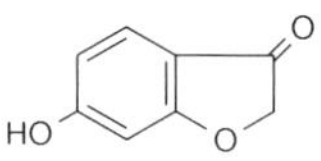
250 mg

S34,632-2
3',4'-DIHYDROXY-2-(ISOPROPYLAMINO)-ACETOPHENONE HYDROCHLORIDE
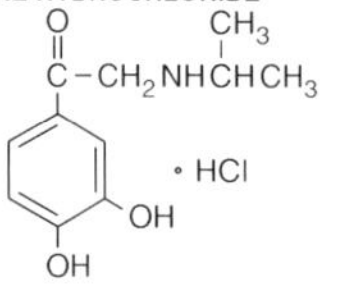
5 gm

S56,080-4
9-[4-(DIMETHYLAMINO)PHENYL]-2,6,7-TRIHYDROXY-3H-XANTHEN-3-ONE
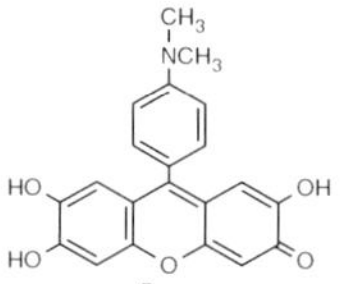
5 gm

S58,677-3
9-[4-(DIMETHYLAMINO)PHENYL]-2,6,7-TRIHYDROXY-3H-XANTHEN-3-ONE SULFATE
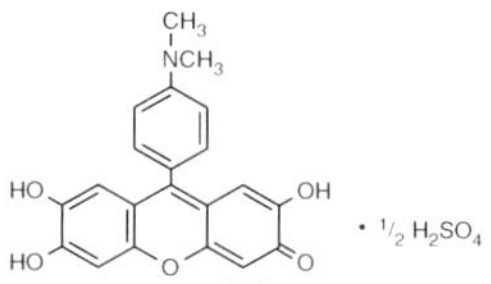
250 mg

S43,927-4
9-METHYL-2,3,7-TRIHYDROXY-6-FLUORONE
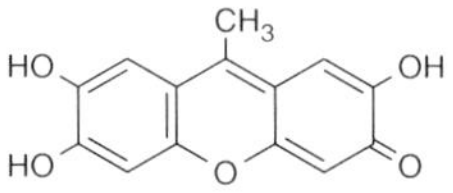
250 mg

S73,945-6
2-HYDROXY-3-PHENYL-3-(2,3,5,6-TETRAMETHYLPHENYL)-2',4',6'-TRIMETHYLPROPIOPHENONE
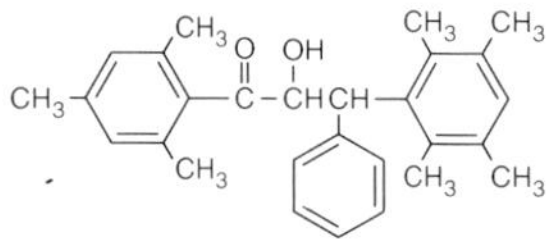
5 gm

S79,206-3
2,4-DIMETHYL-6-HYDROXYCYCLOHEXYL PHENYL KETONE
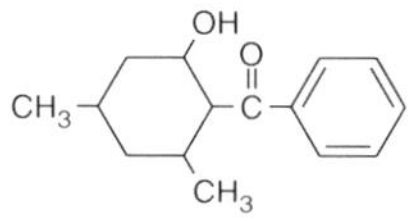
250 mg

S81,268-4
2'-HYDROXY-4'-PHENYLOCTADECANOPHENONE
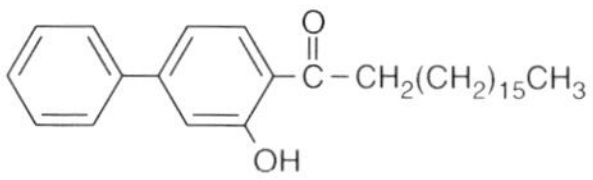
250 mg

S88,301-8
2-(HYDROXYMETHYLENE)-4-PHENYL-1-TETRALONE
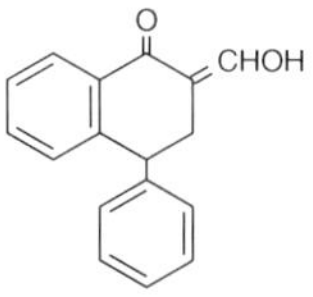
250 mg

S51,481-0
2'-HYDROXYDODECANOPHENONE
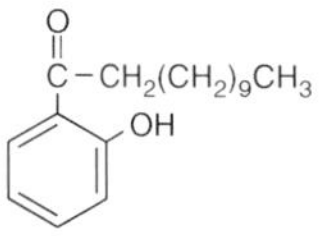
250 mg

S46,228-4
2'-HYDROXY-4',6'-DIMETHYLACETOPHENONE
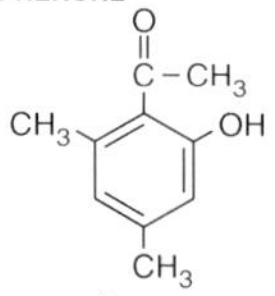
5 gm

S74,046-2
4-HYDROXY-7-METHYL-1-INDANONE
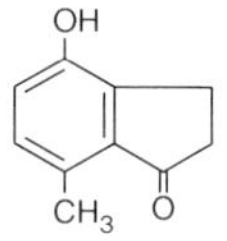
250 mg

S67,930-5
4'-(2-HYDROXYETHYLTHIO)-3'-NITROACETOPHENONE
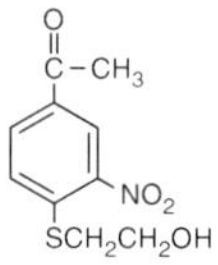
250 mg

S73,975-8
3'-(2-HYDROXYBENZYLAMINO)-ACETOPHENONE
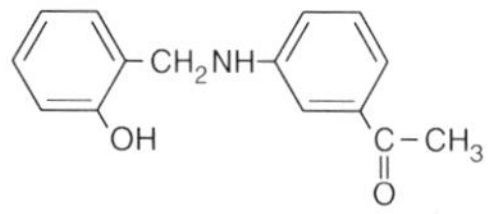
5 gm

S56,462-1
2-BUTYLAMINO-4'-HYDROXYACETOPHENONE HYDROCHLORIDE
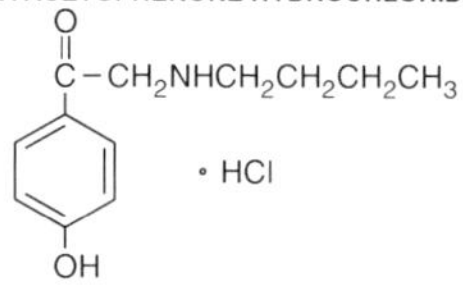
5 gm

S44,997-0
4'-[4-(2-HYDROXYETHYL)PIPERAZINO]-ACETOPHENONE
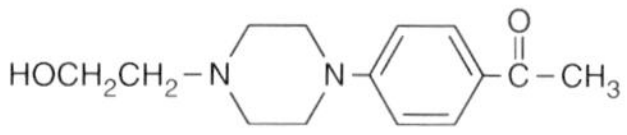
5 gm

S43,313-6
4'-(4-HYDROXYPIPERIDINO)-ACETOPHENONE
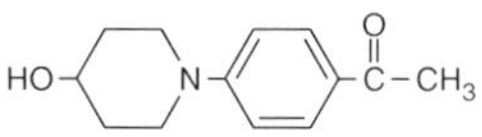
5 gm

S45,053-7
3-BENZOYL-4-HYDROXY-1-METHYL-4-PHENYLPIPERIDINE
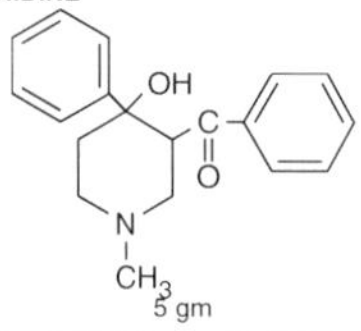
5 gm

S76,755-7
4'-[(5-HYDROXY-1,2,3,4-TETRAHYDRO-2-ISOQUINOLINYL)METHYL]2,2-DIMETHYLPROPIOPHENONE
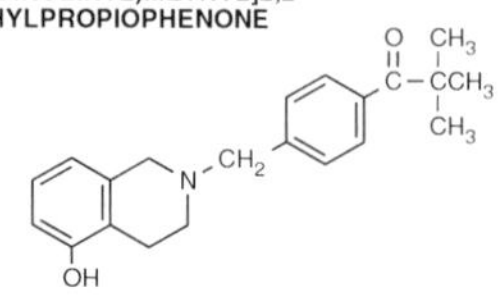
250 mg

S94,367-3
1,2-DIHYDRO-4(3H)CINNOLINONE HYDROCHLORIDE
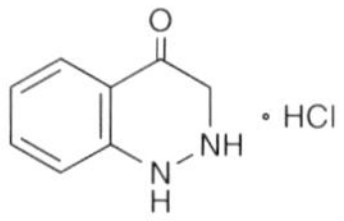
250 mg

S92,841-0
5-[2-(1,3-BENZODIOXOL-5-YL)-2-(HYDROXY)ACETYL]-1,3-BENZODIOXOLE
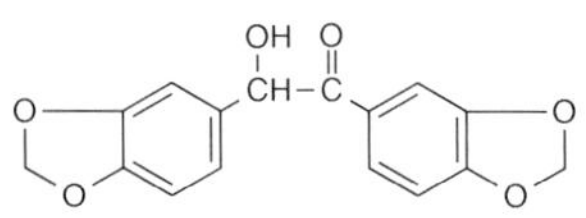
250 mg

S43,978-9
4'-HYDROXYBUTYROPHENONE
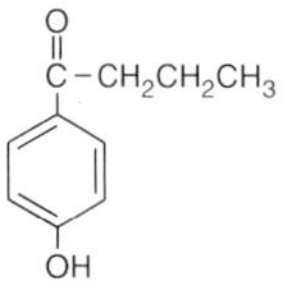
5 gm

S55,837-0
4'-HYDROXYNONANOPHENONE
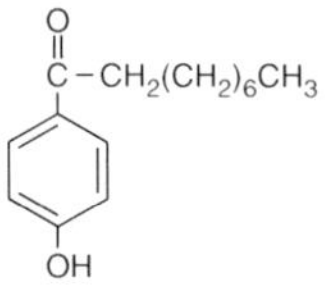
250 mg

S50,989-2
4'-HYDROXYDODECANOPHENONE
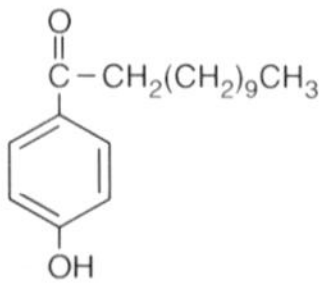
250 mg

S38,656-1
4'-HYDROXYHEXADECANOPHENONE
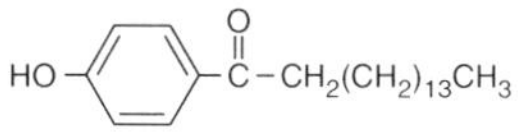
250 mg

S91,423-1
3',5'-DI-TERT-BUTYL-4'-HYDROXYACETOPHENONE
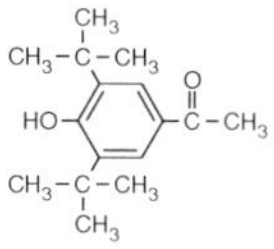
250 mg

S94,622-2
2',5'-DIMETHYL-4'-HYDROXYACETOPHENONE
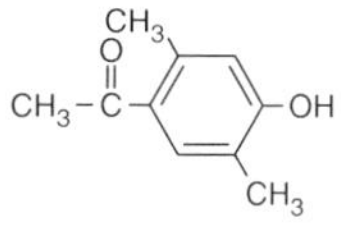
250 mg

S64,862-0
5'-CHLORO-2'-HYDROXY-4'-METHYLHEXANOPHENONE
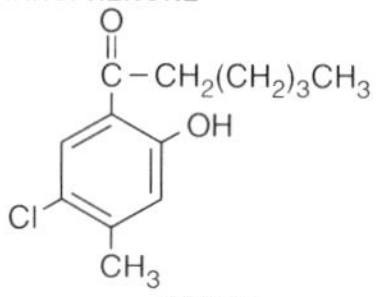
250 mg

S89,450-8
6-HYDROXY-1-TETRALONE
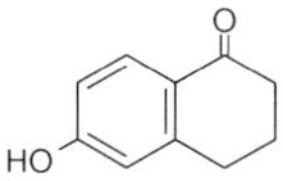
250 mg

S80,893-8
2,2',4'-TRIHYDROXYACETOPHENONE
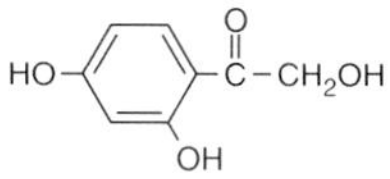
250 mg

S86,399-8
2',4'-BIS(BENZYLOXY)-2-METHOXYACETOPHENONE
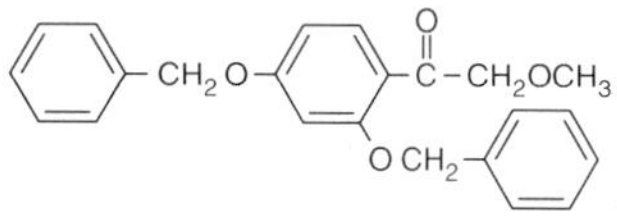
250 mg

S92,392-3
4-(BENZYLOXY)-3,5-DIMETHOXYACETOPHENONE
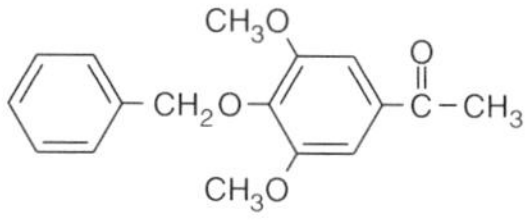
250 mg

S84,995-2
2,3-DIHYDRO-6,7-DIMETHOXY-2,2-DIMETHYL-4H-1-BENZOPYRAN-4-ONE
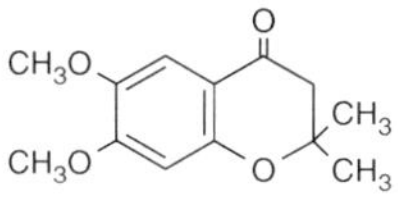
250 mg

S81,210-2
2-AMINO-3',4'-DIMETHOXYACETOPHENONE HYDROCHLORIDE
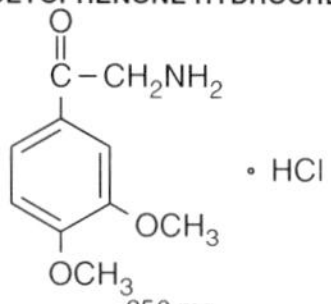
250 mg

S88,316-6
4',5'-DIMETHOXY-2'-METHYLACETOPHENONE
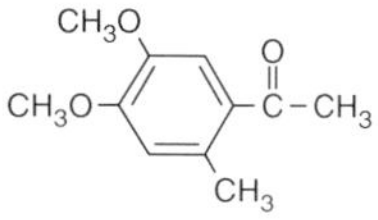
250 mg

S94,910-8
2-(2-METHOXYPHENOXY)ACETOPHENONE
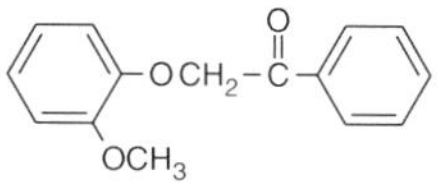
250 mg

S92,882-8
4'-METHOXY-2-PHENOXYACETOPHENONE
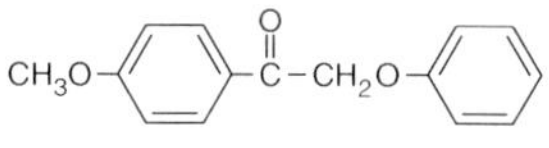
250 mg

S94,911-6
4'-METHOXY-2-(2-METHOXYPHENOXY)-ACETOPHENONE
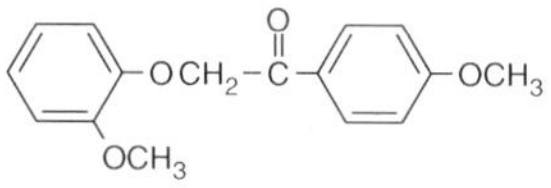
250 mg

S95,525-6
7-HYDROXY-5-METHYL-1-INDANONE
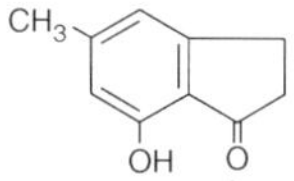
50 mg

S88,830-3
3',5'-DIMETHOXYVALEROPHENONE
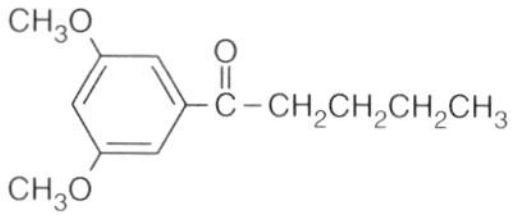
250 mg

S86,069-7
2',5'-DIMETHOXY-4'-METHYLACETOPHENONE
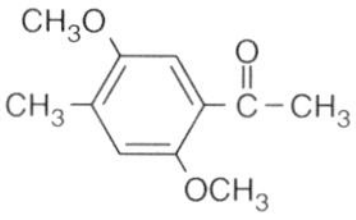
250 mg

S76,323-3
2',4'-DIMETHOXY-6'-METHYLACETOPHENONE
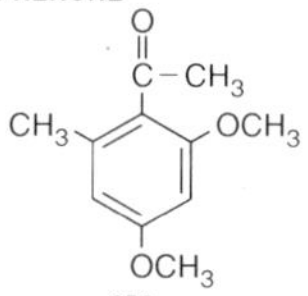
250 mg

S76,233-4
4'-METHOXY-4-[1-(4-METHOXYPHENYL)-3-OXO-1-CYCLOHEXEN-4-YL]-BUTYROPHENONE
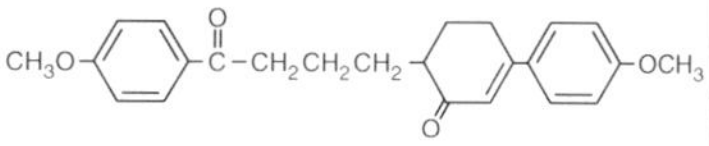
250 mg

S42,500-1
3,3',4,4',5,5'-HEXAMETHOXYLOBELANINE
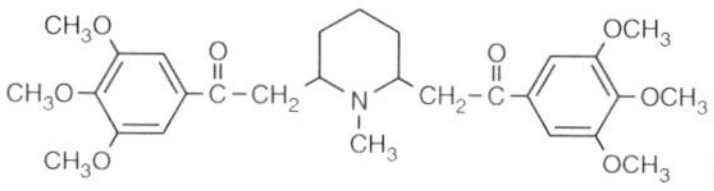
250 mg

S95,525-6
3'-(3,4-DIMETHOXYPHENYL)-1-OXOSPIRO(INDAN-2,2'-OXIRANE)
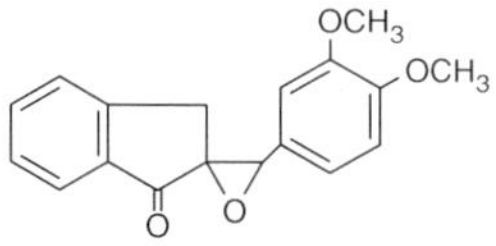
5 gm

S64,290-8
4-BROMO-6,7-DIMETHOXY-1-INDANONE
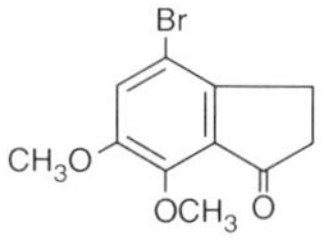
250 mg

S84,987-1
2,3-DIHYDRO-8-METHOXY-4H-1-BENZOPYRAN-4-ONE
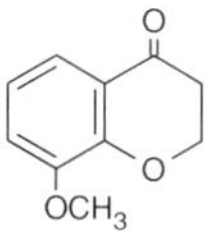
250 mg

S35,439-2
2-PHENOXYPROPIOPHENONE
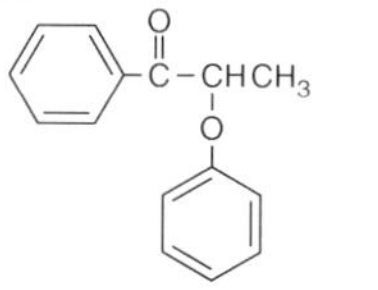
5 gm

S56,623-3
2',4',5'-TRIMETHOXYACETOPHENONE
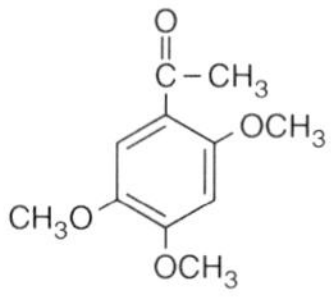
5 gm

S43,370-5
2',4',5'-TRIETHOXYACETOPHENONE
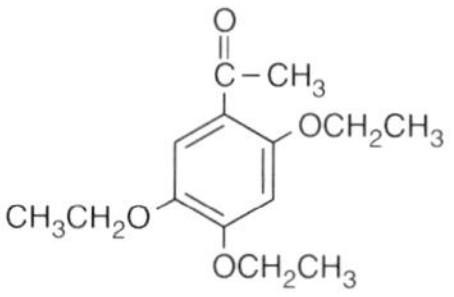
5 gm

S54,126-5
3',4'-DIMETHOXYBUTYROPHENONE
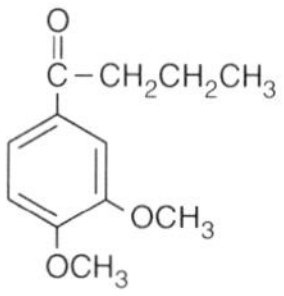
250 mg

S67,150-9
2',5'-DIMETHOXYPROPIOPHENONE
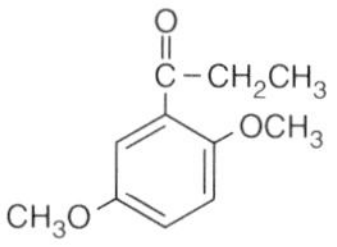
5 gm

S83,897-7
2'-METHOXY-4'-PHENYLOCTADECANOPHENONE
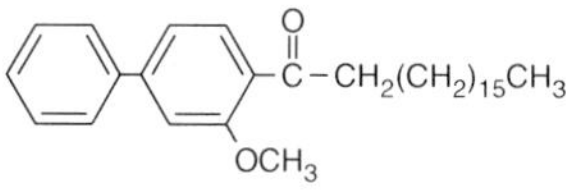
250 mg

S88,336-0
2'-METHOXY-5'-METHYLACETOPHENONE
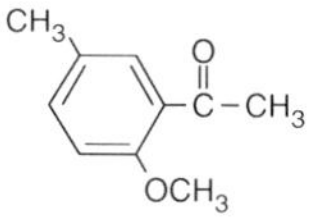
250 mg

S87,690-9
7-PHENOXY-1-TETRALONE
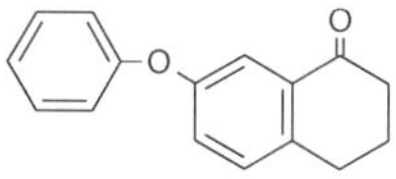
250 mg

S97,328-9
1,2,3,4-TETRAHYDRO-9-METHOXYPHENANTHREN-1-ONE
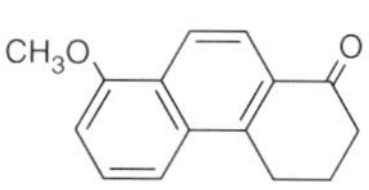
10 mg

S83,418-1
3-[2-(DIETHYLAMINO)ETHYLTHIO]-3-[4-(DIMETHYLAMINO)PHENYL]-4'-METHOXYPROPIOPHENONE HYDROCHLORIDE
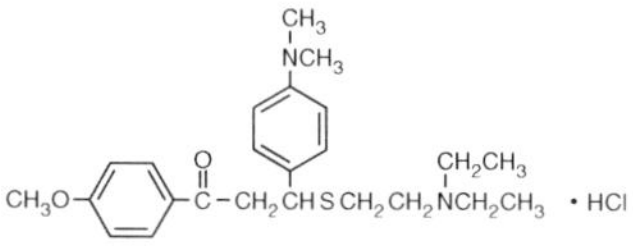
250 mg

S88,485-5
2-(BUTYLTHIOMETHYLENE)-6-METHOXY-1-TETRALONE
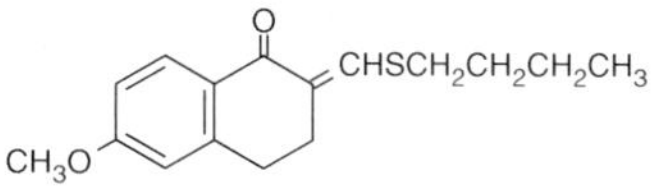
250 mg

S34,115-0
3'-AMINO-4'-METHOXYACETOPHENONE
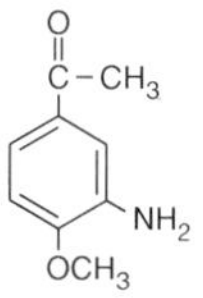
250 mg

S95,524-8
3-AMINO-4-METHOXY-1-INDENONE
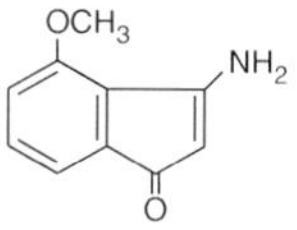
50 mg

S44,781-1
(-)-BETA-NARCOTINE
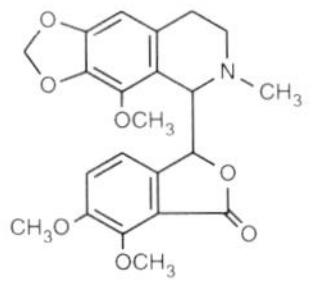
250 mg

S93,521-2
3-[(4-METHOXYPHENYL)AMINO]-PROPIOPHENONE
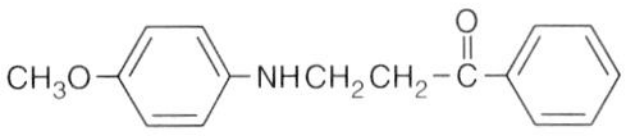
250 mg

S38,905-6
2'-(2-DIETHYLAMINOETHOXY)-DIHYDROCHALCONE HYDROCHLORIDE
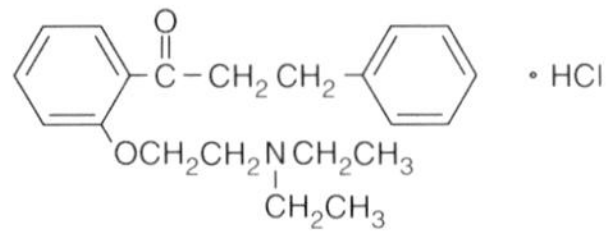
5 gm

S55,481-2
3-(DIMETHYLAMINO)-3'-METHOXYPROPIOPHENONE HYDROCHLORIDE
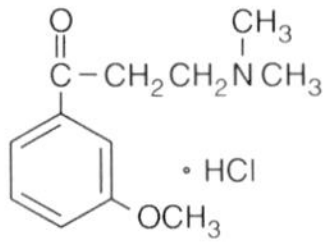
5 gm

S42,589-3
8-(2-OXOPROPYL)NEOOXYBERBERINE
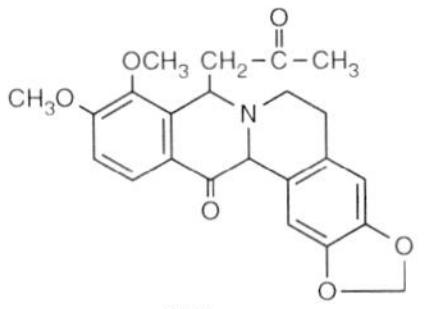
250 mg

S40,352-0
4'-BUTOXY-3-PIPERIDINOPROPIOPHENONE HYDROCHLORIDE
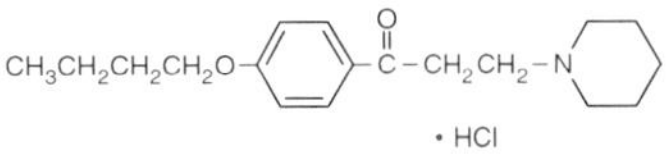
5 gm

S89,361-7
6-METHOXY-2-(4-PHENYL-1,2,5,6-TETRAHYDRO-1-PYRIDYLMETHYL)-1-TETRALONE
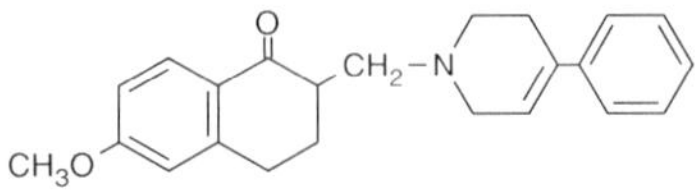
250 mg

S89,357-9
3-(DIMETHYLAMINO)-4'-PHENOXYPROPIOPHENONE HYDROCHLORIDE
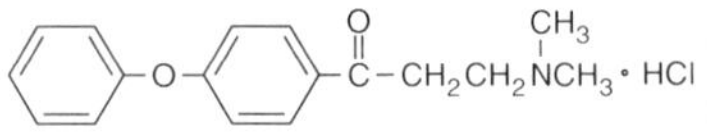
250 mg

S93,523-9
3-(DIMETHYLAMINO)-4'-METHOXYPROPIOPHENONE HYDROCHLORIDE
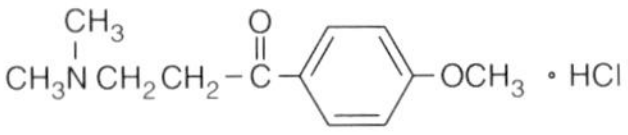
250 mg

S54,284-9
2-DIMETHYLAMINOMETHYL-4'-METHOXY-2-PHENYLPROPIOPHENONE HYDROCHLORIDE
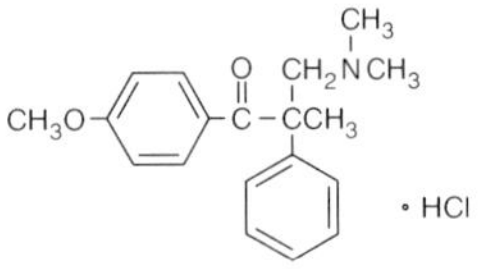
250 mg

S89,342-0
[3-OXO-3-(4-PHENOXYPHENYL)PROPYL]-TRIMETHYLAMMONIUM IODIDE
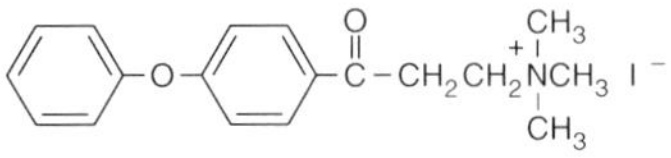
250 mg

S89,355-2
(6-METHOXY-1-OXO-1,2,3,4-TETRAHYDRO-2-NAPHTHYLMETHYL)-TRIMETHYLAMMONIUM IODIDE
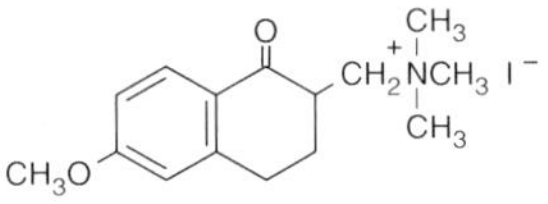
250 mg

S35,221-7
4'-METHOXY-3'-NITROACETOPHENONE
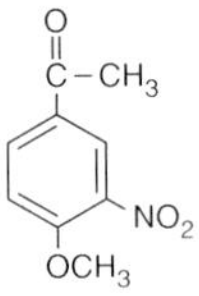
5 gm

S82,712-6
4'-(4-NITROPHENOXY)-OCTADECANOPHENONE
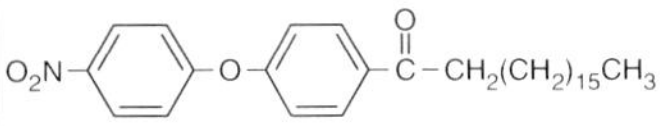
250 mg

S94,673-7
2-(4-METHOXYPHENYL)-4'-NITROACETOPHENONE
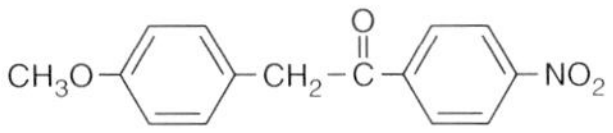
250 mg

S52,497-2
2'-(2-ETHOXY-5-NITROBENZYLOXY)-ACETOPHENONE
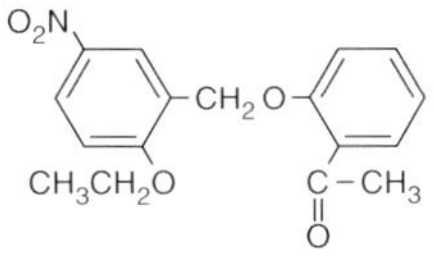
250 mg

S54,905-3
3'-BENZYLOXYACETOPHENONE
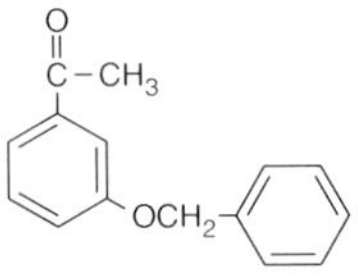
5 gm

S80,465-7
2-BROMO-4'-METHOXY-2-PHENYLACETOPHENONE
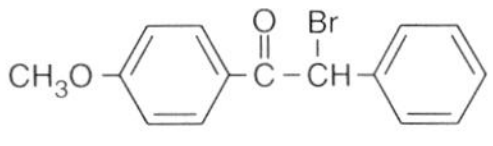
250 mg

S92,986-7
2-PHENOXYACETOPHENONE
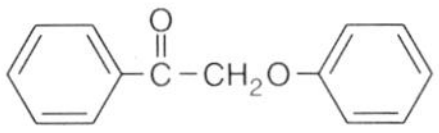
250 mg

S45,774-4
BENZOIN ISOPROPYL ETHER
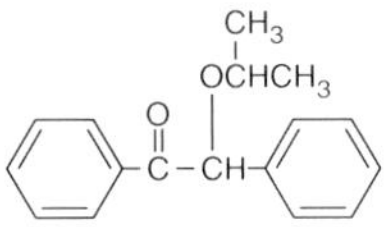
5 gm

S94,621-4
4'-METHOXY-2-PHENYLACETOPHENONE
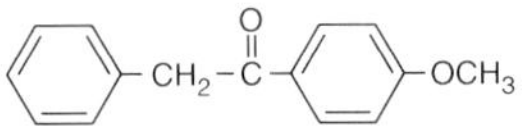
250 mg

S91,964-0
2-(4-METHOXYPHENYL)ACETOPHENONE
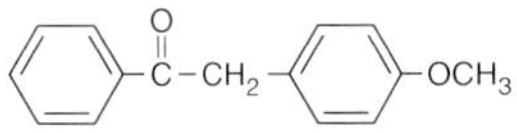
250 mg

S68,319-1
2-METHYL-4'-PHENOXYPROPIOPHENONE
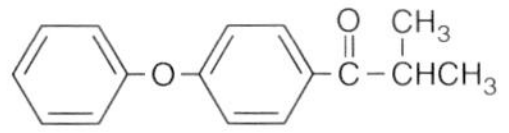
5 gm

S94,563-3
2-PHENYL-4'-PHENOXYACETOPHENONE
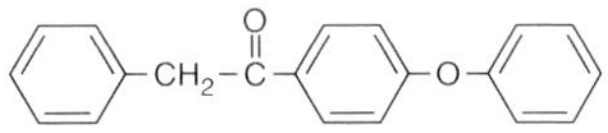
250 mg

S92,960-3
3-CHLORO-3-(4-METHOXYPHENYL)-2-PHENYLPROPIOPHENONE
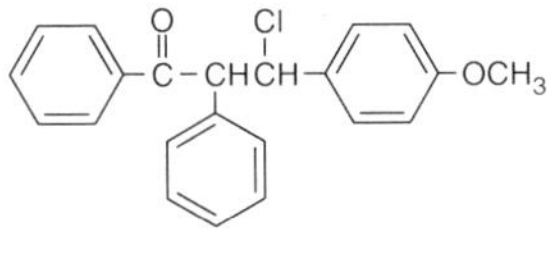
250 mg

S83,892-6
4'-PHENOXYDODECANOPHENONE
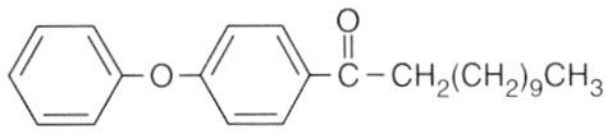
250 mg

S68,155-5
4'-PHENOXYHEXADECANOPHENONE
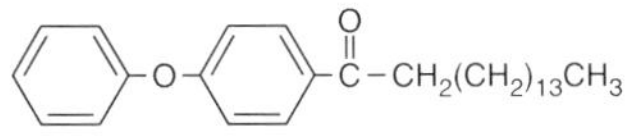
250 mg

S68,153-9
4'-PHENOXYOCTADECANOPHENONE
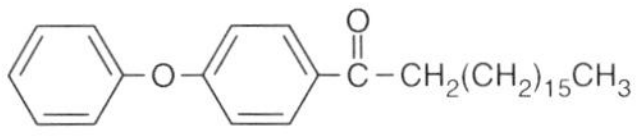
5 gm

S83,600-1
2,2-DIPHENYL-4'-METHOXYACETOPHENONE
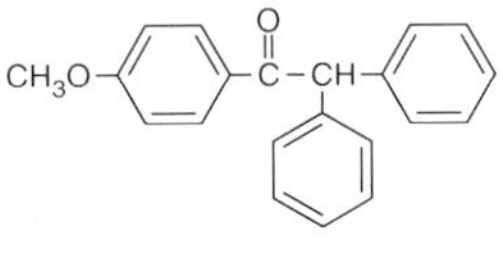
250 mg

S81,725-2
4,4-DIPHENYL-3-(4-METHOXYPHENYL)-BUTYROPHENONE
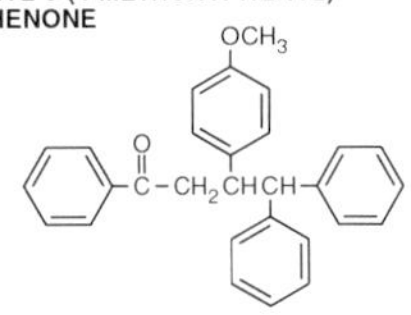
250 mg

S92,613-2
2-PHENOXY-4'-PHENYLACETOPHENONE
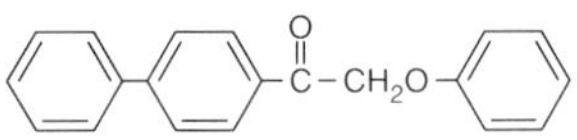
250 mg

S69,928-4
2',4'-DIMETHYL-2-PHENOXYPROPIOPHENONE
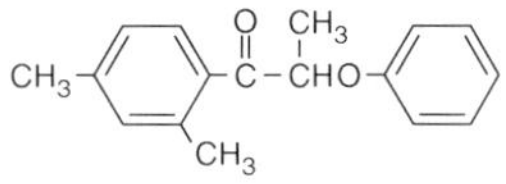
250 mg

S66,316-6
2'-ALLYLOXYACETOPHENONE
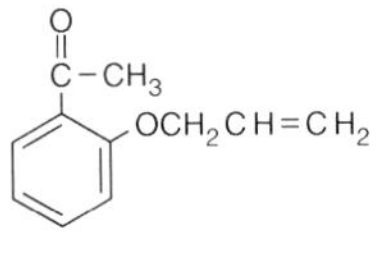
5 gm

S38,303-1
2'-(HEXADECYLOXY)ACETOPHENONE
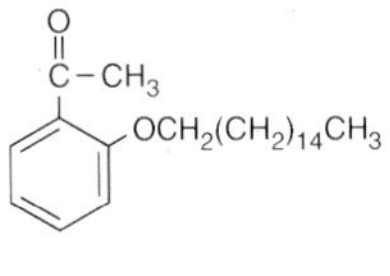
5 gm

S85,034-9
4'-METHOXY-2-METHYLPROPIOPHENONE
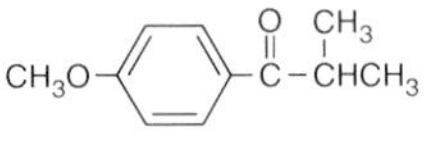
250 mg

S50,525-0
4'-BUTOXYACETOPHENONE
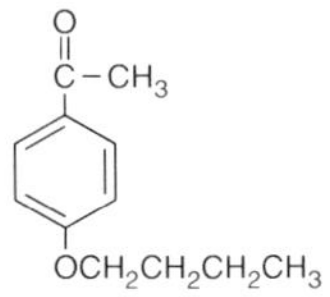
5 gm

S51,620-1
4'-(PENTYLOXY)ACETOPHENONE

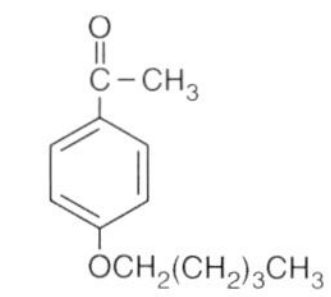

250 mg

S44,497-9
P-ANISYL BENZYL KETONE

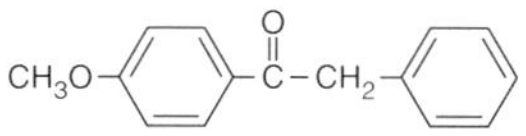

250 mg

S55,761-7
4'-METHOXY-3-PHENYLPROPIOPHENONE

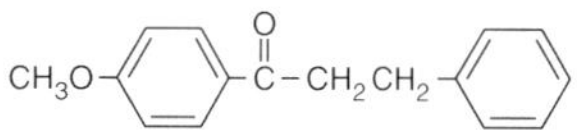

250 mg

S83,889-6
4'-METHOXYDODECANOPHENONE

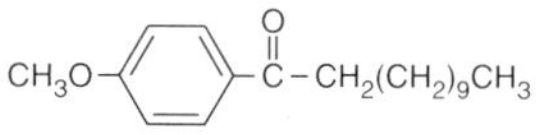

250 mg

S62,915-4
4'-(HEXADECYLOXY)ACETOPHENONE

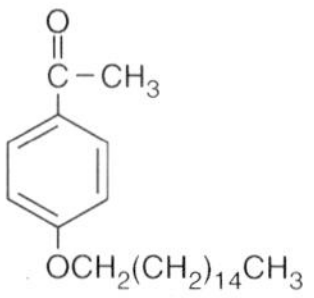

5 gm

S84,087-4
4'-(4-METHYLPHENOXY)-OCTADECANOPHENONE

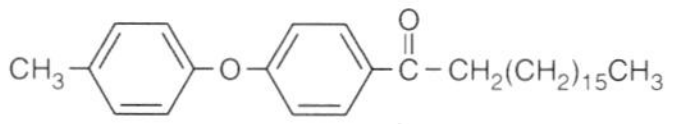

250 mg

S56,016-2
2-(4-METHOXYBENZYLIDENE)-1-TETRALONE

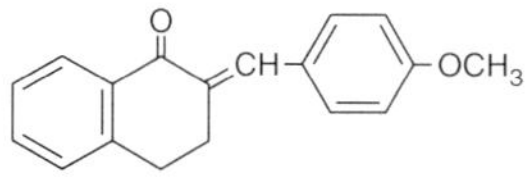

250 mg

S56,229-7
2'-HEXADECYLOXY-5'-METHYLACETOPHENONE

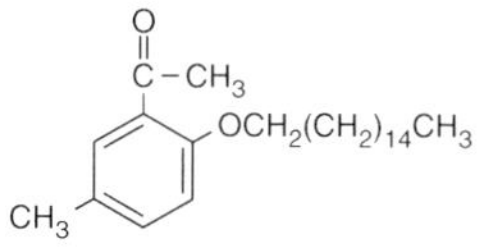

250 mg

S55,623-8
2'-HEXADECYLOXY-4'-METHOXYACETOPHENONE

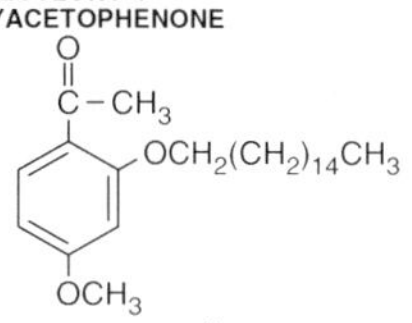

5 gm

S58,208-5
2',5'-DIMETHOXYOCTADECANOPHENONE

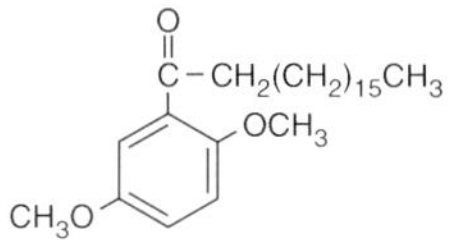

250 mg

S63,916-8
2',4'-DI(DECYLOXY)ACETOPHENONE

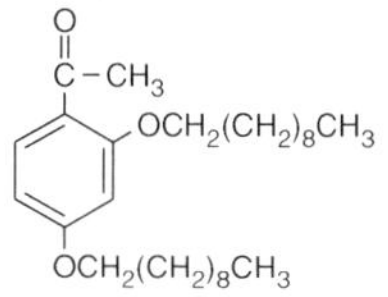

250 mg

S44,937-7
2',4',6'-TRIMETHOXYACETOPHENONE

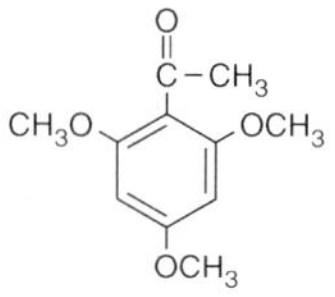

5 gm

S85,998-2
6-METHOXY-5-METHYL-1-TETRALONE

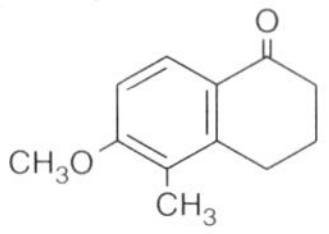

250 mg

S88,445-6
6-METHOXY-2-METHYL-1-TETRALONE

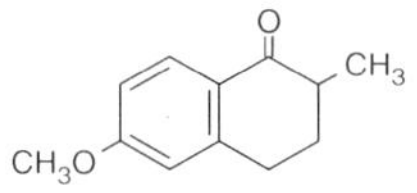

250 mg

S61,974-4
6-ACETYL-2-METHOXY-1-METHYLNAPHTHALENE

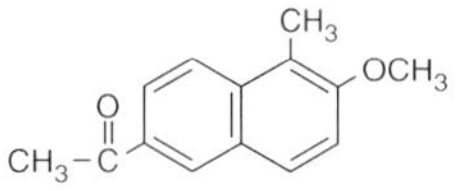

5 gm

S68,020-6
6-METHOXY-2-NAPHTHYL METHYL KETONE

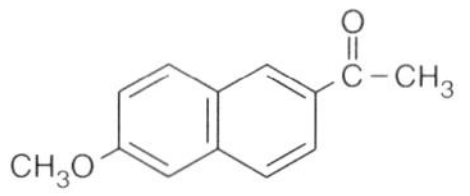

5 gm

S92,884-4
4'-BENZYLOXYACETOPHENONE

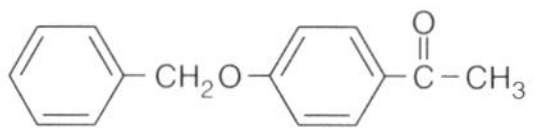

250 mg

S53,575-3
4'-METHOXYBUTYROPHENONE

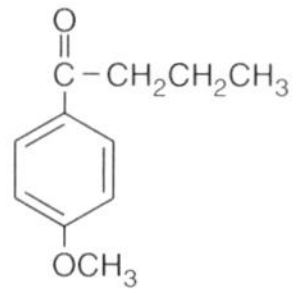

250 mg

S36,960-8
4'-METHOXYVALEROPHENONE

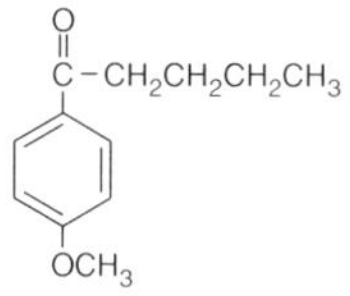

5 gm

S41,298-8
2',4'-DIMETHOXYPROPIOPHENONE

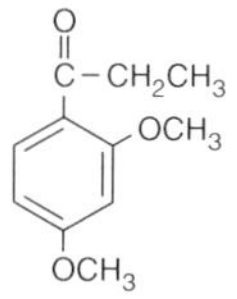

5 gm

S94,161-1
1-ACETYL-4-METHOXYNAPHTHALENE

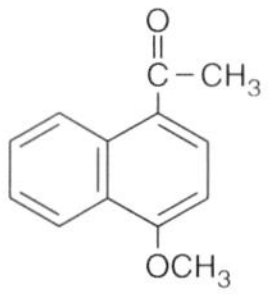

250 mg

S87,417-5
2,6-DIPHENYL-4H-PYRAN-4-THIONE

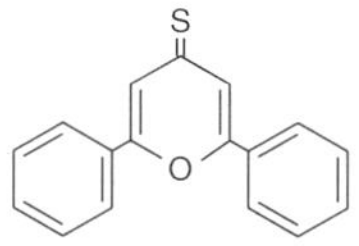

250 mg

S67,252-1
2-HYDROXY-4'-NITROACETOPHENONE, NITRATE ESTER

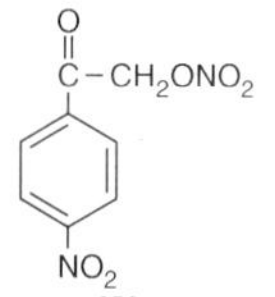

250 mg

S66,256-9
1,3-DIPHENYL-2,3-EPOXY-1-PROPANONE

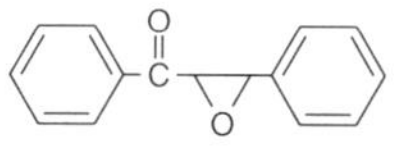

250 mg

S81,836-4
ALPHA,BETA-EPOXYCHALCONE

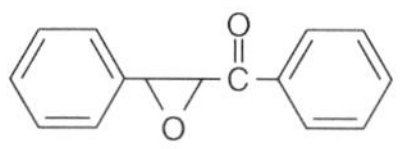

250 mg

S86,119-7
2-(1,3-DIOXOLAN-2-YL)ACETOPHENONE

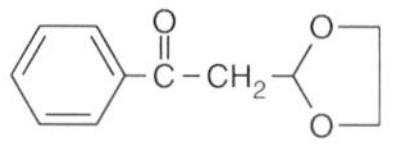

250 mg

S93,579-4
1-CYCLOHEXYL-2,3-DIBENZOYLAZIRIDINE

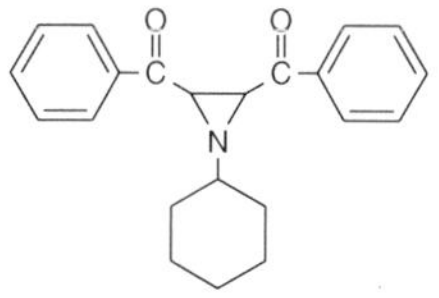

250 mg

S59,688-4
2-(2-NITROPHENYLTHIO)ACETOPHENONE

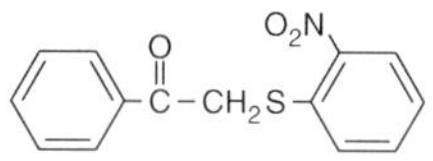

5 gm

S65,607-0
2-(4-NITROPHENYLTHIO)ACETOPHENONE

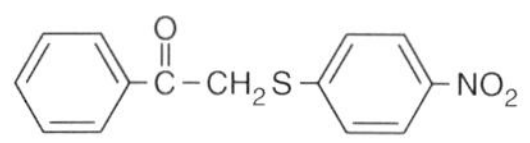

250 mg

S65,634-8
4'-BROMO-2-(4-NITROPHENYLTHIO)-ACETOPHENONE

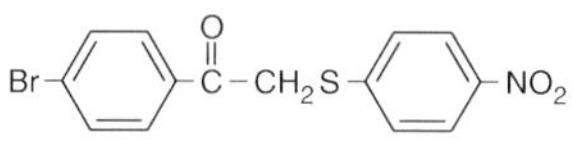

250 mg

S83,401-7
2-(METHYLTHIO)ACETOPHENONE

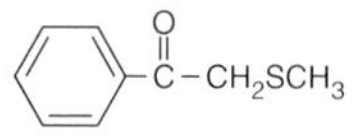

250 mg

S57,703-0
2-(ETHYLTHIO)ACETOPHENONE

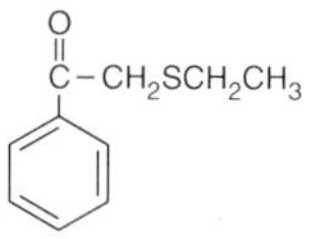

5 gm

S75,464-1
3,3-BIS(METHYLTHIO)-1-PHENYL-2-PROPEN-1-ONE

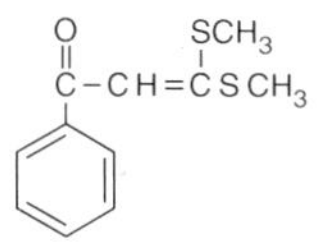

250 mg

S63,370-4
2-(BENZYLTHIO)ACETOPHENONE

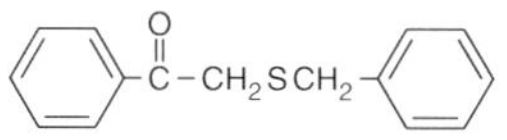

5 gm

S83,417-3
3-PHENYL-3-(PHENYLTHIO)-PROPIOPHENONE

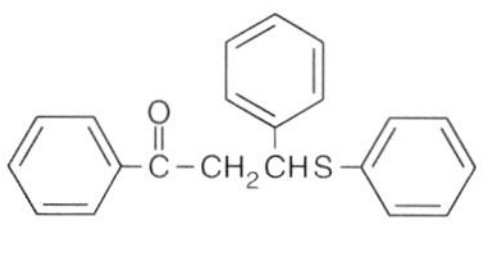

250 mg

S92,620-5
2,2-BIS(PHENYLTHIO)-2-PHENYLACETOPHENONE

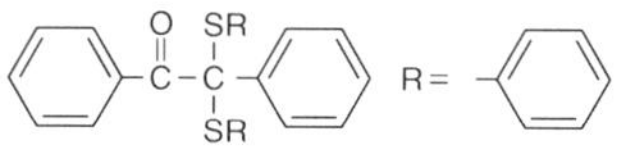

250 mg

S66,203-8
2,2-DIMETHYL-4'-(METHYLTHIO)-PROPIOPHENONE
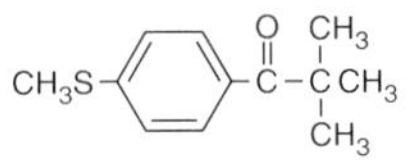
5 gm

S63,124-8
4-ACETYLDIPHENYL SULFIDE
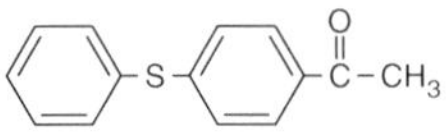
250 mg

S94,717-2
2',5'-DICHLORO-3,3-BIS(METHYLTHIO)-ACRYLOPHENONE
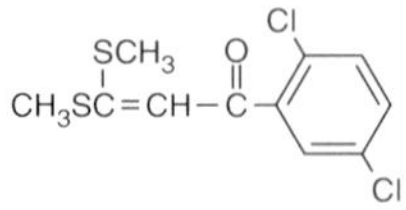
250 mg

S43,414-0
DIMETHYL(PHENACYL)SULFONIUM BROMIDE
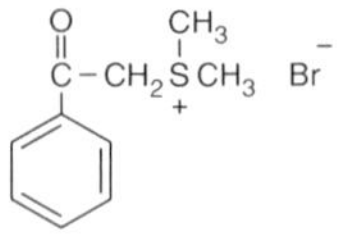
5 gm

S74,246-5
3'-AMINO-4'-METHYL-2'-NITROACETOPHENONE
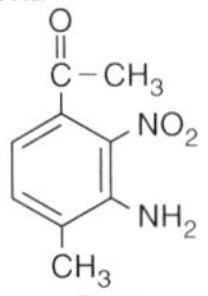
5 gm

S63,996-6
2'-AMINO-4'-NITROACETOPHENONE
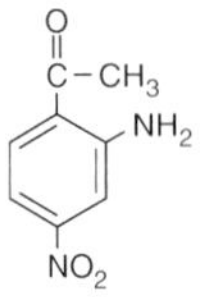
5 gm

S74,244-9
2'-AMINO-5'-NITROACETOPHENONE
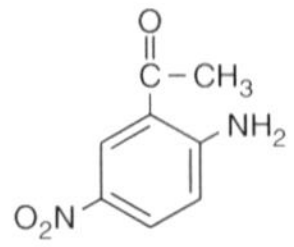
5 gm

S74,513-8
5'-AMINO-2'-NITROACETOPHENONE
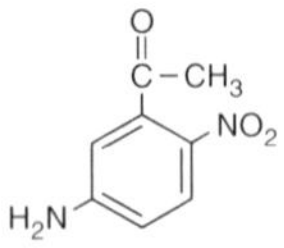
250 mg

S74,286-4
2'-AMINO-5'-NITROPROPIOPHENONE
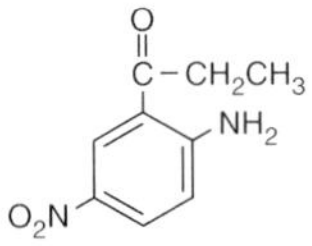
250 mg

S74,253-8
5'-AMINO-4'-METHYL-2'-NITROACETOPHENONE
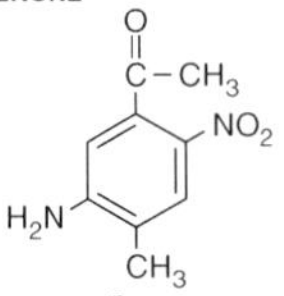
5 gm

S64,219-3
2-AMINO-4'-NITROACETOPHENONE HYDROCHLORIDE
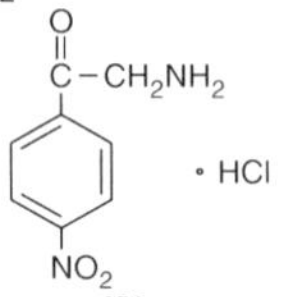
250 mg

S44,661-0
2'-AMINOACETOPHENONE HYDROCHLORIDE
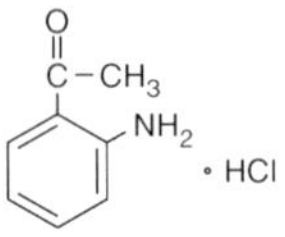
5 gm

S61,804-7
4-AMINOPHENYL PENTADECYL KETONE
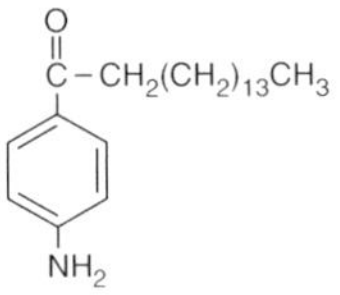
250 mg

S75,945-7
3'-AMINO-2,2,4'-TRIMETHYLPROPIOPHENONE
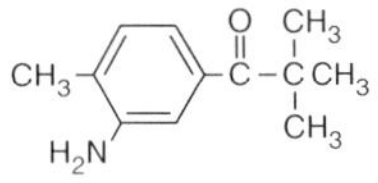
250 mg

S76,475-2
5'-AMINO-2',4'-DIMETHYLACETOPHENONE
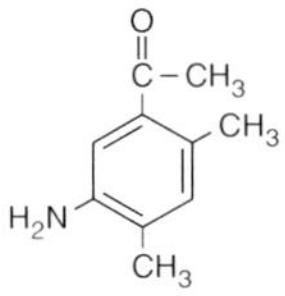
5 gm

S65,350-0
2-AMINO-4'-PHENYLACETOPHENONE HYDROBROMIDE
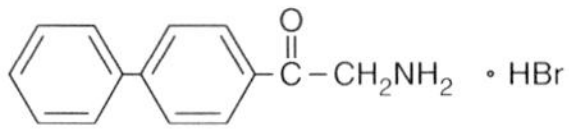
5 gm

S90,725-1
5-AMINO-1-INDANONE
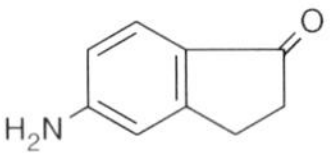
250 mg

S61,965-5
2-(OCTADECANOYL)-1-NAPHTHOL
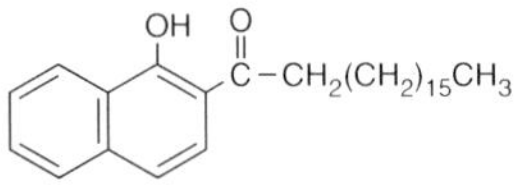
250 mg

S75,914-7
3-AMINO-6,7,8,9-TETRAHYDRO-5H-BENZOCYCLOHEPTEN-5-ONE
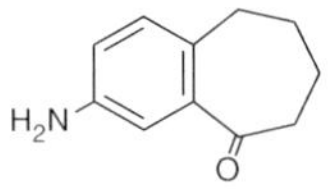
250 mg

S74,663-0
4-ACETYL-2',4'-DINITRODIPHENYLAMINE
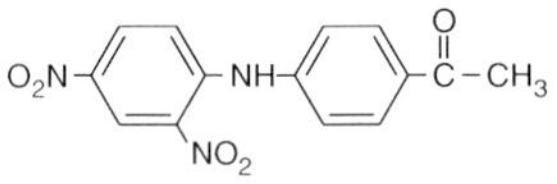
5 gm

S74,662-2
3-ACETYL-2',4'-DINITRODIPHENYLAMINE
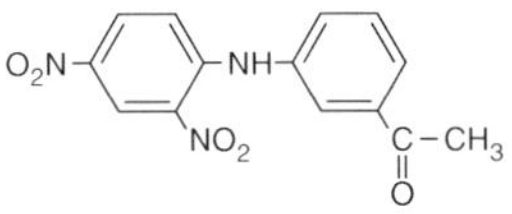
5 gm

S72,061-5
N-(4-ACETYLPHENYL)-2,4,6-TRINITROANILINE
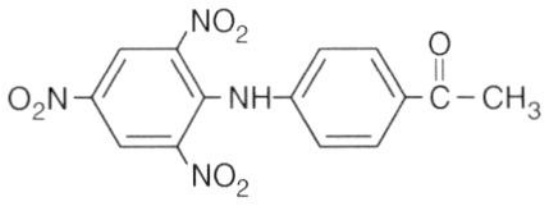
5 gm

S71,914-5
3'-(2,4,6-TRINITROANILINO)-ACETOPHENONE
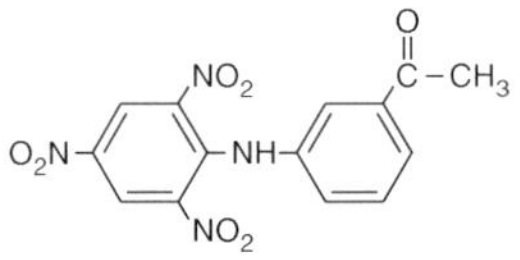
5 gm

S88,771-4
8-(3-ANILINOPROPYLAMINO)-6-PHENYL-2,6-DIAZABICYCLO[5.2.0]NONA-1,7-DIEN-9-ONE PERCHLORATE
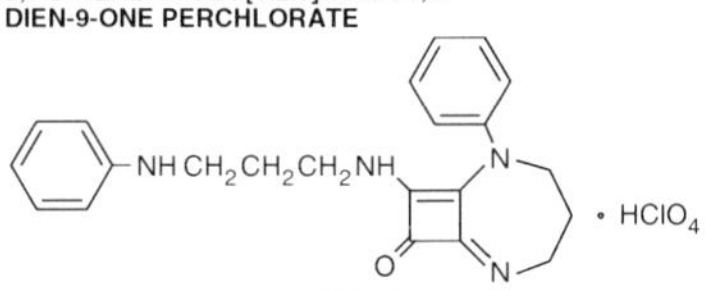
250 mg

S93,336-8
2,3-DIMORPHOLINO-3-PHENYLPROPIOPHENONE
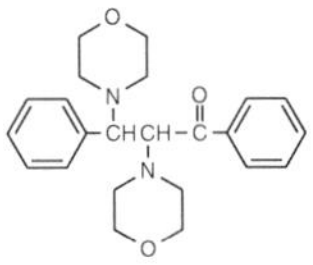
250 mg

S34,296-3
4'-TERT-BUTYL-2',6'-DIMETHYL-3',5'-DINITROACETOPHENONE
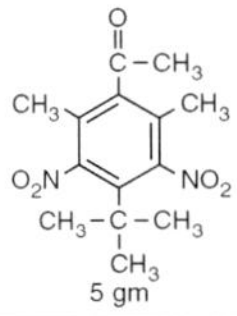
5 gm

S41,469-7
4'-[3-(DIMETHYLAMINO)PROPYLAMINO]-ACETOPHENONE
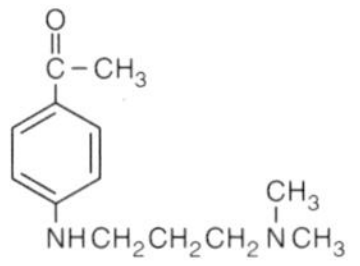
250 mg

S44,578-9
4'-(PHENETHYLAMINO)ACETOPHENONE
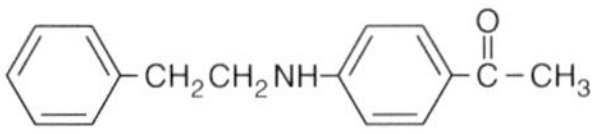
5 gm

S43,307-1
4'-(ALPHA-METHYLPHENETHYLAMINO)-ACETOPHENONE HYDROCHLORIDE
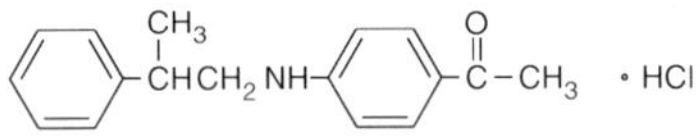
5 gm

S84,626-0
2-(BENZOYLMETHYL)-2-METHYL-1,2,3,4-TETRAHYDROISOQUINOLINIUM CHLORIDE
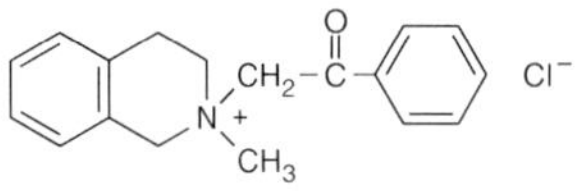
250 mg

S84,625-2
2-(BENZOYLMETHYL)-2-METHYL-1,2,3,4-TETRAHYDROISOQUINOLINIUM BROMIDE
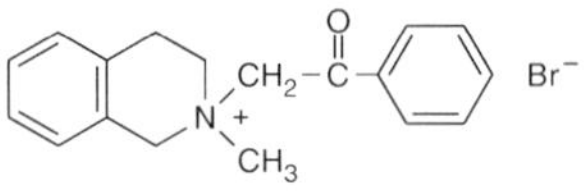
250 mg

S91,467-3
1,4-BIS(BENZOYLMETHYL)-HEXAHYDRO-1H-1,4-DIAZEPINE DIHYDROCHLORIDE
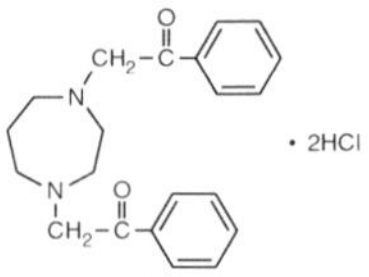
250 mg

S60,214-0
3-ANILINO-3-PHENYLPROPIOPHENONE
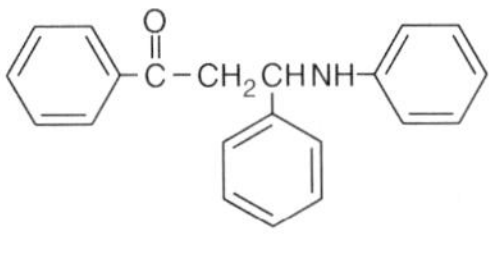
250 mg

S89,672-1
6-(BENZYLAMINO)-3,3-DIMETHYL-1-TETRALONE
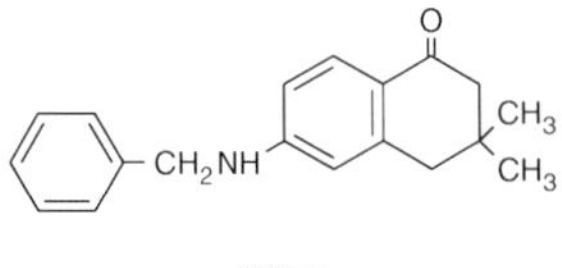
250 mg

S73,237-0
3-(METHYLAMINO)ACRYLOPHENONE
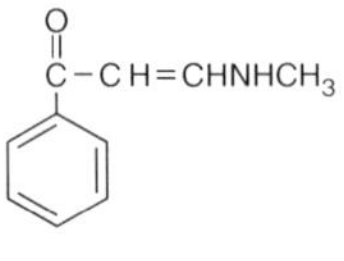
250 mg

S93,719-3
3-(2-NAPHTHYLAMINO)PROPIOPHENONE
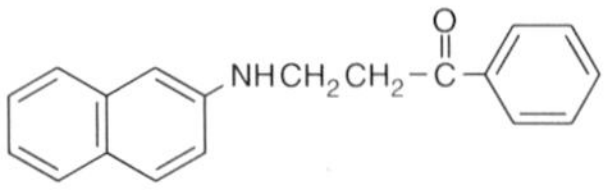
250 mg

S83,603-6
3-[4-(DIMETHYLAMINO)PHENYL]-3-PHENYLPROPIOPHENONE
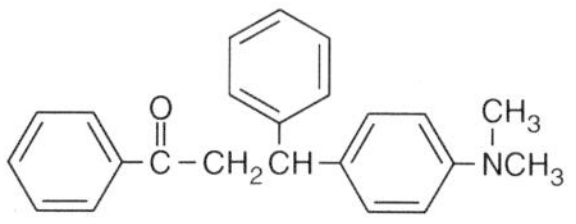
250 mg

S76,176-1
3-(4-METHYLPIPERAZINYL)-3-PHENYLPROPIOPHENONE
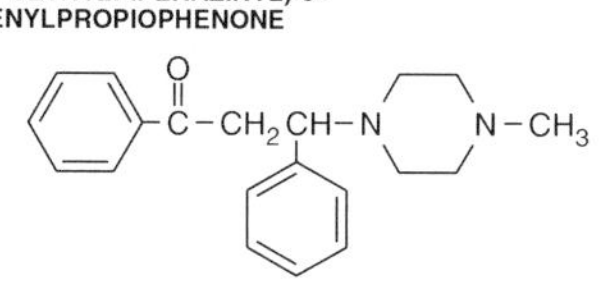
250 mg

S55,480-4
3-(DIMETHYLAMINO)-4'-NITROPROPIOPHENONE HYDROCHLORIDE
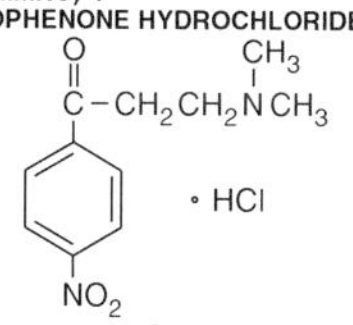
5 gm

S50,647-8
3-DIMETHYLAMINO-2'-NITROACRYLOPHENONE
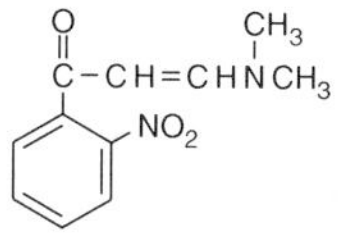
5 gm

S50,648-6
3-DIMETHYLAMINO-3'-NITROACRYLOPHENONE
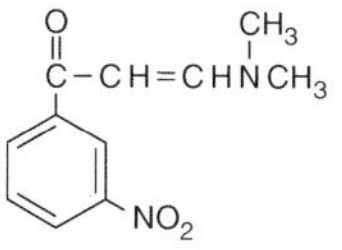
5 gm

S90,724-3
5-(DIMETHYLAMINO)-1-INDANONE
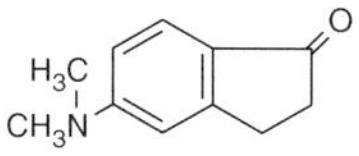
250 mg

S35,540-2
3-PYRROLIDINOPROPIOPHENONE HYDROCHLORIDE
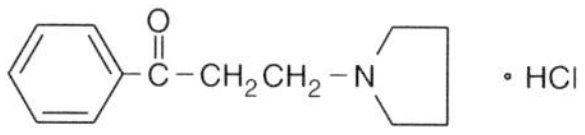
5 gm

S52,919-2
2-METHYL-3-PRROLIDENO-PROPIOPHENONE HYDROCHLORIDE
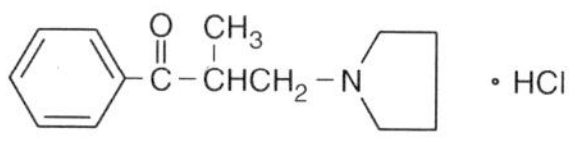
5 gm

S90,837-1
6-(DIMETHYLAMINO)-1-TETRALONE
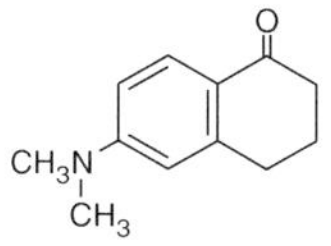
250 mg

S35,493-7
3-PIPERIDINOPROPIOPHENONE HYDROCHLORIDE
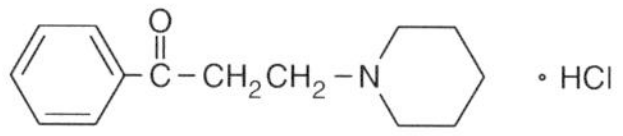
5 gm

S38,337-6
3-PIPERIDINOPROPIOPHENONE
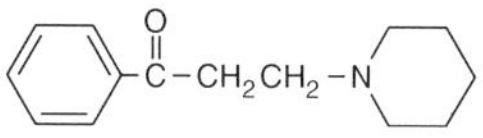
250 mg

S89,360-9
2-(4-PHENYL-1,2,5,6-TETRAHYDRO-1-PYRIDYLMETHYL)-1-TETRALONE
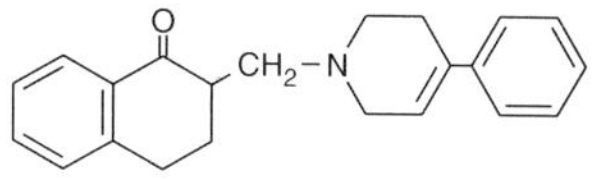
250 mg

S75,878-7
3-(4-METHYLPIPERAZINO)PROPIOPHENONE
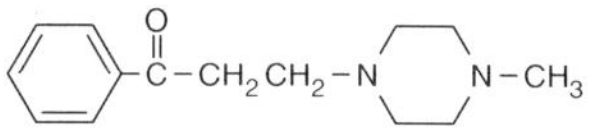
250 mg

S36,398-7
3-MORPHOLINOPROPIOPHENONE HYDROCHLORIDE
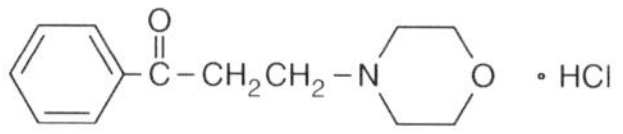
5 gm

S44,767-6
4'-MORPHOLINODEOXYBENZOIN
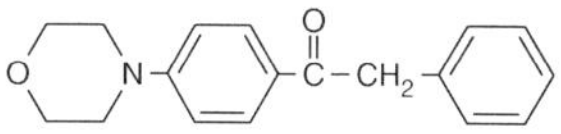
5 gm

S45,520-2
4'-MORPHOLINOPROPIOPHENONE
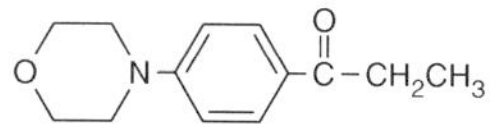
5 gm

S34,666-7
3-DIMETHYLAMINO-2-METHYLPROPIOPHENONE HYDROCHLORIDE
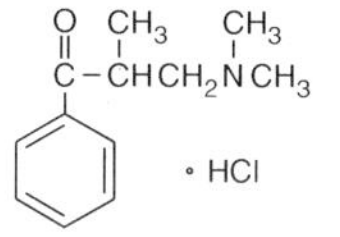
5 gm

S36,583-1
3-DIMETHYLAMINO-2-METHYLPROPIOPHENONE
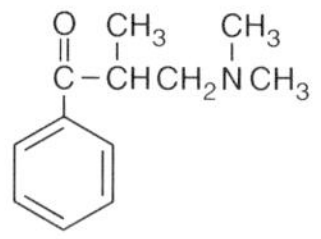
5 gm

S79,278-0
2-(DIMETHYLAMINOMETHYL)-BUTYROPHENONE HYDROCHLORIDE
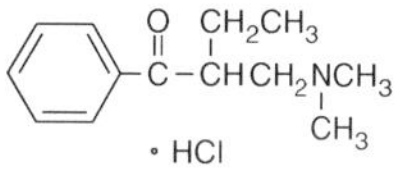
250 mg

S93,522-0
3-(DIMETHYLAMINO)-4'-METHYLPROPIOPHENONE HYDROCHLORIDE
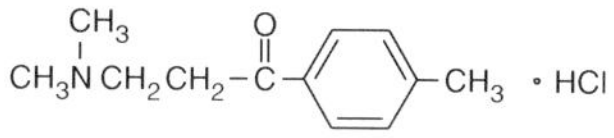
250 mg

S73,496-9
2,2-DIMETHYL-4'-(DIMETHYLAMINOMETHYL)PROPIOPHENONE
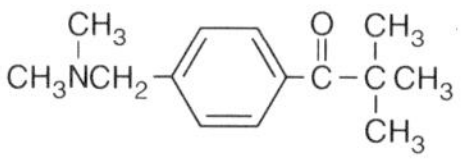
250 mg

S79,355-8
2-(DIMETHYLAMINOMETHYL)-1-INDANONE HYDROCHLORIDE
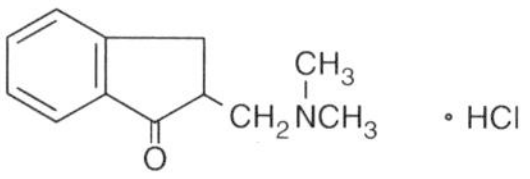
250 mg

S79,277-2
2-(DIMETHYLAMINOMETHYL)-ALPHA-TETRALONE HYDROCHLORIDE
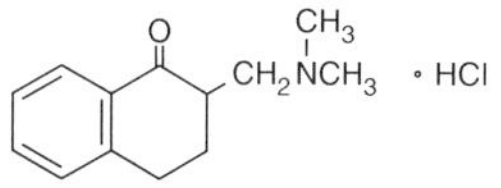
250 mg

S68,603-4
2-DIMETHYLAMINO-1-(2-NAPHTHYL)-1-BUTANONE
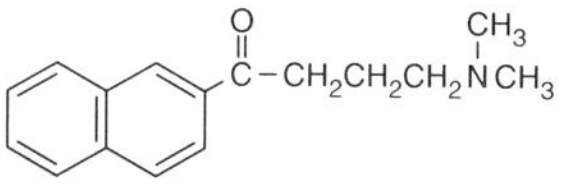
250 mg

S70,337-0
1,2,3,4,4A,9B-HEXAHYDRO-2-METHYL-5H-INDENO(1,2-C)PYRIDIN-5-ONE
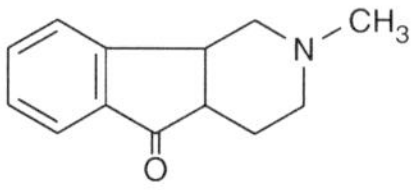
5 gm

S87,412-4
2,6-DIPHENYL-1-METHYL-4(1H)-PYRIDINETHIONE
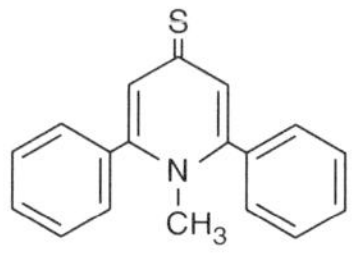
250 mg

S84,614-7
(BENZOYLMETHYL)-BENZYLDIMETHYLAMMONIUM BROMIDE
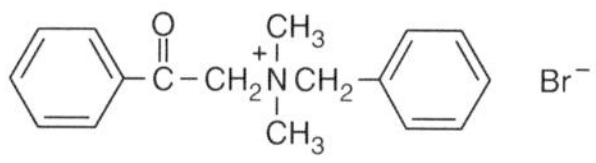
250 mg

S82,967-6
PHENACYLAMMONIUM CHLORIDE
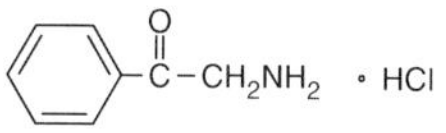
250 mg

S41,606-1
(2-BENZOYLETHYL)-TRIMETHYLAMMONIUM IODIDE
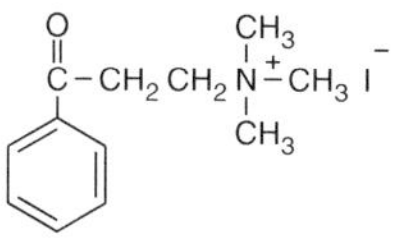
250 mg

S53,194-4
3'-NITROBUTYROPHENONE
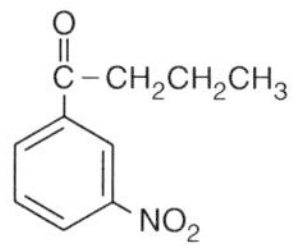
5 gm

S93,980-3
3'-NITROPROPIOPHENONE
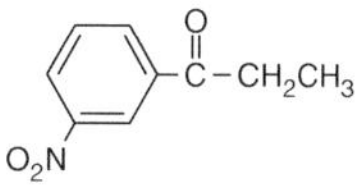
250 mg

S93,225-6
3,4-DIHYDRO-4,4-DIMETHYL-3-(3'-NITROPHENYL)SPIRO[NAPHTHALENE-2(1H)-2'-OXIRANE]-1-ONE
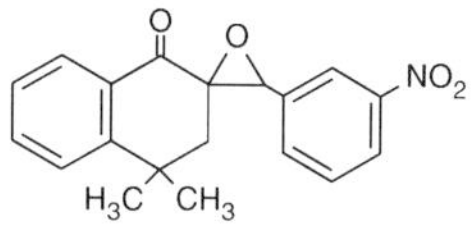
250 mg

S84,891-3
4-NITROVALEROPHENONE
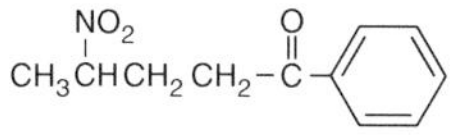
250 mg

S91,924-1
3-NITRO-2,2,3-TRIMETHYLBUTYROPHENONE
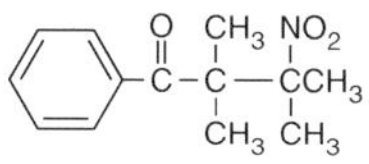
250 mg

S93,358-9
2-(4-NITROBENZYL)-1-INDANONE
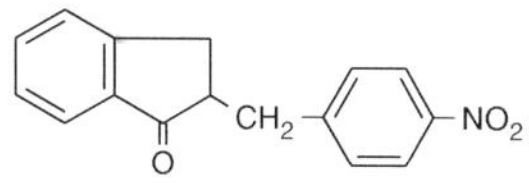
250 mg

S37,013-4
7-NITRO-1-TETRALONE
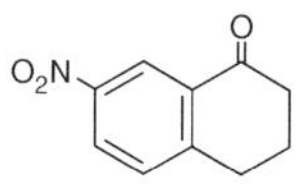
250 mg

S55,911-3
3'-METHYLPROPIOPHENONE

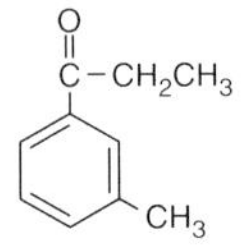

5 gm

S54,978-9
3'-METHYLBUTYROPHENONE

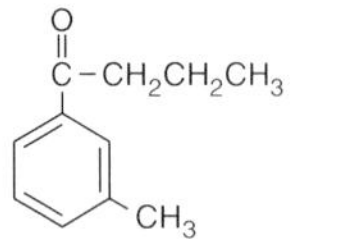

5 gm

S50,001-1
2-BROMOBUTYROPHENONE

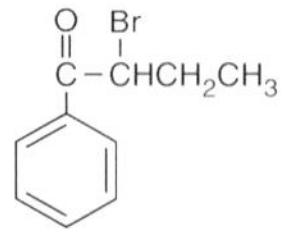

250 mg

S38,721-5
2,3-DIBROMO-3-PHENYLPROPIOPHENONE

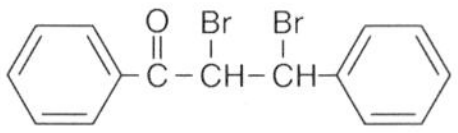

5 gm

S38,816-5
2-BROMOHEXADECANOPHENONE

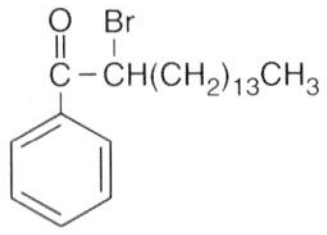

250 mg

S64,977-5
2-BROMO-4'-METHYLISOBUTYROPHENONE

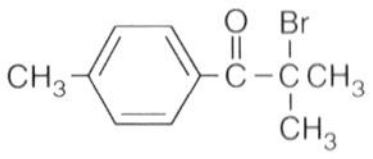

5 gm

S43,512-0
4-BROMOPHENYL CYCLOPROPYL KETONE

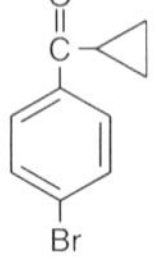

5 gm

S37,244-7
CYCLOPROPYL P-TOLYL KETONE

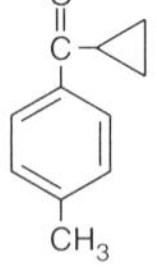

250 mg

S34,519-9
CYCLOPROPYL 4-ETHYLPHENYL KETONE

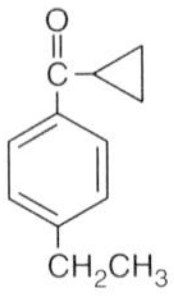

5 gm

S34,521-0
CYCLOPROPYL
4-ISOPROPYLPHENYL KETONE

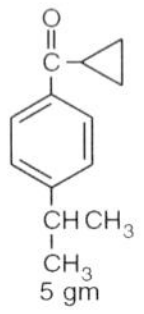

5 gm

S34,310-2
4-TERT-BUTYLPHENYL
CYCLOPROPYL KETONE

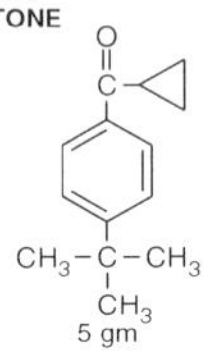

5 gm

S66,285-2
TRANS-1-(4-METHYLBENZOYL)-2-
PHENYLCYCLOPROPANE

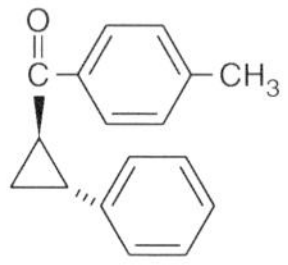

250 mg

S34,516-4
CYCLOPROPYL
2,4-DIMETHYLPHENYL KETONE

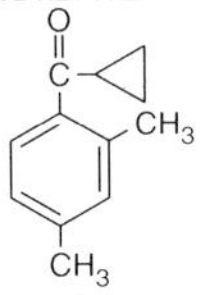

5 gm

S34,517-2
CYCLOPROPYL
2,5-DIMETHYLPHENYL KETONE

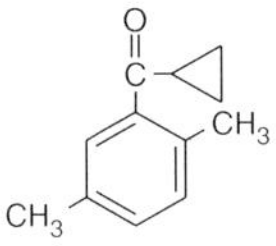

5 gm

S34,518-0
CYCLOPROPYL
3,4-DIMETHYLPHENYL KETONE

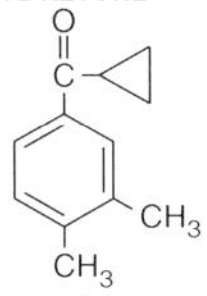

5 gm

S93,395-3
1,2-DIPHENYL-3-(4-METHYLBENZOYL)-
AZIRIDINE

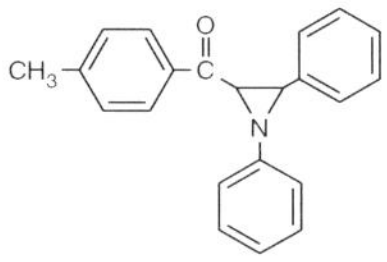

250 mg

S89,436-2
2-CHLORO-3-(TRICHLOROVINYL)-
1-INDANONE

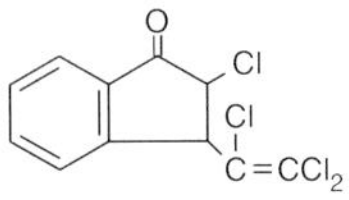

250 mg

S69,001-5
5,7-DICHLORO-4-METHYL-1-
BENZOTHIOPHEN-3(2H)ONE

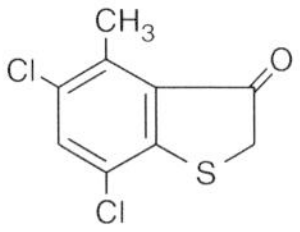

250 mg

S56,013-8
2-PIPERONYLIDENE-1-TETRALONE

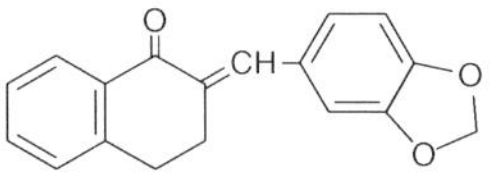

5 gm

S65,743-3
1-METHYL-1-(4-METHYLBENZOYL)-
CYCLOHEXANE

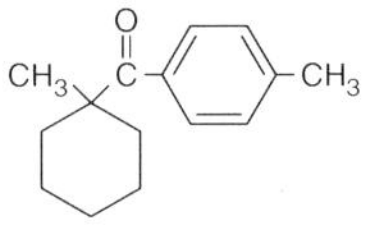

5 gm

S60,333-3
ACETYL-1,2,3,4-
TETRAHYDROANTHRACENE (MIXTURE
OF ISOMERS)

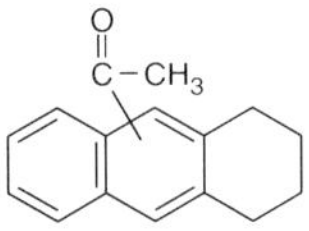

250 mg

S95,701-1
8-METHYL-1,2,3,4-TETRAHYDRO-1-
PHENANTHRENONE

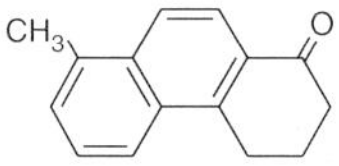

50 mg

S88,835-4
2-PHENYL-2H-1-BENZOTHIOPYRAN-4-
(3H)-ONE

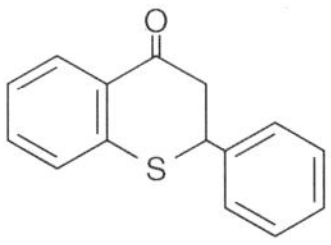

250 mg

S88,825-7
3H-NAPHTHO(2,1-B)THIOPYRAN-1-
(2H)-ONE

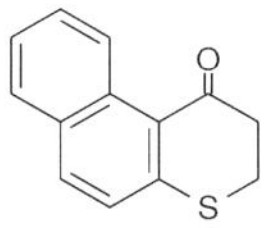

250 mg

S88,826-5
2H-NAPHTHO(1,2-B)THIOPYRAN-4-
(3H)-ONE

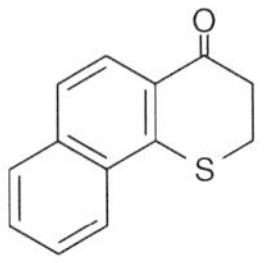

250 mg

S56,115-0
ISOVALEROPHENONE

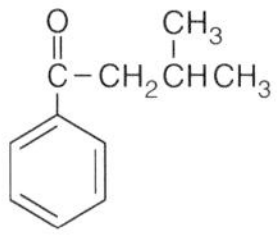

5 gm

S56,104-5
4-METHYLVALEROPHENONE

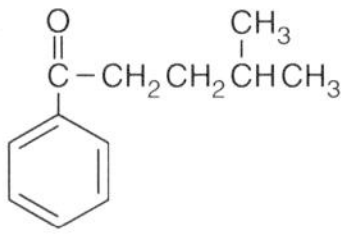

5 gm

S82,419-4
3-PHENYLPROPIOPHENONE

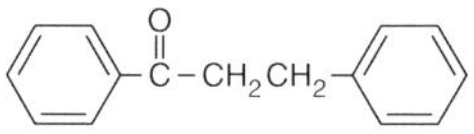

250 mg

S43,348-9
3-PHENYLBUTYROPHENONE

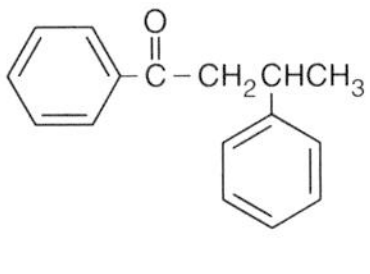

5 gm

S56,144-4
4-PHENYLBUTYROPHENONE

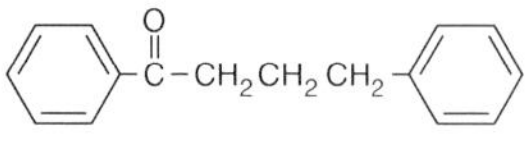

5 gm

S54,737-9
3-METHYL-2-PHENYLBUTYROPHENONE

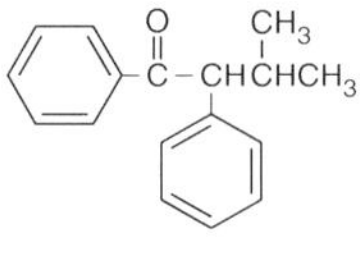

250 mg

S88,841-9
5-PHENYLVALEROPHENONE

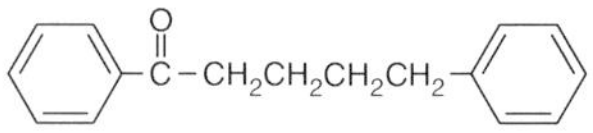

250 mg

S74,804-8
2,2-DIPHENYLACETOPHENONE

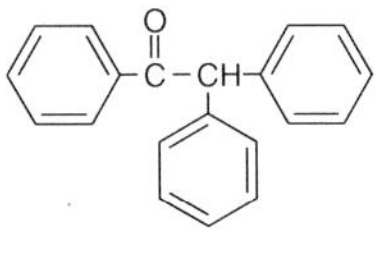

250 mg

S70,549-7
BENZYL TRIETHYLPHENYL KETONE

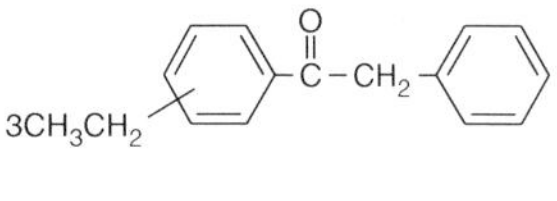

5 gm

S92,958-1
3-CHLORO-2,3-DIPHENYLPROPIOPHENONE

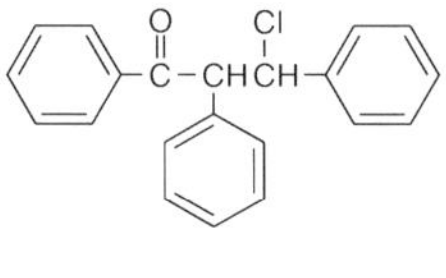

250 mg

S83,475-0
OCTADECANOPHENONE

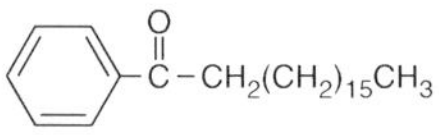

250 mg

S83,932-9
NONADECANOPHENONE
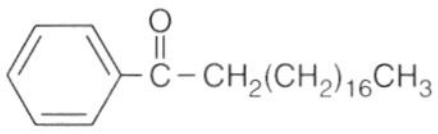
250 mg

S92,619-1
4'-METHYL-2-PHENYLACETOPHENONE
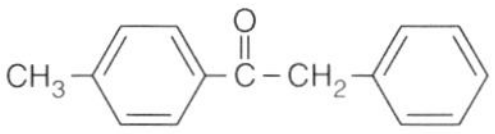
250 mg

S73,944-8
3-PHENYL-3-(P-TOLYL)PROPIOPHENONE
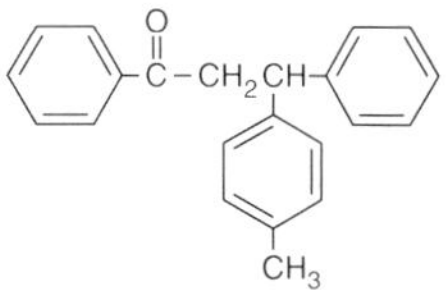
5 gm

S81,351-6
4'-PHENYLHEXADECANOPHENONE
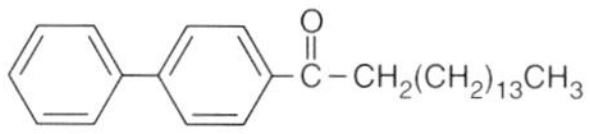
250 mg

S84,086-6
4'-PHENYLOCTADECANOPHENONE
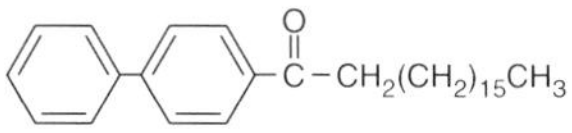
250 mg

S94,564-1
3',4'-DIMETHYL-2-PHENYLACETOPHENONE
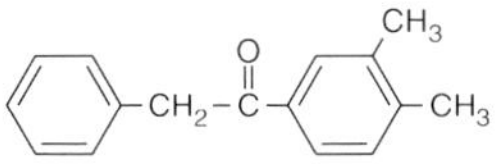
250 mg

S34,770-1
3,3-DIPHENYLPROPIOPHENONE
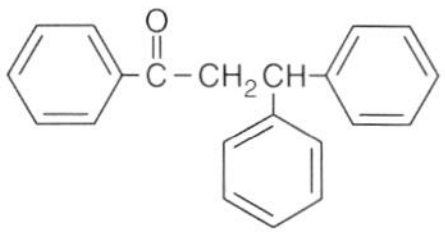
5 gm

S81,381-8
2,2-DIPHENYLPROPIOPHENONE
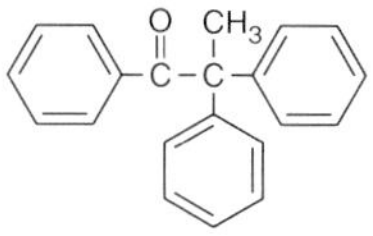
250 mg

S88,674-2
2,2-DIPHENYL-3-METHYLBUTYROPHENONE
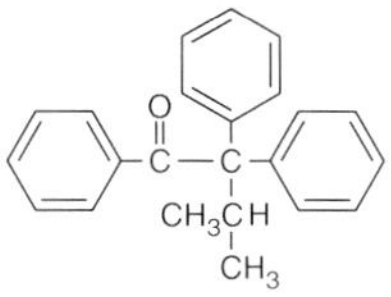
250 mg

S88,676-9
1,2,2-TRIPHENYL-1-HEPTANONE
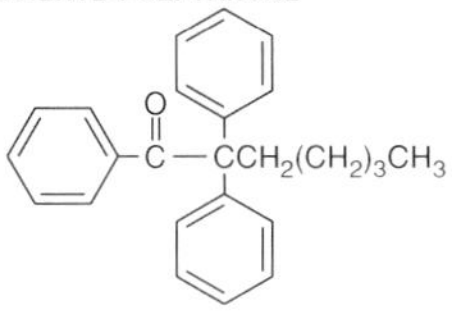
250 mg

S83,476-9
2,3,3-TRIPHENYLPROPIOPHENONE
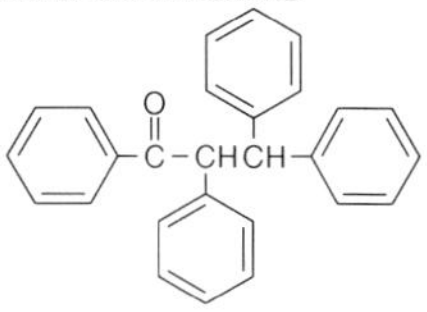
250 mg

S83,995-7
4-BENZYL-3,5,5-TRIPHENYLVALEROPHENONE
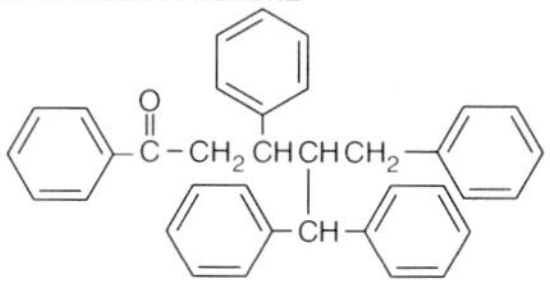
250 mg

S84,207-9
2,2,3,4-TETRAPHENYL-1(2H)-NAPHTHALENONE
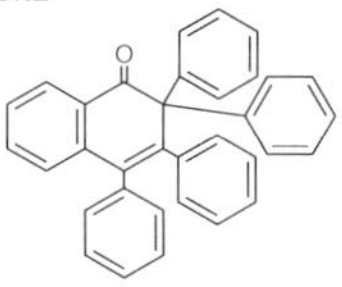
250 mg

S65,949-5
3-PHENYL-1-INDANONE
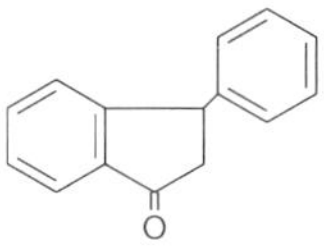
5 gm

S93,226-4
2-BROMO-2-(ALPHA-BROMOBENZYL)-1-TETRALONE
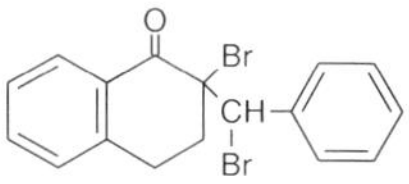
250 mg

S84,234-6
2,3-DIPHENYL-ALPHA-TETRALONE
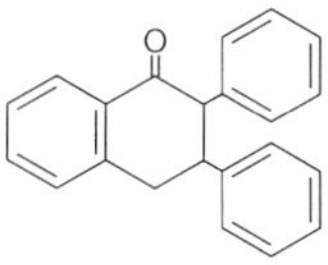
250 mg

S73,273-7
2,2,2'-TRIMETHYLPROPIOPHENONE
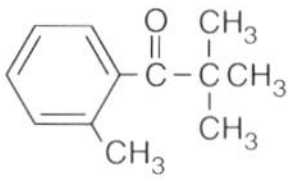
250 mg

S44,687-4
4'-CHLOROBUTYROPHENONE
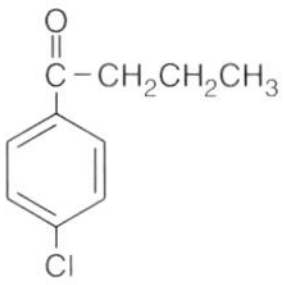
5 gm

S43,281-4
4'-BROMOBUTYROPHENONE
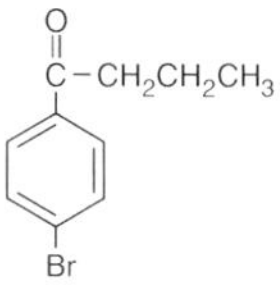
5 gm

S35,270-5
4'-METHYLBUTYROPHENONE
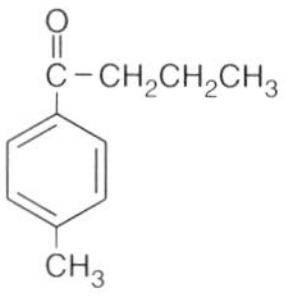
5 gm

S73,251-6
2,4'-DIMETHYLPROPIOPHENONE
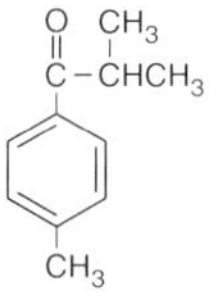
250 mg

S34,991-7
4'-FLUOROVALEROPHENONE
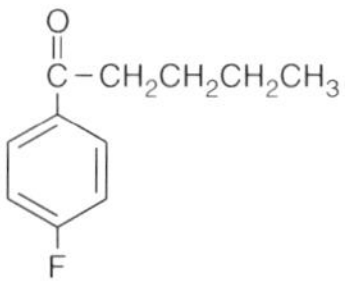
5 gm

S67,241-6
2,2,4'-TRIMETHYLPROPIOPHENONE
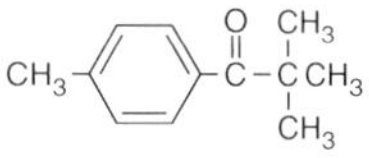
5 gm

S44,327-1
4'-FLUORO-ALPHA-PHENYLACETOPHENONE
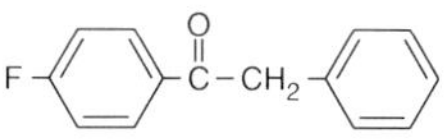
5 gm

S64,751-9
4'-CHLORO-4-METHYLDESOXYBENZOIN
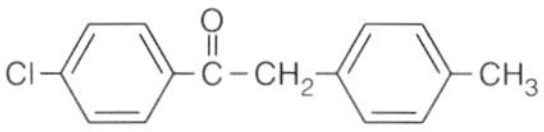
5 gm

S38,659-6
4'-CHLOROHEXADECANOPHENONE
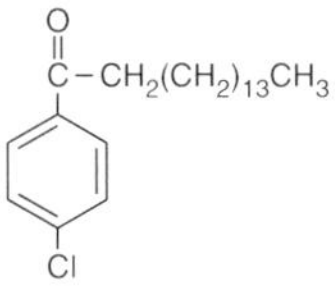
5 gm

S63,262-7
4'-METHYLHEXADECANOPHENONE
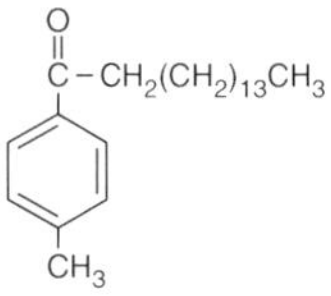
5 gm

S65,649-6
4'-PHENYLPROPIOPHENONE
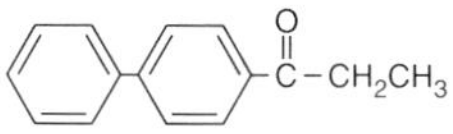
250 mg

S83,896-9
4'-(P-TOLYL)OCTADECANOPHENONE
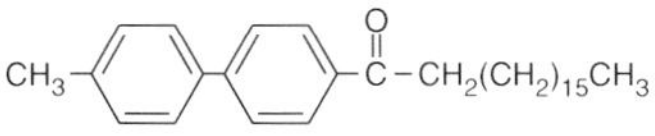
250 mg

S70,586-1
2',4'-DIISOPROPYLPROPIOPHENONE
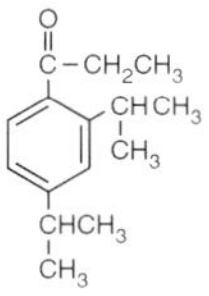
5 gm

S56,019-7
2-(3,4-DICHLOROBENZYLIDENE)-1-TETRALONE
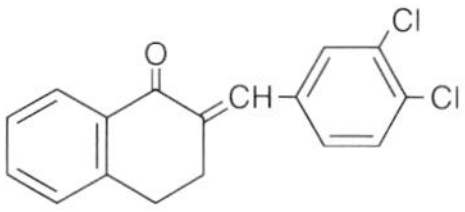
5 gm

S34,295-5
4'-TERT-BUTYL-2',6'-DIMETHYLACETOPHENONE
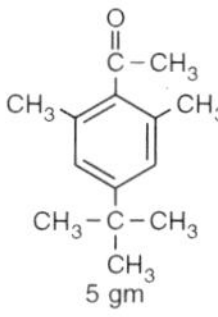
5 gm

S87,747-6
2',4',6'-TRIISOPROPYLACETOPHENONE
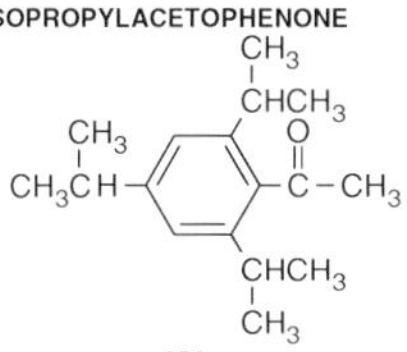
250 mg

S41,360-7
2',4',5'-TRIMETHYLACETOPHENONE
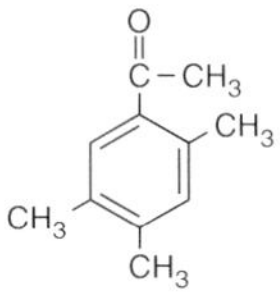
5 gm

S67,101-0
3-METHYL-1-INDANONE
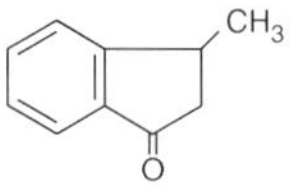
5 gm

S43,784-0
5-ACETYLINDAN
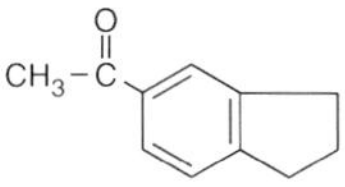
250 mg

S43,547-3
2-INDANYLIDENE-1-INDANONE
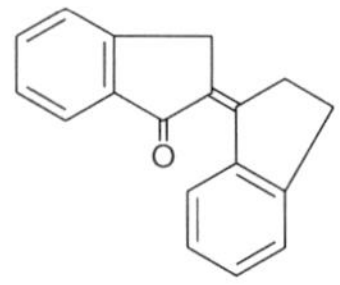
250 mg

S64,017-4
3-ACETYL-9-BROMOPHENANTHRENE
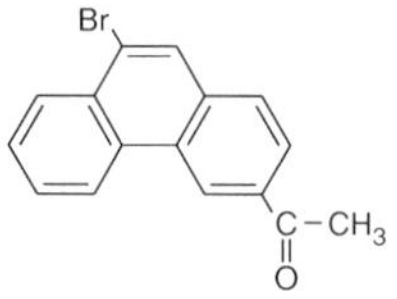

250 mg

S64,007-7
ACETYL-9-BROMOPHENANTHRENE
(MIXTURE OF ISOMERS)
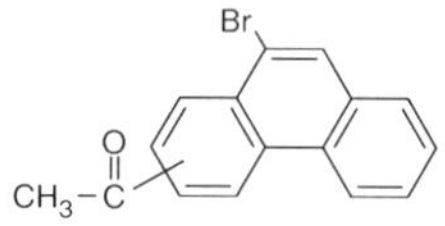

250 mg

S67,211-4
2-BROMOTETRALONE
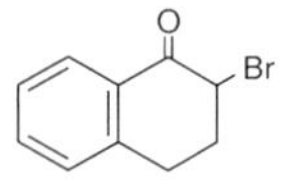

250 mg

S56,838-4
7-METHYL-1-TETRALONE
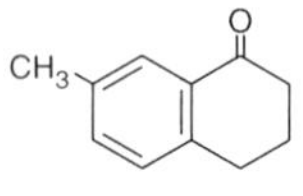

250 mg

S66,236-4
4,4-DIMETHYL-1-TETRALONE
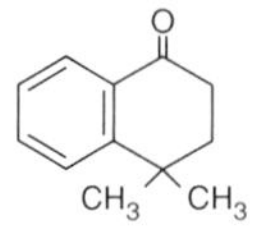

5 gm

S87,671-2
6,7-DIMETHYL-1-TETRALONE
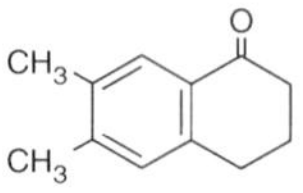

250 mg

S94,790-3
6-ACETYLTETRALIN
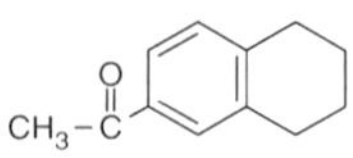

5 gm

S93,217-5
2-BROMO-4,4-DIMETHYL-1-TETRALONE
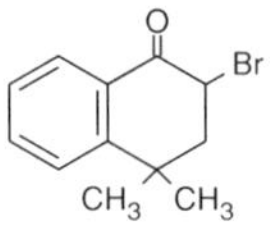

250 mg

S95,628-7
1,2,3,4-TETRAHYDRO-3,4,4-TRIMETHYL-1-NAPHTHALENONE
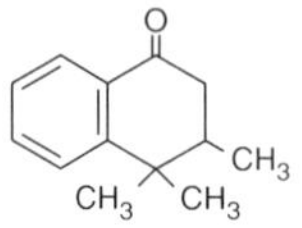

50 mg

S44,809-5
3,3,6,8-TETRAMETHYL-1-TETRALONE
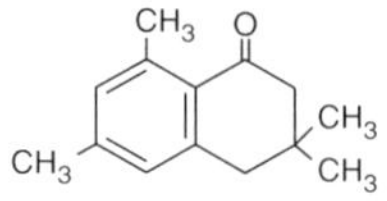

5 gm

S94,796-2
6-ACETYL-1,1,4,4,7-PENTAMETHYL-1,2,3,4-TETRAHYDRONAPHTHALENE
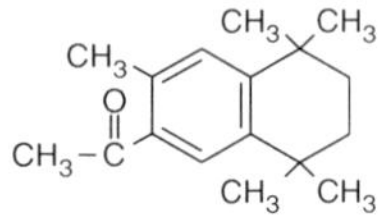

250 mg

S81,316-8
1'-ACETONAPHTHONE PICRATE
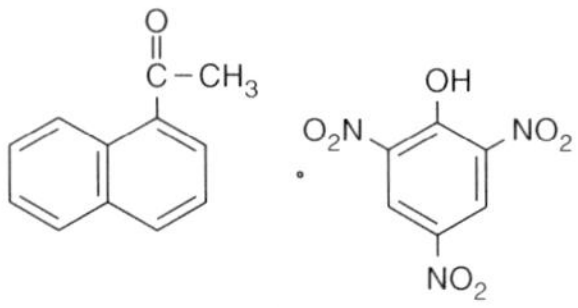

250 mg

S87,994-0
METHYL ALPHA-NAPHTHYL KETONE
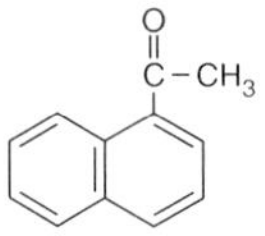

5 gm

S83,891-8
1-NAPHTHYL UNDECYL KETONE
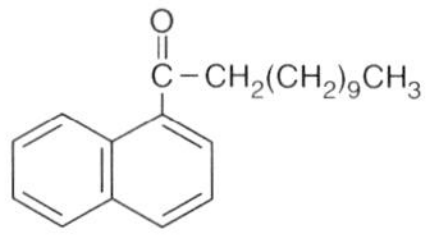

250 mg

S83,893-4
HEPTADECYL 2-NAPHTHYL KETONE
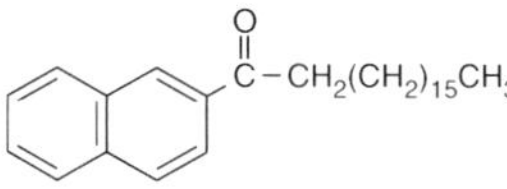

250 mg

S97,543-5
1-ACETYLBENZO(A)PYRENE
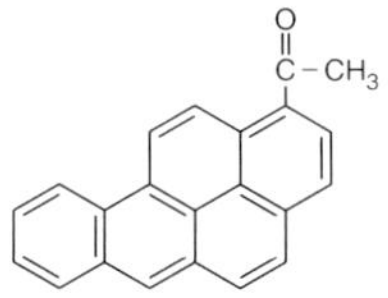

10 mg

S86,447-1
2-ACETYLANTHRACENE
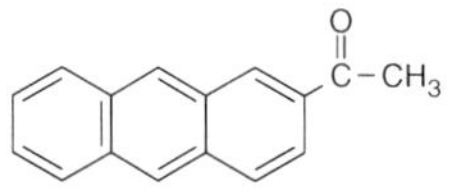

250 mg

S88,270-4
10,10-DIMETHYL-9(10H)-PHENANTHRENONE
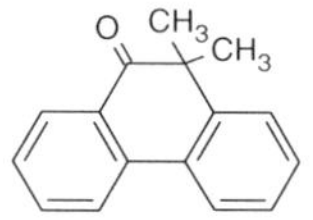

250 mg

S56,347-1
9-ACETYL-1,2,3,4-TETRAHYDROPHENANTHRENE
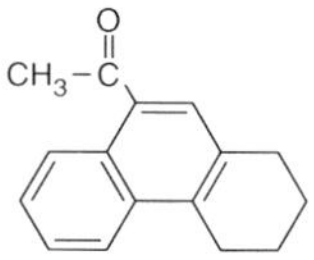

5 gm

S78,695-0
5-ACETYLINDOLINE
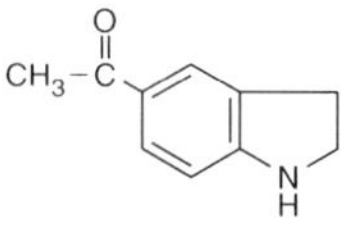

250 mg

S93,787-8
2-BENZOYL-1-BENZYL-3-PHENYLAZIRIDINE
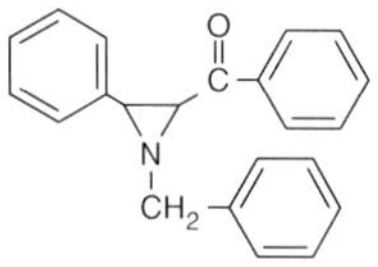

250 mg

S64,310-6
9,10,14,15-TETRAHYDRO-13H-9,10-CYCLOPENTANTHRACEN-13-ONE
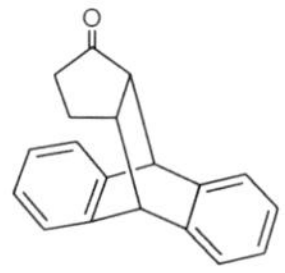

250 mg

S64,303-3
14-(2-METHYLALLYL)-9,10,14,15-TETRAHYDRO-13H-9,10-CYCLOPENTANTHRACEN-13-ONE
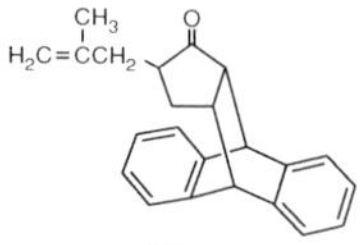

250 mg

S74,242-2
4,7-DICHLOROBENZO(B)THIOPHEN-3(2H)-ONE
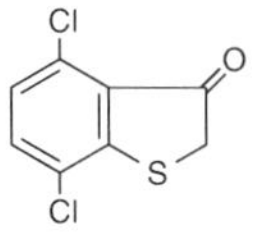

5 gm

S74,167-1
5-CHLORO-7-METHYLBENZO(B)-THIOPHEN-3(2H)-ONE
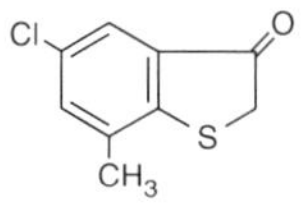

250 mg

S54,684-4
2-BROMO-3,4-DIHYDRO-7-METHOXY-1(2H)-NAPHTHALENONE
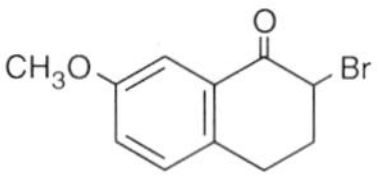

250 mg

S97,349-1
1,2,3,4-TETRAHYDRO-2-PHENYLPHENANTHREN-1-ONE
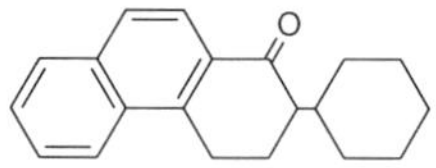

10 mg

S41,545-6
4'-FLUOROBUTYROPHENONE
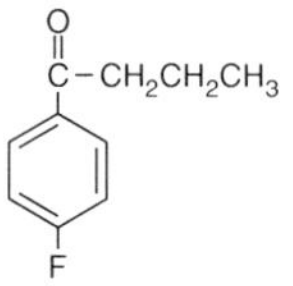

250 mg

S40,948-0
4'-METHYLVALEROPHENONE
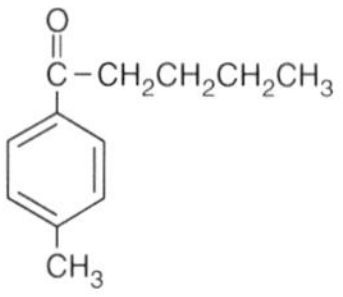

5 gm

S94,155-7
4-METHYL-4(P-TOLYL)-2-PENTANONE
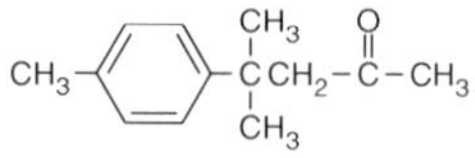

5 gm

S53,263-0
4'-PENTADECYLACETOPHENONE
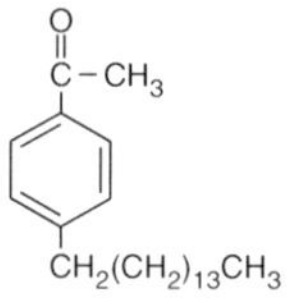

250 mg

S73,363-6
6,16-METHANOBENZOCYCLOPENTADECAN-18-ONE
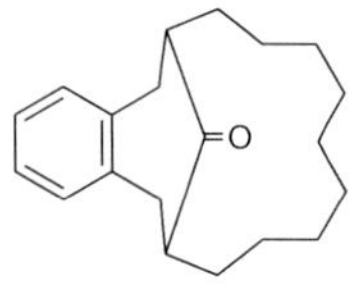

250 mg

S97,541-9
7,8,9,10-TETRAHYDRO-10-METHYLBENZO(A)PYREN-7-ONE
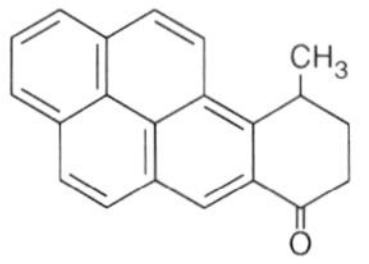

10 mg

S97,360-2
5,6,6A,7,12,12A-HEXAHYDROBENZ(A)ANTHRACEN-5-ONE
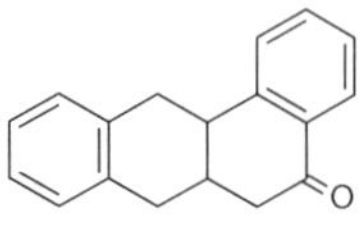

10 mg

S97,316-5
1,2,3,4-TETRAHYDROPHENANTHREN-4-ONE
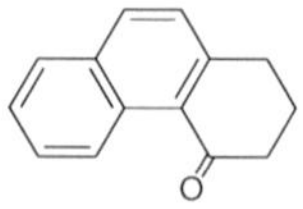

10 mg

S67,254-8
4'-CHLORO-2-HYDROXYACETOPHENONE, NITRATE ESTER
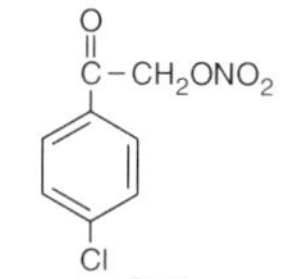
5 gm

S68,124-5
3'-(TRIFLUOROMETHYL)CHALCONE
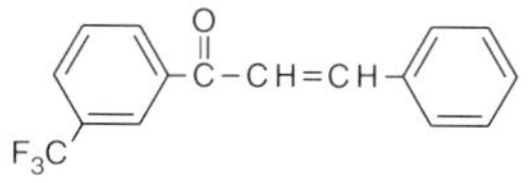
5 gm

S57,611-5
3-FLUORO-2'-HYDROXYCHALCONE
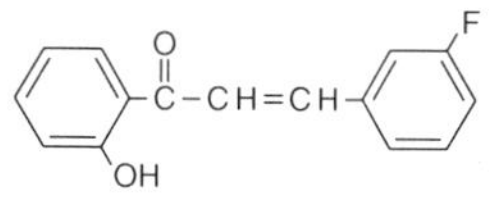
5 gm

S58,937-3
2-CHLORO-2'-HYDROXYCHALCONE
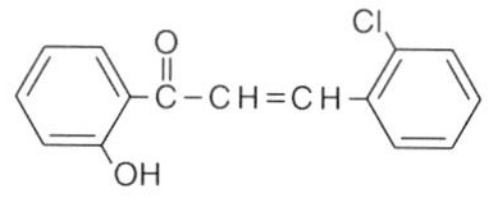
5 gm

S58,962-4
2-BROMO-2'-HYDROXYCHALCONE
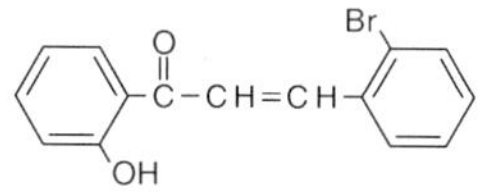
250 mg

S58,929-2
4-FLUORO-2'-HYDROXYCHALCONE
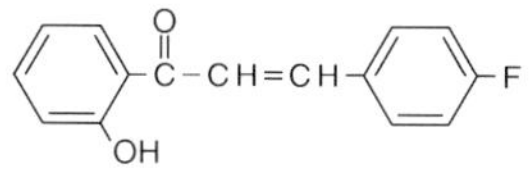
5 gm

S57,381-7
4-BROMO-2'-HYDROXYCHALCONE
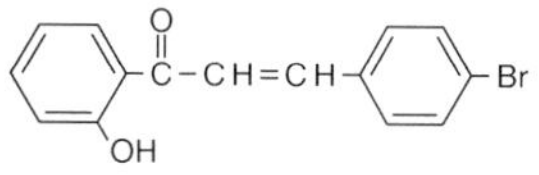
250 mg

S57,238-1
2,6-DICHLORO-2'-HYDROXYCHALCONE
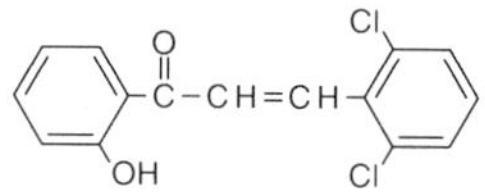
5 gm

S57,241-1
2,4-DICHLORO-2'-HYDROXYCHALCONE
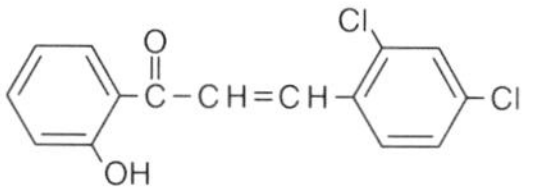
5 gm

S58,972-1
3,4-DICHLORO-2'-HYDROXYCHALCONE
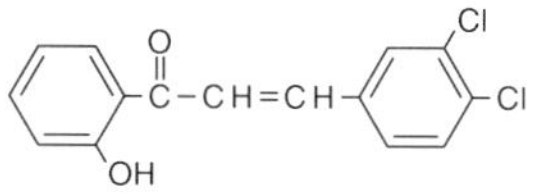
5 gm

S83,335-5
4-CHLORO-4'-HYDROXYCHALCONE
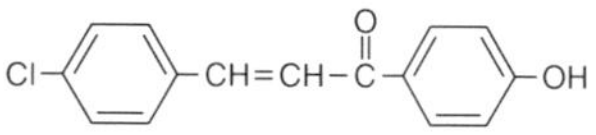
250 mg

S93,118-7
5-BROMO-2,4'-DIMETHOXYCHALCONE
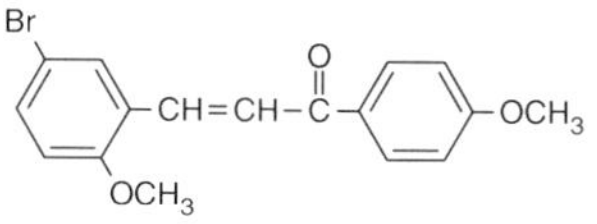
250 mg

S92,165-3
2,6-BIS(5-BROMO-2-METHOXYBENZYLIDENE)-1-CYCLOHEXANONE
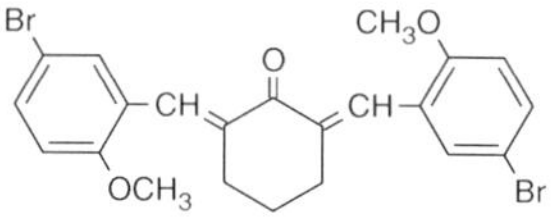
250 mg

S93,121-7
3-(5-BROMO-2-METHOXYPHENYL)-1-CYCLOPROPYL-2-PROPEN-1-ONE
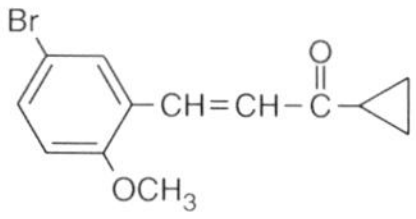
5 gm

S93,120-9
4-CHLORO-4'-METHOXY-3-NITROCHALCONE
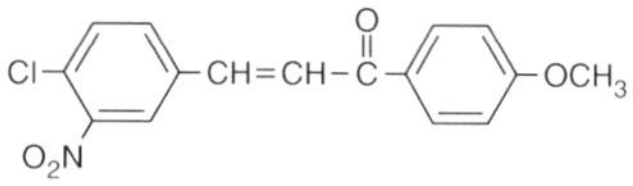
250 mg

S92,147-5
2-(5-BROMO-2-METHOXYBENZYLIDENE)-1-TETRALONE
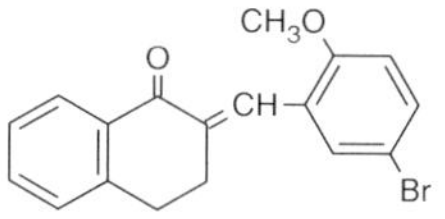
5 gm

S67,000-6
4-CHLORO-4'-METHOXYCHALCONE
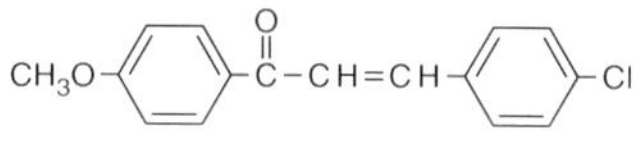
5 gm

S93,117-9
5-BROMO-4'-CHLORO-2-METHOXYCHALCONE
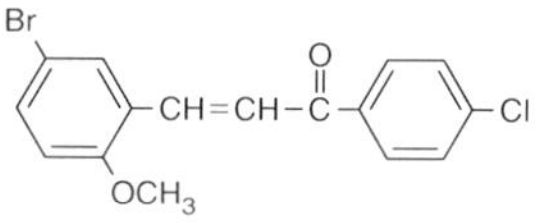
250 mg

S87,876-6
2,4'-DICHLORO-5-NITROCHALCONE
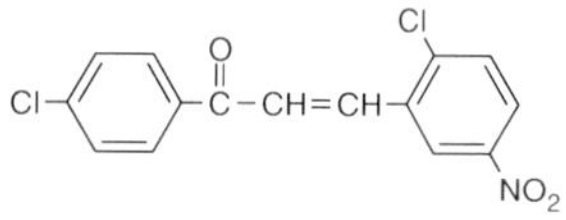
5 gm

S87,877-4
4,4'-DICHLORO-3-NITROCHALCONE
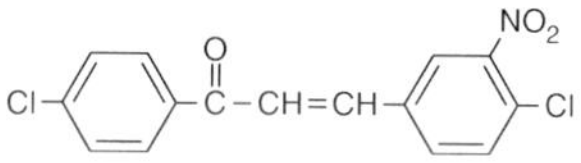
5 gm

S68,518-6
2-CHLORO-4'-METHYL-5-NITROCHALCONE
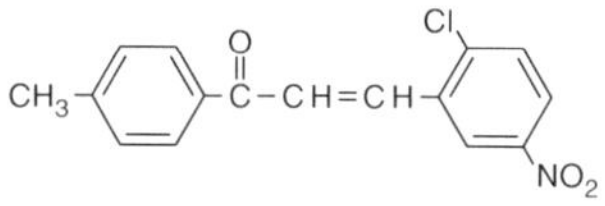
5 gm

S68,519-4
2-CHLORO-5-NITRO-3'-NITROCHALCONE
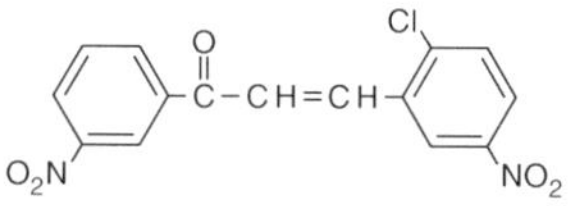
5 gm

S92,144-0
2,6-BIS(4-CHLORO-3-NITROBENZYLIDENE)-1-CYCLOHEXANONE
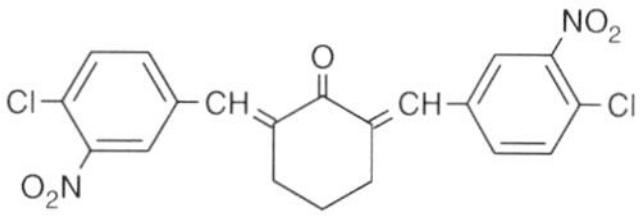
250 mg

S93,122-5
3-(2-CHLORO-5-NITROPHENYL)-1-CYCLOPROPYL-2-PROPEN-1-ONE
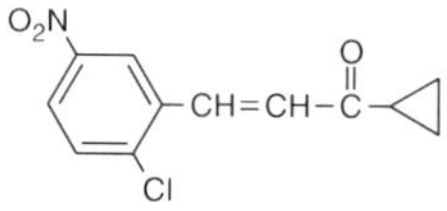
5 gm

S93,116-0
2,6-BIS(2-CHLORO-5-NITROBENZYLIDENE)-1-CYCLOHEXANONE
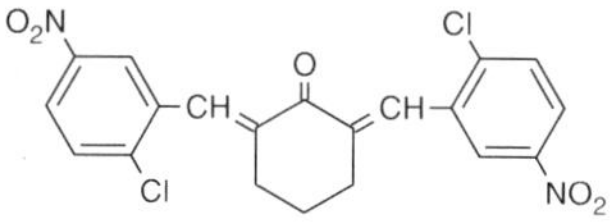
250 mg

S93,114-4
2-(2-CHLORO-5-NITROBENZYLIDENE)-1-TETRALONE
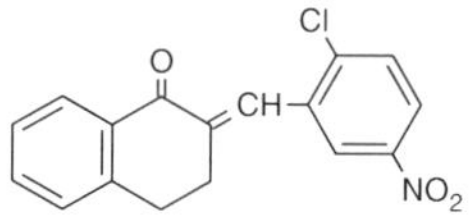
250 mg

S92,161-0
2-(4-CHLORO-3-NITROBENZYLIDENE)-1-TETRALONE
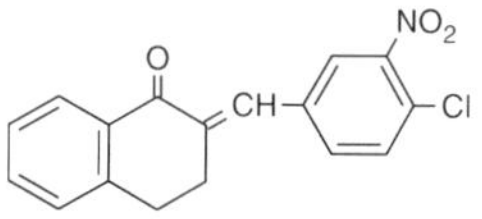
250 mg

S68,278-0
3'-CHLOROCHALCONE
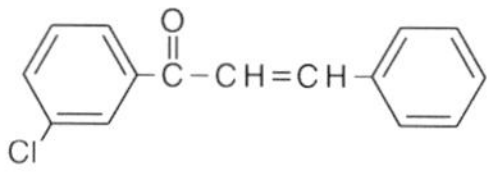
5 gm

S68,103-2
3,4-DICHLOROCHALCONE
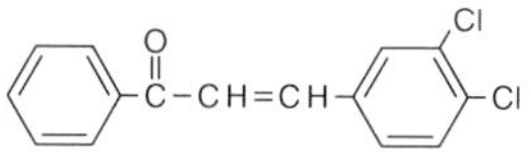
5 gm

S92,158-0
2,6-BIS(2-CHLOROBENZYLIDENE)-1-CYCLOHEXANONE
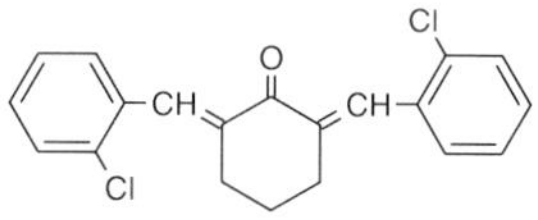
250 mg

S68,109-1
2',4-DICHLOROCHALCONE
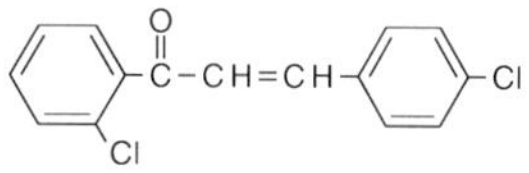
5 gm

S76,663-1
2-(4-CHLOROBENZYLIDENE)-4,4-DIMETHYL-ALPHA-TETRALONE
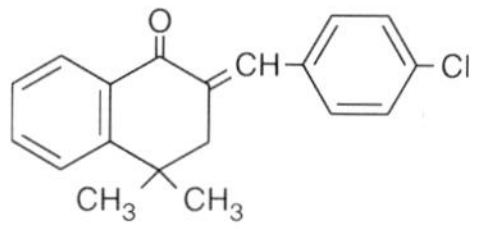
250 mg

S92,162-9
2,6-BIS(2,4-DICHLOROBENZYLIDENE)-1-CYCLOHEXANONE
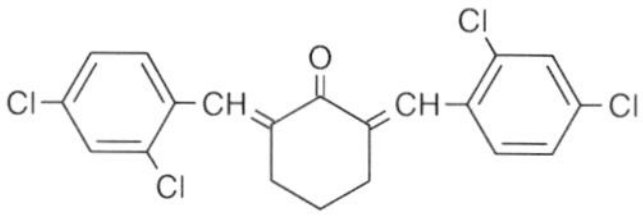
250 mg

S73,996-0
3-(2,4-DICHLOROPHENYL)-2'-ACRYLONAPHTHONE
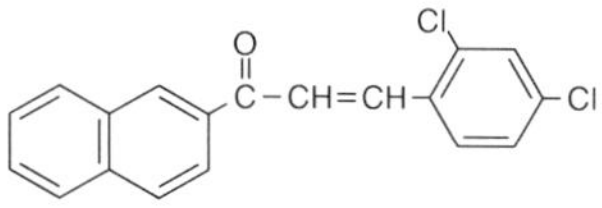
5 gm

S94,731-8
2,5-BIS(4-CHLOROBENZYLIDENE)-1-CYCLOPENTANONE
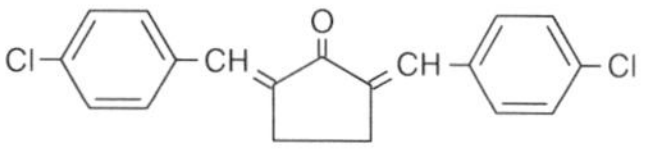
250 mg

S64,181-2
4-BENZYLOXY-2'-HYDROXYCHALCONE
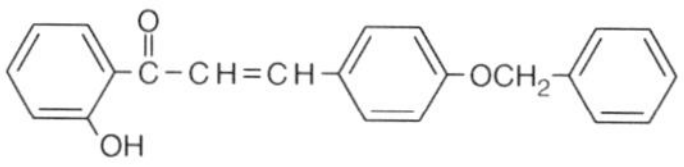
250 mg

S84,996-0
1-(4,5-DIMETHOXY-2-HYDROXYPHENYL)-3-METHYL-2-BUTEN-1-ONE

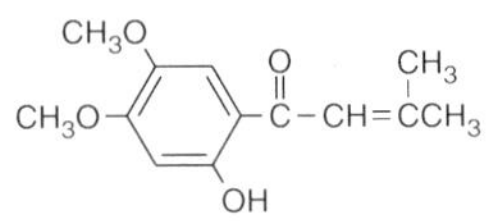

250 mg

S67,002-2
5-BROMO-3,4'-DIMETHOXY-4-HYDROXYCHALCONE

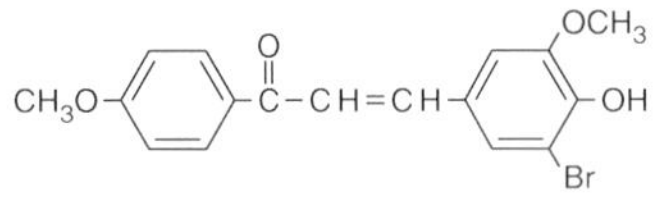

250 mg

S57,264-0
2'-HYDROXY-2-METHOXYCHALCONE

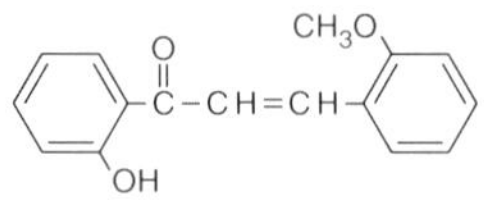

250 mg

S57,380-9
2'-HYDROXY-3-METHOXYCHALCONE

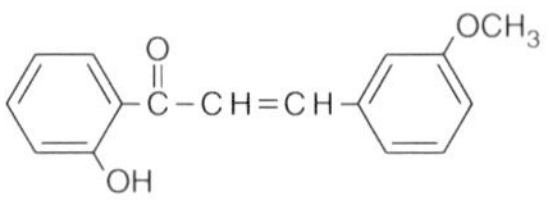

5 gm

S57,267-5
2'-HYDROXY-4-METHOXYCHALCONE

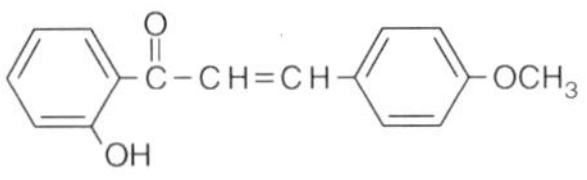

250 mg

S57,601-8
2'-HYDROXY-3,4,5-TRIMETHOXYCHALCONE

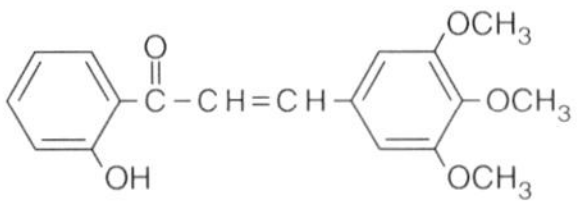

250 mg

S67,001-4
4'-HYDROXY-4-METHOXYCHALCONE

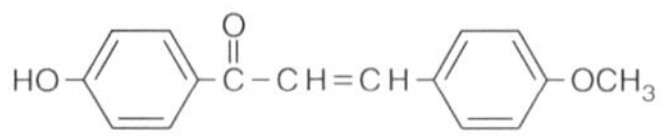

250 mg

S84,219-2
2,6-DI-TERT-BUTYL-4-HYDROXY-4-PHENYL-1(4H)-NAPHTHALENONE

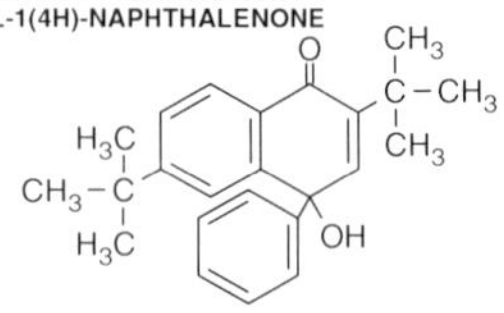

250 mg

S44,972-5
2'-HYDROXYCHALCONE

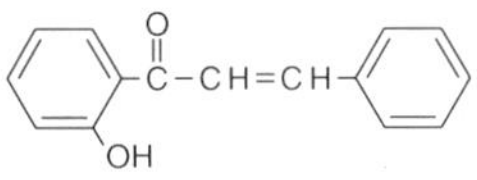

5 gm

S58,934-9
2'-HYDROXY-4-ISOPROPYLCHALCONE

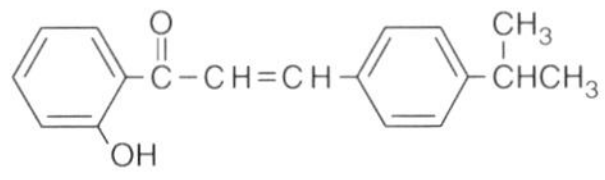

5 gm

S94,749-0
2-(DIETHYLAMINOMETHYL)-CYCLOHEXANONE PICRATE

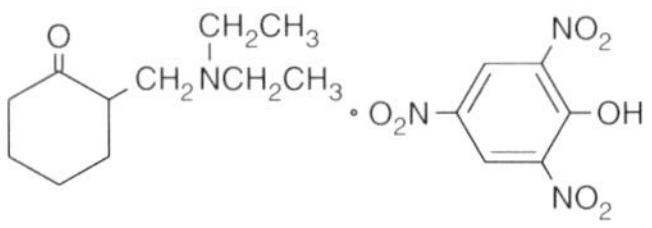

250 mg

S57,607-7
4-(DIETHYLAMINO)-2'-HYDROXYCHALCONE

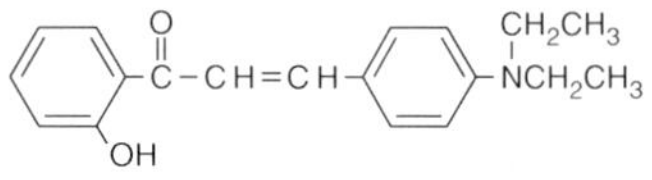

5 gm

S57,603-4
2-CHLORO-4-(DIMETHYLAMINO)-2'-HYDROXYCHALCONE

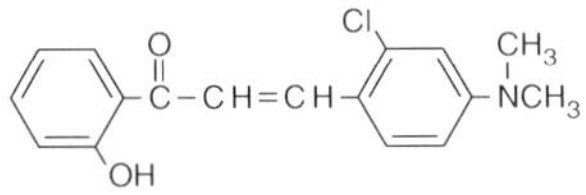

5 gm

S57,263-2
2'-HYDROXY-3-NITROCHALCONE

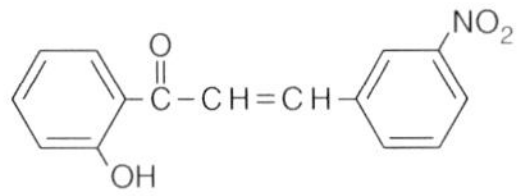

5 gm

S57,268-3
2'-HYDROXY-2-PIPERONYLIDENEACETOPHENONE

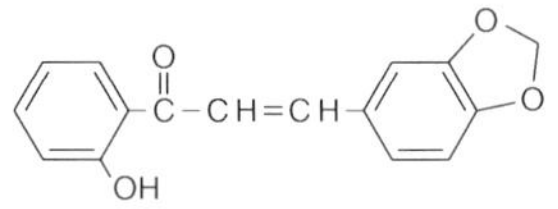

5 gm

S38,881-5
2-HYDROXYCHALCONE

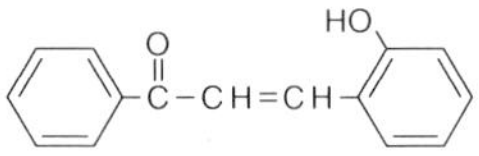

5 gm

S83,336-3
4'-HYDROXYCHALCONE

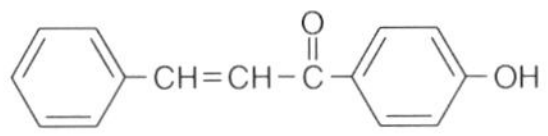

250 mg

S84,953-7
4-(2-HYDROXYPHENYL)-3-BUTEN-2-ONE

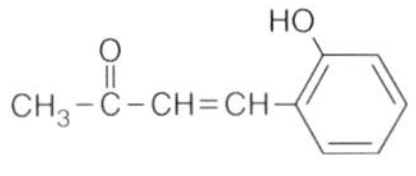

250 mg

S57,237-3
2'-HYDROXY-2-(1-NAPHTHYLMETHYLENE)-ACETOPHENONE

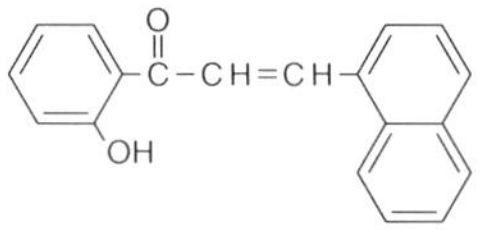

5 gm

S58,448-7
3-(9-ANTHRYL)-2'-HYDROXYACRYLOPHENONE

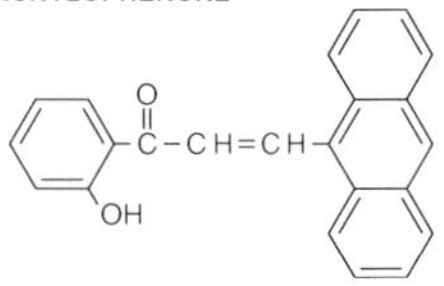

5 gm

S43,184-2
4'-HYDROXY-5'-ISOPROPYL-2'-METHYLACETOPHENONE

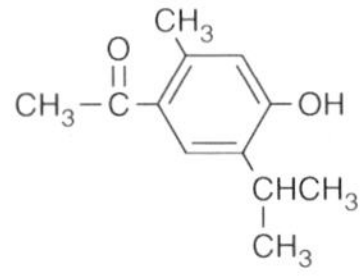

5 gm

S63,976-1
4-BENZYLOXY-4'-CHLOROCHALCONE

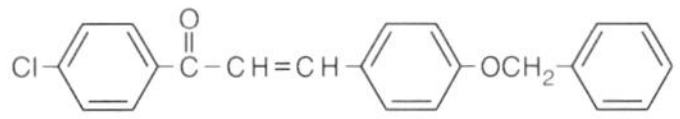

5 gm

S64,179-0
4-BENZYLOXY-4'-METHYLCHALCONE

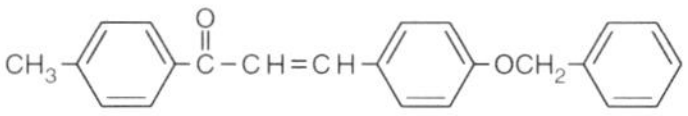

5 gm

S65,317-9
2-(4-BENZYLOXYBENZYLIDENE)-1-TETRALONE

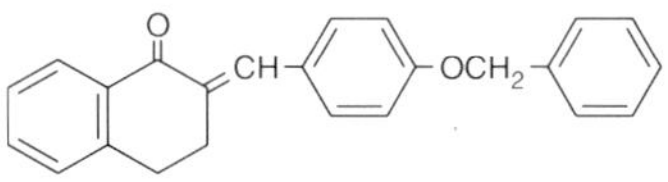

5 gm

S79,812-6
3,4-DIMETHOXYCHALCONE

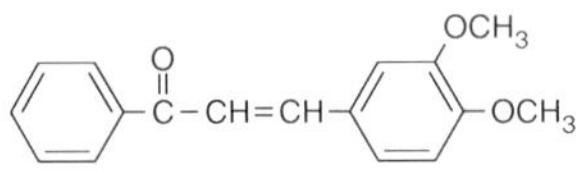

250 mg

S73,323-7
3,4,5-TRIMETHOXYCHALCONE

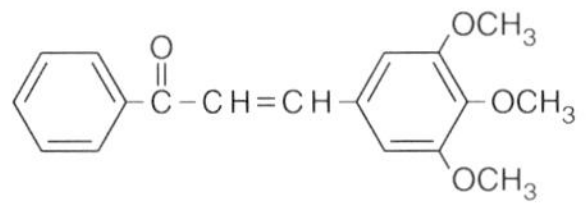

250 mg

S93,115-2
2,6-BIS(2,4-DIMETHOXYBENZYLIDENE)-1-CYCLOHEXANONE

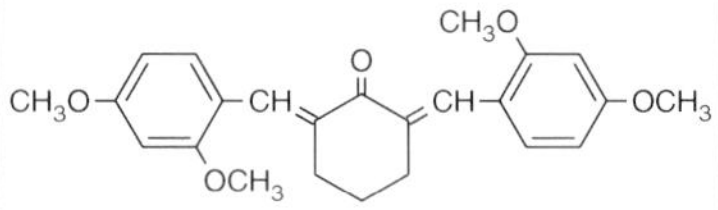

250 mg

S93,119-5
2,4,4'-TRIMETHOXYCHALCONE

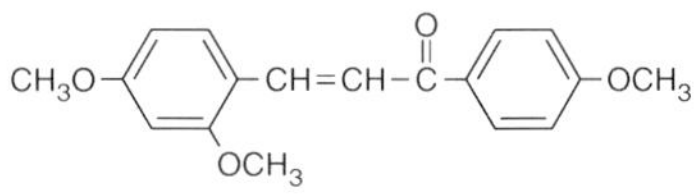

250 mg

S64,880-9
3,4-DIMETHOXY-3'-NITROCHALCONE

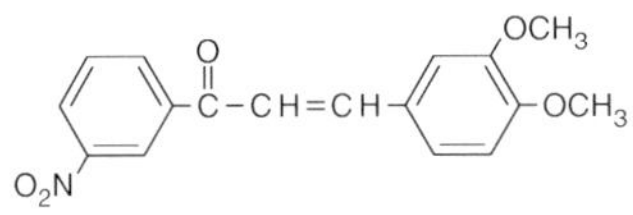

250 mg

S52,449-2
3,4-DIMETHOXY-4'-NITROCHALCONE

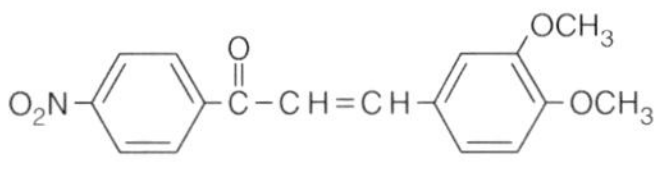

5 gm

S53,651-2
3'-NITRO-3,4,4'-TRIMETHOXYCHALCONE

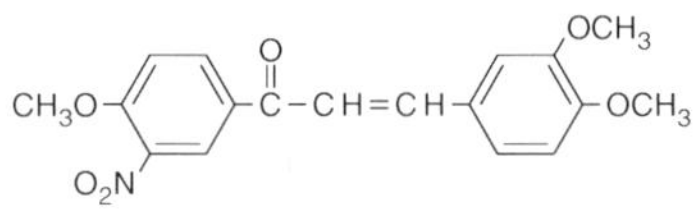

250 mg

S54,837-5
4'-CHLORO-2,3-DIMETHOXYCHALCONE

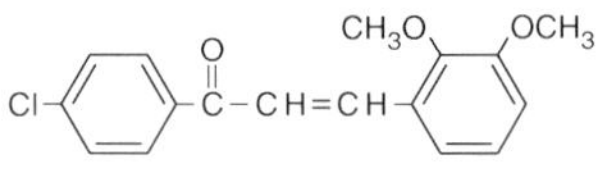

5 gm

S92,982-4
2,3,4'-TRIMETHOXYCHALCONE

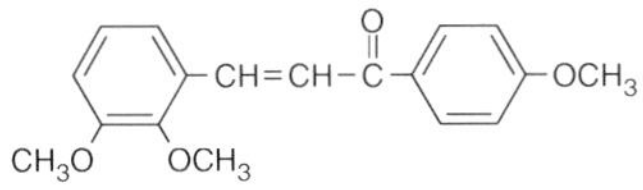

250 mg

S52,447-6
3,4-DIMETHOXY-4'-METHYLCHALCONE

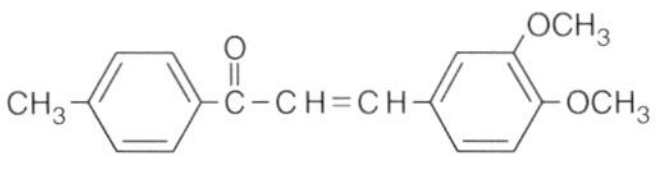

5 gm

S58,417-7
2-(2,3-DIMETHOXYBENZYLIDENE)-1-TETRALONE

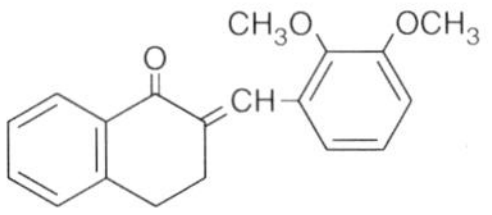

250 mg

S55,573-8
3',4'-DICHLORO-3,4-DIMETHOXYCHALCONE

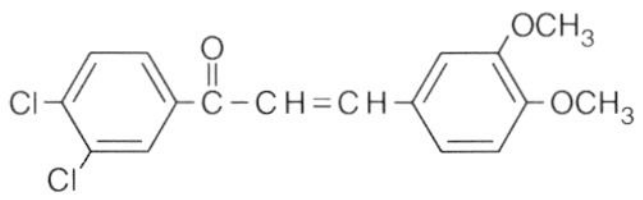

5 gm

S53,603-2
3',4'-DICHLORO-2,3-DIMETHOXYCHALCONE

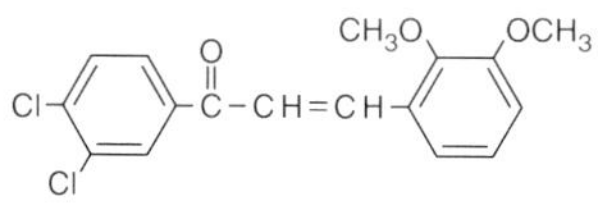

250 mg

S52,577-4
2,3-DIMETHOXY-4'-METHYLCHALCONE

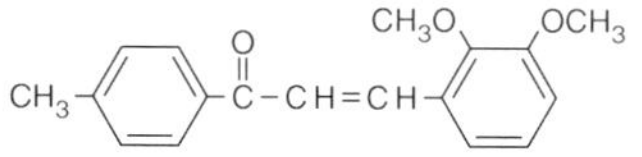

5 gm

S88,329-8
1-(2-METHOXY-4-METHYLPHENYL)-3-METHYL-2-BUTEN-1-ONE

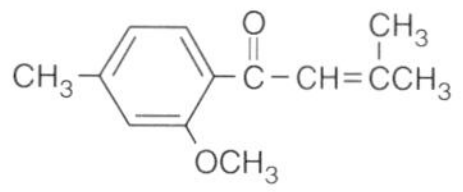

250 mg

S68,125-3
3'-METHOXY-3-NITROCHALCONE

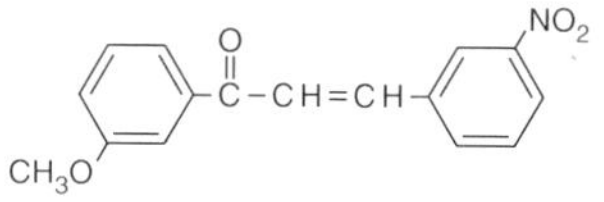

5 gm

S59,995-6
2-(DIMETHYLAMINO)-4'-METHOXYCHALCONE

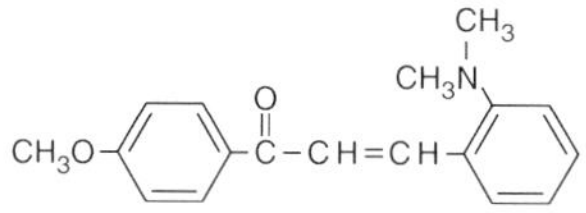

5 gm

S76,622-4
4-(2-DIETHYLAMINOETHOXY)CHALCONE

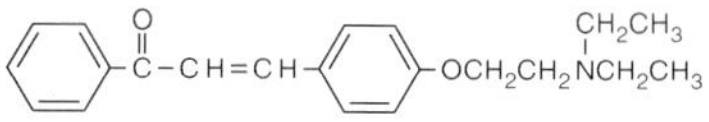

250 mg

S56,930-5
4-(DIMETHYLAMINO)-4'-METHOXYCHALCONE

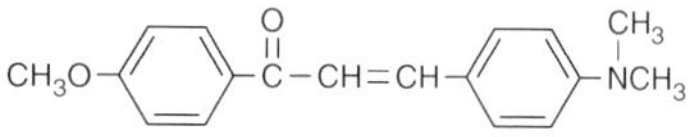

5 gm

S61,495-5
2-METHOXY-4'-NITROCHALCONE

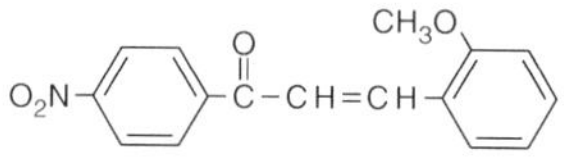

5 gm

S61,383-5
2-ETHOXY-4'-NITROCHALCONE

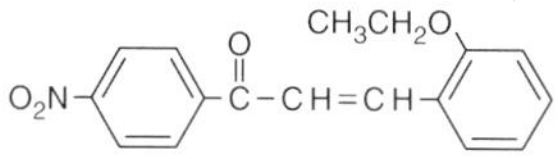

5 gm

S61,376-2
3-METHOXY-4'-NITROCHALCONE

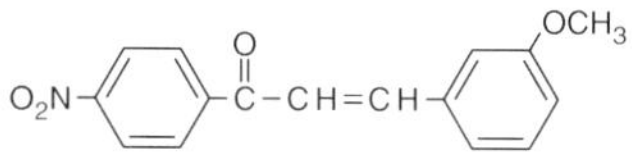

5 gm

S64,379-3
4-METHOXY-3'-NITROCHALCONE

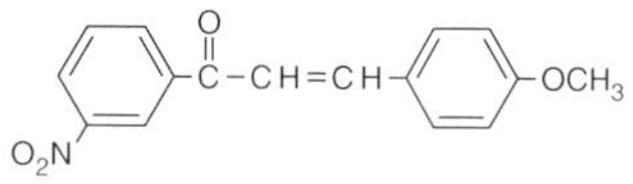

5 gm

S93,221-3
4,4-DIMETHYL-2-(5-METHOXY-2-NITROBENZYLIDENE)-1-TETRALONE

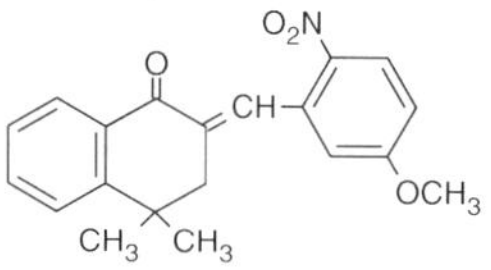

250 mg

S53,964-3
2-CHLORO-4'-METHOXY-3'-NITROCHALCONE

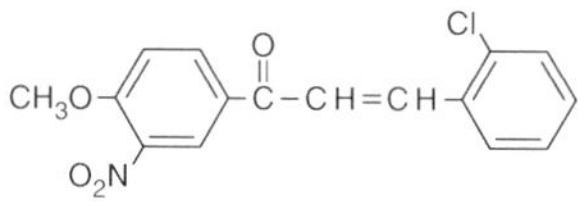

5 gm

S50,953-1
4-METHOXY-4'-NITROCHALCONE

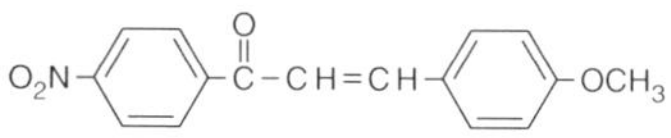

5 gm

S53,469-2
2,6-DICHLORO-4'-METHOXY-3'-NITROCHALCONE

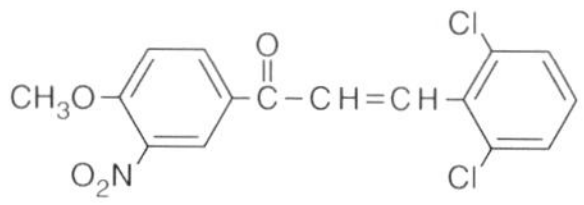

5 gm

S53,660-1
3,3'-DINITRO-4'-METHOXYCHALCONE

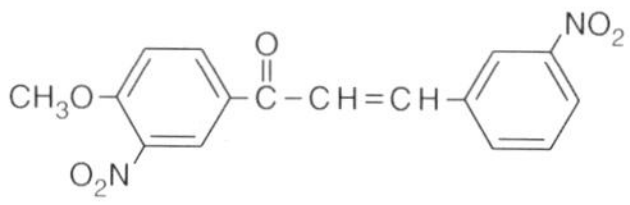

5 gm

S50,458-0
4'-NITRO-5-PHENYL-2,4-PENTADIENOPHENONE

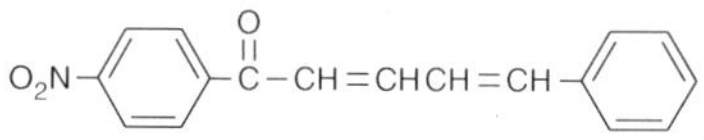

5 gm

S52,430-1
2,3-DIMETHOXY-4'-NITROCHALCONE

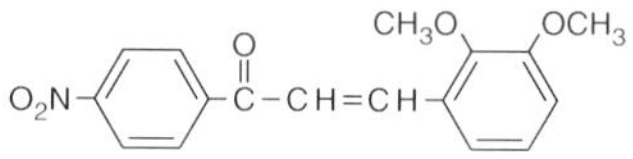

5 gm

S50,450-5
3-(1-NAPHTHYL)-4'-NITROACRYLOPHENONE

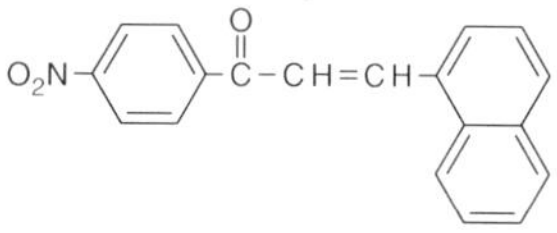

5 gm

S89,433-8
1-(4-METHOXYPHENYL)-2,3,3-TRICHLORO-2-PROPEN-1-ONE

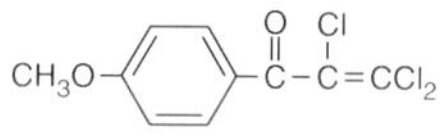

250 mg

S55,979-2
4'-FLUORO-3,4-(METHYLENEDIOXY)-CHALCONE

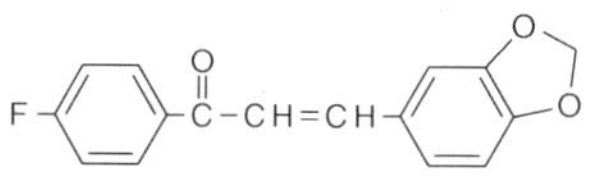

5 gm

S45,674-8
4-METHOXYCHALCONE

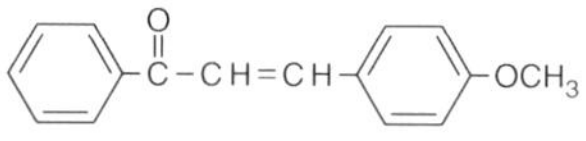

5 gm

S66,274-7
2'-METHOXY-3-PHENYL-1'-ACRYLONAPHTHONE

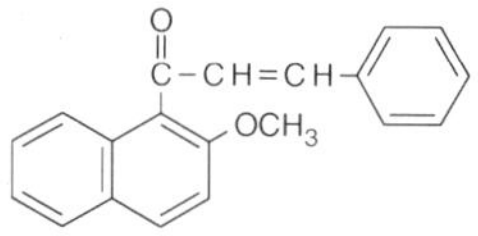

250 mg

S61,497-1
4'-CHLORO-2-METHOXYCHALCONE

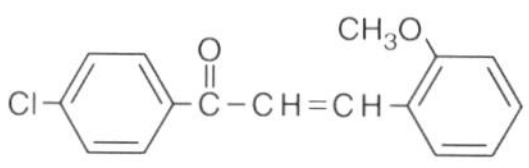

250 mg

S61,645-1
4'-CHLORO-2-ETHOXYCHALCONE

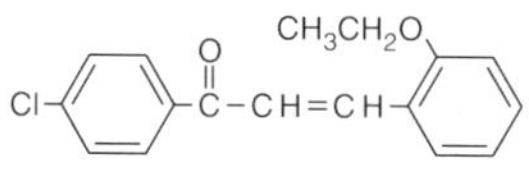

250 mg

S58,660-9
2-ETHOXY-4'-METHYLCHALCONE

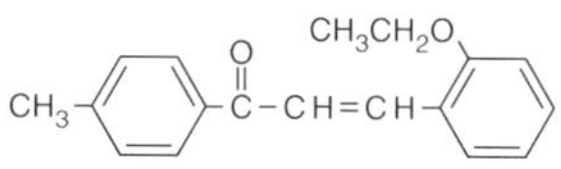

5 gm

S57,724-3
2-(2-ETHOXYBENZYLIDENE)-1-TETRALONE

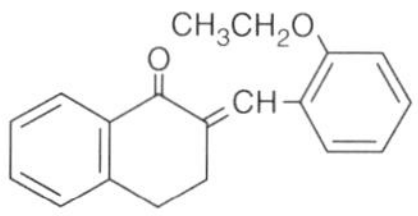

250 mg

S43,399-3
4'-FLUORO-4-METHOXYCHALCONE

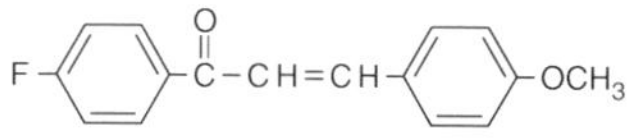

5 gm

S61,382-7
4'-CHLORO-3-METHOXYCHALCONE

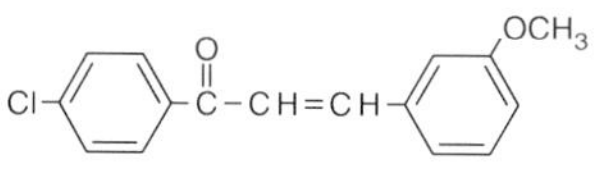

5 gm

S56,921-6
4'-BROMO-4-METHOXYCHALCONE

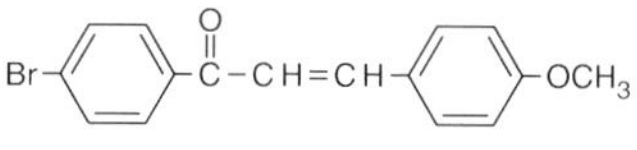

250 mg

S58,983-7
3-METHOXY-4'-METHYLCHALCONE

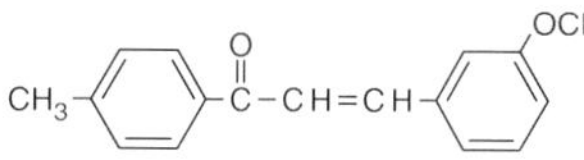

5 gm

S68,039-7
4'-METHOXY-4-METHYLCHALCONE

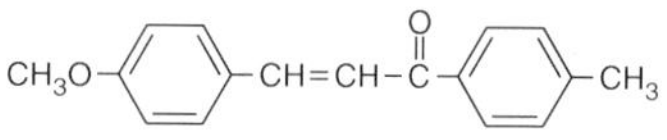

250 mg

S61,723-7
4,4'-DIMETHOXYCHALCONE

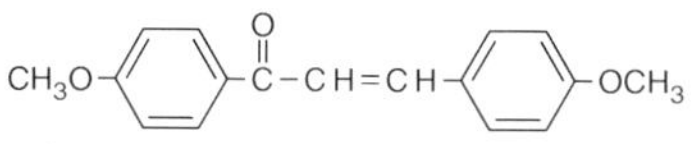

250 mg

S58,416-9
4-ETHOXY-4'-METHYLCHALCONE

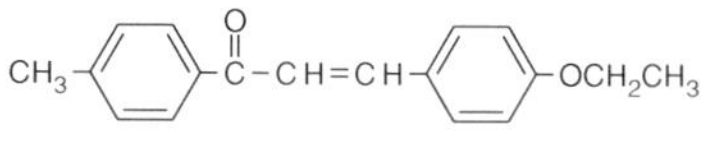

250 mg

S55,959-8
4-METHOXY-4'-PHENYLCHALCONE

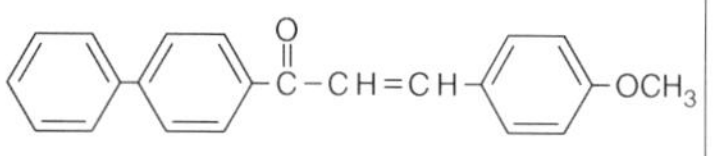

5 gm

S57,717-0
2-(4-ETHOXYBENZYLIDENE)-1-TETRALONE

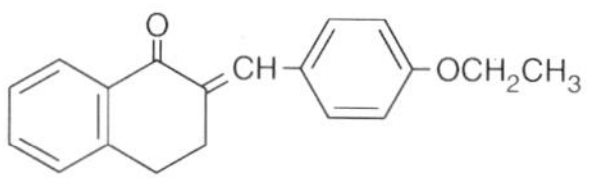

250 mg

S57,681-6
2-(3,4,5-TRIMETHOXYBENZYLIDENE)-1-TETRALONE
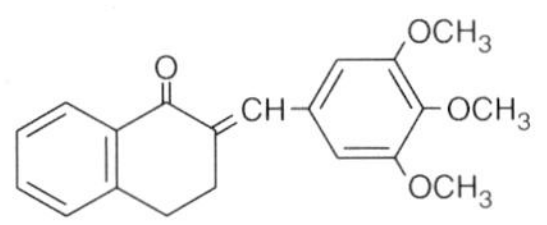
5 gm

S89,350-1
6-METHOXY-2-METHYLENE-1-TETRALONE
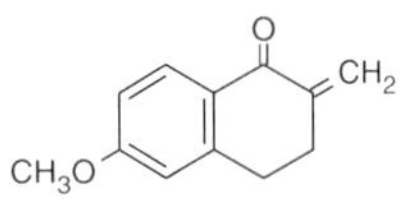
250 mg

S93,589-1
2,5-BIS(4-METHOXYBENZYLIDENE)-CYCLOPENTANONE
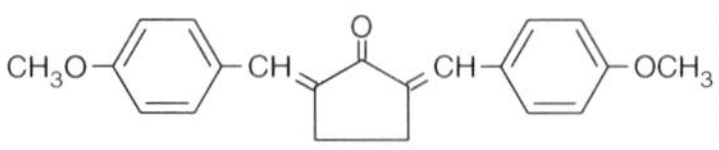
250 mg

S93,587-5
BIS(4-METHOXYBENZYLIDENE)ACETONE
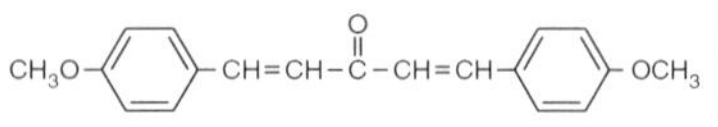
250 mg

S54,766-2
4'-CHLORO-4-METHOXYCHALCONE
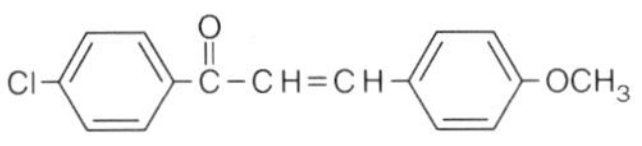
5 gm

S51,102-1
4-METHOXY-4'-METHYLCHALCONE
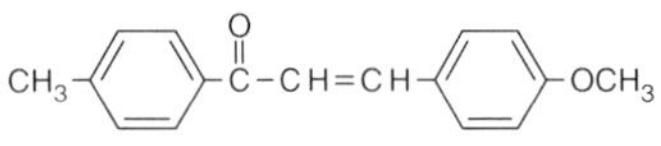
5 gm

S54,813-8
4'-CHLORO-2,4-DIMETHOXYCHALCONE
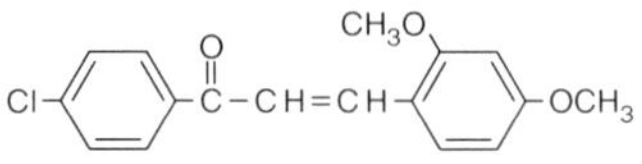
250 mg

S54,883-9
4'-CHLORO-2,5-DIMETHOXYCHALCONE
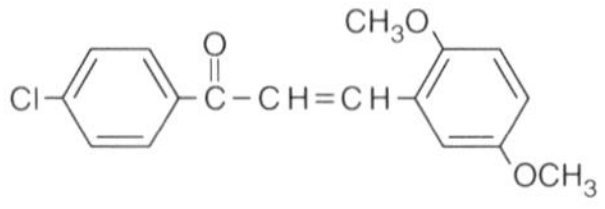
5 gm

S52,429-8
2,4-DIMETHOXY-4'-METHYLCHALCONE
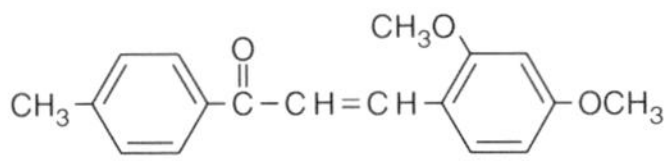
250 mg

S52,446-8
2,5-DIMETHOXY-4'-METHYLCHALCONE
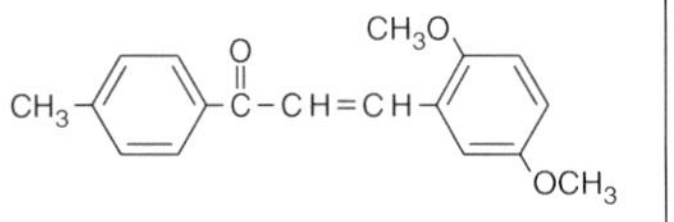
250 mg

S54,696-8
3-BROMO-4-METHOXYCHALCONE
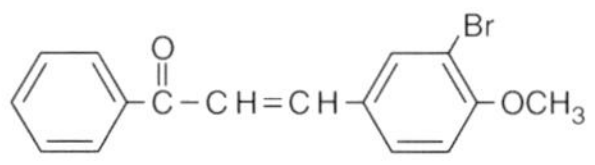
5 gm

S81,255-2
3,4'-DIAMINOCHALCONE
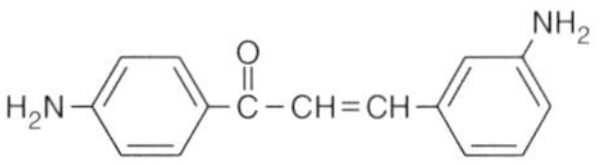
250 mg

S93,220-5
2-(2-AMINOBENZYLIDENE)-4,4-DIMETHYL-1-TETRALONE
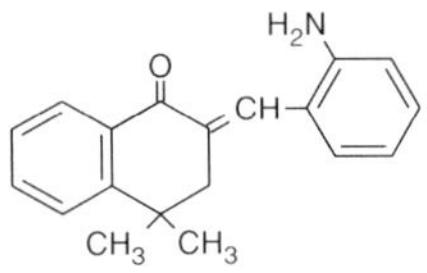
250 mg

S87,375-6
4-[4-(DIMETHYLAMINO)-ALPHA-PHENYLBENZYLIDENE]-1(4H)-NAPHTHALENONE
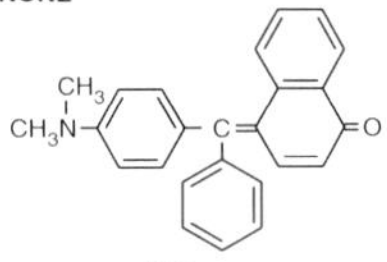
250 mg

S39,340-1
4-DIMETHYLAMINO-4'-METHYLCHALCONE
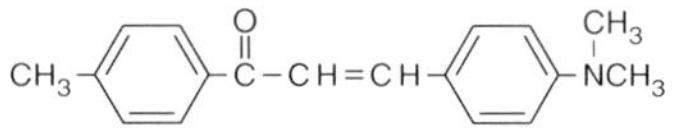
5 gm

S87,370-5
4-(BIS[4-(DIMETHYLAMINO)PHENYL]-METHYLENE)-1(4H)-NAPHTHALENONE
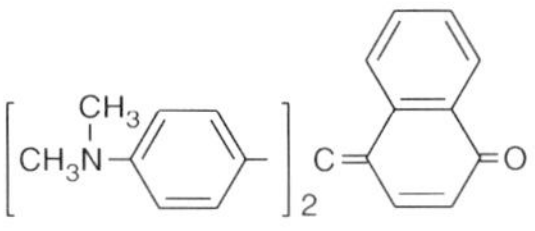
250 mg

S44,786-2
3-DIMETHYLAMINO-1-PHENYL-2-PROPEN-1-ONE
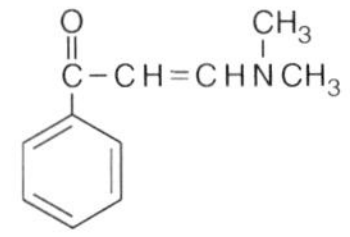
5 gm

S62,316-4
4-DIETHYLAMINO-4'-NITROCHALCONE
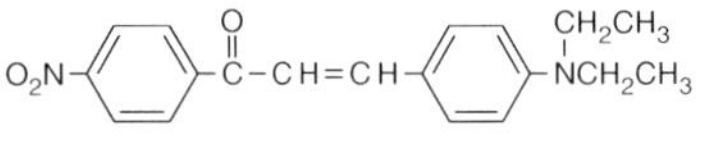
250 mg

S50,959-0
2-CHLORO-4-DIMETHYLAMINO-4'-NITROCHALCONE
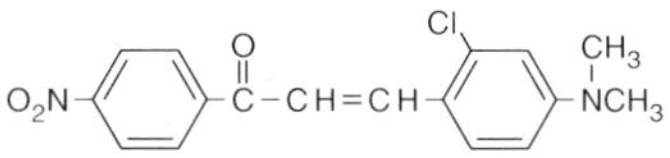
5 gm

S89,896-1
2-(1-BENZYL-3,3-DIMETHYL-2-INDOLINYLIDENE)ACETOPHENONE
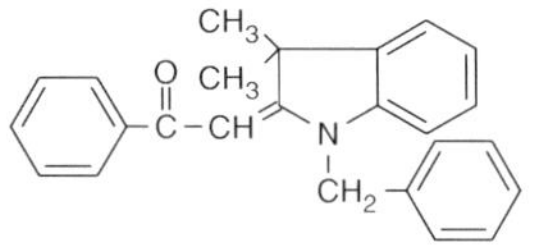
250 mg

S56,002-2
4-DIMETHYLAMINO-4'-FLUOROCHALCONE
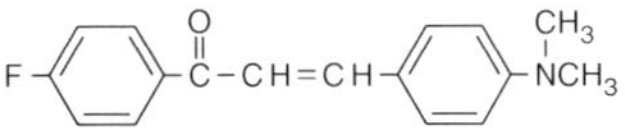
5 gm

S55,976-8
4'-BROMO-4-(DIMETHYLAMINO)CHALCONE
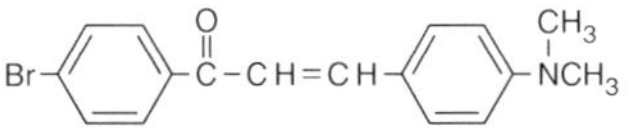
5 gm

S61,490-4
4'-CHLORO-4-DIETHYLAMINOCHALCONE
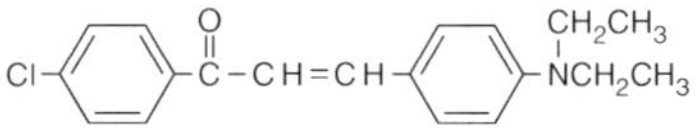
5 gm

S58,974-8
4-DIETHYLAMINO-4'-METHYLCHALCONE
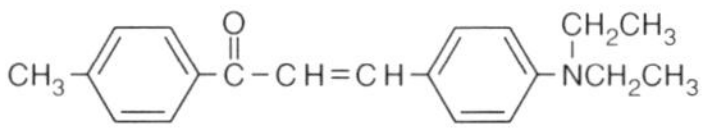
5 gm

S56,919-4
2',4'-DIMETHYL-4-(DIMETHYLAMINO)-CHALCONE
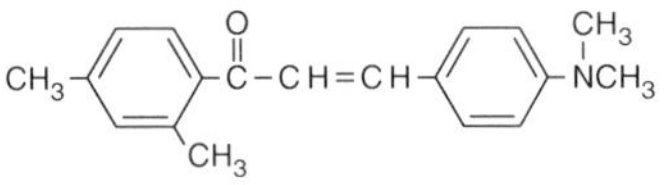
250 mg

S57,652-2
2-[4-(DIETHYLAMINO)BENZYLIDENE]-1-TETRALONE
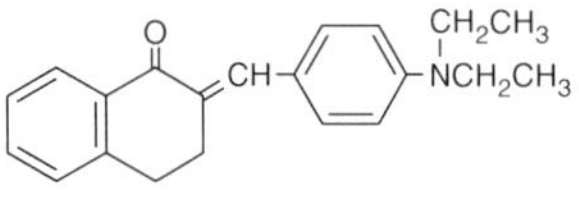
250 mg

S56,922-4
3-[4-(DIMETHYLAMINO)PHENYL]-1'-ACRYLONAPHTHONE
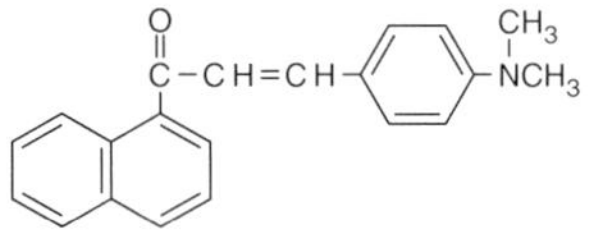
250 mg

S87,351-9
4-([10-METHYL-9(10H)-ACRIDINYLIDENE]ETHYLIDENE)-1(4H)-NAPHTHALENONE
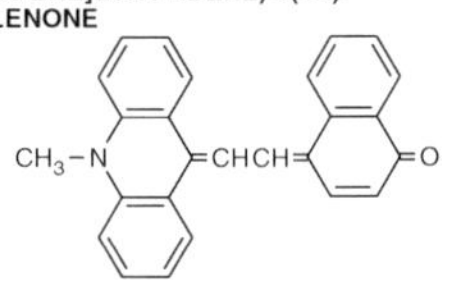
250 mg

S93,586-7
BIS[4-(DIMETHYLAMINO)BENZYLIDENE]-ACETONE
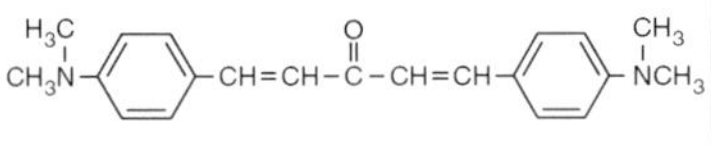
250 mg

S54,769-7
4'-CHLORO-4-(DIMETHYLAMINO)-CHALCONE
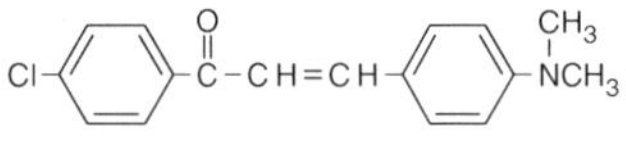
5 gm

S54,838-3
2,4'-DICHLORO-4-(DIMETHYLAMINO)-CHALCONE
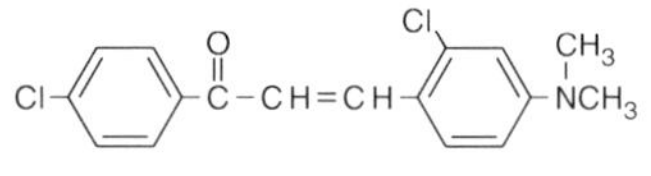
250 mg

S52,416-6
2-CHLORO-4-DIMETHYLAMINO-4'-METHYLCHALCONE
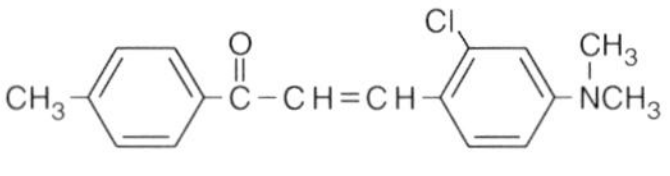
5 gm

S74,536-7
BENZIL MONO(METHYLHYDRAZONE)
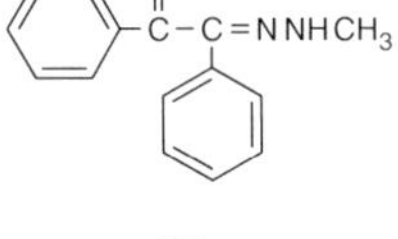
250 mg

S70,684-1
BENZIL MONOPHENYLHYDRAZONE
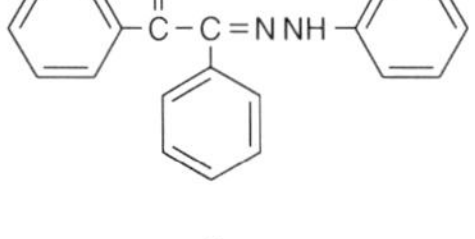
5 gm

S87,281-4
1,4-NAPHTHOQUINONE BIS(P-TOLYL)-HYDRAZONE
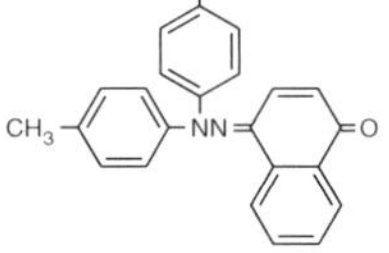
250 mg

S73,987-1
3-(3-NITROPHENYL)-2'-ACRYLONAPHTHONE
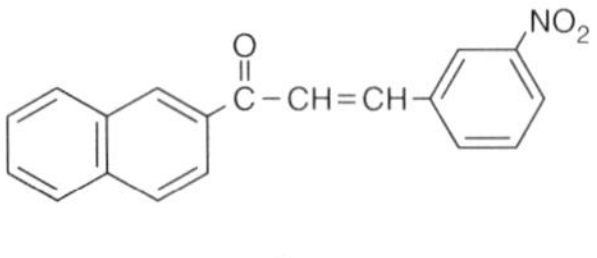
5 gm

S92,197-1
4,4-DIMETHYL-1-(4-NITROPHENYL)-1-PENTEN-3-ONE
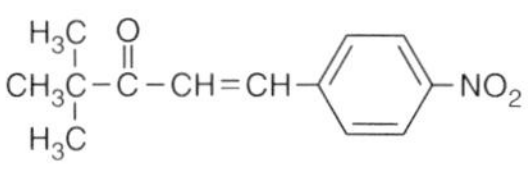
5 gm

S64,375-0
3'-NITROCHALCONE
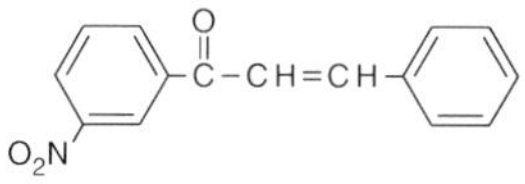
5 gm

S64,373-4
3,3'-DINITROCHALCONE
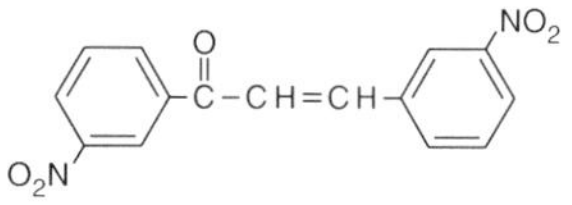
5 gm

S39,903-5
3,4-METHYLENEDIOXY-4'-NITROCHALCONE
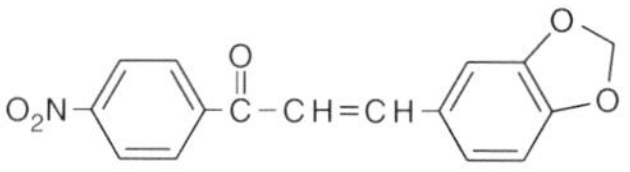
5 gm

S68,126-1
2-NITROCHALCONE
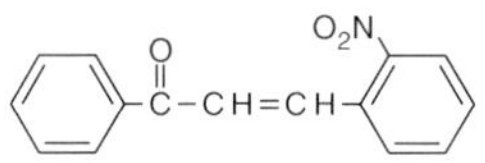
5 gm

S64,372-6
2-CHLORO-3'-NITROCHALCONE
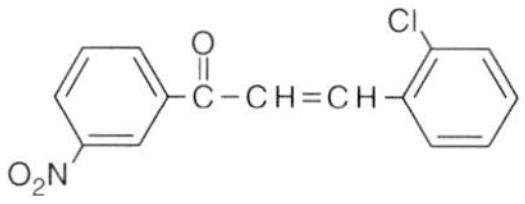
250 mg

S93,227-2
2-(2-NITROBENZYLIDENE)-1-TETRALONE
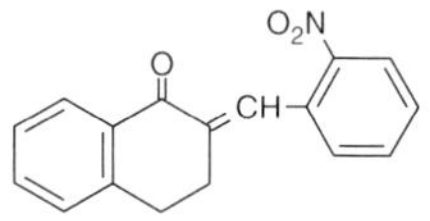
250 mg

S61,380-0
4'-CHLORO-3-NITROCHALCONE
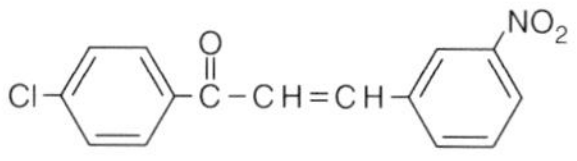
5 gm

S64,374-2
4-CHLORO-3'-NITROCHALCONE
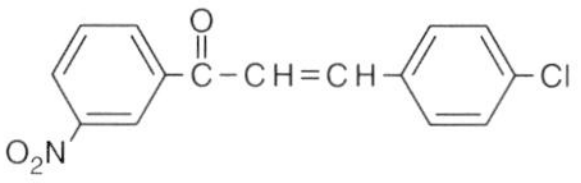
5 gm

S61,638-9
3,4'-DINITROCHALCONE
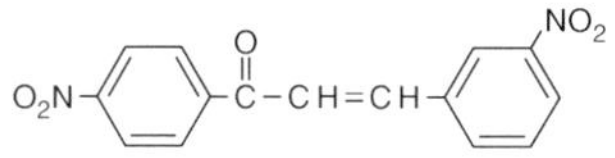
5 gm

S64,377-7
4'-BROMO-3-NITROCHALCONE
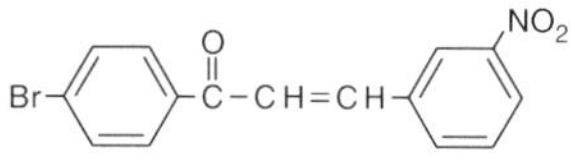
5 gm

S58,918-7
4'-METHYL-3-NITROCHALCONE
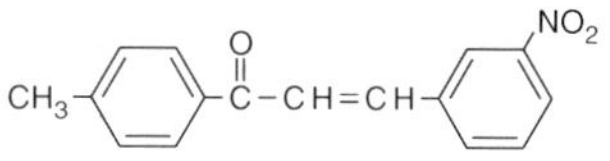
5 gm

S61,752-0
4-ISOPROPYL-4'-NITROCHALCONE
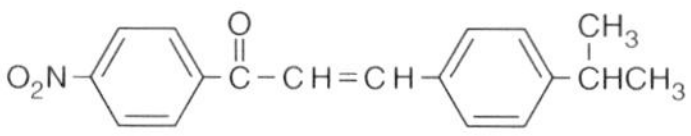
5 gm

S64,556-7
4-ISOPROPYL-3'-NITROCHALCONE
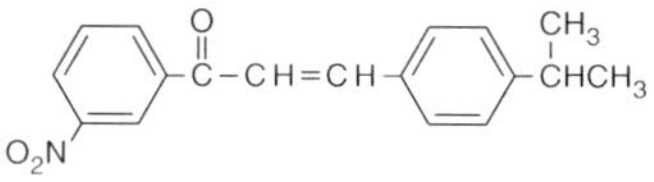
5 gm

S43,412-4
4-CHLORO-4'-FLUORO-3-NITROCHALCONE
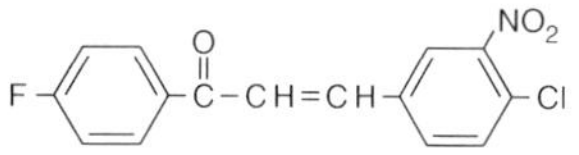
5 gm

S51,465-9
3,4-DICHLORO-4'-NITROCHALCONE
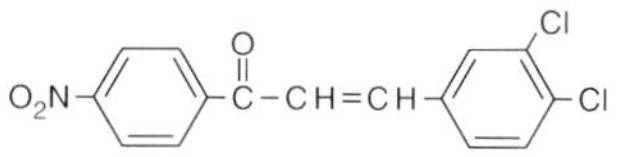
5 gm

S61,499-8
2,4-DICHLORO-4'-NITROCHALCONE
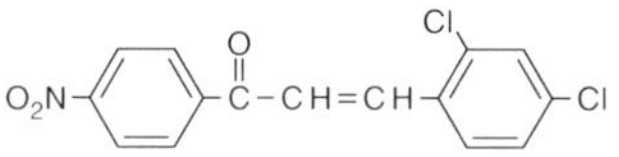
5 gm

S62,124-2
3-(9-ANTHRACENYL)-1-(4-NITROPHENYL)-2-PROPEN-1-ONE
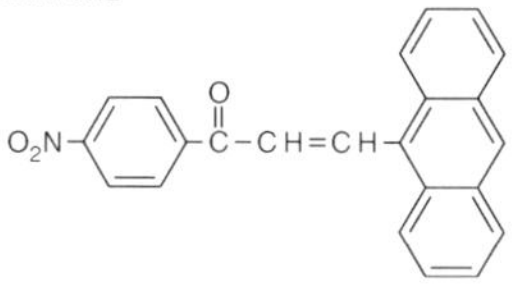
5 gm

S64,565-6
2,4-DICHLORO-3'-NITROCHALCONE
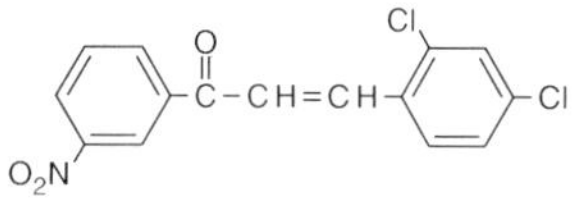
250 mg

S58,960-8
2-(3-NITROBENZYLIDENE)-1-TETRALONE
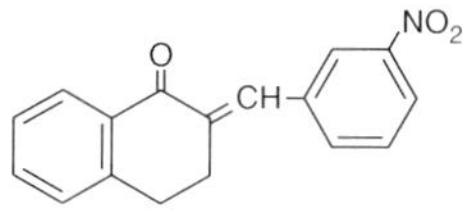
250 mg

S93,222-1
4,4-DIMETHYL-2-(2-NITROBENZYLIDENE)-1-TETRALONE
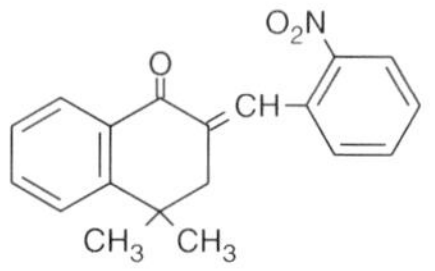
250 mg

S53,896-5
3-NITROCHALCONE
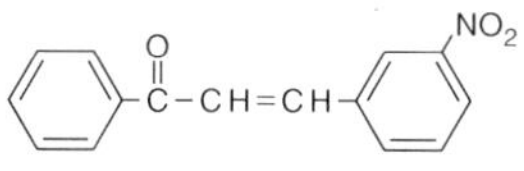
5 gm

S50,955-8
2-CHLORO-4'-NITROCHALCONE
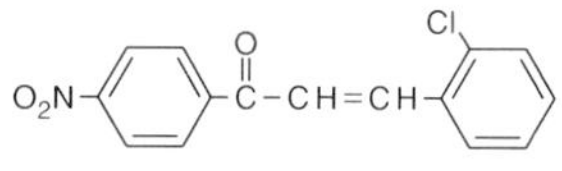
5 gm

S52,263-5
3-FLUORO-4'-NITROCHALCONE
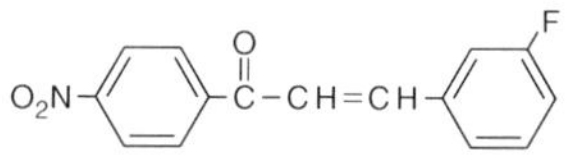
5 gm

S52,450-6
4-FLUORO-4'-NITROCHALCONE
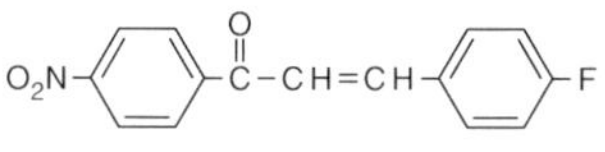
5 gm

S51,466-7
4-CHLORO-4'-NITROCHALCONE
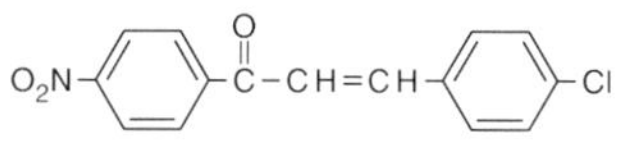
5 gm

S53,872-8
3-CHLORO-4'-NITROCHALCONE
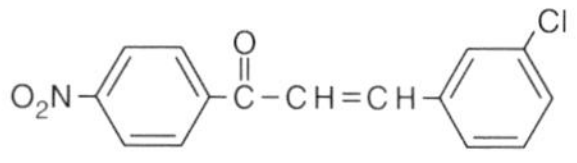
5 gm

S54,798-0
4'-CHLORO-4-NITROCHALCONE
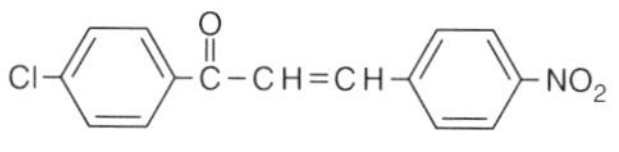
5 gm

S51,467-5
4,4'-DINITROCHALCONE
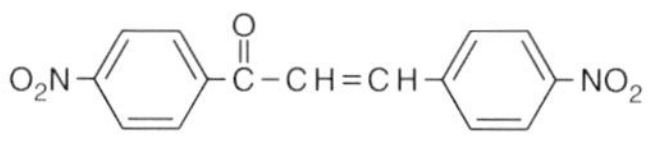
5 gm

S52,259-7
4-BROMO-4'-NITROCHALCONE
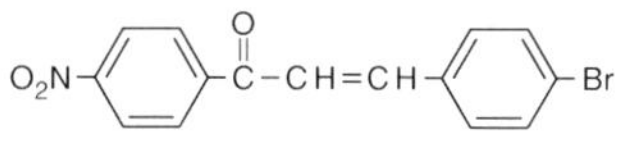
5 gm

S52,262-7
3-BROMO-4'-NITROCHALCONE
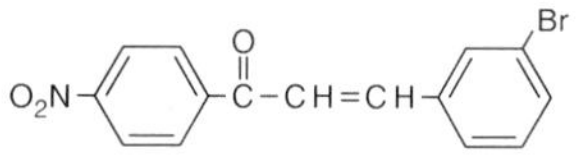
5 gm

S50,956-6
4-METHYL-4'-NITROCHALCONE
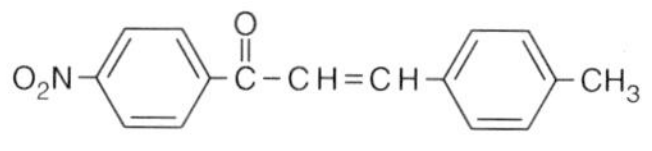
5 gm

S52,427-1
4'-METHYL-4-NITROCHALCONE
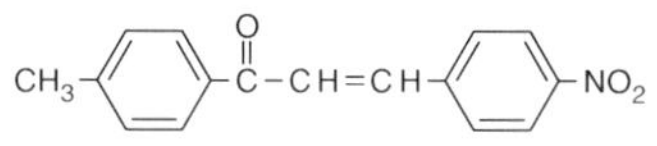
5 gm

S50,963-9
2,6-DICHLORO-4'-NITROCHALCONE
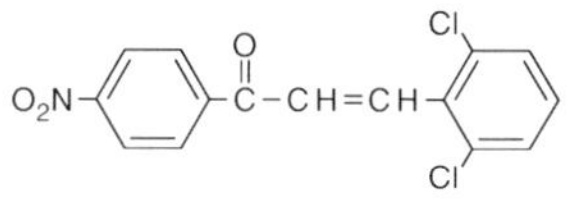
250 mg

S53,898-1
2,6-DICHLORO-3'-NITROCHALCONE
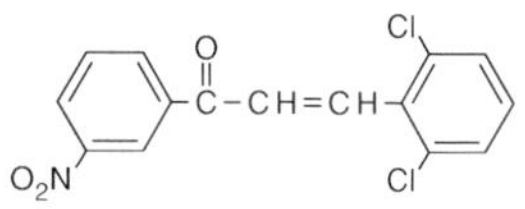
250 mg

S53,916-3
2',4'-DIMETHYL-3-NITROCHALCONE
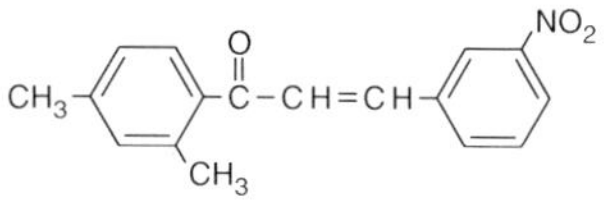
5 gm

S77,176-7
2-BROMO-3-PHENYLACRYLOPHENONE
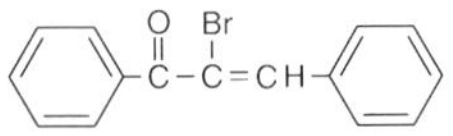
250 mg

S89,435-4
1-(P-TOLYL)-2,3,3-TRICHLORO-2-PROPEN-1-ONE
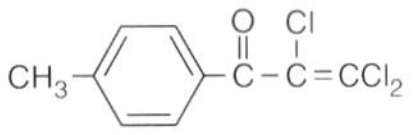
250 mg

S51,725-9
TRANS-3-METHYLCHALCONE
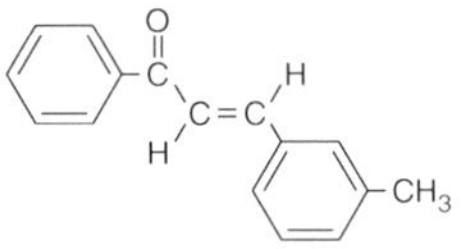
5 gm

S83,710-5
2-BENZYL-5-BENZYLIDENE-1-CYCLOPENTANONE
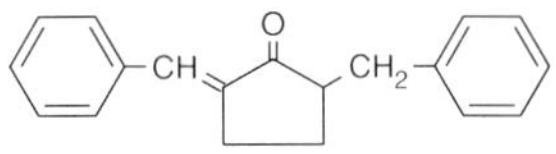
250 mg

S53,390-4
3,4-(METHYLENEDIOXY)CHALCONE
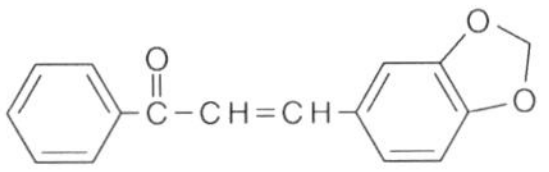
5 gm

S55,963-6
3,4-METHYLENEDIOXY-4'-PHENYLCHALCONE
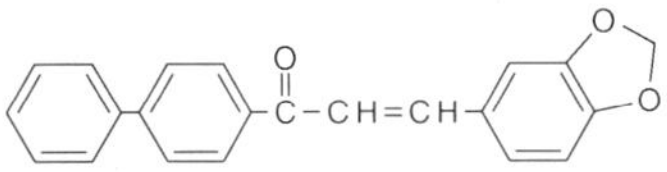
5 gm

S55,972-5
4'-BROMO-3,4-(METHYLENEDIOXY)-CHALCONE
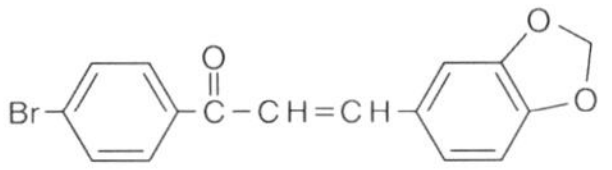
5 gm

S39,185-9
4'-METHYL-3,4-(METHYLENEDIOXY)-CHALCONE
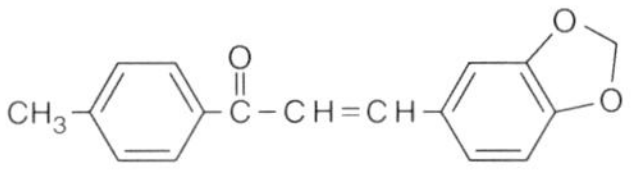
5 gm

S53,607-5
3',4'-DICHLORO-3,4-(METHYLENEDIOXY)-CHALCONE
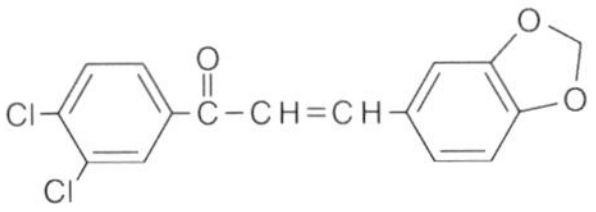
250 mg

S73,997-9
3-[(3,4-METHYLENEDIOXY)PHENYL]-2'-ACRYLONAPHTONE
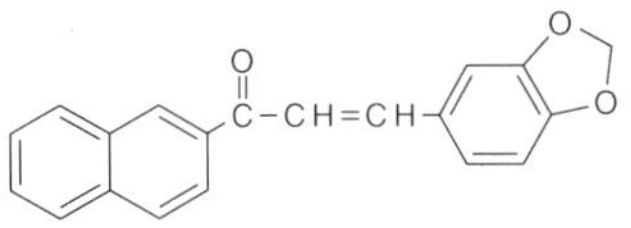
5 gm

S84,629-5
2-BENZYLIDENE-5-METHYL-1-CYCLOHEXANONE
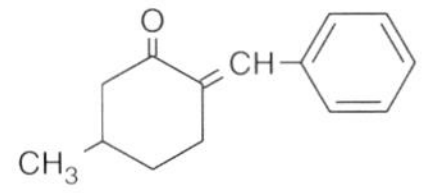
250 mg

S92,622-1
2,6-BIS(BENZYLIDENE)-1-CYCLOHEXANONE
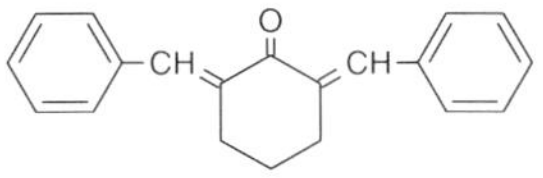
250 mg

S92,152-1
2,6-BIS(CINNAMYLIDENE)-1-CYCLOHEXANONE
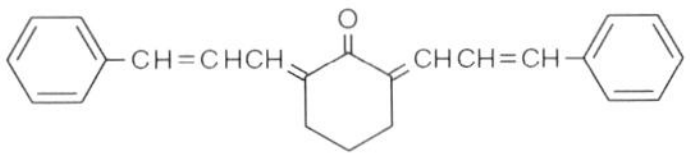
5 gm

S92,621-3
3-(3-BENZYLIDENE-2-OXOCYCLOHEXYL)-3-PHENYLPROPIOPHENONE
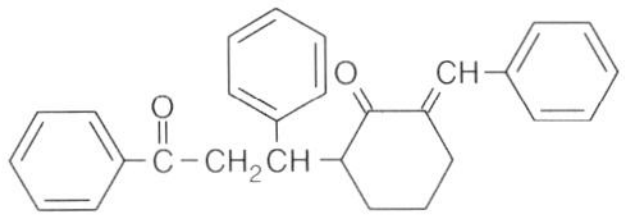
250 mg

S92,156-4
2,6-BIS(4-ISO-PROPYLBENZYLIDENE)-1-CYCLOHEXANONE
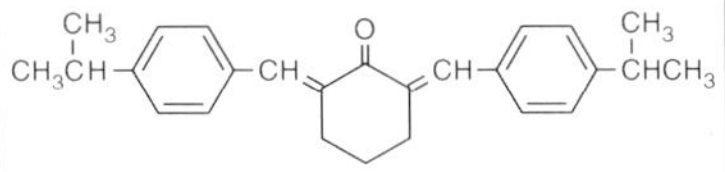
250 mg

S84,951-0
4H-1-BENZOTHIOPYRAN-4-ONE
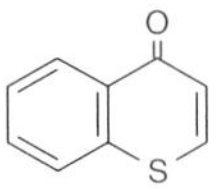
250 mg

S68,297-7
2-BENZOYL-1-PHENYLACETYLENE
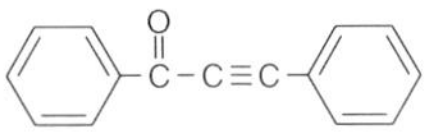
250 mg

S68,251-9
3'-METHYLCHALCONE
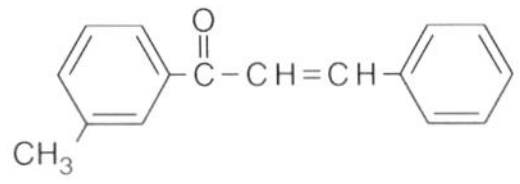
5 gm

S62,189-7
5-PHENYL-2,4-PENTADIENOPHENONE
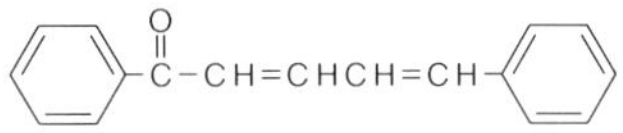
250 mg

S70,553-5
1,5-DIPHENYL-4-PENTEN-1-ONE
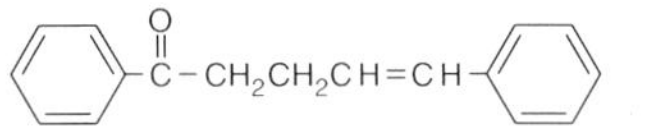
5 gm

S92,983-2
ALPHA-PHENYLCHALCONE
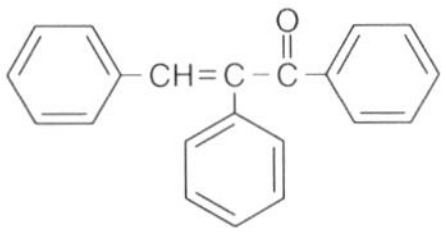
250 mg

S95,450-0
1,3,3-TRIPHENYL-2-PROPEN-1-ONE
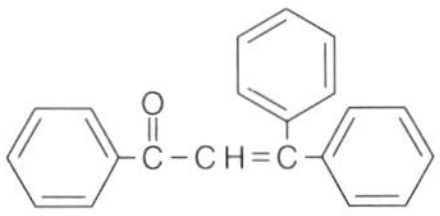
50 mg

S84,065-3
2,7-DIBENZYLIDENE-1-CYCLOHEPTANONE
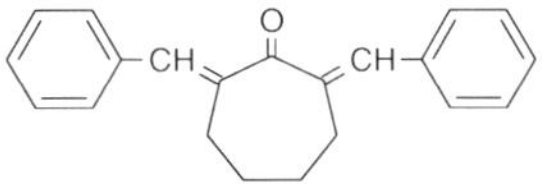
250 mg

S68,307-8
2'-METHYLCHALCONE
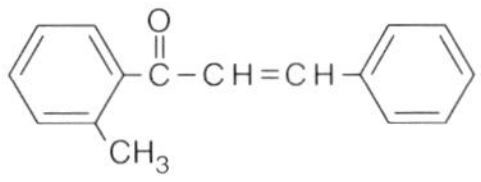
5 gm

S45,539-3
4'-BROMOCHALCONE
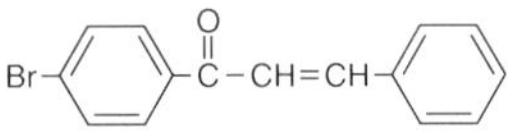
5 gm

S68,023-0
4-METHYLCHALCONE
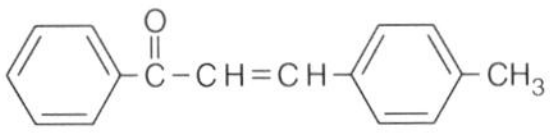
250 mg

S68,123-7
4'-METHYLCHALCONE
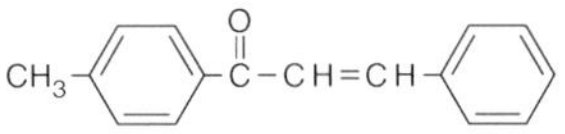
5 gm

S39,187-5
4'-METHYL-5-PHENYL-2,4-PENTADIENOPHENONE
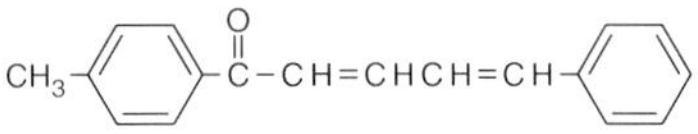
5 gm

S84,662-7
2-(4-METHYLBENZYLIDENE)-3-PHENYL-1-INDANONE
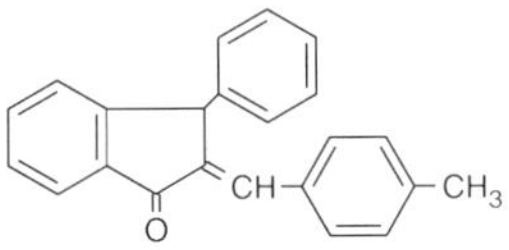
250 mg

S73,501-9
2,6-DIMETHYLCHALCONE
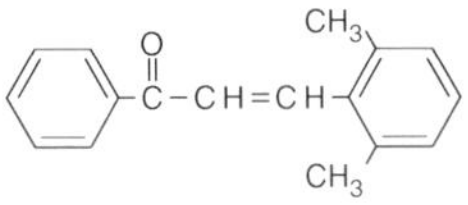
250 mg

S83,993-0
2',4'-DIMETHYLCHALCONE
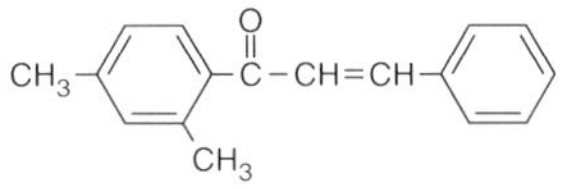
250 mg

S70,697-3
2',4',6'-TRIMETHYLCHALCONE
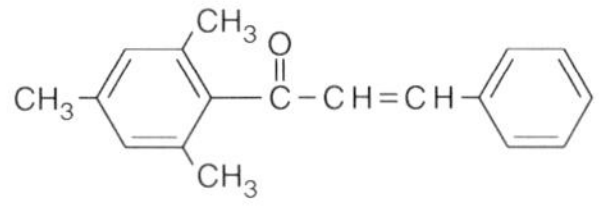
5 gm

S70,698-1
2',4',6'-TRIISOPROPYLCHALCONE
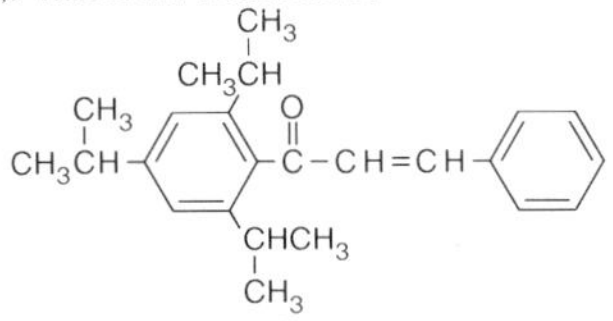
5 gm

S84,658-9
2-BENZYLIDENE-3-PHENYL-1-INDANONE
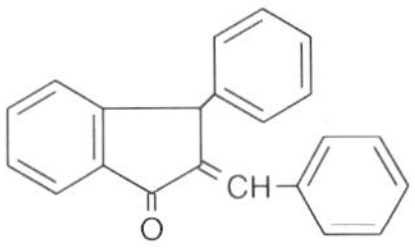
250 mg

S58,977-2
2-BENZYLIDENE-1-TETRALONE
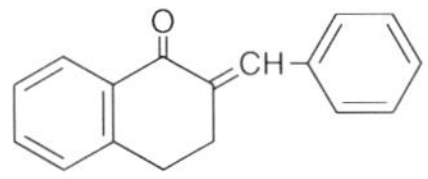
5 gm

S62,193-5
2-(CINNAMYLIDENE)-1-TETRALONE
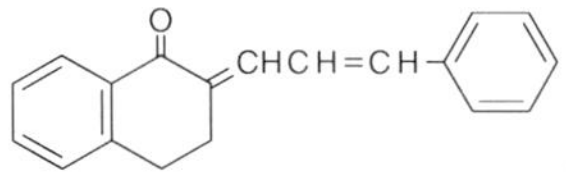
5 gm

S93,219-1
2-BENZYLIDENE-4,4-DIMETHYL-1-TETRALONE
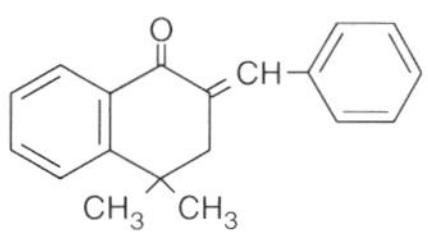
250 mg

S43,397-7
2-CHLORO-4'-FLUOROCHALCONE
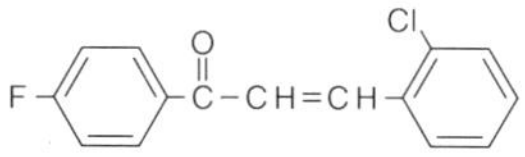
5 gm

S64,376-9
4'-BROMO-2-CHLOROCHALCONE
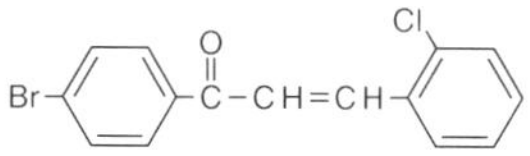
5 gm

S58,935-7
2-(2-CHLOROBENZYLIDENE)-1-TETRALONE
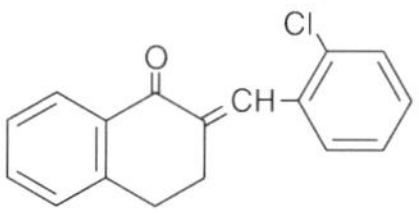
250 mg

S43,398-5
4'-FLUOROCHALCONE
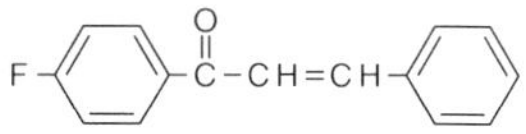
5 gm

S50,696-6
4'-CHLOROCHALCONE
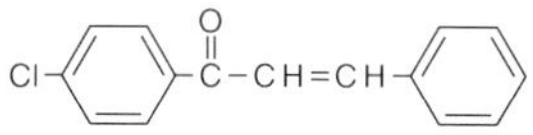
5 gm

S54,291-1
4,4'-DICHLOROCHALCONE
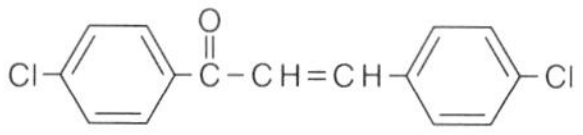
250 mg

S64,378-5
4'-BROMO-4-CHLOROCHALCONE
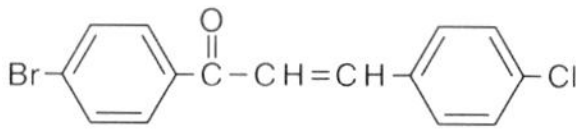
5 gm

S56,008-1
4'-FLUORO-4-METHYLCHALCONE
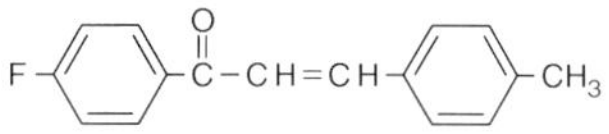
250 mg

S62,626-0
2-METHYL-3-(P-TOLYL)ACRYLOPHENONE
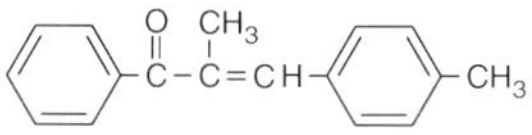
5 gm

S62,471-3
1-(4-FLUOROPHENYL)-5-PHENYL-2,4-PENTADIEN-1-ONE
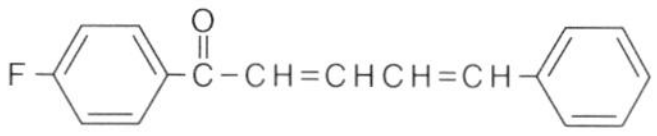
5 gm

S62,192-7
4'-BROMO-5-PHENYL-2,4-PENTADIENOPHENONE
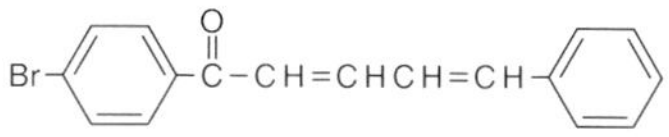
250 mg

S61,371-1
4'-CHLORO-4-ISOPROPYLCHALCONE
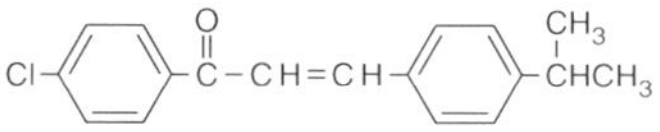
250 mg

S64,371-8
4'-BROMO-4-ISOPROPYLCHALCONE
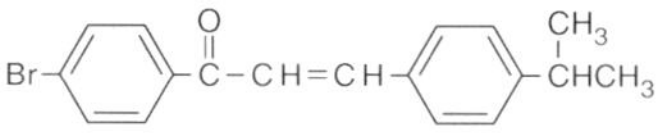
250 mg

S58,418-5
4-ISOPROPYL-4'-METHYLCHALCONE
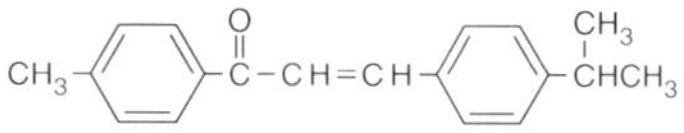
5 gm

S56,005-7
3,4-DICHLORO-4'-FLUOROCHALCONE
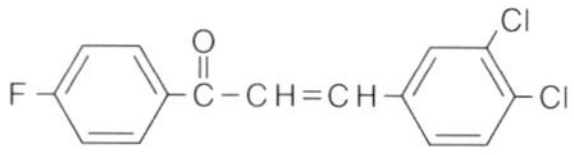
250 mg

S51,107-2
2,4-DICHLORO-4'-METHYLCHALCONE
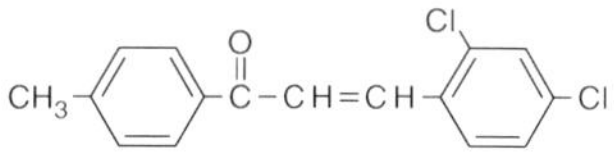
5 gm

S58,415-0
2',4,4'-TRIMETHYLCHALCONE
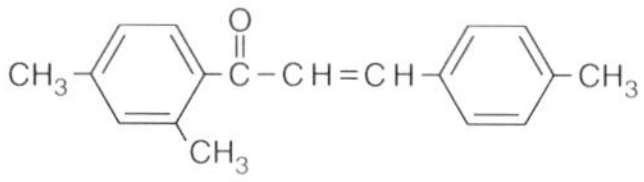
5 gm

S56,476-1
3,4-DICHLORO-4'-PHENYLCHALCONE
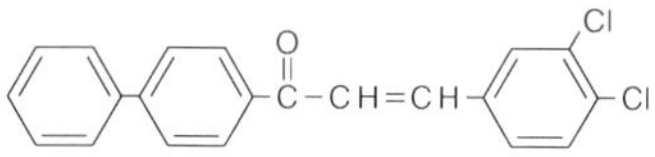
250 mg

S62,194-3
4',5-DIPHENYL-2,4-PENTADIENOPHENONE
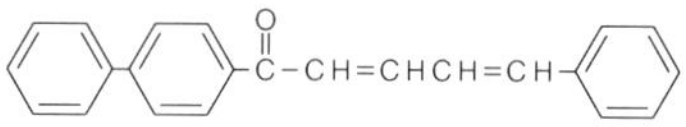
5 gm

S58,944-6
2-(4-CHLOROBENZYLIDENE)-1-TETRALONE
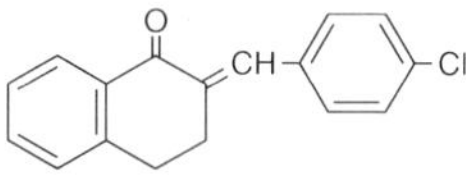
5 gm

S55,826-5
2-(4-METHYLBENZYLIDENE)-1-TETRALONE
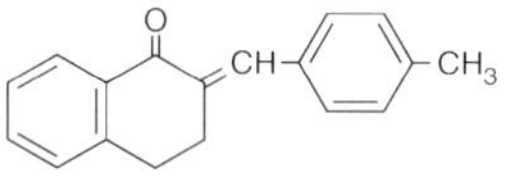
250 mg

S92,130-0
2-(4-METHYLBENZYLIDENE)-1-TETRALONE
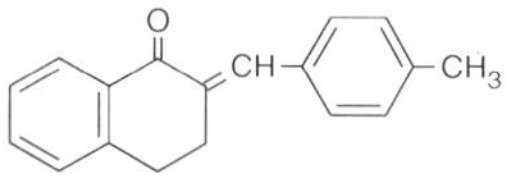
250 mg

S58,968-3
2-(4-ISOPROPYLBENZYLIDENE)-1-TETRALONE
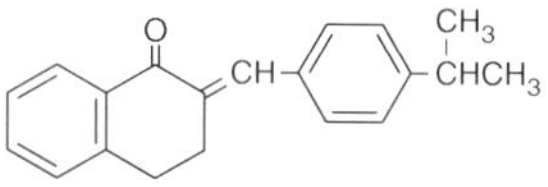
5 gm

S51,709-7
4'-METHYL-3-(1-NAPHTHYL)-ACRYLOPHENONE
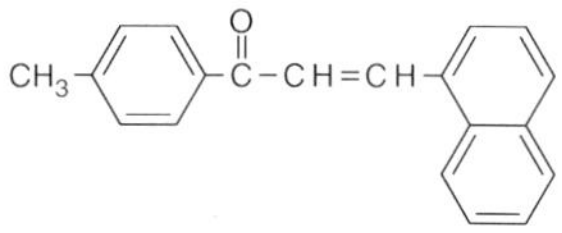
5 gm

S59,304-4
3-(P-TOLYL)-1'-ACRYLONAPHTHONE
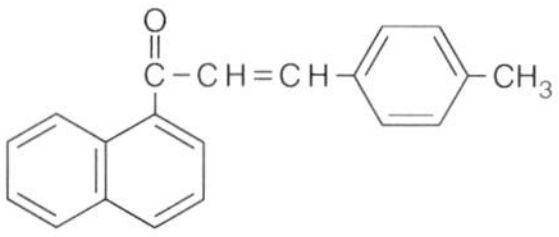
250 mg

S61,961-2
3-(9-ANTHRYL)-4'-CHLOROACRYLOPHENONE
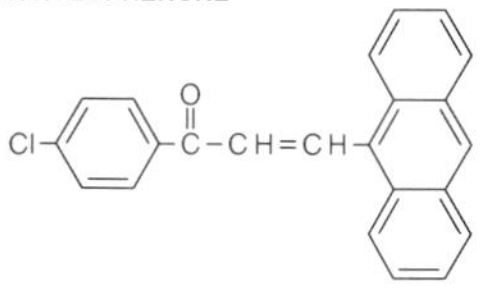
250 mg

S62,186-2
3-(2,4-DICHLOROPHENYL)-2-METHYLACRYLOPHENONE
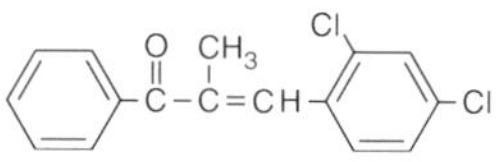
5 gm

S56,070-7
3,4-DICHLORO-2',4'-DIMETHYLCHALCONE
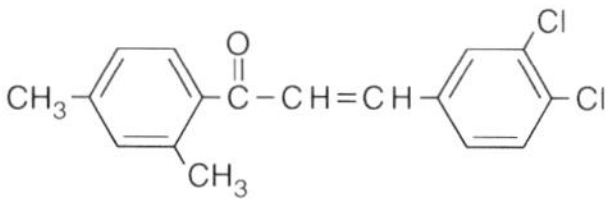
250 mg

S56,484-2
3-(3,4-DICHLOROPHENYL)-1'-ACRYLONAPHTHONE
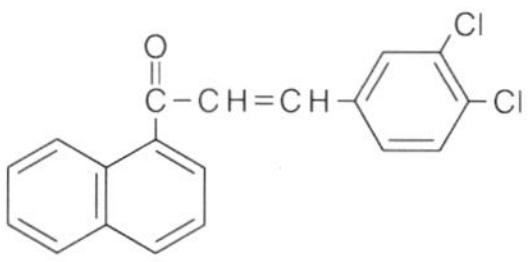
5 gm

S92,664-7
4-MESITYL-3-BUTEN-2-ONE
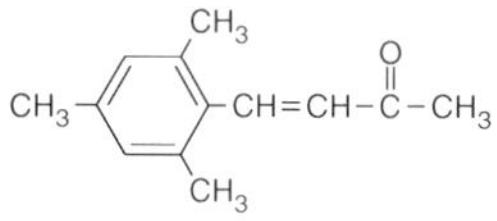
250 mg

S62,402-0
5-PHENYL-1-(2,4,6-TRIMETHYLPHENYL)-2,4-PENTADIEN-1-ONE
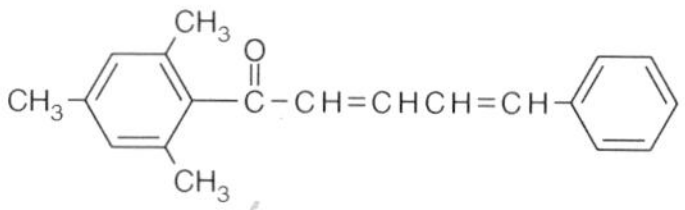
5 gm

S59,996-4
2-(1-NAPHTHYLMETHYLENE)-1-TETRALONE
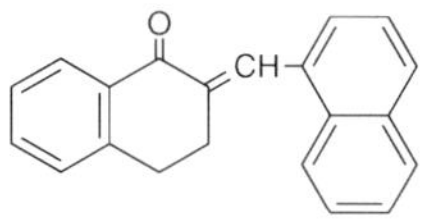
5 gm

S57,714-6
2-(3-CHLOROBENZYLIDENE)-1-TETRALONE
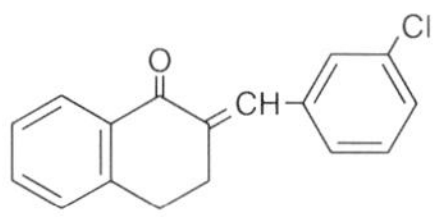
250 mg

S62,543-4
1-(2-NAPHTHYL)-5-PHENYL-2,4-PENTADIEN-1-ONE
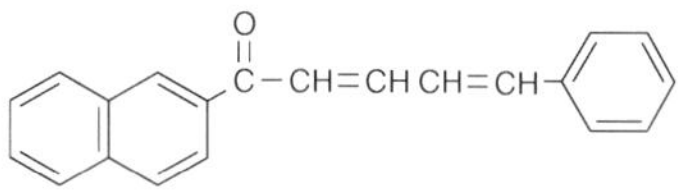
5 gm

S87,383-7
7-METHYL-1H-PHENALEN-1-ONE
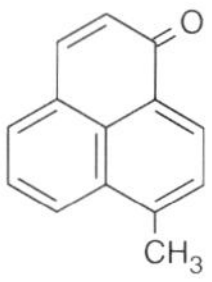
250 mg

S93,588-3
2,5-DICINNAMYLIDENE-1-CYCLOPENTANONE

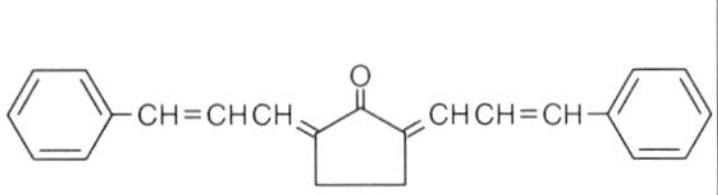

250 mg

S94,562-5
5-(4-BIPHENYLYL)-2-PROPYLCYCLOHEXANONE

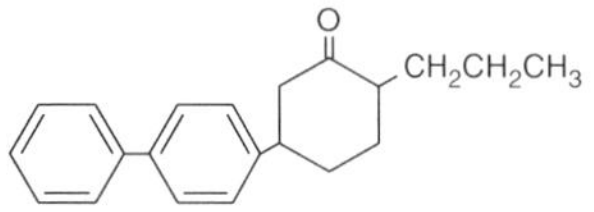

250 mg

S92,344-3
4,4-DIPHENYL-2,5-CYCLOHEXADIEN-1-ONE

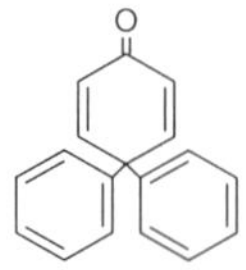

250 mg

S93,359-7
2-BENZYLIDENE-1-INDANONE

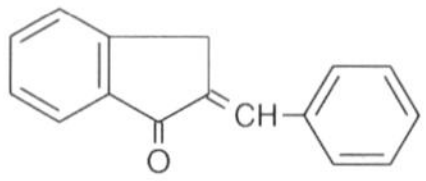

250 mg

S53,394-7
2-CHLOROCHALCONE

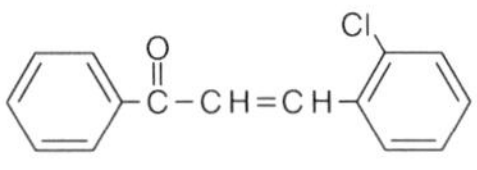

5 gm

S54,844-8
2,4'-DICHLOROCHALCONE

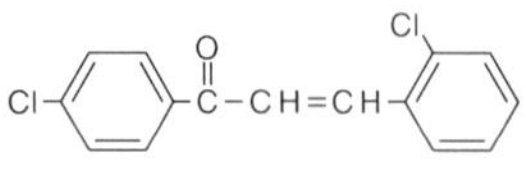

250 mg

S52,575-8
2-CHLORO-4'-METHYLCHALCONE

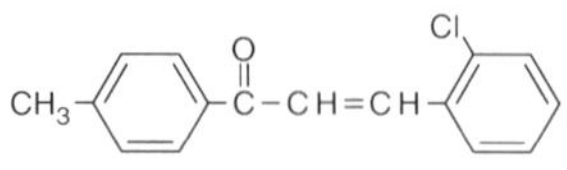

5 gm

S53,895-7
2-CHLORO-2',4'-DIMETHYLCHALCONE

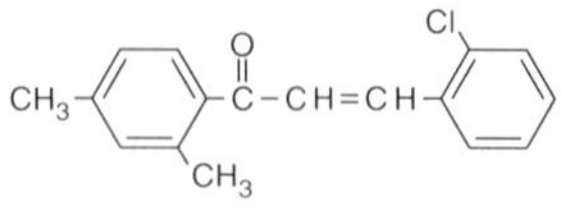

5 gm

S52,571-5
4'-METHYLCROTONOPHENONE

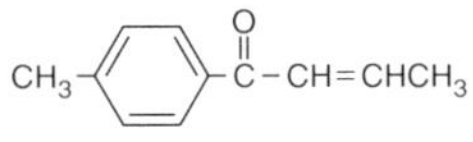

250 mg

S53,392-0
4-CHLOROCHALCONE

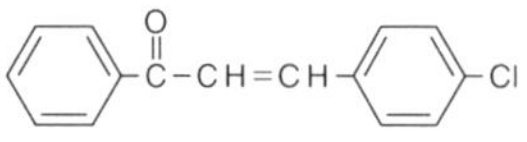

5 gm

S54,880-4
4'-CHLORO-4-FLUOROCHALCONE

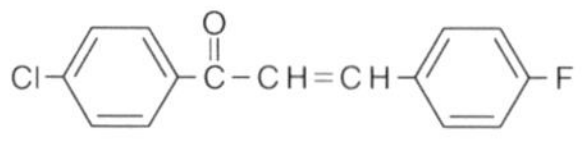

250 mg

S54,877-4
3,4'-DICHLOROCHALCONE

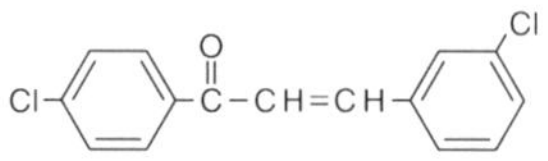

5 gm

S54,799-9
4-BROMO-4'-CHLOROCHALCONE

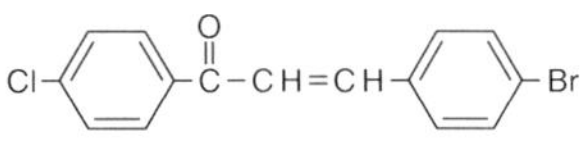

5 gm

S54,800-6
3-BROMO-4'-CHLOROCHALCONE

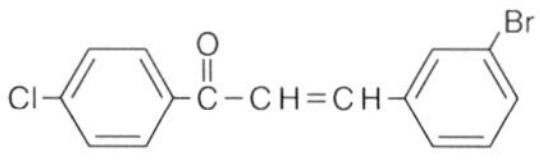

250 mg

S52,448-4
4-FLUORO-4'-METHYLCHALCONE

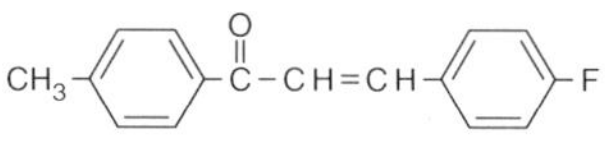

5 gm

S52,572-3
4-CHLORO-4'-METHYLCHALCONE

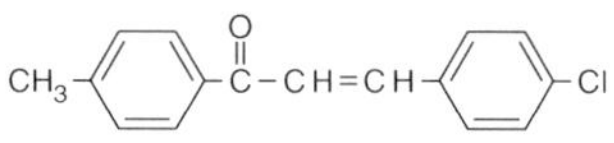

5 gm

S52,576-6
3-CHLORO-4'-METHYLCHALCONE

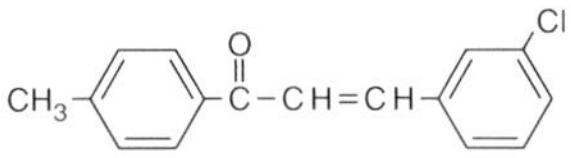

5 gm

S54,885-5
4'-CHLORO-4-METHYLCHALCONE

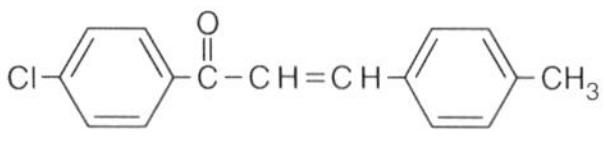

5 gm

S52,579-0
3-BROMO-4'-METHYLCHALCONE

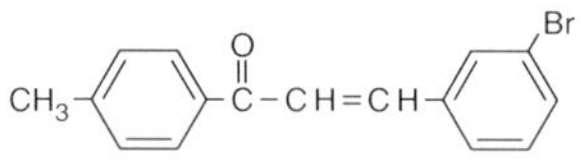

5 gm

S53,873-6
4-BROMO-4'-METHYLCHALCONE

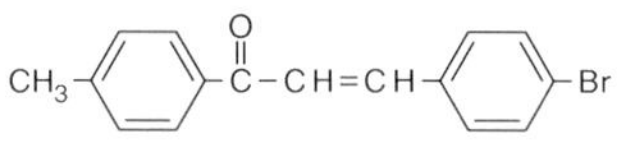

5 gm

S51,095-5
4,4'-DIMETHYLCHALCONE

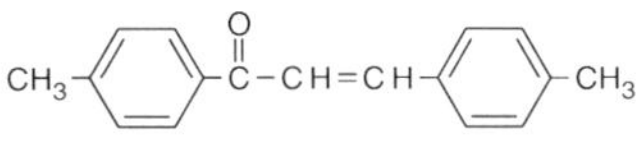

5 gm

S54,768-9
4'-CHLORO-5-PHENYL-2,4-PENTADIENOPHENONE

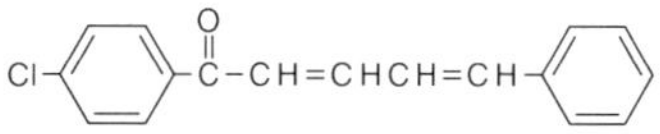

5 gm

S54,879-0
2,4',6-TRICHLOROCHALCONE

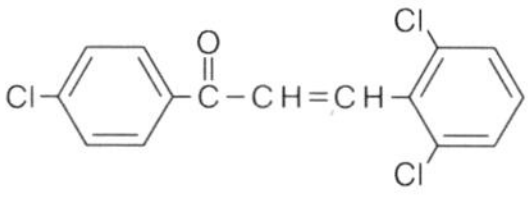

250 mg

S52,342-9
2,6-DICHLORO-4'-METHYLCHALCONE

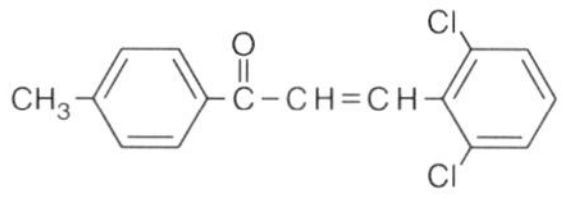

5 gm

S54,881-2
2,4,4'-TRICHLOROCHALCONE

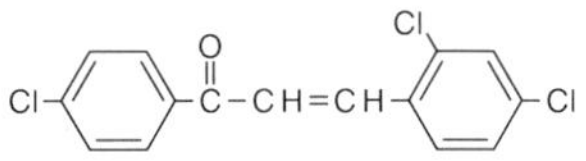

250 mg

S54,876-6
4'-CHLORO-3-(1-NAPHTHYL)-ACRYLOPHENONE

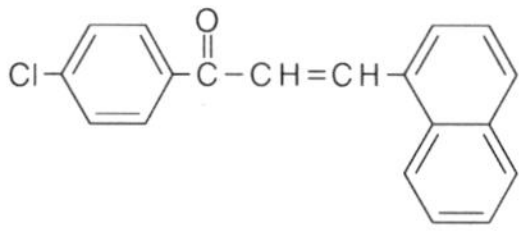

5 gm

S53,393-9
2,6-DICHLOROCHALCONE

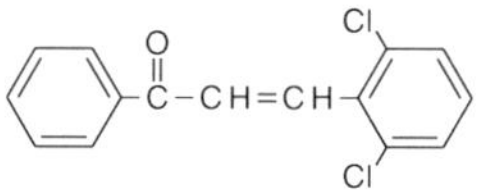

5 gm

S53,281-9
2,3',4',6-TETRACHLOROCHALCONE

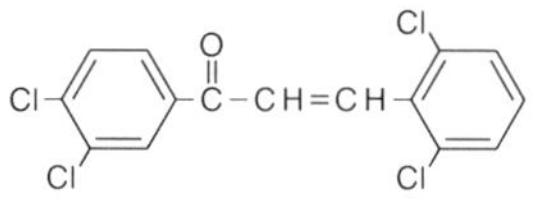

5 gm

S53,897-3
2,6-DICHLORO-2',4'-DIMETHYLCHALCONE

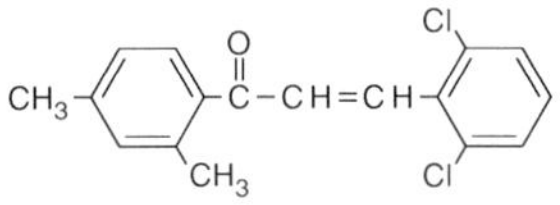

250 mg

S61,627-3
4,4'-BIS(BROMOACETYL)BIPHENYL

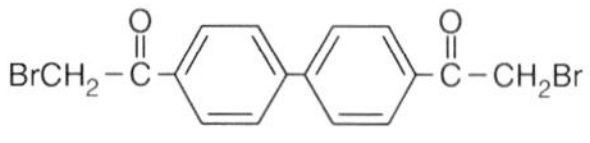

250 mg

S94,930-2
2-(PERFLUOROHEPTANOYL)ACETOPHENONE

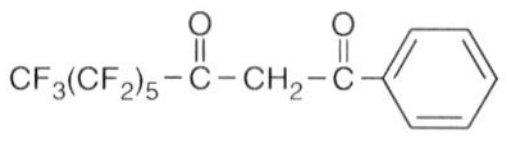

250 mg

S94,612-5
4-(4-CHLOROPHENOXY)BENZIL

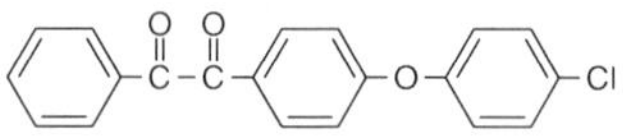

250 mg

S94,613-3
4-(4-CHLOROPHENYLTHIO)BENZIL

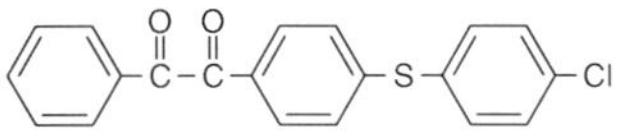

250 mg

S94,715-6
TRANS-1,2-BIS(4-CHLOROBENZOYL)-ETHYLENE

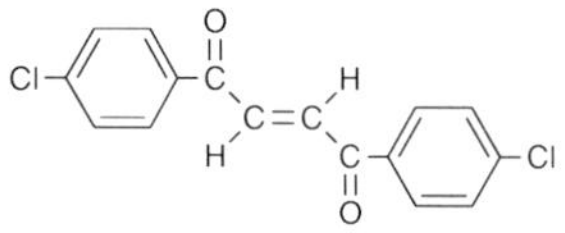

250 mg

S83,173-5
DECAFLUOROBENZIL

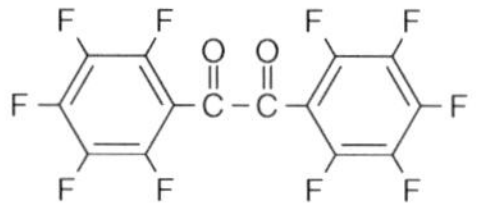

250 mg

S83,172-7
(2,3,4,5,6-PENTACHLORO)-(2',3',4',5',6'-PENTAFLUORO)BENZIL

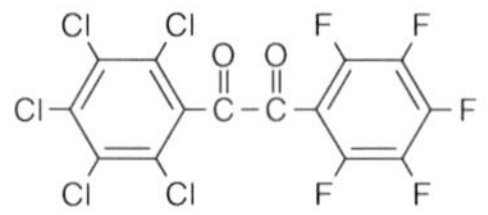

250 mg

S83,009-7
DECACHLOROBENZIL
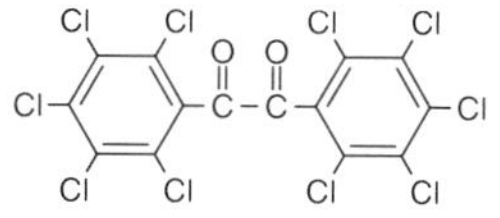
250 mg

S82,644-8
1,3-BIS(PENTACHLOROPHENYL)-2,2-DIMETHYL-1,3-PROPANEDIONE
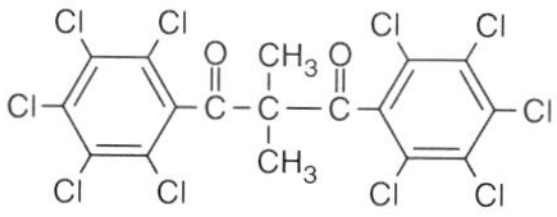
250 mg

S92,978-6
2-(4-CHLOROBENZYLIDENE)-1-PHENYL-1,3-BUTANEDIONE
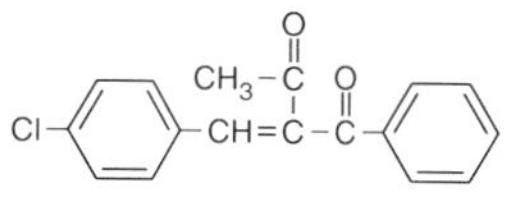
250 mg

S76,337-3
1,3-BIS(4-FLUOROPHENYL)-1,3-PROPANEDIONE
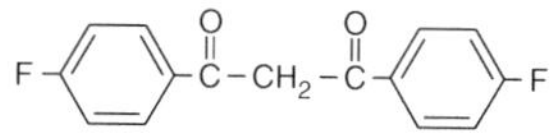
250 mg

S68,345-0
1,3-BIS(4-CHLOROPHENYL)-1,3-PROPANEDIONE
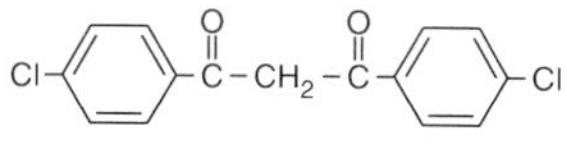
5 gm

S83,012-7
1,4-BIS(PENTAFLUOROPHENYL)-1,4-BUTANEDIONE
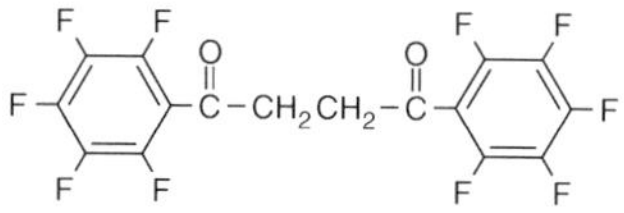
250 mg

S83,175-1
1,4-BIS(PENTACHLOROPHENYL)-2-BUTENE-1,4-DIONE
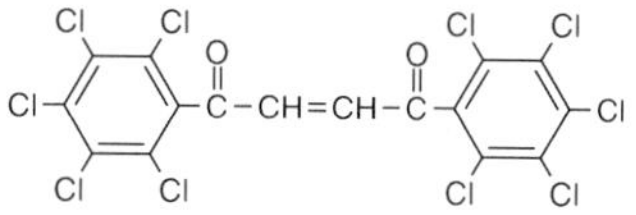
250 mg

S83,013-5
1,4-BIS(PENTACHLOROPHENYL)-1,4-BUTANEDIONE
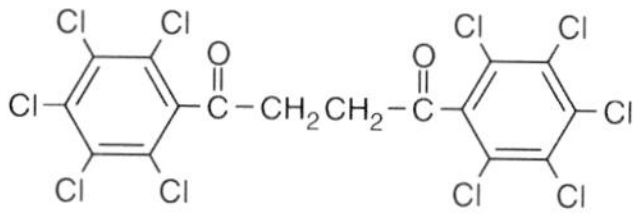
250 mg

S82,642-1
1,5-BIS(PENTACHLOROPHENYL)-1,5-PENTANEDIONE
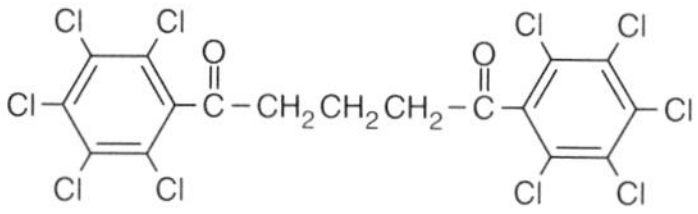
250 mg

S82,637-5
1,6-BIS(PENTAFLUOROPHENYL)-1,6-HEXANEDIONE
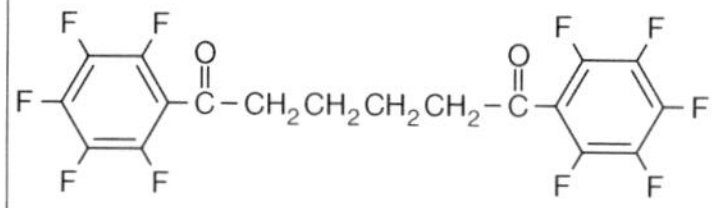
250 mg

S83,006-2
1,6-BIS(PENTACHLOROPHENYL)-1,6-HEXANEDIONE
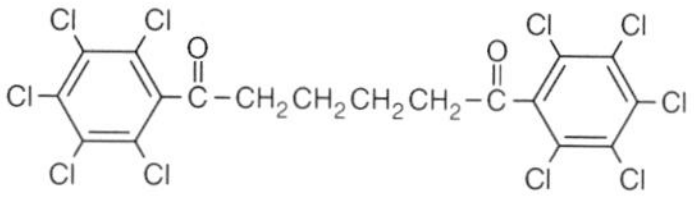
250 mg

S84,118-8
1,4-BIS(PENTAFLUOROBENZOYL)BENZENE
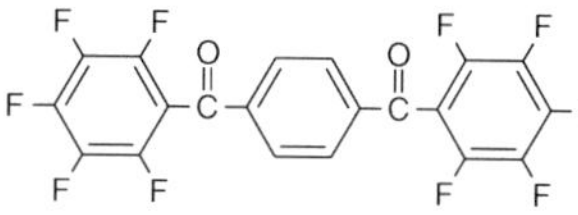
250 mg

S82,828-9
1,4-BIS(PENTACHLOROBENZOYL)BENZENE
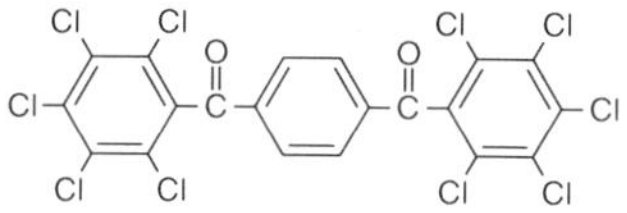
250 mg

S68,121-0
1-(4-CHLOROPHENYL)-3-PHENYL-1,3-PROPANEDIONE
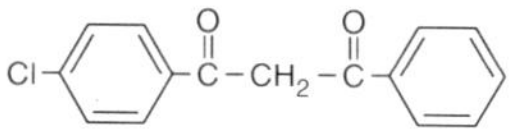
5 gm

S89,894-5
2,2'-(PHENYLMETHYLENE)BIS(5,5-DIMETHYL-3-HYDROXY-2-CYCLOHEXEN-1-ONE)
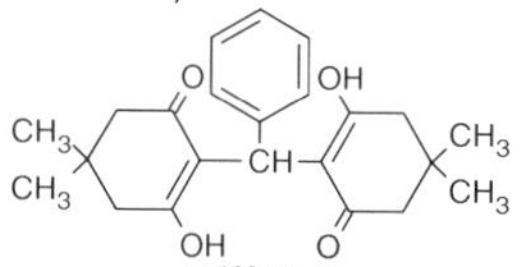
100 mg

S58,452-5
ALPHA,ALPHA'-(1,4-PHENYLENEDIMETHYLIDYNE)BIS(2'-HYDROXYACETOPHENONE)
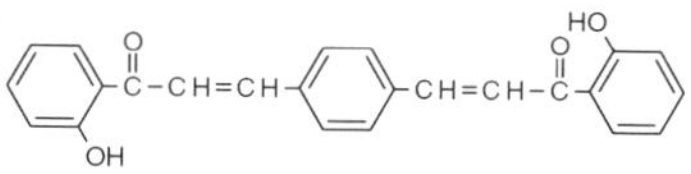
250 mg

S59,777-5
3,3''-(1,4-PHENYLENE)BIS(4'-HYDROXYACRYLOPHENONE)
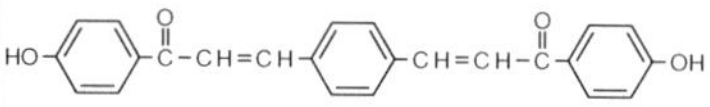
250 mg

S95,522-1
3A,8B-DIHYDRO-8B-HYDROXY-4H-INDENO(1,2-C)-1,2-DIOXOLE-3,4-DIONE
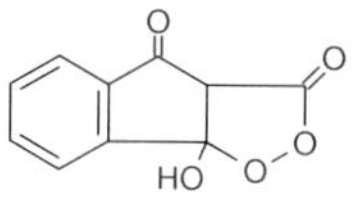
50 mg

S80,928-4
1-(2-HYDROXYPHENYL)-3-(4-METHOXYPHENYL)-1,3-PROPANEDIONE
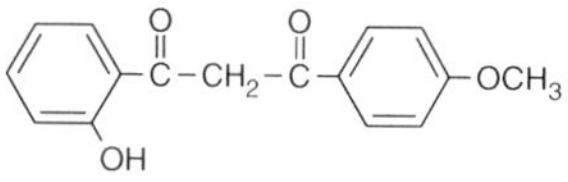
250 mg

S88,327-1
2-BENZYLIDENE-1-(2-HYDROXY-4-METHOXYPHENYL)-1,3-BUTANEDIONE
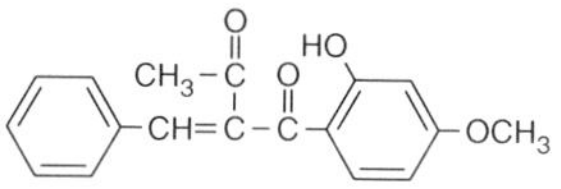
250 mg

S80,887-3
1-(2-HYDROXY-4-METHYLPHENYL)-1,3-BUTANEDIONE
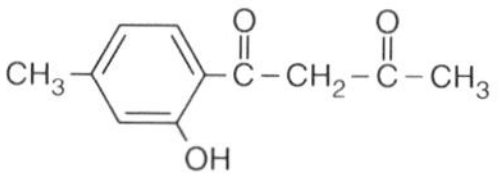
250 mg

S84,333-4
1-(2-HYDROXY-5-METHYLPHENYL)-1,3-BUTANEDIONE
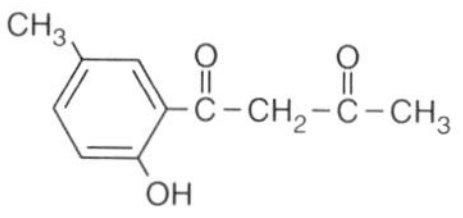
250 mg

S84,329-6
1-(2-HYDROXY-5-METHYLPHENYL)-1,3-PENTANEDIONE
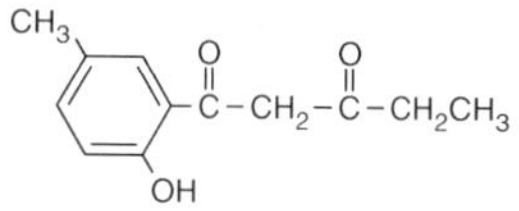
250 mg

S89,899-6
2-(4,4-DIMETHYL-2-HYDROXY-6-OXO-1-CYCLOHEXEN-1-YL)ACETOPHENONE
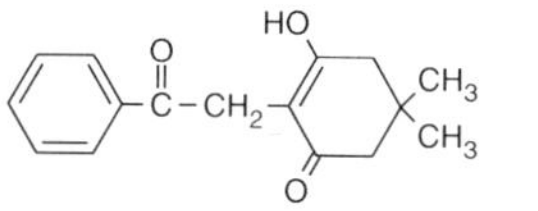
250 mg

S75,646-6
BIS(4-METHOXYBENZOYL)METHANE
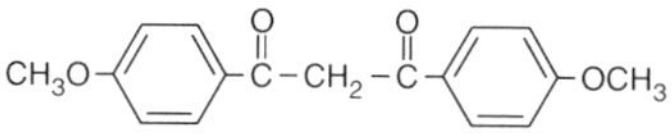
5 gm

S73,955-3
1,5-BIS(4-METHOXYPHENYL)-1,5-PENTANEDIONE
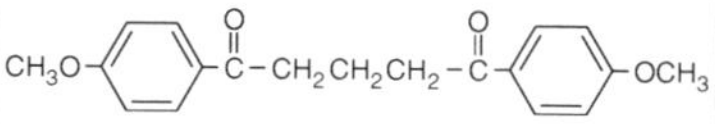
5 gm

S42,473-0
P,P'-DIMETHOXYLOBELANINE HYDROCHLORIDE
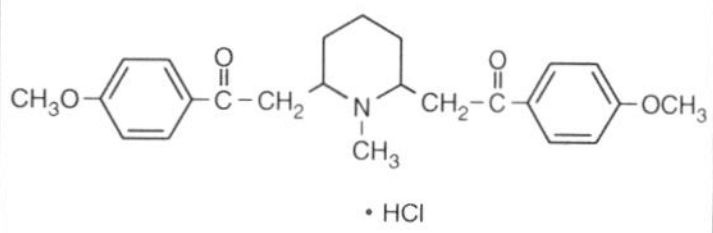
250 mg

S88,517-7
2-(2-METHOXYPHENETHYL)-1,3-CYCLOHEXANEDIONE
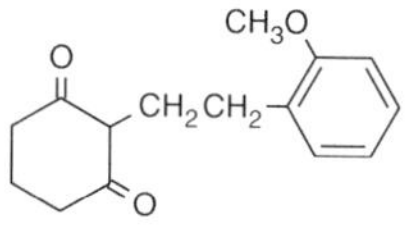
250 mg

S53,397-1
1,5-DIPHENYL-3-(4-METHOXYPHENYL)-1,5-PENTANEDIONE
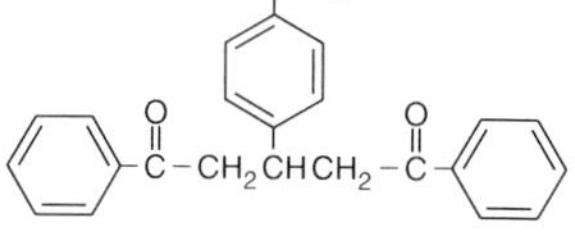
5 gm

S44,866-4
[(2,3,5,6-TETRAMETHYL-1,4-PHENYLENE)DIMETHYLENE]BIS(TERT-BUTYL PEROXIDE)
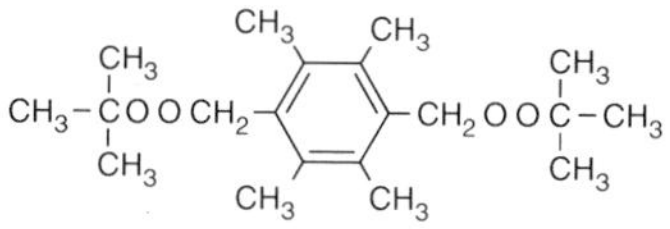
5 gm

S51,767-4
2-(3,4,5-TRIMETHOXYPHENYL)-1,3-INDANDIONE
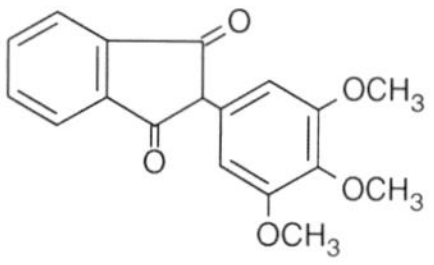
250 mg

S40,788-7
N-(3,4-DIMETHOXYPHENETHYL)-NORLOBELANINE
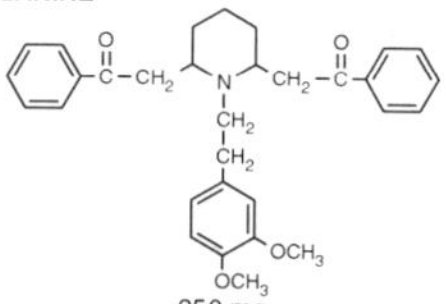
250 mg

S94,922-1
1,1-DIBENZOYLACETONE
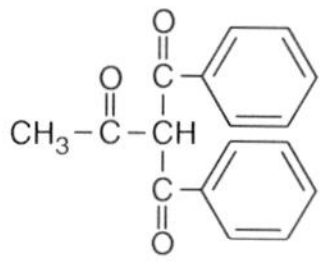
250 mg

S94,925-6
9-(ACETOACETYL)ANTHRACENE
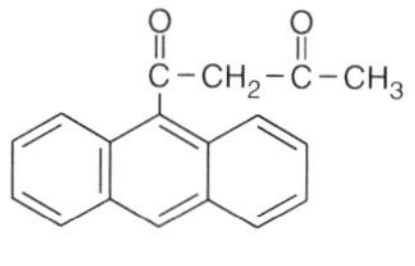
250 mg

S94,591-9
3-(4-BIPHENYLCARBONYL)ACRYLOPHENONE
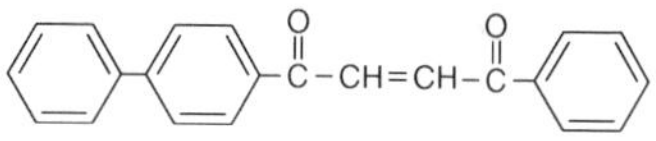
250 mg

S74,009-8
6,7-DIBENZOYL-1,3,5-TRIPHENYLCYCLOPENTA(C)PYRAN
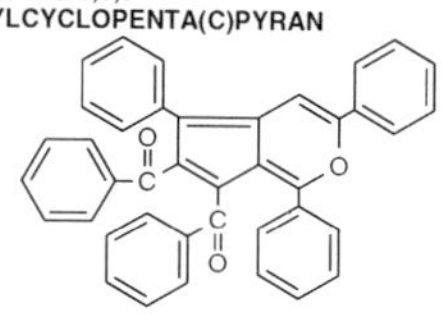
250 mg

S94,476-9
3-MESITOYLACRYLOPHENONE
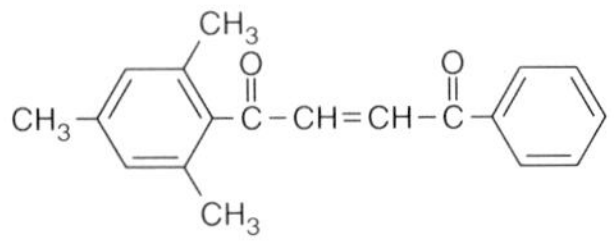
250 mg

S91,492-4
3,3'-THIODIPROPIOPHENONE
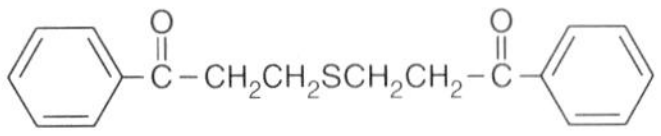
250 mg

S91,511-4
3,3'-(ETHYLENEDITHIO)-DIPROPIOPHENONE
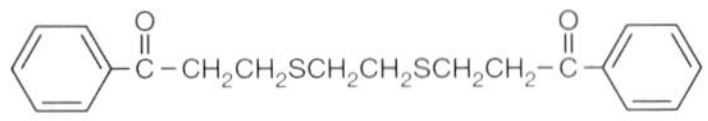
250 mg

S88,708-0
1-PHENYLPHENACYL SULFIDE
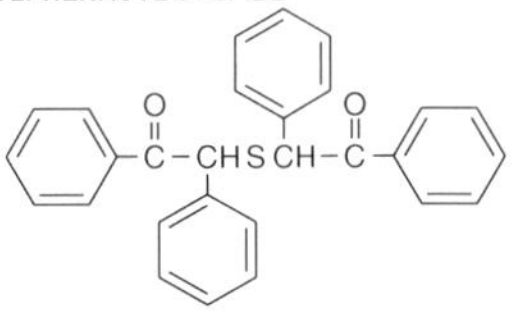
250 mg

S41,285-6
3,3'-THIOBIS[7H-BENZ(DE)-ANTHRACEN-7-ONE]
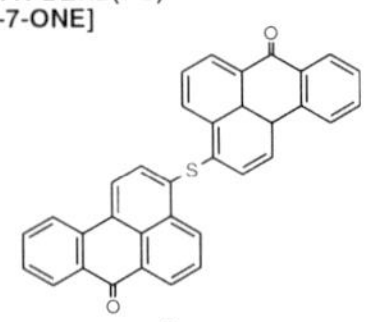
5 gm

S62,207-9
4-BENZOYL-2-NITROPHENYL DISULFIDE
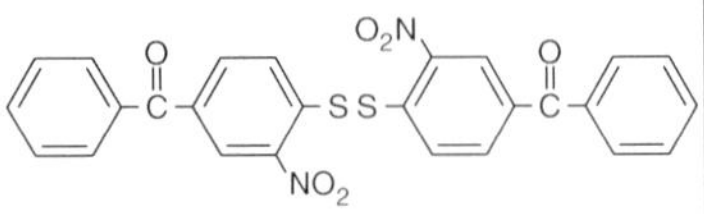
5 gm

S88,914-8
BENZOYLMETHYL DISULFIDE

250 mg

S42,603-2
TRANS-NORLOBELANINE
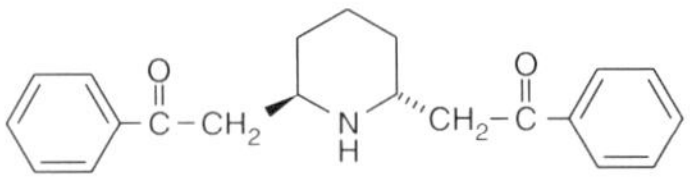
250 mg

S42,604-0
TRANS-NORLOBELANINE HYDROCHLORIDE
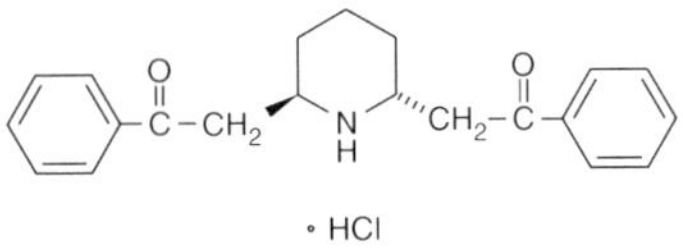
250 mg

S45,910-0
LOBELANINE HYDROCHLORIDE
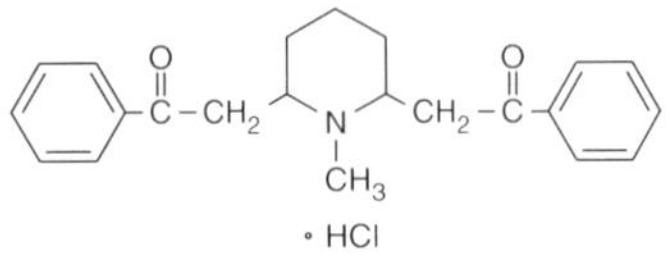
250 mg

S43,364-0
4',4'''-(1,4-PIPERAZINEDIYL)-DIACETOPHENONE
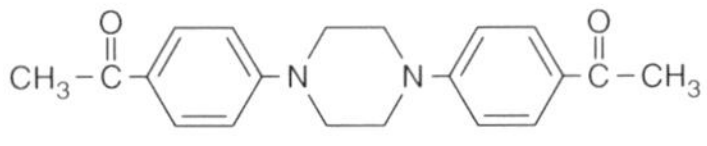
5 gm

S82,900-5
4,4'-DINITROBENZIL
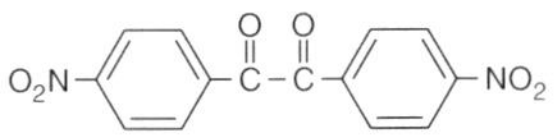
250 mg

S88,701-3
2-(2-NITROBENZYLIDENE)-1-PHENYL-1,3-BUTANEDIONE
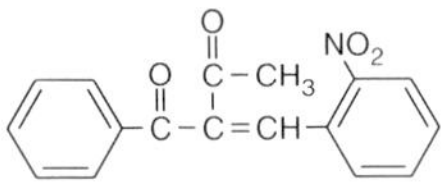
250 mg

S74,516-2
3-(2-NITROBENZYLIDENE)-2,4-PENTANEDIONE
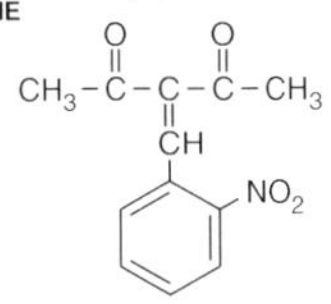
250 mg

S80,582-3
2-NITROPHENANTHRAQUINONE
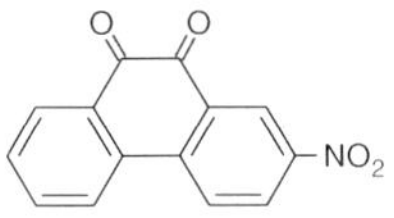
250 mg

S67,649-7
1,7-DIPHENYL-4-ETHYL-4-NITRO-1,7-HEPTANEDIONE
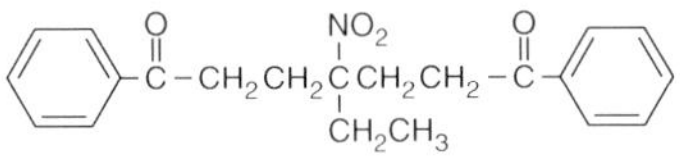
250 mg

S59,576-4
3,3''-(1,4-PHENYLENE)BIS(4'-NITROACRYLOPHENONE)
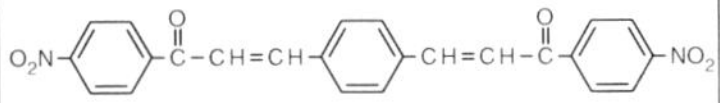
250 mg

S59,781-3
3,3''-(1,4-PHENYLENE)BIS(3'-NITROACRYLOPHENONE)
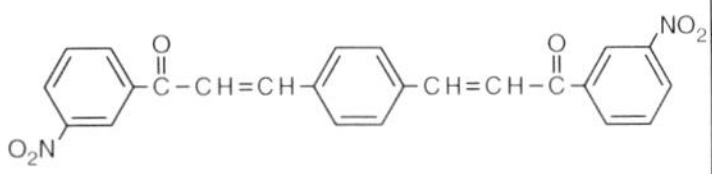
250 mg

S93,229-9
2-(4-NITROBENZYL)-1,3-INDANEDIONE
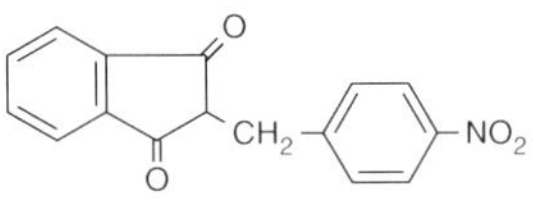
250 mg

S93,312-0
2-CHLORO-2-(ALPHA-CHLORO-4-NITROBENZYL)-1,3-INDANDIONE
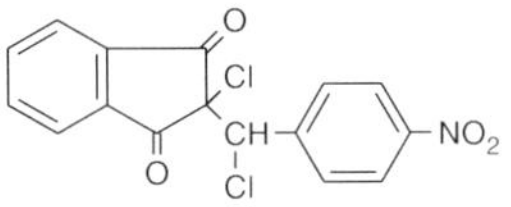
250 mg

S38,922-6
2-BROMO-1,3-DIPHENYL-1,3-PROPANEDIONE
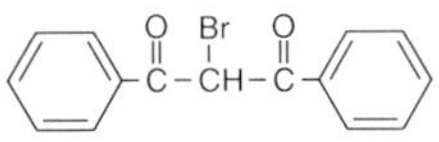
250 mg

S71,011-3
1,2-DIBENZOYL-1-CHLOROETHANE
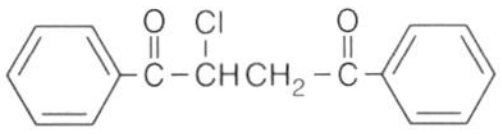
5 gm

S91,967-5
10,11-DIHYDRO-12H-BENZO(C)-PHENOTHIAZINE-10,11-DIONE
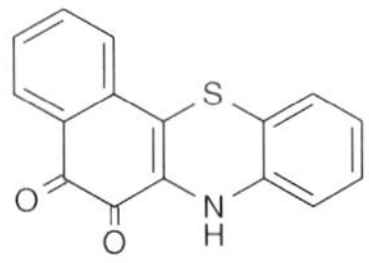
250 mg

S62,033-5
3-BENZYLIDENE-2,4-PENTANEDIONE
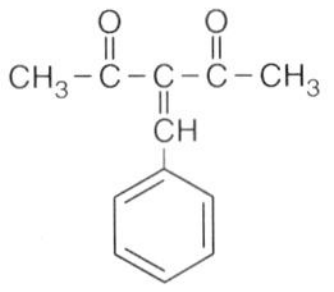
1 gm

S87,405-1
2-BENZOYL-1-CYCLODODECANONE
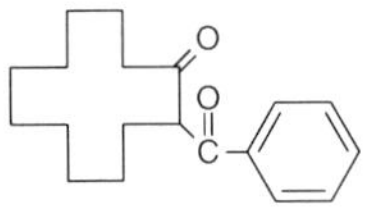
250 mg

S85,985-0
2-BENZYLIDENE-1,3-INDANDIONE
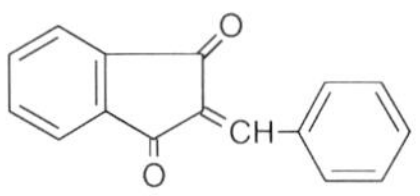
250 mg

S84,718-6
2,2-DIMETHYL-1H-CYCLOPENTA(A)-BIPHENYLENE-1,3(2H,3H)DIONE
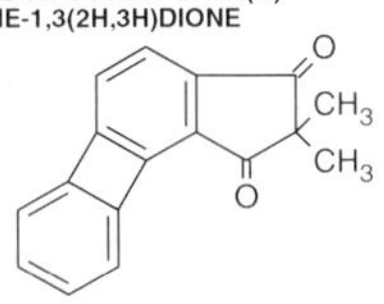
250 mg

S90,619-0
INDENO(2,1A)INDENE-5,10(4BH,9BH)-DIONE
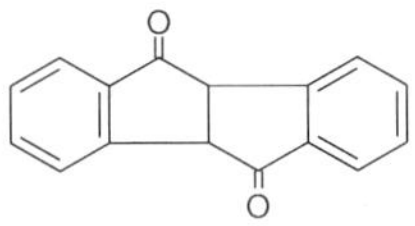
250 mg

S88,353-0
5-PHENYL-2,3,5,6-TETRAMETHYL-2-CYCLOHEXENE-1,4-DIONE
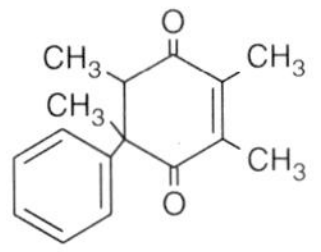
250 mg

S82,026-1
1-PHENYL-1,4-PENTANEDIONE
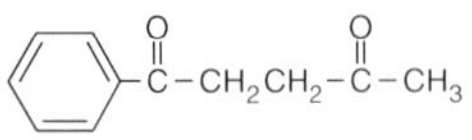
250 mg

S87,641-0
1,2-DIBENZOYLETHANE
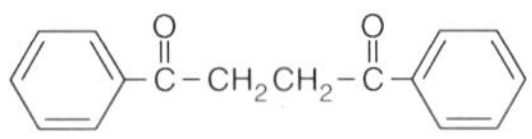
250 mg

S92,626-4
1,2,4-TRIPHENYLBUTAN-1,4-DIONE
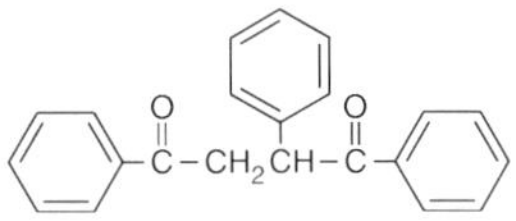
250 mg

S95,753-4
HEXAHYDRO-4'B-PHENYL-SPIRO[INDENE-1,10'-INDENO(1,2-A)INDENE]-3,9'-DIONE
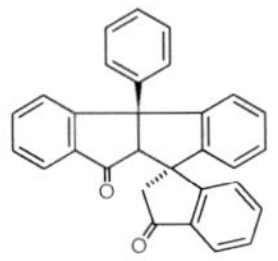
50 mg

S88,003-5
1,4-BIS(4-ACETYLPHENYL)-1,3-BUTADIENE
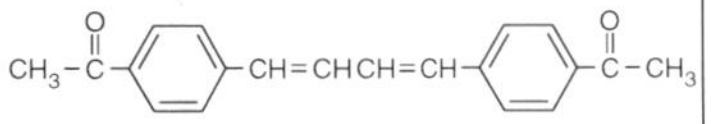
250 mg

S88,002-7
1,6-BIS(4-ACETYLPHENYL)-1,3,5-HEXATRIENE
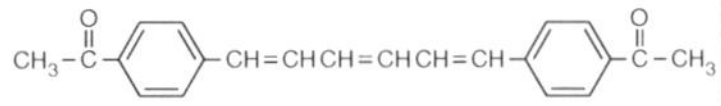
250 mg

S87,672-0
1,3-DIACETYL-5-ETHYLBENZENE
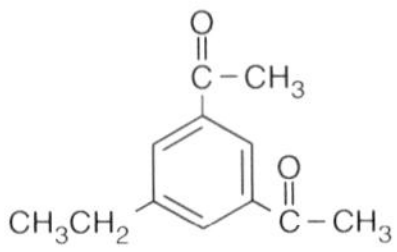
250 mg

S71,442-9
1,6-DIMESITYL-1,6-HEXANEDIONE
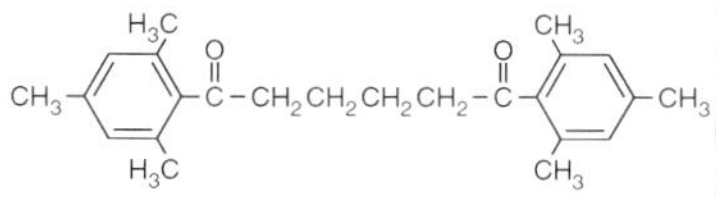
5 gm

S86,228-2
8,16-PYRANTHRENEDIONE
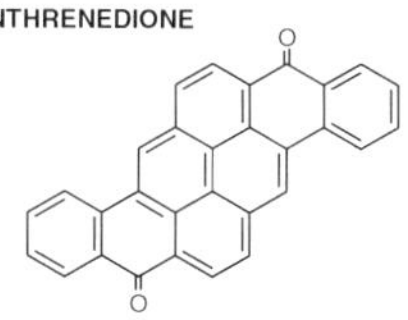
250 mg

S90,612-3
1,6-DIHYDRO-1,6-PYRENEDIONE
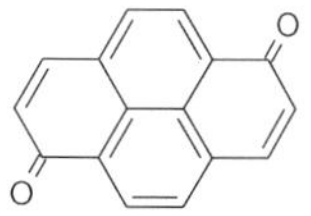
250 mg

S75,956-2
5,5-DIMETHYL-2-(3-MORPHOLINO-1-PHENYL-2-PROPENYLIDENE)-1,3-CYCLOHEXANEDIONE
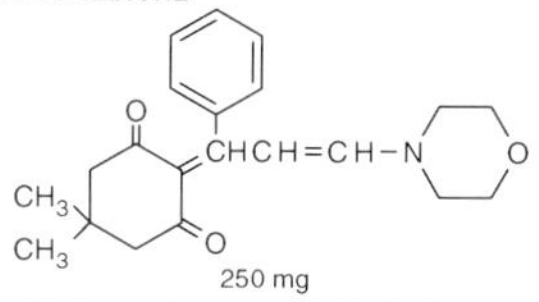
250 mg

S42,445-5
3,9-COEROXONONE
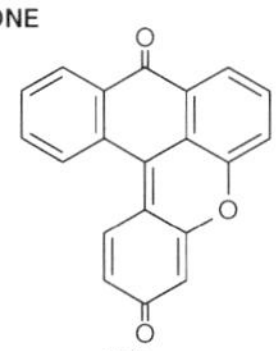
250 mg

S83,111-5
2,8-DIACETYLPHENOXATHIIN
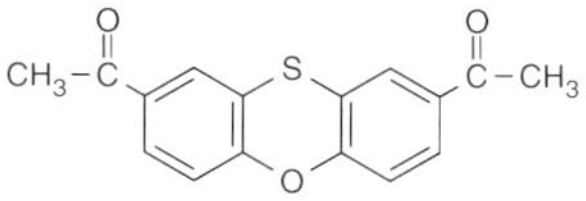
250 mg

S59,784-8
3,3''-(1,4-PHENYLENE)BIS(4'-PHENYLACRYLOPHENONE)
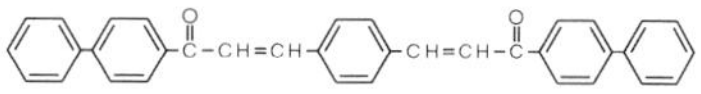
250 mg

S59,649-3
3,3''-(1,4-PHENYLENE)-DIACRYLOPHENONE
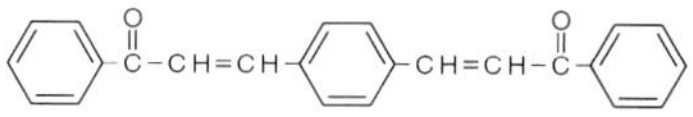
5 gm

S60,033-4
3,3''-(1,4-PHENYLENE)BIS(4'-CHLOROACRYLOPHENONE)
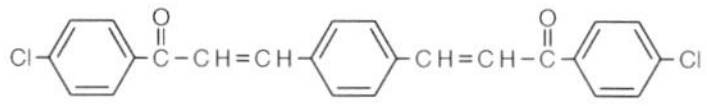
250 mg

S68,243-8
1,10-DIPHENYL-1,3,7,9-DECATETRAENE-5,6-DIONE
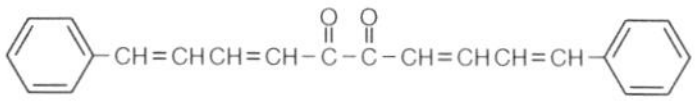
5 gm

S93,228-0
2-BROMO-2-(ALPHA-BROMOBENZYL)-1,3-INDANEDIONE
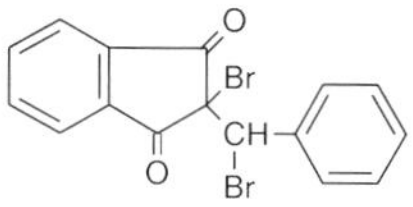
250 mg

S95,754-2
4B-BROMO-4B,5,9B,10-TETRAHYDROINDENO(2,1-A)INDENE-5,10-DIONE
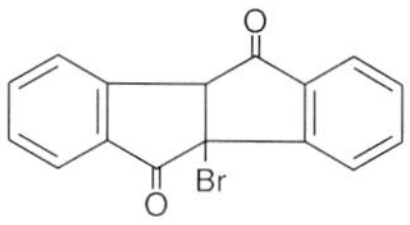
50 mg

S95,756-9
DIBENZO(H,L)TRICYCLO[4.3.3.01,6]-DODEC-3-ENE-7,10-DIONE
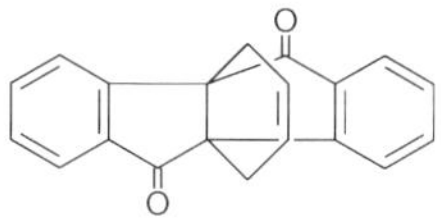
50 mg

S93,343-0
1-(4-BIPHENYLYL)-1,4-HEXANEDIONE
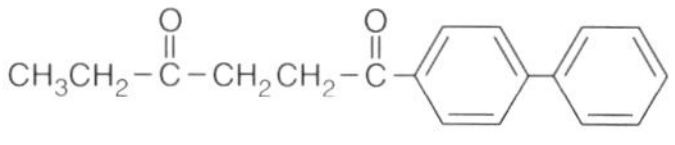
250 mg

S97,464-1
5,6,6A,7,8,12B-HEXAHYDROBENZO(C)-PHENANTHRENE-5,8-DIONE
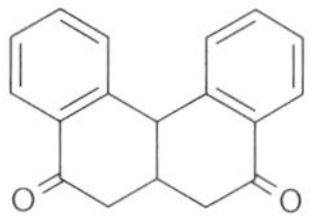
10 mg

S97,478-1
5,6,6A,7,8,12B-HEXAHYDRO-1-METHYLBENZO(C)-PHENANTHRENE-5,8-DIONE
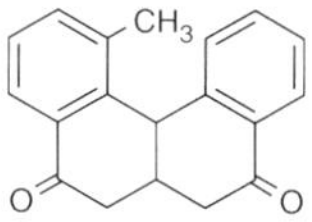
10 mg

S40,739-9
13,13'-DIBENZANTHRONYL
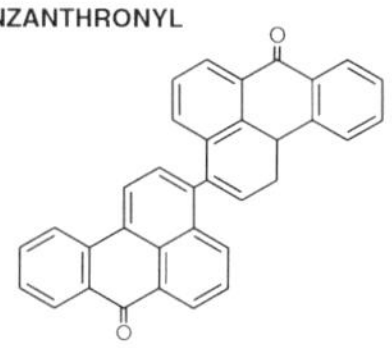
5 gm

S51,441-1
2,2'-(5-INDANYLMETHYLENE)BIS(5,5-DIMETHYL-1,3-CYCLOHEXANEDIONE)
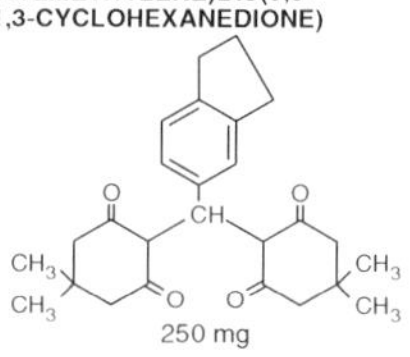
250 mg

S97,412-9
5,6,6A,7,12,12A-HEXAHYDRO-1,12-DIMETHYLBENZ(A)-ANTHRACENE-5,7-DIONE
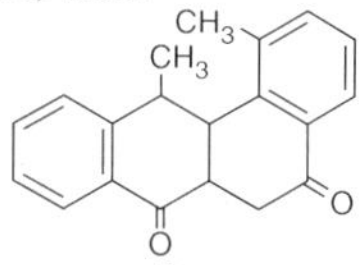
10 mg

S97,414-5
5,6,6A,7,12,12A-HEXAHYDRO-3,9-DIMETHYLBENZ(A)-ANTHRACENE-5,7-DIONE
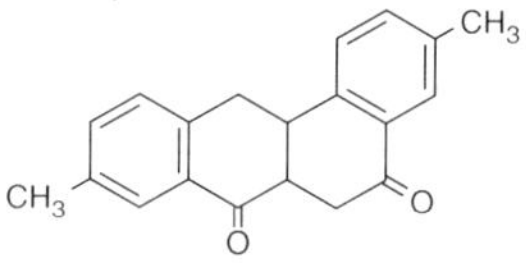
10 mg

S73,478-0
4-(BROMOMETHYL)-3,5-DICHLOROBENZOPHENONE
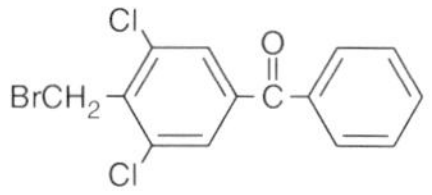
250 mg

S80,689-7
4-(BROMOMETHYL)BENZOPHENONE
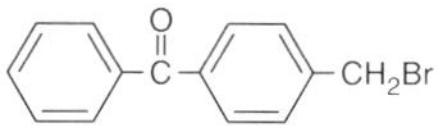
250 mg

S87,675-5
4-BROMO-4'-(TRIFLUOROMETHOXY)-BENZOPHENONE
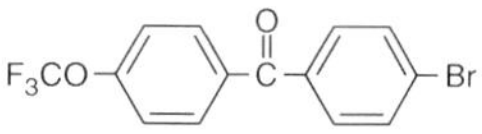
250 mg

S75,726-8
3,4-DICHLORO-4'-(TRIFLUOROMETHYL)-BENZOPHENONE
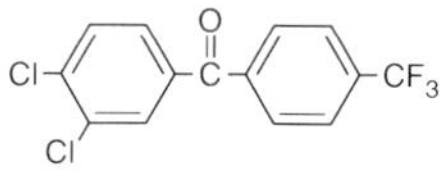
250 mg

S87,676-3
4-CHLORO-4'-(TRIFLUOROMETHYL)-BENZOPHENONE
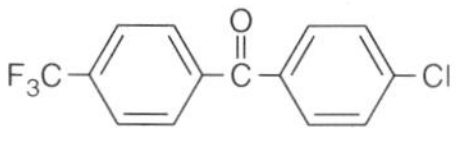
250 mg

S87,677-1
4-IODO-4'-(TRIFLUOROMETHYL)-BENZOPHENONE
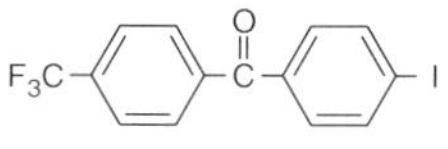
250 mg

S87,960-6
2-AMINO-2',5'-BIS(TRIFLUOROMETHYL)-BENZOPHENONE
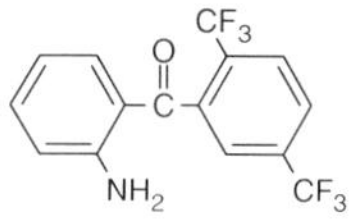
5 gm

S73,446-2
2,5-BIS(TRIFLUOROMETHYL)-BENZOPHENONE
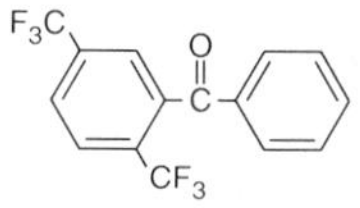
250 mg

S73,978-2
5-CHLORO-2-(2-HYDROXYBENZYLAMINO)-BENZOPHENONE
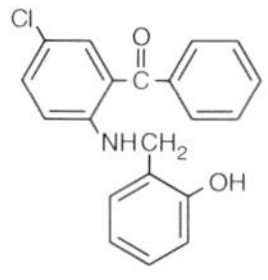
5 gm

S92,291-9
3,4-DIMETHOXY-4'-FLUOROBENZOPHENONE
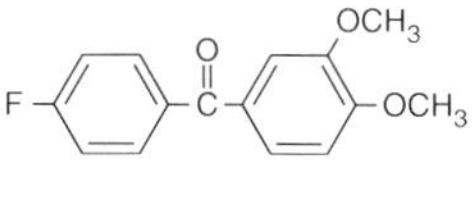
250 mg

S73,351-2
2-(2-AMINOPHENYLTHIOMETHYL)-4'-CHLOROBENZOPHENONE HYDROCHLORIDE
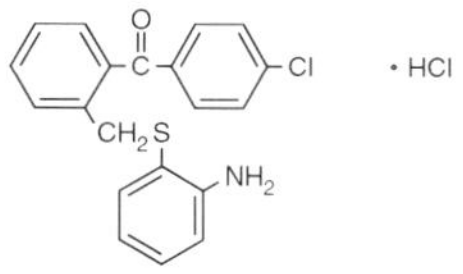
250 mg

S46,081-8
2-AMINO-4'-CHLOROBENZOPHENONE
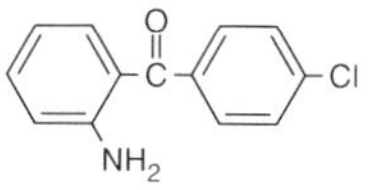
5 gm

S65,144-3
4-AMINO-2,4'-DICHLOROBENZOPHENONE
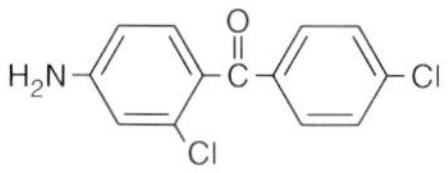
250 mg

S94,623-0
2-AMINO-5-BROMO-2'-CHLOROBENZOPHENONE
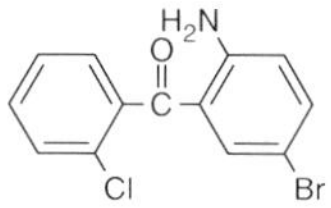
250 mg

S66,239-9
4-CHLORO-2'-(DIMETHYLAMINOMETHYL)-BENZOPHENONE
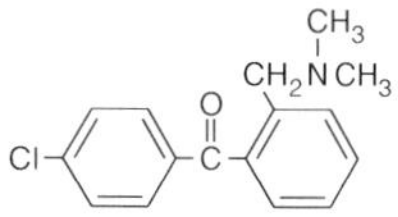
5 gm

S66,259-3
2,2'-DICHLOROBENZIL
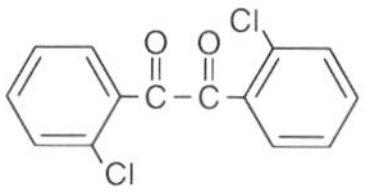
250 mg

S79,330-2
2-CHLORO-4'-PHENYLBENZOPHENONE
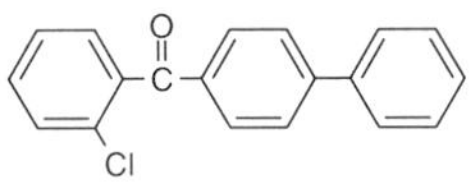
250 mg

S73,281-8
2-BROMOBENZOPHENONE
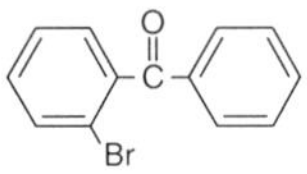
250 mg

S87,629-1
2,6-DICHLOROBENZOPHENONE
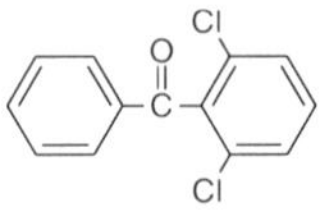
250 mg

S73,514-0
2,4-DICHLOROBENZOPHENONE
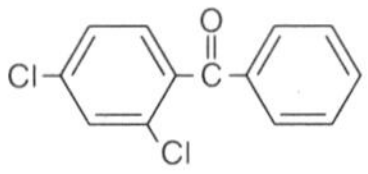
250 mg

S87,630-5
2,4,6-TRICHLOROBENZOPHENONE
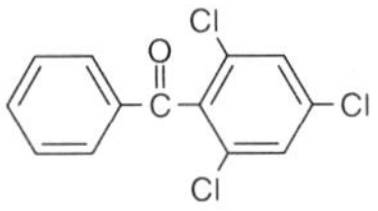
250 mg

S82,871-8
2,3,5,6-TETRACHLOROBENZOPHENONE
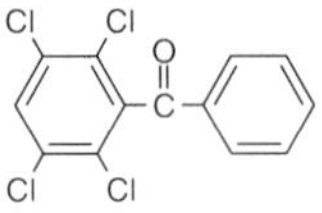
250 mg

S92,334-6
3-BROMO-6-PHENYLBENZANTHRONE
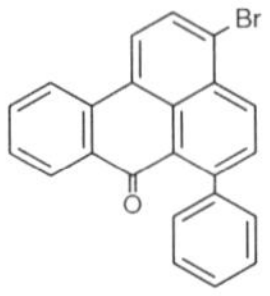
250 mg

S68,312-4
4-CHLORO-2'-METHYLBENZOPHENONE
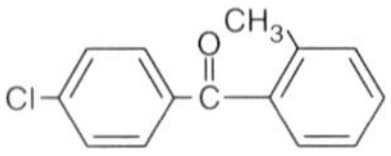
5 gm

S73,382-2
2-[2-(4-CHLOROBENZOYL)BENZYLIDENE]-ALPHA-TETRALONE
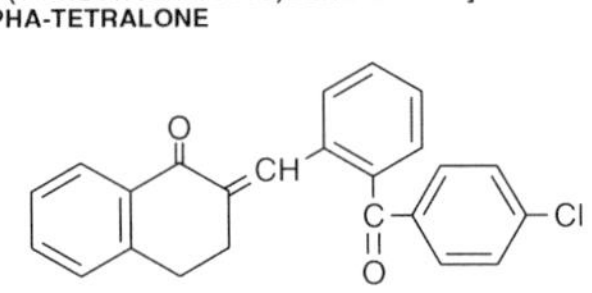
250 mg

S92,695-7
3-IODOBENZANTHRONE
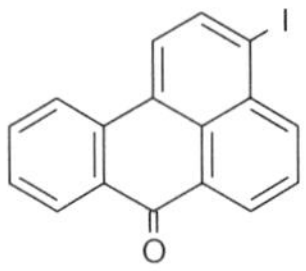
250 mg

S94,395-9
2-CHLORO-2',5'-DIMETHYLBENZOPHENONE
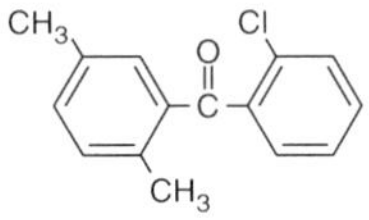
250 mg

S34,635-7
2,2'-DIHYDROXY-4-OCTYLOXYBENZOPHENONE
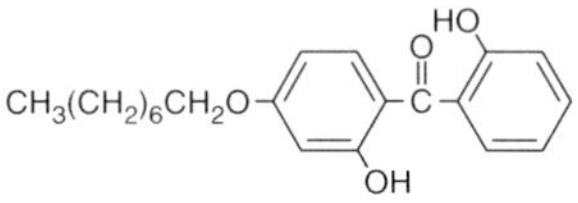
250 mg

S95,698-8
9,10-DIHYDRO-1,4-DIHYDROXY-5-METHOXY-9-METHYL-10-ANTHRACENONE
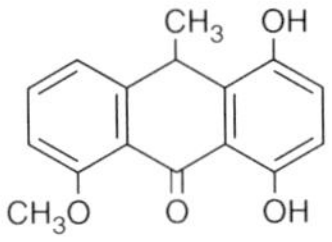
50 mg

S40,905-7
MACLURIN
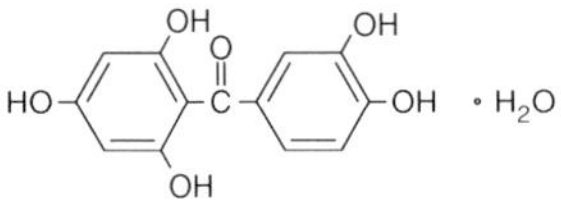
5 gm

S78,759-0
ALOIN
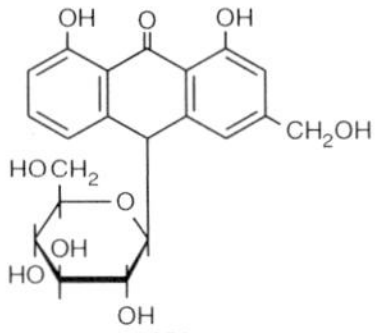
250 mg

S70,686-8
2,4-DIBENZOYLRESORCINOL
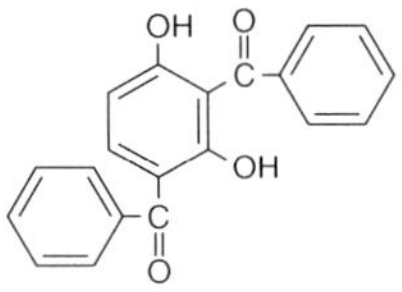
5 gm

S76,288-1
4,4'-DIHYDROXY-3,3'-DINITROBENZOPHENONE
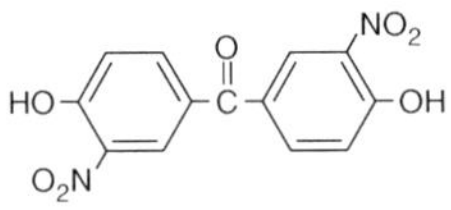
250 mg

S35,112-1
2-HYDROXY-4-OCTYLOXYBENZOPHENONE
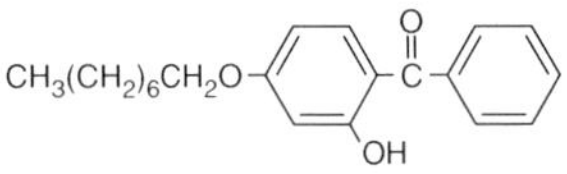
5 gm

S43,833-2
4-DODECYLOXY-2-HYDROXYBENZOPHENONE
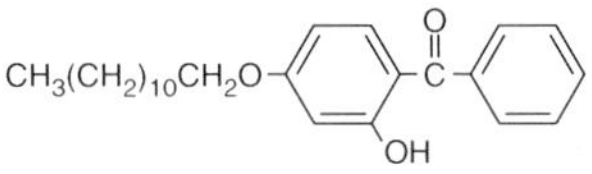
5 gm

S40,888-3
2-HYDROXY-4-METHOXY-4'-METHYLBENZOPHENONE
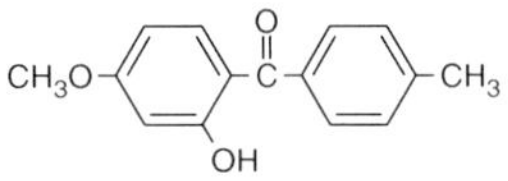
5 gm

S45,008-1
4'-CHLORO-2-HYDROXY-4-METHOXYBENZOPHENONE
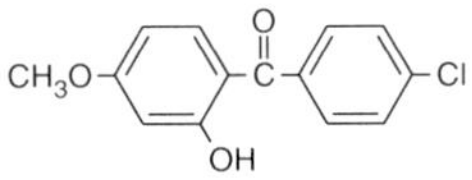
5 gm

S61,736-9
4-HYDROXY-4'-METHOXYBENZOPHENONE
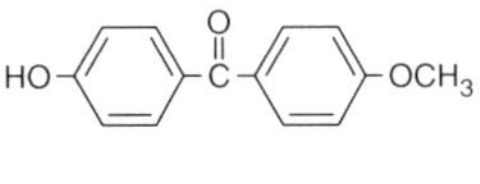
5 gm

S94,144-1
2-AMINO-3-HYDROXY-1H-PHENALEN-1-ONE
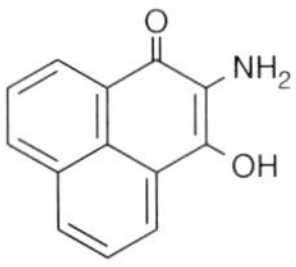
250 mg

S86,588-5
9-TERT-BUTYL-9-HYDROXY-10(9H)-ANTHRONE
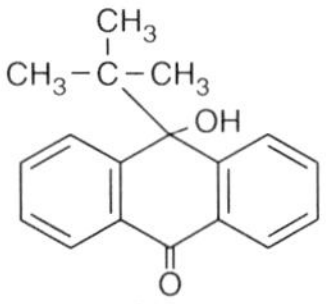
250 mg

S84,255-9
(1,9 OR 2,9 OR 3,9 OR 4,9)-DIISOPROPYL-9-HYDROXY-10(9H)-ANTHRONE
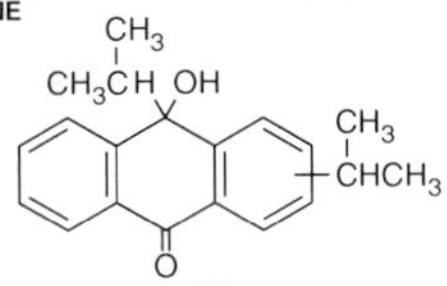
250 mg

S41,436-0
4-HYDROXY-3-METHYLBENZOPHENONE
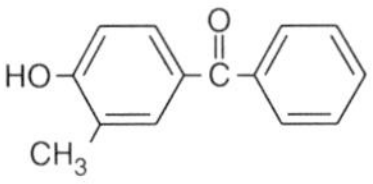
5 gm

S91,451-7
3,5-DI-TERT-BUTYL-4-HYDROXYBENZOPHENONE
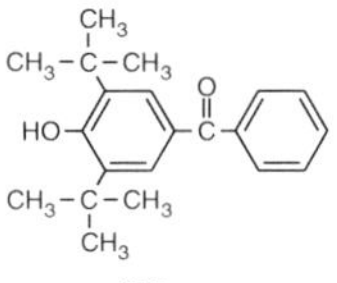
250 mg

S59,622-1
2,3',5',6-TETRACHLORO-4'-HYDROXYBENZOPHENONE
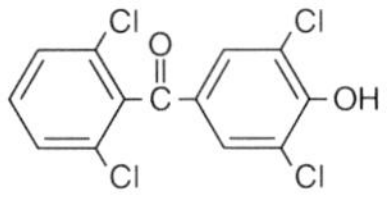
250 mg

S59,724-4
2,2',5',6-TETRACHLORO-4'-HYDROXYBENZOPHENONE
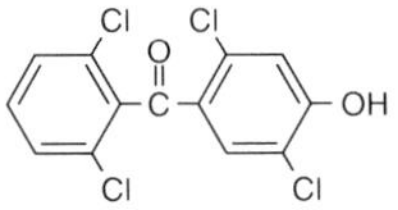
250 mg

S92,863-1
7-HYDROXY-1H-PHENALENE-1-ONE
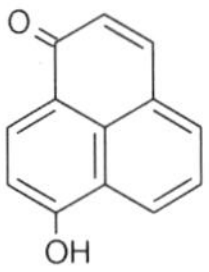
250 mg

S74,943-5
3-HYDROXY-9-FLUORENONE
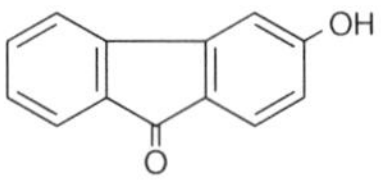
250 mg

S94,086-0
2-HYDROXY-3,5,7-TRIBROMO-2,4,6-CYCLOHEPTATRIENE-1-ONE
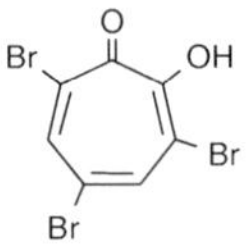
250 mg

S94,087-9
2-HYDROXY-3,5,7-TRIMETHYL-2,4,6-CYCLOHEPTATRIEN-1-ONE
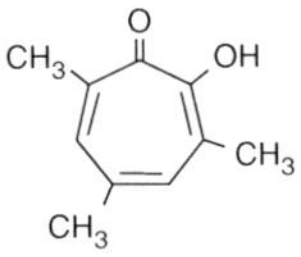
250 mg

S54,685-2
2,4-DIHYDROXYPHENYL 1-NAPHTHYL KETONE
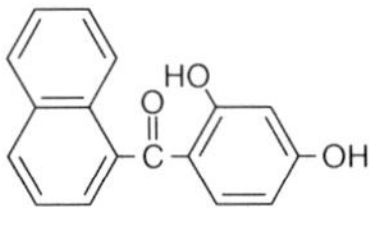
250 mg

S36,319-7
9-METHOXY-9-PHENYLANTHRONE
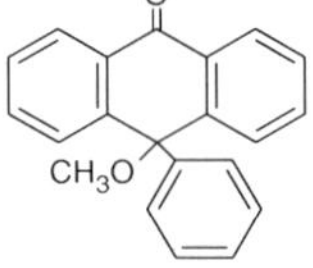
5 gm

S45,693-4
2,4,5-TRIMETHOXYBENZOPHENONE
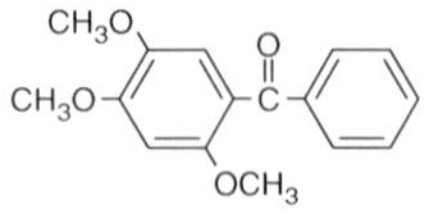
5 gm

S46,302-7
2,3,3',4'-TETRAMETHOXYBENZOPHENONE
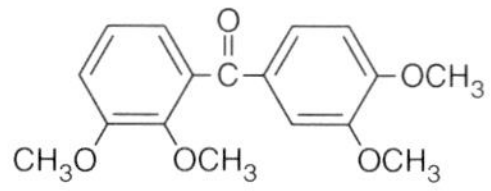

5 gm

S83,473-4
2,6-DIETHOXYBENZOPHENONE
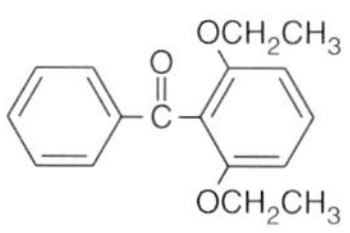

250 mg

S93,322-8
TILORONE
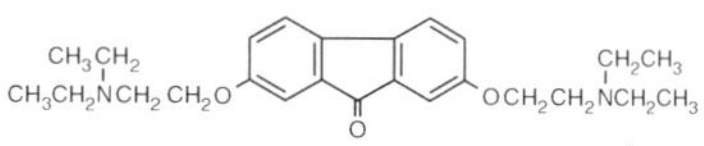

250 mg

S42,555-9
MELICOPINE
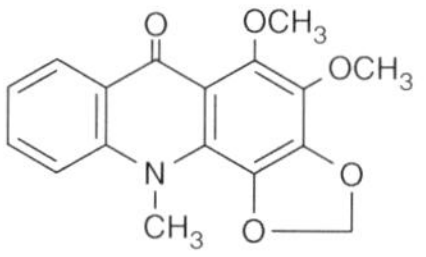

250 mg

S43,371-3
2,4,5-TRIETHOXYBENZOPHENONE
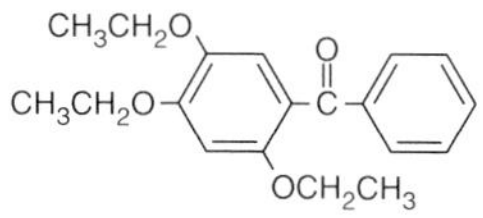

5 gm

S63,514-6
3,4,4'-TRIMETHOXYBENZOPHENONE
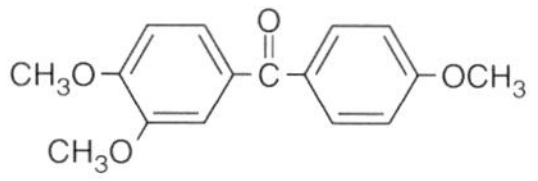

250 mg

S61,957-4
10-(3,4-DIMETHOXYBENZYLIDENE)-9-ANTHRONE
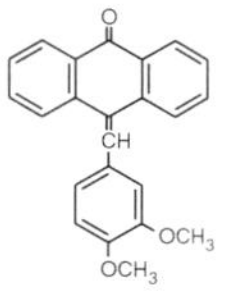

250 mg

S36,435-5
3,3',4,4',5-PENTAMETHOXYBENZOPHENONE
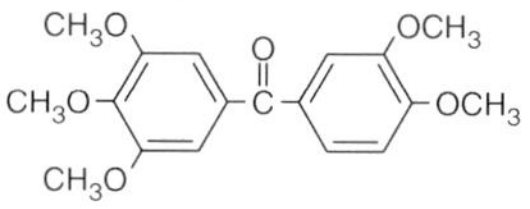

5 gm

S65,687-9
10,11-DIHYDRO-2,3-DIMETHOXY-5H-DIBENZO(A,D)CYCLOHEPTEN-5-ONE
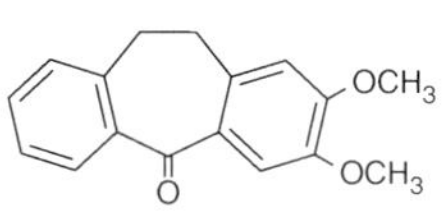

250 mg

S88,691-2
2-METHOXY-7-NITRO-9-THIOXANTHENONE
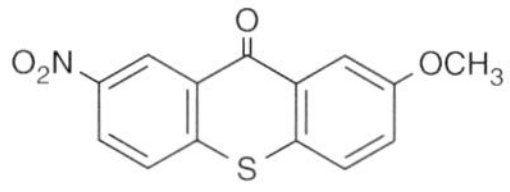

250 mg

S90,745-6
4-(DIMETHYLAMINO)-4'-METHOXYBENZOPHENONE
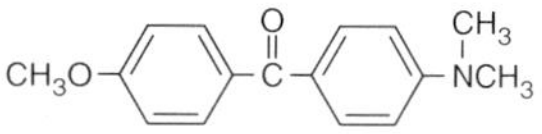

250 mg

S75,608-3
3,7-DINITRO-2-METHOXY-9-FLUORENONE
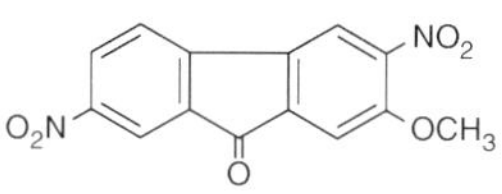

5 gm

S94,141-7
3-(4-METHOXYBENZOYL)BENZANTHRONE
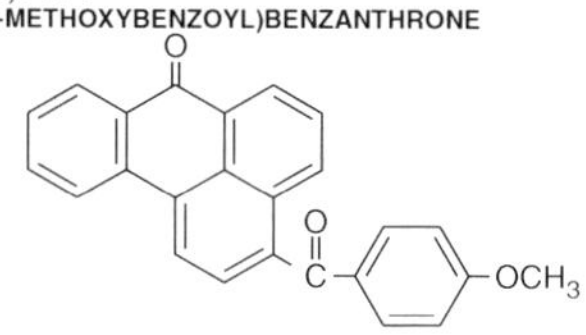

10 mg

S44,760-9
2,4,6-TRIMETHOXYBENZOPHENONE
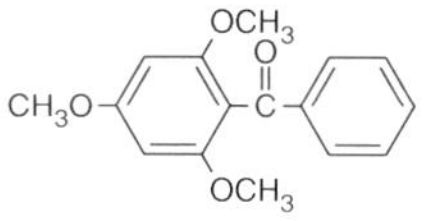

5 gm

S61,757-1
10-(2-ETHOXYBENZYLIDENE)-9-ANTHRONE
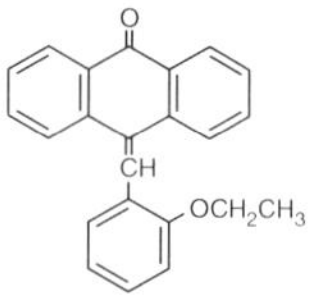

5 gm

S61,370-3
10-(2-METHOXYBENZYLIDENE)-9-ANTHRONE
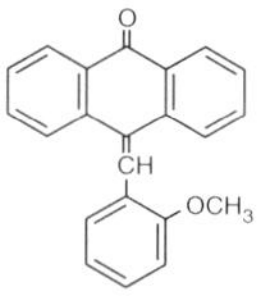

5 gm

S61,753-9
10-(4-ETHOXYBENZYLIDENE)-9-ANTHRONE
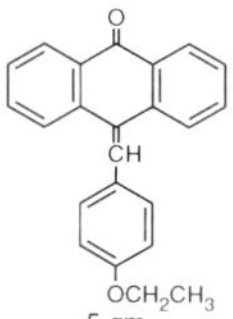

5 gm

S35,214-4
4-METHOXY-4'-METHYLBENZOPHENONE
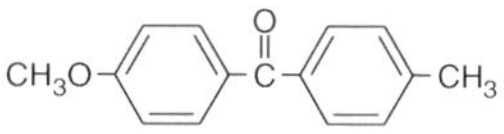

5 gm

S61,373-8
10-(2,5-DIMETHOXYBENZYLIDENE)-9-ANTHRONE
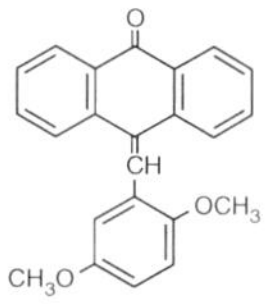

5 gm

S92,865-8
3-METHOXY-1H-PHENALENE-1-ONE
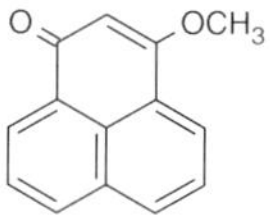

250 mg

S35,213-6
4-METHOXY-9-FLUORENONE
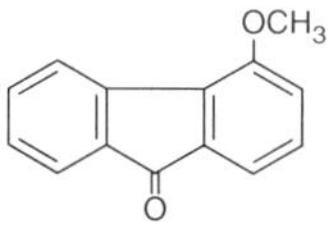

250 mg

S61,378-9
10-(3-METHOXYBENZYLIDENE)-9-ANTHRONE
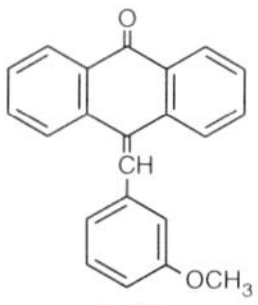

5 gm

S94,142-5
3-METHOXYDIBENZO(B,D,E,F)CHRYSENE-7,14-DIONE
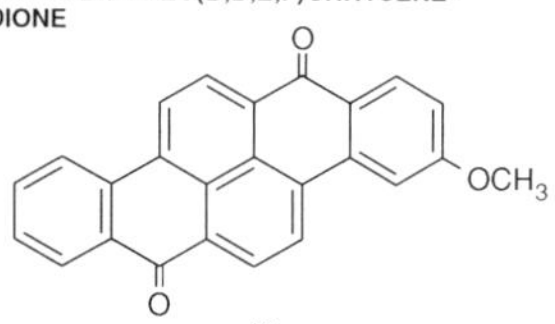

10 mg

S94,143-3
2-METHOXYDIBENZO(B,D,E,F)CHRYSENE-7,14-DIONE
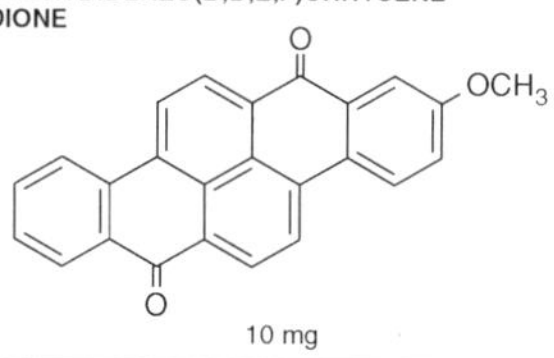

10 mg

S58,435-5
10-(4-METHOXYBENZYLIDENE)-9-ANTHRONE
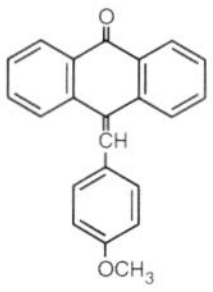

250 mg

S64,864-7
4,4'-DIMETHYLBENZOPHENONE
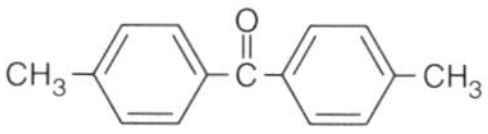

250 mg

S82,059-8
2-BENZOYLDIBENZO-P-DIOXIN
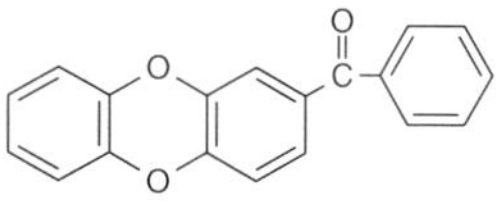

250 mg

S45,661-6
3,4-DIAMINOBENZOPHENONE MONOHYDROCHLORIDE
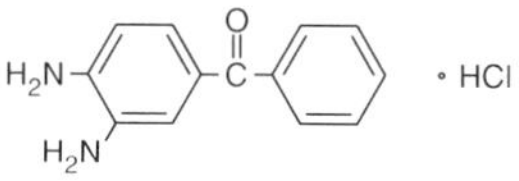

5 gm

S34,120-7
2-AMINO-3-NITRO-9-FLUORENONE
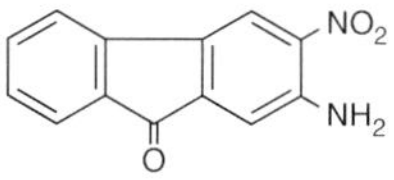

250 mg

S74,992-3
2,4-DIAMINO-9-FLUORENONE
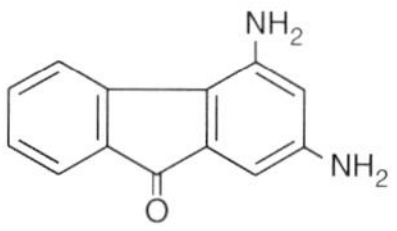

250 mg

S80,611-0
2-AMINO-6-METHYLBENZOPHENONE
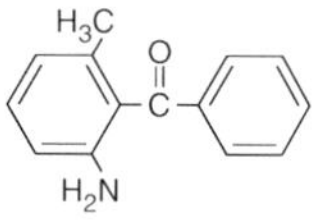

250 mg

S75,310-6
2-AMINO-7-NITRO-9-FLUORENONE
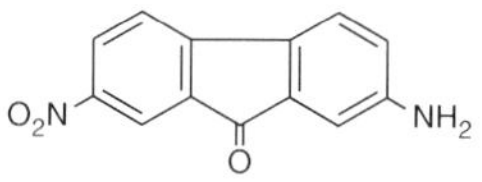

50 mg

S77,422-7
5-AMINO-2-NITRO-9-FLUORENONE
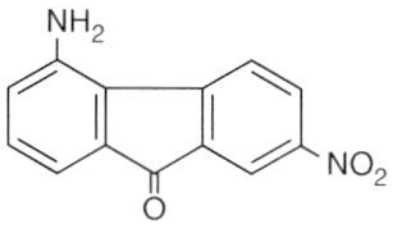

250 mg

S94,624-9
2-AMINO-5-BROMOBENZOPHENONE
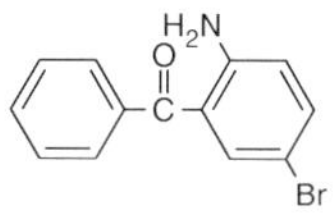

250 mg

S94,625-7
2-AMINO-5-METHYLBENZOPHENONE
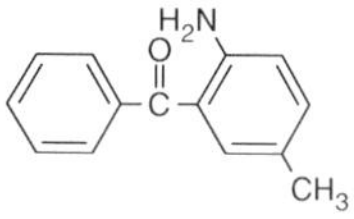

250 mg

S92,690-6
2,7-DIAMINO-9-FLUORENONE
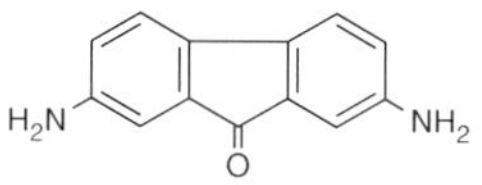

250 mg

S75,607-5
2-AMINO-7-FLUORO-9-FLUORENONE

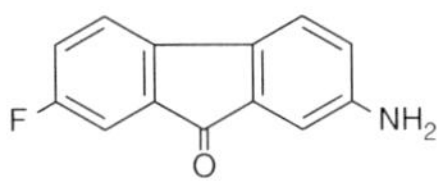

5 gm

S75,407-2
2-AMINO-3-BROMO-7-FLUORO-9-FLUORENONE

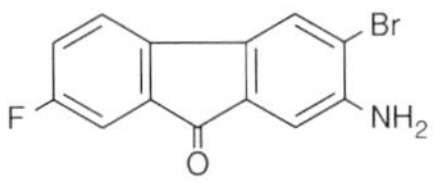

250 mg

S75,419-6
4-AMINO-2,5,7-TRICHLORO-9-FLUORENONE

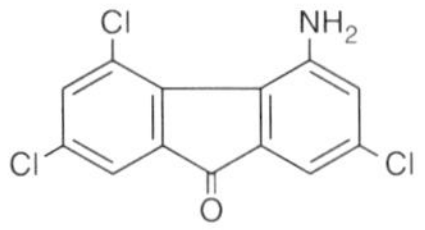

250 mg

S74,899-4
2-AMINO-3-IODO-9-FLUORENONE

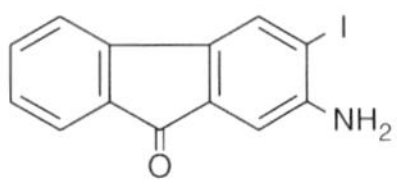

250 mg

S75,437-4
2-AMINO-1,3,4,7-TETRACHLORO-9-FLUORENONE

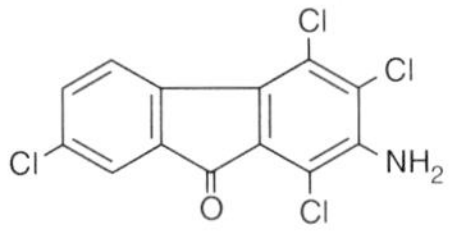

250 mg

S74,900-1
4-AMINO-2,7-DIBROMO-9-FLUORENONE

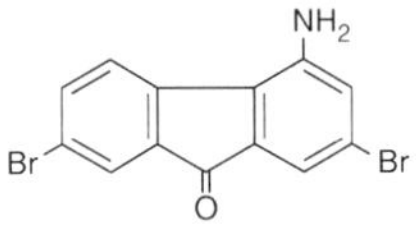

250 mg

S75,477-3
2-AMINO-1,3-DIBROMO-9-FLUORENONE

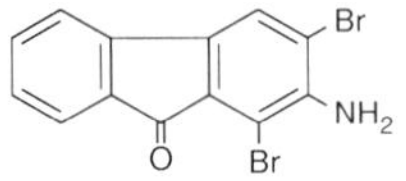

250 mg

S77,406-5
4-AMINO-1,3-DIBROMO-9-FLUORENONE

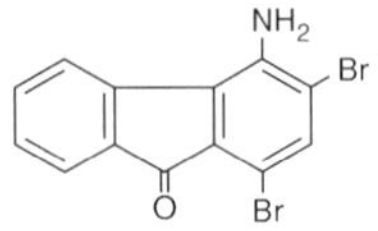

250 mg

S74,893-5
2-AMINO-1,3,7-TRIBROMO-9-FLUORENONE

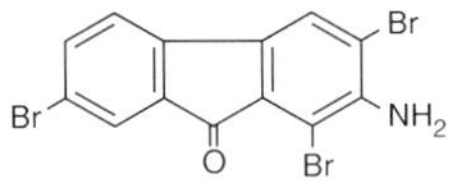

250 mg

S94,139-5
9-AMINO-3-CHLOROBENZANTHRONE

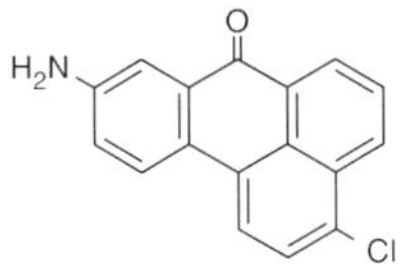

250 mg

S76,644-5
10-AMINOMETHYL-10,11-DIHYDRO-5H-DIBENZO(A,D)CYCLOHEPTEN-5-ONE

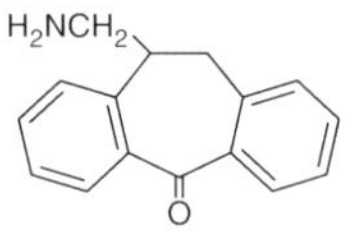

250 mg

S75,824-8
3-AMINO-7-FLUORO-9-FLUORENONE

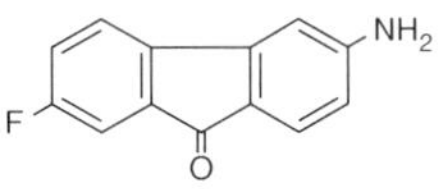

250 mg

S74,593-6
2-AMINO-4,7-DIFLUORO-9-FLUORENONE

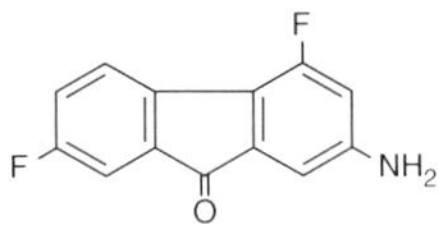

250 mg

S75,551-6
7-AMINO-1,3-DIFLUORO-9-FLUORENONE

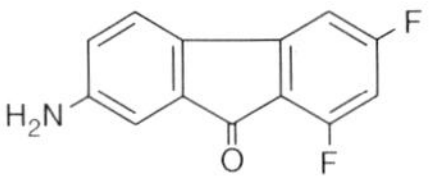

250 mg

S75,524-9
4-AMINO-2,7-DICHLORO-9-FLUORENONE

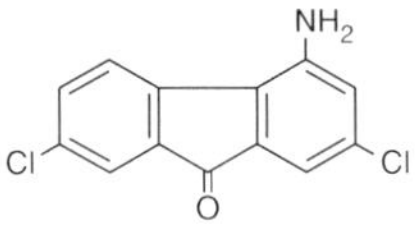

250 mg

S75,168-5
1-AMINO-2,4-DIBROMO-9-FLUORENONE

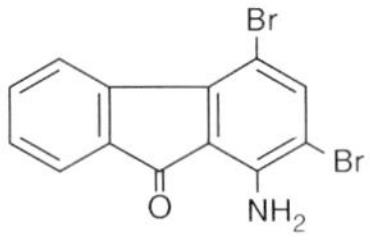

5 gm

S89,469-9
2-(BENZOCYCLOBUTEN-1-YLMETHYLAMINO)-5-NITROBENZOPHENONE

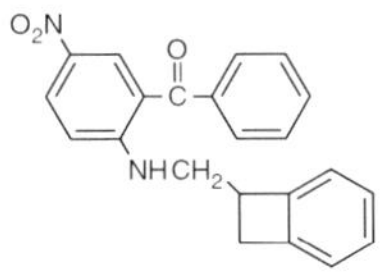

250 mg

S91,879-2
4-(2-NITRO-1,1,2-TRIMETHYLPROPYL)-BENZOPHENONE

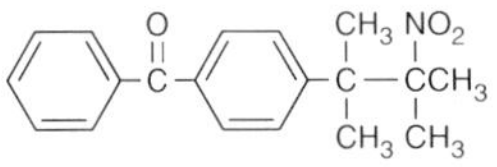

250 mg

S91,878-4
4-(1,2-DIMETHYL-1-ETHYL-2-NITROPROPYL)BENZOPHENONE

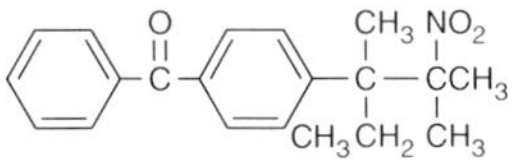

250 mg

S91,884-9
4-(2-NITRO-1,1,2-TRIMETHYLBUTYL)-BENZOPHENONE

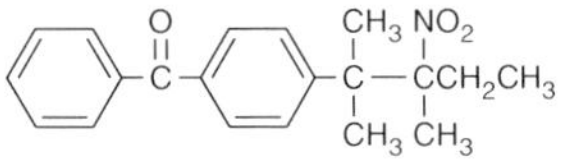

250 mg

S89,647-0
2-(BENZYLAMINO)BENZOPHENONE

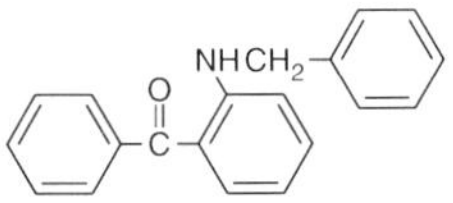

250 mg

S89,636-5
2-(ETHYLAMINO)BENZOPHENONE

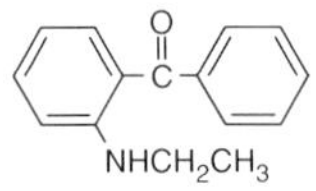

250 mg

S89,646-2
2-ANILINOBENZOPHENONE

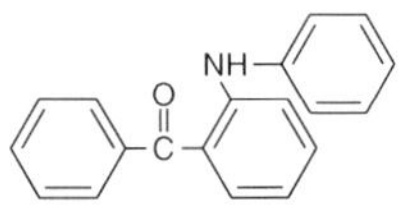

250 mg

S36,335-9
2-METHYLAMINO-9-FLUORENONE

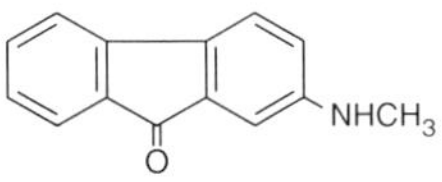

250 mg

S90,736-7
4-BENZOYL-4'-(DIMETHYLAMINO)-BIPHENYL

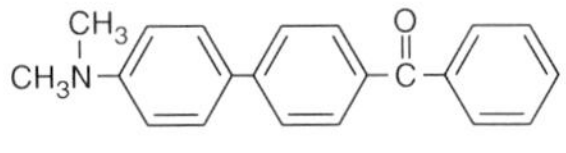

250 mg

S89,629-2
4-(DIMETHYLAMINO)-2'-METHYLBENZOPHENONE

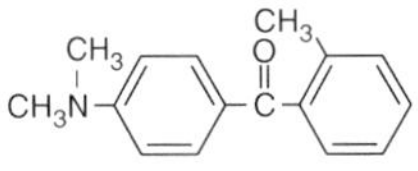

250 mg

S87,327-6
10-[4-(DIMETHYLAMINO)CINNAMYLIDENE]-9-ANTHRONE

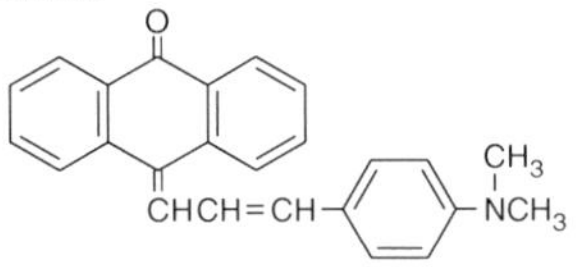

250 mg

S90,755-3
4,4'-BIS(DIMETHYLAMINO)-2,2'-DIMETHYLBENZOPHENONE

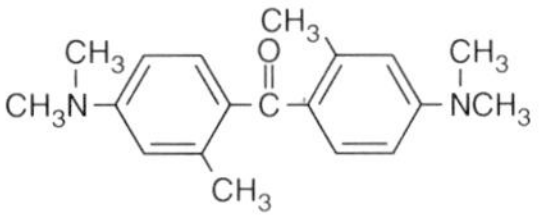

250 mg

S87,369-1
10-[4-(DIMETHYLAMINO)-ALPHA-PHENYLBENZYLIDENE]-9-ANTHRONE

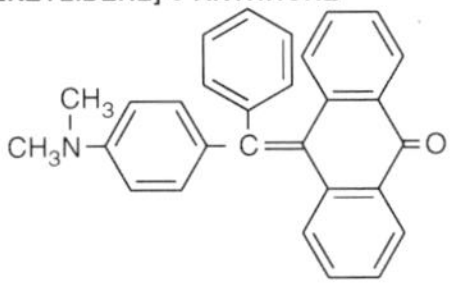

250 mg

S87,328-4
10-(BIS[4-(DIMETHYLAMINO)PHENYL]-METHYLENE)-9-ANTHRONE

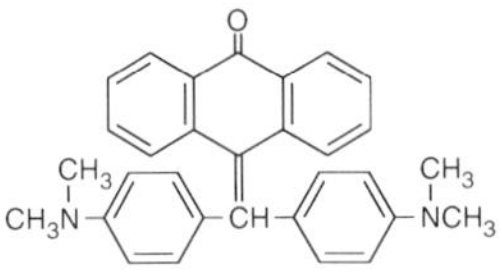

250 mg

S43,992-4
2-MORPHOLINOBENZOPHENONE

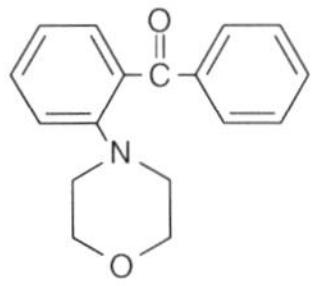

250 mg

S35,861-4
N,N-BIS(2-BENZOYLETHYL)-METHYLAMINE

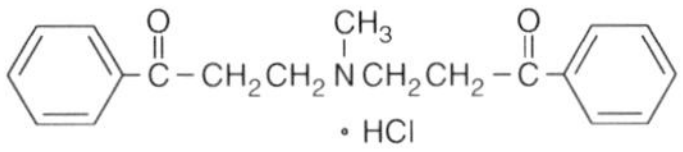

5 gm

S61,959-0
10-(2-CHLORO-4-DIMETHYLAMINOBENZYLIDENE)-9-ANTHRONE

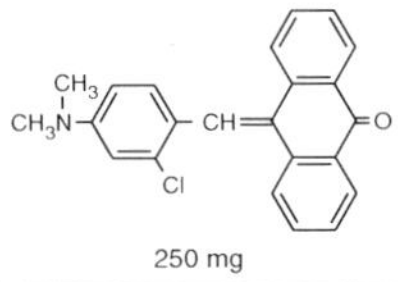

250 mg

S65,744-1
2-DIETHYLAMINO-4-METHYLBENZOPHENONE

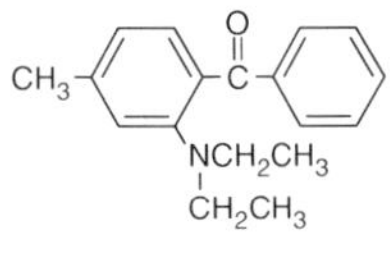

5 gm

S44,190-2
2-DIETHYLAMINO-9-FLUORENONE

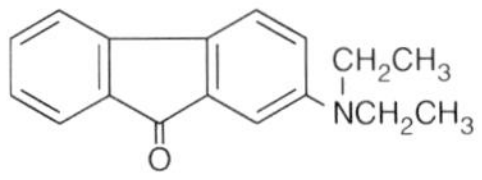

5 gm

S87,329-2
10-(1-METHYL-1,2,3,4-TETRAHYDRO-6-QUINOLYLMETHYLENE)-9-ANTHRONE

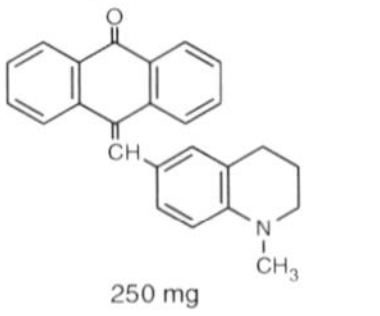

250 mg

S58,450-9
10-[4-(DIMETHYLAMINO)BENZYLIDENE]-9-ANTHRONE
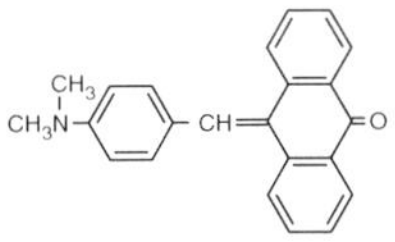
5 gm

S74,840-4
10-(DIMETHYLAMINOMETHYL)-5H-DIBENZO(A,D)CYCLOHEPTEN-5-ONE
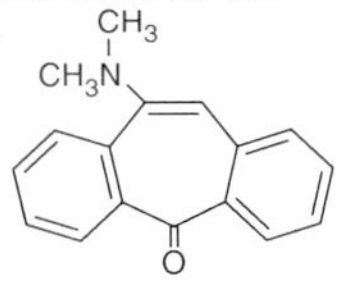
250 mg

S45,479-6
2,4'-DICHLORO-4-NITROBENZOPHENONE
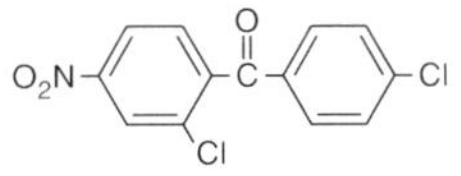
5 gm

S74,518-9
2-NITROBENZOPHENONE
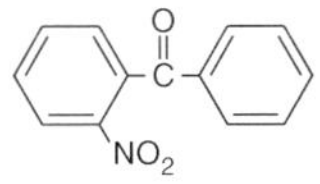
250 mg

S91,602-1
4,4'-DINITROBENZOPHENONE
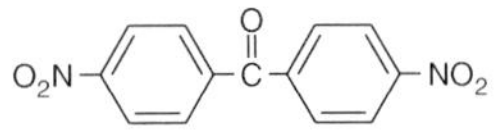
250 mg

S91,870-9
4-(1-METHYL-1-NITROETHYL)-BENZOPHENONE
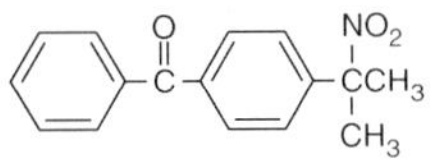
250 mg

S91,874-1
4-(1-METHYL-1-NITROPROPYL)-BENZOPHENONE
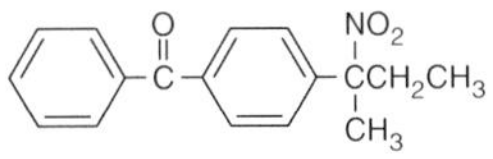
250 mg

S61,756-3
10-(4-NITROBENZYLIDENE)-9-ANTHRONE
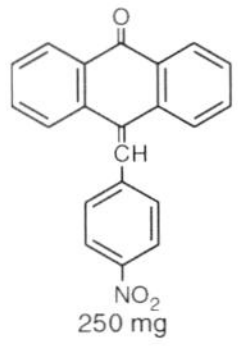
250 mg

S44,763-3
4-FLUORO-4'-NITROBENZOPHENONE
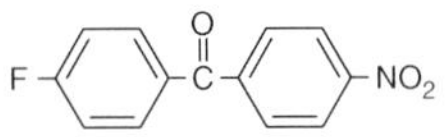
5 gm

S61,489-0
10-(3-NITROBENZYLIDENE)-9-ANTHRONE
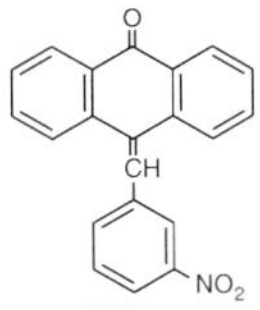
250 mg

S77,402-2
4-NITRO-9-FLUORENONE
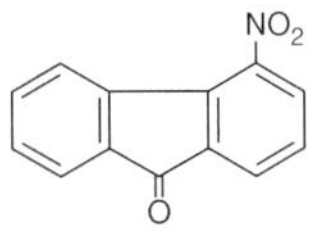
250 mg

S77,086-8
5-FLUORO-2-NITRO-9-FLUORENONE
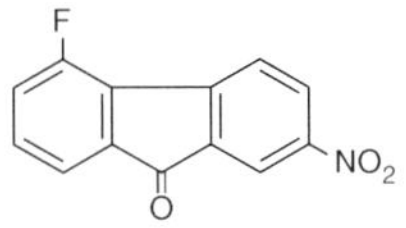
250 mg

S77,408-1
2-FLUORO-6-NITRO-9-FLUORENONE
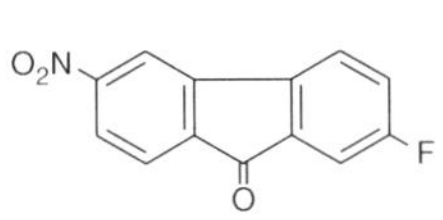
250 mg

S75,600-8
4-CHLORO-2-NITRO-9-FLUORENONE
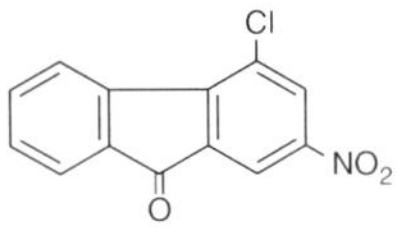
5 gm

S74,892-7
2,3-DICHLORO-7-NITRO-9-FLUORENONE
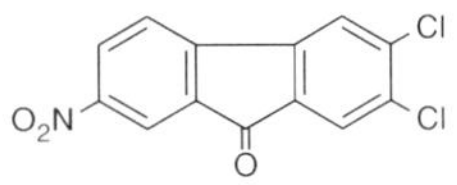
250 mg

S74,938-9
2-BROMO-7-NITRO-9-FLUORENONE
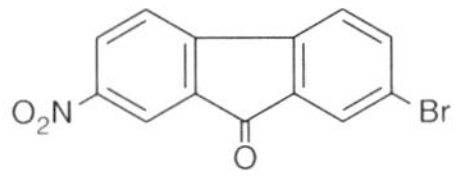
250 mg

S77,083-3
3-BROMO-7-FLUORO-2-NITRO-9-FLUORENONE
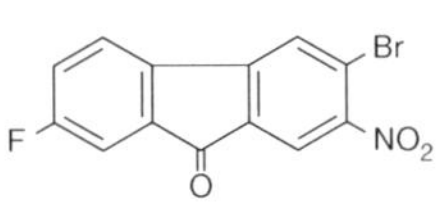
250 mg

S94,138-7
2-NITROBENZANTHRONE
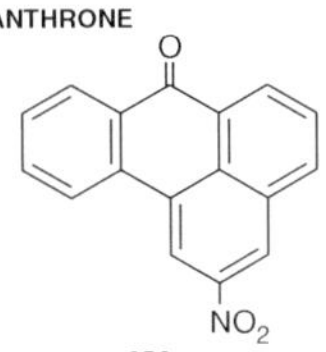
250 mg

S94,140-9
3-NITROBENZANTHRONE
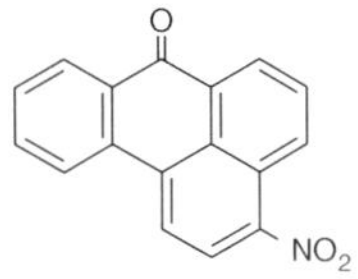
250 mg

S53,701-2
3-BROMO-4'-CHLORO-4-NITROBENZOPHENONE
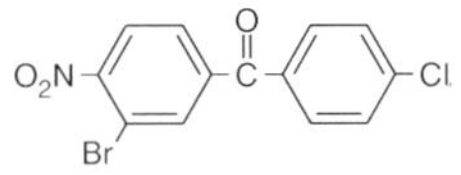
250 mg

S75,533-8
1,3-DIFLUORO-7-NITRO-9-FLUORENONE
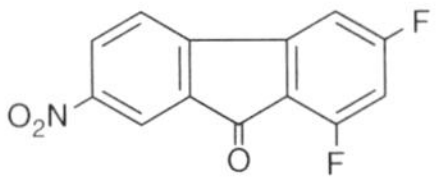
250 mg

S75,402-1
2,5-DINITRO-9-FLUORENONE
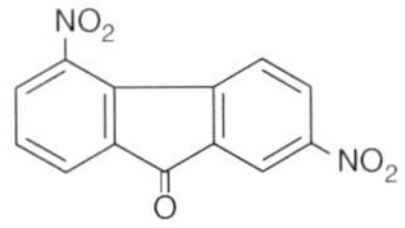
250 mg

S75,154-5
2,7-DICHLORO-4-NITRO-9-FLUORENONE
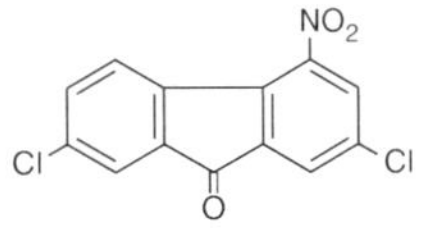
5 gm

S75,821-3
2,7-DIBROMO-4-NITRO-9-FLUORENONE
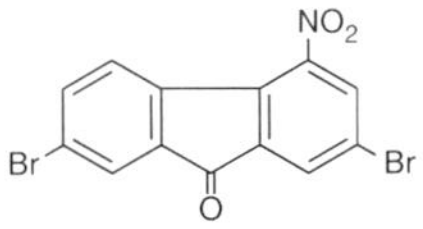
250 mg

S92,987-5
1,2,3,4-TETRAPHENYL-2-BUTEN-1,4-DIONE
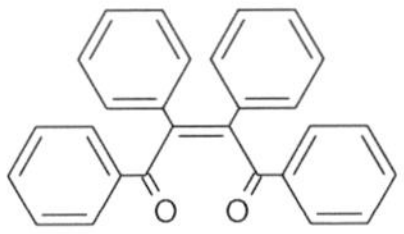
250 mg

S84,963-4
1-BENZOYL-2-(2,4,6-TRIMETHYLBENZOYL)BENZENE
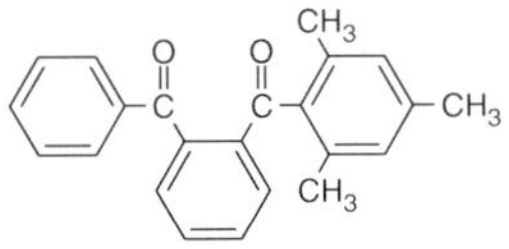
250 mg

S88,917-2
1,3-DIBENZOYLBENZENE
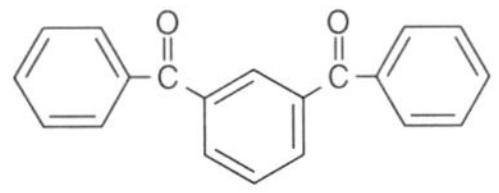
250 mg

S84,962-6
1-(2,4-DIMETHYLBENZOYL)-2-(2,4,6-TRIMETHYLBENZOYL)BENZENE
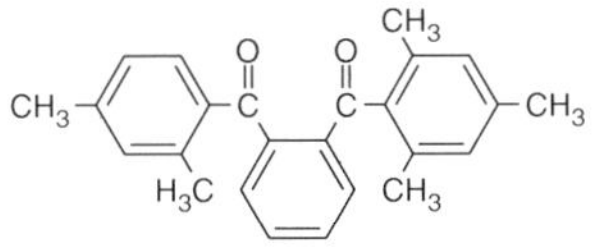
250 mg

S88,847-8
1,2-BIS(2,4,6-TRIMETHYLBENZOYL)-BENZENE
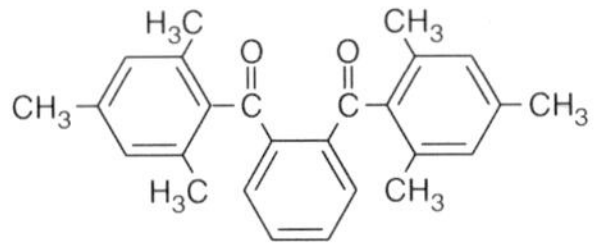
250 mg

S93,372-4
10,10'-BIS(2-DIMETHYLANTHRONE)
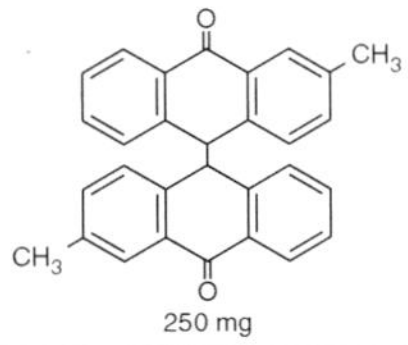
250 mg

S82,922-6
2-BROMO-9-THIOXANTHENONE
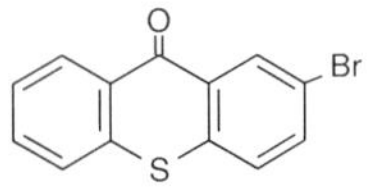
250 mg

S71,322-8
2,3,5,6-TETRAMETHYLBENZOPHENONE
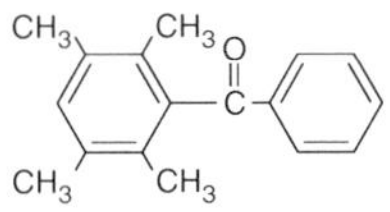
250 mg

S71,747-9
4-ETHYLBENZOPHENONE
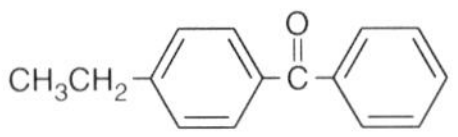
5 gm

S91,871-7
4-ISOPROPYLBENZOPHENONE
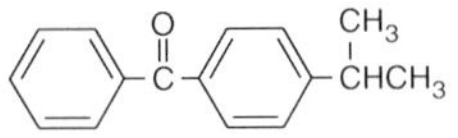
250 mg

S83,201-4
4-(4-BIPHENYLYL)BENZOPHENONE
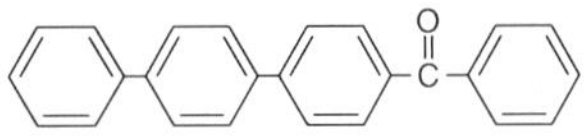
250 mg

S46,153-9
10-BENZHYDRYLIDENE-9-ANTHRONE
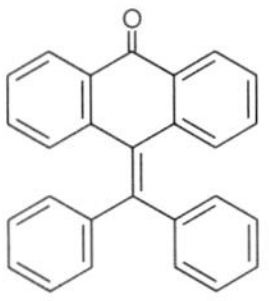
250 mg

S81,196-3
2,4,6-TRIPHENYLBENZOPHENONE
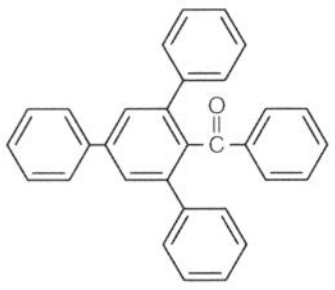
250 mg

S92,696-5
6-PHENYLBENZANTHRONE
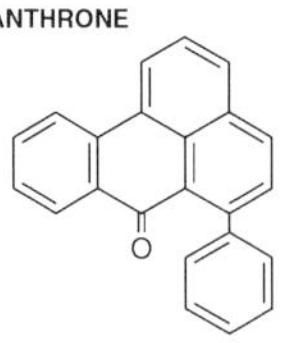
250 mg

S92,227-7
3-BENZOYL BENZANTHRONE
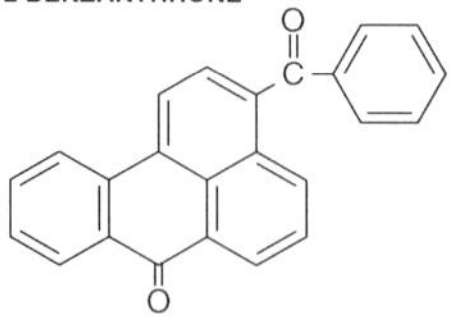
250 mg

S76,353-5
10-PHENYL-5H-DIBENZO(A,D)-CYCLOHEPTEN-5-ONE
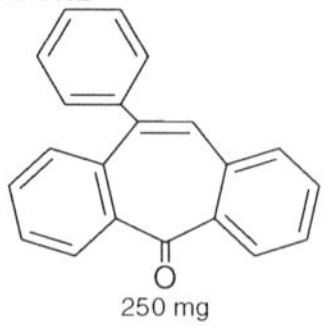
250 mg

S61,751-2
10-(2-CHLOROBENZYLIDENE)-9-ANTHRONE
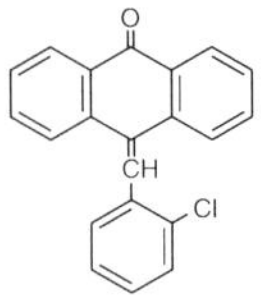
5 gm

S80,610-2
2,3',5'-TRIMETHYLBENZOPHENONE
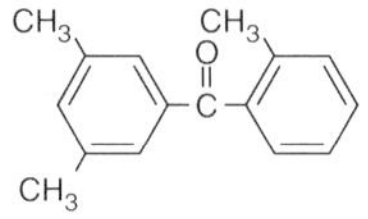
250 mg

S60,387-2
2,3'-DIFLUOROBENZOPHENONE
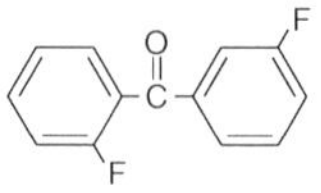
5 gm

S84,688-0
2,2'-DIMETHYLBENZOPHENONE
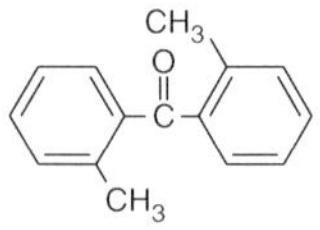
250 mg

S61,750-4
10-(4-CHLOROBENZYLIDENE)-9-ANTHRONE
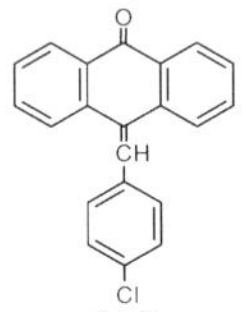
5 gm

S61,755-5
10-(4-ISOPROPYLBENZYLIDENE)-9-ANTHRONE
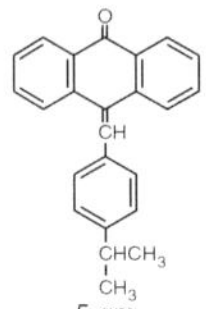
5 gm

S44,929-6
2,4-DICHLORO-4'-FLUOROBENZOPHENONE
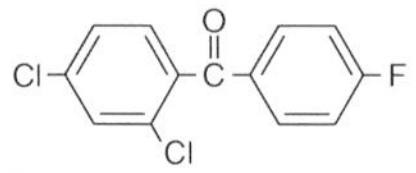
5 gm

S44,984-9
4-BROMO-4'-CHLOROBENZOPHENONE
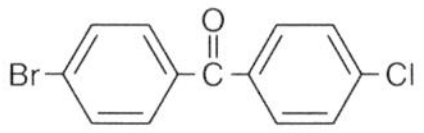
5 gm

S61,492-0
10-(2,4-DICHLOROBENZYLIDENE)-ANTHRONE
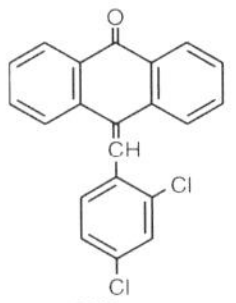
250 mg

S34,248-3
3-BROMOBENZANTHRONE
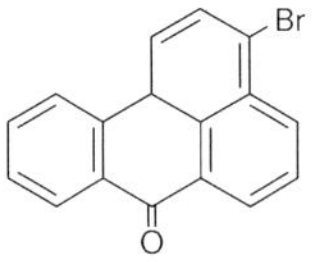
5 gm

S36,016-3
3,9-DIBROMO-7H-BENZ(DE)-ANTHRACEN-7-ONE
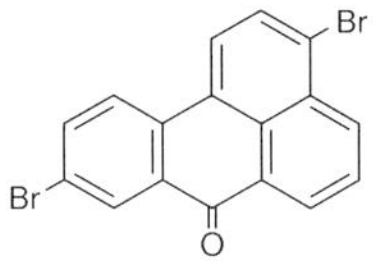
5 gm

S70,635-3
(1-NAPHTHYL)- (2,4,6-TRIMETHYLPHENYL) KETONE
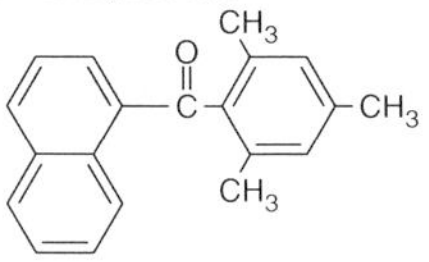
5 gm

S82,116-0
2-(2,4,6-TRIMETHYLBENZOYL)-NAPHTHALENE
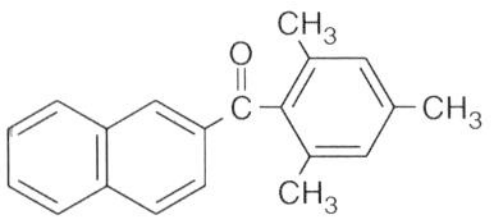
250 mg

S79,248-9
1,8-DIBENZOYLNAPHTHALENE
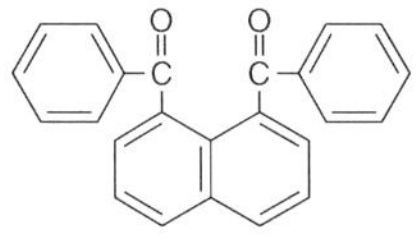
250 mg

S74,911-7
1-FLUORO-9-FLUORENONE
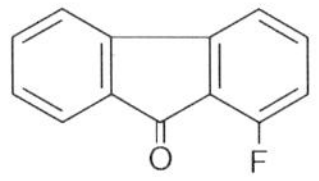
250 mg

S75,474-9
4-FLUORO-9-FLUORENONE
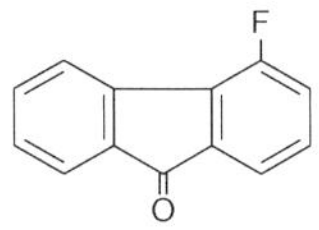
250 mg

S75,819-1
3-FLUORO-9-FLUORENONE
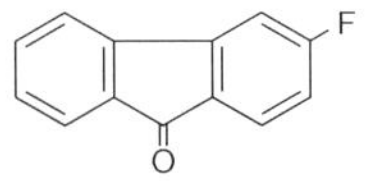
250 mg

S74,891-9
1,4-DIFLUORO-9-FLUORENONE
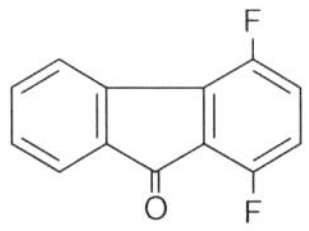
250 mg

S75,435-8
2,5-DIFLUORO-9-FLUORENONE
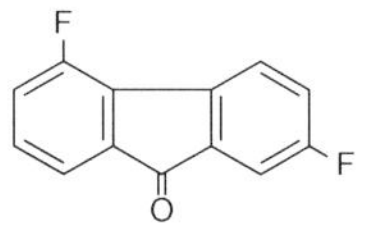
250 mg

S75,478-1
2,4-DIFLUORO-9-FLUORENONE
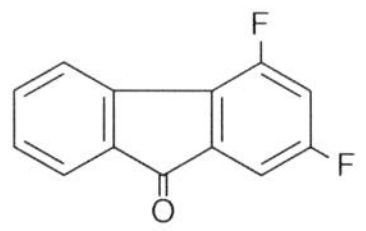
250 mg

S75,414-5
1,3-DICHLORO-9-FLUORENONE
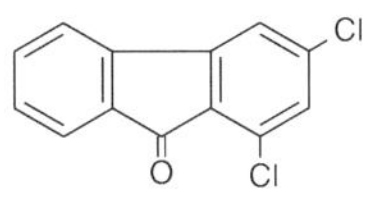
250 mg

S75,438-2
2,4-DICHLORO-9-FLUORENONE
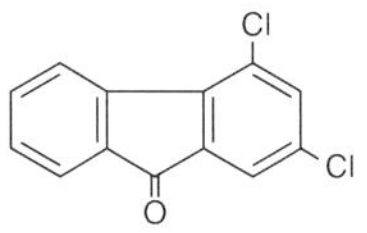
250 mg

S86,558-3
2,3-DIMETHYL-9(10H)-ANTHRACENONE
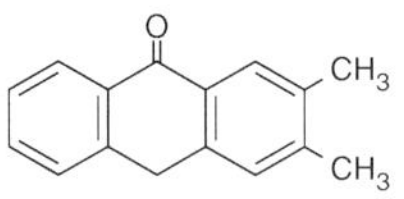
250 mg

S44,201-1
BIANTHRONYL
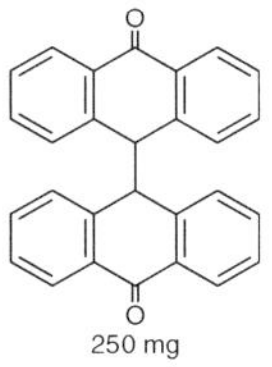
250 mg

S46,191-1
10,11-DIBROMO-5H-DIBENZO(A,D)-CYCLOHEPTEN-5-ONE
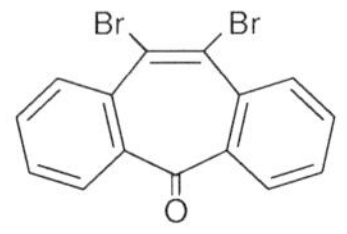
5 gm

S36,298-0
ISOVIOLANTHRONE
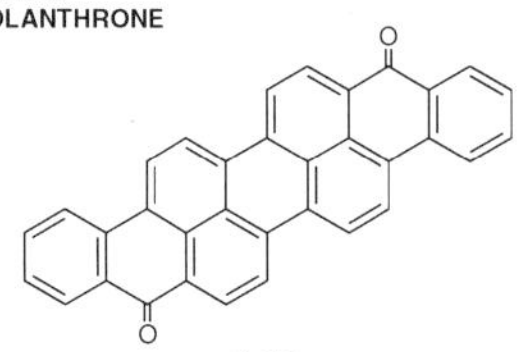
5 gm

S58,434-7
10-PIPERONYLIDENE-9-ANTHRONE
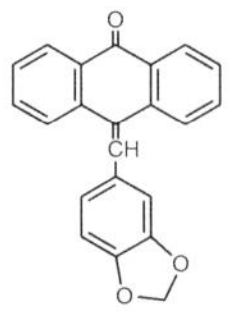
250 mg

S46,213-6
METHYLTHIOXANTHEN-9-ONE, MIXTURE OF 2-AND 4-ISOMERS
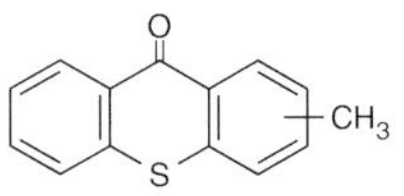
5 gm

S40,667-8
9-BENZYLIDENEANTHRONE
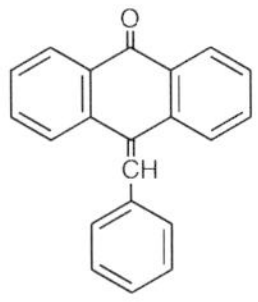
5 gm

S70,701-5
BIS(BIPHENYLYL) KETONE
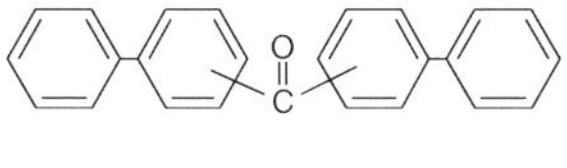
5 gm

S57,383-3
10-(4-METHYLBENZYLIDENE)ANTHRONE
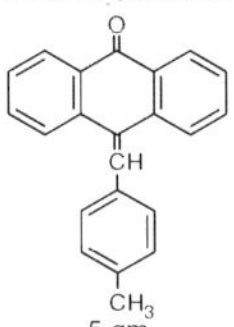
5 gm

S57,382-5
10-(3,4-DICHLOROBENZYLIDENE)-ANTHRONE
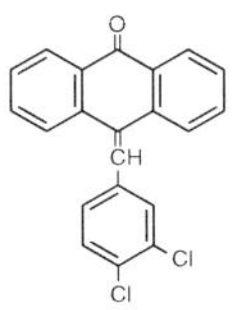
5 gm

S37,316-8
10-BENZHYDRYL-9-ANTHRONE
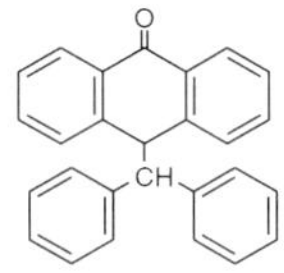
250 mg

S75,820-5
1,4,7-TRIFLUORO-9-FLUORENONE
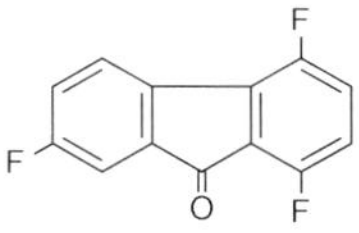
250 mg

S40,897-2
2-IODO-9-FLUORENONE
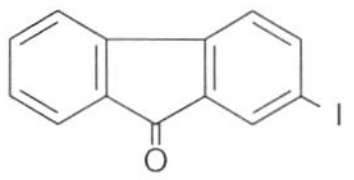
5 gm

S75,552-4
1-IODO-9-FLUORENONE
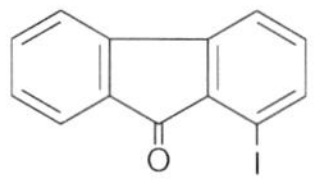
250 mg

S75,558-3
2,7-DICHLORO-4-IODO-9-FLUORENONE
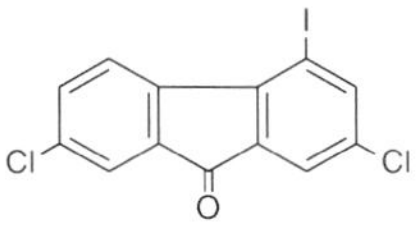
250 mg

S97,406-4
7,12-DIHYDRO-6,8-DIMETHYLBENZ(A)ANTHRACEN-7-ONE
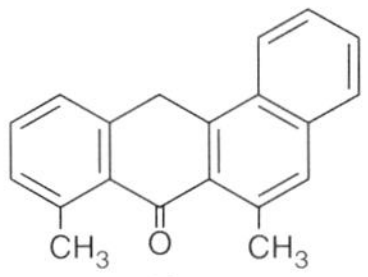
10 mg

S97,433-1
7,12-DIHYDRO-6,8,12-TRIMETHYLBENZ(A)ANTHRACEN-7-ONE
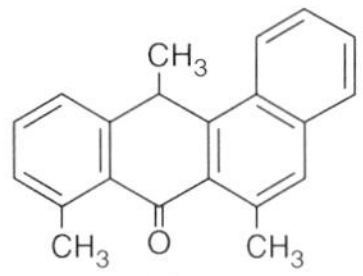
10 mg

S92,318-4
7H-DIBENZ(A,K)ANTHRACEN-7-ONE
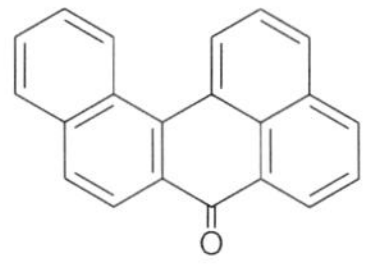
250 mg

S40,798-4
2,2'-DIMETHYLBIANTHRONE
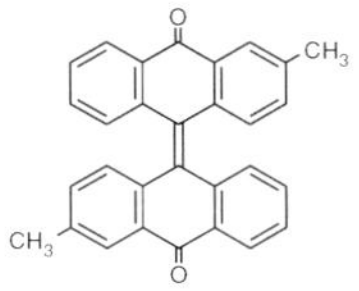
250 mg

S40,799-2
4,4'-DIMETHYLBIANTHRONE
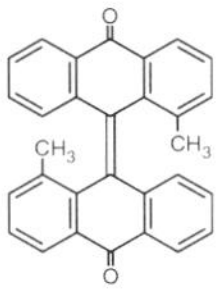
250 mg

S76,393-4
5,9,10,14B-TETRAHYDROBENZO(4,5)-CYCLOHEPT(1,2,3-DE)ANTHRACEN-5-ONE
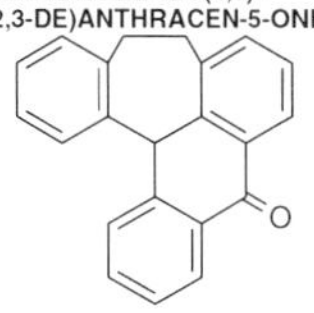
250 mg

S94,135-2
2-CHLORO-3-(DIBROMOMETHYL)-ANTHRAQUINONE
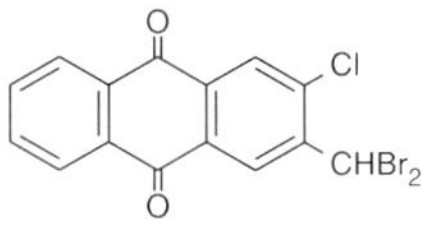
250 mg

S70,439-3
2,5-BIS[3-(TRIFLUOROMETHYL)ANILINO]-3,6-DICHLORO-1,4-BENZOQUINONE
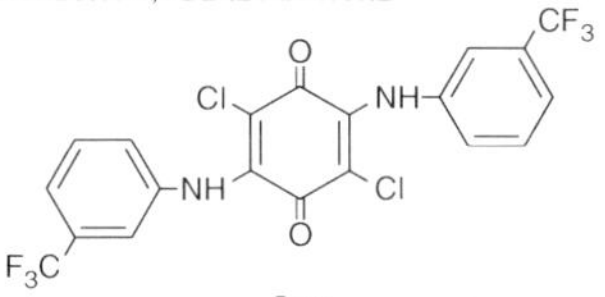
5 gm

S70,358-3
3-CHLORO-2-[3-(TRIFLUOROMETHYL)-ANILINO]-1,4-NAPHTHOQUINONE
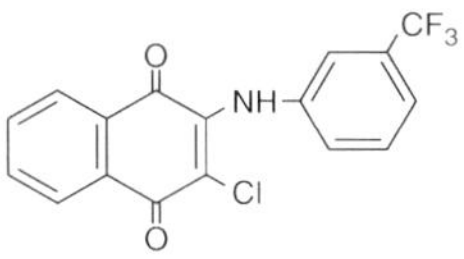
250 mg

S97,206-1
2-[10-(4-CHLOROPHENYLTHIO)-DECYL]-3-HYDROXY-1,4-NAPHTHOQUINONE
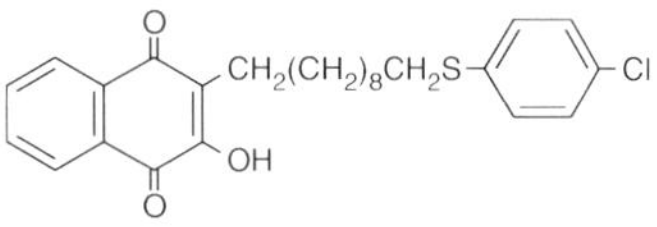
100 mg

S97,266-5
2-(4-BROMOPHENYL)-3-HYDROXY-1,4-NAPHTHOQUINONE
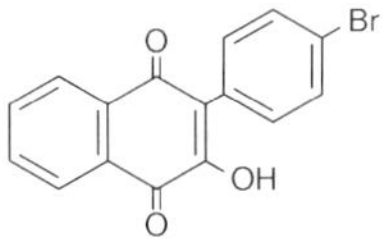
100 mg

S71,737-1
2,5-BIS(3-METHOXYANILINO)-3,6-DICHLORO-1,4-BENZOQUINONE
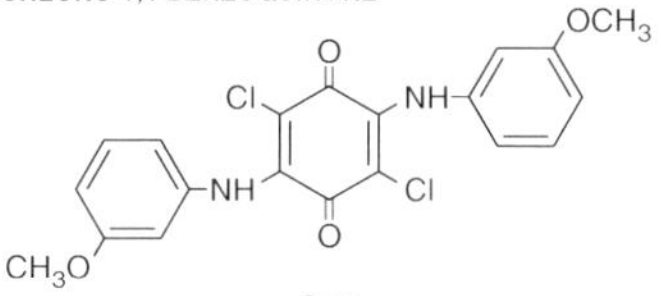
5 gm

S70,437-7
2,5-BIS(3-NITROANILINO)-3,6-DICHLORO-1,4-BENZOQUINONE
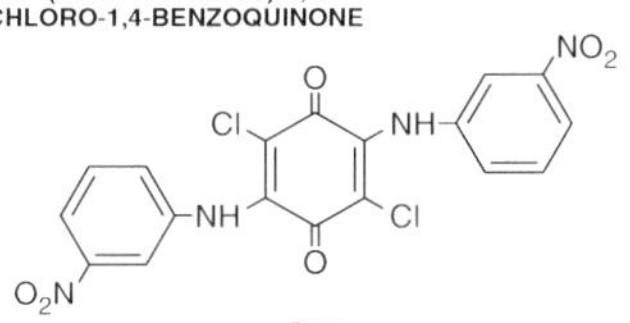
5 gm

S70,432-6
2,5-DIANILINO-3,6-DICHLORO-1,4-BENZOQUINONE
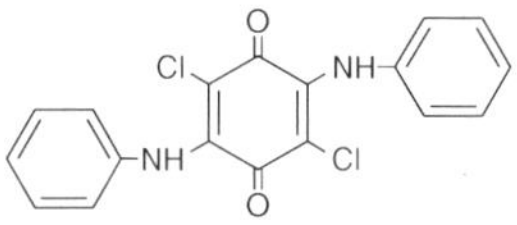
5 gm

S70,485-7
2,5-BIS(4-FLUOROANILINO)-3,6-DICHLORO-1,4-BENZOQUINONE
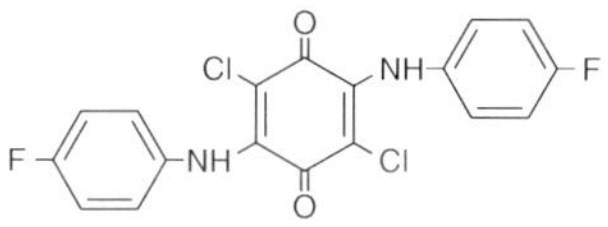
5 gm

S70,389-3
3-CHLORO-2-(4-CHLOROANILINO)-1,4-NAPHTHOQUINONE
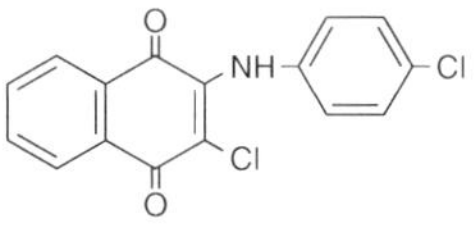
5 gm

S70,391-5
3-CHLORO-2-(3,4-DICHLOROANILINO)-1,4-NAPHTHOQUINONE
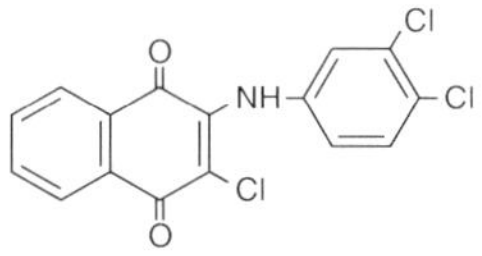
5 gm

S73,983-9
2-(4-BROMO-3-METHYLANILINO)-3-CHLORO-1,4-NAPHTHOQUINONE
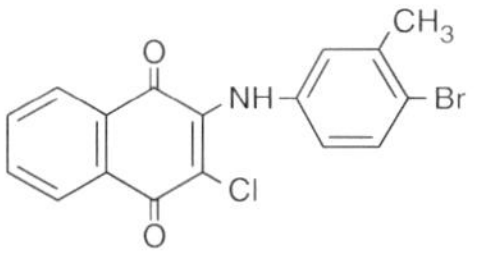
5 gm

S70,482-2
2-CHLORO-3-(4-FLUOROANILINO)-1,4-NAPHTHOQUINONE
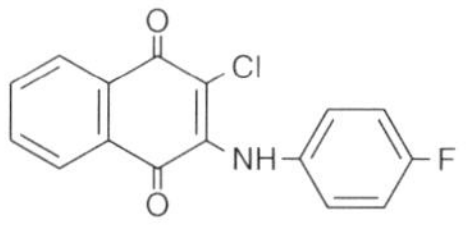
5 gm

S71,963-3
2,5-BIS(4-CHLOROANILINO)-3,6-DICHLORO-1,4-BENZOQUINONE
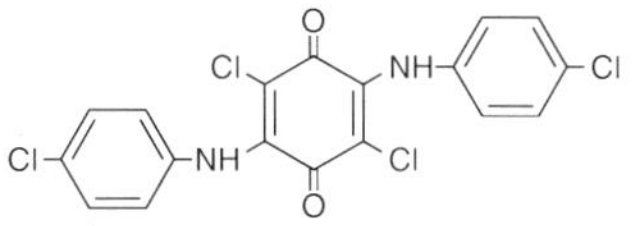
5 gm

S87,815-4
2-CHLORO-3-(4-IODOANILINO)-1,4-NAPHTHOQUINONE
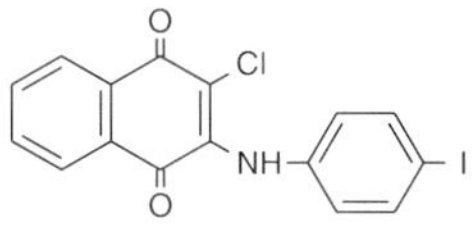
5 gm

S70,436-9
2-CHLORO-3-(2,4-DICHLOROANILINO)-1,4-NAPHTHOQUINONE
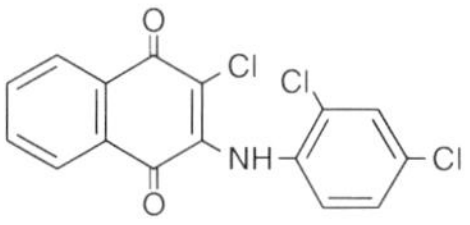
5 gm

S93,106-3
2-BROMO-1-(METHYLAMINO)-ANTHRAQUINONE
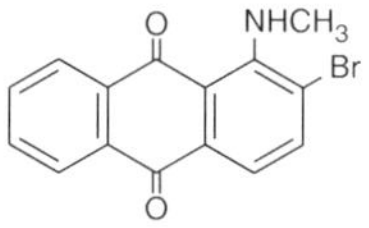
250 mg

S94,496-3
2,5-DICHLORO-1,4-BENZOQUINONE
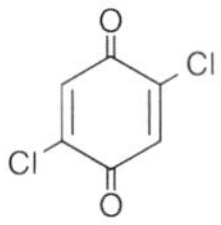
250 mg

S94,495-5
2,5-DIBROMO-1,4-BENZOQUINONE
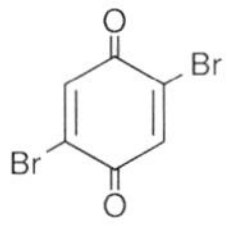
250 mg

S78,692-6
1,2,7-TRIHYDROXYANTHRAQUINONE
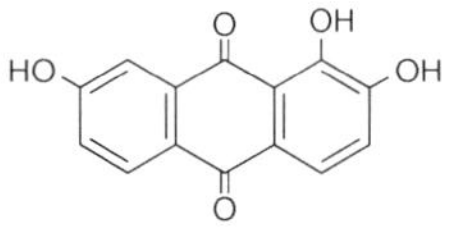
250 mg

S97,197-9
2-HYDROXY-3-(3-HYDROXY-3-METHYLBUTYL)-1,4-NAPHTHOQUINONE
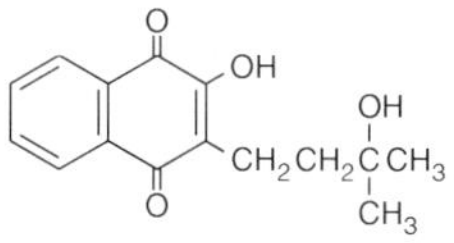
100 mg

S97,223-1
BROMO-2-HYDROXY-3-(3-HYDROXY-3-METHYLBUTYL)-1,4-NAPHTHOQUINONE
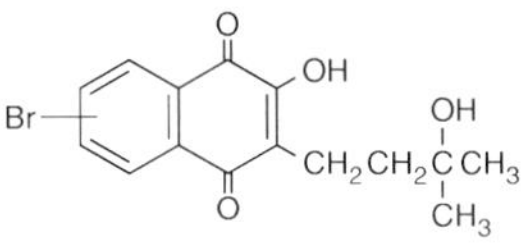
100 mg

S94,137-9
1-AMINO-4,5,8-TRIHYDROXYANTHRAQUINONE
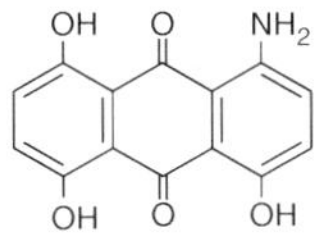
250 mg

S84,981-2
2,7-DIHYDROXYANTHRAQUINONE
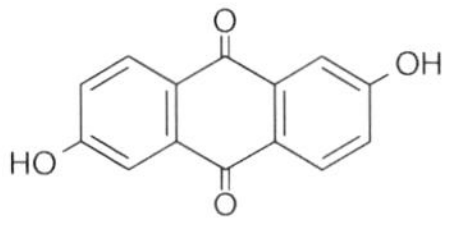
250 mg

S95,696-1
1,4-DIHYDRO-9-HYDROXY-8-METHOXY-10-METHYL-1,4-ANTHRACENEDIONE
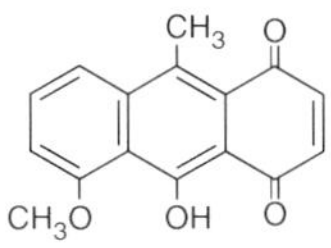
50 mg

S97,261-4
2-[2-(DECAHYDRONAPHTH-2-YL)-ETHYL]-3-HYDROXY-1,4-NAPHTHOQUINONE
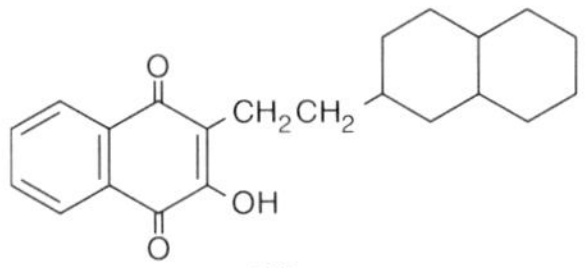
100 mg

S97,262-2
2-HYDROXY-3-(7-METHYLOCTYL)-1,4-NAPHTHOQUINONE
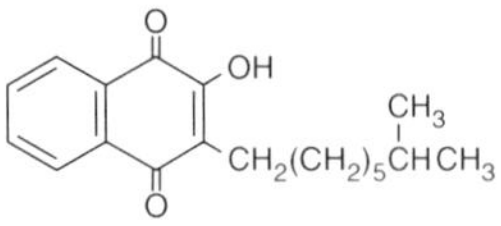
100 mg

S97,204-5
2-HYDROXY-3-[10-(4-METHYLPHENYLTHIO)DECYL]-1,4-NAPHTHOQUINONE
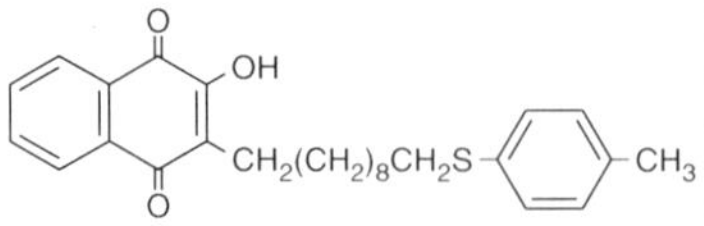
100 mg

S79,804-5
2-HYDROXY-3-[ALPHA-(TETRADECYLAMINO)BENZYL]-1,4-NAPHTHOQUINONE
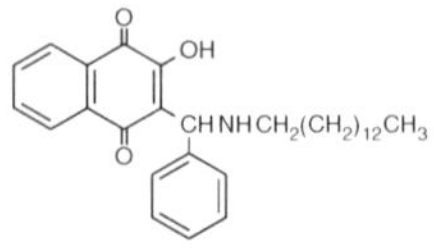
250 mg

S79,809-6
2-DECYLAMINOMETHYL-3-HYDROXY-1,4-NAPHTHOQUINONE
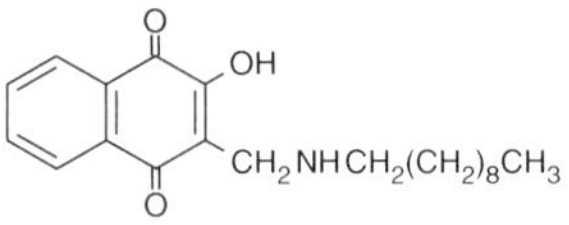
250 mg

S65,958-4
ALIZARIN ORANGE (C.I. 58015)
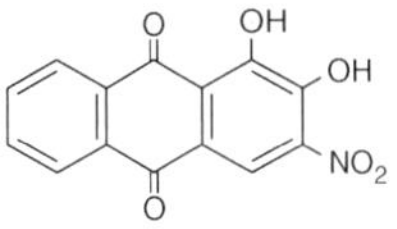
250 mg

S79,818-5
2-CHLORO-3-HYDROXY-1,4-NAPHTHOQUINONE POTASSIUM SALT
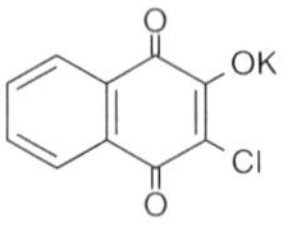
250 mg

S97,268-1
2-HYDROXY-3-[3-(4-TOLYL)-PROPYL]-1,4-NAPHTHOQUINONE
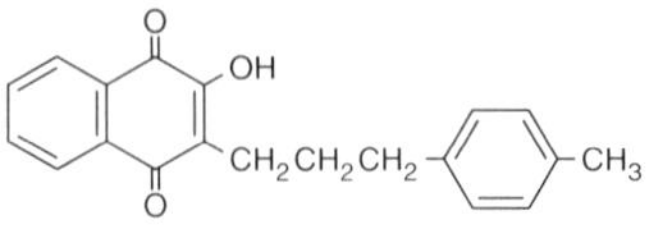
100 mg

S97,265-7
2-[3-(DECAHYDRONAPHTH-1-YL)-PROPYL]-3-HYDROXY-1,4-NAPHTHOQUINONE
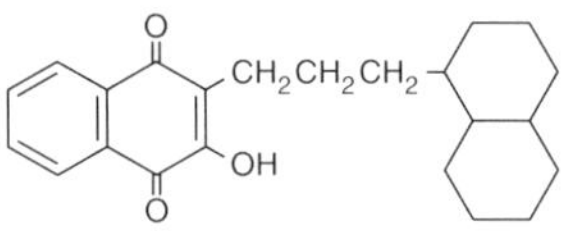
100 mg

S97,271-1
2-HYDROXY-3-(2-METHYLOCTYL)-1,4-NAPHTHOQUINONE SODIUM SALT
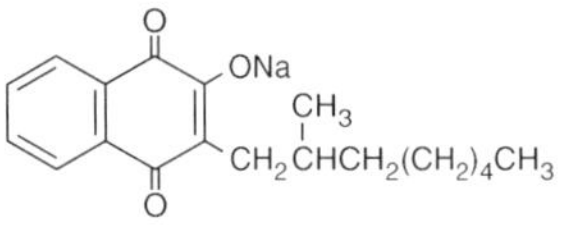
100 mg

S75,769-1
[1-HYDROXY-4-(PARA-TOLUIDINO)ANTHRAQUINONE]
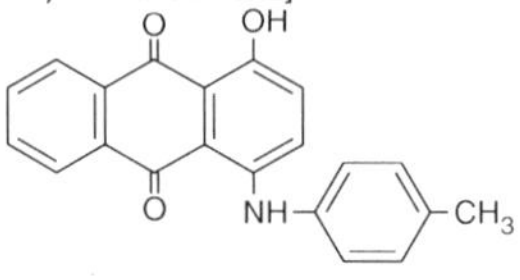
5 gm

S97,302-5
1-HYDROXYBENZO(5,6)-CYCLOHEPTA(1,2,3-DE)NAPHTHALENE-7,12-DIONE
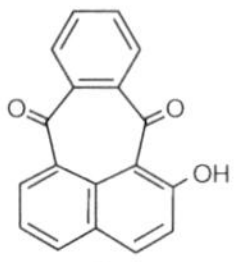
10 mg

S44,878-8
1,8-DIAMINO-4,5-DIHYDROXYANTHRAQUINONE
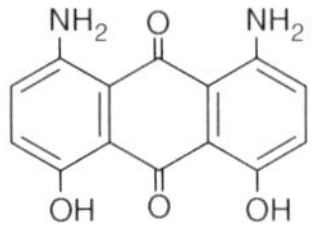
5 gm

S57,197-0
1-AMINO-2-BROMO-4-HYDROXYANTHRAQUINONE
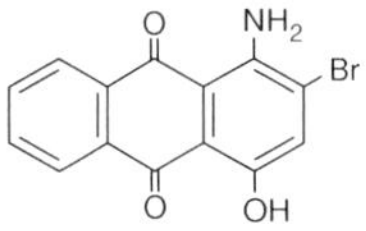
250 mg

S36,063-5
1,4-DIHYDROXY-5,8-DI-P-TOLUIDINOANTHRAQUINONE
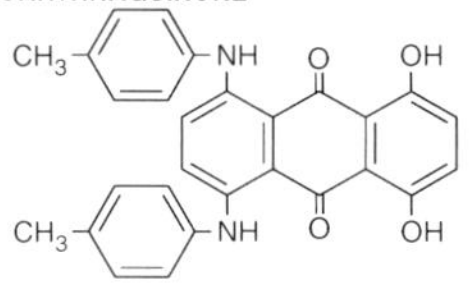
5 gm

S76,436-1
2-CHLORO-3-(2-HYDROXYETHYLAMINO)-1,4-NAPHTHOQUINONE
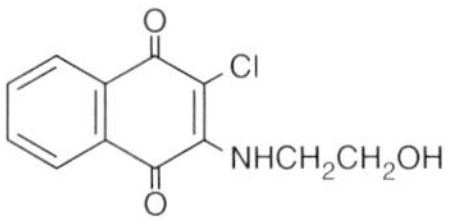
5 gm

S73,979-0
2-CHLORO-3-(2-HYDROXY-5-METHYLANILINO)-1,4-NAPHTHOQUINONE
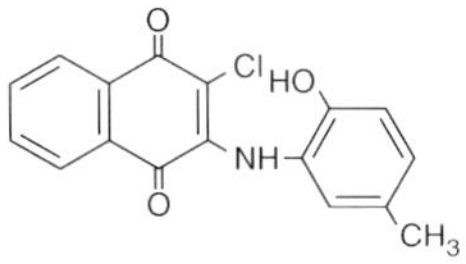
5 gm

S84,685-6
2-(4-HYDROXYPHENYL)-1,4-BENZOQUINONE
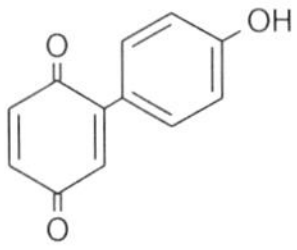
250 mg

S84,245-1
4-HYDROXY-2-PHENYL-2,3,5,6-TETRAMETHYL-3,5-CYCLOHEXADIEN-1-ONE
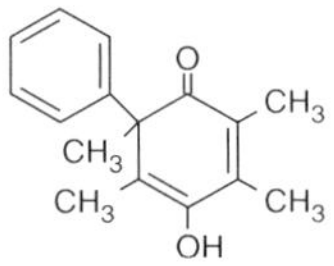
250 mg

S97,199-5
2-HYDROXYBENZ(A)ANTHRACENE-7,12-DIONE
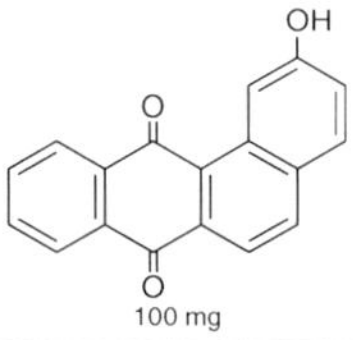
100 mg

S91,984-5
1,4-DIHYDROXY-2-(2-NITROPHENYLTHIO)-ANTHRAQUINONE
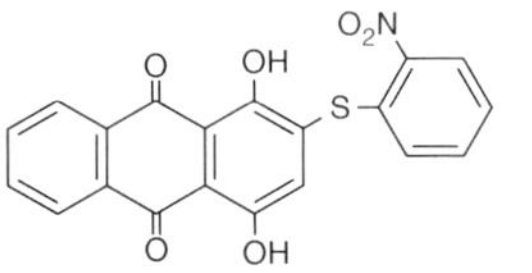
250 mg

S95,616-3
2-METHOXY-3-METHYL-1,4-NAPHTHOQUINONE
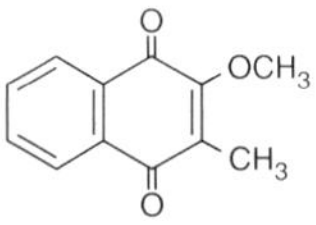
50 mg

S45,960-7
TILORONE ANALOG R 11,043DA
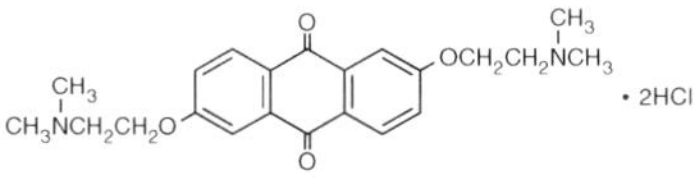
250 mg

S95,700-3
1,8-DIBENZYLOXYANTHRAQUINONE
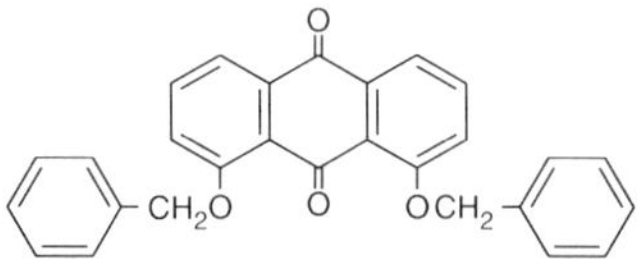
50 mg

S91,985-3
2,5-BIS(4-METHOXYPHENYLTHIO)-1,4-BENZOQUINONE
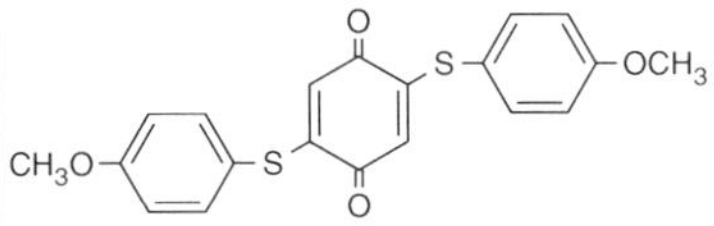
250 mg

S70,487-3
2,5-BIS(2-METHOXYANILINO)-3,6-DICHLORO-1,4-BENZOQUINONE
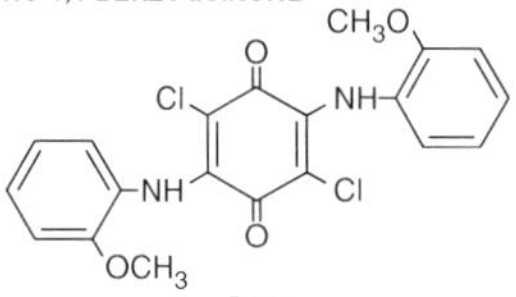
5 gm

S71,962-5
2,5-BIS(4-METHOXYANILINO)-3,6-DICHLORO-1,4-BENZOQUINONE
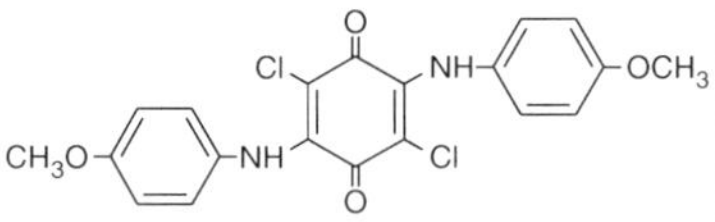
5 gm

S70,445-8
2,5-BIS(4-ETHOXYANILINO)-3,6-DICHLORO-1,4-BENZOQUINONE
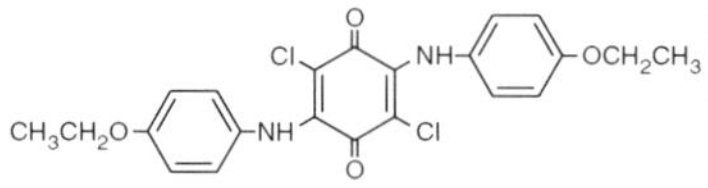
5 gm

S84,645-7
2-METHOXY-5-PHENYL-1,4-BENZOQUINONE
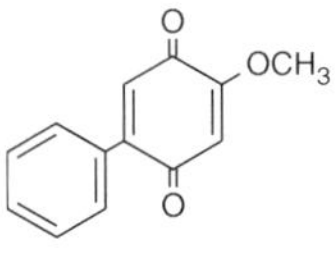
250 mg

S97,260-6
2-BROMO-3-ETHOXY-1,4-NAPHTHOQUINONE
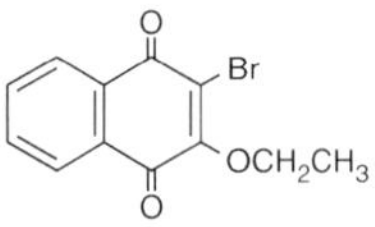
100 mg

S97,248-7
2-PROPOXY-1,4-NAPHTHOQUINONE
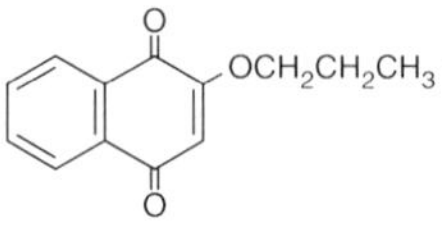
100 mg

S97,258-4
2-BUTOXY-1,4-NAPHTHOQUINONE
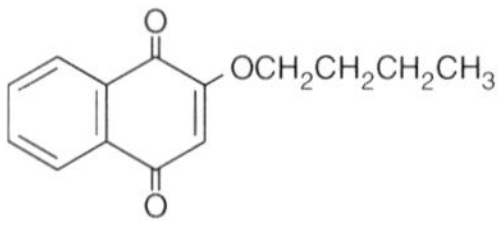
100 mg

S63,334-8
1,4-DIAMINO-2,3-BIS(2-PHENOXYETHOXY)ANTHRAQUINONE
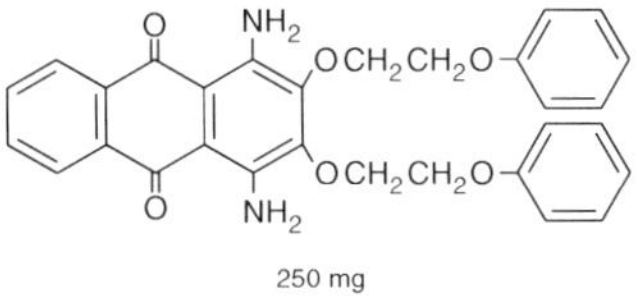
250 mg

S62,307-5
2-METHOXY-3-(4-NITROPHENYL)-1,4-NAPHTHOQUINONE
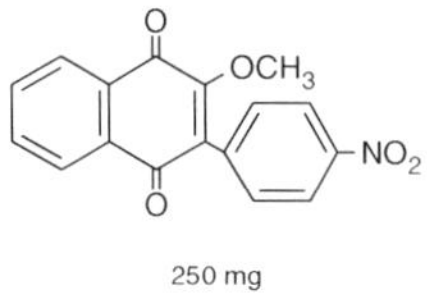
250 mg

S97,208-8
2-(2-BUTEN-1-YLOXY)-1,4-NAPHTHOQUINONE
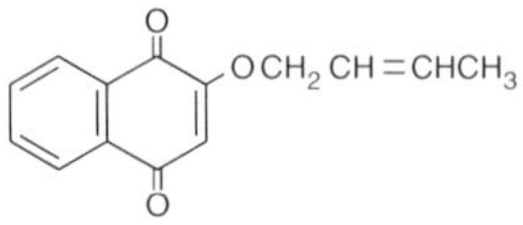
100 mg

S92,698-1
1-AMINO-4-HYDROXY-2-PHENOXYANTHRAQUINONE
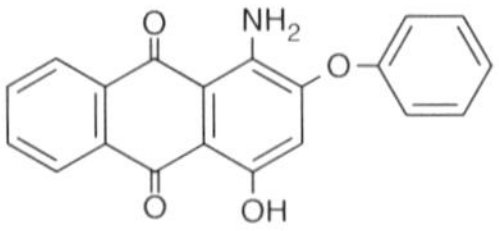
250 mg

S39,635-4
1-AMINO-4-METHOXYANTHRAQUINONE
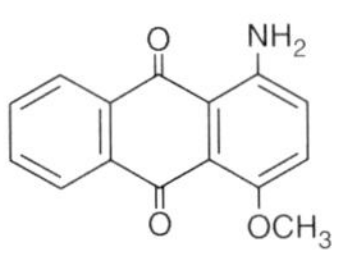
5 gm

S70,392-3
3-CHLORO-2-(2-METHOXYANILINO)-1,4-NAPHTHOQUINONE
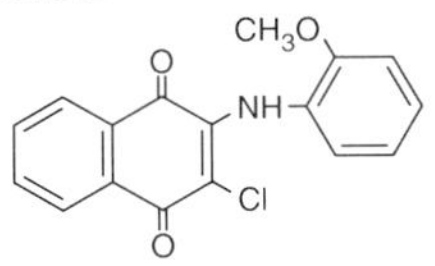
5 gm

S70,393-1
3-CHLORO-2-(4-METHOXYANILINO)-1,4-NAPHTHOQUINONE
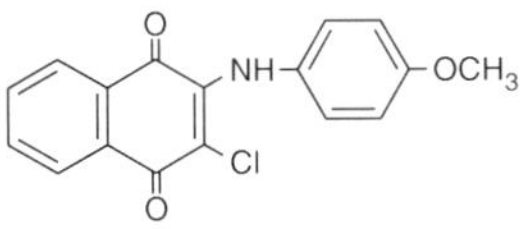
5 gm

S70,357-5
3-CHLORO-2-(4-ETHOXYANILINO)-1,4-NAPHTHOQUINONE
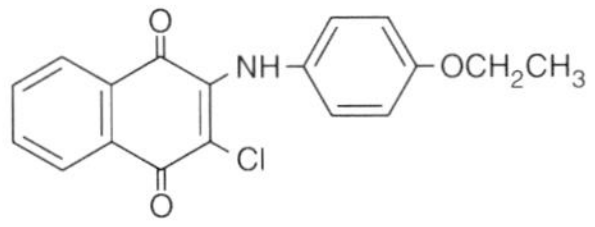
5 gm

S85,043-8
2-(4-METHOXYPHENYL)-1,4-BENZOQUINONE
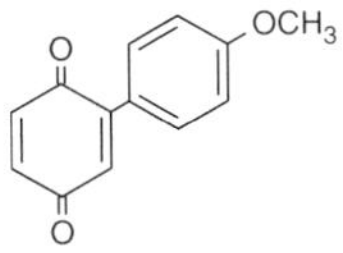
250 mg

S95,695-3
2-METHOXY-3-METHYLANTHRAQUINONE
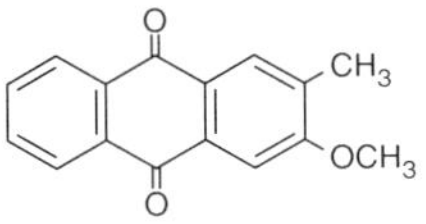
50 mg

S93,357-0
2,5-BIS(4-METHOXYPHENYL)-3,6-DIBROMO-1,4-BENZOQUINONE
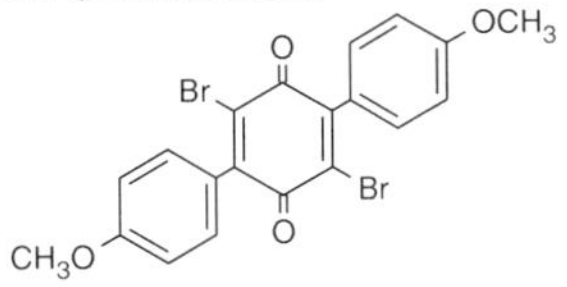
250 mg

S42,484-6
1-(3-METHOXYPHENOXY)ANTHRAQUINONE
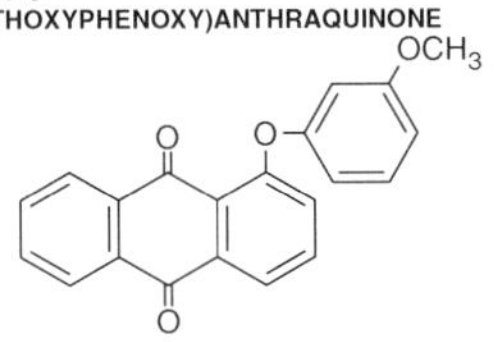
250 mg

S42,419-6
ANTHRARUFIN BIS(3-METHOXYPHENYL)ETHER
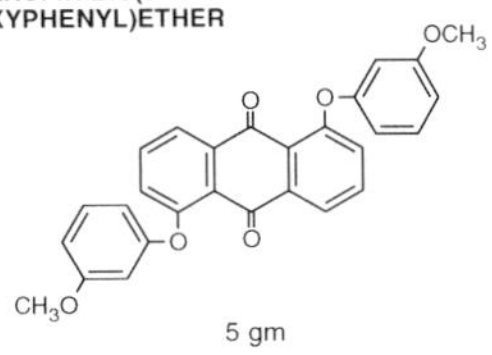
5 gm

S97,363-7
7,12-DIHYDRO-9-METHOXYBENZ(A)ANTHRACENE-7,12-DIONE
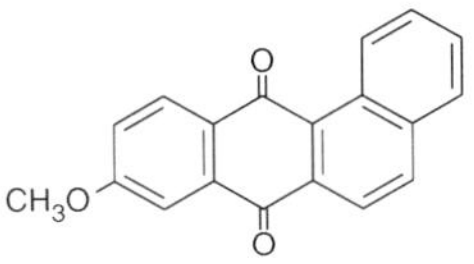
10 mg

S97,364-5
7,12-DIHYDRO-10-METHOXYBENZ(A)ANTHRACENE-7,12-DIONE
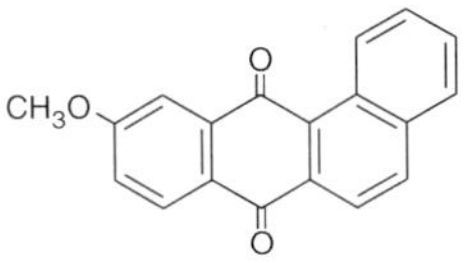
10 mg

S44,336-0
1-AMINO-2,4-DIBROMOANTHRAQUINONE
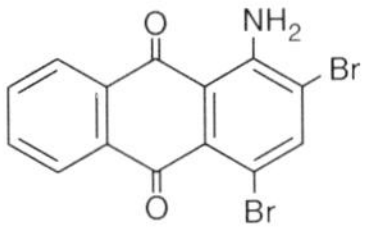
5 gm

S92,127-0
2-(2-NITROPHENYLTHIO)-1,4-BENZOQUINONE
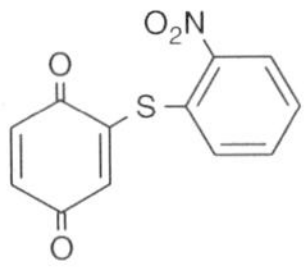
250 mg

S35,305-1
2-METHYL-3-(METHYLTHIO)-1,4-NAPHTHOQUINONE
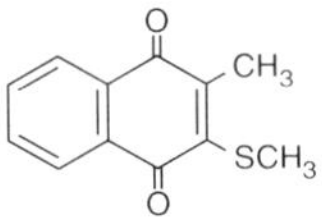
250 mg

S34,963-1
2-(ETHYLTHIO)-3-METHYL-1,4-NAPHTHOQUINONE
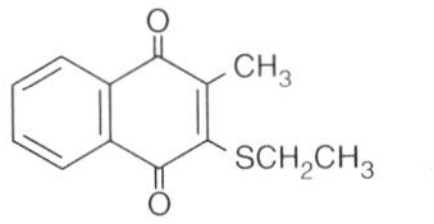
250 mg

S97,214-2
2-METHYL-3-(2-NAPHTHYLTHIO)-1,4-NAPHTHOQUINONE
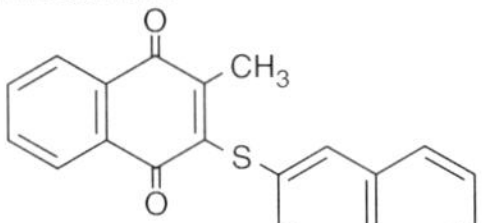
100 mg

S97,229-0
2-(3-METHYLPHENYLTHIO)-1,4-NAPHTHOQUINONE
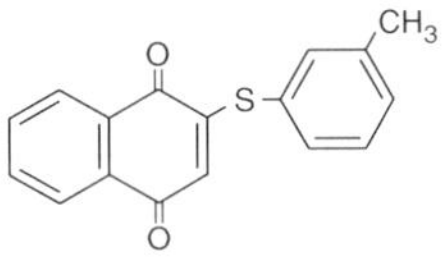
50 mg

S36,774-5
1,1'-IMINOBIS(4-AMINOANTHRAQUINONE)
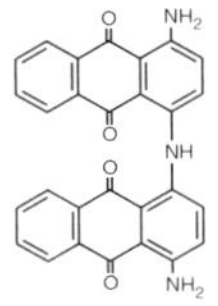
250 mg

S44,195-3
1,4-DIAMINO-5-NITROANTHRAQUINONE
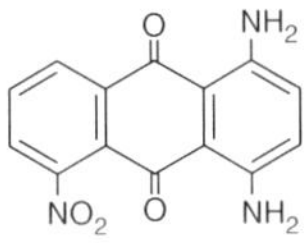
5 gm

S34,095-2
1-AMINO-5-CHLOROANTHRAQUINONE
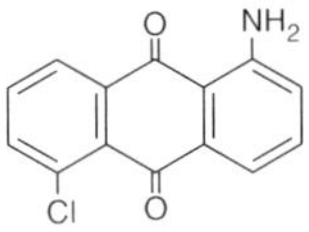
5 gm

S92,276-5
2-AMINO-1-CHLOROANTHRAQUINONE
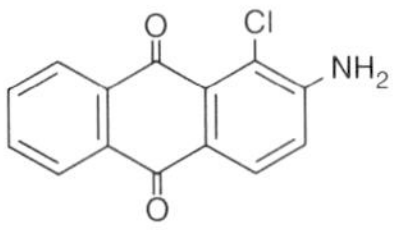
250 mg

S92,275-7
1-AMINO-4-BROMOANTHRAQUINONE
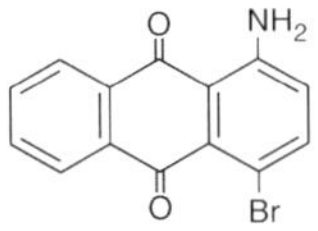
250 mg

S35,748-0
2-AMINO-1-BROMO-3-CHLOROANTHRAQUINONE
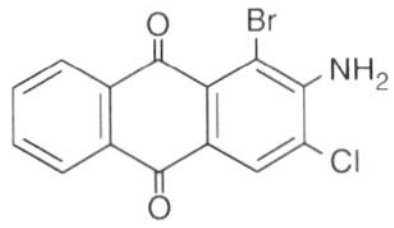
5 gm

S92,338-9
1,4,5-TRIAMINOANTHRAQUINONE
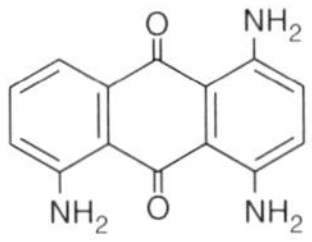
250 mg

S57,193-8
1-AMINO-6,7-DICHLOROANTHRAQUINONE
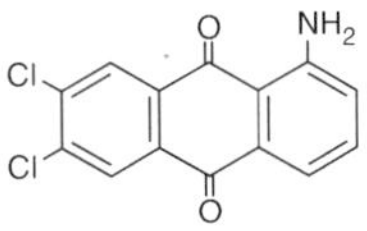
5 gm

S43,788-3
1-AMINO-4-CHLORO-2-METHYLANTHRAQUINONE
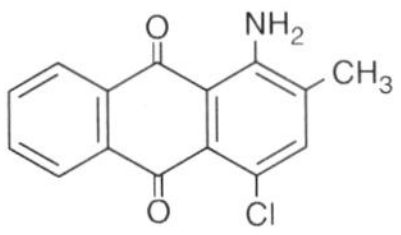
250 mg

S70,433-4
2-CHLORO-3-(3-NITROANILINO)-1,4-NAPHTHOQUINONE
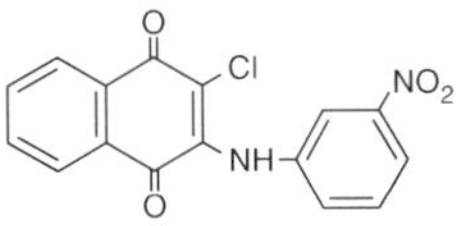
5 gm

S43,611-9
SUDAN GREEN
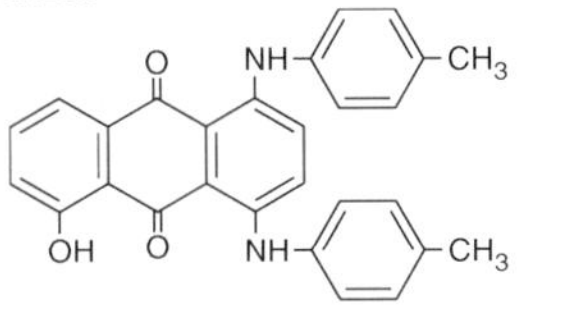
5 gm

S70,390-7
3-CHLORO-2-(4-NITROANILINO)-1,4-NAPHTHOQUINONE
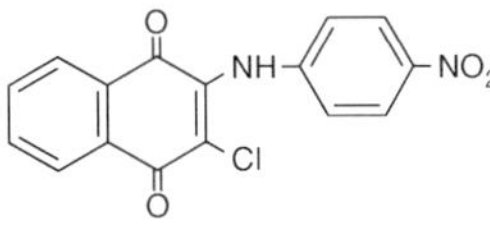
5 gm

S66,025-6
1,4-BIS(P-TOLYLAMINO)-6,7-DICHLOROANTHRAQUINONE
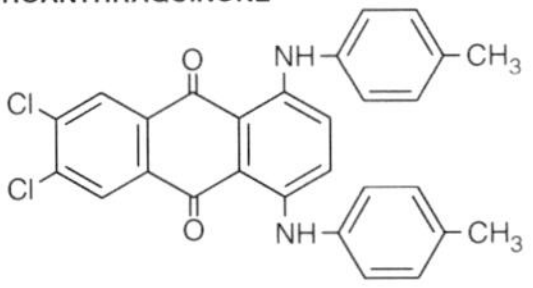
250 mg

S41,261-9
1,4-BIS(1-ANTHRAQUINONYLAMINO)-ANTHRAQUINONE
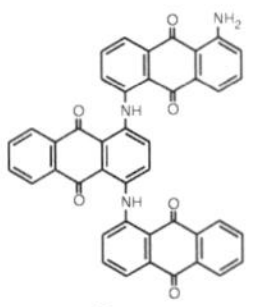
5 gm

S70,483-0
2-CHLORO-3-(PHENETHYLAMINO)-1,4-NAPHTHOQUINONE
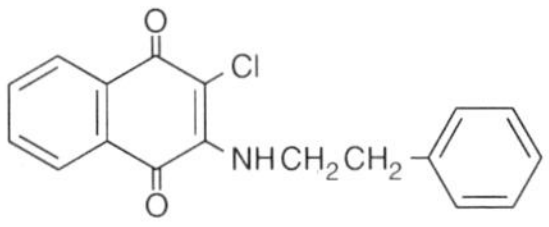
5 gm

S70,481-4
2-CHLORO-3-(4-ETHYLANILINO)-1,4-NAPHTHOQUINONE
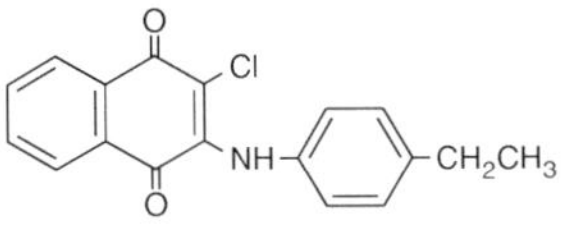
5 gm

S70,396-6
3-CHLORO-2-(3,4-DIMETHYLANILINO)-1,4-NAPHTHOQUINONE
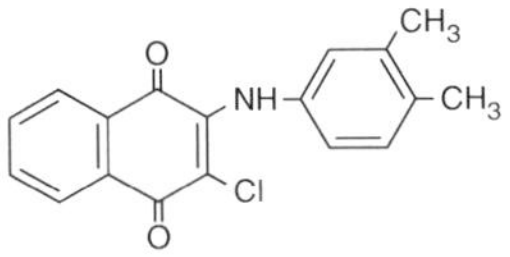
5 gm

S71,961-7
2,5-BIS(2-ETHYLANILINO)-3,6-DICHLORO-1,4-BENZOQUINONE
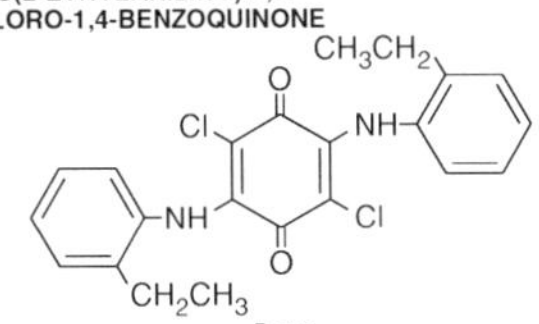
5 gm

S71,960-9
2,5-BIS(4-METHYLANILINO)-3,6-DICHLORO-1,4-BENZOQUINONE
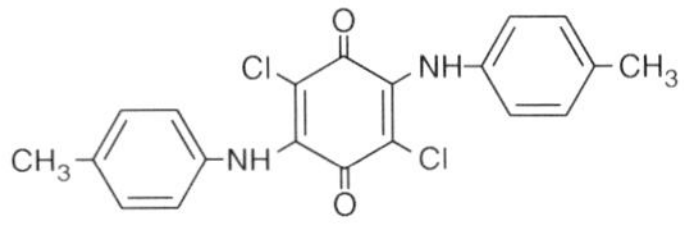
5 gm

S71,959-5
2,5-BIS(3-METHYLANILINO)-3,6-DICHLORO-1,4-BENZOQUINONE
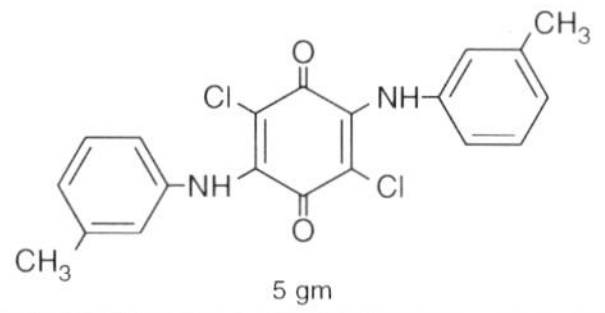
5 gm

S70,423-7
2,5-BIS(4-ETHYLANILINO)-3,6-DICHLORO-1,4-BENZOQUINONE
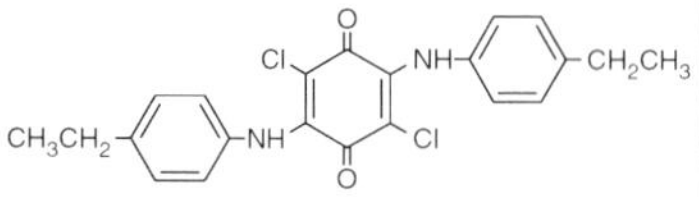
5 gm

S71,957-9
2,5-BIS(2-METHYLANILINO)-3,6-DICHLORO-1,4-BENZOQUINONE
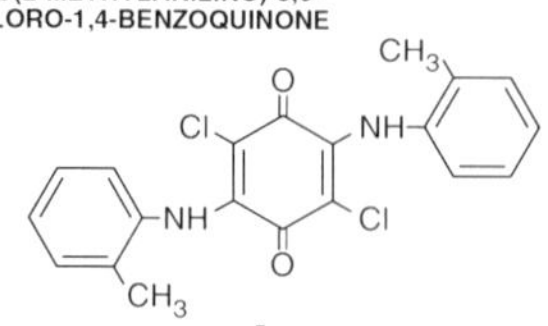
5 gm

S73,981-2
2-(3-ACETYLANILINO)-3-CHLORO-1,4-NAPHTHOQUINONE
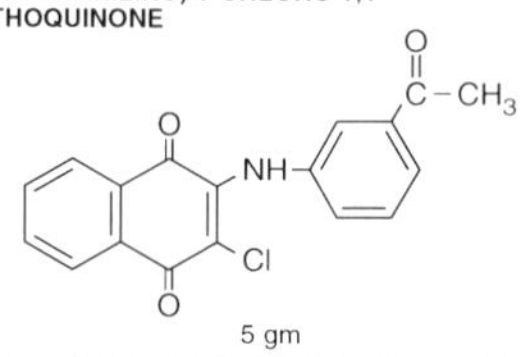
5 gm

S76,435-3
N,N'-DODECAMETHYLENEBIS(2-AMINO-3-CHLORO-1,4-NAPHTHOQUINONE)
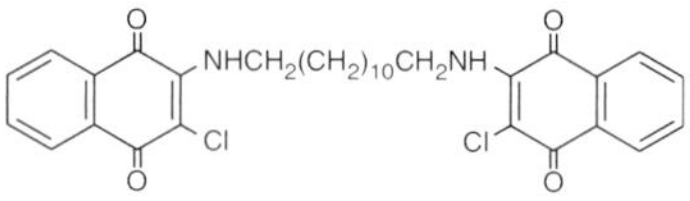
5 gm

S70,356-7
3-CHLORO-2-(2-METHYLANILINO)-1,4-NAPHTHOQUINONE
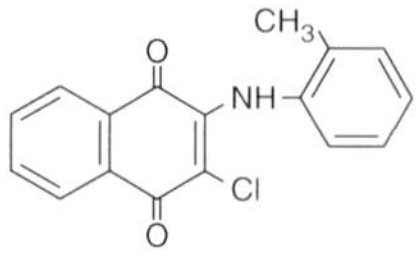
5 gm

S70,395-8
3-CHLORO-2-(2,4-DIMETHYLANILINO)-1,4-NAPHTHOQUINONE
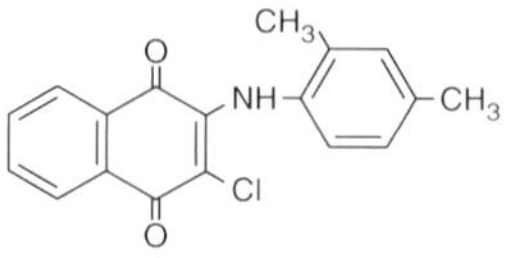
5 gm

S70,426-1
2-CHLORO-3-(2,4,6-TRIMETHYLANILINO)-1,4-NAPHTHOQUINONE
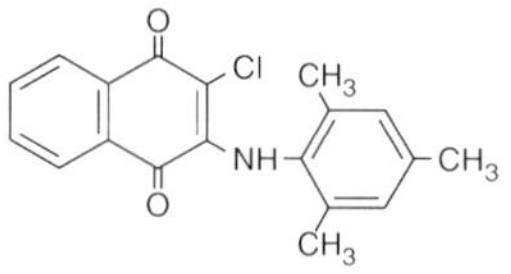
5 gm

S70,359-1
3-CHLORO-2-(3-METHYLANILINO)-1,4-NAPHTHOQUINONE
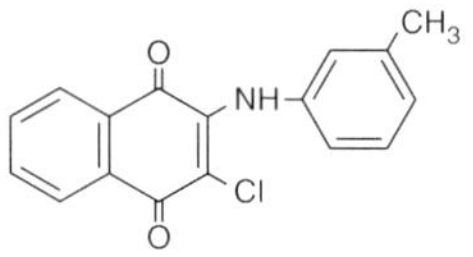
5 gm

S79,805-3
2-OCTADECYLAMINO-1,4-NAPHTHOQUINONE
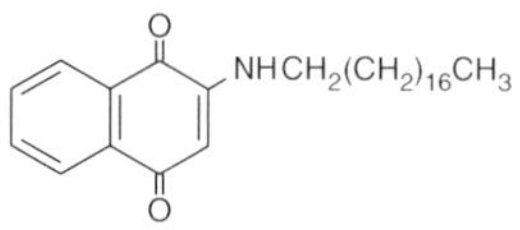
250 mg

S70,441-5
2-CHLORO-3-(CYCLOHEXYLAMINO)-1,4-NAPHTHOQUINONE
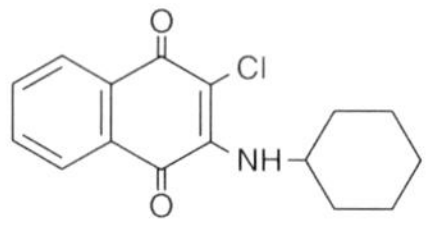
5 gm

S70,372-9
3-CHLORO-2-(4-METHYLANILINO)-1,4-NAPHTHOQUINONE
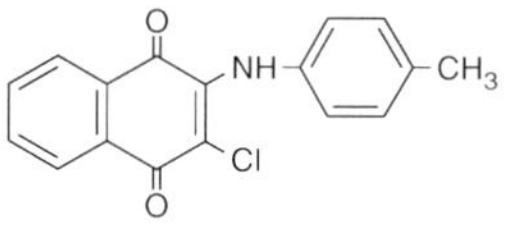
5 gm

S40,737-2
1,1'-DIANTHRIMIDE
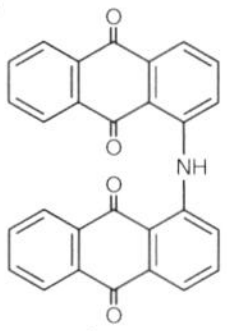
5 gm

S56,855-4
2-[4-(DIMETHYLAMINO)PHENYL]-1,3-INDANDIONE
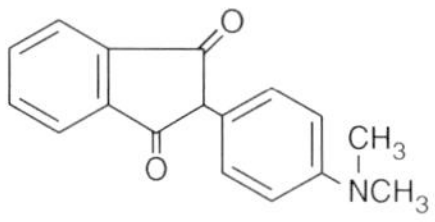
250 mg

S70,434-2
2-CHLORO-3-PIPERIDINO-1,4-NAPHTHOQUINONE
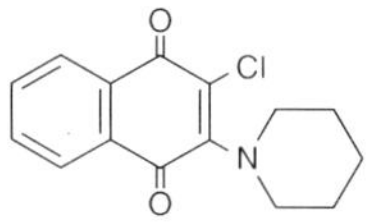
5 gm

S70,484-9
2,5-BIS(3,4-DIMETHYLANILINO)-3,6-DICHLORO-1,4-BENZOQUINONE
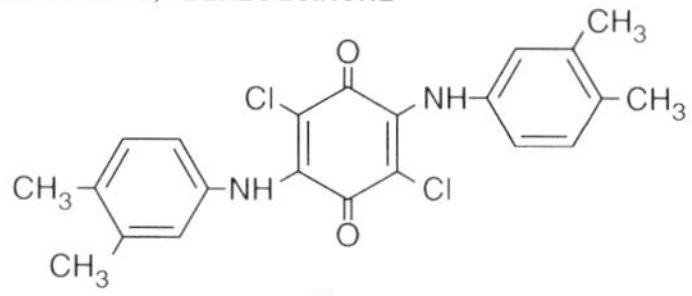
5 gm

S45,704-3
2-NITRO-1,3-INDANEDIONE
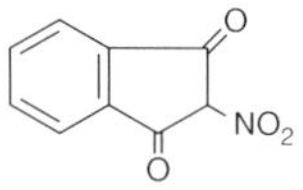
250 mg

S36,365-0
2-METHYL-1-NITROANTHRAQUINONE
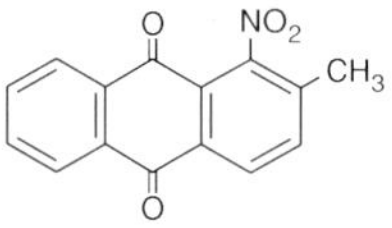
5 gm

S72,240-5
1-NITROANTHRAQUINONE
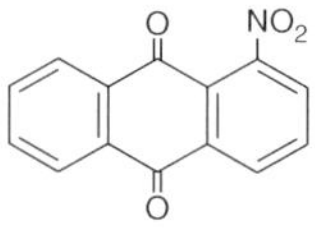
5 gm

S54,777-8
2,6-DICHLORO-3,5-DIMETHYL-1,4-BENZOQUINONE
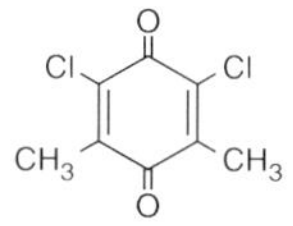
250 mg

S57,280-2
2-ACETONYL-2-PHENYL-1,3-INDANDIONE
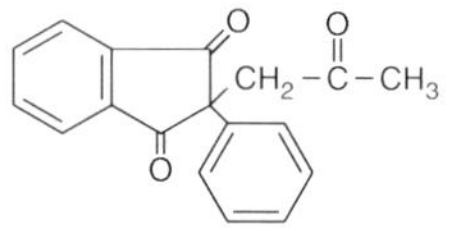
5 gm

S86,364-5
2,3-DIMETHYL-1,4,4A,9A-TETRAHYDROANTHRAQUINONE
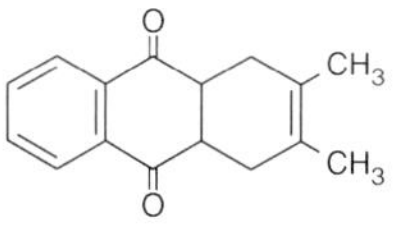
250 mg

S85,987-7
2,2'-BIS(1,4-NAPHTHOQUINONE)
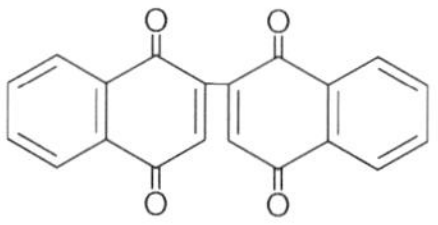
250 mg

S85,035-7
2-BROMO-6-PHENYL-1,4-BENZOQUINONE
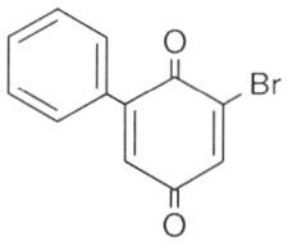
250 mg

S84,239-7
2,3-DIPHENYL-1,4-NAPHTHOQUINONE
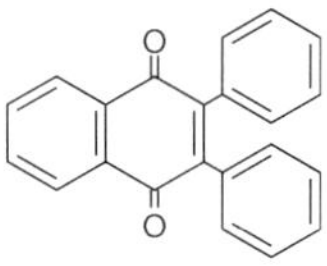
250 mg

S76,092-7
2-(BENZOYLMETHYL)-2-PHENYL-1,3-INDANDIONE
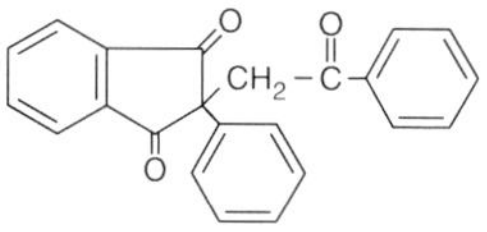
250 mg

S63,902-8
(P-TOLYL)-P-BENZOQUINONE
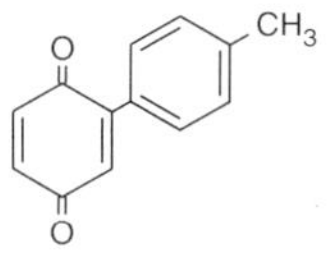
250 mg

S77,353-0
2-CHLORO-3-(2,6-DIMETHYL-3,4,5,6-TETRAHYDRO-2H-1,3,4-OXADIAZIN-4-YL)-1,4-NAPHTHOQUINONE
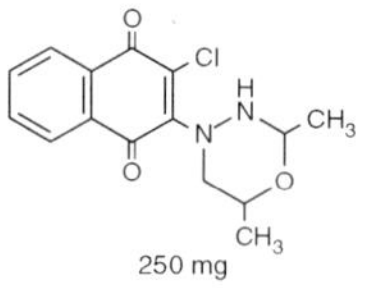
250 mg

S84,243-5
2,7-DI-TERT-BUTYL-1,4-NAPHTHOQUINONE
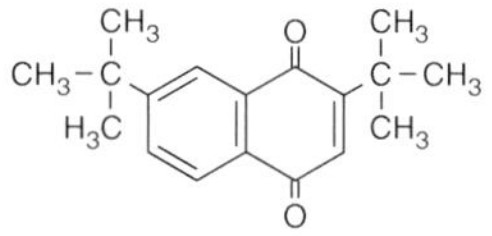
250 mg

S84,244-3
2,6-DI-TERT-BUTYL-1,4-NAPHTHOQUINONE
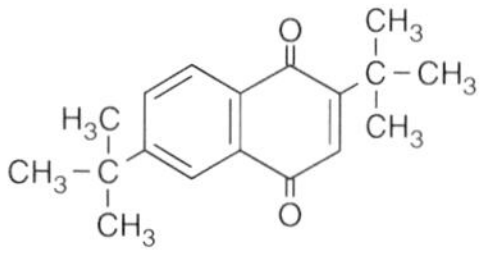
250 mg

S86,358-0
2,3-DICHLOROANTHRAQUINONE
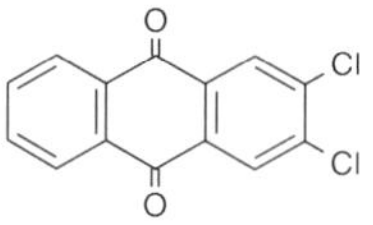
250 mg

S56,203-3
6-CHLORONAPHTHACENEQUINONE
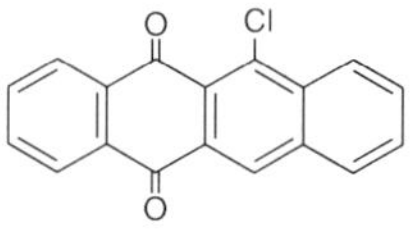
250 mg

S97,348-3
2-PHENYL-9,10-PHENANTHRENEQUINONE
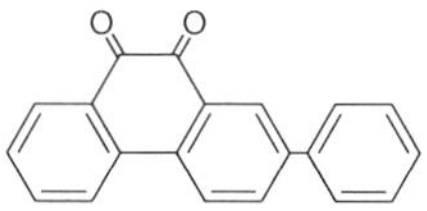
10 mg

S97,333-5
4,5-DIMETHYL-9,10-PHENANTHRENEQUINONE
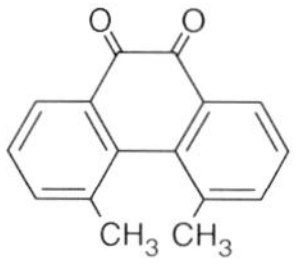
10 mg

S97,337-8
2,7-DIMETHYL-9,10-PHENANTHRENEQUINONE
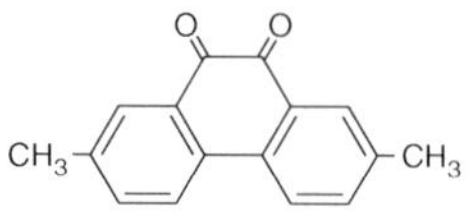
10 mg

S97,303-3
1-CHLOROBENZO(5,6)-CYCLOHEPTA(1,2,3-DE)NAPHTHALENE-7,12-DIONE
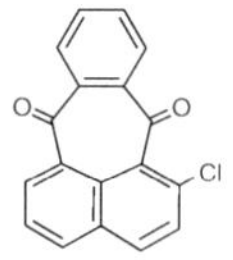
10 mg

S97,230-4
2-BROMO-3-CYCLOHEXYL-1,4-NAPHTHOQUINONE

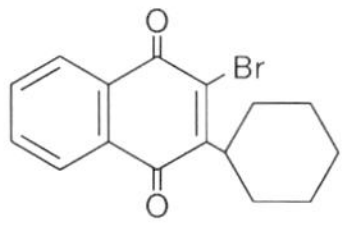

100 mg

S75,875-2
2-CINNAMYLIDENE-1,3-INDANDIONE

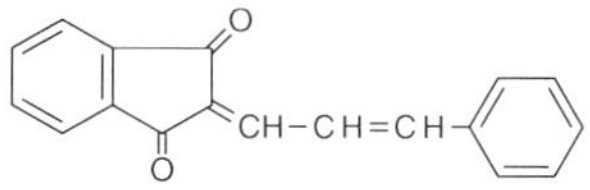

250 mg

S97,213-4
2,3-DICHLORO-2,3-DIHYDRO-1,4-NAPHTHOQUINONE

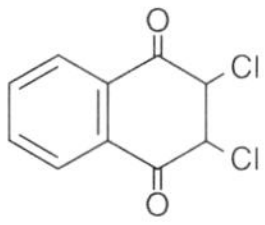

100 mg

S97,351-3
2-CHLORO-7,12-DIHYDROBENZ(A)-ANTHRACENE-7,12-DIONE

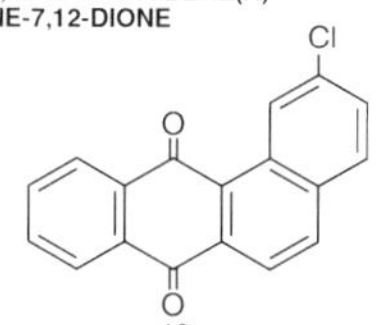

10 mg

S72,238-3
1,4-DIHYDROANTHRAQUINONE

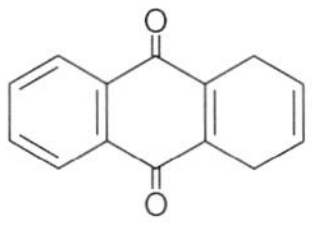

5 gm

S69,699-4
1,4,9,10-ANTHRACENETETRAONE

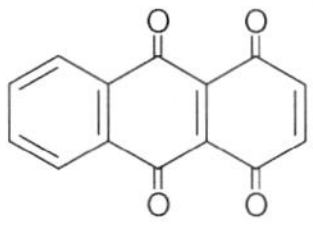

250 mg

S94,136-0
1,3-DIMETHYLANTHRAQUINONE

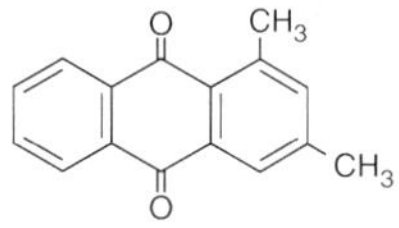

10 mg

S97,443-9
11,12-DIHYDROCHRYSENE-11,12-DIONE

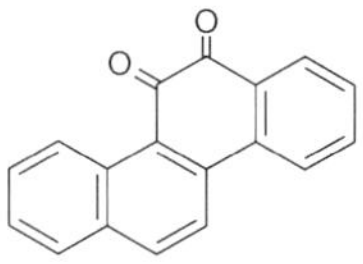

10 mg

S55,082-5
2-(4-BROMOPHENYL)-1,3-INDANDIONE

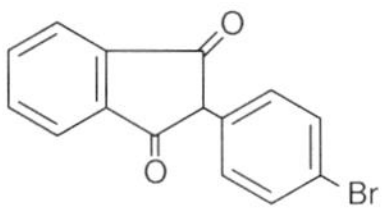

5 gm

S93,584-0
7,12-DIHYDRO-10-METHYLBENZ(A)-ANTHRACENE-7,12-DIONE

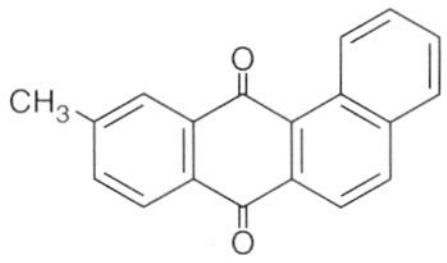

10 mg

S73,321-0
4-(4-CHLOROPHENYL)-2-(2-HYDROXYBUTYL)PHTHALAZIN-1(2H)-ONE

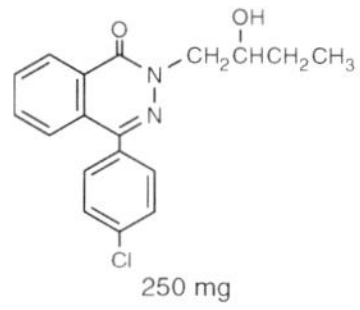

250 mg

S88,925-3
QUERCITRIN

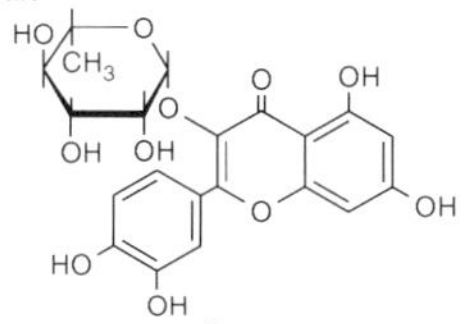

5 gm

S44,218-6
TECHTOCHRYSIN

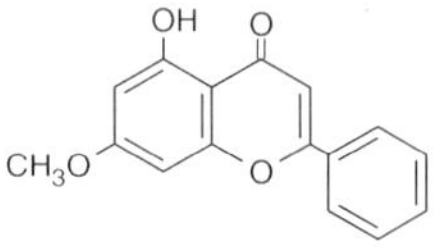

250 mg

S40,447-0
7-HYDROXY-3',4',5'-TRIMETHOXYFLAVONE

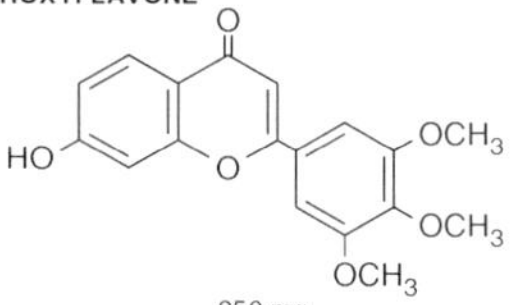

250 mg

S42,508-7
7-HYDROXY-3',4',5'-TRIMETHOXY-ALPHA-NAPHTHOFLAVONE

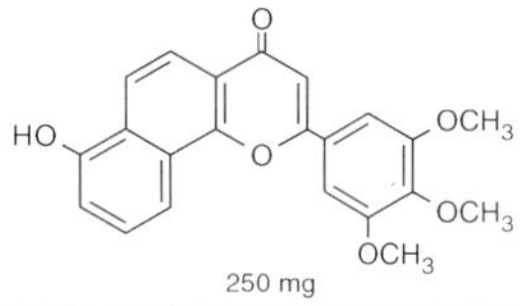

250 mg

S50,084-4
9-(2,4-DICHLOROPHENOXYMETHYL)-6-HYDROXY-3H-XANTHEN-3-ONE

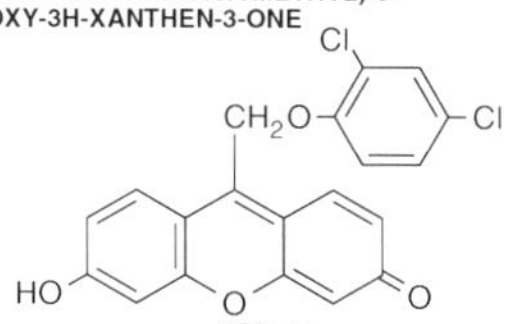

250 mg

S80,936-5
2,3-DIPHENYL-7-HYDROXY-4H-1-BENZOPYRAN-4-ONE

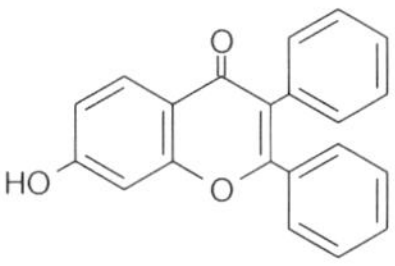

250 mg

S97,242-8
8-BROMO-3,4-DIHYDRO-2,2-DIMETHYL-3-HYDROXYNAPHTHO(1,2-B)-2H-PYRAN-5,6-DIONE

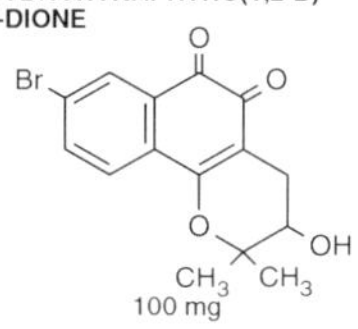

100 mg

S90,685-9
3-ACETYL-5-HYDROXY-2-METHYLCHROMONE

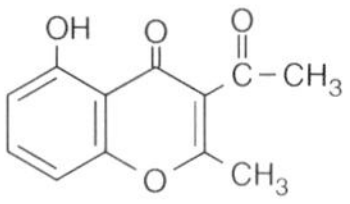

250 mg

S42,520-6
5-HYDROXY-7-METHYLFLAVONE

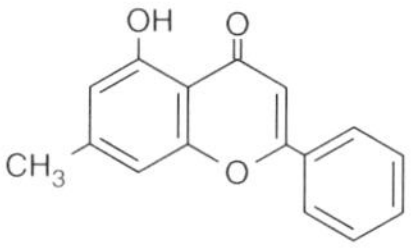

250 mg

S43,945-2
(-)-USNIC ACID

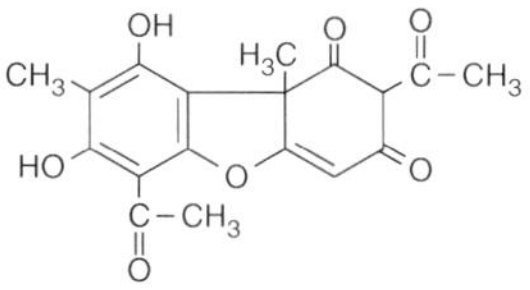

5 gm

S50,077-1
6-HYDROXY-9-(2-HYDROXYPHENYL)-3H-XANTHEN-3-ONE

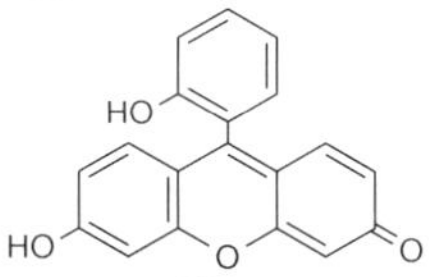

250 mg

S50,083-6
6-HYDROXY-9-METHYL-3H-XANTHEN-3-ONE

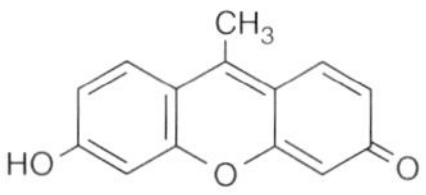

250 mg

S97,236-3
6-(3-HYDROXY-3-METHYLBUTYL)-METHYL-BENZO(A)PHENAZIN-5(7H)-ONE

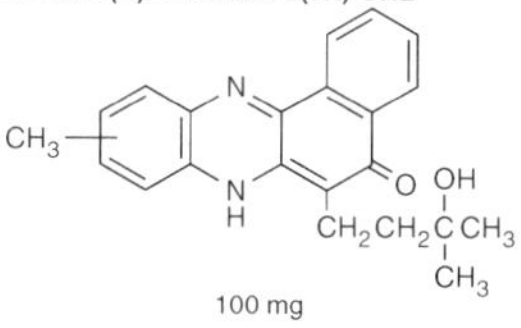

100 mg

S65,847-2
2-(4-BENZYLOXYPHENYL)CHROMONE

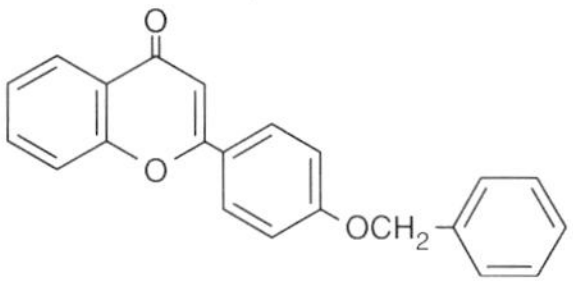

5 gm

S64,184-7
2',5'-DIMETHOXYFLAVONE

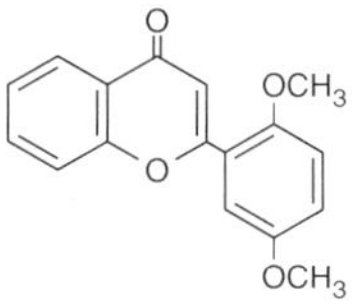

5 gm

S88,834-6
2,2-DIMETHYL-5,7,8-TRIMETHOXY-4-CHROMANONE

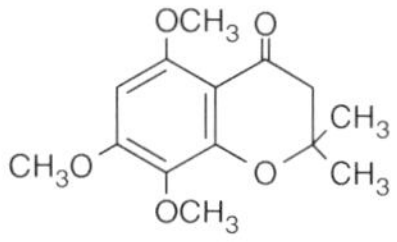

250 mg

S80,941-1
3-BENZOYL-2-(3,4,5-TRIMETHOXYPHENYL)-4H-1-BENZOPYRAN-4-ONE

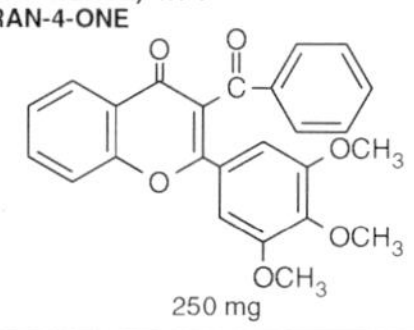

250 mg

S95,840-9
BROMO-METHOXY-METHYL-TETRAHYDRO-IMINOETHANO-PHENANTHRO(4,5-BCD)-FURAN-5-ONE HYDROBROMIDE

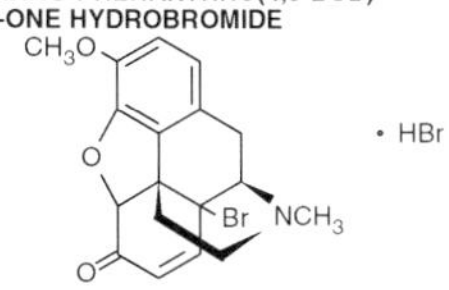

50 mg

S44,919-9
CRYPTOPINE

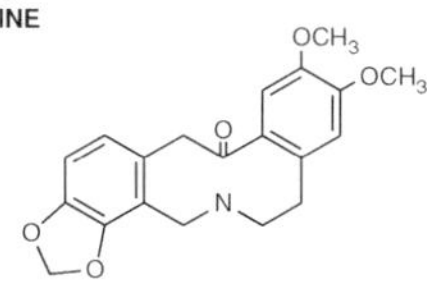

250 mg

S42,657-1
3',4',5'-TRIMETHOXY-ALPHA-NAPHTHOFLAVONE

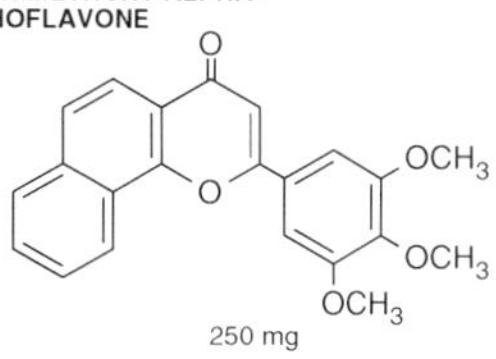

250 mg

S42,659-8
3',4',5'-TRIMETHOXY-BETA-NAPHTHOFLAVONE

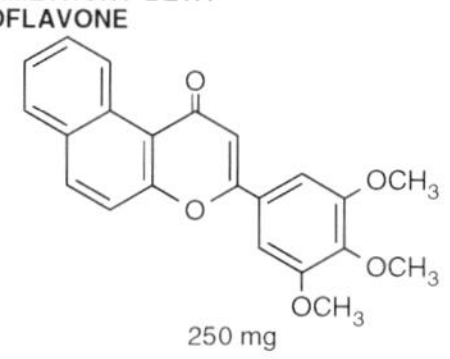

250 mg

S80,937-3
3-BENZOYL-2-(2-METHOXYPHENYL)-4H-1-BENZOPYRAN-4-ONE

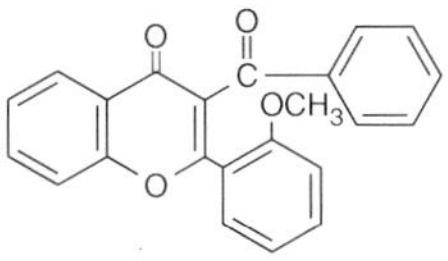

250 mg

S80,938-1
3-(4-METHOXYBENZOYL)-2-PHENYL-4H-1-BENZOPYRAN-4-ONE

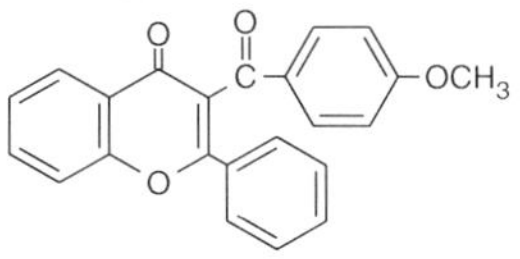

250 mg

S58,038-4
2-METHOXY-9(10H)-ACRIDONE

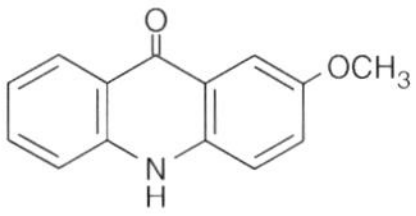

5 gm

S41,614-2
5,8-DIMETHOXY-2-METHYLCHROMONE

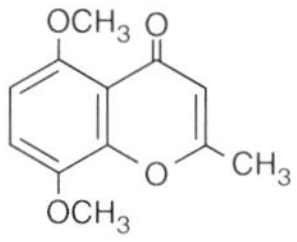

250 mg

S66,805-2
3',4'5'-TRIMETHOXYFLAVONE

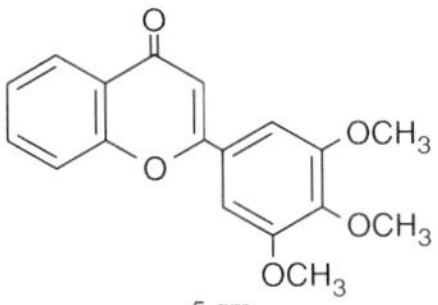

5 gm

S62,630-9
2'-CHLOROFLAVONE

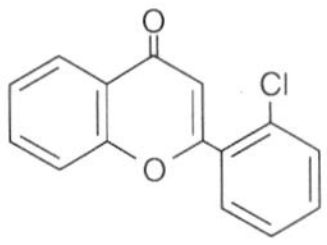

250 mg

S97,244-4
2-ETHYLNAPHTHO(1,2-B)-FURAN-4,5-DIONE

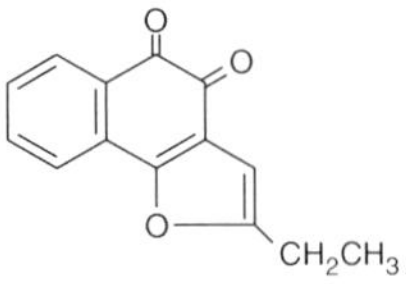

100 mg

S97,218-5
2-ISOPROPYLNAPHTHO(1,2-B)-FURAN-4,5-DIONE

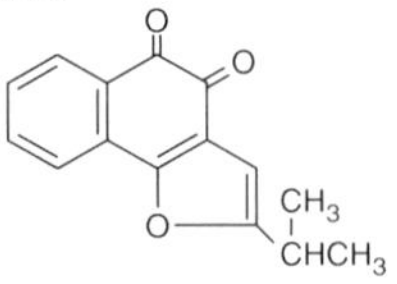

100 mg

S97,231-2
3,8-DIBROMO-3,4-DIHYDRO-2,2-DIMETHYLNAPHTHO(1,2-B)-2H-PYRAN-5,6-DIONE

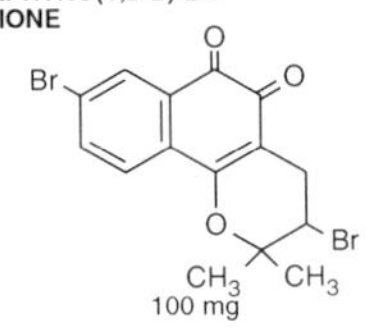

100 mg

S84,317-2
3,7-DIBENZOYL-2,8-DIPHENYL-4H,6H-BENZO(1,2-B:5,4-B')-DIPYRAN-4,6-DIONE

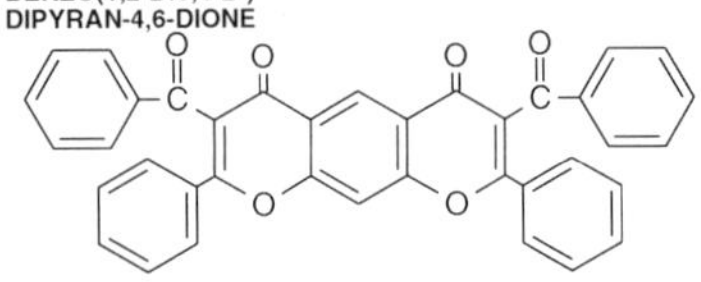

250 mg

S51,559-0
2-METHYL-6-NITRO-4(1H)-QUINOLONE

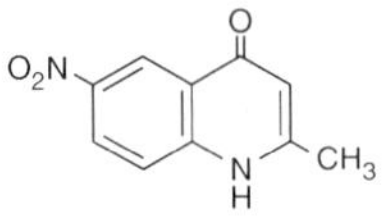

250 mg

S37,386-9
2,6-DIMETHYL-1-PHENYL-4(1H)-PYRIDONE HYDROCHLORIDE

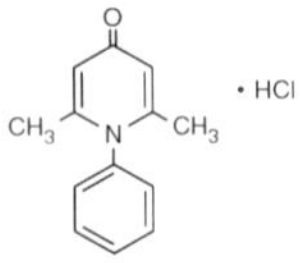

5 gm

S40,440-3
ECHINOPSINE HYDROCHLORIDE

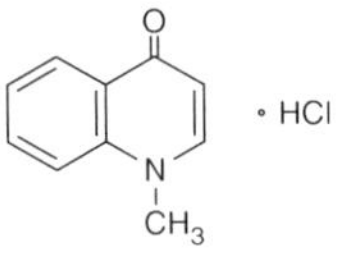

250 mg

S65,482-5
4'-ISOPROPYLISOFLAVONE

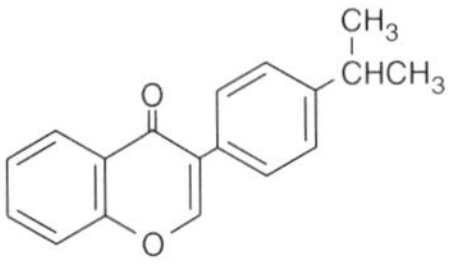

250 mg

S51,769-0
9-(5-INDANYL)-2,3,6,7-TETRAHYDRO-2,2,7,7-TETRAMETHYL-9H-XANTHENE-4,5(1H,8H)-DIONE

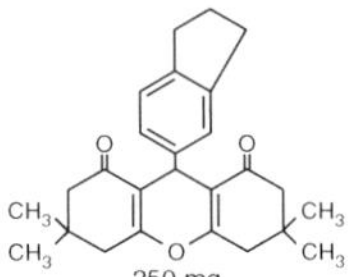

250 mg

S44,439-1
6-METHYL-4-CHROMANONE

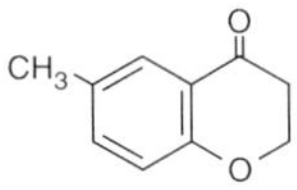

250 mg

S39,261-8
2,6-DIMETHYL-2-CHROMEN-4-ONE

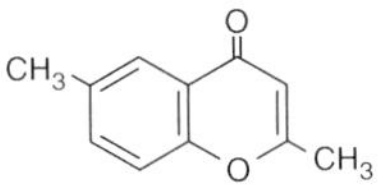

250 mg

S66,186-4
2-BROMO-9-XANTHENONE

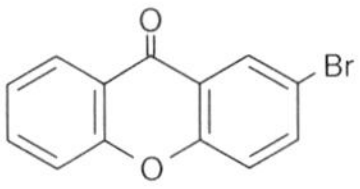

5 gm

S65,120-6
6-METHYLBENZOTHIOPYRAN-4(4H)-ONE

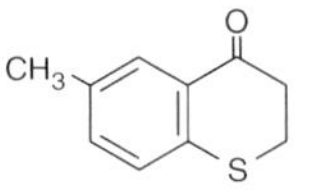

250 mg

S65,191-5
8-METHYLBENZOTHIOPYRAN-4(4H)-ONE

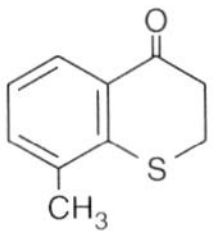

250 mg

S79,873-8
1,2-DIPHENYL-4(1H)-QUINOLINONE

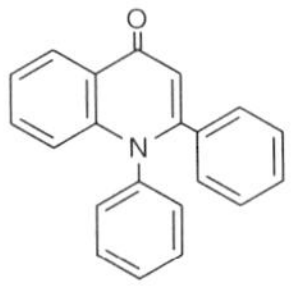

250 mg

S76,126-5
5H-INDENO(1,2-B)QUINOLIN-10(11H)-ONE

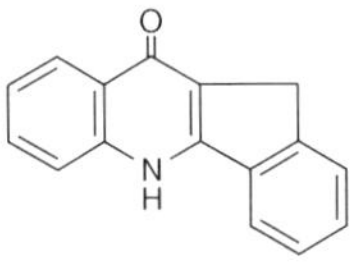

250 mg

S77,917-2
5,6-DIHYDRO-5,5-DIMETHYLBENZ(C)-ACRIDIN-7-(12H)-ONE

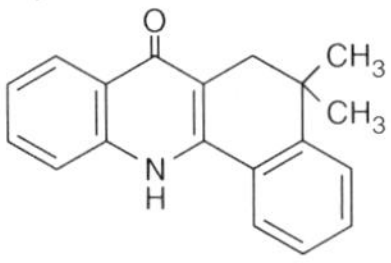

250 mg

S76,187-7
2,3-DIHYDRO-1-BENZOTHIEPIN-5(4H)-ONE

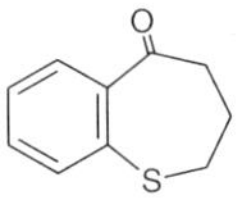

250 mg

S46,218-7
6-CHLORO-4-CHROMANONE

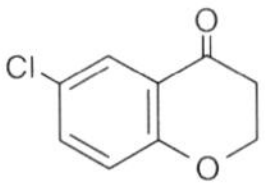

250 mg

S52,633-9
2-METHYLCHROMONE

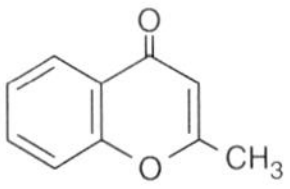

250 mg

S69,923-3
10-CHLORO-10-PHENYL-9(10H)-ANTHRACENONE

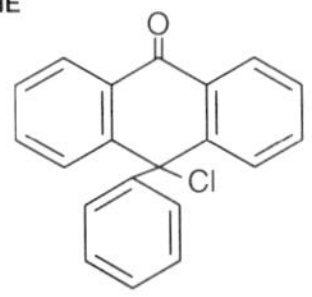

250 mg

S97,256-8
6-(CHLOROMETHYL)BENZO(A)-PHENAZIN-5(7H)-ONE

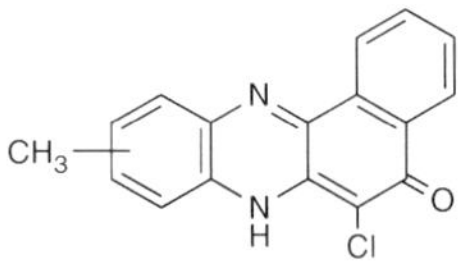

100 mg

S76,122-2
3-CHLORO-6,7,8,9,10,11-HEXAHYDROCYCLOOCTA(B)QUINOLIN-12(5H)-ONE

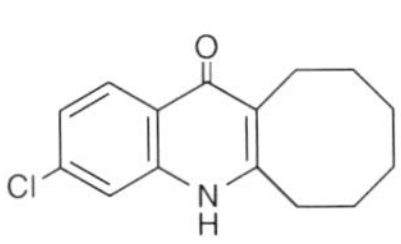

250 mg

S76,045-5
11-CHLORO-6,7-DIHYDRO-5H-BENZO(6,7)-CYCLOHEPTA(1,2-B)QUINOLIN-8(13H)-ONE

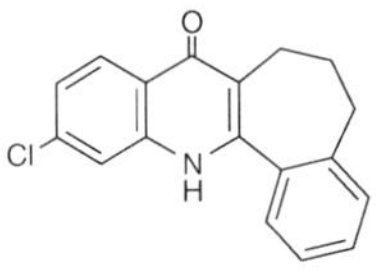

250 mg

S83,526-9
2-(2,4-DICHLOROPHENOXY)ACETALDEHYDE
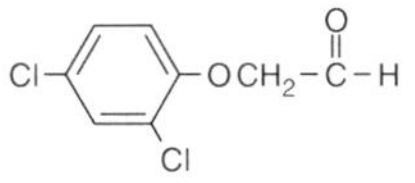
250 mg

S76,604-6
3-(4-CHLOROPHENYL)-2,2-DIMETHYLPROPIONALDEHYDE
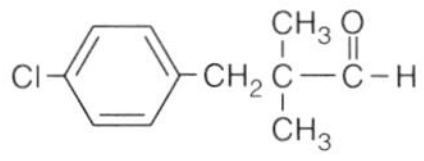
250 mg

S94,996-5
4-CHLOROCINNAMALDEHYDE
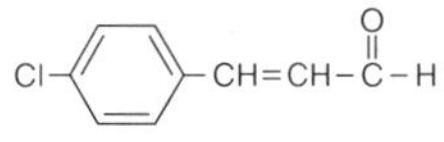
1 gm

S73,269-9
3-(DIMETHYLAMINO)-2-(4-METHOXYPHENYL)ACROLEIN
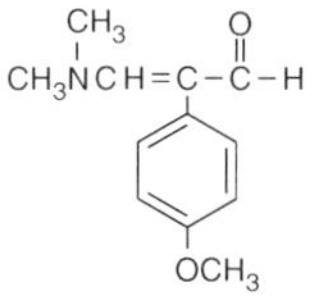
250 mg

S94,787-3
5,5,8,8-TETRAMETHYL-5,6,7,8-TETRAHYDRO-2-NAPHTHOXYACETALDEHYDE
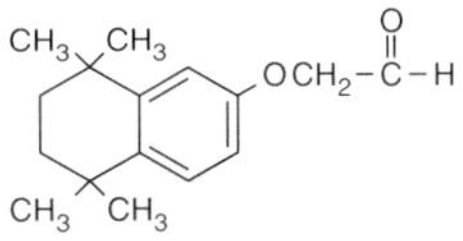
250 mg

S79,930-0
3-(PHENYLTHIO)BUTYRALDEHYDE
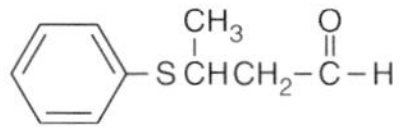
250 mg

S84,345-8
2-CHLORO-3-(3-NITROPHENYL)-PROPIONALDEHYDE
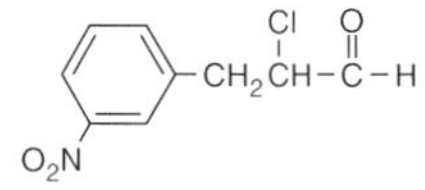
250 mg

S84,346-6
2-CHLORO-3-(4-NITROPHENYL)-PROPIONALDEHYDE
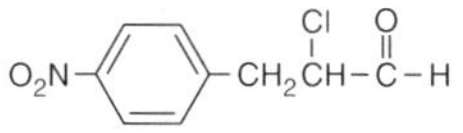
250 mg

S88,110-4
ALPHA-METHYLHYDROCINNAMALDEHYDE
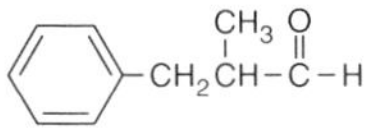
250 mg

S88,055-8
BETA,BETA-DIMETHYLHYDROCINNAMALDEHYDE
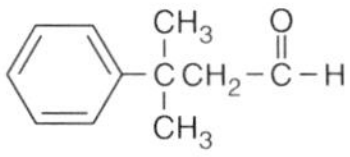
250 mg

S76,302-0
2,2-DIMETHYL-5-PHENYL-4-PENTENAL
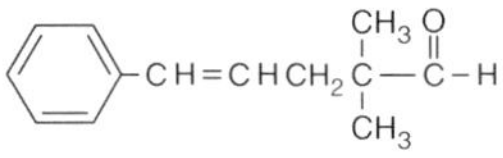
250 mg

S79,226-8
3,3,3-TRIPHENYLPROPIONALDEHYDE
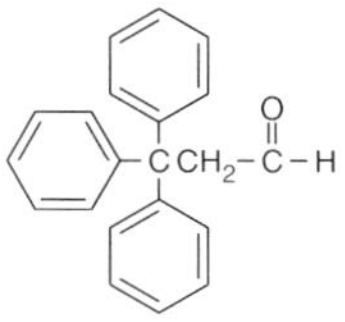
250 mg

S44,397-2
BIS(2-CHLOROPHENYL)ACETALDEHYDE
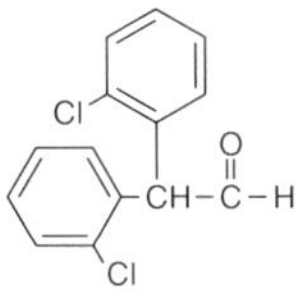
250 mg

S40,536-1
3-(4-ISOPROPYLPHENYL)-2-METHYLPROPIONALDEHYDE
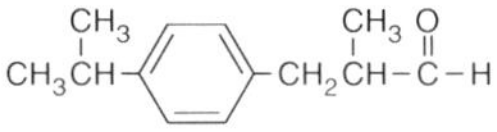
5 gm

S35,928-9
BETA-(4-TERT-BUTYLPHENYL)-ALPHA-METHYLPROPIONALDEHYDE
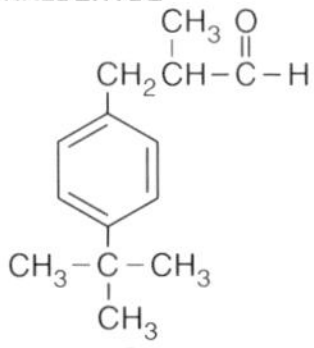
5 gm

S94,156-5
4-METHOXY-ALPHA-METHYLCINNAMALDEHYDE
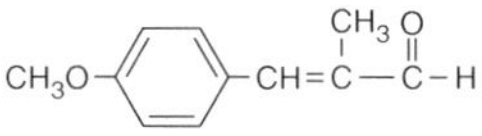
5 gm

S67,953-4
ALPHA-BUTYLCINNAMALDEHYDE
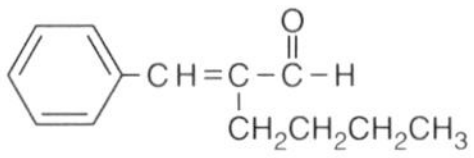
250 mg

S76,605-4
2,2-DIMETHYL-4'-METHYLHYDROCINNAMALDEHYDE
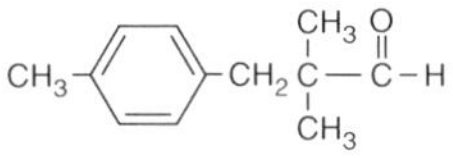
250 mg

S57,369-8
3-[2-(2-CHLOROETHOXY)ETHOXY]-BENZALDEHYDE
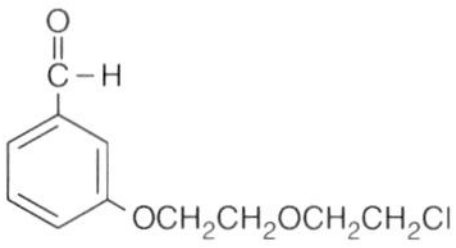
5 gm

S57,467-8
2-[2-(2-CHLOROETHOXY)ETHOXY]-BENZALDEHYDE
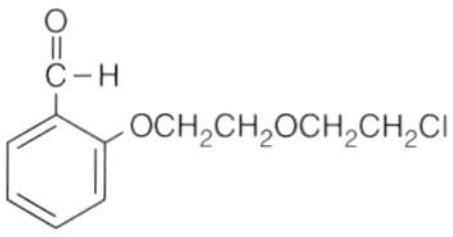
5 gm

S56,931-3
4-[2-(2-CHLOROETHOXY)ETHOXY]-BENZALDEHYDE
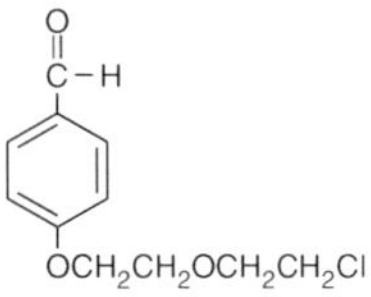
5 gm

S56,943-7
4-(2-BROMOETHOXY)-M-ANISALDEHYDE
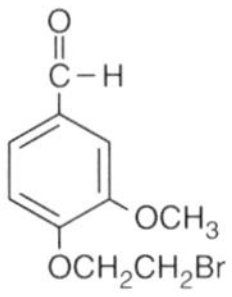
5 gm

S56,934-8
4-(3-CHLOROPROPOXY)-M-ANISALDEHYDE
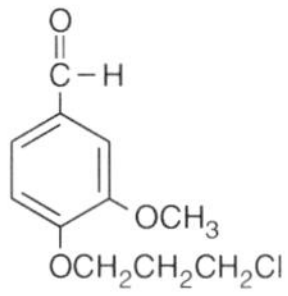
5 gm

S57,370-1
4-[2-(2-CHLOROETHOXY)ETHOXY]-M-ANISALDEHYDE
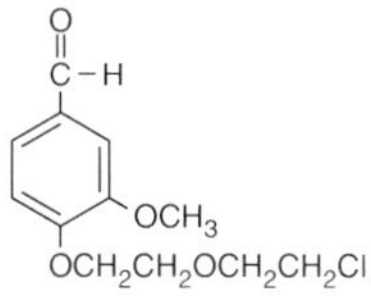
5 gm

S57,585-2
2-(3-CHLOROPROPOXY)BENZALDEHYDE
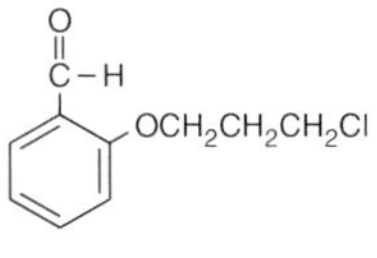
5 gm

S42,862-0
4-(2-BROMOETHOXY)BENZALDEHYDE
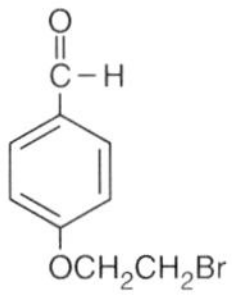
5 gm

S56,938-0
4-(3-CHLOROPROPOXY)-BENZALDEHYDE
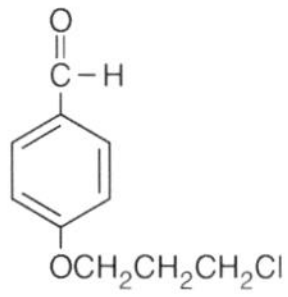
5 gm

S90,847-9
4-[N-(2-CHLOROETHYL)-N-ETHYLAMINO]-2-METHYLBENZALDEHYDE
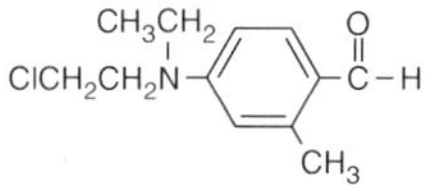
250 mg

S44,384-0
4-[N,N-BIS(2-CHLOROETHYL)AMINO]-BENZALDEHYDE
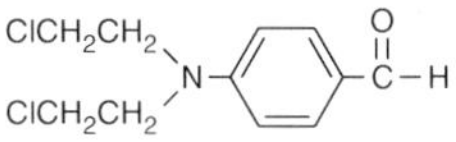
250 mg

S84,378-4
2,3-DIBROMO-4,5-DIHYDROXYBENZALDEHYDE
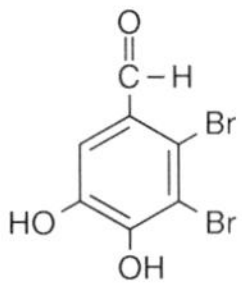
250 mg

S60,049-0
5-BROMO-2-NITROVANILLIN
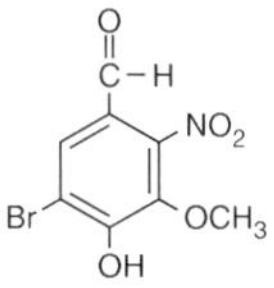
5 gm

S84,393-8
2-BROMOVANILLIN
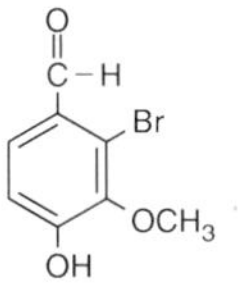
250 mg

S80,430-4
3-BROMO-2-HYDROXY-5-METHOXYBENZALDEHYDE
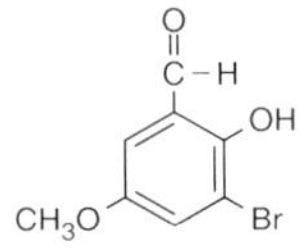
250 mg

S89,926-7
4-HYDROXY-2,3,5,6-TETRACHLOROBENZALDEHYDE
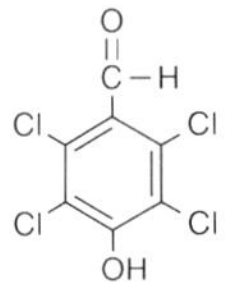
250 mg

S80,695-1
2-BROMO-3-HYDROXYBENZALDEHYDE
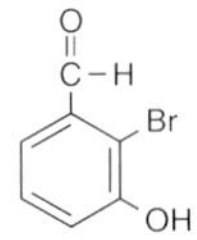
250 mg

S89,924-0
3,5-DIIODO-4-HYDROXYBENZALDEHYDE
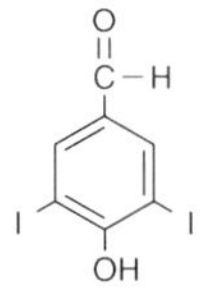
250 mg

S77,033-7
3-BROMO-4,5-DIMETHOXY-2-NITROBENZALDEHYDE
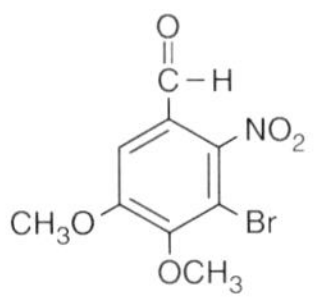
250 mg

S83,942-6
3-BROMO-4,5-DIMETHOXYBENZALDEHYDE
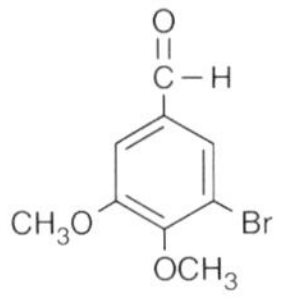
250 mg

S80,349-9
3-BROMO-2,5-DIMETHOXYBENZALDEHYDE
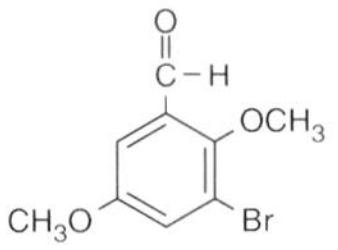
250 mg

S85,004-7
3-BROMO-4-METHOXYBENZALDEHYDE
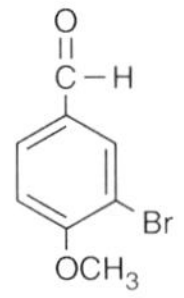
250 mg

S76,153-2
2,6-DICHLORO-3-NITROBENZALDEHYDE
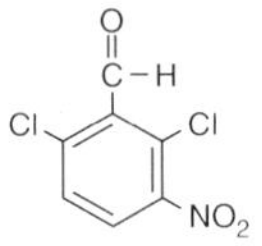
250 mg

S67,063-4
3-BENZOYL-2,4-DICHLOROBENZALDEHYDE
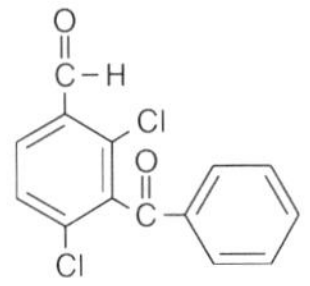
5 gm

S90,818-5
3,5-DIBROMOBENZALDEHYDE
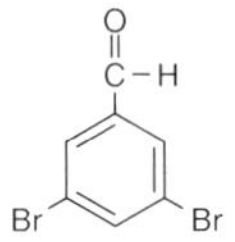
250 mg

S88,577-0
2,2'-DIHYDROXY-3,3'-DIMETHOXY-5,5'-BIPHENYLDICARBOXALDEHYDE
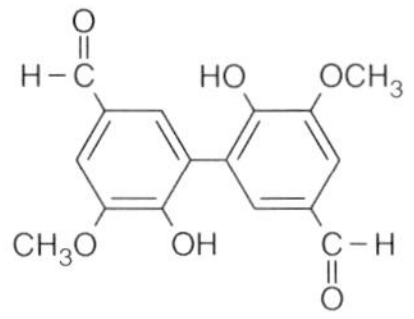
250 mg

S94,779-2
2,4-DIHYDROXY-3-METHYLBENZALDEHYDE
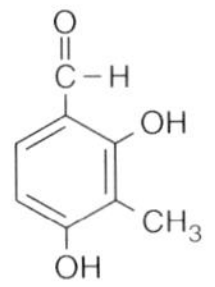
5 gm

S74,970-2
2,4-DIHYDROXY-6-METHYLBENZALDEHYDE
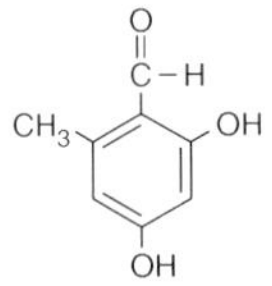
250 mg

S42,953-8
2,4-DIMETHOXY-3-HYDROXYBENZALDEHYDE
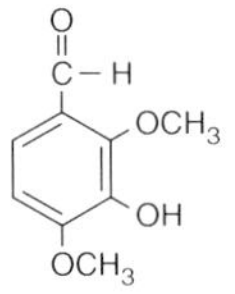
5 gm

S85,005-5
3-HYDROXY-4-METHOXY-6-NITROBENZALDEHYDE, SODIUM SALT
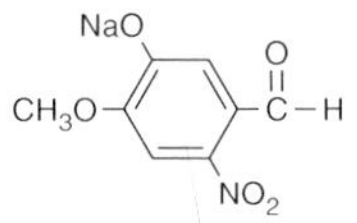
250 mg

S95,316-4
3-HYDROXY-4-METHOXY-2-NITROBENZALDEHYDE
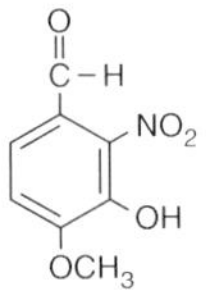
50 mg

S84,394-6
4-HYDROXY-3-METHOXY-2-NITROBENZALDEHYDE
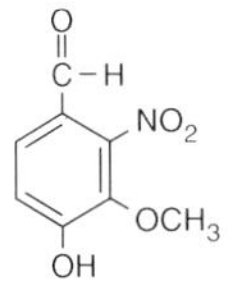
250 mg

S50,133-6
4-HYDROXY-2-NITRO-M-ANISALDEHYDE
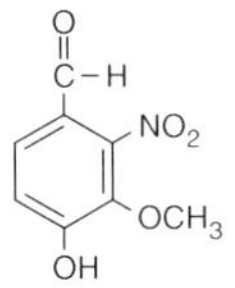
250 mg

S77,013-2
6-NITROVANILLIN
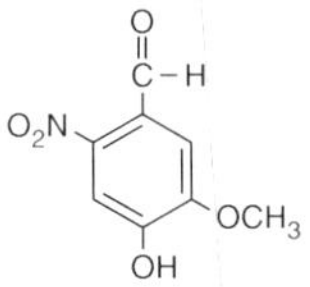
250 mg

S86,086-7
6-METHOXY-4-METHYLSALICYLALDEHYDE
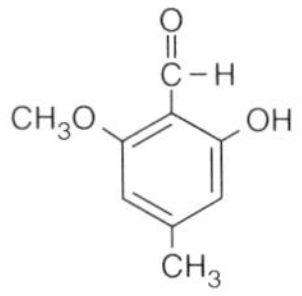
250 mg

S83,544-7
2-HYDROXY-3-NITROBENZALDEHYDE, SODIUM SALT
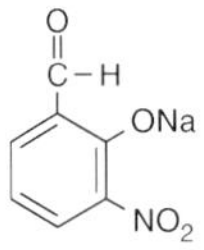
250 mg

S83,545-5
2-HYDROXY-5-NITROBENZALDEHYDE, SODIUM SALT
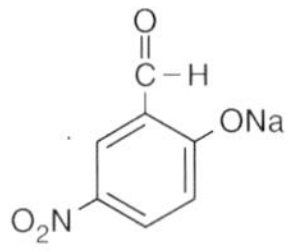
250 mg

S64,282-7
2,4-DIHYDROXY-5-HEXYLBENZALDEHYDE
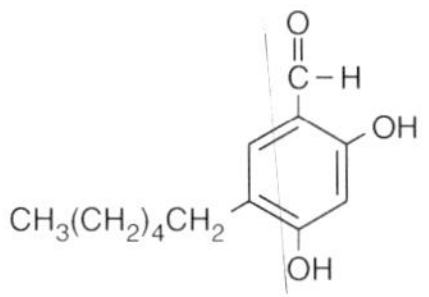
250 mg

S81,070-3
1-HYDROXY-2-NAPHTHALDEHYDE
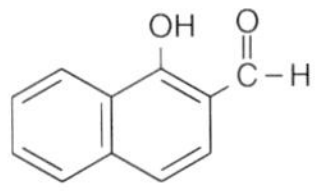
250 mg

S81,529-2
2-HYDROXY-3-NITROBENZALDEHYDE
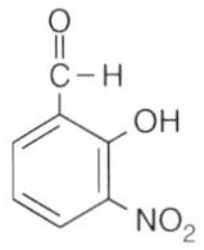
250 mg

S62,074-2
3-CHLORO-4-HYDROXYBENZALDEHYDE
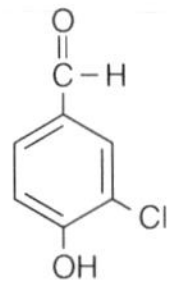
5 gm

S64,354-8
4-BENZYLOXY-3,5-DIMETHOXYBENZALDEHYDE
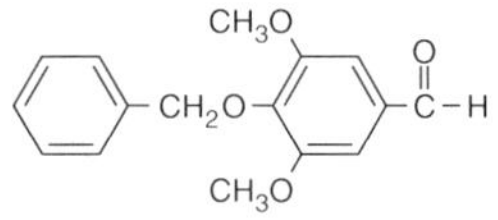
250 mg

S66,995-4
4-(4-NITROBENZYLOXY)BENZALDEHYDE
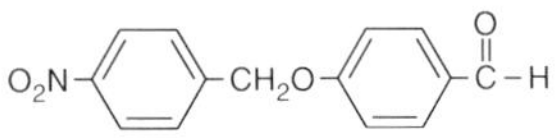
5 gm

S86,088-3
3,4-DIMETHOXY-2-NITROBENZALDEHYDE
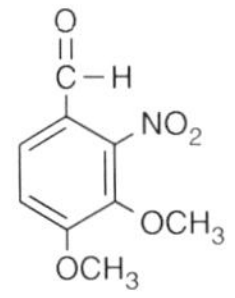
250 mg

S84,302-4
4,5-DIMETHOXY-2-METHYLBENZALDEHYDE
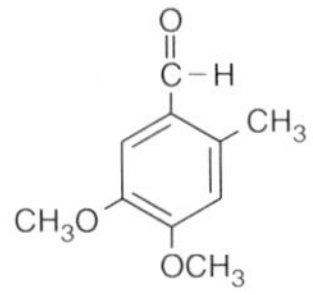
250 mg

S80,350-2
3,6-DIMETHOXY-2-NITROBENZALDEHYDE
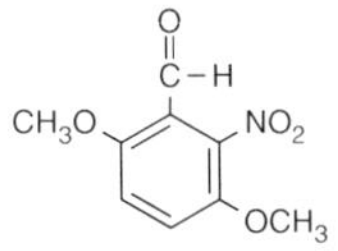
250 mg

S86,005-0
2,5-DIMETHOXYTEREPHTHALDEHYDE
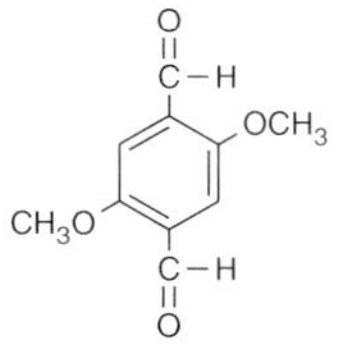
250 mg

S85,000-4
2-FORMYL-1,3,8-TRIMETHOXYANTHRAQUINONE
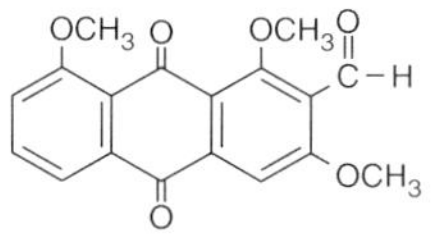
250 mg

S88,361-1
2-(PHENOXYMETHOXY)BENZALDEHYDE
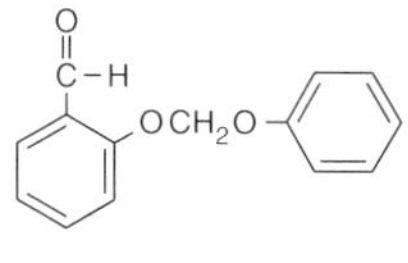
250 mg

S75,192-8
2,4-DIMETHOXY-6-METHYLBENZALDEHYDE
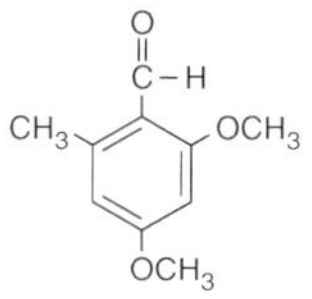
5 gm

S75,193-6
2,4-DIMETHOXY-5-METHYLBENZALDEHYDE
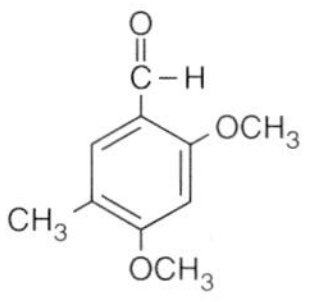
5 gm

S93,735-5
4-(METHOXYMETHOXY)BENZALDEHYDE
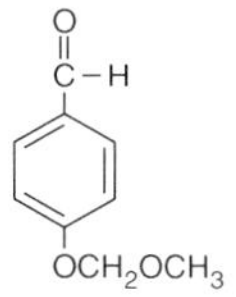
250 mg

S42,776-4
2-ETHOXY-3-METHOXYBENZALDEHYDE
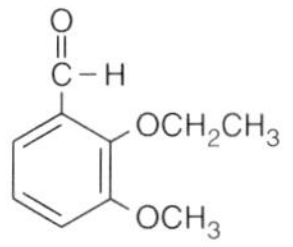
5 gm

S63,092-6
4-ETHOXY-3-METHOXYBENZALDEHYDE
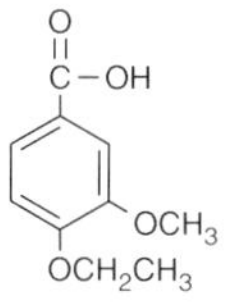
250 mg

S53,077-8
3,4-DIETHOXYBENZALDEHYDE
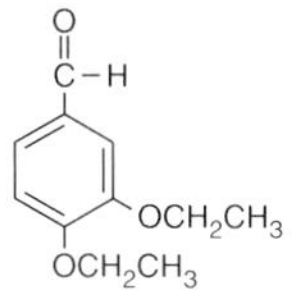
250 mg

S57,579-8
4-(HEXYLOXY)-M-ANISALDEHYDE
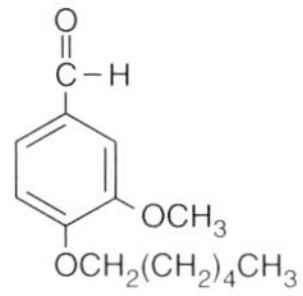
5 gm

S57,584-4
4-(HEPTYLOXY)-M-ANISALDEHYDE
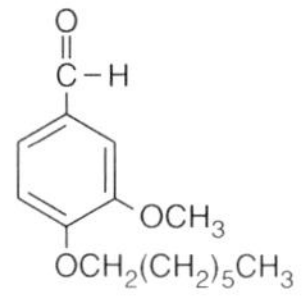
5 gm

S57,538-0
4-(OCTYLOXY)-M-ANISALDEHYDE
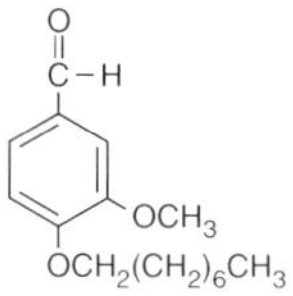
5 gm

S57,570-4
4-(DECYLOXY)-M-ANISALDEHYDE
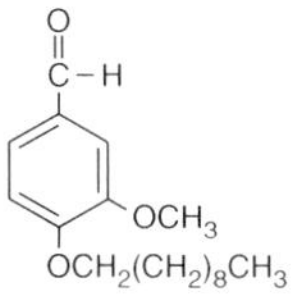
5 gm

S57,498-8
3,3'-(ETHYLENEDIOXY)DI-P-ANISALDEHYDE
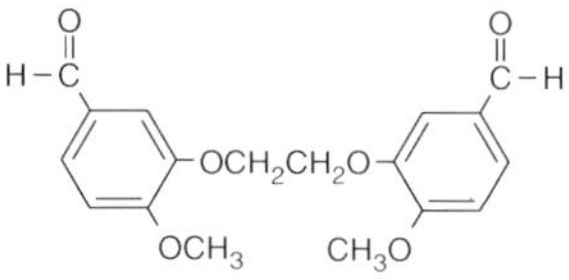
5 gm

S38,590-5
4-HEXADECYLOXY-META-ANISALDEHYDE
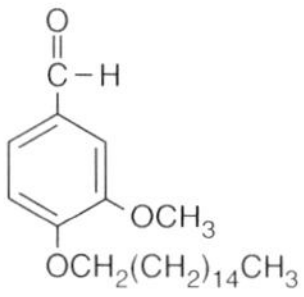
250 mg

S57,456-2
3,3'-(DECAMETHYLENEDIOXY)DI-P-ANISALDEHYDE
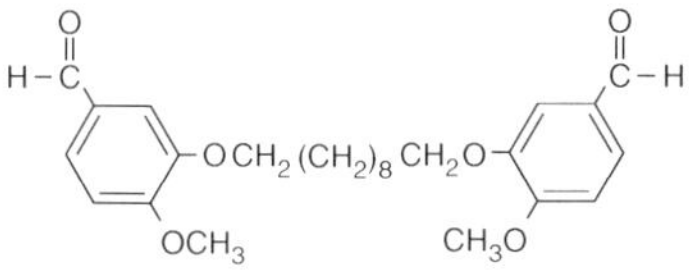
250 mg

S41,872-2
2,4,5-TRIETHOXYBENZALDEHYDE
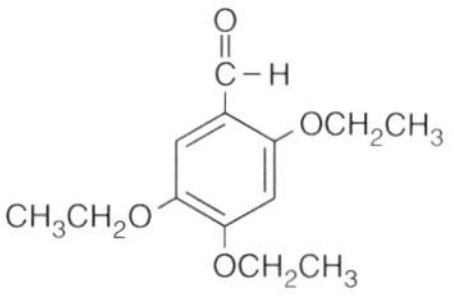
5 gm

S84,049-1
2-METHOXY-3-NITROBENZALDEHYDE
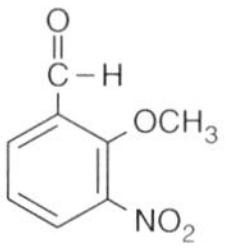
250 mg

S84,608-2
2-ETHOXY-1-NAPHTHALDEHYDE
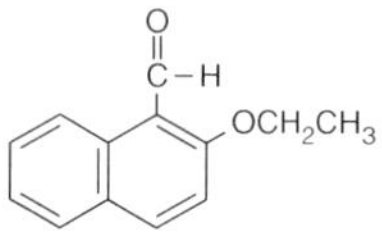
250 mg

S44,886-9
4-[2-(DIETHYLAMINO)-ETHOXY]-BENZALDEHYDE
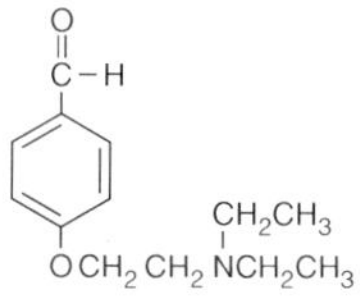
5 gm

S39,316-9
4-HEXADECYLOXY-3-NITROBENZALDEHYDE
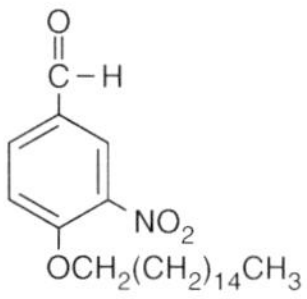
250 mg

S93,348-1
2-(2-NITROPHENOXY)BENZALDEHYDE
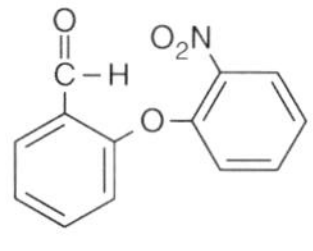
250 mg

S94,199-9
4-METHOXY-3-NITROBENZALDEHYDE
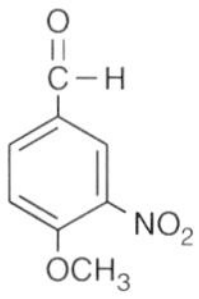
250 mg

S41,789-0
4-METHOXY-2-NITROBENZALDEHYDE
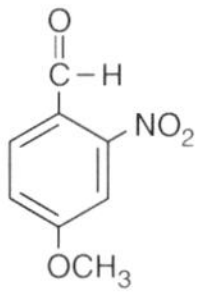
250 mg

S55,356-5
3-(DODECYLOXY)BENZALDEHYDE
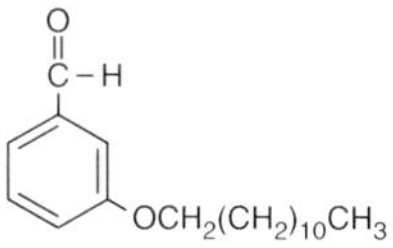
250 mg

S57,545-3
2,2'-(ETHYLENEDIOXY)DIBENZALDEHYDE
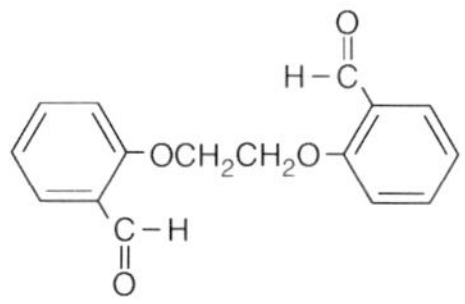
5 gm

S57,548-8
2,2'-(TRIMETHYLENEDIOXY)-DIBENZALDEHYDE
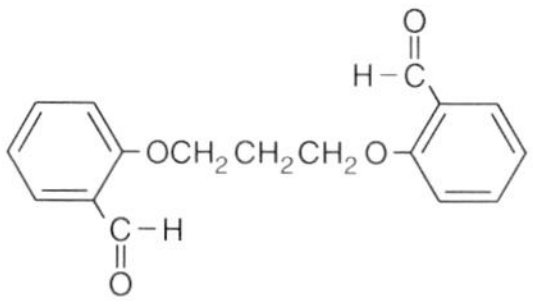
5 gm

S57,543-7
2,2'-(TETRAMETHYLENEDIOXY)-DIBENZALDEHYDE
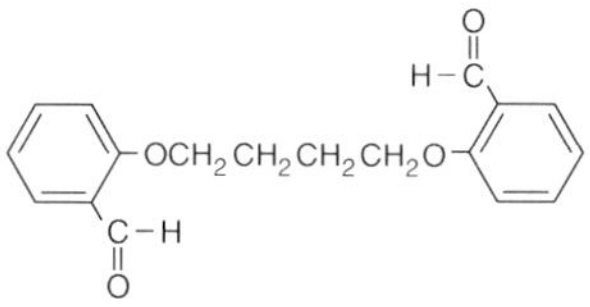
5 gm

S57,347-7
2,2'-(HEXAMETHYLENEDIOXY)-DIBENZALDEHYDE
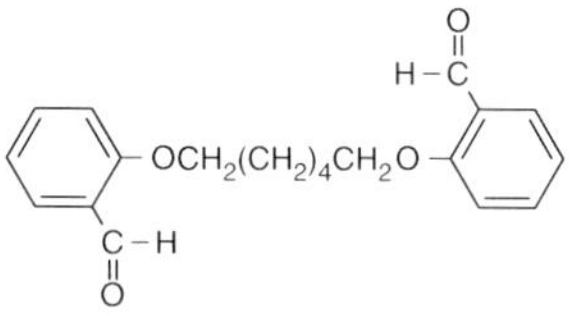
5 gm

S38,722-3
2-(HEXADECYLOXY)BENZALDEHYDE
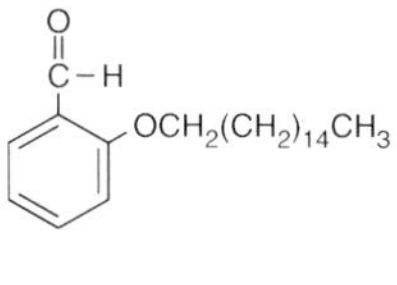
250 mg

S57,460-0
4-(PENTYLOXY)BENZALDEHYDE
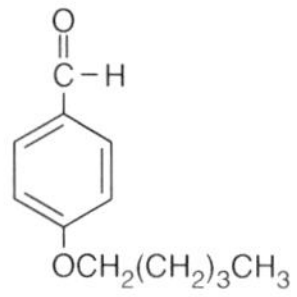
5 gm

S57,473-2
4-(HEPTYLOXY)BENZALDEHYDE
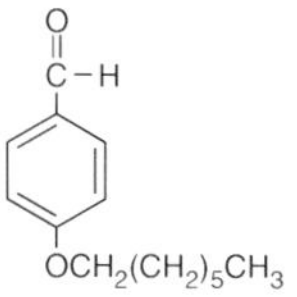
5 gm

S57,536-4
4,4'-(TETRAMETHYLENEDIOXY)-DIBENZALDEHYDE
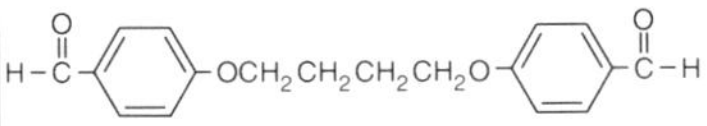
5 gm

S38,584-0
4-(HEXADECYLOXY)BENZALDEHYDE
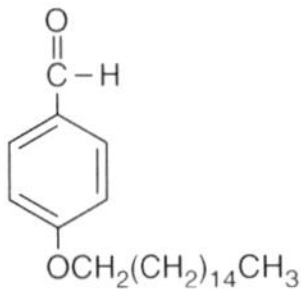
250 mg

S54,529-5
3,5-DIBROMO-2-ETHOXYBENZALDEHYDE
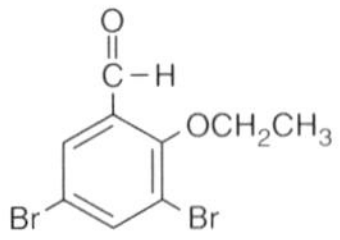
250 mg

S76,130-3
2,3-DIHYDRO-5-METHOXY-3-PHENYL-2-INDOLECARBOXALDEHYDE
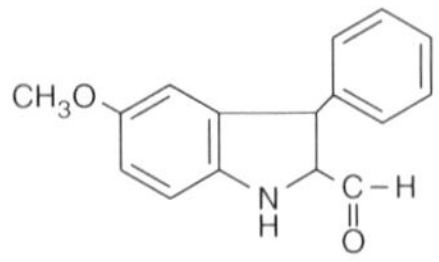
250 mg

S87,962-2
2-(ISOPROPYLAMINO)-4-METHYLBENZALDEHYDE
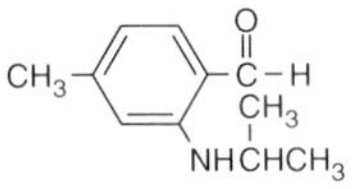
5 gm

S88,616-5
TEREPHTHALDICARBOXALDEHYDE MONO-(PHENYLHYDRAZONE)
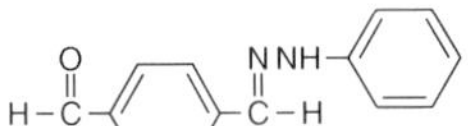
250 mg

S80,516-5
9-OXO-1-FLUORENECARBOXALDEHYDE
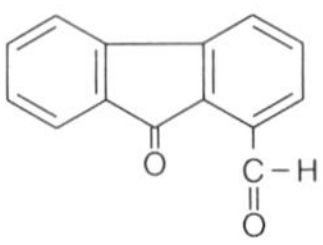
250 mg

S78,239-4
2,4-DIPHENYL-3-OXOBUTYRALDEHYDE
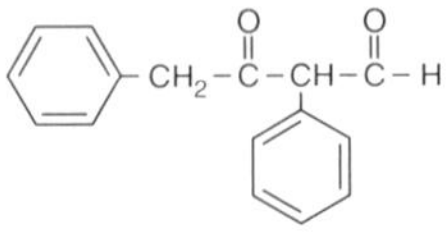
5 gm

S70,552-7
2,3,4,6-TETRAMETHYLBENZALDEHYDE
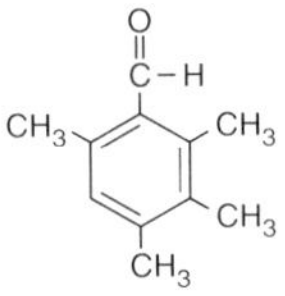
5 gm

S76,310-1
12B-BENZYL-2,3-DIFORMYL-5,6,8,12B-TETRAHYDROISOINDOLO(1,2-A)-ISOQUINOLINE
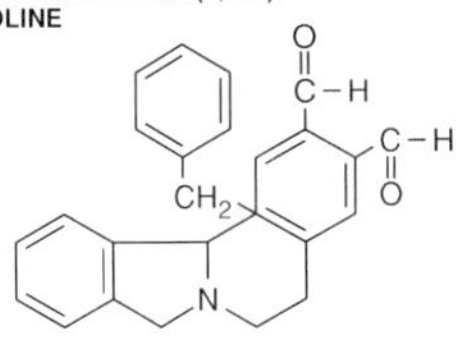
250 mg

S88,230-5
1,6-BIS(4-FORMYLPHENYL)-1,3,5-HEXATRIENE
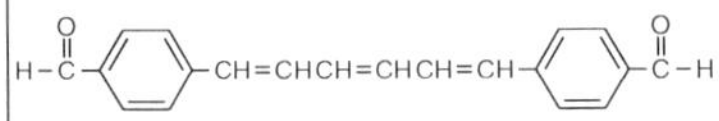
250 mg

S39,024-0
6-CHLOROPIPERONAL
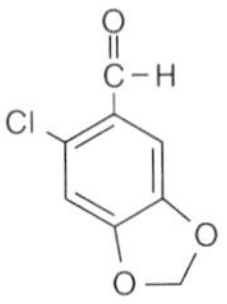
5 gm

S92,008-8
6-BROMO-1,3-BENZODIOXOLE-5-CARBOXALDEHYDE
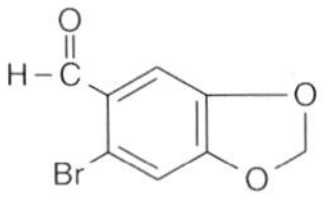
250 mg

S92,273-0
2,3,6,7-TETRAHYDRO-1H,5H-BENZO(IJ)-QUINOLIZIN-9-CARBOXALDEHYDE
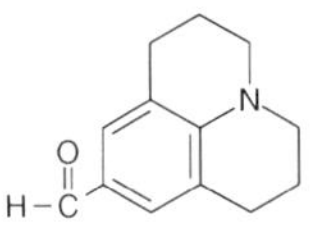
250 mg

S87,782-4
2-STILBENECARBOXALDEHYDE
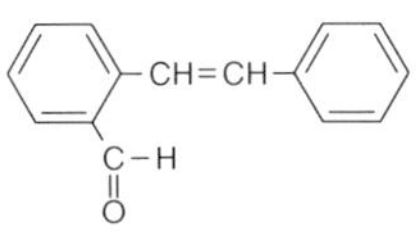
250 mg

S44,445-6
2,4,6-TRIETHOXYBENZALDEHYDE
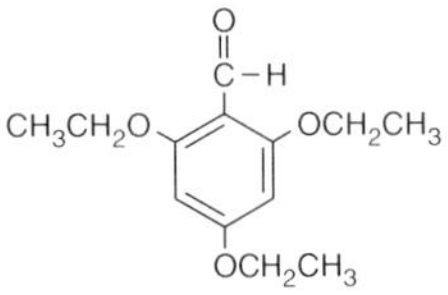
5 gm

S41,654-1
5-INDANCARBOXALDEHYDE
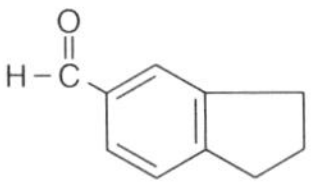
5 gm

S89,462-1
5-BROMO-1-NAPHTHALDEHYDE
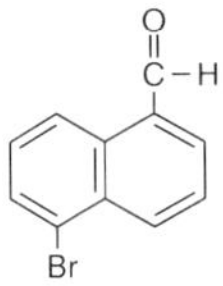
250 mg

S67,125-8
3-METHYL-2-NAPHTHALDEHYDE
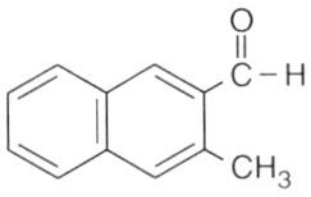
5 gm

S97,534-6
AR-L10 1-BENZO(A)-PYRENECARBOXALDEHYDE
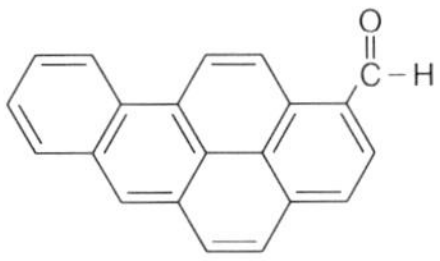
10 mg

S97,535-4
AR-L11 6-BENZO(A)-PYRENECARBOXALDEHYDE
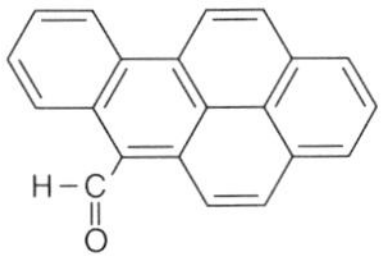
10 mg

S97,509-5
AR-H13 4-PYRENECARBOXALDEHYDE
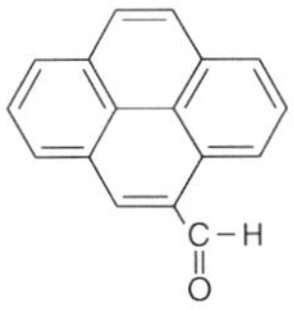
10 mg

S52,210-4
1,2,3,4-TETRAHYDRO-9-PHENANTHRENECARBOXALDEHYDE
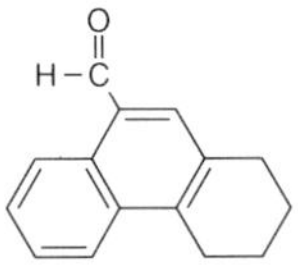
250 mg

S51,131-5
4-PHENYLISONIPECOTALDEHYDE
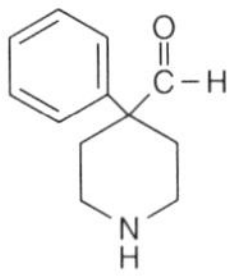
250 mg

S75,598-2
9,10-DIHYDRO-9,10-ETHANOANTHRACENE-11-CARBOXALDEHYDE
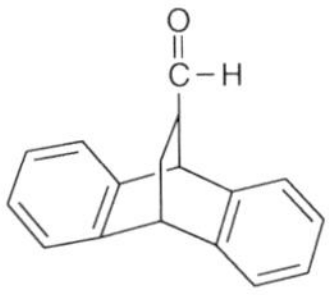
250 mg

S94,163-8
3,5,5,8,8-PENTAMETHYL-5,6,7,8-TETRAHYDRO-2-NAPHTHALENECARBOXALDEHYDE
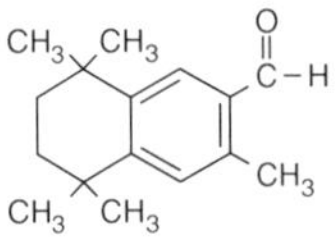
5 gm

S94,791-1
3-ETHYL-5,6,7,8-TETRAHYDRO-5,5,8,8-TETRAMETHYL-2-NAPHTHALDEHYDE
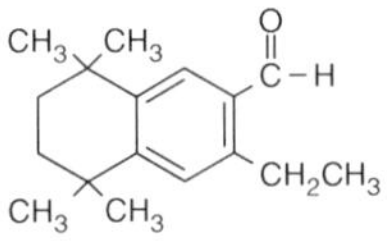
5 gm

S94,378-9
6-METHYL-2-NAPHTHALDEHYDE
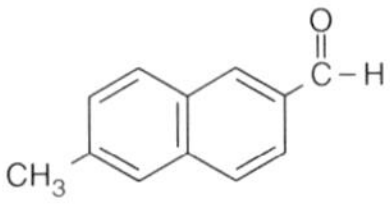
250 mg

S94,337-1
ETHYL 4-CARBOXY-2-(ETHOXYCARBONYL)-5-METHYL-3-PYRROLEACETATE
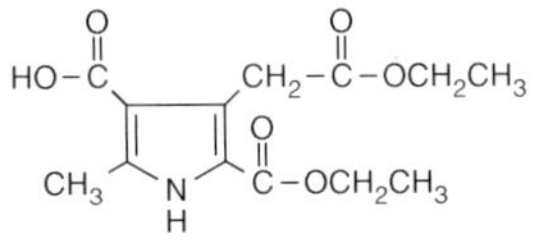
250 mg

S75,665-2
CHLORAMBUCIL
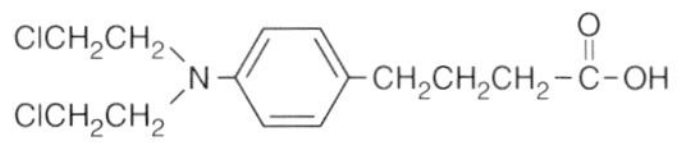
5 gm

S79,357-4
2-(1-PHENYL-2,2,2-TRIFLUOROETHOXY)ACETIC ACID
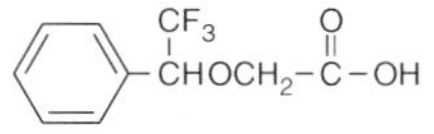
250 mg

S79,846-0
2-[2-NITRO-4-(TRIFLUOROMETHYL)-PHENYL]ACETIC ACID
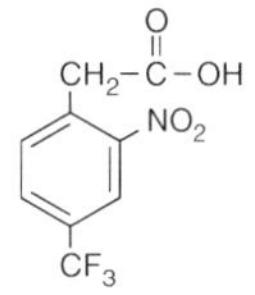
250 mg

S75,470-6
4-OXO-4-(2-TRIFLUOROMETHYLPHENYL)-BUTYRIC ACID
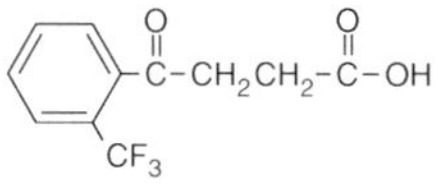
250 mg

S76,611-9
3-(TRIFLUOROMETHYL)-HYDROCINNAMIC ACID
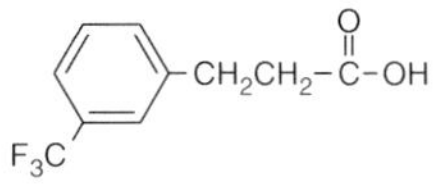
250 mg

S56,711-6
ALPHA-CHLORO-3-(TRIFLUOROMETHYL)-HYDROCINNAMIC ACID
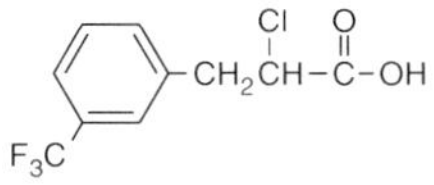
5 gm

S78,885-6
3,5-DIIODO-4-(4-HYDROXYPHENOXY)-HYDROCINNAMIC ACID
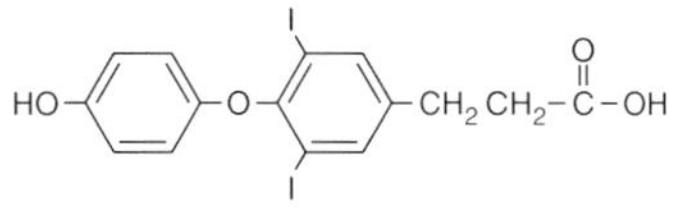
250 mg

S44,688-2
2,2'-DICHLOROBENZILIC ACID
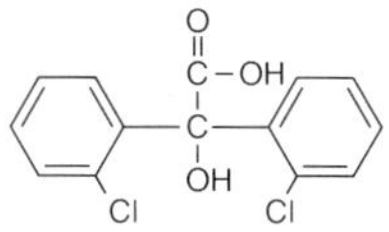
5 gm

S64,870-1
3,4-DICHLORO-ALPHA-METHOXYPHENYLACETIC ACID
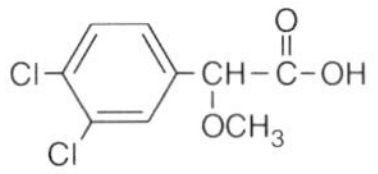
5 gm

S83,527-7
2-(4-CHLORO-2-METHOXYPHENOXY)ACETIC ACID
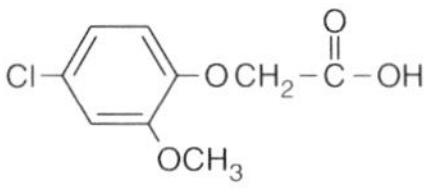
250 mg

S92,094-0
4-(4-BROMO-2,5-DIMETHOXYPHENYL)-BUTYRIC ACID
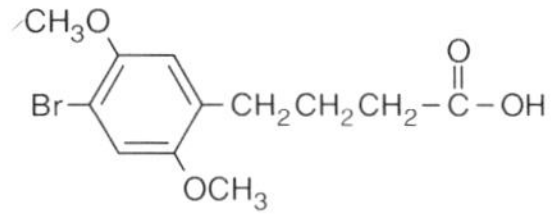
250 mg

S82,328-7
2-(4-BROMOPHENOXY)ACETIC ACID
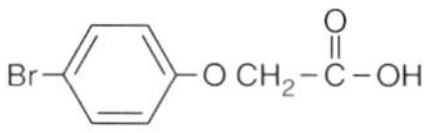
250 mg

S82,487-9
2-(4-FLUOROPHENOXY)-2-METHYLPROPIONIC ACID
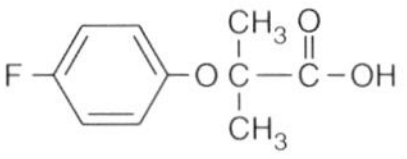
250 mg

S82,488-7
2-(4-IODOPHENOXY)-2-METHYLPROPIONIC ACID
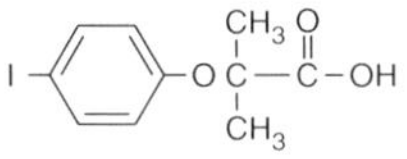
250 mg

S38,406-2
4-AMINO-3,5-DIIODO-ALPHA-(4-METHOXYPHENYL)-HYDROCINNAMIC ACID
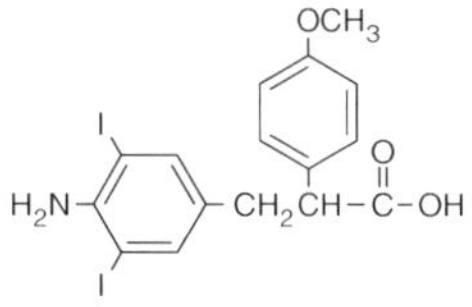
250 mg

S69,714-1
(2-BROMO-4-METHYLPHENOXY)ACETIC ACID
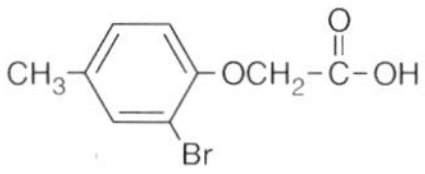
250 mg

S41,232-5
4-(2,4,5-TRICHLOROPHENOXY)-BUTYRIC ACID
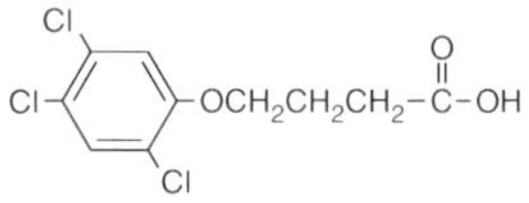
5 gm

S68,059-1
2-(2,4-DICHLORO-5-METHYLPHENYLTHIO)ACETIC ACID
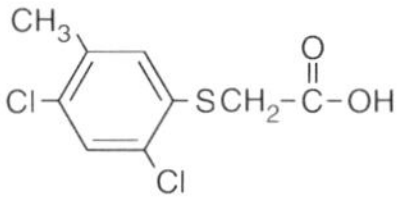
5 gm

S68,048-6
2-(4-CHLORO-2,5-DIMETHYLPHENYLTHIO)ACETIC ACID
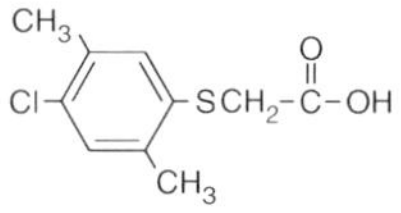
5 gm

S75,468-4
4-(4-CHLOROPHENYL)-3-METHYL-4-OXOBUTYRIC ACID
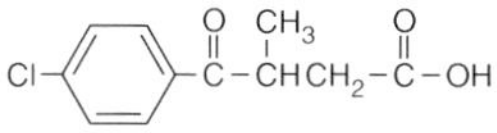
250 mg

S76,658-5
2-(4-CHLOROBENZYL)-1-OXO-2-INDANPROPIONIC ACID
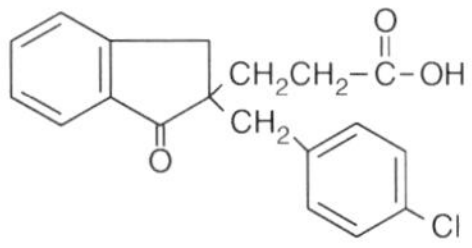
250 mg

S80,865-2
2-(4-CHLOROPHENYL)-3-METHYLSUCCINIC ACID
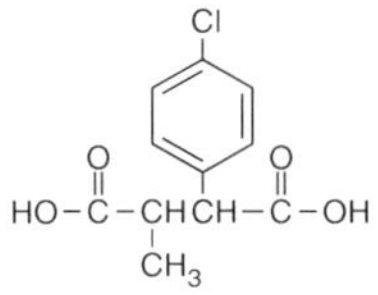
250 mg

S43,121-4
3-(PENTAFLUOROPHENYL)-PROPIONIC ACID
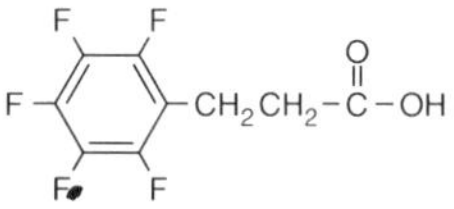
250 mg

S76,113-3
3-(4-CHLOROPHENYL)-5,6-DIHYDROIMIDAZO(2,1-B)-THIAZOLE-2-ACETIC ACID HYDROBROMIDE
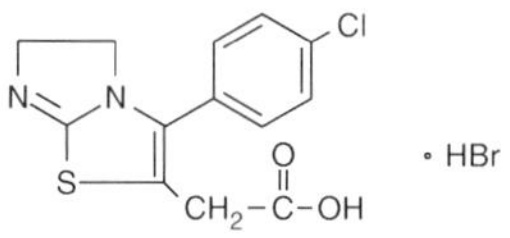
250 mg

S83,020-8
5-(2-CHLOROPHENYL)PENTANOIC ACID
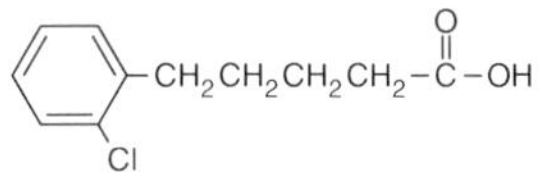
250 mg

S88,193-7
2-(4-IODOPHENYL)ACETIC ACID
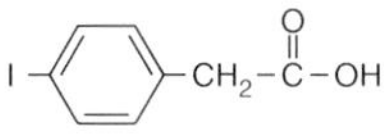
250 mg

S44,752-8
3-HYDROXYMANDELIC ACID
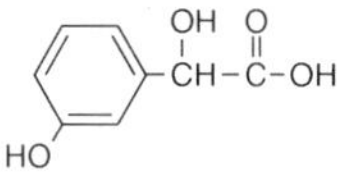
5 gm

S35,860-6
4,4-BIS(3-AMINO-4-HYDROXYPHENYL)-VALERIC ACID
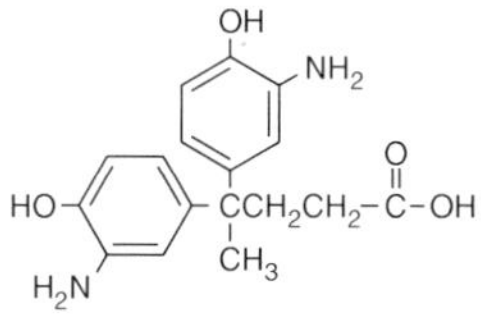
250 mg

S43,280-6
4,4-BIS(4-HYDROXY-3-NITROPHENYL)-VALERIC ACID
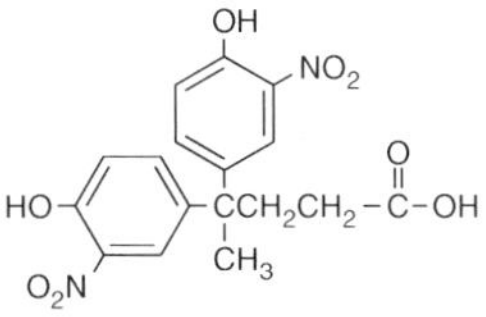
5 gm

S36,686-2
4,4-BIS(3,5-DICHLORO-4-HYDROXYPHENYL)VALERIC ACID
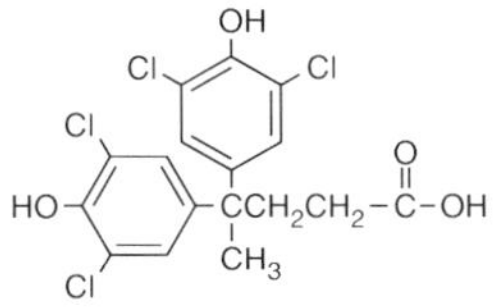
5 gm

S80,918-7
3,4-DIMETHOXYSALICYLIC ACID
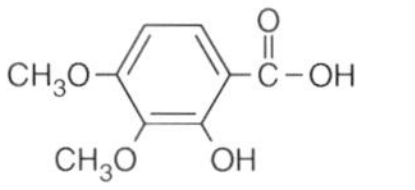
250 mg

S86,079-4
2-(4-HYDROXY-2-METHOXY-3-METHYL-5-NITROPHENOXY)ACETIC ACID
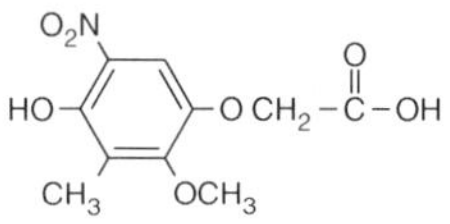
250 mg

S78,557-1
4-NITROPHENYL-BETA-D-GLUCURONIDE

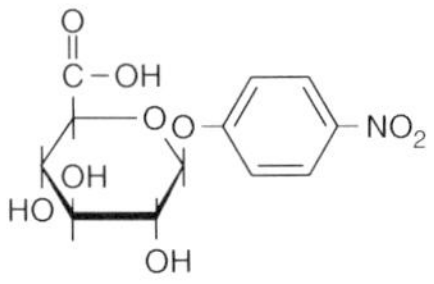

250 mg

S95,328-8
2-(3-HYDROXY-4-METHOXYPHENYL)-GLYOXYLIC ACID

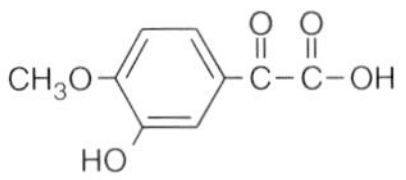

50 mg

S94,905-1
2-(2-HYDROXYPHENOXY)ACETIC ACID

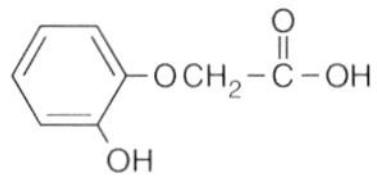

250 mg

S61,674-5
2-(4-HYDROXYPHENOXY)ACETIC ACID

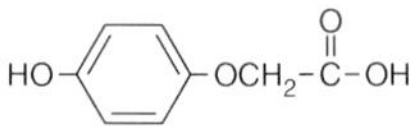

250 mg

S84,115-3
2-[4-(6-BROMO-1-HYDROXY-2-NAPHTHYLAZO)PHENOXY]ACETIC ACID

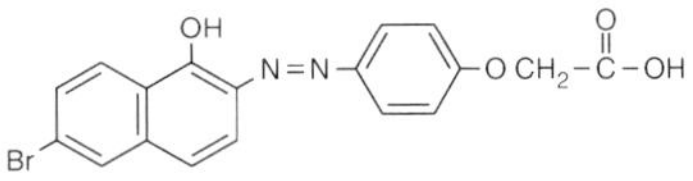

250 mg

S52,743-2
ALPHA-ISOPROPYLMANDELIC ACID

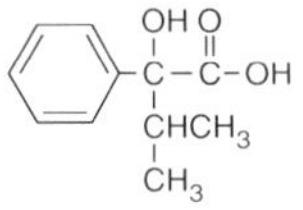

5 gm

S44,848-6
4-FLUOROMANDELIC ACID

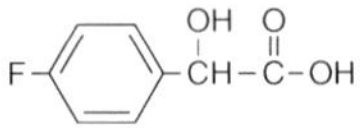

5 gm

S91,485-1
3-PHENYL-DL-SERINE

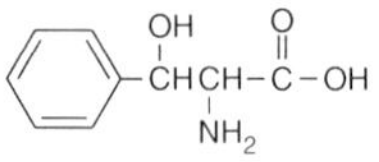

250 mg

S95,429-2
3,4-DIPHENYL-3-HYDROXYBUTYRIC ACID

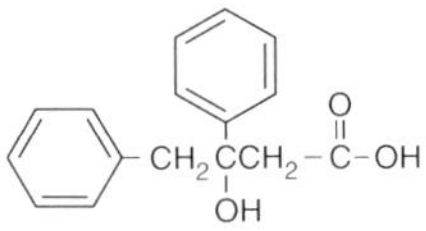

50 mg

S50,373-8
3-OXO-4,5,6-TRIHYDROXY-3H-XANTHENE-9-PROPIONIC ACID

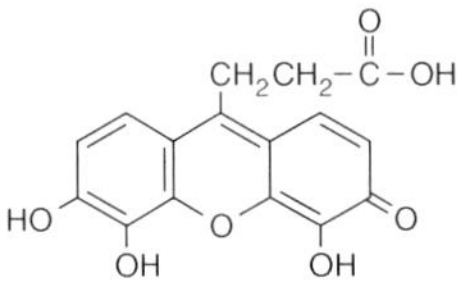

250 mg

S50,079-8
HEXACHLORO-5-(3-OXO-4,5,6-TRIHYDROXY-3H-XANTHEN-9-YL)TRICYCLOUNDEC-9-ENE-4-CARBOXYLIC ACID

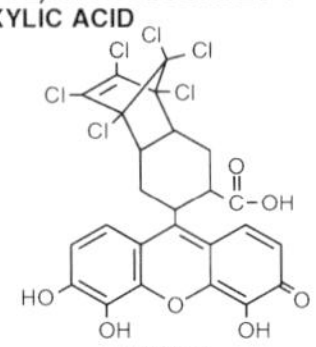

250 mg

S62,894-8
3-(2-HYDROXY-5-METHYLBENZOYL)-PROPIONIC ACID

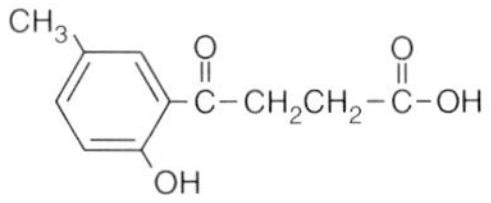

5 gm

S86,568-0
10,11-DIHYDRO-11-HYDROXY-5,10-METHANO-5H-DIBENZO(A,D)-CYCLOHEPTENE-10-CARBOXYLIC ACID

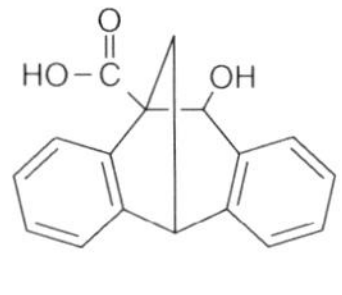

250 mg

S50,070-4
2-(6-HYDROXY-3-OXO-3H-XANTHEN-9-YL)-CYCLOHEXANECARBOXYLIC ACID

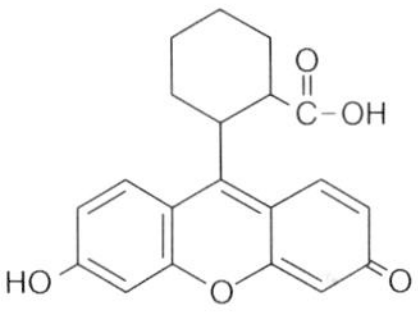

250 mg

S50,071-2
HEXACHLORO-5-(6-HYDROXY-3-OXO-3H-XANTHEN-9-YL)TRICYCLOUNDEC-9-ENE-4-CARBOXYLIC ACID

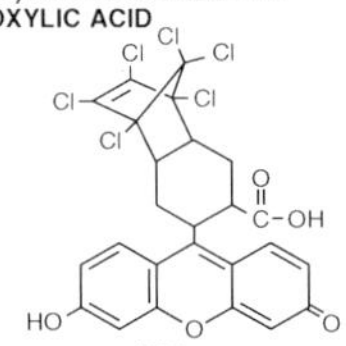

250 mg

S50,072-0
6-(6-HYDROXY-3-OXO-3H-XANTHEN-9-YL)-3-CYCLOHEXENE-1-CARBOXYLIC ACID

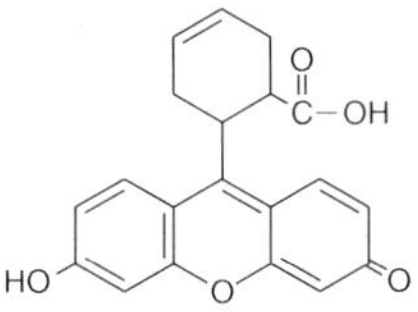

250 mg

S50,082-8
6-HYDROXY-3-OXO-3H-XANTHENE-9-PROPIONIC ACID

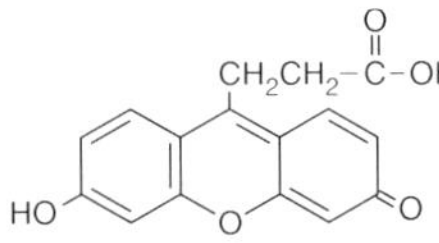

250 mg

S50,367-3
3-(3-OXO-4,5,6-TRIHYDROXY-3H-XANTHEN-9-YL)-5-NORBORNENE-2-CARBOXYLIC ACID

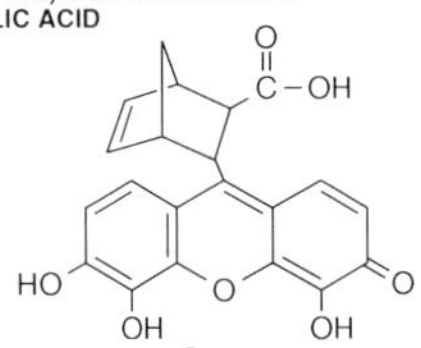

5 gm

S50,368-1
7-METHYL-3-(3-OXO-4,5,6-TRIHYDROXY-3H-XANTHEN-9-YL)-5-NORBORNENE-2-CARBOXYLIC ACID

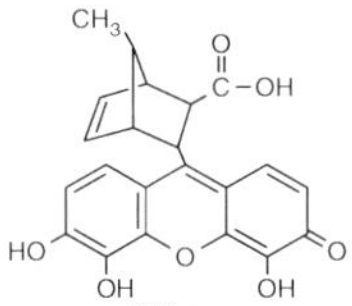

250 mg

S89,478-8
5-(3-HYDROXYPHENYL)VALERIC ACID

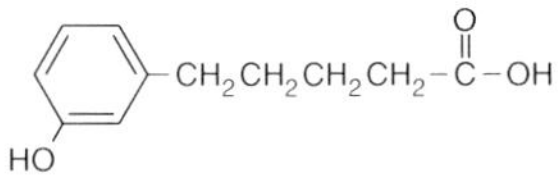

250 mg

S54,360-8
3-(2-HYDROXY-P-TOLYL)GLUTARIC ACID

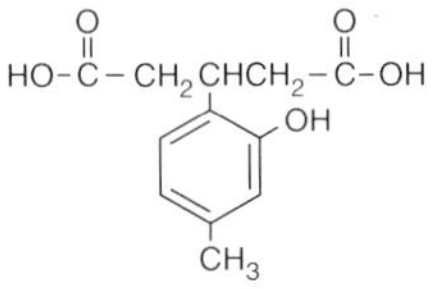

250 mg

S44,477-4
4-CHLOROMANDELIC ACID

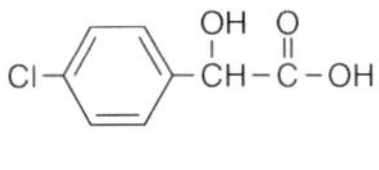

250 mg

S50,075-5
3-(3,6-DIHYDROXY-9H-XANTHEN-9-YL)-HEXACHLORO-5-NORBORNENE-2-CARBOXYLIC ACID

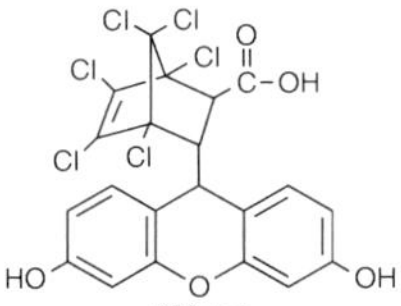

250 mg

S35,119-9
3-(4-HYDROXYPHENYL)-2-PHENYLPROPIONIC ACID

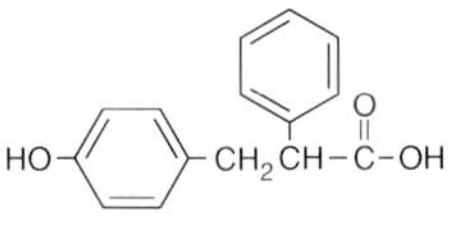

5 gm

S87,680-1
DL-2-(4-HYDROXYPHENYL)-2-PHENYLACETIC ACID

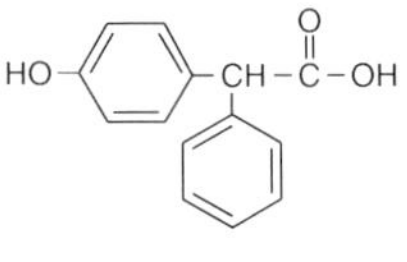

250 mg

S34,643-8
3-(3,5-DIIODO-4-HYDROXYPHENYL)-2-PHENYLPROPIONIC ACID

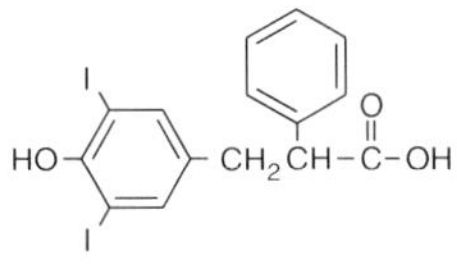

5 gm

S41,989-3
(2-HYDROXYNORBORNAN-2-YL)-PHENYLACETIC ACID

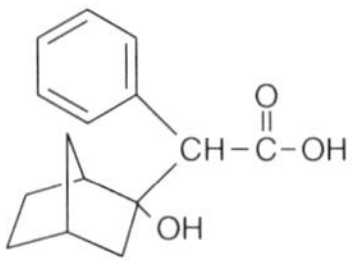

5 gm

S93,821-1
6-(2-HYDROXYBENZYLIDENEAMINO)-HEXANOIC ACID

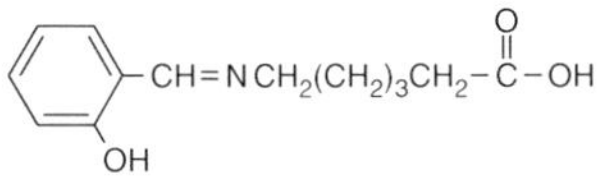

250 mg

S38,404-6
4-HYDROXY-ALPHA-(META-TOLYL)-HYDROCINNAMIC ACID

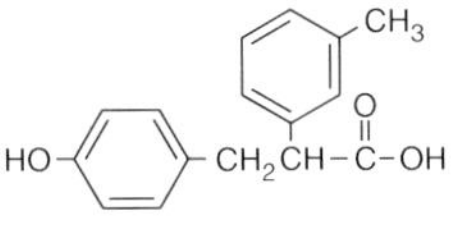

250 mg

S41,748-3
4,4'-DIFLUOROBENZILIC ACID HEMIHYDRATE

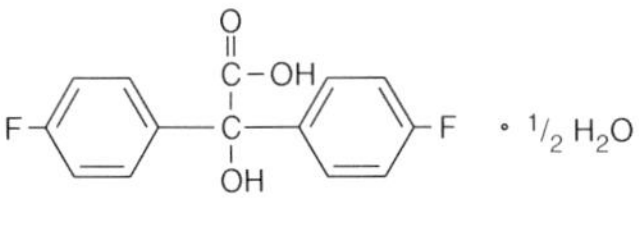

250 mg

S34,218-1
4,4-BIS(3,5-DIBROMO-4-HYDROXYPHENYL)VALERIC ACID

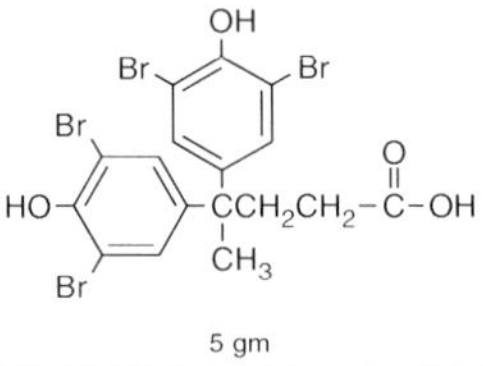

5 gm

S35,106-7
2-HYDROXY-1-NAPHTHALENEACETIC ACID

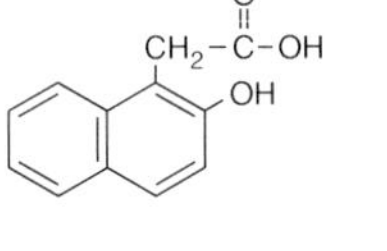

5 gm

S87,945-2
2,2-DIPHENYL-2-[2-(METHYLAMINO)-ETHOXY]ACETIC ACID HYDROBROMIDE
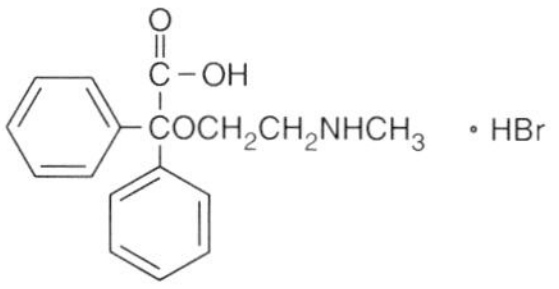
5 gm

S92,612-4
3-(3,4,5-TRIMETHOXYPHENYL)-2-THIOPYRUVIC ACID
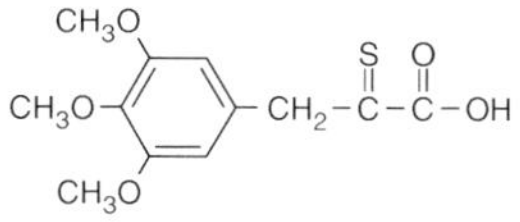
250 mg

S35,542-9
RESORCINOL-O,O'-DIACETIC ACID
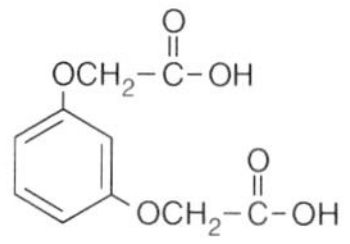
1 gm

S73,142-0
3-(3,4-DIMETHOXYBENZOYL)-PROPIONIC ACID
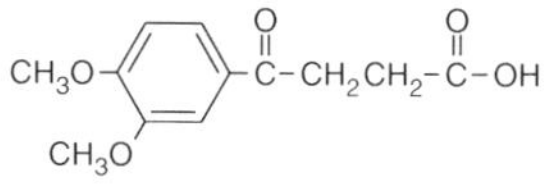
5 gm

S84,104-8
4-(3,4-DIMETHOXYBENZOYL)-BUTYRIC ACID
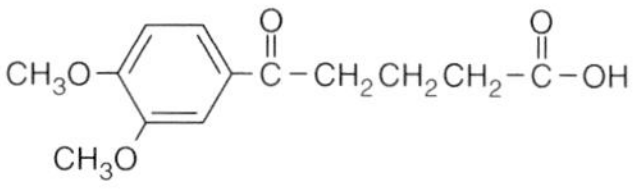
250 mg

S35,062-1
HYDROQUINONE-O,O-DIACETIC ACID
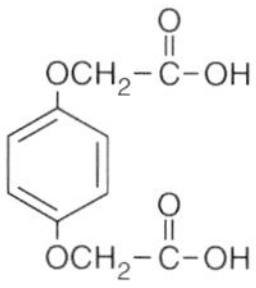
5 gm

S75,810-8
3-(3,4-DIMETHOXY-6-NITROBENZOYL)-PROPIONIC ACID
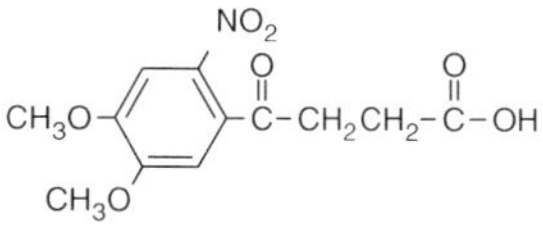
250 mg

S36,347-2
4-METHYLCATECHOL-O,O-DIACETIC ACID
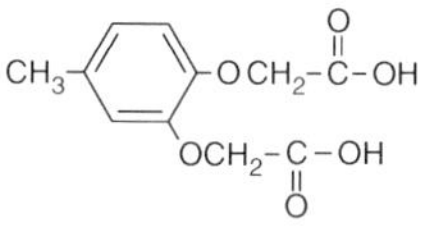
5 gm

S95,380-6
3-(2,3-DIMETHOXYPHENYL)-PROPIONIC ACID
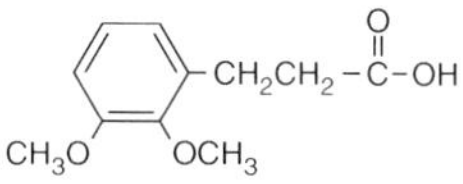
50 mg

S45,870-8
4-ETHOXY-3-METHOXYPHENYLACETIC ACID
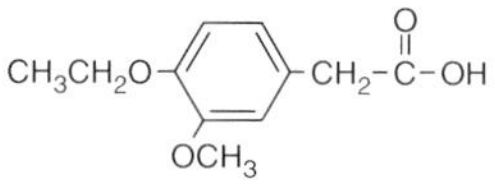
5 gm

S76,349-7
5,7-DIMETHOXY-1-OXO-2-INDANCARBOXYLIC ACID
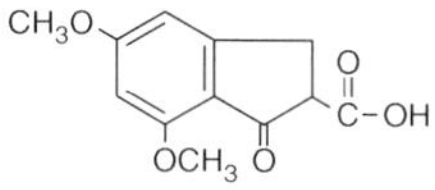
250 mg

S88,856-7
3-(3-METHOXYPHENOXY)PROPIONIC ACID
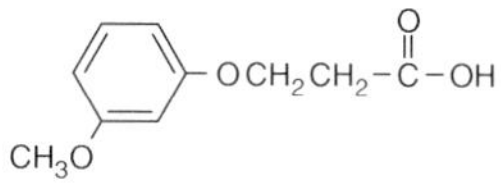
250 mg

S73,436-5
2-(2,4-DIMETHOXYPHENYL)-1-METHYL-5-OXOPYRROLIDINE-3-CARBOXYLIC ACID
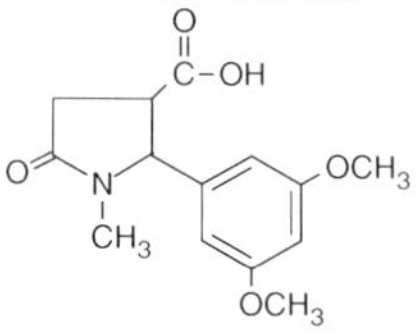
250 mg

S88,700-5
2',5'-DIMETHOXY-2-BIPHENYLACETIC ACID
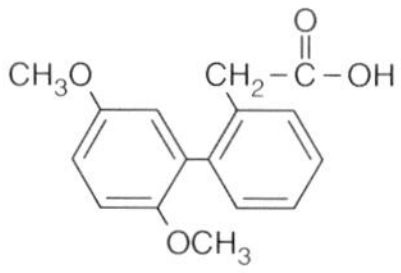
250 mg

S82,484-4
2-(4-AMINOPHENOXY)ACETIC ACID HYDROCHLORIDE
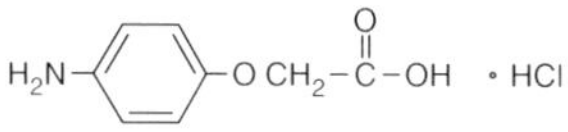
250 mg

S34,996-8
2-FORMYL-4-NITROPHENOXYACETIC ACID
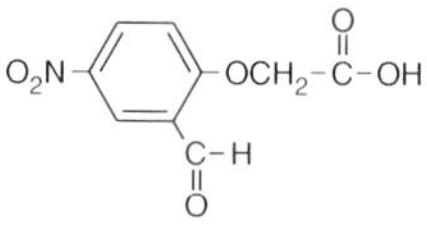
5 gm

S86,940-6
2-(4-FORMYLPHENOXY)ACETIC ACID
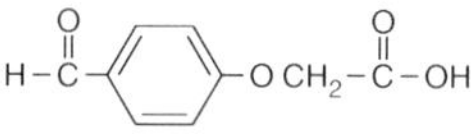
250 mg

S57,178-4
4-FORMYLPHENOXYACETIC ACID
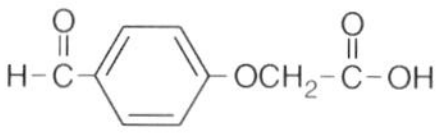
5 gm

S61,962-0
4-(3,4,5-TRIMETHOXYBENZOYL)-BUTYRIC ACID
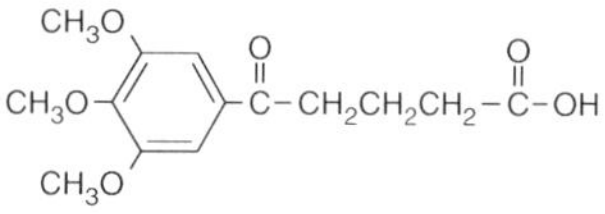
5 gm

S38,362-7
2-(1-ACETYL-2-NAPHTHYLOXY)-PROPIONIC ACID
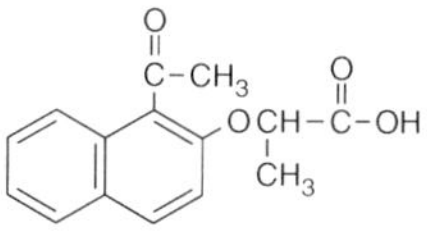
5 gm

S83,339-8
2-(4-CINNAMOYLPHENOXY)ACETIC ACID
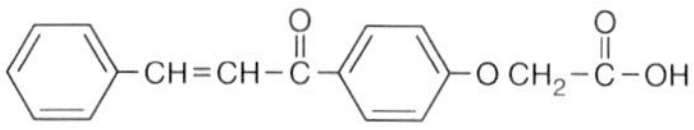
250 mg

S73,354-7
2-(5-OXO-6,7,8,9-TETRAHYDRO-5H-BENZOCYCLOHEPTEN-6-YLIDENE-METHYL)-PHENOXYACETIC ACID
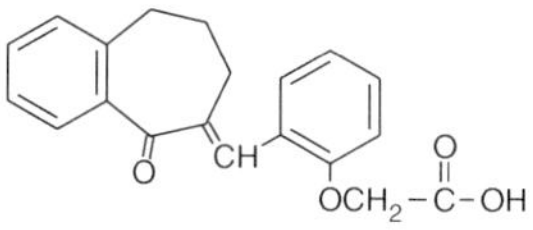
250 mg

S67,285-8
2,3-DIBROMO-3-(3,4-DIMETHOXY-PHENYL)PROPIONIC ACID
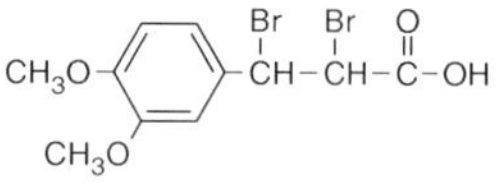
5 gm

S35,319-1
3-METHYLPHENOXYACETIC ACID
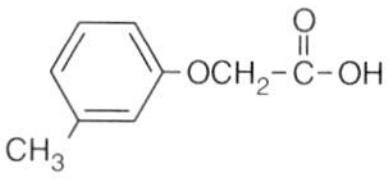
5 gm

S39,135-2
ORTHO-TOLYLOXYACETIC ACID
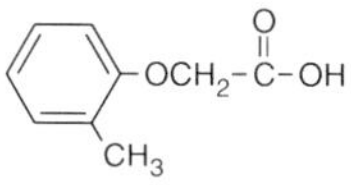
5 gm

S36,318-9
2-(2-METHOXYPHENOXY)ACETIC ACID
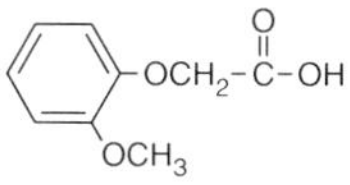
5 gm

S78,681-0
2-(2-ALLYLPHENOXY)ACETIC ACID
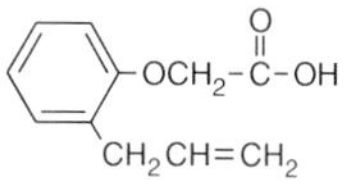
250 mg

S67,290-4
2-(4-TERT-BUTYLPHENOXY)ACETIC ACID
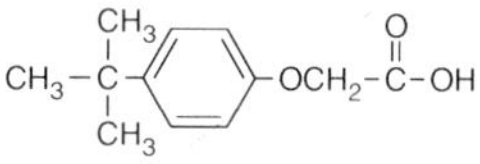
5 gm

S35,856-8
2-(4-BIPHENYLYLOXY)PROPIONIC ACID
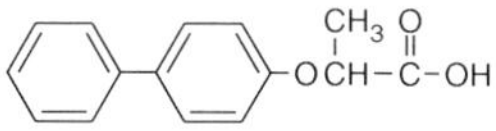
5 gm

S36,110-0
2,3-DIMETHYLPHENOXYACETIC ACID
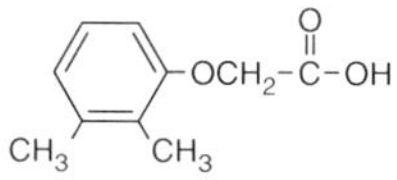
5 gm

S94,786-5
2,6-DIMETHYLPHENOXYACETIC ACID
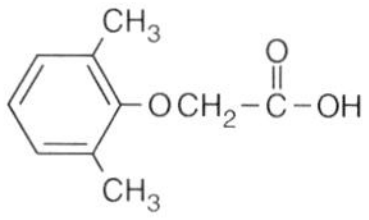
250 mg

S34,552-0
2-(2,5-DICHLOROPHENOXY)-PROPIONIC ACID
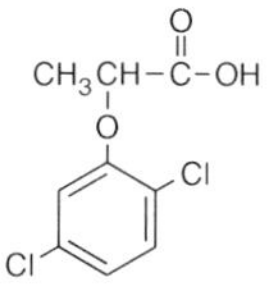
5 gm

S34,879-1
2-(2,4-DIBROMOPHENOXY)-PROPIONIC ACID

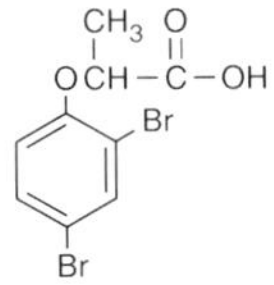

5 gm

S36,507-6
THYMOXYACETIC ACID

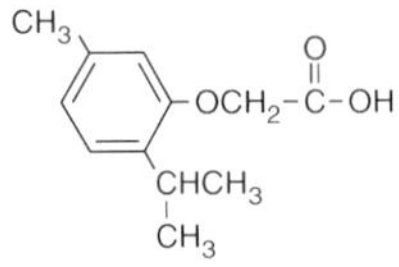

5 gm

S36,521-1
2-(2,4,6-TRICHLOROPHENOXY)-PROPIONIC ACID

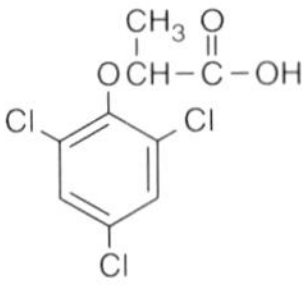

5 gm

S64,265-7
2-BROMO-4,5-DIMETHOXYHYDROCINNAMIC ACID

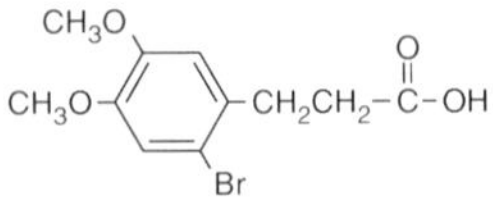

5 gm

S67,218-1
(-)-2-(1-NAPHTHOXY)PROPIONIC ACID

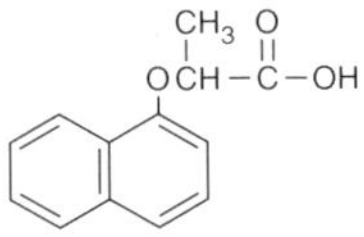

250 mg

S67,272-6
2-(1-NAPHTHYLOXY)PROPIONIC ACID

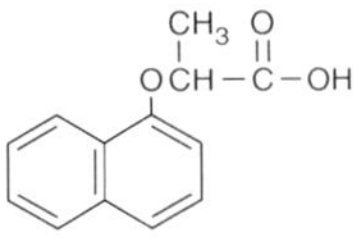

250 mg

S38,172-1
2-(6-BROMO-2-NAPHTHYLOXY)-PROPIONIC ACID

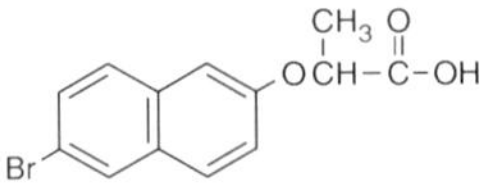

250 mg

S85,261-9
9,10-DIHYDRO-12-ETHOXY-9,10-ETHANOANTHRACENE-11-CARBOXYLIC ACID

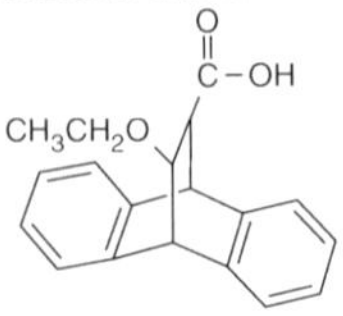

250 mg

S88,837-0
3-(4-BIPHENYLYLOXY)PROPIONIC ACID

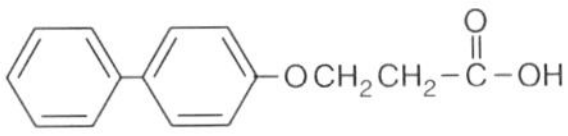

250 mg

S88,828-1
3-(1-NAPHTHOXY)PROPIONIC ACID

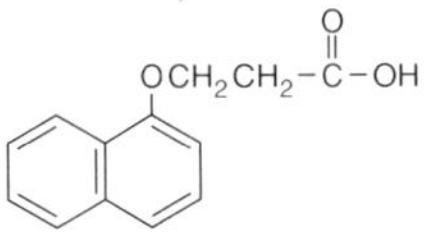

250 mg

S88,827-3
3-(2-NAPHTHOXY)PROPIONIC ACID

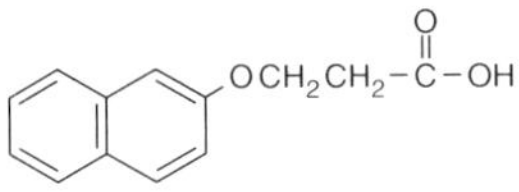

250 mg

S95,415-2
3-(3-METHOXYBENZOYL)ADIPIC ACID

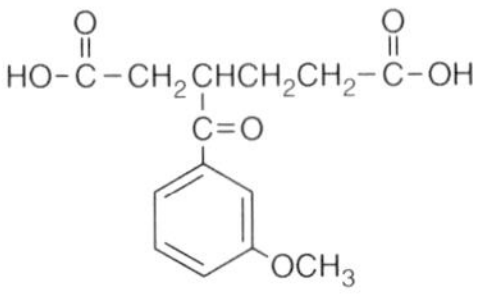

50 mg

S92,584-5
DL-2-(4-PHENOXYPHENYL)-2-PHENYLGLYCINE

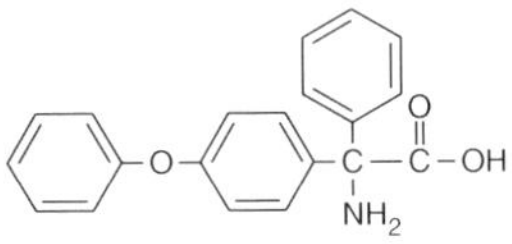

250 mg

S38,002-4
2-(3-AMINOPHENOXY)ACETIC ACID

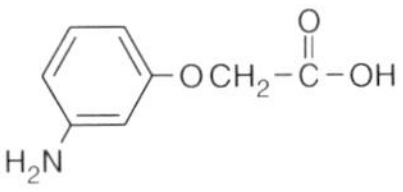

250 mg

S43,459-0
ISOSTRYCHNIC ACID MONOHYDRATE

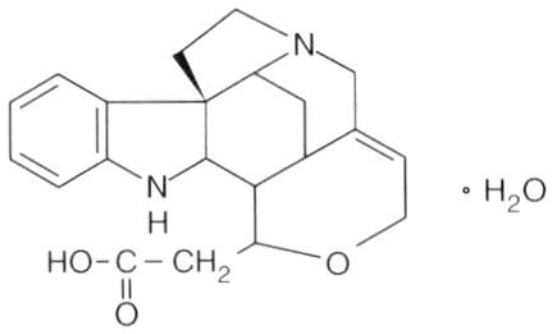

5 gm

S41,416-6
2-(2-NITROPHENOXY)ACETIC ACID

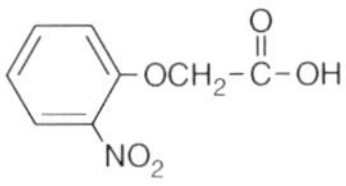

5 gm

S67,066-9
3-(3,6-DIMETHOXY-2-NAPHTHOYL)-PROPIONIC ACID

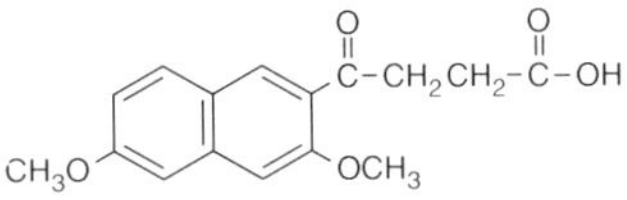

250 mg

S83,337-1
2-[4-(3-OXO-3-PHENYL-1-PROPEN-1-YL)-PHENOXY]ACETIC ACID

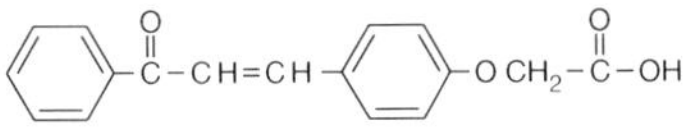

250 mg

S56,587-3
3-(4-PHENOXYBENZOYL)PROPIONIC ACID

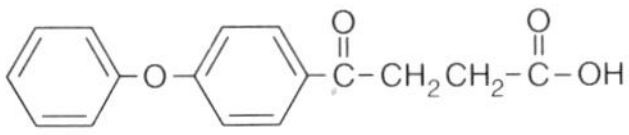

5 gm

S38,672-3
11-(4-ACETYLPHENOXY)-UNDECANOIC ACID

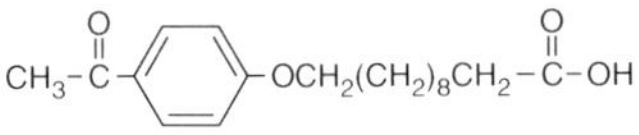

5 gm

S55,862-1
3-[4-(HEXADECYLOXY)BENZOYL]-PROPIONIC ACID

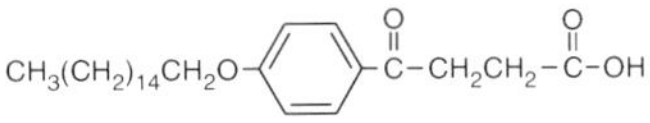

250 mg

S65,114-1
3-(2,5-DIMETHOXYBENZOYL)-PROPIONIC ACID

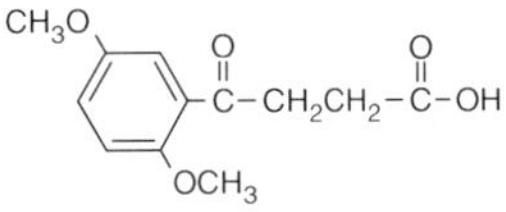

250 mg

S36,317-0
3-(4-METHOXY-1-NAPHTHOYL)-PROPIONIC ACID

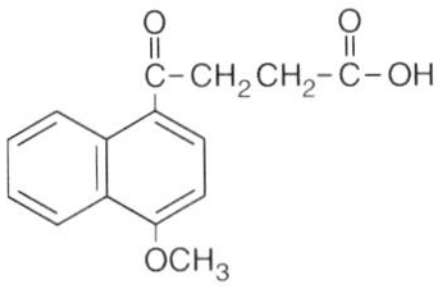

5 gm

S92,611-6
3-(3,4,5-TRIMETHOXYPHENYL)-PYRUVIC ACID

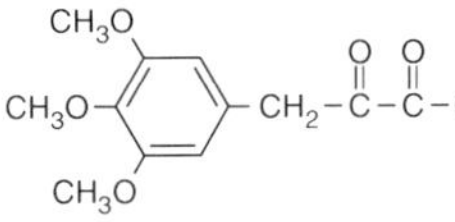

250 mg

S41,276-7
3-(3-FLUORO-P-ANISOYL)-PROPIONIC ACID

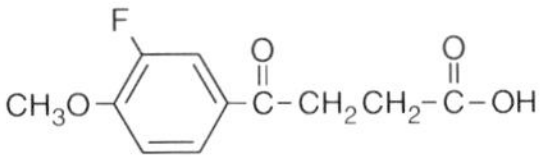

5 gm

S36,959-4
3-(3-METHOXYPHENYL)PROPIONIC ACID

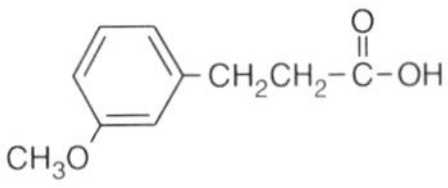

5 gm

S40,816-6
2,5-DIMETHYLPHENOXYACETIC ACID

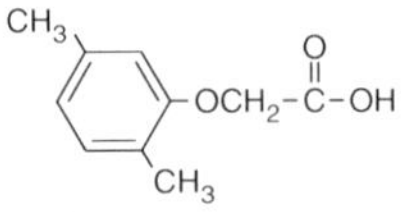

5 gm

S40,817-4
3,5-DIMETHYLPHENOXYACETIC ACID

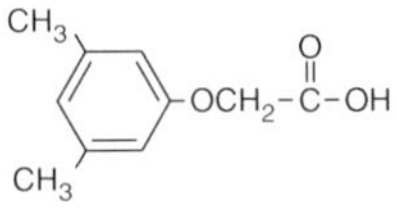

5 gm

S66,075-2
2-(P-ANISYL)-3-PHENYLSUCCINIC ACID

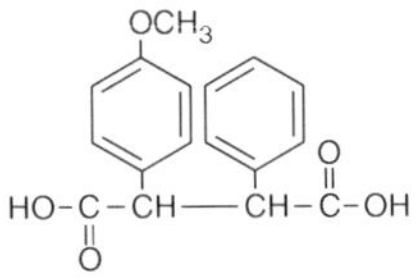

5 gm

S66,151-1
6,7,8,8A,9,10-HEXAHYDRO-2-METHOXYPHENANTHRENE-7,8-DICARBOXYLIC ACID

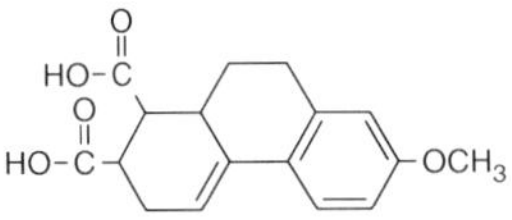

250 mg

S45,046-4
1-(4-METHOXYPHENYL)-1-CYCLOPROPANECARBOXYLIC ACID

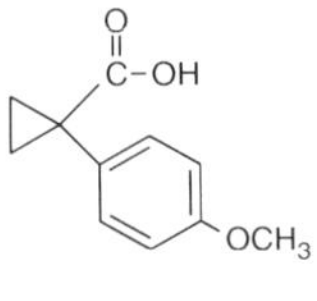

5 gm

S83,690-7
2-(4-METHOXYPHENYL)-2-PHENYLACETIC ACID

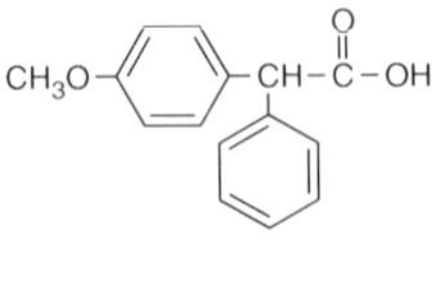

250 mg

S37,661-2
2-ETHOXYPHENYLACETIC ACID

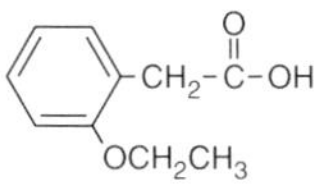

5 gm

S51,576-0
2-(HEXADECYLOXY)HYDROCINNAMIC ACID

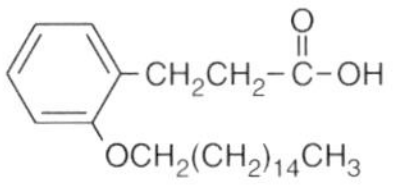

250 mg

S54,293-8
3-(4-METHOXYPHENOXY)PROPIONIC ACID

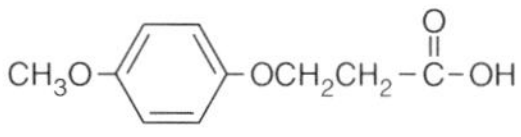

250 mg

S53,264-9
4-(P-TOLYL)BUTYRIC ACID

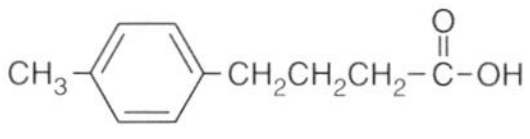

250 mg

S87,660-7
(-)-2-(4-METHOXYPHENYL)BUTYRIC ACID

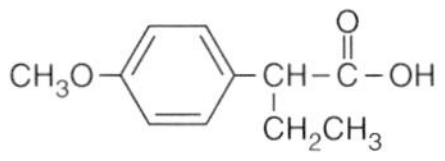

250 mg

S87,949-5
DL-2-(4-METHOXYPHENYL)BUTYRIC ACID

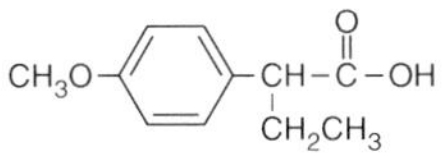

5 gm

S94,059-3
3-(4-ETHOXYPHENYLTELLURO)-PROPIONIC ACID

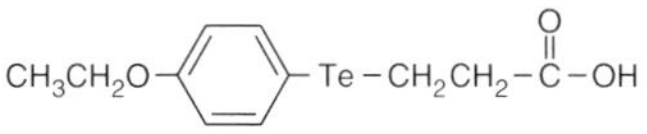

250 mg

S94,065-8
3-(4-ETHOXYPHENYLTELLURO)PROPIONIC ACID, TE,TE-DICHLORIDE

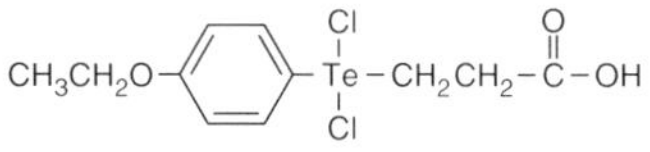

250 mg

S94,063-1
3-(4-ETHOXYPHENYLTELLURO)PROPIONIC ACID, TE,TE-DIBROMIDE

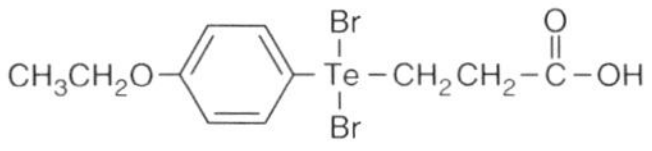

250 mg

S94,066-6
3-(4-ETHOXYPHENYLTELLURO)PROPIONIC ACID, TE,TE-DIIODIDE

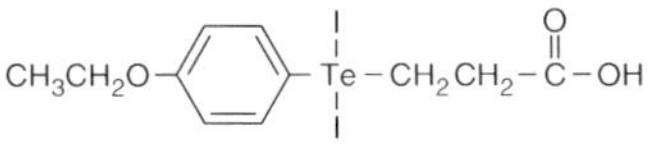

250 mg

S50,420-3
4-BUTOXYPHENYLACETIC ACID

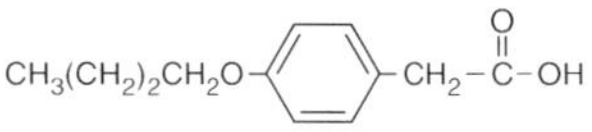

5 gm

S57,842-8
4-[4-(HEXADECYLOXY)PHENYL]-BUTYRIC ACID

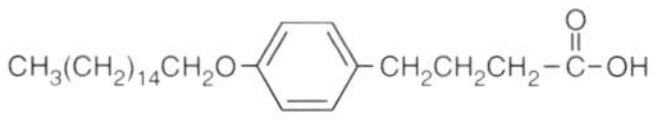

5 gm

S36,809-1
4-(2,4-DICHLOROPHENOXY)BUTYRIC ACID

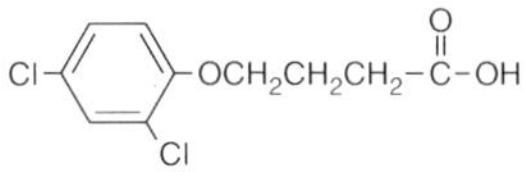

250 mg

S95,379-2
3-(4-METHOXY-3-METHYLPHENYL)-PROPIONIC ACID

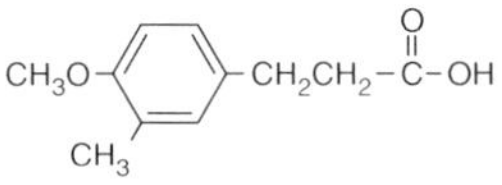

50 mg

S41,231-7
4-(4-CHLORO-M-TOLYLOXY)-BUTYRIC ACID

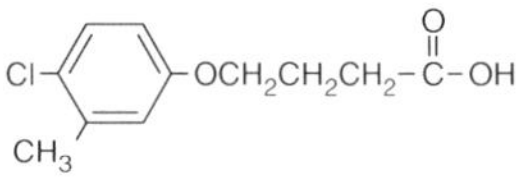

5 gm

S59,799-6
4-[3-BROMO-4-(HEXADECYLOXY)PHENYL]-BUTYRIC ACID

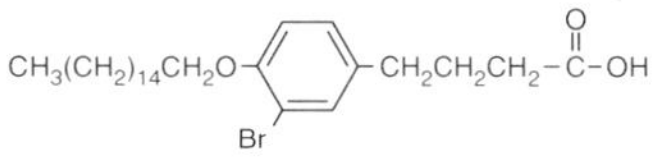

5 gm

S43,251-2
2-PHENYL-2-(2,4,6-TRIIODOPHENOXY)ACETIC ACID

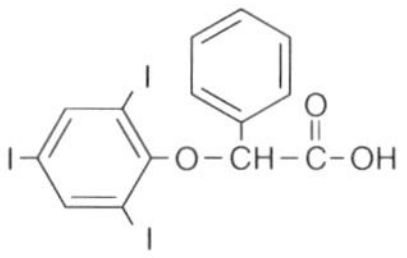

250 mg

S86,156-1
6-METHOXY-2-METHYL-1,2,3,4-TETRAHYDRO-1-NAPHTHOIC ACID

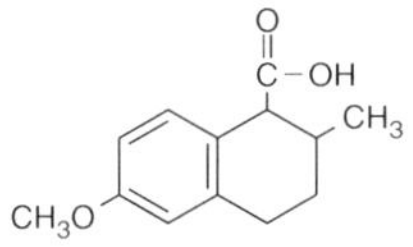

250 mg

S67,162-2
(+)-2-(1-NAPHTHOXY)PROPIONIC ACID

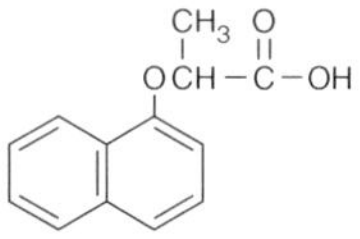

250 mg

S81,689-2
2,3-DIHYDRO-3-PHENYL-2-BENZOFURANCARBOXYLIC ACID

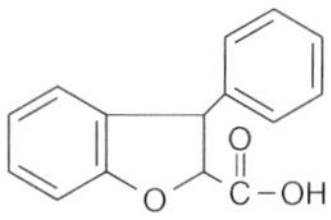

250 mg

S58,871-7
NAPHTHO(2,3-D)-1,3-DIOXOLENE-2-CARBOXYLIC ACID

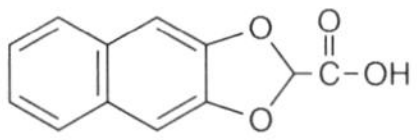

5 gm

S95,343-1
4-(3-AMINOPHENYL)-4-OXOBUTYRIC ACID

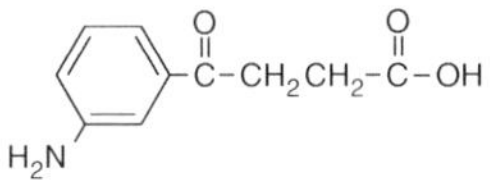

50 mg

S95,371-7
4-OXO-4-(O-TOLYL)BUTYRIC ACID

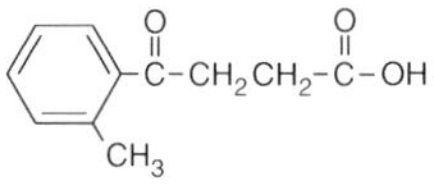

50 mg

S86,509-5
1,2,3,4-TETRAHYDRO-1,4-METHANONAPHTHALENE-2,3-DICARBOXYLIC ACID

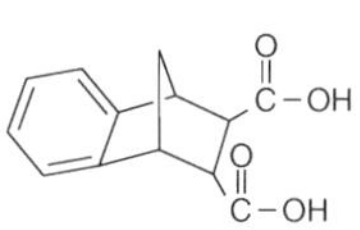

250 mg

S86,573-7
9,10-DIHYDRO-9,10-ETHANOANTHRACENE-11,12-DICARBOXYLIC ACID

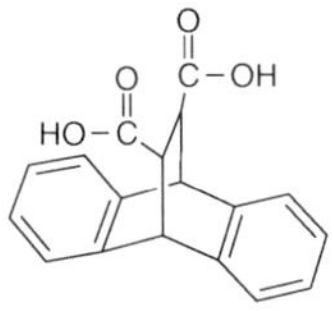

250 mg

S88,354-9
2,3-DIMETHYL-2-PHENYLSUCCINIC ACID

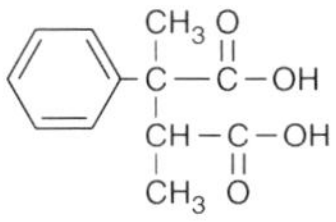

250 mg

S50,195-6
2,2-DIPHENYL-2-MERCAPTOACETIC ACID

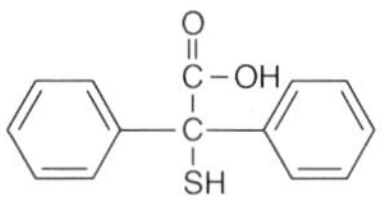

250 mg

S61,488-2
2-(4-AMINOPHENYLTHIO)ACETIC ACID

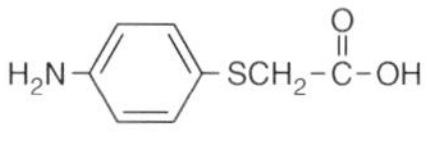

5 gm

S51,328-8
2-(4-NITROPHENYLTHIO)ACETIC ACID

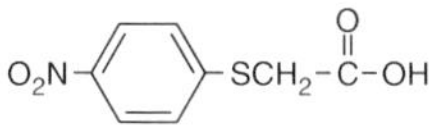

250 mg

S88,693-9
2-[5-NITRO-2-(P-TOLYLTHIO)-PHENYL]ACETIC ACID

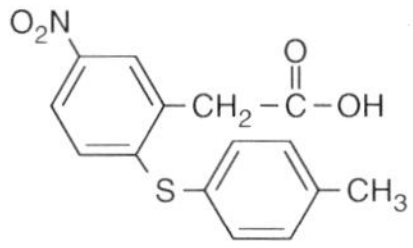

250 mg

S66,221-6
2-(2-NITROBENZYLTHIO)ACETIC ACID

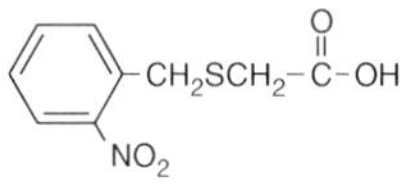

5 gm

S36,431-2
2-[7-OXO-7H-BENZ(DE)ANTHRACEN-3-YLTHIO]ACETIC ACID

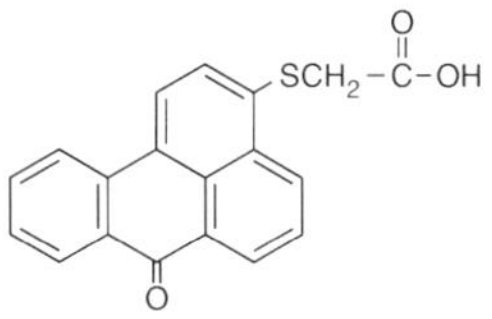

5 gm

S64,752-7
PIPERONAL BIS(CARBOXYMETHYL)-MERCAPTAL

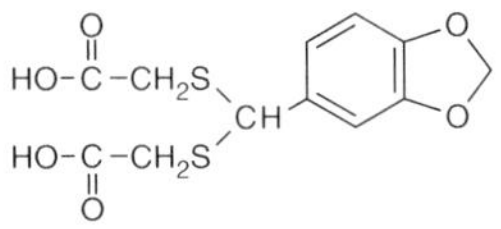

250 mg

S88,416-2
2-(PHENYLTHIO)PROPIONIC ACID

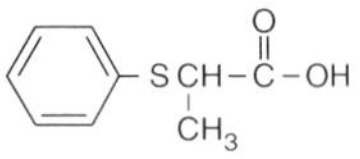

250 mg

S54,156-7
2-(PHENETHYLTHIO)ACETIC ACID

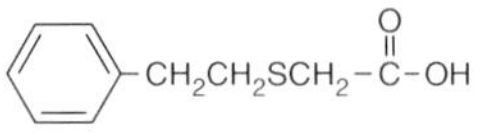

250 mg

S54,157-5
3-(BENZYLTHIO)PROPIONIC ACID

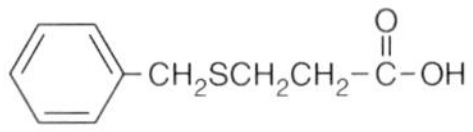

250 mg

S94,769-5
11-(PHENYLTHIO)UNDECANOIC ACID

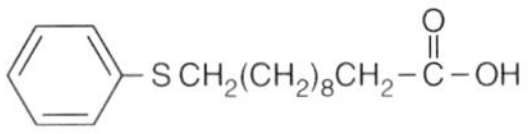

250 mg

S84,956-1
3-(O-TOLYLTHIO)PROPIONIC ACID

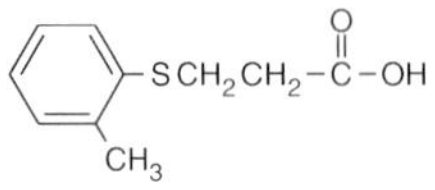

250 mg

S63,790-4
2-(4-CHLOROPHENYLTHIO)ACETIC ACID

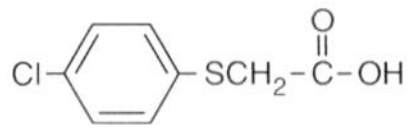

250 mg

S75,571-0
2-(P-TOLYLTHIO)ACETIC ACID

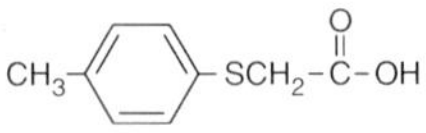

250 mg

S84,958-8
3-(P-TOLYLTHIO)PROPIONIC ACID

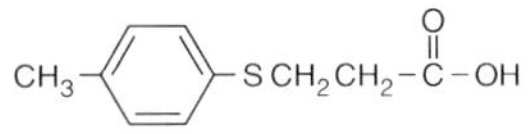

250 mg

S88,824-9
3-(1-NAPHTHYLTHIO)PROPIONIC ACID

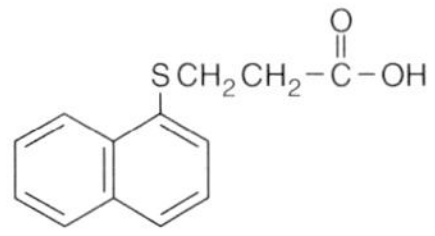

250 mg

S76,185-0
3-AMINO-3-(2-NITROPHENYL)-PROPIONIC ACID

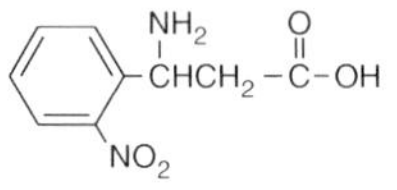

250 mg

S76,203-2
2-AMINO-ALPHA,BETA-DIBROMOHYDROCINNAMIC ACID

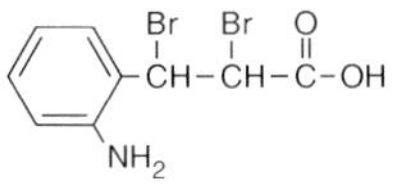

250 mg

S50,658-3
BETA-AMINO-5-INDANPROPIONIC ACID

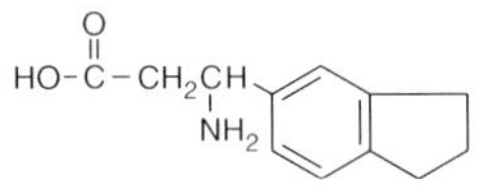

250 mg

S53,127-8
3-(4-CHLORO-2-NITROANILINO)-PROPIONIC ACID

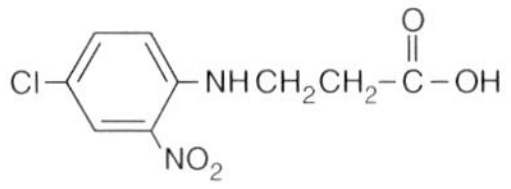

5 gm

S66,828-1
3-[PHENYL(P-TOSYL)AMINO]-PROPIONIC ACID

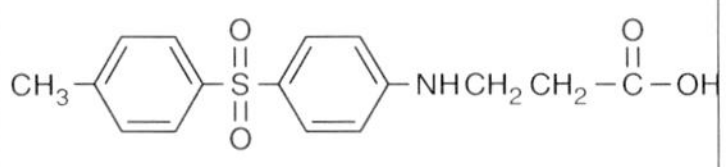

250 mg

S94,602-8
N-BENZYLGLYCINE HYDROCHLORIDE

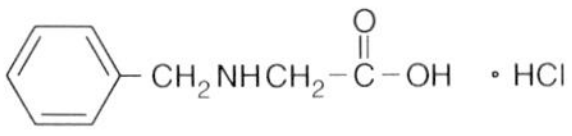

250 mg

S81,877-1
2-[ALPHA-(BENZYLAMINO)BENZYL]-BUTYRIC ACID

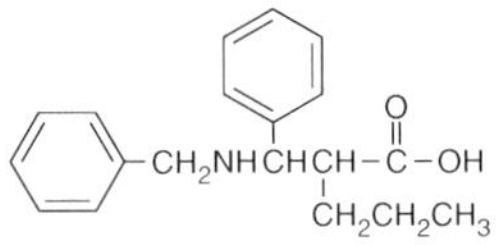

250 mg

S81,065-7
DIBUTYLTETRADECYLAMINE

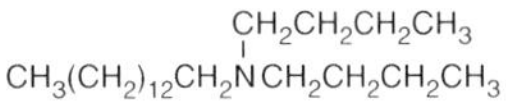

250 mg

S92,044-4
10-PHENOTHIAZINEPROPIONIC ACID

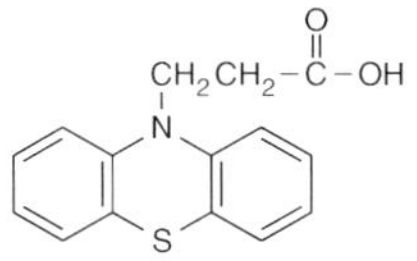

250 mg

S50,671-0
BETA-(HYDROXYAMINO)-HYDROCINNAMIC ACID

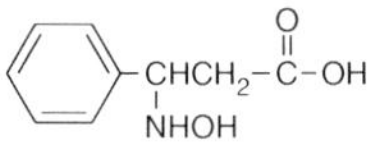

5 gm

S85,250-3
3-BROMO-3-(3-NITROPHENYL)-PROPIONIC ACID

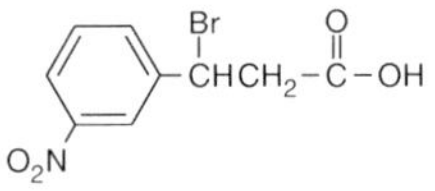

250 mg

S64,004-2
2-(4-NITROPHENYL)SUCCINIC ACID

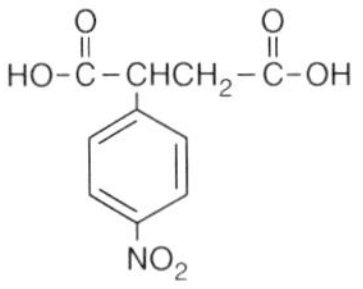

250 mg

S74,508-1
ALPHA,BETA-DIBROMO-2-NITROHYDROCINNAMIC ACID

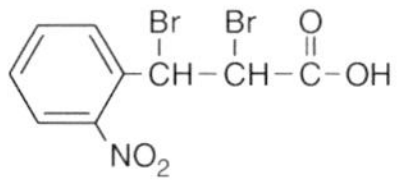

250 mg

S87,721-2
2-(4-NITROPHENYL)BUTYRIC ACID

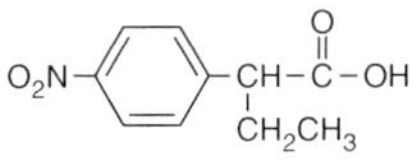

250 mg

S89,590-3
5-NITRO-1,2,3,4-TETRAHYDRO-1,4-ETHENONAPHTHALENE-2,3-DICARBOXYLIC ACID

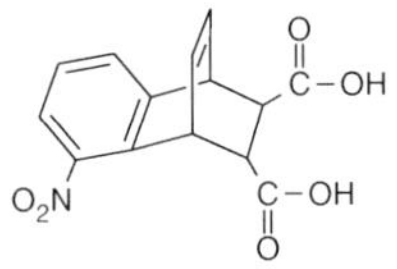

250 mg

S51,497-7
BETA-METHYL-3-NITROHYDROCINNAMIC ACID

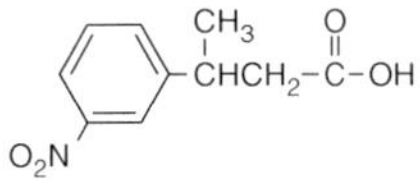

250 mg

S57,427-9
4-(2-NITROPHENYL)BUTYRIC ACID

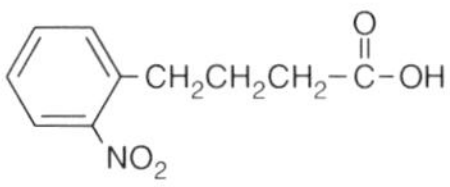

250 mg

S56,805-8
ALPHA,BETA-DIBROMO-4-NITROHYDROCINNAMIC ACID

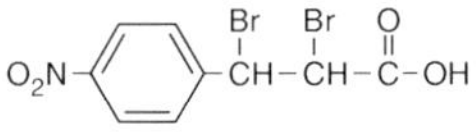

250 mg

S51,493-4
BETA-METHYL-4-NITROHYDROCINNAMIC ACID

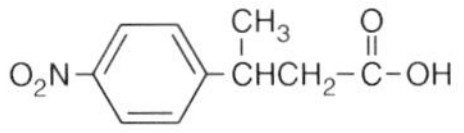

250 mg

S50,051-8
3-BROMO-4-OXO-4-PHENYLBUTYRIC ACID

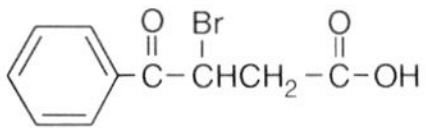

250 mg

S77,825-7
2-BENZOYLPROPIONIC ACID

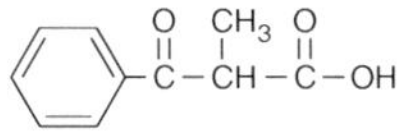

250 mg

S82,651-0
10-OXO-10-PHENYLDECANOIC ACID

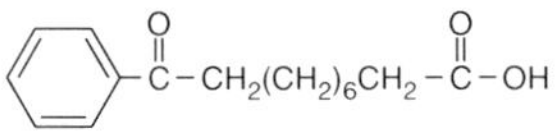

250 mg

S84,228-1
4-OXO-2,3,4-TRIPHENYLBUTYRIC ACID
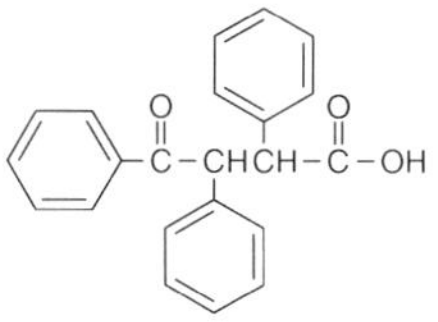
250 mg

S87,691-7
4-(4-BIPHENYLYL)-4-OXOBUTYRIC ACID
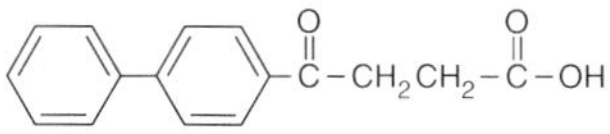
250 mg

S76,641-0
3-(8-BENZYL-9-OXO-6,7,8,9-TETRAHYDRO-5H-BENZOCYCLOHEPTEN-8-YL)PROPIONIC ACID
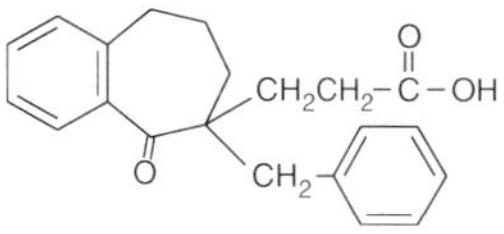
250 mg

S76,181-8
3-[4-(2,2-DIMETHYLPROPIONYL)-BENZOYL]PROPIONIC ACID
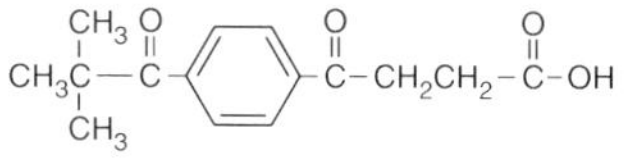
250 mg

S80,919-5
3-OXO-1-INDANCARBOXYLIC ACID
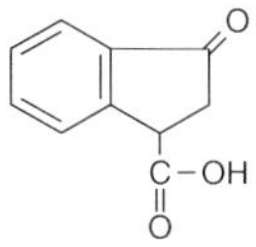
250 mg

S84,609-0
4-OXO-1,2,3,4-TETRAHYDRO-2-NAPHTHALENEPROPIONIC ACID
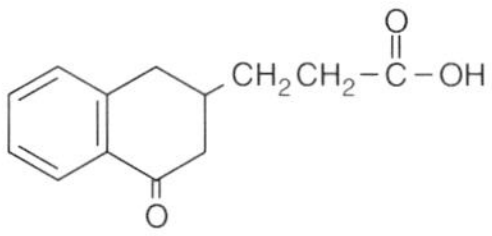
250 mg

S72,246-4
2-METHYL-3-(1-NAPHTHOYL)-PROPIONIC ACID
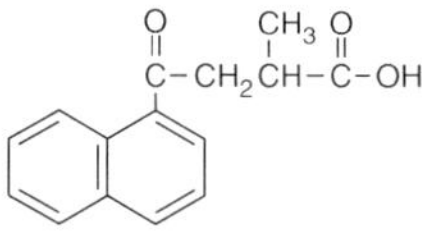
250 mg

S74,255-4
GAMMA-OXO-2-NAPHTHALENEBUTYRIC ACID
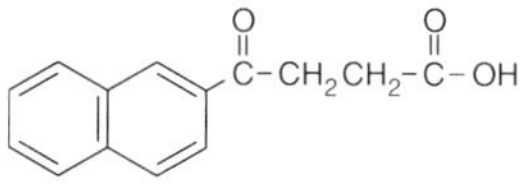
250 mg

S45,055-3
5-BENZOYLVALERIC ACID
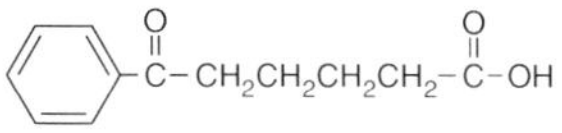
5 gm

S34,648-9
4,7-DIOXO-7-PHENYLHEPTANOIC ACID
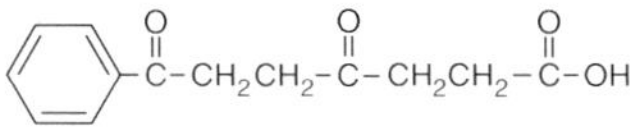
5 gm

S61,722-9
3-BENZOYL-2-PHENYLPROPIONIC ACID
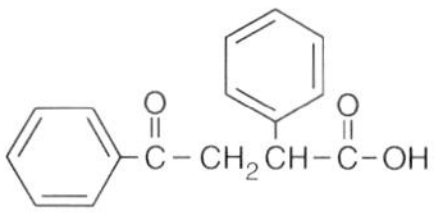
250 mg

S92,618-3
BIS(PHENACYL)ACETIC ACID
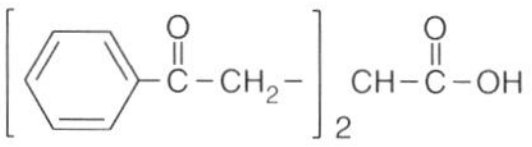
250 mg

S67,248-3
4'-(CARBOXYMETHYL)CHALCONE
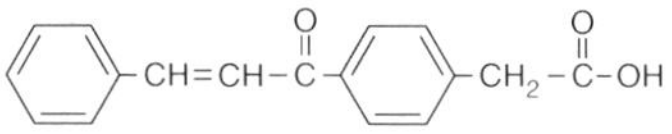
5 gm

S84,624-4
3-OXO-1-PHENYL-1-INDANACETIC ACID
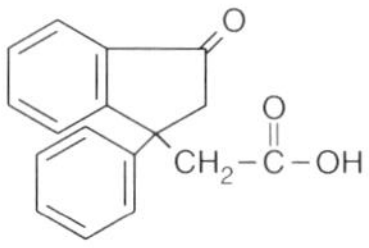
250 mg

S62,064-5
4-(4-ACETYLPHENYL)BUTYRIC ACID
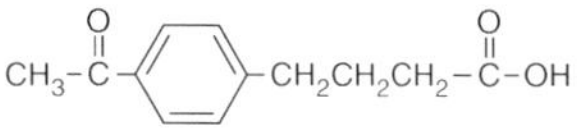
250 mg

S34,681-0
3-(2,4-DIMETHYLBENZOYL)-PROPIONIC ACID
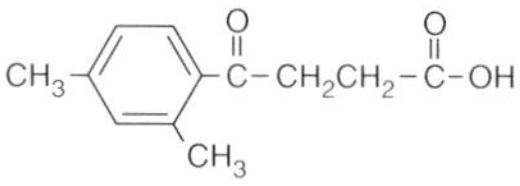
5 gm

S34,682-9
3-(2,5-DIMETHYLBENZOYL)-PROPIONIC ACID
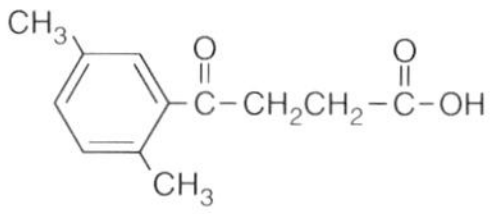
5 gm

S84,603-1
2-(3-OXO-1-INDANYL)ACETIC ACID
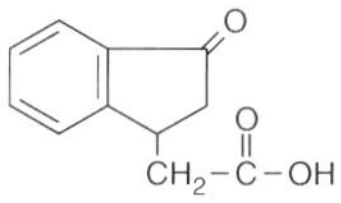
250 mg

S61,969-8
4-(2-ANTHRANOYL)BUTYRIC ACID
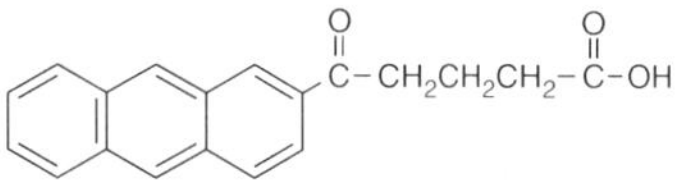
250 mg

S66,248-8
GAMMA-OXO-2-PHENANTHRENEBUTYRIC ACID
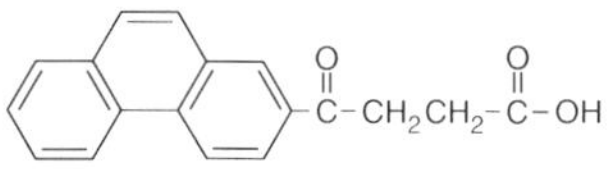
250 mg

S89,518-0
9-OXO-1,2,3,4-TETRAHYDRO-1,4-ETHANONAPHTHALENE-2,3-DICARBOXYLIC ACID
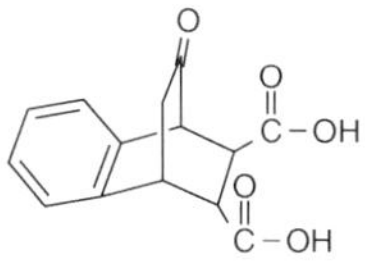
250 mg

S76,197-4
2,6-DIPHENYL-4-OXO-3-PIPERIDINEACETIC ACID
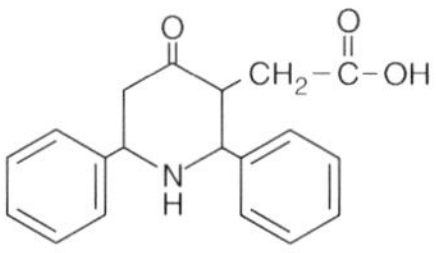
250 mg

S84,642-2
3-OXO-2-PHENYL-1-CYCLOPENTANECARBOXYLIC ACID
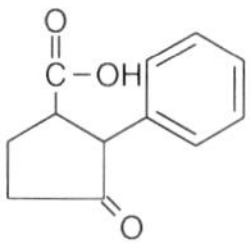
250 mg

S42,088-3
1,4-DIHYDRO-3,5-DIIODO-4-OXO-ALPHA-PHENYL-1-PYRIDINEACETIC ACID
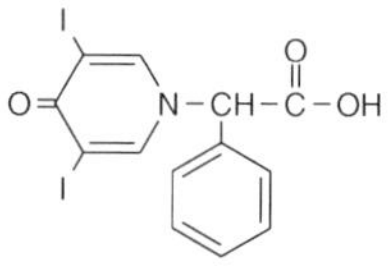
250 mg

S43,559-7
2-PHENYLLEVULINIC ACID
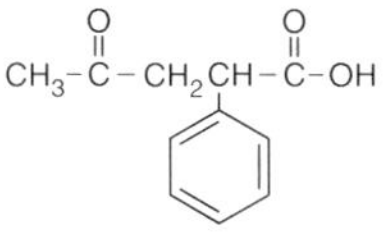
250 mg

S54,480-9
3-(4-BROMOBENZOYL)-PROPIONIC ACID
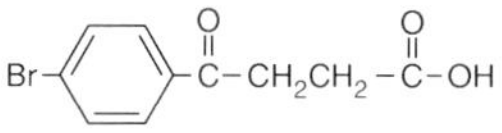
5 gm

S53,267-3
4-OXO-4-(PARA-TOLYL)BUTYRIC ACID
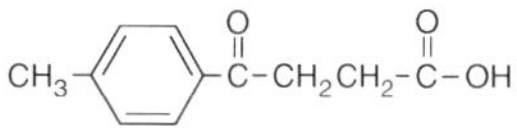
250 mg

S86,239-8
2-(THIOBENZOYL)THIOLACETIC ACID
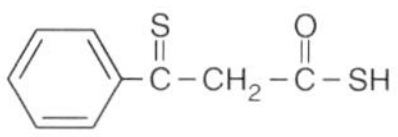
250 mg

S51,042-4
GAMMA-OXO-1-NAPHTHALENEBUTYRIC ACID
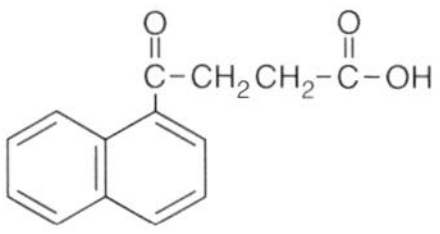
5 gm

S85,213-9
9,10-DIHYDRO-9,10-ETHANOANTHRACENE-11,12-DICARBOXYLIC ACID
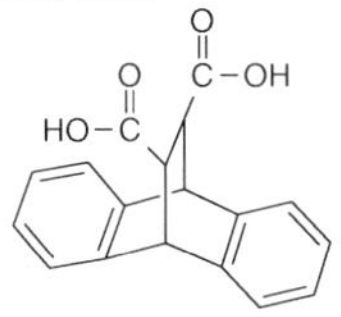
250 mg

S75,791-8
4-CARBOXY-4-CHROMANACETIC ACID
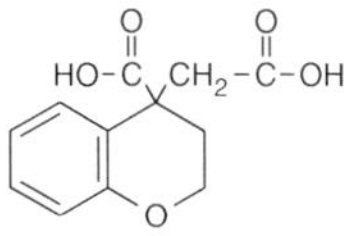
250 mg

S80,831-8
2-METHYL-3-PHENYLSUCCINIC ACID
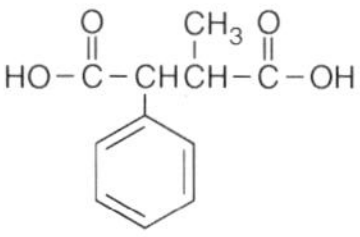
250 mg

S84,605-8
2-BENZYLSUCCINIC ACID

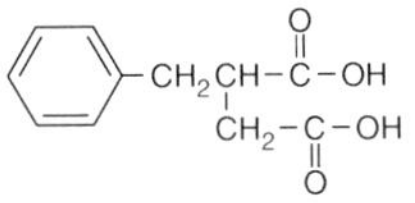

250 mg

S84,611-2
3-BENZYLADIPIC ACID

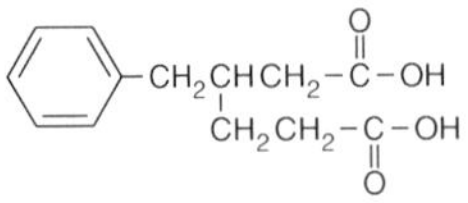

250 mg

S85,018-7
3-PHENYL-1,3,5-PENTANETRICARBOXYLIC ACID

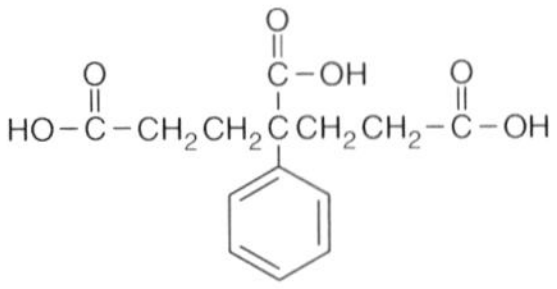

250 mg

S82,625-1
2,3-DIPHENYLSUCCINIC ACID

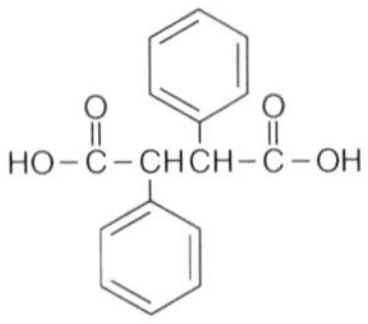

250 mg

S51,911-1
2-ISOPROPYL-2-PHENYLACETIC ACID

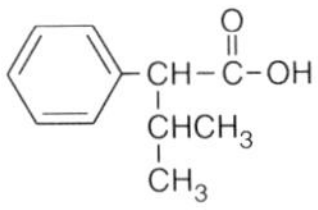

250 mg

S62,059-9
1,3-PHENYLENEDIPROPIONIC ACID

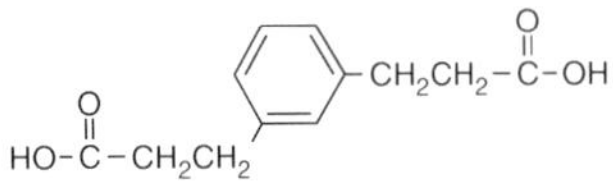

250 mg

S35,322-1
3-METHYL-3-PHENYLGLUTARIC ACID

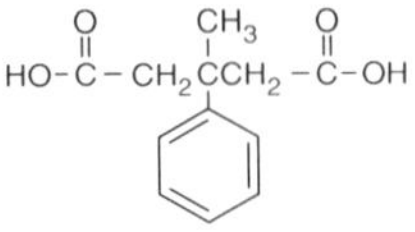

1 gm

S34,955-0
3-ETHYL-3-PHENYLGLUTARIC ACID

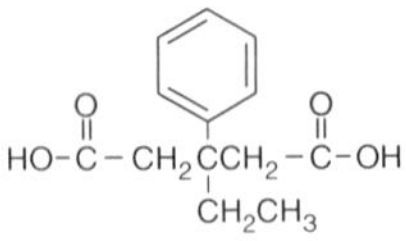

250 mg

S54,122-2
2-PHENYL-2-PROPYLSUCCINIC ACID

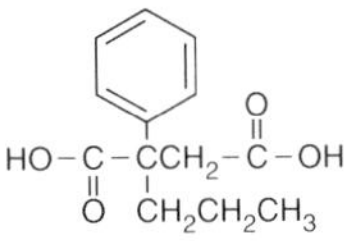

250 mg

S35,472-4
3-PHENYL-3-PROPYLGLUTARIC ACID

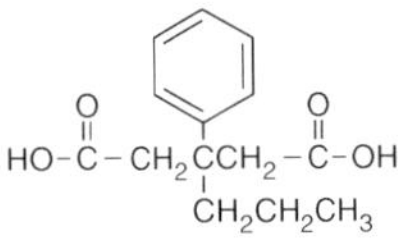

250 mg

S64,302-5
2-(CARBOXYMETHYL)HYDROCINNAMIC ACID

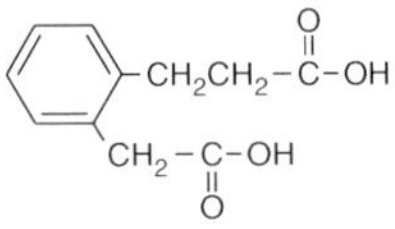

250 mg

S74,260-0
2,2'-BIPHENYLDIACETIC ACID

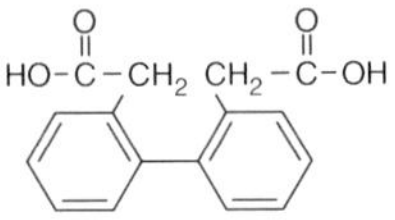

5 gm

S38,372-4
4-ACETOXY-ALPHA-PHENYLHYDROCINNAMIC ACID

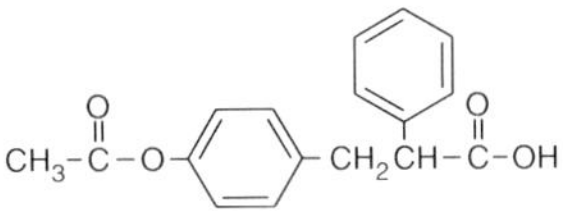

5 gm

S44,820-6
1-(P-TOLYL)-1-CYCLOPROPANECARBOXYLIC ACID

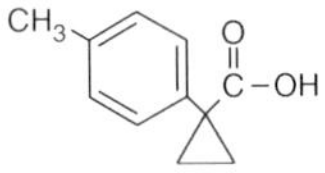

5 gm

S43,389-6
1-(M-TOLYL)-1-CYCLOPENTANECARBOXYLIC ACID

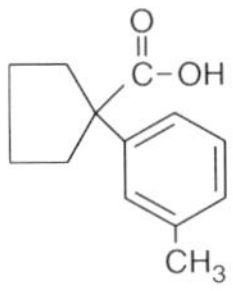

5 gm

S80,924-1
2-PHENYL-1-CYCLOPENTANECARBOXYLIC ACID

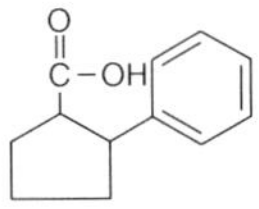

250 mg

S90,638-7
1-INDANACETIC ACID

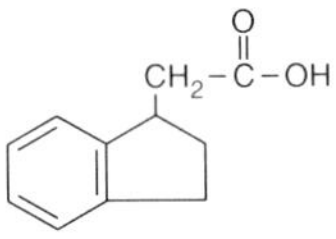

250 mg

S84,315-6
6-BROMO-3,4-(METHYLENEDIOXY)-HYDROCINNAMIC ACID

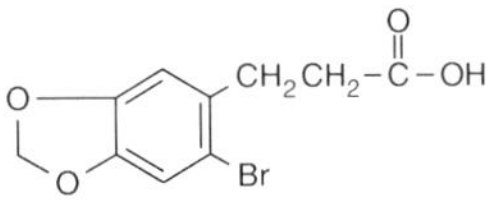

250 mg

S44,818-4
1-(P-TOLYL)-1-CYCLOHEXANECARBOXYLIC ACID

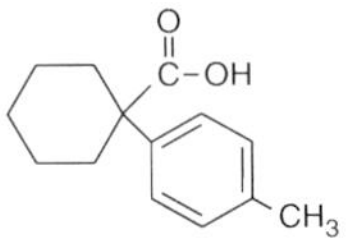

5 gm

S38,351-1
ALPHA-(ORTHO-TOLYL)-CYCLOHEXANEACETIC ACID

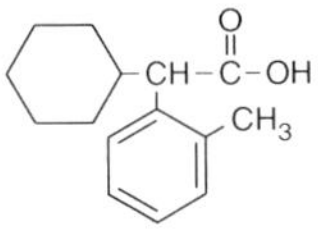

250 mg

S45,047-2
1-(4-CHLOROPHENYL)-1-CYCLOHEXANE-CARBOXYLIC ACID

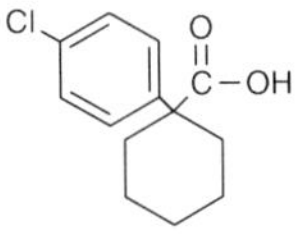

5 gm

S73,441-1
2-(2-PHENYL-1-CYCLOHEXENYL)ACETIC ACID

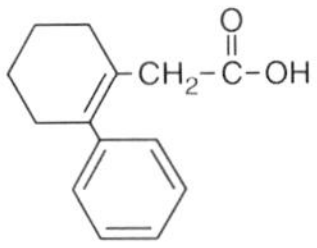

250 mg

S86,427-7
9,10-DIHYDRO-12-PHENYL-9,10-ETHANOANTHRACENE-11-CARBOXYLIC ACID

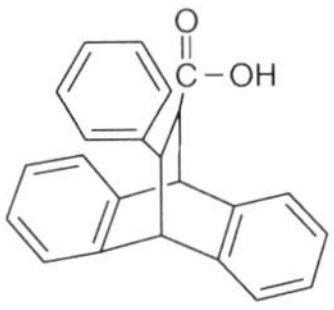

250 mg

S86,416-1
9,10-DIHYDRO-9,10-ETHANOANTHRACENE-11-CARBOXYLIC ACID

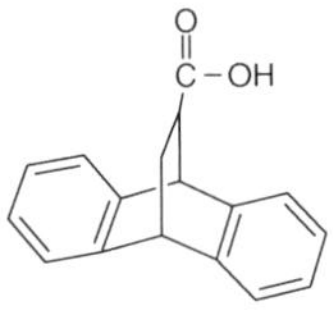

250 mg

S38,154-3
4-PHENYLTETRAHYDROPYRAN-4-CARBOXYLIC ACID

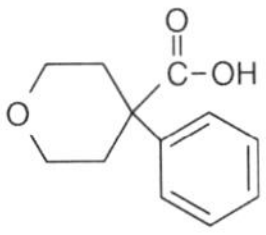

250 mg

S67,781-7
BIS(4-PIVALOYLBENZYL)ACETIC ACID 2,2-DIMETHYLTRIMETHYLENE ACETAL

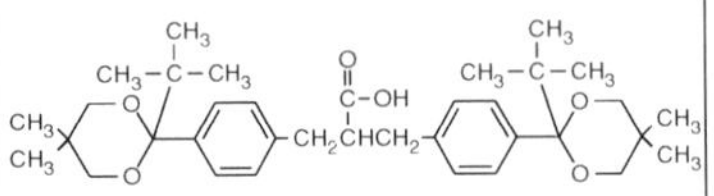

5 gm

S79,252-7
(+/-)-ALPHA-CHLOROPHENYLACETIC ACID

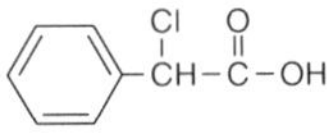

250 mg

S76,484-1
2,3-DIBROMO-3-PHENYLPROPIONIC ACID

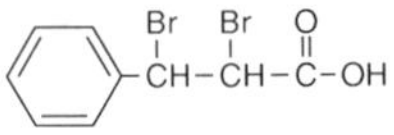

5 gm

S94,062-3
3-(PHENYLTELLURO)PROPIONIC ACID, TE,TE-DIBROMIDE

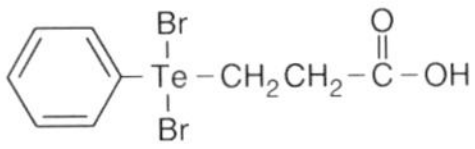

250 mg

S42,731-4
4-PHENYLVALERIC ACID

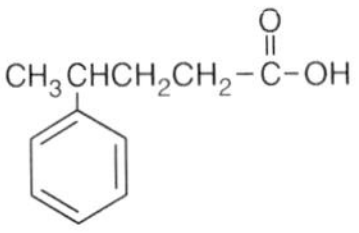

250 mg

S69,721-4
2-METHYL-3-PHENYLBUTYRIC ACID

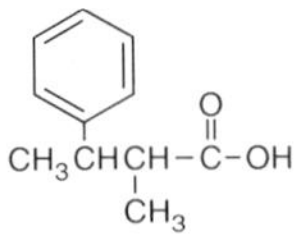

250 mg

S87,623-2
(+)-3-METHYL-2-PHENYLBUTYRIC ACID

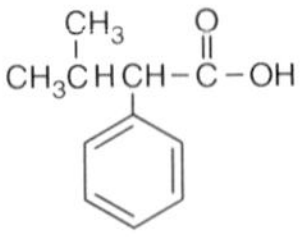

250 mg

S50,313-4
BETA-PROPYLHYDROCINNAMIC ACID
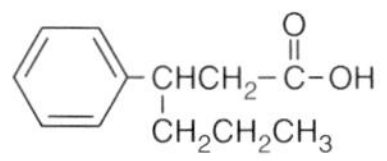
250 mg

S38,197-7
2-PHENYLHEPTANOIC ACID
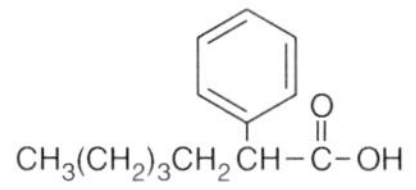
250 mg

S38,329-5
3-METHYL-2-PHENYLHEXANOIC ACID
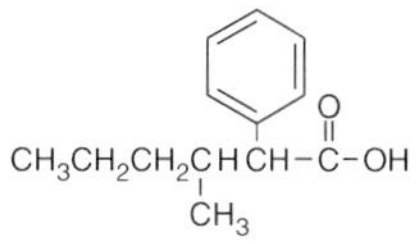
5 gm

S50,302-9
BETA-BUTYLHYDROCINNAMIC ACID
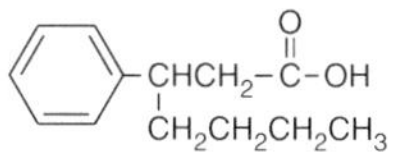
250 mg

S84,247-8
3,4-DIPHENYLBUTYRIC ACID
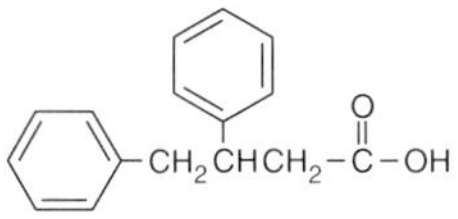
250 mg

S84,248-6
2,4-DIPHENYLBUTYRIC ACID
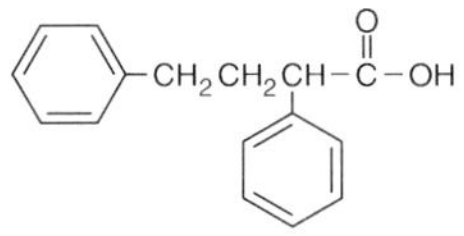
250 mg

S94,594-3
2,3,3-TRIPHENYLPROPIONIC ACID
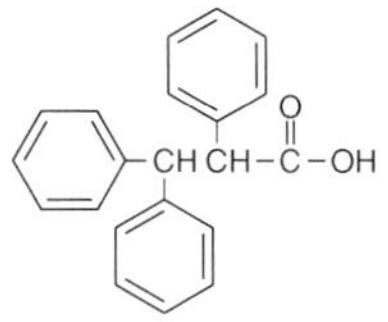
5 gm

S58,157-7
2,3,4-TRIPHENYLBUTYRIC ACID
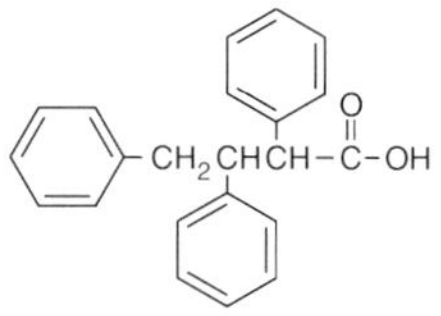
250 mg

S93,566-2
DL-2-(4-BIPHENYLYL)PROPIONIC ACID
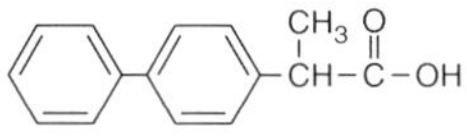
250 mg

S93,775-4
(S)-(+)-2-(4-BIPHENYLYL)-PROPIONIC ACID
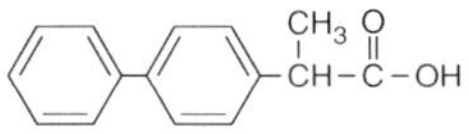
250 mg

S83,389-4
2,2-DIPHENYLHEPTANOIC ACID
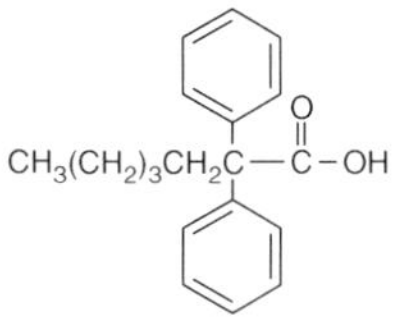
250 mg

S37,100-9
3,3,3-TRIPHENYLPROPIONIC ACID
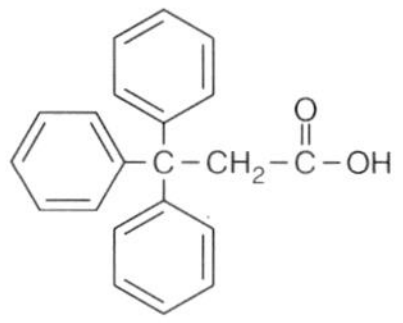
250 mg

S82,627-8
2-(1-NAPHTHYL)-2-PHENYLACETIC ACID
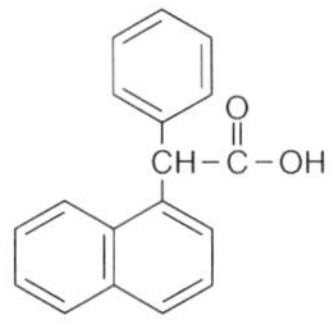
250 mg

S81,876-3
2-[ALPHA-(2-NAPHTHYL)BENZYL]-PROPIONIC ACID
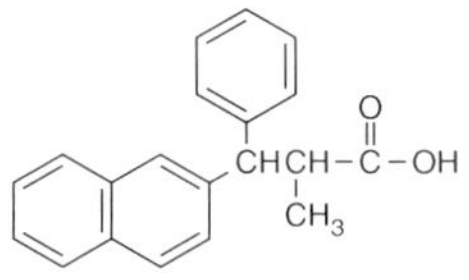
250 mg

S37,677-9
2-(O-TOLYL)-PROPIONIC ACID
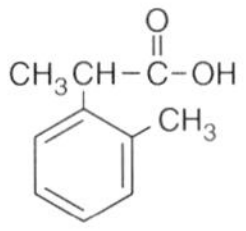
250 mg

S38,149-7
2-(O-TOLYL)-BUTYRIC ACID
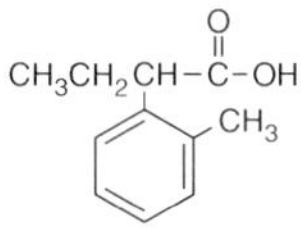
250 mg

S37,730-9
3-METHYL-2-(O-TOLYL)-BUTYRIC ACID
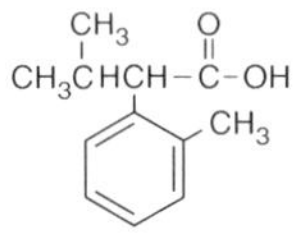
250 mg

S38,165-9
2-(O-TOLYL)VALERIC ACID
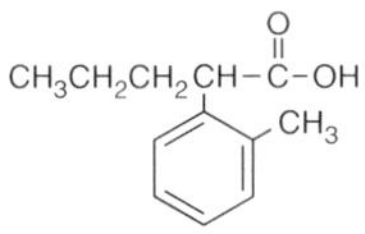
5 gm

S38,181-0
4-METHYL-2-(O-TOLYL)-VALERIC ACID
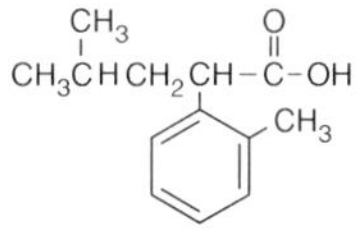
250 mg

S38,239-6
5-METHYL-2-(O-TOLYL)-HEXANOIC ACID
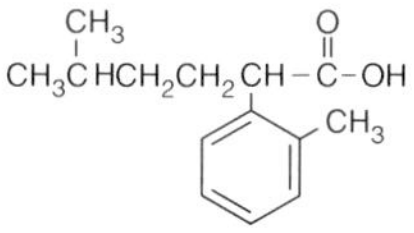
250 mg

S38,241-8
3-METHYL-2-(O-TOLYL)-HEXANOIC ACID
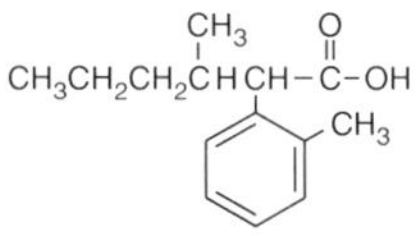
250 mg

S38,247-7
2-(O-TOLYL)HEPTANOIC ACID
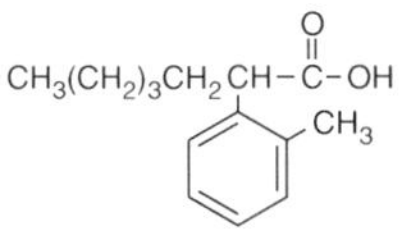
250 mg

S45,035-9
4-FLUORO-ALPHA-METHYLPHENYLACETIC ACID
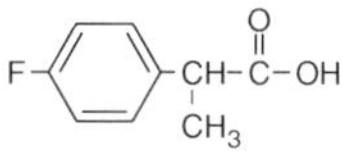
5 gm

S50,309-6
4-CHLORO-BETA-METHYLHYDROCINNAMIC ACID
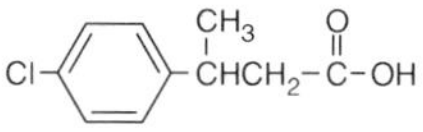
5 gm

S94,060-7
3-(P-TOLYLTELLURO)PROPIONIC ACID
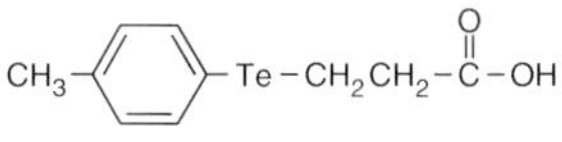
250 mg

S94,061-5
3-(P-TOLYLTELLURO)PROPIONIC ACID, TE,TE-DIBROMIDE
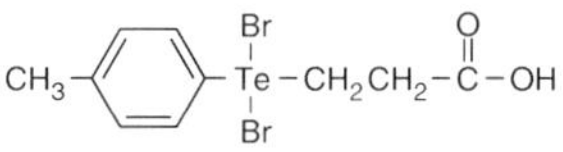
250 mg

S88,861-3
2-(4-ISOPROPYLPHENYL)ACETIC ACID
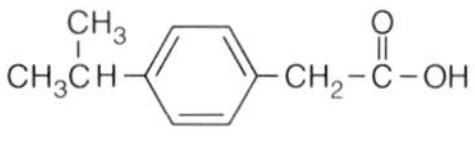
250 mg

S50,315-0
BETA-BUTYL-4-CHLOROHYDROCINNAMIC ACID
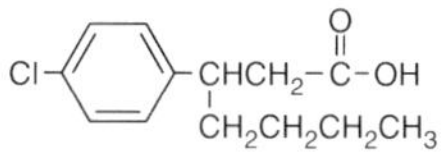
5 gm

S50,251-0
P-(TERT-BUTYL)-BETA-METHYLHYDROCINNAMIC ACID
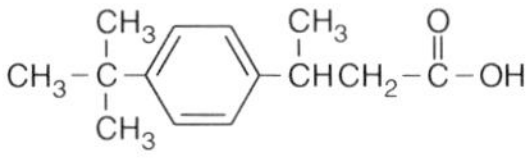
250 mg

S44,275-5
DI-P-TOLYLACETIC ACID
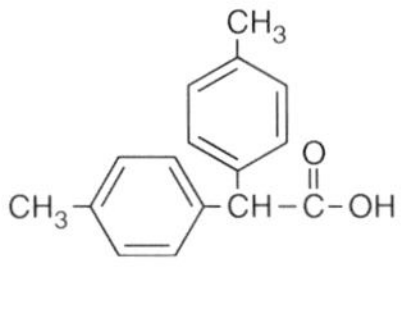
5 gm

S39,042-9
4-(4-HEXADECYLPHENYL)BUTYRIC ACID
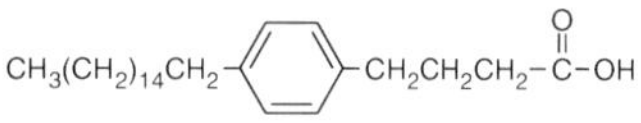
5 gm

S34,720-5
4-(2,4-DIMETHYLPHENYL)BUTYRIC ACID
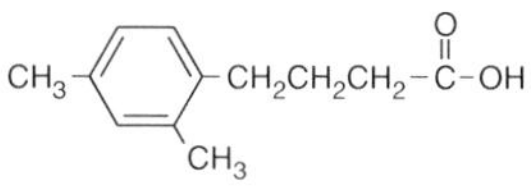
5 gm

S34,721-3
4-(2,5-DIMETHYLPHENYL)BUTYRIC ACID
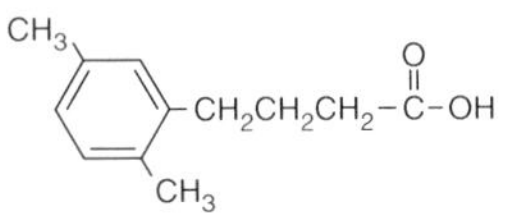
5 gm

S71,446-1
4-(3,4-DIMETHYLPHENYL)BUTYRIC ACID
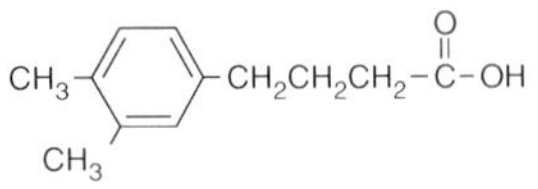
5 gm

S81,811-9
2-(3,5-XYLYL)ACETIC ACID
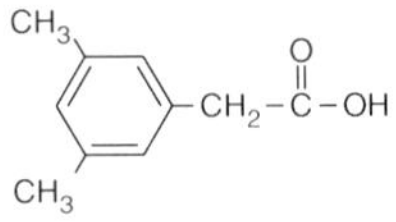
250 mg

S36,307-3
2-MESITYLENEACETIC ACID
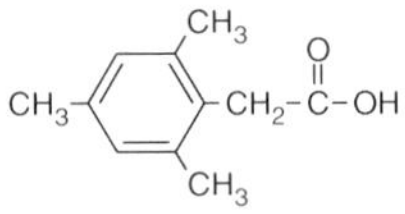
5 gm

S41,604-5
1-INDANCARBOXYLIC ACID
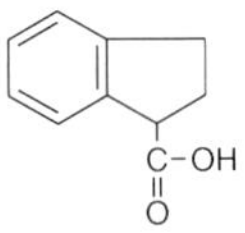
250 mg

S88,274-7
1,2,3,4-TETRAHYDRO-2-NAPHTHALENEPROPIONIC ACID
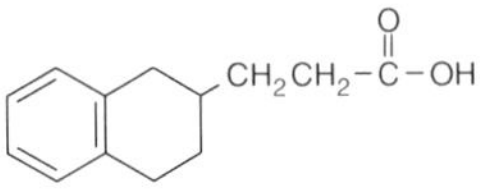
250 mg

S39,882-9
1,2,3,4-TETRAHYDRO-2-NAPHTHALENEBUTYRIC ACID
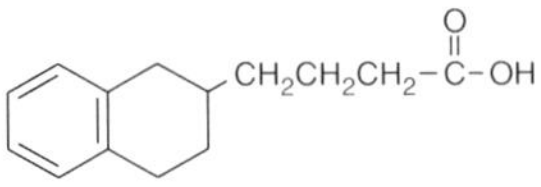
5 gm

S35,309-4
2-METHYL-1-NAPHTHYLACETIC ACID
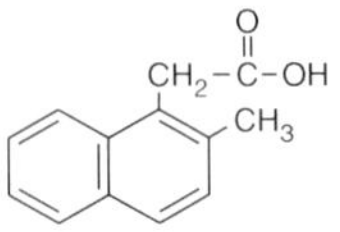
5 gm

S81,775-9
2-NAPHTHALENEPROPIONIC ACID
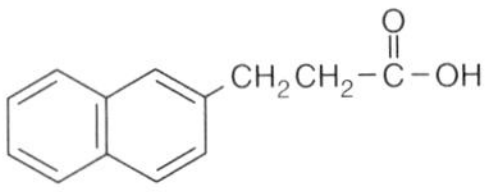
250 mg

S94,581-1
4-(2-NAPHTHYL)BUTYRIC ACID
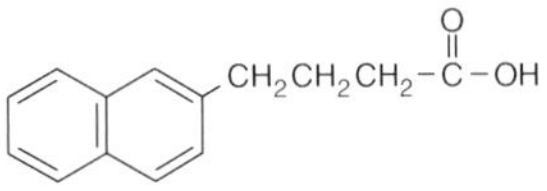
250 mg

S97,522-2
AR-H29 BETA-METHYL-1-PYRENEPROPIONIC ACID
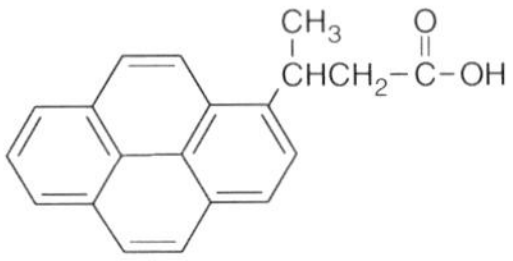
10 mg

S54,486-8
9,10-DIHYDRO-9-ANTHROIC ACID
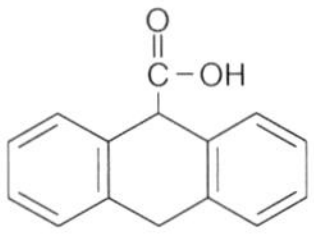
250 mg

S97,335-1
AR-D32A 4-PHENANTHRENEACETIC ACID
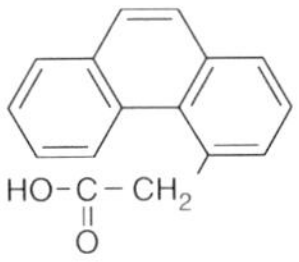
10 mg

S85,295-3
5H-DIBENZO(A,D)CYCLOHEPTENE-5-CARBOXYLIC ACID
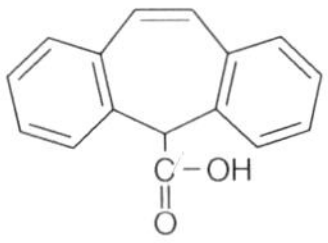
250 mg

S80,870-9
ALPHA-(BROMOMETHYL)CINNAMIC ACID
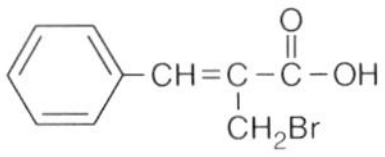
250 mg

S79,851-7
ALPHA[2-AMINO-4-(TRIFLUOROMETHYL)-PHENYL]-4-(TRIFLUOROMETHYL)-CINNAMIC ACID
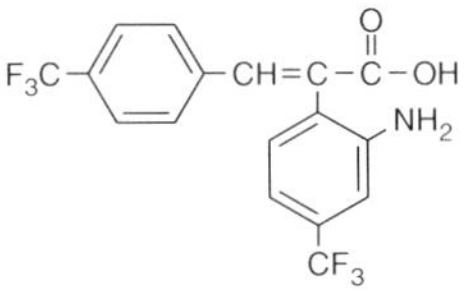
250 mg

S56,713-2
M-(TRIFLUOROMETHYL)CINNAMIC ACID
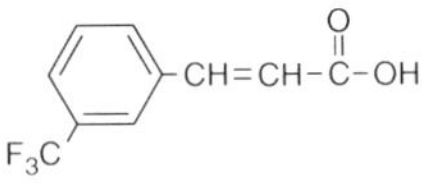
5 gm

S92,173-4
2-BROMO-4,5-(DIMETHOXY)-CINNAMIC ACID
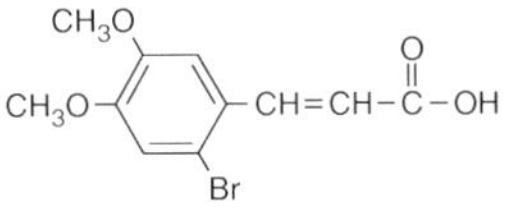
250 mg

S76,260-1
3-CHLORO-ALPHA-MERCAPTOCINNAMIC ACID
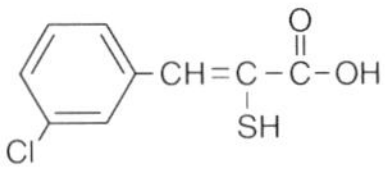
250 mg

S80,869-5
2-CHLORO-ALPHA-METHYLCINNAMIC ACID
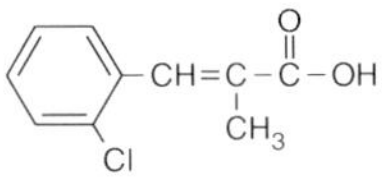
250 mg

S80,868-7
4-CHLORO-ALPHA-METHYLCINNAMIC ACID
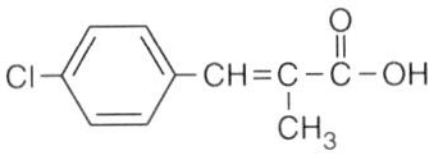
250 mg

S75,588-5
4-HYDROXY-ALPHA-MERCAPTO-3-METHOXYCINNAMIC ACID
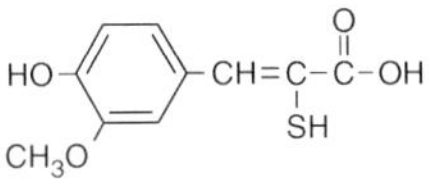
250 mg

S95,526-4
2-(1-INDENYLIDENE)GLYCOLIC ACID
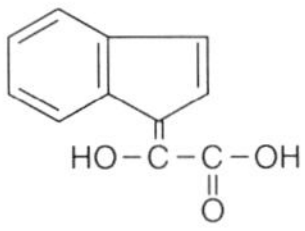
50 mg

S75,659-8
3-HYDROXY-ALPHA-MERCAPTO-BETA-METHYLCINNAMIC ACID
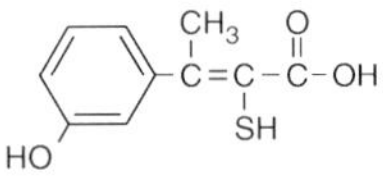
250 mg

S35,117-2
4-HYDROXY-ALPHA-PHENYLCINNAMIC ACID
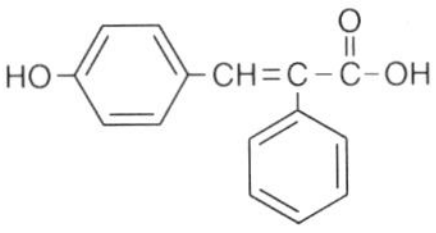
5 gm

S46,085-0
CIS-4-HYDROXYCINNAMIC ACID
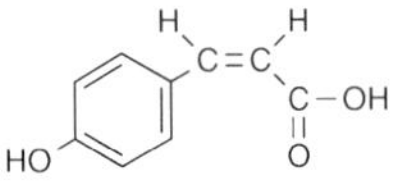
250 mg

S63,997-4
4-BENZYLOXYCINNAMIC ACID
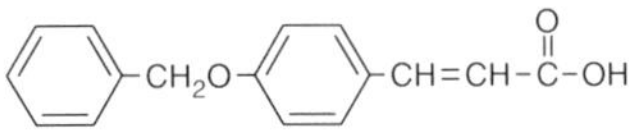
250 mg

S75,656-3
3,4-DIMETHOXY-ALPHA-MERCAPTOCINNAMIC ACID
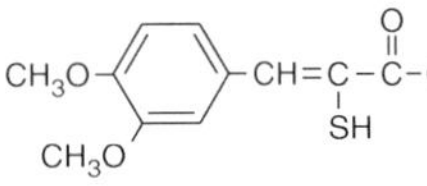
5 gm

S75,687-3
2-(3,4-DIMETHOXYSTYRYL)-GLYOXYLIC ACID
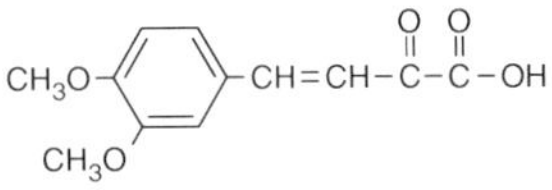
5 gm

S85,016-0
4-BENZYLOXY-3-METHOXYCINNAMIC ACID
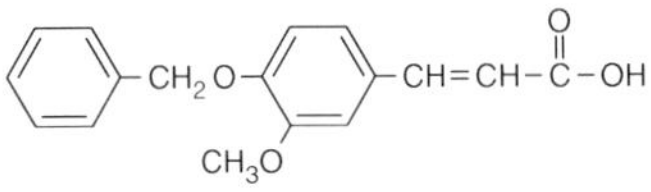
250 mg

S45,584-9
2,3,4-TRIMETHOXYCINNAMIC ACID
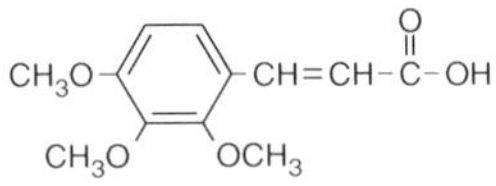
250 mg

S95,391-1
2-(2,3-DIMETHOXYSTYRYL)-GLYOXYLIC ACID
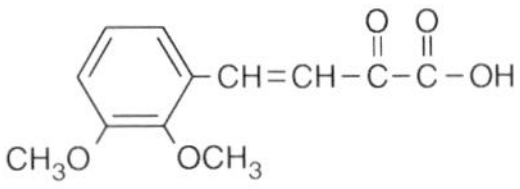
50 mg

S45,908-9
2,4-DIMETHOXYCINNAMIC ACID

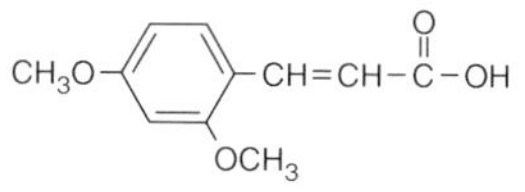

250 mg

S77,018-3
ALPHA-(4-BROMOPHENYL)-3,4-DIMETHOXY-2-NITROCINNAMIC ACID

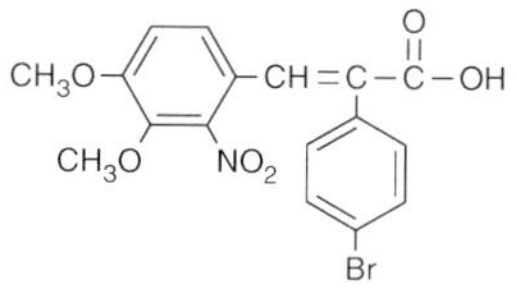

250 mg

S79,331-0
2-(4-METHOXYBENZYLIDENE)-4-(4-METHOXYPHENYL)-3-BUTENOIC ACID

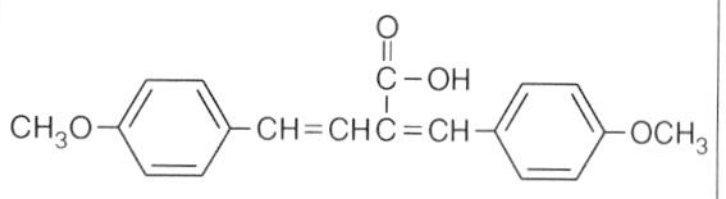

250 mg

S44,785-4
4-ETHOXY-3-METHOXYCINNAMIC ACID

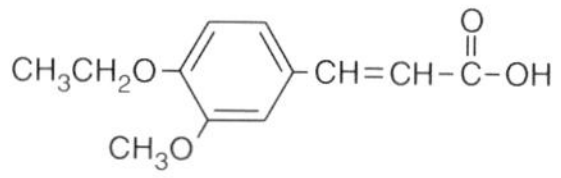

5 gm

S51,386-5
4-HEXADECYLOXY-3-METHOXYCINNAMIC ACID

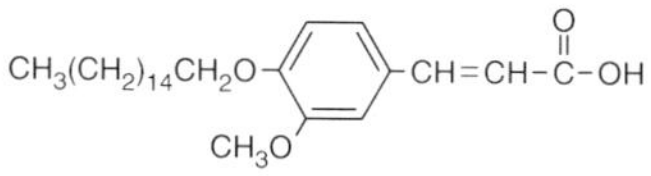

250 mg

S76,253-9
ALPHA-MERCAPTO-2-METHOXYCINNAMIC ACID

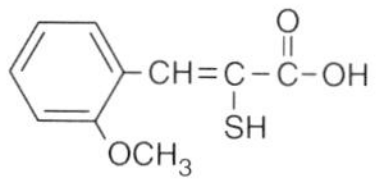

250 mg

S65,167-2
ALPHA-MERCAPTO-2-METHOXY-BETA-METHYLCINNAMIC ACID

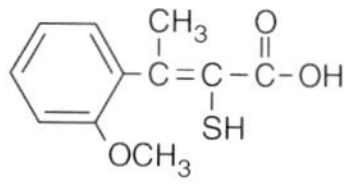

250 mg

S39,804-7
4-AMINO-2-METHOXYCINNAMIC ACID

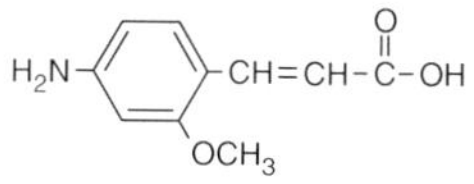

250 mg

S69,636-6
4-(HEXADECYLOXY)-3-NITROCINNAMIC ACID

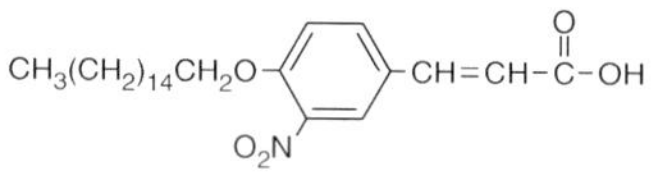

250 mg

S40,001-7
2-METHOXY-4-NITROCINNAMIC ACID

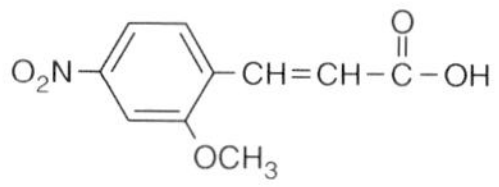

250 mg

S56,040-5
2-METHOXY-ALPHA-METHYLCINNAMIC ACID

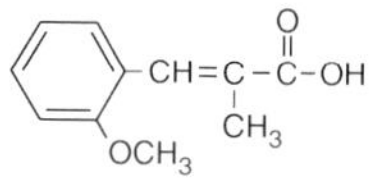

5 gm

S51,220-6
2-(HEXADECYLOXY)CINNAMIC ACID

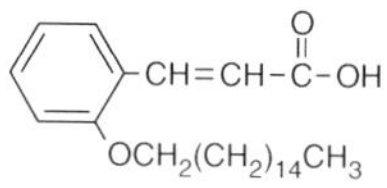

5 gm

S54,958-4
4-ETHOXYCINNAMIC ACID

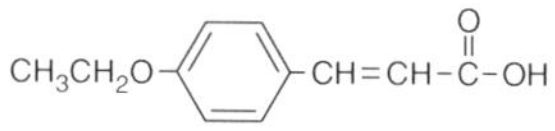

250 mg

S84,711-9
4-METHOXY-BETA-METHYLCINNAMIC ACID

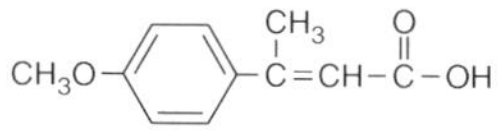

250 mg

S88,333-6
4-METHOXY-ALPHA-METHYLCINNAMIC ACID

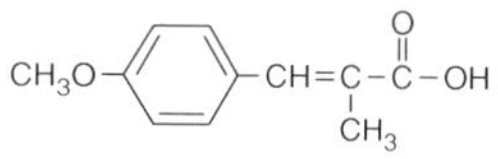

250 mg

S65,950-9
4-METHOXY-ALPHA-PHENYLCINNAMIC ACID

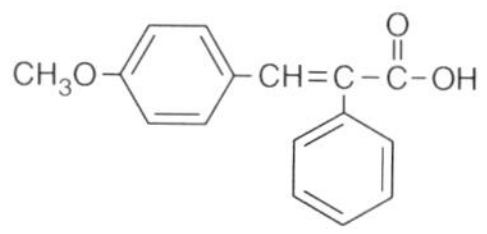

250 mg

S95,366-0
(3-METHOXYSTYRYL)GLYOXYLIC ACID

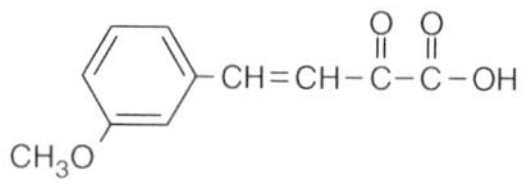

50 mg

S79,215-2
2,4-DIMETHYL-6-METHOXY-ALPHA,BETA-DIPHENYLCINNAMIC ACID

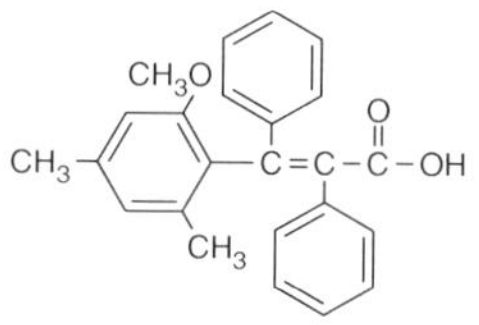

250 mg

S92,910-7
3-(4-METHOXY-1-NAPHTHYL)-ACRYLIC ACID

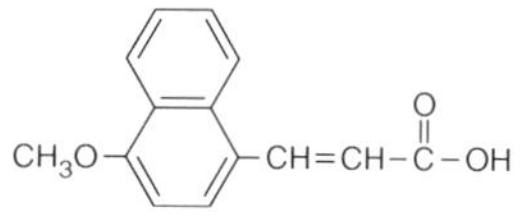

250 mg

S88,860-5
ALPHA-METHYL-2-(METHYLTHIO)-5-NITROCINNAMIC ACID

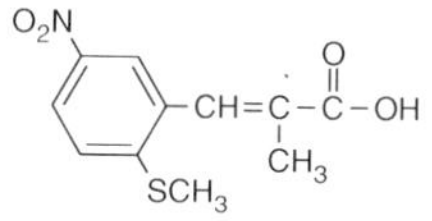

250 mg

S88,810-9
BETA,5-DIMETHYL-2-(METHYLTHIO)-CINNAMIC ACID

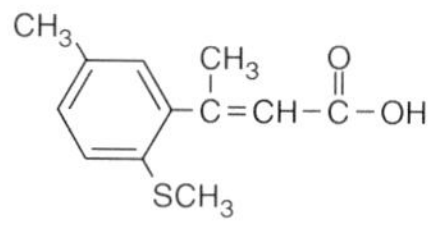

250 mg

S88,079-5
2-AMINOCINNAMIC ACID

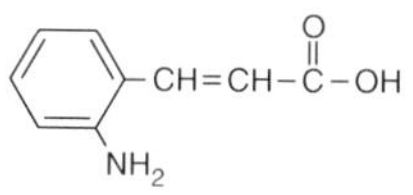

250 mg

S39,801-2
4-AMINO-2-METHYLCINNAMIC ACID

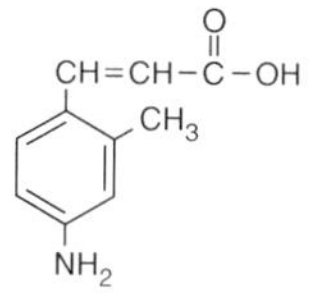

5 gm

S80,867-9
ALPHA-METHYL-4-NITROCINNAMIC ACID

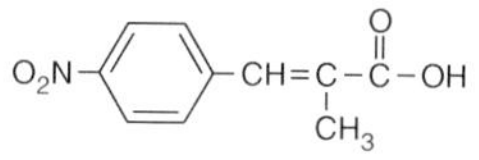

250 mg

S64,818-3
BETA-METHYL-3-NITROCINNAMIC ACID

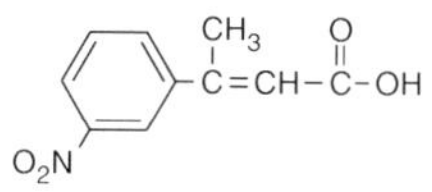

5 gm

S36,185-2
CIS-ALPHA-ETHYL-3-NITROCINNAMIC ACID

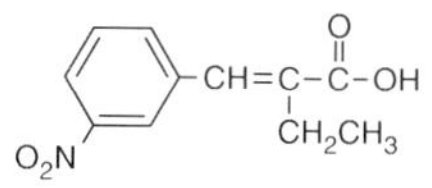

5 gm

S93,168-3
3-NITRO-ALPHA-PHENYLCINNAMIC ACID

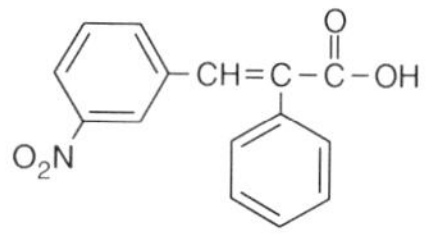

250 mg

S42,142-1
4-NITRO-ALPHA-(O-TOLYL)-CINNAMIC ACID

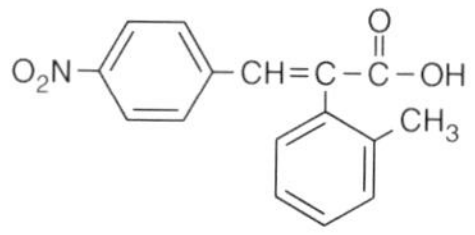

5 gm

S42,137-5
ALPHA-(4-CHLOROPHENYL)-4-NITROCINNAMIC ACID

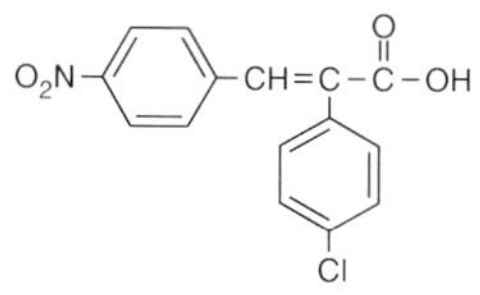

5 gm

S80,876-8
ALPHA-BROMOCINNAMIC ACID

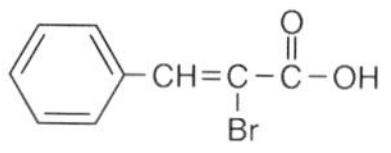

250 mg

S73,137-4
ALPHA, BETA-DIBROMOCINNAMIC ACID

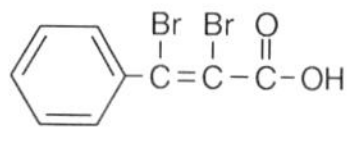

5 gm

S44,634-3
CINNAMYLIDENEMALONIC ACID

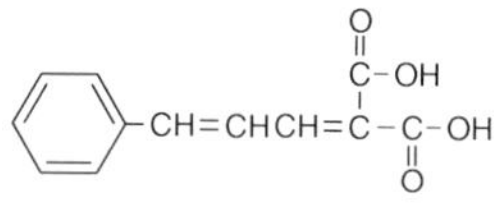

5 gm

S82,606-5
CINNAMIC ACID, MERCURY(II) SALT

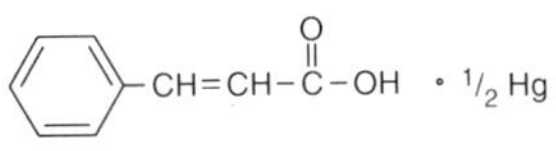

250 mg

S82,833-4
BETA-METHYLCINNAMIC ACID

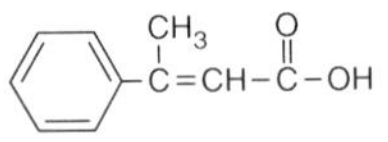

250 mg

S80,832-6
ALPHA-ETHYLCINNAMIC ACID

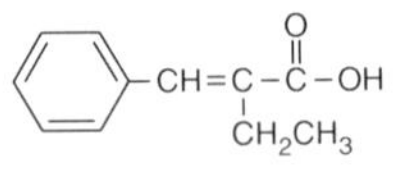

250 mg

S79,319-1
CIS-ALPHA-PHENYLCINNAMIC ACID

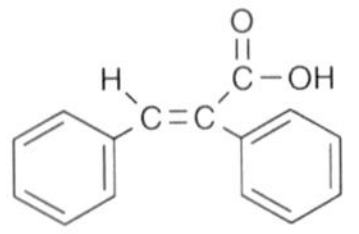

250 mg

S88,143-0
2-PHENYLCINNAMIC ACID

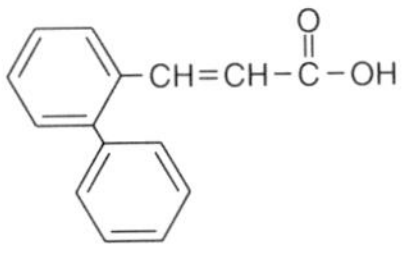

250 mg

S85,255-4
BETA-PHENYLCINNAMIC ACID

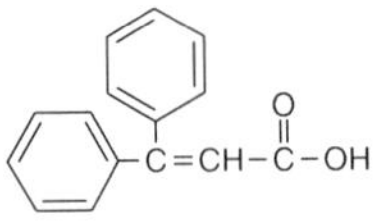

250 mg

S41,964-8
4-ISOPROPYLCINNAMIC ACID

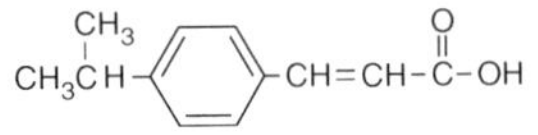

5 gm

S43,404-3
5-INDANACRYLIC ACID

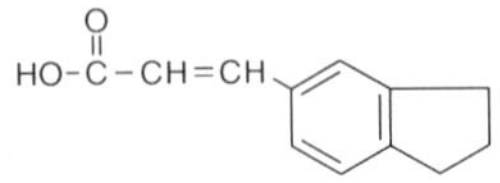

5 gm

S45,582-2
BIPHENYLENEACRYLIC ACID

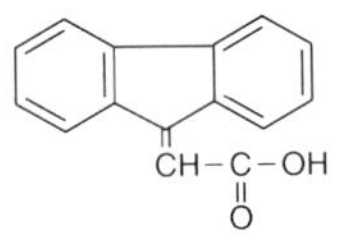

250 mg

S56,940-2
4-(3-CHLOROPROPOXY)BENZOIC ACID

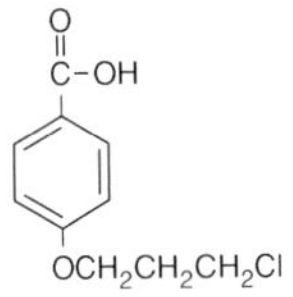

5 gm

S52,409-3
5-CHLOROMETHYL-O-ANISIC ACID

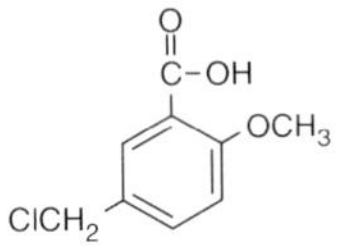

5 gm

S76,335-7
2-(BROMOMETHYL)BENZOIC ACID

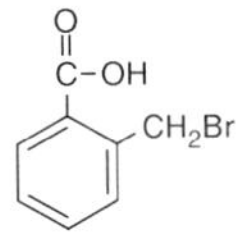

250 mg

S38,324-4
2-(2-CHLORO-1,1,2-TRIFLUOROETHOXY)-BENZOIC ACID

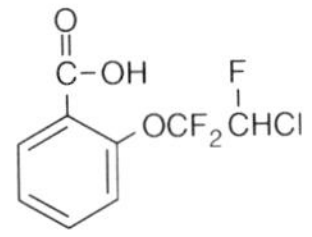

5 gm

S38,597-2
4-(2-CHLORO-1,1,2-TRIFLUOROETHOXY)-BENZOIC ACID

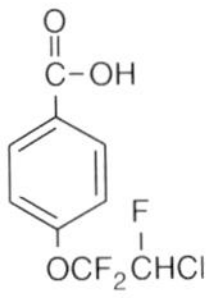

5 gm

S55,898-2
2-(2-CHLORO-1,1,2-TRIFLUOROETHOXY)-P-ANISIC ACID

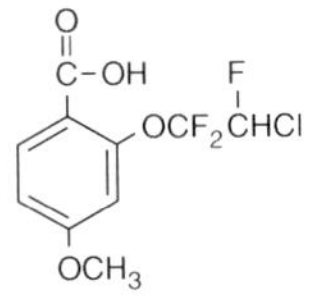

5 gm

S93,350-3
N'-(4-DIFLUOROMETHYLTHIOPHENYL)-ANTHRANILIC ACID

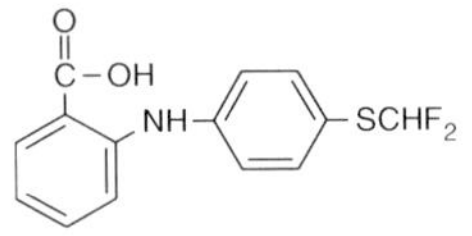

250 mg

S38,308-2
2-(2-CHLORO-1,1,2-TRIFLUOROETHYLTHIO)BENZOIC ACID

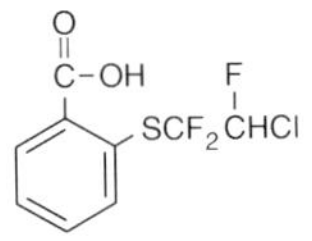

250 mg

S70,324-9
2-CHLORO-5-(TRICHLOROMETHYLTHIO)-BENZOIC ACID

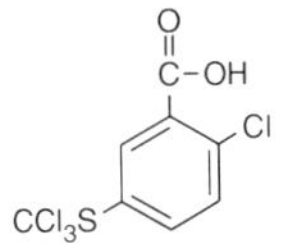

5 gm

S93,825-4
4-(1,1,2,3,3,3-HEXAFLUORO-1-PROPOXY)BENZOIC ACID

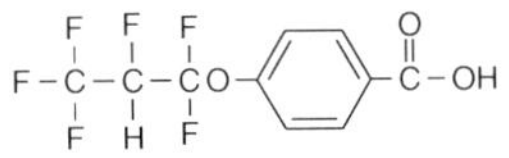

250 mg

S60,067-9
3-NITRO-5-(TRIFLUOROMETHYL)-BENZOIC ACID

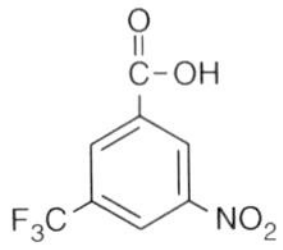

5 gm

S84,379-2
5-BROMO-3,4-DIHYDROXYBENZOIC ACID

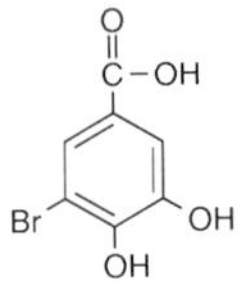

250 mg

S50,116-6
5-AMINO-3-CHLORO-2,4-DIHYDROXYBENZOIC ACID

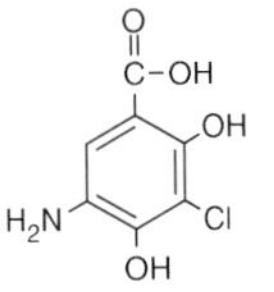

250 mg

S94,777-6
3,5-DIBROMO-2,4-DIHYDROXYBENZOIC ACID

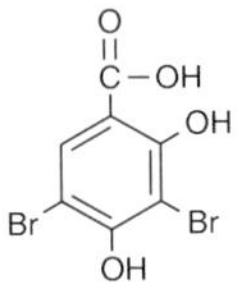

5 gm

S37,968-9
4-AMINO-3,5-DIIODOSALICYLIC ACID

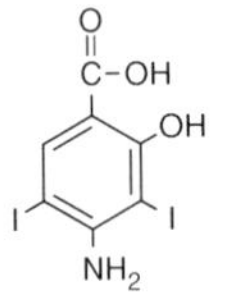

5 gm

S43,545-7
3-HYDROXY-2,4,6-TRIIODOBENZOIC ACID

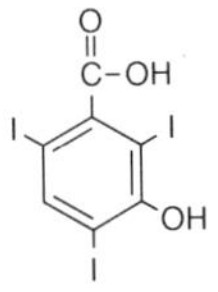

5 gm

S66,306-9
2-BROMO-4,6-DIMETHYL-3-HYDROXYBENZOIC ACID

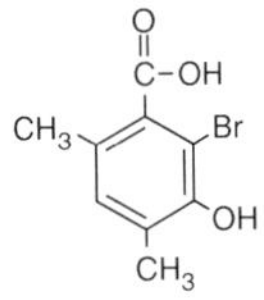

250 mg

S83,943-4
3-BROMO-4,5-DIMETHOXYBENZOIC ACID

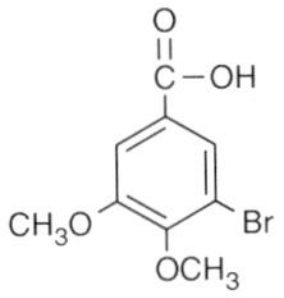

250 mg

S92,015-0
4-BROMO-2,5-DIMETHOXYBENZALDEHYDE

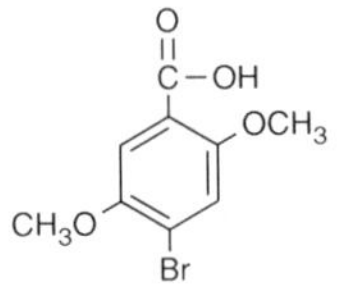

250 mg

S83,202-2
5-BROMO-2-METHOXYBENZOIC ACID

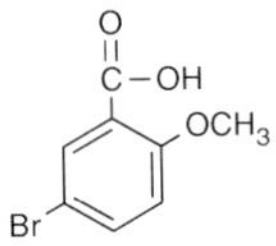

250 mg

S82,584-0
5-BROMO-2-METHOXYISOPHTHALIC ACID

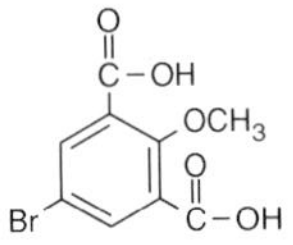

250 mg

S86,057-3
4-BROMO-2-(PHENOXY)BENZOIC ACID

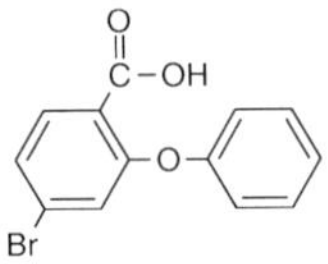

250 mg

S83,574-9
3-CHLORO-2-(4-METHOXYANILINO)-BENZOIC ACID

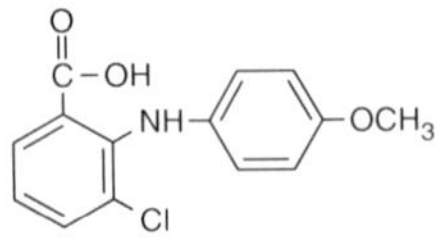

250 mg

S84,109-9
3,5-DIBROMO-4-METHOXYBENZOIC ACID

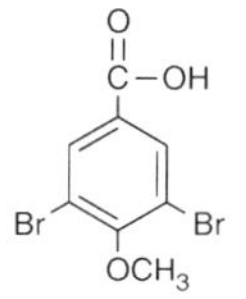

250 mg

S82,585-9
5-BROMO-3-METHYLSALICYLIC ACID

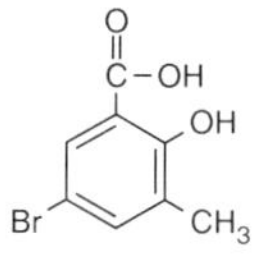

250 mg

S86,405-6
2-BENZOYL-3,4,5,6-TETRACHLOROBENZOIC ACID

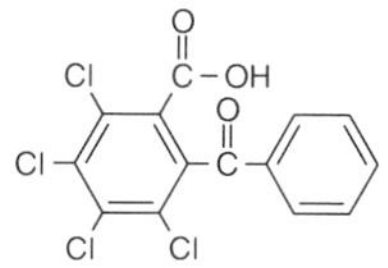

250 mg

S86,382-3
2-(3,4-DICHLOROBENZOYL)BENZOIC ACID

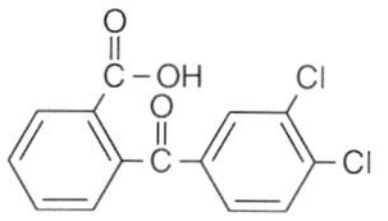

250 mg

S34,127-4
3-AMINO-2,4,6-TRIBROMOBENZOIC ACID

5 gm

S43,787-5
5-AMINO-2-CHLOROBENZOIC ACID SULFATE

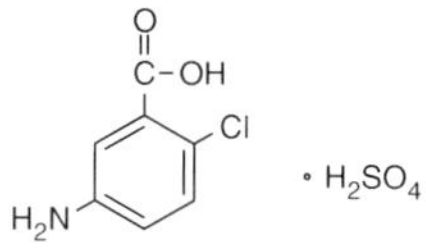

250 mg

S45,864-3
3,5-DICHLOROANTHRANILIC ACID

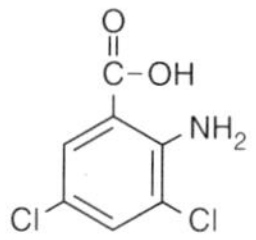

5 gm

S45,830-9
3-AMINO-2,5-DICHLOROBENZOIC ACID

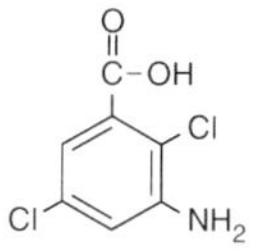

250 mg

S88,871-0
4-AMINO-3,5-DICHLOROBENZOIC ACID

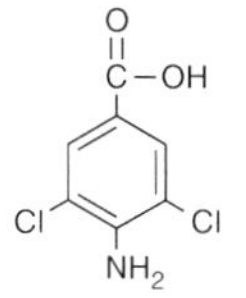

250 mg

S92,206-4
2-AMINO-5-BROMO-4-NITROBENZOIC ACID

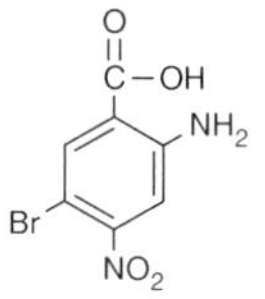

250 mg

S39,140-9
2,6-DICHLORO-3,5-DINITROBENZOIC ACID

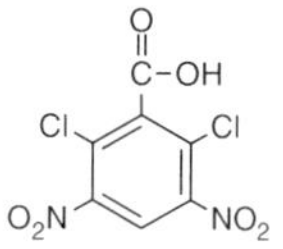

5 gm

S80,232-8
5-CHLORO-2,4-DINITROBENZOIC ACID

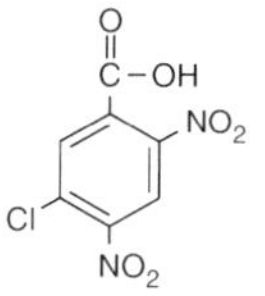

250 mg

S69,303-0
2-ANILINO-5-BROMOTEREPHTHALIC ACID

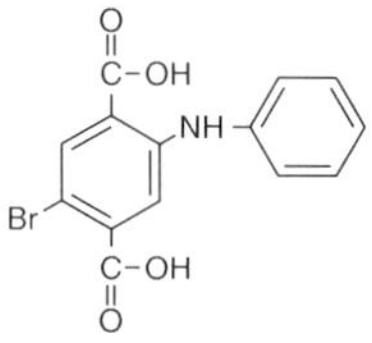

5 gm

S85,236-8
N-(2,5-DICHLOROPHENYL)-ANTHRANILIC ACID

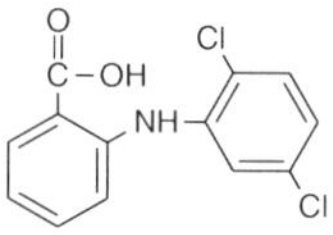

250 mg

S94,385-1
5-CHLORO-N-PHENYLANTHRANILIC ACID

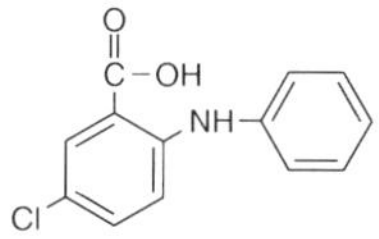

250 mg

S75,127-8
2-(4-METHYLPIPERAZINOCARBONYL)-3,4,5,6-TETRACHLOROBENZOIC ACID

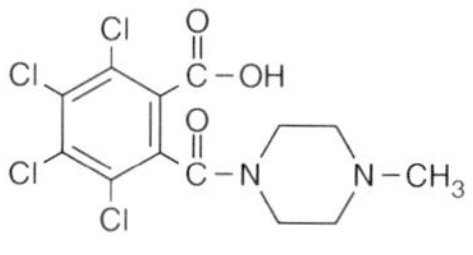

5 gm

S73,624-4
KETOMALONIC ACID 4-CHLOROPHENYLHYDRAZONE

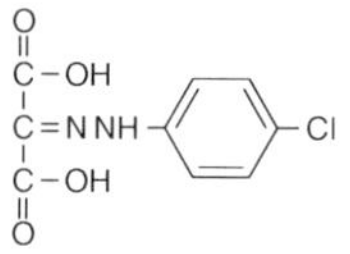

5 gm

S82,230-2
5-BROMO-2-NITROBENZOIC ACID

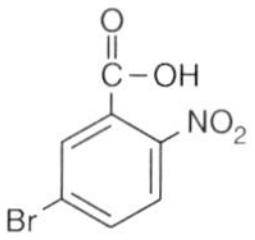

250 mg

S92,658-2
3-BROMO-5-NITROBENZOIC ACID

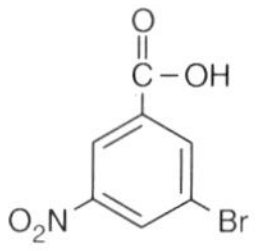

250 mg

S70,694-9
2-(2-CHLOROBENZOYL)BENZOIC ACID

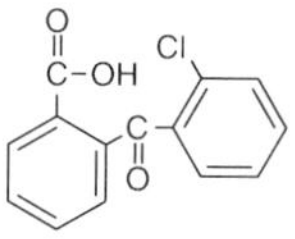

250 mg

S65,625-9
2-ACETYL-3,4,5,6-TETRACHLOROBENZOIC ACID

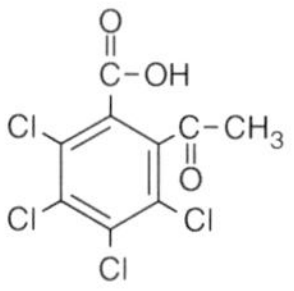

250 mg

S91,563-7
2-BROMOISOPHTHALIC ACID

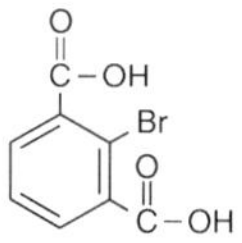

250 mg

S43,717-4
2,5-DICHLOROTEREPHTHALIC ACID

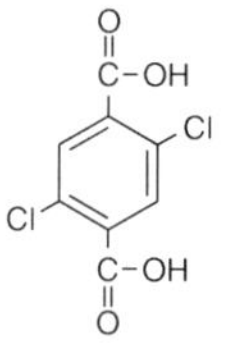

5 gm

S82,913-7
2,3,5,6-TETRACHLOROBENZOIC ACID

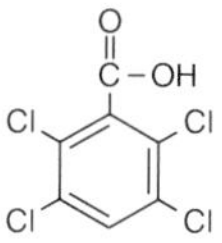

250 mg

S81,223-4
2,3,4,5,6-PENTACHLOROBENZOIC ACID

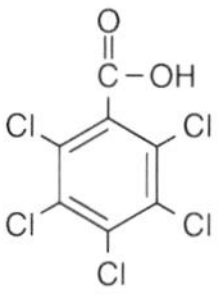

250 mg

S73,517-5
2-(4-CHLOROBENZYL)BENZOIC ACID

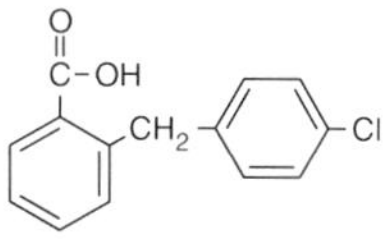

250 mg

S93,720-7
2-CHLORO-5-FLUOROBENZOIC ACID

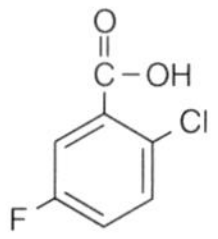

250 mg

S84,253-2
3,4-DIBROMOBENZOIC ACID

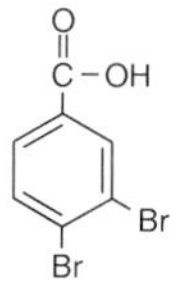

250 mg

S90,817-7
3,5-DIBROMOBENZOIC ACID

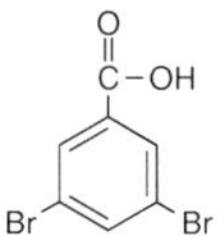

250 mg

S76,654-2
2,3-DICHLORO-6-FLUOROBENZOIC ACID

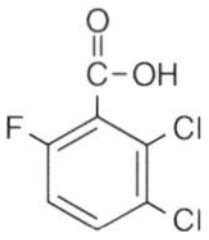

250 mg

S46,224-1
3,4,5-TRIIODOBENZOIC ACID

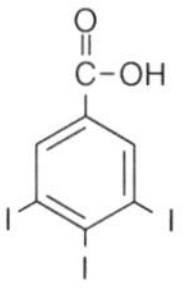

250 mg

S83,314-2
2-(2-CHLOROPHENYL)BENZOIC ACID

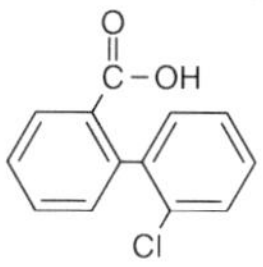

250 mg

S82,517-4
3-(3-BROMOPHENYL)BENZOIC ACID
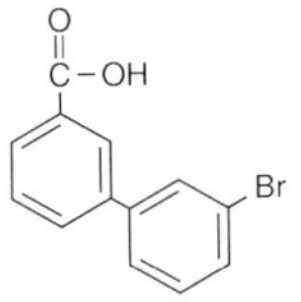
250 mg

S92,311-7
10-BROMO-9-ANTHRACENECARBOXYLIC ACID
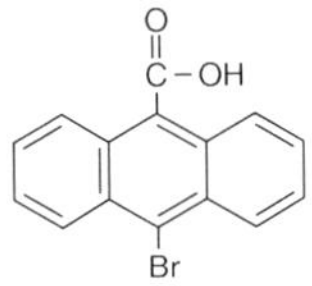
250 mg

S91,561-0
2-CHLOROISOPHTHALIC ACID
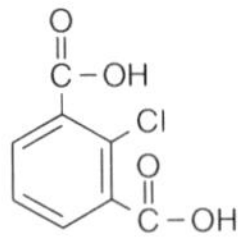
250 mg

S91,559-9
2-IODOISOPHTHALIC ACID
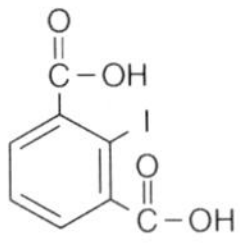
250 mg

S76,455-8
5,5'-THIODISALICYLIC ACID
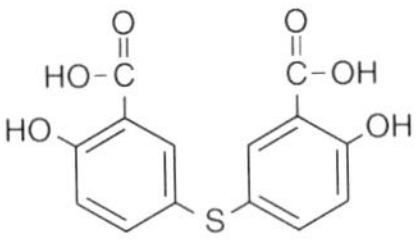
5 gm

S91,947-0
3,6-DIHYDROXYPHTHALIC ACID
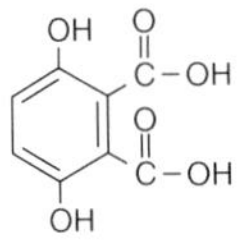
250 mg

S94,781-4
2,4-DIHYDROXY-3-METHYLBENZOIC ACID
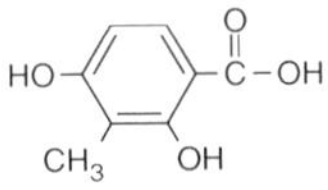
5 gm

S43,452-3
5-(ALPHA,ALPHA-DIMETHYL-4-HYDROXYBENZYL)SALICYLIC ACID, TECH.
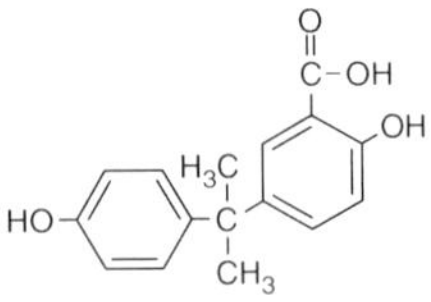
5 gm

S93,193-4
3-BENZOXEPIN-2,4-DICARBOXYLIC ACID
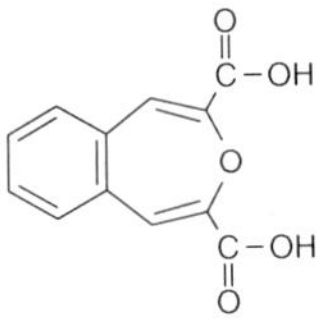
250 mg

S43,808-1
4-ETHOXY-2-HYDROXYBENZOIC ACID
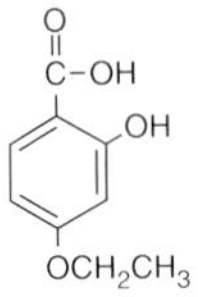
5 gm

S80,891-1
4-ALLYLOXY-2-HYDROXYBENZOIC ACID
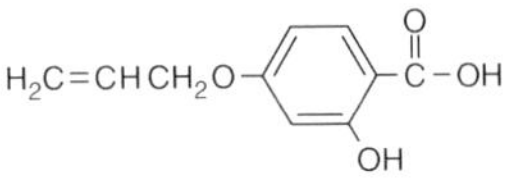
250 mg

S92,655-8
6-METHOXYSALICYLALDEHYDE
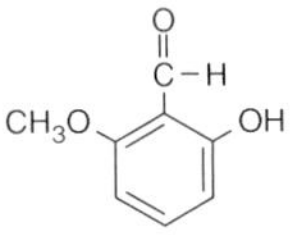
250 mg

S45,059-6
2'-CARBOXY-2-HYDROXY-4-METHOXYBENZOPHENONE
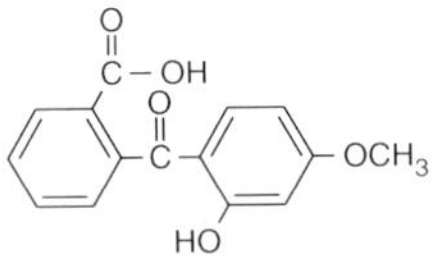
5 gm

S51,758-5
4-(2-HYDROXYETHOXY)BENZOIC ACID
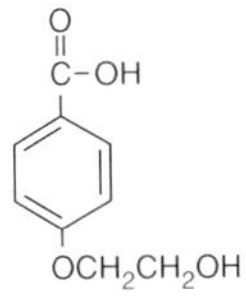
250 mg

S88,831-1
2'-(HYDROXYMETHYL)-2-BIPHENYLCARBOXYLIC ACID
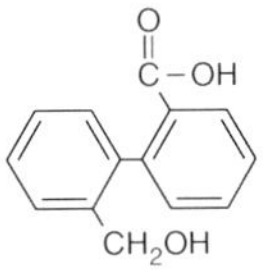
250 mg

S88,317-4
4,5-DIHYDROXY-2-METHYLBENZOIC ACID
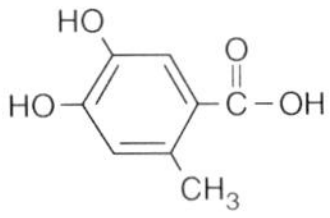
250 mg

S50,076-3
8-(3-OXO-4,5,6-TRIHYDROXY-3H-XANTHEN-9-YL)-1-NAPHTHOIC ACID
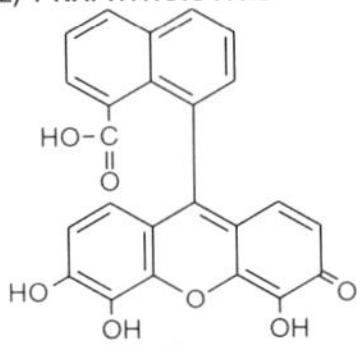
250 mg

S50,073-9
4',5'-DIHYDROXY-4,5,6,7-TETRACHLOROFLUORESCEIN
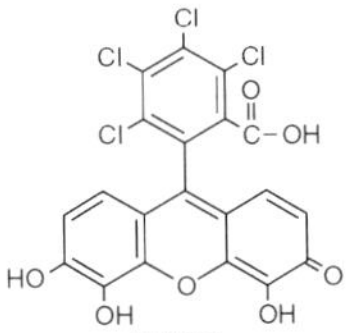
250 mg

S50,074-7
4',5'-DIHYDROXY-4,5,6,7-TETRABROMOFLUORESCEIN
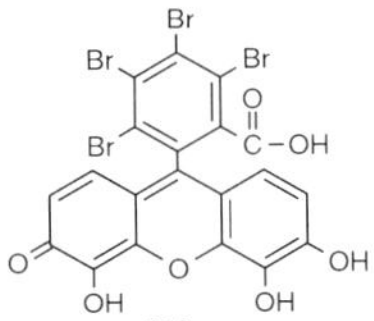
250 mg

S44,938-5
5-(METHYLTHIO)SALICYLIC ACID
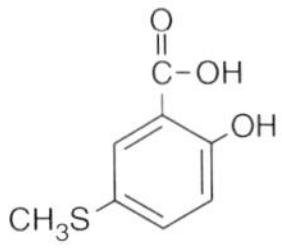
5 gm

S46,219-5
4-DIMETHYLAMINOSALICYLIC ACID
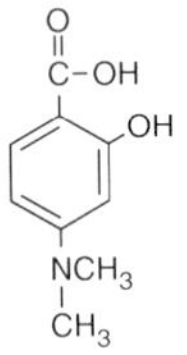
5 gm

S88,872-9
4-NITROSALICYLIC ACID
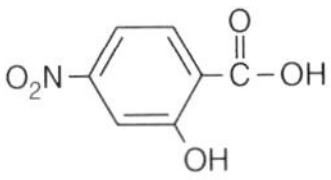
250 mg

S95,311-3
2-HYDROXY-6-NITROBENZOIC ACID
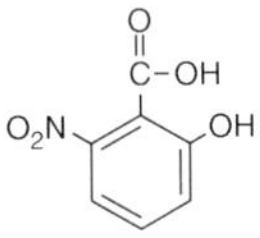
50 mg

S69,301-4
SALICYLIC ACID COMPOUND WITH FORMALDEHYDE
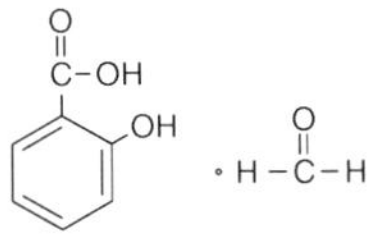
5 gm

S90,684-0
3-ISOPROPYLSALICYLIC ACID
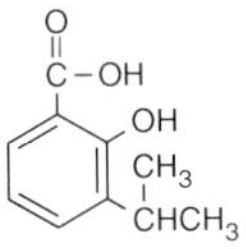
250 mg

S91,454-1
5-TERT-BUTYLSALICYLIC ACID
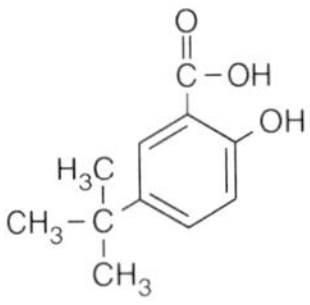
250 mg

S88,337-9
3,6-DIMETHYLSALICYLIC ACID
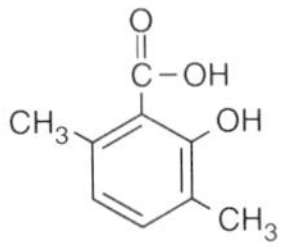
250 mg

S44,557-6
3-TERT-BUTYL-6-METHYLSALICYLIC ACID
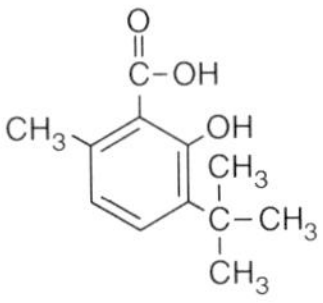
5 gm

S44,429-4
3-TERT-BUTYL-5-METHYLSALICYLIC ACID
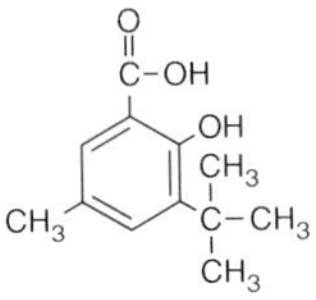
5 gm

S83,824-1
4-BROMO-3-HYDROXY-2-NAPHTHOIC ACID
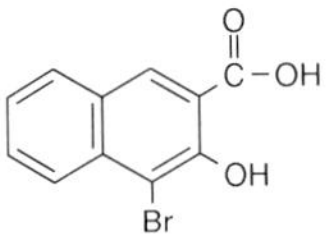
250 mg

S95,529-9
5-HYDROXY-7-METHYL-4,6-INDANDICARBOXYLIC ACID
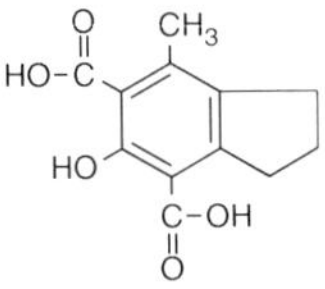
50 mg

S88,318-2
5-HYDROXY-2-METHYL-4-NITROBENZOIC ACID
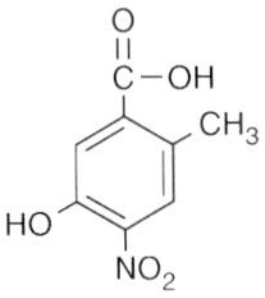
250 mg

S84,993-6
3-(HYDROXYMETHYL)-2-NAPHTHOIC ACID

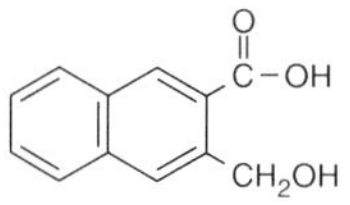

250 mg

S63,030-6
2-(2-HYDROXYETHYLTHIO)BENZOIC ACID

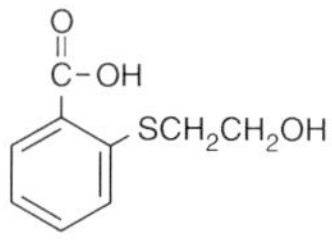

5 gm

S45,146-0
3-AMINO-5-NITROSALICYLIC ACID MONOHYDRATE

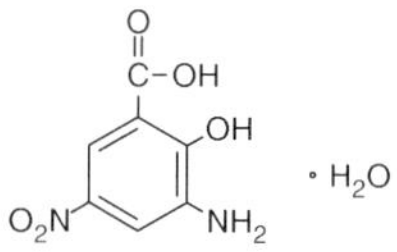

5 gm

S54,705-0
3-HYDROXY-2-OXOBUTYRIC ACID 2,4-DINITROPHENYLHYDRAZONE

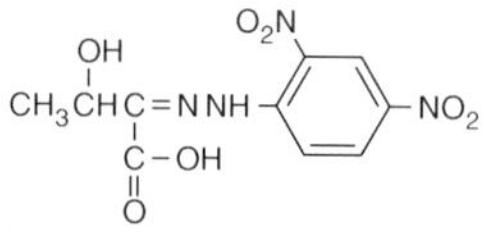

5 gm

S45,269-6
2-(4-HYDROXYBENZOYL)BENZOIC ACID

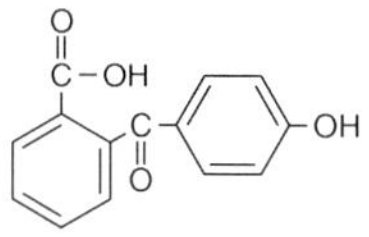

5 gm

S50,069-0
4-(6-HYDROXY-3-OXO-3H-XANTHEN-9-YL)-BENZOIC ACID

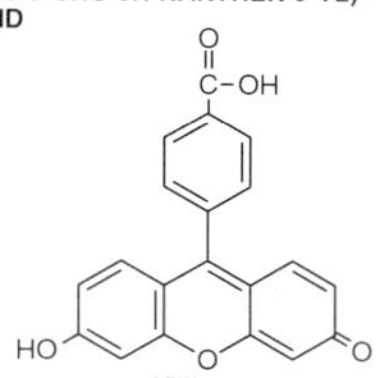

250 mg

S50,085-2
6-HYDROXY-3-OXO-3H-XANTHENE-9-CARBOXYLIC ACID

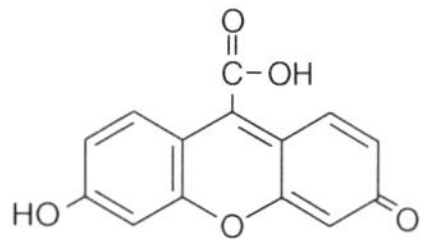

250 mg

S39,962-0
3-(6-HYDROXY-3-OXO-3H-XANTHEN-9-YL)-BENZOIC ACID

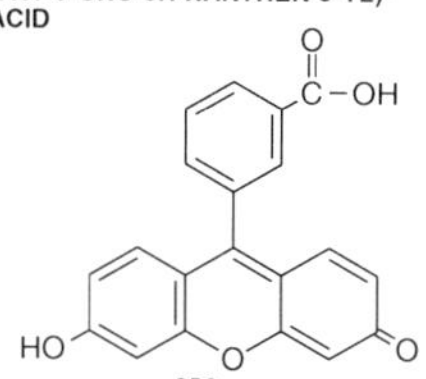

250 mg

S39,774-1
3,4,5,6-TETRABROMOFLUORESCEIN

250 mg

S75,441-2
2-[2-(2-HYDROXY-1-NAPHTHOYL)VINYL]-BENZOIC ACID

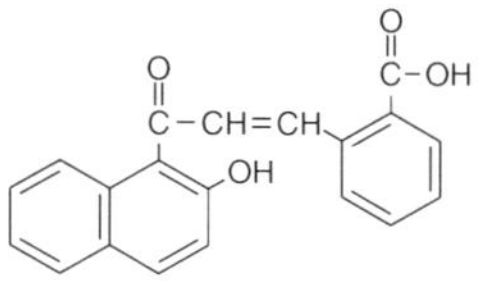

250 mg

S50,371-1
DIMETHYL 4-(6-HYDROXY-3-OXO-3H-XANTHEN-9-YL)ISOPHTHALATE

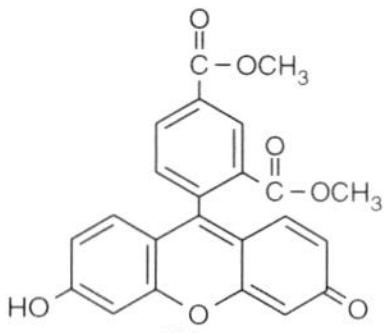

250 mg

S39,965-5
2-(3,6-DIHYDROXY-9H-XANTHEN-9-YL)-3,4,5,6-TETRACHLOROBENZOIC ACID

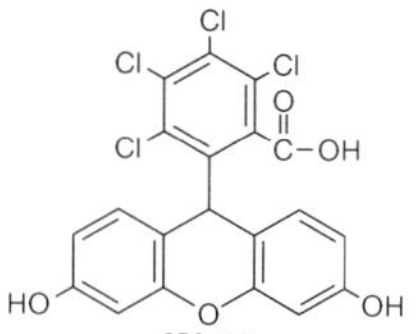

250 mg

S39,964-7
2-(3,6-DIHYDROXY-9H-XANTHEN-9-YL)-3,4,5,6-TETRABROMOBENZOIC ACID

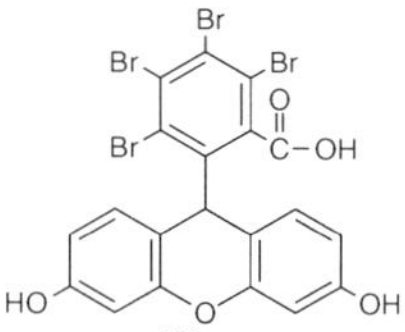

250 mg

S41,567-7
3-PHENYLSALICYLIC ACID

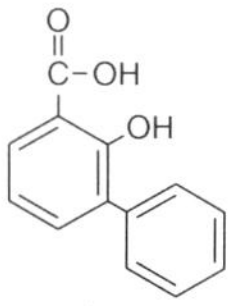

5 gm

S73,303-2
2-[2-(4-HYDROXYPHENYL)PROPYL]-BENZOIC ACID

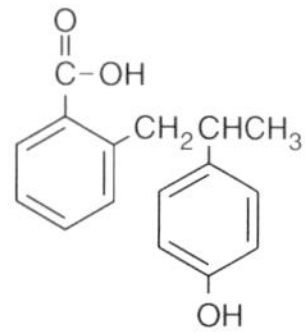

250 mg

S84,011-4
2'-HYDROXY-2-BIPHENYLCARBOXYLIC ACID

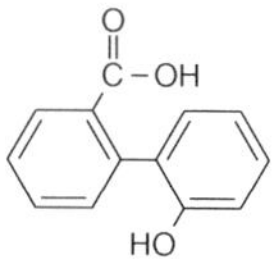

250 mg

S75,867-1
2-(2-HYDROXY-2-ADAMANTYLMETHYL)-BENZOIC ACID

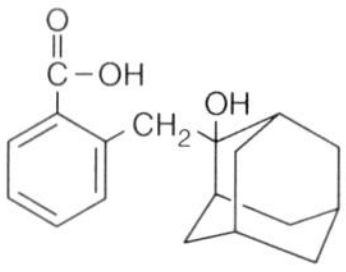

250 mg

S44,452-9
5-TERT-OCTYLSALICYLIC ACID

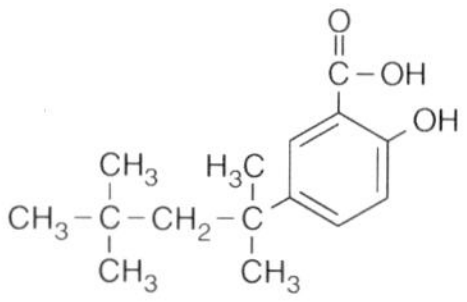

5 gm

S43,836-7
3,5-DIMETHYL-4-HYDROXYBENZOIC ACID

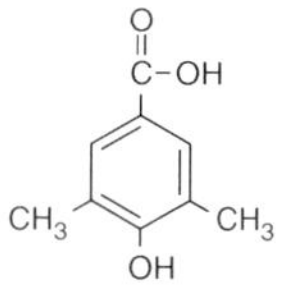

5 gm

S50,690-7
5-(TERT-BUTYL)-4-HYDROXY-M-TOLUIC ACID

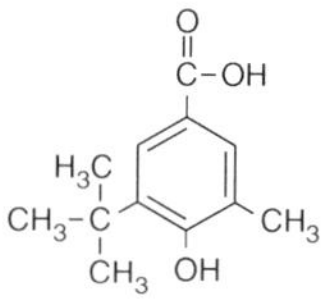

5 gm

S62,969-3
4-BROMO-1-HYDROXY-2-NAPHTHOIC ACID

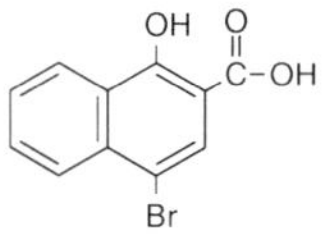

250 mg

S86,136-7
2,5-DI(BENZYLOXY)BENZOIC ACID

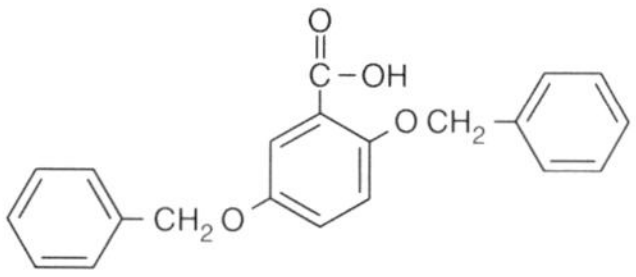

250 mg

S82,788-6
2-(BENZYLOXY)BENZOIC ACID

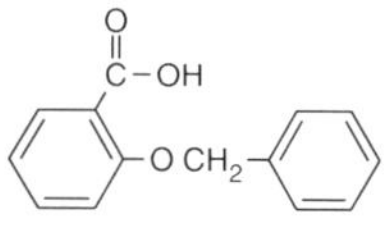

250 mg

S65,453-1
4-BENZYLOXYBENZOIC ACID

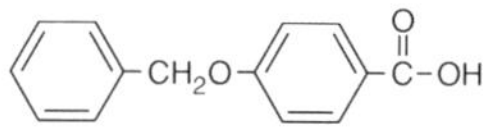

250 mg

S95,462-4
2-(3,4-DIMETHOXY-ALPHA-METHYLBENZYL)-6-METHOXYBENZOIC ACID

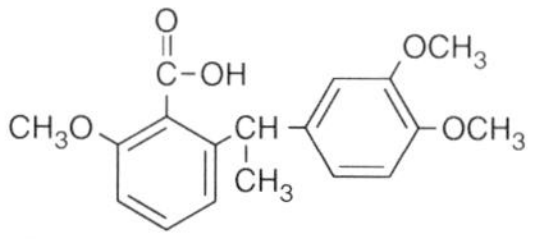

50 mg

S56,557-1
3,4-DIMETHOXYPHTHALIC ACID

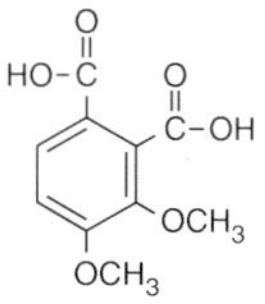

250 mg

S81,495-4
2,3,6-TRIMETHOXYBENZOIC ACID

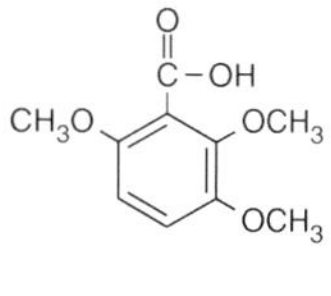

250 mg

S95,349-0
4,5-DIMETHOXY-2-METHYLBENZOIC ACID

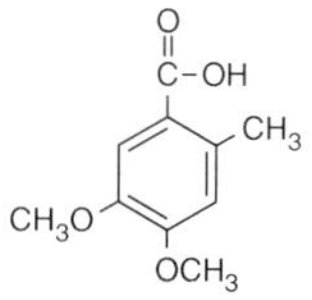

50 mg

S84,284-2
2-(3,4-DIMETHOXYBENZOYL)-BENZOIC ACID

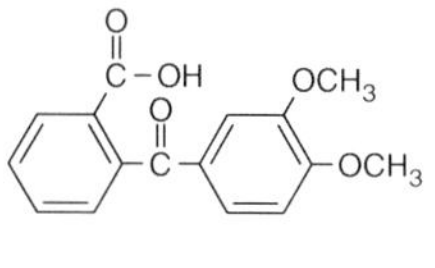

250 mg

S56,942-9
4,4'-(ETHYLENEDIOXY)DIBENZOIC ACID

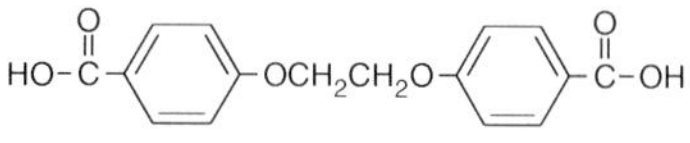

5 gm

S67,194-0
1,3-BIS(4-CARBOXYPHENOXY)PROPANE

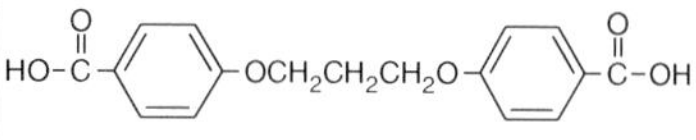

5 gm

S91,555-6
2,6-DIETHOXYBENZOIC ACID

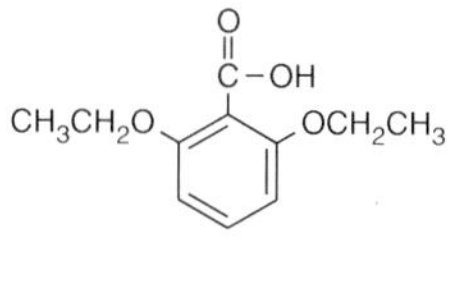

250 mg

S88,702-1
4,4',5,6-TETRAMETHOXY-2-BIPHENYLCARBOXYLIC ACID

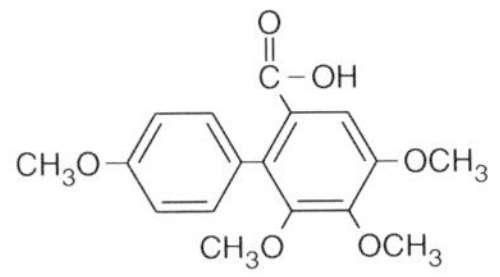

250 mg

S70,726-0
N-(2,5-DIMETHOXY-4-NITROPHENYL)-ANTHRANILIC ACID

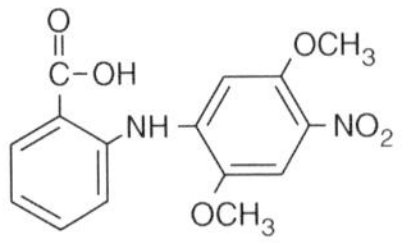

5 gm

S38,020-2
ALPHA-CARBOXY-O-ANISIC ACID

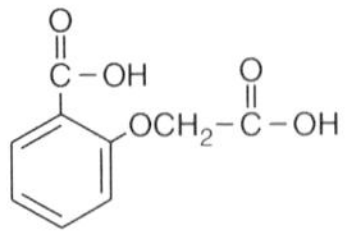

5 gm

S62,875-1
4-(2-PHENOXYETHOXY)BENZOIC ACID

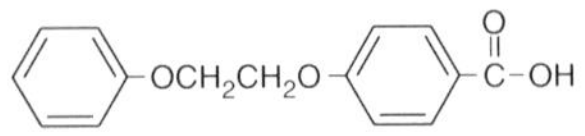

5 gm

S62,967-7
4-[2-(4-TERT-OCTYLPHENOXY)ETHOXY]-BENZOIC ACID

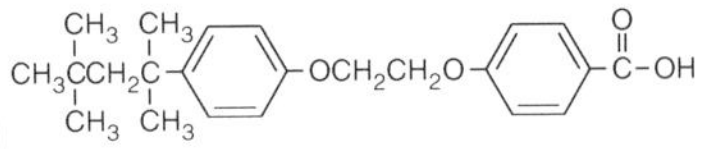

5 gm

S53,751-9
4-HEXADECYLOXY-M-ANISIC ACID

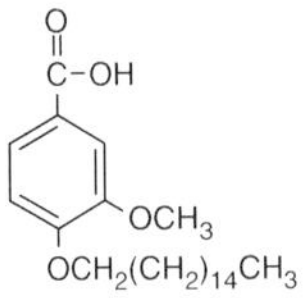

5 gm

S90,831-2
4-NITRO-2-(3-NITROPHENOXY)-BENZOIC ACID

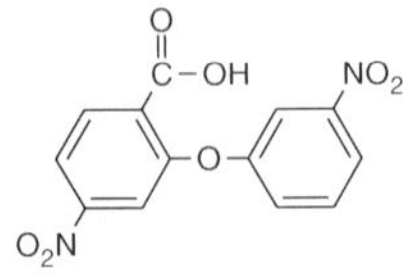

5 gm

S91,558-0
2-ISOPROPOXYBENZOIC ACID

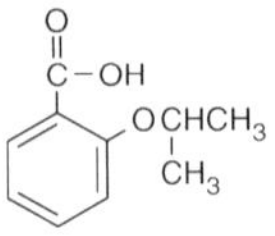

250 mg

S83,828-4
1-METHOXY-2-NAPHTHOIC ACID

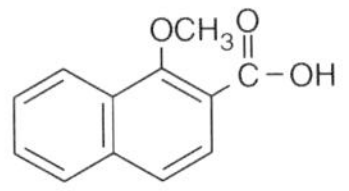

250 mg

S95,388-1
2-NITRO-3-(2,4-PENTADIEN-1-YLOXY)-BENZOIC ACID

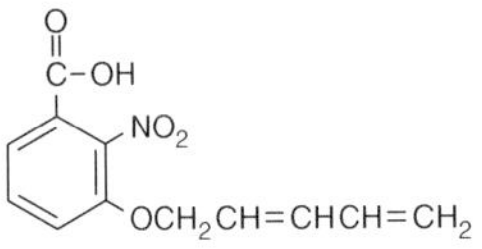

50 mg

S95,390-3
2-NITRO-6-(2,4-PENTADIEN-1-YLOXY)-BENZOIC ACID

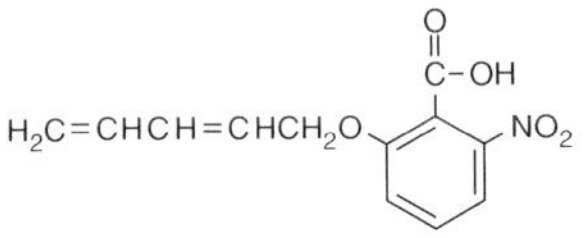

50 mg

S84,707-0
4-METHOXYPHTHALIC ACID

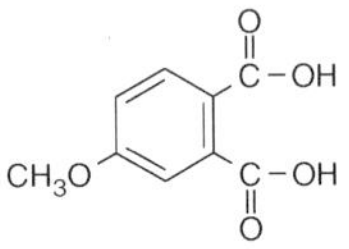

250 mg

S77,064-7
3-METHOXY-P-TOLUIC ACID

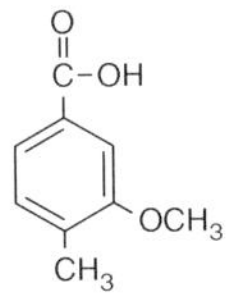

250 mg

S56,273-4
4-AMINO-3-HEXADECYLOXYBENZOIC ACID

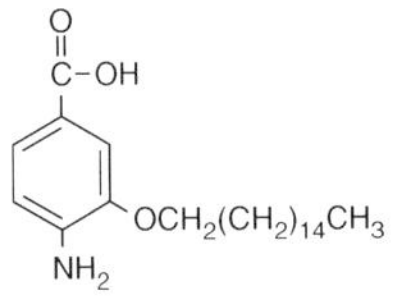

250 mg

S38,140-3
4-AMINO-2-BUTOXYBENZOIC ACID

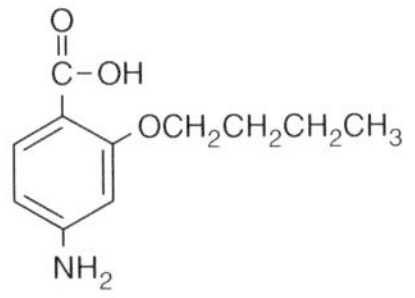

250 mg

S34,801-5
N-(4-ETHOXYPHENYL)-4-NITROANTHRANILIC ACID

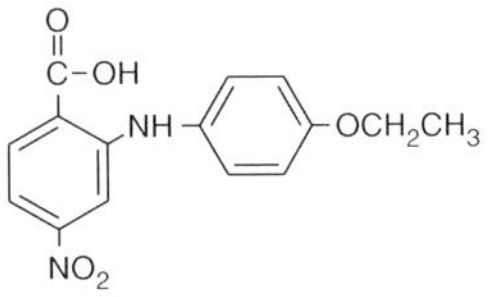

5 gm

S65,612-7
2-(4-METHOXYANILINO)-5-NITROBENZOIC ACID

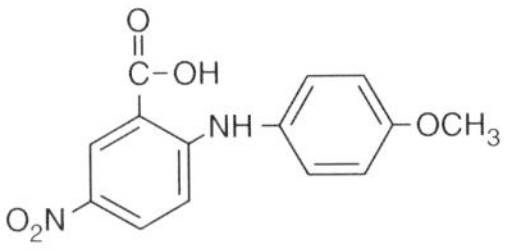

250 mg

S92,913-1
1-(4-ETHOXYANILINO)-2-ANTHRAQUINONECARBOXYLIC ACID

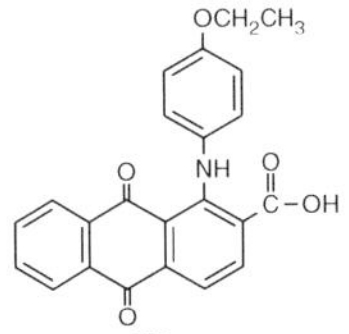

250 mg

S53,653-9
4-(4-METHOXYBENZYLAMINO)-BENZOIC ACID

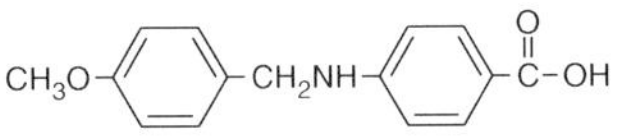

250 mg

S40,678-3
4,4'-BIS(2-CARBOXYPHENYLAMINO)-3,3'-DIMETHOXYBIPHENYL

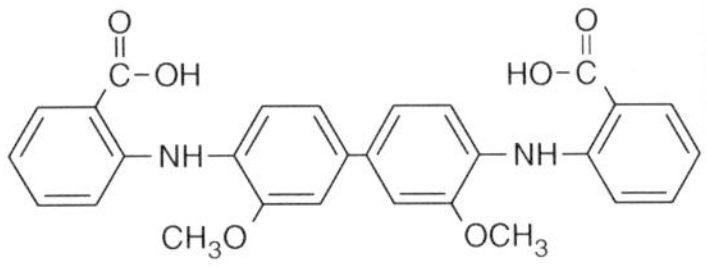

5 gm

S42,610-5
N-OXYNORNARCEINE

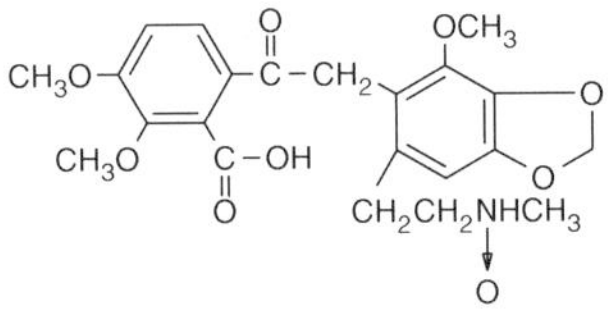

250 mg

S42,611-3
N-OXYNORNARCEINE HYDROCHLORIDE

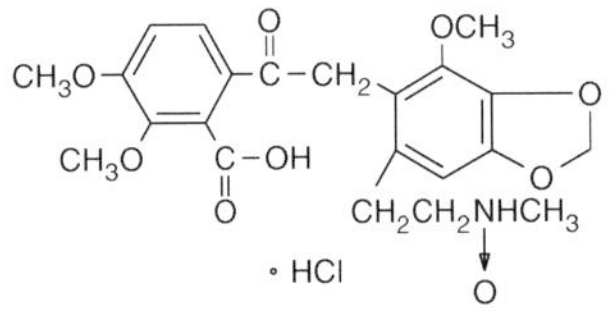

250 mg

S93,782-7
5-METHOXY-N-(1-NAPHTHYL)-ANTHRANILIC ACID

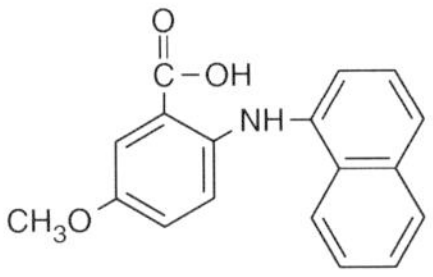

250 mg

S34,287-4
3-BUTOXY-4-NITROBENZOIC ACID

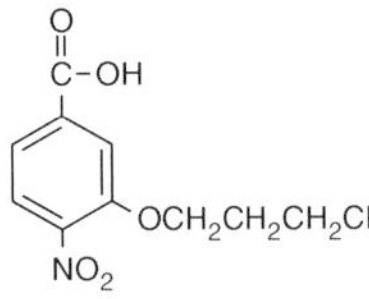

5 gm

S59,287-0
3-HEXADECYLOXY-4-NITROBENZOIC ACID

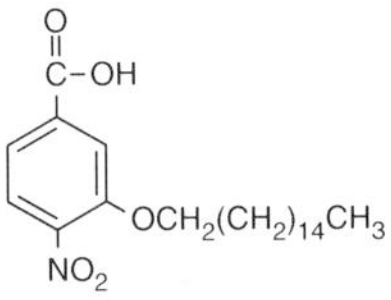

250 mg

S66,275-5
2-(4-METHOXYBENZOYL)-1-NAPHTHOIC ACID

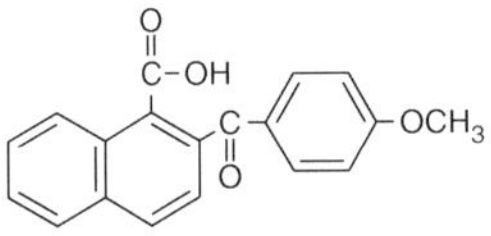

250 mg

S57,457-0
3-(DECYLOXY)BENZOIC ACID

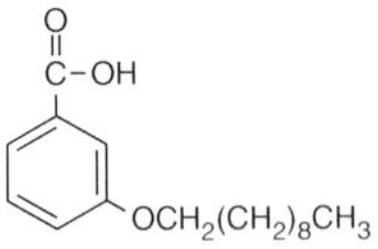

250 mg

S57,471-6
3-(TETRADECYLOXY)BENZOIC ACID

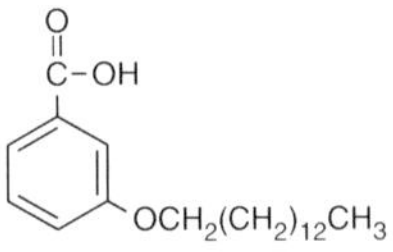

5 gm

S57,488-0
3-(HEXADECYLOXY)BENZOIC ACID

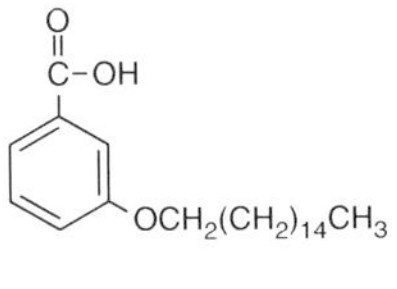

5 gm

S57,574-7
3-(OCTADECYLOXY)BENZOIC ACID

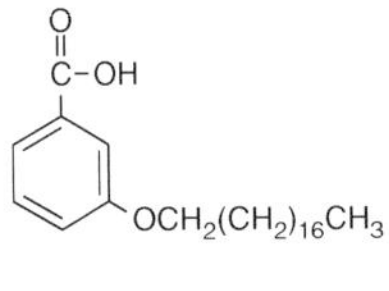

5 gm

S53,114-6
2,2'-OXYDIBENZOIC ACID

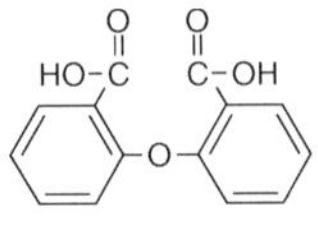

250 mg

S59,506-3
2-(PHENOXYMETHYL)BENZOIC ACID

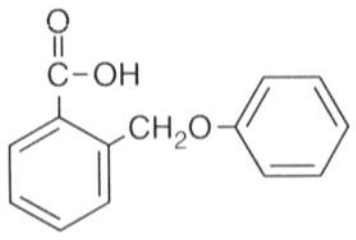

5 gm

S58,261-1
2-(DODECYLOXY)BENZOIC ACID
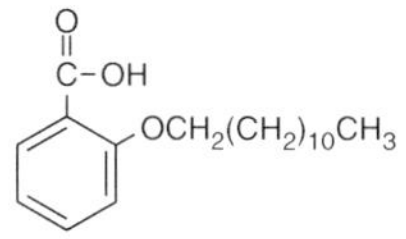
250 mg

S57,823-1
2-(HEXADECYLOXY)BENZOIC ACID
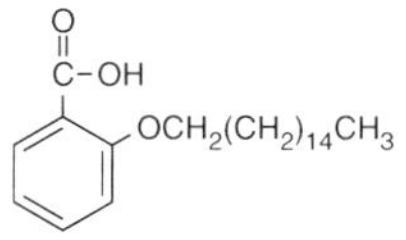
5 gm

S67,154-1
ALPHA-(4-METHOXY-1-NAPHTHYL)-O-TOLUIC ACID
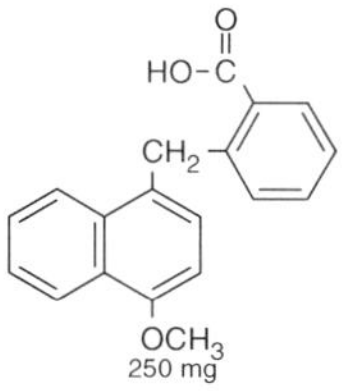
250 mg

S57,492-9
4-(TETRADECYLOXY)BENZOIC ACID
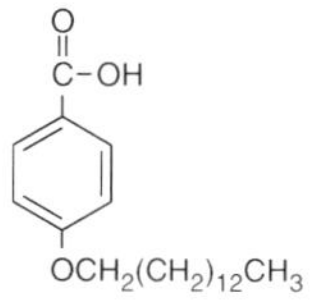
5 gm

S57,485-6
4-(OCTADECYLOXY)BENZOIC ACID
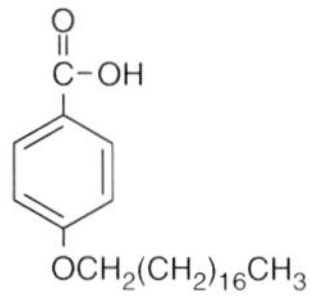
5 gm

S39,053-4
2-CHLORO-PARA-ANISIC ACID
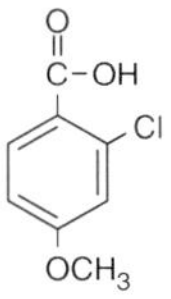
250 mg

S56,168-1
3-BROMO-P-ANISIC ACID
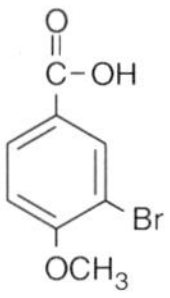
250 mg

S52,326-7
2-METHOXY-4-METHYLBENZOIC ACID
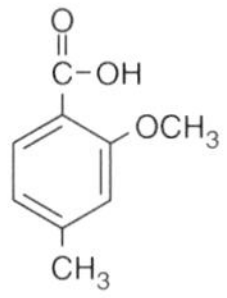
250 mg

S58,263-8
2-(HEXADECYLOXY)-P-ANISIC ACID
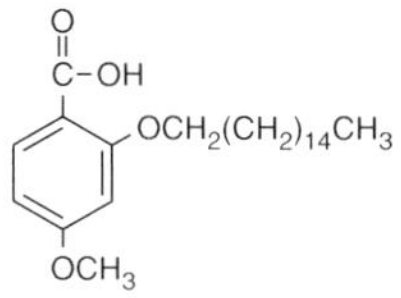
250 mg

S63,917-6
2,4-DI(DECYLOXY)BENZOIC ACID
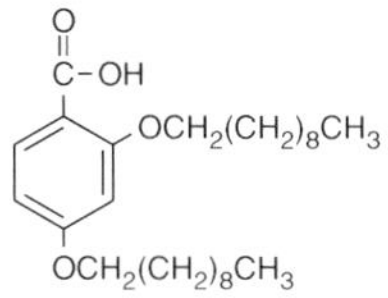
250 mg

S44,624-6
2,4,6-TRIMETHOXYBENZOIC ACID
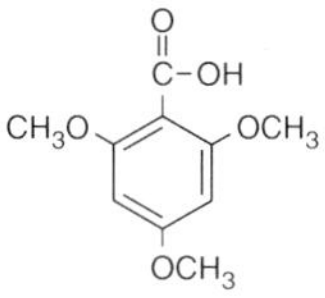
5 gm

S59,954-9
2-METHOXY-1-NAPHTHOIC ACID
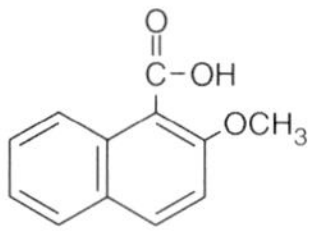
5 gm

S76,152-4
6-METHOXY-1-NAPHTHOIC ACID
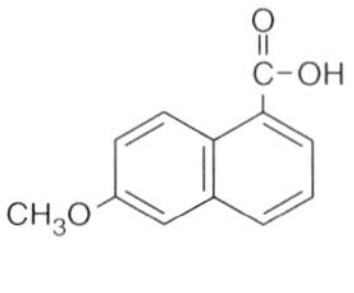
250 mg

S86,008-5
3,4-DIHYDRO-6-METHOXY-1-NAPHTHOIC ACID
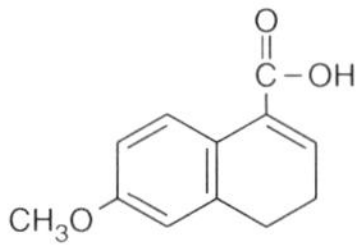
250 mg

S50,691-5
3,5-DIISOPROPOXYBENZOIC ACID
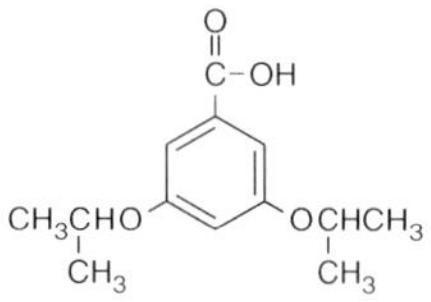
5 gm

S83,117-4
4,6-PHENOXATHIINDICARBOXYLIC ACID
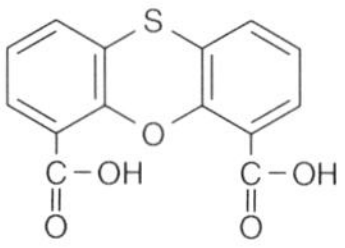
250 mg

S83,827-6
3-METHOXY-2-NAPHTHOIC ACID
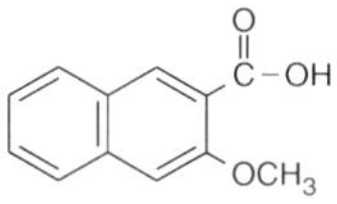
250 mg

S93,783-5
N-(1-NAPHTHYL)-5-NITROANTHRANILIC ACID
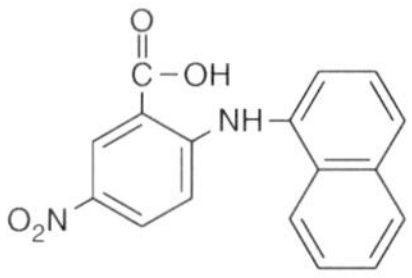
250 mg

S95,435-7
3-BENZOYL-2-BENZYLIDENEPROPIONIC ACID
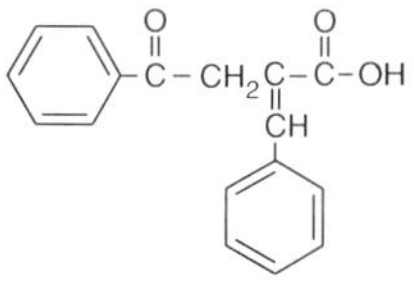
50 mg

S88,821-4
2-ACETYL-3,6-DIMETHYLBENZOIC ACID
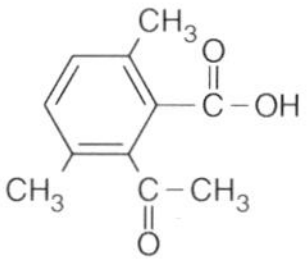
250 mg

S95,434-9
3-BENZOYL-3-BENZYLIDENEPROPIONIC ACID
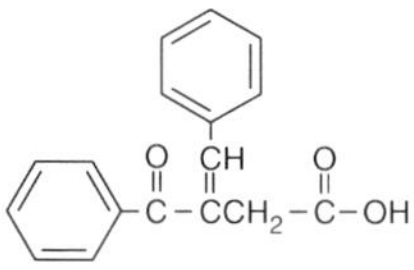
50 mg

S86,575-3
9,10-DIHYDRO-9,10-ETHENOANTHRACENE-11,12-DICARBOXYLIC ACID
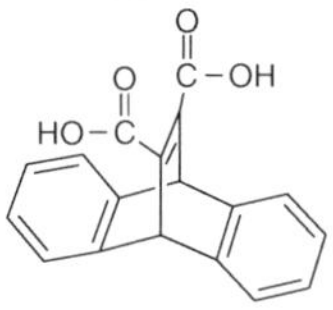
250 mg

S86,011-5
6-BROMO-1,3-BENZODIOXOLE-5-CARBOXYLIC ACID
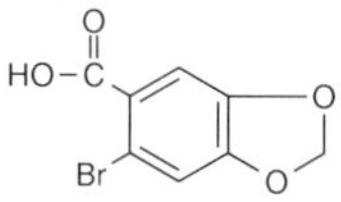
250 mg

S87,252-0
2-(5,6-DIHYDRO-4H-1,3-OXAZIN-2-YL)-BENZOIC ACID HYDROCHLORIDE
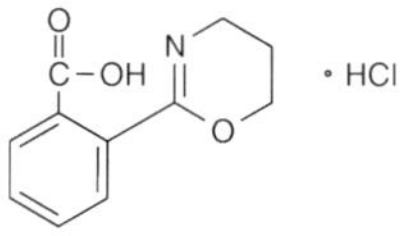
250 mg

S87,253-9
2-(5,6-DIHYDRO-4H-1,3-OXAZIN-2-YL)-BENZOIC ACID
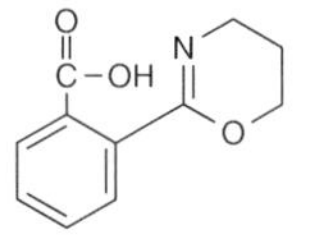
250 mg

S94,653-2
2-CARBOXY-3-PHENYL-3-PYRROLINE-2-PROPIONIC ACID
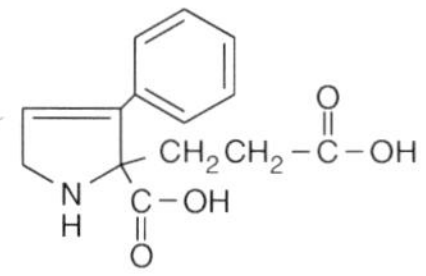
250 mg

S82,045-8
N-(2-MERCAPTOPHENYL)-N-PHENYLANTHRANILIC ACID
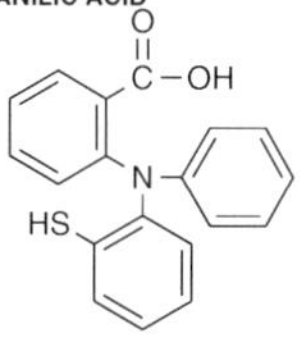
250 mg

S50,118-2
ALPHA-MERCAPTO-P-TOLUIC ACID
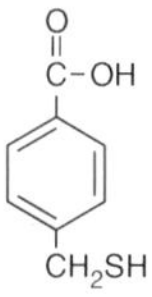
250 mg

S44,856-7
4-METHYLTHIO-3-NITROBENZOIC ACID
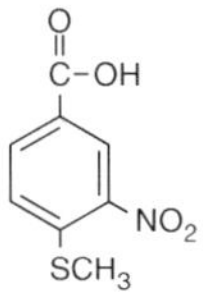
5 gm

S92,582-9
3-(METHYLTHIO)BENZOIC ACID
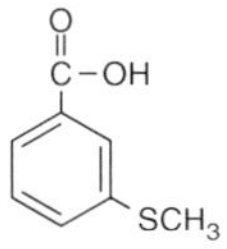
250 mg

S38,697-9
3-(HEXADECYLTHIO)BENZOIC ACID
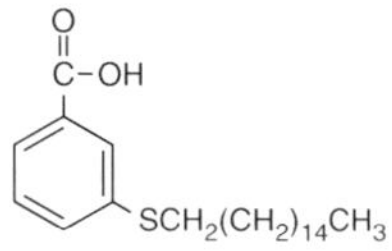
250 mg

S65,886-3
2-(2-CARBOXYPHENYLTHIO)ACETIC ACID
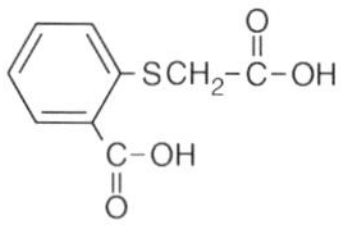
250 mg

S83,652-4
2-(ETHYLTHIO)BENZOIC ACID

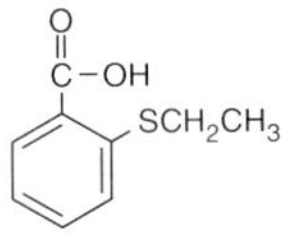

250 mg

S65,677-1
ALPHA-(PHENYLTHIO)-O-TOLUIC ACID

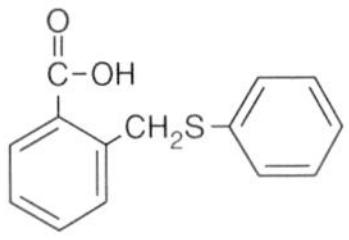

5 gm

S45,021-9
2-(HEXADECYLTHIO)BENZOIC ACID

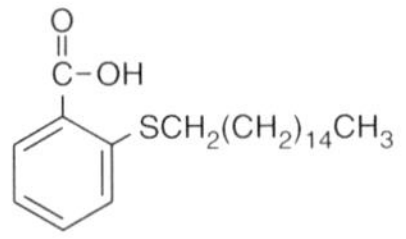

5 gm

S82,400-3
2-[2-(BUTYLTHIO)PHENYL]BENZOIC ACID

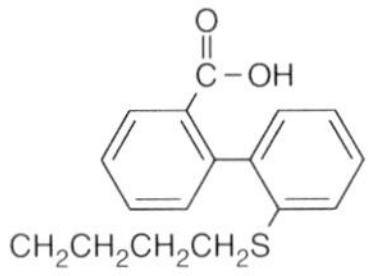

250 mg

S39,046-1
4-(HEXADECYLTHIO)BENZOIC ACID

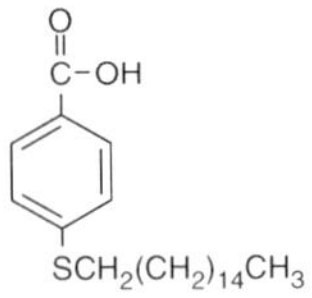

250 mg

S82,277-9
2,2'-(DITHIODI-2,1-PHENYLENE)BIS-(BENZOIC ACID)

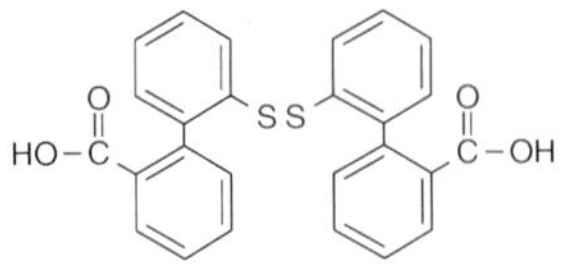

50 mg

S91,554-8
4-AMINO-3,5-DINITROBENZOIC ACID

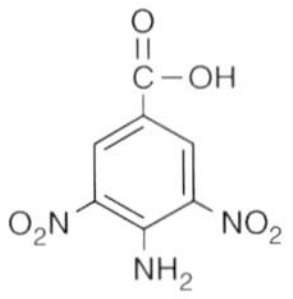

250 mg

S91,588-2
3,5-DINITROANTHRANILIC ACID

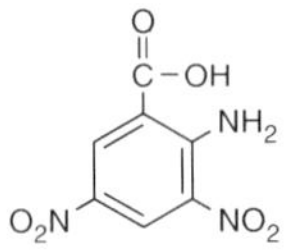

250 mg

S43,089-7
3-NITROANTHRANILIC ACID

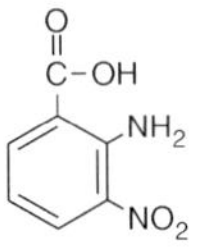

250 mg

S72,235-9
3-AMINO-5-NITROBENZOIC ACID

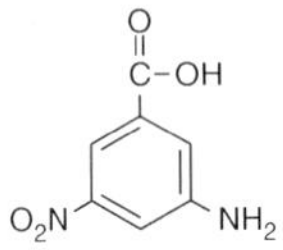

5 gm

S34,845-7
3,5-DIAMINO-4-METHYLBENZOIC ACID

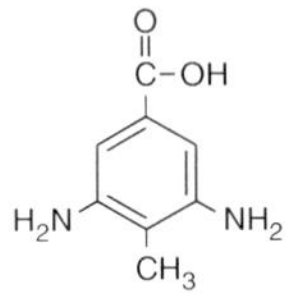

250 mg

S65,593-7
4-AMINO-2-CARBOXYDIPHENYLAMINE

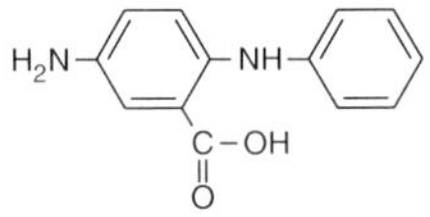

250 mg

S88,315-8
2,2'-BENZIDINEDICARBOXYLIC ACID

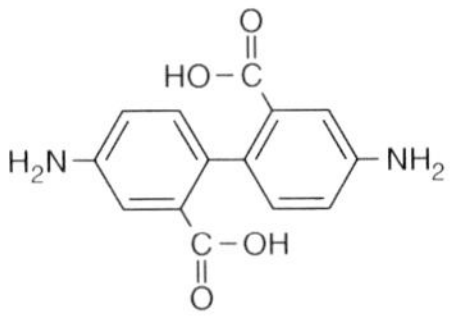

250 mg

S82,008-3
4-AMINOBENZOIC ACID HYDROCHLORIDE

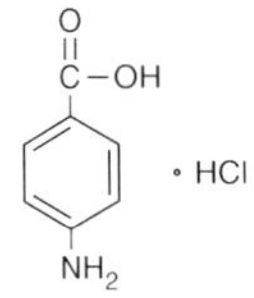

250 mg

S87,715-8
3-AMINO-4-TERT-BUTYLBENZOIC ACID

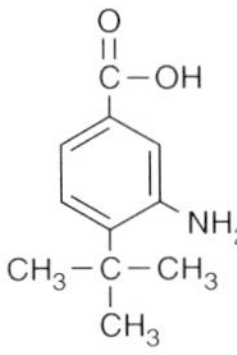

250 mg

S35,759-6
4-AMINO-2,5-DIMETHYLBENZOIC ACID

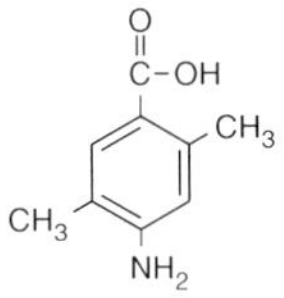

250 mg

S92,662-0
4,5-DIMETHYLANTHRANILIC ACID

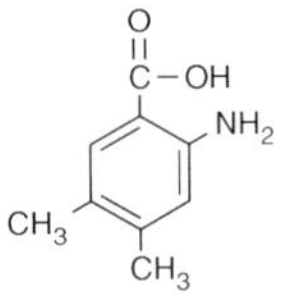

250 mg

S72,254-5
4,6'-BIANTHRANILIC ACID

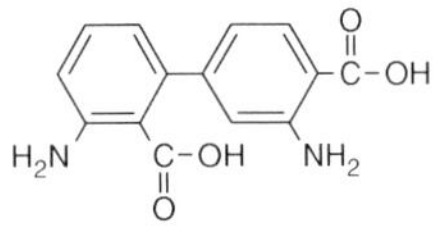

5 gm

S80,464-9
1-AMINO-2-ANTHRAQUINONECARBOXYLIC ACID

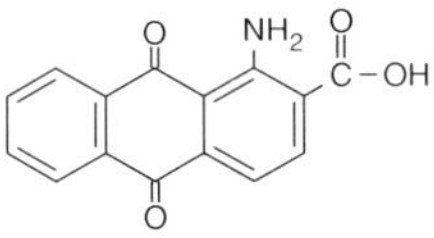

250 mg

S41,435-2
N-(2,4-DINITROPHENYL)-ANTHRANILIC ACID

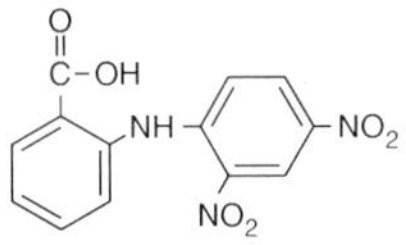

5 gm

S74,665-7
4-CARBOXY-2',4'-DINITRODIPHENYLAMINE

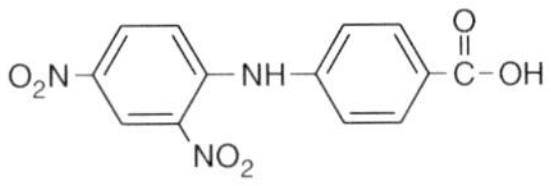

5 gm

S62,400-4
4-(2,4,6-TRINITROPHENYLAMINO)-BENZOIC ACID

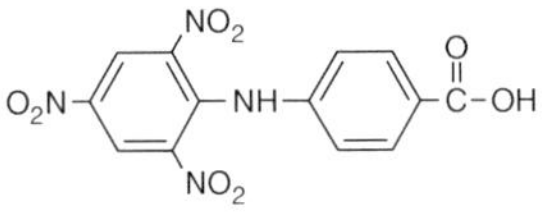

5 gm

S62,405-5
3-(2,4,6-TRINITROPHENYLAMINO)-BENZOIC ACID

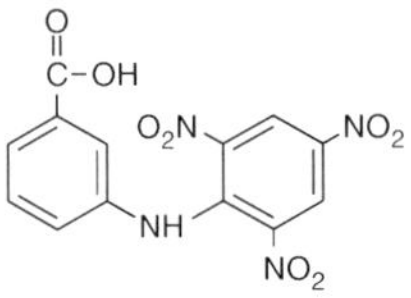

5 gm

S62,320-2
N-PICRYLANTHRANILIC ACID

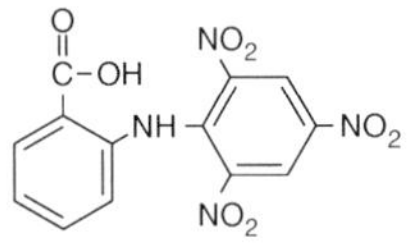

5 gm

S43,395-0
4-DIMETHYLAMINO-3,5-DINITROBENZOIC ACID

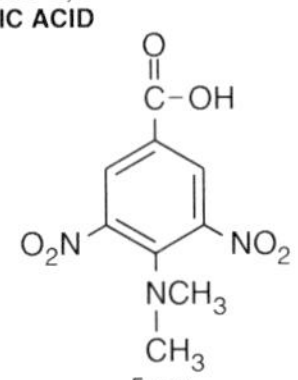

5 gm

S89,466-4
2,3-DICHLORO-4-(2,4-DINITROPHENYLHYDRAZONO)-2-BUTENOIC ACID

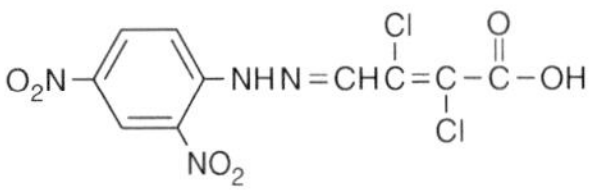

250 mg

S91,556-4
4-ANILINO-3,5-DINITROBENZOIC ACID

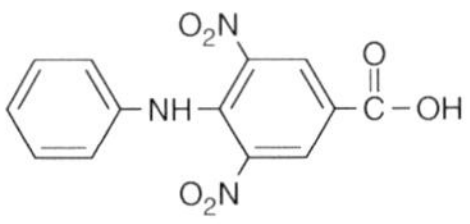

250 mg

S36,417-7
N-(2-NITROPHENYL)ANTHRANILIC ACID

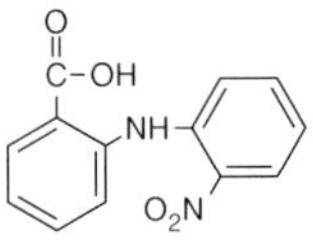

5 gm

S38,261-2
4-(N-HEXADECYL-N-METHYLAMINO)-3-NITROBENZOIC ACID

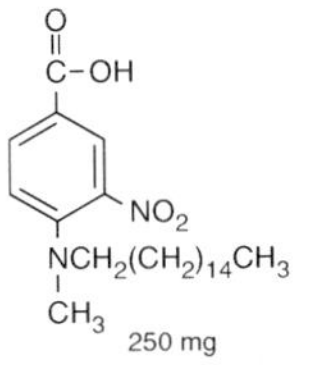

250 mg

S94,238-3
3-NITRO-4-PIPERIDINOBENZOIC ACID

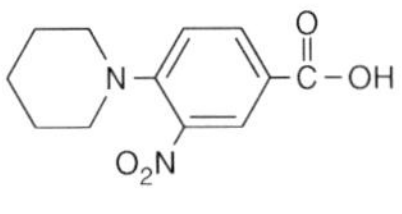

250 mg

S69,700-1
N-(3-NITROPHENYL)ANTHRANILIC ACID

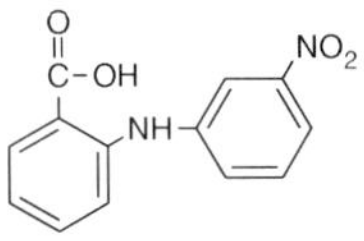

250 mg

S93,823-8
N,N'-(1,4-PHENYLENE)BIS-(ANTHRANILIC ACID)

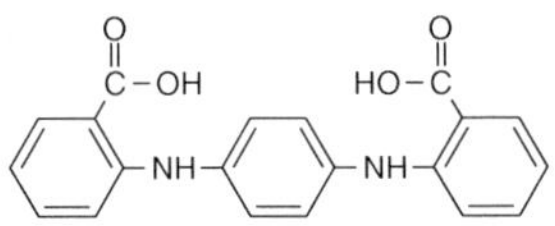

250 mg

S66,171-6
2,5-BIS(P-TOLUIDINO)-TEREPHTHALIC ACID
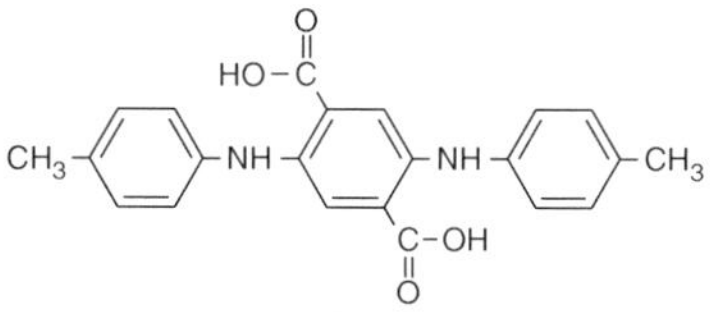
5 gm

S93,542-5
5-NITRO-N-PHENYLANTHRANILIC ACID
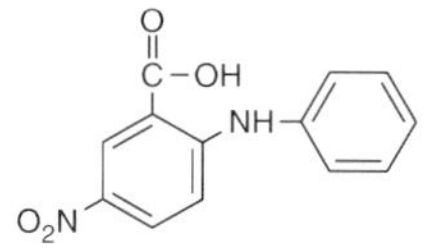
250 mg

S93,539-5
4-CHLORO-N-(4-NITROPHENYL)-ANTHRANILIC ACID
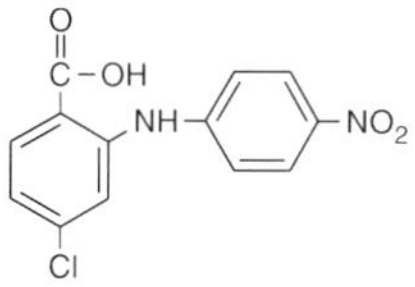
250 mg

S91,435-5
4-(2-NITROVINYL)BENZOIC ACID
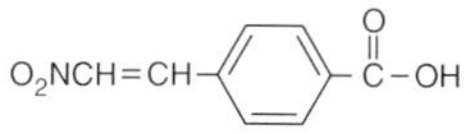
250 mg

S92,697-3
1-(2-METHYLANILINO)-2-ANTHRAQUINONECARBOXYLIC ACID
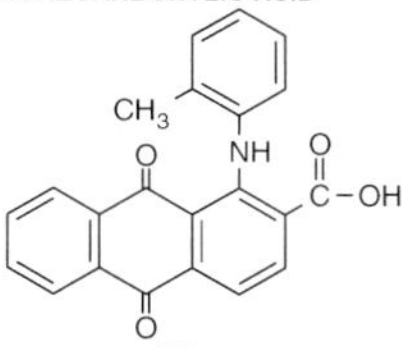
250 mg

S66,852-4
N-ETHYLANTHRANILIC ACID
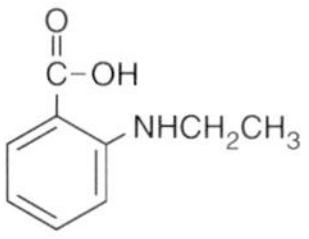
250 mg

S85,238-4
N-(O-TOLYL)ANTHRANILIC ACID
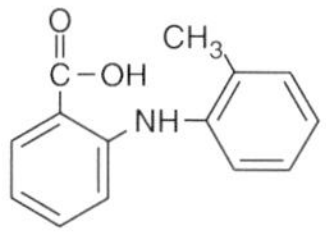
250 mg

S88,836-2
6-ACETYL-N-PHENYLANTHRANILIC ACID
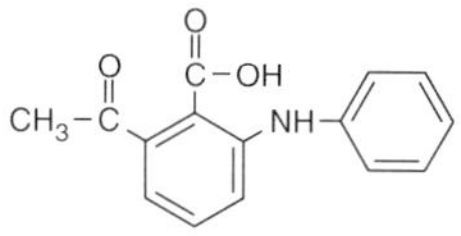
250 mg

S74,028-4
N-ACETONYLANTHRANILIC ACID
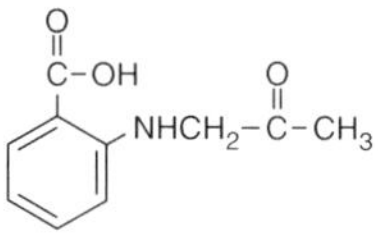
250 mg

S73,980-4
2-(3-CARBOXYANILINO)-3-CHLORO-1,4-NAPHTHOQUINONE
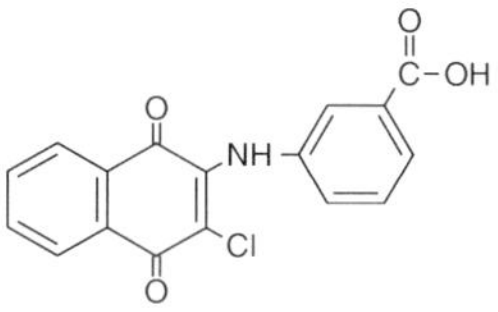
5 gm

S93,824-6
4-(PHENYLAMINO)BENZOIC ACID
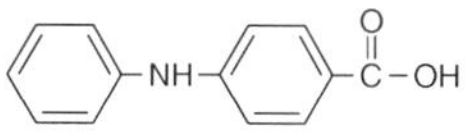
250 mg

S81,040-1
2-(BUTYLAMINO)BENZOIC ACID
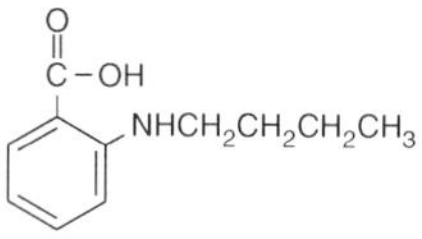
250 mg

S93,541-7
4-CHLORO-N-PHENYLANTHRANILIC ACID
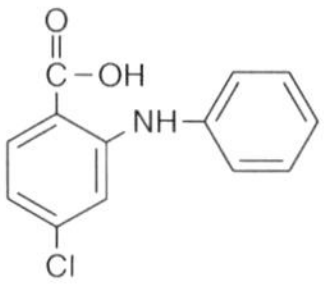
250 mg

S94,921-3
N-(1-NAPHTHYL)ANTHRANILIC ACID
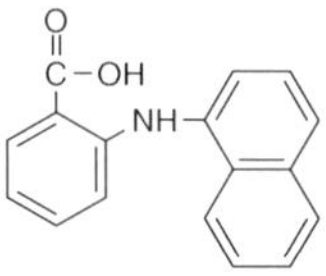
250 mg

S93,540-9
4-CHLORO-N-(4-TOLYL)-ANTHRANILIC ACID
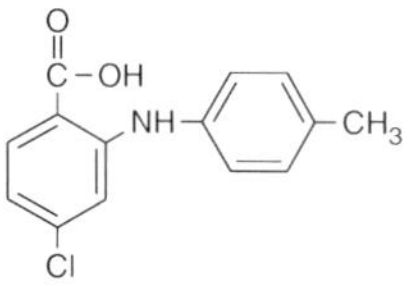
250 mg

S55,215-1
2-CARBOXYPHENYLIMINODIACETIC ACID
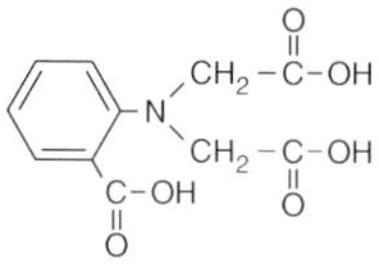
5 gm

S82,028-8
N-METHYL-N-PHENYLANTHRANILIC ACID
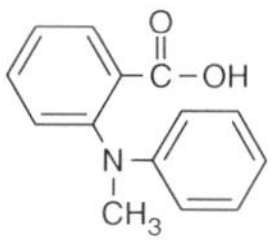
250 mg

S93,550-6
N-BENZYLPROLINE HYDROCHLORIDE
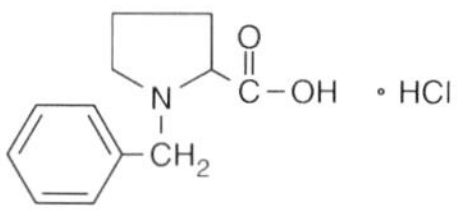
250 mg

S81,924-7
10-ETHYL-4-PHENOTHIAZINECARBOXYLIC ACID
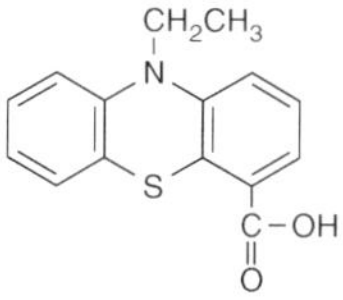
250 mg

S52,083-7
2-[3,6-BIS(DIETHYLAMINO)-9H-XANTHEN-9-YL]BENZOIC ACID SULFATE
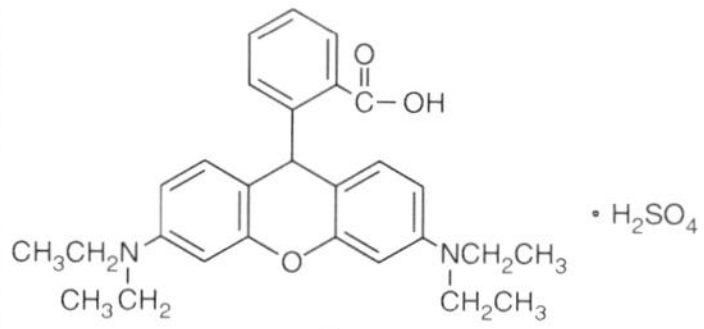
5 gm

S42,733-0
2-DIMETHYLAMINOBENZOIC ACID HYDROCHLORIDE
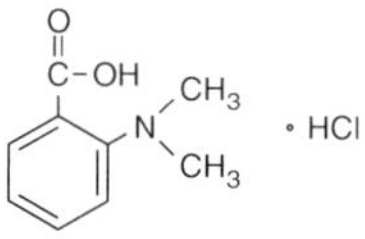
250 mg

S94,573-0
3-ACETYL-2-PHENYL-4-THIAZOLIDINECARBOXYLIC ACID
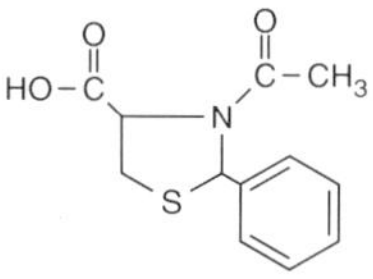
5 gm

S44,934-2
4-CHLORO-2-HYDRAZINOBENZOIC ACID HEMI-HYDROCHLORIDE
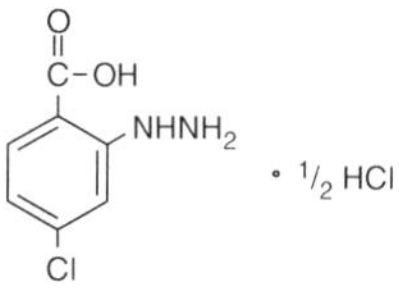
5 gm

S44,306-9
6-HYDRAZINO-M-TOLUIC ACID HYDROCHLORIDE
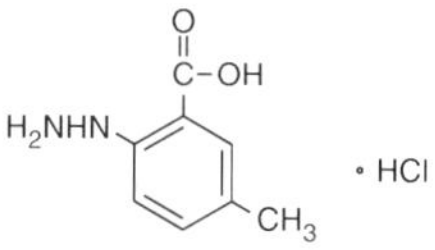
250 mg

S67,924-0
ACETOPHENONE 4-CARBOXYPHENYLHYDRAZONE
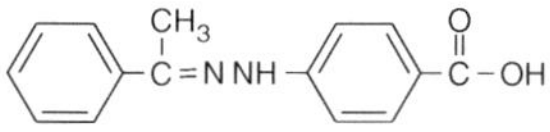
250 mg

S60,540-9
KETOMALONIC ACID PHENYLHYDRAZONE
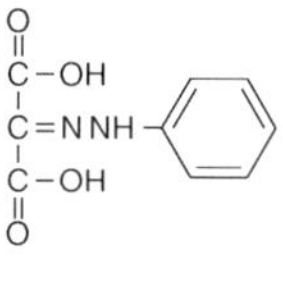
250 mg

S91,846-6
4-NITROISOPHTHALIC ACID
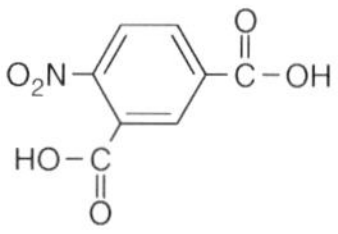
250 mg

S89,475-3
4-(3-NITROPHENYL)-1,3-BUTADIENE-1-CARBOXYLIC ACID
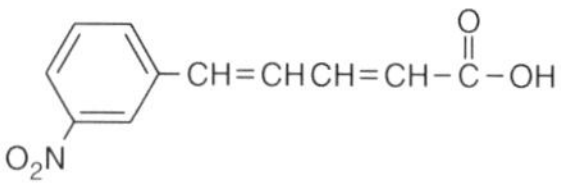
250 mg

S86,534-6
5 OR 6 OR 7 OR 8-NITRO-2-ANTHRAQUINONECARBOXYLIC ACID
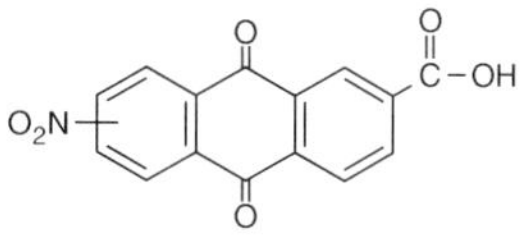
250 mg

S75,164-2
7-NITRO-9-OXO-1-FLUORENECARBOXYLIC ACID
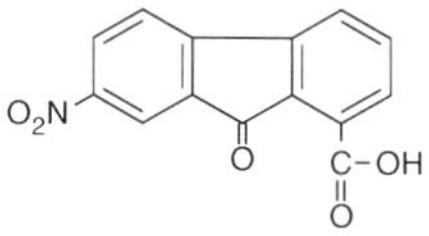
5 gm

S75,412-9
7-NITRO-9-OXO-2-FLUORENECARBOXYLIC ACID
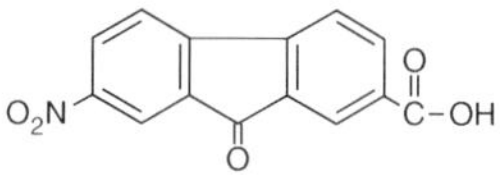
250 mg

S91,456-8
3-(2-NITROVINYL)BENZOIC ACID
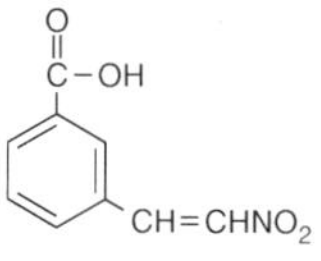
250 mg

S92,656-6
5-METHYL-4-NITROISOPHTHALIC ACID

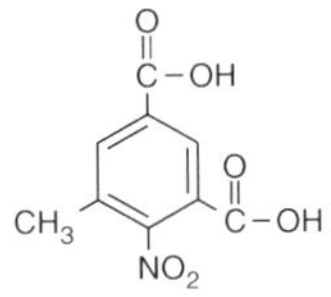

250 mg

S43,821-9
2-NITROISOPHTHALIC ACID

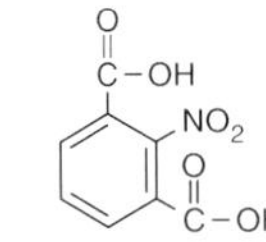

250 mg

S92,258-7
4-TERT-BUTYL-3-NITROBENZOIC ACID

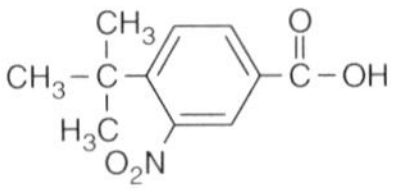

250 mg

S37,204-8
3,5-DIMETHYL-4-NITROBENZOIC ACID

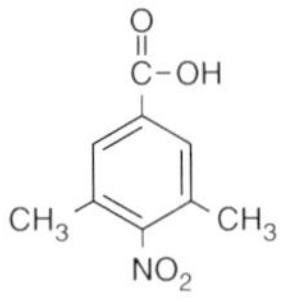

250 mg

S40,810-7
2,5-DIMETHYL-4-NITROBENZOIC ACID

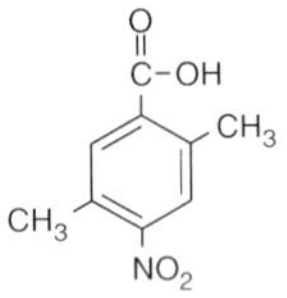

250 mg

S82,215-9
5-NITRO-1-NAPHTHALENECARBOXYLIC ACID

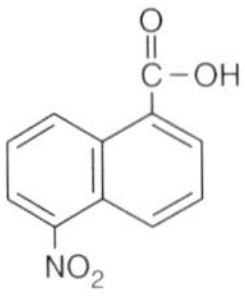

250 mg

S94,502-1
5,5'-DINITRO-2,2'-BIPHENYLDICARBOXYLIC ACID

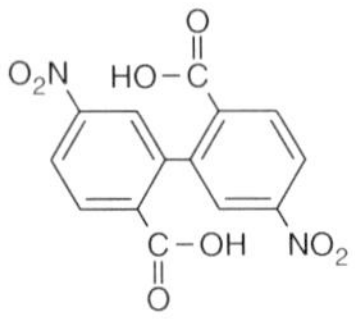

250 mg

S35,390-6
7-NITROFLUORENE-1-CARBOXYLIC ACID

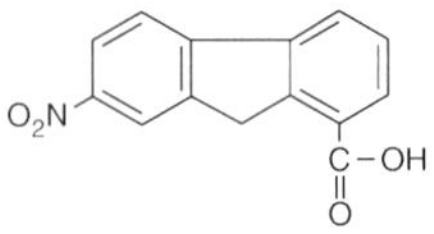

250 mg

S66,311-5
4-CARBOXY-5-PHENANTHRENECARBOXALDEHYDE

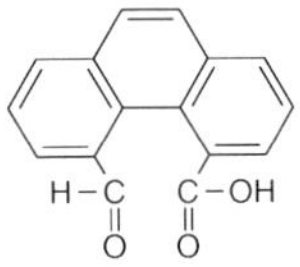

5 gm

S94,377-0
7-METHYL-2-NAPHTHALDEHYDE

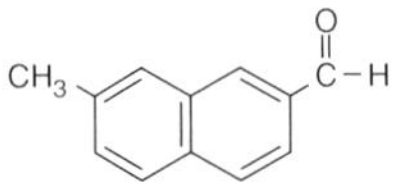

250 mg

S88,742-0
4,4'-(1,4-PHENYLENE)BIS(2-OXO-3-BUTENOIC ACID)

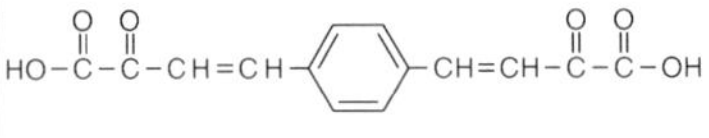

250 mg

S84,965-0
2'-BENZOYL-2-BIPHENYLCARBOXYLIC ACID

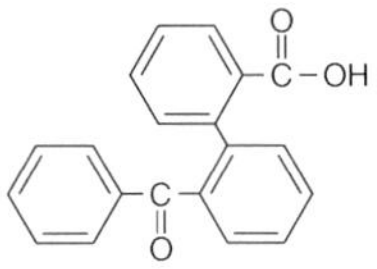

250 mg

S88,902-4
3-(2,4,6-TRIMETHYLBENZOYL)-2-BIPHENYLCARBOXYLIC ACID

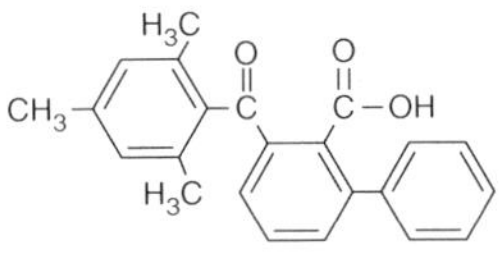

250 mg

S73,447-0
3-BENZOYL-4-METHYLBENZOIC ACID

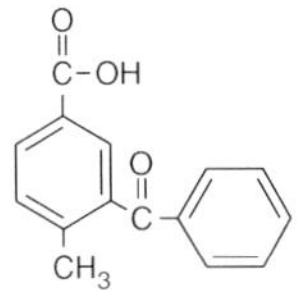

250 mg

S88,901-6
2'-(2,3-DIMETHYLBENZOYL)-2-BIPHENYLCARBOXYLIC ACID

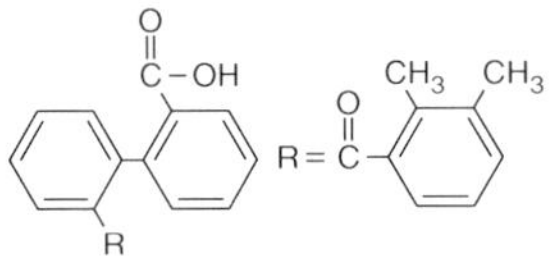

250 mg

S87,953-3
2-(2,4-DIMETHYLBENZOYL)BENZOIC ACID

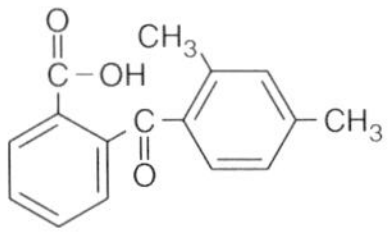

5 gm

S75,802-7
2,2'-(1,3-INDENEDIFORMYL)-DIBENZOIC ACID

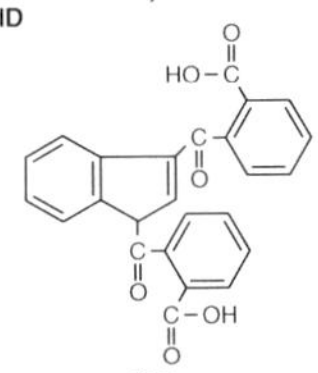

250 mg

S93,146-2
2-(5,6,7,8-TETRAHYDRO-2-NAPHTHOYL)-BENZOIC ACID

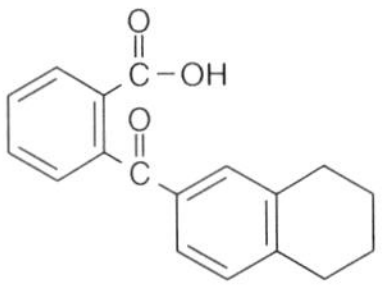

250 mg

S67,164-9
2-(5-FLUORO-1-NAPHTHOYL)-BENZOIC ACID

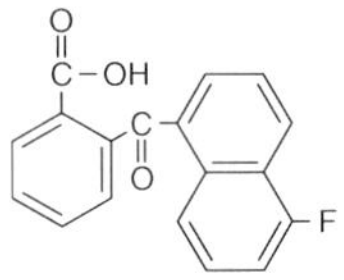

5 gm

S84,968-5
2-(2-NAPHTHOYL)BENZOIC ACID

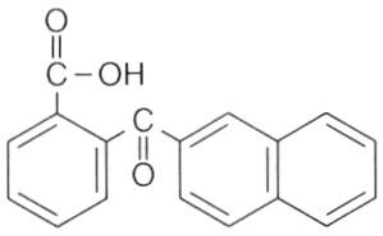

250 mg

S67,247-5
6-(4-TERT-BUTYLPHENYL)-4-OXO-5-HEXENOIC ACID

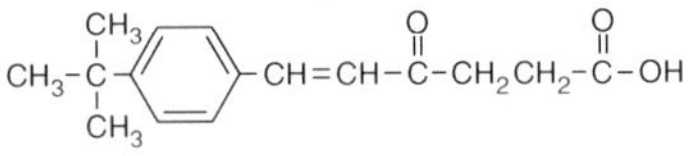

5 gm

S84,970-7
2-(2,4-DIMETHYLBENZOYL)-3-METHYLBENZOIC ACID

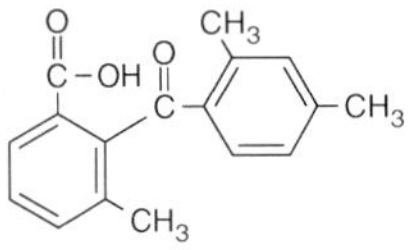

250 mg

S84,967-7
3,6-DIMETHYL-2-(2,4-DIMETHYLBENZOYL)BENZOIC ACID

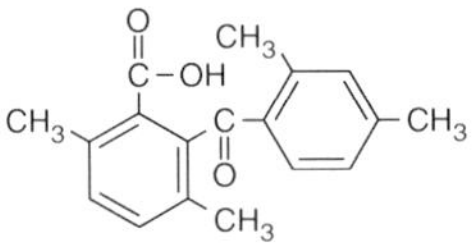

250 mg

S76,465-5
3-ACETYL-2-NAPHTHOIC ACID

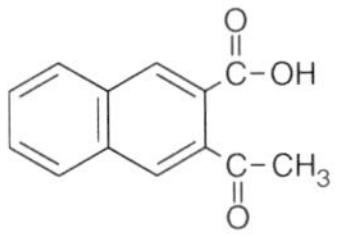

5 gm

S75,439-0
2,7-DIFLUORO-9-OXO-4-FLUORENECARBOXYLIC ACID

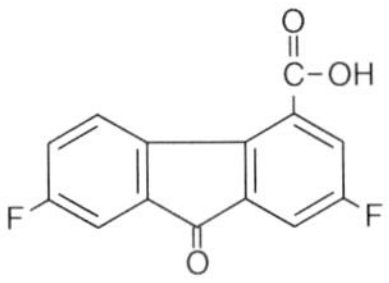

250 mg

S41,353-4
2,4'-CARBONYLDIBENZOIC ACID

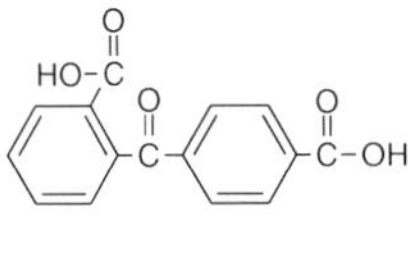

5 gm

S56,210-6
2',3,4-BENZOPHENONETRICARBOXYLIC ACID

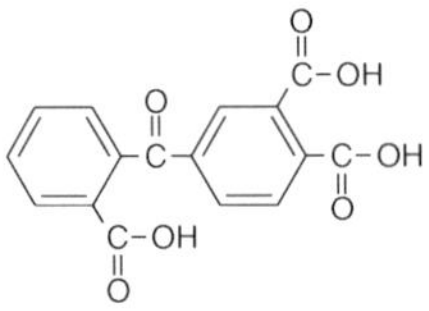

5 gm

S56,208-4
9,10-DIHYDRO-9,10-DIOXO-2,3-ANTHRACENEDICARBOXYLIC ACID

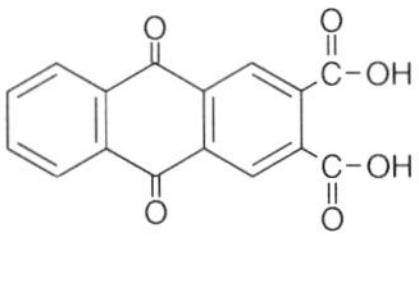

250 mg

S42,706-3
3,3',4,4'-BENZOPHENONETETRACARBOXYLIC ACID

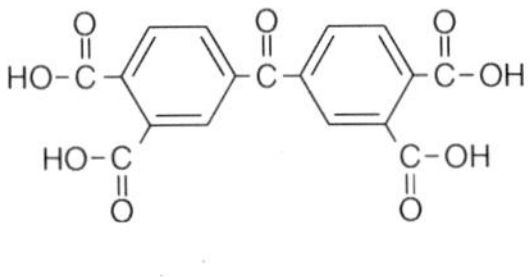

5 gm

S52,765-3
5,5'-CARBONYLBIS(TRIMELLITIC ACID)

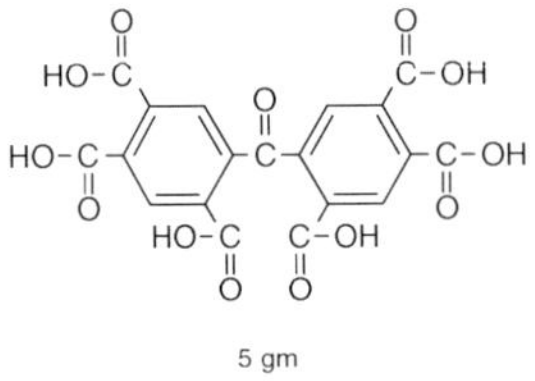

5 gm

S65,668-2
4-OXO-6-PHENYL-5-HEXENOIC ACID

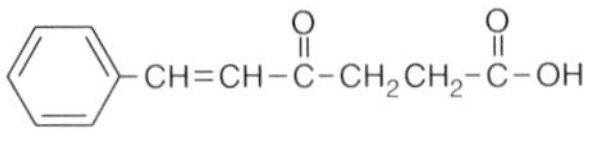

5 gm

S35,227-6
2-(4-ETHYLBENZOYL)BENZOIC ACID

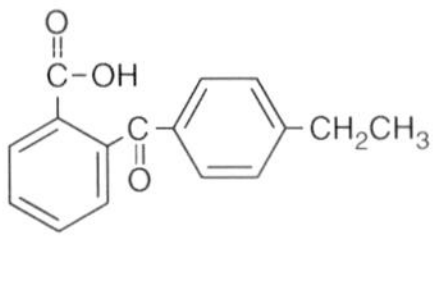

250 mg

S35,171-7
2-(4-ISOPROPYLBENZOYL)BENZOIC ACID

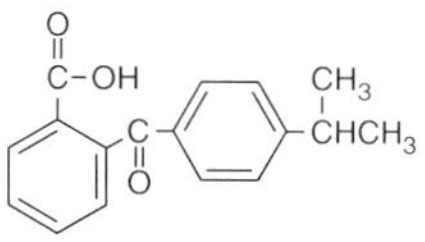

250 mg

S66,284-4
2-(2,4,6-TRIMETHYLBENZOYL)-BENZOIC ACID

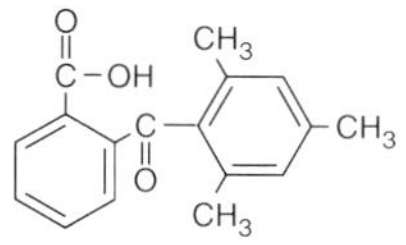

250 mg

S61,964-7
2-(1-NAPHTHOYL)BENZOIC ACID

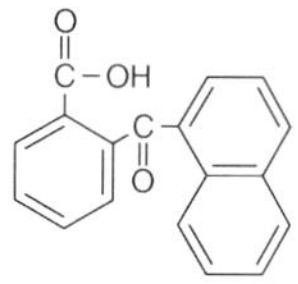

5 gm

S65,738-7
2,4-DICHLORO-3-(3-FLUOROBENZOYL)-BENZOIC ACID

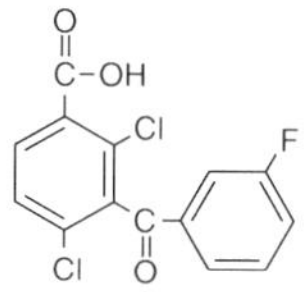

250 mg

S66,287-9
2-BENZOYL-3,6-DIMETHYLBENZOIC ACID

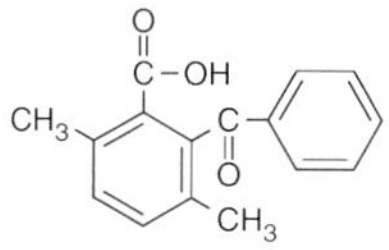

250 mg

S61,977-9
3,4-BENZO-9-OXOFLUORENE-1-CARBOXYLIC ACID

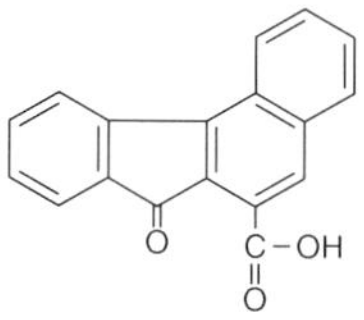

250 mg

S92,362-1
1H-DIBENZ(A,KL)-7-ANTHRACENONE-13-CARBOXYLIC ACID

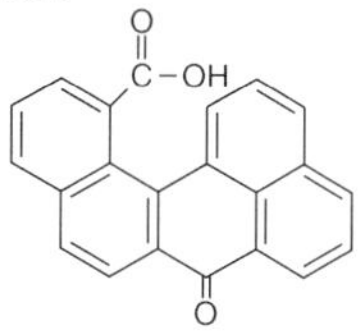

250 mg

S78,674-8
2-CARBOXY-ALPHA-OXOBENZENEACETIC ACID

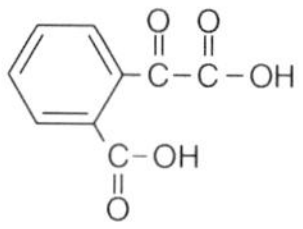

250 mg

S43,710-7
BENZENEPENTACARBOXYLIC ACID

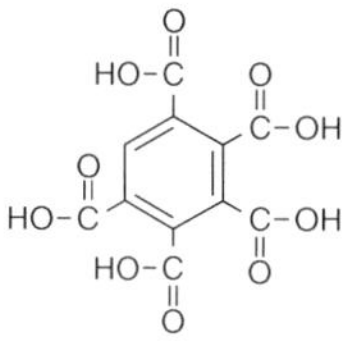

5 gm

S93,355-4
4-PHENYLPHTHALIC ACID

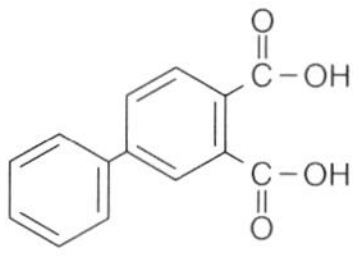

250 mg

S91,562-9
2-METHYLISOPHTHALIC ACID

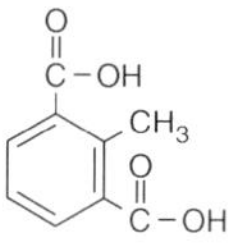

250 mg

S87,761-1
3-VINYLBENZOIC ACID

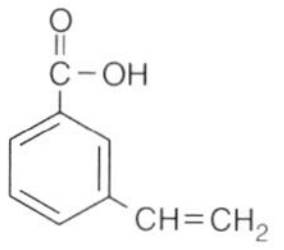

250 mg

S43,476-0
3-PHENYL-1,1,3-TRIMETHYLINDAN-4',5-DICARBOXYLIC ACID

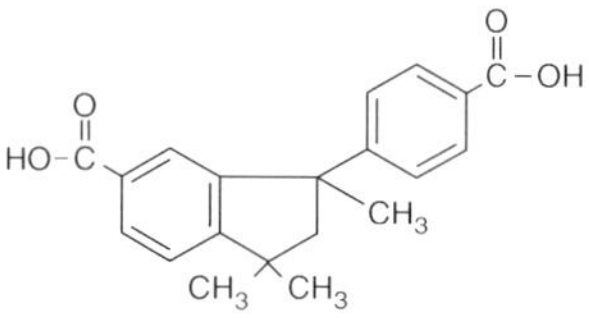

5 gm

S65,873-1
1,6-NAPHTHALENEDICARBOXYLIC ACID

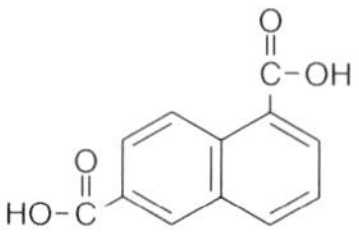

250 mg

S52,240-6
1,8-NAPHTHALENEDICARBOXYLIC ACID

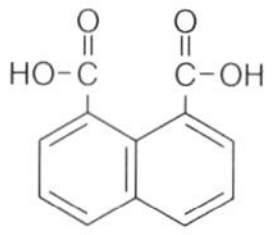

5 gm

S45,700-0
1,1'-BINAPHTHYL-8,8'-DICARBOXYLIC ACID

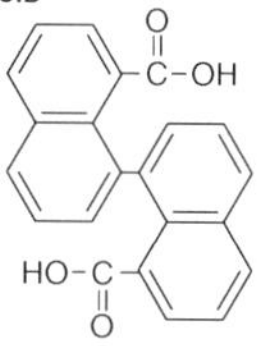

5 gm

S44,348-4
1-METHYLINDENE-2-CARBOXYLIC ACID

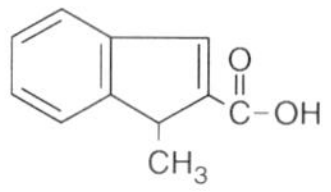

5 gm

S88,048-5
4-[3,4-(METHYLENEDIOXY)PHENYL]-1,3-BUTADIENE-1-CARBOXYLIC ACID

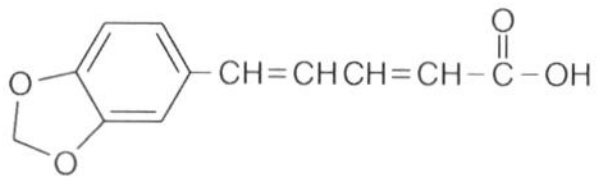

250 mg

S88,279-8
N-CYCLOHEXYLIDENEANTHRANILIC ACID

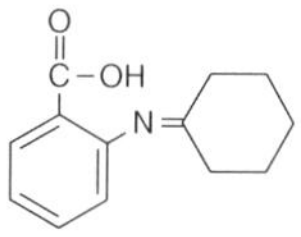

250 mg

S83,998-1
1-PHENOXATHIINCARBOXYLIC ACID

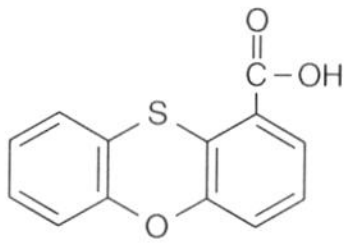

250 mg

S88,083-3
2-PHENYLACRYLIC ACID

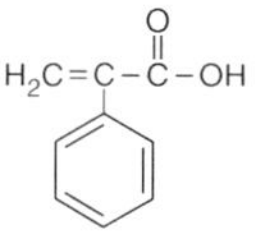

250 mg

S90,617-4
CINNAMYLIDENEACETIC ACID

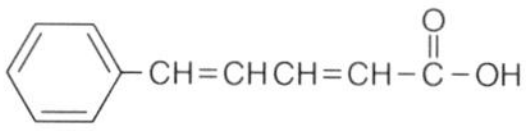

250 mg

S81,482-2
2-(CINNAMYLIDENE)-2-PHENYLACETIC ACID

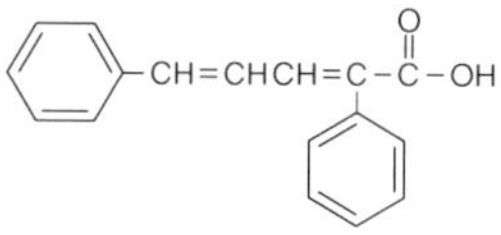

250 mg

S86,570-2
2-(2-METHYL-1-PHENYLPROPYL)-BENZOIC ACID

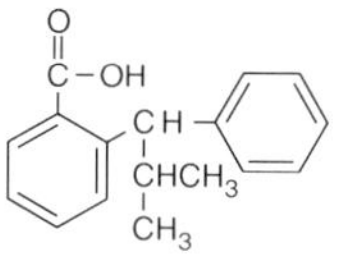

250 mg

S88,019-1
4-STILBENECARBOXYLIC ACID

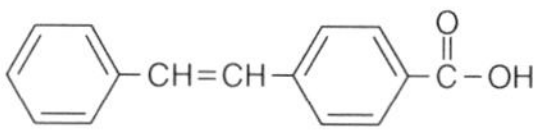

250 mg

S83,205-7
2,4,6-TRIPHENYLBENZOIC ACID

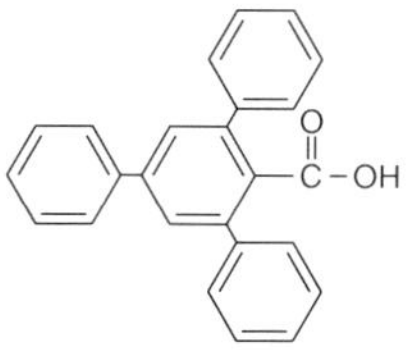

250 mg

S87,408-6
3,5-DIPHENYLBENZOIC ACID

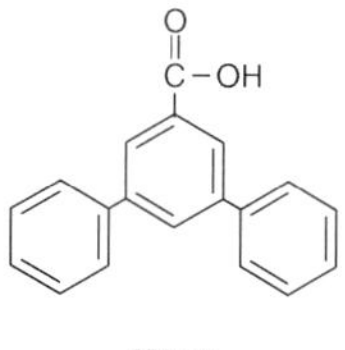

250 mg

S45,958-5
O-TOLUIC ACID, ZONE REFINED

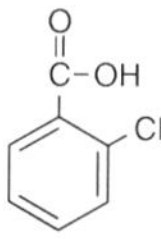

5 gm

S62,281-8
O-ETHYLBENZOIC ACID

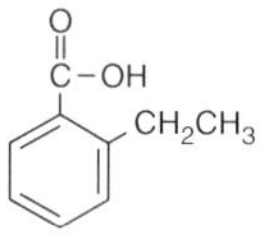

250 mg

S75,926-0
2-(2,2-DIMETHYL-1-INDANYLMETHYL)-BENZOIC ACID

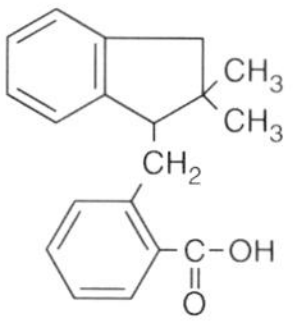

250 mg

S85,284-8
2-(1,2,3,4-TETRAHYDRO-6-NAPHTHYLMETHYL)BENZOIC ACID

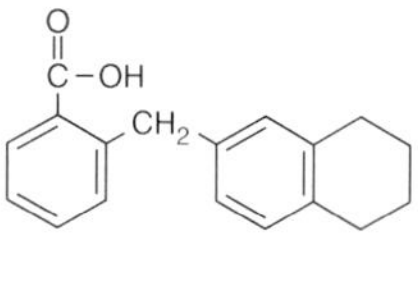

250 mg

S94,557-9
4'-OCTYL-4-BIPHENYLCARBOXYLIC ACID

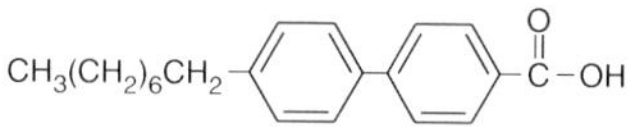

250 mg

S54,697-6
2-BROMO-5-CHLOROBENZOIC ACID
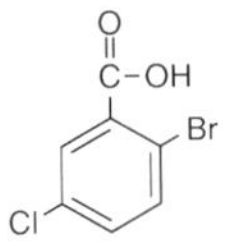

5 gm

S65,665-8
2,6-DICHLORO-3-METHYLBENZOIC ACID
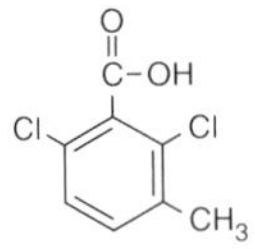

5 gm

S83,232-4
1-INDENECARBOXYLIC ACID
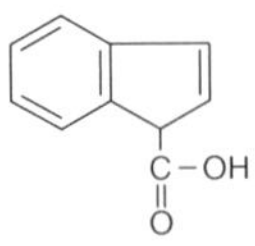

250 mg

S66,290-9
2-METHYL-1-NAPHTHALENECARBOXYLIC ACID
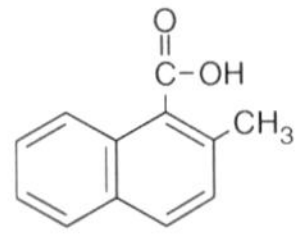

250 mg

S83,919-1
1-METHYL-2-NAPHTHOIC ACID
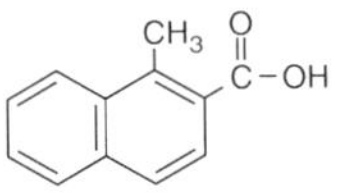

250 mg

S60,303-1
3-(1-NAPHTHYL)ACRYLIC ACID
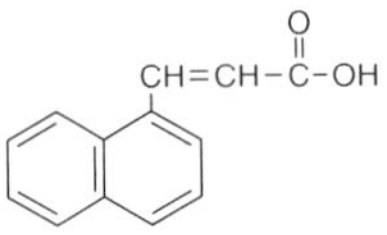

250 mg

S83,902-7
4-BROMO-1-NAPHTHOIC ACID
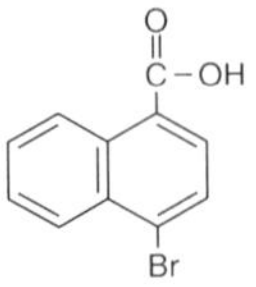

250 mg

S67,119-3
3-BROMO-2-NAPHTHOIC ACID
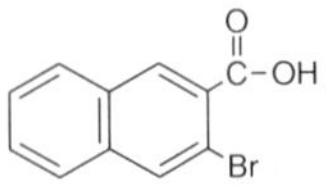

250 mg

S97,466-8
AR-G20 BENZO(C)PHENANTHRENE-6-CARBOXYLIC ACID
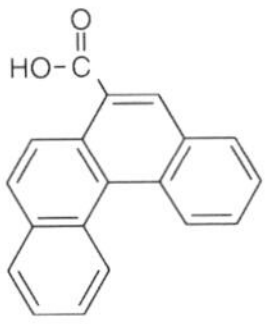

10 mg

S97,467-6
AR-G21 BENZO(C)PHENANTHRENE-5-CARBOXYLIC ACID
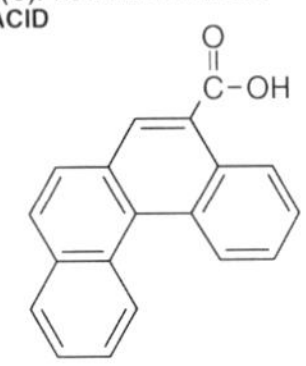

10 mg

S97,469-2
AR-G24 BENZO(C)PHENANTHRENE-3-CARBOXYLIC ACID
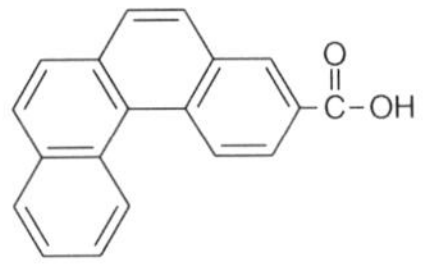

10 mg

S97,471-4
AR-G27 BENZO(C)PHENANTHRENE-1-CARBOXYLIC ACID
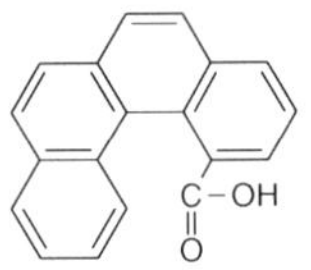

10 mg

S97,391-2
AR-E52 7-METHYLBENZ(A)ANTHRACENE-9-CARBOXYLIC ACID
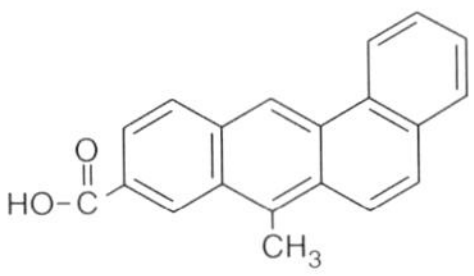

10 mg

S97,392-0
AR-E53 7-METHYLBENZ(A)ANTHRACENE-5-CARBOXYLIC ACID
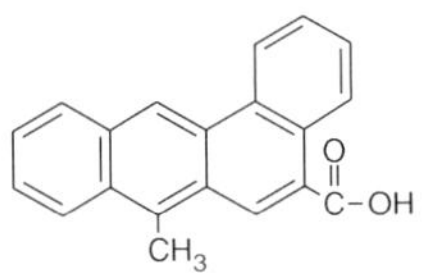

10 mg

S97,423-4
AR-E85 7,12-DIMETHYLBENZ(A)-ANTHRACENE-5-CARBOXYLIC ACID
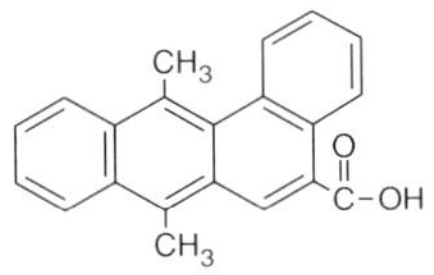

10 mg

S88,311-5
2-ANTHRACENECARBOXYLIC ACID
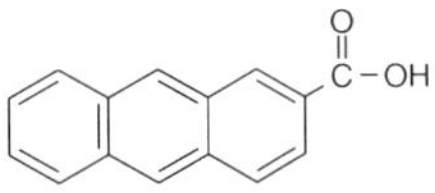

250 mg

S90,732-4
10-METHYL-9-ANTHRACENECARBOXYLIC ACID
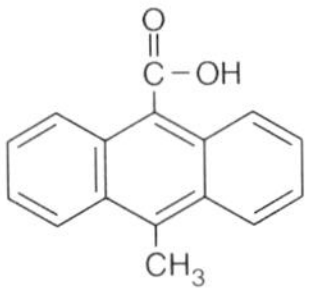

250 mg

S97,318-1
AR-D17
9-PHENANTHRENECARBOXYLIC ACID
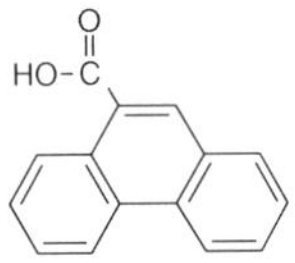

100 mg

S97,326-2
AR-D24 1,2-DIHYDRO-4-PHENANTHROIC ACID
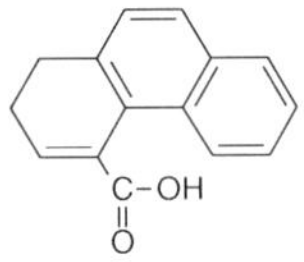

10 mg

S82,280-9
1-THIANTHRENECARBOXYLIC ACID
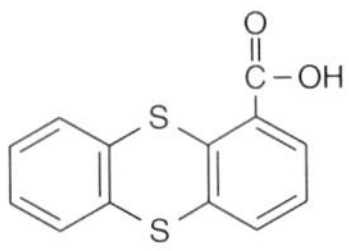

250 mg

S89,335-8
2-PHENYL-1,1-CYCLOPROPANEDICARBOXYLIC ACID HYDRATE
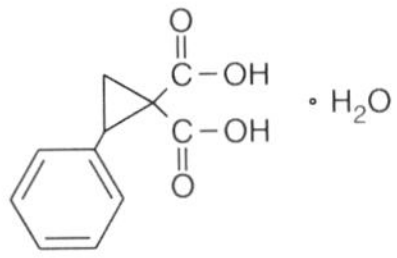

250 mg

S53,196-0
2,4,5-TRIMETHYLBENZOIC ACID
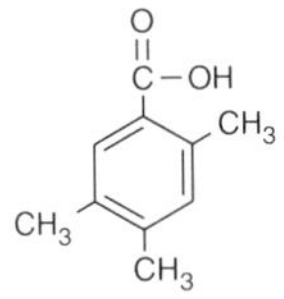

250 mg

S93,734-7
M-CARBORANE-1,7-DICARBOXYLIC ACID
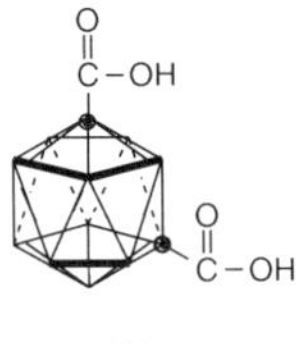

250 mg

S94,494-7
ALPHA-TRUXILLIC ACID
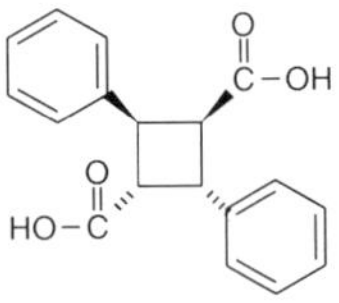

250 mg

S56,500-8
4,4PR-STILBENEDICARBOXYLIC ACID
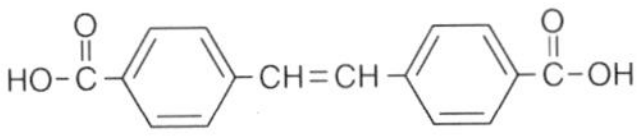

5 gm

S93,583-2
1,2-DIHYDROBENZ(J)ACEANTHRYLENE-12-CARBOXYLIC ACID
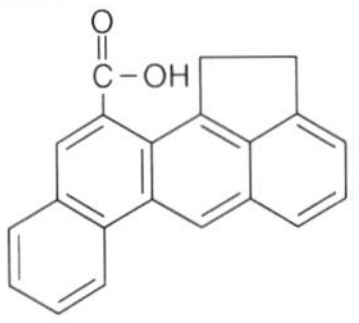

10 mg

S94,374-6
6-METHYL-1,3-BENZODIOXOLE-5-CARBOXYLIC ACID
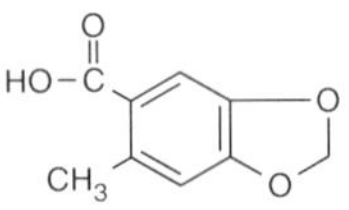

250 mg

S94,571-4
2-(1-ETHYLPROPYL)-4-THIAZOLIDENECARBOXYLIC ACID
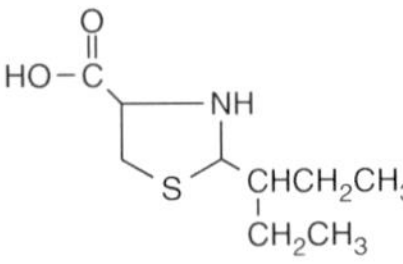

5 gm

S94,657-5
2,3-DIPHENYL-2-CYCLOPROPENE-1-CARBOXYLIC ACID
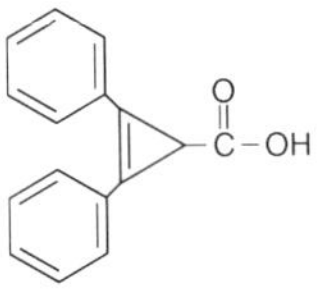

250 mg

S40,824-7
2,2-DIPHENYLBUTYRIC ACID
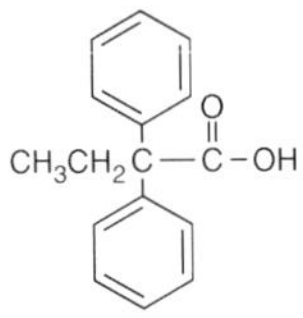

5 gm

S94,639-7
2-(5-METHYL-1,3-DITHIAN-2-YLIDENEHYDRAZINO)BENZOIC ACID
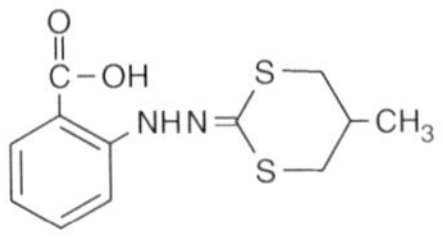

250 mg

S50,170-0
3,5-DIMETHYL-2-IODOBENZOIC ACID
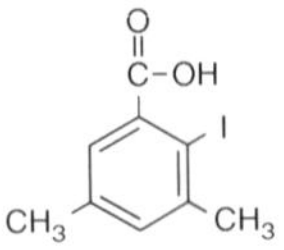

250 mg

S41,605-3
INDENE-3-CARBOXYLIC ACID

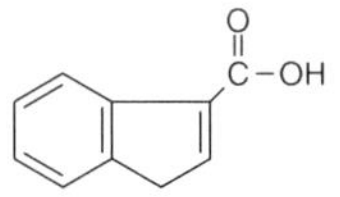

250 mg

S69,685-4
8-BROMO-1-NAPHTHOIC ACID

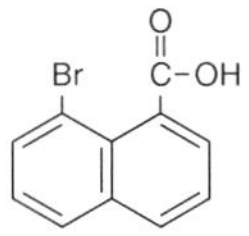

250 mg

S56,195-9
2-NAPHTHOIC ACID

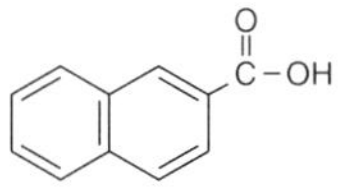

5 gm

S52,227-9
3-PHENANTHRENECARBOXYLIC ACID

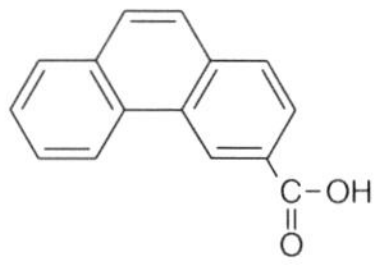

250 mg

S42,788-8
DITHIOTEREPHTHALIC ACID

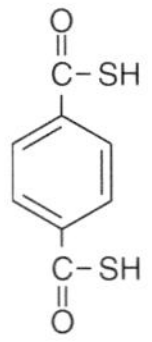

5 gm

S78,427-3
3,5-DIIODO-DL-TYROSINE

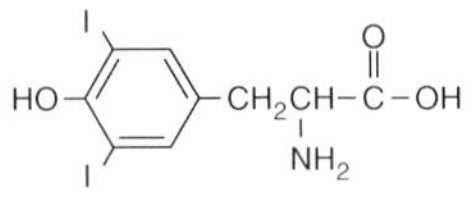

250 mg

S46,296-9
P-FLUORO-D-PHENYLALANINE

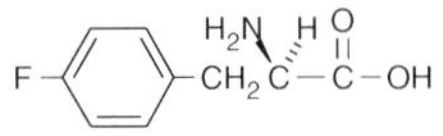

5 gm

S93,968-4
BETA-AMINO-4-CHLOROHYDROCINNAMIC ACID

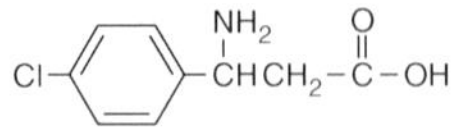

250 mg

S69,684-6
N-(2-BROMOPHENYL)GLYCINE

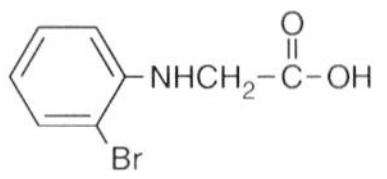

250 mg

S75,670-9
D-3-(3,4-DIHYDROXYPHENYL)ALANINE

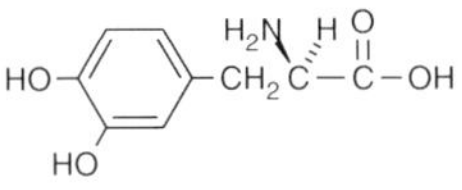

5 gm

S34,867-8
ERYTHRO-3-(3,4-DIBENZYLOXYPHENYL)-SERINE

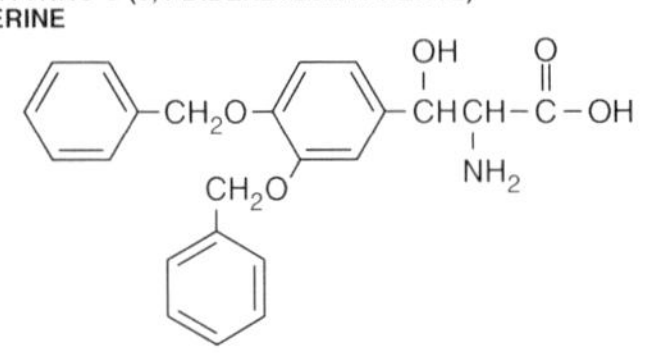

250 mg

S34,868-6
THREO-3-(3,4-DIBENZYLOXYPHENYL)SERINE HYDRATE

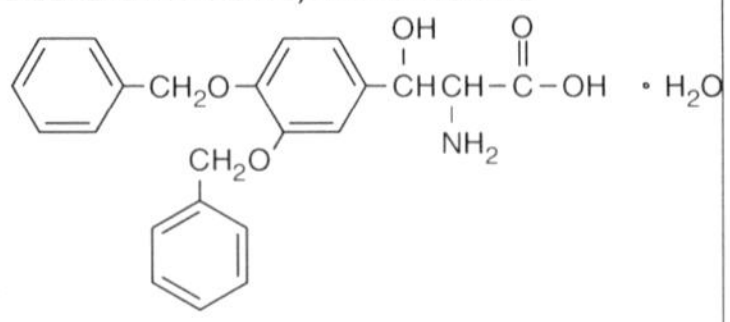

250 mg

S78,322-6
L-3-METHOXYTYROSINE HYDRATE

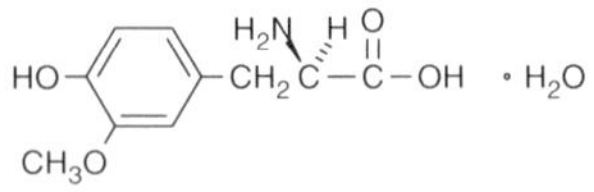

250 mg

S44,517-7
3,5-DIIODO-L-THYRONINE

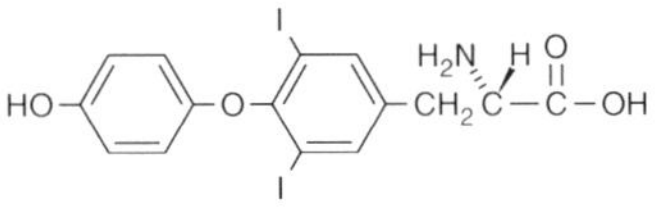

250 mg

S42,850-7
DL-ALPHA-METHYL-M-TYROSINE MONOHYDRATE

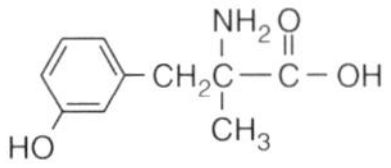

250 mg

S44,610-6
DL-O-TYROSINE

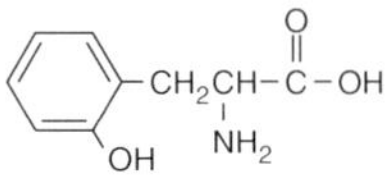

250 mg

S74,975-3
BETA-(2-HYDROXY-3-METHYLPHENYL)-ALANINE

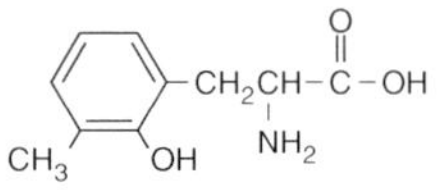

250 mg

S74,966-4
BETA-(2-HYDROXY-5-METHYLPHENYL)-ALANINE

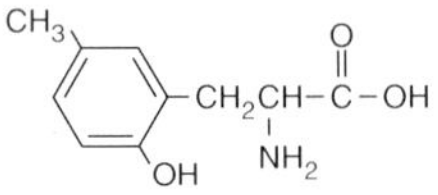

250 mg

S74,974-5
BETA-(2-HYDROXY-4-METHYLPHENYL)-ALANINE

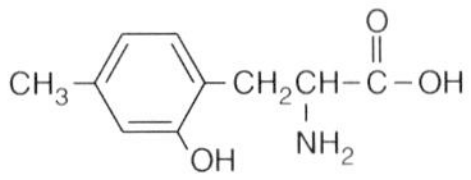

250 mg

S44,892-3
3,5-DIBROMO-L-TYROSINE MONOHYDRATE

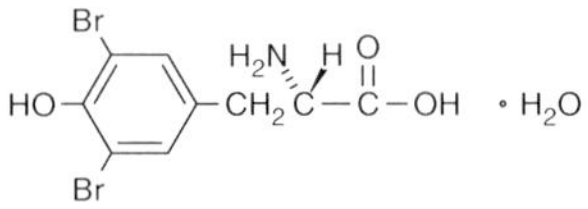

5 gm

S78,321-8
DL-ALPHA-METHYLTYROSINE HYDROCHLORIDE

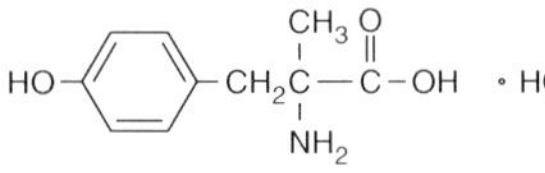

5 gm

S78,197-5
N-(2,4-DINITROPHENYL)-L-SERINE

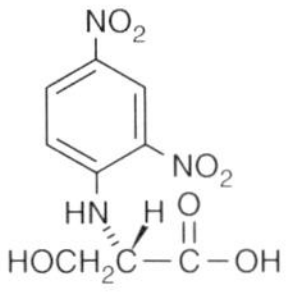

250 mg

S70,504-7
N-(2,4-DINITROPHENYL)-L-THREONINE

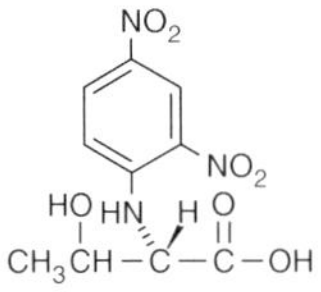

250 mg

S76,584-8
N-(2,4-DINITROPHENYL)-L-HYDROXYPROLINE

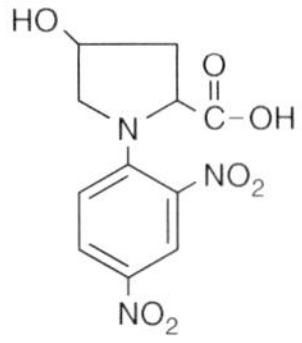

5 gm

S77,692-0
N-BENZYL-L-TYROSINE

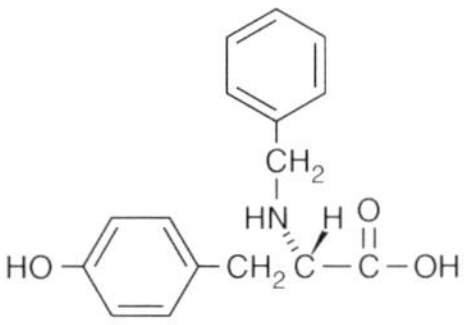

250 mg

S77,706-4
N-BENZYL-L-SERINE

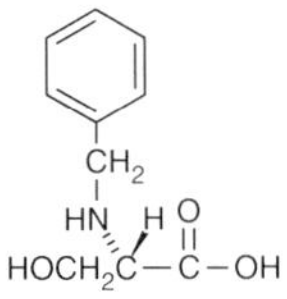

250 mg

S77,754-4
N-BENZYL-L-THREONINE

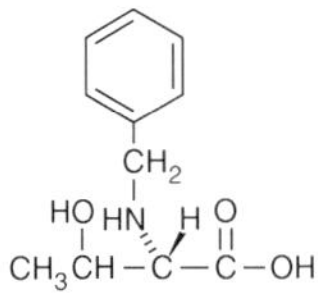

250 mg

S63,502-2
O-BENZYL-DL-SERINE

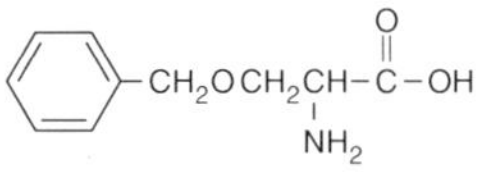

250 mg

S78,337-4
3,4-DIMETHOXY-DL-PHENYLALANINE

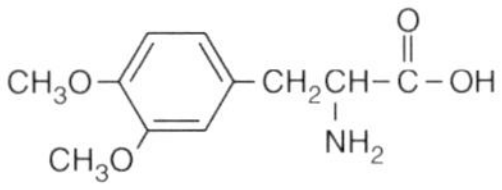

5 gm

S78,743-4
O-(2,4-DINITROPHENYL)-L-TYROSINE MONOHYDRATE

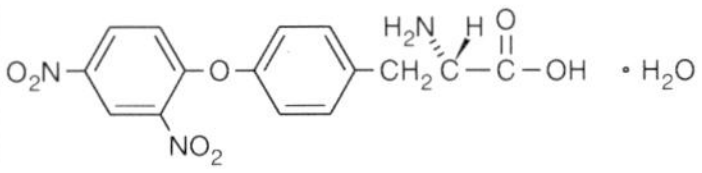

250 mg

S78,732-9
3-METHOXY-ALPHA-METHYLPHENYLALANINE

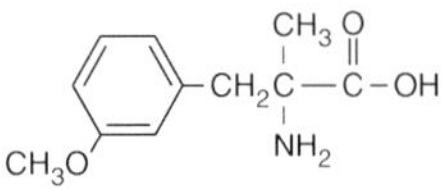

250 mg

S78,931-3
L-VALINE 4-NITROBENZYL ESTER HYDROBROMIDE

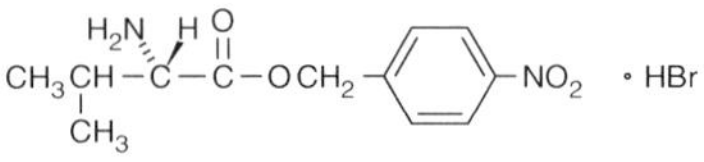

250 mg

S44,728-5
S-DIPHENYLMETHYL-L-CYSTEINE

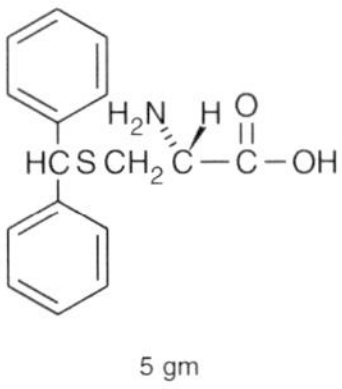

5 gm

S35,840-1
S-BENZYL-DL-HOMOCYSTEINE

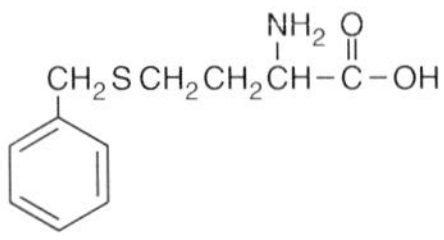

5 gm

S78,924-0
N-(2,4-DINITROPHENYL)-L-METHIONINE DICYCLOHEXYLAMINESALT

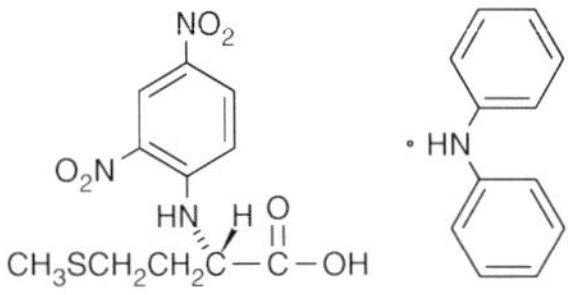

250 mg

S78,644-6
N,N'-BIS(2,4-DINITROPHENYL)-L-CYSTINE

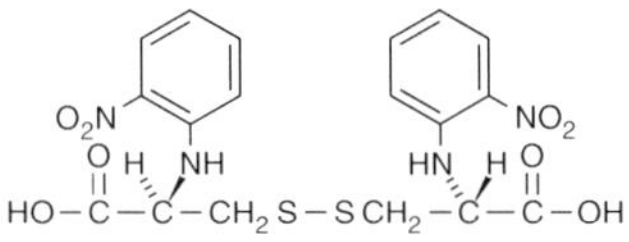

250 mg

S73,154-4
(2,4-DINITROPHENYL)GLYCINE

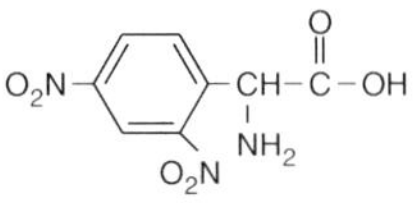

5 gm

S41,704-1
2-CYCLOPROPYL-2-PHENYLGLYCINE
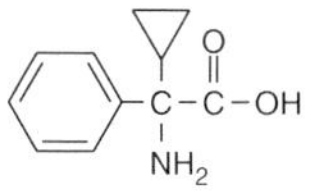
250 mg

S80,281-6
3-BENZOYLALANINE HYDROCHLORIDE
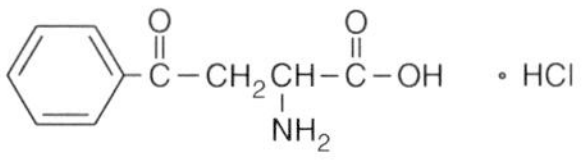
250 mg

S35,194-6
DL-LEUCYLGLYCYL-DL-PHENYLALANINE
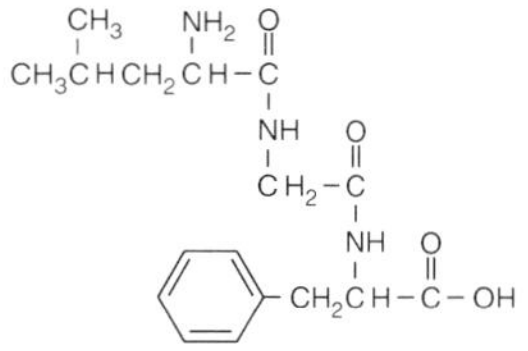
5 gm

S78,884-8
N-(2,4-DINITROPHENYL)-L-ALANINE
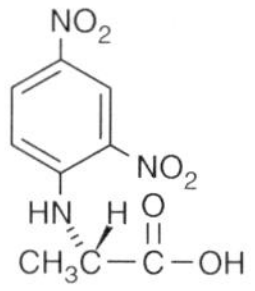
250 mg

S78,914-3
N-(2,4-DINITROPHENYL)-L-2-AMINOBUTYRIC ACID
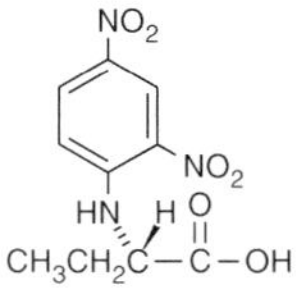
250 mg

S78,167-3
N-(2,4-DINITROPHENYL)-L-ISOLEUCINE
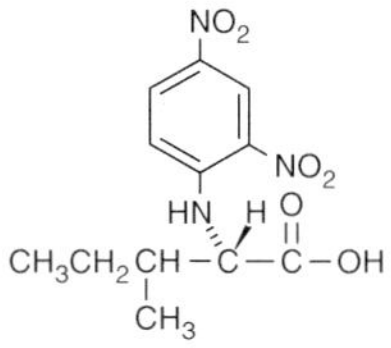
250 mg

S78,659-4
N-(2,4-DINITROPHENYL)-L-LEUCINE
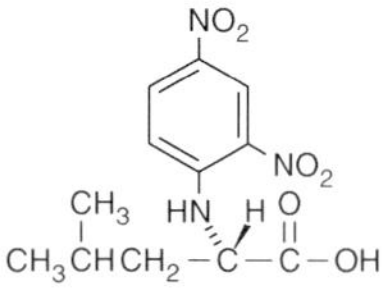
250 mg

S78,921-6
N-(2,4-DINITROPHENYL)-DL-GLUTAMIC ACID HEMIHYDRATE
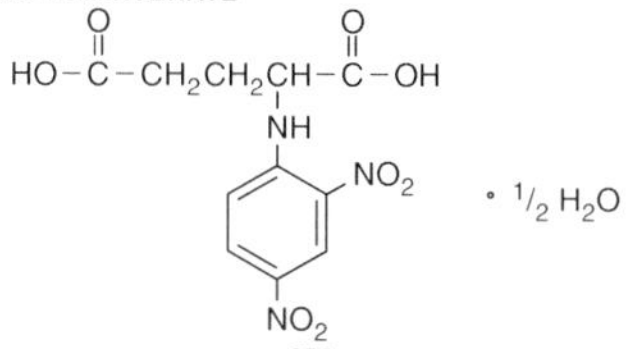
250 mg

S78,730-2
N-(2,4-DINITROPHENYL)-L-VALINE
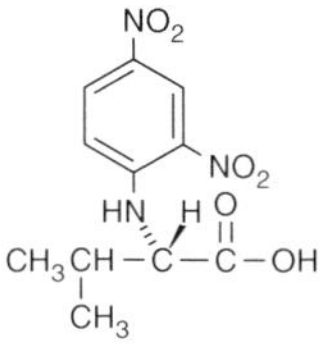
250 mg

S78,177-0
N-(2,4-DINITROPHENYL)-L-ASPARTIC ACID
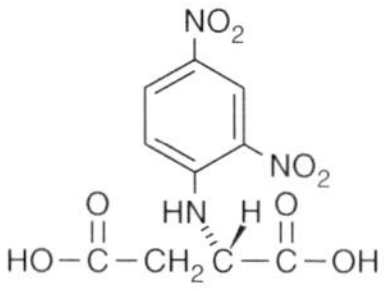
250 mg

S78,660-8
N-(2,4-DINITROPHENYL)-L-PHENYLALANINE
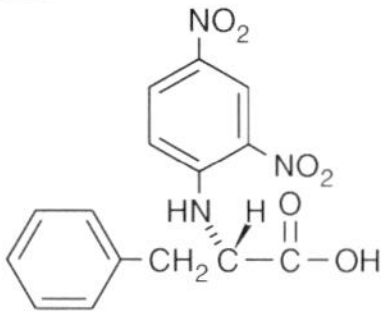
250 mg

S78,380-3
N-BENZYL-L-PHENYLALANINE
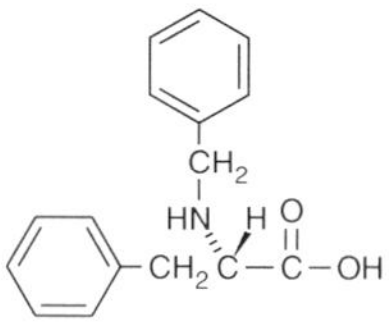
5 gm

S45,230-0
N-BENZYLANTHRANILIC ACID
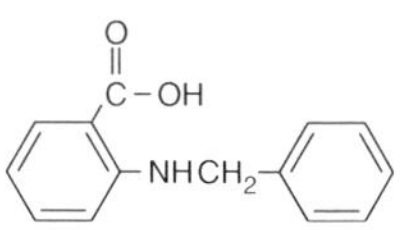
5 gm

S34,159-2
N-BENZYL-DL-ASPARTIC ACID
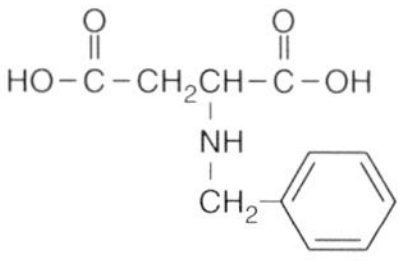
5 gm

S78,441-9
N-BENZYL-L-VALINE
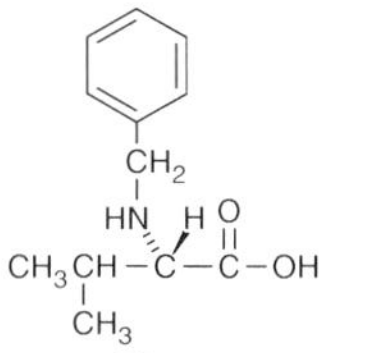
250 mg

S78,939-9
N-BENZYL-DL-LEUCINE
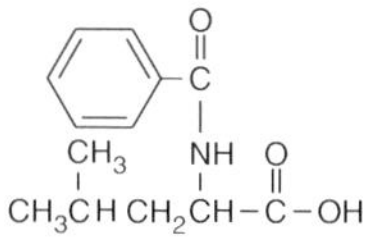
250 mg

S57,949-1
N-TRITYL-L-LEUCINE
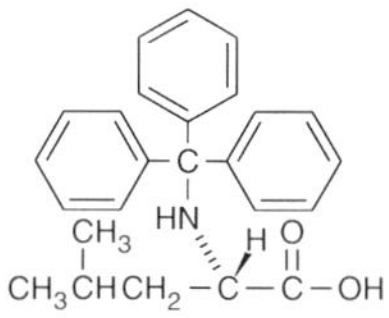
5 gm

S64,011-5
N-TRITYLPHENYLALANINE DIETHYLAMINE SALT
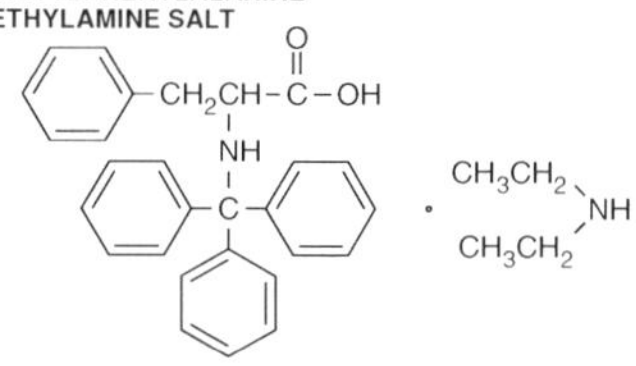
5 gm

S40,651-1
4-ANILINO-1-BENZYLISONIPECOTIC ACID HYDROCHLORIDE
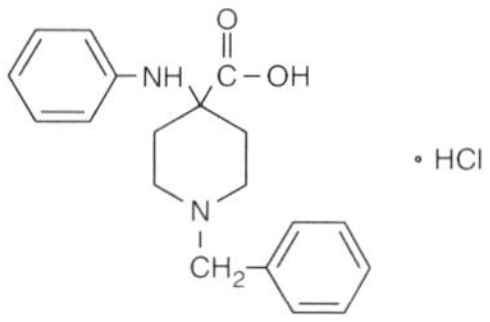
250 mg

S78,582-2
N-BENZYL-L-LEUCINE
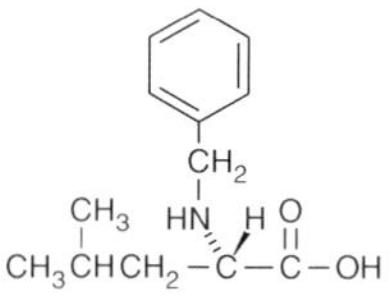
250 mg

S93,471-2
ACRIDONE-10-ACETIC ACID
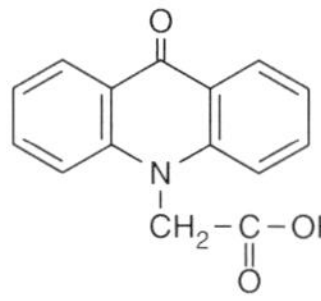
250 mg

S80,470-3
N-ALPHA-BENZOYLASPARAGINE
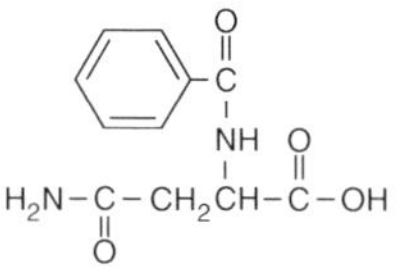
250 mg

S45,179-7
N-BENZOYL-L-ALANINE
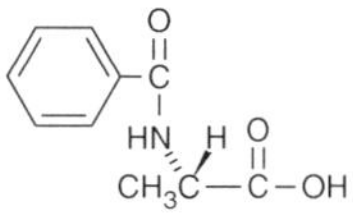
250 mg

S45,506-7
N-BENZOYL-D-ALANINE
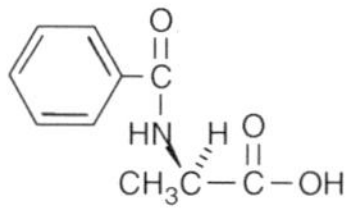
250 mg

S56,001-4
N-BENZOYL-DL-ALANINE
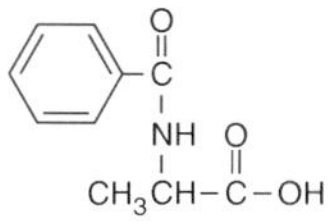
5 gm

S65,820-0
N-(CINNAMOYL)GLYCINE
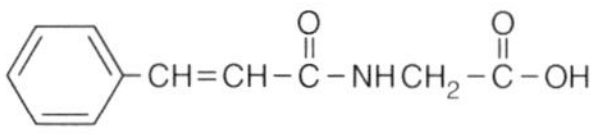
5 gm

S78,439=7
N-BENZOYL-D-VALINE
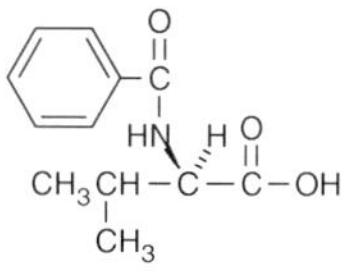
250 mg

S95,428-4
N-BENZOYL-L-PHENYLALANINE
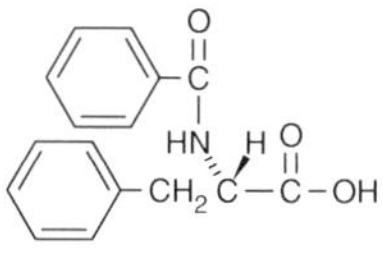
50 mg

S80,428-2
N-ALPHA-[2-(1-NAPHTHYL)ACETYL]-ASPARAGINE
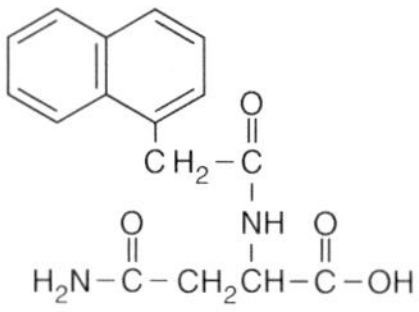
250 mg

S91,464-9
N-(4-CHLOROBENZYL)-L-THREONINE
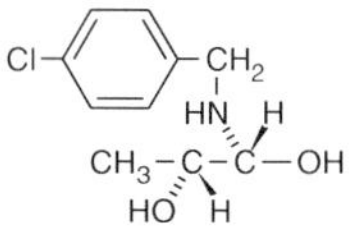
250 mg

S91,478-9
N-(4-METHOXYBENZYL)-L-THREONINE
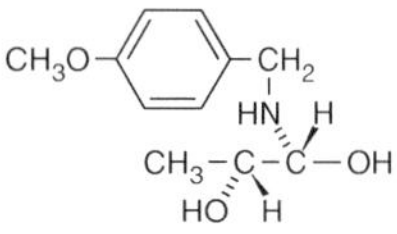
250 mg

S91,481-9
N-(4-NITROBENZYL)-L-THREONINE
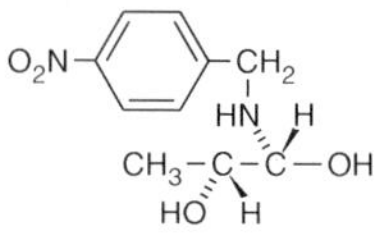
250 mg

S83,725-3
2-AMINO-5-BROMOBENZOIC ACID, POTASSIUM SALT

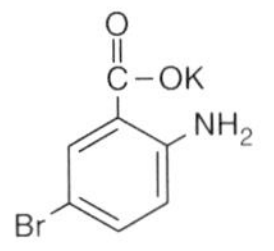

250 mg

S43,479-5
3-AMINO-2,5-DICHLOROBENZOIC ACID, SODIUM SALT, TECH.

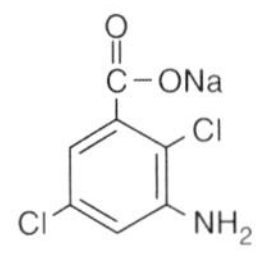

5 gm

S64,369-6
PENTAFLUOROBENZOIC ACID, BARIUM SALT

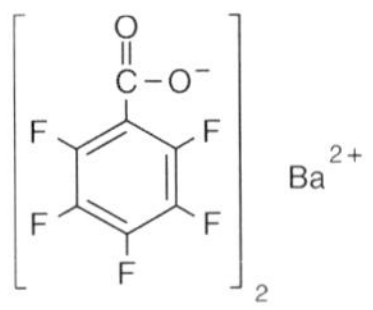

5 gm

S84,089-0
4-IODOBENZOIC ACID, POTASSIUM SALT

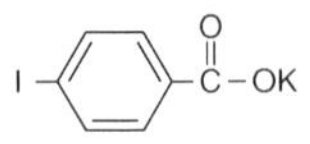

250 mg

S69,556-4
DIPHENYLIODONIUM-2-CARBOXYLATE MONOHYDRATE

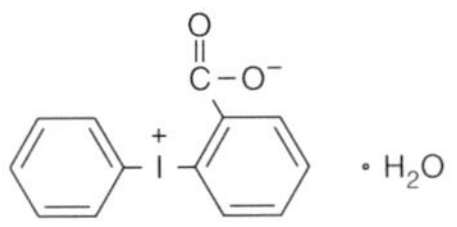

250 mg

S56,061-8
2,6-DIHYDROXYBENZOIC ACID, SODIUM SALT

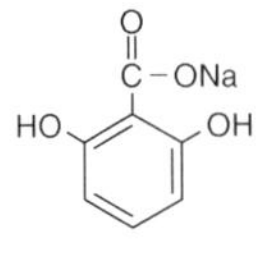

5 gm

S50,416-5
2,4,6-TRIHYDROXYBENZOIC ACID, POTASSIUM SALT

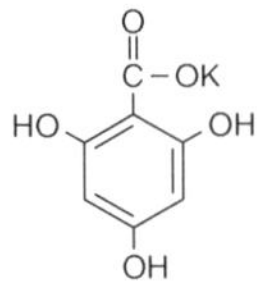

5 gm

S78,908-9
L-3-PHENYLLACTIC ACID SODIUM SALT MONOHYDRATE

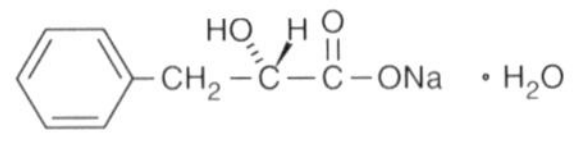

250 mg

S35,458-9
PHENYLMERCURIC LACTATE

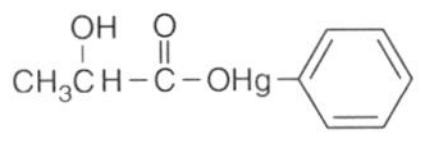

5 gm

S56,106-1
4-AMINOSALICYLIC ACID, POTASSIUM SALT

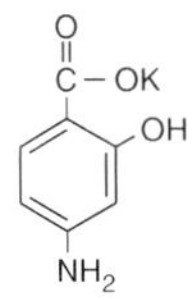

5 gm

S80,482-7
2-HYDROXYBENZOYLFORMIC ACID, SODIUM SALT

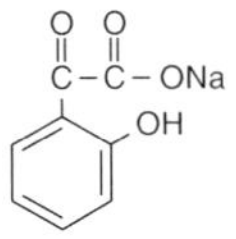

250 mg

S51,918-9
SALICYLIC ACID, POTASSIUM SALT

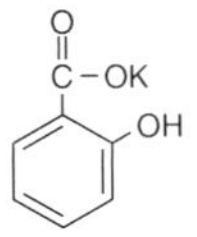

5 gm

S44,217-8
GLYCINECRESOL RED

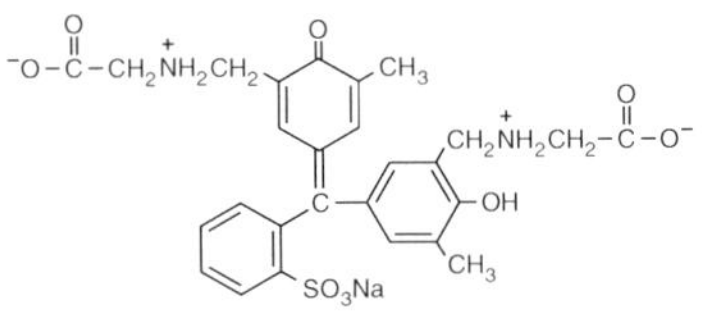

250 mg

S55,391-3
4-HYDROXY-3-NITROPHENYLMERCURI ACETATE

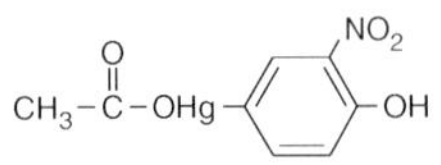

5 gm

S42,136-7
4-HYDROXY-ALPHA-PHENYLCINNAMIC ACID, SODIUM SALT MONOHYDRATE

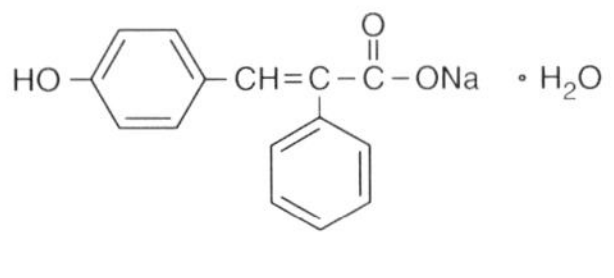

250 mg

S44,469-3
2',4',5',7'-TETRAKIS-(ACETOXYMERCURI)FLUORESCEIN

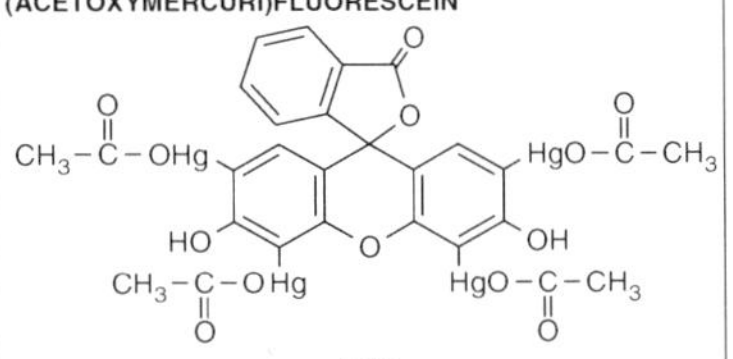

5 gm

S40,128-5
4-HYDROXY-ALPHA-PHENYLHYDROCINNAMIC ACID, SODIUM SALT TRIHYDRATE

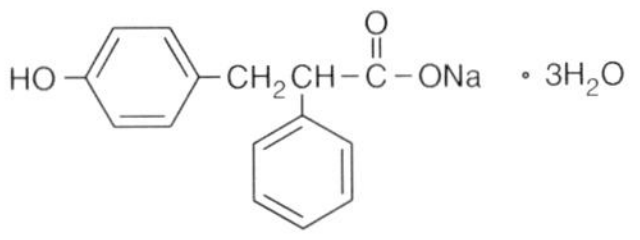

250 mg

S95,044-0
3-(2,3-DIMETHOXYPHENYL)GLYCIDIC ACID, SODIUM SALT

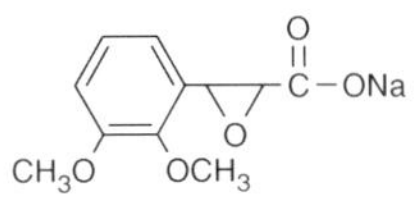

50 mg

S64,005-0
2-AZIDO-3-BENZYLOXYPROPIONIC ACID, BARIUM SALT

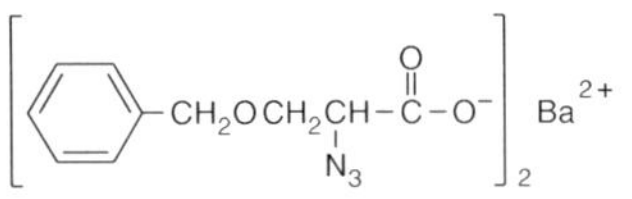

5 gm

S64,300-9
2,3-EPOXY-3-(2-METHOXYPHENYL)-PROPIONIC ACID, SODIUM SALT

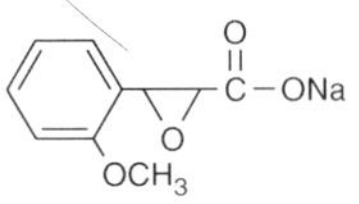

250 mg

S67,893-7
3,3'-DIMETHOXYBENZIDINE-N,N,N',N'-TETRAACETIC ACID, TETRASODIUM SALT

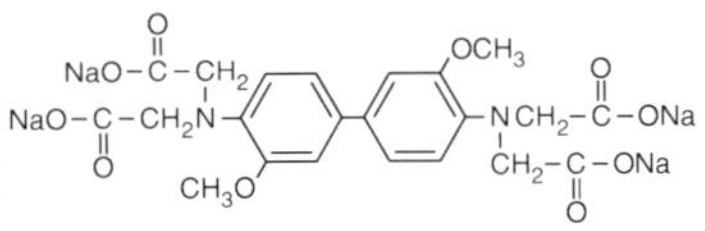

5 gm

S56,579-2
N-(3,4-DIMETHOXYPHENYL)-4-NITROANTHRANILIC ACID, POTASSIUM SALT

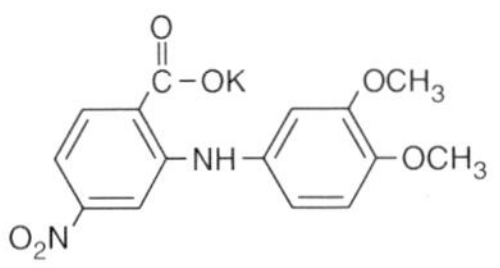

5 gm

S92,883-6
2-(4-FORMYLPHENOXY)ACETIC ACID, SODIUM SALT

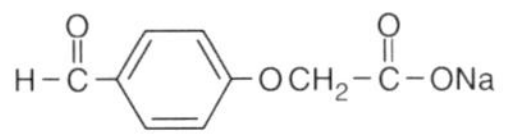

250 mg

S56,400-1
ALPHA-METHOXYPHENYLACETIC ACID, SODIUM SALT

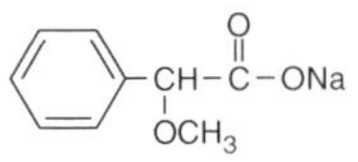

5 gm

S91,589-0
2-ISOPROPOXYBENZOIC ACID, SODIUM SALT

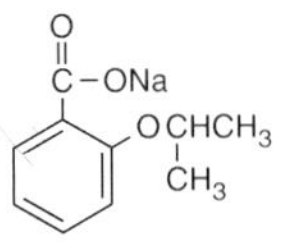

250 mg

S65,420-5
METHYL 3-METHOXYBENZOYLACETATE, COPPER(II) COMPLEX

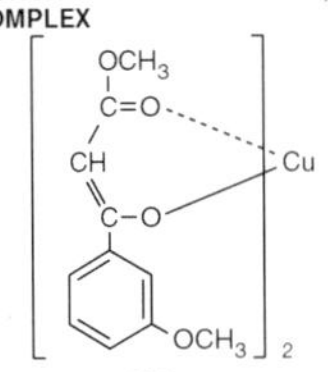

250 mg

S66,046-9
2-(P-ANISYL)-3-PHENYLSUCCINIC ACID, DIAMMONIUM SALT

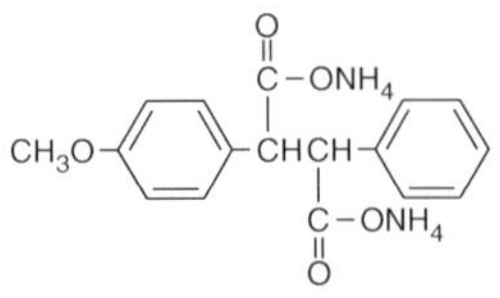

5 gm

S39,061-5
ETHYL ALPHA-ACETYL-3-(METHOXYCARBONYL)BENZOYLACETATE, COPPER(II) DERIVATIVE

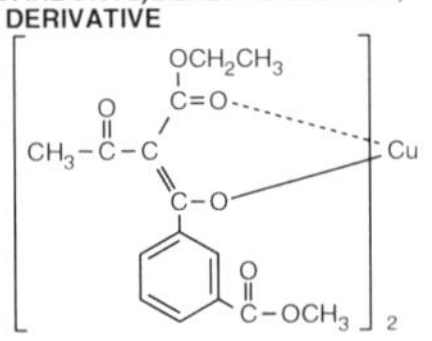

250 mg

S39,057-7
ETHYL ALPHA-ACETYL-4-(METHOXYCARBONYL)BENZOYLACETATE, COPPER(II) DERIVATIVE

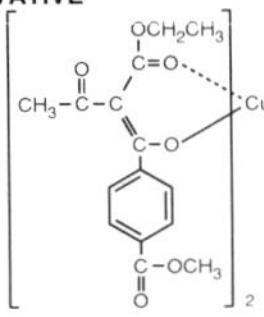

250 mg

S94,674-5
PHTHALIC ACID COPPER(II), SALT

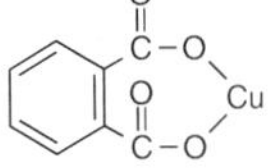

250 mg

S35,456-2
3-PHENYLGLYCIDIC ACID, POTASSIUM SALT

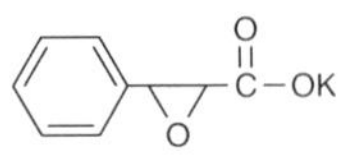

5 gm

S83,322-3
SALICYLIC ACID, OCTADECYLAMINE SALT

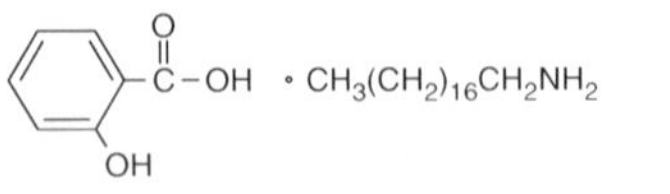

250 mg

S74,217-1
(2-AMINO-1-NAPHTHYLTHIO)ACETIC ACID, SODIUM SALT
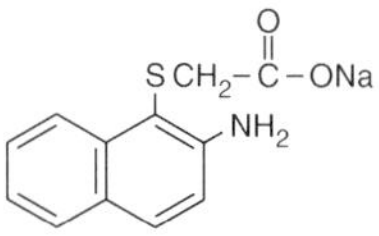
5 gm

S57,102-4
4-AMINOBENZOIC ACID, POTASSIUM SALT
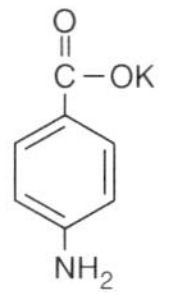
5 gm

S42,407-2
PHENYLACETIC ACID, PIPERAZINE SALT
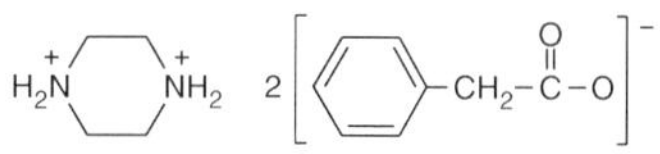
5 gm

S43,895-2
DIHYDROXYTARTARIC ACID OSAZONE, DISODIUM SALT
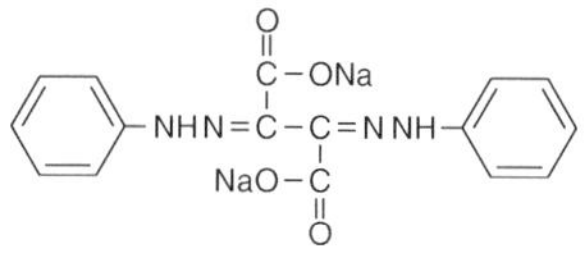
5 gm

S89,318-8
4,5-DINITRO-1,8-NAPHTHALENEDICARBOXYLIC ACID, DISILVER SALT
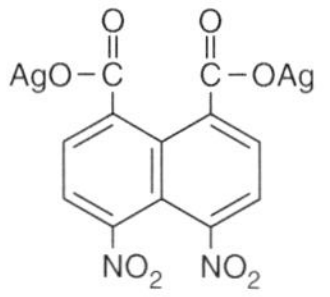
250 mg

S60,518-2
4-(4-HEXADECYLPHENYL)-4-HYDRAZONOBUTYRIC ACID, SODIUM SALT
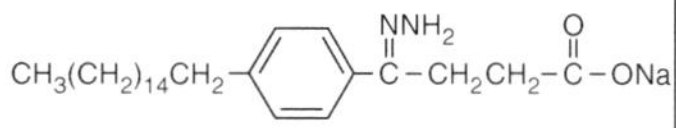
250 mg

S61,542-0
3-NITROPHTHALIC ACID, BARIUM SALT
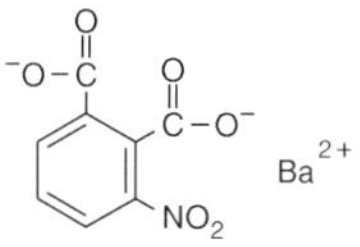
5 gm

S86,432-3
2-NITROBENZOIC ACID, SODIUM SALT
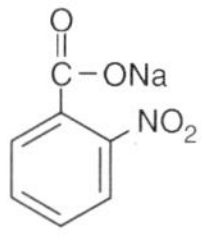
250 mg

S68,382-5
2-(BENZOYLACETYL)BENZOIC ACID DISODIUM DERIVATIVE
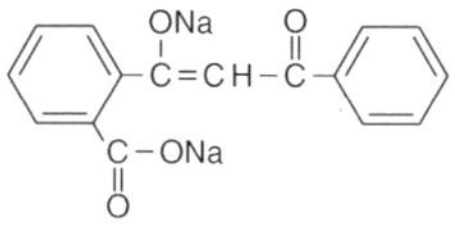
5 gm

S38,594-8
ETHYL ALPHA-ACETYL-3-(FLUOROSULFONYL)BENZOYLACETATE, COPPER(II) DERIVATIVE
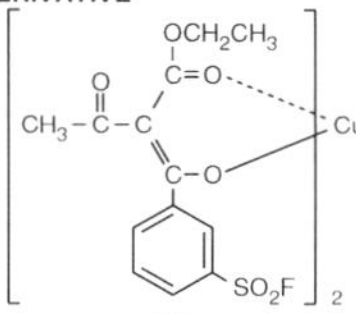
250 mg

S38,586-7
ETHYL BENZOYLACETATE, COPPER(II) DERIVATIVE
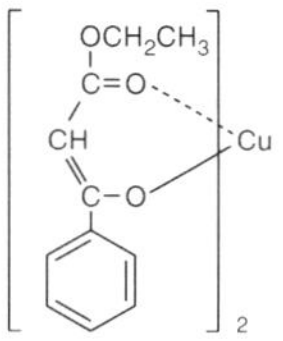
250 mg

S74,495-6
(+)-DIBENZOYL-L-TARTARIC ACID DISILVER SALT
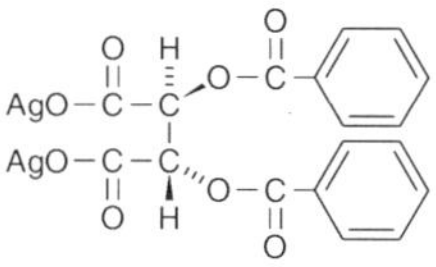
250 mg

S51,817-4
BENZOIC ACID, MAGNESIUM SALT
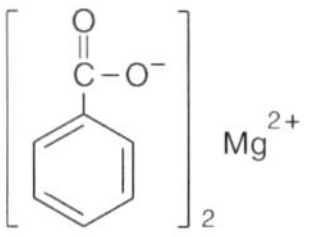
5 gm

S51,856-5
BENZOIC ACID, COPPER (II) SALT DIHYDRATE
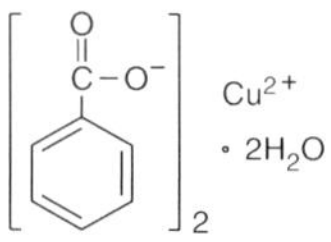
250 mg

S91,551-3
2-TERT-BUTYLBENZOIC ACID, SODIUM SALT
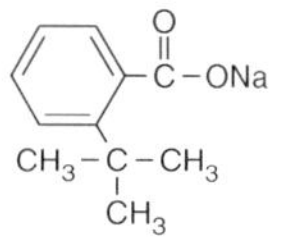
250 mg

S46,263-2
LITHIUM BENZOATE
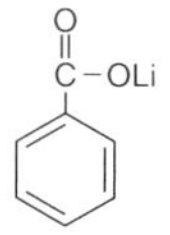
5 gm

S56,635-7
CINNAMIC ACID, POTASSIUM SALT
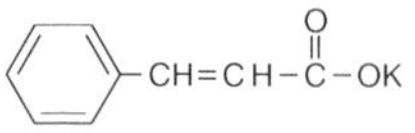
5 gm

S57,310-8
CINNAMIC ACID, SODIUM SALT
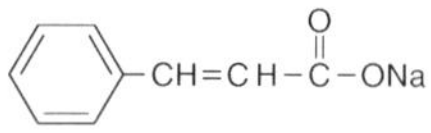
5 gm

S40,190-0
2-PHENYLBUTYRIC ACID, MAGNESIUM SALT MONOHYDRATE
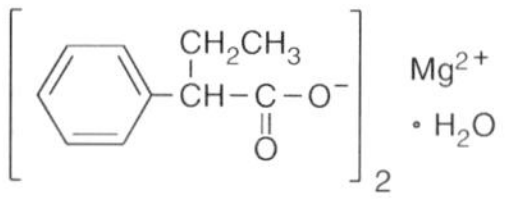
250 mg

S82,591-3
1-NAPHTHOIC ACID, DODECYLAMINE SALT
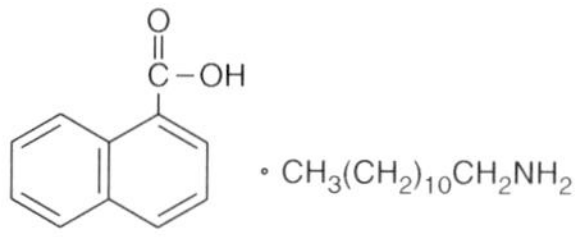
250 mg

S82,592-1
1-NAPHTHOIC ACID, OCTADECYLAMINE SALT
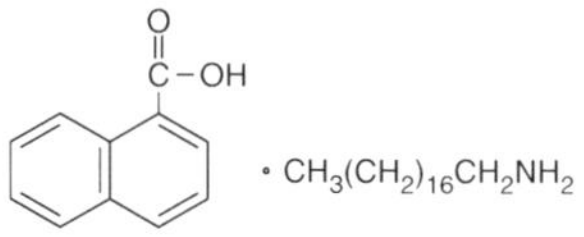
250 mg

S39,043-7
ETHYL 2-CHLOROBENZOYLACETATE, COPPER(II) DERIVATIVE
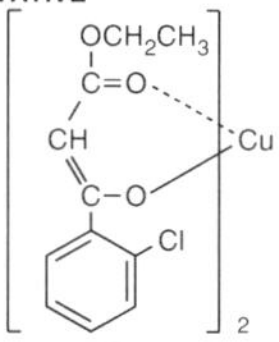
250 mg

S38,311-2
ETHYL 3-(FLUOROSULFONYL)-BENZOYLACETATE, COPPER(II) DERIVATIVE
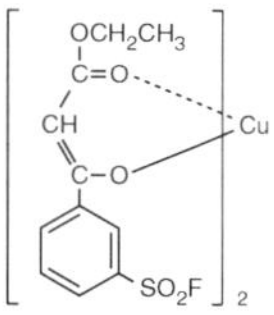
250 mg

S39,062-3
ETHYL 2-FLUOROBENZOYLACETATE, COPPER(II) DERIVATIVE
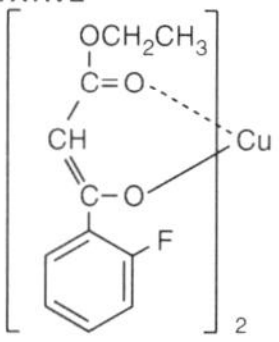
250 mg

S57,636-0
2-(HYDROXYMERCURI)BENZOIC ACID CYCLIC ANHYDRIDE
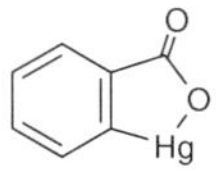
5 gm

S60,831-9
9,10-DIHYDRO-9,10-ETHANOANTHRACENE-11,12-DICARBOXYLIC ACID, DISODIUM SALT
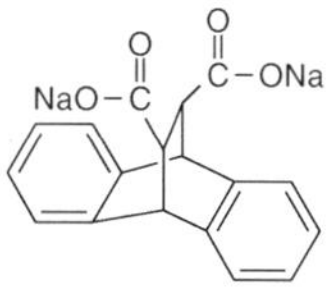
250 mg

S40,307-5
3-METHYL-2-PHENYLVALERIC ACID, PIPERAZINIUM SALT
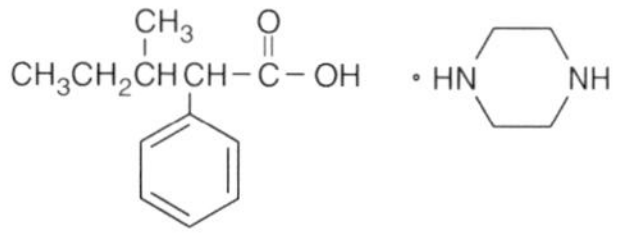
250 mg

S83,323-1
2-PHENYLACETIC ACID, OCTADECYLAMINE SALT
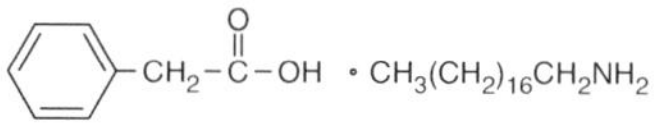
250 mg

S82,596-4
CINNAMIC ACID, DODECYLAMINE SALT
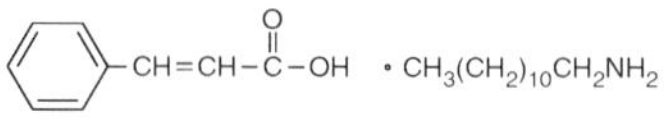
250 mg

S52,247-3
PHENYLMERCURIC PROPIONATE
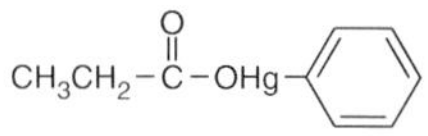
5 gm

S67,936-4
2-PHENYLBUTYRIC ACID, SODIUM SALT
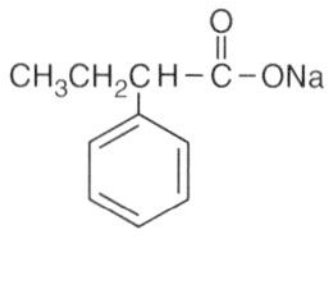
250 mg

S40,030-0
3-METHYL-2-PHENYLBUTYRIC ACID, SODIUM SALT
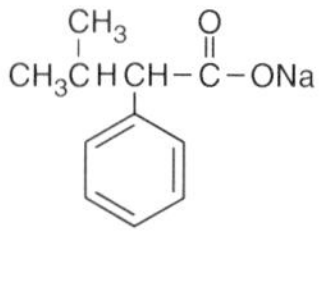
250 mg

S40,080-7
3-METHYL-2-PHENYLVALERIC ACID, SODIUM SALT

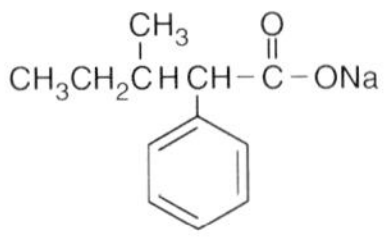

250 mg

S40,081-5
2-PHENYLHEXANOIC ACID, SODIUM SALT

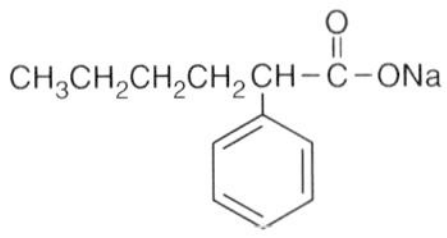

250 mg

S40,272-9
3-METHYL-2-PHENYLVALERIC ACID, MAGNESIUM SALT MONOHYDRATE

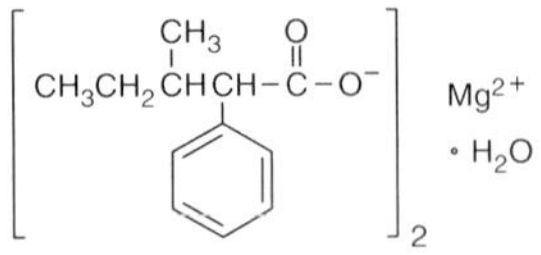

250 mg

S57,289-6
1-NAPHTHALENEACETIC ACID, SODIUM SALT

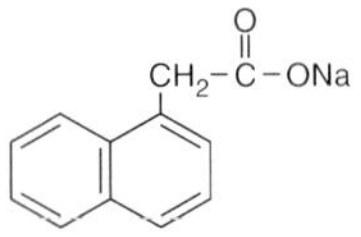

5 gm

S41,762-9
4-ACETAMIDOPHENYLMERCURIC ACETATE

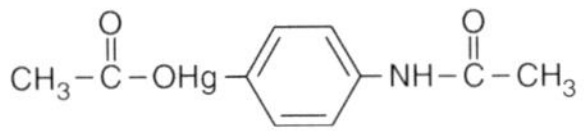

5 gm

S36,614-5
PHENYLMERCURIC DIMETHYLDITHIOCARBAMATE

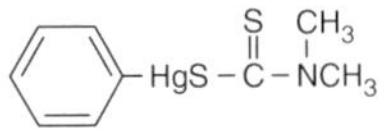

5 gm

S94,732-6
DIPHENYLMETHYL 2-(4-BROMOMETHYLPHENYL)-2-(P-TOLYL)GLYCOLATE
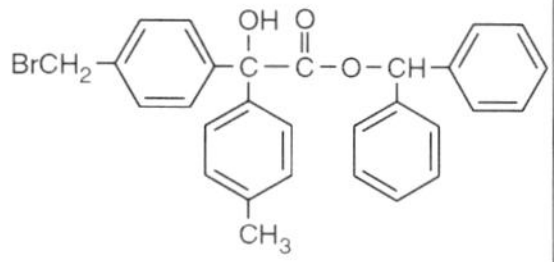
250 mg

S92,885-2
METHYL 2-(4-BENZYLOXYPHENYL)ACETATE
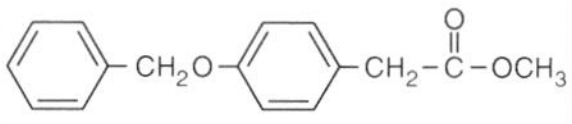
250 mg

S88,196-1
BENZYL 2-CHLOROACETATE
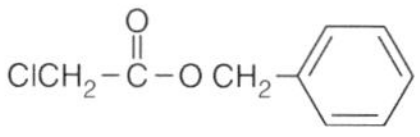
250 mg

S63,055-1
BENZYL 11-BROMOUNDECANOATE
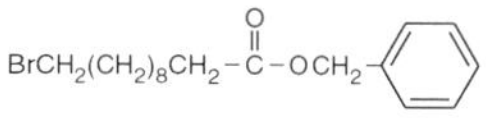
250 mg

S38,356-2
3-CHLOROPROPYL 3-METHYL-2-PHENYLVALERATE
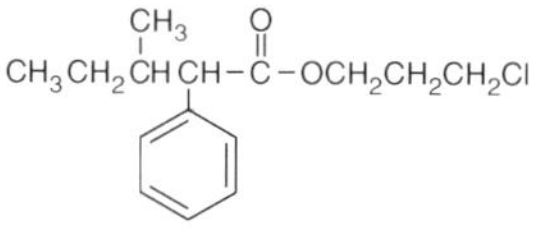
250 mg

S79,255-1
(+/-)-1-(TRIFLUOROMETHYL)ETHYL-PHENYLACETATE
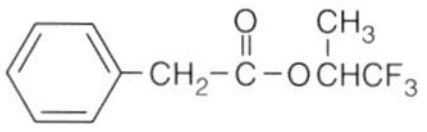
250 mg

S79,272-1
1-(TRIFLUOROMETHYL)ETHYL-2-PHENYLPROPIONATE
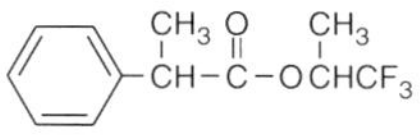
250 mg

S64,128-6
2,4,6-TRIBROMOPHENYL TETRACETYL-BETA-D-GLUCOPYRANOSIDE
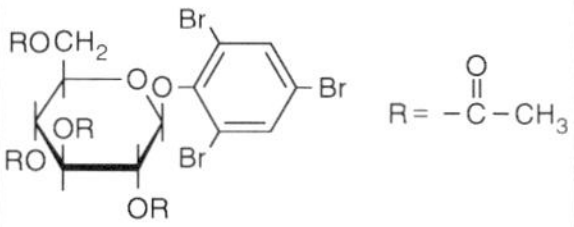
5 gm

S75,853-1
METHYL 2-[2-(4-CHLOROBENZOYLVINYL)-PHENOXY]ACETATE
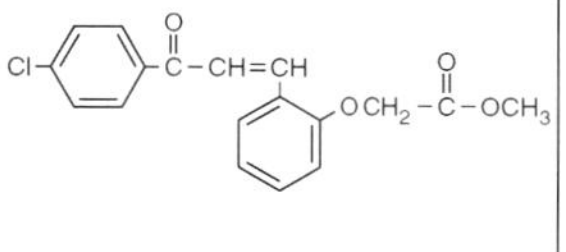
250 mg

S74,030-6
4-CHLOROBENZYL ACETATE
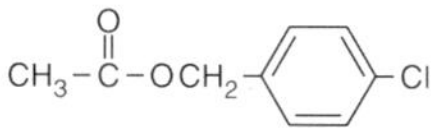
250 mg

S81,949-2
DIETHYL (4-CHLORO-1,2-PHENYLENEDIOXY)DIACETATE
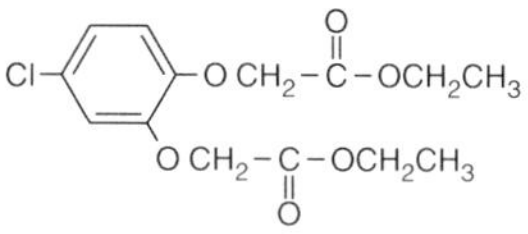
250 mg

S73,440-3
ETHYL 2-(4-CHLOROPHENYL)-1,3-DIOXOLANE-2-PROPIONATE
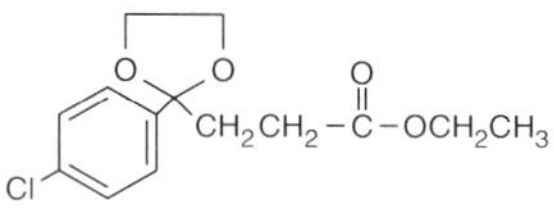
250 mg

S76,131-1
ETHYL 4-(4-CHLOROPHENYL)-2-(1,2,3,4,5,6-HEXAHYDRO-2-PYRIMIDINYL)-4-OXOBUTYRATE
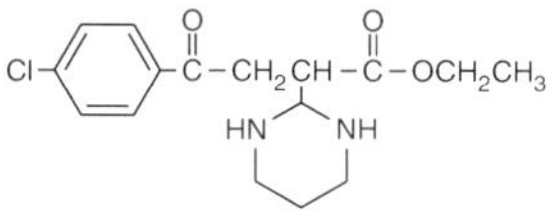
250 mg

S74,802-1
ETHYL 3-(4-CHLOROBENZOYL)PROPIONATE
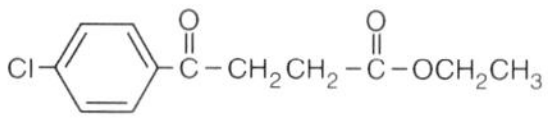
250 mg

S62,090-4
DIETHYL 4-BROMOBENZYLMALONATE
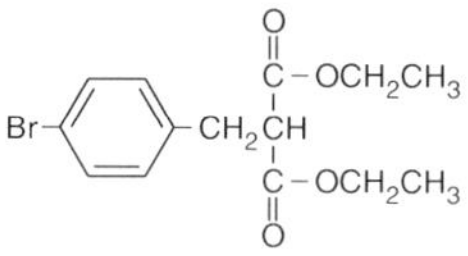
250 mg

S66,816-8
DIMETHYL 2,2-BIS(4-CHLOROBENZYL)-MALONATE
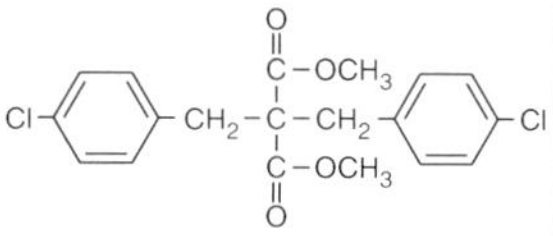
5 gm

S38,443-7
DIETHYL BIS(4-CHLOROBENZYL)-MALONATE
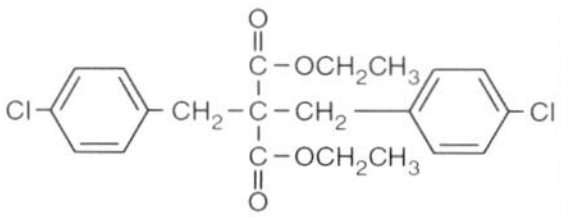
5 gm

S38,442-9
DIETHYL BIS(3,4-DICHLOROBENZYL)-MALONATE
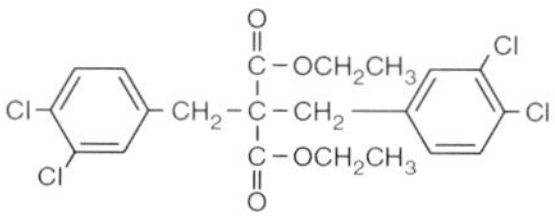
5 gm

S73,457-8
METHYL 2-(4-CHLOROPHENYL)ACETATE
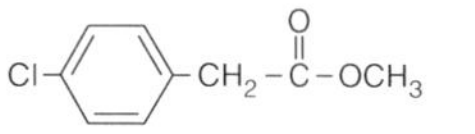
250 mg

S66,874-5
ETHYL 4-CHLOROPHENYLACETATE
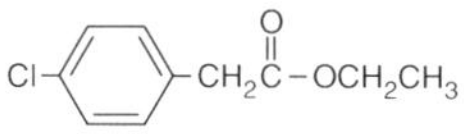
250 mg

S44,570-3
ETHYL 4-HYDROXY-DL-MANDELATE
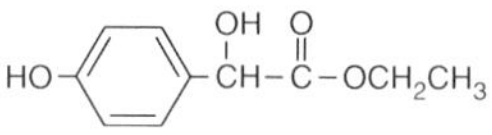
250 mg

S78,176-2
L-3,4-DIHYDROXYPHENYLALANINE METHYL ESTER HYDROCHLORIDE
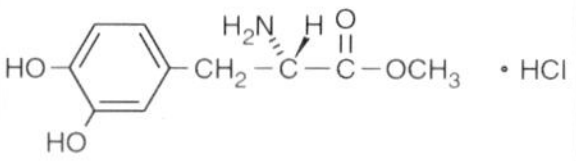
250 mg

S76,052-8
METHYL 3,3-BIS(4,4-DIMETHYL-2-HYDROXY-6-OXO-1-CYCLOHEXENYL)-2-PHENYLPROPIONATE
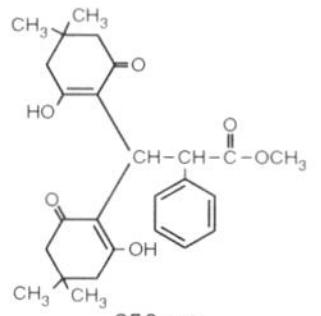
250 mg

S54,488-4
ETHYL 2,5-DIHYDROXYPHENYLACETATE
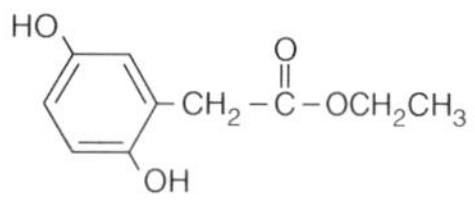
5 gm

S55,112-0
DIETHYL 2-HYDROXY-2-(4-METHOXYPHENYL)MALONATE
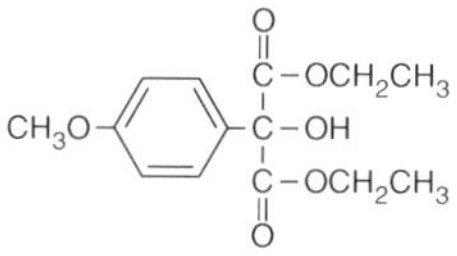
5 gm

S38,046-6
METHYL 3-METHOXYMANDELATE
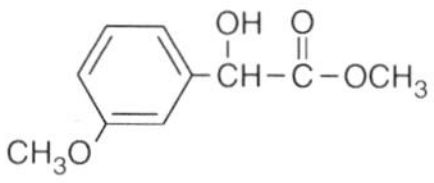
5 gm

S34,550-4
2-(2,4-DICHLOROPHENOXY)-ETHYL LACTATE
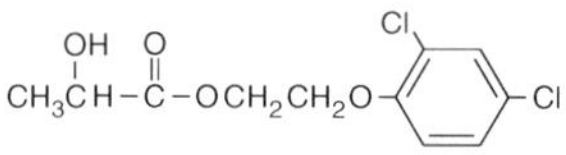
5 gm

S36,379-0
1-METHYL-4-PIPERIDINYL BENZILATE HYDROCHLORIDE
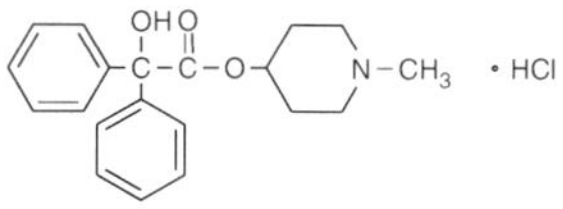
250 mg

S67,939-9
BENZYL MANDELATE
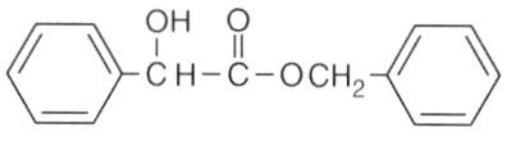
5 gm

S79,389-2
L-MENTHYL DL-MANDELATE
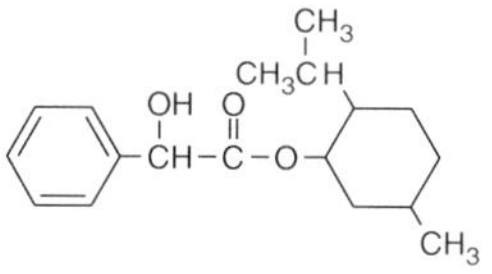
250 mg

S70,567-5
ETHYL 3-HYDROXY-2-METHYL-3-PHENYLPROPIONATE
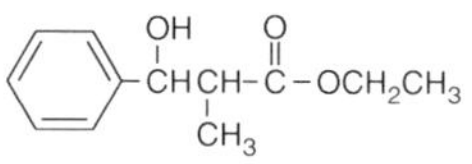
5 gm

S35,151-2
ISOAMYL MANDELATE
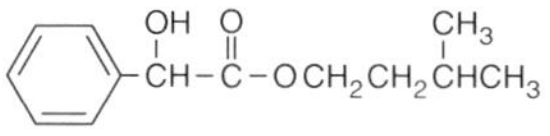
5 gm

S38,229-9
ETHYL ALPHA-(SEC.-BUTYL)MANDELATE
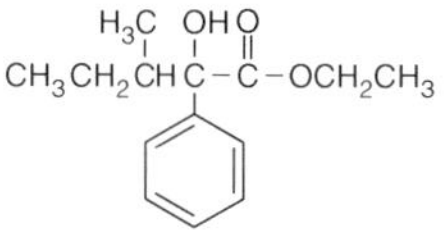
250 mg

S67,938-0
PHENETHYL MANDELATE
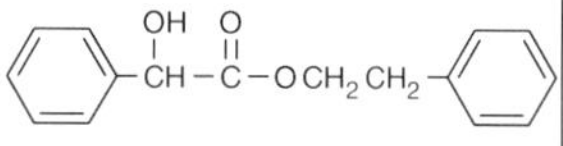
250 mg

S94,720-2
DIPHENYLMETHYL BIS(P-TOLYL)-GLYCOLATE
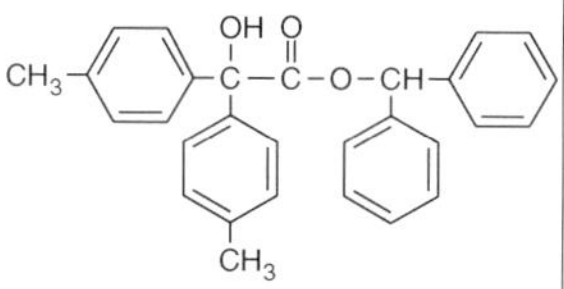
250 mg

S95,445-4
ETHYL 3,4-DIPHENYL-3-HYDROXYBUTYRATE
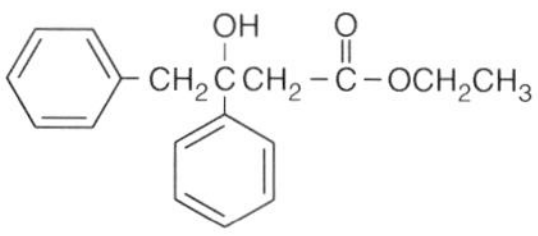
50 mg

S78,595-4
D-TYROSINE BENZYL ESTER P-TOLUENESULFONATE
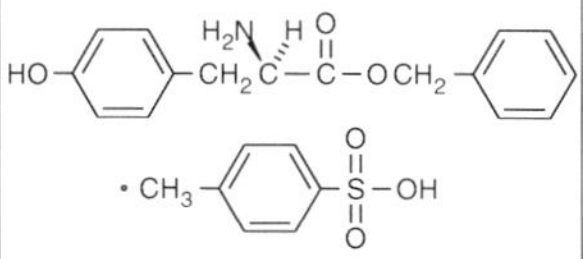
250 mg

S43,965-7
METHYL ALPHA-PHENYLCYCLOPENTANEGLYCOLATE
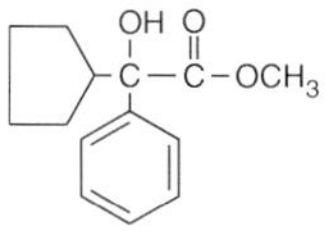
5 gm

S88,838-9
2-ACETOXY-1-HYDROXYANTHRAQUINONE
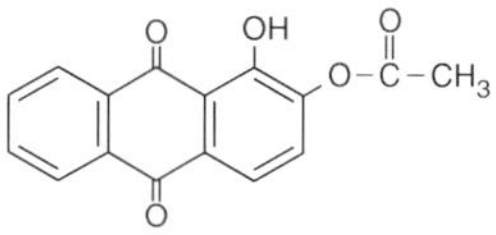
250 mg

S94,990-6
BENZYL 2-HYDROXYPROPIONATE
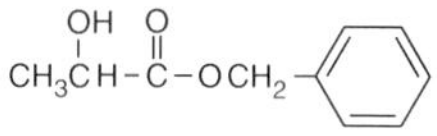
5 gm

S94,988-4
N-TRITYL-DL-SERINE ISOPROPYL ESTER
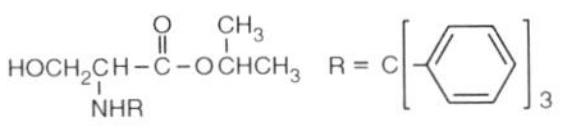
1 gm

S88,150-3
ETHYL 3-HYDROXY-3-METHYL-5-PHENYL-4-PENTENOATE
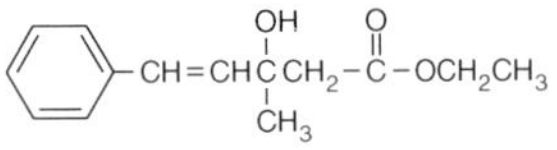
250 mg

S95,461-6
ETHYL 3,3-DIPHENYL-3-HYDROXYPROPIONATE
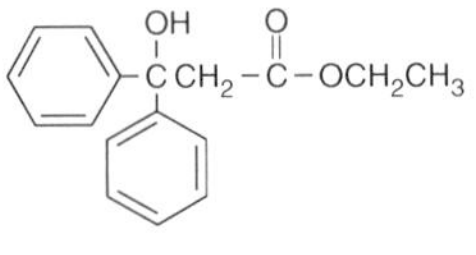
50 mg

S64,762-4
ETHYL 3-(3,5-DIAMINO-4-HYDROXYPHENYL)PROPIONATE
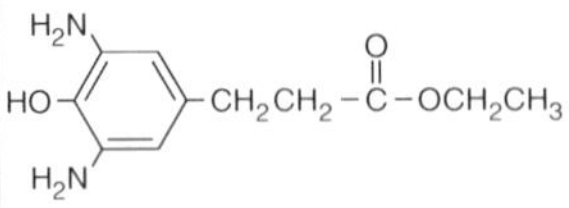
5 gm

S78,727-2
3-NITROTYROSINE METHYL ESTER HYDROCHLORIDE
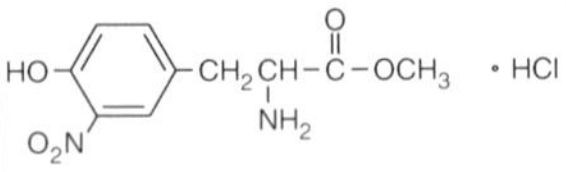
250 mg

S42,772-1
DL-ALPHA-METHYL-M-TYROSINE HYDROCHLORIDE
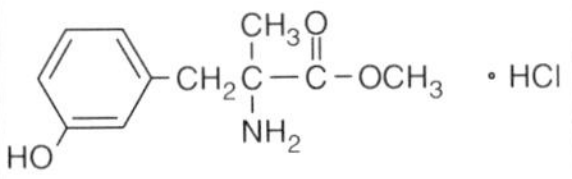
250 mg

S45,532-6
DL-ALPHA,3-DIMETHYLTYROSINE METHYL ESTER HYDROCHLORIDE
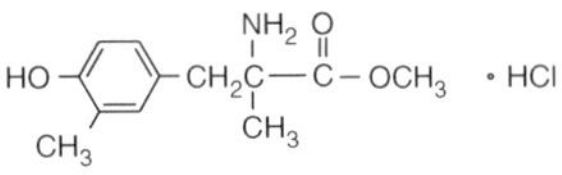
5 gm

S36,073-2
3,5-DIIODO-L-TYROSINE ETHYL ESTER HYDROCHLORIDE
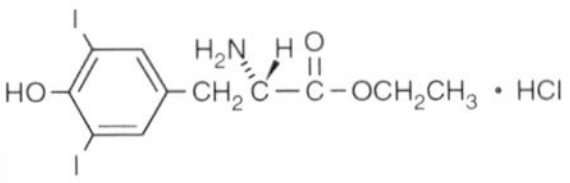
5 gm

S45,778-7
DL-TYROSINE METHYL ESTER HYDROCHLORIDE
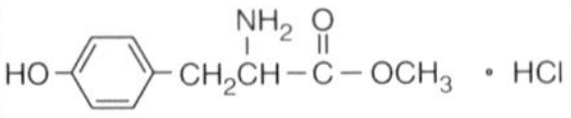
250 mg

S38,126-8
TYROSINE ETHYL ESTER
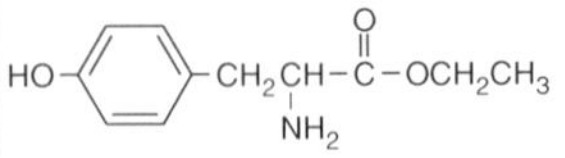
5 gm

S44,455-3
L-TYROSINE ETHYL ESTER HYDROCHLORIDE
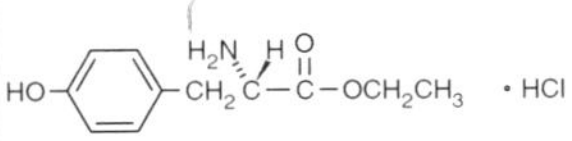
5 gm

S44,566-5
ATROPINE
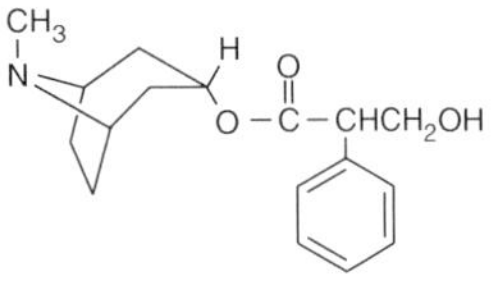
5 gm

S64,757-8
ETHYL 3-(4-HYDROXY-3-NITROPHENYL)-PROPIONATE
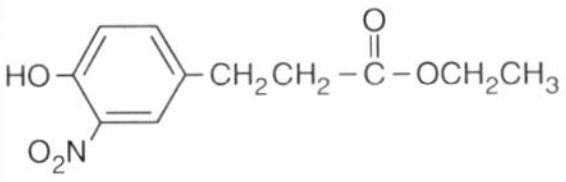
250 mg

S55,109-0
ETHYL 2-HYDROXY-4-OXO-2-PHENYLCYCLOHEXANECARBOXYLATE
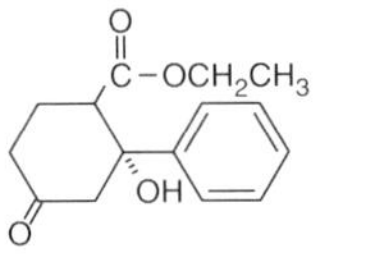
5 gm

S88,386-7
4B-METHYL-4B,5,6,7,8,8A,9,10-OCTAHYDRO-1,7-PHENANTHRENEDIOL 7-ACETATE
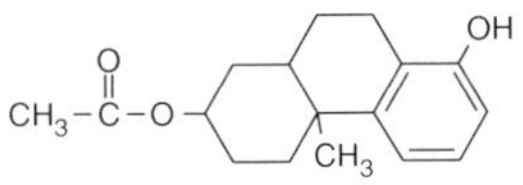
250 mg

S52,100-0
ETHYL BETA-HYDROXYHYDROCINNAMATE
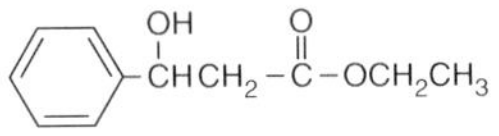
5 gm

S67,517-2
METHYL 2-HYDROXYPHENYLACETATE
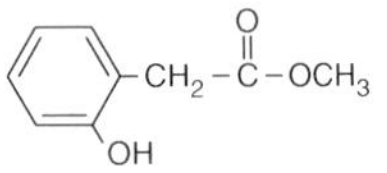
5 gm

S58,213-1
ETHYL 4-HYDROXYHYDROCINNAMATE
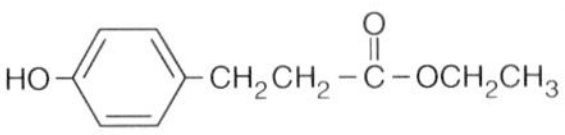
250 mg

S35,267-5
METHYL 4,4-BIS(4-HYDROXYPHENYL)-VALERATE
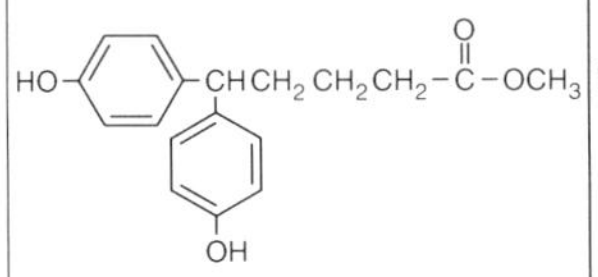
5 gm

S34,813-9
ETHYL 4,4-BIS(4-HYDROXYPHENYL)-VALERATE
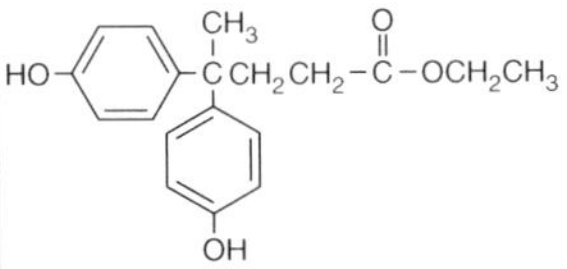
5 gm

S88,432-4
ETHYL 2,3-EPOXY-3-(4-METHOXYPHENYL)-BUTYRATE
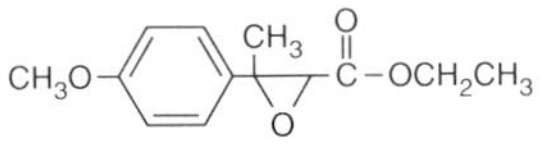
250 mg

S79,341-8
MENTHYL ALPHA-METHOXYPHENYL ACETATE
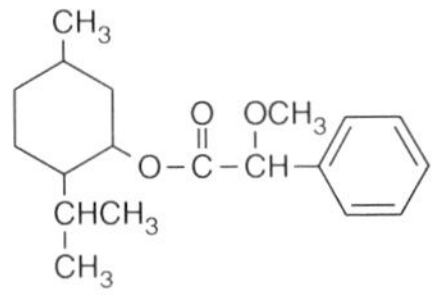
250 mg

S79,286-1
(+/-)-1-METHYLPROPYL ALPHA-METHOXYPHENYLACETATE
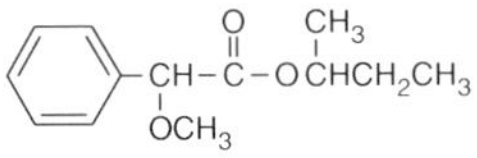
250 mg

S63,053-5
DIETHYLENE GLYCOL BENZYL ETHER ACETATE
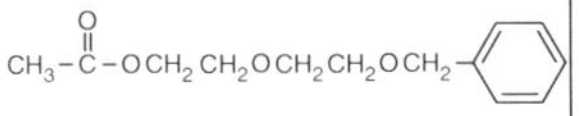
5 gm

S79,360-4
1,2,2-TRIMETHYLPROPYL ALPHA-METHOXYPHENYLACETATE
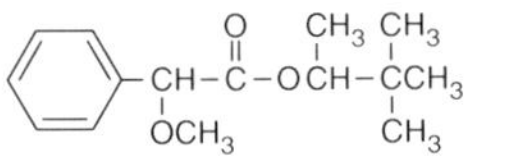
250 mg

S79,299-3
2-METHYL-1-PHENYLPROPYL ALPHA-METHOXYPHENYLACETATE
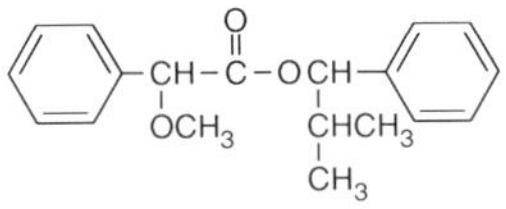
250 mg

S79,258-6
2,2-DIMETHYL-1-PHENYLPROPYL ALPHA-METHOXYPHENYLACETATE
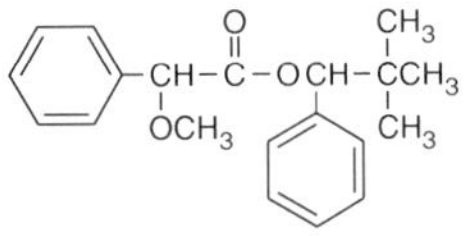
250 mg

S64,301-7
METHYL 3-(5-CHLORO-2-METHOXYPHENYL)-2,3-EPOXYPROPIONATE
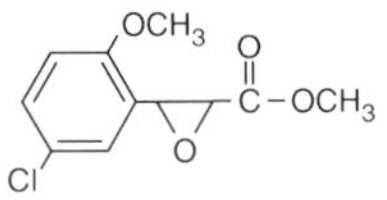
250 mg

S73,805-0
N-(4-BENZYLOXYPHENOXY)GLYCINE ETHYL ESTER
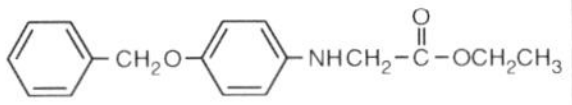
250 mg

S85,301-1
3-METHOXYBENZYL ACETATE
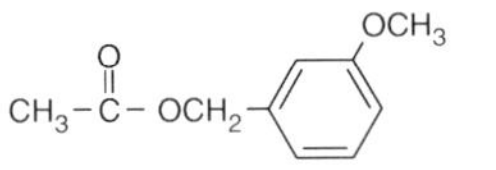
250 mg

S63,501-4
O-BENZYL-DL-SERINE BENZYL ESTER HYDROCHLORIDE
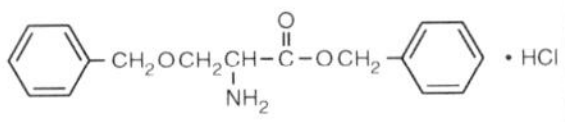
250 mg

S43,248-2
ETHYL ALPHA-(4-BENZYLOXYANILINO)-PHENYLACETATE
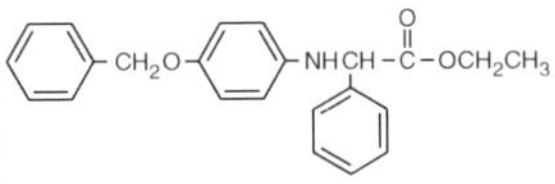
5 gm

S79,380-9
(S)-(-)-ETHYL 2-BENZYLOXYPROPIONATE
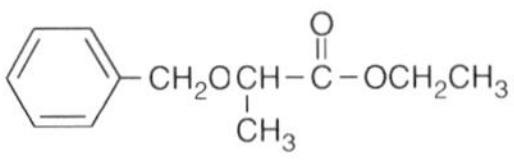
250 mg

S67,948-8
4-METHOXYBENZYL BUTYRATE
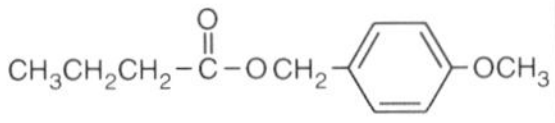
250 mg

S64,350-5
ETHYL 4-BENZYLOXYPHENYLACETATE
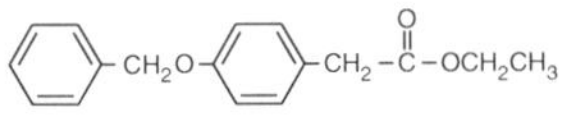
5 gm

S74,612-6
ETHYL 3-(2-AMINO-4,5-DIMETHOXYBENZOYL)PROPIONATE
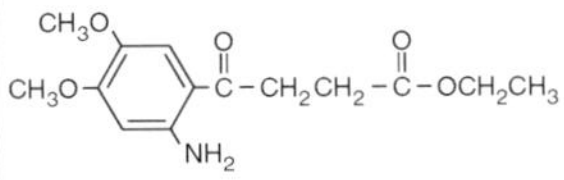
5 gm

S74,456-5
METHYL 1,4-DIHYDRO-6,7-DIMETHOXY-1-METHYL-4-OXO-3-QUINOLINEACETATE
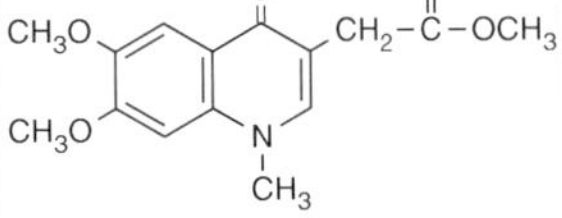
250 mg

S76,077-3
ETHYL 2-(6,7-DIMETHOXY-1,2,3,4-TETRAHYDRO-1-ISOQUINOLYLIDENE)-ACETATE
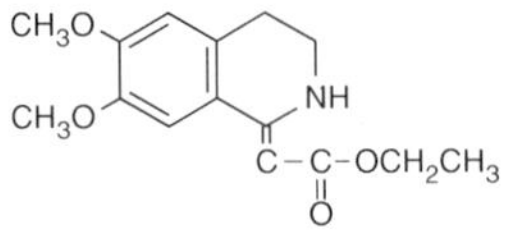
250 mg

S86,012-3
TERT-BUTYL 2-(3,4,5-TRIMETHOXYPHENYL)ACETATE
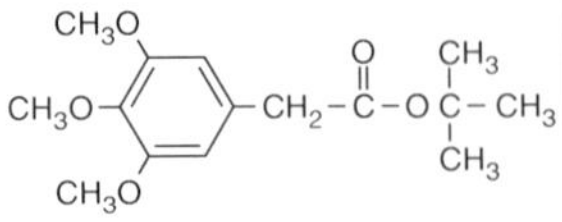
250 mg

S56,891-0
TETRAETHYL 2,2'-(2,5-DIMETHOXY-1,4-PHENYLENEDIMETHYLENE)DIMALONATE
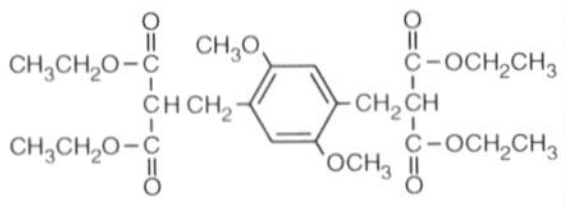
250 mg

S86,163-4
MONO-METHYL 2-(2,5-DIMETHOXYBENZYL)-SUCCINATE
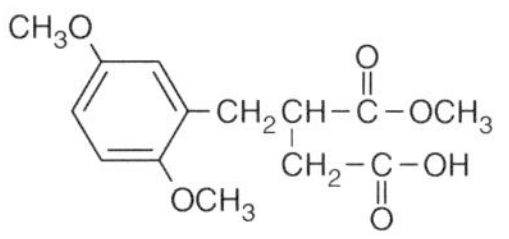
250 mg

S93,356-2
2,5-BIS(3-NITRO-4-METHOXYPHENYL)-3,6-DIACETOXY-1,4-BENZOQUINONE
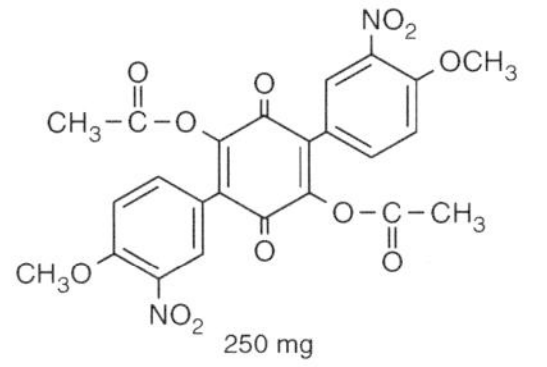
250 mg

S88,550-9
DIMETHYL 3,4-BIS(4-METHOXYPHENYL)-ADIPATE
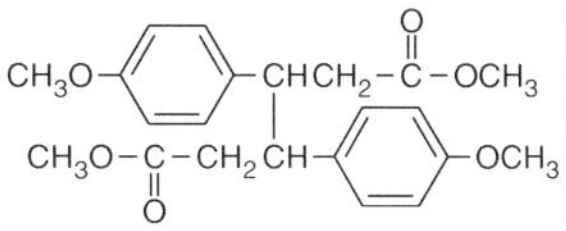
250 mg

S64,158-8
4-ETHYLPHENYLHEPTA-O-ACETYL-BETA-MALTOSIDE
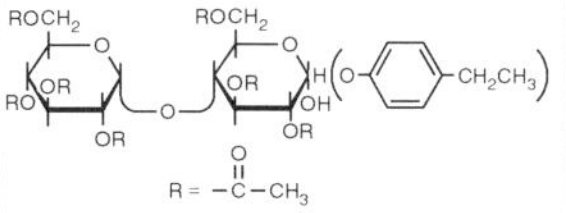
250 mg

S93,974-9
2,3-DI-O-ACETYL-1-O-METHYL-5-O-TRITYL-BETA-D-XYLOFURANOSIDE
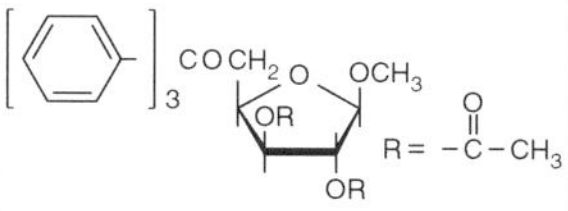
250 mg

S39,848-9
VERATRYL 2-[2-(DIETHYLAMINO)-ETHYLTHIO]-2-PHENYLACETATE, CITRATE SALT
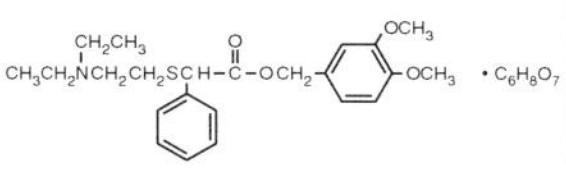
250 mg

S56,118-5
2-(2-DIETHYLAMINOETHOXY)ETHYL-2-ETHYL-2-PHENYLBUTYRATE CITRATE
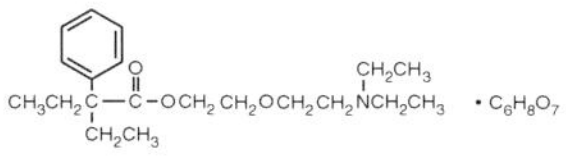
5 gm

S62,935-9
DIETHYL 2-(4-NITROPHENOXY)MALONATE
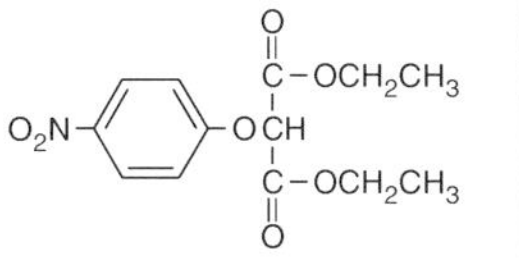
250 mg

S65,299-7
ETHYL 4-NITROPHENOXYACETATE
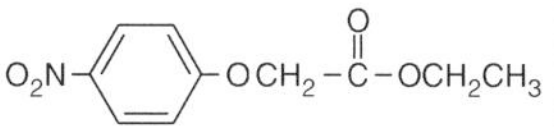
250 mg

S92,355-9
PHENETHYL-2-(2-PHENYLETHOXY)ACETATE
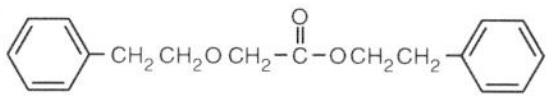
250 mg

S67,276-9
ETHYL 2-PHENOXYACETATE
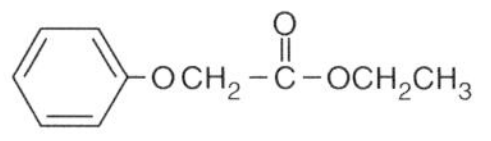
5 gm

S56,336-6
2-[4-(TERT-PENTYL)PHENOXY]-ETHYL ACETATE
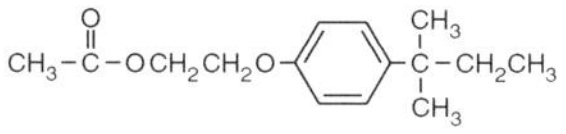
5 gm

S56,626-8
2-[4-(TERT-PENTYL)PHENOXY]-ETHYL LAURATE
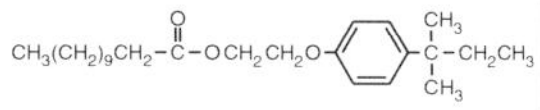
5 gm

S52,928-1
ETHYL 2-(2,3-DICHLOROPHENOXY)-ACETATE
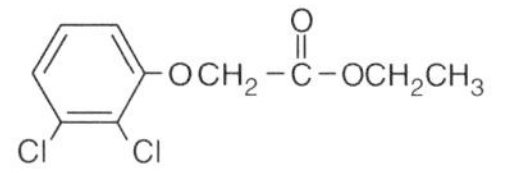
5 gm

S57,598-4
METHYL (2,4-DICHLOROPHENOXY)ACETATE
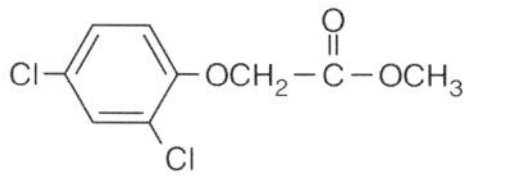
5 gm

S94,294-4
DIETHYL 2-(4-METHOXYBENZYL)MALONATE
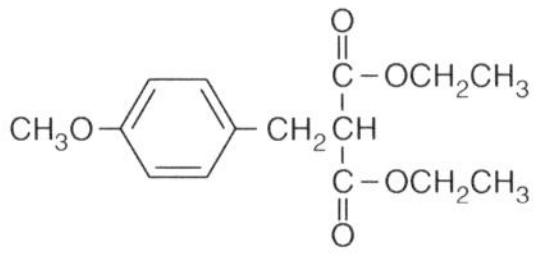
250 mg

S86,078-6
2-METHOXYBENZYL ACETATE
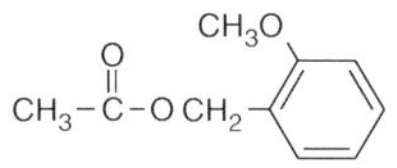
250 mg

S75,664-4
DIMETHYL 3-(3-METHOXYBENZOYL)-ADIPATE
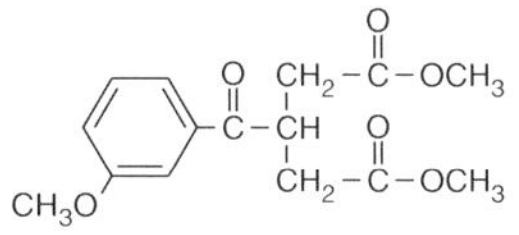
5 gm

S52,673-8
DIETHYL 2-(2-PHENOXYETHYL)MALONATE
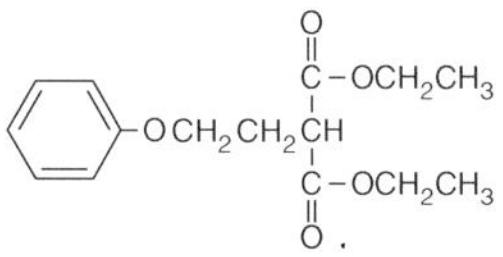
5 gm

S95,717-8
DIMETHYL 1,2,3,4,9,10-HEXAHYDRO-7-METHOXY-1-METHYL-1,2-PHENANTHRENEDICARBOXYLATE
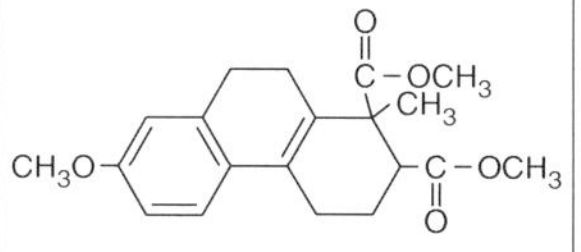
50 mg

S90,694-8
1-O-(3,4-DINITROPHENYL)-2,3,4-TRI-O-ACETYL-BETA-D-XYLOPYRANOSIDE
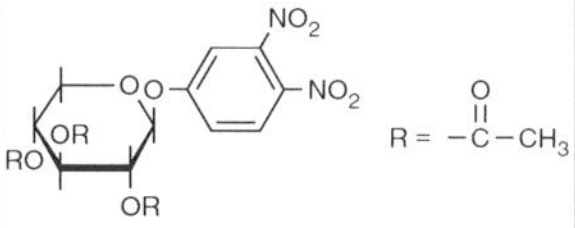
250 mg

S92,879-8
2-(4-METHOXYPHENYL)-DL-GLYCINE TERT-BUTYL ESTER P-TOLUENESULFONATE
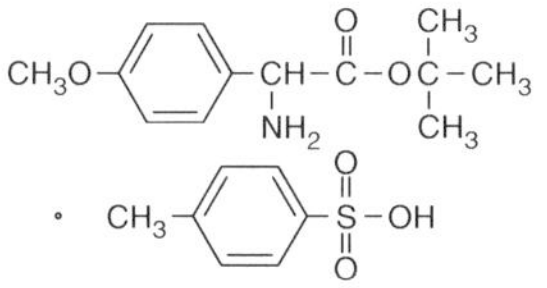
250 mg

S78,813-9
O-(TERT-BUTYL)-L-TYROSINE TERT-BUTYL ESTER HYDROCHLORIDE
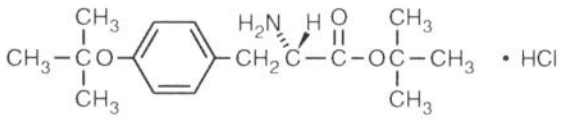
250 mg

S59,263-3
METHYL 3-AMINO-4-(HEXADECYLOXY)-HYDROCINNAMATE
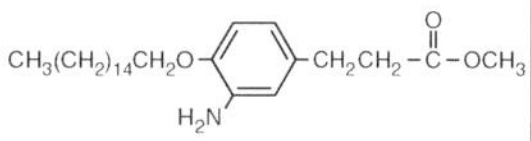
250 mg

S42,668-7
VOACANGINE
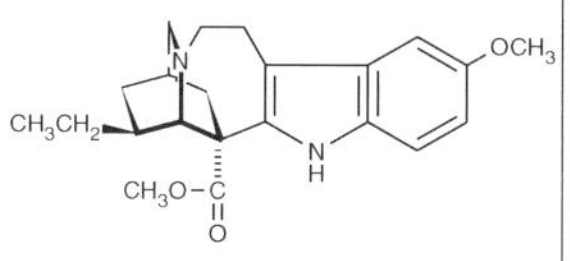
100 mg

S55,364-6
N-[2-(DODECYLOXY)PHENYL]GLYCINE ETHYL ESTER
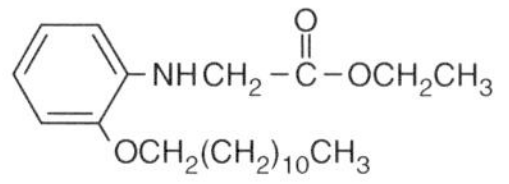
250 mg

S42,280-0
DIETHYLMETHYL[2-(9H-XANTHEN-9-YLCARBONYLOXY)ETHYL]-AMMONIUM BROMIDE
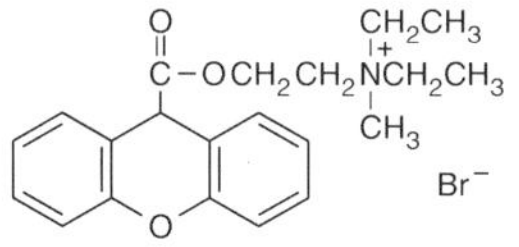
5 gm

S85,017-9
ETHYL 5-(4-METHOXYPHENYL)-5-OXOVALERATE
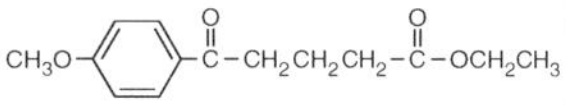
250 mg

S50,236-7
ETHYL 11-(4-ACETYLPHENOXY)-UNDECANOATE
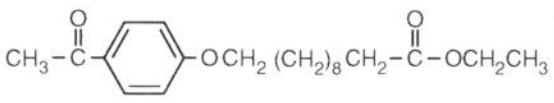
250 mg

S88,561-4
METHYL 3-(5-METHOXY-1-METHYL-2-OXO-1,2,3,4-TETRAHYDRO-1-NAPHTHYL)-PROPIONATE
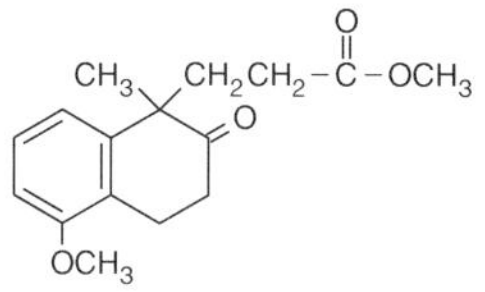
250 mg

S91,757-5
DIMETHYL 2-(3-METHOXYPHENYL)-MALONATE
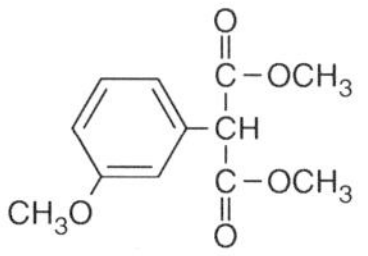
250 mg

S91,758-3
DIMETHYL 2-BROMO-2-(3-METHOXYPHENYL)MALONATE
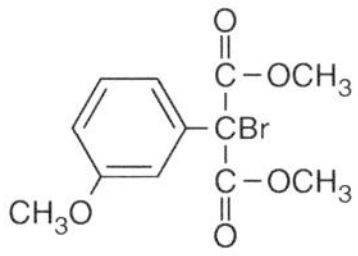
250 mg

S67,118-5
MONOETHYL 2-[1-(3-METHOXYPHENYL)-ETHYL]SUCCINATE, RACEMIC MIXTURE
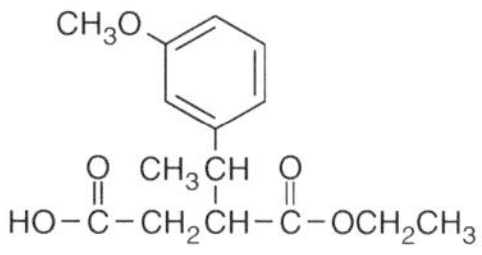
250 mg

S38,113-6
ETHYL 3-(3-CHLOROPHENOXY)PROPIONATE
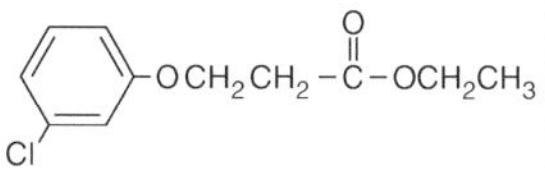
5 gm

S86,137-5
PERMETHRIN
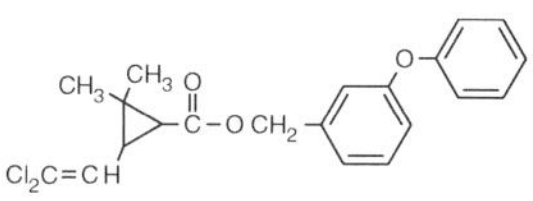
250 mg

S36,372-3
METHYL 3-PHENOXYPROPIONATE
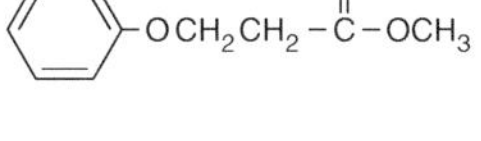
5 gm

S38,079-2
ETHYL 3-PHENOXYPROPIONATE
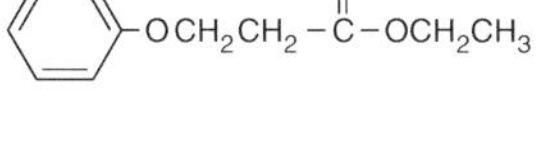
5 gm

S67,983-6
2-PHENOXYETHYL ISOBUTYRATE
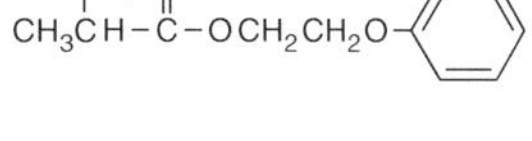
250 mg

S35,209-8
4-METHOXYBENZYL FORMATE
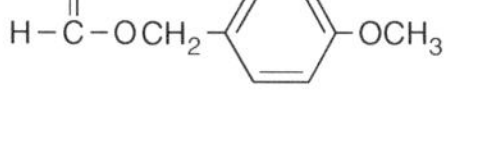
5 gm

S37,782-1
ETHYL 3-(4-CHLOROPHENOXY)-PROPIONATE
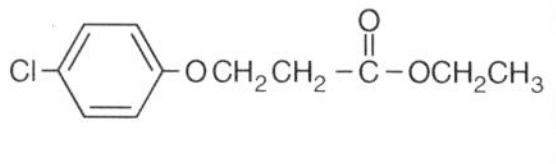
5 gm

S38,104-7
ETHYL ALPHA-CHLORO-4-METHOXYPHENYLACETATE
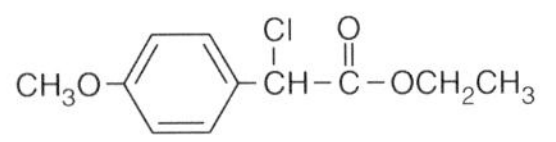
250 mg

S52,814-5
4-ALLYL-2-METHOXYPHENYL PHENYLACETATE
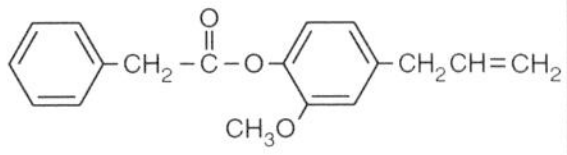
5 gm

S67,108-8
METHYL 3,5-DIMETHOXYPHENYLACETATE
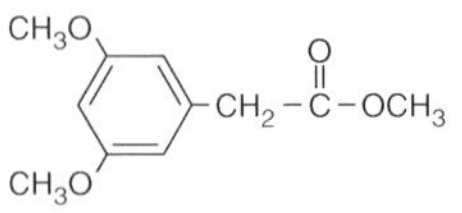
5 gm

S36,611-0
2-NITROTOLUENE-ALPHA,ALPHA-DIOL DIACETATE
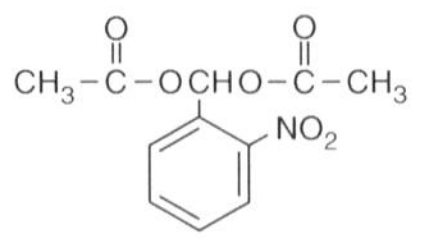
5 gm

S35,398-1
4-NITROTOLUENE-ALPHA,ALPHA-DIOL DIACETATE
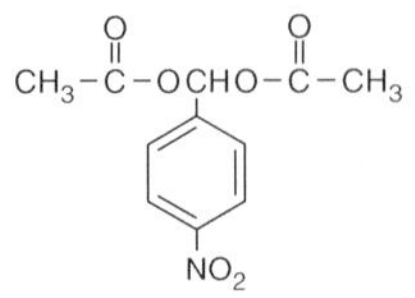
5 gm

S91,406-1
1,6-DI-O-TRITYL-2,3,4,5-TETRA-O-ACETYL-D-MANNITOL
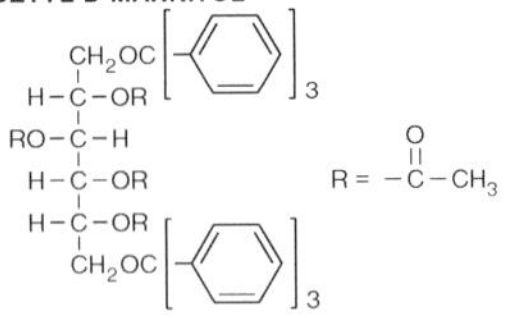
250 mg

S44,637-8
ETHYL BENZILATE
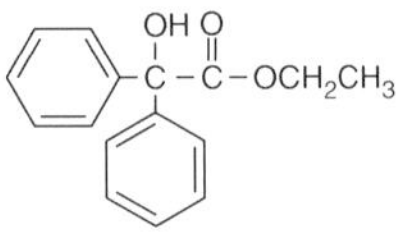
5 gm

S94,120-4
METHYL 2-(2-METHYL-4-PHENYL-1,3-DIOXOLAN-2-YL)ACETATE
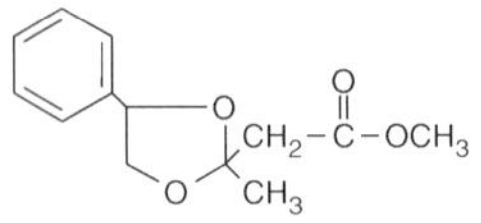
250 mg

S94,122-0
ETHYL 2-(2-METHYL-4-PHENYL-1,3-DIOXOLAN-2-YL)ACETATE
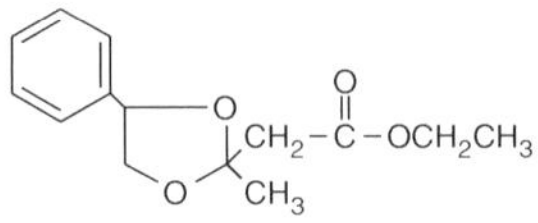
5 gm

S45,645-4
DIBENZYL SUCCINATE
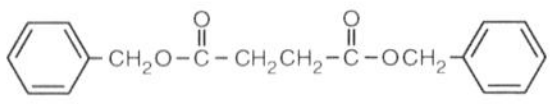
5 gm

S57,920-3
L-CYSTINE DIBENZYL ESTER DITOSYLATE
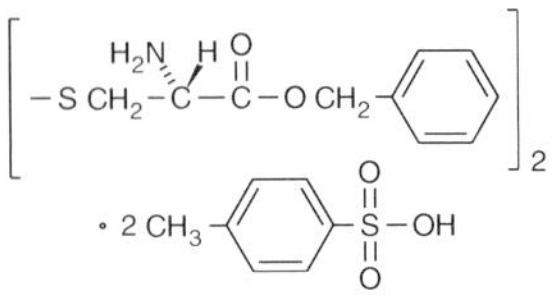
250 mg

S76,505-8
L-ASPARTIC ACID DIBENZYL ESTER-P-TOLUENESULFONATE
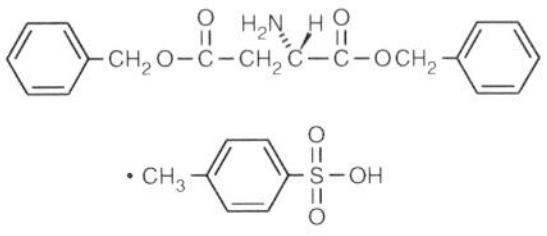
5 gm

S78,734-5
D-GLUTAMIC ACID GAMMA-BENZYL ESTER
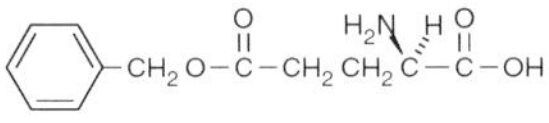
250 mg

S77,629-7
D-ASPARTIC ACID DIBENZYL ESTER-P-TOLUENESULFONATE
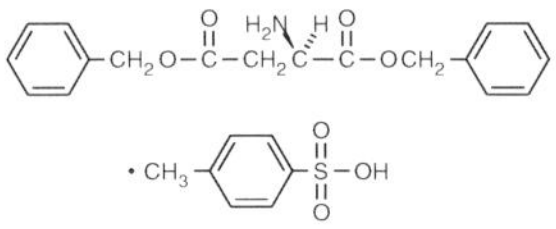
250 mg

S78,362-5
L-NORVALINE BENZYL ESTER-P-TOLUENESULFONATE
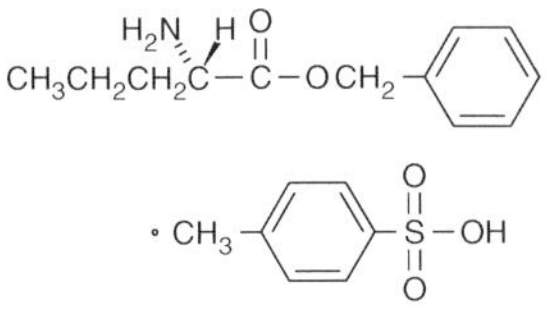
5 gm

S75,805-1
L-GLUTAMIC ACID DIBENZYL ESTER-P-TOLUENESULFONATE
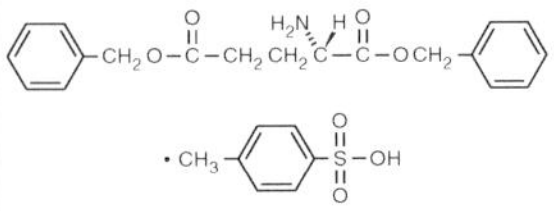
250 mg

S67,836-8
GLYCINE BENZYL ESTER-P-TOLUENESULFONATE
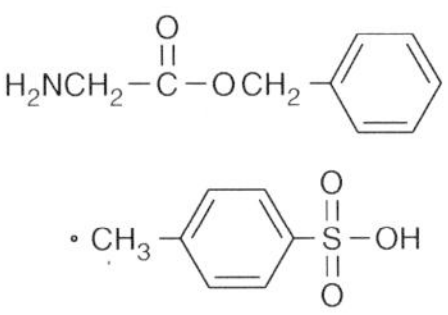
5 gm

S79,013-3
D-2-AMINOBUTYRIC ACID BENZYL ESTER-P-TOLUENESULFONATE
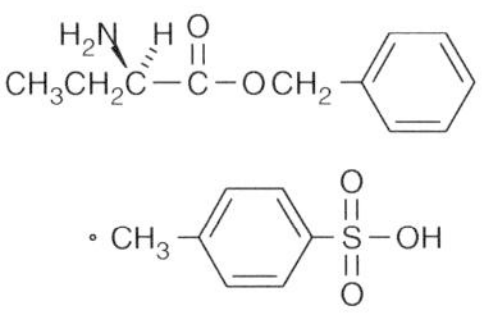
250 mg

S78,325-0
D-VALINE BENZYL ESTER-P-TOLUENESULFONATE
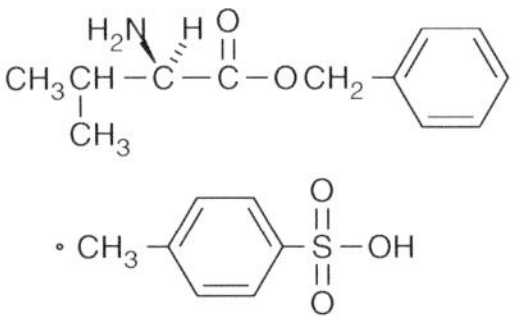
5 gm

S78,391-9
DL-VALINE BENZYL ESTER-P-TOLUENESULFONATE
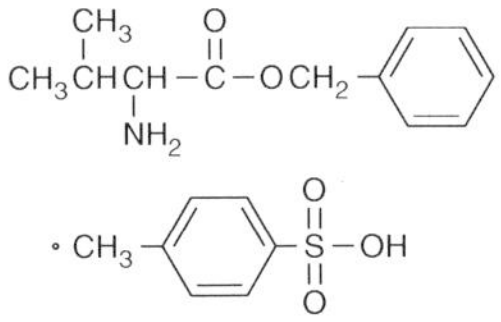
5 gm

S78,389-7
ISOLEUCINE BENZYL ESTER-P-TOLUENESULFONATE
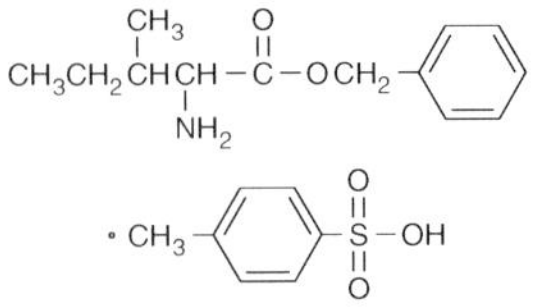
5 gm

S92,872-0
PHENYLALANINE ALPHA-PHENYLPHENACYL ESTER
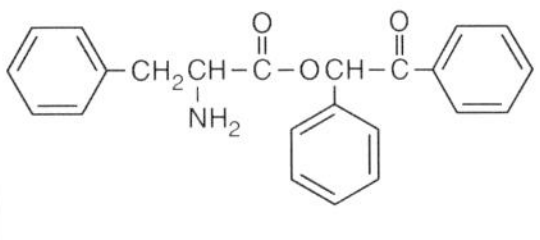
250 mg

S64,340-8
BENZYL LEVULINATE
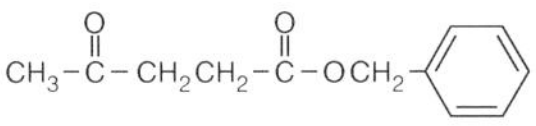
5 gm

S68,067-2
BENZYL 2-BROMOPROPIONATE
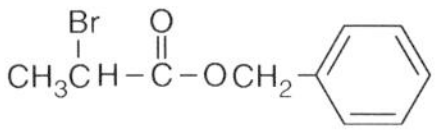
5 gm

S42,912-0
BENZYL FORMATE
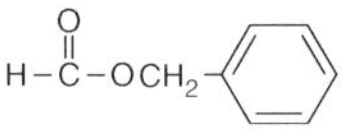
5 gm

S55,757-9
BENZYL PROPIONATE
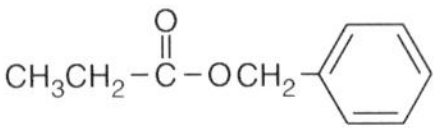
5 gm

S43,170-2
BENZYL BUTYRATE
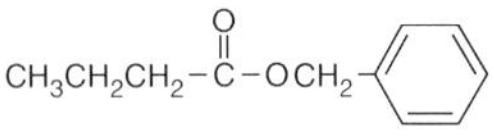
5 gm

S87,706-9
1-METHYL-1-PHENYLPROPYL ACETATE
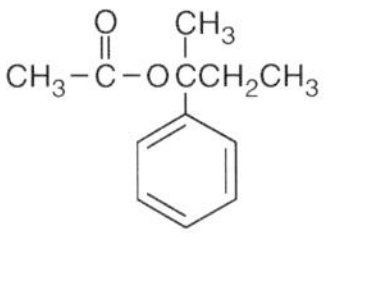
250 mg

S87,707-7
DL-1-PHENYLBUTYL ACETATE
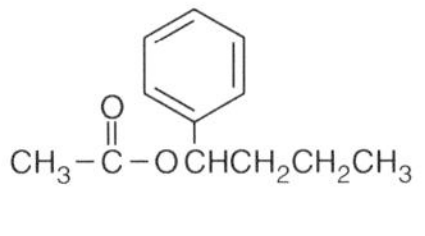
250 mg

S43,278-4
BENZYL PHENYLACETATE
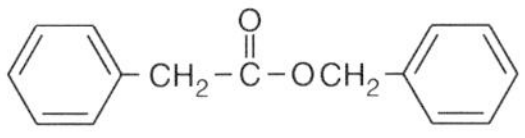
5 gm

S76,767-0
BENZYL ALPHA-CHLOROPHENYLACETATE
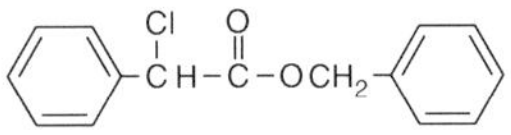
250 mg

S35,846-0
BENZYL LAURATE
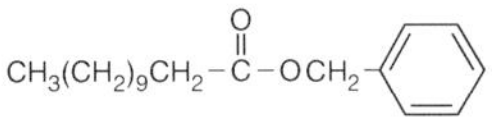
5 gm

S56,904-6
DIPHENYLMETHYL ACETATE
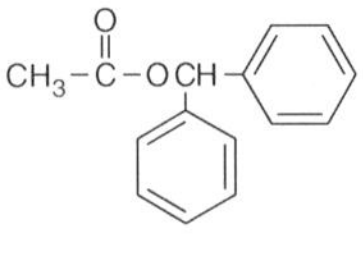
250 mg

S36,419-3
4-NITROPHENYL-ALPHA-D-GALACTOPYRANOSIDE TETRAACETATE
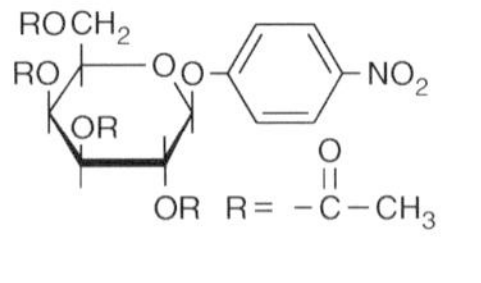
250 mg

S91,547-5
4-(DIACETOXYMETHYL)BENZOIC ACID
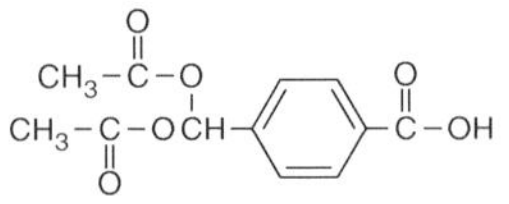
250 mg

S87,524-4
CINNAMYLIDENE DIACETATE
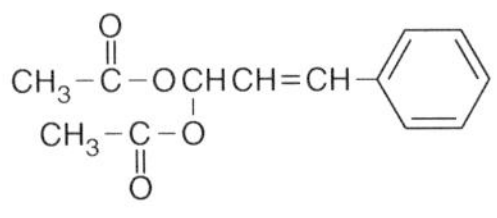
250 mg

S93,975-7
2,3-DI-O-ACETYL-1-O-METHYL-5-O-TRITYL-ALPHA-D-XYLOFURANOSIDE
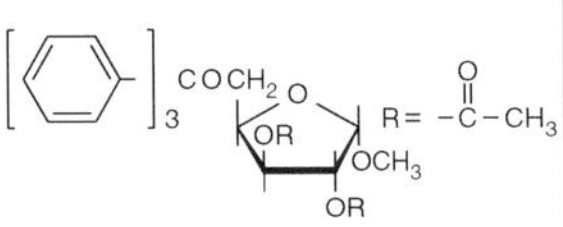
250 mg

S93,931-5
1-S-PHENYL-2,3,4-TRI-O-ACETYL-BETA-D-THIOXYLOSIDE
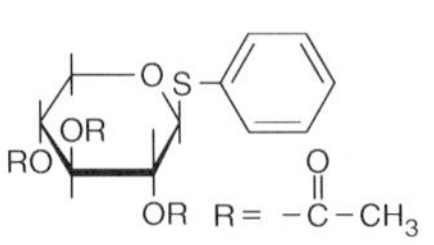
250 mg

S93,804-1
METHYL ALPHA-AMINO-1-ADAMANTANE ACETATE DIBENZOYLTARTRATE
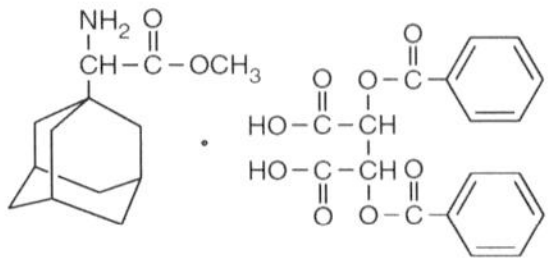
250 mg

S88,685-8
ETHYL 3-(2-NITROPHENYL)-2-OXIRANECARBOXYLATE
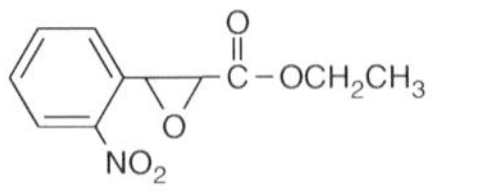
250 mg

S56,426-5
PHENACYL ACETATE
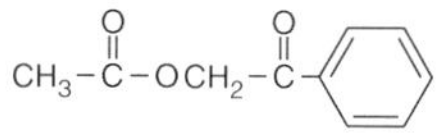
5 gm

S82,121-7
4'-PHENYLPHENACYL DODECANOATE
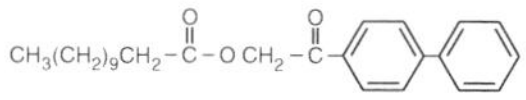
250 mg

S82,122-5
4'-PHENYLPHENACYL TETRADECANOATE
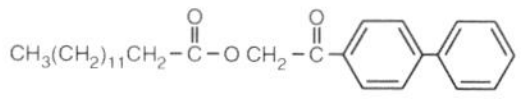
250 mg

S82,123-3
4'-PHENYLPHENACYL HEPTADECANOATE
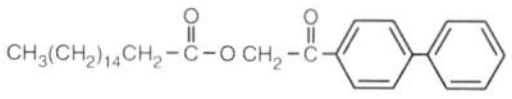
250 mg

S82,124-1
4'-PHENYLPHENACYL STEARATE
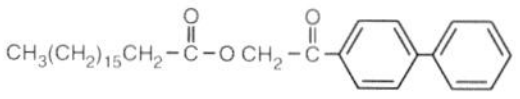
250 mg

S61,564-1
O-(2-METHYLBENZOYL)-DL-BENZOIN
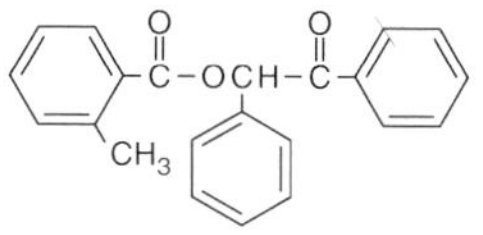
250 mg

S68,195-4
10,11-DIACETOXY-10,11-DIHYDRO-5H-DIBENZO(A,D)CYCLOHEPTEN-5-ONE
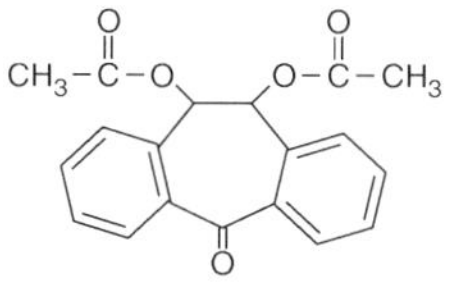
250 mg

S34,936-4
ETHYL 3-METHYL-3-PHENYLGLYCIDATE
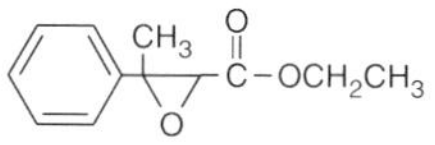
5 gm

S77,109-0
ETHYL 3,3-BIS(4-METHYLPHENETHYL)-2,3-EPOXYPROPIONATE
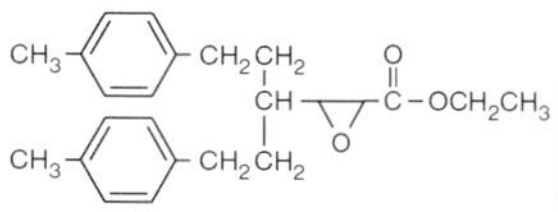
250 mg

S91,752-4
DIETHYL 2-METHYL-2-(PHENYLTHIO)-MALONATE
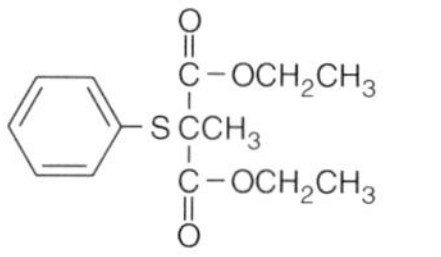
250 mg

S94,474-2
ETHYL METHYL-2,4-DINITROPHENYLMALONATE
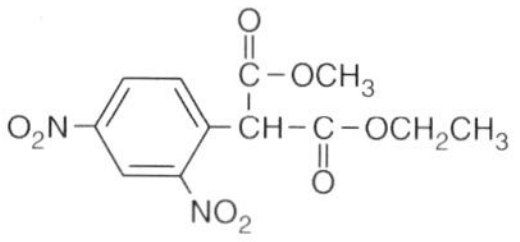
100 mg

S89,468-0
1-BENZYL-4-(PROPIONYLOXY)-3-PYRROLIDINYLMETHYL PROPIONATE
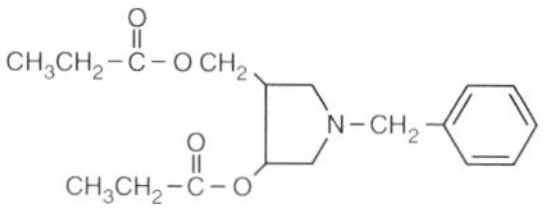
250 mg

S85,278-3
10,11-DIHYDRO-5,11-METHANO-5H-DIBENZO(A,D)CYCLOHEPTEN-10,12-DIYL DIACETATE
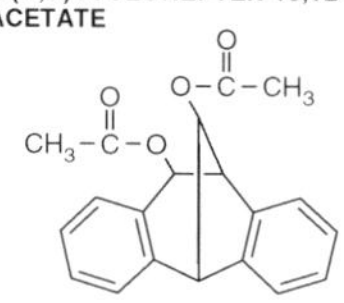
250 mg

S88,352-2
4-OXO-3-PHENYL-2,3,5,6-TETRAMETHYL-1,5-CYCLOHEXADIEN-1-YL ACETATE
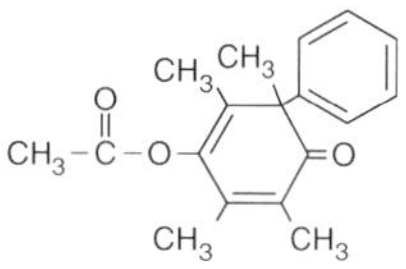
250 mg

S86,426-9
DIETHYL 9,10-DIHYDRO-9,10-ETHANOANTHRACENE-11,12-DICARBOXYLATE
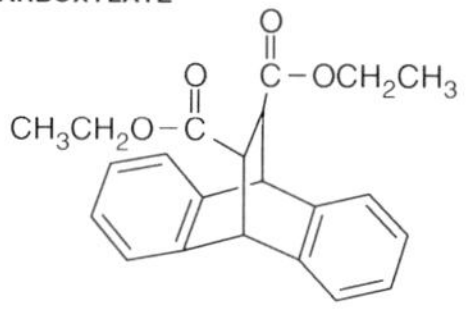
250 mg

S80,512-2
S-BENZYL-L-CYSTEINE ETHYL ESTER HYDROCHLORIDE
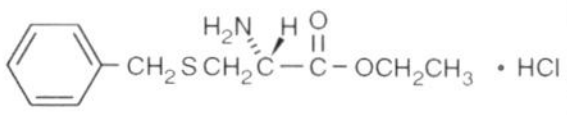
250 mg

S76,487-6
S-BENZYL-L-CYSTEINE METHYL ESTER HYDROCHLORIDE
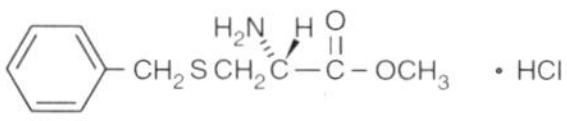
5 gm

S78,378-1
S-BENZYL-L-CYSTEINE TERT-BUTYL ESTER HYDROCHLORIDE
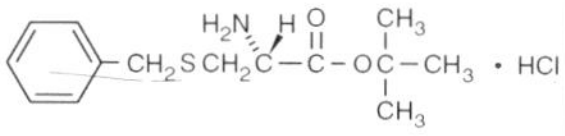
5 gm

S38,373-2
N-[4-(METHYLTHIO)PHENYL]-2-PHENYLGLYCINE ETHYL ESTER
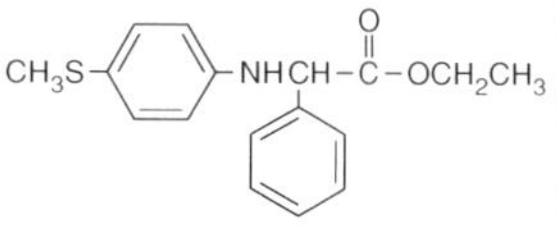
5 gm

S55,361-1
ETHYL (3-PENTADECYLPHENOXY)ACETATE
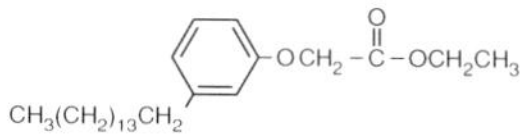
250 mg

S38,435-6
DIETHYL [2-(N-BENZYL-N-METHYL-AMINO)ETHYL][2-(METHYLTHIO)ETHYL]-MALONATE HYDROCHLORIDE
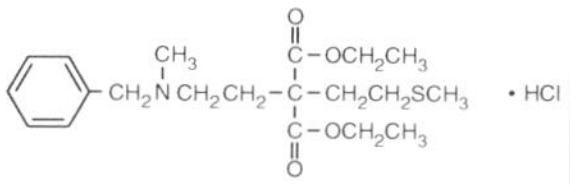
5 gm

S42,293-2
ETHYL ALPHA-[2-(DIETHYLAMINO)-ETHYLTHIO]PHENYLACETATE CITRATE
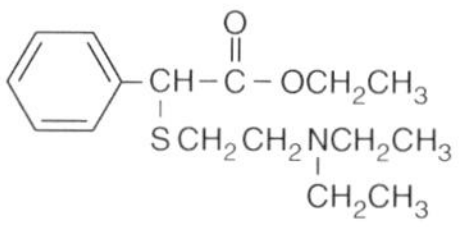
5 gm

S42,262-2
DIETHYLMETHYL(2-[1-PHENYL-2-(PROPIONYLOXY)ETHYLTHIO]ETHYL)-AMMONIUM METHYL SULFATE
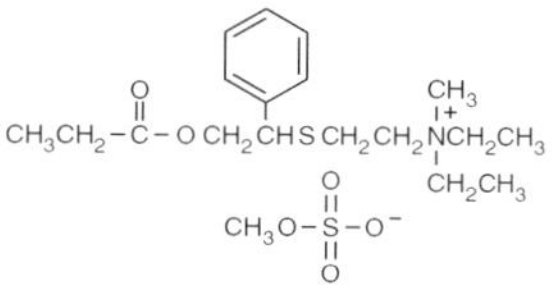
5 gm

S42,274-6
[2-(2-BUTYRYLOXY-1-PHENYLETHYL-THIO)ETHYL]DIETHYLMETHYLAMMONIUM METHYL SULFATE
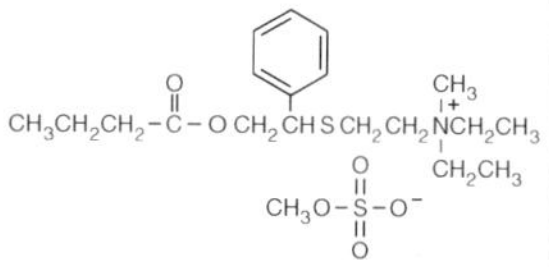
5 gm

S42,298-3
DIETHYLMETHYL(2-[1-PHENYL-2-(PHENYLACETOXY)ETHYLTHIO]ETHYL)-AMMONIUM METHYL SULFATE
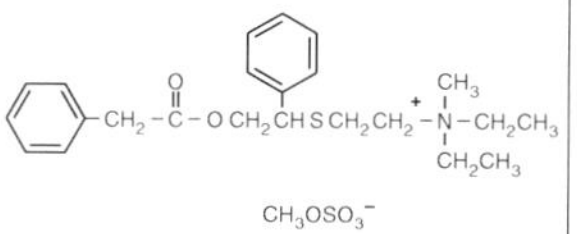
5 gm

S86,626-1
2-(PHENYLTHIO)ETHYL FORMATE
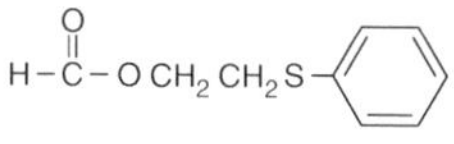
250 mg

S92,614-0
S-(ETHOXYCARBONYLMETHYL)-4-MORPHOLINE-1,1-DITHIOGLYOXYLATE
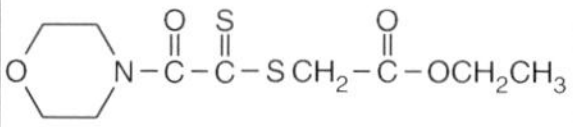
250 mg

S42,297-5
ISOPROPYL ALPHA-(2-PIPERIDINOETHYLTHIO)-PHENYLACETATE CITRATE
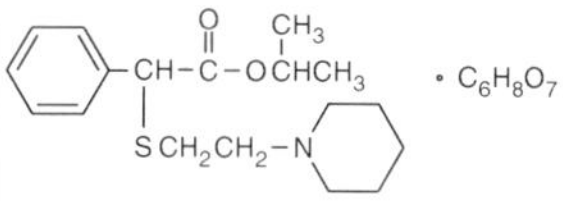
250 mg

S61,414-9
METHYL 2-(BENZYLTHIO)ACETATE
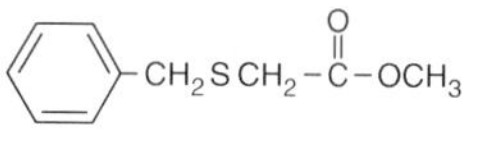
5 gm

S61,437-8
METHYL 3-(PHENYLTHIO)PROPIONATE
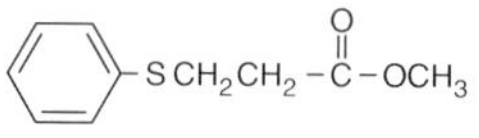
5 gm

S62,020-3
METHYL 2-(PHENYLTHIO)PROPIONATE
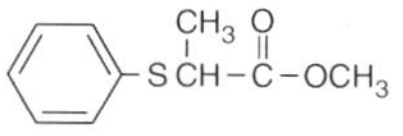
5 gm

S86,639-3
2-(PHENYLTHIO)ETHYL ACETATE
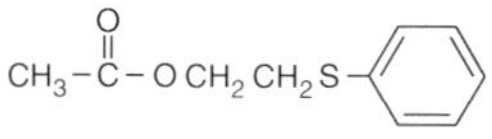
250 mg

S61,403-3
METHYL 2-METHYL-3-(PHENYLTHIO)-PROPIONATE
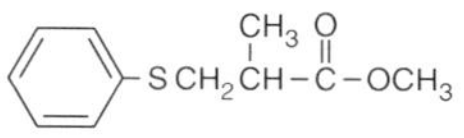
5 gm

S61,416-5
METHYL 3-(BENZYLTHIO)PROPIONATE
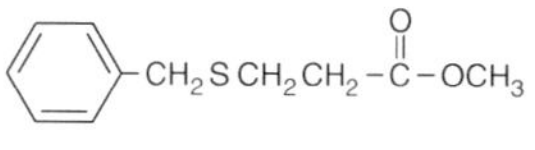
5 gm

S79,991-2
ETHYL 2-(PHENYLTHIO)PROPIONATE
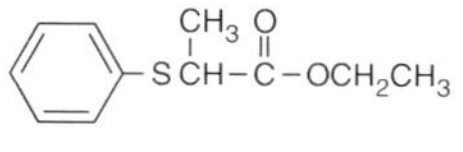
250 mg

S86,656-3
3-(PHENYLTHIO)PROPYL ACETATE
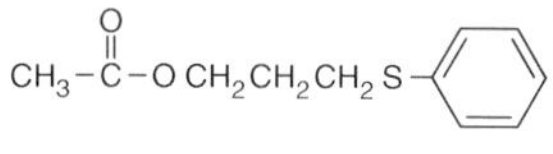
250 mg

S61,415-7
METHYL 3-(BENZYLTHIO)-2-METHYLPROPIONATE
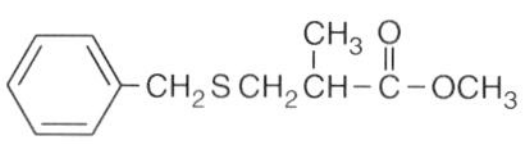
5 gm

S88,422-7
METHYL 2-METHYL-2-(PHENYLTHIO)-4-DECENOATE
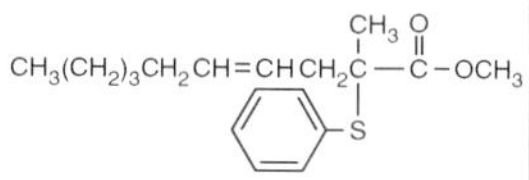
250 mg

S55,360-3
N-[2-(DODECYLTHIO)PHENYL]GLYCINE ETHYL ESTER
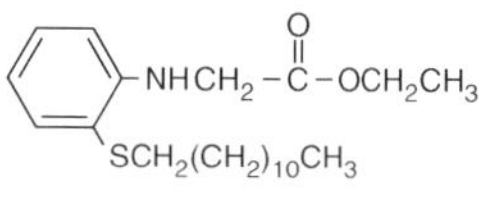
250 mg

S62,017-3
METHYL 2-(4-CHLOROPHENYLTHIO)-ACETATE
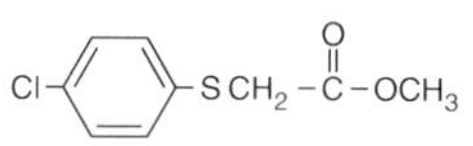
5 gm

S61,417-3
METHYL 3-(4-CHLOROPHENYLTHIO)-2-METHYLPROPIONATE
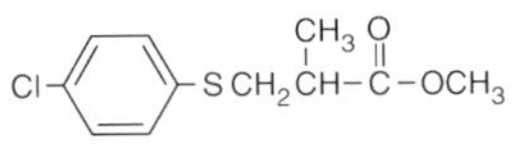
5 gm

S61,432-7
METHYL 3-(4-CHLOROPHENYLTHIO)-PROPIONATE
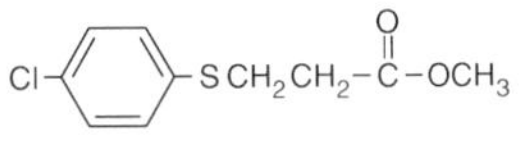
5 gm

S62,028-9
METHYL 2-(4-CHLOROBENZYLTHIO)-ACETATE
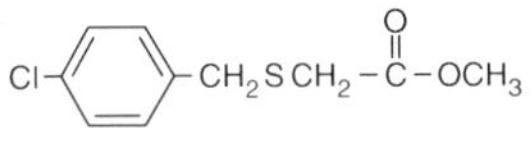
5 gm

S61,418-1
METHYL 3-(4-CHLOROBENZYLTHIO)-PROPIONATE
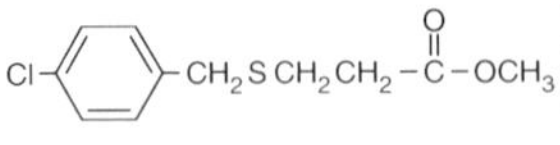
5 gm

S87,968-1
TRYPTOPHAN ETHYL ESTER HYDROCHLORIDE
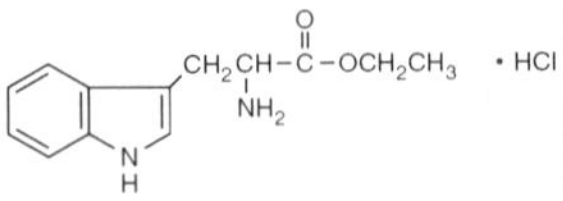
5 gm

S86,006-9
DL-PHENYLALANINE METHYL ESTER HYDROCHLORIDE
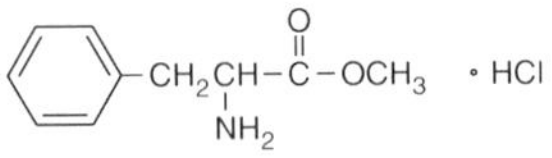
250 mg

S88,806-0
(S)-2-PHENYLGLYCINE METHYL ESTER HYDROCHLORIDE
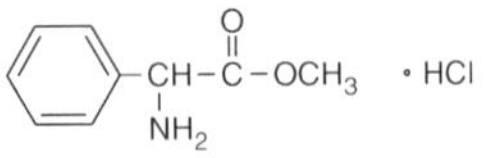
250 mg

S91,952-7
2-PHENYLGLYCINE ETHYL ESTER HYDROCHLORIDE
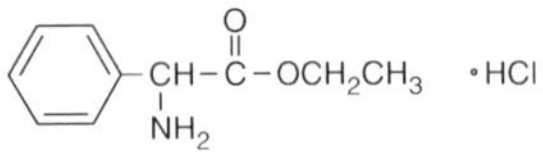
250 mg

S76,474-4
L-ASPARTIC ACID DI-TERT-BUTYL-ESTER HYDROCHLORIDE
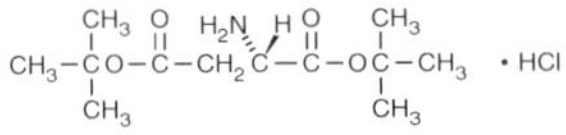
5 gm

S35,190-3
L-LEUCINE BENZYL ESTER HYDROCHLORIDE
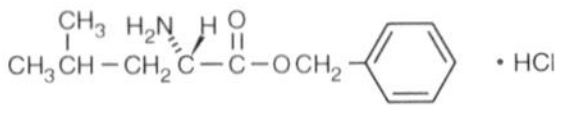
5 gm

S65,706-9
METHYL 2-[N-(2-NITROBENZYL)AMINO]-CYCLOHEXANECARBOXYLATE
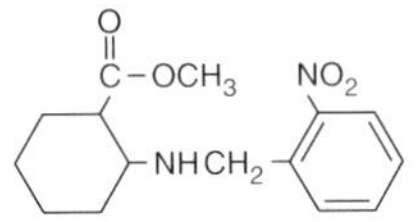
5 gm

S95,498-5
METHYL 4-OXO-3-TETRAHYDROPYRANACETATE-2,4-DINITROPHENYLHYDRAZONE
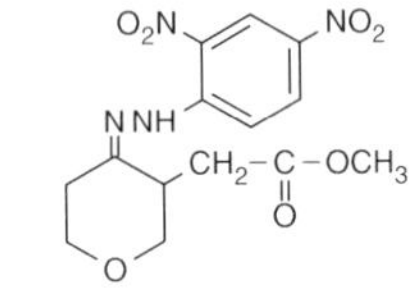
50 mg

S62,947-2
N-(3-NITROPHENYL)GLYCINE-METHYL ESTER
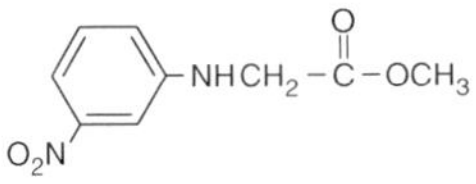
250 mg

S52,616-9
METHYL 3,5-DINITRO-4-HYDROXYHYDROCINNAMATE
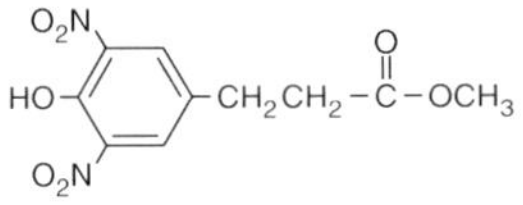
5 gm

S52,328-3
ETHYL 3,5-DINITRO-4-HYDROXYHYDROCINNAMATE
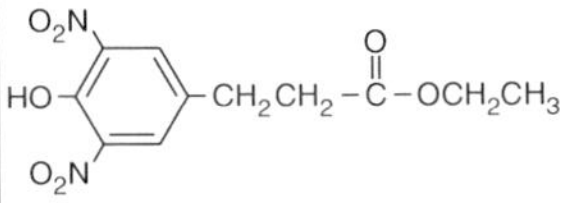
5 gm

S76,470-1
ETHYL 4-NITRO-3-PHENYLBUTYRATE
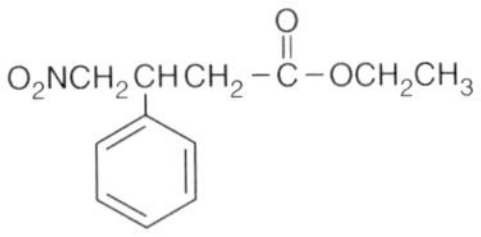
5 gm

S93,330-9
4-BROMO-DL-PHENYLALANINE METHYL ESTER HYDROCHLORIDE
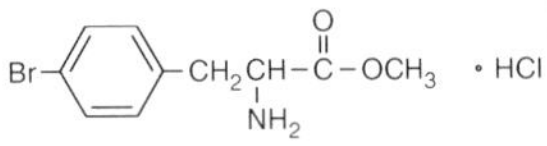
250 mg

S67,139-8
METHYL PHENYLACETIMIDATE HYDROCHLORIDE
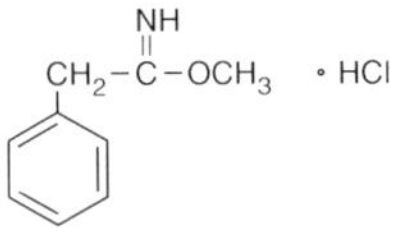
5 gm

S76,773-5
ETHYL 2-CHLORO-2-PHENYLACETIMIDATE HYDROCHLORIDE
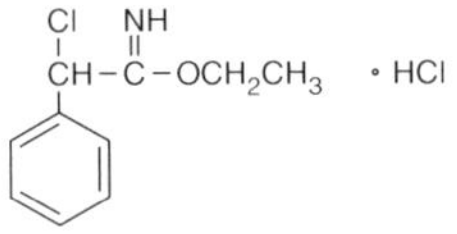
250 mg

S54,872-3
N-(2,2-DICARBETHOXYETHYL)ANILINE
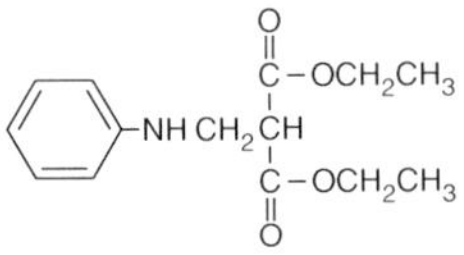
250 mg

S51,480-2
N-PHENYLGLYCINE METHYL ESTER
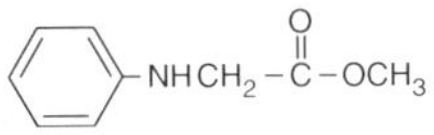
250 mg

S38,374-0
2-PHENYL-N-(PARA-TOLYL)GLYCINE ETHYL ESTER
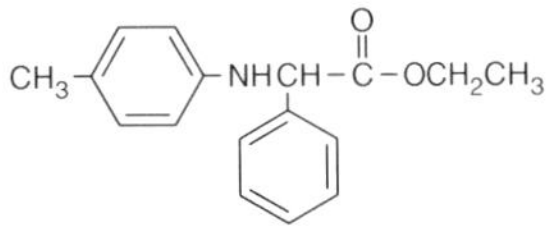
250 mg

S65,301-2
METHYL 3-ANILINOPROPIONATE
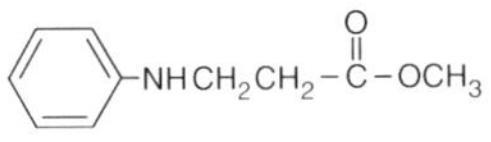
5 gm

S44,831-1
N-(ALPHA-METHYLBENZYL)GLYCINE METHYL ESTER
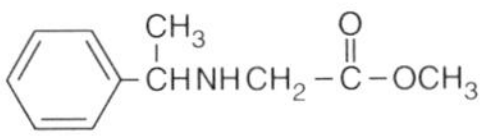
5 gm

S84,606-6
N-BENZYLGLYCINE ETHYL ESTER HYDROBROMIDE
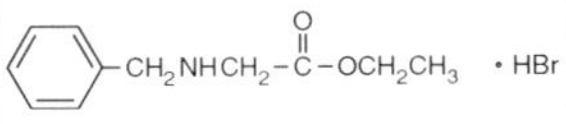
250 mg

S88,332-8
2-PHENYLSARCOSINE ETHYL ESTER
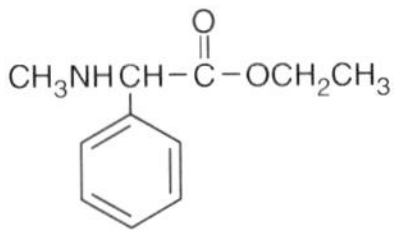
250 mg

S38,175-6
N-BENZYLALANINE ETHYL ESTER
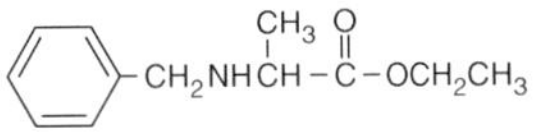
5 gm

S57,051-6
N-TRITYLGLYCINE ETHYL ESTER
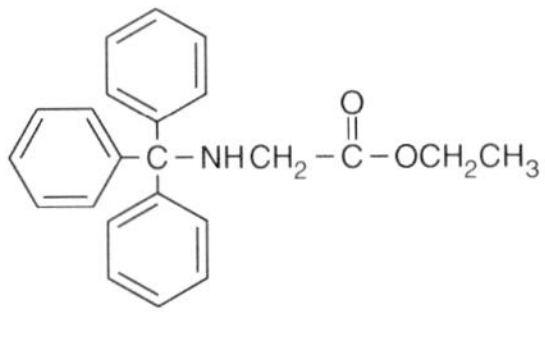
5 gm

S35,979-3
ETHYL 3-BENZYLAMINOPROPIONATE
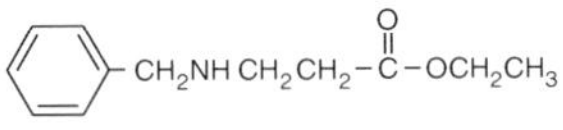
5 gm

S68,199-7
ETHYL 2,6-DIPHENYL-4-OXO-3-PIPERIDINEACETATE
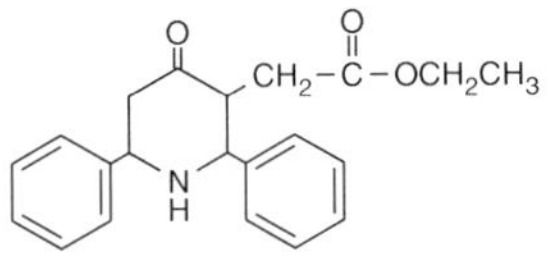
250 mg

S77,436-7
DIETHYL (3-FLUORANTHENYLAMINO)-MALONATE
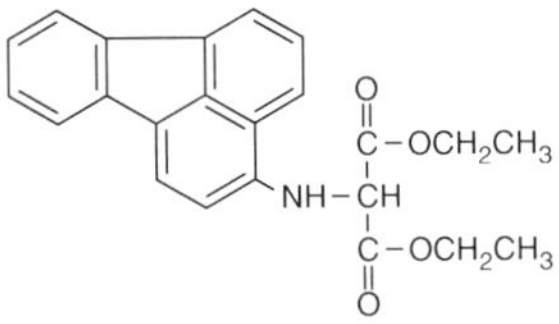
250 mg

S92,799-6
DIETHYL 1,2,3,4,4A,5,10,10A-OCTAHYDROBENZO(B)(1,7)-NAPHTHYRIDINE-1,1DICARBOXYLATE
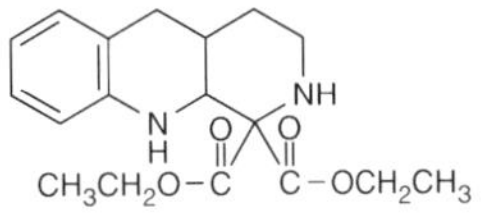
250 mg

S88,482-0
3-(ALLYLAMINO)-1,3-DIMETHYLBUTYL-2-PHENYLACETATE
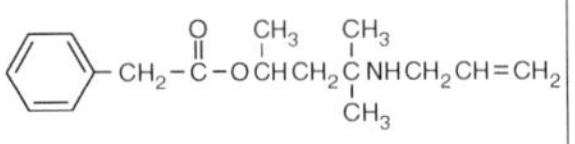
250 mg

S53,929-5
METHYL 3-(2-BIPHENYLYLAMINO)-PROPIONATE HYDROCHLORIDE
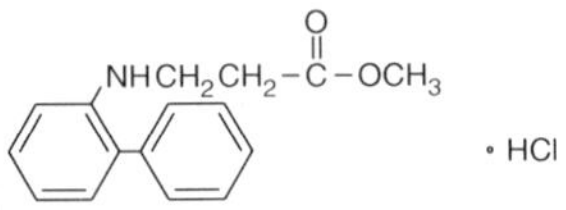
5 gm

S53,937-6
METHYL 3-(2-METHYL-1-NAPHTHYLAMINO)-PROPIONATE HYDROCHLORIDE
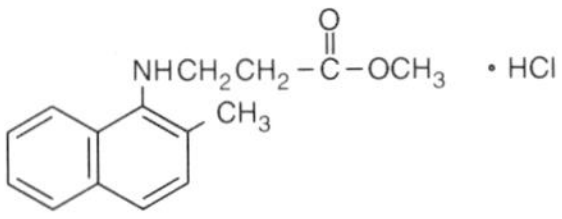
250 mg

S61,860-8
N,N-BIS(2-ACETOXYETHYL)ANILINE
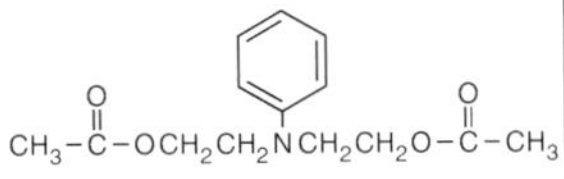
5 gm

S61,760-1
DIISOPENTYL (4-FORMYLPHENYL)-IMINODIACETATE
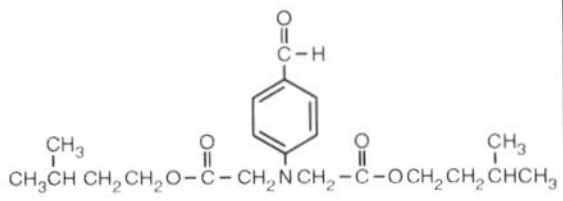
250 mg

S62,206-0
DIISOBUTYL 2,2'-(PHENYLIMINO)-DIACETATE
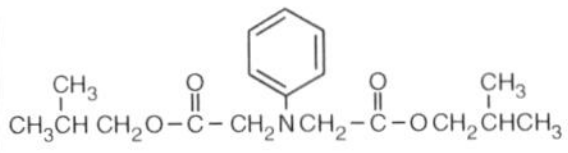
250 mg

S53,561-3
DIMETHYL 3,3'-(BENZYLIMINO)-DIPROPIONATE
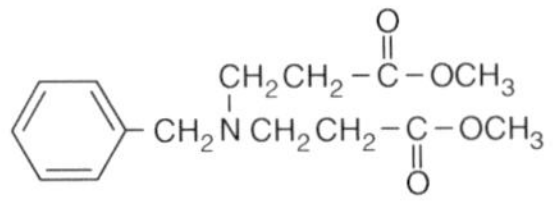
5 gm

S34,213-0
N,N-BIS(2-CARBETHOXYETHYL)-BENZYLAMINE
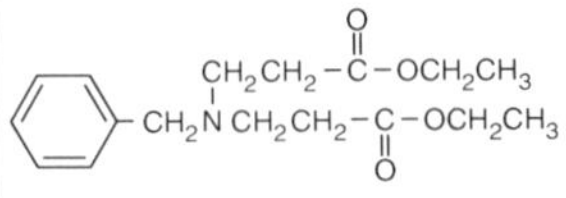
5 gm

S34,340-4
N-(2-CARBETHOXYETHYL)-N-(2-CARBOMETHOXYPROPYL)BENZYLAMINE
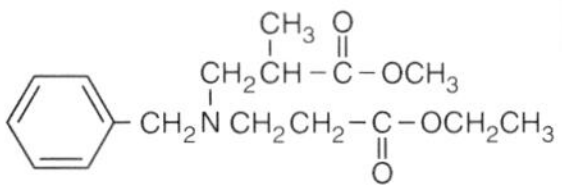
5 gm

S40,207-9
2-(DIETHYLAMINO)ETHYL-ALPHA-PHENYLCYCLOHEXANEACETATE HYDROCHLORIDE
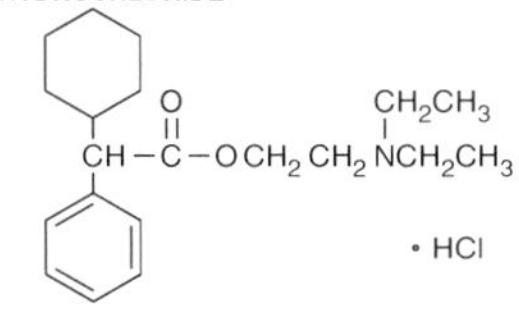
250 mg

S38,423-2
2-(DIETHYLAMINO)ETHYL-9-XANTHENECARBOXYLATE
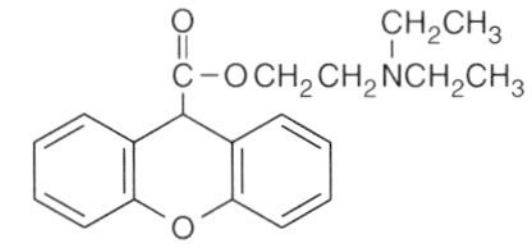
5 gm

S51,227-3
2-(N-ETHYLANILINO)ETHYL ACETATE
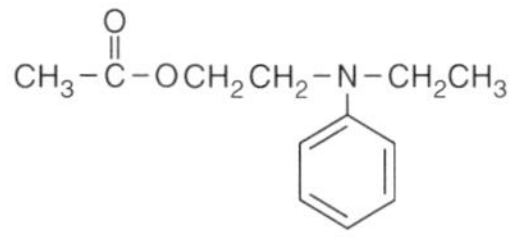
5 gm

S38,441-0
2-(DIETHYLAMINO)ETHYL-PHENYLACETATE
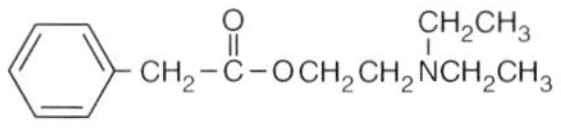
5 gm

S36,044-9
2-(DIETHYLAMINO)ETHYL-2-PHENYLBUTYRATE CITRATE
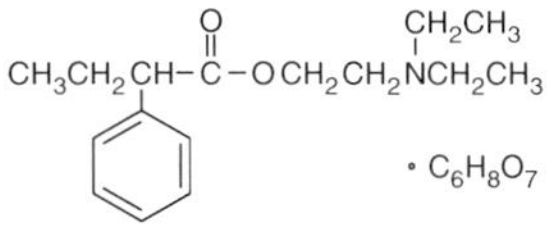
5 gm

S38,529-8
2-(DIMETHYLAMINO)ETHYL 3-METHYL-2-PHENYLVALERATE
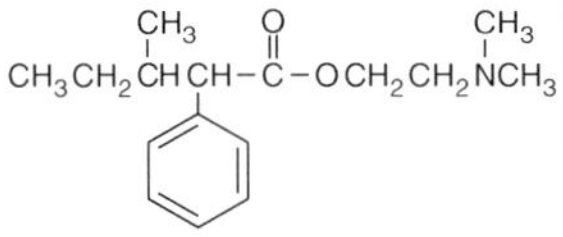
5 gm

S38,388-0
2-(DIETHYLAMINO)ETHYL 3-METHYL-2-PHENYLVALERATE
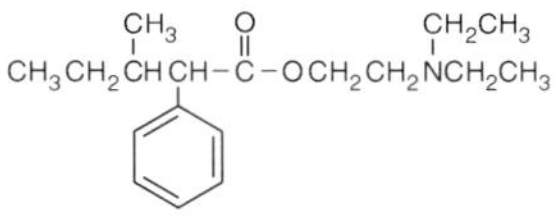
5 gm

S40,162-5
2-(DIMETHYLAMINO)ETHYL-2,2-DIPHENYLACETATE HYDROCHLORIDE
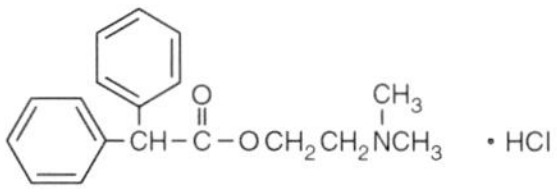
250 mg

S75,920-1
ETHYL 1-TERT-BUTYL-4-PHENYL-1,2,3,4-TETRAHYDRO-2-ISOQUINOLINEACETATE HYDROCHLORIDE
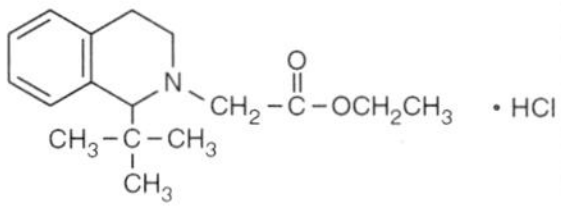
250 mg

S40,242-7
3-(DIETHYLAMINO)PROPYL ALPHA-PHENYLCYCLOHEXANEACETATE HYDROCHLORIDE
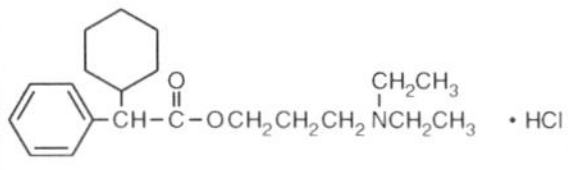
250 mg

S52,562-6
ETHYL 1-BENZYL-4-PIPERIDINEACETATE
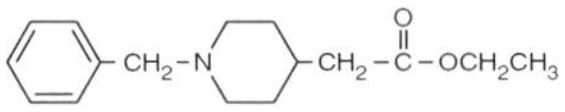
250 mg

S70,342-7
ETHYL 1-METHYL-3-PHENYL-4-PIPERIDINECARBOXYLATE
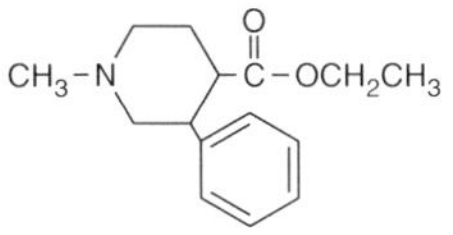
5 gm

S40,480-2
VINDOLININE
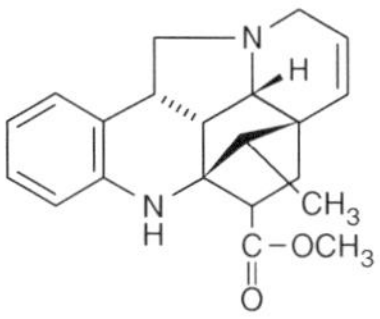
250 mg

S92,048-7
METHYL 10-PHENOTHIAZINEPROPIONATE
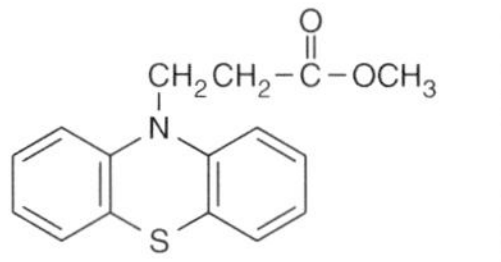
250 mg

S42,265-7
3-(DIETHYLAMINO)PROPYL 3-METHYL-2-PHENYLVALERATE HYDROCHLORIDE
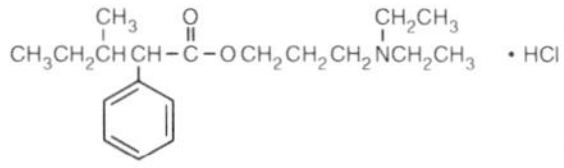
5 gm

S94,771-7
(+)-1-BENZYL-3-(DIMETHYLAMINO)-2-METHYL-1-PHENYLPROPIONATE HYDROCHLORIDE
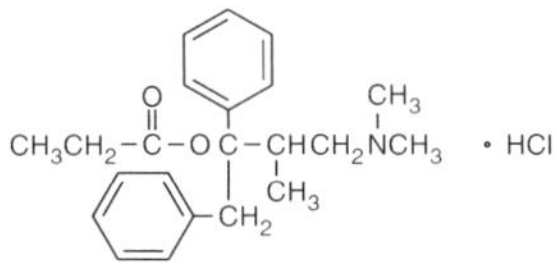
250 mg

S60,512-3
ETHYL 4-[(N-DODECYL)ANILINO]-BUTYRATE
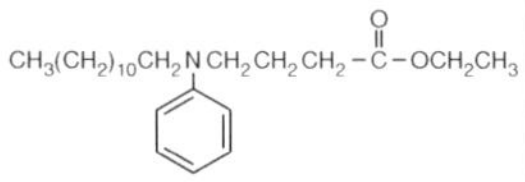
5 gm

S42,108-1
3-(DIETHYLAMINO)PROPYL 3-METHYL-2-(ORTHO-TOLYL)VALERATE HYDROCHLORIDE
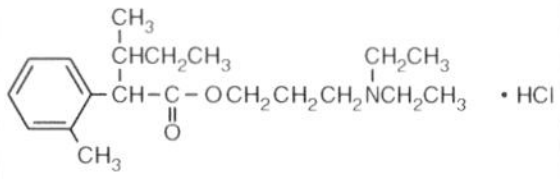
5 gm

S40,181-1
[2-(DIPHENYLACETOXY)ETHYL]-TRIMETHYLAMMONIUM BROMIDE
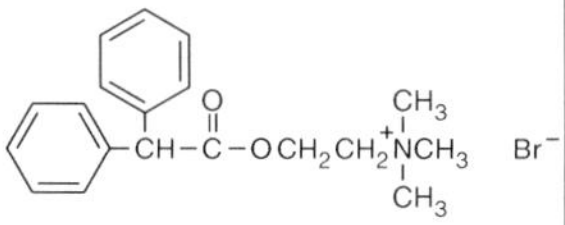
250 mg

S40,249-4
DIETHYL[2-(DIPHENYLACETOXY)ETHYL]-METHYLAMMONIUM BROMIDE
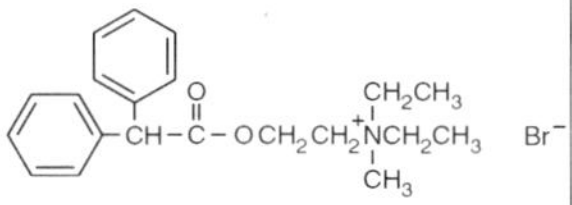
250 mg

S40,237-0
DIETHYLMETHYL(2-[3-METHYL-2-(3,4-XYLYL)VALERYLOXY]ETHYL)-AMMONIUM BROMIDE
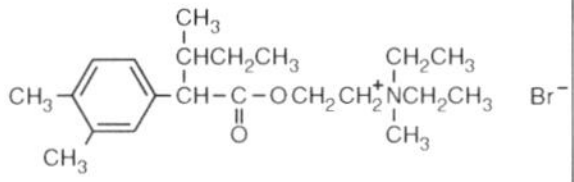
250 mg

S42,245-2
1,1-DIMETHYL-4-(3-METHYL-2-PHENYLVALERYLOXY)-PIPERIDINIUM BROMIDE
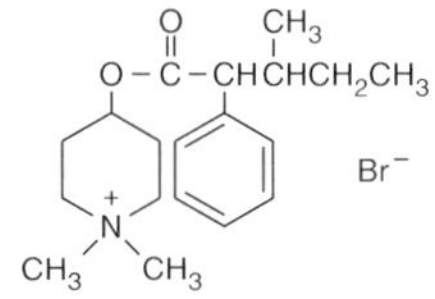
5 gm

S63,179-5
1-METHYL-4-PIPERIDINYL 3-METHYL-2-PHENYLVALERATE, DIMETHYL SULFATE SALT
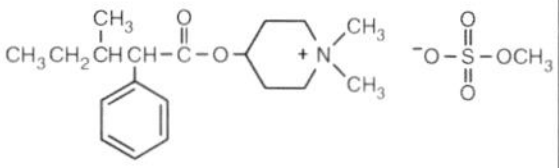
5 gm

S42,264-9
DIETHYLMETHYL[2-(3-METHYL-2-PHENYLVALERYLOXY)ETHYL]-AMMONIUM BROMIDE
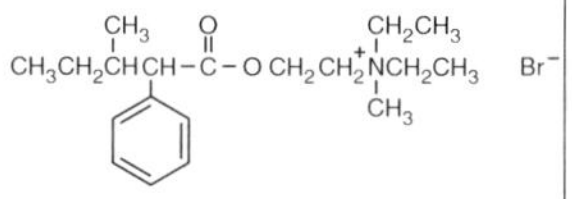
5 gm

S40,236-2
DIETHYLMETHYL[3-(2-PHENYLHEPTANOYLOXY)PROPYL]-AMMONIUM BROMIDE
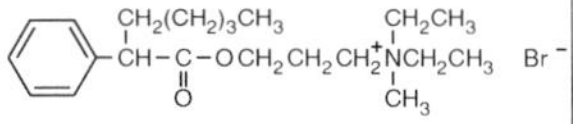
250 mg

S84,620-1
N-TRITYL-L-LEUCINE ETHYL ESTER
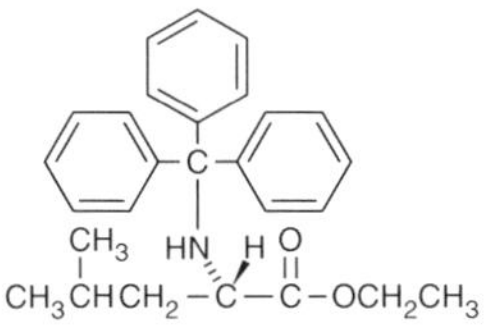
250 mg

S76,763-8
ETHYL 5-PHENYL-1,2,3,9B-TETRAHYDRO-5H-IMIDAZO(2,1A)-ISOINDOLE-5-ACETATE HYDROCHLORIDE
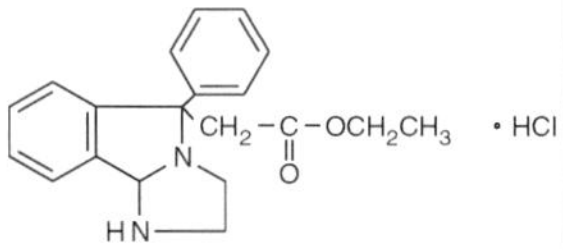
250 mg

S79,387-6
ETHYL 2,2-DIPHENYL-2-PIPERIDINOACETATE
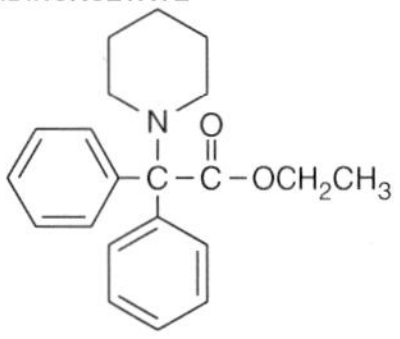
250 mg

S91,506-8
DIETHYL 2-(4-NITROBENZOYL)MALONATE
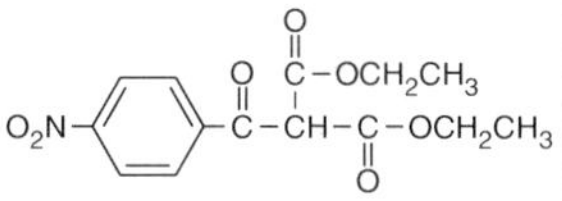
250 mg

S51,495-0
METHYL ALPHA-CHLORO-3-NITROHYDROCINNAMATE
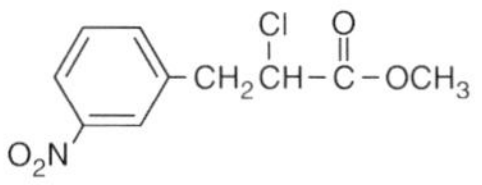
250 mg

S51,494-2
METHYL ALPHA-CHLORO-4-NITROHYDROCINNAMATE
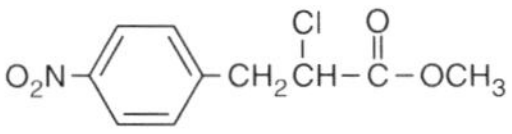
250 mg

S53,982-1
DIETHYL 1,4-DIHYDRO-2,6-DIMETHYL-4-(3-NITROPHENYL)-3,5-PYRIDINEDICARBOXYLATE
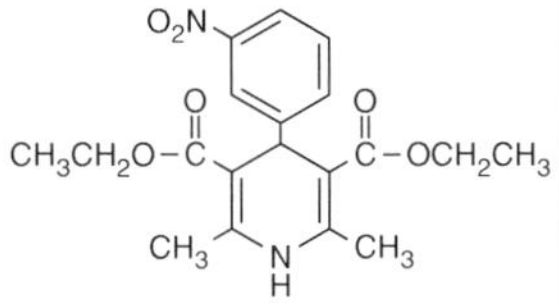
5 gm

S67,251-3
DIETHYL 2-(4-NITROBENZYL)MALONATE
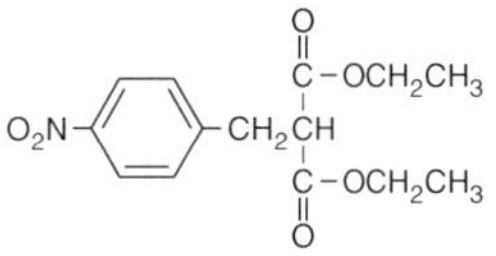
250 mg

S67,265-3
DIETHYL BIS(4-NITROBENZYL)MALONATE
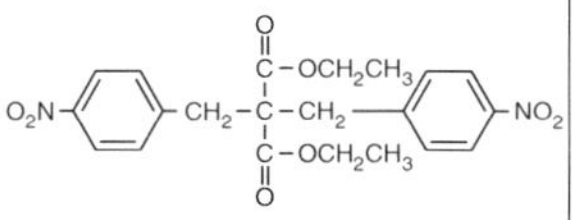
250 mg

S87,698-4
DL-4-NITRODIPHENYLMETHYL ACETATE
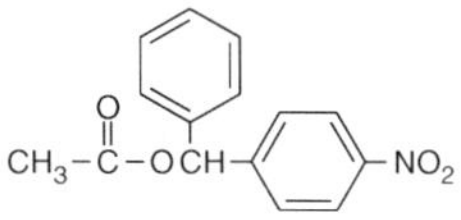
250 mg

S80,921-7
ETHYL 2-NITROPHENYLACETATE
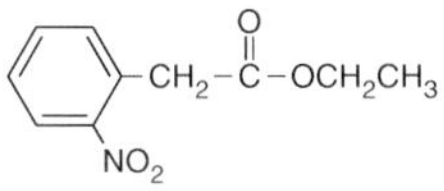
250 mg

S91,773-7
METHYL 9-(ALPHA,ALPHA-DIMETHYL-4-NITROBENZYL)-9-FLUORENECARBOXYLATE
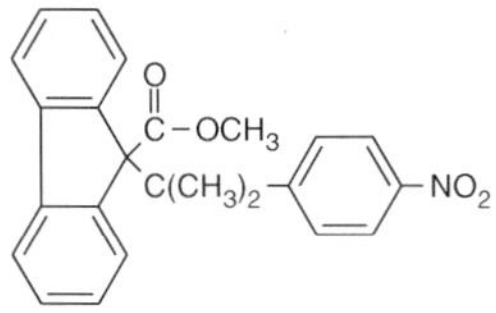
250 mg

S89,523-7
DIMETHYL 5-NITRO-10-OXO-1,2,3,4-TETRAHYDRO-1,4-ETHANO-NAPHTHALENE-2,3-DICARBOXYLATE
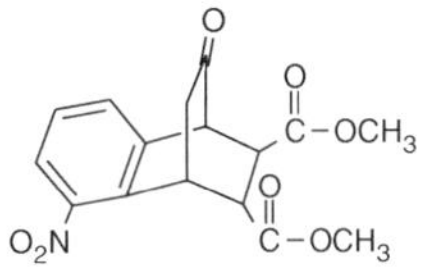
250 mg

S89,612-8
DIETHYL 5-NITRO-1,2,3,4-TETRAHYDRO-1,4-ETHENONAPHTHALENE-2,3-DICARBOXYLATE
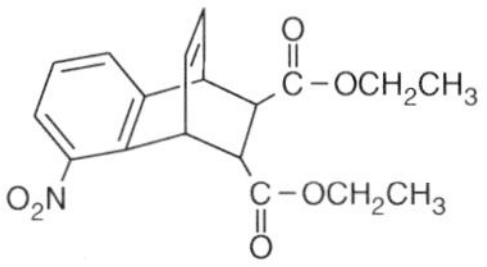
250 mg

S51,496-9
METHYL ALPHA,4-DICHLORO-3-NITROHYDROCINNAMATE
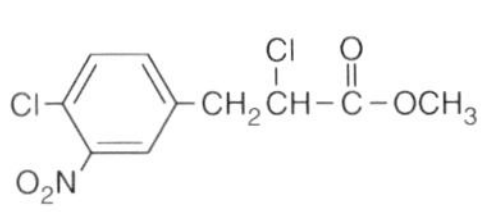
250 mg

S54,123-0
ETHYL 2-(2-NITROPHENOXY)PROPIONATE
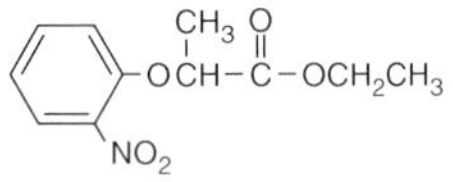
250 mg

S56,226-2
4-NITROBENZYL ACETATE
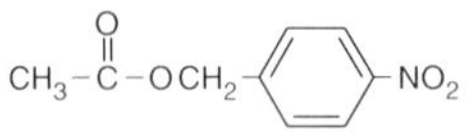
250 mg

S59,560-8
METHYL 4-NITROPHENYLACETATE
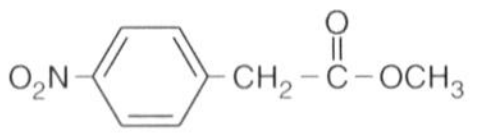
5 gm

S56,014-6
ETHYL 4-NITROPHENYLACETATE
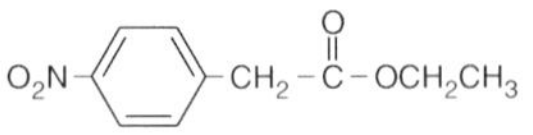
5 gm

S75,554-0
9-ACETOXY-3-NITROFLUORENE
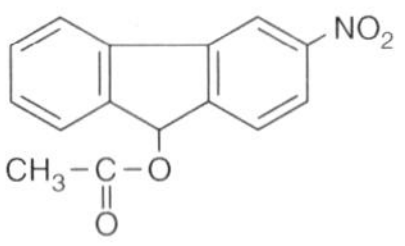
250 mg

S77,424-3
9-ACETOXY-2-NITROFLUORENE
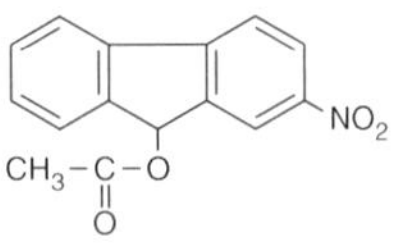
250 mg

S76,357-8
DIBUTYL BIS[4-(2,2-DIMETHYLPROPIONYL)BENZYL]MALONATE
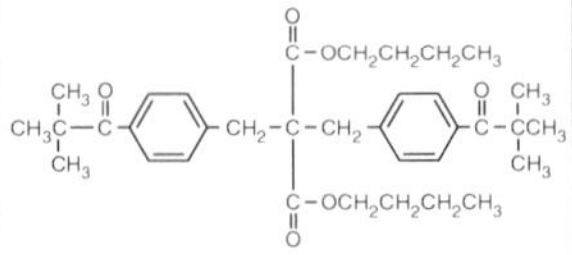
250 mg

S77,834-6
ETHYL 4-OXO-4-PHENYLBUTYRATE
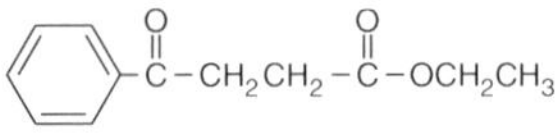
250 mg

S77,103-1
ETHYL 5-OXO-5-PHENYLVALERATE
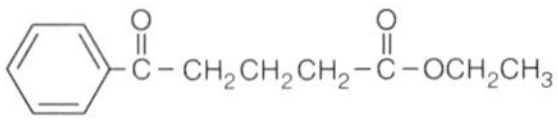
250 mg

S77,115-5
ETHYL 6-OXO-6-PHENYLHEXANOATE
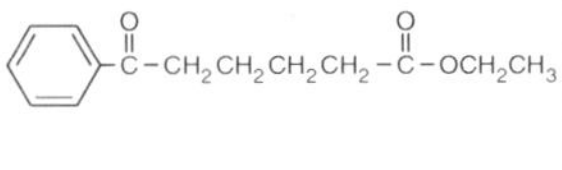
250 mg

S84,224-9
METHYL 4-OXO-2,3,4-TRIPHENYLBUTYRATE
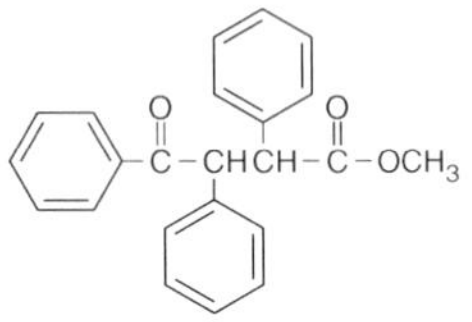
250 mg

S92,887-9
ALPHA-PHENYLPHENACYL ACETATE
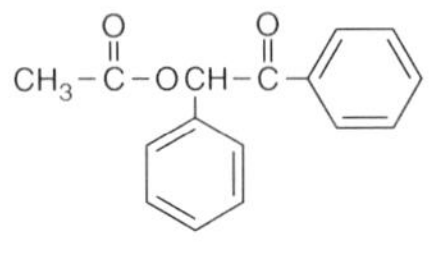
250 mg

S61,966-3
METHYL 3-(9,10-DIHYDRO-2-PHENANTHROYL)PROPIONATE
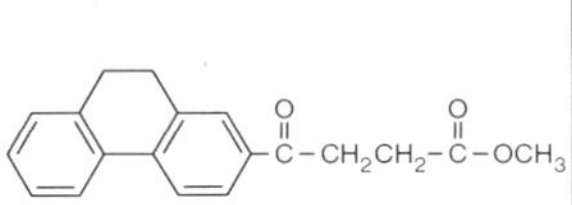
250 mg

S53,168-5
METHYL 2-OXO-3-PHENYLBUTYRATE
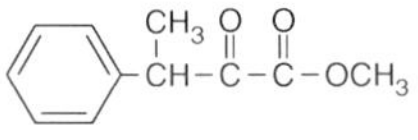
5 gm

S89,524-5
DIMETHYL 9-OXO-1,2,3,4-TETRAHYDRO-1,4-ETHANONAPHTHALENE-2,3-DICARBOXYLATE
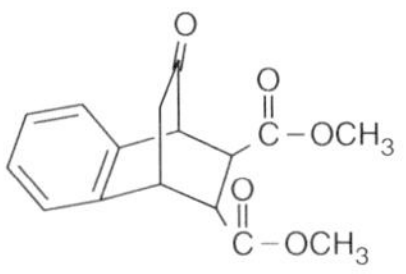
250 mg

S88,544-4
METHYL 2-(1,4A-DIMETHYL-1,2,3,4,4A,9-HEXAHYDRO-2-OXO-1-PHENANTHRYL)ACETATE
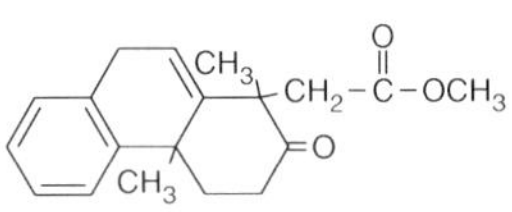
250 mg

S42,135-9
ETHYL 1,4-DIHYDRO-3,5-DIIODO-4-OXO-ALPHA-PHENYL-1-PYRIDINEACETATE
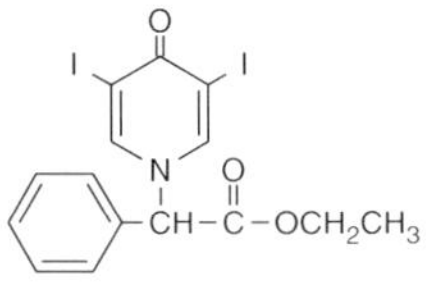
250 mg

S64,341-6
PHENETHYL LEVULINATE
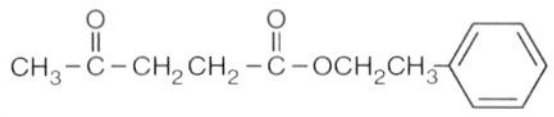
5 gm

S92,617-5
DIETHYL (1,3-DIPHENYL-3-OXOPROPYL)-MALONATE
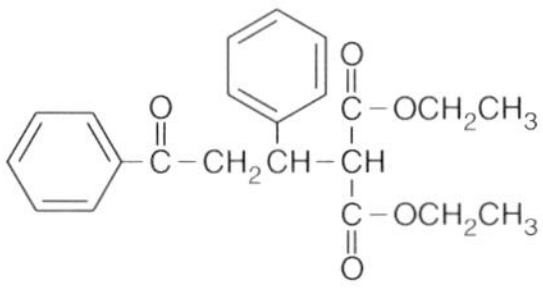
250 mg

S66,317-4
2,6-BIS[2,2-(DIETHOXYCARBONYL)-PROPYL]MESITYLENE
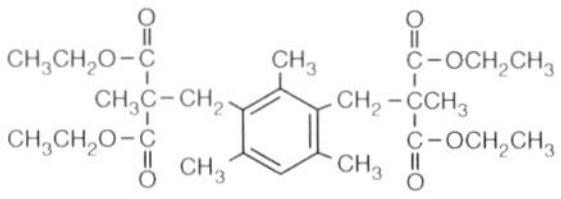
5 gm

S44,845-1
DIETHYL 2-PHENYLCYCLOPROPANE-1,1-DICARBOXYLATE
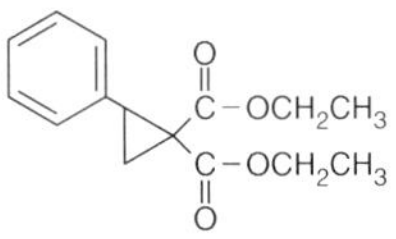
5 gm

S51,872-7
DIMETHYL PHENYLMALONATE
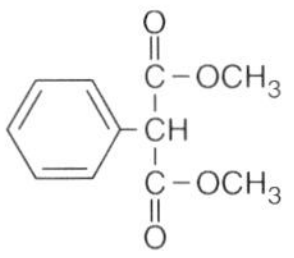
5 gm

S34,605-5
DIETHYL 2-ETHYL-2-PHENYLMALONATE
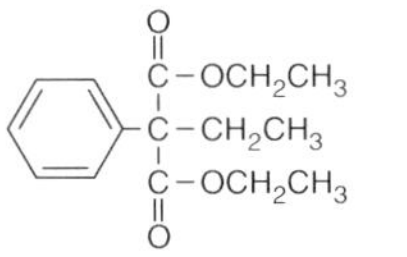
5 gm

S43,799-9
DIETHYL 2-PHENETHYLMALONATE
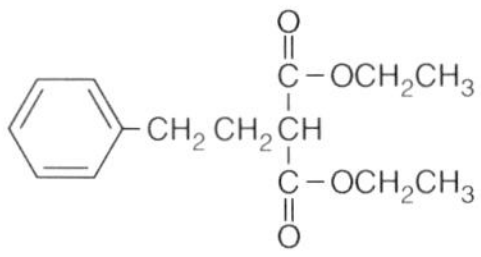
250 mg

S44,772-2
DIETHYL (2-BROMO-2-PHENYLETHYL)-MALONATE
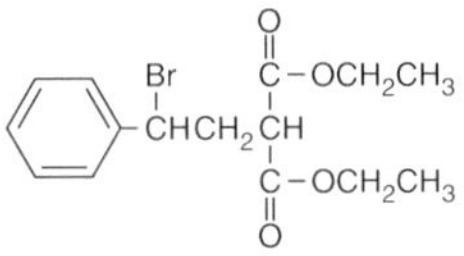
5 gm

S52,164-7
DIETHYL 2-ETHYL-2-(BETA-METHYLPHENETHYL)MALONATE
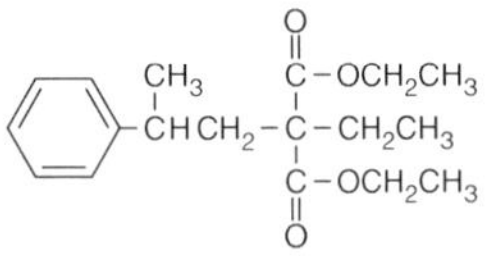
5 gm

S85,260-0
ETHYL 12-CARBOXY-9,10-DIHYDRO-9,10-ETHANOANTHRACENE-11-CARBOXYLATE
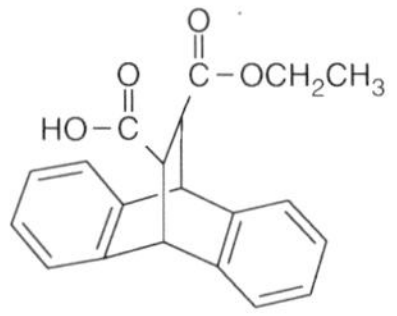
250 mg

S50,934-5
DIMETHYL 3-PHENYLGLUTARATE
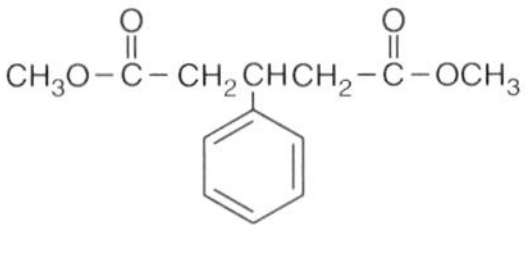
5 gm

S53,303-3
DIETHYL 2-PHENYLSUCCINATE
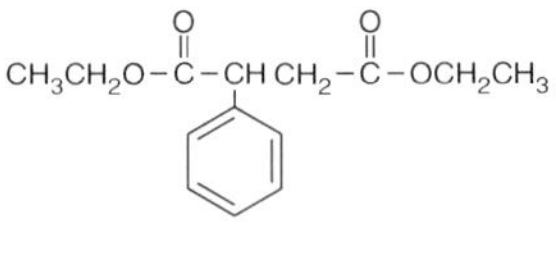
250 mg

S69,765-6
1-ETHYL 5-PROPYL 3-METHYL-2-PHENYLGLUTARATE
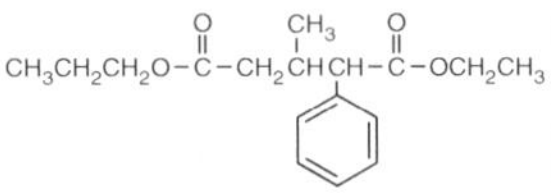
250 mg

S80,642-0
DIETHYL 4-ETHYL-4-PHENYLPIMELATE
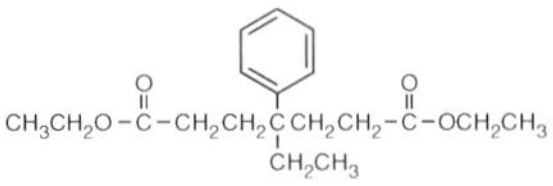
250 mg

S66,083-3
DIETHYL 2,3-DIPHENYLSUCCINATE
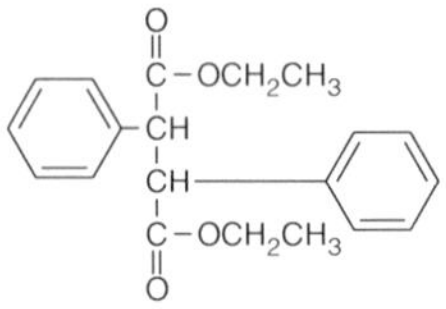
250 mg

S51,615-5
MONO-ETHYL SEBACATE
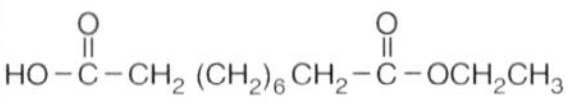
250 mg

S38,440-2
ETHYL ALPHA-(N-HEXANOYLANILINO)-PHENYLACETATE
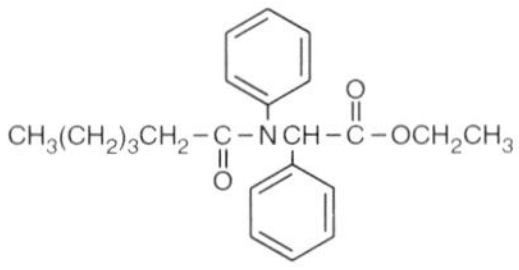
250 mg

S90,930-0
METHYL 1,2,3,4-TETRAHYDRO-3-ISOQUINOLINECARBOXYLATE HYDROCHLORIDE
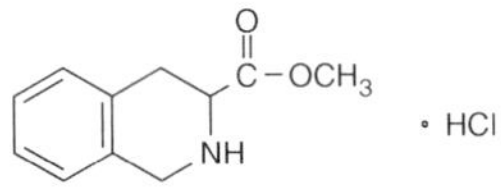
250 mg

S40,858-1
ETHYL META-TOLYLACETATE
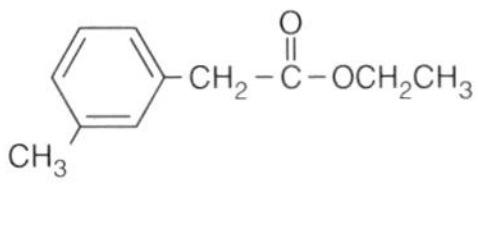
5 gm

S94,768-7
2,3,4,5,6-PENTAMETHYLBENZYL ACETATE
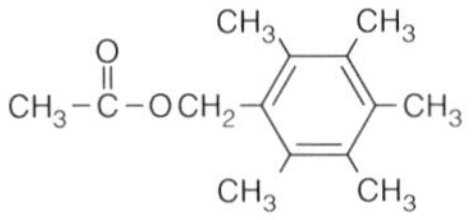
250 mg

S42,425-0
BELLADONNINE BISULFATE
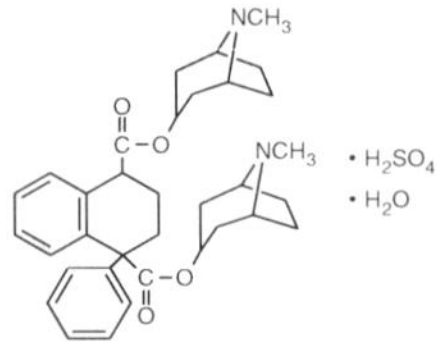
250 mg

S94,289-8
DIETHYL 2-(TRIPHENYLPHOSPHORANYLIDENE)-MALONATE
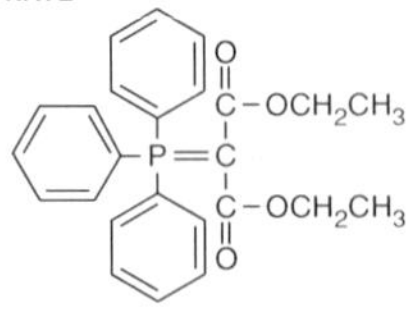
250 mg

S77,450-2
DIETHYL 2,2'-(1,4-PHENYLENE)-DIACETATE
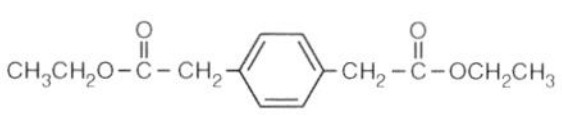
250 mg

S97,498-6
12-FLUORO-HEXAHYDRO-1-METHYLBENZO(C)PHENANTHRENE-5,8-DIOL DIPROPIONATE
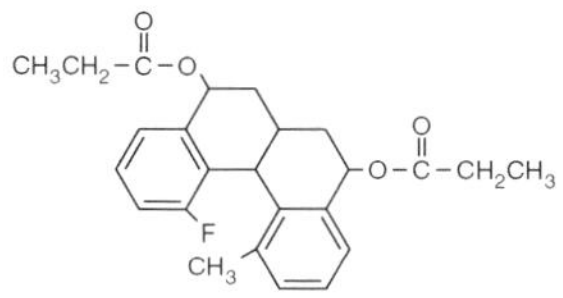
10 mg

S90,632-8
ETHYL 1A,2,7,7A-TETRAHYDRO-1H-CYCLOPROPA(B)NAPHTHALENE-1-CARBOXYLATE
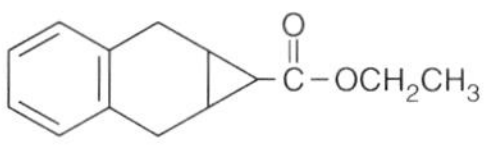
250 mg

S90,641-7
ETHYL 1A,2,3,7B-TETRAHYDRO-1H-CYCLOPROPA(A)NAPHTHALENE-1-CARBOXYLATE
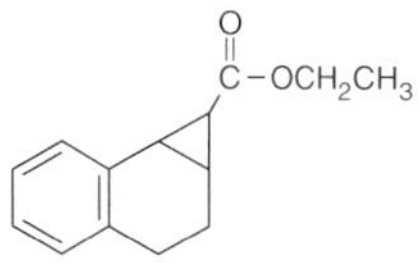
250 mg

S36,724-9
TRANS-1-CARBETHOXY-2-PHENYLCYCLOPROPANE
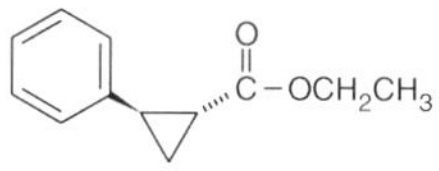
5 gm

S34,812-0
ETHYL 2-BENZYL-1-CYCLOPROPANECARBOXYLATE
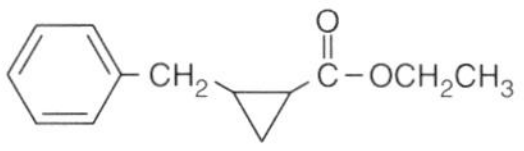
5 gm

S43,560-0
1-PHENYL-2-NORBORNYL ACETATE
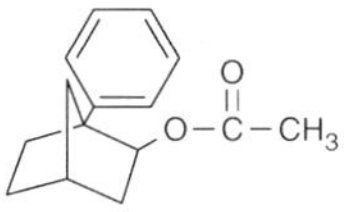
5 gm

S64,359-9
METHYL 1-(2,4-DICHLOROPHENYL)-1-CYCLOPENTANECARBOXYLATE
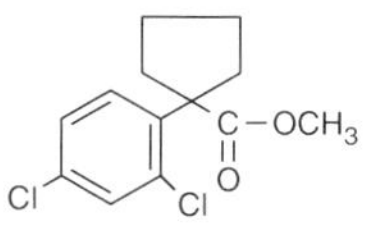
5 gm

S86,583-4
10,11-DIHYDRO-5,10-METHANO-5H-BENZO(A,D)CYCLOHEPTEN-12-YL ACETATE
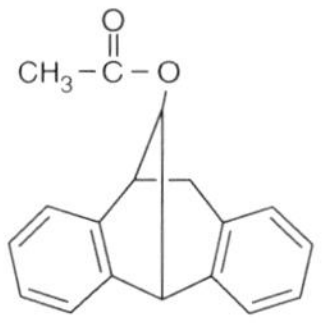
250 mg

S86,584-2
10,11-DIHYDRO-5,10-METHANO-5H-DIBENZO(A,D)CYCLOHEPTEN-11-YL ACETATE
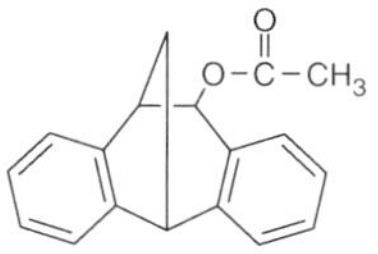
250 mg

S88,898-2
MENTHYL ALPHA,BETA-DIBROMOHYDROCINNAMATE
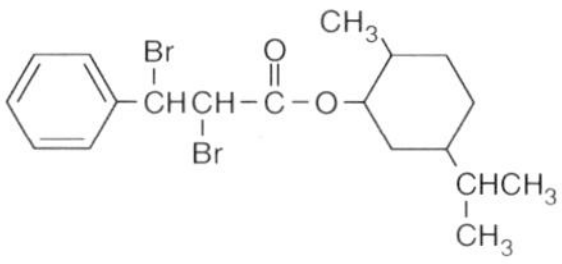
250 mg

S91,748-6
METHYL 9-CYCLOHEXYL-9-FLUORENECARBOXYLATE
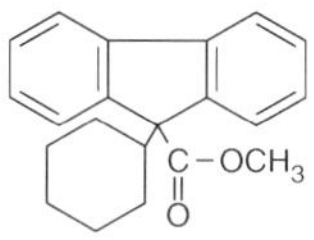
250 mg

S86,539-7
ETHYL 9,10-DIHYDRO-12-PHENYL-9,10-ETHANOANTHRACENE-11-CARBOXYLATE
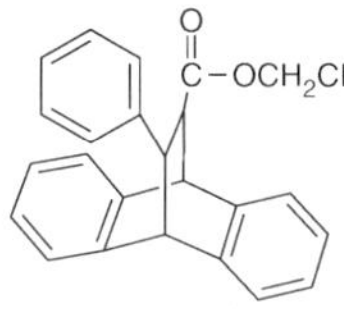
50 mg

S85,264-3
9,10-DIHYDRO-9,10-ETHANOANTHRACEN-11-YL ACETATE
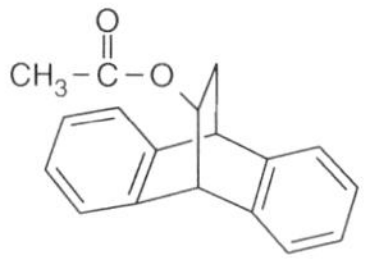
250 mg

S85,290-2
METHYL 9,10-DIHYDRO-9,10-ETHANOANTHRACENE-11-CARBOXYLATE
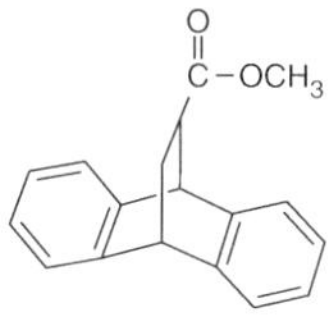
250 mg

S86,582-6
9,10-DIHYDRO-9,10-ETHANOANTHRACEN-9-YL ACETATE
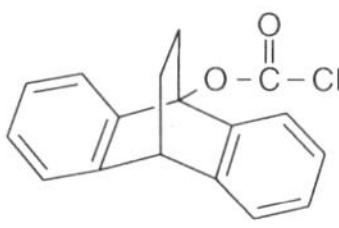
250 mg

S85,319-4
9,10-DIHYDRO-9,10-ETHANOANTHRACENE-11-METHANOL ACETATE
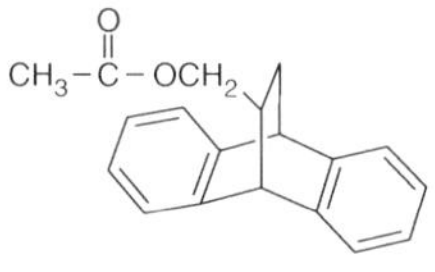
250 mg

S86,593-1
ETHYL 9,10-DIHYDRO-9,10-ETHANOANTHRACENE-11-CARBOXYLATE
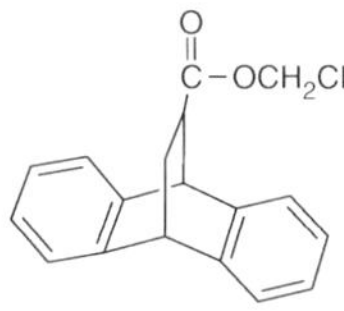
250 mg

S86,595-8
METHYL 9,10-DIHYDRO-11-METHYL-9,10-ETHANOANTHRACENE-11-CARBOXYLATE
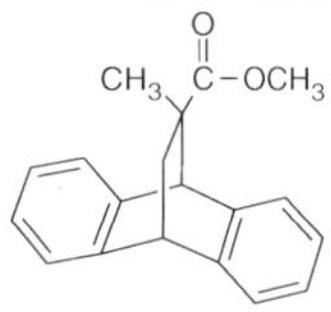
250 mg

S85,288-0
ETHYL 9,10-DIHYDRO-12-METHYL-9,10-ETHANOANTHRACENE-11-CARBOXYLATE
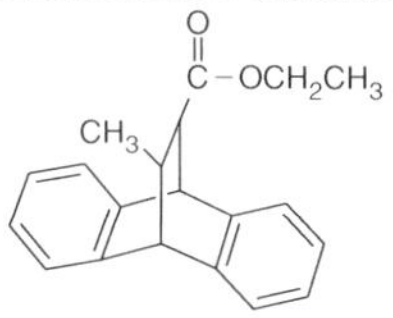
250 mg

S52,635-5
ALLYL 4-PHENYLISONIPECOTATE HYDROCHLORIDE
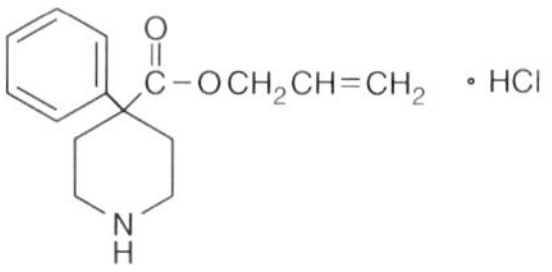
250 mg

S35,434-1
PHENETHYL FORMATE
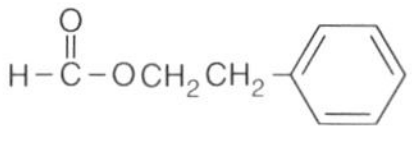
5 gm

S95,334-2
GLYCINE BENZYL ESTER HYDROCHLORIDE
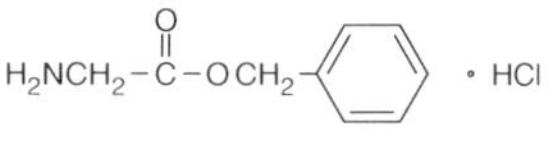
50 mg

S79,253-5
(+/-)-METHYL ALPHA-CHLOROPHENYLACETATE
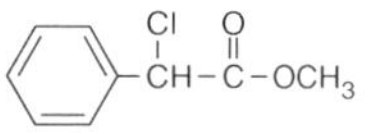
250 mg

S43,287-3
CINNAMYL FORMATE
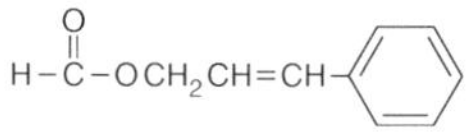
5 gm

S62,046-7
METHYL 3-PHENYLPROPIONATE
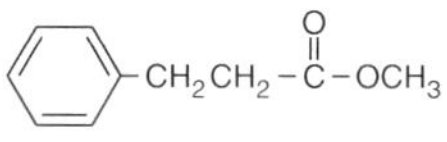
5 gm

S39,941-8
ETHYL ALPHA-BROMOPHENYLACETATE
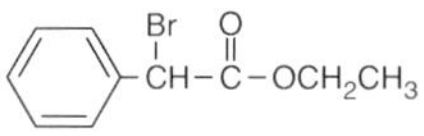
250 mg

S87,718-2
DL-1-PHENYLALLYL ACETATE
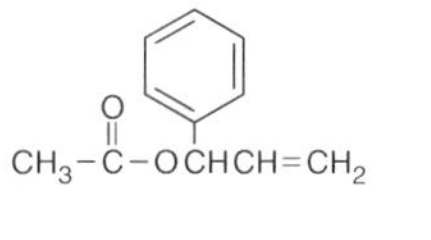
250 mg

S35,437-6
PHENETHYL PROPIONATE
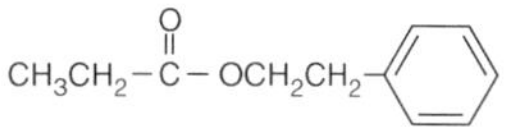
5 gm

S35,470-8
3-PHENYLPROPYL ACETATE
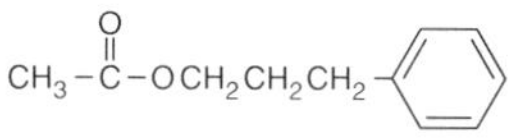
5 gm

S52,832-3
BETA-METHYLPHENETHYL ACETATE
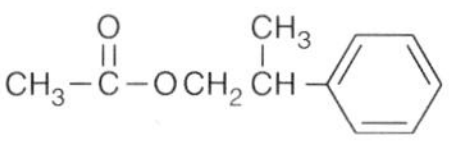
5 gm

S67,792-2
ETHYL 3-PHENYLPROPIONATE
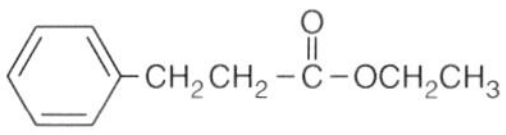
250 mg

S79,885-1
METHYL 2-PHENYLBUTYRATE
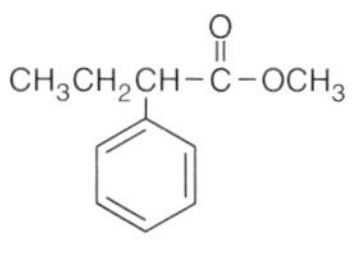
250 mg

S52,846-3
2-PHENOXYETHYL PROPIONATE
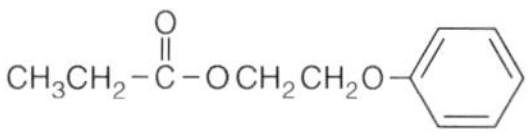
5 gm

S61,672-9
ETHYL 2,3-DIBROMO-3-PHENYLPROPIONATE
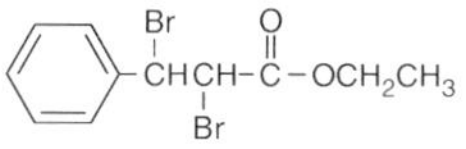
5 gm

S34,954-2
ETHYL 4-PHENYLBUTYRATE
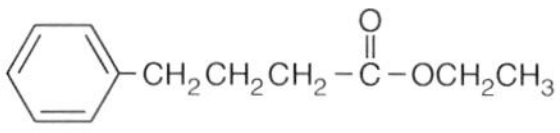
5 gm

S36,187-9
ETHYL 2-PHENYLBUTYRATE
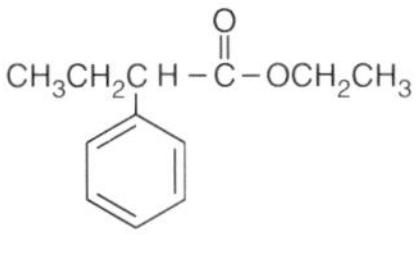
5 gm

S43,134-6
METHYL 5-PHENYLVALERATE
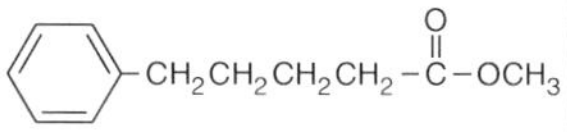
5 gm

S46,057-5
1-METHYL-3-PHENYLPROPYL ACETATE
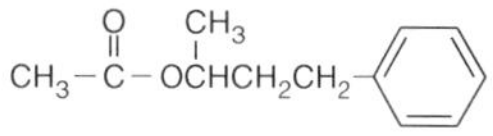
5 gm

S52,802-1
ALPHA,ALPHA-DIMETHYLPHENETHYL ACETATE
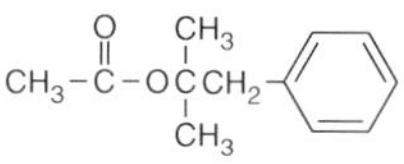
5 gm

S52,852-8
PHENETHYL ISOBUTYRATE
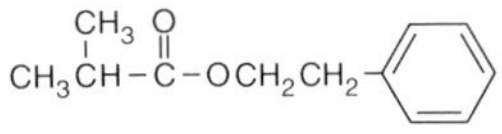
5 gm

S52,854-4
HYDROCINNAMYL PROPIONATE
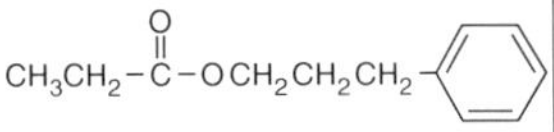
5 gm

S57,081-8
ISOBUTYL PHENYLACETATE
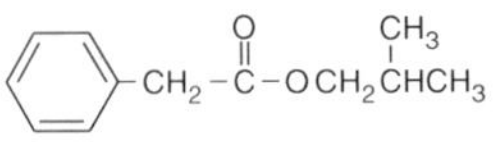
5 gm

S87,627-5
DL-METHYL 3-METHYL-2-PHENYLBUTYRATE
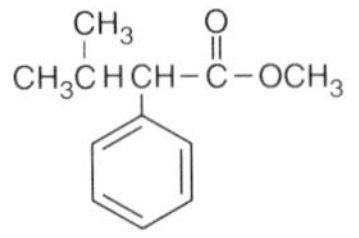
250 mg

S88,891-5
ISOPROPYL ALPHA,BETA-DIBROMOHYDROCINNAMATE
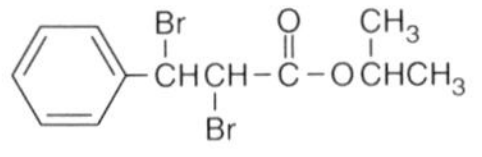
250 mg

S36,754-0
CINNAMYL BUTYRATE
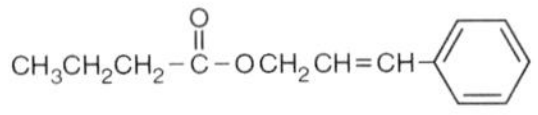
5 gm

S35,165-2
ISOPENTYL PHENYLACETATE
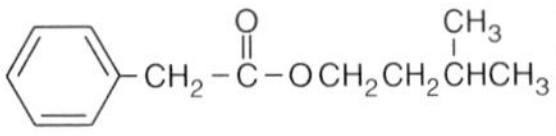
5 gm

S50,935-3
PENTYL PHENYLACETATE
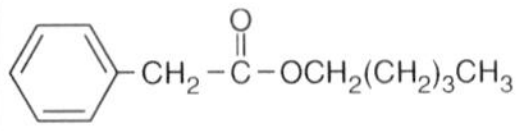
5 gm

S52,826-9
ALPHA,ALPHA-DIMETHYLHYDROCINNAMYL ACETATE
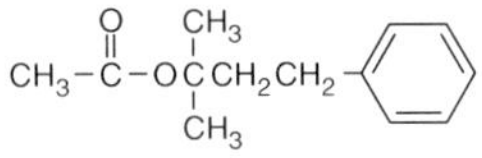
5 gm

S35,150-4
ISOAMYL ALPHA-BROMOPHENYLACETATE
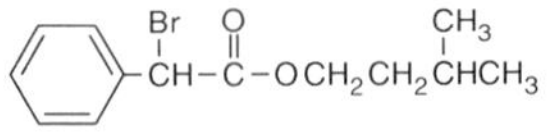
5 gm

S50,941-8
PHENETHYL HEXANOATE
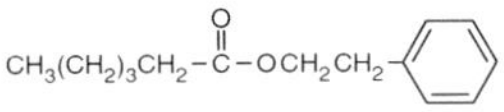
5 gm

S52,811-0
ALPHA,ALPHA-DIMETHYLPHENETHYL BUTYRATE
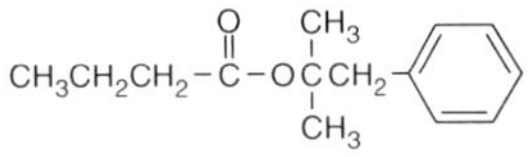
5 gm

S62,041-6
ETHYL 3-PHENYLHEPTANOATE
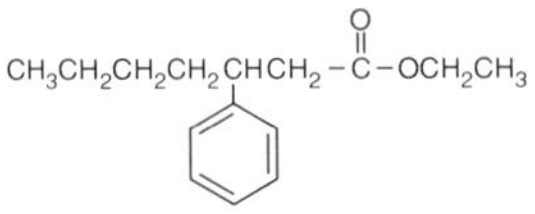
5 gm

S88,424-3
3-PHENYLALLYL 2-BROMOHEXANOATE
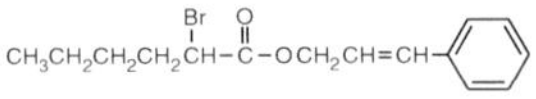
250 mg

S35,436-8
PHENETHYL PHENYLACETATE
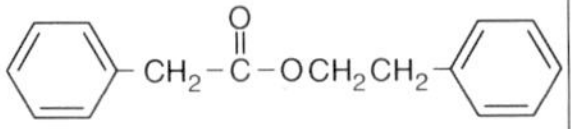
5 gm

S79,254-3
(+/-)-1-PHENYLETHYL PHENYLACETATE
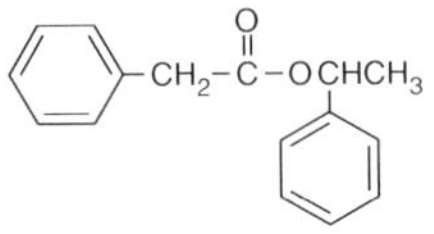
250 mg

S52,816-1
2-BENZYLIDENEHEPTYL ACETATE
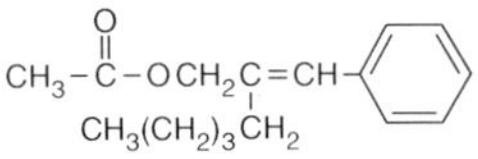
5 gm

S40,149-8
PHENYL 10-UNDECENOATE
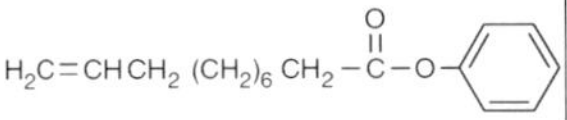
250 mg

S92,999-9
1,2-DIPHENYL-2-OXOETHYL 5-BROMOVALERATE
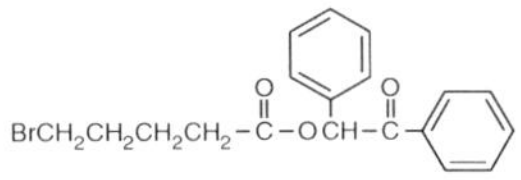
250 mg

S89,322-6
METHYL 2-CHLORO-2,2-DIPHENYLACETATE
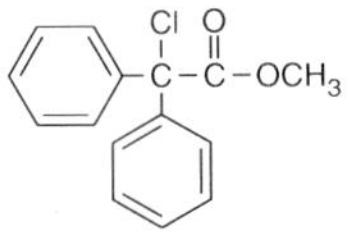
250 mg

S87,687-9
METHYL 2-BROMO-2,2-DIPHENYLACETATE
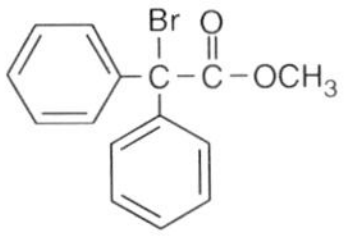
250 mg

S38,369-4
ETHYL DIPHENYLACETATE
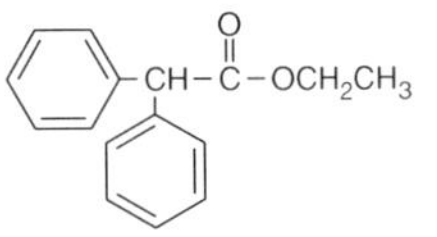
5 gm

S87,751-4
ETHYL 2-BROMO-2,2-DIPHENYLACETATE
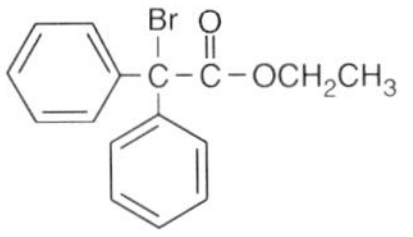
250 mg

S87,740-9
METHYL 2,2,2-TRIPHENYLACETATE
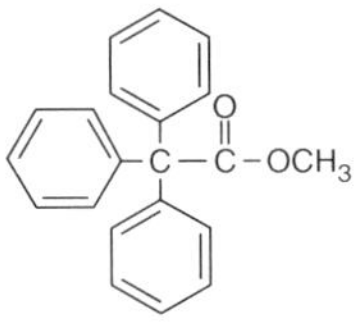
250 mg

S87,739-5
ETHYL 2,2,2-TRIPHENYLACETATE
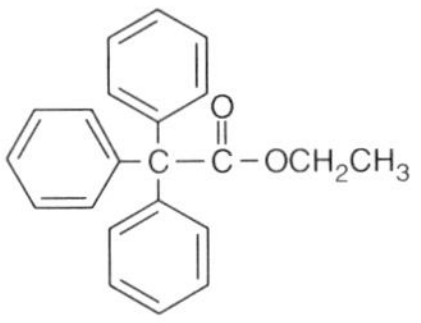
250 mg

S75,941-4
METHYL 1-PHENYL-1,2,3,4-TETRAHYDRO-3-ISOQUINOLINECARBOXYLATE HYDROCHLORIDE
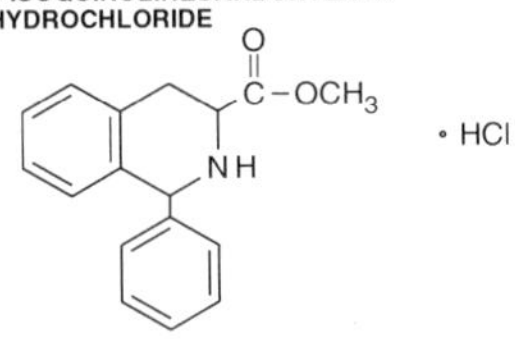
250 mg

S52,849-8
4-METHYLBENZYL ACETATE
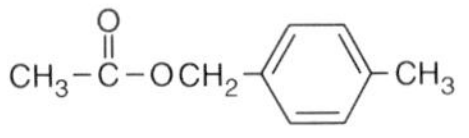
5 gm

S52,808-0
4-ISOPROPYLBENZYL ACETATE
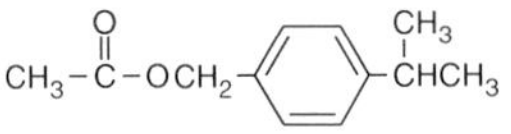
5 gm

S40,150-1
4-CHLOROPHENYL 10-UNDECENOATE
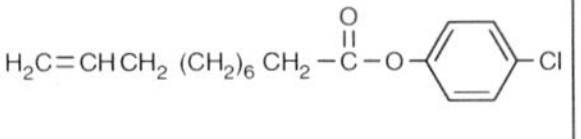
250 mg

S42,315-7
ALPHA,4-DIMETHYLBENZYL 3-METHYL-2-PHENYLVALERATE
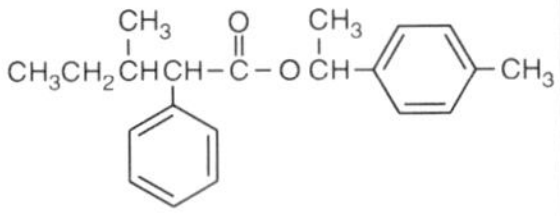
250 mg

S36,551-3
2,6-DICHLOROBENZYL ACETATE
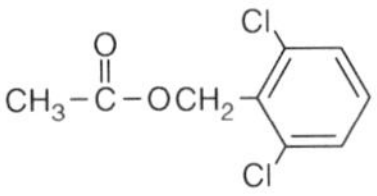
250 mg

S64,368-8
METHYL 2,4-DICHLOROPHENYLACETATE
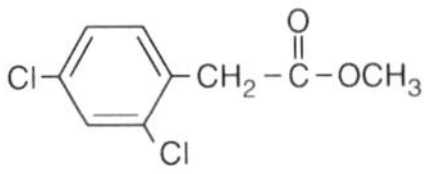
5 gm

S85,276-7
9-CHLORO-1,2,3,4-TETRAHYDRO-1,4-METHANO-2-NAPHTHYL ACETATE
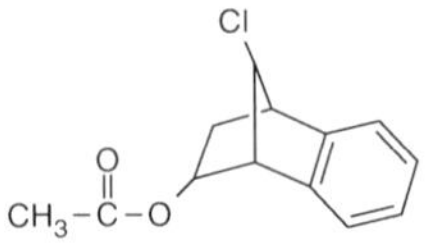
250 mg

S85,234-1
ETHYL 10,11-DIHYDRO-5,10-ETHANO-13-METHYL-5H-DIBENZO(A,D)-CYCLOHEPTENE-1,2-CARBOXYLATE
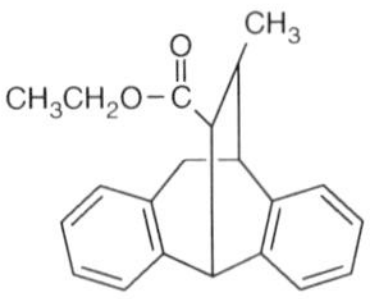
250 mg

S66,324-7
METHYL 5,6,7,8-TETRAHYDRO-2-NAPHTHALENEBUTYRATE
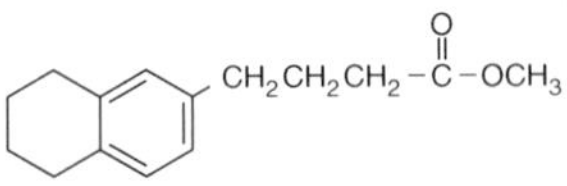
5 gm

S66,326-3
ETHYL 2-(3,4-DIHYDRO-5,8-DIMETHYL-1-NAPHTHYL)PROPIONATE
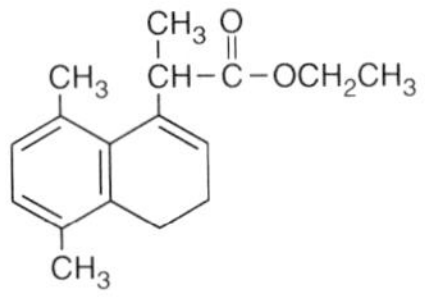
5 gm

S66,340-9
ETHYL 2-(5,8-DIMETHYL-1,2,3,4-TETRAHYDRO-1-NAPHTHYL)PROPIONATE
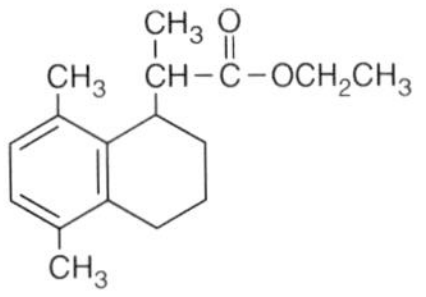
5 gm

S40,853-0
ETHYL 2-NAPHTHYLACETATE
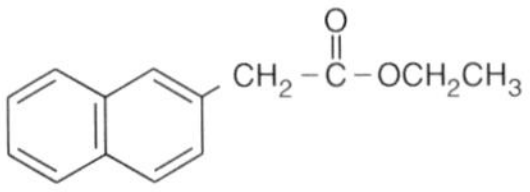
5 gm

S56,358-7
ETHYL 1-NAPHTHALENEACETATE
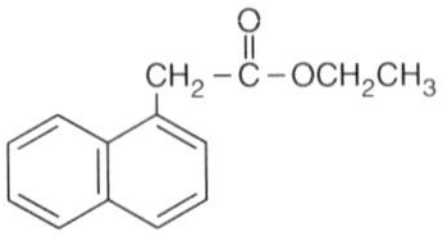
5 gm

S66,465-0
METHYL 4-(2-NAPHTHYL)BUTYRATE
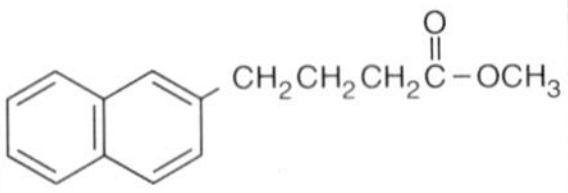
5 gm

S97,439-0
9-FLUORO-12-METHYL-7-BENZ(A)-ANTHRACENYLMETHYL ACETATE
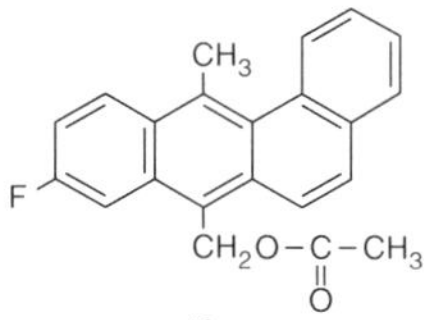
10 mg

S40,852-2
ETHYL 3-METHYL-1,4-BENZODIOXANE-2-CARBOXYLATE
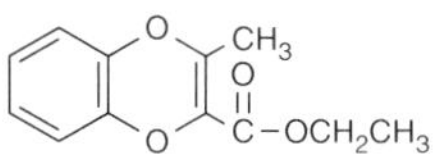
5 gm

S90,627-1
ETHYL 6,7,8,9-TETRAHYDRO-5H-BENZOCYCLOHEPTENE-7-CARBOXYLATE
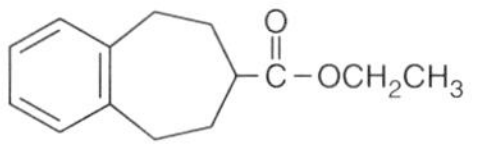
250 mg

S90,634-4
METHYL 4-(6,7,8,9-TETRAHYDRO-5H-BENZOCYCLOHEPTEN-5-YL)BUTYRATE
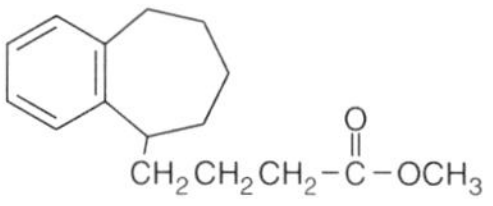
250 mg

S73,500-0
ETHYL 2,2-DIPHENYLACETIMIDATE HYDROCHLORIDE
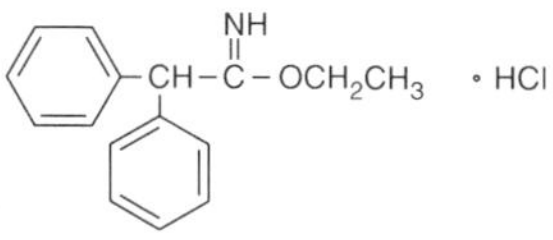
250 mg

S62,220-6
ETHYL 2,4-BIS(2-CHLORO-1,1,2-TRIFLUOROETHOXY)BENZOYLACETATE
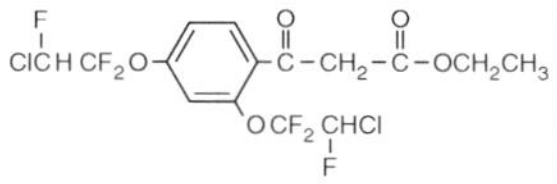
250 mg

S80,674-9
METHYL 7-[3-(4-FLUOROPHENYL)-1-ISOPROPYL-2-INDOLINYL]-5-HYDROXY-3-OXO-6-HEPTENOATE
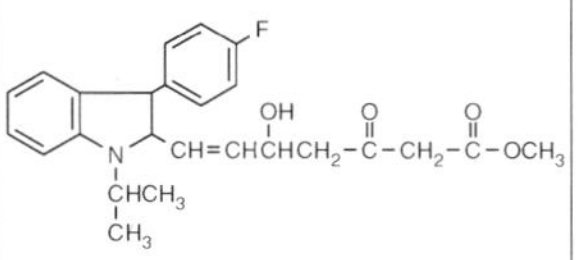
250 mg

S39,163-8
ETHYL O-IODOBENZOYLACETATE
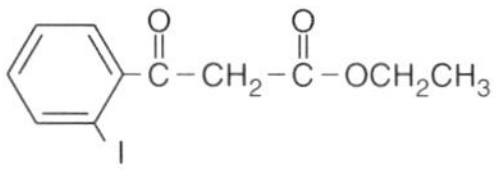
250 mg

S94,570-6
METHYL 4-CHLOROBENZOYLFORMATE
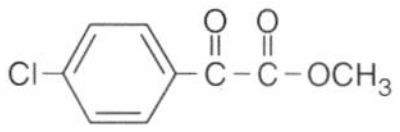
250 mg

S86,154-5
DIMETHYL 2,5-DIMETHOXY-B,B,B',B'-TETRAKIS(ETHOXYCARBONYL)-1,4-BENZENEDIBUTYRATE
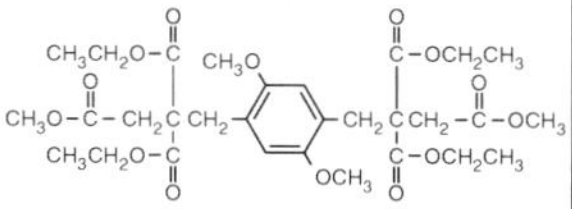
250 mg

S95,555-8
METHYL 2-METHOXY-2-(METHOXYCARBONYL)-3-OXO-1-INDOLINEACETATE
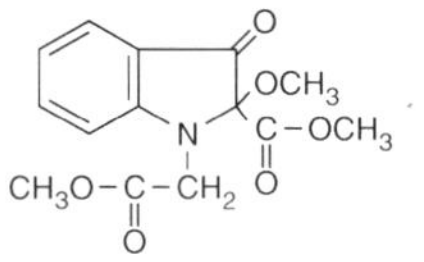
50 mg

S62,878-6
METHYL 2-(3-NITRO-4-[2-(4-TERT-OCTYLPHENOXY)ETHOXY]BENZOYL)-ACETATE
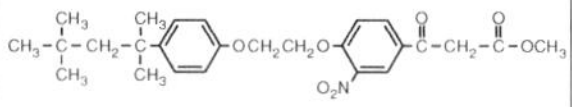
5 gm

S50,545-5
2-METHOXYETHYL-4-NITROBENZOYLACETATE
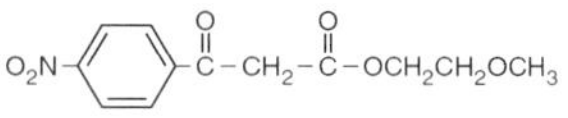
250 mg

S59,801-1
METHYL 4-(2-PHENOXYETHOXY)-BENZOYLACETATE
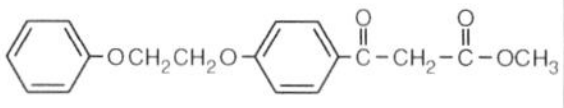
5 gm

S50,539-0
2-METHOXYETHYL 4-(HEXADECYLOXY)-BENZOYLACETATE
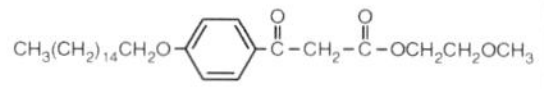
250 mg

S58,062-7
METHYL 4-ETHOXY-3-METHOXYBENZOYLACETATE
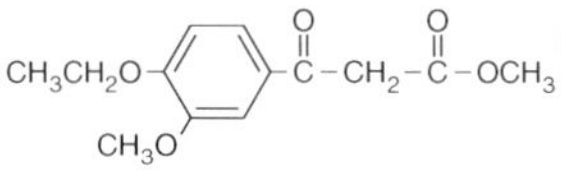
250 mg

S63,098-5
METHYL 2-(4-HEXADECYLOXY-3-METHOXYBENZOYL)ACETATE
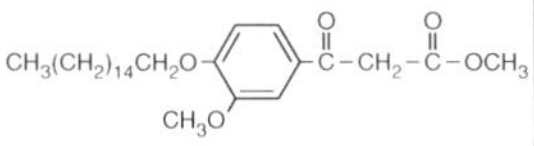
250 mg

S94,468-8
ETHYL 4,4-DIMETHYL-7-METHOXY-1-OXO-1,2,3,4-TETRAHYDRO-2-NAPHTHALENECARBOXYLATE
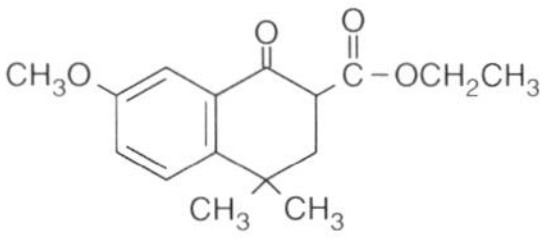
250 mg

S94,469-6
ETHYL 4,4-DIMETHYL-6-METHOXY-1-OXO-1,2,3,4-TETRAHYDRO-2-NAPHTHALENECARBOXYLATE
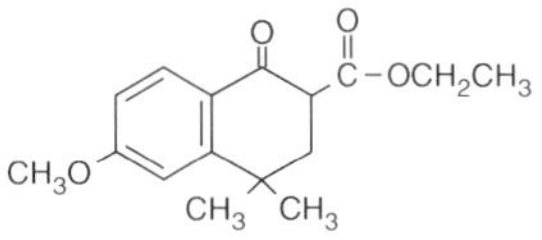
250 mg

S94,208-1
ETHYL 6-METHOXY-1-OXO-1,2,3,4-TETRAHYDRO-2,4,4-TRIMETHYL-2-NAPHTHALENECARBOXYLATE
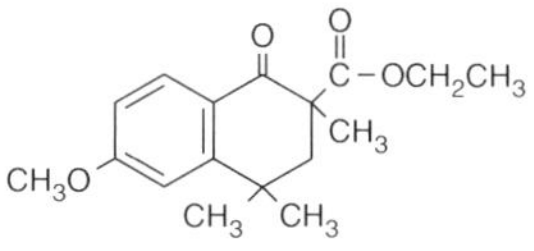
250 mg

S60,400-3
METHYL 2-(4-PENTYLOXYBENZOYL)-ACETATE
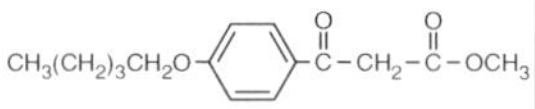
250 mg

S62,851-4
ETHYL 2-[4-(DODECYLOXY)BENZOYL]-ACETATE
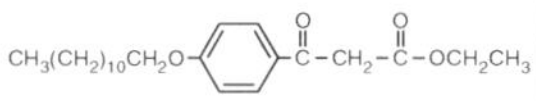
250 mg

S88,428-6
ETHYL 5-(2-METHOXYPHENYL)-3-OXOPENTANOATE
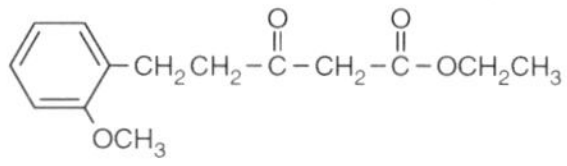
250 mg

S88,423-5
ETHYL 5-(2-METHOXYPHENYL)-2-METHYL-3-OXOPENTANOATE
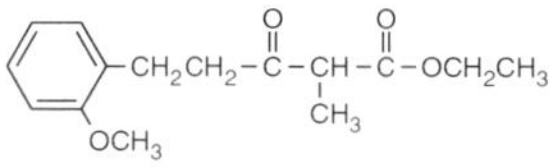
250 mg

S94,578-1
DIETHYL 2,2'-(4-METHOXYBENZYLIDENE)-BISACETOACETATE
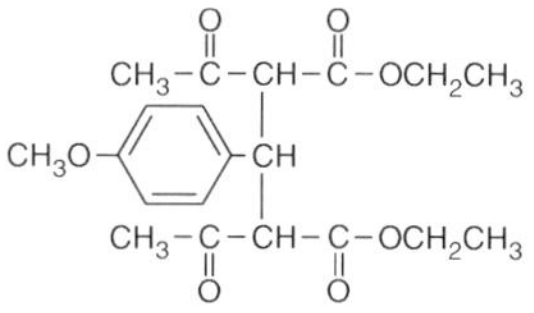
250 mg

S55,004-3
METHYL 2-(ISOHEXADECYLOXY)-BENZOYLACETATE
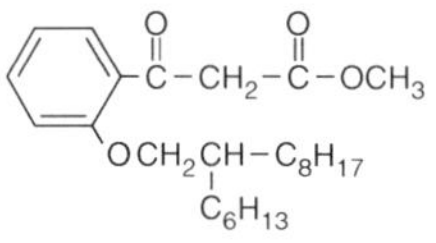
250 mg

S55,007-8
METHYL 2-(HEXADECYLOXY)-BENZOYLACETATE
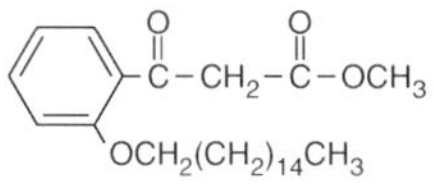
5 gm

S55,455-3
METHYL 4-(DODECYLOXY)-BENZOYLACETATE
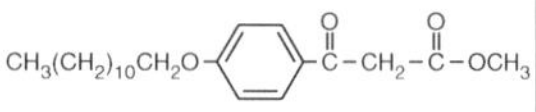
250 mg

S51,436-5
ETHYL M-ANISOYLACETATE
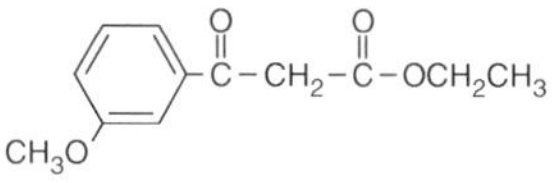
250 mg

S51,628-7
METHYL 4-PHENOXYBENZOYLACETATE
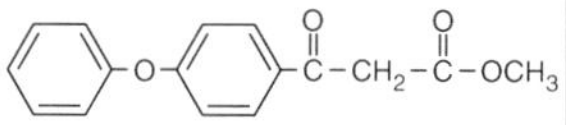
250 mg

S51,074-2
ETHYL 4-(PENTADECYLOXY)-BENZOYLACETATE
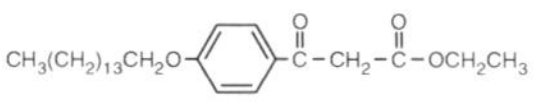
5 gm

S54,629-1
ETHYL 4-(PENTYLOXY)BENZOYLACETATE
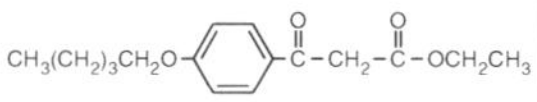
5 gm

S54,627-5
METHYL 4-(HEXADECYLOXY)-BENZOYLACETATE
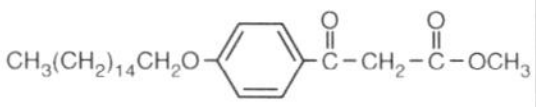
5 gm

S59,171-8
ALPHA-METHYLBENZYL 3-OXOBUTYRATE
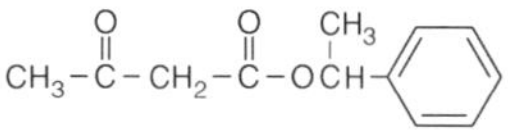
5 gm

S91,838-5
ETHYL 2-(9-ANTHRACENYLMETHYL)-2,3-DIHYDRO-3-OXO-2-BENZOFURANCARBOXYLATE
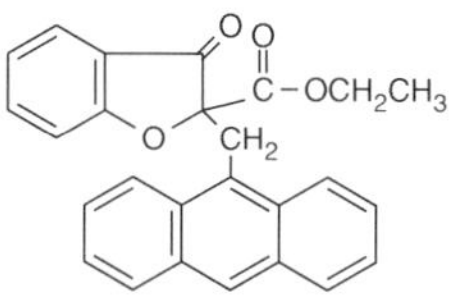
250 mg

S94,215-4
ETHYL 4,4-DIMETHYL-1-OXO-1,2,3,4-TETRAHYDRO-2-NAPHTHALENEGLYOXYLATE
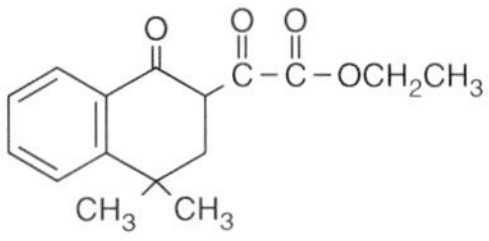
250 mg

S94,214-6
ETHYL 4,4-DIMETHYL-1-OXO-1,2,3,4-TETRAHYDRO-2-NAPHTHALENECARBOXYLATE
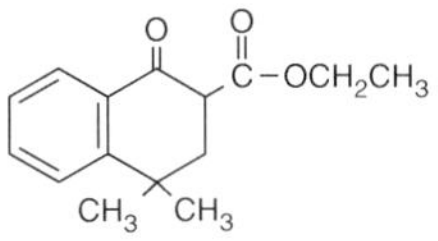
250 mg

S51,753-4
ETHYL (3,4,5-TRIMETHOXYBENZOYL)-ACETATE
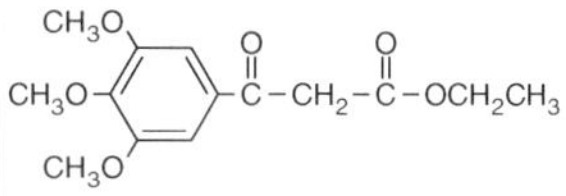
250 mg

S95,433-0
ETHYL 2-ACETYL-3-(BENZYLTHIO)-3,3-DIMETHYLPROPIONATE
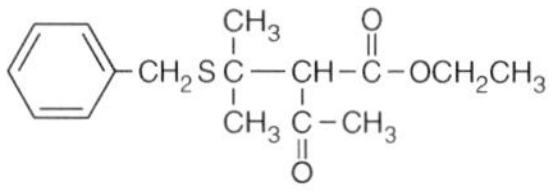
50 mg

S88,224-0
DIETHYL 2-(BENZOYLMETHYL)-2-ETHYLMALONATE
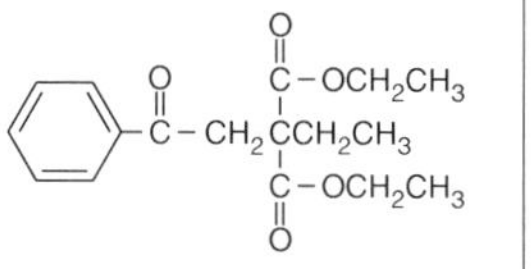
250 mg

S94,270-7
METHYL 5-CHLORO-3(2H)OXOBENZO(B)-THIOPHENE-2-CARBOXYLATE
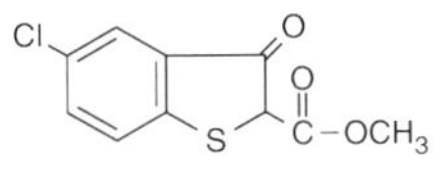
250 mg

S93,473-9
ETHYL 1-(6-BROMOPIPERONYL)-7-CHLORO-1,4-DIHYDRO-6-FLUORO-4-OXO-3-QUINOLINECARBOXYLATE
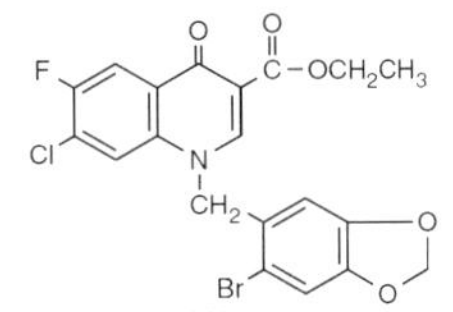
250 mg

S93,474-7
ETHYL 5(6-BR-PIPERONYL)5,8-DIHYDRO-8-OXO-1,3-DIOXOLO(4,5G)QUINOLIN-7-CARBOXYLATE
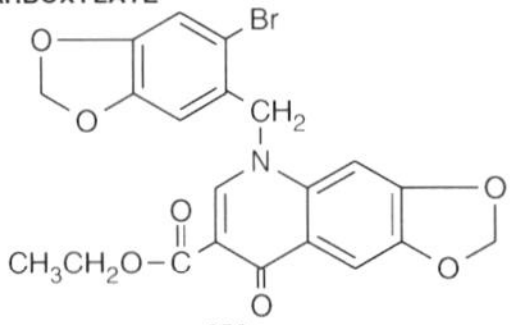
250 mg

S94,210-3
ETHYL 1-OXO-1,2,3,4-TETRAHYDRO-2,4,4-TRIMETHYL-2-NAPHTHALENECARBOXYLATE
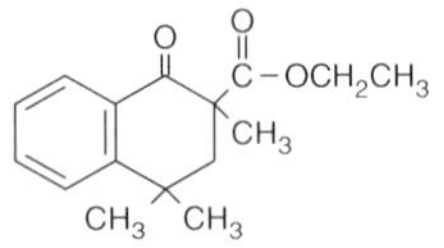
250 mg

S65,606-2
ETHYL 2-BENZOYL-2-(4-NITROPHENYLTHIO)ACETATE
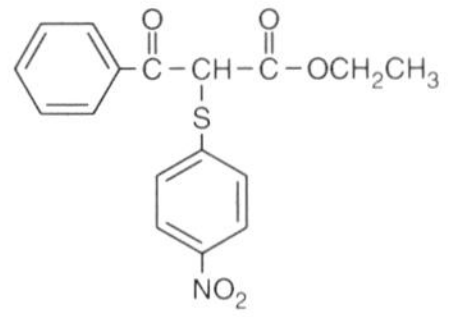
250 mg

S75,651-2
ETHYL 3-OXO-4-(PHENYLTHIO)BUTYRATE
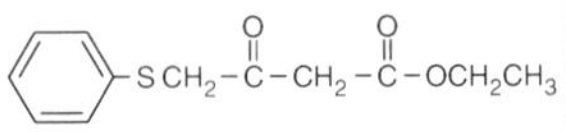
5 gm

S53,726-8
ETHYL 4-CHLORO-3-NITROBENZOYLACETATE
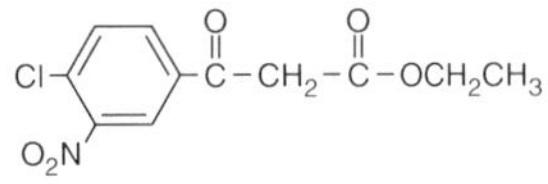
5 gm

S51,680-5
ETHYL 2-ACETYL-2-(4-CYCLOHEXYLBENZOYL)ACETATE
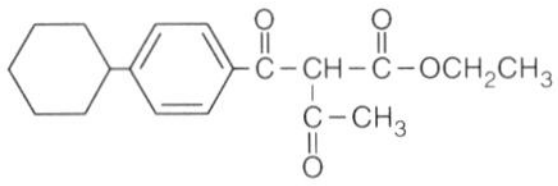
250 mg

S67,214-9
ALLYL BENZOYLACETATE
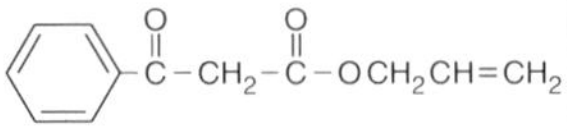
5 gm

S56,663-2
ETHYL 3-OXO-5-PHENYL-4-PENTENOATE
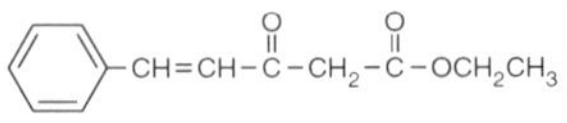
5 gm

S65,472-8
METHYL 4-(TERT-BUTYL)BENZOYLACETATE
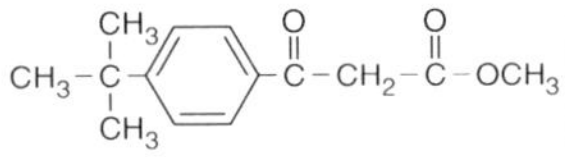
250 mg

S35,559-3
TEREPHTHALOYLBIS(ETHYL ACETATE)
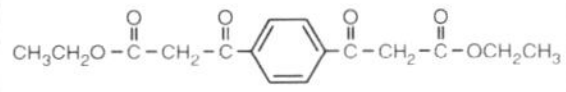
5 gm

S80,691-9
ETHYL 3-OXO-3-(3,5-XYLYL)PROPIONATE
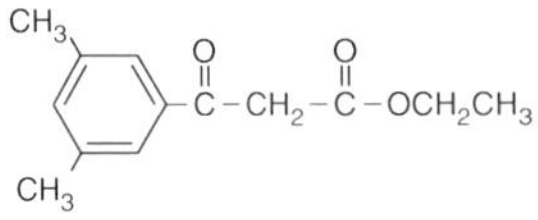
250 mg

S88,480-4
DIETHYL 3-OXO-2-PHENYLADIPATE
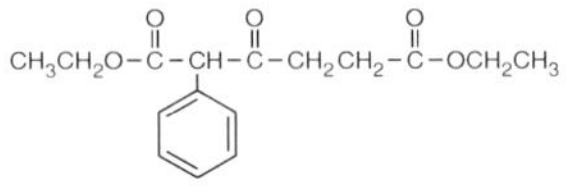
250 mg

S89,726-4
METHYL 2,4-DIOXO-6-PHENYL-1-CYCLOHEXANECARBOXYLATE
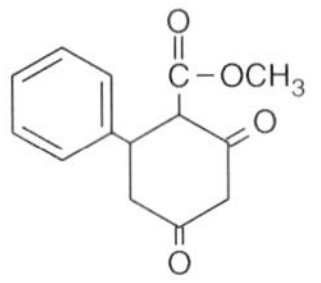
250 mg

S89,720-5
ETHYL 2,4-DIOXO-6-PHENYL-1-CYCLOHEXANECARBOXYLATE
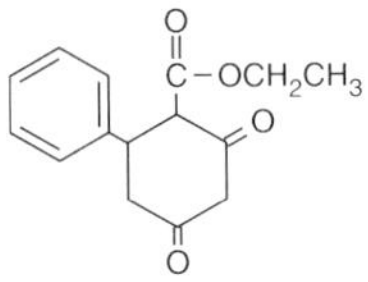
250 mg

S51,432-2
METHYL M-TOLUOYLACETATE
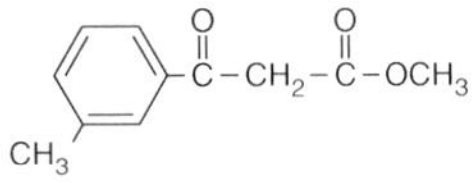
250 mg

S50,947-7
ETHYL 2,2-DIALLYLBENZOYLACETATE
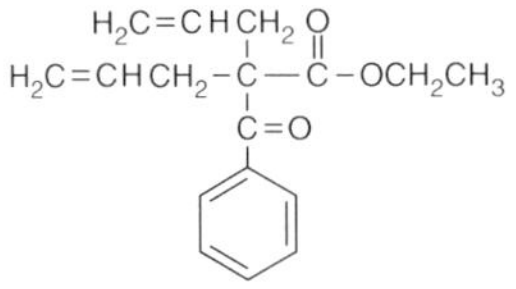
250 mg

S51,278-8
ETHYL 2-BENZOYLOCTADECANOATE
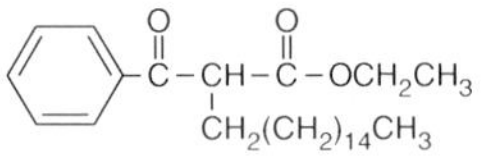
250 mg

S51,282-6
ETHYL 4-CHLOROBENZOYLACETATE
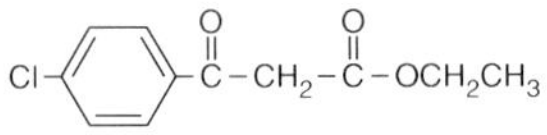
250 mg

S51,689-9
ETHYL 2-ACETYL-2-(5,6,7,8-TETRAHYDRO-2-NAPHTHOYL)ACETATE
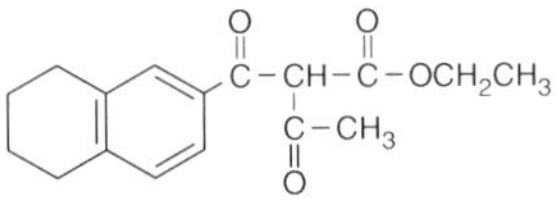
250 mg

S89,338-2
ETHYL 1-CARBOXY-2-PHENYL-1-CYCLOPROPANECARBOXYLATE, POTASSIUM SALT
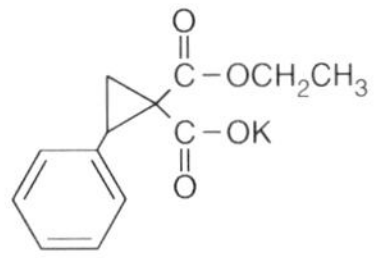
250 mg

S53,767-5
METHYL 4-DODECYLBENZOYLACETATE
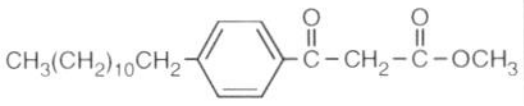
5 gm

S95,386-5
DIMETHYL-4-CHLOROBENZYLIDENEMALONATE
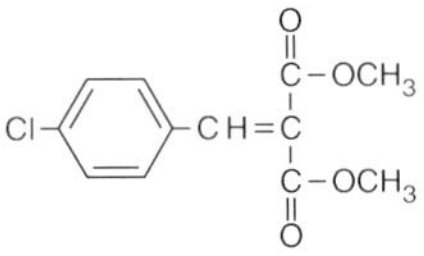
50 mg

S38,230-2
DIETHYL (4-CHLOROBENZYLIDENE)-MALONATE
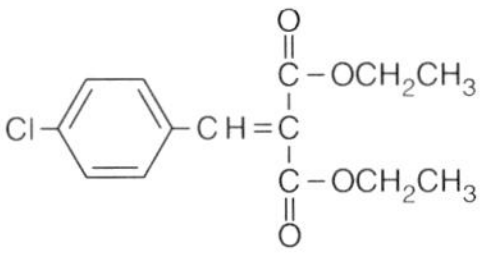
5 gm

S88,067-1
METHYL 4-CHLOROCINNAMATE
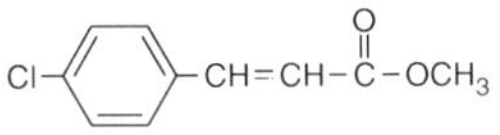
250 mg

S93,167-5
METHYL 4-BROMOCINNAMATE
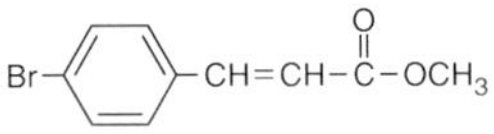
250 mg

S75,653-9
METHYL 3-HYDROXY-ALPHA-MERCAPTO-BETA-METHYLCINNAMATE
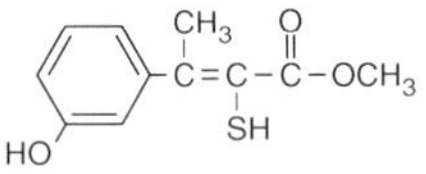
5 gm

S88,070-1
METHYL 4-HYDROXYCINNAMATE
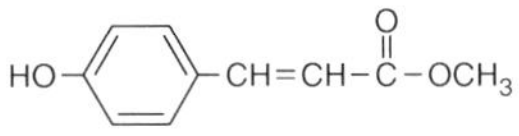
250 mg

S66,110-4
DIETHYL 2-(3,4-DIMETHOXYBENZYLIDENE)MALONATE
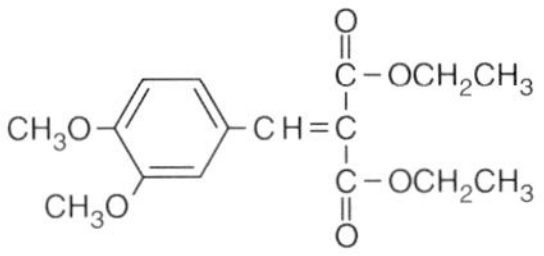
5 gm

S95,417-9
DIMETHYL 2-(DIMETHOXYBENZYLIDENE)-MALONATE
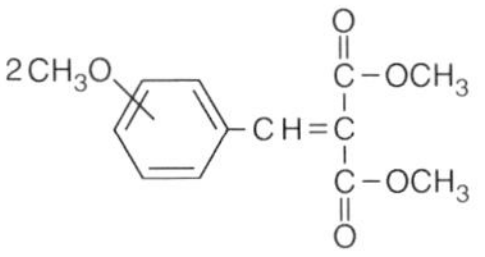
50 mg

S67,106-1
ETHYL 2-[1-(3-METHOXYPHENYL)-ETHYLIDENE]SUCCINATE
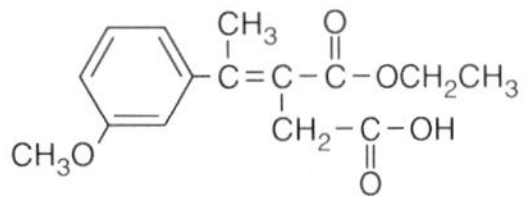
5 gm

S76,561-9
ETHYL 2-(4-METHOXYBENZYLIDENE)-ACETOACETATE
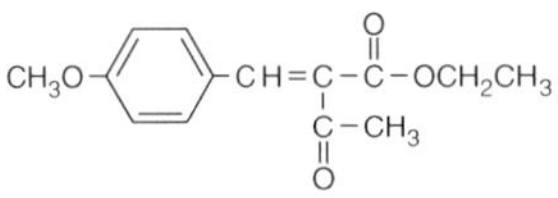
5 gm

S67,375-7
4-METHOXYPHENYL CINNAMATE
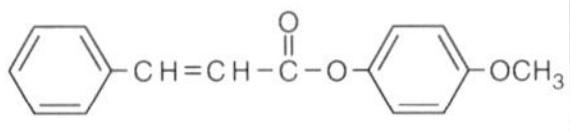
250 mg

S83,875-6
ETHYL ALPHA-(PHENYLTHIO)CINNAMATE
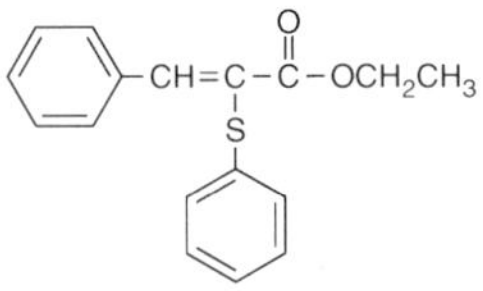
250 mg

S87,771-9
METHYL 4-(METHYLTHIO)CINNAMATE
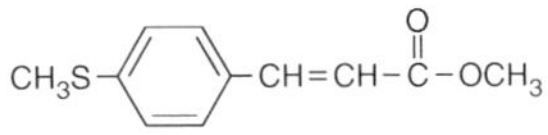
250 mg

S88,059-0
METHYL 4-AMINO-BETA-METHYLCINNAMATE
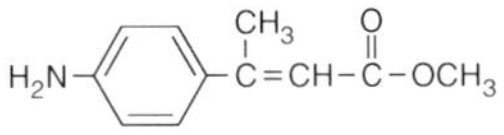
250 mg

S95,342-3
METHYL 2-AMINOCINNAMATE
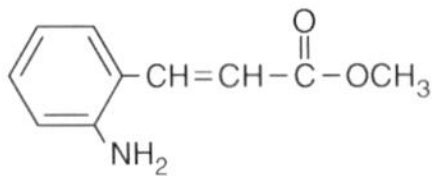
50 mg

S88,215-1
METHYL 4-(DIMETHYLAMINO)CINNAMATE
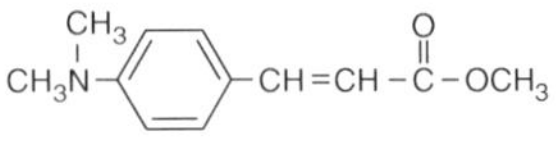
250 mg

S93,361-9
DIMETHYL 2-(1,2-DIPHENYLHYDRAZINO)-MALEATE
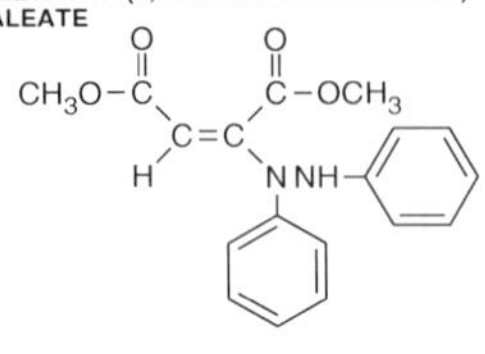
250 mg

S92,972-7
ETHYL ALPHA-BENZOYL-4-NITROCINNAMATE

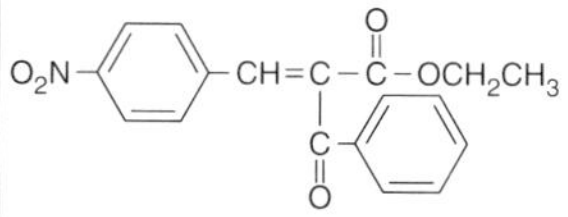

250 mg

S67,910-0
ETHYL 3-NITROCINNAMATE

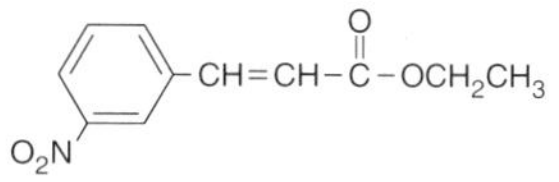

5 gm

S79,913-0
METHYL 4-NITROCINNAMATE

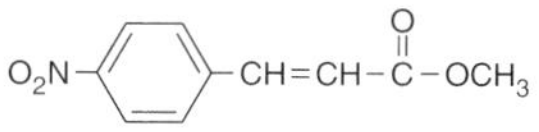

250 mg

S95,340-7
METHYL 2-NITROCINNAMATE

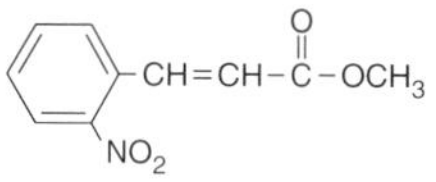

50 mg

S56,288-2
METHYL 3-NITROCINNAMATE

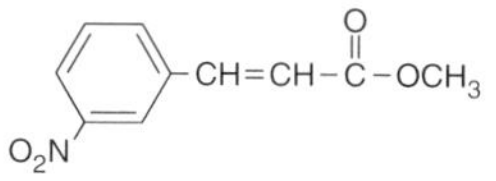

5 gm

S92,747-3
2-(5-NITRO-3-PHTHALIDYLIDENE)-2-PHENYLACETIC ACID

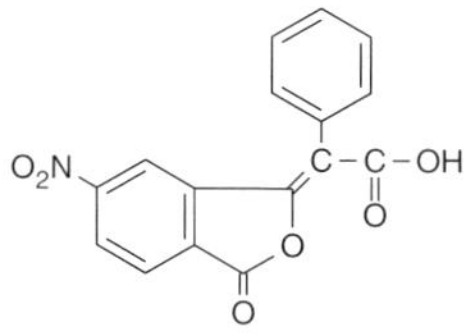

250 mg

S88,062-0
METHYL 3-FORMYLCINNAMATE

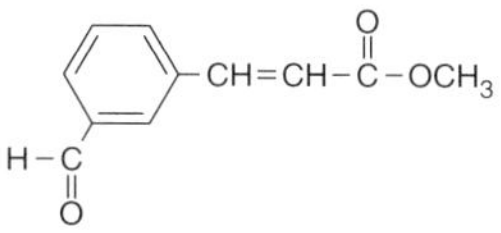

250 mg

S88,163-5
METHYL 4-FORMYLCINNAMATE

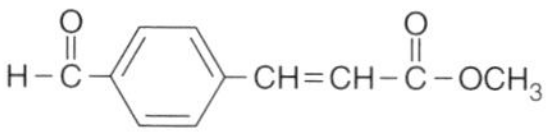

250 mg

S87,765-4
METHYL 2-FORMYLCINNAMATE

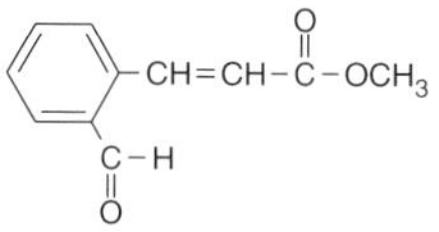

250 mg

S73,147-1
ETHYL 3-CINNAMOYLPYRUVATE

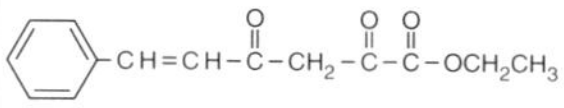

5 gm

S92,964-6
ETHYL 2-(4-METHYLBENZYLIDENE)-ACETOACETATE

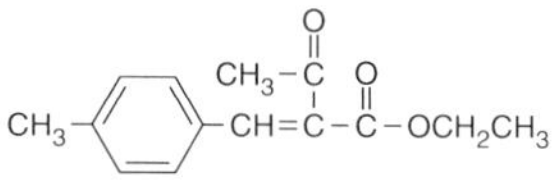

250 mg

S55,418-9
ETHYL 4-METHOXYCINNAMATE

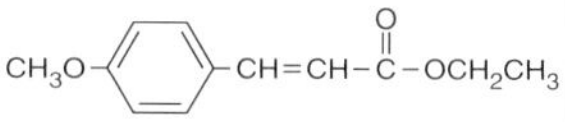

250 mg

S73,952-9
BUTYL 3-CINNAMOYLPYRUVATE

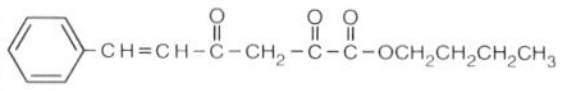

5 gm

S54,548-1
DIMETHYL BENZYLIDENEMALONATE

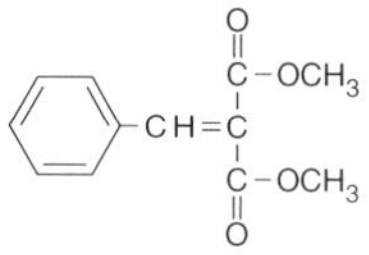

5 gm

S66,129-5
DIETHYL 2-CINNAMYLIDENEMALONATE

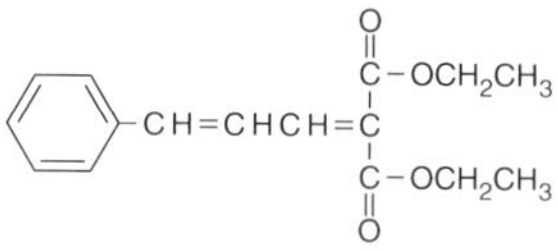

5 gm

S88,047-7
METHYL 2-(METHOXYCARBONYL)-CINNAMATE

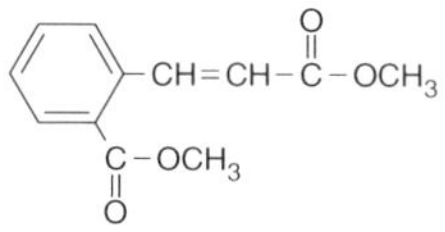

250 mg

S35,719-7
3-ACETOXYCINNAMIC ACID

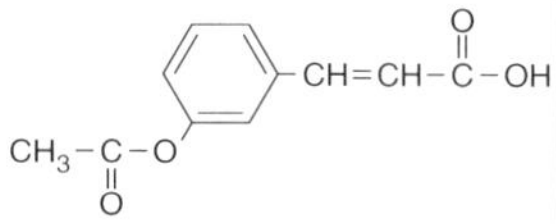

5 gm

S88,148-1
DIETHYL 2-PHENYLMALEATE

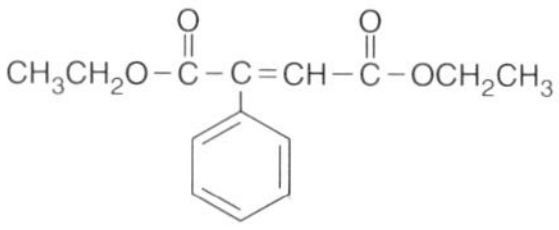

250 mg

S88,061-2
METHYL 4-CYANOCINNAMATE

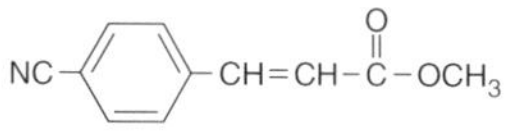

250 mg

S42,444-7
CINNAMOYLTROPINE HYDROCHLORIDE

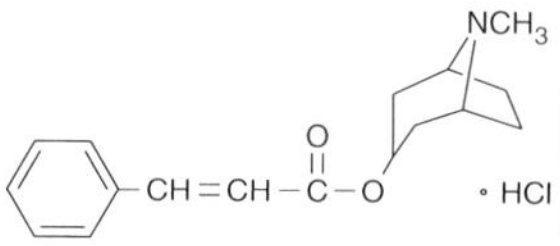

250 mg

S88,916-4
MENTHYL CINNAMATE

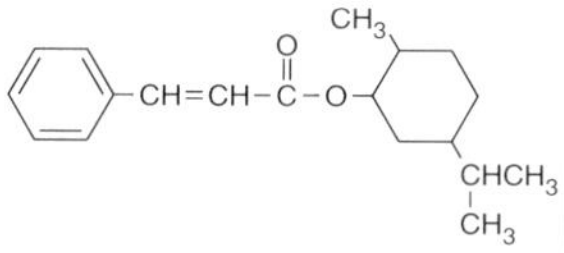

250 mg

S88,208-9
METHYL BETA-METHYLCINNAMATE

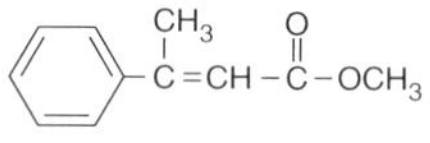

250 mg

S88,918-0
METHYL ALPHA-METHYLCINNAMATE

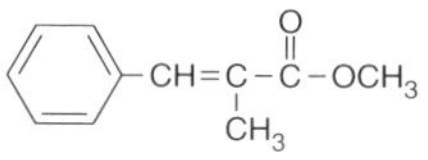

250 mg

S52,819-6
ALLYL CINNAMATE

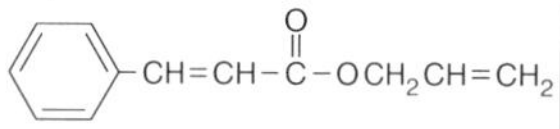

5 gm

S88,890-7
ISOPROPYL CINNAMATE

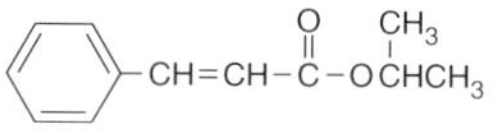

250 mg

S56,140-1
PENTYL CINNAMATE

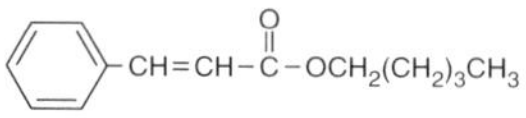

250 mg

S57,054-0
ISOPENTYL CINNAMATE

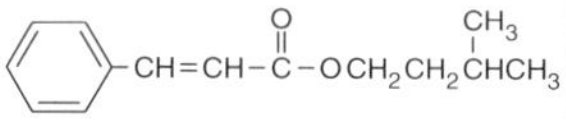

5 gm

S62,045-9
ETHYL 3-PHENYL-2-HEPTENOATE

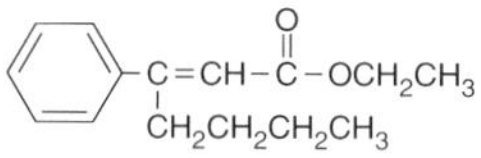

5 gm

S55,657-2
PHENETHYL CINNAMATE

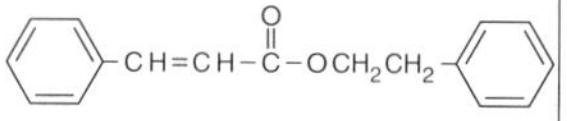

250 mg

S88,007-8
METHYL 4-STYRYLCINNAMATE

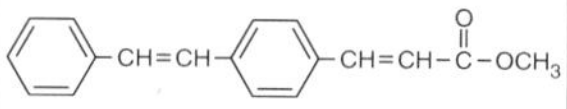

250 mg

S88,012-4
METHYL BETA-PHENYLCINNAMATE

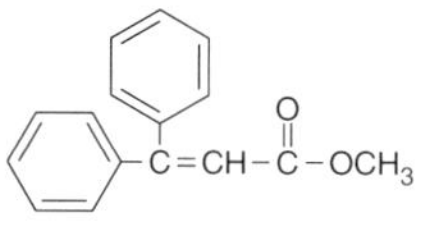

250 mg

S55,079-5
ETHYL 2,6-DICHLOROCINNAMATE

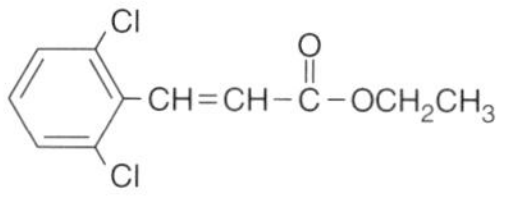

5 gm

S87,798-0
METHYL 2,5-DIISOPROPYLCINNAMATE

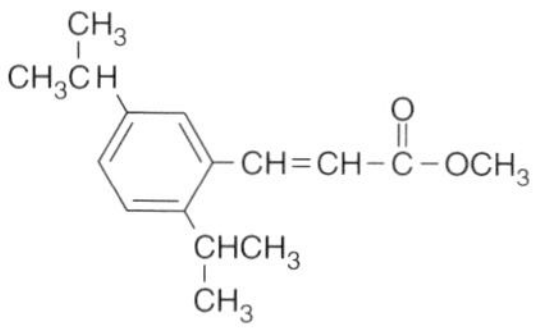

250 mg

S66,241-0
METHYL BETA-METHYL-1-PYRENEACRYLATE

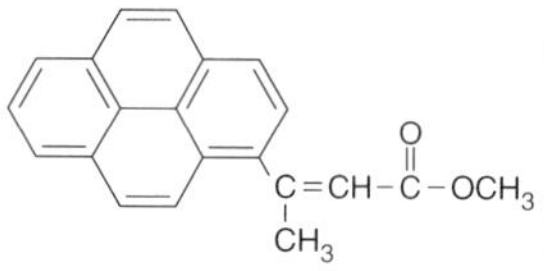

250 mg

S46,147-4
METHYL-4-(DIMETHOXYMETHYL)BENZOATE

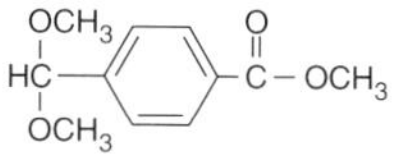

5 gm

S56,936-4
METHYL 4-(3-CHLOROPROPOXY)BENZOATE

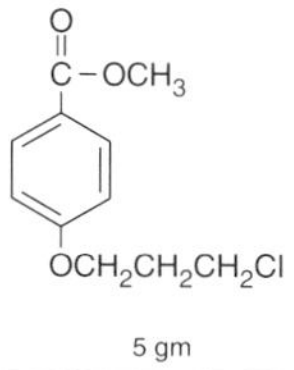

5 gm

S65,169-9
2-CHLOROETHYL 4-NITROBENZOATE

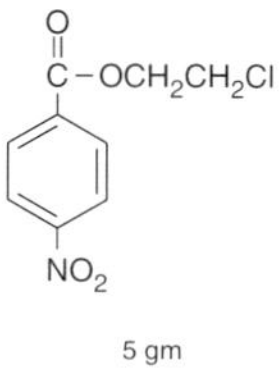

5 gm

S38,096-2
4-CHLOROBUTYL 4-NITROBENZOATE

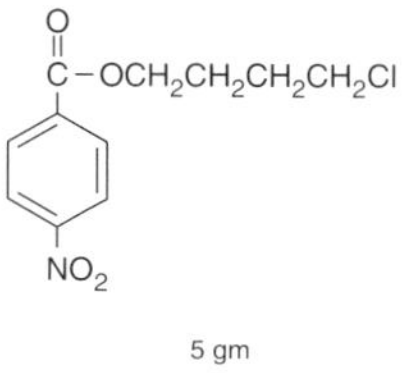

5 gm

S37,727-9
5-CHLOROPENTYL 4-NITROBENZOATE

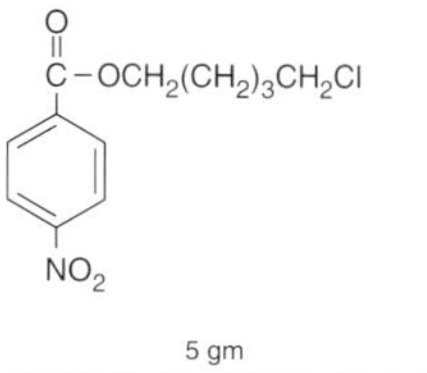

5 gm

S88,198-8
METHYL 3-(CHLOROMETHYL)BENZOATE

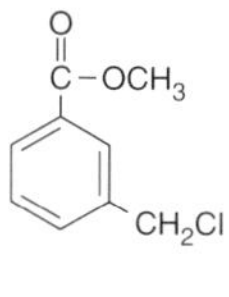

250 mg

S37,687-6
5-CHLOROPENTYL BENZOATE
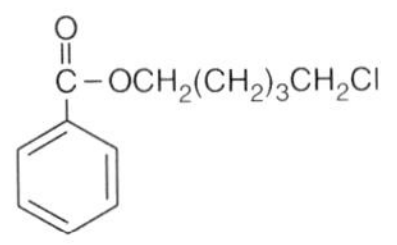
5 gm

S89,627-6
ETHYL 2-(CHLOROMETHYL)BENZOATE
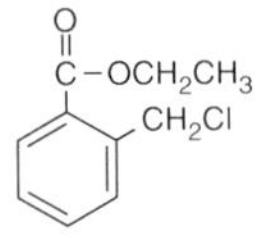
250 mg

S55,449-9
METHYL 2-(2-CHLORO-1,1,2-TRIFLUOROETHOXY)BENZOATE
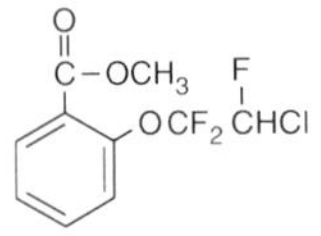
250 mg

S55,583-5
METHYL 4-(2-CHLORO-1,1,2-TRIFLUOROETHOXY)BENZOATE
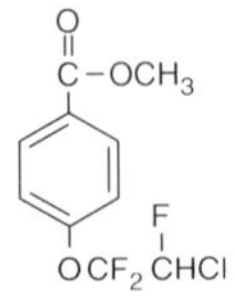
250 mg

S37,342-7
ETHYL ALPHA,ALPHA-DIBROMO-P-TOLUATE
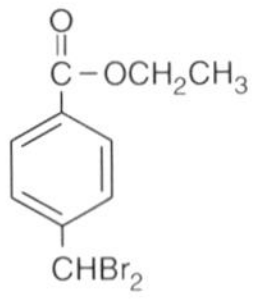
5 gm

S36,054-6
DIETHYL [(ALPHA,ALPHA,ALPHA-TRIFLUORO-M-TOLUIDINO)METHYLENE]-MALONATE
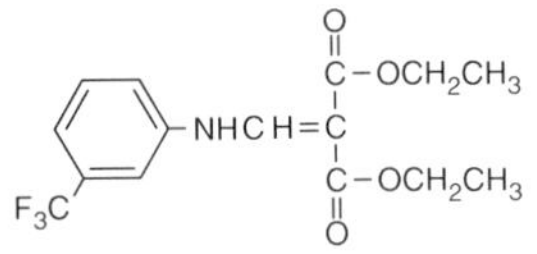
250 mg

S93,723-1
METHYL N-PHENYL-N-(TRIFLUOROACETYL)-ANTHRANILATE
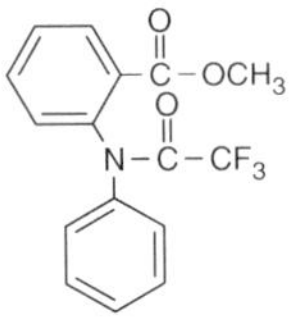
250 mg

S79,377-9
(+)-MONO-1-METHYL-2,2,2-TRIFLUOROETHYL PHTHALATE
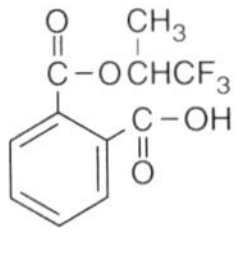
250 mg

S79,376-0
(-)-1-METHYL-2,2,2-TRIFLUOROETHYL BENZOATE
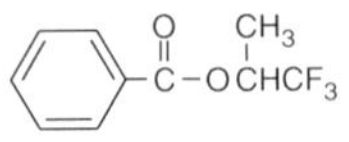
250 mg

S55,493-6
ETHYL A,A,A-TRIFLUORO-M-TOLUATE
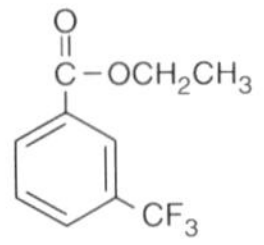
5 gm

S62,203-6
3-HYDROXYBUTYL TETRACHLOROPHTHALATE
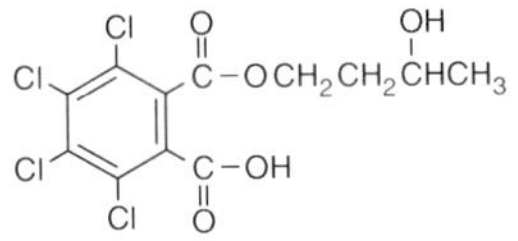
250 mg

S35,297-7
METHYL 5-IODOSALICYLATE
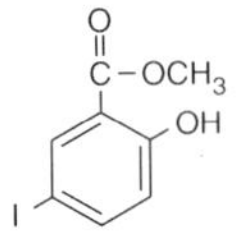
5 gm

S93,145-4
BUTYL 3,5-DICHLOROSALICYLATE
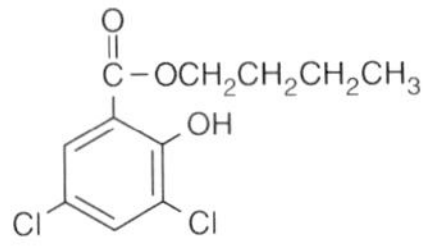
250 mg

S85,033-0
METHYL 2-BROMO-3,4,5-TRIMETHOXYBENZOATE
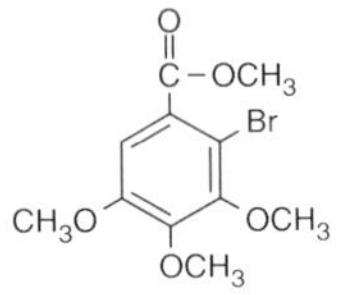
250 mg

S86,066-2
DIMETHYL 2,5-DIBROMO-3,6-DIMETHOXYTEREPHTHALATE
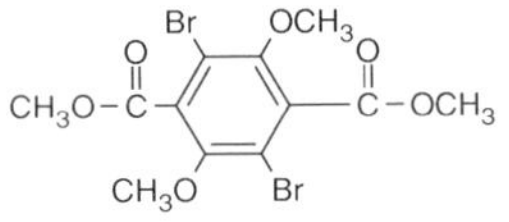
250 mg

S61,791-1
2-METHOXYETHYL TETRACHLOROPHTHALATE
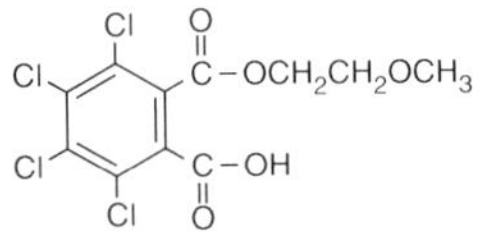
250 mg

S85,292-9
MONO-(3-CHLOROPHENYL)-PHENYLMETHYL PHTHALATE
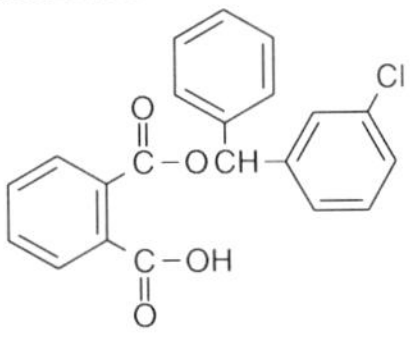
250 mg

S68,924-6
4-CHLOROBENZYL BENZOATE
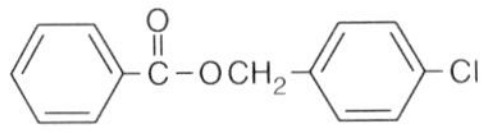
5 gm

S91,868-7
4-METHYLBENZYL PENTACHLOROBENZOATE
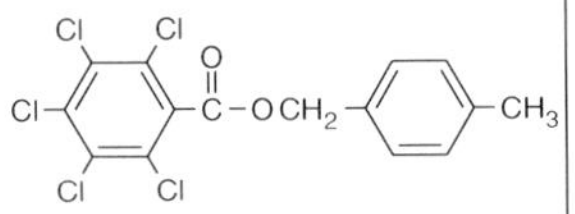
250 mg

S67,428-1
BIS(4'-BROMOPHENACYL) PHTHALATE
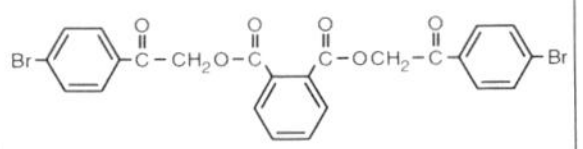
250 mg

S60,107-1
2-CHLOROPROCAINE HYDROCHLORIDE
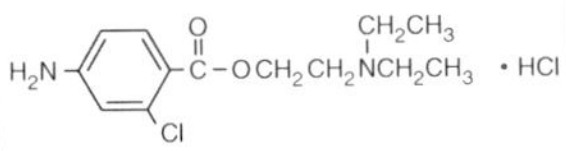
250 mg

S36,333-2
METHYL 5-AMINO-2-BROMOBENZOATE
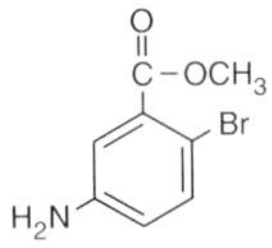
250 mg

S76,344-6
METHYL 4-AMINO-3-IODOBENZOATE
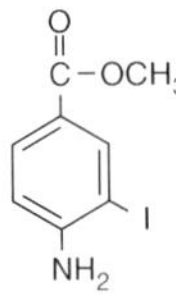
250 mg

S75,406-4
ETHYL 2-CHLORO-3,5-DINITROBENZOATE
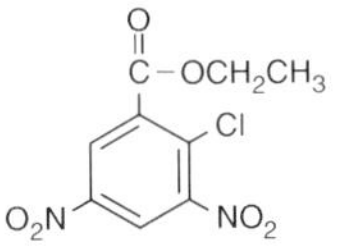
250 mg

S64,769-1
DIETHYL-2-(2-FLUOROANILINOMETHYLENE)-MALONATE
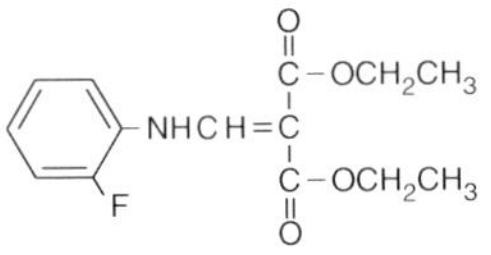
5 gm

S36,562-9
ETHYL 3-(4-CHLOROANILINO)CROTONATE
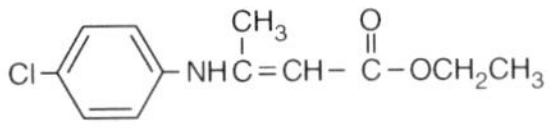
5 gm

S90,145-8
2-(2,6-DINITRO-4-METHYLANILINO)-ETHYL 2-CHLOROBENZOATE
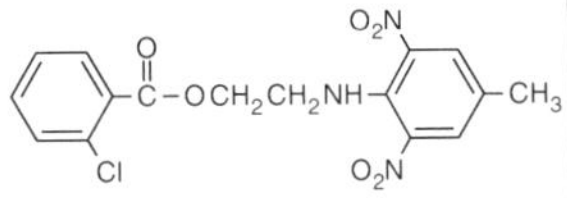
250 mg

S51,434-9
ETHYL 3-(3-CHLOROANILINO)CROTONATE
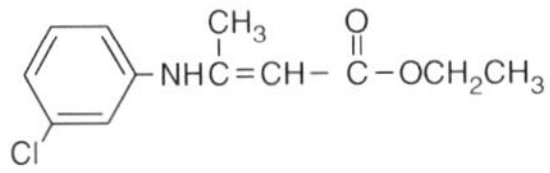
250 mg

S87,770-0
METHYL 4-(2-BROMOANILINO)-2-BUTENOATE
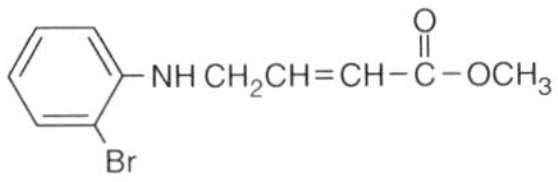
250 mg

S65,978-9
1-DIMETHYLAMINO-2-PROPYL-4-CHLOROBENZOATE HYDROCHLORIDE
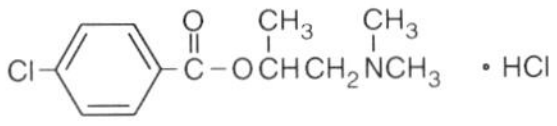
5 gm

S65,984-3
1-DIMETHYLAMINO-2-PROPYL-2,4-DICHLOROBENZOATE HYDROCHLORIDE
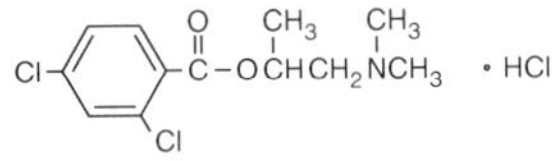
5 gm

S70,674-4
METHYL 4-CHLORO-3-NITROBENZOATE
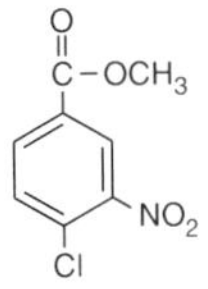
5 gm

S82,021-0
METHYL 5-BROMO-2-NITROBENZOATE
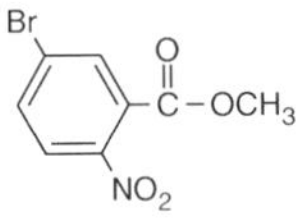
250 mg

S61,853-5
METHYL TETRACHLOROPHTHALATE
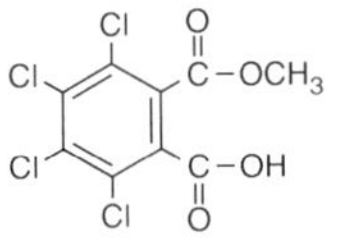
250 mg

S61,849-7
ISOPROPYL TETRACHLOROPHTHALATE
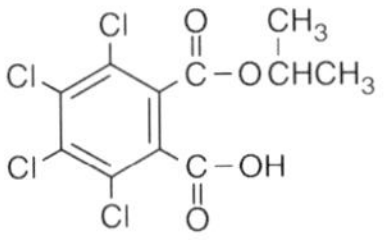
250 mg

S61,826-8
2-METHYLPENTYL TETRACHLOROPHTHALATE
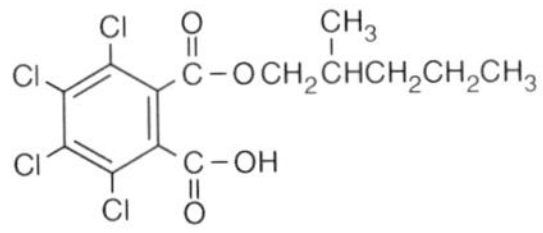
250 mg

S61,837-3
2-ETHYLBUTYL TETRACHLOROPHTHALATE
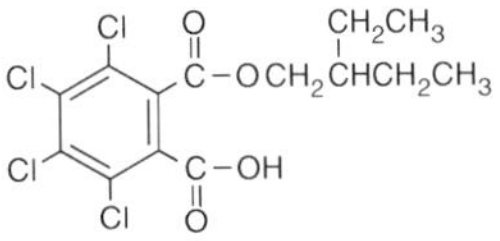
250 mg

S62,210-9
HEXYL TETRACHLOROPHTHALATE
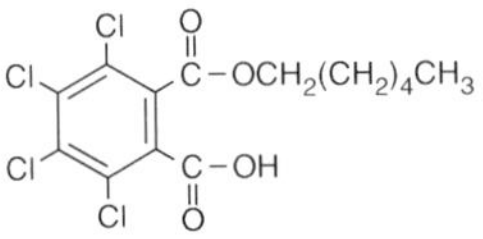
250 mg

S62,198-6
2-OCTYL TETRACHLOROPHTHALATE
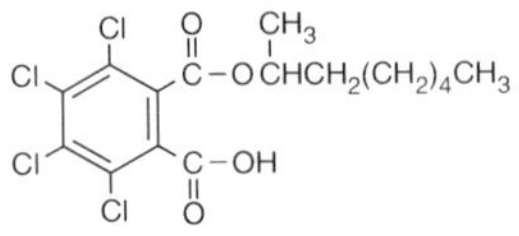
250 mg

S62,199-4
10-UNDECENYL TETRACHLOROPHTHALATE
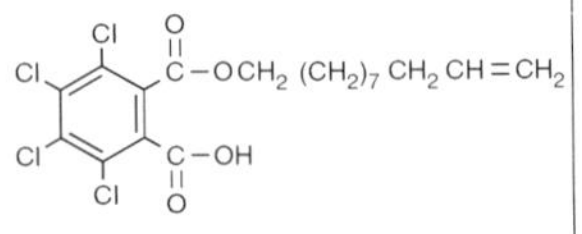
250 mg

S34,736-1
DIMETHYL 2,3,5,6-TETRACHLOROTEREPHTHALATE
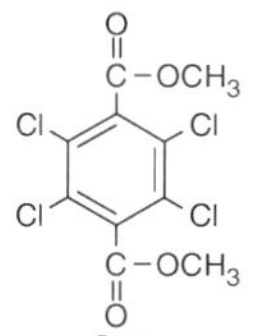
5 gm

S77,088-4
DIETHYL 3,4,5,6-TETRABROMOPHTHALATE
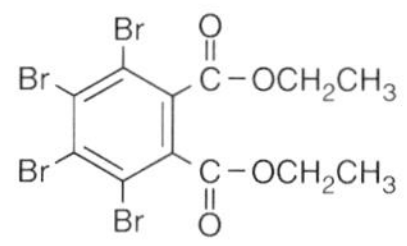
250 mg

S88,072-8
2-BROMOPHENYL METHACRYLATE
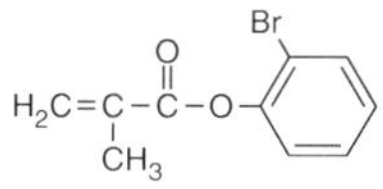
250 mg

S69,768-0
ISOPROPYL 2-BROMOBENZOATE
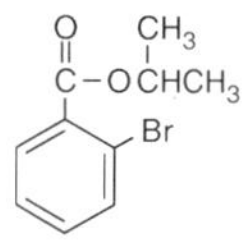
250 mg

S69,767-2
TERT-BUTYL 2-BROMOBENZOATE
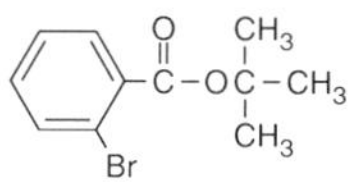
250 mg

S67,433-8
2,4-DICHLOROPHENYL 2-CHLOROBENZOATE
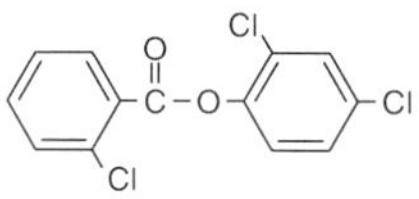
250 mg

S70,576-4
ETHYL 4-CHLOROBENZOATE
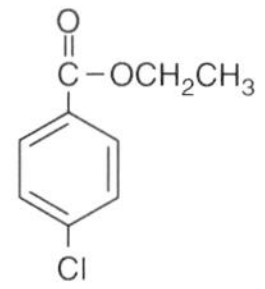
5 gm

S82,582-4
TERT-BUTYL 4-BROMOBENZOATE
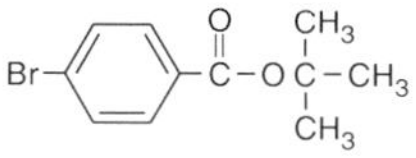
250 mg

S66,341-7
METHYL 3-ETHYL-2-IODOBENZOATE
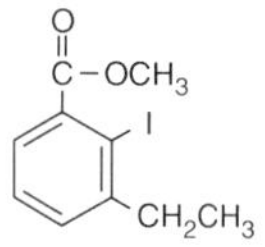
5 gm

S74,626-6
METHYL 3,5-DICHLOROBENZOATE
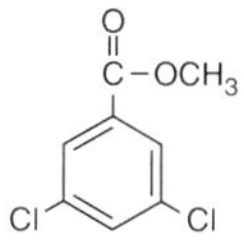
5 gm

S91,893-8
9-ANTHRACENYLMETHYL PENTACHLOROBENZOATE
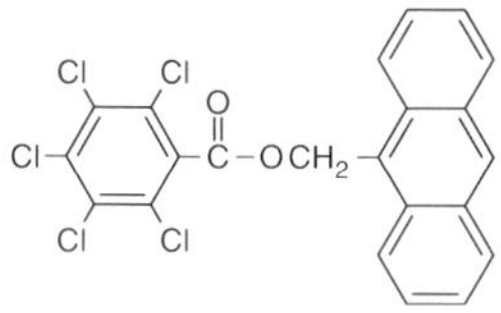
250 mg

S82,578-6
4-BROMOBENZOIC ANHYDRIDE
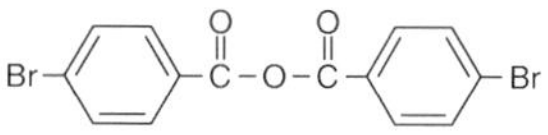
250 mg

S71,372-4
METHYL 3-BROMOBENZOATE
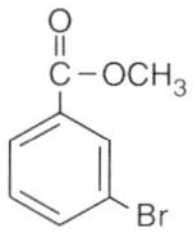
5 gm

S74,850-1
METHYL 2,5-DIHYDROXY-4-METHOXYBENZOATE
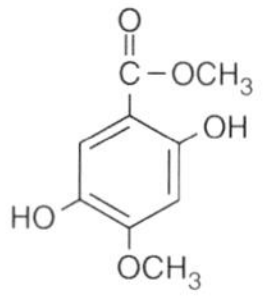
250 mg

S94,716-4
METHYL 3-BENZYLOXY-4,5-DIHYDROXYBENZOATE
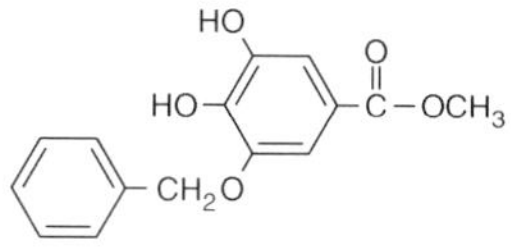
250 mg

S93,906-4
2,3,5-TRI-O-BENZOYL-D-XYLITOL
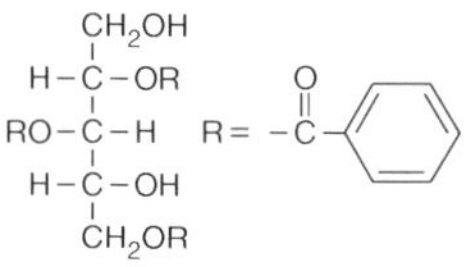
250 mg

S93,973-0
1-O-BENZOYL-D-RIBOSE
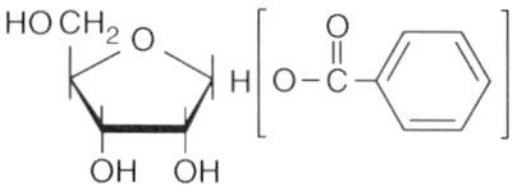
250 mg

S93,925-0
1-O-MESITOYL-BETA-D-GLUCOSE
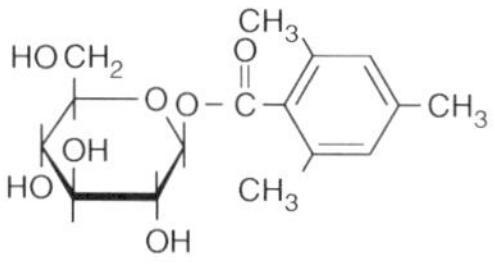
250 mg

S88,403-0
2,4,6-TRIHYDROXYBENZOYLMETHYL BENZOATE
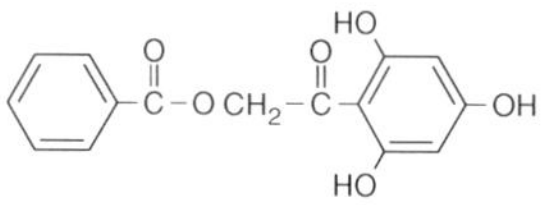
250 mg

S86,018-2
GLYCEROL 2-BENZOATE
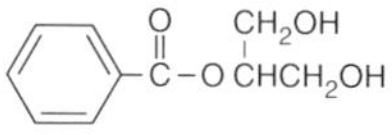
250 mg

S76,486-8
BUTYL GALLATE
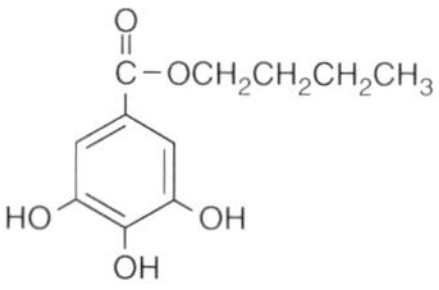
5 gm

S86,081-6
DIMETHYL 2,5-DIHYDROXYTEREPHTHALATE
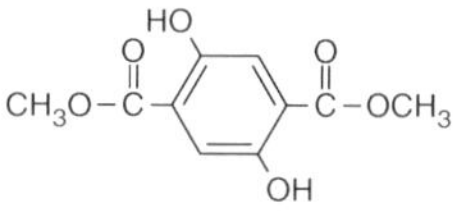
250 mg

S84,986-3
METHYL 2,4-DIHYDROXY-3-FORMYLBENZOATE
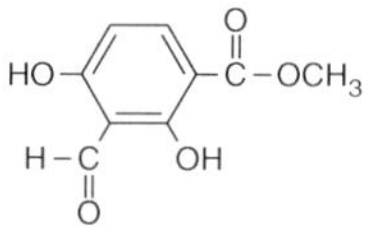
5 gm

S84,286-9
ETHYL [2,4-BIS(ETHOXYCARBONYL)-3,5-DIHYDROXY]PHENYLACETATE
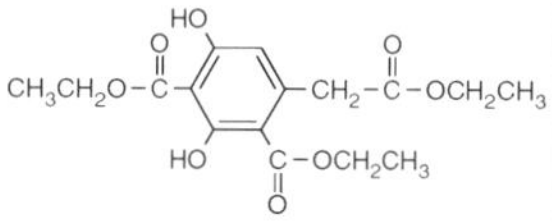
250 mg

S94,782-2
METHYL 2,4-DIHYDROXY-3,6-DIMETHYLBENZOATE
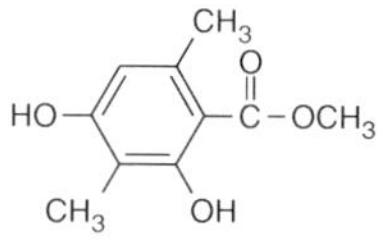
250 mg

S84,399-7
METHYL 2,6-DIHYDROXYBENZOATE
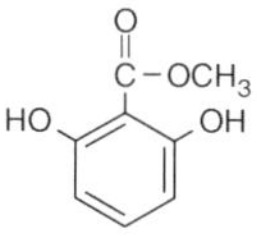
250 mg

S94,778-4
ETHYL 2,4-DIHYDROXYBENZOATE
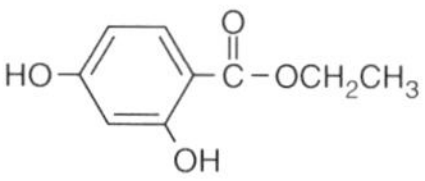
5 gm

S53,042-5
ETHYL 3,5-DIHYDROXYBENZOATE
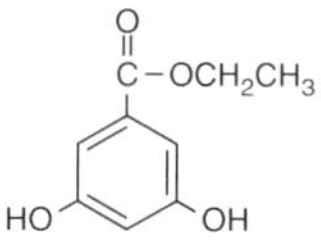
250 mg

S94,780-6
METHYL 2,4-DIHYDROXY-3-METHYLBENZOATE
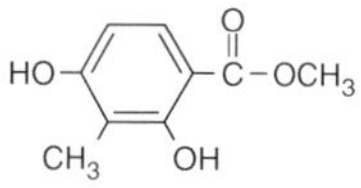
250 mg

S44,303-4
METHYL 2,4,6-TRIHYDROXYBENZOATE
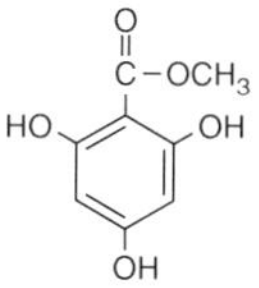
5 gm

S94,096-8
ETHYL 2,4-DIHYDROXY-3-METHYLBENZOATE
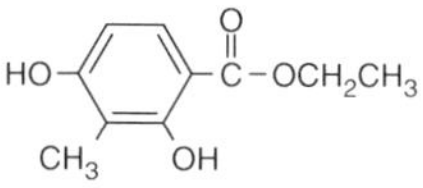
5 gm

S69,653-6
ETHYL 4-(2-HYDROXY-3-METHOXYBENZYLAMINO)BENZOATE
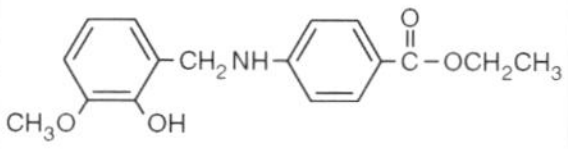
5 gm

S94,097-6
ETHYL 2-(4-HYDROXY-3-ETHOXYBENZYLIDENE)ACETOACETATE
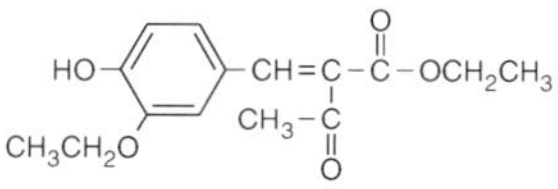
5 gm

S45,926-7
ETHYL VANILLATE
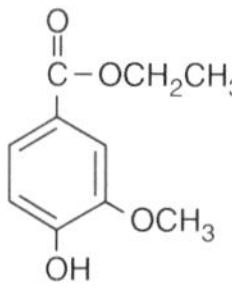
250 mg

S40,944-8
METHYL SYRINGATE
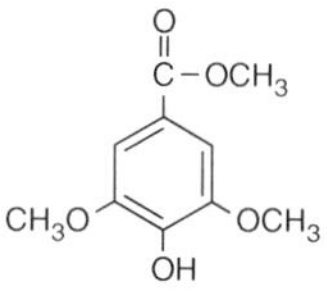
5 gm

S80,886-5
ETHYL 4-(ALLYLOXY)SALICYLATE
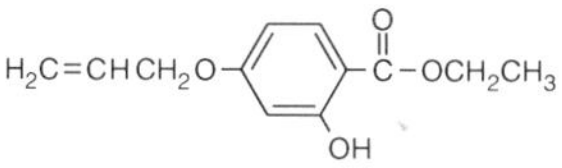
250 mg

S92,997-2
TRIMETHYL 2-HYDROXY-1,3,5-BENZENETRICARBOXYLATE
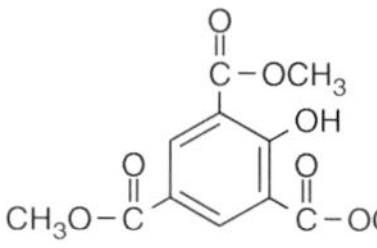
250 mg

S92,629-9
2-HYDROXY-3-PHENOXYPROPYL BENZOATE
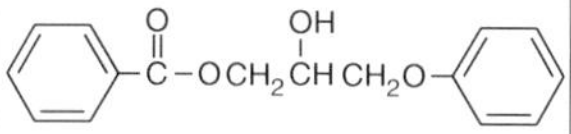
250 mg

S54,151-6
METHYL 4,6-DIMETHOXYSALICYLATE
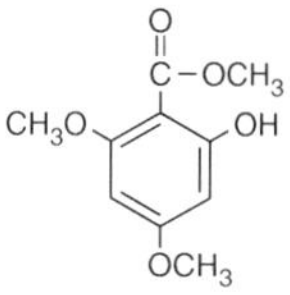
250 mg

S79,313-2
BENZYL 3,5-DIOXO-2-HYDROXY-1-CYCLOPENTENE-1-ACETATE
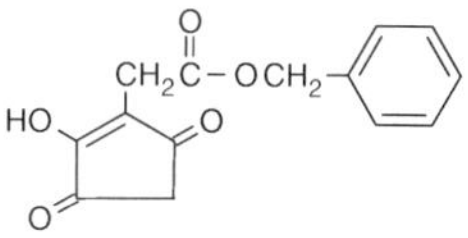
250 mg

S76,089-7
ETHYL 4-BENZOYL-2-HYDROXY-2-PHENYL-1,7-DIAZABICYCLO[4.1.0]HEPT-5-ENE-5-CARBOXYLATE
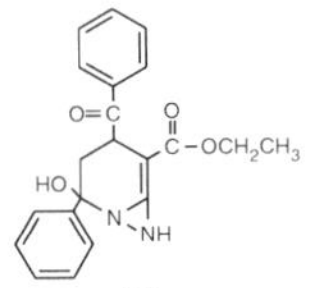
250 mg

S43,753-0
BENZYL SALICYLATE
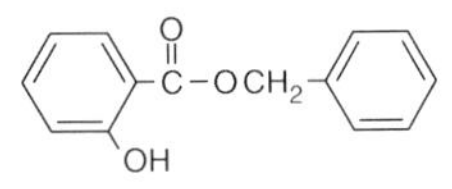
5 gm

S40,127-7
ALPHA-METHYLBENZYL SALICYLATE
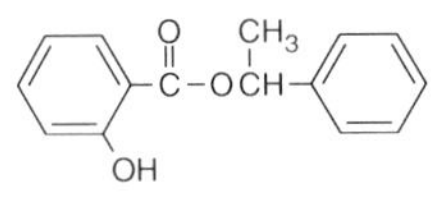
250 mg

S82,654-5
METHYL 4-(DIPHENYLHYDROXYMETHYL)-BENZOATE
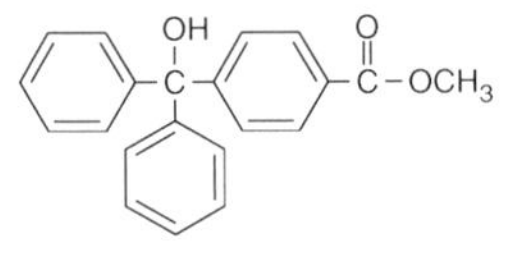
250 mg

S37,530-6
VINYL SALICYLATE
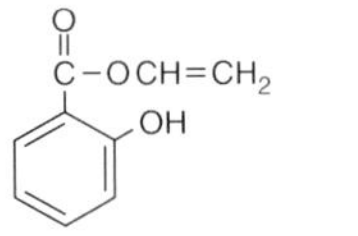
5 gm

S42,624-5
POPULIN DIHYDRATE
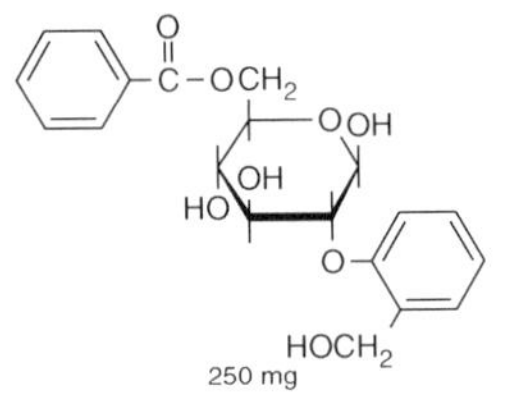
250 mg

S68,065-6
PROPYLENE GLYCOL SALICYLATE
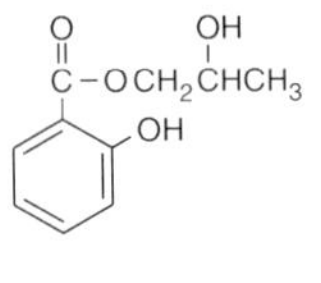
5 gm

S68,036-2
PHENETHYL SALICYLATE
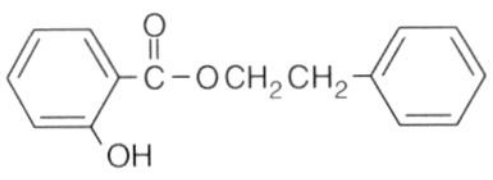
5 gm

S64,123-5
2,3,4-TRIBENZOYL-D-XYLOSE
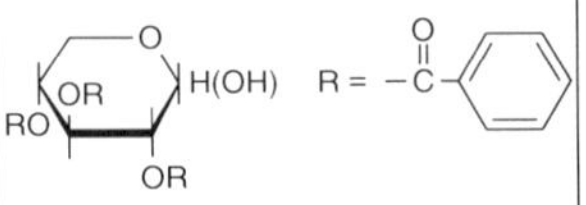
5 gm

S60,605-7
METHYL 4-AMINOSALICYLATE
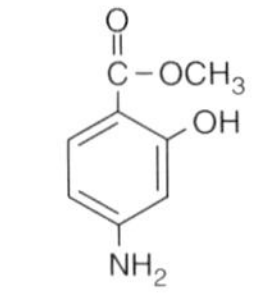
250 mg

S92,634-5
TRANS-2-HYDROXYCYCLOPENTYL-4-NITROBENZOATE
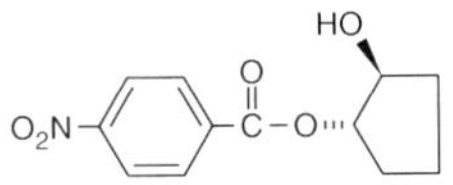
250 mg

S93,561-1
METHYL 1-HYDROXY-2-NAPHTHOATE
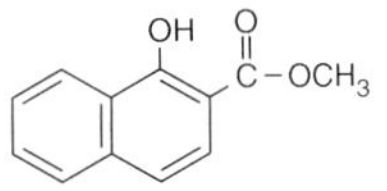
250 mg

S60,489-5
ETHYL 4-(2-HYDROXYBENZYLAMINO)-BENZOATE
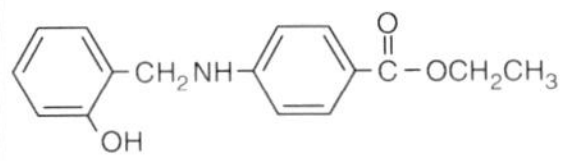
5 gm

S68,897-5
ALPHA-(4-ETHOXYCARBONYLANILINO)-P-CRESOL
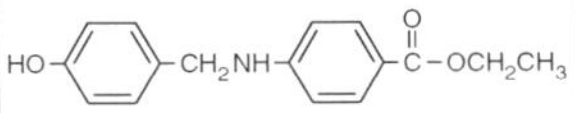
5 gm

S42,511-7
O-(2-HYDROXYBENZOYL)TROPINE HYDROCHLORIDE
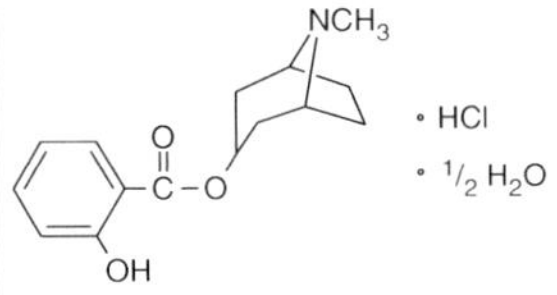
250 mg

S42,509-5
O-(4-HYDROXYBENZOYL)TROPINE
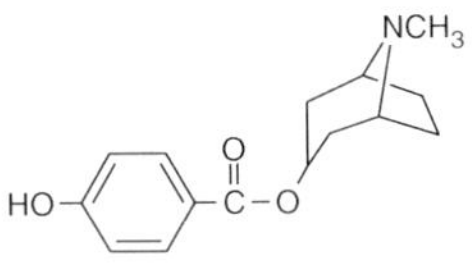
250 mg

S42,510-9
O-(3-HYDROXYBENZOYL)TROPINE HYDROCHLORIDE
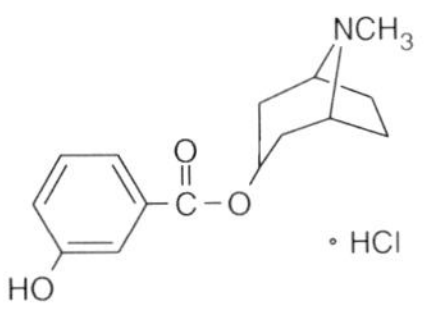
250 mg

S36,974-8
METHYL 3-HYDROXY-4-NITROBENZOATE
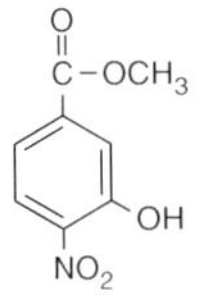
5 gm

S36,975-6
METHYL 4-HYDROXY-3-NITROBENZOATE
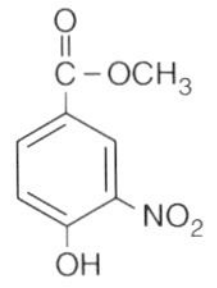
5 gm

S58,823-7
METHYL 4-HYDROXY-3-NITROBENZOATE, SODIUM SALT
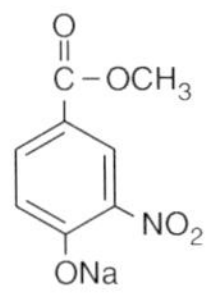
5 gm

S38,279-5
METHYL 5-NITROSALICYLATE
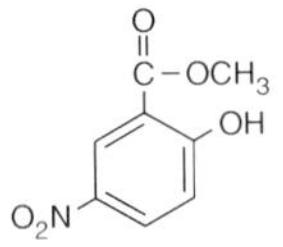
250 mg

S64,019-0
METHYL 3-HYDROXY-4-METHYL-2-NITROBENZOATE
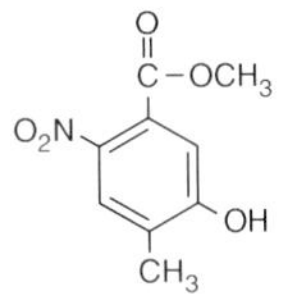
5 gm

S76,316-0
ETHYL 4,4-DIMETHYL-2,6-DIOXO-BETA-(2-HYDROXYPHENYL)-ALPHA-(2-IMIDAZOLIDINYLIDENE)CYCLOHEXANEPROPIONATE
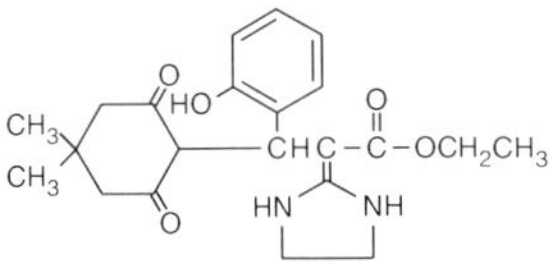
250 mg

S57,901-7
ETHYL 1,4-DIHYDRO-2-HYDROXY-6-METHYL-4-PHENYL-5-PYRIMIDINECARBOXYLATE
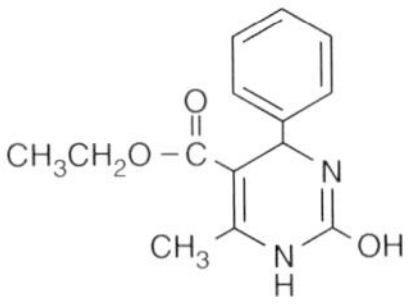
5 gm

S35,666-2
3,3,5-TRIMETHYLCYCLOHEXYL SALICYLATE
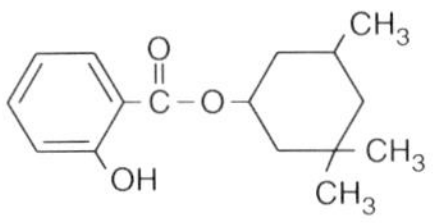
5 gm

S35,156-3
ISOBUTYL SALICYLATE
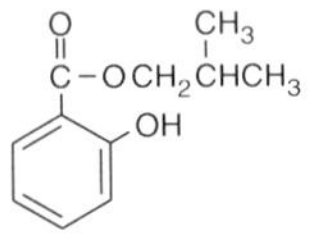
5 gm

S36,720-6
BUTYL SALICYLATE
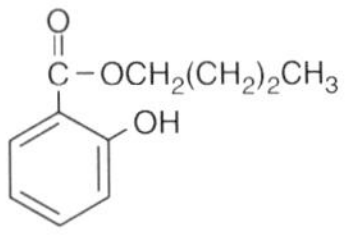
5 gm

S57,106-7
ISOPENTYL SALICYLATE
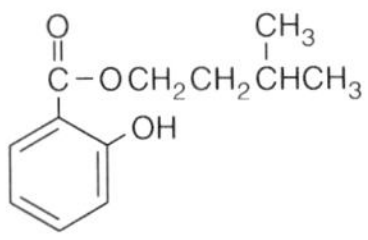
5 gm

S69,555-6
ISOPROPYL 4-HYDROXYBENZOATE
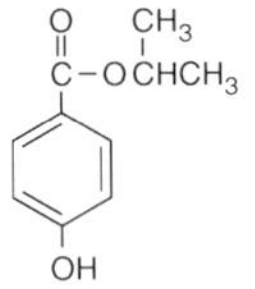
5 gm

S69,552-1
ISOBUTYL 4-HYDROXYBENZOATE
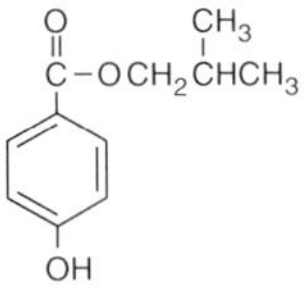
5 gm

S37,814-3
METHYL 3-METHYLSALICYLATE
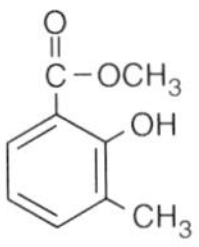
5 gm

S55,083-3
METHYL 5-CHLOROSALICYLATE
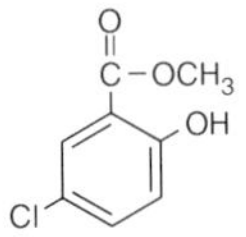
250 mg

S44,155-4
METHYL 5-BROMOSALICYLATE
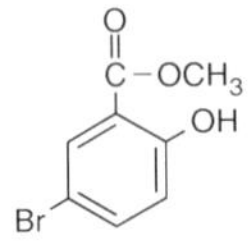
250 mg

S37,785-6
METHYL 5-METHYLSALICYLATE
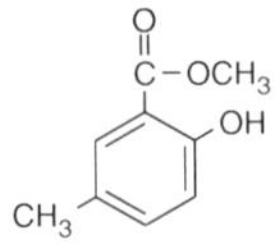
5 gm

S58,024-4
METHYL 4-METHYLSALICYLATE
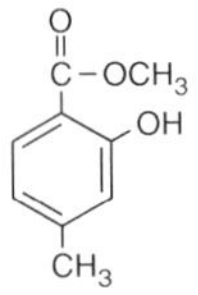
250 mg

S87,384-5
ETHYL 2,6-DIMETHYL-4-HYDROXYBENZOATE
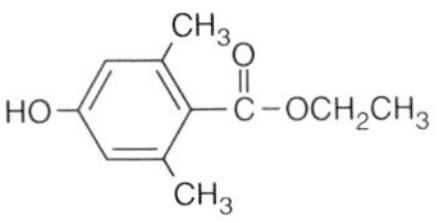
250 mg

S40,696-1
BUTYL 3,5-DIIODO-4-HYDROXYBENZOATE
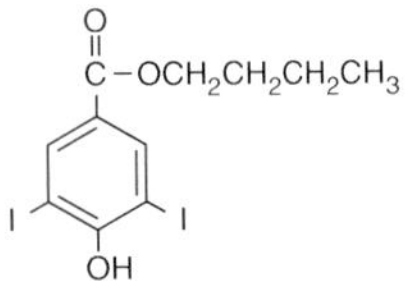
5 gm

S43,461-2
METHYL 3,5-DI-TERT-BUTYL-4-HYDROXYBENZOATE
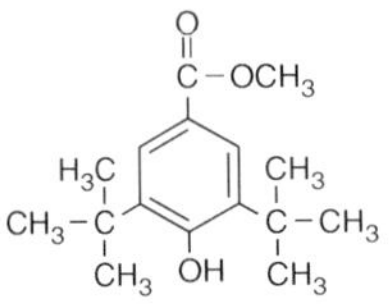
5 gm

S53,927-9
ETHYL 3-HYDROXY-1-PHENYLINDENE-2-CARBOXYLATE
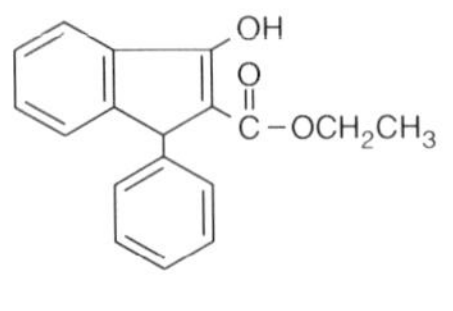
250 mg

S88,868-0
METHYL 3,5-DIBENZYLOXYBENZOATE
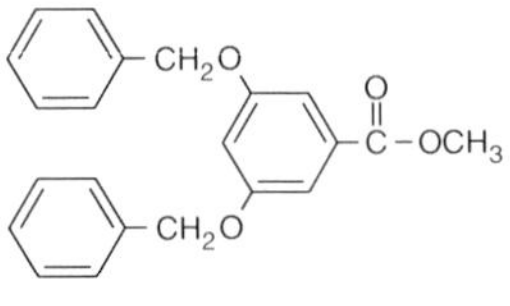
250 mg

S92,633-7
METHYL 4-BENZYLOXYBENZOATE

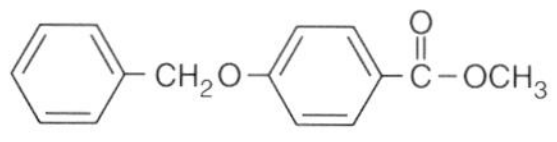

250 mg

S86,150-2
TRANS-2,5-DIHYDRO-2,5-DIMETHOXY-2-METHYL-3-FURYLMETHYL BENZOATE

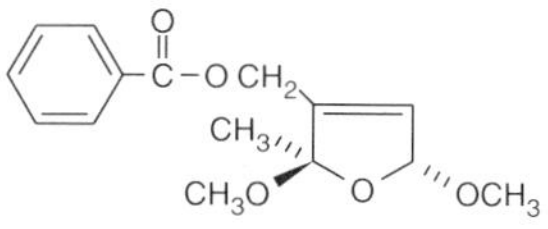

250 mg

S95,414-4
METHYL 4,4-DIMETHOXY-2,3-DIOXOBUTYRATE 2-(4-CARBOXYPHENYLHYDRAZONE)

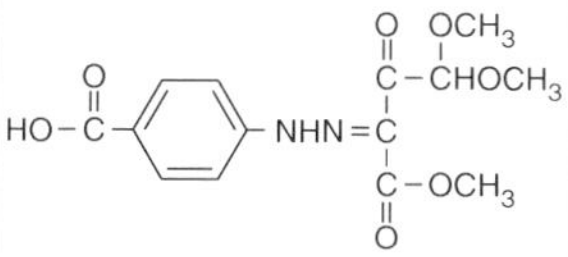

50 mg

S44,544-4
METHYL 4,5-DIMETHOXYANTHRANILATE

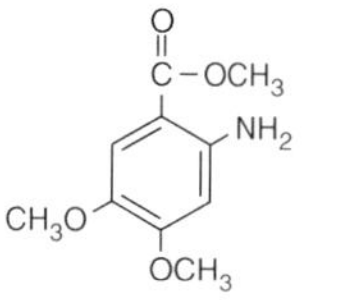

250 mg

S66,177-5
METHYL 2-AMINO-4,5-DIETHOXYBENZOATE

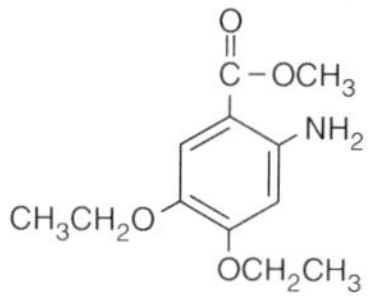

5 gm

S39,671-0
TRIETHYLENE GLYCOL DIBENZOATE

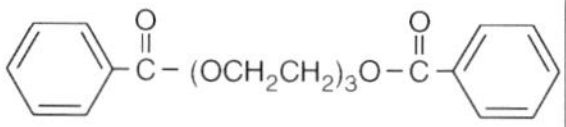

5 gm

S45,975-5
METHYL 4-BENZYLOXY-3-METHOXYBENZOATE

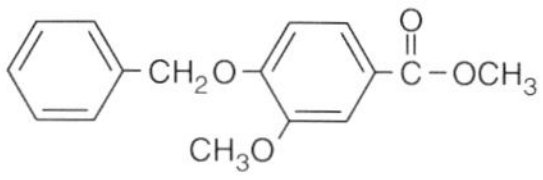

250 mg

S95,403-9
METHYL 2-(2,3-DIMETHOXYSTYRYL)-GLYOXYLATE

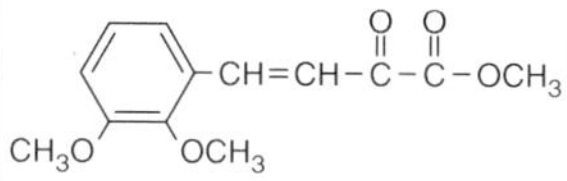

50 mg

S95,348-2
METHYL 2,3-DIMETHOXYBENZOATE

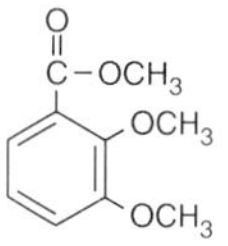

50 mg

S76,032-3
ETHYL 3,4-DIHYDRO-6,7-DIMETHOXY-1-ISOQUINOLINECARBOXYLATE

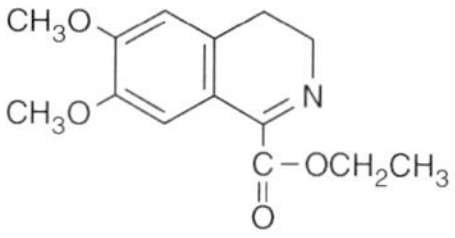

250 mg

S46,118-0
3,5-DIMETHOXY-4-(ETHOXYCARBONYL)-BENZOIC ACID

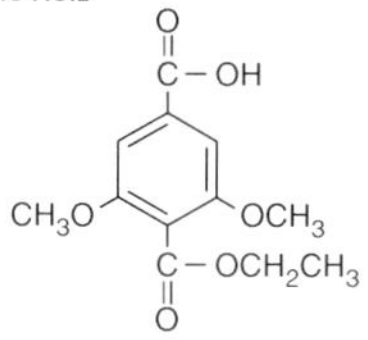

5 gm

S86,077-8
METHYL 2,5-DIMETHOXYBENZOATE

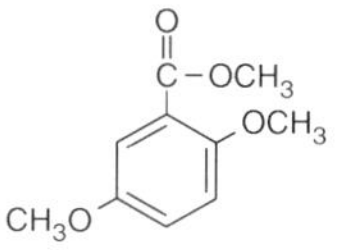

250 mg

S95,405-5
METHYL 4,4-DIMETHOXY-2,3-DIOXOBUTYRATE 2-(4-NITROPHENYLHYDRAZONE)

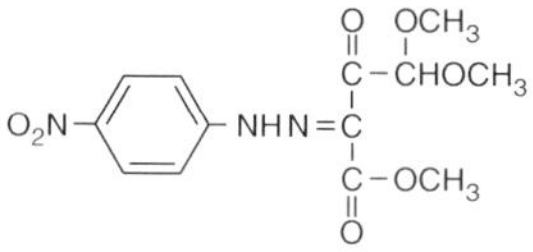

50 mg

S95,406-3
METHYL 4,4-DIMETHOXY-2,3-DIOXOBUTYRATE 2-PHENYLHYDRAZONE

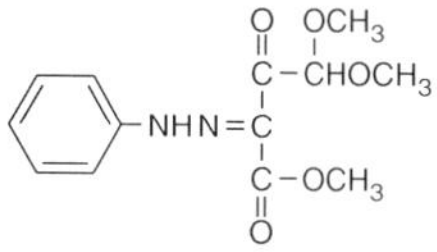

50 mg

S88,399-9
2,7-BIS(BENZYLOXY)-1,4A,4B-TRIMETHYLPERHYDROPHENANTHRENE

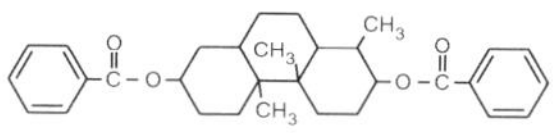

250 mg

S63,035-7
METHYL 3-AMINO-4-[2-(4-TERT-OCTYLPHENOXY)ETHOXY]BENZOATE

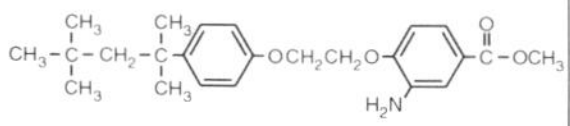

5 gm

S62,445-4
ETHYL 4-(3,4-DIMETHOXYBENZYLAMINO)-BENZOATE

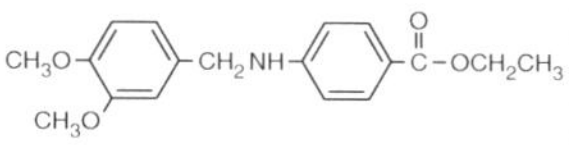

5 gm

S40,438-1
CONVOLVINE

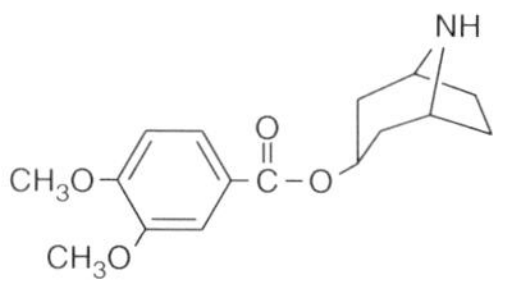

100 mg

S62,964-2
2-DIETHYLAMINOETHYL 3-NITRO-4-[2-(4-TERT-OCTYLPHENOXY)ETHOXY]-BENZOATE

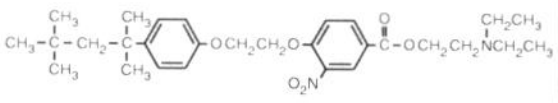

5 gm

S34,593-8
2-DIETHYLAMINOETHYL 3,4,5-TRIMETHOXYBENZOATE HYDROCHLORIDE

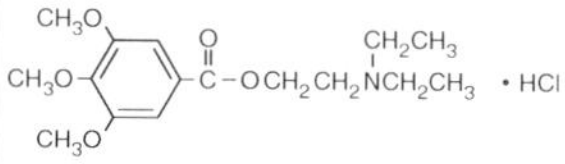

5 gm

S60,074-1
2-DIETHYLAMINOETHYL 3,4,5-TRIMETHOXYBENZOATE

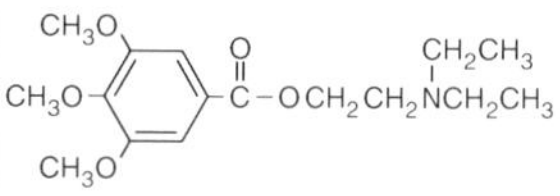

5 gm

S94,714-8
ETHYL 1-(BETA-PHENETHYL)-4-(3,4,5-TRIMETHOXYBENZOYLOXY)-3-PIPERIDINECARBOXYLATE

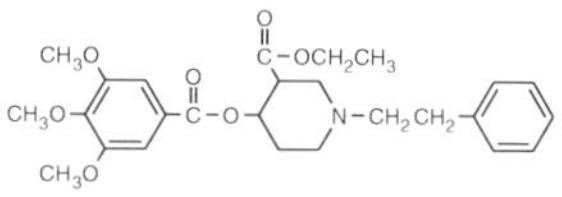

250 mg

S41,134-5
TROXONIUM TOSYLATE

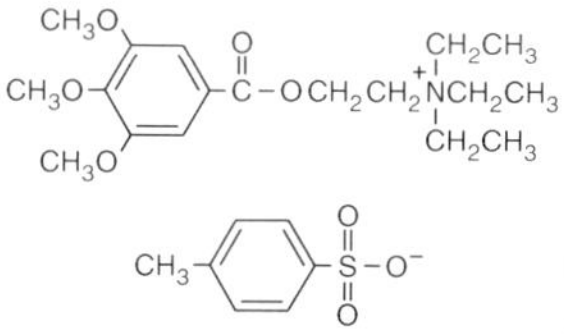

5 gm

S62,908-1
METHYL 3-NITRO-4-[2-(4-TERT-OCTYLPHENOXY)ETHOXY]BENZOATE

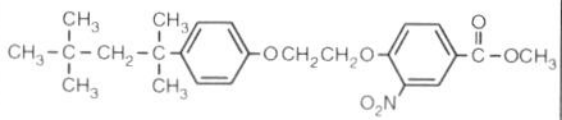

5 gm

S92,628-0
2,3-EPOXYPROPYL 4-(2,3-EPOXYPROPOXY)BENZOATE

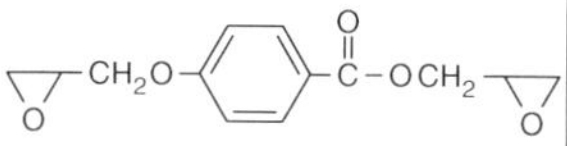

250 mg

S86,155-3
METHYL 2-METHOXY-5-PHENYL-2,4-PENTADIENOATE

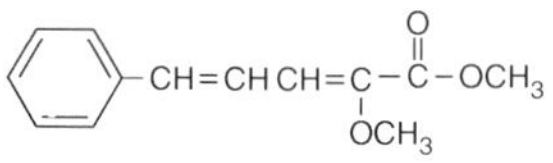

250 mg

S50,004-6
METHYL 4-(2-BUTOXYETHOXY)BENZOATE

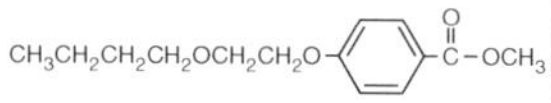

250 mg

S62,906-5
ETHYL 4-[2-(4-TERT-AMYLPHENOXY)-ETHOXY]BENZOATE

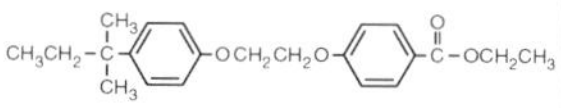

250 mg

S86,454-4
ETHYL 5-METHOXYANTHRAQUINONE-2-CARBOXYLATE

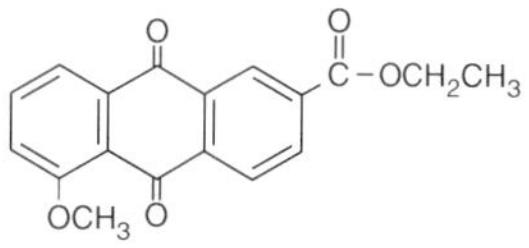

250 mg

S89,489-3
DIMETHYL 2-ETHOXY-3H-1-BENZAZEPINE-3,4-DICARBOXYLATE

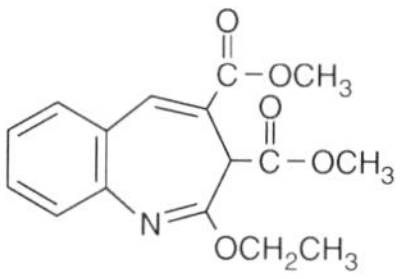

250 mg

S88,853-2
METHYL 3-METHOXY-5-NITROBENZOATE

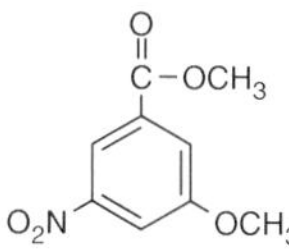

250 mg

S79,979-3
ETHYL 3-(3-METHOXYBENZOYL)PYRUVATE

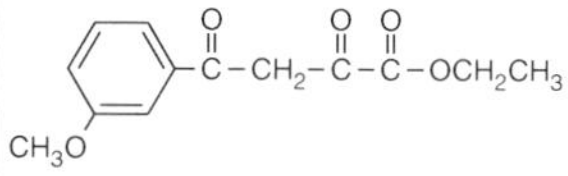

250 mg

S58,243-3
METHYL 3-AMINO-P-ANISATE

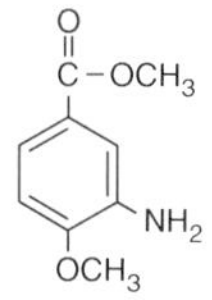

5 gm

S39,894-2
METHYL 3-AMINO-4-HEXADECYLOXYBENZOATE

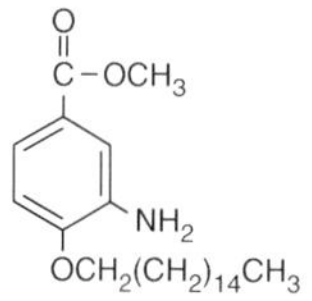

5 gm

S90,078-8
METHYL N-(ALPHA-METHOXY-3-NITROBENZYL)ANTHRANILATE

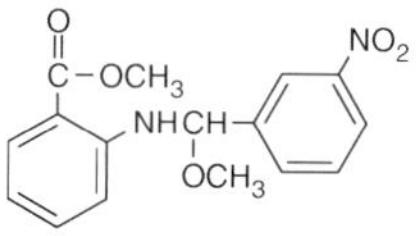

250 mg

S63,023-3
METHYL 4-METHOXY-3-NITROBENZOATE

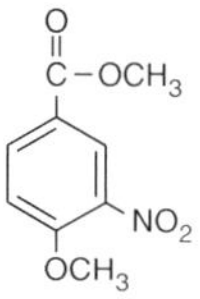

5 gm

S50,725-3
METHYL 4-BENZYLOXY-3-NITROBENZOATE

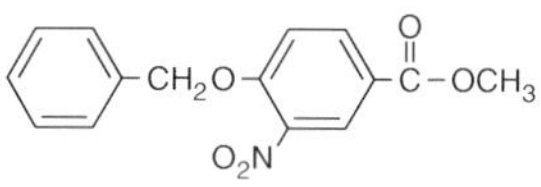

250 mg

S59,817-8
2-(4-NITROPHENOXY)ETHYL METHACRYLATE

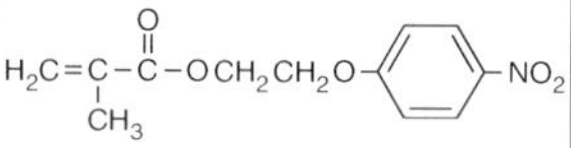

5 gm

S52,645-2
ETHYL 4-(2-DECYLOXY-5-NITROBENZYLOXY)BENZOATE

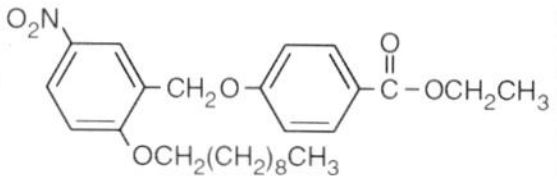

5 gm

S93,179-9
2-ETHYL 3,4-DIMETHOXYPHTHALATE

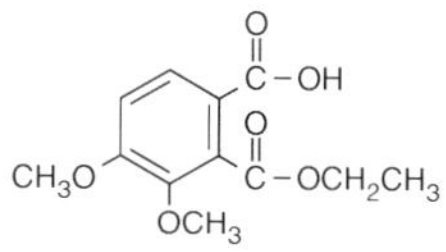

250 mg

S51,691-0
11-[4-(METHOXYCARBONYL)PHENOXY]-UNDECANOIC ACID
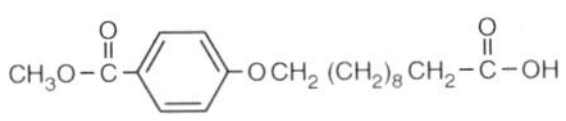
250 mg

S57,551-8
METHYL 3-(DECYLOXY)BENZOATE
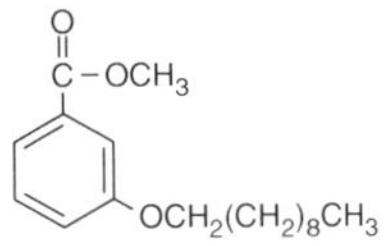
5 gm

S57,540-2
METHYL 3-(DODECYLOXY)BENZOATE
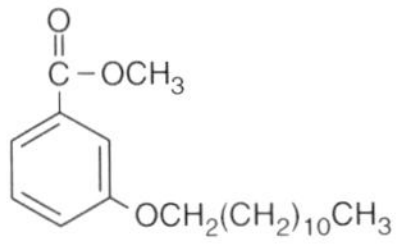
5 gm

S57,506-2
METHYL 3-(TETRADECYLOXY)BENZOATE
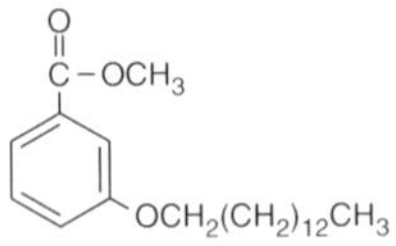
5 gm

S57,463-5
METHYL 3-(HEXADECYLOXY)BENZOATE
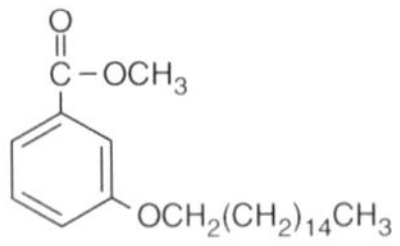
5 gm

S57,590-9
METHYL 3-(OCTADECYLOXY)BENZOATE
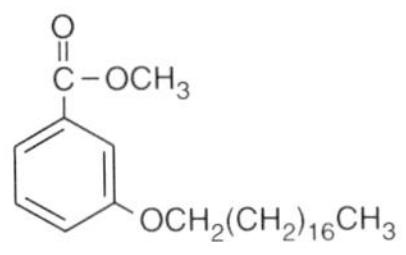
5 gm

S43,541-4
ETHYL 4-ETHOXYBENZOATE
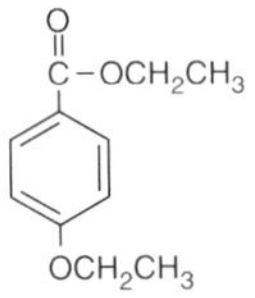
5 gm

S51,625-2
ETHYL 4-(PENTYLOXY)BENZOATE
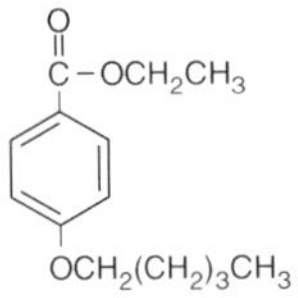
250 mg

S57,501-1
METHYL 4-(DECYLOXY)BENZOATE
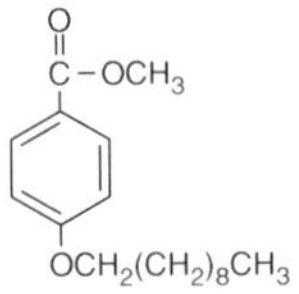
5 gm

S57,447-3
METHYL 4-(DODECYLOXY)BENZOATE
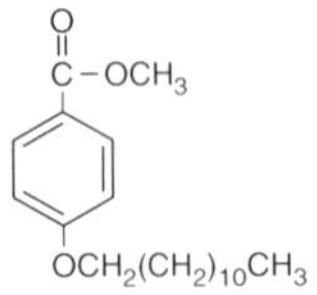
5 gm

S57,507-0
METHYL 4-(TETRADECYLOXY)BENZOATE
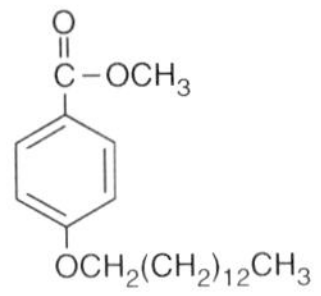
5 gm

S57,444-9
METHYL 4-(HEXADECYLOXY)BENZOATE
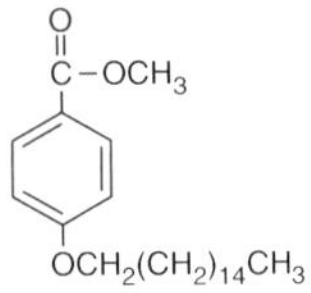
5 gm

S57,589-5
METHYL 4-(OCTADECYLOXY)BENZOATE
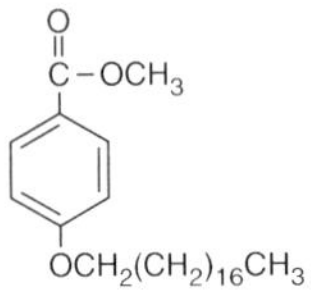
5 gm

S43,366-7
METHYL 2,4,6-TRIMETHOXYBENZOATE
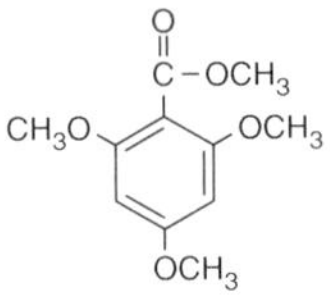
5 gm

S93,907-2
XYLITOL PENTABENZOATE
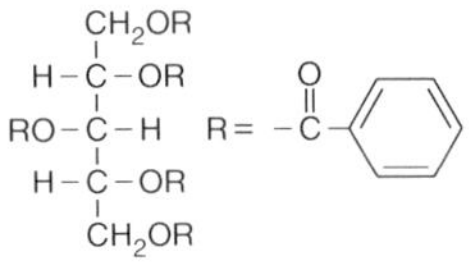
250 mg

S61,218-9
DL-BENZOIN 2,4-DICHLOROBENZOATE
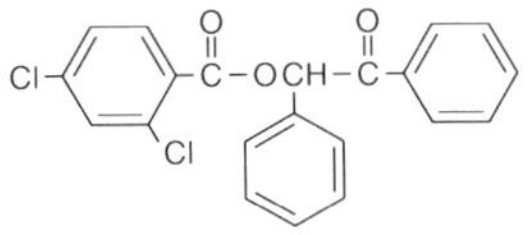
250 mg

S87,561-9
ENDO-2-NORBORNYL PHTHALATE
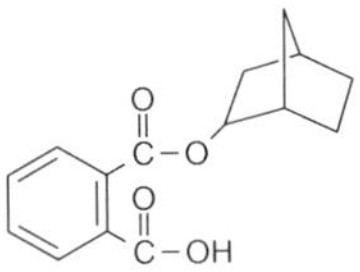
250 mg

S90,646-8
BENZYL 2-(DIPHENYLMETHYLAMINOMETHYL)-ACRYLATE HYDROBROMIDE
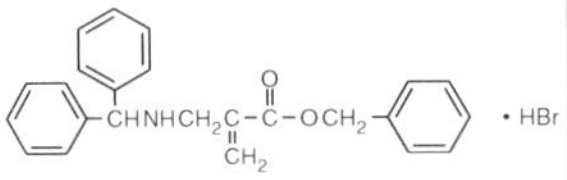
250 mg

S91,915-2
4-NITROBENZYL BENZOATE
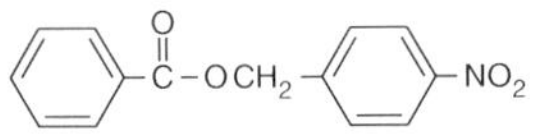
250 mg

S91,831-8
ALPHA,ALPHA-DIMETHYL-4-NITROBENZYL BENZOATE
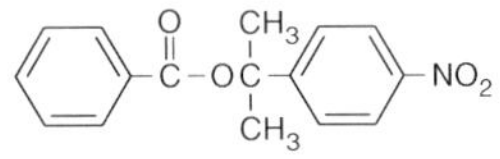
250 mg

S82,579-4
4-NITROBENZYL 4-NITROBENZOATE
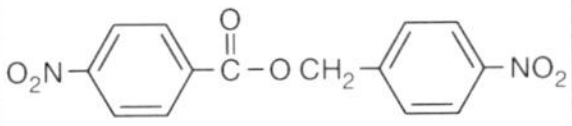
250 mg

S91,869-5
4-NITROBENZYL 4-METHYLBENZOATE
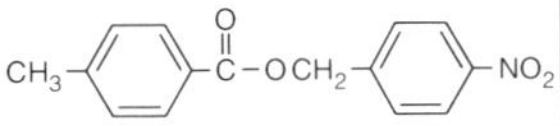
250 mg

S91,916-0
4-NITROBENZYL 2,4,6-TRIMETHYLBENZOATE
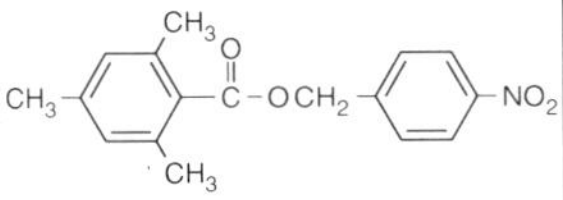
250 mg

S79,285-3
(+/-)-2-METHYL-1-PHENYLPROPYL BENZOYLFORMATE
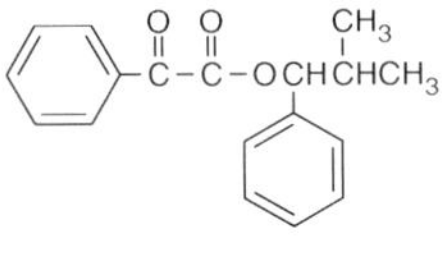
250 mg

S59,906-9
DL-BENZOIN BENZOATE
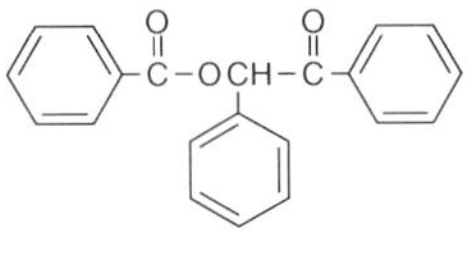
250 mg

S71,008-3
MONO(ALPHA-ETHYL-ALPHA-METHYLBENZYL) PHTHALATE
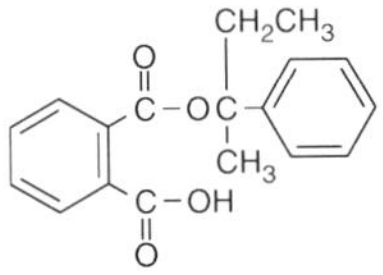
5 gm

S64,808-6
ALPHA-METHYLBENZYL PHTHALATE
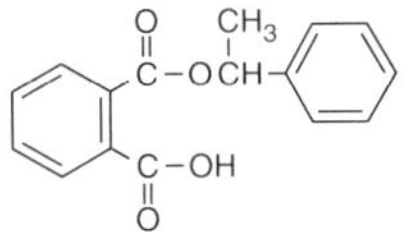
250 mg

S40,334-2
ALPHA-METHYLBENZYL BENZOATE
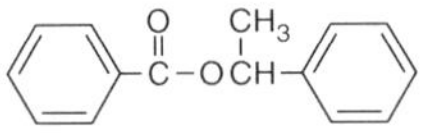
250 mg

S90,608-5
2-BROMO-1,2-DIPHENYLETHYL BENZOATE
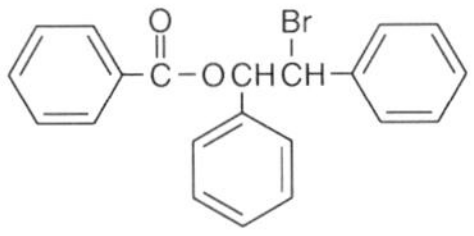
250 mg

S91,876-8
BENZYL 2,4,6-TRIMETHYLBENZOATE
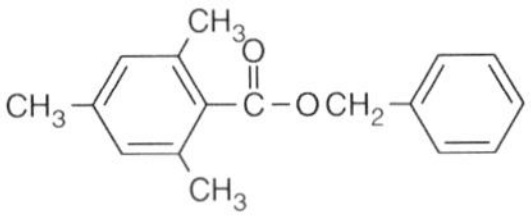
250 mg

S92,080-0
ETHYL 2,4,6-TRIPHENYLBENZOATE
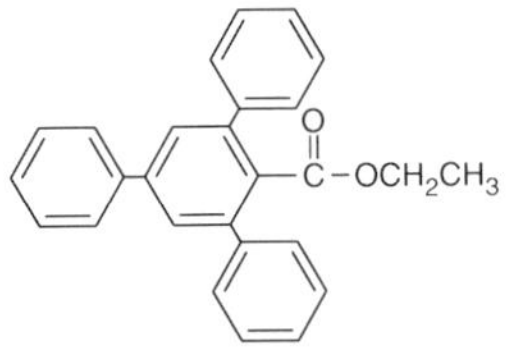
250 mg

S81,385-0
DIPHENYLMETHYL BENZOATE
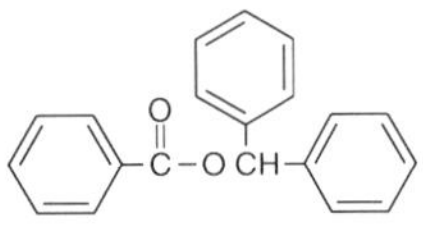
250 mg

S93,976-5
1,2-O-ISOPROPYLIDENE-3,5,6-TRI-O-BENZOYL-ALPHA-D-GLUCOFURANOSIDE
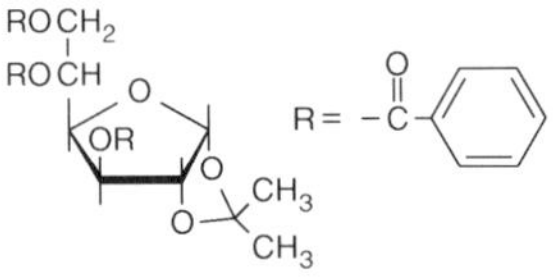
250 mg

S93,919-6
3-O-BENZOYL-1,2:4,6-DI-O-BENZYLIDENE-ALPHA-D-GLUCO-PYRANOSIDE
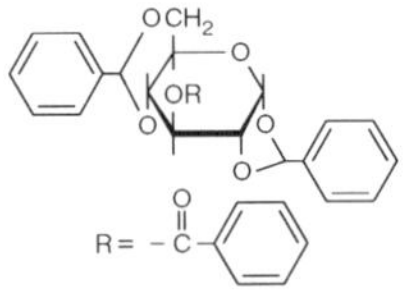
250 mg

S60,875-0
METHYL 2-DEOXY-3,5-BIS-O-(4-NITROBENZOYL)-ALPHA,BETA-D-RIBOSIDE
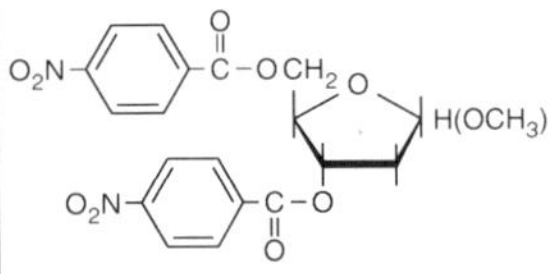
250 mg

S78,131-2
ALPHA-D-XYLOFURANOSE TETRABENZOATE
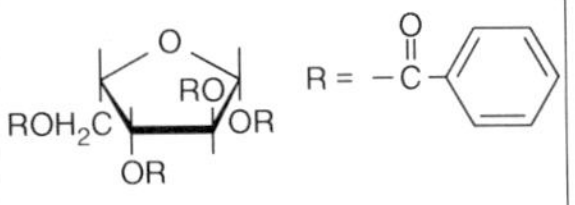
250 mg

S93,924-2
1-O-MESITOYL-2,3,4,6-TETRA-O-ACETYL-ALPHA-D-GLUCOSIDE
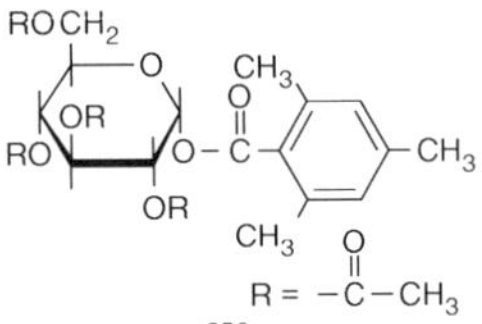
250 mg

S93,972-2
1-O-(4-NITROBENZOYL)-2,3,5-TRI-O-BENZYL-L-ARABINOFURANOSE
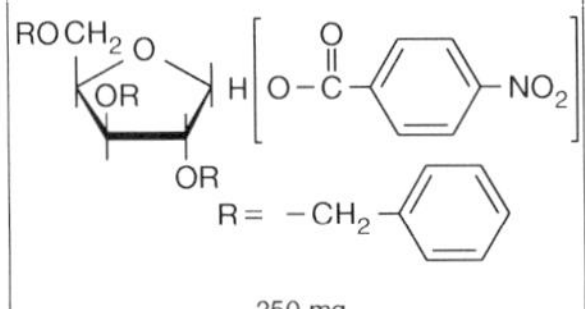
250 mg

S44,895-8
D-LYXOFURANOSE 1-ACETATE 2,3,5-TRIBENZOATE
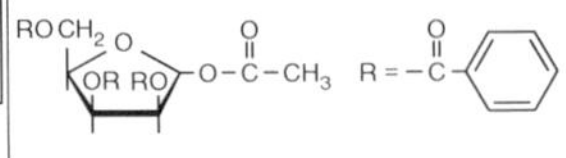
250 mg

S93,908-0
1,5-DI-O-TRITYL-2,3,4-TRI-O-BENZOYLXYLITOL
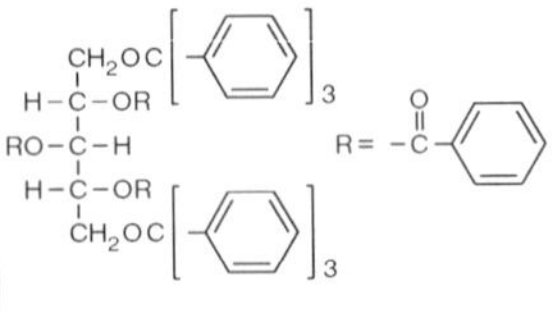
250 mg

S93,928-5
1-O-(4-NITROBENZOYL)-2,3,4,6-TETRA-O-BENZYL-BETA-D-GLUCOSE
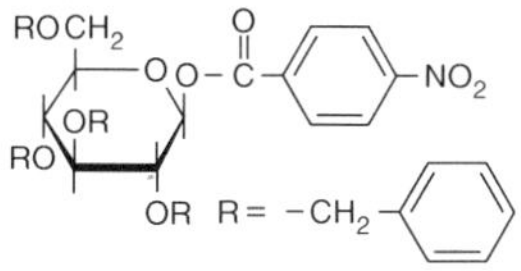
250 mg

S93,927-7
1-O-METHYL-2,3,4,6-TETRA-O-BENZOYL-BETA-D-GLUCOPYRANOSIDE
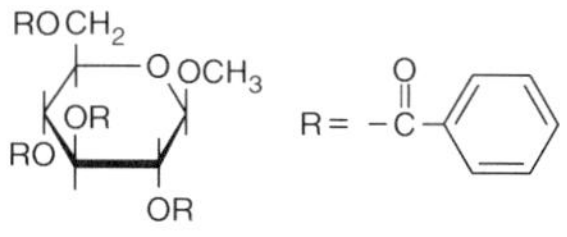
250 mg

S93,915-3
1-O-METHYL-2,3,5-TRI-O-BENZOYL-ALPHA-D-ARABINOFURANOSIDE
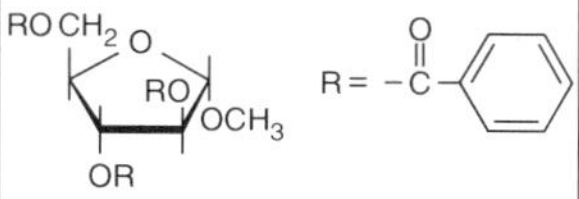
250 mg

S93,938-2
BETA-D-RIBOPYRANOSE TETRABENZOATE
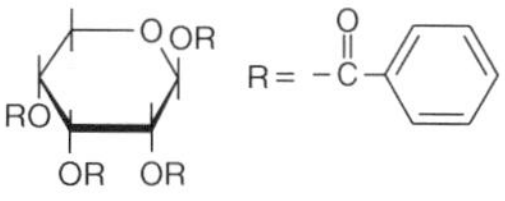
250 mg

S93,926-9
BETA-D-GLUCOSE PENTABENZOATE
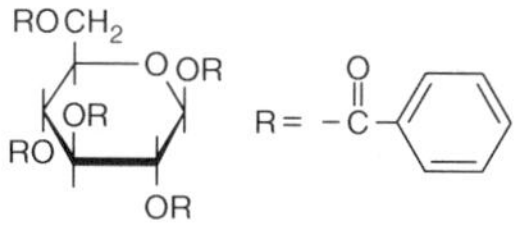
250 mg

S93,935-8
1-O-BENZOYL-2,3,4-TRI-O-BENZYL-BETA-L-ARABINOSE
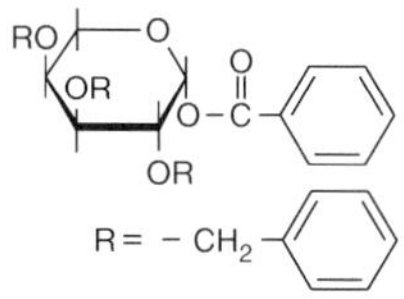
250 mg

S93,904-8
1,6-DI-O-BENZYL-2,3,4,5-TETRA-O-BENZOYLDULCITOL
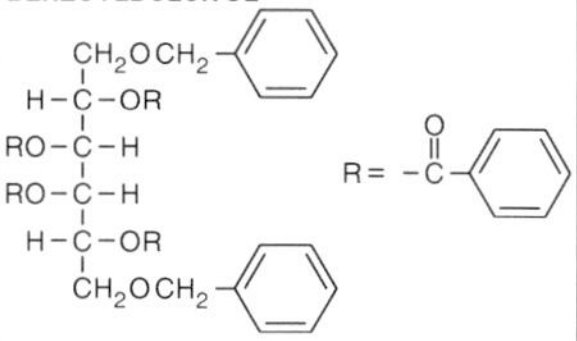
250 mg

S93,803-3
1-(1-ADAMANTYL)ETHYLAMINE DIBENZOYLTARTRATE
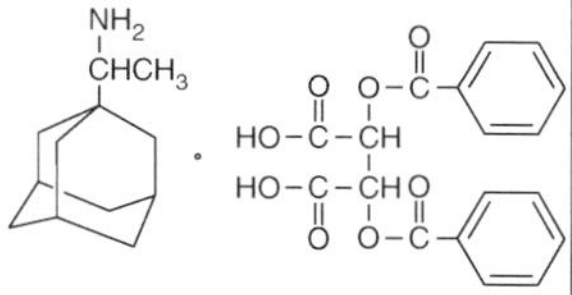
250 mg

S68,604-2
ETHYL 2-(3-NITROPHENYL)-2-BENZOYLOXYACETATE
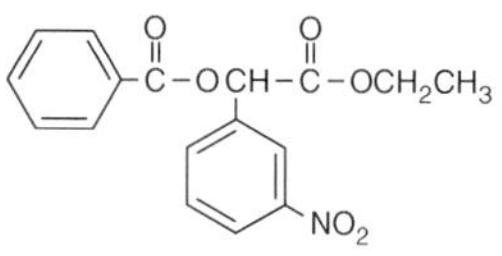
250 mg

S64,311-4
METHYL 4,5-METHYLENEDIOXY-2-NITROBENZOATE
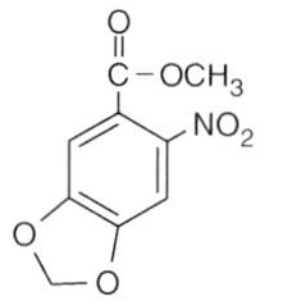
250 mg

S95,528-0
ETHYL 3-HYDROXY-1-OXO-2-INDENECARBOXYLATE, SODIUM SALT
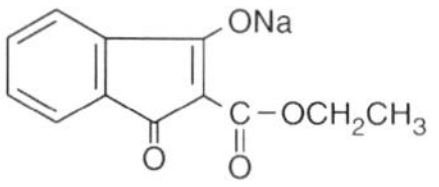
50 mg

S62,934-0
2-(4-TERT-OCTYLPHENOXY)-ETHYL PHTHALATE
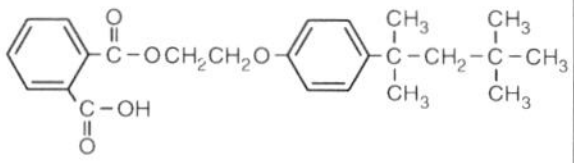
5 gm

S65,651-8
3,4-DIHYDROXY-2,5-DIOXOTETRAHYDROFURAN DI-P-TOLUATE
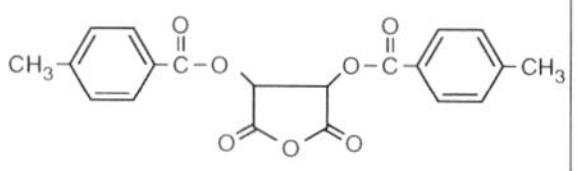
250 mg

S51,161-7
ETHOXYCARBONYLMETHYL BENZOATE
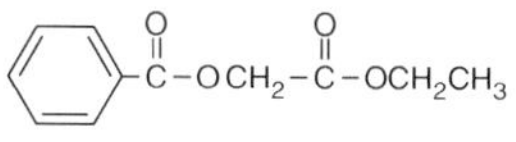
5 gm

S86,032-8
CIS-3,5-CYCLOPENTENEDIOL DIBENZOATE
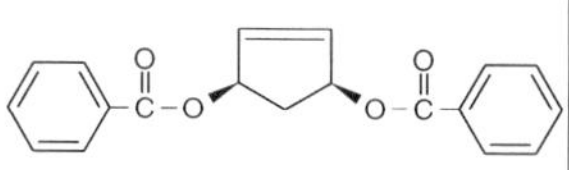
250 mg

S86,022-0
CIS-1,3-CYCLOHEXANEDIOL DIBENZOATE
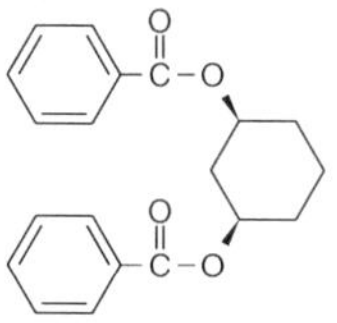
250 mg

S86,023-9
TRANS-1,3-CYCLOHEXANEDIOL DIBENZOATE
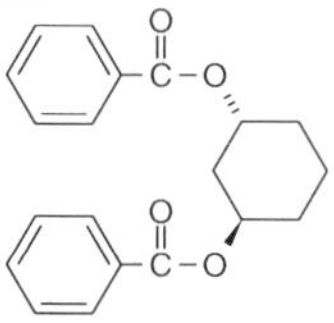
250 mg

S39,674-5
2,2-DIMETHYL-1,3-PROPANEDIOL DIBENZOATE
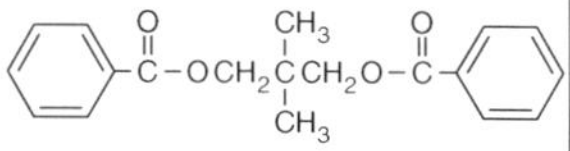
5 gm

S39,675-3
2-HYDROXYMETHYL-2-METHYL-1,3-PROPANEDIOL TRIBENZOATE
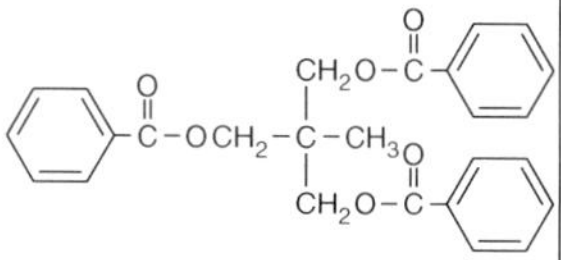
5 gm

S82,610-3
2-TETRAHYDROFURYL SALICYLATE
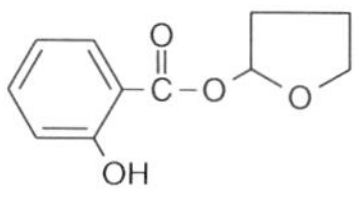
250 mg

S87,775-1
DIMETHYL 2-(4-NITROBENZYLIDENE)-SUCCINATE
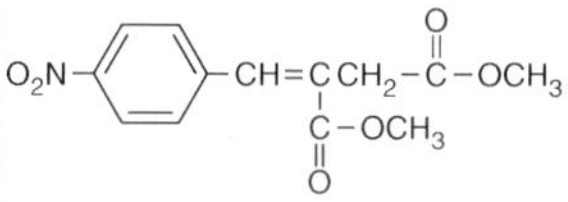
250 mg

S86,535-4
METHYL 2-BENZOYLBENZOATE
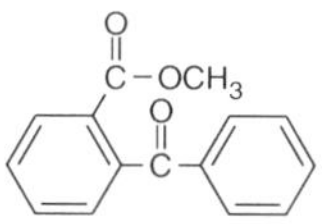
250 mg

S87,743-3
DL-MONO-1-CYCLOHEXYL-3-METHYLBUTYL PHTHALATE
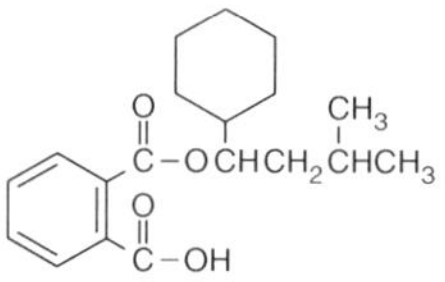
250 mg

S87,635-6
MONO-NEOPENTYL PHTHALATE
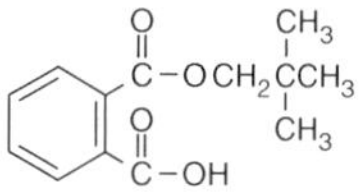
250 mg

S87,688-7
(+)-MONO-1-METHYLHEXYL PHTHALATE
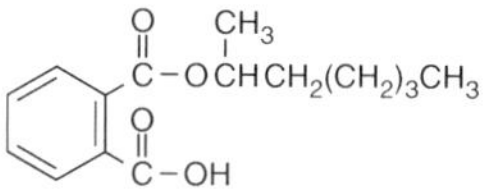
250 mg

S87,947-9
DL-MONO-1-METHYLHEXYL PHTHALATE
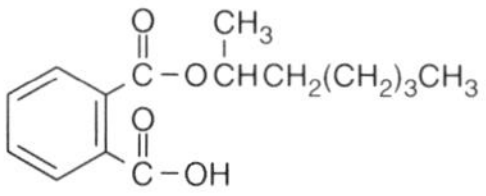
5 gm

S87,694-1
(-)-MONO-2,2-DIMETHYL-1-ISOPROPYLPROPYL PHTHALATE
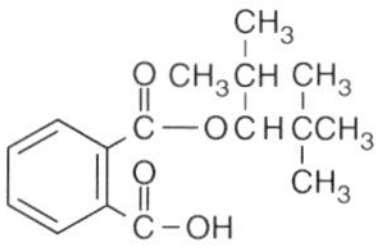
250 mg

S87,946-0
MONO-2,2-DIMETHYL-1-ISOPROPYLPROPYL PHTHALATE
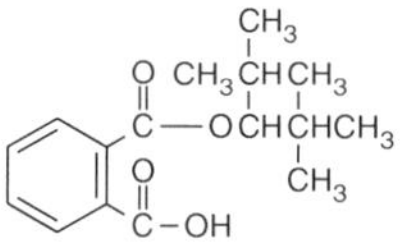
5 gm

S87,693-3
(+)-MONO-1-TERT-BUTYL-3-METHYLBUTYL PHTHALATE
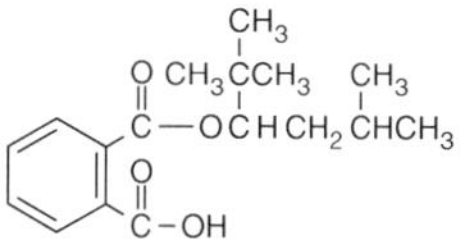
250 mg

S88,797-8
(-)-MONO-1-TERT-BUTYL-3-METHYLBUTYL PHTHALATE, BRUCINE SALT
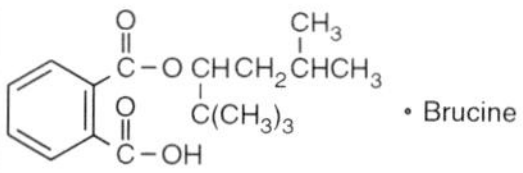
250 mg

S86,138-3
DIMETHYL 5-PYRROLIDINODIBENZO(A,C)-CYCLOOCTENE-6,7-DICARBOXYLATE
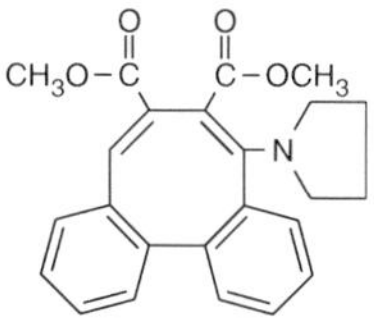
250 mg

S88,153-8
DIMETHYL 2-BENZYLIDENESUCCINATE
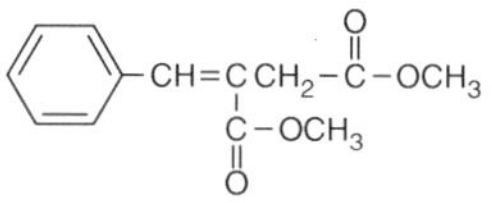
250 mg

S72,192-1
CYCLOHEXANE-1,4-DIOL DIBENZOATE
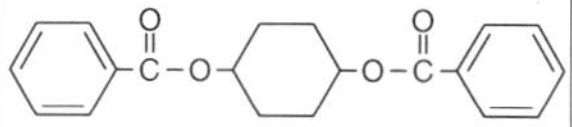
5 gm

S68,560-7
1,4-BUTYLENE GLYCOL DIBENZOATE
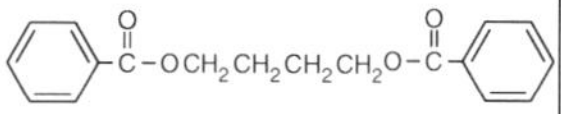
250 mg

S45,971-2
DIISOBUTYL PHTHALATE
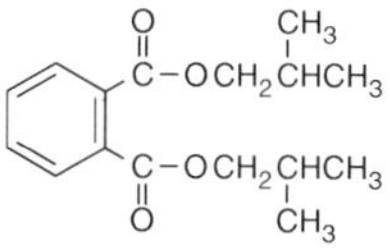
5 gm

S86,598-2
DIMETHYL 9,10-DIHYDRO-9,10-ETHENOANTHRACENE-11,12-DICARBOXYLATE
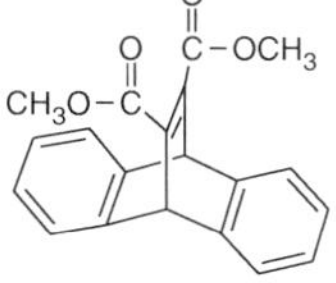
250 mg

S87,748-4
DL-MONO-1-TERT-BUTYLPENTYL PHTHALATE
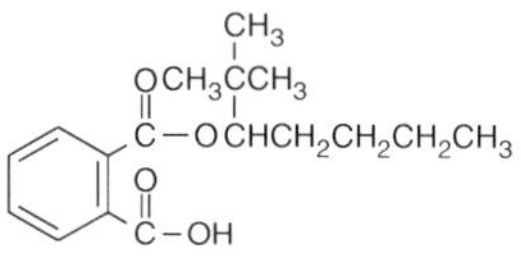
250 mg

S95,865-4
TRANS-TERT-BUTYLCYCLOHEXYL HYDROGEN PHTHALATE
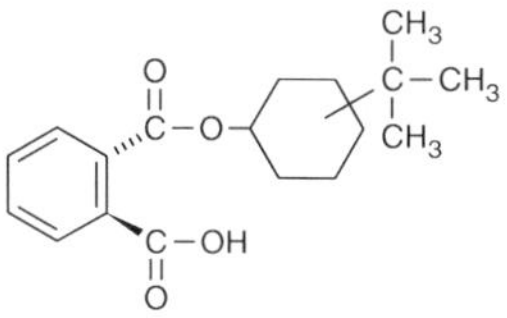
50 mg

S95,648-1
MONO(DECAHYDRONAPHTH-2-YL)-PHTHALATE
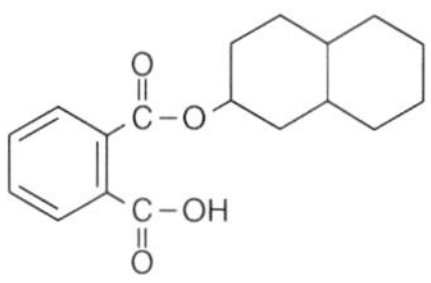
50 mg

S93,937-4
BETA-D-ARABINOSE TETRABENZOATE
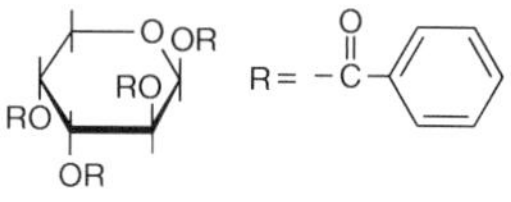
250 mg

S93,929-3
1-O-(4-NITROBENZOYL)-2,3,4,6-TETRA-O-BENZYL-ALPHA-D-GLUCOPYRANOSE

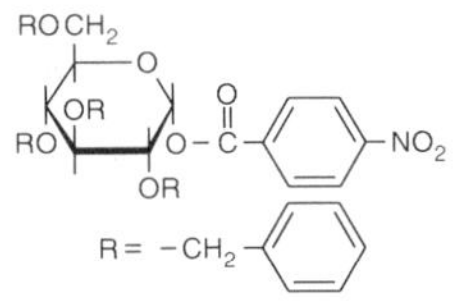

250 mg

S39,672-9
SUCROSE BENZOATE

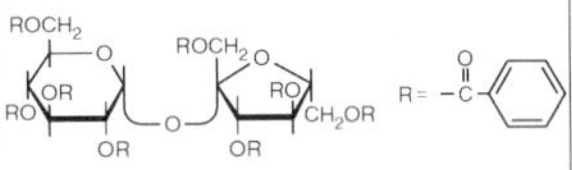

250 mg

S93,936-6
BETA-L-ARABINOPYRANOSE TETRABENZOATE

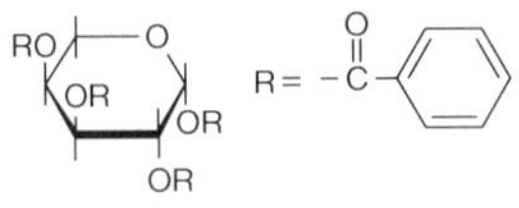

250 mg

S39,611-7
ETHYL 3-[4-(METHYLTHIO)ANILINO]-CROTONATE

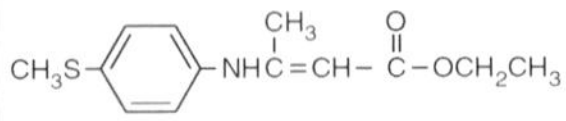

5 gm

S38,189-6
S-(2-DIETHYLAMINOETHYL)-4-NITROTHIOBENZOATE HYDROCHLORIDE

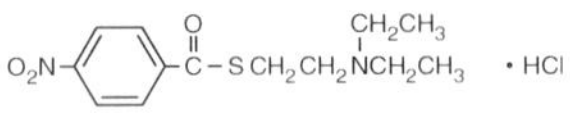

5 gm

S93,538-7
METHYL 2-(4-BROMO-2-NITROPHENYLTHIO)BENZOATE

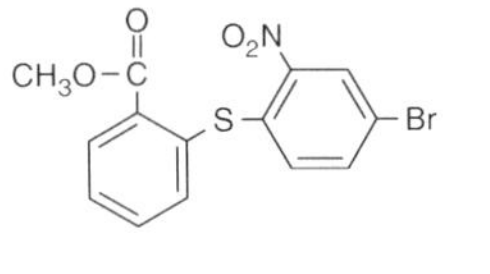

250 mg

S53,262-2
METHYL 4-HEXADECYLTHIO-3-NITROBENZOATE

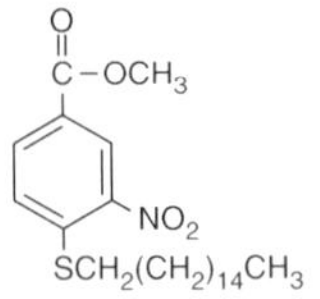

5 gm

S91,787-7
ETHYL 4-(BENZYLTHIO)BENZOATE

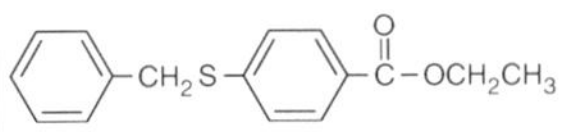

250 mg

S50,734-2
ETHYL 2-(HEXADECYLTHIO)BENZOATE

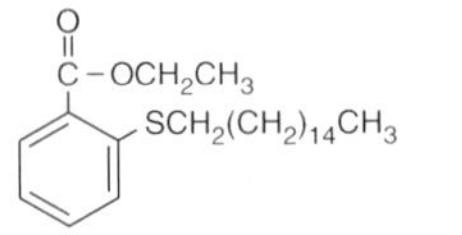

250 mg

S93,993-5
ETHYL 2-(OCTADECYLTHIO)BENZOATE

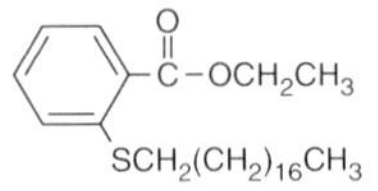

5 gm

S40,660-0
BENZOYL DISULFIDE

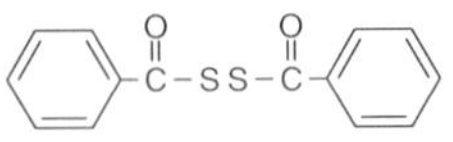

5 gm

S87,979-7
[2-(ETHOXYCARBONYL)PHENYL]-OCTADECYLPROPYLSULFONIUM TETRAFLUOROBORATE

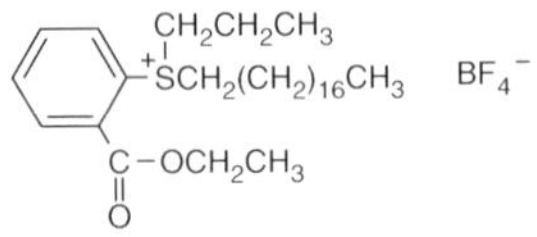

5 gm

S87,553-8
(3H-1,2-BENZODITHIOL-3-YLIDENE)-METHYLSULFONIUM IODIDE

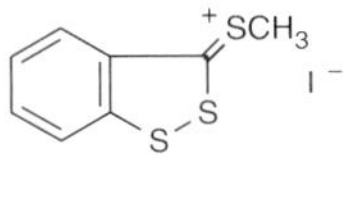

250 mg

S92,896-8
ETHYL 4-AMINO-3-NITROBENZOATE

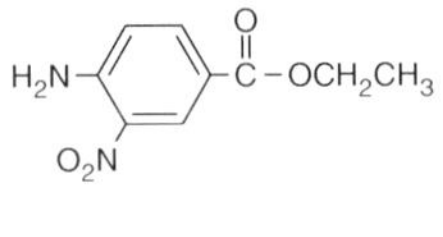

250 mg

S65,548-1
METHYL 4-NITROANTHRANILATE

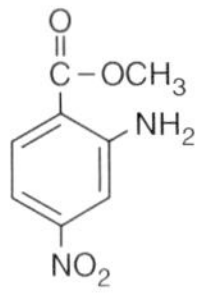

250 mg

S87,708-5
METHYL 3-AMINO-4-TERT-BUTYL-5-NITROBENZOATE

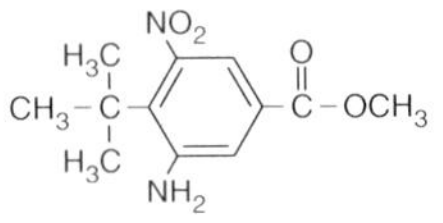

250 mg

S87,705-0
METHYL 4-TERT-BUTYL-3,5-DIAMINOBENZOATE DIHYDROCHLORIDE

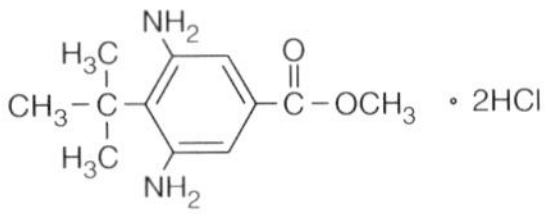

250 mg

S57,982-3
2-(DIMETHYLAMINO)ETHYL-4-AMINOBENZOATE

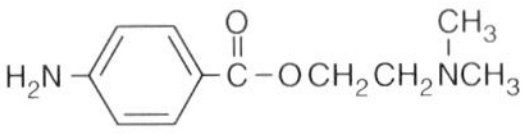

5 gm

S74,486-7
DIMETHYL 4-AMINOPHTHALATE

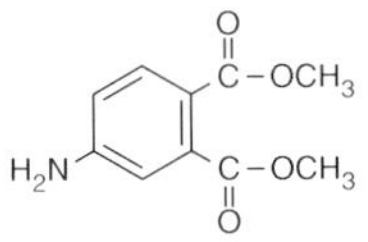

250 mg

S66,222-4
METHYL 3-AMINOBENZOATE

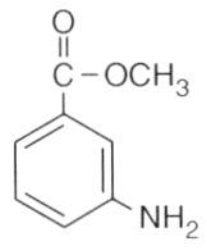

5 gm

S61,772-5
HEPTYL ANTHRANILATE

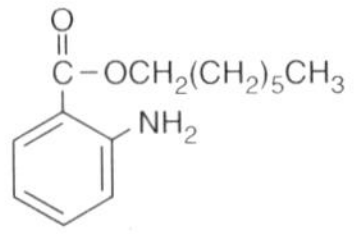

5 gm

S63,045-4
OCTYL ANTHRANILATE

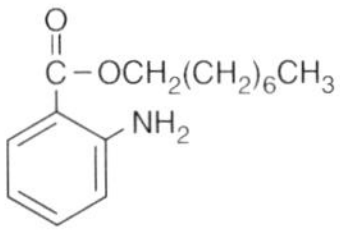

250 mg

S52,060-8
ISOBUTYL 4-AMINOBENZOATE

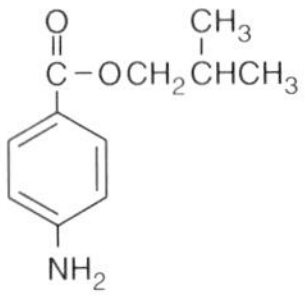

5 gm

S65,286-5
METHYL 3-AMINO-4-METHYLBENZOATE

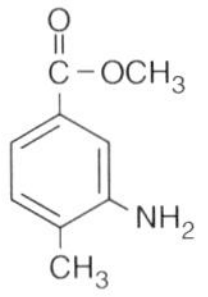

250 mg

S35,433-3
PHENETHYL ANTHRANILATE

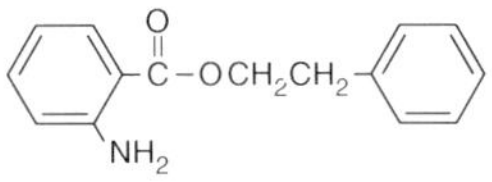

1 gm

S53,404-8
2-AMINOETHYL BENZOATE HYDROBROMIDE

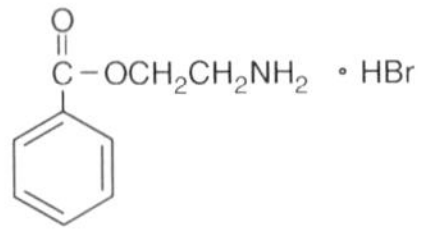

250 mg

S44,506-1
S-BENZOYLCYSTEAMINE HYDROCHLORIDE

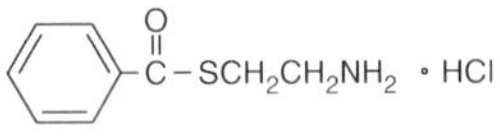

5 gm

S74,987-7
METHYL 7-AMINO-1-FLUORENECARBOXYLATE

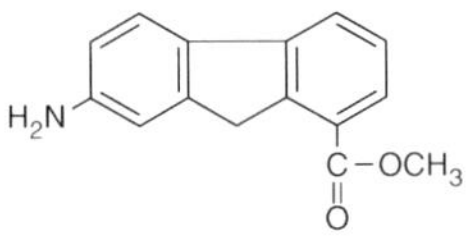

250 mg

S41,224-4
S-BENZOYLCYSTEAMINE

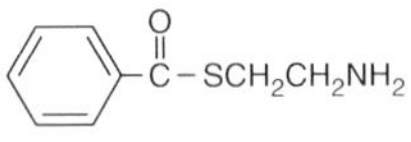

5 gm

S62,406-3
METHYL N-(2,4,6-TRINITROPHENYL)-ANTHRANILATE

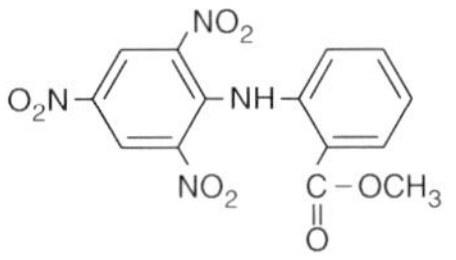

5 gm

S58,961-6
ETHYL 4-(PICRYLAMINO)BENZOATE

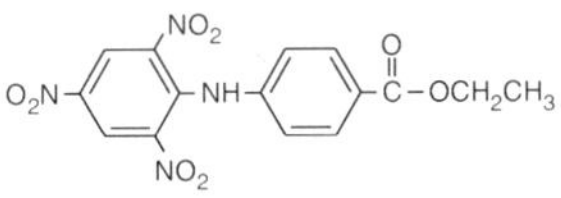

250 mg

S64,729-2
ETHYL BENZOYLFORMATE-2,4-DINITROPHENYLHYDRAZONE

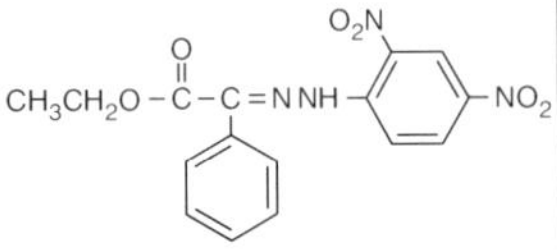

250 mg

S90,146-6
2-(2,6-DINITRO-4-METHYLANILINO)-ETHYL 4-NITROBENZOATE

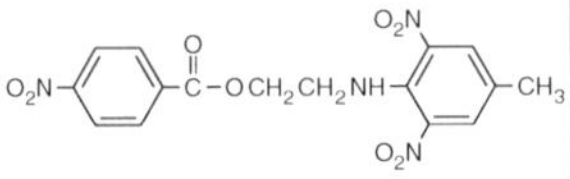

250 mg

S63,021-7
METHYL 4-(N-HEXADECYL-N-METHYLAMINO)-3-NITROBENZOATE

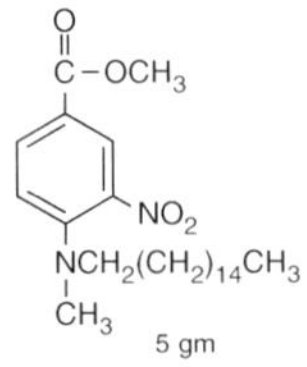

5 gm

S62,460-8
DIETHYL KETOMALONATE, 2-NITROPHENYLHYDRAZONE

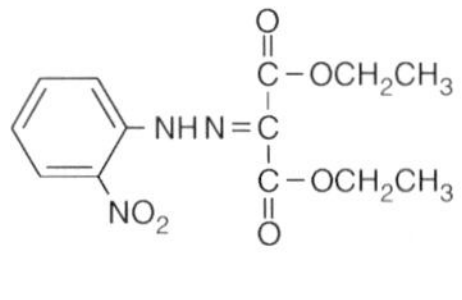

5 gm

S50,041-0
ETHYL 2-CHLORO-4,5-DINITROBENZOATE

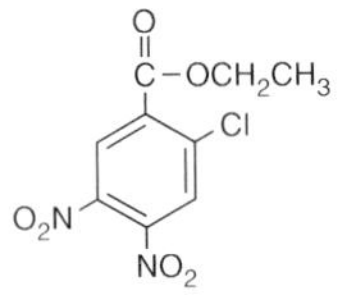

250 mg

S87,749-2
(+)-1-TERT-BUTYLPENTYL-3,5-DINITROBENZOATE

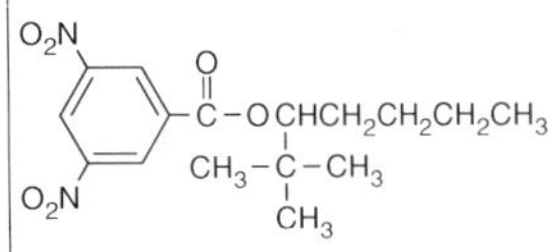

250 mg

S87,709-3
METHYL 4-TERT-BUTYL-3,5-DINITROBENZOATE

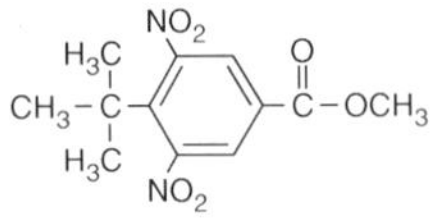

250 mg

S75,428-5
ETHYL 4,6-DINITRO-2-BIPHENYLCARBOXYLATE

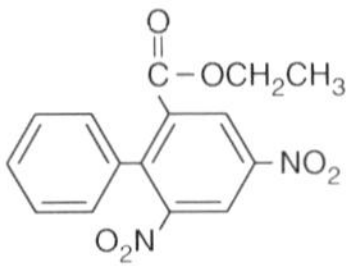

250 mg

S35,274-8
METHYL 2-CHLORO-3,5-DINITROBENZOATE

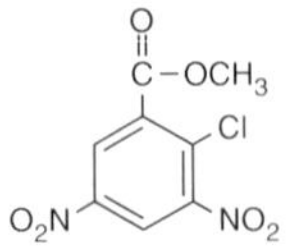

1 gm

S35,275-6
METHYL 4-CHLORO-3,5-DINITROBENZOATE
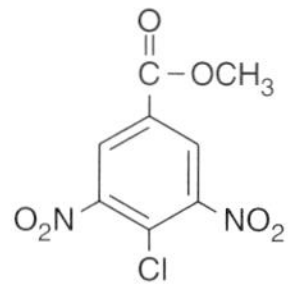
5 gm

S95,393-8
ETHYL 2,3-DIOXOBUTYRATE 2-(4-NITROPHENYLHYDRAZONE)
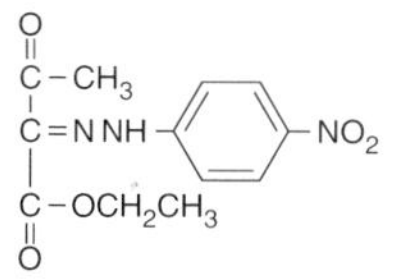
50 mg

S95,444-6
DIETHYL 6,6'-DINITRO-2,2'-BIPHENYLDICARBOXYLATE
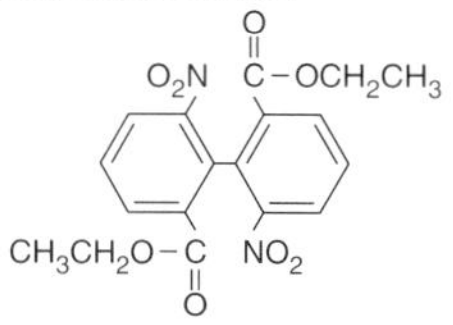
50 mg

S76,397-7
METHYL 4,4-DIBENZYL-4-NITROCROTONATE
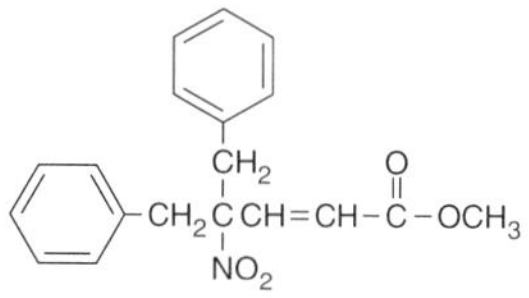
250 mg

S74,676-2
ETHYL 4-[4-(DIMETHYLAMINO)-BENZYLAMINO]BENZOATE
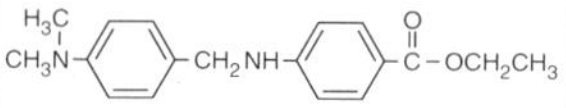
5 gm

S62,123-4
ALPHA,ALPHA'-BIS(4-CARBETHOXYANILINO)-P-XYLENE
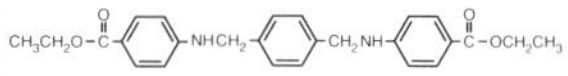
250 mg

S59,372-9
DIETHYL (2,4-XYLIDINOMETHYLENE)-MALONATE
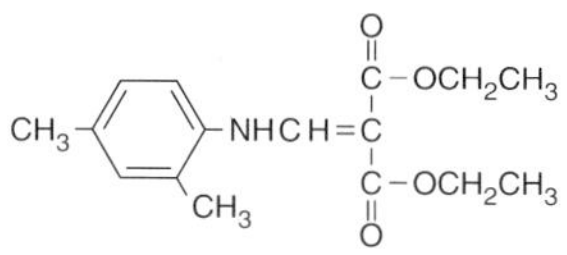
5 gm

S65,855-3
2-(ISOBUTYLAMINO)ETHYL-3-NITROBENZOATE HYDROCHLORIDE
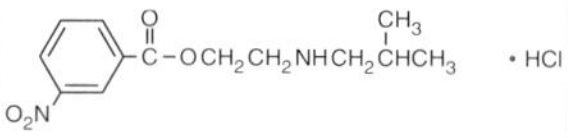
5 gm

S36,148-8
ETHYL 3-ANILINOCROTONATE
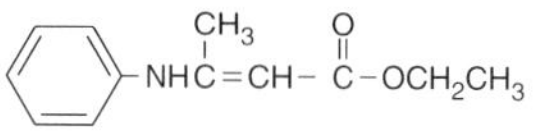
5 gm

S95,376-8
N-[2-(METHOXYCARBONYL)PHENYL]-GLYCINE METHYL ESTER
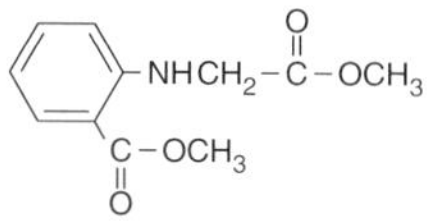
50 mg

S56,383-8
METHYL N-METHYLANTHRANILATE
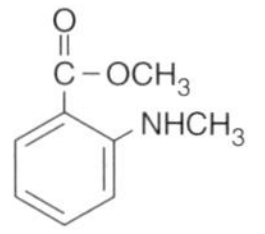
5 gm

S94,326-6
METHYL 4-(METHYLAMINO)BENZOATE
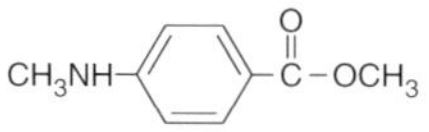
250 mg

S90,759-6
ETHYL 2-[4-(DIMETHYLAMINO)BENZOYL]-BENZOATE
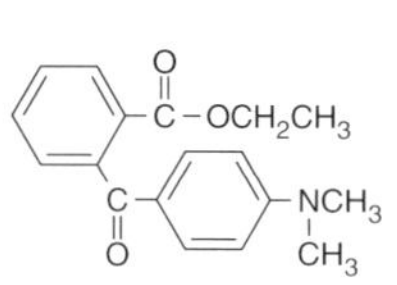
250 mg

S83,312-6
METHYL 3-(DIPHENYLAMINO)BENZOATE
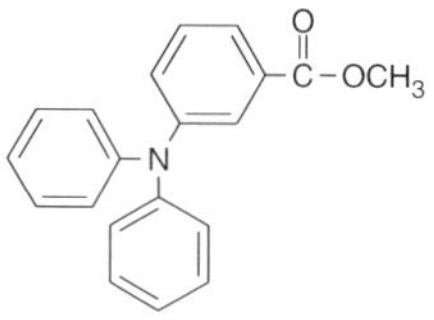
250 mg

S83,313-4
METHYL 4-(DIPHENYLAMINO)BENZOATE
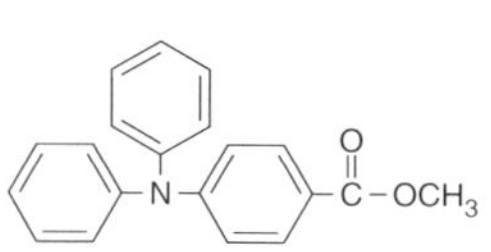
250 mg

S88,682-3
METHYL 4-(DIMETHYLAMINO)BENZOATE
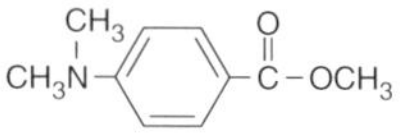
250 mg

S67,115-0
TILORONE ANALOG R 9536-DA
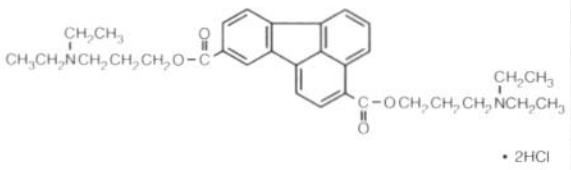
250 mg

S65,976-2
1-DIMETHYLAMINO-2-PROPYL BENZOATE HYDROCHLORIDE
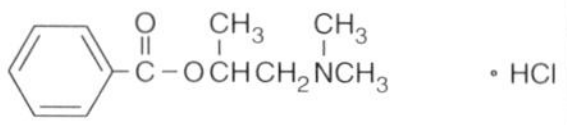
5 gm

S65,980-0
1-DIMETHYLAMINO-2-PROPYL O-TOLUATE HYDROCHLORIDE
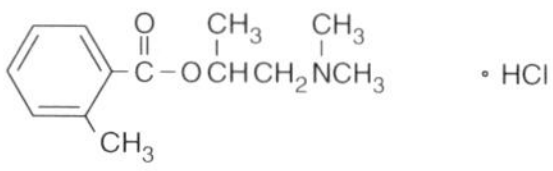
5 gm

S65,997-5
1-DIMETHYLAMINO-2-PROPYL P-TOLUATE HYDROCHLORIDE
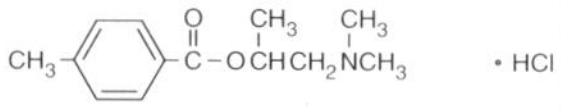
5 gm

S82,535-2
2-(DIETHYLAMINO)ETHYL 2-NAPHTHALENECARBOXYLATE
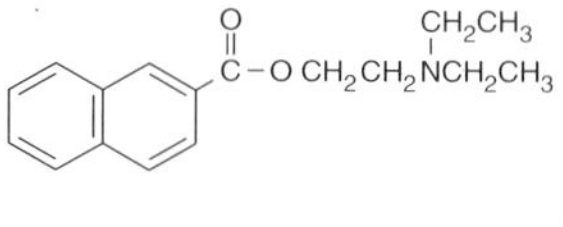
250 mg

S92,715-5
BENZYL 1-(DIPHENYLMETHYL)-2-AZETIDINECARBOXYLATE
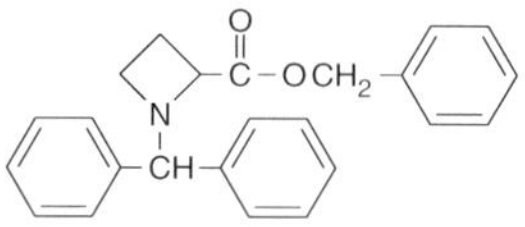
250 mg

S40,325-3
3-(2-METHYLPIPERIDINO)PROPYL-BENZOATE HYDROCHLORIDE
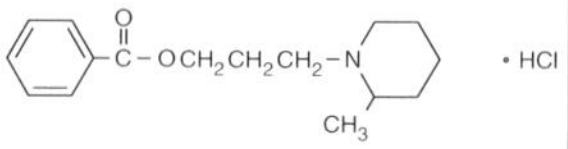
250 mg

S38,398-8
5-(DIETHYLAMINO)PENTYL BENZOATE
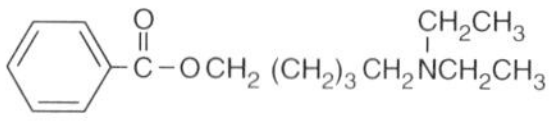
5 gm

S76,305-5
ETHYL 2-[3-(DIMETHYLAMINO)-1-PHENYLPROPYL]BENZOATE
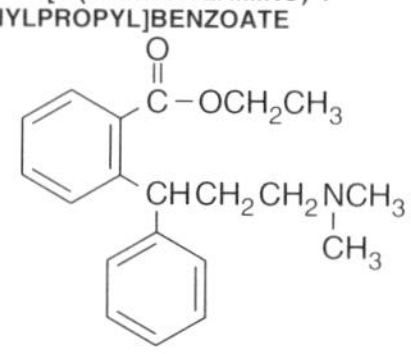
250 mg

S44,793-5
S-PIPERIDINOMETHYL THIOBENZOATE HYDROCHLORIDE
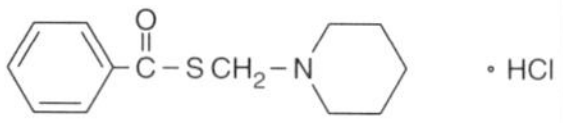
5 gm

S41,868-4
METHYL 2-(DIMETHYLAMINO)BENZOATE
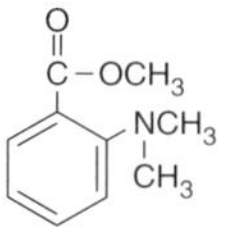
5 gm

S67,669-1
DIMETHYL KETOMALONATE PHENYLHYDRAZONE
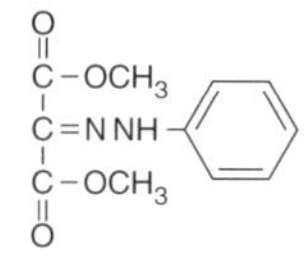
5 gm

S91,922-5
METHYL 4-(1-METHYL-1-NITROETHYL)-BENZOATE
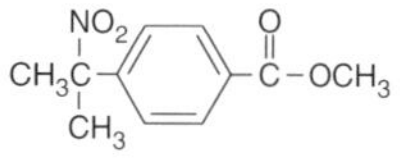
250 mg

S92,087-8
4,4'-DIIMINOBIS(ETHYL BENZOATE)
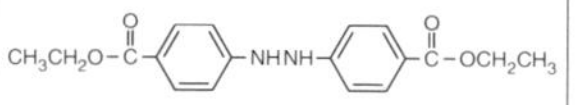
250 mg

S59,227-7
METHYL 2-HYDRAZINOBENZOATE HYDROCHLORIDE
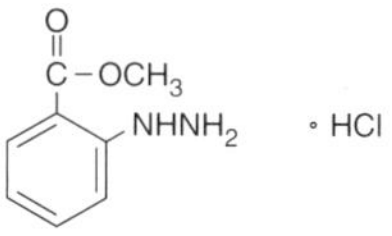
250 mg

S50,204-9
METHYL 4-CHLORO-3-HYDRAZINO-BENZOATE
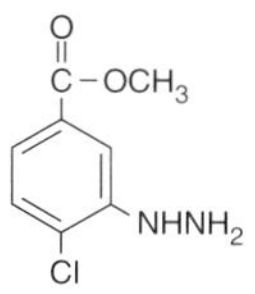

S67,931-3
N-BENZOYLMONOETHANOLAMINE BENZOATE
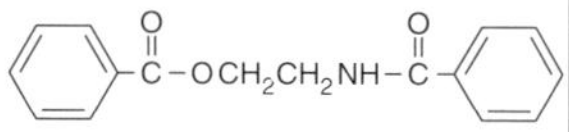
250 mg

S75,808-6
1-METHYL 3-NITROPHTHALATE
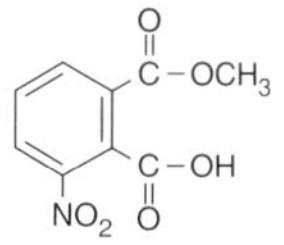
250 mg

S78,213-0
METHYL 2-CARBOXY-6-NITROBENZOATE
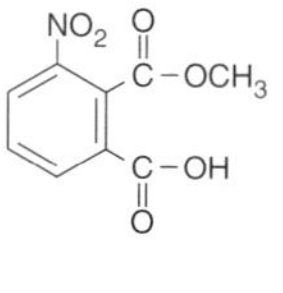
5 gm

S62,946-4
METHYL 4-CARBOXY-2-NITROBENZOATE
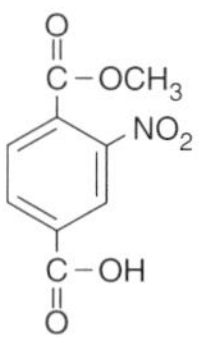
250 mg

S70,660-4
ETHYL 3-NITROBENZOATE
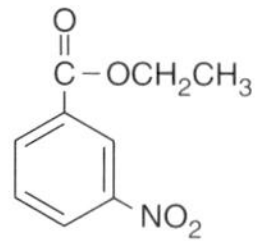
5 gm

S69,402-9
DIMETHYL 2-NITROTEREPHTHALATE
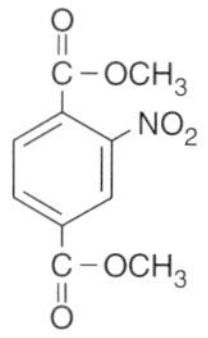
5 gm

S36,601-3
DIMETHYL 5-NITROISOPHTHALATE
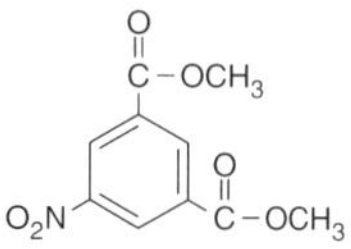
5 gm

S86,706-3
2-(7,7-DIMETHYLBICYCLO[3.1.1]HEPT-2-EN-2-YL)ETHYL 4-NITROBENZOATE
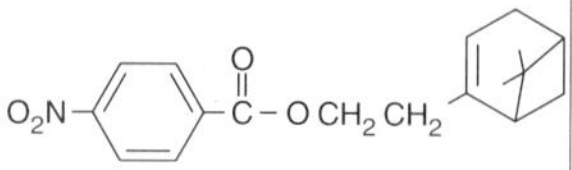
250 mg

S92,638-8
2-TERT-BUTYLCYCLOHEXYL-4-NITROBENZOATE
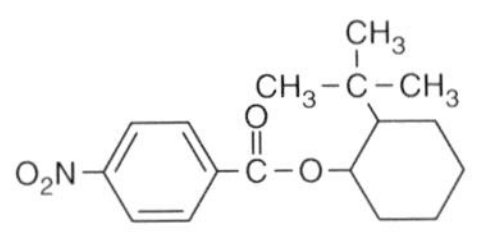
250 mg

S86,554-0
CINNAMYL 4-NITROBENZOATE
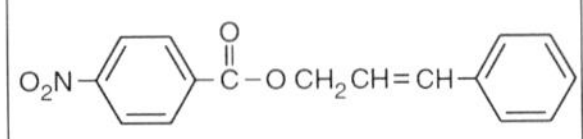
250 mg

S92,635-3
ALPHA-METHYLCINNAMYL-4-NITROBENZOATE
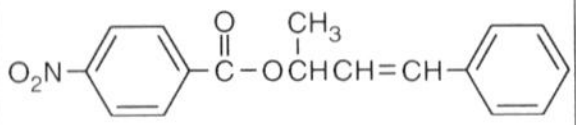
250 mg

S92,636-1
ALPHA-(1-PROPEN-1-YL)BENZYL-4-NITROBENZOATE
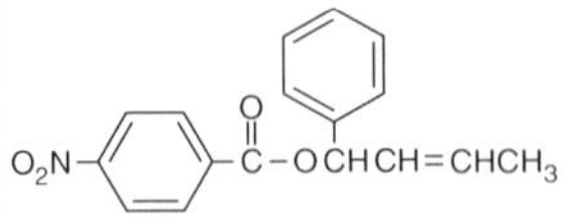
250 mg

S89,459-1
4-NITROBENZYL TRICHLOROACRYLATE
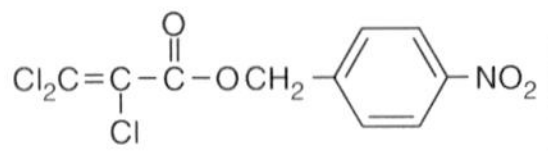
250 mg

S92,631-0
3-METHYL-2-PROPEN-1-YL-4-NITROBENZOATE
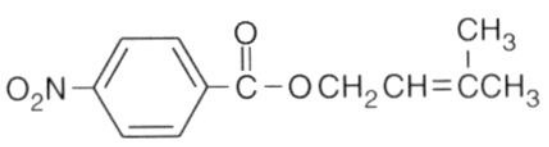
250 mg

S92,632-9
1,1-DIMETHYL-2-PROPEN-1-YL-4-NITROBENZOATE
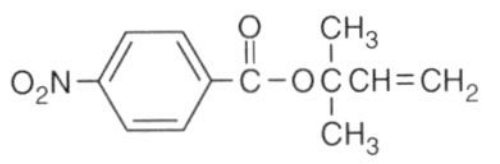
250 mg

S92,646-9
2-METHYL-ENDO-NORBORNYL-4-NITROBENZOATE
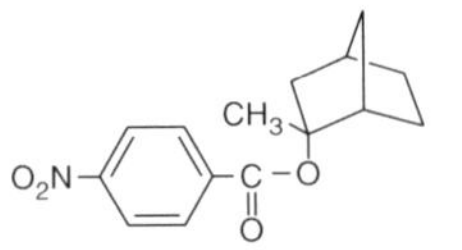
250 mg

S91,466-5
METHYL 5-METHYL-2-NITROBENZOATE
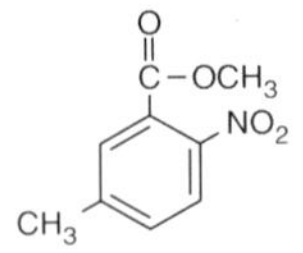
250 mg

S92,268-4
METHYL 10-NITRO-9-ANTHRACENECARBOXYLATE
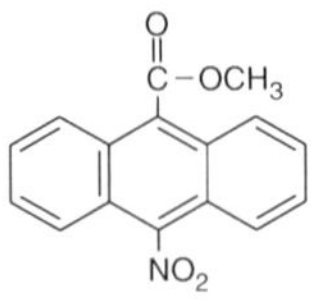
250 mg

S92,563-2
2-(ALPHA-ETHOXYCARBONYL-4-NITROBENZYLIDENE)-3-METHYLBENZOTHIAZOLINE
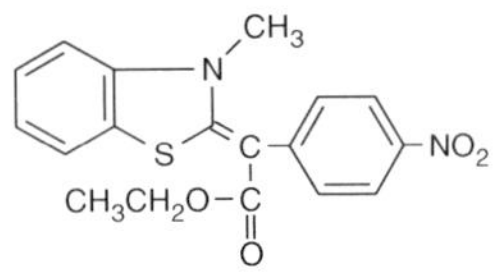
250 mg

S75,643-1
METHYL 7-NITRO-9-OXO-1-FLUORENECARBOXYLATE
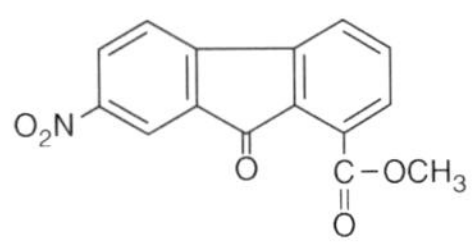
5 gm

S60,496-8
HEXYL 3-NITROPHTHALATE
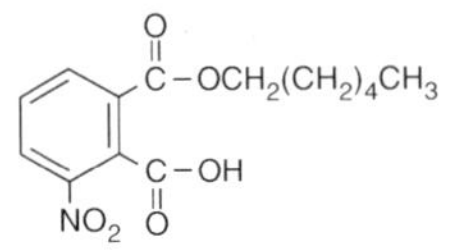
250 mg

S63,014-4
METHYL 4-NITROPHTHALATE
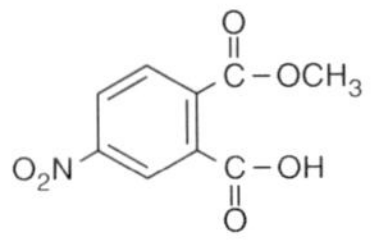
5 gm

S92,645-0
2-METHYL-EXO-NORBORNYL-4-NITROBENZOATE
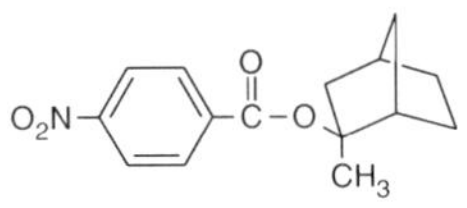
250 mg

S43,500-7
ETHYL 2-NITROBENZOATE
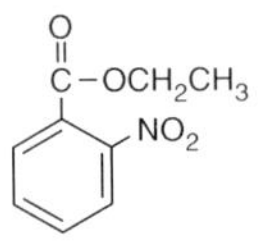
5 gm

S61,978-7
ISOBORNYL 4-NITROBENZOATE
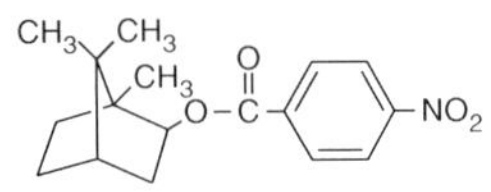
5 gm

S59,299-4
METHYL 2-BROMO-3-NITROBENZOATE
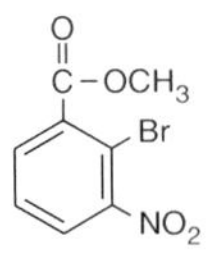
250 mg

S63,872-2
ETHYL 2-BROMO-3-NITROBENZOATE
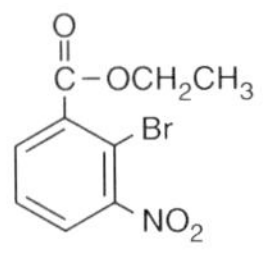
250 mg

S44,838-9
METHYL 2-CHLORO-4-NITROBENZOATE
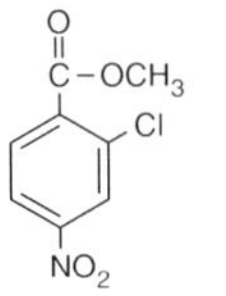
5 gm

S37,465-2
METHYL 4-NITRO-3-METHYLBENZOATE
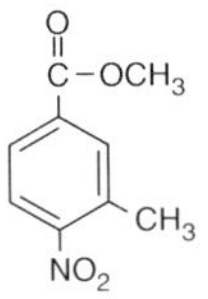
5 gm

S92,644-2
1-METHYL-2-NORBORNYL-3,5-DINITROBENZOATE
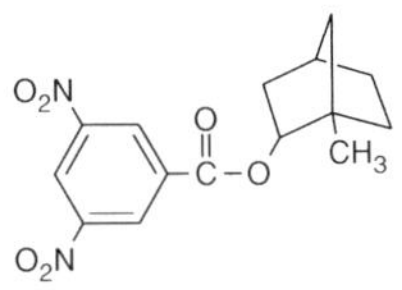
250 mg

S74,887-0
METHYL 7-NITRO-1-FLUORENECARBOXYLATE
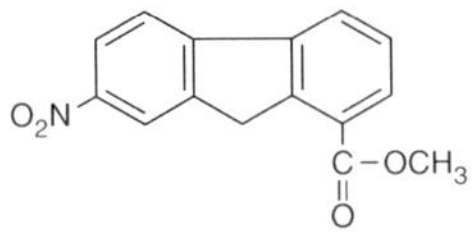
250 mg

S35,311-6
METHYL 3-NITRO-1-NAPHTHOATE
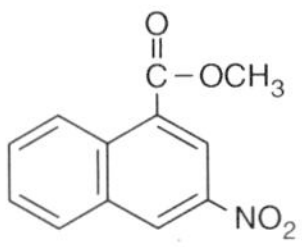
5 gm

S79,361-2
1-METHYLPROPYL BENZOYLFORMATE
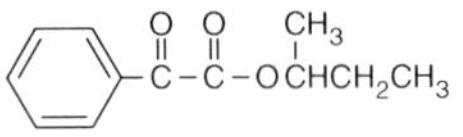
250 mg

S79,276-4
3-METHYL-2-BUTYL BENZOYLFORMATE
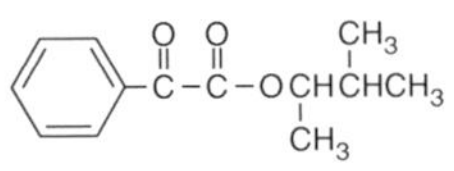
250 mg

S79,345-0
2,2-DIMETHYL-1-PHENYLPROPYL-BENZOYLFORMATE
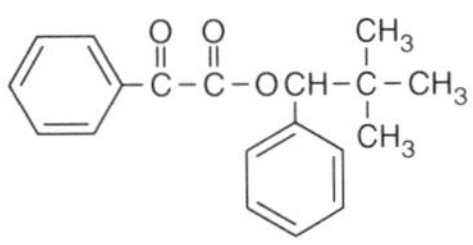
250 mg

S39,169-7
ETHYL 2-(METHOXYCARBONYL)-BENZOYLACETATE
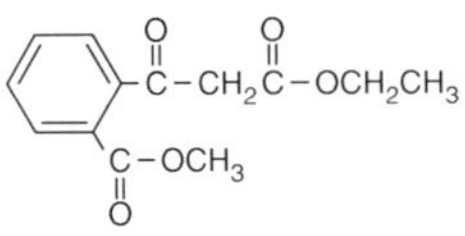
250 mg

S44,366-2
ETHYL BENZOYLACETATE, THALLIUM (I) SALT
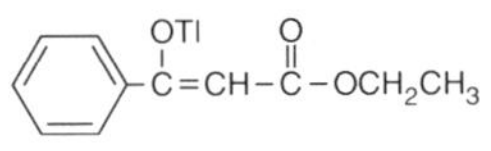
250 mg

S93,373-2
DIMETHYL 10,10'-BIS(2-ANTHRONECARBOXYLATE)
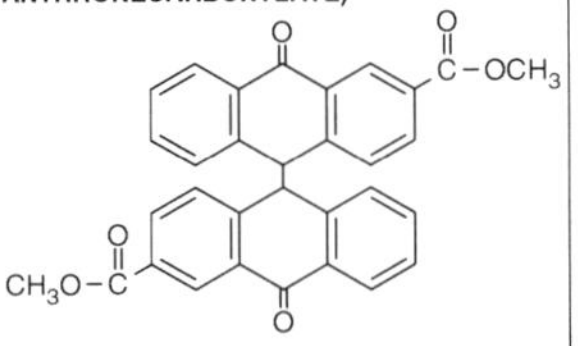
250 mg

S89,531-8
METHYL 6-(9-FLUORENYL)-3-OXO-1-CYCLOHEXENE-1-CARBOXYLATE
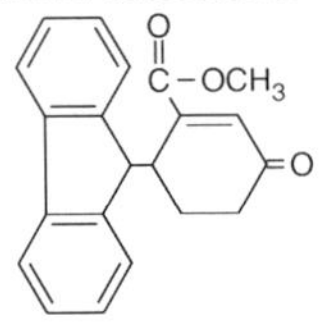
250 mg

S68,136-9
ETHYL 2-BENZOYLBENZOATE
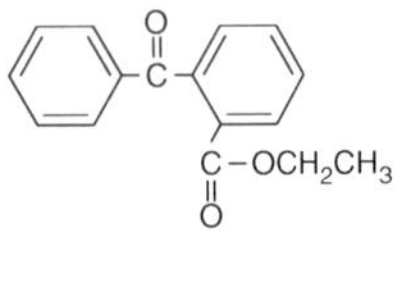
5 gm

S84,975-8
METHYL 9-OXO-1-FLUORENECARBOXYLATE
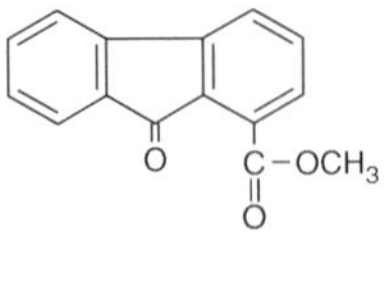
250 mg

S75,645-8
METHYL 7-CHLORO-9-OXO-1-FLUORENECARBOXYLATE
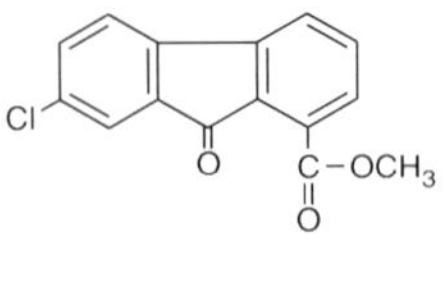
5 gm

S86,017-4
1,3-DIHYDROXYACETONE DIBENZOATE
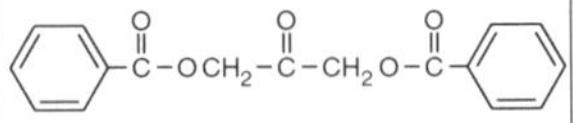
250 mg

S92,630-2
1-BENZOYLETHYL BENZOATE
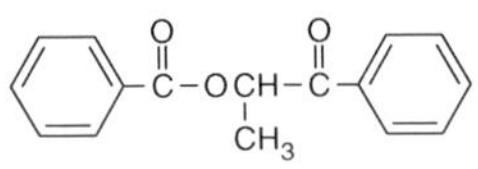
250 mg

S88,844-3
METHYL 7-OXO-7H-BENZO(C)FLUORENE-6-CARBOXYLATE
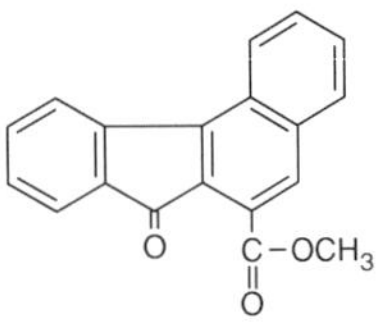
250 mg

S86,063-8
METHYL 2,4-DIOXO-5-PHENYLPENTANOATE
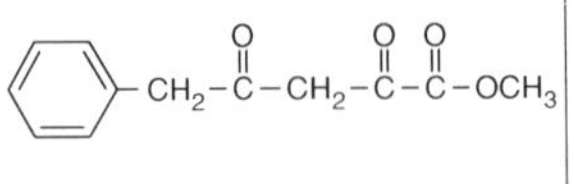
250 mg

S75,443-9
ETHYL 2-(2-IMIDAZOLIDINYLIDENE)-3-OXO-3-PHENYLPROPIONATE
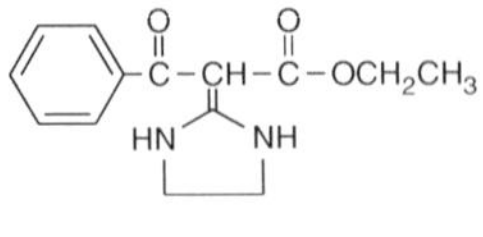
250 mg

S76,629-1
DIMETHYL 1-OXO-2,7-INDANDICARBOXYLATE
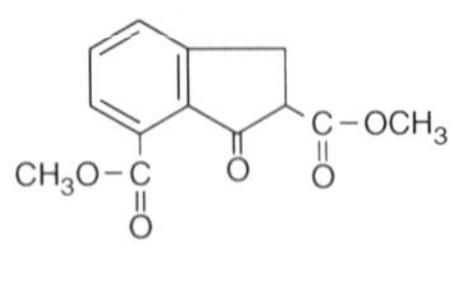
250 mg

S79,382-5
(+)-MONO-(2,2-DIMETHYL-1-PHENYLPROPYL) PHTHALATE
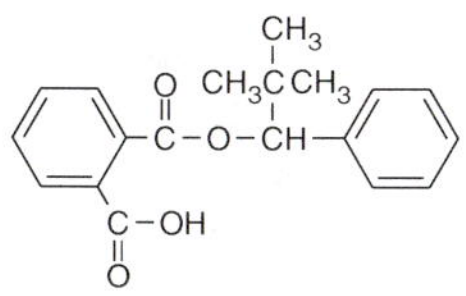
250 mg

S79,388-4
(-)-MONO-(2,2-DIMETHYL-1-PHENYLPROPYL) PHTHALATE
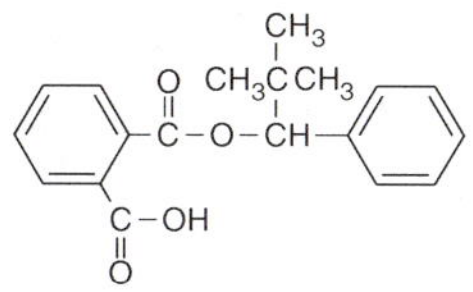
250 mg

S79,349-3
MONO(1,2-DIMETHYLPROPYL) PHTHALATE
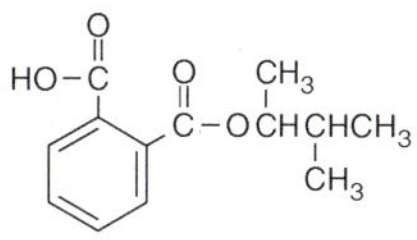
250 mg

S79,280-2
(+)-MONO-(1,2,2-TRIMETHYLPROPYL)-PHTHALATE
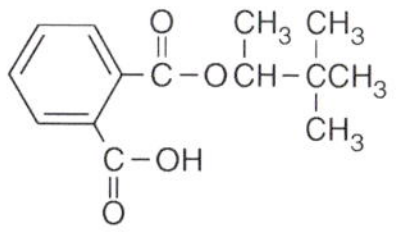
250 mg

S79,392-2
(-)-MONO-(2,2-DIMETHYL-1-ETHYLPROPYL) PHTHALATE
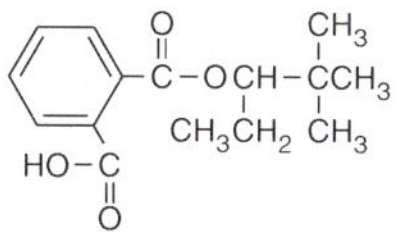
250 mg

S72,188-3
(+)-MONO(2-OCTYL) PHTHALATE
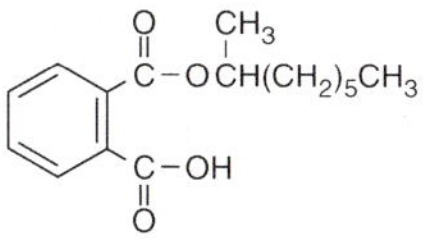
5 gm

S85,282-1
1-METHYLHEPTYL 2-CARBOXYBENZOATE
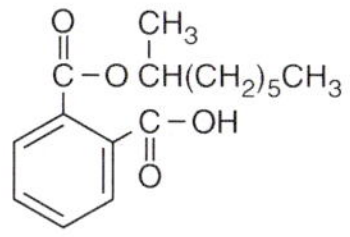
250 mg

S84,260-5
DIMETHYL 2,2'-BIPHENYLDICARBOXYLATE
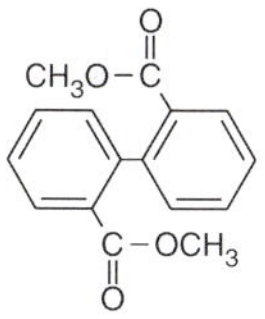
250 mg

S97,345-9
AR-D45 DIMETHYL-4,5-PHENANTHRENEDICARBOXYLATE
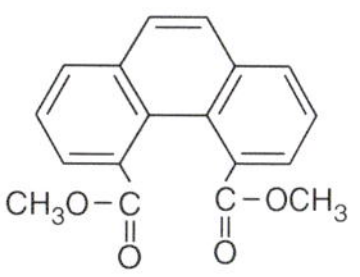
10 mg

S61,868-3
CYCLOHEXYL TETRACHLOROPHTHALATE
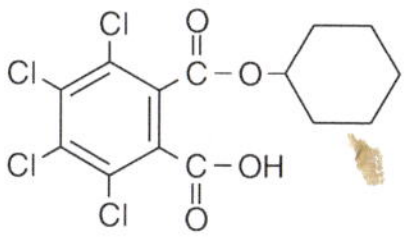
5 gm

S60,408-9
SEC-BUTYL PHTHALATE
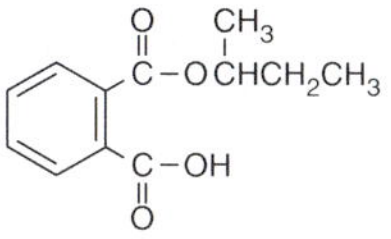
5 gm

S43,279-2
MONO-BENZYL PHTHALATE
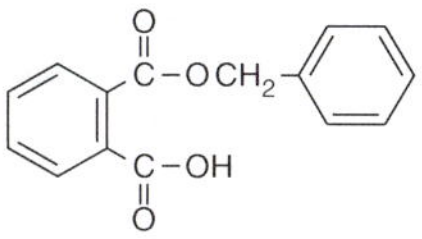
5 gm

S92,643-4
MONO-1-METHYL-2-NORBORNYL-PHTHALATE
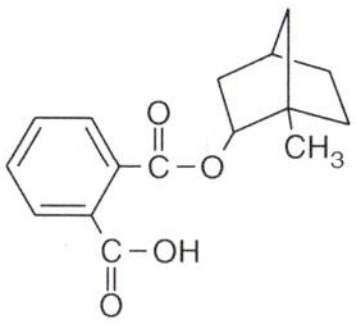
250 mg

S41,777-7
BIS(3,3,5-TRIMETHYLCYCLOHEXYL)-PHTHALATE
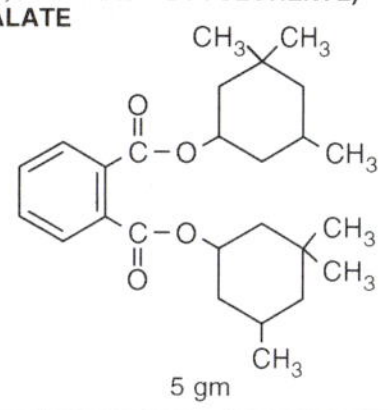
5 gm

S57,324-8
BIS(3,5,5-TRIMETHYLHEXYL)-PHTHALATE
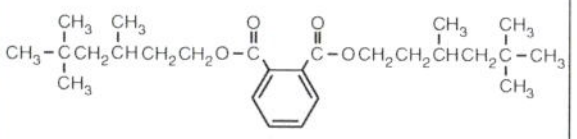
5 gm

S44,253-4
DIMETHYL TEREPHTHALATE, ZONE REFINED
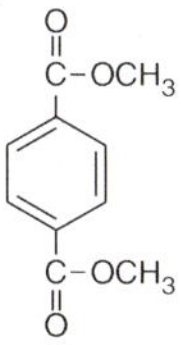
5 gm

S45,733-7
DIETHYL 4,4'-BIPHENYLDICARBOXYLATE
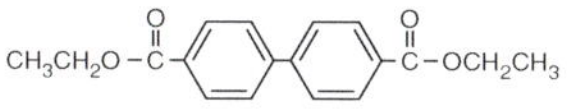
250 mg

S66,243-7
DIMETHYL 6,6'-BIS(1-BROMOETHYL)-2,2'-BIPHENYLDICARBOXYLATE
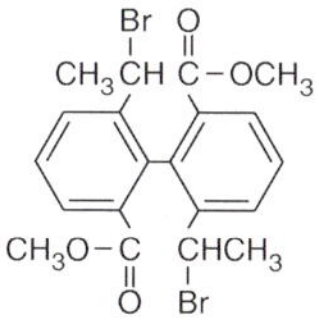
250 mg

S65,321-7
TETRAMETHYL 1,2,4,5-BENZENETETRACARBOXYLATE
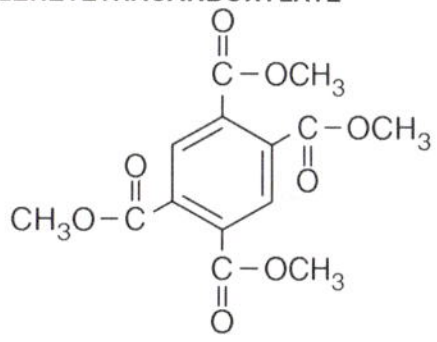
5 gm

S70,974-3
CYCLOPENTYL BENZOATE
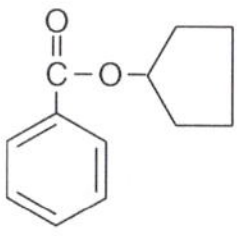
5 gm

S86,383-1
2,5-NORBORNADIEN-7-YL BENZOATE
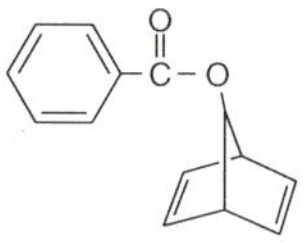
1 gm

S86,538-9
METHYL 3-PHENYL-2,5-NORBORNADIENE-2-CARBOXYLATE
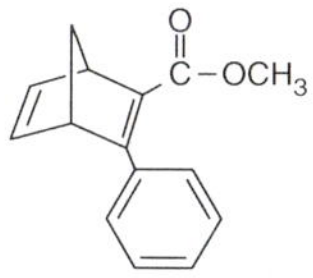
250 mg

S90,613-1
ETHYL 2-(1,2,3,4-TETRAHYDRO-1-FLUORENYLIDENE)ACETATE
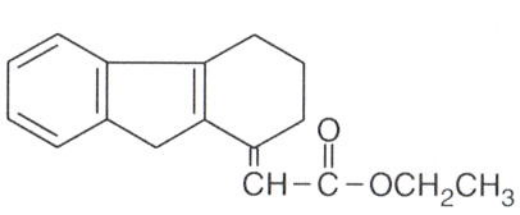
250 mg

S69,677-3
METHYL PIPERONYLATE
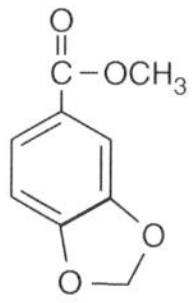
250 mg

S74,807-2
ETHYL 1,2,3,5,6,7,8,9-OCTAHYDRO-5-PHENYLIMIDAZO(1,2-A)QUINOLINE-4-CARBOXYLATE
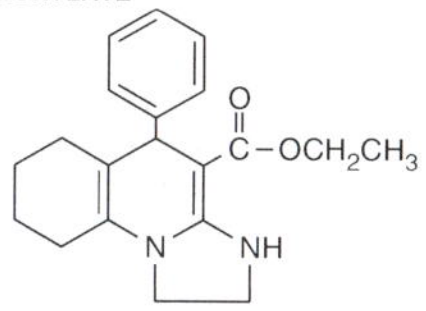
250 mg

S76,796-4
4-METHYLENECYCLOHEXYL BENZOATE
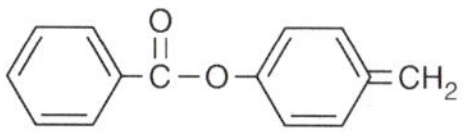
250 mg

S79,205-5
4-BROMO-3,5-DIMETHYLCYCLOHEXYL BENZOATE
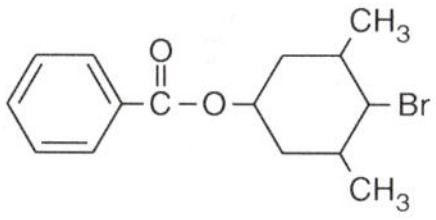
250 mg

S92,834-8
1,3-DIOXAN-5-YL BENZOATE
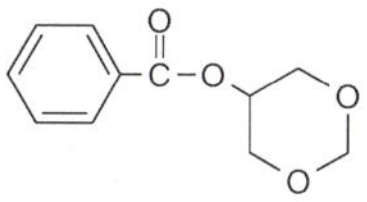
250 mg

S43,786-7
ALLYL BENZOATE
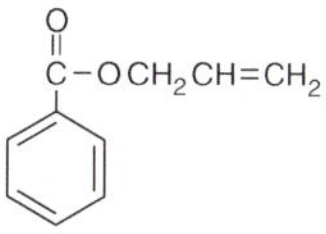
5 gm

S87,717-4
SEC-BUTYL BENZOATE
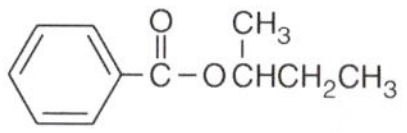
250 mg

S56,583-0
ISOPENTYL BENZOATE
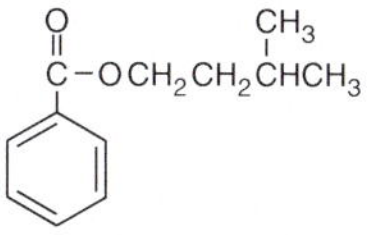
5 gm

S52,804-8
PHENETHYL BENZOATE
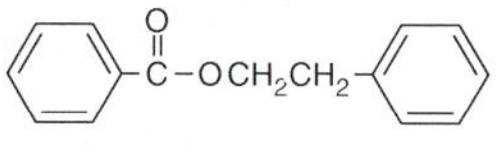
5 gm

S87,797-2
METHYL 4-STILBENECARBOXYLATE
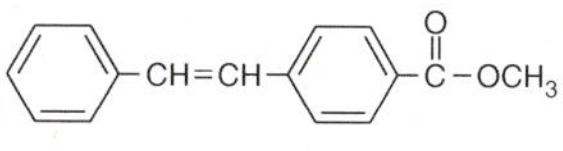
250 mg

S88,225-9
METHYL 4-(4-PHENYL-1,3-BUTADIEN-1-YL)BENZOATE
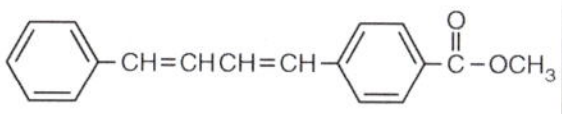
250 mg

S82,656-1
TRIPHENYLMETHYL BENZOATE
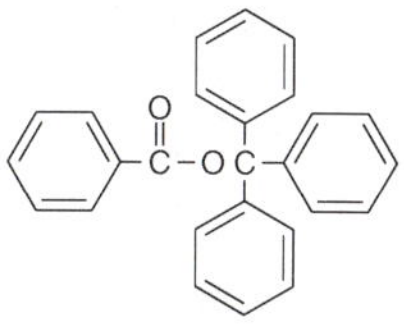
250 mg

S92,278-1
METHYL 4-BIPHENYLCARBOXYLATE
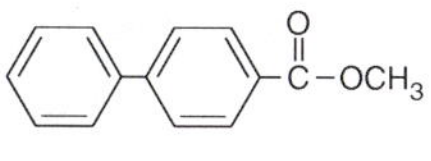
250 mg

S43,645-3
METHYL 2-FLUOROBENZOATE
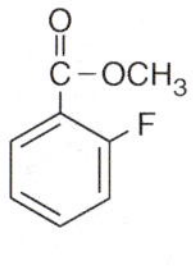
5 gm

S65,623-2
METHYL 2-CHLOROBENZOATE
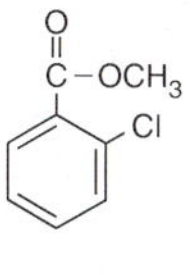
250 mg

S64,636-9
METHYL 4-BROMOBENZOATE
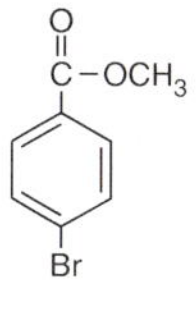
250 mg

S61,686-9
ETHYL P-METHYLBENZOATE
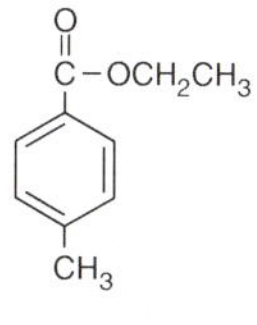
5 gm

S63,497-2
METHYL 3,4-DICHLOROBENZOATE
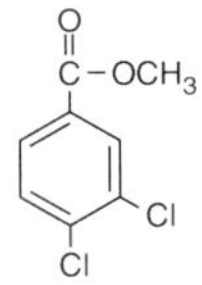
5 gm

S43,424-8
ETHYL 3,4-DIMETHYLBENZOATE
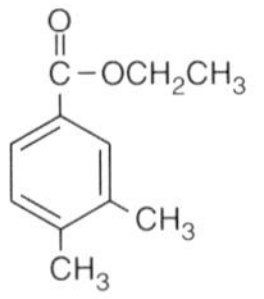
5 gm

S70,148-3
9-ANTHRACENYLMETHYL 2,4,6-TRIMETHYLBENZOATE
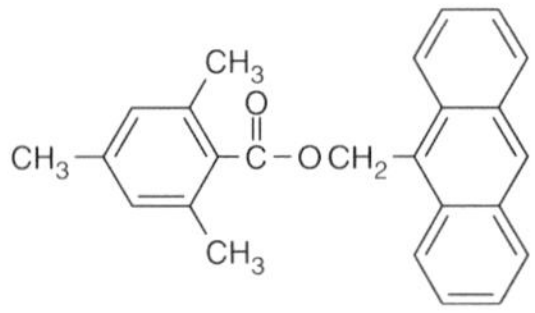
250 mg

S85,206-6
ETHYL 9,10-DIHYDRO-9,10-ETHENOANTHRACENE-11-CARBOXYLATE
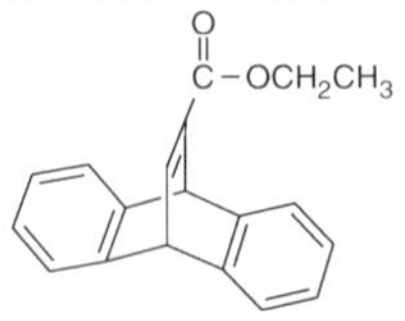
250 mg

S83,934-5
METHYL 1-NAPHTHOATE
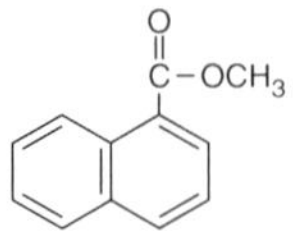
250 mg

S56,146-0
ETHYL 1-NAPHTHOATE
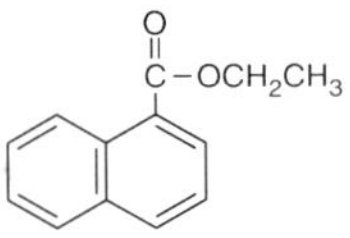
5 gm

S56,148-7
ETHYL 2-NAPHTHOATE
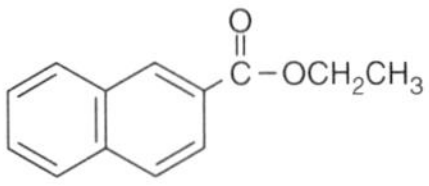
5 gm

S97,485-4
AR-G42 METHYL BENZO(C)PHENANTHRENE-4-CARBOXYLATE
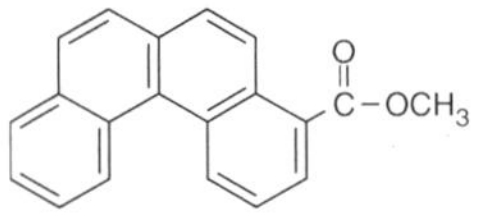
10 mg

S97,486-2
AR-G43 METHYL BENZO(C)PHENANTHRENE-3-CARBOXYLATE
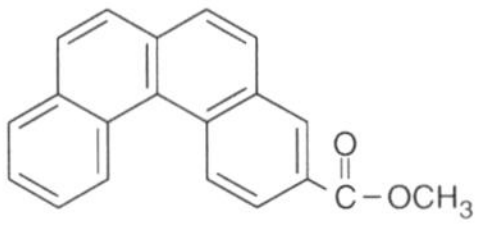
10 mg

S97,440-4
AR-E102 METHYL 7,12-DIMETHYLBENZ(A)-ANTHRACENE-10-CARBOXYLATE
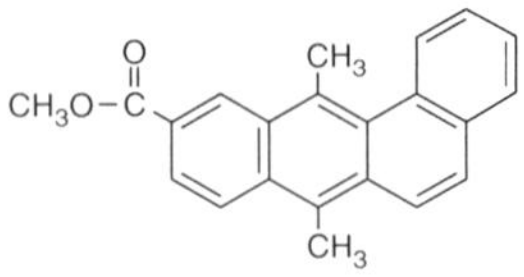
10 mg

S85,281-3
METHYL 9-ANTHRACENECARBOXYLATE
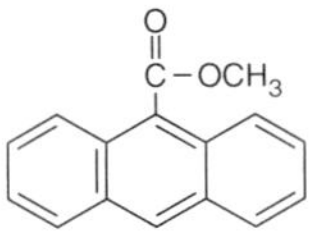
250 mg

S90,742-1
METHYL 10-BROMO-9-ANTHRACENECARBOXYLATE
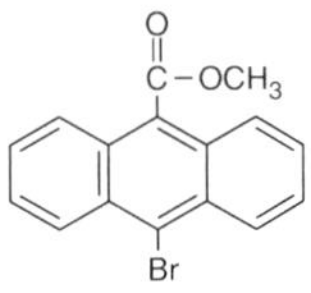
250 mg

S85,246-5
METHYL 3-METHYL-2-ANTHRACENECARBOXYLATE
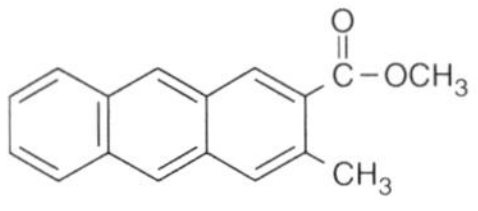
250 mg

S90,764-2
METHYL 10-METHYL-9-ANTHRACENECARBOXYLATE
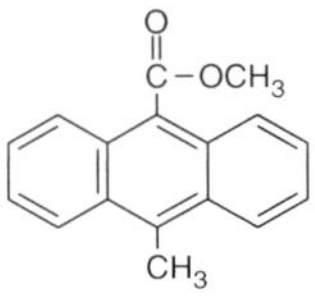
250 mg

S97,334-3
AR-D32 METHYL 4-PHENANTHRENECARBOXYLATE
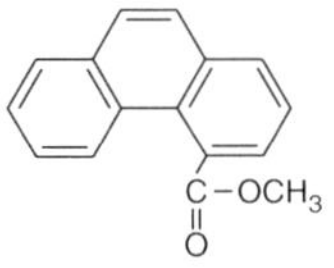
10 mg

S97,340-8
AR-D39 METHYL 1,2-DIHYDRO-4-PHENANTHRENECARBOXYLATE
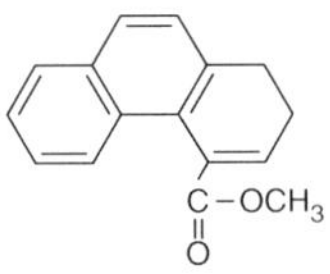
10 mg

S91,805-9
S-METHYL THIOBENZOATE
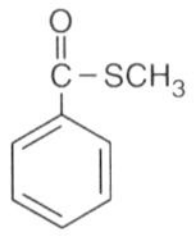
250 mg

S53,386-6
DIETHYL 4-(P-CHLOROPHENYL)-1,4-DIHYDRO-2,6-DIMETHYL-3,5-PYRIDINE DICARBOXYLATE
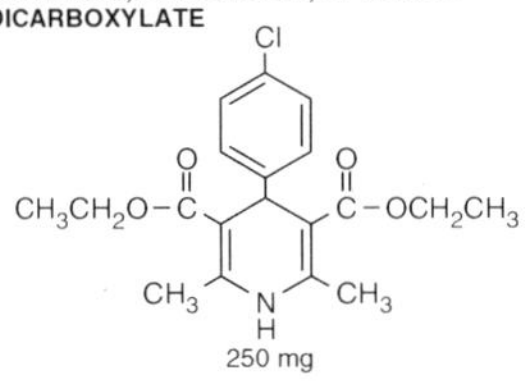
250 mg

S88,140-6
DIMETHYL 4,4'-BIPHENYLDICARBOXYLATE
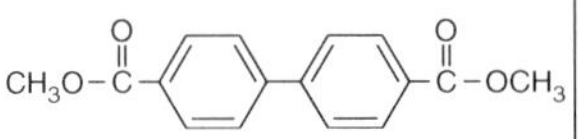
250 mg

S85,966-4
CARBOXYMETHYL DITHIOBENZOATE
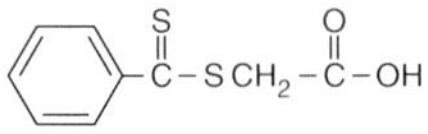
250 mg

S68,580-1
ETHYL ATROPATE
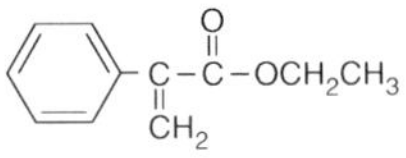
250 mg

S70,627-2
METHYL 2-BENZHYDRYLBENZOATE
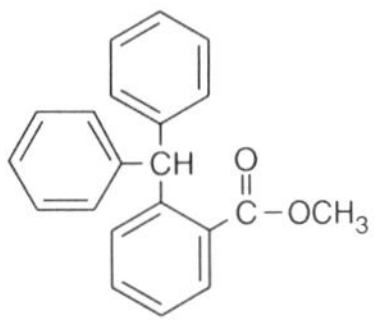
250 mg

S40,920-0
METHYL 2-BROMO-5-FLUOROBENZOATE
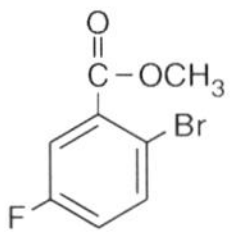
5 gm

S85,310-0
METHYL 9,10-DIHYDRO-9,10-ETHENOANTHRACENE-9-CARBOXYLATE
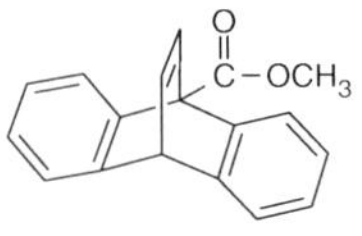
250 mg

S95,372-5
4-BROMOPHENYL 3-CHLORO-2,2-DIMETHYLPROPIONATE
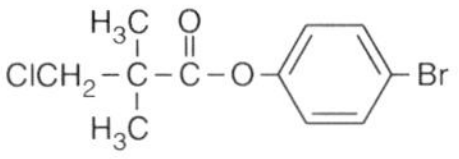
50 mg

S63,245-7
4-(2-BROMOMETHYL)-2,5-DIHYDROXYTOLUENE DIACETATE
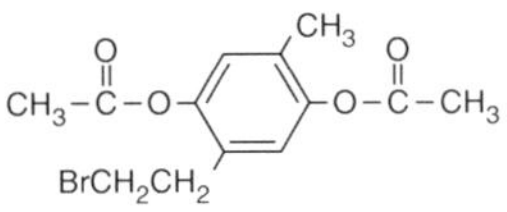
250 mg

S34,156-8
3'-BENZOYLOXY-2-BROMOACETOPHENONE
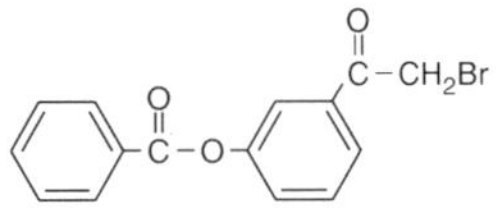
5 gm

S56,395-1
2-(BROMOACETYL)PHENYL BENZOATE
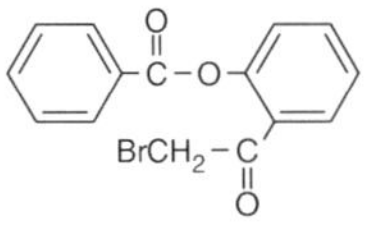
5 gm

S55,970-9
4-(BROMOACETYL)PHENYL BENZOATE
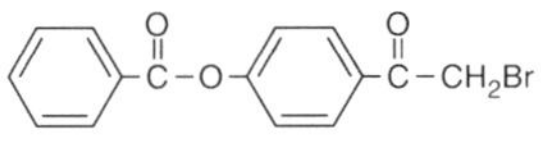
5 gm

S35,453-8
PHENYL 4-CHLOROBUTYRATE
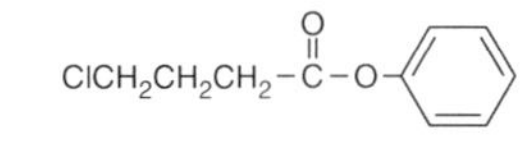
5 gm

S62,250-8
O-TOLYL 10-BROMODECANOATE
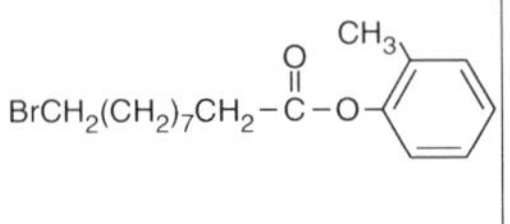
250 mg

S69,601-3
3-(TRIFLUOROMETHYL)PHENYL-3-NITROBENZOATE
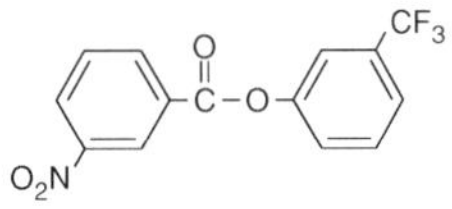
250 mg

S79,372-8
PHENYL 2,2,2-TRIFLUOROACETATE
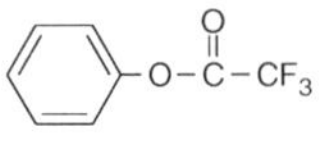
250 mg

S66,107-4
4-CHLOROPHENYL SALICYLATE
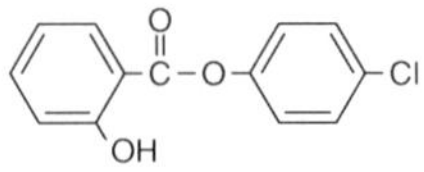
5 gm

S95,339-3
2-BROMO-3-CARBOXY-6-METHOXYPHENYL ACETATE
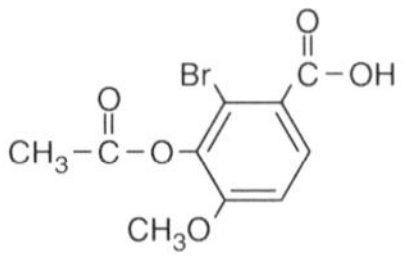
50 mg

S67,376-5
4-METHOXYPHENYL 2-CHLOROBENZOATE
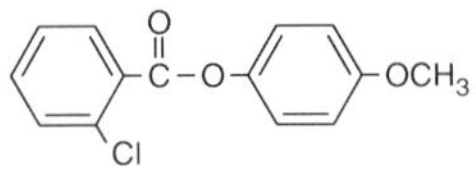
250 mg

S68,738-3
3,4-DICHLOROPHENYL 2-(2,4-DICHLOROPHENOXY)ACETATE
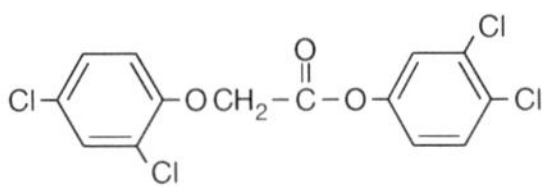
250 mg

S83,342-8
4,6-DICHLORORESORCINOL DIACETATE
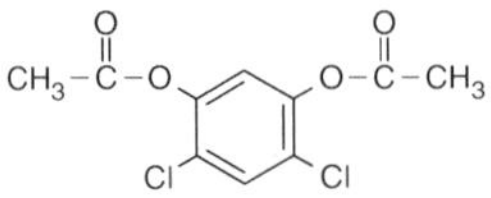
250 mg

S70,242-0
2-CHLOROPHENYL 3-NITROBENZOATE
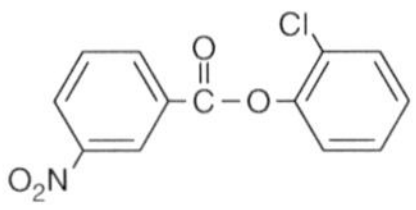
250 mg

S67,660-8
4-FLUOROPHENYL 3-NITROBENZOATE
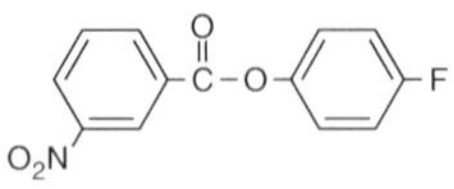
250 mg

S69,602-1
2,4-DICHLOROPHENYL 3-NITROBENZOATE
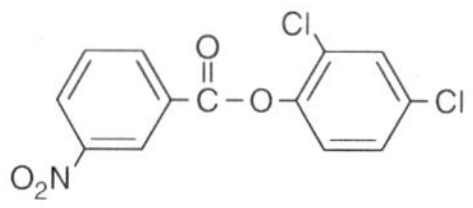
250 mg

S71,837-8
2-CHLORO-4-NITROPHENYL 3-NITROBENZOATE
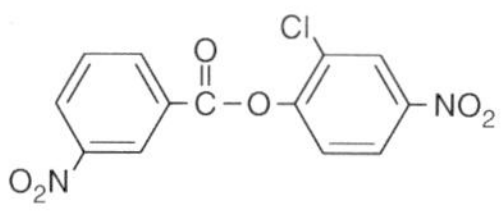
250 mg

S69,585-8
2-CHLORO-4-NITROPHENYL CINNAMATE
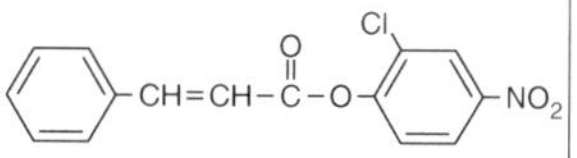
5 gm

S82,581-6
4-BROMOPHENYL BENZOATE
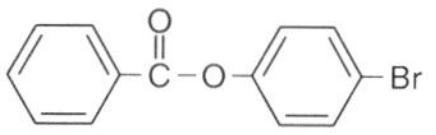
250 mg

S69,507-6
2-CHLOROPHENYL 2-CHLOROBENZOATE
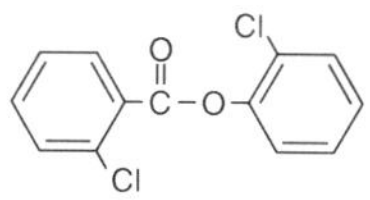
250 mg

S84,972-3
O-TOLYL 2-IODOBENZOATE
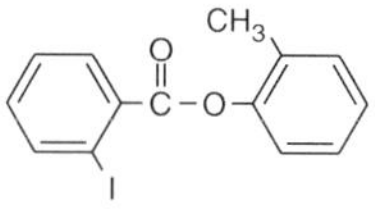
250 mg

S69,980-2
2,4-DICHLOROPHENYL 2-FLUOROBENZOATE
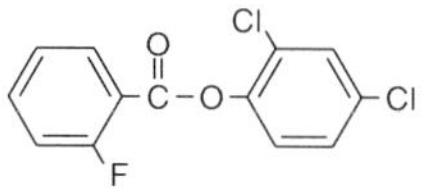
250 mg

S89,378-1
S-PENTACHLOROPHENYL PENTACHLORO-1,3-BUTADIENE-1-THIOCARBOXYLATE
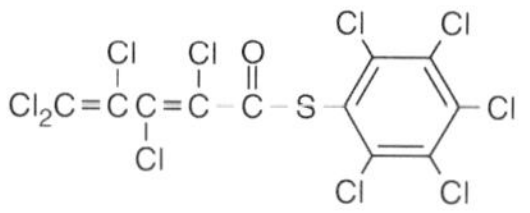
250 mg

S93,141-1
BIS(2-CHLOROPHENYL) OXALATE
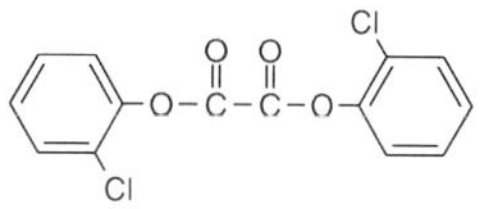
250 mg

S93,143-8
BIS(4-CHLOROPHENYL) OXALATE
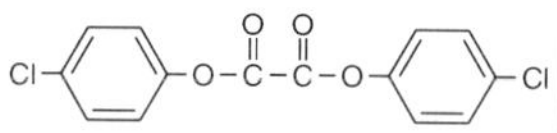
250 mg

S93,144-6
BIS(2,4,5-TRICHLOROPHENYL) OXALATE
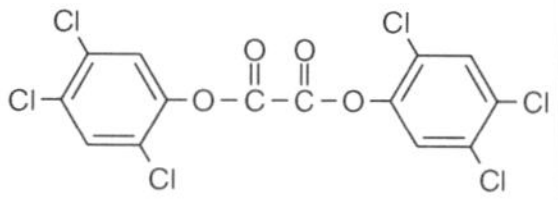
250 mg

S72,941-8
2-CHLORO-4-FLUOROPHENYL N-BUTYLCARBAMATE
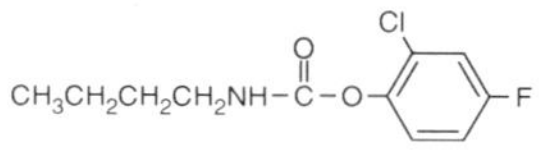
250 mg

S70,000-2
2,4-DICHLOROPHENYL CINNAMATE
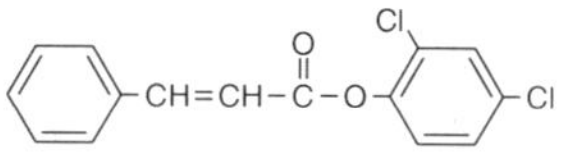
250 mg

S70,180-7
4-FLUOROPHENYL 2-CHLOROBENZOATE
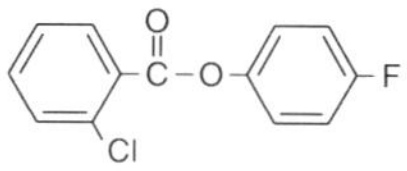
250 mg

S61,636-2
4'-ACETOXY-3',5'-DIMETHOXY-2-HYDROXYACETOPHENONE
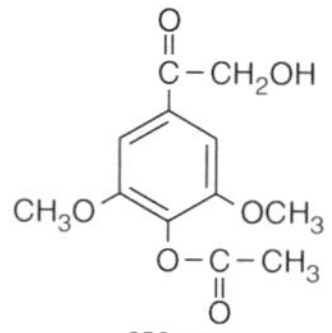
250 mg

S63,265-1
PHENYL 1-HYDROXY-2-ANTHRACENOATE
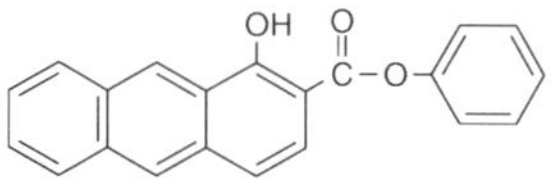
250 mg

S66,421-9
CATECHOL MONOBENZOATE
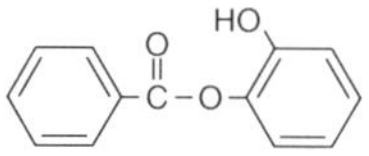
250 mg

S77,042-6
3-FORMYL-4-HYDROXYPHENYL ACETATE
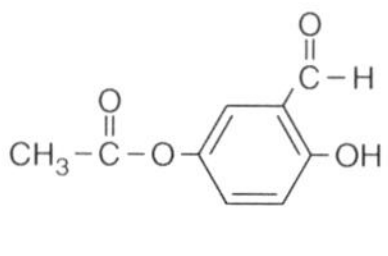
250 mg

S86,151-0
METHYL 5-(BENZOYLOXY)SALICYLATE
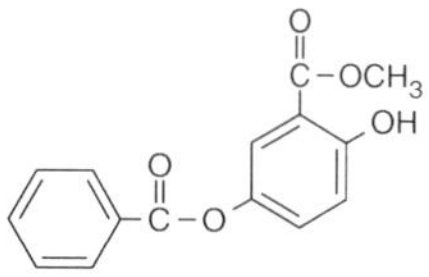
250 mg

S80,920-9
4-ACETYL-3-HYDROXYPHENYL ACETATE
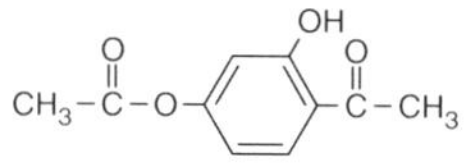
250 mg

S64,192-8
PHENYL 4-(4-CHLOROPHENYLTHIO)-1-HYDROXY-2-NAPHTHOATE
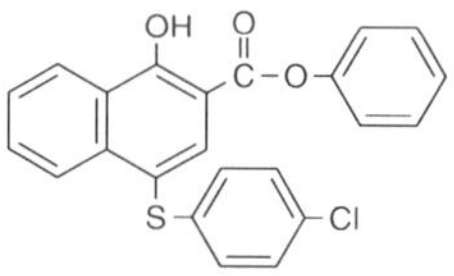
250 mg

S63,863-3
4-AMINO-3-HYDROXYPHENYL BENZOATE
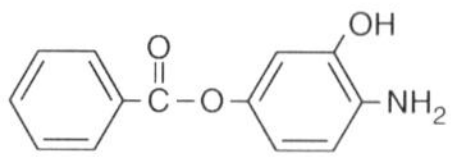
250 mg

S41,126-4
M-TOLYL 4-AMINOSALICYLATE
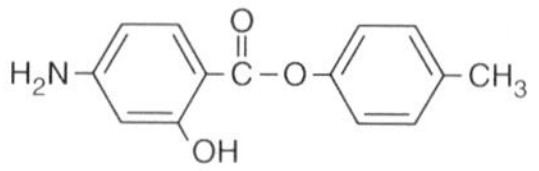
5 gm

S43,105-2
2-CARBOXYPHENYL SALICYLATE
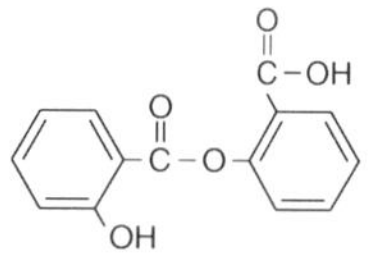
5 gm

S60,574-3
3-HYDROXYPHENYL BENZOATE
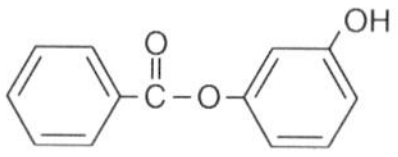
250 mg

S38,316-3
PHENYL 4-CHLORO-1-HYDROXY-2-NAPHTHOATE
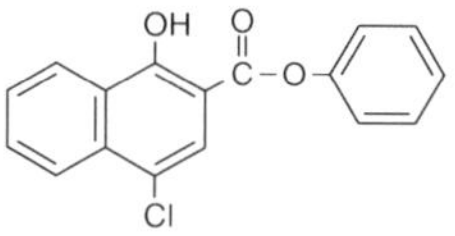
250 mg

S50,009-7
PHENYL 4-BROMO-1-HYDROXY-2-NAPHTHOATE
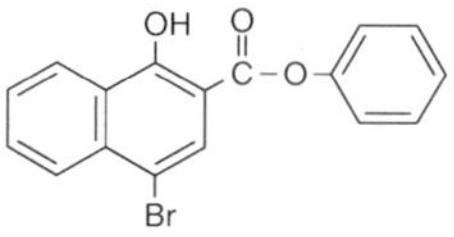
5 gm

S34,193-2
4-BENZYLOXYPHENYL PROPIONATE
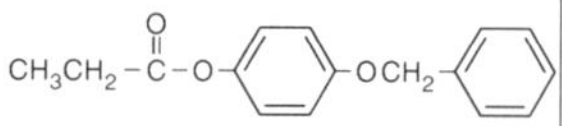
5 gm

S84,314-8
4-ACETYLRESORCINOL BIS(3,4-DIMETHOXYBENZOATE)
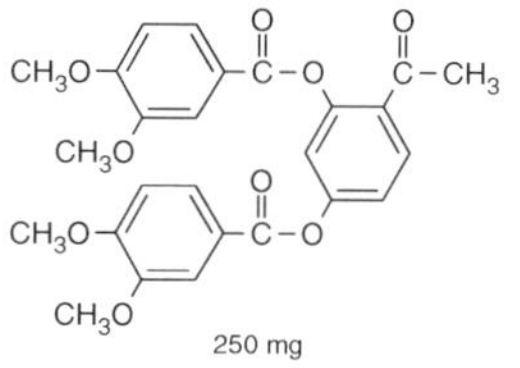
250 mg

S84,320-2
2-ACETYL-5-METHOXYPHENYL 3,4-DIMETHOXYBENZOATE
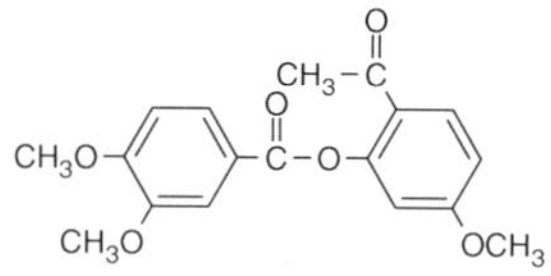
250 mg

S92,907-7
2,6-DIBENZYLOXYHYDROQUINONE DIACETATE
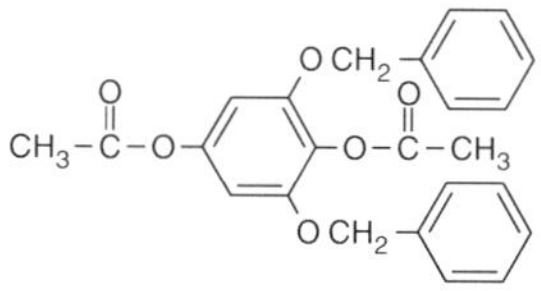
250 mg

S70,414-8
4'-ACETOXY-3',5'-DIMETHOXYACETOPHENONE
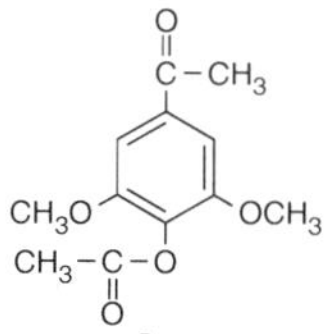
5 gm

S67,890-2
ACETYL SYRINGIC ACID
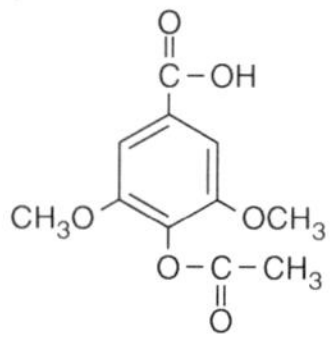
5 gm

S85,036-5
3,3',5,5'-TETRAMETHOXY-2,4,4'-TRIACETOXYBIPHENYL
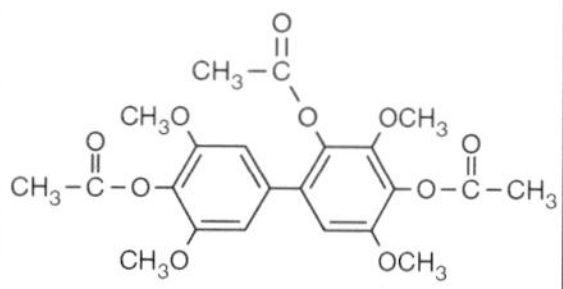
250 mg

S86,507-9
2'-ACETOXY-2,4,4'-TRIS(BENZYLOXY)-CHALCONE
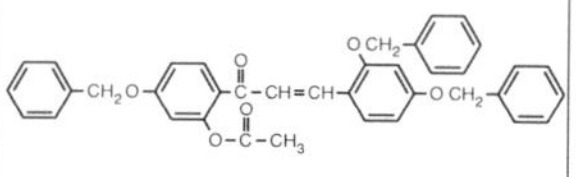
250 mg

S39,447-5
4-FORMYL-2-METHOXY-3-NITROPHENYL ACETATE
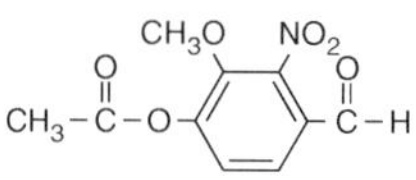
5 gm

S84,663-5
2-METHOXY-4-NITROPHENYL ACETATE
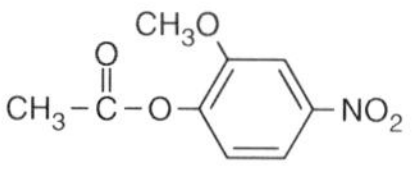
250 mg

S94,913-2
2-(2-HYDROXYPHENOXY)-4'-NITROACETOPHENONE, BENZOATE ESTER
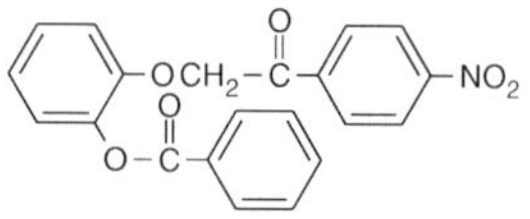
250 mg

S94,915-9
2-(ACETONYLOXY)PHENYL 4-CHLOROBENZOATE
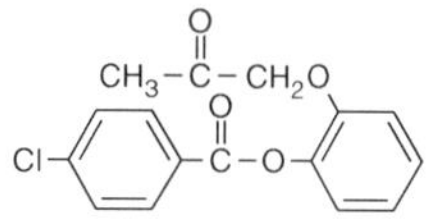
250 mg

S61,652-4
4-ACETOXY-3-METHOXYCINNAMIC ACID
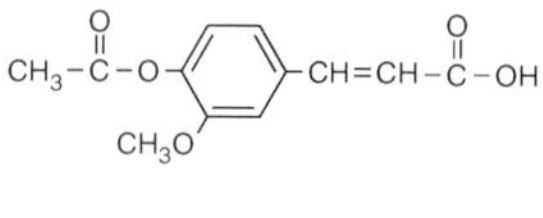
5 gm

S51,933-2
2-METHOXYPHENYL BENZOATE
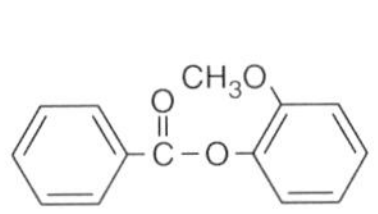
5 gm

S94,914-0
2-(2-BENZOYLOXYPHENOXY)ACETIC ACID
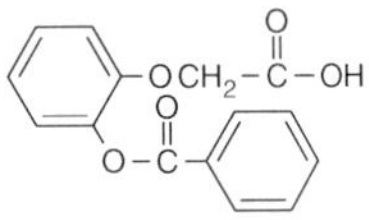
250 mg

S68,055-9
GUAIACOL PHENYLACETATE
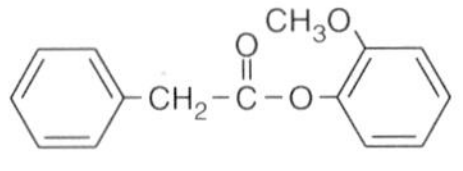
5 gm

S70,687-6
4-ALLYL-2-METHOXYPHENYL CINNAMATE
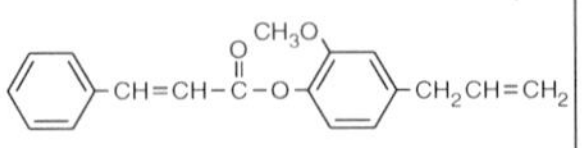
5 gm

S59,190-4
2-(HEXADECYLOXY)PHENYL ACETATE
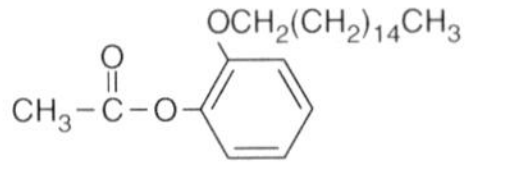
250 mg

S56,364-1
2,6-DIMETHOXYPHENYL ACETATE
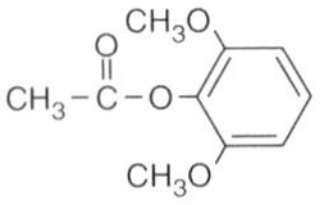
5 gm

S35,163-6
ISOEUGENYL ACETATE
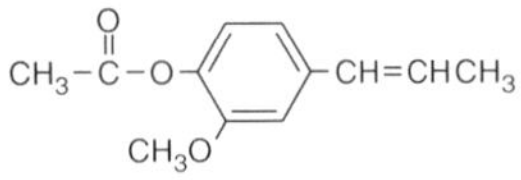
5 gm

S36,197-6
EUGENYL ACETATE
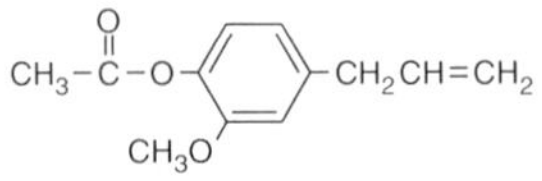
5 gm

S84,324-5
2-ACETYLPHENYL 4-METHOXYBENZOATE
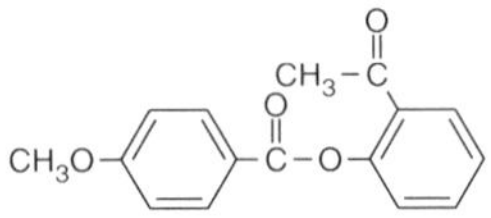
250 mg

S69,909-8
4-METHOXYPHENYL 3-NITROBENZOATE
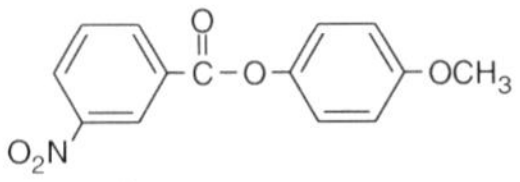
5 gm

S69,662-5
4-BENZOYLOXYPHENOXYACETONE
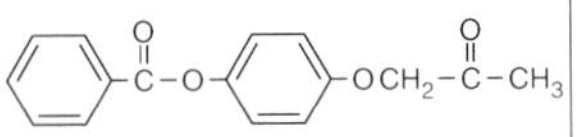
5 gm

S88,811-7
4-METHOXYPHENYL BENZOATE
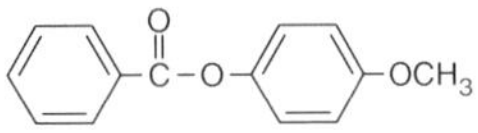
250 mg

S37,673-6
4-ACETOXY-M-ANISIC ACID
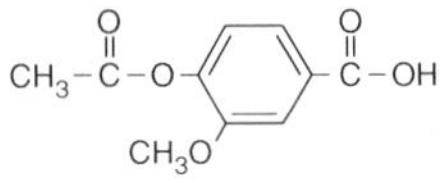
5 gm

S56,120-7
3-METHOXYPHENYL ACETATE
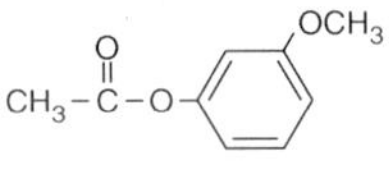
5 gm

S34,194-0
4-BENZYLOXYPHENYL STEARATE
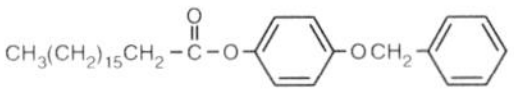
5 gm

S97,426-9
AR-E88 12-ACETOXY-10-METHOXYBENZ(A)-ANTHRACENE
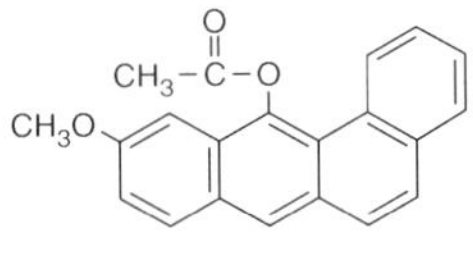
10 mg

S80,947-0
3-BENZYL-7,8-DIACETOXY-2-PHENYL-4H-1-BENZOPYRAN-4-ONE
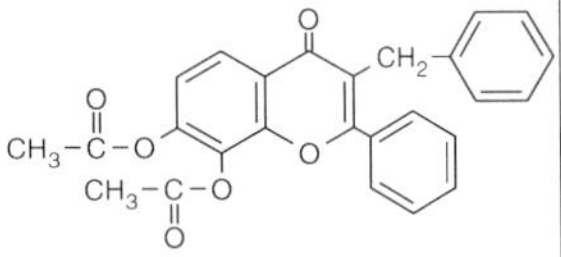
250 mg

S87,692-5
BENZYL 4-(BENZOYLOXY)BENZOATE
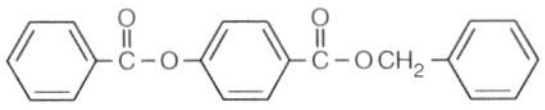
250 mg

S86,147-2
3-METHYL-5-NITRO-1,2,4-BENZENETRIOL TRIACETATE
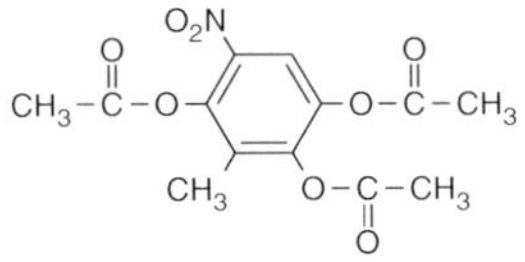
250 mg

S97,216-9
Q269 5,6,8-TRIACETOXY-1,4-NAPHTHOQUINONE
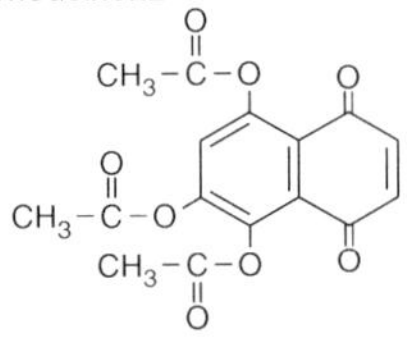
100 mg

S95,416-0
1,4-DIMETHYL-2,3,5-TRIACETOXYBENZENE
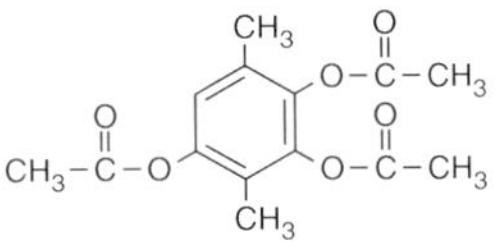
50 mg

S95,404-7
2,3,5-TRIACETOXYTOLUENE
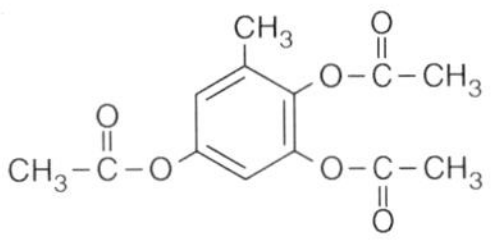
50 mg

S84,673-2
2,4,5-TRIACETOXYTOLUENE
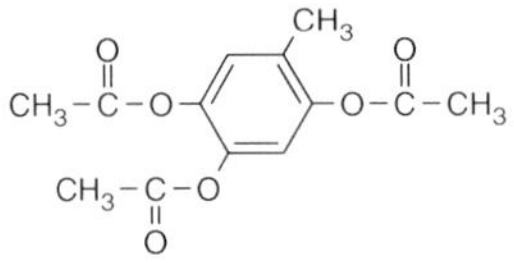
250 mg

S63,749-1
1,2,4-NAPHTHALENETRIOL TRIACETATE
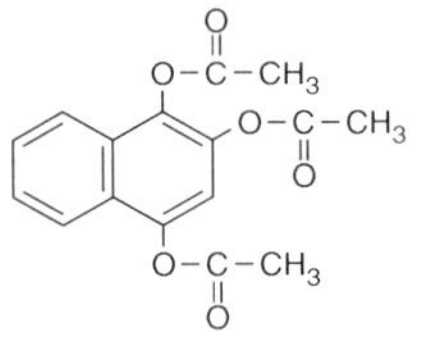
250 mg

S84,654-6
4-NITRO-2',4',5'-TRIACETOXYBIPHENYL
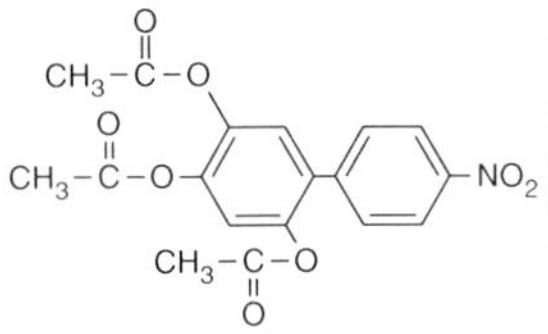
250 mg

S80,949-7
3-BENZYL-7,8-DIACETOXY-2-METHYL-4H-1-BENZOPYRAN-4-ONE
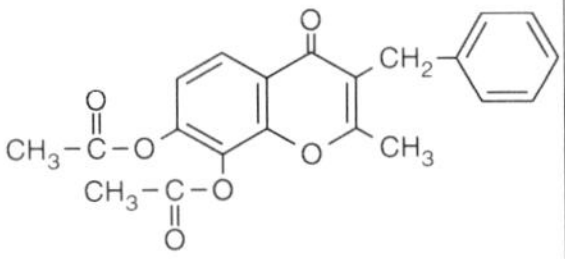
250 mg

S80,948-9
7,8-DIACETOXY-2,3-DIPHENYL-4H-1-BENZOPYRAN-4-ONE
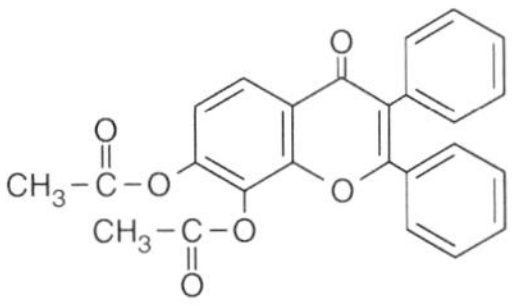
250 mg

S84,694-5
2,4,5-TRIACETOXYBIPHENYL
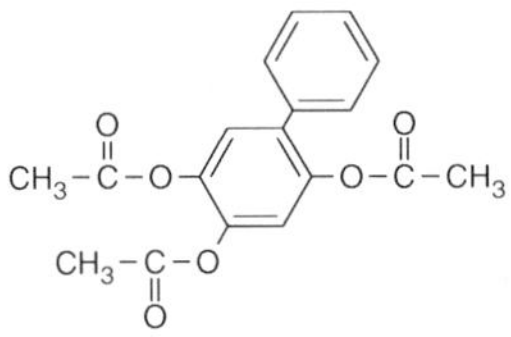
250 mg

S84,692-9
1-TERT-BUTYL-2,4,5-TRIACETOXYBENZENE
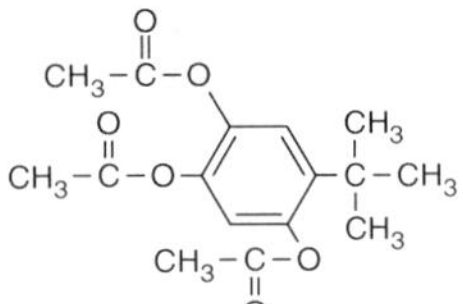
250 mg

S92,915-8
1,2-DIACETOXYANTHRAQUINONE
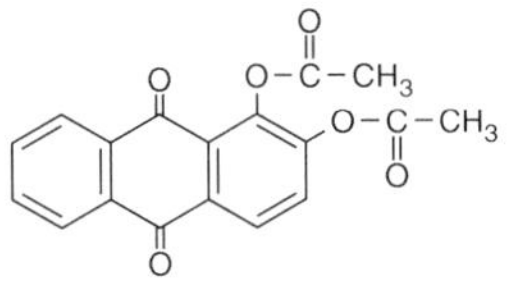
250 mg

S60,284-1
ACETYL-3-ACETOXY-4,5-PHENANTHRYLENE OXIDE (MIXTURE OF ISOMERS)
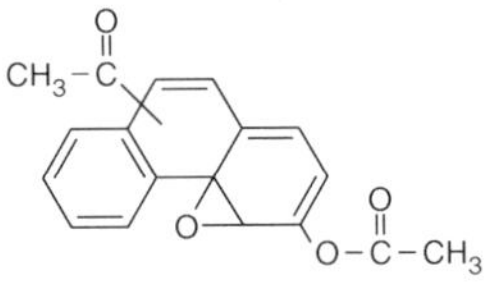
250 mg

S84,204-4
PYROGALLOL TRIACETATE
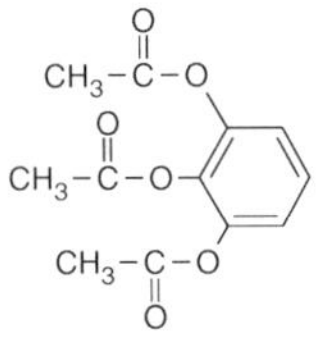
250 mg

S67,199-1
2',5'-DIHYDROXYOCTADECANOPHENONE DIOCTADECANOATE
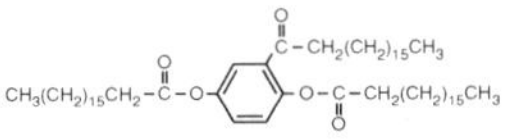
5 gm

S80,952-7
5,7-DIACETOXY-2-METHYL-3-PHENYL-4H-1-BENZOPYRAN-4-ONE
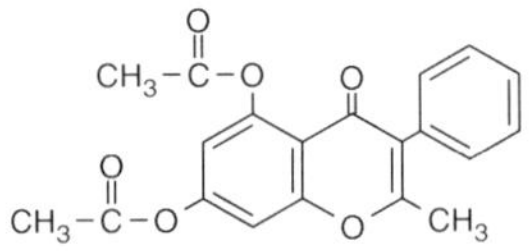
250 mg

S80,950-0
3-BENZYL-5,7-DIACETOXY-2-METHYL-4H-1-BENZOPYRAN-4-ONE
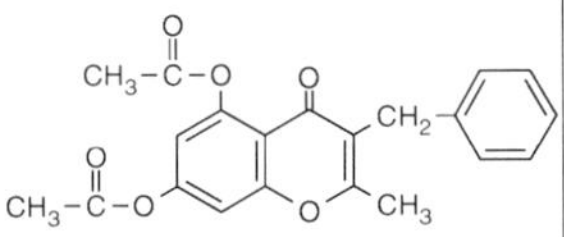
250 mg

S80,945-4
2-ACETYLRESORCINOL DIBENZOATE
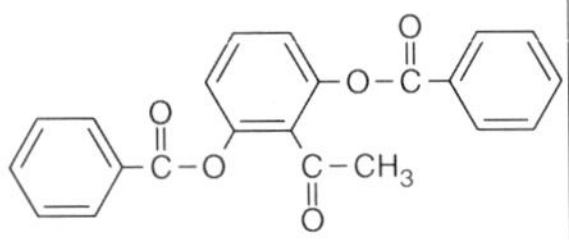
250 mg

S80,951-9
7-ACETOXY-2-METHYL-3-PHENYL-4H-1-BENZOPYRAN-4-ONE
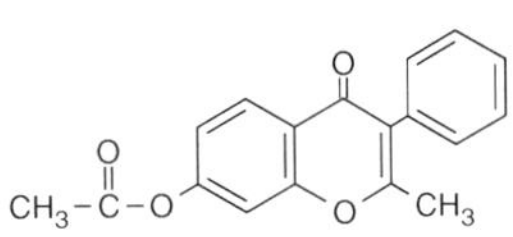
250 mg

S94,721-0
6-ACETYL-2-METHYLRESORCINOL DIACETATE
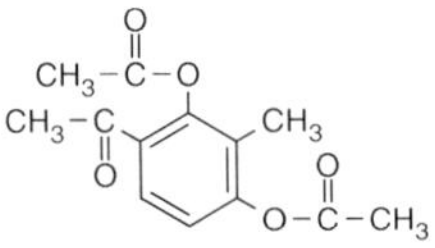
250 mg

S63,300-3
1-(4-AMINOPHENYL)-2-(2,5-DIACETOXYPHENYL)ETHANE HYDROCHLORIDE
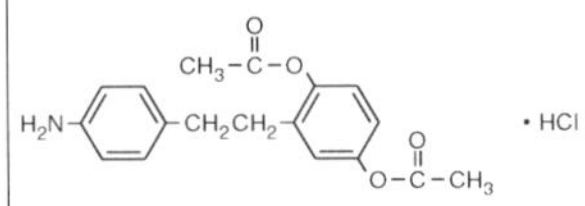
5 gm

S91,432-0
3-ACETOXYBENZOIC ACID
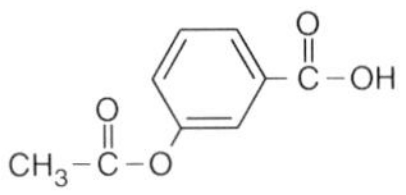
250 mg

S81,904-2
METHYL 3-ACETOXYBENZOATE
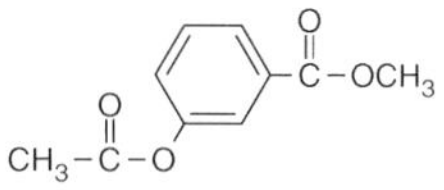
250 mg

S67,202-5
OCTADECYLHYDROQUINONE DIOCTADECANOATE
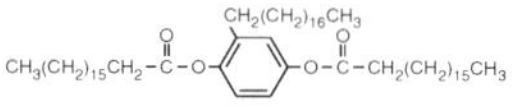
5 gm

S86,061-1
2,5-DIMETHYLHYDROQUINONE DIACETATE
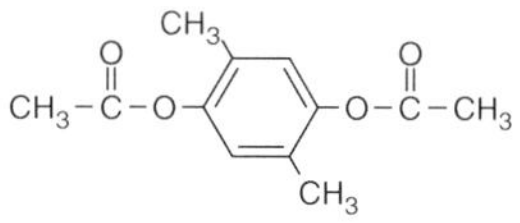
250 mg

S35,307-8
2-METHYL-1,4-NAPHTHALENEDIOL DIACETATE
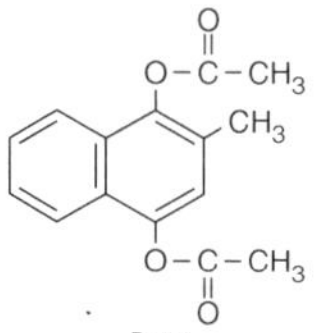
5 gm

S84,203-6
2,3-DIMETHYL-1,4-NAPHTHALENEDIOL DIACETATE
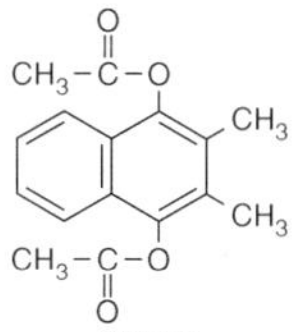
250 mg

S88,350-6
2,6-DI-TERT-BUTYL-1,4-NAPHTHALENEDIOL DIACETATE
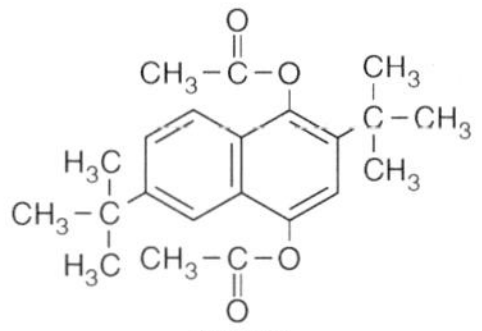
250 mg

S50,327-4
S-(P-TOLYL) 4-NITROTHIOLBENZOATE
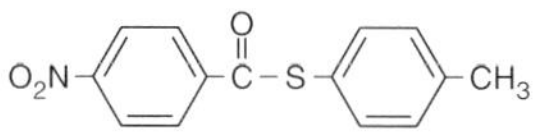
5 gm

S38,904-8
4-GLYCYLPHENYL BENZOATE HYDROCHLORIDE
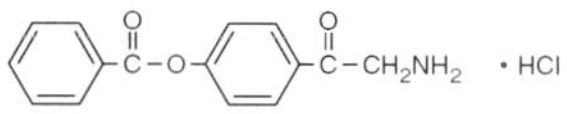
5 gm

S53,492-7
L(+)-LEUCINE 4-NITROPHENYL ESTER HYDROBROMIDE
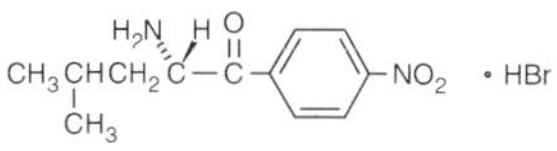
5 gm

S53,094-8
2-(SEC-BUTYL)-4,6-DINITROPHENYL 2-CHLOROBUTYRATE
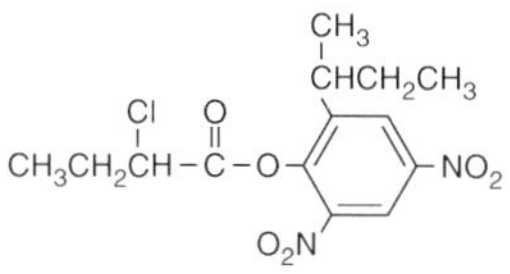
5 gm

S90,749-9
2-(BIS[4-(DIMETHYLAMINO)PHENYL]-METHYL)PHENYL BENZOATE
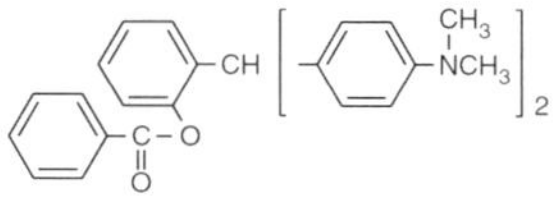
250 mg

S84,644-9
2-NITROPHENYL BENZOATE
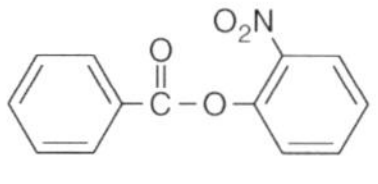
250 mg

S68,021-4
2-METHYL-6-NITROPHENYL ACETATE
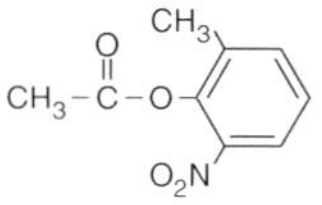
250 mg

S92,627-2
4,4'-ISOPROPYLIDENEDIPHENOL BIS-3,5-DINITROBENZOATE
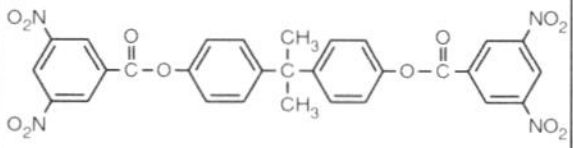
250 mg

S88,365-4
S-PHENYL 4-NITROTHIOBENZOATE
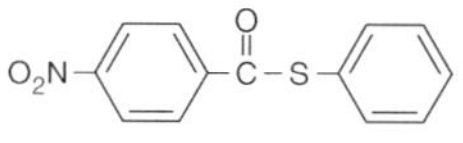
250 mg

S80,874-1
2-METHYL-3-NITROPHENYL ACETATE
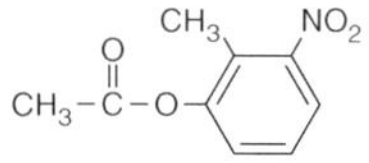
250 mg

S88,852-4
S-(4-NITROPHENYL) THIOACETATE
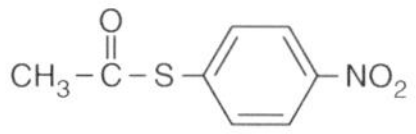
250 mg

S86,034-4
4-NITROFLUORESCEIN
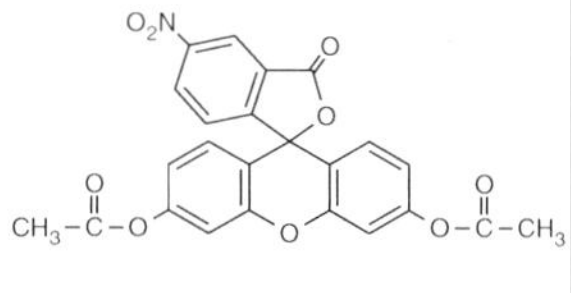
250 mg

S36,413-4
4'-NITRO-4-BIPHENYLYL BENZOATE
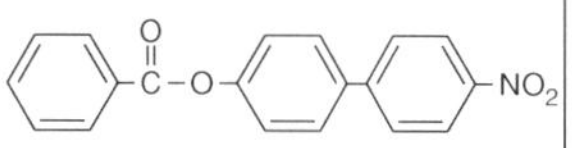
5 gm

S53,317-3
2-NAPHTHYL 4-NITROBENZOATE
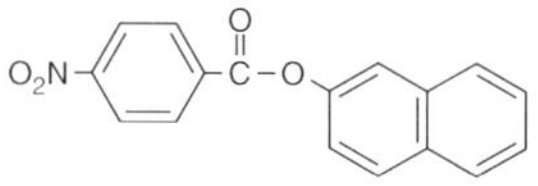
5 gm

S61,655-9
4-NITROPHENYL FORMATE
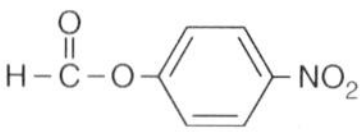
5 gm

S60,320-1
4-NITROPHENYL PROPIONATE
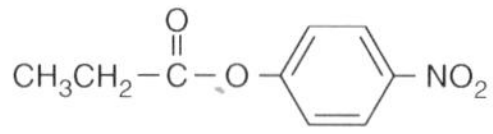
5 gm

S57,412-0
4-NITROPHENYL OCTANOATE
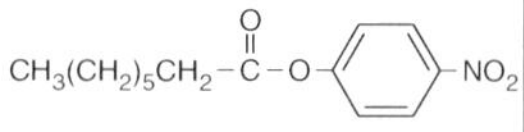
250 mg

S80,720-6
PHENYL 3-OXO-3-PHENYLPROPIONATE
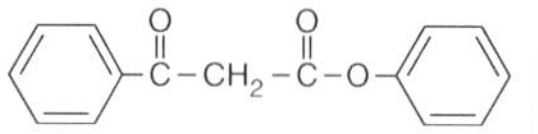
250 mg

S88,903-2
S-PHENYL 2'-(2,4,6-TRIMETHYLBENZOYL)-2-BIPHENYLTHIOCARBOXYLATE
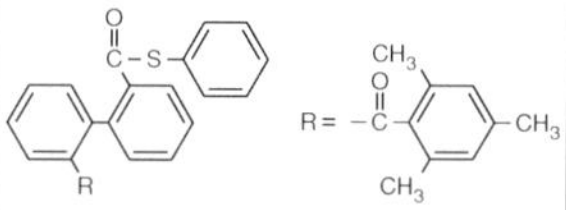
250 mg

S43,100-1
3-ACETYLPHENYL BENZOATE
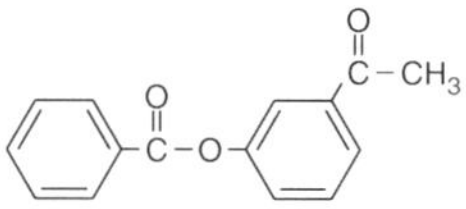
5 gm

S67,003-0
4'-ACETOXY-3,3-DIPHENYLACRYLOPHENONE
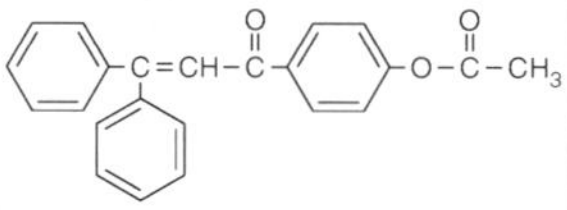
250 mg

S56,003-0
4-ACETYLPHENYL BENZOATE
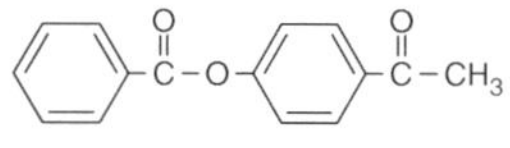
5 gm

S64,016-6
3-ACETOXY-6-ACETYLPHENANTHRENE
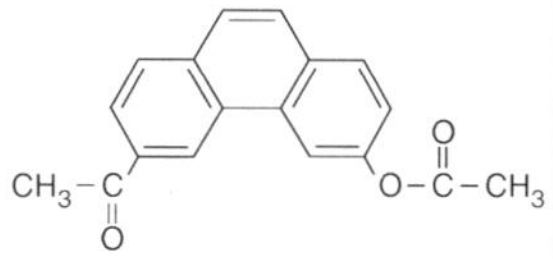
250 mg

S75,507-9
9-OXO-3-FLUORENYL ACETATE
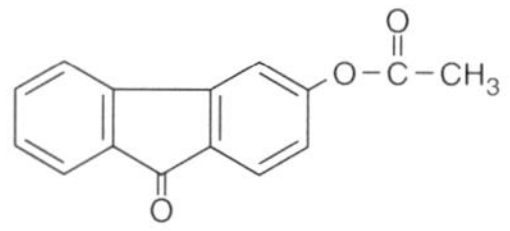
250 mg

S77,032-9
9-ACETYL-7-ISOPROPYL-1-METHYL-3-PHENANTHRYL ACETATE
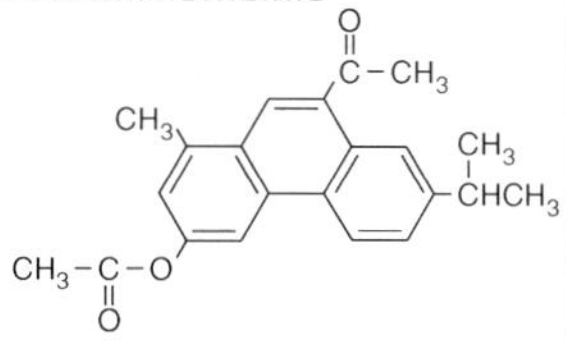
250 mg

S51,751-8
7-ACETOXY-3-ACETYL-2-METHYL-4-CHROMANONE
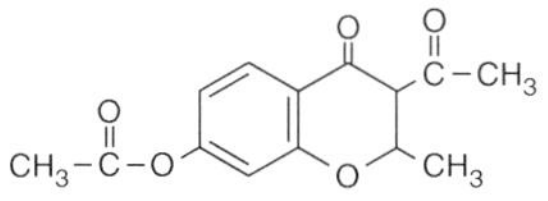
5 gm

S41,236-8
3-ACETYLPHENYL ACETATE
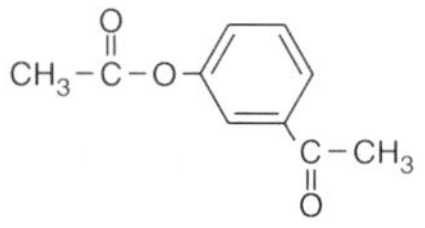
5 gm

S92,888-7
2-ACETYLPHENYL ACETATE
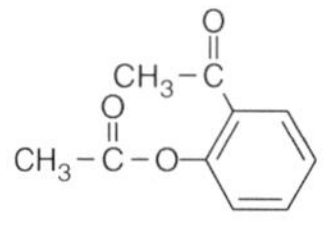
250 mg

S55,371-9
2-ACETYLPHENYL BENZOATE
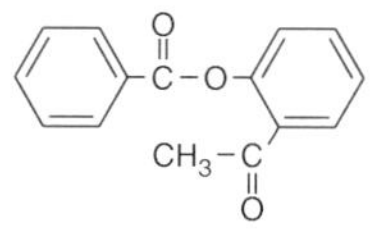
5 gm

S41,234-1
4-ACETYLPHENYL ACETATE
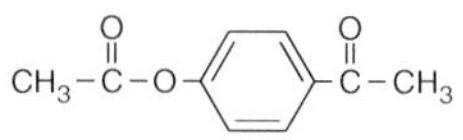
5 gm

S52,205-8
6-ACETYL-3-PHENANTHRYL ACETATE
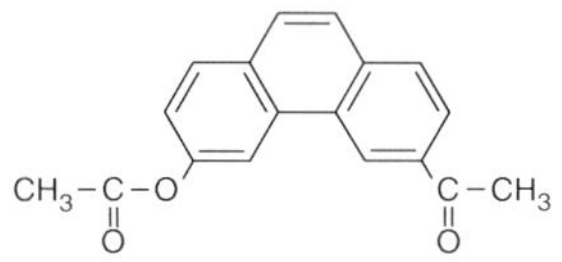
250 mg

S94,387-8
4,4'-DIACETOXYBENZIL
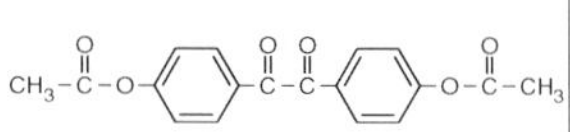
250 mg

S64,127-8
DIETHYL 4-ACETOXYBENZYLIDENEMALONATE
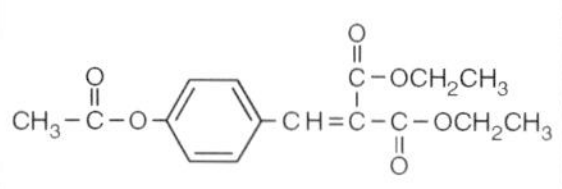
5 gm

S56,637-3
DIPHENYL ISOPHTHALATE
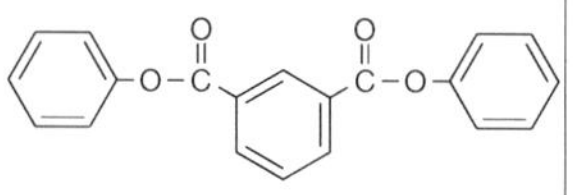
5 gm

S93,354-6
MONO-(1-STYRYLETHYL) PHTHALATE
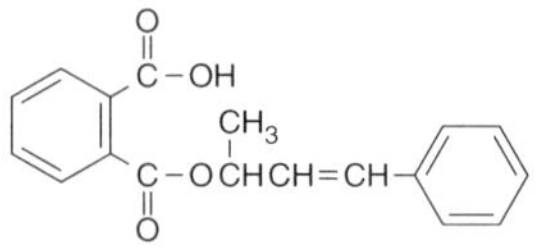
250 mg

S87,702-6
4-CARBOXYPHENYL BENZOATE
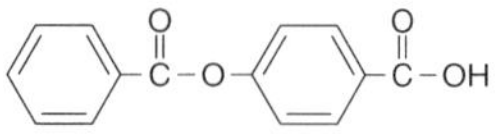
250 mg

S50,511-0
1-ACETOXY-2-NAPHTHOIC ACID
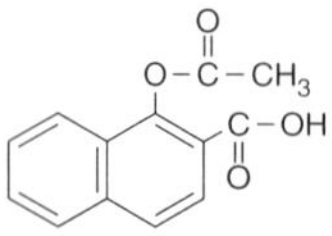
250 mg

S39,245-6
METHYL BENZOYLSALICYLATE
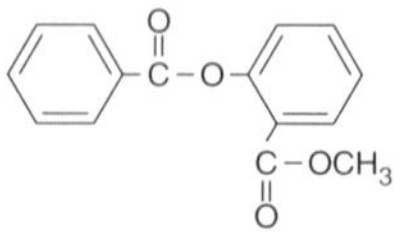
5 gm

S39,167-0
ETHYL ACETYLSALICYLATE
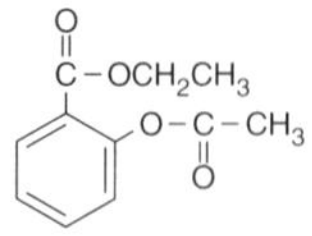
250 mg

S80,743-5
PHENYL 2-METHYLBENZOATE
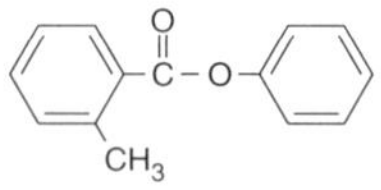
250 mg

S43,305-5
METHYL ACETYLSALICYLATE
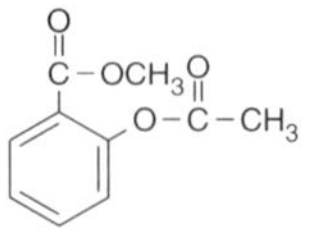
5 gm

S51,147-1
4-(TERT-BUTYL)PHENYL BENZOATE
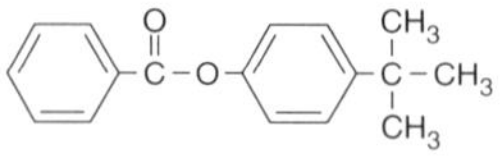
5 gm

S75,534-6
S-(2-FLUORENYL) THIOBENZOATE
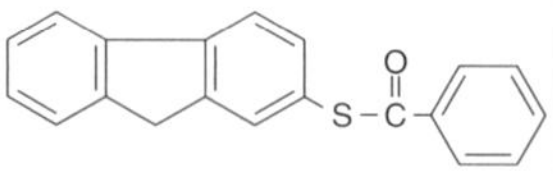
250 mg

S63,114-0
DI(P-TOLYL) MALONATE
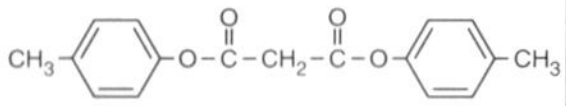
250 mg

S71,449-6
DIPHENYL SUCCINATE
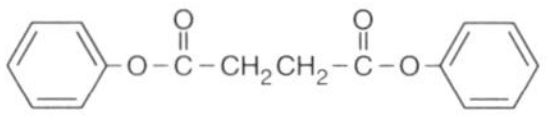
5 gm

S74,027-6
4-ACETOXYPHENYLACETIC ACID
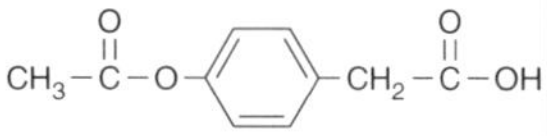
250 mg

S74,535-9
4-ACETOXYHYDROCINNAMIC ACID
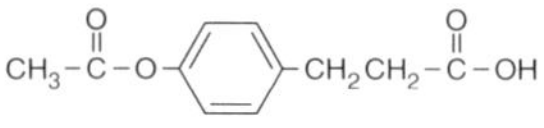
250 mg

S37,664-7
3-METHYLACETYLSALICYLIC ACID
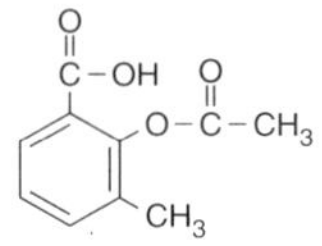
250 mg

S88,881-8
M-TOLYL ACETATE
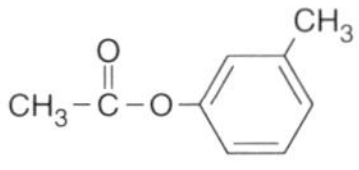
250 mg

S35,612-3
M-TOLYL PHENYLACETATE
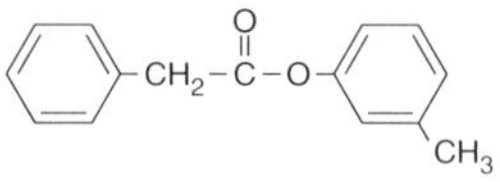
5 gm

S64,863-9
PHENYL STEARATE
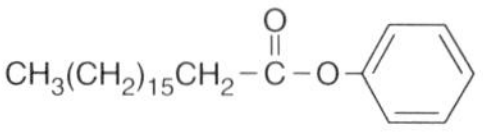
5 gm

S84,205-2
2,3,5,6-TETRAMETHYLHYDROQUINONE DIACETATE
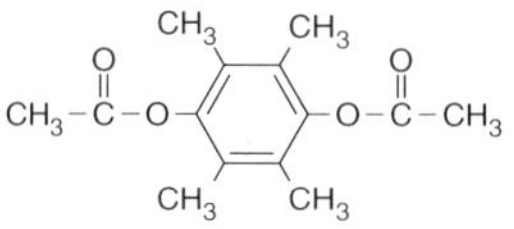
250 mg

S88,390-5
4B,5,6,7,8,8A,9,10-OCTAHYDRO-1,7-PHENANTHRENEDIOL DIACETATE
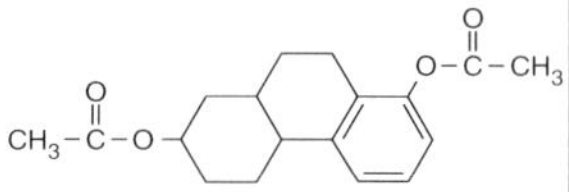
250 mg

S88,100-7
1,4-BIS(4-ACETOXYPHENYL)-1,3-BUTADIENE
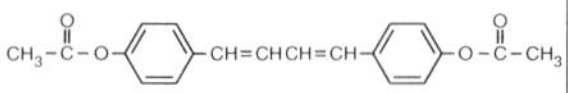
250 mg

S34,210-6
3,4-BIS(4-ACETOXYPHENYL)-2,4-HEXADIENE
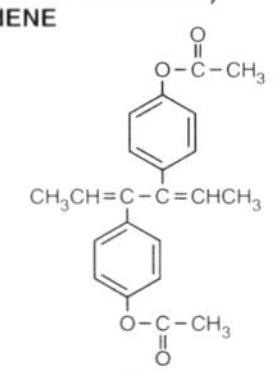
5 gm

S58,013-9
DIETHYLSTILBESTROL DIPROPIONATE
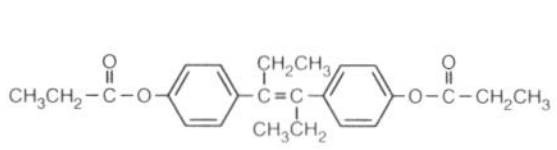
5 gm

S34,616-0
DIETHYLSTILBESTROL DIPALMITATE
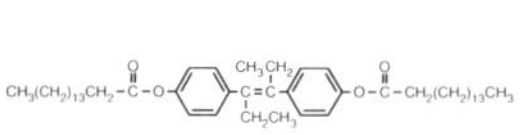
5 gm

S52,757-2
4,4'-ISOPROPYLIDENEDIPHENOL DIACETATE
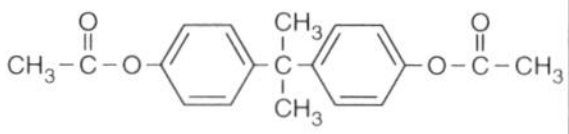
5 gm

S84,269-9
2,5-BIS(2,4-XYLYL) HYDROQUINONE DIACETATE
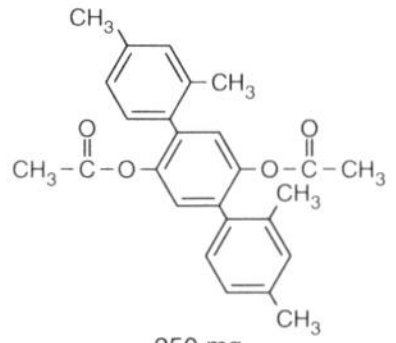
250 mg

S88,583-5
2,2',6,6'-TETRAMETHYL-4,4'-BIPHENOL DIACETATE
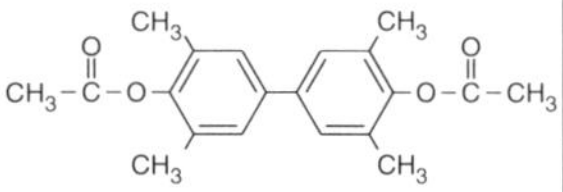
250 mg

S84,693-7
3,3',5,5'-TETRAMETHYL-2,4,4'-TRIACETOXYBIPHENYL
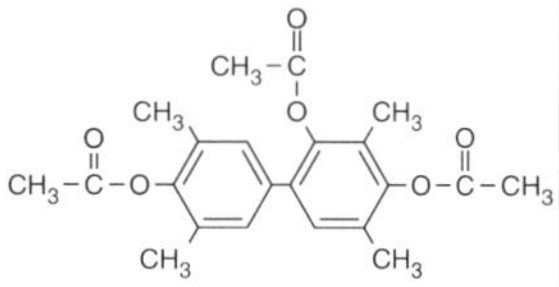
250 mg

S84,695-3
4,4'-DIACETOXY-3,3',5,5'-TETRA-TERT-BUTYLBIPHENYL
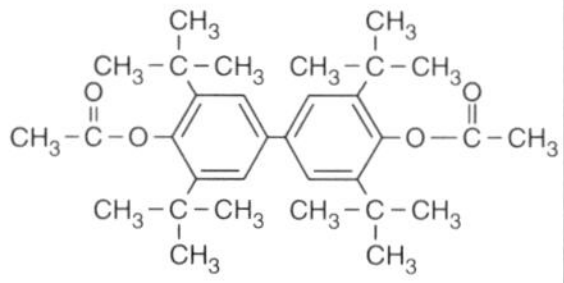
250 mg

S93,977-3
PHENYL ISOBUTYRATE
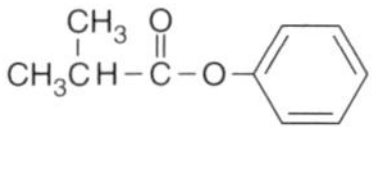
5 gm

S65,789-1
PHENYL HEXANOATE
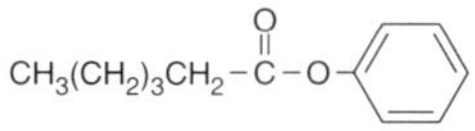
5 gm

S70,695-7
PHENYL CINNAMATE
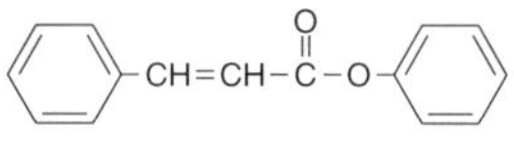
5 gm

S88,223-2
4-STYRYLPHENYL ACETATE
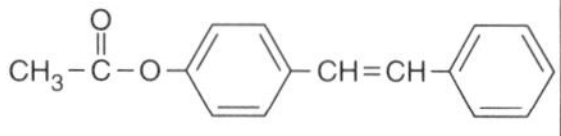
250 mg

S94,221-9
4-(1,1,3,3-TETRAMETHYLBUTYL)PHENYL BENZOATE
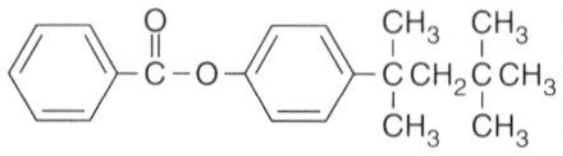
5 gm

S94,559-5
4-BIPHENYLYL ACETATE
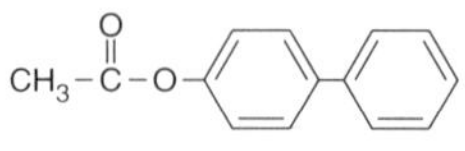
250 mg

S83,895-0
4-BIPHENYLYL OCTADECANOATE
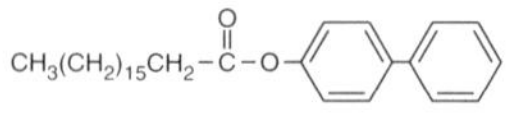
250 mg

S84,217-6
3,4-DIPHENYL-2-METHYL-1-NAPHTHYL ACETATE
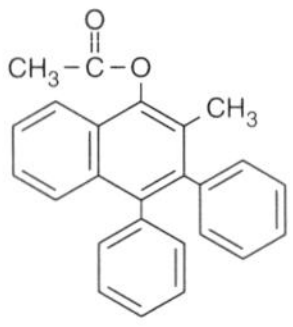
250 mg

S70,675-2
2-NAPHTHYL BENZOATE
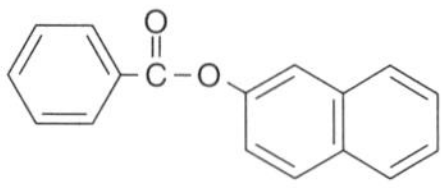
5 gm

S64,001-8
4-ETHYLPHENYL ACETATE
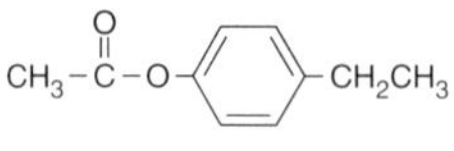
5 gm

S52,850-1
M-TOLYL ISOBUTYRATE
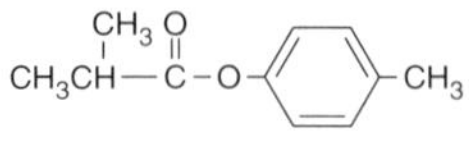
5 gm

S37,412-1
4-TERT-BUTYLPHENYL ACETATE
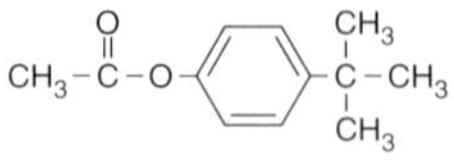
5 gm

S56,341-2
4-(TERT-PENTYL)PHENYL ACETATE
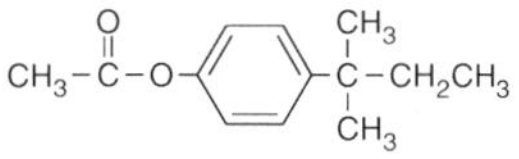
5 gm

S35,613-1
P-TOLYL PHENYLACETATE
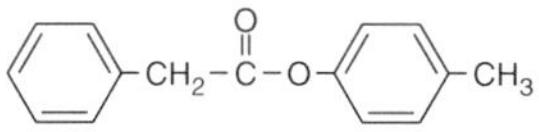
5 gm

S52,847-1
M-TOLYL OCTANOATE
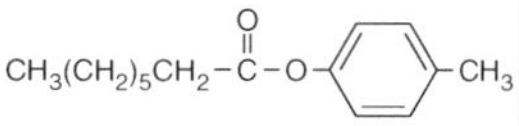
5 gm

S78,353-6
1-NAPHTHYL MYRISTATE
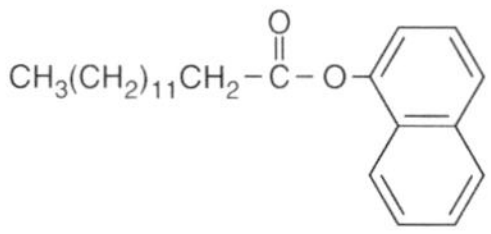
5 gm

S45,181-9
2-NAPHTHYL LAURATE
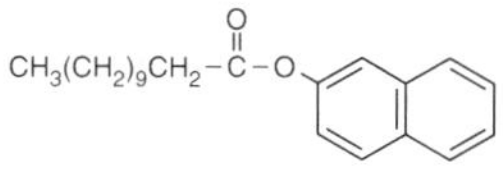
5 gm

S78,550-4
2-NAPHTHYL MYRISTATE
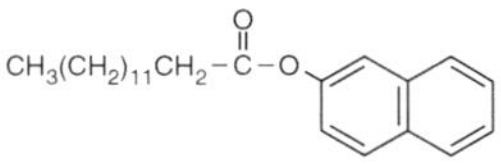
250 mg

S97,544-3
AR-L27 6-BENZO(A)PYRENYL ACETATE
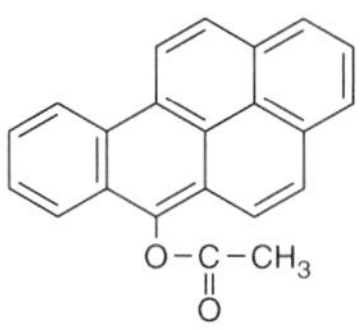
10 mg

S97,547-8
AR-L29 7-BENZO(A)PYRENYL ACETATE
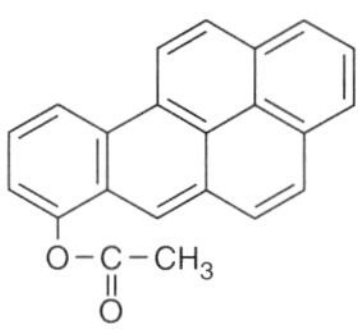
10 mg

S75,397-1
1-FLUORENYL ACETATE
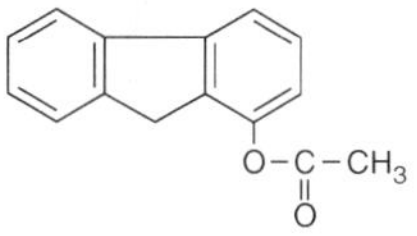
5 gm

S75,500-1
3-FLUORENYL ACETATE
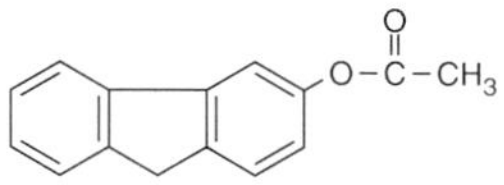
250 mg

S75,822-1
2-FLUORENYL ACETATE
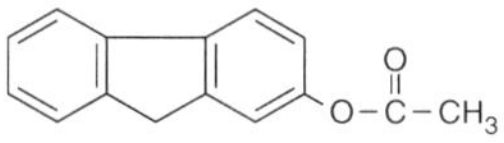
250 mg

S97,390-4
AR-E51 7-BENZ(A)ANTHRYL ACETATE
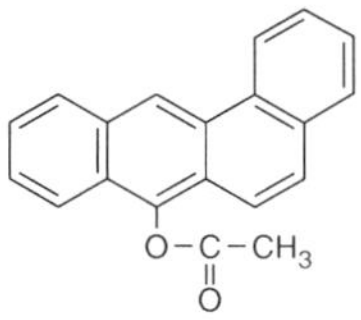
10 mg

S97,393-9
AR-E55 5-ACETOXYBENZ(A)ANTHRACENE
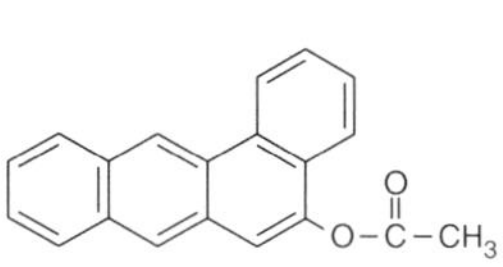
10 mg

S93,613-8
7-METHYL-5-BENZ(A)-ANTHRACENYL ACETATE
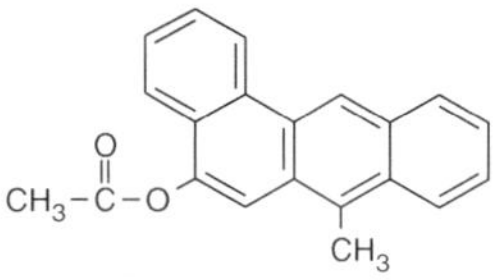
10 mg

S97,422-6
AR-E84 12-METHYL-7-BENZ(A)-ANTHRACENYL ACETATE
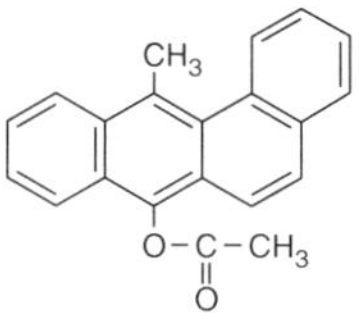
10 mg

S97,438-2
AR-E100 12-ACETOXY-7-METHYLBENZ(A)-ANTHRACENE
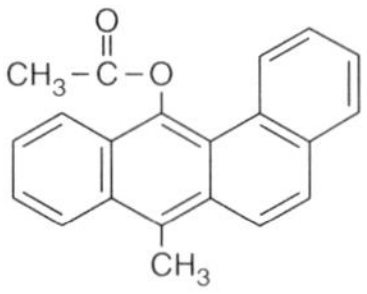
10 mg

S83,651-6
S-(P-TOLYL) THIOACETATE
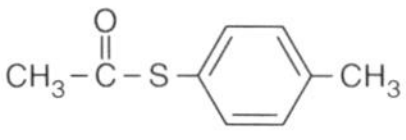
250 mg

S89,887-2
2-CHLOROPHENYL 1,2,2,2-TETRACHLOROETHYL CARBONATE
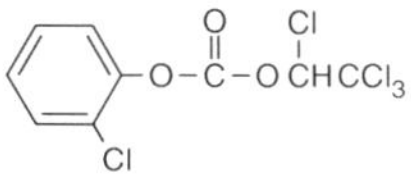
250 mg

S89,889-9
4-BROMOPHENYL 1,2,2,2-TETRACHLOROETHYL CARBONATE
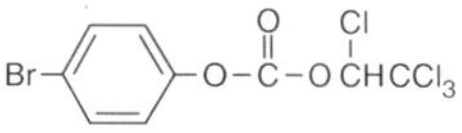
250 mg

S89,888-0
S-(4-CHLOROPHENYL) 1,2,2,2-TETRACHLOROETHYL THIOCARBONATE
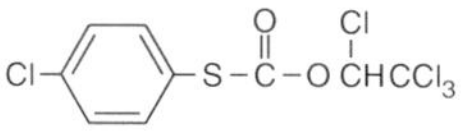
250 mg

S89,882-1
1-CHLOROETHYL-3-(TRIFLUORO-METHYL)PHENYL CARBONATE
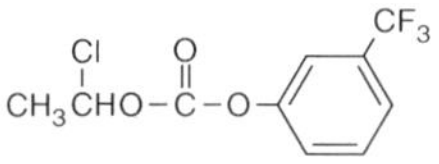
250 mg

S89,890-2
ALPHA,2-DICHLOROBENZYL METHYL CARBONATE
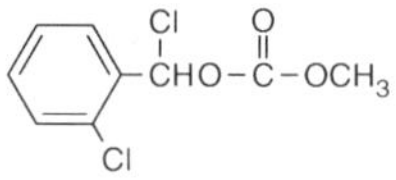
250 mg

S94,588-9
2-CHLORO-4-FLUORO-5-NITROPHENYL ETHYL CARBONATE
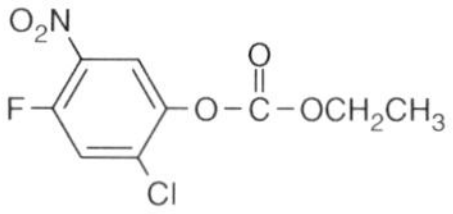
250 mg

S89,877-5
2-CHLOROPHENYL ALPHA,2-DICHLOROBENZYL CARBONATE
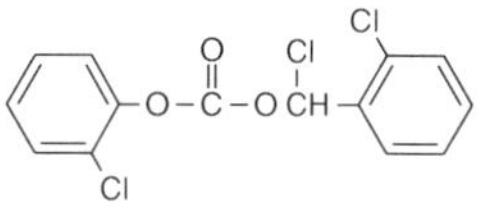
250 mg

S89,870-8
1-CHLOROETHYL-4-CHLOROPHENYL CARBONATE
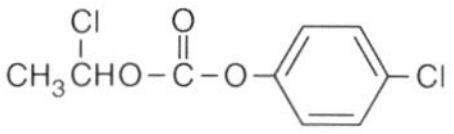
250 mg

S38,763-0
S,S'-BIS(4-CHLOROPHENYL)-DITHIOCARBONATE
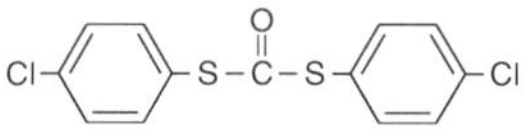
250 mg

S89,871-6
1-CHLOROETHYL S-(4-CHLOROPHENYL)-THIOCARBONATE
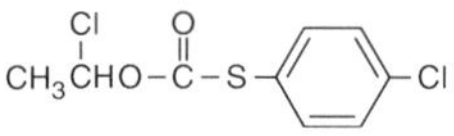
250 mg

S43,492-2
4-METHOXYBENZYL PHENYL CARBONATE
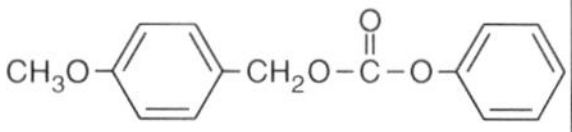
5 gm

S78,745-0
O-(CARBOBENZYLOXY)-L-TYROSINE HYDROCHLORIDE
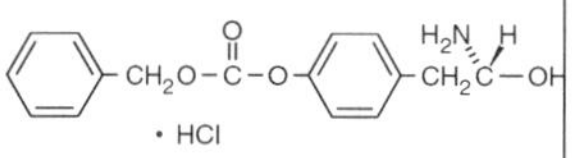
250 mg

S79,026-5
S-CARBOBENZYLOXY-L-CYSTEINE
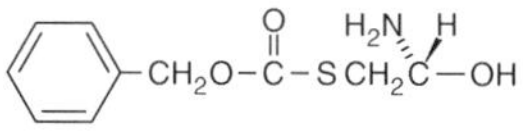
250 mg

S77,117-1
BENZYL S-PHENYL THIOCARBONATE
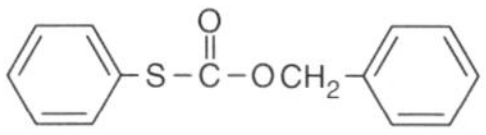
250 mg

S86,347-5
O-BENZYL S-METHYL DITHIOCARBONATE
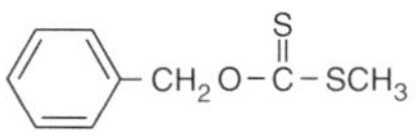
250 mg

S95,436-5
BIS[2-(METHOXYCARBONYL)PHENYL]-CARBONATE
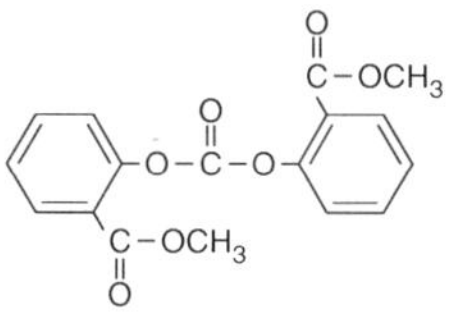
50 mg

S86,537-0
S-DIPHENYLMETHYL S'-METHYL-DITHIOCARBONATE
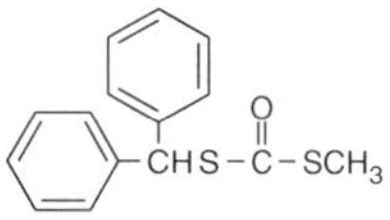
250 mg

S36,592-0
O-(CARBOMETHOXY)SALICYLIC ACID
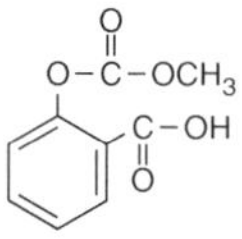
5 gm

S67,112-6
ETHYL 3,5-XYLYL CARBONATE
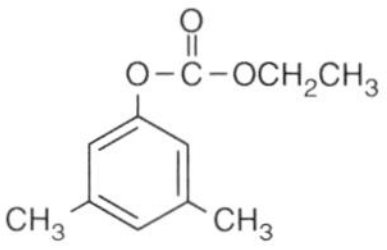
5 gm

S61,854-3
ETHYL 2-NAPHTHYL CARBONATE
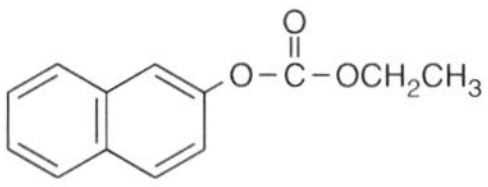
5 gm

S59,078-9
ETHYL M-TOLYL CARBONATE
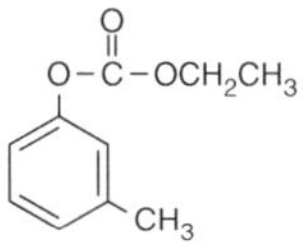
250 mg

S50,904-3
ETHYL PHENYL CARBONATE
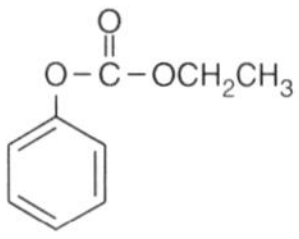
5 gm

S62,130-7
DI-P-TOLYL CARBONATE
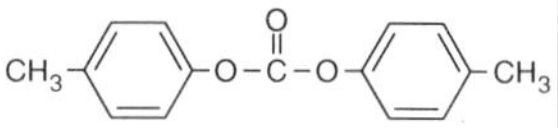
5 gm

S89,878-3
S-(4-CHLOROPHENYL)-ALPHA,2-DICHLOROBENZYL THIOCARBONATE
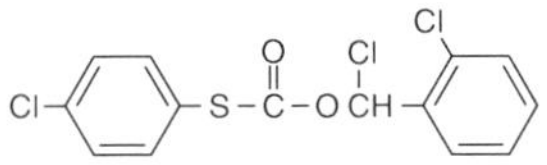
250 mg

S89,891-0
1-CHLOROETHYL S-PHENYL THIOCARBONATE
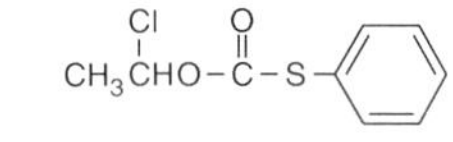
250 mg

S86,592-3
O-(9,10-DIHYDRO-9,10-ETHANOANTHRACEN-11-YLMETHYL)-S-METHYL DITHIOCARBONATE
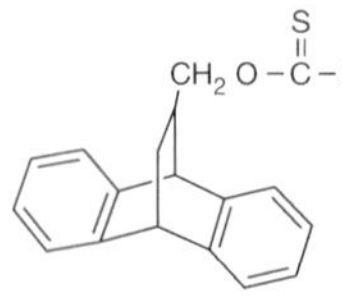
250 mg

S86,494-3
S-METHYL O-PHENETHYL DITHIOCARBONATE
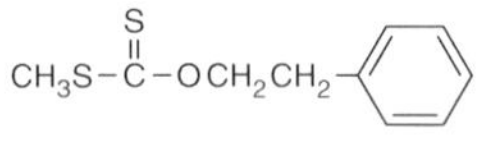
250 mg

S65,265-2
5-(CHLOROMETHYL)MECONIN
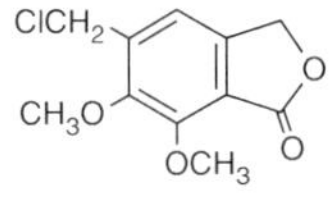
5 gm

S88,859-1
3-(2-BROMOETHYL)COUMARIN
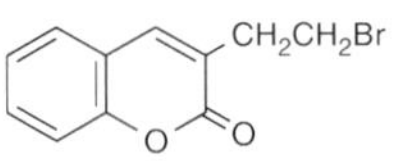
250 mg

S59,221-8
4-CHLOROMETHYL-6-METHYLCOUMARIN
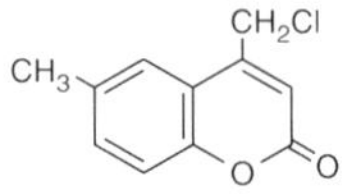
250 mg

S69,427-4
2-(4-CHLOROPHENYL)-4-(3-ETHOXY-2-HYDROXYBENZYLIDENE)-2-OXAZOLIN-5-ONE
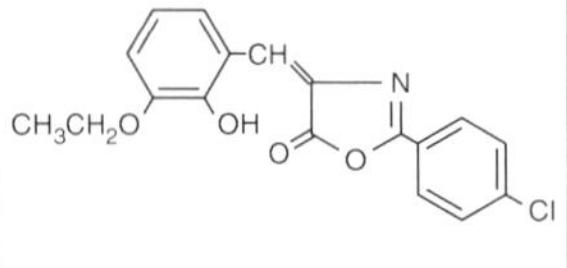
250 mg

S77,399-9
4-(2-CHLOROBENZYLIDENE)-2-(3,4,5-TRIMETHOXYPHENYL)-2-OXAZOLIN-5-ONE
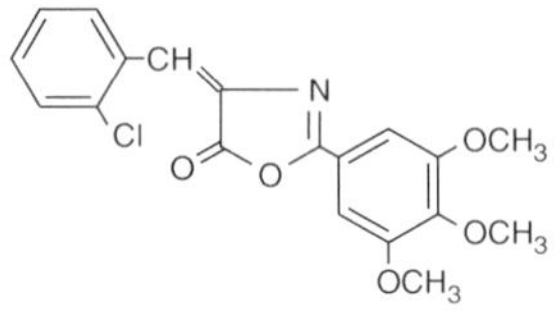
250 mg

S76,327-6
6-CHLORO-3-(4-CHLOROPHENYL)-3-[2-(DIMETHYLAMINO)ETHYL]-1H-2-BENZPYRAN-1-ONE HYDROCHLORIDE
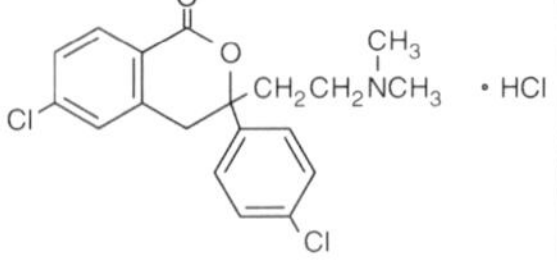
250 mg

S74,868-4
3-[ALPHA-(4-CHLOROPHENYL)-2-(DIMETHYLAMINO)BENZYL]-3-METHYLPHTHALIDE
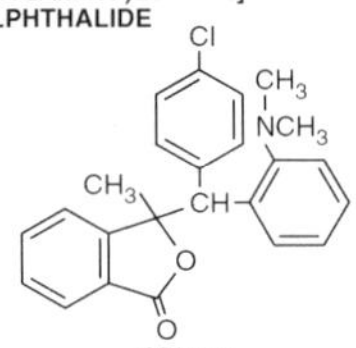
250 mg

S73,380-6
3-(4-CHLOROPHENYL)-3-[2-(DIMETHYLAMINOMETHYL)CYCLOHEXYL]-PHTHALIDE
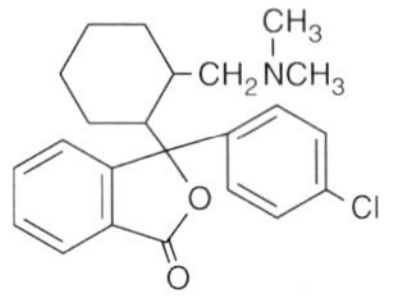
250 mg

S69,234-4
4-(4-CHLOROBENZYLIDENE)-2-(3-NITROPHENYL)-2-OXAZOLIN-5-ONE
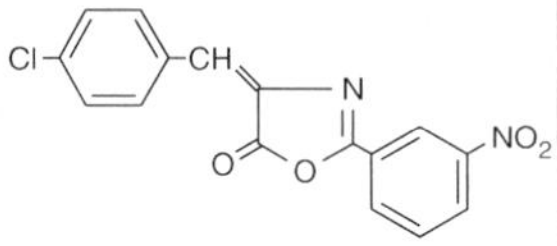
250 mg

S89,923-2
2,6-DIIODO-4-(5-OXO-2-PHENYL-2-OXAZOLIN-4-YLIDENEMETHYL)PHENYL ACETATE
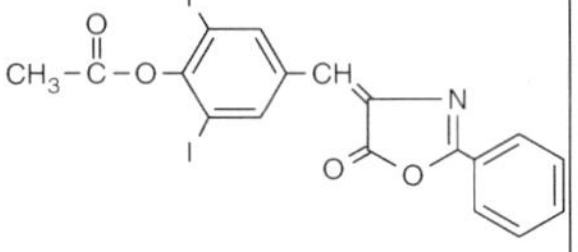
250 mg

S69,233-6
2-(4-TERT-BUTYLPHENYL)-4-(2,4-DICHLOROBENZYLIDENE)-2-OXAZOLIN-5-ONE
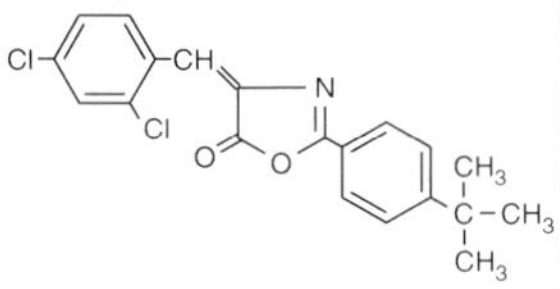
250 mg

S76,165-6
4-(3,4-DICHLOROPHENYL)-1H-2,3-BENZOXAZIN-1-ONE
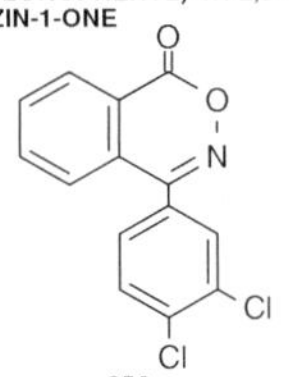
250 mg

S68,104-0
4-(4-CHLOROBENZYLIDENE)-2-METHYL-2-OXAZOLIN-5-ONE
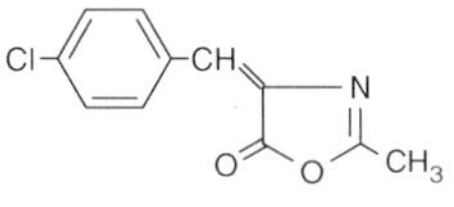
5 gm

S51,788-7
3,3'-(3,4,5-TRIMETHOXYBENZYLIDENE)-BIS(4-HYDROXYCOUMARIN)
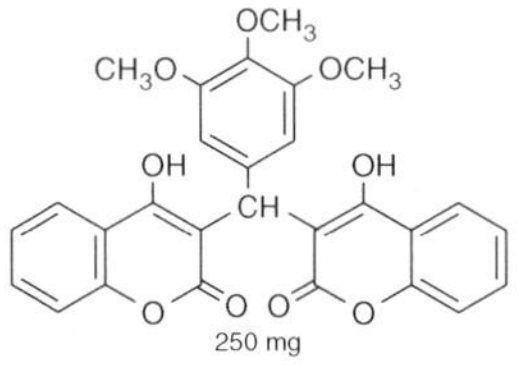
250 mg

S95,677-5
METHYL 2,3-DIHYDRO-5,7-DIHYDROXY-2,4-DIOXOBENZOPYRAN-8-CARBOXYLATE
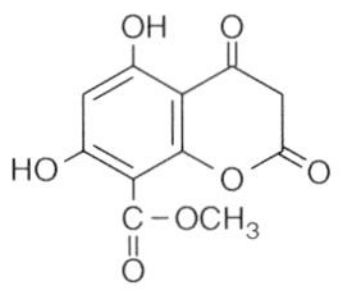
50 mg

S54,238-5
3,3'-(4-ISOPROPYLBENZYLIDENE)BIS(4-HYDROXYCOUMARIN)
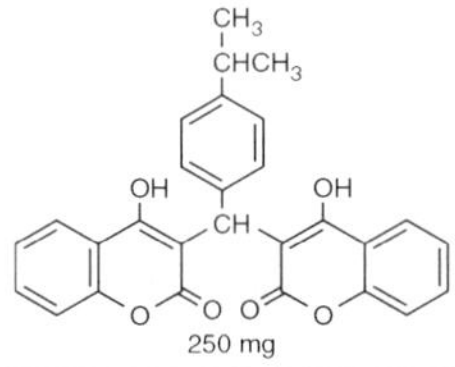
250 mg

S54,234-2
3,3',3'',3'''-(1,4-PHENYLENEDIMETHYLIDYNE)TETRAKIS(4-HYDROXYCOUMARIN)
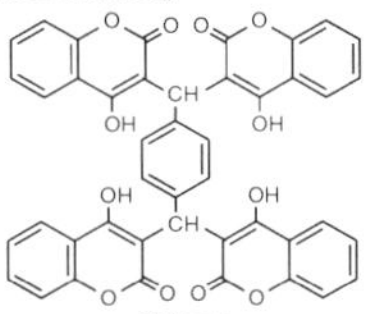
250 mg

S51,211-7
3,3'-(5-INDANYLMETHYLENE)BIS(4-HYDROXYCOUMARIN)
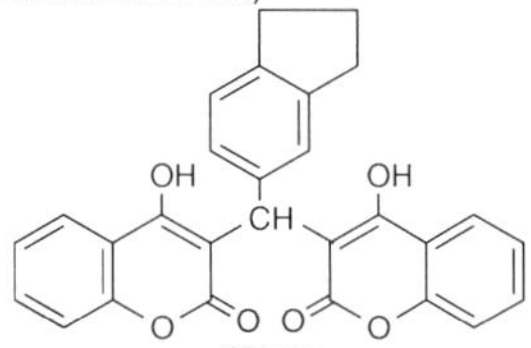
250 mg

S42,428-5
7-BENZYLOXY-6-HYDROXY-4-METHYLCOUMARIN
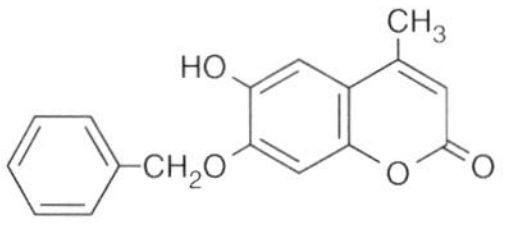
250 mg

S77,391-3
4-(2-HYDROXYBENZYLIDENE)-2-(3,4,5-TRIMETHOXYPHENYL)-2-OXAZOLIN-5-ONE
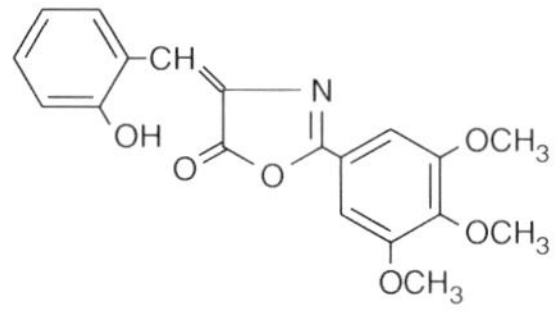
250 mg

S36,625-0
ALPHA-ACETYL-4-HYDROXY-BETA-(HYDROXYMETHYL)-3-METHOXYCINNAMIC ACID GAMMA-LACTONE
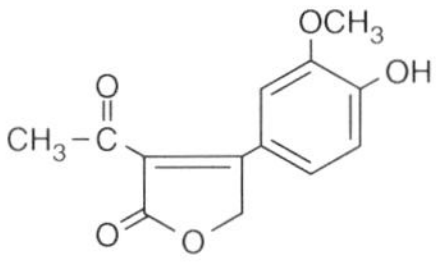
250 mg

S95,594-9
3-HYDROXY-7-METHOXY-3-METHYLPHTHALIDE
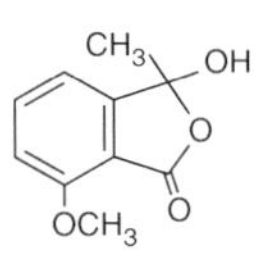
50 mg

S52,525-1
4-(3,5-DICHLOROSALICYLIDENE)-2-METHYL-2-OXAZOLIN-5-ONE
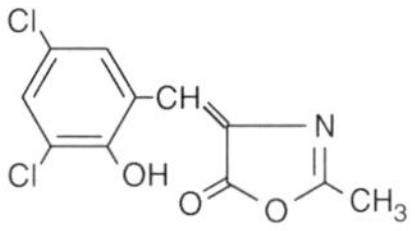
5 gm

S65,655-0
3-(TERT-BUTYL)-3-HYDROXYPHTHALIDE
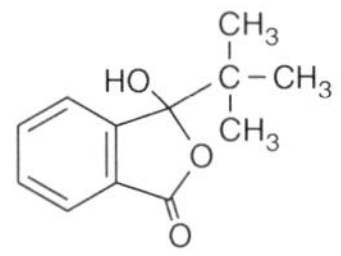
5 gm

S80,671-4
3-TERT-BUTYL-3-HYDROXY-5-METHYLPHTHALIDE
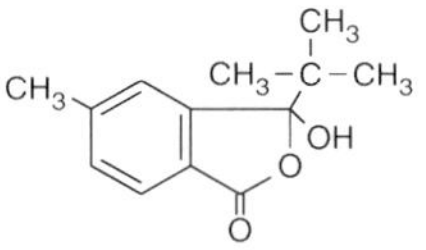
250 mg

S44,860-5
4-METHYLDAPHNETIN
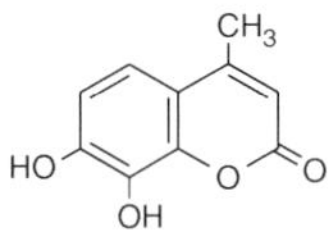
5 gm

S70,325-7
ALPHA-(1-BENZYL-4-HYDROXY-4-PIPERIDYL)-GAMMA-BUTYROLACTONE
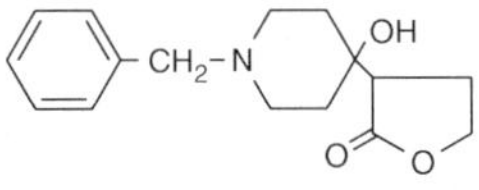
5 gm

S88,843-5
4-HYDROXY-3-(3-OXO-1-PHENYLBUTYL)-COUMARIN
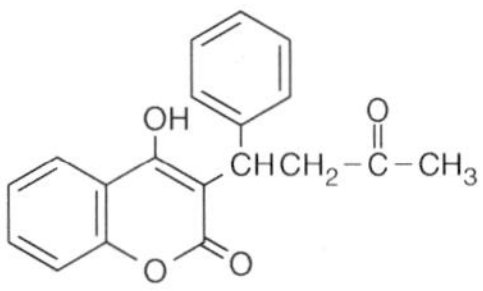
250 mg

S94,352-5
3-CINNAMOYL-4-HYDROXY-6-METHYL-2-PYRONE
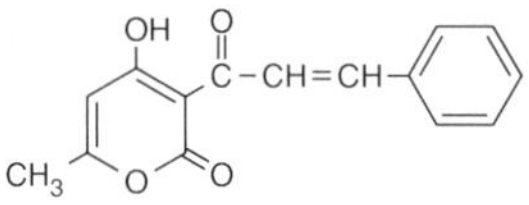
250 mg

S94,353-3
4-HYDROXY-6-METHYL-3-(5-PHENYL-2,4-PENTADIENOYL)-2-PYRONE
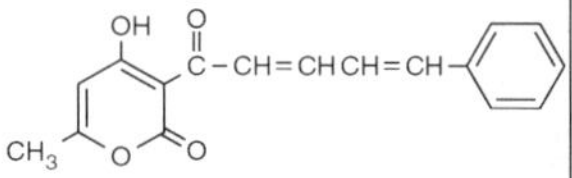
5 gm

S84,307-5
7-HYDROXY-3-COUMARINCARBOXYLIC ACID
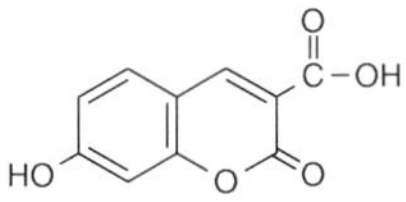
250 mg

S92,775-9
4-HYDROXY-3-PHENYLCOUMARIN
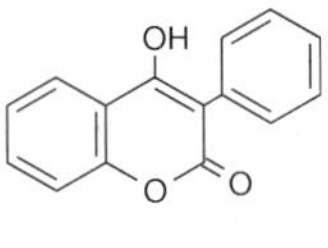
250 mg

S84,998-7
3-[1-(4-ETHYLPHENYL)ETHYL]-4-HYDROXYCOUMARIN
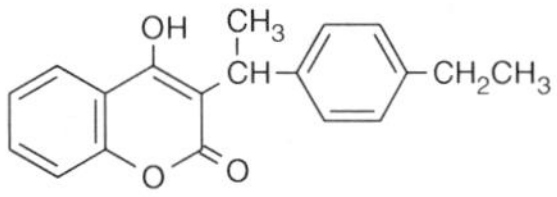
250 mg

S38,212-4
ETHYL 4-HYDROXY-8-METHYLCOUMARIN-3-CARBOXYLATE
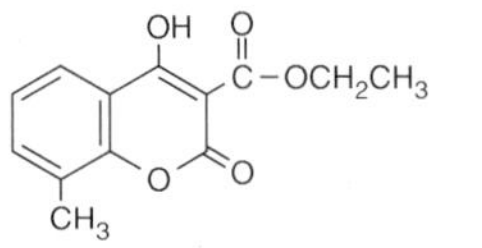
250 mg

S95,591-4
7-HYDROXY-3-METHYLPHTHALIDE MONOHYDRATE
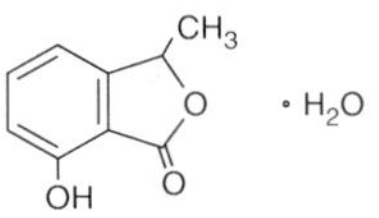
50 mg

S74,912-5
6-HYDROXY-3,4-BENZOCOUMARIN
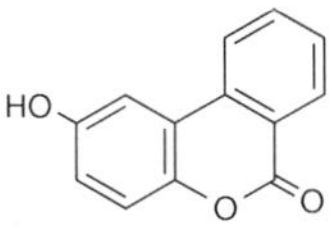
250 mg

S70,084-3
3-HYDROXY-3-PIPERIDINO-4,5,6,7-TETRACHLOROPHTHALIDE
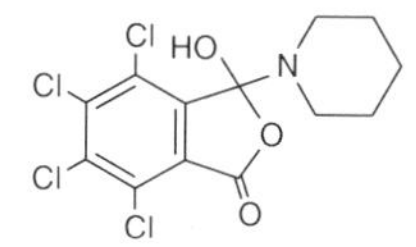
250 mg

S86,092-1
5-HYDROXY-6-METHYL-2(3H)-BENZOFURANONE
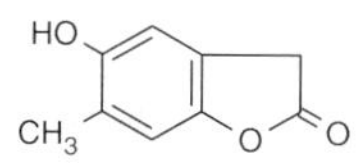
250 mg

S36,265-4
7-HYDROXY-4-PHENYLCOUMARIN
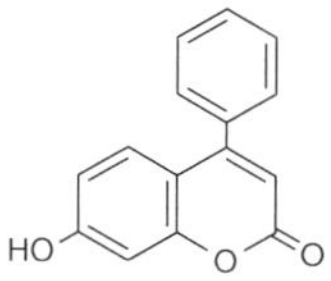
5 gm

S35,295-0
METHYL 5-HYDROXY-4-METHYLCOUMARIN-6-CARBOXYLATE
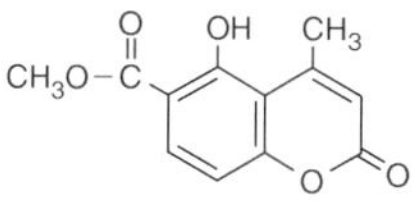
5 gm

S70,316-8
3-HYDROXY-3-PYRROLIDINO-4,5,6,7-TETRACHLOROPHTHALAN-1-ONE
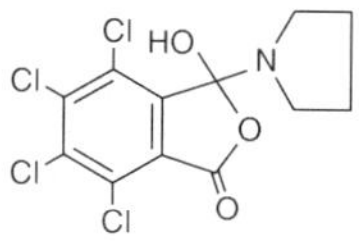
250 mg

S45,294-7
BENGAL RED B
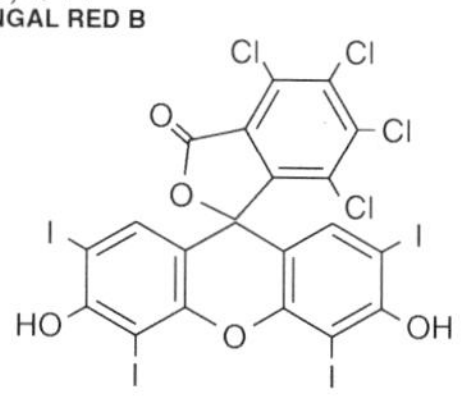
5 gm

S71,958-7
3-HYDROXY-3-MORPHOLINO-4,5,6,7-TETRACHLOROPHTHALIDE
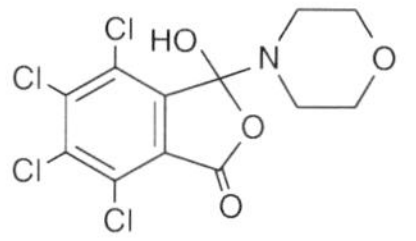
5 gm

S52,711-4
4-(5-BROMOSALICYLIDENE)-2-PHENYL-2-OXAZOLIN-5-ONE
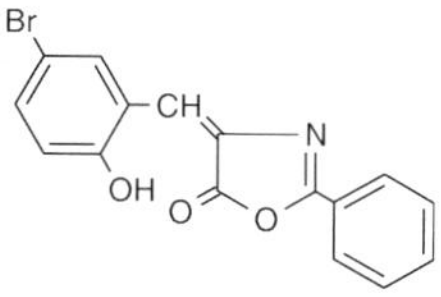
250 mg

S52,727-0
4-(3,5-DIBROMOSALICYLIDENE)-2-PHENYL-2-OXAZOLIN-5-ONE
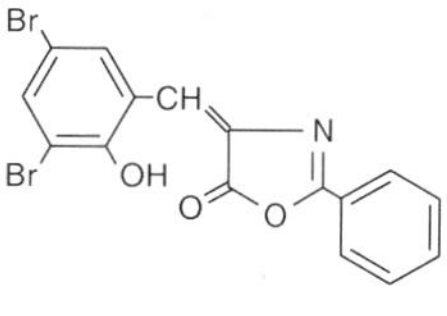
250 mg

S53,824-8
4-(3,5-DIIODOSALICYLIDENE)-2-PHENYL-2-OXAZOLIN-5-ONE
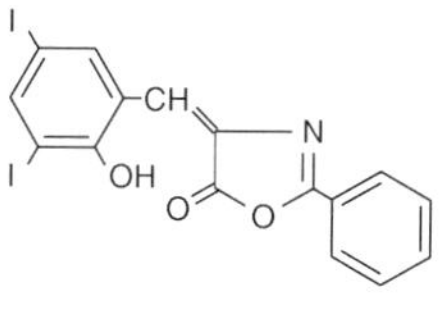
5 gm

S38,072-5
4-HYDROXY-6-METHYLCOUMARIN
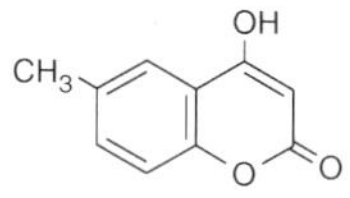
250 mg

S36,101-1
4,8-DIMETHYL-7-HYDROXYCOUMARIN
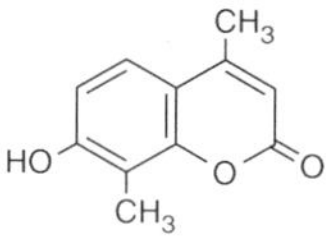
5 gm

S88,887-7
ALPHA-(2-HYDROXYBENZYLIDENE)-GAMMA-BUTYROLACTONE
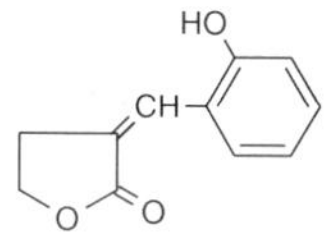
250 mg

S42,902-3
2-BENZYL-3,4-DIHYDROXYCROTONIC ACID GAMMA-LACTONE
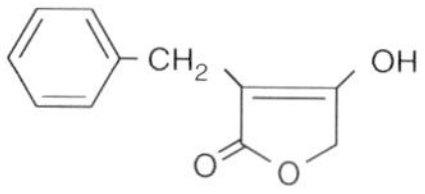
250 mg

S44,690-4
6,6-BIS(4-HYDROXYPHENYL)-3-(1-HYDROXY-1-METHYLETHYL)HEPTANOIC ACID GAMMA-LACTONE
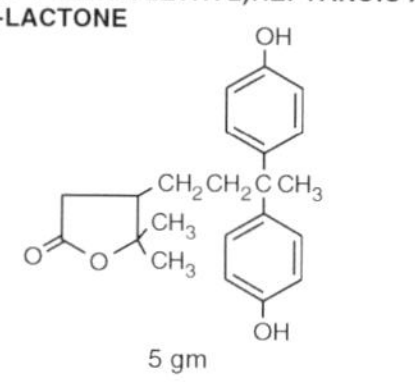
5 gm

S68,245-4
3,5-DI-TERT-BUTYL-3-HYDROXYPHTHALIDE
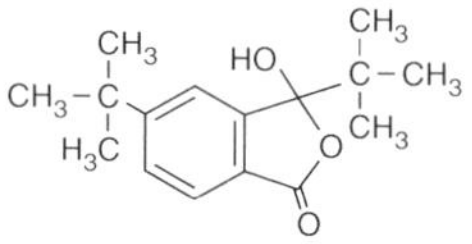
250 mg

S41,631-2
5,7-DIHYDROXY-4-METHYLCOUMARIN MONOHYDRATE
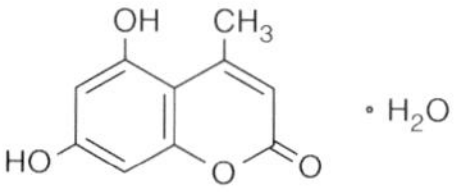
5 gm

S68,247-0
X-METHOXY-3-(4-METHOXYPHENYL)-PHTHALIDE
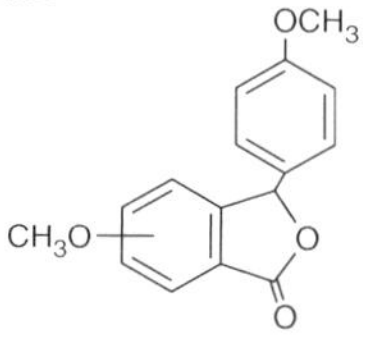
5 gm

S76,094-3
3-[4-(BENZYLOXY)PHENYL]-3-METHYL-1H-2-BENZOPYRAN-1(4H)-ONE
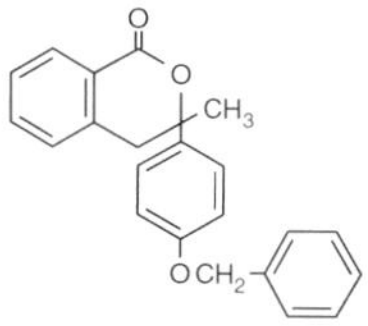
250 mg

S92,977-8
5,6-DIMETHOXYPHTHALIDE
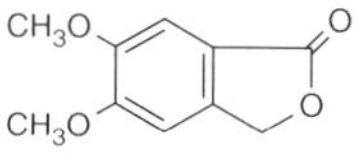
250 mg

S75,208-8
4-(3,5-DIMETHOXYBENZYLIDENE)-2-PHENYL-2-OXAZOLIN-5-ONE
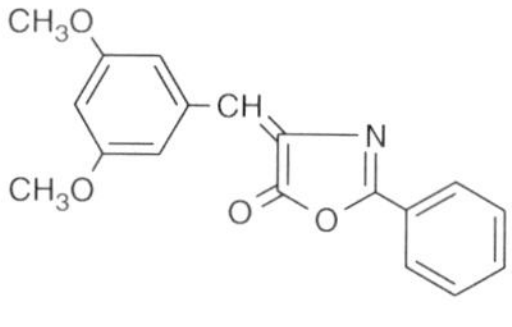
5 gm

S77,394-8
4-(3-NITROBENZYLIDENE)-2-(3,4,5-TRIMETHOXYPHENYL)-2-OXAZOLIN-5-ONE
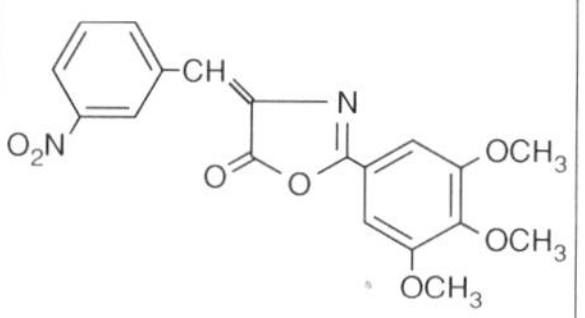
250 mg

S77,051-5
4-(2,4-DIMETHOXYBENZYLIDENE)-2-PHENYL-2-OXAZOLIN-5-ONE
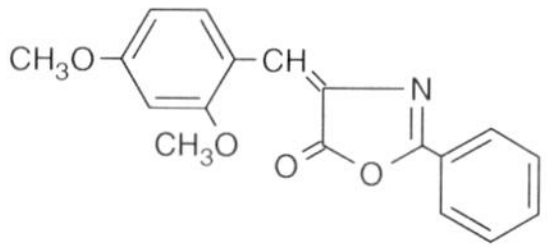
250 mg

S74,973-7
4-(2,6-DIMETHOXYBENZYLIDENE)-2-PHENYL-2-OXAZOLIN-5-ONE
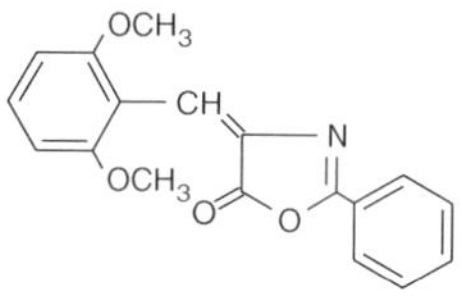
250 mg

S66,228-3
7,8-DIMETHOXY-1-ETHOXY-1-HYDROXY-4-METHYL-3A,4,5,9B-TETRA-H-NAPHTHO-(1,2-C)FURAN-3(1H)-ONE
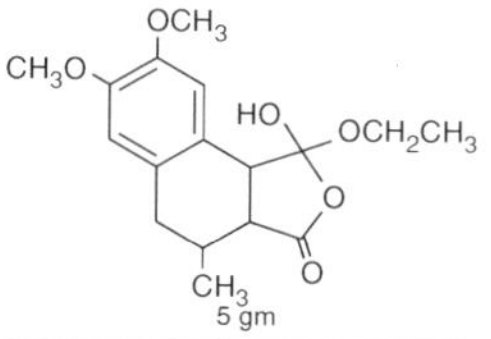
5 gm

S66,301-8
3-ETHOXYPHTHALIDE
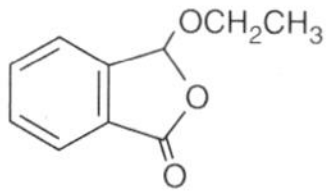
250 mg

S66,262-3
3-ISOPROPOXYPHTHALIDE
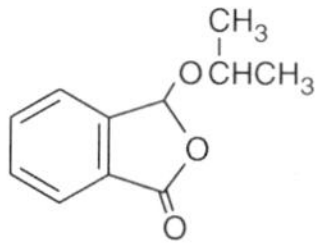
5 gm

S66,296-8
3-PROPOXYPHTHALIDE
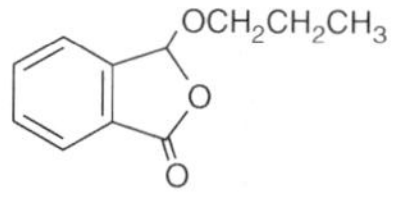
250 mg

S66,818-4
4-(2,4,5-TRIMETHOXYBENZYLIDENE)-2-PHENYL-2-OXAZOLIN-5-ONE
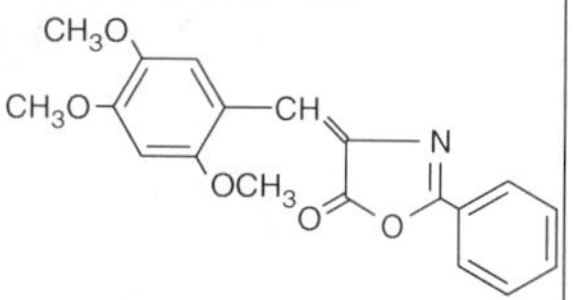
250 mg

S60,034-2
7-(BUTOXYCARBONYLMETHOXY)-3-(2-DIETHYLAMINOETHYL)-4-METHYLCOUMARIN
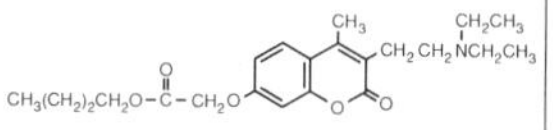
5 gm

S52,916-8
2-PHENYL-4-VERATRYLIDENE-2-OXAZOLIN-5-ONE
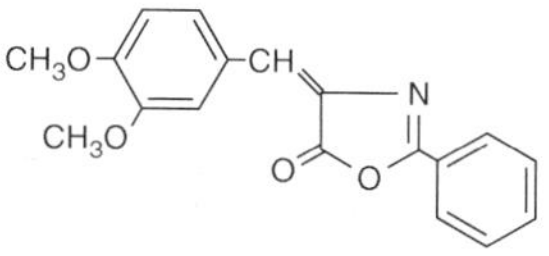
5 gm

S53,829-9
4-(2,3-DIMETHOXYBENZYLIDENE)-2-PHENYL-2-OXAZOLIN-5-ONE
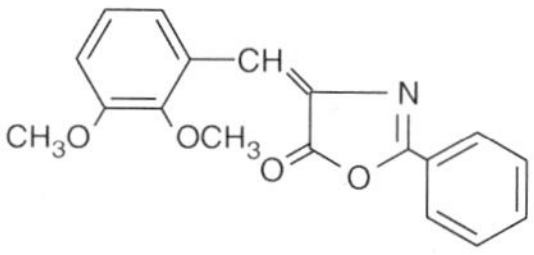
5 gm

S74,979-6
4-(2-ACETOXY-3-METHOXYBENZYLIDENE)-2-PHENYL-2-OXAZOLIN-5-ONE
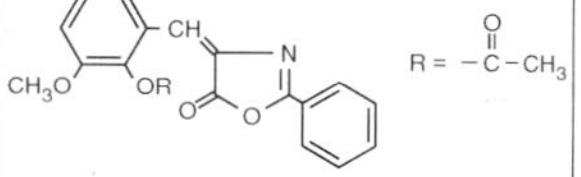
250 mg

S34,079-0
7-ALLYLOXY-4,8-DIMETHYLCOUMARIN
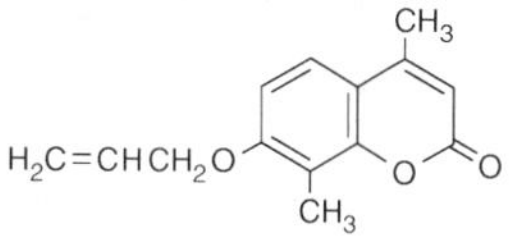
5 gm

S95,593-0
7-METHOXY-3-METHYLPHTHALIDE
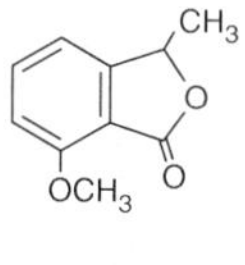
50 mg

S95,592-2
3-CHLORO-7-METHOXY-3-METHYLPHTHALIDE
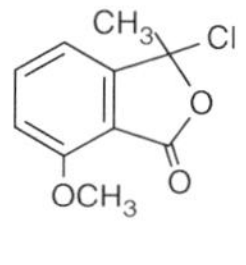
50 mg

S94,349-5
5-(2-METHOXYBENZYLIDENE)-2,2-DIMETHYL-1,3-DIOXANE-4,6-DIONE
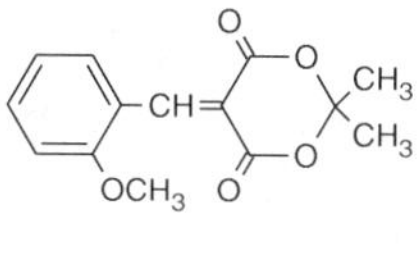
5 gm

S84,715-1
6-METHOXYPHTHALIDE
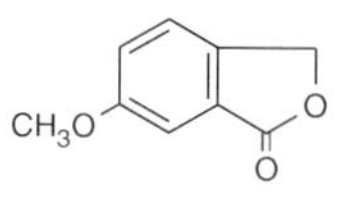
250 mg

S52,911-7
4-(4-BENZYLOXYBENZYLIDENE)-2-PHENYL-2-OXAZOLIN-5-ONE
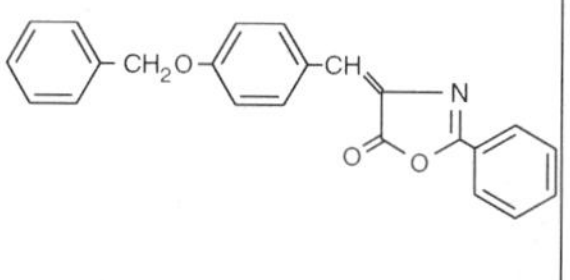
250 mg

S52,917-6
4-(3-METHOXYBENZYLIDENE)-2-PHENYL-2-OXAZOLIN-5-ONE
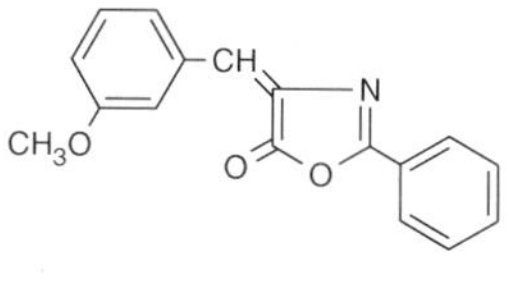
5 gm

S52,712-2
4-(3-ETHOXYBENZYLIDENE)-2-PHENYL-2-OXAZOLIN-5-ONE
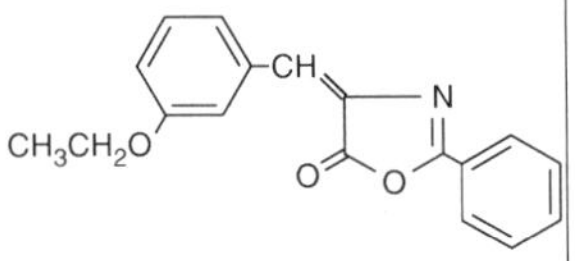
250 mg

S66,277-1
3-(4-METHOXY-1-NAPHTHYL)PHTHALIDE
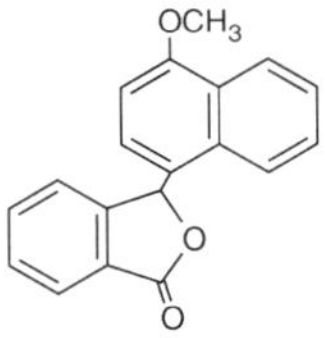
250 mg

S53,917-1
4-(3-BROMO-4-METHOXYBENZYLIDENE)-2-PHENYL-2-OXAZOLIN-5-ONE
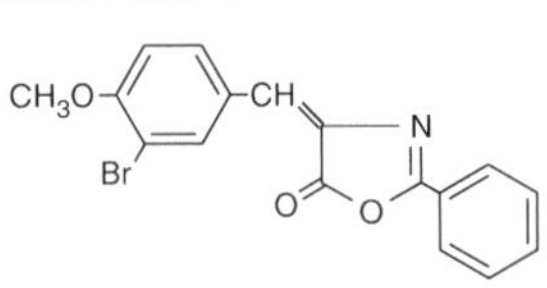
5 gm

S54,948-7
4-(4-ETHOXYBENZYLIDENE)-2-PHENYL-2-OXAZOLIN-5-ONE
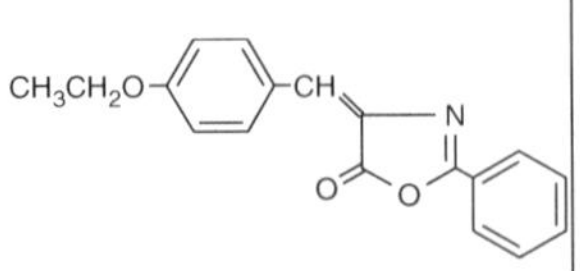
250 mg

S52,708-4
4-(2,5-DIMETHOXYBENZYLIDENE)-2-PHENYL-2-OXAZOLIN-5-ONE
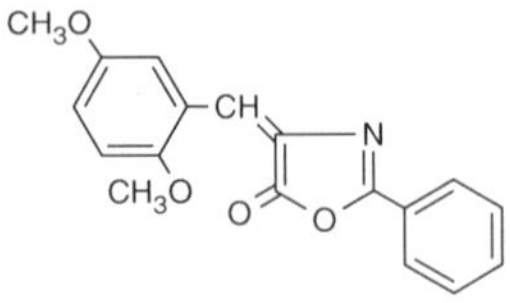
250 mg

S54,516-3
4-(5-BROMO-2-ETHOXYBENZYLIDENE)-2-PHENYL-2-OXAZOLIN-5-ONE
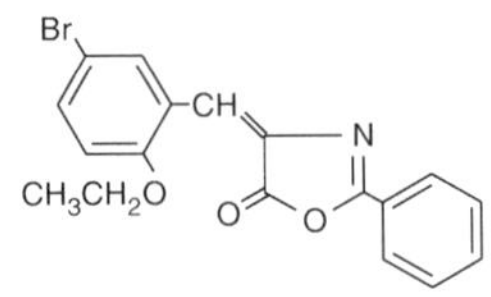
250 mg

S62,689-9
7-ALLYLOXY-4-METHYLCOUMARIN
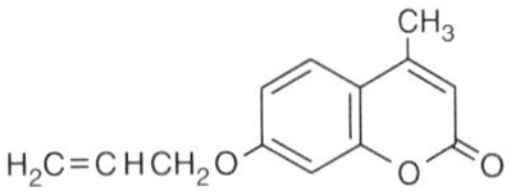
5 gm

S55,062-0
4-CHLORO-N-(3-METHOXYBENZOYL)-ANTHRANILIC ACID
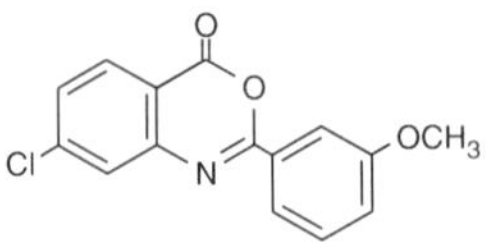
250 mg

S56,483-3
N-(3-METHOXYBENZOYL)-5-METHYLANTHRANILIC ACID
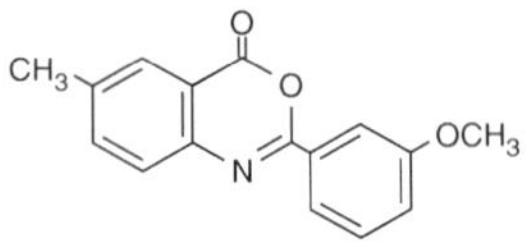
250 mg

S42,413-7
ESCULETIN DIBENZYL ETHER
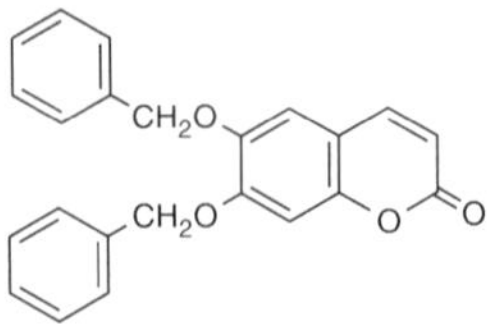
250 mg

S56,194-0
7-ETHOXY-4-METHYLCOUMARIN
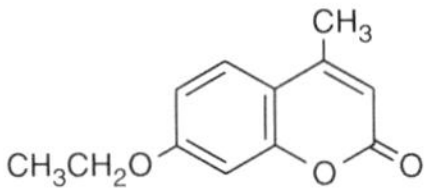
5 gm

S53,886-8
2-(4-ETHOXYPHENYL)-3,1-BENZOXAZIN-4(4H)-ONE
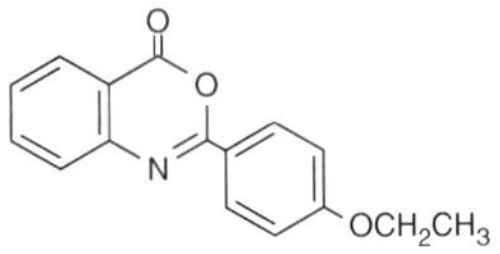
250 mg

S75,953-8
3-(3-AMINOPROPYL)-3-PHENYL-1-ISOCHROMANONE
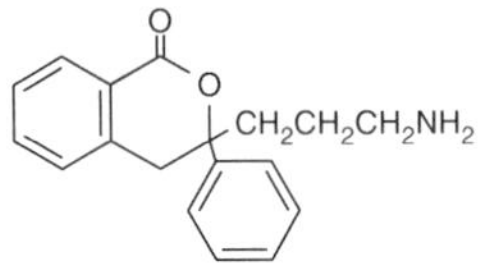
250 mg

S58,932-2
2-(4-CHLORO-2-NITROPHENYL)-3,1-BENZOXAZIN-4(4H)-ONE
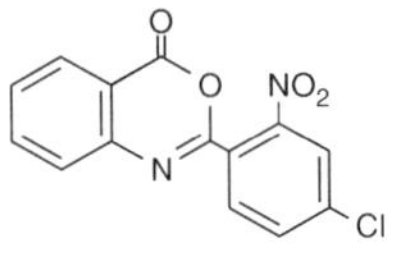
250 mg

S68,381-7
3,4-DIHYDRO-3-[ALPHA-(N-METHYLBENZYLAMINO)BENZYL]-3-PHENYL-1H-2-BENZOPYRAN-1-ONE
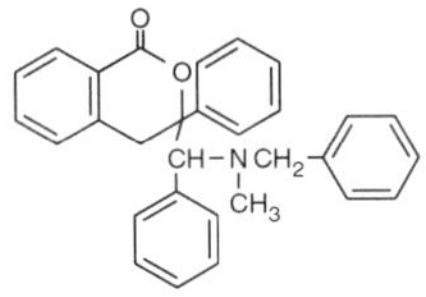
5 gm

S68,364-7
3-[1,1-DIMETHYL-2-(DIMETHYLAMINO)-ETHYL]-3-PHENYLPHTHALIDE
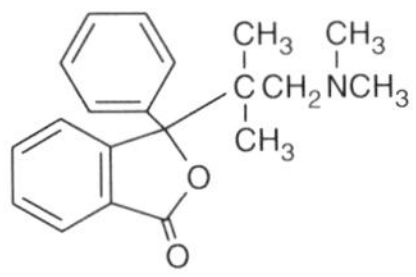
5 gm

S58,926-8
2-(2-CHLORO-4-NITROPHENYL)-4H-3,1-BENZOXAZIN-4-ONE
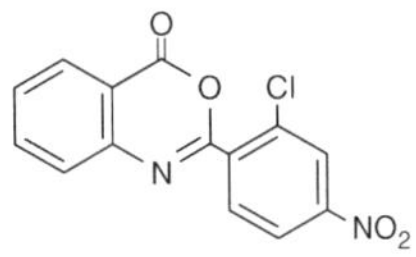
250 mg

S58,910-1
N-(3-CHLOROBENZOYL)ANTHRANILIC ACID
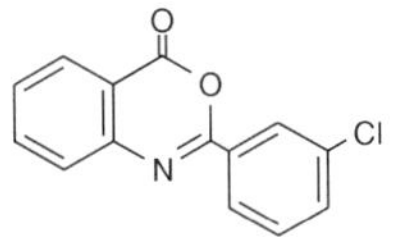
250 mg

S76,271-7
2,4,4-TRIPHENYL-2-OXAZOLIN-5-ONE
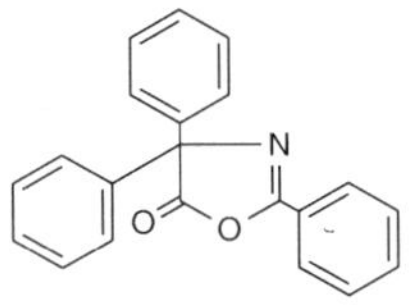
250 mg

S89,380-3
6-PHENYL-3,4,5-TRICHLORO-2H-PYRAN-2-ONE
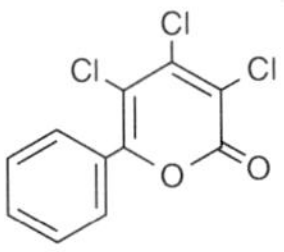
250 mg

S67,772-8
4,6-DIPHENYL-2H-PYRAN-2-ONE
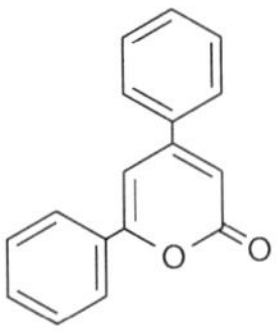
250 mg

S92,042-8
3-PHENYL-1H-2-BENZOPYRAN-1-ONE
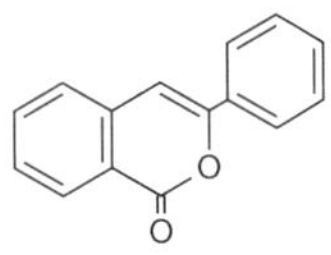
250 mg

S88,867-2
2,2-DIPHENYL-4H-1,3-BENZODIOXIN-4-ONE
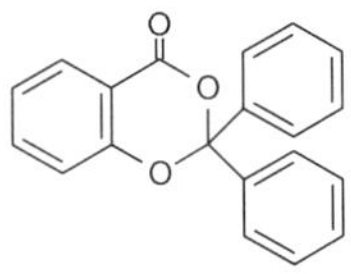
250 mg

S84,291-5
TETRAPHENYL-2-PYRONE
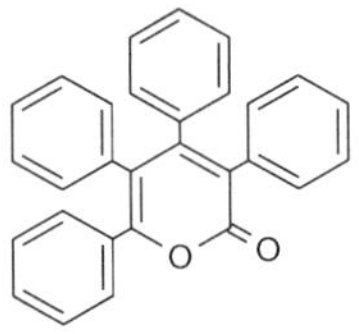
250 mg

S94,724-5
3,6-DIPHENYL-3,6-BIS(O-TOLYL)-1,4-DIOXANE-2,5-DIONE
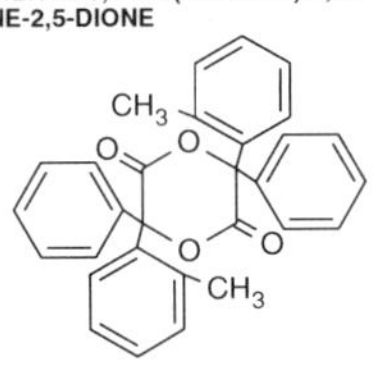
250 mg

S94,725-3
3,3-DIPHENYL-6,6-BIS(P-TOLYL)-1,4-DIOXANE-2,5-DIONE
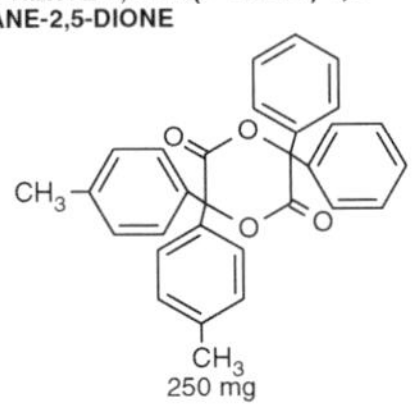
250 mg

S68,230-6
3-METHYL-3-PHENYLPHTHALIDE
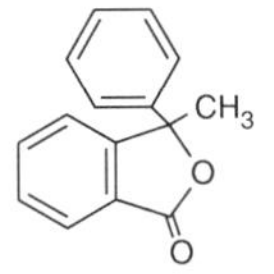
250 mg

S68,249-7
3,4-DIHYDRO-3-PHENYL-1H-2-BENZOPYRAN-1-ONE
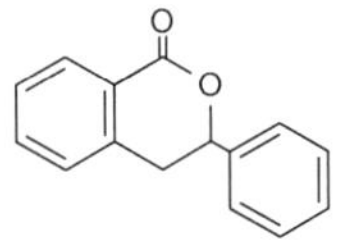
5 gm

S60,036-9
3-BENZAL-6-NITROPHTHALIDE
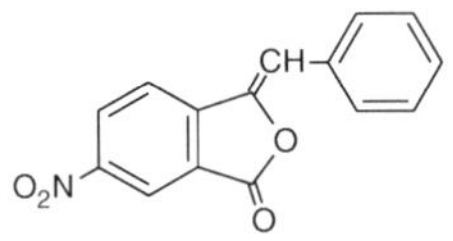
250 mg

S53,792-6
2-PHENYL-4-PIPERONYLIDENE-2-OXAZOLIN-5-ONE
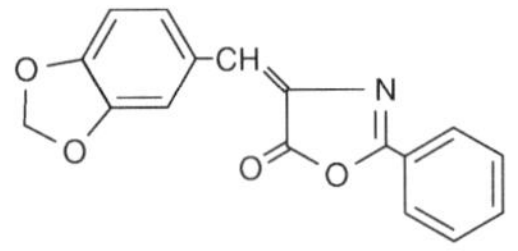
5 gm

S65,764-6
5,6-O-BENZYLIDENE-2,3-DIDEOXY-L-ASCORBIC ACID
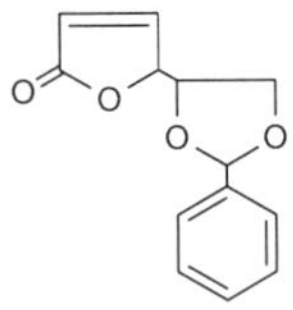
5 gm

S66,211-9
3-BENZOYLPHTHALIDE
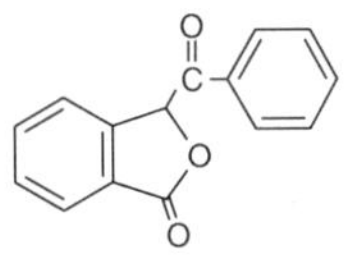
5 gm

S94,350-9
5-(ANILINOMETHYLENE)-2,2-DIMETHYL-1,3-DIOXANE-4,6-DIONE
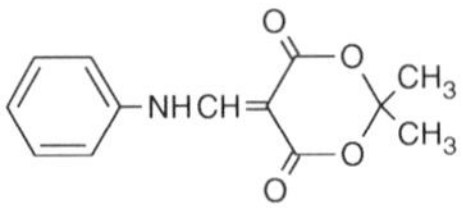
250 mg

S73,471-3
ETHYL 2-(2-IMIDAZOLIDINYLIDENE)-2-PHTHALIDYLACETATE
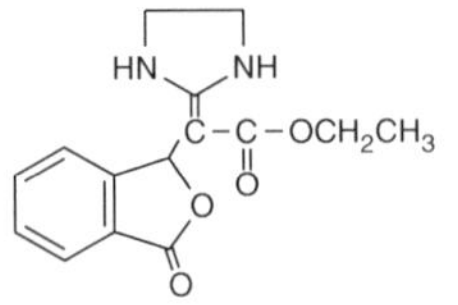
250 mg

S34,033-2
7-ACETOXY-6-ALLYL-4,8-DIMETHYLCOUMARIN
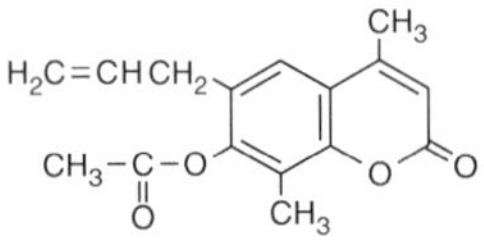
5 gm

S77,441-3
5-METHYL-7-OXO-4-PHENYL-6-OXABICYCLO[3.2.1]OCTAN-2-YLMETHYL ACETATE
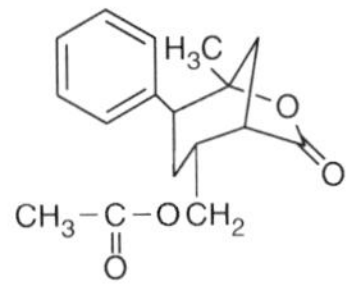
250 mg

S59,268-4
3-ACETYL-4-AMINOCOUMARIN
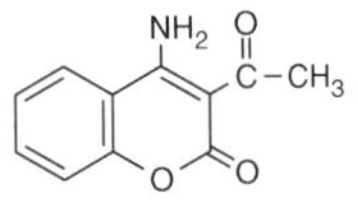
5 gm

S84,050-5
6-AMINOCOUMARIN HYDROCHLORIDE
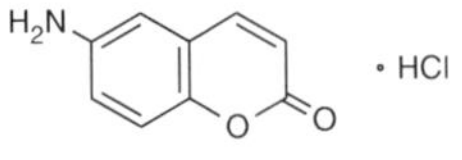
250 mg

S74,519-7
6-AMINOPHTHALIDE
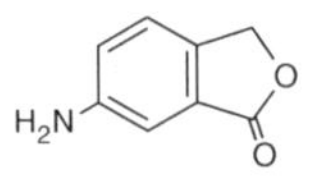
250 mg

S73,218-4
4-AMINOCOUMARIN
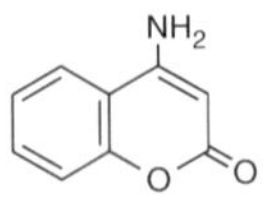
250 mg

S75,540-0
7-AMINO-3,4-BENZOCOUMARIN
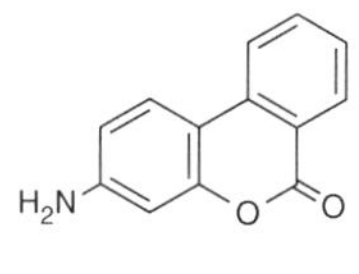
250 mg

S59,992-1
2-(3,5-DINITRO-2-METHYLPHENYL)-3,1-BENZOXAZIN-4(4H)-ONE
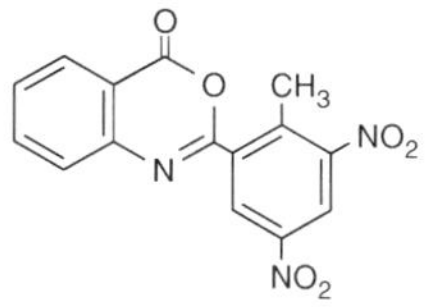
5 gm

S59,650-7
2-(2-CHLORO-3,5-DINITROPHENYL)-4-H-3,1-BENZOXAZIN-4-ONE
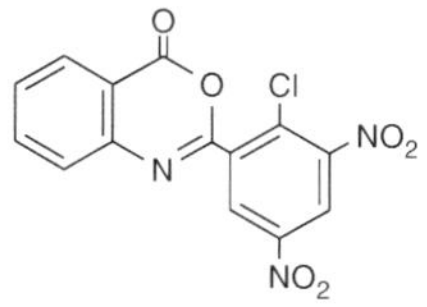
5 gm

S95,270-2
4-BENZYL-2-PHENYL-2-OXAZOLIN-5-ONE
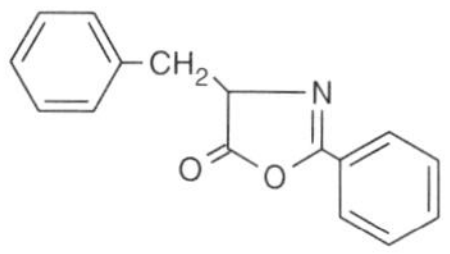
50 mg

S66,485-5
4-BENZYLIDENE-2-METHYL-2-OXAZOLIN-5-ONE
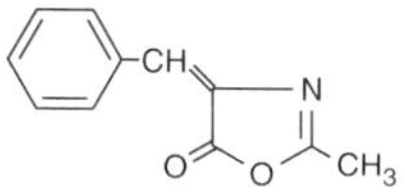
5 gm

S51,133-1
4-(3-ACETOXYBENZYLIDENE)-2-METHYL-2-OXAZOLIN-5-ONE
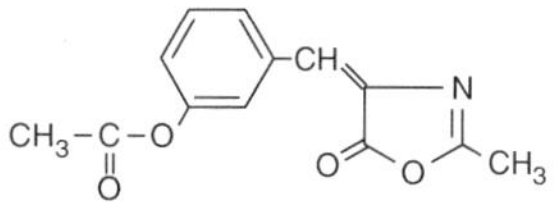
5 gm

S52,075-6
4-(3-ACETOXYBENZYLIDENE)-2-METHYL-5-OXAZOLIDINONE
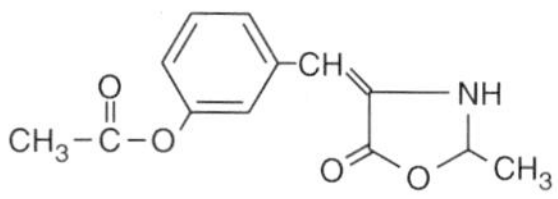
5 gm

S95,678-3
2-IMINO-3-PHENYL-2H-1-BENZOPYRAN P-TOLUENESULFONIC ACID SALT
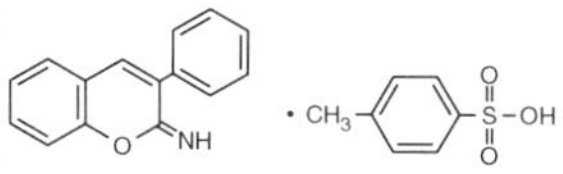
50 mg

S93,472-0
3-(ANILINOMETHYLENE)-2,4-CHROMANEDIONE
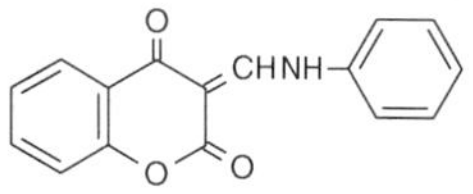
250 mg

S52,997-4
4-BENZYLIDENE-2-PHENYL-5-OXAZOLIDINONE
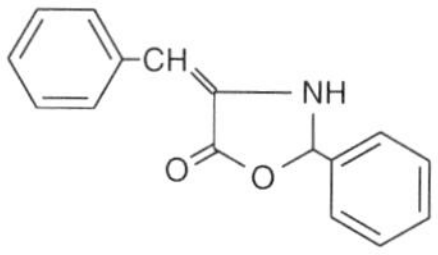
5 gm

S52,435-2
1,2-DIHYDRO-2-PHENYL-4H-3,1-BENZOXAZIN-4-ONE
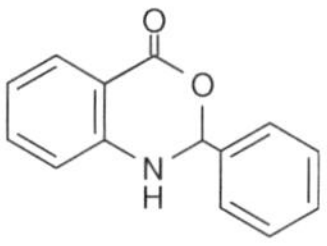
5 gm

S52,434-4
6-CHLORO-1,2-DIHYDRO-2-PHENYL-4H-3,1-BENZOXAZIN-4-ONE
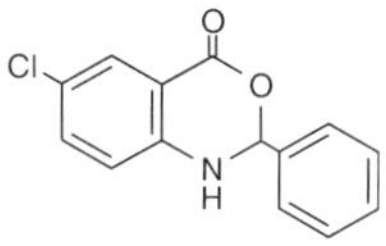
5 gm

S92,570-5
4-[4-(DIMETHYLAMINO)PHENYL]-1,3-DITHIOLE-2-THIONE
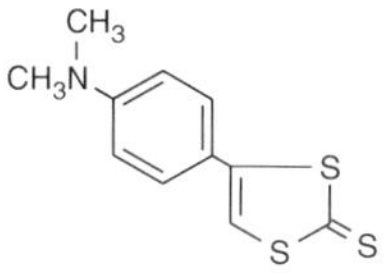
250 mg

S92,801-1
3-(3-METHYL-2-BENZOTHIAZOLINYLIDENE)-6-NITROPHTHALIDE
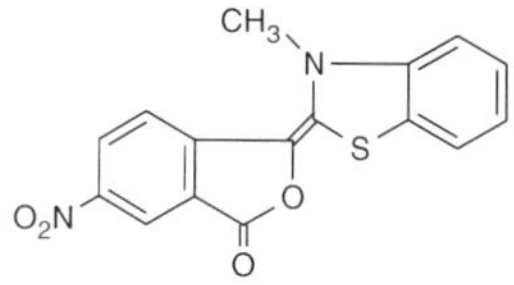
250 mg

S90,729-4
2'-(DIBENZYLAMINO)-6'-(DIETHYLAMINO)SPIRO[ISOBENZOFURAN-1-(3H)-9'-(9H)XANTHENE]-3-ONE
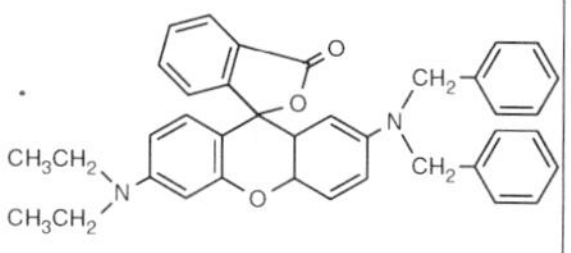
250 mg

S76,765-4
2'-(N-METHYLANILINO)-1-OXO-1,2',3'4-TETRAHYDROSPIRO[3H-2-BENZOPYRAN-3,1'-(1H)INDENE]
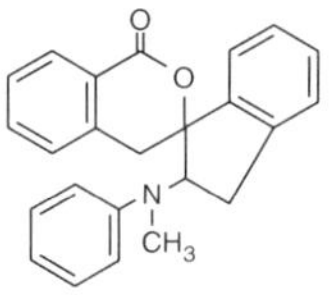
250 mg

S74,903-6
7-(DIMETHYLAMINO)-3,4-BENZOCOUMARIN
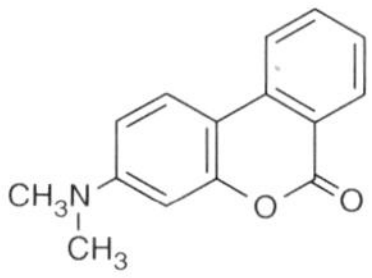
250 mg

S76,623-2
3-[2-(DIMETHYLAMINOMETHYL)PHENYL]-3-PHENYLPHTHALIDE
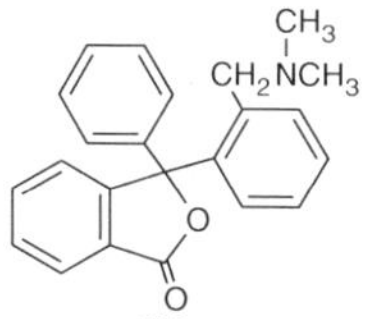
250 mg

S80,514-9
4-[4-(DIMETHYLAMINO)BENZYLIDENE]-2-PHENYL-5(4H)-OXAZOLONE
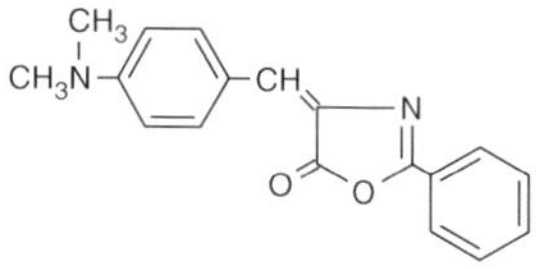
250 mg

S92,462-8
2,3-DIHYDRO-3-[1-(4-ETHYLANILINO)-ETHYLIDENE]-6-METHYL-4H-PYRAN-2,4-DIONE
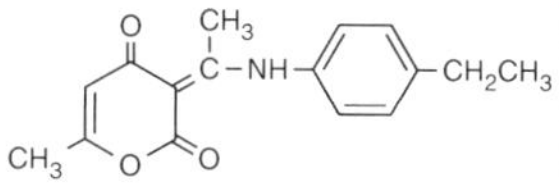
250 mg

S92,463-6
2,3-DIHYDRO-3-[1-(2,5-DIMETHYLANILINO)ETHYLIDENE]-6-METHYL-4H-PYRAN-2,4-DIONE
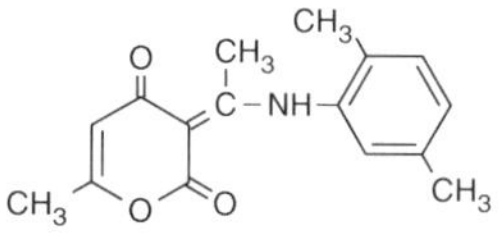
250 mg

S60,129-2
N-(3,4-DICHLOROBENZOYL)-5-METHYLANTHRANILIC ACID
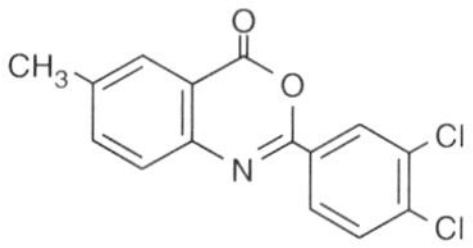
250 mg

S73,431-4
1'-BENZYL-1,4-DIHYDRO-4-PHENYLSPIRO(3H-2-BENZOPYRAN-3,4'-PIPERIDINE)-1-ONE HYDROCHLORIDE
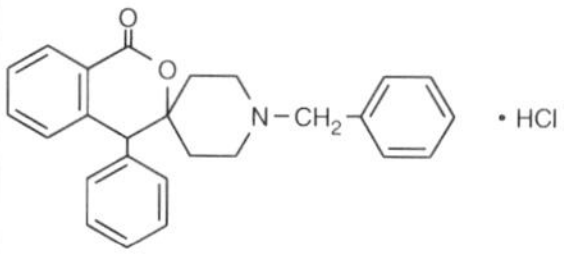
250 mg

S52,494-8
4-(4-NITROBENZYLIDENE)-2-PHENYL-2-OXAZOLIN-5-ONE
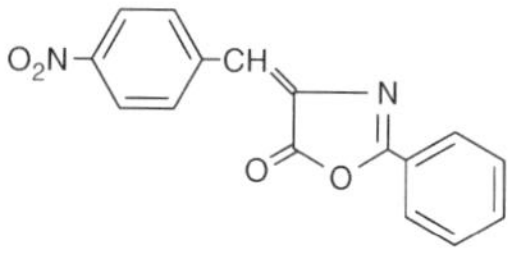
5 gm

S52,912-5
4-BENZYLIDENE-2-PHENYL-2-OXAZOLIN-5-ONE
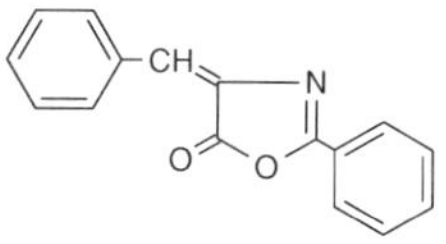
5 gm

S53,612-1
2-(4-CHLORO-3-NITROBENZAMIDO)-BENZOIC ACID
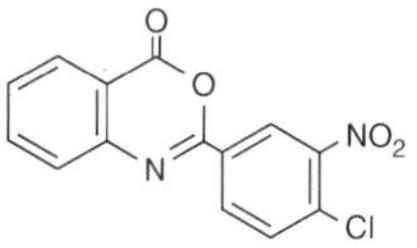
250 mg

S53,115-4
4-(3-NITROBENZYLIDENE)-2-PHENYL-2-OXAZOLIN-5-ONE
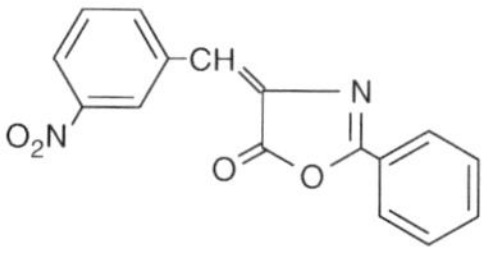
250 mg

S84,054-8
3-BROMO-6-NITROCOUMARIN
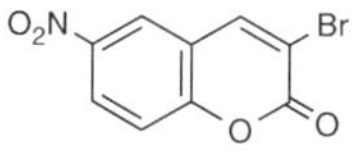
250 mg

S56,474-5
5-METHYL-N-(4-NITROBENZOYL)-ANTHRANILIC ACID
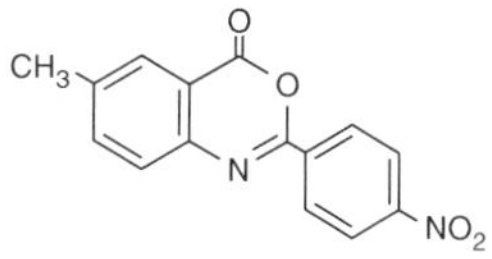
250 mg

S53,312-2
2-(4-NITROPHENYL)-3,1-BENZOXAZIN-4(4H)-ONE
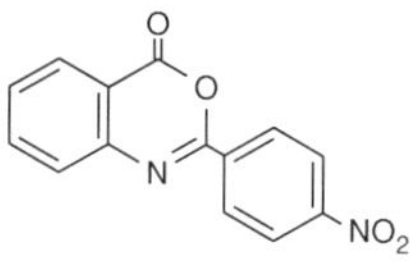
250 mg

S55,332-8
4-CHLORO-N-(4-NITROBENZOYL)-ANTHRANILIC ACID
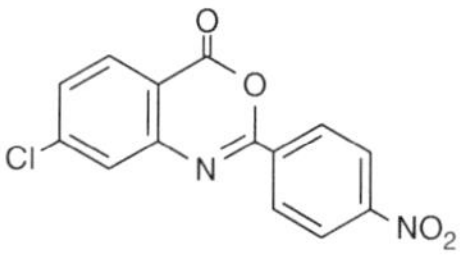
5 gm

S42,722-5
6-NITROCOUMARIN
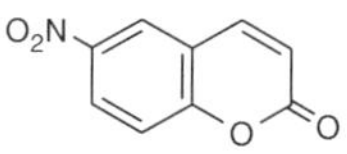
5 gm

S91,503-3
3-ACETYL-4,5-DIPHENYL-2(5H)-FURANONE
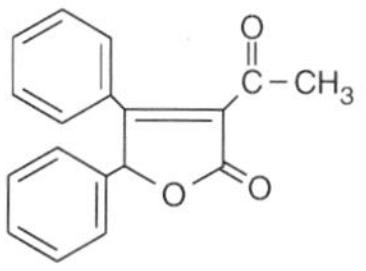
250 mg

S75,943-0
8A-(BENZOYLMETHYL)HEXAHYDRO-3,6,6-TRIMETHYL-3,5-ETHANO-1H,3H-PYRANO(3,4-C)PYRAN-1,8-DIONE
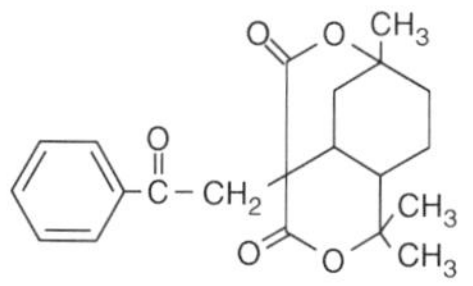
250 mg

S67,132-0
3-(1-NAPHTHYL)PHTHALIDE
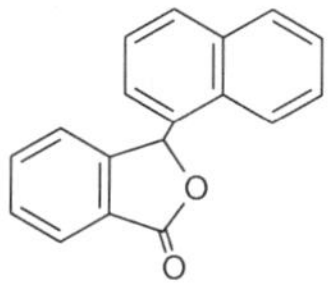
250 mg

S52,735-1
3-ACETYL-6,8-DICHLOROCOUMARIN
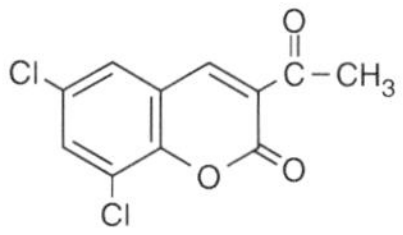
250 mg

S52,734-3
3-ACETYL-6-BROMOCOUMARIN
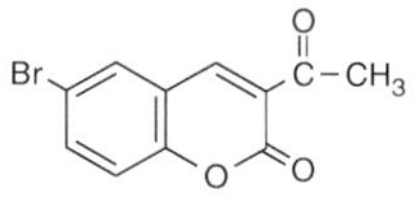
250 mg

S67,571-7
3-(2-CARBOXYPHENYL)PHTHALIDE
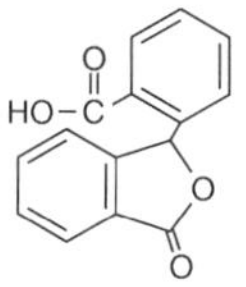
5 gm

S44,950-4
PHTHALIDENEACETIC ACID
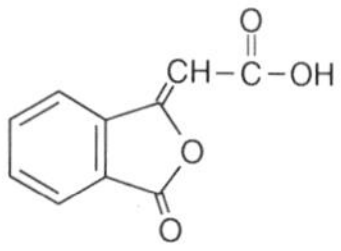
5 gm

S36,202-6
4-(4-FLUOROBENZYLIDENE)-2-PHENYL-2-OXAZOLIN-5-ONE
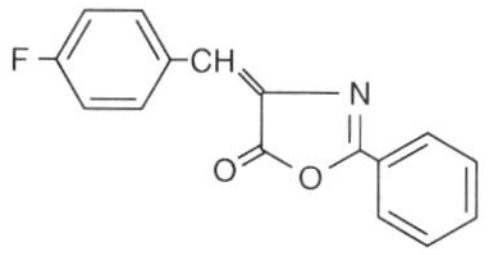
250 mg

S70,564-0
ETHYL 4,6-DIMETHYL-2-OXO-2H-PYRAN-5-CARBOXYLATE
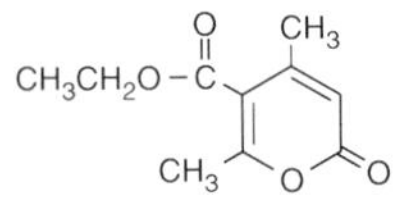
5 gm

S54,570-8
4-(4-CHLOROBENZYLIDENE)-2-PHENYL-2-OXAZOLIN-5-ONE
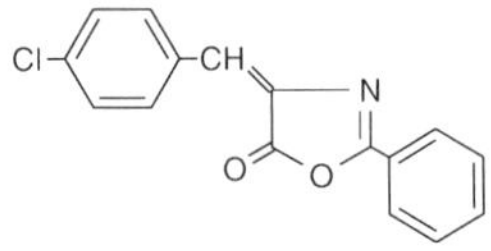
250 mg

S36,678-1
ALPHA-BENZYLIDENE-GAMMA-BUTYROLACTONE
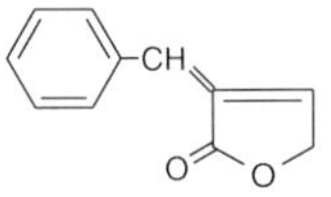
5 gm

S88,539-8
3,4-DIPHENYL-5-METHYL-2(5H)-FURANONE
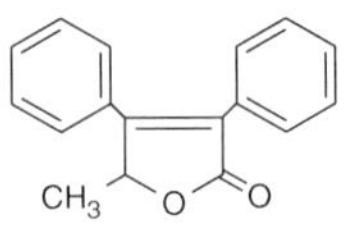
250 mg

S84,964-2
3-PHENYL-3-(2,4-XYLYL)PHTHALIDE
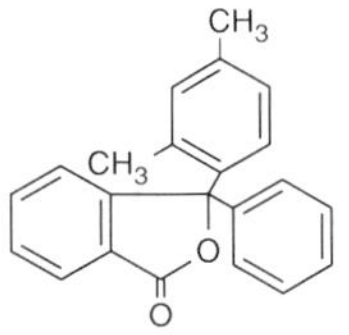
250 mg

S81,143-2
5-(DIPHENYLMETHYLENE)-2(5H)-FURANONE
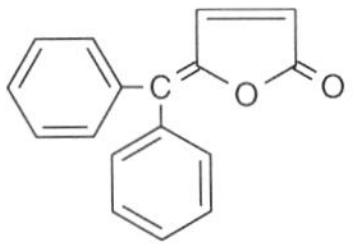
250 mg

S81,144-0
3-CHLORO-5-(DIPHENYLMETHYLENE)-2(5H)-FURANONE
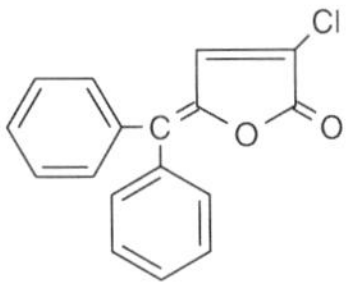
250 mg

S81,146-7
3,4-DICHLORO-5-(DIPHENYLMETHYLENE)-2(5H)-FURANONE
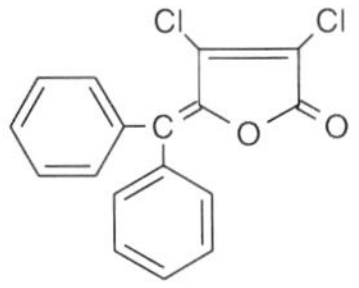
250 mg

S81,145-9
3-BROMO-5-(DIPHENYLMETHYLENE)-2(5H)-FURANONE
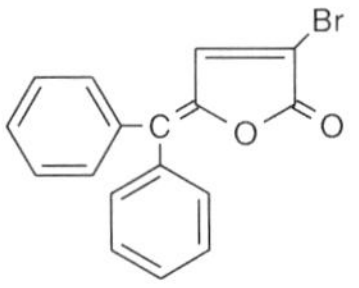
250 mg

S74,241-4
3,3-DIPHENYLPHTHALIDE
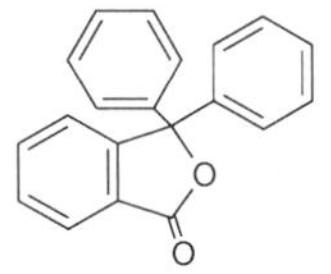
5 gm

S88,310-7
3-(P-TOLYL)PHTHALIDE
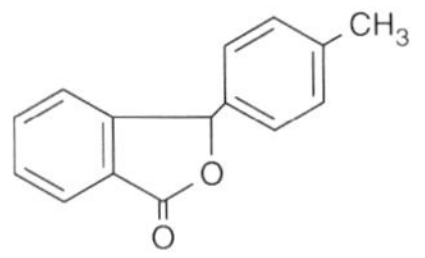
250 mg

S76,600-3
10,10-DIMETHYLSPIRO[ANTHRACENE-9(10H),1'(3'H)-ISOBENZOFURAN]-3'-ONE
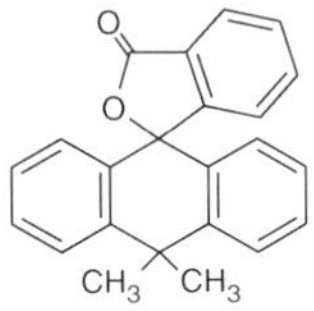
250 mg

S88,275-5
NAPHTHO(1,2-C)FURAN-1(3H)-ONE
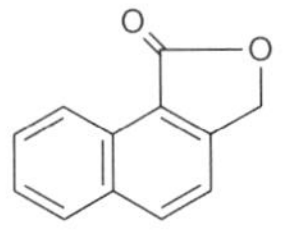
250 mg

S78,683-7
3-BROMOPHTHALIDE
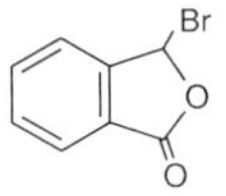
250 mg

S80,922-5
6-METHYL-2H-1,3-DIOXOLO(4,5-H)(1)-BENZOPYRAN
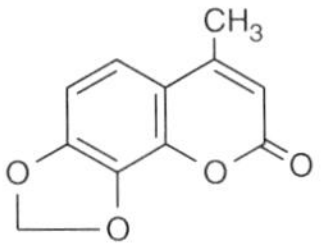
250 mg

S37,347-8
4-(2-FLUOROBENZYLIDENE)-2-PHENYL-2-OXAZOLIN-5-ONE
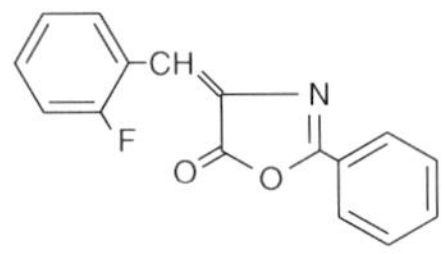
250 mg

S52,715-7
4-(3,4-DIFLUOROBENZYLIDENE)-2-PHENYL-2-OXAZOLIN-5-ONE
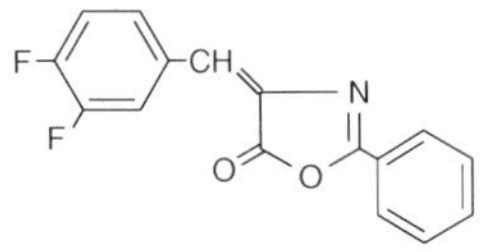
250 mg

S52,723-8
4-(2,4-DIFLUOROBENZYLIDENE)-2-PHENYL-2-OXAZOLIN-5-ONE
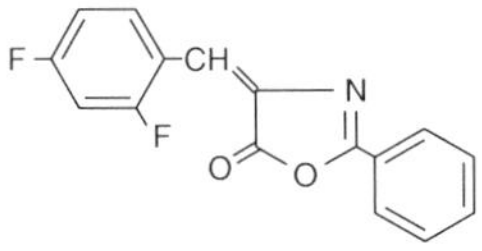
250 mg

S52,724-6
4-(4-METHYLBENZYLIDENE)-2-PHENYL-2-OXAZOLIN-5-ONE
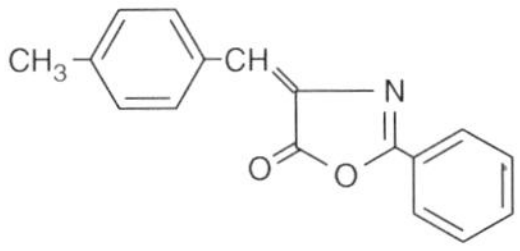
250 mg

S52,487-5
4-(4-METHOXYBENZYLIDENE)-2-PHENYL-2-OXAZOLIN-5-ONE
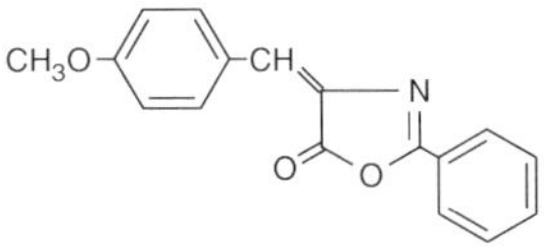
5 gm

S53,925-2
2-PHENYL-4-(4-PHENYLBENZYLIDENE)-2-OXAZOLIN-5-ONE
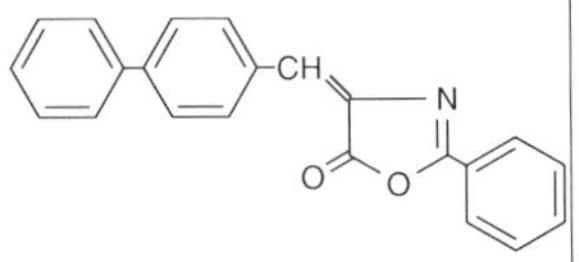
250 mg

S73,946-4
4-ETHYLIDENE-2-PHENYL-2-OXAZOLIN-5-ONE
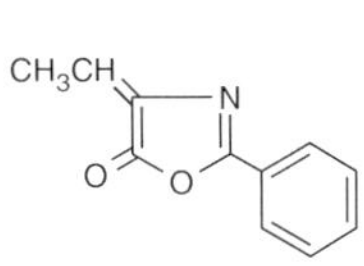
5 gm

S69,443-6
2-(4-TERT-BUTYLPHENYL)-4-CINNAMYLIDENE-2-OXAZOLIN-5-ONE
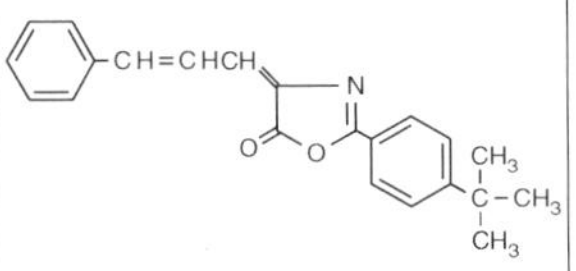
250 mg

S79,249-7
3,3-DIPHENYL-1H,3H-NAPHTHO(1,8-CD)-PYRAN-1-ONE
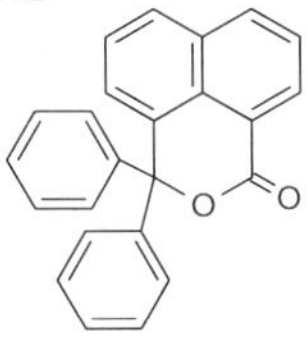
250 mg

S39,326-6
3-PHENYLCOUMARIN
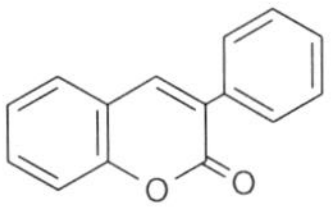
5 gm

S79,260-8
7-CHLORO-4A,5,6,7,8,8A-HEXAHYDRO-4-METHYL-3-PHENYLCOUMARIN
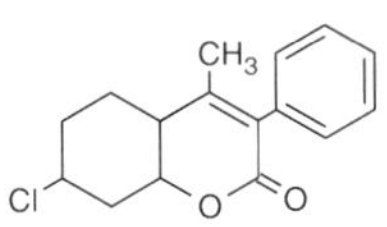
250 mg

S88,273-9
1H,3H-NAPHTHO(1,8-CD)PYRAN-1-ONE
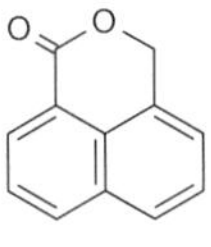
250 mg

S90,718-9
4,6-DIMETHYLCOUMARIN
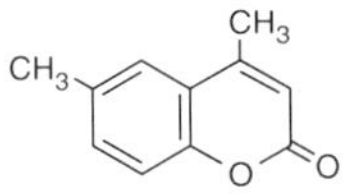
250 mg

S57,244-6
N-CINNAMOYL-5-METHYLANTHRANILIC ACID
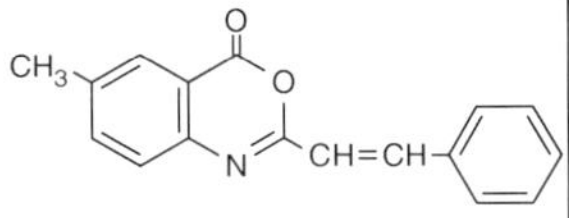
250 mg

S61,975-2
N-(3-CHLOROBENZOYL)-5-METHYLANTHRANILIC ACID
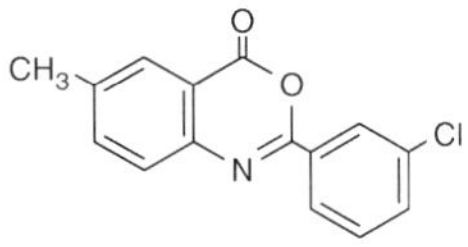
250 mg

S88,272-0
DIBENZ(C,E)OXEPIN-5(7H)-ONE
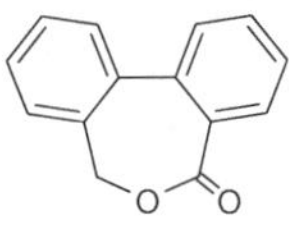
250 mg

S84,949-9
3-CHLORO-2H-1-BENZOTHIOPYRAN-2-ONE
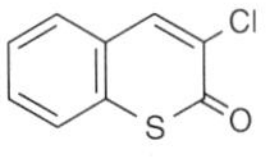
250 mg

S94,245-6
2,2-DIMETHYL-5-(1-NAPHTHYLMETHYLENE)-1,3-DIOXANE-4,6-DIONE
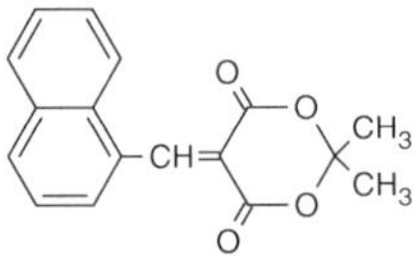
5 gm

S84,714-3
4-PHENYL-1H-2,3-BENZOXAZIN-1-ONE
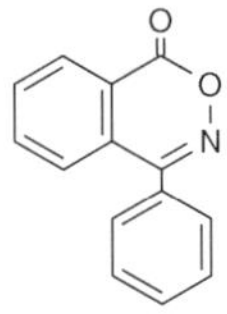
250 mg

S94,242-1
5-CINNAMYLIDENE-2,2-DIMETHYL-1,3-DIOXANE-4,6-DIONE
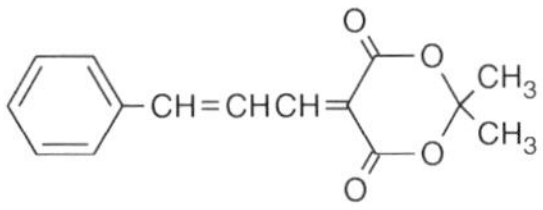
250 mg

S40,557-4
3-CHLOROCOUMARIN
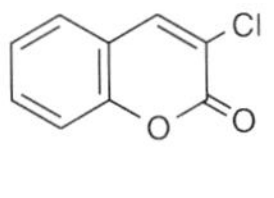
5 gm

S94,492-0
3,6-DIPHENYL-3,6-BIS(M-TOLYL)-1,4-DIOXANE-2,5-DIONE
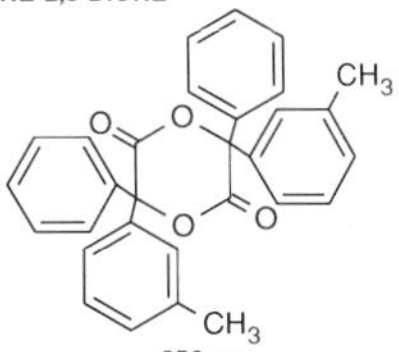
250 mg

S37,473-3
6-HYDROXY-5-IODO-3-PHENYL-2-NORBORNANECARBOXYLIC ACID GAMMA-LACTONE
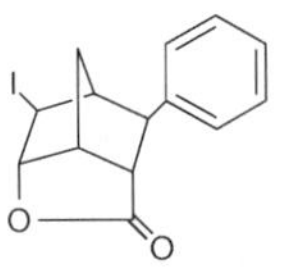
5 gm

S95,252-4
3-BENZYLIDENE-2,3-DIHYDRO-5-PHENYL-2-FURANONE
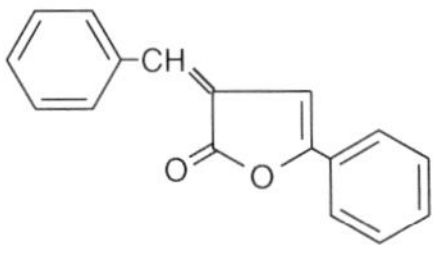
50 mg

S68,116-4
10,11-DIHYDROSPIRO[5H-DIBENZO(A,D)-CYCLOHEPTENE-5,1'-PHTHALAN]-3'-ONE
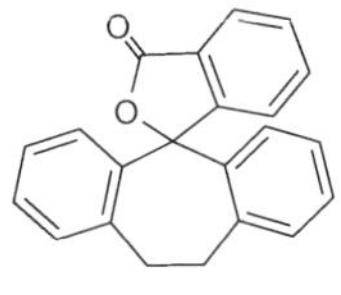
5 gm

S55,424-3
2-PHENYL-2-OXAZOLIN-5-ONE

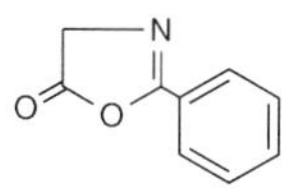

5 gm

S80,521-1
4-CYCLOHEXYLIDENE-2-PHENYL-5(4H)-OXAZOLONE

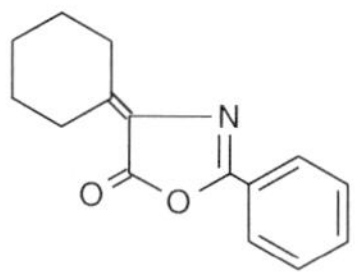

250 mg

S80,515-7
4-ISOPROPYLIDENE-2-PHENYL-5(4H)-OXAZOLE

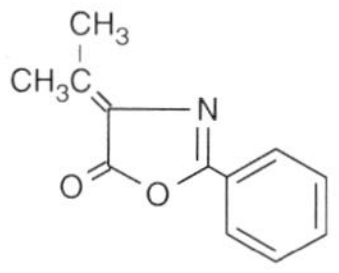

250 mg

S73,397-0
4-BENZYLIDENE-2-PHENYL-2-OXAZOLIN-5-ONE

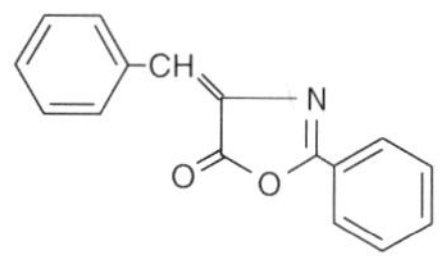

250 mg

S73,337-7
4-(BETA-METHYLCINNAMYLIDENE)-2-PHENYL-2-OXAZOLIN-5-ONE

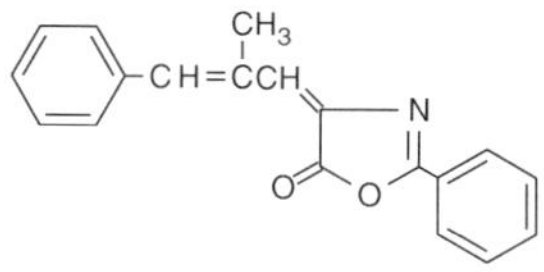

250 mg

S76,734-4
5,5-DIPHENYL-2-(2-PIPERIDINYLETHYL)-1,3-DIOXAN-4-ONE HYDROCHLORIDE

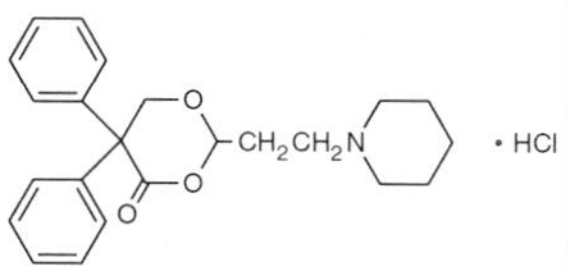

250 mg

S55,942-3
N-BENZOYL-5-METHYLANTHRANILIC ACID

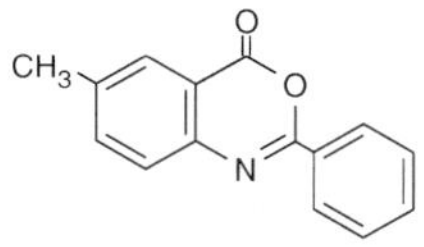

250 mg

S56,283-1
N-(2-METHYLBENZOYL)ANTHRANILIC ACID

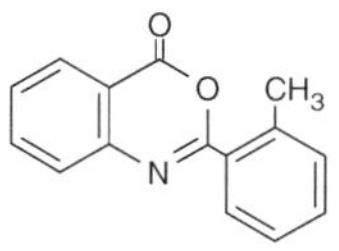

250 mg

S55,936-9
5-METHYL-N-(3-METHYLBENZOYL)-ANTHRANILIC ACID

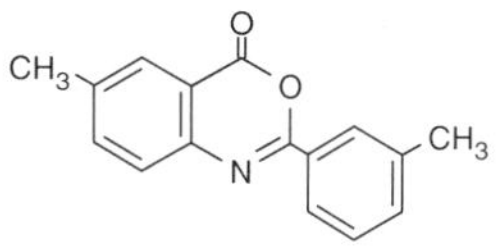

250 mg

S54,148-6
3-PHENYLPHTHALIDE

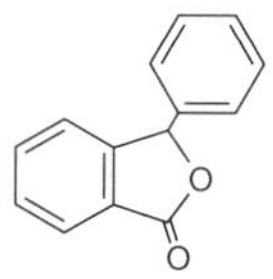

250 mg

S68,166-0
3,4-DIHYDRO-3,3-DIPHENYL-1H-2-BENZOPYRAN-1-ONE

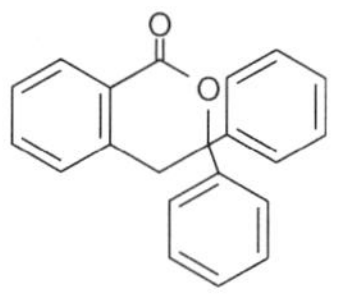

5 gm

S53,972-4
6,7-DIBROMO-2-PHENYL-3,1-BENZOXAZIN-4(4H)-ONE

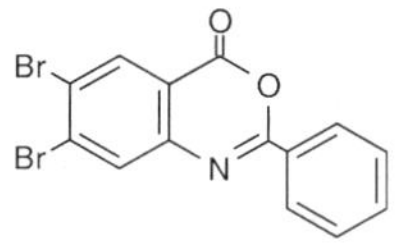

5 gm

S53,879-5
2-(2-CHLOROBENZAMIDO)BENZOIC ACID

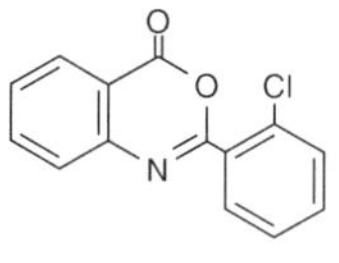

250 mg

S54,368-3
2-(2-IODOPHENYL)-3,1-BENZOXAZIN-4(4H)-ONE

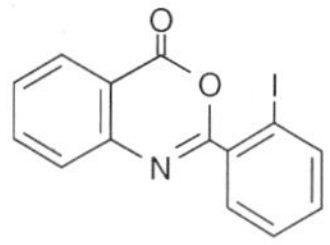

250 mg

S55,061-2
N-(2,4-DIMETHYLBENZOYL)-4-METHYLANTHRANILIC ACID

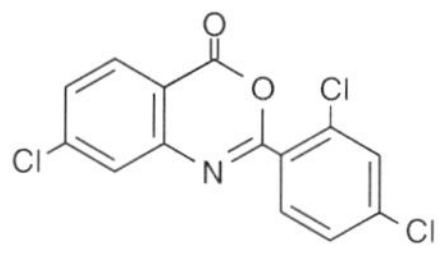

250 mg

S54,811-1
N-(3,4,5-TRIMETHOXYBENZOYL)-ANTHRANILIC ACID

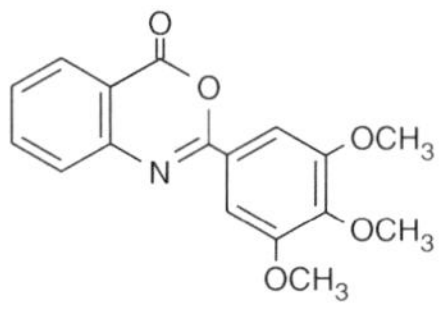

250 mg

S53,876-0
2-(P-FLUOROPHENYL)-3,1-BENZOXAZIN-4(4H)-ONE

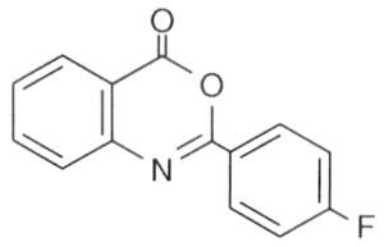

5 gm

S53,151-0
2-(M-TOLYL)-3,1-BENZOXAZIN-4(4H)-ONE

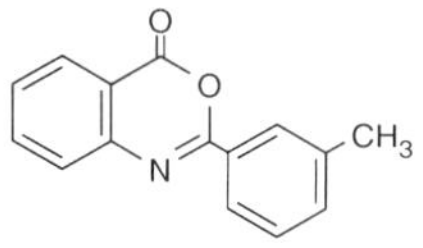

250 mg

S65,663-1
4-BROMOCATECHOL CYCLIC CARBONATE

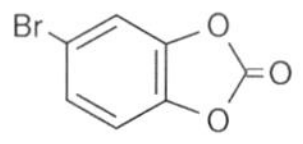

250 mg

S40,976-6
5-PHENYL-3H-1,2-DITHIOL-3-ONE

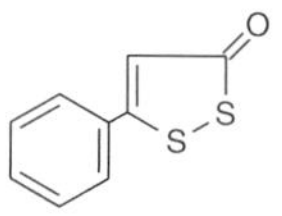

5 gm

S80,848-2
6,7-DIPHENYL-3-OXABICYCLO[3.1.1]-HEPTANE-2,4-DIONE

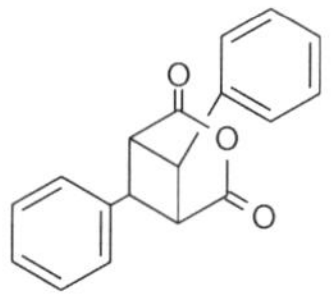

250 mg

S65,807-3
4-(2-HYDROXYETHYLTHIO)-1,8-NAPHTHALIC ANHYDRIDE
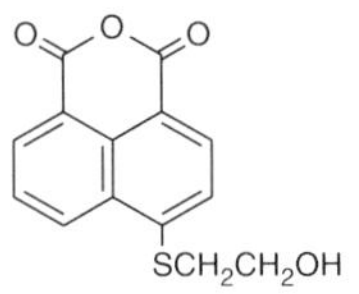
250 mg

S75,655-5
3-METHOXYPHTHALIC ANHYDRIDE
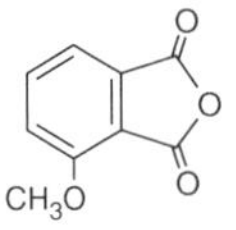
5 gm

S79,819-3
3-METHOXY-5-METHYLHEXAHYDROPHTHALIC ANHYDRIDE
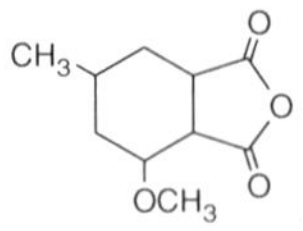
250 mg

S86,226-6
4,4'-OXYDIBENZOIC DIANHYDRIDE
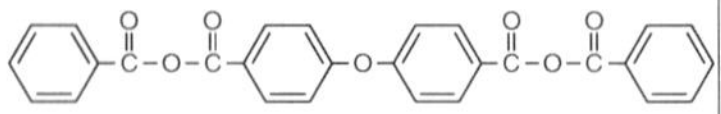
250 mg

S95,709-7
7-METHOXY-1,3,3A,3B,4,5,11,11A-OCTAHYDROPHENANTHRO(1,2-C)-FURAN-1,3-DIONE
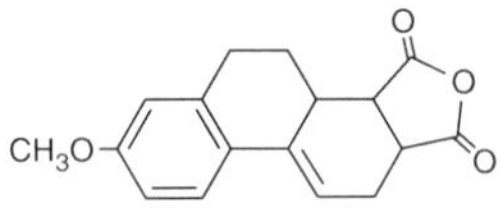
50 mg

S86,703-9
DI-O-BENZOYLTARTARIC ANHYDRIDE
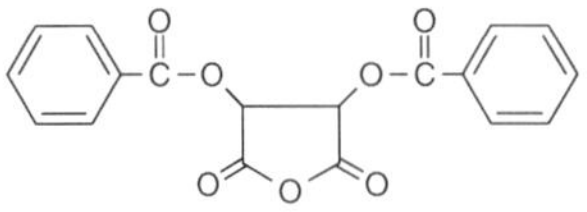
250 mg

S65,825-1
4-(ETHOXYCARBONYLMETHYLTHIO)-1,8-NAPHTHALIC ANHYDRIDE
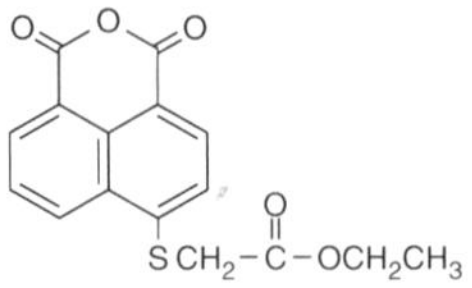
5 gm

S65,817-0
4-PHENYLTHIO-1,8-NAPHTHALIC ANHYDRIDE
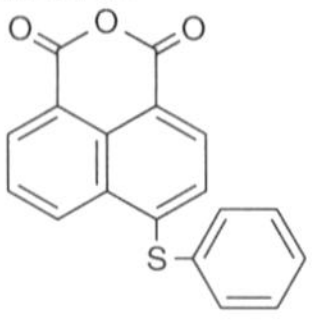
5 gm

S89,588-1
5-NITRO-10-OXO-1,2,3,4-TETRAHYDRO-1,4-ETHANONAPHTHALENE-2,3-DICARBOXYLIC ANHYDRIDE
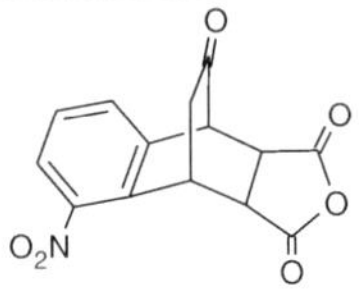
250 mg

S89,587-3
5-NITRO-1,2,3,4-TETRAHYDRO-1,4-ETHENONAPHTHALENE-2,3-DICARBOXYLIC ANHYDRIDE
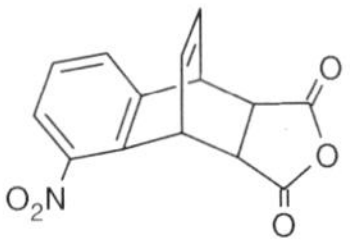
250 mg

S73,335-0
9,10-DIHYDRO-10-FORMYL-9,10-ETHANOANTHRACENE-11,12-DICARBOXYLIC ANHYDRIDE
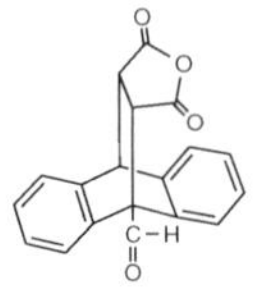
250 mg

S87,578-3
BENZOYL SULFIDE
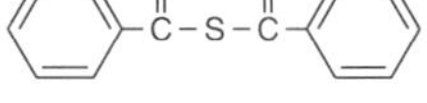
250 mg

S66,292-5
7-FLUORO-1,2-NAPHTHALIC ANHYDRIDE
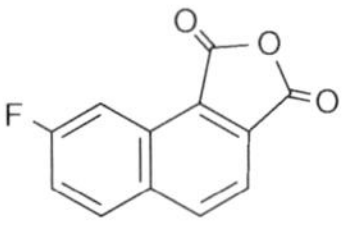
250 mg

S55,059-0
2-(9,10-DIHYDRO-10-OXO-9-ANTHRACENYL)SUCCINIC ANHYDRIDE
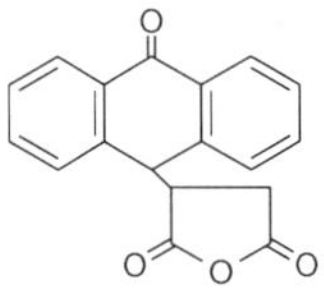
250 mg

S89,604-7
5-(METHOXYCARBONYL)-1,2,3,4-TETRAHYDRO-1,4-ETHENONAPHTHALENE-2,3-DICARBOXYLIC ANHYDRIDE
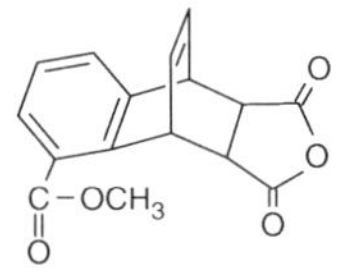
250 mg

S88,596-7
TETRAPHENYLPHTHALIC ANHYDRIDE
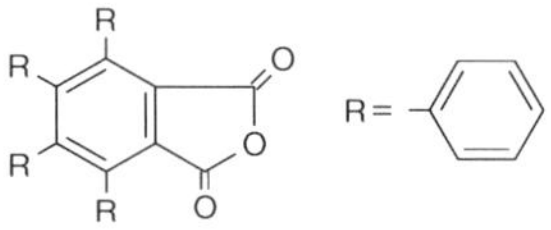
250 mg

S66,307-7
3-CHLOROPHTHALIC ANHYDRIDE
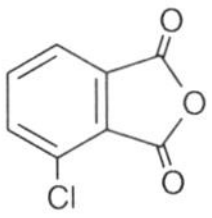
250 mg

S84,952-9
3,6-DIMETHYLPHTHALIC ANHYDRIDE
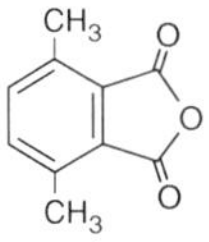
250 mg

S59,621-3
BIS(2,6-DICHLOROBENZOIC) ANHYDRIDE
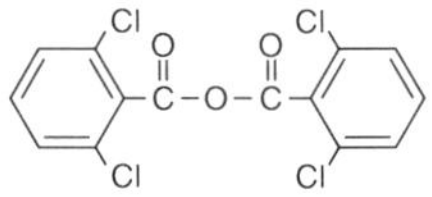
5 gm

S87,668-2
2,3-NAPHTHALIC ANHYDRIDE
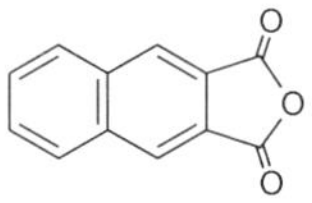
250 mg

S45,372-2
3,4-CORONENEDICARBOXYLIC ANHYDRIDE
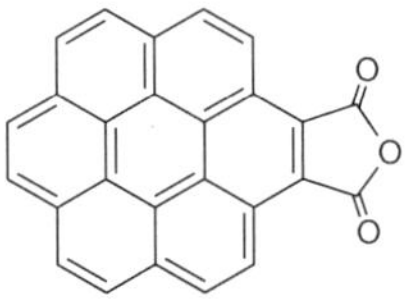
250 mg

S64,125-1
2-(4-ACETOXYPHENYL)-SUCCINIC ANHYDRIDE
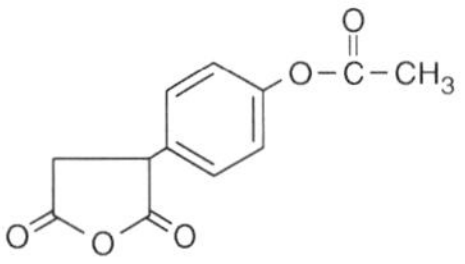
250 mg

S89,613-6
9,10-DIETHYL-1,2,3,4-TETRAHYDRO-1,4-ETHANONAPHTHALENE-2,3-DICARBOXYLIC ANHYDRIDE
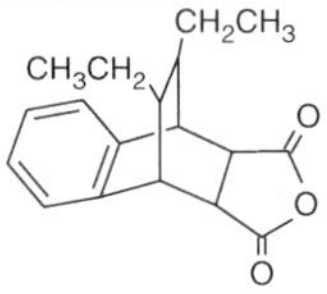
250 mg

S86,424-2
3,6-DIPHENYL-1,2,3,6-TETRAHYDROPHTHALIC ANHYDRIDE
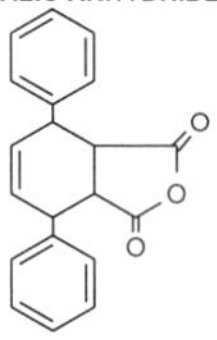
250 mg

S89,516-4
5-BROMO-1,2,3,4-TETRAHYDRO-1,4-ETHENONAPHTHALENE-2,3-DICARBOXYLIC ANHYDRIDE
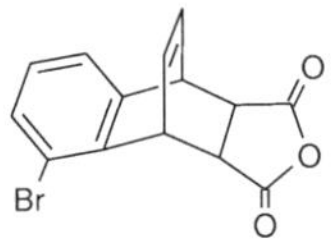
250 mg

S89,522-9
5,8-DIMETHYL-1,2,3,4-TETRAHYDRO-1,4-ETHENONAPHTHALENE-2,3-DICARBOXYLIC ANHYDRIDE
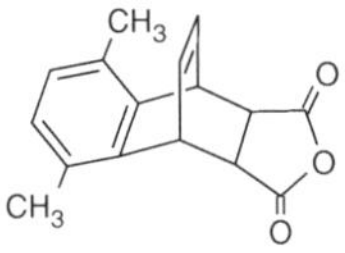
250 mg

S35,596-8
TETRAPHENYL-1,2-DIHYDROPHTHALIC ANHYDRIDE
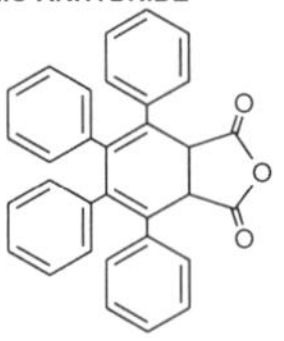
5 gm

S88,277-1
2-BENZYLSUCCINIC ANHYDRIDE
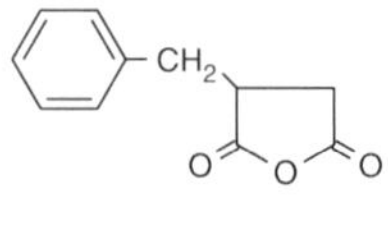
250 mg

S84,690-2
2,3-DIPHENYLSUCCINIC ANHYDRIDE
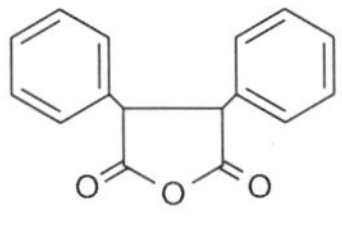
250 mg

S95,646-5
1,3-DIHYDRO-4-PHENYLNAPHTHO(2,3-C)-FURAN-1,3-DIONE
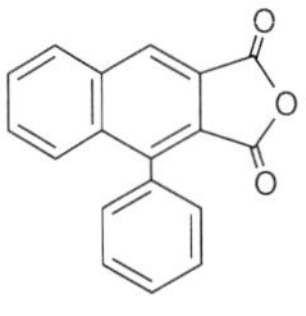
50 mg

S95,816-6
DIBENZO(H,K)-4-OXATRICYCLO(5.2.2.02,6)UNDEC-8,10-DIENE-3,5-DIONE
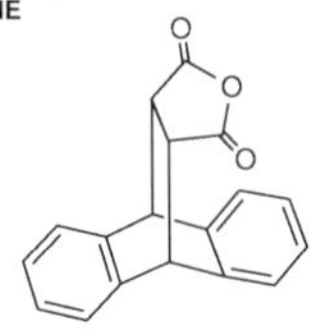
50 mg

S67,848-1
BENZO(GHI)PERYLENE-1,2-DICARBOXYLIC ANHYDRIDE
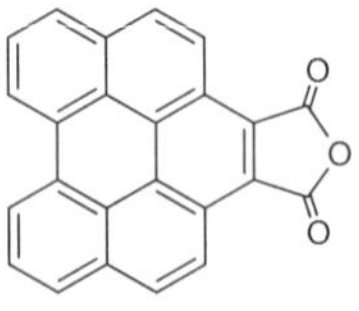
5 gm

S67,828-7
2-(2,4,5-TRICHLOROPHENOXY)ACETYL CHLORIDE

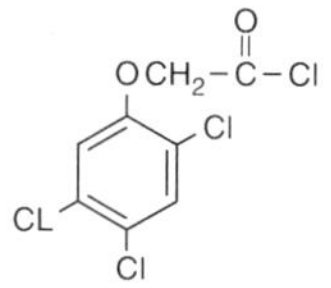

5 gm

S40,503-5
2-PHENOXYPROPIONYL CHLORIDE

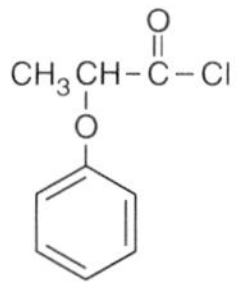

5 gm

S88,415-4
2-(PHENYLTHIO)PROPIONYL CHLORIDE

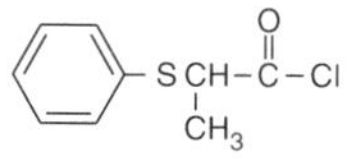

250 mg

S67,595-4
2,3-DIBROMO-3-PHENYLPROPIONYL CHLORIDE

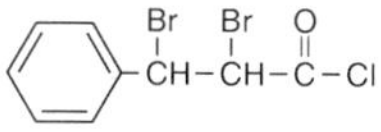

5 gm

S43,722-0
PENTAFLUOROPHENYLACETYL CHLORIDE

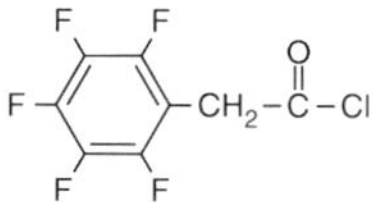

250 mg

S89,632-2
6B,7A-DIHYDRO-7H-CYCLOPROPA(A)-ACENAPHTHYLENE-7-CARBONYL CHLORIDE

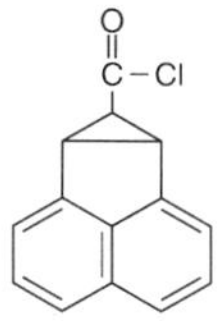

250 mg

S42,860-4
2-PHENYLBUTYRYL CHLORIDE

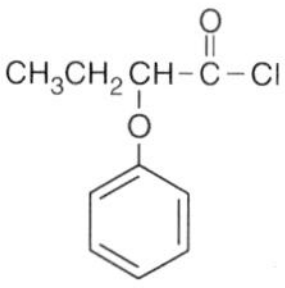

5 gm

S87,742-5
2,2,2-TRIPHENYLACETYL CHLORIDE

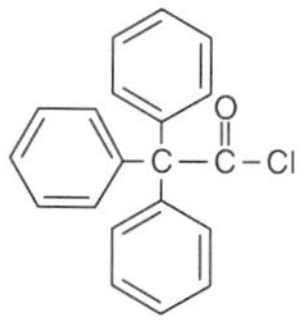

250 mg

S62,043-2
P-TOLYLACETYL CHLORIDE

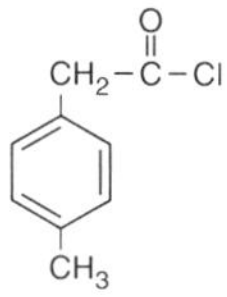

5 gm

S54,188-5
4-(TRICHLOROMETHOXY)-BENZOYL CHLORIDE

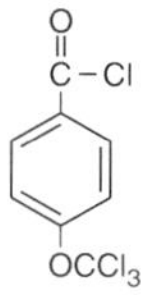

250 mg

S46,137-7
M-TRIFLUOROMETHYLBENZOYL FLUORIDE

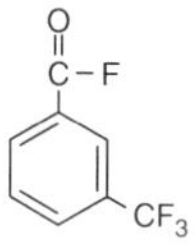

5 gm

S45,663-2
2,3,5,6-TETRACHLOROTEREPHTHALOYL CHLORIDE

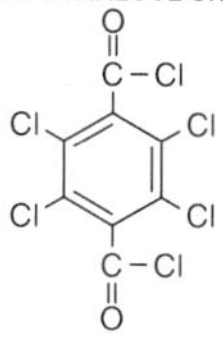

5 gm

S82,975-7
PENTACHLOROBENZOYL CHLORIDE

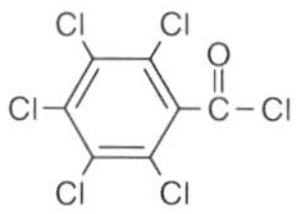

250 mg

S87,294-6
3,4,5-TRIIODOBENZOYL CHLORIDE

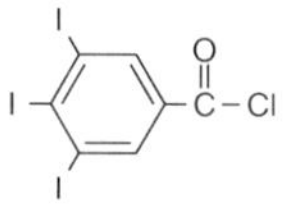

250 mg

S65,659-3
2-(4-CHLOROBENZOYL)BENZOYL CHLORIDE

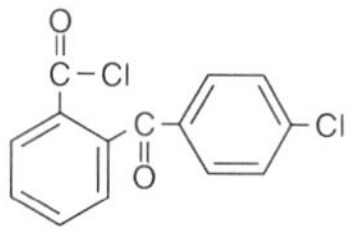

250 mg

S56,448-6
CINNAMYLIDENEMALONYL CHLORIDE

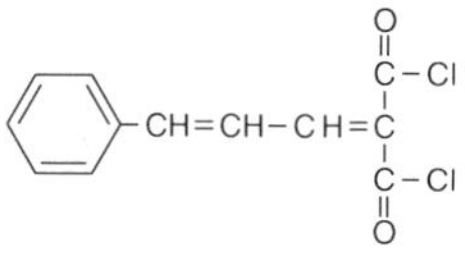

5 gm

S89,465-6
4-PHENYL-1,2,3,4-TETRACHLORO-1,3-BUTADIENE-1-CARBONYL CHLORIDE

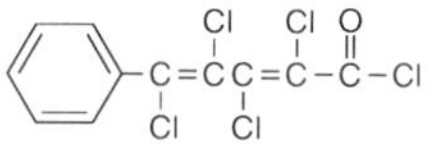

250 mg

S75,893-0
4,4'-STILBENEDICARBONYL CHLORIDE

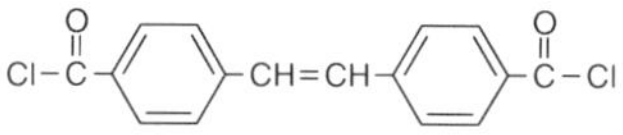

250 mg

S95,625-2
1,4-DIHYDRO-1,4-DIOXO-3-METHYL-2-NAPHTHYLMETHYL CHLOROFORMATE

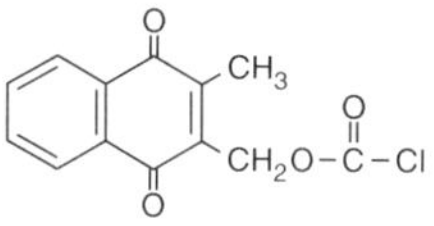

50 mg

S88,924-5
1,4-PHENYLENE BIS(CHLOROFORMATE)

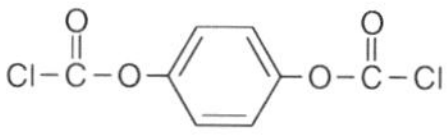

250 mg

S56,123-1
PHENYL CHLOROTHIOLFORMATE

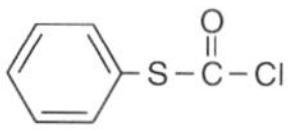

5 gm

S59,277-3
S-(4-CHLOROPHENYL)-CHLOROTHIOLFORMATE

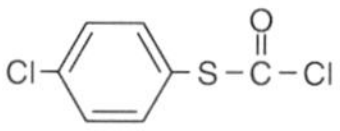

5 gm

S90,507-0
3-BROMO-2'-CHLORO-5'-(TRIFLUOROMETHYL)PROPIONANILIDE

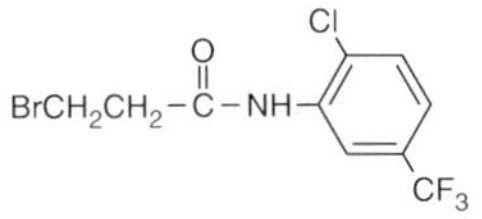

250 mg

S90,503-8
2',4-DICHLORO-5'-(TRIFLUOROMETHYL)-BUTYRANILIDE

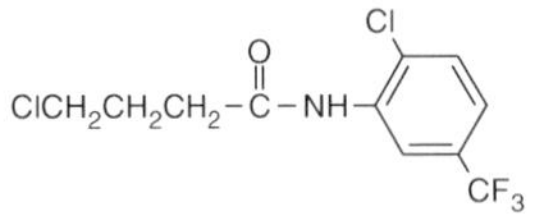

250 mg

S87,844-8
2-CHLORO-3'-(TRIFLUOROMETHYL)-ACETANILIDE

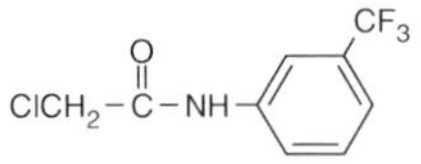

5 gm

S39,747-4
2-CHLORO-2'-(TRIFLUOROMETHYL)-ACETANILIDE

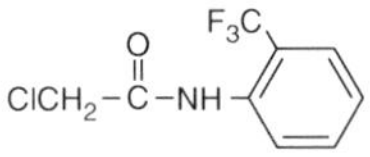

250 mg

S90,508-9
3-BROMO-2'-(TRIFLUOROMETHYL)-PROPIONANILIDE

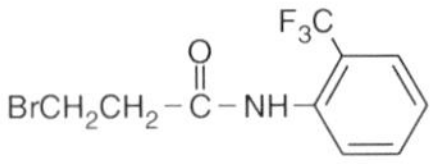

250 mg

S37,817-8
2,2'-DICHLORO-5'-(TRIFLUOROMETHYL)-ACETANILIDE

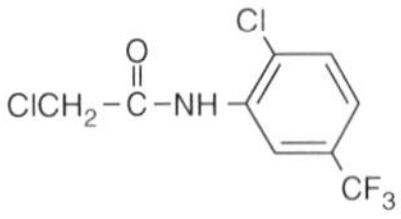

5 gm

S71,810-6
2'-HYDROXY-2,3',5'-TRICHLOROACETANILIDE

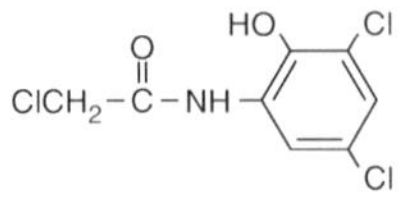

250 mg

S91,666-8
2,5'-DICHLORO-2',4'-DIMETHOXYACETANILIDE

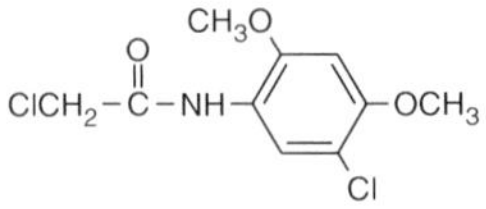

250 mg

S90,495-3
4,5'-DICHLORO-2',4'-DIMETHOXYBUTYRANILIDE

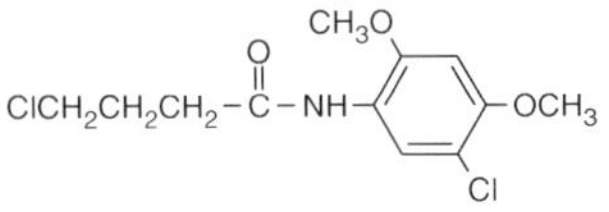

250 mg

S90,505-4
4,4'-DICHLORO-2',5'-DIMETHOXYBUTYRANILIDE

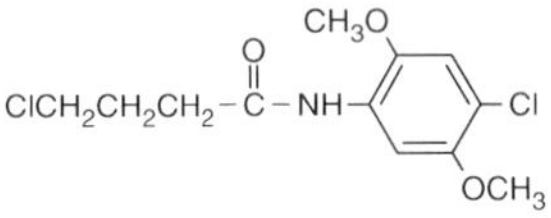

250 mg

S91,669-2
2,3'-DICHLORO-2'-METHOXYACETANILIDE

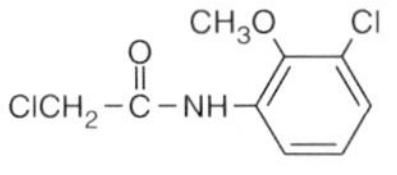

250 mg

S91,380-4
2,5'-DICHLORO-2'-METHOXYACETANILIDE

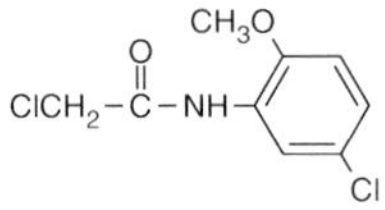

250 mg

S91,386-3
2,4'-DICHLORO-2'-METHOXYACETANILIDE

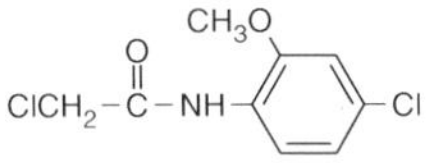

250 mg

S71,804-1
2,4'-DICHLORO-2'-PHENOXYACETANILIDE

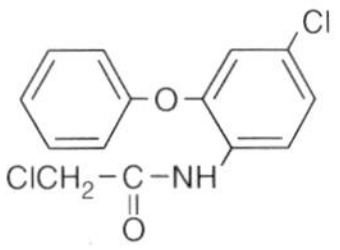

250 mg

S91,664-1
2'-NITRO-2,4',6'-TRICHLOROACETANILIDE

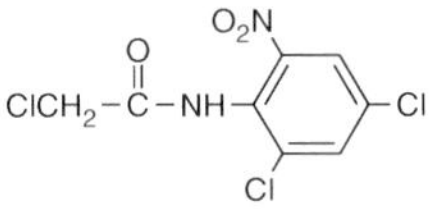

250 mg

S91,668-4
N,N'-(4,5-DICHLORO-1,2-PHENYLENE)-BIS(2-CHLOROACETAMIDE)

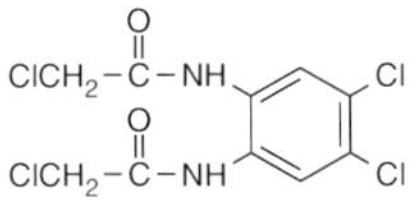

250 mg

S90,454-6
2-CHLORO-4'-FLUORO-3'-NITROACETANILIDE

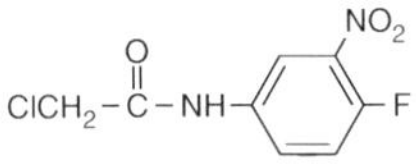

250 mg

S91,661-7
2,4'-DICHLORO-3'-NITROACETANILIDE

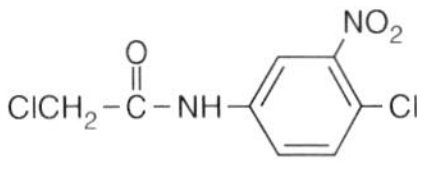

250 mg

S92,160-2
2,4'-CHLORO-3'-NITROACETANILIDE

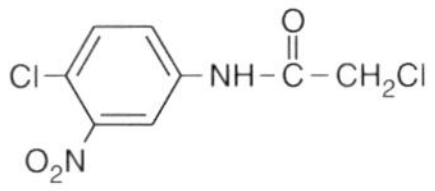

250 mg

S71,808-4
2,2'-DICHLORO-4'-NITROACETANILIDE

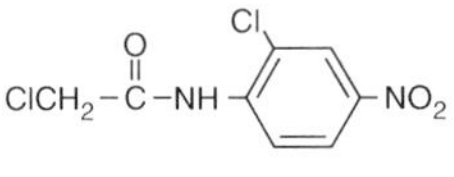

250 mg

S38,016-4
2-CHLORO-N-(2-CHLOROBENZYL)-ACETAMIDE

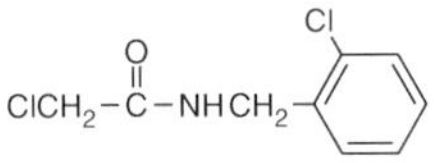

250 mg

S73,484-5
2'-BENZOYL-2,5'-DICHLOROACETANILIDE

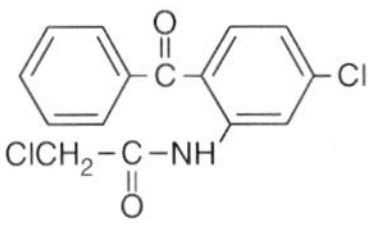

250 mg

S91,667-6
5-CHLORO-N-(2-CHLOROACETYL)-ANTHRANILIC ACID

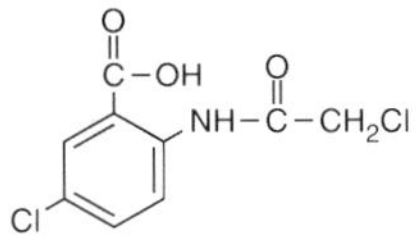

250 mg

S72,037-2
2'-CARBOXY-2-CHLORO-4'-IODOACETANILIDE

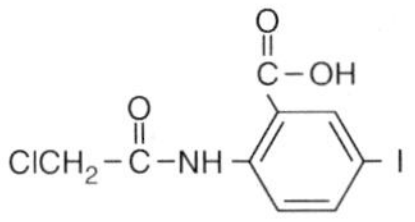

250 mg

S71,596-4
3'-BROMO-2-CHLOROACETANILIDE

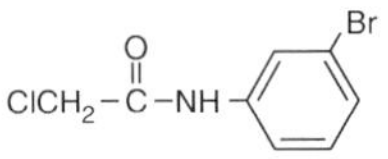

250 mg

S90,511-9
3-BROMO-3'-FLUOROPROPIONANILIDE

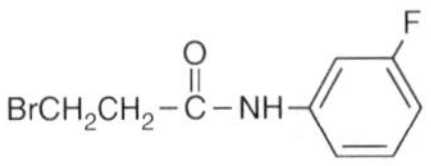

250 mg

S90,519-4
4-BROMO-3'-FLUOROBUTYRANILIDE

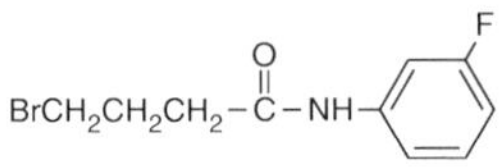

250 mg

S72,895-0
2-CHLORO-4'-FLUOROACETANILIDE

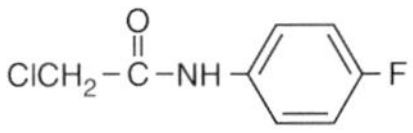

5 gm

S87,821-9
2-CHLORO-4'-IODOACETANILIDE

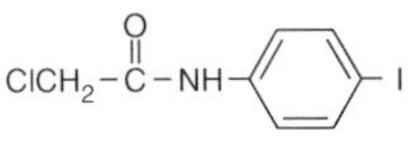

5 gm

S39,927-2
4'-BROMO-2-IODOACETANILIDE

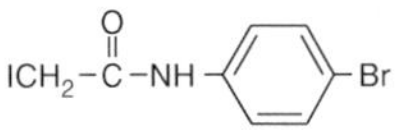

5 gm

S52,954-0
3-CHLORO-4'-FLUOROPROPIONANILIDE

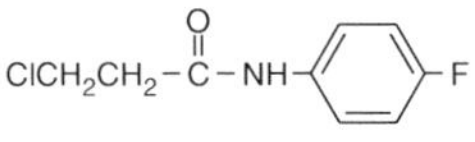

5 gm

S71,803-3
2,2',3'-TRICHLOROACETANILIDE

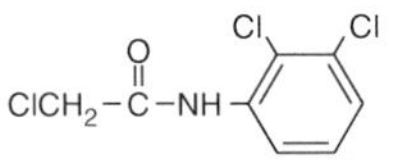

5 gm

S89,332-3
3-CHLORO-2',6'-DIBROMOPROPIONANILIDE

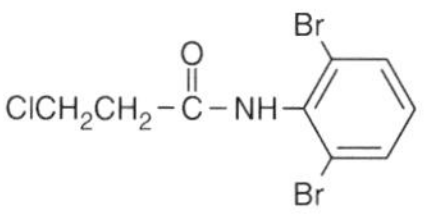

250 mg

S90,506-2
2',3',4-TRICHLOROBUTYRANILIDE

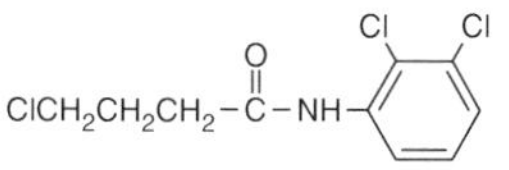

250 mg

S91,659-5
2-CHLORO-2',5'-DIFLUOROACETANILIDE

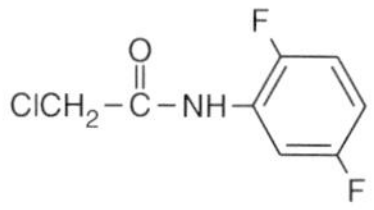

250 mg

S74,658-4
2,3',4'-TRICHLOROACETANILIDE

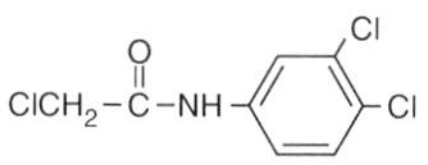

5 gm

S92,146-7
4'-BROMO-2,2'-CHLOROACETANILIDE

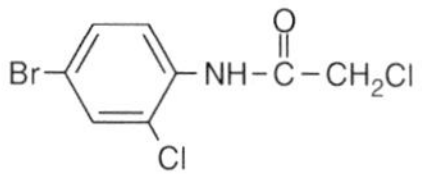

5 gm

S74,659-2
2,3'-DICHLORO-P-ACETOTOLUIDIDE

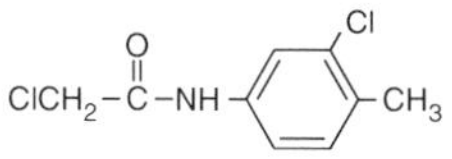

5 gm

S87,845-6
4'-BROMO-2-CHLORO-O-ACETOTOLUIDIDE

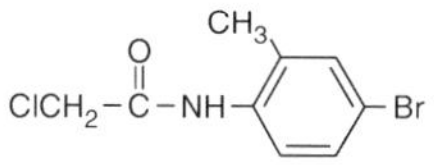

5 gm

S90,455-4
2,3',5'-TRICHLOROACETANILIDE

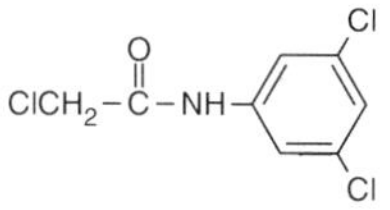

250 mg

S67,393-5
N-(2,6-DIBROMO-4-FLUOROPHENYL)-2-BROMOACETAMIDE

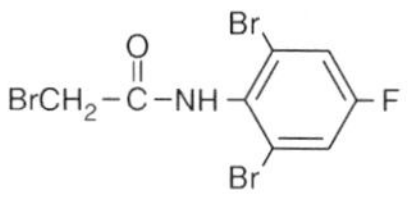

250 mg

S68,942-4
2-BROMO-N-(4-CHLORO-2,6-DIBROMOPHENYL)ACETAMIDE

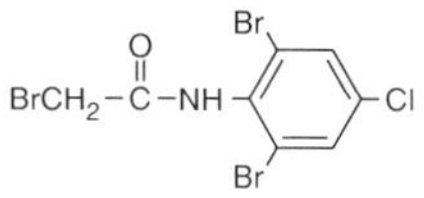

250 mg

S69,170-4
2-BROMO-N-(2-CHLORO-4,6-DIBROMOPHENYL)ACETAMIDE

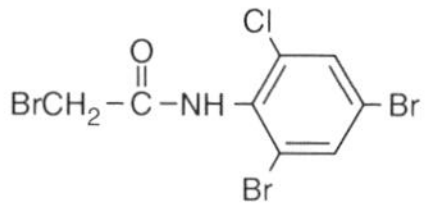

250 mg

S71,806-8
4'-BROMO-2-CHLORO-2',6'-DIMETHYLACETANILIDE

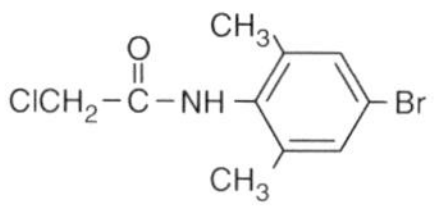

250 mg

S68,784-7
2-BROMO-N-(2,6-DIBROMO-4-ETHYLPHENYL)ACETAMIDE

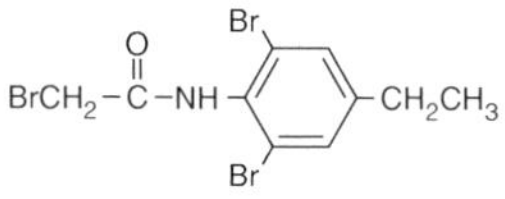

250 mg

S90,504-6
N-METHYL-2',3',4-TRICHLOROBUTYRANILIDE

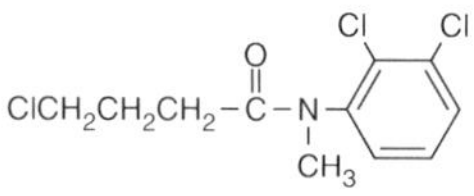

250 mg

S45,353-6
N-CHLOROACETYL-L-TYROSINE ETHYL ESTER

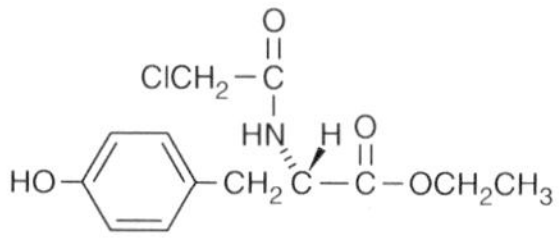

250 mg

S50,106-9
2-CHLORO-2'-HYDROXYACETANILIDE

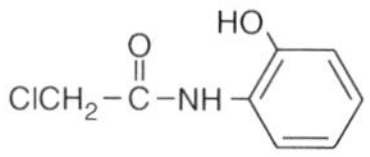

5 gm

S85,482-4
2-CHLORO-2',5'-DIMETHOXYACETANILIDE

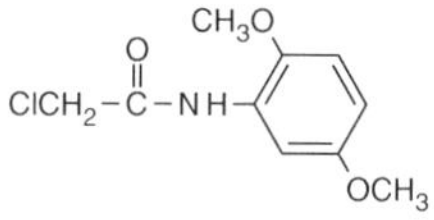

250 mg

S91,383-9
2-CHLORO-4'-METHOXY-2'-NITROACETANILIDE

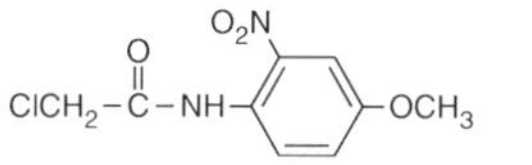

250 mg

S68,934-3
2-BROMO-N-(4-METHOXY-2-NITROPHENYL)-ACETAMIDE

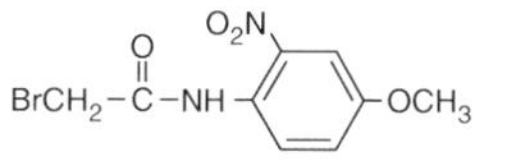

250 mg

S91,660-9
2-CHLORO-4'-METHOXY-3'-NITROACETANILIDE

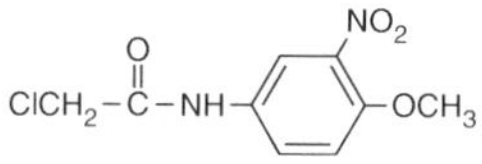

250 mg

S71,807-6
2-CHLORO-2'-METHOXY-4'-NITROACETANILIDE

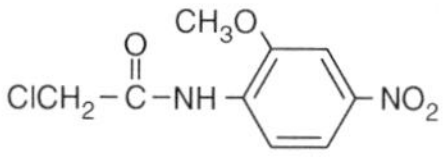

250 mg

S68,961-0
N-(4-ACETAMIDO-3-METHOXYPHENYL)-2-BROMOACETAMIDE

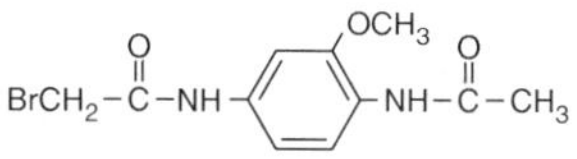

250 mg

S90,509-7
4'-ACETAMIDO-3-BROMO-3'-METHOXYPROPIONANILIDE

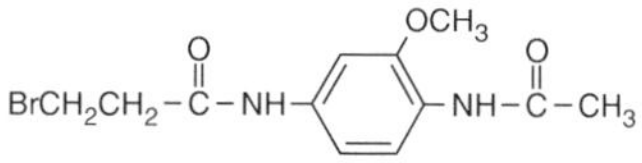

250 mg

S90,518-6
4'-ACETAMIDO-4-BROMO-2'-METHOXYBUTYRANILIDE

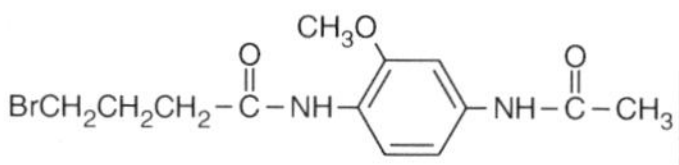

250 mg

S85,494-8
2-CHLORO-2'-METHOXYACETANILIDE

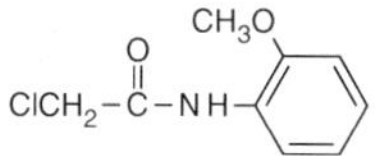

250 mg

S85,481-6
2-CHLORO-2'-METHOXY-5'-METHYLACETANILIDE

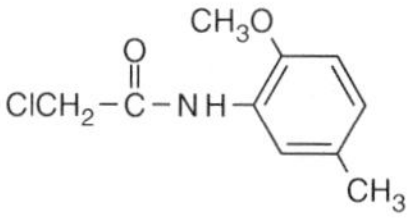

250 mg

S90,510-0
3-BROMO-2'-METHOXY-5'-METHYLPROPIONANILIDE

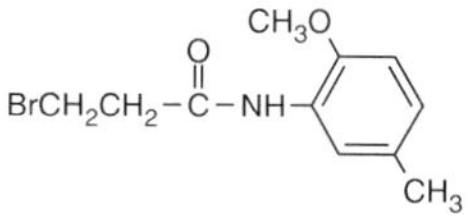

250 mg

S90,520-8
4-BROMO-2'-METHOXY-5'-METHYLBUTYRANILIDE

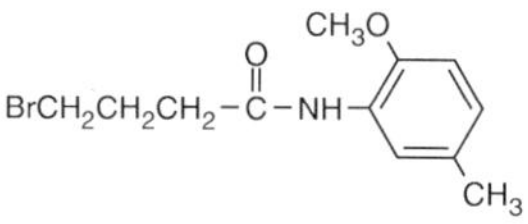

250 mg

S52,956-7
3-CHLORO-PARA-PROPIONOPHENETIDIDE

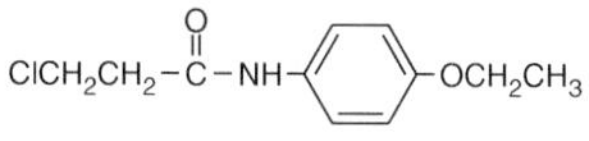

250 mg

S52,032-2
2-CHLORO-META-ACETANISIDIDE

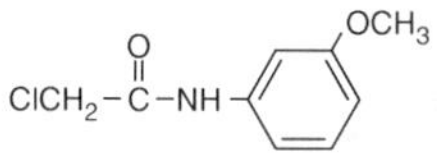

250 mg

S92,155-6
2-CHLORO-2'-ETHOXYACETANILIDE

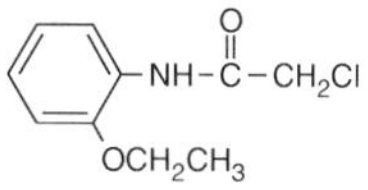

250 mg

S93,827-0
4'-(FLUOROMETHOXY)ACETANILIDE

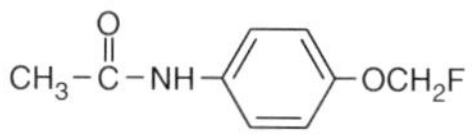

250 mg

S38,012-1
2-CHLORO-PARA-ACETANISIDIDE
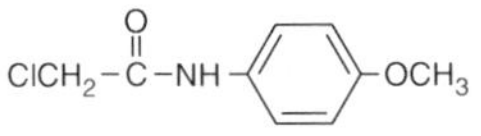
5 gm

S35,950-5
2-CHLORO-4'-ETHOXYACETANILIDE
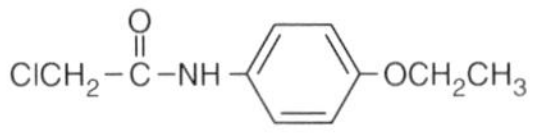
5 gm

S89,319-6
4-CHLORO-4'-ETHOXYBUTYRANILIDE
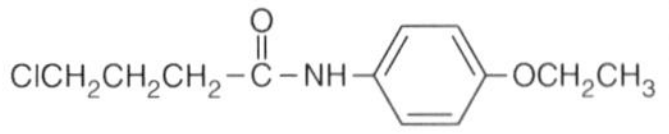
250 mg

S91,672-2
2-CHLORO-3'-(METHYLTHIO)ACETANILIDE
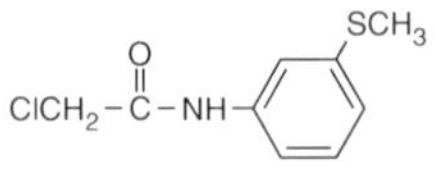
250 mg

S91,663-3
2-CHLORO-4',5'-DIMETHYL-2'-NITROACETANILIDE
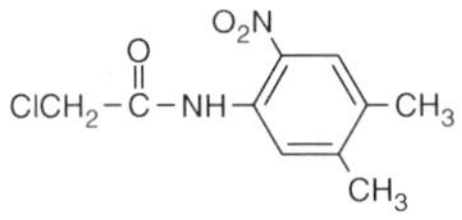
250 mg

S41,732-7
2-CHLORO-2'-NITROACETANILIDE
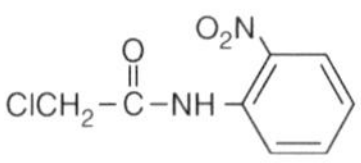
5 gm

S63,608-8
2,4'-DICHLORO-2'-NITROACETANILIDE
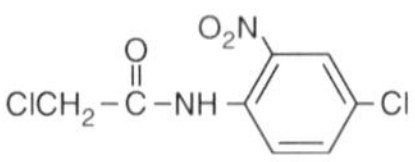
5 gm

S73,625-2
2'-CARBOXY-2-CHLORO-5'-NITROACETANILIDE
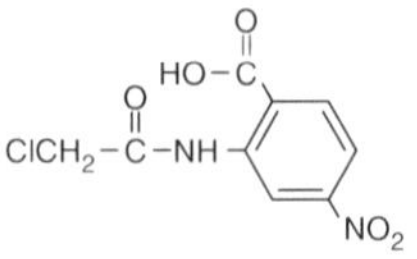
5 gm

S41,733-5
2-CHLORO-3'-NITROACETANILIDE
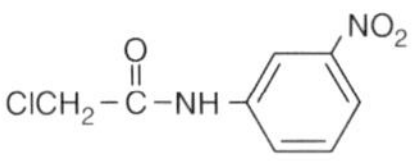
5 gm

S38,200-0
N,N'-(2,4,6-TRIMETHYL-1,3-PHENYLENE)BIS(2-CHLOROACETAMIDE)
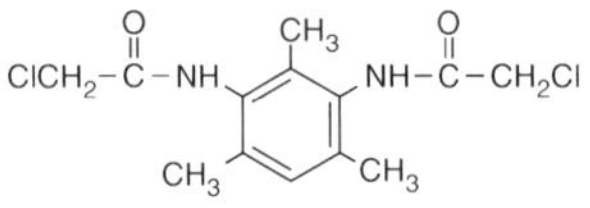
250 mg

S71,812-2
2-CHLORO-4'-NITRO-O-ACETOTOLUIDIDE
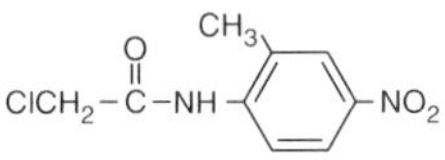
250 mg

S38,403-8
2-CHLORO-N-(1,2-DIPHENYLETHYL)-ACETAMIDE
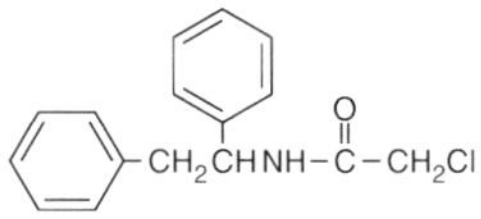
250 mg

S38,386-4
3-CHLORO-N-(1,2-DIPHENYLETHYL)-PROPIONAMIDE
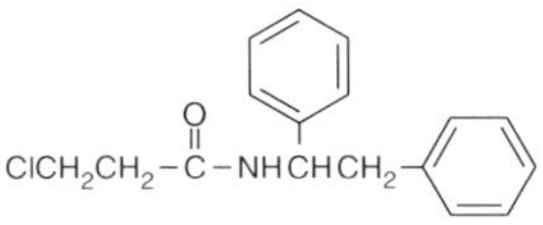
5 gm

S34,392-7
CHLOROACETYLGLYCYL-DL-PHENYLALANINE
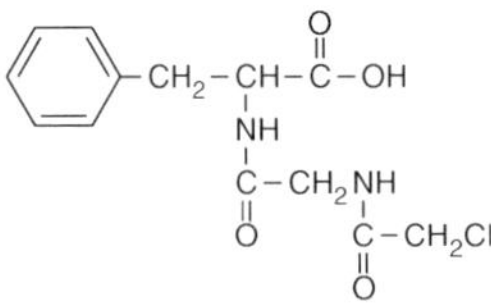
5 gm

S78,923-2
N-(CHLOROACETYL)-DL-PHENYLALANINE
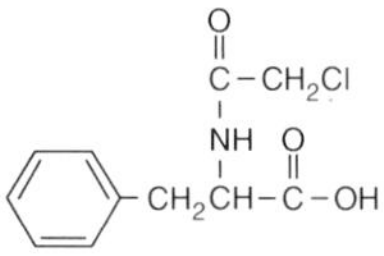
250 mg

S78,697-7
N-(2-CHLOROACETYL)-L-PHENYLALANINE
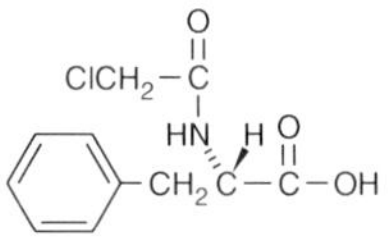
250 mg

S52,951-6
3-CHLORO-N-PHENETHYLPROPIONAMIDE
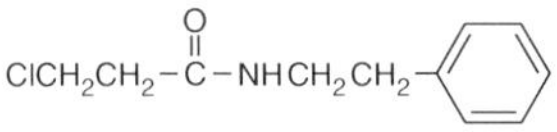
5 gm

S52,955-9
(-)-3-CHLORO-N-(ALPHA-METHYLPHENETHYL)PROPIONAMIDE
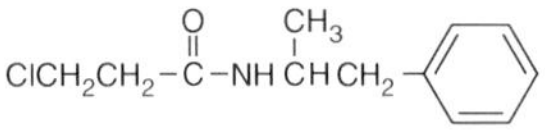
5 gm

S60,099-7
N-(CHLOROACETYL)-N,ALPHA,ALPHA-DIMETHYLPHENETHYLAMINE
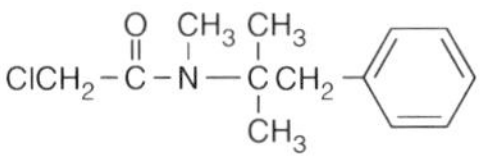
250 mg

S91,662-5
4'-ANILINO-2-CHLOROACETANILIDE
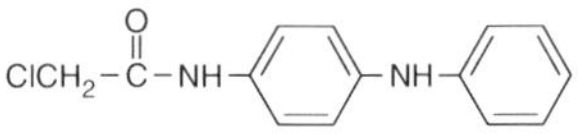
250 mg

S95,332-6
2-CHLORO-N-METHYLACETANILIDE
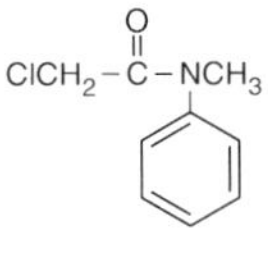
50 mg

S35,828-2
N-BENZYL-3-CHLOROPROPIONAMIDE
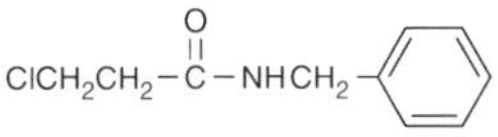
5 gm

S52,949-4
N-BENZYL-4-CHLOROBUTYRAMIDE
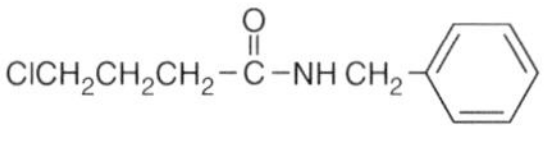
5 gm

S38,011-3
2-CHLORO-N-(4-CHLOROBENZYL)-ACETAMIDE
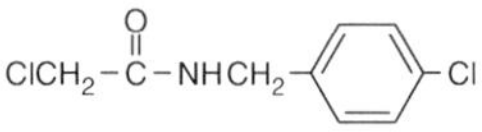
5 gm

S74,207-4
2-CHLORO-N,N-DIBENZYLACETAMIDE
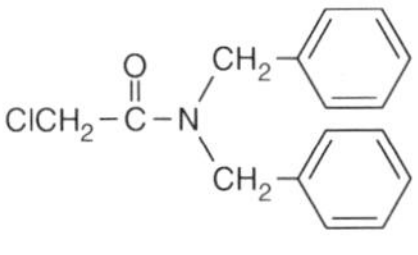
5 gm

S67,861-9
2-CHLORO-5'-NITRO-O-ACETOTOLUIDIDE
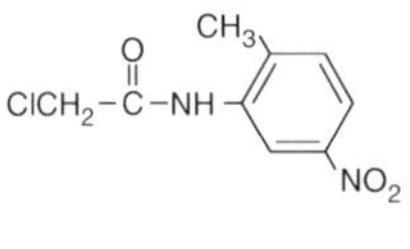
5 gm

S57,988-2
2'-BENZOYL-2,4'-DICHLOROACETANILIDE
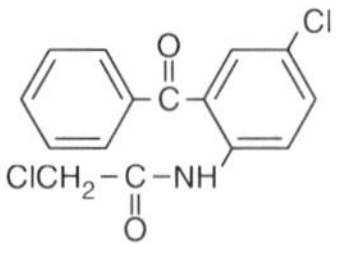
5 gm

S57,407-4
2'-BENZOYL-4'-CHLORO-2-IODOACETANILIDE
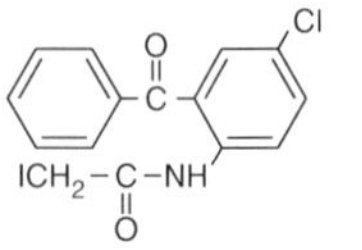
250 mg

S92,149-1
4'-ACETYL-2-CHLOROACETANILIDE
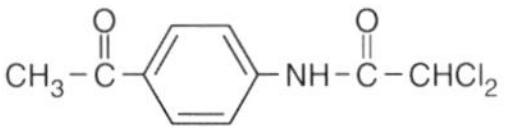
250 mg

S92,137-8
4-(FORMAMIDO)ACETOPHENONE
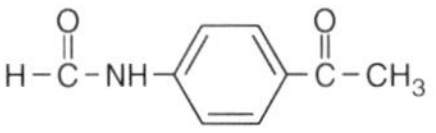
5 gm

S91,671-4
DIMETHYL 5-(2-CHLOROACETAMIDO)-ISOPHTHALATE
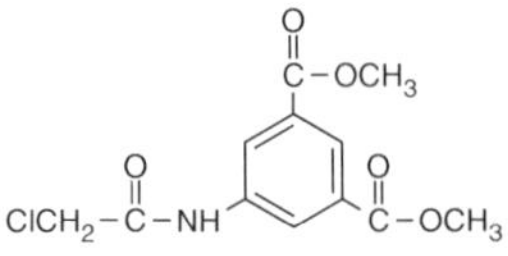
250 mg

S70,032-0
2-(2-CHLOROACETAMIDO)BENZOIC ACID
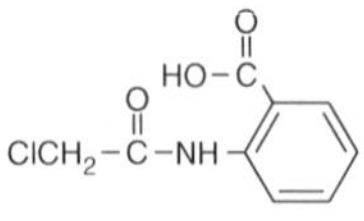
250 mg

S37,660-4
METHYL N-(CHLOROACETYL)-ANTHRANILATE

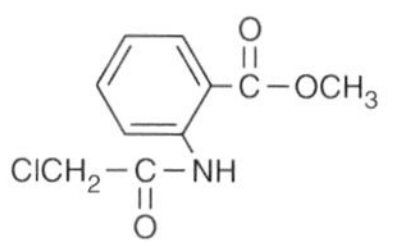

5 gm

S38,124-1
ETHYL N-(2-CHLOROACETYL)-ANTHRANILATE

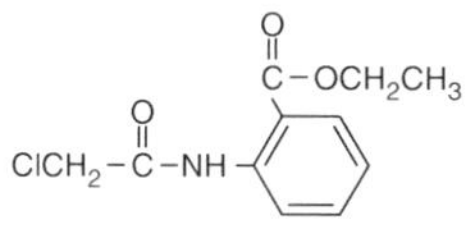

5 gm

S39,093-3
2,2',4',6'-TETRACHLOROACETANILIDE

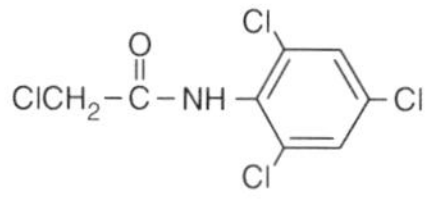

250 mg

S52,960-5
3-CHLORO-2',6'-PROPIONOXYLIDIDE

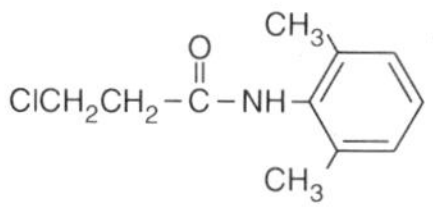

5 gm

S37,639-6
2,3'-DICHLOROACETANILIDE

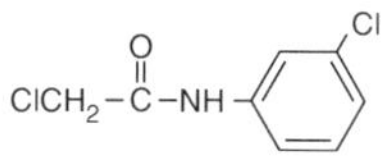

5 gm

S39,027-5
2-CHLORO-META-ACETOTOLUIDIDE

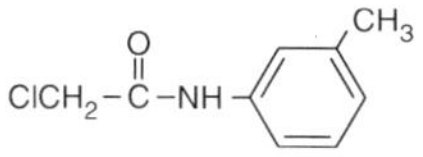

5 gm

S52,569-3
4-CHLORO-META-BUTYROTOLUIDIDE

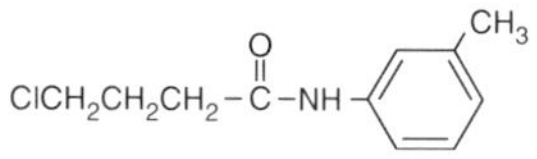

250 mg

S52,933-8
3-CHLOROPROPIONANILIDE

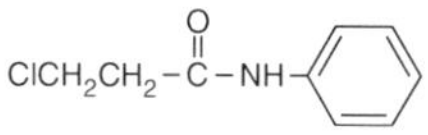

250 mg

S74,657-6
2,2'-DICHLOROACETANILIDE

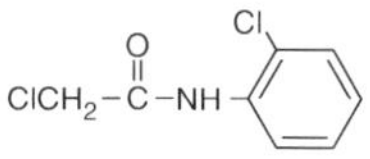

5 gm

S37,820-8
2-CHLORO-ORTHO-ACETOTOLUIDIDE

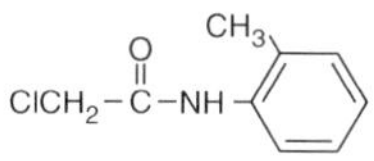

5 gm

S40,748-8
2,4'-DICHLOROACETANILIDE

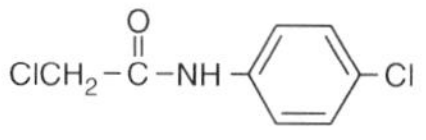

5 gm

S41,731-9
4'-BROMO-2-CHLOROACETANILIDE

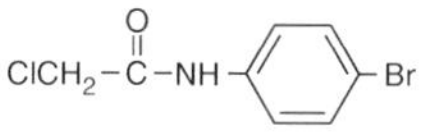

5 gm

S39,030-5
2-CHLORO-PARA-ACETOTOLUIDIDE

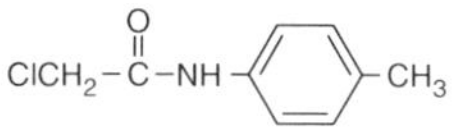

5 gm

S67,860-0
2-CHLORO-4'-ETHYLACETANILIDE

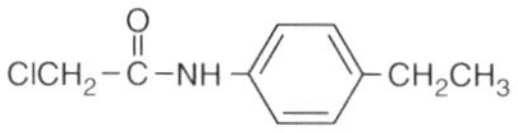

5 gm

S52,878-1
3-CHLORO-N-(4-CHLORO-ALPHA,ALPHA-DIMETHYLPHENETHYL)PROPIONAMIDE

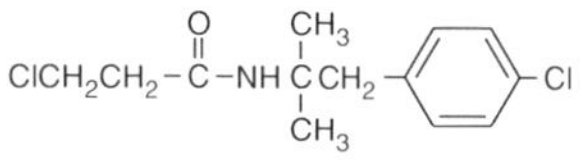

5 gm

S92,168-8
4'-BUTYL-2-CHLOROACETANILIDE

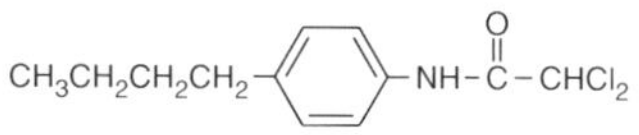

250 mg

S69,016-3
4',4''-METHYLENEBIS(2-CHLOROACETANILIDE)

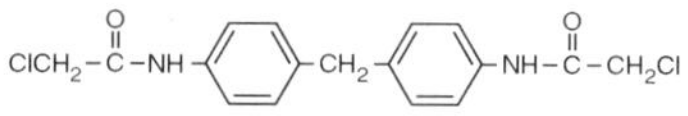

5 gm

S37,803-8
2,3'-DICHLORO-ORTHO-ACETOTOLUIDIDE

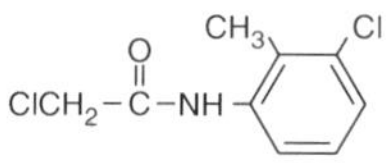

5 gm

S39,172-7
2-CHLORO-2',3'-ACETOXYLIDIDE

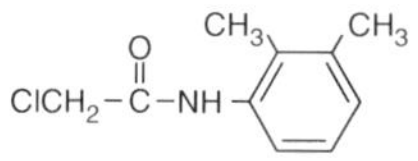

5 gm

S38,530-1
2,2',4'-TRICHLOROACETANILIDE

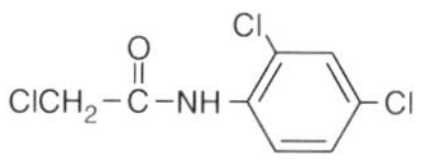

250 mg

S92,139-4
2,2',5'-TRICHLOROACETANILIDE

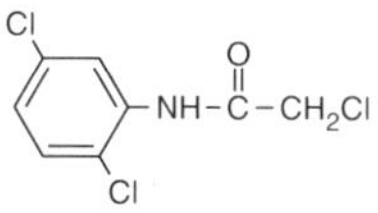

5 gm

S37,800-3
2,6'-DICHLORO-META-ACETOTOLUIDIDE

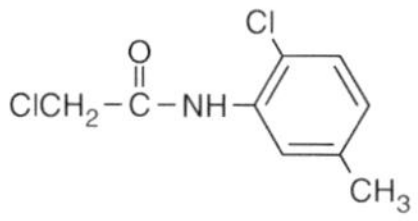

250 mg

S37,807-0
2,5'-DICHLORO-ORTHO-ACETOTOLUIDIDE

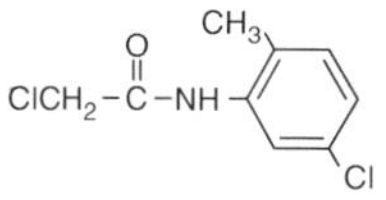

5 gm

S37,809-7
2,4'-DICHLORO-ORTHO-ACETOTOLUIDIDE

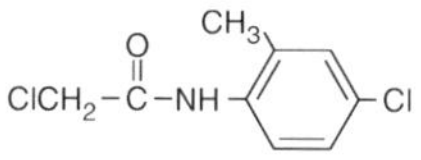

5 gm

S39,173-5
2-CHLORO-3',4'-ACETOXYLIDIDE

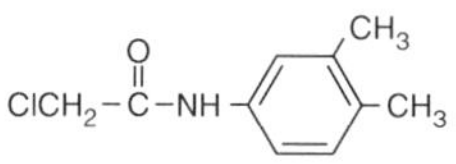

5 gm

S74,656-8
2-CHLORO-2',5'-ACETOXYLIDIDE

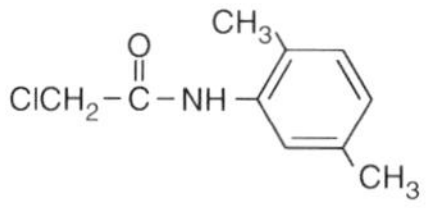

5 gm

S74,616-9
N,N'-(3,3'-DIMETHYL-1,1'-BIPHENYL-4,4'-DIYL)BIS(3-BROMOPROPIONAMIDE)

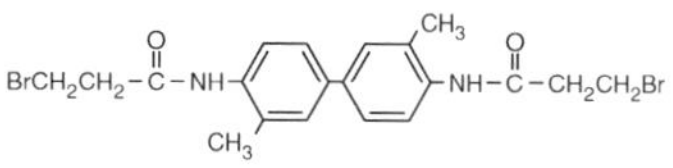

5 gm

S92,164-5
2-CHLORO-2',4',6'-TRIMETHYLACETANILIDE

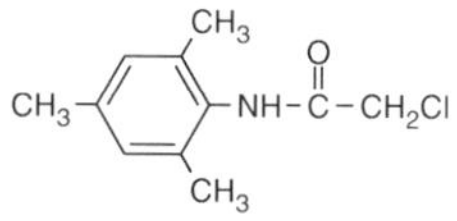

250 mg

S37,633-7
2,2',4',5'-TETRACHLOROACETANILIDE

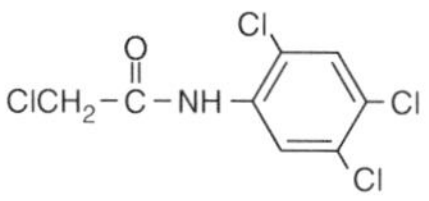

5 gm

S52,948-6
3-CHLORO-N-(1-PIPERONYLETHYL)-PROPIONAMIDE

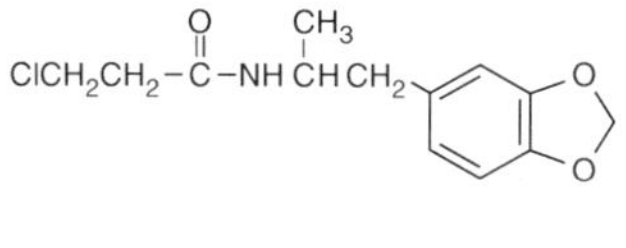

250 mg

S38,117-9
2-CHLORO-N-ETHYL-ORTHO-ACETOTOLUIDIDE

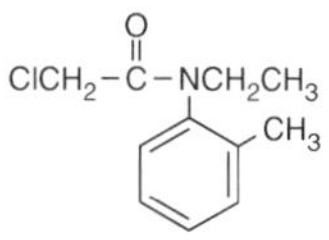

5 gm

S90,095-8
2,2-DICHLORO-3'-(TRIFLUOROMETHYL)-ACETANILIDE

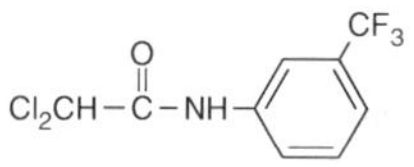

250 mg

S91,681-1
N-(2,2-DICHLOROACETYL)-5-IODOANTHRANILIC ACID

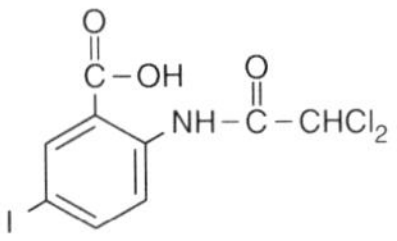

250 mg

S79,554-2
2,2,3'-TRICHLOROACETANILIDE

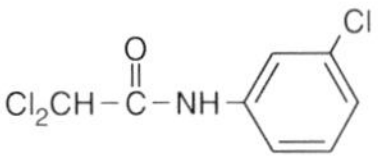

250 mg

S91,683-8
2,2-DICHLORO-2',3',5',6'-TETRAFLUOROACETANILIDE

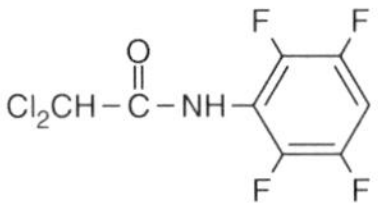

250 mg

S89,233-5
2'-METHYL-2,2,6'-TRICHLOROACETANILIDE

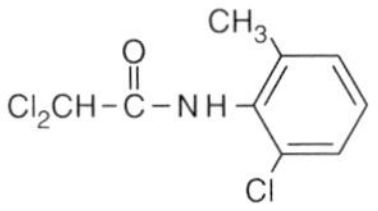

250 mg

S91,676-5
2,2-DICHLORO-2',5'-DIFLUOROACETANILIDE

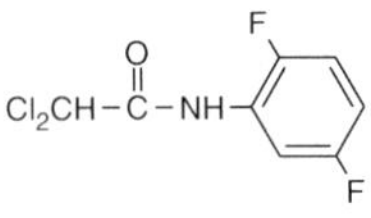

250 mg

S89,232-7
2,2,3',4'-TETRACHLOROACETANILIDE

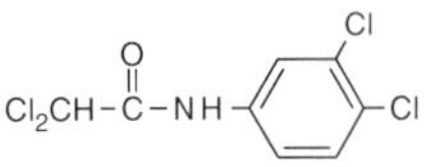

250 mg

S79,193-8
4'-METHYL-2,2,3'-TRICHLOROACETANILIDE

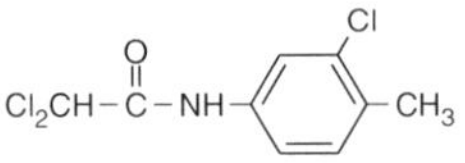

250 mg

S90,093-1
2'-METHYL-2,2,5'-TRICHLOROACETANILIDE

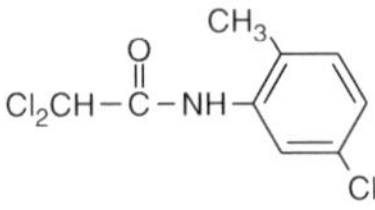

250 mg

S75,517-6
2,2,3',5'-TETRACHLOROACETANILIDE

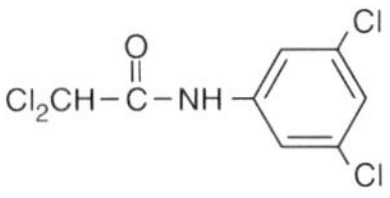

250 mg

S73,784-4
2',6'-DIBROMO-2,2-DICHLORO-4'-FLUOROACETANILIDE

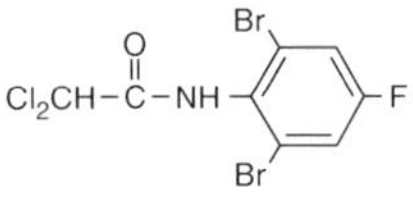

5 gm

S68,979-3
N-(2-BROMO-4,5-DIMETHYLPHENYL)-2,2-DICHLOROACETAMIDE

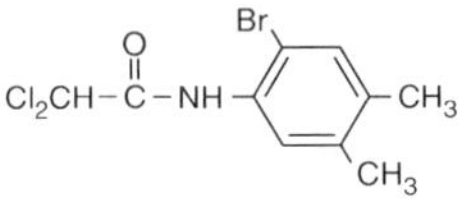

250 mg

S70,073-8
2,2-DICHLORO-4'-METHOXY-2'-NITROACETANILIDE

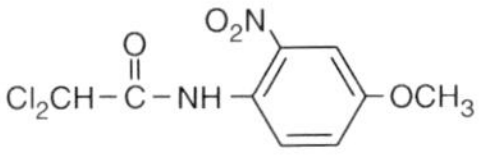

250 mg

S91,678-1
2,2-DICHLORO-4'-METHOXY-3'-NITROACETANILIDE

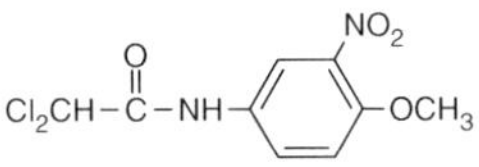

250 mg

S72,352-5
N-(4-ACETAMIDO-3-METHOXYPHENYL)-2,2-DICHLOROACETAMIDE

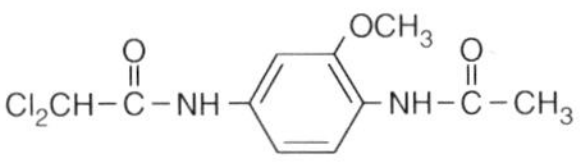

5 gm

S42,041-7
N-[2-(4-CHLOROPHENOXY)-1-METHYLETHYL]-2,2-DICHLOROACETAMIDE

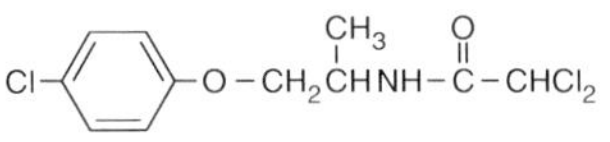

250 mg

S42,042-5
N-[2-(4-BROMOPHENOXY)-1-METHYLETHYL]-2,2-DICHLOROACETAMIDE

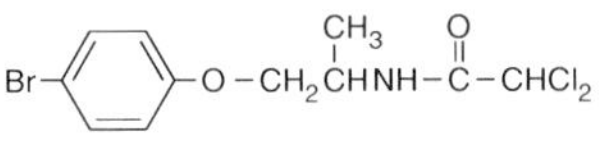

5 gm

S69,572-6
2,2-DICHLORO-2'-METHOXY-5'-METHYLACETANILIDE

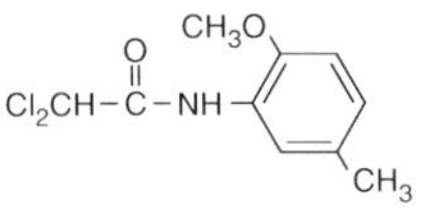

250 mg

S79,194-6
2,2-DICHLORO-2'-METHOXYACETANILIDE

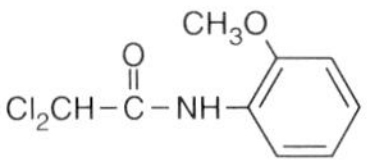

250 mg

S38,930-7
4'-(DIFLUOROMETHYLTHIO)ACETANILIDE

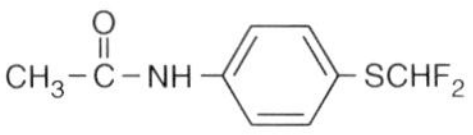

250 mg

S68,828-2
2,2-DICHLORO-4'-NITRO-O-ACETOTOLUIDIDE

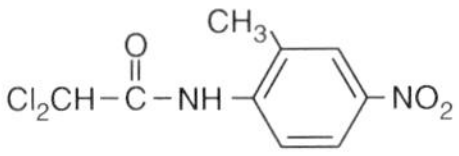

250 mg

S84,404-7
2,2-DICHLORO-N-PHENETHYLACETAMIDE

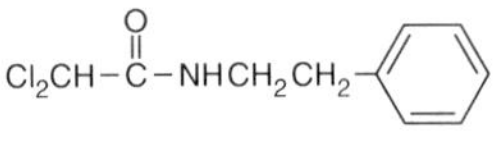

250 mg

S79,197-0
2,2-DICHLORO-2'-ETHYLACETANILIDE

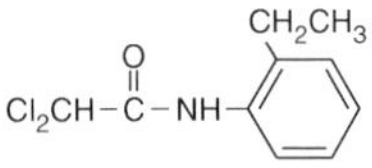

250 mg

S79,556-9
2,2-DICHLORO-4'-ETHYLACETANILIDE

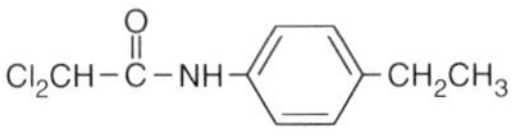

250 mg

S79,557-7
2,2-DICHLORO-2',3'-DIMETHYLACETANILIDE

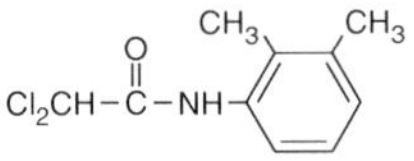

250 mg

S77,212-7
2,2-DICHLORO-5'-ISOPROPYL-2'-METHYLACETANILIDE

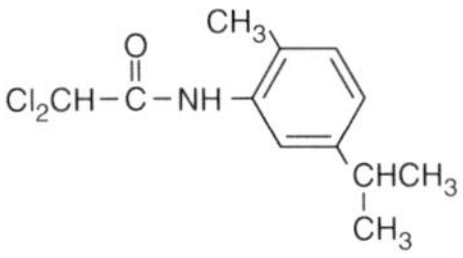

250 mg

S77,986-5
2',5'-DIMETHOXY-2,2,2,4'-TETRACHLOROACETANILIDE

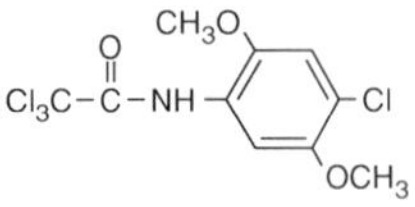

250 mg

S89,037-5
2-(4-CHLOROPHENOXY)-2'-(TRIFLUOROMETHYL)ACETANILIDE

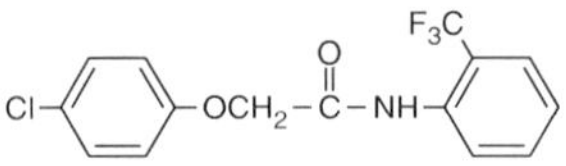

250 mg

S89,158-4
2'-CHLORO-2-(2-CHLOROPHENOXY)-5'-(TRIFLUOROMETHYL)ACETANILIDE

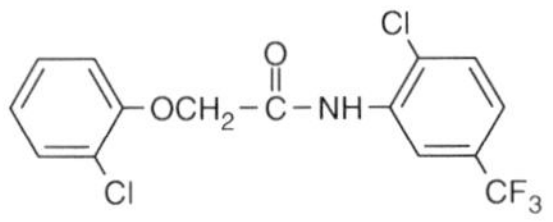

250 mg

S89,157-6
2-(2,4,5-TRICHLOROPHENOXY)-2'-(TRIFLUOROMETHYL)ACETANILIDE

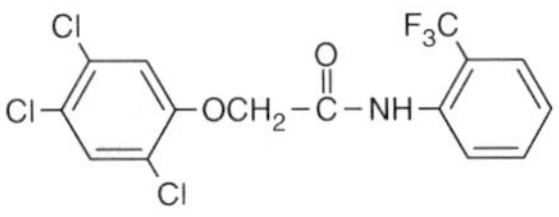

250 mg

S72,509-9
2-(2,4,5-TRICHLOROPHENOXY)-3'-(TRIFLUOROMETHYL)ACETANILIDE

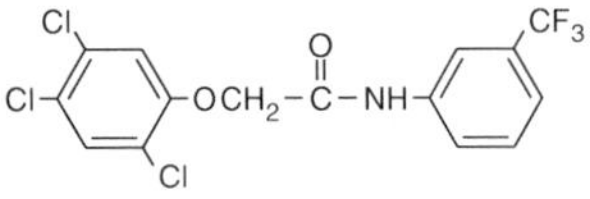

5 gm

S75,342-4
2-(2-CHLOROPHENOXY)-3'-(TRIFLUOROMETHYL)ACETANILIDE

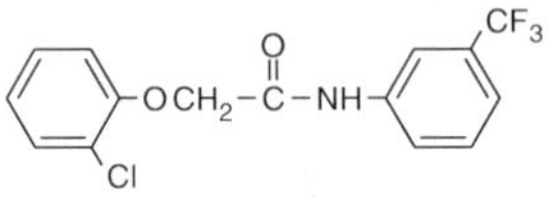

250 mg

S72,507-2
2'-CHLORO-2-(2,4,5-TRICHLOROPHENOXY)-5'-(TRIFLUOROMETHYL)ACETANILIDE

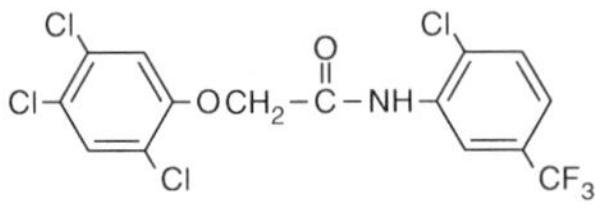

5 gm

S75,352-1
2-(4-CHLORO-2-METHYLPHENOXY)-3'-(TRIFLUOROMETHYL)ACETANILIDE

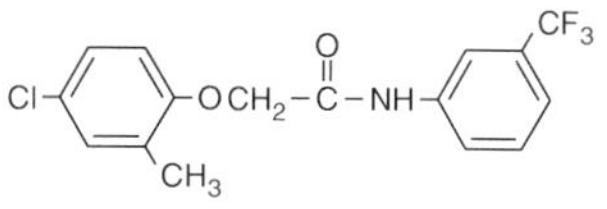

5 gm

S72,506-4
2'-CHLORO-2-(4-CHLORO-2-METHYLPHENOXY)-5'-(TRIFLUOROMETHYL)-ACETANILIDE

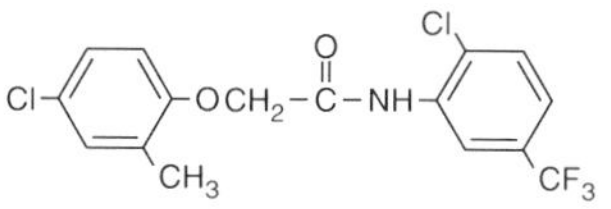

5 gm

S79,517-8
2-(2-CHLOROPHENOXY)-2'-(TRIFLUOROMETHYL)ACETANILIDE

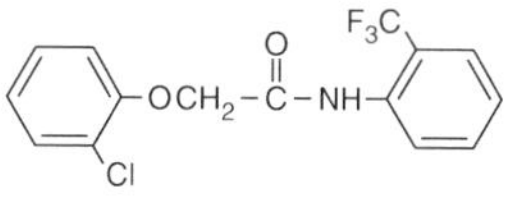

250 mg

S72,608-7
2-(4-CHLORO-2-METHYLPHENOXY)-2'-(TRIFLUOROMETHYL)ACETANILIDE

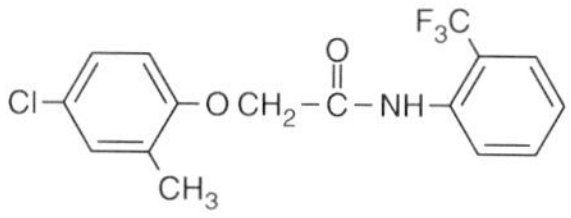

250 mg

S78,085-5
2'-CHLORO-2-(4-CHLOROPHENYL)-5'-(TRIFLUOROMETHYL)ACETANILIDE

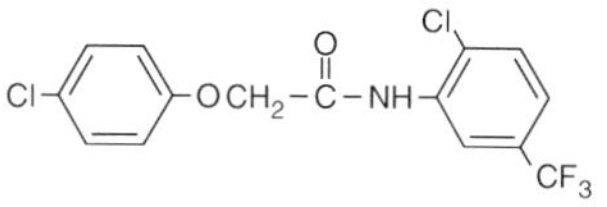

250 mg

S79,119-9
2'-CHLORO-2-(2,4-DICHLOROPHENOXY)-5'-(TRIFLUOROMETHYL)ACETANILIDE

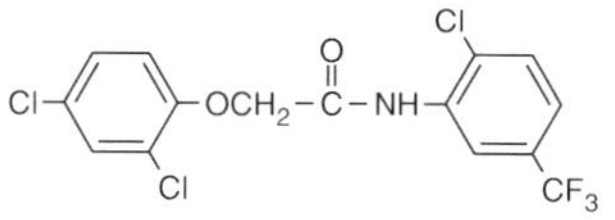

250 mg

S87,822-7
2'-CHLORO-5'-(TRIFLUOROMETHYL)-SUCCINANILIC ACID

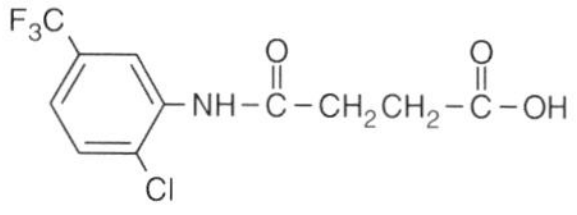

5 gm

S93,592-1
2-(4-CHLOROPHENYLTHIO)-3'-TRIFLUOROMETHYLACETANILIDE

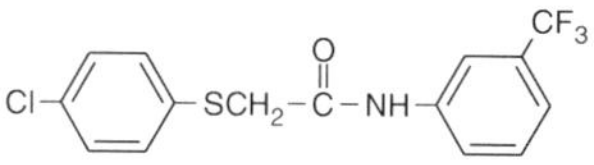

5 gm

S90,477-5
4'-FLUORO-3'-NITRO-2,2,2-TRICHLOROACETANILIDE

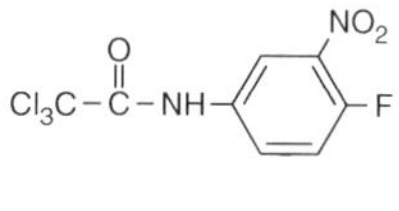

250 mg

S79,601-8
3'-NITRO-2,2,3,3,4,4,4,4'-OCTAFLUOROBUTYRANILIDE

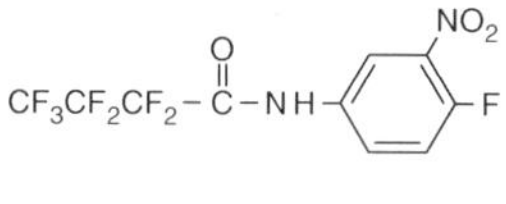

250 mg

S79,603-4
2'-CHLORO-2,2,3,3,4,4,4-HEPTAFLUORO-5'-NITROBUTYRANILIDE

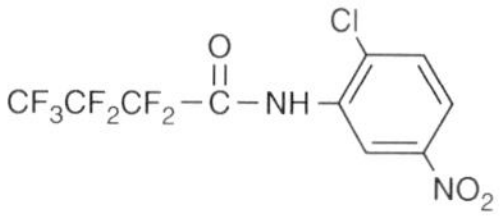

250 mg

S79,140-7
6'-CHLORO-ALPHA,ALPHA,ALPHA-TRIFLUORO-M-OCTANOTOLUIDIDE

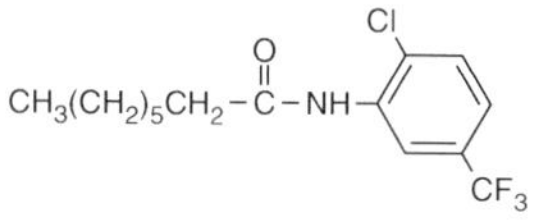

250 mg

S90,525-9
ETHYL 2-CHLORO-5-(TRIFLUOROMETHYL)-OXANILATE

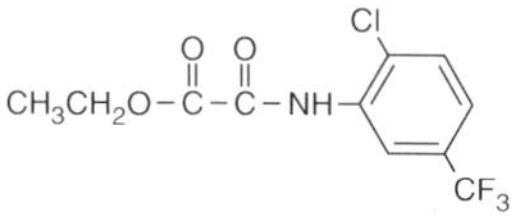

250 mg

S91,273-5
2'-CHLORO-HEXAHYDRO-5'-(TRIFLUOROMETHYL)PHTHALANILIC ACID

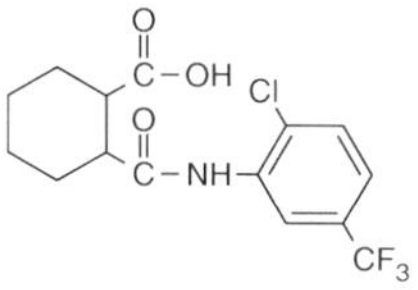

250 mg

S72,917-5
2'-CHLORO-2,3-DIBROMO-5'-(TRIFLUOROMETHYL)SUCCINANILIC ACID

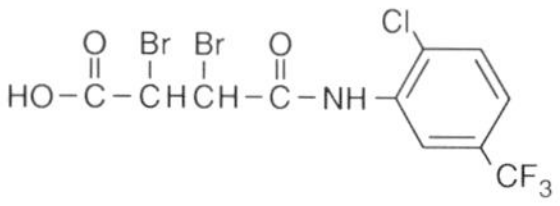

5 gm

S72,097-6
2'-CHLORO-5'-(TRIFLUOROMETHYL)-GLUTARANILIC ACID

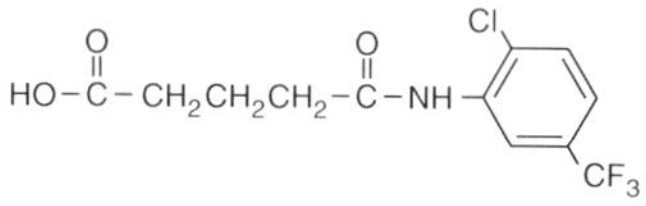

5 gm

S69,283-2
2,2,2,3'-TETRACHLOROACETANILIDE

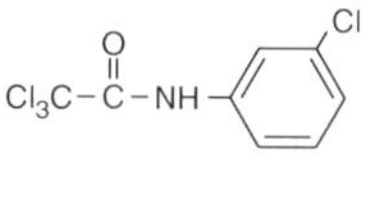

5 gm

S90,494-5
4'-CHLORO-2'-(TRIFLUOROMETHYL)-1-CYCLOHEXANECARBOXANILIDE

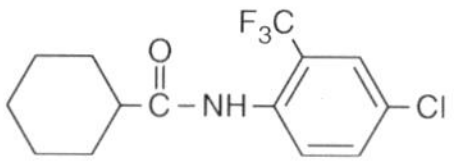

250 mg

S90,499-6
2'-CHLORO-5'-(TRIFLUOROMETHYL)-1-CYCLOHEXANECARBOXANILIDE

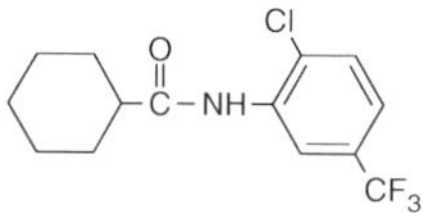

250 mg

S90,043-5
2'-CHLORO-2-PHENYL-5'-(TRIFLUOROMETHYL)ACETANILIDE

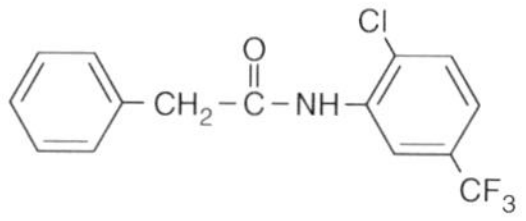

250 mg

S79,519-4
2'-BENZOYL-2,2,2,4'-TETRACHLOROACETANILIDE

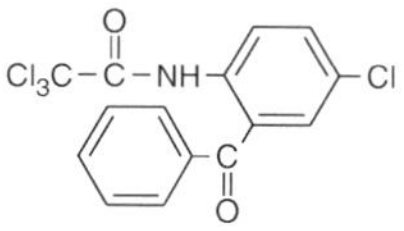

250 mg

S69,284-0
4'-FLUORO-2,2,2-TRICHLOROACETANILIDE

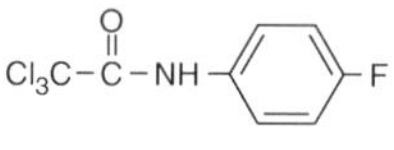

5 gm

S70,258-7
2,2,2,4'-TETRACHLOROACETANILIDE

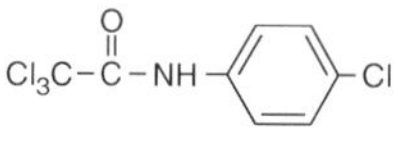

5 gm

S72,026-7
4'-BROMO-2,2,2-TRICHLOROACETANILIDE

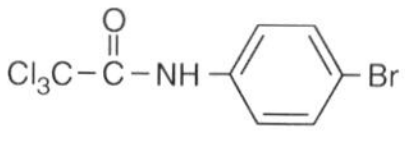

5 gm

S72,054-2
N-(4-IODOPHENYL)-2,2,2-TRICHLOROACETAMIDE

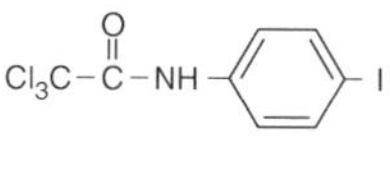

5 gm

S67,536-9
N-(2,5-DICHLOROPHENYL)-2,2,2-TRICHLOROACETAMIDE

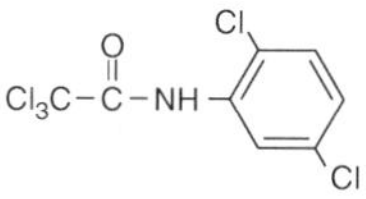

5 gm

S69,059-7
N-(2,6-DICHLOROPHENYL)-2,2,2-TRICHLOROACETAMIDE

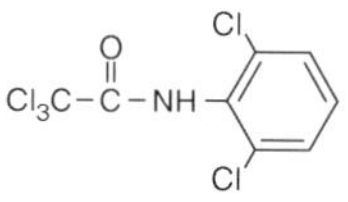

250 mg

S79,624-7
N-(2,3-DICHLOROPHENYL)-2,2,3,3,4,4,5,5,6,6,7,7,8,8,8-PENTADECYLFLUOROOCTANILIDE

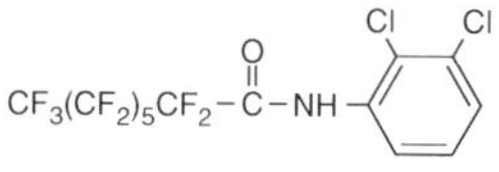

250 mg

S89,243-2
2,2,2,3',4'-PENTACHLOROACETANILIDE

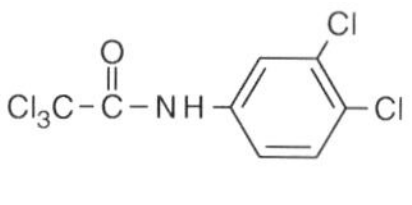

250 mg

S84,400-4
4'-METHYL-2,2,2,3'-TETRAFLUOROACETANILIDE

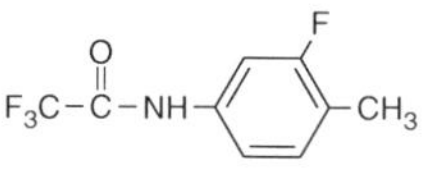

250 mg

S73,024-6
3'-CHLORO-4'-METHYL-2,2,2-TRICHLOROACETANILIDE

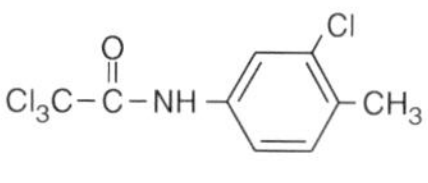

5 gm

S92,143-2
4'-BROMO-3'-METHYL-2-TRICHLOROACETANILIDE

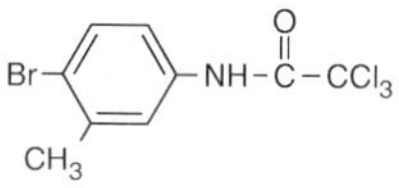

5 gm

S90,476-7
2,2,2,4'-TETRACHLORO-2'-(TRIFLUOROMETHYL)ACETANILIDE

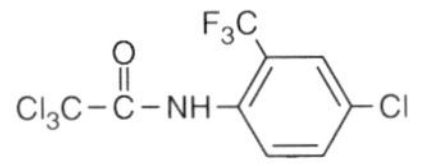

250 mg

S85,453-0
2',5'-DIBROMO-2,2,3,3,3-PENTAFLUOROPROPIONANILIDE

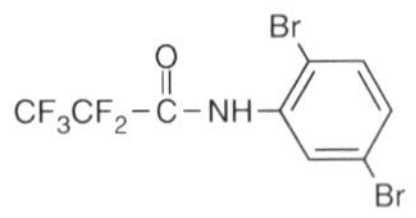

250 mg

S89,027-8
2'-CHLORO-5'-(TRIFLUOROMETHYL)-PROPIONANILIDE

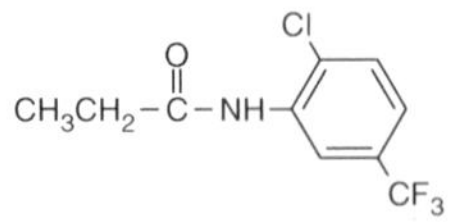

250 mg

S90,473-2
4'-CHLORO-2'-(TRIFLUOROMETHYL)-BUTYRANILIDE

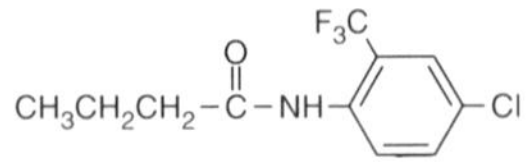

250 mg

S90,521-6
2-BROMO-4'-CHLORO-2'-(TRIFLUOROMETHYL)BUTYRANILIDE

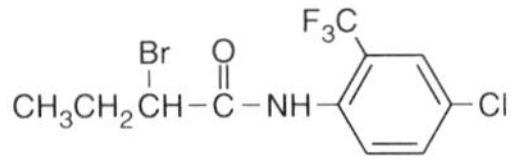

250 mg

S67,551-2
N-(4-CHLORO-2,6-DIBROMOPHENYL)-2,2,2-TRICHLOROACETAMIDE

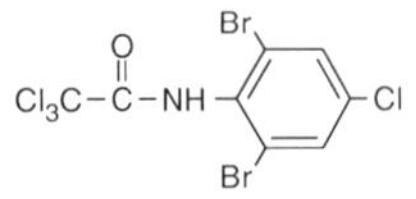

250 mg

S67,397-8
N-(2,4,6-TRIBROMOPHENYL)-2,2,2-TRICHLOROACETAMIDE

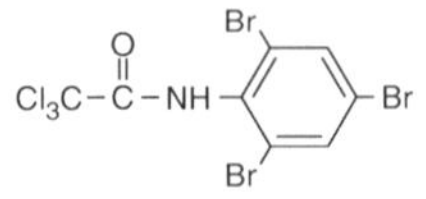

250 mg

S69,951-9
2,2,2,2',4',5'-HEXACHLOROACETANILIDE

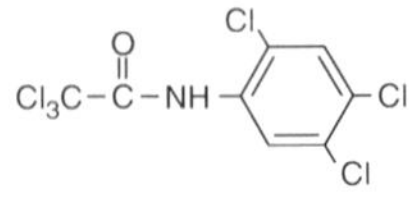

250 mg

S67,552-0
N-(2-BROMO-4,5-DIMETHYLPHENYL)-2,2,2-TRICHLOROACETAMIDE

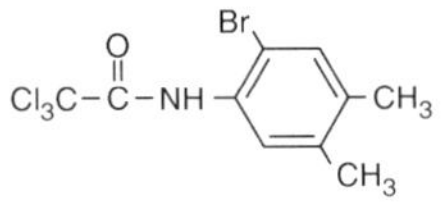

250 mg

S88,908-3
4-(2,2,2-TRICHLOROACETAMIDO)-SALICYLIC ACID

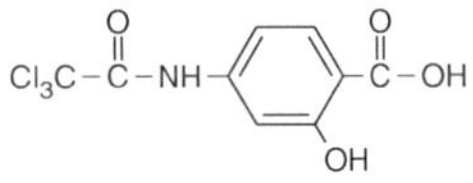

250 mg

S76,395-0
N-TERT-BUTYL-3-HYDROXY-3-(1,3-DICHLORO-6-TRIFLUOROMETHYL-9-PHENANTHRYL)PROPIONAMIDE HYDRATE

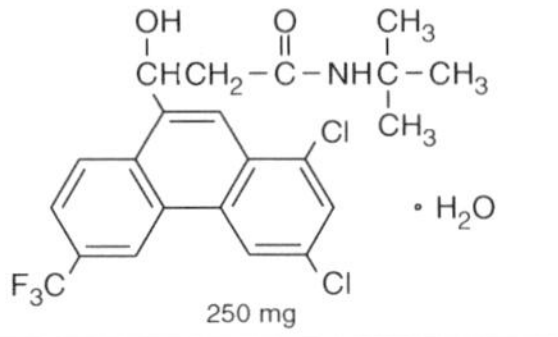

250 mg

S75,267-3
4'-(2,5-DIMETHOXYSTYRYL)-2,2,2-TRICHLOROACETANILIDE

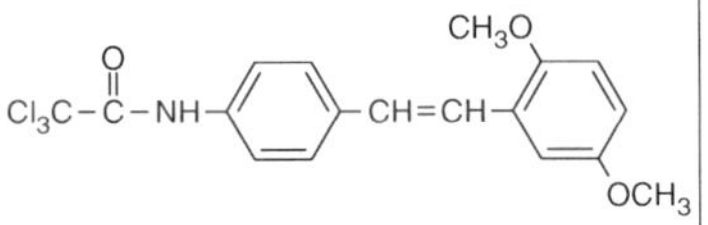

250 mg

S90,895-9
4'-METHOXY-3'-NITRO-2,2,2-TRIFLUOROACETANILIDE

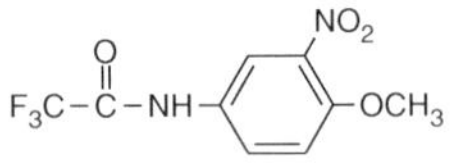

250 mg

S72,177-8
N-(2-ETHOXYPHENYL)-2,2,2-TRICHLOROACETAMIDE

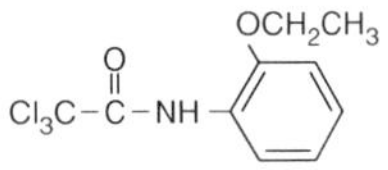

5 gm

S73,021-1
3'-METHOXY-2,2,2-TRICHLOROACETANILIDE

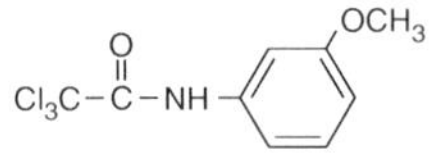

5 gm

S69,538-6
4'-METHOXY-2,2,2-TRICHLOROACETANILIDE

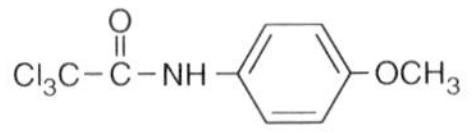

5 gm

S69,530-0
4'-ETHOXY-2,2,2-TRICHLOROACETANILIDE

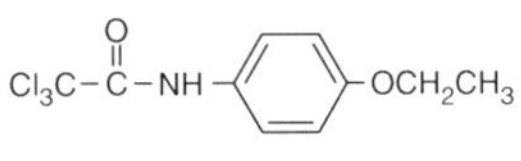

5 gm

S92,181-5
2'-(TRIFLUOROMETHYL)MALEANILIC ACID

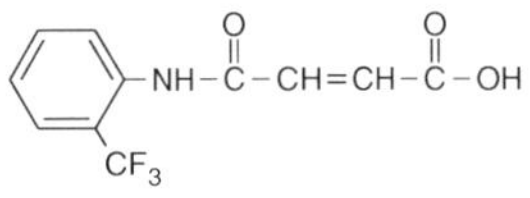

250 mg

S87,823-5
2'-(TRIFLUOROMETHYL)-SUCCINANILIC ACID

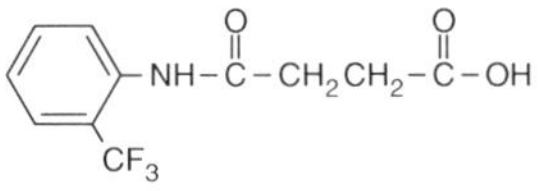

5 gm

S64,764-0
2'-AMINO-4'-TRIFLUOROMETHYLACETANILIDE

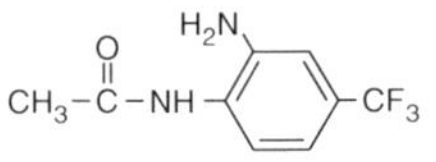

5 gm

S54,784-0
2',4'-DINITRO-2,2,2-TRICHLOROACETANILIDE

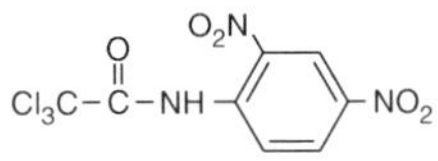

250 mg

S91,653-6
2'-NITRO-4'-(TRIFLUOROMETHYL)-PROPIONANILIDE

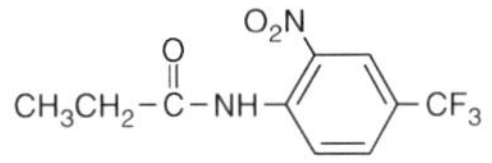

250 mg

S85,842-0
2',6'-DINITRO-4'-(TRIFLUOROMETHYL)-PROPIONANILIDE

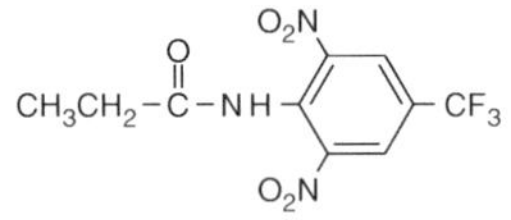

250 mg

S85,843-9
2',6'-DINITRO-4'-(TRIFLUOROMETHYL)-BUTYRANILIDE

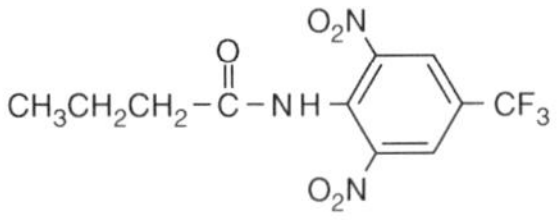

250 mg

S85,844-7
2',6'-DINITRO-N-METHYL-4'-(TRIFLUOROMETHYL)ACETANILIDE

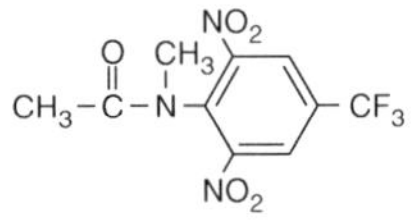

250 mg

S85,845-5
2',6'-DINITRO-N-ETHYL-4'-(TRIFLUOROMETHYL)ACETANILIDE

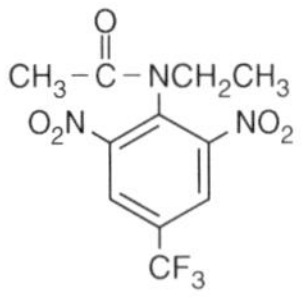

250 mg

S85,849-8
2',6'-DINITRO-N-METHYL-4'-(TRIFLUOROMETHYL)PROPIONANILIDE

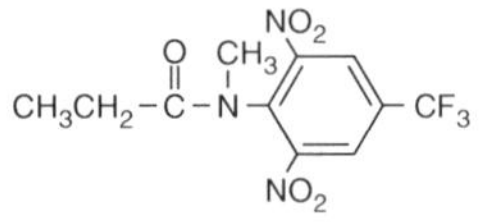

250 mg

S85,846-3
2',6'-DINITRO-N-PROPYL-4'-(TRIFLUOROMETHYL)ACETANILIDE

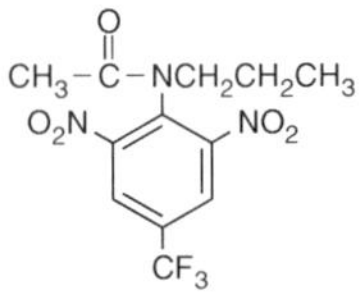

250 mg

S85,850-1
2',6'-DINITRO-N-ETHYL-4'-(TRIFLUOROMETHYL)PROPIONANILIDE

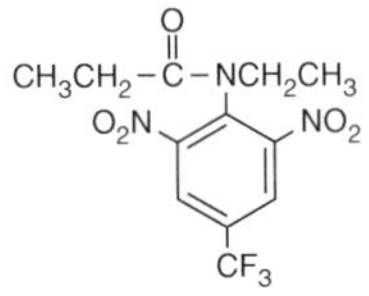

250 mg

S85,856-0
2',6'-DINITRO-N-METHYL-4'-(TRIFLUOROMETHYL)BUTYRANILIDE

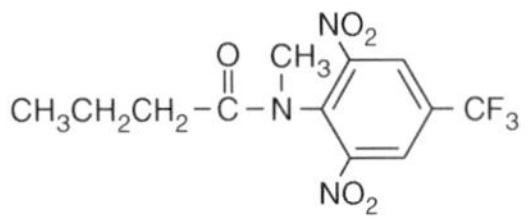

250 mg

S85,847-1
N-BUTYL-2',6'-DINITRO-4'-(TRIFLUOROMETHYL)ACETANILIDE

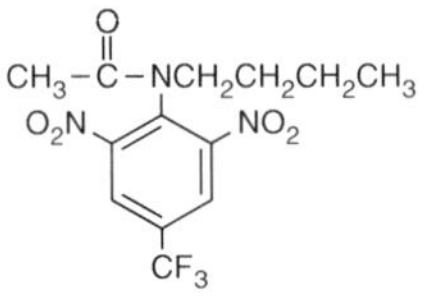

250 mg

S85,852-8
2',6'-DINITRO-N-PROPYL-4'-(TRIFLUOROMETHYL)PROPIONANILIDE

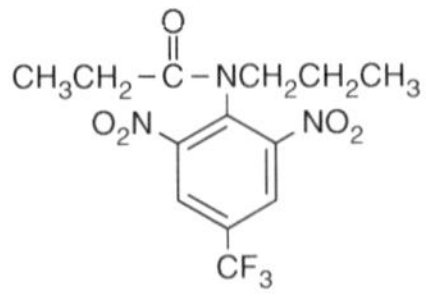

250 mg

S85,857-9
2',6'-DINITRO-N-ETHYL-4'-(TRIFLUOROMETHYL)BUTYRANILIDE

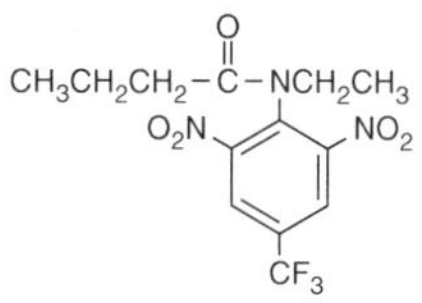

250 mg

S85,853-6
N-BUTYL-2',6'-DINITRO-4'-(TRIFLUOROMETHYL)PROPIONANILIDE

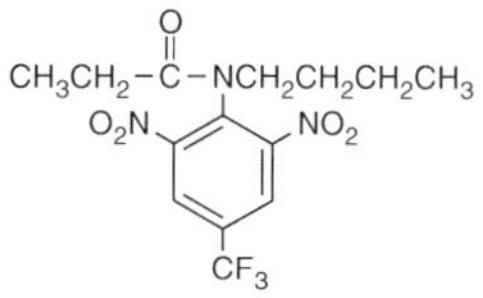

250 mg

S85,854-4
2',6'-DINITRO-N-PROPYL-4'-(TRIFLUOROMETHYL)BUTYRANILIDE

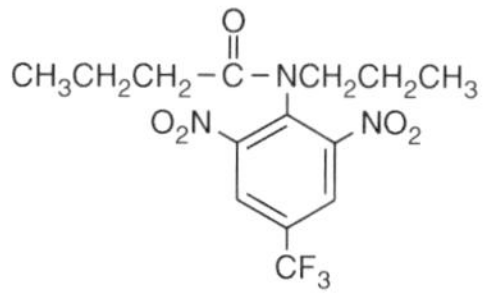

250 mg

S85,855-2
N-BUTYL-2',6'-DINITRO-4'-(TRIFLUOROMETHYL)BUTYRANILIDE

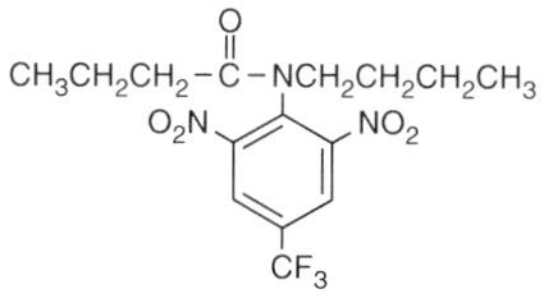

250 mg

S79,604-2
2,2,3,3,4,4,4-HEPTAFLUORO-3'-NITROBUTYRANILIDE

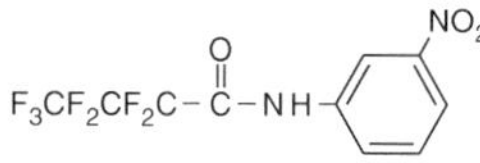

250 mg

S90,800-2
N-ACETYL-N'-(2,2,2-TRIFLUOROACETYL)-1,3-PHENYLENEDIAMINE

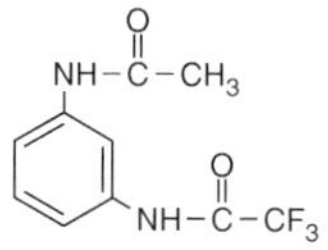

250 mg

S70,257-9
4'-ANILINO-2,2,2-TRICHLOROACETANILIDE

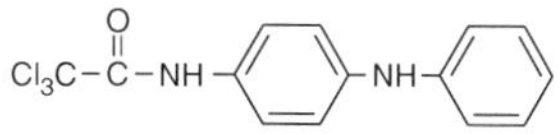

250 mg

S36,694-3
5-ACETAMIDO-2-NITROBENZOTRIFLUORIDE

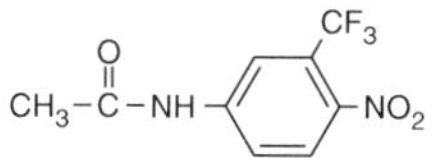

5 gm

S79,611-5
N-ACETYL-N'-(2,2,3,3,4,4,4-HEPTAFLUOROBUTYRYL)-1,4-PHENYLENEDIAMINE

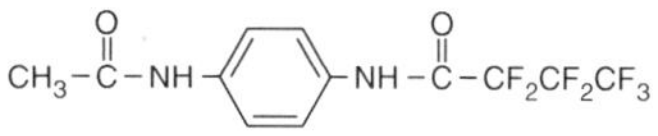

250 mg

S72,024-0
N-PHENETHYL-2,2,2-TRICHLOROACETAMIDE

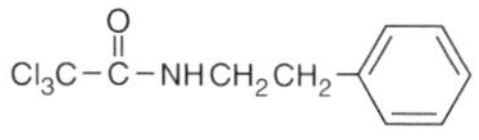

5 gm

S39,868-3
BIS(3,3'-TRIFLUOROMETHYLCARBAMOYL-METHYL)DIMETHYLAMMONIUM CHLORIDE

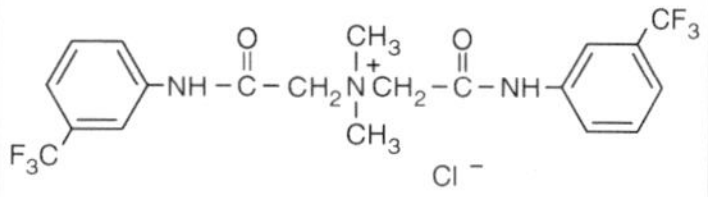

250 mg

S39,866-7
(3,4-DICHLOROPHENYLCARBAMOYLMETHYL)-DIMETHYL-(3'-TRIFLUOROMETHYLTOLYL-CARBAMOYLMETHYL)AMMONIUM CHLORIDE

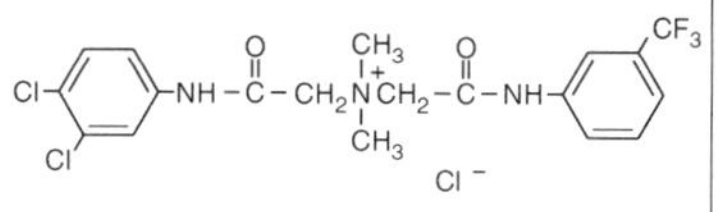

5 gm

S95,364-4
N-BENZYL-N-(TRIFLUOROACETYL)GLYCINE

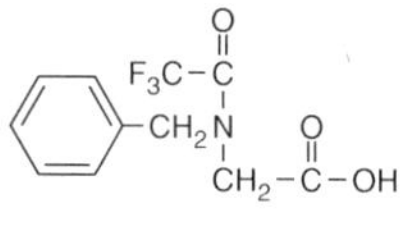

50 mg

S73,018-1
N-BENZYL-2,2,2-TRICHLOROACETAMIDE

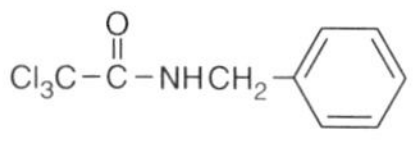

5 gm

S40,829-8
N-(1,1-DIPHENYL-2,2,2-TRIFLUOROETHYL)ACETAMIDE

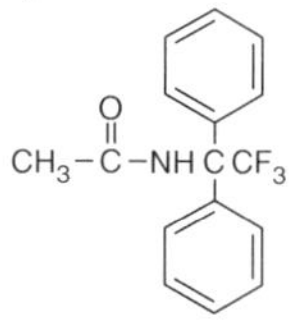

5 gm

S77,475-8
N-BENZYL-N-ISOPROPYL-2,2,2-TRICHLOROACETAMIDE

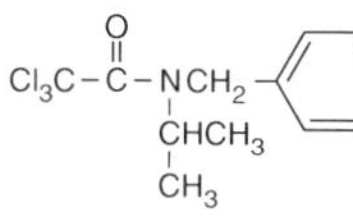

250 mg

S73,020-3
3'-ACETYL-2,2,2-TRICHLOROACETANILIDE

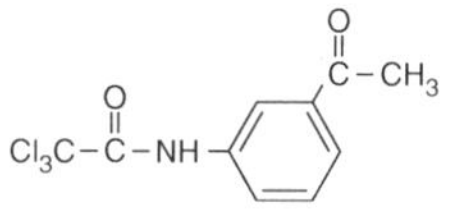

5 gm

S85,577-4
2,3-DIBROMO-3-PHENYL-3'-(TRIFLUOROMETHYL)PROPIONANILIDE

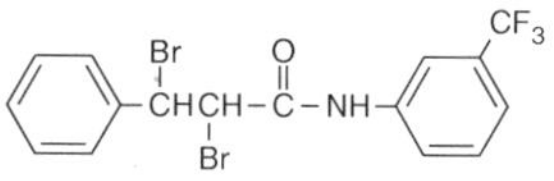

250 mg

S74,997-4
6-(2,2,2-TRIFLUOROACETAMIDO)-3,4-BENZOCOUMARIN

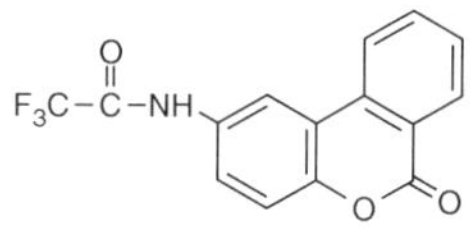

250 mg

S74,998-2
7-(2,2,2-TRIFLUOROACETAMIDO)-3,4-BENZOCOUMARIN

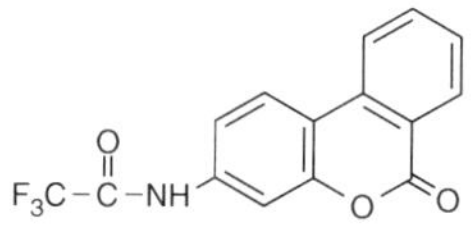

250 mg

S70,312-5
3'-(TRIFLUOROMETHYL)-SUCCINANILIC ACID

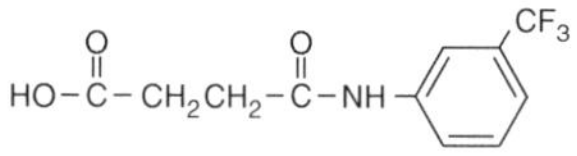

5 gm

S70,295-1
CIS-HEXAHYDRO-N-[3-(TRIFLUOROMETHYL)PHENYL]-PHTHALAMIC ACID

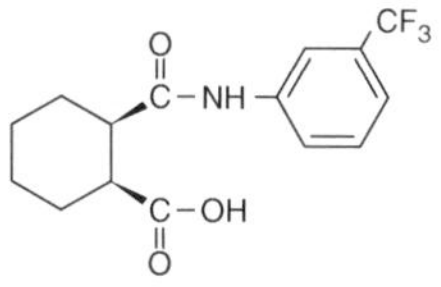

5 gm

S71,926-9
CIS-HEXAHYDRO-N-[2-(TRIFLUOROMETHYL)PHENYL]-PHTHALAMIC ACID

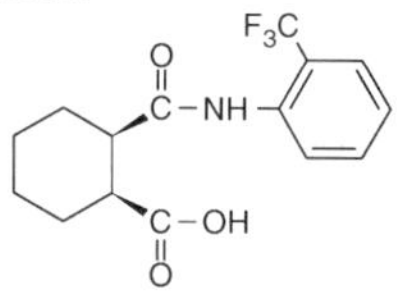

5 gm

S74,631-2
3'-(TRIFLUOROMETHYL)-1,2,3,6-TETRAHYDROPHTHALANILIC ACID

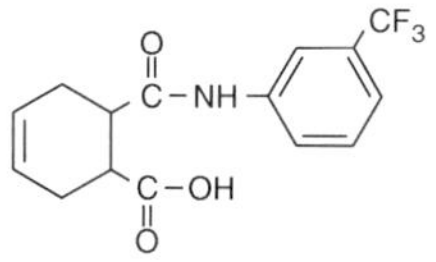

5 gm

S85,461-1
N,N'-BIS[3-(TRIFLUOROMETHYL)PHENYL]-1,10-DECANEDIAMIDE

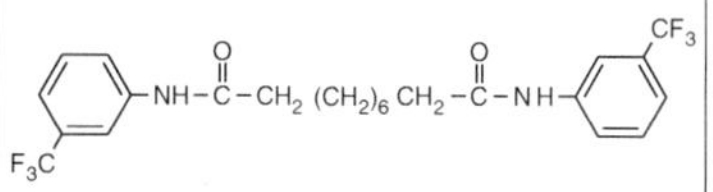

250 mg

S57,720-0
ALPHA,ALPHA,ALPHA-TRIFLUORO-M-FORMOTOLUIDIDE

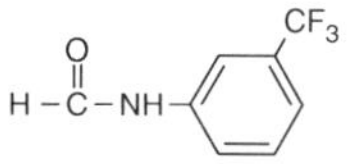

250 mg

S69,525-4
3'-METHYL-2,2,2-TRICHLOROACETANILIDE

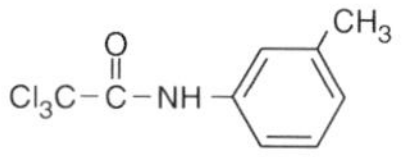

5 gm

S71,921-8
2,2,2-TRICHLORO-3'-(TRIFLUOROMETHYL)ACETANILIDE

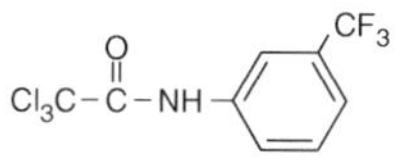

5 gm

S90,271-3
2,2-DIMETHYL-3'-(TRIFLUOROMETHYL)-PROPIONANILIDE

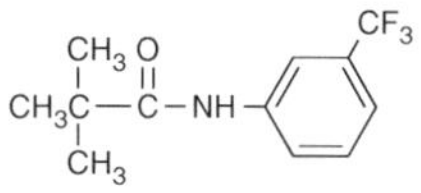

250 mg

S69,529-7
2,2,2-TRICHLOROACETANILIDE

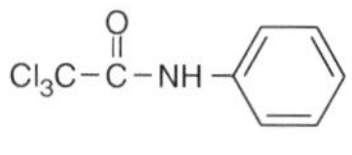

5 gm

S90,047-8
2-PHENYL-3'-(TRIFLUOROMETHYL)-ACETANILIDE

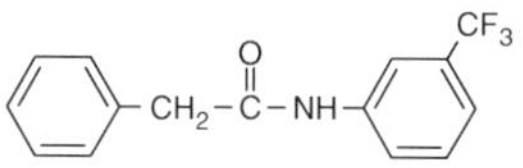

250 mg

S76,865-0
2,2-DIPHENYL-2'-(TRIFLUOROMETHYL)-ACETANILIDE

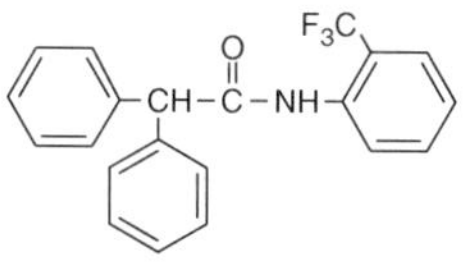

250 mg

S57,655-7
ALPHA,ALPHA,ALPHA-TRIFLUORO-O-FORMOTOLUIDIDE

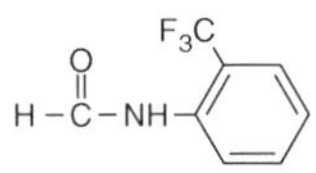

250 mg

S55,307-7
A,A,A-TRIFLUORO-O-ACETOTOLUIDIDE

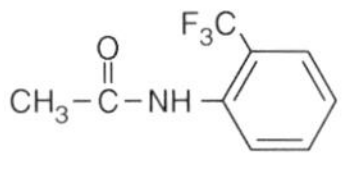

250 mg

S69,527-0
2'-METHYL-2,2,2-TRICHLOROACETANILIDE

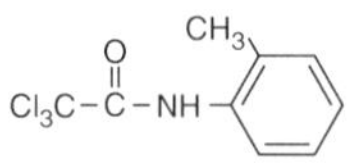

5 gm

S73,994-4
2'-ETHYL-2,2,2-TRICHLOROACETANILIDE

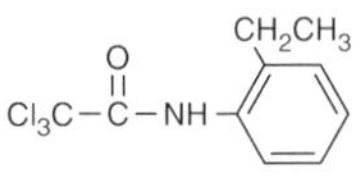

5 gm

S78,214-9
ALPHA,ALPHA,ALPHA-TRIFLUORO-P-ACETOTOLUIDIDE

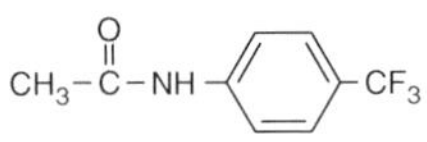

5 gm

S69,282-4
4'-METHYL-2,2,2-TRICHLOROACETANILIDE

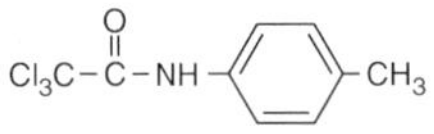

5 gm

S71,922-6
4'-ETHYL-2,2,2-TRICHLOROACETANILIDE

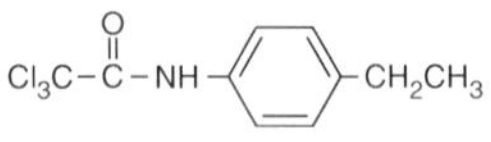

5 gm

S92,150-5
4'-BUTYL-2,2,2-TRICHLOROACETANILIDE

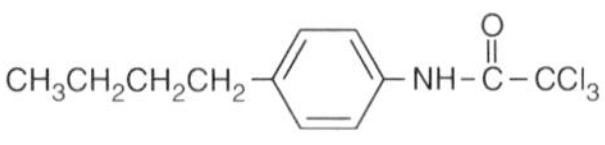

5 gm

S40,679-1
N-[1,1-BIS(4-FLUOROPHENYL)-2,2,2-TRIFLUOROETHYL]ACETAMIDE

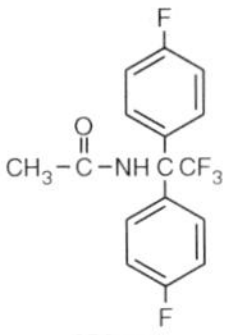

250 mg

S69,526-2
2',3'-DIMETHYL-2,2,2-TRICHLOROACETANILIDE

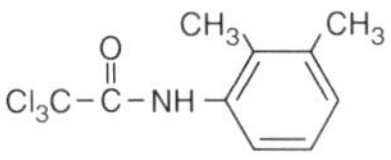

5 gm

S72,011-9
3',5'-DIMETHYL-2,2,2-TRICHLOROACETANILIDE

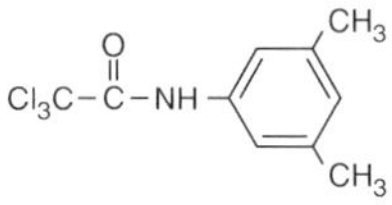

5 gm

S72,178-6
2,2,2-TRICHLORO-N-(2,6-XYLYL)-ACETAMIDE

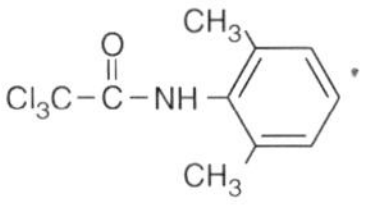

5 gm

S57,656-5
6'-CHLORO-ALPHA,ALPHA,ALPHA-TRIFLUORO-M-FORMOTOLUIDIDE

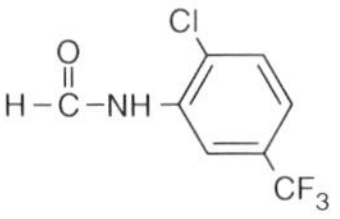

5 gm

S69,801-6
3',4'-DIMETHYL-2,2,2-TRICHLOROACETANILIDE

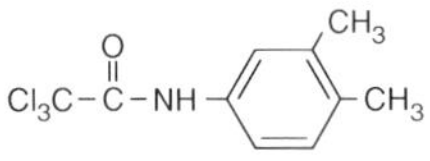

5 gm

S72,025-9
2',4'-DIMETHYL-2,2,2-TRICHLOROACETANILIDE

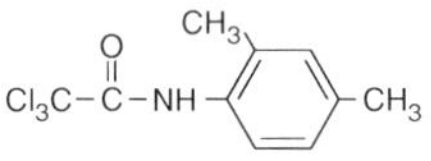

5 gm

S73,023-8
2',5'-DIMETHYL-2,2,2-TRICHLOROACETANILIDE

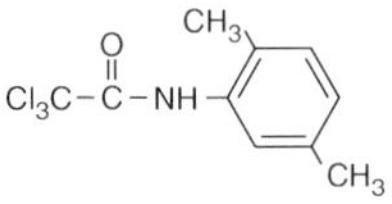

5 gm

S65,707-7
3',5'-BIS(TRIFLUOROMETHYL)-ACETANILIDE

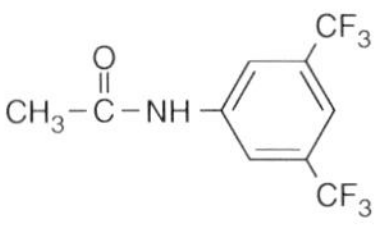

5 gm

S71,983-8
2,2,2-TRICHLORO-2',4',6'-TRIMETHYLACETANILIDE

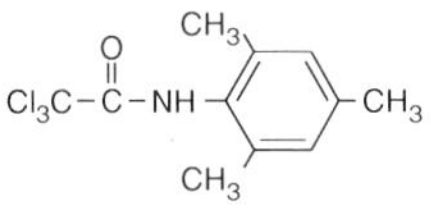

5 gm

S90,478-3
N-(1-NAPHTHYL)-2,2,2-TRICHLOROACETAMIDE

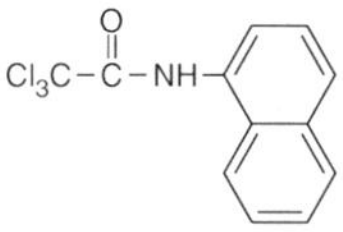

250 mg

S89,241-6
N-BENZYL-2,2,2-TRICHLOROACETANILIDE

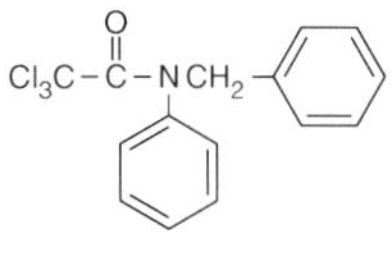

250 mg

S90,472-4
N-ETHYL-2'-(TRIFLUOROMETHYL)-BUTYRANILIDE

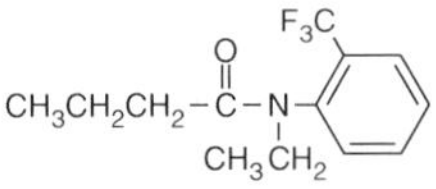

250 mg

S38,250-7
ALPHA,ALPHA,ALPHA-TRIFLUORO-1-PIPERIDINEACETO-META-TOLUIDIDE

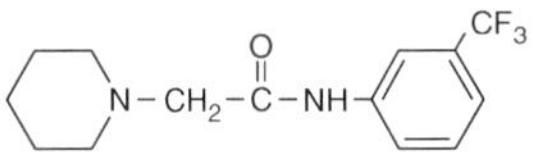

5 gm

S89,451-6
N-(2,4-DICHLOROPHENYL)-2,3-DIHYDROXYSUCCINAMIC ACID

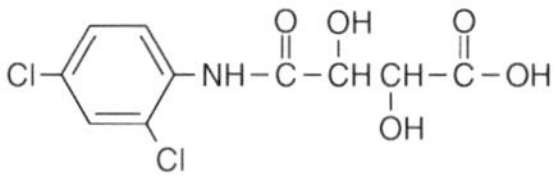

250 mg

S64,209-6
2-(CARBOXYMETHOXY)-3',5'-DICHLORO-2'-HYDROXY-4'-METHYLACETANILIDE

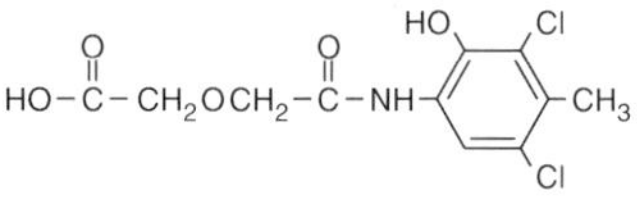

250 mg

S68,552-6
2-(2,4-DICHLOROPHENOXY)-N-(4-HYDROXYPHENYL)ACETAMIDE

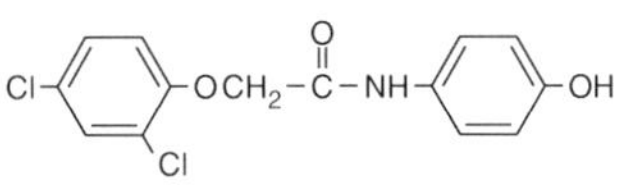

250 mg

S78,182-7
N-ACETYL-3-IODO-L-TYROSINE

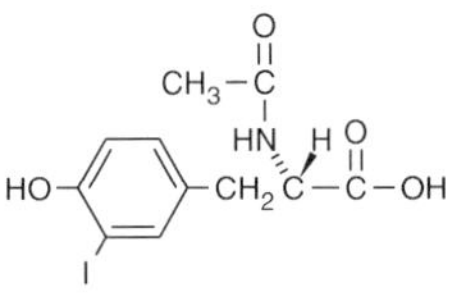

250 mg

S78,260-2
N-ACETYL-3,5-DIBROMO-L-TYROSINE HEMIHYDRATE

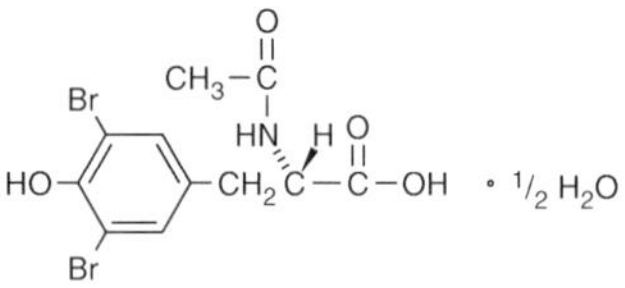

5 gm

S78,872-4
N-ACETYL-D-PHENYLALANYL-3,5-DIIODO-L-TYROSINE

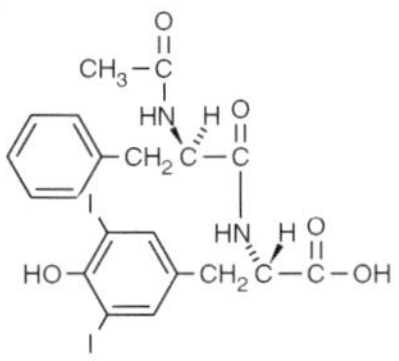

250 mg

S79.017-6
N-ACETYL-3,5-DIIODO-L-TYROSINE ETHYL ESTER

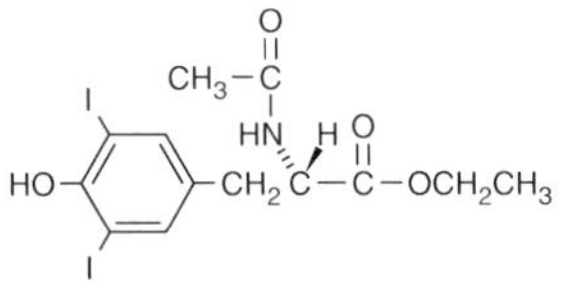

250 mg

S80,659-5
2'-BENZOYL-4'-CHLORO-2-(2-HYDROXYETHYLAMINO)ACETANILIDE

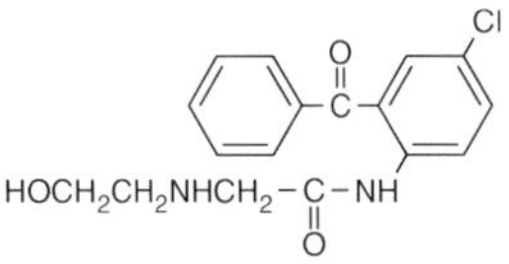

250 mg

S70,190-4
5'-CHLORO-2'-HYDROXYPROPIONANILIDE

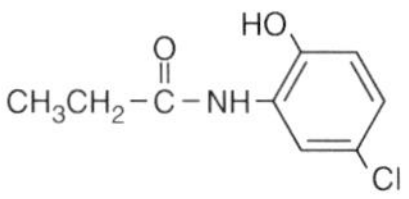

250 mg

S89,103-7
N,N-BIS(2-[2-(2,4,5-TRICHLOROPHENOXY)ACETOXY]ETHYL)-2-(4-CHLOROPHENOXY)ACETAMIDE

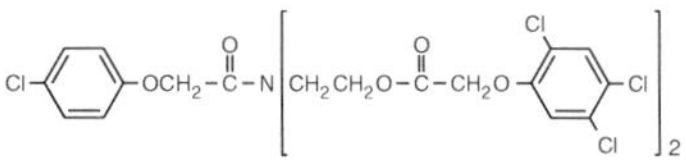

250 mg

S89,297-1
4'-CHLORO-2-(4-CHLOROPHENOXY)-2'-METHOXYACETANILIDE

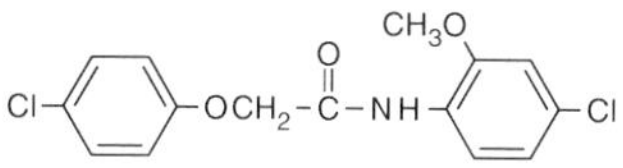

250 mg

S73,045-9
N-(5-CHLORO-2,4-DIMETHOXYPHENYL)-GLUTARAMIC ACID

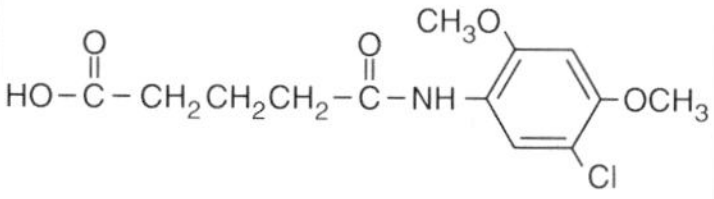

5 gm

S90,496-1
5'-CHLORO-2',4'-DIMETHOXY-1-CYCLOHEXANECARBOXANILIDE

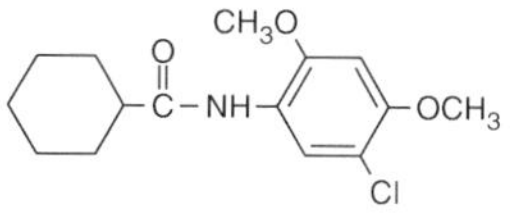

250 mg

S73,620-1
5'-CHLORO-2',4'-DIMETHOXYPROPIONANILIDE

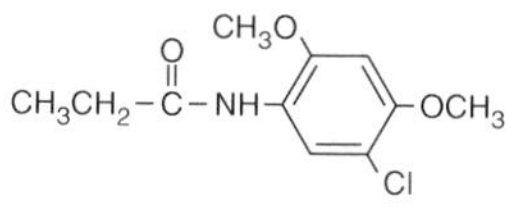

5 gm

S70,082-7
5'-CHLORO-2',4'-DIMETHOXYBUTYRANILIDE

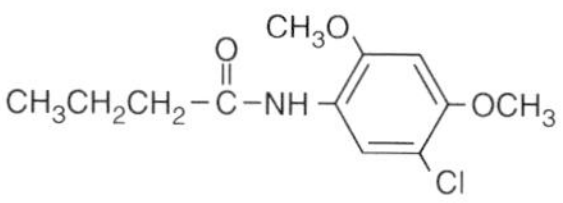

250 mg

S73,915-4
4'-CHLORO-2',5'-DIMETHOXYDODECANANILIDE

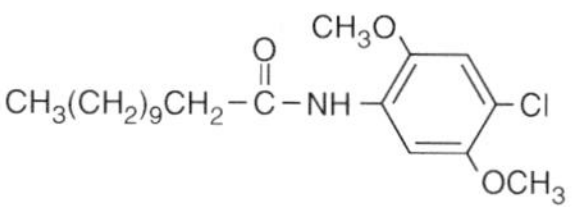

5 gm

S67,610-1
N-(2-METHOXY-5-NITROPHENYL)-2-(2,4,5-TRICHLOROPHENOXY)ACETAMIDE

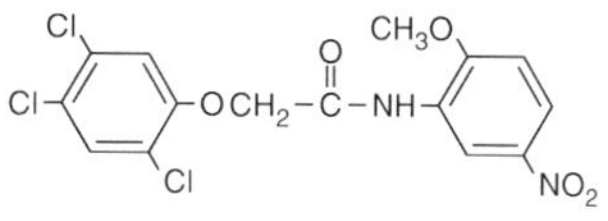

250 mg

S75,019-0
4'-CHLORO-2',5'-DIMETHOXYHEXAHYDROPHTHALANILIC ACID

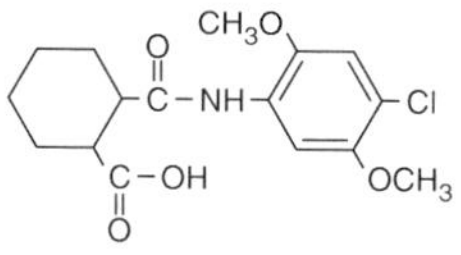

5 gm

S67,579-2
N-(2-ETHOXYPHENYL)-2,4,5-TRICHLOROPHENOXYACETAMIDE

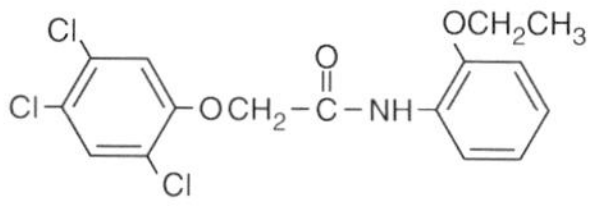

250 mg

S67,612-8
N-(5-CHLORO-2-METHOXYPHENYL)-2-(2,4,5-TRICHLOROPHENOXY)ACETAMIDE

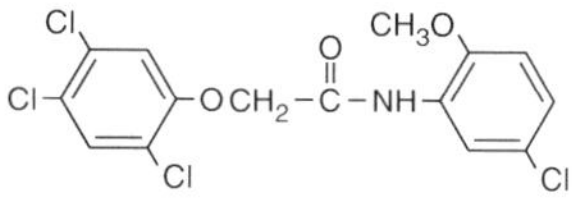

5 gm

S90,524-0
4'-CHLORO-2',5'-DIMETHOXY-2-PROPOXYACETANILIDE

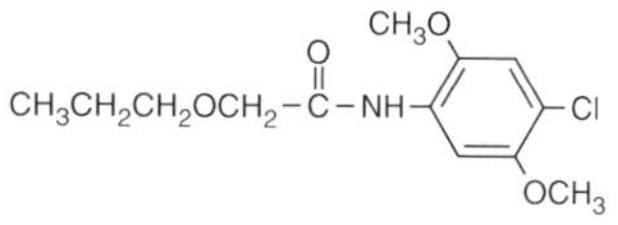

250 mg

S77,928-8
2-(4-CHLOROPHENOXY)-4'-ETHOXYACETANILIDE

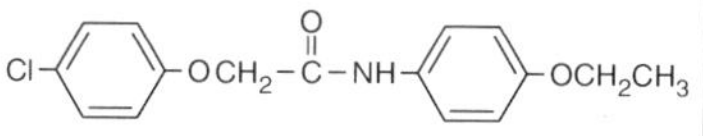

250 mg

S73,619-8
4'-CHLORO-3',5'-DIMETHOXYPROPIONANILIDE

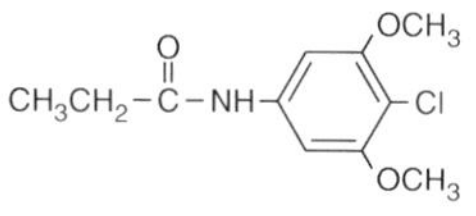

250 mg

S85,483-2
4'-CHLORO-3',5'-DIMETHOXYISOBUTYRANILIDE

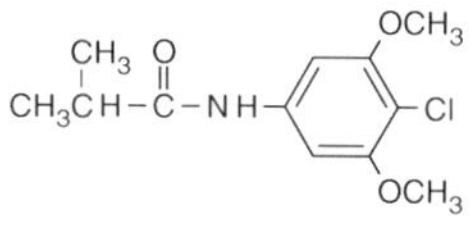

250 mg

S89,299-8
2-(4-CHLOROPHENOXY)-4'-(METHYLTHIO)-ACETANILIDE

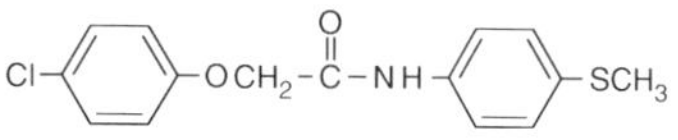

250 mg

S88,940-7
2-(4-CHLOROPHENOXY)-4'-METHYL-2'-NITROACETANILIDE

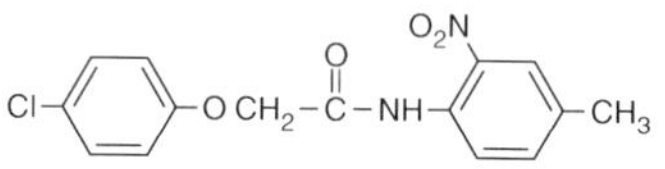

250 mg

S85,175-2
4'-METHYL-2'-NITRO-2-(2,4,5-TRICHLOROPHENOXY)ACETANILIDE

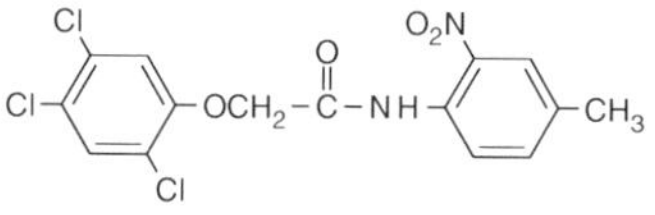

250 mg

S89,300-5
2-(4-CHLOROPHENOXY)-3'-NITROACETANILIDE

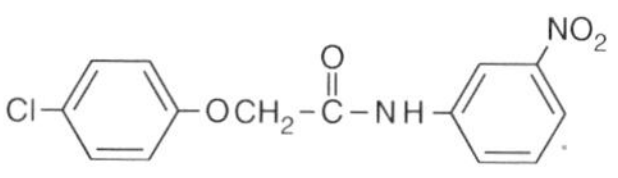

250 mg

S88,954-7
2-(4-CHLOROPHENXOY)-2'-METHYL-5'-NITROACETANILIDE

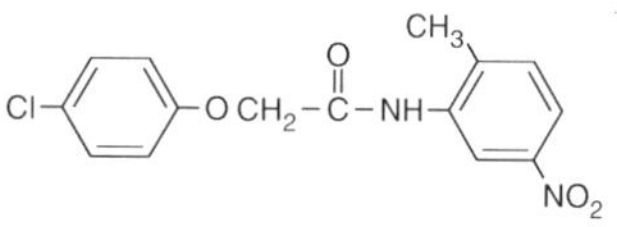

250 mg

S88,953-9
2-(2,4-DICHLOROPHENOXY)-2'-METHYL-5'-NITROACETANILIDE

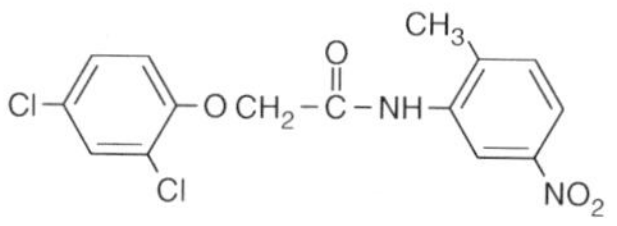

250 mg

S88,952-0
2'-METHYL-5'-NITRO-2-(2,4,5-TRICHLOROPHENOXY)ACETANILIDE

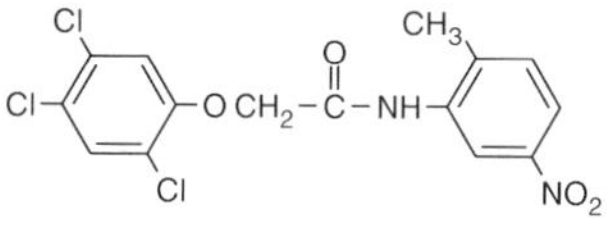

250 mg

S76,979-7
4'-CHLORO-2-(2,4-DICHLOROPHENOXY)-3'-NITROACETANILIDE

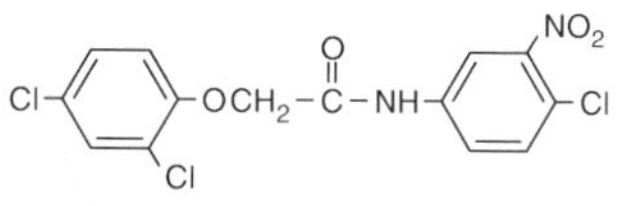

250 mg

S67,605-5
N-[4-(ETHOXYCARBONYL)PHENYL]-2-(2,4,5-TRICHLOROPHENOXY)ACETAMIDE

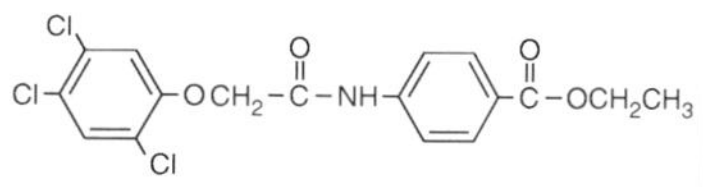

5 gm

S41,110-8
2-(PENTAFLUOROPHENOXY)ACETAMIDE

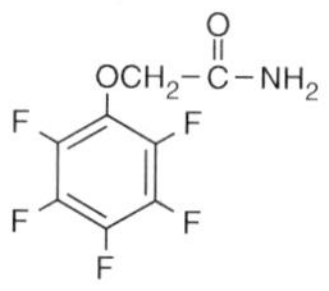

5 gm

S89,032-4
2-(4-CHLOROPHENOXY)-4'-FLUOROACETANILIDE

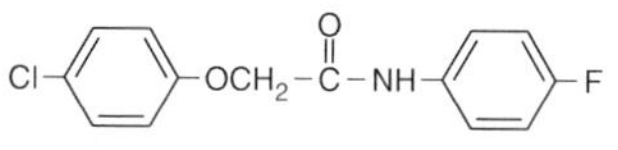

250 mg

S89,237-8
5'-CHLORO-2'-METHYL-2-(PHENOXY)-PROPIONANILIDE

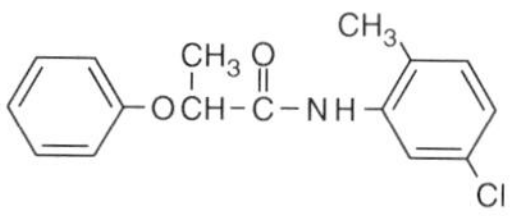

250 mg

S89,163-0
2-(2-CHLOROPHENOXY)-4'-FLUOROACETANILIDE

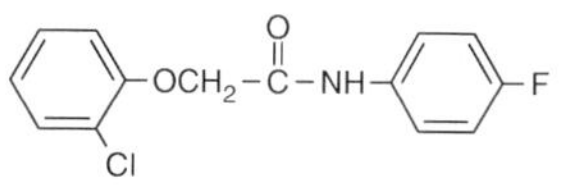

250 mg

S89,150-9
2-(4-CHLORO-2-METHYLPHENOXY)-2'-METHYLACETANILIDE

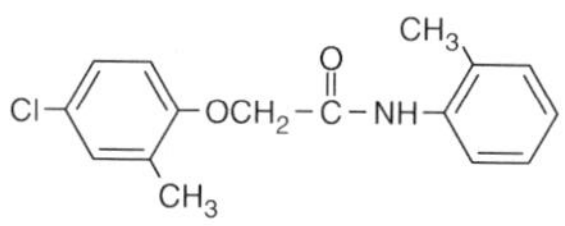

250 mg

S89,039-1
3'-CHLORO-2-(4-CHLOROPHENOXY)-2'-METHYLACETANILIDE

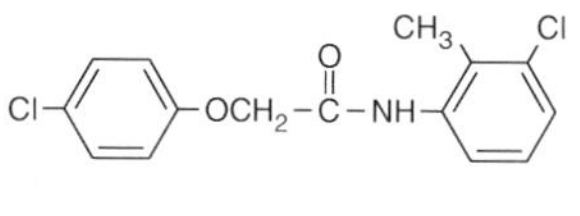

250 mg

S89,301-3
2-(4-CHLOROPHENOXY)-2',5'-DICHLOROACETANILIDE

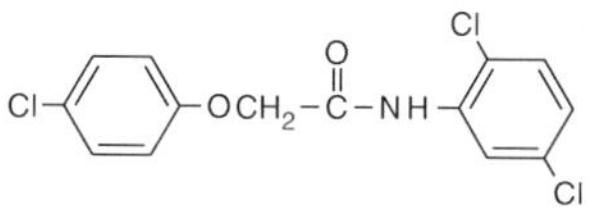

250 mg

S89,040-5
5'-CHLORO-2-(4-CHLOROPHENOXY)-2'-METHYLACETANILIDE

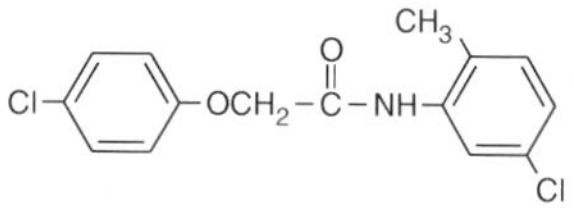

250 mg

S89,041-3
4'-CHLORO-2-(4-CHLOROPHENOXY)-2'-METHYLACETANILIDE

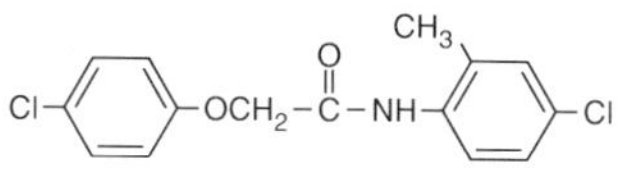

250 mg

S89,044-8
2-(4-CHLOROPHENOXY)-2',4'-DIMETHYLACETANILIDE

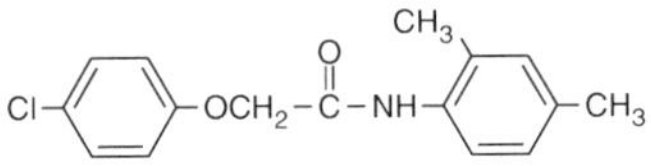

250 mg

S77,249-6
5'-CHLORO-2-(4-CHLOROPHENOXY)-2'-METHOXYACETANILIDE

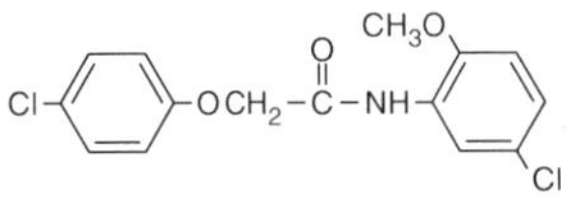

250 mg

S89,083-9
3'-CHLORO-2-(2,4-DICHLOROPHENOXY)-2'-METHYLACETANILIDE

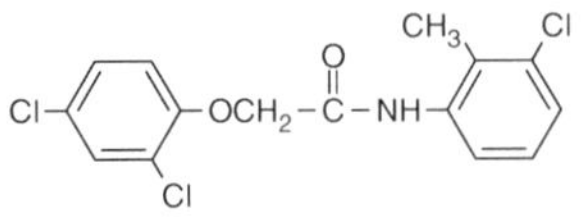

250 mg

S72,598-6
5'-CHLORO-2-(2,4-DICHLOROPHENOXY)-2'-METHYLACETANILIDE

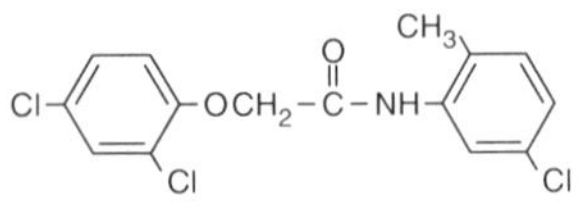

5 gm

S89,082-0
3'-CHLORO-2-(2,4-DICHLOROPHENOXY)-4'-METHYLACETANILIDE

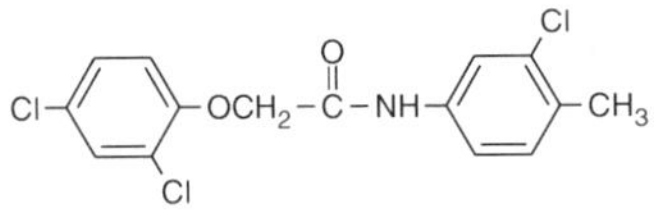

250 mg

S89,085-5
2-(2,4-DICHLOROPHENOXY)-2',4'-DIMETHYLACETANILIDE

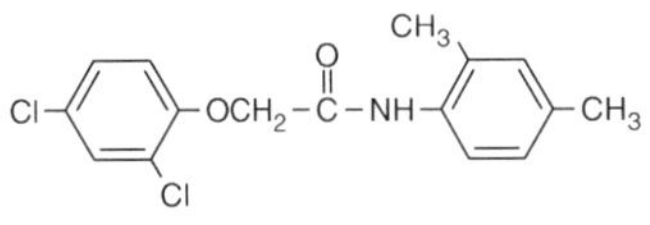

250 mg

S89,148-7
2-(4-CHLORO-2-METHYLPHENOXY)-2',4'-DIMETHYLACETANILIDE

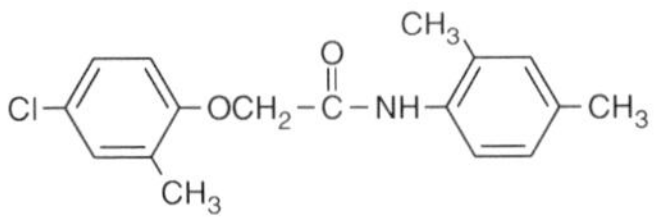

250 mg

S89,149-5
2',4'-DIMETHYL-2-(2,4,5-TRICHLOROPHENOXY)ACETANILIDE

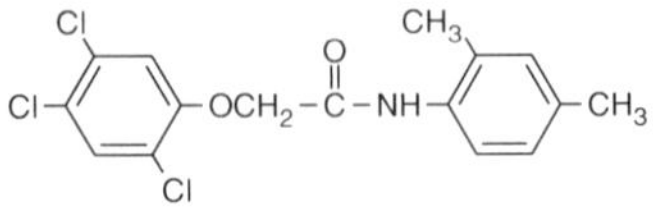

250 mg

S82,491-7
2-(4-CHLOROPHENOXY)-2-METHYLPROPIONAMIDE

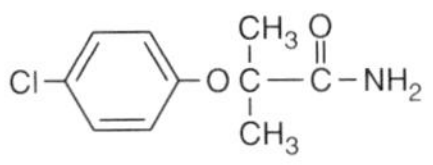

250 mg

S82,490-9
2-(4-BROMOPHENOXY)-2-METHYLPROPIONAMIDE

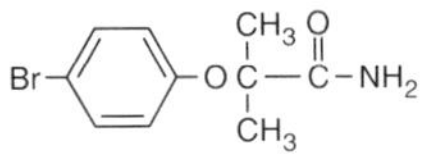

250 mg

S75,334-3
2-(2-CHLOROPHENOXY)-2'-FLUOROACETANILIDE

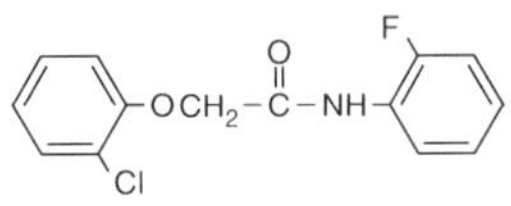

5 gm

S75,338-6
2'-CHLORO-2-(2-CHLOROPHENOXY)-ACETANILIDE

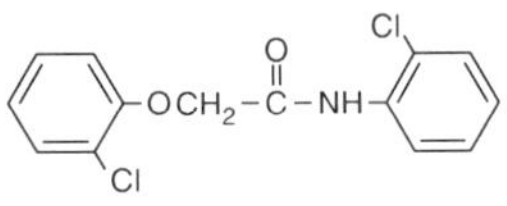

5 gm

S72,483-1
2-(2-CHLOROPHENOXY)-2'-METHYLACETANILIDE

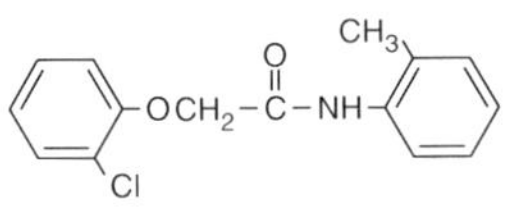

5 gm

S72,446-7
2'-METHYL-2-(2,4,5-TRICHLOROPHENOXY)ACETANILIDE

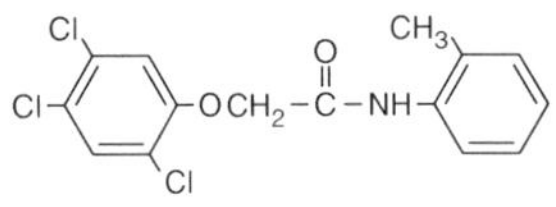

5 gm

S75,362-9
2-(2-CHLOROPHENOXY)-2'-ETHYLACETANILIDE

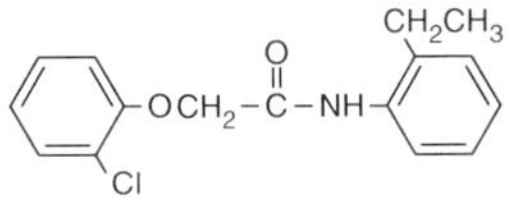

5 gm

S75,337-8
4'-CHLORO-2-(2-CHLOROPHENOXY)-ACETANILIDE

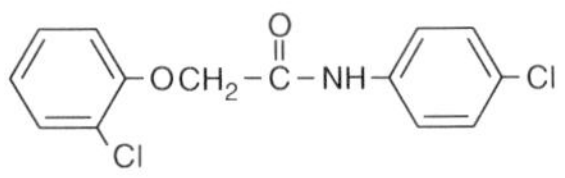

250 mg

S72,490-4
2-(4-CHLOROPHENOXY)-2'-ETHYLACETANILIDE

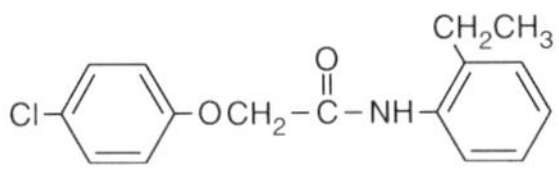

5 gm

S75,348-3
3'-CHLORO-2-(2-CHLOROPHENOXY)-2'-METHYLACETANILIDE

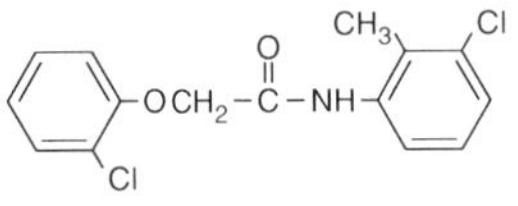

5 gm

S75,229-0
2-(2-CHLOROPHENOXY)-2',5'-DICHLOROACETANILIDE

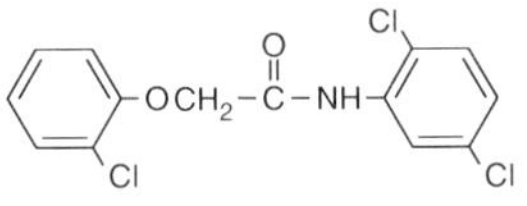

5 gm

S75,240-1
2-(4-CHLORO-2-METHYLPHENOXY)-2'-FLUOROACETANILIDE

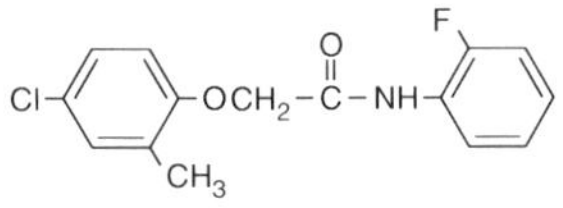

5 gm

S75,371-8
2-(4-CHLORO-2-METHYLPHENOXY)-2'-ETHYLACETANILIDE

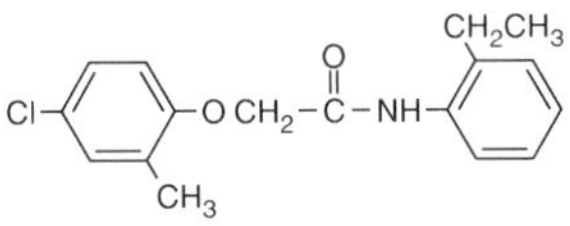

250 mg

S75,376-9
2-(2-CHLOROPHENOXY)-5'-ISOPROPYL-2'-METHYLACETANILIDE

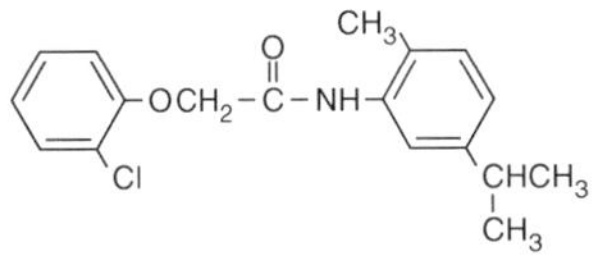

5 gm

S75,332-7
2'-FLUORO-2-(2,4,5-TRICHLOROPHENOXY)ACETANILIDE

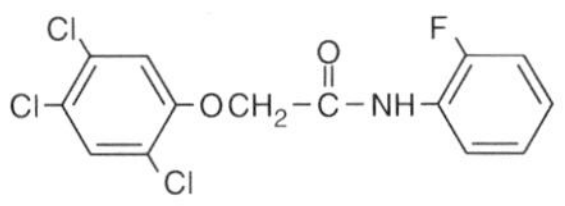

5 gm

S75,354-8
2'-ETHYL-2-(2,4,5-TRICHLOROPHENOXY)-ACETANILIDE

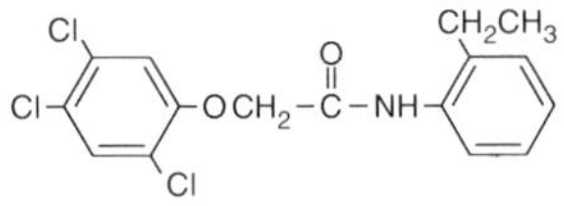

5 gm

S72,476-9
2-(4-CHLOROPHENOXY)-4',5'-DICHLORO-2'-METHYLACETANILIDE

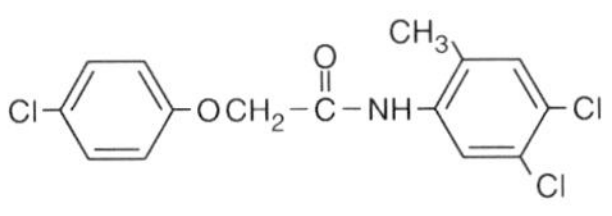

50 mg

S75,147-2
2'-CHLORO-2-(4-CHLOROPHENOXY)-6'-METHYLACETANILIDE

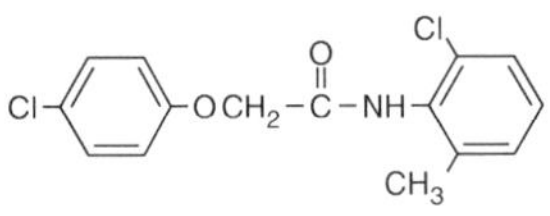

5 gm

S76,954-1
2-(4-CHLOROPHENOXY)-2',6'-ACETOXYLIDIDE

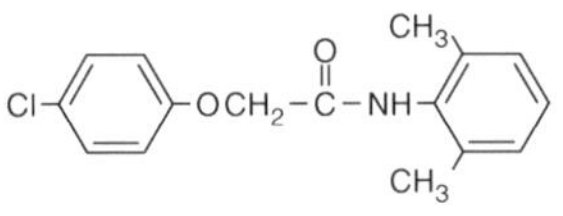

250 mg

S72,599-4
2-(4-CHLORO-2-METHYLPHENOXY)-4'-FLUOROACETANILIDE

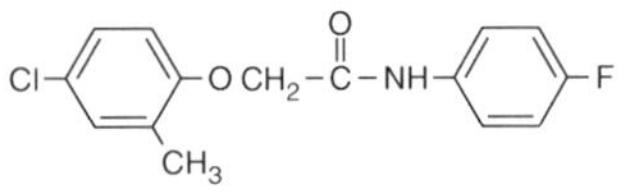

250 mg

S72,455-6
4'-CHLORO-2-(4-CHLORO-2-METHYLPHENOXY)ACETANILIDE

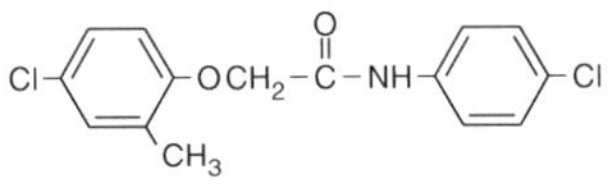

5 gm

S79,126-1
3'-CHLORO-2-(4-CHLOROPHENOXY)-4'-METHYLACETANILIDE

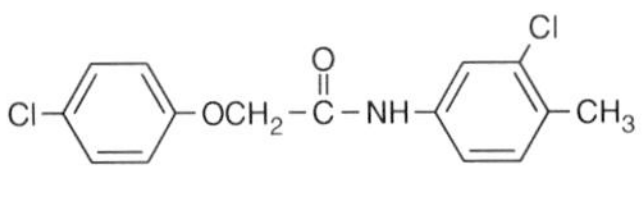

250 mg

S75,333-5
4'-CHLORO-2-(2,4,5-TRICHLOROPHENOXY)ACETANILIDE

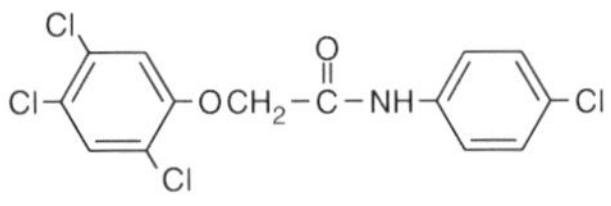

5 gm

S72,434-3
2',3'-DIMETHYL-2-(2,4,5-TRICHLOROPHENOXY)ACETANILIDE

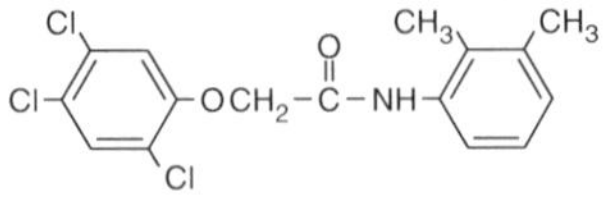

5 gm

S75,356-4
3'-CHLORO-2-(4-CHLORO-2-METHYLPHENOXY)-2'-METHYLACETANILIDE

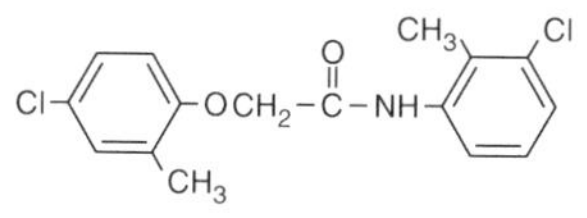

5 gm

S75,381-5
2-(4-CHLORO-2-METHYLPHENOXY)-2',6'-DIETHYLACETANILIDE

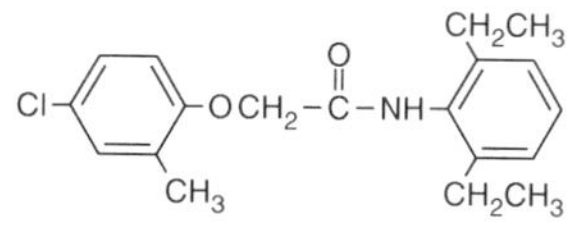

5 gm

S75,344-0
3'-CHLORO-2'-METHYL-2-(2,4,5-TRICHLOROPHENOXY)ACETANILIDE

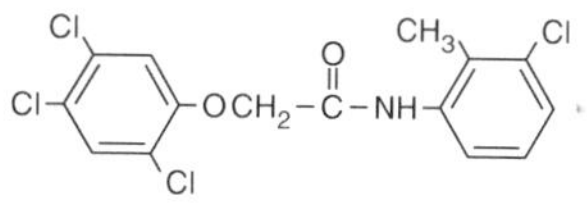

5 gm

S72,487-4
2',5'-DICHLORO-2-(2,4,5-TRICHLOROPHENOXY)ACETANILIDE

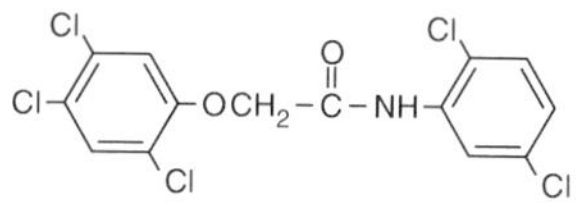

250 mg

S75,347-5
2',5'-DICHLORO-2-(4-CHLORO-2-METHYLPHENOXY)ACETANILIDE

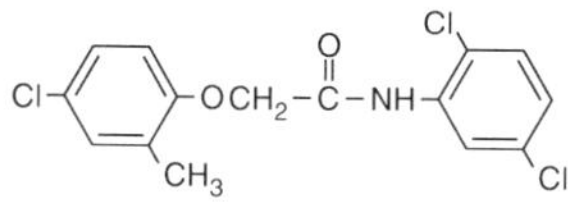

5 gm

S75,355-6
3'-CHLORO-2-(4-CHLORO-2-METHYLPHENOXY)-4'-METHYLACETANILIDE

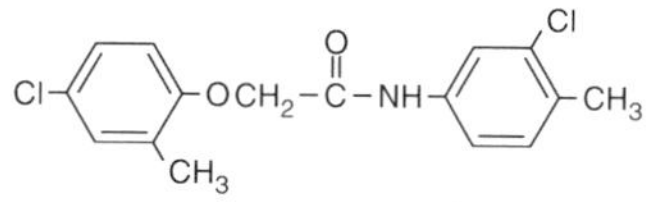

5 gm

S75,343-2
3'-CHLORO-4'-METHYL-2-(2,4,5-TRICHLOROPHENOXY)ACETANILIDE

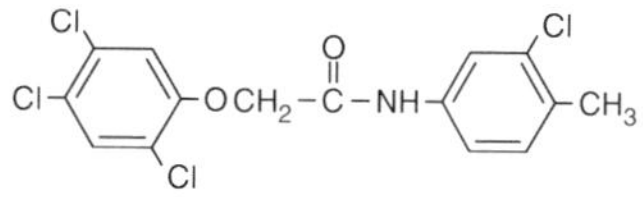

5 gm

S75,375-0
5'-ISOPROPYL-2'-METHYL-2-(2,4,5-TRICHLOROPHENOXY)ACETANILIDE

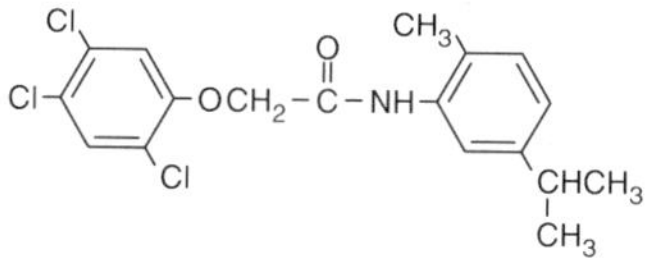

5 gm

S68,403-1
2-(2,4-DICHLOROPHENOXY)-N-(DIPHENYLMETHYL)ACETAMIDE

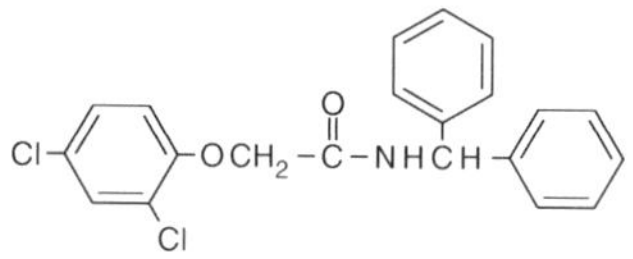

250 mg

S72,421-1
2-(4-CHLOROPHENOXY)-N-CYCLOHEXYLACETANILIDE

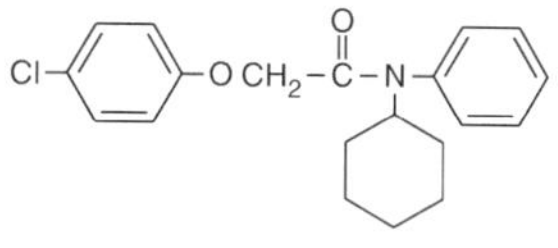

250 mg

S90,691-3
1-O-(4-BROMOPHENYL)-2,3,4,6-TETRAACETYL-BETA-D-GLUCOSAMINE

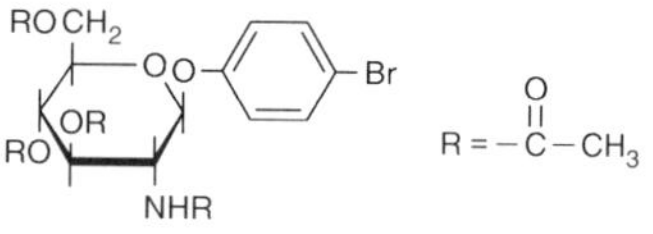

250 mg

S93,594-8
2-(4-CHLOROPHENYLTHIO)-4'-ETHOXYACETANILIDE

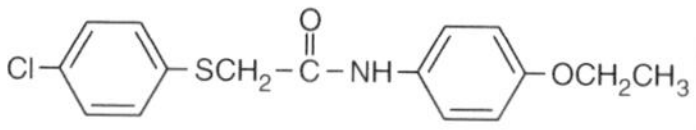

5 gm

S93,599-9
2-(4-CHLOROPHENYLTHIO)-4'-METHOXYACETANILIDE

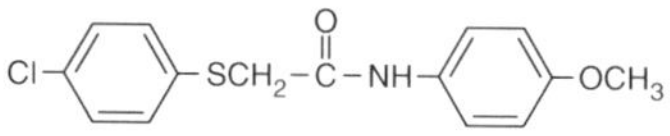

5 gm

S68,638-7
N-(4,5-DIMETHYL-2-NITROPHENYL)-2-(2,4-DICHLOROPHENOXY)ACETAMIDE

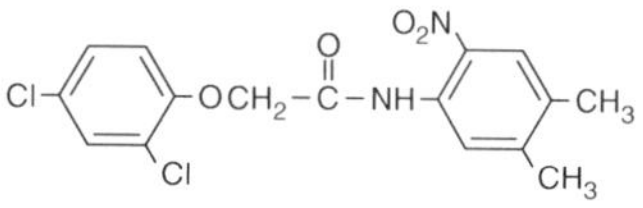

250 mg

S72,858-6
2'-CHLORO-4'-NITRO-2-(2,4,5-TRICHLOROPHENOXY)ACETANILIDE

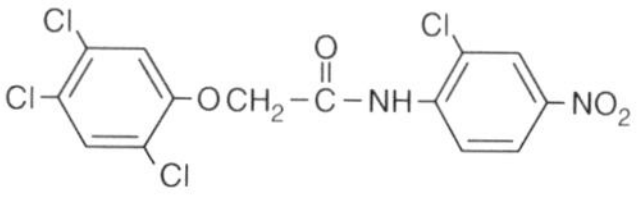

5 gm

S71,845-9
2'-CHLORO-2-(4-CHLORO-2-METHYLPHENOXY)-4'-NITROACETANILIDE MONOHYDRATE

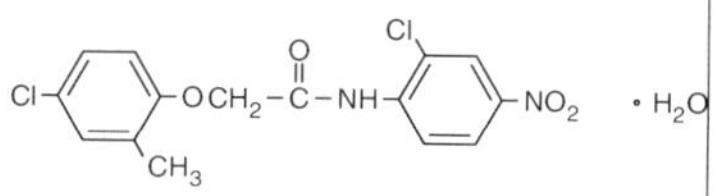

250 mg

S67,631-4
N-(3-NITROPHENYL)-2-(2,4,5-TRICHLOROPHENOXY)ACETAMIDE

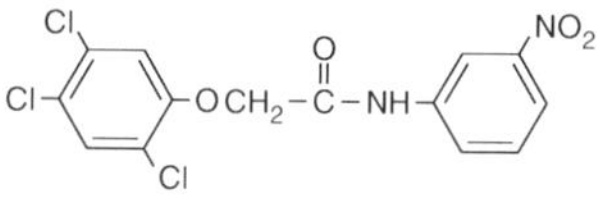

5 gm

S67,607-1
N-(3-ACETYLPHENYL)-2-(2,4,5-TRICHLOROPHENOXY)ACETAMIDE

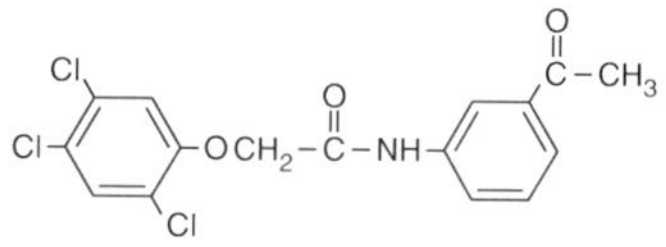

5 gm

S67,584-9
N-[2-(METHOXYCARBONYL)PHENYL]-2,4,5-TRICHLOROPHENOXYACETAMIDE

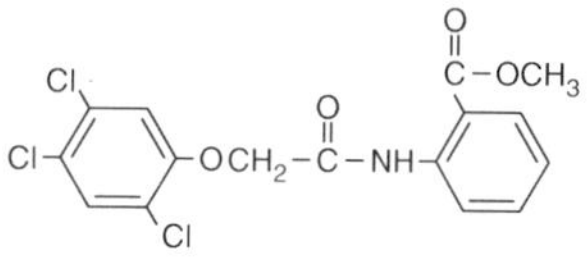

5 gm

S72,918-3
5'-CHLORO-2,3-DIBROMO-2'-METHOXYSUCCINANILIC ACID

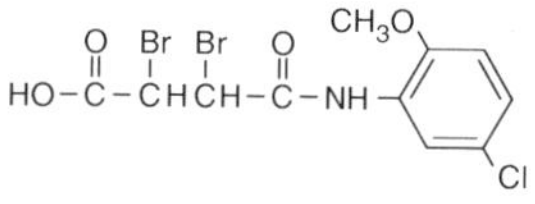

5 gm

S72,912-4
CIS-N-(5-CHLORO-2-METHOXYPHENYL)-HEXAHYDROPHTHALAMIC ACID

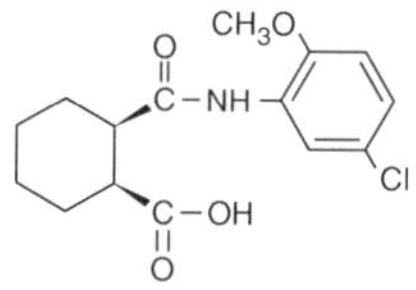

5 gm

S67,427-3
N-(3-CHLOROPHENYL)-2,4,5-TRICHLOROPHENOXYACETAMIDE

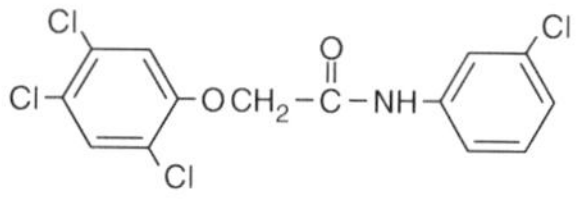

5 gm

S72,169-7
4'-CHLORO-2'-METHOXYCYCLOPROPANECARBOXANILIDE

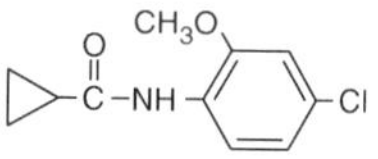

250 mg

S69,573-4
N-(4-CHLORO-2-PHENOXYPHENYL)-2,2-DIPHENYLACETAMIDE

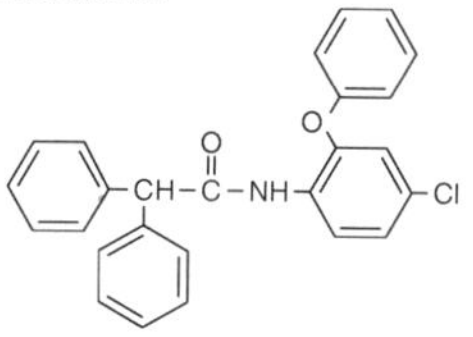

250 mg

S90,501-1
4'-IODO-2-METHOXYACETANILIDE

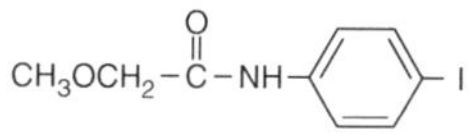

250 mg

S67,622-5
N-(2,6-DIETHYLPHENYL)-2-(2,4,5-TRICHLOROPHENOXY)ACETAMIDE

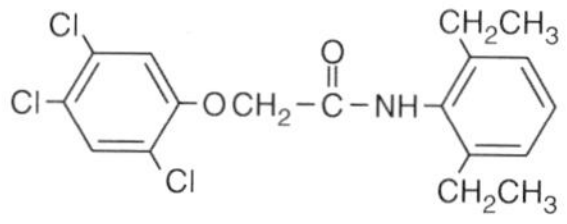

5 gm

S70,278-1
5'-CHLORO-2'-METHOXYFORMANILIDE

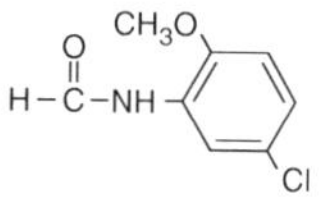

5 gm

S79,198-9
4'-CHLORO-2'-METHOXYPROPIONANILIDE

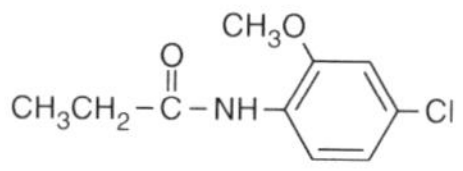

250 mg

S89,053-7
5'-CHLORO-2'-METHOXYPROPIONANILIDE

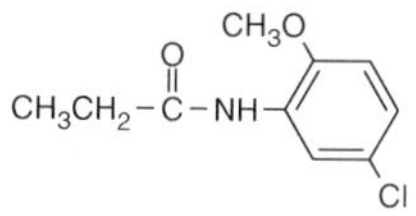

250 mg

S69,960-8
4'-CHLORO-2'-METHOXYBUTYRANILIDE

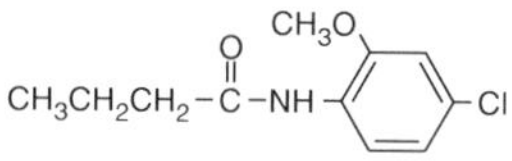

250 mg

S91,389-8
5'-CHLORO-2,2-DIMETHYL-2'-METHOXYPROPIONANILIDE

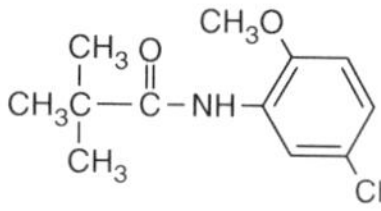

250 mg

S76,983-5
4'-CHLORO-2,2-DIMETHYL-2'-METHOXYPROPIONANILIDE

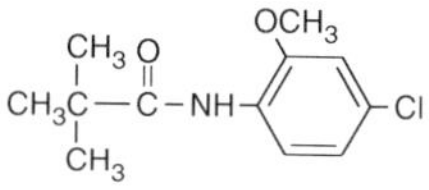

250 mg

S89,781-7
2-(2,4-DICHLOROPHENOXY)-N-METHYL-N-(TETRAHYDROFURFURYL)ACETAMIDE

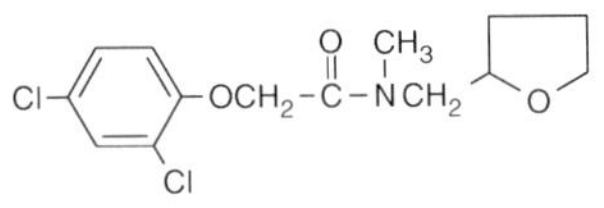

250 mg

S89,783-3
N-CYCLOHEXYL-2-(2,4-DICHLOROPHENOXY)-N-METHYLACETAMIDE

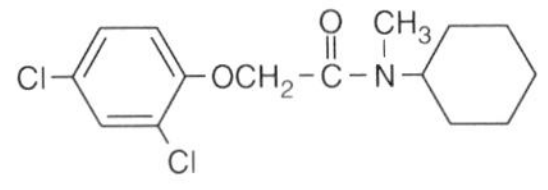

250 mg

S72,477-7
2'-BENZOYL-4'-CHLORO-2-(4-CHLOROPHENOXY)ACETANILIDE

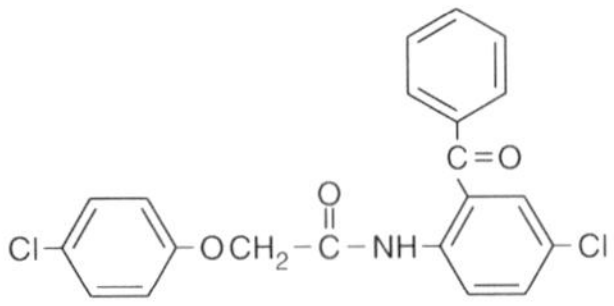

5 gm

S68,716-2
N-(4-ACETYLPHENYL)-2-(2,4-DICHLOROPHENOXY)ACETAMIDE

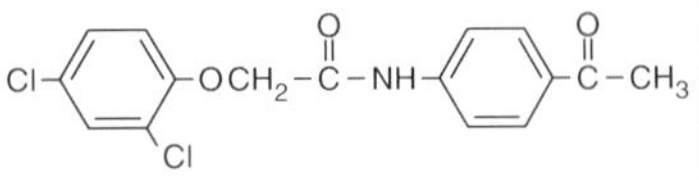

250 mg

S85,579-0
2'-CHLORO-2,3-DIBROMO-4'-METHOXY-3-PHENYLPROPIONANILIDE

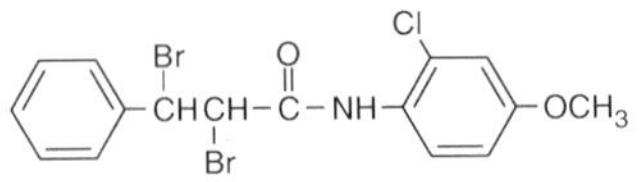

250 mg

S85,460-3
METHYL 2-[2-(4-CHLORO-2-METHYLPHENXOY)ACETAMIDO]BENZOATE

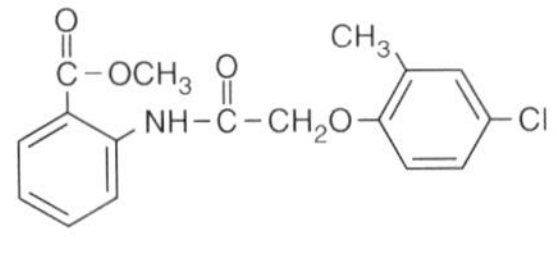

250 mg

S72,439-4
2-(2-CHLOROPHENOXY)-2',4'-DIMETHYLACETANILIDE

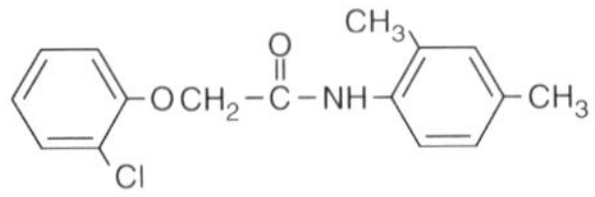

5 gm

S76,850-2
4'-CHLORO-2-(4-CHLOROPHENOXY)-ACETANILIDE

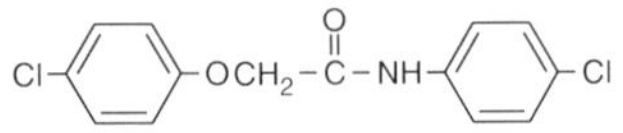

250 mg

S79,146-6
2-(4-CHLOROPHENOXY)-2',3'-DIMETHYLACETANILIDE

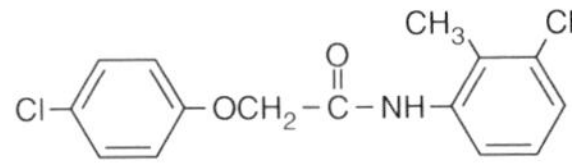

250 mg

S68,831-2
2,4-DICHLOROPHENOXY-N-(2,3-DICHLOROPHENYL)ACETAMIDE

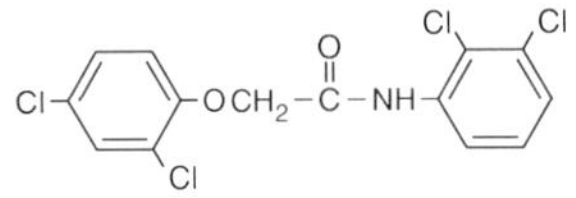

250 mg

S71,851-3
2-(2,4,5-TRICHLOROPHENOXY)-2',4',6'-TRIMETHYLACETANILIDE

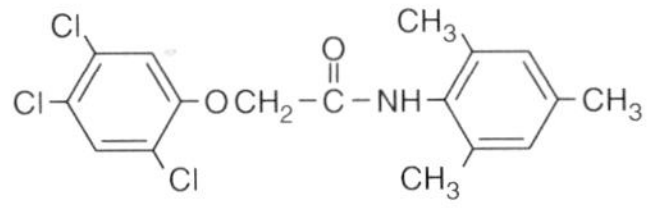

5 gm

S87,835-9
4'-BROMO-2'-CHLOROHEXAHYDRO-PHTHALANILIC ACID

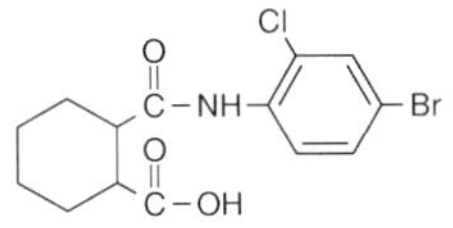

5 gm

S87,811-1
N-(2,5-DICHLOROPHENYL)-1,2,3,6-TETRAHYDROPHTHALAMIC ACID

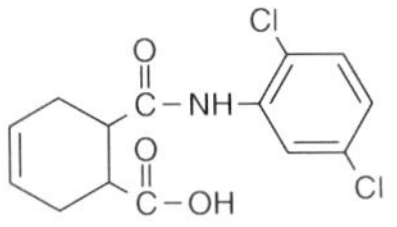

5 gm

S92,163-7
N-(4-BROMOPHENYL)GLUTARANILIC ACID

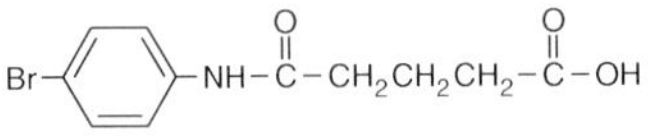

250 mg

S87,802-2
3'-CHLORO-2'-METHYLSUCCINANILIC ACID

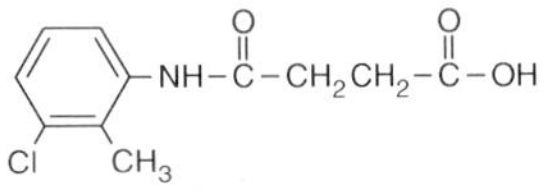

5 gm

S92,175-0
4'-BROMO-2'-CHLOROSUCCINANILIC ACID

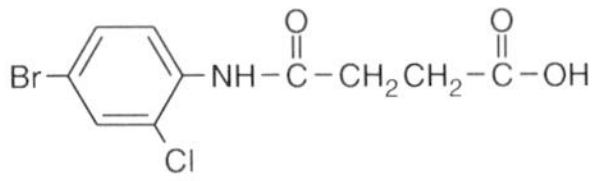

5 gm

S87,803-0
3'-CHLORO-4'-METHYLSUCCINANILIC ACID

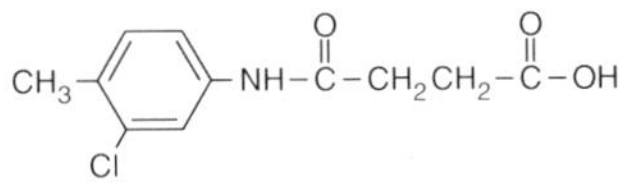

5 gm

S92,183-1
4'-BROMO-3'-METHYLSUCCINANILIC ACID

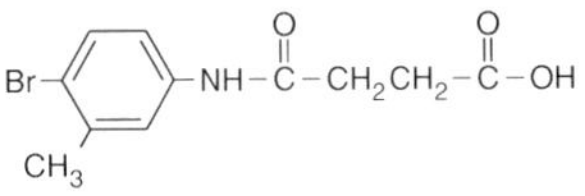

5 gm

S92,176-9
4'-BROMO-2'-CHLOROGLUTARANILIC ACID

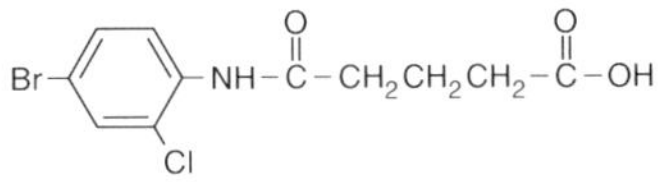

5 gm

S92,177-7
4'-BROMO-3'-METHYLGLUTARANILIC ACID

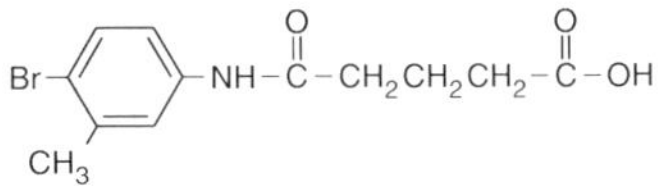

5 gm

S67,174-6
4'-CHLORO-2-MERCAPTOSTEARANILIDE

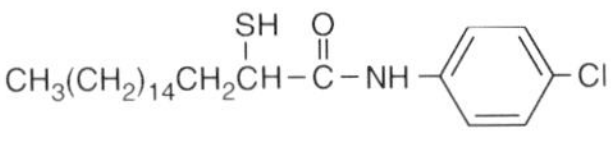

250 mg

S93,597-2
2-(4-CHLOROPHENYLTHIO)-2'-NITROACETANILIDE

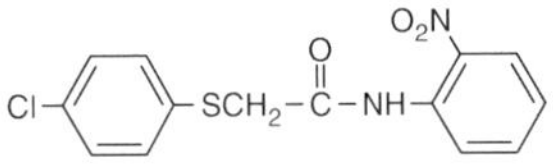

5 gm

S93,595-6
2-(4-CHLOROPHENYLTHIO)-4'-CHLORO-3'-NITROACETANILIDE

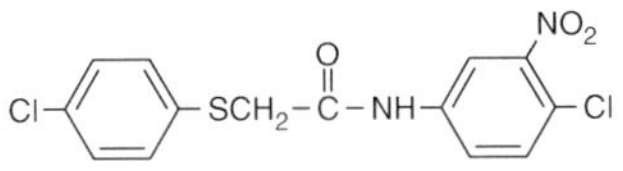

5 gm

S93,380-5
2-(4-CHLOROPHENYLTHIO)-3'-NITROACETANILIDE

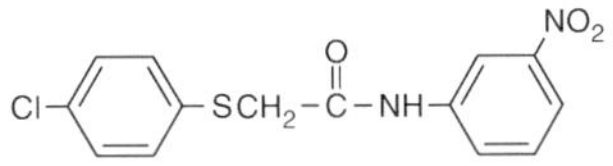

250 mg

S94,605-2
2-(BENZYLTHIO)-4'-CHLORO-3'-NITROACETANILIDE

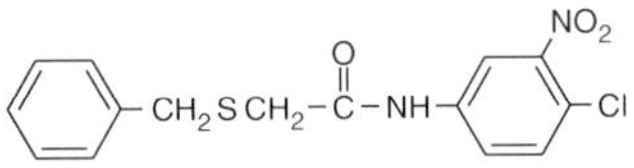

5 gm

S93,676-6
2-(BENZYLTHIO)-4'-FLUOROACETANILIDE

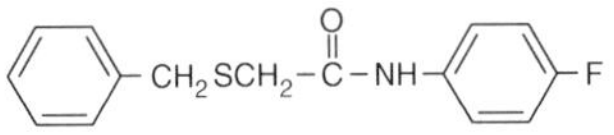

5 gm

S94,604-4
2-(BENZYLTHIO)-4'-CHLOROACETANILIDE

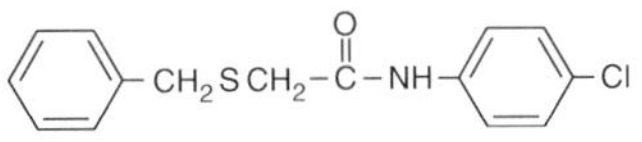

5 gm

S93,680-4
2-(BENZYLTHIO)-2',5'-DICHLOROACETANILIDE

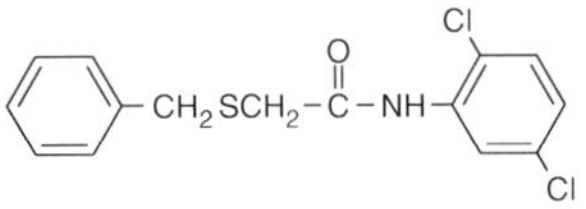

5 gm

S93,677-4
2-(BENZYLTHIO)-4'-BROMOACETANILIDE

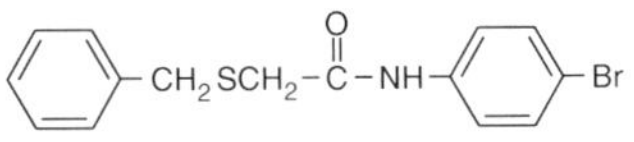

5 gm

S93,684-7
2-(BENZYLTHIO)-4'-IODOACETANILIDE

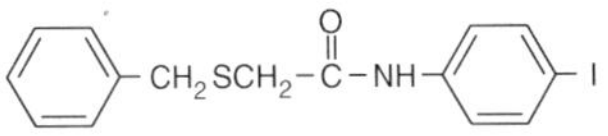

5 gm

S94,607-9
2-(BENZYLTHIO)-4'-BROMO-2'-CHLOROACETANILIDE

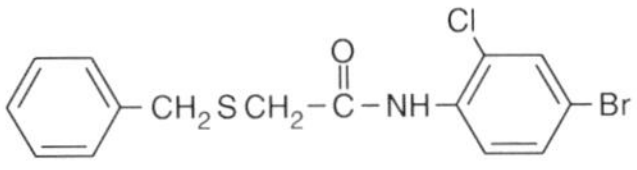

5 gm

S94,182-4
2-(BENZYLTHIO)-3'-CHLORO-4'-METHYLACETANILIDE

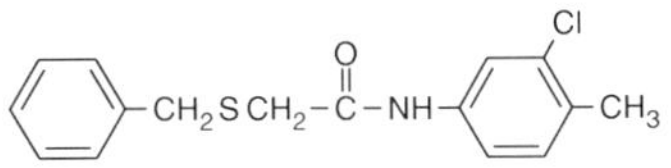

5 gm

S94,606-0
2-(BENZYLTHIO)-4'BROMO-3'-METHYLACETANILIDE

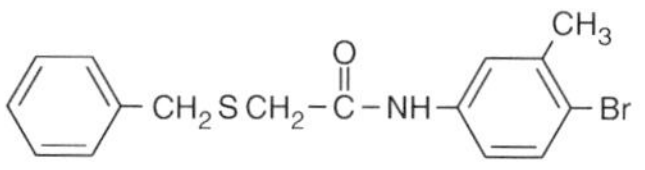

5 gm

S94,183-2
2-(BENZYLTHIO)-2',4',5'-TRICHLOROACETANILIDE

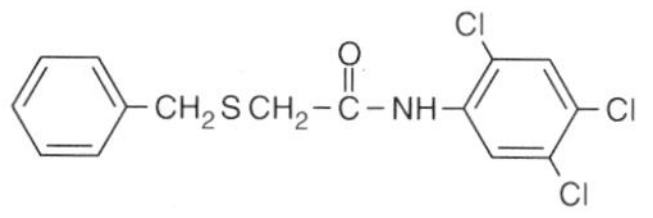

5 gm

S94,609-5
4'-CHLORO-2-(4-CHLOROPHENYLTHIO)-ACETANILIDE

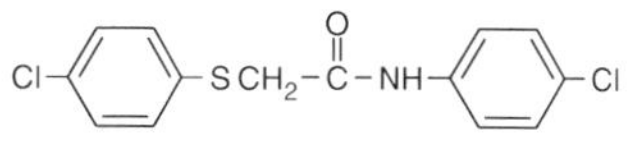

5 gm

S93,646-4
4'-BROMO-2-(4-CHLOROPHENYLTHIO)-ACETANILIDE

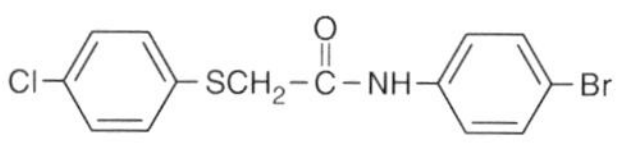

5 gm

S93,647-2
2-(4-CHLOROPHENYLTHIO)-4'-METHYLACETANILIDE

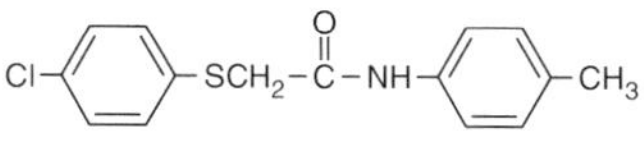

5 gm

S93,653-7
4'-BUTYL-2-(4-CHLOROPHENYLTHIO)-ACETANILIDE

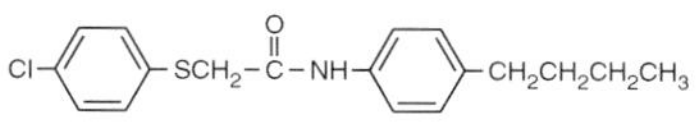

5 gm

S93,648-0
2-(4-CHLOROPHENYLTHIO)-2',3'-DIMETHYLACETANILIDE

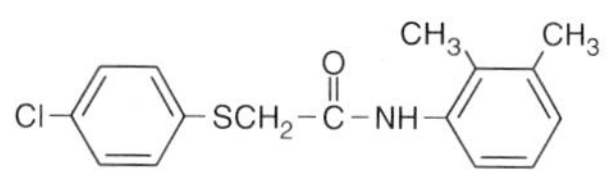

5 gm

S93,591-3
2-(4-CHLOROPHENYLTHIO)-2',5'-DICHLOROACETANILIDE

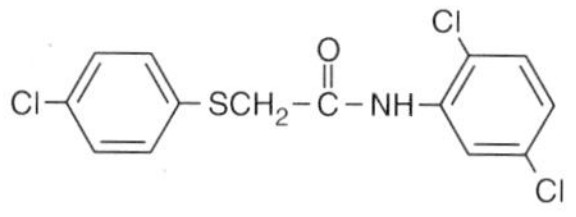

5 gm

S93,596-4
2-(4-CHLOROPHENYLTHIO)-3',4'-DICHLOROACETANILIDE

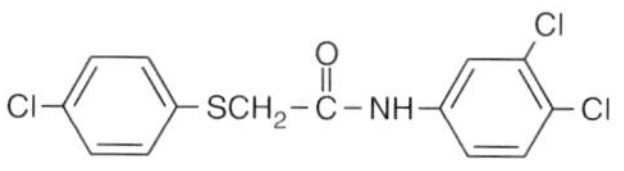

5 gm

S93,645-6
4'-BROMO-2'-CHLORO-2-(4-CHLOROPHENYLTHIO)ACETANILIDE

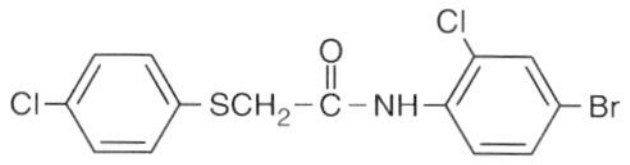

5 gm

S93,598-0
2-(4-CHLOROPHENYLTHIO)-4'-CHLORO-2'-METHYLACETANILIDE

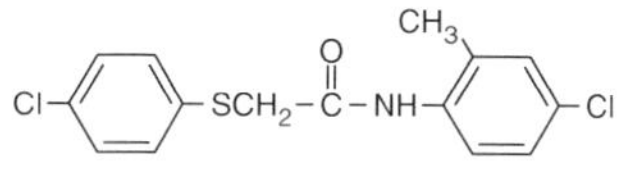

5 gm

S93,650-2
5'-CHLORO-2-(4-CHLOROPHENYLTHIO)-2'-METHYLACETANILIDE

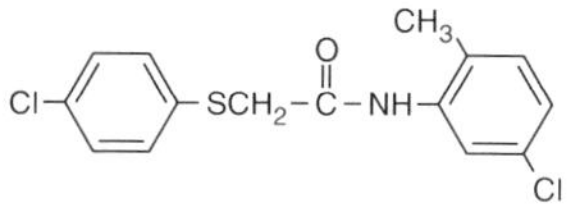

5 gm

S93,655-3
3'-CHLORO-2-(4-CHLOROPHENYLTHIO)-4'-METHYLACETANILIDE

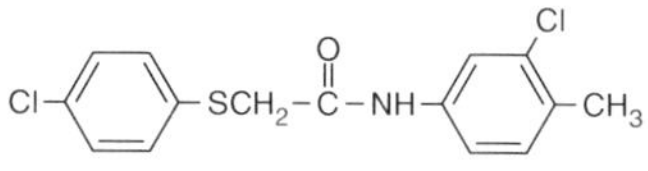

5 gm

S93,652-9
4'-BROMO-2-(4-CHLOROPHENYLTHIO)-3'-METHYLACETANILIDE

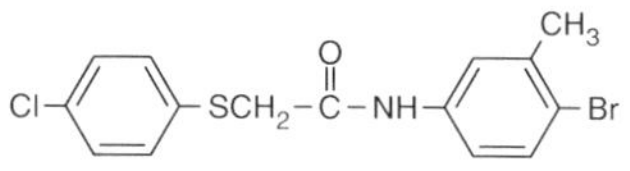

5 gm

S94,179-4
4'-BROMO-2-(4-CHLOROPHENYLTHIO)-2'-METHYLACETANILIDE

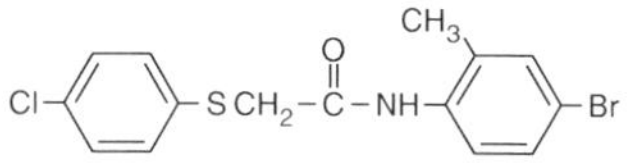

5 gm

S93,651-0
2-(4-CHLOROPHENYLTHIO)-3',4'-DIMETHYLACETANILIDE

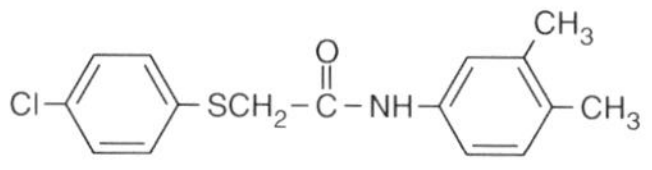

5 gm

S93,654-5
2-(4-CHLOROPHENYLTHIO)-2',4',5'-TRICHLOROACETANILIDE

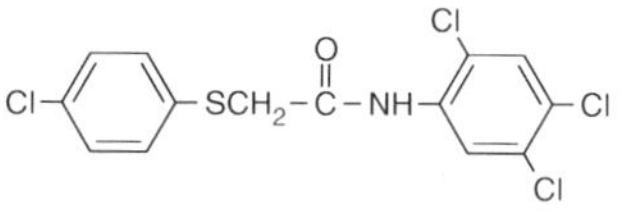

5 gm

S93,649-9
2-(4-CHLOROPHENYLTHIO)-2',4',6'-TRIMETHYLACETANILIDE

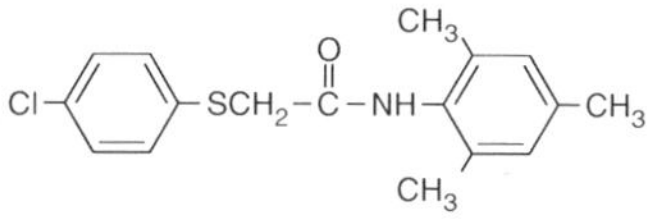

5 gm

S94,180-8
2-(BENZYLTHIO)-3',4'-DICHLOROACETANILIDE

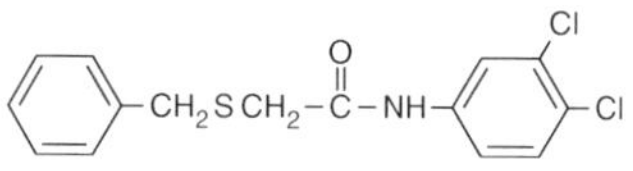

5 gm

S75,487-0
4'-AMINO-2'-CHLOROACETANILIDE

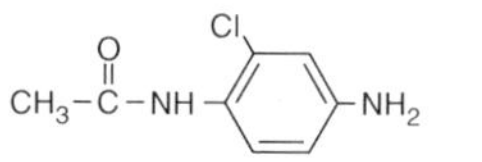

250 mg

S37,305-2
4'-AMINO-2',6'-DICHLOROACETANILIDE

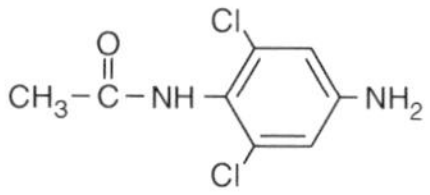

5 gm

S76,794-8
4'-AMINO-3',5'-DIFLUOROACETANILIDE

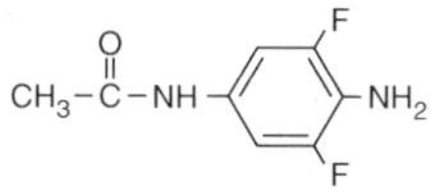

250 mg

S67,448-6
N,N'-BIS(4-CHLORO-3-NITROPHENYL)-OXAMIDE

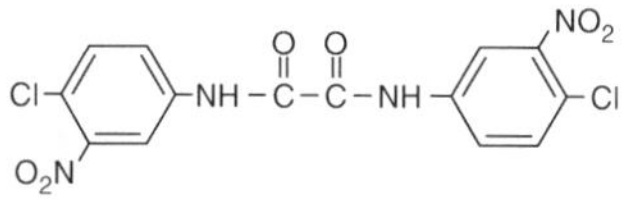

5 gm

S42,208-8
6'-CHLORO-1-PYRROLIDINEACETO-ORTHO-TOLUIDIDE SUCCINATE

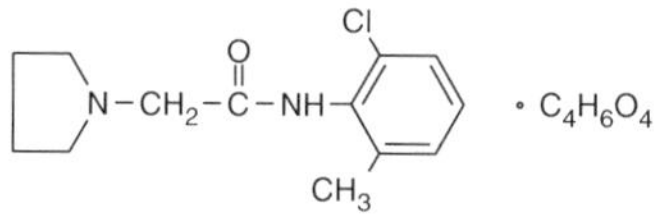

250 mg

S90,275-6
4'-CHLORO-2,2-DIMETHYL-2'-NITROPROPIONANILIDE

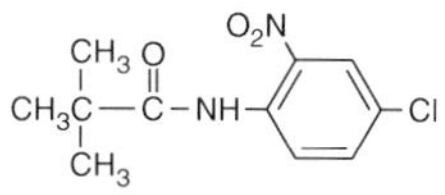

250 mg

S74,995-8
2',4'-DIFLUORO-6'-NITROACETANILIDE

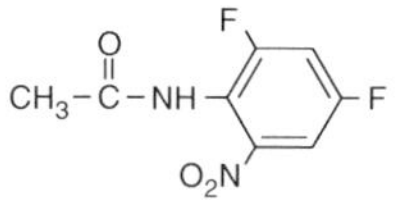

250 mg

S91,655-2
2',4'-DICHLORO-6'-NITROPROPIONANILIDE

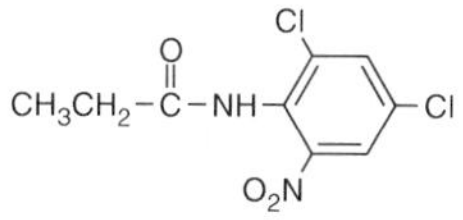

250 mg

S75,198-7
4'-CHLORO-6'-NITRO-M-ACETOTOLUIDIDE

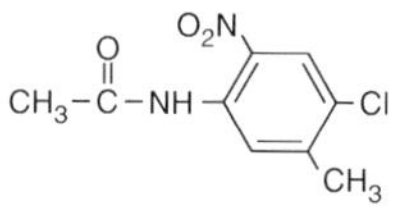

5 gm

S77,041-8
5'-CHLORO-4'-METHYL-2'-NITROACETANILIDE

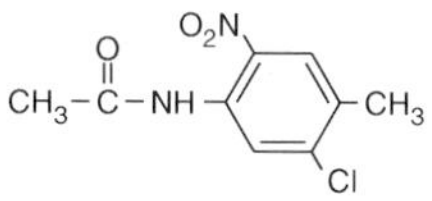

250 mg

S77,044-2
5'-BROMO-4'-ETHYL-2'-NITROACETANILIDE

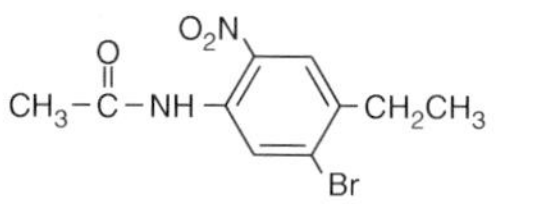

250 mg

S91,654-4
4'-CHLORO-N-ETHYL-2'-NITROPROPIONANILIDE

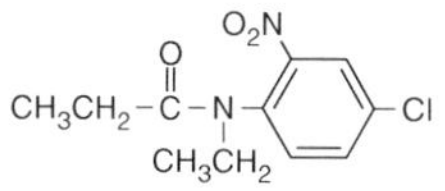

250 mg

S67,416-8
N,N'-BIS(4-FLUOROPHENYL)OXAMIDE

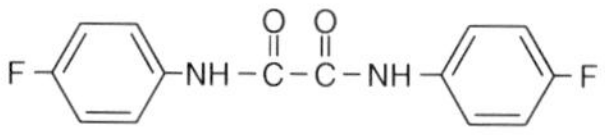

250 mg

S80,365-0
N,N'-(4-FLUORO-1,2-PHENYLENE)-BISACETAMIDE
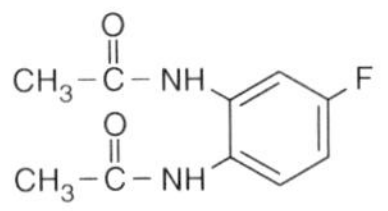

250 mg

S91,674-9
N,N'-(4,5-DICHLORO-1,2-PHENYLENE)-BISPROPIONAMIDE
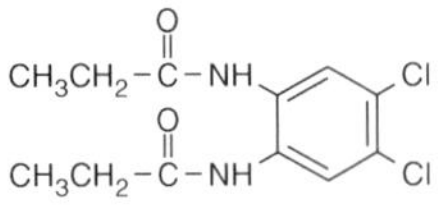
250 mg

S73,658-9
4'-FLUORO-3'-NITROHEXAHYDROPHTHALANILIC ACID
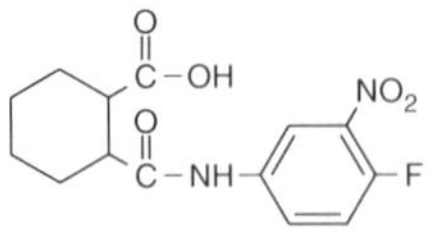
5 gm

S77,207-0
4'-FLUORO-3'-NITROGLUTARANILIC ACID
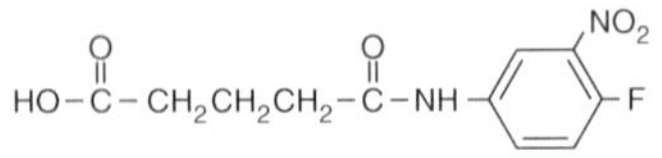
250 mg

S72,922-1
2'-CHLORO-5'-NITROGLUTARANILIC ACID
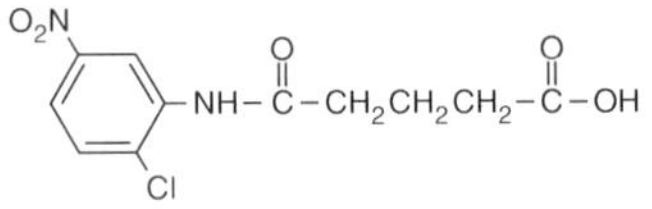
250 mg

S76,431-0
3-[N-(4-CHLORO-3-NITROPHENYL)-CARBAMOYL]-5-NORBORNENE-2-CARBOXYLIC ACID
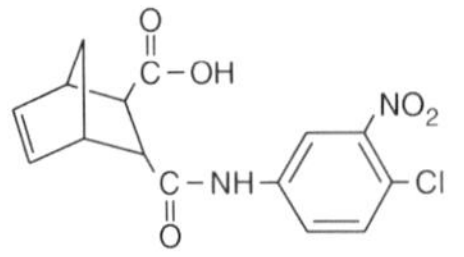
5 gm

S79,620-4
4'-FLUORO-3'-NITROCYCLOHEXANECARBOXANILIDE
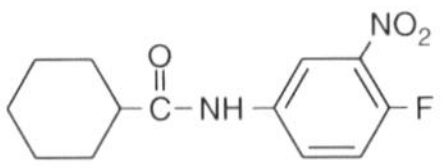
250 mg

S76,429-9
4'-CHLORO-3'-NITROFORMANILIDE
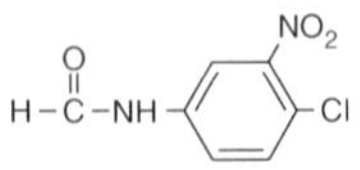
5 gm

S69,866-0
2'-CHLORO-5'-NITROPROPIONANILIDE
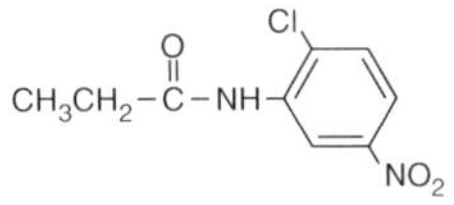
250 mg

S70,077-0
4'-CHLORO-3'-NITROPROPIONANILIDE
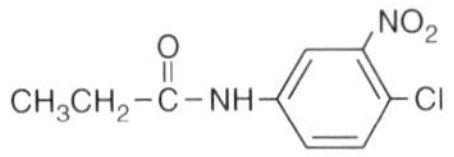
250 mg

S71,814-9
2'-CHLORO-5'-NITROBUTYRANILIDE
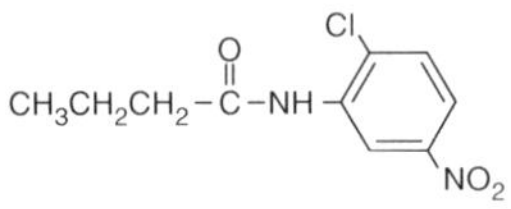
250 mg

S74,935-4
3'-IODO-5'-NITROACETANILIDE
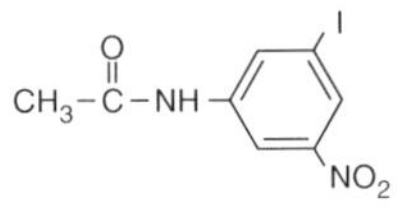
250 mg

S89,132-0
2'-BROMO-4'-NITRO-1-CYCLOPROPANECARBOXANILIDE
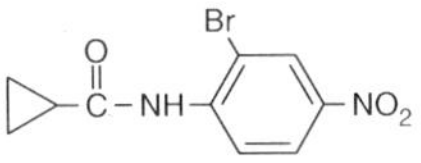
250 mg

S90,016-8
2'-CHLORO-4'-NITRO-2-PHENYLACETANILIDE
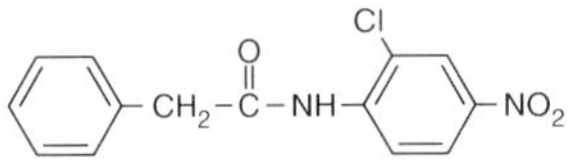
250 mg

S77,418-9
2'-FLUORO-4'-NITROACETANILIDE
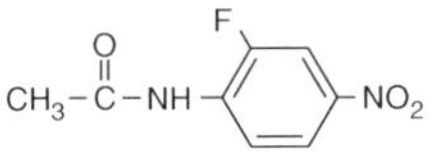
250 mg

S73,626-0
2'-CHLORO-4'-NITROPROPIONANILIDE
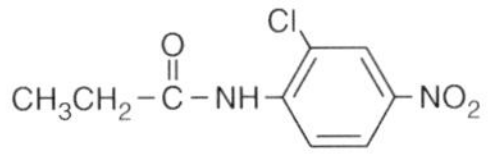
5 gm

S70,030-4
2'-CHLORO-4'-NITROBUTYRANILIDE
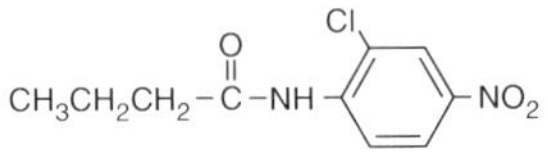
250 mg

S75,103-0
2'-BROMO-2,2-DIMETHYL-4'-NITROACETANILIDE
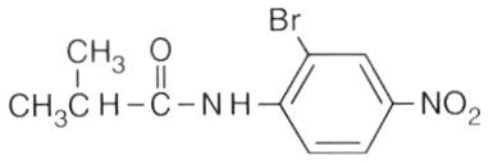
5 gm

S91,392-8
2'-BROMO-2,2-DIMETHYL-4'-NITROPROPIONANILIDE
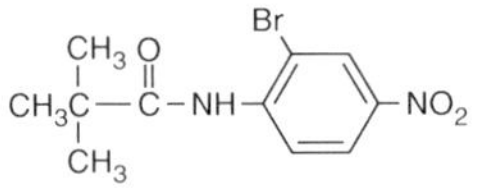
250 mg

S74,384-4
2'-CHLORO-4'-NITRODECANANILIDE
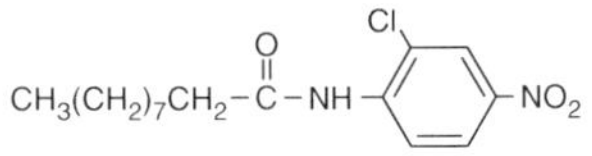
5 gm

S78,158-4
N-ACETYL-4-CHLORO-DL-PHENYLALANINE
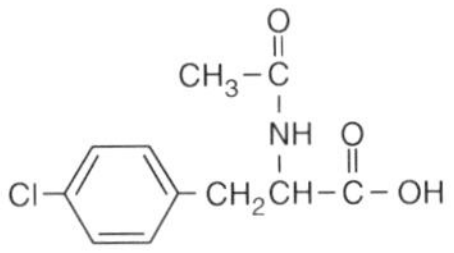
250 mg

S79.045-1
N-ACETYL-4-IODO-L-PHENYLALANINE
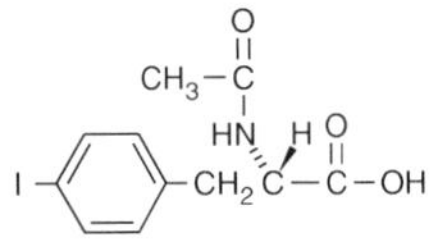
250 mg

S90,017-6
2',4'-DICHLORO-2-PHENYLACETANILIDE
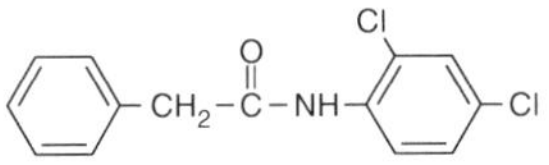
250 mg

S89,049-9
5'-CHLORO-2,2-DIPHENYL-2'-METHYLACETANILIDE
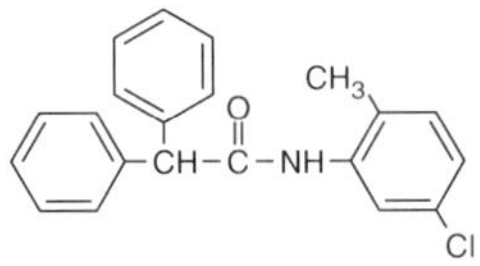
250 mg

S89,050-2
3'-CHLORO-2,2-DIPHENYL-4'-METHYLACETANILIDE
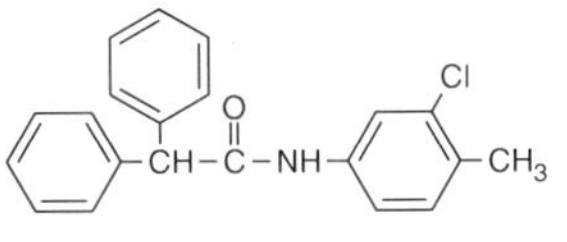
250 mg

S39,228-6
N-(PARA-CHLORO-ALPHA,ALPHA-DIMETHYLPHENETHYL)FORMAMIDE
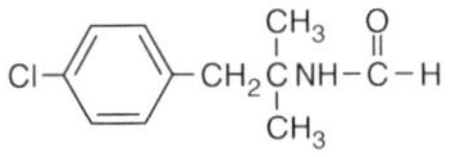
250 mg

S87,913-4
4'-BROMO-3'-METHYL-2-(TETRAHYDROFURFURYLAMINO)-ACETANILIDE
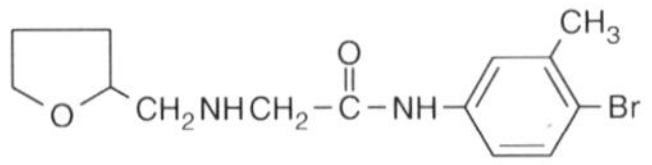
5 gm

S87,879-0
2-(BENZYLAMINO)-4'-BROMO-3'-METHYLACETANILIDE
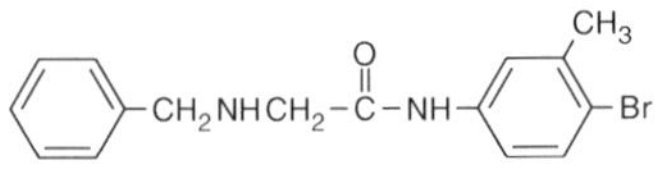
5 gm

S93,377-5
2-ANILINO-4'-IODOACETANILIDE
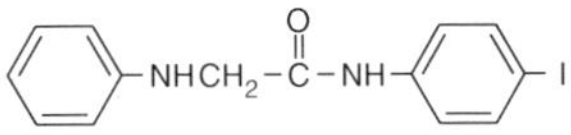
250 mg

S87,875-8
4'-BROMO-3'-METHYL-2-PIPERIDINOACETANILIDE
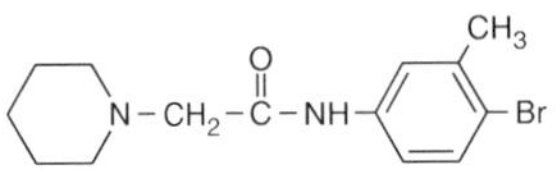
5 gm

S77,466-9
N-BENZYL-2'-CHLORO-2-METHYLPROPIONANILIDE
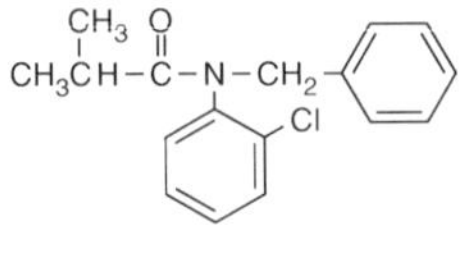
250 mg

S75,449-8
4'-CHLORO-4-OXO-4-(4-PHENYL-2-PYRAZOLIN-5-YL)BUTYRANILIDE
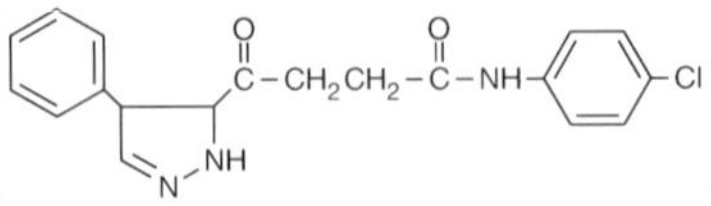
250 mg

S75,049-2
2',5'-DICHLORO-4-NITROBENZANILIDE

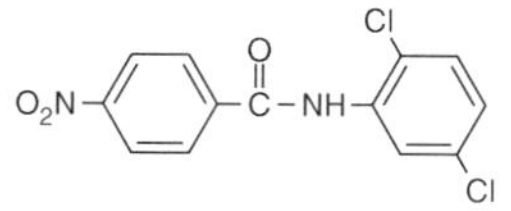

250 mg

S71,927-7
CIS-HEXAHYDRO-N-(4-CHLORO-3-NITROPHENYL)PHTHALAMIC ACID

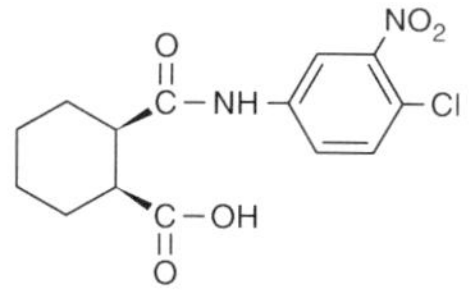

5 gm

S71,520-4
5'-CHLORO-2'-NITROACETANILIDE

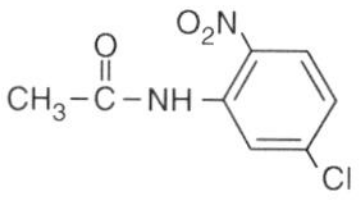

5 gm

S90,272-1
2,2-DIMETHYL-2'-FLUORO-5'-NITROPROPIONANILIDE

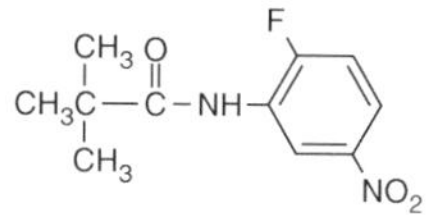

250 mg

S90,039-7
2'-BENZOYL-4'-CHLORO-2-PHENYLACETANILIDE

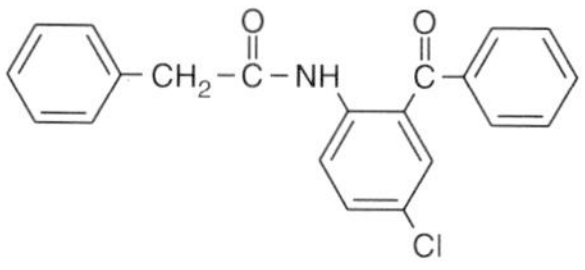

250 mg

S75,287-8
2'-BENZOYL-4'-CHLORO-3-PHENYLPROPIONANILIDE

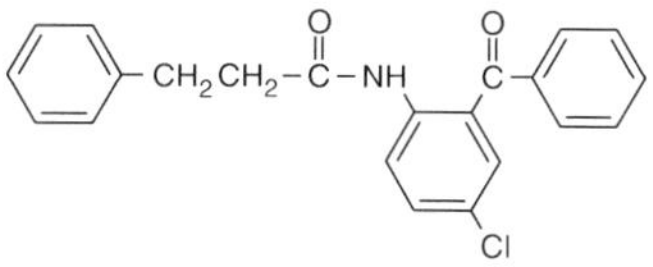

250 mg

S75,295-9
2'-BENZOYL-4'-CHLORO-2,2-DIPHENYLACETANILIDE

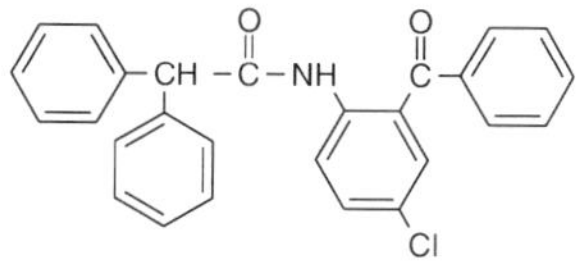

250 mg

S70,034-7
2'-BENZOYL-4'-CHLOROBUTYRANILIDE

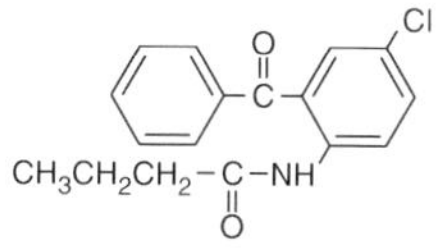

250 mg

S74,469-7
5'-ACETYL-2'-CHLOROACETANILIDE

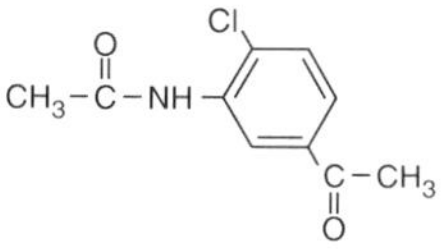

250 mg

S74,511-1
2'-ACETYL-4'-CHLOROACETANILIDE

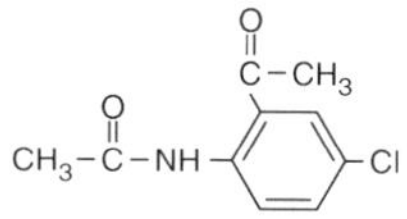

250 mg

S73,356-3
2-(4-CHLOROBENZYL)-1-OXO-1,2,3,4-TETRAHYDRO-2-NAPHTHALENEPROPIONAMIDE

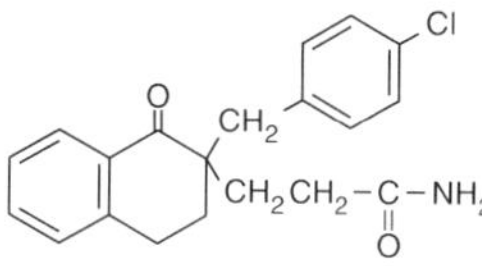

250 mg

S73,855-7
3'-CHLORO-2,3-DIBROMOSUCCINANILIC ACID

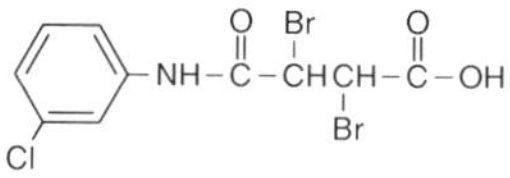

250 mg

S85,559-6
2,3-DIBROMO-4'-FLUOROSUCCINANILIC ACID

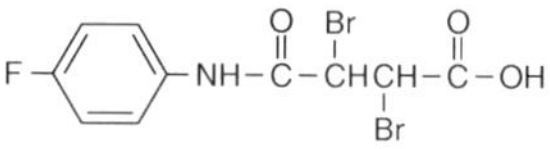

250 mg

S73,888-3
2,3,4'-TRIBROMOSUCCINANILIC ACID

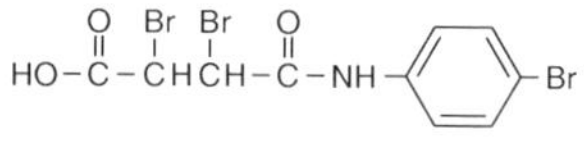

5 gm

S85,558-8
2,3-DIBROMO-3',4'-DICHLOROSUCCINANILIC ACID

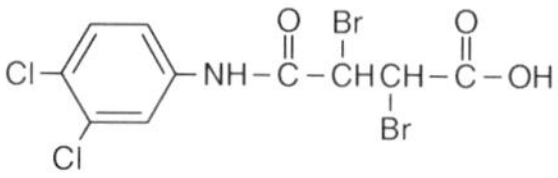

250 mg

S72,921-3
2,3-DIBROMO-4',5'-DICHLORO-2'-METHYLSUCCINANILIC ACID

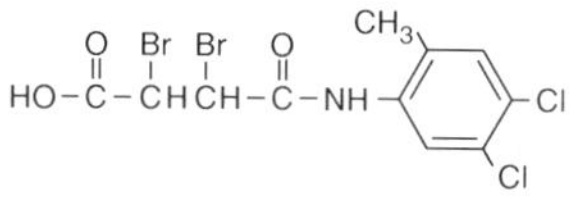

5 gm

S85,572-3
2,3-DIBROMO-2'-FLUORO-3-PHENYLPROPIONANILIDE

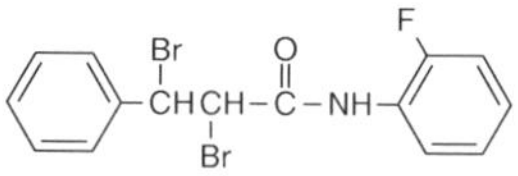

250 mg

S80,774-5
ALPHA,BETA-DIBROMO-4'-FLUOROHYDROCINNAMANILIDE

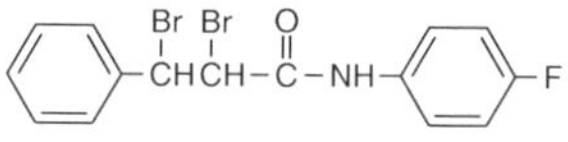

250 mg

S85,571-5
3-PHENYL-2,3,4'-TRIBROMOPROPIONANILIDE

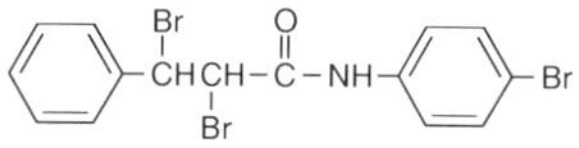

250 mg

S80,778-8
ALPHA,BETA-DIBROMO-3',4'-DICHLOROHYDROCINNAMANILIDE

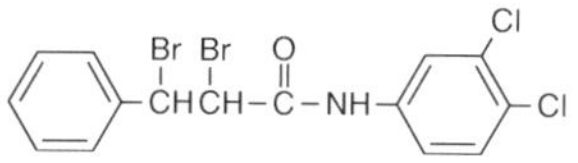

250 mg

S42,256-8
N,N'-ADIPOYLBIS(3-AMINO-2,4,6-TRIIODOBENZOIC ACID, SODIUM SALT)

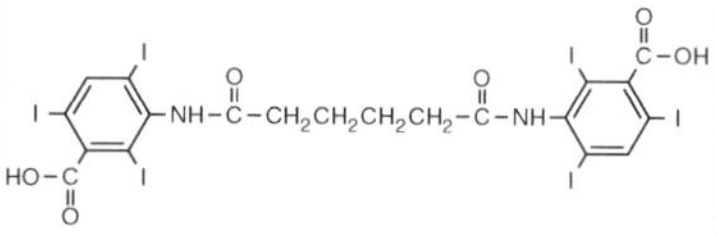

5 gm

S92,255-2
2'-CHLORO-6'-METHYLSUCCINANILIC ACID

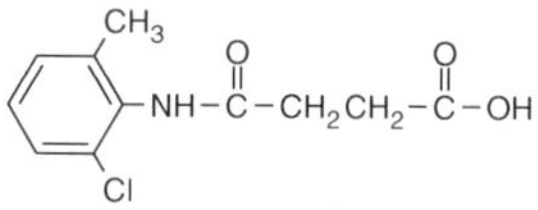

250 mg

S40,387-3
3-ACETAMIDO-2,4,6-TRIIODOBENZOIC ACID, SODIUM SALT DIHYDRATE

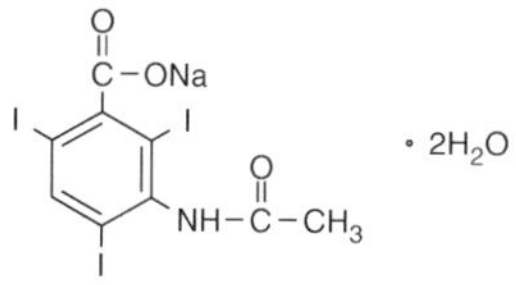

250 mg

S42,079-4
3-ACETAMIDO-2,4,6-TRIIODOBENZOIC ACID, BIS(2-HYDROXYETHYL)-AMMONIUM SALT

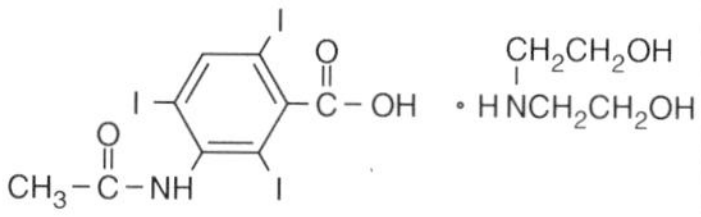

5 gm

S40,398-9
3-ACETAMIDO-2,4,6-TRIIODOBENZOIC ACID, COMPOUND WITH 1-DEOXY-1-(METHYL-AMINO)GLUCITOL

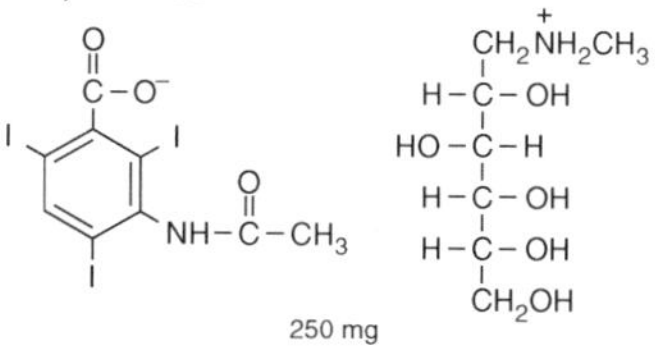

250 mg

S88,686-6
N-BUTYRYL-5-CHLOROANTHRANILIC ACID

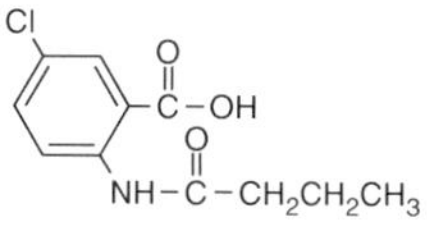

250 mg

S88,909-1
4-ACETAMIDO-3,5-DICHLOROBENZOIC ACID

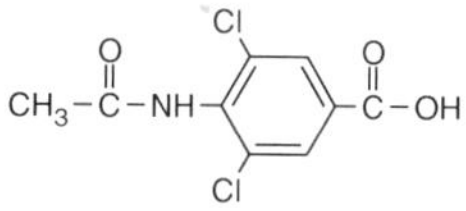

250 mg

S75,173-1
2-CHLORO-5'-ISOPROPYL-2'-METHYLBENZANILIDE

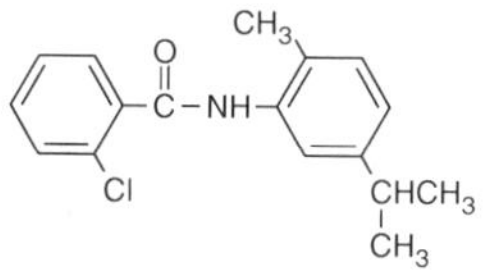

250 mg

S91,271-9
3'-FLUOROGLUTARANILIC ACID

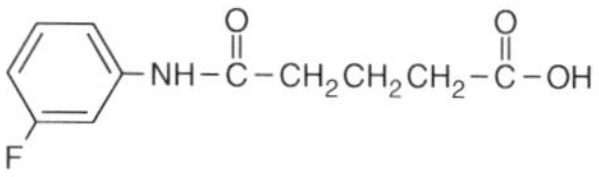

250 mg

S67,508-3
4-[N-(3-CHLOROPHENYL)CARBAMOYL]-BUTYRIC ACID

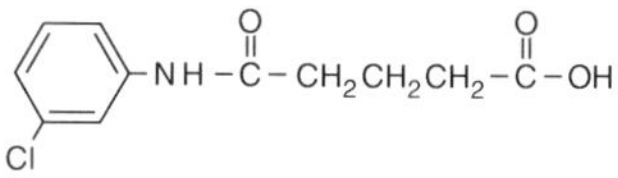

5 gm

S75,022-0
3'-BROMOGLUTARANILIC ACID

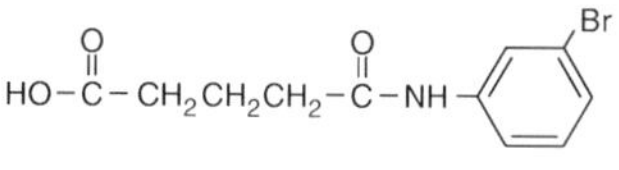

5 gm

S42,307-6
N,N'-ADIPOYLBIS(3-AMINO-ALPHA-ETHYL-2,4,6-TRIIODOHYDRO-CINNAMIC ACID)

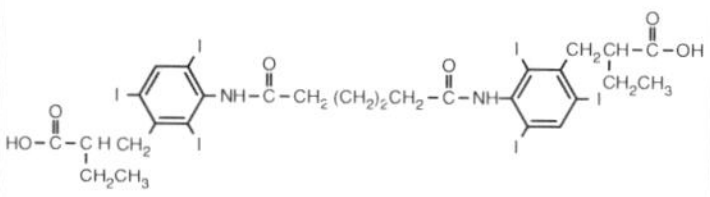

5 gm

S73,059-9
3-[N-(4-FLUOROPHENYL)CARBAMOYL]-5-NORBORNENE-2-CARBOXYLIC ACID

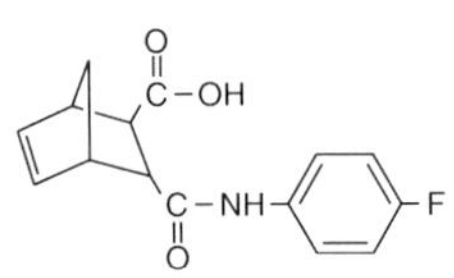

5 gm

S72,023-2
3-[N-(4-CHLOROPHENYL)CARBAMOYL]-5-NORBORNENE-2-CARBOXYLIC ACID

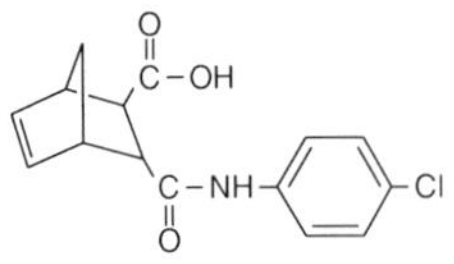

5 gm

S76,863-4
3-[N-(2,4-DIFLUOROPHENYL)CARBAMOYL]-5-NORBORNENE-2-CARBOXYLIC ACID

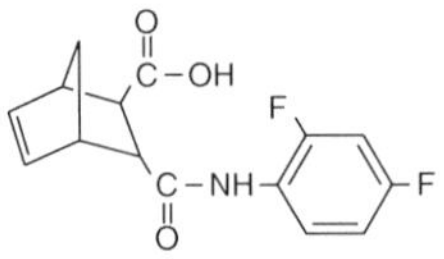

250 mg

S72,022-4
3-[N-(3,4-DICHLOROPHENYL)CARBAMOYL]-5-NORBORNENE-2-CARBOXYLIC ACID

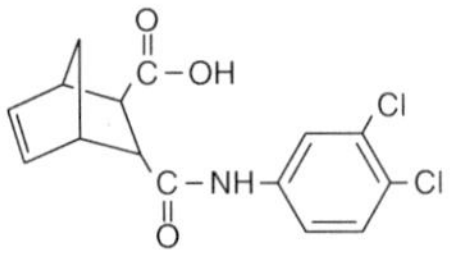

5 gm

S71,987-0
3-[N-(3-CHLORO-4-METHYLPHENYL)-CARBAMOYL]-1,4,5,6,7,7-HEXACHLORO-5-NORBORNENE-2-CARBOXYLIC ACID

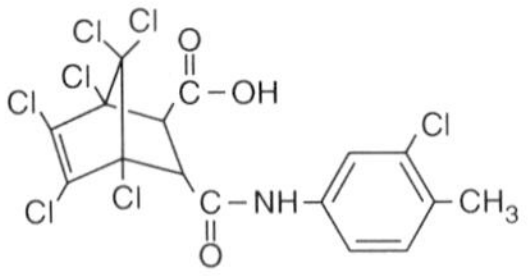

5 gm

S91,275-1
3'-FLUOROHEXAHYDROPHTHALANILIC ACID

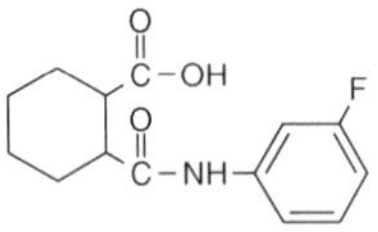

250 mg

S70,287-0
CIS-N-(3-CHLOROPHENYL)-HEXAHYDROPHTHALAMIC ACID

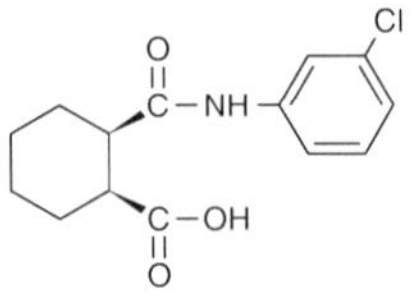

5 gm

S72,055-0
CIS-N-(2-FLUOROPHENYL)-HEXAHYDROPHTHALAMIC ACID

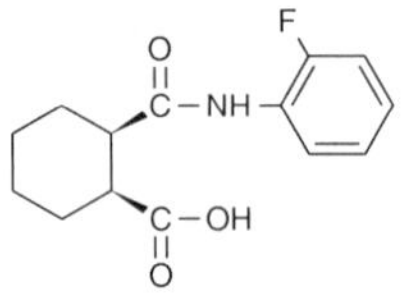

5 gm

S73,071-8
CIS-N-(2-CHLOROPHENYL)-HEXAHYDROPHTHALAMIC ACID

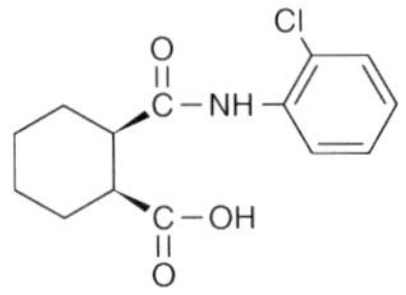

5 gm

S70,302-8
CIS-N-(4-FLUOROPHENYL)-HEXAHYDROPHTHALAMIC ACID

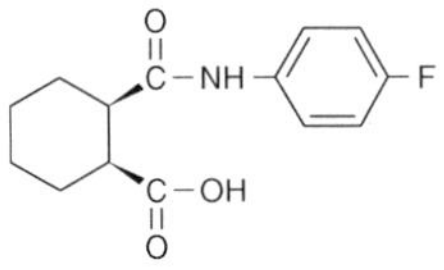

5 gm

S72,019-4
CIS-HEXAHYDRO-N-(4-CHLOROPHENYL)-PHTHALAMIC ACID

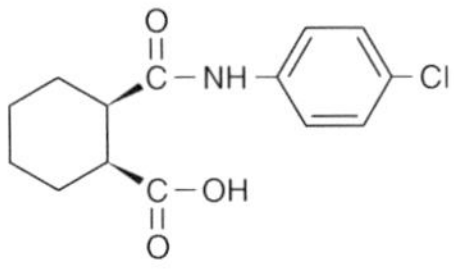

5 gm

S70,440-7
4'-BROMOHEXAHYDROPHTHALANILIC ACID

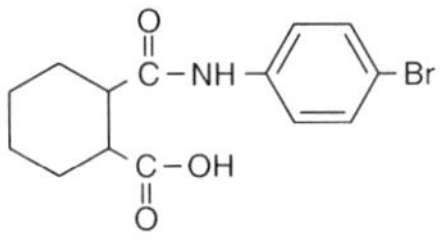

5 gm

S72,010-0
CIS-HEXAHYDRO-N-(4-IODOPHENYL)-PHTHALAMIC ACID

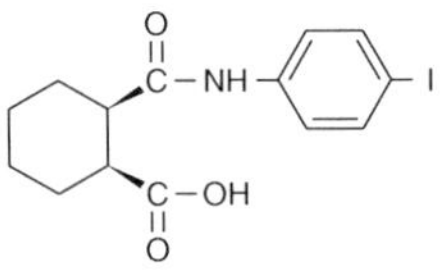

5 gm

S70,369-9
CIS-N-(2,3-DICHLOROPHENYL)-HEXAHYDROPHTHALAMIC ACID

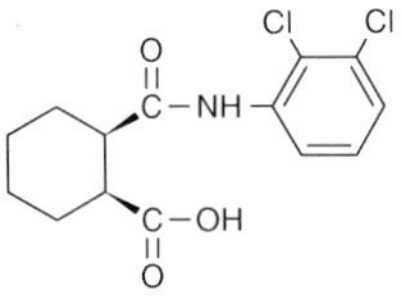

5 gm

S70,307-9
CIS-N-(3-CHLORO-2-METHYLPHENYL)-HEXAHYDROPHTHALAMIC ACID

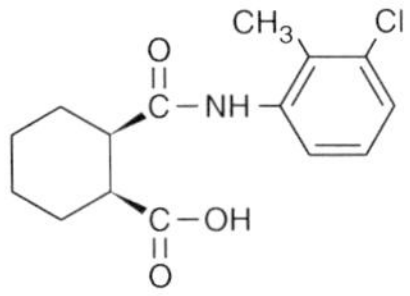

5 gm

S71,644-8
CIS-N-(2,5-DICHLOROPHENYL)-HEXAHYDROPHTHALAMIC ACID

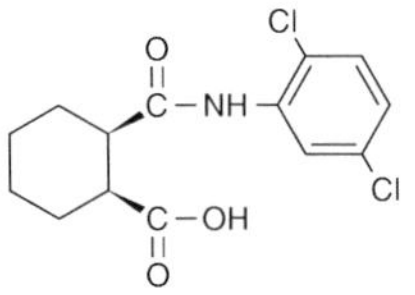

50 mg

S71,928-5
CIS-HEXAHYDRO-N-(2,4-DICHLOROPHENYL)PHTHALAMIC ACID

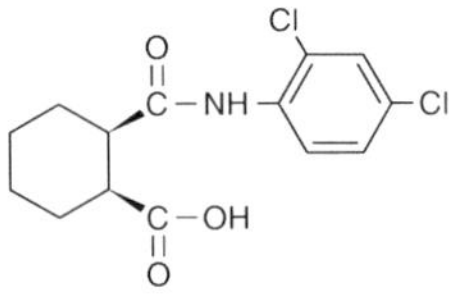

5 gm

S72,020-8
CIS-HEXAHYDRO-N-(3-CHLORO-4-METHYLPHENYL)PHTHALAMIC ACID

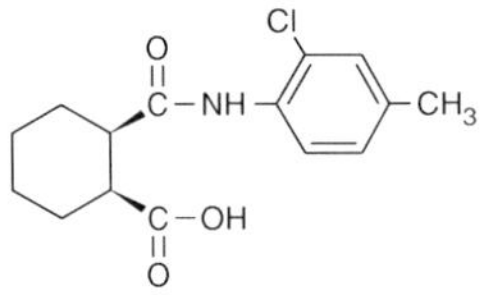

5 gm

S67,637-3
2-(3,4-DICHLOROPHENYLCARBAMOYL)-CYCLOHEXANECARBOXYLIC ACID

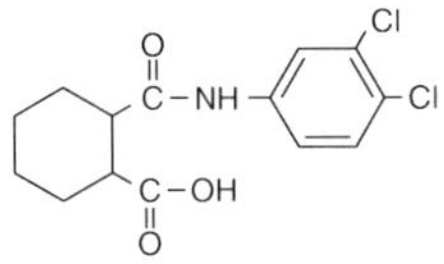

5 gm

S71,931-5
N-(3-CHLOROPHENYL)-1,2,5,6-TETRAHYDROPHTHALAMIC ACID

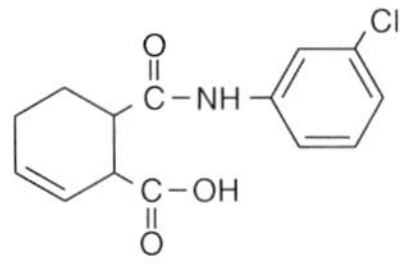

5 gm

S93,112-8
4'-FLUORO-1,2,3,6-TETRAHYROPHTHALANILIC ACID

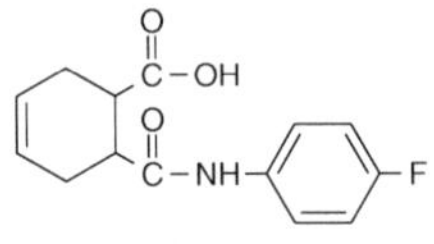

250 mg

S93,113-6
4'-BROMO-1,2,3,6-TETRAHYDROPHTHALANILIC ACID

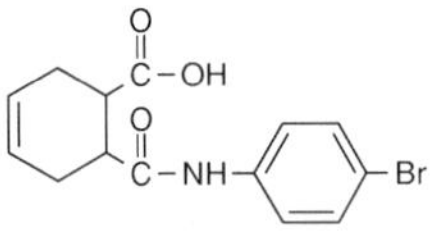

250 mg

S74,633-9
2',4'-DICHLORO-1,2,3,6-TETRAHYDROPHTHALANILIC ACID

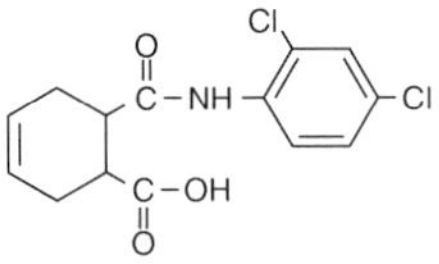

5 gm

S92,180-7
3',4'-DICHLORO-1,2,3,6-TETRAHYDROPHTHALANILIC ACID

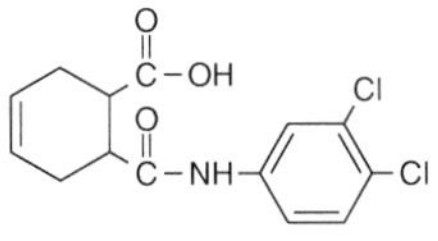

5 gm

S90,000-1
N-(2,4-DICHLOROBENZYL)-1,2,3,6-TETRAHYDROPHTHALAMIC ACID

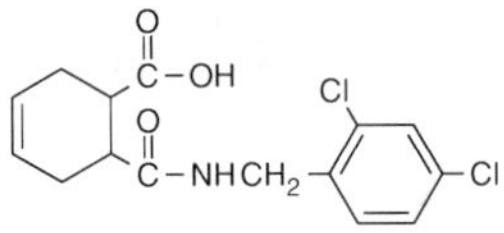

250 mg

S70,373-7
2-FLUOROSUCCINANILIC ACID

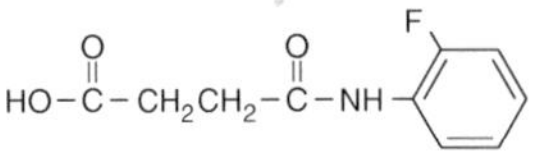

5 gm

S72,068-2
N-(2-FLUOROPHENYL)GLUTARAMIC ACID

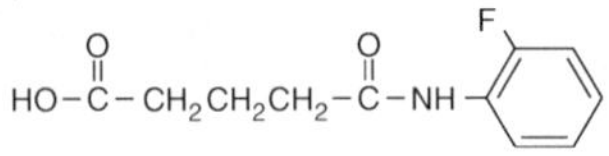

5 gm

S73,033-5
N-(2-CHLOROPHENYL)GLUTARAMIC ACID

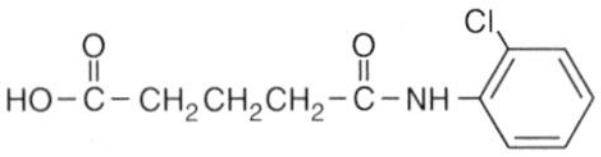

5 gm

S70,281-1
4'-FLUOROSUCCINANILIC ACID

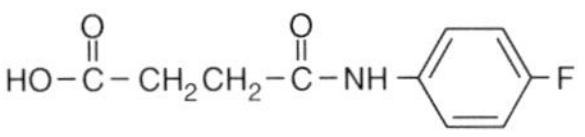

5 gm

S70,109-2
4'-CHLOROSUCCINANILIC ACID

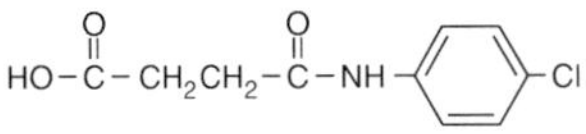

5 gm

S70,443-1
4'-BROMOSUCCINANILIC ACID

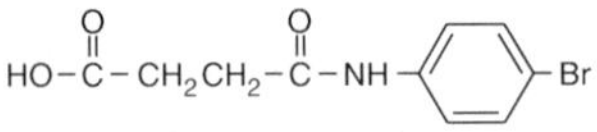

5 gm

S71,942-0
4'-IODOSUCCINANILIC ACID

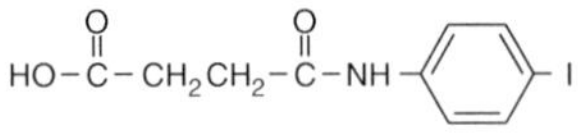

5 gm

S70,385-0
4'-FLUOROGLUTARANILIC ACID

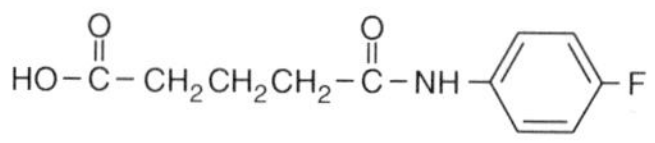

5 gm

S73,039-4
N-(4-CHLOROPHENYL)GLUTARAMIC ACID

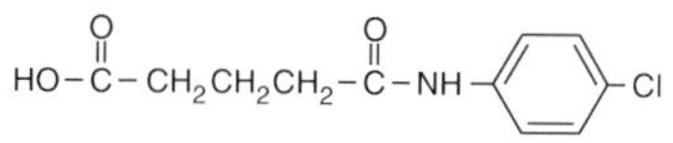

5 gm

S73,034-3
N-(4-IODOPHENYL)GLUTARAMIC ACID

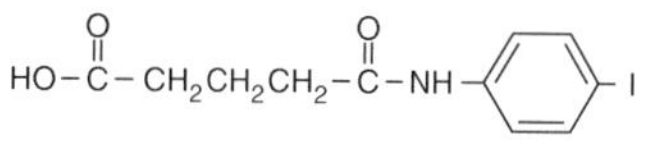

5 gm

S70,374-5
2',3'-DICHLOROSUCCINANILIC ACID

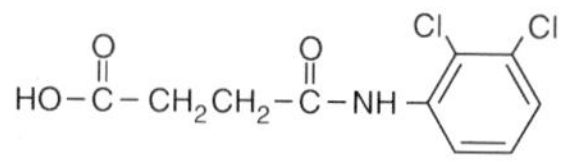

5 gm

S74,629-0
2',3'-DICHLOROGLUTARANILIC ACID

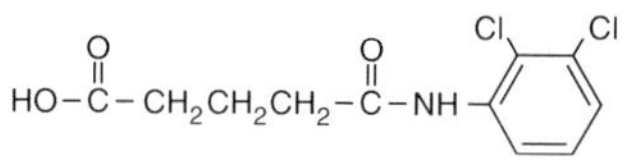

5 gm

S87,804-9
N-(3-CHLORO-2-METHYLPHENYL)-GLUTARAMIC ACID

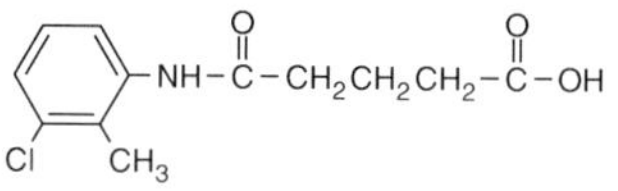

5 gm

S70,362-1
2',5'-DICHLOROSUCCINANILIC ACID

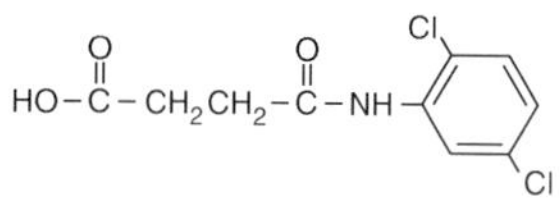

5 gm

S70,375-3
3',4'-DICHLOROSUCCINANILIC ACID

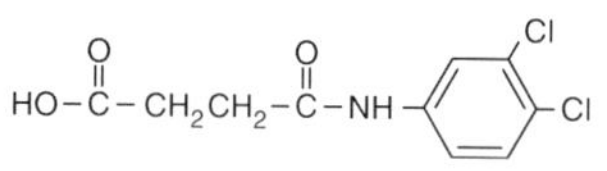

5 gm

S70,376-1
2',4'-DICHLOROSUCCINANILIC ACID

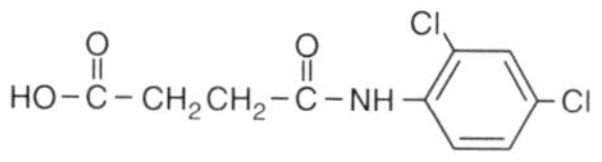

5 gm

S74,628-2
2',4'-DIFLUOROGLUTARANILIC ACID

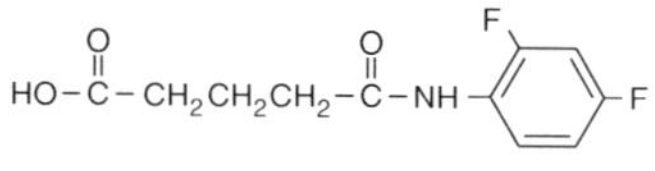

5 gm

S70,386-9
2',4'-DICHLOROGLUTARANILIC ACID

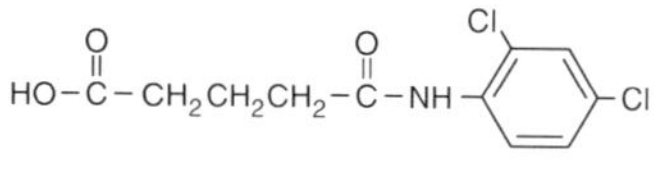

5 gm

S70,387-7
3',4'-DICHLOROGLUTARANILIC ACID

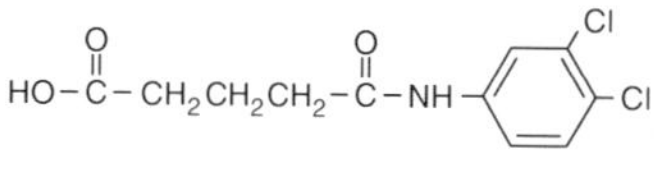

5 gm

S70,388-5
2',5'-DICHLOROGLUTARANILIC ACID

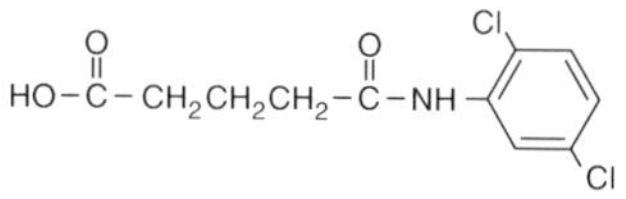

5 gm

S74,651-7
4'-BROMO-2'-METHYLSUCCINANILIC ACID

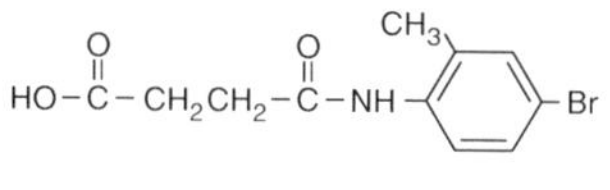

5 gm

S72,919-1
3'-CHLORO-2,3-DIBROMO-4'-METHYLSUCCINANILIC ACID

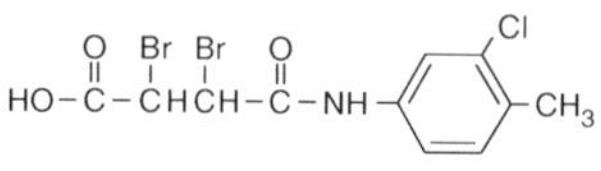

5 gm

S73,041-6
N-(3-CHLORO-4-METHYLPHENYL)-GLUTARAMIC ACID

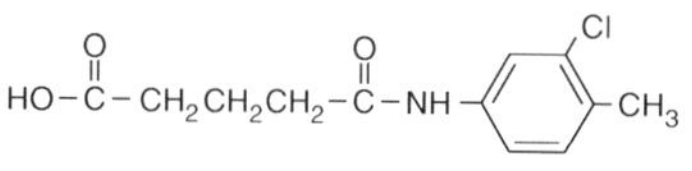

5 gm

S74,650-9
4'-BROMO-2'-METHYLGLUTARANILIC ACID

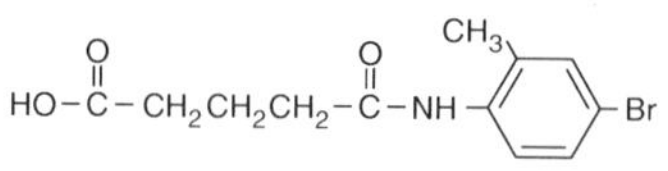

5 gm

S74,634-7
4'-BROMO-2'-CHLORO-1,2,3,6-TETRAHYDROPHTHALANILIC ACID

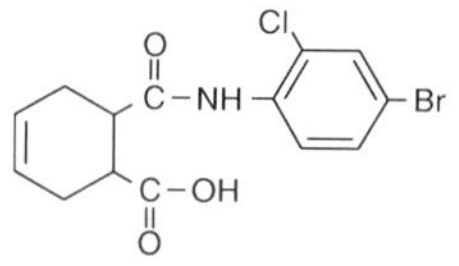

5 gm

S73,885-9
4'-BROMO-2',6'-DIMETHYLSUCCINANILIC ACID

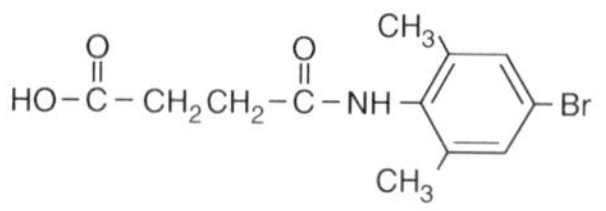

5 gm

S70,378-8
2',4',5'-TRICHLOROSUCCINANILIC ACID

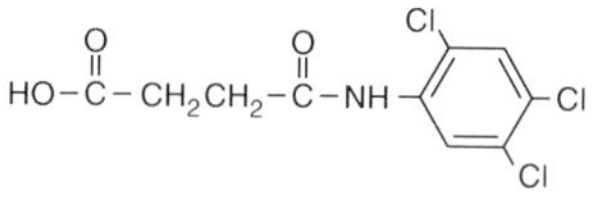

5 gm

S91,398-7
2',4',5'-TRICHLOROGLUTARANILIC ACID

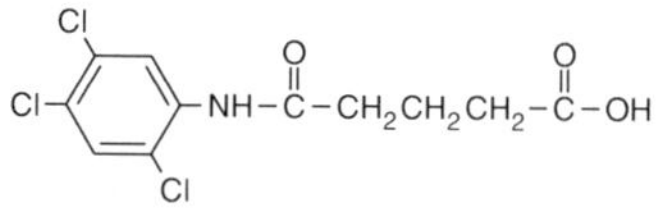

250 mg

S90,546-1
3-[N-(4-FLUOROPHENYL)-N-METHYL-CARBAMOYL]-1,4,5,6,7,7-HEXACHLORO-5-NORBORNENE-2-CARBOXYLIC ACID

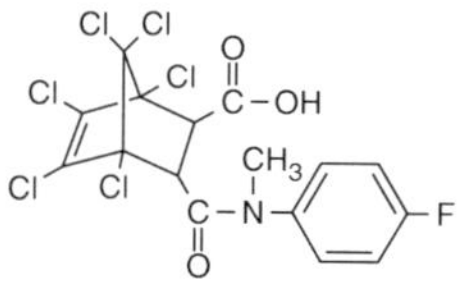

250 mg

S74,394-1
2-[N-(3-CHLOROBENZYL)-N-METHYLCARBAMOYL)-CYCLOHEXANECARBOXYLIC ACID

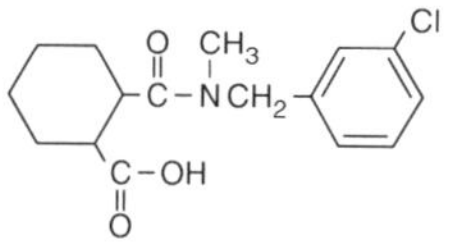

5 gm

S92,298-6
4'-BROMO-3'-METHYLHEXAHYDROPHTHALANILIC ACID

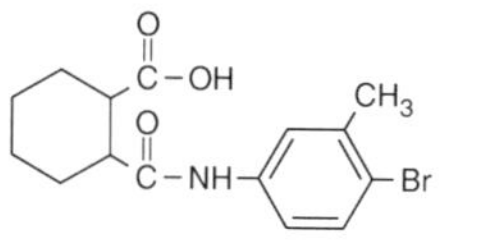

250 mg

S92,283-8
4'-IODO-1,2,3,6-TETRAHYDROPHTHALANILIC ACID

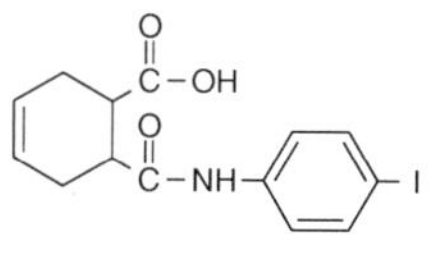

250 mg

S93,109-8
3'-CHLORO-2'-METHYL-1,2,3,6-TETRAHYDROPHTHALANILIC ACID

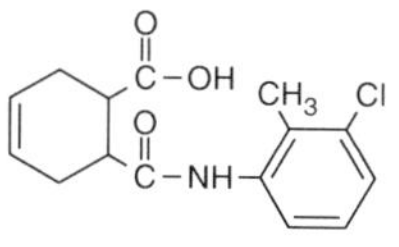

250 mg

S41,117-5
3-(PENTAFLUOROPHENYL)PROPIONAMIDE

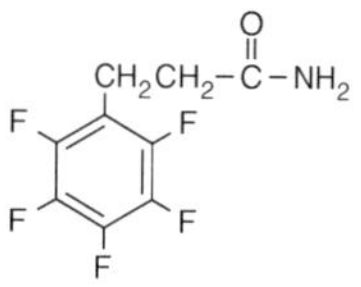

5 gm

S75,697-0
2',3',4',5'-TETRAFLUOROACETANILIDE

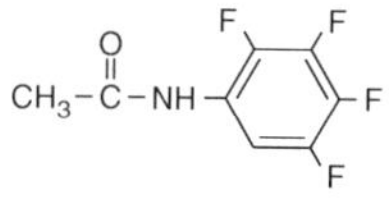

5 gm

S69,975-6
3'-FLUOROPROPIONANILIDE

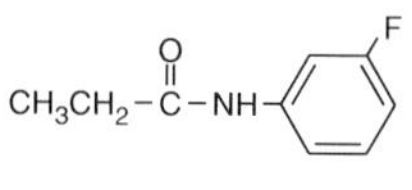

250 mg

S70,029-0
3'-BROMOPROPIONANILIDE

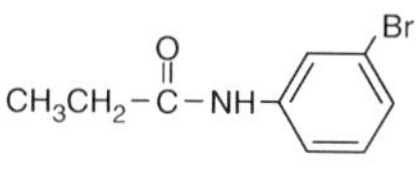

250 mg

S73,238-9
N-(3-FLUOROPHENYL)BUTYRAMIDE

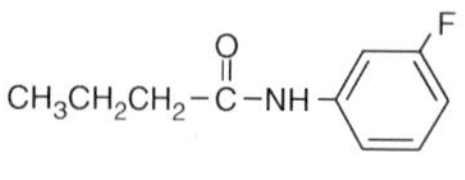

250 mg

S71,597-2
3'-BROMO-2,2-DIMETHYLPROPIONANILIDE

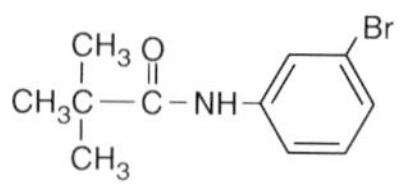

5 gm

S87,909-6
4'-BROMO-3'-METHYL-2-MORPHOLINOACETANILIDE

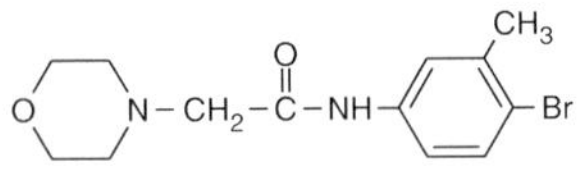

5 gm

S72,175-1
3'-BROMOCYCLOPROPANECARBOXANILIDE

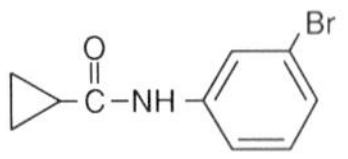

250 mg

S72,786-5
N-(CYCLOPROPYLCARBONYL)-2,4,5-TRICHLOROANILINE

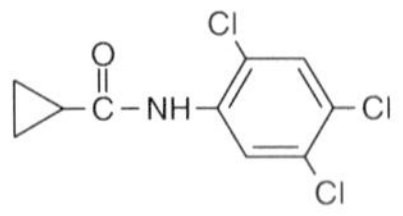

5 gm

S89,146-0
2'-CHLORO-1-CYCLOPROPANECARBOXANILIDE

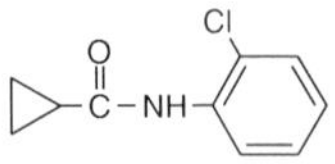

250 mg

S84,403-9
4'-FLUOROCYCLOPROPANECARBOXANILIDE

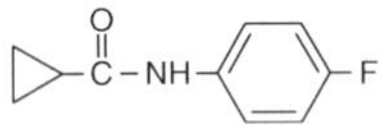

250 mg

S89,145-2
4'-CHLORO-1-CYCLOPROPANECARBOXANILIDE

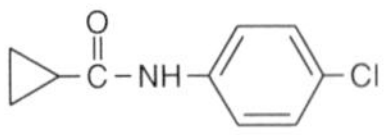

250 mg

S75,225-8
4'-BROMOCYCLOPROPANECARBOXANILIDE

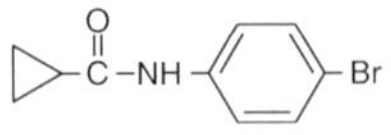

5 gm

S72,652-4
N-(2,3-DICHLOROPHENYL)-CYCLOPROPANECARBOXAMIDE

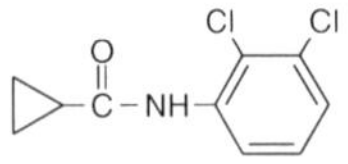

250 mg

S72,642-7
N-(3-CHLORO-2-METHYLPHENYL)-CYCLOPROPANECARBOXAMIDE

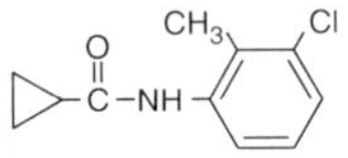

250 mg

S89,134-7
2'-CHLORO-6'-METHYL-1-CYCLOPROPANECARBOXANILIDE

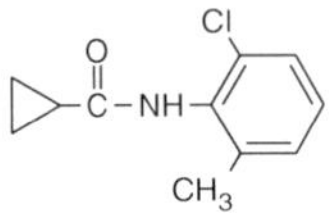

250 mg

S89,144-4
2',5'-DICHLORO-1-CYCLOPROPANECARBOXANILIDE

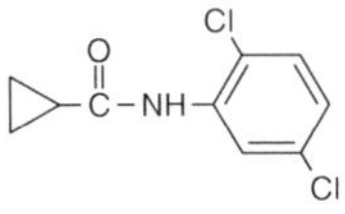

250 mg

S89,142-8
3'-CHLORO-4'-METHYL-1-CYCLOPROPANECARBOXANILIDE

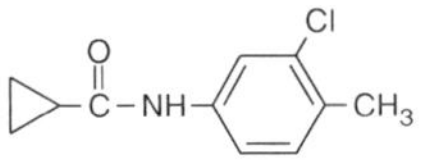

250 mg

S89,143-6
5'-CHLORO-2'-METHYL-1-CYCLOPROPANECARBOXANILIDE

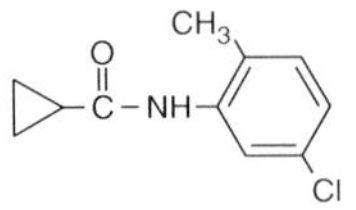

250 mg

S72,618-4
N-(4,5-DICHLORO-2-METHYLPHENYL)-CYCLOPROPANECARBOXAMIDE

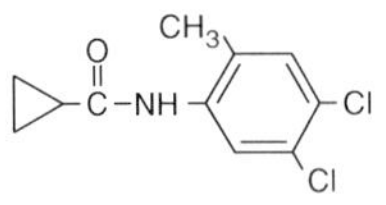

5 gm

S90,500-3
4'-IODO-1-CYCLOHEXANECARBOXANILIDE

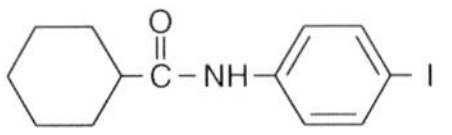

250 mg

S90,042-7
3'-CHLORO-2-PHENYLACETANILIDE

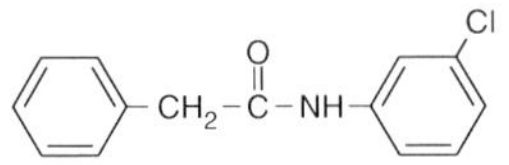

250 mg

S78,062-6
2,2-DIPHENYL-2'-FLUOROACETANILIDE

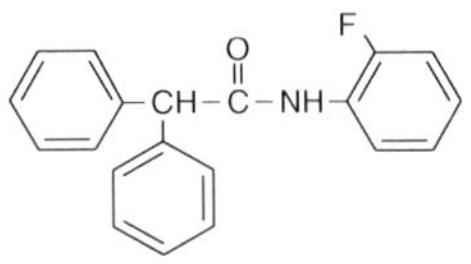

250 mg

S72,082-8
4'-FLUORO-3-PHENYLPROPIONANILIDE

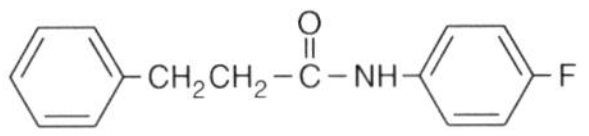

5 gm

S90,048-6
N-(4-CHLOROBENZYL)-2-PHENYLACETAMIDE

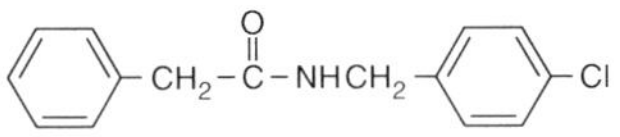

250 mg

S69,849-0
4'-BROMO-2,2-DIPHENYLACETANILIDE

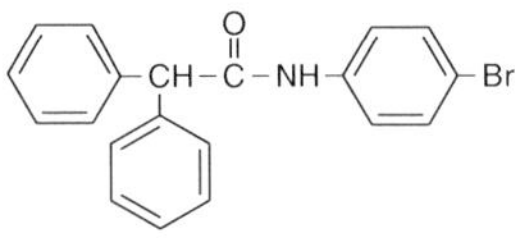

250 mg

S77,365-4
2,2-DIPHENYL-4'-FLUOROACETANILIDE

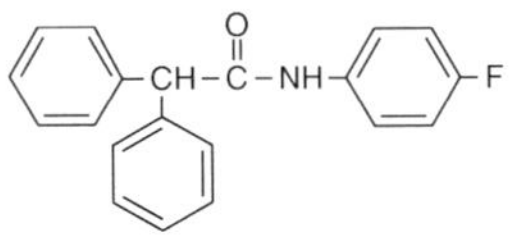

250 mg

S73,391-1
4'-CHLORO-2,2-DIPHENYLACETANILIDE

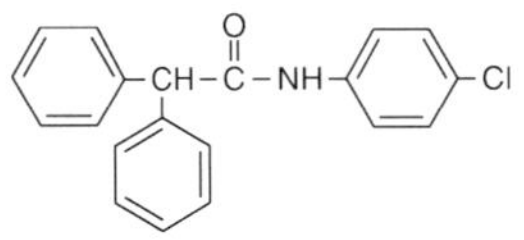

250 mg

S76,976-2
2',3'-DICHLORO-2,2-DIPHENYLACETANILIDE

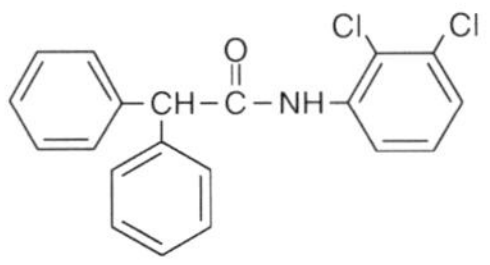

250 mg

S78,066-9
3'-CHLORO-2,2-DIPHENYL-2'-METHYLACETANILIDE

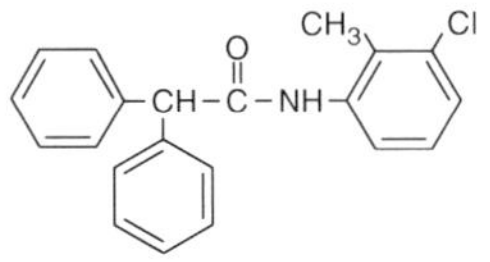

250 mg

S72,682-6
5'-CHLORO-2'-METHYL-3-PHENYLPROPIONANILIDE

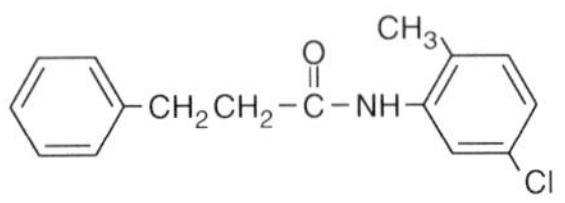

5 gm

S70,261-7
3',4'-DICHLORO-2,2-DIPHENYLACETANILIDE

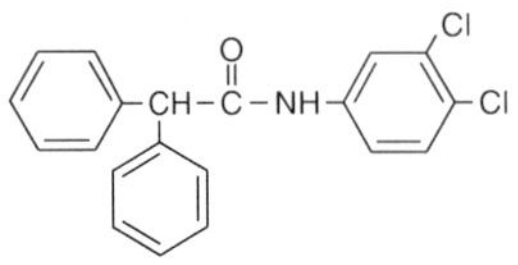

250 mg

S77,367-0
4'-CHLORO-2,2-DIPHENYL-2'-METHYLACETANILIDE

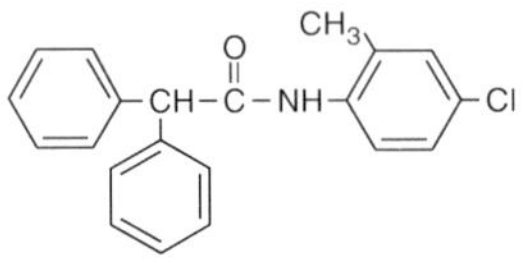

250 mg

S77,509-6
2'-CHLOROFORMANILIDE

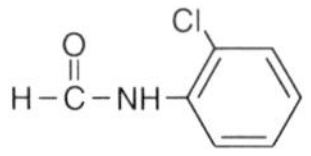

250 mg

S91,550-5
2'-BROMOACETANILIDE

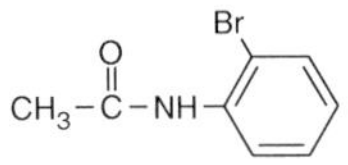

250 mg

S87,760-3
2'-IODOACETANILIDE

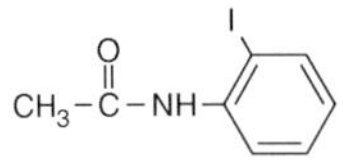

250 mg

S39,763-6
2'-FLUOROOCTANANILIDE

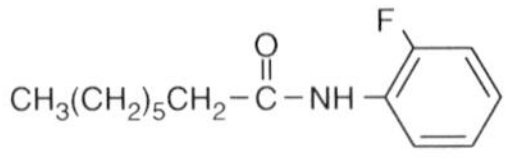

250 mg

S87,297-0
4'-IODOACETANILIDE

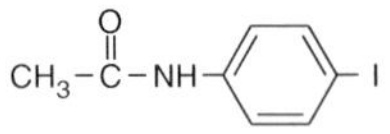

250 mg

S79,095-8
4'-CHLOROPROPIONANILIDE

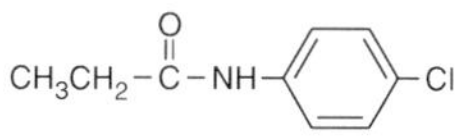

250 mg

S69,511-4
N-(4-BROMOPHENYL)PROPIONAMIDE

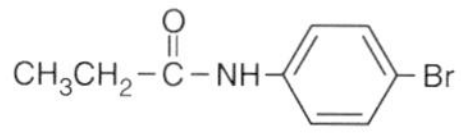

250 mg

S91,656-0
4'-IODOPROPIONANILIDE

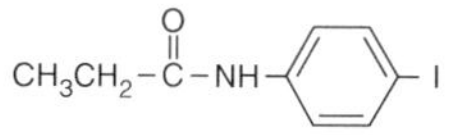

250 mg

S90,088-5
2,2-DIMETHYL-4'-FLUOROPROPIONANILIDE

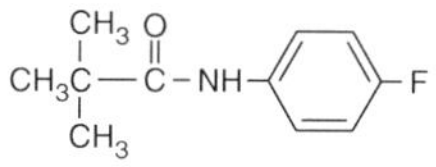

250 mg

S90,264-0
4'-CHLORO-2,2-DIMETHYLPROPIONANILIDE

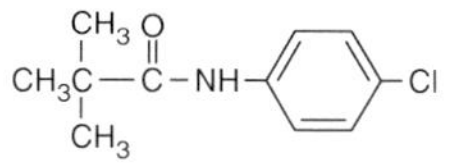

250 mg

S67,478-8
N-(4-CHLOROPHENYL)HEXADECANAMIDE

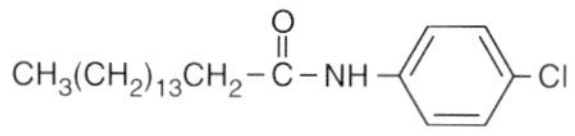

5 gm

S74,373-9
2',4'-DIMETHYL-2-(4-FLUOROPHENYL)-ACETANILIDE

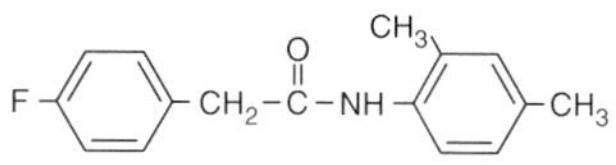

5 gm

S94,560-9
4'-(4-IODOPHENYL)ACETANILIDE

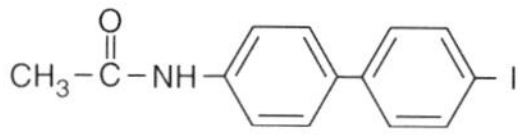

250 mg

S69,518-1
2',3'-DICHLOROBUTYRANILIDE

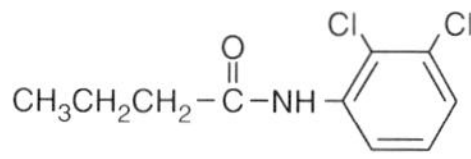

250 mg

S90,522-4
2-BROMO-3'-CHLORO-2'-METHYLBUTYRANILIDE

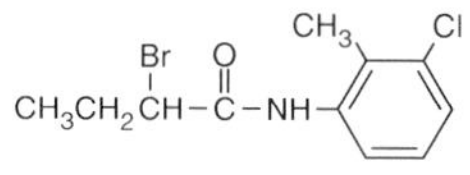

250 mg

S90,268-3
3'-CHLORO-2,2,2'-TRIMETHYLPROPIONANILIDE

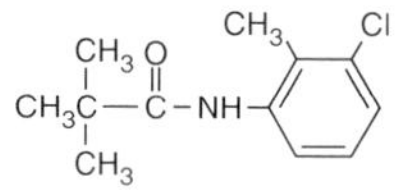

250 mg

S67,594-6
N-(2-CHLORO-6-METHYLPHENYL)-HEXADECANAMIDE

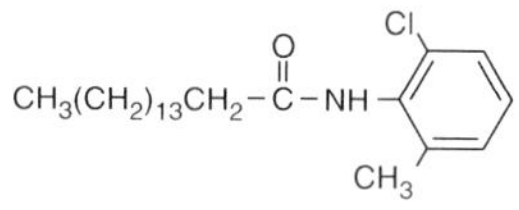

5 gm

S76,441-8
2',4'-DIFLUOROFORMANILIDE

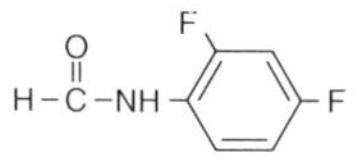

5 gm

S74,635-5
4'-BROMO-2'-CHLOROFORMANILIDE

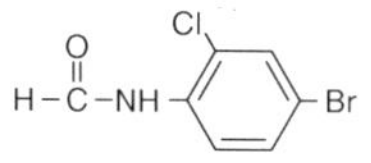

5 gm

S71,406-2
3'-CHLORO-4'-METHYLFORMANILIDE

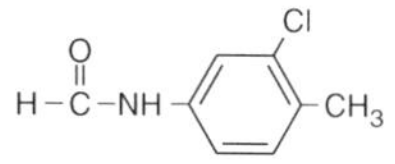

5 gm

S75,150-2
2',5'-DIFLUOROACETANILIDE

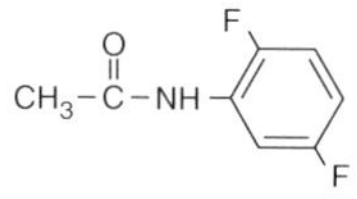

5 gm

S75,475-7
2',4'-DIFLUOROACETANILIDE

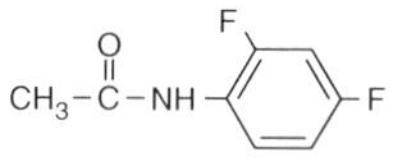

250 mg

S93,367-8
2'-BROMO-4'-CHLOROACETANILIDE

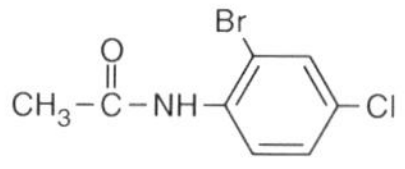

250 mg

S77,821-4
2',4'-DIBROMOACETANILIDE

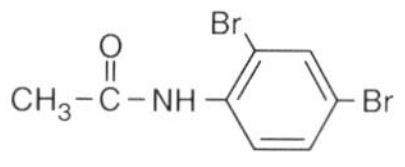

250 mg

S74,484-0
4'-CHLORO-2'-METHYLACETANILIDE

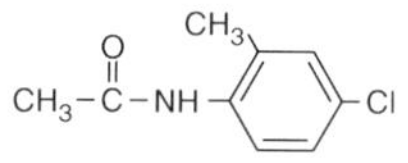

250 mg

S75,206-1
4'-CHLORO-M-ACETOTOLUIDIDE

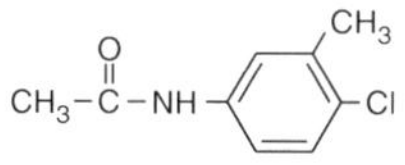

5 gm

S75,308-4
3'-CHLORO-P-ACETOTOLUIDIDE

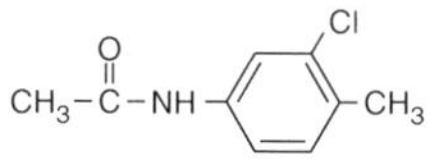

5 gm

S69,479-7
N-(2,5-DIFLUOROPHENYL)PROPIONAMIDE

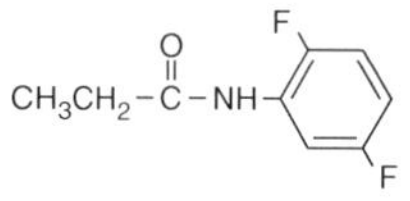

250 mg

S75,212-6
3'-CHLORO-4'-ETHYLACETANILIDE

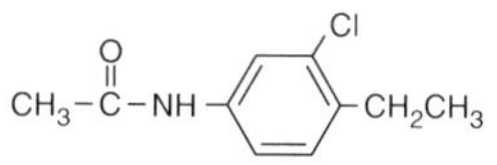

5 gm

S77,276-3
4'-CHLORO-O-PROPIONOTOLUIDIDE

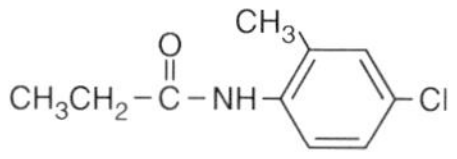

250 mg

S77,872-9
5'-CHLORO-O-PROPIONOTOLUIDIDE

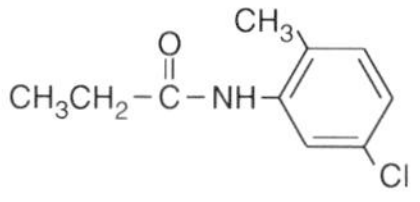

250 mg

S79,099-0
3'-CHLORO-P-PROPIONOTOLUIDIDE

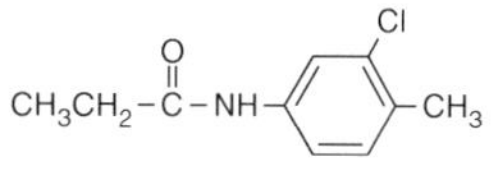

250 mg

S84,405-5
6'-CHLORO-M-PROPIONOTOLUIDIDE

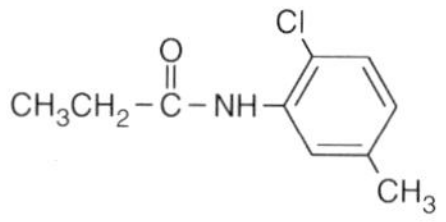

250 mg

S75,185-5
3'-BROMO-4'-ETHYLACETANILIDE

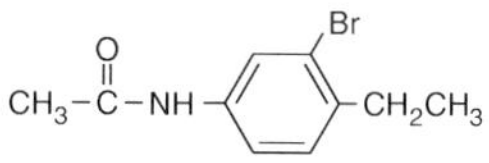

5 gm

S75,186-3
4'-BROMO-3'-ETHYLACETANILIDE

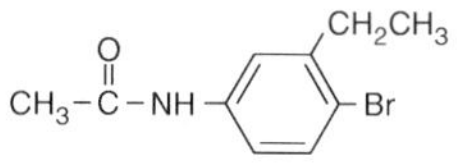

5 gm

S90,265-9
2',5'-DICHLORO-2,2-DIMETHYLPROPIONANILIDE

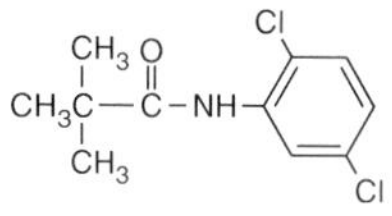

250 mg

S90,266-7
3',4'-DICHLORO-2,2-DIMETHYLPROPIONANILIDE

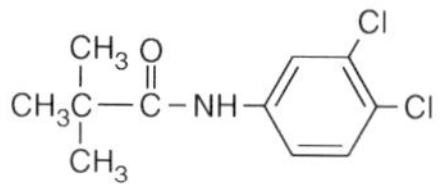

250 mg

S92,625-6
2'-BROMO-4'-ISOPROPYLACETANILIDE
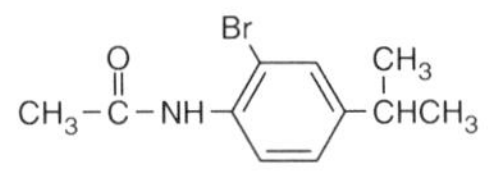
250 mg

S84,992-8
2'-BROMO-4'-CHLORO-2,2-DIMETHYLPROPIONANILIDE
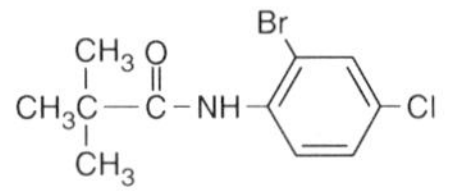
250 mg

S92,457-1
2-BROMO-3'-CHLORO-4'-METHYLBUTYRANILIDE
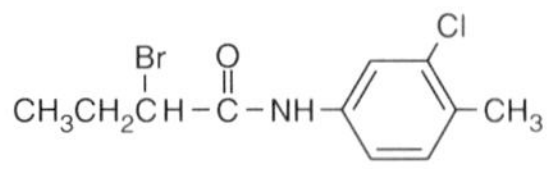
250 mg

S90,267-5
3'-CHLORO-2,2,4'-TRIMETHYLPROPIONANILIDE
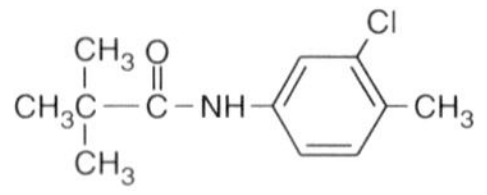
250 mg

S90,269-1
4'-CHLORO-2,2,2'-TRIMETHYLPROPIONANILIDE
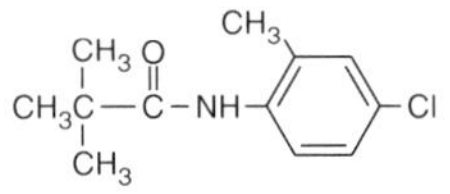
250 mg

S71,820-3
3',4'-DICHLORODECANANILIDE
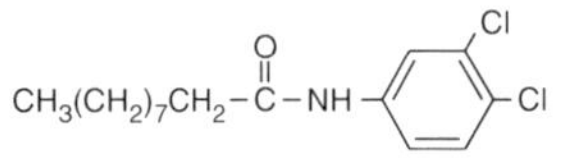
250 mg

S67,600-4
N-(3,4-DICHLOROPHENYL)-HEXADECANAMIDE
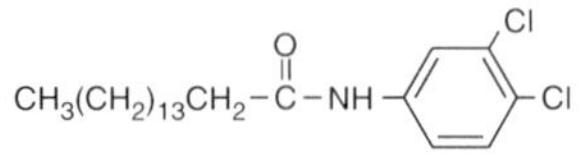
250 mg

S76,782-4
3',5'-DIFLUOROFORMANILIDE
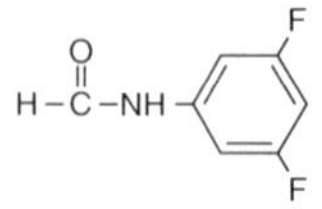
250 mg

S90,270-5
3',5'-DICHLORO-2,2-DIMETHYLPROPIONANILIDE
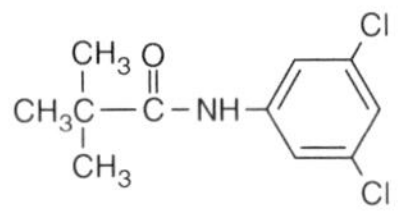
250 mg

S71,816-5
3',5'-DICHLORODECANANILIDE
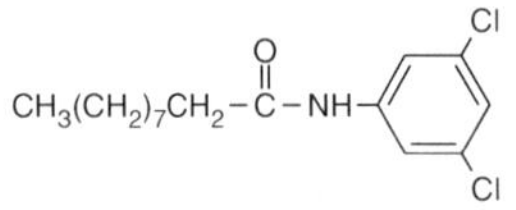
5 gm

S93,192-6
3'-BROMO-2',4'-DIMETHYLACETANILIDE
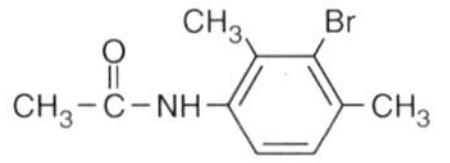
250 mg

S71,541-7
2',4',6'-TRICHLOROACETANILIDE
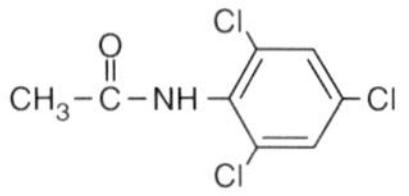
5 gm

S68,787-1
N-(4-CHLORO-2,6-DIBROMOPHENYL)-BUTANAMIDE
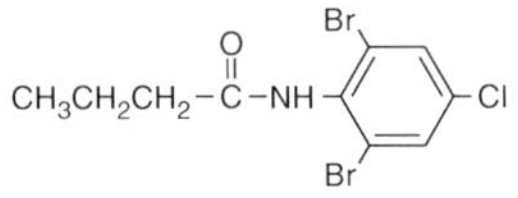
250 mg

S91,675-7
4'-BROMO-2',6'-DIMETHYLPROPIONANILIDE
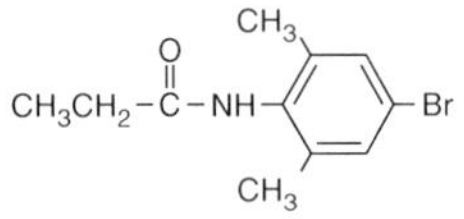
250 mg

S68,943-2
N-(4-CHLORO-2,6-DIBROMOPHENYL)-2,2,2-TRIMETHYLACETAMIDE
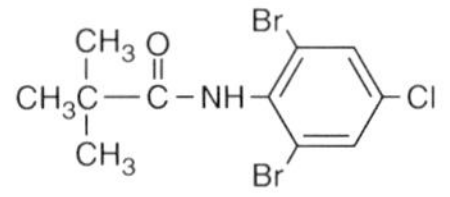
250 mg

S69,847-4
4'-BROMO-2',6'-DIMETHYLBUTYRANILIDE
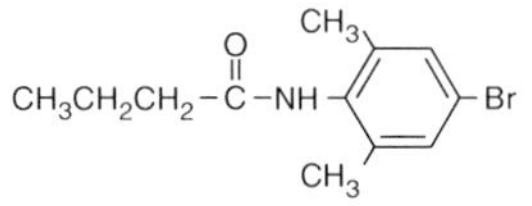
250 mg

S73,815-8
4'-BROMO-2',6'-DIMETHYLDODECANANILIDE
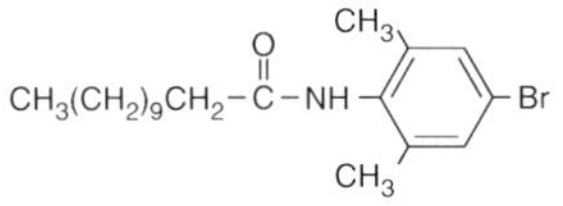
5 gm

S73,887-5
4'-BROMO-2',6'-DIMETHYLSTEARANILIDE
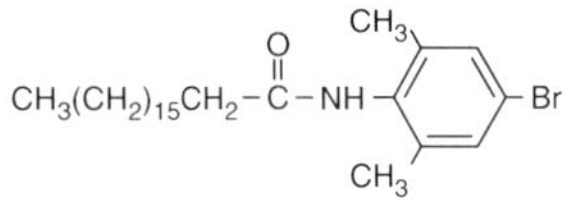
5 gm

S70,020-7
2',4',5'-TRICHLOROPROPIONANILIDE
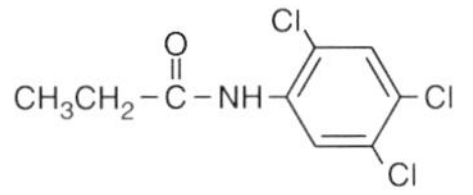
250 mg

S80,368-5
4'-BROMO-2',5'-ACETOXYLIDIDE
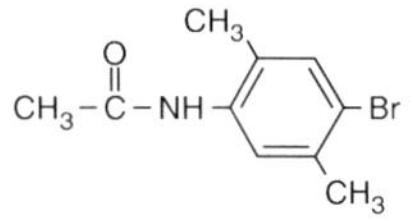
250 mg

S74,379-8
2,2-DIMETHYL-2',4',5'-TRICHLOROPROPIONANILIDE
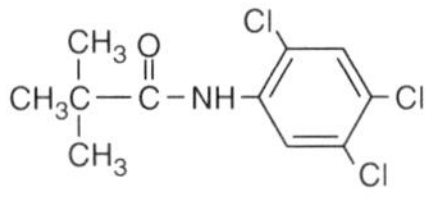
5 gm

S74,383-6
2',4',5'-TRICHLORODECANANILIDE
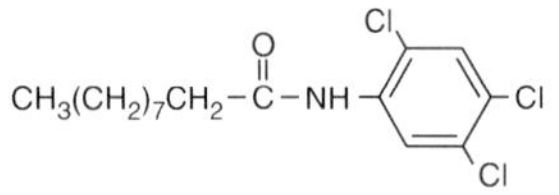
5 gm

S77,525-8
6'-BROMO-N-METHYL-2-PHENYL-O-PROPIONOTOLUIDIDE
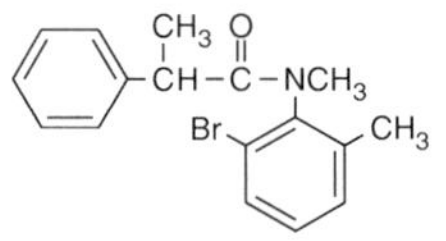
250 mg

S77,494-4
2-CHLORO-4',6'-DIBROMO-N-ETHYL-O-ACETOTOLUIDIDE
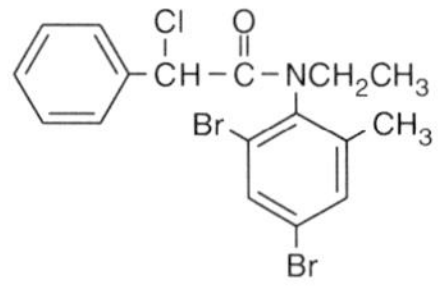
250 mg

S83,869-1
4'-BROMO-N-METHYLACETANILIDE
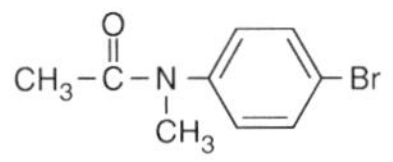
250 mg

S90,523-2
2-BROMO-2',3'-DICHLORO-N-METHYLBUTYRANILIDE
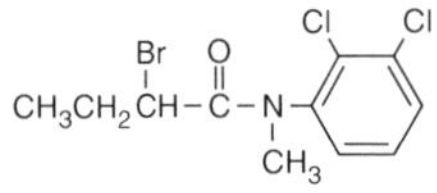
250 mg

S42,080-8
3,4-DICHLORO-N,N-DIETHYLHYDROCINNAMAMIDE
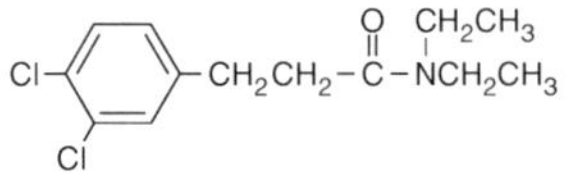
5 gm

S94,908-6
N-(2-HYDROXYETHYL)-2-(2-HYDROXYPHENOXY)ACETAMIDE
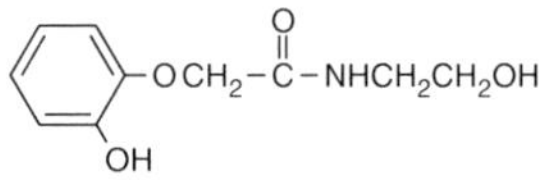
250 mg

S77,691-2
VALYLTYROSYLSERINE
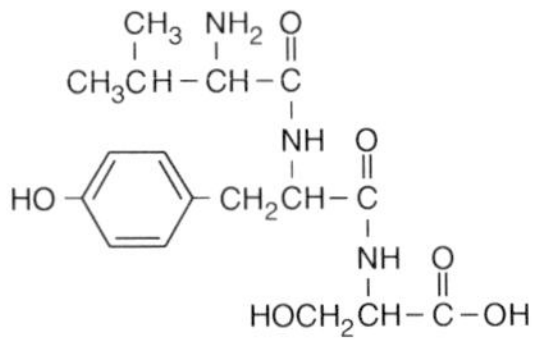
250 mg

S78,473-7
L-TYROSYL-L-TYROSINAMIDE HYDROCHLORIDE
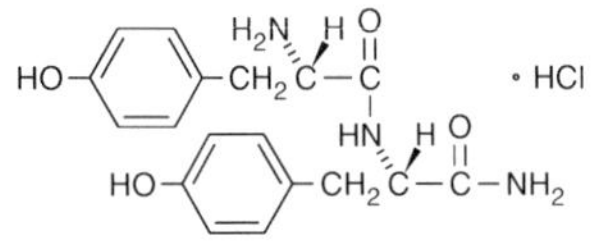
250 mg

S86,939-2
2-(ETHOXYHYDROXYMETHYL)BUTYRANILIDE
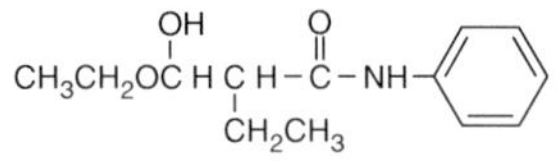
250 mg

S93,690-1
4'-(3-AMINO-2-HYDROXY-1-PROPOXY)-ACETANILIDE
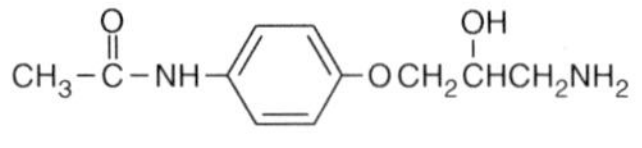
250 mg

S67,876-7
N-VANILLYL-10-UNDECENOAMIDE

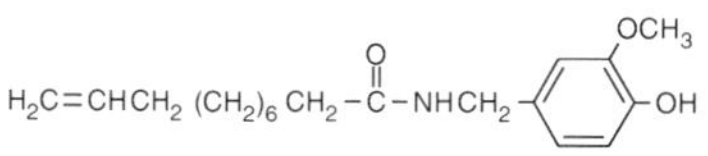

5 gm

S35,188-1
N-LACTOYL-P-PHENETIDINE

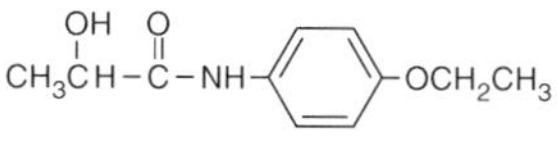

5 gm

S94,909-4
2-(2-HYDROXYPHENOXY)ACETANILIDE

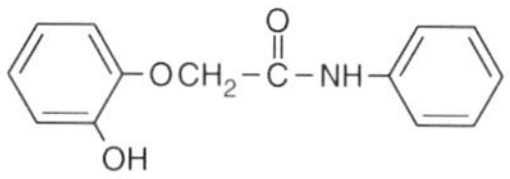

100 mg

S94,907-8
2-(2-HYDROXYPHENOXY)-N-METHYLACETAMIDE

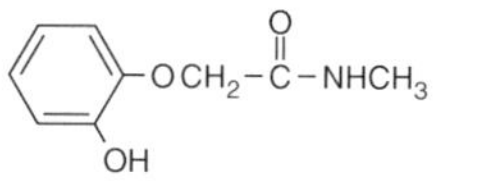

250 mg

S93,970-6
4'-(2-HYDROXYETHOXY)ACETANILIDE

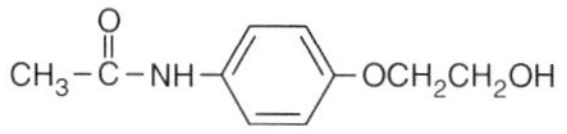

5 gm

S56,786-8
N-VANILLYLNONANOAMIDE

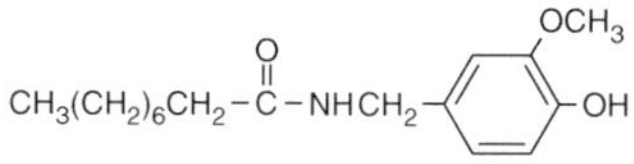

5 gm

S89,471-0
N-[1-(1-HYDROXY-6-METHOXY-1,2,3,4-TETRAHYDRO-2-NAPHTHYLMETHYL)-4-PIPERIDINYL]PROPIONANILIDE

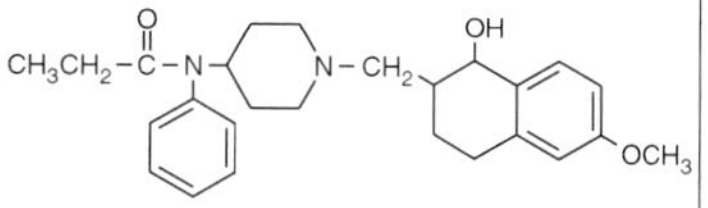

250 mg

S35,079-6
N-(2-HYDROXYETHYL)PHENOXYACETAMIDE

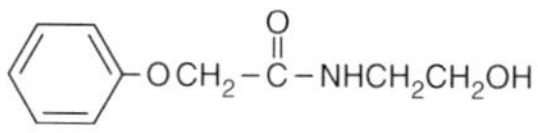

5 gm

S67,181-9
2'-HYDROXY-4'-METHOXYACETANILIDE

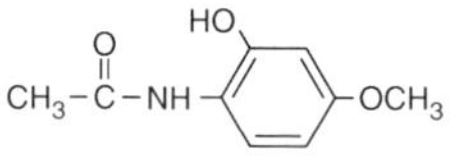

250 mg

S51,502-7
N-(2-HYDROXYPROPYL)-N-[2-(META-TOLYLOXY)ETHYL]ACETAMIDE

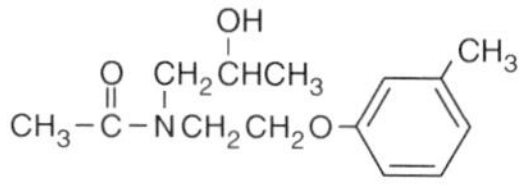

250 mg

S51,500-0
N-(2-HYDROXYPROPYL)-N-(2-PHENOXYETHYL)ACETAMIDE

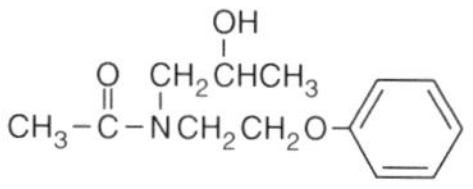

250 mg

S51,501-9
N-(2-HYDROXYPROPYL)-N-[2-(ORTHO-TOLYLOXY)ETHYL]ACETAMIDE

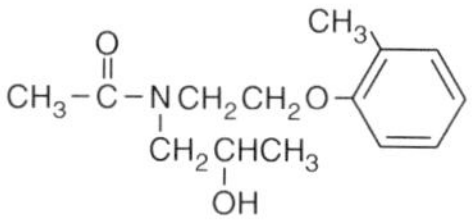

5 gm

S51,503-5
N-(2-HYDROXYPROPYL)-N-[2-(PARA-TOLYLOXY)ETHYL]ACETAMIDE

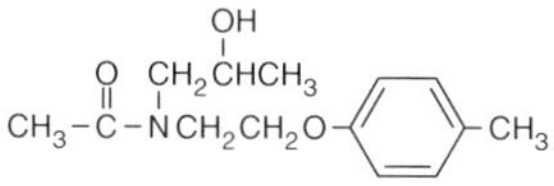

250 mg

S67,184-3
N-ALLYL-2'-HYDROXY-4'-METHOXYACETANILIDE

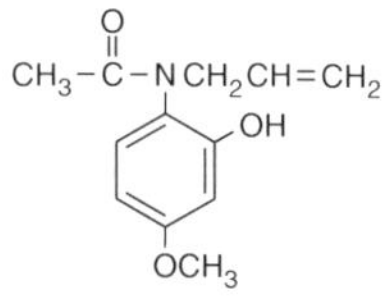

5 gm

S53,866-3
BETA-HYDROXY-PARA-ACETOPHENETIDIDE

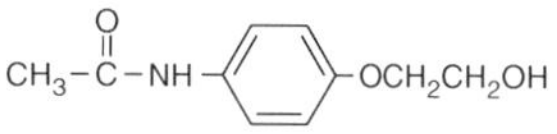

5 gm

S59,107-6
3-HYDROXY-P-BUTYROPHENETIDIDE

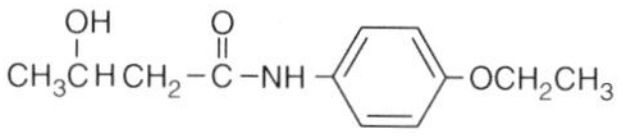

250 mg

S88,854-0
2-HYDROXY-2-PHENYLPROPIONAMIDE

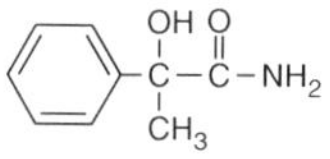

250 mg

S71,017-2
MANDELANILIDE

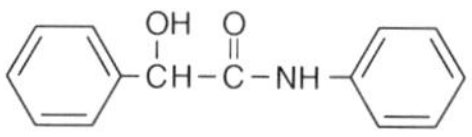

5 gm

S40,908-1
MANDELAMIDE

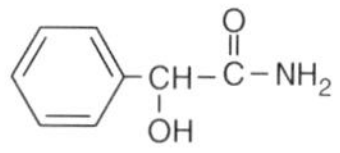

5 gm

S90,671-9
ETHYL 3-ACETAMIDO-2-(HYDROXYMETHYL)-2-PHENYLPROPIONATE

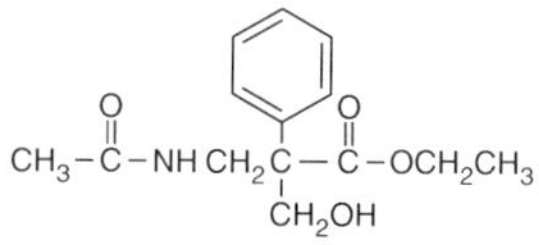

250 mg

S52,548-0
4'-ACETYL-3'-HYDROXYACETANILIDE

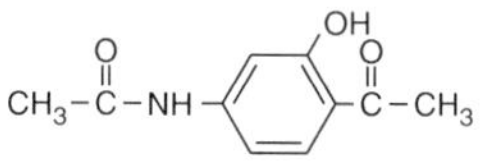

250 mg

S40,012-2
4-ACETAMIDOSALICYLIC ACID, SODIUM SALT DIHYDRATE

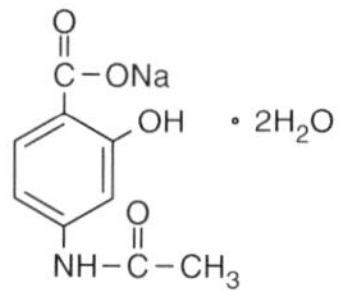

250 mg

S37,823-2
4-ACETAMIDOSALICYLIC ACID

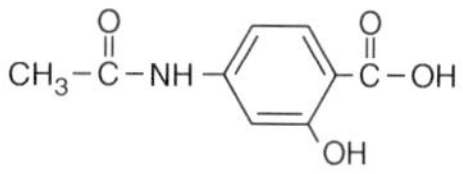

250 mg

S40,013-0
4-ACETAMIDOSALICYLIC ACID, CALCIUM SALT TRIHYDRATE

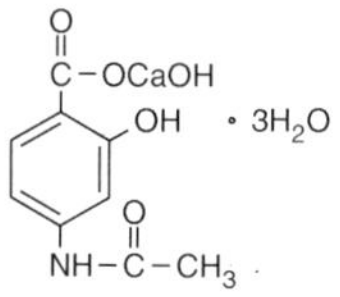

250 mg

S58,325-1
4'-AMINO-2'-HYDROXYISONONANANILIDE

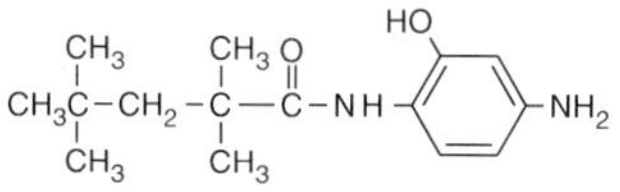

250 mg

S61,832-2
4'-AMINO-2'-HYDROXYSUCCINANILIC ACID

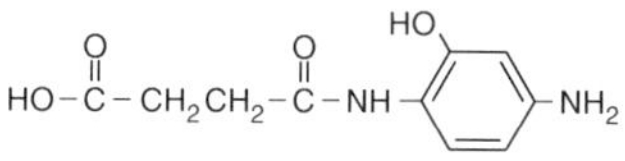

250 mg

S77,657-2
L-SERYL-L-PHENYLALANINAMIDE HYDROCHLORIDE

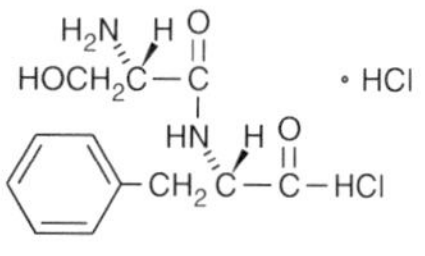

250 mg

S78,932-1
L-PHENYLALANYL-L-SERINAMIDE HYDROCHLORIDE

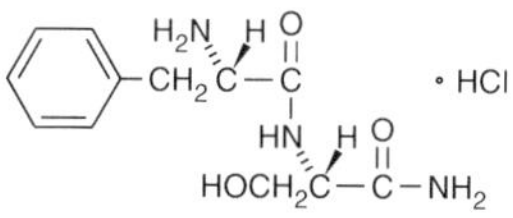

250 mg

S78,850-3
N-(6-AMINOHEXANOYL)-L-PHENYLALANYL-L-TYROSINE

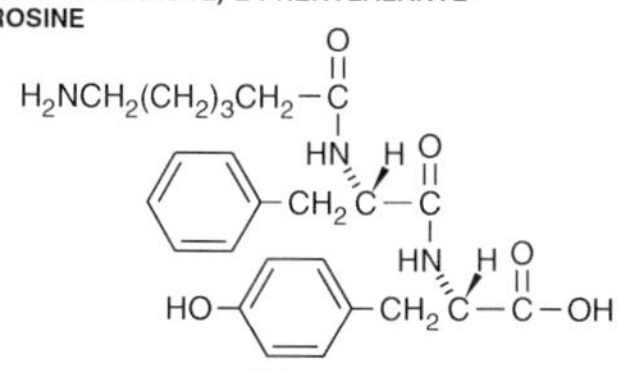

250 mg

S45,081-2
L-TYROSYL-L-ALANINE

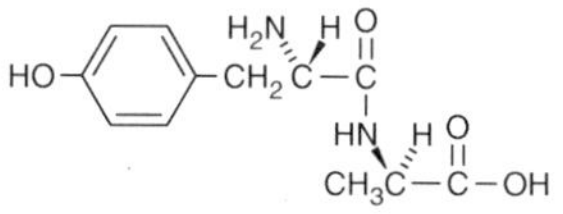

250 mg

S77,715-3
L-TYROSYL-BETA-ALANINE

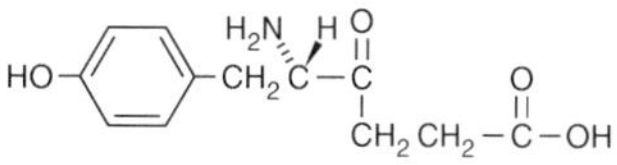

250 mg

S78,524-5
L-2-AMINOBUTYRYL-L-TYROSINE

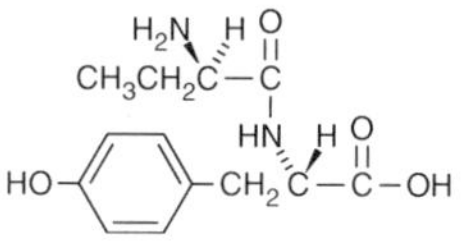

250 mg

S45,273-4
GLYCYL-L-TYROSINAMIDE ACETATE
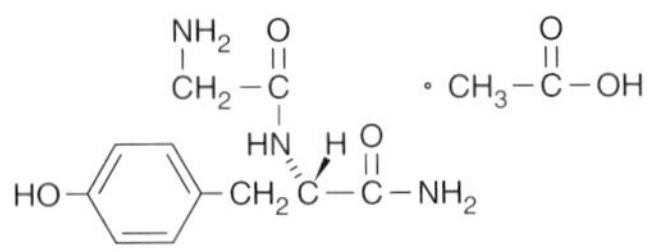
250 mg

S45,275-0
GLYCYL-L-TYROSINAMIDE HYDROCHLORIDE
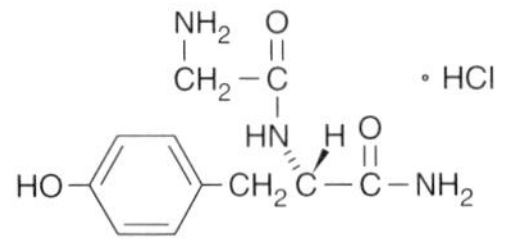
250 mg

S78,152-5
L-TYROSYL-L-GLYCINAMIDE HYDROCHLORIDE
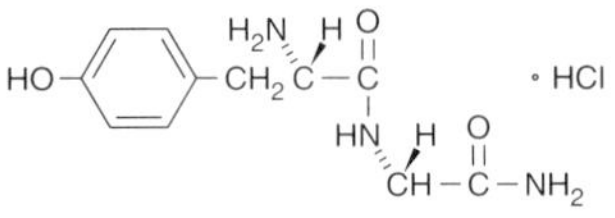
250 mg

S78,829-5
NEPSILON-(2,4-DINITROPHENYL)-L-LYSYL-L-TYROSINE HYDROBROMIDE MONOHYDRATE
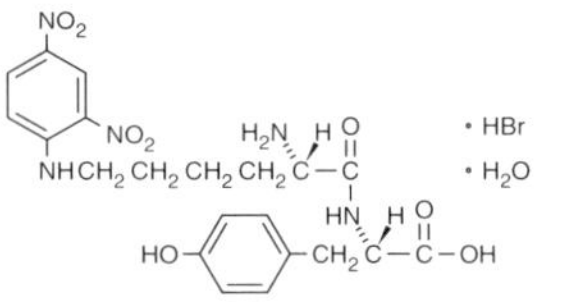
250 mg

S67,911-9
D-LEUCYL-L-TYROSINE
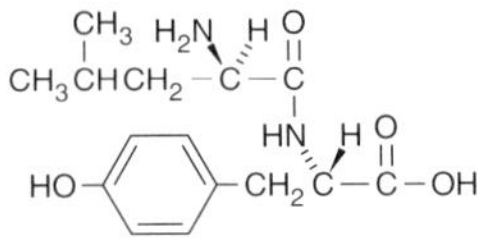
5 gm

S37,103-3
L-VALYL-L-TYROSINE METHYL ESTER HYDROBROMIDE
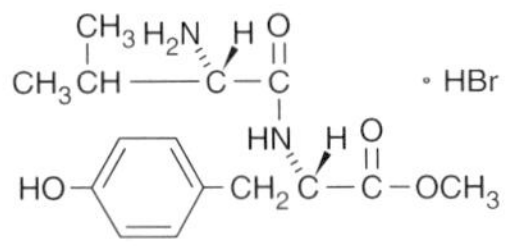
250 mg

S78,190-8
L-ALANYL-L-TYROSYL-L-PHENYLALANINAMIDE HYDROCHLORIDE
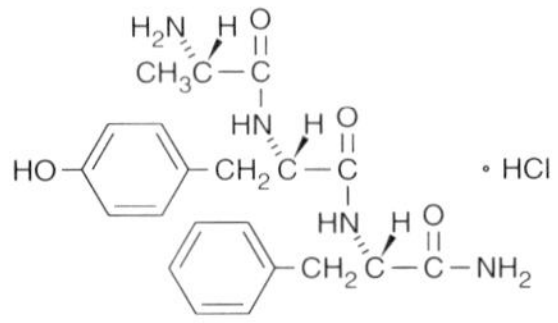
250 mg

S77,626-2
L-NORLEUCYL-L-TYROSINAMIDE
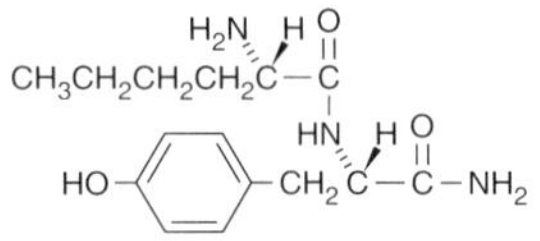
250 mg

S42,312-2
2-[BIS(2-HYDROXYETHYL)AMINO]-6'-CHLORO-ORTHO-ACETOTOLUIDIDE HYDROCHLORIDE
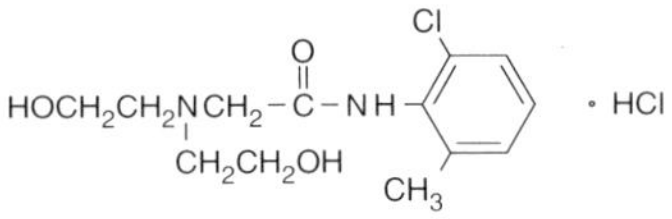
5 gm

S85,506-5
2'-HYDROXY-5'-NITROBUTYRANILIDE
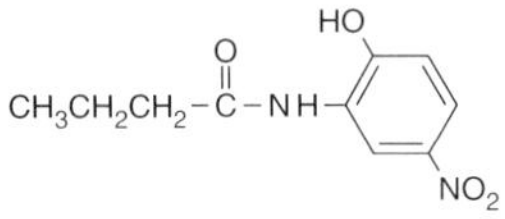
250 mg

S34,046-4
N-ACETYL-3,5-DINITRO-L-TYROSINE ETHYL ESTER
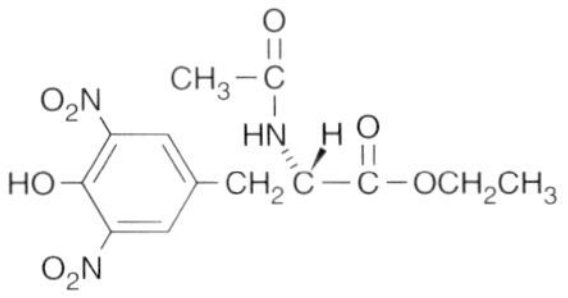
250 mg

S65,526-0
2'-HYDROXY-3'-NITROACETANILIDE
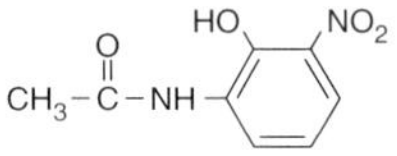
250 mg

S65,162-1
2-ACETAMIDO-4-NITROPHENOL, SODIUM SALT
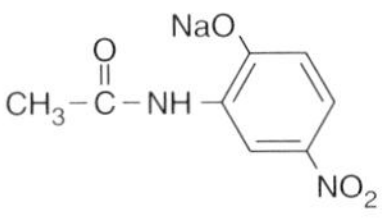
250 mg

S41,402-6
4-ACETAMIDO-2-DIETHYLAMINOMETHYLPHENOL
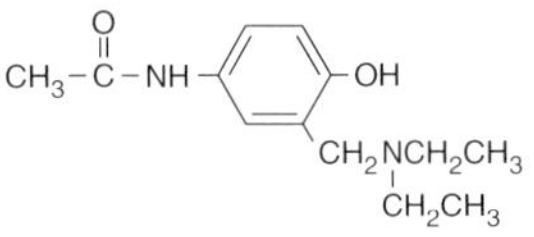
5 gm

S65,515-5
2'-HYDROXY-4'-NITRO-2-OCTADECYLSUCCINANILIC ACID
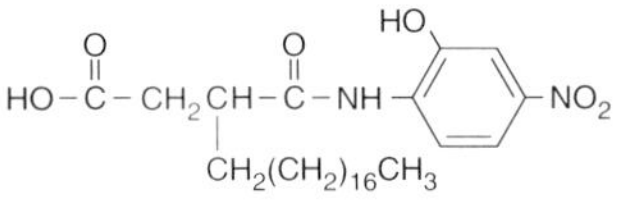
250 mg

S34,090-1
2-ACETAMIDO-1-HYDROXY-1-(4-NITROPHENYL)-2-PROPENE
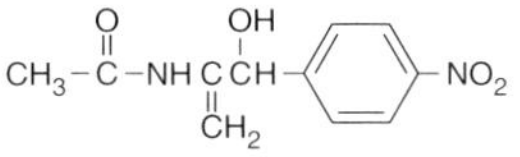
250 mg

S34,024-3
DL-2-ACETAMIDO-1-(4-NITROPHENYL)-1,3-PROPANEDIOL
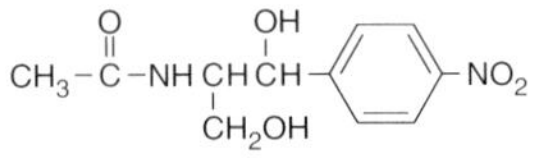
250 mg

S78,528-8
L-PROPYL-L-TYROSINAMIDE HYDROCHLORIDE
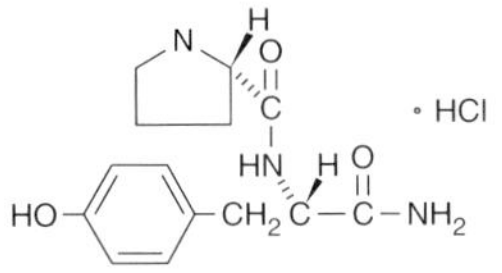
250 mg

S35,806-1
L-ALPHA-BENZAMIDO-BETA-(4-HYDROXYPHENYL)PROPIONAMIDE
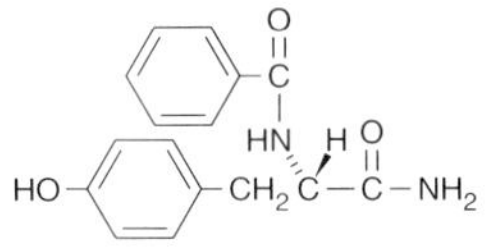
5 gm

S78,543-1
N-ACETYL-L-TYROSYL-L-PHENYLALANINE
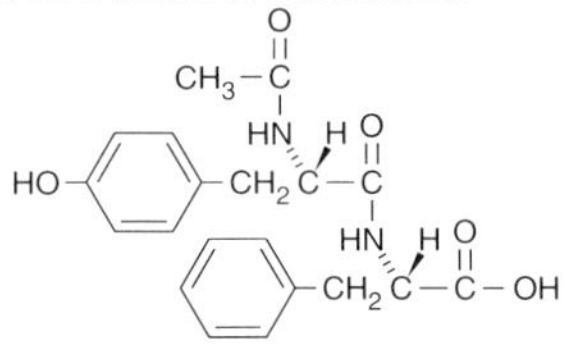
250 mg

S78,452-4
L-VALYL-L-TYROSINE
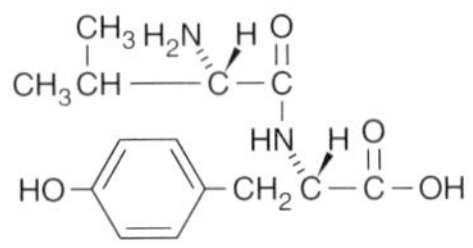
250 mg

S35,007-9
N-FORMYL-L-TYROSINE
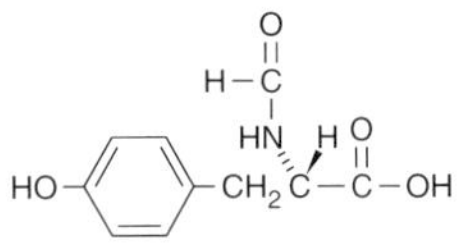
5 gm

S78,173-8
N-ACETYL-DL-TYROSINE
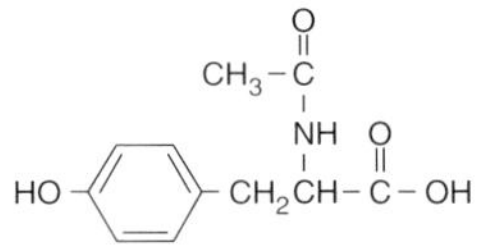
250 mg

S77,639-4
N-ACETYL-L-PHENYLALANYL-L-TYROSINE METHYL ESTER
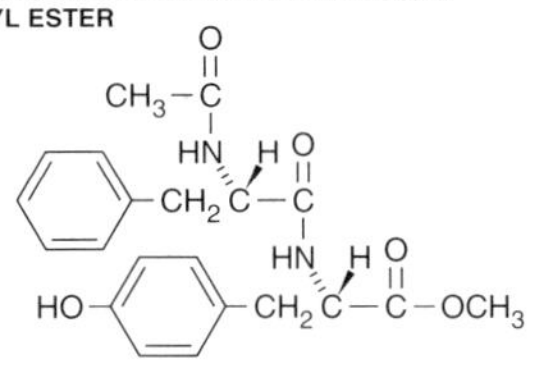
250 mg

S78,136-3
N-ACETYL-DL-TYROSINE METHYL ESTER
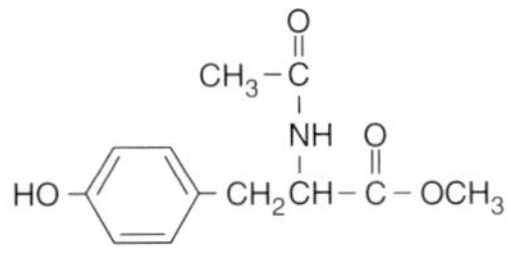
250 mg

S76,585-6
N-ACETYL-D-TYROSINE ETHYL ESTER MONOHYDRATE
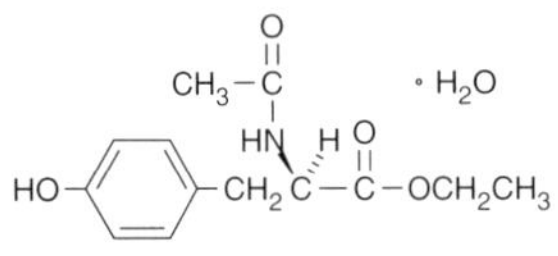
5 gm

S78,360-9
N-ACETYL-D-TYROSINE ETHYL ESTER
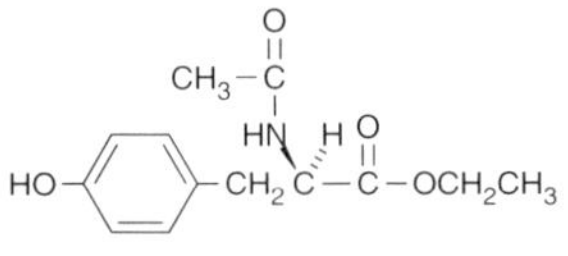
5 gm

S78,377-3
N-ACETYL-DL-TYROSINE ETHYL ESTER
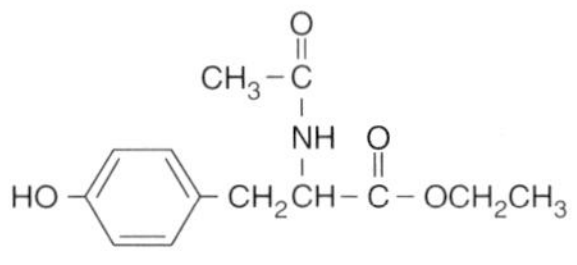
5 gm

S77,645-9
N-ACETYL-L-PHENYLALANYL-L-TYROSINAMIDE
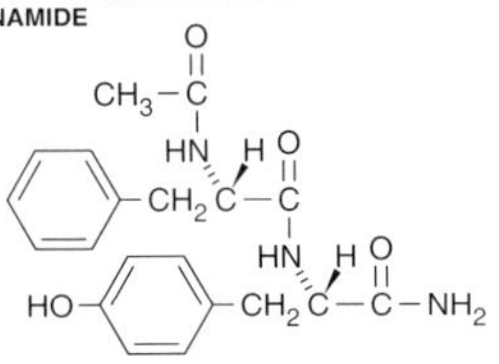
250 mg

S77,621-1
N-ACETYL-DL-TYROSINAMIDE
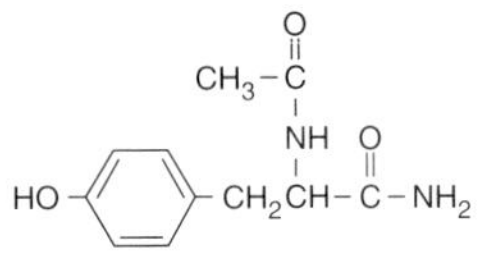
250 mg

S77,623-8
N-ACETYL-D-TYROSINAMIDE
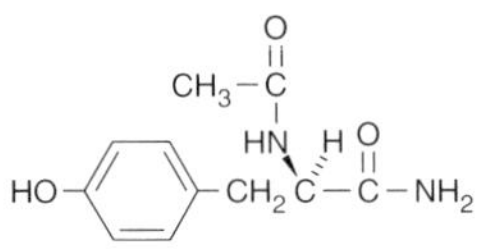
250 mg

S78,445-1
GLYCYLGLYCYL-L-TYROSINAMIDE HYDROCHLORIDE
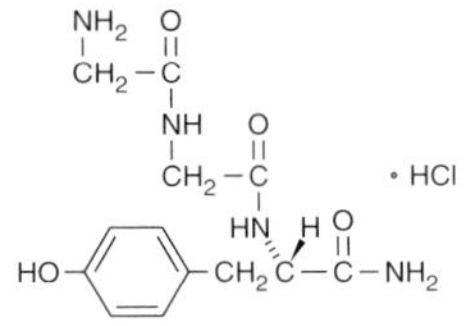
250 mg

S78,269-6
N-ACETYL-L-TYROSINAMIDE HYDROCHLORIDE
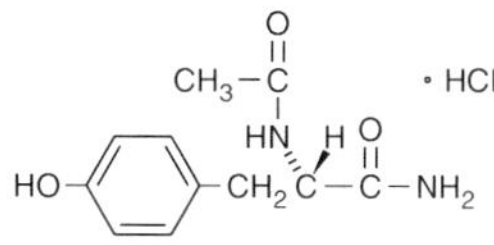
5 gm

S78,593-8
ISOLEUCYLTYROSINAMIDE
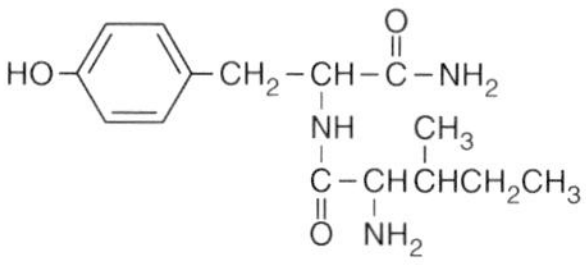
250 mg

S66,898-2
2-(2-HYDROXYPHENYL)ACETAMIDE
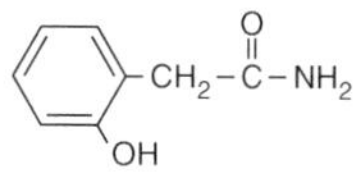
5 gm

S40,011-4
N-HYDROXYMETHYL-2-PHENYLACETAMIDE
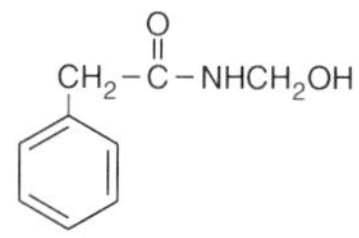
250 mg

S42,246-0
N,N-DIETHYL-3,5-DIIODO-4-HYDROXY-ALPHA-PHENYLHYDROCINNAMAMIDE
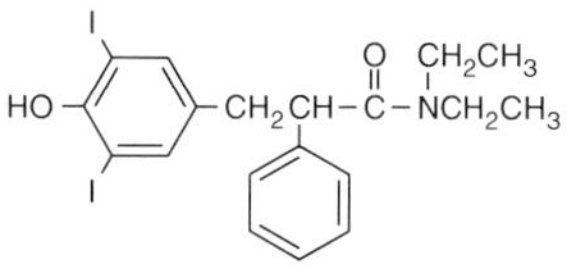
250 mg

S35,724-3
N-ACETYL-3,5-DINITRO-L-TYROSINE
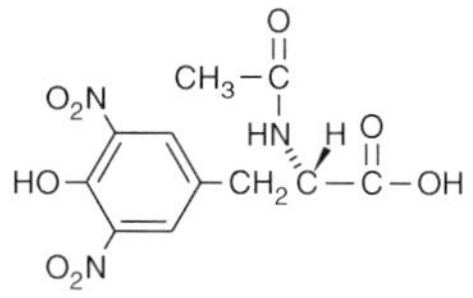
250 mg

S38,931-5
ETHYL 3-ACETAMIDO-4-HYDROXYBENZOATE
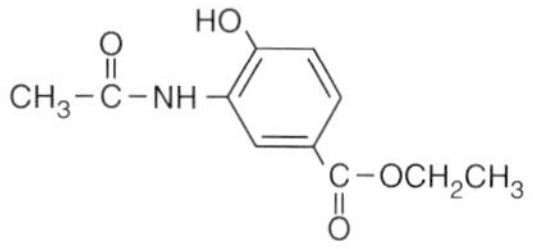
5 gm

S36,247-6
2'-HYDROXYFORMANILIDE
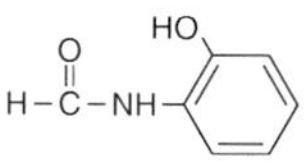
5 gm

S81,455-5
3'-ALLYL-2'-HYDROXYACETANILIDE
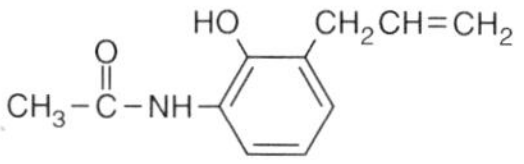
250 mg

S57,198-9
6'-HYDROXY-M-ACETOTOLUIDIDE
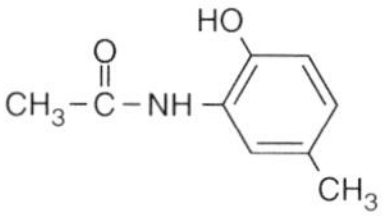
250 mg

S58,251-4
5'-CHLORO-2'-HYDROXY-P-ACETOTOLUIDIDE
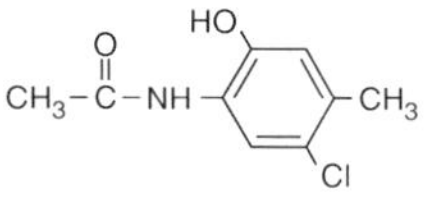
5 gm

S59,803-8
4-ACETAMIDOPHENYL SALICYLATE
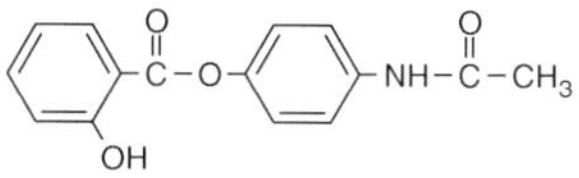
5 gm

S93,138-1
ETHYL 4-HYDROXYOXANILATE
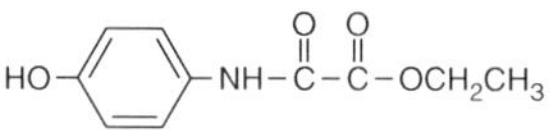
250 mg

S78,034-0
3'-HYDROXYHEXAHYDROPHTHALANILIC ACID
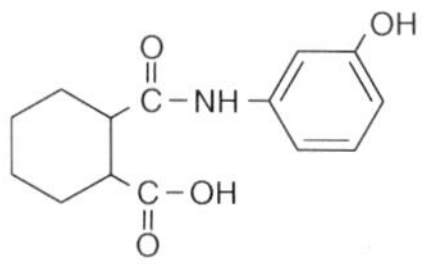
250 mg

S73,616-3
4'-HYDROXYGLUTARANILIC ACID
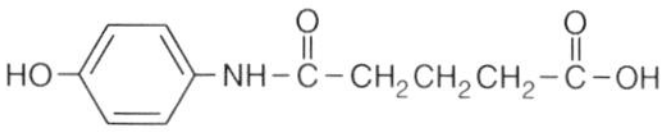
5 gm

S91,270-0
4'-(2-HYDROXYETHYL)-GLUTARANILIC ACID
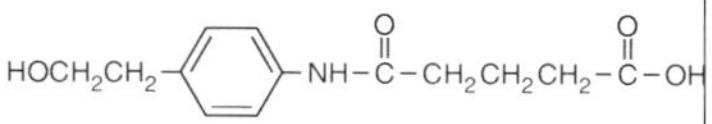
250 mg

S52,654-1
N-ACETYL-3,5-DIIODO-L-TYROSINE
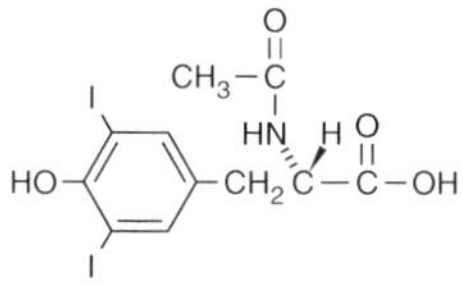
250 mg

S71,818-1
3'-HYDROXYDECANANILIDE
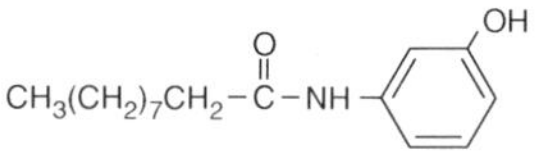
250 mg

S54,056-0
4'-HYDROXYFORMANILIDE
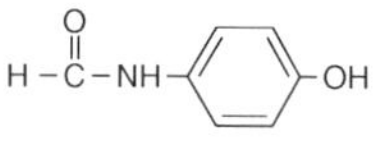
5 gm

S82,921-8
4'-HYDROXYACETANILIDE, SODIUM SALT
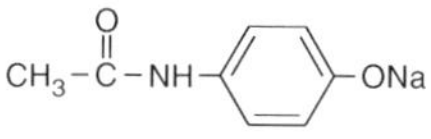
250 mg

S38,906-4
4'-HYDROXYBUTYRANILIDE
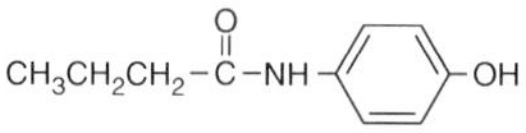
5 gm

S65,288-1
3-HYDROXY-4'-METHYLPROPIONANILIDE
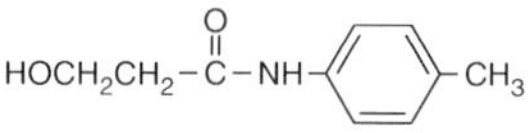
5 gm

S90,279-9
2,2-DIMETHYL-4'-(2-HYDROXYETHYL)-PROPIONANILIDE
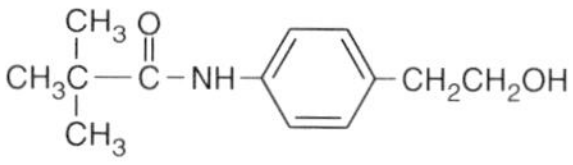
250 mg

S38,907-2
4'-HYDROXYNONANANILIDE
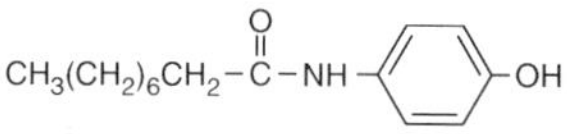
5 gm

S38,908-0
4'-HYDROXYDODECANANILIDE
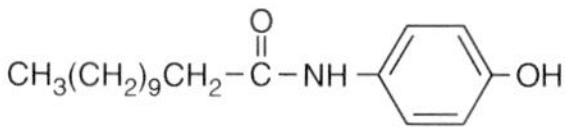
5 gm

S51,029-7
4'-HYDROXYPALMITANILIDE
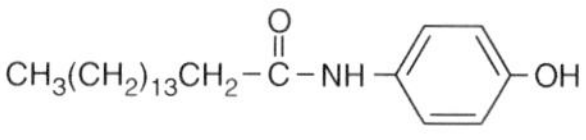
250 mg

S81,349-4
4'-HYDROXYEICOSANANILIDE
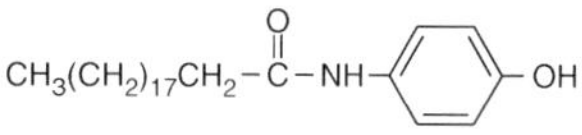
250 mg

S95,452-7
N-(3,4-DIBENZYLOXYPHENYL)ACETAMIDE
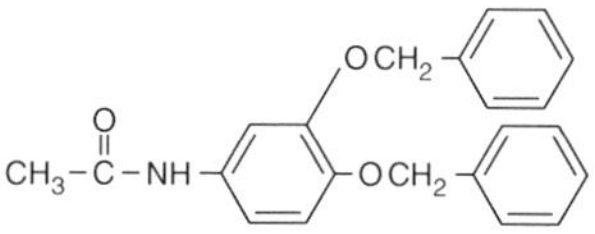
50 mg

S78,814-7
N-ACETYL-O-BENZYL-L-TYROSINE
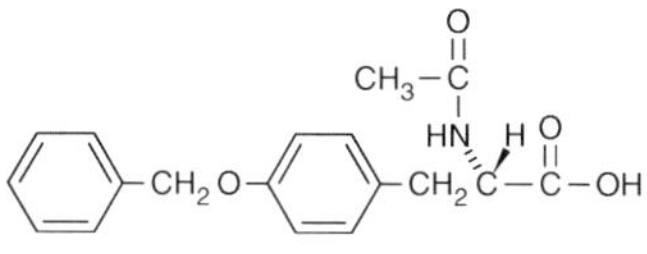
250 mg

S75,869-8
2-(BENZYLAMINO)-N-[2-(3,4-DIMETHOXYPHENYL)ETHYL]ACETAMIDE HYDROCHLORIDE
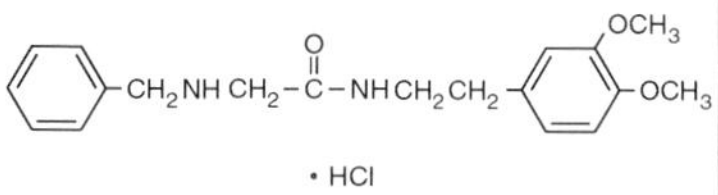
250 mg

S70,340-0
N,N''-HEXAMETHYLENEBIS(N'-[2-(3,4-DIMETHOXYPHENYL)ETHYL]-HEXANEDIAMIDE)
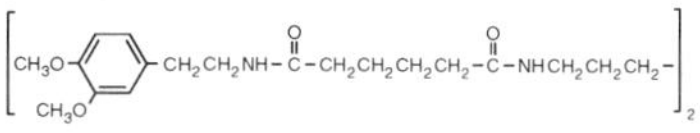
5 gm

S70,339-7
N,N'-HEXAMETHYLENEBIS[2-(4,5-DIMETHOXY-2-METHYLPHENYL)ACETAMIDE]-
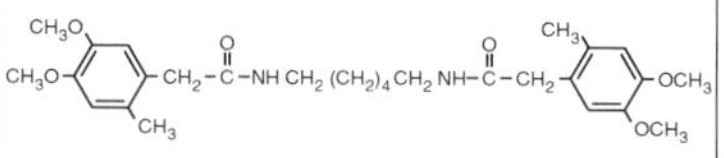
5 gm

S68,607-7
METHYL 6-[3-(3,4-DIMETHOXYCINNAMOYL)PROPIONAMIDO]-HEXANOATE
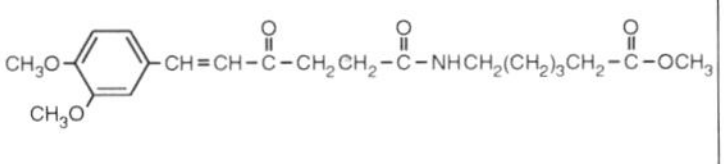
250 mg

S84,332-6
2-(3,4,5-TRIMETHOXYPHENYL)ACETAMIDE
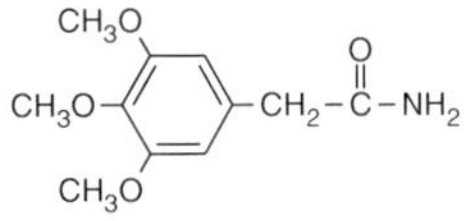
250 mg

S73,714-3
2',4'-DIMETHOXYHEXAHYDROPHTHALANILIC ACID
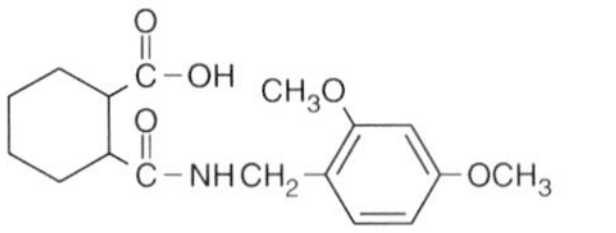
5 gm

S95,419-5
N-(2,3-DIMETHOXYPHENYLACETYL)-GLYCINE ETHYL ESTER
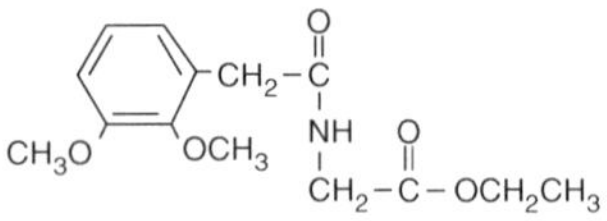
50 mg

S90,889-4
3'-(N-ETHYL-N-[2-(4-METHOXYPHENOXY)-ETHYL]AMINO)ACETANILIDE
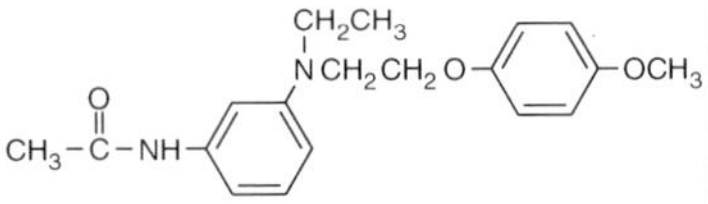
250 mg

S78,045-6
2',5'-DIMETHOXYPROPIONANILIDE
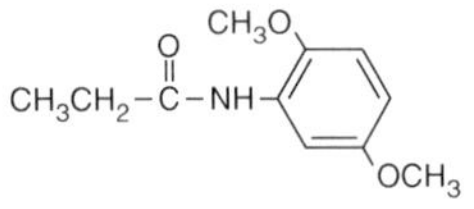
250 mg

S80,855-5
4-(2,5-DIMETHOXYPHENYL)BUTYRAMIDE
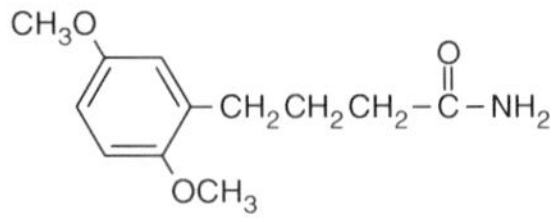
250 mg

S73,709-7
1,4-BIS(2-METHOXY-4-NITROPHENYL)-OXAMIDE
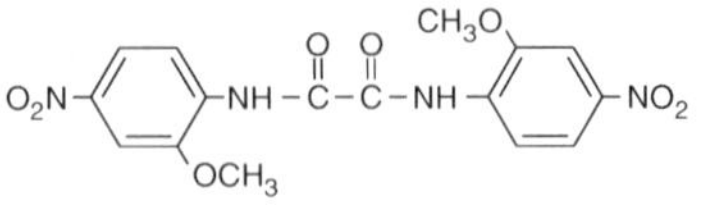
5 gm

S71,866-1
N,N'-BIS(2-ETHOXYPHENYL)-DECANEDIAMIDE
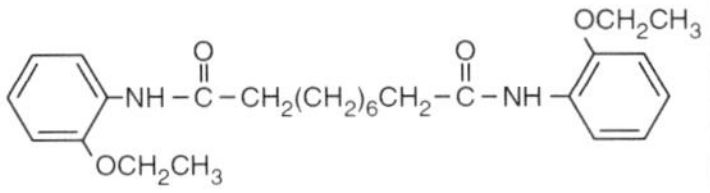
5 gm

S84,655-4
2',2''-DIMETHOXY-4',4''-BIACETANILIDE
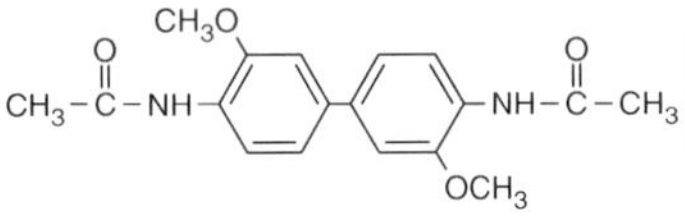
250 mg

S81,808-9
2',5'-DIMETHOXY-4'-METHYLACETANILIDE
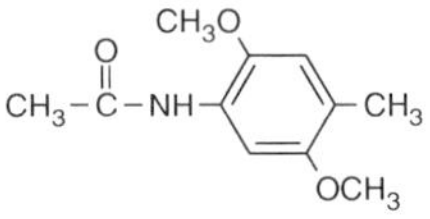
250 mg

S85,464-6
N,N'-BIS(4-ETHOXYPHENYL)-1,10-DECANEDIAMIDE
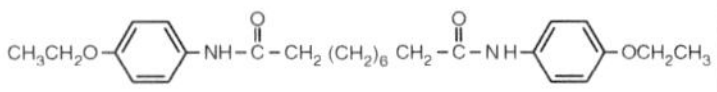
250 mg

S72,491-2
4'-(METHYLTHIO)-2-PHENOXYACETANILIDE
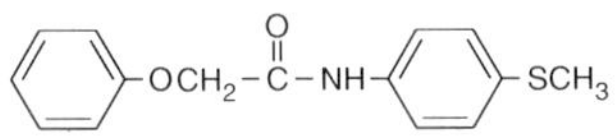
5 gm

S59,193-9
2-(4-AMINOPHENOXY)ACETANILIDE
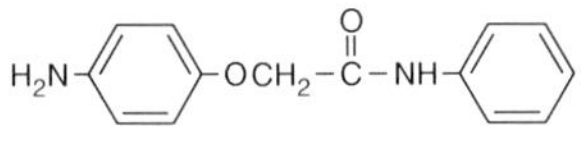
5 gm

S58,710-9
2',5'-DIETHOXY-4'-NITRO-2-PHENOXYACETANILIDE
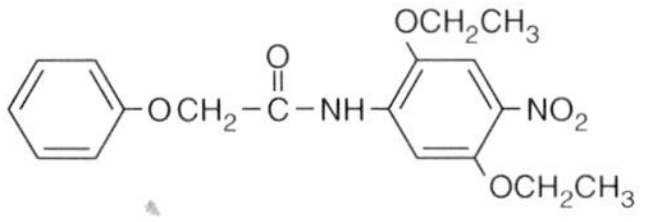
250 mg

S93,742-8
(R)-2-METHOXY-2-PHENYLACETAMIDE
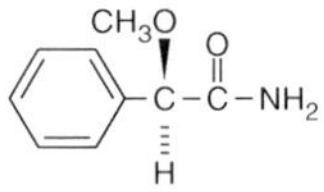
250 mg

S42,559-1
METHYLAMINOCOLCHICINE HYDRATE
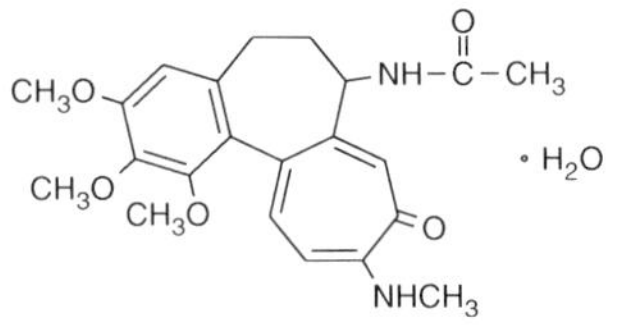
250 mg

S43,344-6
6'-CHLORO-2-(2-DIETHYLAMINOETHOXY)-O-ACETOTOLUIDIDE HYDROCHLORIDE
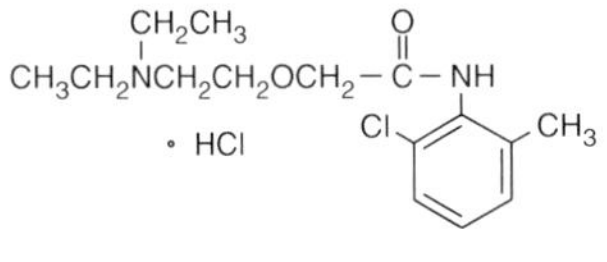
5 gm

S59,189-0
3-[2-(2,4-DI-TERT-PENTYLPHENOXY)-ACETAMIDO]BENZOIC ACID
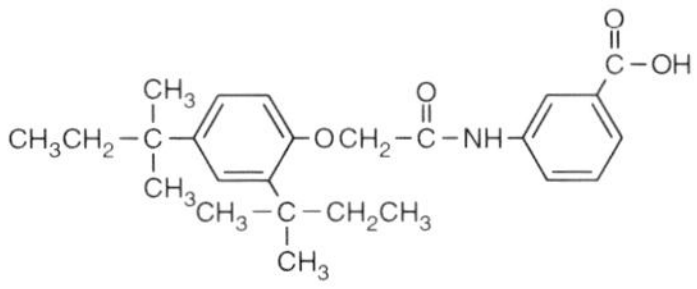
250 mg

S58,708-7
2',5'-DIETHOXY-2-PHENOXYACETANILIDE
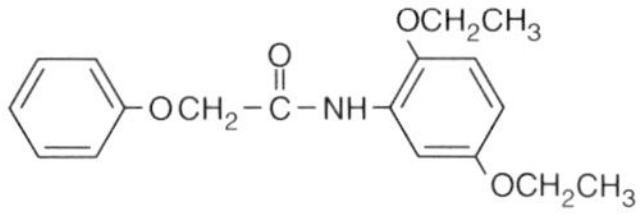
250 mg

S68,376-0
4'-(N,N-DIMETHYLCARBAMOYLMETHOXY)-CHALCONE
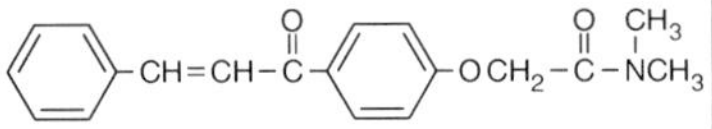
5 gm

S57,443-0
3',3''-(ETHYLENEDIOXY)DIACETANILIDE
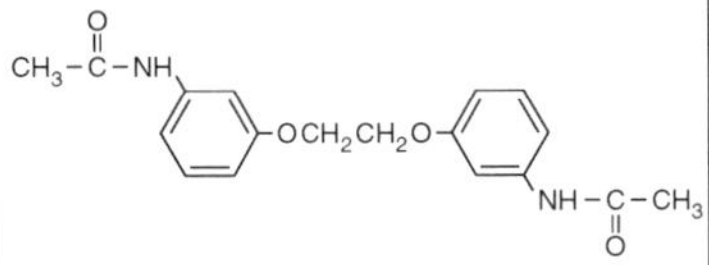
5 gm

S57,366-3
4',4''-(ETHYLENEDIOXY)DIACETANILIDE
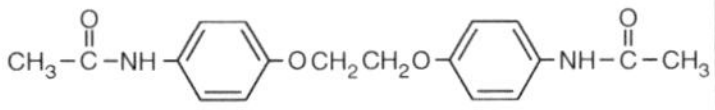
5 gm

S56,888-0
3',4'-DIMETHOXYACETANILIDE
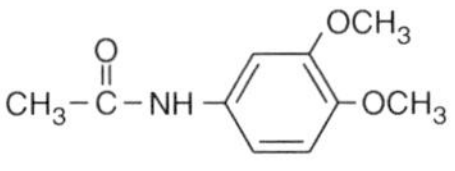
250 mg

S79,768-5
ETHYL 4-(3-PHENOXYPROPIONAMIDO)-BENZOATE
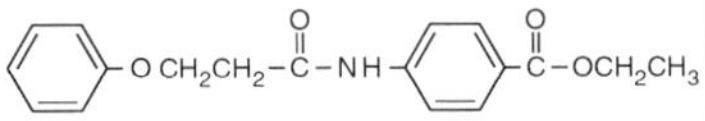
250 mg

S76,252-0
2-(3-METHOXYPHENYLTHIO)ACETAMIDE
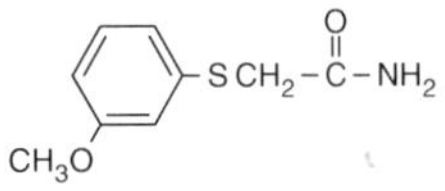
250 mg

S93,673-1
2-(BENZYLTHIO)-4'-METHOXYACETANILIDE
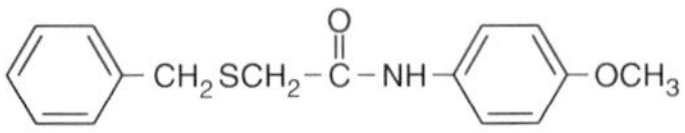
5 gm

S93,682-0
2-(BENZYLTHIO)-4'-ETHOXYACETANILIDE
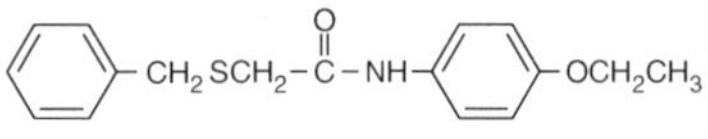
5 gm

S64,741-1
2'-AMINO-4'-METHOXYACETANILIDE
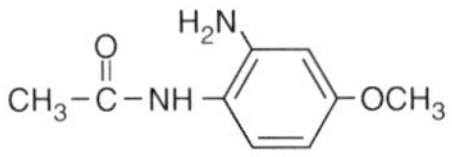
5 gm

S57,201-2
3'-AMINO-P-ACETOPHENETIDIDE
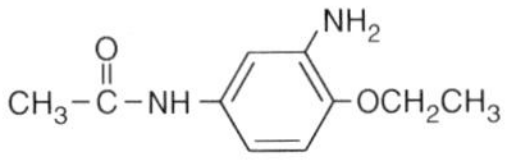
5 gm

S90,803-7
2',4'-DINITRO-5'-METHOXYACETANILIDE
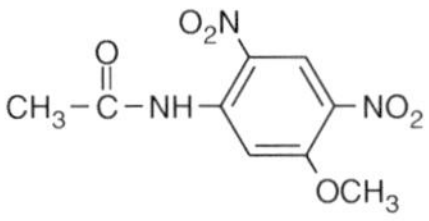
250 mg

S42,361-0
4'-BUTOXY-2-(BUTYLAMINO)-ACETANILIDE HYDROCHLORIDE
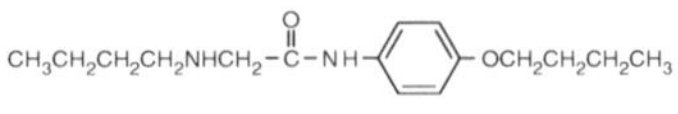
5 gm

S42,170-7
2'-PROPOXY-1-PIPERIDINEACETANILIDE HYDROCHLORIDE

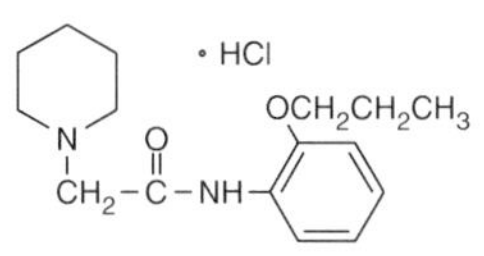

250 mg

S42,168-5
N-(1-METHYL-2-PHENOXYETHYL)-1-PYRROLIDINEACETAMIDE

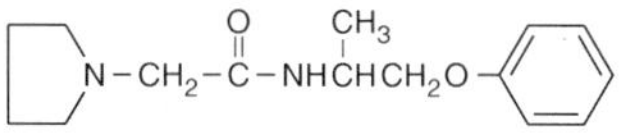

250 mg

S42,121-9
1-PYRROLIDINEACETO-ORTHO-PHENETIDIDE HYDROCHLORIDE

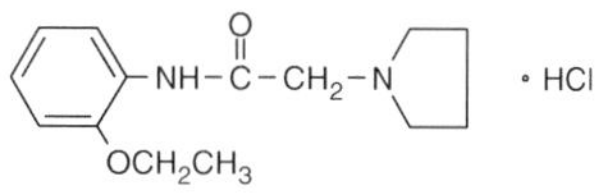

250 mg

S42,169-3
1-PIPERIDINEACETO-PARA-PHENETIDIDE HYDROCHLORIDE

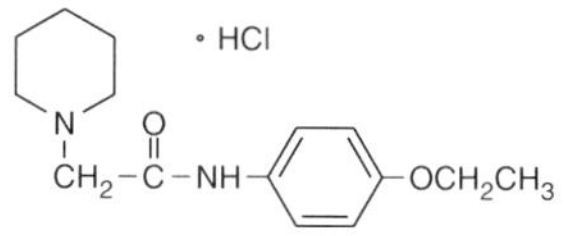

250 mg

S42,402-1
2-DIETHYLAMINO-PARA-ACETOPHENETIDIDE HYDROCHLORIDE

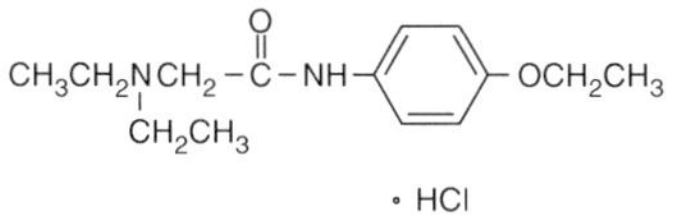

250 mg

S42,199-5
4'-BUTOXY-2-(DIETHYLAMINO)-ACETANILIDE HYDROCHLORIDE

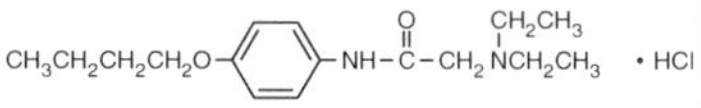

5 gm

S42,021-2
(2,4,6-TRIIODOPHENOXY)ACETAMIDE

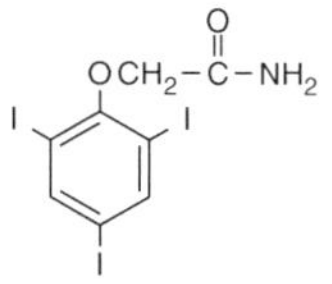

250 mg

S35,438-4
(+)-(PHENOXYMETHYL)PENICILLOIC ACID HYDRATE

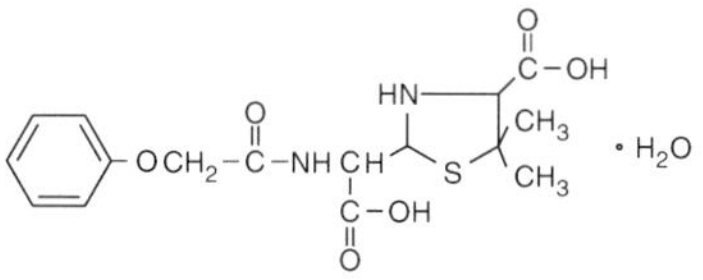

5 gm

S91,387-1
4'-METHOXY-2'-NITRO-1-CYCLOPROPANECARBOXANILIDE

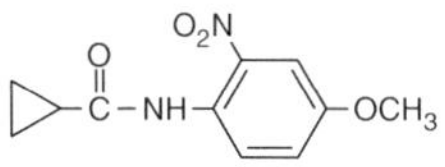

250 mg

S74,377-1
2,2-DIMETHYL-4'-METHOXY-2'-NITROPROPIONANILIDE

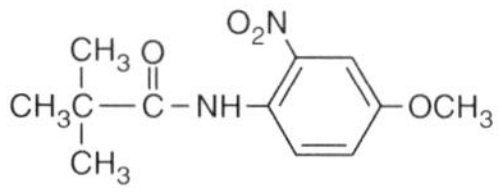

5 gm

S35,708-1
4-ACETAMIDO-3-NITROANISOLE

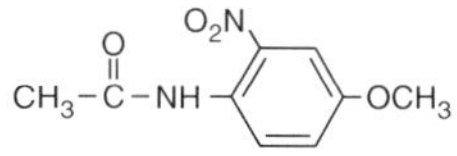

5 gm

S94,153-0
5'-CHLORO-4'-METHOXY-2'-NITROACETANILIDE

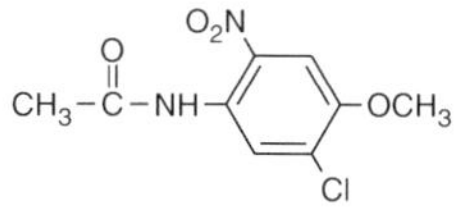

250 mg

S42,268-1
2'-(ORTHO-TOLYLOXY)-1-PIPERIDINEACETANILIDE HYDROCHLORIDE

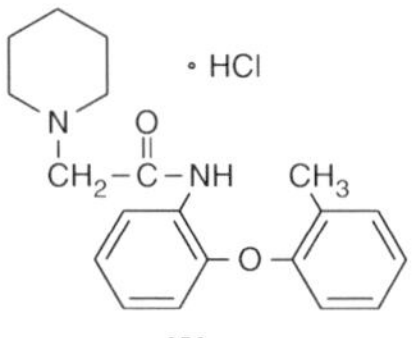

250 mg

S42,204-5
4'-BUTOXY-1-PIPERIDINEACETANILIDE HYDROCHLORIDE

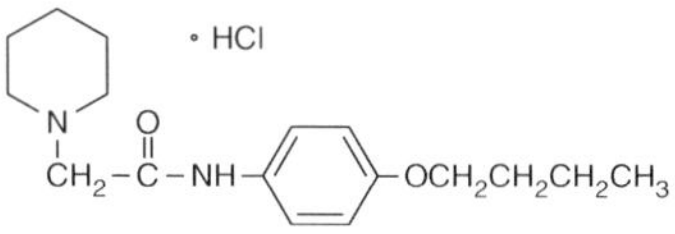

250 mg

S92,875-5
ETHYL 3-[5-(ACETAMIDO)-2-METHOXYANILINO]BUTYRATE

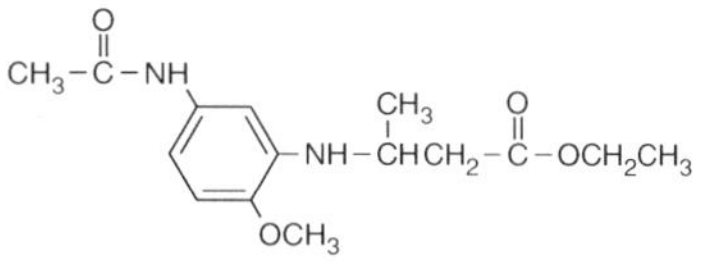

250 mg

S90,893-2
3'-(CYCLOHEXYLAMINO)-4'-METHOXYACETANILIDE

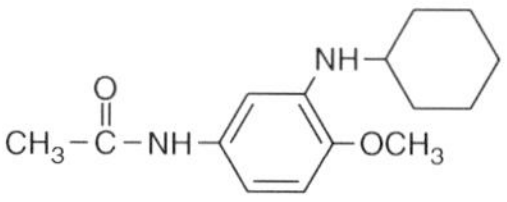

250 mg

S90,892-4
4'-METHOXY-3'-(1-METHYLPENTYLAMINO)-ACETANILIDE

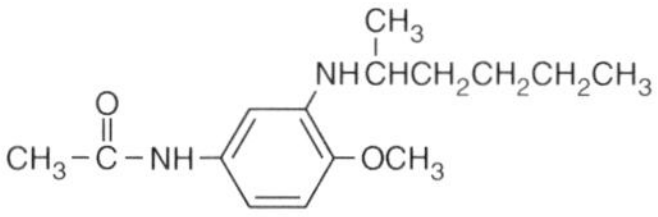

250 mg

S90,880-0
4'-METHOXY-3'-(1-METHYLHEXYLAMINO)-ACETANILIDE

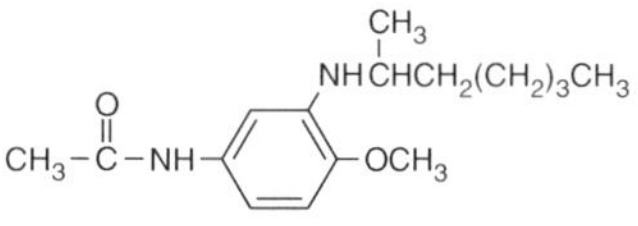

250 mg

S92,393-1
3'-ANILINO-4'-METHOXYACETANILIDE

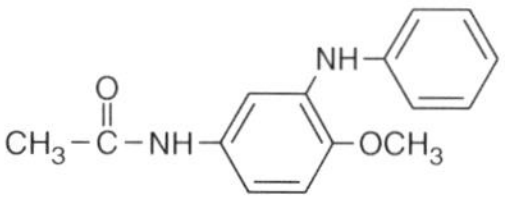

250 mg

S90,902-5
3'-(ISOPROPYLAMINO)-4'-METHOXYACETANILIDE

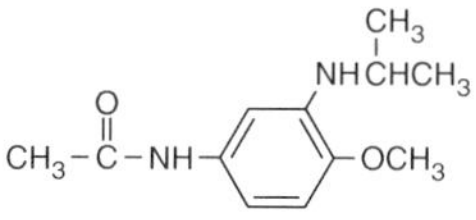

250 mg

S90,852-5
N-(5-ACETAMIDO-2-METHOXYPHENYL)-N-METHYLGLYCINE ETHYL ESTER

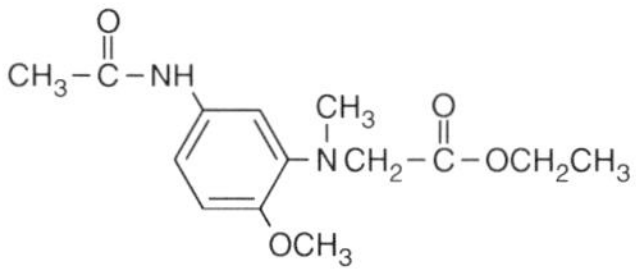

250 mg

S90,885-1
3'-[N-ETHYL-N-(2-PHENOXYETHYL)-AMINO]ACETANILIDE

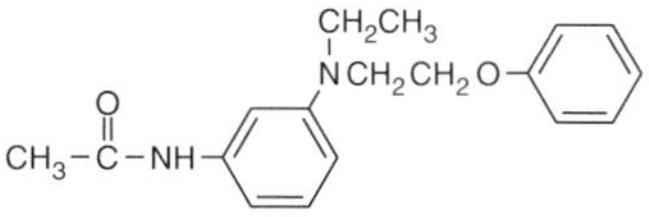

250 mg

S42,401-3
3-(N-ETHYL-N-METHYLAMINO)-N-(2-PHENOXYETHYL)PROPIONAMIDE

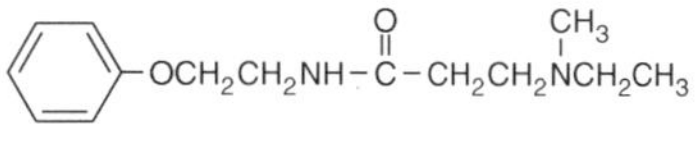

250 mg

S89,290-4
2'-METHOXY-3'-NITROPROPIONANILIDE

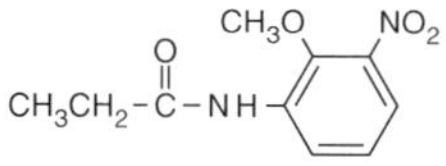

250 mg

S73,614-7
2'-METHOXY-5'-NITROPROPIONANILIDE

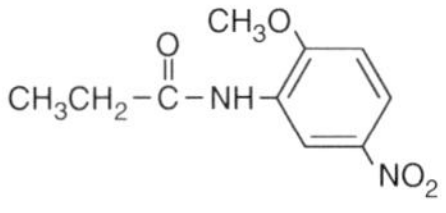

5 gm

S72,545-5
2'-METHOXY-5'-NITROBUTYRANILIDE

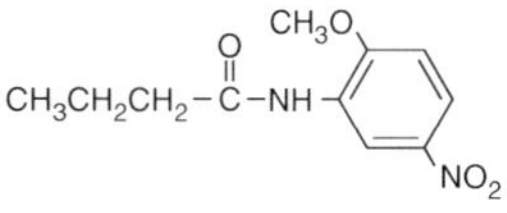

5 gm

S62,609-0
2'-METHOXY-5'-NITROFORMANILIDE

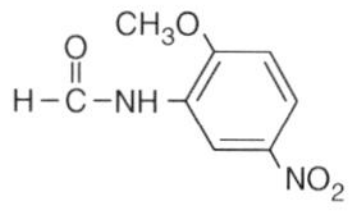

5 gm

S72,920-5
2,3-DIBROMO-2'-METHOXY-4'-NITROSUCCINANILIC ACID

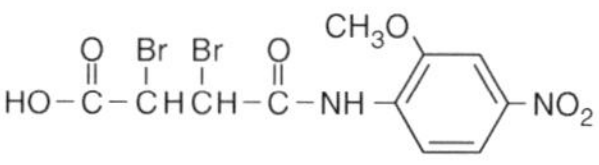

5 gm

S67,998-4
2'-METHOXY-4'-NITROBUTYRANILIDE

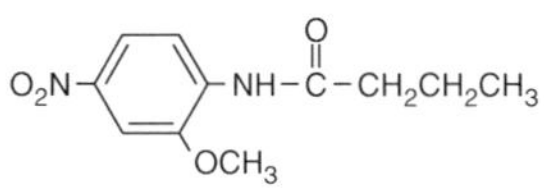

5 gm

S65,246-6
2'-METHOXY-4'-NITROFORMANILIDE

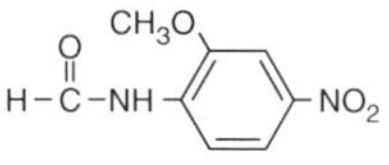

5 gm

S65,462-0
2'-METHOXY-4'-NITROACETANILIDE

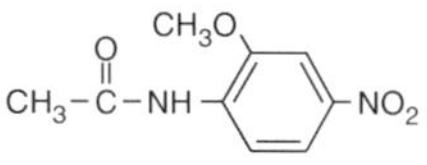

250 mg

S71,599-9
4'-(CYCLOPROPYLCARBOXAMIDO)-2'-METHOXYACETANILIDE

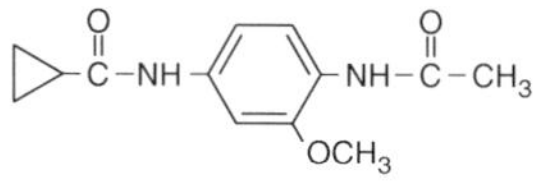

50 mg

S71,587-5
4'-ACETAMIDO-3'-METHOXYPROPIONANILIDE
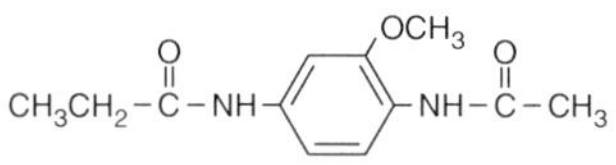
5 gm

S91,390-1
4'-ACETAMIDO-2,2-DIMETHYL-3'-METHOXYPROPIONANILIDE
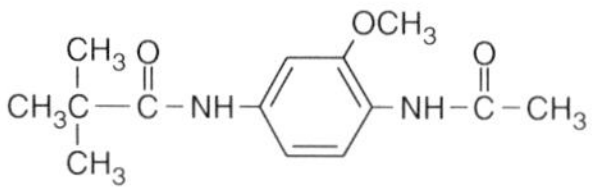
250 mg

S74,385-2
4'-ACETAMIDO-3'-METHOXYDECANANILIDE
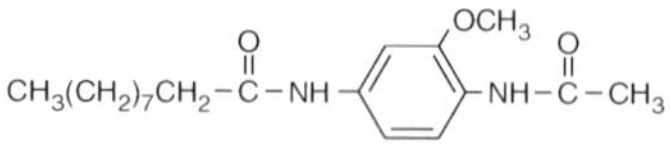
5 gm

S53,359-9
2,5-DIETHOXY-4-(2-PHENYLACETAMIDO)-BENZENEDIAZONIUM CHLORIDE
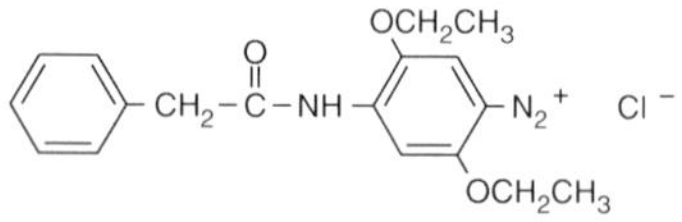
250 mg

S77,634-3
N-ACETYL-O-METHYL-L-TYROSINE
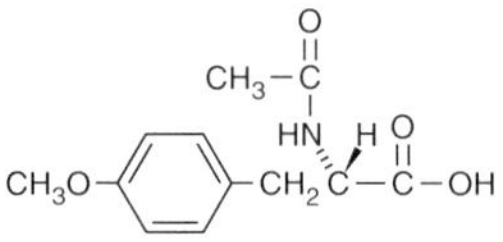
250 mg

S77,602-5
N-ACETYL-O-METHYL-DL-TYROSINE METHYL ESTER
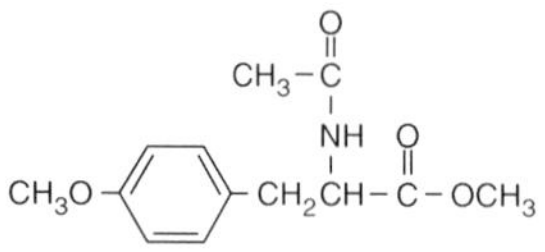
250 mg

S58,081-3
2-(4-METHOXYPHENYL)ACETAMIDE
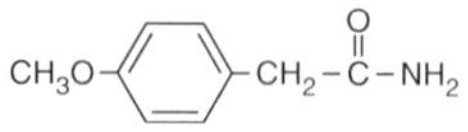
5 gm

S50,421-1
2-PHENYL-P-BUTYROPHENETIDIDE
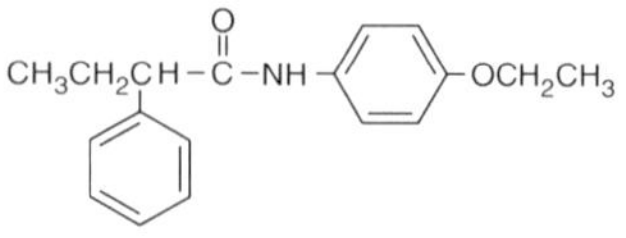
5 gm

S53,152-9
ALPHA-(PARA-ANISIDINO)-PARA-ACETOTOLUIDIDE
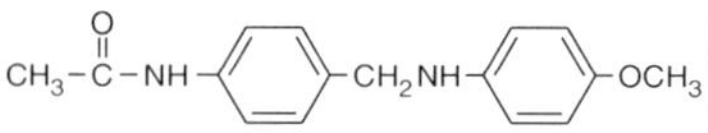
250 mg

S42,231-2
2,BETA-BIS(DIETHYLAMINO)-PARA-ACETOPHENETIDIDE
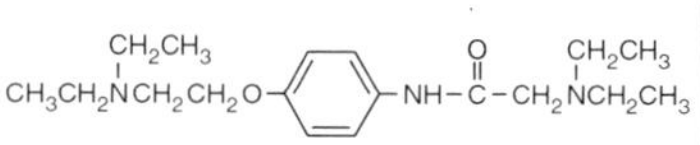
5 gm

S35,250-0
ALPHA-(4-METHOXYPHENYL)-1-PYRROLIDINEACETAMIDE
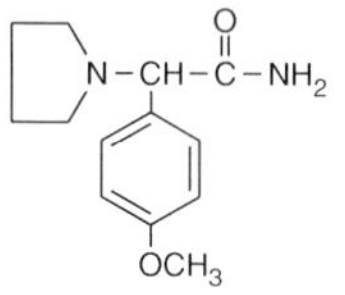
5 gm

S35,245-4
ALPHA-(4-METHOXYPHENYL)-1-PIPERIDINEACETAMIDE
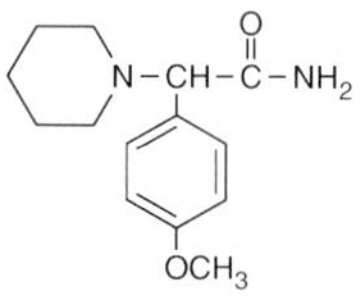
5 gm

S35,237-3
ALPHA-(4-METHOXYPHENYL)-4-MORPHOLINEACETAMIDE
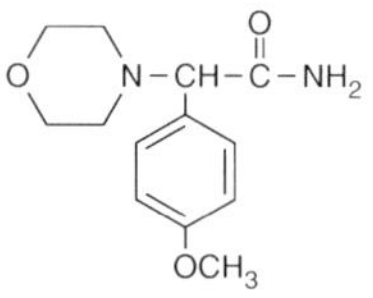
5 gm

S70,101-7
4'-METHOXY-2-PIPERIDINOACETANILIDE
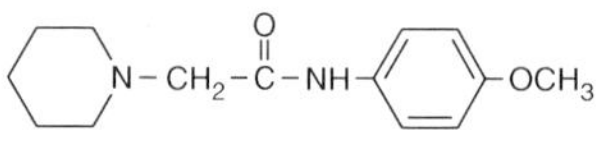
5 gm

S85,561-8
2,3-DIBROMO-2'-METHOXY-3'-NITROSUCCINANILIC ACID
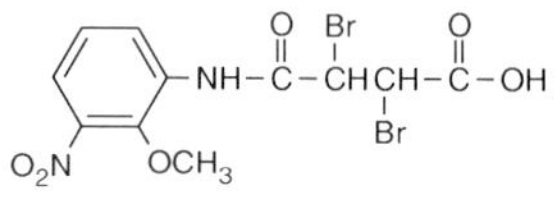
250 mg

S80,775-3
ALPHA,BETA-DIBROMO-2'-METHOXY-3'-NITROHYDROCINNAMANILIDE
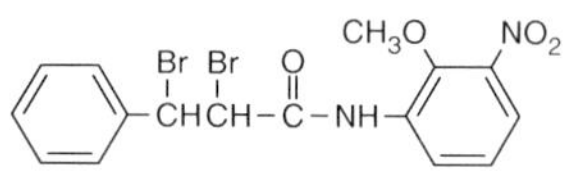
250 mg

S67,651-9
N-(2-METHOXY-4-NITROPHENYL)-2-PHENYLACETAMIDE
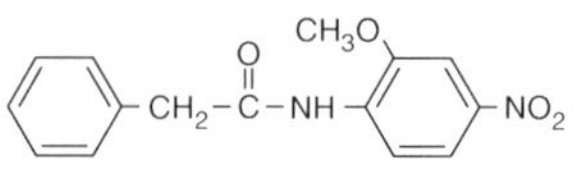
5 gm

S67,652-7
N-(2-METHOXY-5-NITROPHENYL)-2-PHENYLACETAMIDE
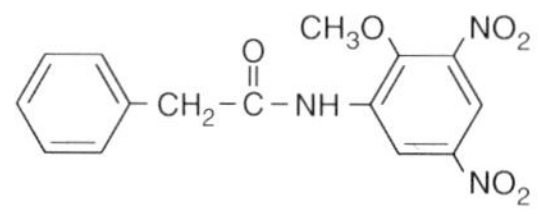
250 mg

S94,154-9
5'-CHLORO-4'-NITRO-2'-PROPOXYACETANILIDE
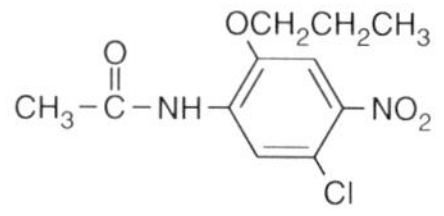
250 mg

S38,110-1
ALPHA'-ACETYL-ORTHO-ACETANISIDIDE
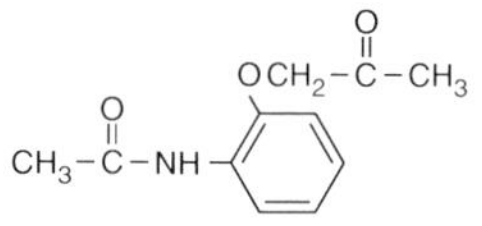
5 gm

S85,563-4
2,3-DIBROMO-2'-ETHOXYSUCCINANILIC ACID
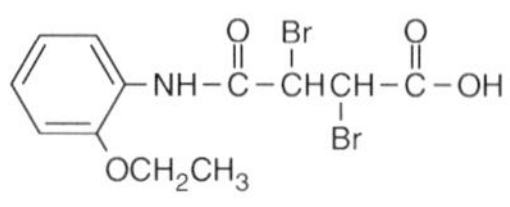
250 mg

S85,564-2
2,3-DIBROMO-2'-METHOXY-5'-METHYLSUCCINANILIC ACID
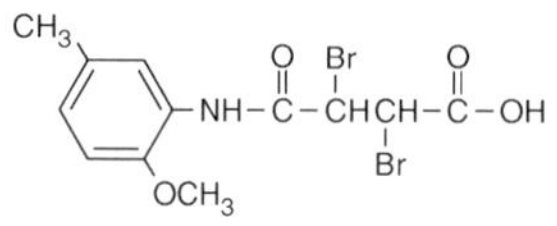
250 mg

S80,776-1
ALPHA,BETA-DIBROMO-2'-ETHOXYHYDROCINNAMANILIDE
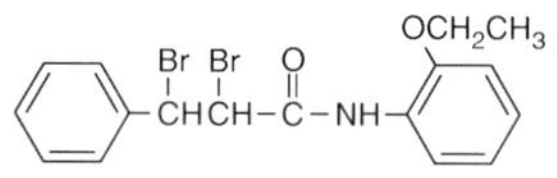
250 mg

S61,874-8
3-ACETAMIDO-4-HEXADECYLOXYBENZOIC ACID
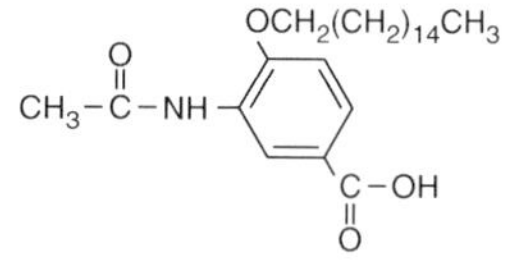
5 gm

S67,604-7
2-(2-ETHOXYPHENYLCARBAMOYL)-CYCLOHEXANECARBOXYLIC ACID
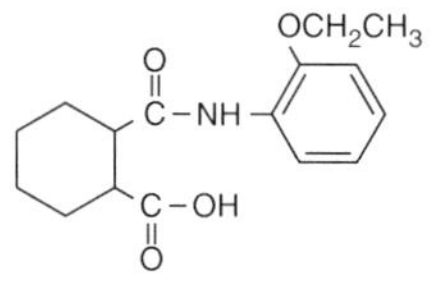
250 mg

S72,910-8
CIS-HEXAHYDRO-N-(2-METHOXY-5-METHYLPHENYL)PHTHALAMIC ACID
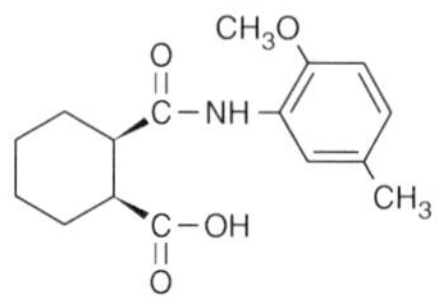
5 gm

S70,401-6
2'-ETHOXYSUCCINANILIC ACID
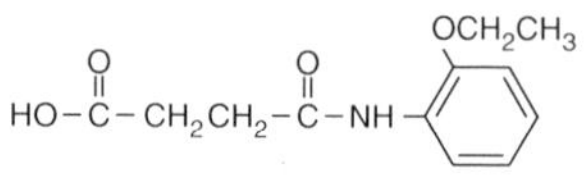
5 gm

S70,424-5
2'-METHOXY-5'-METHYLSUCCINANILIC ACID
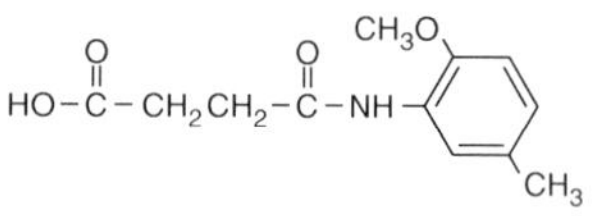
5 gm

S72,925-6
2'-METHOXY-5'-METHYLGLUTARANILIC ACID
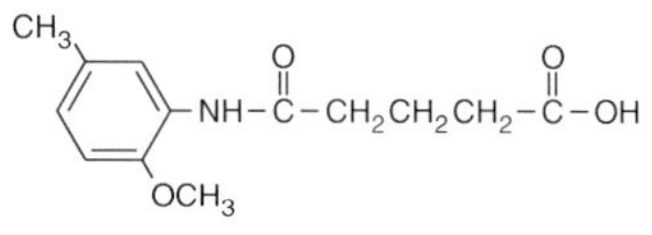
5 gm

S72,105-0
1,4-BENZODIOXAN-2-YLMETHYL N-(2-METHOXY-5-METHYLPHENYL)-CARBAMATE
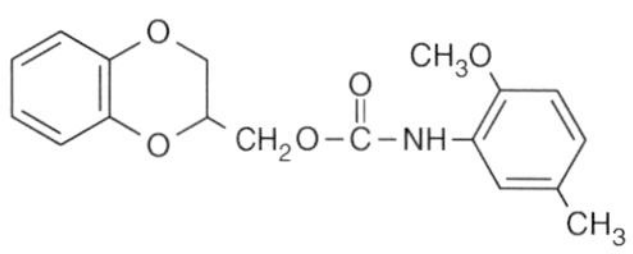
5 gm

S74,375-5
2'-METHOXY-2-PHENYLACETANILIDE
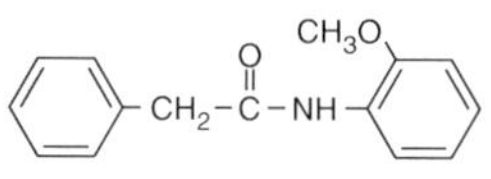
5 gm

S78,054-5
2,2-DIPHENYL-2'-ETHOXYACETANILIDE
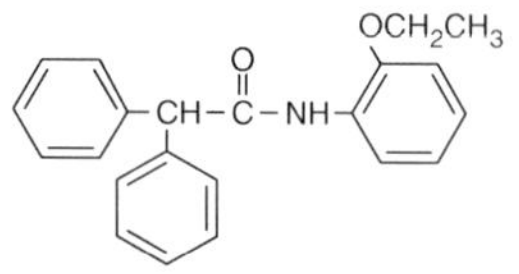
250 mg

S51,668-6
ORTHO-FORMANISIDIDE
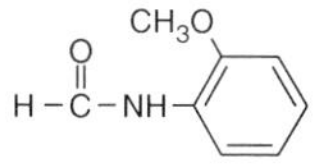
5 gm

S51,665-1
ORTHO-FORMOPHENETIDIDE
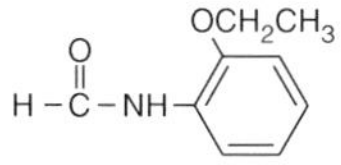
5 gm

S59,088-6
O-ACETANISIDIDE
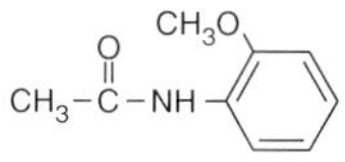
250 mg

S81,526-8
2'-(ALLYLOXY)ACETANILIDE
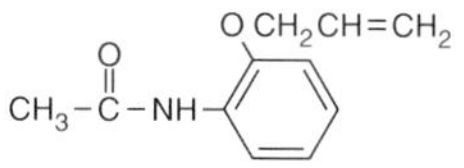
250 mg

S79,104-0
2'-ETHOXYPROPIONANILIDE
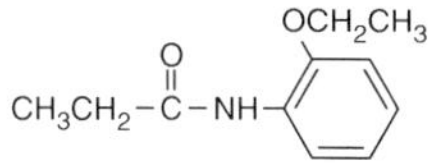
250 mg

S57,355-8
2'-(OCTYLOXY)ACETANILIDE
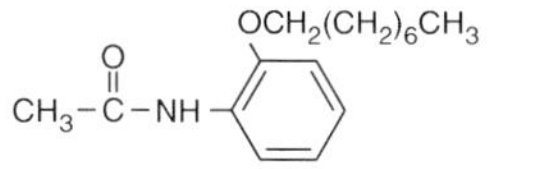
5 gm

S57,482-1
2'-(DECYLOXY)ACETANILIDE
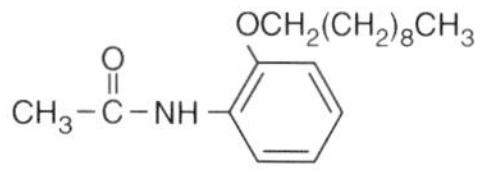
5 gm

S57,557-7
2',2'''-(TRIMETHYLENEDIOXY)-DIACETANILIDE
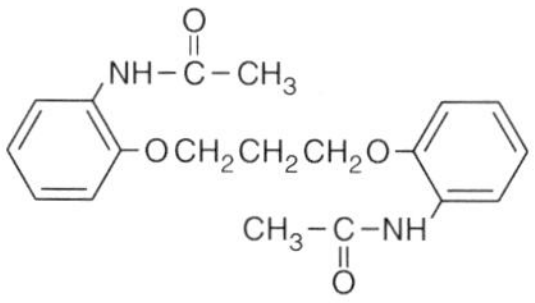
5 gm

S57,515-1
2'-(DODECYLOXY)ACETANILIDE
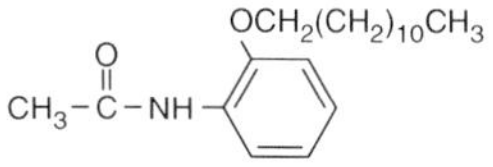
5 gm

S57,349-3
2',2'''-(TETRAMETHYLENEDIOXY)-DIACETANILIDE
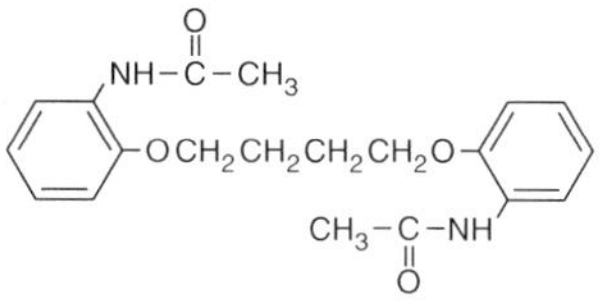
5 gm

S57,350-7
2',2''-(PENTAMETHYLENEDIOXY)-DIACETANILIDE
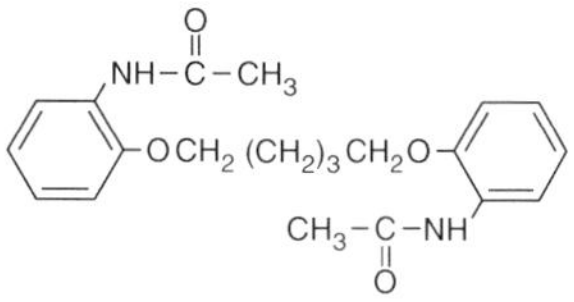
5 gm

S57,495-3
2'-(TETRADECYLOXY)ACETANILIDE
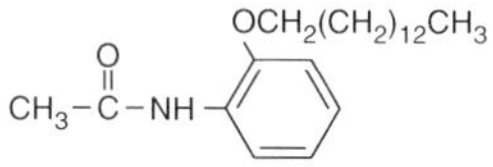
5 gm

S57,373-6
2',2''-(HEXAMETHYLENEDIOXY)-DIACETANILIDE
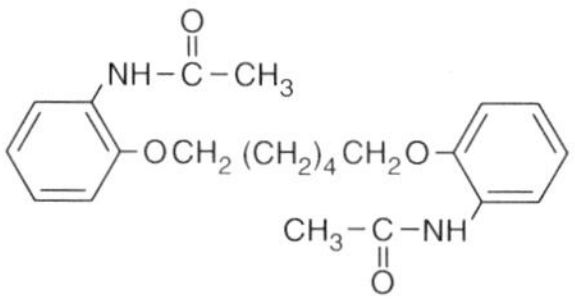
5 gm

S60,403-8
2'-(HEXADECYLOXY)ACETANILIDE
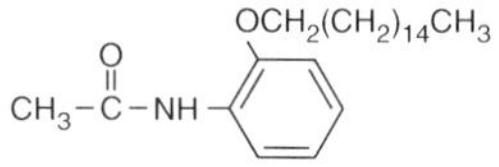
5 gm

S57,519-4
2'-(OCTADECYLOXY)ACETANILIDE
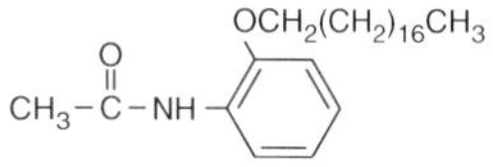
5 gm

S57,517-8
2',2''-(DECAMETHYLENEDIOXY)-DIACETANILIDE
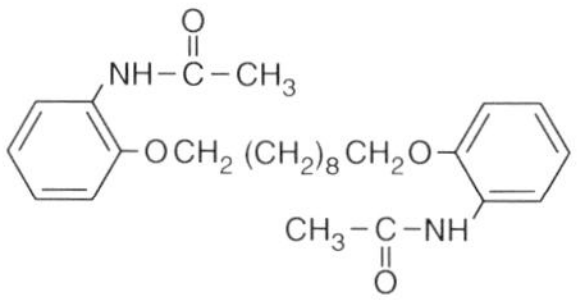
5 gm

S62,333-4
3'-CHLORO-2'-METHOXYFORMANILIDE
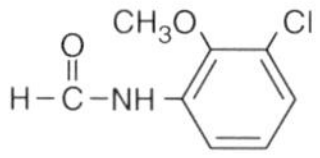
5 gm

S62,326-1
2'-METHOXY-5'-METHYLFORMANILIDE
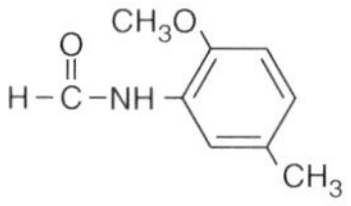
5 gm

S57,670-0
2',5'-DIMETHOXYFORMANILIDE
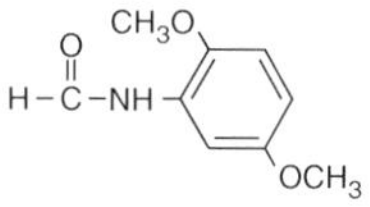
5 gm

S65,467-1
2',4'-DIMETHOXYACETANILIDE
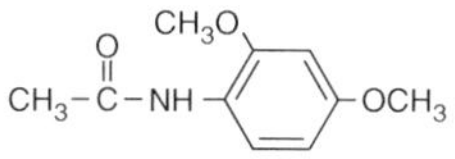
250 mg

S85,503-0
2'-METHOXY-5'-METHYLBUTYRANILIDE
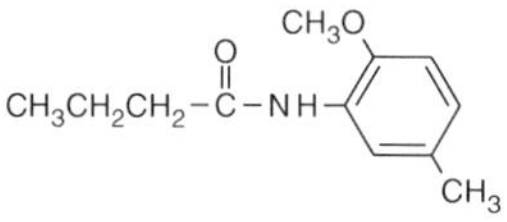
250 mg

S91,395-2
2'-METHOXY-5'-METHYLDECANANILIDE
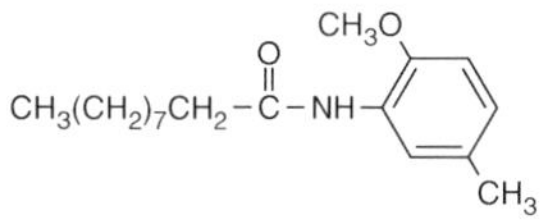
250 mg

S67,629-2
N-(2-METHOXY-5-METHYLPHENYL)-DODECANAMIDE
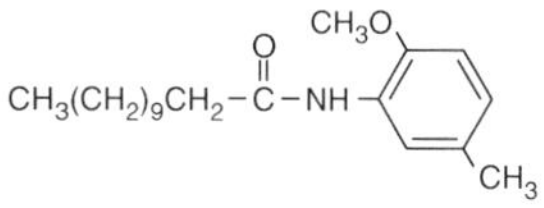
5 gm

S57,638-7
5'-CHLORO-2',4'-DIMETHOXYFORMANILIDE
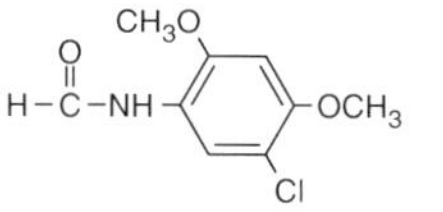
250 mg

S77,087-6
4-METHOXYHYDROCINNAMOHYDROXAMIC ACID
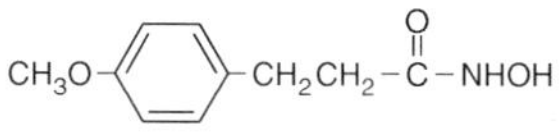
250 mg

S90,964-5
4'-METHOXYFORMANILIDE
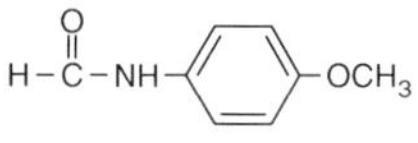
250 mg

S94,481-5
7-METHOXY-1-METHYL-4-OXO-1,2,3,4-TETRAHYDRO-1-NAPHTHALENECARBOXAMIDE
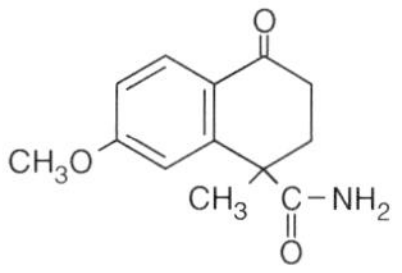
250 mg

S73,854-9
2,3-DIBROMO-4'-ETHOXYSUCCINANILIC ACID
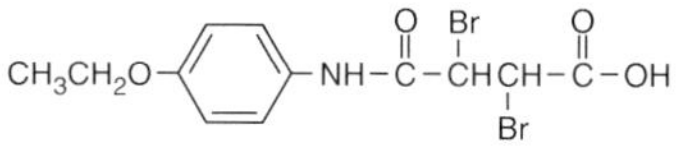
250 mg

S44,799-4
METHYL 4-ACETAMIDO-2-METHOXYBENZOATE
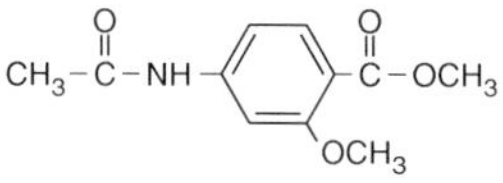
5 gm

S44,803-6
METHYL 4-ACETAMIDO-5-CHLORO-2-METHOXY-BENZOATE
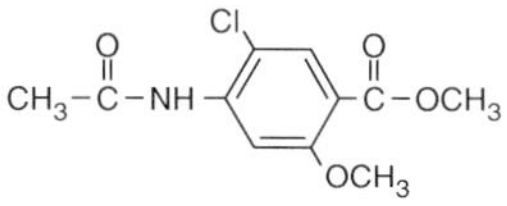
5 gm

S42,098-0
N-ETHYL-N-(2-PHENOXYETHYL)-CROTONAMIDE
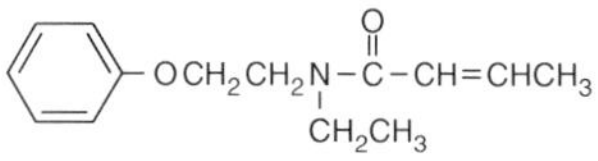
5 gm

S42,130-8
N-ETHYL-3-METHYL-N-(2-PHENOXYETHYL)-CROTONAMIDE
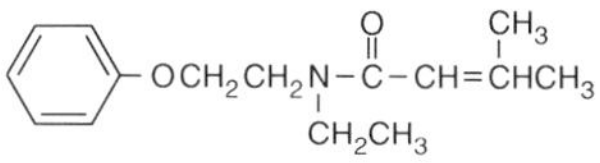
5 gm

S42,244-4
N,3-DIMETHYL-N-[1-METHYL-2-(2,4,6-TRIMETHYLPHENOXY)ETHYL]-CROTONAMIDE
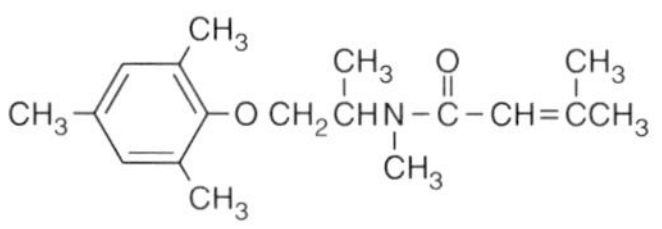
250 mg

S93,140-3
ETHYL 4-METHOXYOXANILATE
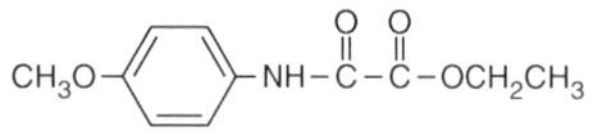
250 mg

S54,141-9
PARA-GLYOXYLOPHENETIDIDE
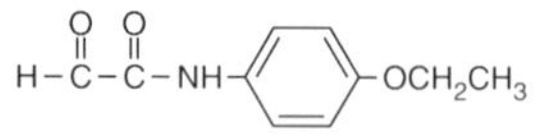
250 mg

S72,070-4
N-(3-METHOXYPHENYL)GLUTARAMIC ACID
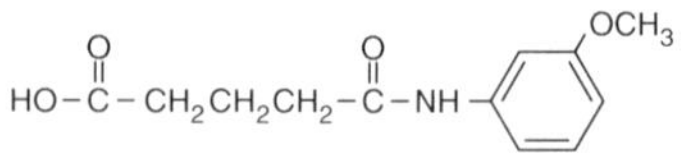
5 gm

S71,935-8
3-[N-(4-METHOXYPHENYL)CARBAMOYL]-5-NORBORNENE-2-CARBOXYLIC ACID
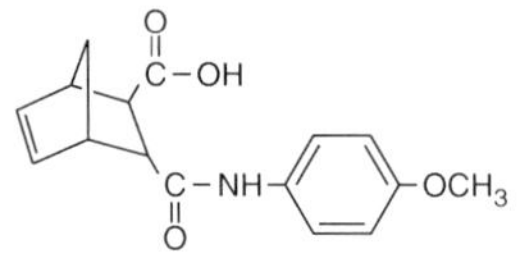
5 gm

S72,015-5
1,4,5,6,7,7-HEXACHLORO-3-[N-(4-METHOXYPHENYL)CARBAMOYL]-5-NORBORNEN-2-CARBOXYLIC ACID
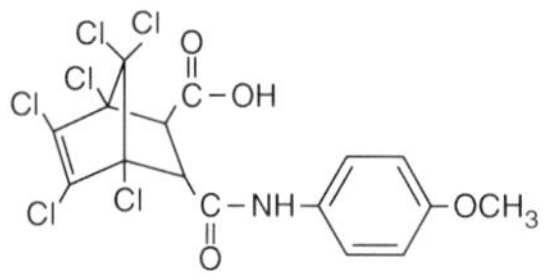
5 gm

S71,743-6
3-[N-(4-ETHOXYPHENYL)CARBAMOYL]-BICYCLO[2.2.1]HEPT-5-ENE-2-CARBOXYLIC ACID
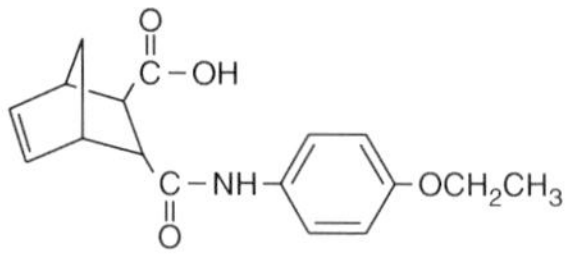
5 gm

S71,989-7
3-[N-(4-ETHOXYPHENYL)CARBAMOYL]-1,4,5,6,7,7-HEXACHLORO-5-NORBORNENE-2-CARBOXYLIC ACID
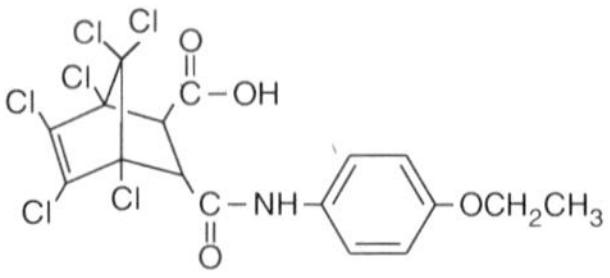
5 gm

S70,365-6
CIS-HEXAHYDRO-N-(4-METHOXYPHENYL)-PHTHALAMIC ACID
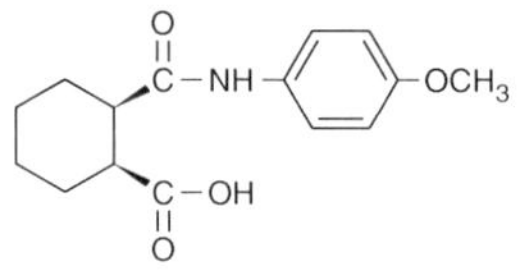
5 gm

S72,014-3
CIS-HEXAHYDRO-N-(4-ETHOXYPHENYL)-PHTHALAMIC ACID
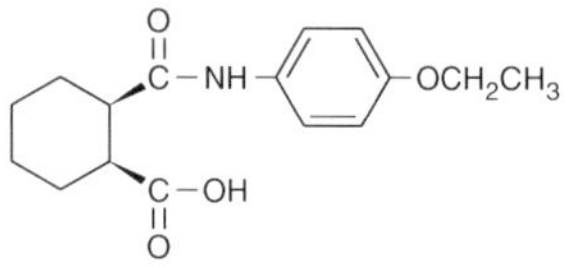
5 gm

S73,044-0
N-(4-PHENOXYPHENYL)GLUTARAMIC ACID
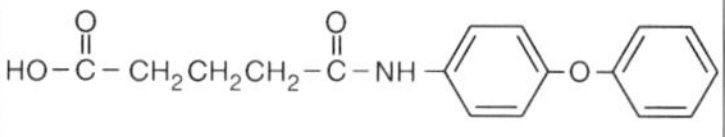
5 gm

S70,102-5
4'-METHOXYSUCCINANILIC ACID
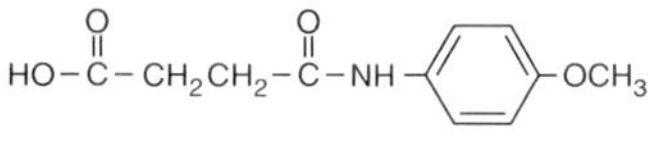
5 gm

S67,471-0
3-(4-ETHOXYPHENYLCARBAMOYL)-PROPIONIC ACID
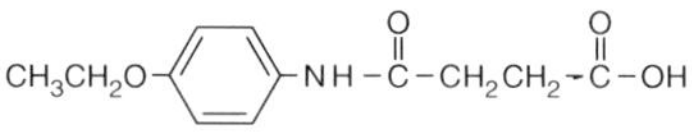
5 gm

S87,805-7
N-(4-METHOXYPHENYL)GLUTARAMIC ACID
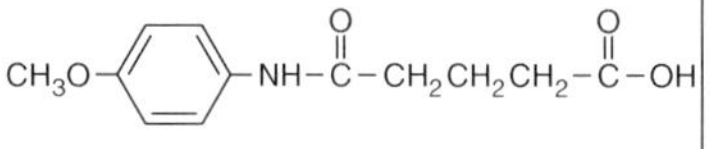
5 gm

S70,364-8
4'-ETHOXYGLUTARANILIC ACID
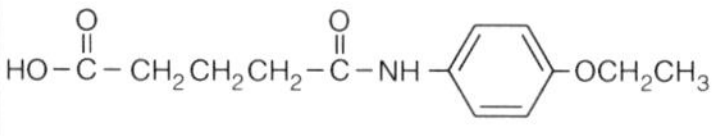
5 gm

S34,318-8
DEXTRO-CAMPHORIC P-PHENETIDIDE
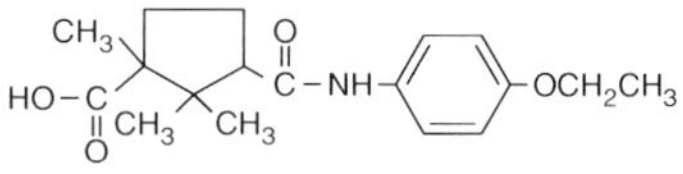
5 gm

S92,998-0
N-(4-ETHOXYPHENYL)-1,2,3,6-TETRAHYDROPHTHALAMIC ACID
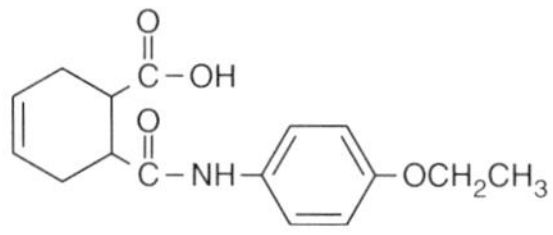
250 mg

S71,929-3
N-(4-METHOXYPHENYL)-1,2,5,6-TETRAHYDROPHTHALAMIC ACID
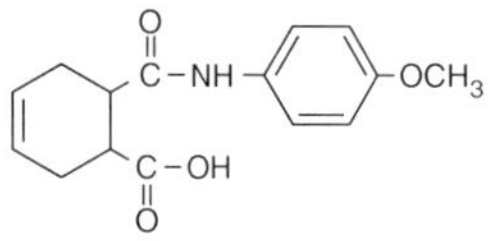
5 gm

S92,580-2
S-(ETHOXYCARBONYLMETHYL)-1,1-DITHIO-4'-METHOXYOXANILATE
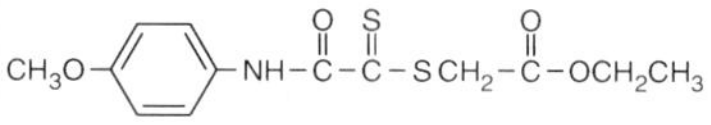
250 mg

S85,009-8
3-(4-METHOXYPHENYL)PROPIONAMIDE
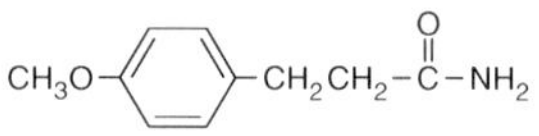
250 mg

S51,660-0
META-FORMANISIDIDE
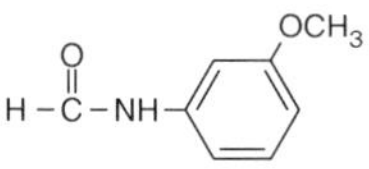
5 gm

S38,968-4
META-ACETANISIDIDE
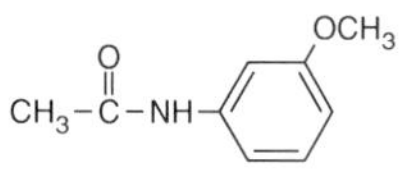
5 gm

S80,615-3
3'-ETHOXYACETANILIDE
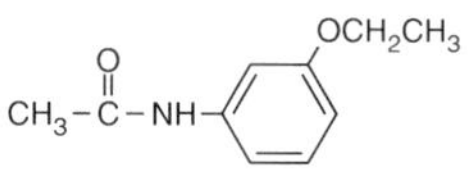
250 mg

S57,357-4
3'-BENZYLOXYACETANILIDE
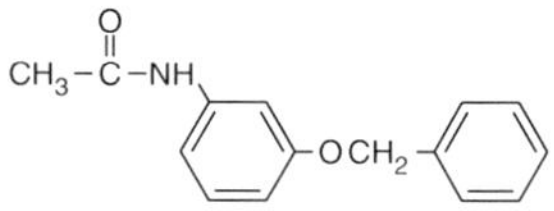
5 gm

S57,480-5
3'-(OCTYLOXY)ACETANILIDE
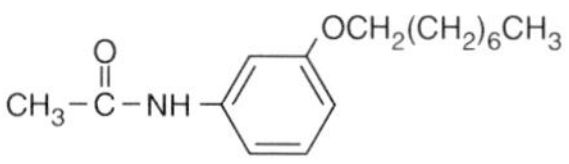
5 gm

S57,569-0
3'-(DECYLOXY)ACETANILIDE
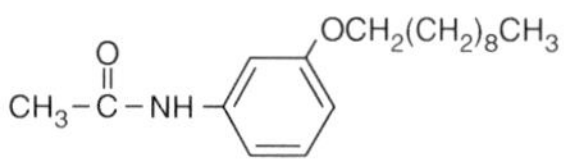
5 gm

S57,562-3
3',3'''-(TRIMETHYLENEDIOXY)-DIACETANILIDE
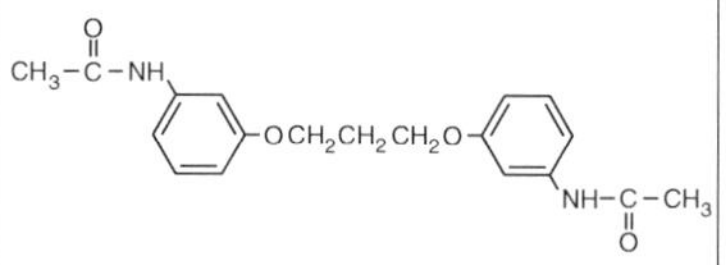
5 gm

S57,361-2
3'-(DODECYLOXY)ACETANILIDE
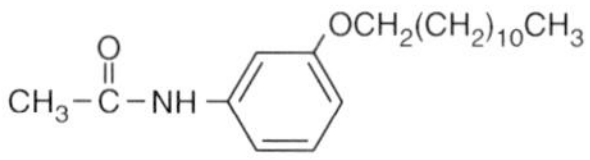
5 gm

S57,558-5
3',3'''-(TETRAMETHYLENEDIOXY)-DIACETANILIDE
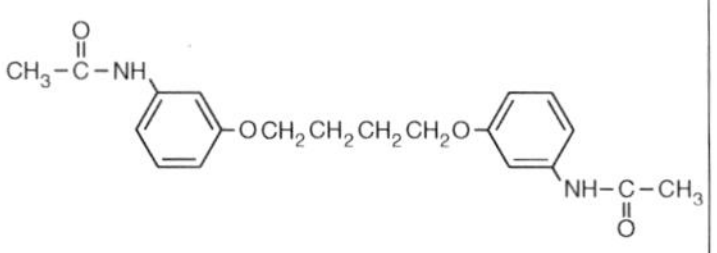
5 gm

S57,359-0
3',3'''-(PENTAMETHYLENEDIOXY)-DIACETANILIDE
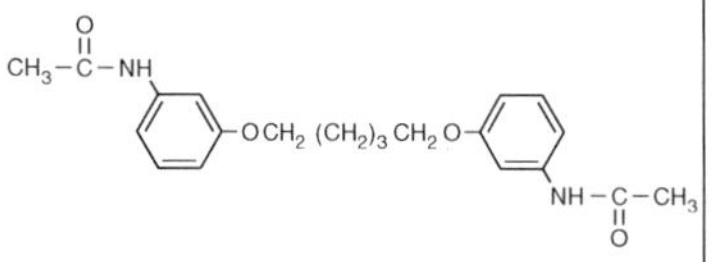
5 gm

S57,530-5
3'-(TETRADECYLOXY)ACETANILIDE
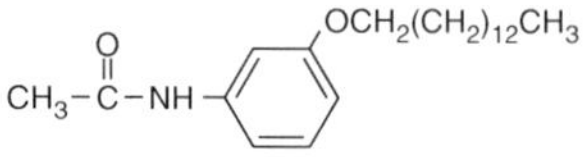
5 gm

S57,479-1
3'-(HEXADECYLOXY)ACETANILIDE
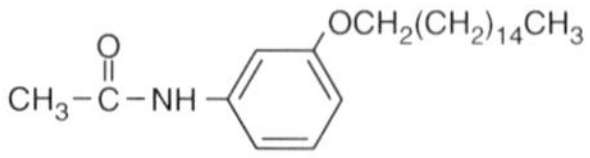
5 gm

S57,372-8
3'-(OCTADECYLOXY)ACETANILIDE

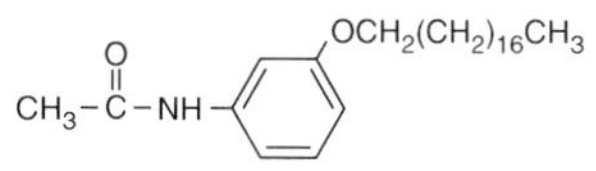

5 gm

S57,486-4
3',3''-(DECAMETHYLENEDIOXY)-DIACETANILIDE

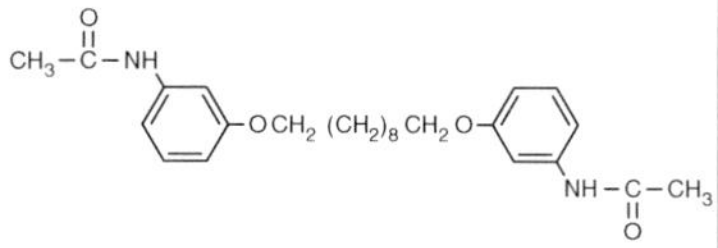

5 gm

S68,898-3
N-(4-ETHOXYPHENYL)-2-PIPERIDINOACETAMIDE

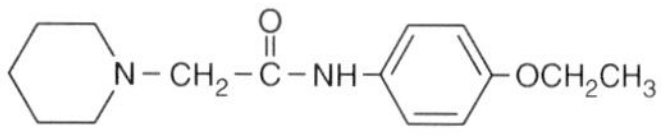

5 gm

S74,723-8
4'-METHOXY-2-PHENYLACETANILIDE

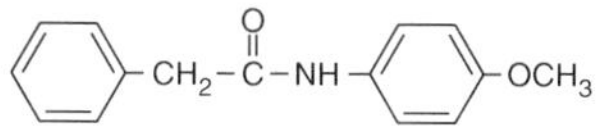

5 gm

S78,055-3
2,2-DIPHENYL-4'-ETHOXYACETANILIDE

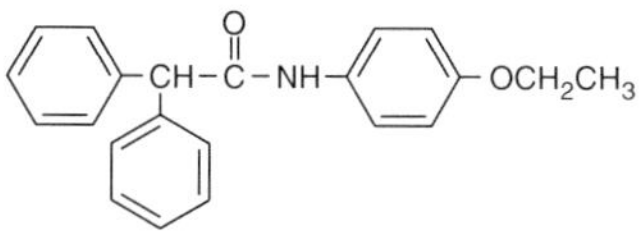

250 mg

S93,705-3
N-(4-BIPHENYLYL)-2-METHOXYACETAMIDE

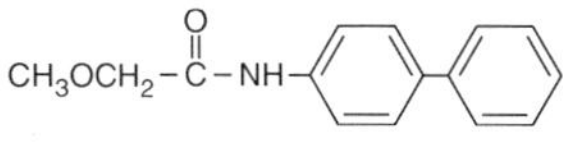

250 mg

S51,662-7
PARA-FORMANISIDIDE

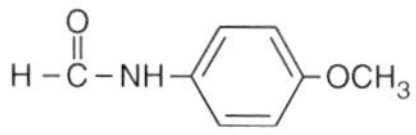

5 gm

S51,667-8
PARA-FORMOPHENETIDIDE

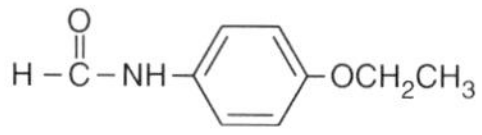

5 gm

S59,089-4
P-ACETANISIDIDE

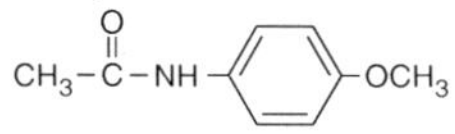

5 gm

S89,069-3
4'-ETHOXYPROPIONANILIDE

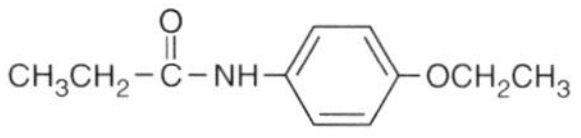

250 mg

S57,617-4
4'-BUTOXYACETANILIDE

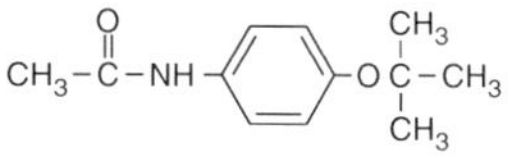

5 gm

S90,274-8
2,2-DIMETHYL-4'-METHOXYPROPIONANILIDE

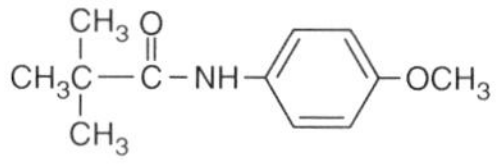

250 mg

S90,091-5
2,2-DIMETHYL-4'-ETHOXYPROPIONANILIDE

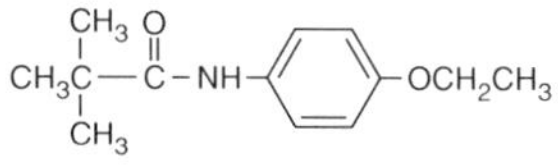

250 mg

S57,526-7
4'-BENZYLOXYACETANILIDE

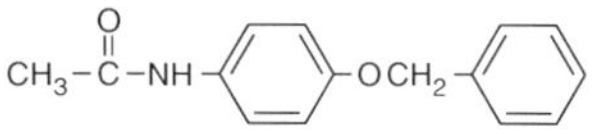

5 gm

S51,266-4
4'-(OCTYLOXY)ACETANILIDE

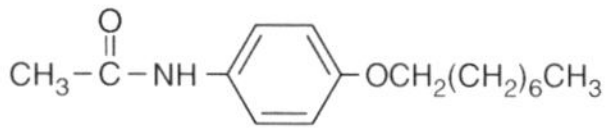

5 gm

S89,086-3
4'-ETHOXYOCTANANILIDE

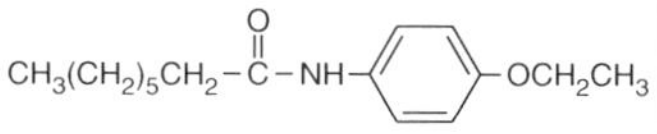

250 mg

S53,153-7
ALPHA-(5-CHLORO-2,4-DIMETHOXYANILINO)-PARA-ACETOTOLUIDIDE

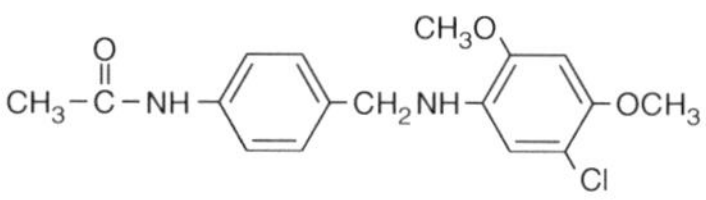

5 gm

S57,561-5
4'-(DECYLOXY)ACETANILIDE

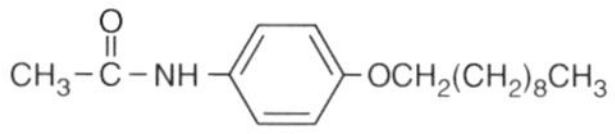

5 gm

S71,817-3
4'-ETHOXYDECANANILIDE

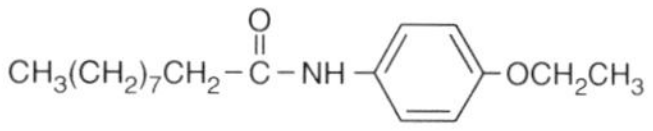

250 mg

S91,657-9
4',4''-OXYBIS(PROPIONANILIDE)

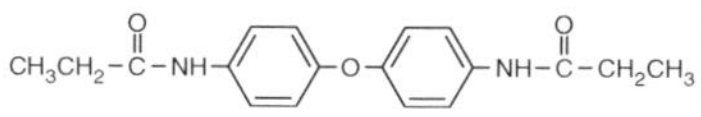

250 mg

S57,514-3
4',4'''-(TRIMETHYLENEDIOXY)-DIACETANILIDE

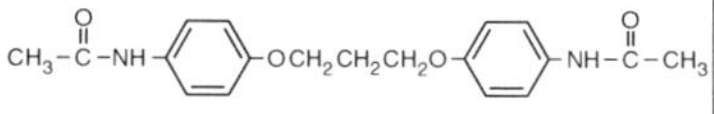

5 gm

S57,566-6
4'-(DODECYLOXY)ACETANILIDE

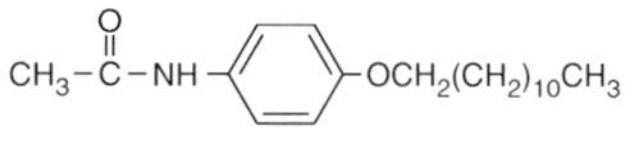

5 gm

S57,528-3
4',4'''-(TETRAMETHYLENEDIOXY)-DIACETANILIDE

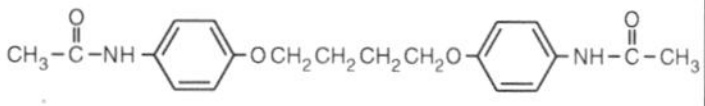

5 gm

S57,534-8
4',4''-(PENTAMETHYLENEDIOXY)-DIACETANILIDE

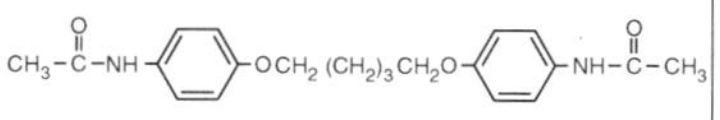

5 gm

S57,474-0
4'-(TETRADECYLOXY)ACETANILIDE

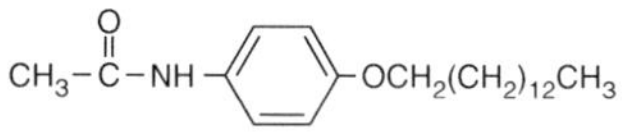

5 gm

S57,353-1
4',4''-(HEXAMETHYLENEDIOXY)-DIACETANILIDE

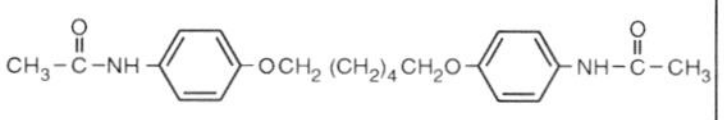

5 gm

S61,844-6
HEXADECA-P-ANISIDIDE

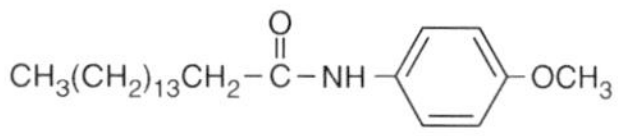

5 gm

S57,560-7
4'-(HEXADECYLOXY)ACETANILIDE

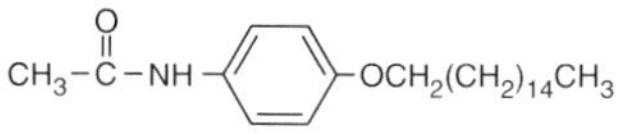

5 gm

S82,136-5
4'-ETHOXYHEPTADECANANILIDE

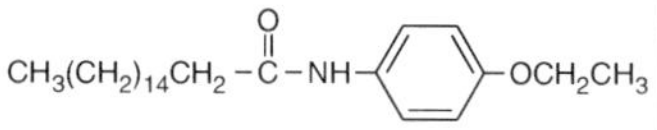

250 mg

S57,518-6
4'-(OCTADECYLOXY)ACETANILIDE

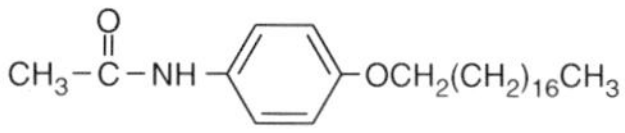

5 gm

S57,352-3
4',4''-(DECAMETHYLENEDIOXY)-DIACETANILIDE

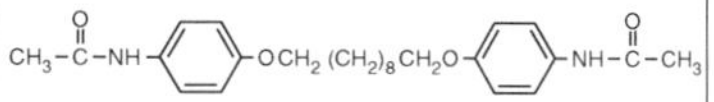

5 gm

S62,608-2
4'-METHOXY-2'-METHYLFORMANILIDE

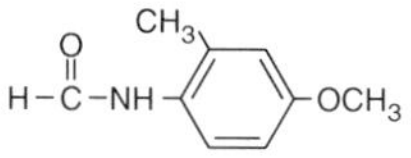

5 gm

S94,152-2
3'-CHLORO-4'-METHOXYACETANILIDE
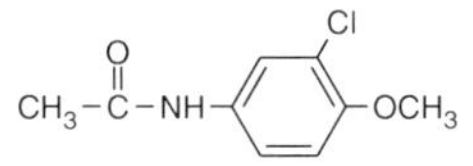
250 mg

S78,930-5
L-GLUTAMINE 4-NITROBENZYL ESTER HYDROBROMIDE
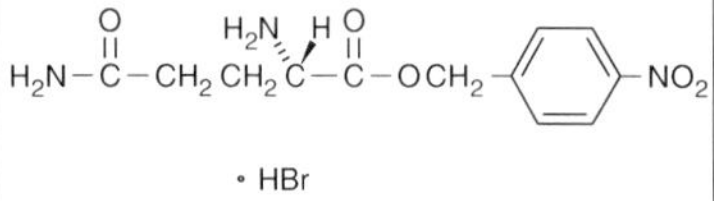
250 mg

S78,809-0
D-ALANYL-D-ALANYL-D-ALANINE BENZYL ESTER
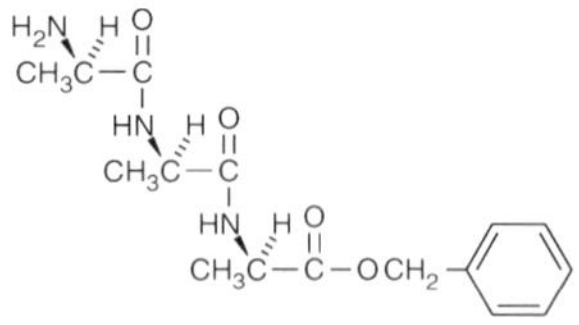
250 mg

S92,454-7
METHYL 2-(CYCLOPROPYLCARBOXAMIDO)-BENZOATE
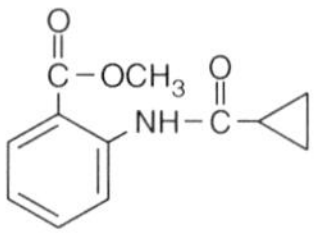
250 mg

S35,422-8
GAMMA-PENTABENZYL N-CARBOBENZYLOXY-L-GLUTAMYL-L-GLUTAMYL-L-GLUTAMYL-L-GLUTAMATE
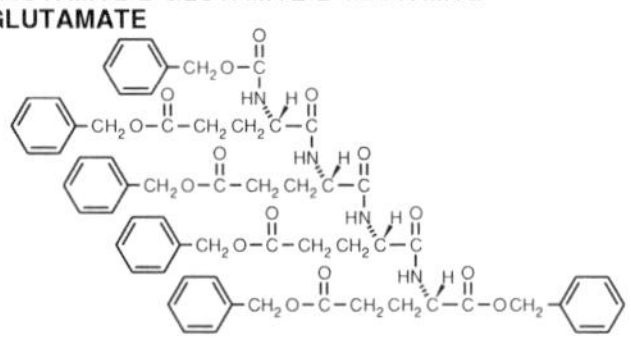
250 mg

S78,338-2
GLYCYLGLYCINE BENZYL ESTER P-TOLUENESULFONATE
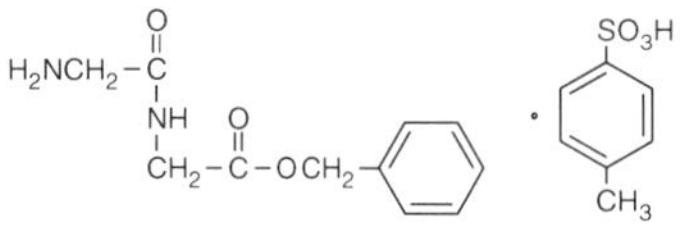
5 gm

S73,015-7
N-(2-PHENETHYL)GLUTARAMIC ACID
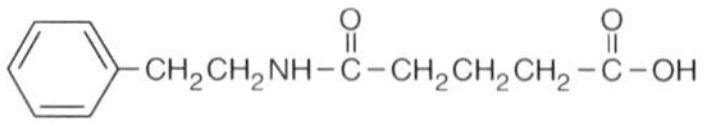
5 gm

S80,503-3
4'-(DIACETOXYMETHYL)-2'-NITROACETANILIDE
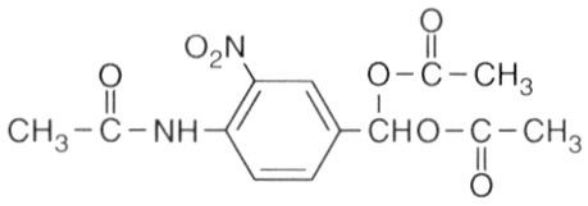
250 mg

S67,833-3
N-(3,4-DIACETOXYPHENETHYL)ACETAMIDE
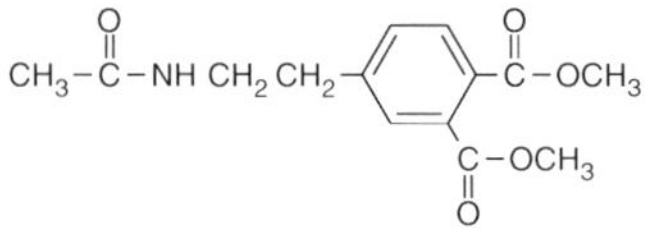
5 gm

S56,621-7
3-METHYL-1,4-BENZODIOXAN-2-CARBOXAMIDE
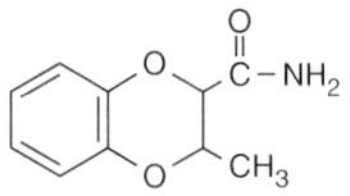
5 gm

S65,748-4
2-(4-BROMOPHENOXY)-N,N,2-TRIMETHYLPROPIONAMIDE
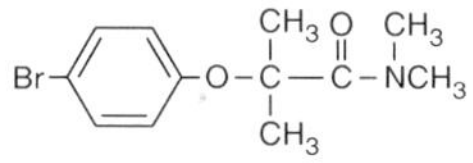
5 gm

S95,374-1
4'-(METHYLTHIO)SUCCINANILIC ACID
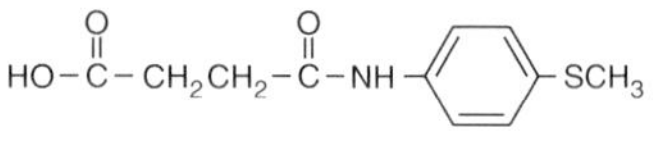
50 mg

S87,816-2
4'-ANILINOSUCCINANILIC ACID
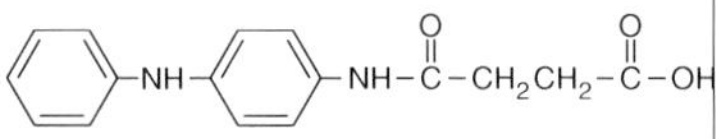
5 gm

S92,182-3
4'-METHYL-3'-NITROSUCCINANILIC ACID
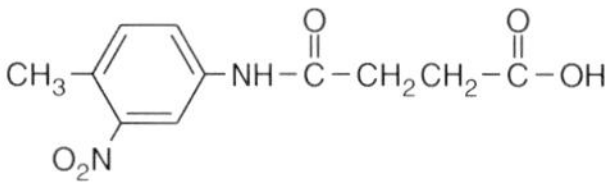
5 gm

S87,808-1
4'-(ETHOXYCARBONYL)-SUCCINANILIC ACID
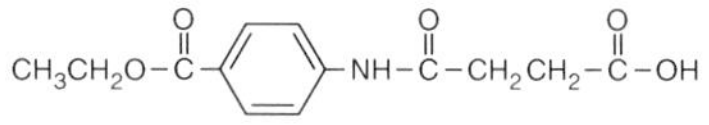
5 gm

S87,842-1
4'-BUTYLHEXAHYDROPHTHALANILIC ACID
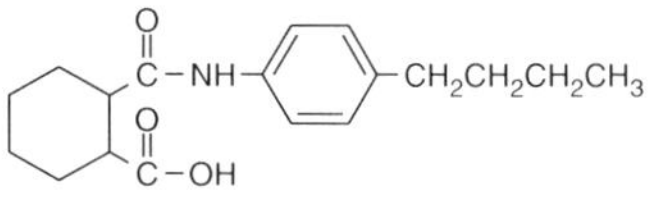
5 gm

S72,526-9
CIS-2',5'-DIMETHYLHEXAHYDROPHTHALANILIC ACID
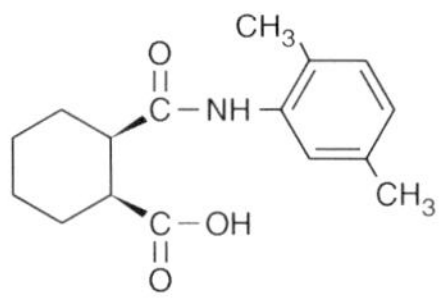
5 gm

S87,818-9
N-(2,3-DIMETHYLPHENYL)-1,2,3,6-TETRAHYDROPHTHALAMIC ACID
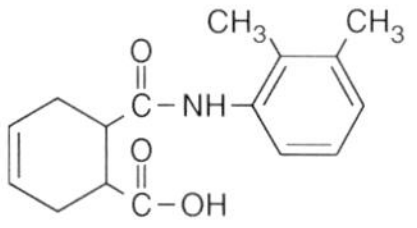
5 gm

S87,819-7
N-(2,6-DIMETHYLPHENYL)-1,2,3,6-TETRAHYDROPHTHALAMIC ACID
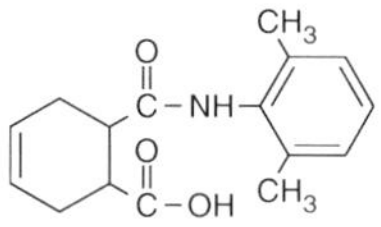
5 gm

S87,817-0
N-(2,4-DIMETHYLPHENYL)-1,2,3,6-TETRAHYDROPHTHALAMIC ACID
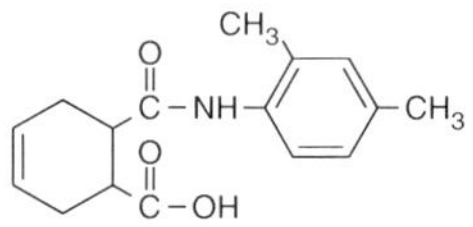
5 gm

S95,368-7
2'-CARBOXYSUCCINANILIC ACID
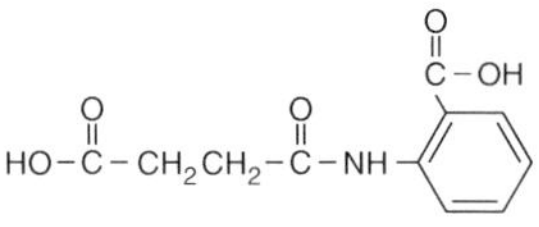
50 mg

S37,634-5
2-MERCAPTOACETANILIDE
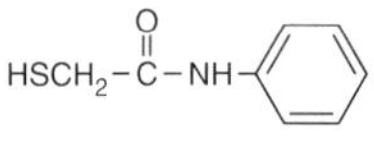
5 gm

S37,379-6
2-MERCAPTO-PARA-PROPIONOTOLUIDIDE
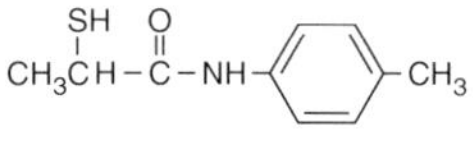
5 gm

S38,024-5
6'-CHLORO-2-MERCAPTO-ORTHO-ACETOTOLUIDIDE
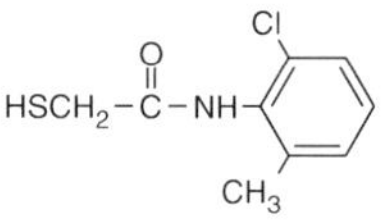
5 gm

S36,305-7
3-MERCAPTO-PARA-PROPIONOTOLUIDIDE
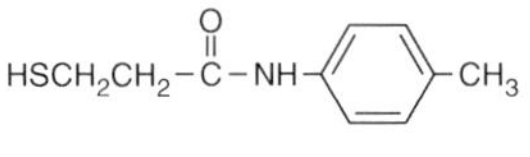
5 gm

S86,052-2
2-MERCAPTO-N-(2-NAPHTHYL)ACETAMIDE
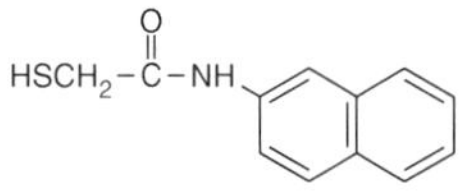
250 mg

S65,523-6
2'-METHYLTHIO-5'-NITROACETANILIDE
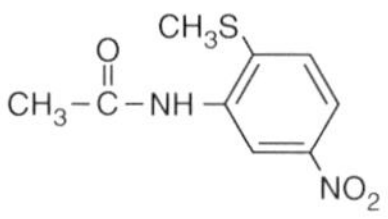
5 gm

S93,384-8
2-(BENZYLTHIO)-3'-NITROACETANILIDE
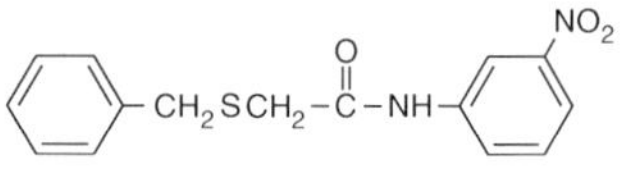
250 mg

S93,385-6
2-(BENZYLTHIO)-2'-METHYL-5'-NITROACETANILIDE
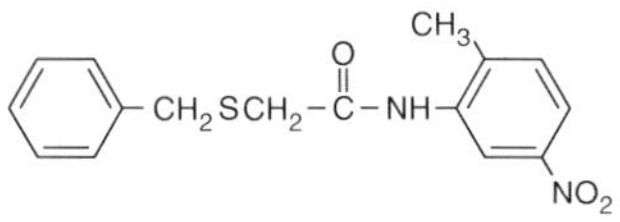
250 mg

S38,360-0
3-METHYL-N-[2-(METHYLTHIO)ETHYL]-2-PHENYLVALERAMIDE
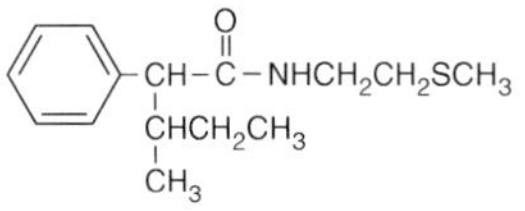
250 mg

S53,142-1
ALPHA-[4-(METHYLTHIO)ANILINO]-PARA-ACETOTOLUIDIDE
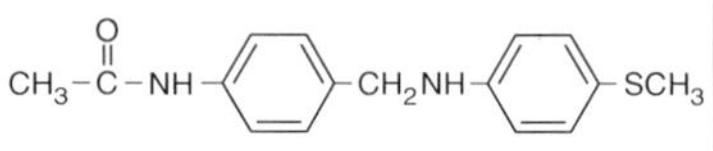
250 mg

S40,173-0
6'-CHLORO-[2-(2-DIBUTYLAMINO)-ETHYLTHIO]-ORTHO-ACETOTOLUIDIDE HYDROCHLORIDE
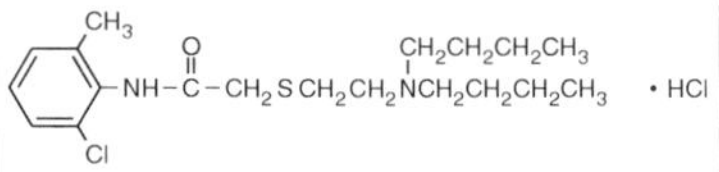
250 mg

S42,065-4
2'-CHLORO-2-[2-(DIMETHYLAMINO)-ETHYLTHIO]-ACETANILIDE HYDROCHLORIDE

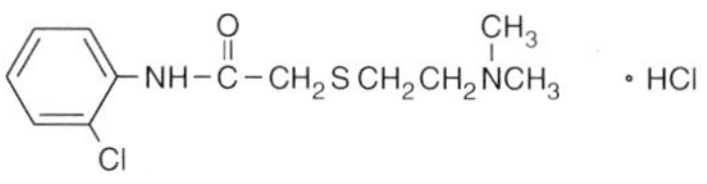

5 gm

S42,237-1
2'-CHLORO-2-[2-(DIBUTYLAMINO)-ETHYLTHIO]-ACETANILIDE HYDROCHLORIDE

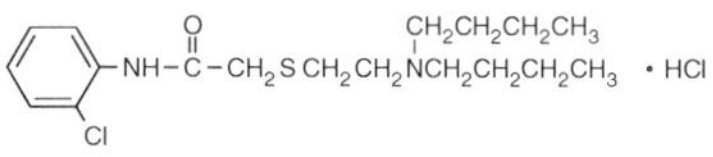

5 gm

S42,122-7
2',5'-DICHLORO-2-[2-(DIETHYLAMINO)-ETHYLTHIO]-ACETANILIDE HYDROCHLORIDE

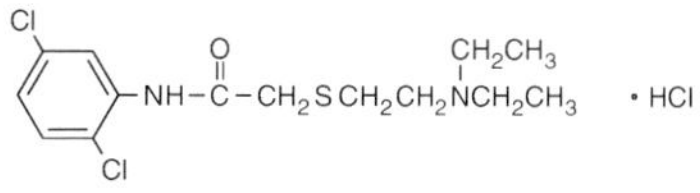

250 mg

S42,123-5
2',4'-DICHLORO-2-[2-(DIETHYLAMINO)-ETHYLTHIO]-ACETANILIDE HYDROCHLORIDE

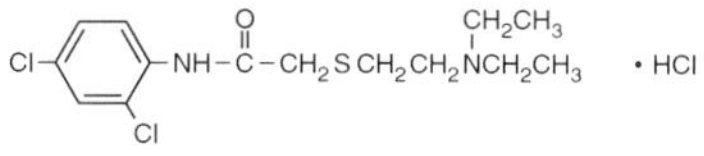

250 mg

S42,128-6
3',4'-DICHLORO-2-[2-(DIETHYLAMINO)-ETHYLTHIO]-PROPIONANILIDE

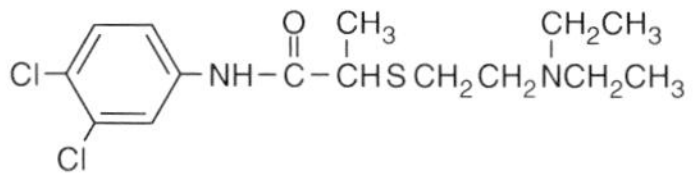

250 mg

S42,172-3
2',5'-DICHLORO-2-[2-(DIPROPYLAMINO)-ETHYLTHIO]-ACETANILIDE HYDROCHLORIDE

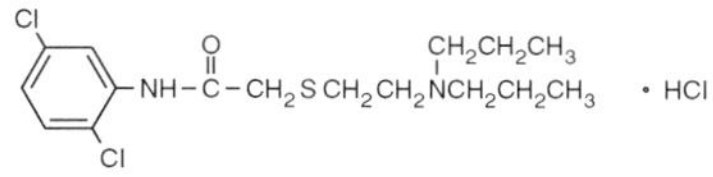

250 mg

S42,242-8
2-[2-(DIBUTYLAMINO)ETHYLTHIO]-2',5'-DICHLOROACETANILIDE HYDROCHLORIDE

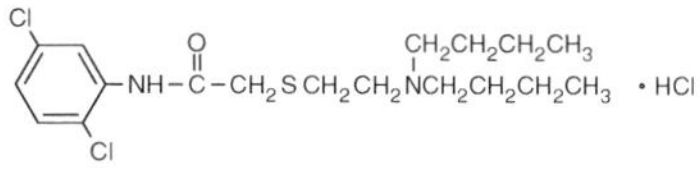

5 gm

S63,792-0
2-(4-CHLOROPHENYLTHIO)ACETAMIDE

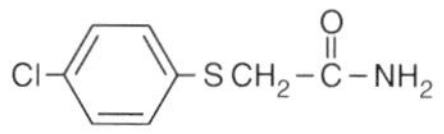

5 gm

S93,722-3
2'-(4-BROMO-2-NITROPHENYLTHIO)-ACETANILIDE

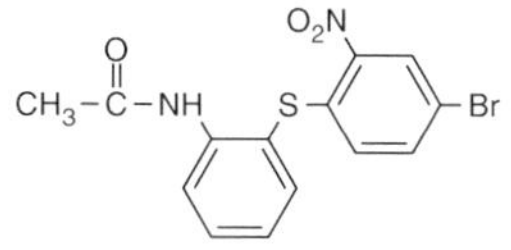

250 mg

S85,578-2
2,3-DIBROMO-4'-(METHYLTHIO)-3-PHENYLPROPIONANILIDE

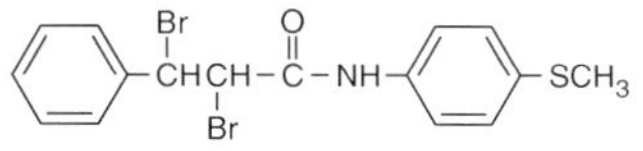

250 mg

S68,379-5
2-(CARBAMOYLMETHYLTHIO)BENZOIC ACID

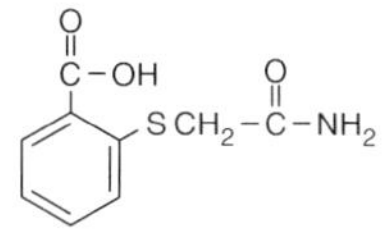

250 mg

S90,024-9
HEXAHYDRO-N-[4-(METHYLTHIO)PHENYL]-PHTHALAMIC ACID

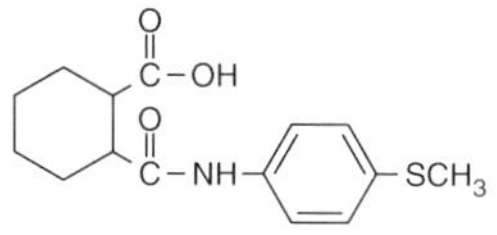

250 mg

S78,268-8
N-ACETYL-S-BENZYL-L-CYSTEINE

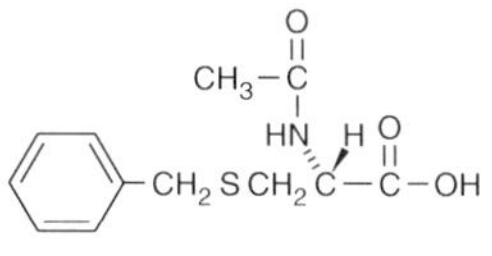

5 gm

S61,438-6
2-(BENZYLTHIO)ACETAMIDE

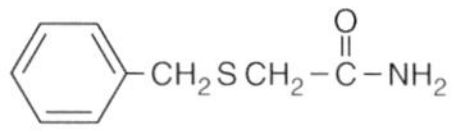

5 gm

S62,015-7
3-(BENZYLTHIO)PROPIONAMIDE

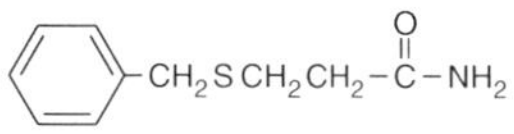

5 gm

S64,618-0
2-METHYL-3-(PHENYLTHIO)PROPIONAMIDE

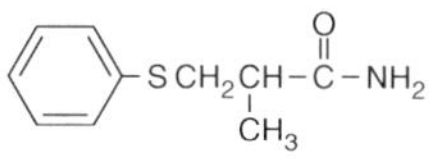

250 mg

S54,557-0
3'-(METHYLTHIO)ACETANILIDE

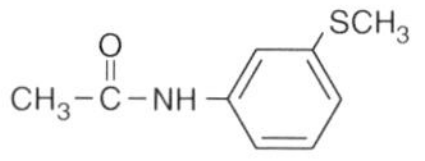

5 gm

S89,238-6
4'-(METHYLTHIO)-3-PHENYLPROPIONANILIDE

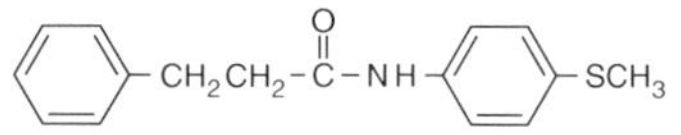

250 mg

S93,675-8
2-(BENZYLTHIO)-4'-METHYLACETANILIDE

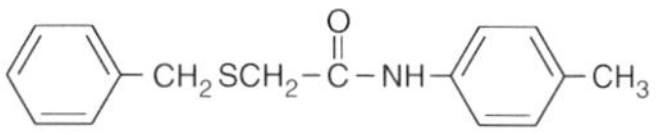

5 gm

S93,681-2
2-(BENZYLTHIO)-4'-BUTYLACETANILIDE

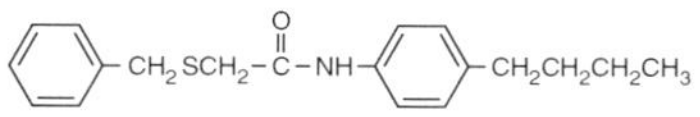

5 gm

S79,535-6
2,2-DIPHENYL-4'-(METHYLTHIO)-ACETANILIDE

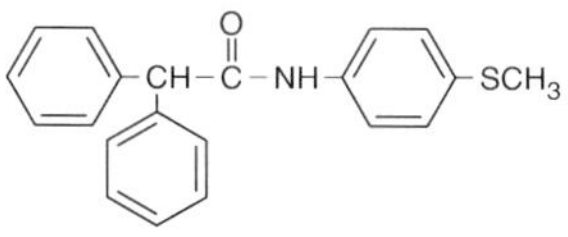

250 mg

S94,608-7
2-(BENZYLTHIO)-2',3'-DIMETHYLACETANILIDE

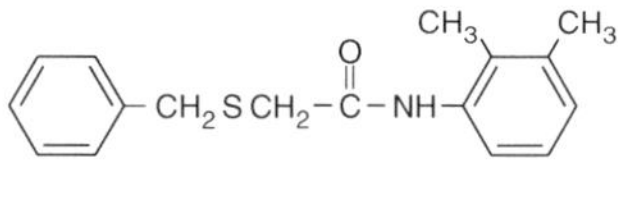

5 gm

S94,181-6
2-(BENZYLTHIO)-4'-CHLORO-2'-METHYLACETANILIDE

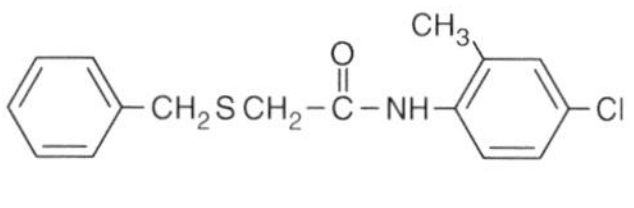

5 gm

S93,678-2
4'-BROMO-2-(BENZYLTHIO)-2'-METHYLACETANILIDE

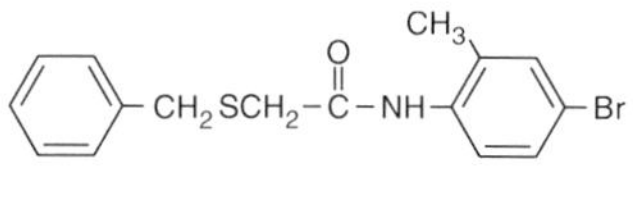

5 gm

S93,679-0
2-(BENZYLTHIO)-3',4'DIMETHYLACETANILIDE

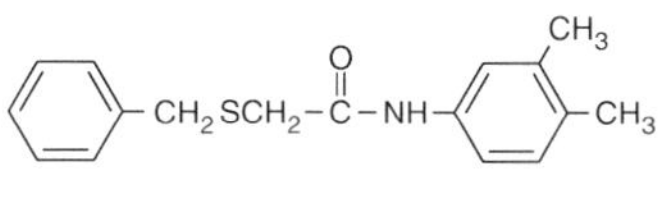

5 gm

S93,683-9
2-(BENZYLTHIO)-2',4',6'-TRIMETHYLACETANILIDE

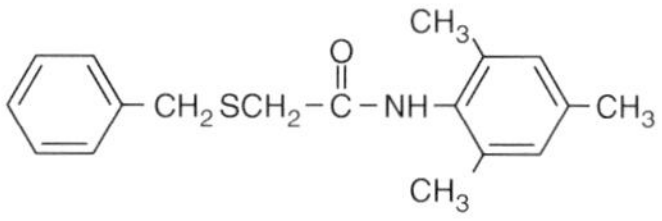

5 gm

S39,028-3
4'-(METHYLTHIO)ACETANILIDE

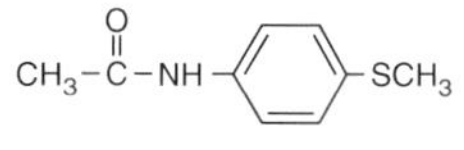

5 gm

S90,277-2
2,2-DIMETHYL-4'-(METHYLTHIO)-PROPIONANILIDE

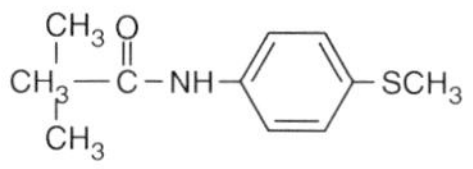

250 mg

S82,904-8
4'-(M-TOLYLTHIO)FORMANILIDE

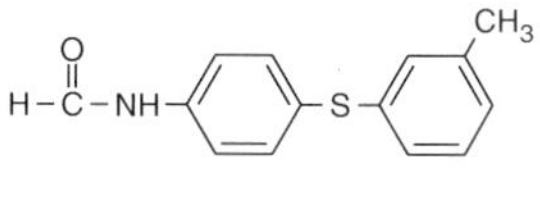

250 mg

S75,275-4
4'-(METHYLTHIO)-10-UNDECENANILIDE

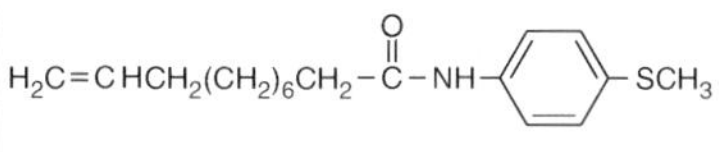

250 mg

S64,567-2
4'-(HEXADECYLTHIO)ACETANILIDE

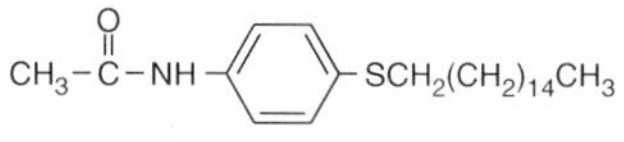

5 gm

S93,390-2
2',4'-DIFLUORO-2-(4-TOLYLTHIO)-ACETANILIDE

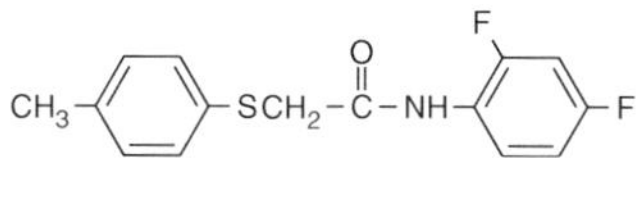

250 mg

S35,713-8
4-ACETAMIDOPHENYL DISULFIDE

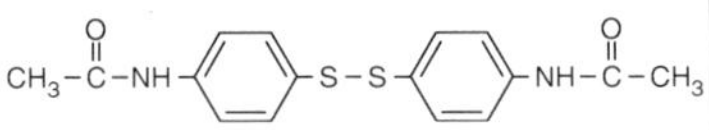

5 gm

S73,136-6
4'-AMINO-2'-NITROACETANILIDE

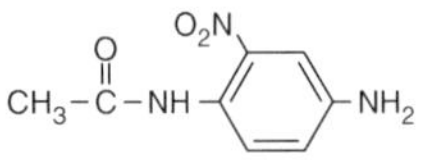

5 gm

S36,003-1
2',4'-DIAMINOACETANILIDE

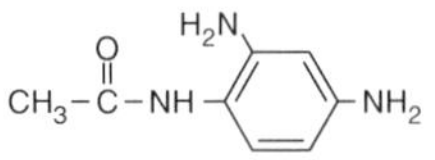

5 gm

S45,163-0
GLYCYL-L-PHENYLALANINE

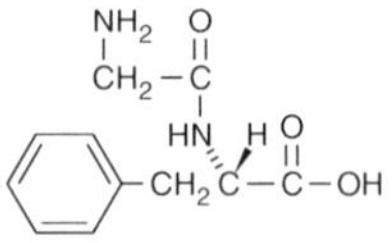

250 mg

S85,952-4
N,N'-BIS(2-AMINOPHENYL)OXAMIDE

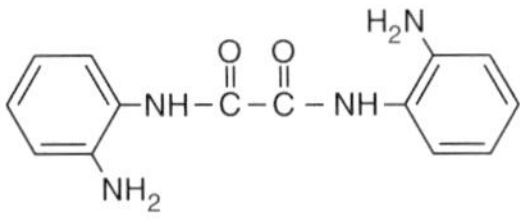

250 mg

S60,037-7
DL-2-PHENYLGLYCINAMIDE

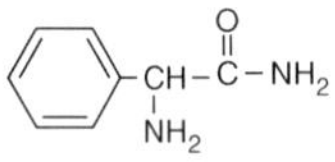

250 mg

S89,359-5
N-[1-(4-AMINOPHENYL)-4-PIPERIDINYL]-PROPIONANILIDE

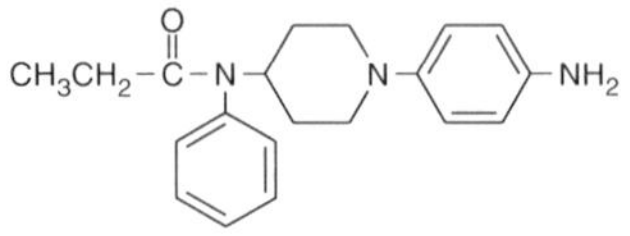

250 mg

S37,304-4
3'-AMINOACETANILIDE

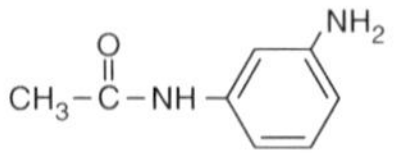

5 gm

S56,239-4
3'-AMINOHEXADECANANILIDE

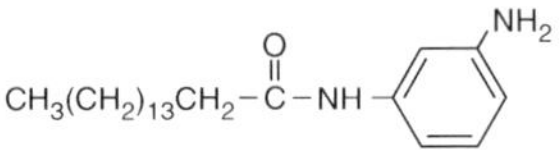

5 gm

S61,504-8
3'-AMINO-4'-METHYLACETANILIDE

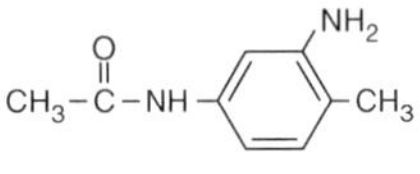

5 gm

S56,266-1
4'-AMINOHEXADECANANILIDE

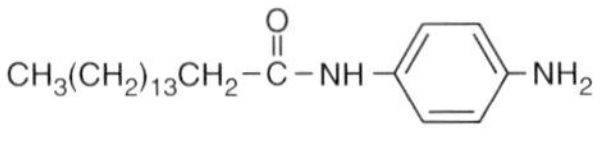

5 gm

S36,396-0
ALPHA-PHENYL-4-MORPHOLINEACETAMIDE

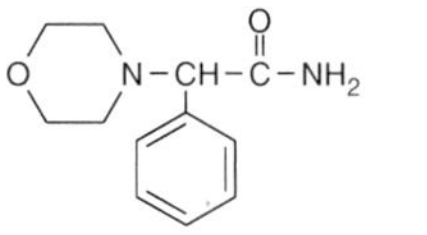

5 gm

S78,522-9
L-PHENYLALANYLGLYCYLGLYCINAMIDE

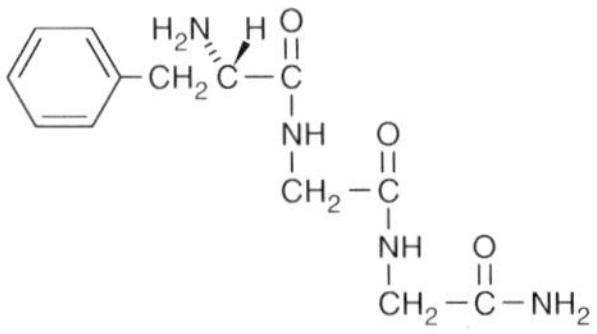

250 mg

S45,157-6
GLYCYL-DL-PHENYLALANINE

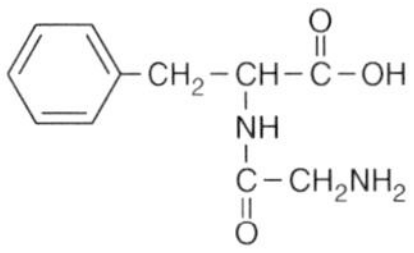

250 mg

S45,190-8
L-PHENYLALANYLGLYCINE MONOHYDRATE

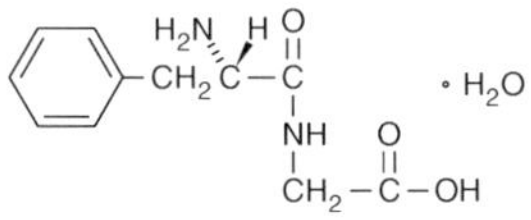

250 mg

S77,735-8
L-ALANYL-D-PHENYLALANINE

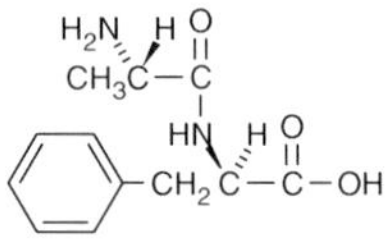

250 mg

S79,014-1
D-ALANYL-L-PHENYLALANINE

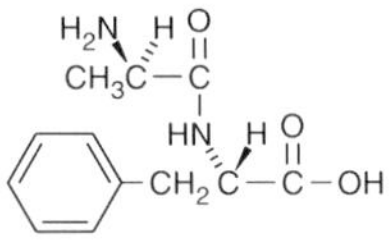

250 mg

S77,617-3
L-2-AMINOBUTYRYL-L-PHENYLALANINE DIHYDRATE

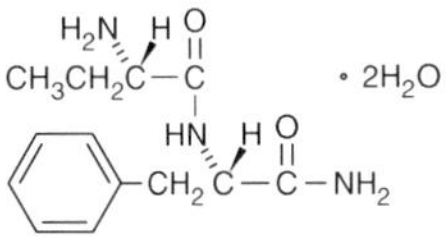

250 mg

S45,142-8
L-PHENYLALANYL-L-VALINE

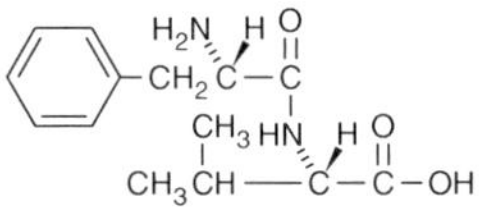

250 mg

S77,736-6
L-VALYL-L-PHENYLALANINE

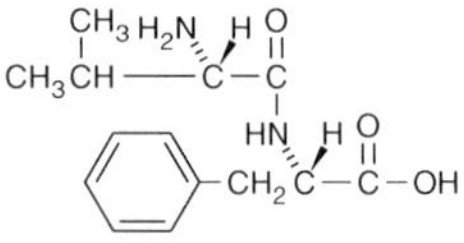

250 mg

S40,973-1
DL-PHENYLALANYL-DL-PHENYLALANINE

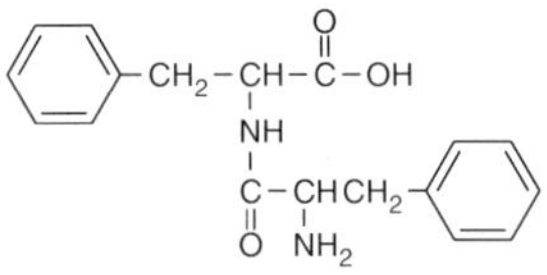

250 mg

S45,277-7
L-PHENYLALANYL-L-PHENYLALANINE

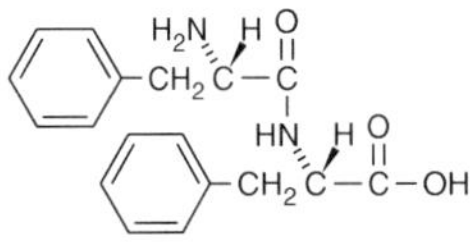

250 mg

S35,020-6
GLYCYLGLYCYL-DL-PHENYLALANINE

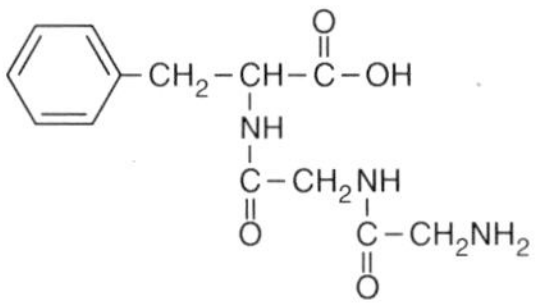

5 gm

S45,188-6
GLYCYL-L-PHENYLALANINAMIDE ACETATE

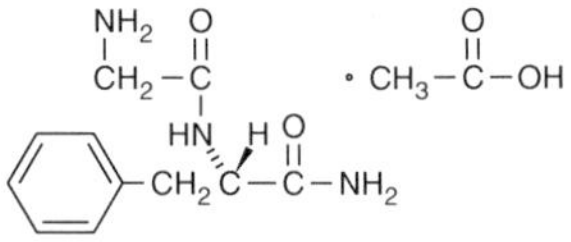

250 mg

S77,733-1
L-ALANYL-L-PHENYLALANINAMIDE HYDROCHLORIDE

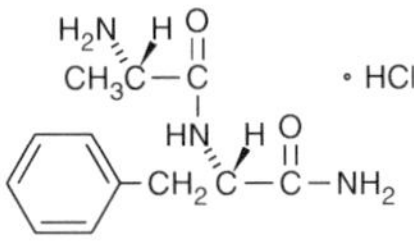

250 mg

S77,618-1
GLYCYLGLYCYL-L-PHENYLALANINAMIDE HYDROCHLORIDE

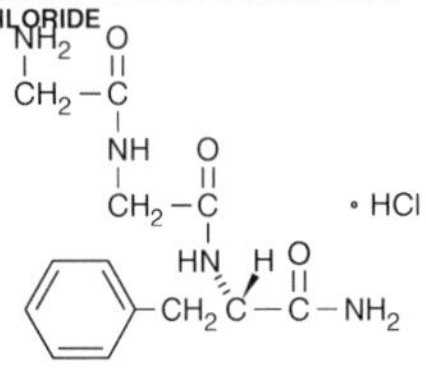

250 mg

S77,781-1
L-VALYL-L-PHENYLALANINAMIDE ACETATE

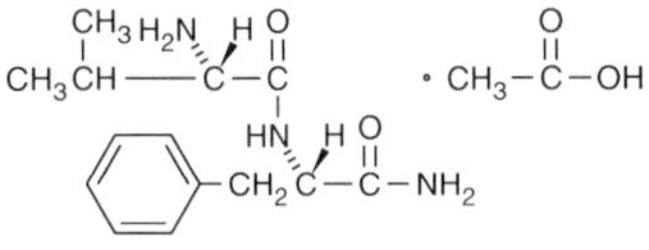

250 mg

S78,525-3
L-NORVALYL-L-PHENYLALANINAMIDE HYDROCHLORIDE

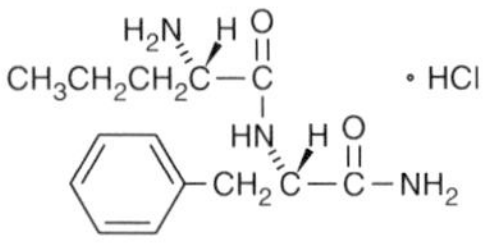

250 mg

S77,672-6
L-PHENYLALANYL-L-ISOLEUCINAMIDE

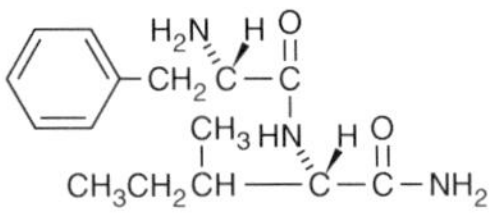

250 mg

S78,148-7
L-ISOLEUCYL-L-PHENYLALANINAMIDE HYDROCHLORIDE

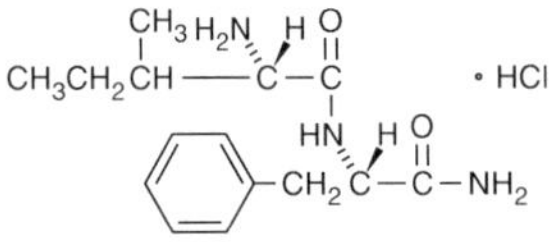

250 mg

S88,076-0
2-AMINOCINNAMAMIDE

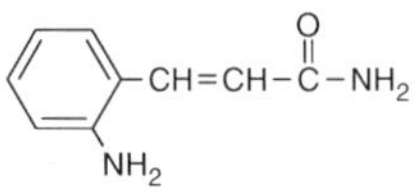

250 mg

S75,671-7
GLYCYL-D-PHENYLALANINE

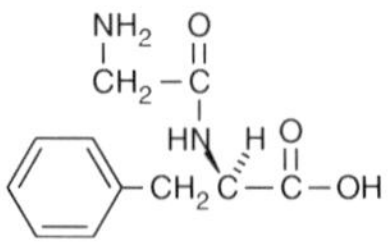

5 gm

S38,015-6
2-AMINO-6'-CHLORO-ORTHO-ACETOTOLUIDIDE HYDROCHLORIDE
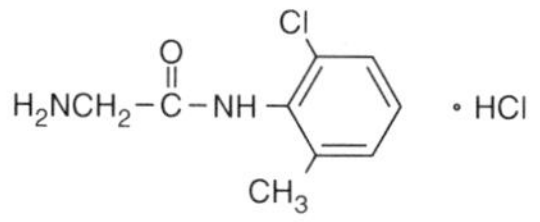
250 mg

S53,550-8
1-BENZYL-4-(PARA-TOLUIDINO)-ISONIPECOTAMIDE
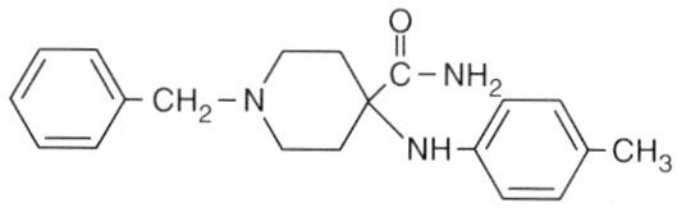
250 mg

S74,661-4
4-ACETAMIDO-2',4'-DINITRODIPHENYLAMINE
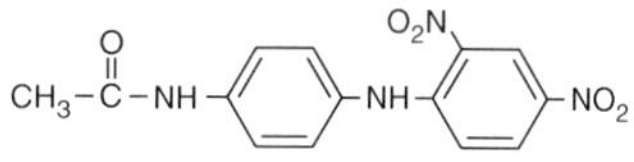
5 gm

S53,549-4
1-BENZYL-4-(DIMETHYLAMINO)-ISONIPECOTAMIDE
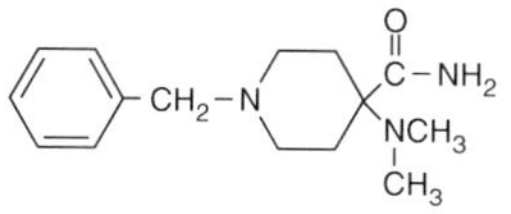
250 mg

S80,406-1
N,N'-DIACETYL-2-NITRO-P-PHENYLENEDIAMINE
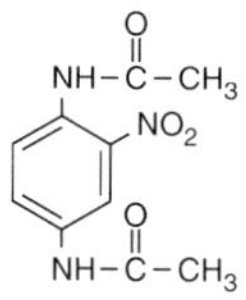
250 mg

S54,062-5
2',4'-DINITROACETANILIDE
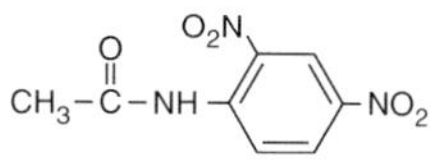
5 gm

S46,249-7
N,N'-(2,3-DINITRO-1,4-PHENYLENE)-BISACETAMIDE
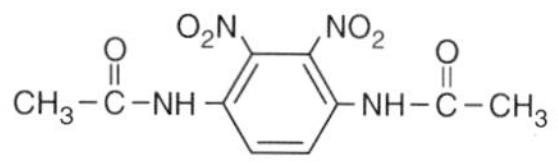
250 mg

S78,291-2
N-(2,4-DINITROPHENYL)-L-ASPARAGINE
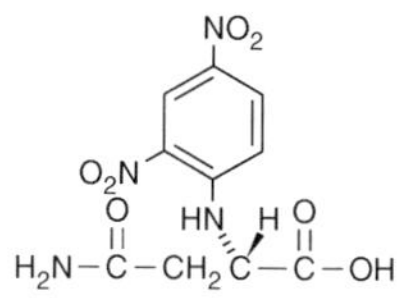
5 gm

S80,927-6
N-(2,4-DINITROPHENYLHYDRAZONOETHYL)-HEXANAMIDE
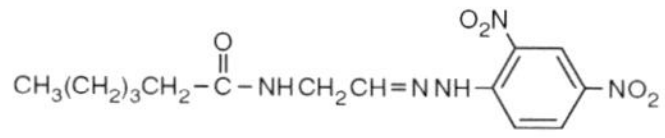
250 mg

S80,246-8
2-ACETAMIDO-4,5-DINITROBENZOIC ACID
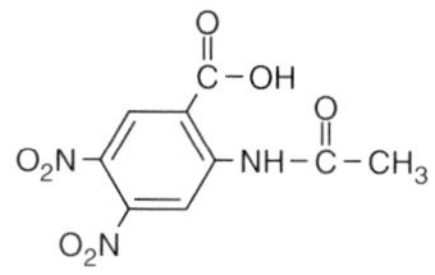
250 mg

S80,248-4
2',5'-DINITRO-P-ACETOTOLUIDIDE
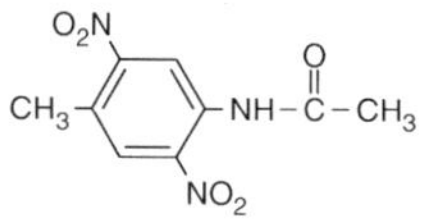
250 mg

S80,436-3
4',5'-DINITRO-O-ACETOTOLUIDIDE
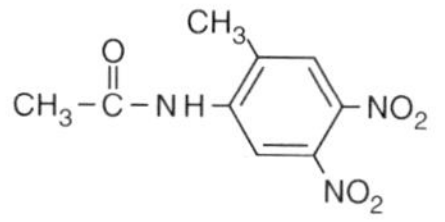
250 mg

S65,600-3
1-PHENYL-3-(N-PHENYLCARBAMOYL)-2-PYRAZOLIN-4-ONE
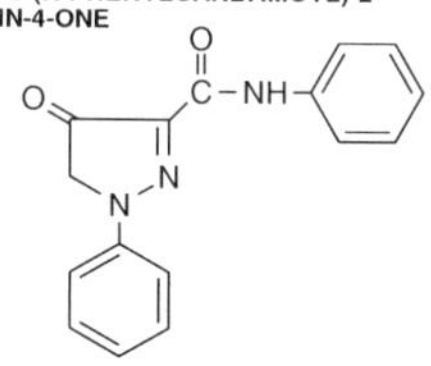
5 gm

S40,160-9
N-[2-(DIMETHYLAMINO)ETHYL]-2,2-DIPHENYLACETAMIDE HYDROCHLORIDE
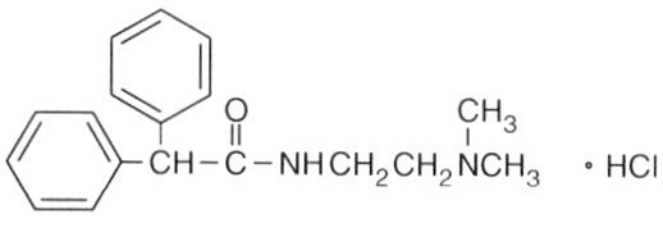
250 mg

S40,178-1
N-[2-(DIMETHYLAMINO)ETHYL]-2,2-DIPHENYL-N-METHYLACETAMIDE HYDROCHLORIDE
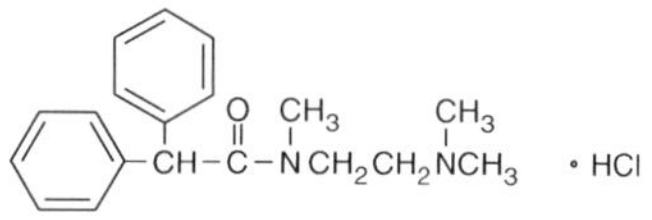
250 mg

S40,209-5
N-[2-(DIMETHYLAMINO)ETHYL]-2,2-DIPHENYL-N-ETHYLACETAMIDE HYDROCHLORIDE
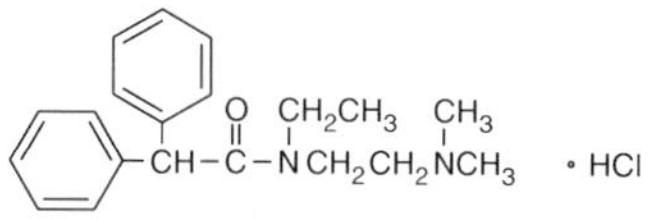
250 mg

S40,247-8
N-[2-(DIETHYLAMINO)ETHYL]-2,2-DIPHENYL-N-METHYLACETAMIDE HYDROCHLORIDE
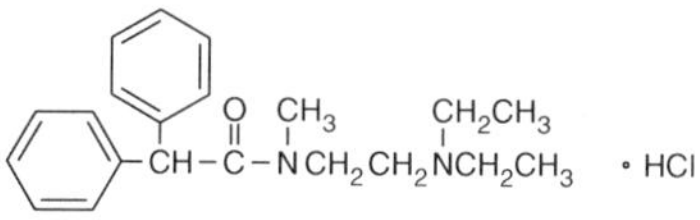
250 mg

S40,175-7
[2-(DIPHENYLACETAMIDO)ETHYL]-TRIMETHYLAMMONIUM BROMIDE
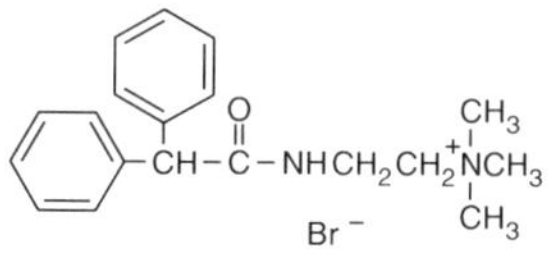
250 mg

S40,244-3
DIETHYL-[2-(DIPHENYLACETAMIDO)-ETHYL]METHYLAMMONIUM BROMIDE
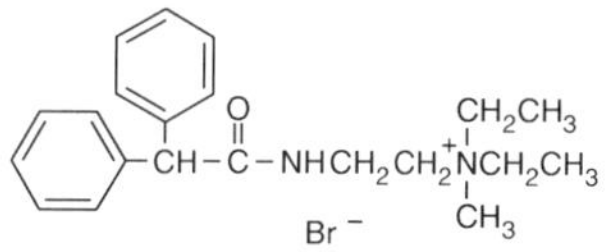
250 mg

S73,050-5
N,N'-BIS(2-PHENYLETHYL)OXAMIDE
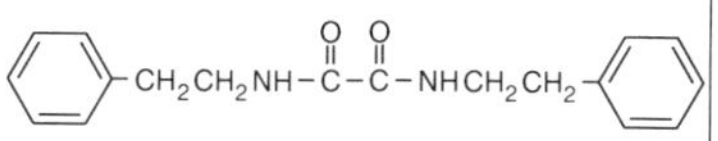
5 gm

S42,359-9
6'-CHLORO-2-DIETHYLAMINO-ORTHO-ACETOTOLUIDIDE HYDROCHLORIDE
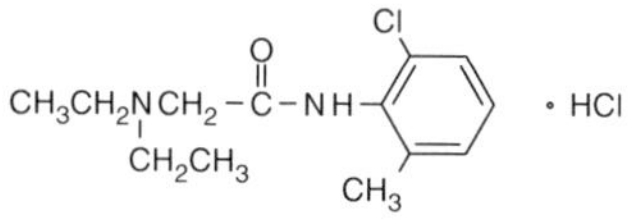
5 gm

S92,403-2
4,4'-BIS(DIMETHYLAMINO)OXANILIDE
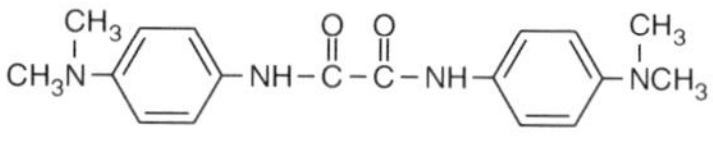
250 mg

S90,184-9
N-[2-(N-METHYLANILINO)ETHYL]-1,2,3,6-TETRAHYDROPHTHALAMIC ACID
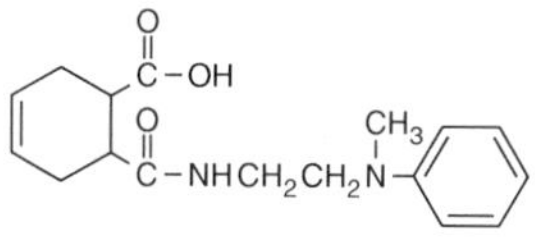
250 mg

S42,295-9
2-[N-BENZYL-N-(2-DIETHYLAMINOETHYL]-AMINO)-6'-CHLORO-ORTHO-ACETOTOLUIDIDE
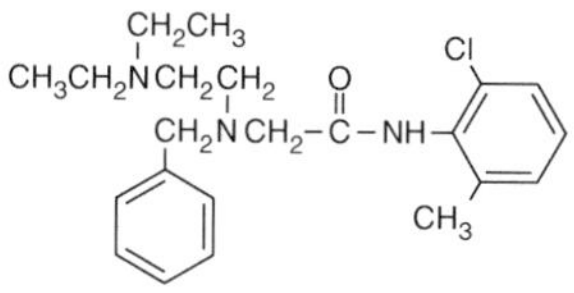
5 gm

S38,211-6
N-BENZYL-N-(2-DIMETHYLAMINOETHYL)-ACETAMIDE
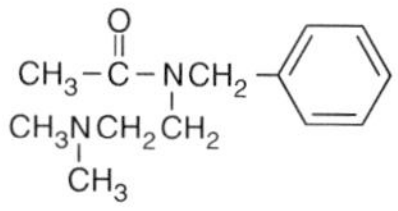
5 gm

S42,234-7
2-DIETHYLAMINO-N-METHYL-N-(2-METHYL-1-PHENYLBUTYL)ACETAMIDE HYDROCHLORIDE
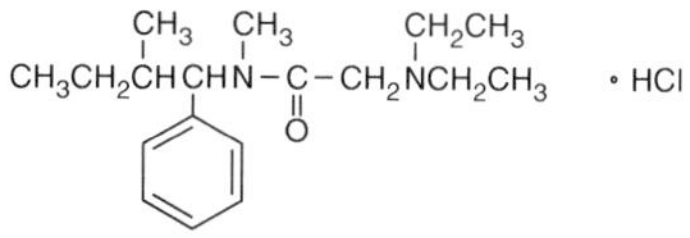
250 mg

S74,783-1
2-(N-[2-(N-METHYLANILINO)ETHYL]-CARBAMOYL)-CYCLOHEXANECARBOXYLIC ACID
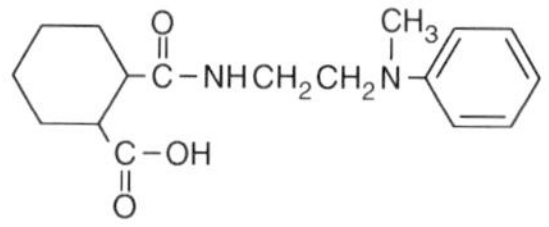
5 gm

S61,733-4
N-[2-(DIETHYLAMINO)ETHYL]-N-(ETHYL)-GLYCYL-2-CHLOROANILIDE
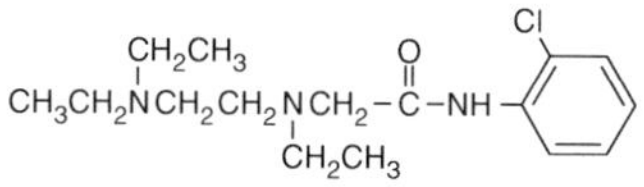
5 gm

S42,176-6
2'-CHLORO-2-[N-ETHYL-N-(2-PYRROLIDINOETHYL)AMINO]-ACETANILIDE DIHYDROCHLORIDE
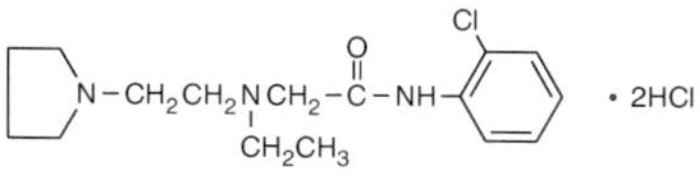
250 mg

S42,081-6
6'-CHLORO-1-PYRROLIDINEACETO-ORTHO-TOLUIDIDE NITRATE
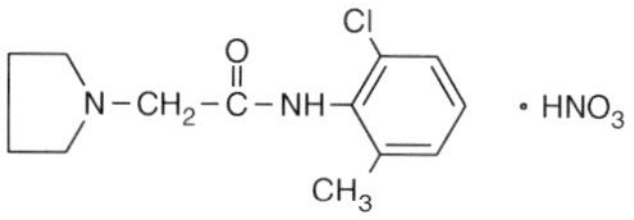
250 mg

S42,324-6
6'-CHLORO-1-PYRROLIDINEACETO-ORTHO-TOLUIDIDE PERCHLORATE
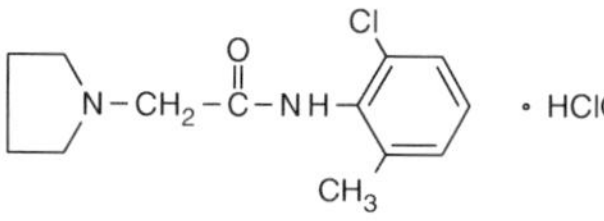
250 mg

S42,366-1
6'-CHLORO-1-PYRROLIDINEACETO-ORTHO-TOLUIDIDE METHANESULFONATE
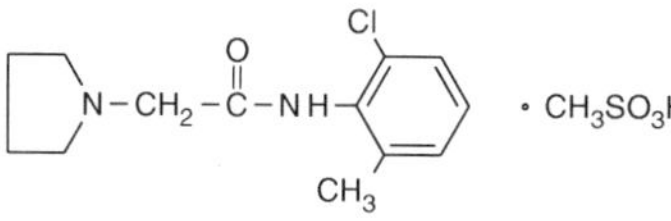
250 mg

S42,178-2
6'-CHLORO-2-(N-CYCLOHEXYL-N-METHYLAMINO)-ORTHO-ACETOTOLUIDIDE HYDROCHLORIDE

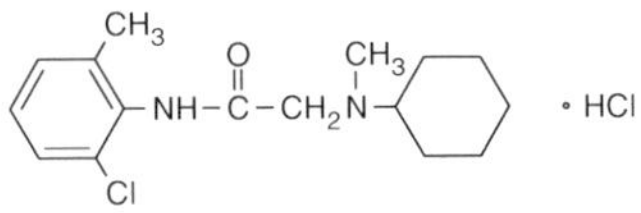

5 gm

S42,177-4
6'-CHLORO-2-(2,6-DIMETHYLPIPERIDINO)-ORTHO-ACETOTOLUIDIDE HYDROCHLORIDE

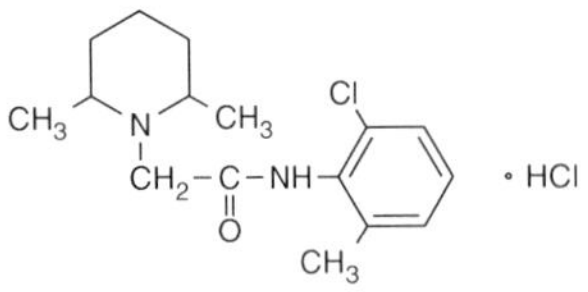

250 mg

S90,823-1
2'-(HEXAHYDRO-1H-AZEPIN-1-YL)-ACETANILIDE

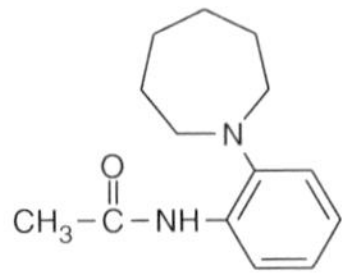

250 mg

S42,163-4
2'-CHLORO-2-[N-(2-DIETHYLAMINOETHYL)-N-METHYLAMINO]-ACETANILIDE DIHYDROCHLORIDE

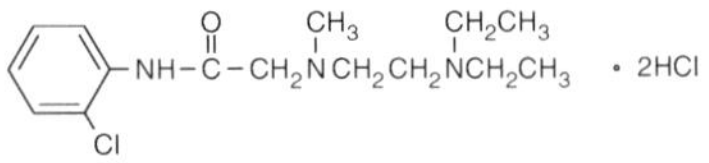

5 gm

S42,201-0
2'-CHLORO-2-[N-(2-DIETHYLAMINOETHYL)-N-METHYLAMINO]-PROPIONANILIDE DIHYDROCHLORIDE

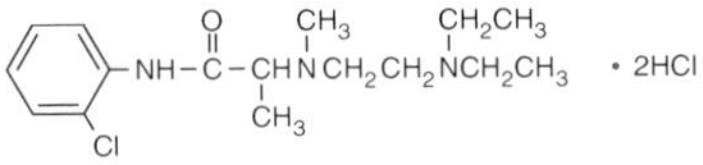

5 gm

S42,202-9
6'-CHLORO-2-[N-(2-DIETHYLAMINOETHYL)-N-METHYLAMINO]-O-ACETOTOLUIDIDE DIHYDROCHLORIDE

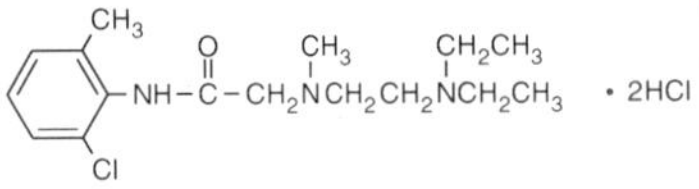

5 gm

S42,218-5
2-[N-(2-DIETHYLAMINOETHYL)-N-METHYLAMINO]-2',6'-ACETOXYLIDIDE DIHYDROCHLORIDE

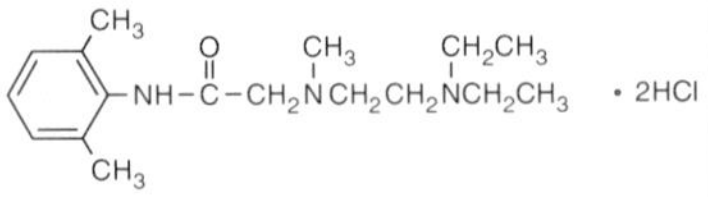

250 mg

S42,070-0
6'-CHLORO-2-DIETHYLAMINO-META-ACETOTOLUIDIDE HYDROCHLORIDE

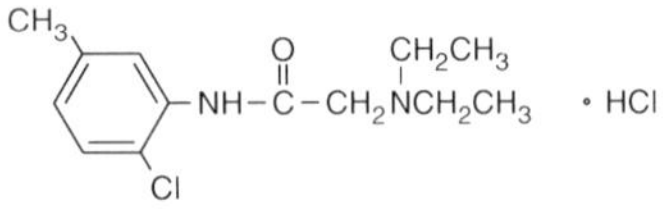

5 gm

S42,038-7
6'-CHLORO-2-DIMETHYLAMINO-ORTHO-ACETOTOLUIDIDE HYDROCHLORIDE

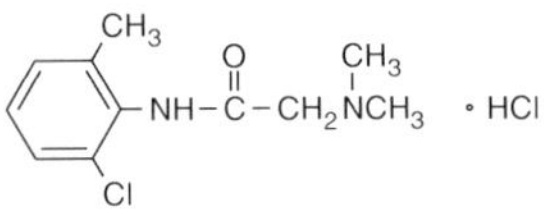

5 gm

S40,306-7
BENZYLDIMETHYL (HEXADECYLCARBAMOYLMETHYL)-AMMONIUM CHLORIDE

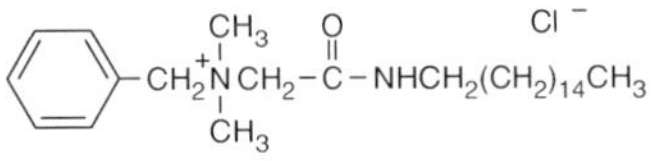

250 mg

S73,048-3
N,N'-DIBENZYLOXAMIDE

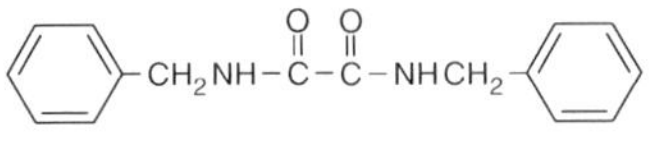

5 gm

S77,469-3
N,N'-DIBENZYL-N,N'-DI-TERT-BUTYLOXAMIDE

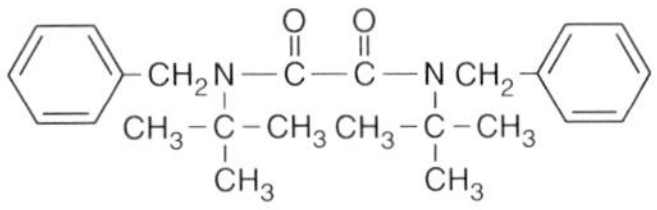

250 mg

S77,467-7
N,N'-DIBENZYL-N,N'-DI(2,6-XYLYL)-OXAMIDE

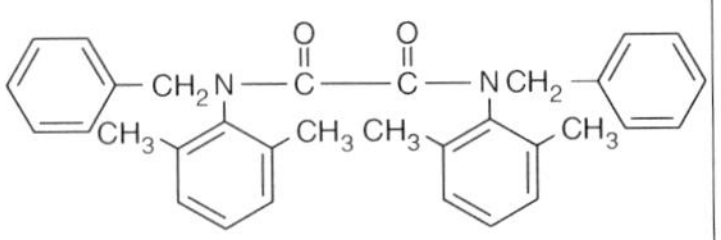

250 mg

S90,119-9
4-ACETAMIDOBENZALDEHYDE (2,6-DINITRO-4-METHYLPHENYL)-HYDRAZONE

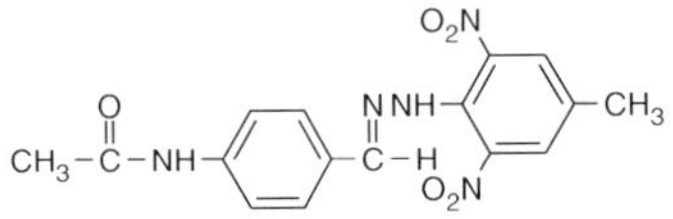

250 mg

S34,196-7
DEXTRO-(-)BENZYLPENICILLOIC ACID HYDRATE

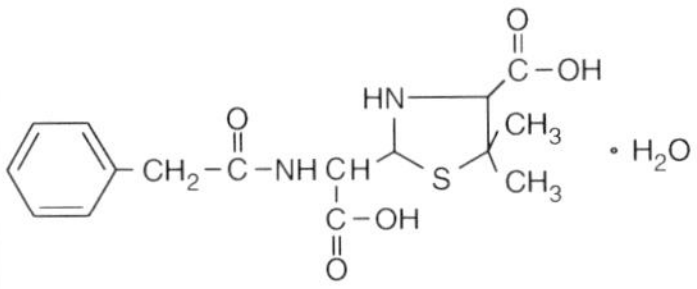

5 gm

S73,132-3
2'-ACETYL-6'-NITROACETANILIDE

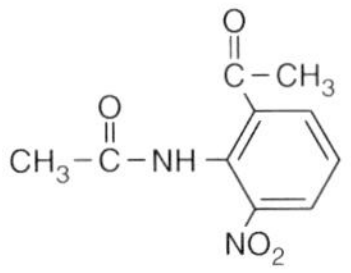

5 gm

S74,690-8
2'-NITRO-6'-PROPIONYLACETANILIDE

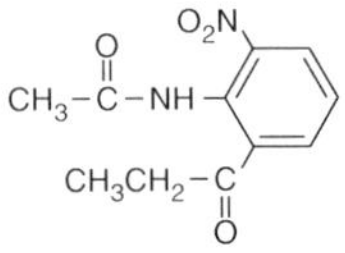

5 gm

S85,964-8
N,N'-BIS(2-NITROPHENYL)OXAMIDE

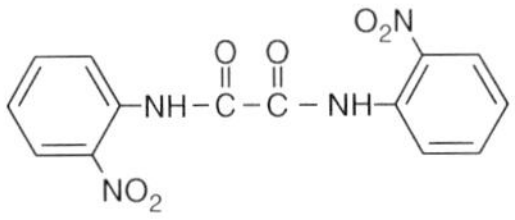

250 mg

S69,574-2
4'-METHYL-2'-NITROPROPIONANILIDE

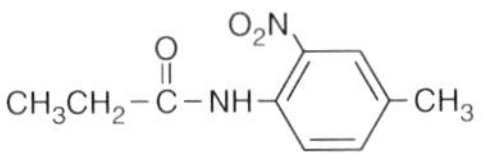

250 mg

S84,056-4
4'-ISOPROPYL-2'-NITROACETANILIDE

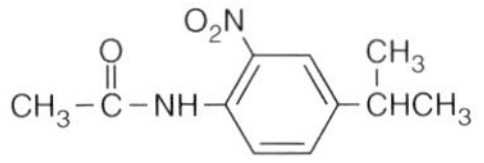

250 mg

S80,351-0
5',6'-DINITRO-O-ACETOTOLUIDIDE

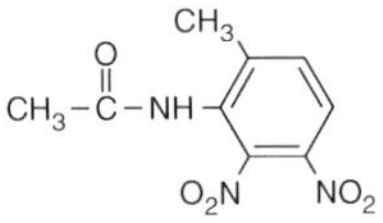

250 mg

S93,364-3
2',3'-DIMETHYL-6'-NITROACETANILIDE

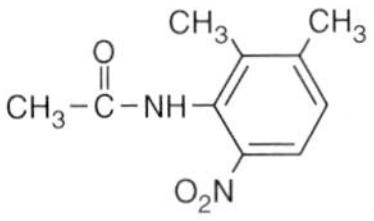

250 mg

S72,166-2
4',5'-DIMETHYL-2'-NITROPROPIONANILIDE

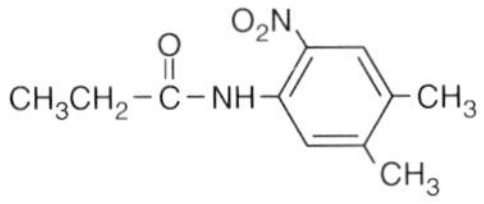

250 mg

S77,046-9
4'-ETHYL-5'-METHYL-2'-NITROACETANILIDE

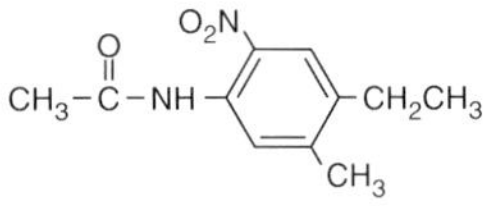

250 mg

S80,559-9
4-ACETAMIDO-3-NITROBENZALDEHYDE

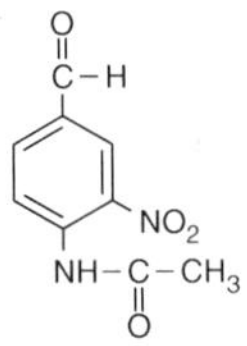

250 mg

S94,285-5
METHYL N-ACETYL-3-NITROANTHRANILATE

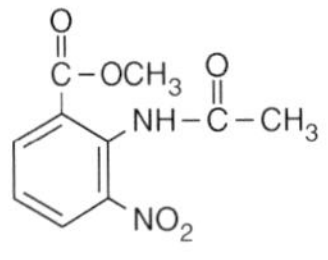

250 mg

S58,833-4
4'-ACETYLTHIO-2'-NITROACETANILIDE

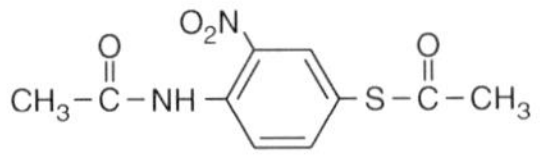

250 mg

S63,614-2
2'-NITROFORMANILIDE

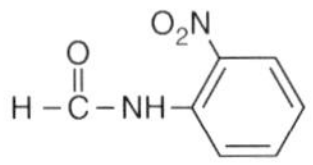

5 gm

S50,052-6
6'-NITRO-ORTHO-ACETOTOLUIDIDE

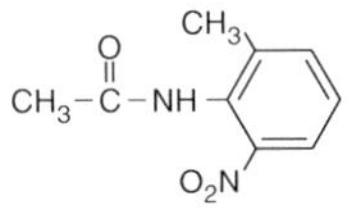

250 mg

S63,421-2
4-CHLORO-2-NITROFORMANILIDE

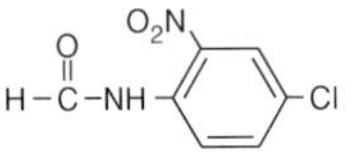

5 gm

S61,471-8
4'-BROMO-2'-NITROACETANILIDE

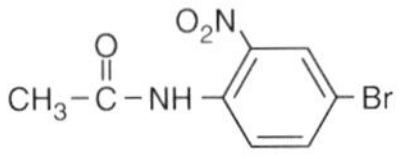

5 gm

S61,506-4
4'-METHYL-2'-NITROACETANILIDE

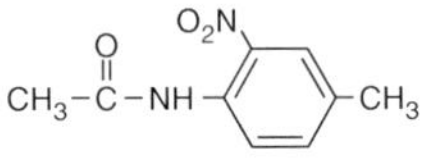

5 gm

S94,603-6
4',6'-DIMETHYL-2'-NITROACETANILIDE

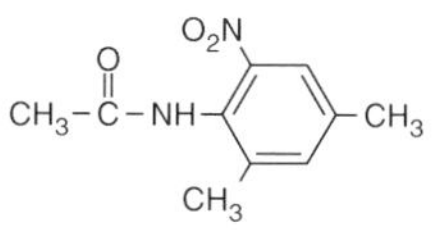

250 mg

S38,745-2
5'-CHLORO-4'-FLUORO-2'-NITROACETANILIDE

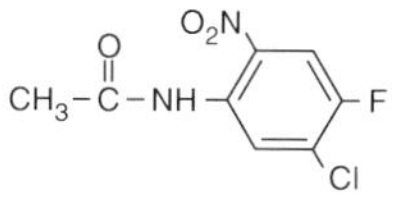

5 gm

S36,034-1
4',5'-DICHLORO-2'-NITROACETANILIDE

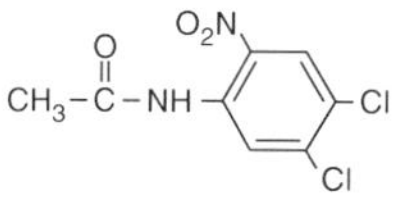

5 gm

S34,715-9
4',5'-DIMETHYL-2'-NITROACETANILIDE

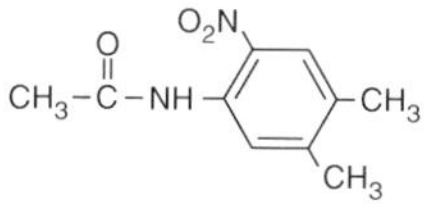

5 gm

S77,470-7
N,N'-DIMETHYL-N,N'-DI(O-TOLYL)-OXAMIDE

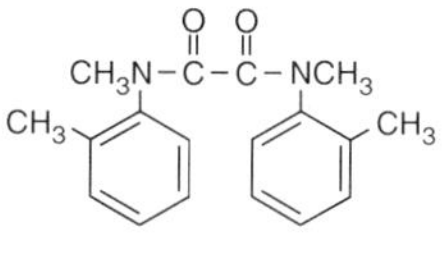

250 mg

S77,471-5
N,N'-DIETHYL-N,N'-DI(O-TOLYL)-OXAMIDE

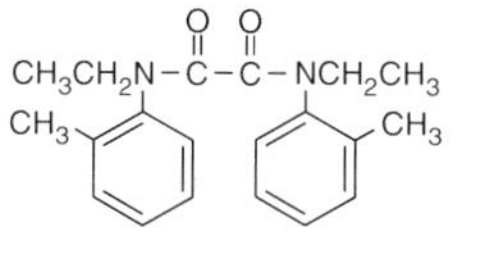

250 mg

S77,539-8
N,N'-DIALLYL-N,N'-DI(O-TOLYL)-OXAMIDE

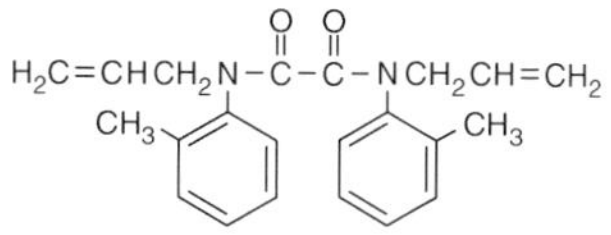

250 mg

S77,468-5
N,N'-DIISOPROPYL-N,N'-DI(O-TOLYL)-OXAMIDE

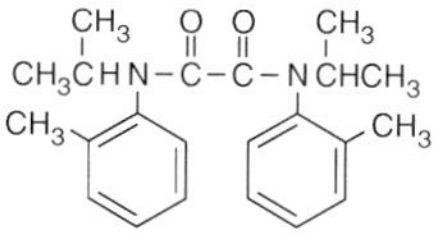

250 mg

S42,209-6
6'-CHLORO-1-PYRROLIDINEACETO-ORTHO-TOLUIDIDE MALEATE

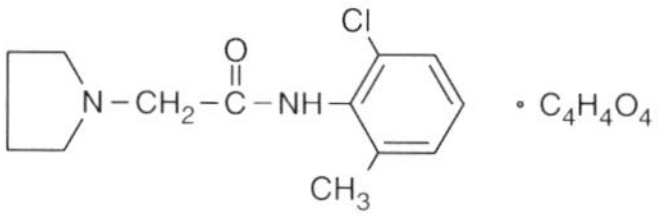

250 mg

S81,801-1
1,2-DIACETAMIDOBENZENE

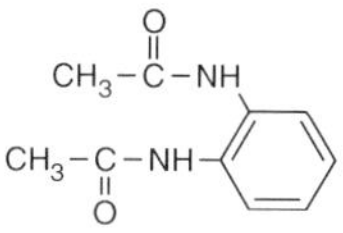

250 mg

S84,111-0
4'-TERT-BUTYL-2'-NITROACETANILIDE

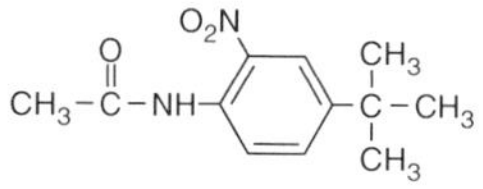

250 mg

S51,401-2
4'-CHLORO-6'-FORMAMIDO-ORTHO-ACETOTOLUIDIDE

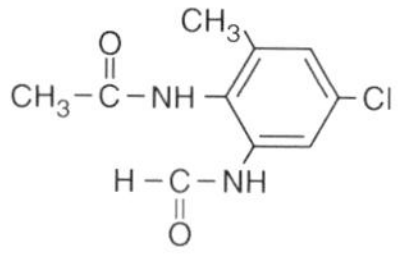

250 mg

S51,404-7
6'-FORMAMIDO-2',4'-ACETOXYLIDIDE

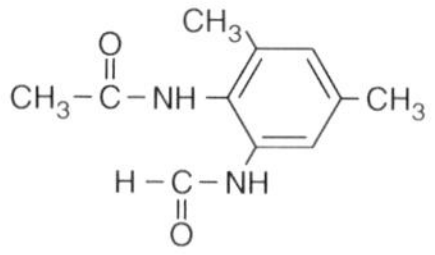

250 mg

S77,508-8
N,N'-(1,2-PHENYLENE)BIS(N-ETHYLFORMAMIDE)

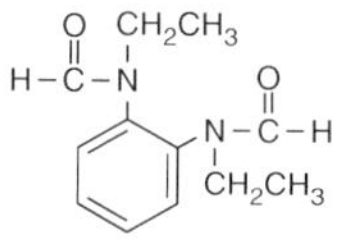

250 mg

S35,800-2
4-ANILINO-4-CARBAMYLPIPERIDINE

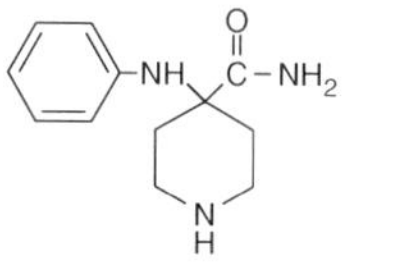

250 mg

S40,243-5
[3-(2,2-DIPHENYL-N-METHYLACETAMIDO)-PROPYL]TRIMETHYLAMMONIUM BROMIDE

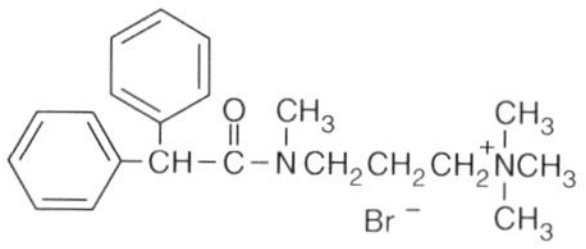

250 mg

S40,278-8
DIETHYL[3-(2,2-DIPHENYL-N-METHYLACETAMIDO)PROPYL]-METHYLAMMONIUM BROMIDE

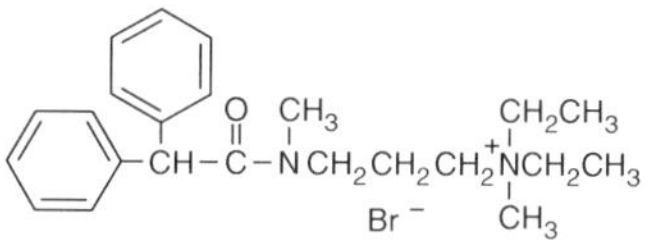

250 mg

S34,171-1
1-BENZYL-4-CARBAMYL-4-PYRROLIDINOPIPERIDINE

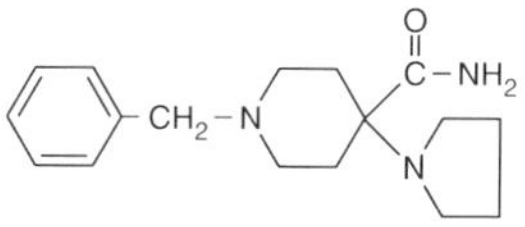

250 mg

S34,162-2
1-BENZYL-4-CYCLOHEXYLAMINO-4-PIPERIDINECARBOXAMIDE

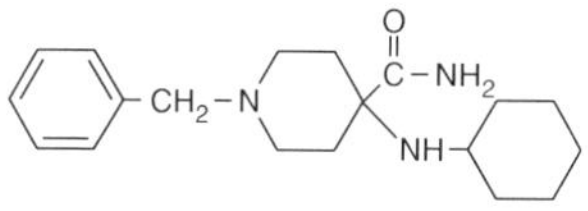

5 gm

S34,166-5
1-BENZYL-4-METHYLAMINO-4-PIPERIDINECARBOXAMIDE

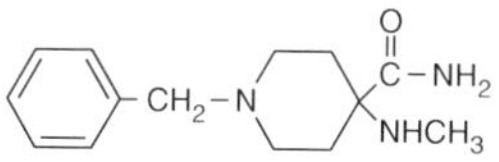

5 gm

S34,164-9
1-BENZYL-4-ETHYLAMINO-4-PIPERIDINECARBOXAMIDE

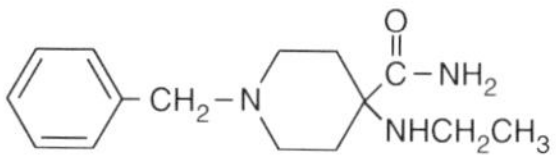

5 gm

S34,165-7
1-BENZYL-4-ISOPROPYLAMINO-4-PIPERIDINECARBOXAMIDE

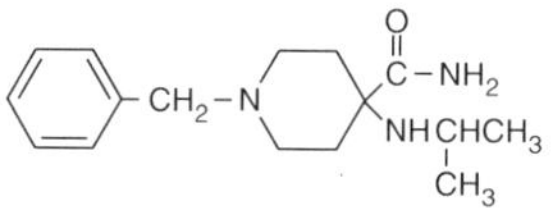

5 gm

S34,170-3
1-BENZYL-4-PROPYLAMINO-4-PIPERIDINECARBOXAMIDE

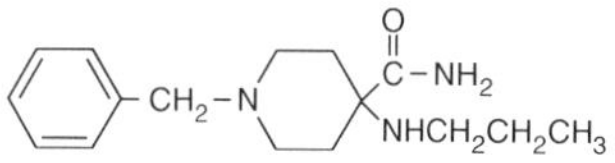

5 gm

S90,871-1
3'-(ETHYLAMINO)ACETANILIDE

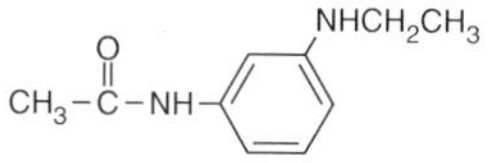

5 gm

S90,900-9
3'-(ETHYLAMINO)-4'-METHYLACETANILIDE

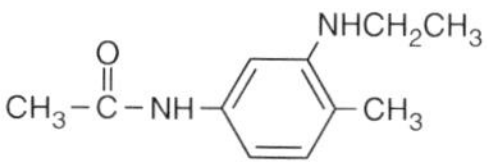

5 gm

S53,874-4
ALPHA-(3-NITROANILINO)-PARA-ACETOTOLUIDIDE

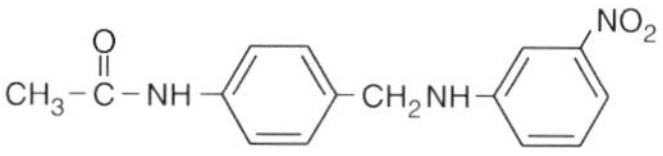

5 gm

S90,883-5
3'-(N-BENZYL-N-ETHYLAMINO)-ACETANILIDE

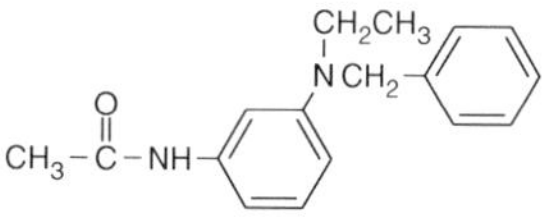

5 gm

S90,879-7
3'-(DIPROPYLAMINO)SUCCINANILIC ACID

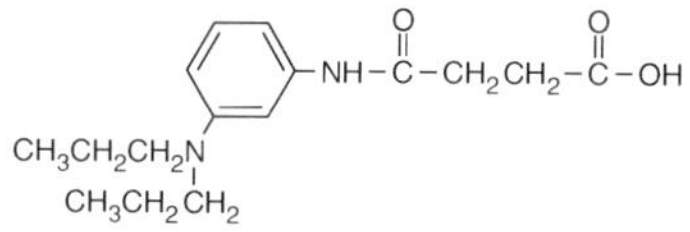

250 mg

S90,849-5
3'-(DIETHYLAMINO)ACETANILIDE

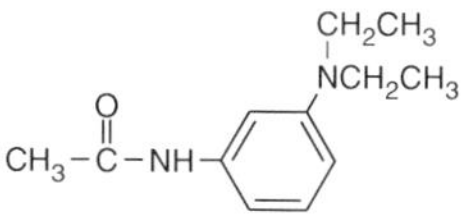

5 gm

S90,853-3
3'-(DIPROPYLAMINO)ACETANILIDE

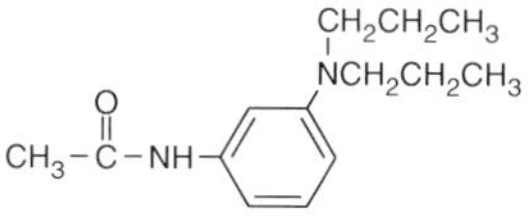

5 gm

S42,313-0
N-BENZYL-3-(N-BUTYL-N-METHYLAMINO)-PROPIONAMIDE

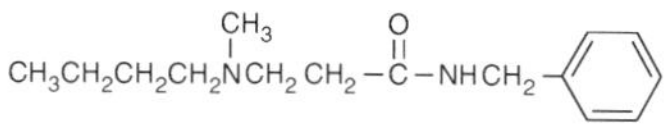

250 mg

S65,853-7
2-CHLORO-3'-(DIETHYLAMINO)-PROPIONANILIDE

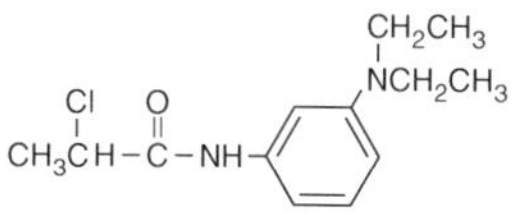

5 gm

S42,171-5
3-(2-METHYLPIPERIDINO)-BUTYRANILIDE HYDROCHLORIDE

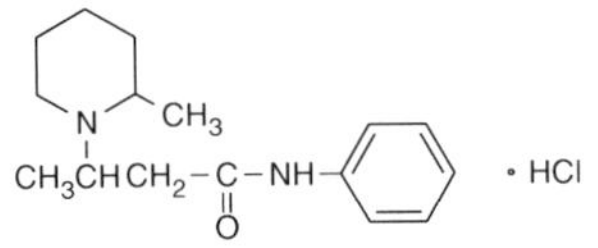

250 mg

S38,575-1
6'-CHLORO-3-DIETHYLAMINO-ORTHO-BUTYROTOLUIDIDE

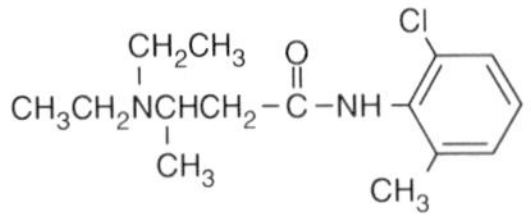

5 gm

S90,878-9
N-(1-ETHYL-1,2,3,4-TETRAHYDRO-2,2,4-TRIMETHYL-7-QUINOLYL)-ACETAMIDE

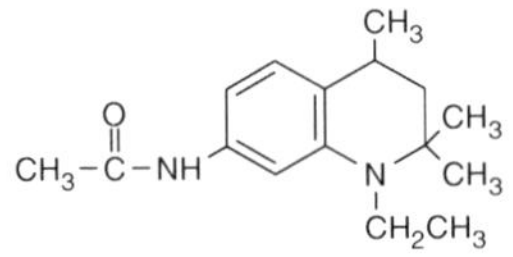

250 mg

S90,884-3
N-(1-ETHYL-1,2,3,4-TETRAHYDRO-2,2,4-TRIMETHYL-7-QUINOLYL)-PROPIONAMIDE

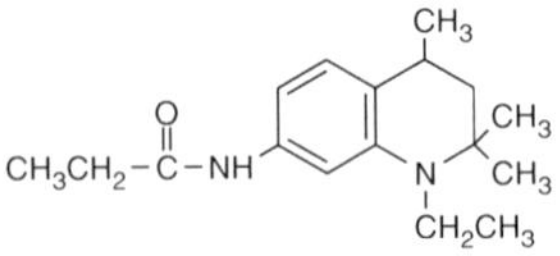

250 mg

S35,722-7
4-(N-ACETYLANILINO)-1-BENZYLPIPERIDINE

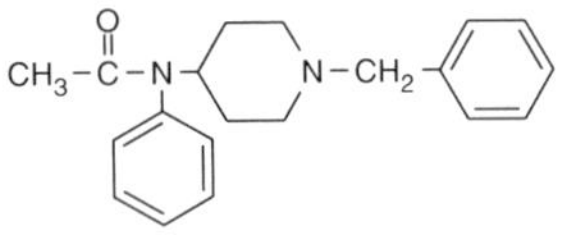

250 mg

S34,199-1
1-BENZYL-4-(N-PROPIONYLANILINO)-PIPERIDINE

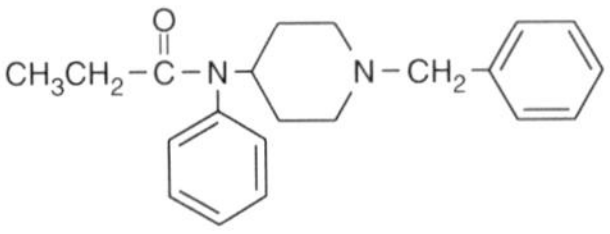

5 gm

S39,191-3
N,N'-(2-NITROBENZYLIDENE)-BISFORMAMIDE

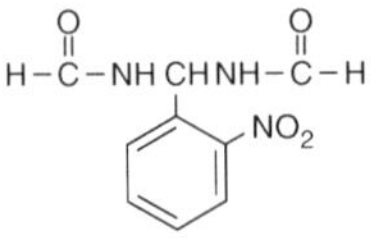

5 gm

S70,309-5
3'-NITROSUCCINANILIC ACID

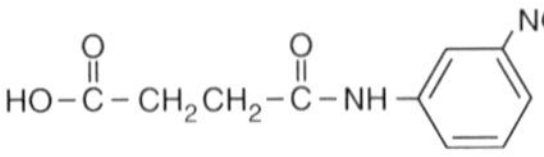

5 gm

S70,442-3
3'-NITROGLUTARANILIC ACID

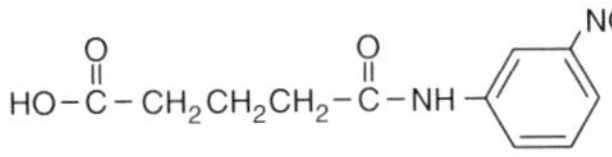

5 gm

S70,313-3
CIS-HEXAHYDRO-N-(3-NITROPHENYL)-PHTHALAMIC ACID

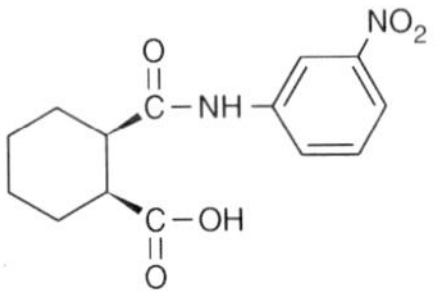

5 gm

S71,901-3
CIS-HEXAHYDRO-N-(4-METHYL-3-NITROPHENYL)PHTHALAMIC ACID

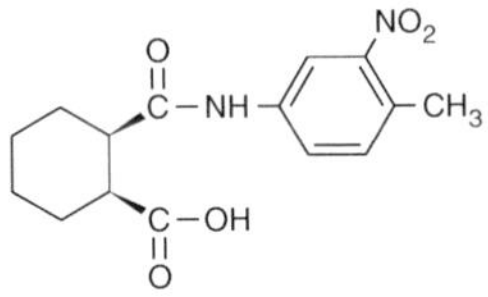

5 gm

S71,930-7
N-(3-NITROPHENYL)-1,2,5,6-TETRAHYDROPHTHALAMIC ACID

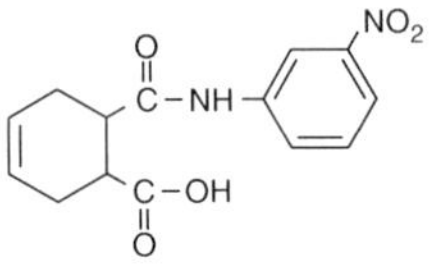

5 gm

S73,042-4
N-(4-METHYL-3-NITROPHENYL)-GLUTARAMIC ACID

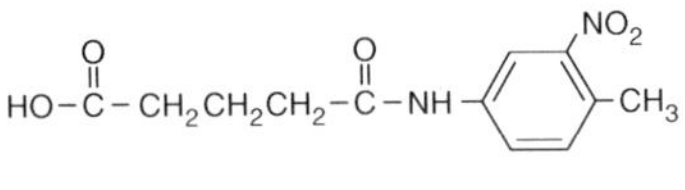

5 gm

S85,586-3
4'-METHYL-3'-NITROSUCCINANILIC ACID

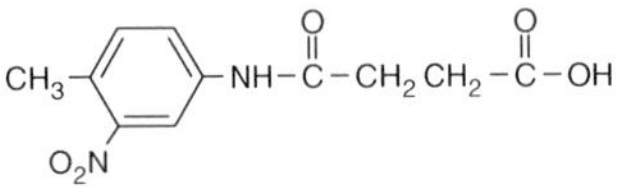

250 mg

S71,731-2
3-[N-(3-NITROPHENYL)CARBAMOYL]-BICYCLO[2.2.1]HEPT-5-ENE-2-CARBOXYLIC ACID

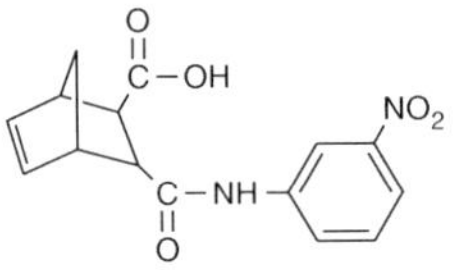

5 gm

S81,347-8
N-(3-NITROPHENYL)OCTADECANAMIDE

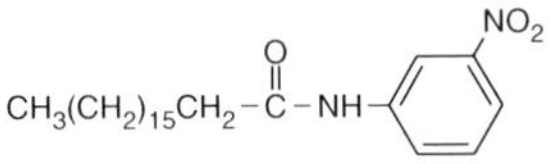

250 mg

S89,133-9
3'-NITRO-1-CYCLOPROPANECARBOXANILIDE

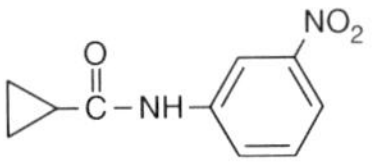

250 mg

S90,021-4
2'-METHYL-5'-NITRO-2-PHENYLACETANILIDE

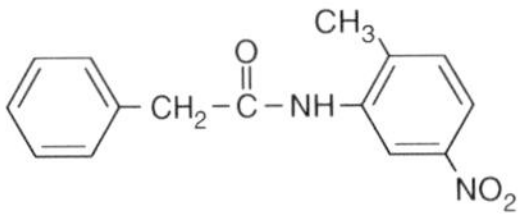

250 mg

S69,958-6
4'-METHYL-3'-NITROPROPIONANILIDE

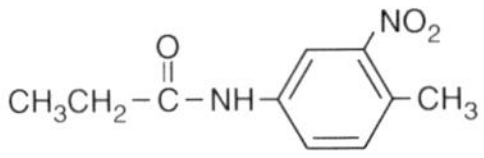

250 mg

S38,291-4
N-FORMYL-4-NITROANTHRANILIC ACID

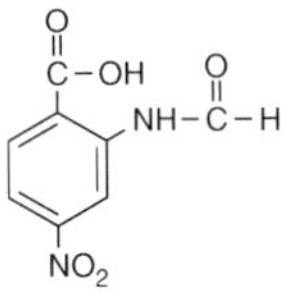

250 mg

S40,623-6
2-ACETAMIDO-4-NITROBENZOIC ACID

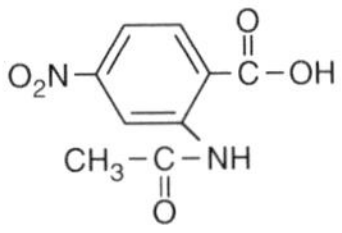

5 gm

S41,239-2
3'-NITROFORMANILIDE

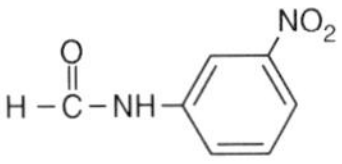

5 gm

S44,764-1
3'-NITROPALMITANILIDE

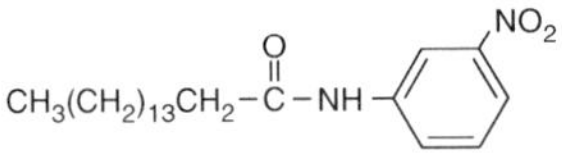

5 gm

S55,063-9
4'-(2,4-DINITROSTYRYL)ACETANILIDE

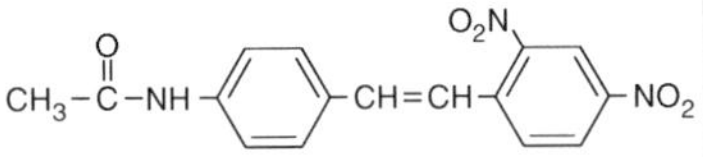

250 mg

S57,672-7
5'-NITRO-O-FORMOTOLUIDIDE

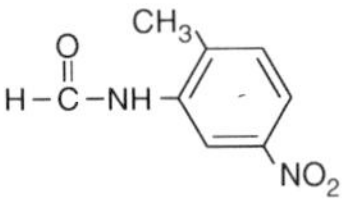

5 gm

S57,673-5
3'-NITRO-P-FORMOTOLUIDIDE

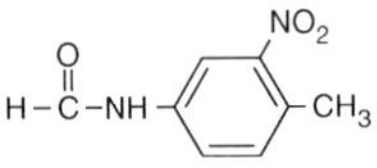

5 gm

S36,408-8
5'-NITRO-ORTHO-ACETOTOLUIDIDE

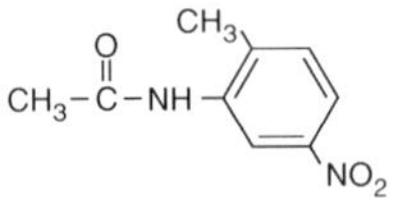

5 gm

S61,502-1
4'-METHYL-3'-NITROACETANILIDE

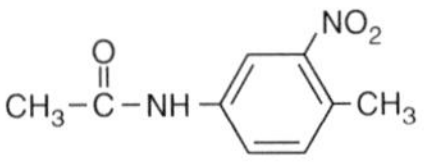

5 gm

S57,765-0
N-ETHYL-3'-NITROACETANILIDE

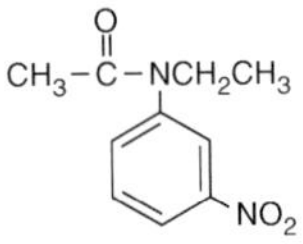

5 gm

S36,002-3
3,5-DIACETAMIDOBENZOIC ACID

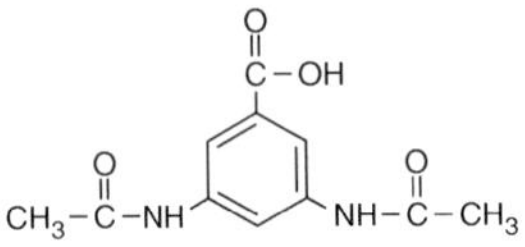

250 mg

S84,416-0
3'-ACETAMIDOCYCLOPROPANECARBOXANILIDE MONOHYDRATE

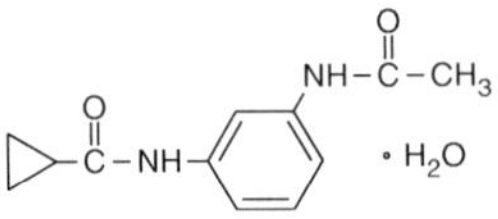

250 mg

S42,131-6
6'-CHLORO-3-PYRROLIDINO-ORTHO-BUTYROTOLUIDIDE HYDROCHLORIDE

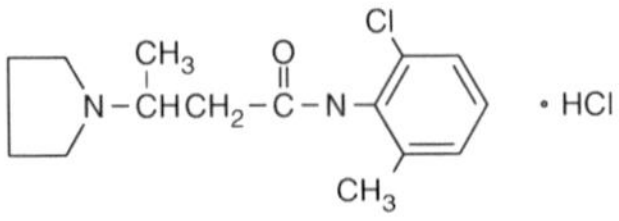

5 gm

S52,898-6
N-BENZYL-3-(4-PHENYL-1,2,3,6-TETRAHYDROPYRIDINO)PROPIONAMIDE

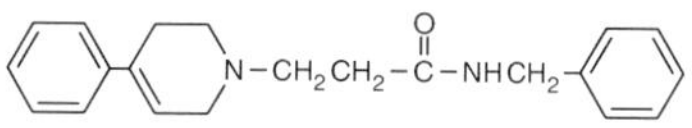

250 mg

S93,103-9
7-ACETAMIDO-1-ETHYL-1,2,3,4-TETRAHYDRO-2,2,4-TRIMETHYLQUINOLINE

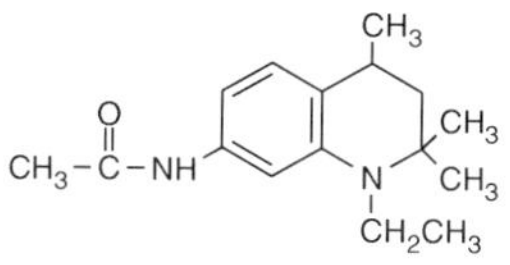

250 mg

S45,441-9
4-(N-PROPIONYLANILINO)PIPERIDINE

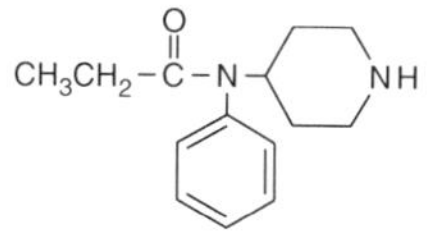

250 mg

S35,839-8
1-BENZYL-4-FORMAMIDOMETHYL-4-PHENYLPIPERIDINE

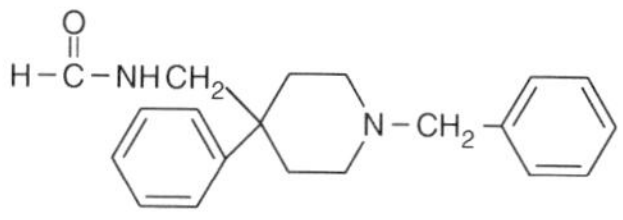

5 gm

S66,070-1
2-PHENYL-3-(M-TOLYL)SUCCINAMIDE

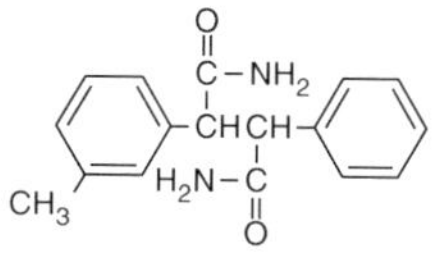

250 mg

S91,400-2
4'-ANILINOHEXAHYDROPHTHALANILIC ACID

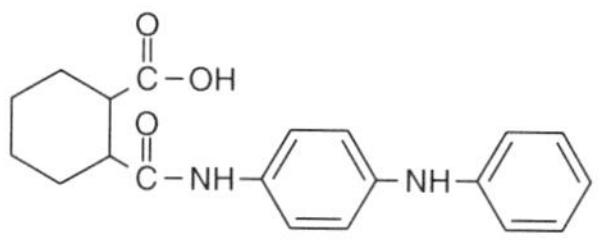

250 mg

S76,438-8
4'-ANILINOGLUTARANILIC ACID

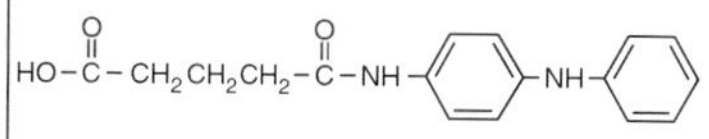

5 gm

S53,554-0
4-PHENYLISONIPECOTAMIDE HYDROCHLORIDE

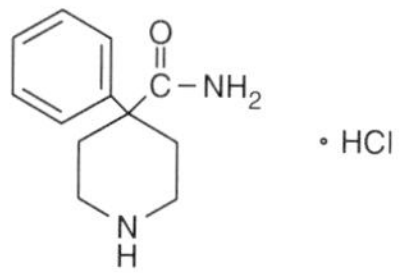

250 mg

S34,020-0
4-ACETAMIDOMETHYL-4-PHENYLPIPERIDINE

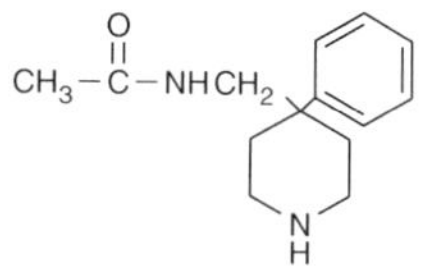

250 mg

S69,850-4
4'-ANILINOPROPIONANILIDE

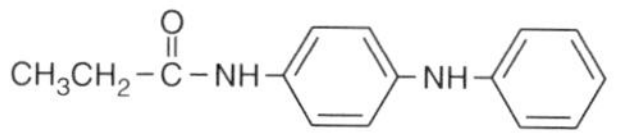

250 mg

S91,277-8
4'-(DIMETHYLAMINO)GLUTARANILIC ACID

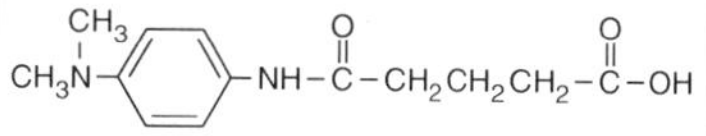

250 mg

S87,542-2
4'-(DIMETHYLAMINO)ACETANILIDE

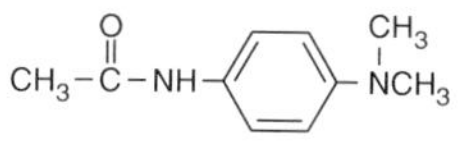

250 mg

S90,280-2
2,2-DIMETHYL-4'-(DIMETHYLAMINO)-PROPIONANILIDE

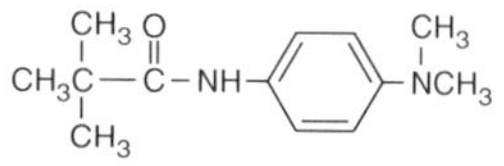

250 mg

S74,699-1
4'-(DIMETHYLAMINO)-HEXAHYDROPHTHALANILIC ACID

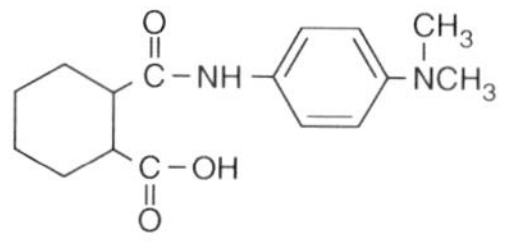

5 gm

S35,706-5
4-ACETAMIDOMETHYL-1-BENZYL-4-PHENYLPIPERIDINE

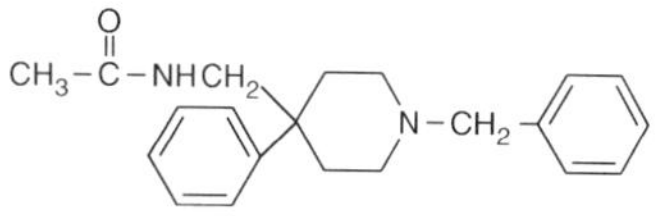

250 mg

S43,372-1
4'-DIETHYLAMINOACETANILIDE

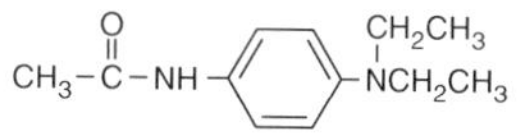

5 gm

S71,598-0
2,2-DIMETHYL-4'-(DIMETHYLAMINO)-PROPIONANILIDE HYDROCHLORIDE

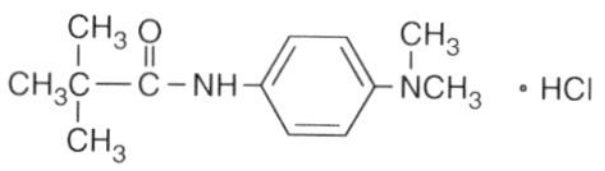

250 mg

S73,133-1
2'-ACETYL-4'-NITROACETANILIDE

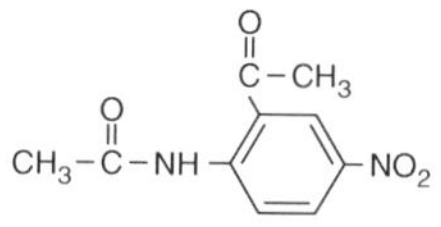

5 gm

S85,562-6
2,3-DIBROMO-2'-METHYL-4'-NITROSUCCINANILIC ACID

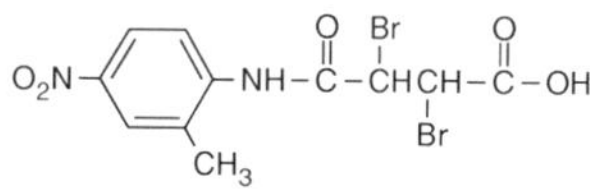

250 mg

S69,609-9
5-NITRO-2-PROPIONAMIDOBENZOIC ACID

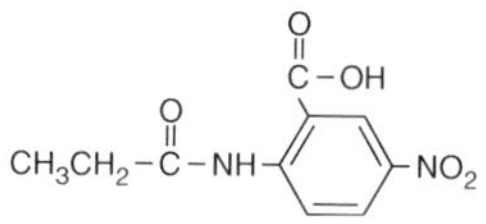

250 mg

S70,290-0
CIS-HEXAHYDRO-N-(4-NITROPHENYL)-PHTHALAMIC ACID

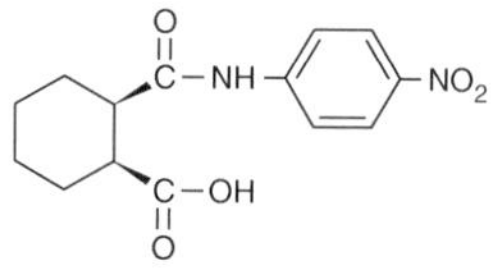

5 gm

S70,093-2
4'-NITROSUCCINANILIC ACID

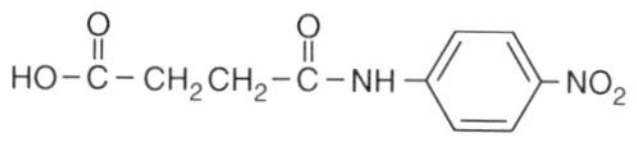

5 gm

S78,042-1
4'-NITROGLUTARANILIC ACID

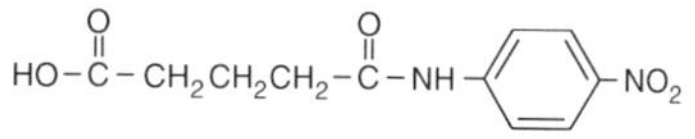

250 mg

S89,130-4
2'-METHYL-4'-NITRO-1-CYCLOPROPANECARBOXANILIDE

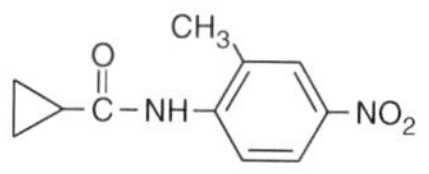

250 mg

S74,261-9
4'-NITRO-2'-PHENYLACETANILIDE

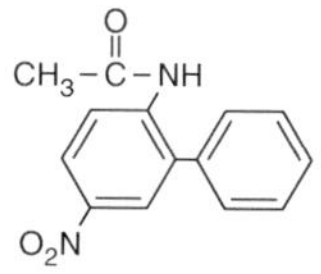

5 gm

S75,055-7
4'-NITRODECANANILIDE

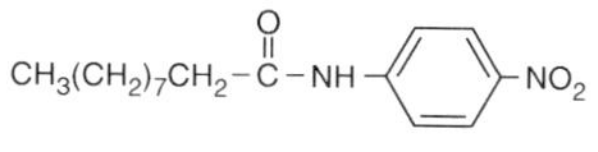

5 gm

S82,132-2
4'-NITROHEXADECANANILIDE

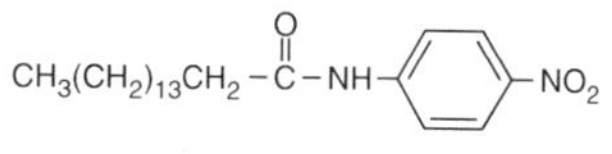

250 mg

S75,190-1
4'-NITRO-O-ACETOTOLUIDIDE

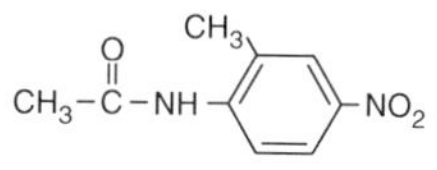

5 gm

S69,957-8
2'-METHYL-4'-NITROPROPIONANILIDE

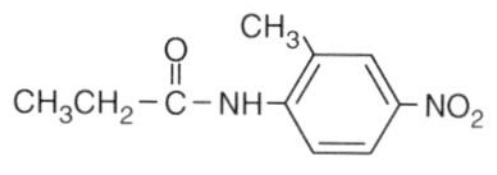

250 mg

S50,238-3
ETHYL 9-(2-HYDROXY-4-NITROPHENYLCARBAMOYL)NONANOATE

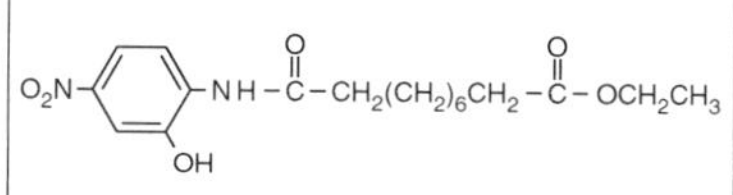

250 mg

S94,286-3
METHYL N-ACETYL-5-NITROANTHRANILATE

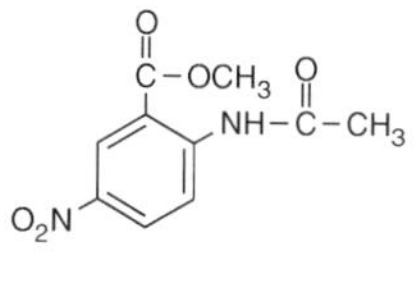

250 mg

S54,055-2
4'-NITROFORMANILIDE

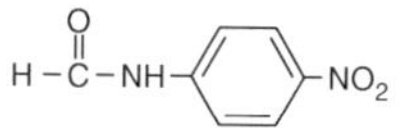

5 gm

S74,170-1
2-(4-ACETAMIDOPHENYL)-7-NITRO-1,2,3,4-TETRAHYDROISOQUINOLINE
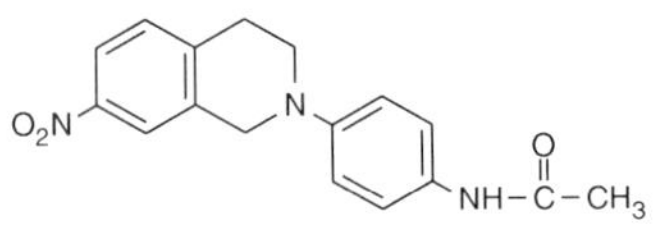
250 mg

S62,334-2
2'-CHLORO-4'-NITROFORMANILIDE
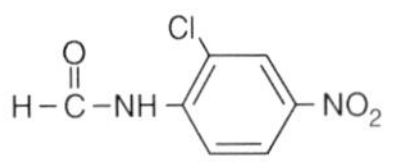
5 gm

S57,640-9
4'-NITRO-O-FORMOTOLUIDIDE
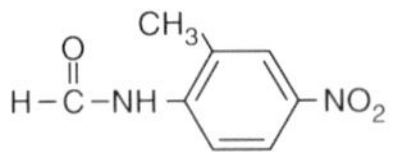
5 gm

S50,542-0
2',6'-DICHLORO-4'-NITROACETANILIDE
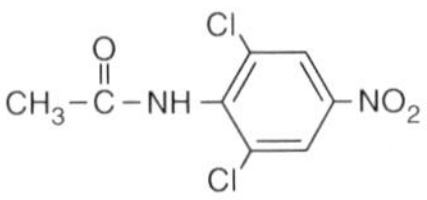
5 gm

S53,041-7
2',5'-DICHLORO-4'-NITROACETANILIDE
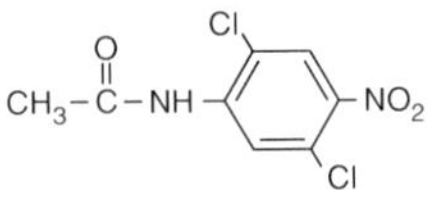
5 gm

S67,464-8
4'-ACETAMIDOSUCCINANILIC ACID
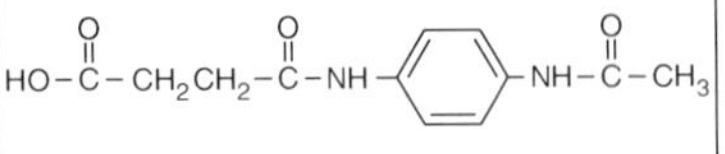
250 mg

S71,595-6
4'-(N-METHYLACETAMIDO)-GLUTARANILIC ACID
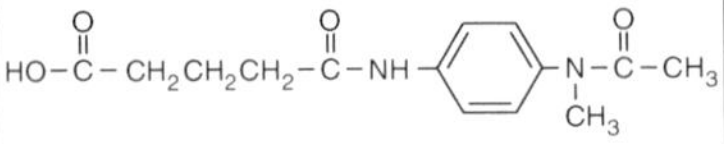
5 gm

S90,030-3
N-ACETYL-N-METHYL-N'-PHENYLACETYL-1,4-PHENYLENEDIAMINE
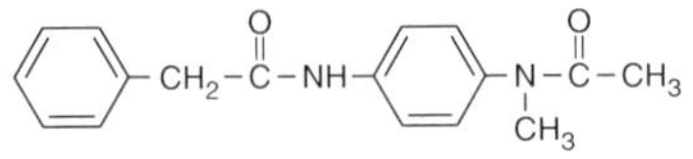
250 mg

S73,899-9
4'-(N-METHYLACETAMIDO)-DODECANANILIDE
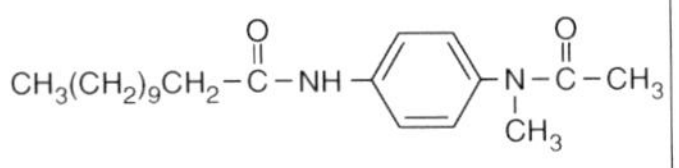
5 gm

S66,949-0
N,N'-DIACETYL-1,4-PHENYLENEDIAMINE
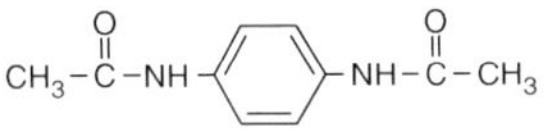
5 gm

S69,612-9
N-ACETYL-N'-PROPIONYL-1,4-PHENYLENEDIAMINE
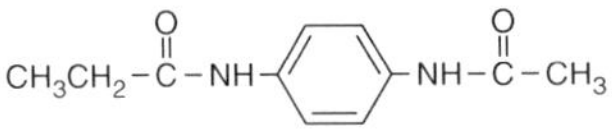
250 mg

S85,504-9
4'-ACETAMIDOBUTYRANILIDE
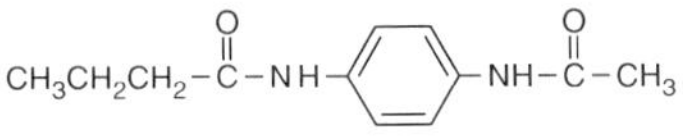
250 mg

S73,860-3
4'-ACETAMIDODODECANANILIDE
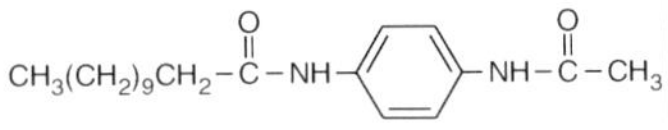
250 mg

S81,355-9
N,N'-(1,4-PHENYLENE)-BISOCTADECANAMIDE
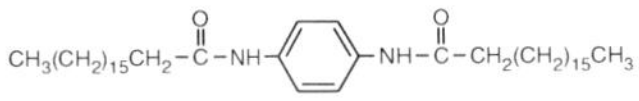
250 mg

S77,752-8
SARCOSYL-L-PHENYLALANINE
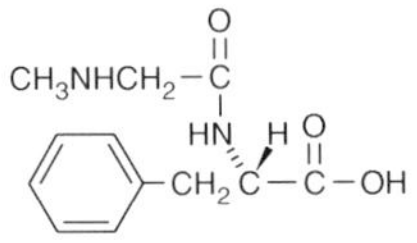
250 mg

S78,738-8
SARCOSYL-L-PHENYLALANINAMIDE ACETATE
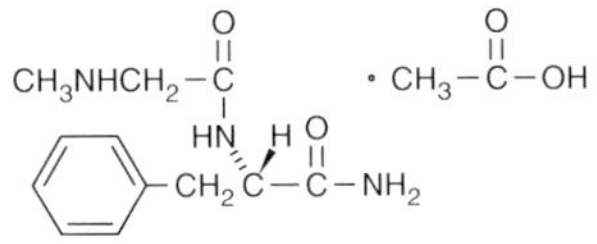
250 mg

S77,145-7
N-[1-(BUTYLAMINO)-2-PHENYLETHYL]-ACETAMIDE
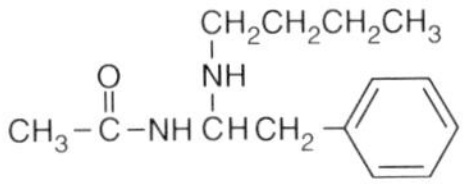
250 mg

S38,418-6
N-(1,2-DIPHENYLETHYL)-2-(N-ETHYL-N-METHYLAMINO)ACETAMIDE
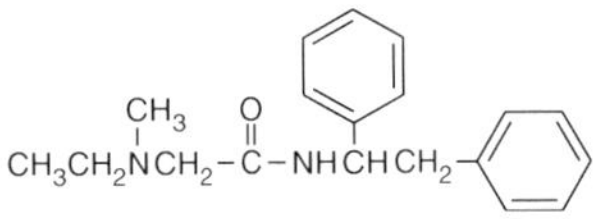
250 mg

S38,422-4
N-(1,2-DIPHENYLETHYL)-2-(N-METHYL-N-PROPYLAMINO)ACETAMIDE
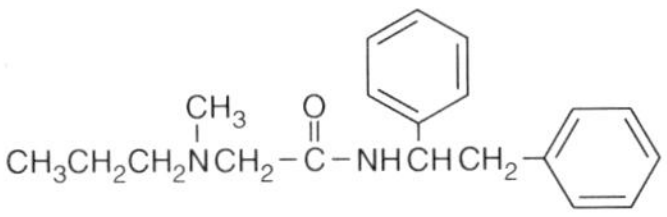
250 mg

S38,445-3
2-(N-BUTYL-N-METHYLAMINO)-N-(1,2-DIPHENYLETHYL)ACETAMIDE
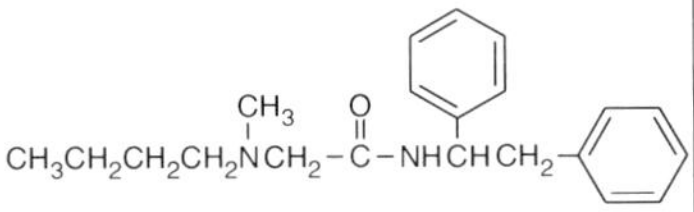
250 mg

S35,473-2
ALPHA-PHENYL-1-PYRROLIDINEACETAMIDE
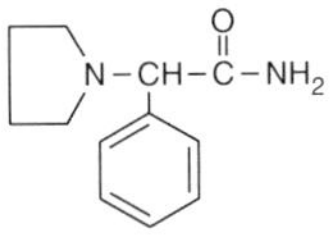
5 gm

S35,464-3
ALPHA-PHENYL-1-PIPERIDINEACETAMIDE
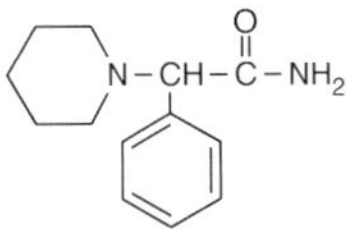
5 gm

S38,424-0
N-(1,2-DIPHENYLETHYL)-1-PYRROLIDINEACETAMIDE
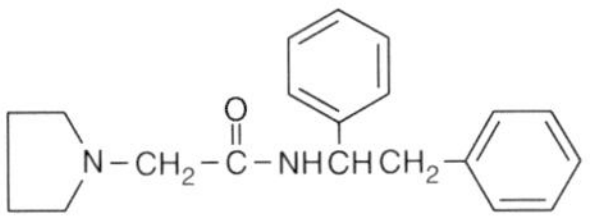
250 mg

S40,135-8
2-DIETHYLAMINO-N-(ALPHA-METHYLPHENETHYL)ACETAMIDE
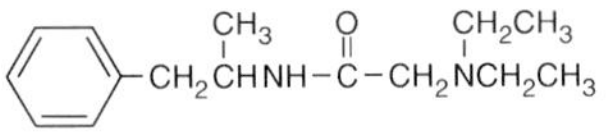
250 mg

S40,138-2
N-METHYL-N-(ALPHA-METHYLPHENETHYL)-1-PYRROLIDINEACETAMIDE
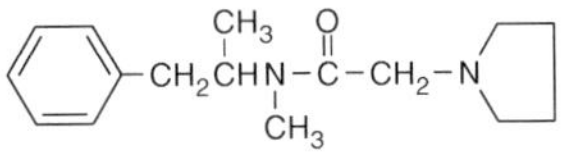
250 mg

S40,145-5
2-DIETHYLAMINO-N-METHYL-N-(ALPHA-METHYLPHENETHYL)ACETAMIDE
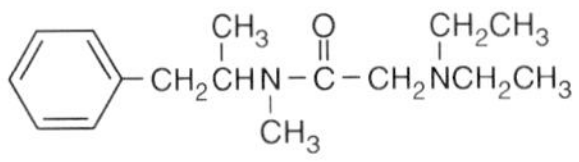
250 mg

S42,263-0
2-DIETHYLAMINO-N-METHYL-N-(3-METHYL-2-PHENYLPENTYL)ACETAMIDE HYDROCHLORIDE
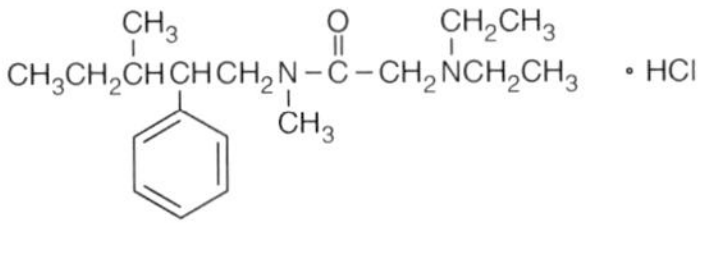
250 mg

S79,021-4
N-BENZYLGLYCYLGLYCYLPHENYLALANINE
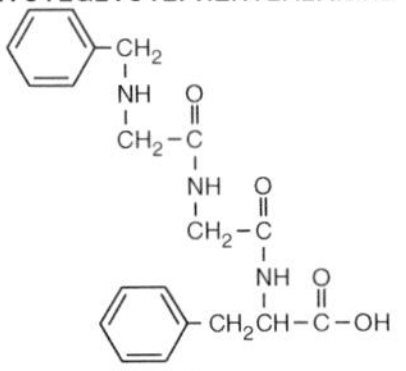
250 mg

S77,465-0
N-BENZYL-2,2-DIMETHYL-N-(2-PHENYLETHYL)PROPIONAMIDE
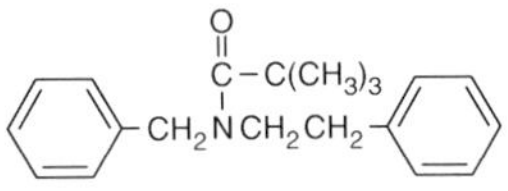
250 mg

S77,464-2
N-BENZYL-2,2-DIMETHYL-N-(1-METHYL-2-PHENYLETHYL)PROPIONAMIDE
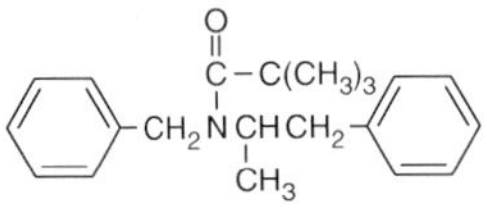
250 mg

S56,985-2
N-(4-NITROPHENETHYL)ACETAMIDE
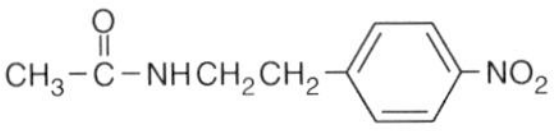
250 mg

S90,979-3
N-FORMYL-2-PHENYLACETAMIDE
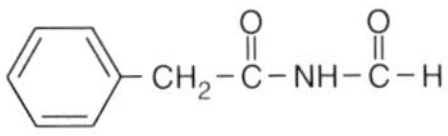
250 mg

S68,142-3
DIETHYL 2-ACETAMIDO-2-[4-(TRIMETHYLACETYL)BENZYL]MALONATE

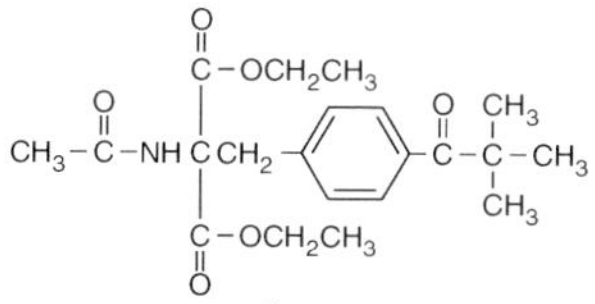

5 gm

S84,971-5
DIETHYL 2-ACETAMIDO-2-BENZYLMALONATE

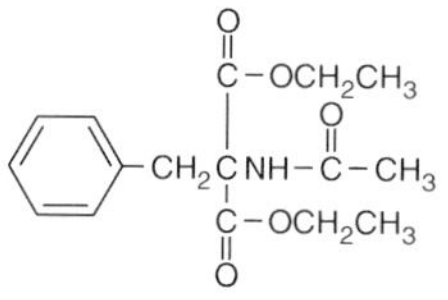

250 mg

S78,166-5
L-PHENYLALANYL-L-GLUTAMIC ACID

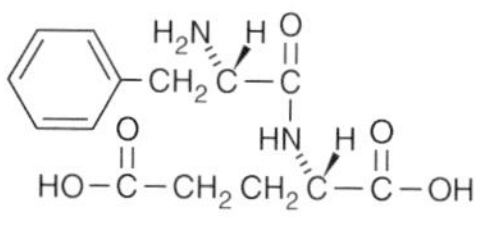

250 mg

S73,060-2
3-[N-(2-PHENYLETHYL)CARBAMOYL]-5-NORBORNENE-2-CARBOXYLIC ACID

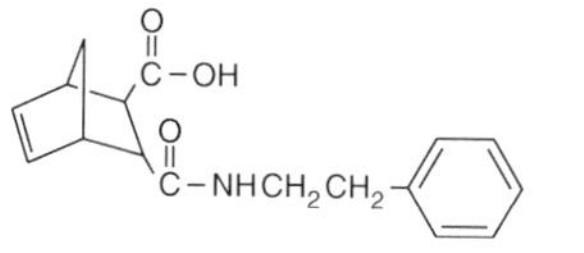

5 gm

S73,029-7
CIS-HEXAHYDRO-N-(2-PHENETHYL)-PHTHALAMIC ACID

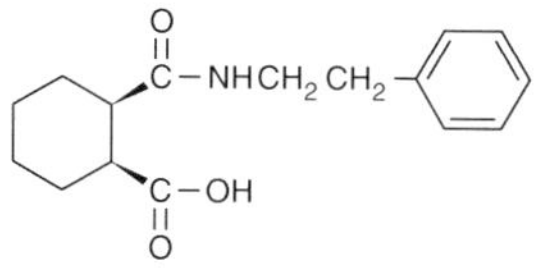

5 gm

S78,247-5
N-(PHENYLETHYL)SUCCINAMIC ACID

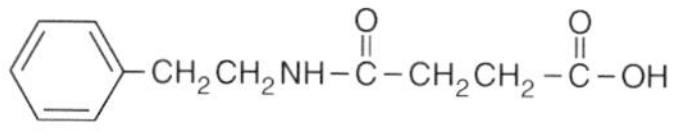

5 gm

S76,589-9
N,O-DIACETYL-L-TYROSINE

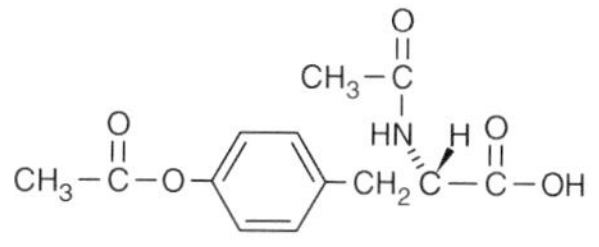

5 gm

S78,318-8
N,O-DIACETYL-DL-TYROSINE

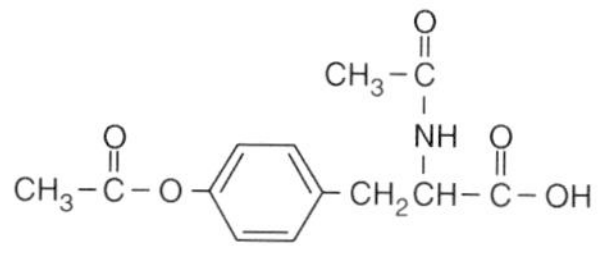

5 gm

S34,537-7
N,O-DIACETYL-3,5-DIIODO-L-TYROSINE

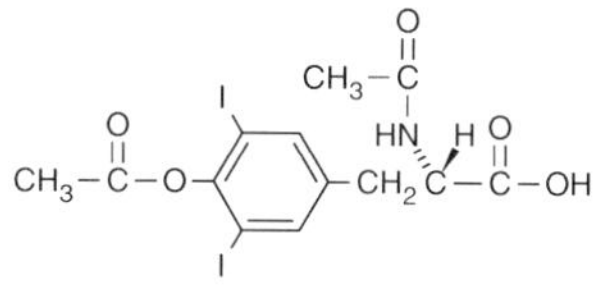

5 gm

S78,453-2
L-NORVALYLPHENYLALANINE

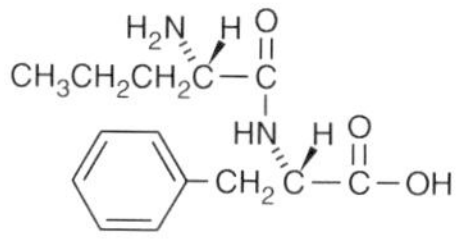

250 mg

S78,591-1
L-NORLEUCYL-L-PHENYLALANINE HYDRATE

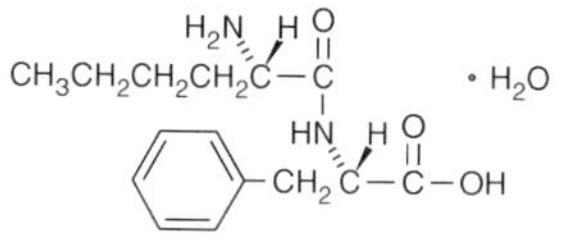

250 mg

S34,258-0
DL-2-BROMOISOCAPROYLGLYCYL-DL-PHENYLALANINE

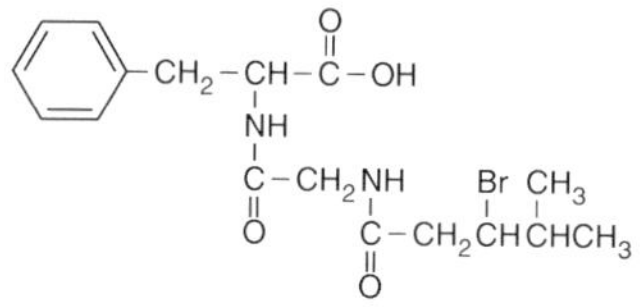

5 gm

S62,075-0
DL-N-FORMYLPHENYLALANINE

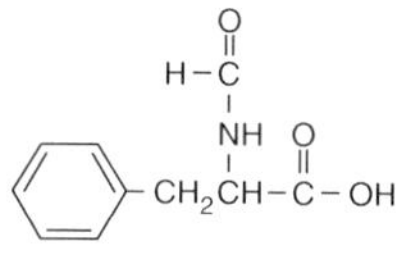

5 gm

S63,306-2
N-(PHENYLACETYL)GLYCINE

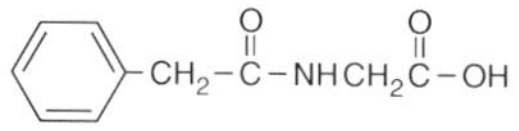

250 mg

S77,755-2
N-ACETYL-D-PHENYLALANINE

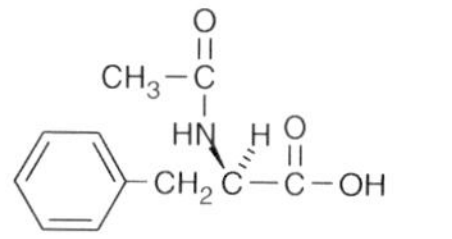

250 mg

S78,356-0
N-ACETYL-DL-PHENYLALANINE

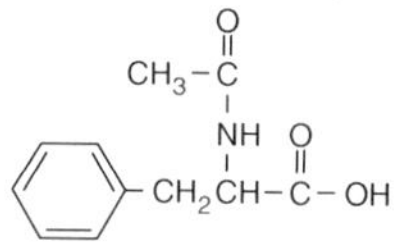

5 gm

S34,277-7
2-BROMOPROPIONYL-DL-PHENYLALANINE

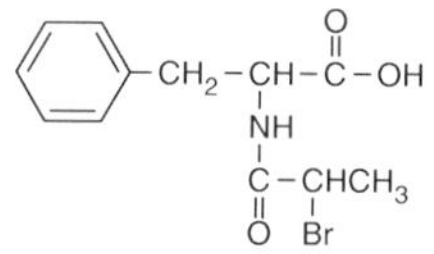

5 gm

S34,262-9
DL-2-BROMOISOVALERYL-DL-PHENYLALANINE

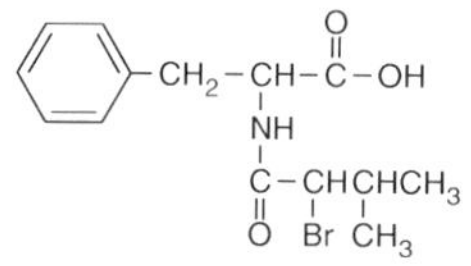

250 mg

S42,287-8
3-(N-METHYL-N-[1-(4-METHYLBENZYL)-ETHYL]CARBAMOYL)-2,3,3-TRIMETHYL-CYCLOPENTANECARBOXYLIC ACID, SODIUM SALT

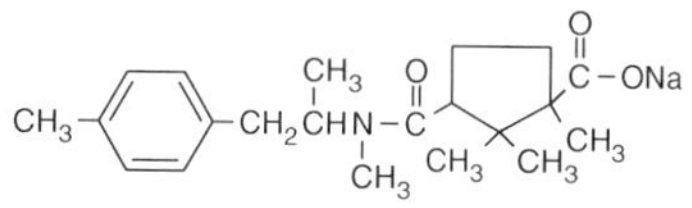

250 mg

S42,179-0
5-[N-(1-BENZYLETHYL)-N-METHYLCARBAMOYL]VALERIC ACID, SODIUM SALT

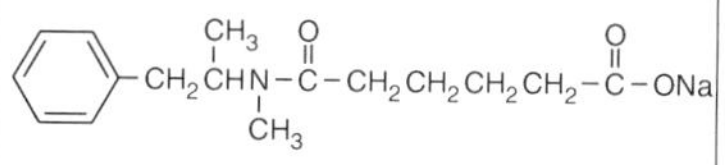

250 mg

S77,641-6
N-ACETYLGLYCYL-L-PHENYLALANINE METHYL ESTER MONOHYDRATE

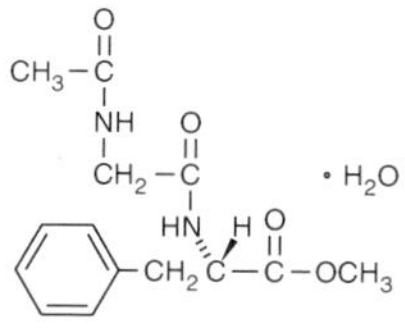

250 mg

S35,440-6
DL-N-(PHENYLACETYL)ALANINE METHYL ESTER

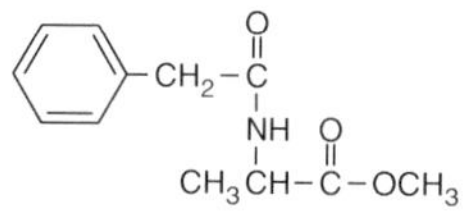

5 gm

S76,586-4
N-ACETYL-L-PHENYLALANINE METHYL ESTER

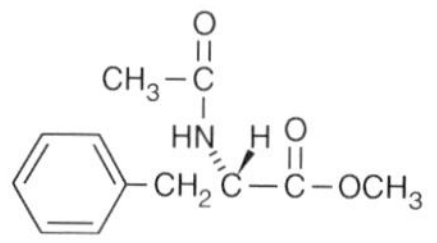

5 gm

S77,624-6
N-ACETYL-D-PHENYLALANINAMIDE

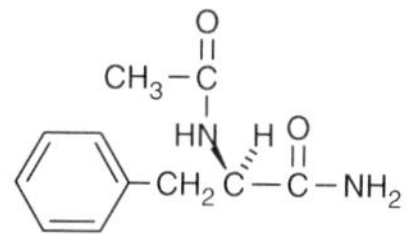

250 mg

S78,435-4
GLYCYL-D-PHENYLALANINAMIDE

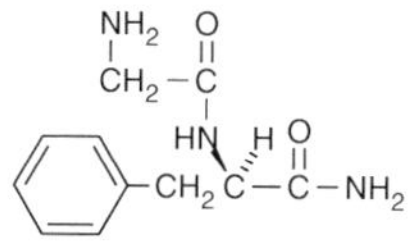

250 mg

S78,519-9
N-ACETYL-DL-PHENYLALANINAMIDE

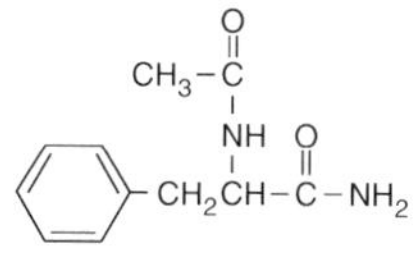

250 mg

S44,075-2
2-PHENYLACETAMIDE

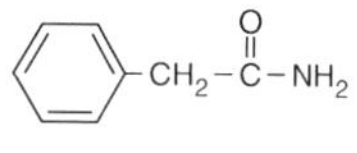

5 gm

S36,440-1
2-PHENYLBUTYRAMIDE

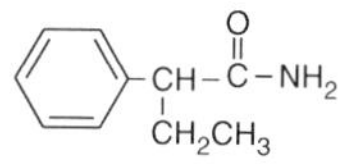

5 gm

S40,028-9
3-METHYL-2-PHENYLBUTYRAMIDE

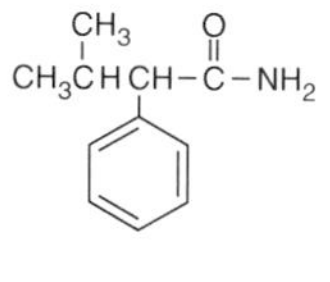

250 mg

S37,762-7
ALPHA-(SEC-BUTYL)PHENYLACETAMIDE

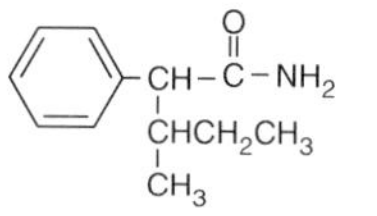

5 gm

S63,167-1
2,2-DIPHENYLACETAMIDE

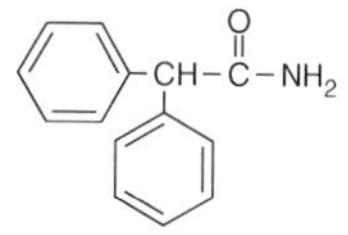

250 mg

S41,813-7
1-INDANCARBOXAMIDE

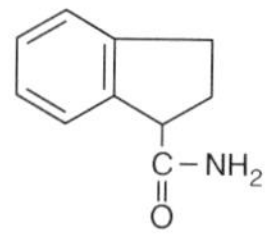

250 mg

S89,225-4
N-CYCLOHEXYL-2,2-DIPHENYLACETAMIDE
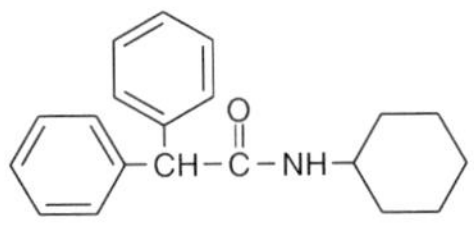
250 mg

S58,188-7
N-PHENETHYLACETAMIDE
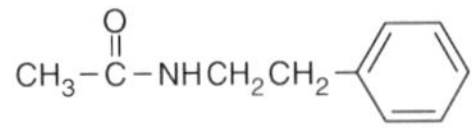
250 mg

S52,161-2
2-PHENYLACETANILIDE
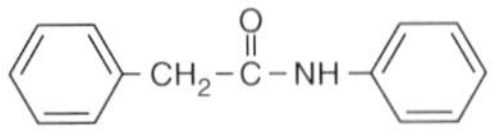
5 gm

S34,764-7
N-(1-BENZYL-2-PHENYLETHYL)FORMAMIDE
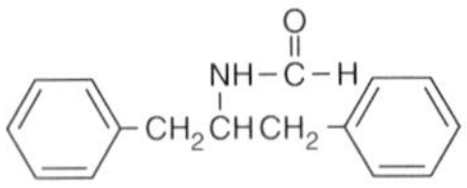
5 gm

S79,390-6
N-PHENETHYL-2-PHENYLACETAMIDE
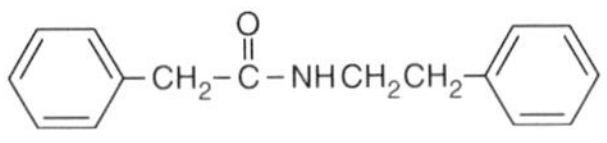
250 mg

S89,052-9
2',3'-DIMETHYL-2,2-DIPHENYLACETANILIDE
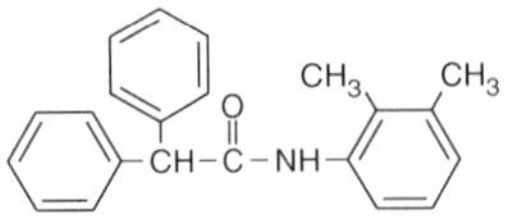
250 mg

S52,122-1
2,2-DIPHENYL-N-METHYLACETAMIDE
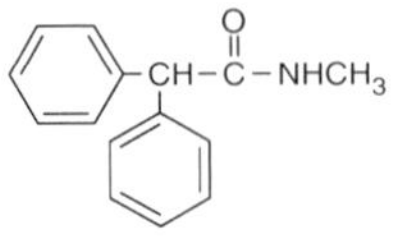
5 gm

S52,661-4
2,2-DIPHENYL-N-ETHYLACETAMIDE
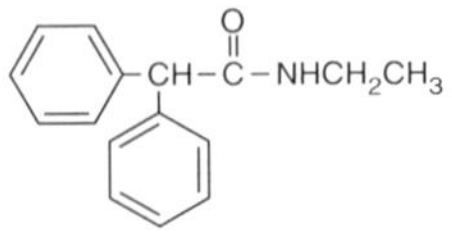
5 gm

S52,685-1
2,2-DIPHENYLACETANILIDE
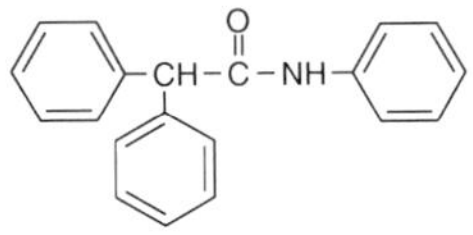
5 gm

S89,224-6
N-BENZYL-2,2-DIPHENYLACETAMIDE
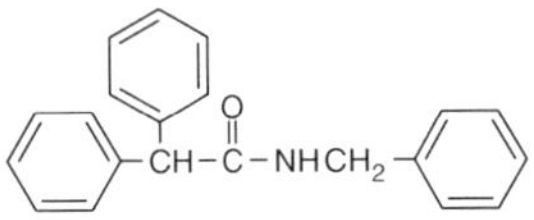
250 mg

S51,946-4
4'-BROMO-2-PHENYLACETANILIDE
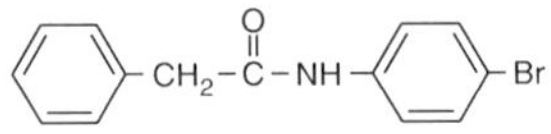
5 gm

S38,368-6
2-CHLORO-2-PHENYL-2',6'-ACETOXYLIDIDE
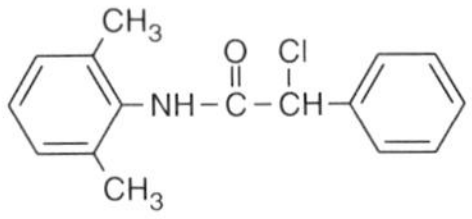
5 gm

S40,114-5
N-METHYL-N-PHENETHYL-1-PYRROLIDINEACETAMIDE
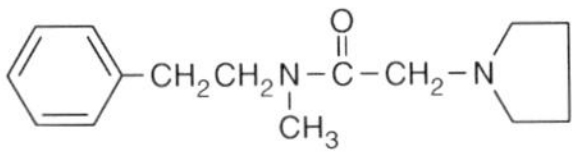
250 mg

S38,157-8
N-METHYL-N-(ALPHA-METHYLPHENETHYL)-ACETAMIDE
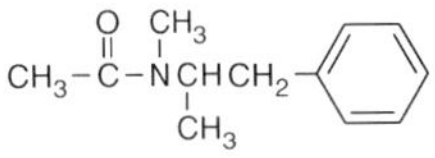
250 mg

S42,120-0
2-PHENYL-N,N,3-TRIMETHYLVALERAMIDE
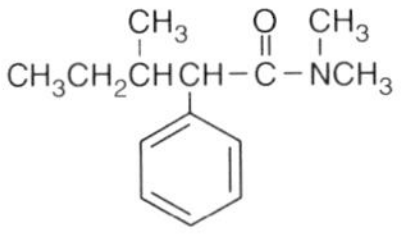
5 gm

S42,317-3
N,N-DIETHYL-3-METHYL-2-PHENYLVALERAMIDE
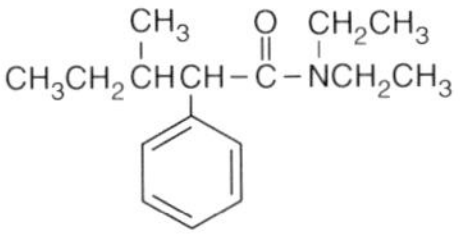
5 gm

S77,455-3
2,2-DIMETHYL-N-ETHYL-N-(1-METHYL-2-PHENYLETHYL)PROPIONAMIDE
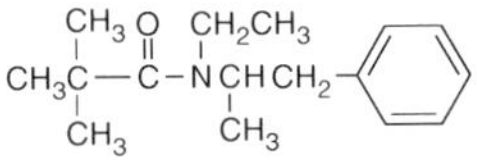
250 mg

S52,578-2
ALPHA-(4-FLUOROANILINO)-PARA-ACETOTOLUIDIDE
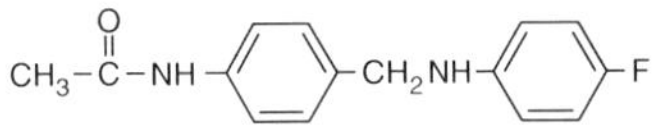
5 gm

S53,147-2
ALPHA-(4-CHLOROANILINO)-PARA-ACETOTOLUIDIDE
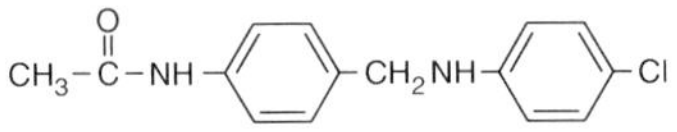
5 gm

S53,884-1
ALPHA-(4-BROMOANILINO)-PARA-ACETOTOLUIDIDE
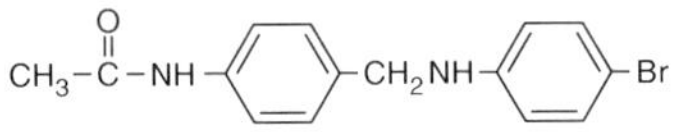
250 mg

S53,314-9
ALPHA-(4-IODOANILINO)-PARA-ACETOTOLUIDIDE
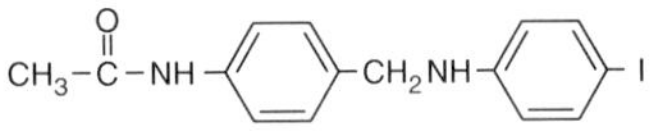
250 mg

S53,146-4
ALPHA-(PARA-TOLUIDINO)-PARA-ACETOTOLUIDIDE
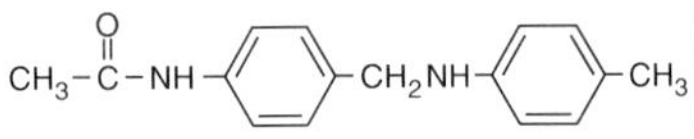
250 mg

S53,883-3
ALPHA-(3-CHLORO-PARA-TOLUIDINO)-PARA-ACETOTOLUIDIDE
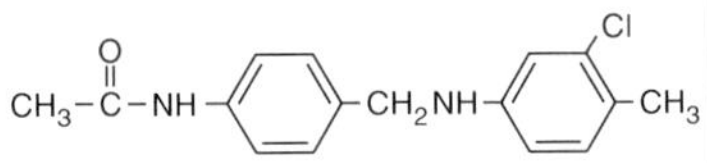
5 gm

S53,285-1
ALPHA-(2,4-XYLIDINO)-PARA-ACETOTOLUIDIDE
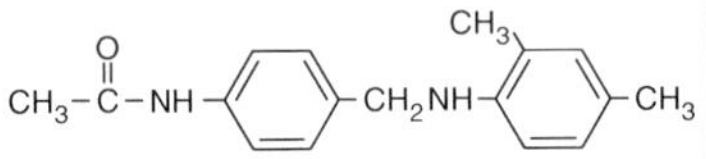
250 mg

S39,820-9
6'-CHLORO-2-PROPYLAMINO-ORTHO-ACETOTOLUIDIDE HYDROCHLORIDE
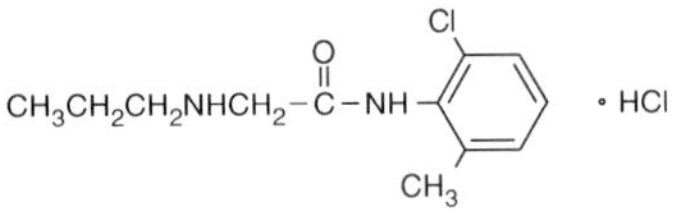
250 mg

S62,261-3
4'-ANILINOSTEARANILIDE
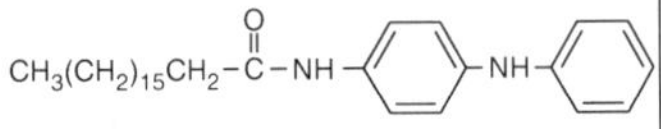
250 mg

S42,071-9
2-BUTYLAMINO-6'-CHLORO-ORTHO-ACETOTOLUIDIDE HYDROCHLORIDE
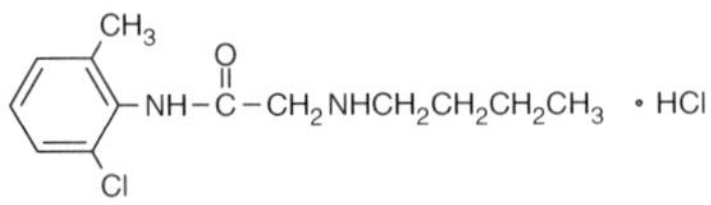
250 mg

S38,071-7
N-ETHYL-N-(ORTHO-TOLYL)-CARBAMOYL CHLORIDE
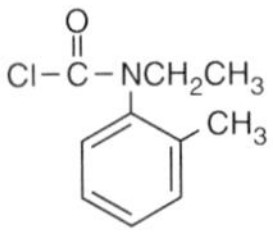
250 mg

S42,133-2
N-ACETYL-6'-CHLORO-1-PYRROLIDINEACETO-ORTHO-TOLUIDIDE HYDROCHLORIDE
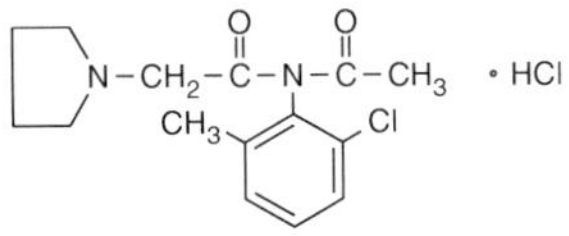
250 mg

S39,818-7
6'-CHLORO-ALPHA-METHYL-1-PYRROLIDINE-O-ACETOTOLUIDIDE HYDROCHLORIDE
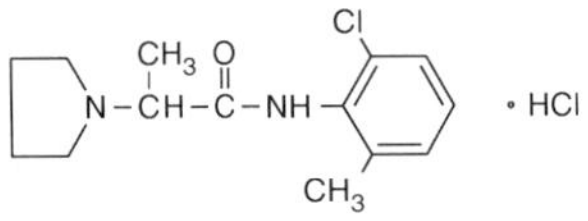
5 gm

S40,180-3
2-(DIETHYLAMINO)-N-(DIPHENYLMETHYL)-ACETAMIDE HYDROCHLORIDE
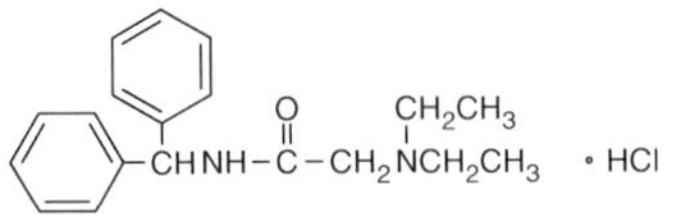
250 mg

S38,086-5
6'-CHLORO-2-DIMETHYLAMINO-ORTHO-ACETOTOLUIDIDE
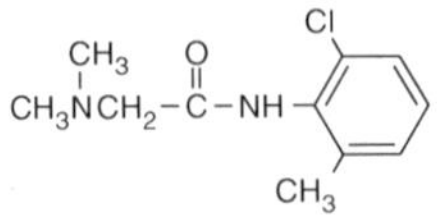
5 gm

S38,255-8
2-DIETHYLAMINO-2',6'-ACETOXYLIDIDE

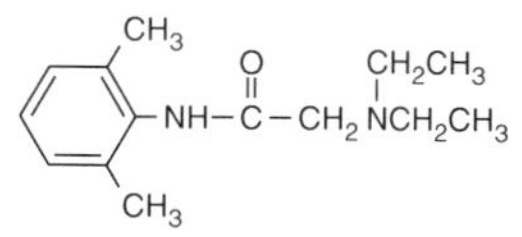

5 gm

S39,819-5
2',6'-DICHLORO-2-(DIPROPYLAMINO)-ACETANILIDE HYDROCHLORIDE

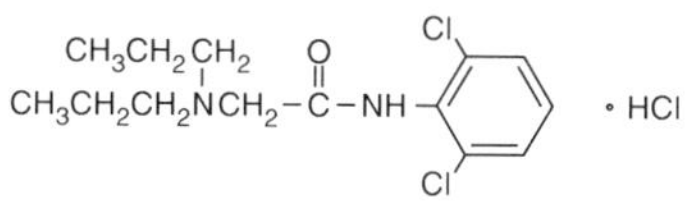

250 mg

S38,053-9
3',4'-DICHLORO-2-(DIMETHYLAMINO)-ACETANILIDE HYDROCHLORIDE

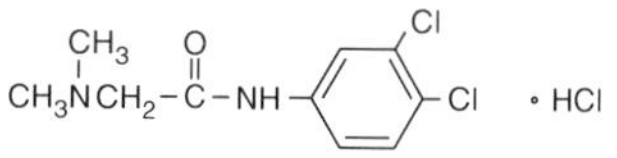

250 mg

S42,061-1
4'-CHLORO-1-PYRROLIDINEACETANILIDE HYDROCHLORIDE

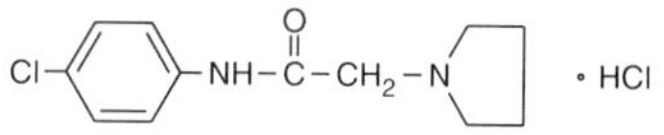

5 gm

S76,277-6
N,N'-DIPHENYL-1-METHYL-2,6-PIPERIDINEDIACETAMIDE

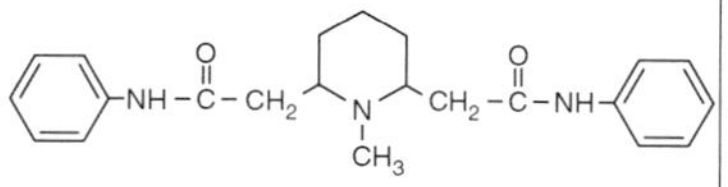

250 mg

S40,293-1
BENZYLDIMETHYL (TETRADECYLCARBAMOYLMETHYL)-AMMONIUM CHLORIDE

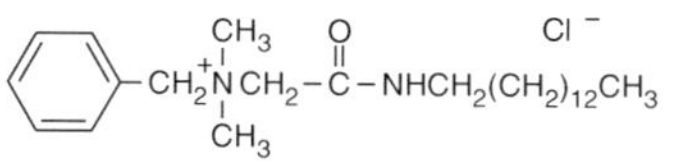

250 mg

S65,263-6
(4'-ACETAMIDOPHENACYL)DIETHYL(10-ETHOXYCARBONYLDECYL)AMMONIUM IODIDE

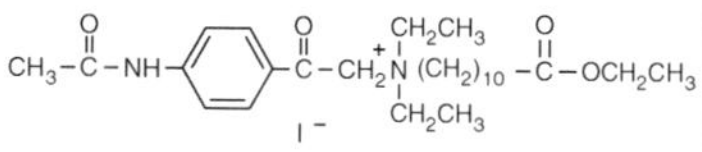

5 gm

S39,897-7
(4-CHLOROPHENYLCARBAMOYLMETHYL)-DIMETHYL(TETRAHYDRO-2-NAPHTHYLMETHYL)-AMMONIUM CHLORIDE

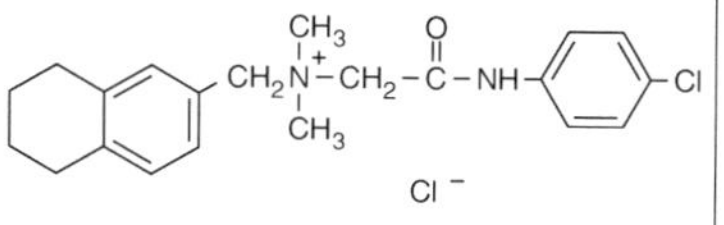

250 mg

S40,304-0
DIMETHYL(DODECYLCARBAMOYLMETHYL)-(5,6,7,8-TETRAHYDRO-2-NAPHTHYLMETHYL)-AMMONIUM CHLORIDE

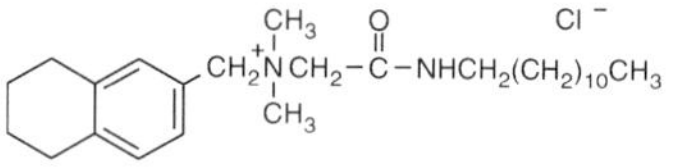

5 gm

S36,373-1
3-METHYL-2-[2-(N-PHENYLACETAMIDO)-VINYL]-2-THIAZOLINIUM IODIDE

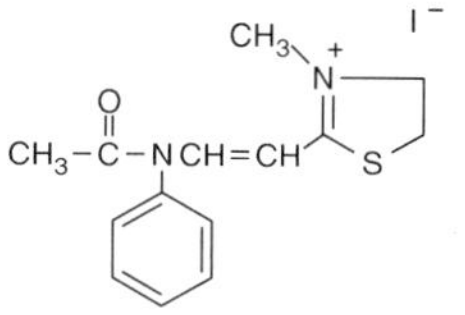

250 mg

S67,659-4
2-[N-(2,4,6-TRIMETHYLPHENYL)-CARBAMOYL]-CYCLOHEXANECARBOXYLIC ACID

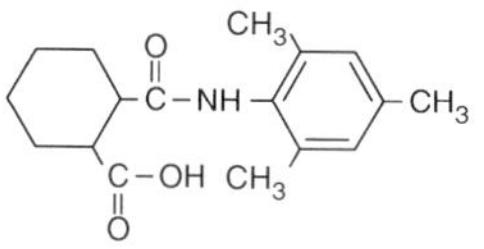

5 gm

S67,678-0
2-(4-METHYLBENZYLCARBAMOYL)-BENZOIC ACID

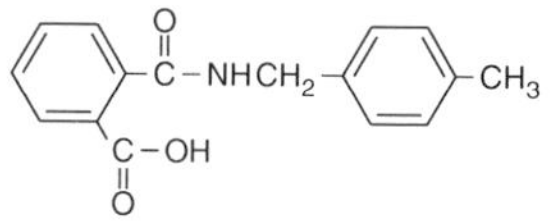

250 mg

S65,145-1
N,N'-DIBENZYL-N,N'-DIMETHYLOXAMIDE

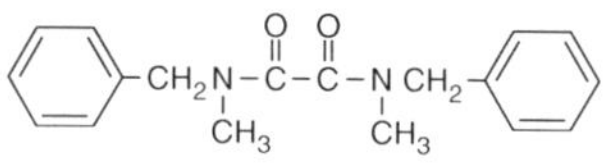

5 gm

S77,487-1
N,N'-DIBENZYL-N,N'-DIISOPROPYLMALONAMIDE

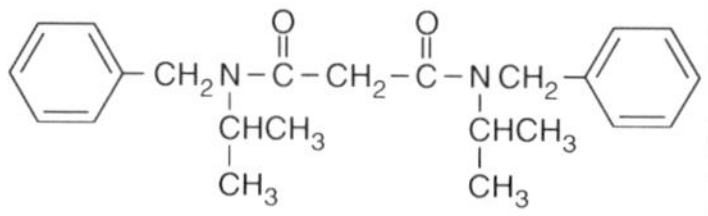

250 mg

S73,061-0
3-(N-BENZYLCARBAMOYL)-5-NORBORNENE-2-CARBOXYLIC ACID

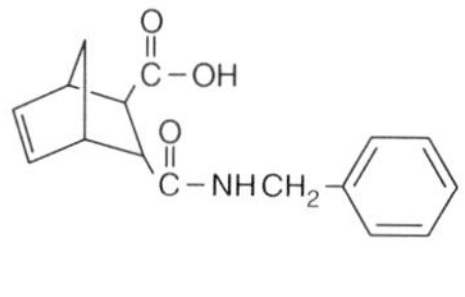

5 gm

S72,007-0
CIS-N-BENZYLHEXAHYDRO-PHTHALAMIC ACID

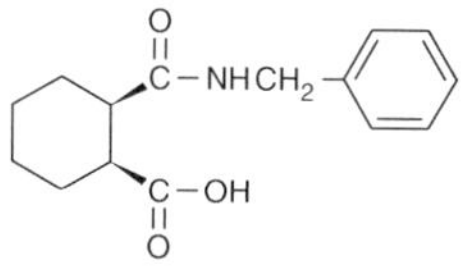

5 gm

S73,031-9
N-BENZYLGLUTARAMIC ACID

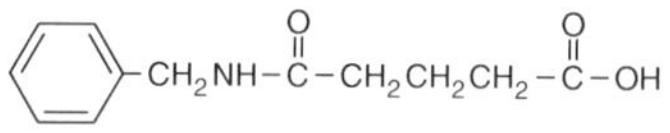

5 gm

S75,021-2
N,N-DIBENZYLHEXAHYDROPHTHALAMIC ACID

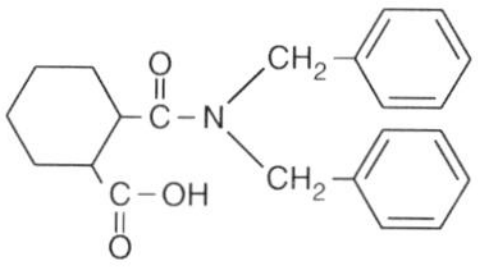

5 gm

S35,735-9
2-ALLYL-2-BENZYLACETAMIDE

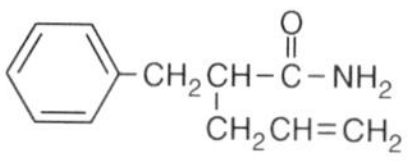

5 gm

S35,592-5
N,N'-(2,3,5,6-TETRAMETHYL-1,4-XYLYLENE)BISACETAMIDE

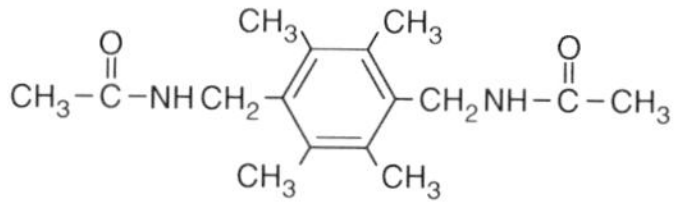

5 gm

S52,034-9
N-(ALPHA-METHYLBENZYL)FORMAMIDE

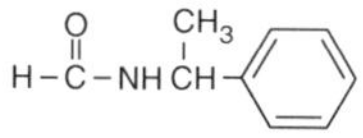

250 mg

S95,034-3
N-BENZYLDODECANAMIDE

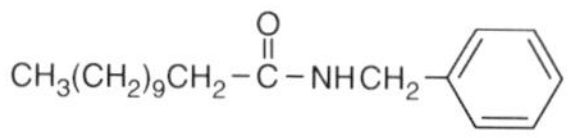

50 mg

S35,665-4
N-(2,4,5-TRIMETHYLBENZYL)ACETAMIDE

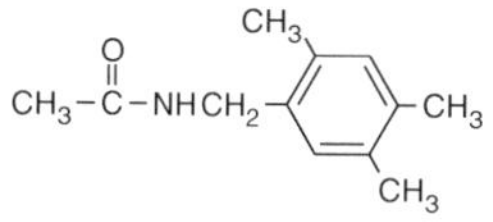

5 gm

S77,461-8
N-BENZYL-2,2-DIMETHYL-N-PROPYLPROPIONAMIDE

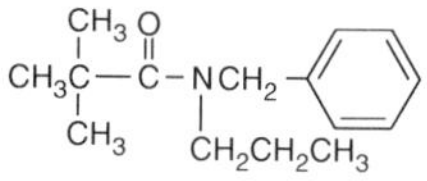

250 mg

S77,493-6
N-BENZYL-2-CHLORO-2-PHENYL-O-ACETOTOLUIDIDE

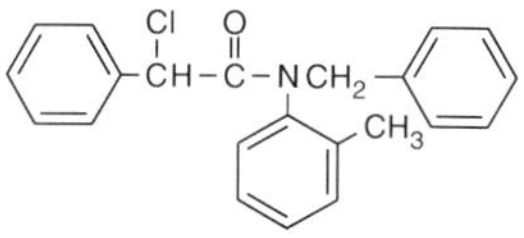

250 mg

S77,492-8
N-BENZYL-2,2-DIPHENYL-O-PROPIONOTOLUIDIDE

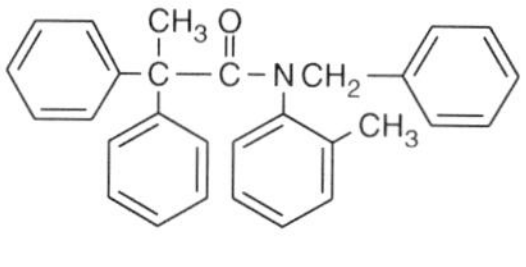

250 mg

S37,808-9
N-BENZYLACETAMIDE

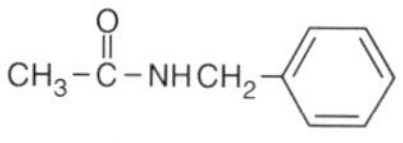

250 mg

S89,252-1
4-ACETAMIDOBENZALDEHYDE PHENYLHYDRAZONE

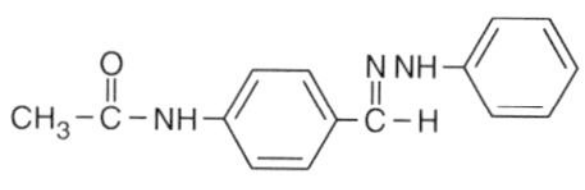

250 mg

S80,256-5
2-ACETAMIDO-3-CHLORO-1,4-NAPHTHOQUINONE

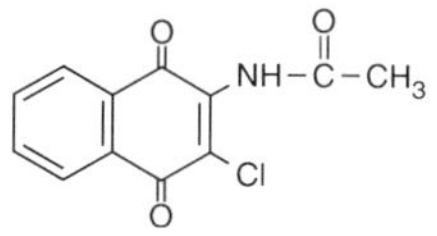

250 mg

S50,027-5
OXANILIC ACID

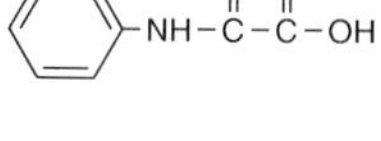

250 mg

S58,688-9
2-ACETAMIDOPHENYL ACETATE

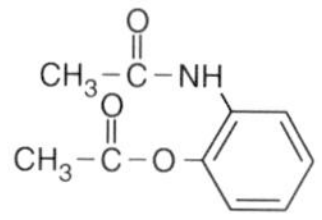

5 gm

S35,702-2
4-ACETAMIDO-3,5-DICHLOROBENZENEDIAZONIUM TETRAFLUOROBORATE

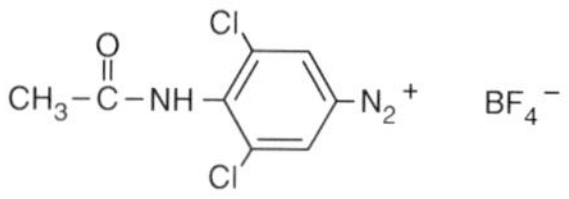

250 mg

S81,253-6
4'-ACETAMIDO-3-NITROCHALCONE
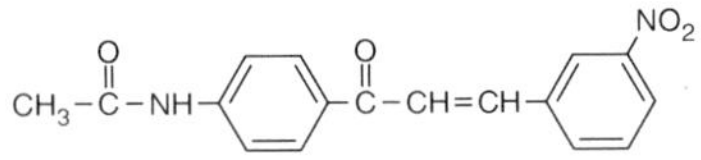
250 mg

S75,140-5
5'-ISOPROPYL-2'-METHYL-3-NITROBENZANILIDE
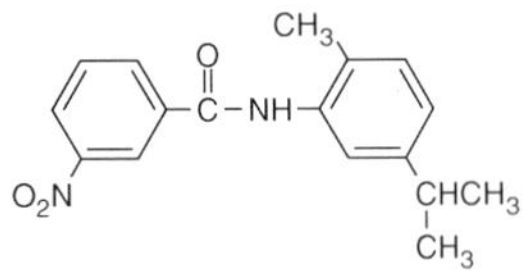
250 mg

S88,712-9
DIPROPYL 2-(ALPHA-CARBAMOYL-2-NITROBENZYL)MALONATE
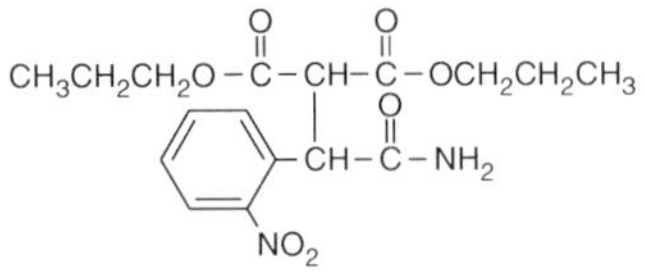
250 mg

S71,988-9
1,4,5,6,7,7-HEXACHLORO-3-[N-(3-NITROPHENYL)CARBAMOYL]-5-NORBORNENE-2-CARBOXYLIC ACID
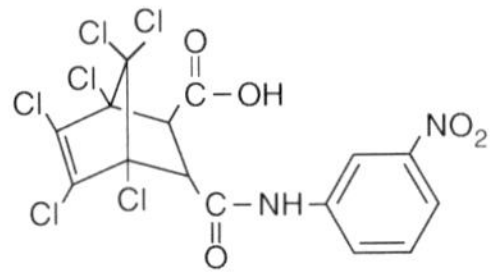
5 gm

S76,192-3
4-METHYL-3-PHENYL-N-(O-TOLYL)-2-ISOXAZOLINE-4-CARBOXAMIDE
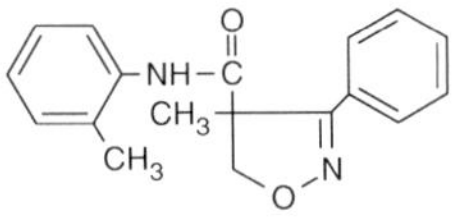
250 mg

S77,458-8
N-BENZYL-N-TERT-BUTYL-2,2-DIPHENYLPROPIONAMIDE
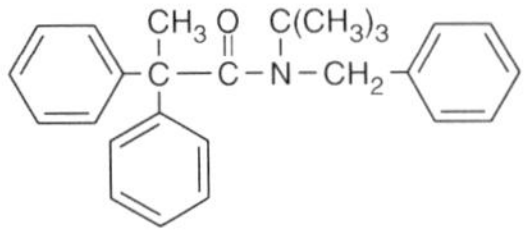
250 mg

S50,091-7
2'-BROMO-2-(3-NITROBENZOYL)-ACETANILIDE
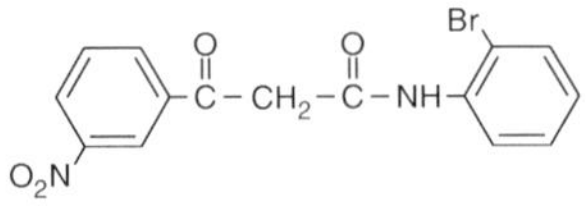
5 gm

S53,247-9
2'-NITRO-PARA-FORMOTOLUIDIDE
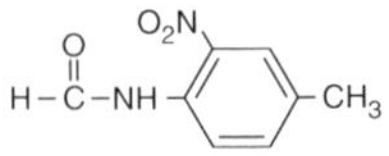
250 mg

S50,733-4
2',6'-DIETHYL-3'-NITROACETANILIDE
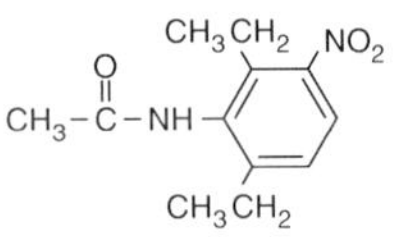
250 mg

S88,845-1
DIETHYL 2-ACETAMIDO-2-(4-ACETYLPHENETHYL)MALONATE
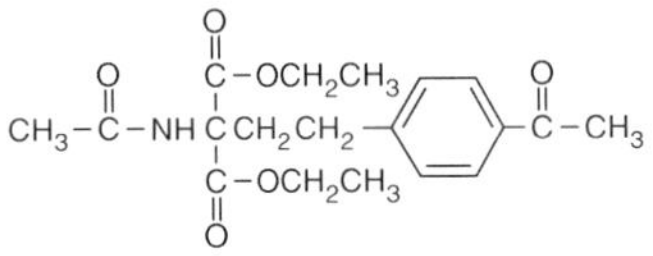
250 mg

S84,999-5
DIETHYL 2-ACETAMIDO-2-(3-OXO-3-PHENYLPROPYL)MALONATE
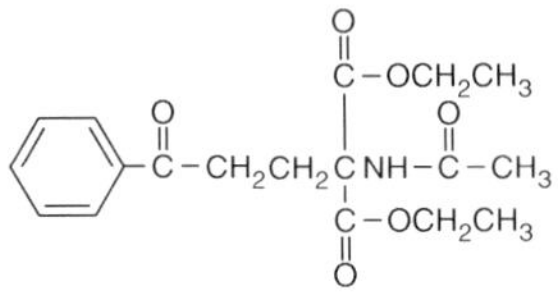
250 mg

S80,480-0
DIETHYL 2-ACETAMIDO-2-(BENZOYLMETHYL)MALONATE
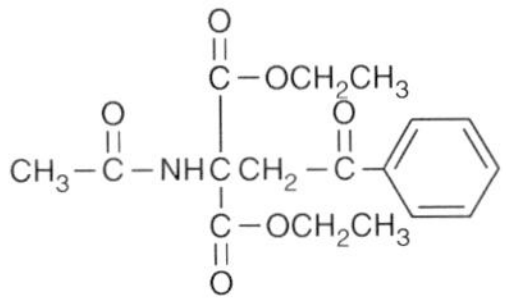
250 mg

S87,807-3
N-(3-ACETYLPHENYL)GLUTARAMIC ACID
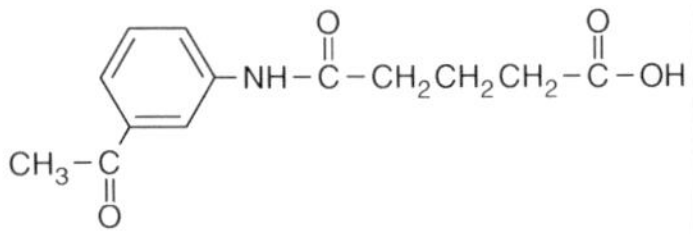
5 gm

S71,966-8
CIS-N-(4-ACETYLPHENYL)-HEXAHYDROPHTHALAMIC ACID
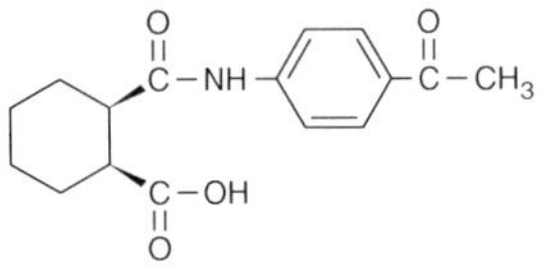
5 gm

S93,110-1
3'-ACETYL-1,2,3,6-TETRAHYDROPHTHALANILIC ACID
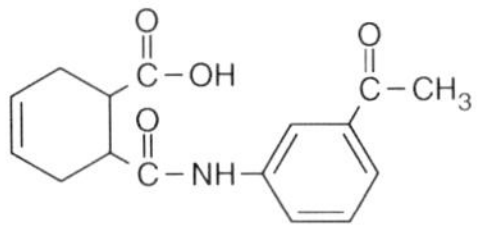
250 mg

S74,630-4
4'-ACETYLGLUTARANILIC ACID
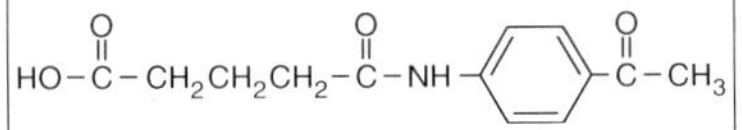
5 gm

S73,342-3
N-(1-INDANYL)-4-OXO-4-PHENYLBUTYRAMIDE
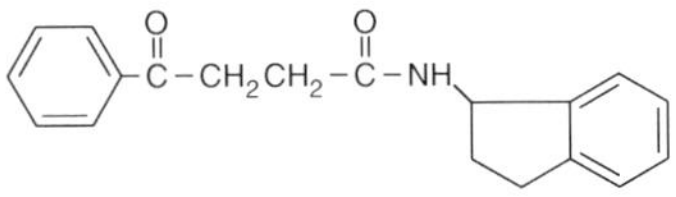
250 mg

S74,251-1
3'-ACETYLACETANILIDE
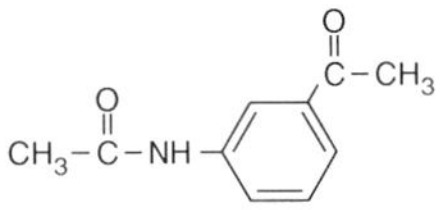
5 gm

S74,691-6
3'-PROPIONYLACETANILIDE
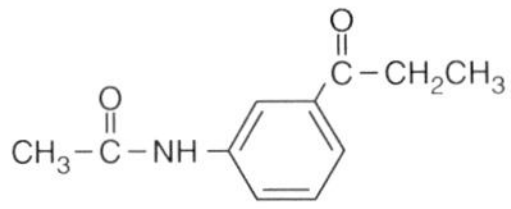
5 gm

S55,823-0
ALPHA-(2-FUROYLMETHYLENE)-P-ACETOTOLUIDIDE
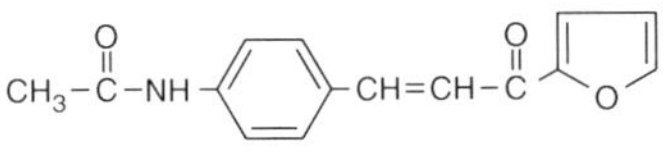
5 gm

S87,884-7
2,6-BIS(4-ACETAMIDOBENZYLIDENE)-1-CYCLOHEXANONE
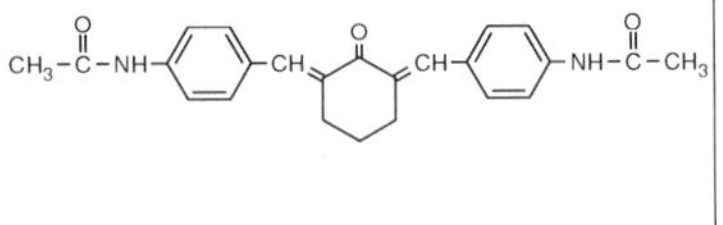
5 gm

S90,843-6
6-ACETAMIDO-1-TETRALONE
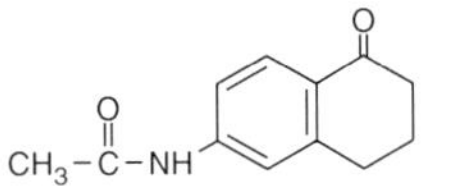
250 mg

S68,238-1
N-(2-BENZOYLPHENYL)-3,3-DIMETHYLBUTYRAMIDE
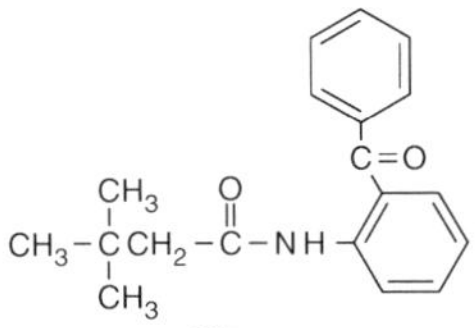
250 mg

S76,173-7
N-(2-ACETYLPHENYL)-5-METHYL-3-PHENYL-4-ISOXAZOLIN-4-CARBOXAMIDE
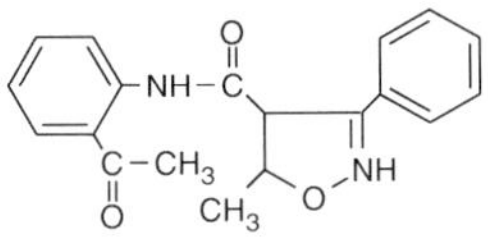
250 mg

S74,282-1
2'-PROPIONYLACETANILIDE
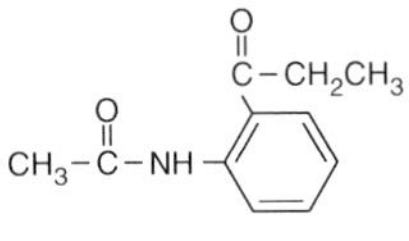
5 gm

S73,135-8
2'-ACETAMIDOBUTYROPHENONE
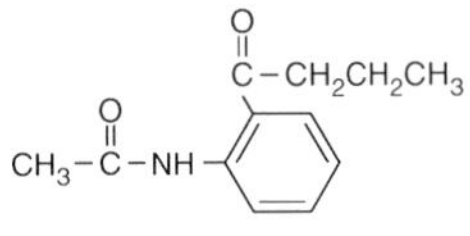
5 gm

S74,254-6
4'-ACETYLACETANILIDE
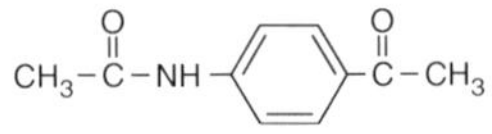
5 gm

S62,412-8
4'-PROPIONYLACETANILIDE
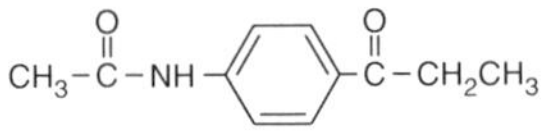
250 mg

S72,982-5
4'-ACETYLPROPIONANILIDE
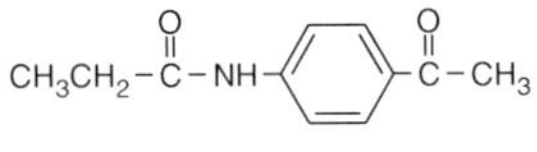
5 gm

S69,962-4
4'-ACETYLBUTYRANILIDE
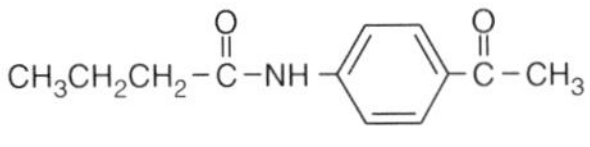
250 mg

S59,305-2
4-ACETAMIDO-4'-FLUOROCHALCONE
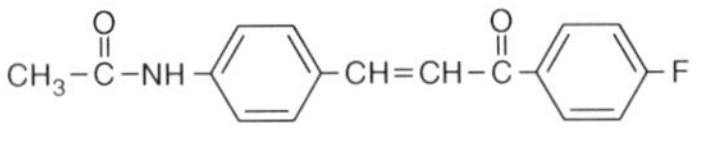
250 mg

S50,246-4
4'-ACETYLPALMITANILIDE
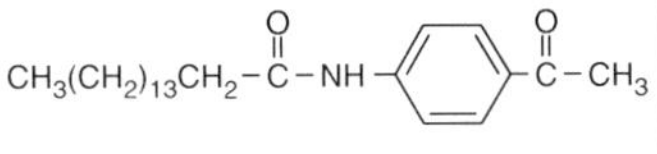
5 gm

S55,966-0
ALPHA-(4'-PHENYLPHENACYLIDENE)-P-ACETOTOLUIDIDE

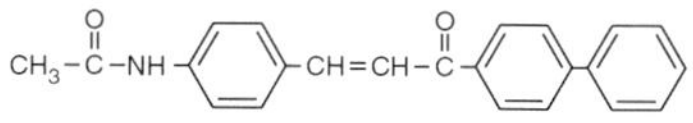

250 mg

S55,825-7
2-(4-ACETAMIDOBENZYLIDENE)-1-TETRALONE

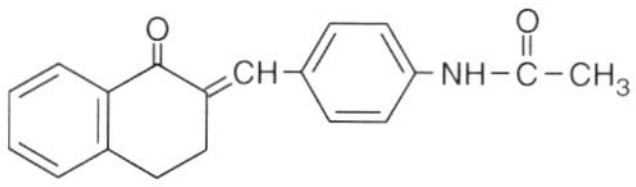

5 gm

S56,923-2
ALPHA-(1-NAPHTHOYLMETHYLENE)-P-ACETOTOLUIDIDE

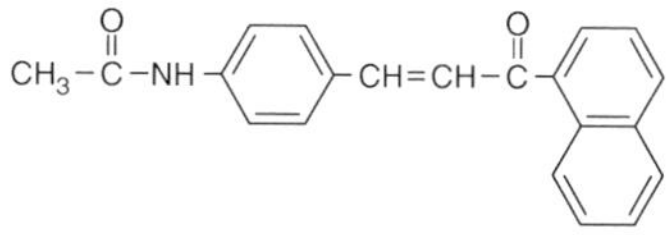

250 mg

S74,250-3
5'-ACETYL-2'-METHYLACETANILIDE

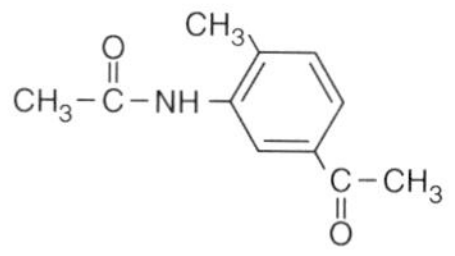

5 gm

S90,882-7
3,6-DIACETAMIDO-9-FLUORENONE

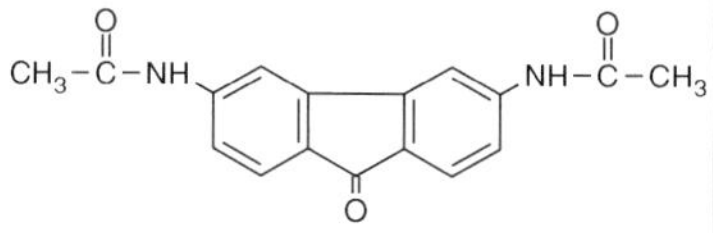

250 mg

S59,303-6
2,4-DIACETYLGLUTARANILIDE

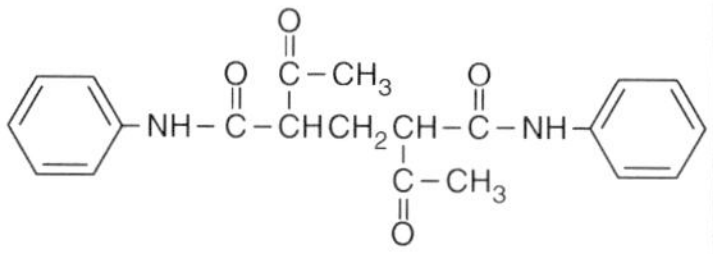

250 mg

S92,166-1
3'-ACETYLFORMANILIDE

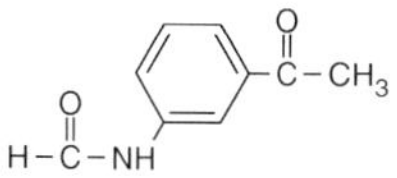

250 mg

S58,449-5
10-(4-ACETAMIDOBENZYLIDENE)-9-ANTHRONE

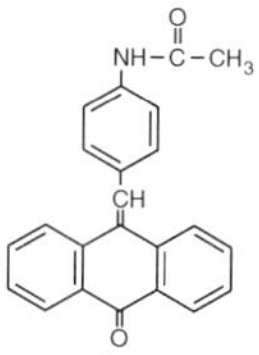

5 gm

S92,340-0
N-(1-ETHYL-1,2,3,4-TETRAHYDRO-2,2,4-TRIMETHYL-7-QUINOLYL)PROPIONAMIDE

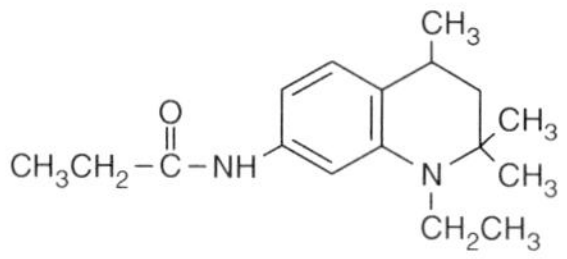

250 mg

S67,485-0
2,3-DIBROMO-3-[2-(METHOXYCARBONYL)-PHENYLCARBAMOYL]PROPIONIC ACID

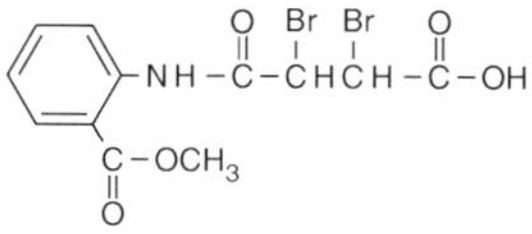

250 mg

S85,565-0
2,3-DIBROMO-4'-(ETHOXYCARBONYL)-SUCCINANILIC ACID

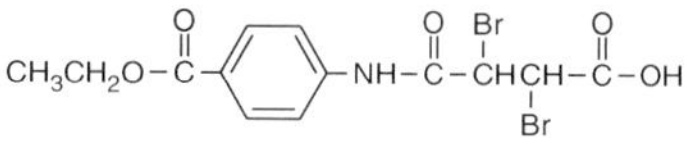

250 mg

S85,582-0
METHYL 2-(2,3-DIBROMO-3-PHENYLPROPIONAMIDO)BENZOATE

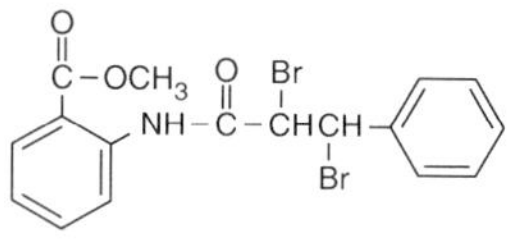

250 mg

S85,567-7
2,3-DIBROMO-N-(1-NAPHTHYL)-SUCCINAMIC ACID

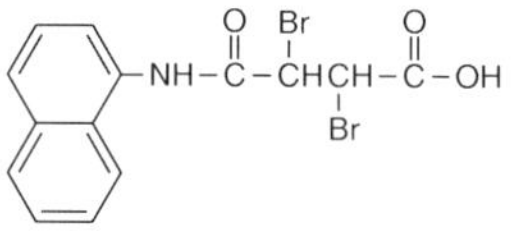

250 mg

S74,397-6
METHYL 2-[(2-CARBOXYCYCLOHEXYL)-CARBOXAMIDO]BENZOATE

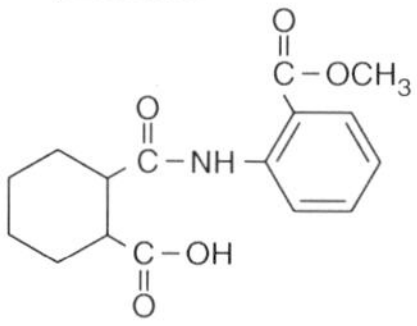

5 gm

S72,056-9
CIS-N-[4-(ETHOXYCARBONYL)PHENYL]-HEXAHYDROPHTHALAMIC ACID

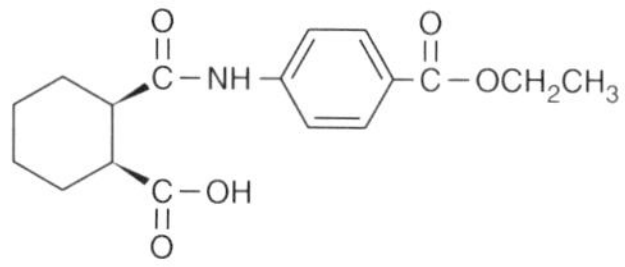

5 gm

S71,912-9
4'-(ETHOXYCARBONYL)-GLUTARANILIC ACID

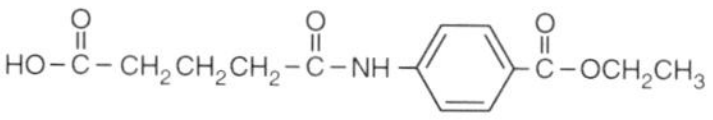

5 gm

S79,275-6
METHYL 2'-(METHOXYCARBONYL)-SUCCINANILATE

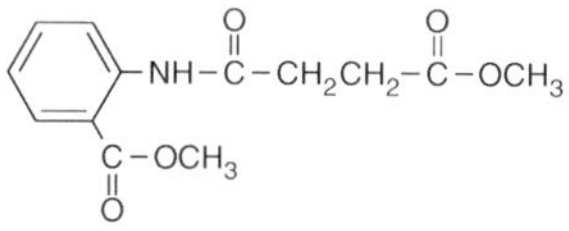

250 mg

S91,673-0
DIMETHYL 5-(PROPIONAMIDO)-ISOPHTHALATE

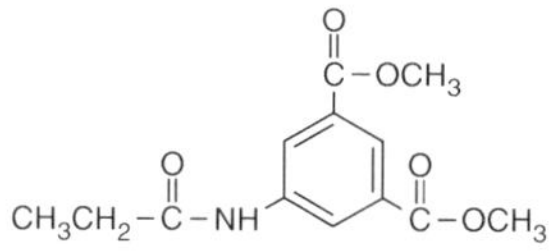

250 mg

S79,402-3
N,N'-BIS[4-(ETHOXYCARBONYL)PHENYL]-1,10-DECANEDIAMIDE

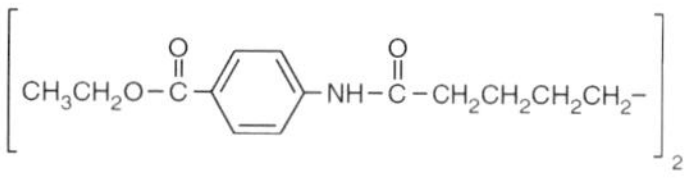

250 mg

S38,150-0
4-(3-CARBOXYPROPIONAMIDO)-BENZOIC ACID

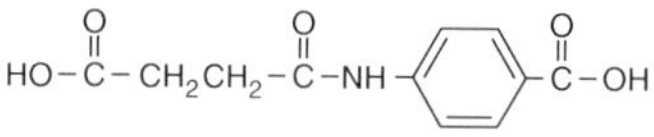

250 mg

S91,560-2
2-ACETAMIDOISOPHTHALIC ACID

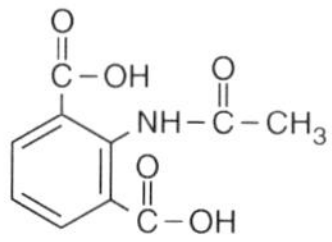

250 mg

S90,276-4
3-(2,2-DIMETHYLPROPIONAMIDO)-BENZOIC ACID

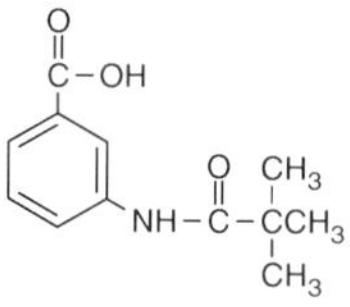

250 mg

S61,866-7
3-(NONYLAMIDO)BENZOIC ACID, BRANCHED ISOMERS

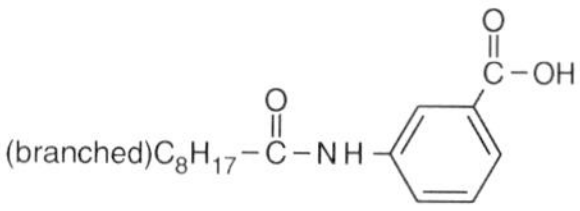

5 gm

S56,278-5
3-(HEXADECANAMIDO)BENZOIC ACID

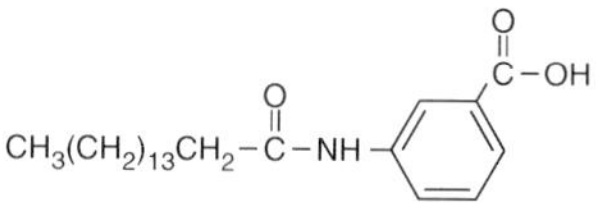

250 mg

S63,688-6
N-FORMYLANTHRANILIC ACID

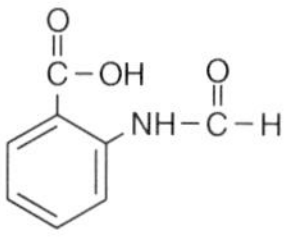

5 gm

S79,952-1
2-ACETAMIDOBENZOIC ACID

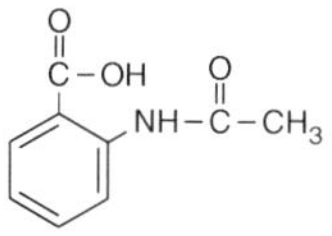

250 mg

S87,767-0
2-ACETAMIDOCINNAMIC ACID

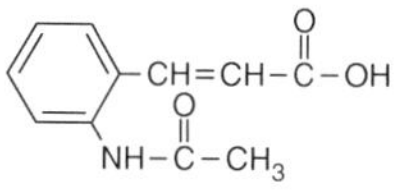

250 mg

S62,403-9
4-FORMAMIDOBENZOIC ACID

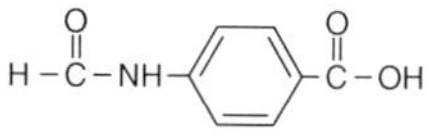

5 gm

S44,559-2
4-ACETAMIDOCINNAMIC ACID

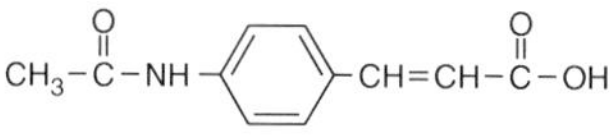

5 gm

S82,129-2
4-(TETRADECANAMIDO)BENZOIC ACID

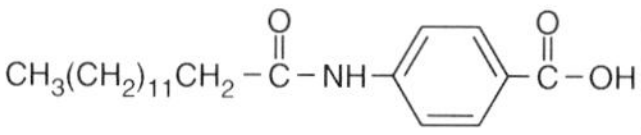

250 mg

S82,130-6
4-(OCTADECANAMIDO)BENZOIC ACID

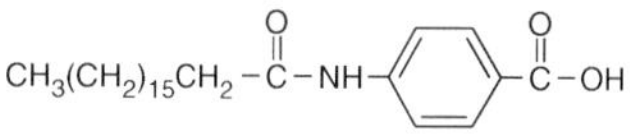

250 mg

S45,489-3
3-ACETAMIDO-P-TOLUIC ACID

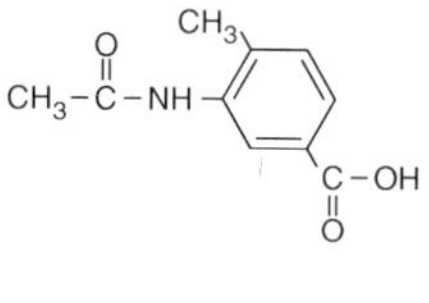

5 gm

S76,576-7
(N-BENZOYL)GLYCYLGLYCINE METHYL ESTER

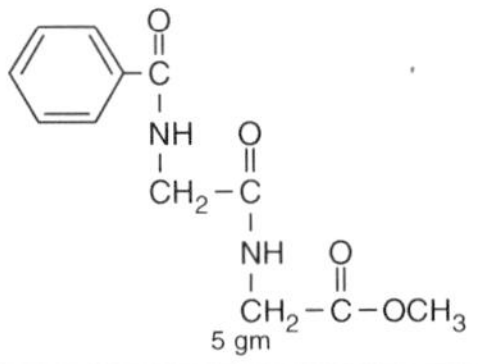

5 gm

S66,407-3
METHYL 3-(PIVALOYLAMINO)BENZOATE

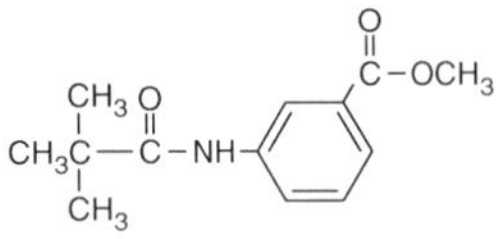

250 mg

S37,295-1
6-ACETAMIDOCOUMARIN

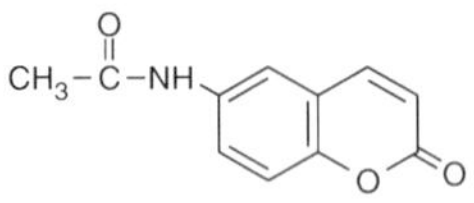

5 gm

S77,435-9
7-ACETAMIDO-3,4-BENZOCOUMARIN

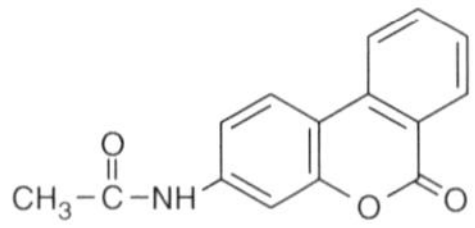

250 mg

S74,396-8
METHYL 2-(2-PHENYLACETAMIDO)-BENZOATE

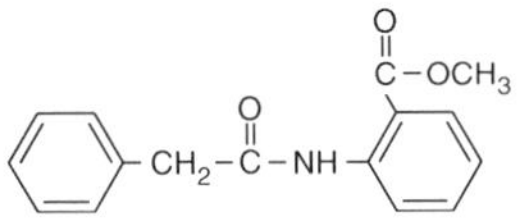

250 mg

S79,562-3
METHYL 2-BUTYRAMIDOBENZOATE

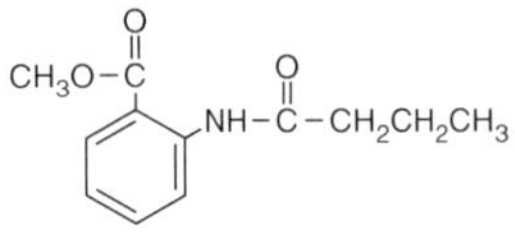

250 mg

S62,444-6
ETHYL 4-FORMAMIDOBENZOATE

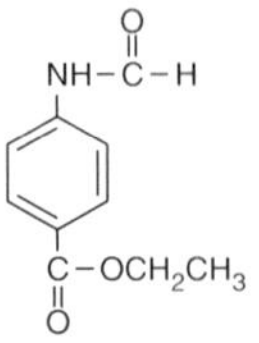

5 gm

S94,916-7
ETHYL 4-(ACETAMIDO)BENZOATE

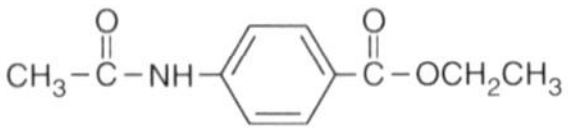

250 mg

S69,857-1
4'-ETHOXYCARBONYLPROPIONANILIDE

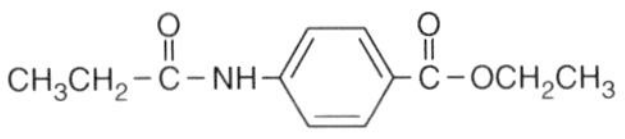

250 mg

S85,485-9
ETHYL 4-(2,2-DIMETHYLACETAMIDO)-BENZOATE

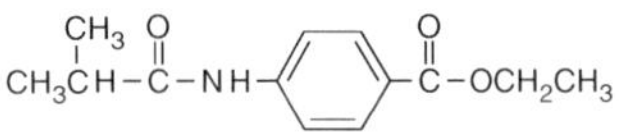

250 mg

S85,501-4
ETHYL 4-BUTYRAMIDOBENZOATE

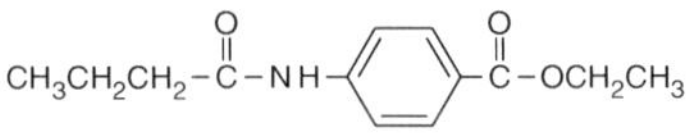

250 mg

S63,001-2
ETHYL 4-(DECANOYLAMINO)BENZOATE

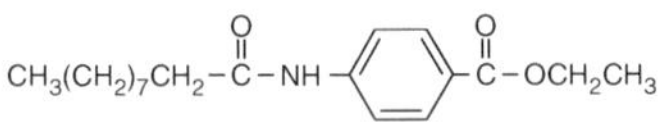

250 mg

S54,767-0
4'-[2-(4-CHLOROBENZOYL)VINYL]-ACETANILIDE

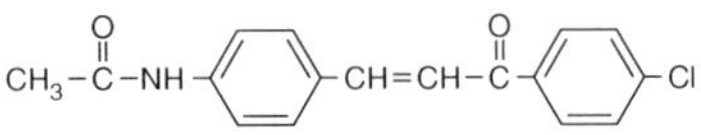

5 gm

S51,109-9
4'-[2-(P-TOLUOYL)VINYL]ACETANILIDE

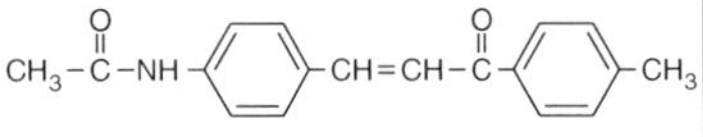

5 gm

S53,774-8
4',4'''-(3-OXO-1,4-PENTADIEN-1,5-DIYL)BISACETANILIDE

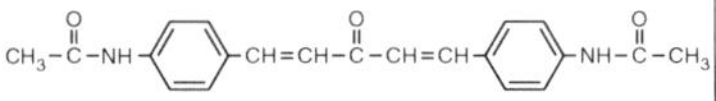

5 gm

S75,934-1
ETHYL 2-(CHLOROCARBONYL)OXANILATE

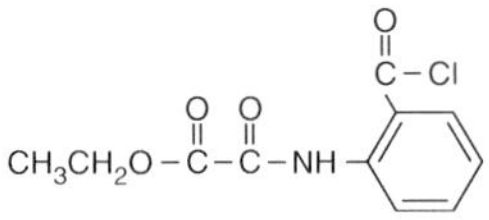

250 mg

S44,837-0
ETHYL 4'-METHYLOXANILATE

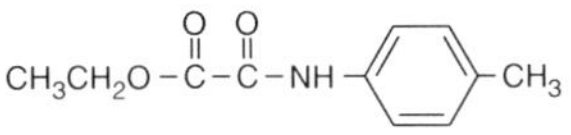

5 gm

S92,607-8
OXANILIDE

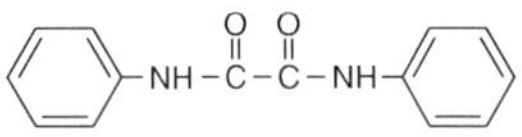

250 mg

S95,329-6
PYRUVANILIDE

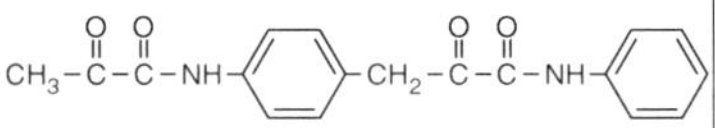

50 mg

S70,683-3
N,N'-DI(O-TOLYL)OXAMIDE

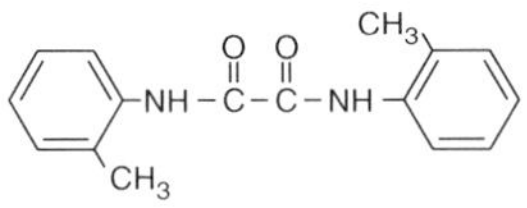

5 gm

S80,411-8
N-ALPHA-(2-PHENYLACETYL)-L-ASPARAGINE

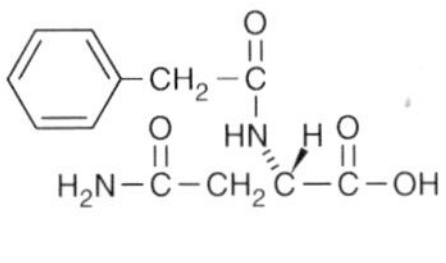

250 mg

S70,299-4
N-(M-TOLYL)SUCCINAMIC ACID

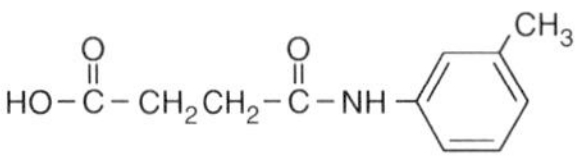

5 gm

S70,383-4
3'-METHYLGLUTARANILIC ACID

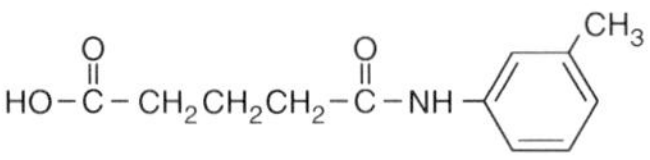

5 gm

S71,936-6
3-(N-PHENYLCARBAMOYL)-5-NORBORNENE-2-CARBOXYLIC ACID

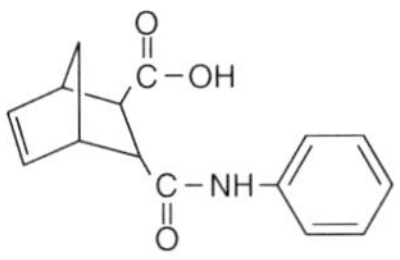

5 gm

S71,937-4
1,4,5,6,7,7-HEXACHLORO-3-(N-PHENYLCARBAMOYL)-5-NORBORNENE-2-CARBOXYLIC ACID

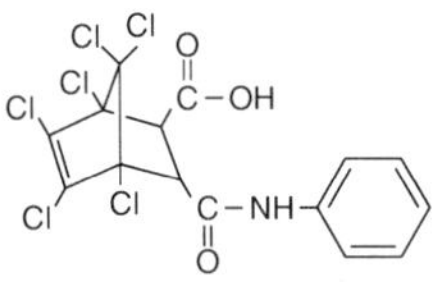

5 gm

S71,986-2
3-(4-CHLOROPHENYLCARBAMOYL)-1,4,5,6,7,7-HEXACHLORO-5-NORBORNENE-2-CARBOXYLIC ACID

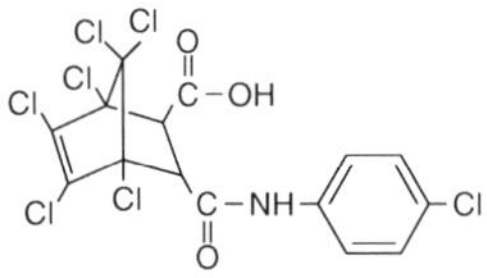

5 gm

S72,013-5
3-[N-(4-METHYLPHENYL)CARBAMOYL]-5-NORBORNENE-2-CARBOXYLIC ACID

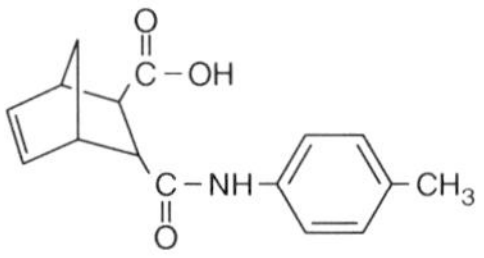

5 gm

S72,017-8
3-[N-(3,4-DIMETHYLPHENYL)CARBAMOYL]-5-NORBORNENE-2-CARBOXYLIC ACID

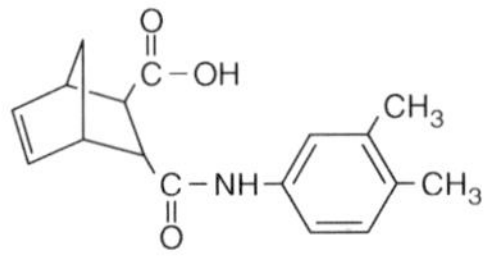

5 gm

S73,058-0
3-[N-(2,5-XYLYL)CARBAMOYL]-5-NORBORNENE-2-CARBOXYLIC ACID

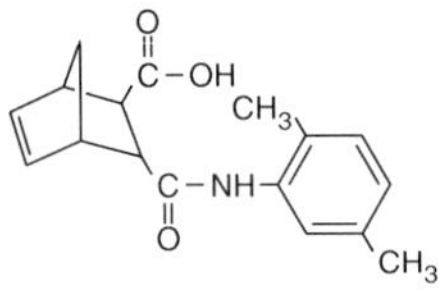

5 gm

S71,985-4
1,4,5,6,7,7-HEXACHLORO-3-[N-(2,4-XYLYL)CARBAMOYL]-5-NORBORNENE-2-CARBOXYLIC ACID

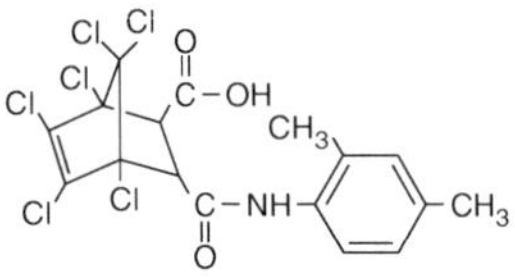

5 gm

S73,062-9
3-[N-(2,4,6-TRIMETHYLPHENYL)-CARBAMOYL]-5-NORBORNENE-2-CARBOXYLIC ACID

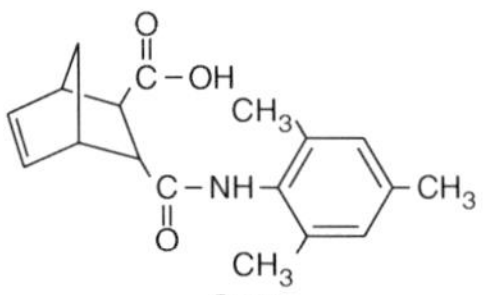

5 gm

S71,984-6
1,4,5,6,7,7-HEXACHLORO-3-[N-(M-TOLYL)CARBAMOYL]-5-NORBORNENE-2-CARBOXYLIC ACID

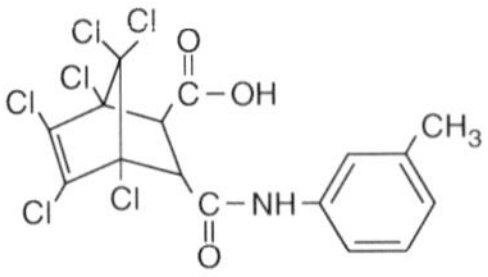

5 gm

S70,402-4
CIS-HEXAHYDRO-N-(3-METHYLPHENYL)-PHTHALAMIC ACID
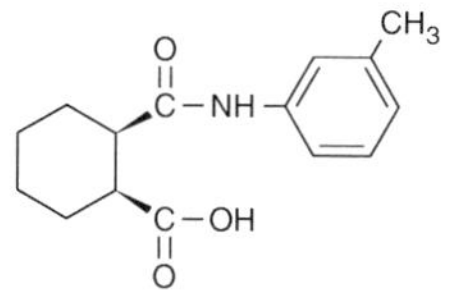
5 gm

S70,289-7
CIS-HEXAHYDRO-N-PHENYLPHTHALAMIC ACID
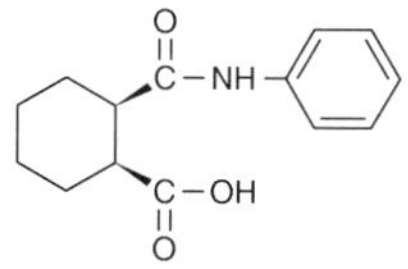
5 gm

S70,370-2
CIS-HEXAHYDRO-N-(2-METHYLPHENYL)-PHTHALAMIC ACID
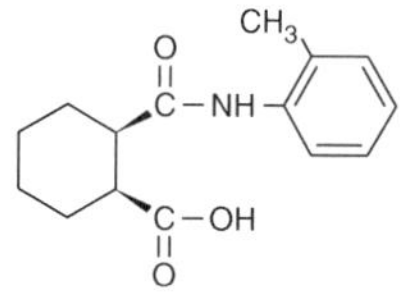
5 gm

S70,291-9
CIS-HEXAHYDRO-N-(4-METHYLPHENYL)-PHTHALAMIC ACID
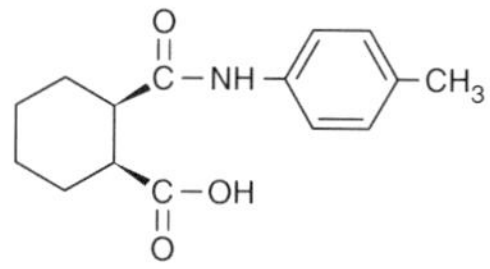
5 gm

S71,330-9
CIS-4'-ETHYLHEXAHYDROPHTHALANILIC ACID
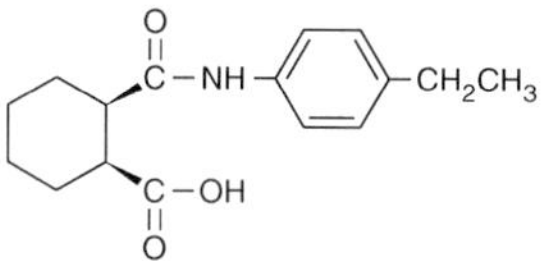
5 gm

S71,900-5
CIS-2',6'-DIMETHYLHEXAHYDROPHTHALANILIC ACID
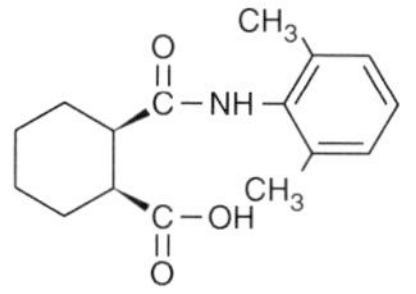
5 gm

S70,366-4
CIS-2',4'-DIMETHYLHEXAHYDROPHTHALANILIC ACID
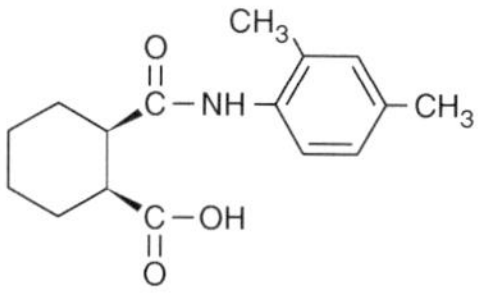
5 gm

S71,734-7
CIS-3',4'-DIMETHYLHEXAHYDROPHTHALANILIC ACID
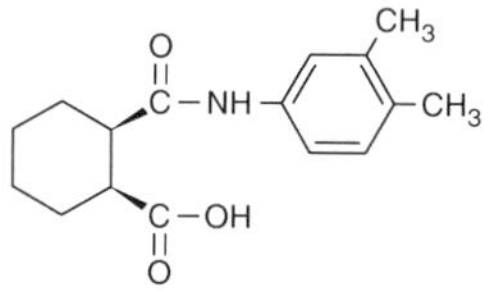
5 gm

S72,009-7
CIS-3',5'-DIMETHYLHEXAHYDROPHTHALANILIC ACID
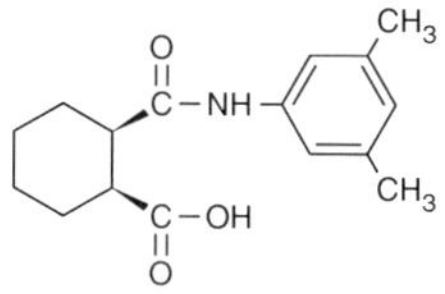
5 gm

S90,028-1
HEXAHYDRO-N-(2,3-XYLYL)-PHTHALAMIC ACID
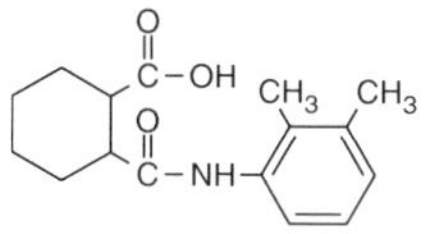
250 mg

S74,632-0
4'-BUTYL-1,2,3,6-TETRAHYDROPHTHALANILIC ACID
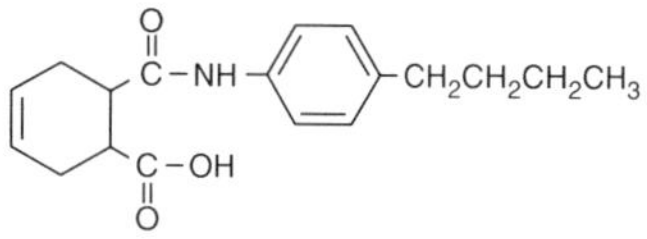
5 gm

S74,653-3
4'-BROMO-2'-METHYL-1,2,3,6-TETRAHYDROPHTHALANILIC ACID
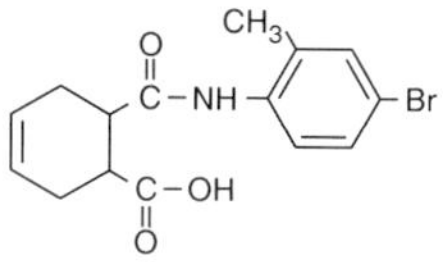
5 gm

S71,932-3
3',4'-DIMETHYL-1,2,3,6-TETRAHYDROPHTHALANILIC ACID
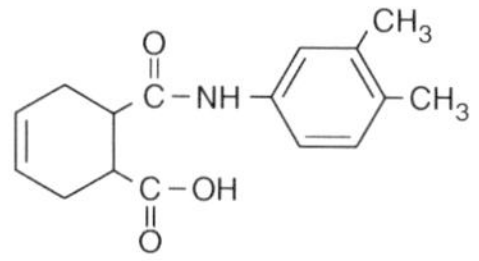
5 gm

S70,098-3
SUCCINANILIC ACID
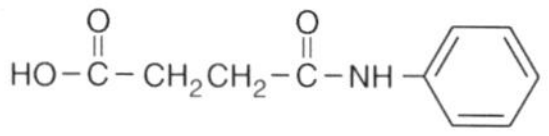
5 gm

S70,382-6
GLUTARANILIC ACID
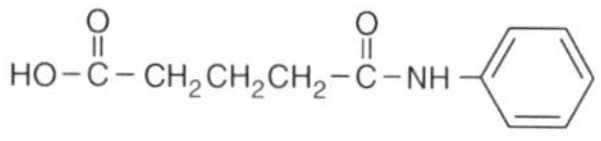
5 gm

S78,007-3
N-(DIPHENYLMETHYL)GLUTARAMIC ACID
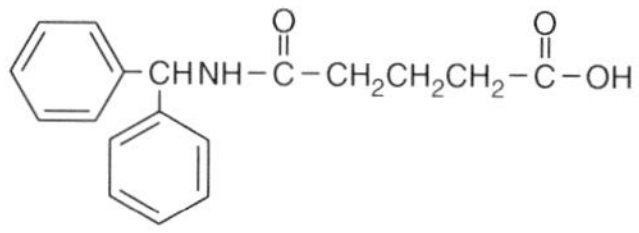
250 mg

S70,360-5
2'-METHYLGLUTARANILIC ACID
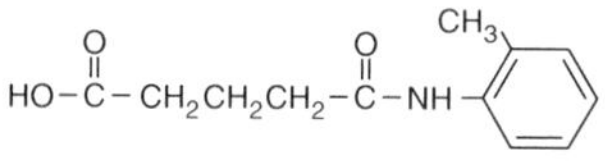
5 gm

S73,040-8
N-(2-ETHYLPHENYL)GLUTARAMIC ACID
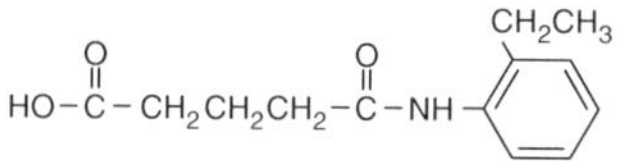
5 gm

S70,280-3
4'-ETHYLSUCCINANILIC ACID
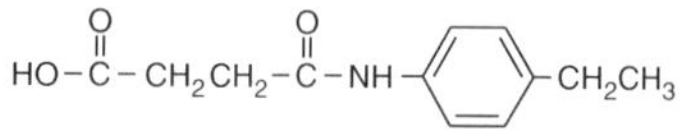
5 gm

S70,361-3
4'-METHYLGLUTARANILIC ACID
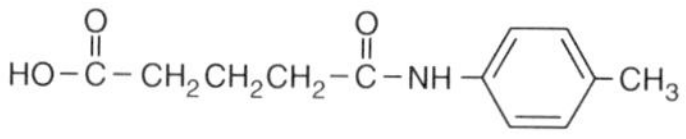
5 gm

S70,368-0
4'-ETHYLGLUTARANILIC ACID
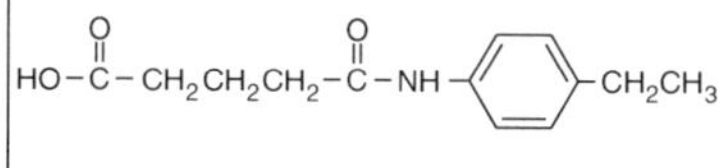
5 gm

S74,627-4
4'-BUTYLGLUTARANILIC ACID
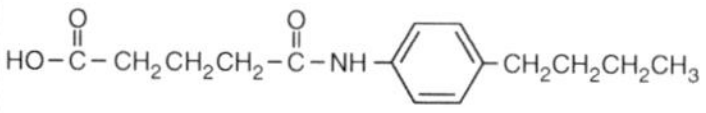
5 gm

S72,067-4
N-(2,3-XYLYL)GLUTARAMIC ACID
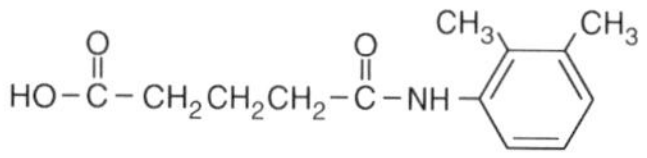
5 gm

S73,037-8
N-(2,6-XYLYL)GLUTARAMIC ACID
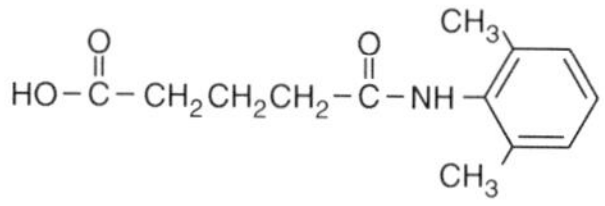
5 gm

S91,397-9
2',6'-DIISOPROPYLGLUTARANILIC ACID
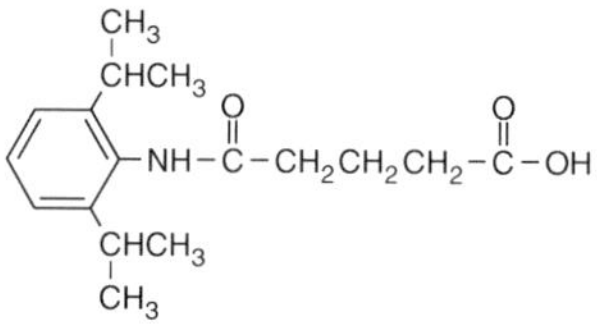
250 mg

S70,114-9
2',4'-DIMETHYLSUCCINANILIC ACID
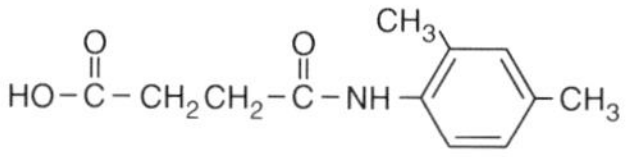
5 gm

S70,308-7
2',5'-DIMETHYLSUCCINANILIC ACID
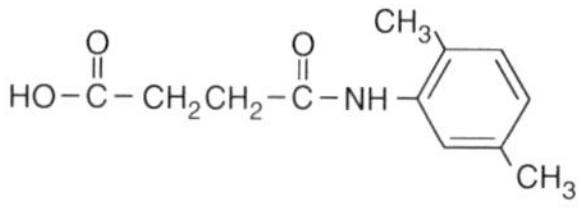
5 gm

S70,318-4
3',4'-DIMETHYLSUCCINANILIC ACID
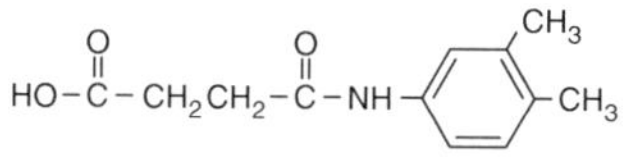
5 gm

S70,352-4
2',4'-DIMETHYLGLUTARANILIC ACID
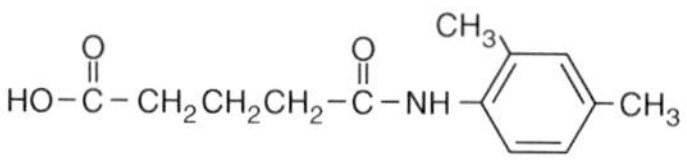
5 gm

S70,384-2
3',4'-DIMETHYLGLUTARANILIC ACID
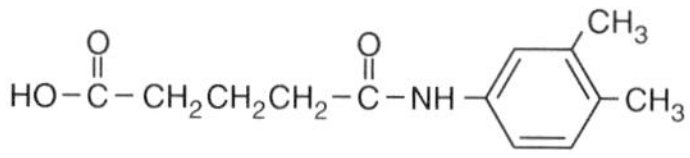
5 gm

S73,035-1
N-(2,5-XYLYL)GLUTARAMIC ACID
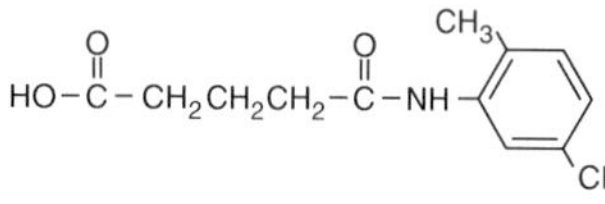
5 gm

S73,038-6
N-(3,5-XYLYL)GLUTARAMIC ACID
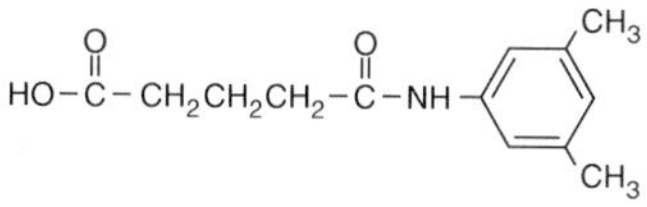
5 gm

S70,284-6
2',4',6'-TRIMETHYLSUCCINANILIC ACID

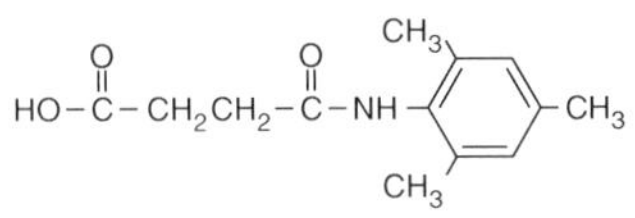

5 gm

S70,480-6
N-(2,4,6-TRIMETHYLPHENYL)-GLUTARAMIC ACID

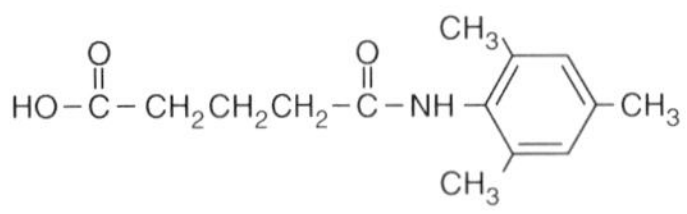

5 gm

S76,442-6
3-[N-(M-TOLYL)CARBAMOYL]-5-NORBORNENE-2-CARBOXYLIC ACID

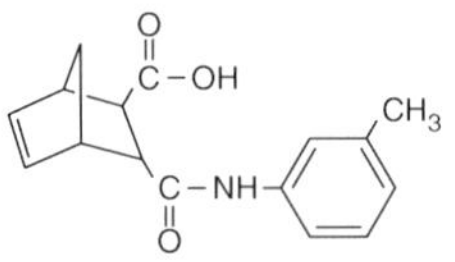

5 gm

S84,612-0
3-BENZYLADIPAMIDE

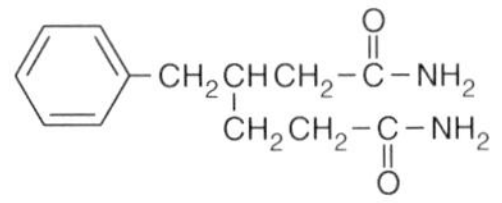

250 mg

S88,326-3
N,N'-DIPHENYLPIMELAMIDE

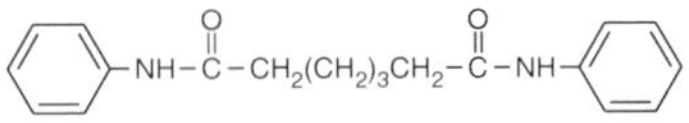

250 mg

S88,762-5
N,N'-DIPHENYL-N,N',3,3-TETRAMETHYLGLUTARAMIDE

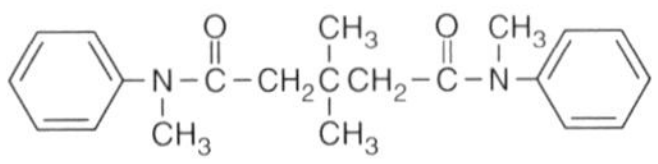

250 mg

S34,271-8
(2-BROMO-3-PHENYLPROPIONYL)-GLYCYLGLYCINE

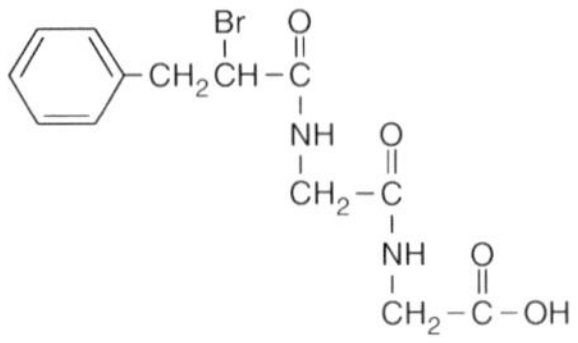

5 gm

S56,758-2
N-FORMYL-2-PHENYLALANINE

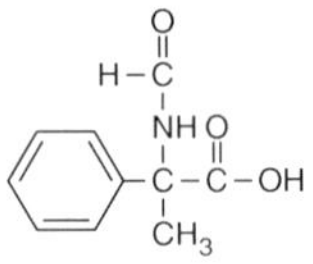

5 gm

S34,269-6
DL-N-(2-BROMO-3-PHENYLPROPIONYL)-GLYCINE

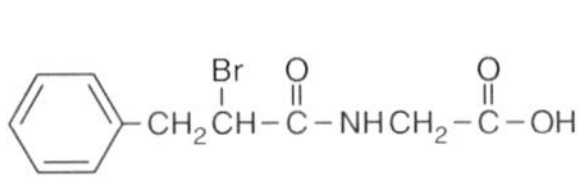

5 gm

S68,169-5
3-ACETAMIDOPHENYLACETIC ACID

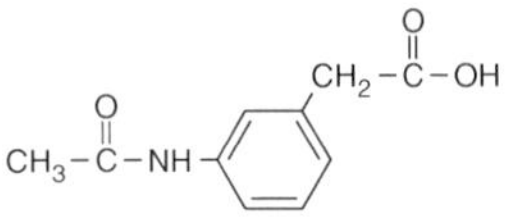

5 gm

S34,321-8
DEXTRO-CAMPHORIC M-TOLUIDIDE

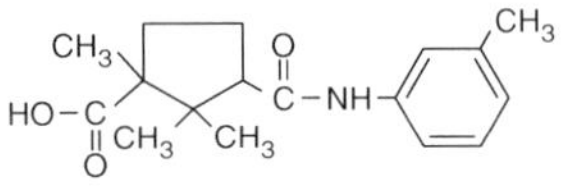

5 gm

S34,319-6
DEXTRO-CAMPHORIC O-TOLUIDIDE

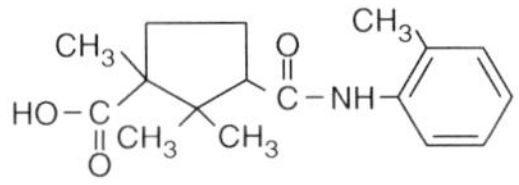

5 gm

S34,322-6
DEXTRO-CAMPHORIC P-TOLUIDIDE

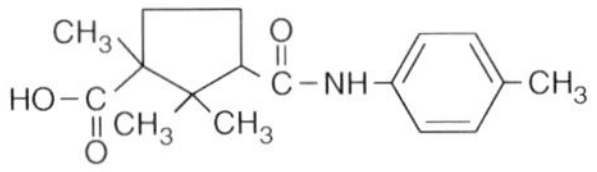

5 gm

S73,030-0
CIS-2'-ETHYLHEXAHYDROPHTHALANILIC ACID

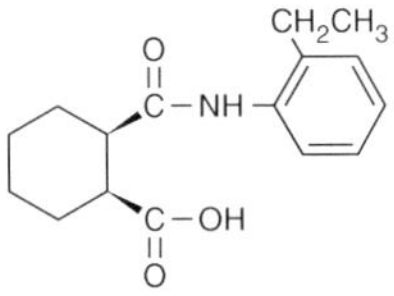

5 gm

S92,322-2
N-(3,5-DIMETHYLPHENYL)-1,2,3,6-TETRAHYDROPHTHALAMIC ACID

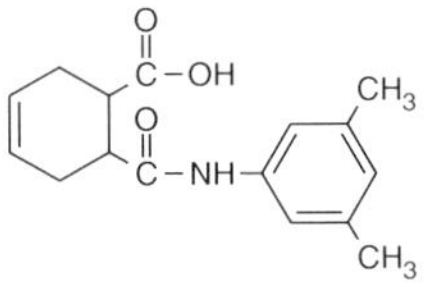

250 mg

S66,157-0
N-(O-TOLYL)SUCCINAMIC ACID

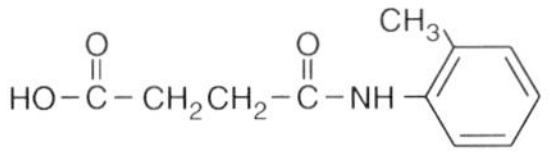

250 mg

S50,223-5
4-PALMITAMIDOBENZOIC ACID

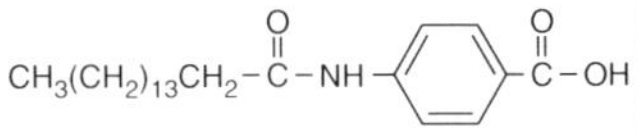

250 mg

S38,143-8
3-[N-(6-CHLORO-ORTHO-TOLYL]-CARBAMOYL)PROPIONIC ACID, SODIUM SALT

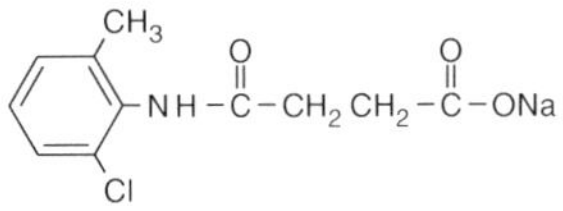

5 gm

S38,170-5
3-[N-(2,6-XYLYL)CARBAMOYL]-PROPIONIC ACID

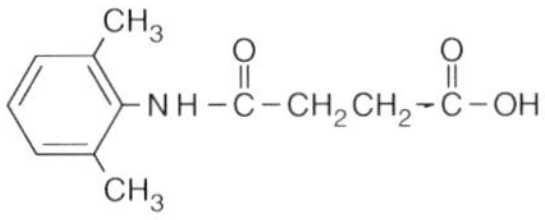

250 mg

S76,588-0
N-ACETYLGLYCYL-L-PHENYLALANINE

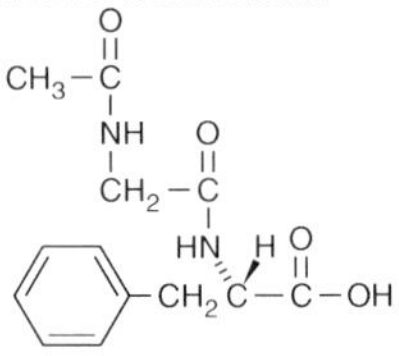

5 gm

S92,581-0
S-(ETHOXYCARBONYLMETHYL)-1,1-DITHIO-4'-METHYLOXANILATE

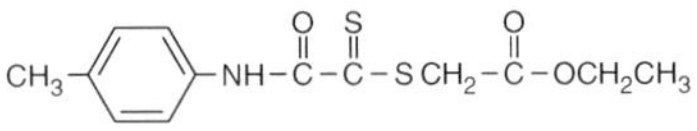

250 mg

S50,211-1
METHYL 3-(PHENYLCARBAMOYL)-PROPIONATE

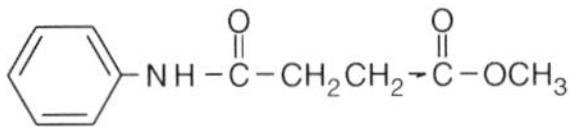

5 gm

S94,284-7
METHYL N-ACETYLANTHRANILATE

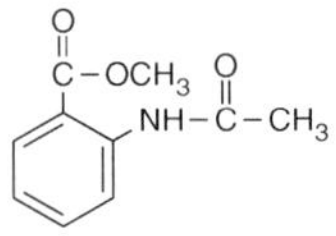

5 gm

S52,860-9
2-PROPIONAMIDOPHENYL ACETATE

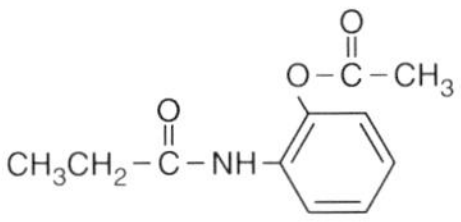

5 gm

S55,984-9
S-(4-ACETAMIDOPHENYL) THIOACETATE

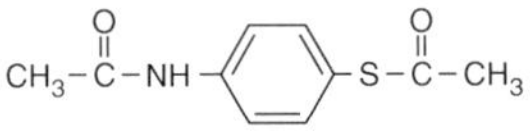

250 mg

S92,602-7
4-(4-ACETAMIDOPHENYL)-1,3-DITHIOLE-2-THIONE

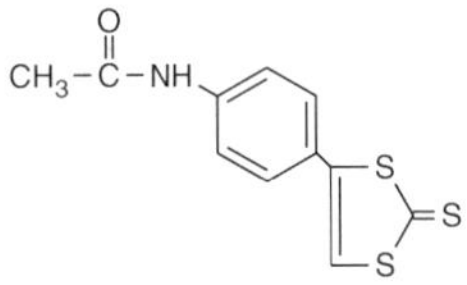

250 mg

S80,931-4
6-BROMO-1,3-BENZODIOXOLE-5-PROPIONAMIDE

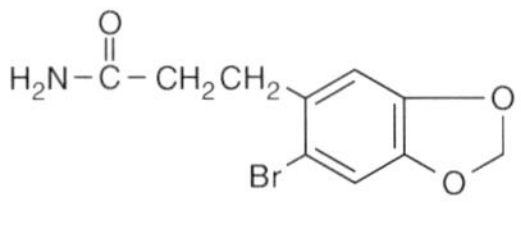

250 mg

S86,414-5
10-CHLORO-9,10-DIHYDRO-9,10-ETHANOANTHRACENE-11-CARBOXAMIDE

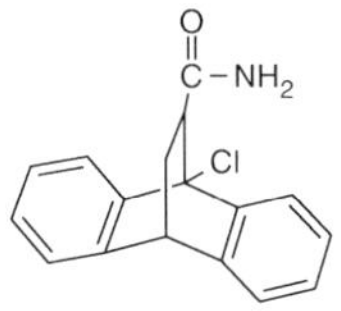

250 mg

S63,148-5
3-PHENYLPROPIONAMIDE

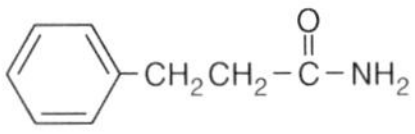

250 mg

S59,531-4
3-PHENYLGLUTARAMIDE

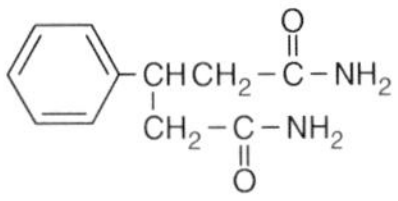

5 gm

S84,226-5
2,3,4-TRIPHENYLBUTYRAMIDE

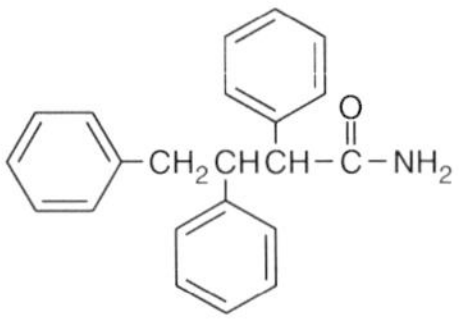

250 mg

S53,676-8
2-(PARA-TOLYL)ACETAMIDE

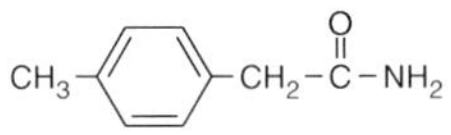

5 gm

S56,122-3
4-(4-BROMOPHENYL)BUTYRAMIDE

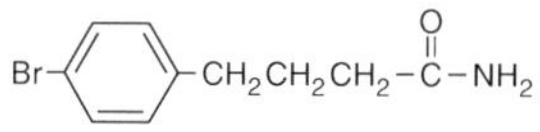

5 gm

S38,097-0
1,2,3,4-TETRAHYDRO-1-NAPHTHALENECARBOXAMIDE

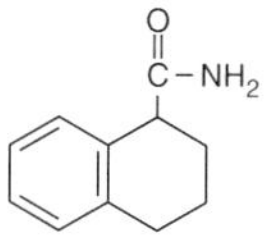

250 mg

S50,415-7
3'-CHLOROFORMANILIDE

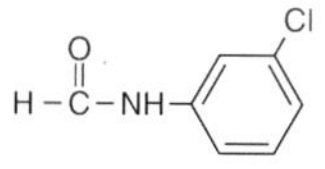

5 gm

S35,005-2
M-FORMOTOLUIDIDE

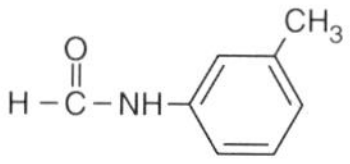

5 gm

S89,131-2
3'-METHYL-1-CYCLOPROPANECARBOXANILIDE

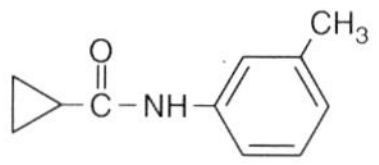

250 mg

S71,590-5
N-BENZHYDRYLCYCLOPROPANECARBOXAMIDE

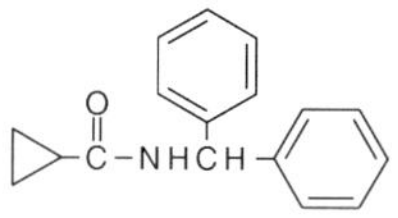

5 gm

S89,137-1
2'-ETHYL-1-CYCLOPROPANECARBOXANILIDE

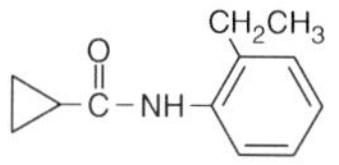

250 mg

S84,410-1
4'-METHYLCYCLOPROPANECARBOXANILIDE

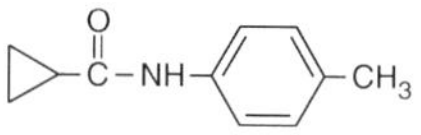

250 mg

S89,139-8
4'-ETHYL-1-CYCLOPROPANECARBOXANILIDE

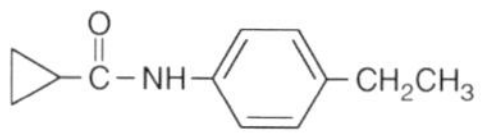

250 mg

S84,417-9
2',6'-DIMETHYLCYCLOPROPANECARBOXANILIDE

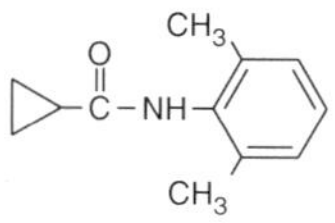

250 mg

S89,135-5
2',6'-DIETHYL-1-CYCLOPROPANECARBOXANILIDE

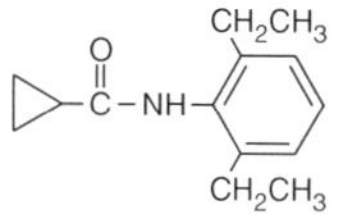

250 mg

S89,140-1
2',4'-DIMETHYL-1-CYCLOPROPANECARBOXANILIDE

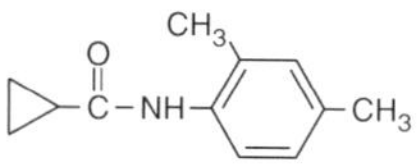

250 mg

S89,136-3
5'-ISOPROPYL-2'-METHYL-1-CYCLOPROPANECARBOXANILIDE

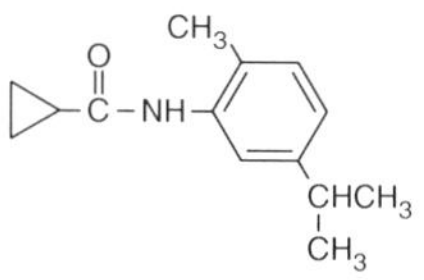

250 mg

S76,114-1
N-METHYL-1-PHENYLCYCLOPENTANECARBOXAMIDE

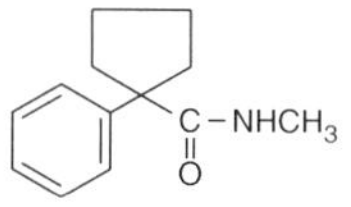

250 mg

S84,323-7
N,3-DIPHENYLCYCLOPENTANECARBOXAMIDE

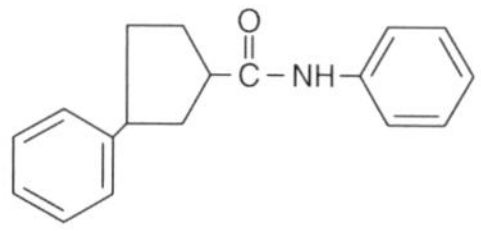

250 mg

S41,668-1
N-(1-PHENYLCYCLOHEXYL)FORMAMIDE

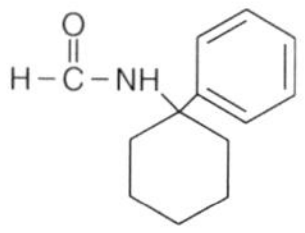

250 mg

S90,498-8
2',3'-DICHLORO-1-CYCLOHEXANECARBOXANILIDE

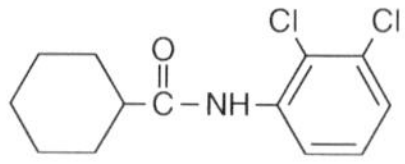

250 mg

S85,215-5
9,10-DIHYDRO-9,10-ETHANOANTHRACENE-11-CARBOXANILIDE

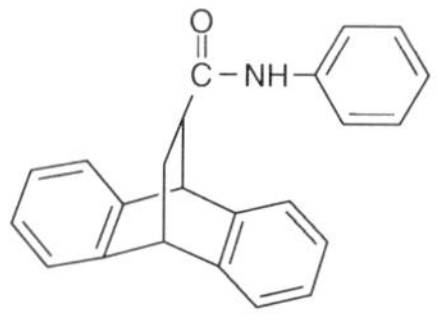

250 mg

S81,399-0
PROPIONANILIDE

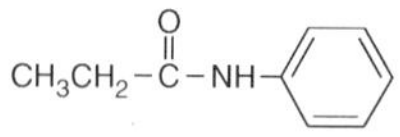

250 mg

S66,920-2
4-PHENYLBUTYRAMIDE

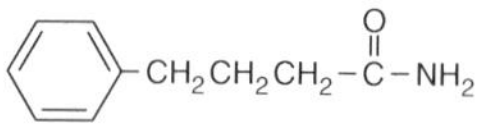

5 gm

S79,987-4
N-PHENYLISOBUTYRAMIDE

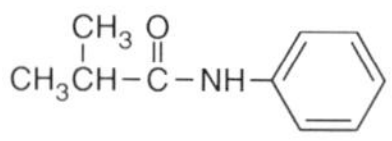

250 mg

S87,645-3
(-)-N,2-DIPHENYL-3-METHYLBUTYRAMIDE

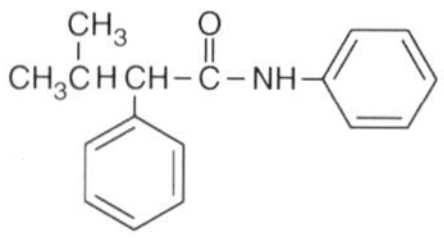

250 mg

S82,134-9
OCTADECANANILIDE

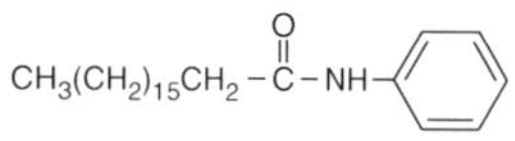

250 mg

S77,984-9
2,2-DIPHENYL-2'-ETHYLACETANILIDE

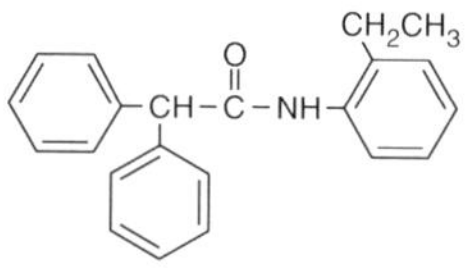

250 mg

S74,729-7
2-PHENYLACETO-2',3'-XYLIDIDE

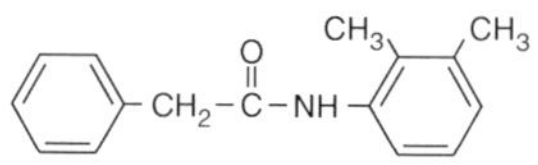

5 gm

S90,050-8
2',3'-DIMETHYL-2-PHENYLACETANILIDE

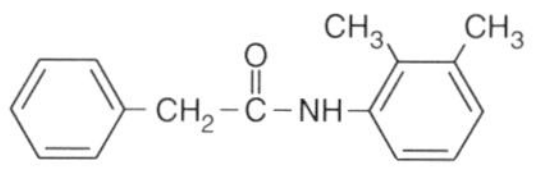

250 mg

S75,053-0
2',6'-DIETHYL-2-PHENYLACETANILIDE

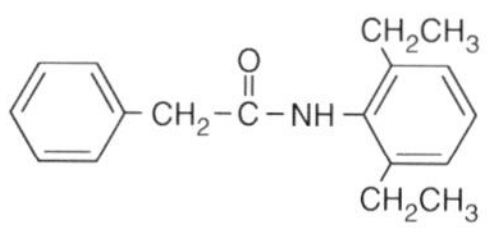

5 gm

S76,806-5
2,2-DIPHENYL-2',6'-ACETOXYLIDIDE

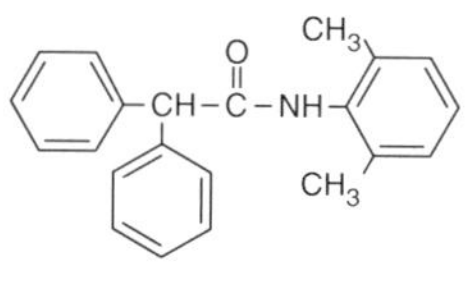

250 mg

S74,380-1
2-PHENYLACETO-2',5'-XYLIDIDE

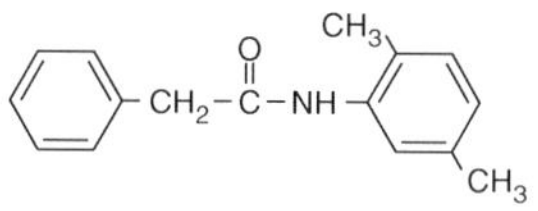

5 gm

S90,049-4
2',5'-DIMETHYL-2-PHENYLACETANILIDE

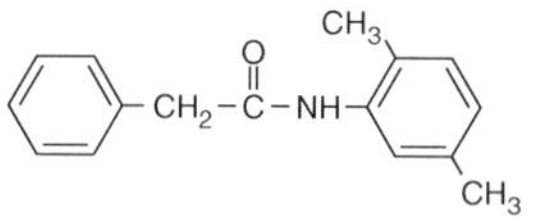

250 mg

S77,985-7
2,2-DIPHENYL-2',5'-ACETOXYLIDIDE

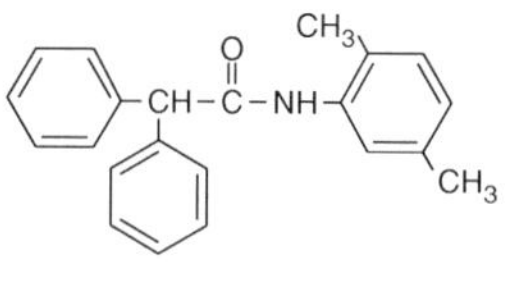

250 mg

S78,053-7
2,2-DIPHENYL-2',4'-ACETOXYLIDIDE

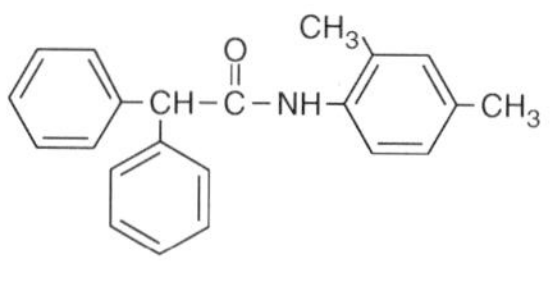

250 mg

S75,054-9
2-PHENYL-2',4',6'-TRIMETHYLACETANILIDE

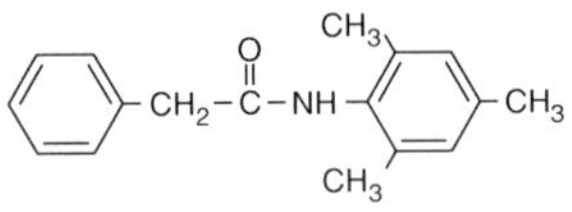

5 gm

S90,053-2
2-PHENYL-2',4',5'-TRIMETHYLACETANILIDE

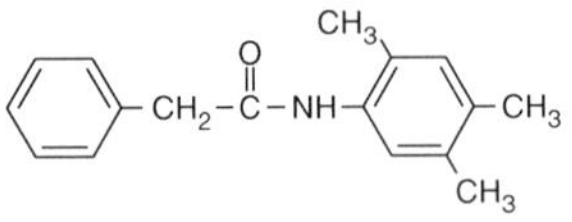

250 mg

S69,562-9
N-DIPHENYLMETHYLPROPIONAMIDE

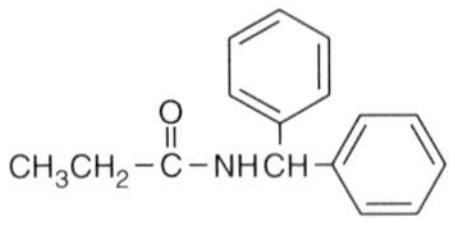

250 mg

S91,393-6
2,2-DIMETHYL-N-(DIPHENYLMETHYL)-PROPIONAMIDE

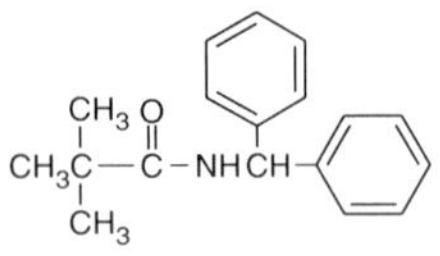

250 mg

S35,006-0
N-TRITYLFORMAMIDE

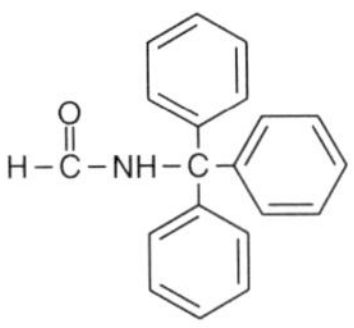

5 gm

S74,382-8
N-(DIPHENYLMETHYL)DECANAMIDE

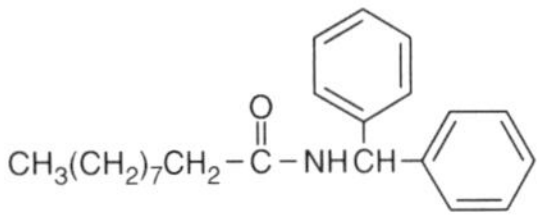

5 gm

S81,049-5
N,3,4,4-TETRAPHENYLBUTYRAMIDE

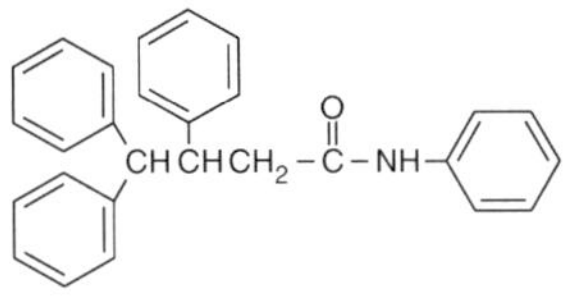

250 mg

S34,999-2
O-FORMOTOLUIDIDE

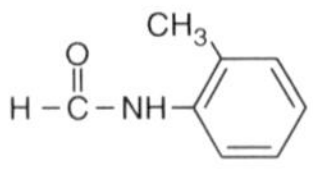

5 gm

S55,438-3
O-ACETOTOLUIDIDE

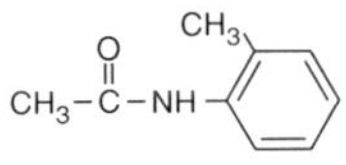

5 gm

S62,335-0
2'-ETHYLFORMANILIDE

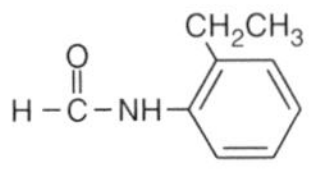

5 gm

S37,774-0
2-BROMO-2'-CHLOROPROPIONANILIDE

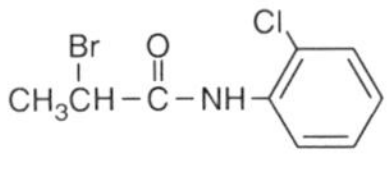

250 mg

S76,491-4
2'-ETHYLACETANILIDE

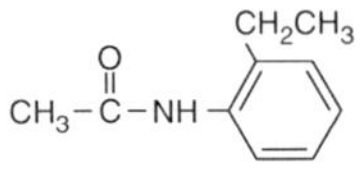

5 gm

S75,217-7
2'-PROPYLACETANILIDE

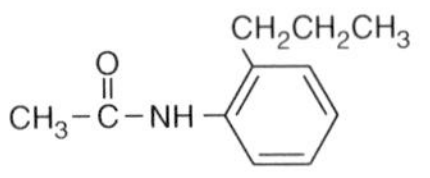

5 gm

S76,934-7
2'-ETHYLOCTANANILIDE

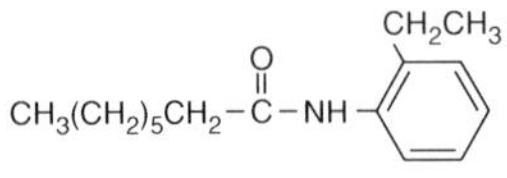

250 mg

S75,379-3
2'-ETHYLDECANANILIDE

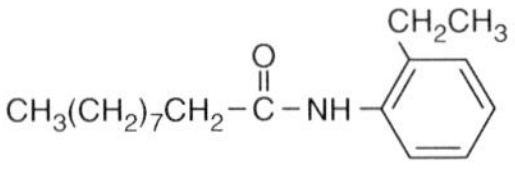

5 gm

S63,595-2
N-(2-CHLOROBENZYLIDENE)-TETRADECANAMIDE

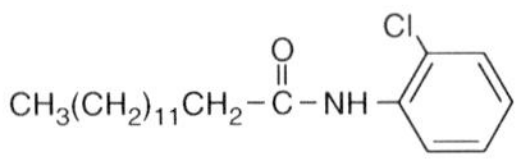

5 gm

S75,391-2
2'-ETHYLHEXADECANANILIDE

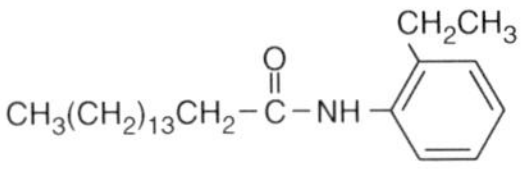

5 gm

S57,653-0
4'-FLUOROFORMANILIDE

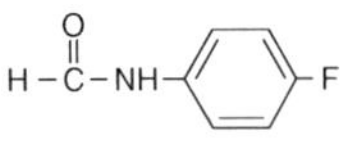

5 gm

S50,322-3
4-CHLOROFORMANILIDE

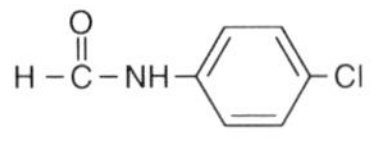

5 gm

S62,450-0
4'-BROMOFORMANILIDE

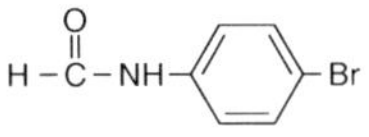

5 gm

S50,326-6
4'-IODOFORMANILIDE

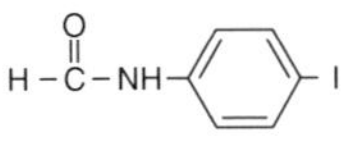

5 gm

S58,017-1
P-FORMOTOLUIDIDE

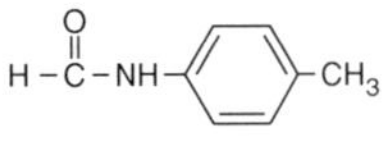

5 gm

S62,311-3
4'-ETHYLFORMANILIDE

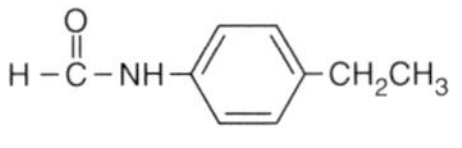

5 gm

S39,751-2
4'-FLUOROPROPIONANILIDE

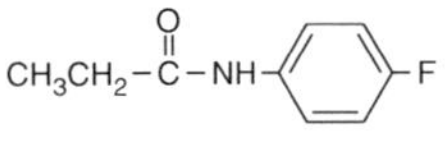

250 mg

S75,200-2
4'-ETHYLACETANILIDE

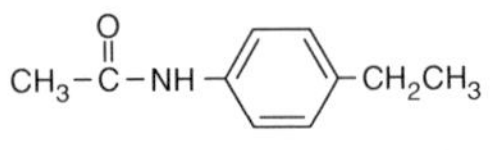

5 gm

S82,525-5
P-PROPIONOTOLUIDIDE

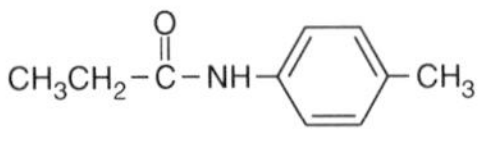

250 mg

S54,541-4
4'-ISOPROPYLACETANILIDE

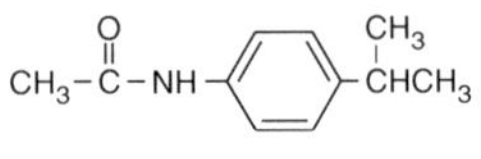

5 gm

S94,567-6
4'-(HEPTAFLUOROPROPYL)ACETANILIDE

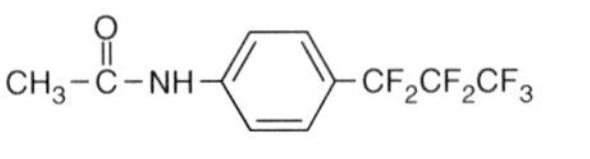

250 mg

S90,262-4
2,2,4'-TRIMETHYLPROPIONANILIDE

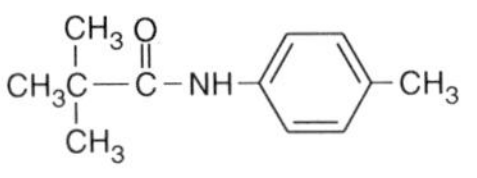

250 mg

S91,448-7
4'-TERT-BUTYLACETANILIDE

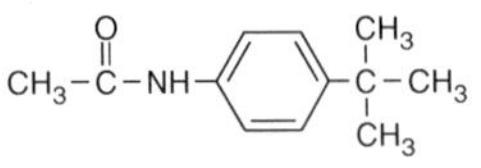

250 mg

S90,087-7
2,2-DIMETHYL-4'-ETHYLPROPIONANILIDE

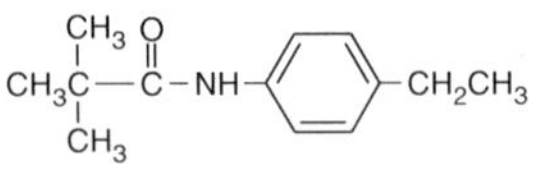

250 mg

S75,278-9
4'-ETHYLDECANANILIDE

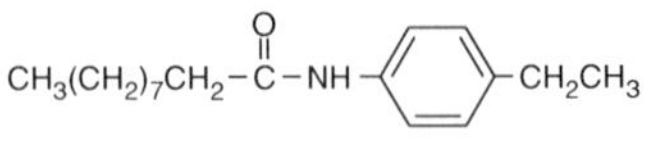

5 gm

S75,392-0
4'-ETHYLHEXADECANANILIDE
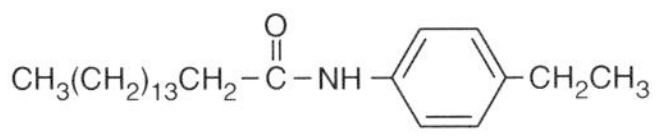
5 gm

S69,017-1
4',4''-METHYLENEBIS(FORMANILIDE)
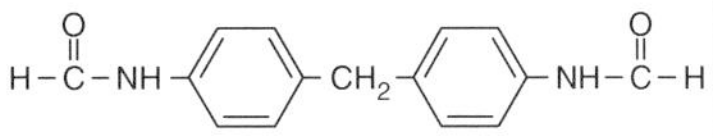
5 gm

S68,568-2
4',4''-METHYLENEBIS(ACETANILIDE)
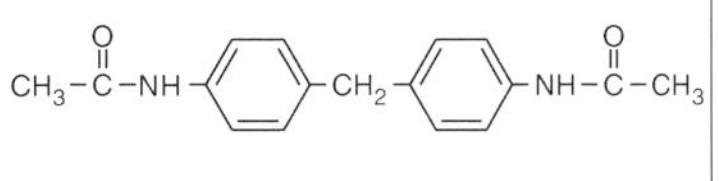
5 gm

S62,329-6
2',3'-DICHLOROFORMANILIDE
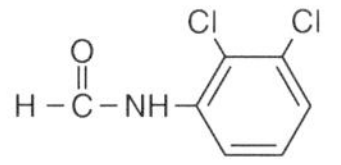
5 gm

S57,679-4
3'-CHLORO-O-FORMOTOLUIDIDE
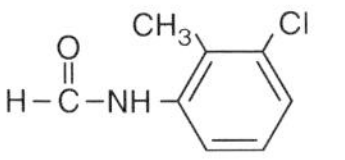
5 gm

S62,547-7
6'-CHLORO-O-FORMOTOLUIDIDE
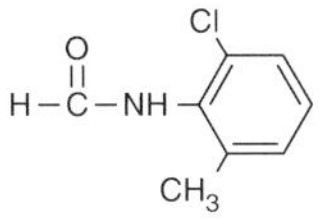
5 gm

S39,665-6
2',6'-FORMOXYLIDIDE
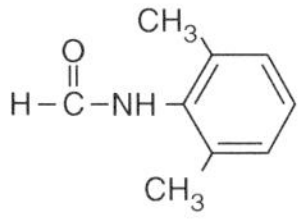
5 gm

S51,565-5
3'-CHLORO-ORTHO-ACETOTOLUIDIDE
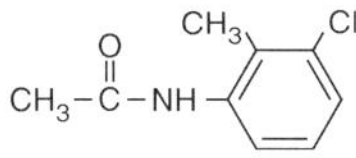
5 gm

S90,787-1
2',3'-DIMETHYLACETANILIDE
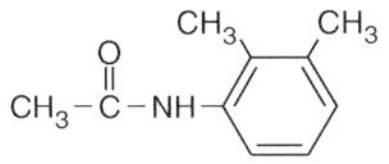
250 mg

S55,535-5
3'-CHLORO-O-PROPIONOTOLUIDIDE
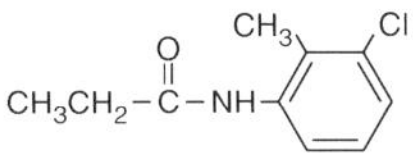
5 gm

S79,103-2
2',6'-PROPIONOXYLIDIDE
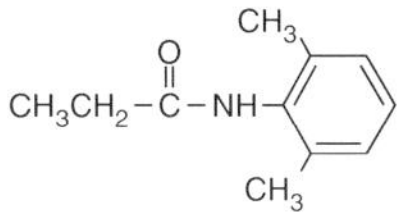
250 mg

S89,029-4
2',3'-DIMETHYLPROPIONANILIDE
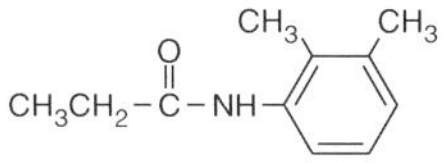
250 mg

S38,091-1
6'-CHLORO-ORTHO-ISOBUTYROTOLUIDIDE
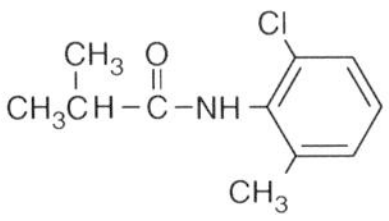
5 gm

S38,144-6
2-BROMO-6'-CHLORO-ORTHO-BUTYROTOLUIDIDE
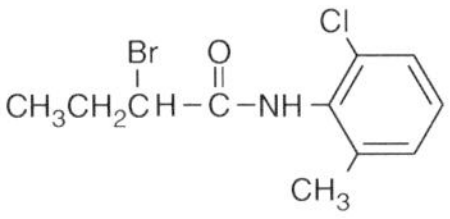
5 gm

S37,729-5
2',6'-ISOBUTYROXYLIDIDE
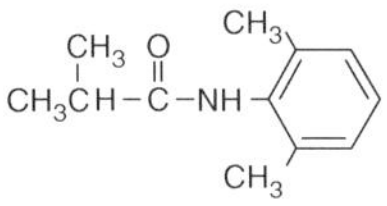
5 gm

S77,251-8
2',3'-OCTANOXYLIDIDE
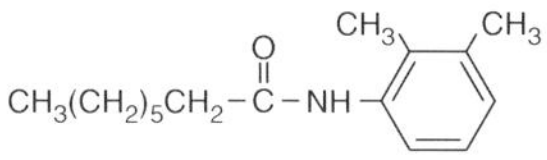
250 mg

S89,007-3
2',6'-DIMETHYLOCTANANILIDE
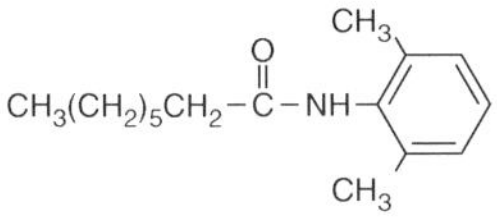
250 mg

S72,659-1
2',6'-DIMETHYLDECANANILIDE
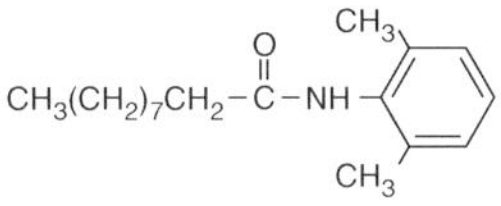
5 gm

S75,276-2
2',3'-DECANOXYLIDIDE
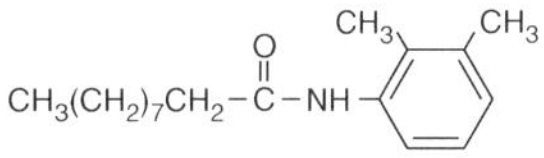
5 gm

S63,580-4
N-(2,3-DIMETHYLPHENYL)-TETRADECANAMIDE
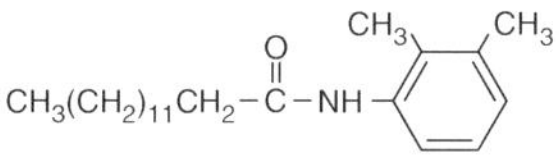
250 mg

S75,292-4
2',6'-HEXADECANOXYLIDIDE
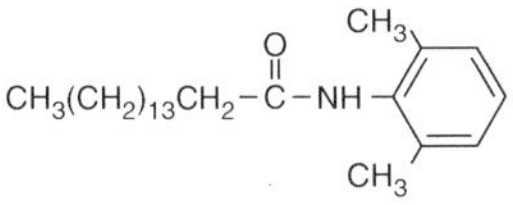
5 gm

S75,393-9
2',3'-HEXADECANOXYLIDIDE
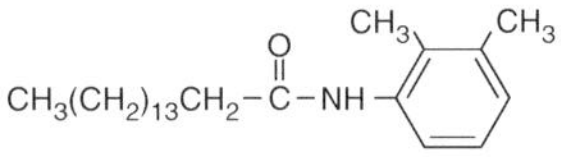
5 gm

S57,671-9
2',5'-DICHLOROFORMANILIDE
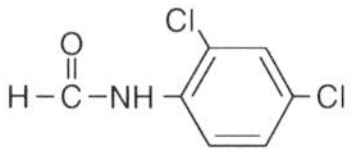
5 gm

S57,678-6
2',4'-DICHLOROFORMANILIDE
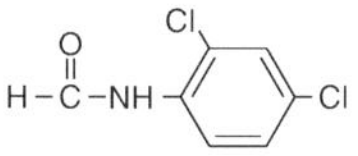
5 gm

S58,016-3
3',4'-DICHLOROFORMANILIDE
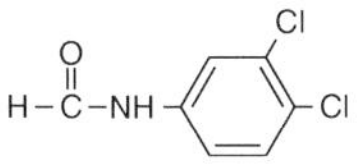
5 gm

S57,639-5
4'-CHLORO-O-FORMOTOLUIDIDE
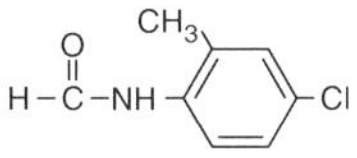
5 gm

S62,328-8
5'-CHLORO-O-FORMOTOLUIDIDE
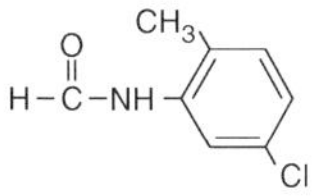
5 gm

S50,068-2
3',4'-DIFLUOROACETANILIDE
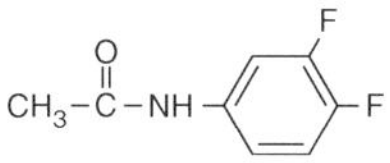
250 mg

S38,825-4
3'-CHLORO-4'-FLUOROACETANILIDE
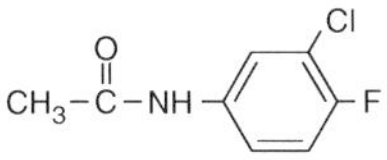
5 gm

S36,024-4
3',4'-DICHLOROACETANILIDE
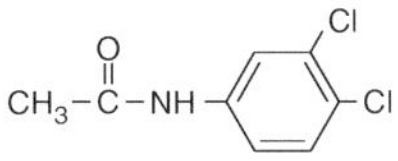
5 gm

S39,011-9
2',5'-DICHLOROACETANILIDE
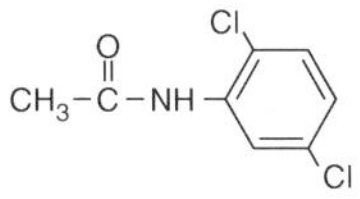
5 gm

S62,458-6
4'-BROMO-O-FORMOTOLUIDIDE
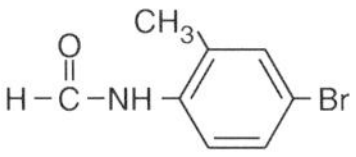
5 gm

S39,919-1
3',4'-FORMOXYLIDIDE
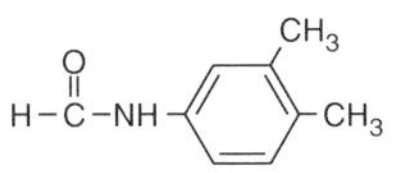
5 gm

S57,654-9
2',5'-FORMOXYLIDIDE
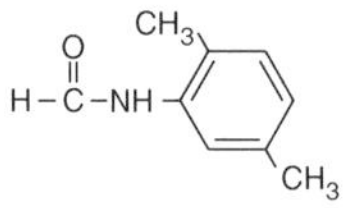
5 gm

S57,680-8
2',4'-FORMOXYLIDIDE
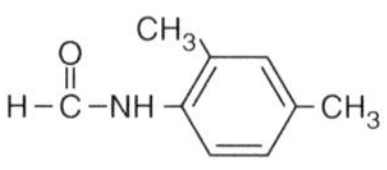
5 gm

S34,431-1
5'-CHLORO-2'-METHYLACETANILIDE
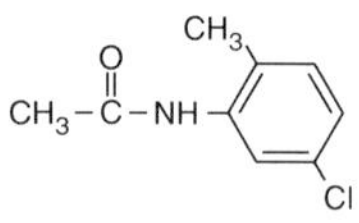
5 gm

S65,676-3
3',4'-DICHLOROPROPIOANILIDE
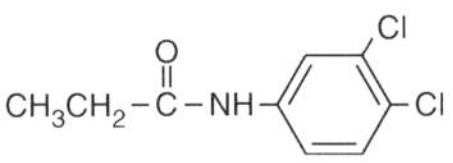
5 gm

S40,690-2
2'-BROMO-4'-METHYLACETANILIDE
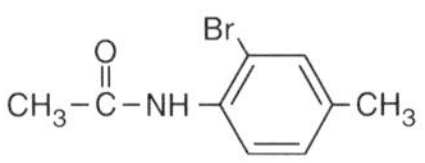
5 gm

S50,741-5
4'-BROMO-O-ACETOTOLUIDIDE
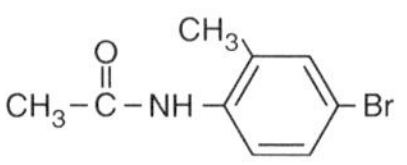
5 gm

S56,372-2
3',4'-ACETOXYLIDIDE
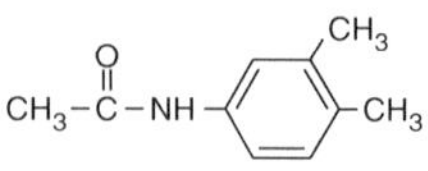
5 gm

S74,227-9
2',4'-DIMETHYLACETANILIDE
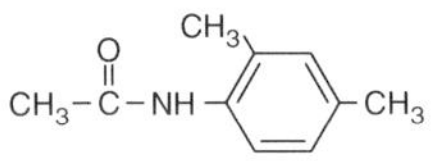
5 gm

S72,164-6
5'-ISOPROPYL-2'-METHYLFORMANILIDE
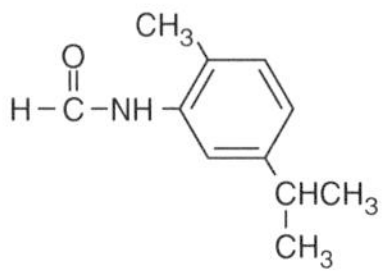
5 gm

S89,030-8
2',4'-DIMETHYLPROPIONANILIDE
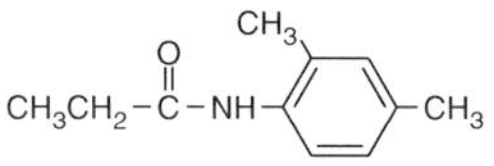
250 mg

S75,307-6
3',4'-DIETHYLACETANILIDE
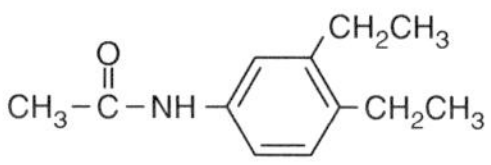
5 gm

S89,286-6
5'-ISOPROPYL-2'-METHYLPROPIONANILIDE
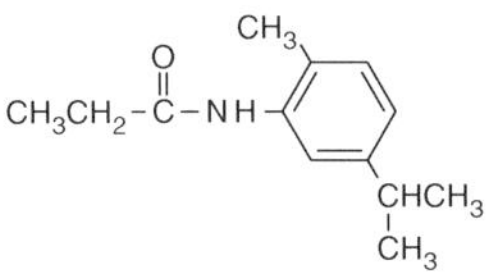
250 mg

S90,263-2
2,2,3',4'-TETRAMETHYLPROPIONANILIDE
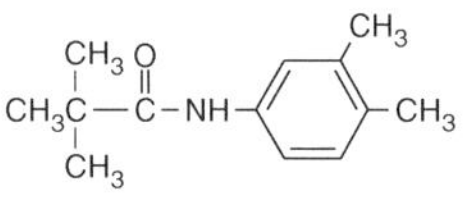
250 mg

S76,923-1
2',4'-OCTANOXYLIDIDE
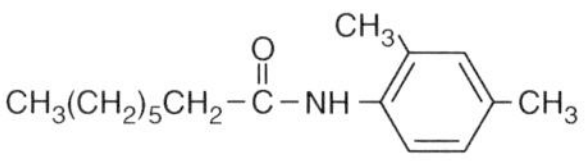
250 mg

S79,155-5
3',4'-OCTANOXYLIDIDE
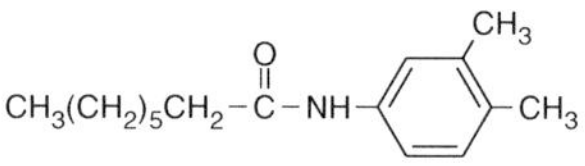
250 mg

S89,008-1
2',5'-DIMETHYLOCTANANILIDE
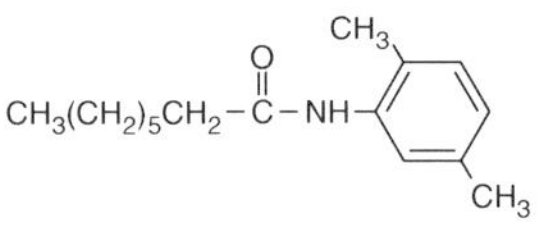
250 mg

S39,424-6
5'-ISOPROPYL-ORTHO-OCTANOTOLUIDIDE
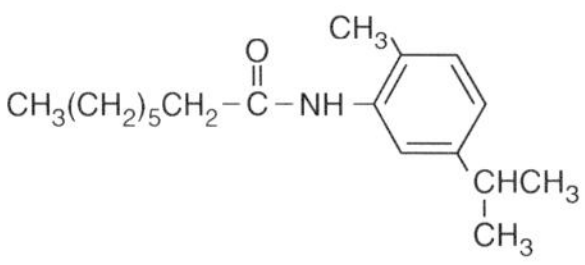
250 mg

S75,277-0
2',4'-DECANOXYLIDIDE
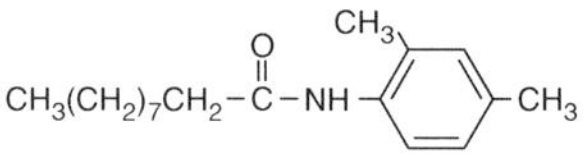
5 gm

S75,274-6
5'-CHLORO-2'-METHYL-10-UNDECENANILIDE
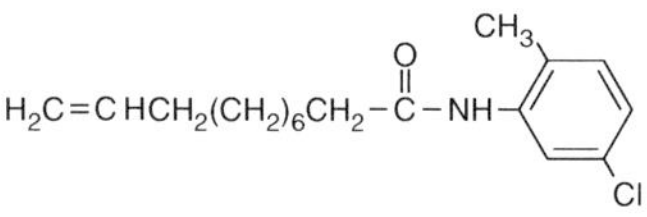
250 mg

S39,426-2
5'-ISOPROPYL-ORTHO-LAUROTOLUIDIDE
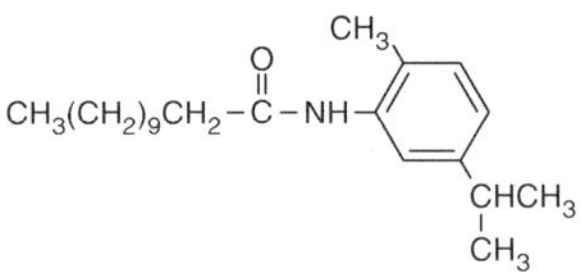
250 mg

S75,293-2
2',4'-HEXADECANOXYLIDIDE
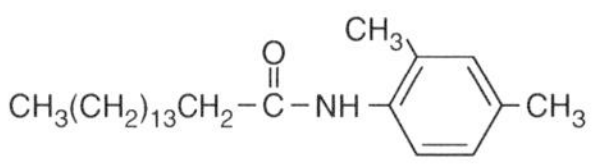
5 gm

S53,177-4
4',4''-METHYLENEBIS(2'-CHLOROFORMANILIDE)
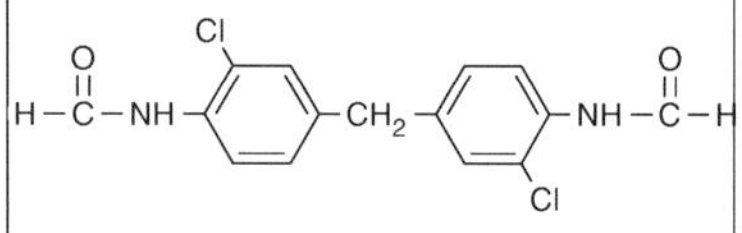
5 gm

S62,541-8
3',5'-DIMETHYLFORMANILIDE
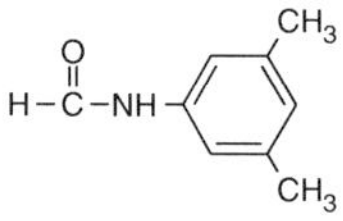
5 gm

S50,634-6
3',5'-ACETOXYLIDIDE
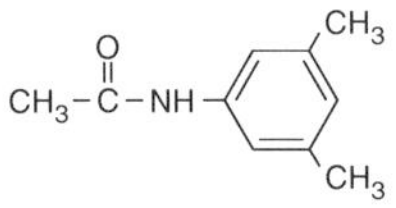
250 mg

S63,579-0
N-(3,5-DIMETHYLPHENYL)-TETRADECANAMIDE
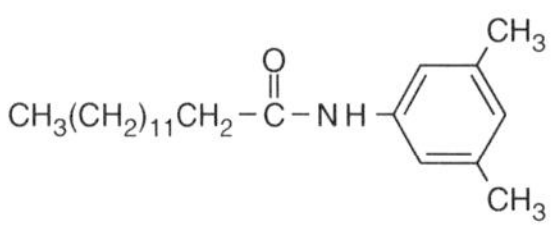
250 mg

S55,845-1
2',3',4'-TRICHLOROACETANILIDE
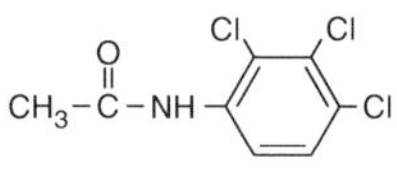
5 gm

S70,275-7
2',4',6'-TRIMETHYLFORMANILIDE
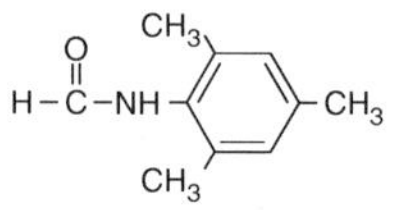
5 gm

S89,006-5
2',4',6'-TRIMETHYLOCTANANILIDE
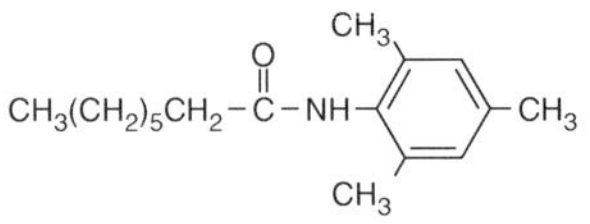
250 mg

S89,166-5
2',4',6'-TRIMETHYLHEXADECANANILIDE
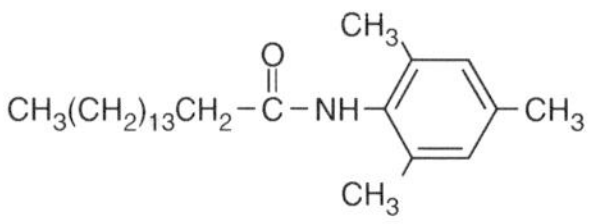
250 mg

S75,294-0
2',4',6'-TRIMETHYLHEXADECANANILIDE
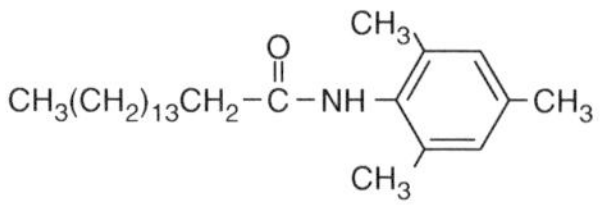
5 gm

S53,182-0
2',4',5'-TRICHLOROFORMANILIDE
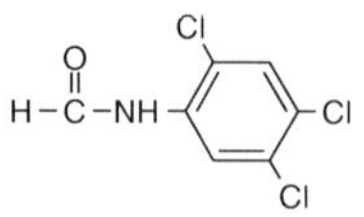
5 gm

S55,838-9
2',4',5'-TRICHLOROACETANILIDE

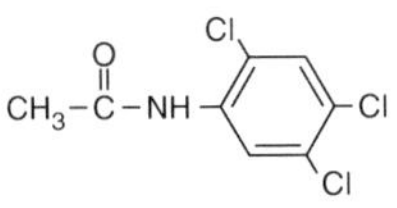

5 gm

S94,362-2
2',4',5'-TRIMETHYLACETANILIDE

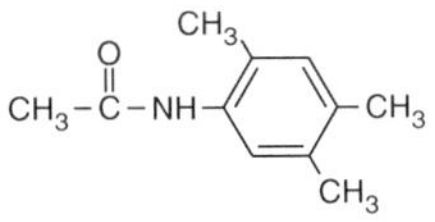

250 mg

S55,536-3
2',4',5'-TRIMETHYLPROPIONANILIDE

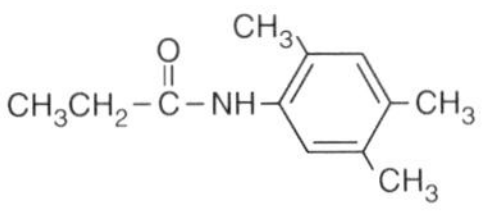

5 gm

S34,742-6
N,N'-(4,6-DIMETHYL-1,3-XYLYLENE)-BISACETAMIDE

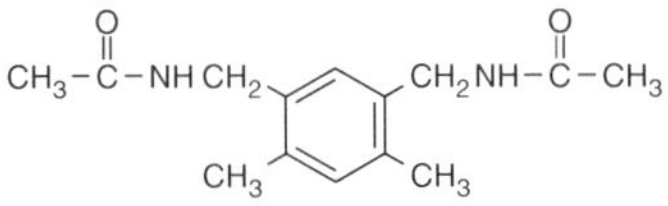

5 gm

S52,029-2
N-(1,2,3,4-TETRAHYDRO-1-NAPHTHYL)-FORMAMIDE

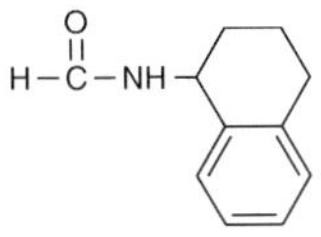

250 mg

S34,027-8
1-ACETAMIDO-5,6,7,8-TETRAHYDRONAPHTHALENE

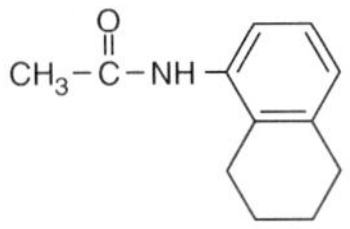

5 gm

S66,413-8
2-ACETAMIDO-5,6,7,8-TETRAHYDRONAPHTHALENE

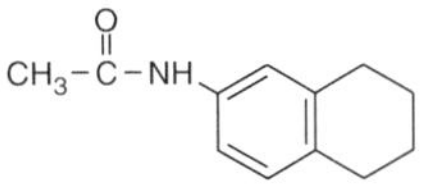

250 mg

S75,426-9
3-ACETAMIDOFLUORENE

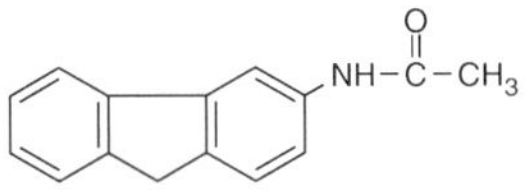

250 mg

S90,734-0
N-(3-PHENANTHRYL)ACETAMIDE

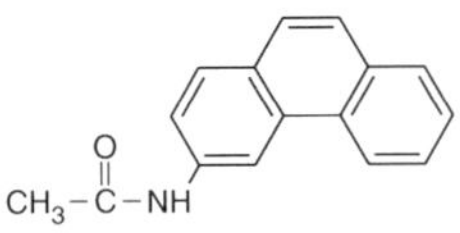

250 mg

S94,595-1
N-(9-PHENANTHRYL)ACETAMIDE

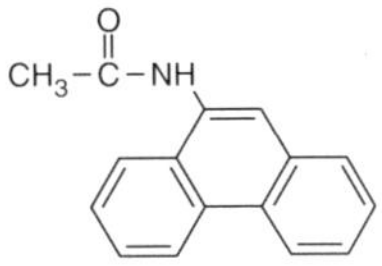

250 mg

S66,455-3
N-ACETYLDEHYDROABIETYLAMINE

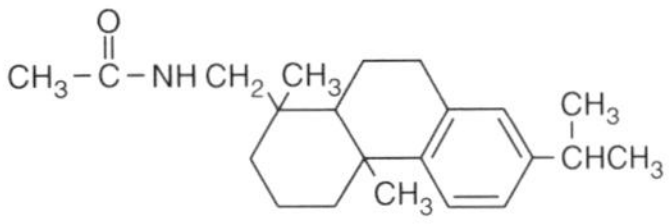

5 gm

S55,902-4
3'-CHLORO-N-METHYLACETANILIDE

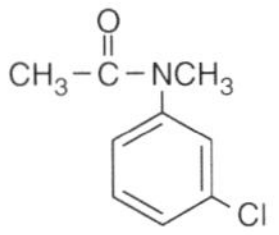

5 gm

S61,384-3
3,3-DIPHENYL-N,N,N',N'-TETRAMETHYL-1,2-CYCLOPROPANEDICARBOXAMIDE

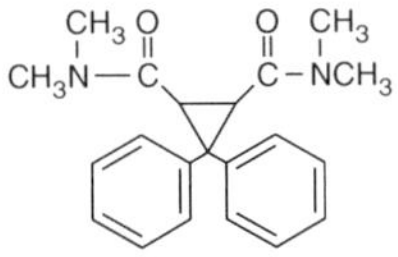

250 mg

S89,277-7
N-CYCLOHEXYLPROPIONANILIDE

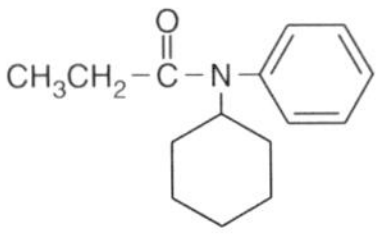

250 mg

S87,172-9
N-(3-CYCLOHEXEN-1-YL)ACETANILIDE

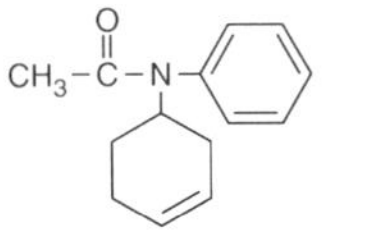

250 mg

S39,214-6
N-METHYLACETANILIDE

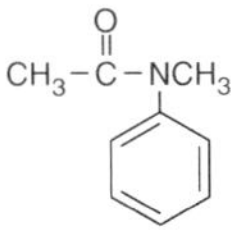

5 gm

S56,081-2
N-ETHYLFORMANILIDE

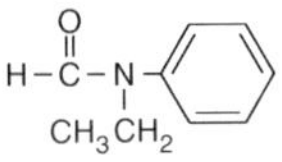

5 gm

S42,827-2
N-ETHYLACETANILIDE

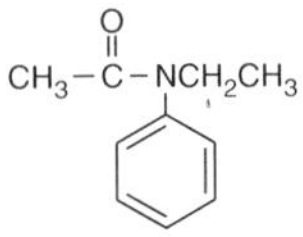

5 gm

S59,340-0
N-METHYLPROPIONANILIDE

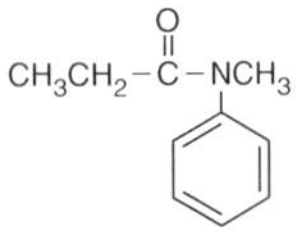

5 gm

S88,910-5
N,N-DIMETHYL-2-PHENYLACETAMIDE

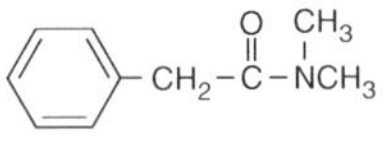

250 mg

S86,127-8
N-BENZYL-N-METHYLISOBUTYRAMIDE

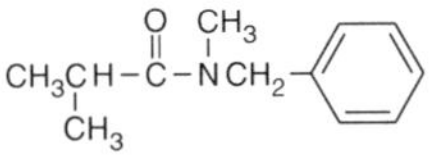

250 mg

S88,402-2
N-ALLYL-N-PHENYLHEXANAMIDE

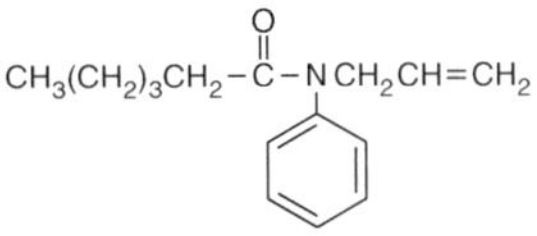

250 mg

S77,523-1
2-CHLORO-N-METHYL-2-PHENYL-O-ACETOTOLUIDIDE

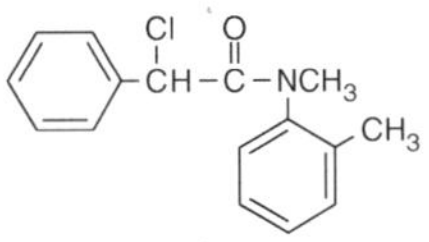

250 mg

S77,495-2
N-ETHYL-2-PHENYL-O-BUTYROTOLUIDIDE

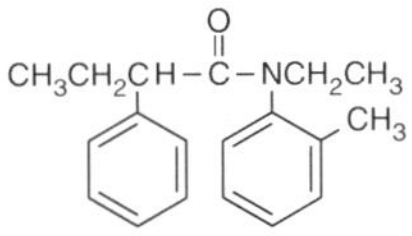

250 mg

S77,522-3
2-CHLORO-N-METHYL-2-PHENYL-2',6'-ACETOXYLIDIDE

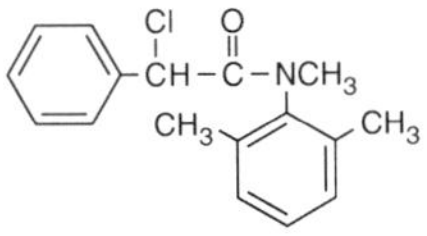

250 mg

S77,521-5
2-CHLORO-N-ISOPROPYL-2-PHENYL-2',5'-ACETOXYLIDIDE

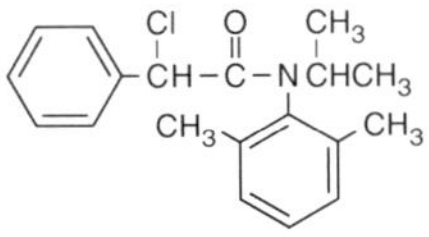

250 mg

S77,533-9
N,N-DIETHYL-2,2-DIPHENYLBUTYRAMIDE

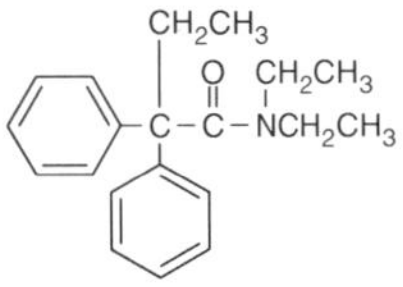

250 mg

S42,099-9
N-ETHYL-ORTHO-ISOBUTYROTOLUIDIDE

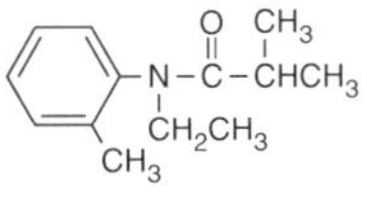

250 mg

S77,506-1
N-(SEC-BUTYL)-O-ACETOTOLUIDIDE

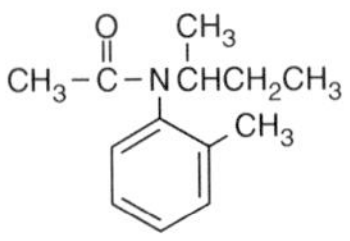

250 mg

S40,323-7
2-CHLORO-N,N-DIETHYLHYDROCINNAMAMIDE

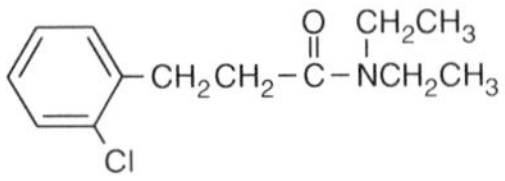

5 gm

S77,484-7
N-ISOPROPYL-2',6'-ACETOXYLIDIDE

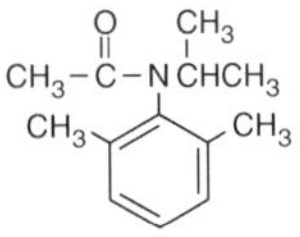

250 mg

S34,538-5
N,N-DIACETYL-5,6,7,8-TETRAHYDRO-1-NAPHTHYLAMINE

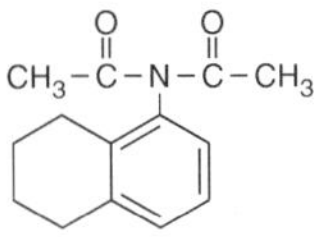

250 mg

S92,199-8
4'-BROMO-3'-METHYL-1,2,3,6-TETRAHYDROPHTHALANILIC ACID

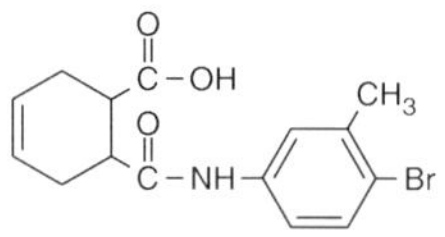

5 gm

S91,384-7
N-ACETYL-N'-(2-CHLOROACETYL)-2-METHOXY-1,4-PHENYLENEDIAMINE

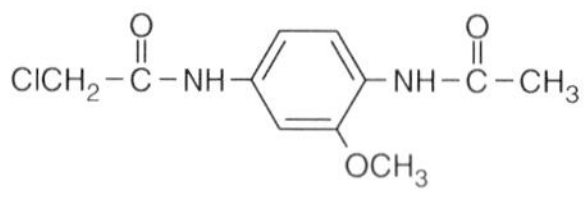

250 mg

S60,145-4
2',4-DICHLOROACETOACETANILIDE

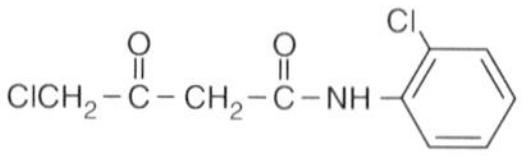

5 gm

S60,146-2
4-CHLORO-O-ACETOACETOTOLUIDIDE

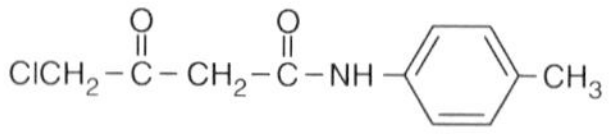

250 mg

S93,721-5
4'-METHOXY-L-TARTRANILIC ACID

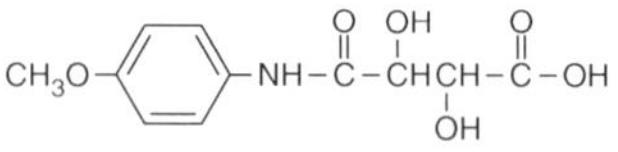

250 mg

S38,983-8
(3-HYDROXYBENZOYL)-ORTHO-ACETANISIDIDE

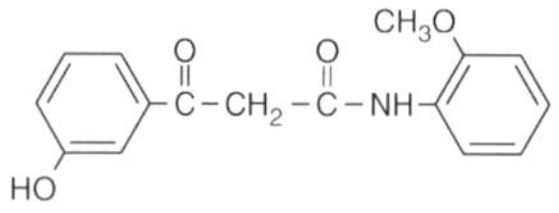

5 gm

S92,605-1
2',5'-DIMETHOXYACETOACETANILIDE

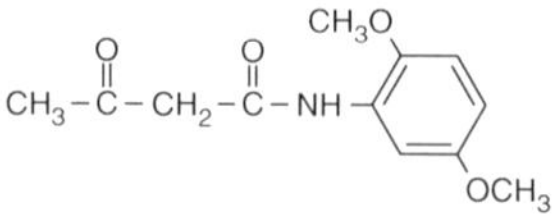

250 mg

S62,971-5
2-(3-NITROBENZOYL)-2'-[2-(4-TERT-OCTYLPHENOXY)ETHOXY]ACETANILIDE

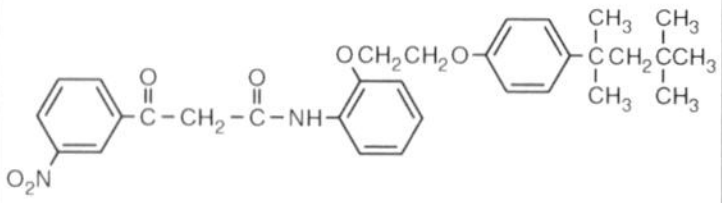

250 mg

S62,956-1
4'-CARBETHOXY-3-[2-(3-NITROBENZOYL)-ACETAMIDO]-4-[2-(4-TERT-OCTYLPHENOXY)-ETHOXY]BENZANILIDE

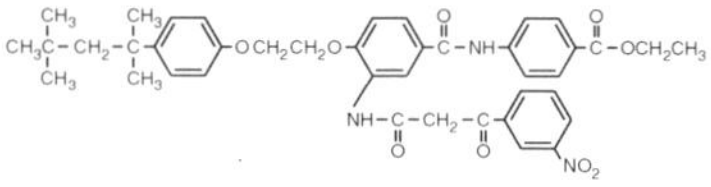

250 mg

S63,050-6
2-(3-NITROBENZOYL)-4'-[2-(P-TOLYLOXY)ETHOXY]ACETANILIDE

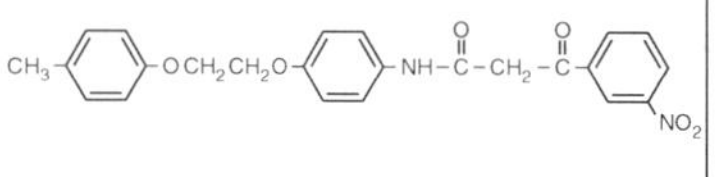

5 gm

S60,580-8
2-(3-AMINOBENZOYL)-4'-HEXADECYLOXYACETANILIDE

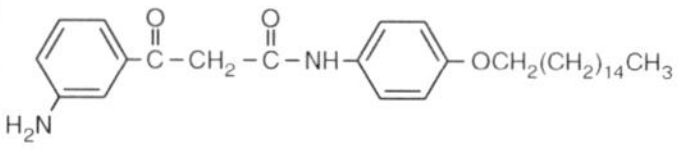

250 mg

S94,975-2
4-METHOXY-2-METHYL-3-NITROBENZAMIDE

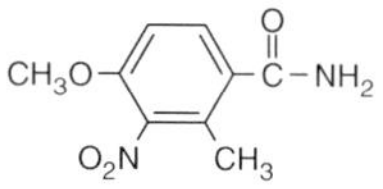

100 mg

S60,492-5
4'-[2-(4-METHOXYBENZOYL)ACETAMIDO]-SUCCINANILIC ACID

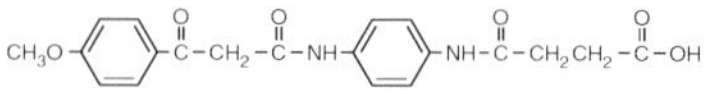

250 mg

S63,060-8
2-(4-NITROBENZOYL)ACET-O-ANISIDIDE

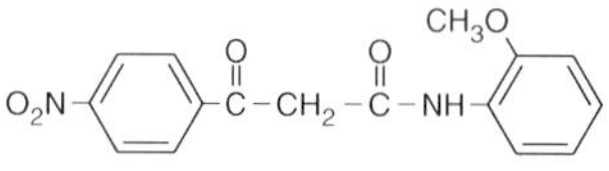

250 mg

S64,223-1
2-[4-(METHACRYLAMIDO)BENZOYL]-ACET-O-ANISIDE

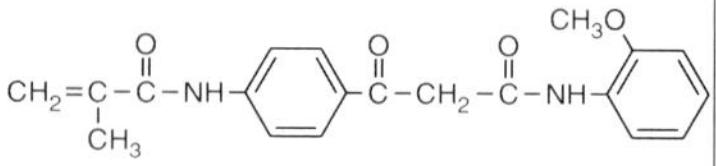

250 mg

S38,463-1
ORTHO-ACETOACETOPHENETIDIDE

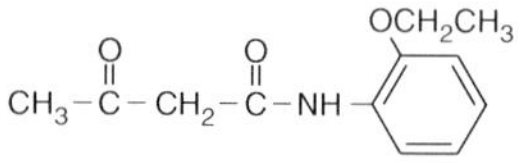

5 gm

S58,820-2
2-HEXADECANOYL-O-ACETANISIDIDE

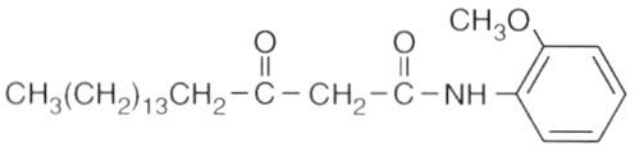

250 mg

S38,488-7
5'-CHLORO-ORTHO-ACETOACETANISIDIDE

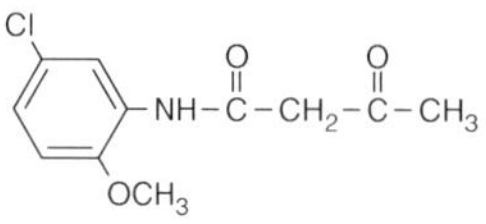

5 gm

S62,284-2
5'-CHLORO-2'-METHOXYACETOACETANILIDE

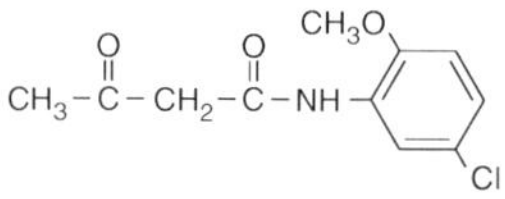

250 mg

S35,945-9
4'-CHLORO-2',5'-DIMETHOXYACETOACETANILIDE

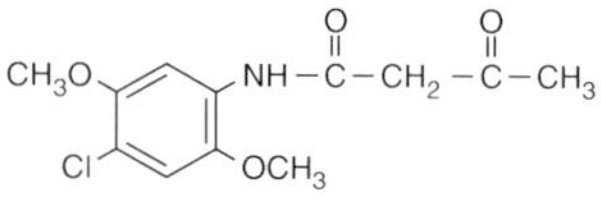

5 gm

S50,996-5
3-ACETOACETAMIDO-4-(HEXADECYLOXY)-BENZOIC ACID

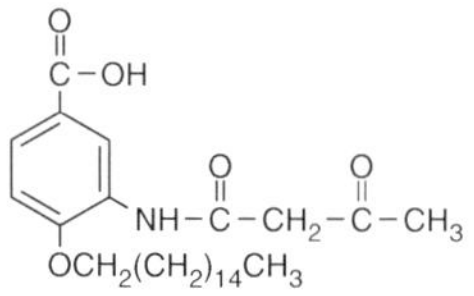

250 mg

S50,024-0
2'-(HEXADECYLOXY)ACETOACETANILIDE

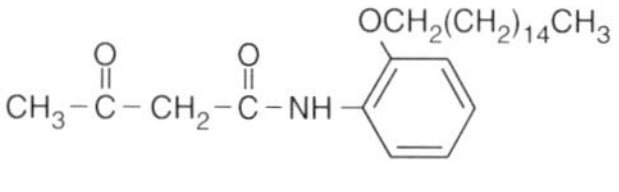

250 mg

S37,548-9
N-(2-METHOXY-5-NITROPHENYL)-BENZOYLACETAMIDE

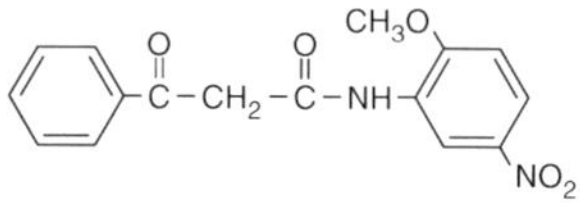

5 gm

S60,589-1
4'-CARBOXY-2-(P-ANISOYL)ACETANILIDE

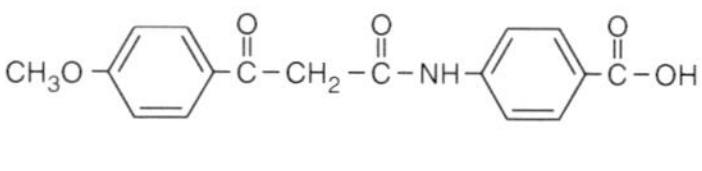

250 mg

S41,173-6
N-(4-METHOXYBENZYL)ACETOACETAMIDE

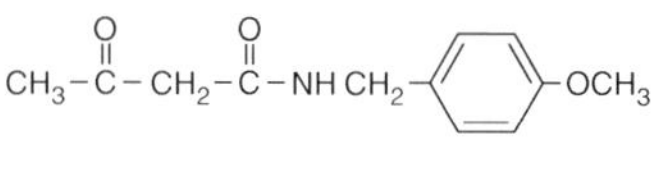

5 gm

S37,111-4
N-(4-METHOXYPHENETHYL)-ACETOACETAMIDE

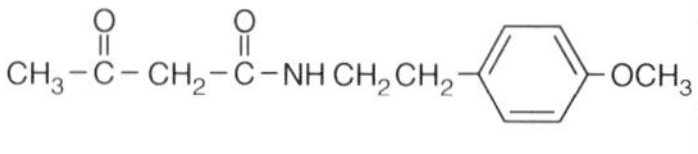

250 mg

S56,846-5
P-HEXADECANOACETANISIDIDE

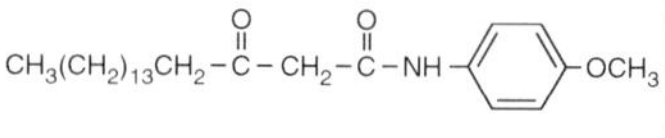

250 mg

S38,485-2
2',5'-DIETHOXYACETOACETANILIDE

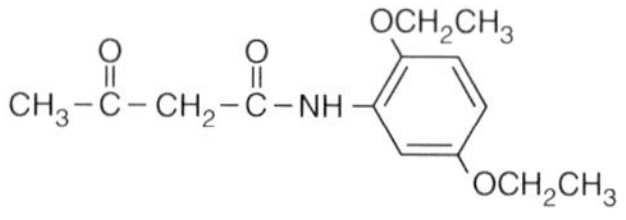

5 gm

S36,076-7
3',5'-DIMETHOXYACETOACETANILIDE,

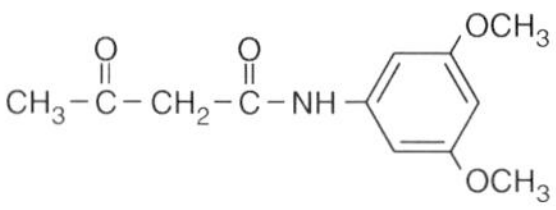

250 mg

S34,032-4
3',4',5'-TRIMETHOXY-ACETOACETANILIDE

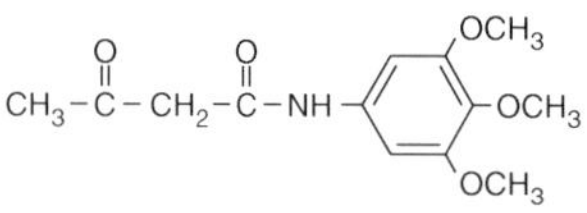

5 gm

S42,124-3
N-ETHYL-N-(2-PHENOXYETHYL)-ACETOACETAMIDE

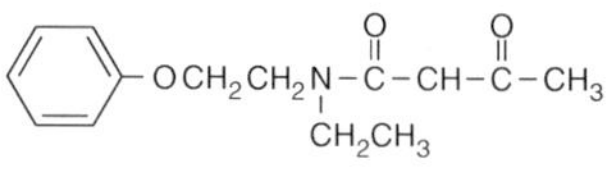

250 mg

S59,882-8
4'-CARBOXY-2-(O-ANISOYL)ACETANILIDE

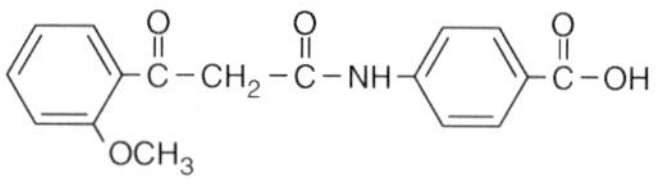

250 mg

S55,365-4
2'-BROMO-2-[4-(HEXADECYLOXY)-BENZOYL]ACETANILIDE

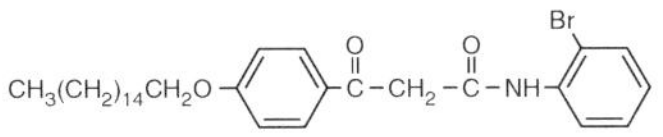

250 mg

S95,443-8
DIETHYL BETA-OXOBENZYLIMINODIPROPIONATE

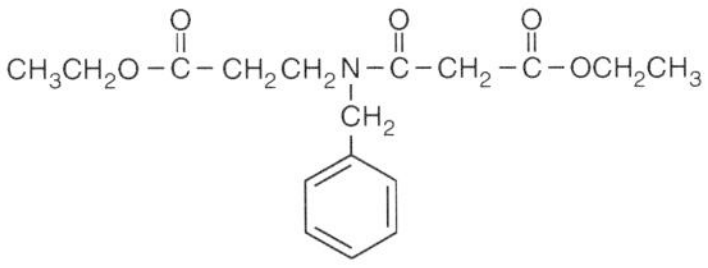

50 mg

S53,066-2
2-BENZOYL-2-(PHENYLTHIO)-ACETANILIDE

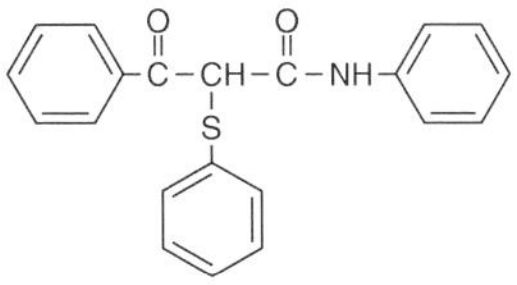

250 mg

S60,569-7
4-(2-BENZOYLACETAMIDO)ANILINE

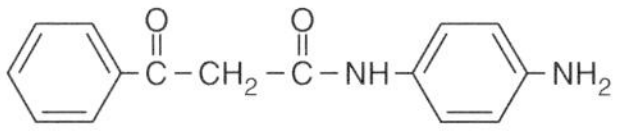

5 gm

S62,963-4
DIMETHYL 5-[2-(3-AMINOBENZOYL)-ACETAMIDO]ISOPHTHALATE

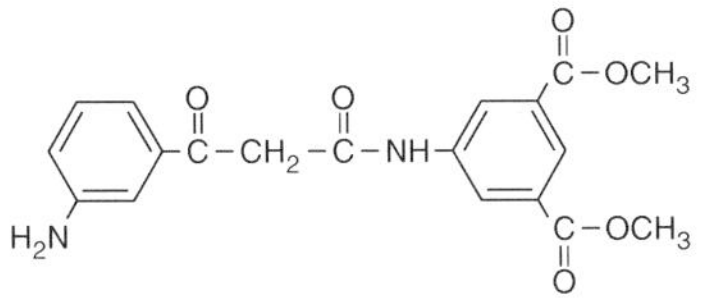

5 gm

S51,520-5
2-(4-AMINOBENZOYL)PROPIONANILIDE

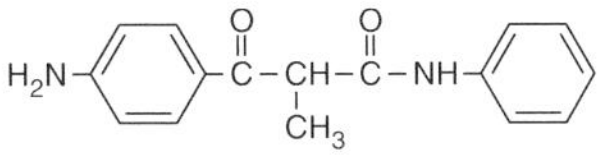

5 gm

S34,594-6
2-(DIETHYLAMINO)MALONDIANILIDE

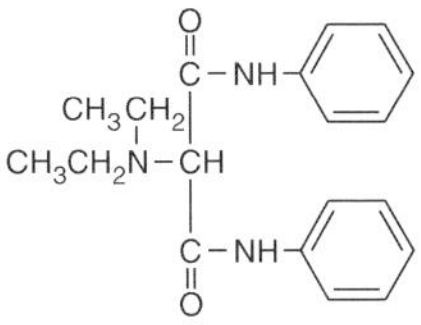

5 gm

S50,200-6
3'-NITRO-2-BENZOYLACETANILIDE

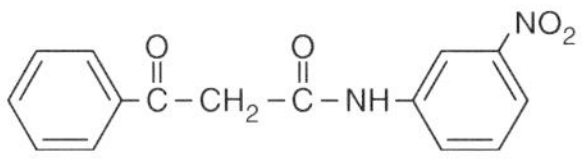

250 mg

S59,192-0
N,N'-DIPHENYLMALONAMIDE

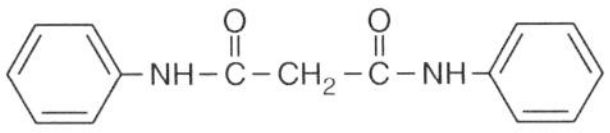

250 mg

S38,417-8
2-BUTYLMALONANILIDE

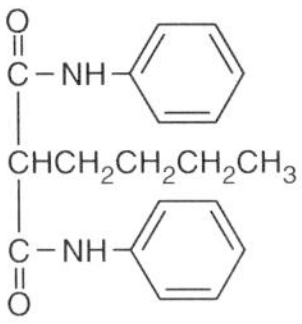

100 mg

S38,270-1
2-BENZOYL-4'-NITROACETANILIDE

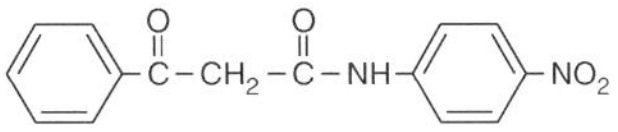

5 gm

S54,659-3
2-(2-FLUOROBENZOYL)-4'-NITROACETANILIDE

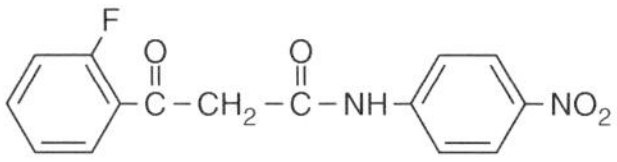

250 mg

S59,246-3
4'-(2-BENZOYLACETAMIDO)-SUCCINANILIC ACID

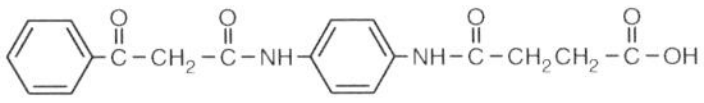

250 mg

S36,444-4
N,N'-(1,4-PHENYLENE)BIS-(ACETOACETAMIDE)

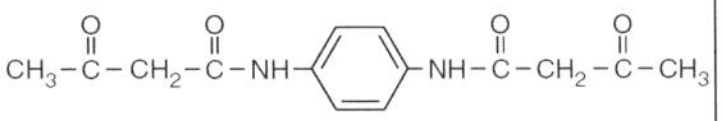

5 gm

S38,493-3
N,N'-(2,5-DIMETHYL-1,4-PHENYLENE)-BIS(ACETOACETAMIDE)

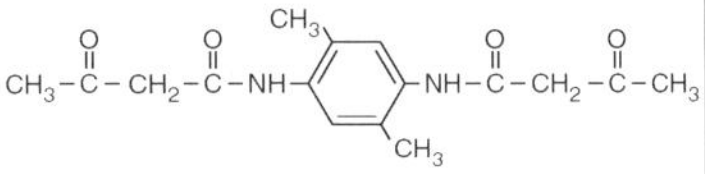

5 gm

S89,778-7
6,6-DIMETHYL-2-OXO-N-PHENETHYL-4-(PHENETHYLAMINO)-3-CYCLOHEXENE-1-CARBOXAMIDE

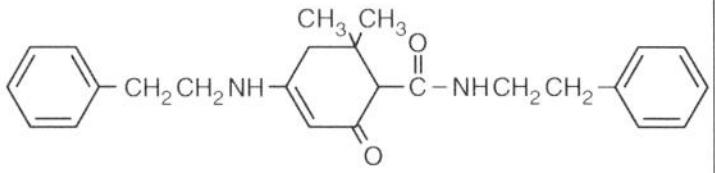

250 mg

S89,773-6
2-OXO-N-PHENETHYL-4-(PHENETHYLAMINO)-6-PHENYL-3-CYCLOHEXENE-1-CARBOXAMIDE

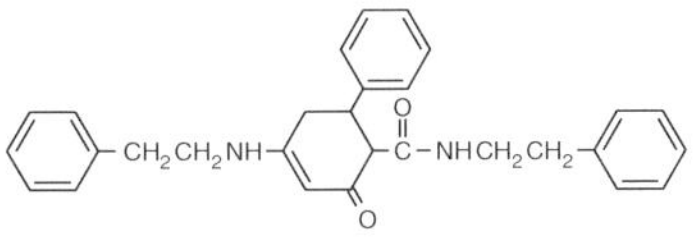

250 mg

S78,255-6
N-(2-PHENYLETHYL)MALONAMIC ACID

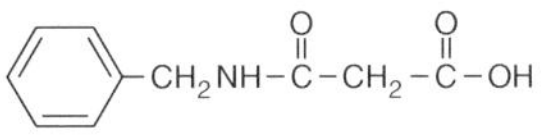

5 gm

S38,327-9
DIETHYL 2-(PHENYLCARBAMOYL)-MALONATE

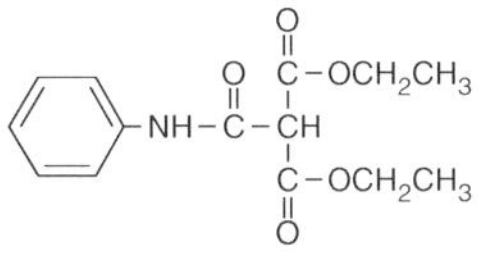

5 gm

S35,717-0
ACETOACETOBENZYLAMIDE

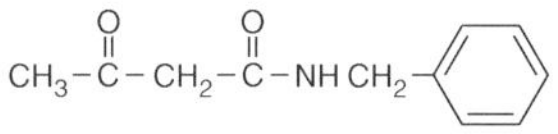

5 gm

S56,489-3
N-(DIPHENYLMETHYL)ACETOACETAMIDE

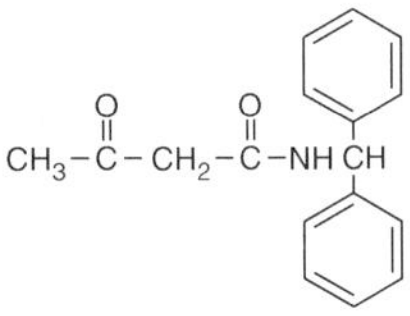

5 gm

S41,169-8
N-BENZYL-N-METHYLACETOACETAMIDE

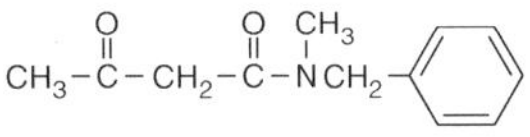

5 gm

S62,940-5
DIMETHYL [2-(4-NITROBENZOYL)-ACETAMIDO]TEREPHTHALATE

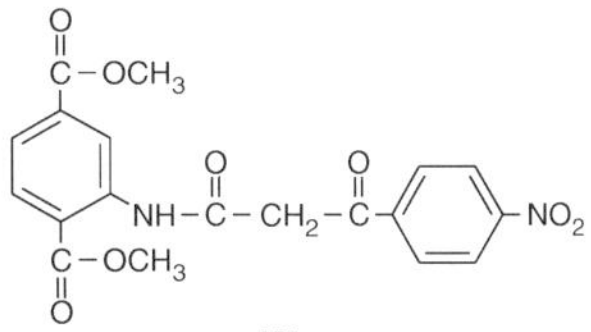

250 mg

S63,048-9
2-(4-NITROBENZOYL)PROPIONANILIDE

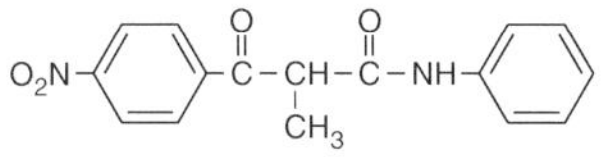

250 mg

S39,354-1
2-CHLORO-2-BENZOYLACETANILIDE

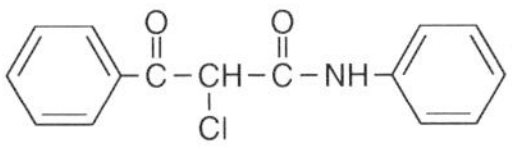

250 mg

S53,409-9
2-BENZOYL-2-BROMOACETANILIDE

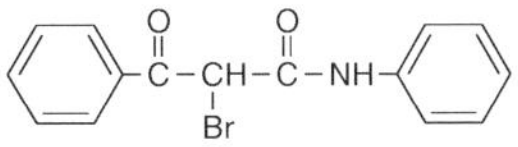

250 mg

S58,179-8
4-[2-(2-CHLOROBENZOYL)ACETAMIDO]-BENZOIC ACID

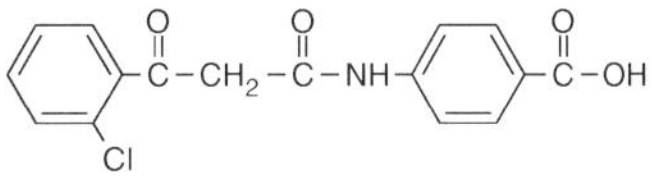

5 gm

S60,597-2
4'-CARBOXY-2-PIVALOYLACETANILIDE

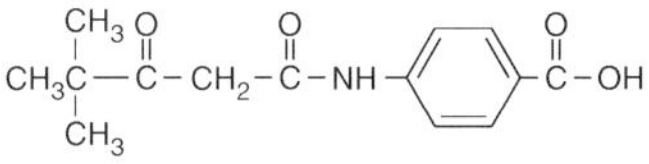

250 mg

S58,201-8
4-(2-BENZOYLACETAMIDO)BENZOIC ACID

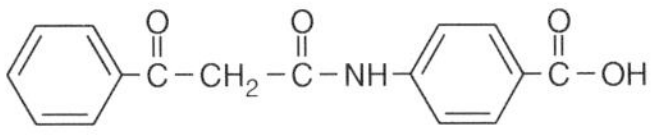

5 gm

S60,549-2
4-[2-(4-CHLOROBENZOYL)ACETAMIDO]-BENZOIC ACID

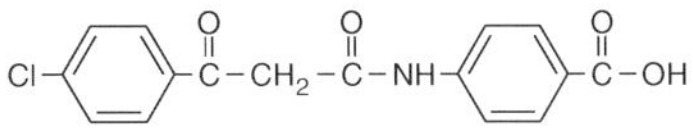

5 gm

S67,177-0
N-(ACETOACETYL)ANTHRANILIC ACID

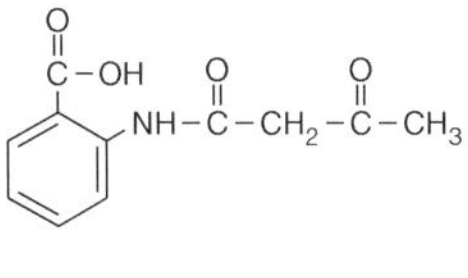

5 gm

S59,808-9
2-(P-TOLUOYL)ACETANILIDE

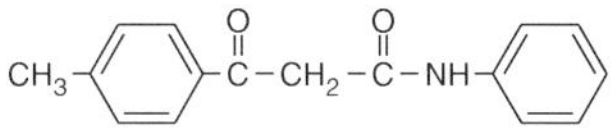

250 mg

S55,730-7
N-(O-TOLYL)DIACETOACETAMIDE
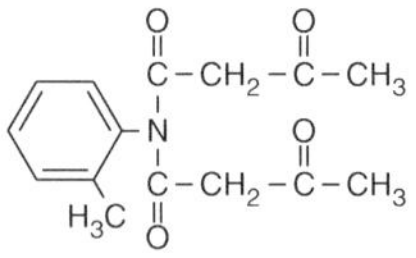
250 mg

S38,461-5
3'-CHLOROACETOACETANILIDE
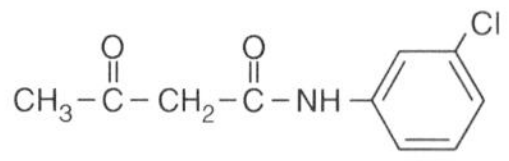
5 gm

S38,487-9
META-ACETOACETOTOLUIDIDE
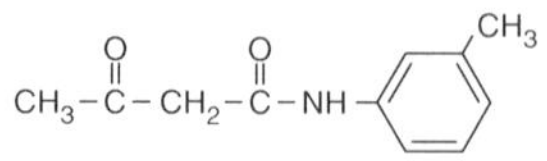
5 gm

S86,985-6
2-OXO-1-CYCLOPENTANECARBOXANILIDE
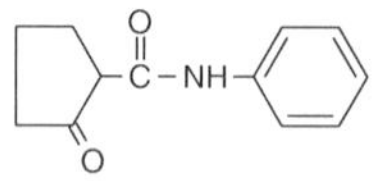
250 mg

S87,001-3
2-OXO-1-CYCLOHEXANECARBOXANILIDE
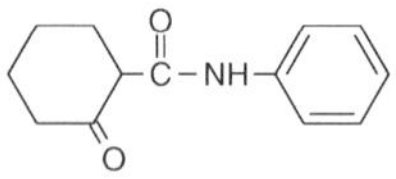
250 mg

S35,471-6
N-(3-PHENYLPROPYL)ACETOACETAMIDE
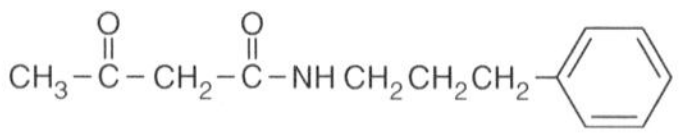
5 gm

S37,601-9
N-(1-METHYL-3-PHENYLPROPYL)-ACETOACETAMIDE
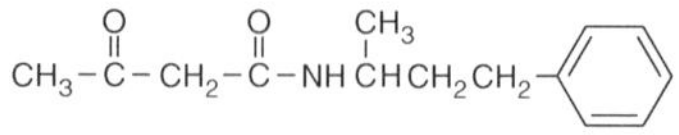
5 gm

S59,310-9
2,4-DIACETYL-2',2''-DICHLOROGLUTARANILIDE
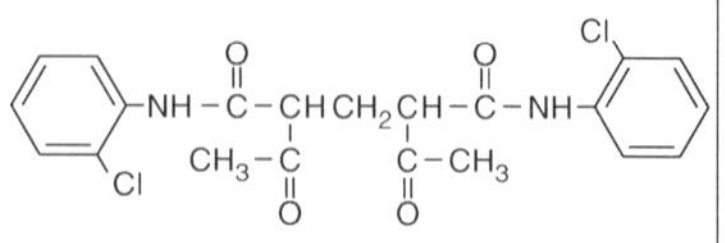
250 mg

S59,312-5
2,4-DIACETYL-4',4''-DIMETHYLGLUTARANILIDE
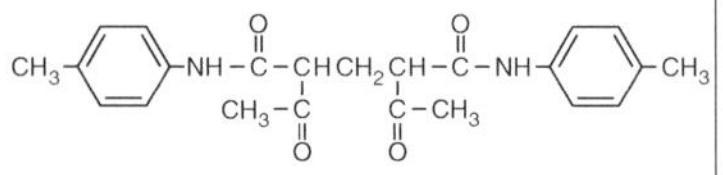
5 gm

S59,162-9
4',4'''-METHYLENEBIS-(ACETOACETANILIDE)
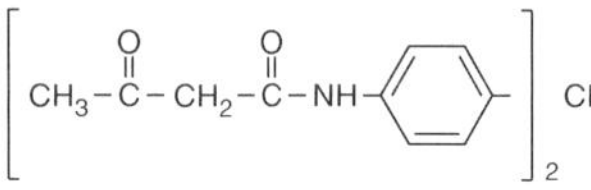
5 gm

S38,459-3
3'-CHLORO-ORTHO-ACETOACETOTOLUIDIDE
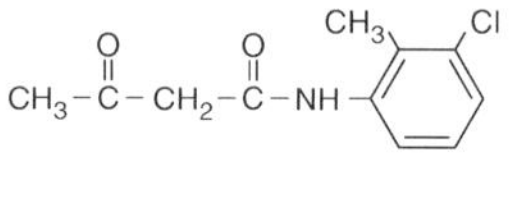
5 gm

S38,477-1
2',3'-ACETOACETOXYLIDIDE
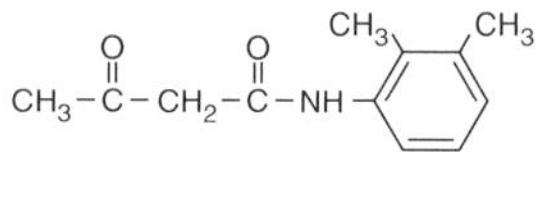
5 gm

S34,030-8
ACETOACETO-2,4-DICHLOROANILIDE
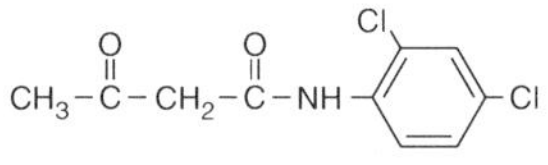
5 gm

S38,465-8
2',5'-DICHLOROACETOACETANILIDE
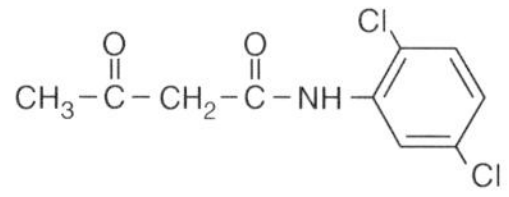
5 gm

S36,593-9
4'-CHLORO-ORTHO-ACETOACETOTOLUIDIDE
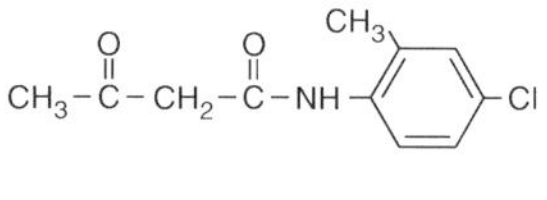
5 gm

S38,476-3
5'-CHLORO-ORTHO-ACETOACETOTOLUIDIDE
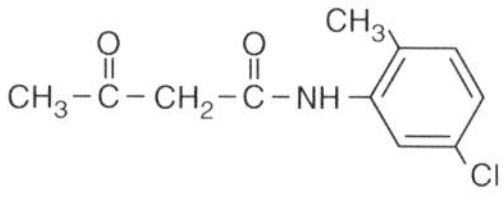
5 gm

S38,468-2
3',4'-ACETOACETOXYLIDIDE
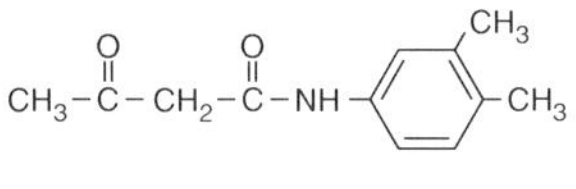
5 gm

S38,475-5
2',5'-ACETOACETOXYLIDIDE
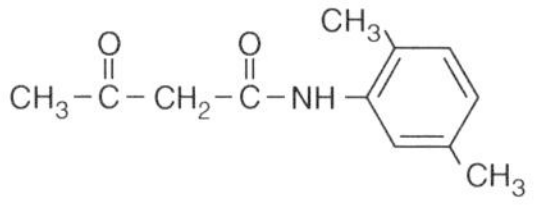
5 gm

S35,803-7
4',4''-BIS(ORTHO-ACETOACETOTOLUIDIDE)
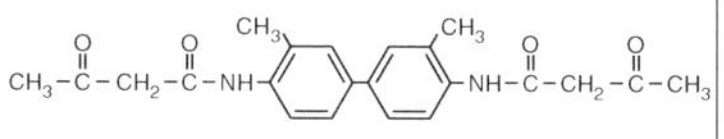
5 gm

S36,580-7
N-1-INDANYLACETOACETAMIDE
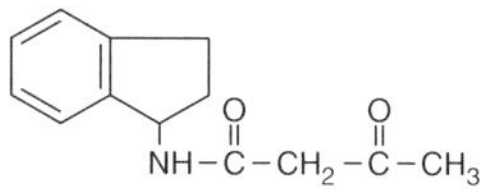
250 mg

S42,327-0
ALPHA-ACETYL-2-CHLORO-N,N-DIETHYLHYDROCINNAMAMIDE
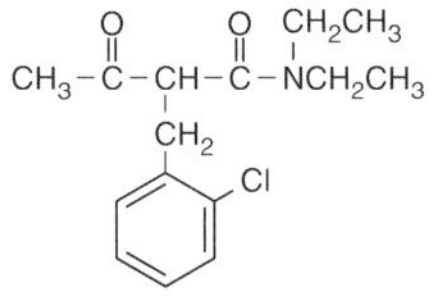
5 gm

S55,454-5
METHYL 3-ACETOACETAMIDO-4-(HEXADECYLOXY)BENZOATE
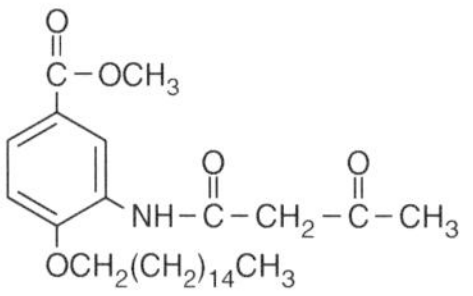
250 mg

S42,338-6
2'-CHLOROACETOACETANILIDE
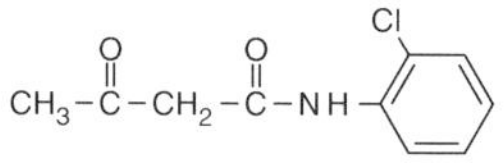
5 gm

S42,002-6
4'-BROMOACETOACETANILIDE
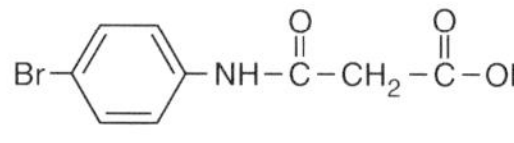
5 gm

S42,067-0
2',6'-ACETOACETOXYLIDIDE
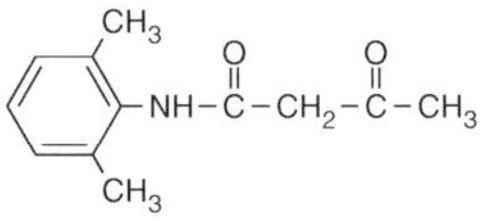
5 gm

S42,075-1
N-ETHYL-ORTHO-ACETOACETOTOLUIDIDE
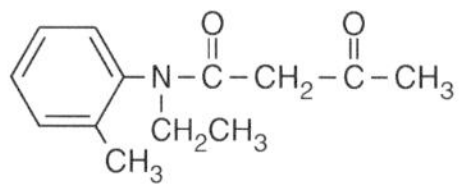
5 gm

S51,136-6
2,5-DIOXO-3-TETRAHYDROFURANDOCOSA-BETA,DELTA-DIENOIC ACID, ALPHA-SUBSTITUTED
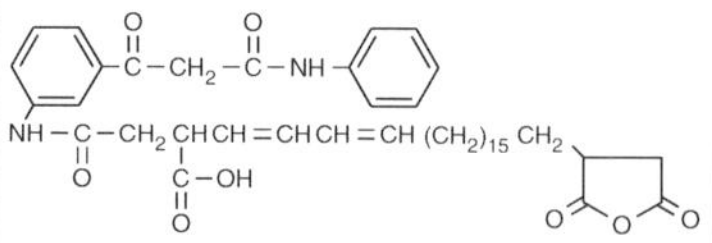
5 gm

S51,355-5
2-[4-(DECANOYLAMINO)BENZOYL]-PROPIONANILIDE
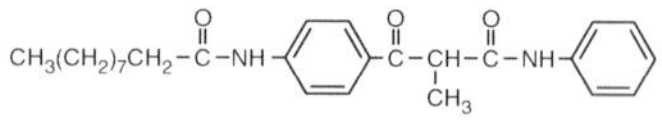
250 mg

S51,692-9
4'-[2-(N-PHENYLCARBAMOYL)PROPIONYL]-HEXADECANANILIDE
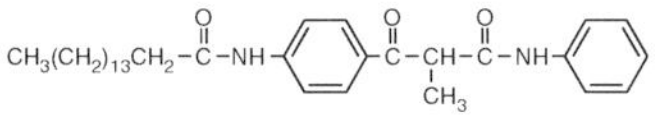
250 mg

S51,348-2
4'-CHLORO-N-METHYLBENZOYLACETANILIDE
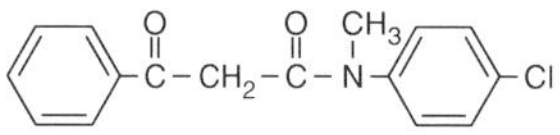
5 gm

S41,692-4
N-PIPERONYLACETOACETAMIDE
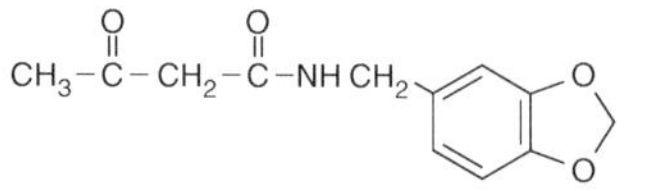
5 gm

S51,476-4
N-ETHYLACETOACETANILIDE
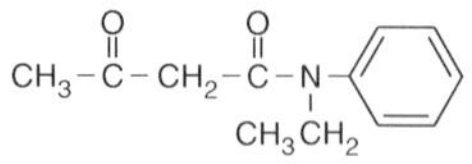
250 mg

S65,262-8
N-METHYL-3-OXOOCTADECANANILIDE

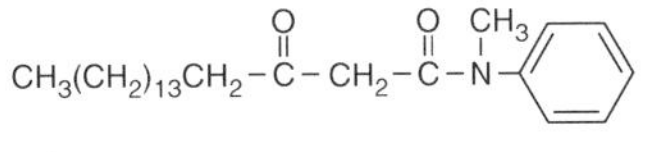

250 mg

S94,997-3
1-PHENYL-1-CYCLOPENTANECARBOXAMIDE

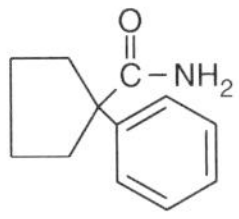

250 mg

S90,986-6
3-(4-BROMOPHENYL)THIAZOLO(3,2-A)-PYRIDINIUM TETRAFLUOROBORATE

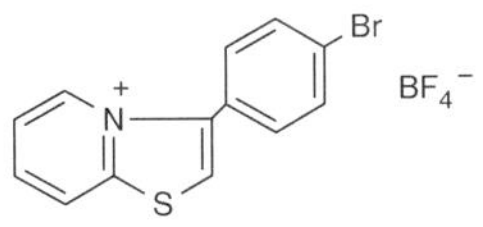

250 mg

S77,502-9
N,N'-ETHYLENEBIS[N-(CHLOROMETHYL)-2,6-DIMETHOXYBENZAMIDE]

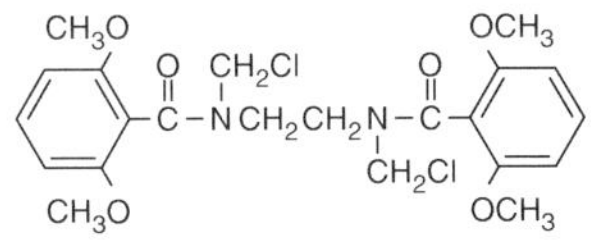

250 mg

S68,859-2
N-[3-(2-CHLOROACETAMIDO)-4-METHOXYPHENYL]BENZAMIDE

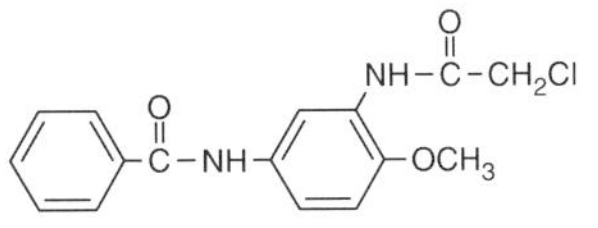

250 mg

S71,809-2
2'-CARBAMOYL-2-CHLOROACETANILIDE

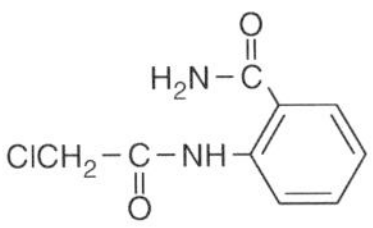

250 mg

S72,290-1
3-(2,2-DICHLOROACETAMIDO)-4,4'-DIMETHYLBENZANILIDE

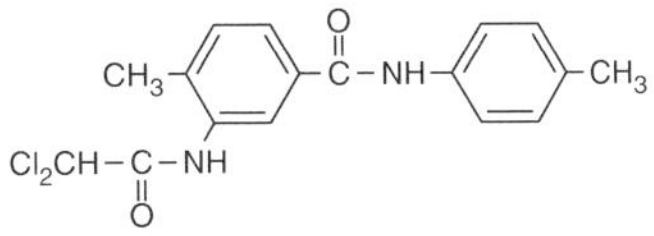

5 gm

S79,586-0
2'-CHLORO-5'-(TRIFLUOROMETHYL)-3,4,5-TRIMETHOXYBENZANILIDE

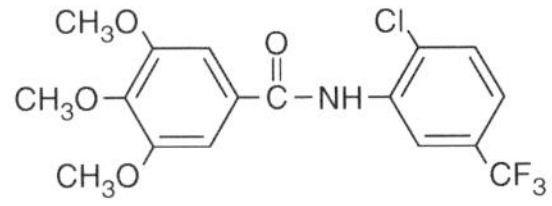

250 mg

S90,491-0
4'-CHLORO-2'-METHOXY-N-METHYL-4-(TRIFLUOROMETHYL)BENZANILIDE

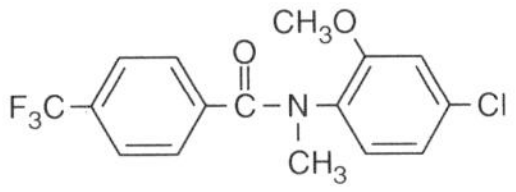

250 mg

S88,931-8
2'-CHLORO-3-NITRO-5'-(TRIFLUOROMETHYL)BENZANILIDE

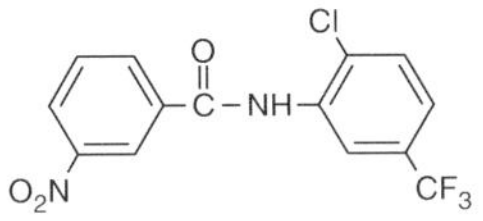

250 mg

S70,398-2
3,4,5,6-TETRACHLORO-3'-(TRIFLUOROMETHYL)PHTHALANILIC ACID

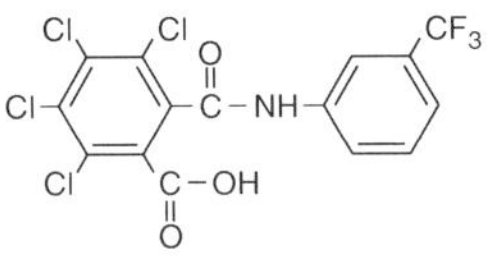

5 gm

S73,858-1
3,4,5,6-TRIBROMO-3'-(TRIFLUOROMETHYL)PHTHALANILIC ACID

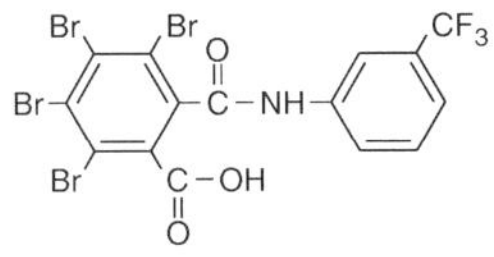

5 gm

S73,898-0
2'-(TRIFLUOROMETHYL)-3,4,5,6-TETRABROMOPHTHALANILIC ACID

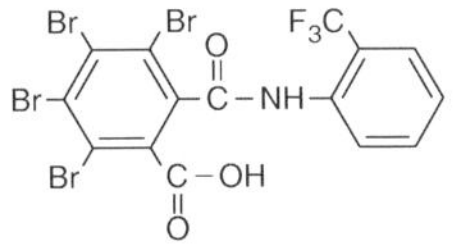

5 gm

S85,568-5
4'-CHLORO-3,4,5,6-TETRABROMO-2'-(TRIFLUOROMETHYL)PHTHALANILIC ACID

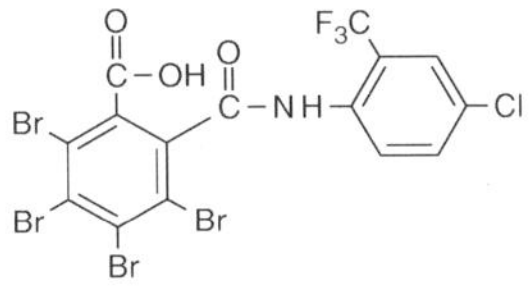

250 mg

S79,498-8
METHYL N-[2-CHLORO-5-(TRIFLUOROMETHYL)PHENYL]PHTHALAMATE

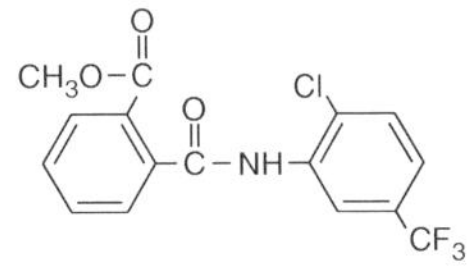

250 mg

S77,369-7
N,N'-BIS[2-CHLORO-5-(TRIFLUOROMETHYL)PHENYL]-ISOPHTHALAMIDE

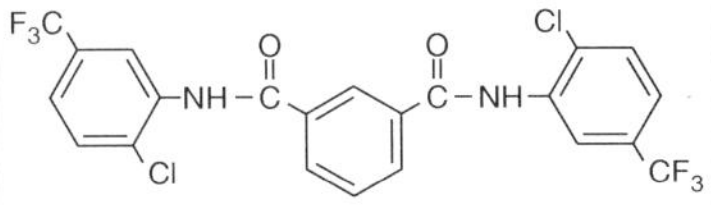

250 mg

S89,015-4
3-CHLORO-3'-(TRIFLUOROMETHYL)-BENZANILIDE

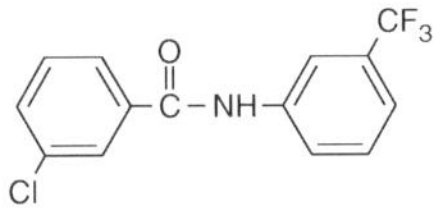

250 mg

S71,779-7
3-BROMO-3'-(TRIFLUOROMETHYL)-BENZANILIDE

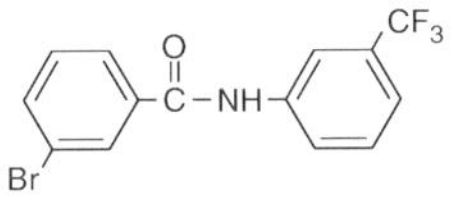

5 gm

S79,176-8
6'-CHLORO-ALPHA,ALPHA,ALPHA-TRIFLUORO-M-BENZOTOLUIDIDE

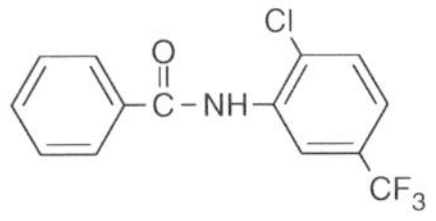

250 mg

S78,234-3
2'-FLUORO-3-(TRIFLUOROMETHYL)-BENZANILIDE

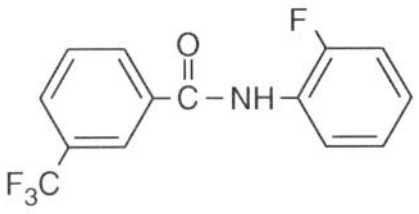

5 gm

S79,177-6
3-CHLORO-ALPHA,ALPHA,ALPHA-TRIFLUORO-O-BENZOTOLUIDIDE

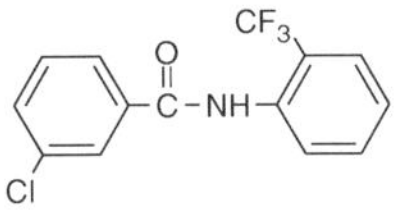

250 mg

S78,235-1
2-FLUORO-4'-(TRIFLUOROMETHYL)-BENZANILIDE

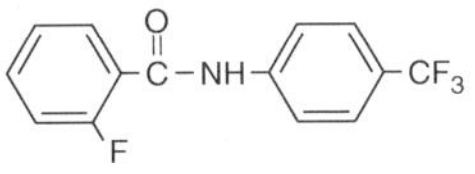

5 gm

S80,148-8
4-FLUORO-2'-(TRIFLUOROMETHYL)-BENZANILIDE

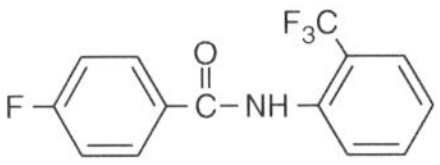

250 mg

S39,722-9
4-CHLORO-ALPHA,ALPHA,ALPHA-TRIFLUORO-ORTHO-BENZOTOLUIDIDE

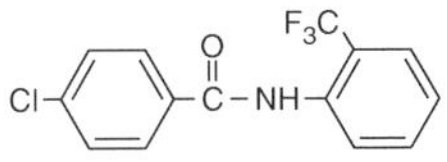

250 mg

S79,174-4
2,6'-DICHLORO-ALPHA,ALPHA,ALPHA-TRIFLUORO-M-BENZOTOLUIDIDE

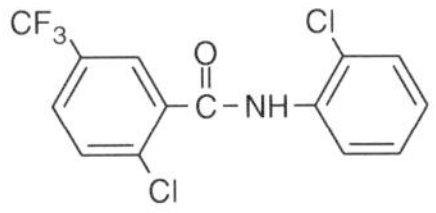

250 mg

S78,237-8
4-FLUORO-4'-(TRIFLUOROMETHYL)-BENZANILIDE

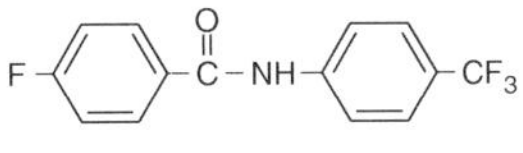

5 gm

S78,238-6
4'-FLUORO-3-(TRIFLUOROMETHYL)-BENZANILIDE

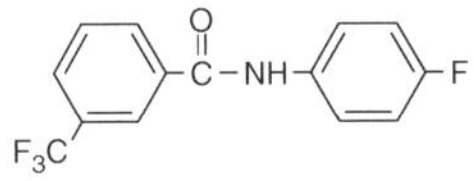

5 gm

S80,149-6
4-FLUORO-3'-(TRIFLUOROMETHYL)-BENZANILIDE

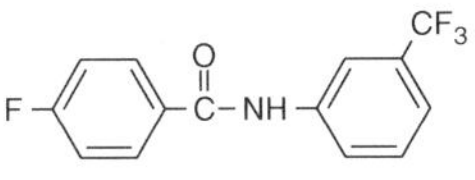

250 mg

S39,752-0
4-CHLORO-ALPHA,ALPHA,ALPHA-TRIFLUORO-META-BENZOTOLUIDIDE

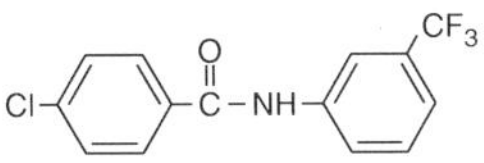

250 mg

S79,173-3
4,6'-DICHLORO-ALPHA,ALPHA,ALPHA-TRIFLUORO-M-BENZOTOLUIDIDE

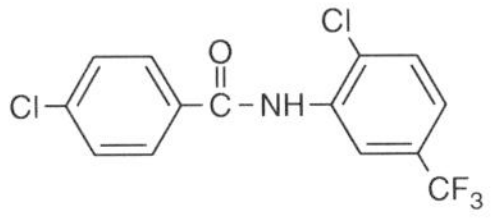

250 mg

S91,014-7
2'-CHLORO-4-METHYL-5'-(TRIFLUOROMETHYL)BENZANILIDE

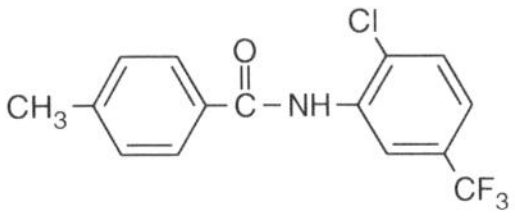

250 mg

S39,740-7
2,4-DICHLORO-ALPHA,ALPHA,ALPHA-TRIFLUORO-META-BENZOTOLUIDIDE

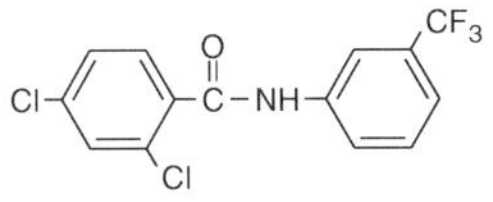

250 mg

S89,016-2
2',3-DICHLORO-5'-(TRIFLUOROMETHYL)-BENZANILIDE
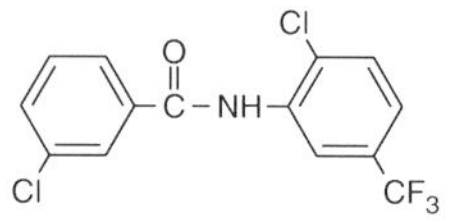
250 mg

S90,555-0
3,4,4'-TRICHLORO-2'-(TRIFLUOROMETHYL)BENZANILIDE
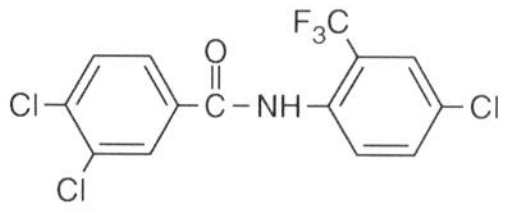
250 mg

S91,002-3
2'-CHLORO-3-METHYL-5'-(TRIFLUOROMETHYL)BENZANILIDE
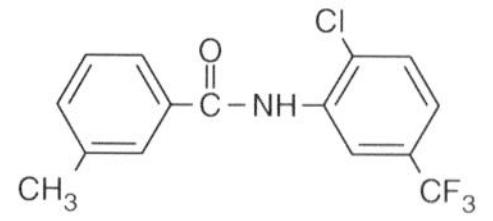
250 mg

S90,457-0
4-CHLORO-N-ETHYL-2'-(TRIFLUOROMETHYL)BENZANILIDE
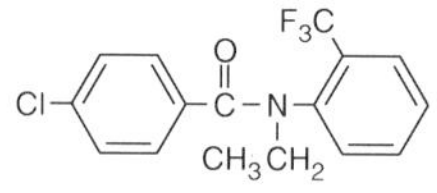
250 mg

S89,269-6
3-[N-(2-CHLORO-5-TRIFLUOROMETHYLPHENYL)-CARBAMOYL]HEXACHLORO-5-NORBORNENE-2-CARBOXYLIC ACID
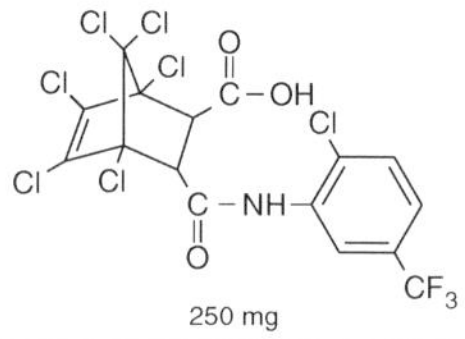
250 mg

S90,098-2
4-METHOXY-3'-(TRIFLUOROMETHYL)-BENZANILIDE
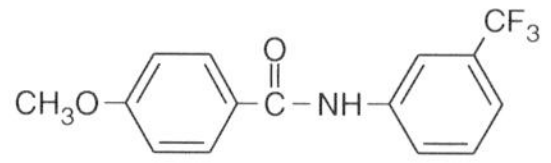
250 mg

S88,932-6
3-NITRO-2'-(TRIFLUOROMETHYL)-BENZANILIDE
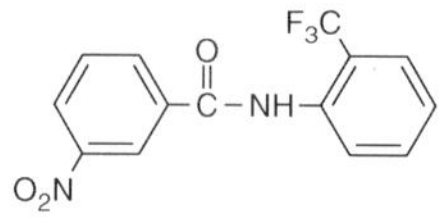
250 mg

S94,568-4
4-(TRIFLUOROMETHOXY)-BENZOHYDROXAMIC ACID
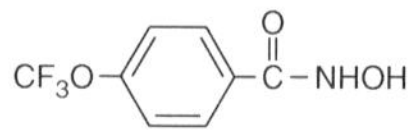
250 mg

S94,569-2
2-(TRIFLUOROMETHYL)-BENZOHYDROXAMIC ACID
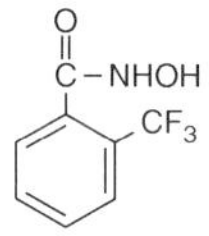
250 mg

S77,333-6
3-NITRO-3'-(TRIFLUOROMETHYL)-BENZANILIDE
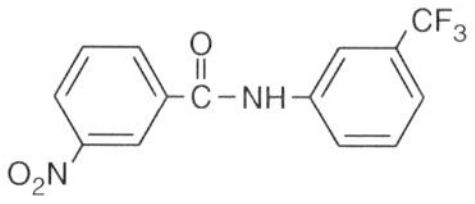
250 mg

S80,791-5
N,N'-(2,2,2-TRICHLOROETHYLIDENE)-BIS(4-NITROBENZAMIDE)
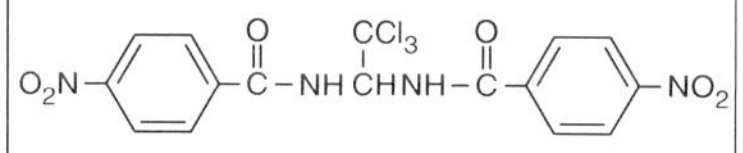
250 mg

S69,289-1
3'-(TRIFLUOROMETHYL)MALEANILIC ACID
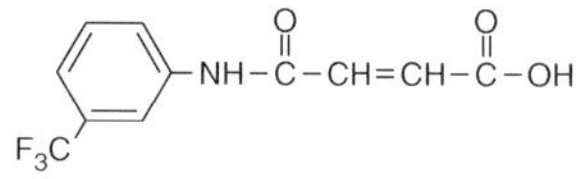
5 gm

S70,129-7
3'-(TRIFLUOROMETHYL)-PHTHALANILIC ACID
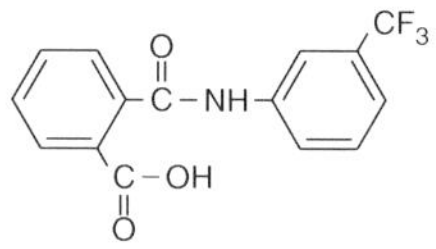
5 gm

S77,243-7
N,N'-BIS[2-(TRIFLUOROMETHYL)PHENYL]-ISOPHTHALAMIDE
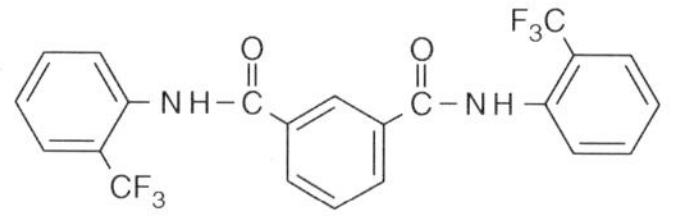
250 mg

S68,246-2
N-[3-(TRIFLUOROMETHYL)BENZOYL]-GLYCINE
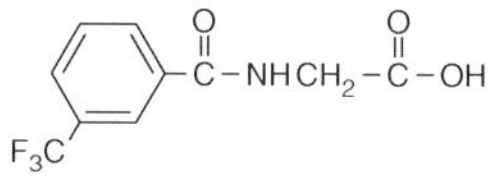
5 gm

S78,223-8
3-(TRIFLUOROMETHYL)BENZAMIDE
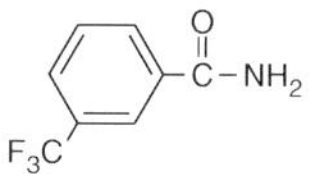
5 gm

S78,224-6
2-(TRIFLUOROMETHYL)BENZAMIDE
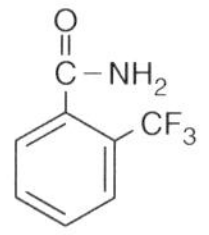
5 gm

S76,631-3
N-METHYL-3-TRIFLUOROMETHYLBENZAMIDE
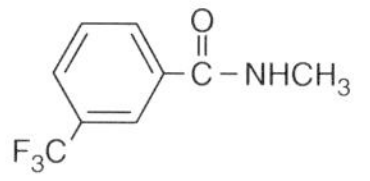
250 mg

S90,092-3
3'-(TRIFLUOROMETHYL)-2-BUTENANILIDE
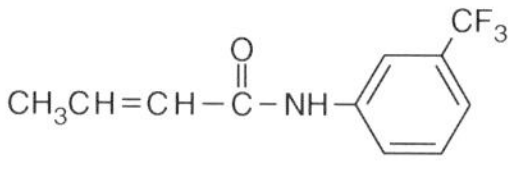
250 mg

S78,231-9
3'-(TRIFLUOROMETHYL)BENZANILIDE
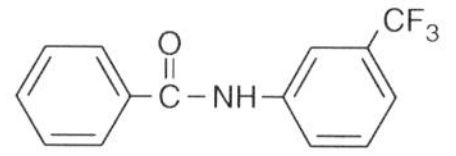
5 gm

S39,734-2
ALPHA,ALPHA,ALPHA-TRIFLUORO-ORTHO-BENZOTOLUIDIDE
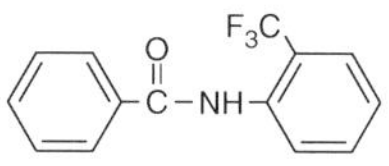
250 mg

S78,232-7
2-(TRIFLUOROMETHYL)BENZANILIDE
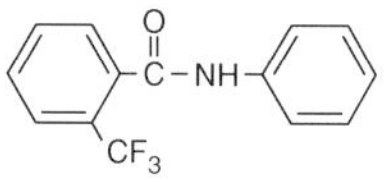
5 gm

S89,254-8
2'-(TRIFLUOROMETHYL)CINNAMANILIDE
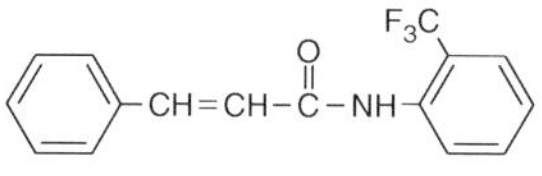
250 mg

S39,715-6
2,4-DICHLORO-ALPHA,ALPHA,ALPHA-TRIFLUORO-ORTHO-BENZOTOLUIDIDE
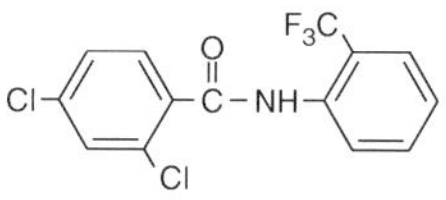
250 mg

S78,228-9
4,4'-BIS(TRIFLUOROMETHYL)-BENZANILIDE
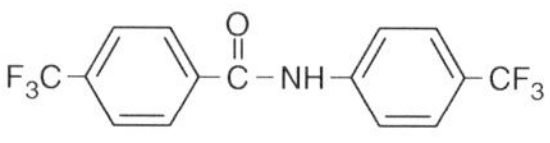
5 gm

S78,229-7
3,4'-BIS(TRIFLUOROMETHYL)-BENZANILIDE
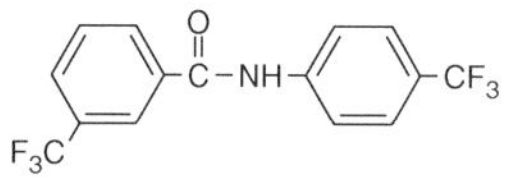
5 gm

S55,526-6
2'-(TRIFLUOROMETHYL)-PHTHALANILIC ACID
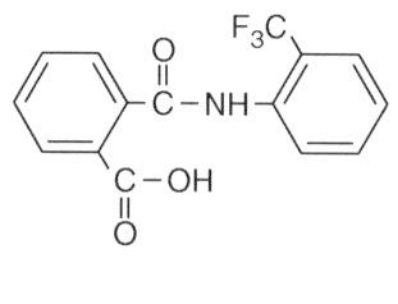
5 gm

S55,523-1
3,4-DICHLORO-A,A,A-TRIFLUORO-O-BENZOTOLUIDIDE
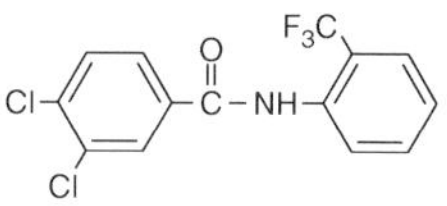
5 gm

S91,274-3
N-[1,1-BIS(HYDROXYMETHYL)ETHYL]-3,4,5,6-TETRACHLOROPHTHALAMIC ACID
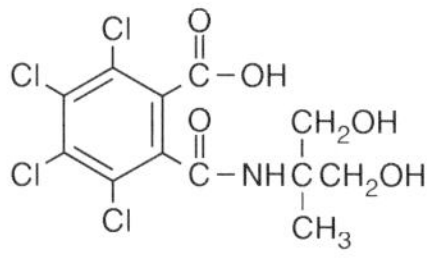
250 mg

S69,652-8
N-(4-CHLOROPHENYL)-2-[(2,2-DIPHENYL-2-HYDROXY)ETHYL]BENZAMIDE
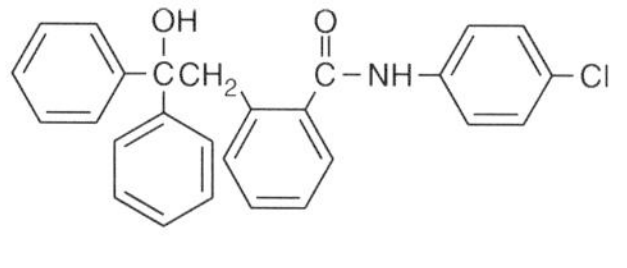
250 mg

S73,492-6
2-[2-(4-CHLOROPHENYL)-2-HYDROXYPROPYL]-N-METHYLBENZAMIDE
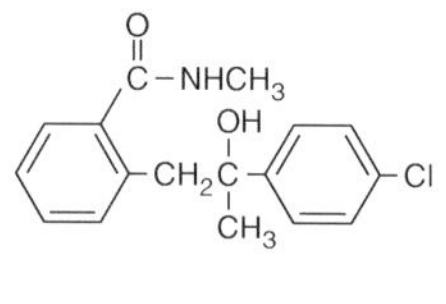
250 mg

S86,114-6
5-BROMO-2-HYDROXYBENZAMIDE
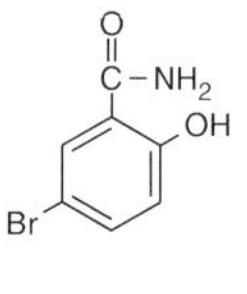
250 mg

S68,223-3
2-AMINO-5-CHLORO-N-(2-HYDROXYETHYL)-BENZAMIDE

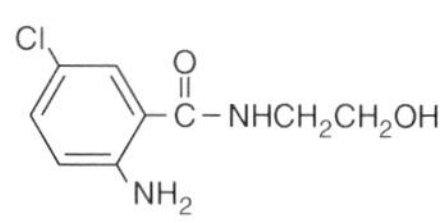

250 mg

S60,583-2
4'-AMINO-4-CHLORO-2'-HYDROXYBENZANILIDE

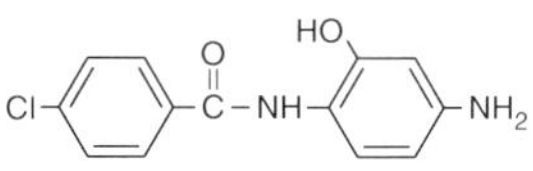

250 mg

S68,176-8
2-AMINO-5-CHLORO-N-(3-HYDROXYPROPYL)BENZAMIDE

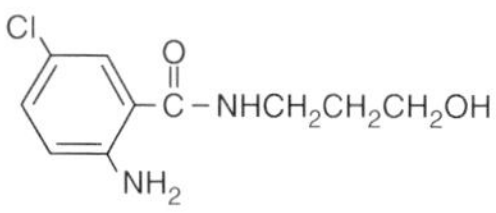

5 gm

S64,227-4
3-AMINO-3',5'-DICHLORO-2'-HYDROXY-4'-METHYLBENZANILIDE

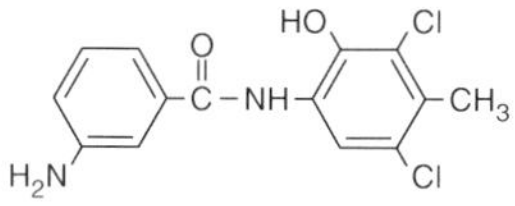

250 mg

S85,502-2
3,4-DICHLORO-2'-HYDROXY-5'-NITROBENZANILIDE

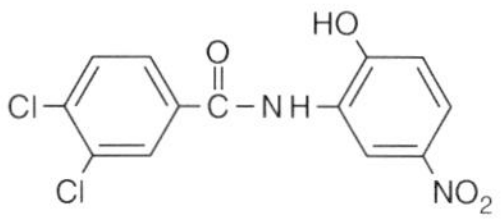

250 mg

S35,817-7
N-BENZOYL-3,5-DIIODO-L-TYROSINE ETHYL ESTER

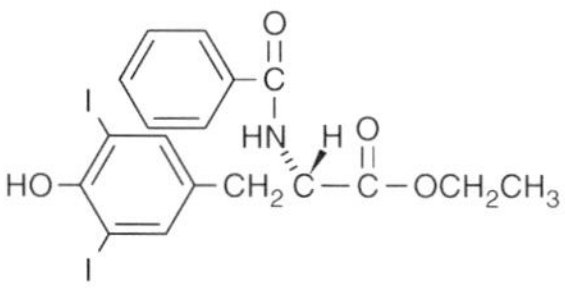

250 mg

S85,484-0
2-CHLORO-2'-HYDROXY-5'-NITROBENZANILIDE

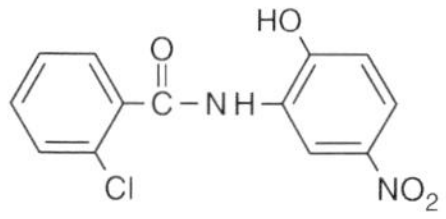

250 mg

S70,260-9
5'-CHLORO-2'-HYDROXY-4-NITROBENZANILIDE

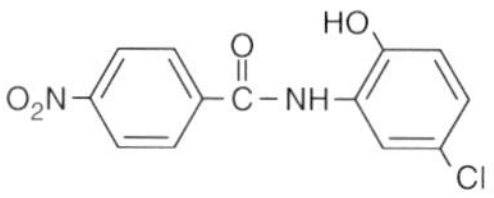

250 mg

S73,891-3
4-CHLORO-2'-HYDROXY-5'-NITROBENZANILIDE

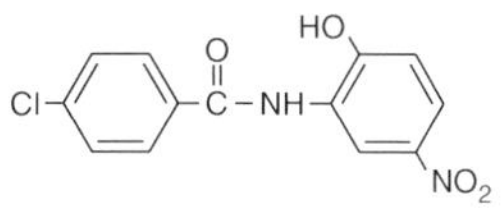

5 gm

S64,196-0
3',5'-DICHLORO-2'-HYDROXY-4'-METHYLACRYLANILIDE

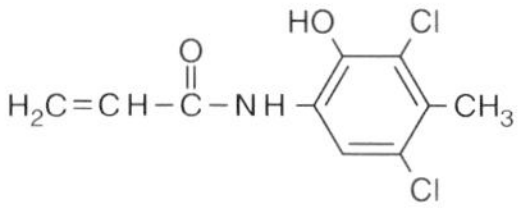

250 mg

S90,850-9
5'-CHLORO-2'-HYDROXYBENZANILIDE

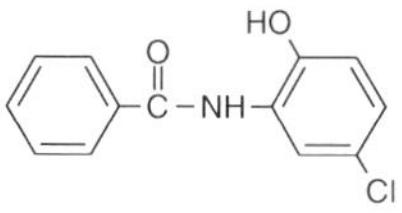

250 mg

S72,472-6
3,4-DICHLORO-2'-HYDROXYBENZANILIDE

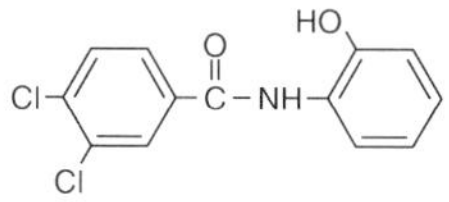

5 gm

S51,776-3
2'-CHLORO-3,5-DI-TERT-BUTYL-4-HYDROXYBENZANILIDE

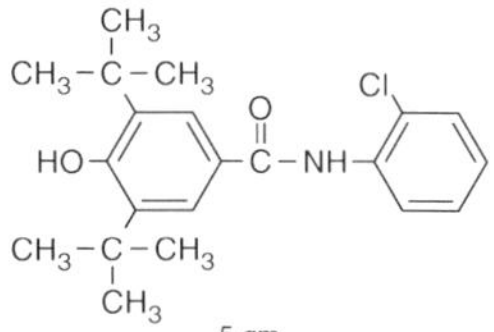

5 gm

S53,754-3
3',5'-DICHLORO-2'-HYDROXY-4'-METHYLMALEANILIC ACID

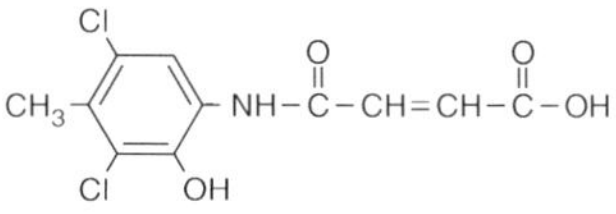

5 gm

S71,636-7
3'-HYDROXY-3,4,5,6-TETRABROMOPHTHALANILIC ACID

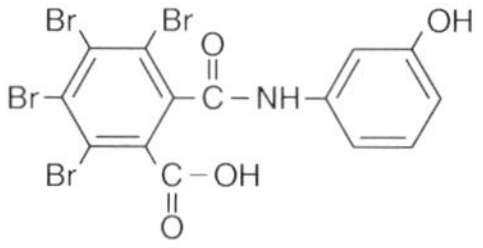

50 mg

S80,146-1
2-FLUORO-4'-HYDROXYBENZANILIDE

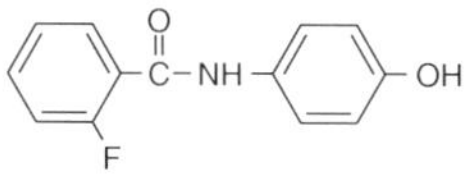

250 mg

S64,932-5
3',4'-DICHLORO-2-HYDROXYBENZANILIDE

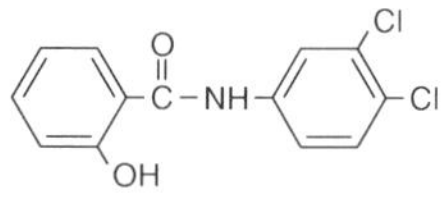

5 gm

S42,089-1
2',4',5'-TRICHLOROSALICYLANILIDE

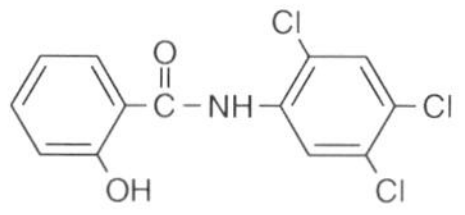

5 gm

S80,136-4
3-FLUORO-4'-(2-HYDROXYETHYL)-BENZANILIDE

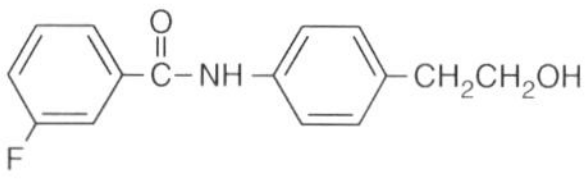

250 mg

S80,157-7
4-FLUORO-4'-(2-HYDROXYETHYL)-BENZANILIDE

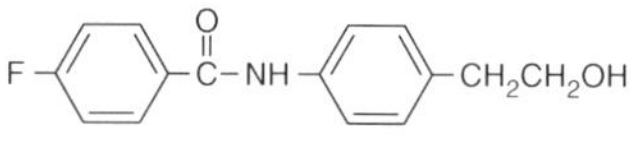

250 mg

S67,844-9
3,4-DICHLORO-3'-HYDROXYBENZANILIDE

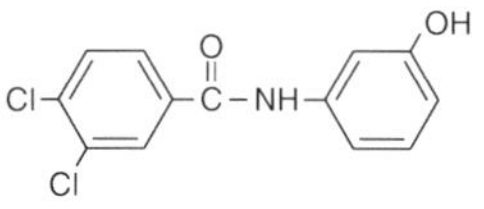

5 gm

S75,137-5
2,4-DICHLORO-3'-HYDROXYBENZANILIDE

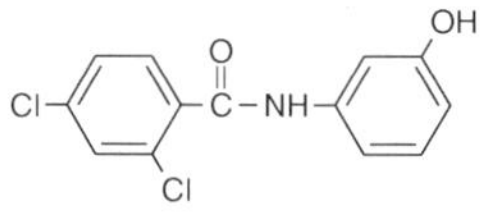

250 mg

S90,553-4
3,4-DICHLORO-N-(5-HYDROXY-1-NAPHTHYL)BENZAMIDE

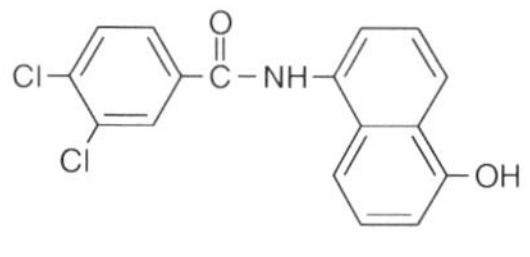

250 mg

S79,585-2
4'-FLUORO-3,4,5-TRIMETHOXYBENZANILIDE

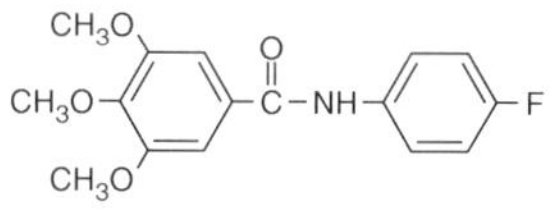

250 mg

S73,583-3
2',4',6'-TRICHLORO-3,4,5-TRIMETHOXYBENZANILIDE

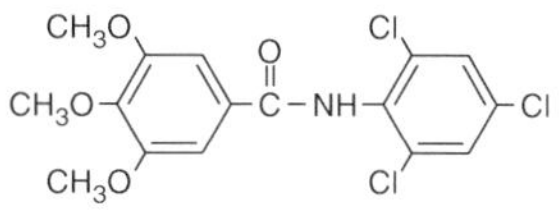

250 mg

S71,532-8
2',6'-DICHLORO-3,4,5-TRIMETHOXYBENZANILIDE

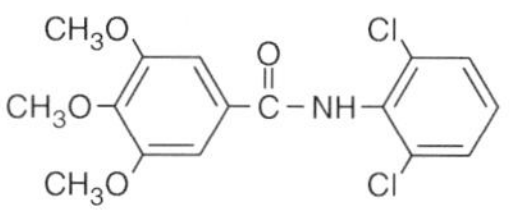

250 mg

S86,133-2
2,5-DIBROMO-N,N'-DI-TERT-BUTYL-3,6-DIMETHOXYTEREPHTHALAMIDE

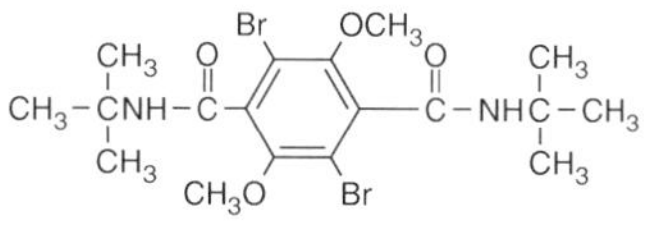

250 mg

S74,352-6
2',4'-DIMETHOXY-3,4,5,6-TETRABROMOPHTHALANILIC ACID

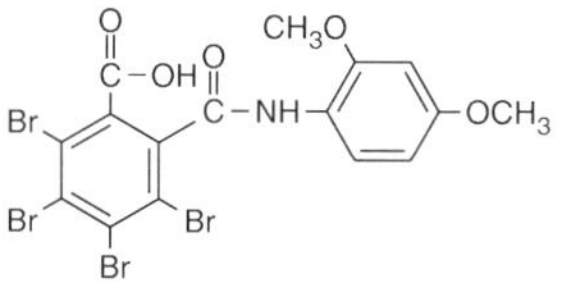

250 mg

S75,149-9
2-CHLORO-2',4'-DIMETHOXYBENZANILIDE

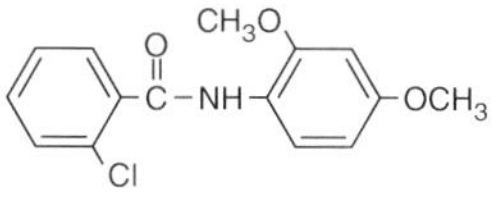

5 gm

S72,326-6
5'-CHLORO-2',4'-DIMETHOXY-4-FLUOROBENZANILIDE

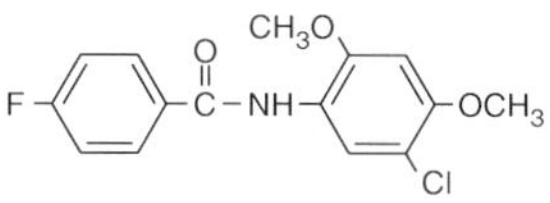

5 gm

S69,494-0
N-(5-CHLORO-2,4-DIMETHOXYPHENYL)-4-TERT-BUTYLBENZAMIDE

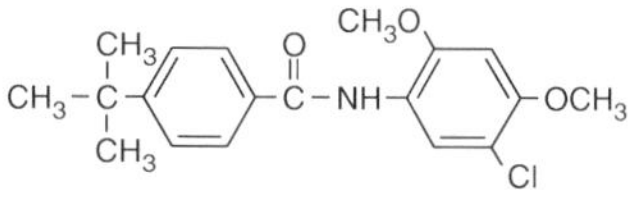

250 mg

S71,594-8
5'-CHLORO-2',4'-DIMETHOXYSORBANILIDE

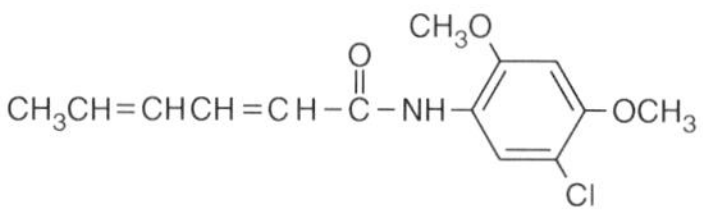

250 mg

S70,330-3
5'-CHLORO-2',4'-DIMETHOXYMALEANILIC ACID

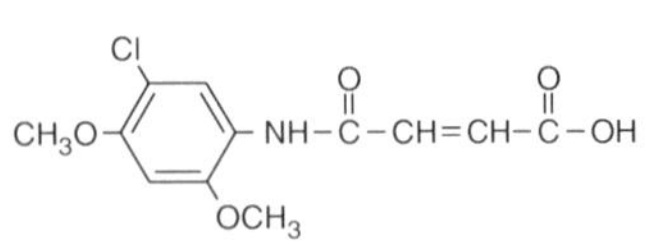

5 gm

S73,591-4
2'-BROMO-2,6-DIMETHOXYBENZANILIDE

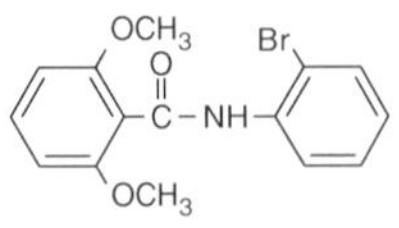

250 mg

S71,134-9
2'-CHLORO-2,6-DIMETHOXY-6'-METHYLBENZANILIDE

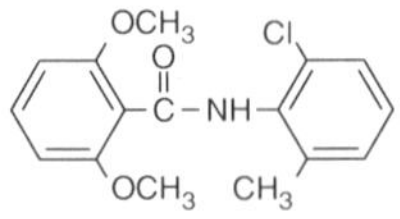

250 mg

S78,039-1
5'-CHLORO-2',4'-DIMETHOXY-2-BUTENANILIDE

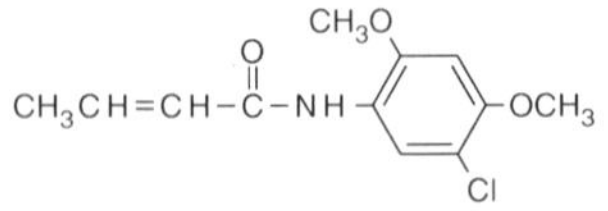

250 mg

S71,601-4
2,4-DICHLORO-2',5'-DIMETHOXYBENZANILIDE

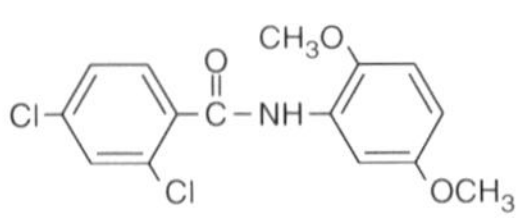

250 mg

S85,499-9
3,4-DICHLORO-2',5'-DIMETHOXYBENZANILIDE

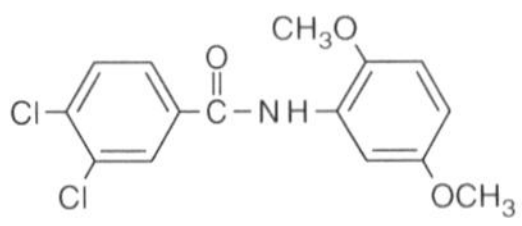

250 mg

S92,459-8
2,4-DICHLORO-2',4'-DIMETHOXYBENZANILIDE

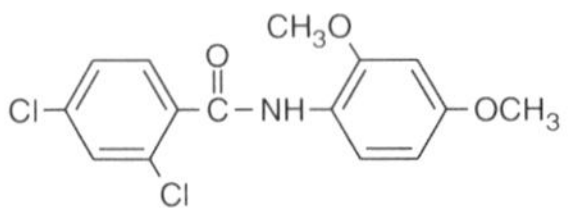

250 mg

S69,898-9
4'-CHLORO-2',5'-DIMETHOXY-2-BUTENANILIDE

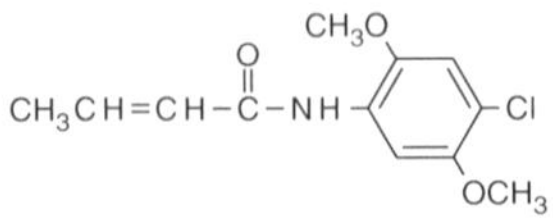

250 mg

S85,468-9
2-CHLORO-2',5'-DIMETHOXYBENZANILIDE

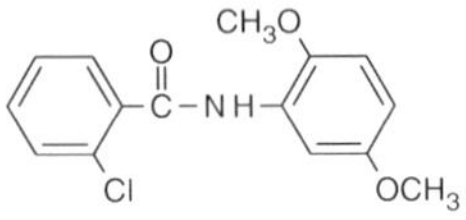

250 mg

S67,613-6
2',5'-DIMETHOXY-2,4,4'-TRICHLOROBENZANILIDE

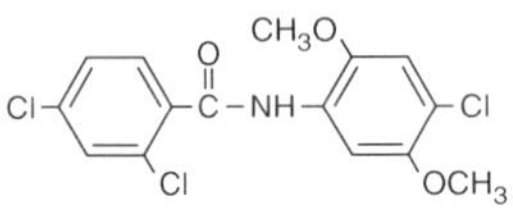

5 gm

S69,475-4
N-(2,5-DIMETHOXYPHENYL)-2-FLUOROBENZAMIDE

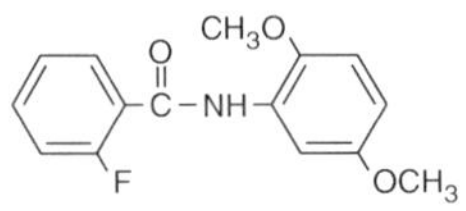

250 mg

S74,711-4
4'-CHLORO-2',5'-DIMETHOXY-3,4,5,6-TETRABROMOPHTHALANILIC ACID

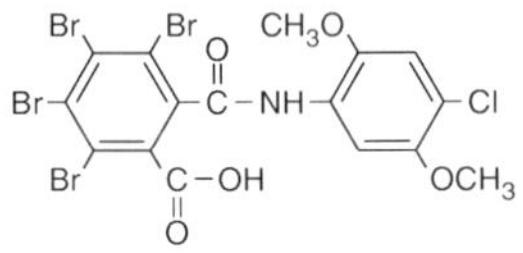

5 gm

S89,248-3
5'-CHLORO-2',4'-DIMETHOXYCINNAMANILIDE

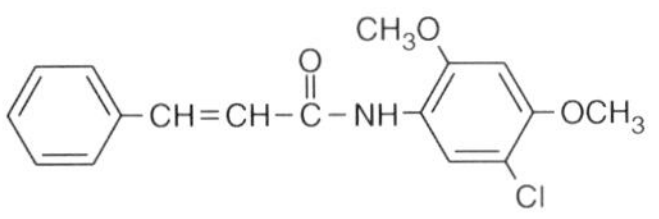

250 mg

S80,159-3
2',5'-DIMETHOXY-4-FLUOROBENZANILIDE

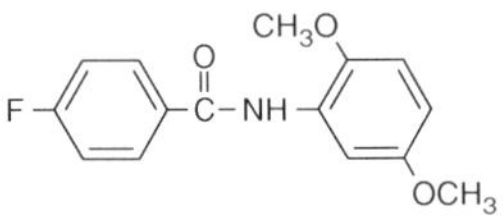

250 mg

S79,580-1
5'-CHLORO-2',4-DIMETHOXYBENZANILIDE

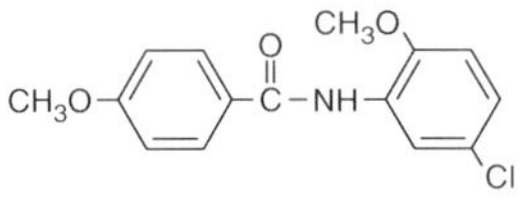

250 mg

S80,135-6
2',5'-DIMETHOXY-3-FLUOROBENZANILIDE

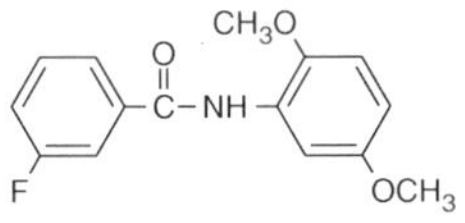

250 mg

S68,611-5
N-(4-CHLORO-3-NITROPHENYL)-3,4,5-TRIMETHOXYBENZAMIDE

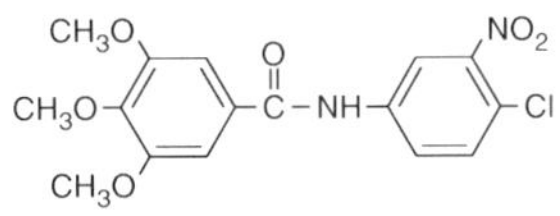

250 mg

S91,363-4
5'-CHLORO-2',4'-DIMETHOXYPHTHALANILIC ACID

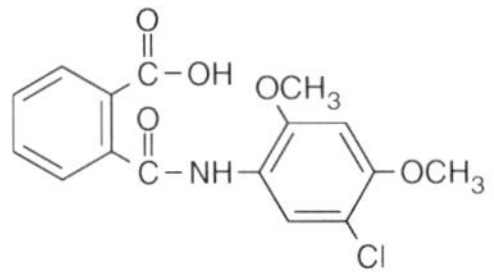

250 mg

S73,564-7
2',6'-DICHLORO-2,6-DIMETHOXYBENZANILIDE

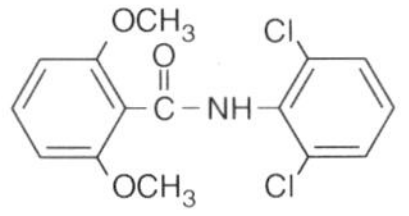

250 mg

S69,252-2
N-(4-IODOPHENYL)-2-(3,4,5-TRIMETHOXYBENZYLIDENE)ACETAMIDE

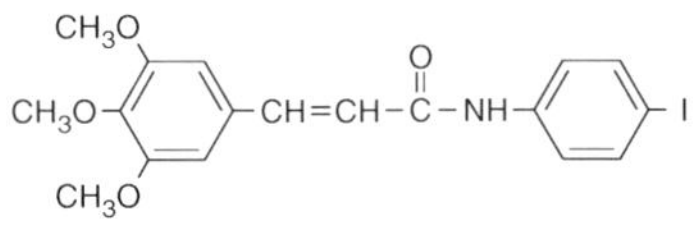

250 mg

S71,680-4
2-METHOXY-2',4',6'-TRIBROMOBENZANILIDE

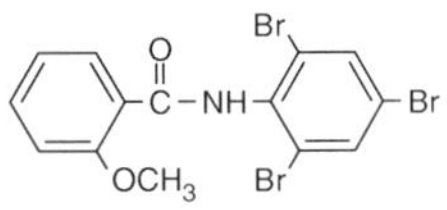

250 mg

S70,971-9
2',6'-DICHLORO-3-METHOXYBENZANILIDE

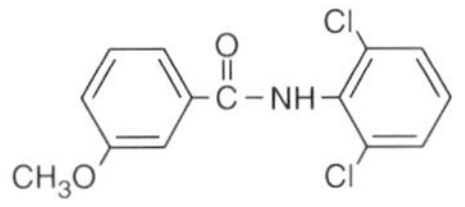

5 gm

S70,810-0
2'-CHLORO-3-METHOXY-6-METHYLBENZANILIDE

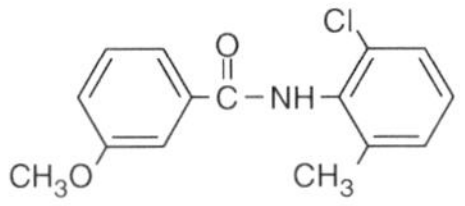

250 mg

S85,500-6
4-CHLORO-2'-METHOXY-5'-NITROBENZANILIDE

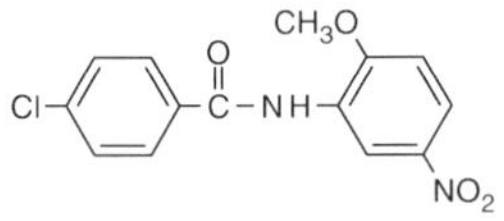

250 mg

S71,886-6
3,4-DICHLORO-2'-METHOXY-5'-NITROBENZANILIDE

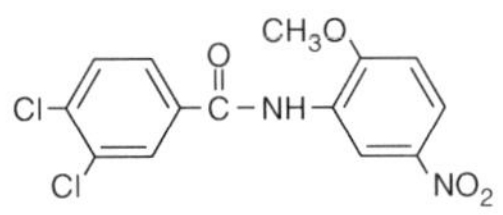

5 gm

S72,300-2
4'-ACETAMIDO-4-FLUORO-3'-METHOXYBENZANILIDE

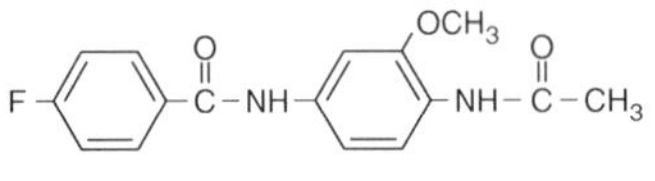

250 mg

S73,857-3
2'-METHOXY-3'-NITRO-3,4,5,6-TETRABROMOPHTHALANILIC ACID

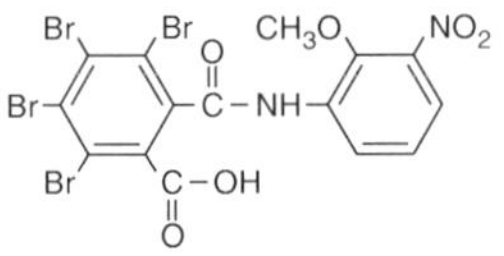

5 gm

S75,114-6
2,4-DICHLORO-3'-NITRO-O-BENZANISIDIDE

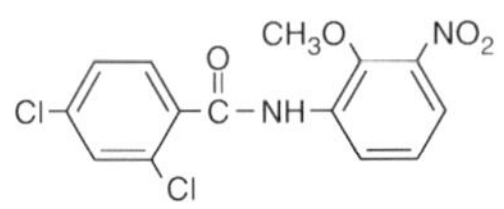

5 gm

S67,606-3
3-CHLORO-5'-NITRO-O-BENZANISIDIDE

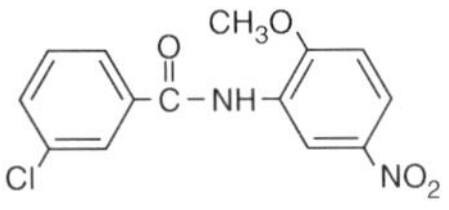

5 gm

S89,283-1
5'-CHLORO-2'-METHOXY-3-NITROBENZANILIDE

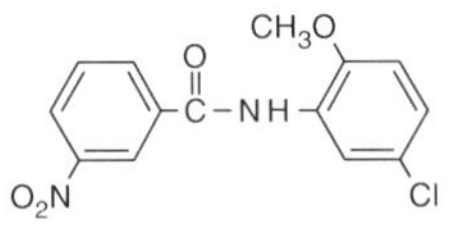

250 mg

S72,909-4
2'-ETHOXY-3,4,5,6-TETRABROMOPHTHALANILIC ACID

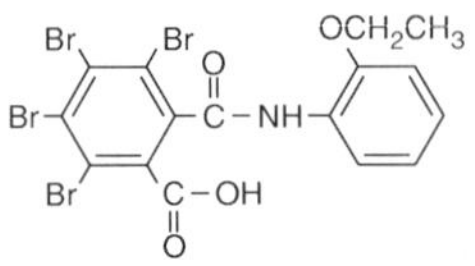

5 gm

S74,752-1
4'-CHLORO-2'-METHOXYPHTHALANILIC ACID

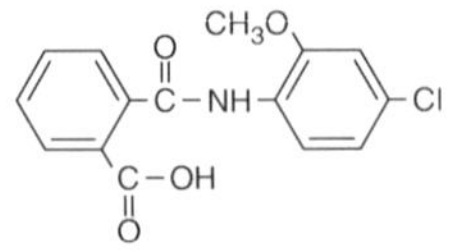

5 gm

S69,295-6
3'-CHLORO-6'-METHOXYMALEANILIC ACID

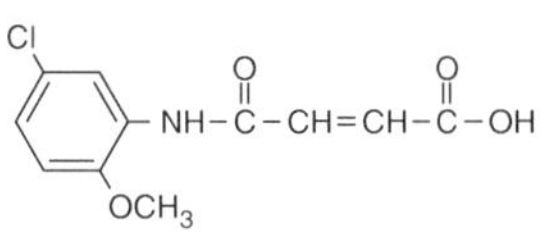

5 gm

S71,640-5
2'-METHOXY-3,4,4',5,6-PENTACHLOROPHTHALANILIC ACID

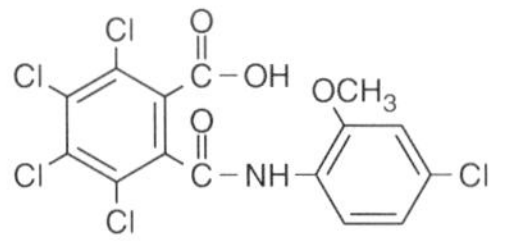

5 gm

S91,365-0
2'-METHOXY-3,4,5,5',6-PENTACHLOROPHTHALANILIC ACID

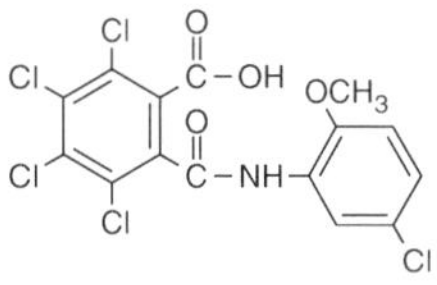

250 mg

S91,609-9
4'-CHLORO-2'-METHOXY-3,4,5,6-TETRABROMOPHTHALANILIC ACID

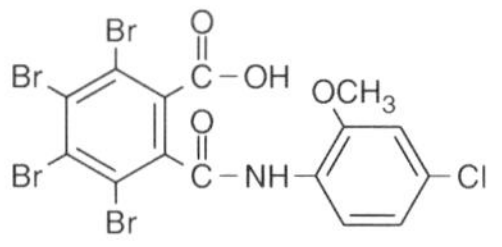

250 mg

S80,160-7
4-[N-(5-CHLORO-2-METHOXYPHENYL)-CARBAMOYL]PHTHALIC ANHYDRIDE

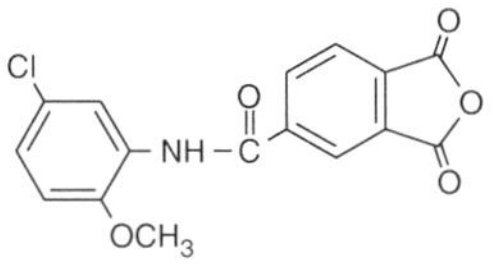

250 mg

S74,789-0
N,N'-BIS(5-CHLORO-2-METHOXYPHENYL)-TEREPHTHALAMIDE

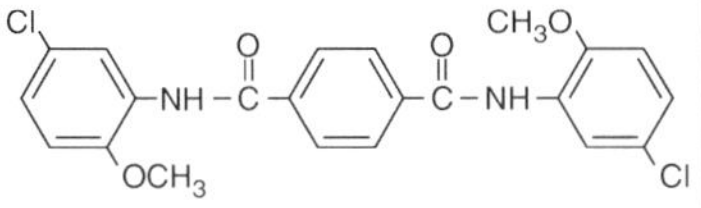

250 mg

S69,599-8
4-TERT-BUTYL-N-(4-CHLORO-2-PHENOXYPHENYL)BENZAMIDE

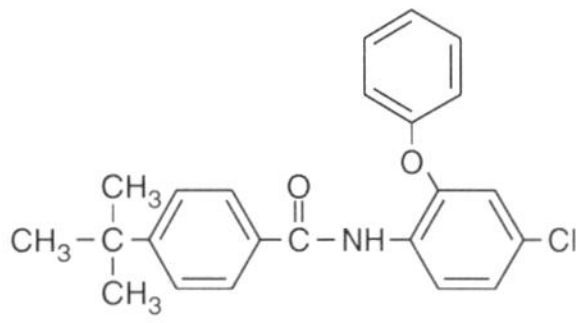

250 mg

S90,069-9
2-FLUORO-2'-METHOXYBENZANILIDE

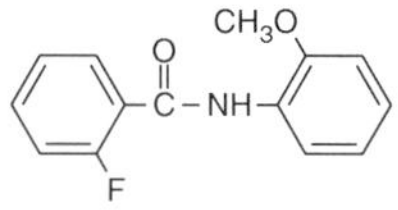

250 mg

S71,600-6
3-CHLORO-2'-METHOXYBENZANILIDE

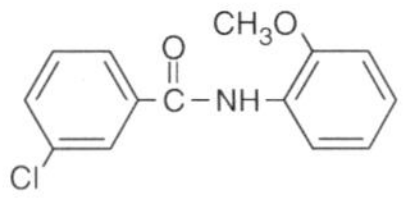

250 mg

S71,524-7
3-BROMO-2'-METHOXYBENZANILIDE

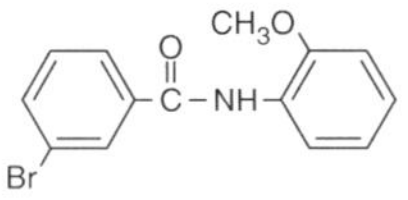

250 mg

S80,138-0
2'-ETHOXY-3-FLUOROBENZANILIDE

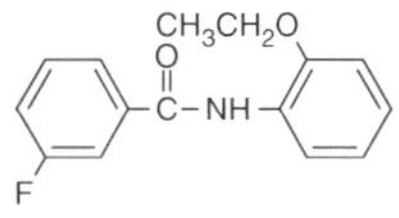

250 mg

S79,132-6
2-CHLORO-2'-ETHOXYBENZANILIDE

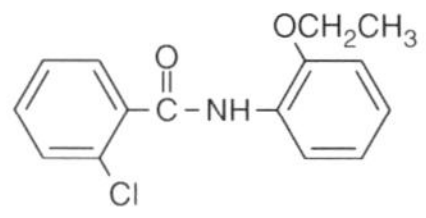

250 mg

S75,880-9
2,6-DICHLOROBENZO-O-ANISIDIDE

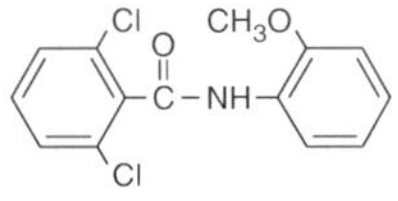

250 mg

S77,957-1
2,5'-DICHLORO-2'-METHOXYBENZANILIDE

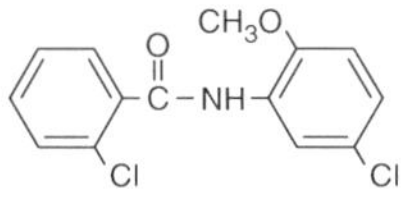

250 mg

S85,469-7
2-CHLORO-2'-METHOXY-5'-METHYLBENZANILIDE

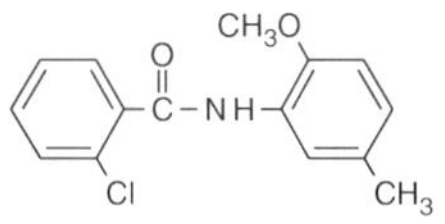

250 mg

S80,152-6
5'-CHLORO-4-FLUORO-2'-METHOXYBENZANILIDE

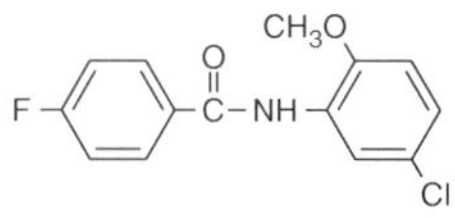

250 mg

S80,158-5
4-FLUORO-2'-METHOXY-5'-METHYLBENZANILIDE

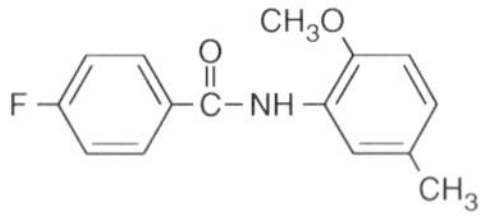

250 mg

S69,473-8
4-TERT-BUTYL-N-(5-CHLORO-2-METHOXYPHENYL)BENZAMIDE

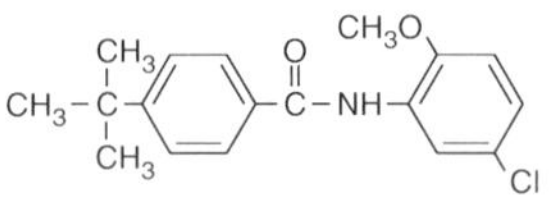

250 mg

S78,076-6
2'-METHOXY-2,4,5'-TRICHLOROBENZANILIDE

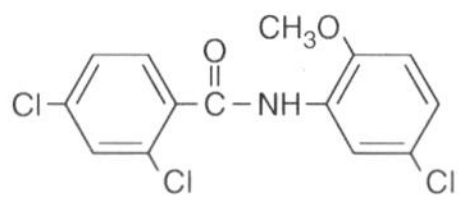

250 mg

S80,137-2
3-FLUORO-2'-METHOXY-5'-METHYLBENZANILIDE

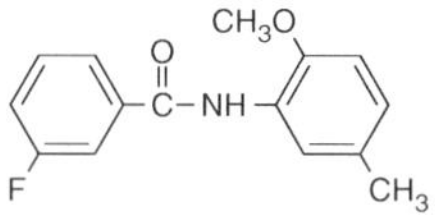

250 mg

S79,664-6
4-CHLORO-N,5'-DIMETHYL-2'-METHOXYBENZANILIDE

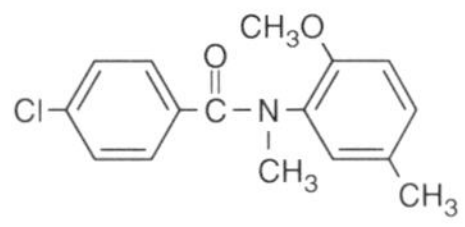

250 mg

S90,468-6
3,4-DICHLORO-N,5'-DIMETHYL-2'-METHOXYBENZANILIDE

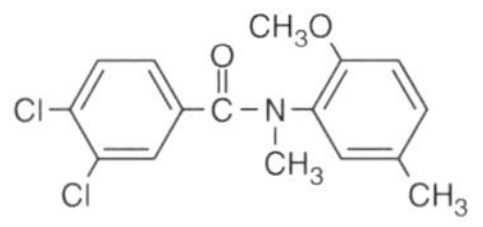

250 mg

S72,121-2
2'-METHOXY-5'-METHYL-3,4,5,6-TETRACHLOROPHTHALANILIC ACID

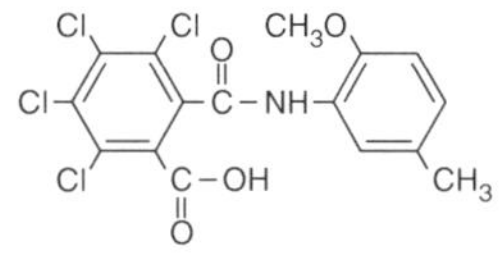

250 mg

S79,522-4
5'-CHLORO-2'-METHOXYCINNAMANILIDE

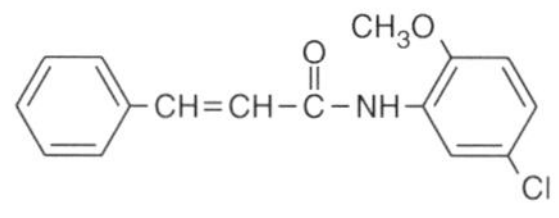

250 mg

S85,570-7
2'-METHOXY-4'-NITRO-3,4,5,6-TETRABROMOPHTHALANILIC ACID

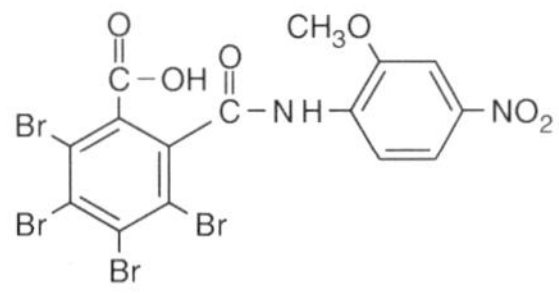

250 mg

S75,115-4
2'-CHLORO-4-NITRO-P-BENZANISIDIDE

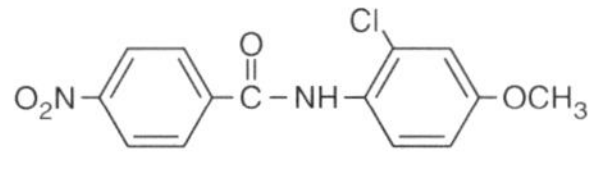

5 gm

S72,132-8
N-(3-METHOXYPHENYL)-TETRACHLOROPHTHALAMIC ACID

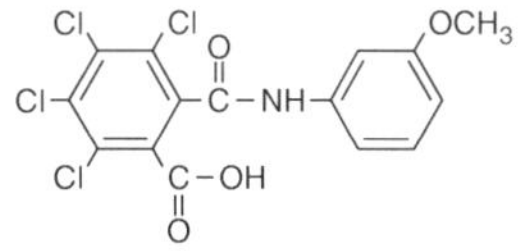

5 gm

S70,355-9
4'-ETHOXY-3,4,5,6-TETRACHLOROPHTHALANILIC ACID

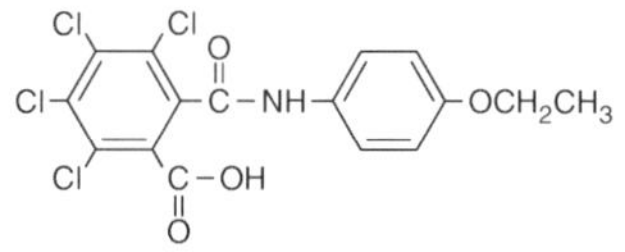

5 gm

S70,294-3
N-(4-METHOXYPHENYL)-TETRACHLOROPHTHALAMIC ACID

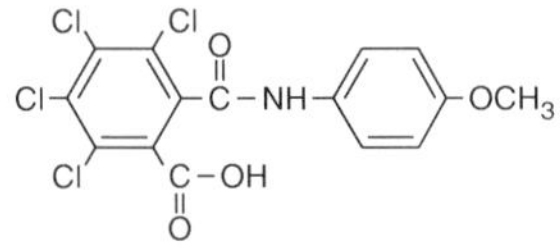

5 gm

S89,067-7
2-CHLORO-4'-ETHOXYBENZANILIDE

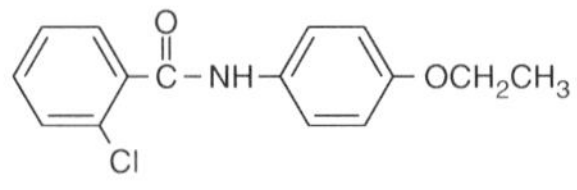

250 mg

S90,102-4
2'-FLUORO-3,4,5-TRIMETHOXYBENZANILIDE

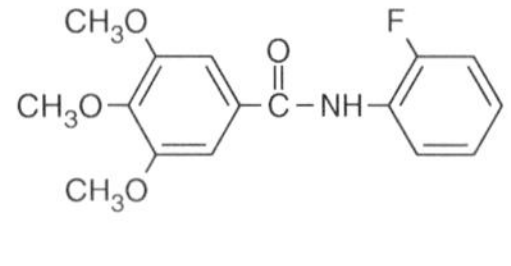

250 mg

S75,233-9
4'-FLUORO-4-METHOXYBENZANILIDE

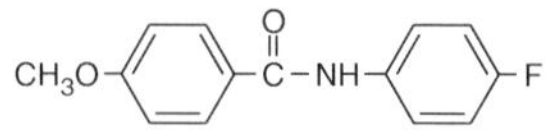

5 gm

S90,096-6
3'-CHLORO-4-METHOXYBENZANILIDE
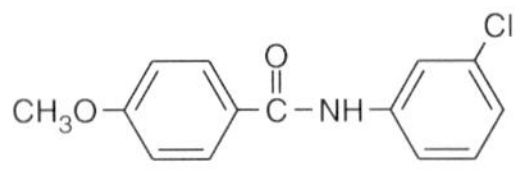
250 mg

S90,141-5
3-CHLORO-4'-ETHOXYBENZANILIDE
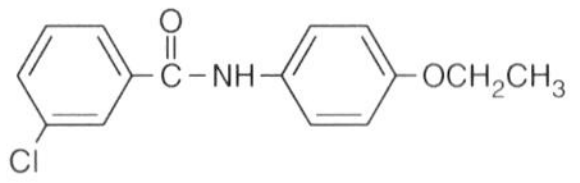
250 mg

S70,826-7
2',6'-DICHLORO-4-METHOXYBENZANILIDE
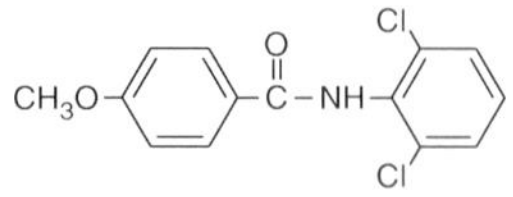
5 gm

S72,698-2
3',4'-DICHLORO-4-METHOXYBENZANILIDE
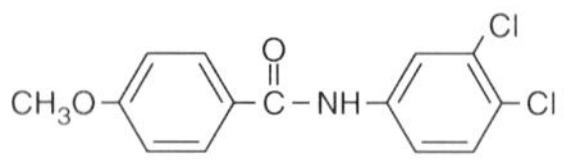
5 gm

S89,221-1
3'-CHLORO-4-METHOXY-4'-METHYLBENZANILIDE
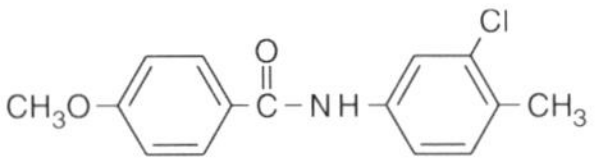
250 mg

S71,235-3
4-METHOXY-2',4',6'-TRICHLOROBENZANILIDE
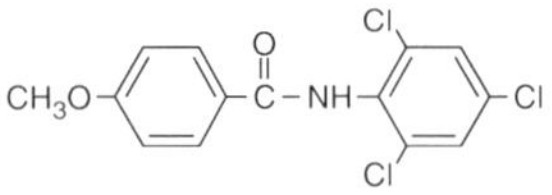
250 mg

S70,896-8
4-METHOXY-2',4',6'-TRIBROMOBENZANILIDE
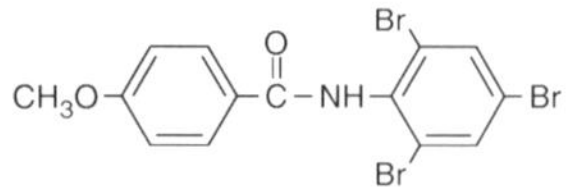
250 mg

S89,937-2
4'-METHOXY-2,2',4-TRICHLOROBENZANILIDE
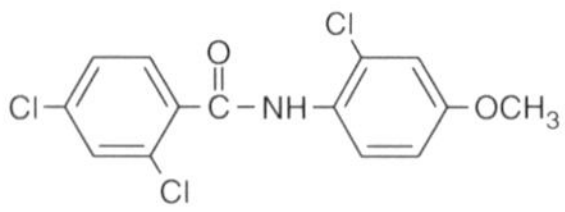
250 mg

S67,578-4
3,4-DICHLORO-5'-METHYL-M-BENZANISIDIDE
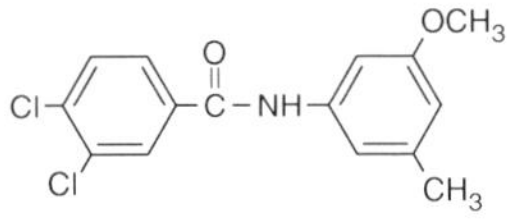
5 gm

S68,593-3
N-(2-FLUORO-5-NITROPHENYL)-3,4,5-TRIMETHOXYBENZAMIDE
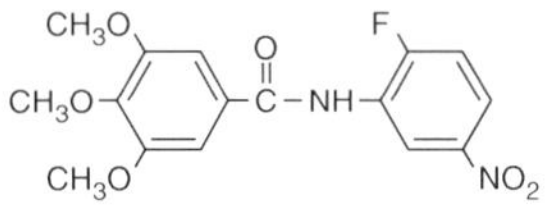
250 mg

S92,179-3
2',4'-DIFLUOROMALEANILIC ACID
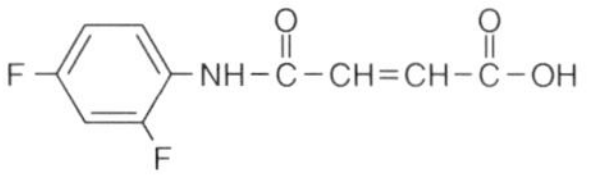
250 mg

S87,800-6
2',5'-DICHLOROMALEANILIC ACID
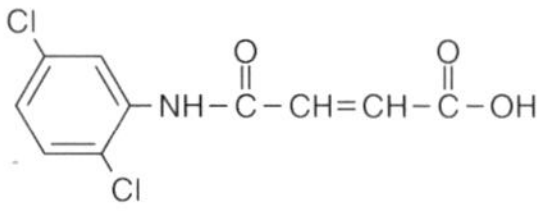
5 gm

S72,441-6
3,3',4,4',5,6-HEXACHLOROPHTHALANILIC ACID
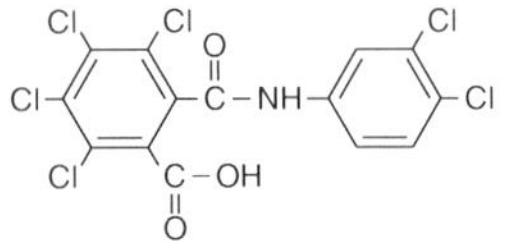
5 gm

S72,643-5
2',6'-DIMETHYL-3,4,4',5,6-PENTABROMOPHTHALANILIC ACID
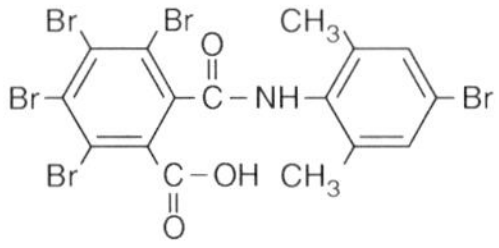
5 gm

S92,140-8
N-(BENZYL)-3,4,5,6-TETRACHLOROPHTHALAMIC ACID
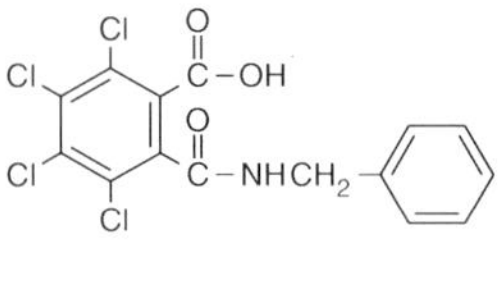
5 gm

S89,273-4
4'-(METHYLTHIO)-3,4,5,6-TETRACHLOROPHTHALANILIC ACID
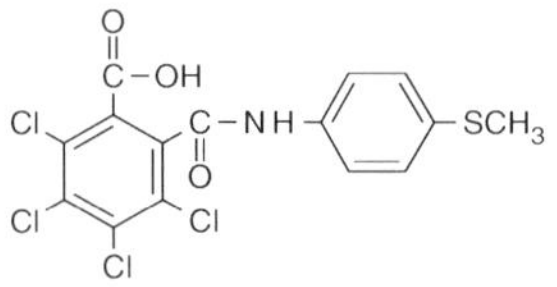
250 mg

S89,239-4
4-CHLORO-4'-(METHYLTHIO)BENZANILIDE
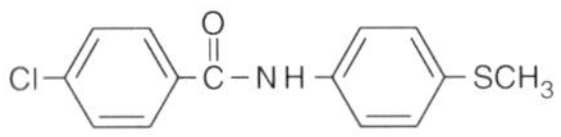
250 mg

S72,141-7
2,4-DICHLORO-4'-(METHYLTHIO)-BENZANILIDE
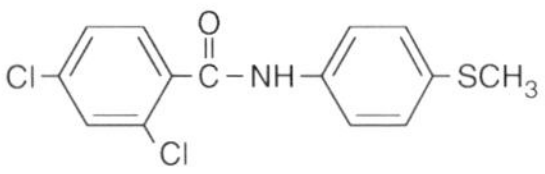
250 mg

S90,813-4
3,5-DICHLOROANTHRANILAMIDE
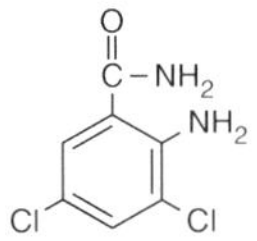
250 mg

S43,503-1
4-AMINO-N-CYCLOPROPYL-3,5-DICHLOROBENZAMIDE
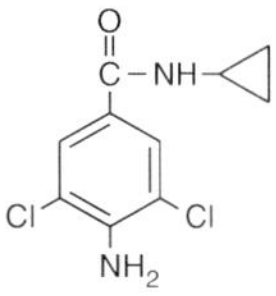
5 gm

S59,791-0
4-CHLORO-N1-HEXADECYL-N1-METHYL-3-AMINOBENZAMIDE HYDROCHLORIDE
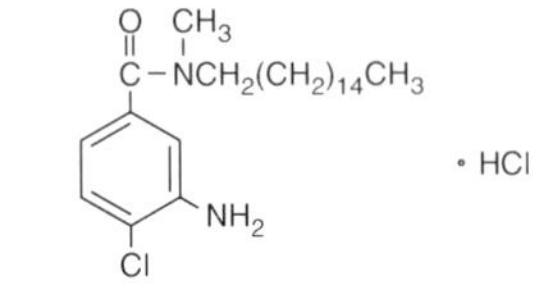
250 mg

S76,100-1
N-[1,2-BIS(2-FLUOROPHENYL)-2-(2-FLUOROBENZYLIDENEAMINO)ETHYL]-2-FLUOROBENZAMIDE
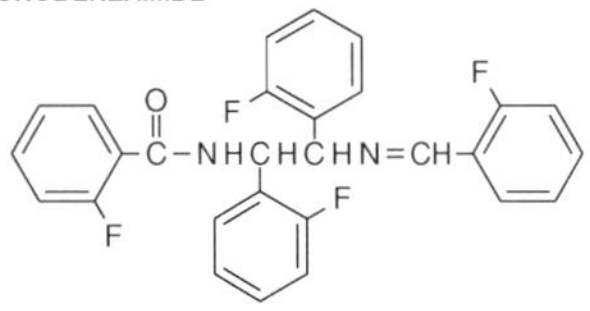
250 mg

S71,642-1
4'-METHYL-2'-NITRO-3,4,5,6-TETRACHLOROPHTHALANILIC ACID
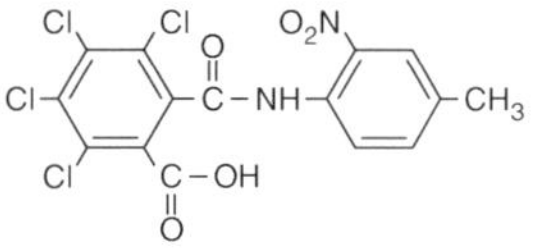
5 gm

S59,439-3
2-CHLORO-2'-NITROBENZANILIDE
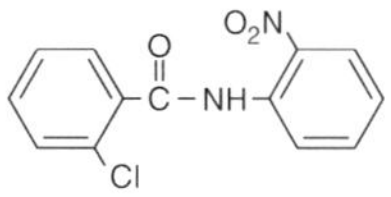
250 mg

S73,594-9
4-CHLORO-2'-METHYL-6'-NITROBENZANILIDE
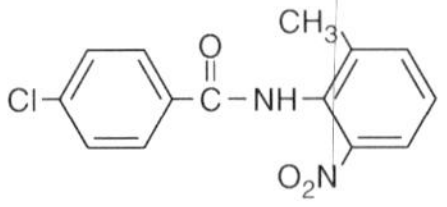
250 mg

S85,050-0
4-CHLORO-2',6'-DINITRO-4'-METHYLBENZANILIDE
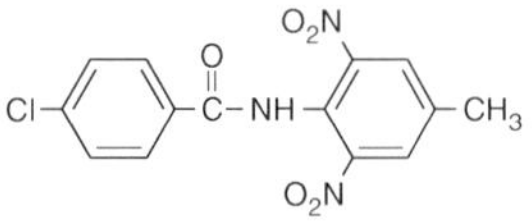
250 mg

S73,598-1
3-CHLORO-2'-METHYL-6'-NITROBENZANILIDE
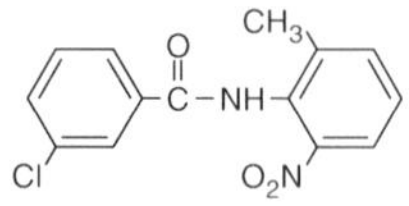
250 mg

S73,574-4
3-BROMO-2'-METHYL-6'-NITROBENZANILIDE
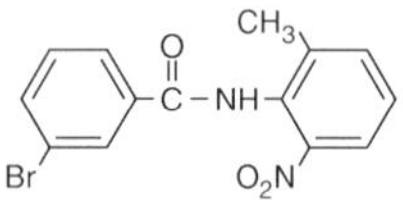
250 mg

S73,566-3
3,4-DICHLORO-2'-METHYL-6'-NITROBENZANILIDE
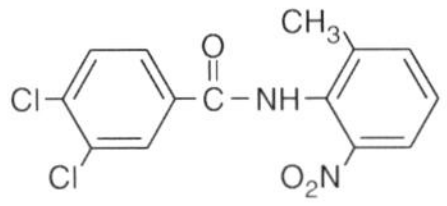
250 mg

S77,875-3
2,5-DICHLORO-2'-METHYL-6'-NITROBENZANILIDE
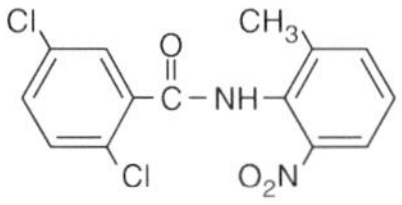
250 mg

S68,119-9
5-CHLORO-2-(METHYLAMINO)BENZAMIDE
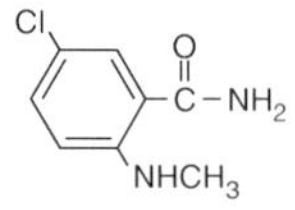
5 gm

S70,428-8
3'-NITRO-3,4,5,6-TETRACHLOROPHTHALANILIC ACID
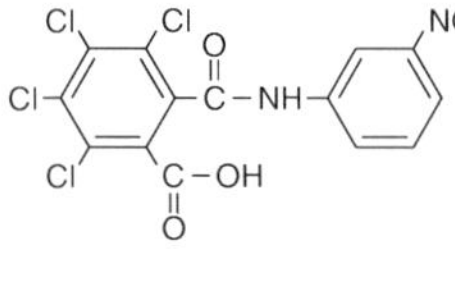
5 gm

S90,557-7
3'-NITRO-3,4,5,6-TETRABROMOPHTHALANILIC ACID

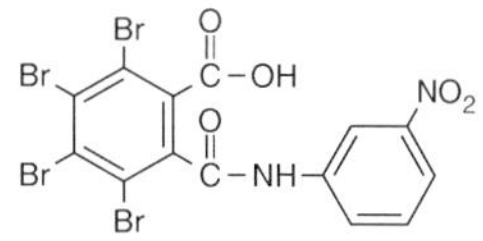

250 mg

S73,990-1
4'-CHLORO-3'-NITROPHTHALANILIC ACID

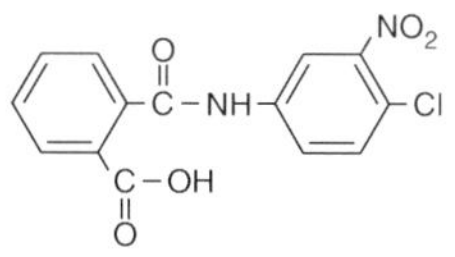

5 gm

S91,371-5
2'-METHYL-3'-NITRO-3,4,5,6-TETRACHLOROPHTHALANILIC ACID

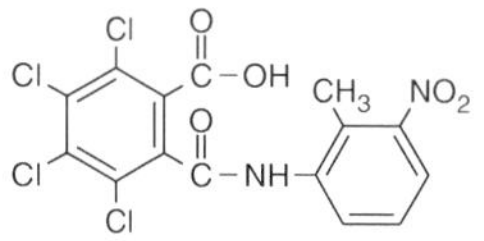

250 mg

S91,611-0
3'-NITRO-3,4,4',5,6-PENTACHLOROPHTHALANILIC ACID

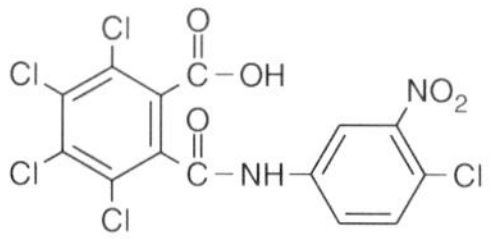

250 mg

S77,954-7
4'-FLUORO-3'-NITRO-3,4,5,6-TETRABROMOPHTHALANILIC ACID

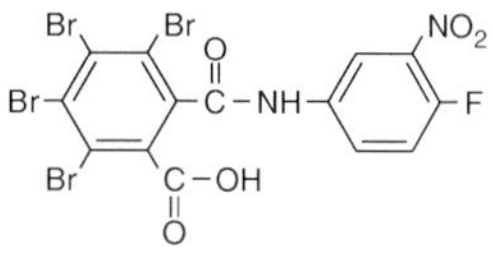

250 mg

S74,354-2
4'-CHLORO-3'-NITRO-3,4,5,6-TETRABROMOPHTHALANILIC ACID

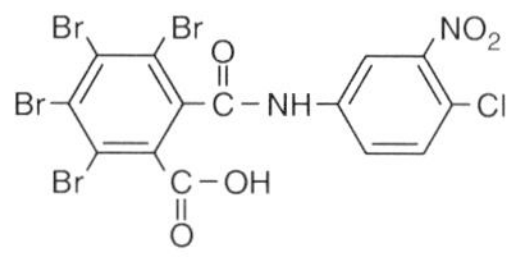

5 gm

S78,029-4
2'-CHLORO-5'-NITRO-3,4,5,6-TETRABROMOPHTHALANILIC ACID

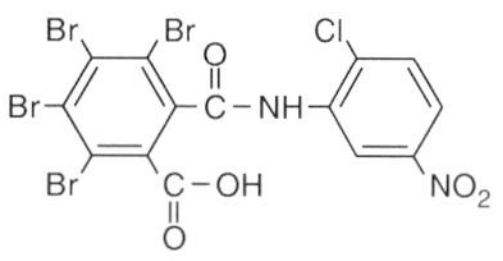

250 mg

S74,712-2
2-FLUORO-3'-NITROBENZANILIDE

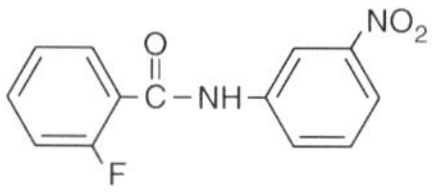

5 gm

S76,817-0
2,2'-DIFLUORO-5'-NITROBENZANILIDE

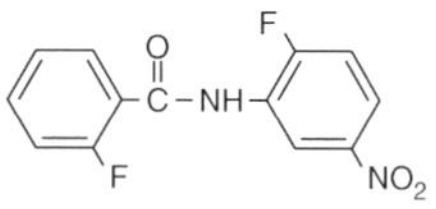

250 mg

S69,913-6
N-(4-CHLORO-3-NITROPHENYL)-2-FLUOROBENZAMIDE

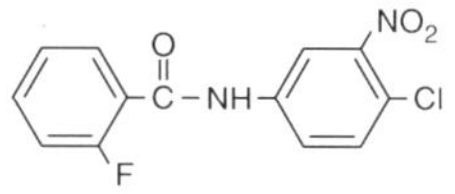

250 mg

S72,976-0
4-CHLORO-4'-FLUORO-3'-NITROBENZANILIDE

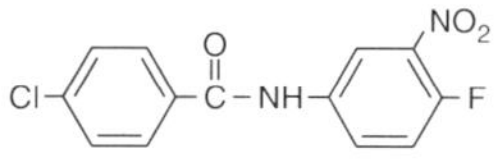

5 gm

S80,145-3
2'-CHLORO-4-FLUORO-5'-NITROBENZANILIDE

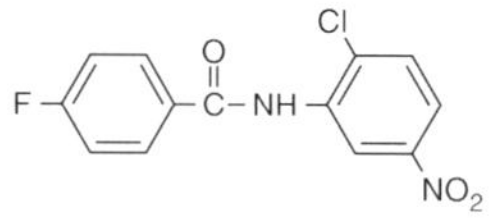

250 mg

S70,222-6
4-ETHYL-2'-FLUORO-5'-NITROBENZANILIDE

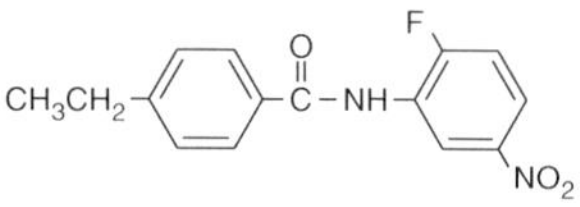

250 mg

S69,969-1
4-TERT-BUTYL-2'-FLUORO-5'-NITROBENZANILIDE

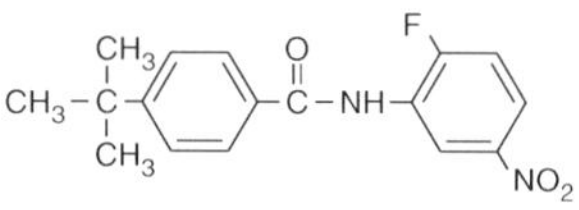

250 mg

S70,016-9
4-TERT-BUTYL-4'-CHLORO-3'-NITROBENZANILIDE

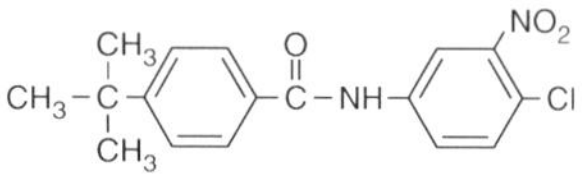

250 mg

S76,869-3
3,4-DICHLORO-4'-FLUORO-3'-NITROBENZANILIDE

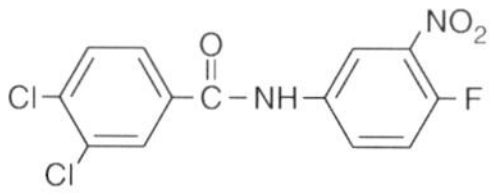

250 mg

S84,457-8
4,4'-DICHLORO-3-METHYL-3'-NITROBENZANILIDE

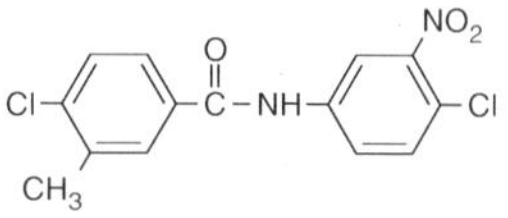

250 mg

S84,455-1
3'-CHLORO-3,5-DINITRO-4'-METHYLBENZANILIDE

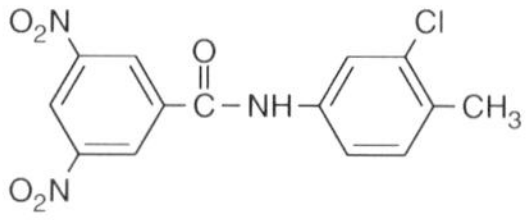

250 mg

S70,233-1
4'-BROMO-2',6'-DIMETHYL-3,5-DINITROBENZANILIDE

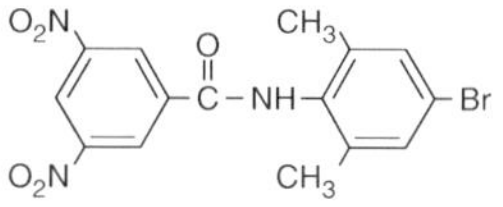

250 mg

S70,047-9
4-TERT-BUTYL-4'-FLUORO-3'-NITROBENZANILIDE

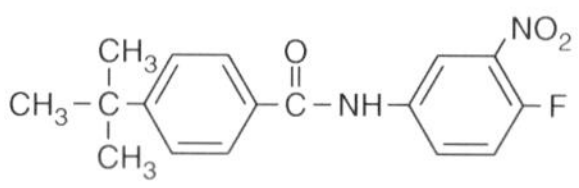

250 mg

S91,272-7
2'-CARBAMOYL-3,4,5,6-TETRACHLOROPHTHALANILIC ACID

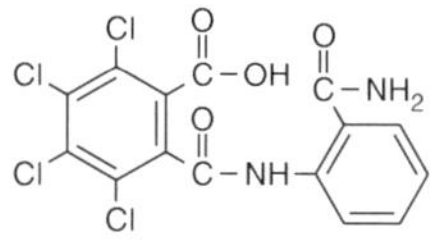

250 mg

S75,020-4
4'-ANILINO-3,4,5,6-TETRACHLOROPHTHALANILIC ACID

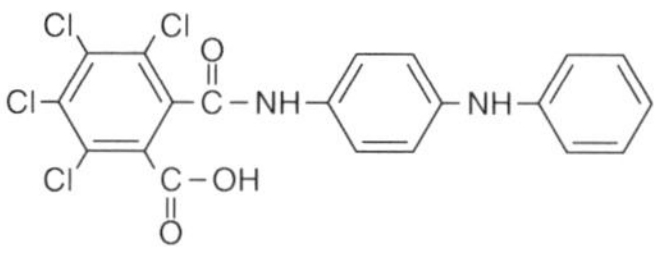

5 gm

S71,638-3
4'-ANILINO-3,4,5,6-TETRABROMOPHTHALANILIC ACID

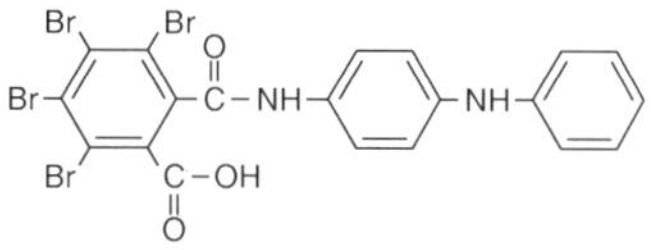

5 gm

S74,614-2
N-(4-ANILINOPHENYL)-4-FLUOROBENZAMIDE

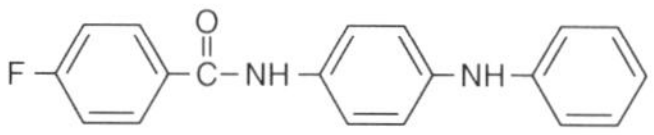

250 mg

S80,131-3
4'-ANILINO-3-FLUOROBENZANILIDE

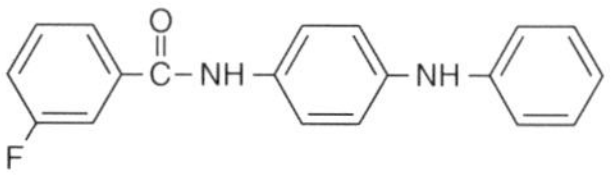

250 mg

S72,119-0
4'-(DIMETHYLAMINO)-3,4,5,6-TETRACHLOROPHTHALANILIC ACID

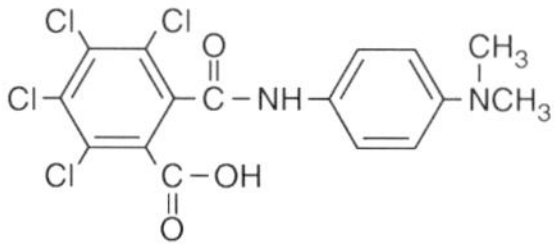

250 mg

S72,931-0
4'-(DIMETHYLAMINO)-3,4,5,6-TETRABROMOPHTHALANILIC ACID

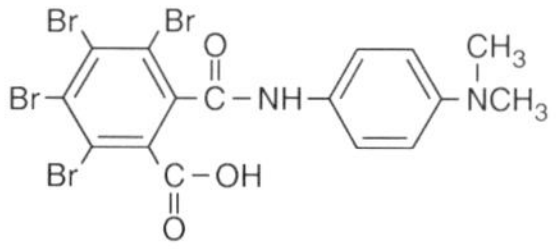

5 gm

S91,610-2
2'-CHLORO-4'-NITRO-3,4,5,6-TETRABROMOPHTHALANILIC ACID

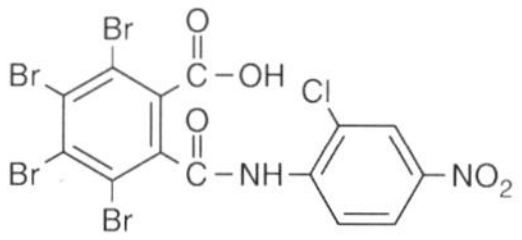

250 mg

S88,929-6
3'-CHLORO-3-NITROBENZANILIDE

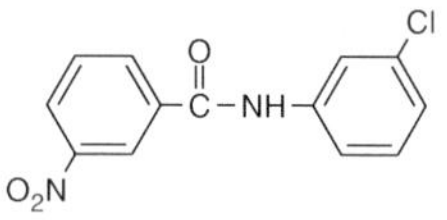

250 mg

S87,309-8
4'-NITRO-3,4,5-TRIIODOBENZANILIDE

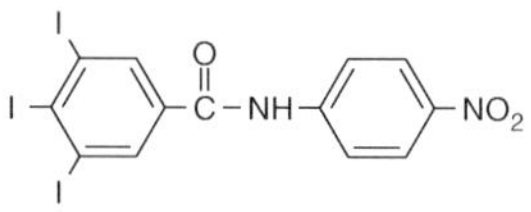

250 mg

S88,936-9
3'-CHLORO-2'-METHYL-3-NITROBENZANILIDE

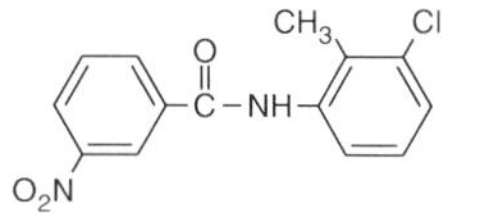

250 mg

S79,107-5
4'-NITRO-2,2',4-TRICHLOROBENZANILIDE

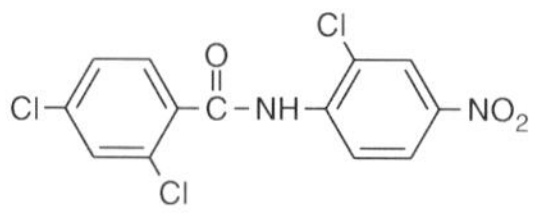

250 mg

S88,934-2
3'-CHLORO-4'-METHYL-3-NITROBENZANILIDE

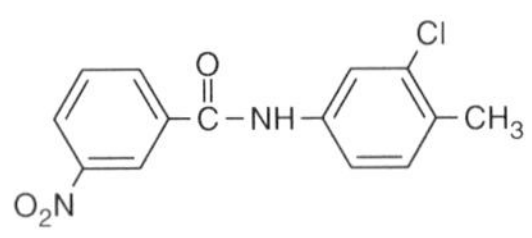

250 mg

S85,001-2
2-IODO-N-METHYL-4'-NITROBENZANILIDE

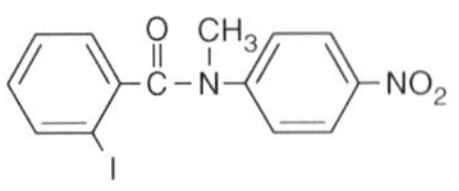

250 mg

S85,490-5
4'-(N-METHYLACETAMIDO)-3,4,5,6-TETRABROMOPHTHALANILIC ACID

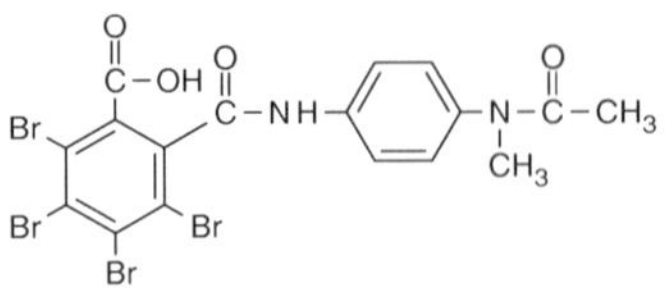

250 mg

S74,353-4
2'-BROMO-4'-NITROPHTHALANILIC ACID

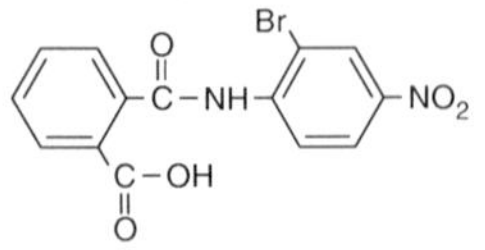

5 gm

S80,139-9
N-ACETYL-N'-(3-FLUOROBENZOYL)-1,4-PHENYLENEDIAMINE

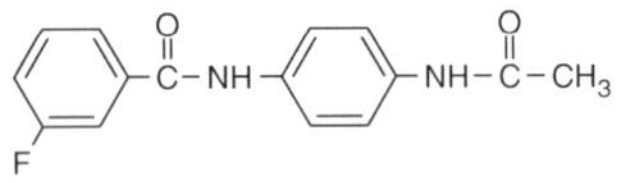

250 mg

S80,156-9
N-ACETYL-N'-(4-FLUOROBENZOYL)-1,4-PHENYLENEDIAMINE

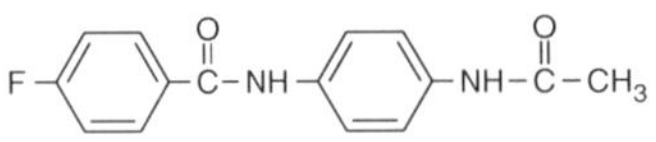

250 mg

S72,913-2
N-(2-PHENYLETHYL)-3,4,5,6-TETRACHLOROPHTHALAMIC ACID

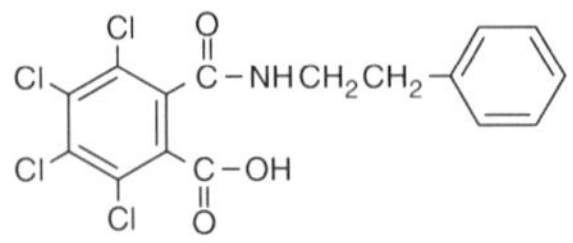

5 gm

S67,621-7
N-[3,5-BIS(METHOXYCARBONYL)PHENYL]-3,4,5,6-TETRABROMOPHTHALAMIC ACID

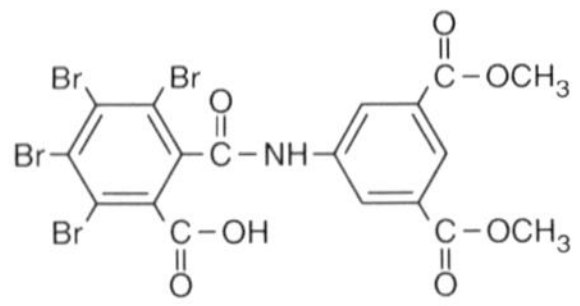

250 mg

S67,619-5
N-[(4-ETHOXYCARBONYL)PHENYL]-3,4,5,6-TETRABROMOPHTHALAMIC ACID

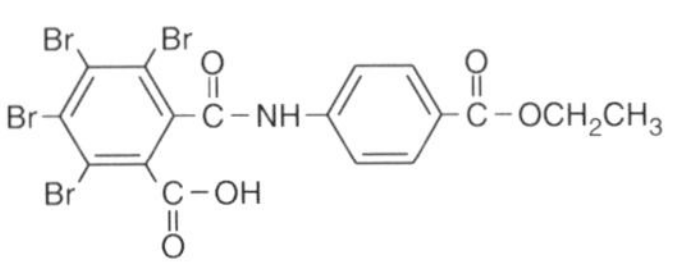

5 gm

S76,111-7
2,6-DICHLORO-3-(MORPHOLINOMETHYL)-BENZAMIDE

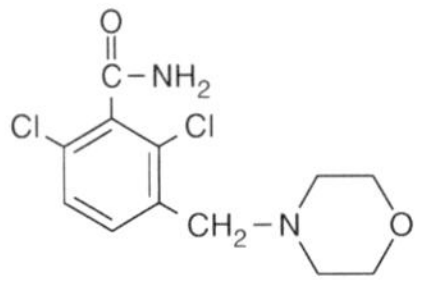

250 mg

S76,971-1
4-CHLORO-N-(2,4-DICHLOROBENZYL)-BENZAMIDE

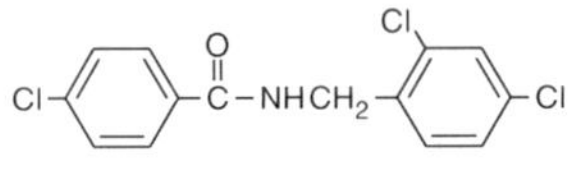

250 mg

S90,333-7
ETHYL 4-CHLOROHIPPURATE

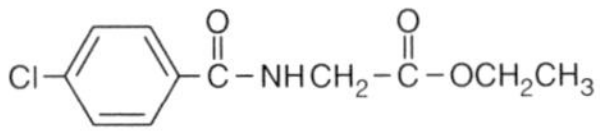

250 mg

S85,498-0
4-CHLORO-2'-METHOXY-5'-METHYLBENZANILIDE

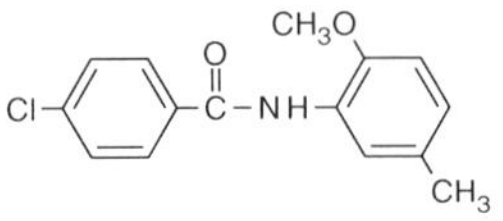

250 mg

S77,080-9
4-CHLORO-N-PHENYLBENZOHYDROXAMIC ACID

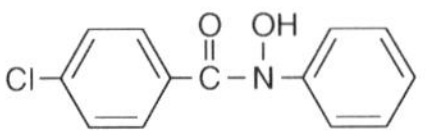

250 mg

S74,256-2
2-(4-CHLORO-3-NITROBENZOYL)-BENZAMIDE

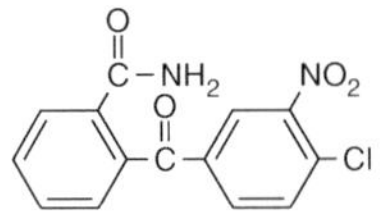

5 gm

S75,124-3
2'-BENZOYL-4'-CHLORO-4-NITROBENZANILIDE

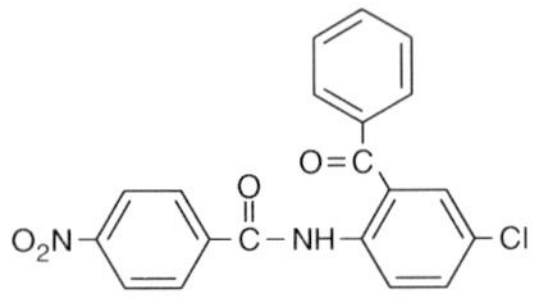

5 gm

S71,641-3
2',5'-DICHLORO-4'-NITRO-3,4,5,6-TETRACHLOROPHTHALANILIC ACID

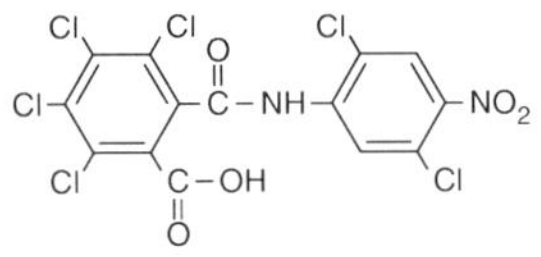

5 gm

S72,146-8
4'-FLUORO-3'-NITROPHTHALANILIC ACID

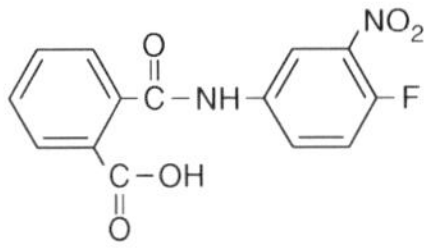

5 gm

S77,962-8
2'-FLUORO-3-NITROBENZANILIDE

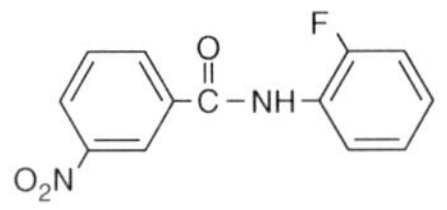

250 mg

S76,896-0
2'-CHLORO-3-NITROBENZANILIDE

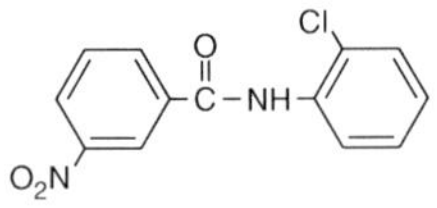

250 mg

S88,145-7
N-(2-BROMOPHENYL)-2-NITROCINNAMAMIDE

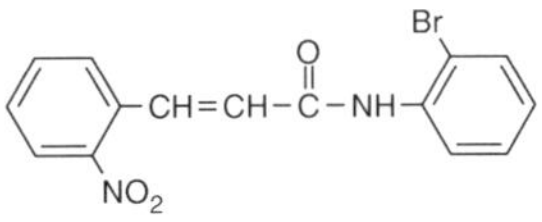

250 mg

S88,928-8
2'-CHLORO-4-NITROBENZANILIDE

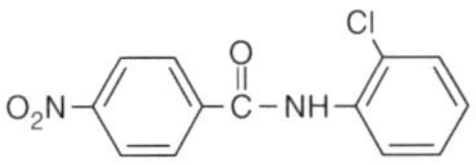

250 mg

S76,912-6
4'-FLUORO-3-NITROBENZANILIDE

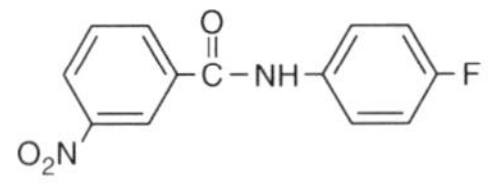

250 mg

S79,594-1
3'-FLUORO-4-NITROBENZANILIDE

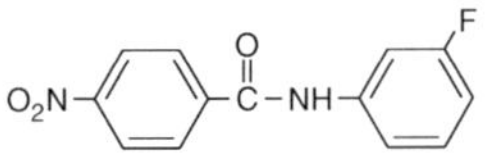

250 mg

S89,930-5
4'-FLUORO-4-NITROBENZANILIDE

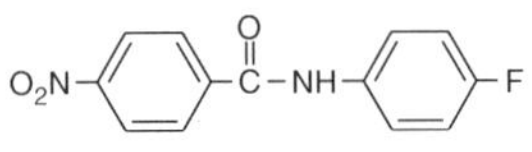

250 mg

S70,831-3
2'-CHLORO-6'-METHYL-4-NITROBENZANILIDE

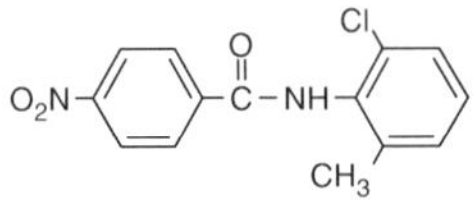

250 mg

S88,938-5
3'-CHLORO-2'-METHYL-4-NITROBENZANILIDE

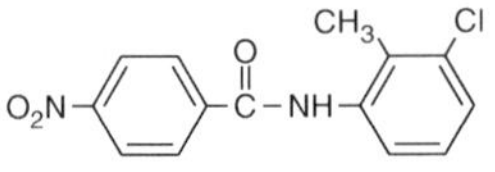

250 mg

S88,933-4
5'-CHLORO-2'-METHYL-4-NITROBENZANILIDE

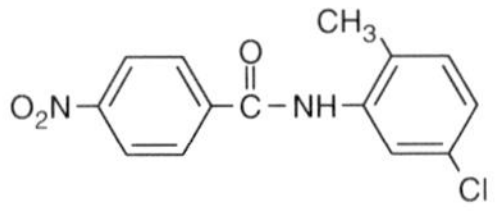

250 mg

S88,935-0
4'-CHLORO-2'-METHYL-4-NITROBENZANILIDE

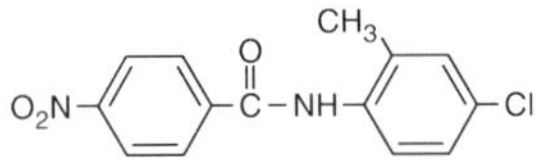

250 mg

S88,937-7
3'-CHLORO-4'-METHYL-4-NITROBENZANILIDE

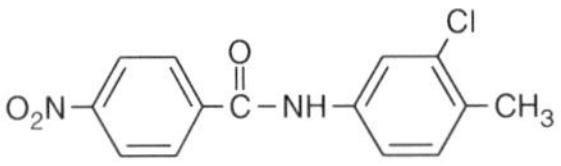

250 mg

S72,480-7
2'-CHLORO-6'-METHYL-3-NITROBENZANILIDE

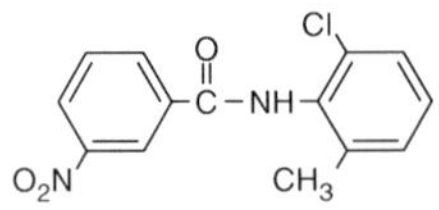

5 gm

S69,605-6
2'-CHLORO-4'-NITRO-2-BUTENANILIDE

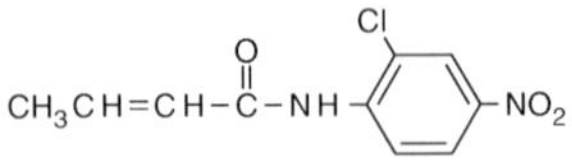

250 mg

S76,964-9
5'-CHLORO-3-NITROBENZO-O-TOLUIDIDE

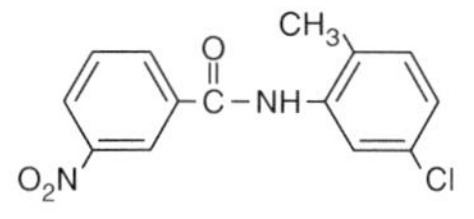

250 mg

S77,340-9
4'-CHLORO-3-NITROBENZO-O-TOLUIDIDE

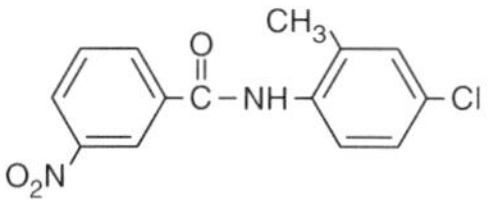

250 mg

S73,921-9
4'-BROMO-2',6'-DIMETHYL-3-NITROBENZANILIDE

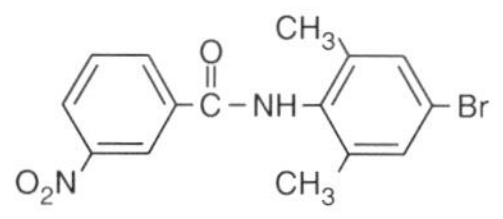

5 gm

S89,285-8
4',5'-DICHLORO-2'-METHYL-3-NITROBENZANILIDE

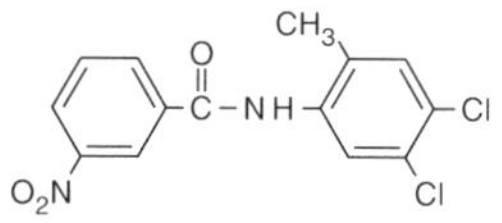

250 mg

S79,882-7
2-BROMO-N-(3-FORMYL-1-METHYLPROPYL)-BENZAMIDE

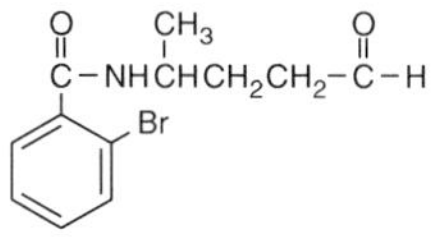

250 mg

S85,487-5
3'-ACETYL-3,4,5,6-TETRABROMOPHTHALANILIC ACID

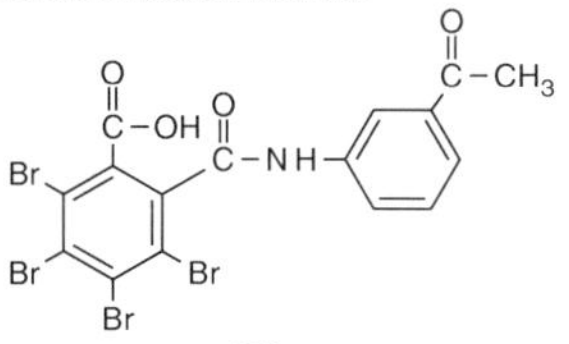

250 mg

S80,141-0
3'-ACETYL-3-FLUOROBENZANILIDE

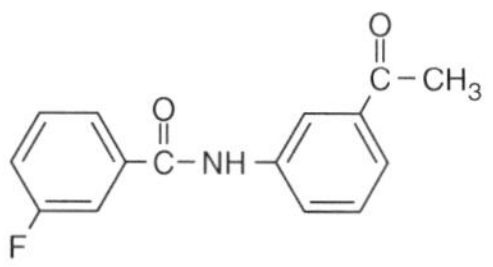

250 mg

S90,138-5
3'-ACETYL-3-CHLOROBENZANILIDE

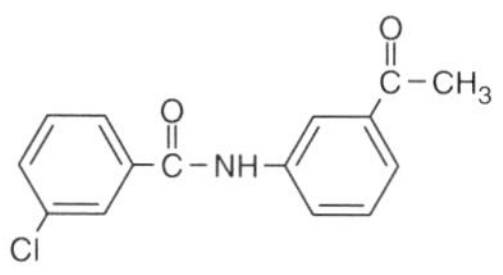

250 mg

S75,279-7
2'-BENZOYL-4,4'-DICHLOROBENZANILIDE

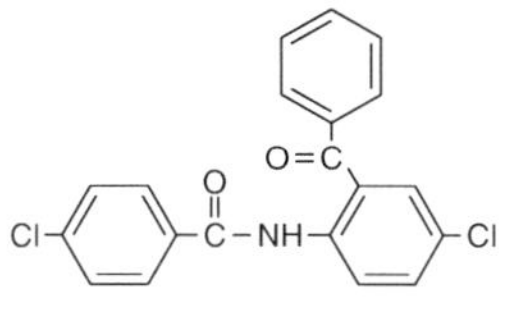

5 gm

S75,291-6
2'-BENZOYL-4'-CHLORO-10-UNDECENANILIDE

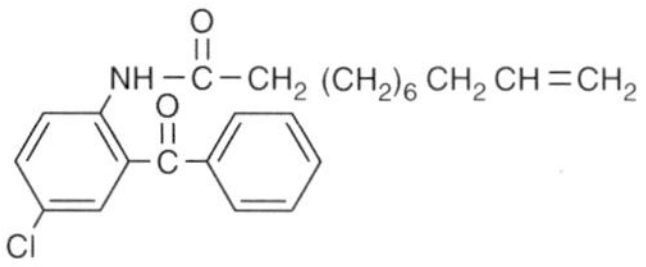

5 gm

S89,245-9
2'-BENZOYL-4'-CHLOROCINNAMANILIDE

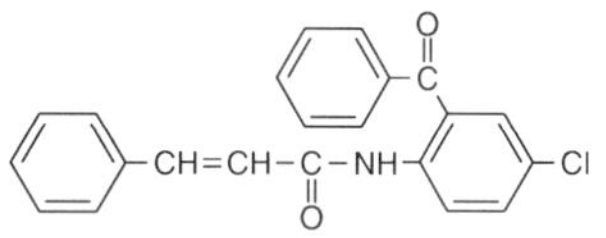

250 mg

S90,067-2
2'-ACETYL-2-FLUOROBENZANILIDE

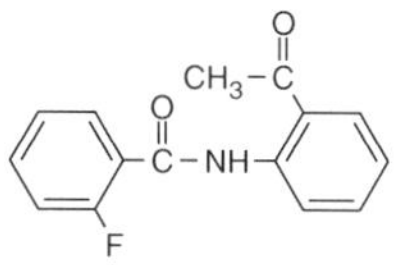

250 mg

S72,780-6
3'-ACETYL-4-FLUOROBENZANILIDE

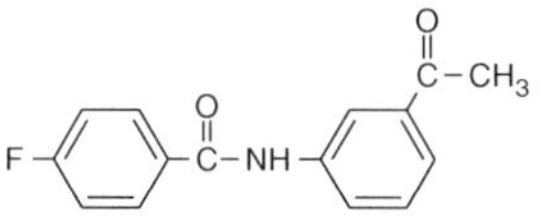

250 mg

S71,652-9
N-(4-ETHOXYCARBONYLPHENYL)-3,4,5,6-TETRACHLOROPHTHALAMIC ACID

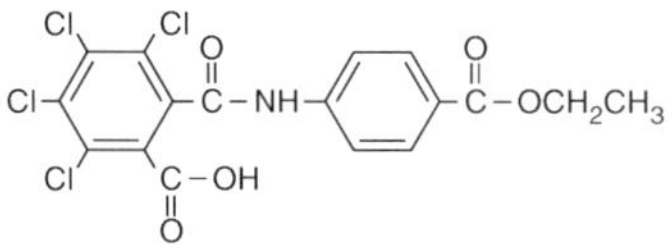

250 mg

S73,861-1
4'-(BUTOXYCARBONYL)-3,4,5,6-TETRABROMOPHTHALANILIC ACID

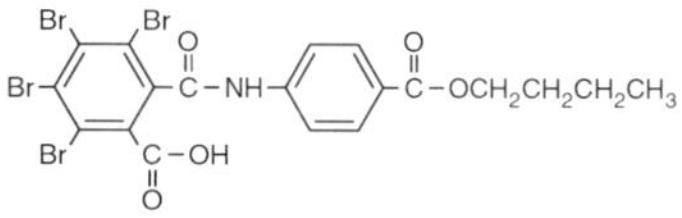

5 gm

S77,946-6
2'-CARBAMOYL-3,4,5,6-TETRABROMOPHTHALANILIC ACID

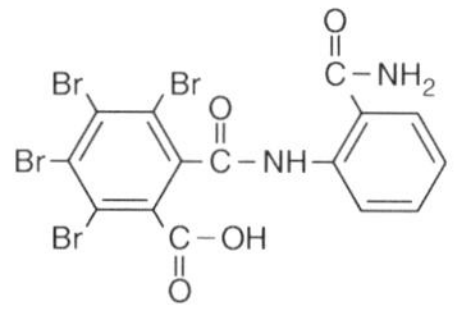

250 mg

S73,231-1
3'-FLUOROMALEANILIC ACID

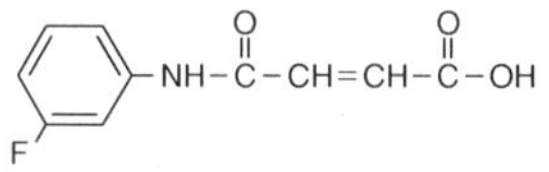

250 mg

S70,089-4
N-TERT-BUTYL-3,4,5,6-TETRACHLOROPHTHALAMIC ACID

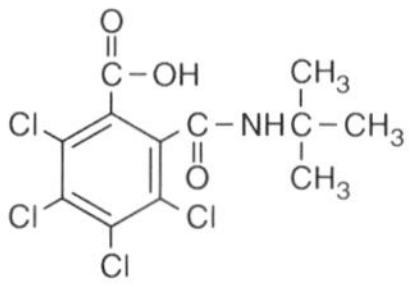

5 gm

S72,911-6
3'-FLUORO-3,4,5,6-TETRACHLOROPHTHALANILIC ACID

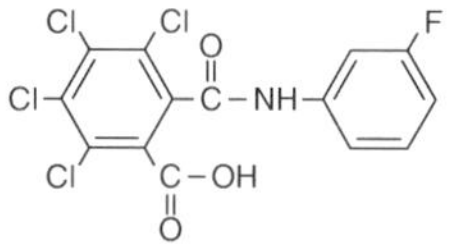

5 gm

S92,157-2
3'-CHLORO-3,4,5,6-TETRACHLOROPHTHALANILIC ACID

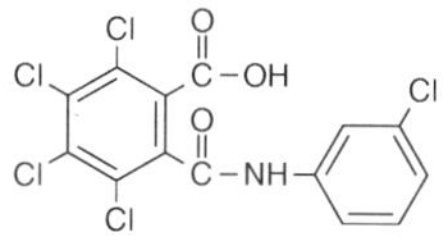

250 mg

S73,653-8
3'-BROMO-3,4,5,6-TETRACHLOROPHTHALANILIC ACID

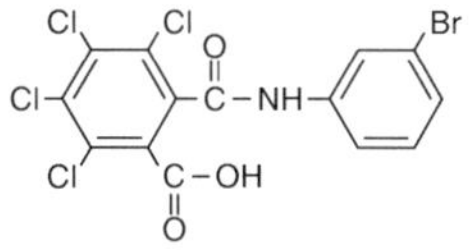

5 gm

S75,025-5
3'-FLUORO-3,4,5,6-TETRABROMOPHTHALANILIC ACID

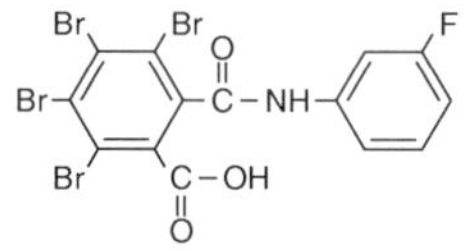

5 gm

S72,845-4
3,3',4,5,6-PENTABROMOPHTHALANILIC ACID

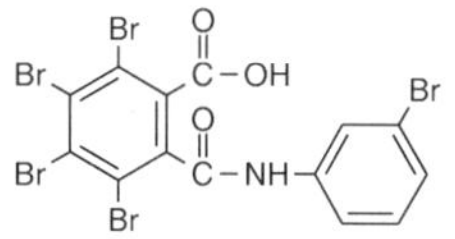

250 mg

S70,087-8
3,4,5,6-TETRACHLOROPHTHALANILIC ACID

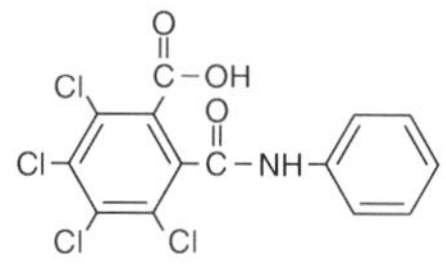

5 gm

S70,085-1
3,4,5,6-TETRABROMOPHTHALANILIC ACID

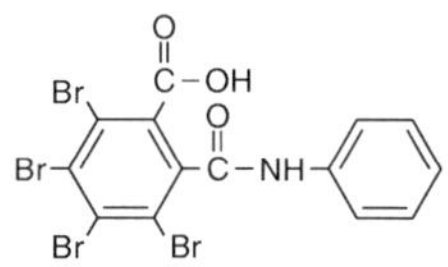

5 gm

S91,269-7
N-(DIPHENYLMETHYL)-3,4,5,6-TETRACHLOROPHTHALAMIC ACID

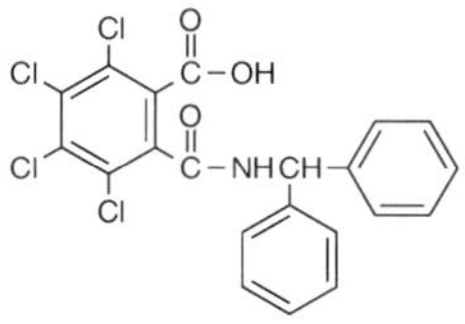

250 mg

S88,174-0
N-(2-BROMOPHENYL)MALEAMIC ACID

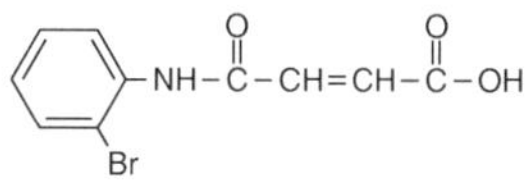

250 mg

S91,364-2
3'-FLUOROPHTHALANILIC ACID

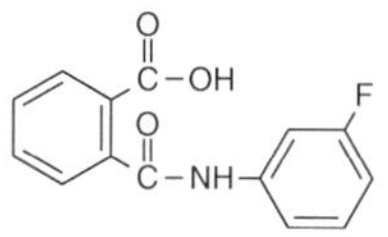

250 mg

S71,924-2
N-(3-CHLOROPHENYL)PHTHALAMIC ACID

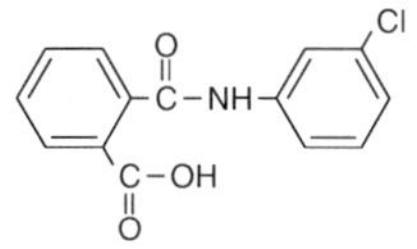

5 gm

S71,954-4
2'-CHLOROPHTHALANILIC ACID

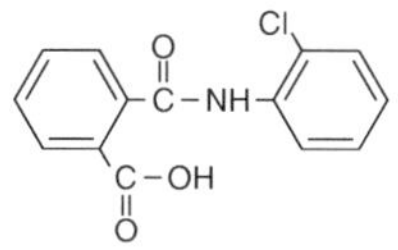

5 gm

S91,361-8
3'-BROMOPHTHALANILIC ACID

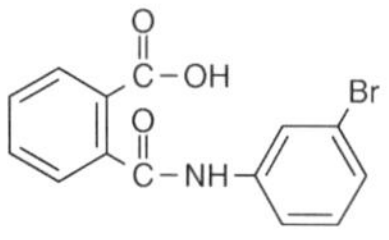

250 mg

S74,755-6
2'-FLUORO-3,4,5,6-TETRACHLOROPHTHALANILIC ACID

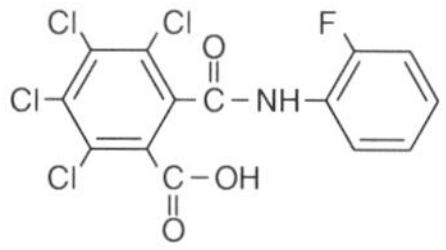

5 gm

S92,159-9
2'-CHLORO-3,4,5,6-TETRACHLOROPHTHALANILIC ACID

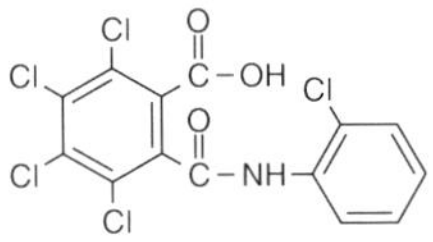

250 mg

S71,832-7
2'-FLUORO-3,4,5,6-TETRABROMOPHTHALANILIC ACID

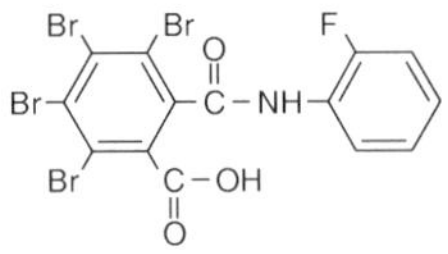

250 mg

S70,399-0
2'-METHYL-3,4,5,6-TETRACHLOROPHTHALANILIC ACID

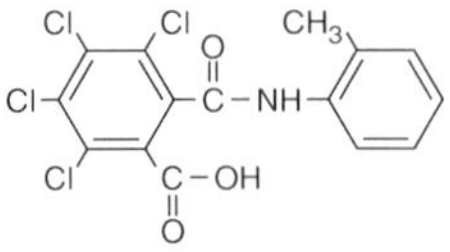

5 gm

S70,486-5
4'-FLUOROPHTHALANILIC ACID

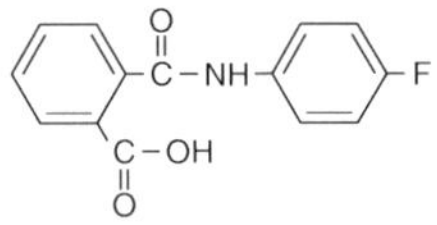

5 gm

S70,108-4
4'-CHLOROPHTHALANILIC ACID

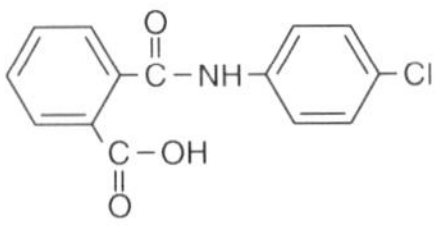

5 gm

S71,925-0
N-(4-IODOPHENYL)PHTHALAMIC ACID

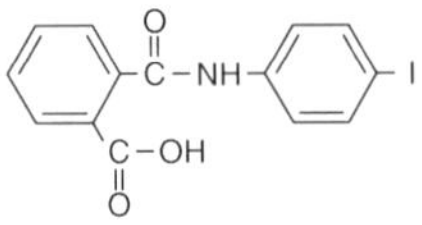

5 gm

S76,432-9
2',3'-DICHLOROPHTHALANILIC ACID

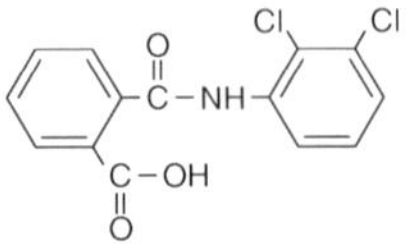

5 gm

S73,986-3
2'-CHLORO-6'-METHYLPHTHALANILIC ACID

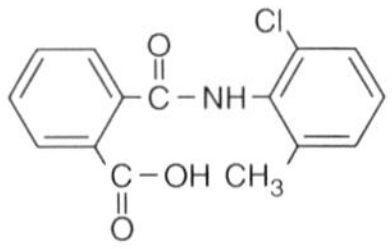

5 gm

S73,985-5
2',4'-DIFLUOROPHTHALANILIC ACID

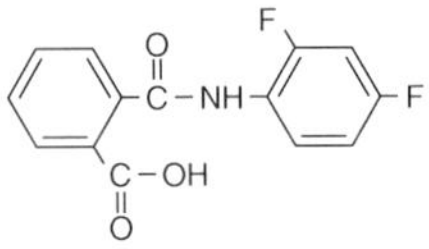

5 gm

S73,989-8
3'-CHLORO-4'-FLUOROPHTHALANILIC ACID

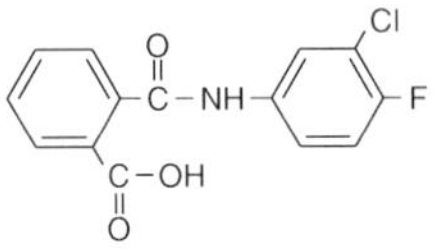

5 gm

S70,490-3
3',4'-DICHLOROPHTHALANILIC ACID

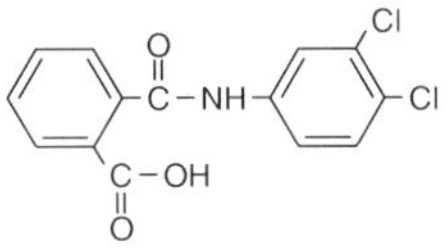

5 gm

S91,355-3
3',4'-DIBROMOPHTHALANILIC ACID

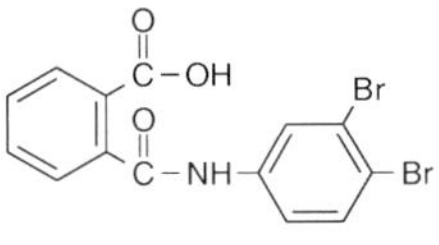

250 mg

S72,147-6
4'-FLUORO-3,3',4,5,6-PENTACHLOROPHTHALANILIC ACID

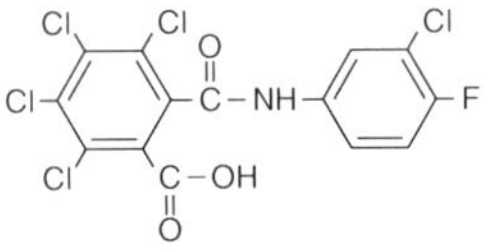

250 mg

S71,730-4
3'-CHLORO-4'-METHYLPHTHALANILIC ACID

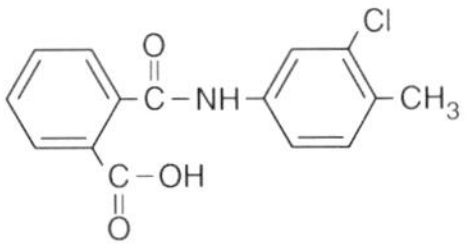

5 gm

S79,120-2
4'-CHLORO-2'-METHYLPHTHALANILIC ACID

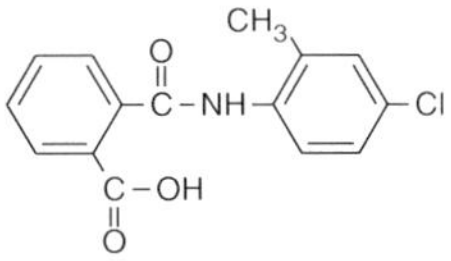

250 mg

S89,021-9
N-(5-CHLORO-2-METHYLPHENYL)-PHTHALAMIC ACID

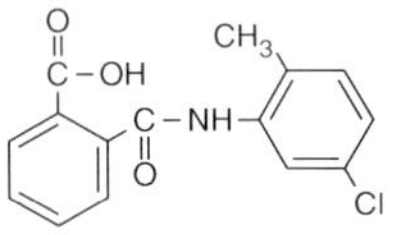

250 mg

S73,984-7
4'-BROMO-3'-METHYLPHTHALANILIC ACID

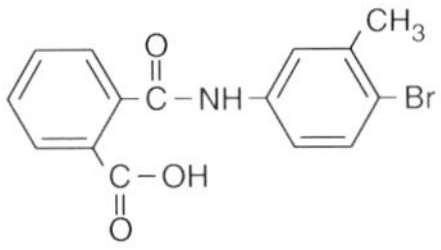

5 gm

S74,649-5
4'-BROMO-2'-METHYLPHTHALANILIC ACID

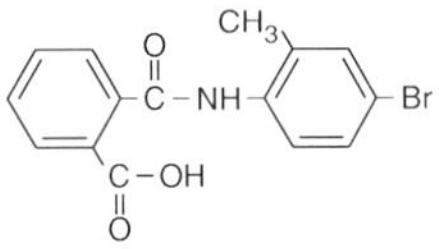

5 gm

S74,754-8
3',5'-DICHLOROPHTHALANILIC ACID

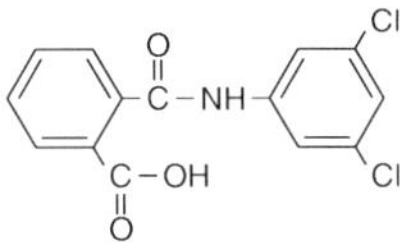

5 gm

S72,304-5
4'-BROMO-2',6'-DIMETHYLPHTHALANILIC ACID

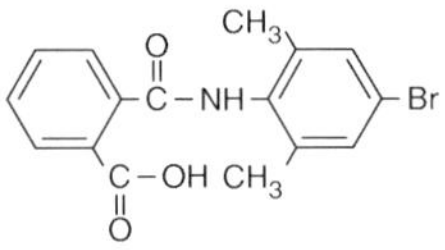

250 mg

S69,290-5
4'-FLUOROMALEANILIC ACID

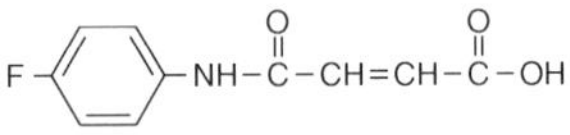

5 gm

S69,544-0
4'-BROMOMALEANILIC ACID

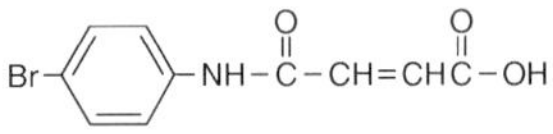

5 gm

S70,422-9
4'-FLUORO-3,4,5,6-TETRACHLOROPHTHALANILIC ACID

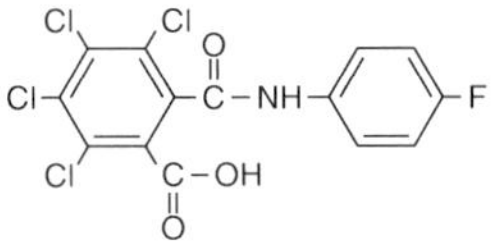

5 gm

S70,421-0
4'-CHLORO-3,4,5,6-TETRACHLOROPHTHALANILIC ACID

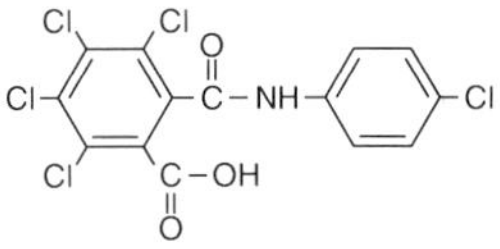

5 gm

S70,429-6
4'-ETHYL-3,4,5,6-TETRACHLOROPHTHALANILIC ACID

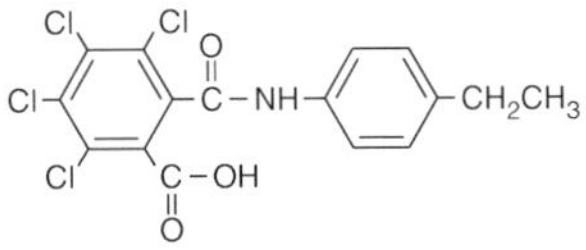

5 gm

S85,575-8
4'-ETHYL-3,4,5,6-TETRABROMOPHTHALANILIC ACID

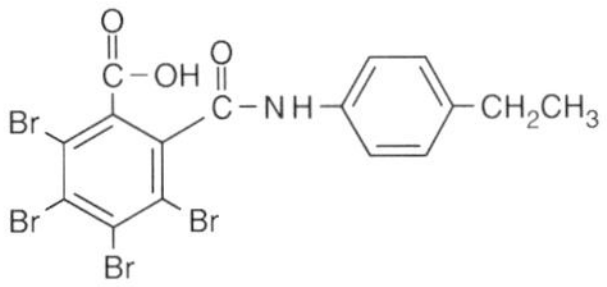

250 mg

S92,230-7
2',3'-DICHLOROMALEANILIC ACID

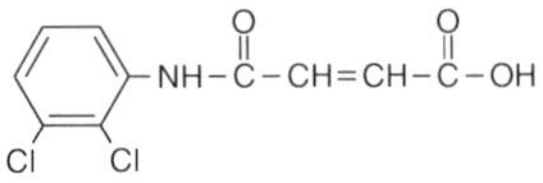

250 mg

S74,351-8
2',3,3',4,5,6-HEXACHLOROPHTHALANILIC ACID

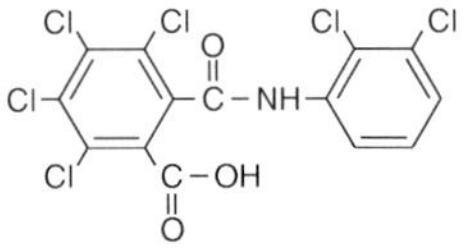

5 gm

S75,026-3
2',3'-DICHLORO-3,4,5,6-TETRABROMOPHTHALANILIC ACID

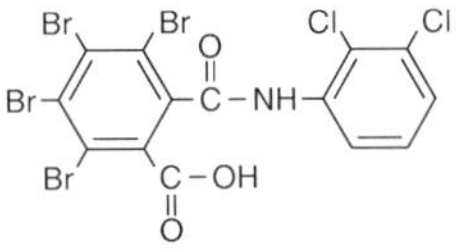

5 gm

S75,144-8
2'-METHYL-3,4,5,6,6'-PENTACHLOROPHTHALANILIC ACID

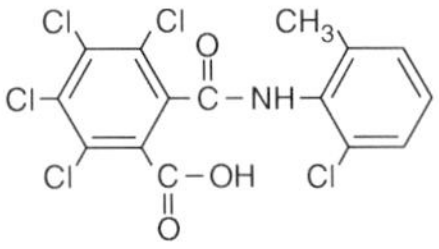

5 gm

S75,145-6
2'-METHYL-3,3',4,5,6-PENTACHLOROPHTHALANILIC ACID

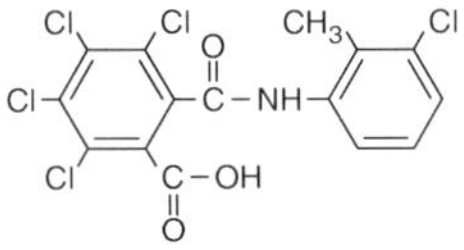

5 gm

S74,764-5
2',6'-DIMETHYL-3,4,5,6-TETRACHLOROPHTHALANILIC ACID

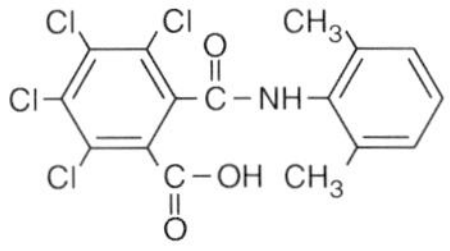

5 gm

S85,488-3
2',3'-DIMETHYL-3,4,5,6-TETRABROMOPHTHALANILIC ACID
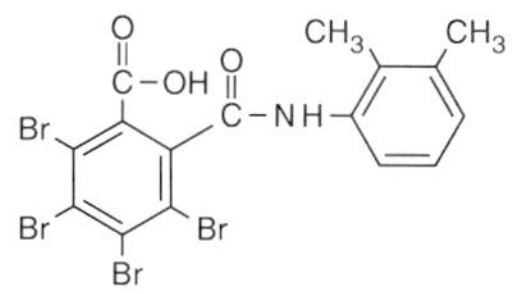
250 mg

S73,188-9
2',6'-DIISOPROPYL-3,4,5,6-TETRACHLOROPHTHALANILIC ACID
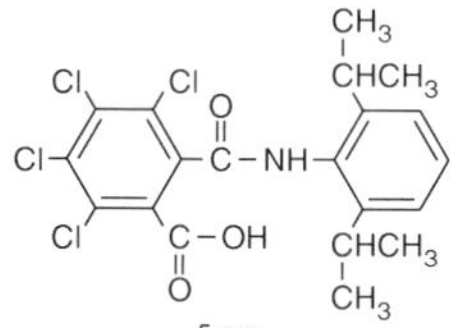
5 gm

S92,229-3
3'-CHLORO-4'-FLUOROMALEANILIC ACID
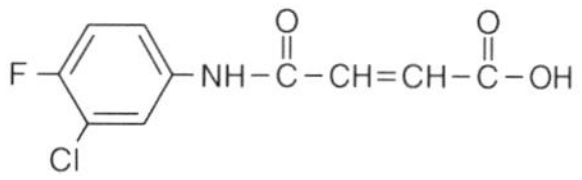
250 mg

S69,291-3
2',4'-DICHLOROMALEANILIC ACID
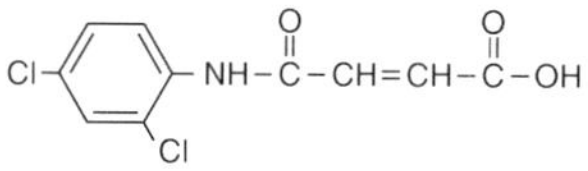
5 gm

S74,652-5
4'-BROMO-2'-METHYLMALEANILIC ACID
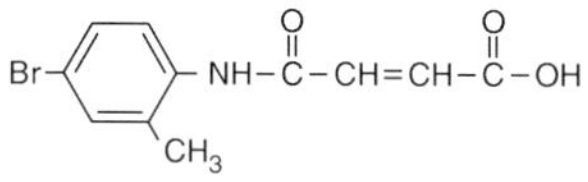
5 gm

S92,171-8
4'-BROMO-2'-CHLOROPHTHALANILIC ACID
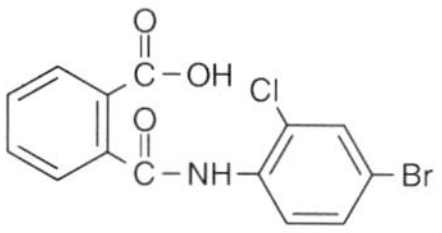
250 mg

S92,172-6
4'-BROMO-2'-CHLORO-3,4,5,6-TETRACHLOROPHTHALANILIC ACID
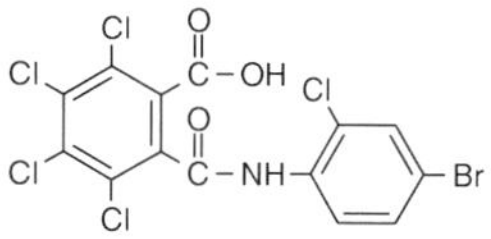
250 mg

S71,637-5
2',5'-DICHLORO-3,4,5,6-TETRABROMOPHTHALANILIC ACID
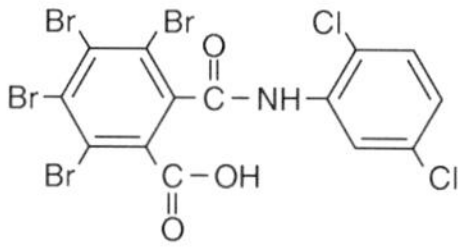
5 gm

S85,566-9
3',4'-DICHLORO-3,4,5,6-TETRABROMOPHTHALANILIC ACID
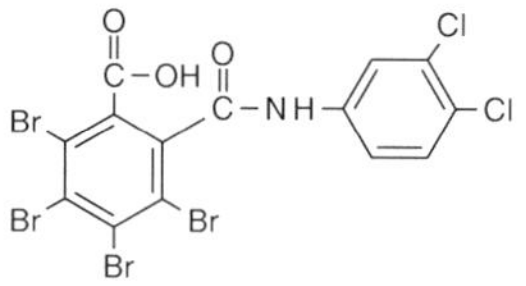
250 mg

S75,142-1
4'-METHYL-3,3',4,5,6-PENTACHLOROPHTHALANILIC ACID
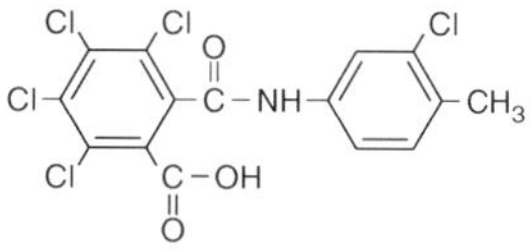
5 gm

S92,169-6
4'-BROMO-3'-METHYL-3,4,5,6-TETRACHLOROPHTHALANILIC ACID
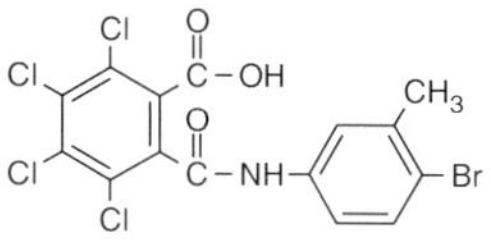
250 mg

S85,569-3
5'-CHLORO-2'-METHYL-3,4,5,6-TETRABROMOPHTHALANILIC ACID
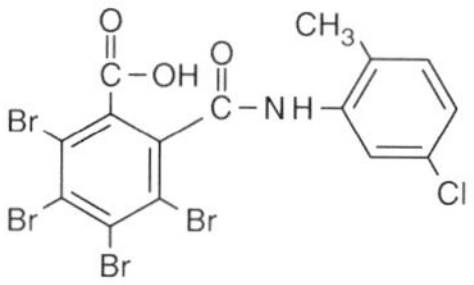
250 mg

S84,505-1
2'-CHLORO-4'-METHOXY-3,4,5,6-TETRABROMOPHTHALANILIC ACID
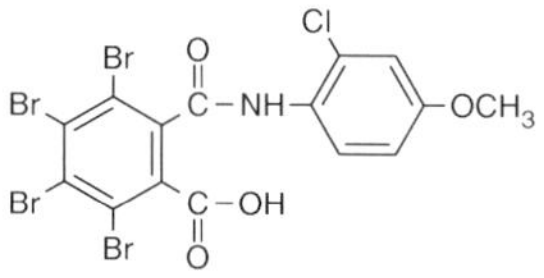
250 mg

S70,400-8
3',4'-DIMETHYL-3,4,5,6-TETRACHLOROPHTHALANILIC ACID
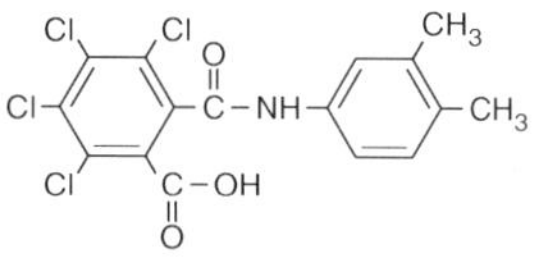
5 gm

S89,263-7
5'-ISOPROPYL-2'-METHYL-3,4,5,6-TETRACHLOROPHTHALANILIC ACID
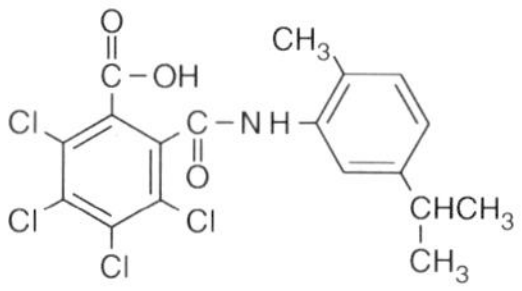
250 mg

S71,635-9
3,3',4,5,5',6-HEXACHLOROPHTHALANILIC ACID
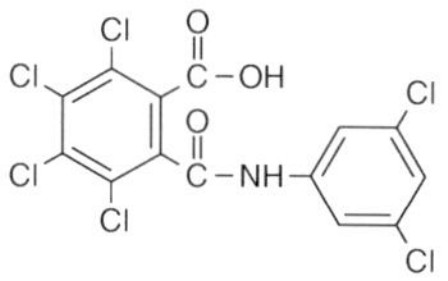
5 gm

S73,907-3
4'-BROMO-2',6'-DIMETHYL-3,4,5,6-TETRACHLOROPHTHALANILIC ACID
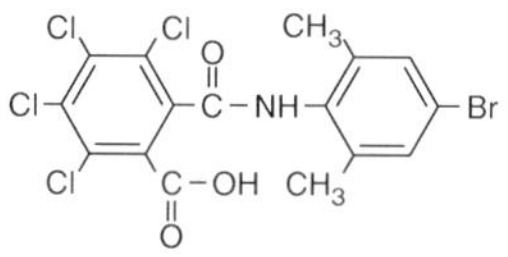
5 gm

S89,264-5
3,4,5,6-TETRACHLORO-2',4',6'-TRIMETHYLPHTHALANILIC ACID
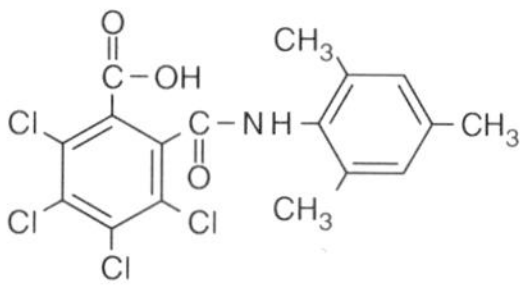
250 mg

S91,370-7
2',3,4,4',5,5',6-HEPTACHLOROPHTHALANILIC ACID
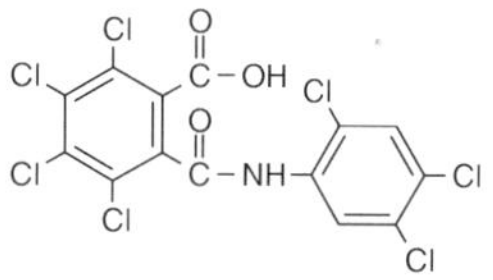
250 mg

S74,388-7
3,4,5,6-TETRABROMO-2',4',5'-TRICHLOROPHTHALANILIC ACID
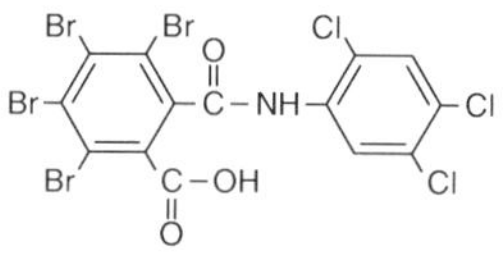
5 gm

S91,372-3
3,4,5,6-TETRACHLORO-2',4',5'-TRIMETHYLPHTHALANILIC ACID
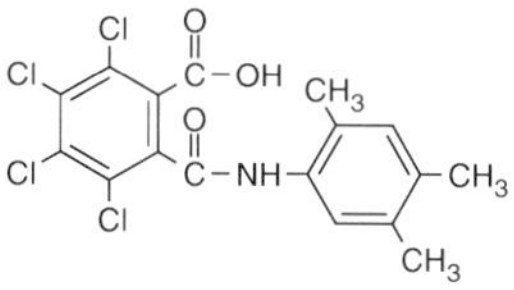
250 mg

S71,639-1
N-(1-NAPHTHYL)-3,4,5,6-TETRACHLOROPHTHALAMIC ACID
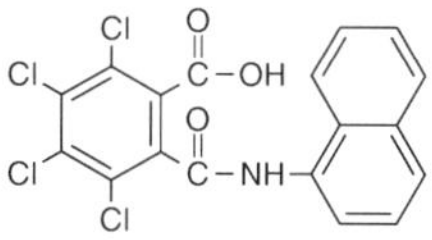
5 gm

S72,868-3
DIMETHYL 5-(2-FLUOROPHENYLCARBOXAMIDO)-ISOPHTHALATE
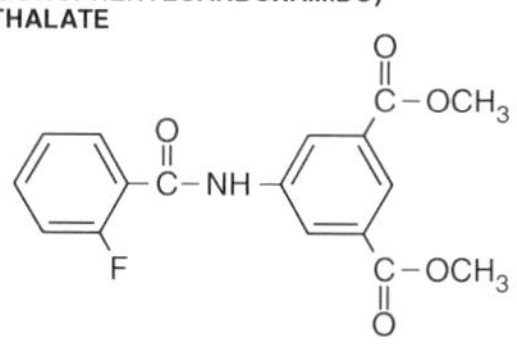
250 mg

S80,153-4
4-[N-(2,4-DICHLOROPHENYL)CARBAMOYL]-PHTHALIC ANHYDRIDE
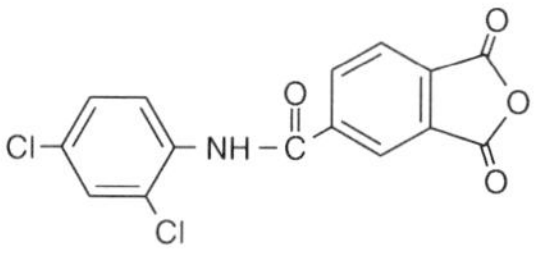
250 mg

S80,154-2
4-[N-(3,4-DICHLOROPHENYL)CARBAMOYL]-PHTHALIC ANHYDRIDE
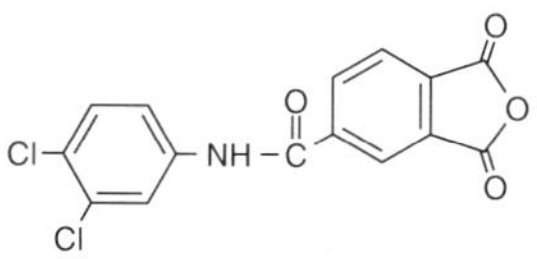
250 mg

S76,836-7
N,N'-BIS(2-FLUOROPHENYL)-ISOPHTHALAMIDE
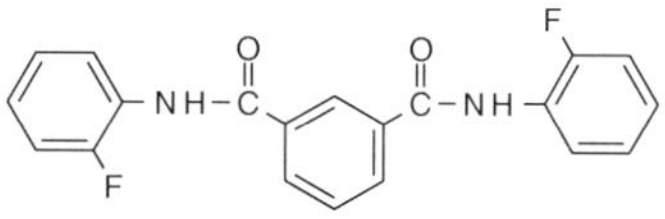
250 mg

S77,218-6
N,N'-BIS(4-FLUOROPHENYL)-ISOPHTHALAMIDE
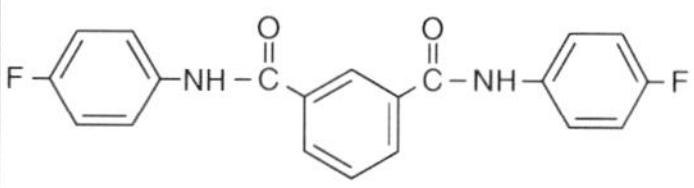
250 mg

S77,983-0
N,N'-BIS(2,5-DICHLOROPHENYL)-TEREPHTHALAMIDE
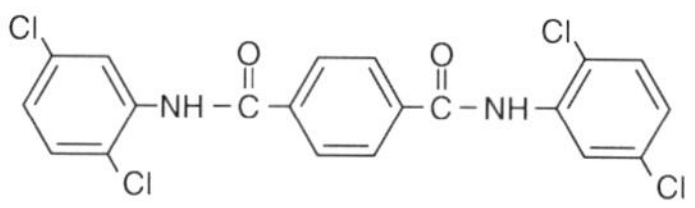
250 mg

S78,051-0
N,N'-BIS(3-CHLORO-4-METHYLPHENYL)-TEREPHTHALAMIDE
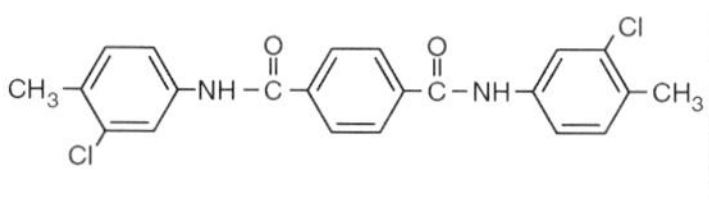
250 mg

S89,075-8
N,N'-BIS(3-CHLORO-2-METHYLPHENYL)-ISOPHTHALAMIDE
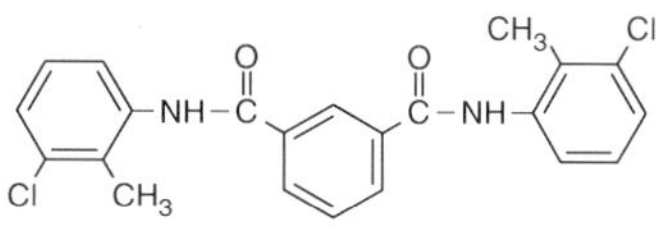
250 mg

S74,791-2
N,N'-BIS(2-CHLORO-5-METHYLPHENYL)-ISOPHTHALAMIDE
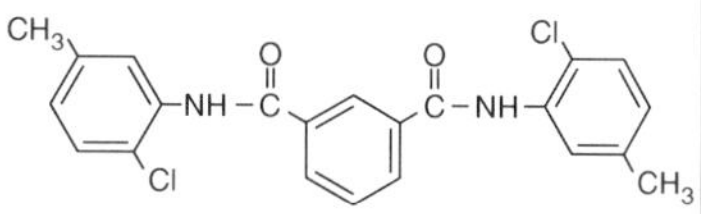
5 gm

S74,797-1
N,N'-BIS(2-CHLORO-5-METHYLPHENYL)-TEREPHTHALAMIDE
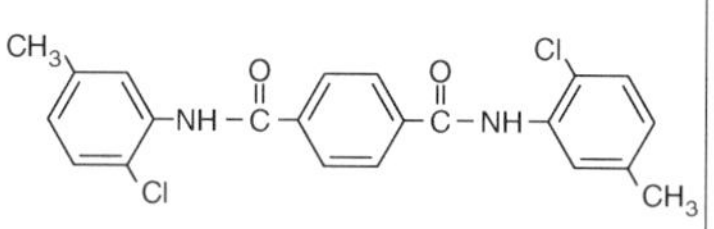
5 gm

S89,076-6
N,N'-BIS(5-CHLORO-2-METHYLPHENYL)-ISOPHTHALAMIDE

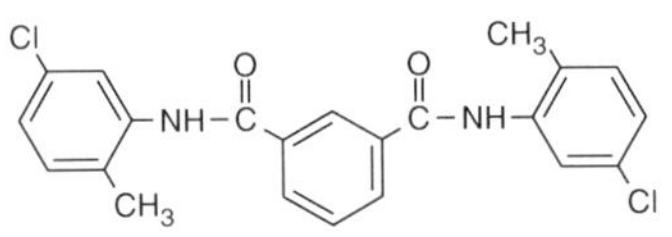

250 mg

S89,295-5
N,N'-BIS(4,5-DICHLORO-2-METHYLPHENYL)ISOPHTHALAMIDE

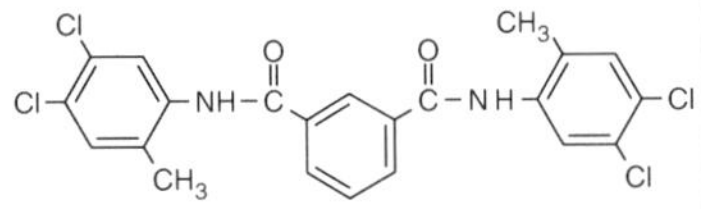

250 mg

S92,138-6
3'-METHYL-3,4,5,6-TETRACHLOROPHTHALANILIC ACID

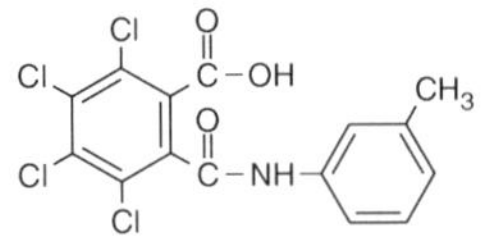

5 gm

S67,695-0
2-CHLOROHIPPURIC ACID

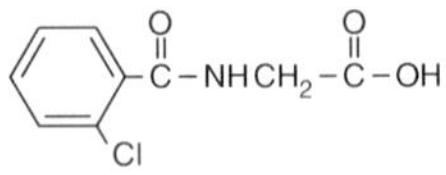

250 mg

S80,144-5
4-(3-FLUOROBENZAMIDO)BENZOIC ACID

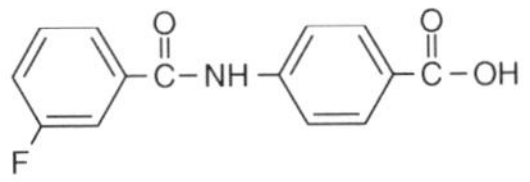

250 mg

S92,154-8
4'-BROMOPHTHALANILIC ACID

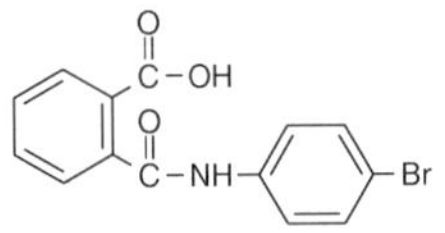

250 mg

S92,141-6
4'-BROMO-3,4,5,6-TETRACHLOROPHTHALANILIC ACID

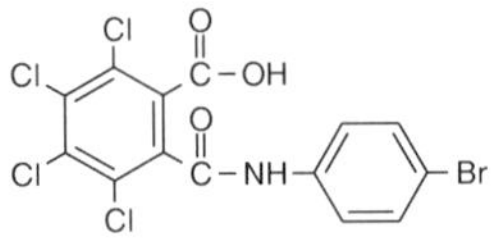

250 mg

S92,151-3
2',5'-DIMETHYL-3,4,5,6-TETRACHLOROPHTHALANILIC ACID

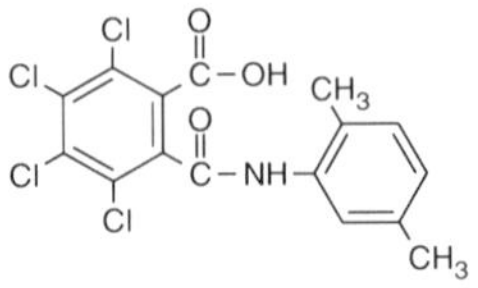

250 mg

S72,525-0
3-CHLORO-3'-(METHOXYCARBONYL)-BENZANILIDE

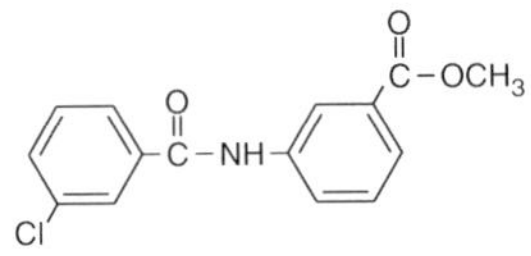

5 gm

S80,140-2
METHYL 2-(3-FLUOROBENZAMIDO)-BENZOATE

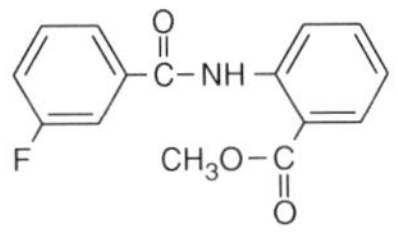

250 mg

S90,139-3
METHYL N-(3-CHLOROBENZOYL)-ANTHRANILATE

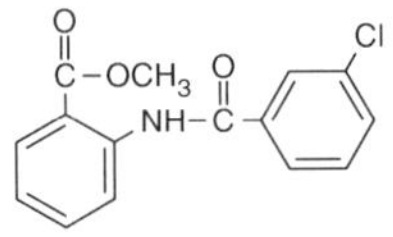

250 mg

S72,782-2
METHYL 2-(4-FLUOROPHENYLCARBOXAMIDO)-BENZOATE

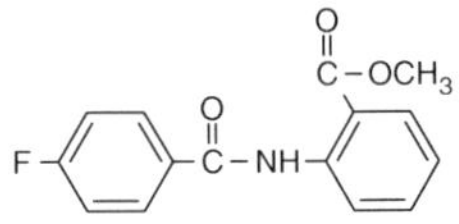

5 gm

S73,710-0
4'-(ETHOXYCARBONYL)-4-FLUOROBENZANILIDE

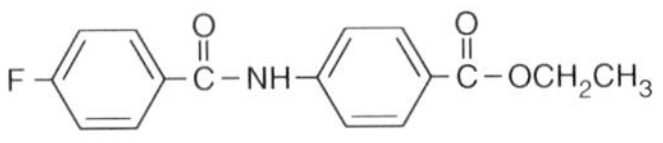

5 gm

S80,134-8
ETHYL 4-(3-FLUOROBENZAMIDO)BENZOATE

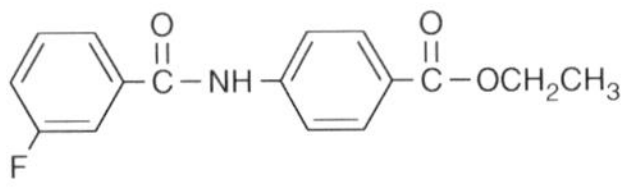

250 mg

S73,711-9
ETHYL 4-(4-CHLOROBENZAMIDO)BENZOATE

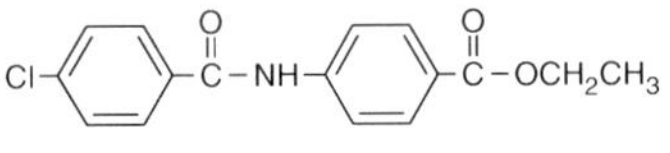

5 gm

S75,009-3
ETHYL 4-(3-CHLOROBENZAMIDO)BENZOATE

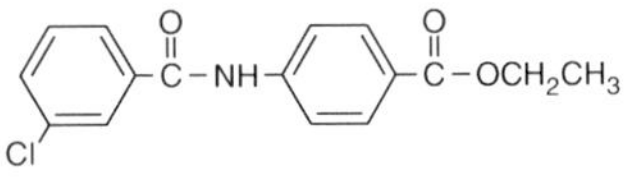

5 gm

S71,678-2
3-CHLORO-4-METHYLBENZAMIDE

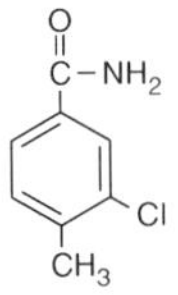

5 gm

S68,120-2
2-2'-FLUOROBENZOIC IMIDE

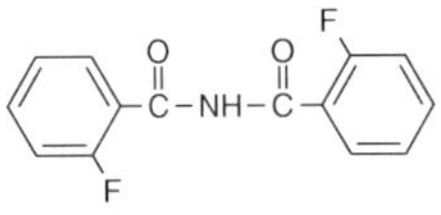

5 gm

S75,051-4
3-CHLORO-3'-FLUOROBENZANILIDE

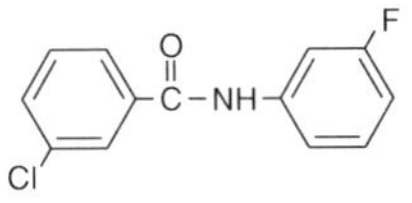

5 gm

S90,126-1
3,3'-DICHLOROBENZANILIDE

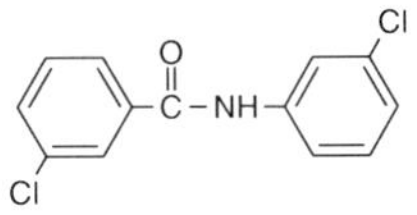

250 mg

S72,492-0
4-CHLORO-N-CYCLOHEXYLBENZAMIDE

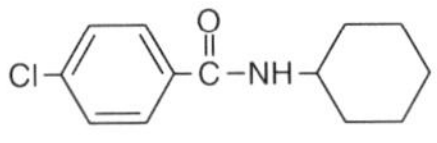

5 gm

S89,240-8
N-CYCLOHEXYL-2,4-DICHLOROBENZAMIDE

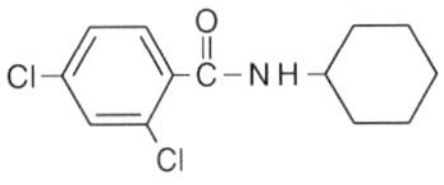

250 mg

S72,227-8
3'-CHLOROBENZANILIDE

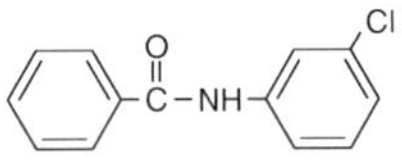

5 gm

S88,304-2
3-CHLOROBENZANILIDE

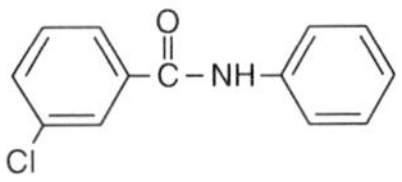

250 mg

S80,771-0
3'-FLUOROCINNAMANILIDE

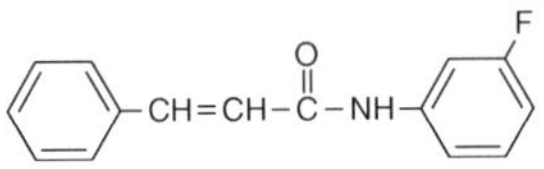

250 mg

S39,741-5
2'-FLUOROBENZANILIDE

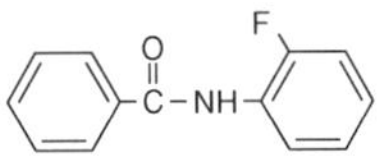

250 mg

S85,014-4
2'-CHLOROBENZANILIDE

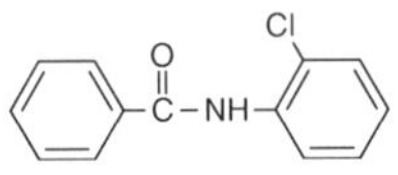

250 mg

S89,175-4
2'-CHLOROCINNAMANILIDE

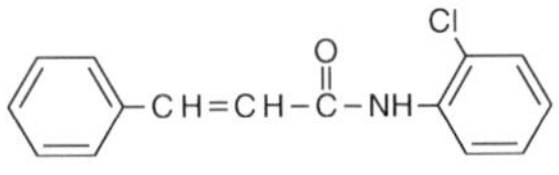

250 mg

S87,781-6
N-(2-BROMOPHENYL)CINNAMAMIDE

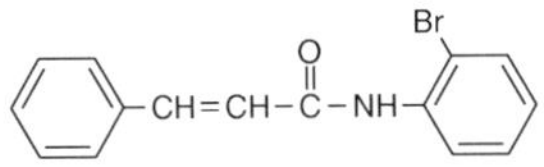

250 mg

S89,062-6
3'-CHLORO-2'-METHYLBENZANILIDE

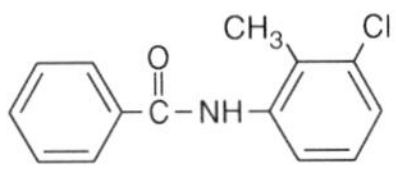

250 mg

S89,061-8
4'-CHLORO-2'-METHYLBENZANILIDE

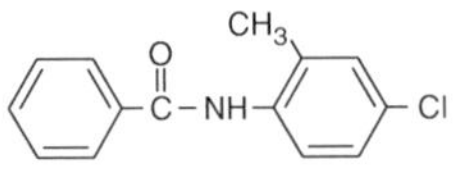

250 mg

S92,598-5
4'-CHLORO-2'-METHYLBENZANILIDE

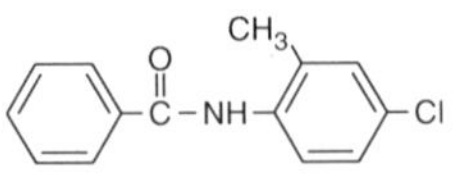

250 mg

S89,255-6
2',5'-DICHLOROCINNAMANILIDE

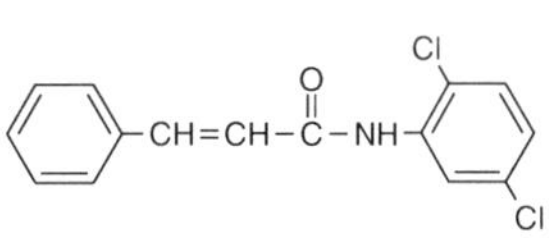

250 mg

S72,486-6
3'-CHLORO-4'-METHYLCINNAMANILIDE

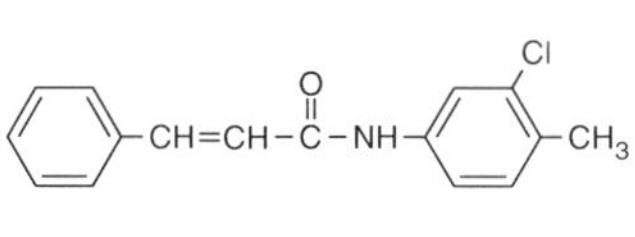

5 gm

S89,251-3
5'-CHLORO-2'-METHYLCINNAMANILIDE

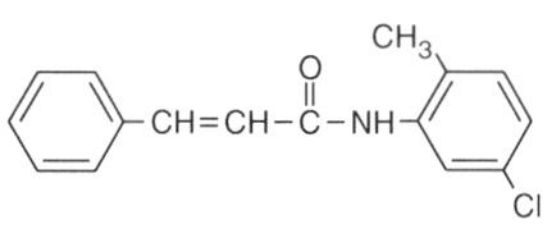

250 mg

S70,856-9
2',4',6'-TRICHLOROBENZANILIDE

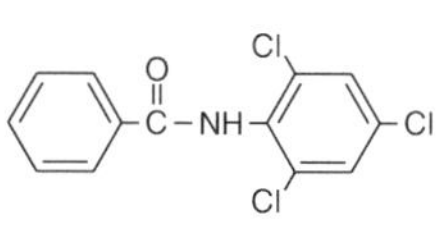

250 mg

S70,728-7
2',4',6'-TRIBROMOBENZANILIDE

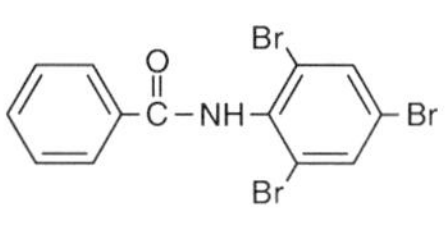

250 mg

S70,223-4
4'-BROMO-2',6'-DIMETHYLBENZANILIDE

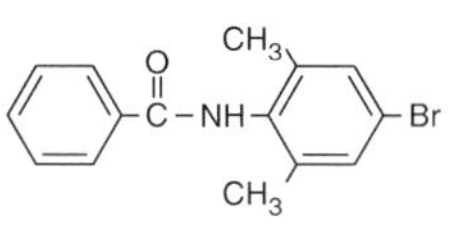

250 mg

S77,486-3
2-CHLORO-N-METHYLBENZAMIDE

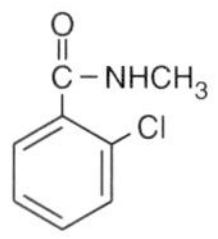

250 mg

S79,558-5
2'-FLUORO-2-BUTENANILIDE

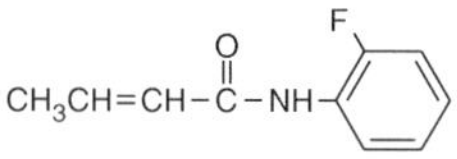

250 mg

S88,069-8
N-ALLYL-2-BROMOBENZAMIDE

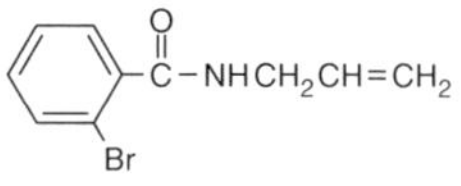

250 mg

S88,046-9
N-(2-BROMOPHENYL)-1,3-PENTADIENE-1-CARBOXAMIDE

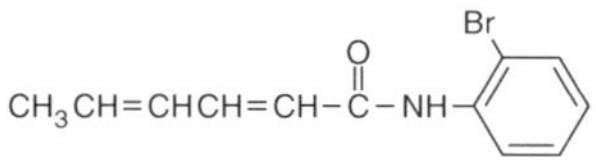

250 mg

S79,108-3
3-CHLORO-2'-FLUOROBENZANILIDE

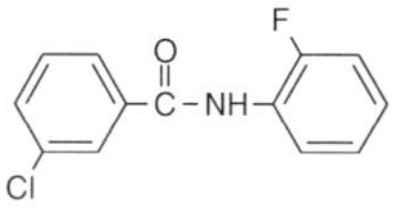

250 mg

S71,408-9
2-BROMO-2',4',5'-TRICHLOROBENZANILIDE

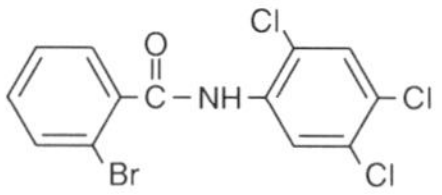

5 gm

S90,058-3
2-FLUORO-2'-METHYLBENZANILIDE

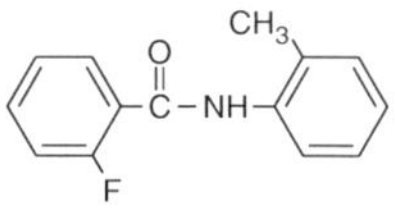

250 mg

S71,781-9
3-BROMO-2'-METHYLBENZANILIDE

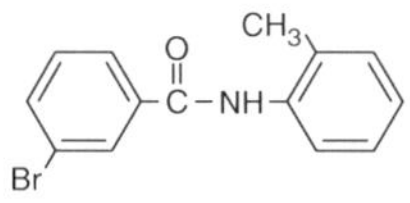

5 gm

S79,129-6
2-CHLORO-2'-ETHYLBENZANILIDE

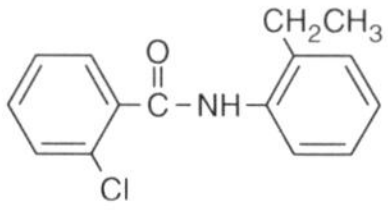

250 mg

S89,011-1
3-CHLORO-2'-ETHYLBENZANILIDE

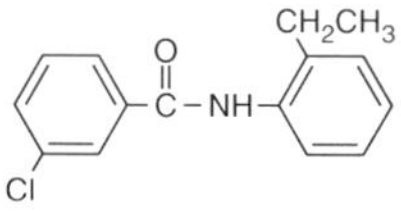

250 mg

S69,749-4
4'-CHLORO-O-TOLUANILIDE

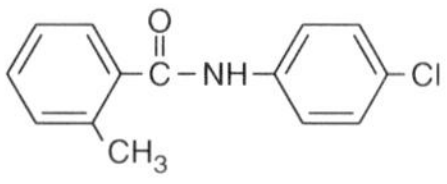

5 gm

S79,131-8
2-CHLORO-4'-ETHYLBENZANILIDE

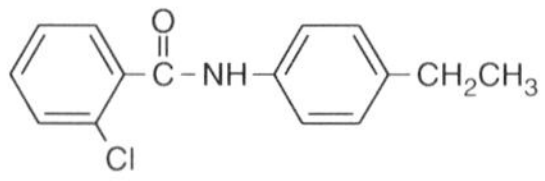

250 mg

S90,284-5
4-CHLORO-2'-(N-METHYLCARBAMOYL)-BENZANILIDE

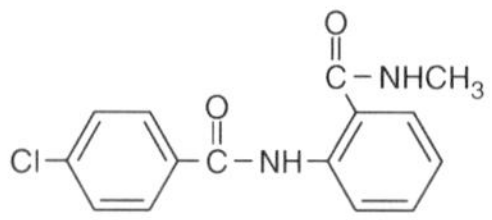

250 mg

S39,743-1
4-CHLORO-2'-FLUOROBENZANILIDE

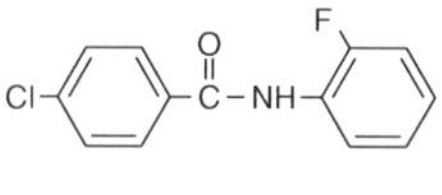

250 mg

S69,495-9
2',3'-DICHLORO-2-FLUOROBENZANILIDE

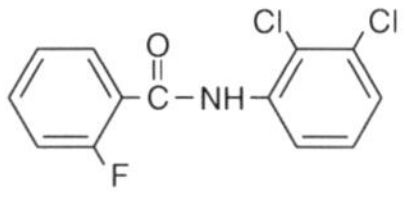

250 mg

S71,409-7
2,2',6'-TRICHLOROBENZANILIDE

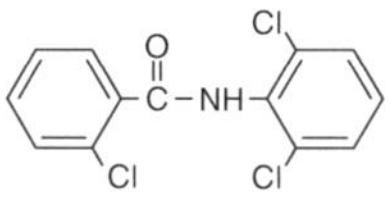

5 gm

S70,853-4
2-BROMO-2',6'-DICHLOROBENZANILIDE

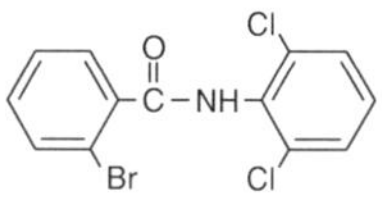

250 mg

S71,399-6
2,2'-DICHLORO-6'-METHYLBENZANILIDE

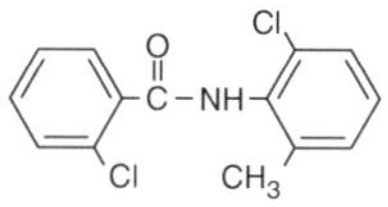

5 gm

S70,832-1
2-BROMO-2'-CHLORO-6'-METHYLBENZANILIDE

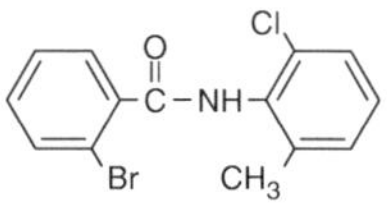

250 mg

S70,814-3
2-CHLORO-2',6'-DIMETHYLBENZANILIDE

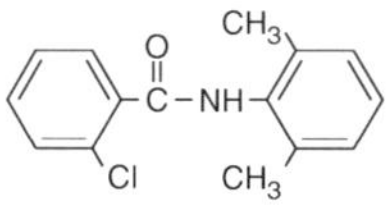

5 gm

S71,542-5
2-BROMO-2',6'-DIMETHYLBENZANILIDE

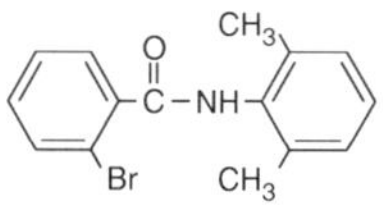

250 mg

S72,457-2
2-CHLORO-2',6'-DIETHYLBENZANILIDE

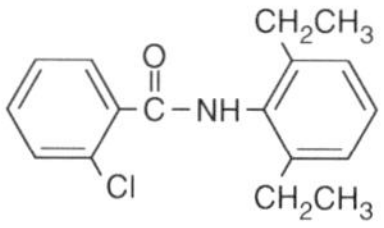

5 gm

S70,922-0
2-BROMO-2',6'-DIETHYLBENZANILIDE

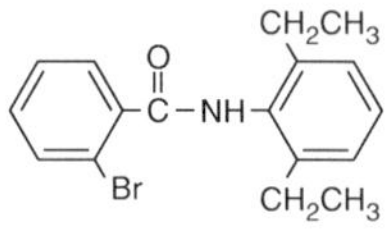

250 mg

S71,396-1
2-CHLORO-2',6'-DIISOPROPYLBENZANILIDE

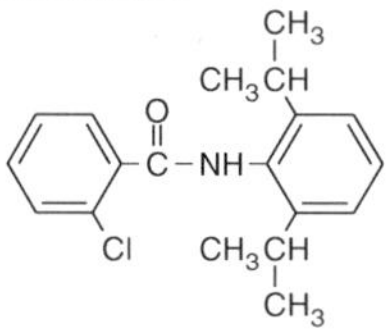

5 gm

S70,754-6
2-BROMO-2',6'-DIISOPROPYLBENZANILIDE

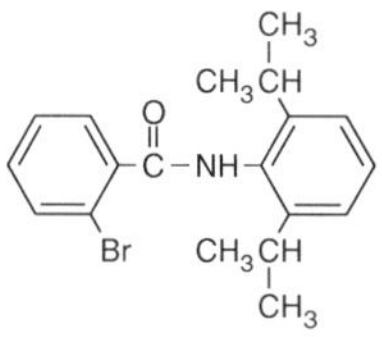

5 gm

S71,664-2
2,3',4'-TRICHLOROBENZANILIDE

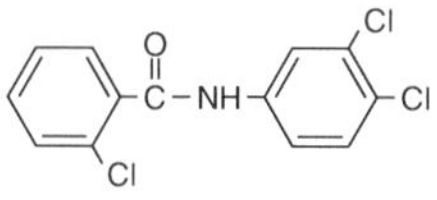

5 gm

S79,117-2
2,3'-DICHLORO-4'-METHYLBENZANILIDE
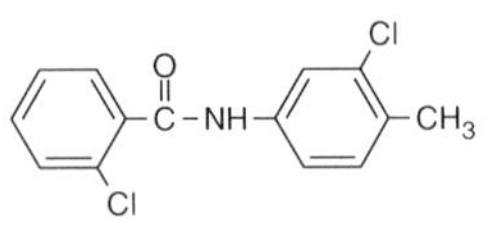
250 mg

S84,469-1
2,2'-DICHLORO-5'-METHYLBENZANILIDE
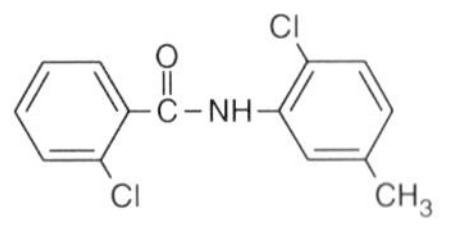
250 mg

S89,059-6
2,5'-DICHLORO-2'-METHYLBENZANILIDE
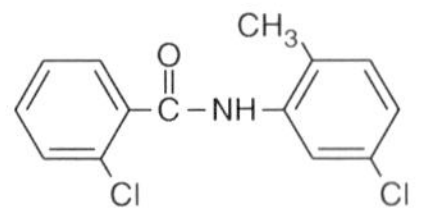
250 mg

S70,888-7
2,3',5'-TRICHLOROBENZANILIDE
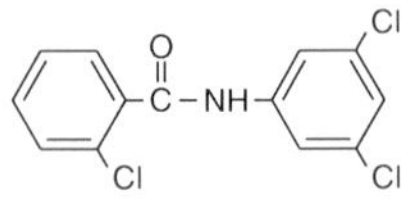
250 mg

S70,897-6
2,2',4',6'-TETRACHLOROBENZANILIDE
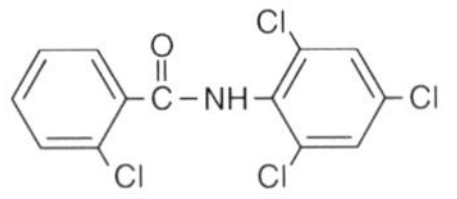
250 mg

S70,835-6
2-METHYL-2',4',6'-TRICHLOROBENZANILIDE
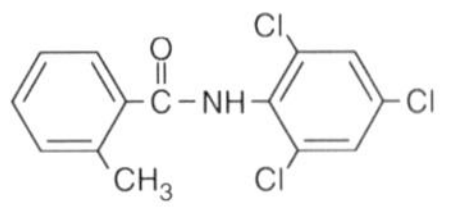
5 gm

S71,690-1
2-METHYL-2',4',6'-TRIBROMOBENZANILIDE
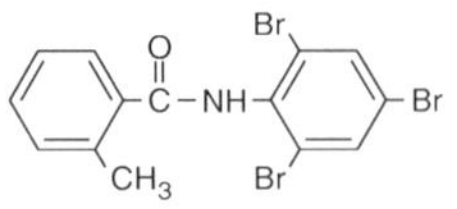
5 gm

S71,405-4
2-CHLORO-2',4',6'-TRIMETHYLBENZANILIDE
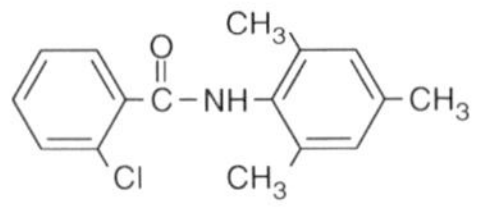
5 gm

S71,151-9
2-BROMO-2',4',6'-TRIMETHYLBENZANILIDE
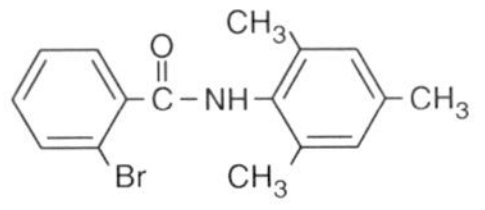
5 gm

S89,306-4
2'-METHYL-2,4',5'-TRICHLOROBENZANILIDE
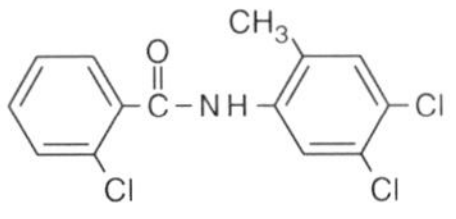
250 mg

S73,213-3
4-FLUORO-N-METHYLBENZAMIDE
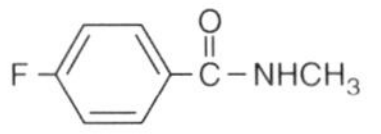
250 mg

S89,017-0
3-CHLORO-4'-FLUOROBENZANILIDE
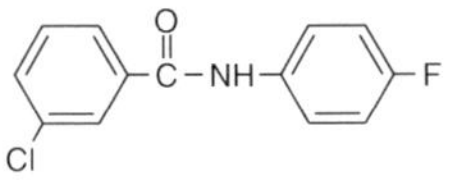
250 mg

S71,780-0
3,4'-DIBROMOBENZANILIDE
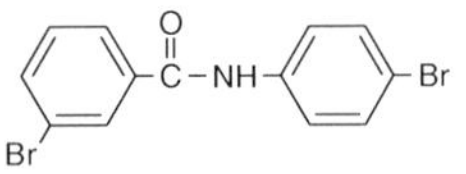
5 gm

S91,001-5
4'-FLUORO-3-METHYLBENZANILIDE
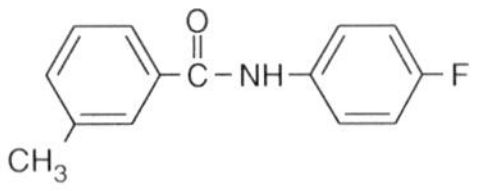
250 mg

S91,013-9
4'-FLUORO-4-METHYLBENZANILIDE
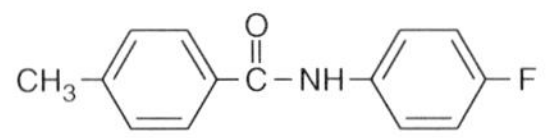
250 mg

S89,013-8
3-CHLORO-4'-ETHYLBENZANILIDE
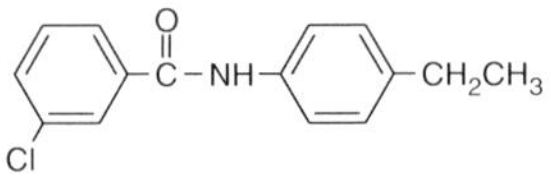
250 mg

S71,034-2
2',4,6'-TRICHLOROBENZANILIDE
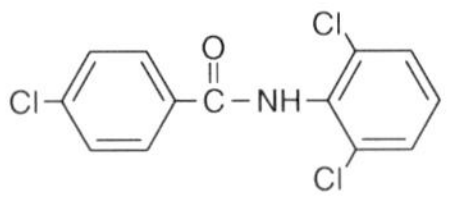
5 gm

S71,685-5
2',3',4-TRICHLOROBENZANILIDE
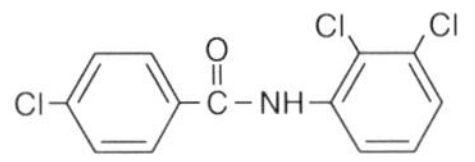
5 gm

S70,823-2
2',4-DICHLORO-6'-METHYLBENZANILIDE
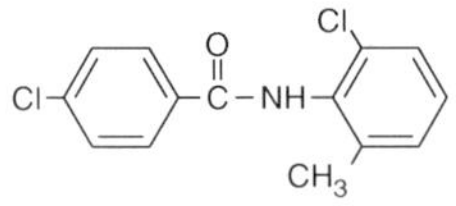
5 gm

S70,828-3
2',6'-DICHLORO-4-METHYLBENZANILIDE
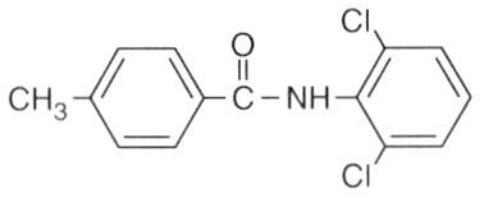
250 mg

S71,236-1
4-BROMO-2'-CHLORO-6'-METHYLBENZANILIDE
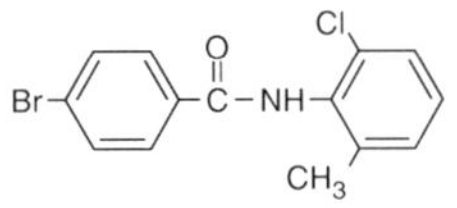
250 mg

S70,813-5
2'-CHLORO-4,6'-DIMETHYLBENZANILIDE
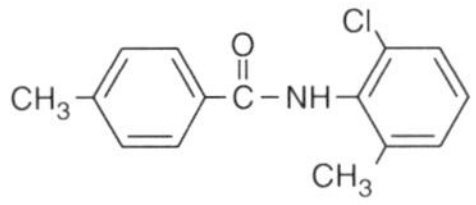
250 mg

S71,466-6
4-CHLORO-2',6'-DIMETHYLBENZANILIDE
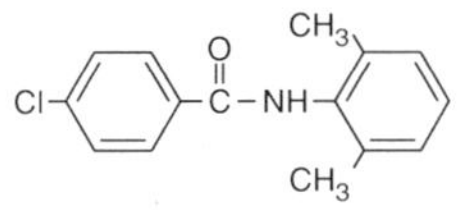
5 gm

S70,784-8
4-CHLORO-2'-ETHYL-6'-METHYLBENZANILIDE
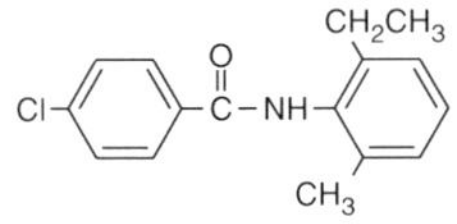
5 gm

S71,511-5
4-CHLORO-2',6'-DIETHYLBENZANILIDE
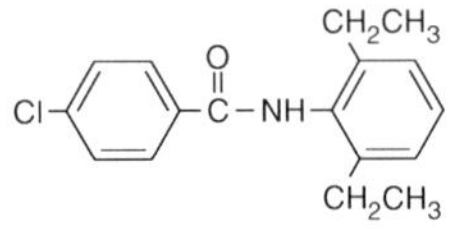
250 mg

S71,663-4
4-BROMO-2',6'-DIETHYLBENZANILIDE
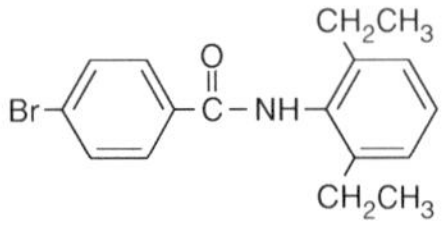
5 gm

S73,580-9
4-BROMO-2',6'DIISOPROPYLBENZANILIDE
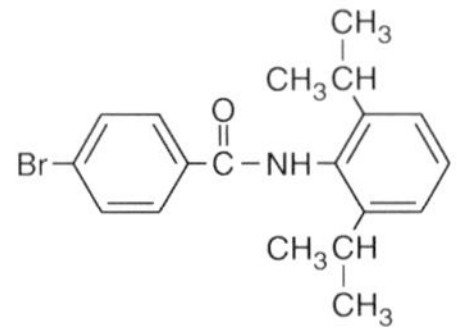
250 mg

S74,701-7
N,N'-BIS(2-CHLORO-6-METHYLPHENYL)-TEREPHTHALAMIDE
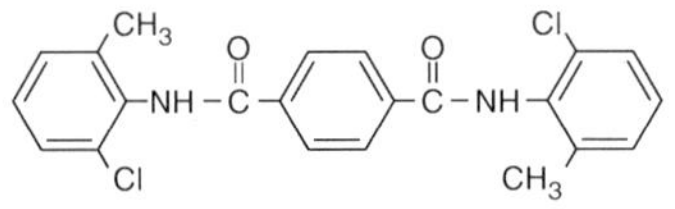
5 gm

S89,055-3
3',4-DICHLORO-4'-METHYLBENZANILIDE
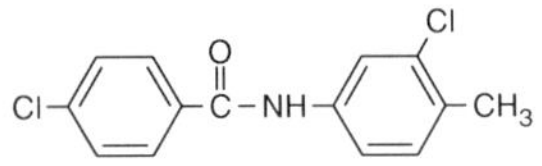
250 mg

S89,056-1
4,4'-DICHLORO-2'-METHYLBENZANILIDE
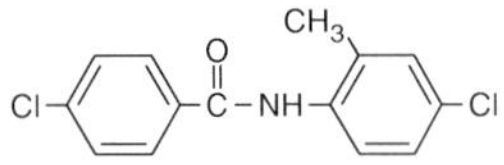
250 mg

S89,058-8
4,5'-DICHLORO-2'-METHYLBENZANILIDE
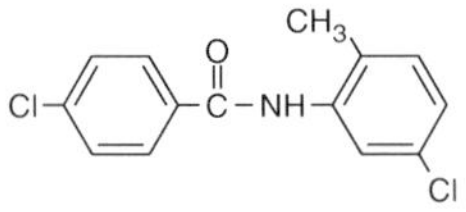
250 mg

S92,641-8
2'-CHLORO-4,4'-DIMETHYLBENZANILIDE
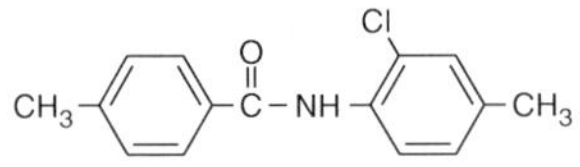
250 mg

S72,456-4
3,4-DICHLORO-4'-ETHYLBENZANILIDE

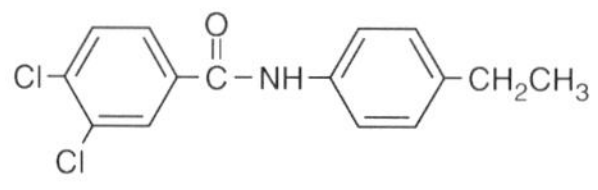

5 gm

S72,471-8
2,4-DICHLORO-4'-ETHYLBENZANILIDE

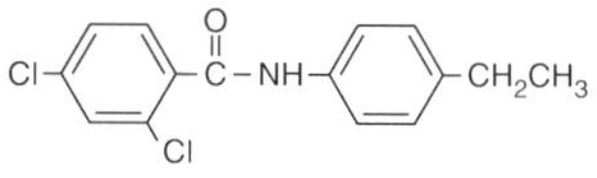

5 gm

S39,720-2
2,4-DICHLORO-4'-FLUOROBENZANILIDE

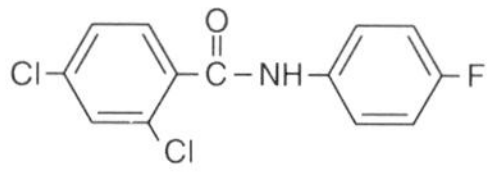

250 mg

S90,459-7
3',4,5'-TRICHLOROBENZANILIDE

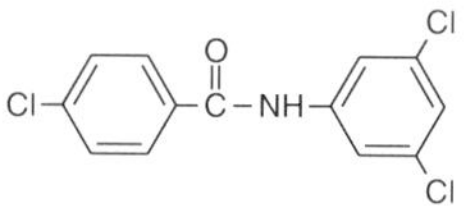

250 mg

S73,771-2
4'-BROMO-2',6'-DIMETHYL-4-FLUOROBENZANILIDE

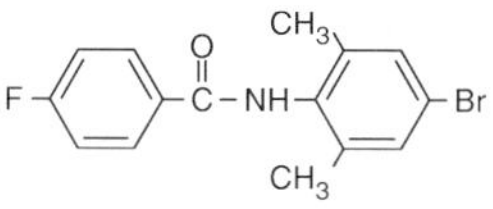

5 gm

S71,404-6
4-CHLORO-2',4',6'-TRIMETHYLBENZANILIDE

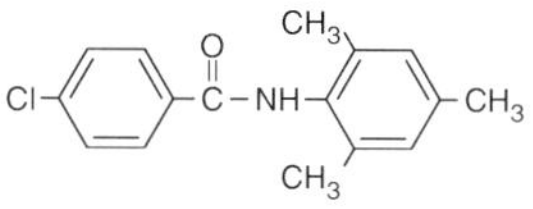

5 gm

S69,509-2
N-(4-BROMO-2,6-DIMETHYLPHENYL)-4-TERT-BUTYLBENZAMIDE

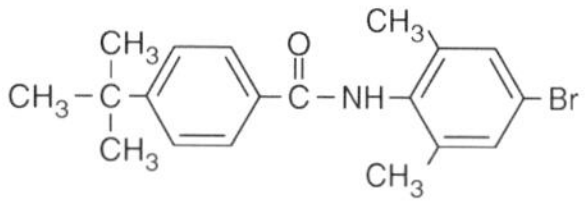

250 mg

S70,857-7
2',3,6'-TRICHLOROBENZANILIDE

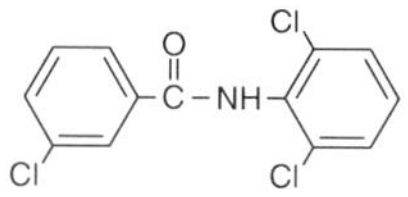

250 mg

S70,827-5
2',3-DICHLORO-6'-METHYLBENZANILIDE

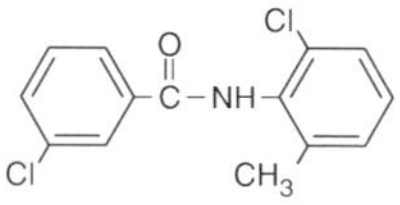

250 mg

S70,966-2
2,2',6-TRICHLORO-O-BENZOTOLUIDIDE

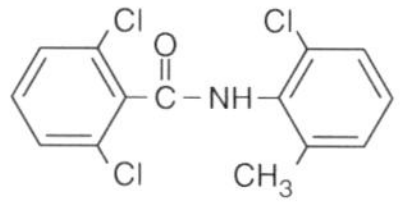

250 mg

S73,573-6
3-BROMO-2'-CHLORO-6'-METHYLBENZANILIDE

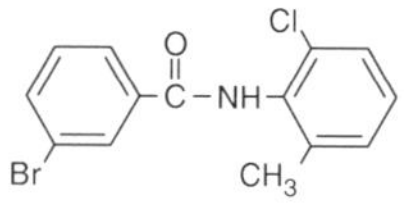

250 mg

S80,127-5
2',3'-DIMETHYL-3-FLUOROBENZANILIDE

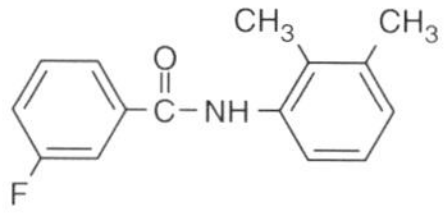

250 mg

S70,811-9
2'-CHLORO-3,6'-DIMETHYLBENZANILIDE

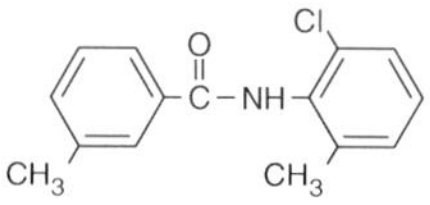

5 gm

S70,812-7
3-CHLORO-2',6'-DIMETHYLBENZANILIDE

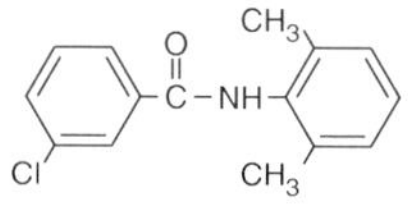

5 gm

S77,371-9
3-CHLORO-2',3'-DIMETHYLBENZANILIDE

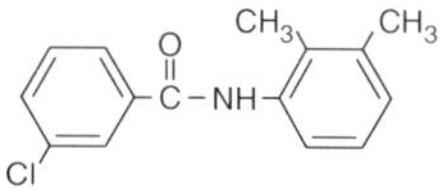

250 mg

S70,898-4
2,6-DICHLOROBENZO-2',6'-XYLIDIDE

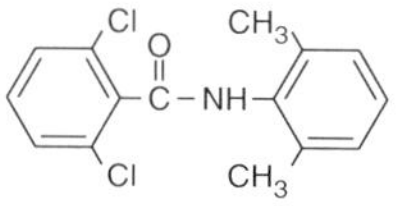

250 mg

S73,581-7
3-BROMO-2',6'-DIMETHYLBENZANILIDE

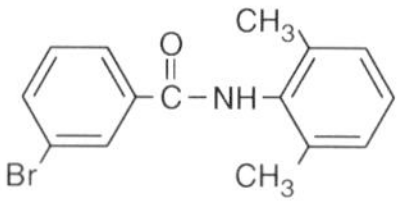

250 mg

S70,782-1
3-CHLORO-2'-ETHYL-6'-METHYLBENZANILIDE

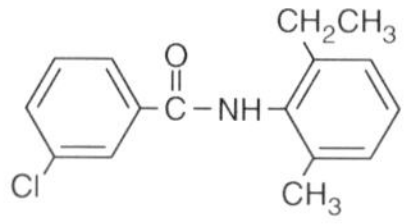

5 gm

S70,788-0
3-BROMO-2'-ETHYL-6'-METHYLBENZANILIDE

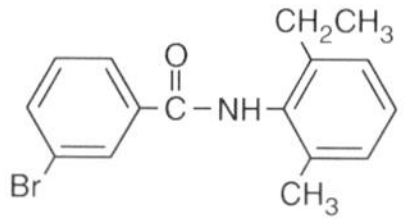

250 mg

S70,774-0
3-CHLORO-2',6'-DIETHYLBENZANILIDE

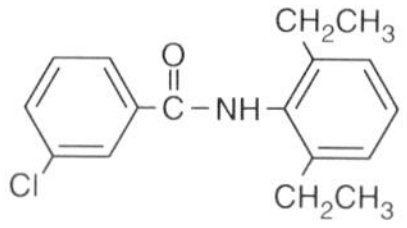

5 gm

S71,776-2
2,6-DICHLORO-2',6'-DIETHYLBENZANILIDE

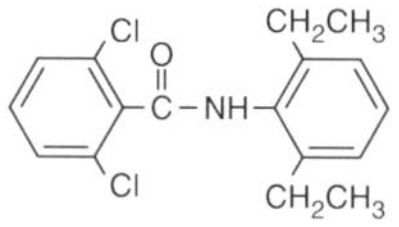

5 gm

S71,284-1
3-BROMO-2',6'-DIETHYLBENZANILIDE

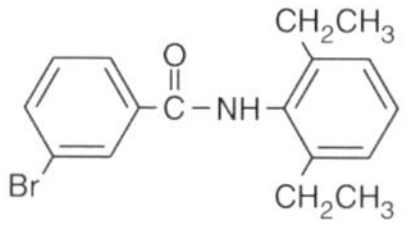

5 gm

S70,753-8
3-CHLORO-2',6'-DIISOPROPYLBENZANILIDE

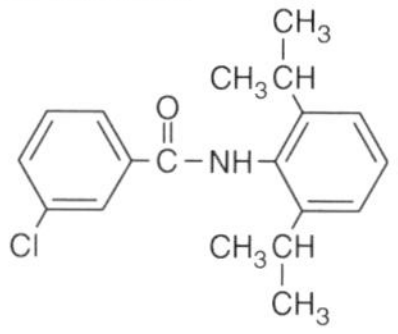

250 mg

S71,692-8
3-BROMO-2',6'-DIISOPROPYLBENZANILIDE

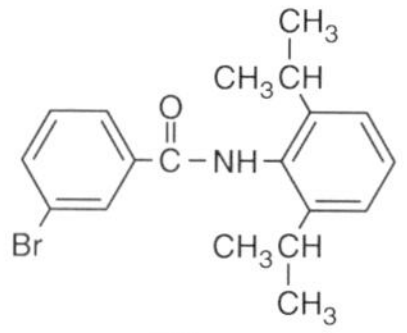

250 mg

S70,845-3
2,2',5,6'-TETRACHLOROBENZANILIDE

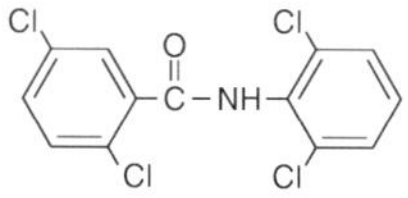

250 mg

S73,572-8
2',3,4,6'-TETRACHLOROBENZANILIDE

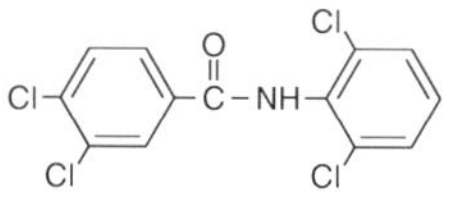

250 mg

S77,411-1
2',6'-DIBROMO-3,4-DICHLOROBENZANILIDE

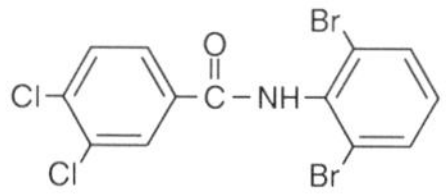

250 mg

S70,836-4
2,4,6'-TRICHLORO-O-BENZOTOLUIDIDE

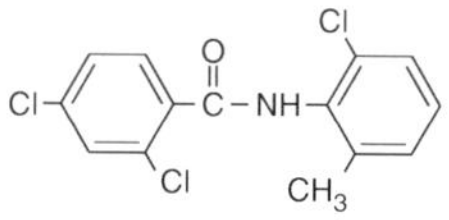

5 gm

S70,837-2
3,4,6'-TRICHLORO-O-BENZOTOLUIDIDE

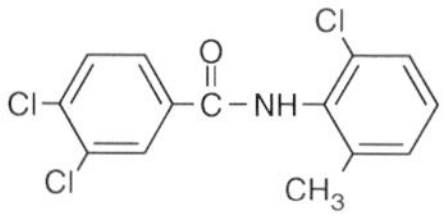

5 gm

S70,799-6
2,4-DICHLORO-2',6'-DIMETHYLBENZANILIDE

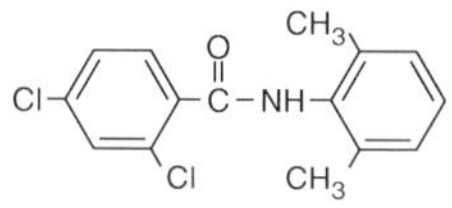

250 mg

S70,800-3
2,5-DICHLORO-2',6'-DIMETHYLBENZANILIDE

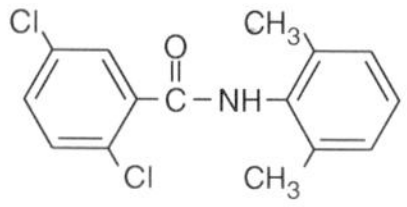

250 mg

S70,776-7
2,4-DICHLORO-2',6'-DIETHYLBENZANILIDE

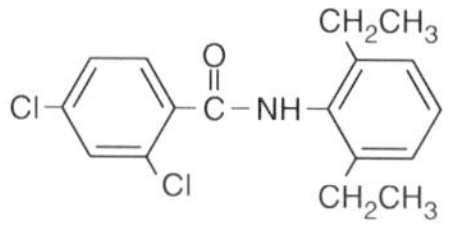

250 mg

S71,529-8
2,5-DICHLORO-2',6'-DIETHYLBENZANILIDE
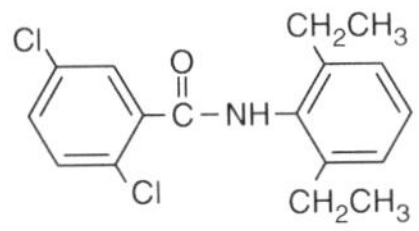
100 mg

S70,756-2
2,4-DICHLORO-2',6'-DIISOPROPYLBENZANILIDE
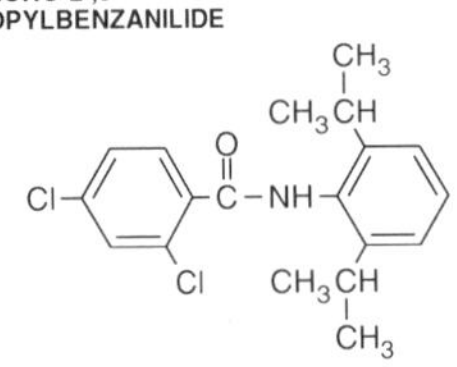
5 gm

S71,401-1
3,4-DICHLORO-2',6'-DIISOPROPYLBENZANILIDE
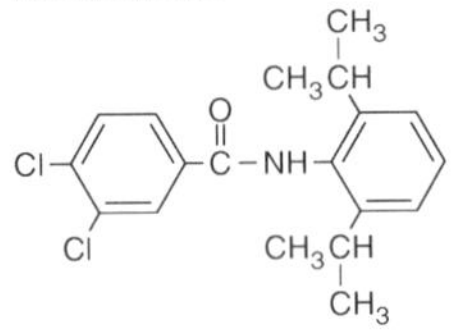
5 gm

S71,470-4
2,5-DICHLORO-2',6'-DIISOPROPYLBENZANILIDE
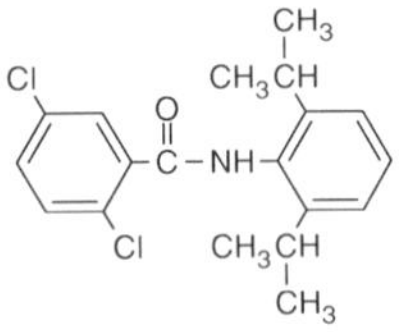
5 gm

S72,916-7
3'-CHLORO-4'-METHYL-2,4-HEXADIENANILIDE
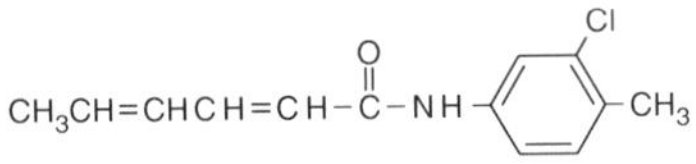
5 gm

S80,143-7
2',5'-DICHLORO-3-FLUOROBENZANILIDE
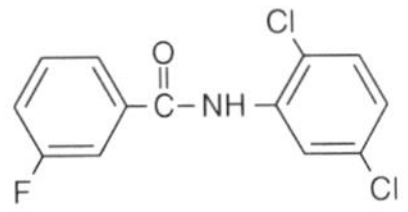
250 mg

S90,134-2
2',3,5'-TRICHLOROBENZANILIDE
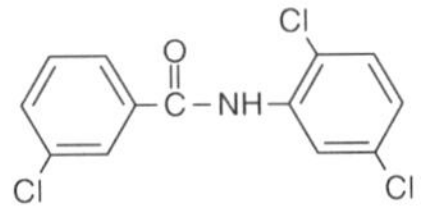
250 mg

S70,944-1
3-BROMO-2',5'-DICHLOROBENZANILIDE
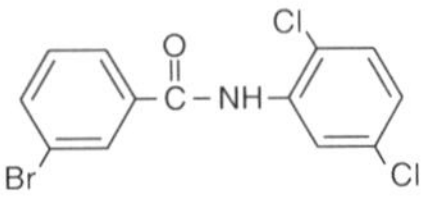
5 gm

S71,407-0
3-BROMO-3',4'-DICHLOROBENZANILIDE
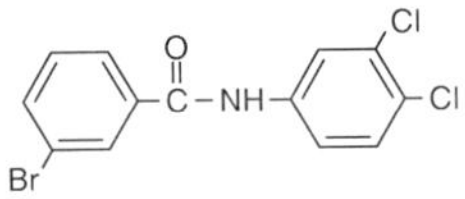
5 gm

S80,142-9
3'-CHLORO-3-FLUORO-4'-METHYLBENZANILIDE
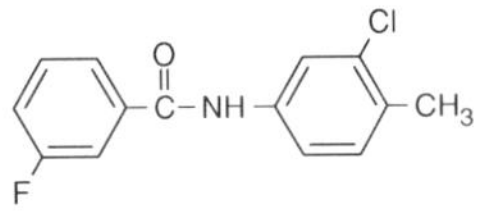
250 mg

S75,136-7
3,4,6'-TRICHLORO-M-BENZOTOLUIDIDE
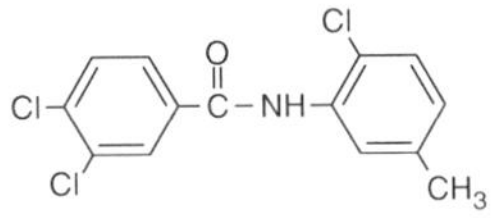
5 gm

S77,321-2
2,4,5'-TRICHLORO-O-BENZOTOLUIDIDE
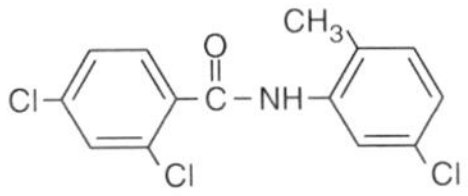
250 mg

S79,115-6
2'-METHYL-2,4,4'-TRICHLOROBENZANILIDE
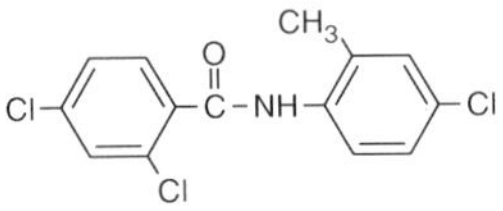
250 mg

S89,018-9
4'-METHYL-2,3',4-TRICHLOROBENZANILIDE
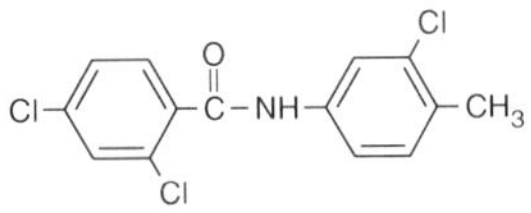
250 mg

S89,935-6
5'-METHYL-2',3,4-TRICHLOROBENZANILIDE
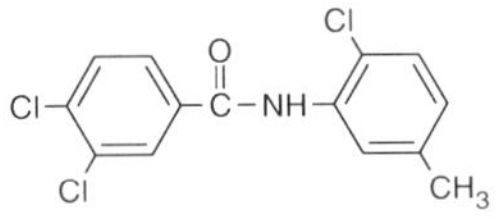
250 mg

S89,936-4
5'-METHYL-2,2',4-TRICHLOROBENZANILIDE
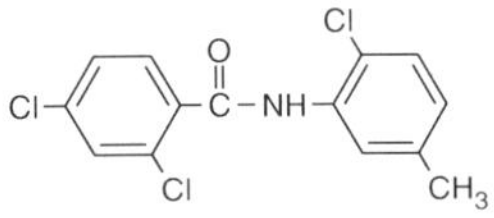
250 mg

S90,463-5
3,4-DICHLORO-N-(2,4-DICHLOROBENZYL)-BENZAMIDE
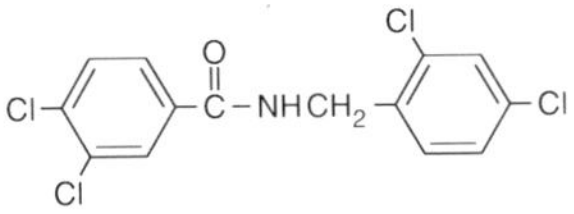
250 mg

S76,827-8
3-CHLORO-2',4'-DIMETHYLBENZANILIDE
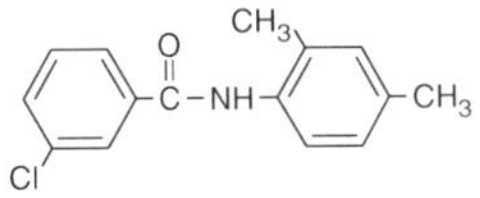
250 mg

S89,014-6
3-CHLORO-2',5'-DIMETHYLBENZANILIDE
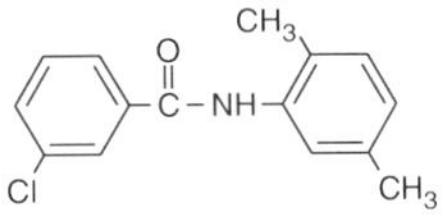
250 mg

S79,124-5
3,4-DICHLORO-2',5'-DIMETHYLBENZANILIDE
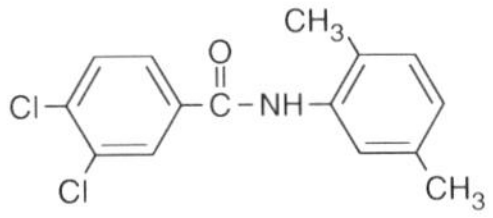
250 mg

S90,465-1
3,3',4,5'-TETRACHLOROBENZANILIDE
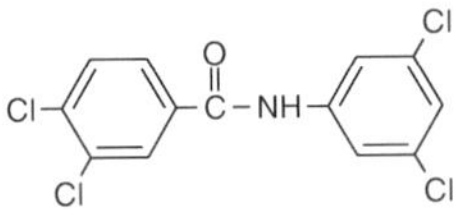
250 mg

S73,797-6
4'-BROMO-2,4-DICHLORO-2',6'-DIMETHYLBENZANILIDE
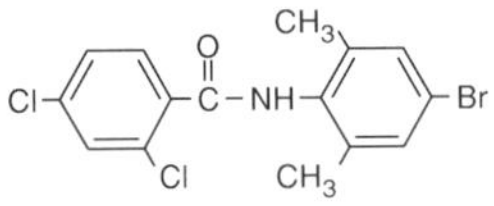
5 gm

S89,943-7
2,4-DICHLORO-2',4',6'-TRIMETHYLBENZANILIDE
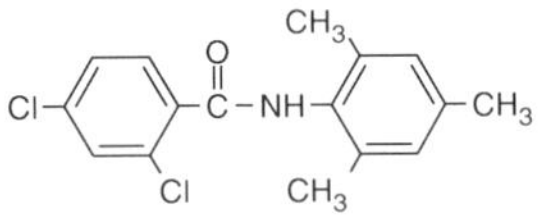
250 mg

S89,934-8
2'-METHYL-2,4,4',5'-TETRACHLOROBENZANILIDE
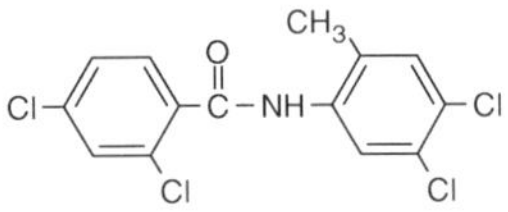
250 mg

S71,778-9
2',3,4',6'-TETRACHLOROBENZANILIDE
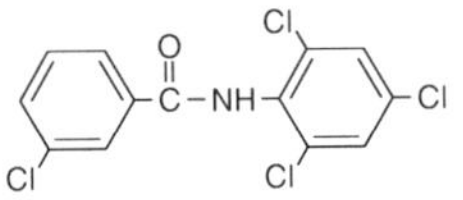
250 mg

S73,561-2
3-CHLORO-2',4',6'-TRIBROMOBENZANILIDE
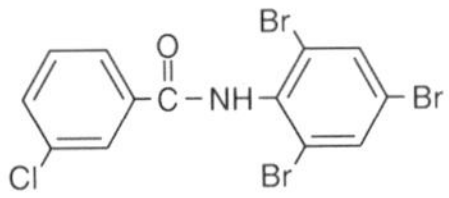
250 mg

S71,465-8
3-METHYL-2',4',6'-TRICHLOROBENZANILIDE
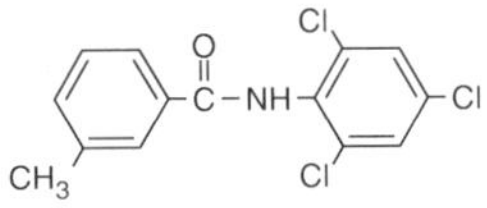
250 mg

S71,233-7
3-METHYL-2',4',6'-TRIBROMOBENZANILIDE
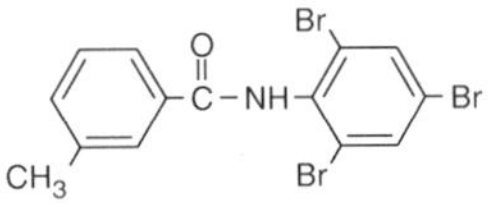
250 mg

S71,403-8
3-CHLORO-2',4',6'-TRIMETHYLBENZANILIDE
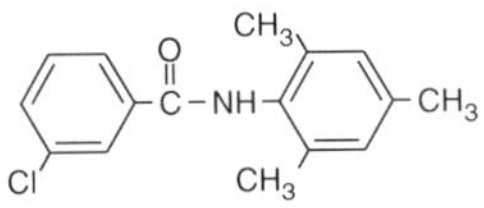
5 gm

S73,576-0
3-BROMO-2',4',6'-TRIMETHYLBENZANILIDE
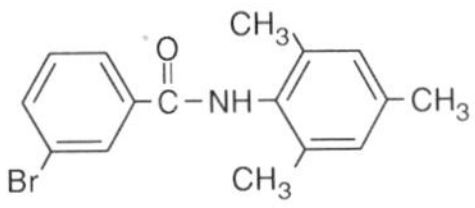
250 mg

S78,040-5
2',4',5'-TRICHLORO-2,4-HEXADIENANILIDE
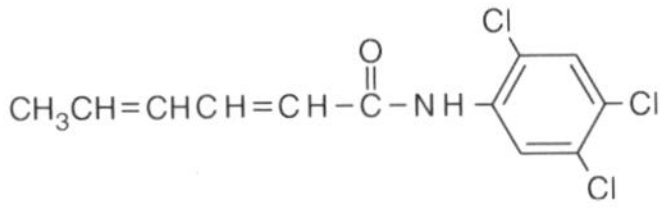
250 mg

S80,130-5
3-FLUORO-2',4',5'-TRIMETHYLBENZANILIDE
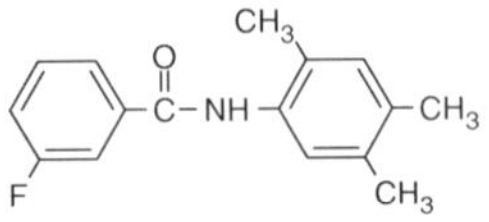
250 mg

S77,478-2
4-CHLORO-N,N-DIISOPROPYLBENZAMIDE

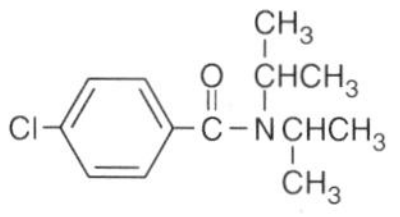

250 mg

S70,113-0
2'-FLUOROMALEANILIC ACID

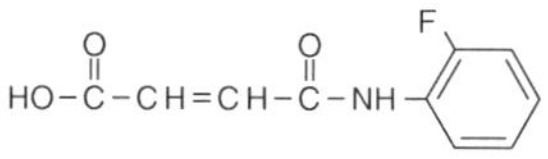

5 gm

S70,104-1
2',4'-DICHLOROPHTHALANILIC ACID

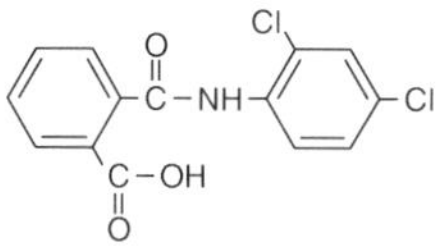

5 gm

S94,390-8
2'-BROMO-3'-METHYLBENZANILIDE

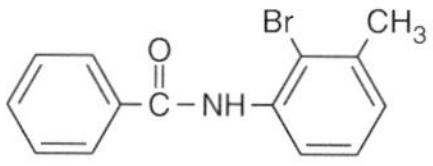

250 mg

S79,521-6
3'-CHLORO-2'-METHYLCINNAMANILIDE

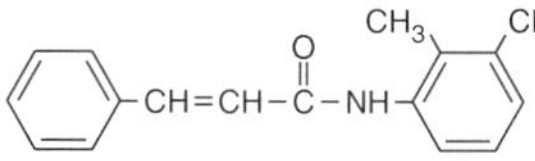

250 mg

S79,523-2
4',5'-DICHLORO-2'-METHYLCINNAMANILIDE

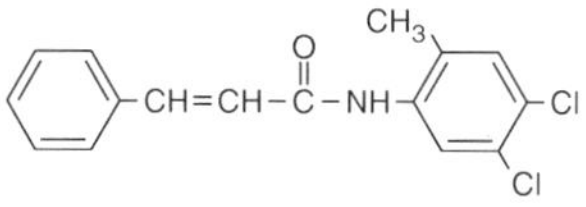

250 mg

S71,815-7
4'-BROMO-2-BUTENANILIDE

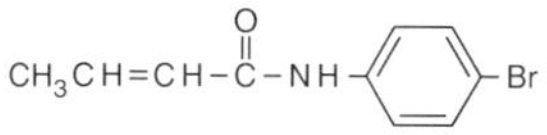

250 mg

S89,010-3
2,4-DICHLORO-N-ETHYL-2'-METHYLBENZANILIDE

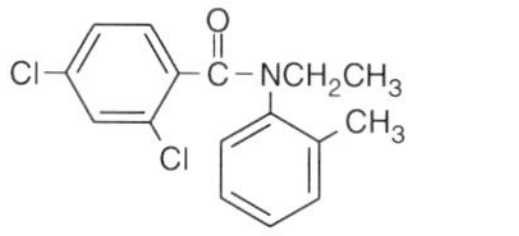

250 mg

S56,593-8
N-(2-HYDROXYETHYL)-3,4,5-TRIHYDROXYBENZAMIDE

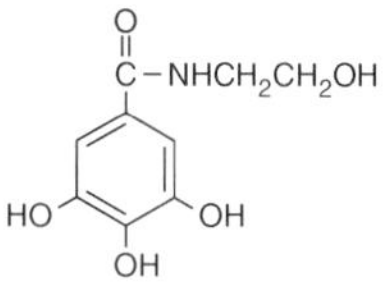

5 gm

S36,060-0
3,5-DIHYDROXYBENZAMIDE

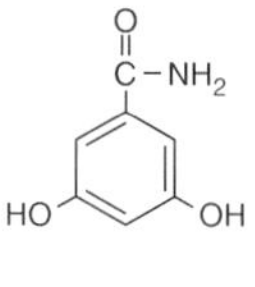

5 gm

S40,892-1
4'-HYDROXYSALICYLANILIDE

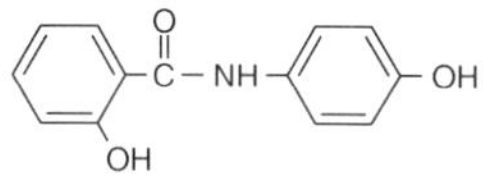

5 gm

S68,156-3
4-[2-(DIETHYLAMINO)ETHOXY]-A-METHYL-A-[2-(METHYLCARBAMOYL)-BENZYL]BENZYL ALCOHOL

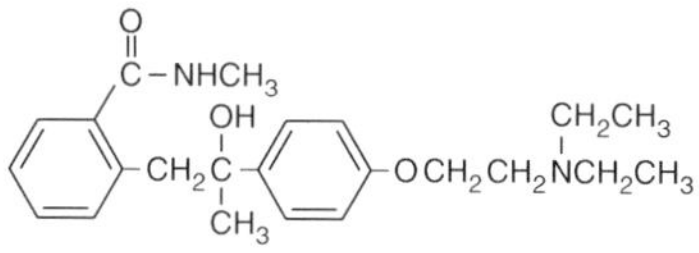

250 mg

S73,533-7
2-[2-(4-BENZYLOXYPHENYL)-2-HYDROXYPROPYL]-N-METHYLBENZAMIDE

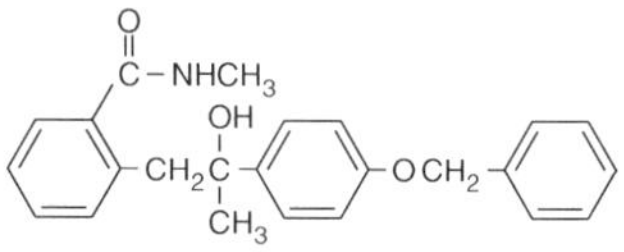

250 mg

S75,947-3
2-[2-(4-BENZYLOXYPHENYL)-2-HYDROXY-1-PHENYLETHYL]-N-METHYLBENZAMIDE

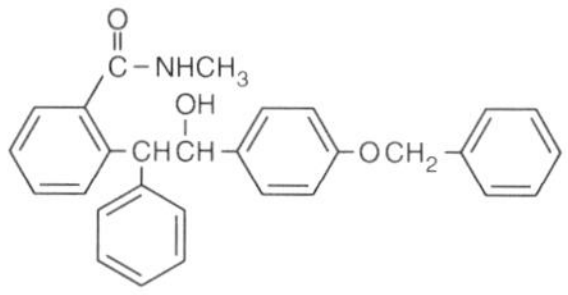

250 mg

S76,854-5
3'-HYDROXY-3,4,5-TRIMETHOXYBENZANILIDE

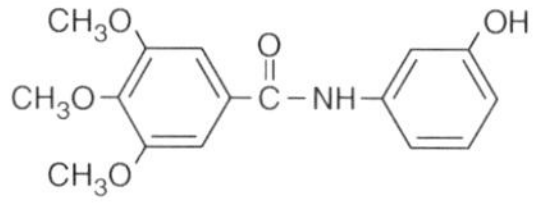

250 mg

S60,399-6
4'-AMINO-4-HEXADECYLOXY-2'-HYDROXY-3-METHOXYBENZANILIDE

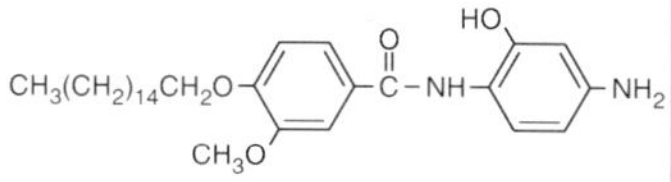

250 mg

S40,394-6
2-(2-HYDROXYETHOXY)BENZAMIDE

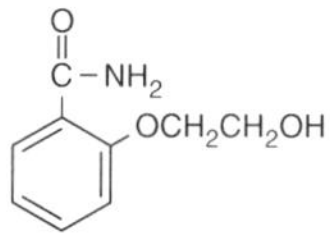

250 mg

S35,554-2
SYRINGAMIDE

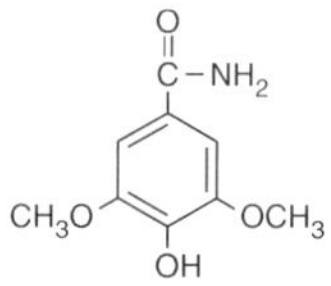

5 gm

S36,055-4
N,N-DIETHYLVANILLAMIDE

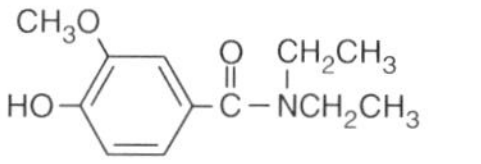

5 gm

S68,370-1
5-[5-METHOXY-2-(N-METHYLCARBAMOYL)-PHENYL]DIBENZOSUBEROL

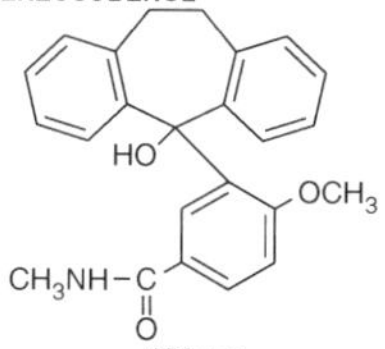

250 mg

S59,880-1
4'-AMINO-2'-HYDROXY-O-ANISANILIDE

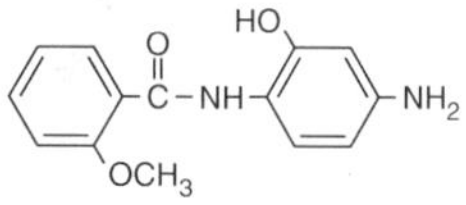

5 gm

S59,186-6
4'-AMINO-2-HEXADECYLOXY-2'-HYDROXY-M-TOLUANILIDE

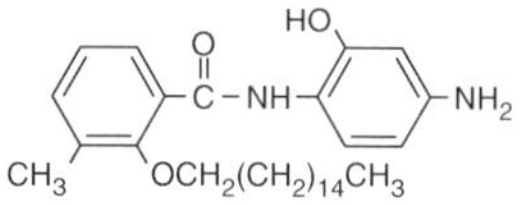

250 mg

S59,194-7
4'-AMINO-3-HEXADECYLOXY-2'-HYDROXYBENZANILIDE

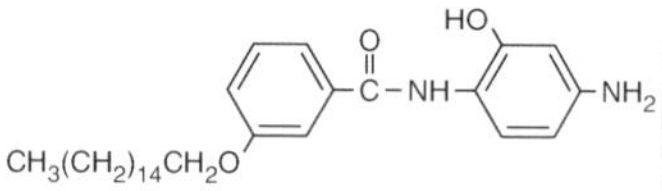

250 mg

S50,237-5
2'-HYDROXY-4-METHOXY-4'-NITROBENZANILIDE

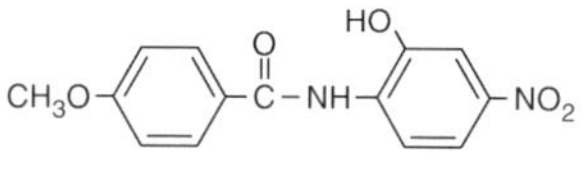

250 mg

S76,098-6
2-[10,11-DIHYDRO-5-HYDROXY-5H-DIBENZO(A,D)CYCLOHEPTEN-5-YL]4,N-DIMETHYLBENZAMIDE

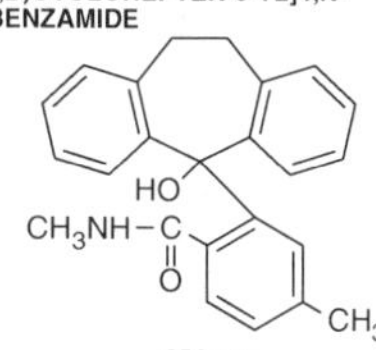

250 mg

S65,709-3
4-METHYL-2'-[N-(METHYL)CARBAMOYL]-BENZHYDRYL ALCOHOL

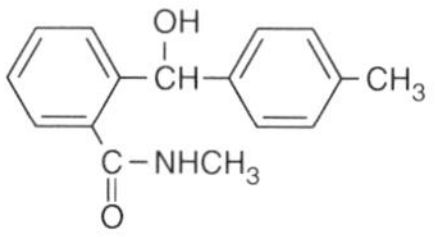

5 gm

S86,969-4
N-(2-HYDROXYBENZOYL)-L-ASPARTIC ACID

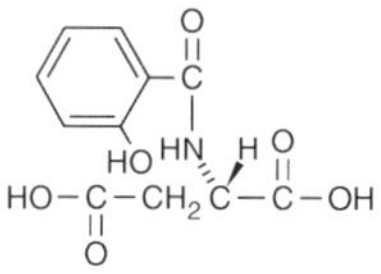

250 mg

S80,487-8
N-(2-HYDROXYBENZOYL)GLUTAMIC ACID

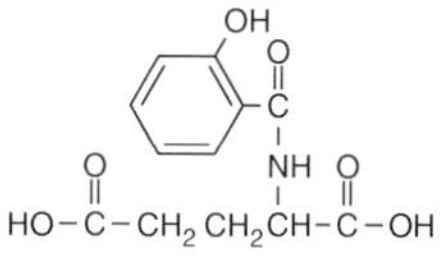

250 mg

S86,993-7
N-ACETYL-3-HYDROXY-2-NAPHTHALENECARBOXAMIDE

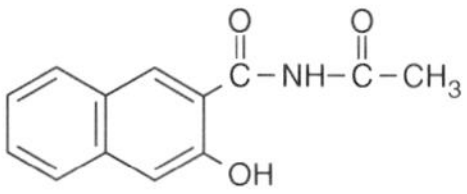

250 mg

S69,287-5
2'-HYDROXY-5'-METHYLMALEANILIC ACID

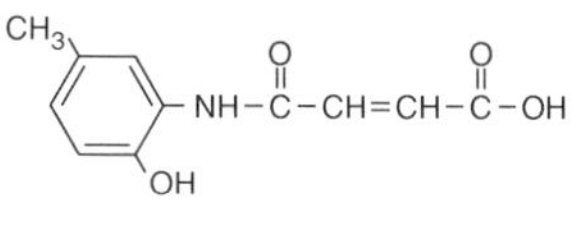

5 gm

S87,226-1
2-(HYDROXYMETHYL)-N-METHYLBENZAMIDE

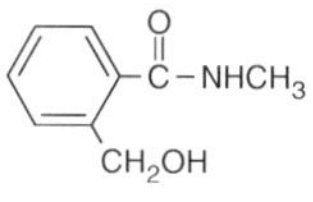

250 mg

S76,666-6
2-[10,11-DIHYDRO-5-HYDROXY-5H-DIBENZO(A,D)CYCLOHEPTEN-5-YL]-N-ETHYLBENZAMIDE

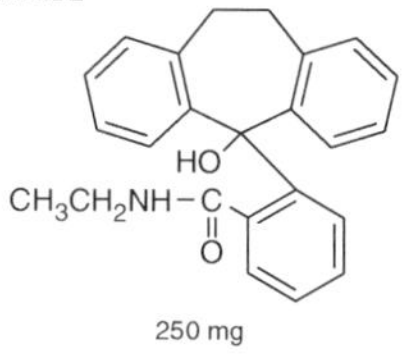

250 mg

S68,150-4
2-(ALPHA-HYDROXY-2-METHYL-ALPHA-PHENYLBENZYL)-N-METHYLBENZAMIDE

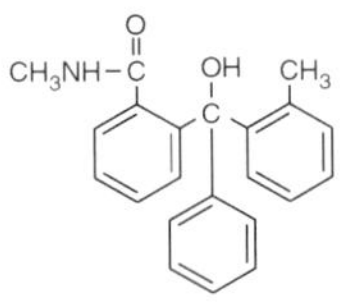

5 gm

S58,315-4
4'-AMINO-2'-HYDROXY-P-TOLUANILIDE

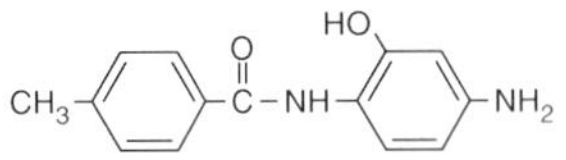

5 gm

S62,058-0
N-(2-DIETHYLAMINOETHYL)-P-HYDROXYBENZAMIDE HYDROCHLORIDE

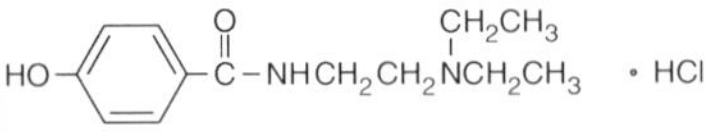

250 mg

S56,233-5
4-(TERT-BUTYL)-2'-HYDROXY-4'-NITROBENZANILIDE

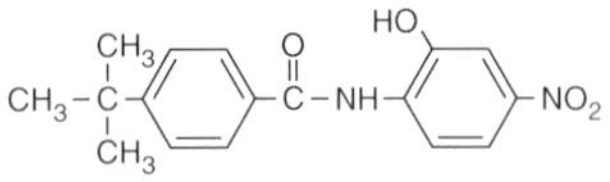

250 mg

S50,188-3
2'-HYDROXY-4'-NITROBENZANILIDE

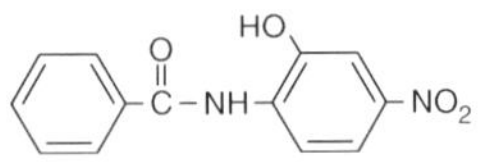

250 mg

S60,592-1
4'-BENZAMIDO-3'-HYDROXYSUCCINANILATE

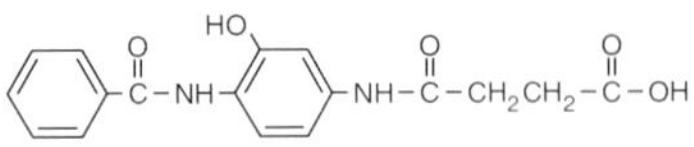

250 mg

S50,235-9
4'-AMINO-2'-HYDROXYBENZANILIDE

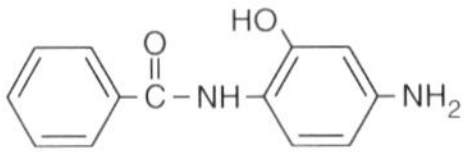

5 gm

S40,680-5
N,N'-BIS(2-HYDROXYETHYL)-TEREPHTHALAMIDE

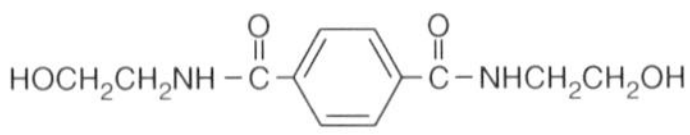

5 gm

S45,174-6
N-BENZOYL-L-TYROSINAMIDE

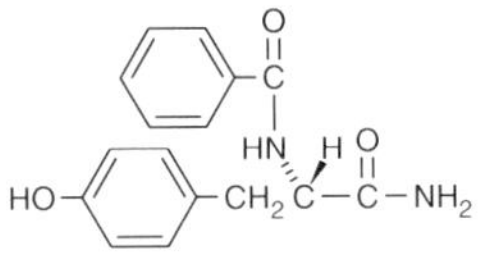

250 mg

S36,589-0
N-BENZOYL-3,5-DIIODO-DL-TYROSINAMIDE

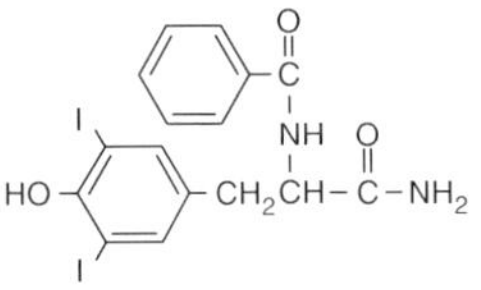

250 mg

S86,380-7
N-(2-HYDROXYCYCLOHEXYL)-4-NITROBENZAMIDE

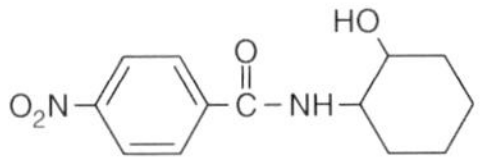

250 mg

S89,931-3
2'-HYDROXY-4-NITROBENZANILIDE

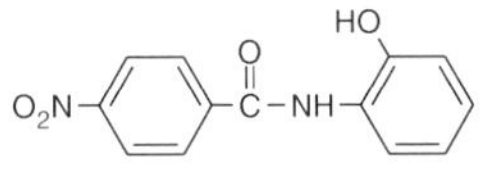

250 mg

S85,486-7
2'-HYDROXY-5'-NITROCINNAMANILIDE

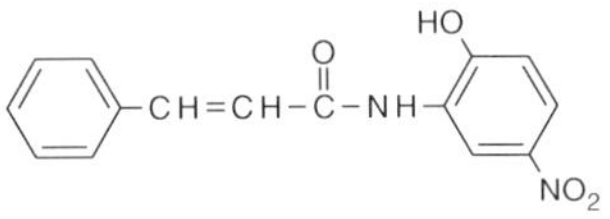

250 mg

S89,996-8
3'-HYDROXYPHTHALANILIC ACID

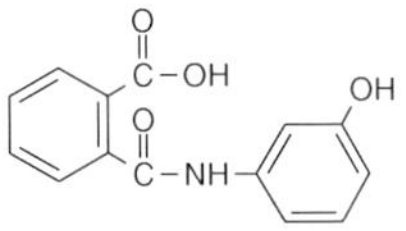

250 mg

S74,389-5
4'-(2-HYDROXYETHYL)-3,4,5,6-TETRABROMOPHTHALANILIC ACID

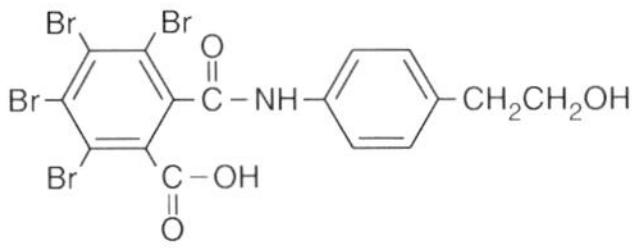

5 gm

S43,231-8
4-BENZAMIDOSALICYLIC ACID, CALCIUM SALT

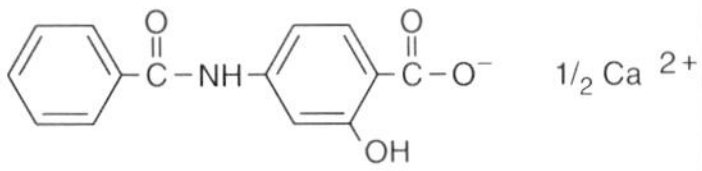

5 gm

S40,333-4
N-ACETYLSALICYLAMIDE

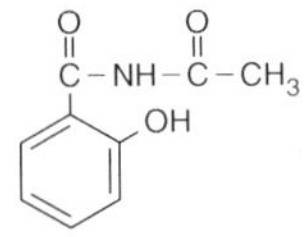

250 mg

S40,346-6
N-(HYDROXYMETHYL)BENZAMIDE

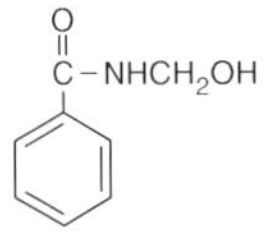

250 mg

S79,857-6
N-(1,1-DIMETHYL-2-HYDROXYETHYL)-BENZAMIDE

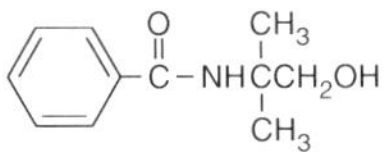

250 mg

S54,283-0
N-ALLYLSALICYLAMIDE

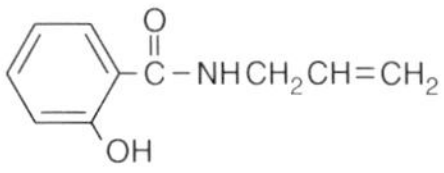

250 mg

S58,077-5
N-PROPYLSALICYLAMIDE

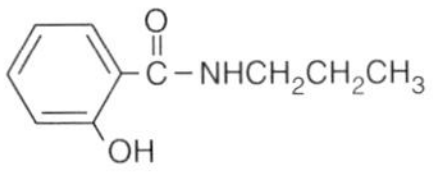

5 gm

S55,045-0
2-CHLORO-2'-HYDROXYBENZANILIDE

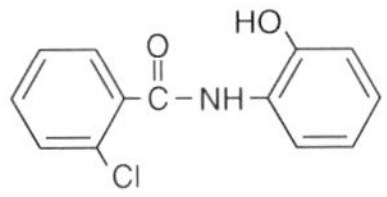

250 mg

S55,046-9
2-CHLORO-3'-HYDROXYBENZANILIDE

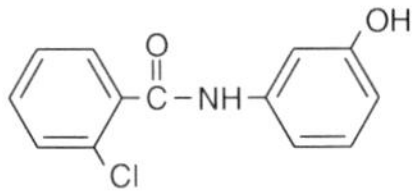

250 mg

S55,047-7
2-CHLORO-4'-HYDROXYBENZANILIDE

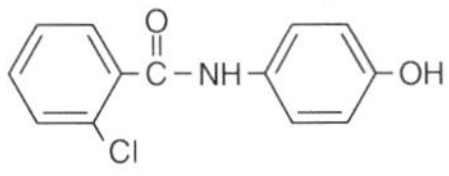

250 mg

S55,048-5
4-CHLORO-2'-HYDROXYBENZANILIDE

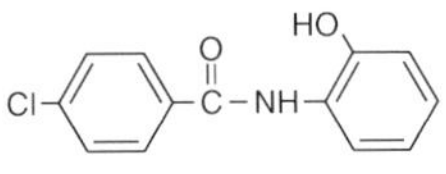

250 mg

S39,853-5
2-HYDROXY-2',6'-DIMETHYLBENZANILIDE

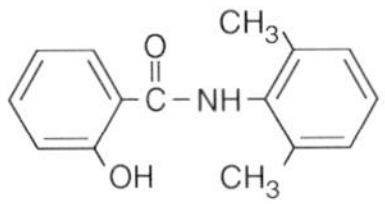

250 mg

S55,051-5
2,4-DICHLORO-2'-HYDROXYBENZANILIDE

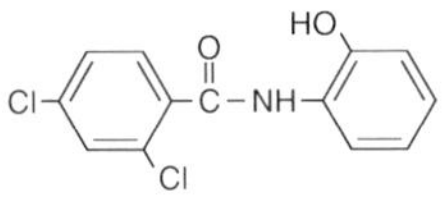

250 mg

S42,105-7
4'-CHLORO-O-SALICYLOTOLUIDIDE

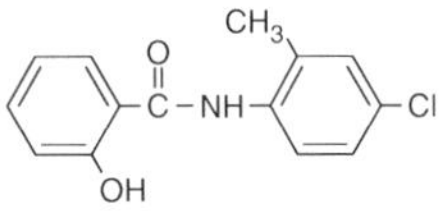

5 gm

S68,128-8
2-[5-HYDROXY-5H-DIBENZO(A,D)-CYCLOHEPTEN-5-YL]-N-METHYLBENZAMIDE

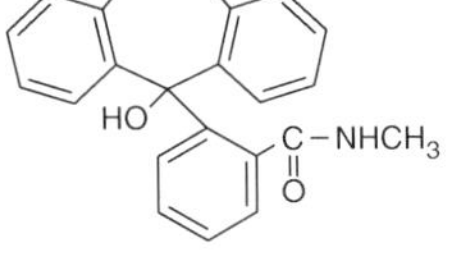

250 mg

S67,243-2
O-[10,11-DIHYDRO-5-HYDROXYDIBENZO-(A,D)CYCLOHEPTEN-5-YL]-N-METHYLBENZAMIDE

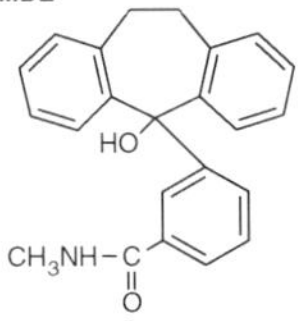

5 gm

S68,380-9
9-[2-(N-METHYLCARBAMOYL)BENZYL]-XANTHEN-9-OL

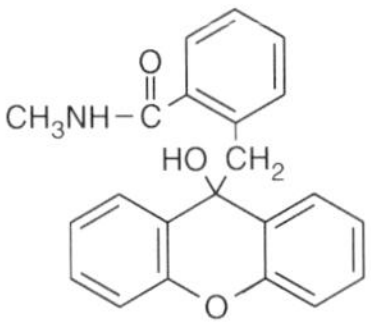

5 gm

S55,049-3
4-CHLORO-3'-HYDROXYBENZANILIDE

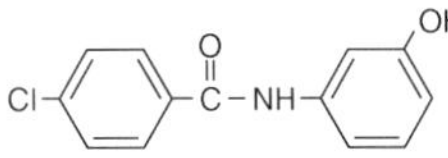

250 mg

S55,050-7
4-CHLORO-4'-HYDROXYBENZANILIDE
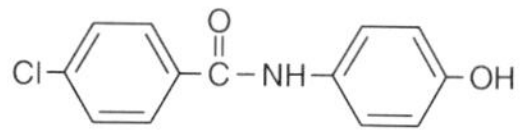
250 mg

S55,052-3
2,4-DICHLORO-4'-HYDROXYBENZANILIDE
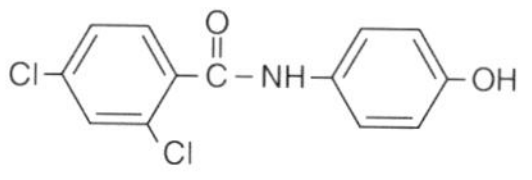
250 mg

S42,340-8
5-BROMO-4'-CHLOROSALICYLANILIDE
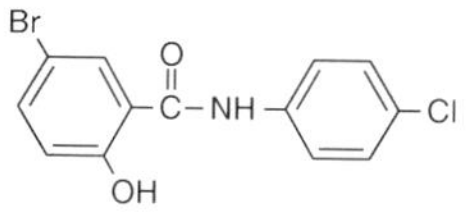
5 gm

S38,505-0
N-(HYDROXYMETHYL)SALICYLAMIDE
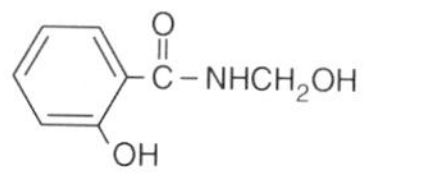
5 gm

S86,419-6
2-BENZAMIDO-3-METHOXY-3-PHENYLPROPIONIC ACID
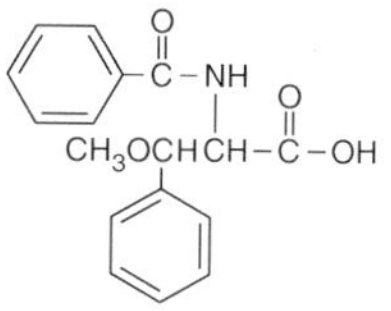
250 mg

S38,228-0
2-BENZYLOXYBENZAMIDE
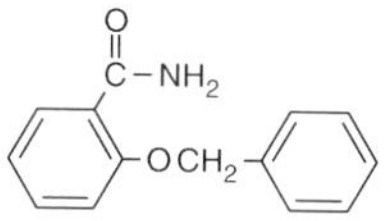
5 gm

S76,106-0
N-METHYL-3,4,5-TRIMETHOXYBENZAMIDE
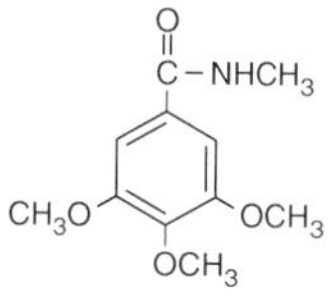
250 mg

S74,213-9
3,4,5-TRIMETHOXYBENZANILIDE
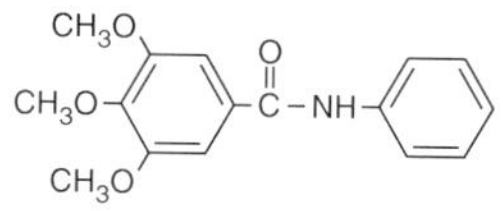
5 gm

S79,588-7
4'-ETHYL-3,4,5-TRIMETHOXYBENZANILIDE
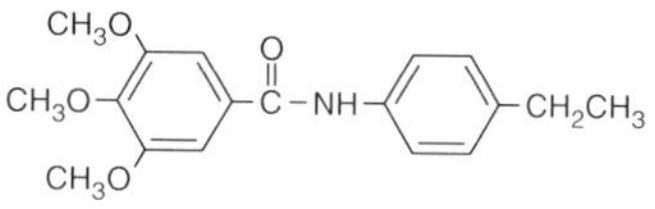
250 mg

S79,587-9
4'-ETHOXY-3,4,5-TRIMETHOXYBENZANILIDE
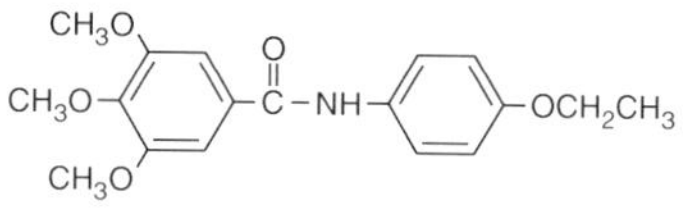
250 mg

S70,323-0
N-[3-(3,4-DIMETHOXYPHENYL)-1-METHYL-4-PIPERIDYL]-4-NITROBENZAMIDE
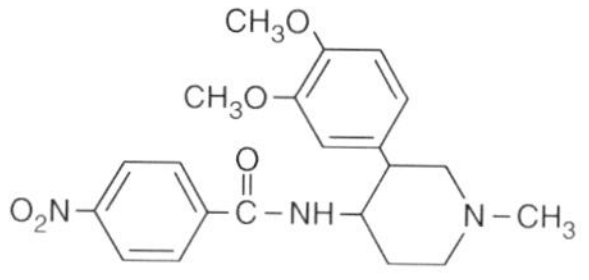
5 gm

S75,189-8
2,3-DIMETHOXYBENZAMIDE
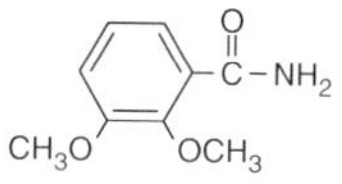
5 gm

S71,539-5
3,5-DIMETHOXY-2',6'-DIMETHYLBENZANILIDE
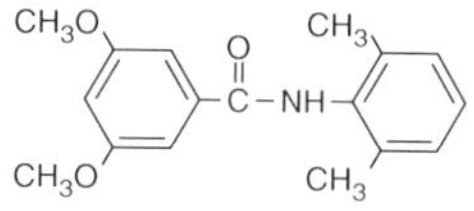
100 mg

S86,015-8
N,N'-DI-TERT-BUTYL-2,5-DIMETHOXYTEREPHTHALAMIDE
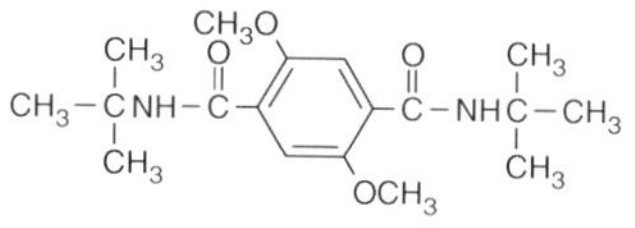
250 mg

S68,642-5
N-(4-METHYL-3-NITROPHENYL)-3,4,5-TRIMETHOXYBENZAMIDE
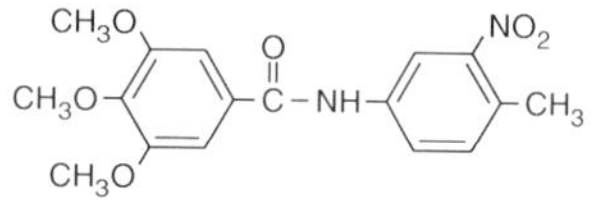
250 mg

S73,498-5
2',4'-DIMETHOXYMALEANILIC ACID
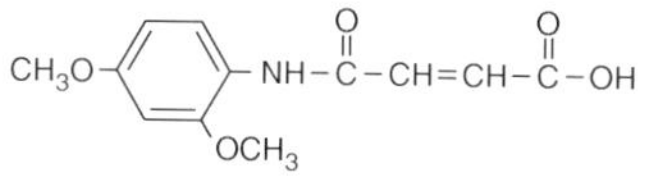
250 mg

S75,222-3
N-BENZOYL-3-(2,4-DIMETHOXYPHENYL)-ALANINE
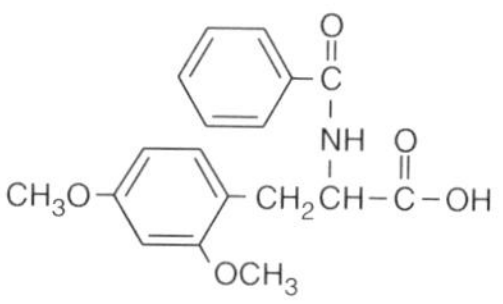
5 gm

S73,266-4
3',5'-DIMETHOXYMALEANILIC ACID
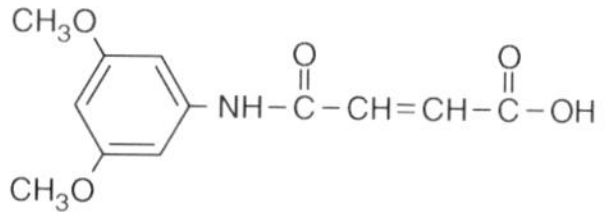
250 mg

S69,280-8
2',5'-DIMETHOXYMALEANILIC ACID
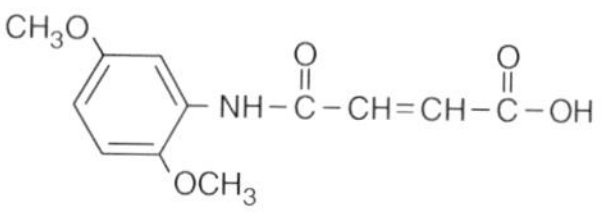
5 gm

S85,489-1
2',5'-DIMETHOXYCINNAMANILIDE
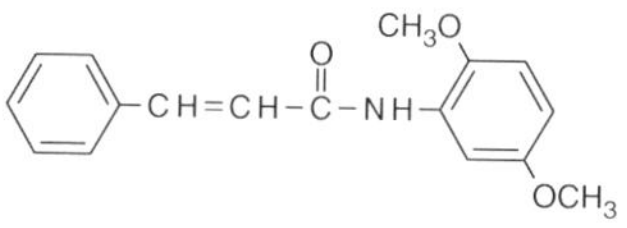
250 mg

S79,552-6
N,N'-BIS(2-METHOXY-3-NITROPHENYL)-TEREPHTHALAMIDE
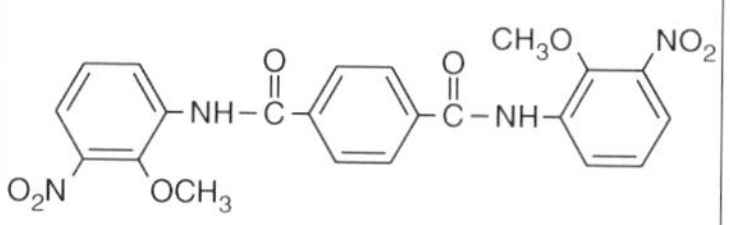
250 mg

S74,793-9
N,N'-BIS(2-METHOXY-3-NITROPHENYL)-ISOPHTHALAMIDE
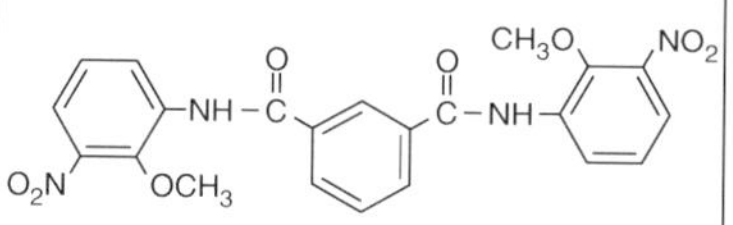
5 gm

S67,816-3
2',5'-DIBUTOXY-4'-NITROBENZANILIDE
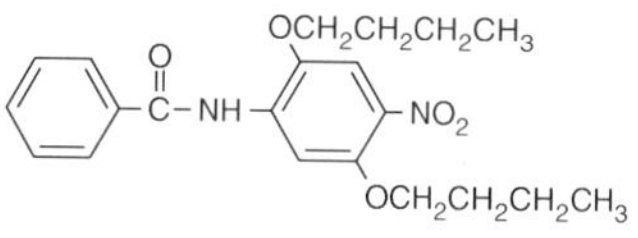
250 mg

S89,220-3
2',4-DIMETHOXYBENZANILIDE
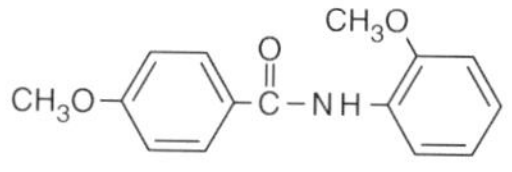
250 mg

S92,470-9
2,5-BIS[N-(2-METHOXY-5-NITROPHENYL)-CARBAMOYL]TEREPHTHALIC ACID
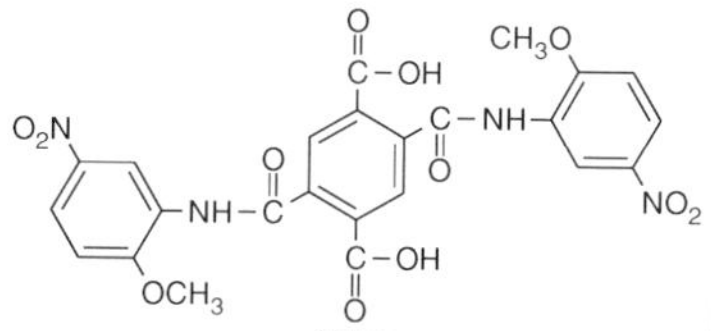
250 mg

S68,610-7
N-(4-METHOXY-2-NITROPHENYL)-3,4,5-TRIMETHOXYBENZAMIDE
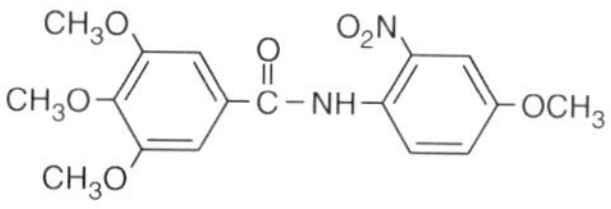
250 mg

S76,638-0
2-BENZOYL-N-[2-(3,4-DIMETHOXYPHENYL)ETHYL]BENZAMIDE
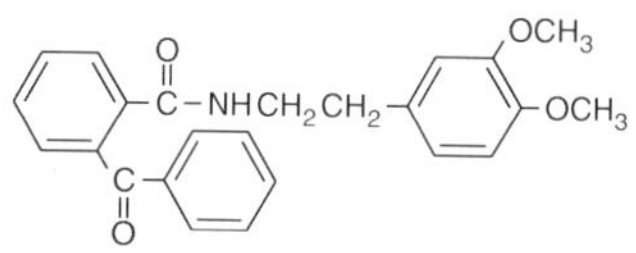
250 mg

S90,181-4
2,5-BIS[N-(4-ETHOXYPHENYL)-CARBAMOYL]TEREPHTHALIC ACID HYDRATE
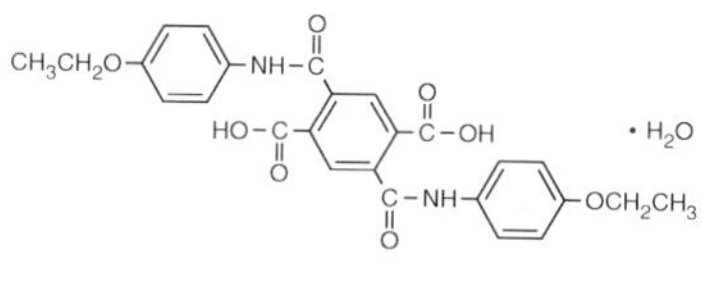
250 mg

S74,799-8
N,N'-BIS(3-METHOXYPHENYL)-TEREPHTHALAMIDE
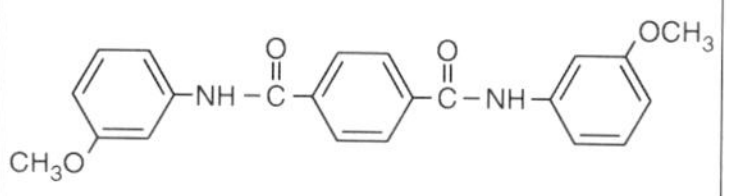
5 gm

S77,988-1
N,N'-BIS(4-METHOXYPHENYL)-ISOPHTHALAMIDE
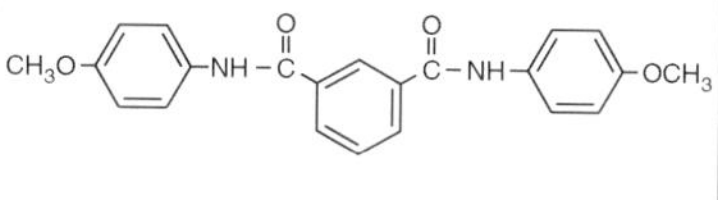
250 mg

S77,990-3
N,N'-BIS(4-ETHOXYPHENYL)-ISOPHTHALAMIDE
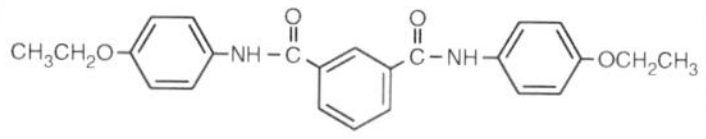
250 mg

S74,956-7
ALPHA-(2,6-DIMETHOXYBENZYLIDENE)-HIPPURIC ACID
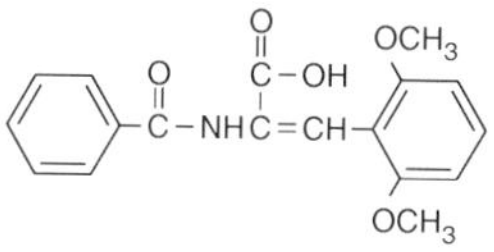
250 mg

S74,945-1
N-BENZOYL-3-(2,6-DIMETHOXYPHENYL)-ALANINE

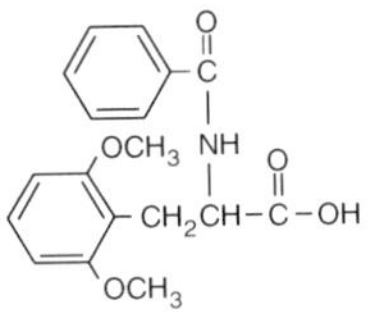

250 mg

S74,961-3
ALPHA-(2,4-DIMETHOXYBENZYLIDENE)-HIPPURIC ACID

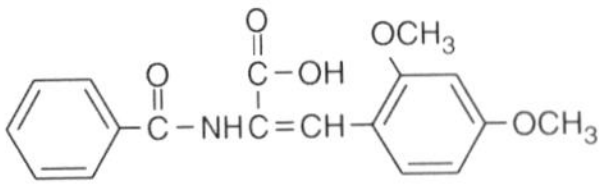

250 mg

S77,001-9
ALPHA-(2,4-DIMETHOXY-3-METHYLBENZYLIDENE)HIPPURIC ACID

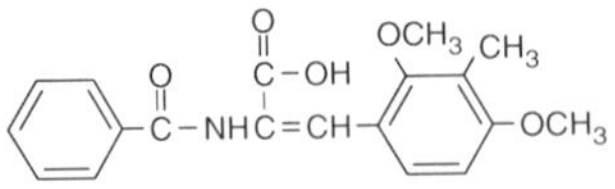

250 mg

S77,000-0
ALPHA-(2,4-DIMETHOXY-6-METHYLBENZYLIDENE)HIPPURIC ACID

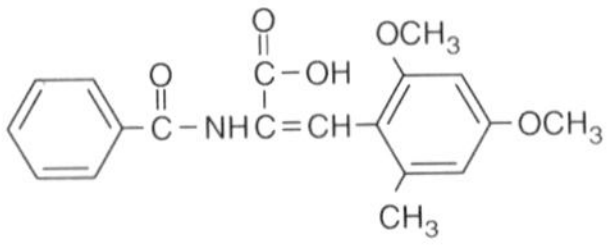

250 mg

S72,101-8
N,N'-BIS(3-METHOXYPHENYL)-ISOPHTHALAMIDE

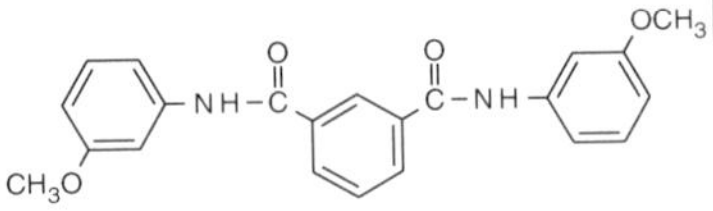

5 gm

S34,072-3
N-ALLYL-2-(CARBOXYMETHOXY)-BENZAMIDE

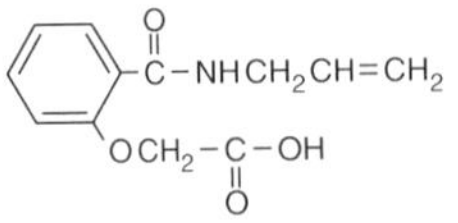

5 gm

S62,941-3
4'-AMINO-4-[2-(4-TERT-OCTYLPHENOXY)-ETHOXY]BENZANILIDE

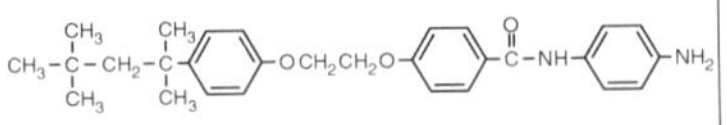

5 gm

S62,921-9
DIMETHYL 5-(3-AMINO-4-[2-(4-TERT-OCTYLPHENOXY)ETHOXY]BENZAMIDO)-ISOPHTHALATE

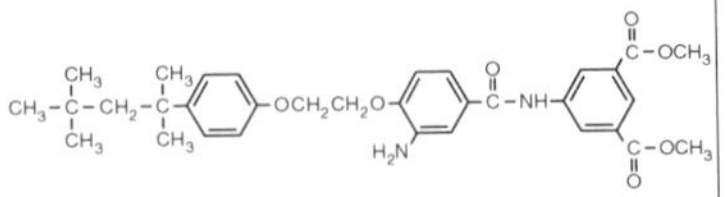

5 gm

S65,080-3
4-CHLORO-N-(3,4-DIMETHOXYPHENETHYL)-BENZAMIDE

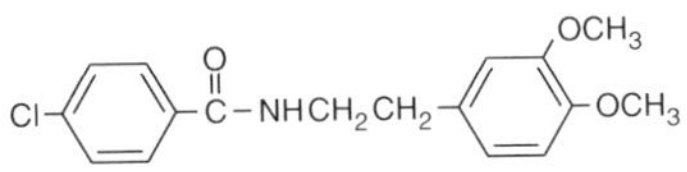

5 gm

S64,734-9
N-(3,4-DIMETHOXYPHENETHYL)-4-NITROBENZAMIDE

5 gm

S65,125-7
N-(3,4-DIMETHOXYPHENETHYL)BENZAMIDE

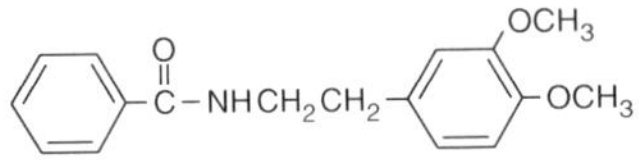

5 gm

S65,085-4
3-CHLORO-N-(3,4-DIMETHOXYPHENETHYL)-BENZAMIDE

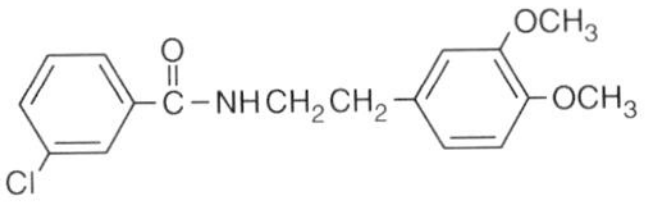

5 gm

S65,078-1
3,4-DICHLORO-N-(3,4-DIMETHOXYPHENETHYL)BENZAMIDE

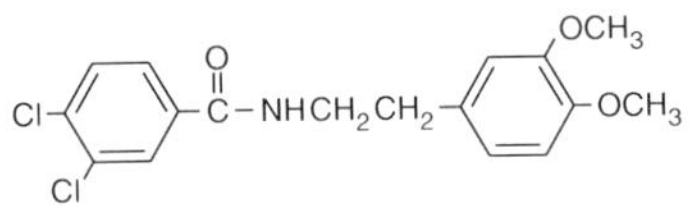

5 gm

S42,005-0
2-CARBAMOYLPHENOXYACETIC ACID, SODIUM SALT

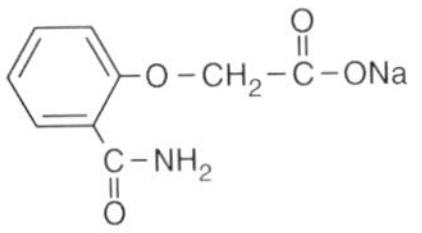

5 gm

S54,279-2
N-(3,4-DIMETHOXYPHENETHYL)-MALEAMIC ACID

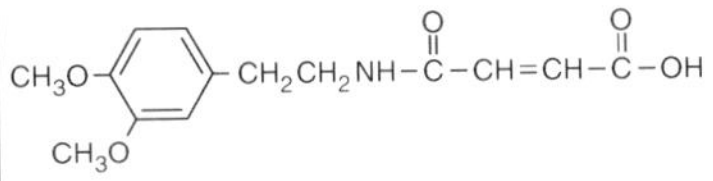

250 mg

S74,969-9
3-BENZAMIDO-8-METHOXYCOUMARIN

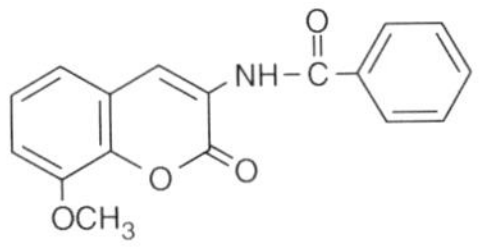

250 mg

S44,857-5
3,4,5-TRIMETHOXYBENZAMIDE

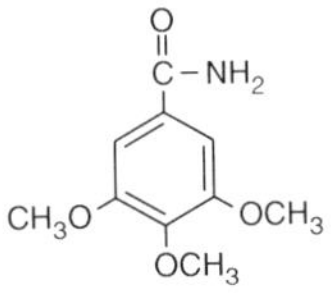

5 gm

S70,906-9
2',6'-DIMETHYL-2-METHOXYBENZANILIDE

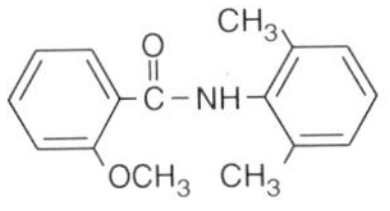

5 gm

S71,398-8
2',6'-DIISOPROPYL-2-METHOXYBENZANILIDE

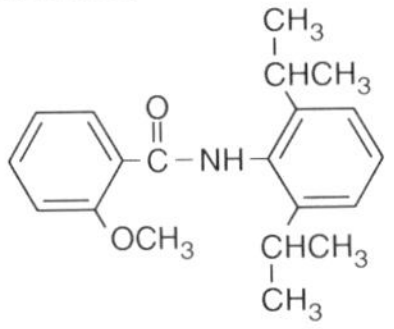

5 gm

S70,767-8
2-METHOXY-2',4',6'-TRIMETHYLBENZANILIDE

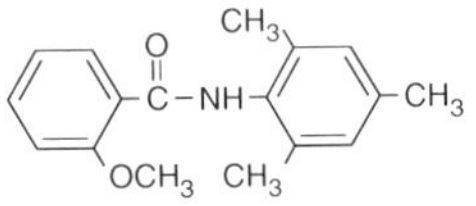

5 gm

S73,597-3
3-METHOXY-2'-METHYL-6'-NITROBENZANILIDE

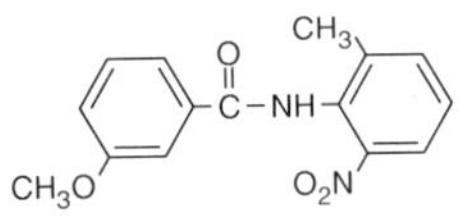

250 mg

S68,825-8
N-(2-METHOXY-5-NITROPHENYL)-3,4,5-TRIMETHOXYBENZAMIDE

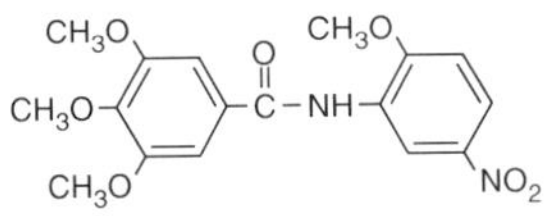

250 mg

S85,999-0
N-TERT-BUTYL-3-METHOXYBENZAMIDE

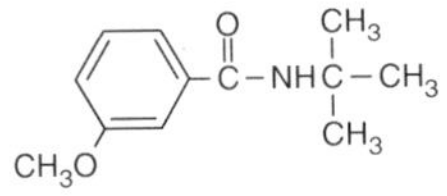

250 mg

S71,278-7
2',6'-DIMETHYL-3-METHOXYBENZANILIDE

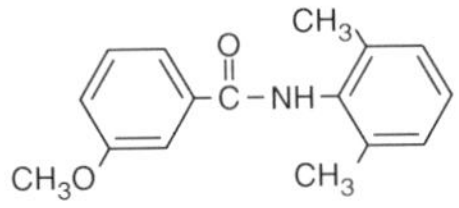

5 gm

S70,758-9
2',6'-DIETHYL-3-METHOXYBENZANILIDE

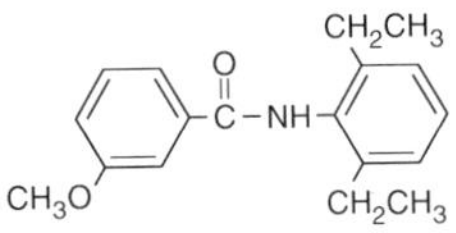

5 gm

S71,683-9
2',6'-DIISOPROPYL-3-METHOXYBENZANILIDE

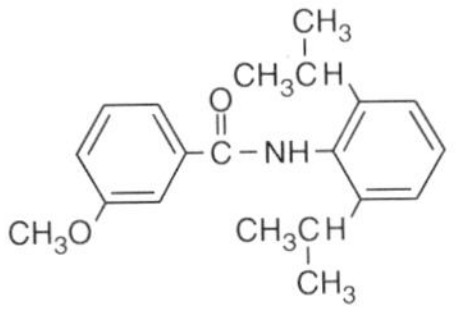

5 gm

S86,158-8
N-TERT-BUTYL-3-METHOXY-N-METHYLBENZAMIDE

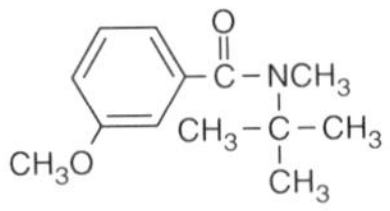

250 mg

S92,145-9
N-(3-ACETYLPHENYL)MALEANILIC ACID

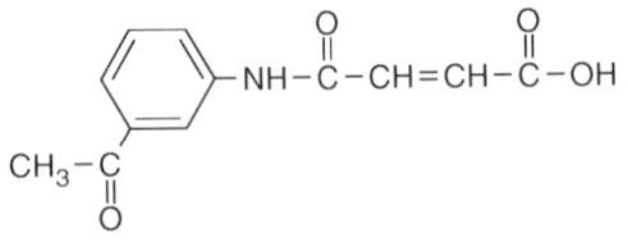

5 gm

S92,148-3
N-(4-ACETYLPHENYL)MALEANILIC ACID

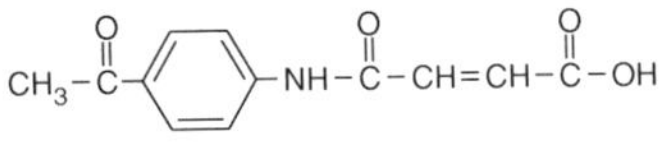

250 mg

S64,733-0
2,2'-DIAMINO-4'-METHOXY-4-METHYLBENZANILIDE

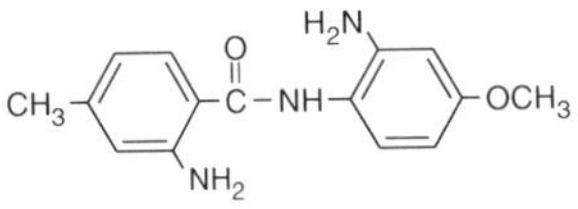

5 gm

S63,231-7
4'-AMINO-2,2'-DIMETHOXYBENZANILIDE

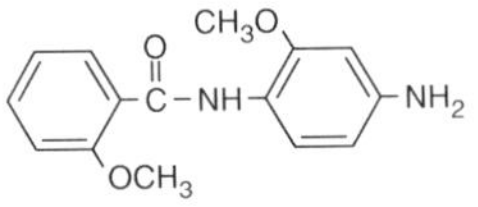

250 mg

S58,698-6
3-AMINO-N-[3-(DODECYLOXY)PROPYL]-BENZAMIDE HYDROCHLORIDE

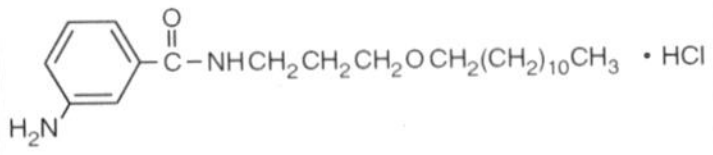

5 gm

S74,277-5
N-[(3-AMINO-4-ETHOXY-2-METHYL-5-PROPYL)PHENYL]BENZAMIDE

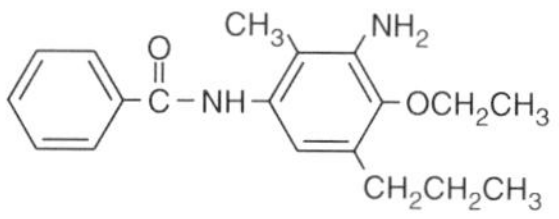

5 gm

S58,775-3
3-AMINO-2'-(DODECYLOXY)BENZANILIDE HYDROCHLORIDE

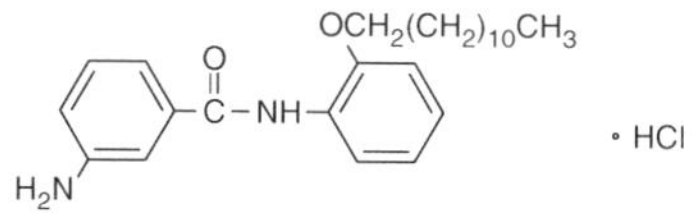

5 gm

S37,307-9
3-AMINO-4-METHOXYBENZANILIDE

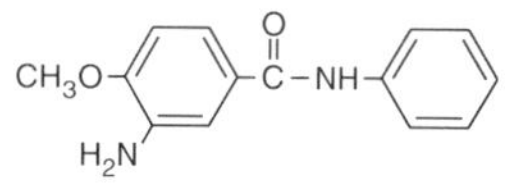

5 gm

S58,694-3
4-AMINO-N-[3-(DODECYLOXY)PROPYL]-BENZAMIDE HYDROCHLORIDE

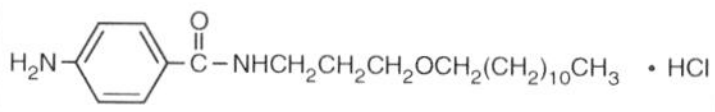

250 mg

S51,478-0
N-(2-DIETHYLAMINOETHYL)-4-(HEXADECYLOXY)BENZANILIDE

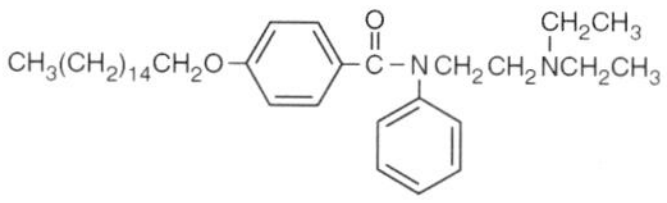

250 mg

S91,040-6
2'-METHOXY-4-METHYL-5'-NITROBENZANILIDE

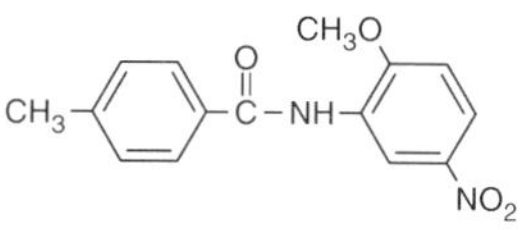

250 mg

S69,503-3
2'-METHOXY-5'-NITRO-2-BUTENANILIDE

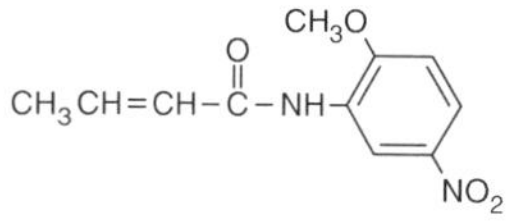

250 mg

S73,088-2
2'-METHOXY-4-METHYL-4'-NITROBENZANILIDE

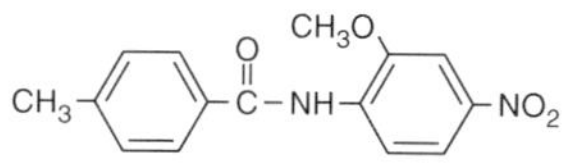

5 gm

S91,003-1
2'-METHOXY-3-METHYL-4'-NITROBENZANILIDE

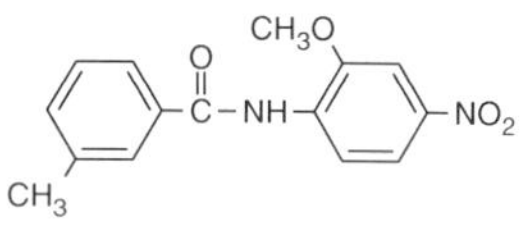

250 mg

S35,222-5
4-METHOXY-3-NITROBENZANILIDE

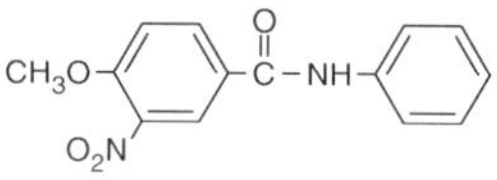

5 gm

S67,515-6
N-(4-ACETAMIDO-3-METHOXYPHENYL)-PHTHALAMIC ACID

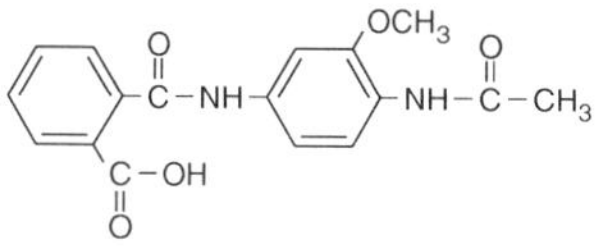

250 mg

S69,862-8
4'-ACETAMIDO-3'-METHOXY-2-BUTENANILIDE

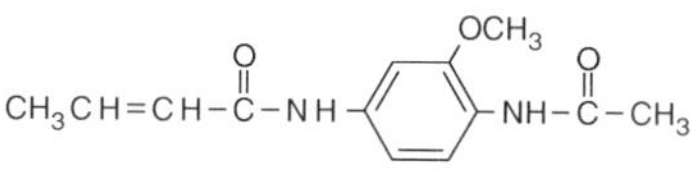

250 mg

S89,282-3
4'-METHOXY-3-NITROBENZANILIDE

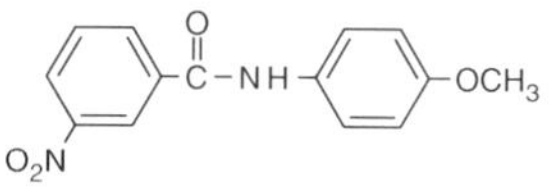

250 mg

S84,976-6
2-METHOXY-N-METHYL-4'-NITROBENZANILIDE

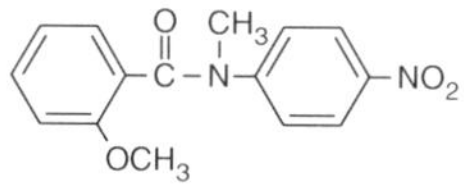

250 mg

S70,285-4
2'-METHOXY-4'-NITROMALEANILIC ACID

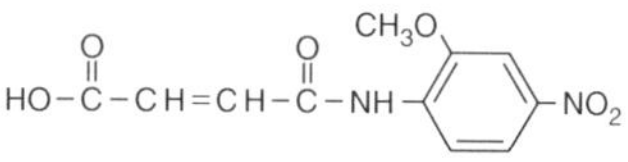

5 gm

S75,123-5
3',4-DINITRO-2'-METHOXYBENZANILIDE

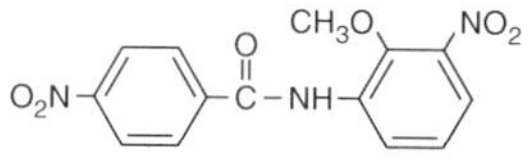

5 gm

S76,314-4
4-(2-METHOXYANILINO)-4-OXO-2-BUTENOIC ACID

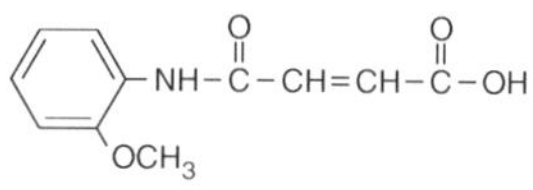

250 mg

S69,281-6
2'-ETHOXYMALEANILIC ACID

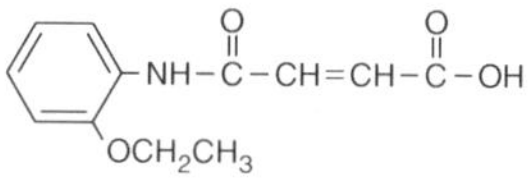

5 gm

S70,435-0
2'-METHOXYPHTHALANILIC ACID

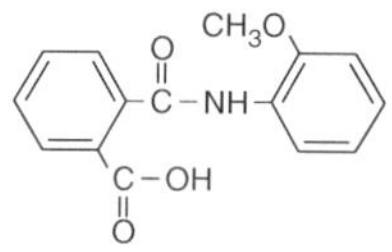

5 gm

S76,408-6
2'-ETHOXYPHTHALANILIC ACID

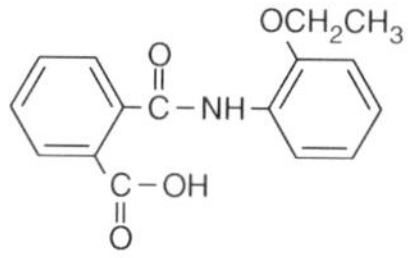

5 gm

S69,539-4
2'-METHOXY-5'-METHYLMALEANILIC ACID

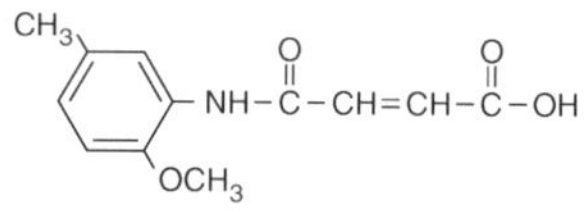

5 gm

S80,165-8
4-[N-(2-ETHOXYPHENYL)CARBAMOYL]-PHTHALIC ANHYDRIDE

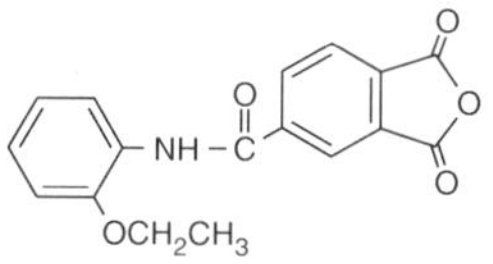

250 mg

S71,258-2
2'-METHOXYBENZANILIDE

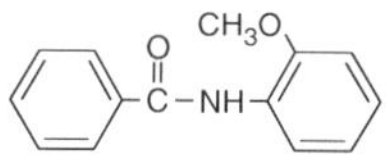

250 mg

S76,842-1
2'-ETHOXYBENZANILIDE

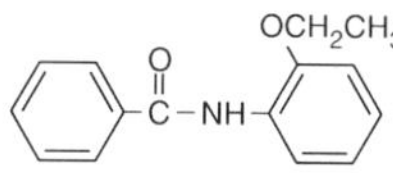

250 mg

S59,449-0
2-CHLORO-2'-METHOXYBENZANILIDE

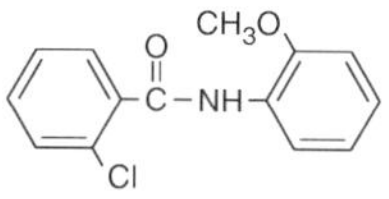

250 mg

S61,572-2
2'-METHOXY-O-TOLUANILIDE

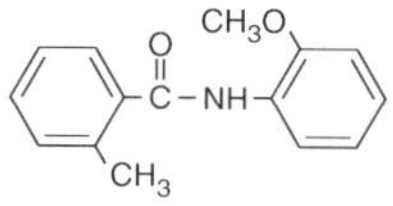

250 mg

S59,897-6
4-CHLORO-O-BENZANISIDIDE

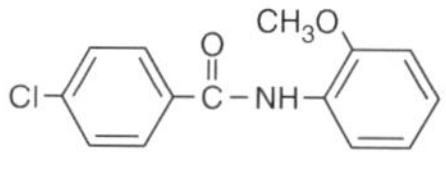

5 gm

S61,227-8
2,4-DICHLOROBENZO-O-ANISIDIDE

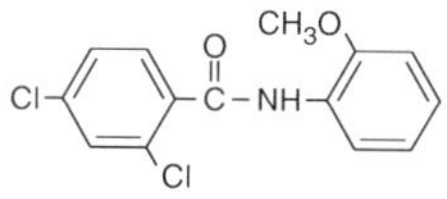

5 gm

S69,964-0
2'-METHOXY-5'-METHYL-2-BUTENANILIDE

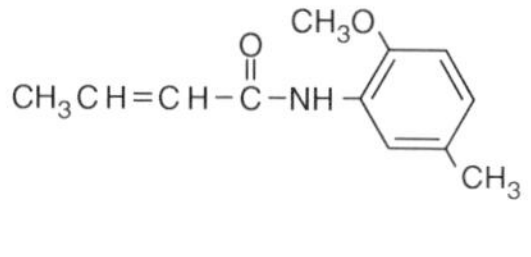

250 mg

S72,915-9
2'-METHOXY-5'-METHYL-2,4-HEXADIENANILIDE

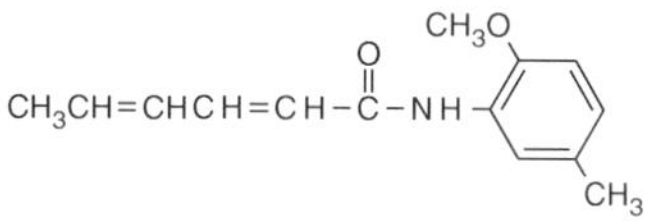

5 gm

S34,585-7
2',5'-DIETHOXYBENZANILIDE

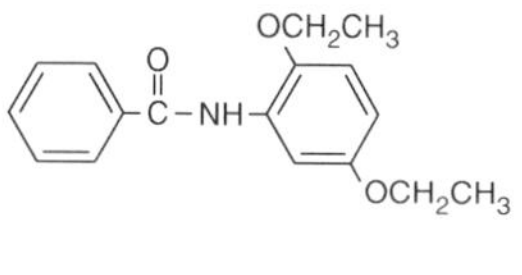

5 gm

S73,605-8
2'-METHOXY-5'-METHYL-1-NAPHTHANILIDE

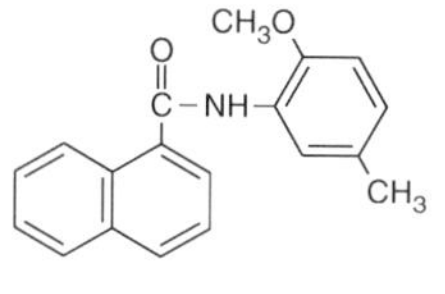

5 gm

S77,113-9
4-METHOXY-N-PHENYLBENZOHYDROXAMIC ACID
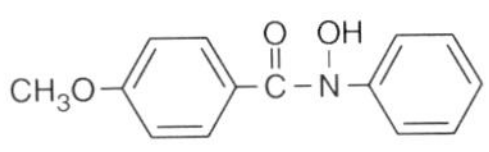
250 mg

S90,045-1
4',6-DINITRO-2'-METHOXYPHTHALANILIC ACID
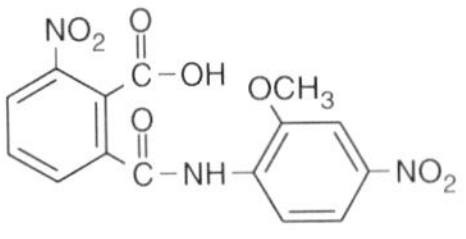
250 mg

S80,161-5
4-[N-(2-METHOXY-4-NITROPHENYL)-CARBAMOYL]PHTHALIC ANHYDRIDE
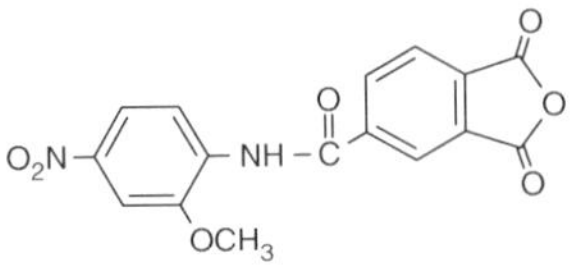
250 mg

S80,162-3
4-[N-(2-METHOXY-5-NITROPHENYL)-CARBAMOYL]PHTHALIC ANHYDRIDE
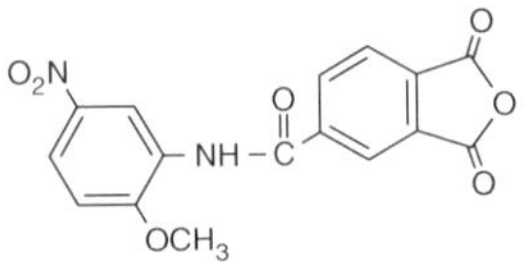
250 mg

S94,278-2
4'-METHOXY-4-NITROBENZANILIDE
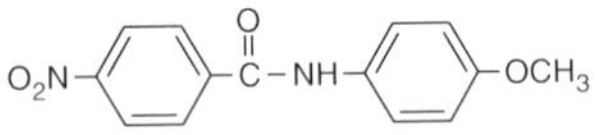
250 mg

S76,149-4
2-[(3,3-DIMETHYL-2-OXO-1-PHENYL)-BUTYL]-4-METHOXY-N-METHYLBENZAMIDE
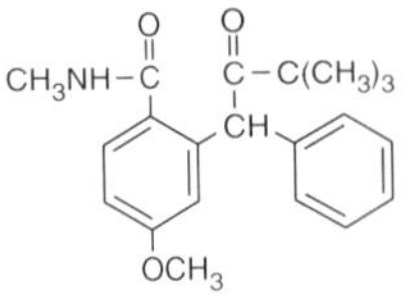
250 mg

S90,157-1
4-[N-(4-METHOXYPHENYL)CARBAMOYL]-ISOPHTHALIC ACID, 4-METHOXYANILINE SALT
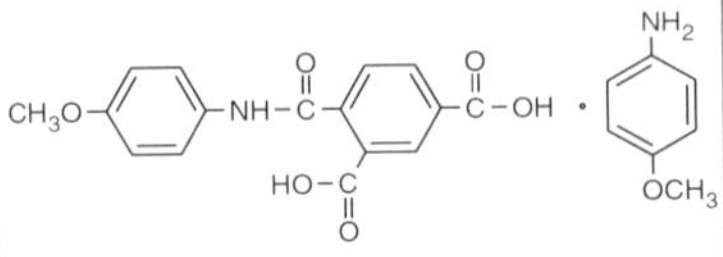
250 mg

S76,439-6
3'-METHOXYPHTHALANILIC ACID
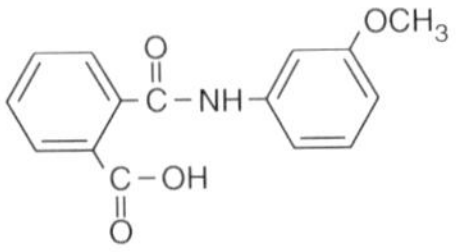
5 gm

S70,120-3
4'-METHOXYPHTHALANILIC ACID
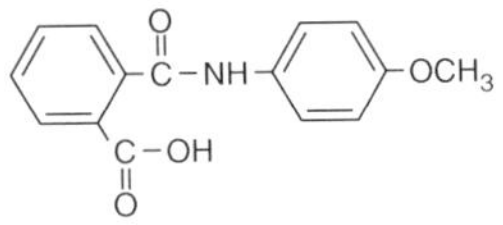
5 gm

S55,389-1
3'-METHOXYMALEANILIC ACID
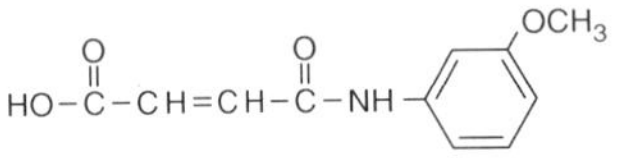
250 mg

S53,257-6
4'-METHOXYMALEANILIC ACID
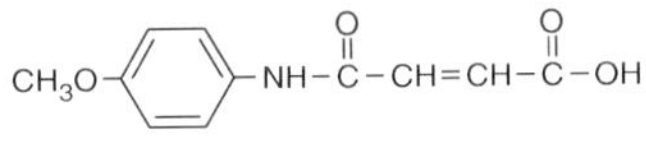
5 gm

S53,822-1
4'-ETHOXYMALEANILIC ACID
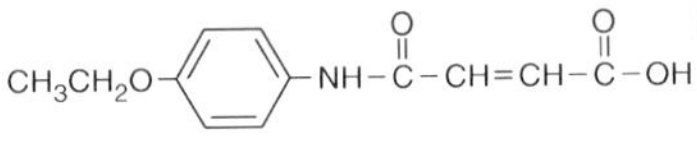
5 gm

S93,699-5
METHYL 4-(4-METHOXY-N-PHENYLBENZAMIDO)BENZOATE
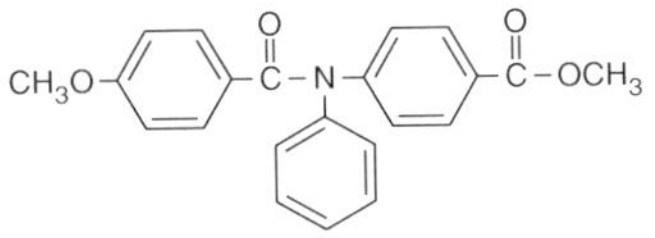
250 mg

S38,055-5
2-ISOPROPOXYBENZAMIDE
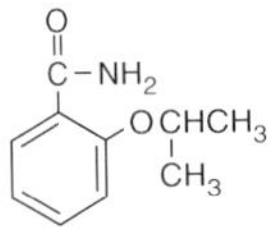
250 mg

S38,101-2
2-BUTOXYBENZAMIDE
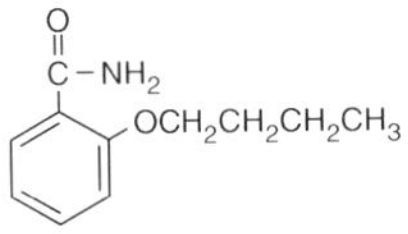
5 gm

S59,213-7
M-BENZANISIDIDE
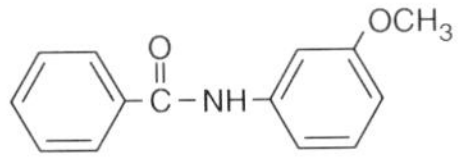
5 gm

S89,070-7
4'-ETHOXYBENZANILIDE
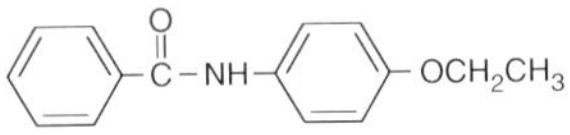
250 mg

S93,694-4
N-(5-CHLORO-2-METHOXYPHENYL)-BENZANILIDE
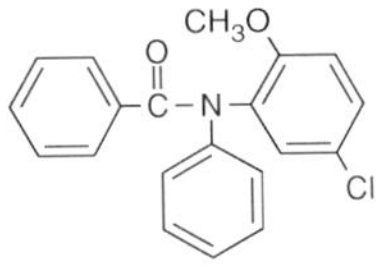
250 mg

S59,200-5
2-CHLORO-M-BENZANISIDIDE
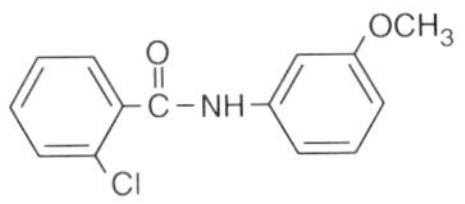
250 mg

S59,434-2
2-CHLORO-4'-METHOXYBENZANILIDE
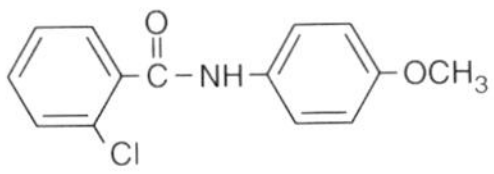
250 mg

S60,922-6
4'-METHOXY-O-TOLUANILIDE
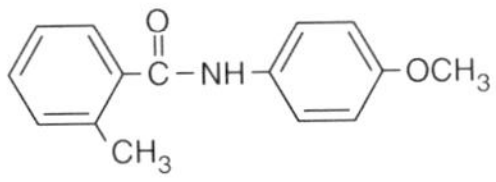
250 mg

S61,563-3
O-TOLU-P-PHENETIDIDE
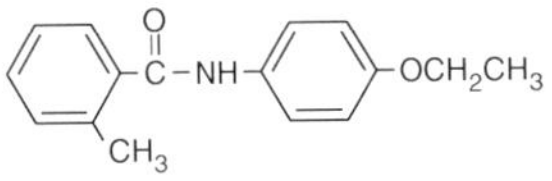
250 mg

S55,512-6
2-CHLORO-5'-METHYL-O-BENZANISIDIDE
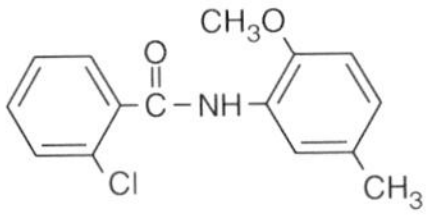
5 gm

S59,075-4
P-BENZANISIDIDE
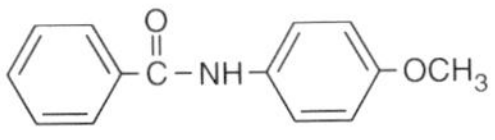
250 mg

S66,486-3
4-METHOXY-N-PHENYLBENZAMIDE
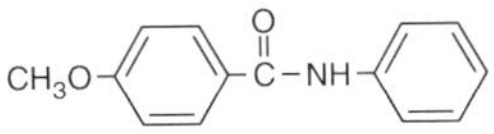
5 gm

S74,386-0
4'-ETHOXY-2,4-HEXADIENANILIDE
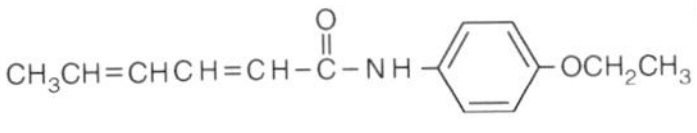
250 mg

S59,068-1
P-BENZOPHENETIDIDE
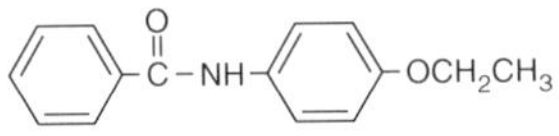
250 mg

S60,236-1
P-TOLU-M-ANISIDIDE
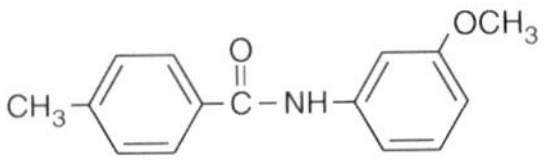
250 mg

S59,905-0
4-CHLORO-4'-ETHOXYBENZANILIDE
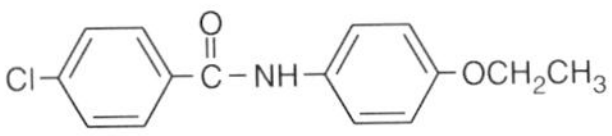
250 mg

S60,060-1
N-(CINNAMYL)-P-ANISIDINE
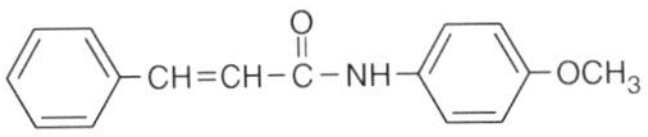
250 mg

S60,250-7
P-TOLU-P-PHENETIDIDE
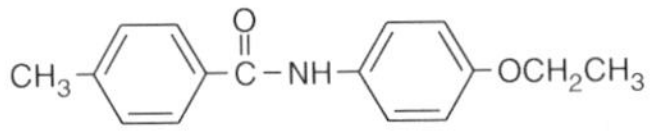
250 mg

S71,602-2
2',3'-DIMETHYL-4-METHOXYBENZANILIDE
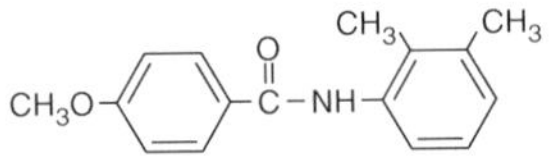
5 gm

S73,584-1
4-METHOXY-2',6'-DIMETHYLBENZANILIDE

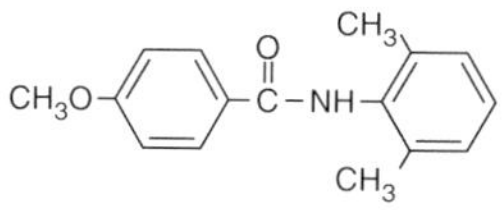

250 mg

S89,218-1
2',6'-DIMETHYL-4-METHOXYBENZANILIDE

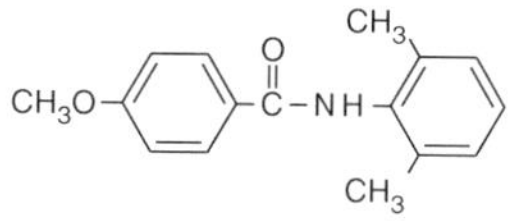

250 mg

S70,890-9
2'-ETHYL-4-METHOXY-6'-METHYLBENZANILIDE

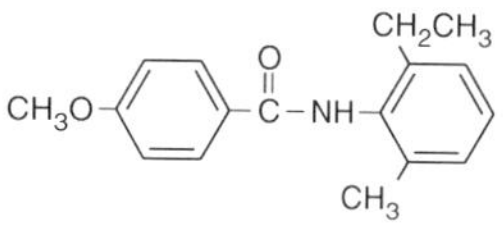

5 gm

S70,757-0
2',6'-DIETHYL-4-METHOXYBENZANILIDE

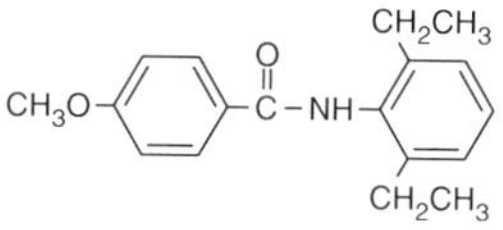

5 gm

S61,214-6
2,4-DICHLOROBENZO-P-ANISIDIDE

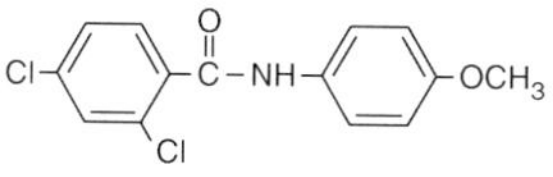

250 mg

S61,225-1
2,4-DICHLOROBENZO-P-PHENETIDIDE

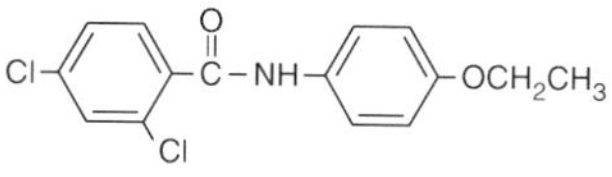

5 gm

S70,768-6
4-METHOXY-2',4',6'-TRIMETHYLBENZANILIDE

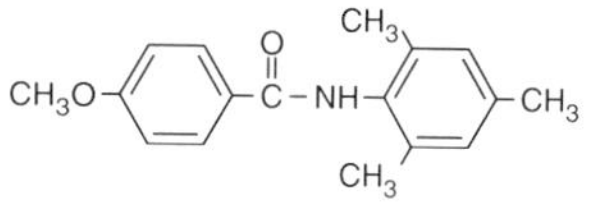

5 gm

S61,229-4
2,4-DICHLOROBENZO-M-ANISIDIDE

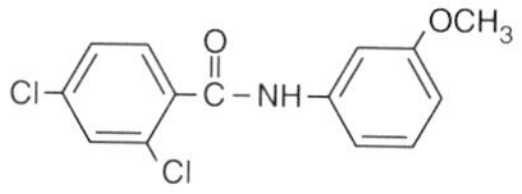

250 mg

S51,477-2
N-ETHYL-2-METHOXYBENZO-M-TOLUIDIDE

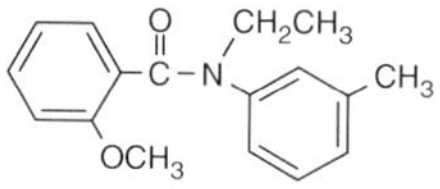

250 mg

S55,522-3
4'-ETHOXYPHTHALANILIC ACID

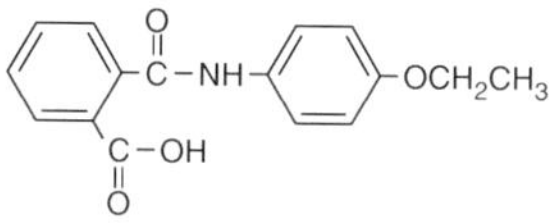

5 gm

S55,525-8
2'-METHOXY-5'-METHYLPHTHALANILIC ACID

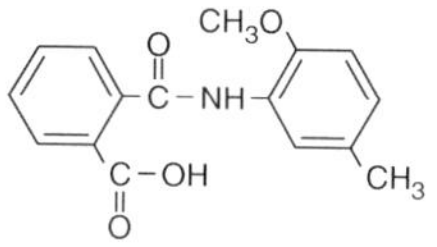

5 gm

S53,493-5
2-METHOXYBENZAMIDE

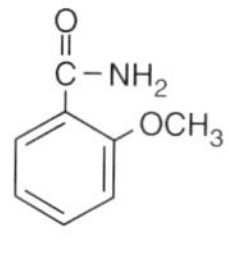

5 gm

S38,032-6
2-ALLYLOXYBENZAMIDE

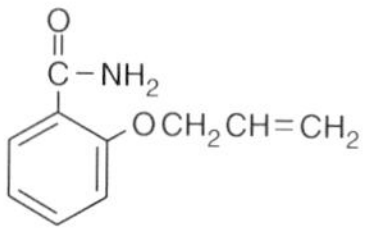

250 mg

S55,558-4
4-CHLORO-5'-METHYL-O-BENZANISIDIDE

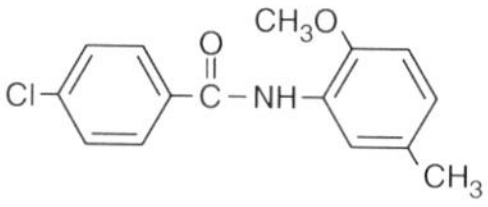

5 gm

S55,501-0
3,4-DICHLORO-5'-METHYL-O-BENZANISIDIDE

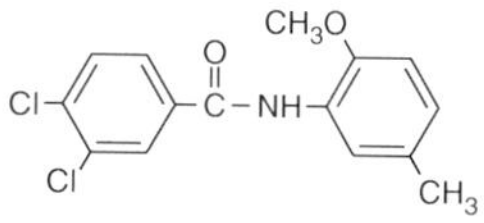

5 gm

S51,629-5
N,N-DIETHYL-P-ANISAMIDE

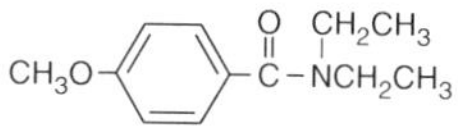

5 gm

S51,233-8
N-METHYL-P-ANISANILIDE

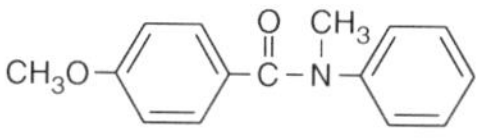

250 mg

S51,483-7
2,5-DIMETHOXY-N-HEXADECYL-N-METHYLBENZAMIDE

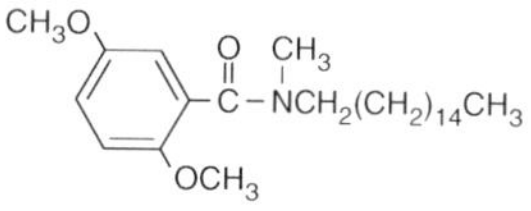

250 mg

S64,606-7
2'-BENZOYL-4'-CHLOROMALEANILIC ACID

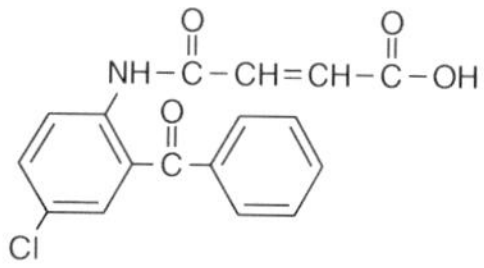

5 gm

S69,288-3
2'-CARBOXYMALEANILIC ACID

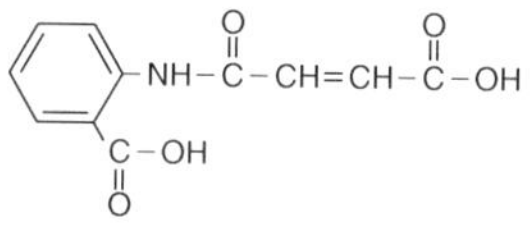

5 gm

S90,183-0
2,5-BIS(N-MESITYLCARBAMOYL)-TEREPHTHALIC ACID

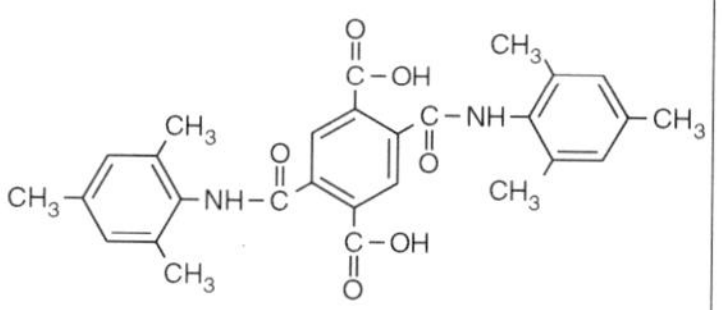

250 mg

S50,056-9
3-NITRO-4-(2-[4-(1,1,3,3-TETRAMETHYLBUTYL)PHENOXY]ETHOXY)-BENZAMIDE

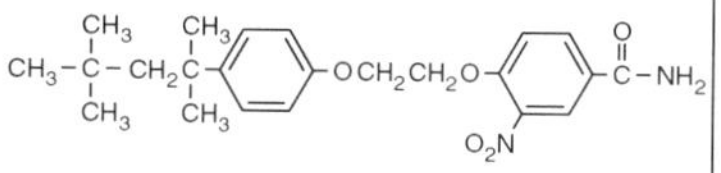

250 mg

S94,611-7
4-BENZILCARBOXAMIDE

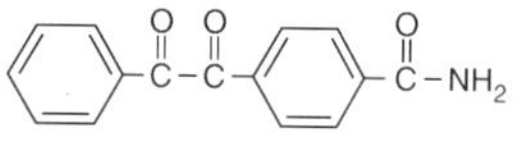

250 mg

S37,828-3
2-CARBAMOYLPHENOXYACETIC ACID

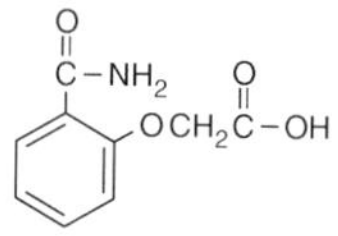

5 gm

S77,027-2
7-ACETOXY-3-(BENZAMIDO)COUMARIN

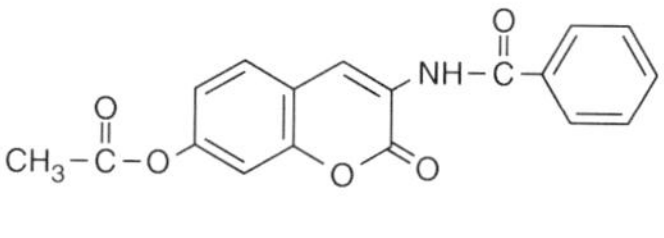

250 mg

S80,575-0
N-(2-ACETOXYBENZOYL)GLUTAMIC ACID

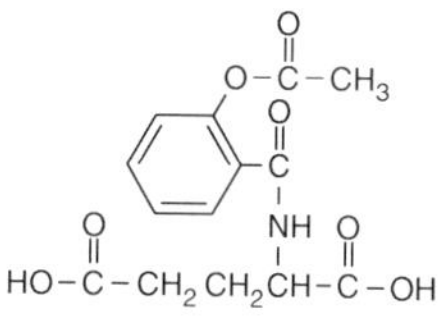

250 mg

S95,420-9
2-(PHENYLACETYL)BENZAMIDE

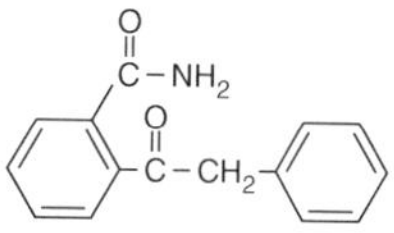

50 mg

S92,178-5
4'-BUTYLMALEANILIC ACID

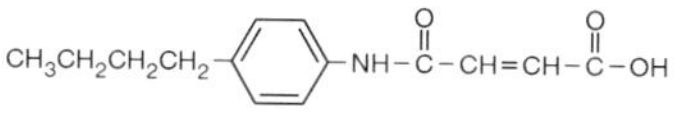

5 gm

S87,167-2
METHYL N-ETHYLPHTHALAMATE

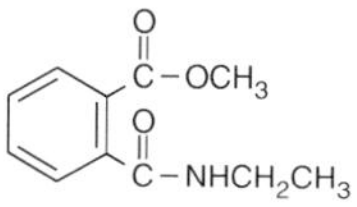

250 mg

S66,209-7
2-MERCAPTO-N-METHYLBENZAMIDE

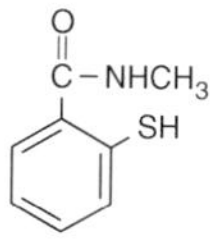

5 gm

S75,118-9
3,5-DINITRO-4'-(METHYLTHIO)-BENZANILIDE

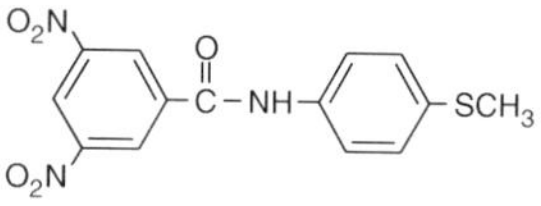

5 gm

S61,547-1
2'-(METHYLTHIO)BENZANILIDE

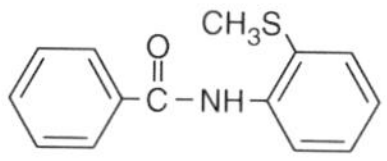

250 mg

S75,139-1
4'-(METHYLTHIO)-3-NITROBENZANILIDE

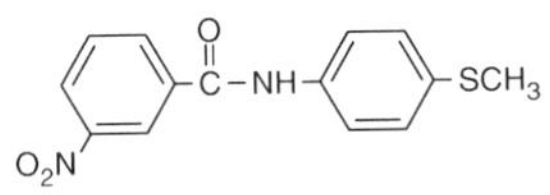

5 gm

S89,938-0
4'-(METHYLTHIO)-4-NITROBENZANILIDE

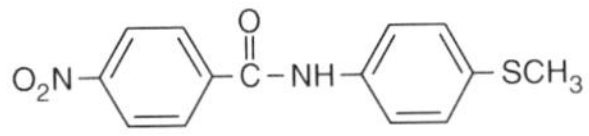

250 mg

S72,469-6
N,N'-BIS[4-(METHYLTHIO)PHENYL]-ISOPHTHALAMIDE

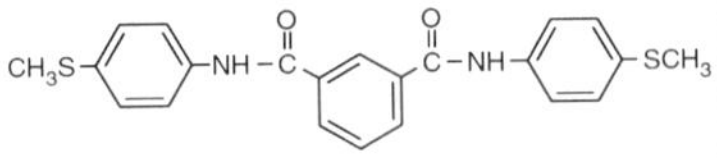

5 gm

S74,790-4
N,N'-BIS[4-(METHYLTHIO)PHENYL]-TEREPHTHALAMIDE

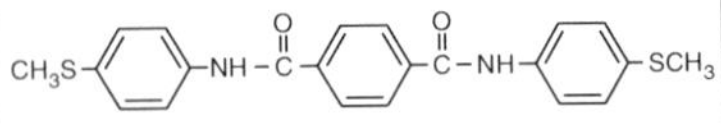

5 gm

S45,136-3
N-BENZOYL-DL-METHIONINE

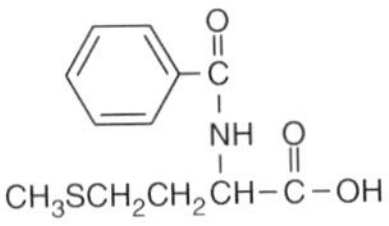

250 mg

S91,005-8
2-[4-(3-METHYLBENZAMIDO)-PHENYLTHIO]ACETIC ACID

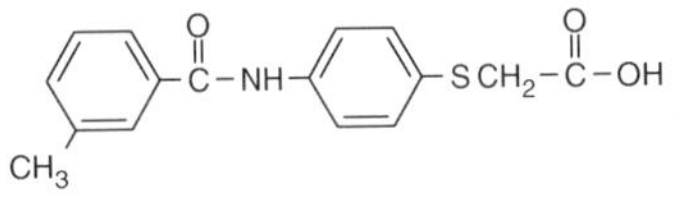

250 mg

S78,658-6
N-ALPHA-BENZOYL-DL-METHIONINE AMIDE

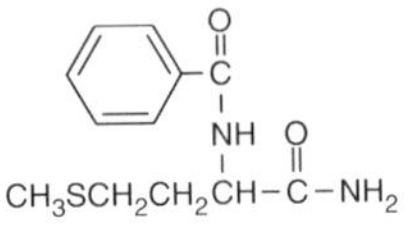

250 mg

S93,315-5
N-ALPHA-BENZOYL-L-METHIONINE AMIDE

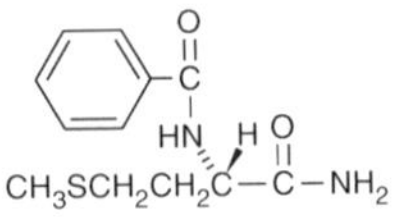

250 mg

S60,440-2
3'-(METHYLTHIO)BENZANILIDE

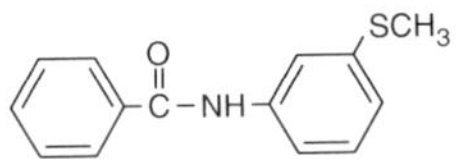

250 mg

S76,215-6
N-METHYL-4-(METHYLTHIO)BENZAMIDE

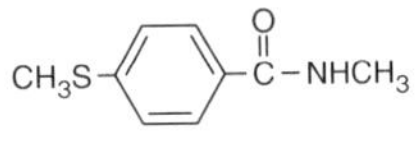

250 mg

S62,954-5
2-ACETAMIDOPHENYL DISULFIDE

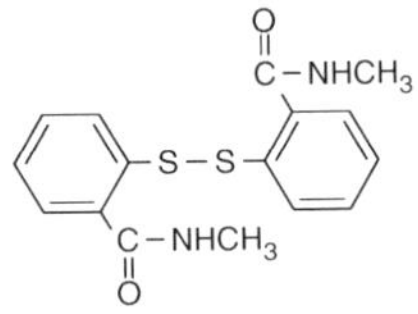

5 gm

S60,053-9
4-AMINO-N-[2-(DIETHYLAMINO)ETHYL]-BENZAMIDE SULFATE

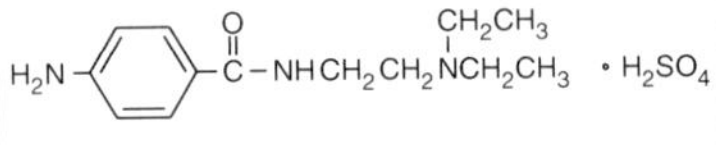

5 gm

S76,496-5
2'-AMINOBENZANILIDE

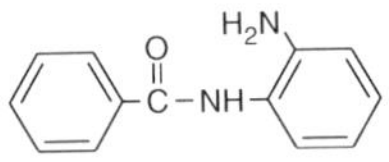

250 mg

S69,764-8
N-(2-AMINOETHYL)BENZAMIDE

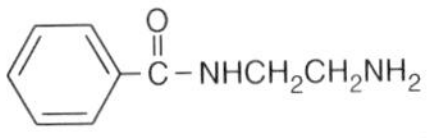

250 mg

S68,244-6
4-(2-AMINOBENZAMIDO)-PHENYLACETIC ACID

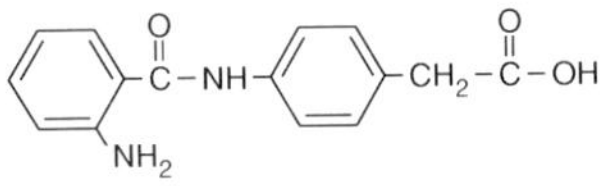

250 mg

S76,023-4
2-AMINO-5-NITROBENZAMIDE

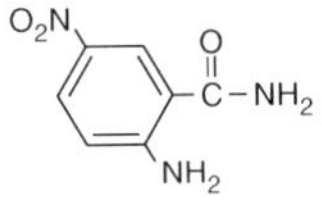

250 mg

S63,230-9
3-AMINO-N-BUTYLBENZAMIDE

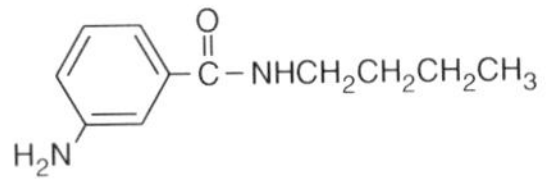

250 mg

S58,700-1
3-AMINO-N-HEXADECYLBENZAMIDE HYDROCHLORIDE

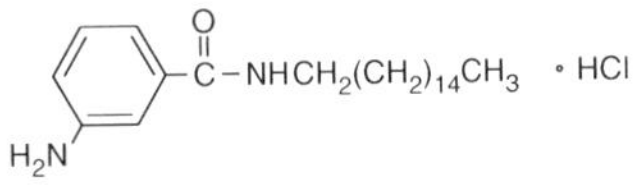

5 gm

S57,206-3
4'-AMINOBENZANILIDE

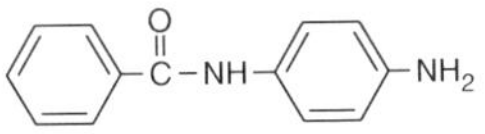

5 gm

S38,123-3
3-AMINO-N,N-DIETHYLBENZAMIDE

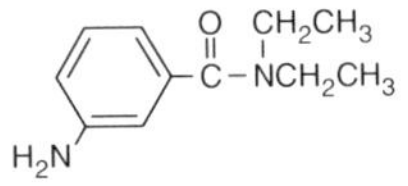

5 gm

S58,689-7
3-AMINO-N,N-DIETHYLBENZAMIDE HYDROCHLORIDE

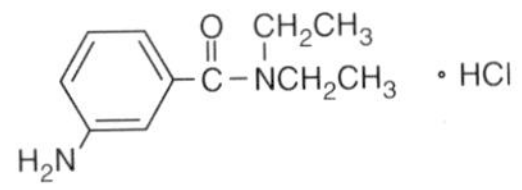

250 mg

S58,348-0
3-AMINO-N,N-BIS(2-ETHYLHEXYL)-BENZAMIDE

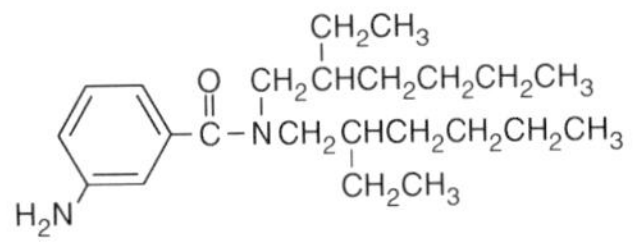

250 mg

S58,219-0
3-AMINO-N-HEXADECYL-N-METHYLBENZAMIDE HYDROCHLORIDE

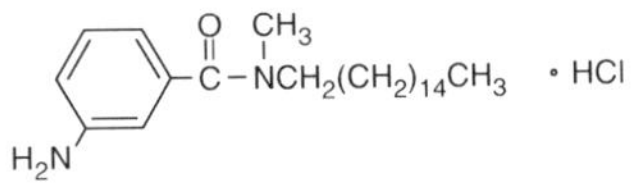

5 gm

S58,683-8
3-AMINO-N,N-DIDODECYLBENZAMIDE

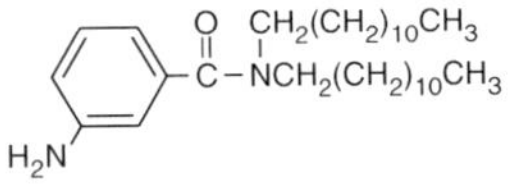

250 mg

S52,105-1
N-(4-AMINOBENZOYL)GLUTAMIC ACID, ZINC SALT

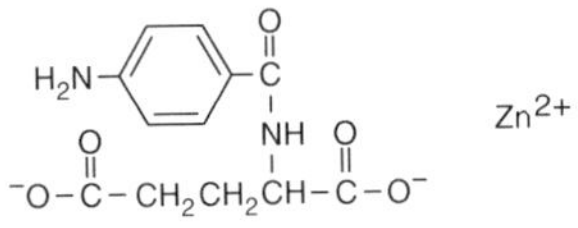

5 gm

S90,848-7
5-(4-AMINOBENZAMIDO)VALERIC ACID

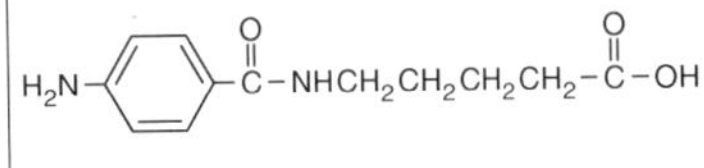

250 mg

S59,885-2
4-(3-AMINOPHENYL)-3-BUTENANILIDE

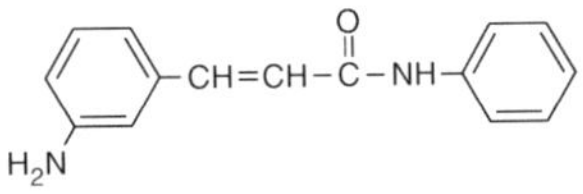

250 mg

S73,495-0
2-AMINOBENZANILIDE

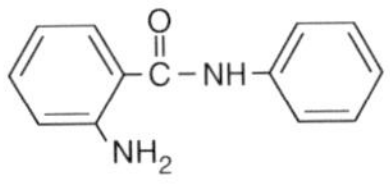

250 mg

S64,216-9
4-AMINO-N-HEXADECYLBENZAMIDE

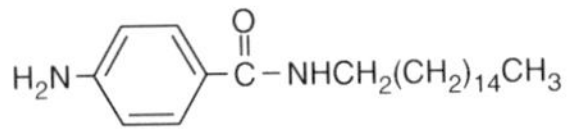

5 gm

S58,678-1
P-AMINO-N,N-BIS(2-ETHYLHEXYL)-BENZAMIDE

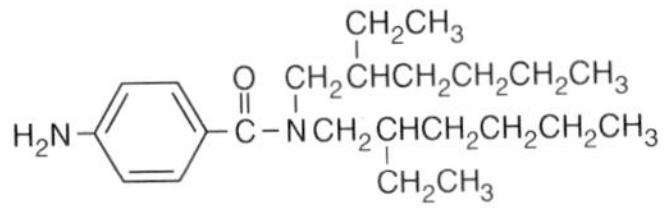

5 gm

S58,706-0
4-AMINO-N,N-DIDODECYLBENZAMIDE

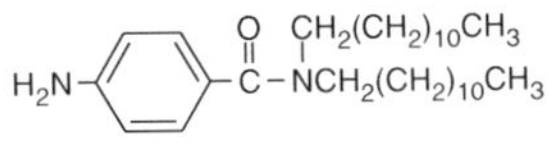

250 mg

S34,106-1
3-AMINO-2',4'-DIMETHYL-P-TOLUANILIDE

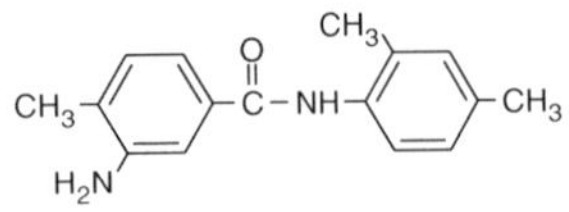

5 gm

S78,710-8
NEPSILON-BENZOYL-DL-LYSINE

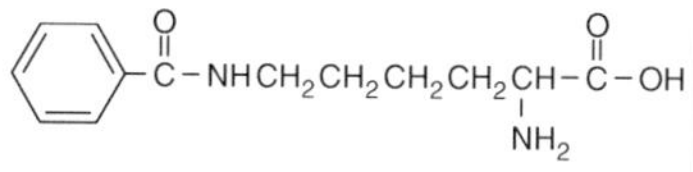

250 mg

S78,447-8
N-ALPHA-BENZOYL-L-ORNITHINE METHYL ESTER HYDROCHLORIDE

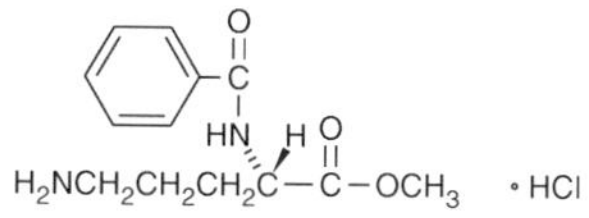

250 mg

S38,607-3
N-HEXADECYLANTHRANILAMIDE

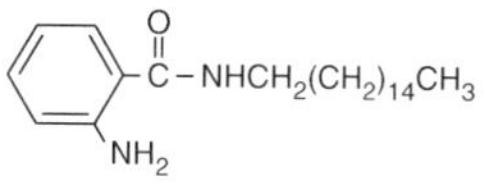

250 mg

S51,545-0
4-ETHYLAMINO-3-NITROBENZAMIDE

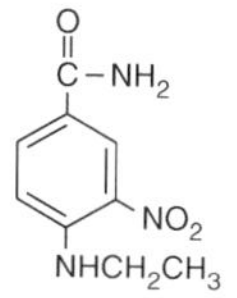

250 mg

S91,008-2
3-METHYL-N-(2-MORPHOLINOETHYL)-BENZAMIDE HYDROCHLORIDE

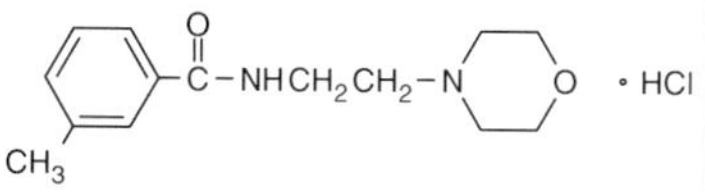

250 mg

S76,189-3
N-[2-(DIMETHYLAMINOETHYL)]-2-METHYLBENZAMIDE

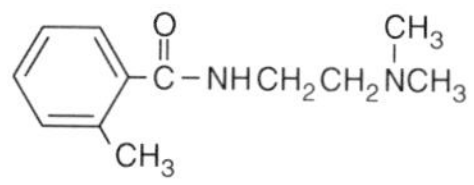

250 mg

S87,210-5
1-(2-BENZAMIDOETHYL)-1-PHENYLPYRROLIDINIUM BROMIDE

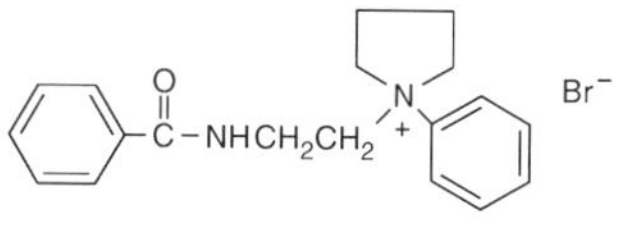

250 mg

S76,407-8
2'-NITROPHTHALANILIC ACID

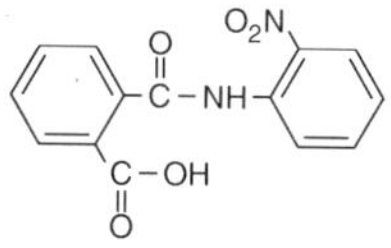

5 gm

S74,861-7
2',6'-DINITROBENZANILIDE

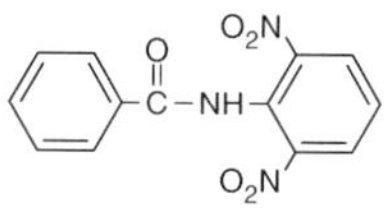

250 mg

S85,013-6
2',4-DINITROBENZANILIDE

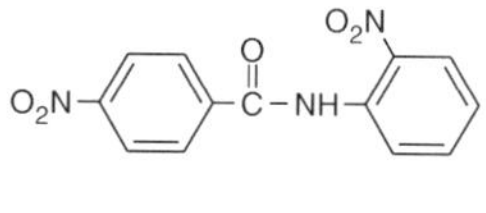

250 mg

S39,379-7
2'-NITROMALEANILIC ACID

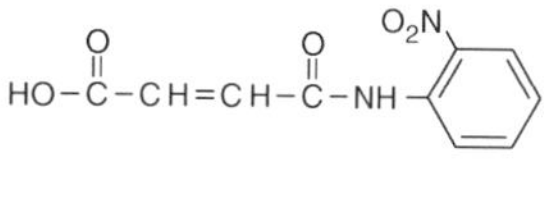

5 gm

S53,827-2
4'-METHYL-2'-NITROMALEANILIC ACID

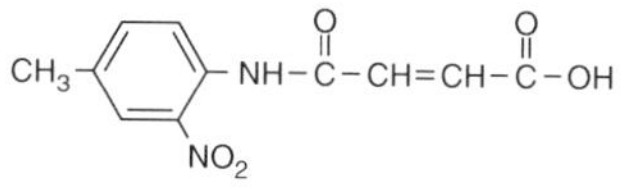

5 gm

S86,422-6
METHYL ALPHA-BENZAMIDO-BETA-NITROMETHACRYLATE

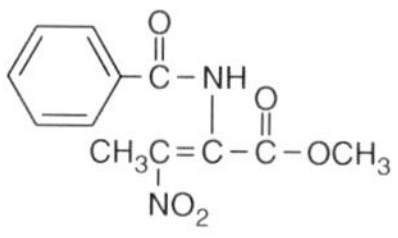

250 mg

S59,062-2
2'-NITROBENZANILIDE

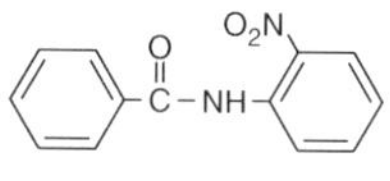

250 mg

S60,244-2
4-CHLORO-2'-NITROBENZANILIDE

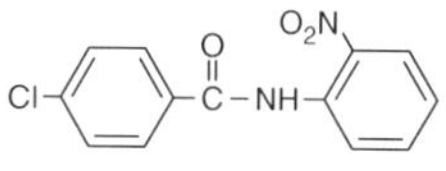

250 mg

S60,213-2
4,4'-DICHLORO-2'-NITROBENZANILIDE

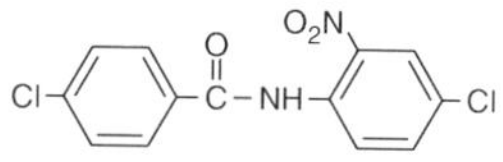

250 mg

S34,416-8
4-CHLORO-3,5-DINITROBENZAMIDE

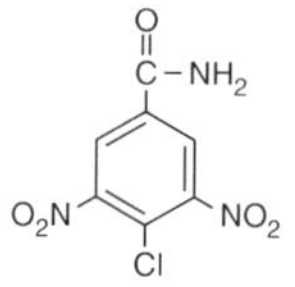

5 gm

S54,785-9
N2-METHYLANTHRANILAMIDE

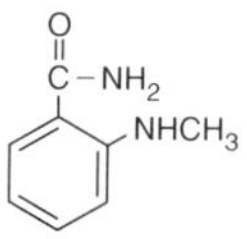

250 mg

S90,888-6
3'-(DIPROPYLAMINO)BENZANILIDE

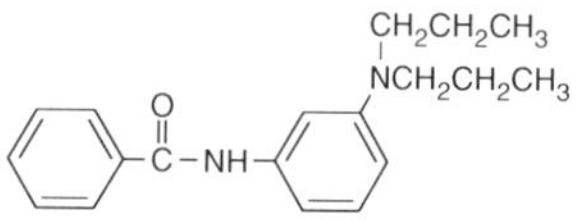

5 gm

S70,131-9
3'-NITROMALEANILIC ACID

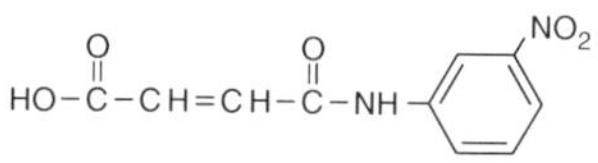

5 gm

S70,123-8
3'-NITROPHTHALANILIC ACID

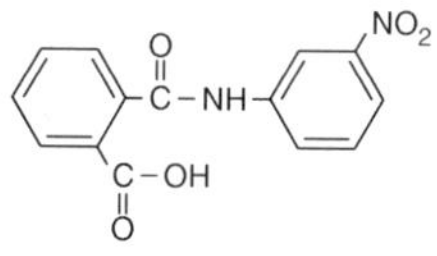

5 gm

S71,733-9
4'-METHYL-3'-NITROPHTHALANILIC ACID

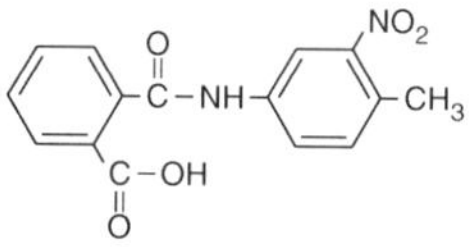

5 gm

S85,534-0
METHYL 4-[N-(3-NITROPHENYL)-CARBAMOYL]BENZOATE

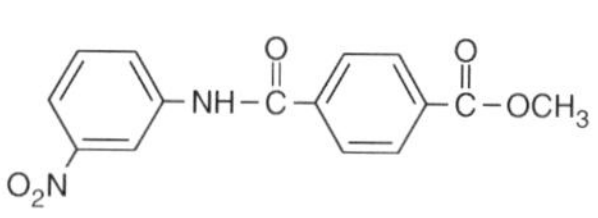

250 mg

S80,155-0
4-[N-(3-NITROPHENYL)CARBAMOYL]-PHTHALIC ANHYDRIDE

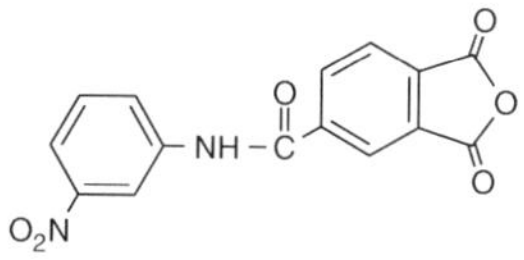

250 mg

S74,794-7
N,N'-BIS(3-NITROPHENYL)-ISOPHTHALAMIDE

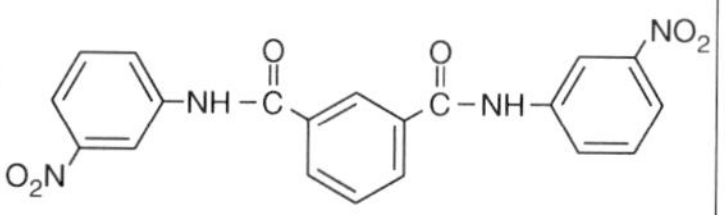

5 gm

S72,468-8
N,N'-BIS(3-NITROPHENYL)-TEREPHTHALAMIDE

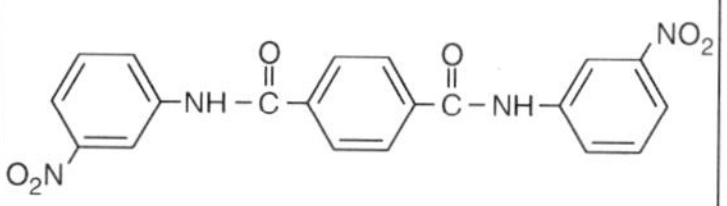

5 gm

S88,943-1
2',3'-DIMETHYL-3-NITROBENZANILIDE

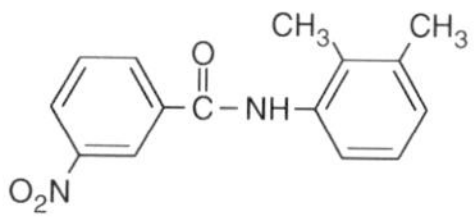

250 mg

S70,775-9
2',6'-DIETHYL-3,5-DINITROBENZANILIDE

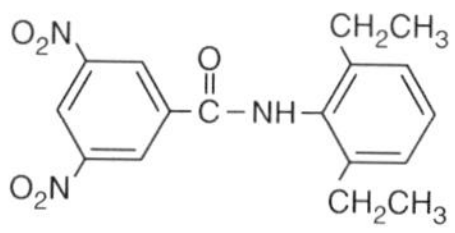

250 mg

S73,617-1
4'-METHYL-3'-NITRO-2-BUTENANILIDE

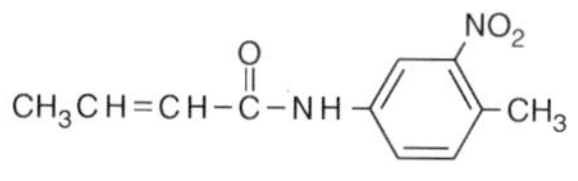

5 gm

S55,513-4
2-CHLORO-3'-NITROBENZANILIDE

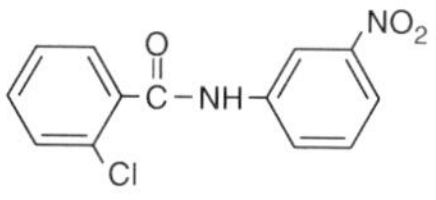

5 gm

S61,790-3
2-NITROBENZO-M-TOLUIDIDE

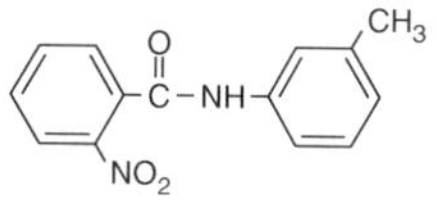

5 gm

S90,802-9
N-ACETYLANTHRANILAMIDE

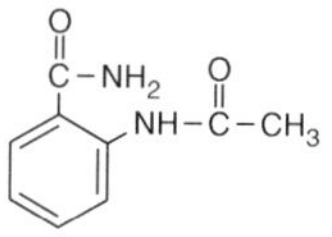

250 mg

S69,865-2
2'-CARBAMOYL-2-BUTENANILIDE
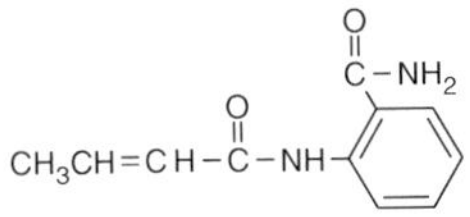
250 mg

S75,056-5
2'-CARBAMOYL-2,4-HEXADIENANILIDE
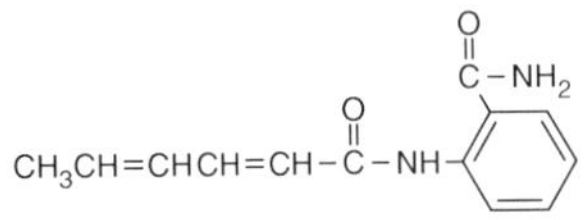
5 gm

S91,010-4
4'-(DIETHYLAMINO)-3-METHYLBENZANILIDE HYDROCHLORIDE
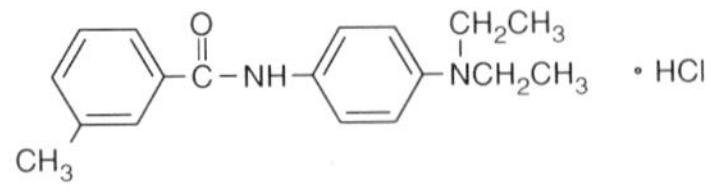
250 mg

S91,019-8
4'-(DIETHYLAMINO)-4-METHYLBENZANILIDE HYDROCHLORIDE
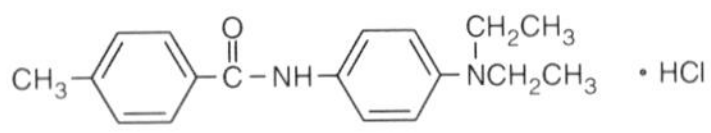
250 mg

S69,285-9
4'-ANILINOMALEANILIC ACID
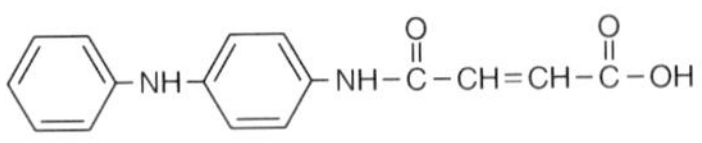
5 gm

S70,106-8
4'-NITROPHTHALANILIC ACID
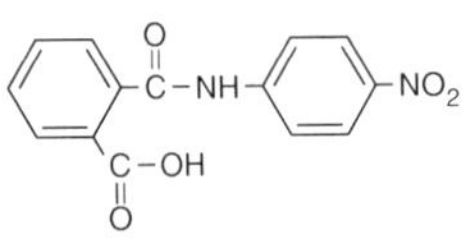
5 gm

S71,910-2
N-(2-METHYL-4-NITROPHENYL)-PHTHALAMIC ACID
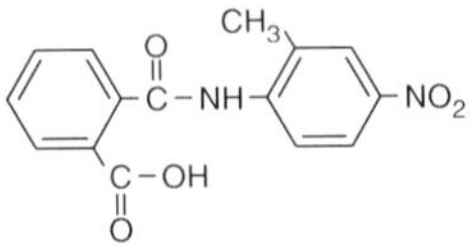
5 gm

S72,148-4
4'-NITRO-2',3,4,5,6-PENTACHLOROPHTHALANILIC ACID
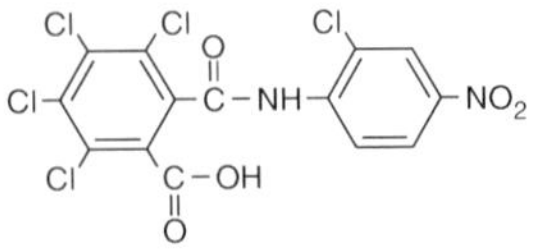
250 mg

S75,050-6
3-METHYL-4'-NITROBENZANILIDE
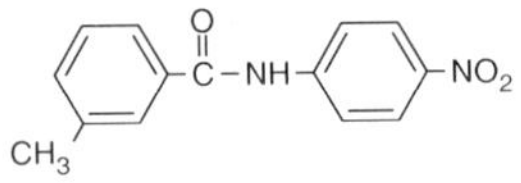
5 gm

S91,016-3
4-METHYL-4'-NITROBENZANILIDE
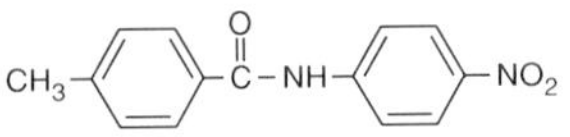
250 mg

S91,015-5
2',4-DIMETHYL-4'-NITROBENZANILIDE
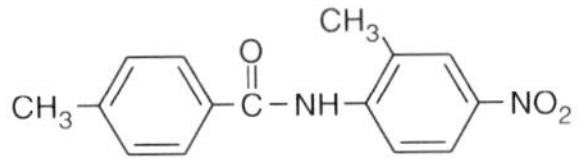
250 mg

S69,497-5
2'-METHYL-4'-NITRO-2-BUTENANILIDE
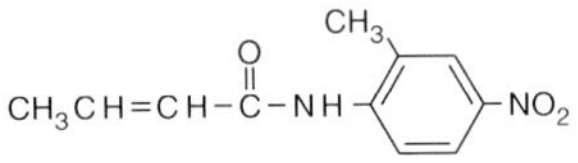
250 mg

S91,006-6
2',3-DIMETHYL-4'-NITROBENZANILIDE
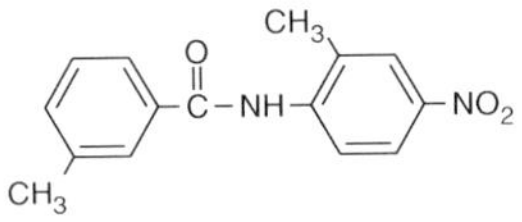
250 mg

S89,280-7
N-BENZYL-N-METHYL-3-NITROBENZAMIDE
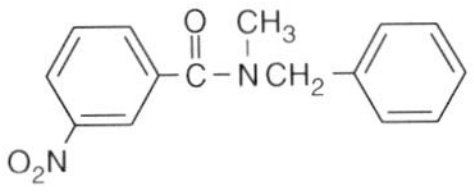
250 mg

S84,973-1
N-METHYL-4'-NITROBENZANILIDE
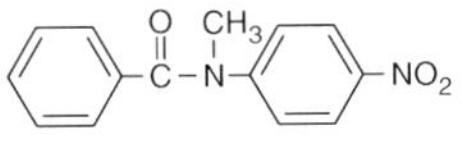
250 mg

S70,112-2
4'-NITROMALEANILIC ACID
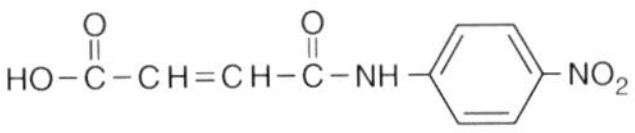
5 gm

S64,114-6
2-CARBAMOYL-6-NITROBENZOIC ACID
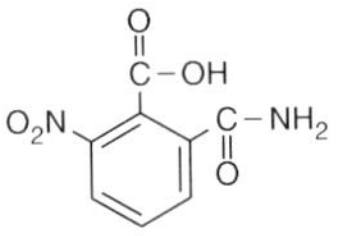
250 mg

S65,803-0
3-NITROHIPPURIC ACID
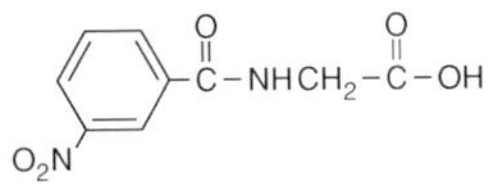
5 gm

S51,016-5
2-HEXADECYLCARBAMOYL-4-NITROBENZOIC ACID
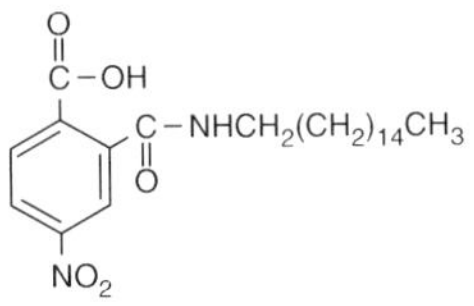
250 mg

S50,025-9
2-BROMO-3-NITROBENZAMIDE
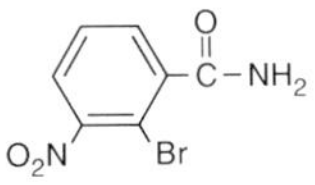
250 mg

S61,395-9
2-CHLORO-5-NITROBENZAMIDE
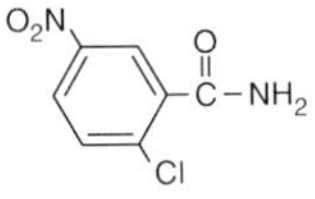
5 gm

S61,396-7
4-CHLORO-3-NITROBENZAMIDE
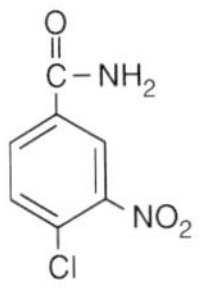
5 gm

S59,445-8
2-CHLORO-4'-NITROBENZANILIDE
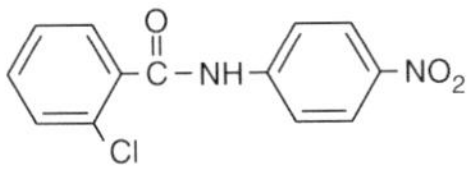
250 mg

S59,061-4
4'-NITROBENZANILIDE
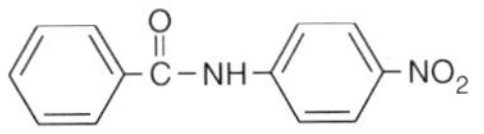
250 mg

S60,247-7
4-CHLORO-4'-NITROBENZANILIDE
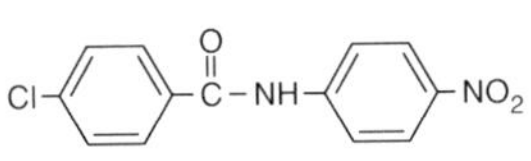
250 mg

S34,441-9
4-CHLORO-3-NITROBENZANILIDE
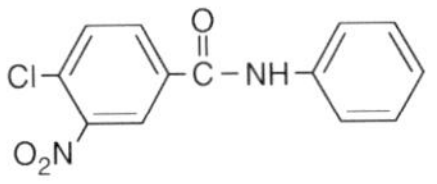
5 gm

S34,717-5
2',4'-DIMETHYL-3-NITRO-4-TOLUANILIDE
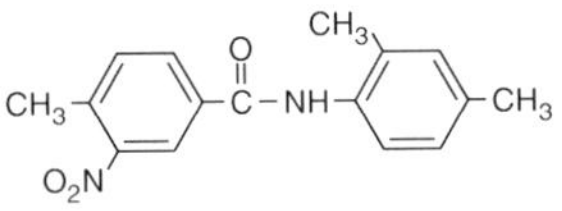
5 gm

S58,212-3
N,N-BIS(2-ETHYLHEXYL)-3-NITROBENZAMIDE
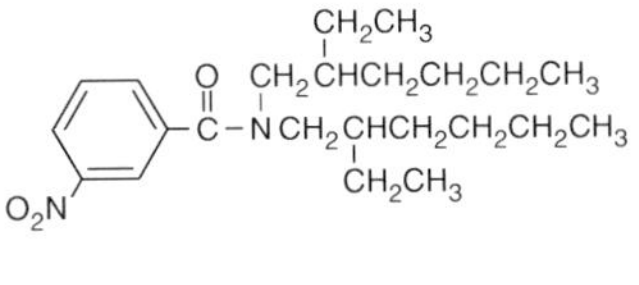
250 mg

S55,549-5
2,2'-DICHLORO-4'-NITROBENZANILIDE
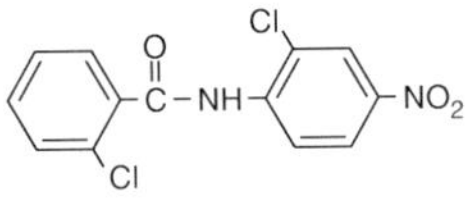
5 gm

S80,163-1
4-[N-(4-ACETAMIDOPHENYL)CARBAMOYL]-PHTHALIC ANHYDRIDE
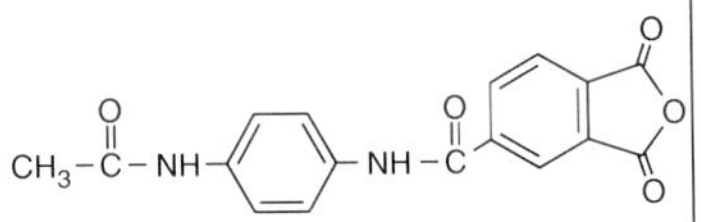
250 mg

S35,479-1
PHTHALIC DIAMIDE
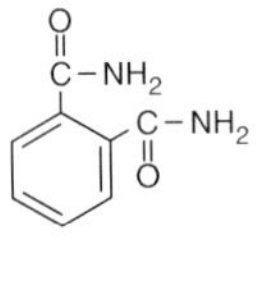
5 gm

S85,491-3
4'-(N-METHYLACETAMIDO)CINNAMANILIDE
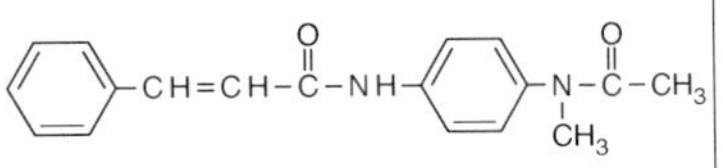
250 mg

S84,615-5
N,N'-DIBENZOYL-1,4-BUTANEDIAMINE

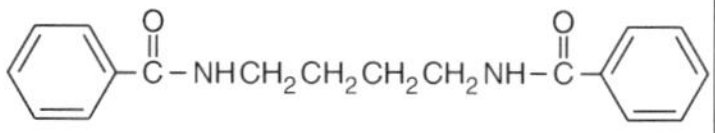

250 mg

S84,616-3
2-BENZYL-N,N'-DIBENZOYL-1,4-BUTANEDIAMINE

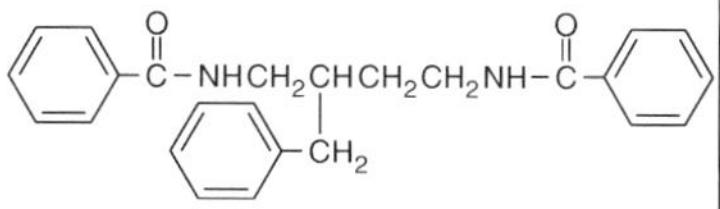

250 mg

S51,525-6
N,N'-DIMETHYLPHTHALAMIDE

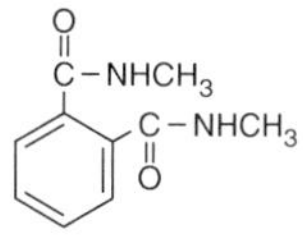

250 mg

S55,544-4
2-CHLORO-4'-(N-METHYLACETAMIDO)-BENZANILIDE

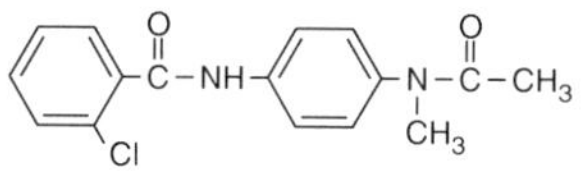

5 gm

S55,542-8
4'-ACETAMIDO-2-CHLOROBENZANILIDE

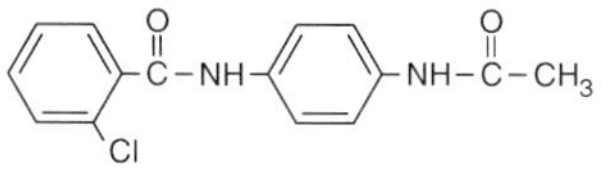

5 gm

S55,554-1
4'-ACETAMIDO-4-CHLOROBENZANILIDE

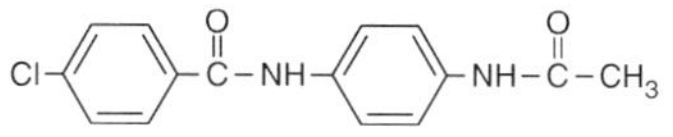

5 gm

S55,537-1
4'-ACETAMIDO-3,4-DICHLOROBENZANILIDE

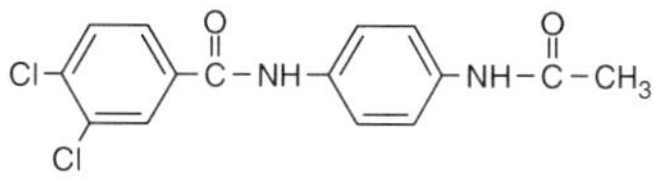

5 gm

S55,562-2
4'-ACETAMIDO-2,4-DICHLOROBENZANILIDE

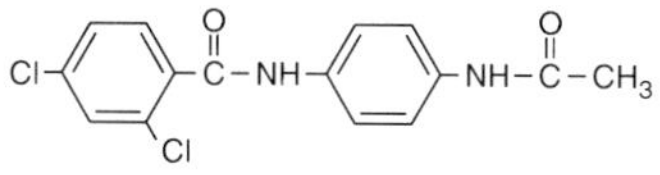

5 gm

S80,843-1
2-(BENZOYLPHENYLMETHYLAMINO)-BENZAMIDE

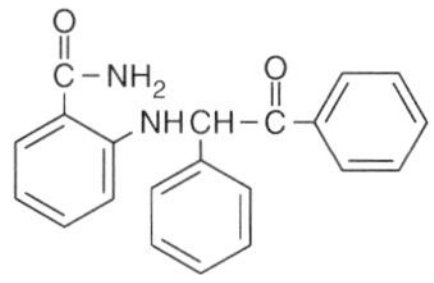

250 mg

S68,599-2
N-(2-TERT-BUTYL-1,2-DIPHENYL-2-NITROSOETHYL)BENZAMIDE, MIXTURE OF ISOMERS

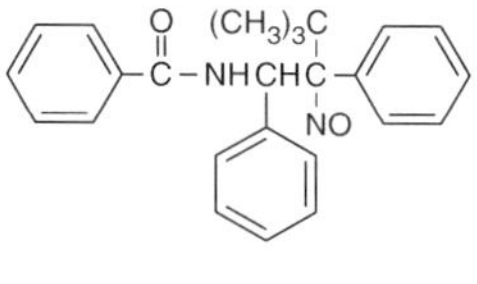

250 mg

S41,782-3
N-PHENACYLBENZAMIDE

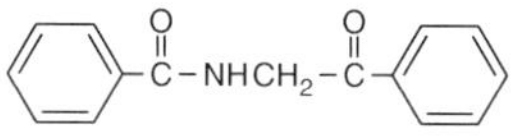

5 gm

S73,055-6
2-(PHENYLACETYL)-N-(2-PHENYLETHYL)-BENZAMIDE

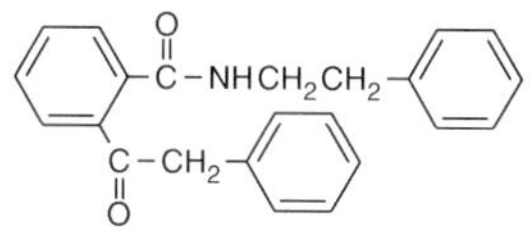

5 gm

S73,074-2
N-(2-PHENETHYL)PHTHLALAMIC ACID

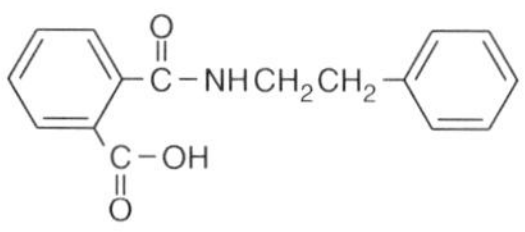

5 gm

S78,297-1
N-BENZOYL-D-PHENYLALANINE

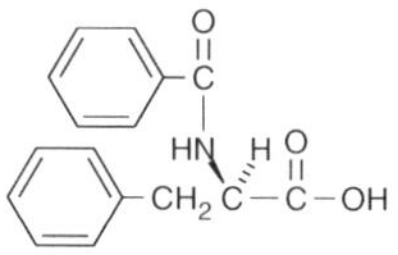

5 gm

S78,299-8
N-BENZOYL-DL-PHENYLALANINE

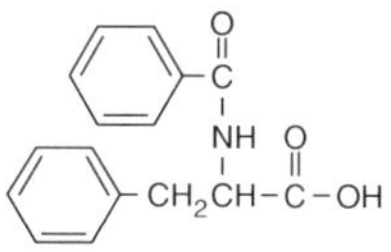

5 gm

S64,595-8
N-PHENETHYLMALEAMIC ACID

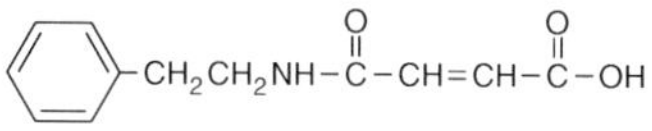

5 gm

S78,594-6
N-BENZOYL-DL-PHENYLALANINAMIDE

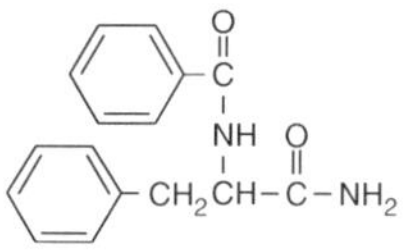

250 mg

S77,517-7
N-METHYL-N-(1-METHYL-2-PHENYLETHYL)-BENZAMIDE

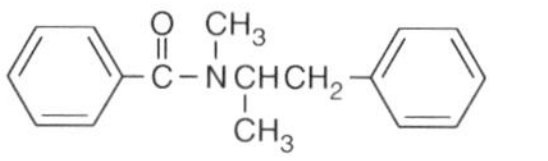

250 mg

S45,137-1
N-ACETYL-L-PHENYLALANINE ETHYL ESTER

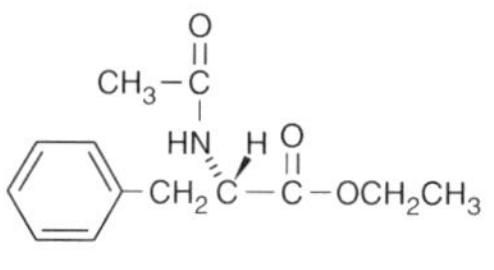

250 mg

S55,548-7
2,4-DICHLORO-N-PHENETHYLBENZAMIDE

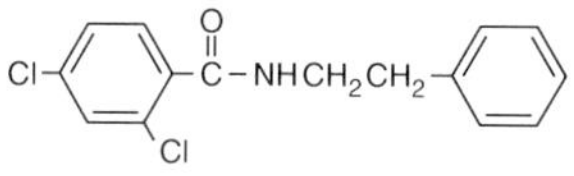

5 gm

S65,963-0
DIETHYL N-(4-METHYLAMINOBENZOYL)-L-GLUTAMATE

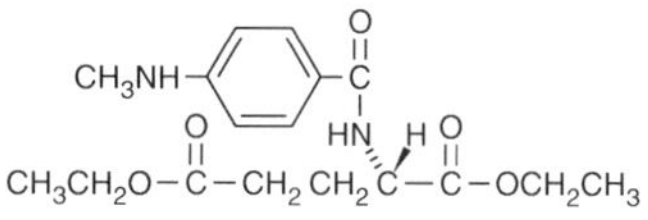

250 mg

S89,648-9
BETA-(CYCLOHEXYLAMINO)CINNAMANILIDE

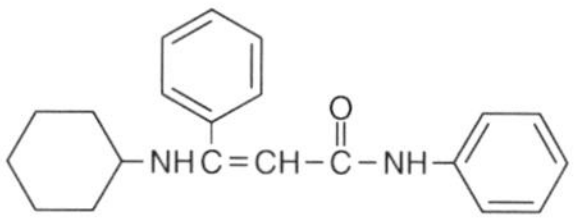

250 mg

S92,661-2
N,N-DIETHYLPHTHALAMIC ACID

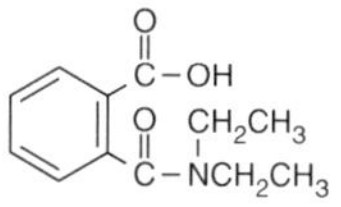

250 mg

S87,085-4
BETA-MORPHOLINO-N-PHENYLCINNAMAMIDE

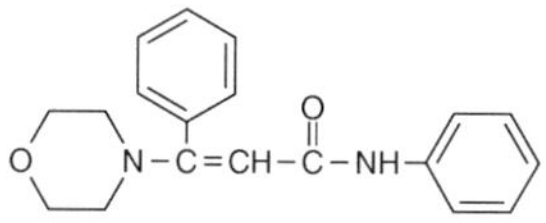

250 mg

S76,632-1
2-[(4-DIMETHYLAMINO-1-PHENYL)BUTYL]-N-METHYLBENZAMIDE

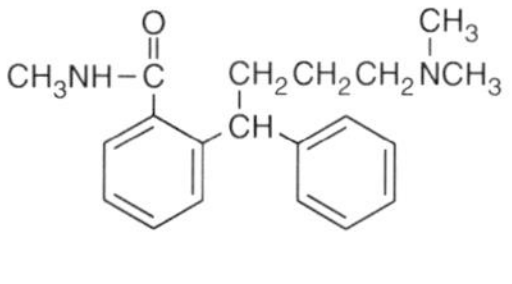

250 mg

S93,728-2
N-BENZOYL-N-PHENYLANTHRANILIC ACID

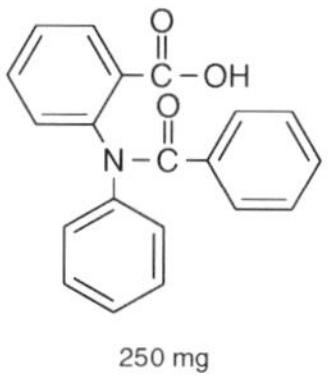

250 mg

S93,727-4
N-4-TERT-BUTYLBENZOYL-N-PHENYLANTHRANILIC ACID

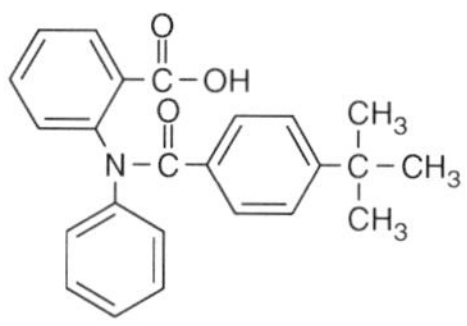

250 mg

S93,729-0
METHYL N-PHENYL-N-TOLUOYLANTHRANILATE

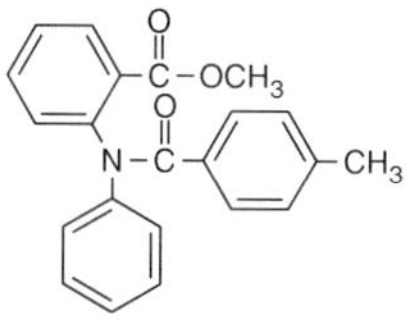

250 mg

S73,596-5
(3-HYDROXYCARBAMOYLPHENYL)-TRIMETHYLAMMONIUM IODIDE

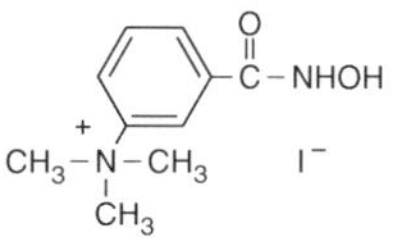

250 mg

S73,054-8
N-BENZYL-2-(PHENYLACETYL)BENZAMIDE

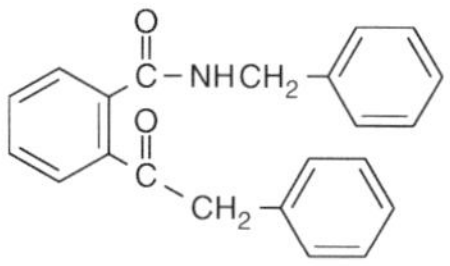

5 gm

S71,902-1
N-BENZYLPHTHALAMIC ACID

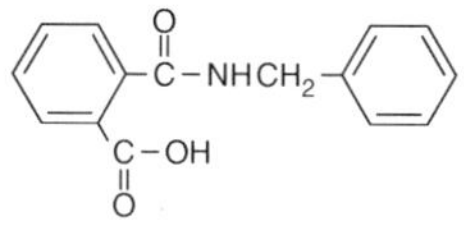

5 gm

S72,098-4
N,N'-DIBENZYL-N,N'-DIMETHYLISOPHTHALAMIDE

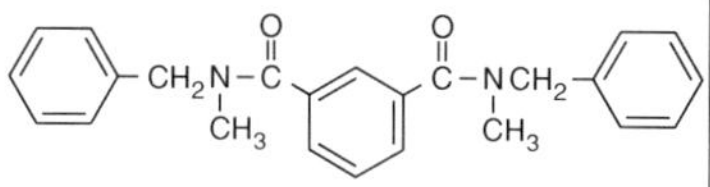

5 gm

S77,511-8
N,N'-DIBENZYL-N,N'-DIMETHYLPHTHALAMIDE

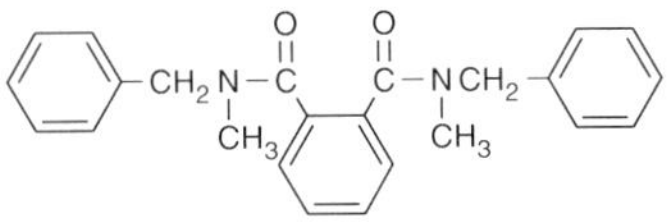

250 mg

S77,501-0
N,N'-DIBENZYL-N,N'-DIMETHYL-2,2'-BIPHENYLDICARBOXAMIDE

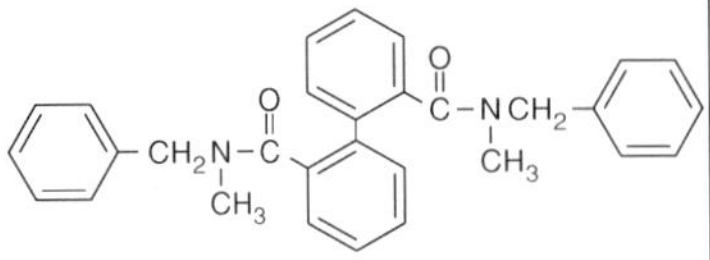

250 mg

S77,512-6
N,N'-DIBENZYL-N,N'-DI(O-TOLYL)-PHTHALAMIDE

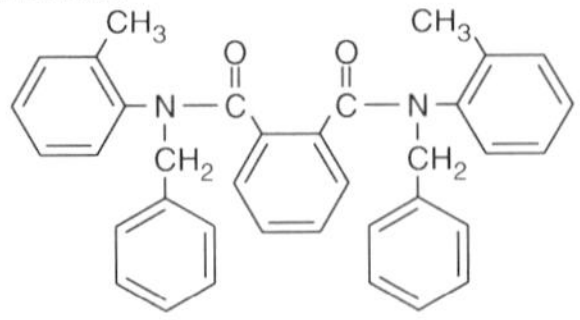

250 mg

S39,383-5
N-BENZYLMALEAMIC ACID

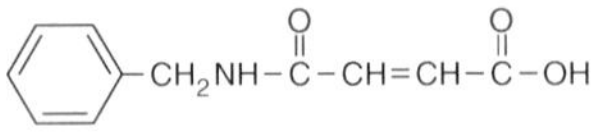

5 gm

S64,743-8
METHYL N-BENZOYLANTHRANILATE

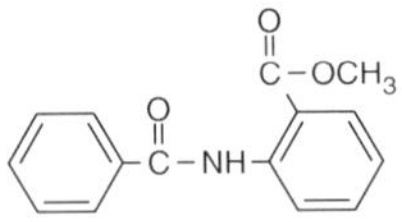

5 gm

S64,759-4
METHYL N-(2,4-DICHLOROBENZOYL)-ANTHRANILATE

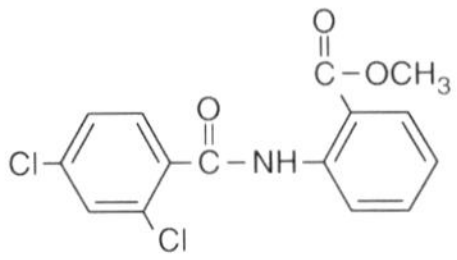

5 gm

S61,823-3
ETHYL 3-(4-CHLOROBENZAMIDO)-CROTONATE

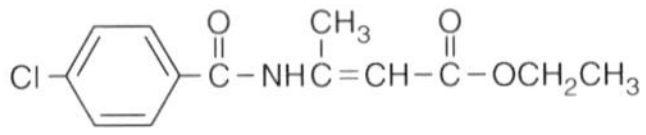

250 mg

S52,959-1
N-BENZYLMETHACRYLAMIDE

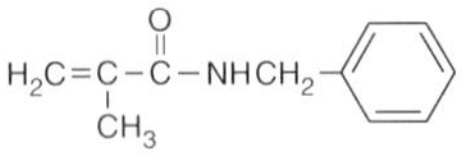

5 gm

S79,496-1
4-(N-BENZYLCARBAMOYL)-PHTHALIC ANHYDRIDE

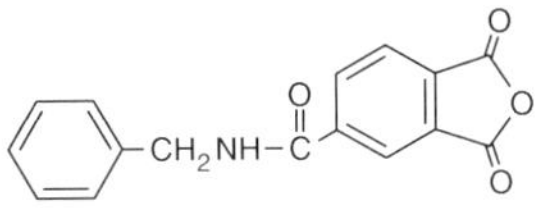

250 mg

S52,794-7
N-BENZYLCROTONAMIDE

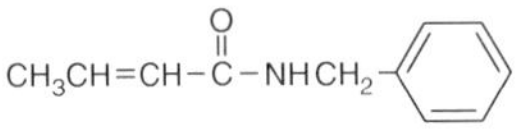

250 mg

S62,343-1
4'-CHLORO-M-TOLUANILIDE

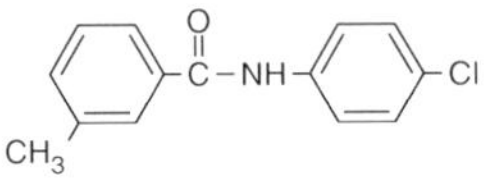

250 mg

S59,894-1
N-BENZYL-4-METHYLBENZAMIDE

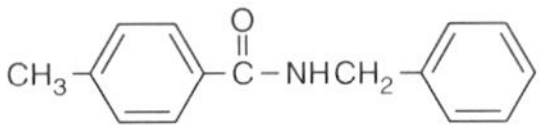

250 mg

S51,485-3
N-BENZYL-N-METHYLBENZAMIDE

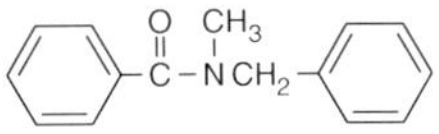

250 mg

S77,535-5
N-BENZYL-N-TERT-BUTYLBENZAMIDE

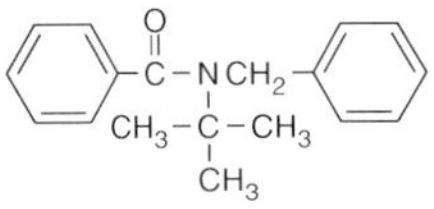

250 mg

S51,368-7
N-BENZYLBENZANILIDE

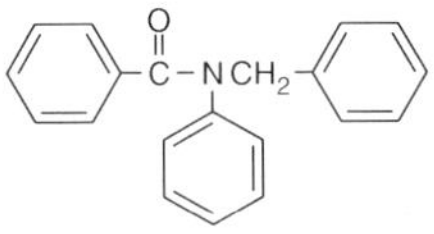

250 mg

S51,557-4
N,N-DIBENZYLBENZAMIDE

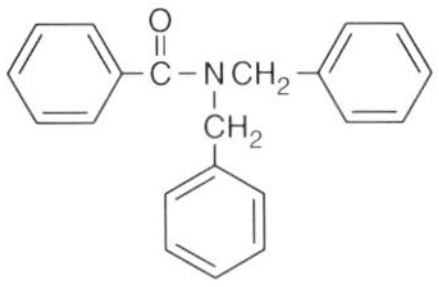

250 mg

S77,514-2
N-BENZYL-N-TERT-BUTYL-O-TOLUAMIDE

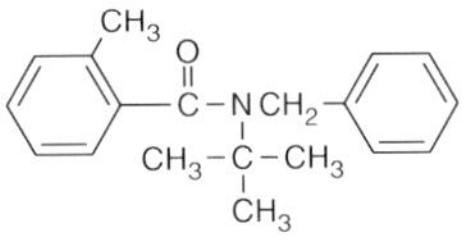

250 mg

S76,785-9
4-NITRO-N-PHENYLBENZOHYDROXAMIC ACID

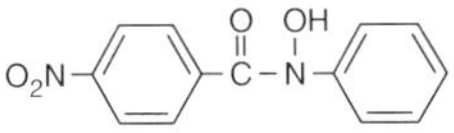

250 mg

S77,081-7
N-METHYL-4-NITROBENZOHYDROXAMIC ACID

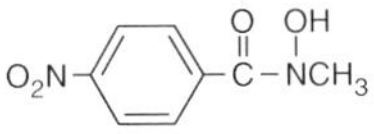

250 mg

S94,600-1
4-NITROBENZOHYDROXAMIC ACID

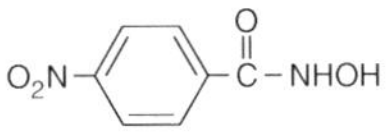

250 mg

S64,789-6
N,O-BIS(4-NITROBENZOYL)-METHYLHYDROXYLAMINE

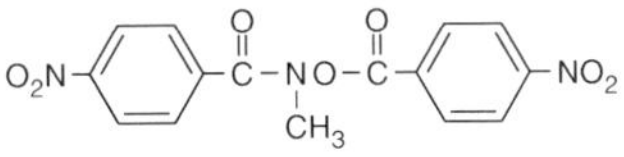

250 mg

S45,308-0
BENZOHYDROXAMIC ACID, POTASSIUM SALT

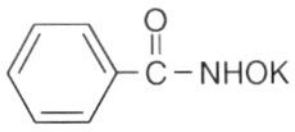

5 gm

S80,576-9
O-ALLYLBENZOHYDROXAMIC ACID

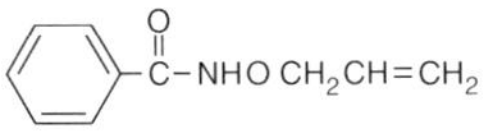

250 mg

S90,054-0
N-(1-NAPHTHYL)-6-NITROPHTHALAMIC ACID

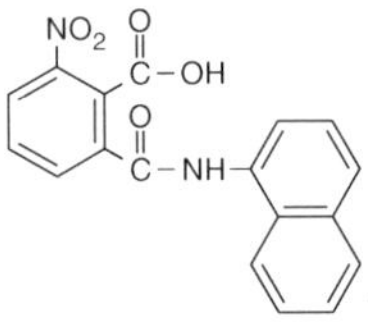

250 mg

S93,698-7
METHYL N-(4-NITROBENZOYL)-N-PHENYLANTHRANILATE

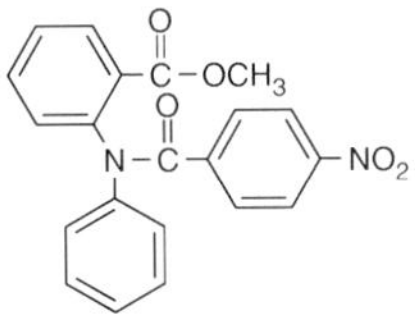

250 mg

S67,775-2
3-METHYL-4-NITROBENZAMIDE

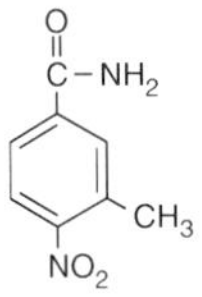

5 gm

S88,036-1
4-NITROBENZANILIDE

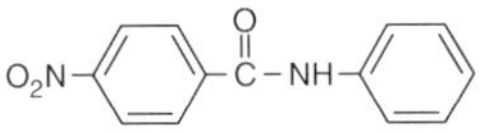

250 mg

S89,940-2
2'-ETHYL-4-NITROBENZANILIDE

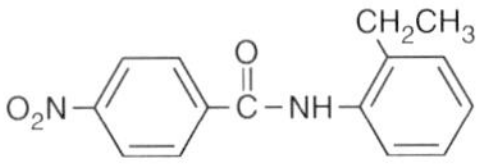

250 mg

S72,125-5
4-NITROBENZO-2',3'-XYLIDIDE

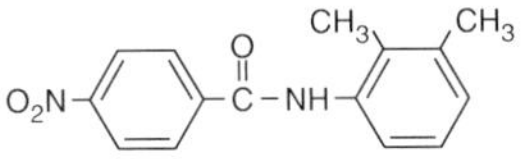

250 mg

S73,569-8
4-NITROBENZO-2',6'-XYLIDIDE

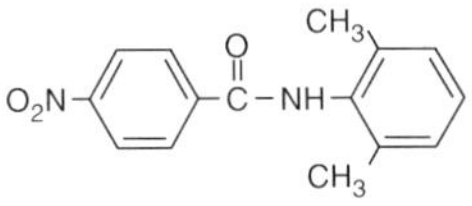

250 mg

S70,771-6
2',6'-DIETHYL-4-NITROBENZANILIDE

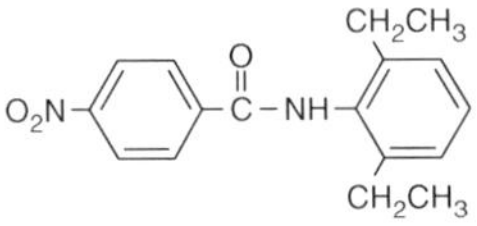

5 gm

S75,138-3
4-NITROBENZO-2',5'-XYLIDIDE

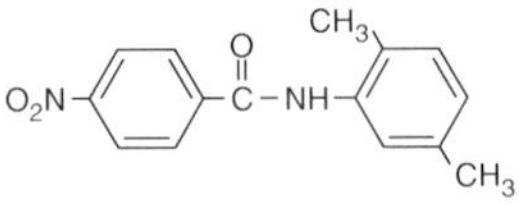

5 gm

S72,111-5
5'-ISOPROPYL-2'-METHYL-4-NITROBENZANILIDE

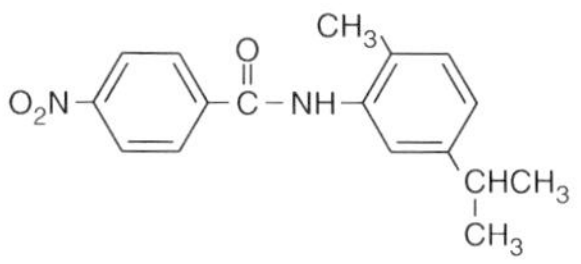

250 mg

S70,787-2
4-NITRO-2',4',6'-TRIMETHYLBENZANILIDE

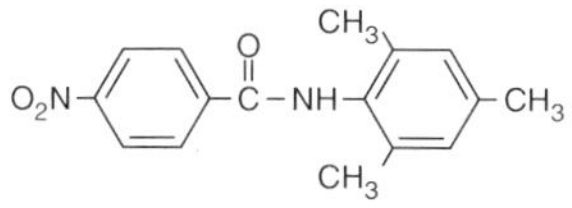

250 mg

S70,818-6
3-NITROBENZO-2',6'-XYLIDIDE

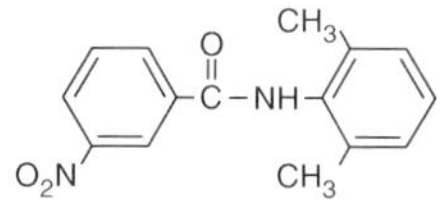

250 mg

S70,773-2
2',6'-DIETHYL-3-NITROBENZANILIDE

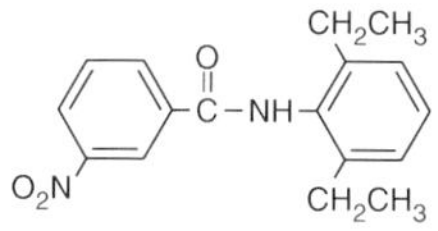

5 gm

S77,952-0
3-NITROBENZO-2',4'-XYLIDIDE

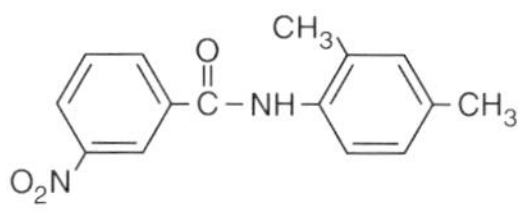

250 mg

S78,081-2
3-NITROBENZO-3',4'-XYLIDIDE

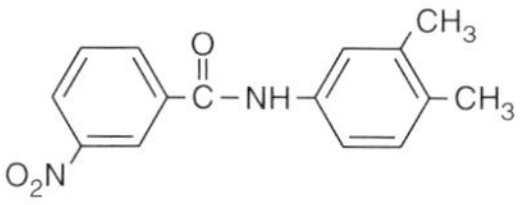

250 mg

S87,193-1
4-NITROBENZOIC N2-(AZACYCLODEC-1-EN-2-YL)HYDRAZIDE

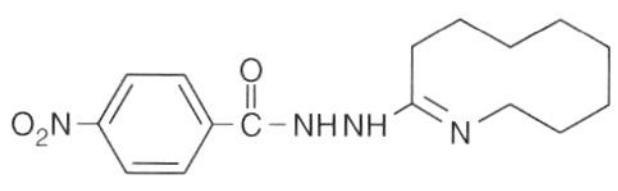

250 mg

S77,871-0
2'-CHLORO-5'-NITROPHTHALANILIC ACID

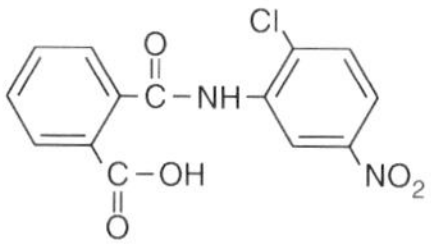

250 mg

S71,586-7
4'-ETHYL-4-NITROBENZANILIDE

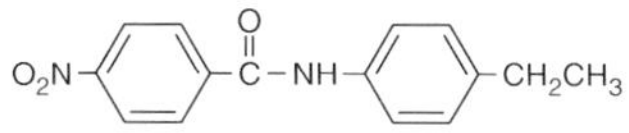

250 mg

S56,886-4
N-(4-NITROBENZOYL)ANTHRANILIC ACID

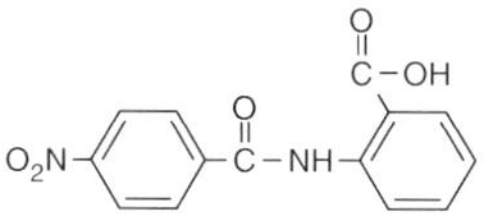

5 gm

S53,240-1
4'-METHYL-3'-NITROMALEANILIC ACID

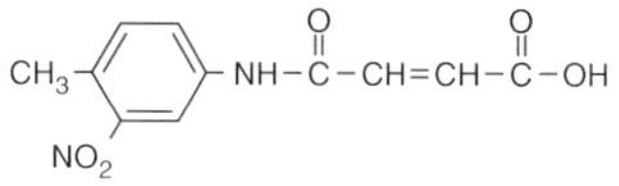

5 gm

S53,246-0
2'-METHYL-5'-NITROMALEANILIC ACID

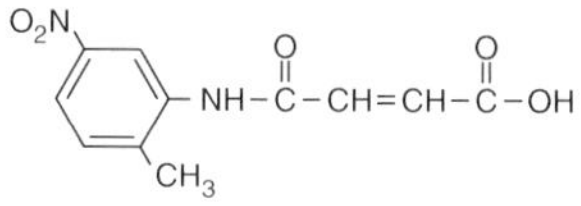

5 gm

S56,881-3
2'-CARBAMOYL-4-NITROBENZANILIDE

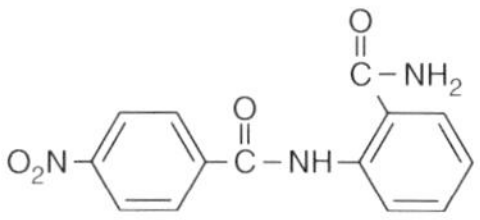

250 mg

S94,198-0
N-BUTYL-4-NITROBENZAMIDE

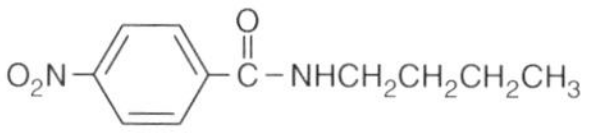

250 mg

S61,862-4
N,N-DIDODECYL-3-NITROBENZAMIDE

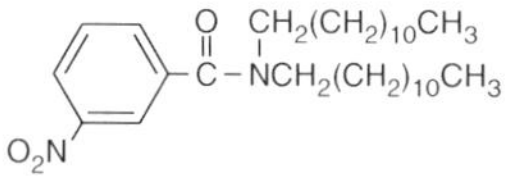

250 mg

S38,103-9
N-(4-NITROBENZOYL)GLYCINE ETHYL ESTER

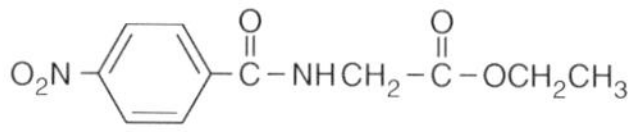

250 mg

S55,539-8
3,4-DICHLORO-3'-NITROBENZANILIDE

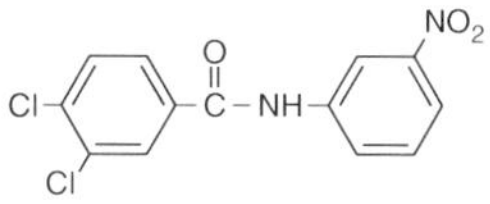

5 gm

S88,839-7
2-BENZAMIDOCINNAMALDEHYDE

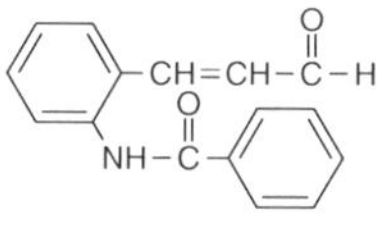

250 mg

S90,625-5
N-(2-FORMYL-2-METHYLPROPYL)-N-METHYLBENZAMIDE

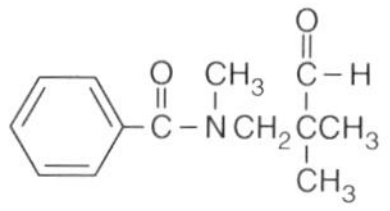

250 mg

S76,402-7
3'-ACETYLPHTHALANILIC ACID

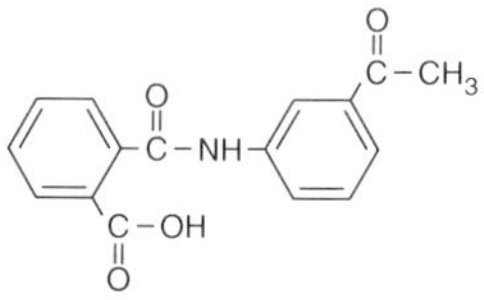

5 gm

S73,992-8
4'-ACETYLPHTHALANILIC ACID

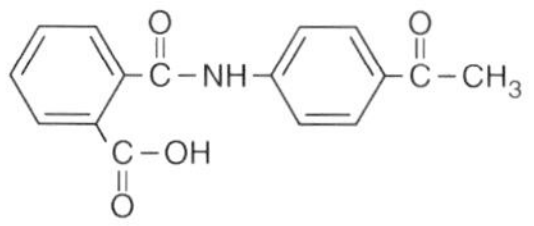

5 gm

S69,661-7
2-BENZOYL-N-METHYLBENZAMIDE

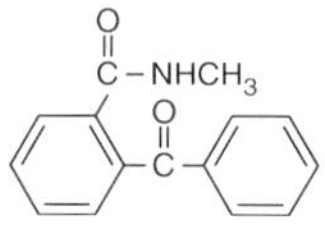

250 mg

S38,222-1
2-BENZOYLBENZAMIDE

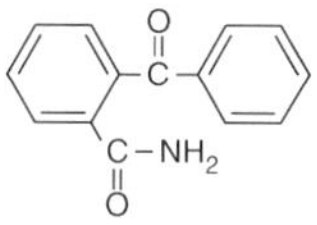

5 gm

S73,368-7
5-BENZAMIDO-6-BENZOYL-1,3-BENZODIOXOLE

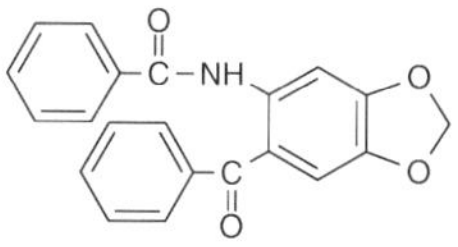

250 mg

S61,741-5
1-BENZAMIDO-4-CHLOROANTHRAQUINONE

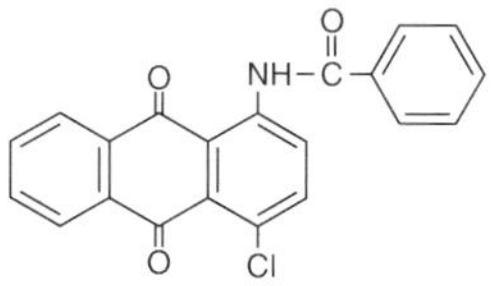

5 gm

S55,531-2
2'-BENZOYL-2,4,4'-TRICHLOROBENZANILIDE

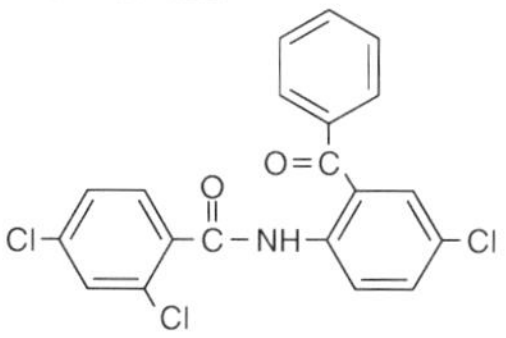

5 gm

S93,499-2
N-BENZOYL-2-(4-OXOCYCLOHEXYL)-GLYCINE

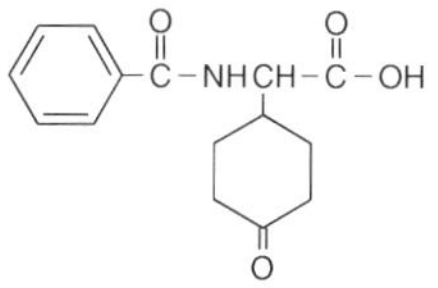

250 mg

S94,024-0
N-(1-ACETYL-2-METHYL-1-PROPYL)-BENZAMIDE

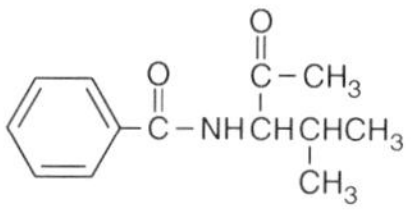

250 mg

S55,545-2
2'-BENZOYL-2,4-DICHLOROBENZANILIDE

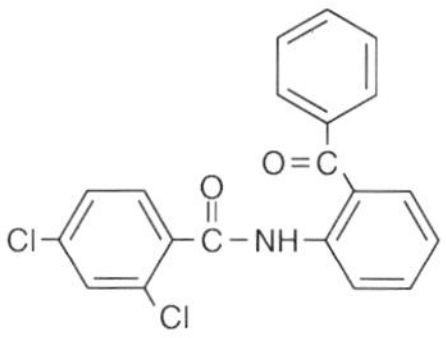

250 mg

S34,600-4
DIETHYL BENZAMIDOMALONATE

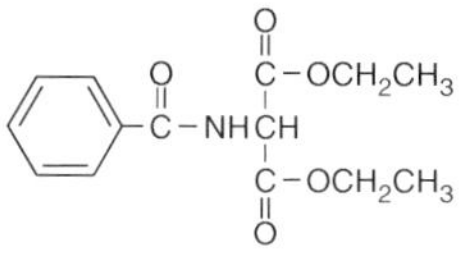

5 gm

S78,920-8
N-BENZOYL-D-ASPARTIC ACID

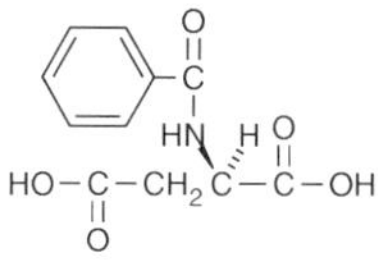

250 mg

S92,471-7
2,5-BIS[N-(2-METHOXYCARBONYLPHENYL)-CARBAMOYL]TEREPHTHALIC ACID

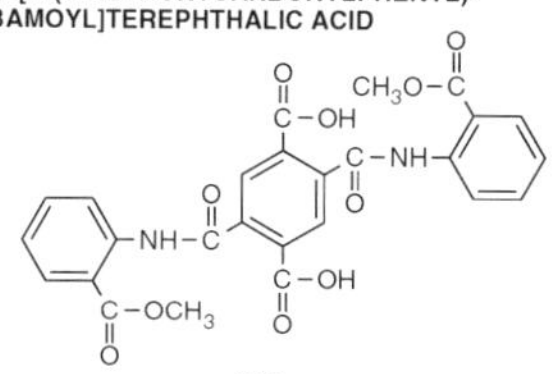

250 mg

S90,156-3
4-[N-(2,3-XYLYL)CARBAMOYL]-ISOPHTHALIC ACID, 2,3-DIMETHYLANILINE SALT

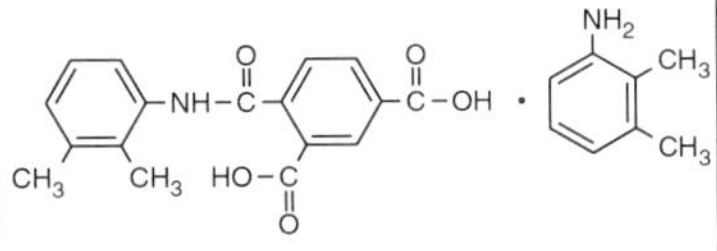

250 mg

S90,155-5
4-(N-MESITYLCARBAMOYL)ISOPHTHALIC ACID, 2,4,6-TRIMETHYLANILINE SALT

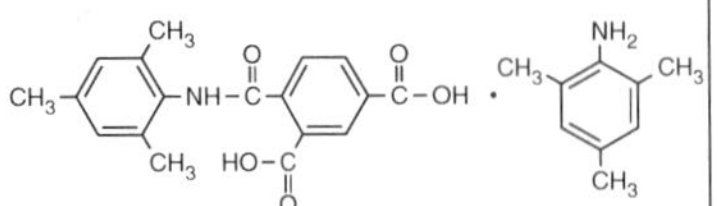

250 mg

S90,182-2
2,5-BIS[N-(2,5-XYLYL)CARBAMOYL]-TEREPHTHALIC ACID

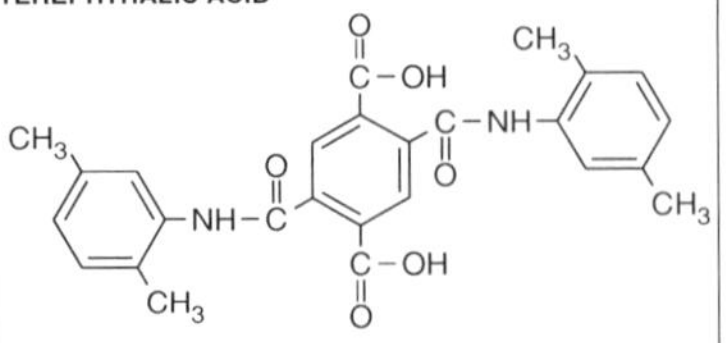

250 mg

S90,162-8
4-[N-(1-NAPHTHYL)CARBAMOYL]-ISOPHTHALIC ACID

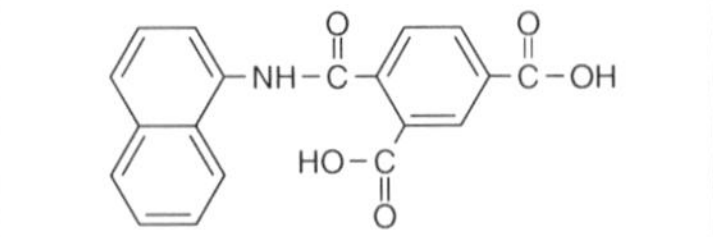

250 mg

S76,958-4
2'-CARBAMOYLGLUTARANILIC ACID

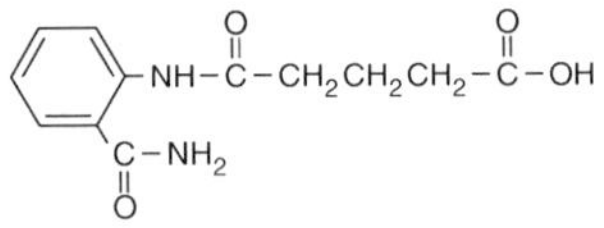

250 mg

S71,904-8
N-CYCLOHEXYLPHTHALAMIC ACID

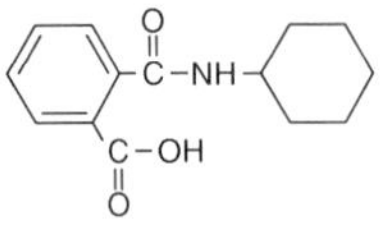

5 gm

S70,130-0
2'-ETHYLMALEANILIC ACID

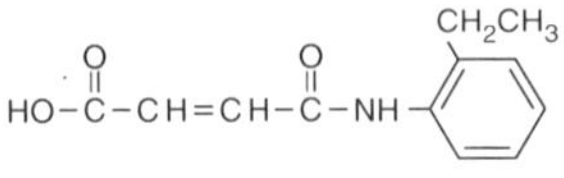

5 gm

S70,107-6
N-TERT-BUTYLPHTHALAMIC ACID

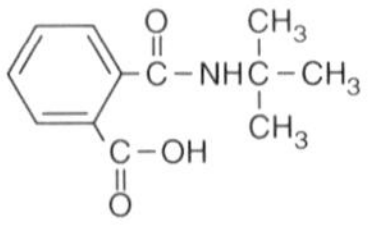

5 gm

S71,953-6
3'-METHYLPHTHALANILIC ACID

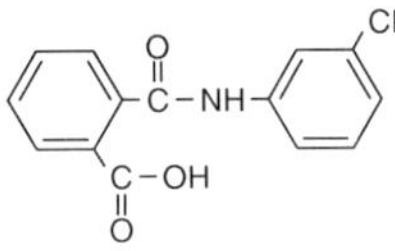

5 gm

S92,142-4
2'-METHYLPHTHALANILIC ACID

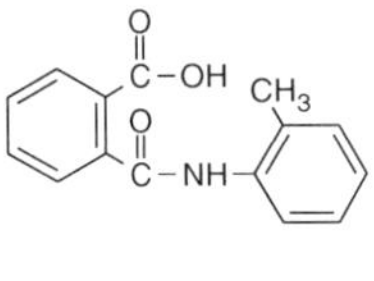

250 mg

S73,085-8
N-(2-ETHYLPHENYL)PHTHALAMIC ACID

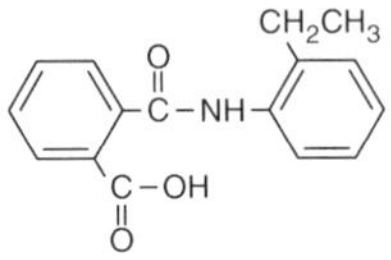

5 gm

S70,100-9
4'-METHYLPHTHALANILIC ACID

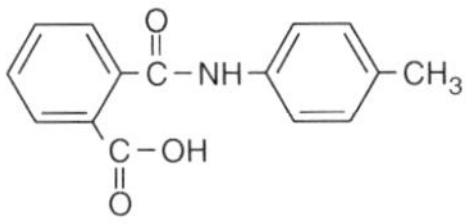

5 gm

S89,022-7
N-(P-TOLYL)PHTHALAMIC ACID

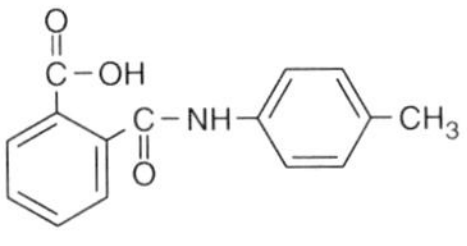

250 mg

S71,955-2
4'-ETHYLPHTHALANILIC ACID

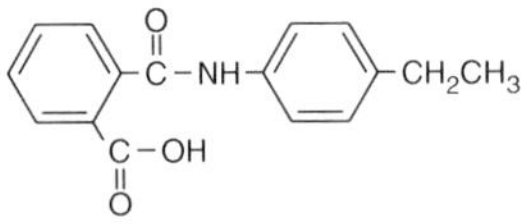

5 gm

S71,905-6
N-(2,6-XYLYL)PHTHALAMIC ACID

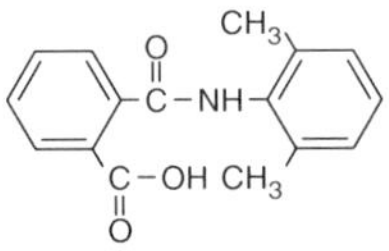

5 gm

S71,908-0
N-(2,3-XYLYL)PHTHALAMIC ACID

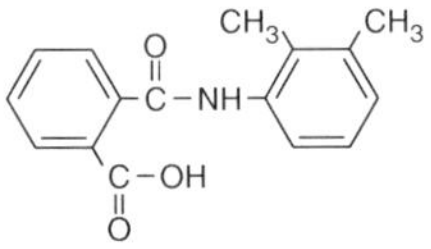

5 gm

S71,742-8
2',4'-DIMETHYLPHTHALANILIC ACID

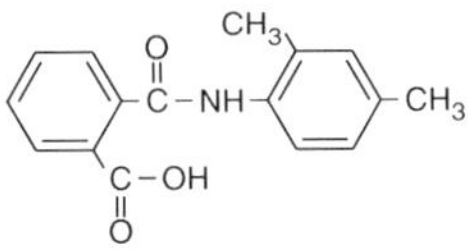

5 gm

S71,907-2
N-(3,4-XYLYL)PHTHALAMIC ACID

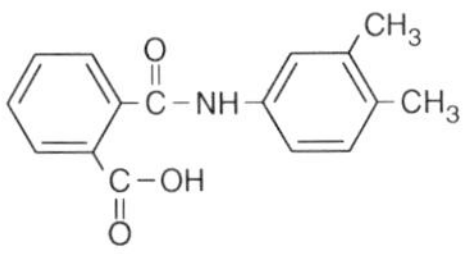

5 gm

S71,956-0
2',5'-DIMETHYLPHTHALANILIC ACID

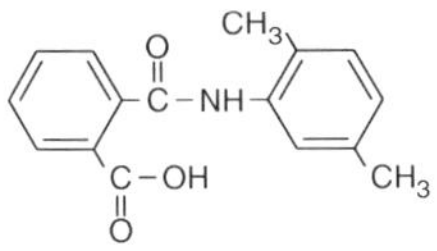

5 gm

S76,405-1
2',4',6'-TRIMETHYLPHTHALANILIC ACID

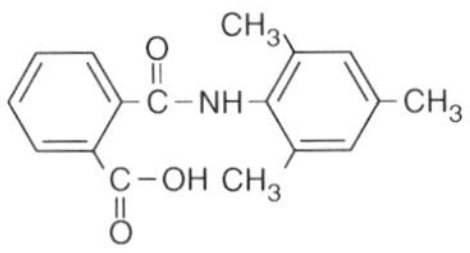

5 gm

S67,354-4
N-(4-ETHYLPHENYL)MALEAMIC ACID

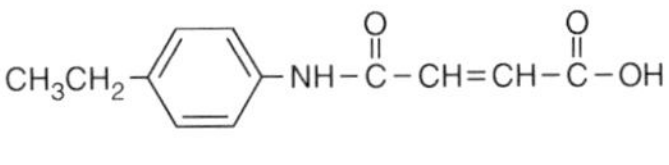

250 mg

S69,286-7
2',3'-DIMETHYLMALEANILIC ACID

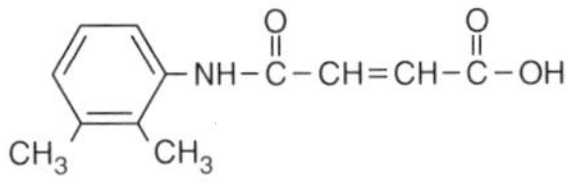

5 gm

S69,532-7
2',4'-DIMETHYLMALEANILIC ACID

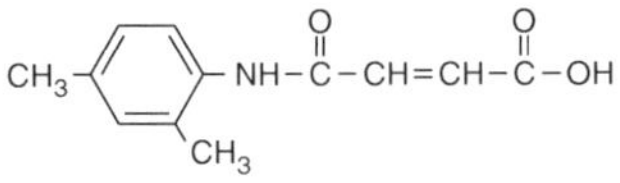

5 gm

S70,397-4
2',4'-DIMETHYL-3,4,5,6-TETRACHLOROPHTHALANILIC ACID

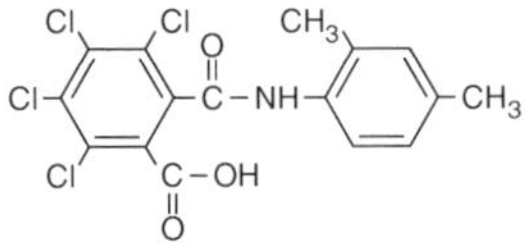

5 gm

S92,264-1
3',5'-DIMETHYLMALEANILIC ACID

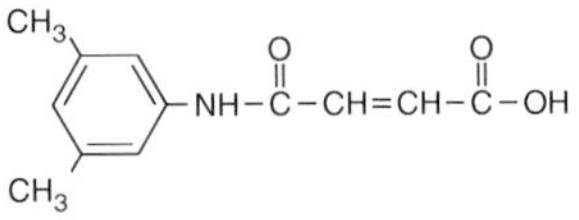

5 gm

S70,103-3
2',4',6'-TRIMETHYLMALEANILIC ACID

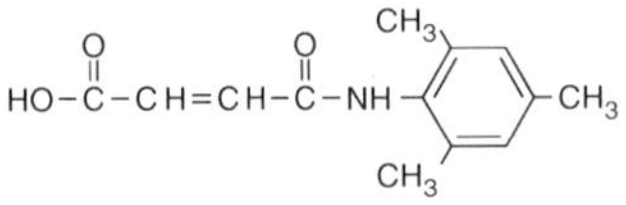

5 gm

S89,616-0
DIETHYL 5-BENZAMIDO-1,2,3,4-TETRAHYDRO-1,4-ETHENONAPHTHALENE-2,3-DICARBOXYLATE

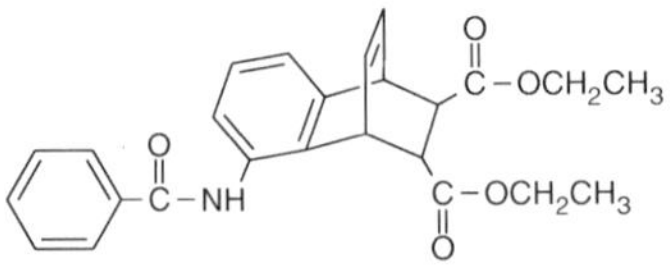

250 mg

S89,993-3
N,N'-BIS[2-(METHOXYCARBONYL)PHENYL]-TEREPHTHALAMIDE

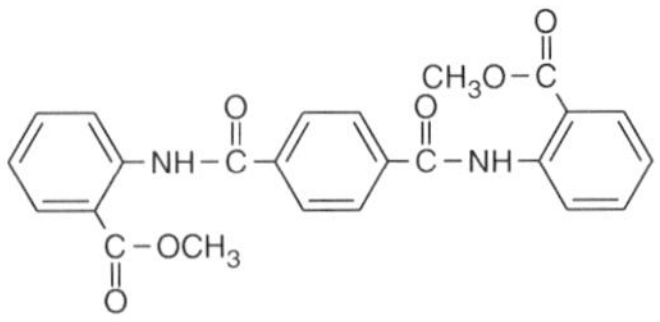

250 mg

S88,016-7
METHYL 4-(N-PHENYLCARBAMOYL)-BENZOATE

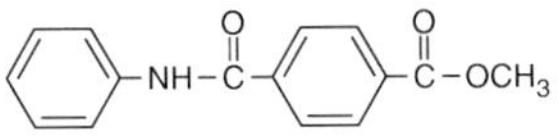

250 mg

S73,582-5
2-(METHOXYCARBONYL)-2',6'-DIMETHYLBENZANILIDE

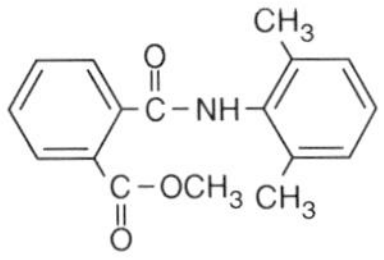

250 mg

S90,579-8
4-[N-(2-NAPHTHYL)CARBAMOYL]-PHTHALIC ANHYDRIDE

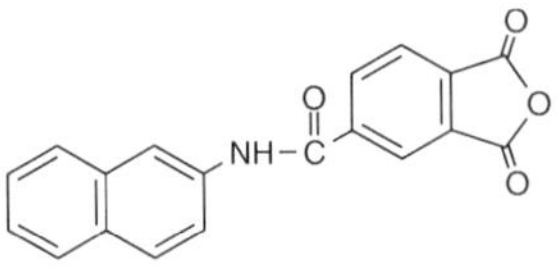

250 mg

S79,546-1
N,N'-DICYCLOHEXYLTEREPHTHALAMIDE

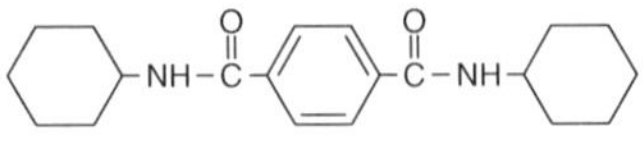

250 mg

S79,169-5
N,N'-BIS(2-ETHYLPHENYL)-ISOPHTHALAMIDE

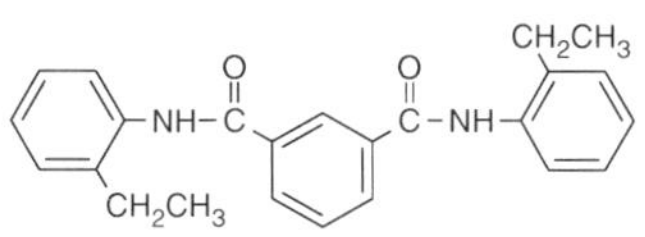

250 mg

S89,296-3
N,N'-DIHEXYLTEREPHTHALAMIDE

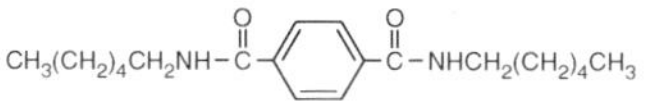

250 mg

S89,077-4
N,N'-BIS(2,3-XYLYL)ISOPHTHALAMIDE

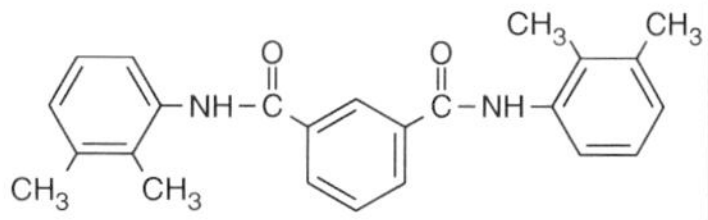

250 mg

S77,242-9
N,N'-BIS(2,4-XYLYL)ISOPHTHALAMIDE

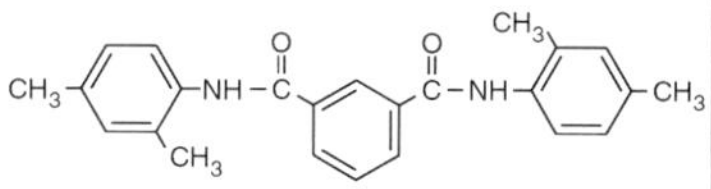

250 mg

S89,078-2
N,N'-BIS(3,4-XYLYL)ISOPHTHALAMIDE

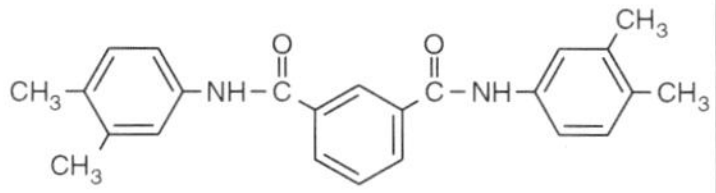

250 mg

S77,536-3
N,N,N',N'-TETRAISOBUTYLISOPHTHALAMIDE

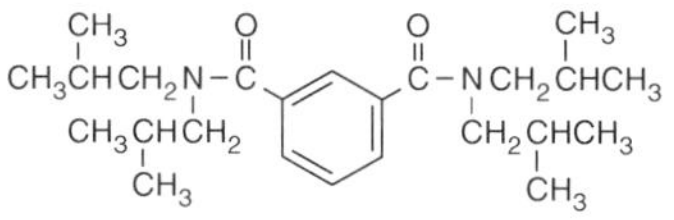

250 mg

S79,555-0
N,N'-DICYCLOHEXYL-N,N'-DIPHENYLISOPHTHALAMIDE

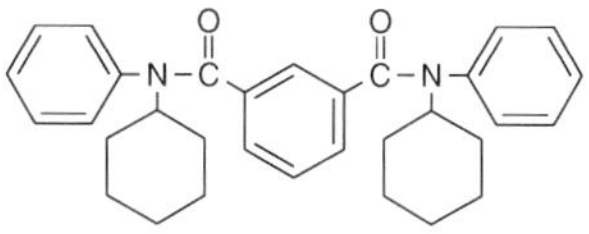

250 mg

S75,170-7
N,N'-DICYCLOHEXYL-N,N'-DIPHENYLTEREPHTHALAMIDE

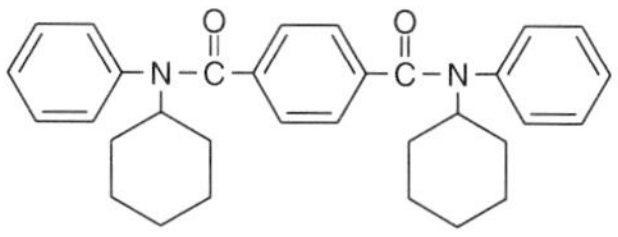

5 gm

S78,067-7
N,N'-DIETHYL-N,N'-DIPHENYLTEREPHTHALAMIDE

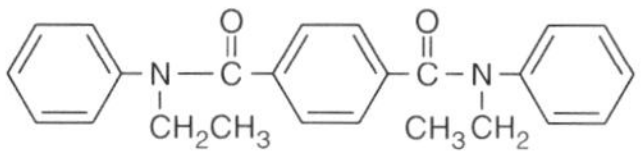

250 mg

S89,294-7
N,N'-DIBENZYL-N,N'-DIMETHYLTEREPHTHALAMIDE

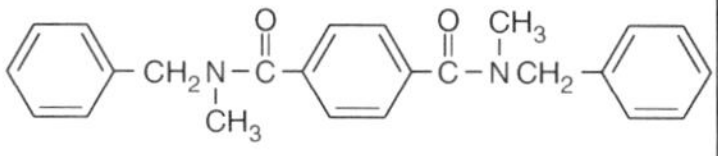

250 mg

S74,608-8
N,N,N',N'-TETRAMETHYLPHTHALIC DIAMIDE

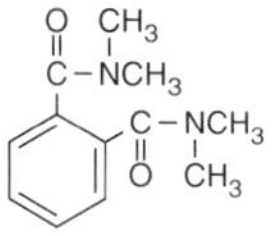

5 gm

S77,490-1
N,N'-DIETHYL-N,N'-BIS(O-TOLYL)-ISOPHTHALAMIDE

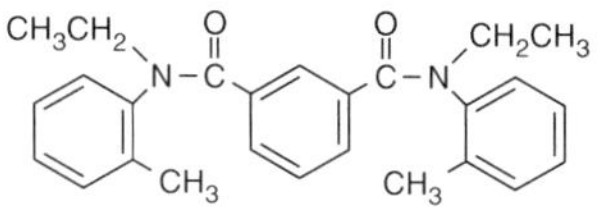

250 mg

S77,513-4
N,N',N''-TRIETHYL-N,N',N''-TRI(O-TOLYL)-1,3,5-BENZENETRICARBOXAMIDE

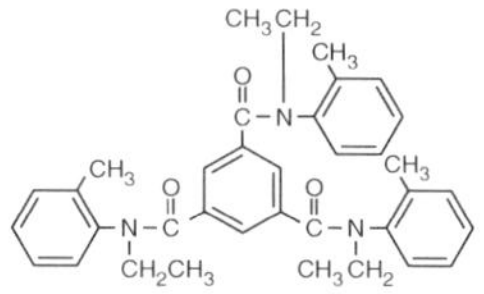

250 mg

S74,796-3
N,N'-DIETHYL-N,N'-BIS(M-TOLYL)-TEREPHTHALAMIDE

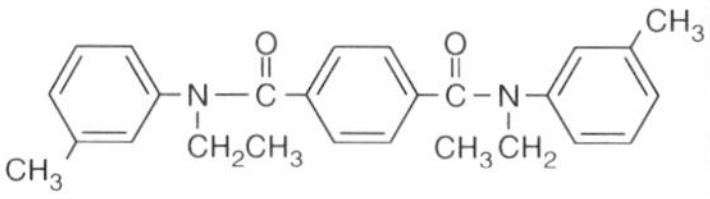

5 gm

S50,603-6
N,N'-(1,3-PHENYLENE)DIMALEAMIC ACID

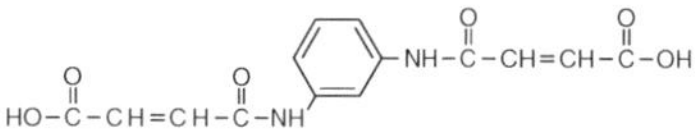

5 gm

S50,602-8
N,N'-(1,2-PHENYLENE)DIMALEAMIC ACID

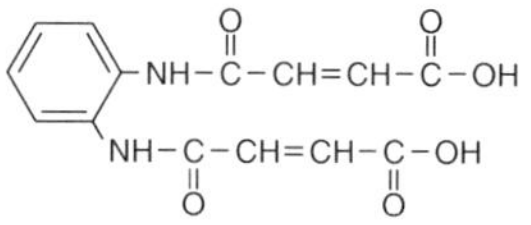

250 mg

S64,605-9
4,4'-METHYLENE-BIS-(MALEANILIC ACID)

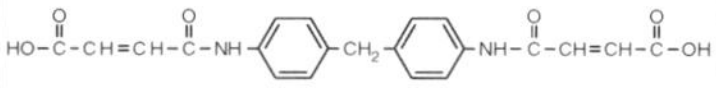

5 gm

S80,286-7
2-BENZAMIDO-3-METHYLCROTONIC ACID

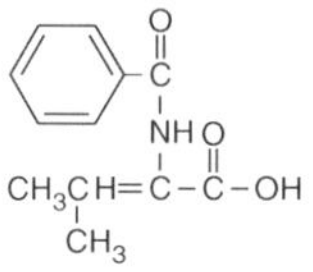

250 mg

S65,804-9
4-CHLOROHIPPURIC ACID

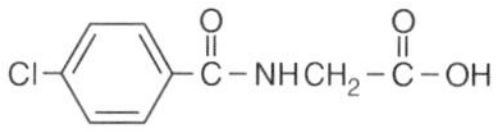

5 gm

S67,537-7
4-(TERT-BUTYL)BENZAMIDOACETIC ACID

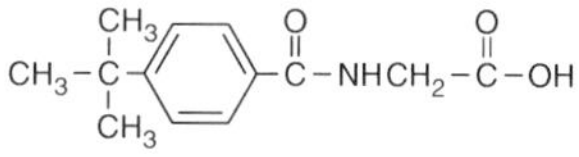

5 gm

S65,814-6
N-(3,4,5-TRIMETHOXYBENZOYL)GLYCINE

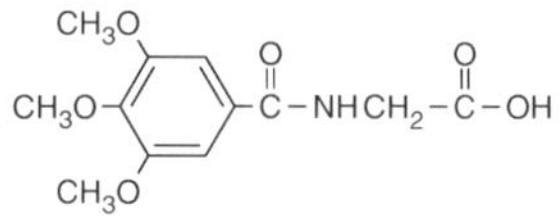

5 gm

S67,514-8
N-(4-ETHOXYCARBONYLPHENYL)-PHTHALAMIC ACID

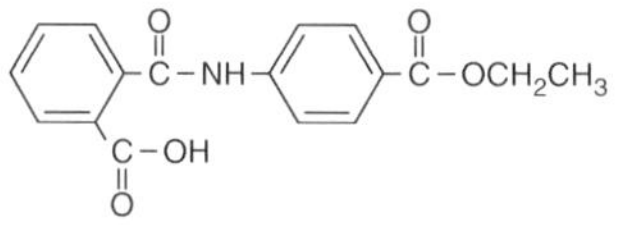

250 mg

S55,468-5
4'-(ETHOXYCARBONYL)MALEANILIC ACID

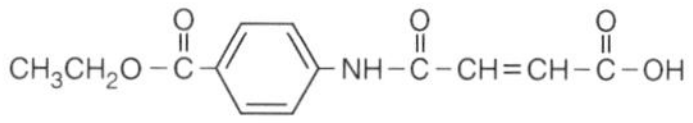

5 gm

S38,879-3
O-BENZAMIDOGLYCOLIC ACID

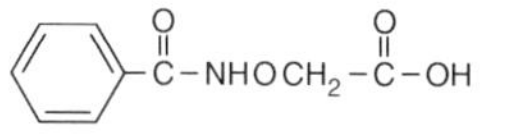

5 gm

S54,757-3
PHTHALAMIC ACID

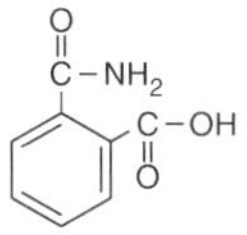

250 mg

S53,969-4
3'-CHLOROMALEANILIC ACID

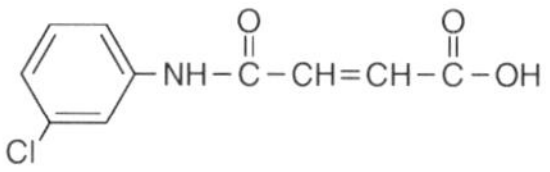

5 gm

S39,381-9
3'-METHYLMALEANILIC ACID

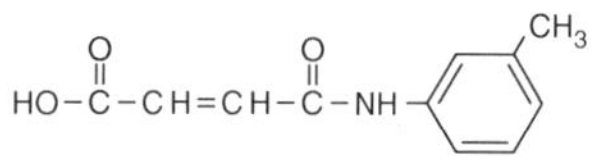

5 gm

S72,127-1
3'-CARBOXY-3-METHYLBENZANILIDE

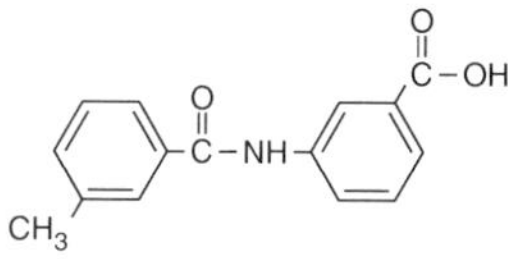

250 mg

S89,521-0
3-BENZAMIDO-2-NORBORNANECARBOXYLIC ACID

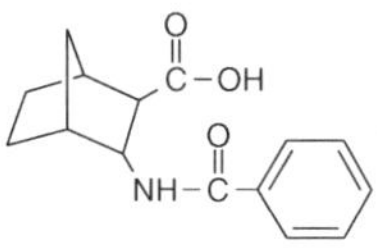

250 mg

S36,445-2
N-PHENYLMALEAMIC ACID

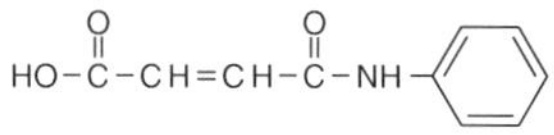

5 gm

S70,682-5
2-BENZAMIDOCROTONIC ACID

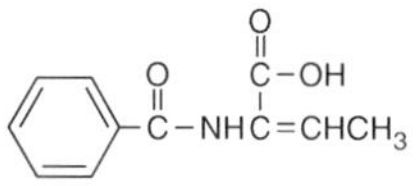

5 gm

S45,084-7
5-BENZAMIDOVALERIC ACID

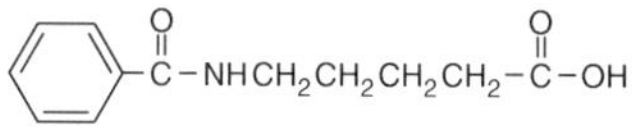

5 gm

S70,688-4
6-BENZAMIDOHEXANOIC ACID

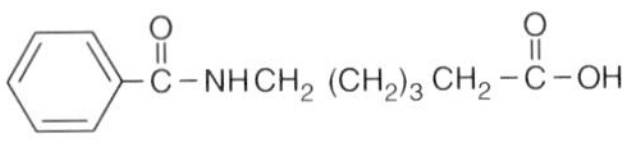

5 gm

S73,956-1
6-(BENZOYLAMINO)-2-BROMOHEXANOIC ACID

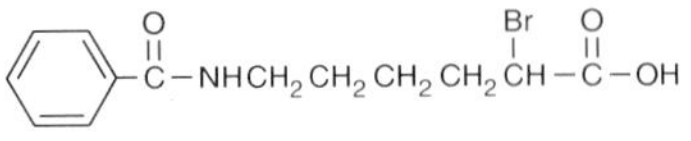

5 gm

S90,996-3
3-BENZAMIDOBENZOIC ACID

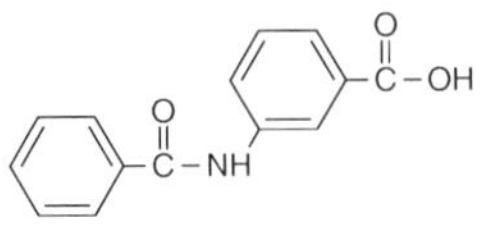

250 mg

S94,718-0
ALPHA-(BENZAMIDO)CINNAMIC ACID

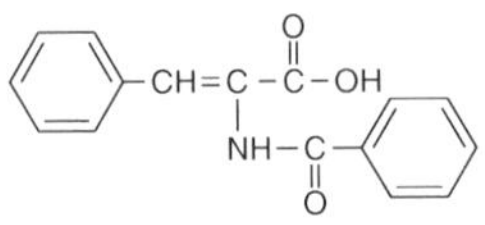

100 mg

S39,385-1
2'-CHLOROMALEANILIC ACID

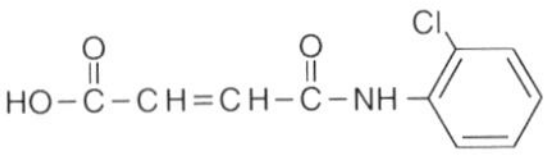

5 gm

S62,244-3
2-(BUTYLCARBAMOYL)BENZOIC ACID

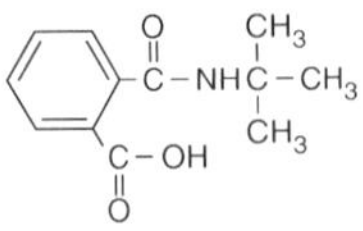

250 mg

S55,533-9
2'-FLUOROPHTHALANILIC ACID

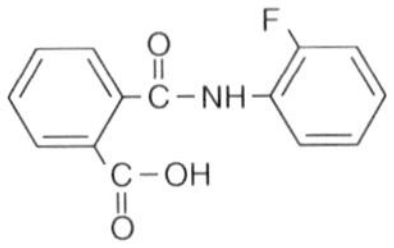

5 gm

S55,527-4
3'-CHLORO-2'-METHYLPHTHALANILIC ACID

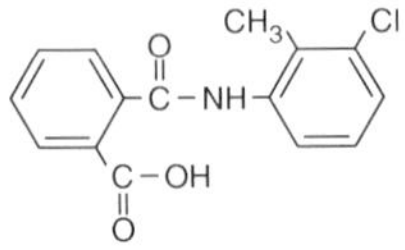

5 gm

S67,516-4
N-(DIPHENYLMETHYL)PHTHALAMIC ACID

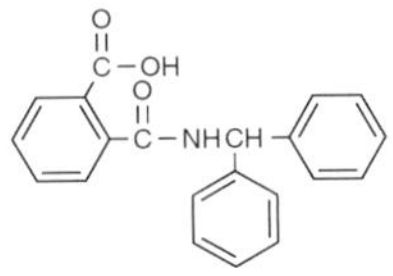

250 mg

S39,384-3
4'-CHLOROMALEANILIC ACID

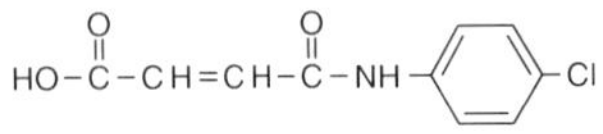

5 gm

S54,280-6
4'-IODOMALEANILIC ACID

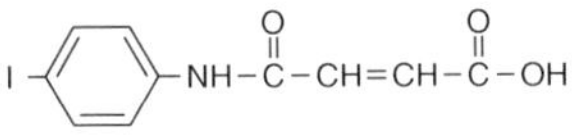

250 mg

S39,380-0
4'-METHYLMALEANILIC ACID

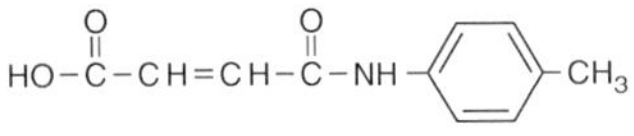

5 gm

S91,011-2
3-(4-METHYLBENZAMIDO)BENZOIC ACID

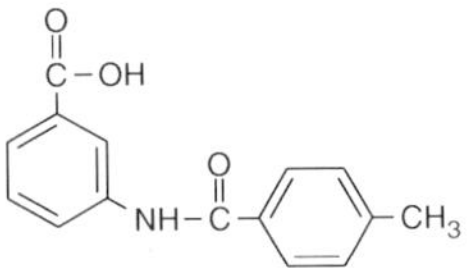

250 mg

S53,831-0
2',6'-DICHLOROMALEANILIC ACID

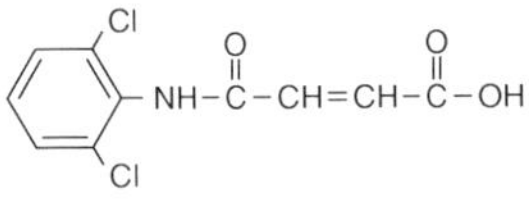

250 mg

S38,085-7
3-[N-(6-CHLORO-ORTHO-TOLYL)-CARBAMOYL]ACRYLIC ACID, SODIUM SALT

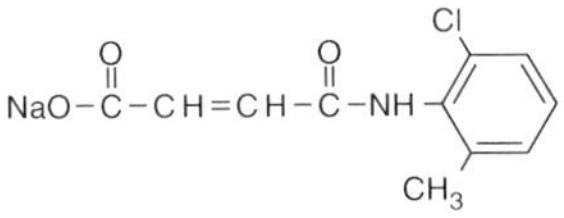

250 mg

S54,832-4
3'-CHLORO-2'-METHYLMALEANILIC ACID

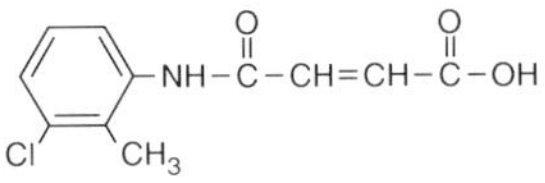

250 mg

S54,866-9
2',6'-DIMETHYLMALEANILIC ACID

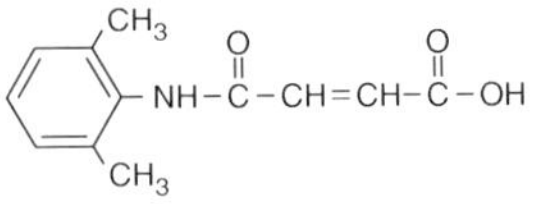

250 mg

S53,237-1
3',4'-DICHLOROMALEANILIC ACID

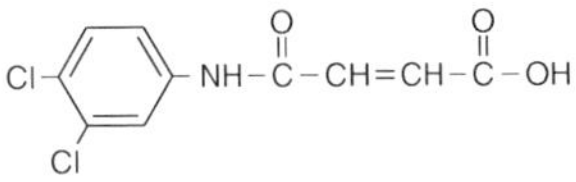

5 gm

S54,509-0
4'-BROMO-2'-CHLOROMALEANILIC ACID

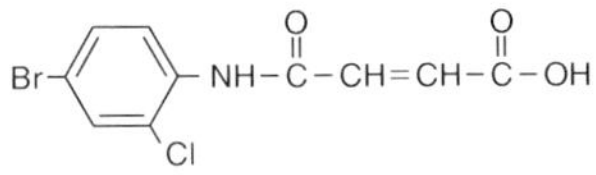

5 gm

S53,261-4
3'-CHLORO-4'-METHYLMALEANILIC ACID

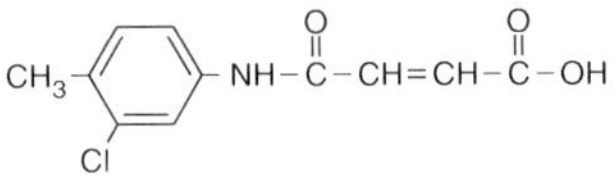

5 gm

S53,913-9
3',4'-DIMETHYLMALEANILIC ACID

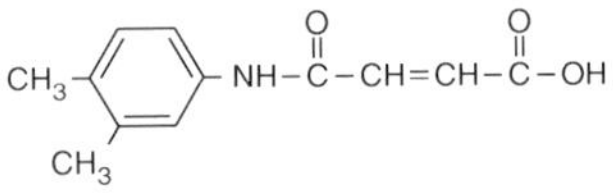

5 gm

S75,011-5
METHYL HIPPURATE

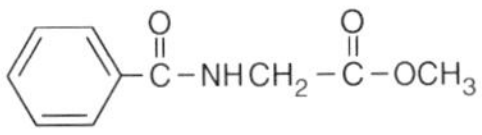

5 gm

S46,082-6
N-BENZOYL-L-ALANINE METHYL ESTER

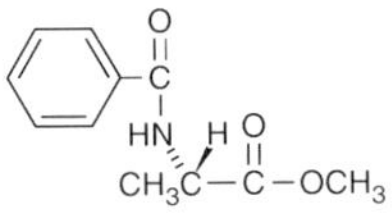

5 gm

S67,761-2
ETHYL HIPPURATE

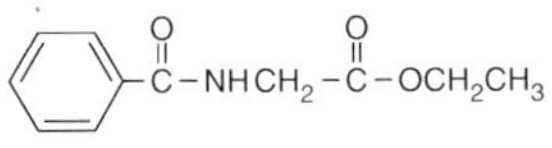

250 mg

S78,580-6
N-BENZOYL-L-VALINE METHYL ESTER

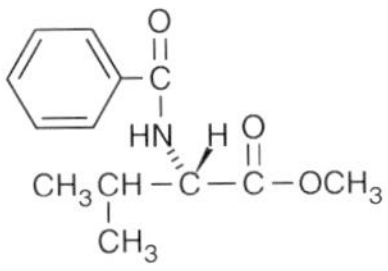

250 mg

S78,296-3
N-BENZOYL-L-LEUCINE METHYL ESTER

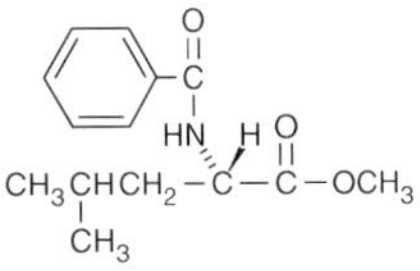

5 gm

S52,054-3
N-BENZOYL-2,2-DIMETHYLGLYCINE ETHYL ESTER

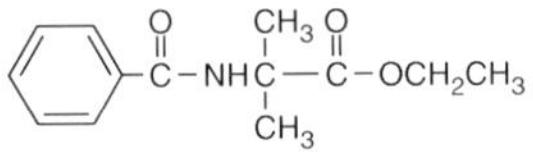

5 gm

S43,204-0
ETHYL 6-BENZAMIDOHEXANOATE

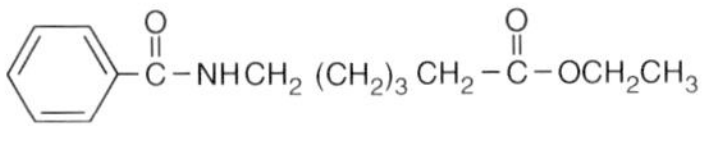

5 gm

S93,695-2
METHYL N-(4-CHLOROBENZOYL)-N-PHENYLANTHRANILATE

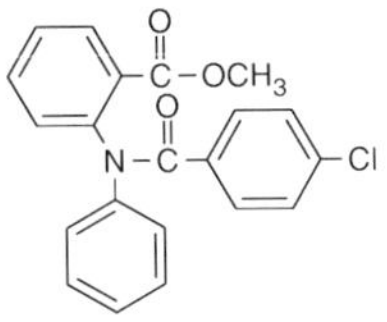

250 mg

S93,696-0
ETHYL N-PHENYL-N-P-TOLUOYLANTHRANILATE

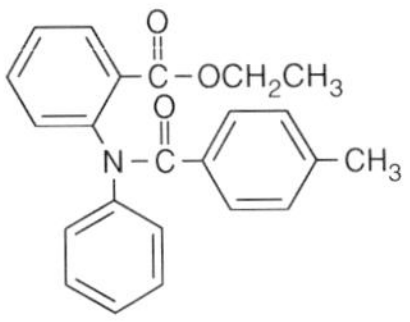

250 mg

S93,697-9
METHYL N-(4-TERT-BUTYLBENZOYL)-N-PHENYLANTHRANILATE

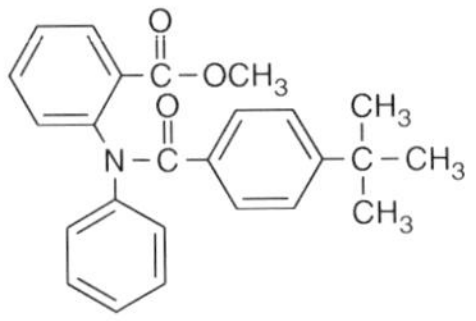

250 mg

S59,909-3
METHYL N-(4-CHLOROBENZOYL)-ANTHRANILATE

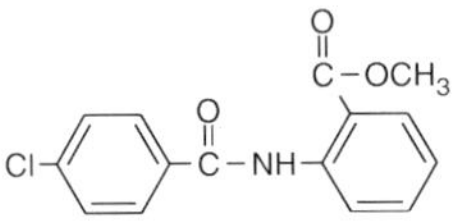

250 mg

S88,368-9
N-BENZOYL-BETA-ALANINE METHYL ESTER

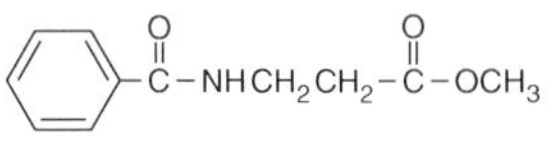

250 mg

S71,591-3
4'-ETHOXYCARBONYL-2-BUTENANILIDE

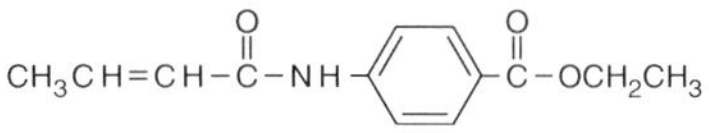

5 gm

S61,570-6
METHYL 4-BENZAMIDOBENZOATE

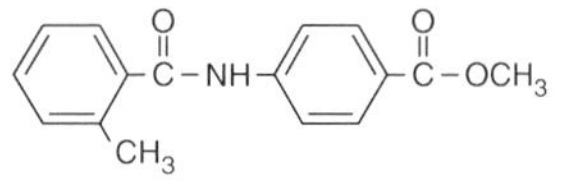

250 mg

S72,973-6
ETHYL 4-(2,4-HEXADIENAMIDO)BENZOATE

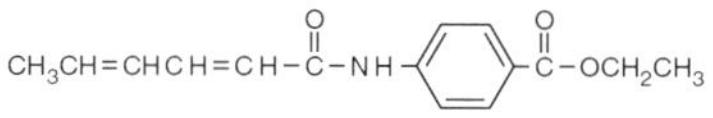

5 gm

S55,500-2
ETHYL 4-(2,4-DICHLOROBENZAMIDO)-BENZOATE

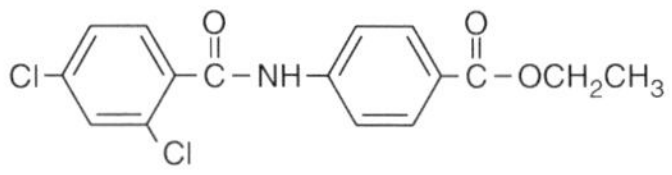

5 gm

S55,518-5
ETHYL 4-(3,4-DICHLOROBENZAMIDO)-BENZOATE

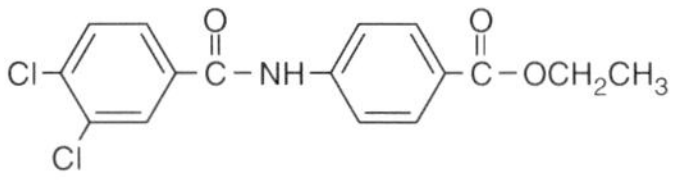

5 gm

S88,866-4
S-(3-BENZAMIDOPROPYL) THIOBENZOATE

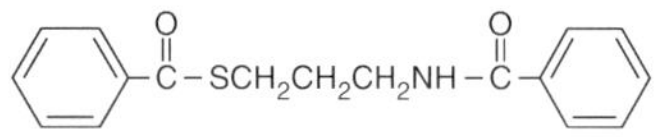

250 mg

S55,978-4
3-CHLOROBENZAMIDE

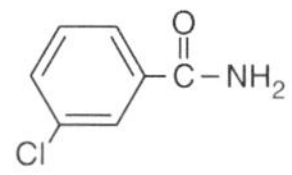

5 gm

S35,882-7
3-BROMOBENZAMIDE

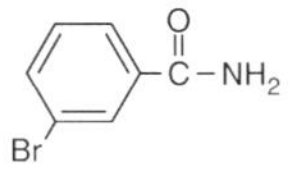

5 gm

S77,620-3
N-BENZOYL-DL-LEUCINAMIDE

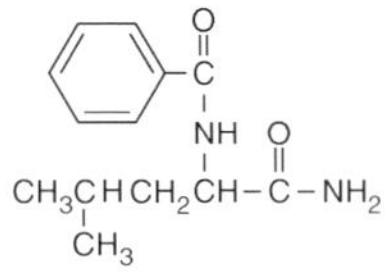

250 mg

S93,329-5
N-(BENZOYLGLYCYL)GLYCINE AMIDE

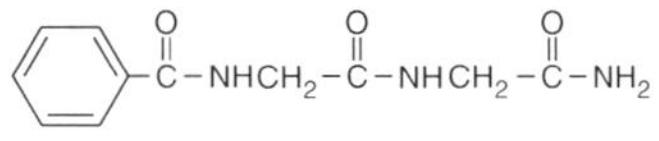

250 mg

S78,938-0
N-BENZOYL-L-LEUCINAMIDE

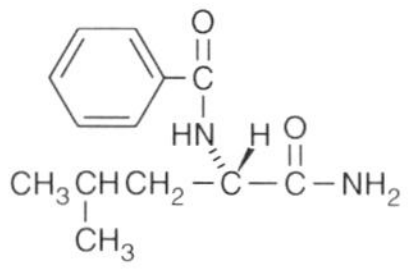

250 mg

S56,412-5
TEREPHTHALAMIDE

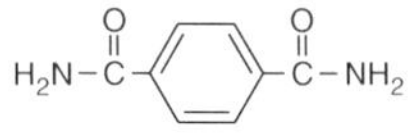

5 gm

S61,703-2
4-(TERT-BUTYL)BENZAMIDE

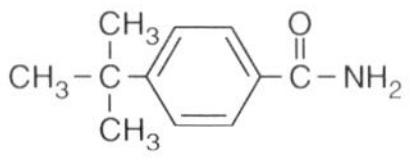

250 mg

S35,661-1
2,4,6-TRIMETHYLBENZAMIDE

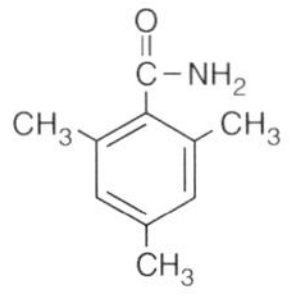

5 gm

S73,305-9
2-[10,11-DIHYDRO-5H-DIBENZO(A,D)-CYCLOHEPTEN-5-YLIDENE]ACETAMIDE

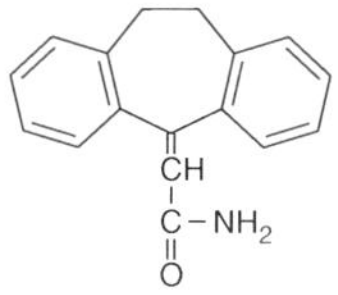

250 mg

S81,032-0
N-PHENYLDIBENZAMIDE

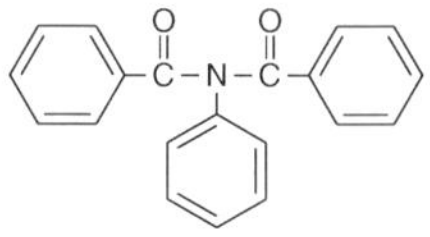

250 mg

S59,070-3
M-BENZOTOLUIDIDE

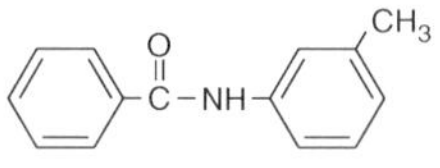

5 gm

S62,387-3
M-TOLUANILIDE

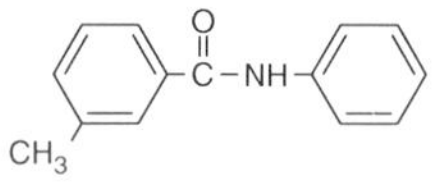

5 gm

S62,383-0
3'-CHLORO-M-TOLUANILIDE

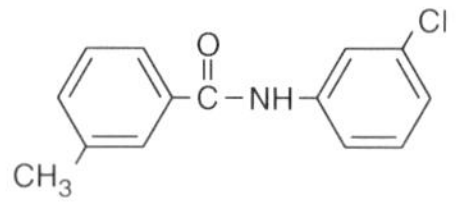

5 gm

S55,511-8
2,4-DICHLORO-4'-(N-METHYLACETAMIDO)-ENZANILIDE

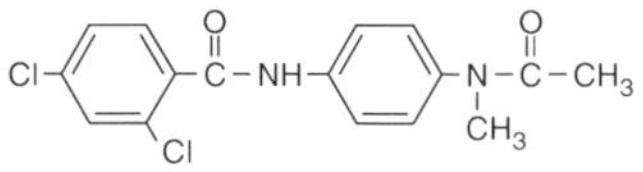

250 mg

S72,914-0
4'-SULFAMOYL-2,4-HEXADIENANILIDE

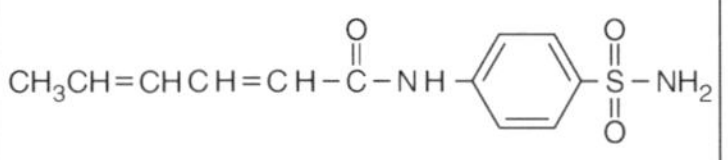

250 mg

S42,485-4
FAGARAMIDE

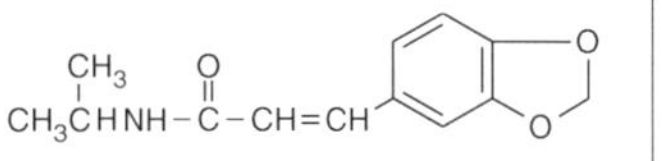

250 mg

S76,200-8
N-(3,5-DIMETHYL-1-PHENYL-2-PYRAZOLIN-4-YL)BENZAMIDE

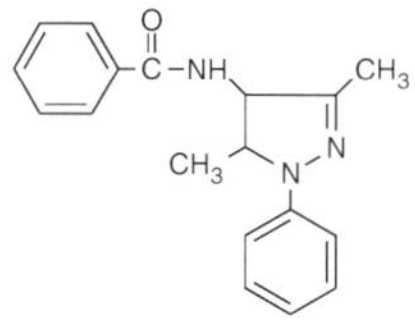

250 mg

S85,247-3
N-(2-PHENYLCYCLOHEXYL)BENZAMIDE

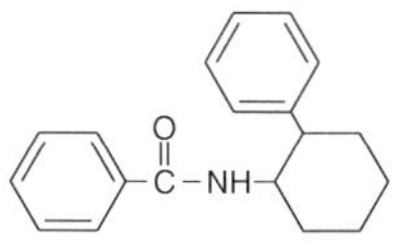

250 mg

S80,818-0
2-(CYCLOHEXYLIDENEAMINO)-N-PHENYLBENZAMIDE

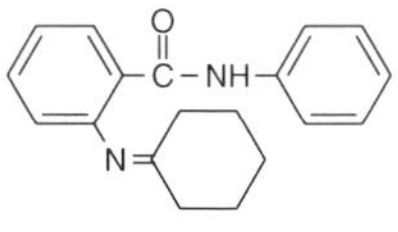

250 mg

S95,458-6
N-(4-CYCLOHEXYLCYCLOHEXYL)BENZAMIDE

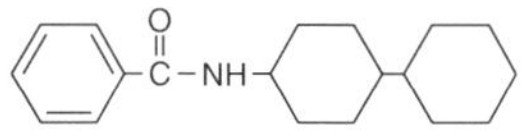

50 mg

S73,270-2
N-(TETRAHYDROPYRAN-2-YL)BENZAMIDE

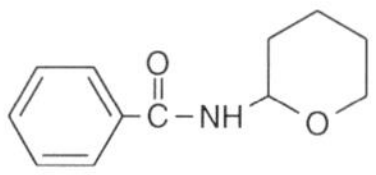

250 mg

S75,946-5
N-[2-METHYL-5-(2,5,5-TRIMETHYL-1,3-DIOXAN-2-YL)PHENYL]BENZAMIDE

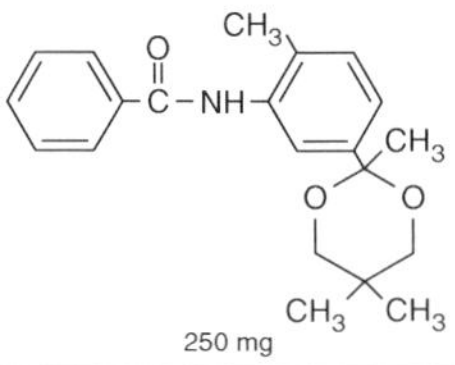

250 mg

S74,500-6
N-METHYL-2-(4-METHYL-2H-1,3-OXAZIN-6-YL)BENZAMIDE

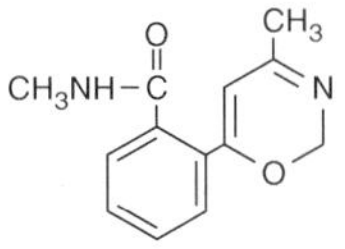

250 mg

S68,105-9
N-ETHYLBENZAMIDE

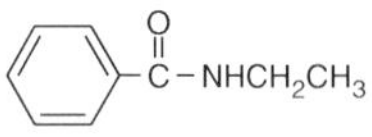

5 gm

S87,725-5
N-(2-BROMOETHYL)BENZAMIDE

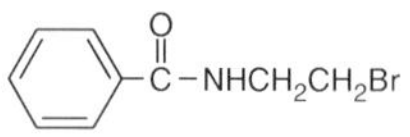

250 mg

S88,173-2
N-PHENYLMETHACRYLAMIDE

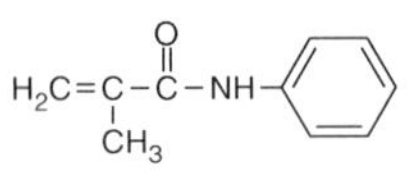

250 mg

S88,455-3
N-ALLYLBENZAMIDE

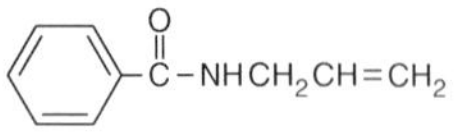

250 mg

S77,912-1
N-PROPYLBENZAMIDE

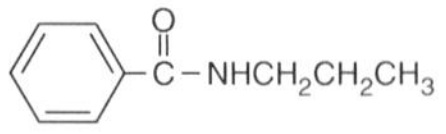

250 mg

S89,461-3
1,2,3,4,4-PENTACHLORO-1,3-BUTADIENE-1-CARBOXANILIDE

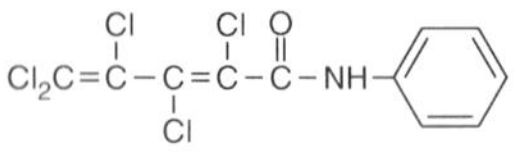

250 mg

S89,437-0
4-BROMO-1,2,3,4-TETRACHLORO-1,3-BUTADIENE-1-CARBOXANILIDE

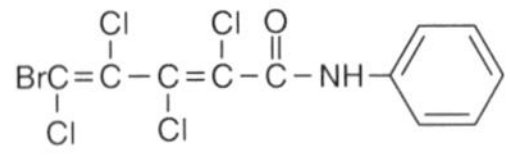

250 mg

S87,777-8
2-ETHYL-2-PENTENANILIDE

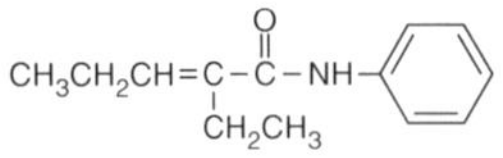

250 mg

S63,157-4
CINNAMANILIDE

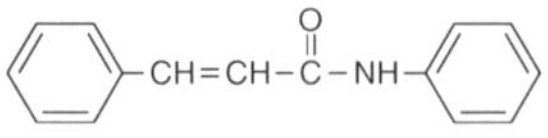

250 mg

S86,215-0
N-STYRYLBENZAMIDE

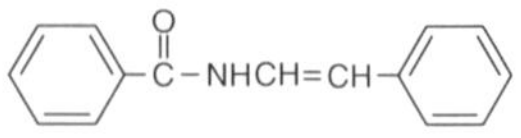

250 mg

S88,011-6
BETA-METHYL-N-PHENYLCINNAMAMIDE

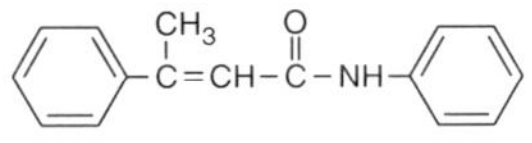

250 mg

S56,431-1
N,N'-HEXAMETHYLENEBIS(BENZAMIDE)

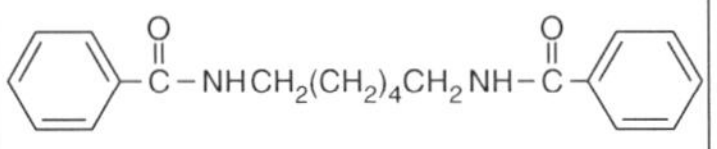

5 gm

S82,708-8
N-OCTADECYLBENZAMIDE

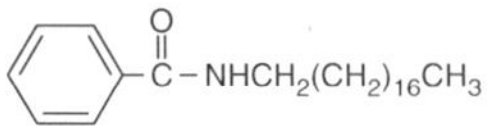

250 mg

S77,310-7
2'-ETHYLCINNAMANILIDE

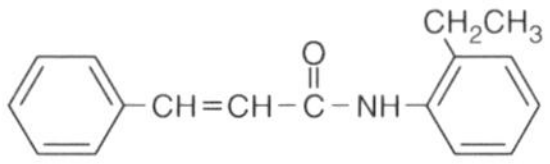

250 mg

S78,245-9
2'-PHENYLCINNAMANILIDE

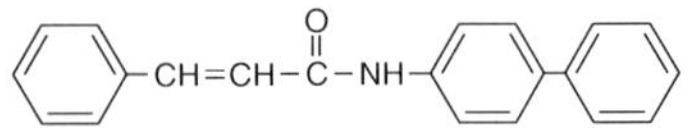

5 gm

S90,115-6
N-(4-METHYLBENZYLIDENE)BENZAMIDE

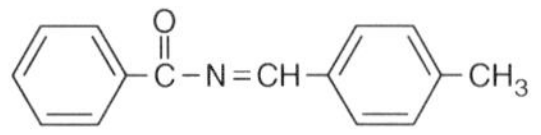

250 mg

S51,543-4
4',4'''-VINYLENEDIBENZANILIDE

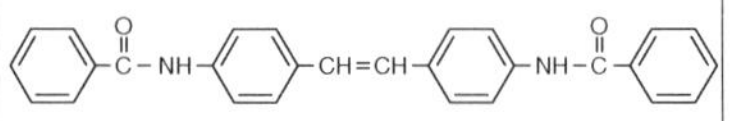

250 mg

S77,874-5
2,2',6'-TRIBROMOBENZANILIDE

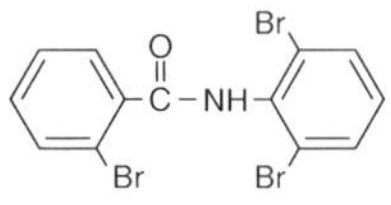

250 mg

S71,147-0
2'-ETHYL-6'-METHYLBENZANILIDE

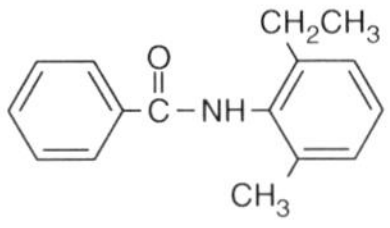

5 gm

S70,792-9
2'-CHLORO-6'-METHYLCINNAMANILIDE

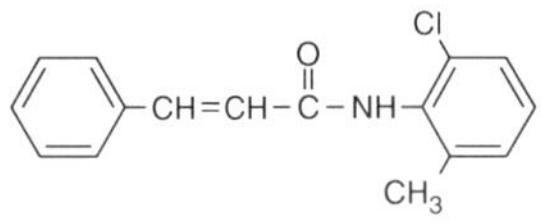

250 mg

S70,950-6
2'-ETHYL-6'-METHYLCINNAMANILIDE

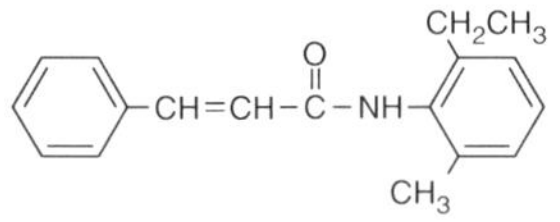

250 mg

S93,371-6
2-BENZOYL-4-FLUORO-N-METHYLBENZAMIDE

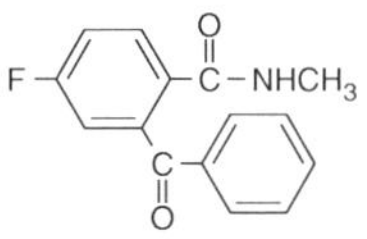

250 mg

S89,246-7
5'-ISOPROPYL-2'-METHYLCINNAMANILIDE

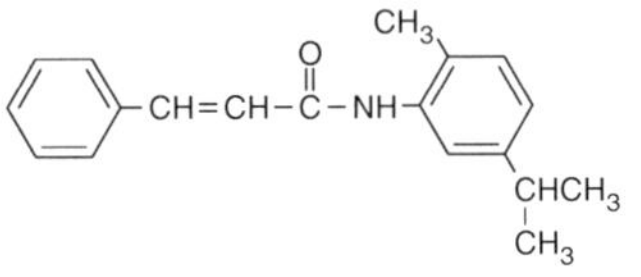

250 mg

S70,921-2
2',4',6'-TRIMETHYLBENZANILIDE

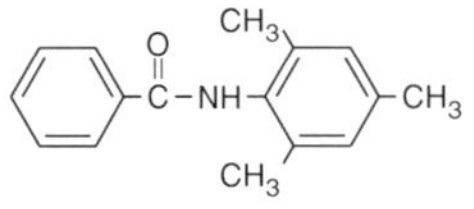

5 gm

S76,191-5
N-(5,6,7,8-TETRAHYDRO-1-NAPHTHYL)-BENZAMIDE

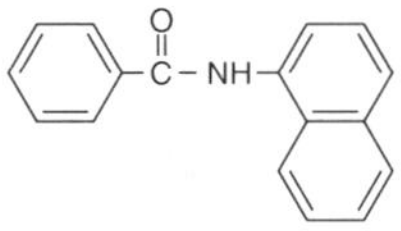

250 mg

S89,167-3
N-(1-NAPHTHYL)CINNAMAMIDE

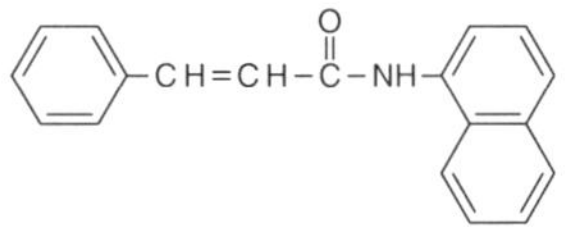

250 mg

S57,120-2
2-CHLOROBENZANILIDE

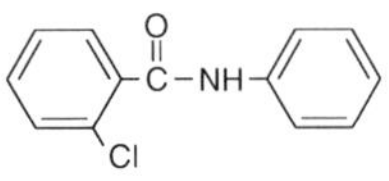

5 gm

S59,214-5
2,2'-DICHLOROBENZANILIDE

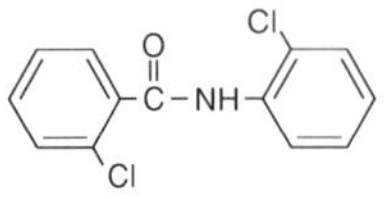

250 mg

S59,073-8
O-BENZOTOLUIDIDE

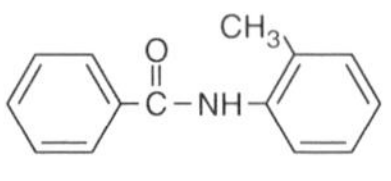

5 gm

S61,550-1
O-TOLUANILIDE

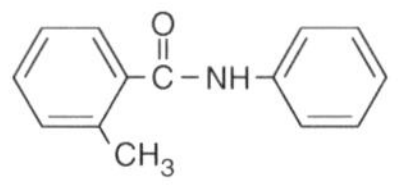

250 mg

S59,199-8
2-CHLORO-M-BENZOTOLUIDIDE

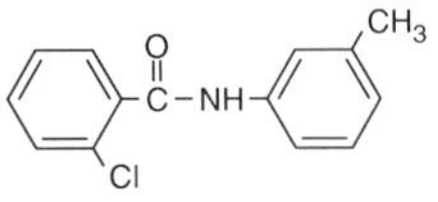

5 gm

S59,218-8
2-CHLORO-O-BENZOTOLUIDIDE

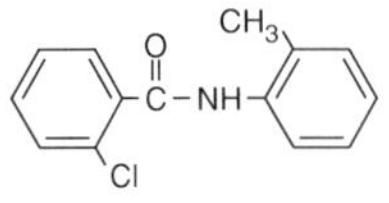

250 mg

S61,560-9
2'-CHLORO-O-TOLUANILIDE

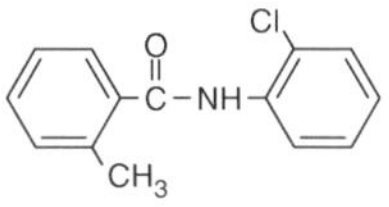

250 mg

S62,396-2
2'-CHLORO-M-TOLUANILIDE

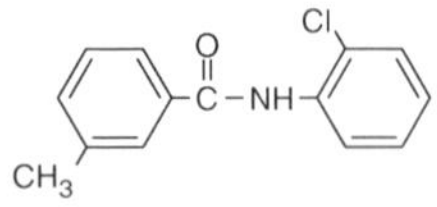

5 gm

S62,393-8
2'-METHYL-M-TOLUANILIDE

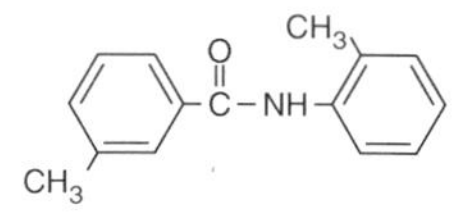

5 gm

S65,662-3
2-BENZYL-N-METHYLBENZAMIDE

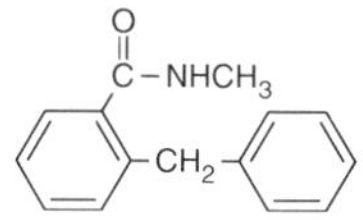

5 gm

S68,148-2
2,2'-ETHYLENEBIS(N-METHYLBENZAMIDE)

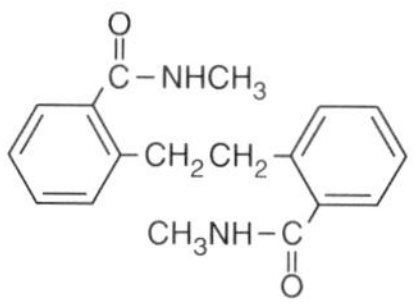

5 gm

S59,208-0
2,4'-DICHLOROBENZANILIDE

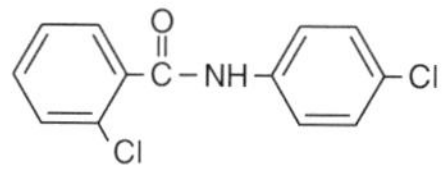

250 mg

S59,898-4
2',4-DICHLOROBENZANILIDE

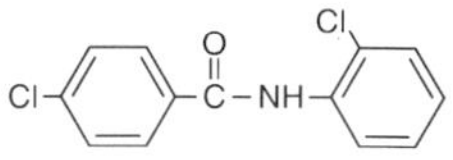

250 mg

S59,207-2
4'-BROMO-2-CHLOROBENZANILIDE

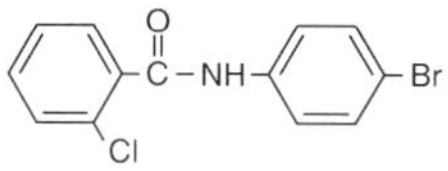

250 mg

S59,209-9
2-CHLORO-P-BENZOTOLUIDIDE

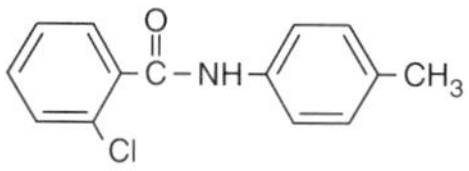

250 mg

S59,901-8
4-CHLORO-O-BENZOTOLUIDIDE

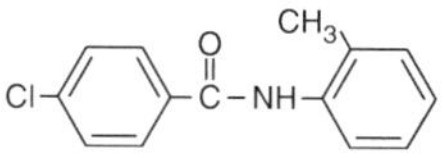

5 gm

S60,242-6
2'-CHLORO-P-TOLUANILIDE

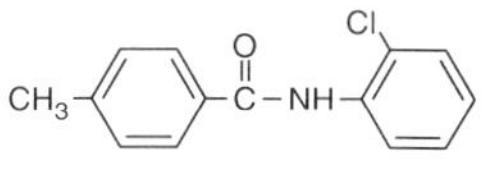

250 mg

S61,553-6
4'-BROMO-2-METHYLBENZANILIDE

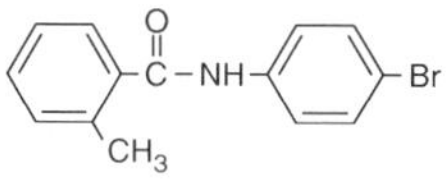

250 mg

S60,933-1
4'-METHYL-O-TOLUANILIDE

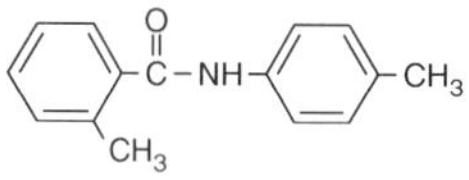

250 mg

S68,114-8
N-(TERT-BUTYL)-2-(4-METHYLBENZYL)-BENZAMIDE

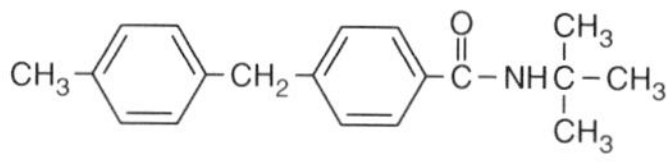

5 gm

S59,212-9
2-CHLORO-2',3'-BENZOXYLIDIDE

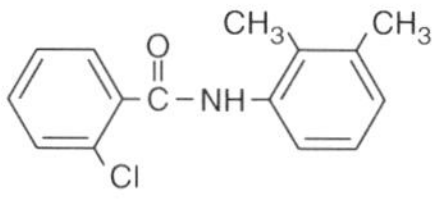

250 mg

S61,574-9
2'-CHLORO-2,6'-DIMETHYLBENZANILIDE

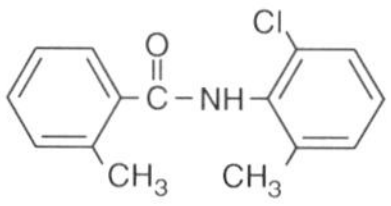

250 mg

S61,566-8
2-METHYLBENZO-2',3'-XYLIDIDE

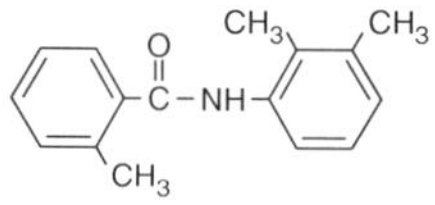

250 mg

S70,781-3
2,2',6'-TRIMETHYLBENZANILIDE

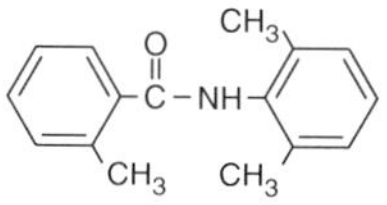

5 gm

S71,777-0
2,2'-DIMETHYL-6'-ETHYLBENZANILIDE

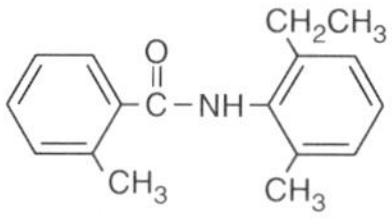

5 gm

S71,136-5
2',6'-DIETHYL-2-METHYLBENZANILIDE

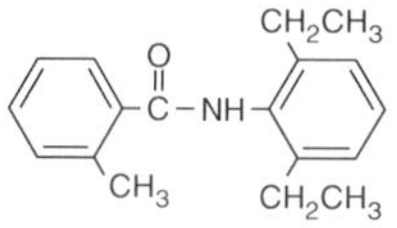

5 gm

S70,920-4
2',6'-DIISOPROPYL-2-METHYLBENZANILIDE

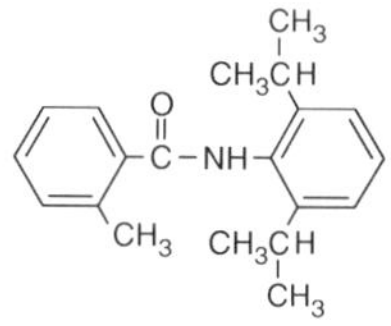

250 mg

S55,506-1
3,4-DICHLORO-2'-FLUOROBENZANILIDE

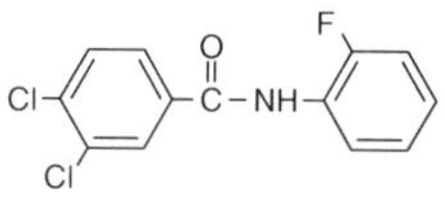

5 gm

S59,436-9
2,2',4'-TRICHLOROBENZANILIDE

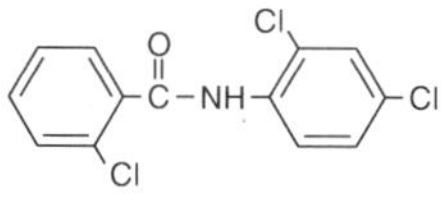

250 mg

S61,211-1
2,2',4-TRICHLOROBENZANILIDE

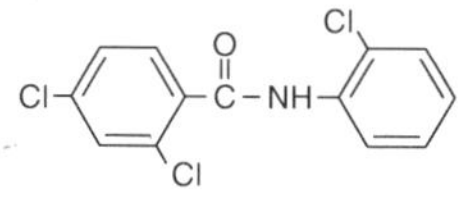

250 mg

S61,220-0
2,4-DICHLOROBENZO-O-TOLUIDIDE

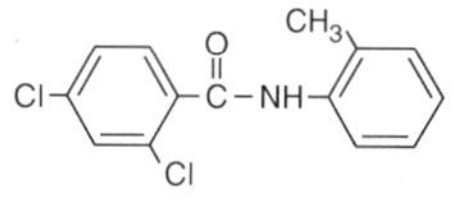

5 gm

S61,554-4
2',4'-DICHLORO-2-METHYLBENZANILIDE

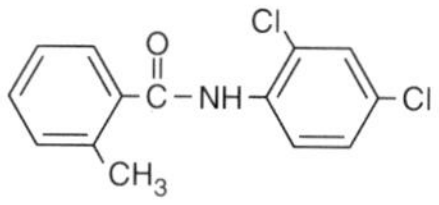

250 mg

S61,555-2
2',5'-DICHLORO-2-METHYLBENZANILIDE

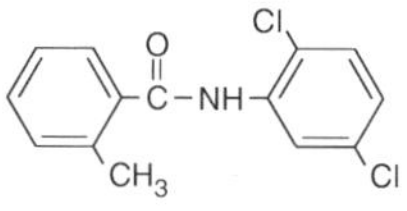

250 mg

S59,197-1
2-CHLORO-2',4'-BENZOXYLIDIDE

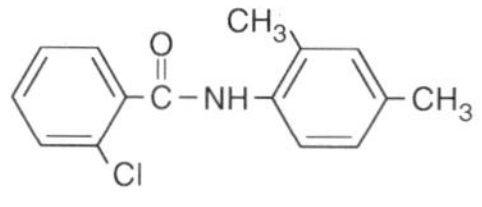

5 gm

S59,211-0
2-CHLORO-3',4'-BENZOXYLIDIDE

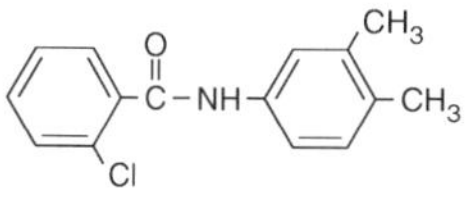

250 mg

S59,216-1
2-CHLORO-2',5'-BENZOXYLIDIDE

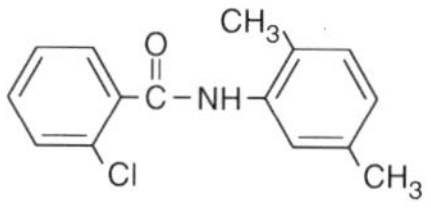

5 gm

S61,575-7
3'-CHLORO-2,4'-DIMETHYLBENZANILIDE

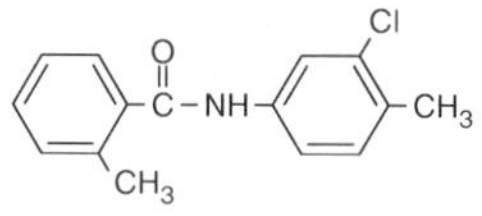

250 mg

S61,567-6
2-METHYLBENZO-2',5'-XYLIDIDE

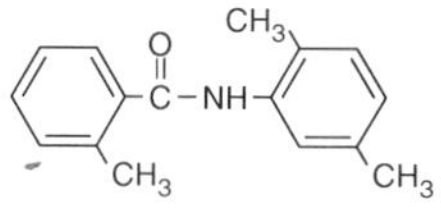

250 mg

S61,568-4
2-METHYLBENZO-2',4'-XYLIDIDE

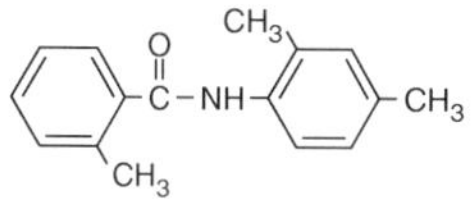

250 mg

S59,448-2
2-CHLORO-2',4',6'-TRIBROMOBENZANILIDE

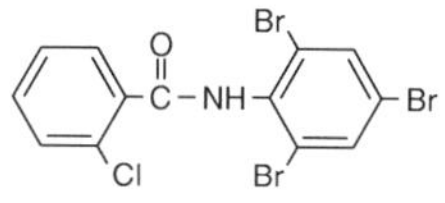

250 mg

S71,543-3
2,2',4',6'-TETRAMETHYLBENZANILIDE

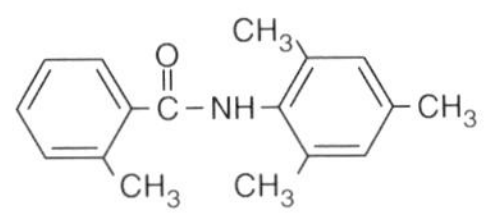

250 mg

S73,485-3
10,11-DIHYDRO-5-[2-(N-METHYLCARBAMOYL)PHENYL]-5H-DIBENZO (A,D)CYCLOHEPTENE

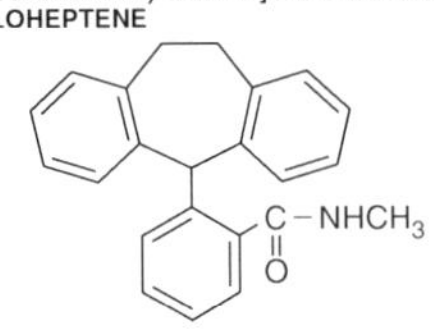

250 mg

S34,735-3
N,N'-DIMETHYLTEREPHTHALAMIDE

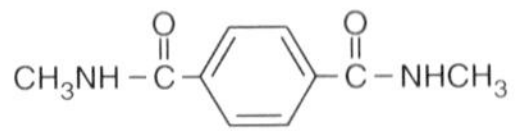

5 gm

S65,940-1
N-CROTONYL-P-TOLUIDINE

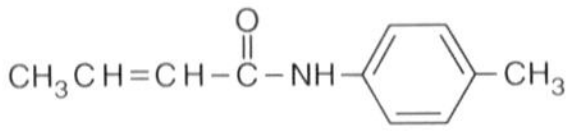

5 gm

S90,089-3
4'-ETHYL-2-BUTENANILIDE

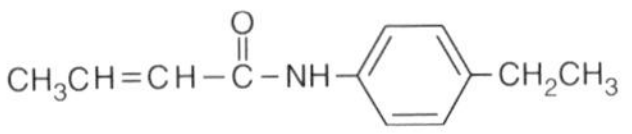

250 mg

S59,076-2
4'-BROMOBENZANILIDE

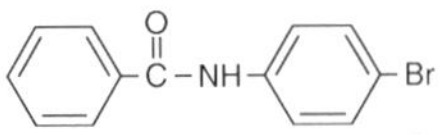

5 gm

S59,907-7
4'-BROMO-4-CHLOROBENZANILIDE

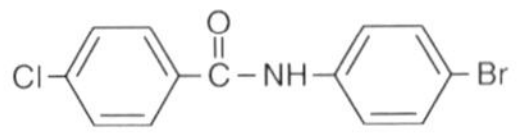

250 mg

S59,074-6
P-BENZOTOLUIDIDE

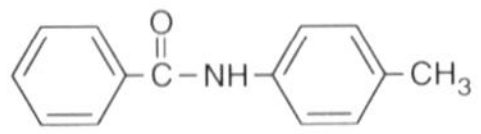

250 mg

S59,899-2
P-TOLUANILIDE

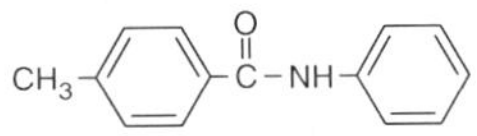

250 mg

S55,505-3
N-BENZYL-4-CHLOROBENZAMIDE

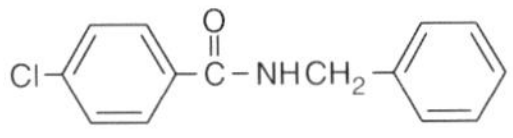

5 gm

S59,910-7
4'-CHLORO-P-TOLUANILIDE

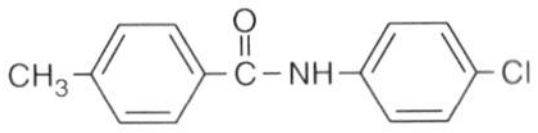

250 mg

S60,249-3
4-CHLORO-M-BENZOTOLUIDIDE

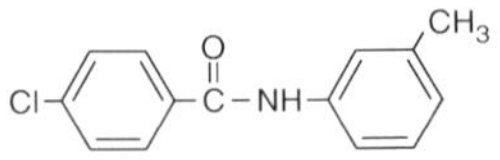

250 mg

S59,903-4
4'-BROMO-4-METHYLBENZANILIDE

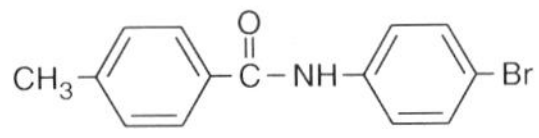

250 mg

S62,570-1
4'-BROMO-3-METHYLBENZANILIDE

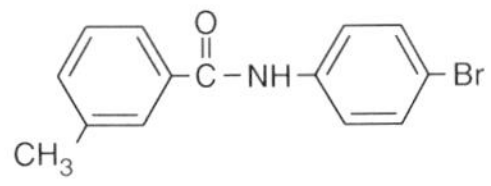

250 mg

S60,245-0
P-TOLU-M-TOLUIDIDE

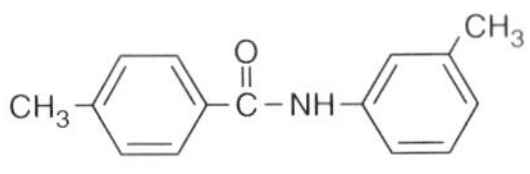

250 mg

S60,365-1
P-TOLU-P-TOLUIDIDE

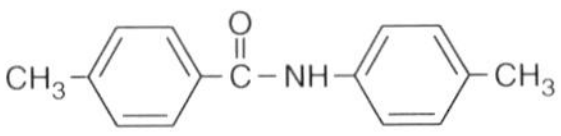

250 mg

S55,504-5
4-CHLORO-N-PHENETHYLBENZAMIDE

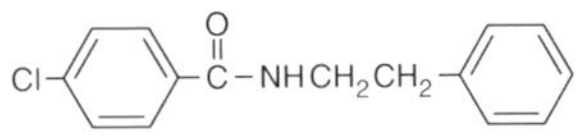

5 gm

S60,232-9
4-CHLORO-2',3'-BENZOXYLIDIDE

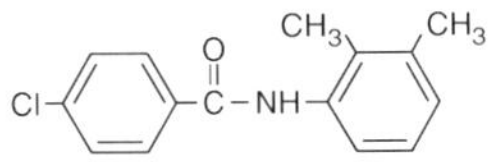

250 mg

S70,779-1
2',4,6'-TRIMETHYLBENZANILIDE

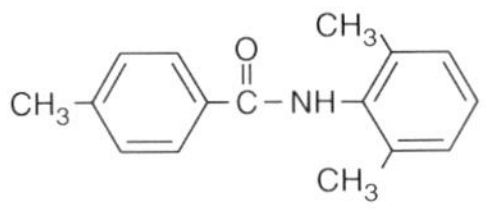

5 gm

S70,904-2
2',4-DIMETHYL-6'-ETHYLBENZANILIDE

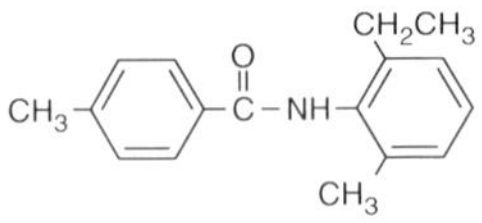

5 gm

S70,760-0
2',6'-DIETHYL-4-METHYLBENZANILIDE

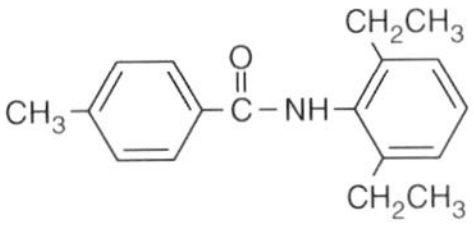

250 mg

S60,241-8
2',4,5'-TRICHLOROBENZANILIDE

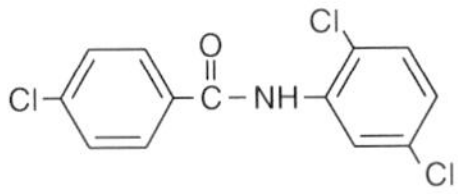

250 mg

S61,223-5
2,4,4'-TRICHLOROBENZANILIDE

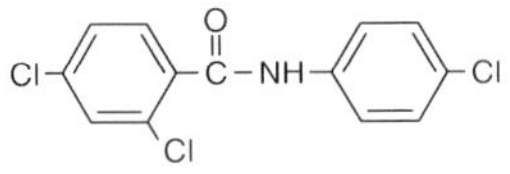

250 mg

S61,222-7
4'-BROMO-2,4-DICHLOROBENZANILIDE

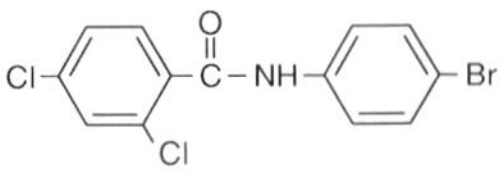

250 mg

S61,224-3
2,4-DICHLOROBENZO-P-TOLUIDIDE

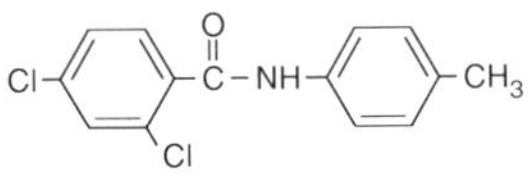

5 gm

S60,233-7
4-CHLORO-2',5'-BENZOXYLIDIDE

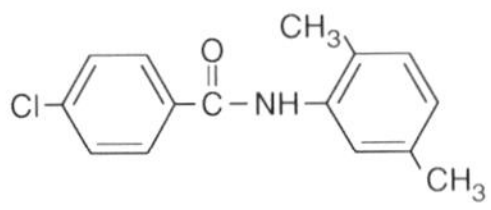

250 mg

S60,248-5
4-CHLORO-2',4'-BENZOXYLIDIDE

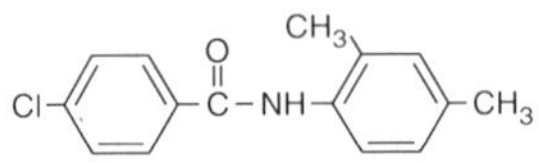

5 gm

S59,893-3
4-METHYLBENZO-2',4'-XYLIDIDE

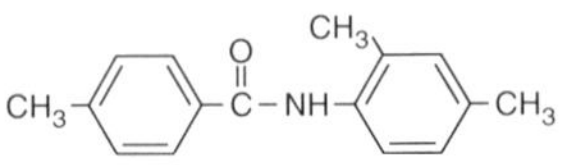

250 mg

S59,896-8
2',4,5'-TRIMETHYLBENZANILIDE

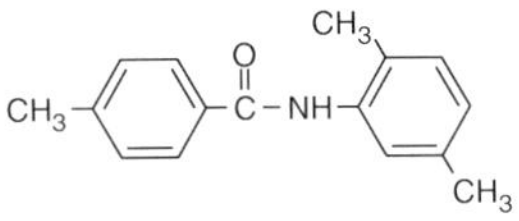

250 mg

S63,694-0
N,N'_BIS(CARVACRYL)TEREPHTHALAMIDE

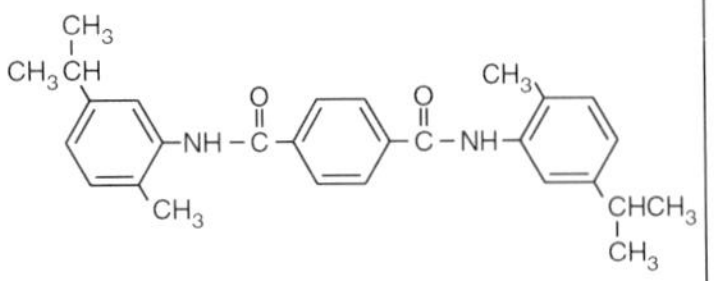

250 mg

S60,251-5
4-CHLORO-2',4',6'-TRIBROMOBENZANILIDE

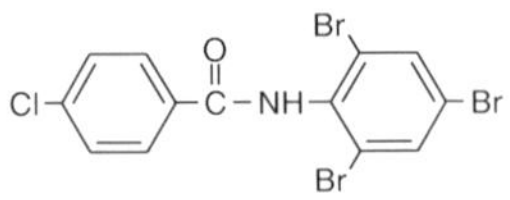

250 mg

S59,902-6
4-METHYL-2',4',6'-TRIBROMOBENZANILIDE

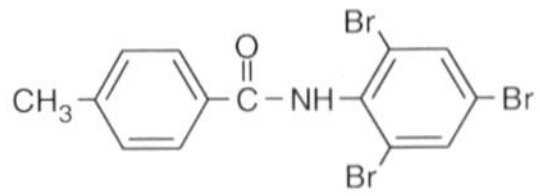

250 mg

S71,462-3
2',4,4',6'-TETRAMETHYLBENZANILIDE
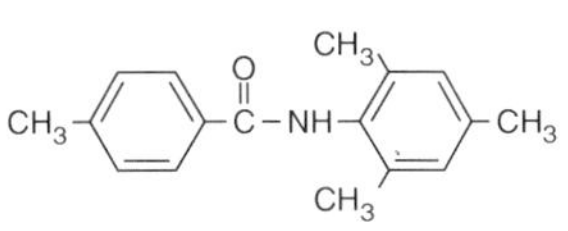
5 gm

S43,236-9
6'-CHLORO-ORTHO-CROTONOTOLUIDIDE
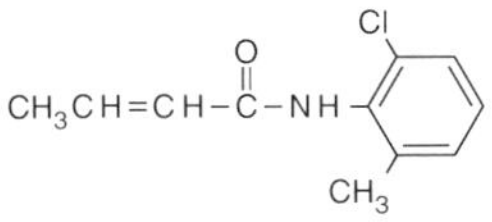
5 gm

S38,169-1
2',6'-METHACRYLOXYLIDIDE
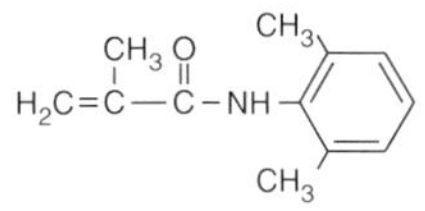
5 gm

S38,177-2
2',6'-CROTONOXYLIDIDE
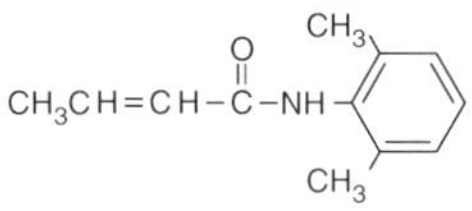
250 mg

S90,090-7
2',3'-DIMETHYL-2-BUTENANILIDE
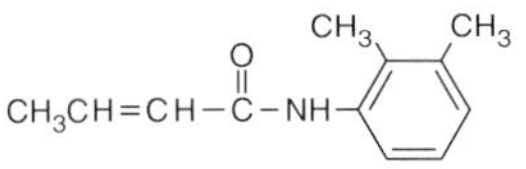
250 mg

S60,442-9
2',6'-DICHLOROBENZANILIDE
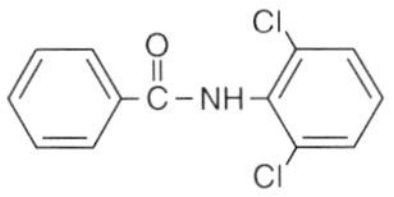
250 mg

S60,444-5
2',3'-DICHLOROBENZANILIDE
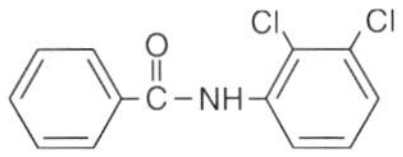
250 mg

S60,425-9
2'-CHLORO-6'-METHYLBENZANILIDE
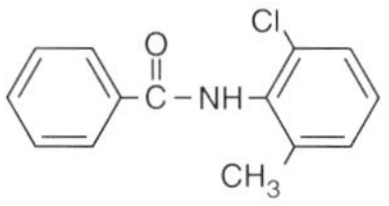
250 mg

S34,670-5
2',6'-BENZOXYLIDIDE
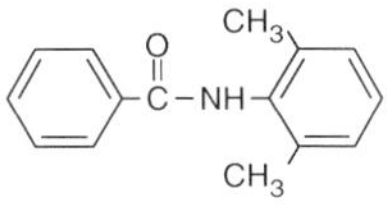
5 gm

S59,205-6
2',3'-BENZOXYLIDIDE
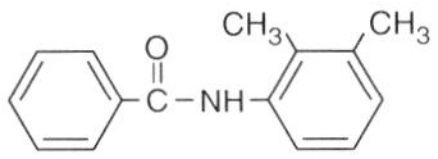
250 mg

S60,428-3
2',6'-DIETHYLBENZANILIDE
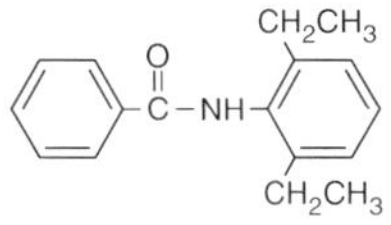
250 mg

S70,769-4
2',3-DIMETHYL-6'-ETHYLBENZANILIDE
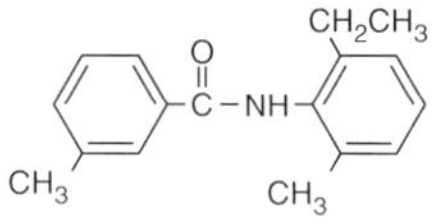
5 gm

S61,216-2
2,2',3',4-TETRACHLOROBENZANILIDE
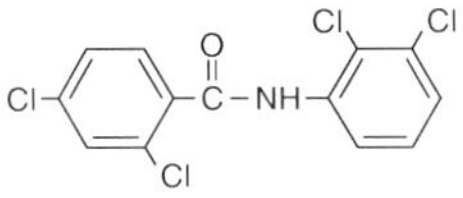
250 mg

S61,228-6
2,2',4,6'-TETRACHLOROBENZANILIDE
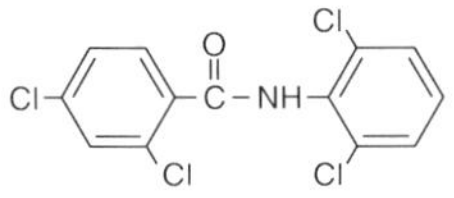
5 gm

S55,508-8
3,3',4-TRICHLORO-O-BENZOTOLUIDIDE
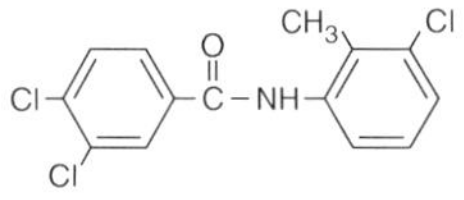
5 gm

S61,219-7
2,4-DICHLOROBENZO-2',3'-XYLIDIDE
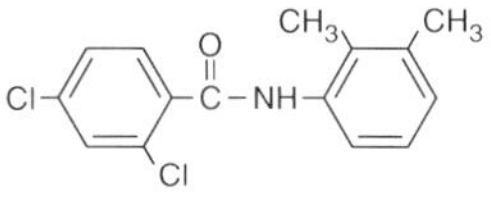
250 mg

S59,063-0
2',4'-DICHLOROBENZANILIDE
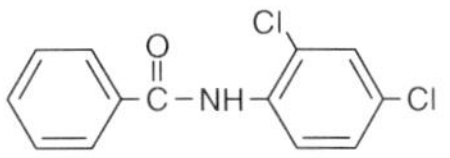
250 mg

S60,426-7
2',5'-DICHLOROBENZANILIDE
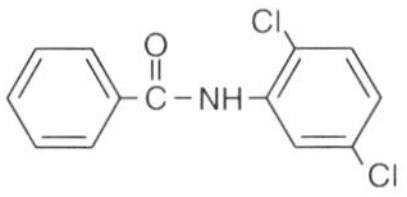
250 mg

S60,427-5
3',4'-DICHLOROBENZANILIDE
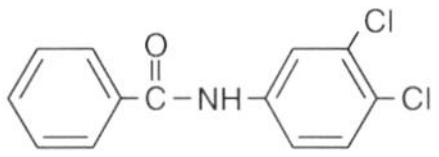
250 mg

S61,230-8
2,4-DICHLOROBENZANILIDE
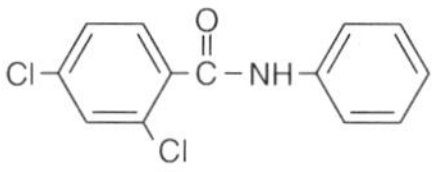
250 mg

S50,869-1
3,3',4,4'-TETRACHLOROBENZANILIDE
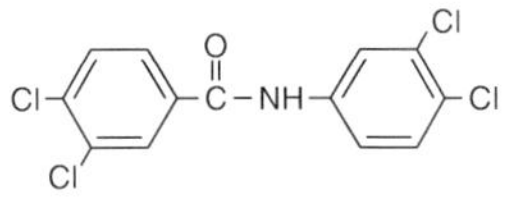
250 mg

S60,429-1
3'-CHLORO-4'-METHYLBENZANILIDE
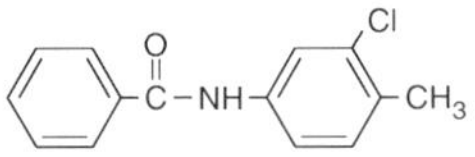
250 mg

S60,443-7
5'-CHLORO-2'-METHYLBENZANILIDE
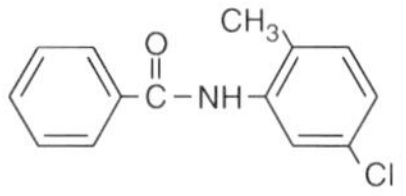
250 mg

S62,566-3
2',4'-DICHLORO-3-METHYLBENZANILIDE
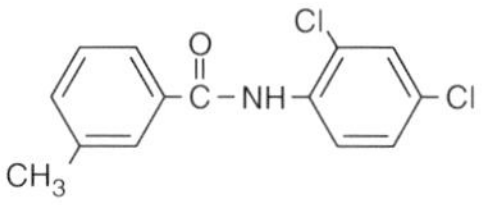
250 mg

S64,079-4
2',5'-DICHLORO-3-METHYLBENZANILIDE
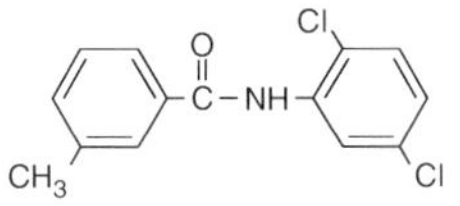
250 mg

S64,080-8
3',4'-DICHLORO-3-METHYLBENZANILIDE
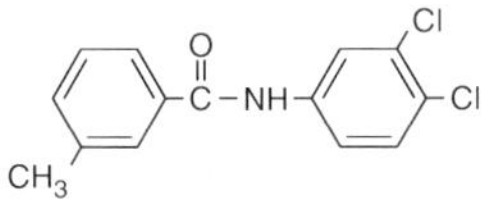
250 mg

S59,060-6
3',4'-BENZOXYLIDIDE
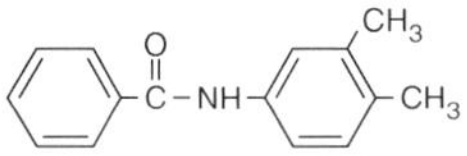
5 gm

S59,204-8
2',5'-BENZOXYLIDIDE
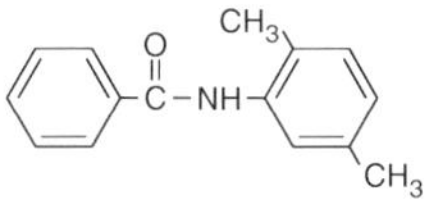
250 mg

S59,206-4
2',4'-BENZOXYLIDIDE
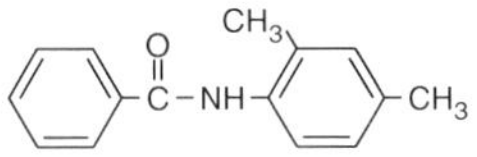
250 mg

S62,582-5
3'-CHLORO-3,4'-DIMETHYLBENZANILIDE
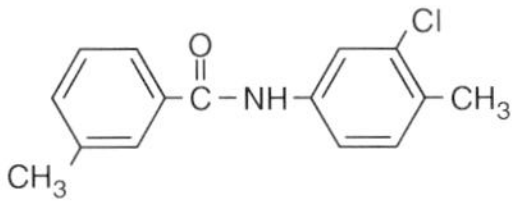
250 mg

S61,213-8
2,4-DICHLOROBENZO-3',4'-XYLIDIDE
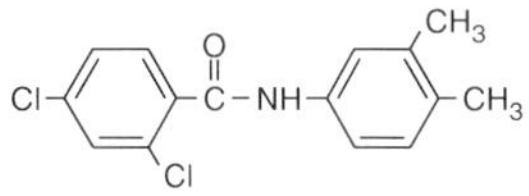
250 mg

S61,217-0
2,4-DICHLOROBENZO-2',4'-XYLIDIDE
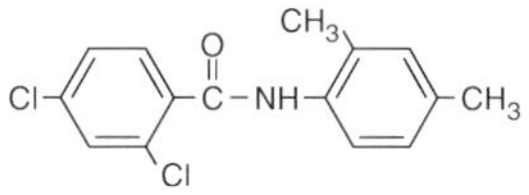
250 mg

S61,221-9
2,4-DICHLOROBENZO-2',5'-XYLIDIDE

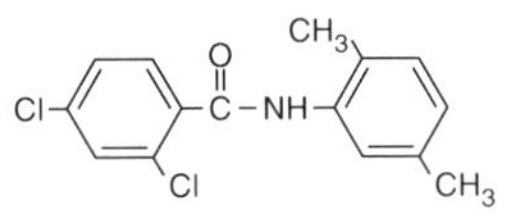

5 gm

S62,594-9
3-METHYLBENZO-2',4'-XYLIDIDE

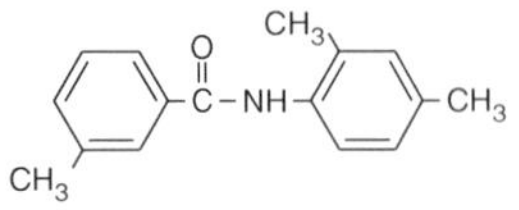

250 mg

S62,595-7
3-METHYLBENZO-2',5'-XYLIDIDE

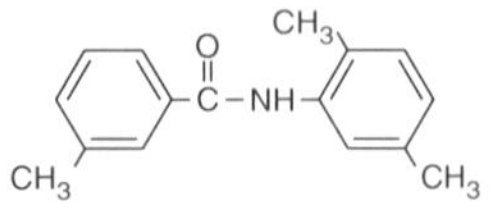

250 mg

S71,467-4
2',3,4',6'-TETRAMETHYLBENZANILIDE

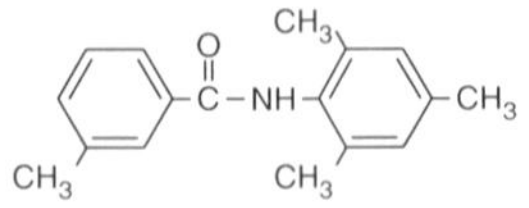

250 mg

S85,477-8
N-(2-NAPHTHYL)-1-NAPHTHALENECARBOXAMIDE

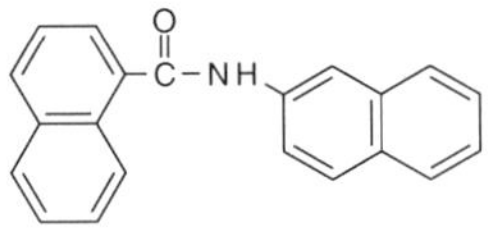

250 mg

S42,380-7
N-ETHYLACRYLANILIDE

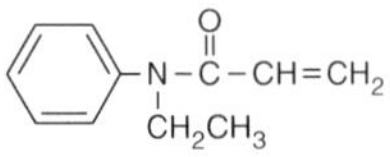

250 mg

S42,077-8
N-ETHYL-META-METHACRYLOTOLUIDIDE

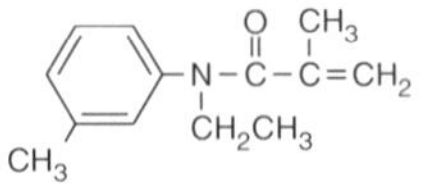

5 gm

S51,235-4
N-CYCLOHEXYLBENZANILIDE

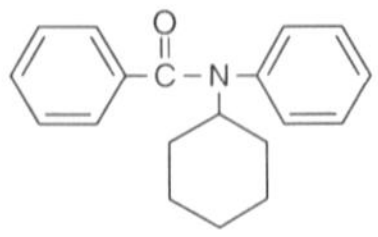

250 mg

S89,249-1
N-CYCLOHEXYL-N-METHYLCINNAMAMIDE

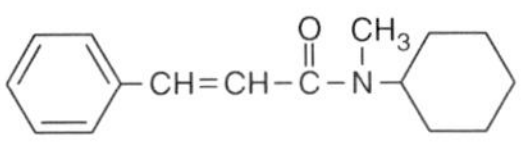

250 mg

S51,374-1
1,4-DIBENZOYLPIPERAZINE

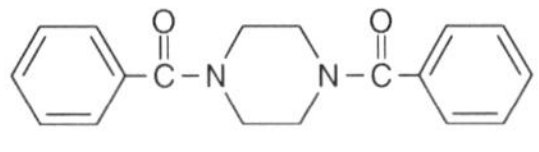

250 mg

S42,379-3
N-METHYLMETHACRYLANILIDE

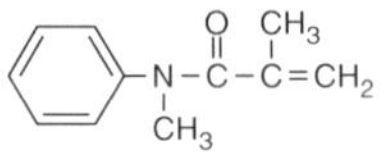

250 mg

S38,087-3
N,N-DIETHYLBENZAMIDE

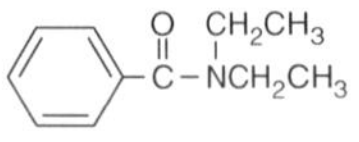

5 gm

S42,388-2
N-ETHYLMETHACRYLANILIDE

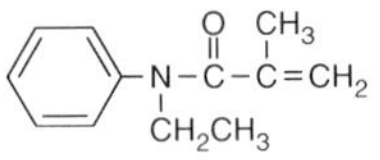

250 mg

S57,075-3
N-ETHYLCROTONANILIDE

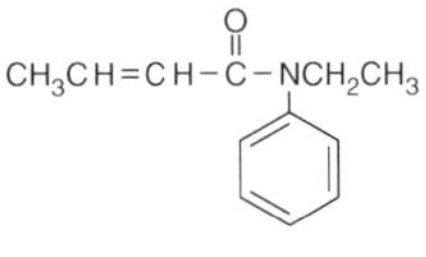

5 gm

S77,479-0
N,N-DIISOPROPYLBENZAMIDE

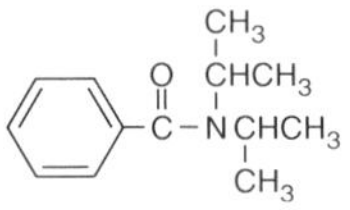

250 mg

S82,641-3
N-METHYLBENZANILIDE

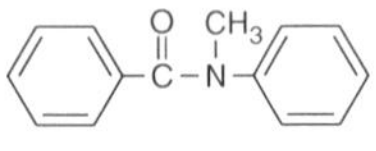

250 mg

S37,455-5
N,N-DIMETHALLYLBENZAMIDE

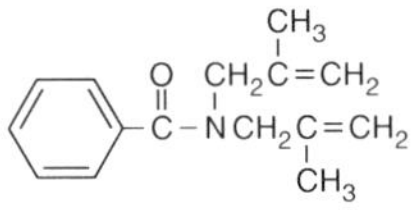

5 gm

S82,947-1
N,N-DIISOBUTYLBENZAMIDE

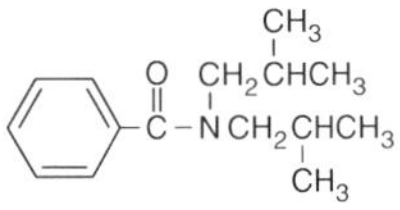

250 mg

S88,481-2
N-ALLYL-N-HEXYLBENZAMIDE

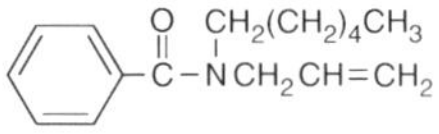

250 mg

S89,247-5
N-BENZYL-N-METHYLCINNAMAMIDE

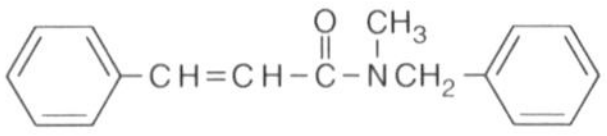

250 mg

S89,644-6
N,N-DIPHENYLSORBAMIDE

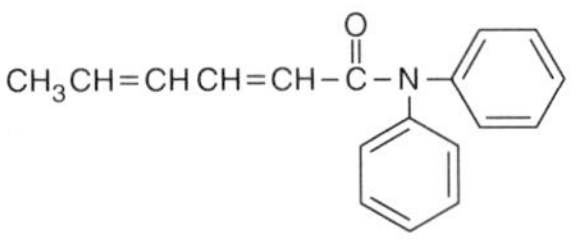

250 mg

S81,335-4
N,N-DIPHENYLBENZAMIDE

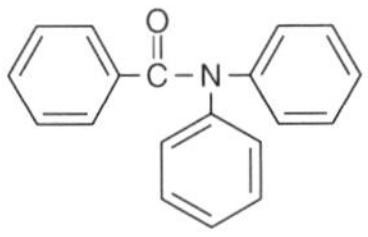

250 mg

S89,244-0
N-BENZYLCINNAMANILIDE

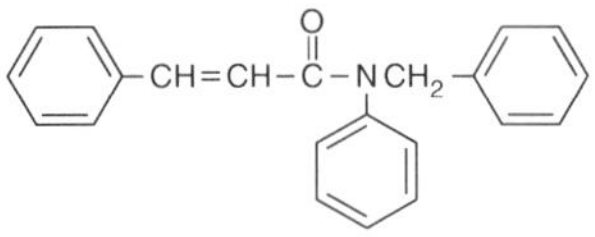

250 mg

S77,534-7
N,N-BIS(ALPHA-METHYLBENZYL)-BENZAMIDE

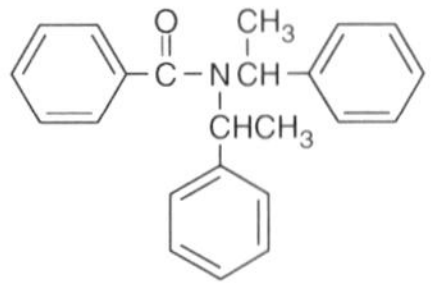

250 mg

S80,724-9
N-METHYL-O-TOLUANILIDE

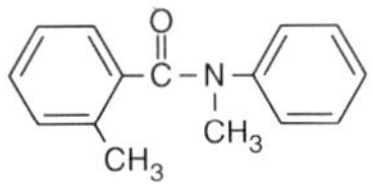

250 mg

S77,987-3
N,N'-DIISOPROPYL-N,N'-DIPHENYLTEREPHTHALAMIDE

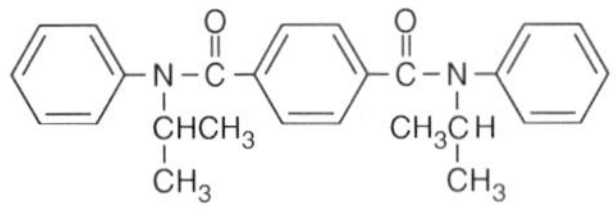

250 mg

S42,069-7
N-METHYL-ORTHO-METHACRYLOTOLUIDIDE

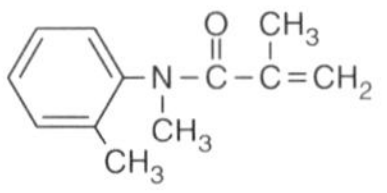

5 gm

S42,078-6
N-ETHYL-ORTHO-METHACRYLOTOLUIDIDE

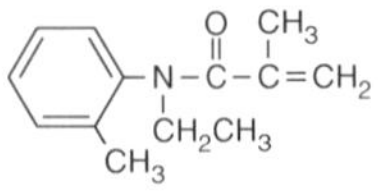

5 gm

S77,480-4
N,N-DIISOPROPYL-O-TOLUAMIDE

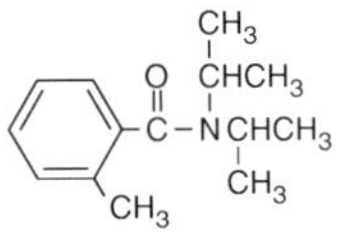

250 mg

S88,603-3
4,N,N-TRIMETHYLBENZAMIDE

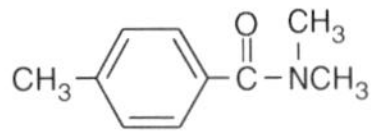

250 mg

S42,068-9
N-METHYL-PARA-METHACRYLOTOLUIDIDE

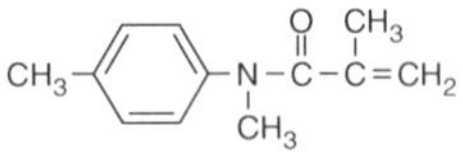

5 gm

S42,385-8
N-ETHYL-PARA-METHACRYLOTOLUIDIDE

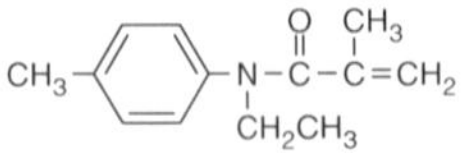

5 gm

S70,099-1
4'-ETHYLMALEANILIC ACID
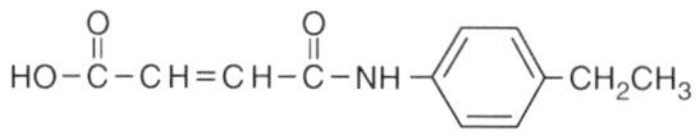
5 gm

S70,305-2
N-(4-METHYLPHENYL)-TETRACHLOROPHTHALAMIC ACID
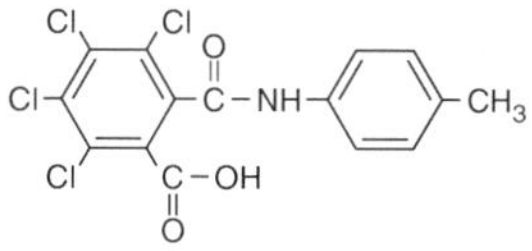
5 gm

S72,120-4
N,N'-DIETHYL-N,N'-DI(M-TOLYL)-ISOPHTHALAMIDE
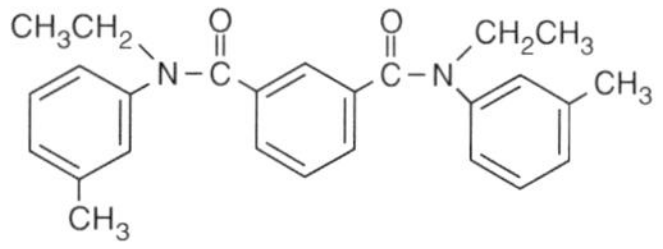
250 mg

S55,534-7
2',5'-DICHLOROPHTHALANILIC ACID
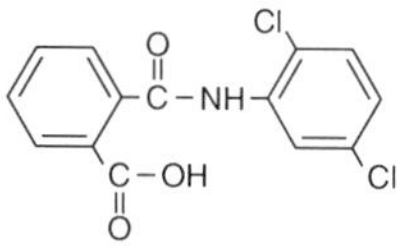
5 gm

S42,149-9
6-BROMO-3-(2,3-DICHLOROPHENYLCARBAMOYL)COUMARIN
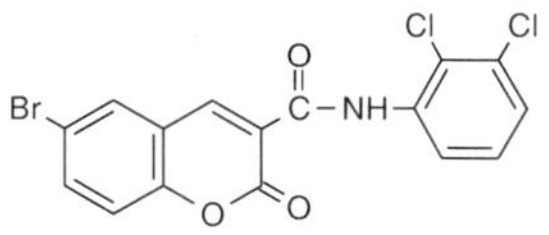
250 mg

S42,146-4
6-BROMO-3-(3,4-DICHLOROPHENYLCARBAMOYL)COUMARIN
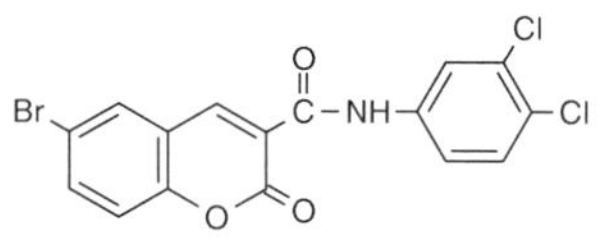
250 mg

S42,147-2
6-BROMO-3-(2,5-DICHLOROPHENYLCARBAMOYL)COUMARIN
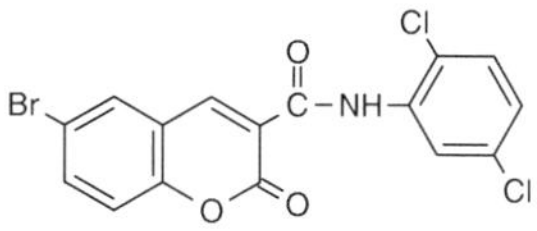
250 mg

S42,148-0
6-BROMO-3-(2,4-DICHLOROPHENYLCARBAMOYL)COUMARIN
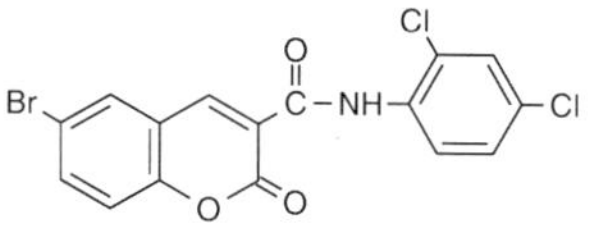
250 mg

S55,556-8
ETHYL 4-(2-CHLOROBENZAMIDO)BENZOATE
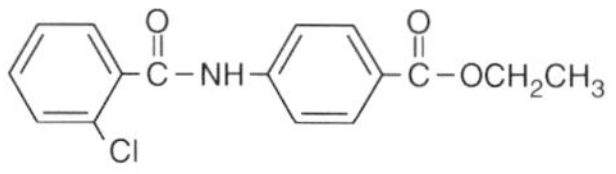
5 gm

S52,327-5
ETHYL 4-BENZAMIDOBENZOATE
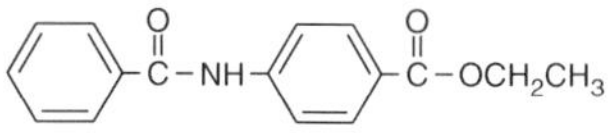
250 mg

S55,767-6
1-BENZOYL-3-(3-CHLORO-ORTHO-TOLYL)UREA
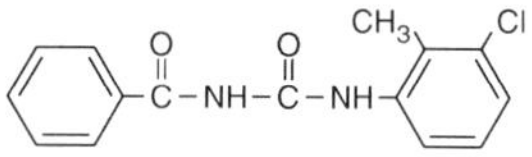
250 mg

S59,557-8
3-BROMOBENZANILIDE
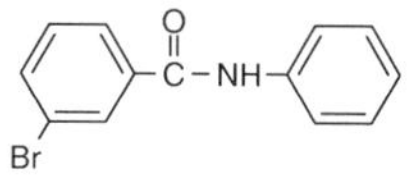
5 gm

S54,022-6
N-DODECYLBENZAMIDE
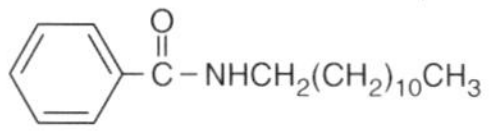
5 gm

S79,526-7
4'-ETHYLCINNAMANILIDE
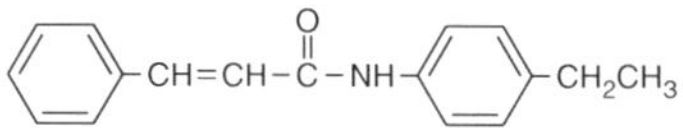
250 mg

S72,087-9
N,N'-(3,3'-DIMETHYLBIPHENYL-4,4'-DIYL)BISCINNAMAMIDE
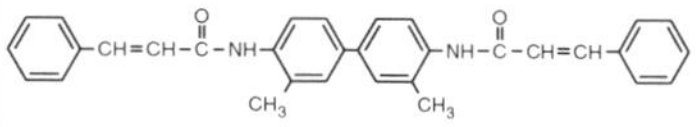
250 mg

S55,553-3
2,3'-DICHLORO-O-BENZOTOLUIDIDE
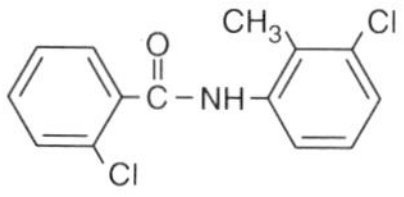
5 gm

S55,528-2
2,2',5'-TRICHLOROBENZANILIDE
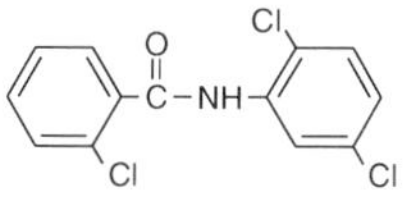
5 gm

S55,547-9
4-CHLORO-P-BENZOTOLUIDIDE
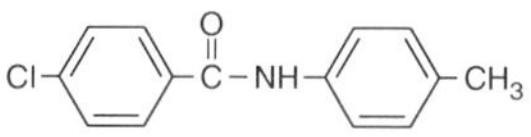
5 gm

S55,559-2
3',4-DICHLORO-O-BENZOTOLUIDIDE
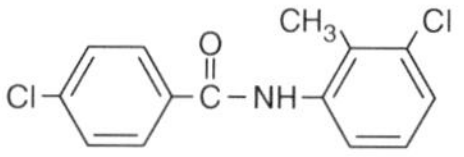
5 gm

S55,560-6
3',4,4'-TRICHLOROBENZANILIDE
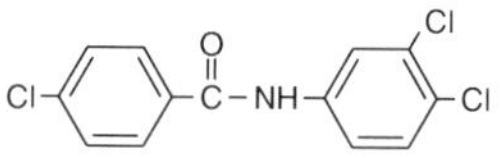
5 gm

S59,553-5
N-(2,3-DICHLOROPHENYL)-3-BROMOBENZAMIDE
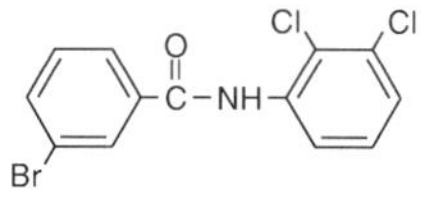
5 gm

S55,532-0
2,3',4-TRICHLORO-O-BENZOTOLUIDIDE
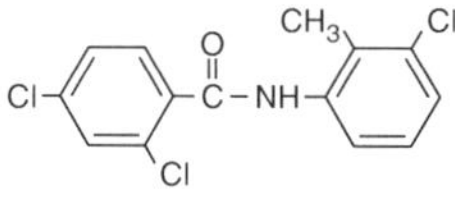
5 gm

S55,497-9
2,3',4,4'-TETRACHLOROBENZANILIDE
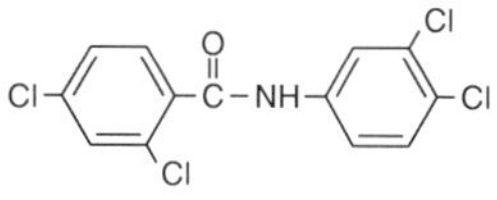
5 gm

S55,529-0
2,2',4,5'-TETRACHLOROBENZANILIDE
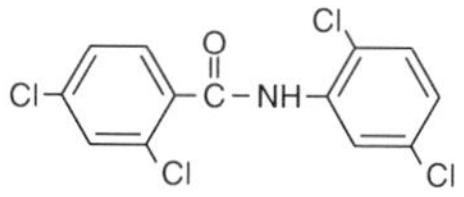
5 gm

S59,546-2
N-(2,4-DICHLOROPHENYL)-3-BROMOBENZAMIDE
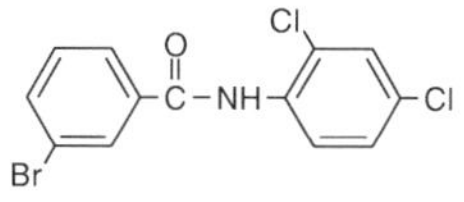
5 gm

S55,552-5
N-BENZYL-2,4-DICHLOROBENZAMIDE
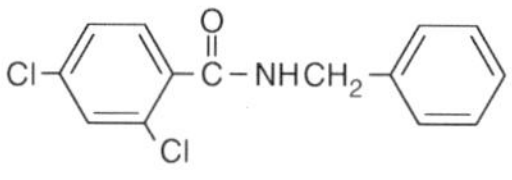
5 gm

S55,498-7
2,4-DICHLORO-5'-METHYL-O-BENZANISIDIDE
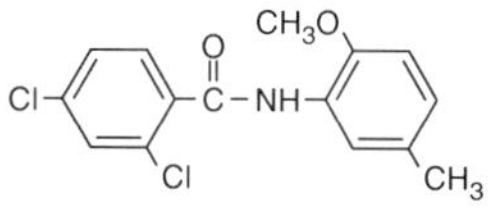
5 gm

S55,515-0
2,4-DICHLORO-2',4',5'-TRIMETHYLBENZANILIDE
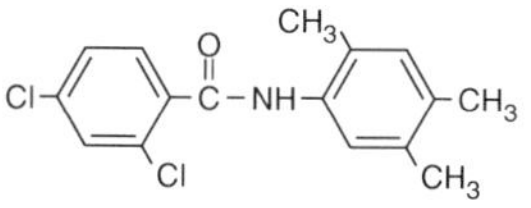
250 mg

S79,536-4
N-PHENYLCINNAMANILIDE
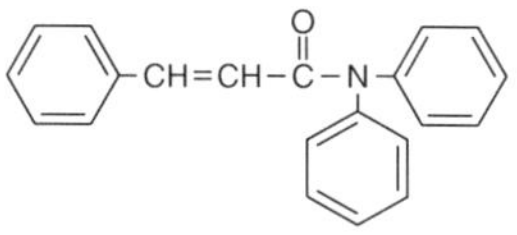
250 mg

S42,405-6
N-METHYL-ORTHO-TOLUANILIDE
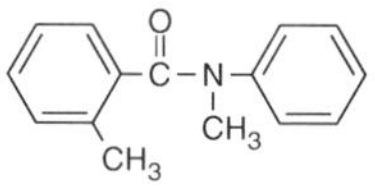
5 gm

S55,550-9
N-BENZYL-4-CHLORO-N-METHYLBENZAMIDE
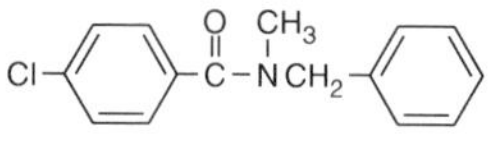
250 mg

S50,814-4
4'-(4-CHLOROPHENOXY)-2,3'-DICHLOROACETANILIDE
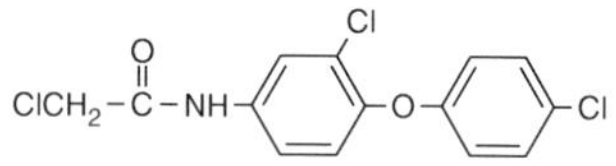
5 gm

S39,188-3
2-CHLORO-N-(1-NAPHTHYL)ACETAMIDE

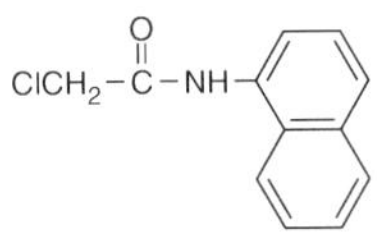

5 gm

S75,300-9
2-(2,2-DICHLOROACETAMIDO)-7-NITROFLUORENE

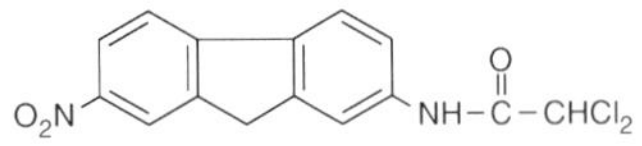

5 gm

S37,543-8
N-(2,2-DICHLOROACETYL)-2-FLUORENAMINE

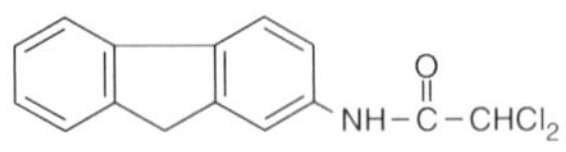

250 mg

S75,548-6
2,5-BIS(2,2-DICHLOROACETAMIDO)-FLUORENE

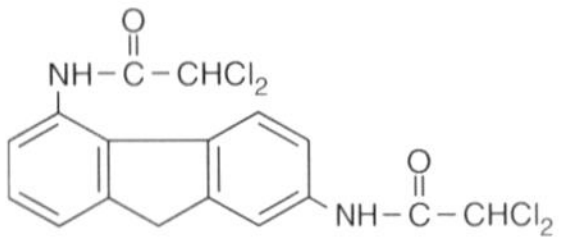

250 mg

S75,502-8
6-(2,2-DICHLOROACETAMIDO)CHRYSENE

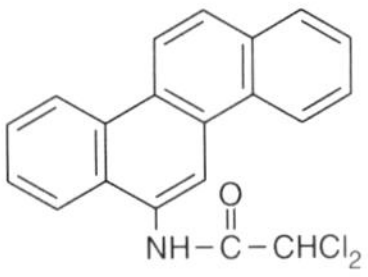

250 mg

S92,437-7
2'-CHLORO-5'-TRIFLUOROMETHYL-1-NAPHTHALENECARBOXANILIDE

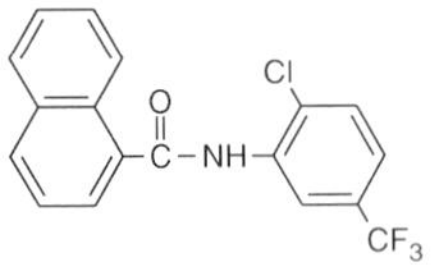

250 mg

S75,526-5
3,7-DINITRO-2-(2,2,2-TRIFLUOROACETAMIDO)FLUORENE

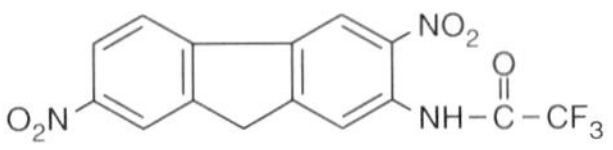

250 mg

S75,167-7
3-CHLORO-7-NITRO-2-(2,2,2-TRIFLUOROACETAMIDO)-9-FLUORENONE

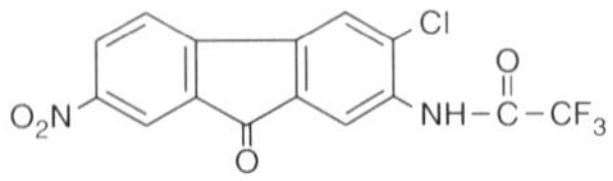

5 gm

S75,299-1
2-(2,2,2-TRIFLUOROACETAMIDO)-9-FLUORENONE

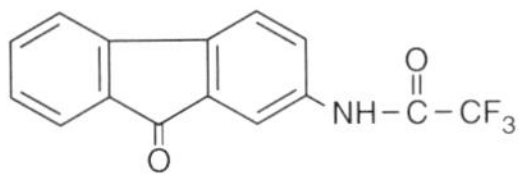

5 gm

S75,704-7
3-BROMO-2-(2,2,2-TRIFLUOROACETAMIDO)-9-FLUORENONE

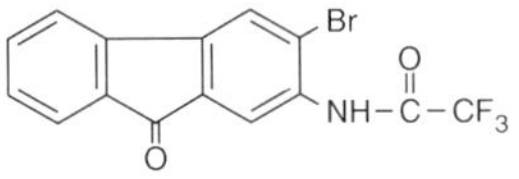

5 gm

S75,151-0
1,3,7-TRIBROMO-2-(2,2,2-TRIFLUOROACETAMIDO)-9-FLUORENONE

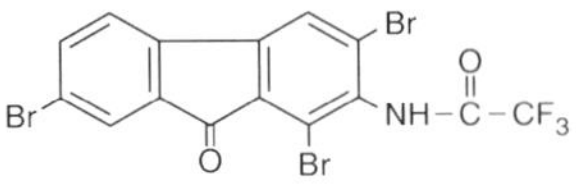

5 gm

S35,655-7
2-(2,2,2-TRIFLUOROACETAMIDO)-FLUORENE

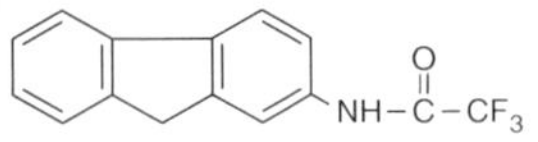

5 gm

S75,162-6
9-(2,2,2-TRIFLUOROACETAMIDO)-FLUORENE

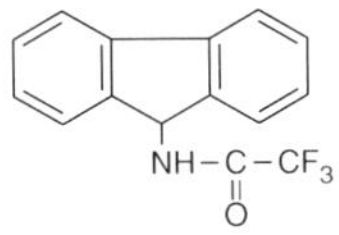

5 gm

S34,279-3
9-BROMO-2-TRIFLUOROACETAMIDO-FLUORENE

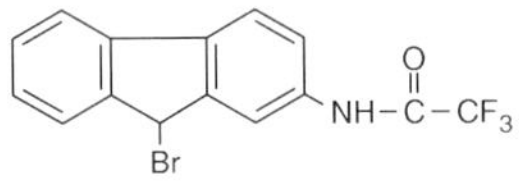

5 gm

S77,427-8
2-ACETAMIDO-7-(2,2,2-TRIFLUOROACETAMIDO)FLUORENE

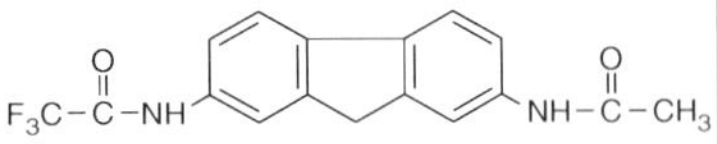

250 mg

S77,400-6
3-(2,2,2-TRIFLUOROACETAMIDO)-FLUORANTHENE

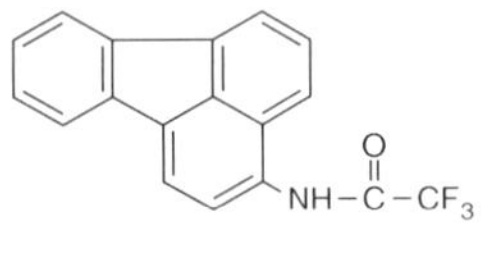

250 mg

S72,978-7
4-CHLORO-N-(5-HYDROXY-1-NAPHTHYL)-BENZAMIDE

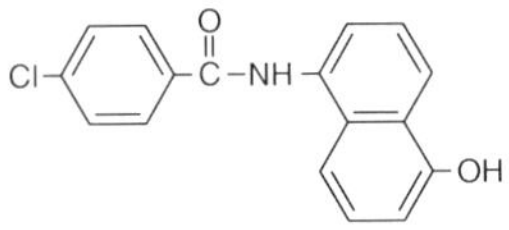

5 gm

S75,047-6
4'-CHLORO-2',5'-DIMETHOXY-1-NAPHTHANILIDE

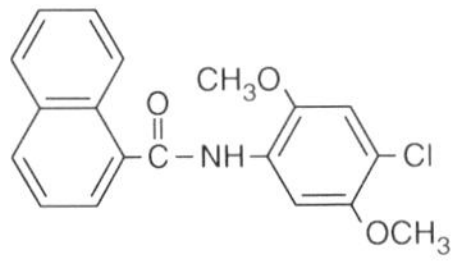

5 gm

S67,586-5
N-(1-NAPHTHYL)-2,4,5-TRICHLOROPHENOXYACETAMIDE

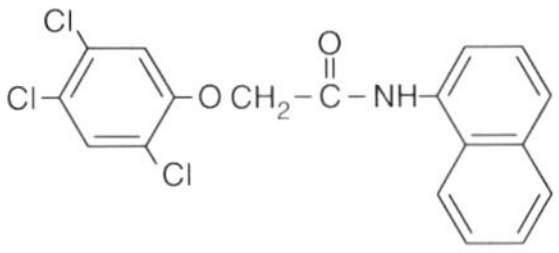

5 gm

S80,649-8
2-(4-CHLOROBENZOYL)-N-METHYL-1-NAPHTHAMIDE

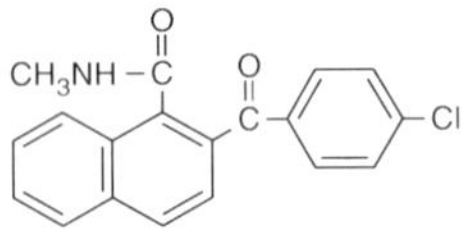

250 mg

S92,700-7
N-(5-CHLORO-1-ANTHRAQUINONYL)-BENZAMIDE

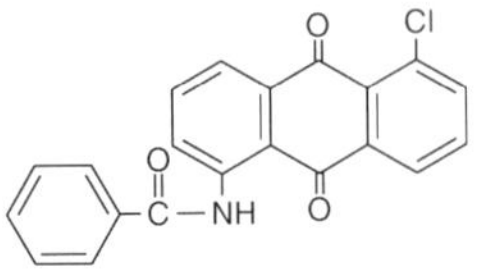

250 mg

S76,420-5
N,N'-DODECAMETHYLENEBIS(3,4,5,6-TETRABROMOPHTHALAMIC ACID)

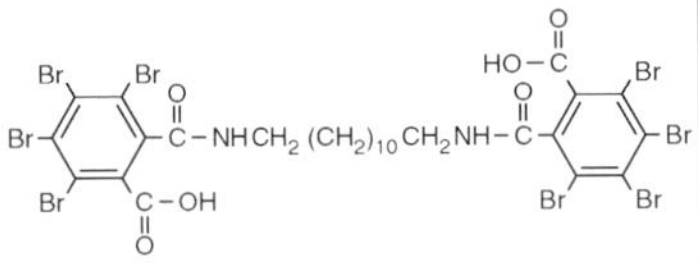

5 gm

S79,400-7
N-(1-NAPHTHYL)-TETRABROMOPHTHALAMIC ACID

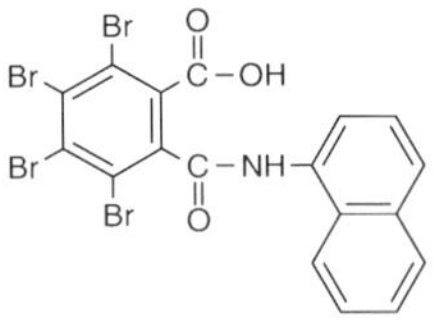

250 mg

S72,944-2
N-(5-HYDROXY-1-NAPHTHYL)-3,4,5,6-TETRABROMOPHTHALAMIC ACID

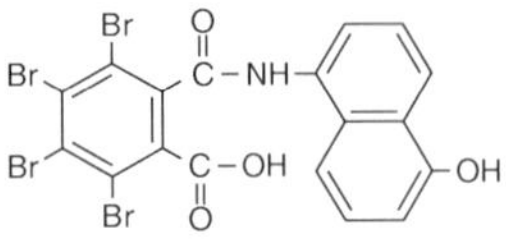

5 gm

S85,474-3
2-CHLORO-N-(2-NAPHTHYL)BENZAMIDE

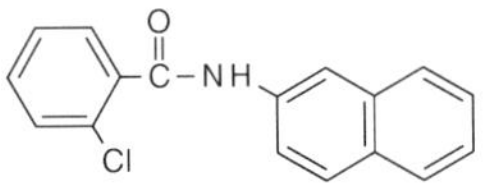

250 mg

S80,132-1
4-FLUORO-N-(2-NAPHTHYL)BENZAMIDE

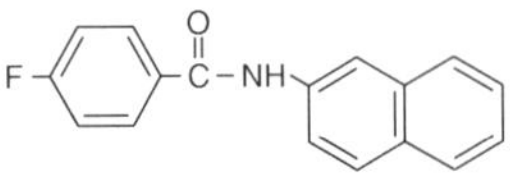

250 mg

S90,534-8
3'-CHLORO-4'-FLUORO-1-NAPHTHANILIDE

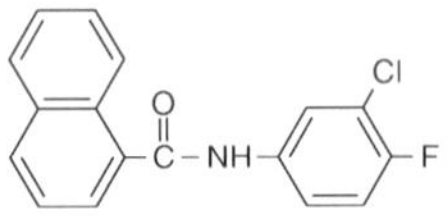

250 mg

S73,812-3
3,4-DICHLORO-N-(2-NAPHTHYL)-BENZAMIDE

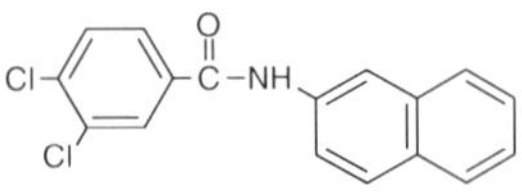

250 mg

S85,473-5
2,4-DICHLORO-N-(2-NAPHTHYL)-BENZAMIDE

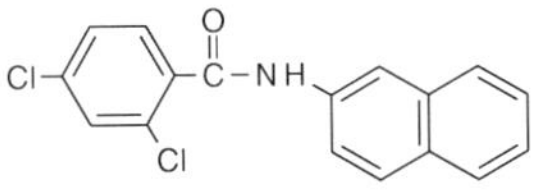

250 mg

S80,129-1
3-FLUORO-N-(2-NAPHTHYL)BENZAMIDE

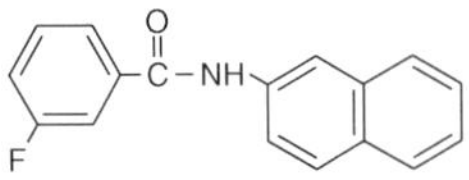

250 mg

S72,109-3
N-(2-NAPHTHYL)-3,4,5,6-TETRACHLOROPHTHALAMIC ACID

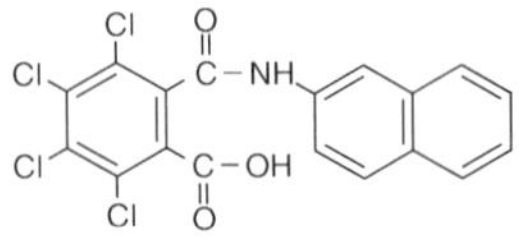

250 mg

S84,105-6
4'-BROMO-2'-PHENYLACETANILIDE

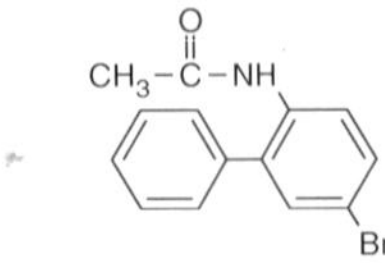

250 mg

S93,781-9
N-(4-FLUORO-1-NAPHTHYL)BENZAMIDE

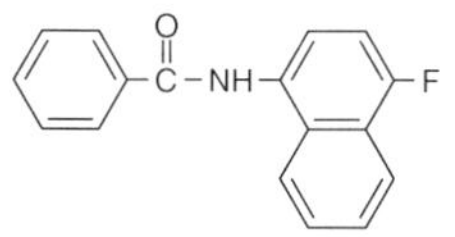

250 mg

S74,398-4
2-FLUORO-N-(1-NAPHTHYL)BENZAMIDE

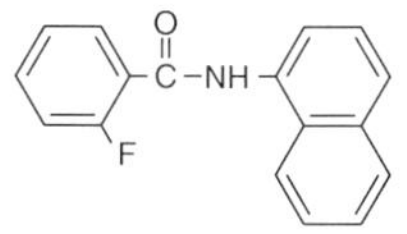

5 gm

S75,566-4
4'-(4-FLUOROPHENYL)ACETANILIDE

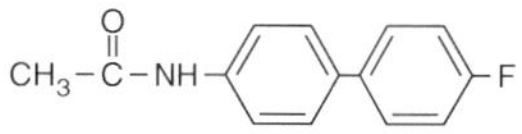

250 mg

S70,012-6
4'-BROMO-2-(1-NAPHTHYL)ACETANILIDE

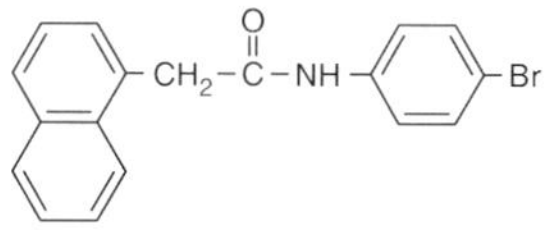

250 mg

S73,745-3
4'-BROMO-2',6'-DIMETHYL-2-(1-NAPHTHYL)ACETANILIDE

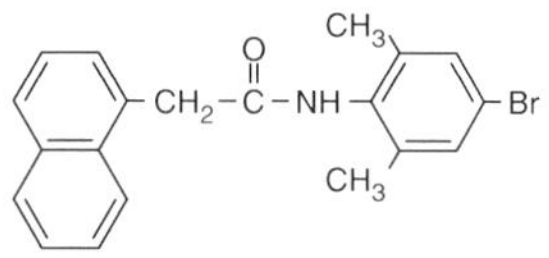

5 gm

S75,048-4
3-CHLORO-N-(1-NAPHTHYL)BENZAMIDE

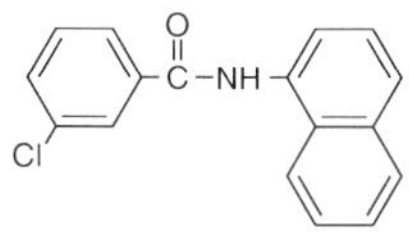

5 gm

S75,493-5
2-ACETAMIDO-7-[N,N-BIS(2-HYDROXYETHYL)AMINO]-1,3-DIBROMOFLUORENE

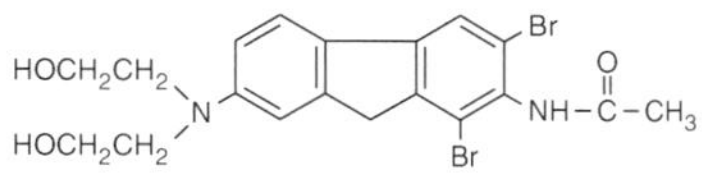

250 mg

S62,916-2
N-(3-AMINO-4-[2(4-TERT-OCTYL-PHENOXY)ETHOXY])BENZOYL-N'-(1-HYDROXY-2-NAPHTHOYL)ETHYLENEDIAMINE

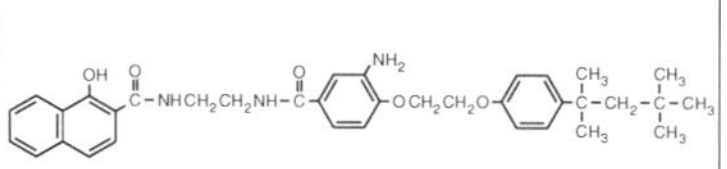

5 gm

S80,609-9
N-(1,1-DIMETHYL-2-HYDROXYETHYL)-2-METHOXY-1-NAPHTHALENECARBOXAMIDE

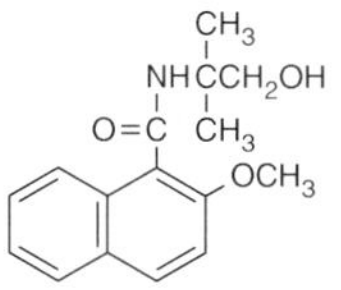

250 mg

S34,410-9
5'-CHLORO-2',4'-DIMETHOXY-2-HYDROXY-3-NAPHTHOYLANILIDE

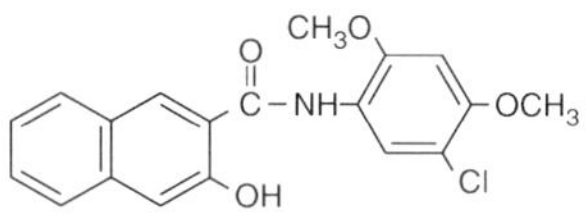

5 gm

S35,110-5
3-HYDROXY-2-NAPHTHOYL-ORTHO-PHENETIDIDE

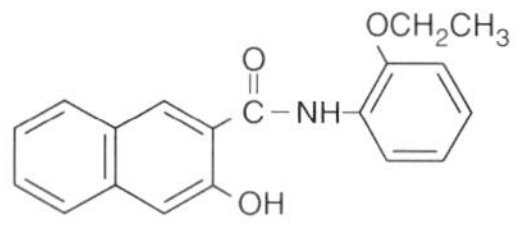

5 gm

S50,707-5
4-CHLORO-1-HYDROXY-4'-NITRO-2-NAPHTH-O-ANISIDIDE

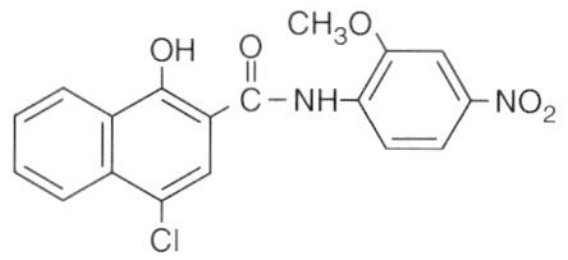

5 gm

S50,705-9
1-HYDROXY-4'-NITRO-2-NAPHTHO-ORTHO-ANISIDIDE

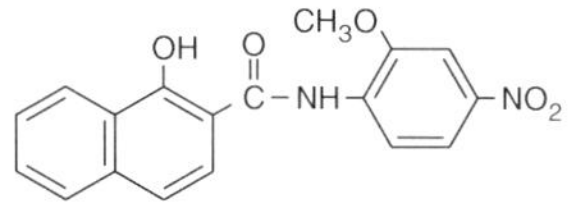

5 gm

S62,979-0
1-HYDROXY-5'-NITRO-2'-[2-(4-TERT-OCTYLPHENOXY)ETHOXY]-2-NAPHTHANILIDE

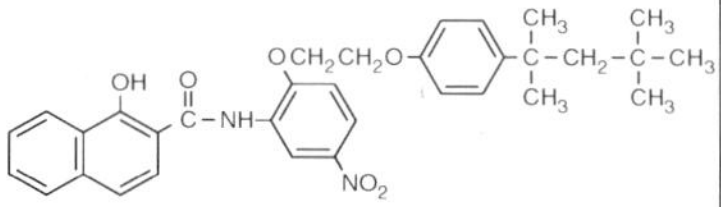

5 gm

S50,412-2
3-HYDROXY-2-NAPHTHOYL-P-ANISIDIDE

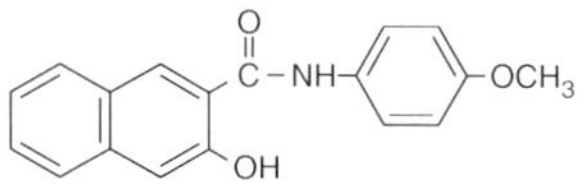

5 gm

S75,561-3
N-(2-FLUORENYL)MANDELAMIDE

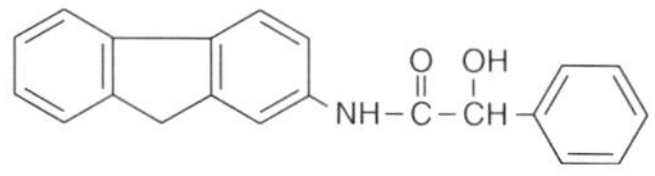

250 mg

S58,792-3
N-(2-AMINOETHYL)-1-HYDROXY-2-NAPHTHAMIDE

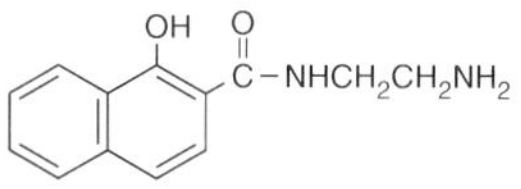

250 mg

S58,255-7
N-(2-AMINOETHYL)-4-CHLORO-1-HYDROXY-2-NAPHTHAMIDE

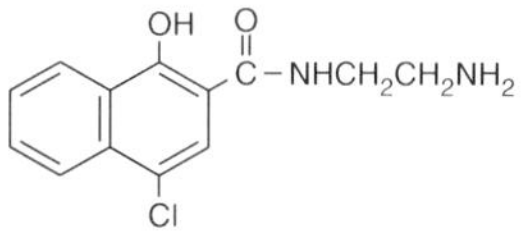

5 gm

S64,234-7
3'-AMINO-4-CHLORO-1-HYDROXY-2-NAPHTHANILIDE

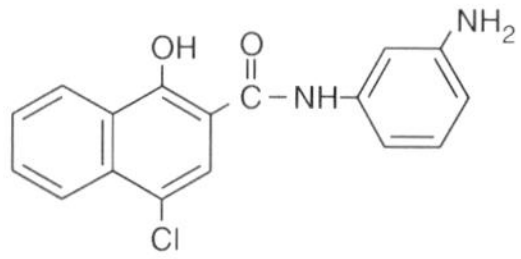

250 mg

S51,551-5
4'-AMINO-4-CHLORO-1-HYDROXY-2-NAPHTHANILIDE

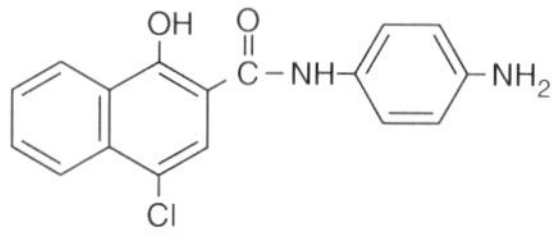

5 gm

S51,515-9
4'-AMINO-1-HYDROXY-2-NAPHTHANILIDE

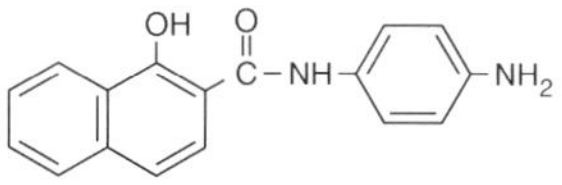

5 gm

S62,939-1
4'-AMINO-1-HYDROXY-2-NAPHTHANILIDE SULFUROUS ACID SALT

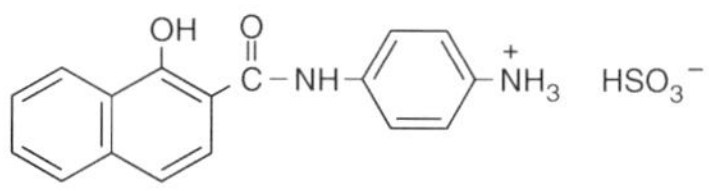

250 mg

S50,043-7
4'-AMINO-2'-CHLORO-1-HYDROXY-2-NAPHTHANILIDE

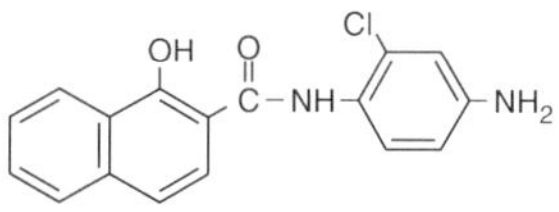

5 gm

S63,139-6
3'-AMINO-1-HYDROXY-2-NAPHTHANILIDE

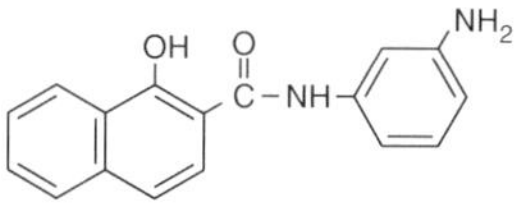

5 gm

S62,945-6
N-(6-AMINOHEXYL)-1-HYDROXY-2-NAPHTHANILIDE

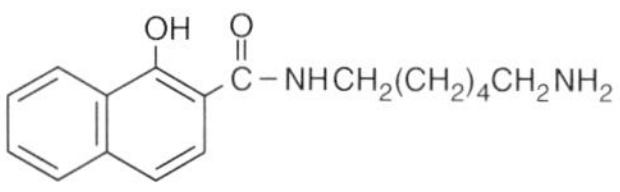

250 mg

S51,711-9
3-HYDROXY-2'-NITRO-2-NAPHTHANILIDE

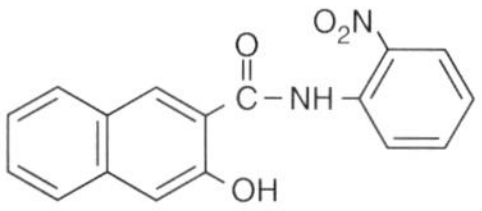

5 gm

S75,495-1
2-ACETAMIDO-9-HYDROXY-3-NITROFLUORENE

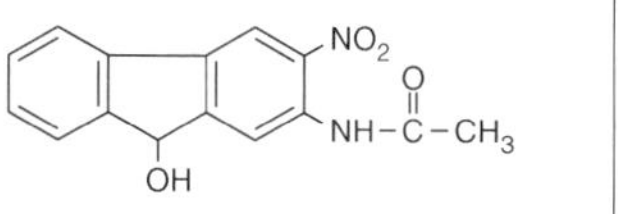

250 mg

S64,231-2
N-[N-(ACRYLAMIDO)ETHYL]-4-CHLORO-1-HYDROXY-2-NAPHTHAMIDE

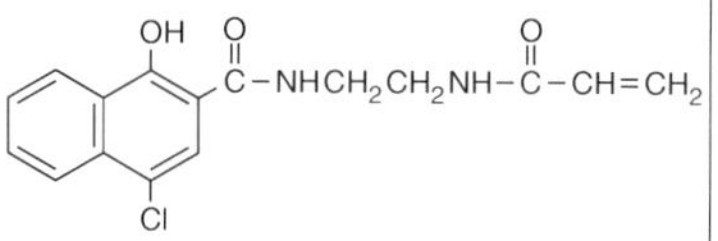

250 mg

S64,220-7
4-CHLORO-1-HYDROXY-N-[N-(METHACRYLAMIDO)ETHYL]-2-NAPHTHAMIDE

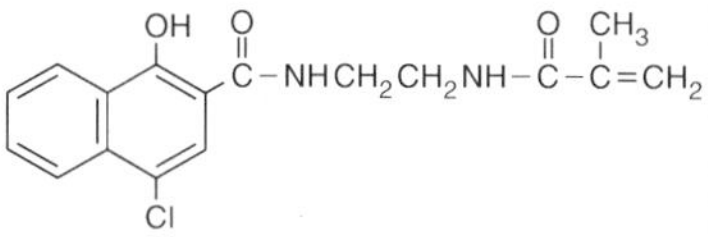

250 mg

S53,065-4
4-CHLORO-1-HYDROXY-3'-NITRO-2-NAPHTHANILIDE

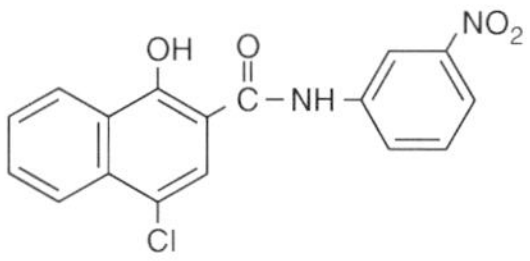

5 gm

S35,109-1
3-HYDROXY-3'-NITRO-2-NAPHTHANILIDE

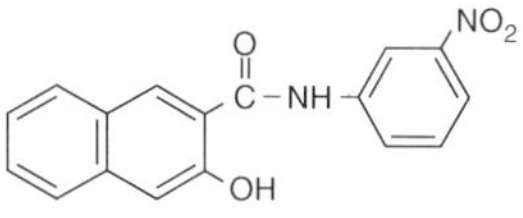

5 gm

S62,936-7
4-CHLORO-1-HYDROXY-4'-NITRO-2-NAPHTHANILIDE

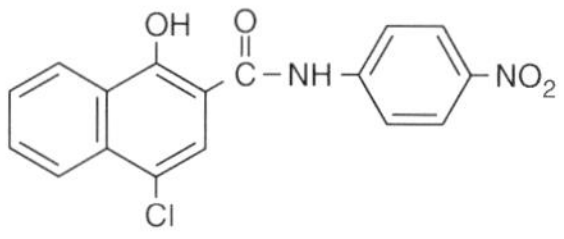

5 gm

S35,107-5
2-HYDROXY-3-NAPHTHOIC ACID ETHANOLAMIDE

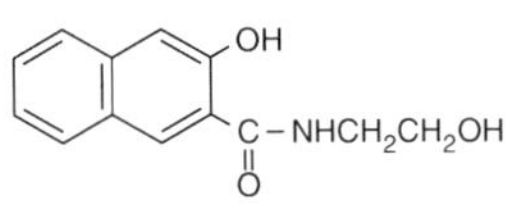

5 gm

S74,890-0
2-ACETAMIDO-3-HYDROXYFLUORENE

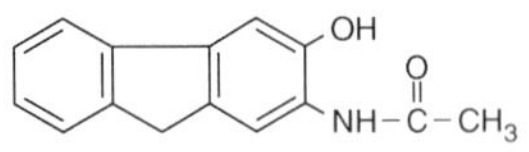

250 mg

S50,427-0
3-HYDROXY-4'-NITRO-2-NAPHTHANILIDE

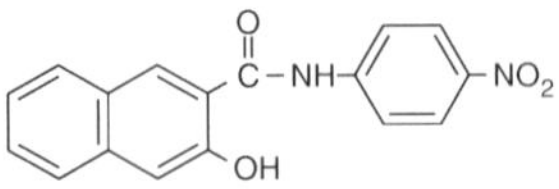

5 gm

S50,721-0
1-HYDROXY-4'-NITRO-2-NAPHTHANILIDE

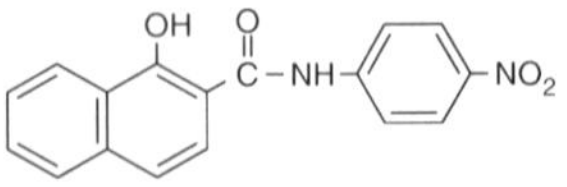

5 gm

S62,932-4
4'-CHLORO-1-HYDROXY-2'-NITRO-2-NAPHTHANILIDE

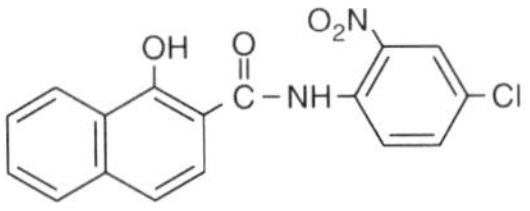

5 gm

S62,938-3
2'-CHLORO-1-HYDROXY-4'-NITRO-2-NAPHTHANILIDE

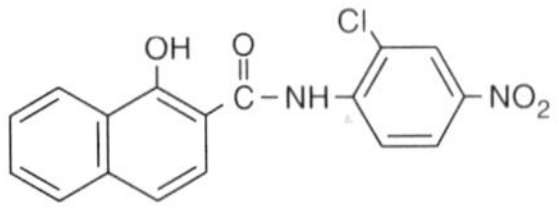

250 mg

S50,401-7
3-HYDROXY-2-NAPHTHO-2',3'-XYLIDIDE

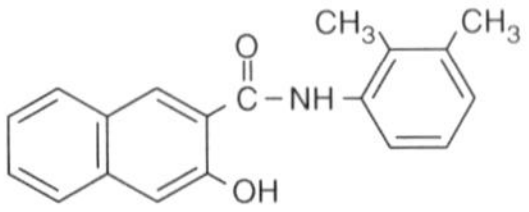

5 gm

S42,907-4
ALGOL RED (1,5-DIBENZAMIDO-4-HYDROXYANTHRAQUINONE)

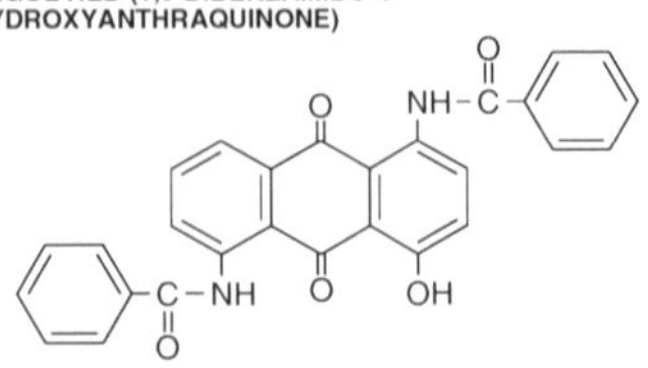

5 gm

S62,927-8
2'-CHLORO-5'-FLUOROSULFONYL-1-HYDROXY-2-NAPHTHANILIDE

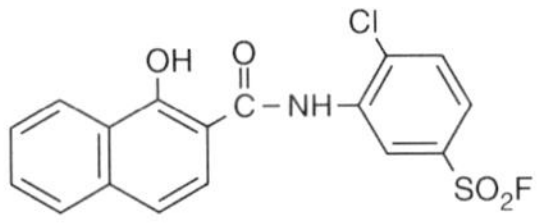

5 gm

S50,110-7
N-CYCLOHEXYL-1-HYDROXY-2-NAPHTHAMIDE

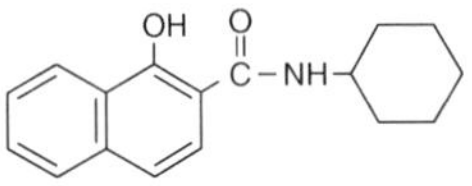

250 mg

S59,114-9
2-HYDROXY-1-NAPHTHO-O-TOLUIDIDE

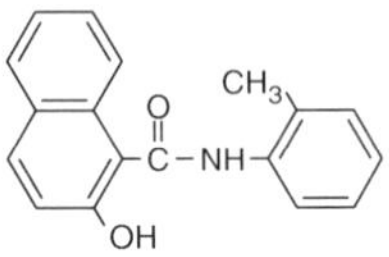

250 mg

S35,954-8
5'-CHLORO-3-HYDROXY-2'-METHYL-2-NAPHTHANILIDE

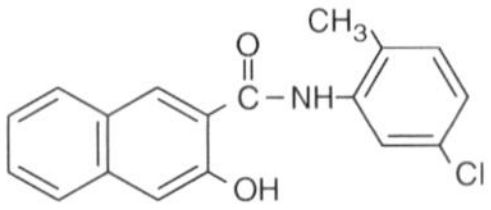

5 gm

S34,705-1
2',4'-DIMETHYL-2-HYDROXY-3-NAPHTHOYLANILIDE

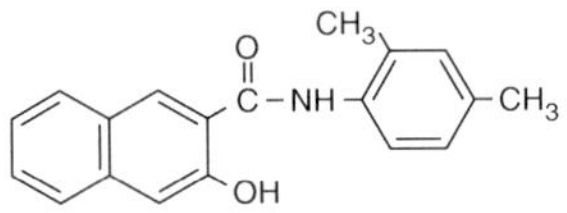

5 gm

S59,116-5
2-HYDROXY-N-(1-NAPHTHYL)-1-NAPHTHAMIDE

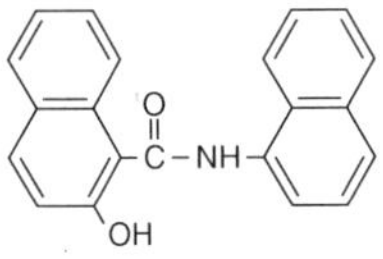

250 mg

S35,100-8
3-HYDROXY-2'-METHYL-2-NAPHTHANILIDE

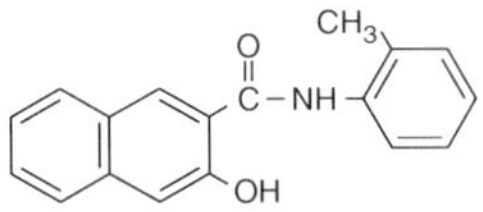

5 gm

S50,404-1
4'-CHLORO-3-HYDROXY-2-NAPHTHANILIDE

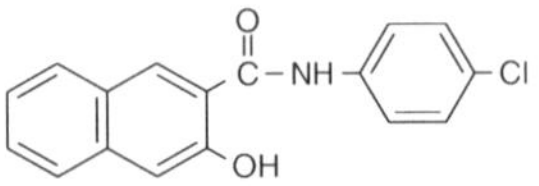

5 gm

S37,324-9
4'-CHLORO-3-HYDROXY-2'-METHYL-2-NAPHTHANILIDE

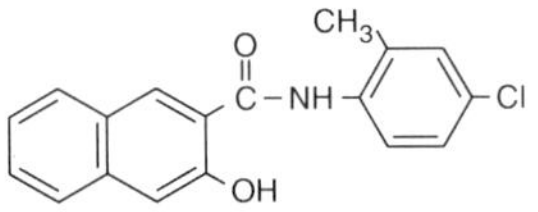

5 gm

S37,562-4
1-ACETAMIDO-6-NAPHTHOL

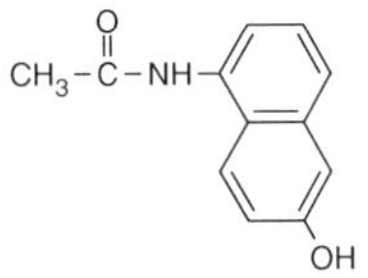

5 gm

S79,633-6
N-(5-HYDROXY-1-NAPHTHYL)BUTYRAMIDE

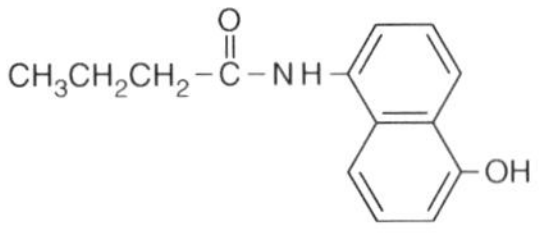

250 mg

S42,748-9
2-ACETAMIDO-7-HYDROXYFLUORENE

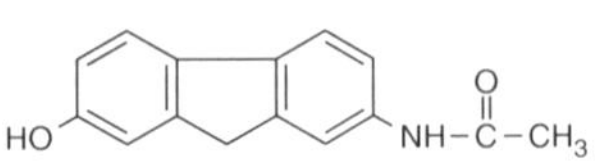

250 mg

S74,585-5
2-ACETAMIDO-3-CHLORO-9-HYDROXYFLUORENE

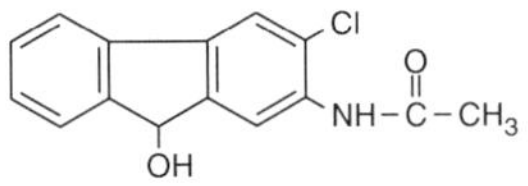

250 mg

S85,463-8
N,N'-BIS(2-METHOXY-5-NITROPHENYL)-1,10-DECANEDIAMIDE

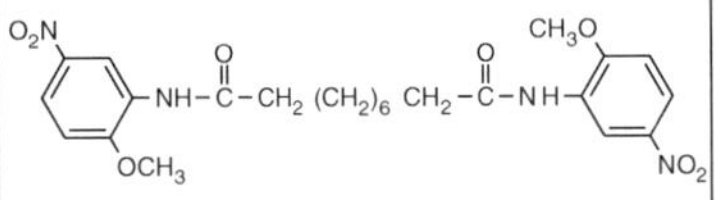

250 mg

S85,479-4
2',5'-DIMETHOXY-1-NAPHTHALENECARBOXANILIDE

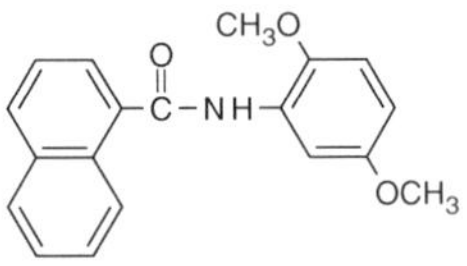

250 mg

S53,333-5
4'-(2-METHOXYPHENOXY)-MALEANILIC ACID

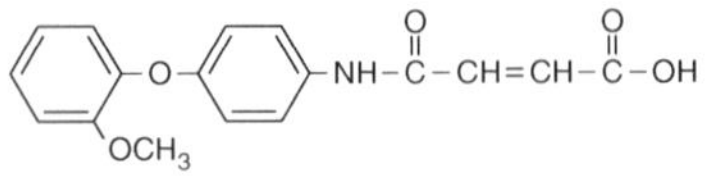

5 gm

S75,266-5
N-(1-NAPHTHYL)-2-PHENOXYACETAMIDE

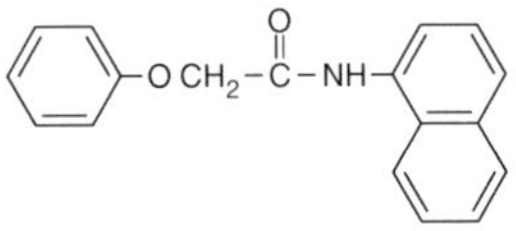

250 mg

S77,722-6
L-METHIONINE-4-METHOXY-2-NAPHTHYLAMIDE

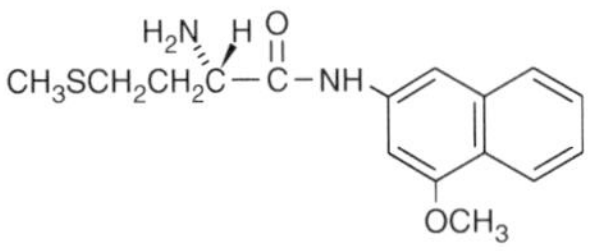

250 mg

S78,123-1
L-VALINE-4-METHOXY-2-NAPHTHYLAMIDE

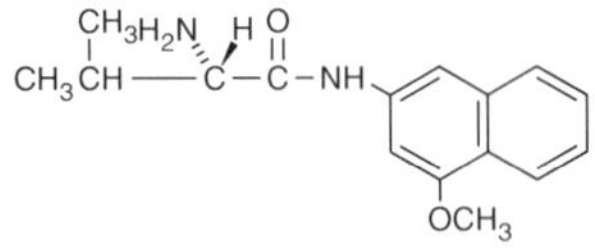

250 mg

S78,458-3
L-ASPARTIC ACID 4-METHOXY-2-NAPHTHYLAMIDE

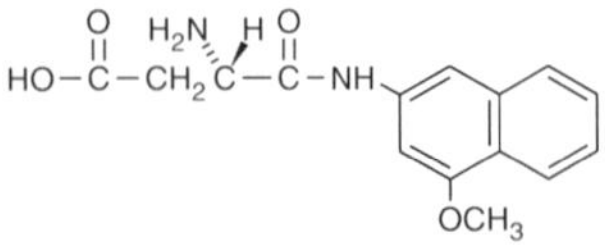

250 mg

S45,276-9
L-LEUCINE-4-METHOXY-2-NAPHTHYLAMIDE HYDROCHLORIDE

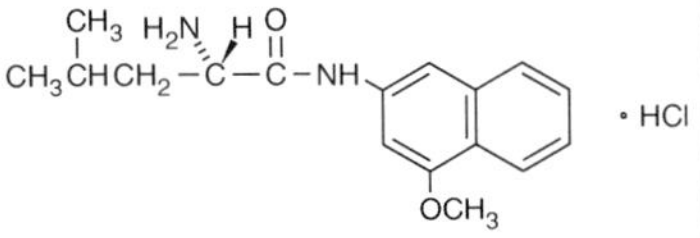

250 mg

S79,008-7
L-ALANINE-4-METHOXY-2-NAPHTHYLAMIDE HYDROCHLORIDE

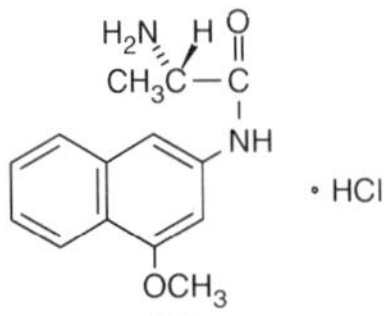

250 mg

S39,841-1
5'-CHLORO-2'-(4-CHLOROPHENOXY)-2-(DIMETHYLAMINO)ACETANILIDE HYDROCHLORIDE

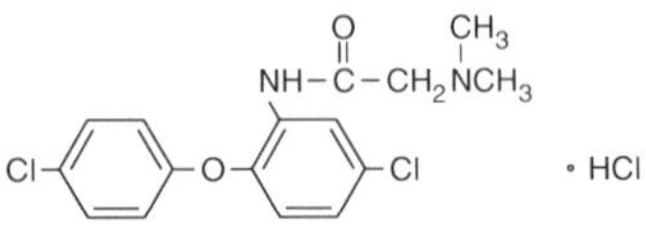

5 gm

S34,533-4
4,4'-BIS(ACETOACETAMIDO)-3,3'-DIMETHOXYBIPHENYL

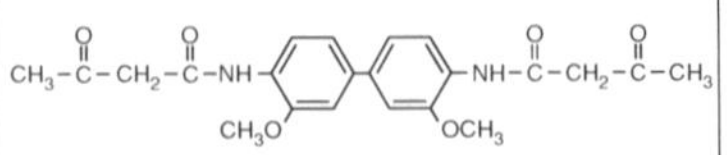

5 gm

S53,330-0
4'-PHENOXYMALEANILIC ACID
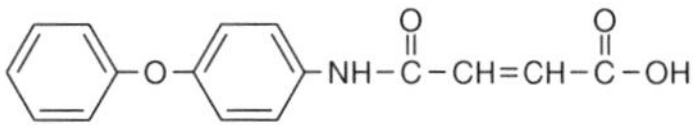
5 gm

S75,262-2
4-METHOXY-N-(1-NAPHTHYL)BENZAMIDE
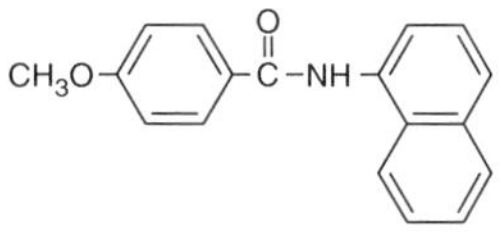
250 mg

S94,671-0
(L)-2,3-DIACETOXY-N-[1-(1-NAPHTHYL)-ETHYL]SUCCINAMIC ACID
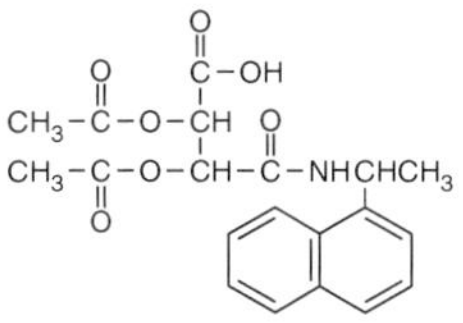
250 mg

S95,852-2
ISOINDIGO
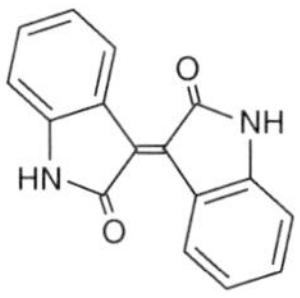
50 mg

S46,259-4
S-BENZYL-L-CYSTEINE-BETA-NAPHTHYLAMIDE
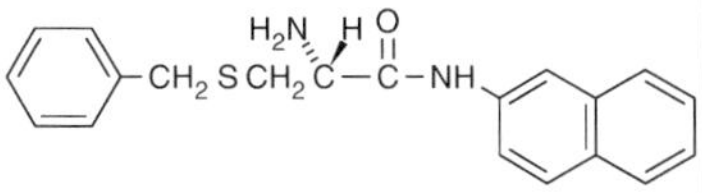
250 mg

S75,527-3
2'-(METHYLTHIO)-4'-PHENYLACETANILIDE
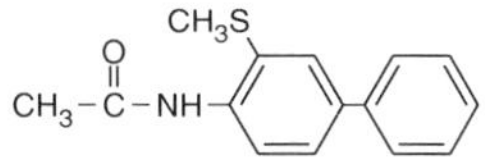
250 mg

S40,820-4
L-CYSTINE BIS[N-(2-NAPHTHYL)AMIDE]
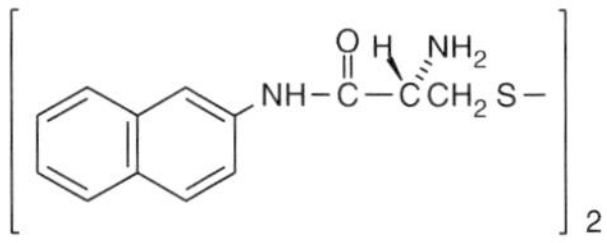
250 mg

S40,821-2
L-CYSTINE BIS[N-(2-NAPHTHYL)AMIDE]-DIHYDROCHLORIDE
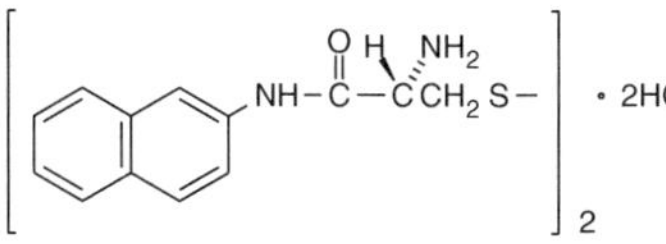
250 mg

S75,516-8
2-ACETAMIDO-4-AMINOFLUORENE
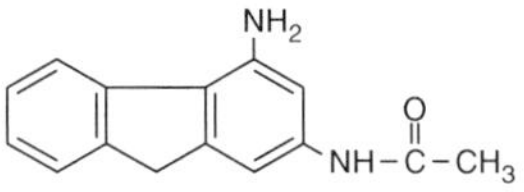
250 mg

S35,372-8
N-GAMMA-(1-NAPHTHYL)-DL-GLUTAMINE
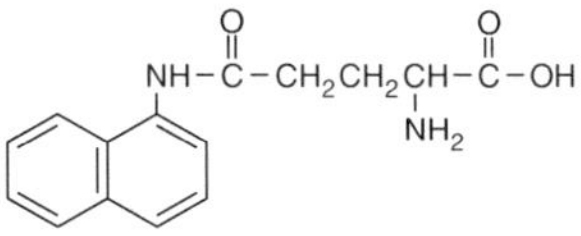
250 mg

S45,189-4
GLYCYL-L-PHENYLALANINE-2-NAPHTHYLAMIDE, HEMIHYDRATE
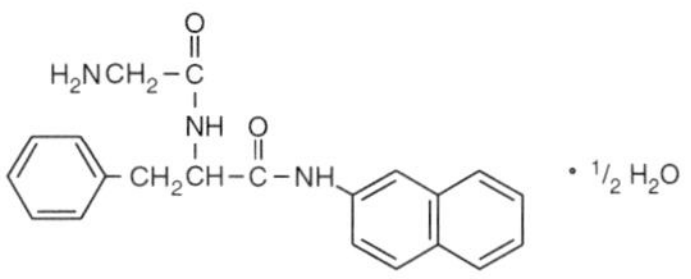
250 mg

S79,002-8
ALANYLPHENYLALANINE BETA-NAPHTHYLAMIDE
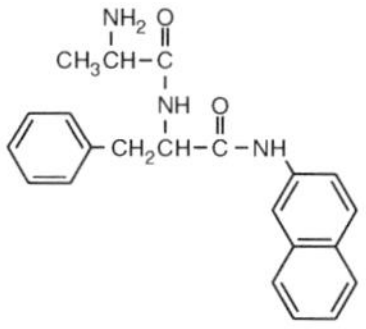
250 mg

S40,640-6
N-(2-AMINOETHYL)-1-NAPHTHALENEACETAMIDE HYDROCHLORIDE
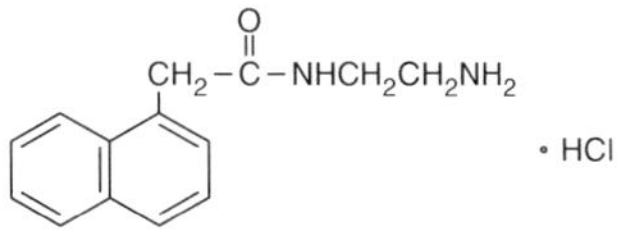
5 gm

S34,086-3
1-AMINO-5-BENZAMIDOANTHRAQUINONE
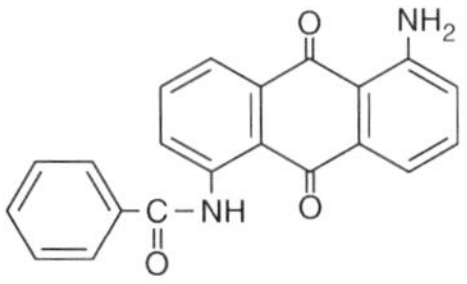
5 gm

S45,112-6
L-ALANINE-2-NAPHTHYLAMIDE
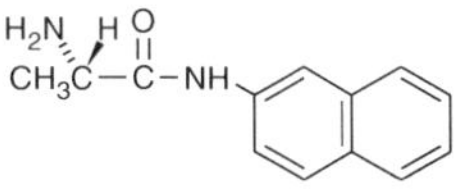
5 gm

S45,140-1
DL-ALANINE-2-NAPHTHYLAMIDE
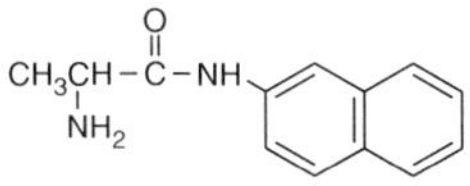
5 gm

S78,476-1
N-ACETYLGLYCYL-L-LYSINE-BETA-NAPHTHYLAMIDE
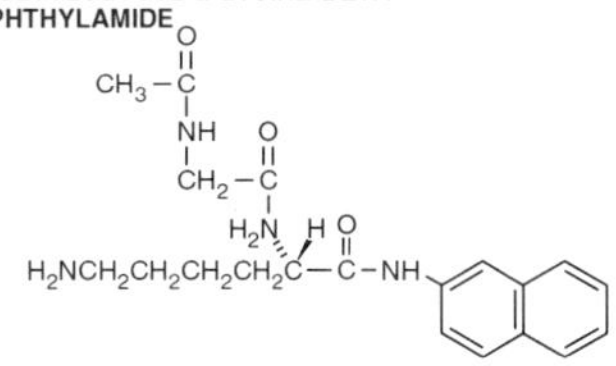
250 mg

S78,875-9
GLYCINE-2-NAPHTHYLAMIDE MONOHYDRATE
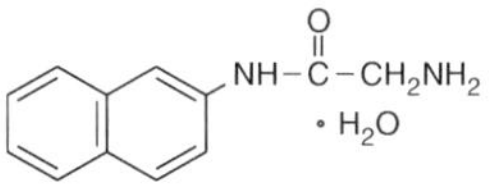
250 mg

S78,523-7
BETA-ALANINE 2-NAPHTHYLAMIDE
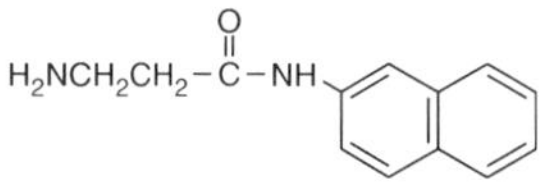
250 mg

S35,701-4
2-ACETAMIDO-7-AMINOFLUORENE
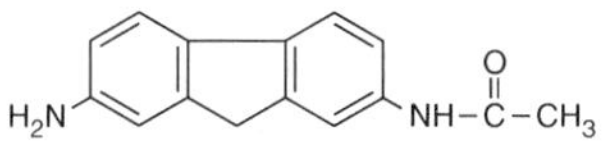
250 mg

S36,404-5
N-(2-NAPHTHYL)-L-LEUCINAMIDE HYDROCHLORIDE
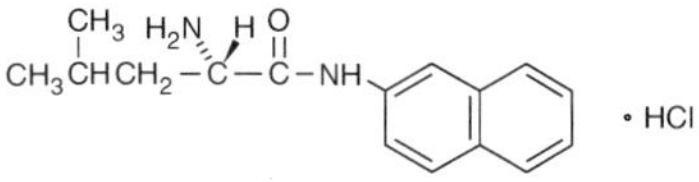
250 mg

S77,101-5
2-ACETAMIDO-7-FLUORO-3-NITRO-9-FLUORENONE
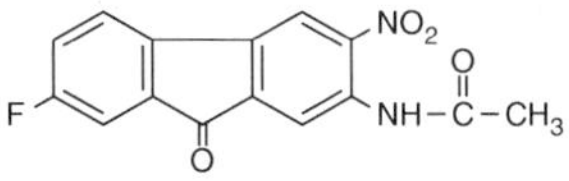
250 mg

S35,704-9
2-ACETAMIDO-3,7-DINITRO-9-FLUORENONE
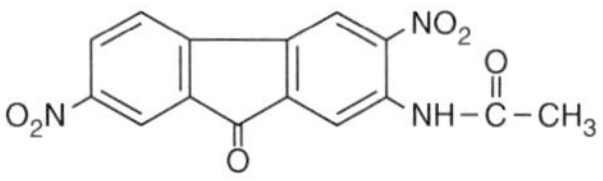
250 mg

S74,586-3
2,7-DIACETAMIDO-3,6-DINITRO-9-FLUORENONE
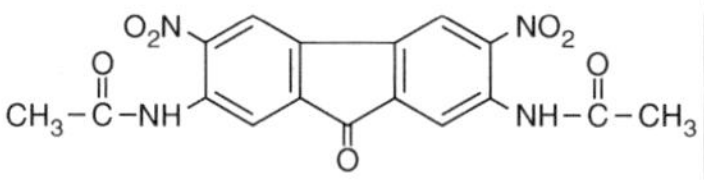
250 mg

S35,707-3
5-ACETAMIDO-4-NITROACENAPHTHENE
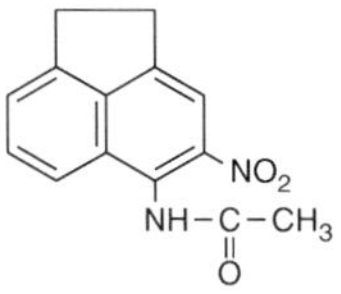
250 mg

S41,676-2
2'-NITRO-4'-PHENYLACETANILIDE
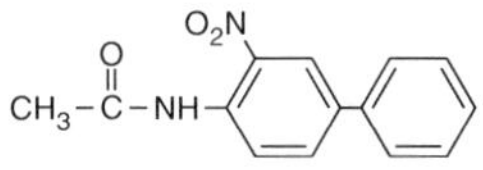
5 gm

S75,549-4
3-ACETAMIDO-2-NITROFLUORENE
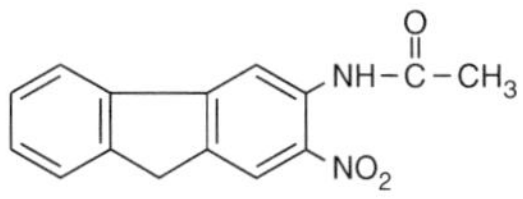
250 mg

S75,410-2
2-ACETAMIDO-7-FLUORO-3-NITROFLUORENE
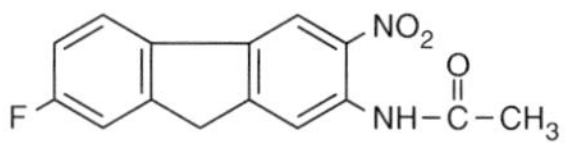
250 mg

S34,016-2
2-ACETAMIDO-3,7-DINITROFLUORENE
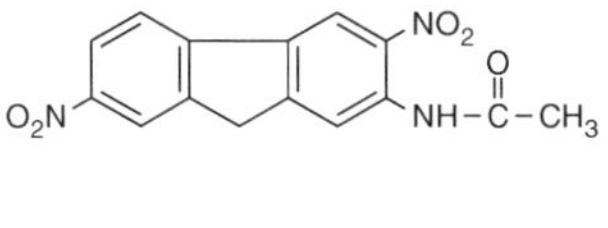
5 gm

S34,859-7
4,5'-DIBENZAMIDO-1,1'-DIANTHRIMIDE
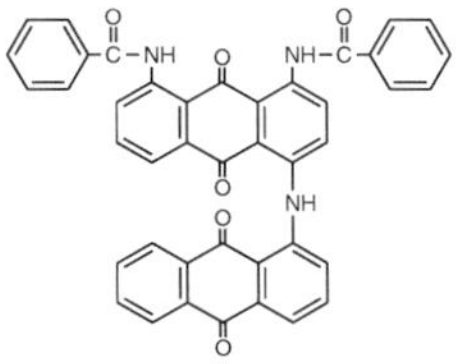
5 gm

S34,858-9
4,4'-DIBENZAMIDO-1,1'-DIANTHRIMIDE
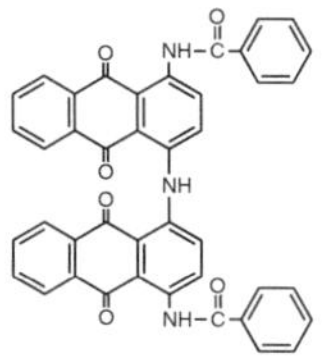
5 gm

S35,871-1
9,10-BIS(FORMAMIDO)ANTHRACENE
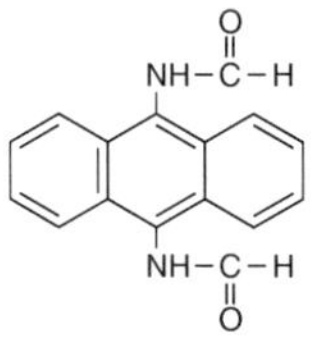
250 mg

S44,597-5
N-BENZOYL-DL-PHENYLALANINE BETA-NAPHTHYL ESTER
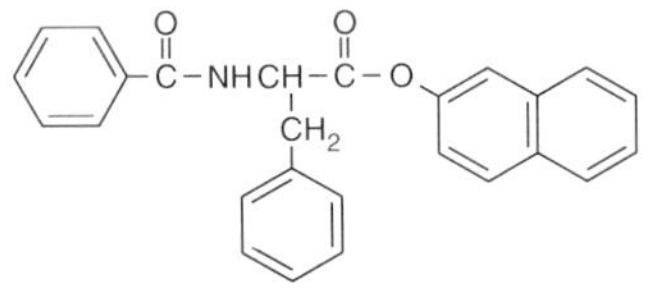
250 mg

S45,128-2
N-BENZOYL-DL-PHENYLALANINE-2-NAPHTHYLAMIDE
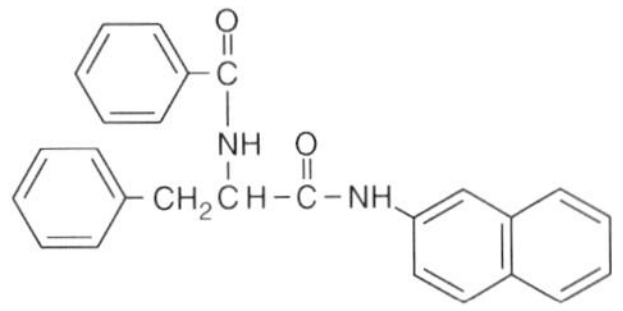
5 gm

S77,531-2
N-METHYL-N-(2-PHENYLETHYL)-2-NAPHTHALENECARBOXAMIDE
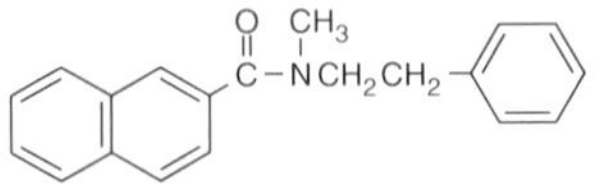
250 mg

S75,815-9
2-ACETAMIDO-9-(DIMETHYLAMINO)-FLUORENE
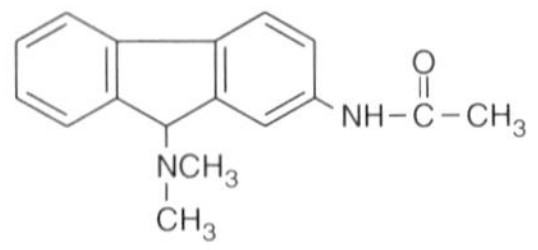
250 mg

S77,459-6
N-BENZYL-2,2-DIMETHYL-2'-PHENYLPROPIONANILIDE
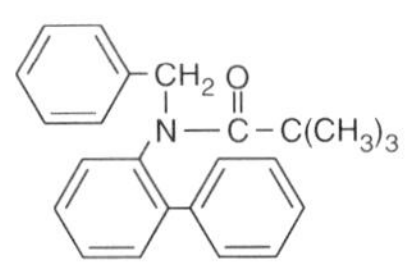
250 mg

S68,101-6
N,2-DIMETHYL-1-NAPHTHAMIDE
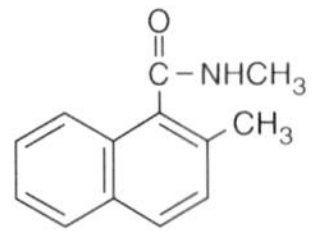
5 gm

S80,523-8
DIETHYL 2-ACETAMIDO-2-(9-FLUORENYL)-MALONATE
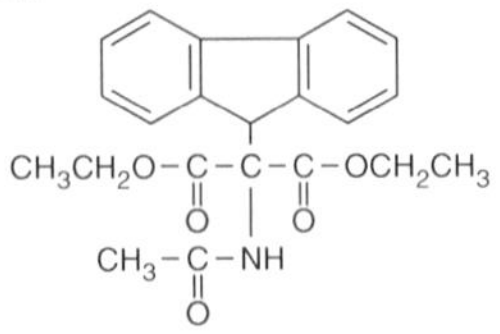
250 mg

S41,291-0
2-FORMAMIDO-7-NITROFLUORENE
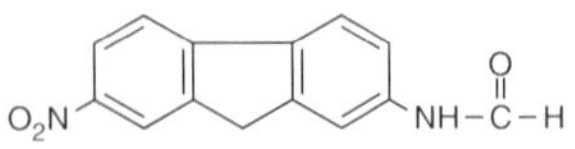
250 mg

S75,316-5
2-ACETAMIDO-1,3-DICHLORO-7-NITROFLUORENE
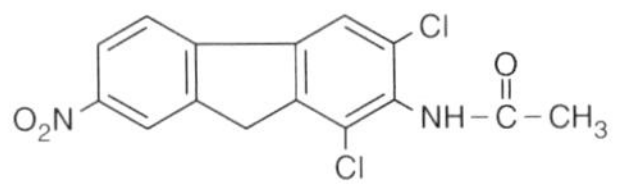
5 gm

S61,541-2
9-OXO-4-FLUORENECARBOXAMIDE
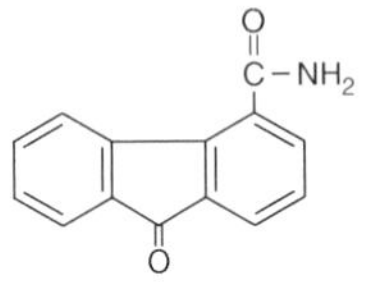
5 gm

S92,395-8
1,8-BIS(BENZAMIDO)ANTHRAQUINONE
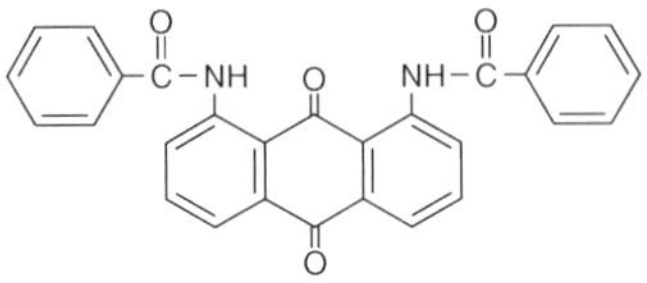
5 gm

S94,134-4
N-(2-ANTHRAQUINONYL)BENZAMIDE
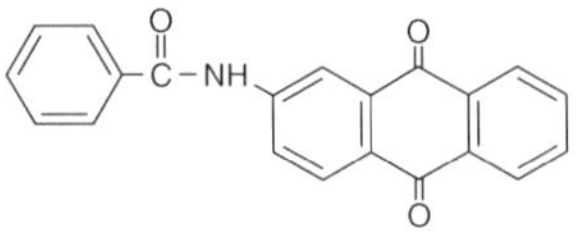
250 mg

S92,389-3
1-CHLORO-4-BENZAMIDOANTHRAQUINONE
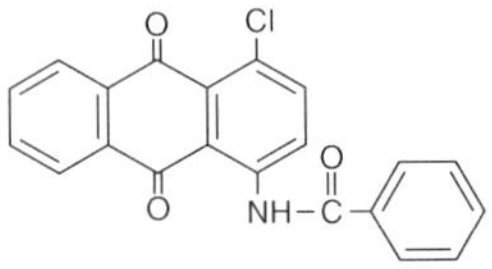
250 mg

S34,993-3
2-FORMAMIDO-9-FLUORENONE
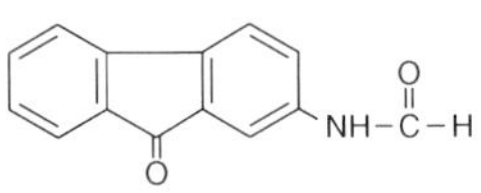
250 mg

S75,492-7
2-ACETAMIDO-9-FLUORENONE
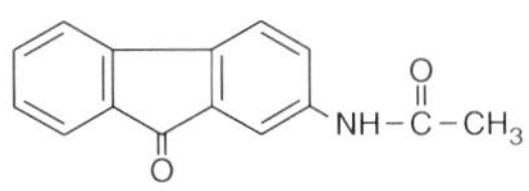
250 mg

S75,616-4
3-ACETAMIDO-9-FLUORENONE
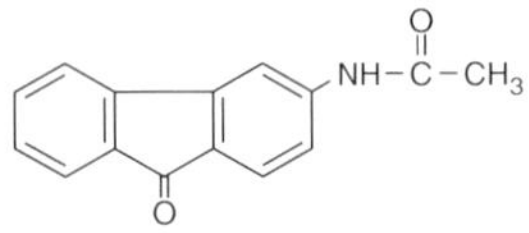
5 gm

S77,102-3
1-ACETAMIDO-9-FLUORENONE
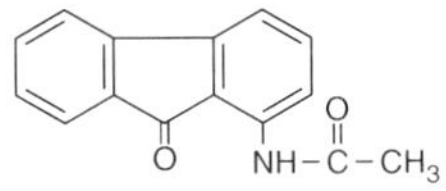
250 mg

S75,610-5
2-ACETAMIDO-7-CHLORO-9-FLUORENONE
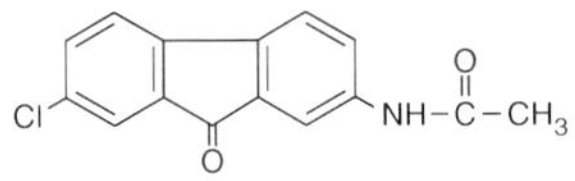
5 gm

S74,886-2
2-ACETAMIDO-3-BROMO-9-FLUORENONE
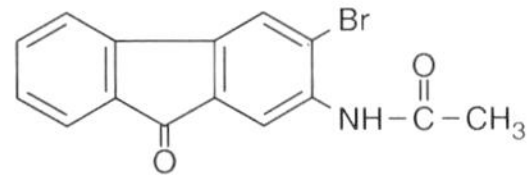
250 mg

S75,315-7
2-ACETAMIDO-1,3,4,7-TETRACHLORO-9-FLUORENONE
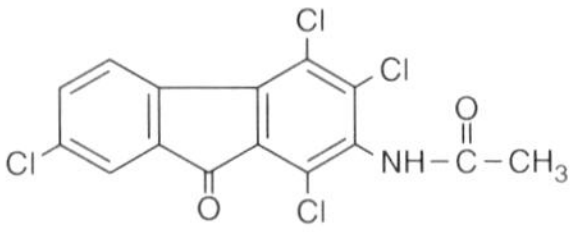
5 gm

S75,901-5
9-(2,2-DIMETHYLPROPIONYL)-N-METHYL-1-FLUORENECARBOXAMIDE
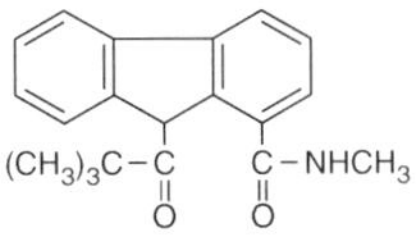
250 mg

S75,513-3
2-(N-METHYLACETAMIDO)-9-FLUORENONE
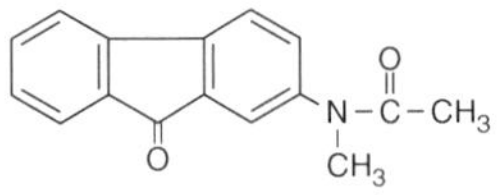
250 mg

S74,991-5
2-(N-ETHYLACETAMIDO)-9-FLUORENONE
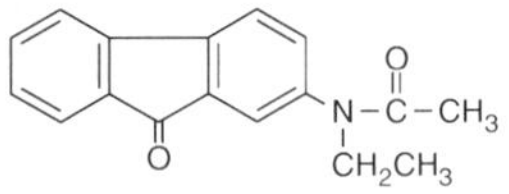
250 mg

S92,293-5
2-(FORMAMIDO)ANTHRAQUINONE
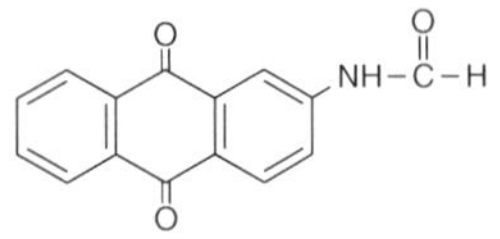
250 mg

S71,646-4
2-PROPIONAMIDOANTHRAQUINONE
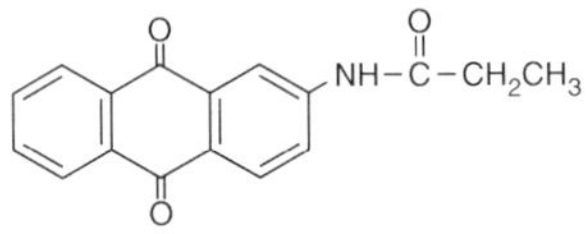
5 gm

S34,529-6
1,4-DIACETAMIDOANTHRAQUINONE
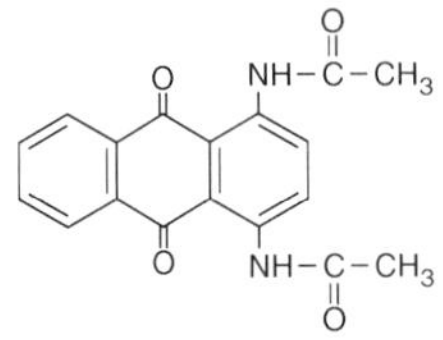
5 gm

S67,648-9
2-(2,2,2-TRIMETHYLACETAMIDO)-ANTHRAQUINONE
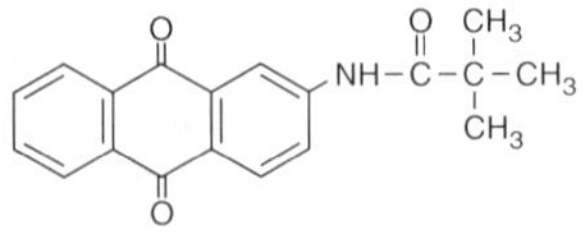
250 mg

S75,705-5
2-ACETAMIDO-3-BROMO-1-CHLORO-9-FLUORENONE
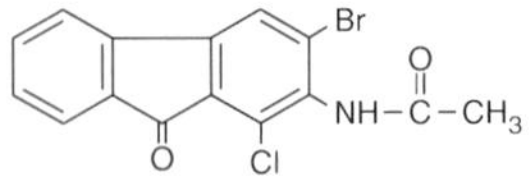
5 gm

S37,359-1
2-(N-METHYLFORMAMIDO)-9-FLUORENONE
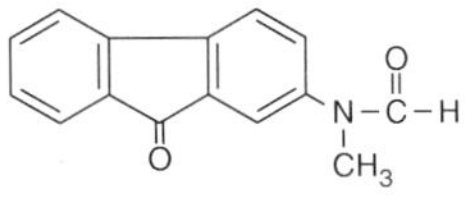
250 mg

S85,478-6
DIMETHYL 5-[(1-NAPHTHYL)-CARBOXAMIDO]ISOPHTHALATE
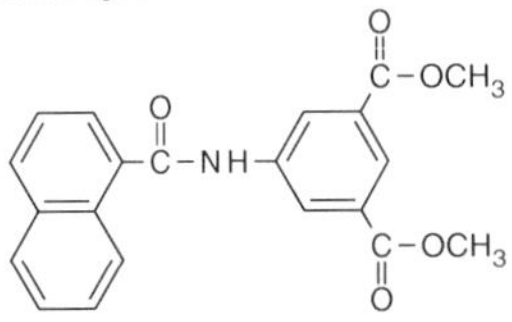
250 mg

S77,530-4
N,N'-DIETHYL-N,N'-DI(2-NAPHTHYL)-PHTHALAMIDE
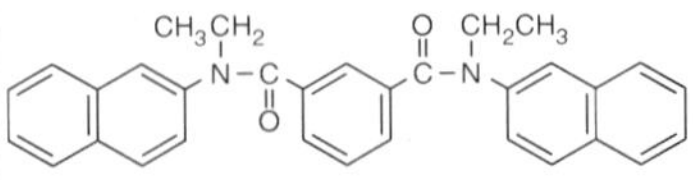
250 mg

S44,707-2
4-PHENYLHIPPURIC ACID
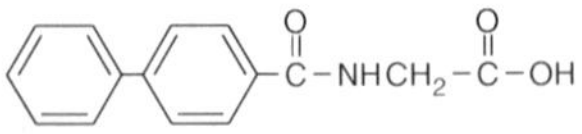
5 gm

S37,540-3
N-(7-BROMO-2-FLUORENYL)-PHTHALAMIC ACID
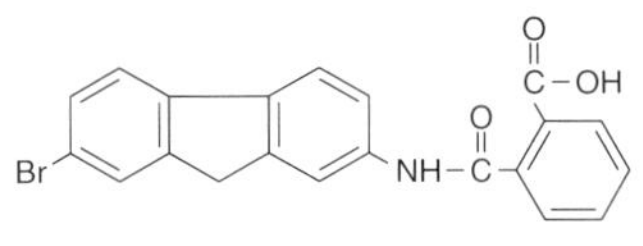
250 mg

S39,387-8
N-(1-NAPHTHYL)MALEAMIC ACID
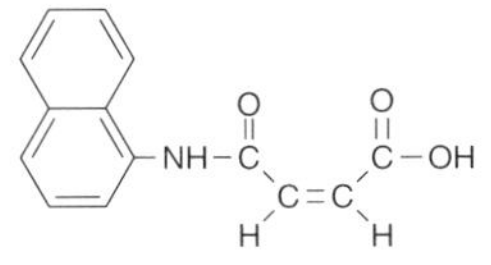
5 gm

S67,598-9
3-(METHYLCARBAMOYLOXY)-2-NAPHTHANILIDE
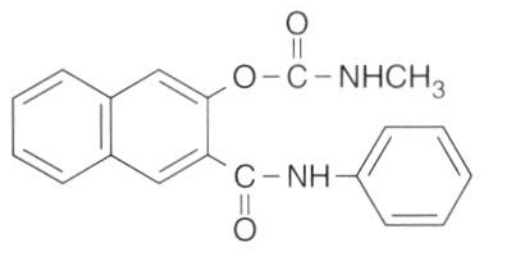
250 mg

S97,473-0
AR-G29 BENZO(C)PHENANTHRENE-1-CARBOXAMIDE
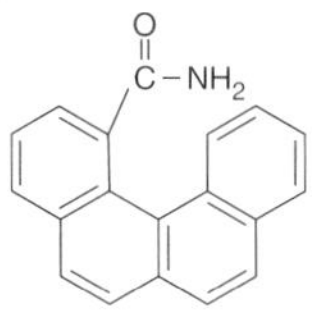
10 mg

S45,126-6
N-BENZOYL-DL-LEUCINE-2-NAPHTHYLAMIDE
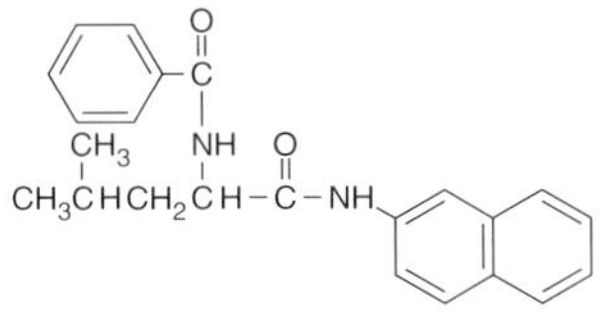
5 gm

S45,127-4
N-BENZOYL-L-LEUCINE-2-NAPHTHYLAMIDE
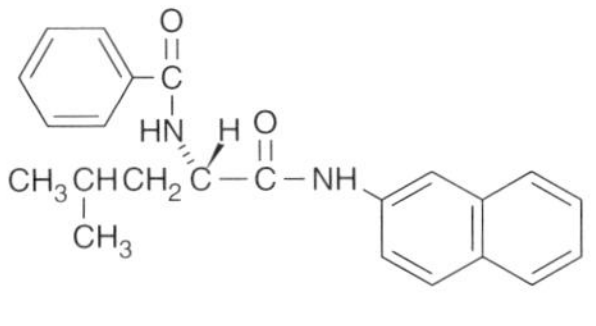
5 gm

S52,006-3
N-(1-NAPHTHYL)BENZAMIDE
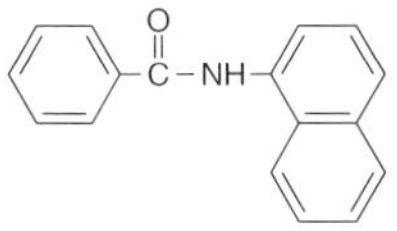
250 mg

S67,581-4
N-(2-NAPHTHYL)CINNAMAMIDE
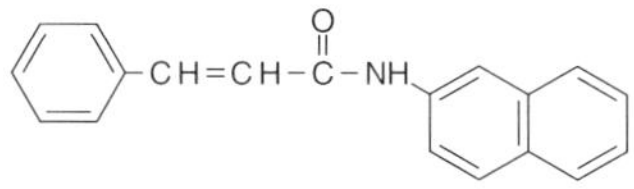
250 mg

S69,707-9
N-(2-BIPHENYLYL)-1-NAPHTHALENECARBOXAMIDE
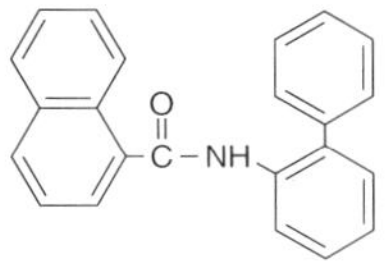
250 mg

S55,514-2
2-CHLORO-N-(1-NAPHTHYL)BENZAMIDE
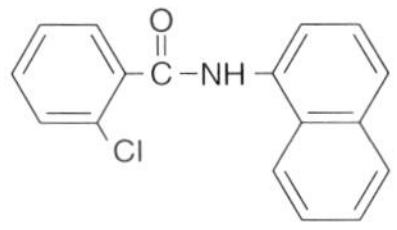
5 gm

S43,200-8
N-(2-FLUORENYL)-PARA-TOLUAMIDE
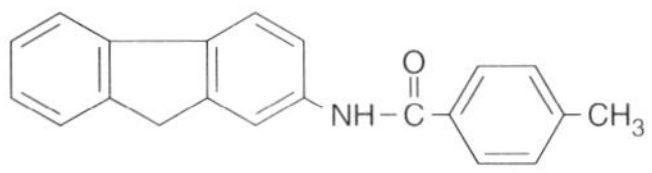
5 gm

S55,509-6
3,4-DICHLORO-N-(1-NAPHTHYL)-BENZAMIDE
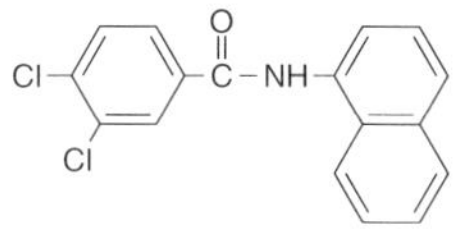
250 mg

S55,516-9
2,4-DICHLORO-N-(1-NAPHTHYL)-BENZAMIDE
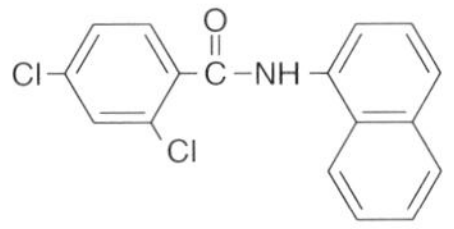
5 gm

S77,482-0
N-ETHYL-N-(1-NAPHTHYL)BENZAMIDE
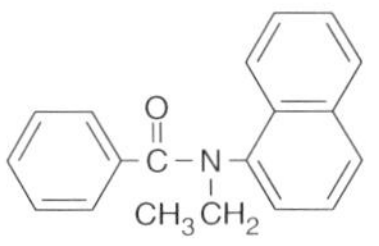
250 mg

S77,481-2
N-METHYL-N-(2-NAPHTHYL)-O-TOLUAMIDE
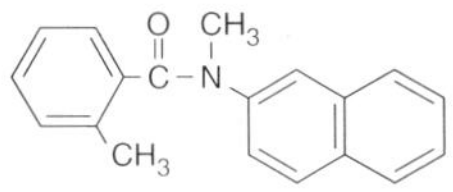
250 mg

S44,824-9
ETHYL 4'-PHENYLOXANILATE
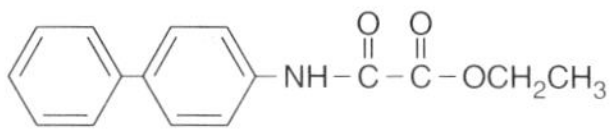
5 gm

S97,520-6
AR-H27 ETHYL 1-PYRENEOXAMATE
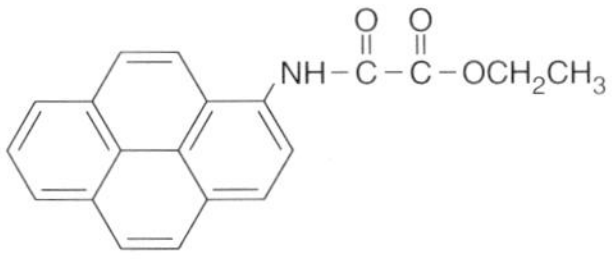
10 mg

S40,952-9
N-(1-NAPHTHYL)ACETOACETAMIDE
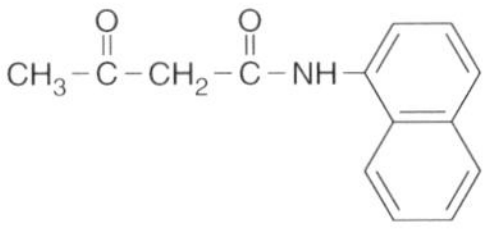
5 gm

S67,650-0
2-[N-(1-NAPHTHYL)CARBAMOYL]-CYCLOHEXANECARBOXYLIC ACID
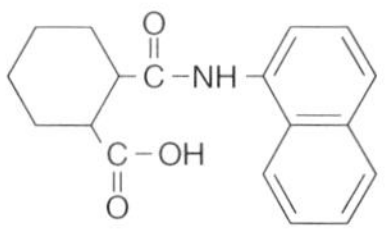
250 mg

S90,006-0
N-(1-NAPHTHYL)-1,2,3,6-TETRAHYDROPHTHALAMIC ACID
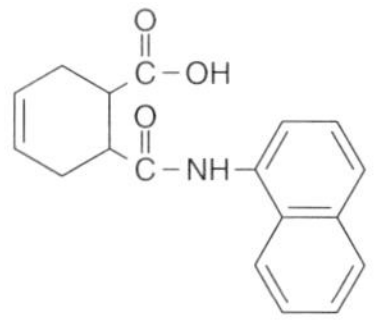
250 mg

S72,924-8
N-(1-NAPHTHYL)GLUTARAMIC ACID
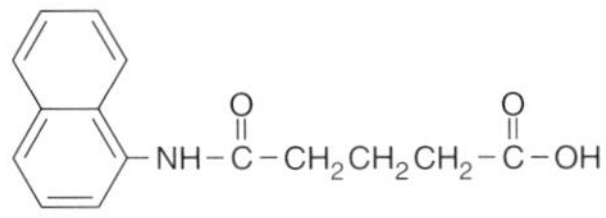
5 gm

S77,695-5
N-ACETYL-L-LEUCINE 1-NAPHTHYL ESTER
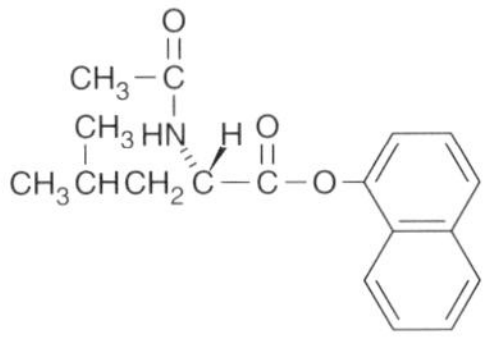
250 mg

S75,401-3
2-ACETAMIDO-9-FLUORENYL ACETATE
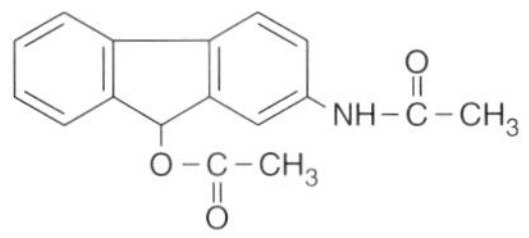
250 mg

S44,196-1
2-DIACETYLAMINOFLUORENE
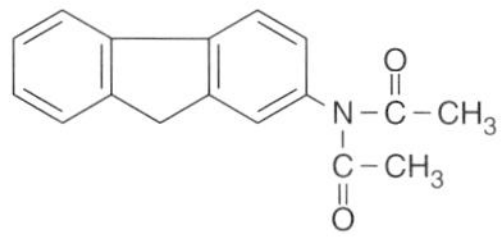
5 gm

S78,459-1
L-PROLINE 2-NAPHTHYLAMIDE HYDROBROMIDE
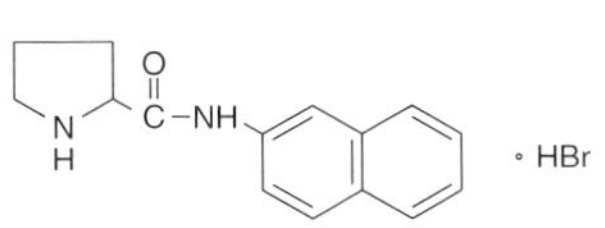
250 mg

S84,102-1
2'-PHENYLFORMANILIDE
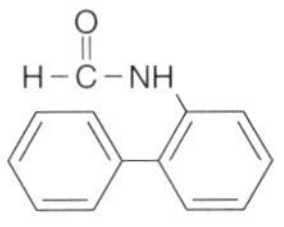
250 mg

S81,126-2
2'-PHENYLACETANILIDE
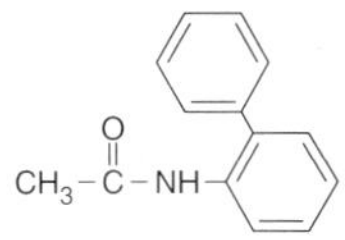
250 mg

S82,138-1
4'-PHENYLHEXADECANANILIDE
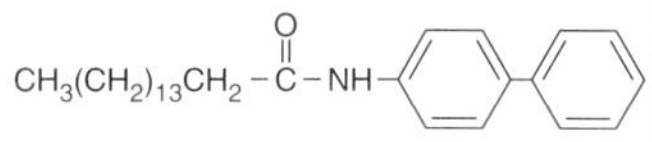
250 mg

S82,140-3
4'-PHENYLOCTADECANANILIDE
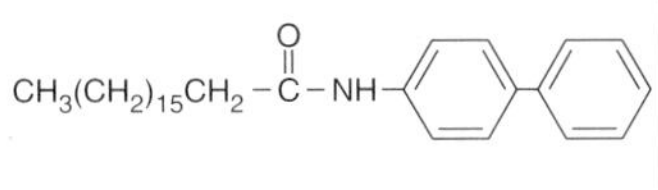
250 mg

S67,655-1
N-(1-NAPHTHYL)-2-PHENYLACETAMIDE
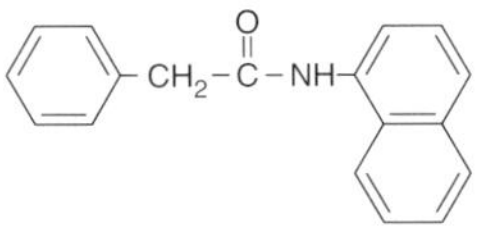
5 gm

S63,880-3
N-(1-NAPHTHYL)-3-PHENYLPROPIONAMIDE
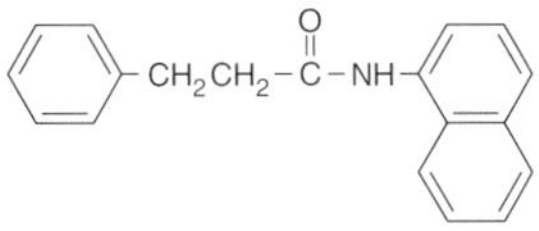
250 mg

S81,354-0
N,N'-(1,1'-BIPHENYL-4,4'-DIYL)-BISOCTADECAMIDE
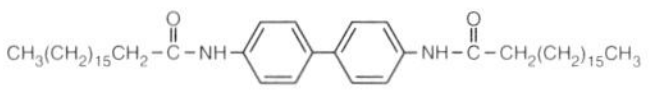
250 mg

S55,546-0
4-CHLORO-N-(2-NAPHTHYL)BENZAMIDE
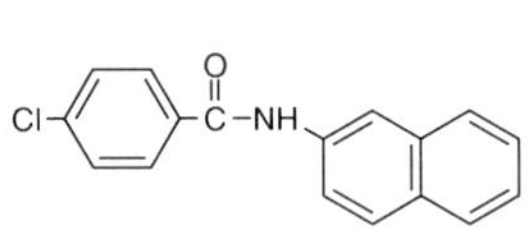
5 gm

S50,408-4
N-(1-NAPHTHYL)FORMAMIDE
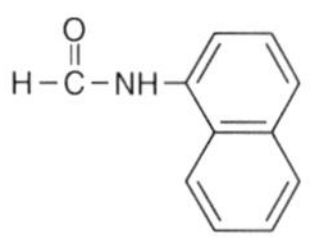
5 gm

S55,561-4
N-(1-NAPHTHYL)PROPIONAMIDE
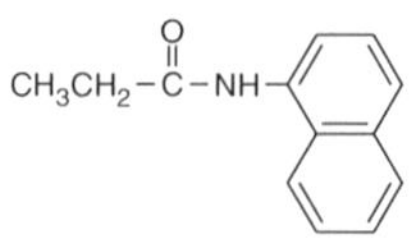
5 gm

S77,463-4
2-METHYL-N-(1-NAPHTHYL)PROPIONAMIDE
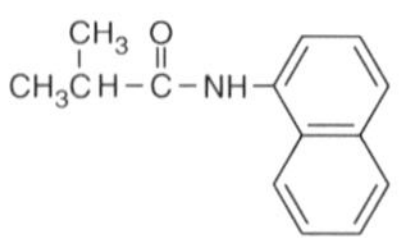
250 mg

S81,442-3
N-(1-NAPHTHYLMETHYL)HEXANOAMIDE
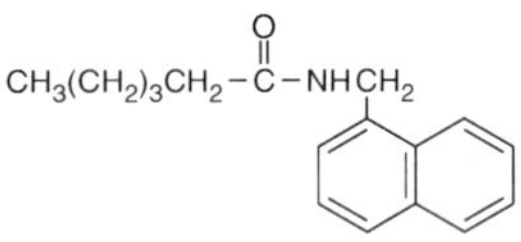
250 mg

S81,441-5
N-(1-NAPHTHYLMETHYL)HEPTANOAMIDE
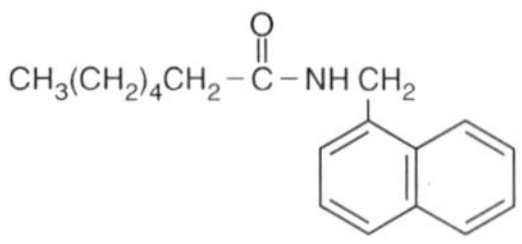
250 mg

S75,046-8
N-(1-NAPHTHYL)DECANAMIDE
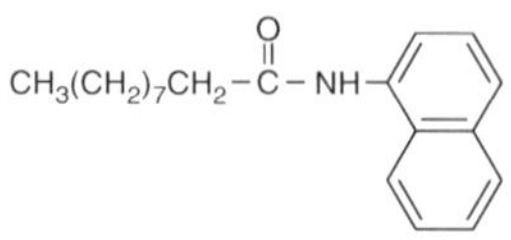
5 gm

S75,394-7
N-(1-NAPHTHYL)HEXADECANAMIDE
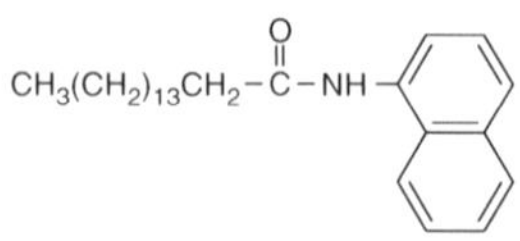
5 gm

S39,134-4
N-(4-BROMO-1-NAPHTHYL)ACETAMIDE
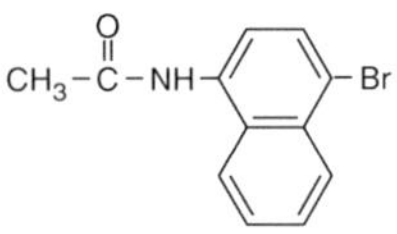
5 gm

S71,592-1
N-(1-NAPHTHYL)SORBAMIDE
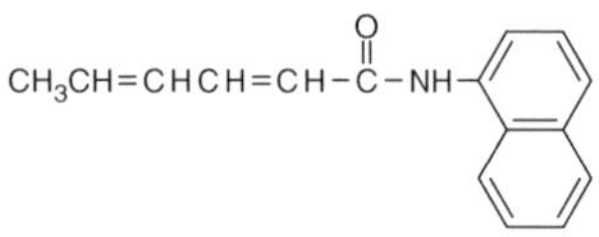
5 gm

S45,844-9
2-ACETAMIDONAPHTHALENE
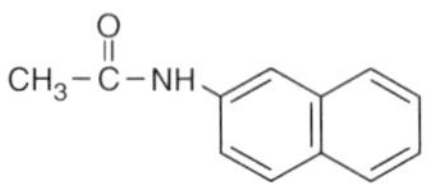
5 gm

S73,901-4
N-(2-NAPHTHYL)BUTYRAMIDE
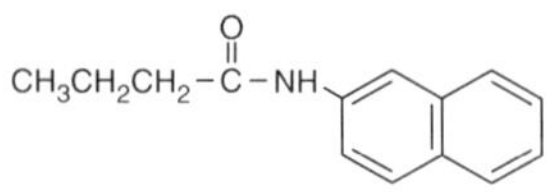
5 gm

S67,615-2
N-(2-NAPHTHYL)DODECANAMIDE
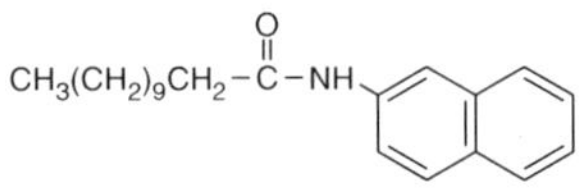
5 gm

S81,352-4
N-(2-NAPHTHYL)OCTADECANAMIDE
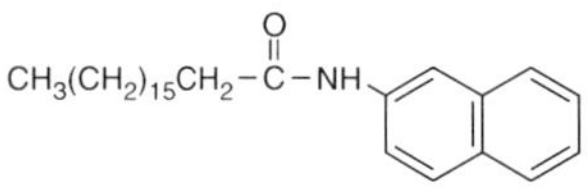
250 mg

S97,488-9
AR-G45 3-ACETAMIDOBENZO(C)-PHENANTHRENE
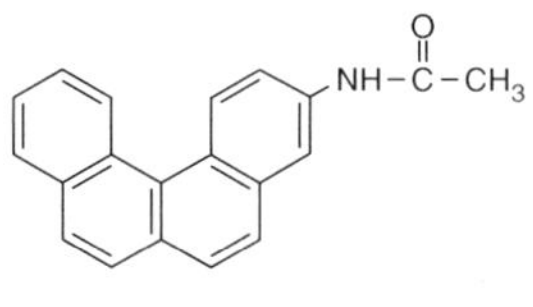
10 mg

S97,489-7
AR-G46 2-ACETAMIDOBENZO(C)-PHENANTHRENE
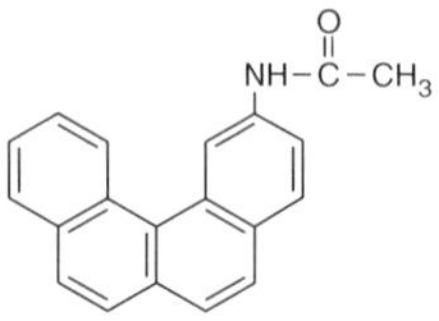
10 mg

S97,490-0
AR-G47 4-ACETAMIDOBENZO(C)-PHENANTHRENE
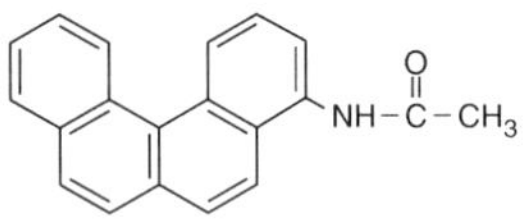
10 mg

S97,491-9
AR-G48 1-ACETAMIDOBENZO(C)-PHENANTHRENE
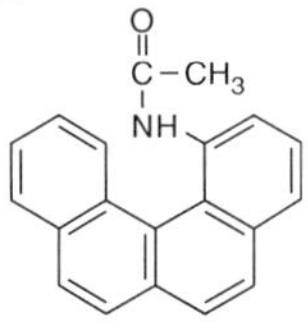
10 mg

S97,492-7
AR-G49 5-ACETAMIDOBENZO(C)-PHENANTHRENE
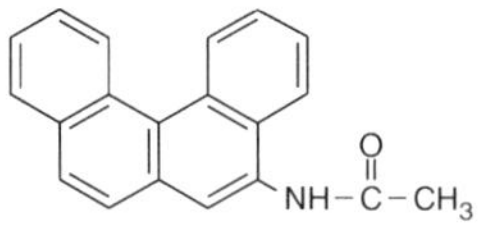
10 mg

S97,493-5
AR-G50 6-ACETAMIDOBENZO(C)-PHENANTHRENE
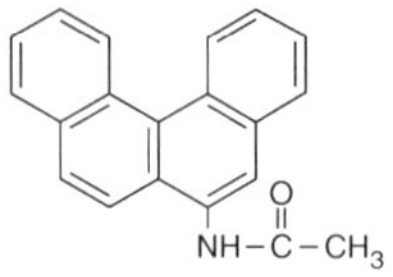
10 mg

S93,672-3
N-(2-NAPHTHYL)-2,2,2-TRIMETHYLACETAMIDE
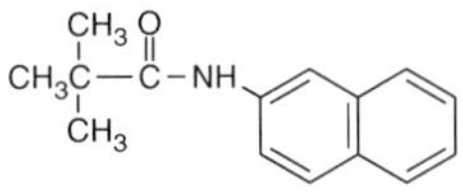
250 mg

S34,992-5
2-FORMAMIDOFLUORENE
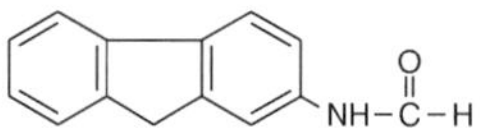
250 mg

S75,490-0
9-BROMO-2-FORMAMIDOFLUORENE
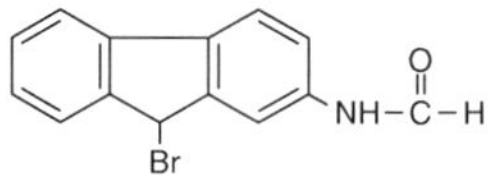
250 mg

S75,433-1
4-ACETAMIDOFLUORENE
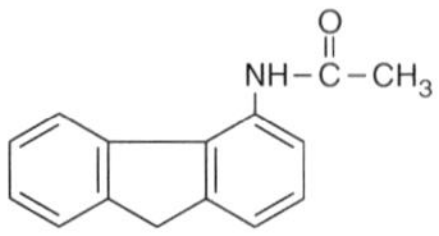
250 mg

S77,100-7
2-ACETAMIDO-5-FLUOROFLUORENE
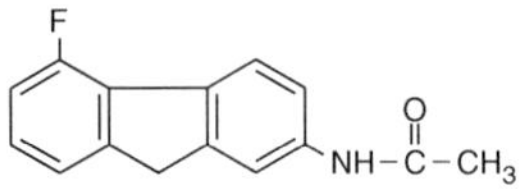
250 mg

S34,622-5
2,7-DIFORMAMIDOFLUORENE
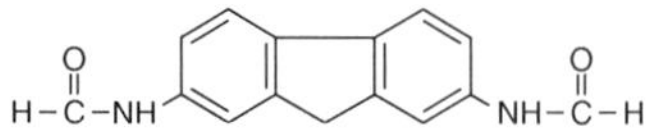
250 mg

S74,581-2
2-ACETAMIDO-6-CHLOROFLUORENE
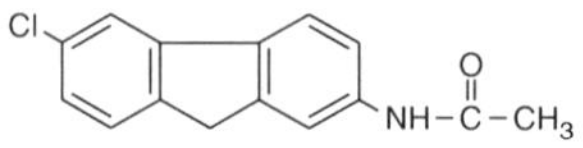
250 mg

S77,066-3
2-ACETAMIDO-4-CHLOROFLUORENE
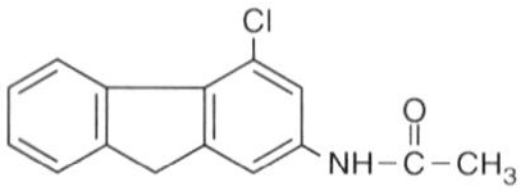
250 mg

S75,514-1
7-ACETAMIDO-1,4-DIFLUOROFLUORENE
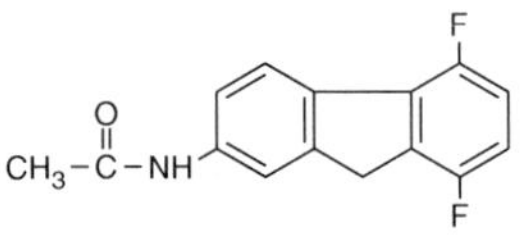
250 mg

S75,480-3
2-ACETAMIDO-1,4,7-TRIFLUOROFLUORENE
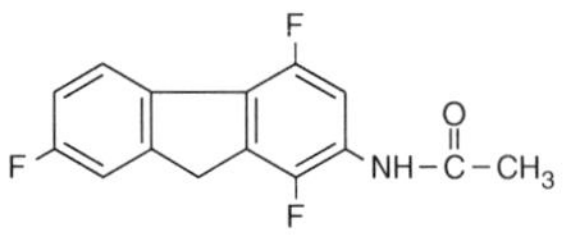
250 mg

S77,067-1
4-ACETAMIDO-2,7-DICHLOROFLUORENE
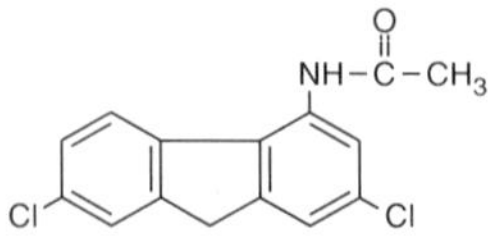
250 mg

S75,528-1
7-ACETAMIDO-1,2,3,4-TETRAFLUOROFLUORENE
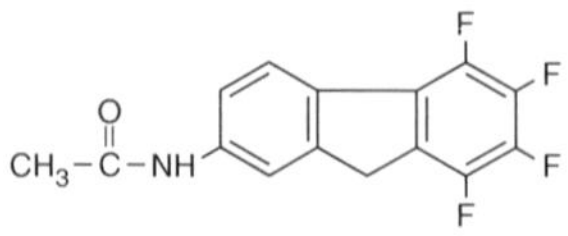
250 mg

S34,014-6
2-ACETAMIDO-9-BROMOFLUORENE
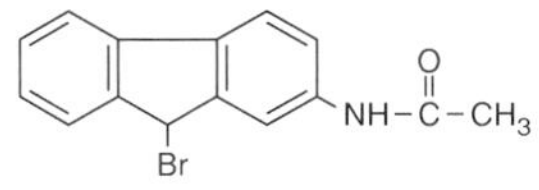
1 gm

S75,499-4
2-ACETAMIDO-3-BROMO-7-FLUOROFLUORENE
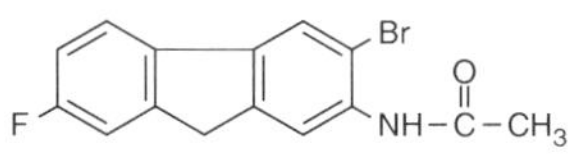
250 mg

S75,485-4
2-ACETAMIDO-7-FLUORO-1,3,4-TRICHLOROFLUORENE
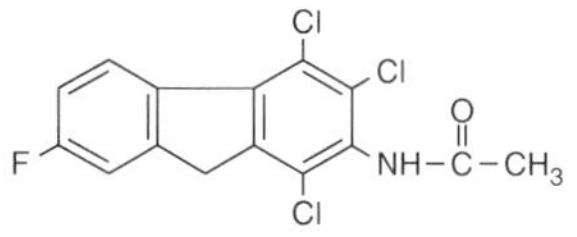
250 mg

S74,983-4
2-ACETAMIDO-7-IODOFLUORENE
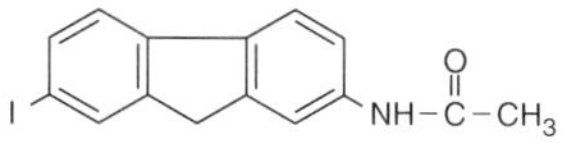
250 mg

S74,924-9
2-ACETAMIDO-1,3,4,7-TETRACHLOROFLUORENE
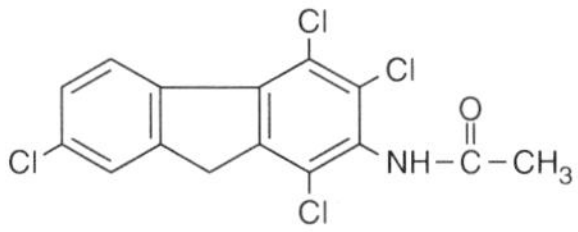
250 mg

S75,163-4
2-ACETAMIDO-1,3-DIBROMOFLUORENE
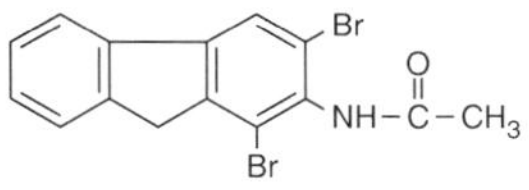
5 gm

S75,538-9
2,7-DIACETAMIDOFLUORENE
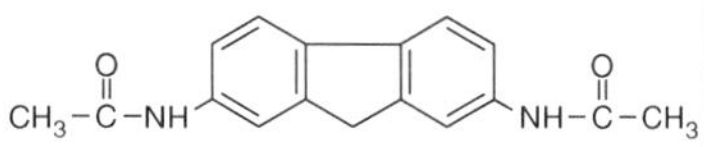
250 mg

S75,559-1
2,5-DIACETAMIDOFLUORENE
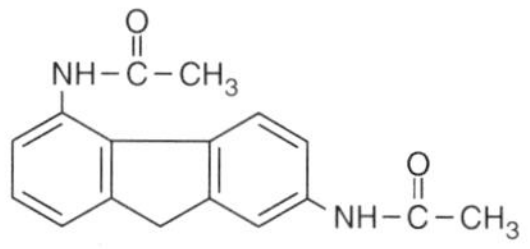
250 mg

S75,814-0
2,7-DIACETAMIDO-1,3-DICHLOROFLUORENE
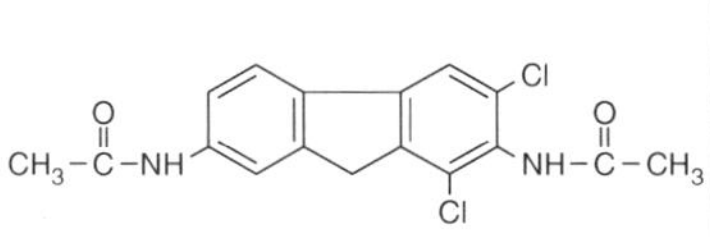
250 mg

S80,565-3
1-ACETAMIDOANTHRACENE
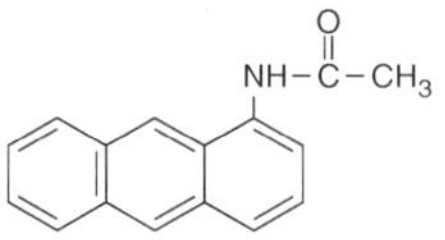
250 mg

S94,295-2
N-(2-ANTHRACENYL)ACETAMIDE
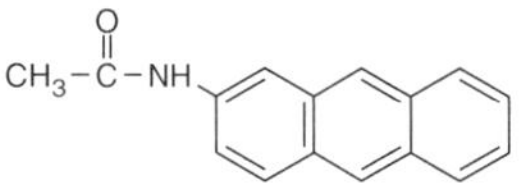
250 mg

S45,622-5
5-ACETAMIDOACENAPHTHENE
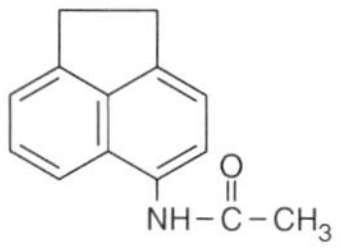
250 mg

S75,504-4
2-(N-METHYLFORMAMIDO)FLUORENE
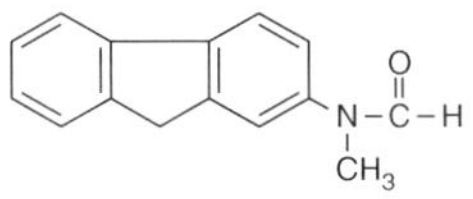
250 mg

S74,985-0
2-(N-METHYLACETAMIDO)FLUORENE
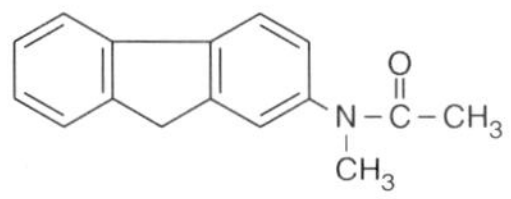
250 mg

S68,696-4
1-(4-CHLOROBENZOYL)-2-(DICHLOROACETYL)HYDRAZINE
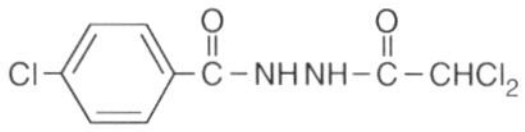
250 mg

S90,480-5
2,4-DICHLOROBENZOIC N2-(2,2,2-TRICHLOROACETYL)HYDRAZIDE
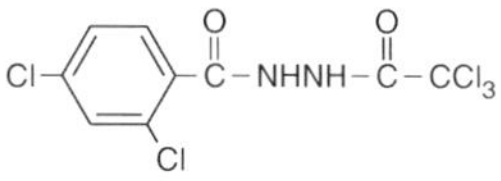
250 mg

S90,479-1
1-(2-PHENYLACETYL)-2-(2,2,2-TRICHLOROACETYL)HYDRAZINE
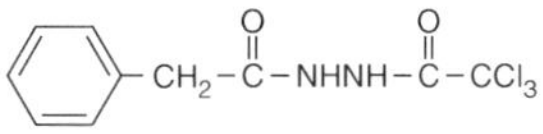
250 mg

S90,403-1
4-FLUOROBENZOIC (4-HYDROXY-3-METHOXYBENZYLIDENE)HYDRAZIDE
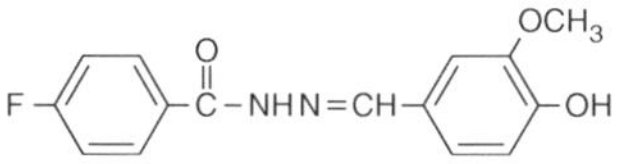
250 mg

S90,335-3
4-CHLOROBENZOIC (4-HYDROXY-3-METHOXYBENZYLIDENE)HYDRAZIDE
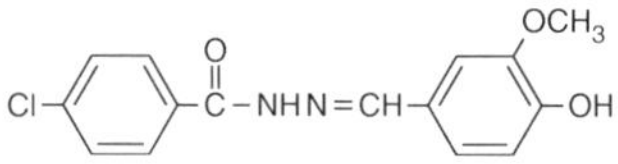
250 mg

S90,140-7
2-METHOXYBENZOIC (5-BROMO-2-HYDROXYBENZYLIDENE)HYDRAZIDE
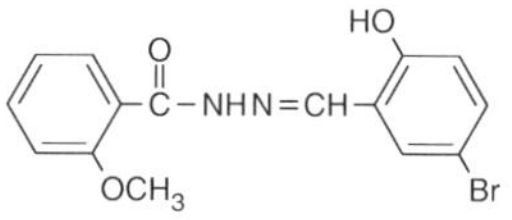
250 mg

S79,695-6
BENZILIC (4-CHLOROBENZYLIDENE)-HYDRAZIDE
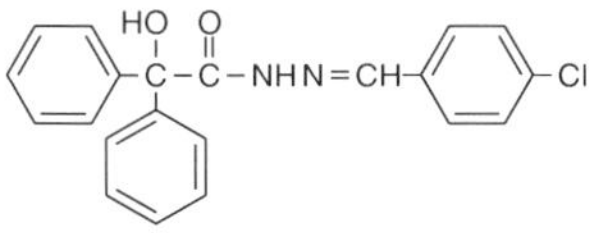
250 mg

S72,317-7
4-(3-CHLORO-2-METHYLPHENYL)-1-(2-HYDROXYBENZOYL)SEMICARBAZIDE
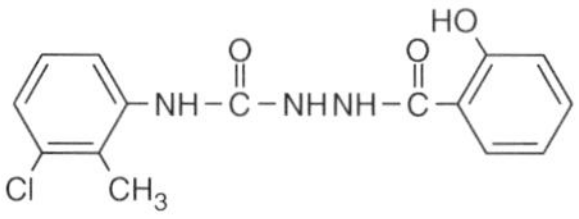
5 gm

S72,316-9
4-(4-CHLORO-2-METHYLPHENYL)-1-(2-HYDROXYBENZOYL)SEMICARBAZIDE
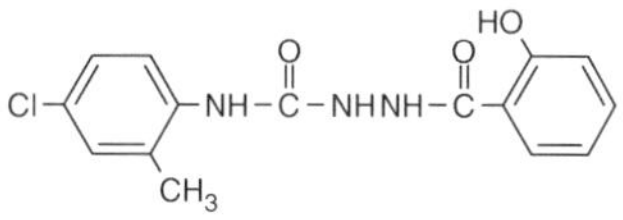
5 gm

S75,092-1
SALICYLIC (2-CHLOROBENZYLIDENE)-HYDRAZIDE
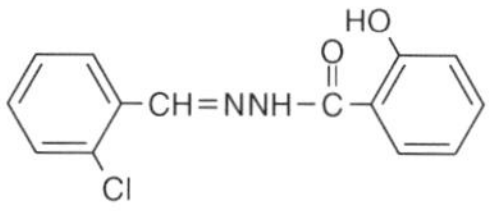
5 gm

S67,469-9
N-(4-CHLORO-3-NITROPHENYL)-N'-(2-HYDROXYBENZYLIDENEAMINO)OXAMIDE
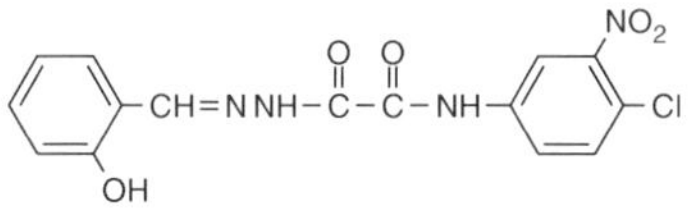
5 gm

S90,338-8
4-CHLOROHIPPURIC (2-HYDROXYBENZYLIDENE)HYDRAZIDE
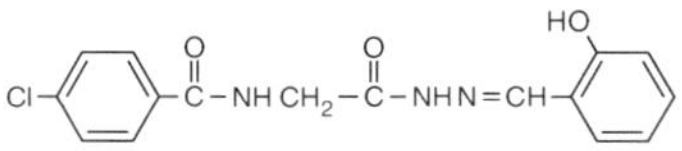
250 mg

S90,386-8
2-HYDROXYISOBUTYRIC (2,4-DICHLOROBENZYLIDENE)HYDRAZIDE
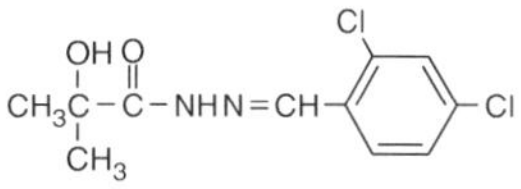
250 mg

S90,349-3
2-CHLOROHIPPURIC (2-HYDROXYBENZYLIDENE)HYDRAZIDE
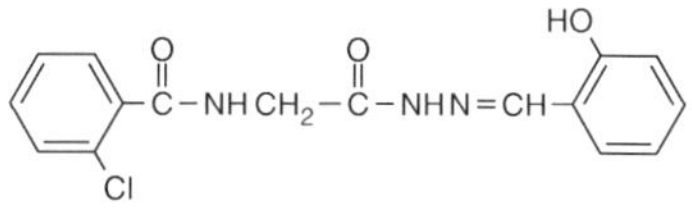
250 mg

S90,344-2
3,4-DICHLOROHIPPURIC (2-HYDROXYBENZYLIDENE)HYDRAZIDE
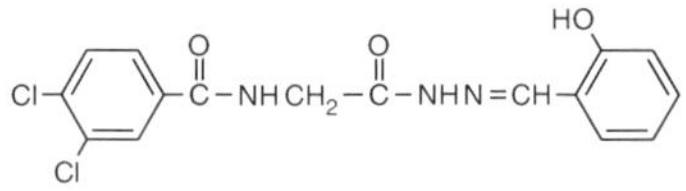
250 mg

S90,345-0
3,4-DICHLOROHIPPURIC (4-HYDROXYBENZYLIDENE)HYDRAZIDE
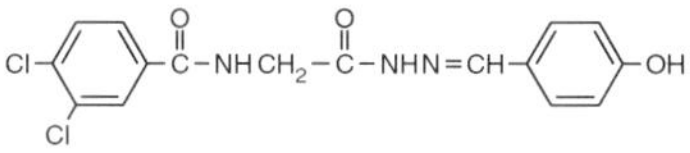
250 mg

S86,883-3
N-(2-CHLOROBENZOYL)ANTHRANILIC (2-HYDROXYBENZYLIDENE)HYDRAZIDE
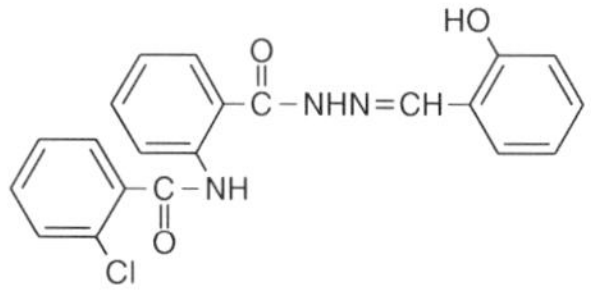
250 mg

S75,090-5
4-CHLOROBENZOIC (2-HYDROXYBENZYLIDENE)HYDRAZIDE
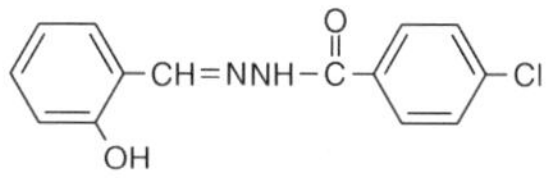
5 gm

S72,561-7
3,4-DICHLOROBENZOIC ACID N2-(3-HYDROXYPHENYL)HYDRAZIDE

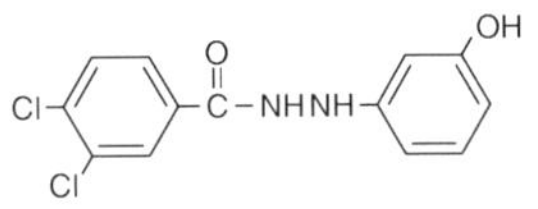

5 gm

S68,843-6
2-(3,4-DICHLOROPHENYLUREIDO)ACETIC ACID (2,3-DIMETHOXYBENZYLIDENE)-HYDRAZIDE

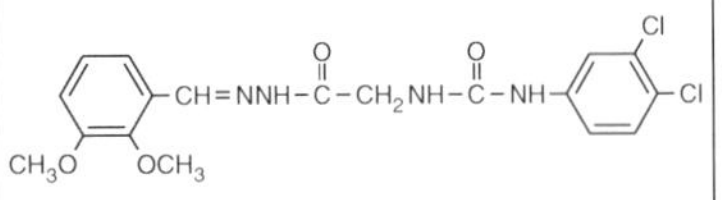

250 mg

S67,370-6
N-(5-CHLORO-2,4-DIMETHOXYBENZAMIDO)-N'-(3,4-DICHLOROPHENYL)OXAMIDE

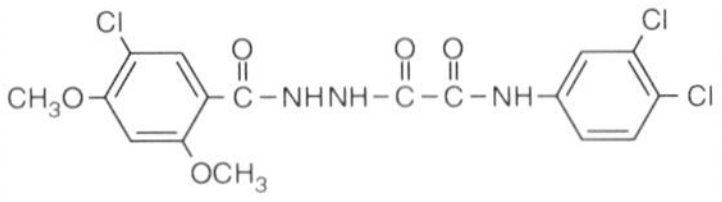

250 mg

S74,359-3
4-CHLOROBENZOIC (2,5-DIMETHOXYBENZYLIDENE)HYDRAZIDE

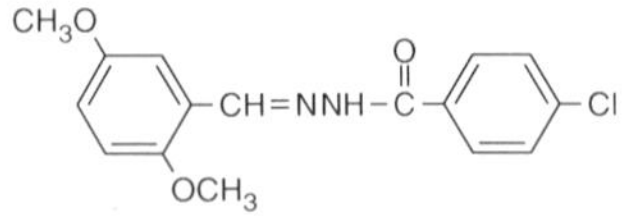

5 gm

S68,999-8
1-(4-CHLOROBENZOYL)-2-(3,4,5-TRIMETHOXYBENZOYL)HYDRAZINE

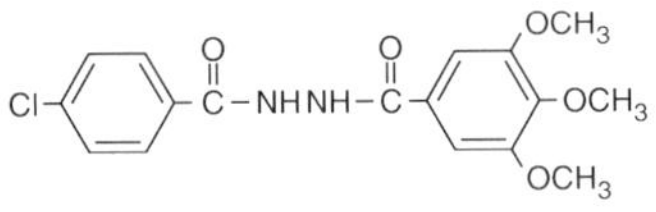

250 mg

S74,757-2
2-METHOXYBENZOIC [2-CHLORO-4-(DIETHYLAMINO)BENZYLIDENE]HYDRAZIDE

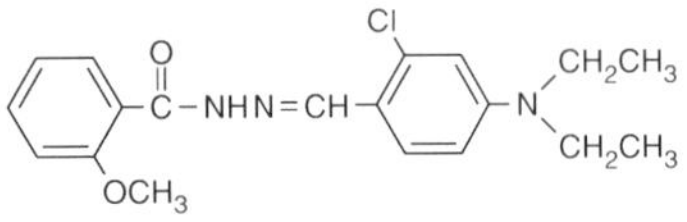

250 mg

S74,748-3
1-(3,4-DICHLOROPHENYL)-4-(2-METHOXYBENZYLIDENEAMINO)OXAMIDE

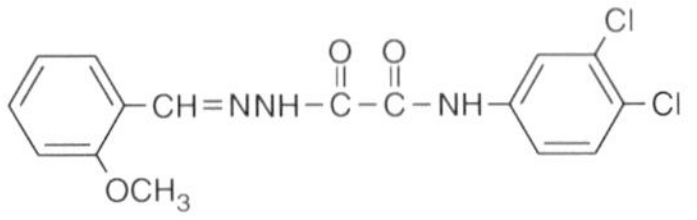

5 gm

S72,956-6
1-(2,4-DICHLOROBENZOYL)-4-(2-METHOXY-5-METHYLPHENYL)-SEMICARBAZIDE

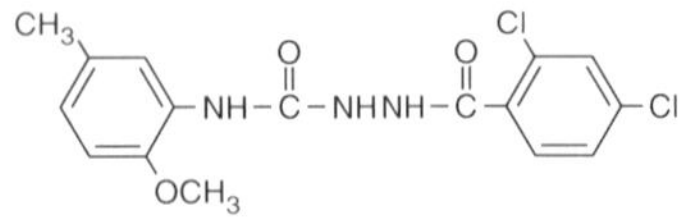

5 gm

S86,892-2
N-(2-CHLOROBENZOYL)ANTHRANILIC (4-METHOXYBENZYLIDENE)HYDRAZIDE

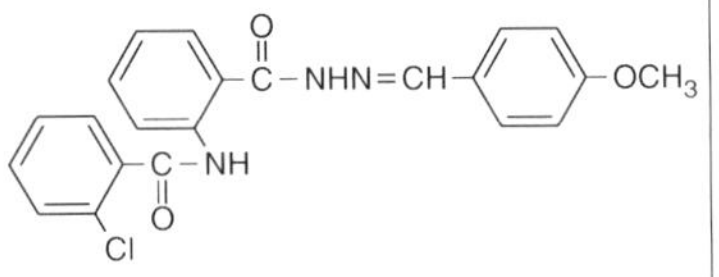

250 mg

S90,334-5
2-CHLOROBENZOIC (4-METHOXYBENZYLIDENE)HYDRAZIDE

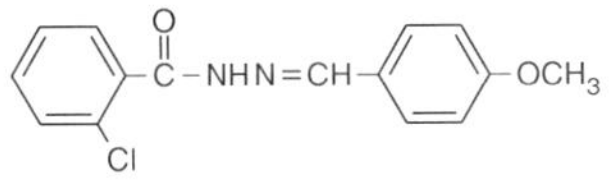

250 mg

S69,576-9
1-(2,4-DICHLOROPHENOXYACETYL)-2-(4-METHYLBENZOYL)HYDRAZINE

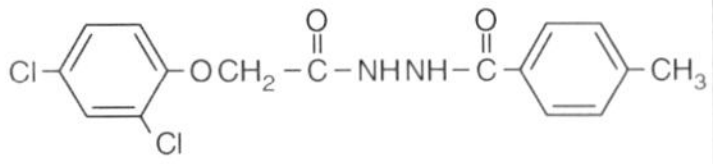

250 mg

S67,472-9
N-(2-AMINOBENZYLIDENEAMINO)-N'-(4-CHLORO-3-NITROPHENYL)OXAMIDE

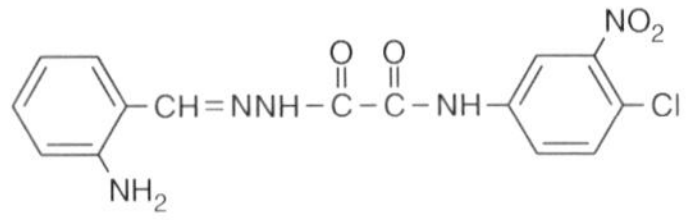

250 mg

S74,965-6
DIMETHYL 2-(2-AMINO-5-CHLOROBENZOYLHYDRAZONO)SUCCINATE

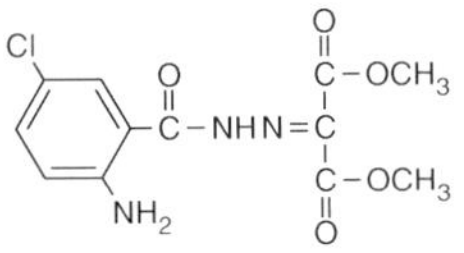

250 mg

S74,358-5
4-AMINOBENZOIC (2-CHLOROBENZYLIDENE)HYDRAZIDE

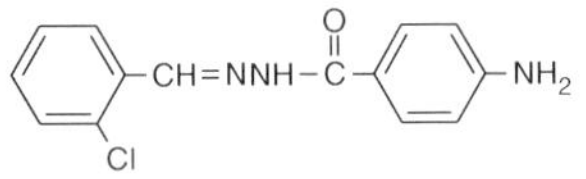

5 gm

S77,150-3
ANTHRANILIC (2,6-DICHLOROBENZYLIDENE)HYDRAZIDE

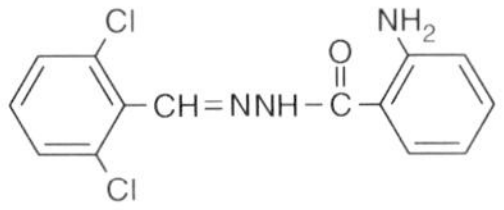

250 mg

S67,361-7
N-(3-CHLOROPHENYL)-N'-(2,4-DICHLOROBENZYLIDENEAMINO)OXAMIDE

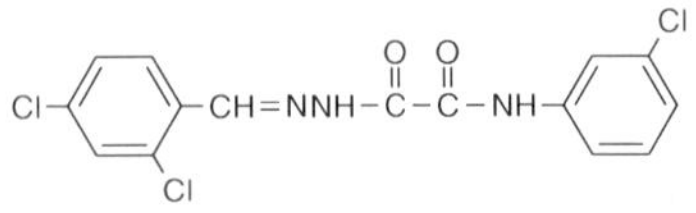

5 gm

S67,383-8
N-(3-CHLOROPHENYL)-N'-(PIPERONYLIDENEAMINO)OXAMIDE

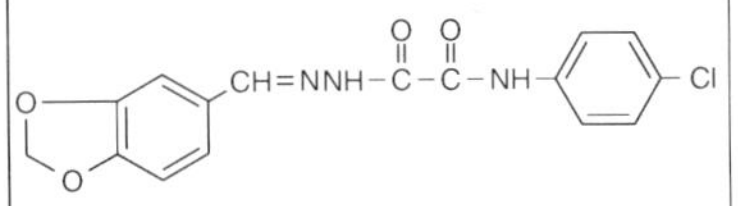

250 mg

S90,339-6
4-CHLOROHIPPURIC (3,4-DICHLOROBENZYLIDENE)HYDRAZIDE

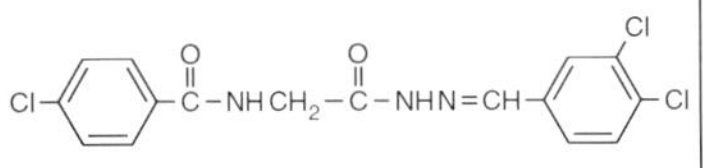

250 mg

S73,104-8
4-CHLOROBENZOIC N2-(4-NITROPHENYL)-HYDRAZIDE

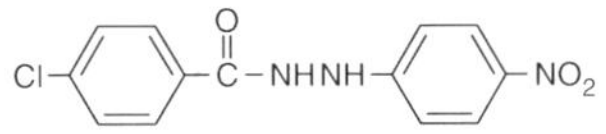

5 gm

S79,615-8
3,4-DICHLOROBENZOIC N2-(4-NITROPHENYL)HYDRAZIDE

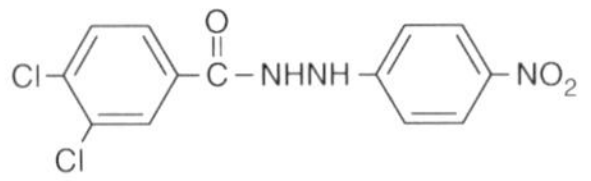

250 mg

S90,460-0
4-CHLOROBENZOIC N2-(2-PHENYLACETYL)-HYDRAZIDE

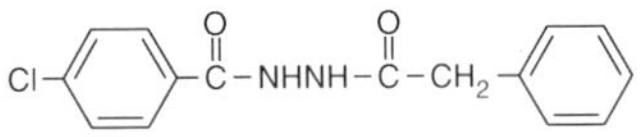

250 mg

S90,469-4
3,4-DICHLOROBENZOIC N2-(2-PHENYLACETYL)HYDRAZIDE

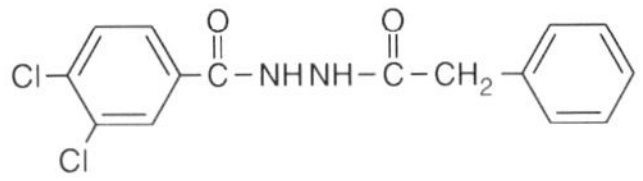

250 mg

S73,688-0
N-METHYLANTHRANILIC (2-CHLOROBENZYLIDENE)HYDRAZIDE

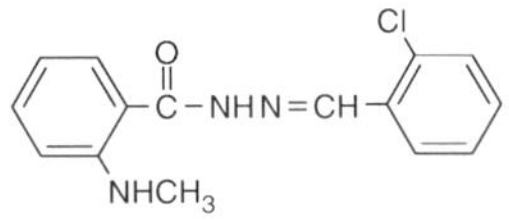

5 gm

S90,350-7
2-CHLOROHIPPURIC [4-(DIMETHYLAMINO)-BENZYLIDENE]HYDRAZIDE

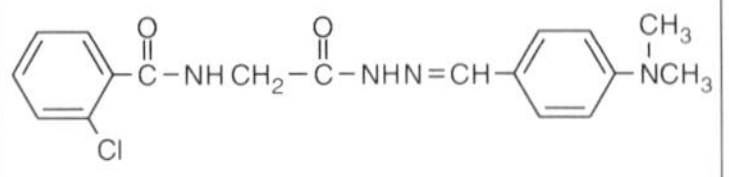

250 mg

S85,596-0
2-(2-CHLOROBENZAMIDO)BENZOIC [2-CHLORO-4-(DIMETHYLAMINO)-BENZYLIDENE]HYDRAZIDE

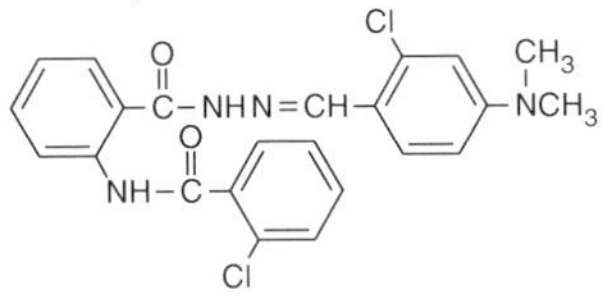

250 mg

S90,330-2
4-(DIMETHYLAMINO)BENZOIC (2-CHLOROBENZYLIDENE)HYDRAZIDE

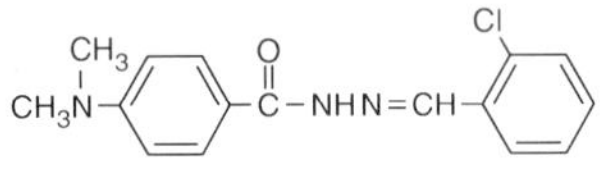

250 mg

S90,336-1
4-CHLOROBENZOIC [4-(DIMETHYLAMINO)-BENZYLIDENE]HYDRAZIDE

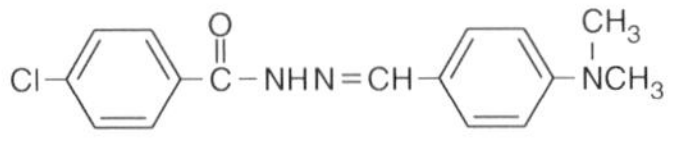

250 mg

S66,843-5
N1-(4-CHLOROBENZOYL)-N1-PHENYLHYDRAZINE HYDROCHLORIDE

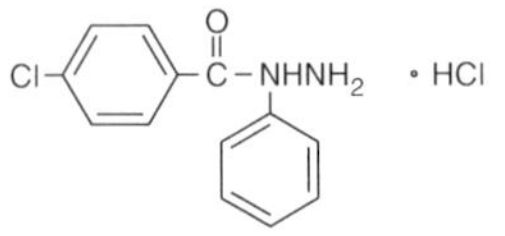

5 gm

S72,547-1
STEARIC ACID (3,4-DICHLOROBENZYLIDENE)HYDRAZIDE

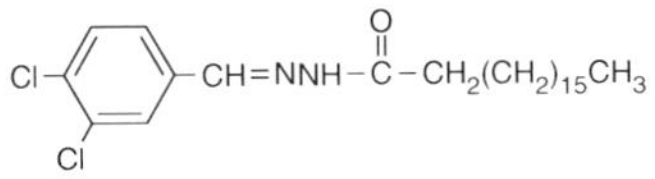

5 gm

S75,013-1
METHYL N-(2,4-DICHLOROBENZYLIDENEAMINO)CARBAMATE

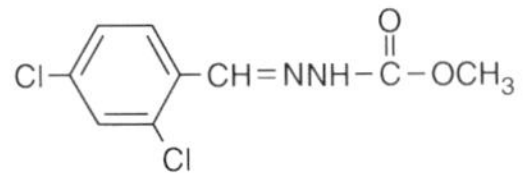

5 gm

S84,917-0
4-CHLOROBENZOIC N2-(4-CHLOROBENZYL)-HYDRAZIDE

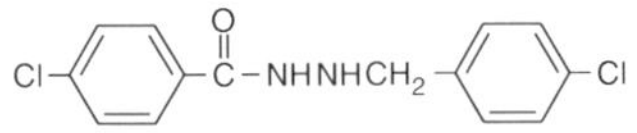

250 mg

S73,810-7
STEARIC (2,4-DICHLOROBENZYLIDENE)-HYDRAZIDE

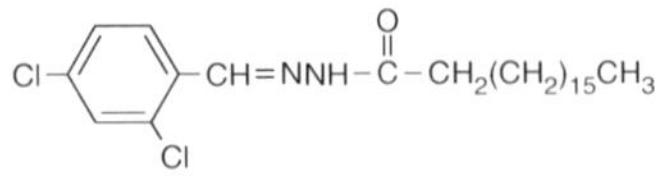

5 gm

S69,484-3
1-(4-CHLOROBENZOYL)-2-(2-FLUOROBENZOYL)HYDRAZINE
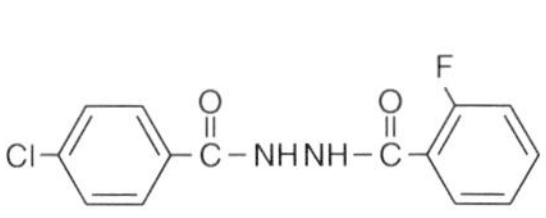
250 mg

S69,587-4
1-(2-CHLOROBENZOYL)-2-(1-NAPHTHYLACETYL)HYDRAZINE
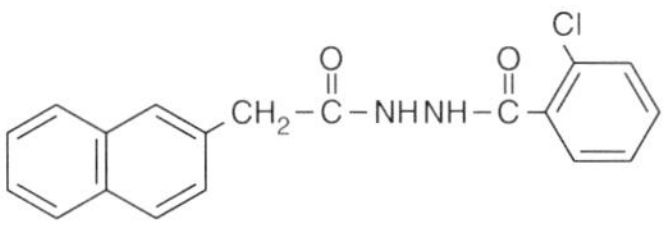
250 mg

S72,996-5
1-(4-CHLOROBENZOYL)-2-(3,4-DICHLOROBENZOYL)HYDRAZIDE
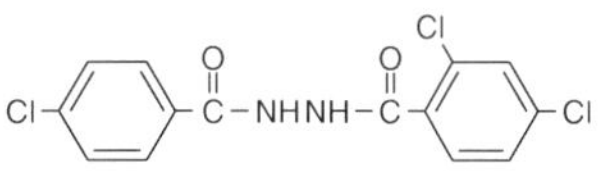
250 mg

S79,608-5
1-BUTYRYL-2-(2,4-DICHLOROBENZOYL)-HYDRAZINE
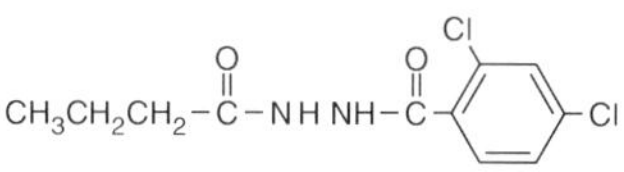
250 mg

S72,934-5
1-(2,4-DICHLOROBENZOYL)-2-(3,4-DICHLOROBENZOYL)HYDRAZINE
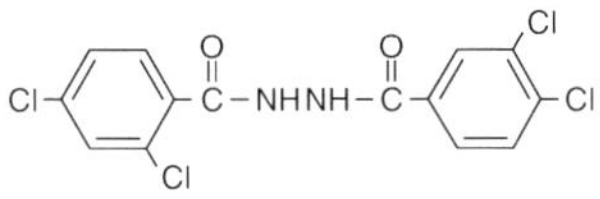
5 gm

S90,563-1
2,2'-OXYBIS[ACETIC (3,4-DICHLOROBENZYLIDENE)HYDRAZIDE]
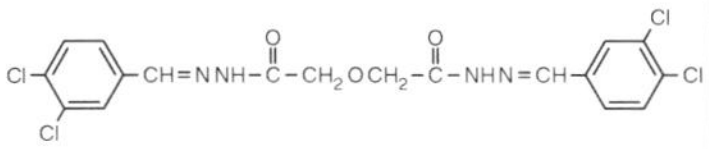
250 mg

S90,346-9
3,4-DICHLOROHIPPURIC (3-NITROBENZYLIDENE)HYDRAZIDE
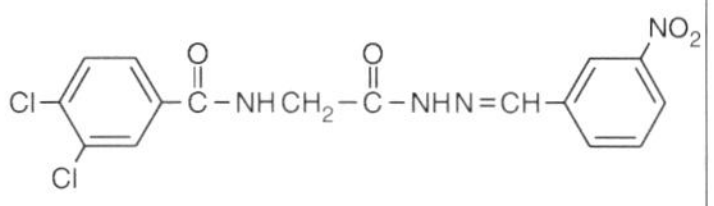
250 mg

S90,389-2
HIPPURIC (4-CHLOROBENZYLIDENE)-HYDRAZIDE
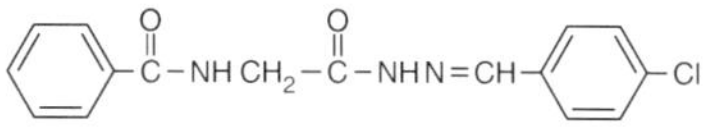
250 mg

S90,395-7
HIPPURIC (2,4-DICHLOROBENZYLIDENE)-HYDRAZIDE
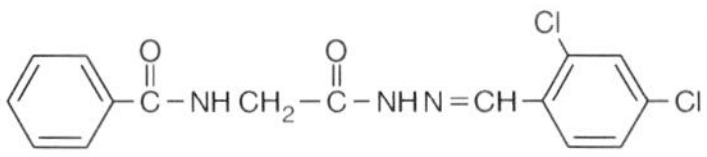
250 mg

S90,342-6
2,4-DICHLOROHIPPURIC (3-PHENYLALLYLIDENE)HYDRAZIDE
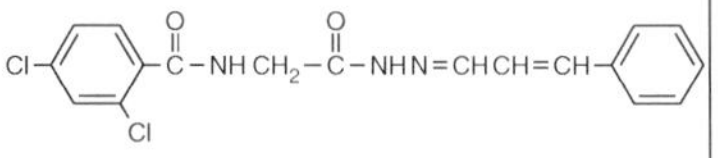
250 mg

S90,351-5
2-CHLOROHIPPURIC (4-FLUOROBENZYLIDENE)HYDRAZIDE
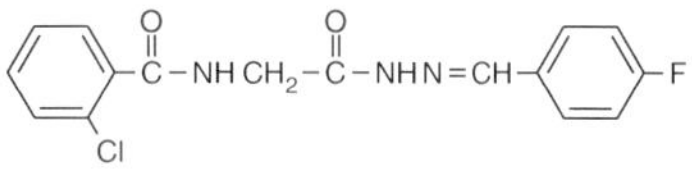
250 mg

S90,347-7
3,4-DICHLOROHIPPURIC (2-FLUOROBENZYLIDENE)HYDRAZIDE
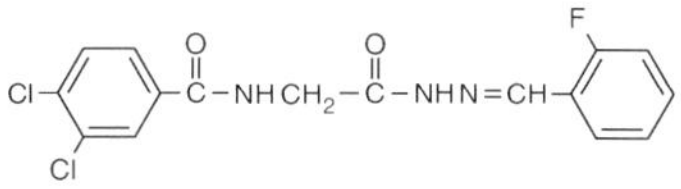
250 mg

S90,100-8
N-(4-CHLOROBENZOYL)GLYCINE (3-FLUOROBENZYLIDENE)HYDRAZIDE
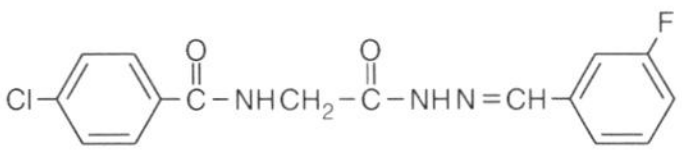
250 mg

S90,348-5
3,4-DICHLOROHIPPURIC (4-FLUOROBENZYLIDENE)HYDRAZIDE
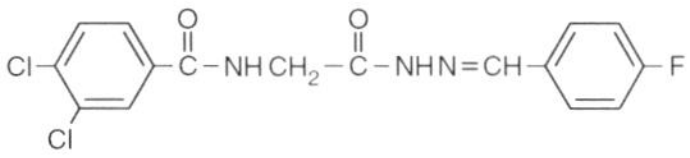
250 mg

S90,341-8
2,4-DICHLOROHIPPURIC (3-FLUOROBENZYLIDENE)HYDRAZIDE
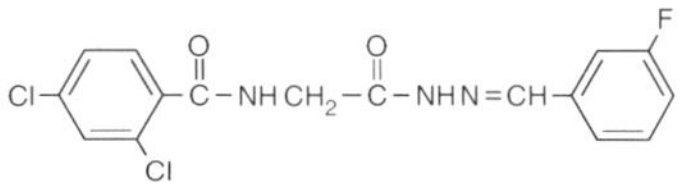
250 mg

S90,343-4
3,4-DICHLOROHIPPURIC (3-FLUOROBENZYLIDENE)HYDRAZIDE
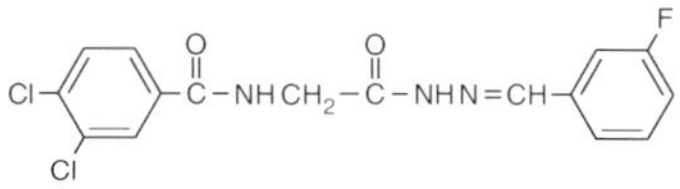
250 mg

S85,594-4
2-(CYCLOPROPYLCARBOXAMIDO)BENZOIC (2,4-DICHLOROBENZYLIDENE)HYDRAZIDE
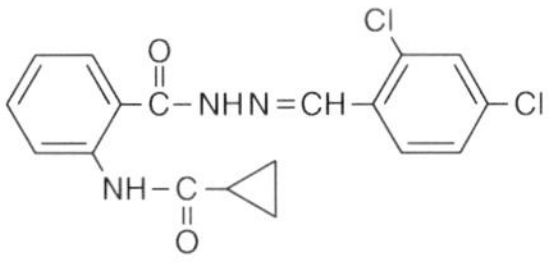
250 mg

S86,901-5
N-(2-CHLOROBENZOYL)ANTHRANILIC (3-PHENYLPROPYLIDENE)HYDRAZIDE
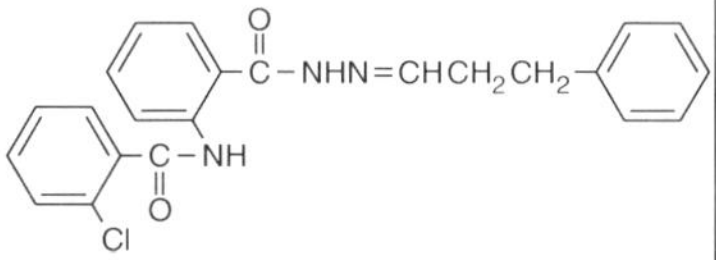
250 mg

S77,997-0
3,4-DICHLOROHIPPURIC (3-PHENYL-2-PROPENYLIDENE)HYDRAZIDE
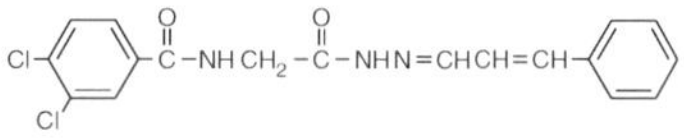
250 mg

S77,996-2
2,4-DICHLOROHIPPURIC (4-FLUOROBENZYLIDENE)HYDRAZIDE
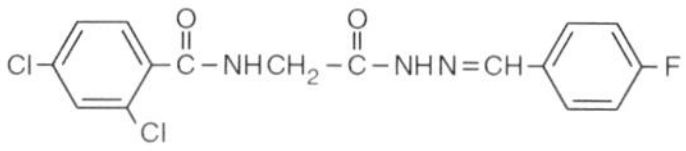
250 mg

S75,068-9
BENZOIC (2-CHLOROBENZYLIDENE)-HYDRAZIDE
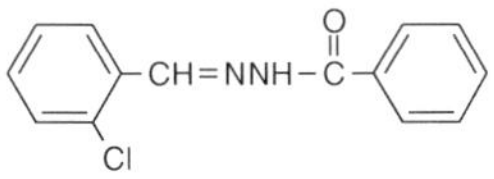
5 gm

S75,082-4
CINNAMIC (2-CHLOROBENZYLIDENE)-HYDRAZIDE
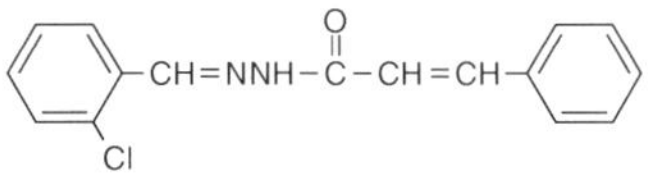
5 gm

S79,625-5
2-CHLOROBENZOIC (3,4-DICHLOROBENZYLIDENE)HYDRAZIDE
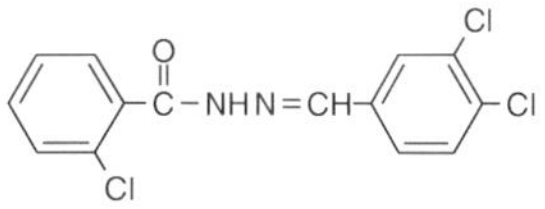
250 mg

S75,067-0
P-TOLUIC (4-CHLOROBENZYLIDENE)-HYDRAZIDE
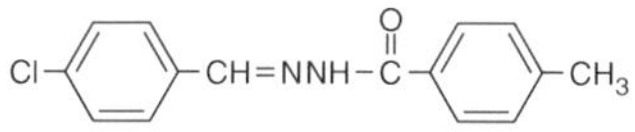
5 gm

S79,579-8
4-METHYLBENZOIC (2,4-DICHLOROBENZYLIDENE)HYDRAZIDE
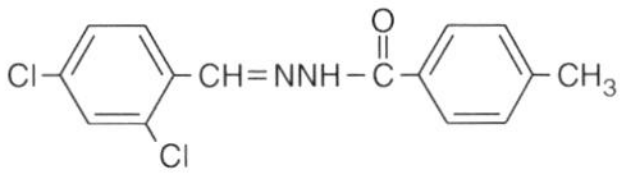
250 mg

S71,623-5
4-PYRIDINECARBOXYLIC ACID N-(4-CHLOROBENZYLIDENE)HYDRAZIDE
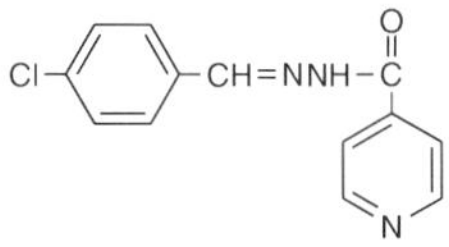
5 gm

S90,286-1
2,4-DICHLOROBENZOIC (3-FLUOROBENZYLIDENE)HYDRAZIDE
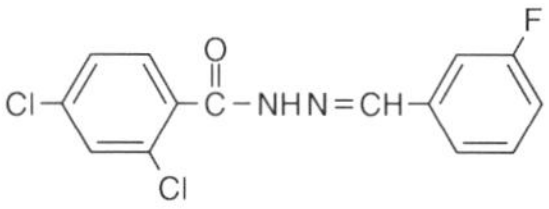
250 mg

S90,289-6
3,4-DICHLOROBENZOIC (3,4-DICHLOROBENZYLIDENE)HYDRAZIDE
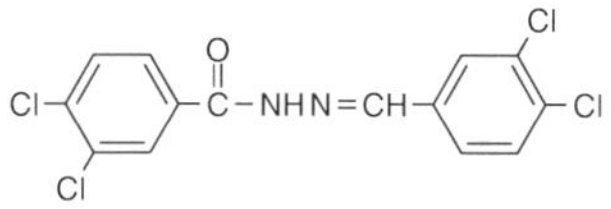
250 mg

S73,723-2
OXALIC BIS[(2,4-DICHLOROBENZYLIDENE)HYDRAZIDE]
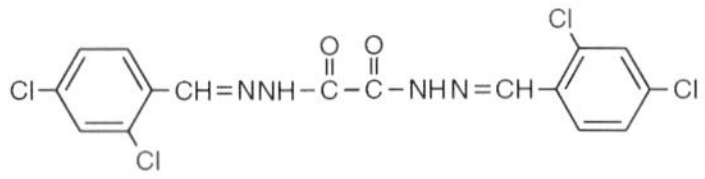
5 gm

S89,992-5
ADIPIC BIS(2-CHLOROBENZYLIDENE)-HYDRAZIDE
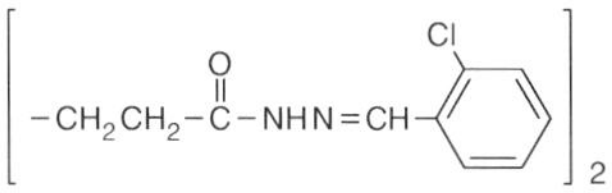
250 mg

S86,911-2
SEBACIC BIS[(4-CHLOROBENZYLIDENE)-HYDRAZIDE]
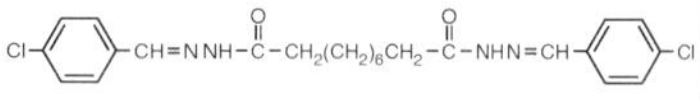
250 mg

S86,910-4
SEBACIC BIS[(3,4-DICHLOROBENZYLIDENE)HYDRAZIDE]
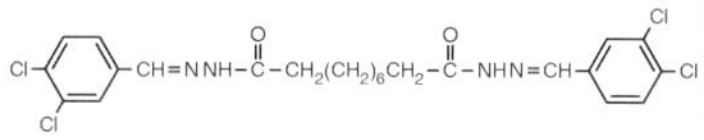
250 mg

S72,522-6
2-(4-FLUOROPHENYLUREIDO)ACETIC ACID (2-CHLOROBENZYLIDENE)HYDRAZIDE
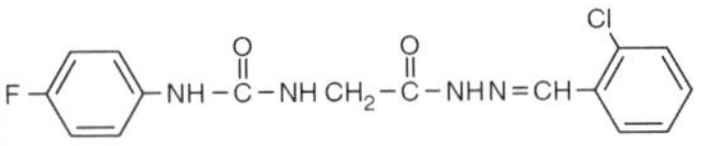
250 mg

S68,867-3
2-(3,4-DICHLOROPHENYLUREIDO)ACETIC ACID PIPERONYLIDENE HYDRAZIDE
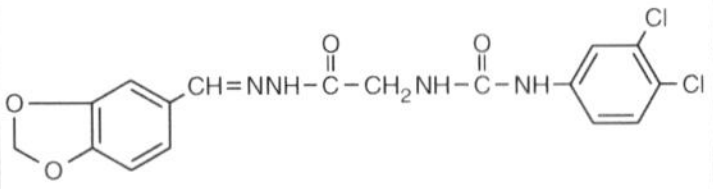
250 mg

S74,698-3
10-UNDECENOIC [ALPHA-(3-CHLOROPHENYL)BENZYLIDENE]HYDRAZIDE
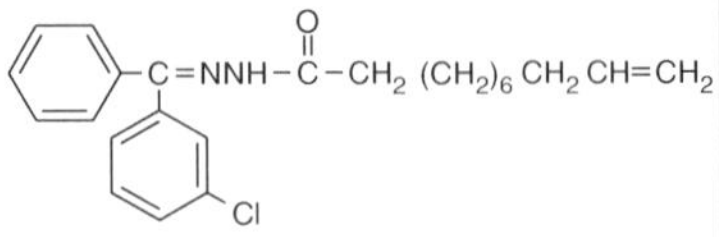
5 gm

S74,759-9
LAURIC (4-CHLORO-ALPHA-PHENYLBENZYLIDENE)HYDRAZIDE
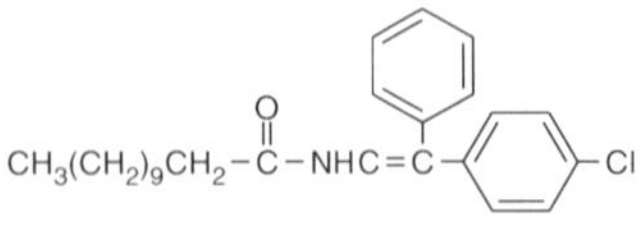
250 mg

S73,863-8
STEARIC (2-CHLOROBENZYLIDENE)-HYDRAZIDE
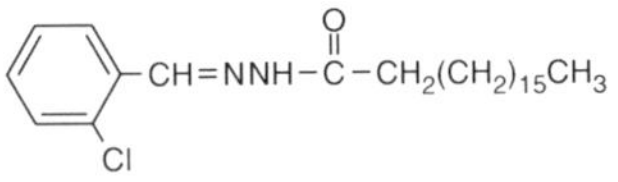
5 gm

S79,413-9
2-(1-NAPHTHYL)ACETIC 2-CHLOROBENZYLIDENEHYDRAZIDE
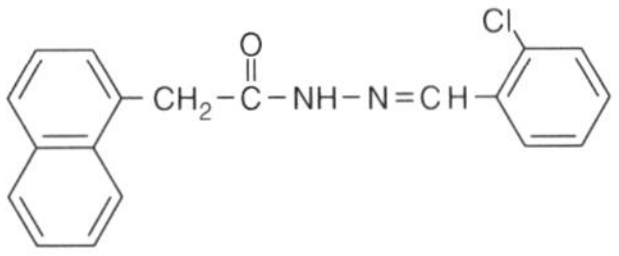
250 mg

S73,101-3
2-(1-NAPHTHYL)ACETIC 4-CHLOROBENZYLIDENEHYDRAZIDE
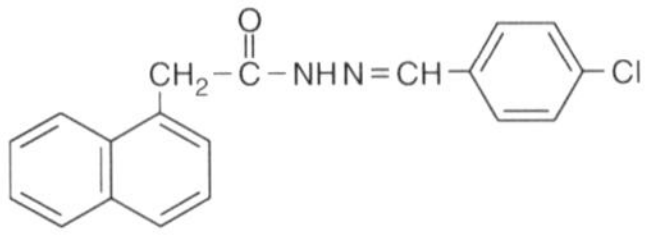
5 gm

S77,194-5
BUTYRIC (2,6-DICHLOROBENZYLIDENE)-HYDRAZIDE
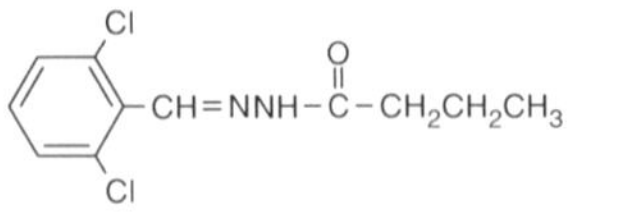
250 mg

S74,361-5
10-UNDECENOIC (2,6-DICHLOROBENZYLIDENE)HYDRAZIDE
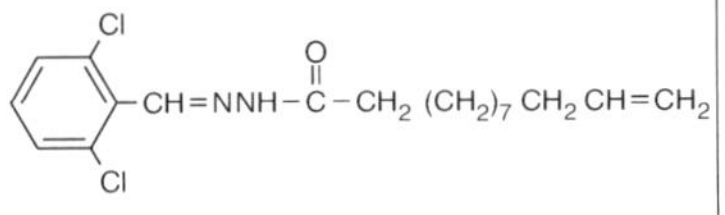
5 gm

S90,406-6
2-(1-NAPHTHYL)ACETIC (2,4-DICHLOROBENZYLIDENE)HYDRAZIDE
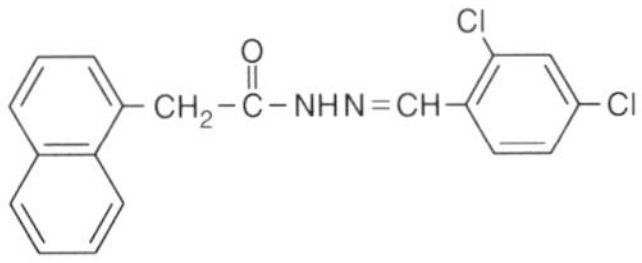
250 mg

S92,666-3
ACETIC 2-(4-CHLORO-2-NITROPHENYL)-HYDRAZIDE
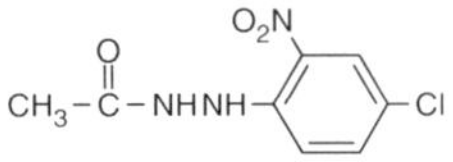
250 mg

S90,358-2
PHTHALIC MONO-(3,4-DICHLOROBENZOYL)-HYDRAZIDE
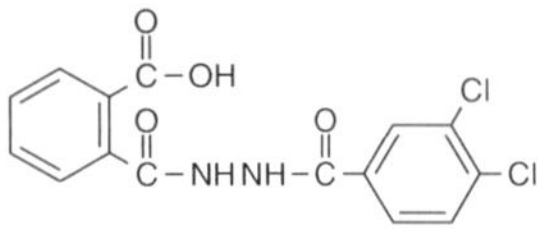
250 mg

S94,504-8
2-BROMOBENZOIC ACID HYDRAZIDE
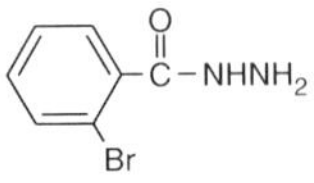
250 mg

S90,461-9
1-(2-CHLOROBENZOYL)-2-(4-CHLOROBENZOYL)HYDRAZINE
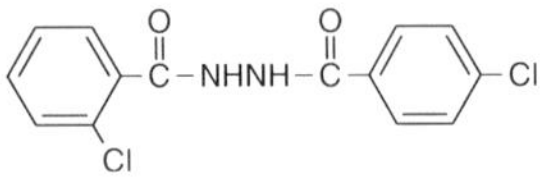
250 mg

S90,470-8
1-(2-CHLOROBENZOYL)-2-(3,4-DICHLOROBENZOYL)HYDRAZINE
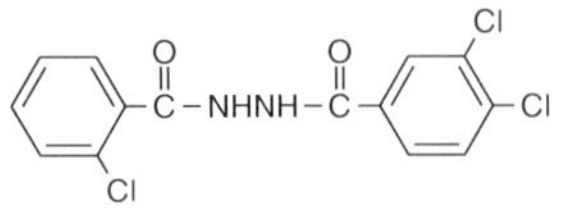
250 mg

S90,474-0
4-CHLOROBENZOIC N2-BUTYRYLHYDRAZIDE
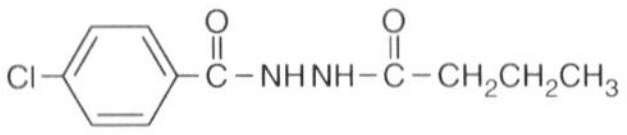
250 mg

S90,462-7
1-(4-CHLOROBENZOYL)-2-(2,4-DICHLOROBENZOYL)HYDRAZINE
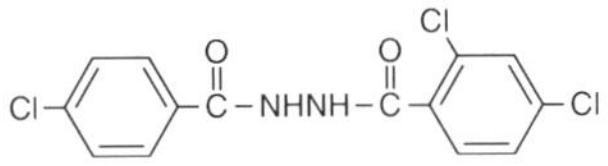
250 mg

S69,153-4
3,4-DICHLOROBENZOIC ACID TERT-BUTYLHYDRAZIDE
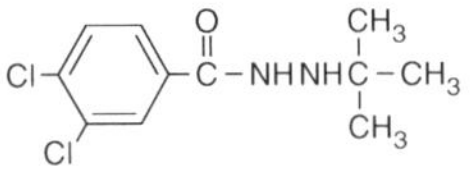
250 mg

S93,362-7
4-CHLOROPHENYLACETIC HYDRAZIDE
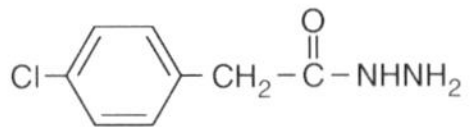
250 mg

S90,475-9
BUTYRIC N2-(4-CHLOROPHENYL)-HYDRAZIDE
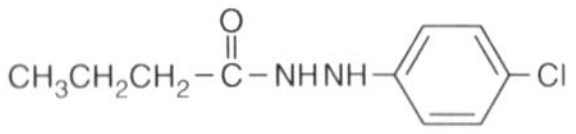
250 mg

S90,412-0
BENZILIC (4-HYDROXY-3-METHOXYBENZYLIDENE)HYDRAZIDE
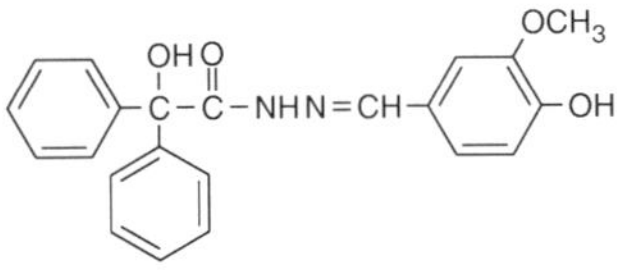
250 mg

S77,214-3
1-(3,4,5-TRIHYDROXYBENZOYL)-2-(3,4,5-TRIMETHOXYBENZOYL)HYDRAZINE
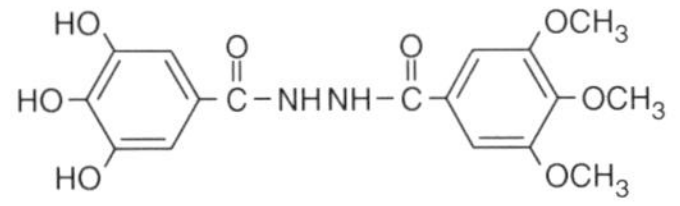
250 mg

S73,811-5
GLUCONIC (4-METHOXYBENZYLIDENE)-HYDRAZIDE
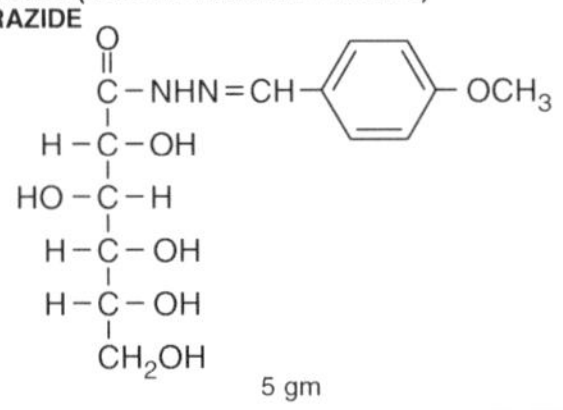
5 gm

S75,072-7
2-METHOXYBENZOIC (2,4-DIHYDROXYBENZYLIDENE)HYDRAZIDE
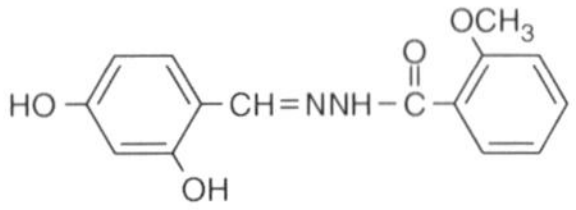
250 mg

S71,620-0
MANDELIC ACID N-(2-HYDROXYBENZYLIDENE)HYDRAZIDE
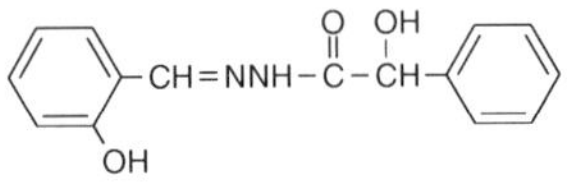
5 gm

S72,385-1
GLUCONIC PHENYLHYDRAZIDE
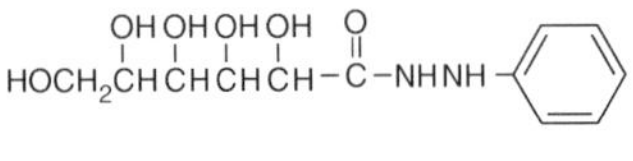
5 gm

S75,071-9
BENZOIC (2,4-DIHYDROXYBENZYLIDENE)-HYDRAZIDE
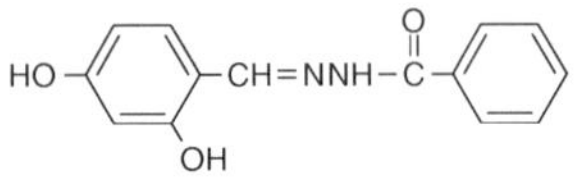
250 mg

S90,142-3
P-TOLUIC (2,4-DIHDYROXYBENZYLIDENE)-HYDRAZIDE
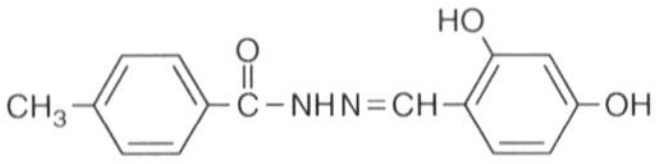
250 mg

S74,761-0
10-UNDECENOIC (2,4-DIHYDROXYBENZYLIDENE)HYDRAZIDE
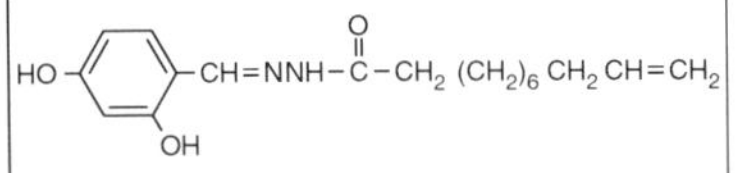
250 mg

S91,294-8
2,4-DIHYDROXYBENZOIC HYDRAZIDE
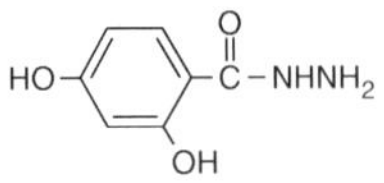
250 mg

S53,622-9
3-HYDROXYBENZOIC ACID (3,5-DINITRO-4-HYDROXYBENZYLIDENE)HYDRAZIDE
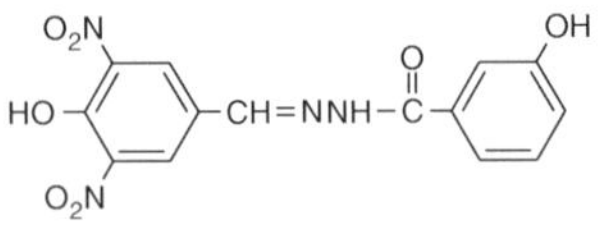
5 gm

S75,079-4
4-HYDROXYBENZOIC (2-HYDROXYBENZYLIDENE)HYDRAZIDE
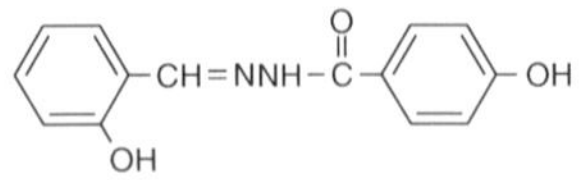
250 mg

S90,381-7
3-HYDROXY-2-NAPHTHOIC (2-HYDROXYBENZYLIDENE)HYDRAZIDE

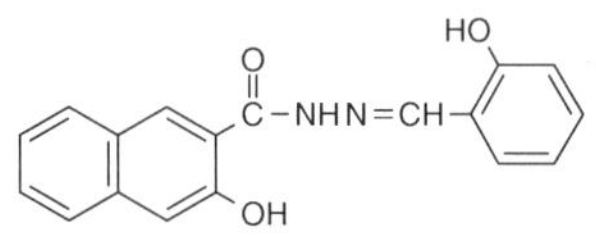

250 mg

S52,615-0
4-HYDROXYHYDROCINNAMIC ACID (5-BROMOSALICYLIDENE)HYDRAZIDE

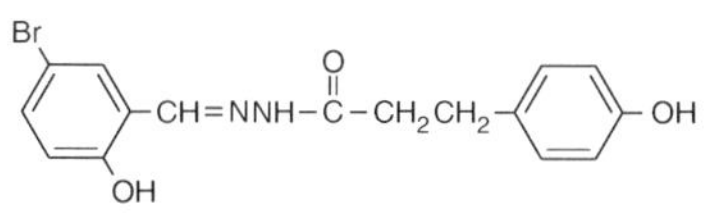

5 gm

S52,351-8
4-HYDROXYHYDROCINNAMIC ACID (3,5-DICHLOROSALICYLIDENE)HYDRAZIDE

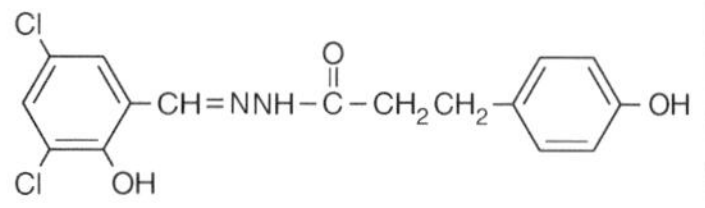

5 gm

S52,443-3
3,5-DIBROMO-4-HYDROXYHYDROCINNAMIC ACID (3,5-DICHLOROSALICYLIDENE)-HYDRAZIDE

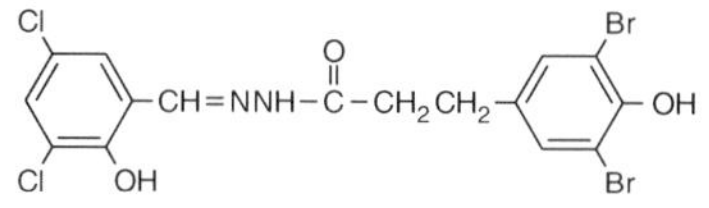

5 gm

S71,619-7
MANDELIC ACID N-(4-METHOXYBENZYLIDENE)HYDRAZIDE

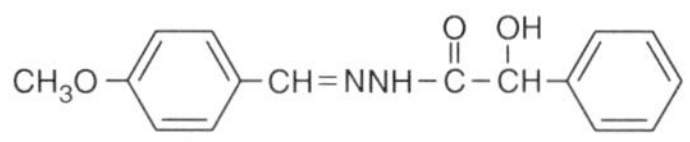

5 gm

S90,411-2
BENZILIC (4-METHOXYBENZYLIDENE)-HYDRAZIDE

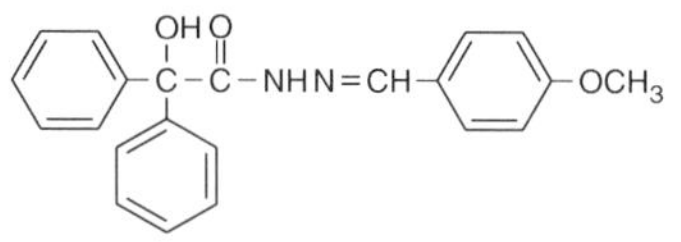

250 mg

S90,379-5
2-(PHENOXY)ACETIC (4-HYDROXY-3-METHOXYBENZYLIDENE)HYDRAZIDE

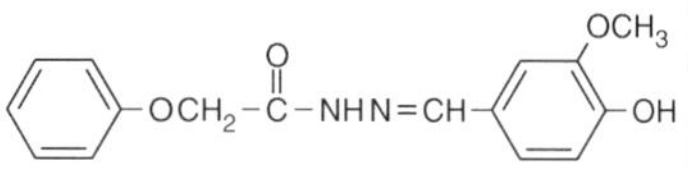

250 mg

S90,147-4
2-METHOXYBENZOIC (4-HYDROXY-3-METHOXYBENZYLIDENE)HYDRAZIDE

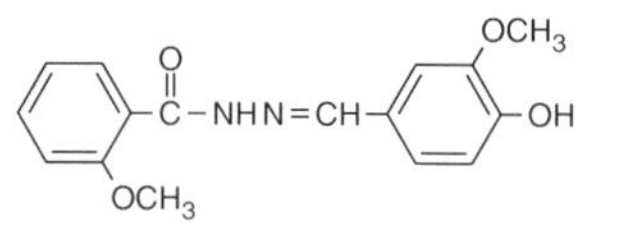

250 mg

S73,708-9
2-(METHYLAMINO)BENZOIC VANILLYLIDENEHYDRAZIDE

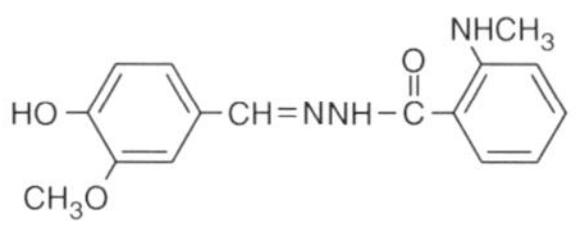

5 gm

S52,598-7
4-HYDROXYHYDROCINNAMIC ACID VANILLYLIDENEHYDRAZIDE

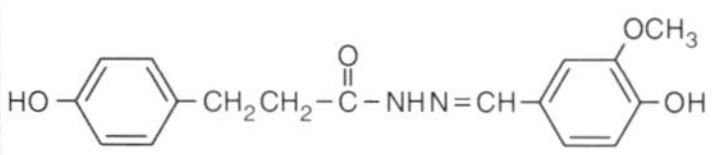

5 gm

S73,818-2
STEARIC (4-HYDROXY-3-METHOXYBENZYLIDENE)HYDRAZIDE

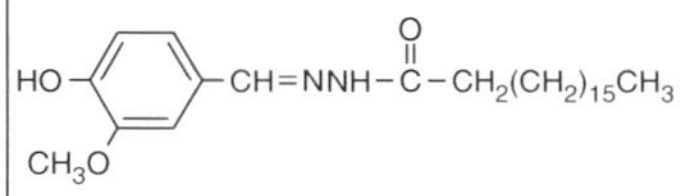

5 gm

S76,895-2
4-HYDROXY-3-METHOXYBENZOIC (3-ETHOXY-4-HYDROXYBENZYLIDENE)-HYDRAZIDE

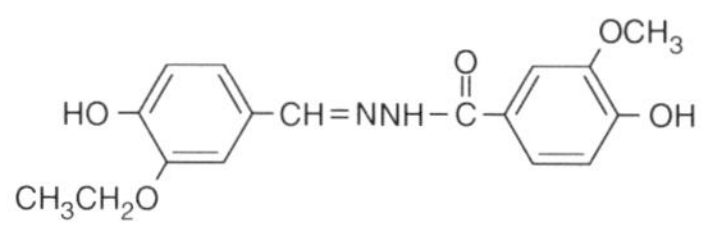

250 mg

S90,380-9
METHYL (4-HYDROXY-3-METHOXYBENZYLIDENE)CARBAZATE

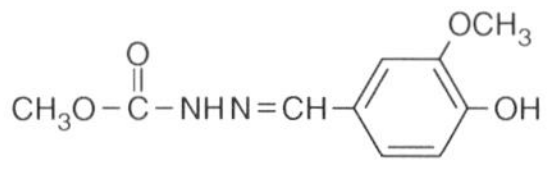

250 mg

S533750
4-HYDROXYBENZOIC ACID (3,4,5-TRIMETHOXYBENZYLIDENE)HYDRAZIDE

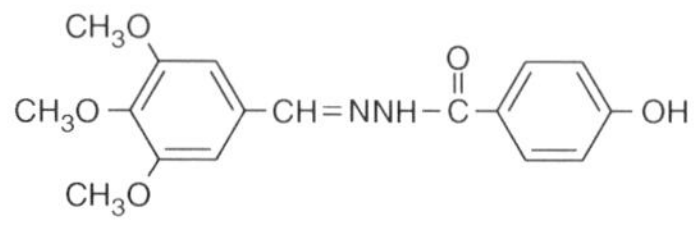

5 gm

S90,328-0
ACETIC (4-HYDROXY-3-METHOXYBENZYLIDENE)HYDRAZIDE

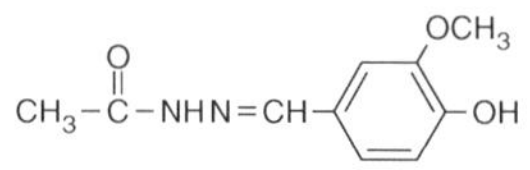

250 mg

S78,050-2
STEARIC (3-ETHOXY-4-HYDROXYBENZYLIDENE)HYDRAZIDE

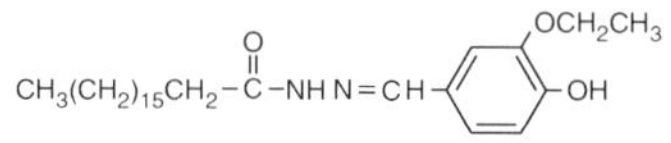

250 mg

S75,087-5
SALICYLIC (2-METHOXYBENZYLIDENE)-HYDRAZIDE

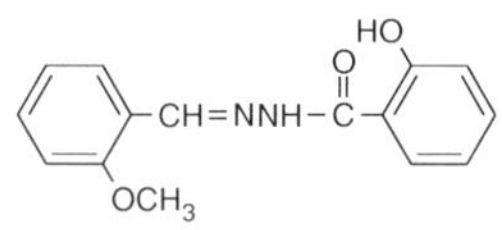

5 gm

S75,095-6
SALICYLIC (4-METHOXYBENZYLIDENE)-HYDRAZIDE

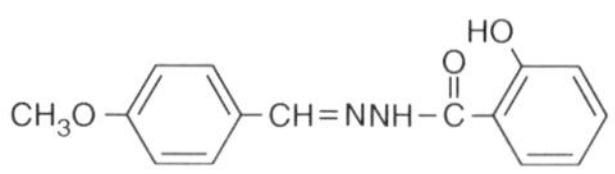

5 gm

S73,696-1
ANTHRANILIC (4-HYDROXY-3-METHOXYBENZYLIDENE)HYDRAZIDE

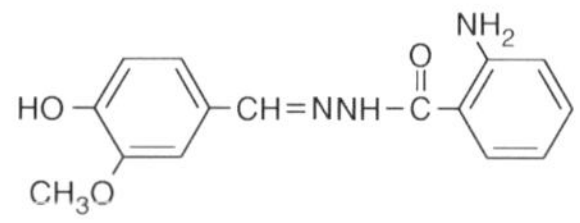

250 mg

S90,332-9
4-(DIMETHYLAMINO)BENZOIC (4-HYDROXY-3-METHOXYBENZYLIDENE)-HYDRAZIDE

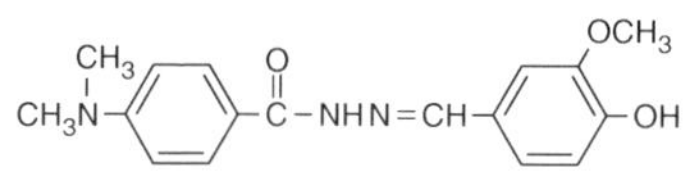

250 mg

S53,624-5
3-HYDROXYBENZOIC ACID (5-BROMO-2-METHOXYBENZYLIDENE)HYDRAZIDE

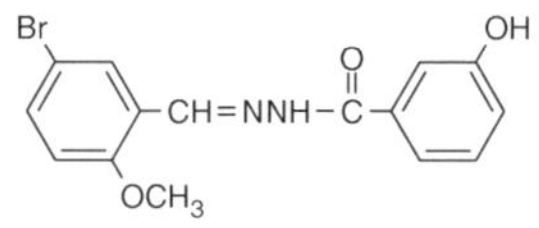

5 gm

S68,971-8
1-(4-HYDROXYBENZOYL)-2-(3,4,5-TRIMETHOXYCINNAMOYL)HYDRAZINE

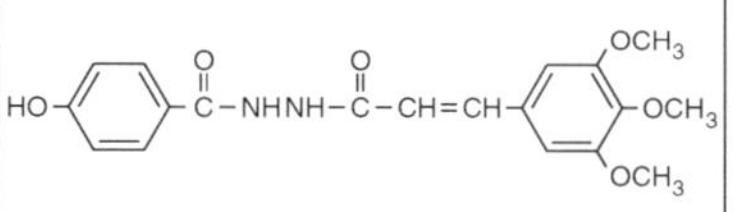

250 mg

S67,567-9
3-HYDROXY-2-NAPHTHOIC ACID N2-(3,4,5-TRIMETHOXYCINNAMOYL)-HYDRAZIDE

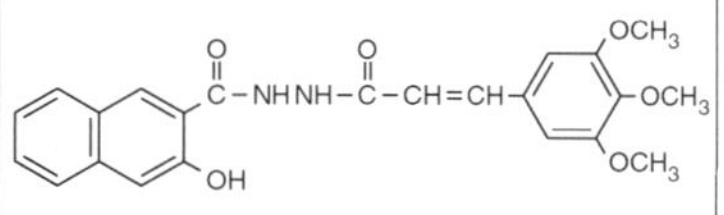

250 mg

S73,681-3
4-HYDROXY-3-METHOXYBENZOIC 3-NITROBENZYLIDENEHYDRAZIDE

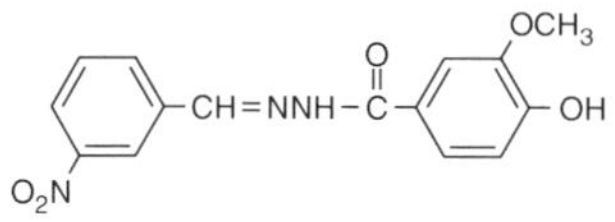

5 gm

S90,394-9
HIPPURIC (4-HYDROXY-3-METHOXYBENZYLIDENE)HYDRAZIDE

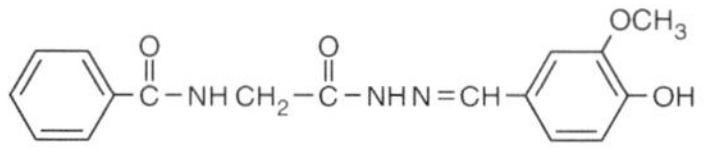

250 mg

S86,894-9
N-BENZOYLANTHRANILIC (4-HYDROXY-3-METHOXYBENZYLIDENE)HYDRAZIDE

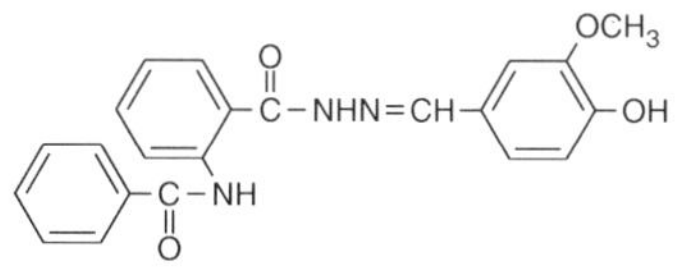

250 mg

S52,451-4
PIPERONYLIC ACID VANILLYLIDENEHYDRAZIDE

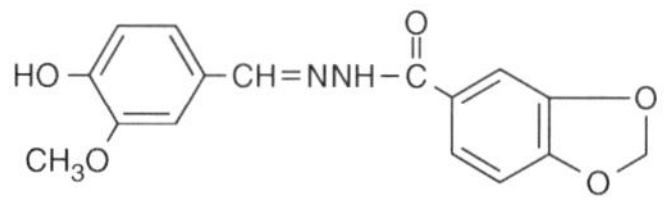

5 gm

S73,825-5
2-METHOXYBENZOIC (2-HYDROXY-1-NAPHTHYLMETHYLENE)HYDRAZIDE

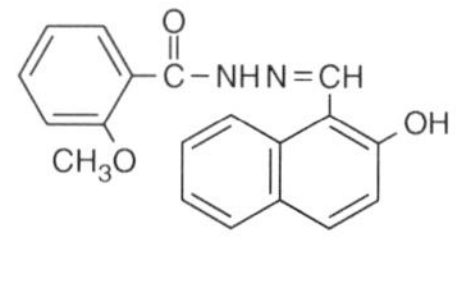

250 mg

S73,102-1
2-(1-NAPHTHYL)ACETIC 4-HYDROXY-3-METHOXYBENZYLIDENEHYDRAZIDE

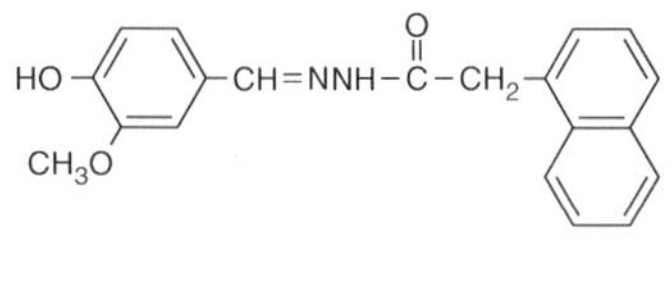

5 gm

S41,970-2
BENZILIC ACID HYDRAZIDE

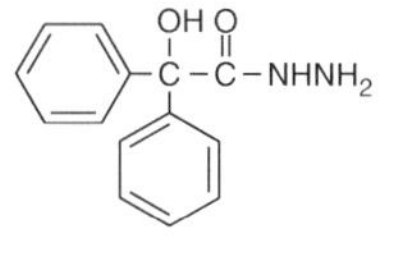

5 gm

S67,131-2
3-(3,5-DI-TERT-BUTYLPHENYL)-3-HYDROXYPROPIONIC ACID HYDRAZIDE

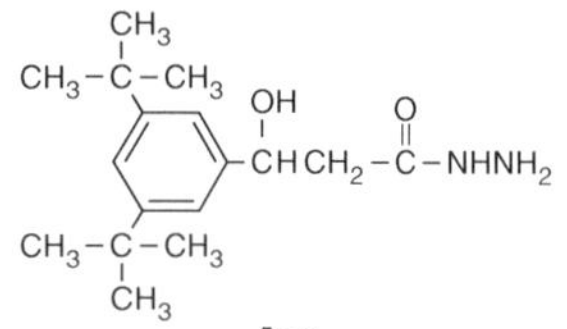

5 gm

S90,409-0
MANDELIC (1-NAPHTHYLMETHYLENE)-HYDRAZIDE

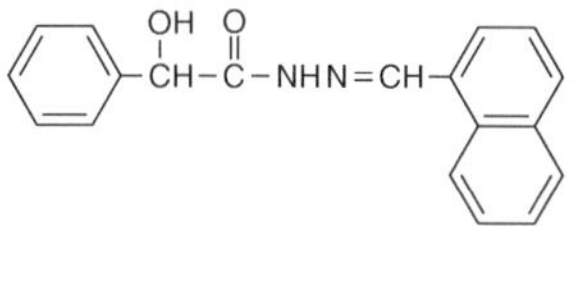

250 mg

S50,659-1
MANDELIC ACID (5-INDANYLMETHYLENE)-HYDRAZIDE

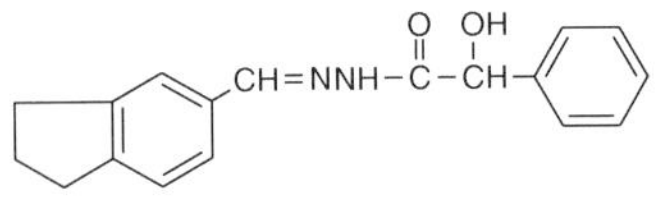

250 mg

S73,099-8
MANDELIC 3-NITROBENZYLIDENEHYDRAZIDE

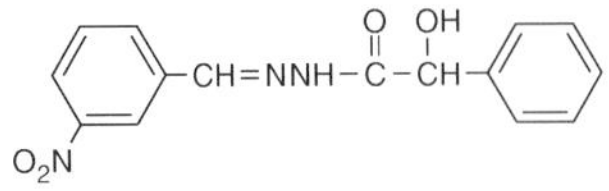

5 gm

S75,089-1
SALICYLIC (2-PHENYLETHYLIDENE)-HYDRAZIDE

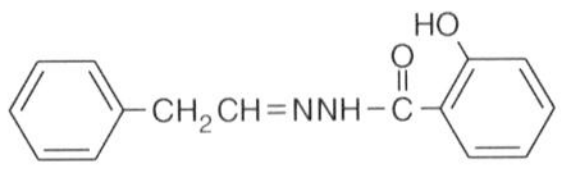

5 gm

S75,091-3
SALICYLIC (ALPHA-METHYLBENZYLIDENE)-HYDRAZIDE

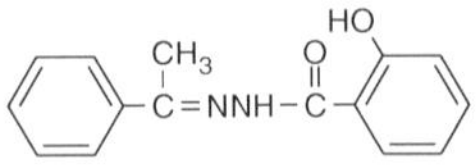

5 gm

S76,930-4
4-AMINOBENZOIC (2-HYDROXYBENZYLIDENE)HYDRAZIDE

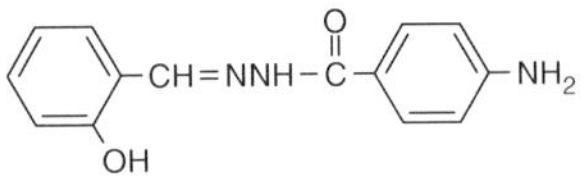

250 mg

S73,656-2
ANTHRANILIC (2-HYDROXYBENZYLIDENE)-HYDRAZIDE

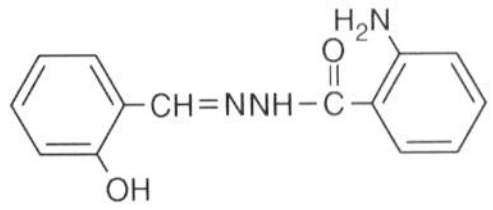

250 mg

S73,664-3
OXALIC BIS[(2-HYDROXY-1-NAPHTHYLMETHYLENE)HYDRAZIDE]

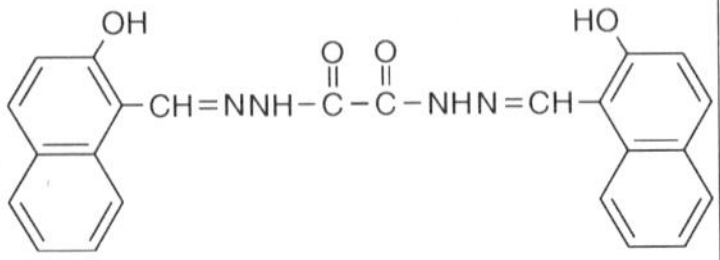

5 gm

S90,382-5
3-HYDROXY-2-NAPHTHOIC (2-PHENYLETHYLIDENE)HYDRAZIDE

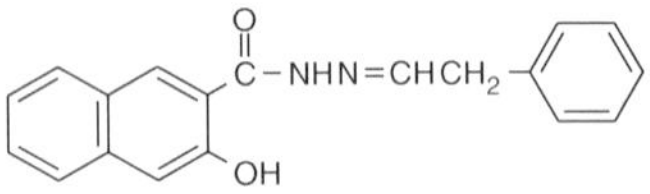

250 mg

S45,139-0
N-ACETYL-L-TYROSINE HYDRAZIDE

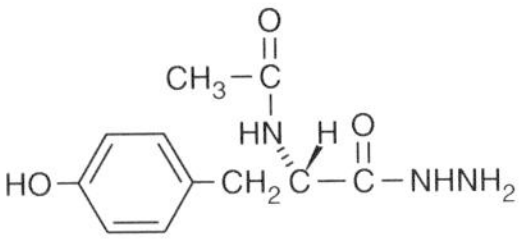

250 mg

S73,602-3
N-METHYLANTHRANILIC (2-HYDROXY-1-NAPHTHYLMETHYLENE)HYDRAZIDE

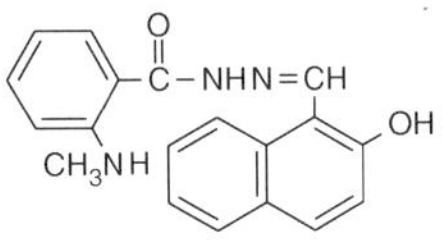

250 mg

S90,329-9
4-(DIMETHYLAMINO)BENZOIC (2-HYDROXYBENZYLIDENE)HYDRAZIDE

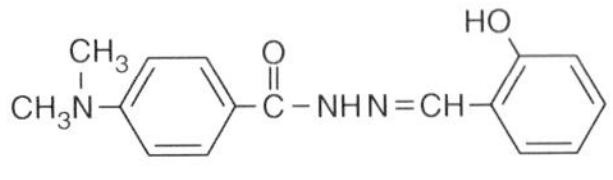

250 mg

S53,623-7
3-HYDROXYBENZOIC ACID [2-CHLORO-4-(DIMETHYLAMINO)BENZYLIDENE]-HYDRAZIDE

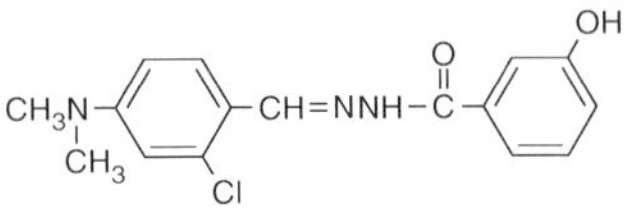

5 gm

S53,625-3
3-HYDROXYBENZOIC ACID (2,4-DICHLOROBENZYLIDENE)HYDRAZIDE

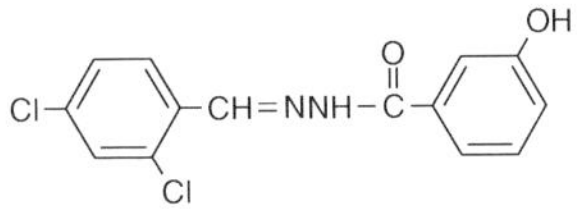

5 gm

S64,296-7
MALEIC ACID (3-HYDROXYBENZYLIDENE)-HYDRAZIDE

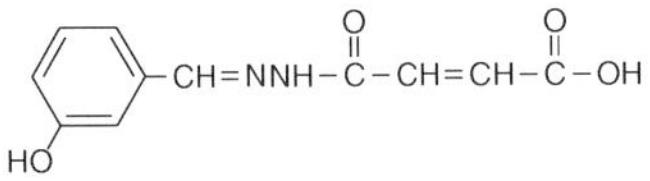

250 mg

S94,919-1
DIPHENYLPHOSPHINIC (2-HYDROXYBENZYLIDENE)HYDRAZIDE

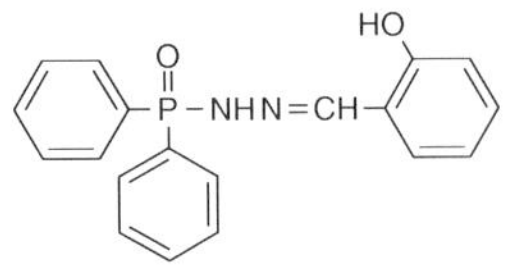

250 mg

S73,095-5
STEARIC 2-HYDROXYBENZYLIDENEHYDRAZIDE

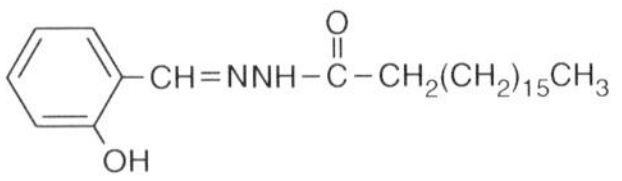

5 gm

S72,871-3
4-BUTYL-1-(4-HYDROXYBENZOYL)-SEMICARBAZIDE

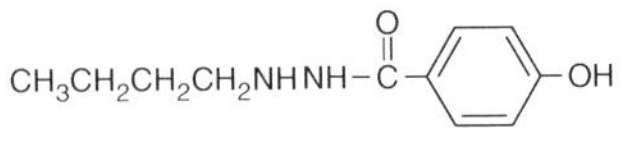

250 mg

S90,387-6
2-HYDROXYISOBUTYRIC (1-NAPHTHYLMETHYLENE)HYDRAZIDE

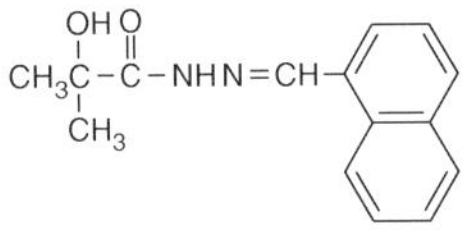

250 mg

S90,392-2
HIPPURIC (2-HYDROXYBENZYLIDENE)-HYDRAZIDE

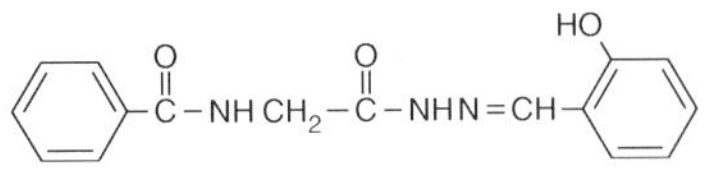

250 mg

S86,884-1
N-BENZOYLANTHRANILIC (3-HYDROXYBENZYLIDENE)HYDRAZIDE

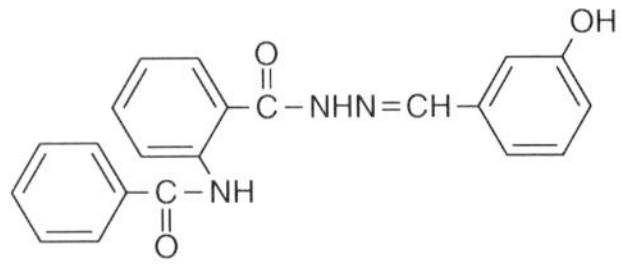

250 mg

S90,400-7
HIPPURIC (2-HYDROXY-1-NAPHTHYLMETHYLENE)HYDRAZIDE

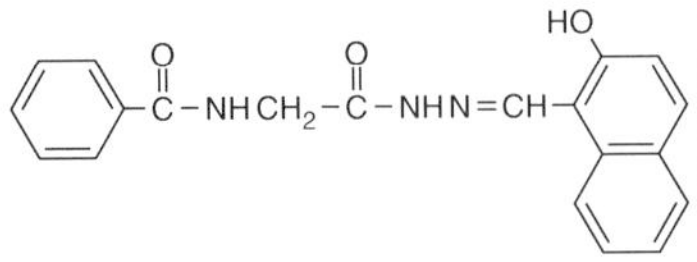

250 mg

S90,117-2
BENZOIC (7-HYDROXY-1-NAPHTHYLMETHYLENE)HYDRAZIDE

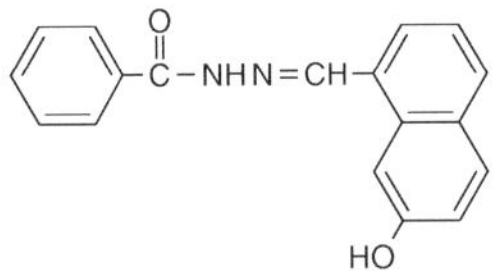

250 mg

S75,066-2
N,N'-TEREPHTHALYLIDENEBIS(4-HYDROXYBENZOIC HYDRAZIDE)

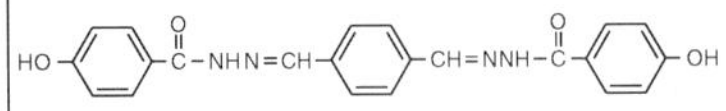

5 gm

S50,661-3
4-HYDROXYBENZOIC ACID (5-INDANYLMETHYLENE)HYDRAZIDE

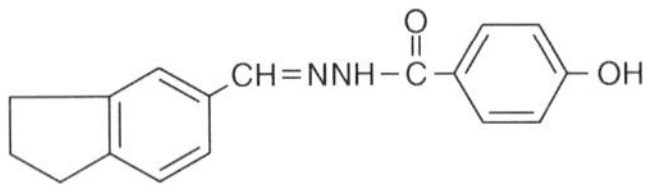

250 mg

S73,601-5
4-METHYLBENZOIC (2-HYDROXY-1-NAPHTHYLMETHYLENE)HYDRAZIDE

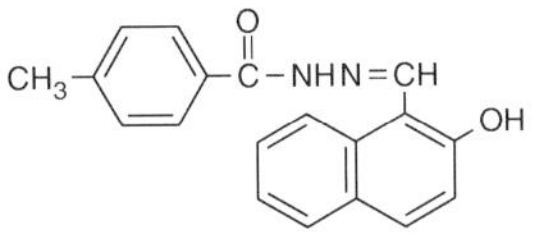

250 mg

S76,949-5
4-HYDROXYBENZOIC (1-NAPHTHYLMETHYLENE)HYDRAZIDE

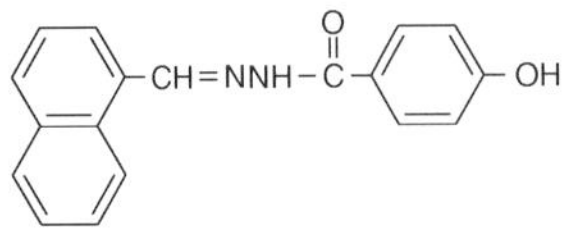

250 mg

S86,926-0
SEBACIC BIS[(3,5-DI-TERT-BUTYL-4-HYDROXYBENZYLIDENE)HYDRAZIDE]

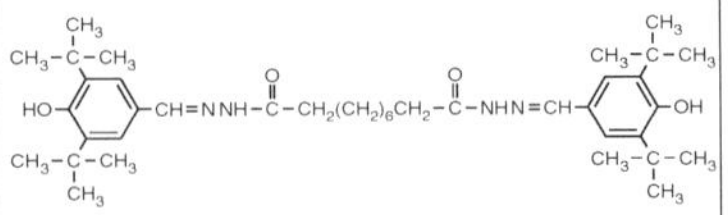

250 mg

S52,358-5
4-HYDROXYHYDROCINNAMIC ACID PIPERONYLIDENEHYDRAZIDE

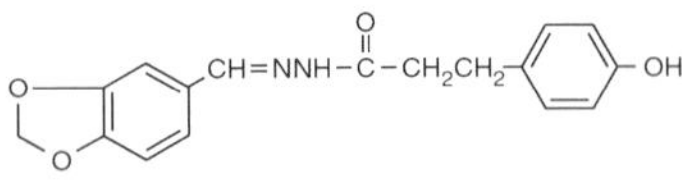

5 gm

S52,614-2
PIPERONYLIC ACID (5-BROMOSALICYLIDENE)HYDRAZIDE

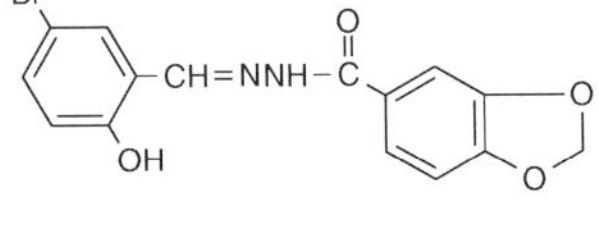

5 gm

S52,452-2
PIPERONYLIC ACID (3,5-DICHLOROSALICYLIDENE)-HYDRAZIDE

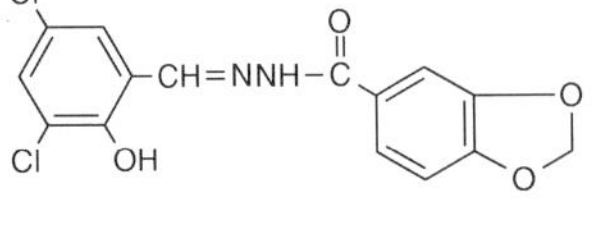

250 mg

S77,968-7
ACETIC (2-HYDROXYBENZYLIDENE)-HYDRAZIDE

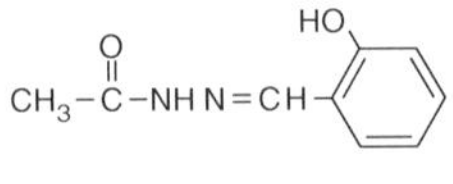

250 mg

S86,915-5
OCTADECANOIC (4-HYDROXYBENZYLIDENE)-HYDRAZIDE

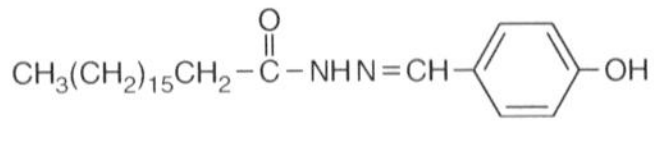

250 mg

S77,964-4
ACETIC (2-HYDROXY-1-NAPHTHYLMETHYLENE)HYDRAZIDE
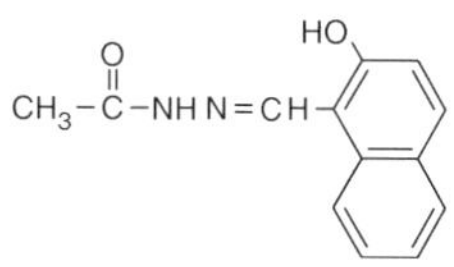
250 mg

S73,873-5
STEARIC (2-HYDROXY-1-NAPHTHYLMETHYLENE)HYDRAZIDE
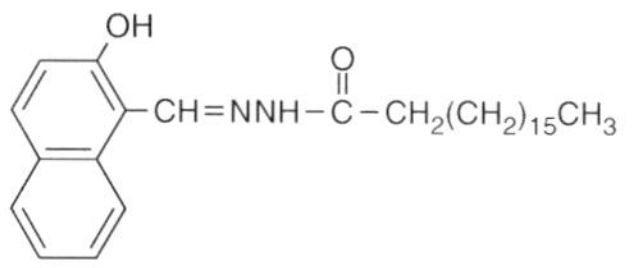
5 gm

S50,342-8
3-HYDROXY-2-NAPHTHOIC ACID (5-INDANYLMETHYLENE)HYDRAZIDE
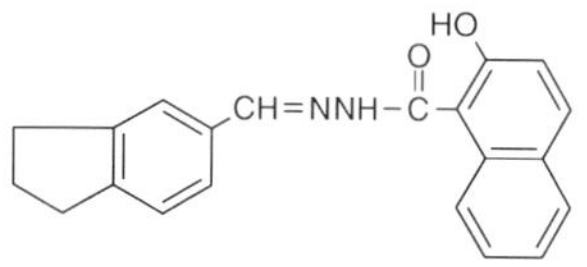
250 mg

S37,313-3
N-BENZOYL-L-TYROSINE HYDRAZIDE
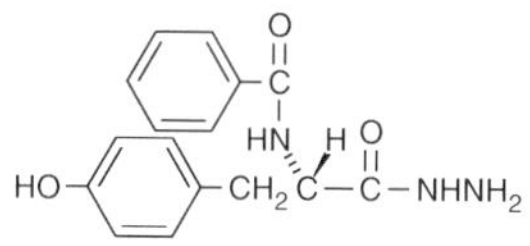
5 gm

S66,885-0
2-(2-HYDROXYPHENYL)ACETHYDRAZIDE
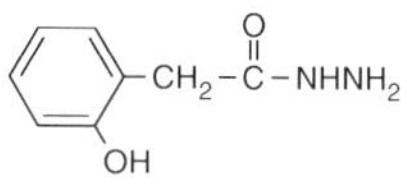
5 gm

S52,388-7
4-HYDROXYHYDROCINNAMIC ACID HYDRAZIDE
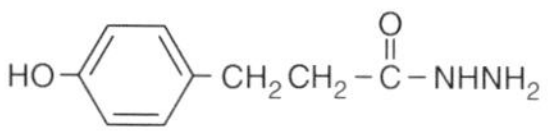
5 gm

S76,890-1
4-AMINOBENZOIC (2,3-DIMETHOXYBENZYLIDENE)HYDRAZIDE
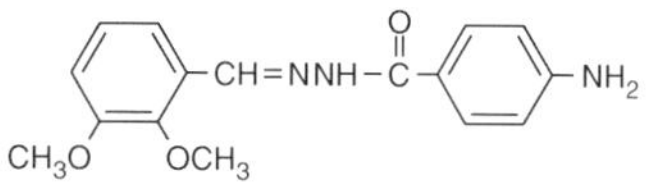
250 mg

S71,613-8
HIPPURIC ACID N-(4-ETHOXY-3-METHOXYBENZYLIDENE)HYDRAZIDE
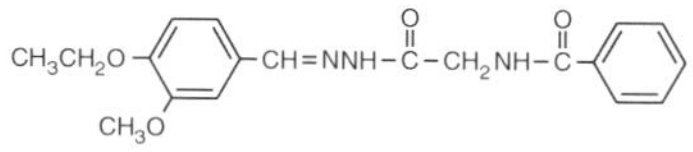
250 mg

S67,519-9
BENZOIC ACID (4-ETHOXY-3-METHOXYBENZYLIDENE)HYDRAZIDE
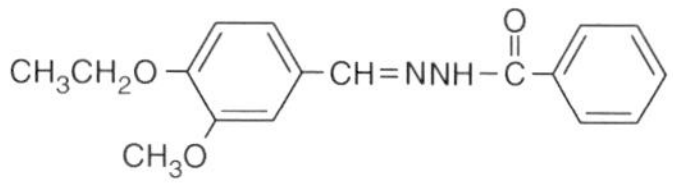
5 gm

S71,610-3
2,2'-OXYDIACETIC ACID BIS[(3,4-DIMETHOXYBENZYLIDENE)HYDRAZIDE]
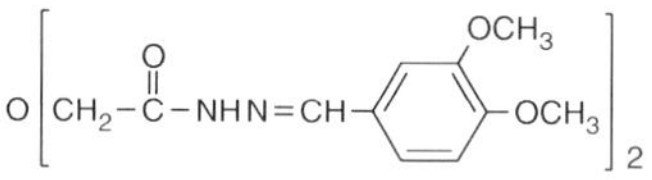
5 gm

S87,612-7
DIGLYCOLIC BIS[(4-ETHOXY-3-METHOXYBENZYLIDENE)HYDRAZIDE]
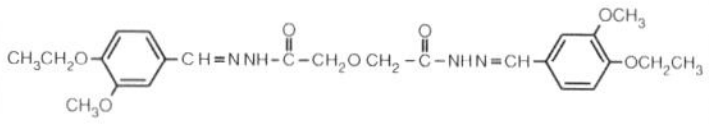
250 mg

S73,802-6
STEARIC (3,4-DIMETHOXYBENZYLIDENE)-HYDRAZIDE
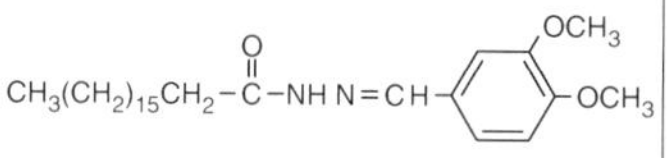
5 gm

S86,909-0
N-BENZOYLANTHRANILIC (3-ETHOXY-4-METHOXYBENZYLIDENE)HYDRAZIDE
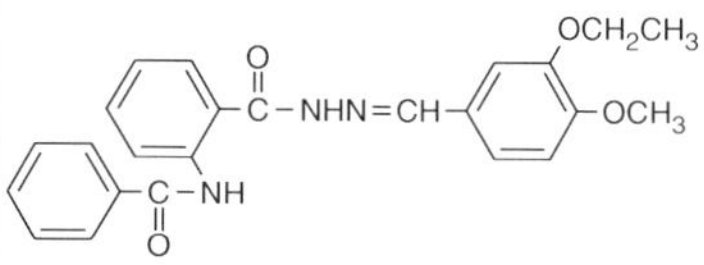
250 mg

S86,923-6
SEBACIC BIS[(3,4-DIMETHOXYBENZYLIDENE)HYDRAZIDE]
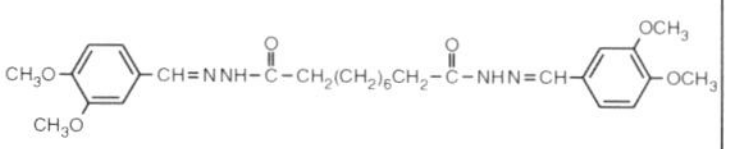
250 mg

S86,925-2
SEBACIC BIS[(4-ETHOXY-3-METHOXYBENZYLIDENE)HYDRAZIDE]
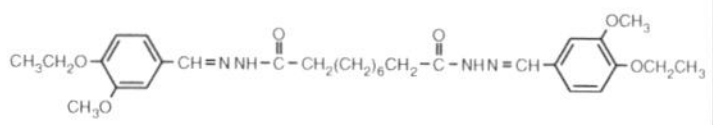
250 mg

S75,083-2
10-UNDECENOIC (2,3-DIMETHOXYBENZYLIDENE)HYDRAZIDE
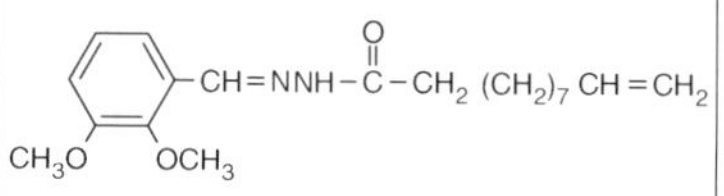
5 gm

S75,080-8
CINNAMIC (2,4-DIMETHOXYBENZYLIDENE)-HYDRAZIDE
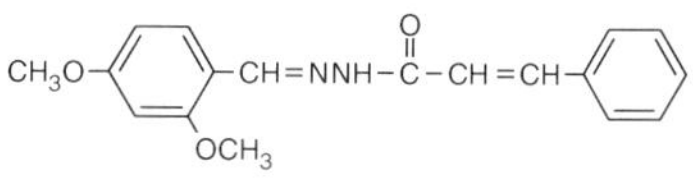
5 gm

S75,084-0
2-METHOXYBENZOIC (2-METHOXYBENZYLIDENE)HYDRAZIDE
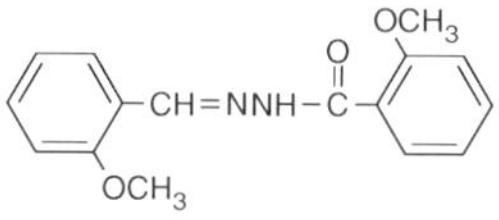
5 gm

S76,900-2
1-[2-(METHYLAMINO)BENZOYL]-2-(3,4,5-TRIMETHOXYBENZOYL)HYDRAZINE
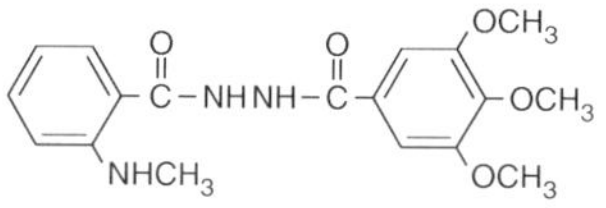
250 mg

S73,853-0
STEARIC (2,5-DIMETHOXYBENZYLIDENE)-HYDRAZIDE
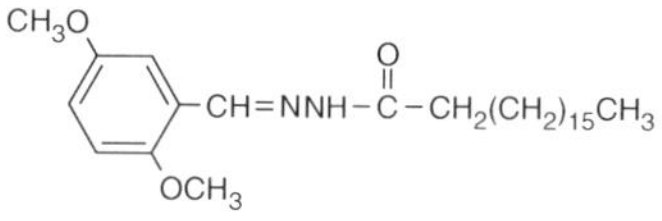
5 gm

S90,004-4
4-AMINOBENZOIC (2,4-DIMETHOXYBENZYLIDENE)HYDRAZIDE
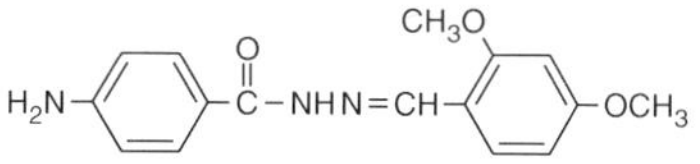
250 mg

S89,970-4
MALONIC BIS(4-METHOXYBENZYLIDENEHYDRAZIDE)
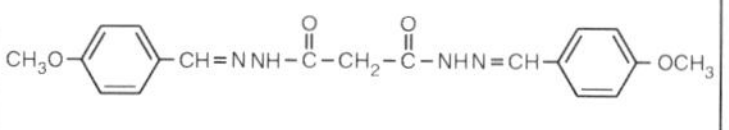
250 mg

S86,916-3
SEBACIC BIS[(2-METHOXYBENZYLIDENE)-HYDRAZIDE]
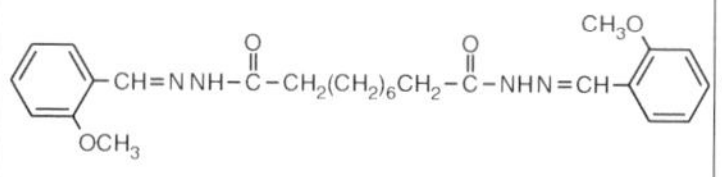
250 mg

S86,917-1
SEBACIC BIS[(4-METHOXYBENZYLIDENE)-HYDRAZIDE]
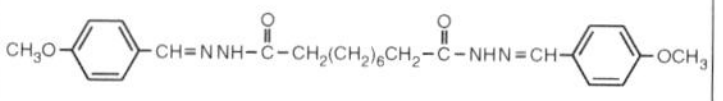
250 mg

S78,001-4
2-[3-(2-ETHOXYPHENYL)UREIDO]ACETIC (4-METHOXYBENZYLIDENE)HYDRAZIDE
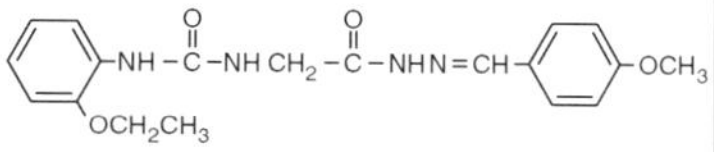
250 mg

S64,295-9
METHYL MALEATE VERATRYLIDENEHYDRAZIDE
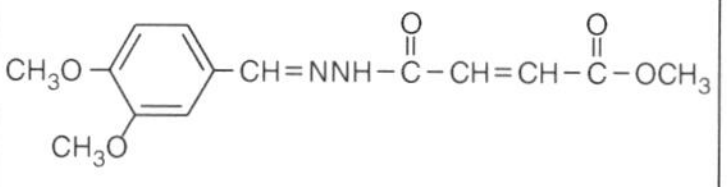
250 mg

S59,260-9
ACETIC ACID N2-(4-HEXADECYLOXY-3-METHOXYBENZYL)HYDRAZIDE
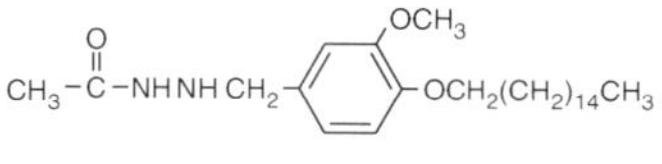
250 mg

S50,986-8
4-BUTOXYBENZOIC ACID (3,4,5-TRIMETHOXYBENZYLIDENE)HYDRAZIDE
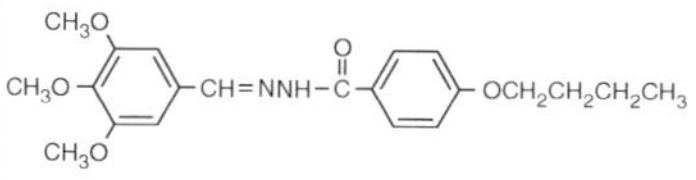
5 gm

S56,299-8
3,4,5-TRIMETHOXYBENZOIC ACID HYDRAZIDE
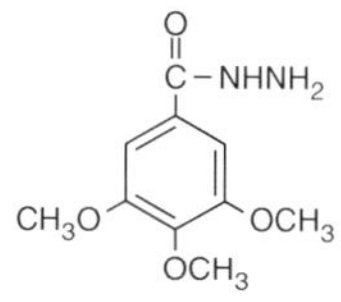
5 gm

S74,356-9
2-METHOXYBENZOIC (3-NITROBENZYLIDENE)HYDRAZIDE
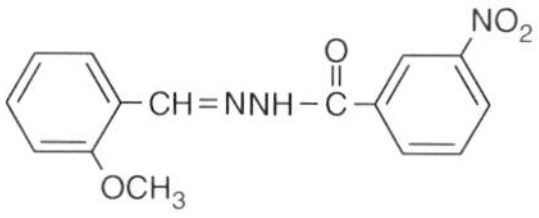
5 gm

S75,086-7
2-METHOXYBENZOIC (4-ACETAMIDOBENZYLIDENE)HYDRAZIDE
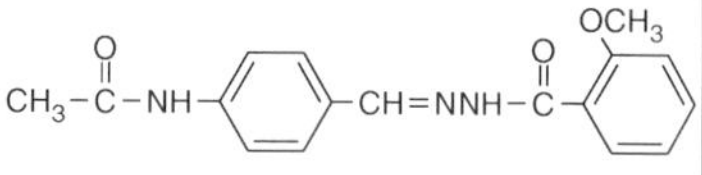
5 gm

S75,096-4
2-METHOXYBENZOIC BENZYLIDENEHYDRAZIDE
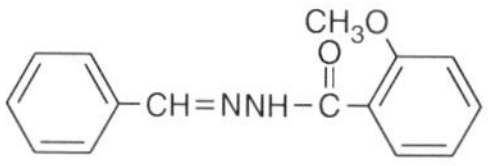
5 gm

S75,074-3
2-METHOXYBENZOIC (4-METHYLBENZYLIDENE)HYDRAZIDE

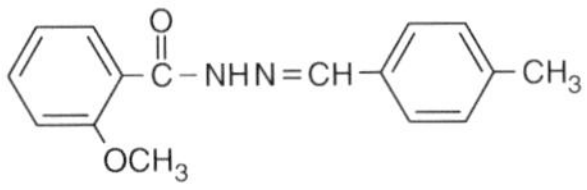

250 mg

S95,858-1
4-PHENOXYBUTYRIC PHENYLHYDRAZIDE

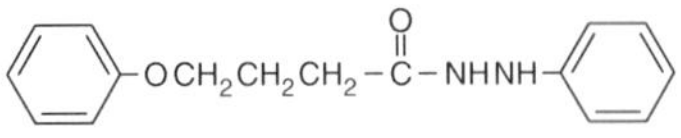

50 mg

S90,002-8
4-AMINOBENZOIC (4-METHOXYBENZYLIDENE)HYDRAZIDE

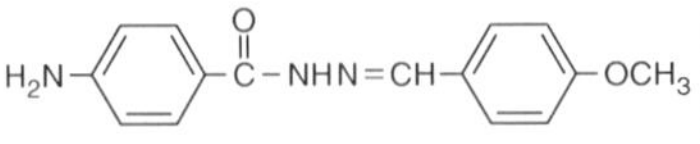

250 mg

S90,378-7
2-(PHENOXY)ACETIC (3-NITROBENZYLIDENE)HYDRAZIDE

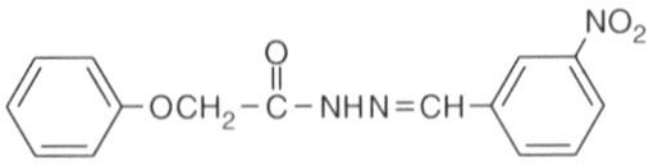

250 mg

S66,996-2
PHENYLACETIC ACID (2-METHOXYBENZYLIDENE)HYDRAZIDE

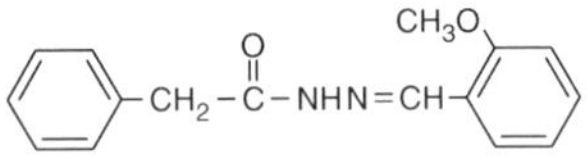

250 mg

S90,376-0
2-PHENYLACETIC (4-METHOXYBENZYLIDENE)HYDRAZIDE

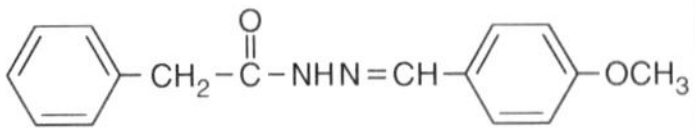

250 mg

S90,377-9
2-(PHENOXY)ACETIC [4-(DIMETHYLAMINO)BENZYLIDENE]-HYDRAZIDE

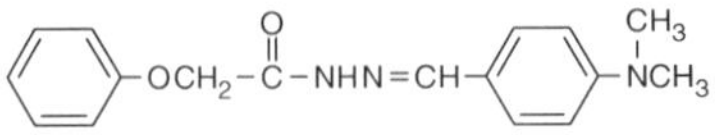

250 mg

S87,614-3
DIGLYCOLIC BIS([4-(DIMETHYLAMINO)-BENZYLIDENE]HYDRAZIDE)

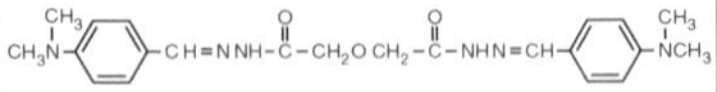

250 mg

S71,622-7
10-UNDECENOIC ACID N-(2-METHOXYBENZYLIDENE)HYDRAZIDE

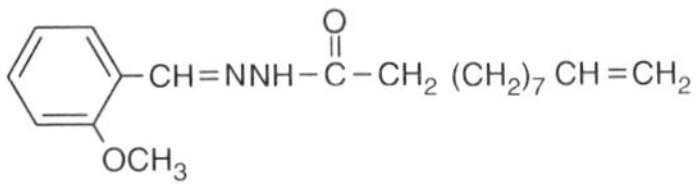

250 mg

S35,821-5
1-BENZOYL-2-(3-METHOXYBENZOYL)-HYDRAZINE

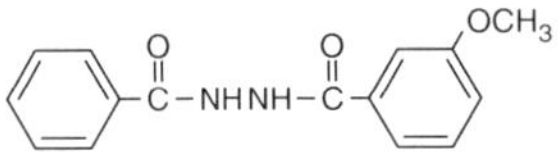

5 gm

S36,311-1
1-(4-METHOXYBENZOYL)-2-(1-NAPHTHOYL)HYDRAZINE

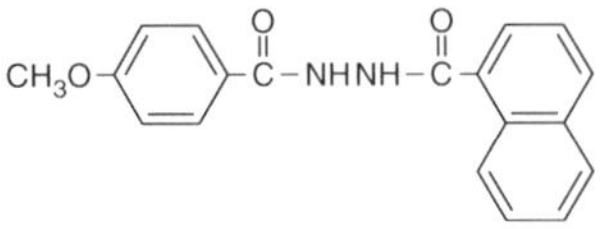

250 mg

S36,310-3
1-(3-METHOXYBENZOYL)-2-(1-NAPHTHOYL)HYDRAZINE

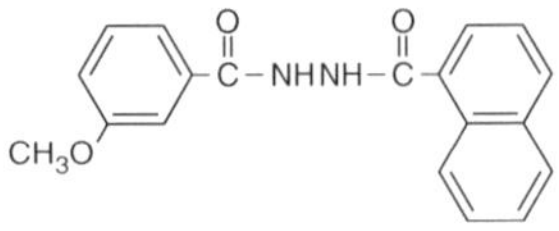

250 mg

S90,390-6
HIPPURIC (4-METHOXYBENZYLIDENE)-HYDRAZIDE

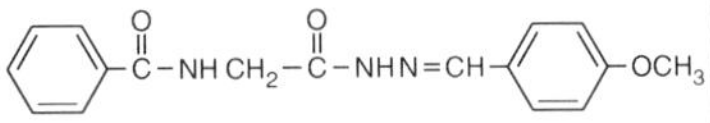

250 mg

S85,595-2
2-(CYCLOPROPYLCARBOXAMIDO)BENZOIC (4-METHOXYBENZYLIDENE)HYDRAZIDE

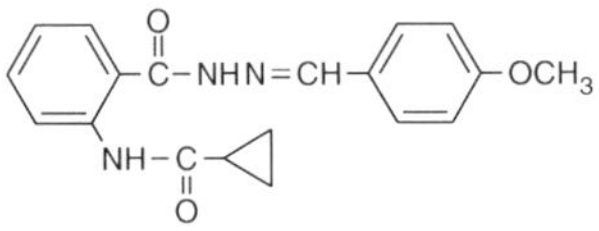

250 mg

S75,077-8
BENZOIC (2-METHOXYBENZYLIDENE)-HYDRAZIDE

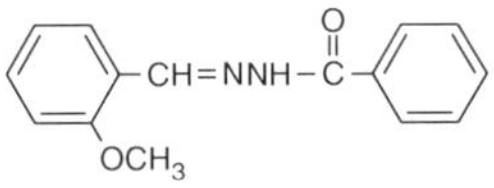

5 gm

S75,073-5
BENZOIC (4-METHOXYBENZYLIDENE)-HYDRAZIDE

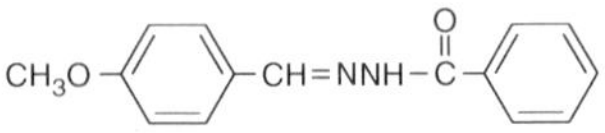

5 gm

S90,149-0
2-METHOXYBENZOIC [1-(1-NAPHTHYL)-ETHYLIDENE]HYDRAZIDE

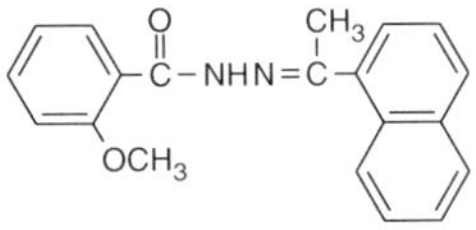

250 mg

S76,892-8
4-METHYLBENZOIC (4-METHOXYBENZYLIDENE)HYDRAZIDE

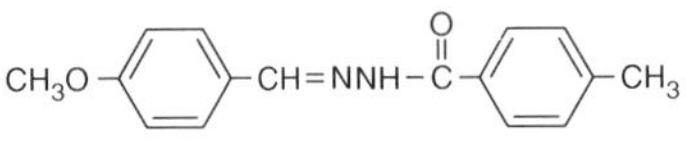

250 mg

S51,449-7
4-ETHOXYBENZOIC ACID (5-INDANYLMETHYLENE)HYDRAZIDE

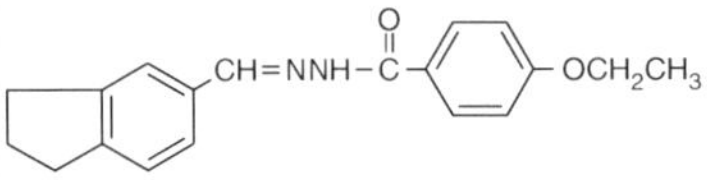

250 mg

S51,445-4
4-BUTOXYBENZOIC ACID (5-INDANYLMETHYLENE)HYDRAZIDE

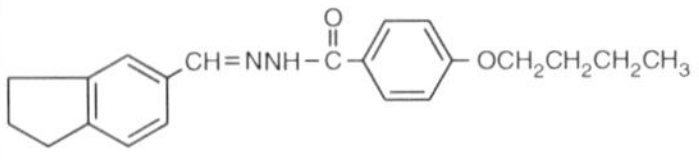

5 gm

S54,915-0
ETHYL (3,4,5-TRIMETHOXYBENZYLIDENE)-ARBAZATE

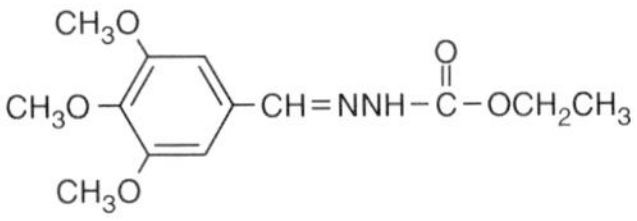

5 gm

S54,916-9
TERT-BUTYL (3,4,5-TRIMETHOXYBENZYLIDENE)CARBAZATE

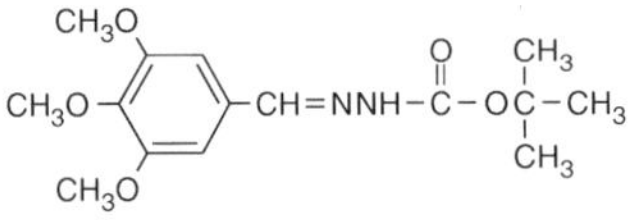

5 gm

S56,944-5
ACETIC ACID (3-ISOOCTADECYLOXYBENZYLIDENE)-HYDRAZIDE

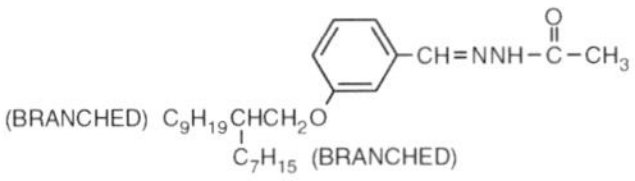

5 gm

S92,536-5
4-METHOXYBENZOIC (CYCLOPENTYLIDENE)-HYDRAZIDE

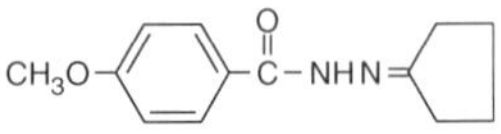

250 mg

S87,611-9
DIGLYCOLIC BIS[(3-PHENYLPROPYLIDENE)HYDRAZIDE]

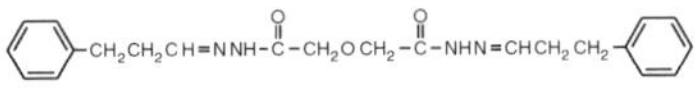

250 mg

S54,929-0
ACETIC ACID (2-METHOXYBENZYLIDENE)-HYDRAZIDE

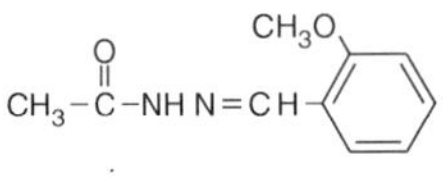

250 mg

S73,874-3
STEARIC (2-METHOXYBENZYLIDENE)-HYDRAZIDE

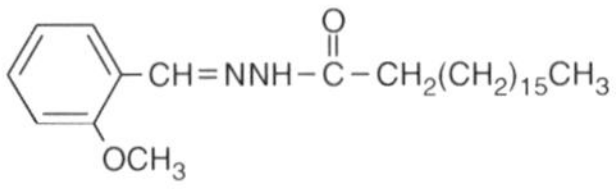

5 gm

S73,732-1
STEARIC (4-METHOXYBENZYLIDENE)-HYDRAZIDE

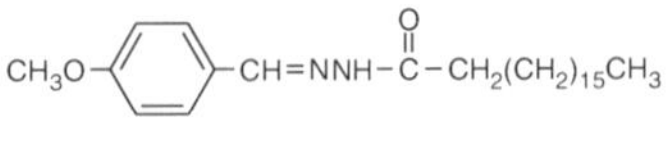

5 gm

S54,824-3
BENZOIC ACID (3,4,5-TRIMETHOXYBENZYLIDENE)HYDRAZIDE

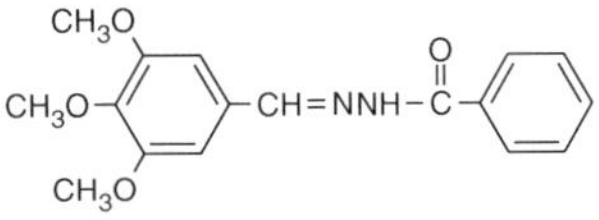

5 gm

S52,704-1
5-INDANCARBOXALDEHYDE (PHENYLACETYL)HYDRAZONE

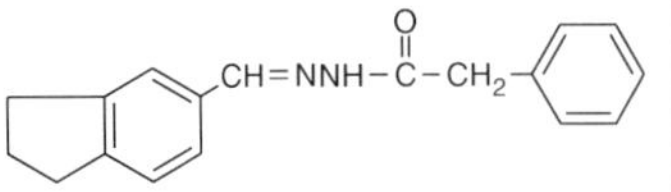

250 mg

S58,754-0
ACETIC ACID N2-[3-(DODECYLOXY)-BENZYL]HYDRAZIDE

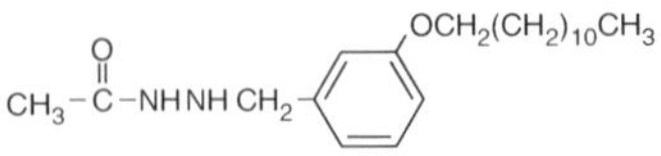

5 gm

S55,038-8
O-ANISIC ACID HYDRAZIDE HYDROCHLORIDE

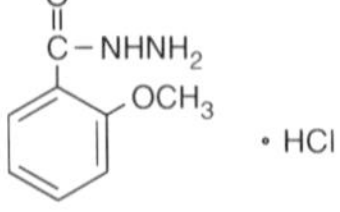

5 gm

S55,039-6
P-ANISIC ACID HYDRAZIDE
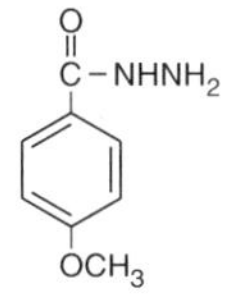
5 gm

S55,781-1
3,5-DIMETHOXYBENZOIC ACID HYDRAZIDE
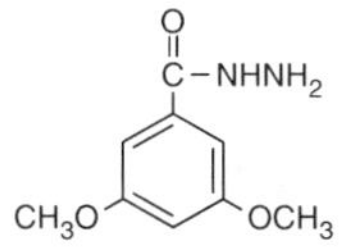
250 mg

S95,437-3
N-(BENZYLIDENEAMINO)-N-(BENZYLOXYCARBONYL)GLYCINE
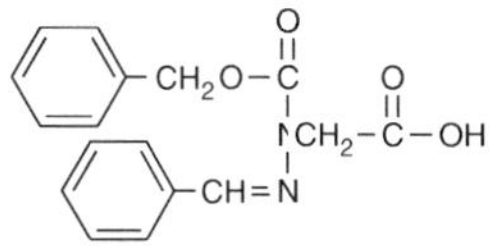
50 mg

S51,210-9
N,N'-BIS(3,4,5-TRIMETHOXYBENZYLIDENE)ANTHRANILIC ACID HYDRAZIDE
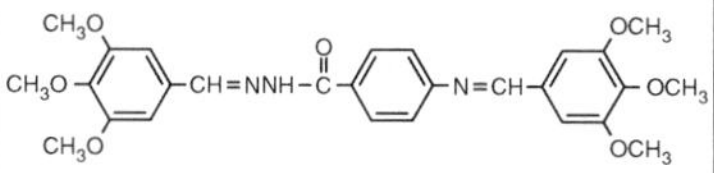
5 gm

S54,216-4
TERT-BUTYL (2,3,4-TRIMETHOXYBENZYLIDENE)CARBAZATE
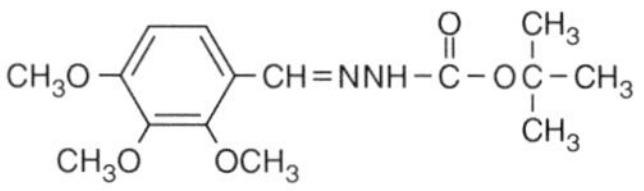
5 gm

S93,560-3
DIETHYL 2-PHTHALIMIDOGLUTARATE
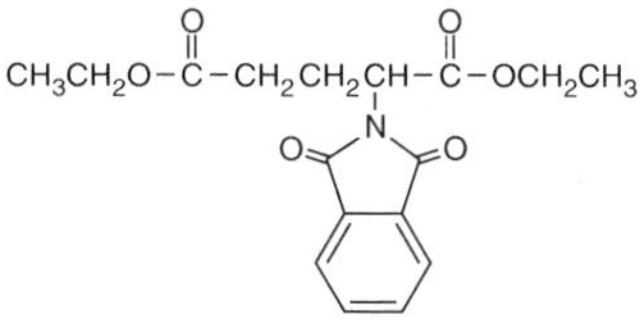
250 mg

S50,976-0
OXAMIC ACID (5-INDANYLMETHYLENE)-HYDRAZIDE
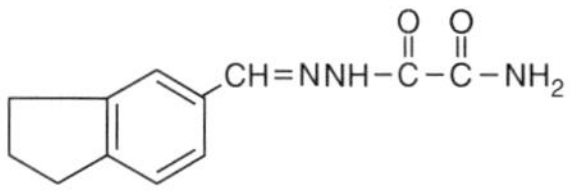
250 mg

S90,375-2
ANTHRANILIC [4-(DIMETHYLAMINO)-BENZYLIDENE]HYDRAZIDE
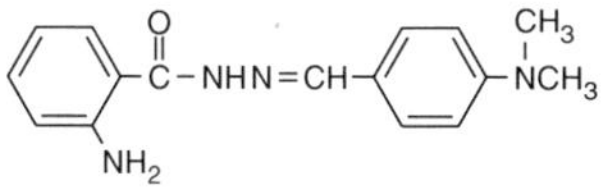
250 mg

S67,784-1
1-(2-AMINOBENZOYL)-2-BENZOYLHYDRAZINE
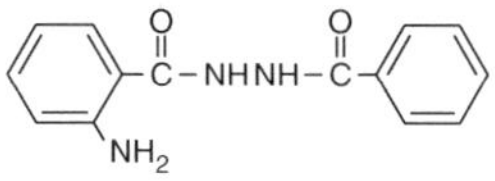
5 gm

S73,670-8
ANTHRANILIC (3-NITROBENZYLIDENE)-HYDRAZIDE
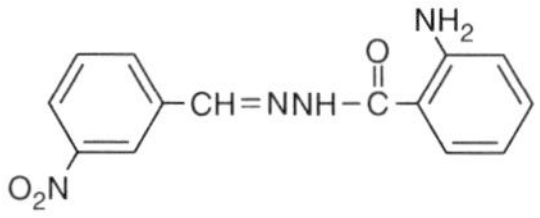
5 gm

S74,760-2
ANTHRANILIC (1-NAPHTHYLMETHYLENE)-HYDRAZIDE
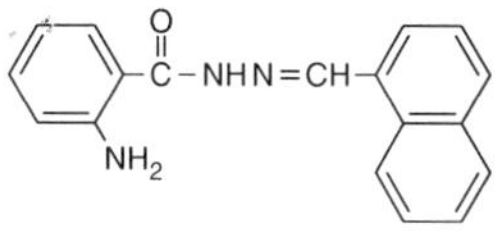
5 gm

S75,081-6
4-AMINOBENZOIC (4-METHYLBENZYLIDENE)HYDRAZIDE
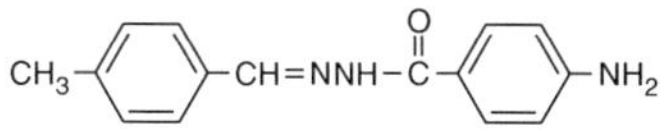
5 gm

S92,639-6
ACETIC 2-(2-AMINO-4-METHYLPHENYL)-HYDRAZIDE
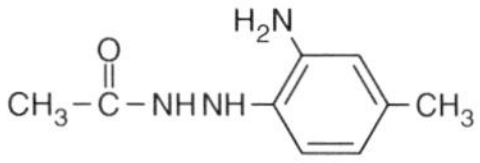
250 mg

S92,519-5
4-AMINOBENZOIC (ALPHA,4-DIMETHYLBENZYLIDENE)HYDRAZIDE
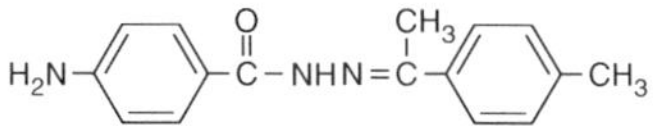
250 mg

S82,154-3
HEXADECANOIC N2-(2,4-DINITROPHENYL)-HYDRAZIDE
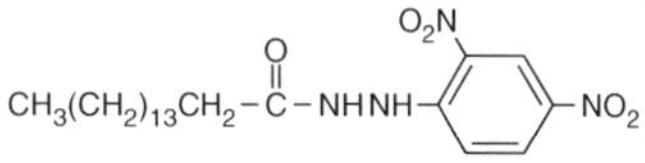
250 mg

S82,157-8
STEARIC N2-(2,4-DINITROPHENYL)-HYDRAZIDE
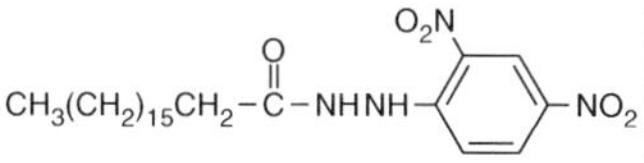
250 mg

S75,098-0
OXALIC BIS[(ALPHA-PROPYLBENZYLIDENE)HYDRAZIDE]
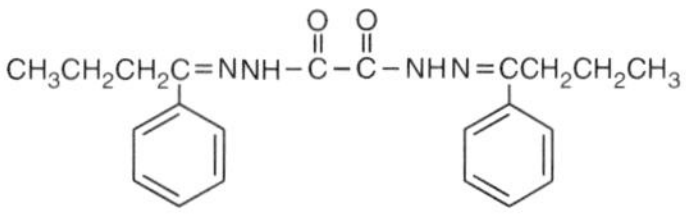
5 gm

S87,321-7
BENZOIC [2,6-BIS(ETHOXYCARBONYL)-1,4-DIHYDRO-1-METHYL-4-PYRIDYLIDENE]HYDRAZIDE
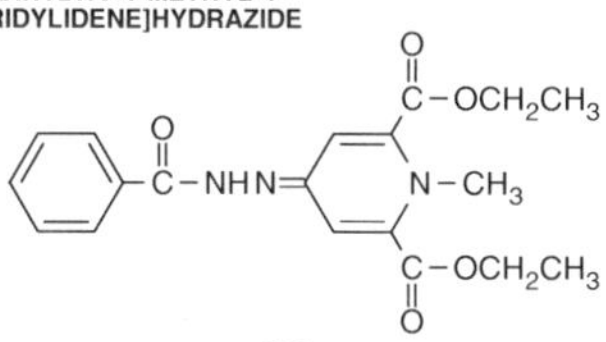
250 mg

S34,153-3
4-(BENZOYLHYDRAZINO)-1-BENZYLPIPERIDINE
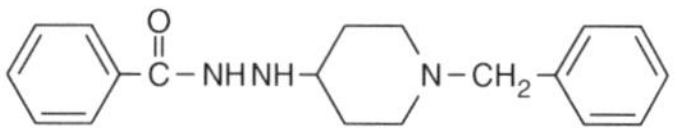
5 gm

S53,948-1
O-(DIMETHYLAMINO)BENZOIC ACID HYDRAZIDE
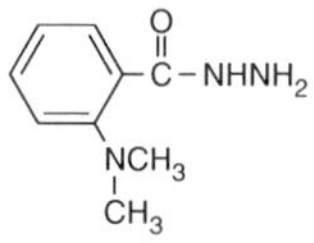
5 gm

S35,140-7
INDOLE-3-CARBOXALDEHYDE CYANOACETYLHYDRAZONE
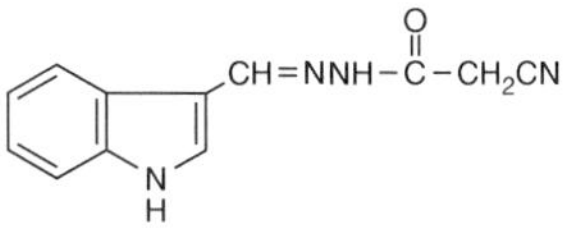
250 mg

S50,337-1
2-NITROBENZOIC ACID (5-INDANYLMETHYLENE)HYDRAZIDE
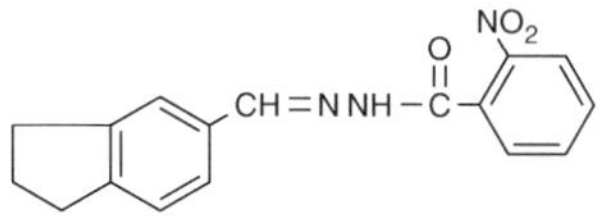
250 mg

S34,746-9
3,5-DINITROBENZHYDRAZIDE
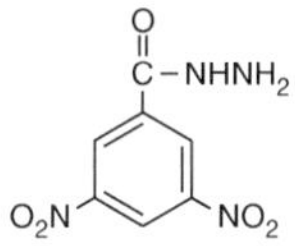
5 gm

S64,750-0
N-ACETYLANTHRANILIC ACID HYDRAZIDE
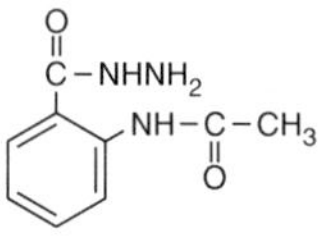
5 gm

S90,331-0
4-(DIMETHYLAMINO)BENZOIC (3-NITROBENZYLIDENE)HYDRAZIDE
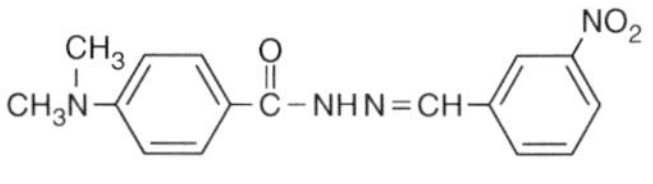
250 mg

S38,892-0
ACETIC ACID 2-(4-NITROPHENYL)-HYDRAZIDE
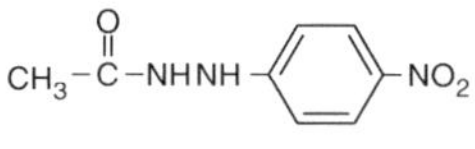
5 gm

S90,393-0
HIPPURIC (3-NITROBENZYLIDENE)-HYDRAZIDE
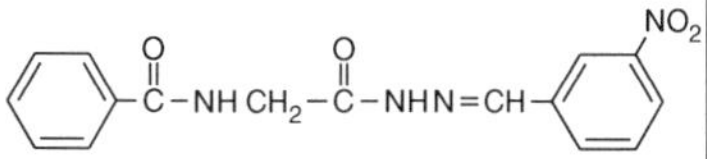
250 mg

S90,384-1
ADIPIC BIS[(3-NITROBENZYLIDENE)-HYDRAZIDE]
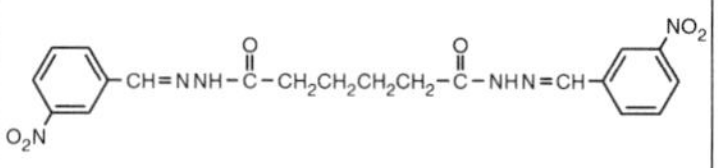
250 mg

S55,044-2
3-NITROBENZOIC ACID HYDRAZIDE
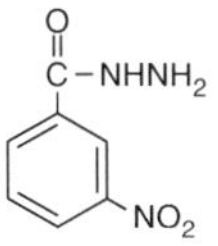
250 mg

S90,674-3
9,10-DIHYDRO-9,10-ETHANOANTHRACENE-11,12-DICARBOXYLIC DIHYDRAZIDE
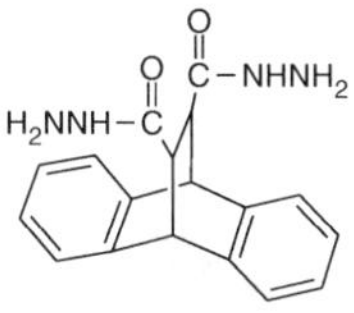
250 mg

S66,998-9
1-ACETYL-2-PHENYLACETYLHYDRAZINE
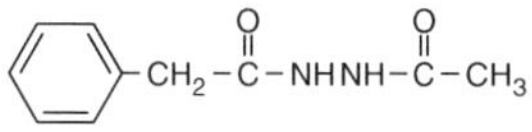
5 gm

S90,402-3
HIPPURIC (2-PHENYLETHYLIDENE)-HYDRAZIDE
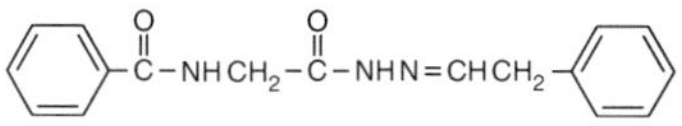
250 mg

S73,691-0
N-METHYLANTHRANILIC (3-NITROBENZYLIDENE)HYDRAZIDE
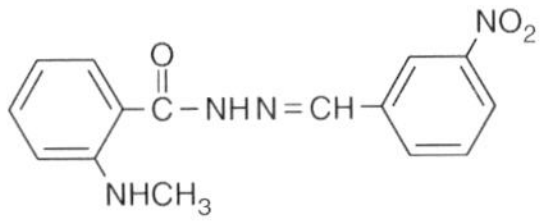
5 gm

S55,034-5
2-(METHYLAMINO)BENZOIC ACID HYDRAZIDE
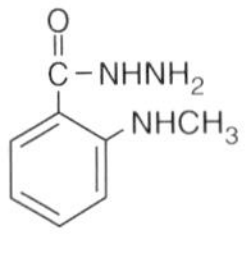
250 mg

S86,906-6
N-BENZOYLANTHRANILIC [4-(DIMETHYLAMINO)BENZYLIDENE]-HYDRAZIDE
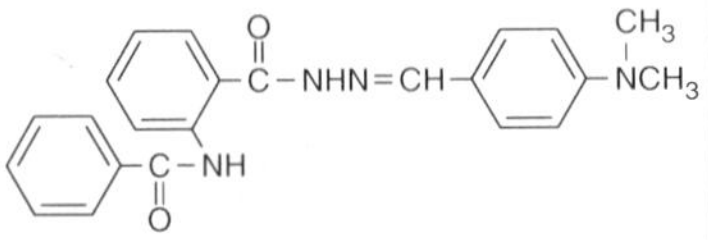
250 mg

S90,405-8
2-(1-NAPHTHYL)ACETIC [4-(DIMETHYLAMINO)BENZYLIDENE]-HYDRAZIDE
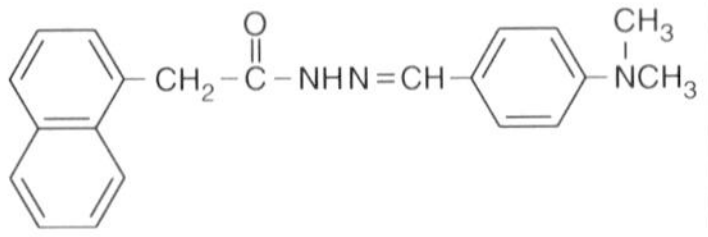
250 mg

S76,948-7
4-(3-ALLYLUREIDO)BENZOIC [4-(DIMETHYLAMINO)BENZYLIDENE]-HYDRAZIDE
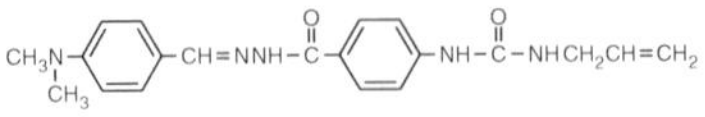
250 mg

S79,696-4
ADIPIC BIS[4-(DIMETHYLAMINO)-BENZYLIDENEHYDRAZIDE]
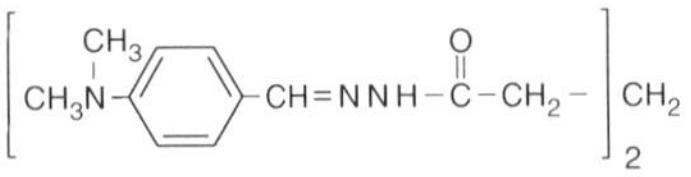
250 mg

S73,184-6
10-UNDECENOIC 4-(DIMETHYLAMINO)-BENZYLIDENEHYDRAZIDE
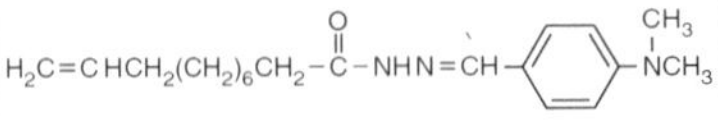
250 mg

S94,902-7
6-(DIETHYLAMINO)-1,4-(2H,3H)-PHTHALAZINEDIONE
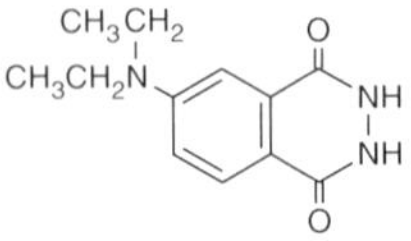
250 mg

S60,204-3
4-DIMETHYLAMINOBENZHYDRAZIDE
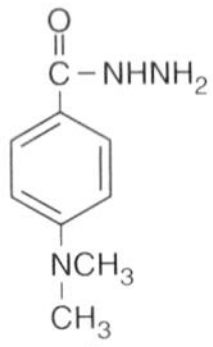
5 gm

S66,999-7
BENZOIC ACID BENZYLIDENEHYDRAZIDE
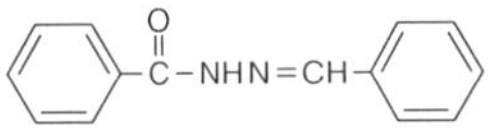
5 gm

S71,621-9
10-UNDECENOIC ACID N-BENZYLIDENEHYDRAZIDE
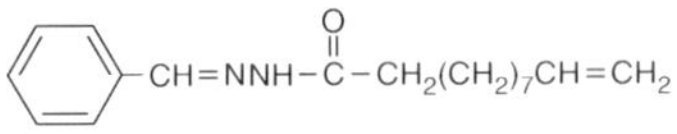
250 mg

S62,221-4
4-ISOPROPYLBENZALDEHYDE ACETYLHYDRAZONE
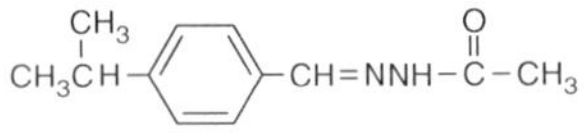
250 mg

S64,278-9
MALEIC ACID (3-NITROBENZYLIDENE)-HYDRAZIDE
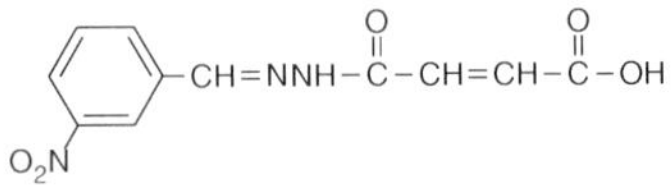
250 mg

S92,624-8
ACETIC 2-(2,4-DINITROPHENYL)-HYDRAZIDE
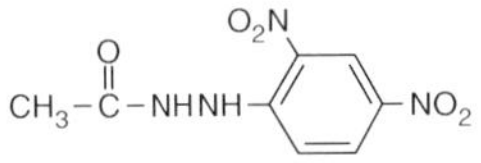
250 mg

S35,819-3
1-BENZOYL-2-(3-FLUOROBENZOYL)-HYDRAZINE
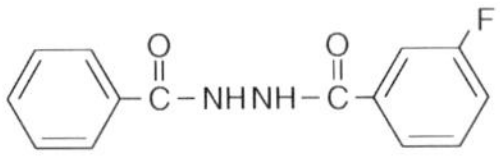
5 gm

S35,815-0
1-BENZOYL-2-(3-CHLOROBENZOYL)-HYDRAZINE
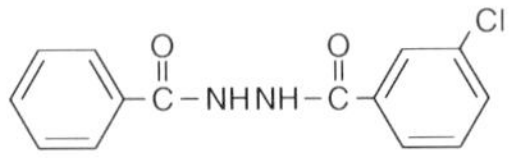
5 gm

S35,812-6
1-BENZOYL-2-(3-BROMOBENZOYL)-HYDRAZINE
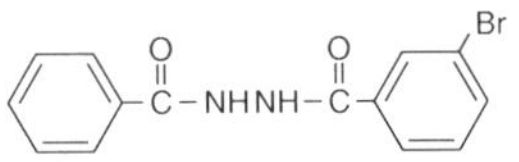
5 gm

S66,997-0
1-ACETYL-2-BENZOYLHYDRAZINE
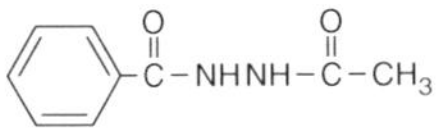
5 gm

S36,011-2
1,2-DIBENZOYL-1-METHYLHYDRAZINE
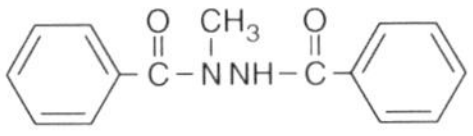
5 gm

S82,273-6
1,2-DIBENZOYL-1,2-DIPHENYLHYDRAZINE
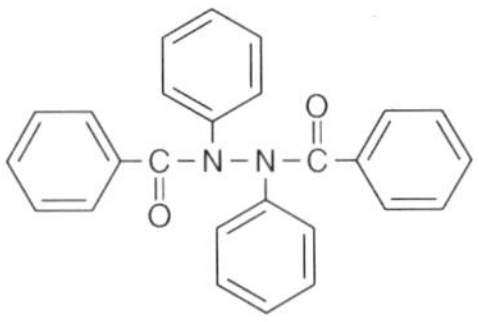
250 mg

S35,818-5
1-BENZOYL-2-(2-FLUOROBENZOYL)-HYDRAZINE
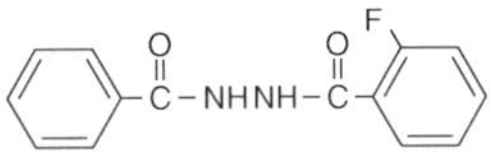
250 mg

S35,814-2
1-BENZOYL-2-(2-CHLOROBENZOYL)-HYDRAZINE
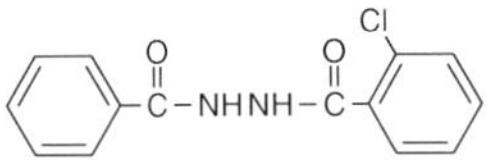
5 gm

S35,822-3
1-BENZOYL-2-(2-METHYLBENZOYL)-HYDRAZINE
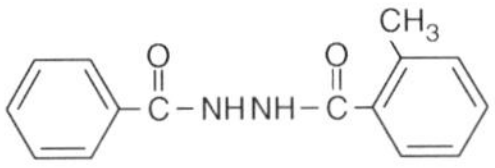
5 gm

S36,199-2
1-(2-FLUOROBENZOYL)-2-(2-NAPHTHOYL)-HYDRAZINE
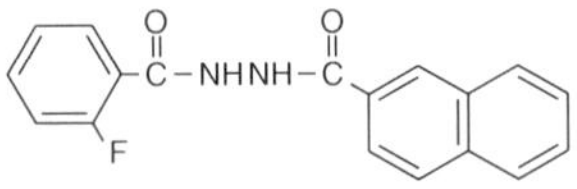
250 mg

S72,634-6
1-(2,2-DIMETHYLPROPIONYL)-2-(4-METHYLBENZOYL)HYDRAZINE
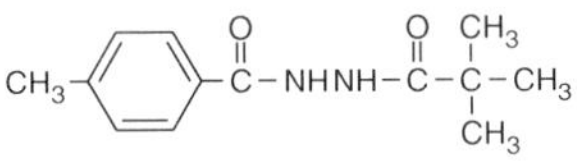
5 gm

S35,820-7
1-BENZOYL-2-(4-FLUOROBENZOYL)-HYDRAZINE
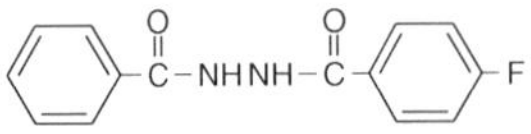
5 gm

S35,816-9
1-BENZOYL-2-(4-CHLOROBENZOYL)-HYDRAZINE
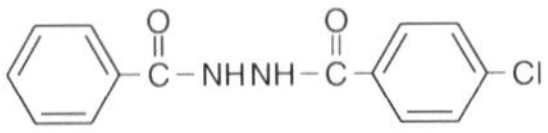
5 gm

S34,906-2
1,2-BIS(4-CHLOROBENZOYL)-HYDRAZINE
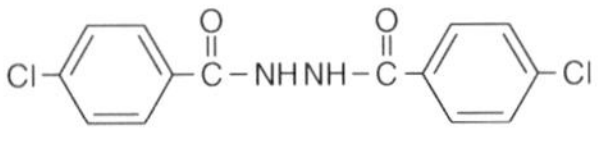
250 mg

S36,586-6
1-(4-BROMOBENZOYL)-2-(1-NAPHTHOYL)-HYDRAZINE
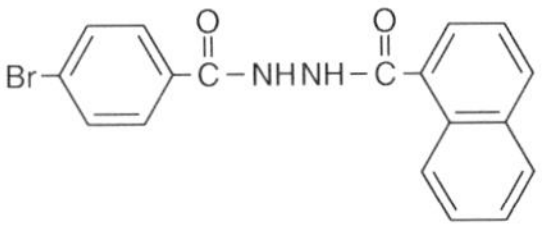
5 gm

S36,340-5
1-(4-METHYLBENZOYL)-2-(1-NAPHTHOYL)-HYDRAZINE
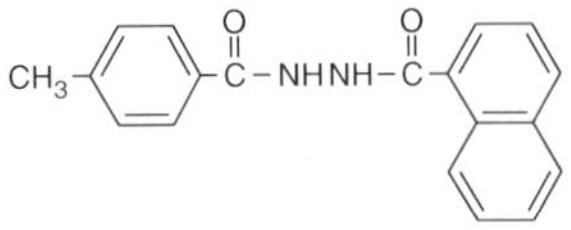
250 mg

S36,341-3
1-(4-METHYLBENZOYL)-2-(2-NAPHTHOYL)-HYDRAZINE
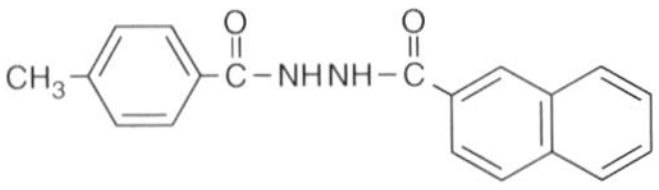
250 mg

S36,585-8
1-BENZOYL-2-(2,4-DICHLOROBENZOYL)-HYDRAZINE
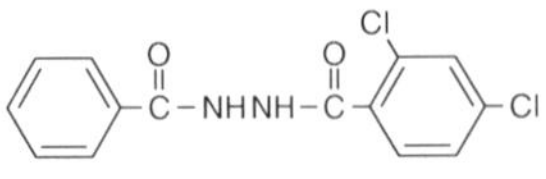
250 mg

S35,867-3
1,2-BIS(3,4-DICHLOROBENZOYL)-HYDRAZINE

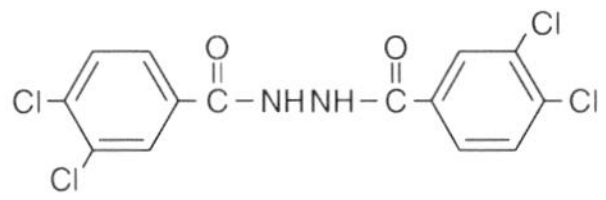

250 mg

S36,041-4
1,2-BIS(2,5-DIBROMOBENZOYL)-HYDRAZINE

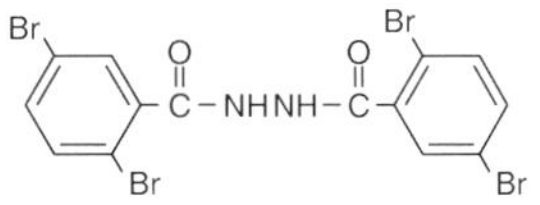

250 mg

S35,884-3
1-(3-BROMOBENZOYL)-2-(1-NAPHTHOYL)-HYDRAZINE

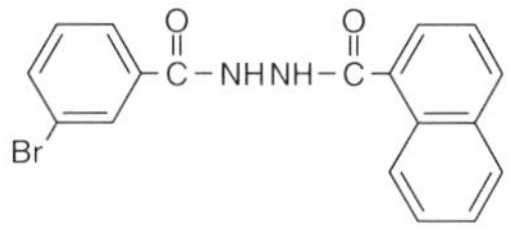

5 gm

S35,885-1
1-(3-BROMOBENZOYL)-2-(2-NAPHTHOYL)-HYDRAZINE

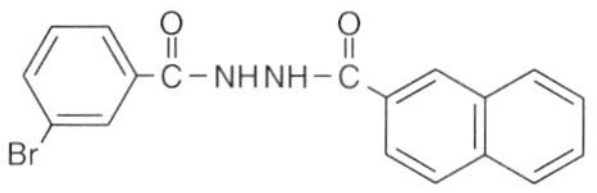

250 mg

S55,718-8
PROPIONIC ACID 2-PHENYLHYDRAZIDE

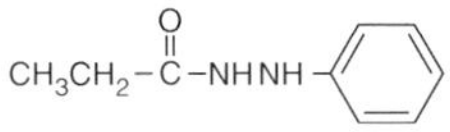

5 gm

S84,923-5
1,2-BIS(2-PHENYL-4-PENTENOYL)-HYDRAZINE

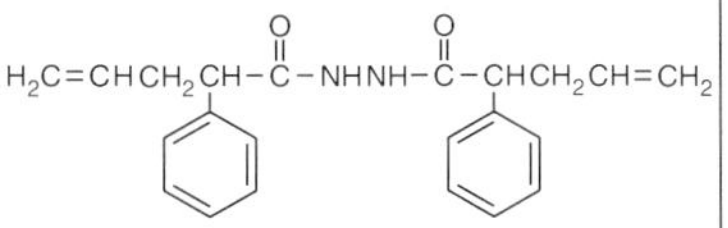

250 mg

S41,267-8
1,2-BIS(2-FLUOROBENZOYL)HYDRAZINE

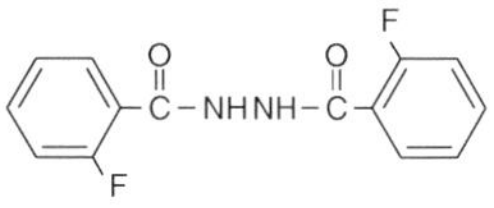

250 mg

S56,992-5
HEXADECANOIC ACID (2,4,6-TRICHLOROPHENYL)HYDRAZIDE

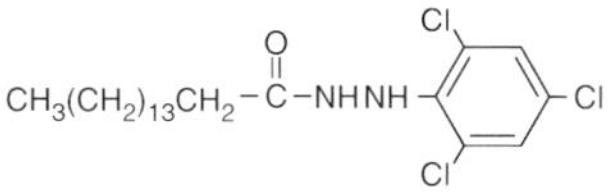

250 mg

S68,003-6
1-ACETYL-1-PHENYLHYDRAZINE

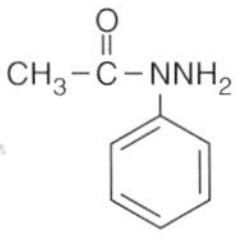

5 gm

S74,763-7
CINNAMIC (3-NITROBENZYLIDENE)-HYDRAZIDE

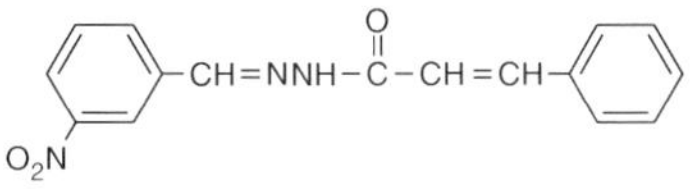

5 gm

S73,100-5
2-(1-NAPHTHYL)ACETIC 3-NITROBENZYLIDENEHYDRAZIDE

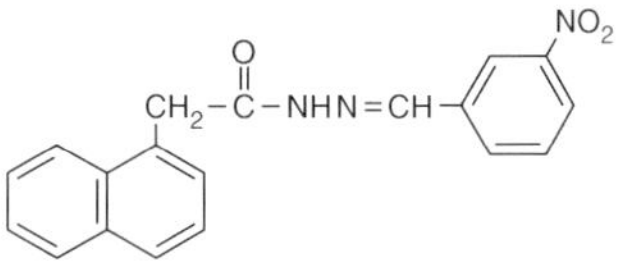

5 gm

S69,425-8
2-(CARBOXYMETHYLENE)ACETIC ACID N2-BENZYLIDENEHYDRAZIDE

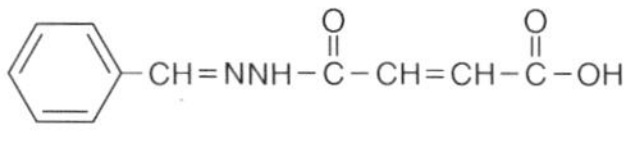

250 mg

S90,401-5
HIPPURIC [3,4-(METHYLENEDIOXY)-BENZYLIDENE]HYDRAZIDE

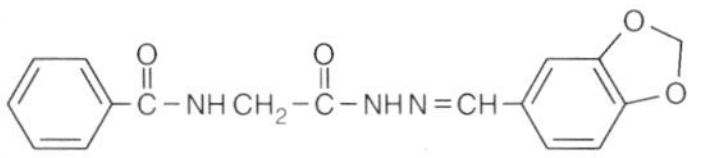

250 mg

S90,397-3
HIPPURIC (4-ACETAMIDOBENZYLIDENE)-HYDRAZIDE

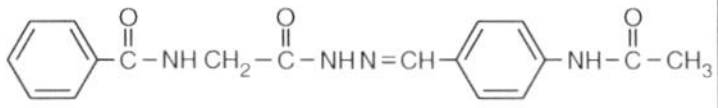

250 mg

S90,396-5
HIPPURIC (1-NAPHTHYLMETHYLENE)-HYDRAZIDE

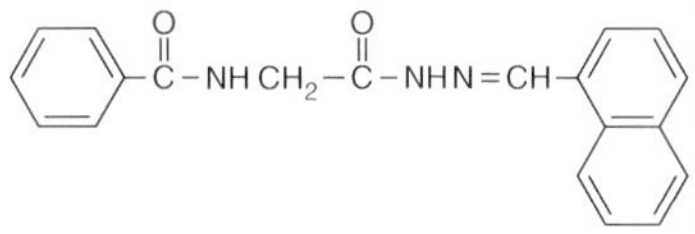

250 mg

S90,398-1
HIPPURIC [1-(2-NAPHTHYL)ETHYLIDENE]-HYDRAZIDE

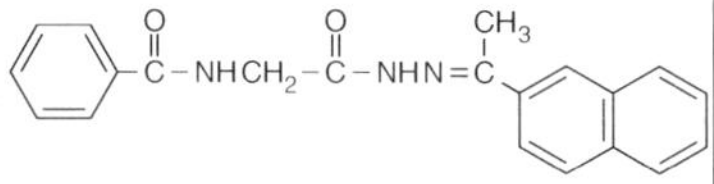

250 mg

S71,614-6
HIPPURIC ACID N-CYCLOPENTYLIDENEHYDRAZIDE

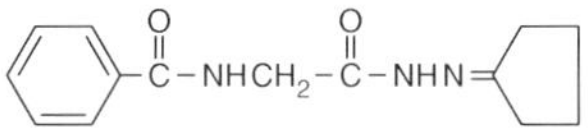

5 gm

S79,689-1
HIPPURIC (1-PHENYL-1-BUTYLIDENE)-HYDRAZIDE

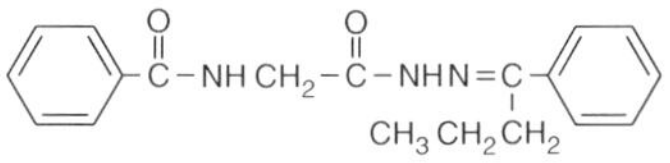

250 mg

S76,943-6
2-BENZAMIDOBENZOIC (3-PHENYLPROPYLIDENE)HYDRAZIDE

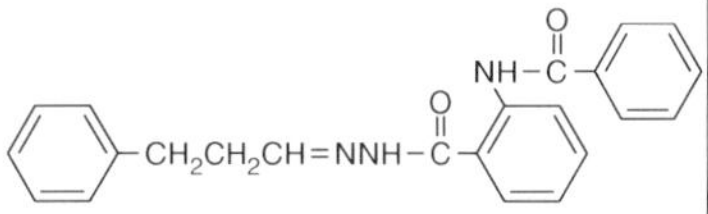

250 mg

S75,085-9
CINNAMIC (4-ACETAMIDOBENZYLIDENE)-HYDRAZIDE

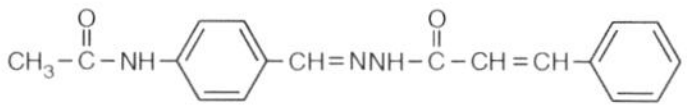

5 gm

S75,078-6
BENZOIC (CYCLOHEXYLMETHYLENE)-HYDRAZIDE

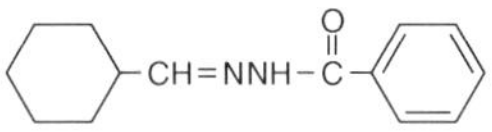

5 gm

S86,214-2
BENZOIC BENZYLIDENEHYDRAZIDE

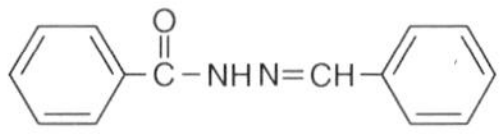

250 mg

S73,715-1
BENZOIC (1-METHYLOCTYLIDENE)-HYDRAZIDE

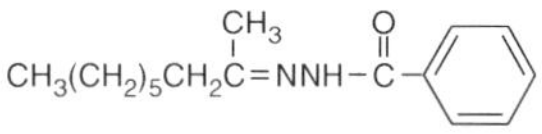

5 gm

S73,164-1
4-METHYLBENZOIC CINNAMYLIDENEHYDRAZIDE

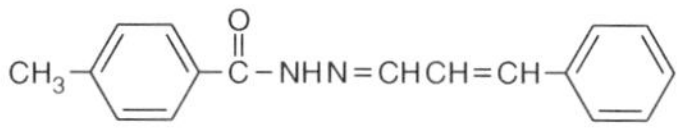

5 gm

S75,075-1
BENZOIC (1-NAPHTHYLMETHYLENE)-HYDRAZIDE

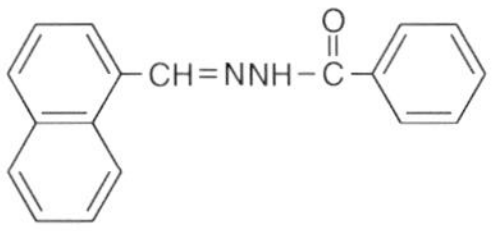

5 gm

S50,633-8
P-TOLUIC ACID (5-INDANYLMETHYLENE)-HYDRAZIDE

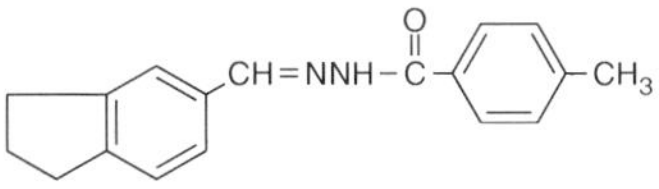

250 mg

S50,656-7
4-(5-INDANYLMETHYLENEAMINO)BENZOIC ACID (5-INDANYLMETHYLENE)HYDRAZIDE

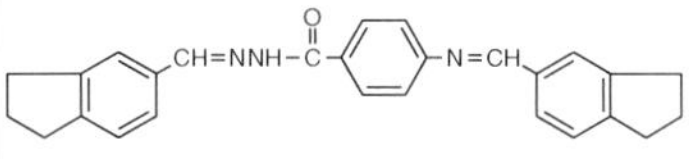

250 mg

S74,360-7
4-METHYLBENZOIC (1-NAPHTHYLMETHYLENE)HYDRAZIDE

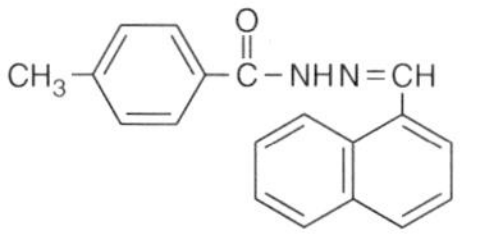

250 mg

S90,150-4
OXALIC BIS[(3-PHENYLALLYLIDENE)-HYDRAZIDE]

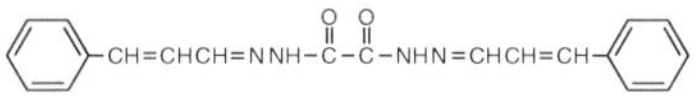

250 mg

S73,611-2
OXALIC BIS([1-(1-NAPHTHYL)-ETHYLIDENE]HYDRAZIDE)

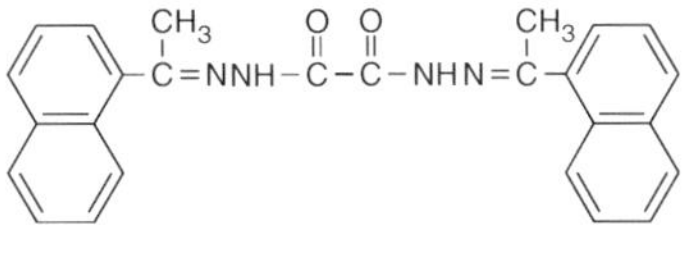

5 gm

S86,922-8
SEBACIC BIS(CINNAMYLIDENEHYDRAZIDE)

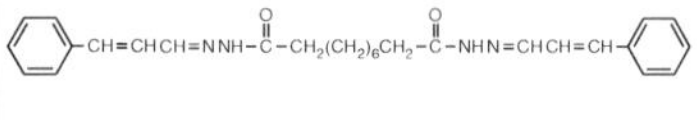

250 mg

S73,613-9
ADIPIC BIS[(1-NAPHTHYLMETHYLENE)-HYDRAZIDE]

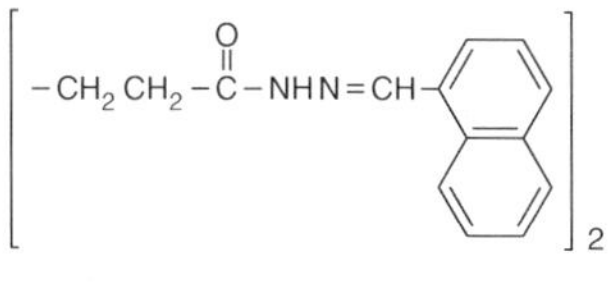

5 gm

S95,345-8
ETHYL N-(BENZYLIDENEAMINO)CARBAMATE
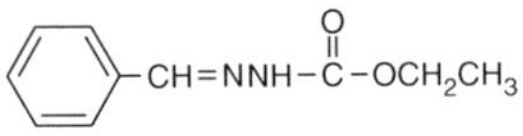
50 mg

S90,408-2
2-CYCLOHEXYLACETIC (4-ACETAMIDOBENZYLIDENE)HYDRAZIDE
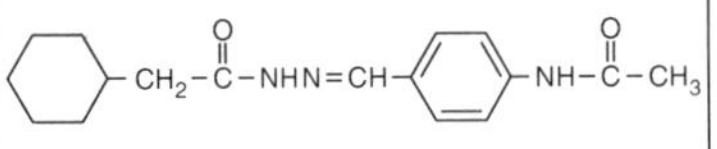
250 mg

S73,872-7
STEARIC (4-ACETAMIDOBENZYLIDENE)-HYDRAZIDE
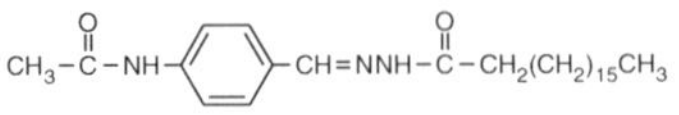
5 gm

S77,318-2
TETRAHYDRO-2-FUROIC (ALPHA-PROPYLBENZYLIDENE)HYDRAZIDE
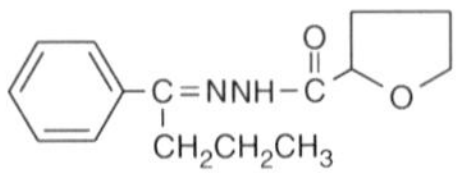
250 mg

S52,604-5
PIPERONYLIC ACID PIPERONYLIDENEHYDRAZIDE
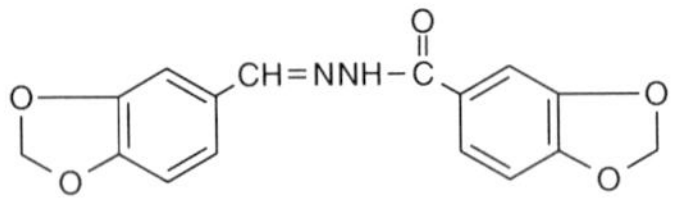
5 gm

S54,285-7
BENZOIC ACID CYCLOHEXYLIDENEHYDRAZIDE
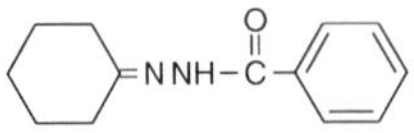
250 mg

S87,224-5
ACETIC BENZYLIDENEHYDRAZIDE
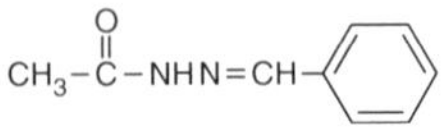
250 mg

S42,125-1
VALERIC ACID CINNAMYLIDENEHYDRAZIDE
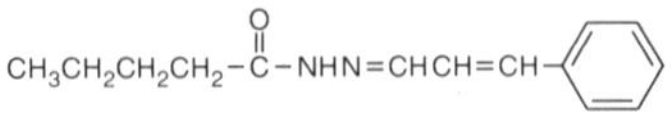
250 mg

S88,624-6
ACETIC N2-BENZYLIDENE-N1-PHENYLHYDRAZIDE
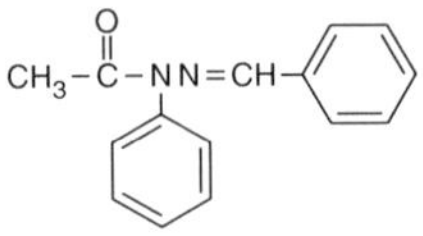
250 mg

S76,909-6
DODECANOIC (ALPHA-PROPYLBENZYLIDENE)HYDRAZIDE
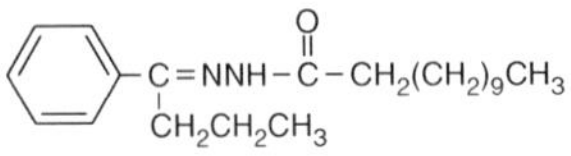
250 mg

S89,792-2
OCTADECANOIC BENZYLIDENEHYDRAZIDE
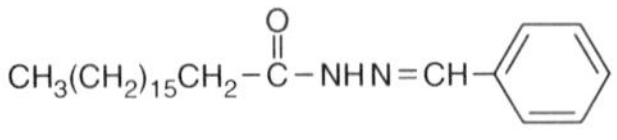
250 mg

S86,920-1
OCTADECANOIC CINNAMYLIDENEHYDRAZIDE
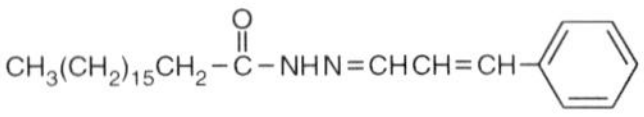
250 mg

S73,866-2
STEARIC (3-PHENYLPROPYLIDENE)-HYDRAZIDE
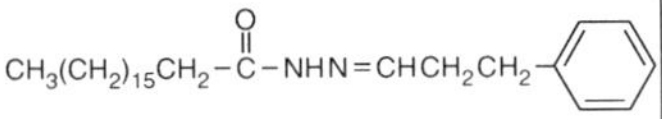
5 gm

S73,916-2
STEARIC (BETA-HEXYLCINNAMYLIDENE)-HYDRAZIDE
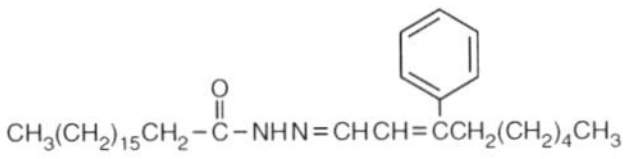
5 gm

S87,613-5
SEBACIC BIS[(ALPHA-HEXYLCINNAMYLIDENE)HYDRAZIDE]
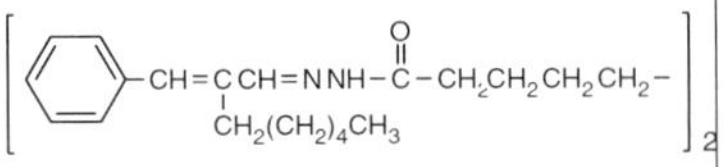
250 mg

S59,868-2
ACETIC ACID DIPHENYLMETHYLENEHYDRAZIDE
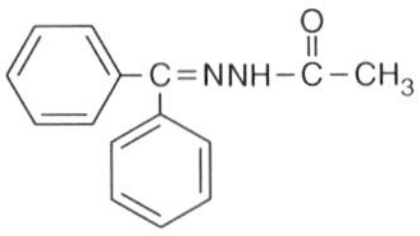
250 mg

S90,324-8
ACETIC (4-ISOPROPYLBENZYLIDENE)-HYDRAZIDE
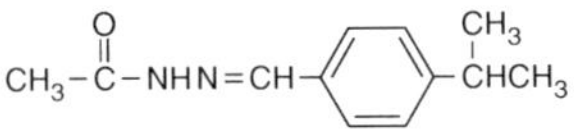
250 mg

S74,357-7
STEARIC (4-METHYLBENZYLIDENE)-HYDRAZIDE
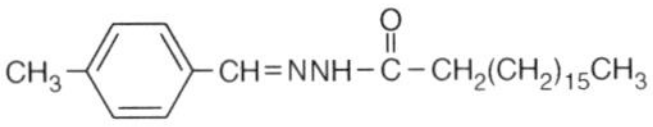
5 gm

S74,785-8
N,N'-TEREPHTHALYLIDENEBIS(10-UNDECENOIC HYDRAZIDE)
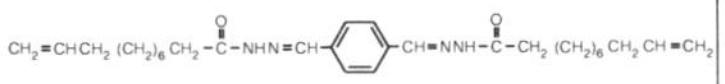
5 gm

S54,815-4
ACETIC ACID (3,4,5-TRIMETHOXYBENZYLIDENE)HYDRAZIDE
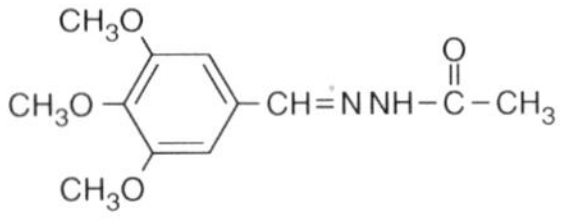
5 gm

S71,624-3
STEARIC ACID (4-PYRIDYLMETHYLENE)-HYDRAZIDE
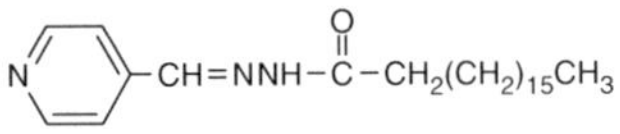
5 gm

S50,339-8
1-NAPHTHALENEACETIC ACID (5-INDANYLMETHYLENE)HYDRAZIDE
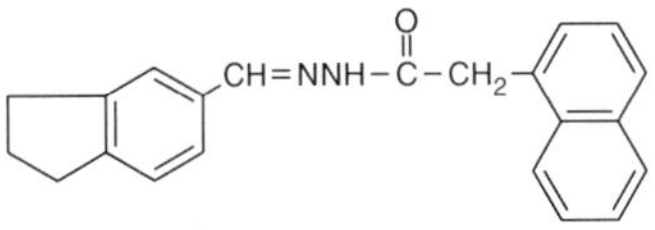
250 mg

S90,325-6
ACETIC (1-NAPHTHYLMETHYLENE)-HYDRAZIDE
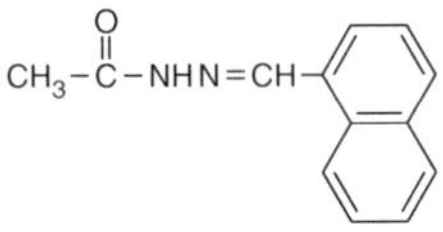
250 mg

S75,088-3
STEARIC (1-NAPHTHYLMETHYLENE)-HYDRAZIDE
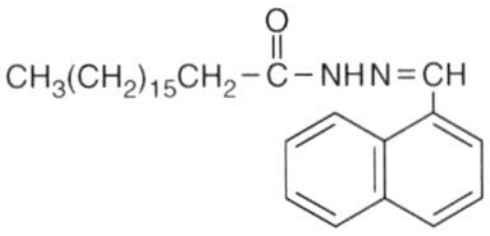
5 gm

S50,627-3
ETHYL (5-INDANYLMETHYLENE)-CARBAZATE
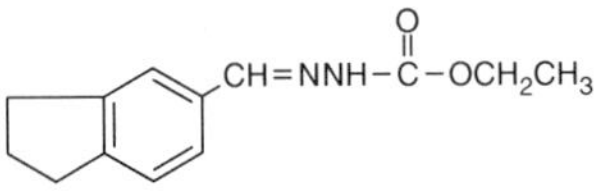
250 mg

S54,868-5
1,2-DIACETYL-1-(2,4-DINITROPHENYL)-2-(1-METHYLPROPENYL)HYDRAZINE
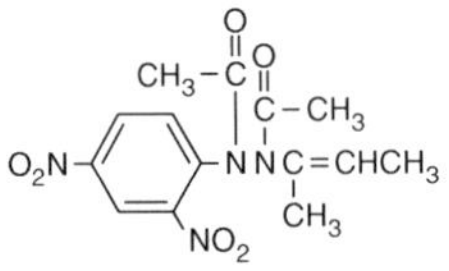
250 mg

S67,742-6
2-(4-METHYLPHENYLCARBOXAMIDO)-BENZOIC HYDRAZIDE
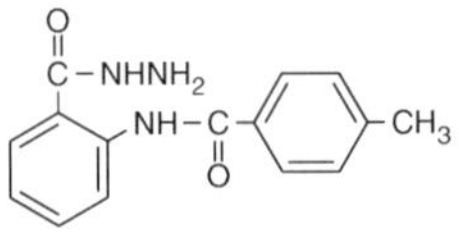
250 mg

S62,791-7
4-(DECANOYLAMINO)BENZOIC HYDRAZIDE
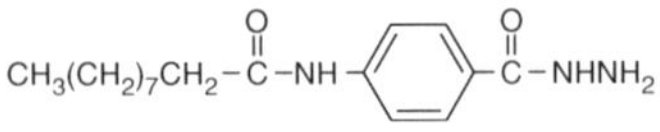
250 mg

S50,183-2
BENZOIC ACID (5-INDANYLMETHYLENE)-HYDRAZIDE
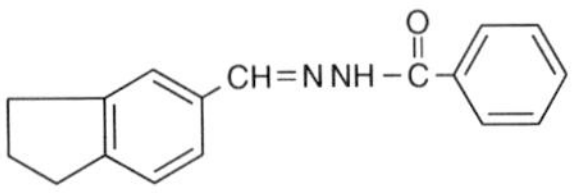
250 mg

S85,026-8
6-BROMO-1,4(2H,3H)-PHTHALAZINEDIONE
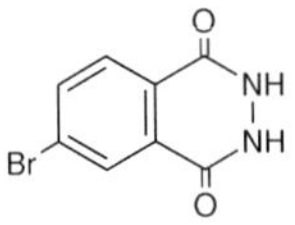
250 mg

S84,899-9
2-BUTYL-1,4(2H,3H)-PHTHALAZINEDIONE
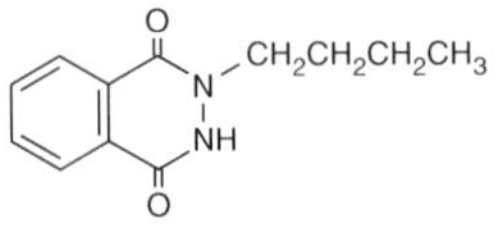
250 mg

S55,032-9
CINNAMIC ACID HYDRAZIDE
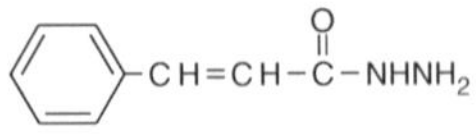
250 mg

S68,227-6
BENZOIC ACID N2-PHENYLHYDRAZIDE
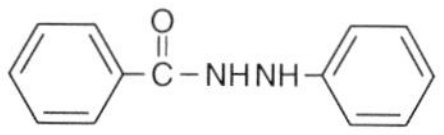
250 mg

S34,860-0
1,2-DIBENZOYL-1,2-DIMETHYLHYDRAZINE
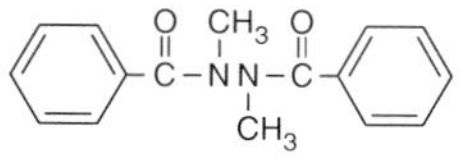
5 gm

S81,781-3
N1,N2-DIPHENYLBENZOIC HYDRAZIDE
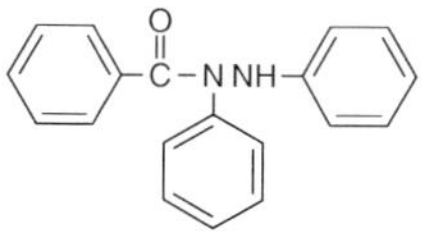
250 mg

S61,898-5
O-TOLUIC ACID HYDRAZIDE
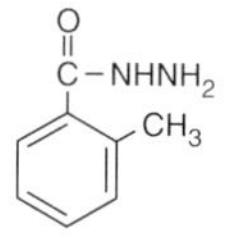
5 gm

S36,587-4
1-(2,4-DICHLOROBENZOYL)-2-(1-NAPHTHOYL)HYDRAZINE
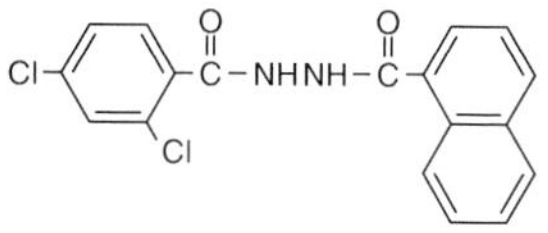
250 mg

S97,431-5
AR-E94 7,12-DIMETHYLBENZ(A)-ANTHRACENE-5-CARBOXYLIC ACID HYDRAZIDE
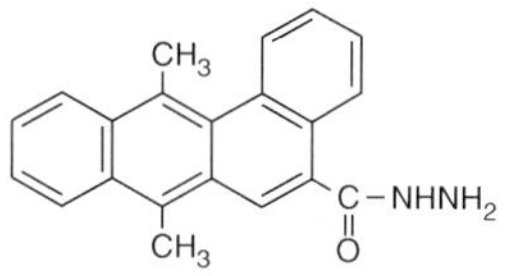
10 mg

S50,621-4
OXALIC ACID (3,4,5-TRIMETHOXYBENZYLIDENE)HYDRAZIDE
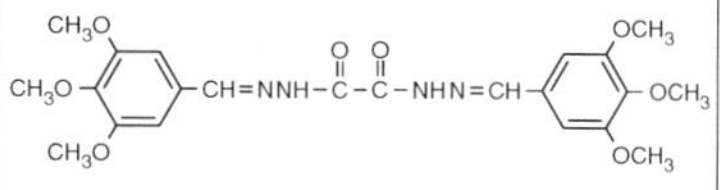
250 mg

S50,620-6
OXALIC ACID (5-INDANYLMETHYLENE)-HYDRAZIDE
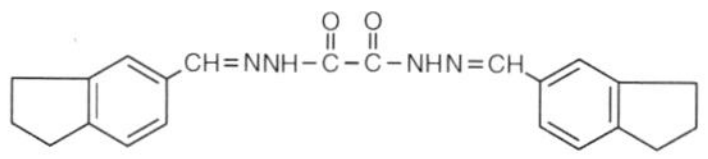
250 mg

S50,184-0
MALONIC ACID (5-INDANYLMETHYLENE)-HYDRAZIDE
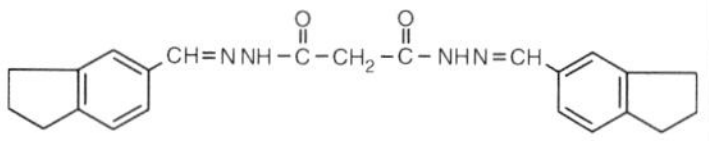
250 mg

S74,737-8
2-[N-(2,5-DIMETHYLBENZYL)CARBAMOYL]-CYCLOHEXANECARBOXYLIC ACID
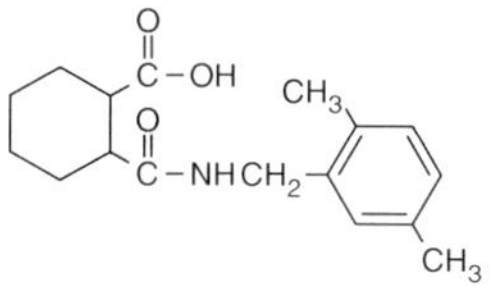
250 mg

S70,348-6
N-ANILINOSUCCINAMIC ACID
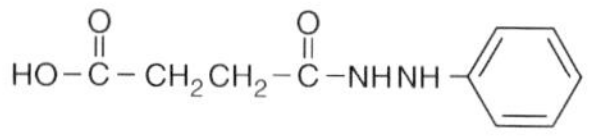
5 gm

S70,379-6
N-ANILINOGLUTARAMIC ACID
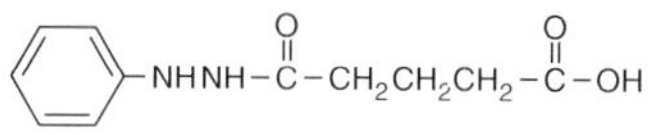
5 gm

S84,924-3
3-(METHOXYCARBONYL)PROPIONIC N1,N2-BIS(P-TOLYL)HYDRAZIDE
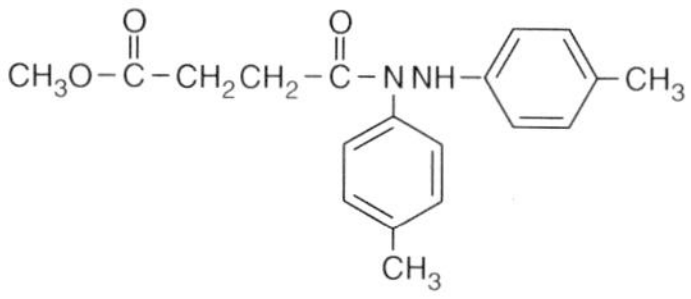
250 mg

S50,185-9
TERT-BUTYL (5-INDANYLMETHYLENE)-CARBAZATE
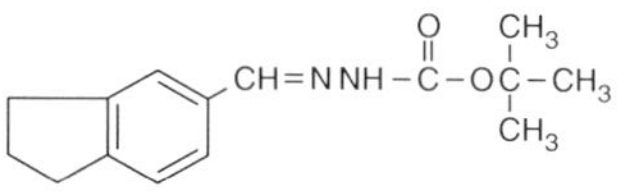
5 gm

S54,791-3
ISOPHTHALIC ACID DIHYDRAZIDE
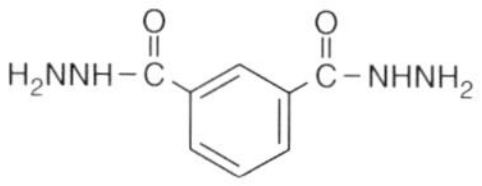
250 mg

S51,268-0
ETHYL 3-PHENYLCARBAZATE
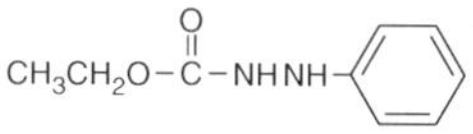
250 mg

S85,287-2
9,10-DIHYDRO-9,10-ETHANOANTHRACENE-11-CARBOXYLIC HYDRAZIDE
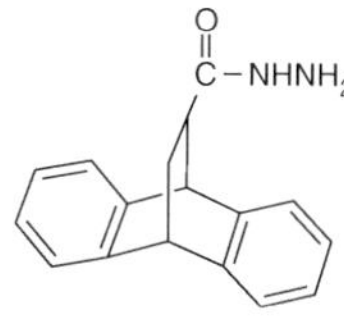
250 mg

S86,586-9
9,10-DIHYDRO-11-METHYL-9,10-ETHANOANTHRACENE-11-CARBOXYLIC HYDRAZIDE
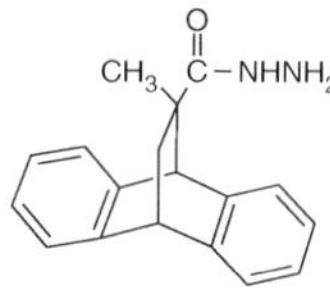
250 mg

S54,892-8
ACETIC ACID (1-METHYL-2-PHENYL)-HYDRAZIDE
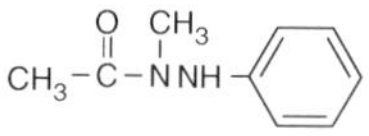
250 mg

S80,834-2
2-PHENYLACETIC N2-METHYLHYDRAZIDE
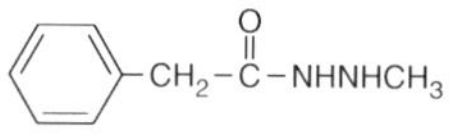
250 mg

S93,365-1
HYDROCINNAMIC HYDRAZIDE
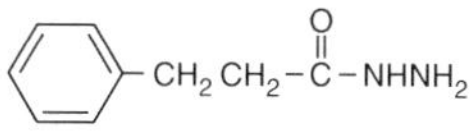
250 mg

S82,152-7
TRIDECANOIC N2-PHENYLHYDRAZIDE
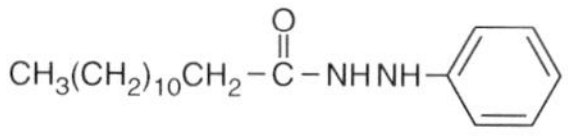
250 mg

S82,155-1
HEXADECANOIC N2-PHENYLHYDRAZIDE
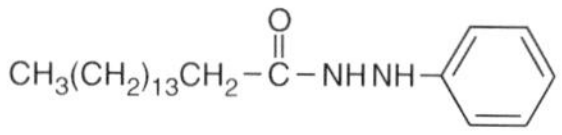
250 mg

S82,158-6
STEARIC N2-PHENYLHYDRAZIDE
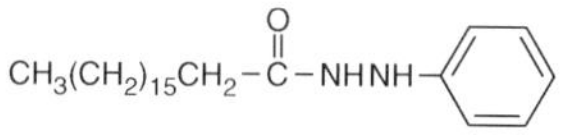
250 mg

S55,036-1
2-IODOBENZOIC ACID HYDRAZIDE
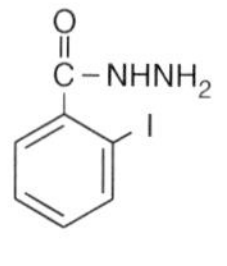
250 mg

S41,773-4
1-(2-CHLOROBENZOYL)-2-(1-NAPHTHOYL)-HYDRAZINE
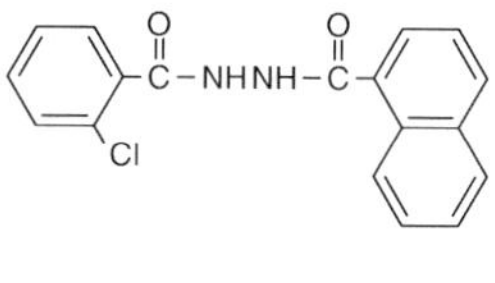
5 gm

S76,391-8
PALMITIC N2-(2,6-XYLYL)HYDRAZIDE
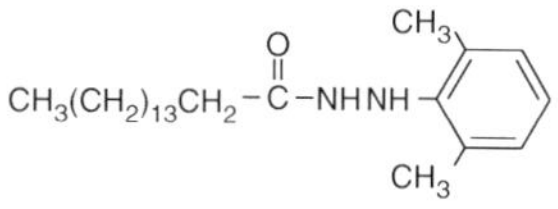
250 mg

S55,043-4
2,4-DICHLOROBENZOIC ACID HYDRAZIDE
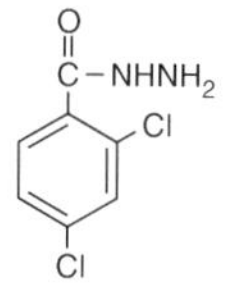
250 mg

S50,346-0
FORMIC ACID (5-INDANYLMETHYLENE)-HYDRAZIDE
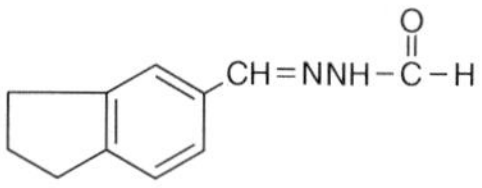
250 mg

S50,623-0
STEARIC ACID (5-INDANYLMETHYLENE)-HYDRAZIDE
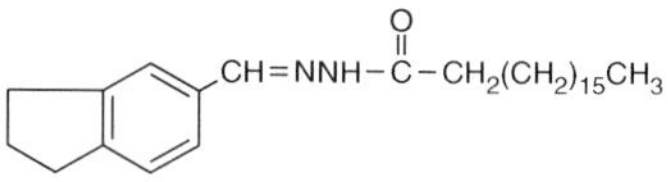
250 mg

S43,589-9
1-NAPHTHYLACETIC HYDRAZIDE
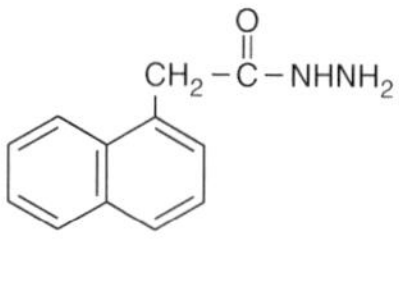
5 gm

S91,290-5
2-(1-NAPHTHYL)ACETIC HYDRAZIDE
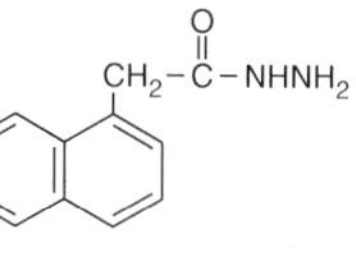
250 mg

S36,403-7
2-NAPHTHYLACETIC HYDRAZIDE

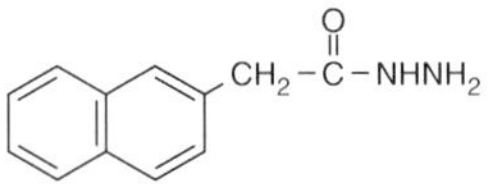

5 gm

S40,918-9
1-(3-METHYLBENZOYL)-2-(2-NAPHTHOYL)-HYDRAZINE

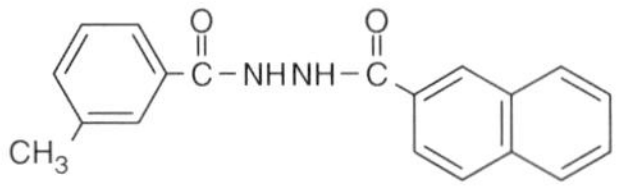

250 mg

S90,486-4
3-(2-CHLOROETHYL)-1-(2,3-DICHLOROPHENYL)-1-METHYLUREA

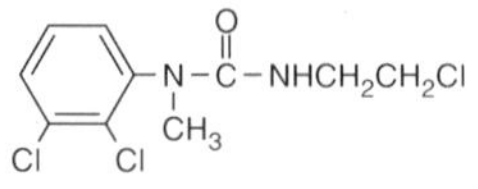

250 mg

S50,297-9
1,3-BIS(2-CHLOROETHYL)-1,3-DIBENZOYLUREA

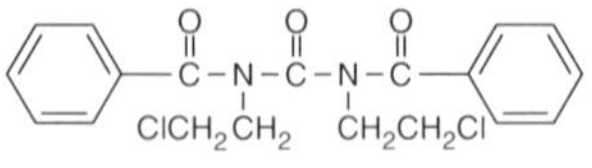

5 gm

S72,173-5
2-(CHLOROMETHYL)-2,3-DIHYDRO-S-TRIAZOLO(4,3-A)PYRIDIN-3-ONE

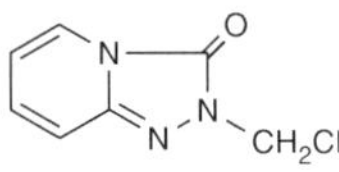

5 gm

S75,298-3
1-(2-CHLOROETHYL)-3-(P-TOLYL)UREA

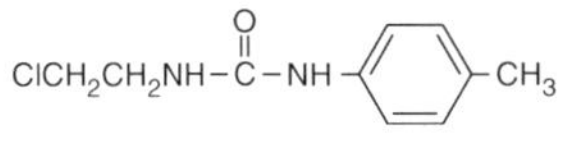

5 gm

S74,972-9
2-(P-TOLYLUREIDO)ETHYL 2,2-DICHLOROACETATE

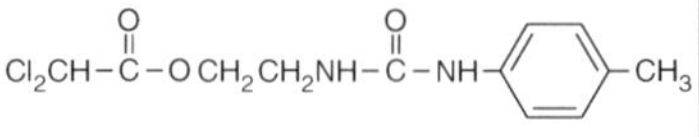

250 mg

S68,968-8
4-[4-CHLORO-2-(TRIFLUOROMETHYL)-PHENYL]-1-(2-HYDROXYBENZOYL)-SEMICARBAZIDE

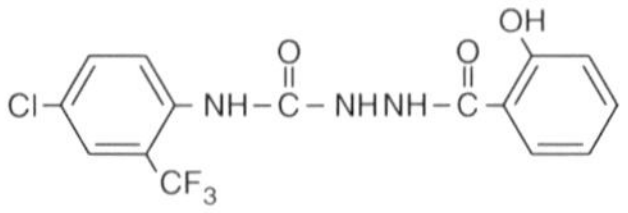

250 mg

S77,969-5
1-[4-CHLORO-2-(TRIFLUOROMETHYL)-PHENYL]-3-(2-HYDROXYETHYL)UREA

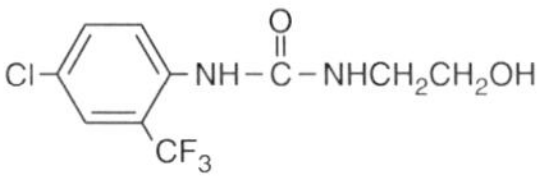

250 mg

S68,744-8
4-[4-CHLORO-2-(TRIFLUOROMETHYL)-PHENYL]-1-(4-HYDROXYBENZOYL)-SEMICARBAZIDE

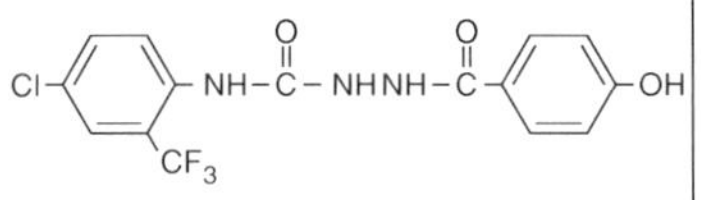

250 mg

S69,320-0
4-[4-BROMO-2-(TRIFLUOROMETHYL)-PHENYL]-1-(4-HYDROXYBENZOYL)-SEMICARBAZIDE

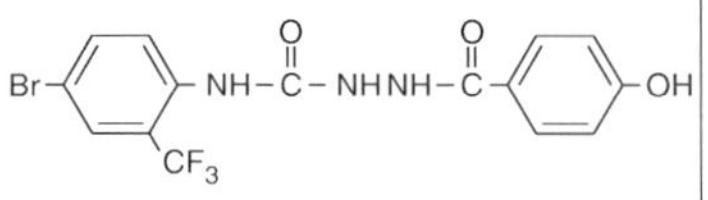

250 mg

S80,061-9
1-[2-CHLORO-5-(TRIFLUOROMETHYL)-PHENYL]-3-(4-HYDROXYPHENYL)UREA

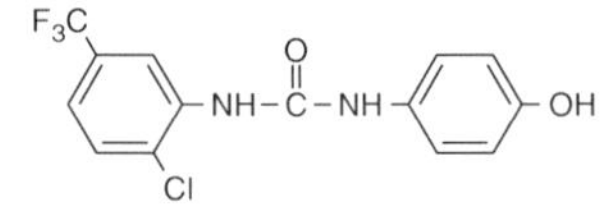

250 mg

S69,566-1
1-(5-CHLORO-2,4-DIMETHOXYPHENYL)-3-[2-(TRIFLUOROMETHYL)PHENYL]UREA

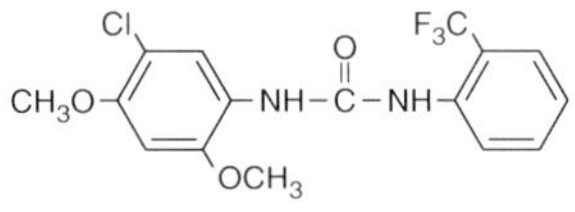

250 mg

S91,268-9
1-(5-CHLORO-2-METHOXYPHENYL)-3-[4-NITRO-3-(TRIFLUOROMETHYL)-PHENYL]UREA

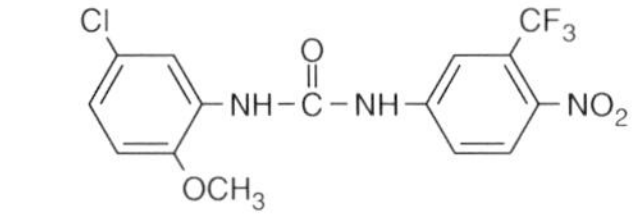

250 mg

S79,668-9
1-(4-ACETAMIDO-3-METHOXYPHENYL)-3-[4-CHLORO-3-(TRIFLUOROMETHYL)-PHENYL]UREA

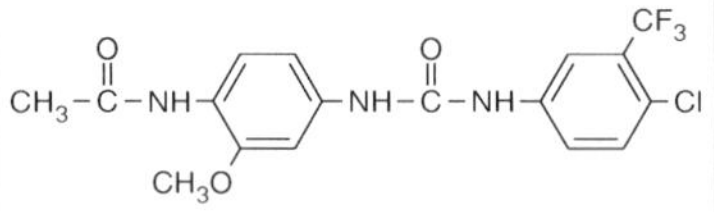

250 mg

S72,281-2
1-[2-CHLORO-5-(TRIFLUOROMETHYL)-PHENYL]-6-(2-METHOXY-5-METHYL-PHENYL)-3-METHYLBIUREA

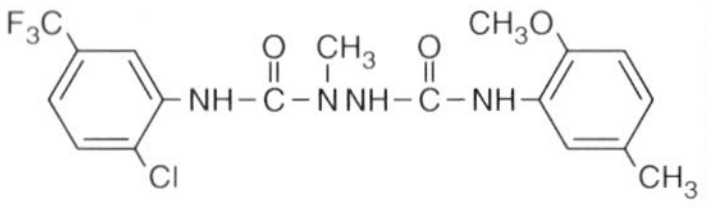

250 mg

S77,950-4
1-(3-BUTOXYPROPYL)-3-[4-CHLORO-2-(TRIFLUOROMETHYL)PHENYL]UREA

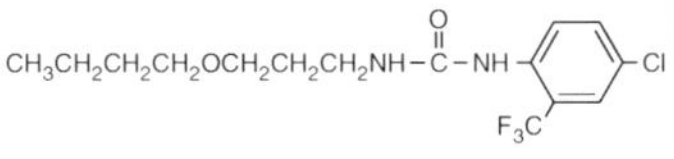

250 mg

S74,777-7
1-[4-CHLORO-2-(TRIFLUOROMETHYL)-PHENYL]-3-[2-(2,4-DINITROPHENYLTHIO)PHENYL]UREA

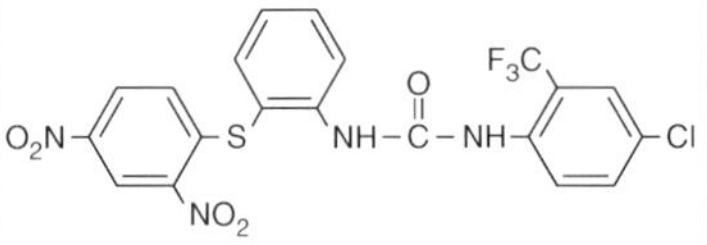

5 gm

S77,306-9
1-(4-CHLORO-2-NITROPHENYL)-3-[2-CHLORO-5-(TRIFLUOROMETHYL)-PHENYL]UREA

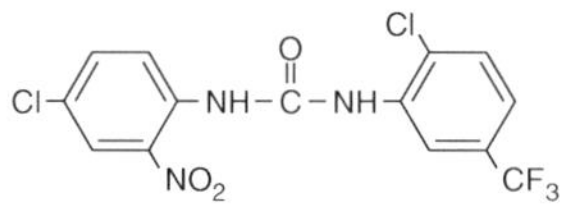

250 mg

S84,454-3
1-(2-FLUORO-5-NITROPHENYL)-3-[2-(TRIFLUOROMETHYL)PHENYL]UREA

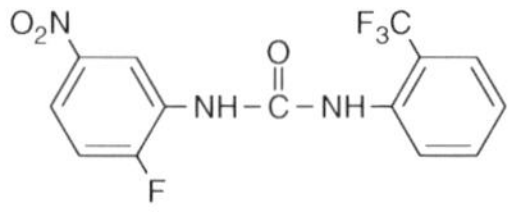

250 mg

S72,822-5
1-(4-FLUORO-3-NITROPHENYL)-3-[3-(TRIFLUOROMETHYL)PHENYL]UREA

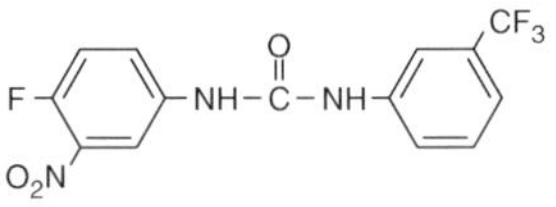

5 gm

S86,743-8
1-[2-CHLORO-5-(TRIFLUOROMETHYL)-PHENYL]-3-(4-FLUORO-3-NITROPHENYL)UREA

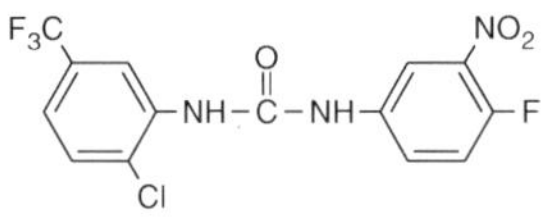

250 mg

S71,561-1
1-(2-CARBAMOYLPHENYL)-3-[4-CHLORO-3-(TRIFLUOROMETHYL)PHENYL]UREA

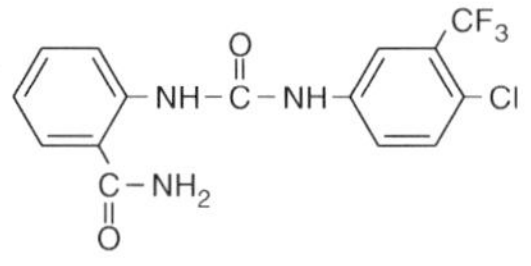

250 mg

S91,614-5
1-(2,5-DIFLUOROPHENYL)-3-[4-NITRO-2-(TRIFLUOROMETHYL)PHENYL]UREA

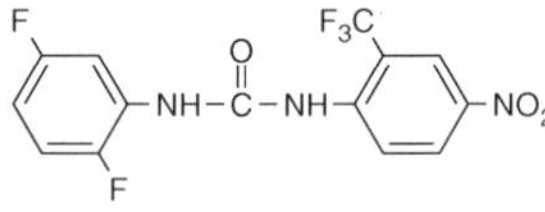

250 mg

S66,498-7
1-BENZYL-3-(3-TRIFLUOROMETHYLPHENYL)UREA

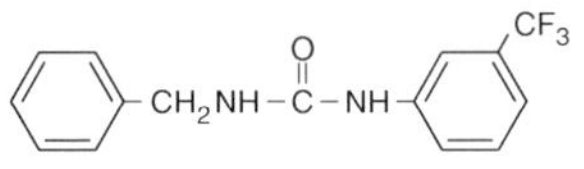

250 mg

S80,063-5
1-BENZYL-3-[2-CHLORO-5-(TRIFLUOROMETHYL)PHENYL]UREA

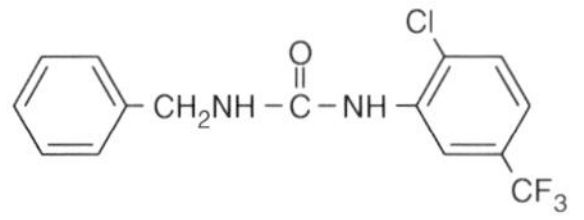

250 mg

S79,634-4
1-(2,4-DICHLOROBENZYL)-3-[2-(TRIFLUOROMETHYL)PHENYL]UREA

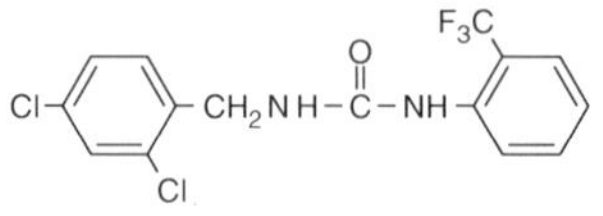

250 mg

S76,927-4
1-(2,4-DICHLOROBENZOYL)-4-(2-TRIFLUOROMETHYLPHENYL)SEMICARBAZIDE

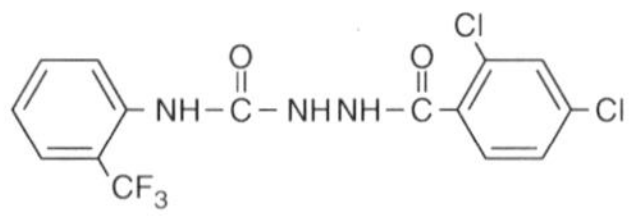

250 mg

S69,091-0
1-(4-CHLOROBENZOYL)-4-[4-CHLORO-2-(TRIFLUOROMETHYL)PHENYL]-SEMICARBAZIDE

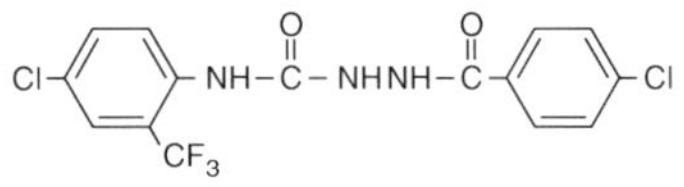

250 mg

S69,434-7
4-[4-CHLORO-2-(TRIFLUOROMETHYL)-PHENYL]-1-(2,4-DICHLOROBENZOYL)-SEMICARBAZIDE

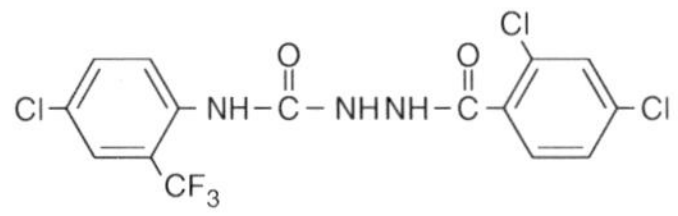

250 mg

S69,408-8
1-ACETYL-4-[4-CHLORO-2-(TRIFLUOROMETHYL)PHENYL]-SEMICARBAZIDE

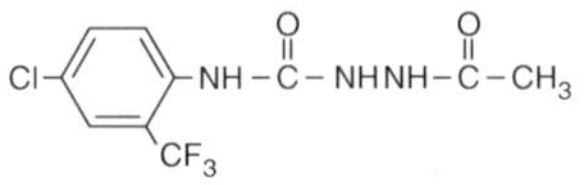

250 mg

S67,424-9
4-[4-BROMO-2-(TRIFLUOROMETHYL)-PHENYL]-1-(4-FLUOROPHENYL-UREIDOACETYL)SEMICARBAZIDE

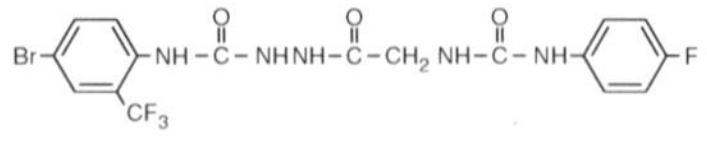

250 mg

S84,774-7
4-(3-[3-(TRIFLUOROMETHYL)PHENYL]-UREIDO)BENZOIC (2-FLUOROBENZYLIDENE)HYDRAZIDE

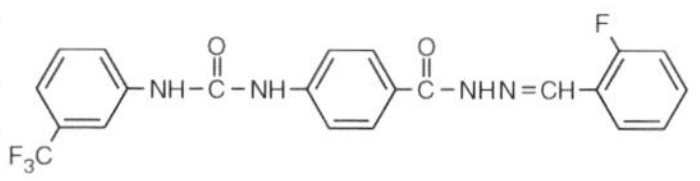

250 mg

S79,484-8
DIMETHYL 5-(3-[2-CHLORO-5-(TRIFLUOROMETHYL)PHENYL]UREIDO)-ISOPHTHALATE

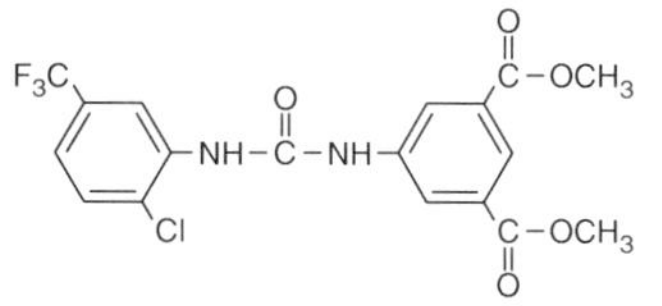

250 mg

S91,297-2
1-(4-CARBOXYPHENYL)-4-[2-CHLORO-5-(TRIFLUOROMETHYL)PHENYL]-SEMICARBAZIDE

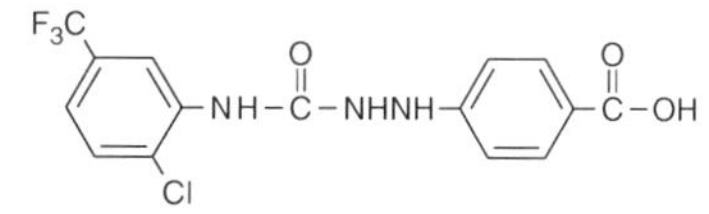

250 mg

S66,946-6
1-(4-CARBOXYPHENYL)-3-(2-CHLORO-5-TRIFLUOROMETHYLPHENYL)UREA

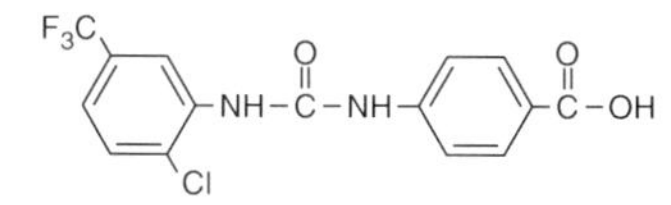

5 gm

S90,199-7
3-(3-[4-CHLORO-3-(TRIFLUOROMETHYL)-PHENYL]UREIDO)BENZOIC ACID

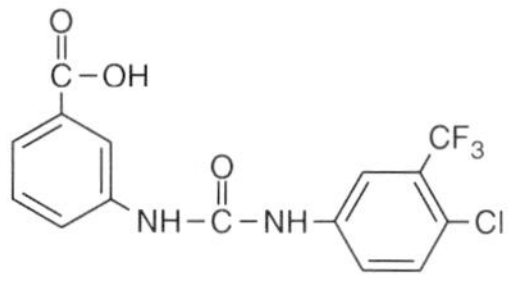

250 mg

S66,978-4
1-(2-CARBOXYPHENYL)-3-(2-CHLORO-5-TRIFLUOROMETHYLPHENYL)UREA

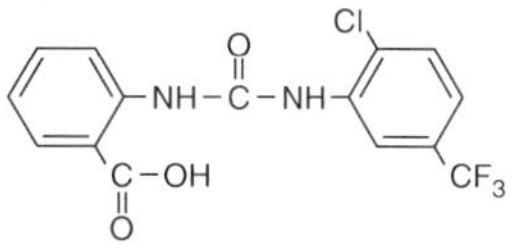

5 gm

S72,675-3
ETHYL 4-[2-CHLORO-5-(TRIFLUOROMETHYL)PHENYLUREIDO]-BENZOATE

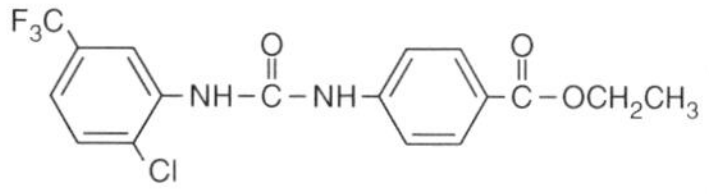

5 gm

S69,119-4
1-BENZOYL-4-[4-CHLORO-2-(TRIFLUOROMETHYL)PHENYL]-SEMICARBAZIDE

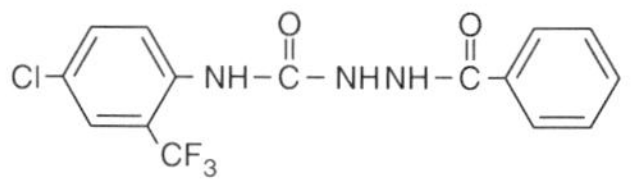

250 mg

S90,446-5
1-(2-CHLOROBENZOYL)-4-[3-(TRIFLUOROMETHYL)PHENYL]-SEMICARBAZIDE

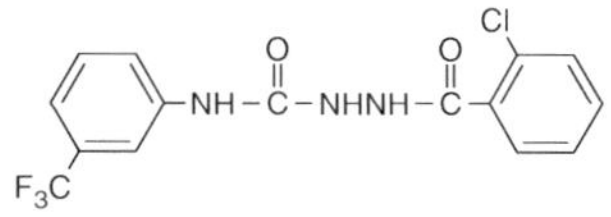

250 mg

S91,644-7
OXAMIC N2-(N-[2-CHLORO-5-(TRIFLUOROMETHYL)PHENYL]CARBAMOYL)-HYDRAZIDE

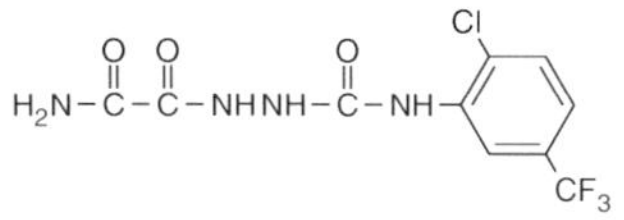

250 mg

S68,856-8
N-(3-CHLOROPHENYLUREIDO)-N'-[3-(TRIFLUOROMETHYL)PHENYL]OXAMIDE

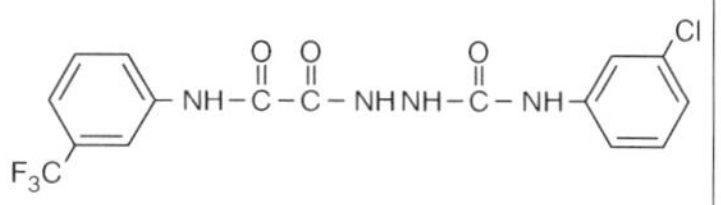

250 mg

S68,875-4
N-[4-BROMO-2-(TRIFLUOROMETHYL)-PHENYLUREIDO]-N'-(4-FLUOROPHENYL)-OXAMIDE

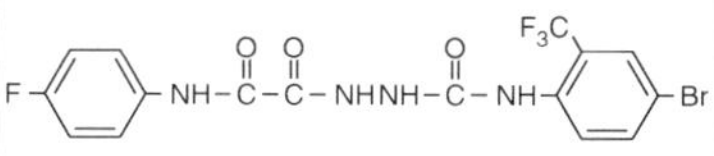

250 mg

S68,870-3
N-[4-BROMO-2-(TRIFLUOROMETHYL)-PHENYLUREIDO]-N'-(3,4-DICHLOROPHENYL)OXAMIDE

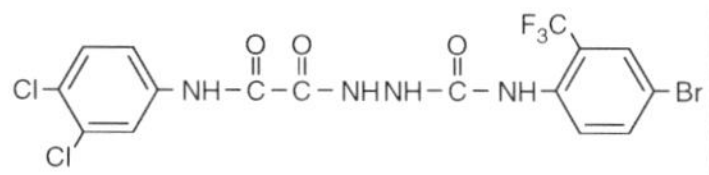

250 mg

S91,299-9
1-ACETYL-4-[2-CHLORO-5-(TRIFLUOROMETHYL)PHENYL]-SEMICARBAZIDE

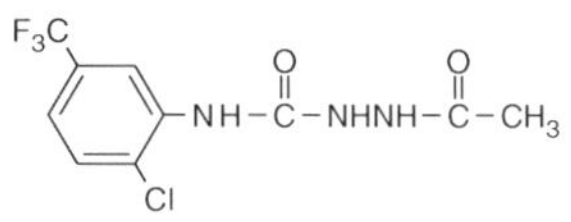

250 mg

S77,279-8
6-(4-CHLORO-3-METHYLPHENYL)-3-PHENYL-1-[2-(TRIFLUOROMETHYL)-PHENYL]BIUREA

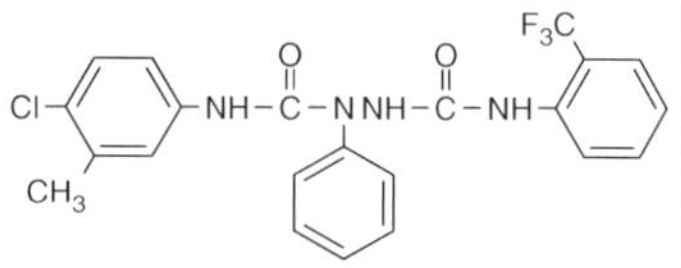

250 mg

S72,282-0
1-[2-CHLORO-5-(TRIFLUOROMETHYL)-PHENYL]-6-(2-ETHYLPHENYL)-3-METHYLBIUREA

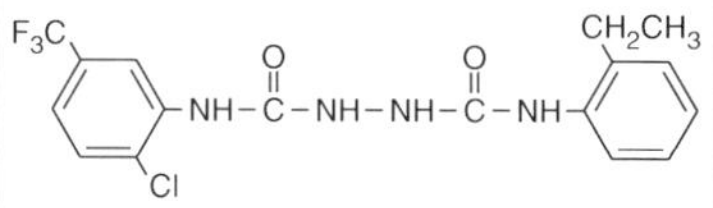

250 mg

S91,646-3
1-[2-CHLORO-5-(TRIFLUOROMETHYL)-PHENYL]-3-PIPERIDINOUREA

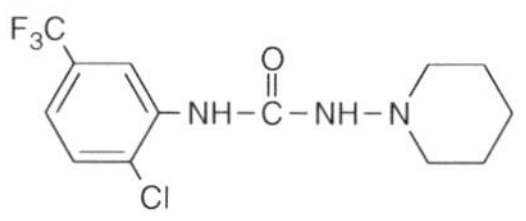

250 mg

S91,298-0
4-[2-CHLORO-5-(TRIFLUOROMETHYL)-PHENYL]-1-METHYL-1-PHENYLSEMICARBAZIDE

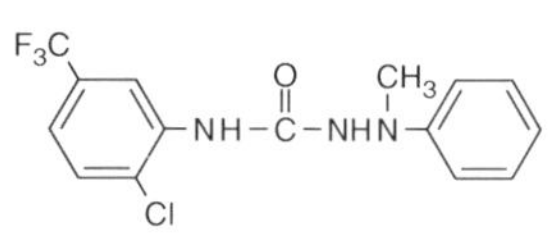

250 mg

S68,844-4
1,6-BIS[4-CHLORO-2-(TRIFLUOROMETHYL)PHENYL]BIUREA

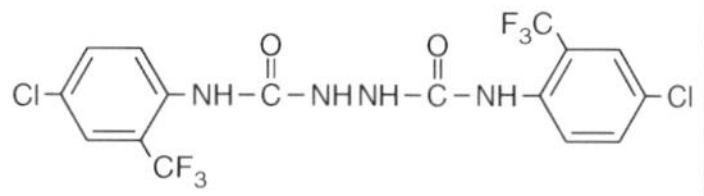

250 mg

S90,420-1
1-(3-CHLOROPHENYL)-3-(2,2,2-TRIFLUOROACETYL)UREA

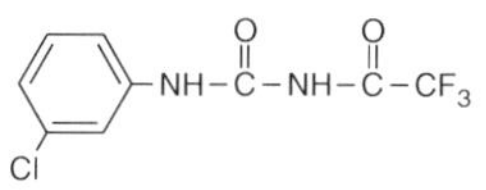

250 mg

S70,197-1
1,1'-(2-BROMO-6-METHYLPHENYLENE)-BIS(3-[3-(TRIFLUOROMETHYL)-PHENYL]UREA)

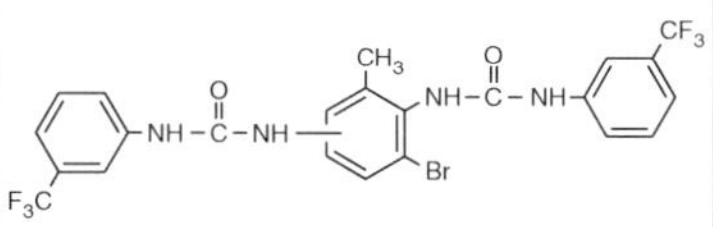

250 mg

S77,304-2
1-[4-CHLORO-2-(TRIFLUOROMETHYL)-PHENYL]-3-CYCLOPENTYLUREA

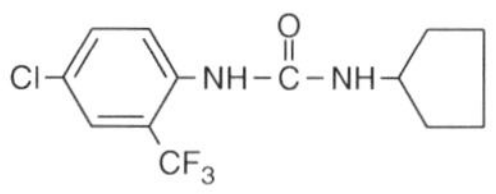

250 mg

S90,576-3
3-[2-CHLORO-5-(TRIFLUOROMETHYL)-PHENYL]-1,1-DICYCLOHEXYLUREA

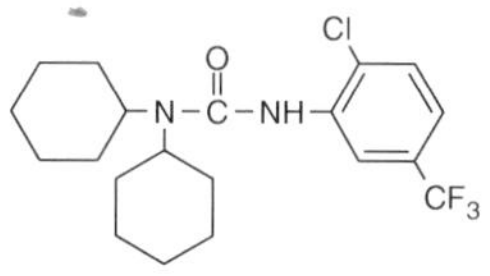

250 mg

S91,302-2
1-[2-CHLORO-5-(TRIFLUOROMETHYL)-PHENYL]-3-(DIPHENYLMETHYL)UREA

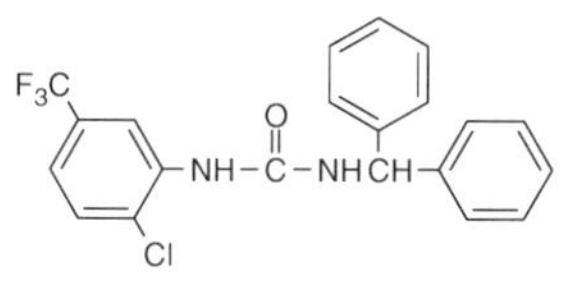

250 mg

S89,786-8
1-(2-FLUOROPHENYL)-3-[3-(TRIFLUOROMETHYL)PHENYL]UREA

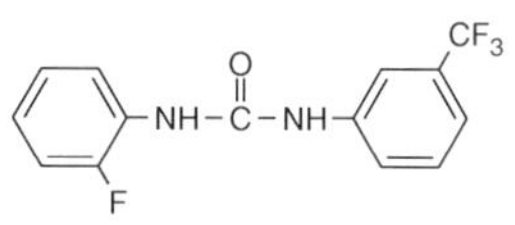

250 mg

S79,114-8
1-(3-CHLOROPHENYL)-3-[2-(TRIFLUOROMETHYL)PHENYL]UREA

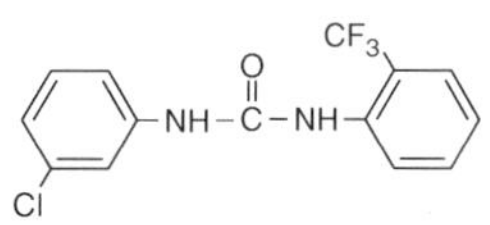

250 mg

S39,721-0
1-(2-CHLOROPHENYL)-3-TRIFLUORO(M-TOLYL)UREA

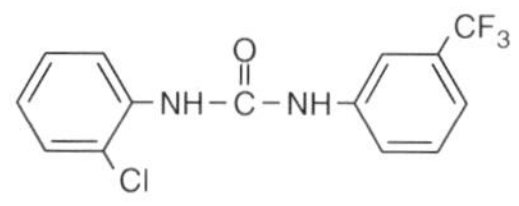

250 mg

S79,503-8
1-(4-FLUOROPHENYL)-3-[2-(TRIFLUOROMETHYL)PHENYL]UREA

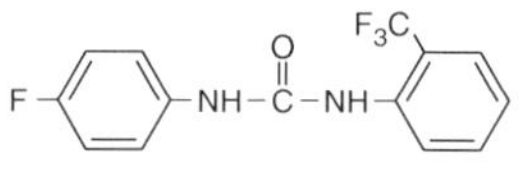

250 mg

S77,308-5
1-(4-CHLOROPHENYL)-3-[2-(TRIFLUOROMETHYL)PHENYL]UREA

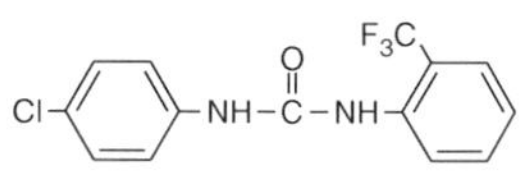

250 mg

S90,428-7
1-ETHYL-3-(4-FLUOROPHENYL)-1-[2-(TRIFLUOROMETHYL)PHENYL]UREA

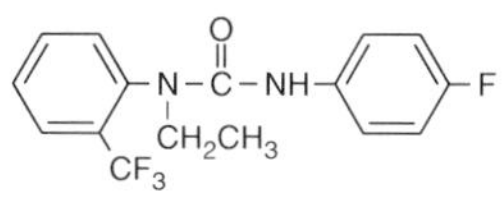

250 mg

S75,346-7
1-(3-CHLORO-2-METHYLPHENYL)-3-[2-(TRIFLUOROMETHYL)PHENYL]UREA

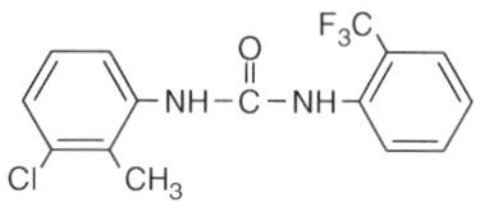

5 gm

S79,112-1
1-(2-CHLOROPHENYL)-3-[2-CHLORO-5-(TRIFLUOROMETHYL)PHENYL]UREA

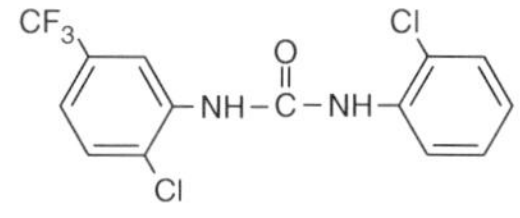

250 mg

S75,345-9
1-(3-CHLORO-4-METHYLPHENYL)-3-[2-(TRIFLUOROMETHYL)PHENYL]UREA
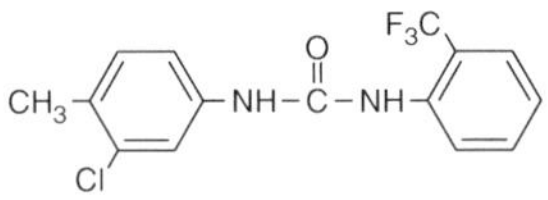
5 gm

S76,939-8
1-(5-CHLORO-2-METHYLPHENYL)-3-[2-(TRIFLUOROMETHYL)PHENYL]UREA
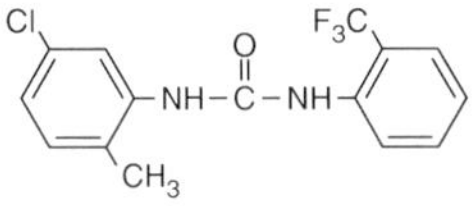
250 mg

S69,890-3
1-(4-BROMO-2,6-DIMETHYLPHENYL)-3-[2-(TRIFLUOROMETHYL)PHENYL]UREA
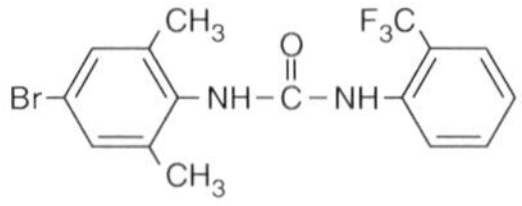
250 mg

S80,024-4
1-(4-FLUOROPHENYL)-3-[3-(TRIFLUOROMETHYL)PHENYL]UREA
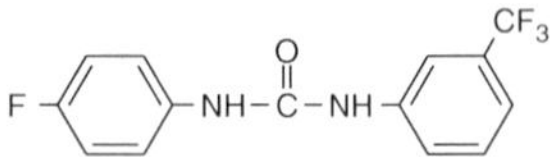
250 mg

S71,506-9
1-(4-CHLOROPHENYL)-3-[3-(TRIFLUOROMETHYL)PHENYL]UREA
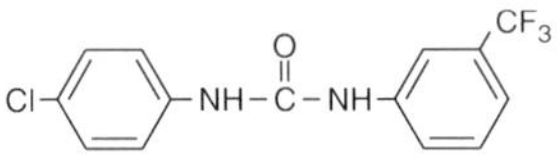
5 gm

S71,232-9
1-(4-BROMOPHENYL)-3-[3-TRIFLUOROMETHYLPHENYL]UREA
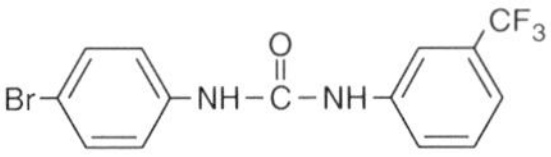
5 gm

S79,657-3
1-(4-CHLOROPHENYL)-1-ETHYL-3-[3-(TRIFLUOROMETHYL)PHENYL]UREA
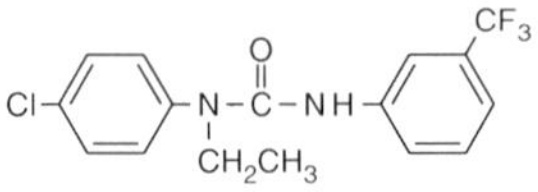
250 mg

S79,121-0
1-(4-CHLOROPHENYL)-3-[2-METHYL-3-(TRIFLUOROMETHYL)PHENYL]UREA
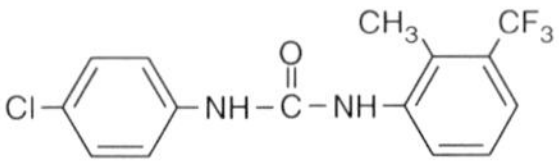
250 mg

S80,022-8
1-[2-CHLORO-5-(TRIFLUOROMETHYL)-PHENYL]-3-(4-FLUOROPHENYL)UREA
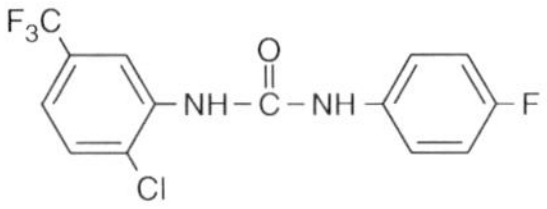
250 mg

S77,330-1
1-(4-CHLOROPHENYL)-3-[2-CHLORO-5-(TRIFLUOROMETHYL)PHENYL]UREA
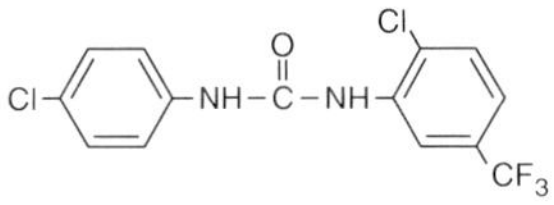
250 mg

S80,060-0
1-[2-CHLORO-5-(TRIFLUOROMETHYL)-PHENYL]-3-(4-IODOPHENYL)UREA
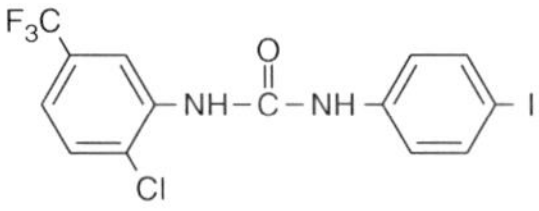
250 mg

S86,769-1
1-(3-CHLORO-4-METHYLPHENYL)-3-[4-(TRICHLOROMETHYL)PHENYL]UREA
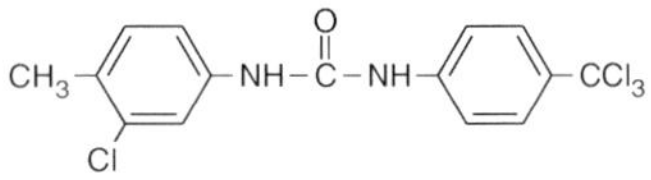
250 mg

S71,381-3
1-(2,3-DICHLOROPHENYL)-3-[3-(TRIFLUOROMETHYL)PHENYL]UREA
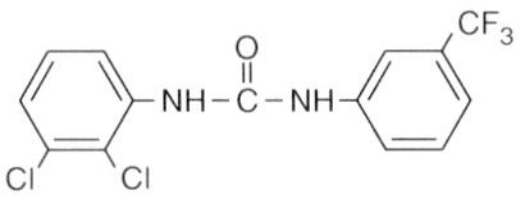
5 gm

S71,507-7
1-(2,6-DICHLOROPHENYL)-3-[3-(TRIFLUOROMETHYL)PHENYL]UREA
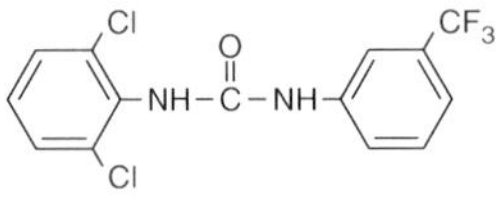
5 gm

S71,386-4
1-(2-CHLORO-6-METHYLPHENYL)-3-[3-(TRIFLUOROMETHYL)PHENYL]UREA
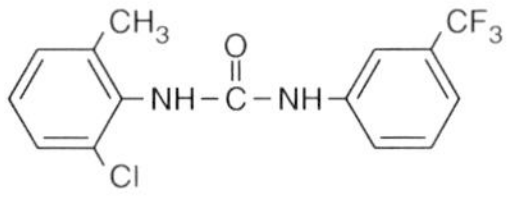
5 gm

S86,767-5
1-(3-CHLORO-2-METHYLPHENYL)-3-[3-(TRIFLUOROMETHYL)PHENYL]UREA
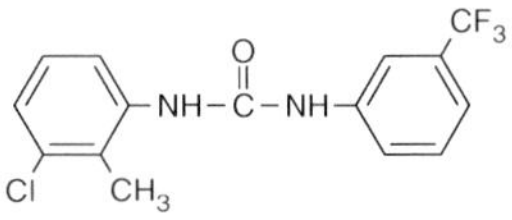
250 mg

S71,556-5
1-[2-CHLORO-5-(TRIFLUOROMETHYL)-PHENYL]-3-(2,3-DICHLOROPHENYL)UREA
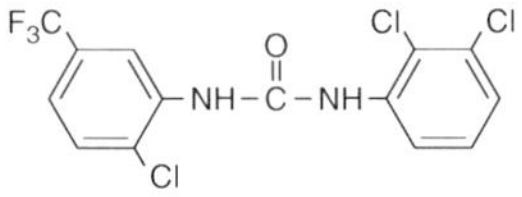
250 mg

S66,612-2
1-(2-CHLORO-5-TRIFLUOROMETHYLPHENYL)-3-METHYLUREA
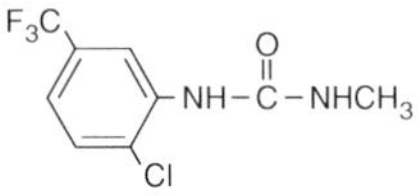
5 gm

S66,420-0
3-(2-CHLORO-5-TRIFLUOROMETHYLPHENYL)-1,1-DIMETHYLUREA
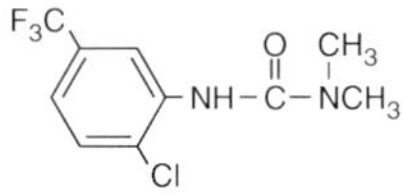
5 gm

S66,456-1
1-(2-CHLORO-5-TRIFLUOROMETHYLPHENYL)-3-ETHYLUREA
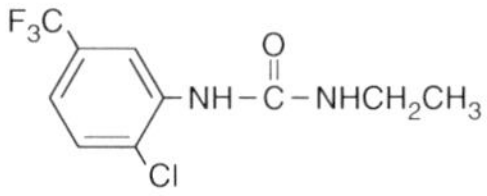
5 gm

S66,434-0
1-(2-CHLORO-5-TRIFLUOROMETHYLPHENYL)-3-ISOPROPYLUREA
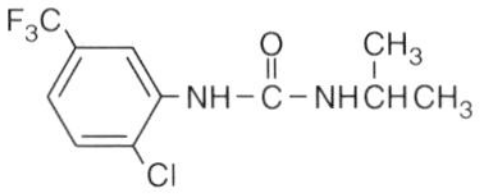
5 gm

S92,416-4
1-(4-CHLORO-2-TRIFLUOROMETHYLPHENYL)-3-N-PROPYLUREA
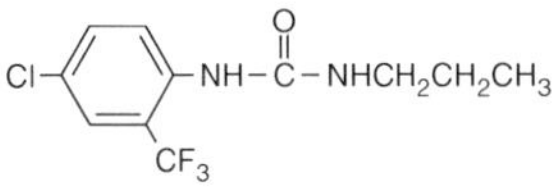
250 mg

S84,453-5
1-[2-CHLORO-5-(TRIFLUOROMETHYL)-PHENYL]-3-(3-FLUOROPHENYL)UREA
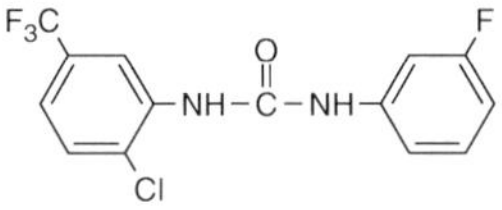
250 mg

S70,840-2
1-(2,4-DICHLOROPHENYL)-3-(3-TRIFLUOROMETHYLPHENYL)UREA
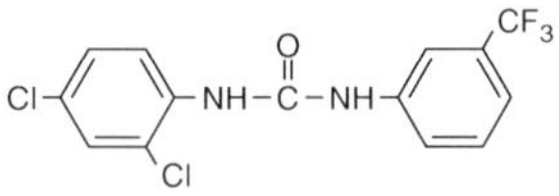
5 gm

S71,225-6
1-(2,5-DICHLOROPHENYL)-3-(3-TRIFLUOROMETHYLPHENYL)UREA
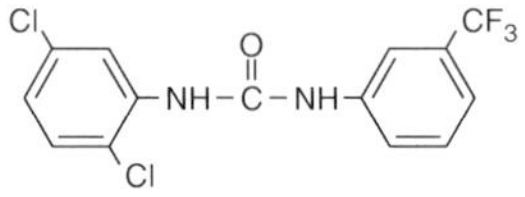
5 gm

S73,548-5
1-(3,4-DICHLOROPHENYL)-3-(3-TRIFLUOROMETHYLPHENYL)UREA
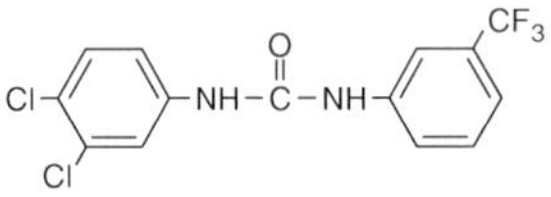
250 mg

S76,849-9
1-(3-CHLOROPHENYL)-3-[2-CHLORO-5-(TRIFLUOROMETHYL)PHENYL]UREA
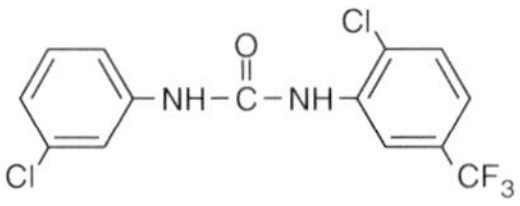
250 mg

S91,612-9
1-[2-CHLORO-5-(TRIFLUOROMETHYL)-PHENYL]-3-(2,5-DIFLUOROPHENYL)UREA
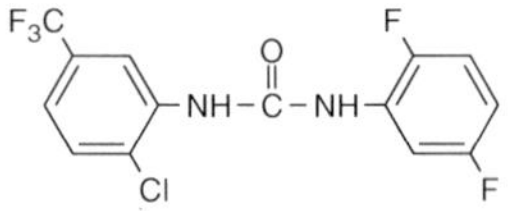
250 mg

S92,409-1
1-(2-CHLORO-4-NITROPHENYL)-3-(3-TRIFLUOROMETHYLPHENYL)UREA
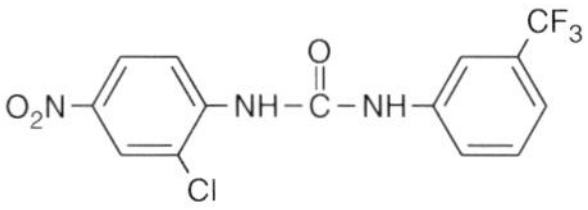
250 mg

S72,819-5
1-[2-CHLORO-5-(TRIFLUOROMETHYL) PHENYL]-3-(2,4-DICHLOROPHENYL)UREA
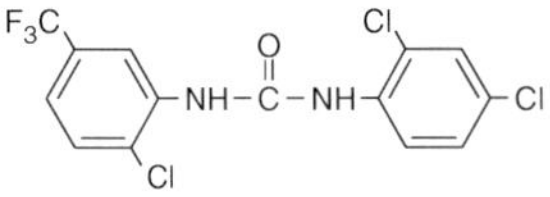
5 gm

S71,574-3
1-(5-CHLORO-2-METHYLPHENYL)-3-[3-(TRIFLUOROMETHYL)PHENYL]UREA
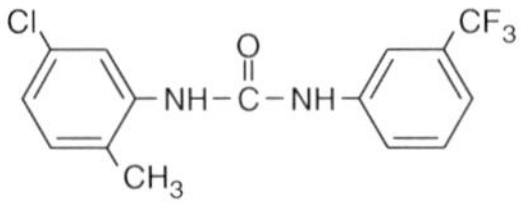
5 gm

S73,673-2
1-(3-CHLORO-4-METHYLPHENYL)-3-[3-(TRIFLUOROMETHYL)PHENYL]UREA
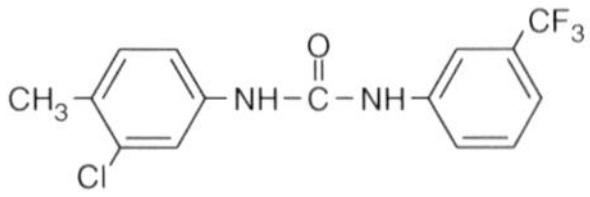
5 gm

S90,447-3
1-(2,4-DICHLOROBENZYL)-3-[3-(TRIFLUOROMETHYL)PHENYL]UREA
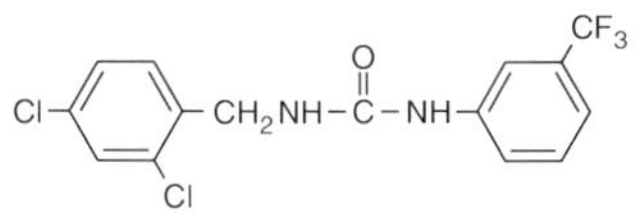
250 mg

S75,341-6
1-[2-CHLORO-5-(TRIFLUOROMETHYL)-PHENYL]-3-[3-(TRIFLUOROMETHYL)-PHENYL]UREA
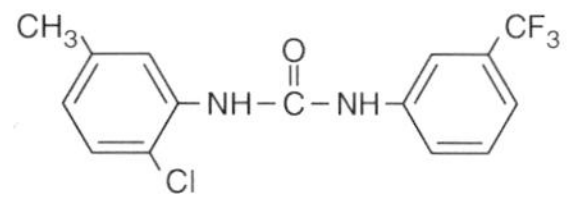
5 gm

S80,070-8
1-[2-CHLORO-5-(TRIFLUOROMETHYL)-PHENYL]-3-(3,4-XYLYL)UREA
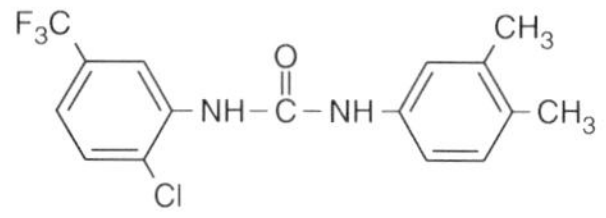
250 mg

S67,068-5
1-(2-CHLORO-5-TRIFLUOROMETHYLPHENYL)-3-OCTADECYLUREA
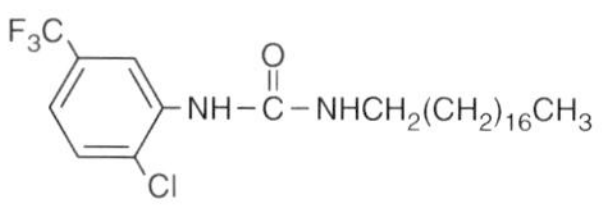
5 gm

S67,331-5
4,4'-BIS[(4-CHLORO-3-TRIFLUOROMETHYL)PHENYLUREIDO]-3,3'-DIMETHYLBIPHENYL

250 mg

S71,682-0
1-(2,4,6-TRICHLOROPHENYL)-3-[3-(TRIFLUOROMETHYL)PHENYL]UREA
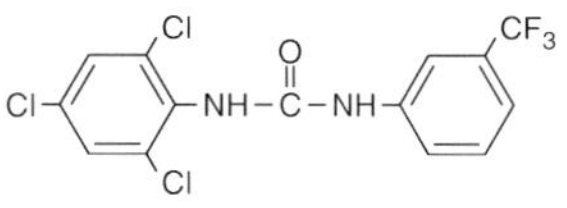
5 gm

S71,224-8
1-(2,4,5-TRICHLOROPHENYL)-3-[3-(TRIFLUOROMETHYL)PHENYL]UREA
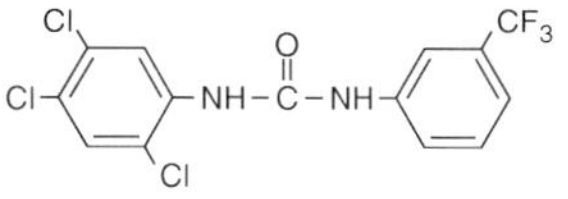
5 gm

S80,085-6
1-(4,5-DICHLORO-2-METHYLPHENYL)-3-[3-(TRIFLUOROMETHYL)PHENYL]UREA
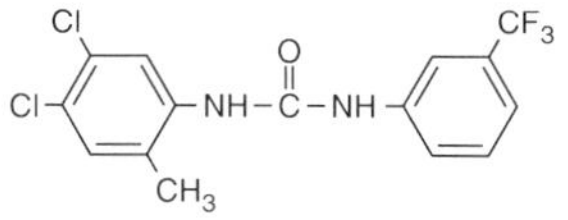
250 mg

S80,086-4
1-(2-CHLORO-5-METHYLPHENYL)-3-[3-(TRIFLUOROMETHYL)PHENYL]UREA
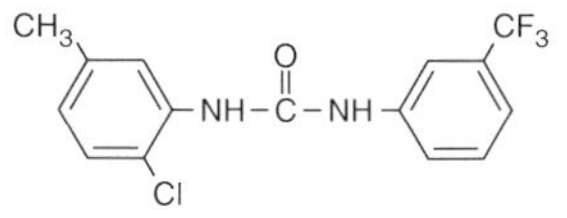
250 mg

S90,217-9
2-(3-[3-(TRIFLUOROMETHYL)PHENYL]-UREIDO)ACETIC (3-HYDROXY-4-METHOXYBENZYLIDENE)HYDRAZIDE
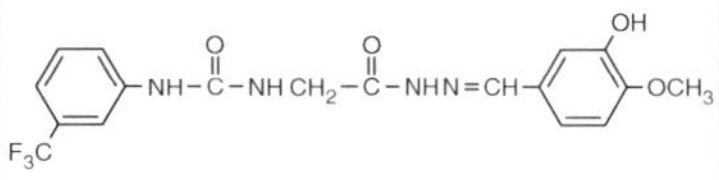
250 mg

S80,114-3
1-[3-(1-HYDROXYETHYL)PHENYL]-3-[3-(TRIFLUOROMETHYL)PHENYL]UREA
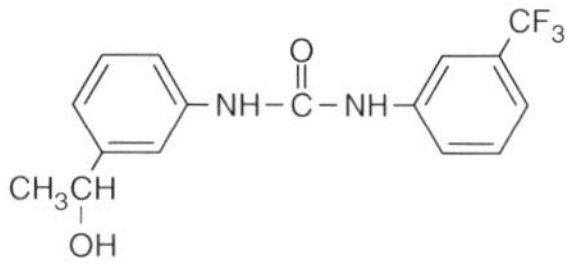
250 mg

S73,767-4
1-(2-HYDROXYBENZOYL)-4-[2-(TRIFLUOROMETHYL)PHENYL]-SEMICARBAZIDE
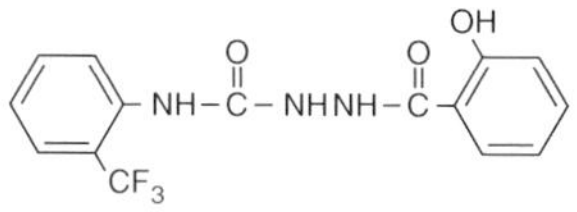
5 gm

S73,731-3
1-(6-HYDROXYHEXANOYL)-4-[3-(TRIFLUOROMETHYL)PHENYL]-SEMICARBAZIDE
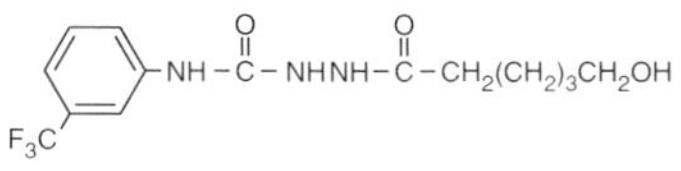
5 gm

S79,487-2
1-CYCLOHEXYL-1-(2-HYDROXYETHYL)-3-[2-(TRIFLUOROMETHYL)PHENYL]UREA
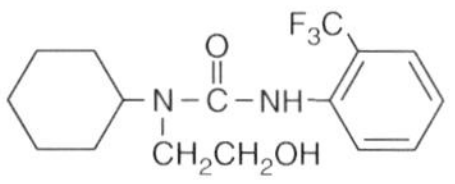
250 mg

S69,160-7
1-(4-HYDROXYBENZOYL)-4-[2-(TRIFLUOROMETHYL)PHENYL]-SEMICARBAZIDE
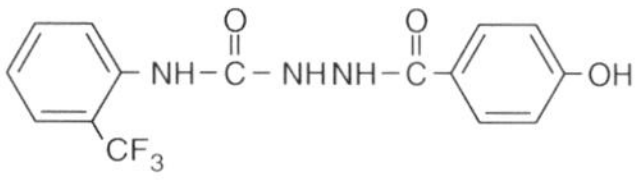
250 mg

S80,168-2
1-(3-HYDROXYPHENYL)-3-[2-(TRIFLUOROMETHYL)PHENYL]UREA
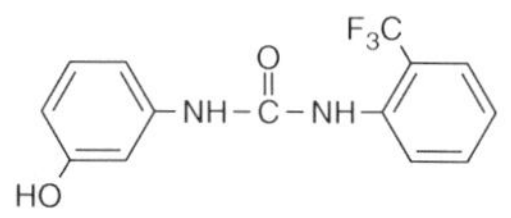
250 mg

S80,169-0
1-(4-HYDROXYPHENYL)-3-[2-(TRIFLUOROMETHYL)PHENYL]UREA
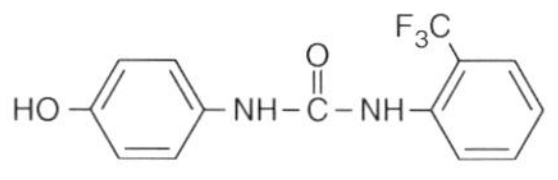
250 mg

S79,490-2
1-[4-(2-HYDROXYETHYL)PHENYL]-3-[2-(TRIFLUOROMETHYL)PHENYL]UREA
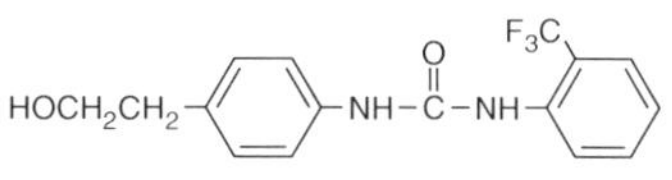
250 mg

S73,719-4
1-[4-(2-HYDROXYETHYL)PHENYL]-3-[3-(TRIFLUOROMETHYL)PHENYL]UREA
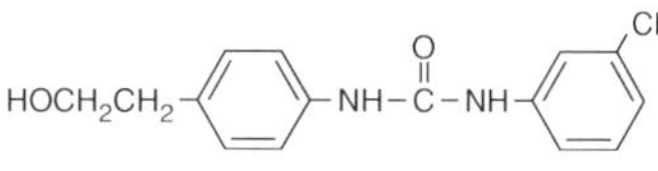
5 gm

S80,090-2
1-(2,4-DIMETHOXYPHENYL)-3-[3-(TRIFLUOROMETHYL)PHENYL]UREA
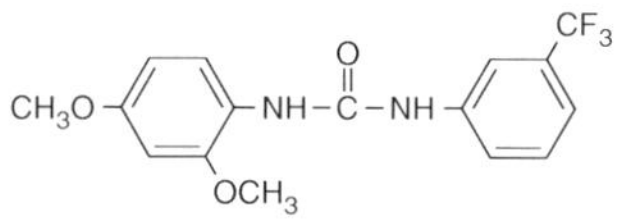
250 mg

S80,186-0
1-(2,5-DIMETHOXYPHENYL)-3-[2-(TRIFLUOROMETHYL)PHENYL]UREA
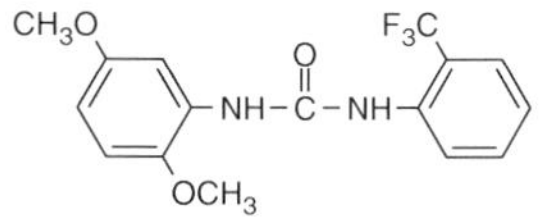
250 mg

S80,105-4
1-(2,5-DIMETHOXYPHENYL)-3-[3-(TRIFLUOROMETHYL)PHENYL]UREA
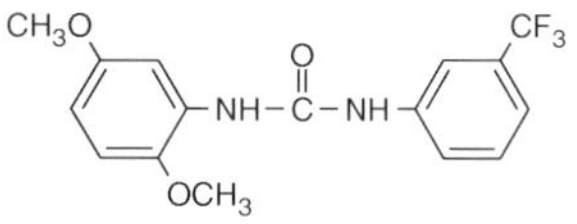
250 mg

S84,780-1
4-(3-[3-(TRIFLUOROMETHYL)PHENYL]-UREIDO)BENZOIC (3,4-DIMETHOXY-BENZYLIDENE)HYDRAZIDE
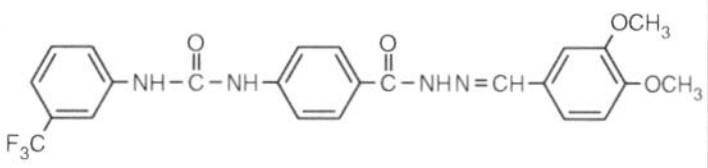
250 mg

S69,340-5
1-(2-METHOXYBENZOYL)-4-[3-(TRIFLUOROMETHYL)PHENYL]-SEMICARBAZIDE
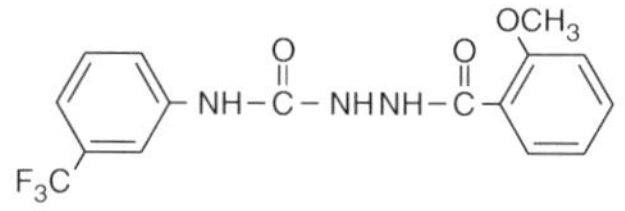
250 mg

S90,216-0
2-(3-[3-(TRIFLUOROMETHYL)PHENYL]-UREIDO)ACETIC (4-BUTOXYBENZYLIDENE)-HYDRAZIDE
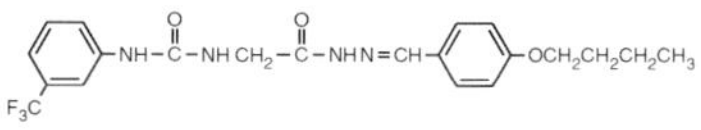
250 mg

S90,431-7
1-[2,6-DINITRO-4-(TRIFLUOROMETHYL)-PHENYL]-4-(2-METHOXY-5-METHYL)-SEMICARBAZIDE
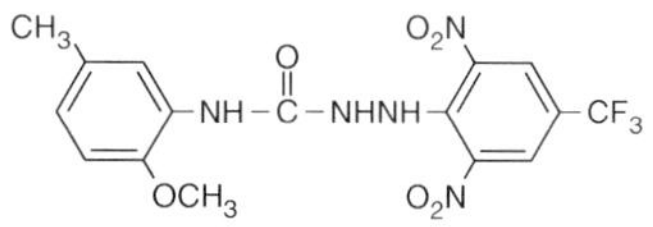
250 mg

S80,184-4
1-(2-METHOXY-5-NITROPHENYL)-3-[2-(TRIFLUOROMETHYL)PHENYL]UREA
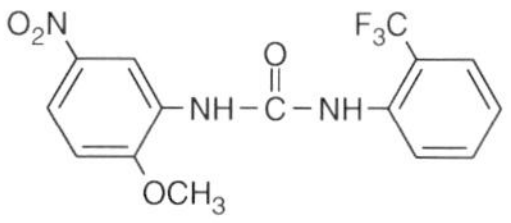
250 mg

S80,030-9
1-(2-METHOXY-5-NITROPHENYL)-3-[3-(TRIFLUOROMETHYL)PHENYL]UREA
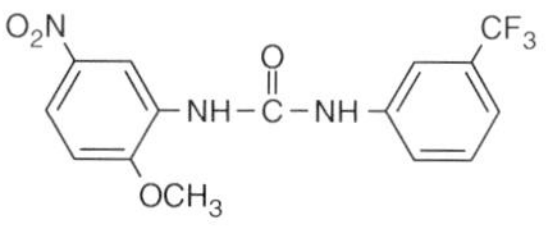
250 mg

S50,843-8
1,1'-(1,3-PHENYLENE)BIS(3-[6-(3,4-DICHLOROPHENOXY)-A,A,A-TRIFLUORO-M-TOLYL]UREA)
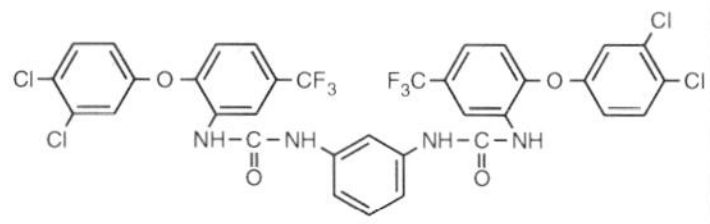
5 gm

S50,837-3
1,1'-(1,4-PHENYLENE)BIS(3-[6-(3,4-DICHLOROPHENOXY)-A,A,A-TRIFLUORO-M-TOLYL]UREA)
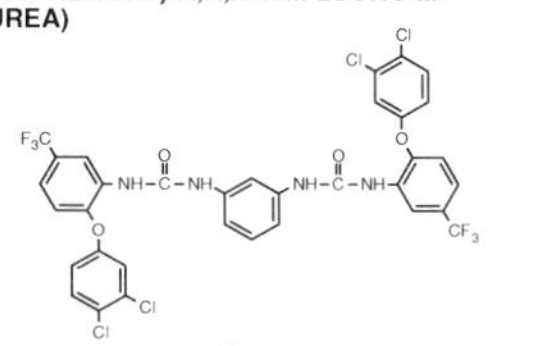
5 gm

S72,936-1
1-(4-METHOXYBENZOYL)-4-[2-(TRIFLUOROMETHYL)PHENYL]-SEMICARBAZIDE
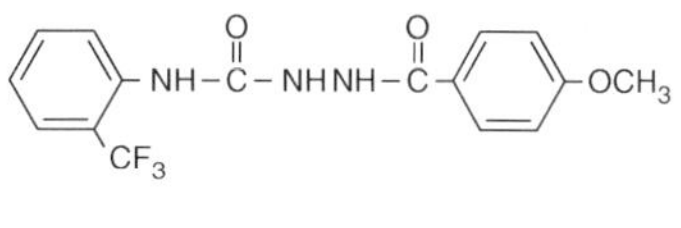
5 gm

S84,783-6
4-(3-[3-(TRIFLUOROMETHYL)PHENYL]-UREIDO)BENZOIC (4-BUTOXYBENZYLIDENE)HYDRAZIDE
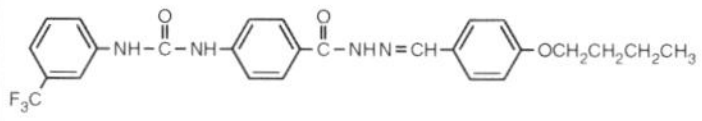
250 mg

S84,786-0
4-(3-[3-(TRIFLUOROMETHYL)PHENYL]-UREIDO)BENZOIC(4-HYDROXY-3-METHOXY-BENZYLIDENE)HYDRAZIDE

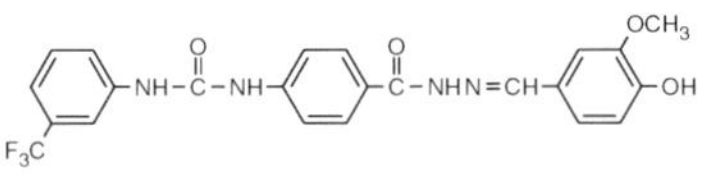

250 mg

S80,188-7
1-[2-METHOXY-5-(N-PHENYLCARBAMOYL)-PHENYL]-3-[2-(TRIFLUOROMETHYL)-PHENYL]UREA

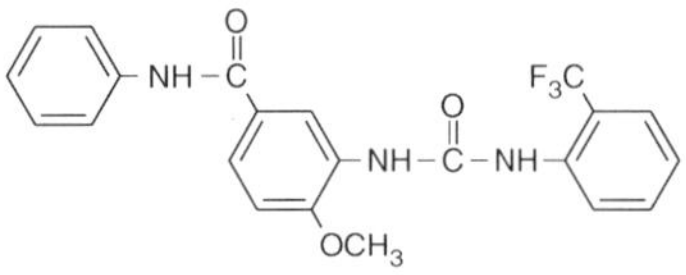

250 mg

S90,440-6
1-(2-METHOXY-5-METHYLPHENYL)-3-(2,2,2-TRIFLUOROACETYL)UREA

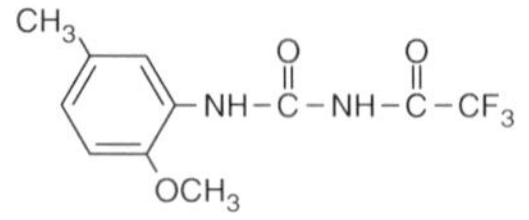

250 mg

S50,818-7
1-[5-CHLORO-2-(3,4-DICHLOROPHENOXY)-PHENYL]-3-(6-CHLOROTRIFLUORO-M-TOLYL)UREA

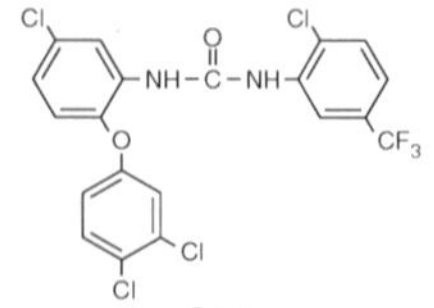

5 gm

S61,182-4
1-(2-METHOXYETHYL)-3-[3-(TRIFLUOROMETHYL)PHENYL]UREA

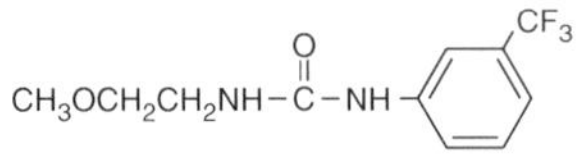

5 gm

S79,491-0
1-(2-ETHOXYPHENYL)-3-[3-(TRIFLUOROMETHYL)PHENYL]UREA

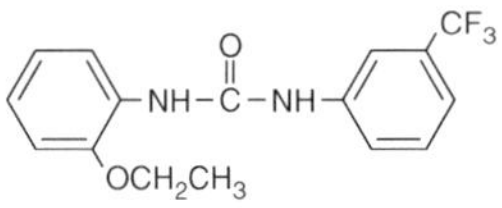

250 mg

S79,492-9
1-(2-ETHOXYPHENYL)-3-[2-(TRIFLUOROMETHYL)PHENYL]UREA

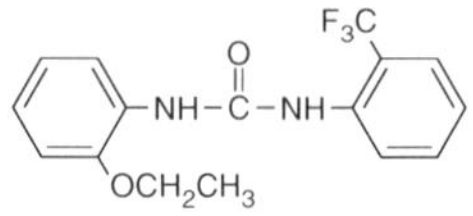

250 mg

S80,185-2
1-(2-METHOXY-5-METHYLPHENYL)-3-[2-(TRIFLUOROMETHYL)PHENYL]UREA

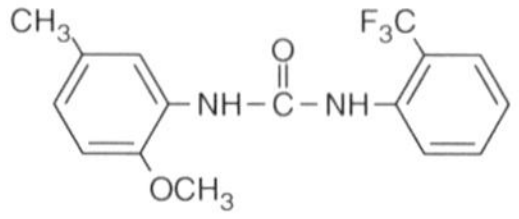

250 mg

S79,489-9
1-(2-METHOXY-5-METHYLPHENYL)-3-[3-(TRIFLUOROMETHYL)PHENYL]UREA

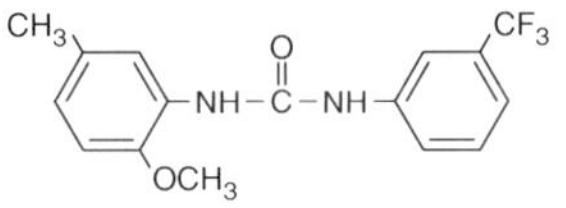

250 mg

S92,408-3
1-(2-METHOXY-4-NITROPHENYL)-3-[3-(TRIFLUOROMETHYL)PHENYL]UREA

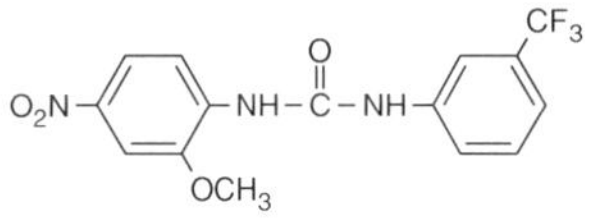

250 mg

S71,262-0
1-(3-METHOXYPHENYL)-3-[3-(TRIFLUOROMETHYL)PHENYL]UREA

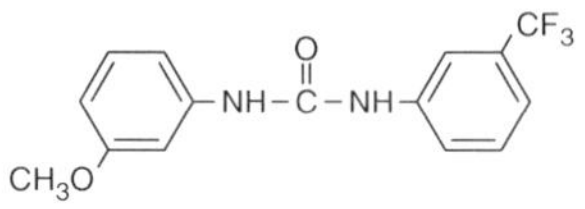

5 gm

S78,084-7
1-(3-BUTOXYPROPYL)-3-[3-(TRIFLUOROMETHYL)PHENYL]UREA

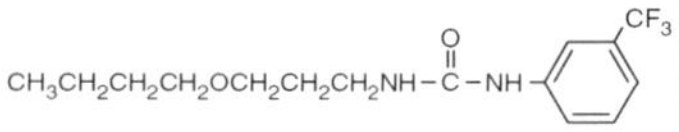

250 mg

S79,713-8
1-[4-(METHYLTHIO)PHENYL]-3-[4-NITRO-2-(TRIFLUOROMETHYL)-PHENYL]UREA

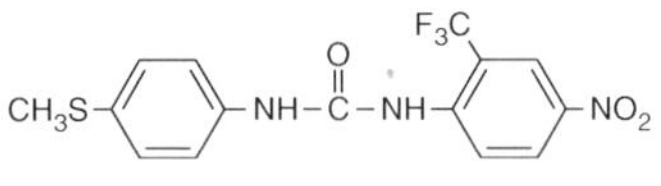

250 mg

S79,714-6
1-[4-(METHYLTHIO)PHENYL]-3-[2-(TRIFLUOROMETHYL)PHENYL]UREA

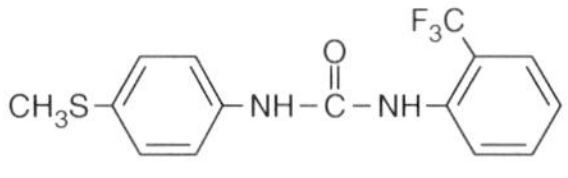

250 mg

S79,715-4
1-[4-(METHYLTHIO)PHENYL]-3-[3-(TRIFLUOROMETHYL)PHENYL]UREA

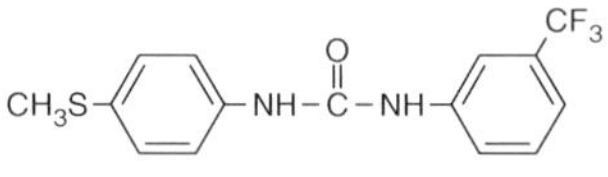

250 mg

S74,766-1
1-(2,4-DINITROPHENYL)-4-[3-(TRIFLUOROMETHYL)PHENYL]-SEMICARBAZIDE

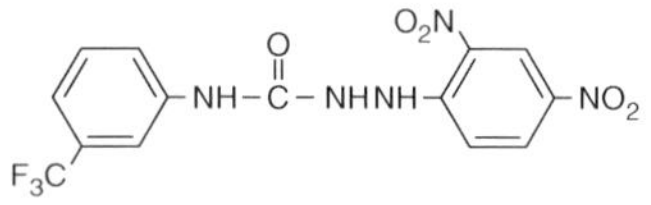

5 gm

S90,215-2
2-(3-[3-(TRIFLUOROMETHYL)PHENYL]-UREIDO)ACETIC [4-(DIMETHYLAMINO)-BENZYLIDENE]HYDRAZIDE

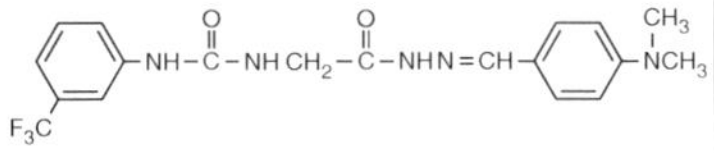

250 mg

S90,214-4
2-(3-[3-(TRIFLUOROMETHYL)PHENYL]-UREIDO)ACETIC (4-METHYLBENZYLIDENE)-HYDRAZIDE

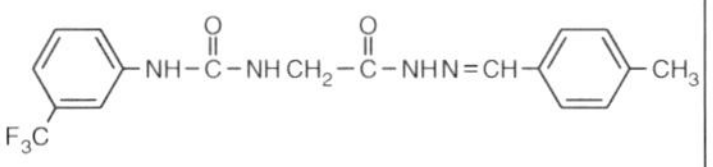

250 mg

S91,224-7
ETHYL 2-(3-[2-NITRO-4-(TRIFLUOROMETHYL)PHENYL]UREIDO)-ACETATE

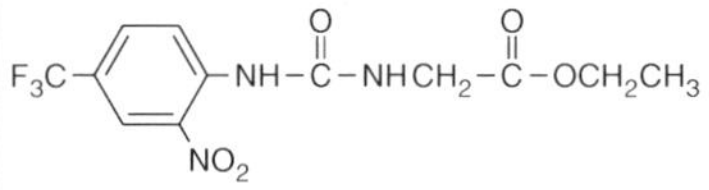

250 mg

S79,504-6
1-(3-NITROPHENYL)-3-[2-(TRIFLUOROMETHYL)PHENYL]UREA

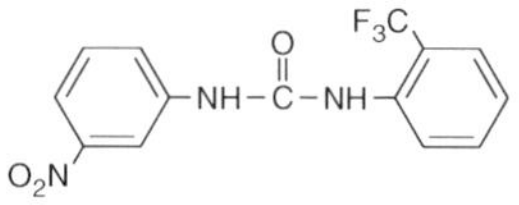

250 mg

S80,194-1
1-(2-METHYL-5-NITROPHENYL)-3-[2-(TRIFLUOROMETHYL)PHENYL]UREA

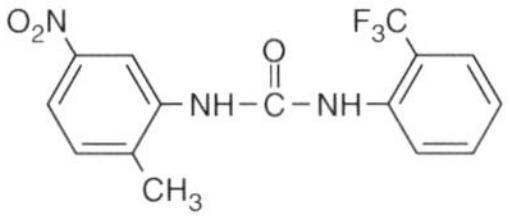

250 mg

S90,577-1
1-(2-METHYL-5-NITROPHENYL)-3-[3-(TRIFLUOROMETHYL)PHENYL]UREA

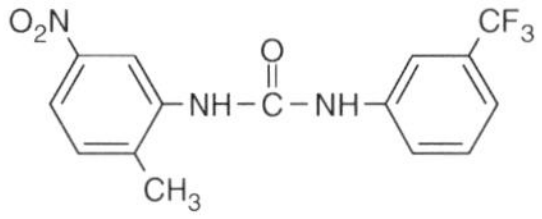

250 mg

S80,117-8
1-(2-CYANOETHYL)-1-CYCLOHEXYL-3-[3-(TRIFLUOROMETHYL)PHENYL]UREA

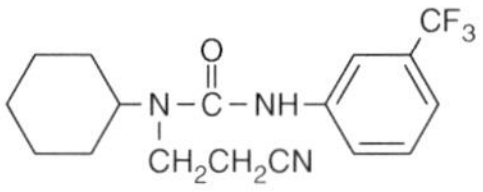

250 mg

S66,018-3
1-(3-MORPHOLINOPROPYL)-3-[3-(TRIFLUOROMETHYL)PHENYL]UREA

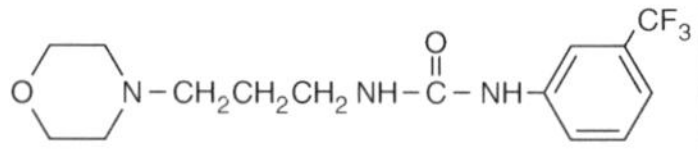

5 gm

S89,090-1
1,1'-(4-METHYL-1,3-PHENYLENE)BIS(3-[3-(TRIFLUOROMETHYL)PHENYL]UREA)

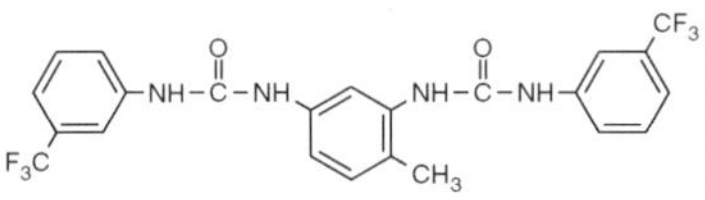

250 mg

S79,480-5
1-(4-ANILINOPHENYL)-3-[2-(TRIFLUOROMETHYL)PHENYL]UREA

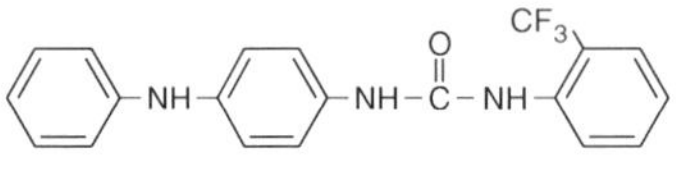

250 mg

S80,120-8
1-(4-ANILINOPHENYL)-3-[3-(TRIFLUOROMETHYL)PHENYL]UREA

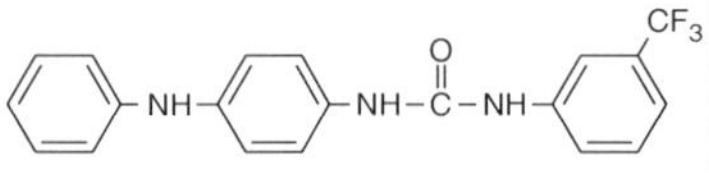

250 mg

S79,661-1
1-[4-(DIMETHYLAMINO)PHENYL]-3-[2-(TRIFLUOROMETHYL)PHENYL]UREA

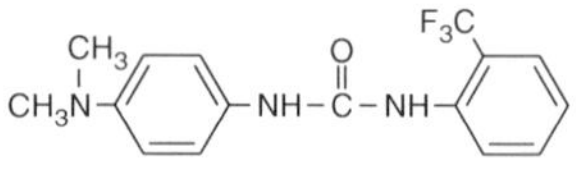

250 mg

S85,339-9
1-[4-(DIMETHYLAMINO)PHENYL]-3-[3-(TRIFLUOROMETHYL)PHENYL]UREA

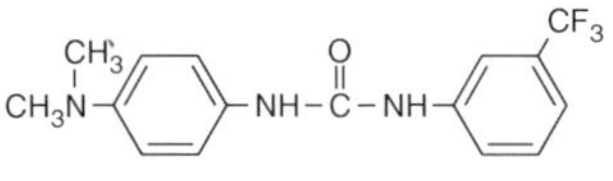

250 mg

S90,442-2
1-(4-NITROPHENYL)-4-[2-(TRIFLUOROMETHYL)PHENYL]-SEMICARBAZIDE

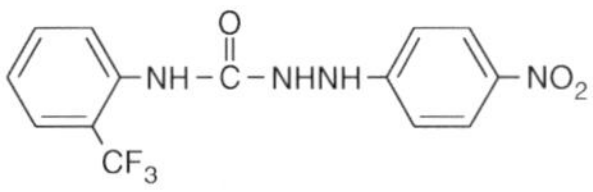

250 mg

S68,672-7
1,3-BIS[4-NITRO-2-(TRIFLUOROMETHYL)-PHENYL]UREA

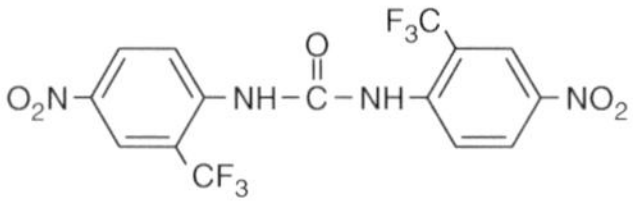

5 gm

S79,495-3
1-(4-ACETAMIDOPHENYL)-3-[2-(TRIFLUOROMETHYL)PHENYL]UREA

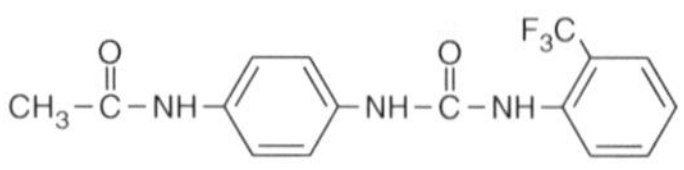

250 mg

S60,176-4
1-(4-ACETAMIDOPHENYL)-3-[3-TRIFLUOROMETHYLPHENYL]UREA

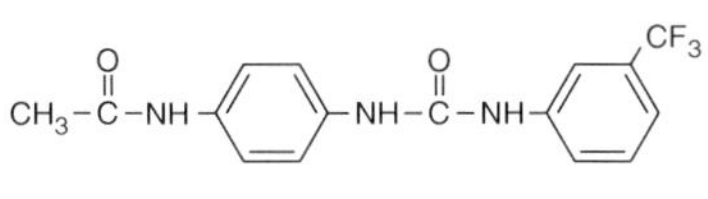

5 gm

S90,448-1
1-(2-PHENYLACETYL)-4-[3-(TRIFLUOROMETHYL)PHENYL]-SEMICARBAZIDE

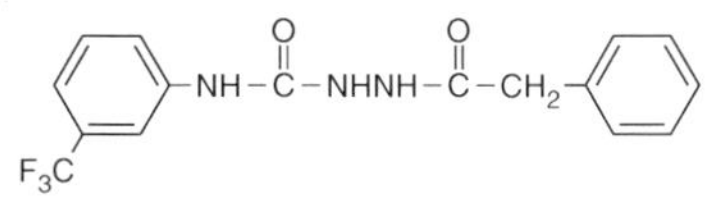

250 mg

S84,782-8
4-(3-[3-(TRIFLUOROMETHYL)PHENYL]-UREIDO)BENZOIC [4-(4-DIMETHYLAMINO)-BENZYLIDENE]HYDRAZIDE

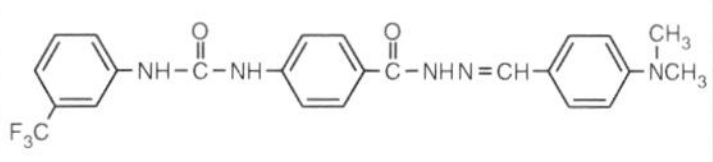

250 mg

S70,940-9
1,1-DIMETHYL-3-[2-(TRIFLUOROMETHYL)PHENYL]UREA

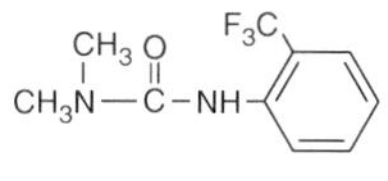

5 gm

S59,918-2
1,1-DIBENZYL-3-[3-(TRIFLUOROMETHYL)PHENYL]UREA

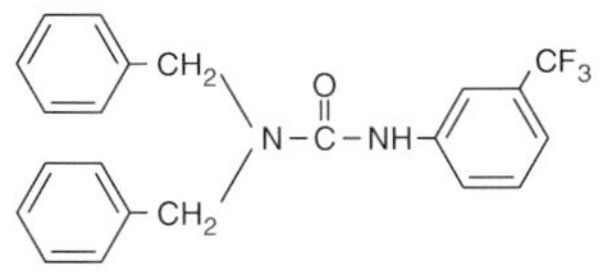

5 gm

S80,103-8
1-BENZYL-1-PHENYL-3-[3-(TRIFLUOROMETHYL)PHENYL]UREA

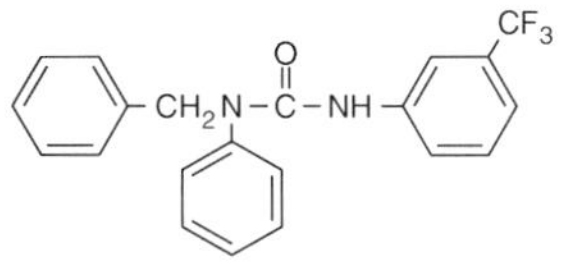

250 mg

S79,542-9
1-BENZYL-1-METHYL-3-[2-(TRIFLUOROMETHYL)PHENYL]UREA

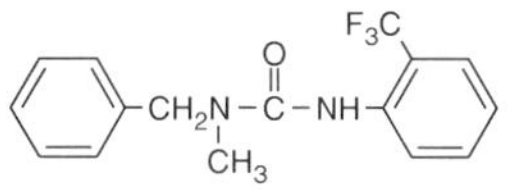

250 mg

S80,173-9
1-(3,4-DIMETHYLBENZYL)-3-[2-(TRIFLUOROMETHYL)PHENYL]UREA

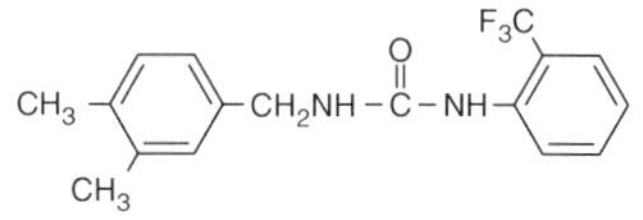

250 mg

S80,115-1
1-(3,4-DIMETHYLBENZYL)-3-[3-(TRIFLUOROMETHYL)PHENYL]UREA

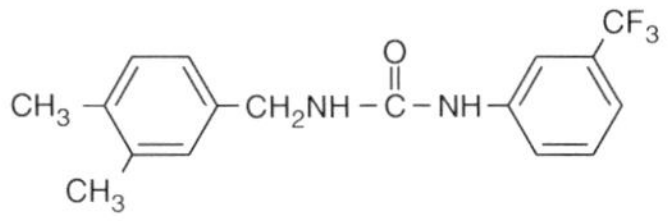

250 mg

S76,901-0
1-ACETYL-4-[3-(TRIFLUOROMETHYL)PHENYL]SEMICARBAZIDE

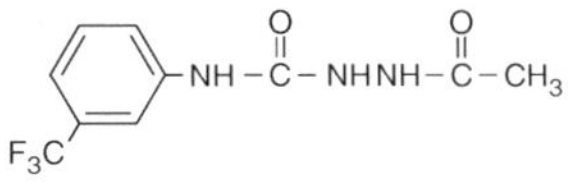

250 mg

S78,083-9
1-[2-(3,3-DIETHYLUREIDO)ACETYL]-4-[3-(TRIFLUOROMETHYL)PHENYL]-SEMICARBAZIDE

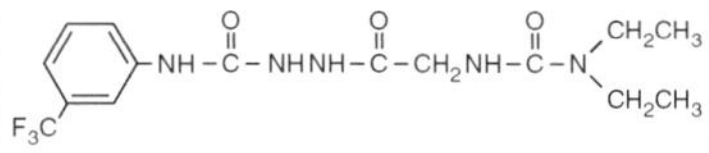

250 mg

S79,656-5
1-(2-PHENYLACETYL)-4-[2-(TRIFLUOROMETHYL)PHENYL]-SEMICARBAZIDE

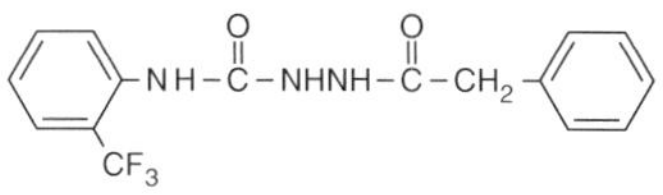

250 mg

S84,773-9
4-(3-[3-(TRIFLUOROMETHYL)PHENYL]-UREIDO)BENZOIC (3-NITROBENZYLIDENE)-HYDRAZIDE

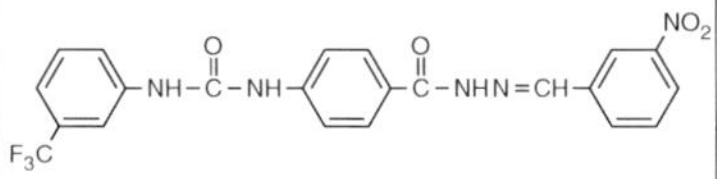

250 mg

S84,779-8
4-(3-[3-(TRIFLUOROMETHYL)PHENYL]-UREIDO)BENZOIC (4-ACETAMIDOBENZYLIDENE)HYDRAZIDE

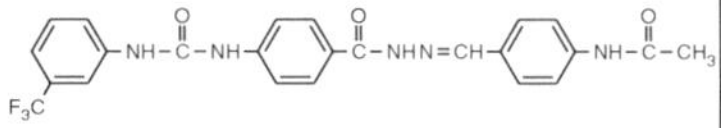

250 mg

S84,785-2
4-(3-[3-(TRIFLUOROMETHYL)PHENYL]-UREIDO)BENZOIC (4-METHYLBENZYLIDENE)HYDRAZIDE

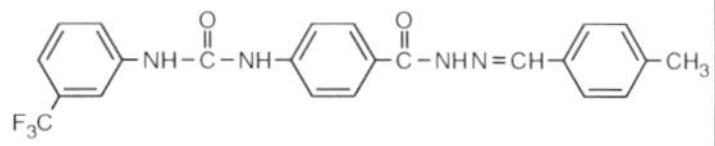

250 mg

S90,202-0
ETHYL 2-(3-[3-(TRIFLUOROMETHYL)-PHENYL]UREIDO)ACETATE

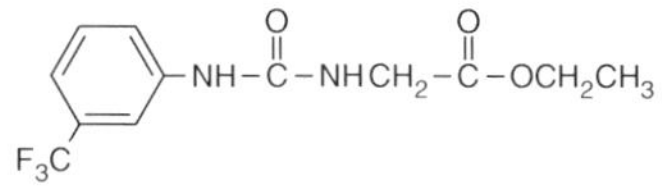

250 mg

S64,959-7
1-HYDROXY-3-[3-(TRIFLUOROMETHYL)PHENYL]UREA

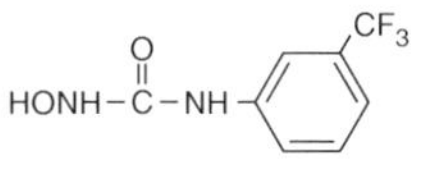

5 gm

S69,903-9
1-(4-ACETYLPHENYL)-3-[2-(TRIFLUOROMETHYL)PHENYL]UREA

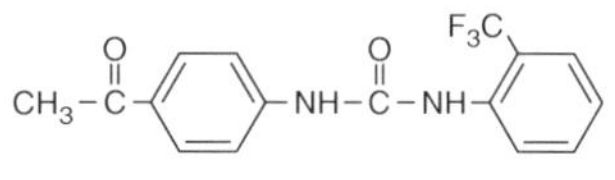

250 mg

S64,654-7
1-(2-BENZOYL-4-CHLOROPHENYL)-3-[3-(TRIFLUOROMETHYL)PHENYL]UREA

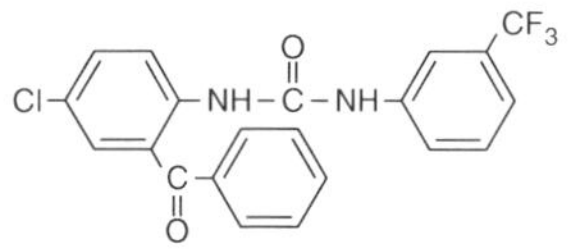

5 gm

S80,107-0
DIMETHYL 5-(3-[3-(TRIFLUOROMETHYL)-PHENYL]UREIDO)ISOPHTHALATE

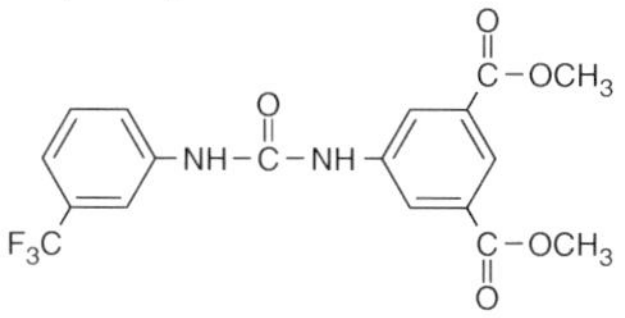

250 mg

S91,322-7
1-(4-CARBOXYPHENYL)-4-[2-(TRIFLUOROMETHYL)PHENYL]-SEMICARBAZIDE

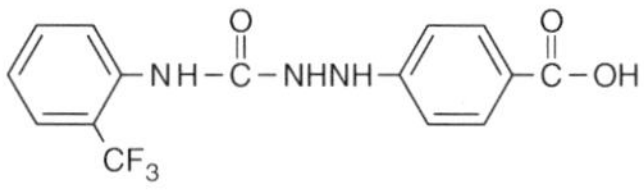

250 mg

S80,111-9
3-(3-[3-(TRIFLUOROMETHYL)PHENYL]-UREIDO)BENZOIC ACID

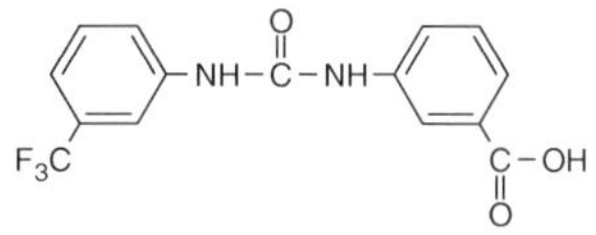

250 mg

S80,112-7
4-(3-[3-(TRIFLUOROMETHYL)PHENYL]-UREIDO)BENZOIC ACID

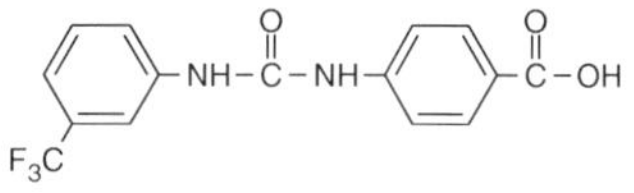

250 mg

S72,589-7
METHYL 2-[3-(TRIFLUOROMETHYL)-PHENYLUREIDO]BENZOATE

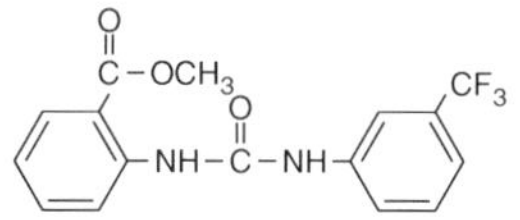

5 gm

S79,486-4
ETHYL 4-(3-[2-(TRIFLUOROMETHYL)-PHENYL]UREIDO)BENZOATE

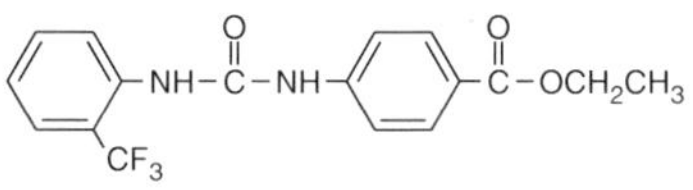

250 mg

S80,106-2
ETHYL 4-(3-[3-(TRIFLUOROMETHYL)-PHENYL]UREIDO)BENZOATE

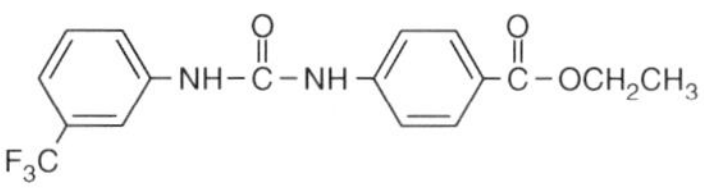

250 mg

S84,524-8
4-(3-[3-(TRIFLUOROMETHYL)PHENYL]-UREIDO)BENZOIC HYDRAZIDE

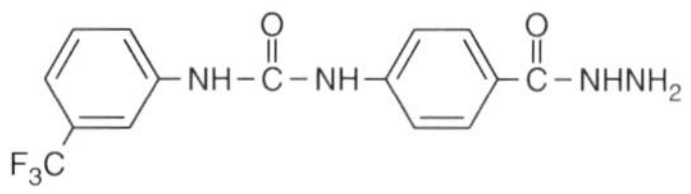

250 mg

S79,482-1
11-(3-[3-(TRIFLUOROMETHYL)PHENYL]-UREIDO)UNDECANOIC ACID

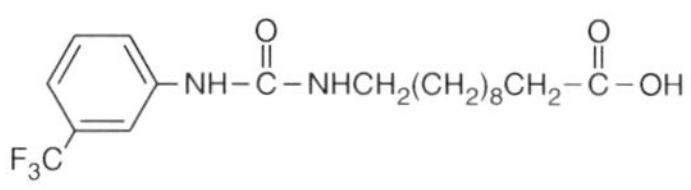

250 mg

S61,206-5
1,1-DIMETHYL-4-[3-(TRIFLUOROMETHYL)PHENYL]SEMICARBAZIDE

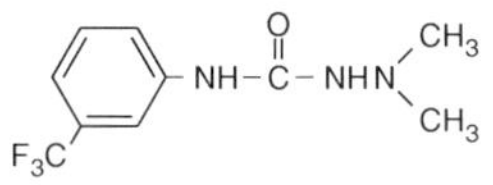

5 gm

S71,056-3
1-PHENYL-4-[3-(TRIFLUOROMETHYL)-PHENYL]SEMICARBAZIDE

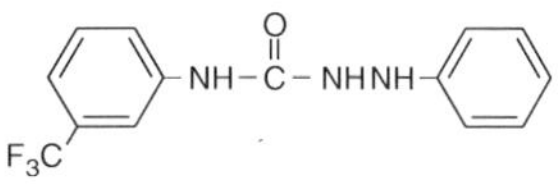

250 mg

S90,450-3
1-BUTYRYL-3-[3-(TRIFLUOROMETHYL)-PHENYL]UREA

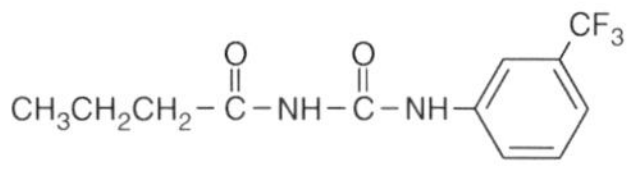

250 mg

S90,445-7
1-(2,2,2-TRIFLUOROACETYL)-3-[2-(TRIFLUOROMETHYL)PHENYL]UREA

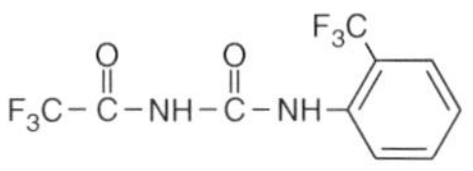

250 mg

S79,685-9
4'-[4-(3-METHYLUREIDO)BENZYL]-2,2,3,3,3-PENTAFLUOROPROPIONANILIDE
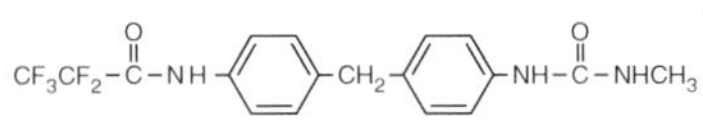
250 mg

S66,043-4
1-METHYL-3-(3-TRIFLUOROMETHYLPHENYL)UREA
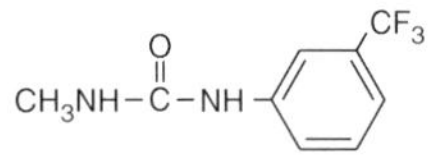
5 gm

S59,926-3
1,1-DIMETHYL-3-(3-TRIFLUOROMETHYLPHENYL)UREA
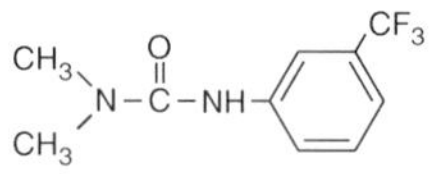
5 gm

S73,545-0
1-ETHYL-3-(3-TRIFLUOROMETHYLPHENYL)UREA
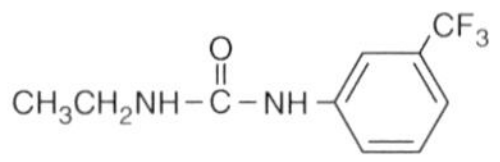
250 mg

S66,041-8
1-ALLYL-3-(3-TRIFLUOROMETHYLPHENYL)UREA
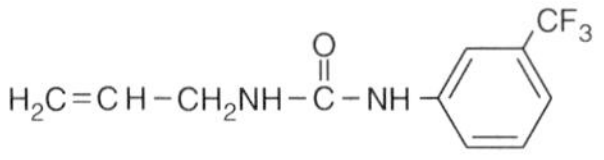
5 gm

S61,142-5
1-ISOPROPYL-3-(3-TRIFLUOROMETHYLPHENYL)UREA
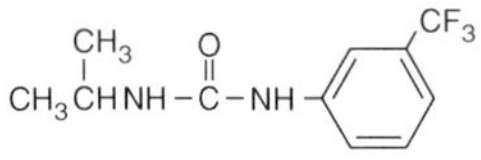
5 gm

S61,190-5
1-PROPYL-3-[3-(TRIFLUOROMETHYL)-PHENYL]UREA
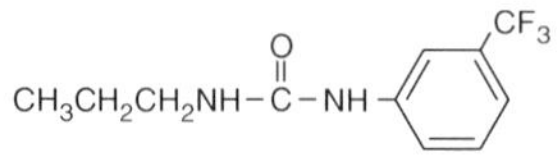
5 gm

S59,923-9
1,1-DIETHYL-3-(3-TRIFLUOROMETHYLPHENYL)UREA
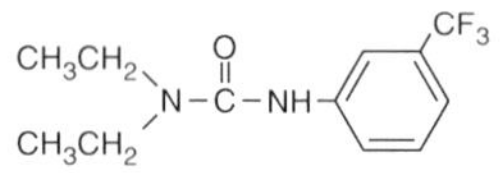
5 gm

S61,140-9
1-BUTYL-3-(3-TRIFLUOROMETHYLPHENYL)UREA
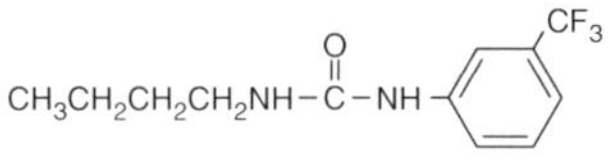
5 gm

S64,655-5
1-ISOBUTYL-3-(3-TRIFLUOROMETHYLPHENYL)UREA
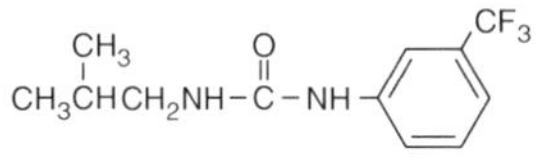
5 gm

S89,115-0
1-TERT-BUTYL-3-[3-(TRIFLUOROMETHYL)-PHENYL]UREA
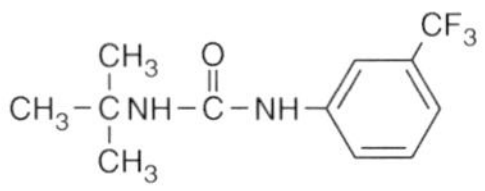
250 mg

S61,166-2
1-BUTYL-1-METHYL-3-(3-TRIFLUOROMETHYLPHENYL)UREA
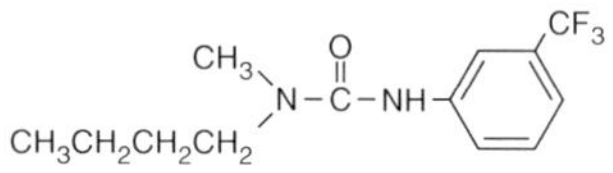
5 gm

S59,934-4
1-PHENYL-3-[3-(TRIFLUOROMETHYL)-PHENYL]UREA
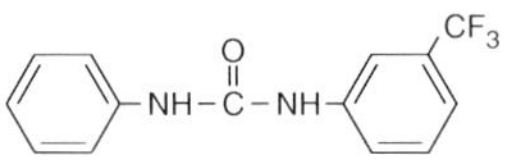
5 gm

S59,912-3
1,1-DIALLYL-3-(3-TRIFLUOROMETHYLPHENYL)UREA
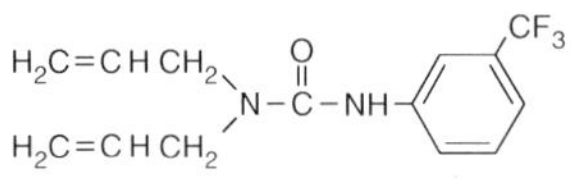
5 gm

S59,921-2
1,1-DIPROPYL-3-(3-TRIFLUOROMETHYLPHENYL)UREA
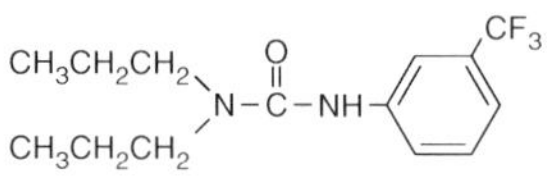
5 gm

S71,431-3
1,1-DIISOPROPYL-3-(3-TRIFLUOROMETHYLPHENYL)UREA
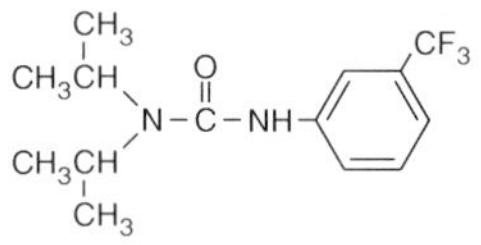
250 mg

S59,933-6
1-(3-CHLOROPHENYL)-3-(3-TRIFLUOROMETHYLPHENYL)UREA
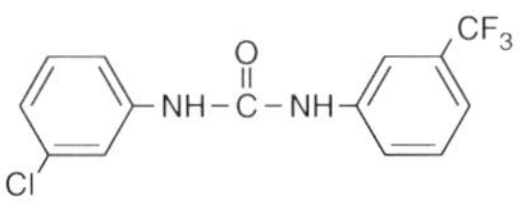
5 gm

S60,083-0
1,3-BIS(3-TRIFLUOROMETHYLPHENYL)UREA
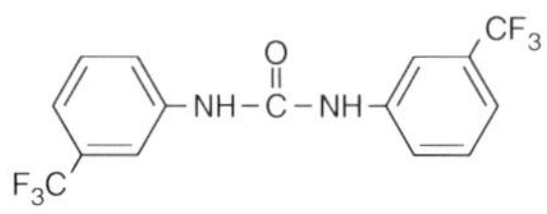
5 gm

S59,930-1
1,1-DIBUTYL-3-(3-TRIFLUOROMETHYLPHENYL)UREA
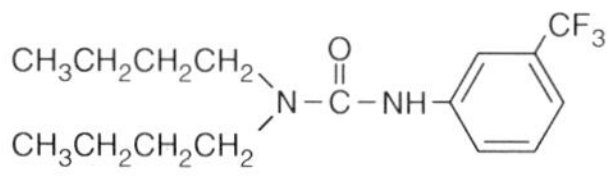
250 mg

S59,913-1
1-BUTYL-1-PHENYL-3-(3-TRIFLUOROMETHYLPHENYL)UREA
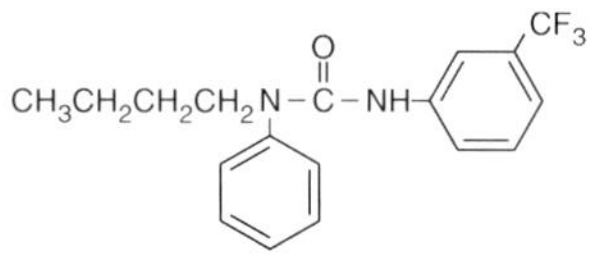
5 gm

S79,409-0
1,1'-HEXAMETHYLENEBIS(3-[3-(TRIFLUOROMETHYL)PHENYL]UREA)
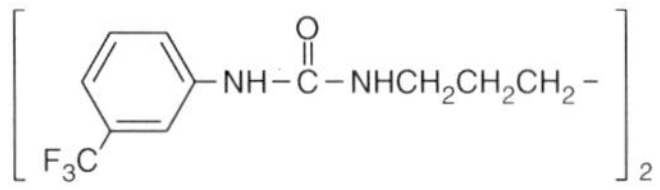
250 mg

S67,056-1
1-OCTADECYL-3-[3-(TRIFLUOROMETHYL)-PHENYL]UREA
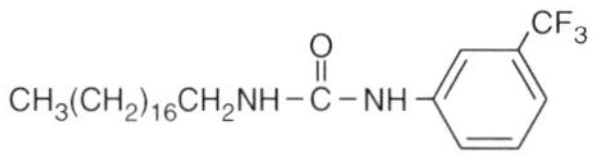
5 gm

S77,245-3
1-CYCLOPENTYL-3-[3-(TRIFLUOROMETHYL)PHENYL]UREA
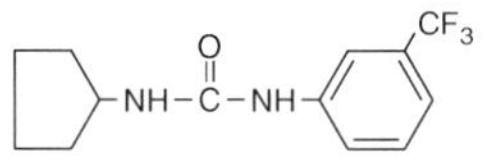
250 mg

S61,197-2
1-CYCLOHEXYL-3-(3-TRIFLUOROMETHYLPHENYL)UREA
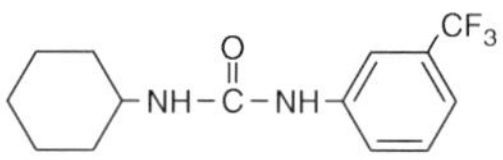
5 gm

S61,189-1
1-CYCLOHEXYL-1-METHYL-3-(3-TRIFLUOROMETHYLPHENYL)UREA
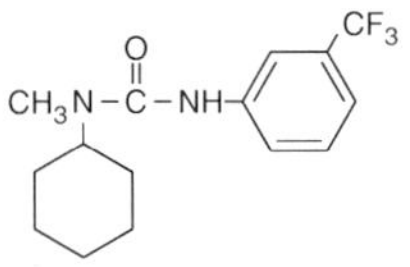
5 gm

S80,118-6
1-CYCLOHEXYL-1-ISOPROPYL-3-[3-(TRIFLUOROMETHYL)PHENYL]UREA
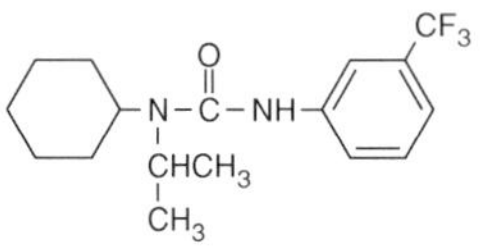
250 mg

S61,137-9
1,1-DICYCLOHEXYL-3-(3-TRIFLUOROMETHYLPHENYL)UREA
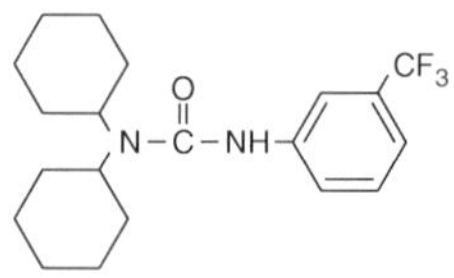
5 gm

S66,470-7
1-(2-CHLORO-5-TRIFLUOROMETHYLPHENYL)-3-CYCLOHEXYLUREA
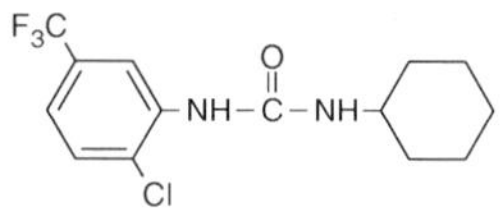
5 gm

S80,087-2
1-METHYL-1-PHENYL-3-[3-(TRIFLUOROMETHYL)PHENYL]UREA
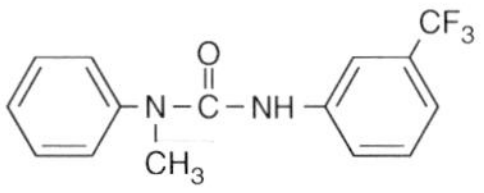
250 mg

S71,695-2
1-PHENETHYL-3-(3-TRIFLUOROMETHYLPHENYL)UREA
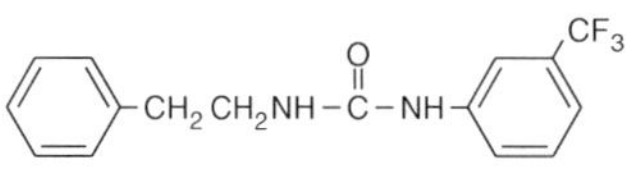
5 gm

S72,952-3
1-(DIPHENYLMETHYL)-3-[2-(TRIFLUOROMETHYL)PHENYL]UREA
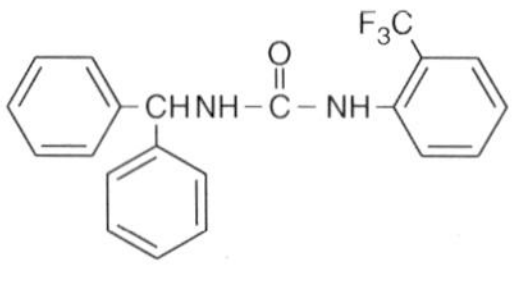
5 gm

S91,332-4
1-(DIPHENYLMETHYL)-3-[3-(TRIFLUOROMETHYL)PHENYL]UREA
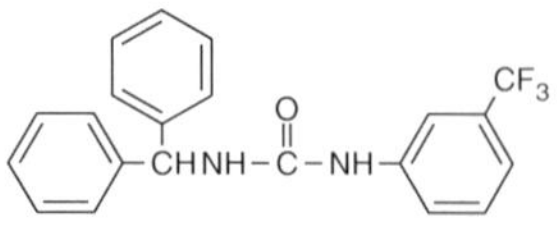
250 mg

S71,115-2
1-METHYL-3-(2-TRIFLUOROMETHYLPHENYL)UREA

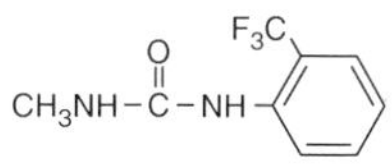

250 mg

S71,655-3
1-ISOPROPYL-3-(2-TRIFLUOROMETHYLPHENYL)UREA

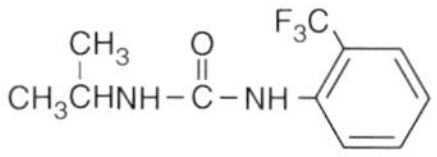

5 gm

S79,101-6
1-PROPYL-3-[2-(TRIFLUOROMETHYL)-PHENYL]UREA

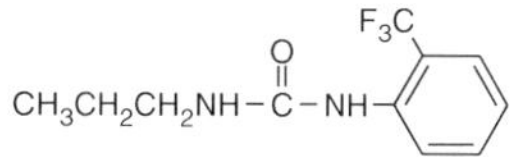

250 mg

S75,227-4
1-TERT-BUTYL-3-(2-TRIFLUOROMETHYLPHENYL)UREA

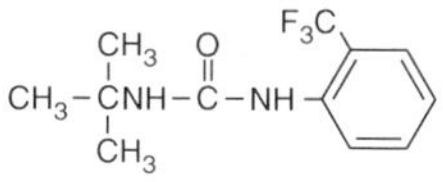

5 gm

S79,106-7
1-ISOBUTYL-3-[2-(TRIFLUOROMETHYL)-PHENYL]UREA

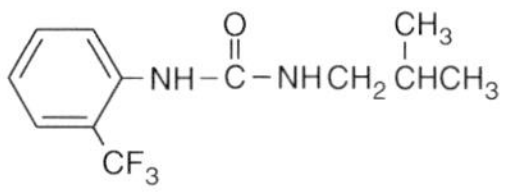

250 mg

S80,084-8
1-[2-(TRIFLUOROMETHYL)PHENYL]-3-[3-(TRIFLUOROMETHYL)PHENYL]UREA

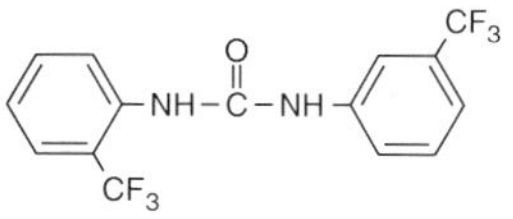

250 mg

S68,821-5
1-(3-ETHYLPHENYL)-3-[2-(TRIFLUOROMETHYL)PHENYL]UREA

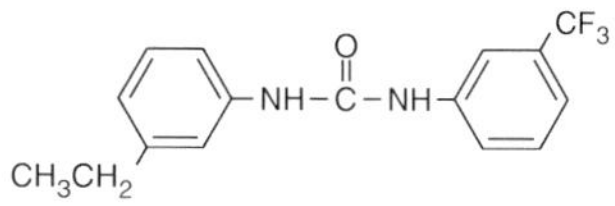

250 mg

S75,361-0
1-(2-ETHYLPHENYL)-3-[2-(TRIFLUOROMETHYL)PHENYL]UREA

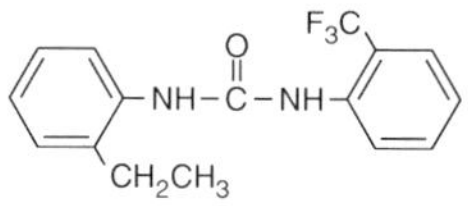

5 gm

S88,968-7
1-(2-ETHYLPHENYL)-3-[3-(TRIFLUOROMETHYL)PHENYL]UREA

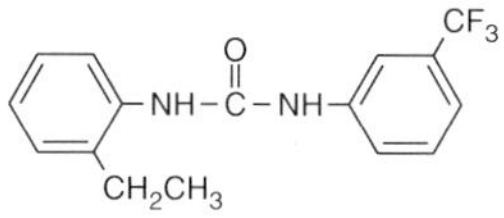

250 mg

S79,670-0
1-ETHYL-1-[2-(TRIFLUOROMETHYL)-PHENYL]-3-[3-(TRIFLUOROMETHYL)-PHENYL]UREA

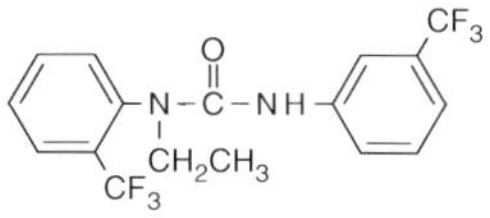

250 mg

S90,444-9
1,3-BIS[2-(TRIFLUOROMETHYL)PHENYL]-1-ETHYLUREA

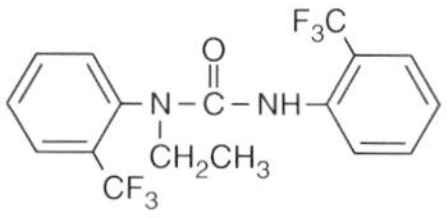

250 mg

S67,046-4
1-OCTADECYL-3-[2-(TRIFLUOROMETHYL)-PHENYL]UREA

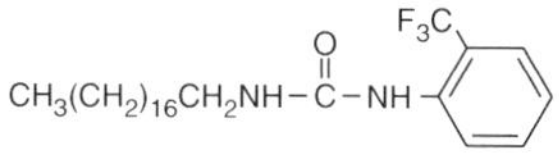

5 gm

S75,360-2
1-(4-ETHYLPHENYL)-3-[2-(TRIFLUOROMETHYL)PHENYL]UREA

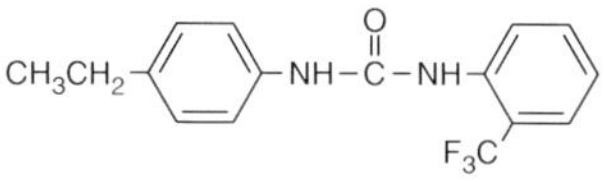

5 gm

S77,977-6
1-(4-BUTYLPHENYL)-3-[2-(TRIFLUOROMETHYL)PHENYL]UREA

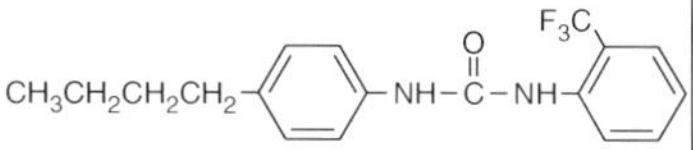

250 mg

S75,357-2
1-[2-(TRIFLUOROMETHYL)PHENYL]-3-(2,3-XYLYL)UREA

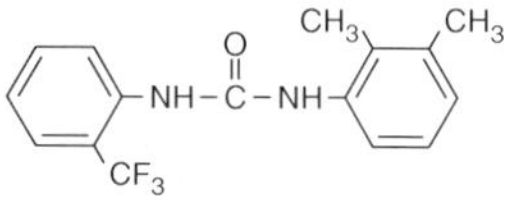

5 gm

S75,358-0
1-[2-(TRIFLUOROMETHYL)PHENYL]-3-(2,6-XYLYL)UREA

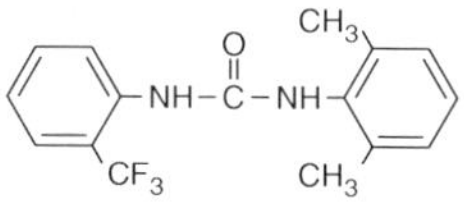

5 gm

S75,359-9
1-[2-(TRIFLUOROMETHYL)PHENYL]-3-(2,4-XYLYL)UREA

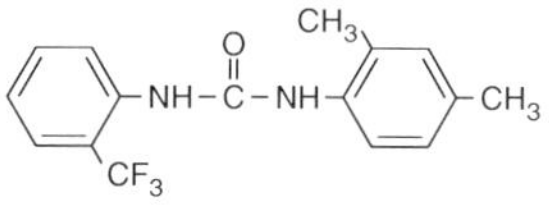

5 gm

S67,352-8
4,4'-BIS(3-[2-(TRIFLUOROMETHYL)-PHENYL]UREIDO)-3,3'-DIMETHYLBIPHENYL

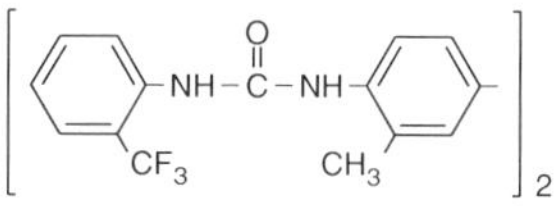

250 mg

S79,499-6
1-[3,5-BIS(TRIFLUOROMETHYL)PHENYL]-3-[2-(TRIFLUOROMETHYL)PHENYL]UREA

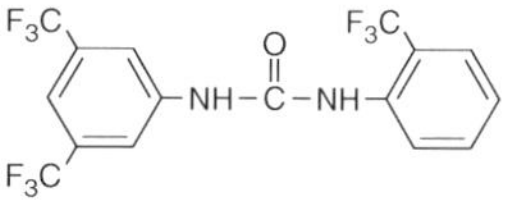

250 mg

S75,369-6
1-[2-(TRIFLUOROMETHYL)PHENYL]-3-(2,4,6-TRIMETHYLPHENYL)UREA

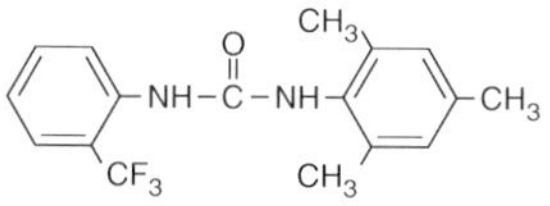

5 gm

S71,150-0
1-(4-TOLYL)-3-(3-TRIFLUOROMETHYLPHENYL)UREA

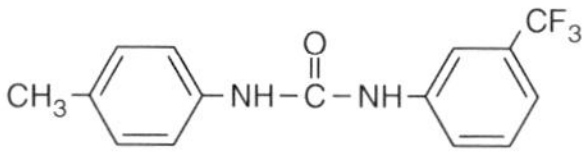

5 gm

S88,966-0
1-(4-ETHYLPHENYL)-3-[3-(TRIFLUOROMETHYL)PHENYL]UREA

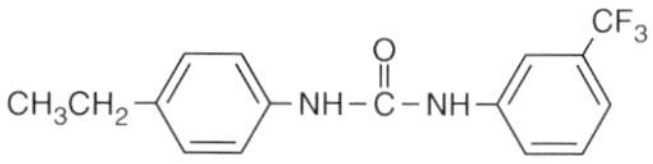

250 mg

S73,550-7
1,1-DIBUTYL-3-(4-TRIFLUOROMETHYLPHENYL)UREA

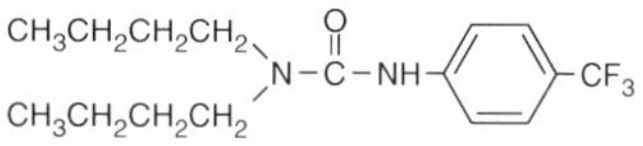

250 mg

S78,008-1
1-(4-BUTYLPHENYL)-3-[3-(TRIFLUOROMETHYL)PHENYL]UREA

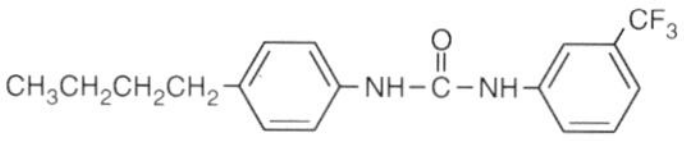

250 mg

S66,010-8
1-(2-ETHYL-6-METHYLPHENYL)-3-(3-TRIFLUOROMETHYLPHENYL)UREA

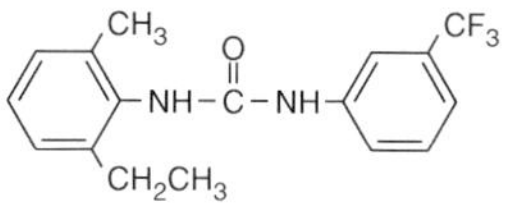

5 gm

S86,804-3
1-[(4-TRICHLOROMETHYL)PHENYL]-3-(2,5-XYLYL)UREA

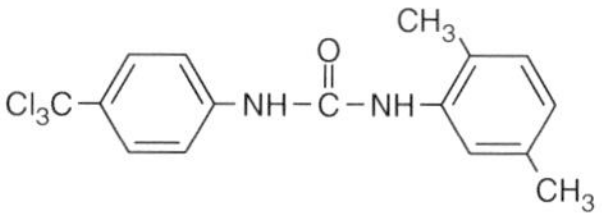

250 mg

S59,932-8
1-(2,3-DICHLOROPHENYL)UREA

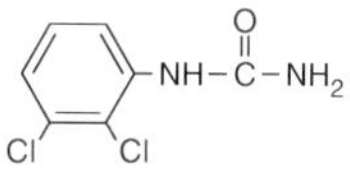

5 gm

S70,909-3
1-[3-(TRIFLUOROMETHYL)PHENYL]-3-(2,6-XYLYL)UREA

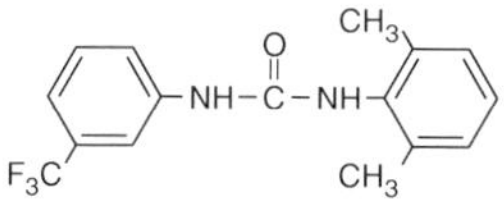

5 gm

S88,967-9
1-[3-(TRIFLUOROMETHYL)PHENYL]-3-(2,3-XYLYL)UREA

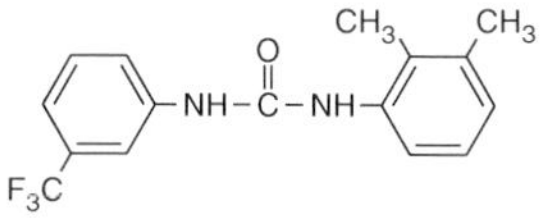

250 mg

S71,505-0
1-(2,6-DIETHYLPHENYL)-3-(3-TRIFLUOROMETHYLPHENYL)UREA

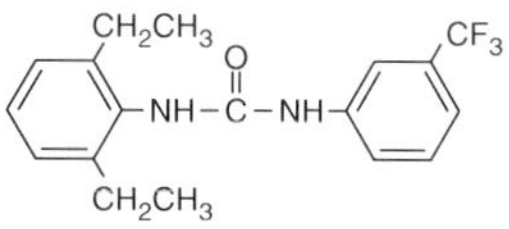

5 gm

S70,741-4
1-(2,6-DIISOPROPYLPHENYL)-3-(3-TRIFLUOROMETHYLPHENYL)UREA

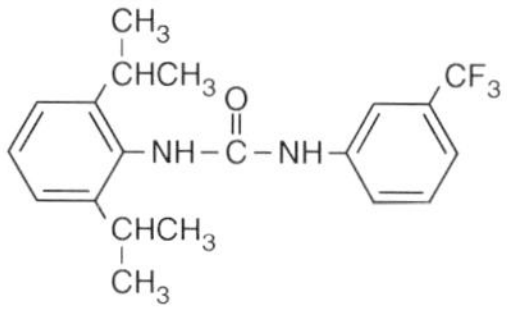

5 gm

S60,193-4
1-(4-CHLORO-2-METHYLPHENYL)-3-(3-TRIFLUOROMETHYLPHENYL)UREA

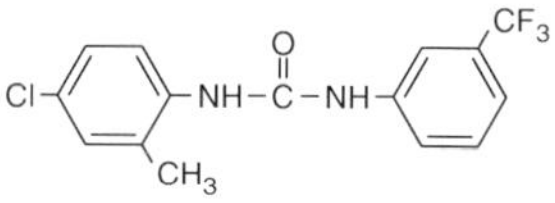

5 gm

S79,494-5
1-[3-(TRIFLUOROMETHYL)PHENYL]-3-(3,4-XYLYL)UREA

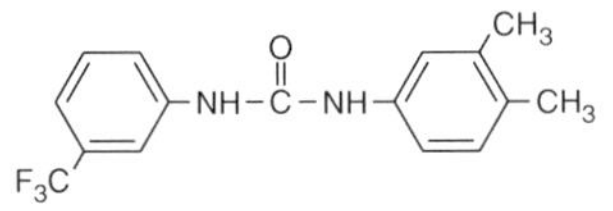

250 mg

S88,965-2
1-[3-(TRIFLUOROMETHYL)PHENYL]-3-(2,4-XYLYL)UREA

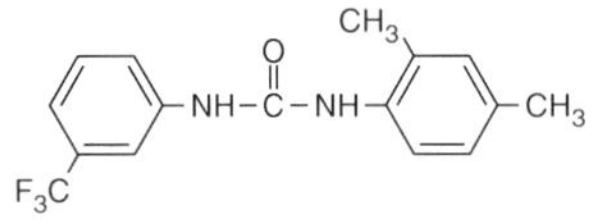

250 mg

S89,118-5
1-[3,5-BIS(TRIFLUOROMETHYL)PHENYL]-3-(1-NAPHTHYL)UREA

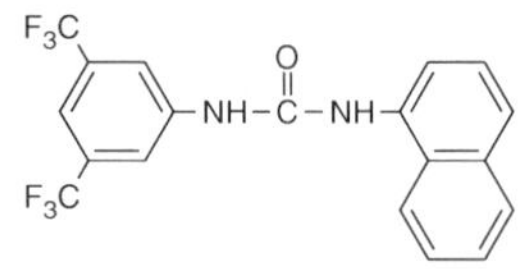

250 mg

S88,981-4
1-MESITYL-3-[3-(TRIFLUOROMETHYL)-PHENYL]UREA

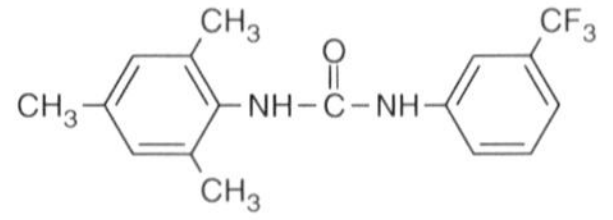

250 mg

S88,980-6
1-[3-(TRIFLUOROMETHYL)PHENYL]-3-(2,4,5-TRIMETHYLPHENYL)UREA

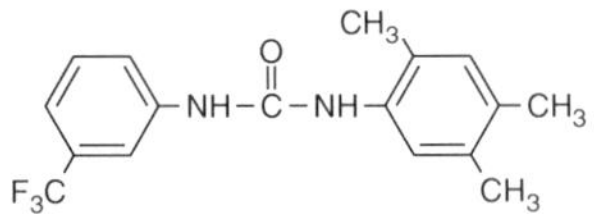

250 mg

S74,336-4
1,1'-(4,6-DIBROMO-1,3-PHENYLENE)-BIS(3-[1,1-BIS(HYDROXYMETHYL)-2-HYDROXYETHYL]UREA)

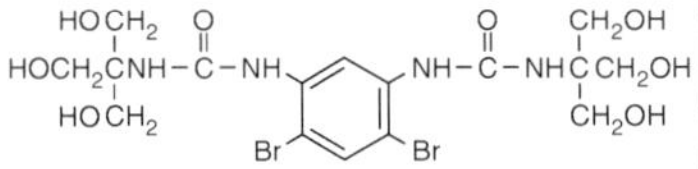

5 gm

S85,421-2
1-(5-CHLORO-2,4-DIMETHOXYPHENYL)-3-[4-(2-HYDROXYETHYL)PHENYL]UREA

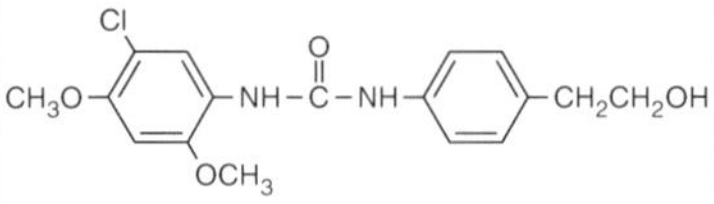

250 mg

S90,209-8
2-[3-(4-FLUOROPHENYL)UREIDO]ACETIC (4-HYDROXY-3-METHOXYBENZYLIDENE)-HYDRAZIDE

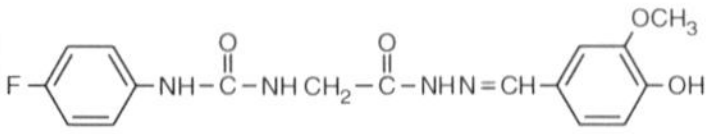

250 mg

S90,213-6
2-[3-(4-FLUOROPHENYL)UREIDO]ACETIC (3-HYDROXY-4-METHOXYBENZYLIDENE)-HYDRAZIDE

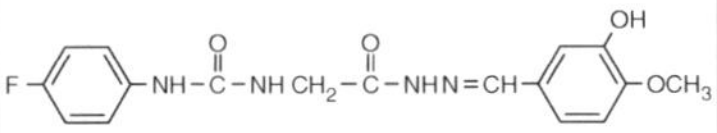

250 mg

S69,517-3
4-(5-CHLORO-2-METHOXYPHENYL)-1-(6-HYDROXYHEXANOYL)SEMICARBAZIDE

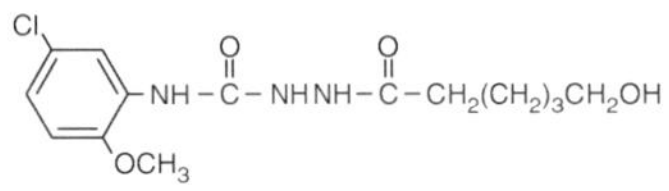

5 gm

S69,055-4
4-(5-CHLORO-2-METHOXYPHENYL)-1-(4-HYDROXYBENZOYL)SEMICARBAZIDE

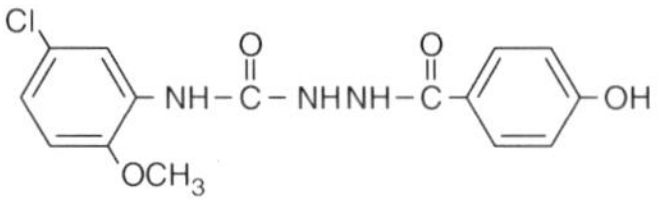

250 mg

S90,319-1
2-[3-(2-CHLORO-4-NITROPHENYL)-UREIDO]ACETIC(4-HYDROXY-3-METHOXY-BENZYLIDENE)HYDRAZIDE

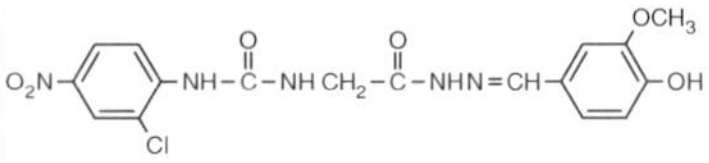

250 mg

S86,902-3
4-[3-(3-CHLOROPHENYL)UREIDO]-BENZOIC (3-ETHOXY-4-HYDROXYBENZYLIDENE)HYDRAZIDE

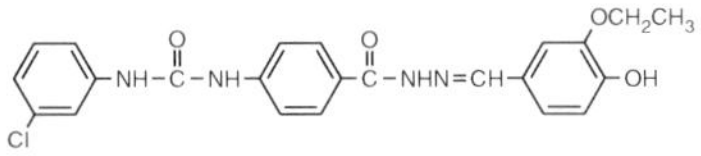

250 mg

S72,315-0
4-(4-CHLORO-2-METHOXYPHENYL)-1-(2-HYDROXYBENZOYL)SEMICARBAZIDE

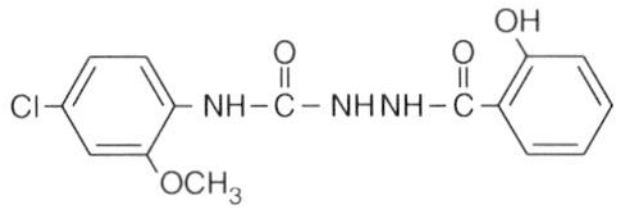

250 mg

S91,728-1
4-(4-CHLORO-2-METHOXYPHENYL)-1-(4-HYDROXYBENZOYL)SEMICARBAZIDE

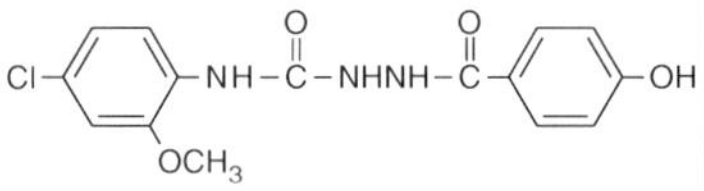

250 mg

S85,093-4
1-(4-CHLORO-2-METHOXYPHENYL)-3-CYCLOHEXYL-3-(2-HYDROXYETHYL)UREA

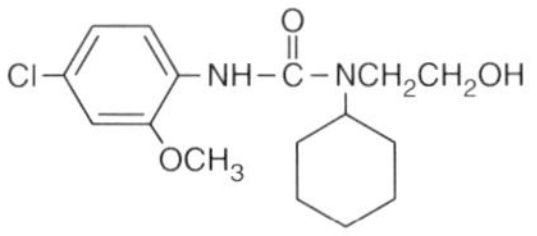

250 mg

S85,430-1
1-(4-CHLORO-2-METHOXYPHENYL)-3-(2-HYDROXYPHENYL)UREA

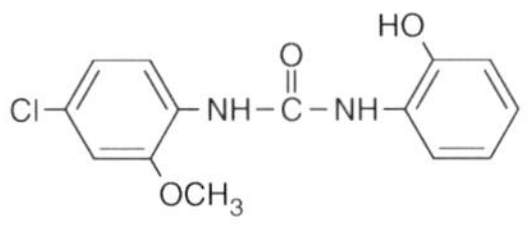

250 mg

S79,444-9
1-(5-CHLORO-2-METHOXYPHENYL)-3-[4-(2-HYDROXYETHYL)PHENYL]UREA

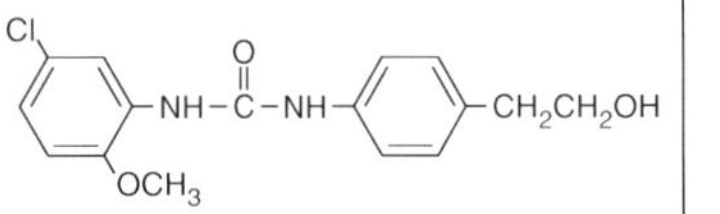

250 mg

S79,470-8
1-(5-CHLORO-2-METHOXYPHENYL)-3-(3-HYDROXYPHENYL)UREA

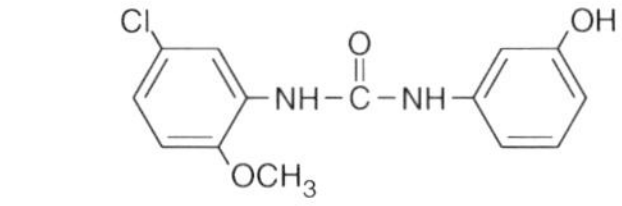

250 mg

S79,763-4
1-(5-CHLORO-2-METHOXYPHENYL)-3-(5-HYDROXY-1-NAPHTHYL)UREA

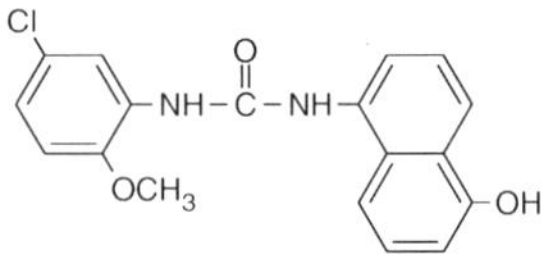

250 mg

S85,065-9
1-(2-CHLORO-4-METHOXYPHENYL)-3-(3-HYDROXYPHENYL)UREA

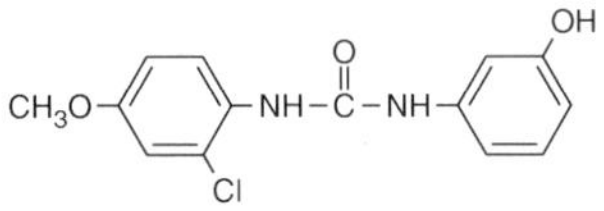

250 mg

S69,249-2
PHENYLGLYCOLIC ACID N2-[N-(3,4-DICHLOROPHENYL)CARBAMOYL]HYDRAZIDE

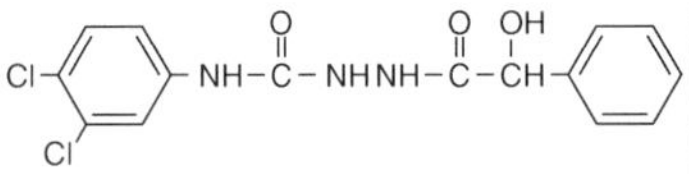

250 mg

S84,752-6
1-(2-CHLORO-4-NITROPHENYL)-3-CYCLOHEXYL-3-(2-HYDROXY-2-PHENYLETHYL)UREA

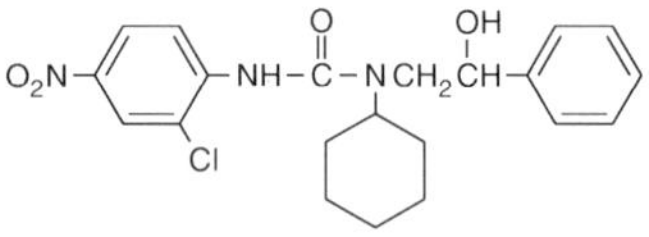

250 mg

S71,570-0
1-(3,4-DICHLOROPHENYL)-3-(1-HYDROXYETHYL)PHENYLUREA

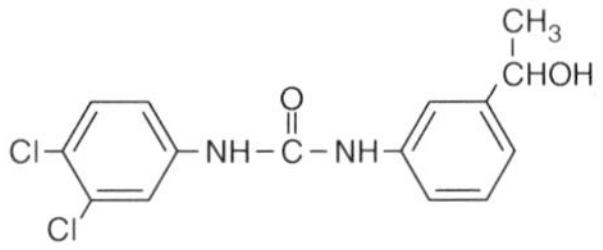

5 gm

S72,378-9
4-(2,5-DICHLOROPHENYL)-1-(2-HYDROXYBENZOYL)SEMICARBAZIDE

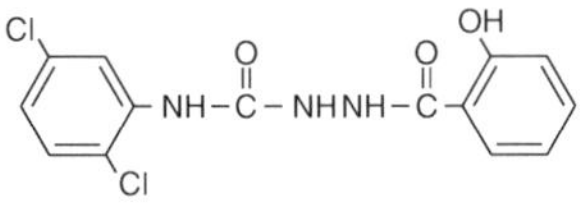

5 gm

S72,379-7
4-(3,4-DICHLOROPHENYL)-1-(2-HYDROXYBENZOYL)SEMICARBAZIDE

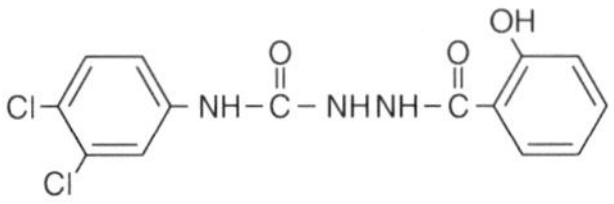

5 gm

S72,764-4
1-(4-CARBOXY-3-HYDROXYPHENYL)-3-(3-FLUOROPHENYL)UREA

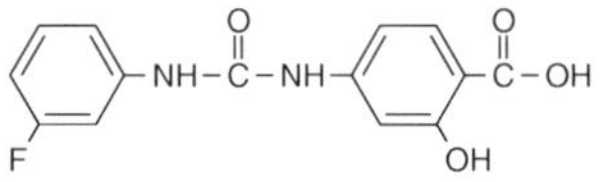

250 mg

S90,237-3
2-[3-(3-CHLOROPHENYL)UREIDO]ACETIC (3-HYDROXYBENZYLIDENE)HYDRAZIDE

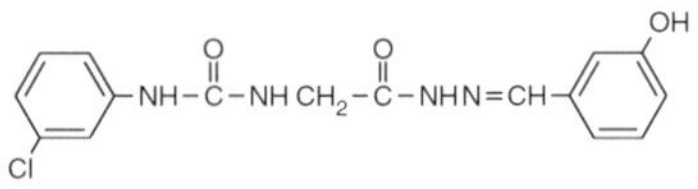

250 mg

S90,205-5
2-[3-(4-FLUOROPHENYL)UREIDO]ACETIC (2-HYDROXYBENZYLIDENE)HYDRAZIDE

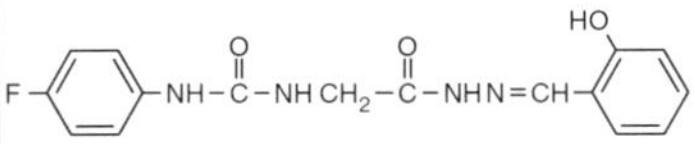

250 mg

S90,292-6
2-[3-(2,3-DICHLOROPHENYL)-UREIDO]ACETIC (3-HYDROXYBENZYLIDENE)HYDRAZIDE

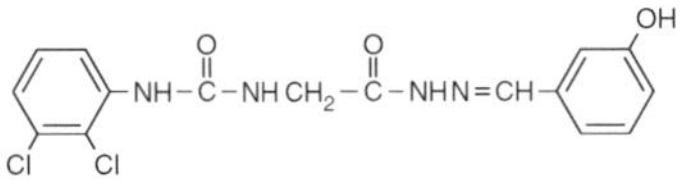

250 mg

S86,783-7
1-(4-CHLORO-2-NITROPHENYL)-3-[4-(2-HYDROXYETHYL)PHENYL]UREA

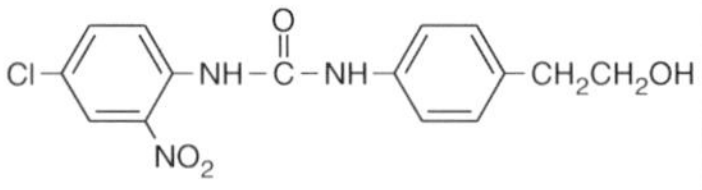

250 mg

S86,733-0
1-(4-CHLORO-2-NITROPHENYL)-3-(3-HYDROXYPHENYL)UREA

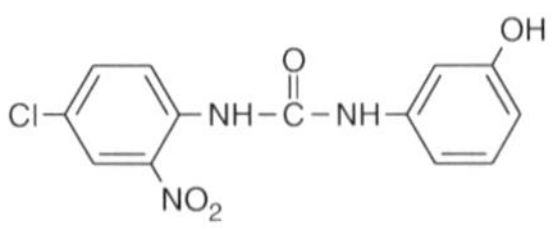

250 mg

S84,541-8
1-CYCLOHEXYL-3-(2-FLUORO-5-NITROPHENYL)-1-(2-HYDROXYETHYL)UREA
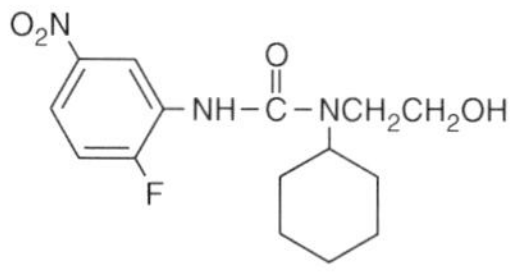
250 mg

S85,150-7
1-(2,3-DICHLOROPHENYL)-3-(2-HYDROXY-5-NITROPHENYL)UREA
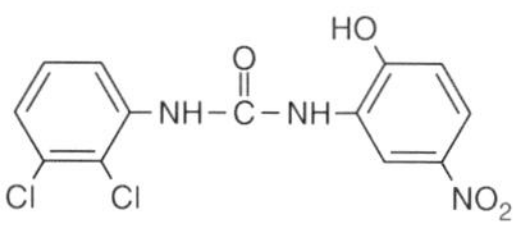
250 mg

S71,571-9
1-(3-CHLORO-2-METHYLPHENYL)-3-(2-HYDROXY-5-NITROPHENYL)UREA
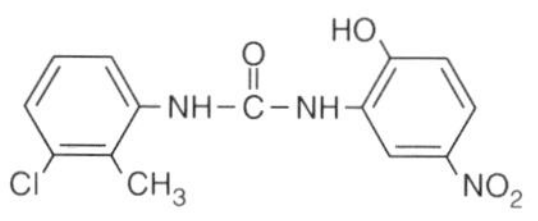
5 gm

S84,432-2
1-(5-CHLORO-2-HYDROXYPHENYL)-3-(2-FLUORO-5-NITROPHENYL)UREA
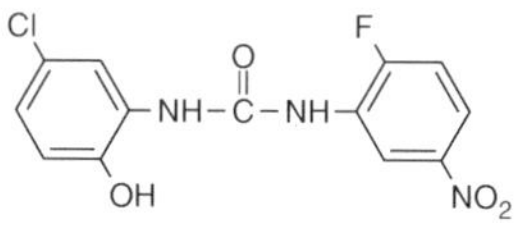
250 mg

S84,435-7
1-(2-FLUORO-5-NITROPHENYL)-3-(2-HYDROXY-5-NITROPHENYL)UREA
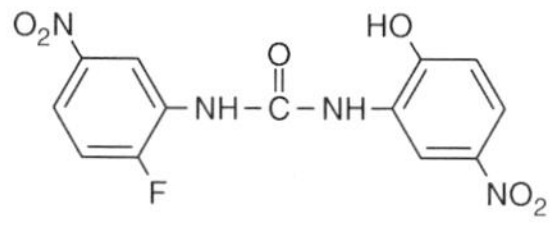
250 mg

S84,789-5
1-(5-CHLORO-2-HYDROXYPHENYL)-3-(2-METHYL-5-NITROPHENYL)UREA
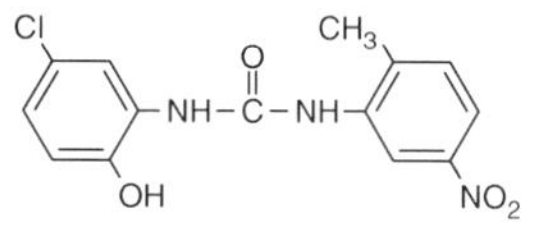
250 mg

S69,338-3
4-(4-FLUOROPHENYL)-1-(4-HYDROXYBENZOYL)SEMICARBAZIDE
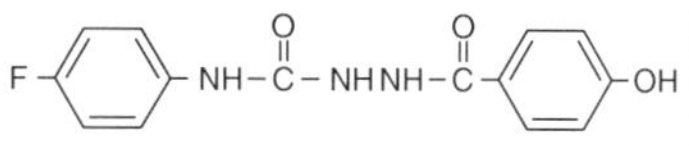
250 mg

S69,414-2
4-(2,4-DIFLUOROPHENYL)-1-(4-HYDROXYBENZOYL)SEMICARBAZIDE
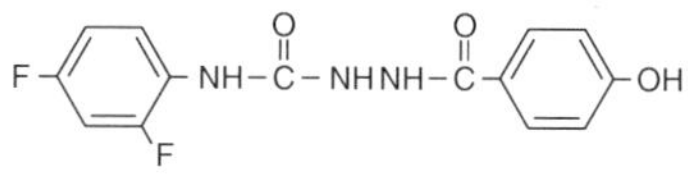
250 mg

S69,324-3
4-(3,4-DICHLOROPHENYL)-1-(4-HYDROXYBENZOYL)SEMICARBAZIDE
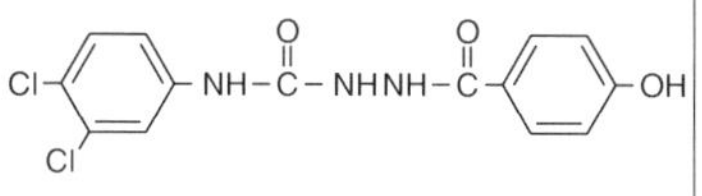
250 mg

S72,928-0
1-(6-HYDROXYHEXANOYL)-4-(3-CHLOROPHENYL)SEMICARBAZIDE
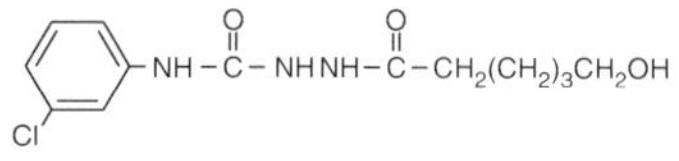
5 gm

S68,978-5
1-(6-HYDROXYHEXANOYL)-4-(2-CHLOROPHENYL)SEMICARBAZIDE
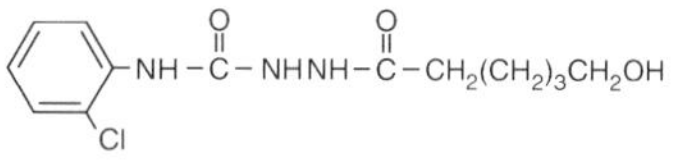
250 mg

S72,418-1
4-(3,4-DICHLOROPHENYL)-1-(6-HYDROXYHEXANOYL)SEMICARBAZIDE
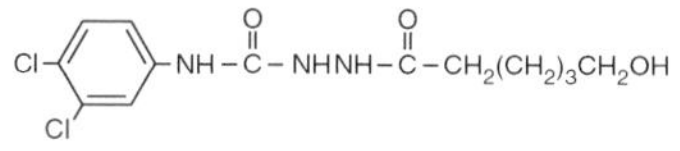
5 gm

S78,020-0
2-[3-(3-CHLOROPHENYL)UREIDO]ACETIC (2-HYDROXYBENZYLIDENE)HYDRAZIDE
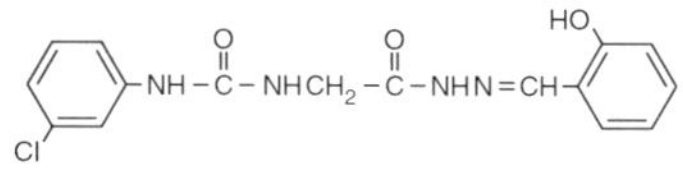
250 mg

S79,660-3
2-[3-(4-FLUOROPHENYL)UREIDO]ACETIC (3-HYDROXYBENZYLIDENE)HYDRAZIDE
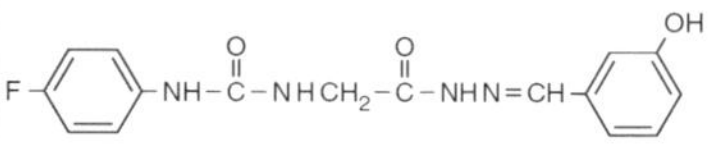
250 mg

S77,938-5
2-[3-(3,4-DICHLOROPHENYL)-UREIDO]ACETIC (5-BROMO-2-HYDROXYBENZYLIDENE)HYDRAZIDE
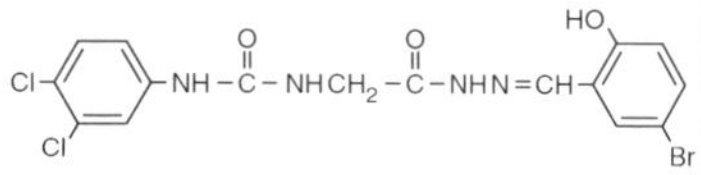
250 mg

S90,243-8
2-[3-(3-CHLOROPHENYL)UREIDO]ACETIC (2-HYDROXY-1-NAPHTHYLMETHYLENE)-HYDRAZIDE
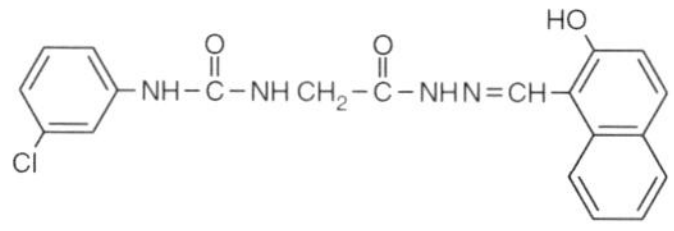
250 mg

S84,542-6
1-CYCLOHEXYL-3-(3-FLUOROPHENYL)-1-(2-HYDROXYETHYL)UREA
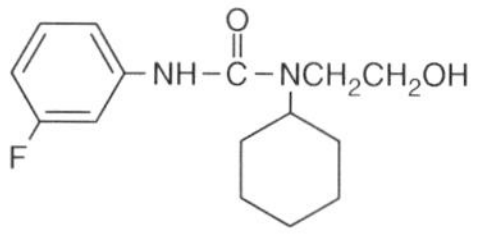
250 mg

S85,079-9
1-CYCLOHEXYL-3-(2,3-DICHLOROPHENYL)-1-(2-HYDROXYETHYL)UREA
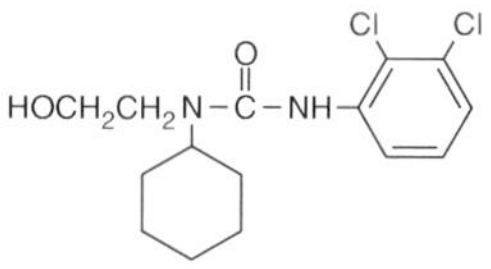
250 mg

S85,059-4
1-(3-CHLORO-4-FLUOROPHENYL)-3-CYCLOHEXYL-3-(2-HYDROXYETHYL)UREA
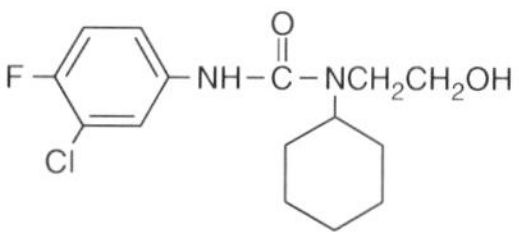
250 mg

S85,146-9
1-(2-CHLORO-6-METHYLPHENYL)-3-(2-HYDROXYPHENYL)UREA
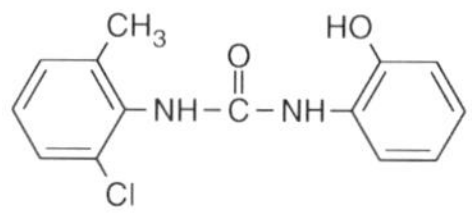
250 mg

S86,760-8
1-(5-CHLORO-2-METHYLPHENYL)-3-(2-HYDROXYPHENYL)UREA
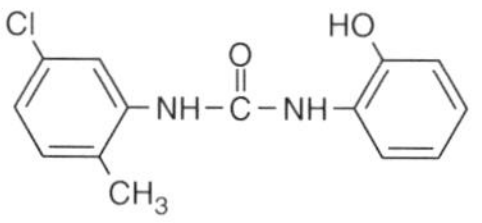
250 mg

S89,794-9
1-(5-CHLORO-2-HYDROXYPHENYL)-3-(3,4-DICHLOROPHENYL)UREA
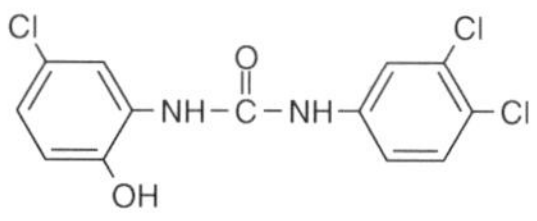
250 mg

S69,148-8
4-(4-CHLORO-3-NITROPHENYL)-1-(4-HYDROXYBENZOYL)SEMICARBAZIDE
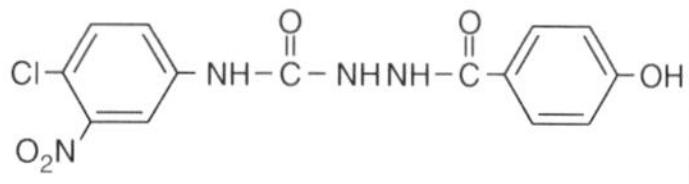
250 mg

S92,406-7
1-(5-CHLORO-2-HYDROXYPHENYL)-3-(4-FLUORO-3-NITROPHENYL)UREA
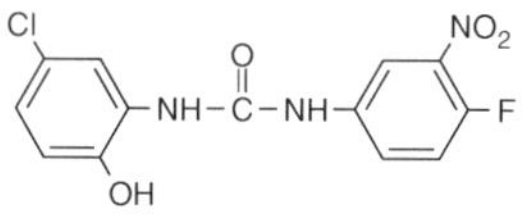
250 mg

S86,819-1
1-(4-FLUORO-3-NITROPHENYL)-3-(5-HYDROXY-1-NAPHTHYL)UREA
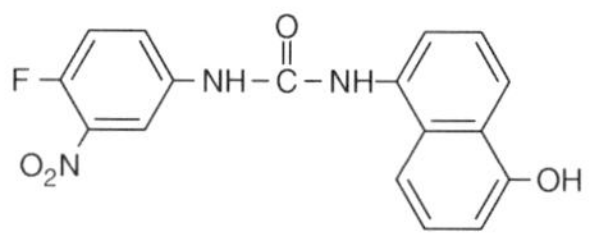
250 mg

S85,143-4
4-[3-(2-CHLORO-6-METHYLPHENYL)-UREIDO]SALICYLIC ACID
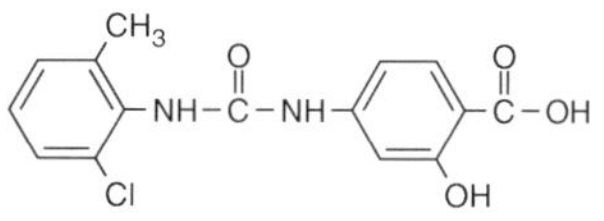
250 mg

S80,072-4
2-HYDROXYPROPYL 4-[3-(4-FLUOROPHENYL)UREIDO]BENZOATE
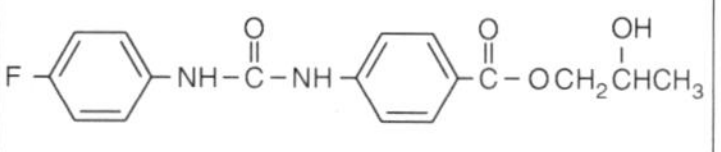
250 mg

S73,841-7
4-(2-FLUOROPHENYL)-1-(4-HYDROXYBENZOYL)SEMICARBAZIDE
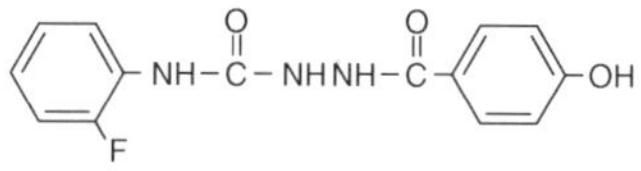
5 gm

S91,731-1
4-(2,6-DICHLOROPHENYL)-1-(4-HYDROXYBENZOYL)SEMICARBAZIDE
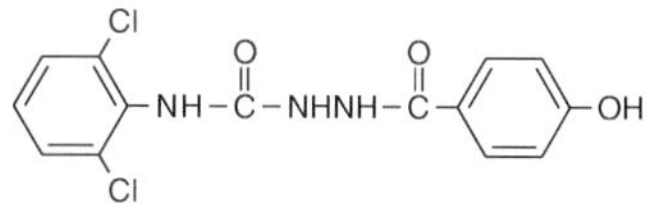
250 mg

S91,727-3
4-(3-CHLORO-2-METHYLPHENYL)-1-(4-HYDROXYBENZOYL)SEMICARBAZIDE
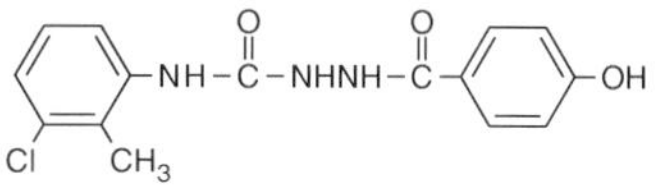
250 mg

S68,798-7
4-(4-CHLORO-2-METHYLPHENYL)-1-(6-HYDROXYHEXANOYL)SEMICARBAZIDE
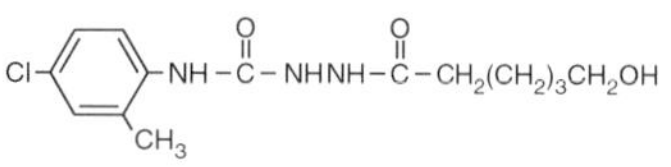
250 mg

S79,506-2
1-(3-FLUOROPHENYL)-3-(3-HYDROXYPHENYL)UREA
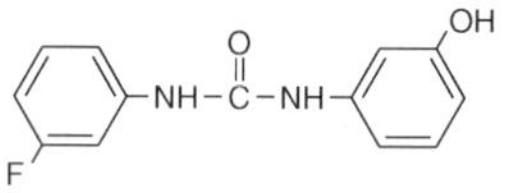
250 mg

S71,880-7
1-(2-FLUOROPHENYL)-3-[4-(2-HYDROXYETHYL)PHENYL]UREA
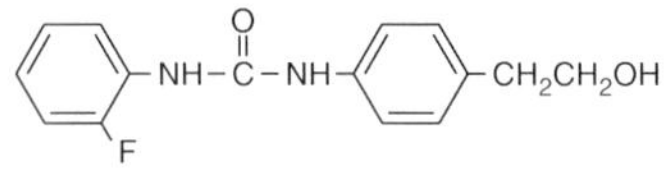
5 gm

S73,159-5
1-(2-FLUOROPHENYL)-3-(7-HYDROXY-1-NAPHTHYL)UREA
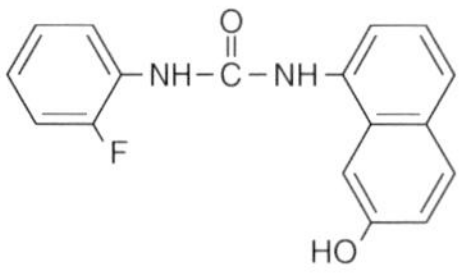
5 gm

S80,059-7
1-(4-FLUOROPHENYL)-3-(3-HYDROXYPHENYL)UREA
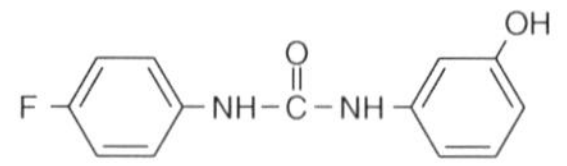
250 mg

S72,756-3
1-(4-FLUOROPHENYL)-3-[4-(2-HYDROXYETHYL)PHENYL]UREA
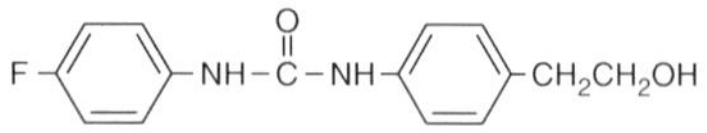
5 gm

S72,757-1
1-(3-FLUOROPHENYL)-3-[4-(2-HYDROXYETHYL)PHENYL]UREA
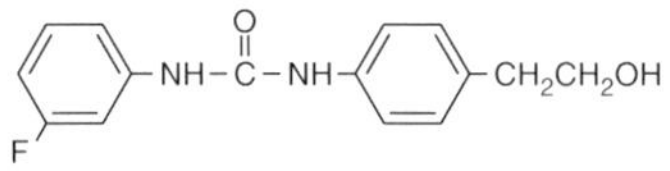
5 gm

S85,142-6
1-(2,3-DICHLOROPHENYL)-3-[4-(2-HYDROXYETHYL)PHENYL]UREA
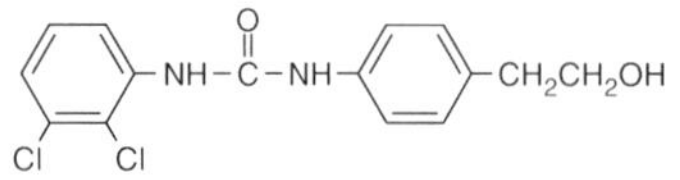
250 mg

S85,139-6
1-(2-CHLORO-6-METHYLPHENYL)-3-[4-(2-HYDROXYETHYL)PHENYL]UREA
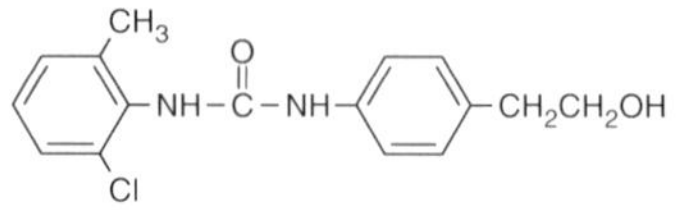
250 mg

S72,661-3
1-(2,5-DIFLUOROPHENYL)-3-(4-HYDROXYPHENYL)UREA
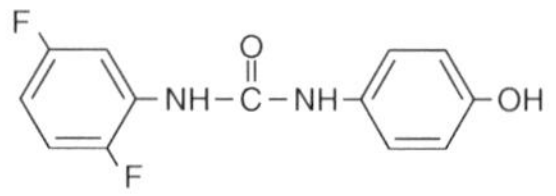
5 gm

S91,627-7
1-(2,5-DIFLUOROPHENYL)-3-[4-(2-HYDROXYETHYL)PHENYL]UREA
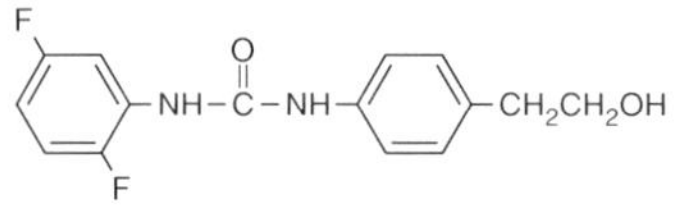
250 mg

S85,061-6
1-(3-CHLORO-4-FLUOROPHENYL)-3-[4-(2-HYDROXYETHYL)PHENYL]UREA
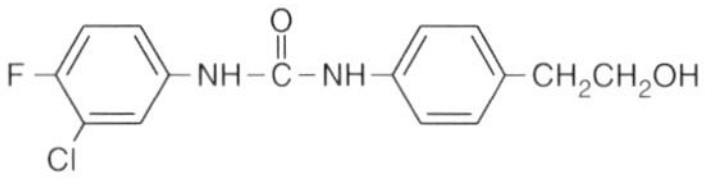
250 mg

S72,964-7
1-(2,5-DIFLUOROPHENYL)-3-[4-(2-HYDROXYETHYL)PHENYL]UREA
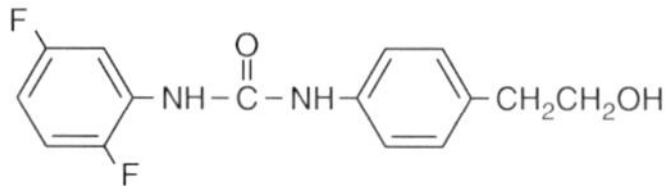
5 gm

S85,145-0
1-(2-CHLORO-6-METHYLPHENYL)-3-(3-HYDROXYPHENYL)UREA
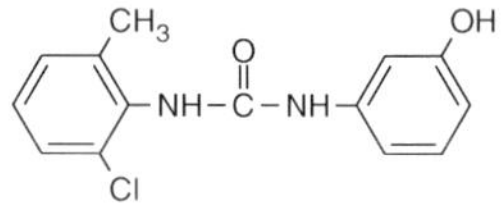
250 mg

S86,761-6
1-(3-CHLORO-2-METHYLPHENYL)-3-(3-HYDROXYPHENYL)UREA
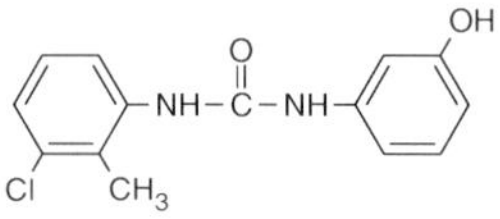
250 mg

S86,844-2
1-(3-CHLORO-2-METHYLPHENYL)-3-(5-HYDROXY-1-NAPHTHYL)UREA
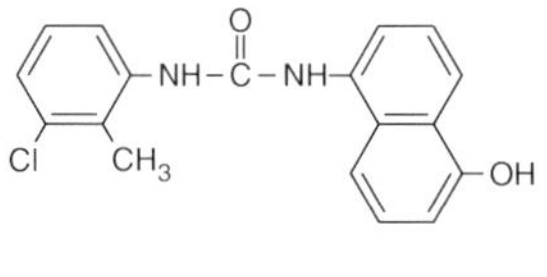
250 mg

S73,636-8
1-(2,5-DIFLUOROPHENYL)-3-(3-HYDROXYPHENYL)UREA
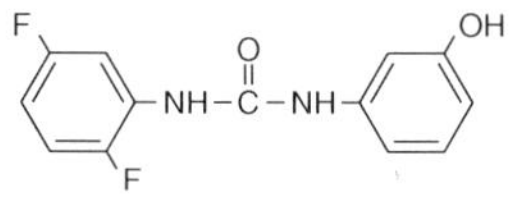
5 gm

S86,842-6
1-(5-CHLORO-2-METHYLPHENYL)-3-(7-HYDROXY-1-NAPHTHYL)UREA
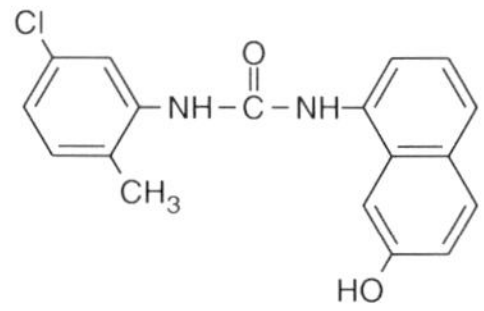
250 mg

S86,843-4
1-(5-CHLORO-2-METHYLPHENYL)-3-(5-HYDROXY-1-NAPHTHYL)UREA
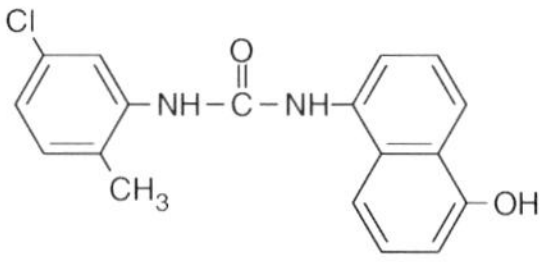
250 mg

S84,874-3
1-(5-CHLORO-2,4-DIMETHOXYPHENYL)-3-(2,5-DIMETHOXYPHENYL)UREA
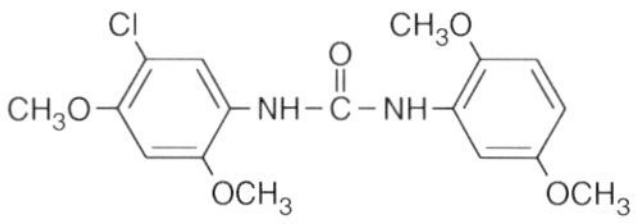
250 mg

S85,420-4
2-(4-[3-(5-CHLORO-2,4-DIMETHOXYPHENYL)UREIDO]-PHENYLTHIO)ACETIC ACID
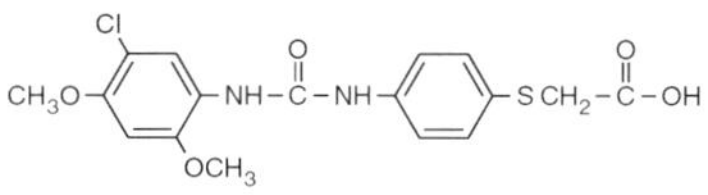
250 mg

S85,397-6
1-(5-CHLORO-2,4-DIMETHOXYPHENYL)-3-[3-(METHYLTHIO)PHENYL]UREA
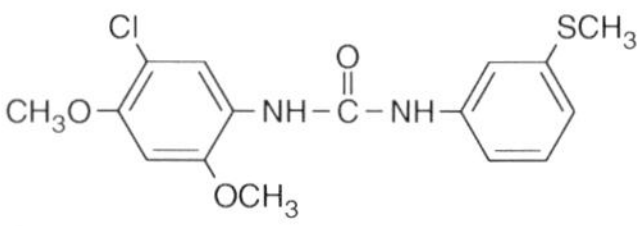
250 mg

S69,226-3
1-(5-CHLORO-2,4-DIMETHOXYPHENYL)-3-[2-(4-NITROPHENYLTHIO)PHENYL]UREA
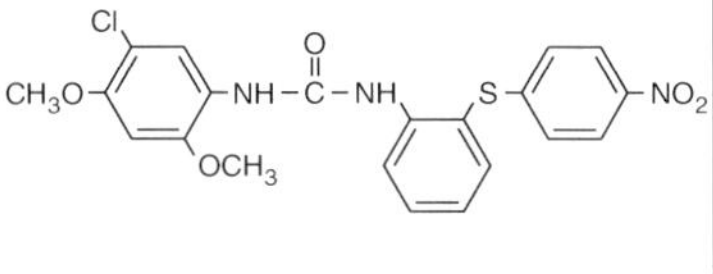
250 mg

S69,591-2
1-(5-CHLORO-2,4-DIMETHOXYPHENYL)-3-(4-CHLORO-3-NITROPHENYL)UREA
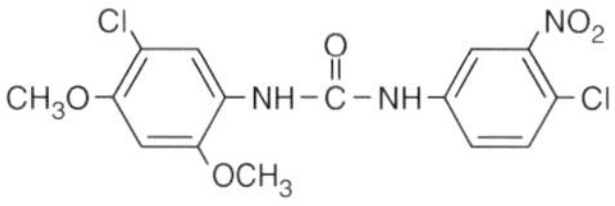
250 mg

S86,774-8
1-(5-CHLORO-2,4-DIMETHOXYPHENYL)-3-(4-FLUORO-3-NITROPHENYL)UREA
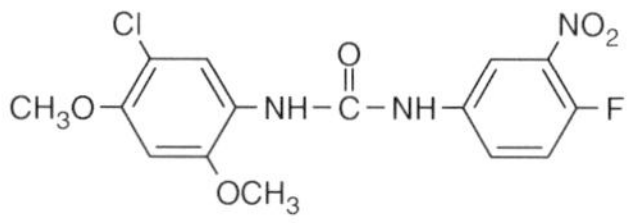
250 mg

S71,557-3
1-(4-ACETAMIDOPHENYL)-3-(5-CHLORO-2,4-DIMETHOXYPHENYL)UREA
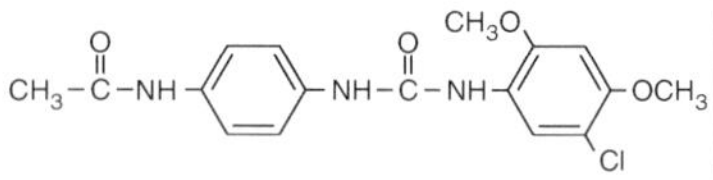
250 mg

S85,388-7
1-(5-CHLORO-2,4-DIMETHOXYPHENYL)-3-[4-(N-METHYLACETAMIDO)PHENYL]UREA
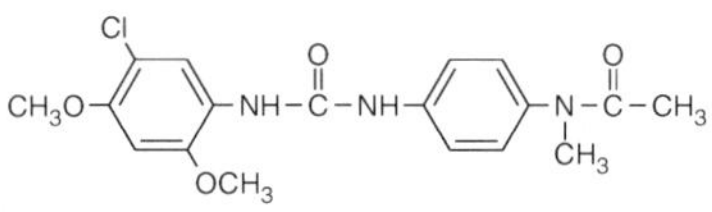
250 mg

S85,389-5
1-(5-CHLORO-2,4-DIMETHOXYPHENYL)-3-[4-(DIMETHYLAMINO)PHENYL]UREA
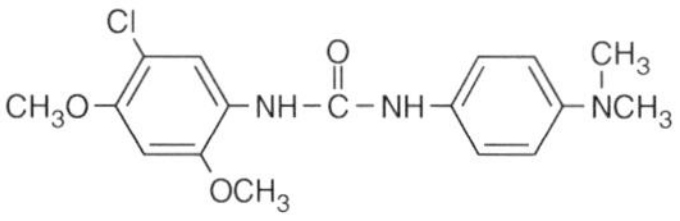
250 mg

S85,419-0
1-BENZYL-3-(5-CHLORO-2,4-DIMETHOXYPHENYL)UREA
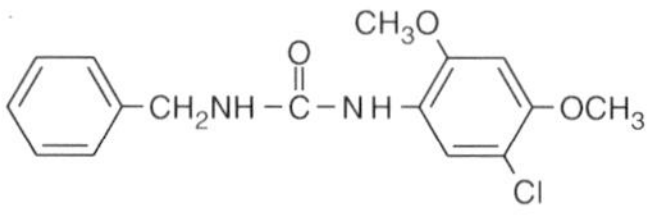
250 mg

S85,422-0
1-BENZYL-3-(5-CHLORO-2,4-DIMETHOXYPHENYL)-1-METHYLUREA
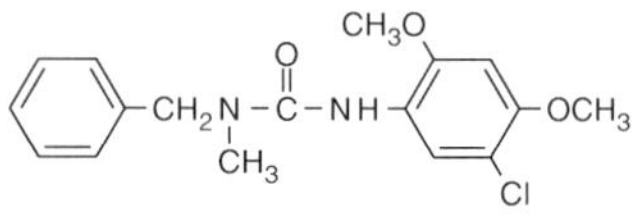
250 mg

S69,222-0
1-[N-(5-CHLORO-2,4-DIMETHOXYPHENYL)-CARBAMOYL]-4-(3-METHYL-PHENYL)-1-PHENYLSEMICARBAZIDE
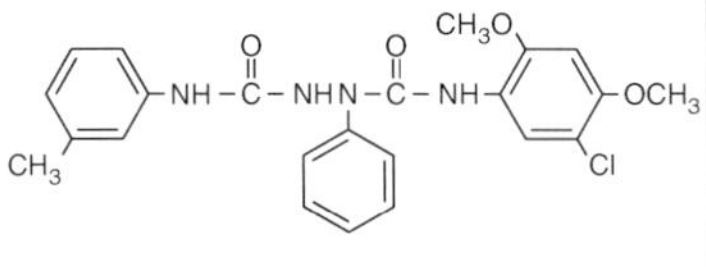
250 mg

S80,012-0
1-(5-CHLORO-2,4-DIMETHOXYPHENYL)-3-(2-METHOXY-4-NITROPHENYL)UREA
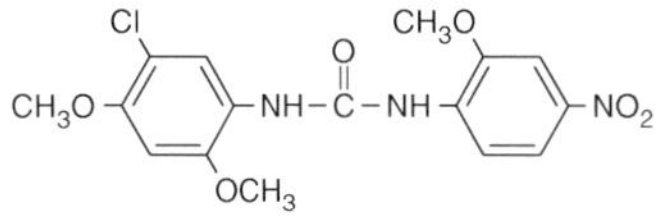
250 mg

S85,417-4
4-[3-(5-CHLORO-2,4-DIMETHOXYPHENYL)-UREIDO]BENZOIC ACID
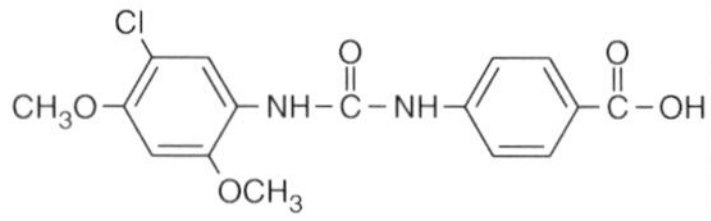
250 mg

S85,399-2
3-[3-(5-CHLORO-2,4-DIMETHOXYPHENYL)-UREIDO]BENZOIC ACID
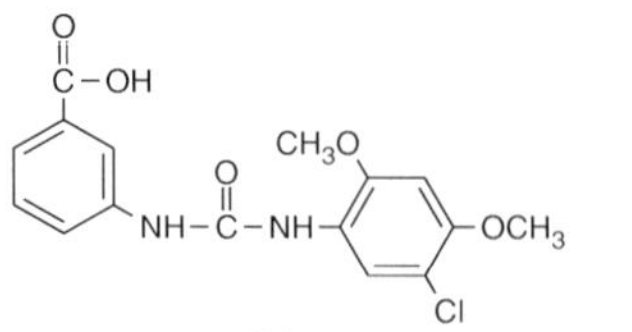
250 mg

S73,193-5
1-(5-CHLORO-2,4-DIMETHOXYPHENYL)-3-[4-(ETHOXYCARBONYL)PHENYL]UREA

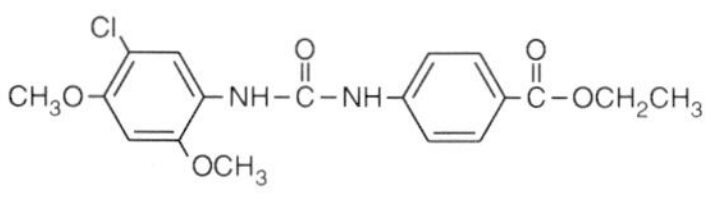

5 gm

S85,377-1
1-(2-CARBAMOYLPHENYL)-3-(5-CHLORO-2,4-DIMETHOXYPHENYL)UREA

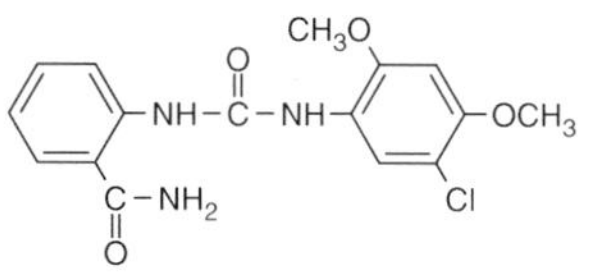

250 mg

S73,695-3
1-(2,4-DIMETHOXYPHENYL)-3-(2-FLUOROPHENYL)UREA

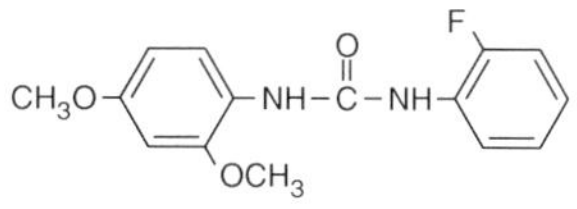

5 gm

S69,401-0
1-(2-CHLOROPHENYL)-3-(2,4-DIMETHOXYPHENYL)UREA

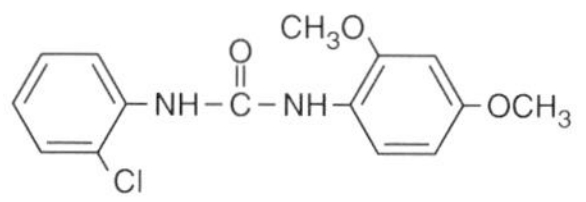

5 gm

S80,038-4
1-(2,4-DIMETHOXYPHENYL)-3-(4-FLUOROPHENYL)UREA

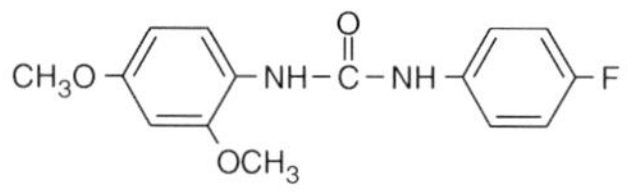

250 mg

S89,170-3
1-(5-CHLORO-2,4-DIMETHOXYPHENYL)-3-(4-METHOXYPHENYL)UREA

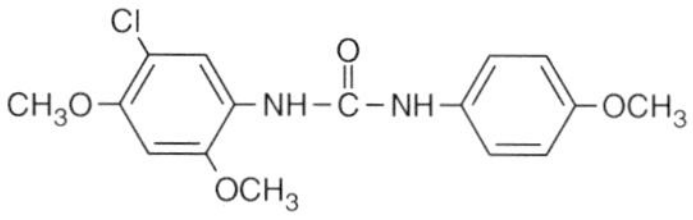

250 mg

S69,485-1
1-(5-CHLORO-2,4-DIMETHOXYPHENYL)-3-(3-CHLORO-2-METHYLPHENYL)UREA

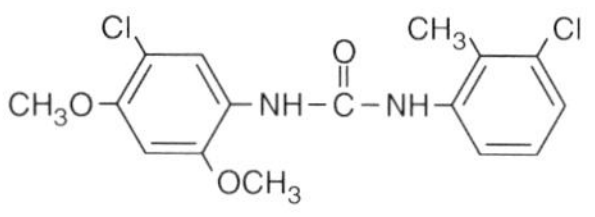

250 mg

S80,784-2
1-(3-CHLORO-2-METHYLPHENYL)-3-(2,4-DIMETHOXYPHENYL)UREA

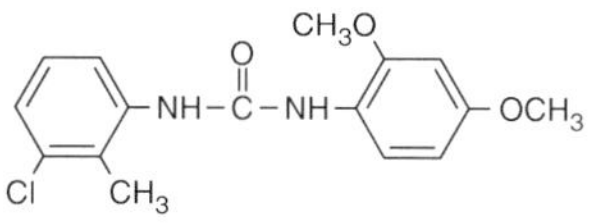

250 mg

S85,401-8
1-(5-CHLORO-2,4-DIMETHOXYPHENYL)-3-(2,3-DICHLOROPHENYL)UREA

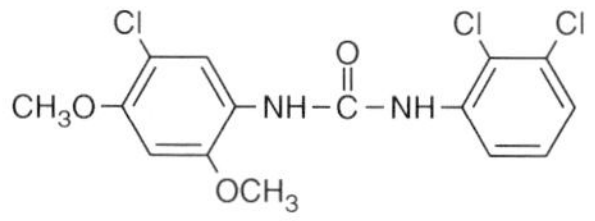

250 mg

S75,246-0
1-(3-CHLOROPHENYL)-3-(2,4-DIMETHOXYPHENYL)UREA

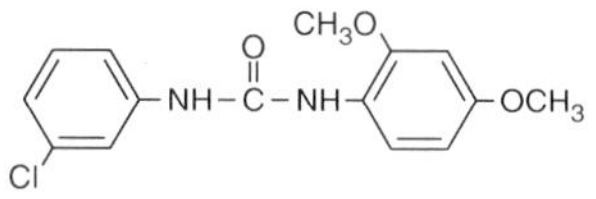

5 gm

S74,718-1
1-(2,5-DIFLUOROPHENYL)-3-(2,4-DIMETHOXYPHENYL)UREA

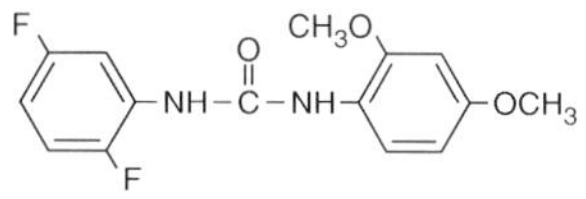

5 gm

S69,592-0
1-(5-CHLORO-2,4-DIMETHOXYPHENYL)-3-(2,4-DIFLUOROPHENYL)UREA

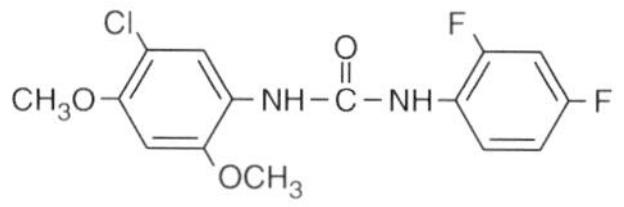

250 mg

S85,060-8
1-(3-CHLORO-4-FLUOROPHENYL)-3-(2,4-DIMETHOXYPHENYL)UREA

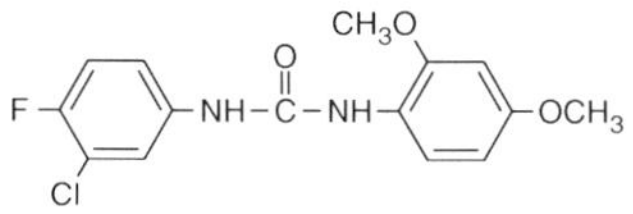

250 mg

S85,125-6
1-(3,4-DICHLOROPHENYL)-3-(2,4-DIMETHOXYPHENYL)UREA

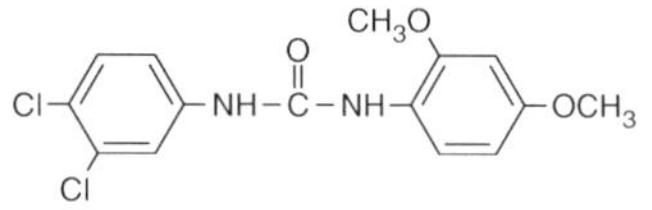

250 mg

S74,318-6
1-(5-CHLORO-2,4-DIMETHOXYPHENYL)-3-(2,5-DIFLUOROPHENYL)UREA

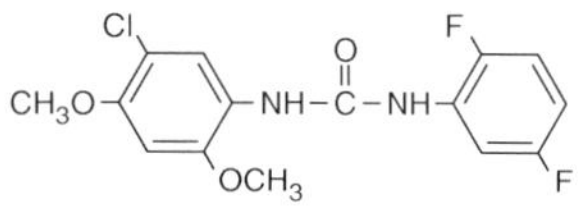

5 gm

S72,530-7
1-(5-CHLORO-2-METHOXYPHENYL)-3-(2,4-DIMETHOXYPHENYL)UREA

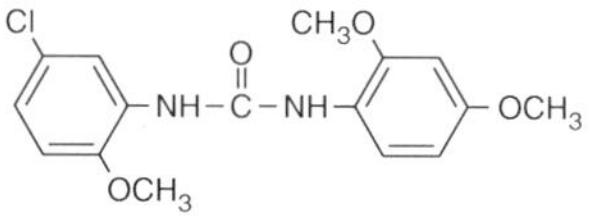

250 mg

S69,499-1
1-(5-CHLORO-2,4-DIMETHOXYPHENYL)-3-(2-METHOXY-5-METHYLPHENYL)UREA

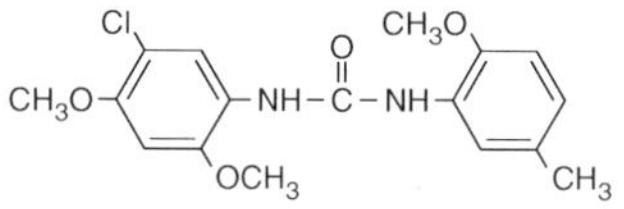

5 gm

S91,710-9
1-(5-CHLORO-2,4-DIMETHOXYPHENYL)-3-METHYLUREA

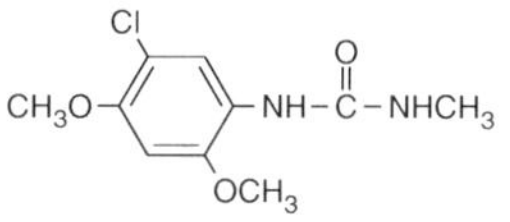

250 mg

S71,788-6
1-(5-CHLORO-2,4-DIMETHOXYPHENYL)-3-ETHYLUREA

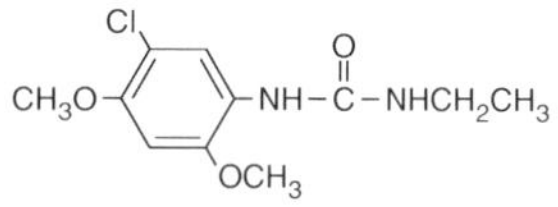

250 mg

S69,379-0
1-(3-BROMOPHENYL)-3-(3,5-DIMETHOXYPHENYL)UREA

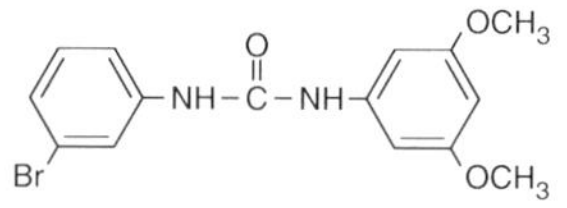

250 mg

S86,830-2
1-(2,5-DIETHOXYPHENYL)-3-(4-FLUORO-3-NITROPHENYL)UREA

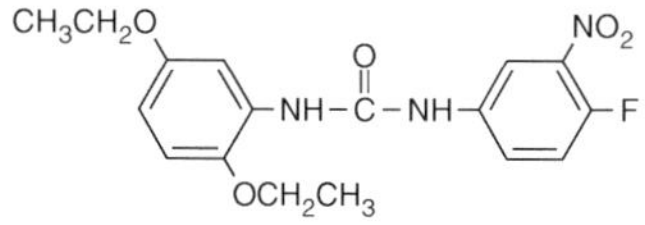

250 mg

S90,369-8
1-(3,4-DICHLOROBENZOYL)-4-(2,5-DIMETHOXYPHENYL)SEMICARBAZIDE

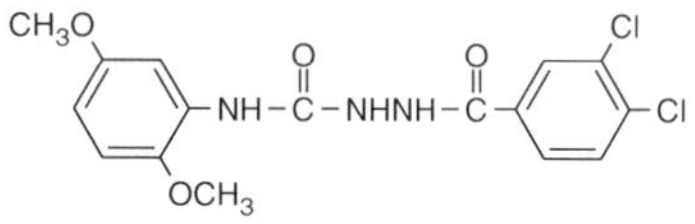

250 mg

S71,877-7
1-(4-CHLORO-2-PHENOXYPHENYL)-3-(2,5-DIMETHOXYPHENYL)UREA

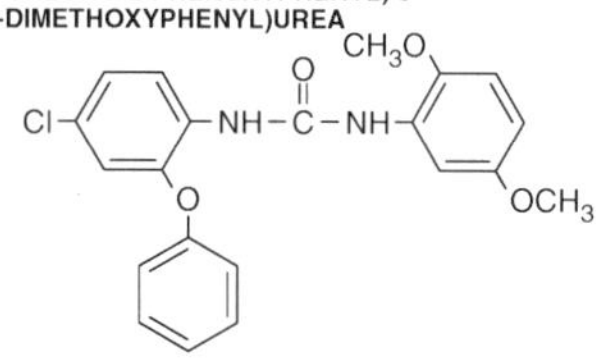

250 mg

S89,147-9
1-(2-CHLOROPHENYL)-3-(2,5-DIETHOXYPHENYL)UREA

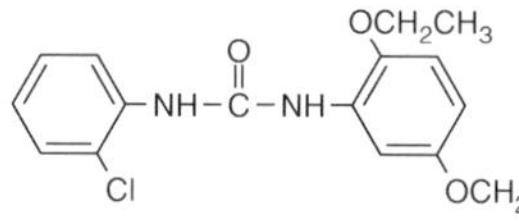

250 mg

S80,037-6
1-(2,5-DIMETHOXYPHENYL)-3-(4-FLUOROPHENYL)UREA

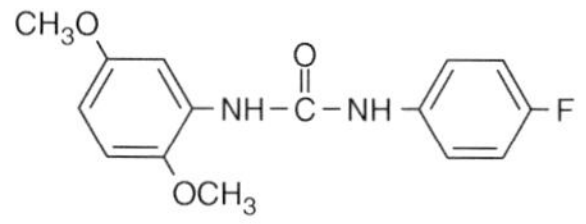

250 mg

S84,867-0
1-(2,5-DIMETHOXYPHENYL)-3-(4-IODOPHENYL)UREA

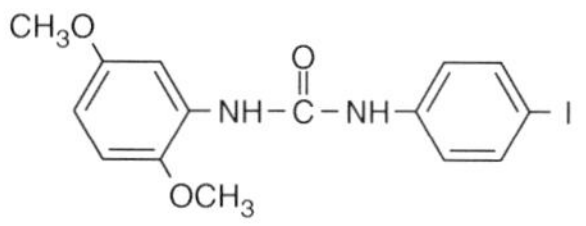

250 mg

S84,866-2
1-(2,3-DICHLOROPHENYL)-3-(2,5-DIMETHOXYPHENYL)UREA

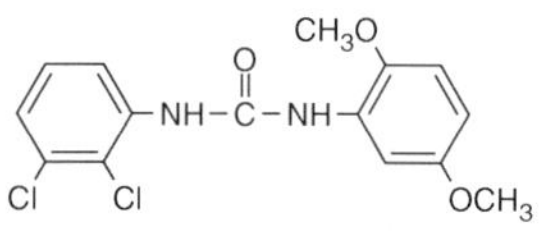

250 mg

S71,488-7
1-(6-CHLORO-2-METHYLPHENYL)-3-(2,5-DIMETHOXYPHENYL)UREA

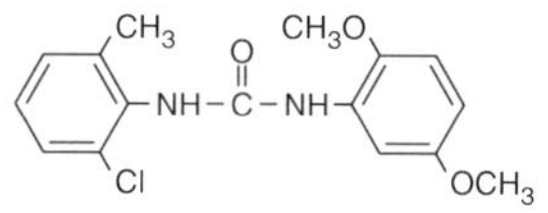

5 gm

S78,086-3
1-(3-CHLORO-2-METHYLPHENYL)-3-(2,5-DIMETHOXYPHENYL)UREA

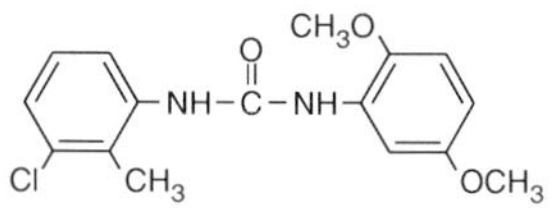

250 mg

S84,535-3
1-(2,5-DIMETHOXYPHENYL)-3-(3-FLUOROPHENYL)UREA

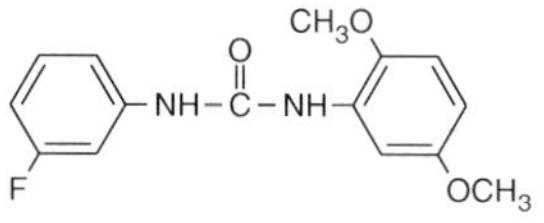

250 mg

S67,521-0
1-(5-CHLORO-2-METHYLPHENYL)-3-(2,5-DIMETHOXYPHENYL)UREA

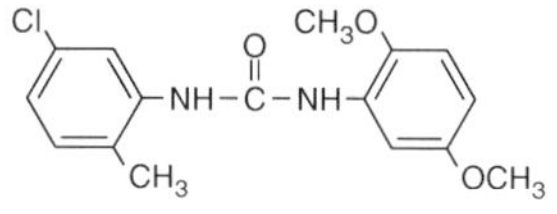

5 gm

S79,433-3
1-(5-CHLORO-2-METHOXYPHENYL)-3-(2,5-DIETHOXYPHENYL)UREA

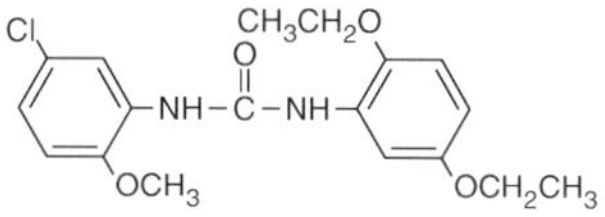

250 mg

S69,504-1
1-(4-BROMO-2,6-DIMETHYLPHENYL)-3-(2,5-DIMETHOXYPHENYL)UREA

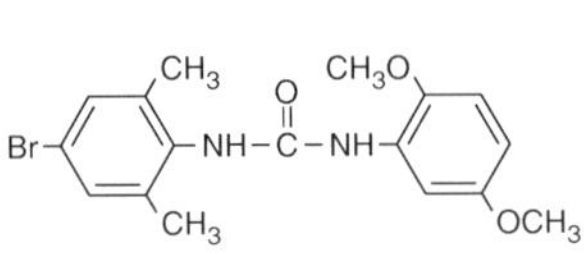

250 mg

S84,873-5
1-(4-CHLORO-2,5-DIMETHOXYPHENYL)-3-(2,5-DIMETHOXYPHENYL)UREA

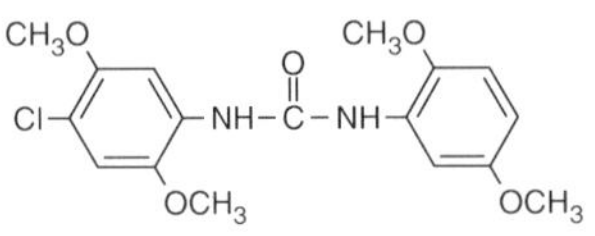

250 mg

S89,174-6
1-(4-CHLORO-2,5-DIMETHOXYPHENYL)-3-(3-CHLOROPHENYL)UREA

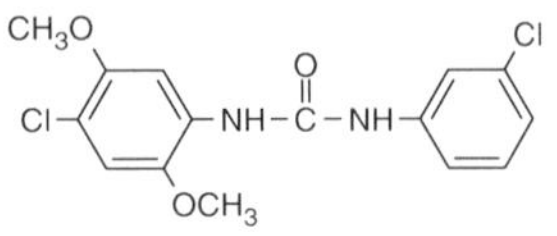

250 mg

S75,288-6
1-(2-CHLOROPHENYL)-3-[4-(2,5-DIMETHOXYSTYRYL)PHENYL]UREA

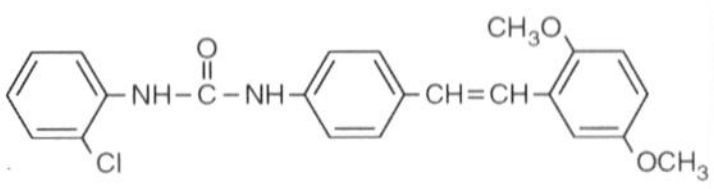

5 gm

S90,293-4
2-[3-(2,3-DICHLOROPHENYL)-UREIDO]ACETIC (3,4-DIMETHOXYBENZYLIDENE)HYDRAZIDE

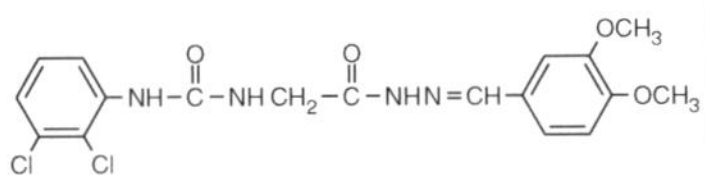

250 mg

S86,777-2
1-(4-CHLORO-2,5-DIMETHOXYPHENYL)-3-(4-CHLORO-2-NITROPHENYL)UREA

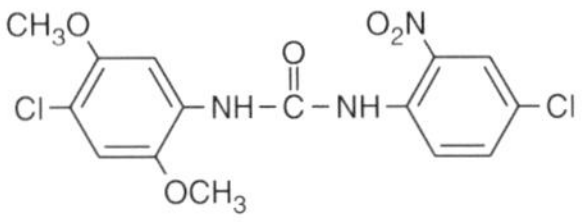

250 mg

S68,850-9
N-[3-(5-CHLORO-2,4-DIMETHOXYPHENYL)UREIDO]-N'-(4-CHLORO-3-NITROPHENYL)OXAMIDE

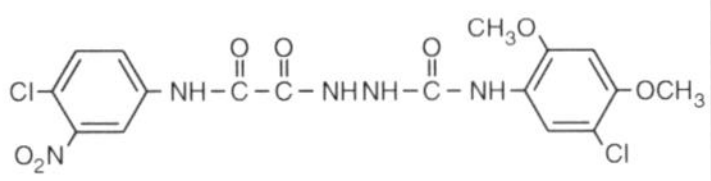

5 gm

S84,532-9
1-(2,5-DIMETHOXYPHENYL)-3-(2-FLUORO-5-NITROPHENYL)UREA

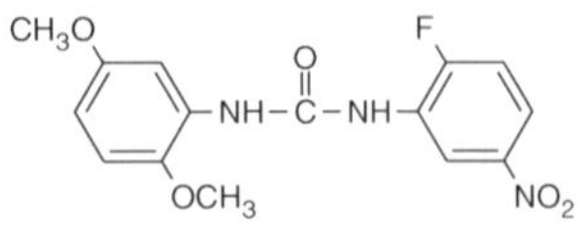

250 mg

S84,832-8
1-(4-CHLORO-2,5-DIMETHOXYPHENYL)-3-(2-METHYL-5-NITROPHENYL)UREA

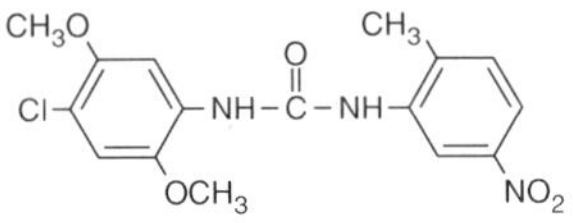

250 mg

S85,376-3
1-(5-CHLORO-2,4-DIMETHOXYPHENYL)-3-(4-METHYL-3-NITROPHENYL)UREA

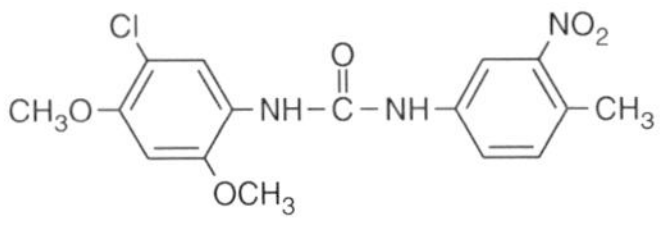

250 mg

S85,425-5
1-(4-ANILINOPHENYL)-3-(5-CHLORO-2,4-DIMETHOXYPHENYL)UREA

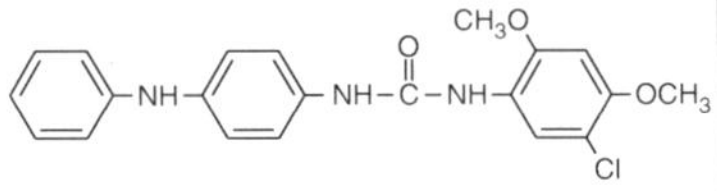

250 mg

S69,396-0
1-(4-CHLOROBENZOYL)-4-(3,4,5-TRIMETHOXYPHENYL)SEMICARBAZIDE

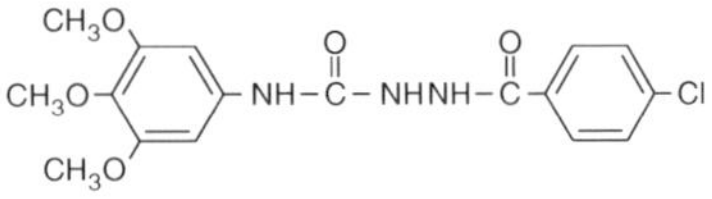

250 mg

S84,788-7
4-[3-(2-FLUOROPHENYL)UREIDO]-BENZOIC (3,4-DIMETHOXYBENZYLIDENE)-HYDRAZIDE

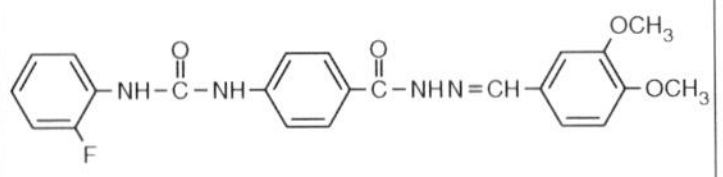

250 mg

S86,908-2
4-[3-(3-CHLOROPHENYL)UREIDO]-BENZOIC (4-ETHOXY-3-METHOXYBENZYLIDENE)HYDRAZIDE

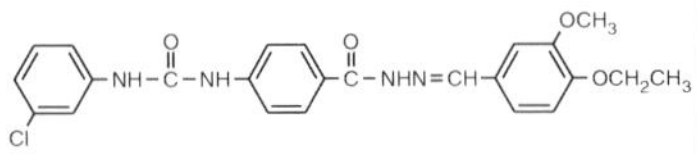

250 mg

S73,689-9
1-(2-CHLORO-4-METHOXYPHENYL)-3-(2-METHOXY-4-NITROPHENYL)UREA

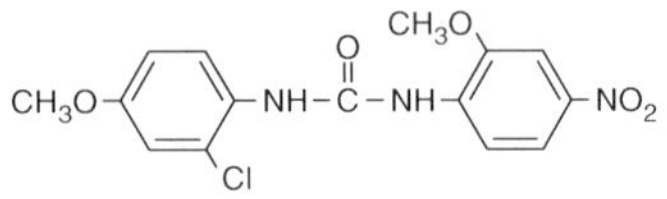

5 gm

S80,782-6
1-(3-ACETYLPHENYL)-3-(4-CHLORO-2,5-DIMETHOXYPHENYL)UREA

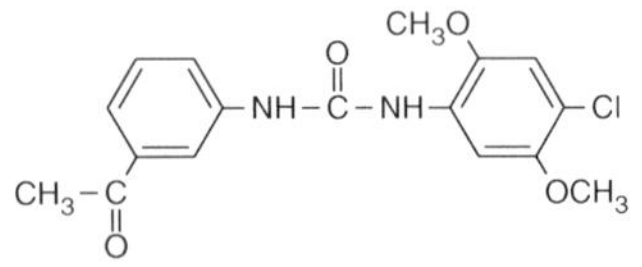

250 mg

S79,438-4
ETHYL 4-[3-(4-CHLORO-2,5-DIMETHOXYPHENYL)UREIDO]BENZOATE

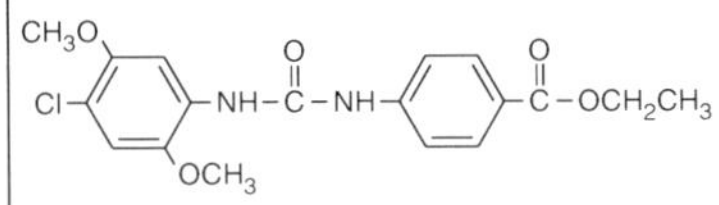

250 mg

S90,370-1
1-(4-CHLOROBENZOYL)-4-(2,5-DIMETHOXYPHENYL)SEMICARBAZIDE

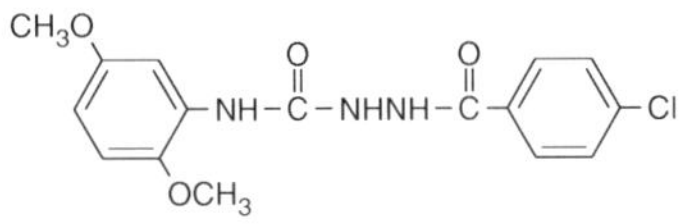

250 mg

S90,365-5
4-(2,5-DIMETHOXYPHENYL)-1-[3-(4-FLUOROPHENYL)UREIDOACETYL]-SEMICARBAZIDE

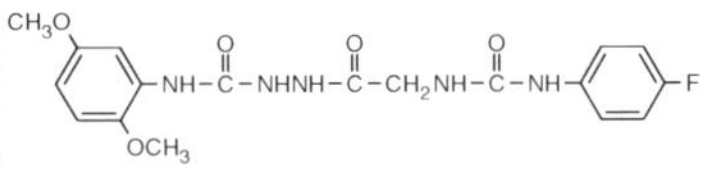

250 mg

S72,655-9
1-(4-CHLORO-2-METHOXYPHENYL)-3-(2-METHOXYPHENYL)UREA

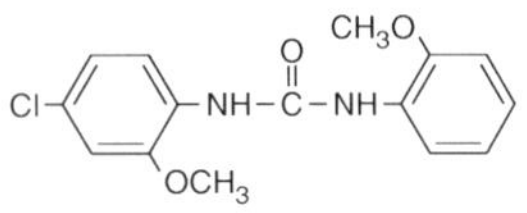

5 gm

S75,247-9
1-(2-CHLORO-4-METHOXYPHENYL)-3-(2-METHOXYPHENYL)UREA

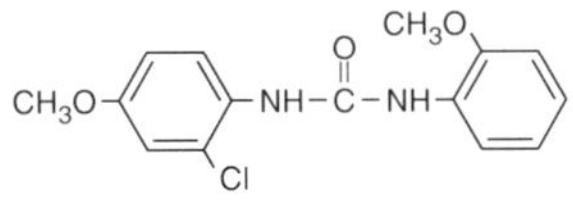

5 gm

S75,244-4
1-(2-CHLOROPHENYL)-3-(4-CHLORO-2,5-DIMETHOXYPHENYL)UREA

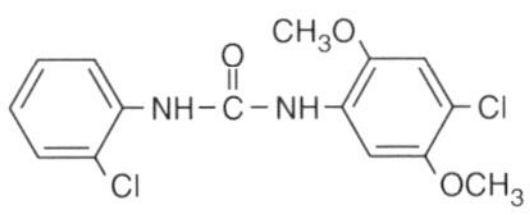

5 gm

S75,251-7
1-(4-CHLORO-2,5-DIMETHOXYPHENYL)-3-(O-TOLYL)UREA

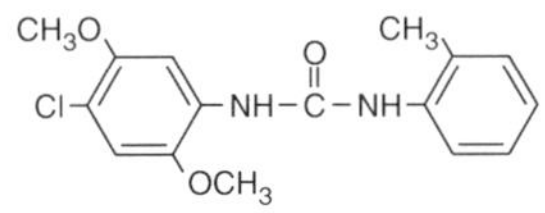

5 gm

S79,136-9
1-(5-CHLORO-2-METHOXYPHENYL)-3-(4-METHOXYPHENYL)UREA

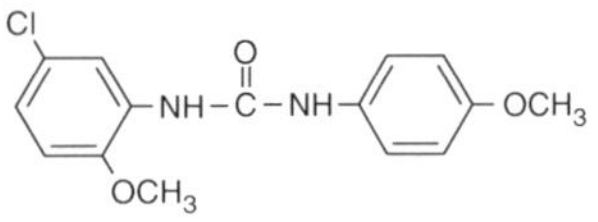

250 mg

S89,171-1
1-(4-CHLORO-2-METHOXYPHENYL)-3-(4-METHOXYPHENYL)UREA

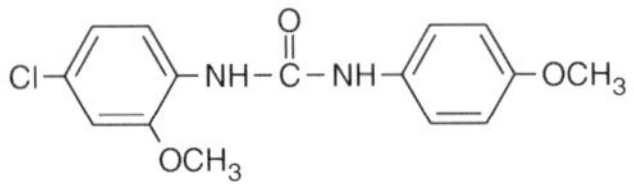

250 mg

S75,256-8
1-(4-CHLORO-2,5-DIMETHOXYPHENYL)-3-(4-METHOXYPHENYL)UREA

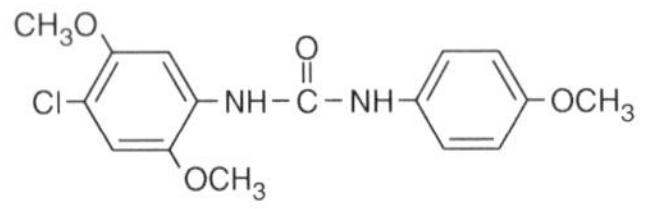

5 gm

S90,993-9
1-(3-CHLORO-4-METHYLPHENYL)-3-(2,5-DIMETHOXYPHENYL)-1-ETHYLUREA

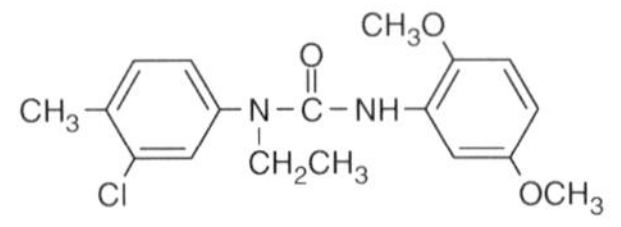

250 mg

S78,025-1
1-(4-CHLORO-2,5-DIMETHOXYPHENYL)-3-(2,5-DIFLUOROPHENYL)UREA

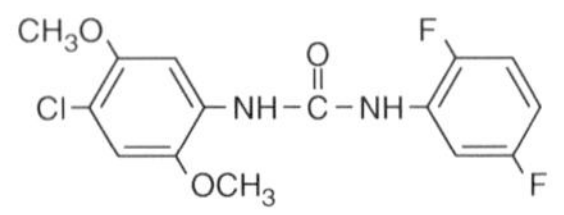

250 mg

S85,454-9
1-(4-CHLORO-2,5-DIMETHOXYPHENYL)-3-METHYLUREA

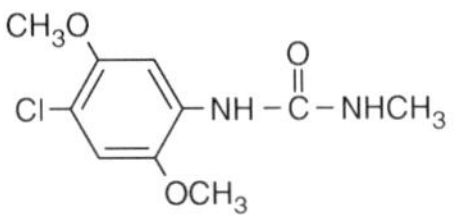

250 mg

S72,609-5
ETHYL 2-(4-CHLORO-2,5-DIMETHOXYPHENYLUREIDO)ACETATE

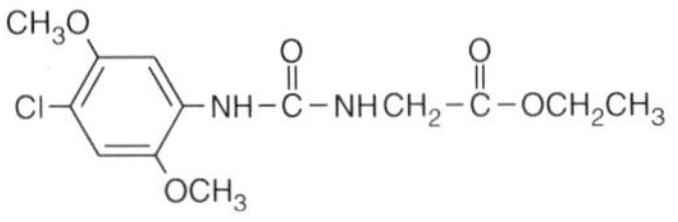

5 gm

S75,248-7
1-(2-CHLORO-4-METHOXYPHENYL)-3-(4-METHOXYPHENYL)UREA

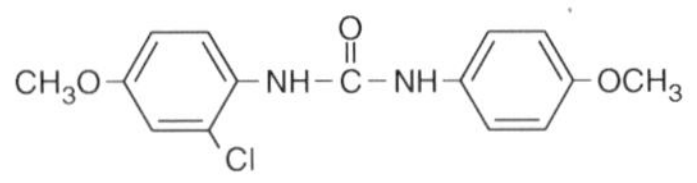

5 gm

S41,158-2
5-CHLORO-2,4-DIMETHOXYPHENYLUREA

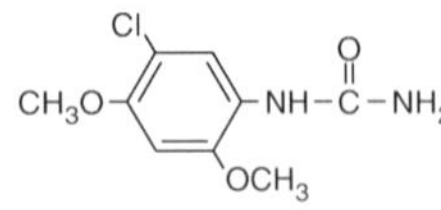

5 gm

S72,958-2
4-(3-CHLOROPHENYL)-1-(2-METHOXYBENZOYL)SEMICARBAZIDE

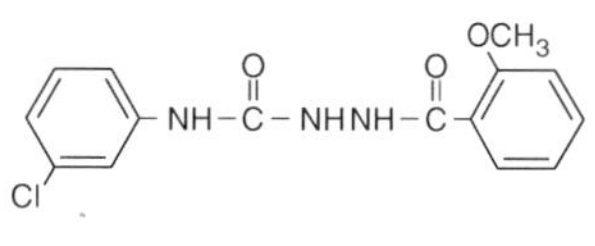

5 gm

S76,808-1
4-(2-CHLOROPHENYL)-1-(2-METHOXYBENZOYL)SEMICARBAZIDE

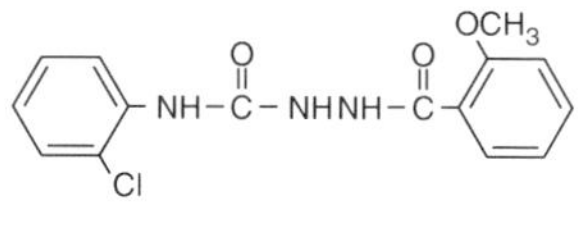

250 mg

S72,658-3
4-(4-FLUOROPHENYL)-1-(2-METHOXYBENZOYL)SEMICARBAZIDE

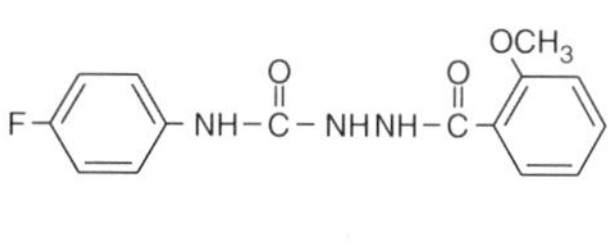

5 gm

S69,090-2
4-(4-CHLORO-2-METHYLPHENYL)-1-(2-METHOXYBENZOYL)SEMICARBAZIDE

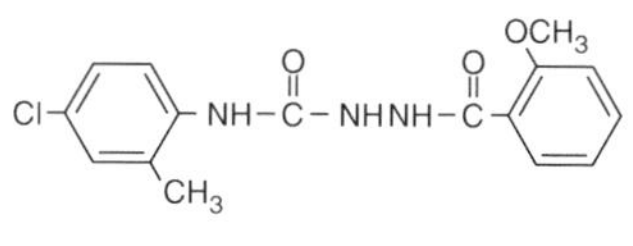

250 mg

S68,847-9
6-(5-CHLORO-2,4-DIMETHOXYPHENYL)-1,1-DIPHENYLBIUREA

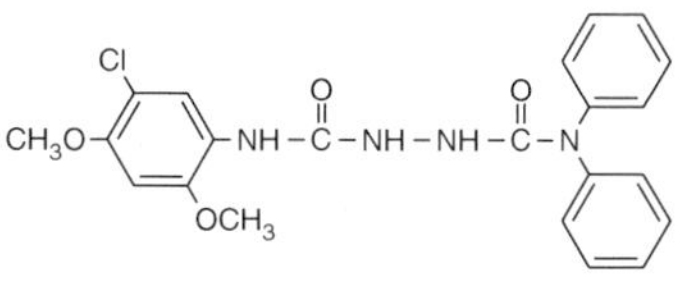

250 mg

S69,174-7
4-(2,3-DICHLOROPHENYL)-1-(2-METHOXYBENZOYL)SEMICARBAZIDE

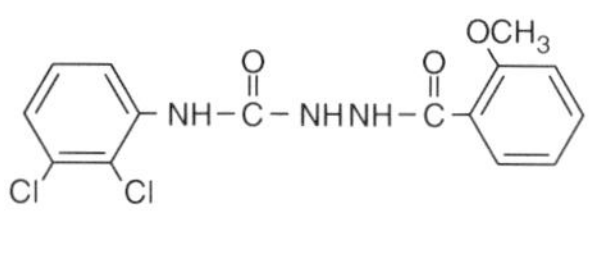

250 mg

S68,758-8
4-(2,4-DIFLUOROPHENYL)-1-(2-METHOXYBENZOYL)SEMICARBAZIDE

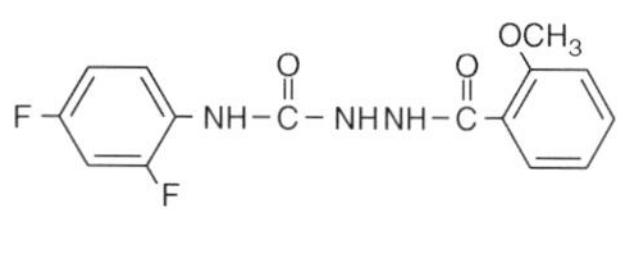

250 mg

S79,727-8
1-(5-CHLORO-2-METHOXYPHENYL)-3-[4-(METHYLTHIO)PHENYL]UREA

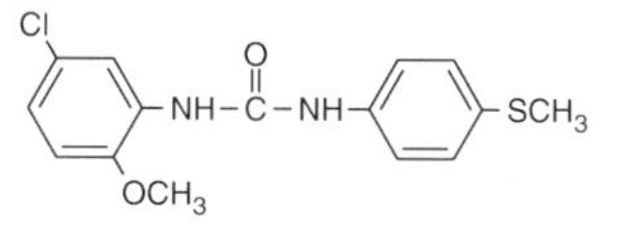

250 mg

S79,725-1
1-(2-CHLORO-4-METHOXYPHENYL)-3-[4-(METHYLTHIO)PHENYL]UREA

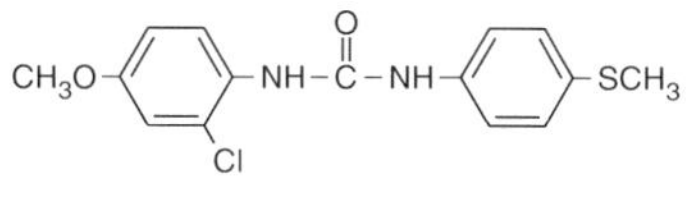

250 mg

S90,296-9
2-[3-(2-ETHOXYPHENYL)UREIDO]ACETIC [2-CHLORO-4-(DIMETHYLAMINO)-BENZYLIDENE]HYDRAZIDE

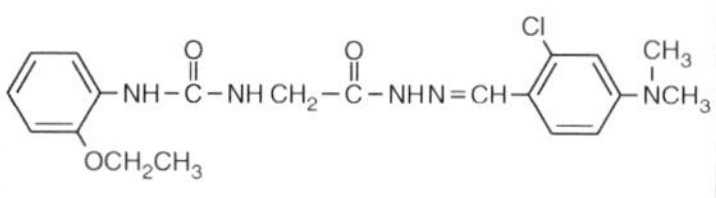

250 mg

S90,318-3
2-[3-(2-CHLORO-4-NITROPHENYL)-UREIDO]ACETIC (4-METHOXYBENZYLIDENE)HYDRAZIDE

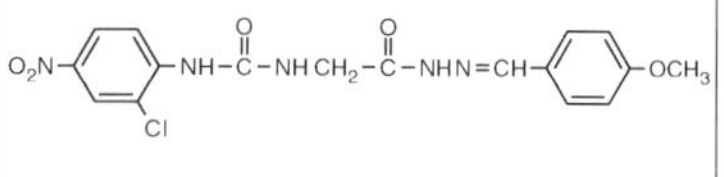

250 mg

S90,206-3
2-[3-(4-FLUOROPHENYL)UREIDO]ACETIC (4-METHOXYBENZYLIDENE)HYDRAZIDE

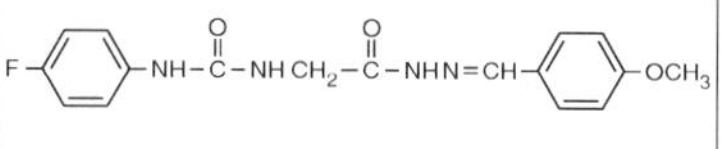

250 mg

S90,244-6
2-[3-(3-CHLOROPHENYL)UREIDO]ACETIC (4-BUTOXYBENZYLIDENE)HYDRAZIDE

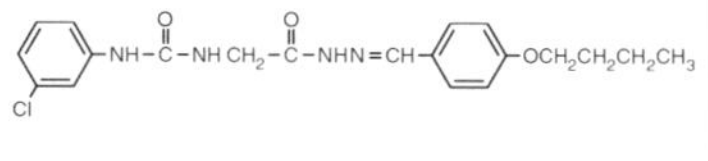

250 mg

S85,187-6
1-[2-(4-BROMOPHENOXY)ETHYL]-3-(4-METHYL-2-NITROPHENYL)UREA

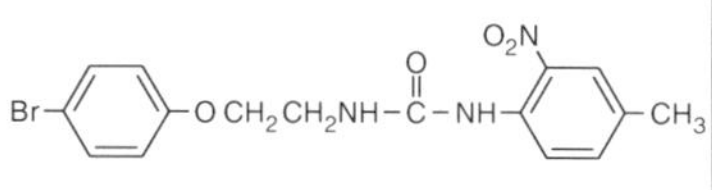

250 mg

S86,754-3
1-(5-CHLORO-2-METHOXYPHENYL)-3-(4-CHLORO-2-NITROPHENYL)UREA

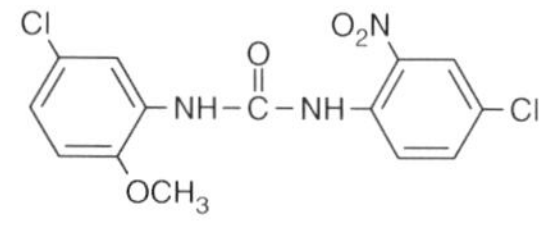

250 mg

S67,455-9
N-[3-(5-CHLORO-2-METHOXYPHENYL)-UREIDO]-N'-(4-FLUOROPHENYL)OXAMIDE

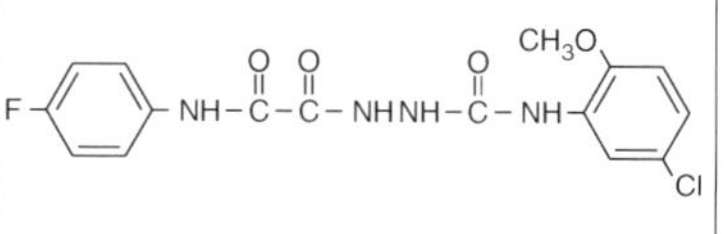

250 mg

S89,161-4
1-(2-CHLOROPHENYL)-3-(2-METHOXY-3-NITROPHENYL)UREA

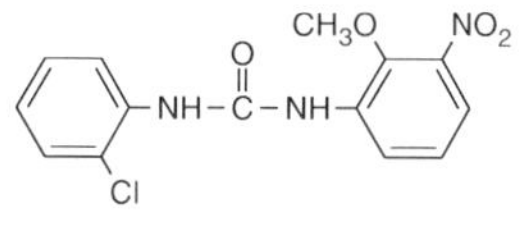

250 mg

S84,483-7
1-(2-FLUORO-5-NITROPHENYL)-3-(2-METHOXYPHENYL)UREA

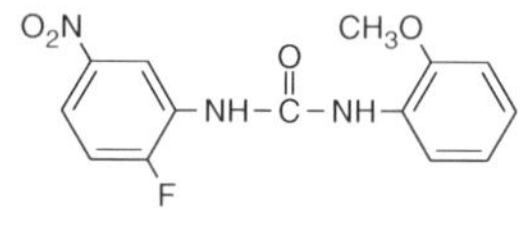

250 mg

S75,235-5
1-(2-CHLOROPHENYL)-3-(2-METHOXY-5-NITROPHENYL)UREA

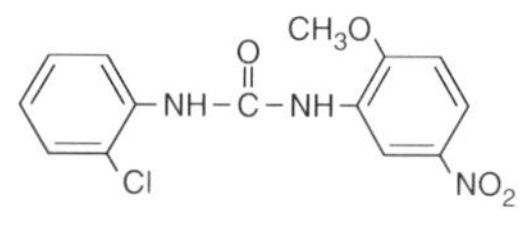

5 gm

S79,501-1
1-(4-IODOPHENYL)-3-(2-METHOXY-5-NITROPHENYL)UREA

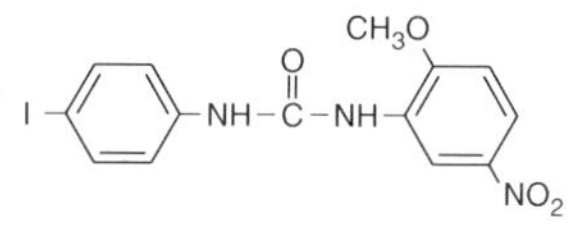

250 mg

S72,824-1
1-(4-FLUORO-3-NITROPHENYL)-3-(2-METHOXY-3-NITROPHENYL)UREA

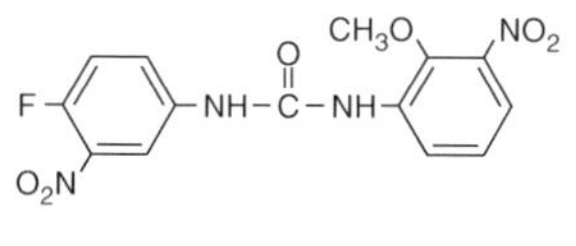

5 gm

S86,784-5
1-(3-CHLORO-2-METHYLPHENYL)-3-(2-METHOXY-5-NITROPHENYL)UREA

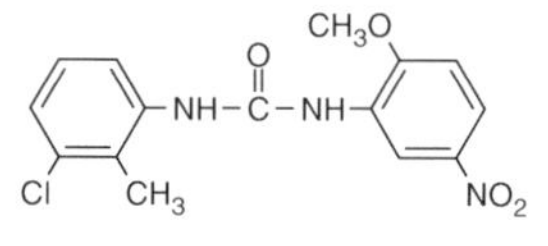

250 mg

S84,485-3
1-(3-FLUOROPHENYL)-3-(2-METHOXY-5-NITROPHENYL)UREA

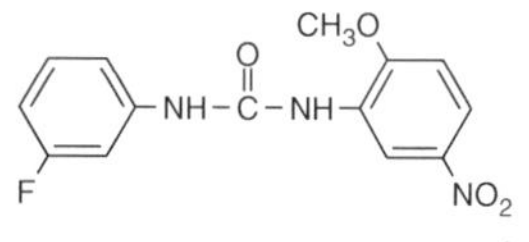

250 mg

S84,478-0
1-(3-CHLOROPHENYL)-3-(4-METHOXY-3-NITROPHENYL)UREA

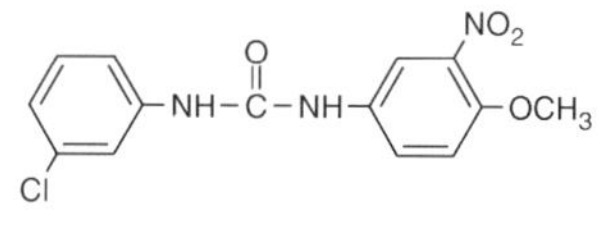

250 mg

S74,321-6
1-(2,5-DIFLUOROPHENYL)-3-(2-METHOXY-5-NITROPHENYL)UREA

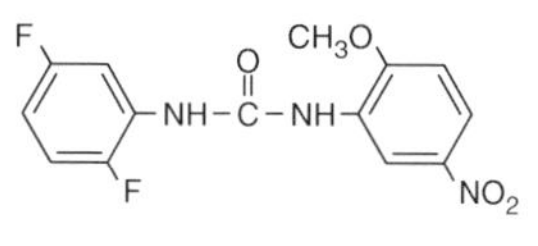

5 gm

S86,752-7
1-(5-CHLORO-2-METHOXYPHENYL)-3-(4-FLUORO-3-NITROPHENYL)UREA

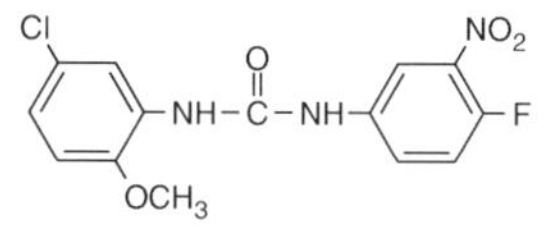

250 mg

S73,645-7
1-(4-FLUORO-3-NITROPHENYL)-3-(2-METHOXY-5-NITROPHENYL)UREA

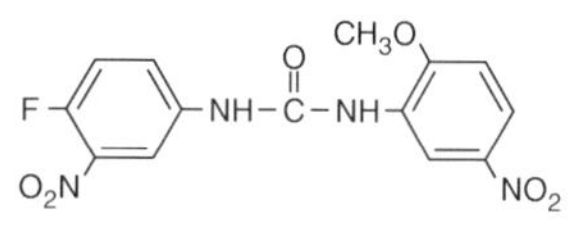

5 gm

S84,531-0
1-(2-FLUORO-5-NITROPHENYL)-3-(2-METHOXY-5-METHYLPHENYL)UREA

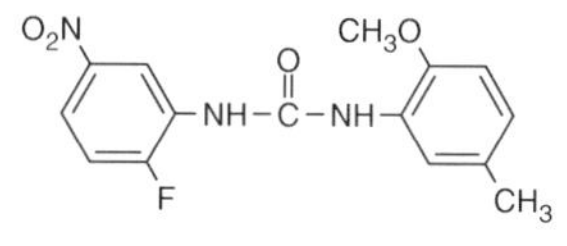

250 mg

S86,785-3
1-(5-CHLORO-2-METHYLPHENYL)-3-(2-METHOXY-5-NITROPHENYL)UREA

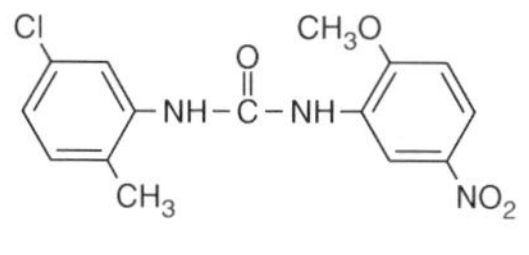

250 mg

S75,041-7
1-(2,5-DIFLUOROPHENYL)-3-(4-METHOXY-3-NITROPHENYL)UREA

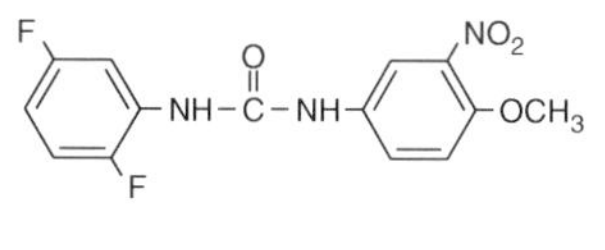

5 gm

S91,280-8
1-(5-CHLORO-2-METHOXYPHENYL)-3-(3,5-DINITROPHENYL)UREA

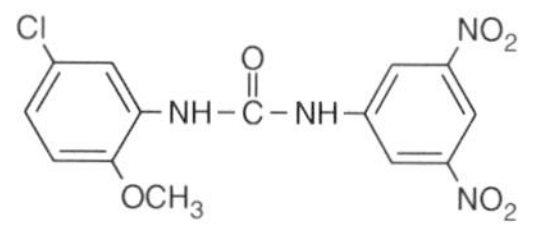

250 mg

S84,484-5
1-(2-FLUORO-5-NITROPHENYL)-3-(4-METHOXYPHENYL)UREA

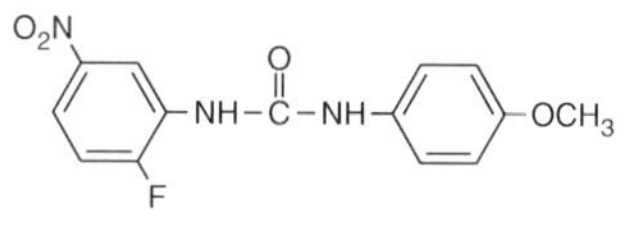

250 mg

S86,753-5
1-(2-CHLORO-4-METHOXYPHENYL)-3-(4-FLUORO-3-NITROPHENYL)UREA
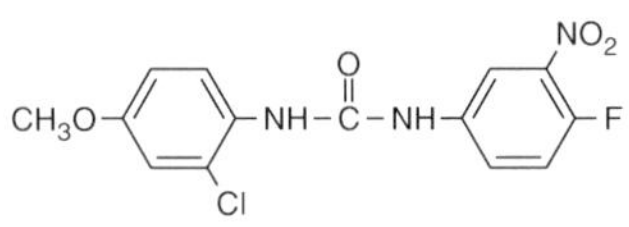
250 mg

S72,535-8
1-(4-ANILINOPHENYL)-3-(5-CHLORO-2-METHOXYPHENYL)UREA
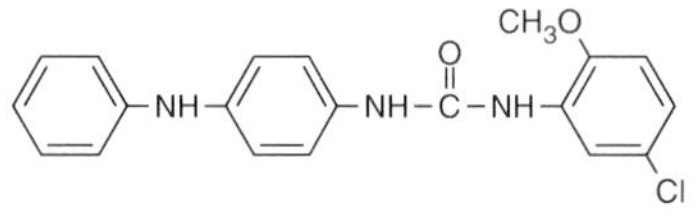
250 mg

S92,235-8
1-(5-CHLORO-2-METHOXYPHENYL)-3-[4-(DIMETHYLAMINO)PHENYL]UREA
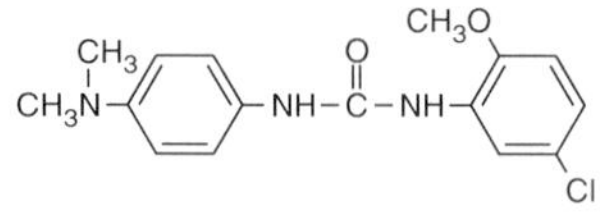
250 mg

S68,864-9
1-(4-CHLORO-2-PHENOXYPHENYL)-3-(4-NITROPHENYL)UREA
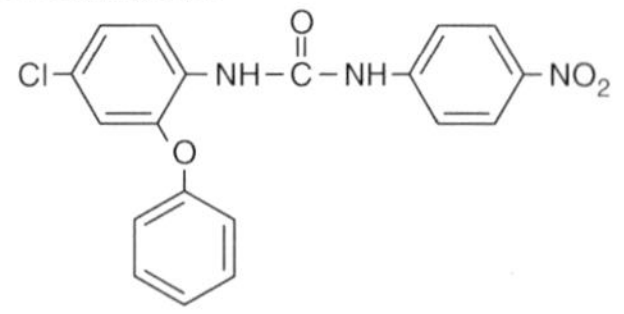
250 mg

S78,072-3
1-(2-FLUOROPHENYL)-3-(2-METHOXY-4-NITROPHENYL)UREA
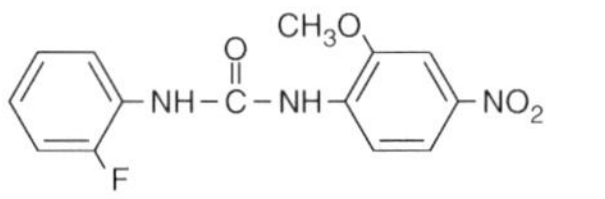
250 mg

S89,162-2
1-(2-BROMO-4-NITROPHENYL)-3-(2-METHOXYPHENYL)UREA
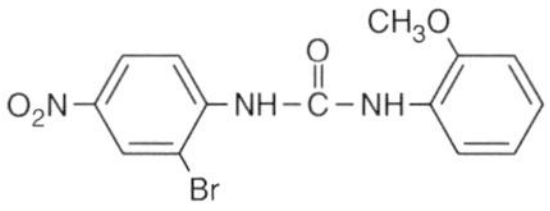
250 mg

S84,486-1
1-(3-FLUOROPHENYL)-3-(2-METHOXY-4-NITROPHENYL)UREA
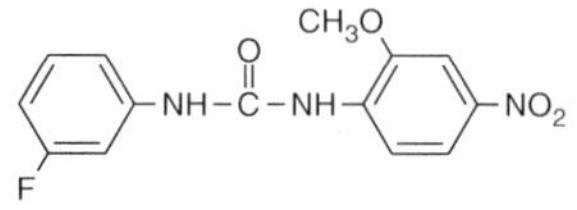
250 mg

S90,991-2
1-(4-CHLORO-2-METHOXYPHENYL)-3-(2-CHLORO-4-NITROPHENYL)-1-METHYLUREA
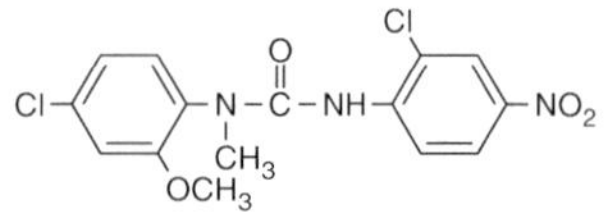
250 mg

S79,462-7
1-(4,5-DICHLORO-2-METHYLPHENYL)-3-(2-METHOXY-4-NITROPHENYL)UREA
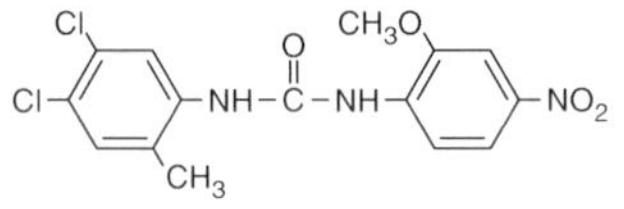
250 mg

S72,115-8
1-(4-ACETAMIDO-3-METHOXYPHENYL)-3-(3-FLUOROPHENYL)UREA
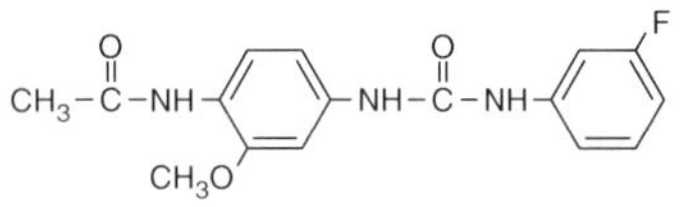
250 mg

S84,474-8
1-(2-BROMO-4-NITROPHENYL)-3-(4-METHOXYPHENYL)UREA
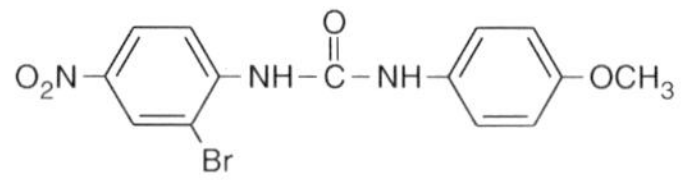
250 mg

S79,796-0
1-(5-CHLORO-2-METHOXYPHENYL)-3-(1,2-DIPHENYLETHYL)UREA
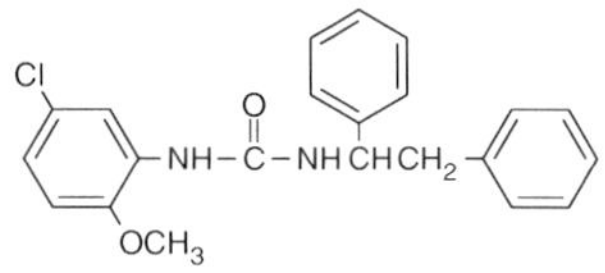
250 mg

S69,044-9
4-(5-CHLORO-2-METHOXYPHENYL)-1-[4-(DIMETHYLAMINO)BENZOYL]-SEMICARBAZIDE
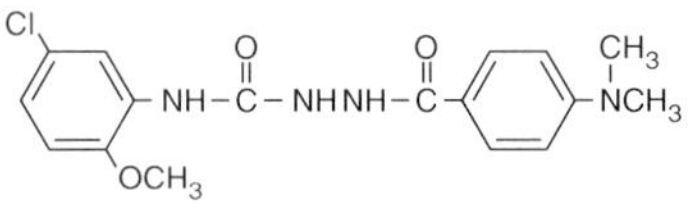
250 mg

S85,432-8
1-BENZYL-3-(4-CHLORO-2-METHOXYPHENYL)-1-METHYLUREA
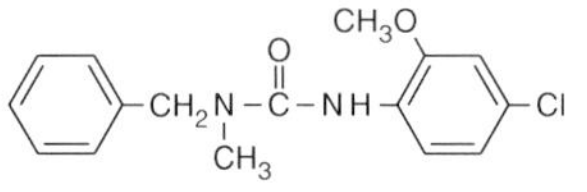
250 mg

S74,299-6
1-(2-CHLOROBENZOYL)-4-(2-METHOXY-5-METHYLPHENYL)SEMICARBAZIDE
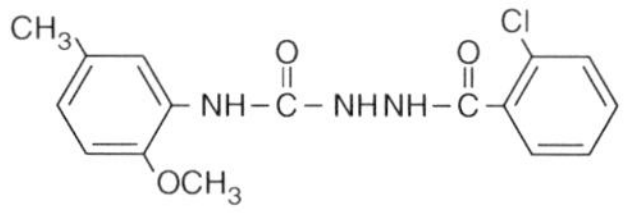
5 gm

S69,516-5
4-(5-CHLORO-2-METHOXYPHENYL)-1-(2,4-DICHLOROBENZOYL)SEMICARBAZIDE
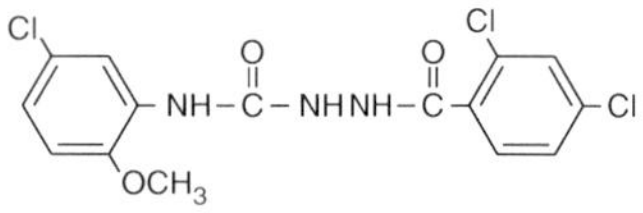
250 mg

S72,966-3
1-ACETYL-4-(5-CHLORO-2-METHOXYPHENYL)SEMICARBAZIDE
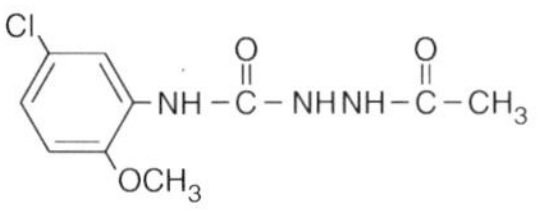
5 gm

S68,473-2
1-(5-CHLORO-2-ETHOXYPHENYL)-6-(3-CHLOROPHENYL)-3-PHENYLBIUREA
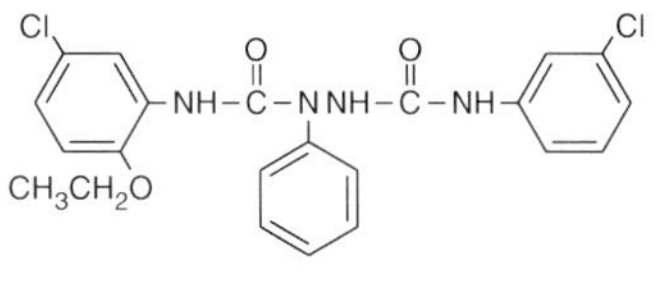
250 mg

S69,412-6
1,6-BIS(4-CHLORO-2-METHOXYPHENYL)-BIUREA
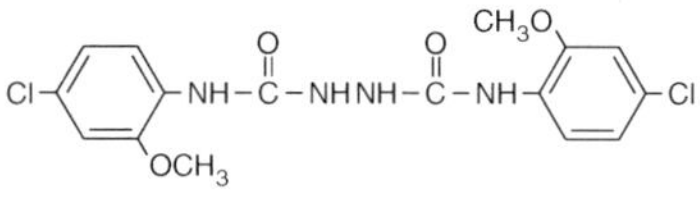
250 mg

S77,939-3
4-(2,5-DIFLUOROPHENYL)-1-(4-METHOXYBENZOYL)SEMICARBAZIDE
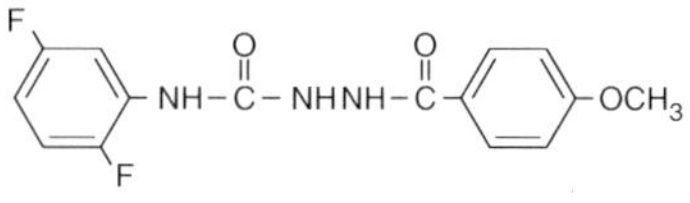
250 mg

S77,229-1
6-(4-CHLORO-3-METHOXYPHENYL)-3-PHENYL-1-(O-TOLYL)BIUREA
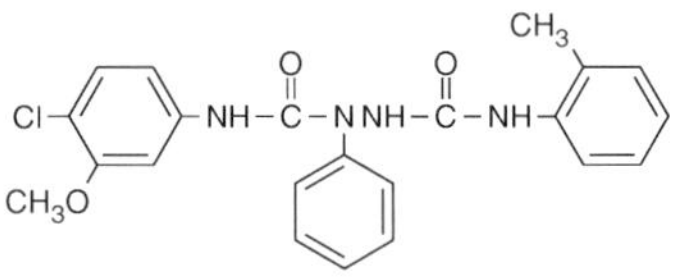
250 mg

S76,801-4
6-(4-CHLORO-3-METHYLPHENYL)-1-(2-METHOXY-5-METHYLPHENYL)-3-PHENYLBIUREA
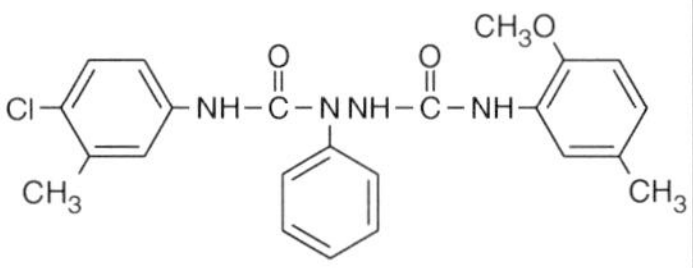
250 mg

S68,596-8
4-(5-CHLORO-2-METHOXYPHENYL)-1-(9-FLUORENYLIDENE)SEMICARBAZIDE
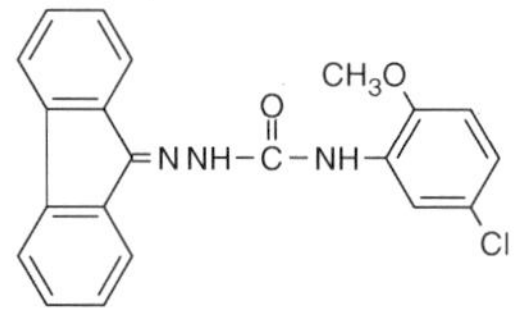
250 mg

S77,978-4
2-[3-(2-ETHOXYPHENYL)UREIDO]ACETIC (4-CHLOROBENZYLIDENE)HYDRAZIDE
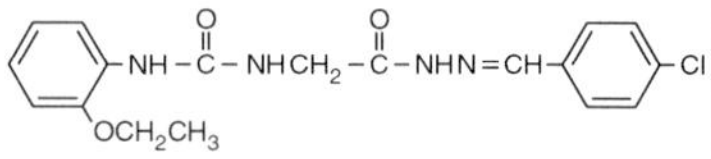
250 mg

S84,776-3
4-[3-(2-FLUOROPHENYL)UREIDO]-BENZOIC (4-METHOXYBENZYLIDENE)-HYDRAZIDE
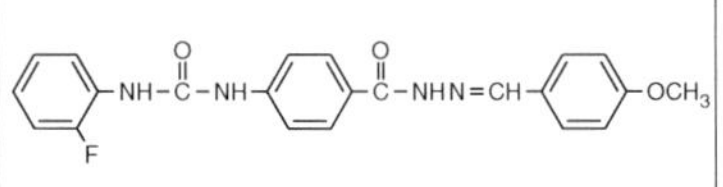
250 mg

S84,777-1
4-[3-(2-FLUOROPHENYL)UREIDO]-BENZOIC (4-HYDROXY-3-METHOXYBENZYLIDENE)HYDRAZIDE
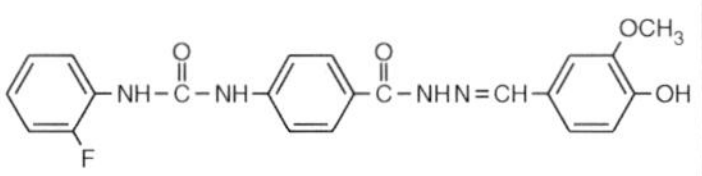
250 mg

S72,657-5
1-(3-CHLOROPHENYL)-3-(2-METHOXY-3-NITROPHENYL)UREA
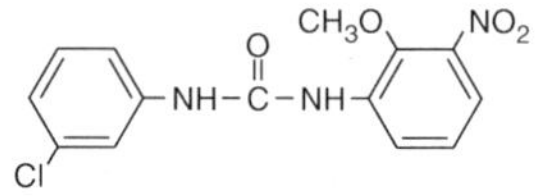
5 gm

S92,407-5
1-(4-CHLORO-2-NITROPHENYL)-3-(2-METHOXY-4-NITROPHENYL)UREA
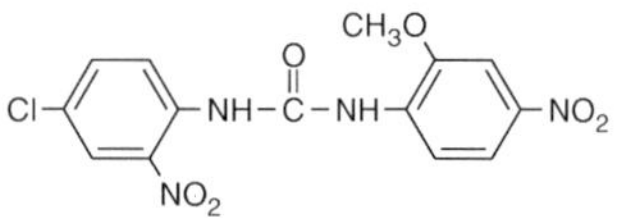
250 mg

S66,733-1
1-(P-ANISYL)-3-(2-BENZOYL-4-CHLOROPHENYL)UREA
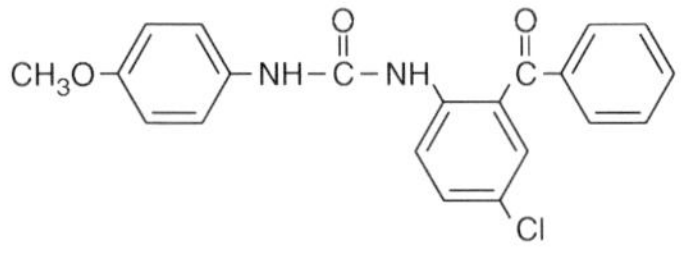
5 gm

S66,737-4
1-(O-ANISYL)-3-(2-BENZOYL-4-CHLOROPHENYL)UREA
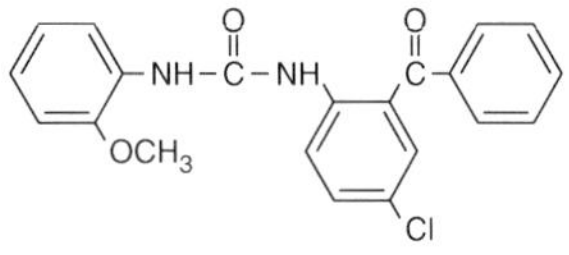
5 gm

S66,727-7
1-(M-ANISYL)-3-(2-BENZOYL-4-CHLOROPHENYL)UREA
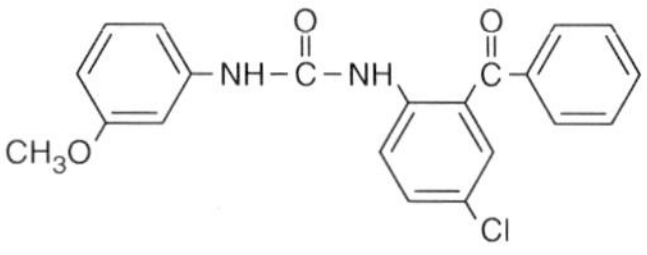
5 gm

S80,788-5
1-(3-ACETYLPHENYL)-3-(5-CHLORO-2-METHOXYPHENYL)UREA
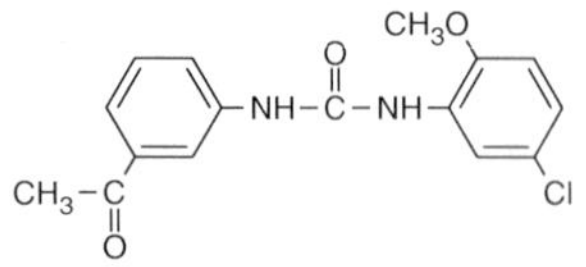
250 mg

S91,258-1
4-[3-(5-CHLORO-2-METHOXYPHENYL)-UREIDO]BENZOIC ACID

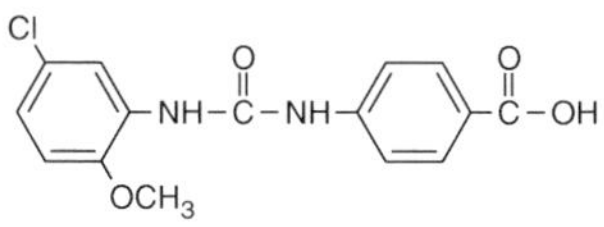

250 mg

S73,377-6
ETHYL 4-[3-(5-CHLORO-2-PHENOXYPHENYL)UREIDO]BENZOATE

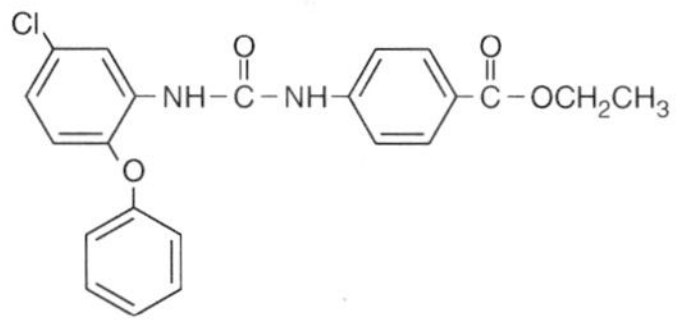

250 mg

S79,454-6
METHYL 2-[3-(5-CHLORO-2-METHOXYPHENYL)UREIDO]BENZOATE

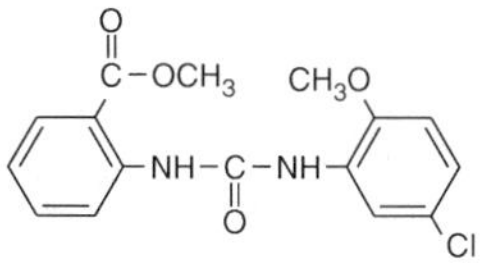

250 mg

S72,892-6
ETHYL 4-(5-CHLORO-2-METHOXYPHENYLUREIDO)BENZOATE

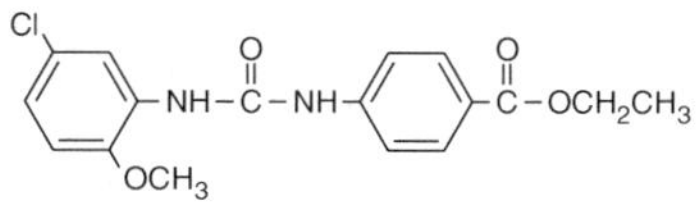

5 gm

S79,430-9
BUTYL 4-[3-(5-CHLORO-2-METHOXYPHENYL)UREIDO]BENZOATE

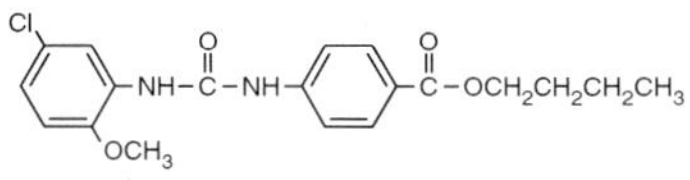

250 mg

S71,651-0
1-(4-CHLOROBENZOYL)-4-(2-METHOXY-5-METHYLPHENYL)SEMICARBAZIDE

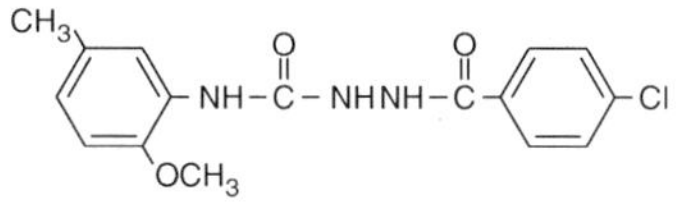

5 gm

S90,432-5
1-(2,4-DICHLOROBENZOYL)-3-(2-METHOXY-5-METHYLPHENYL)UREA

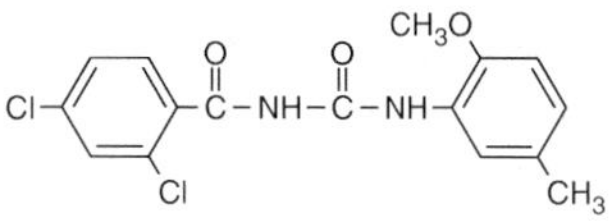

250 mg

S91,261-1
1-(TERT-BUTOXYCARBONYL)-4-(5-CHLORO-2-METHOXYPHENYL)-SEMICARBAZIDE

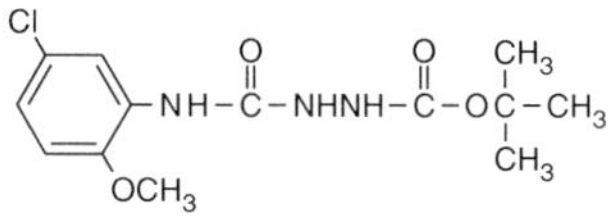

250 mg

S91,260-3
4-(5-CHLORO-2-METHOXYPHENYL)-1-METHYL-1-PHENYLSEMICARBAZIDE

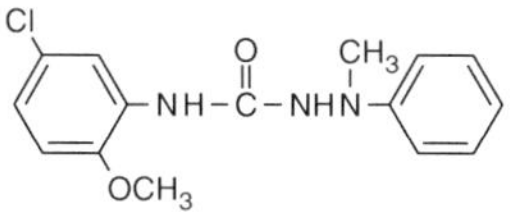

250 mg

S66,610-6
4-(3-CHLORO-2-METHOXYPHENYL)-1,1-DIMETHYLSEMICARBAZIDE

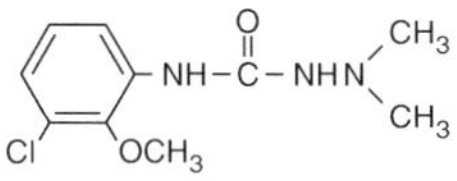

5 gm

S91,265-4
1-(5-CHLORO-2-METHOXYPHENYL)-3-(2-METHYLCYCLOHEXYL)UREA

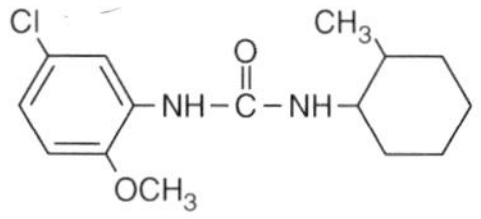

250 mg

S72,840-3
1-(5-CHLORO-2-METHOXYPHENYL)-3,3-DICYCLOHEXYLUREA

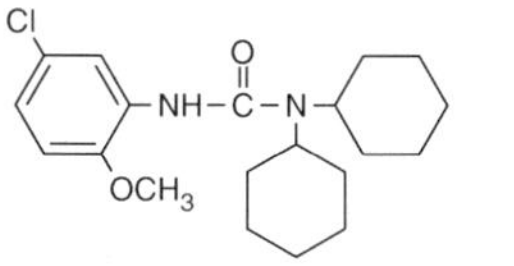

250 mg

S89,020-0
1-(5-CHLORO-2-METHOXYPHENYL)-3-PHENYLUREA

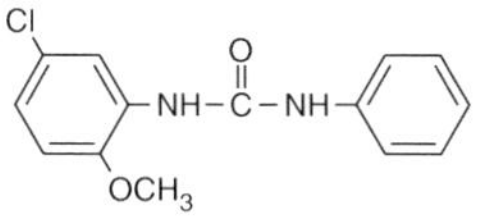

250 mg

S69,502-5
1-(4-CHLORO-2-PHENOXYPHENYL)-3-(2,5-DIFLUOROPHENYL)UREA

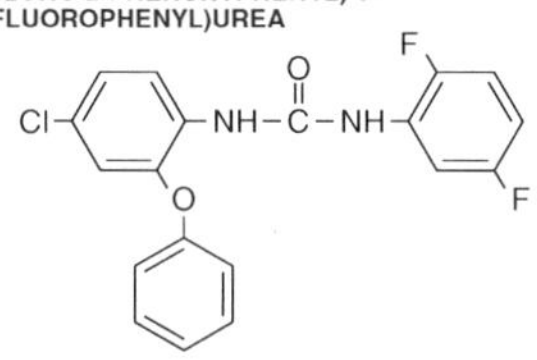

250 mg

S91,281-6
1-(5-CHLORO-2-METHOXYPHENYL)-3-(DIPHENYLMETHYL)UREA

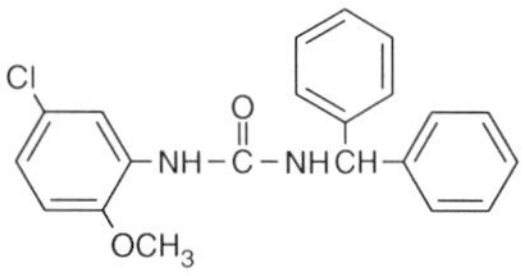

250 mg

S89,080-4
1-(3-CHLOROPHENYL)-3-(2-METHOXYPHENYL)UREA

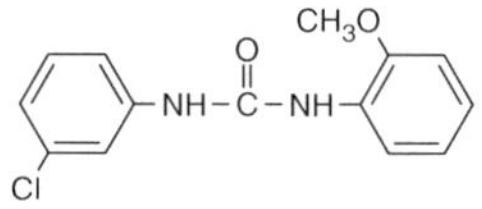

250 mg

S80,193-3
1-(2-ETHOXYPHENYL)-3-(3-FLUOROPHENYL)UREA

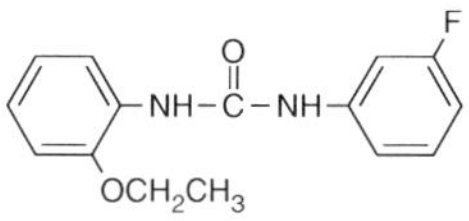

250 mg

S89,063-4
1-(2-CHLOROPHENYL)-3-(2-ETHOXYPHENYL)UREA

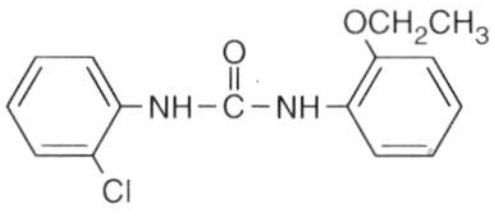

250 mg

S76,862-6
1-(4-FLUOROPHENYL)-3-(2-METHOXYPHENYL)UREA

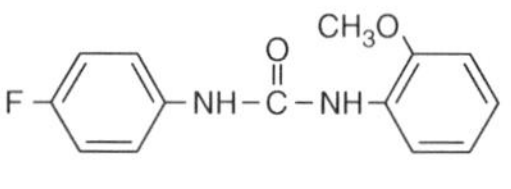

250 mg

S75,234-7
1-(4-BROMOPHENYL)-3-(2-METHOXYPHENYL)UREA

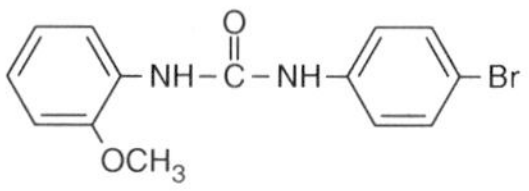

250 mg

S71,251-5
1-(2,6-DICHLOROPHENYL)-3-(2-METHOXYPHENYL)UREA

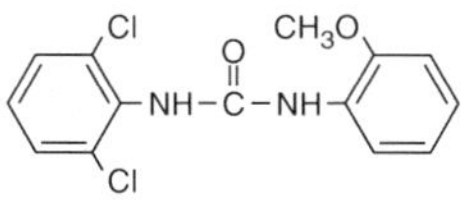

5 gm

S89,160-6
1-(2,3-DICHLOROPHENYL)-3-(2-METHOXYPHENYL)UREA

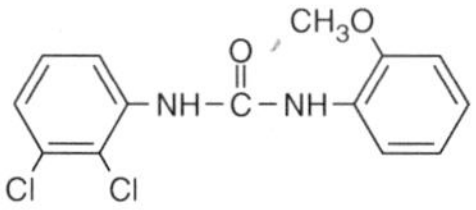

250 mg

S89,151-7
1-(2-CHLORO-6-METHYLPHENYL)-3-(2-METHOXYPHENYL)UREA

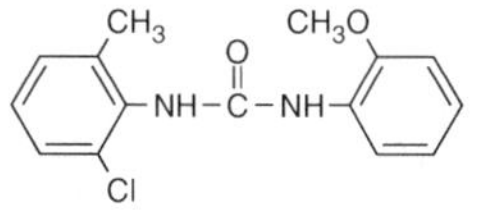

250 mg

S67,593-8
1-(4-CHLORO-2-METHOXYPHENYL)-3-(2-CHLOROPHENYL)UREA

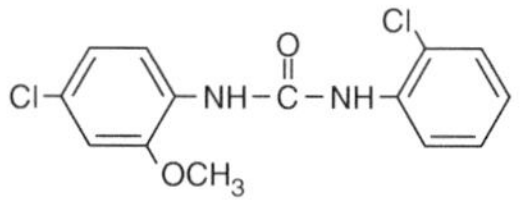

5 gm

S73,694-5
1-(2-FLUOROPHENYL)-3-(2-METHOXY-5-METHYLPHENYL)UREA

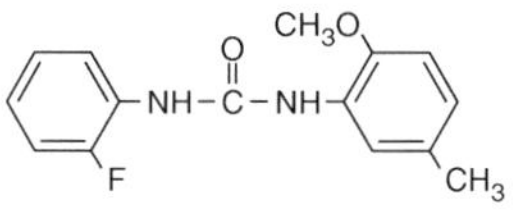

5 gm

S74,320-8
1-(2,5-DIFLUOROPHENYL)-3-(2-ETHOXYPHENYL)UREA

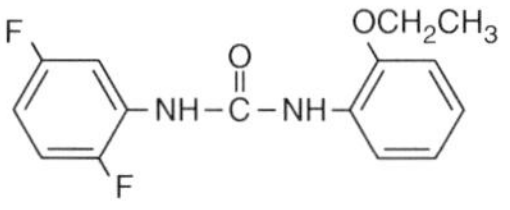

5 gm

S90,020-6
1-(3,4-DICHLOROBENZYL)-3-(2-METHOXYPHENYL)UREA

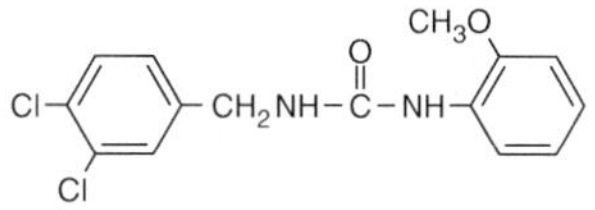

250 mg

S71,239-6
1-(2-METHOXYPHENYL)-3-(2,4,5-TRICHLOROPHENYL)UREA

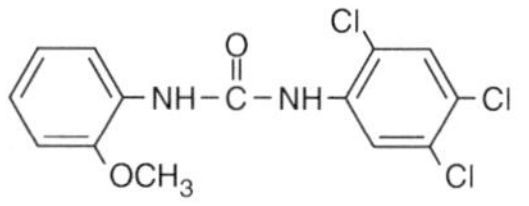

5 gm

S71,112-8
1-(4-CHLOROPHENYL)-3-(2-METHOXYETHYL)UREA

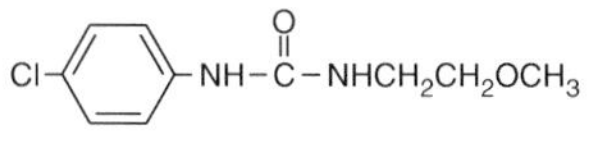

5 gm

S90,439-2
1-(4-CHLOROPHENYL)-1-ETHYL-3-(2-METHOXY-5-METHYLPHENYL)UREA

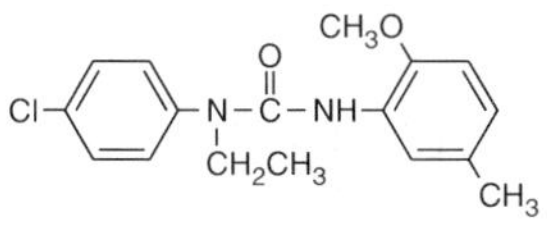

250 mg

S92,424-5
1-(4-CHLOROPHENYL)-1-ETHYL-3-(2-METHOXY-5-METHYLPHENYL)UREA

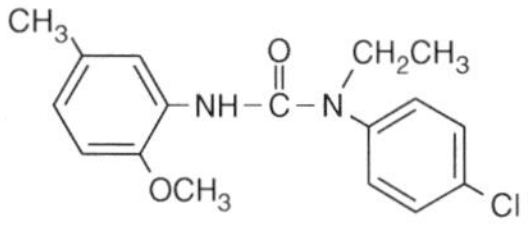

250 mg

S91,711-7
1-(3-CHLORO-2-METHOXYPHENYL)-3-METHYLUREA

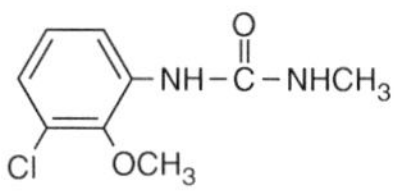

250 mg

S71,087-3
1-(2,6-DICHLOROPHENYL)-3-(2-METHOXYETHYL)UREA

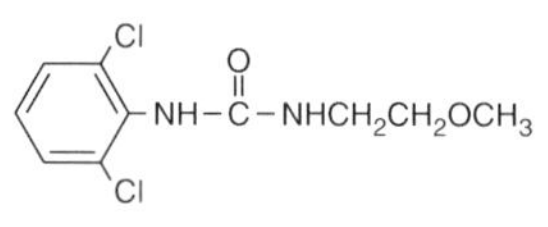

5 gm

S66,089-2
1-ALLYL-3-(3-CHLORO-2-METHOXYPHENYL)UREA

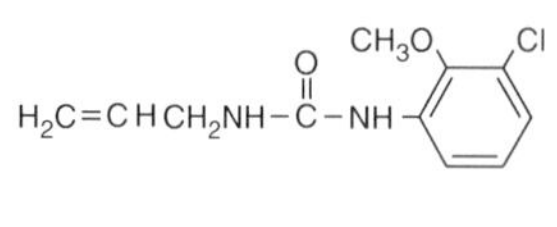

5 gm

S80,799-0
1-(5-CHLORO-2-METHOXYPHENYL)-3-(3-CHLORO-2-METHYLPHENYL)UREA

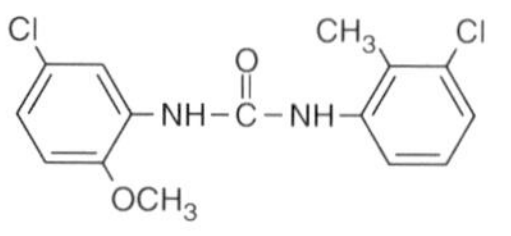

250 mg

S89,079-0
1-(5-CHLORO-2-METHOXYPHENYL)-3-PROPYLUREA

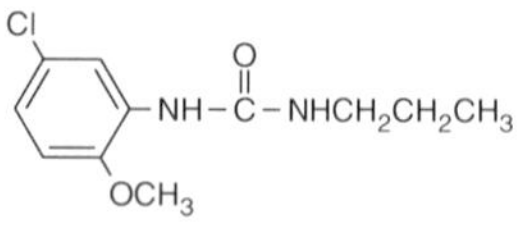

250 mg

S84,476-4
1-(5-CHLORO-2-METHOXYPHENYL)-3-(3-FLUOROPHENYL)UREA

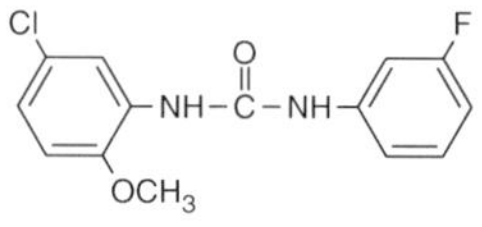

250 mg

S74,727-0
1-(5-CHLORO-2-METHOXYPHENYL)-3-(2,5-DIFLUOROPHENYL)UREA

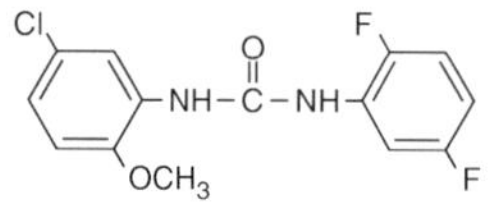

5 gm

S91,621-8
1-(4-CHLORO-2-METHOXYPHENYL)-3-(2,5-DIFLUOROPHENYL)UREA

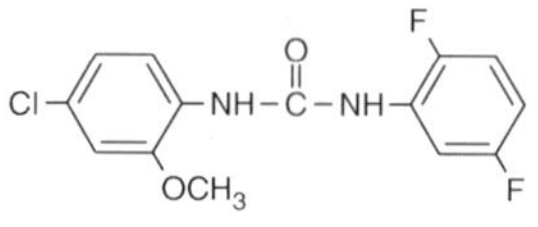

250 mg

S91,283-2
1-(5-CHLORO-2-METHOXYPHENYL)-3-(2,5-DIBROMOPHENYL)UREA

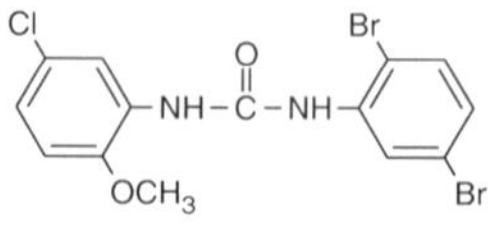

250 mg

S91,285-9
1-(5-CHLORO-2-METHOXYPHENYL)-3-(3,4-DIBROMOPHENYL)UREA

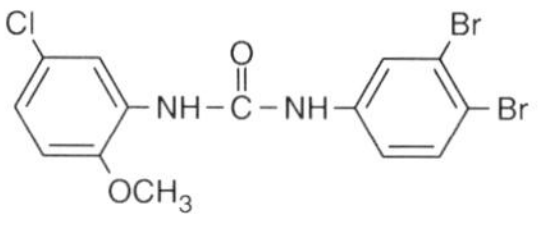

250 mg

S84,534-5
1-(3-FLUOROPHENYL)-3-(2-METHOXY-5-METHYLPHENYL)UREA

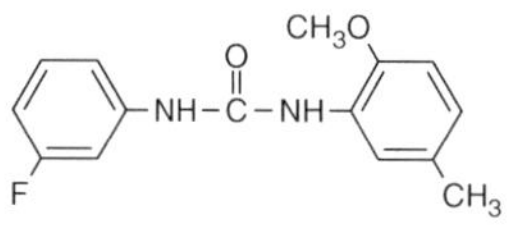

250 mg

S91,616-1
1-(2,5-DIFLUOROPHENYL)-3-(2-METHOXY-5-METHYLPHENYL)UREA

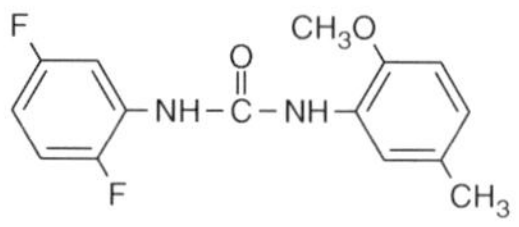

250 mg

S79,724-3
1-(5-CHLORO-2-METHOXYPHENYL)-3-(2-CHLORO-5-METHYLPHENYL)UREA

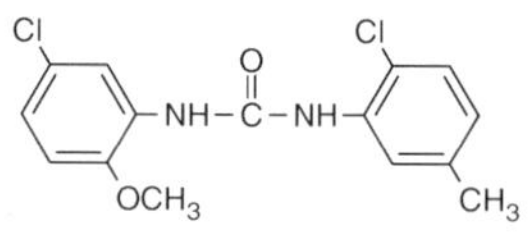

250 mg

S80,798-2
1-(5-CHLORO-2-METHOXYPHENYL)-3-(5-CHLORO-2-METHYLPHENYL)UREA

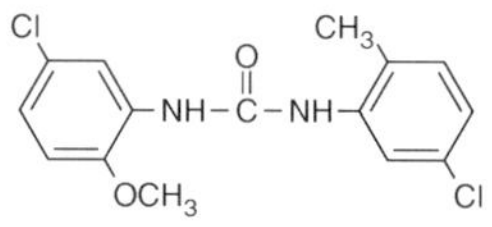

250 mg

S90,434-1
1-(2,4-DICHLOROBENZYL)-3-(2-METHOXY-5-METHYLPHENYL)UREA

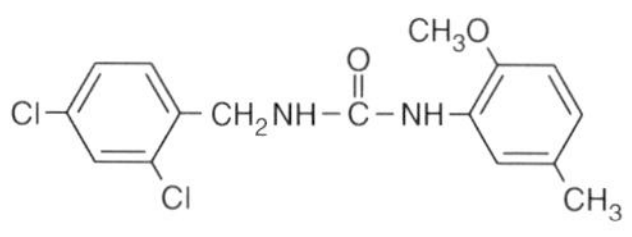

250 mg

S90,552-6
1-(3-CHLORO-4-METHYLPHENYL)-1-ETHYL-3-(2-METHOXY-5-METHYLPHENYL)UREA

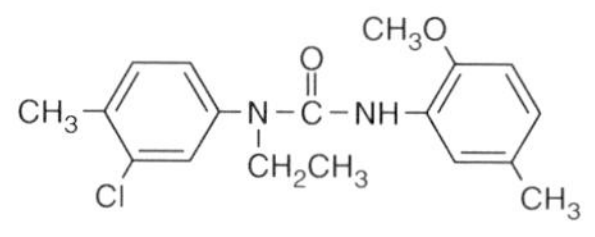

250 mg

S72,611-7
1-(5-CHLORO-2-METHOXYPHENYL)-3-(1-NAPHTHYL)UREA

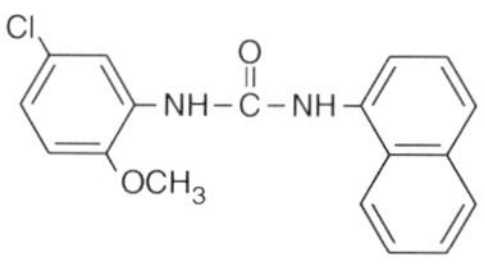

5 gm

S91,618-8
1-(4-ACETYL-3-METHOXYPHENYL)-3-(2,5-DIFLUOROPHENYL)UREA

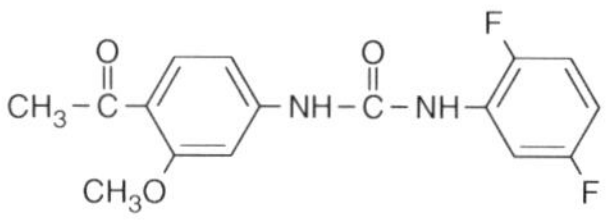

250 mg

S79,747-2
ETHYL 4-[3-(2-CHLORO-4-METHOXYPHENYL)UREIDO]BENZOATE

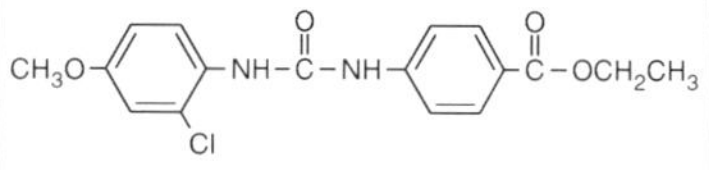

250 mg

S90,426-0
4-(4-FLUOROPHENYL)-1-(4-METHOXYBENZOYL)SEMICARBAZIDE

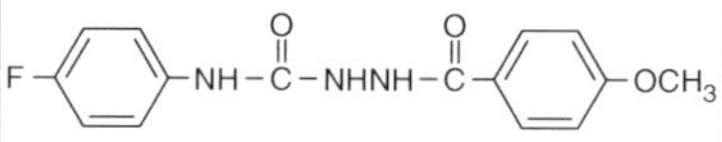

250 mg

S79,642-5
4-(3-CHLOROPHENYL)-1-(4-METHOXYBENZOYL)SEMICARBAZIDE

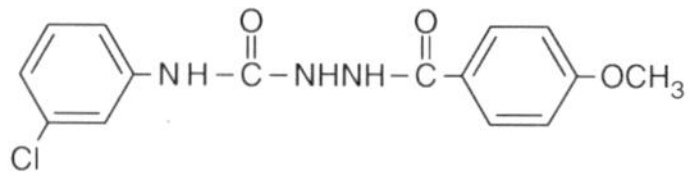

250 mg

S71,417-8
4-(2,6-DICHLOROPHENYL)-1-(4-METHOXYBENZOYL)SEMICARBAZIDE

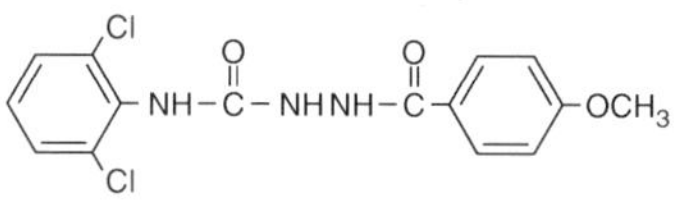

5 gm

S72,930-2
1-(3,4-DICHLOROBENZOYL)-3-(4-ETHOXYPHENYL)-2-THIOUREA

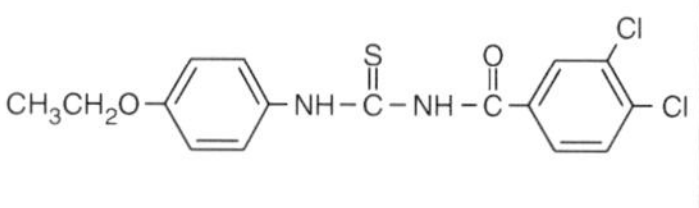

250 mg

S68,863-0
N-[3-(3-CHLOROPHENYL)UREIDO]-N'-(4-ETHOXYPHENYL)OXAMIDE

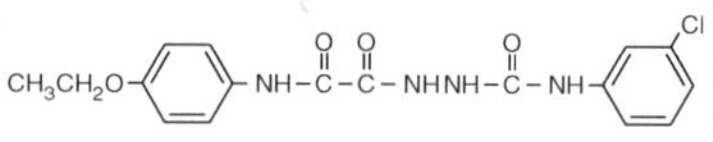

250 mg

S90,481-3
ETHYL 2-[3-(4-CHLORO-2-METHOXYPHENYL)-3-METHYLUREIDO]-ACETATE

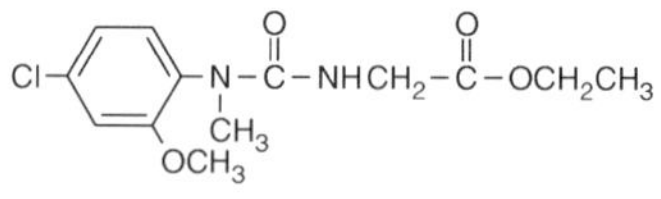

250 mg

S68,795-2
4-(4-CHLORO-2-METHYLPHENYL)-1-[3-(4-METHOXYPHENYL)UREIDOACETYL]-SEMICARBAZIDE

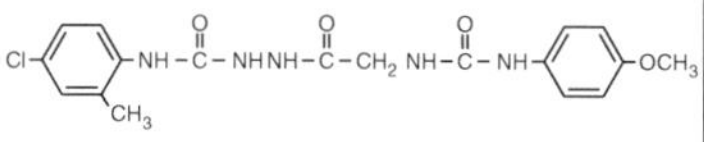

250 mg

S78,032-4
1-(3-BUTOXYPROPYL)-3-(3-CHLOROPHENYL)UREA

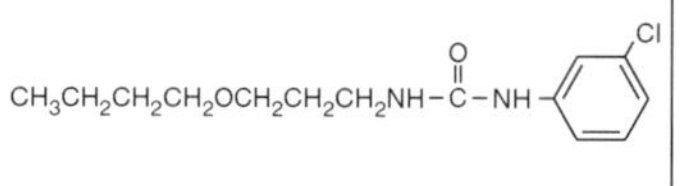

250 mg

S89,939-9
1-(2-CHLORO-4-METHOXYPHENYL)-3-CYCLOHEXYLUREA

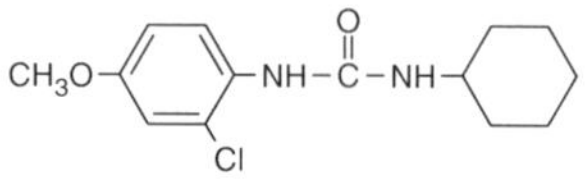

250 mg

S77,335-2
1-(2-CHLOROPHENYL)-3-(4-ETHOXYPHENYL)UREA

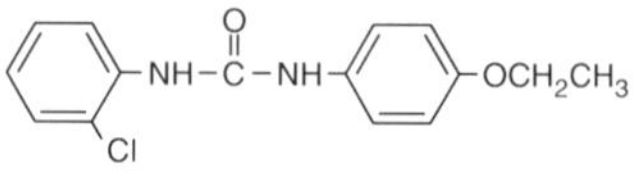

250 mg

S75,231-2
1-(2-CHLORO-4-METHOXYPHENYL)-3-(2-CHLOROPHENYL)UREA

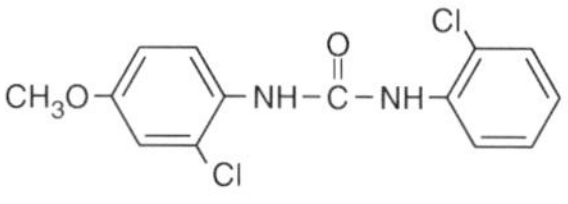

5 gm

S75,245-2
1-(2-CHLORO-4-METHOXYPHENYL)-3-(O-TOLYL)UREA

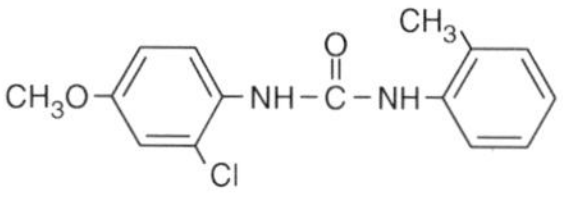

5 gm

S77,956-3
1-(4-CHLOROPHENYL)-3-(4-METHOXYPHENYL)UREA

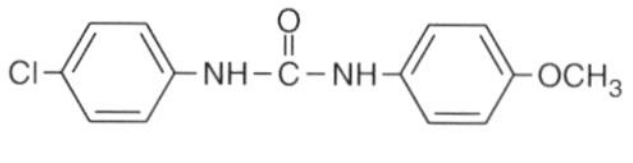

250 mg

S78,075-8
1-(4-CHLOROPHENYL)-3-(3-METHOXYPHENYL)UREA

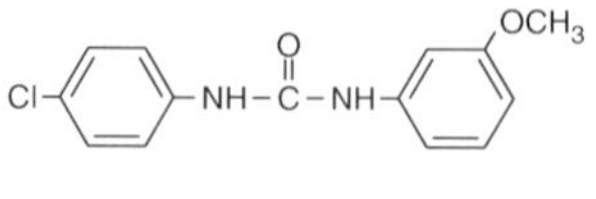

250 mg

S67,683-7
1-(4-BROMOPHENYL)-3-(4-METHOXYPHENYL)UREA
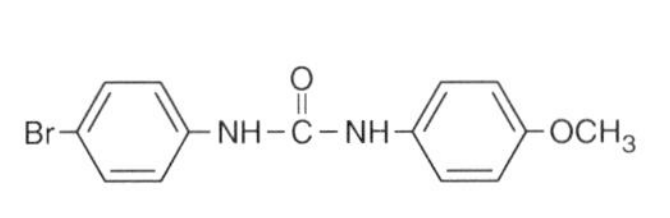
5 gm

S70,824-0
1-(4-METHOXYPHENYL)-3-(2,4,5-TRICHLOROPHENYL)UREA
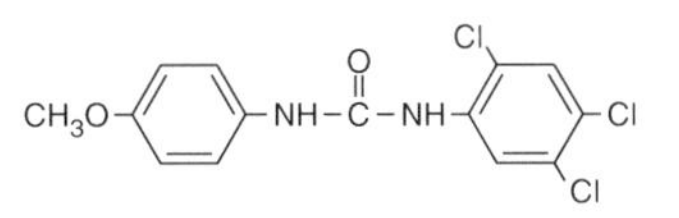
5 gm

S80,192-5
1-(4-ETHOXYPHENYL)-3-(3-FLUOROPHENYL)UREA
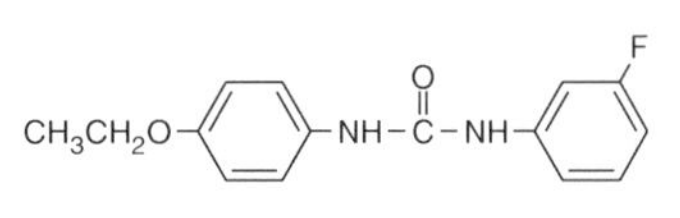
250 mg

S90,023-0
1-(3-CHLOROBENZYL)-3-(4-METHOXYPHENYL)UREA
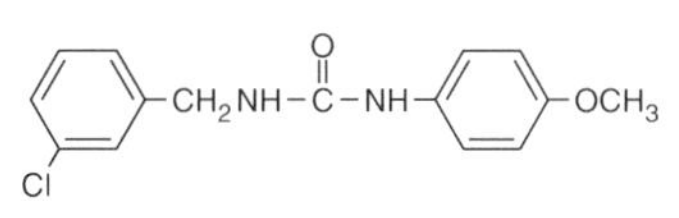
250 mg

S71,498-4
1-(2,6-DICHLOROPHENYL)-3-(4-METHOXYPHENYL)UREA
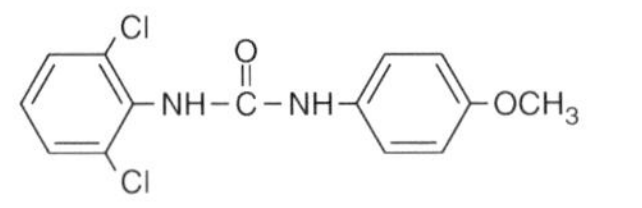
5 gm

S84,479-9
1-(2,3-DICHLOROPHENYL)-3-(4-METHOXYPHENYL)UREA
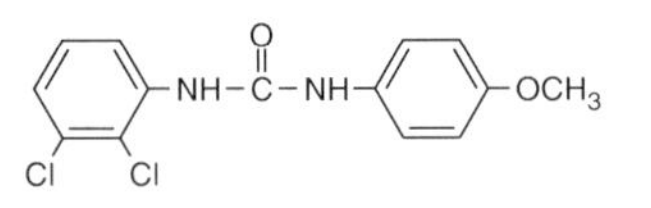
250 mg

S67,484-2
1-(2-CHLORO-6-METHYLPHENYL)-3-(4-METHOXYPHENYL)UREA
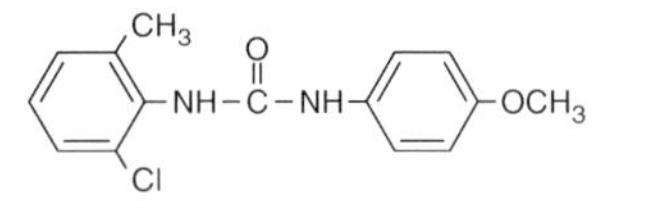
250 mg

S77,317-4
1-(3-CHLORO-2-METHYLPHENYL)-3-(4-METHOXYPHENYL)UREA
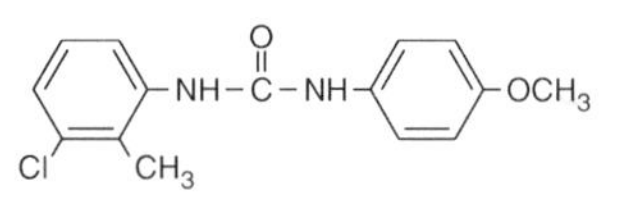
250 mg

S76,866-9
1-(3-CHLORO-4-METHYLPHENYL)-3-(4-METHOXYPHENYL)UREA
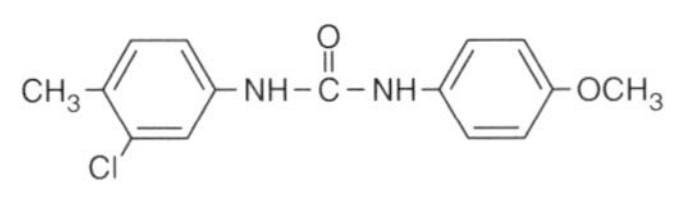
250 mg

S76,968-1
1-(4-CHLORO-2-METHYLPHENYL)-3-(4-METHOXYPHENYL)UREA
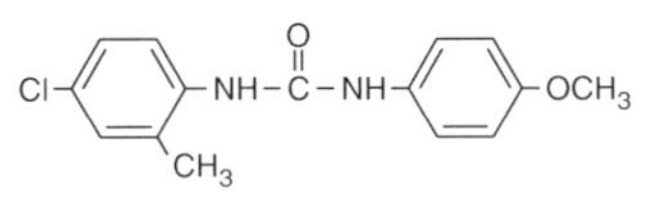
250 mg

S75,243-6
1-(4,5-DICHLORO-2-METHYLPHENYL)-3-(4-METHOXYPHENYL)UREA
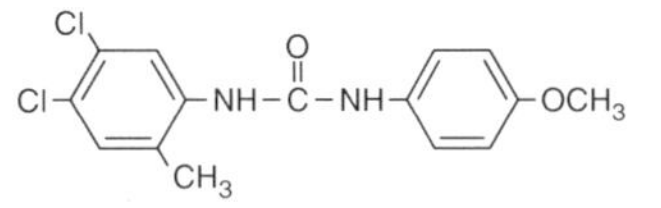
5 gm

S71,376-7
1-(2-CHLORO-6-METHYLPHENYL)-3-(2-METHOXYETHYL)UREA
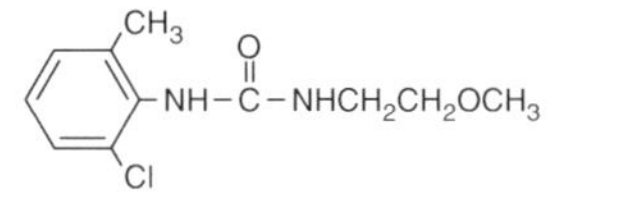
5 gm

S70,821-6
1-(2,6-DICHLOROPHENYL)-3-(3-METHOXYPHENYL)UREA
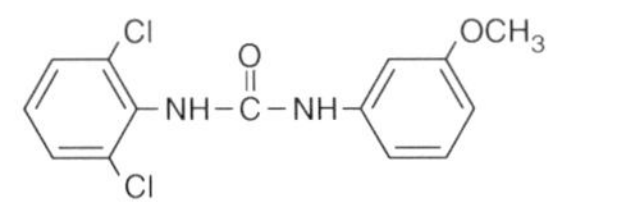
5 gm

S75,035-2
1-(2-CHLORO-4-METHOXYPHENYL)-3-PROPYLUREA
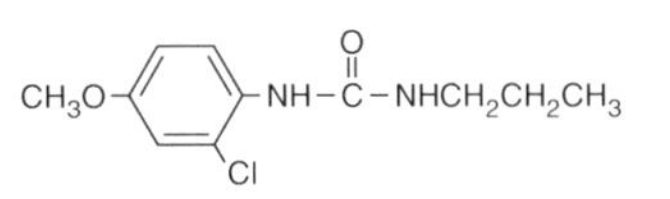
5 gm

S89,275-0
1-BUTYL-3-(2-CHLORO-4-METHOXYPHENYL)UREA
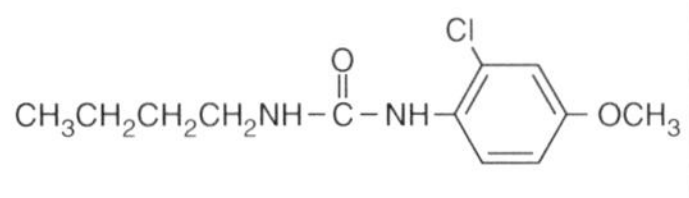
250 mg

S66,347-6
3-(3-CHLORO-2-METHOXYPHENYL)-1,1-DIALLYLUREA
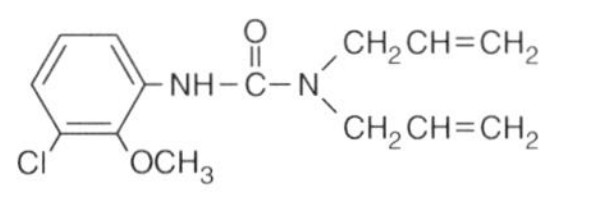
5 gm

S78,031-6
1-(3-BUTOXYPROPYL)-3-(3-CHLORO-4-FLUOROPHENYL)UREA
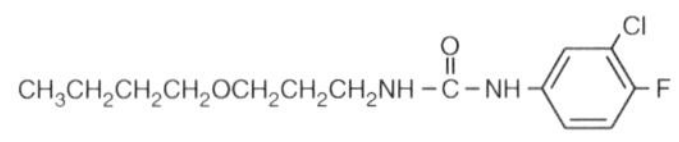
250 mg

S86,786-1
1-(2-CHLORO-4-METHOXYPHENYL)-3-(5-CHLORO-2-METHYLPHENYL)UREA
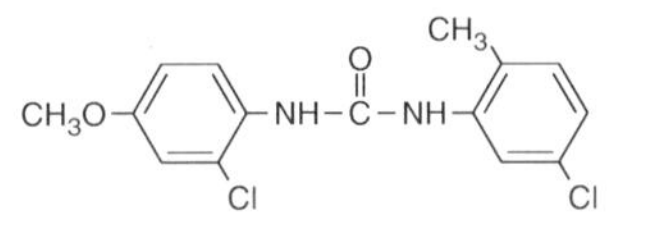
250 mg

S92,236-6
1-(2-CHLORO-4-METHOXYPHENYL)-3-(5-CHLORO-2-METHOXYPHENYL)UREA
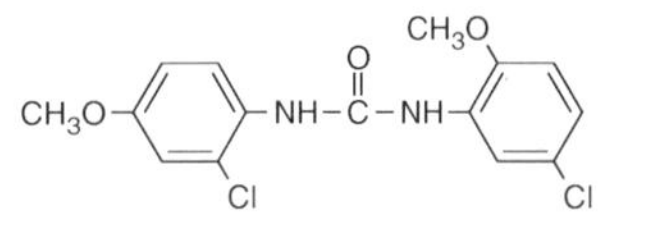
250 mg

S90,362-0
3-(3-CHLORO-4-FLUOROPHENYL)-1-ETHYL-1-(2-METHOXY-5-METHYLPHENYL)UREA
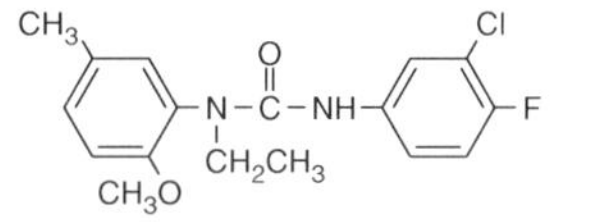
250 mg

S71,241-8
1-(3-METHOXYPHENYL)-3-(2,4,5-TRICHLOROPHENYL)UREA
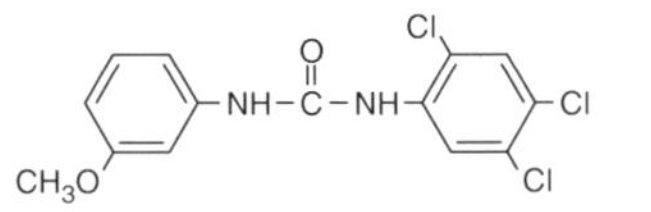
250 mg

S77,993-8
2-(METHACRYLOYLOXY)ETHYL N-[N-(3,5-DICHLOROPHENYL)CARBAMOYL]CARBAMATE
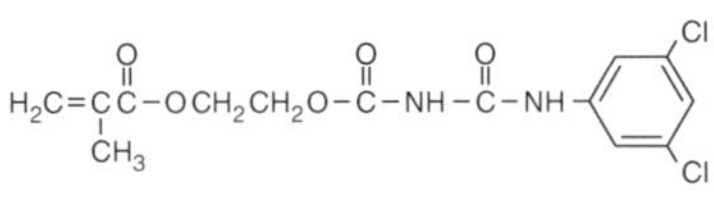
250 mg

S68,963-7
4-(3,4-DICHLOROPHENYL)-1-(2-METHOXYBENZOYL)SEMICARBAZIDE
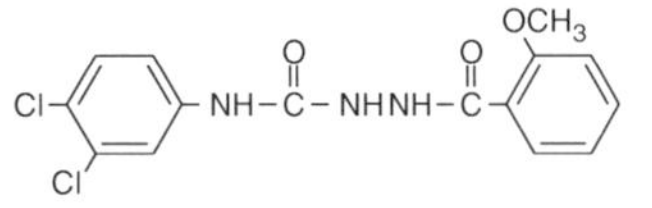
250 mg

S68,411-2
4-(2,4-DIFLUOROPHENYL)-1-(2-HYDROXYBENZOYL)SEMICARBAZIDE
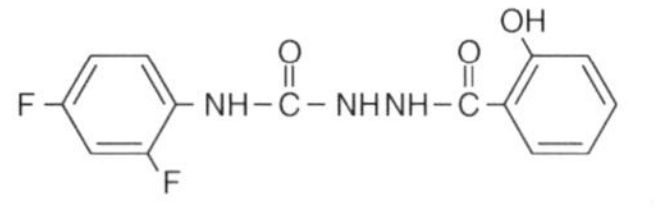
250 mg

S94,510-2
1-(4-CHLOROPHENYL)-5-OXO-3-PYRROLIDINECARBOXYLIC ACID
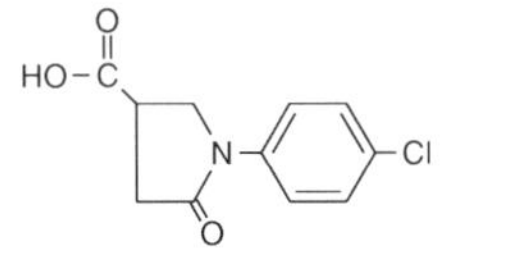
250 mg

S72,374-6
1-(3-CHLORO-2-METHYLPHENYL)-3-(2-MERCAPTOPHENYL)UREA
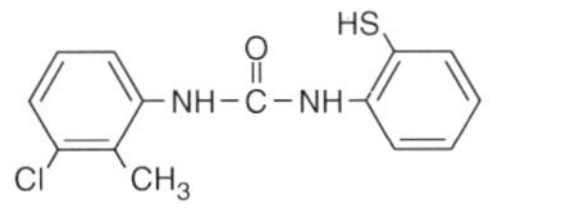
250 mg

S69,072-4
1-(4-CHLORO-2-METHYLPHENYL)-3-(2-MERCAPTOPHENYL)UREA
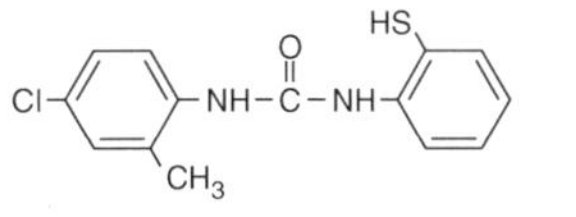
250 mg

S86,758-6
1-(4-CHLORO-2-NITROPHENYL)-3-[4-(METHYLTHIO)PHENYL]UREA
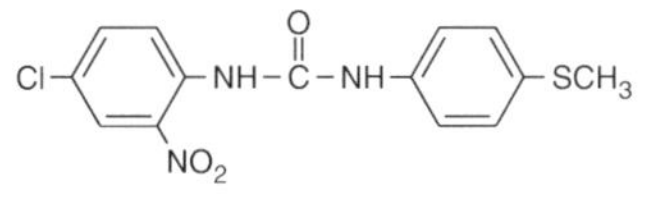
250 mg

S67,444-3
N-(4-FLUOROPHENYL)-N'-(3-[4-(METHYLTHIO)PHENYL]UREIDO)OXAMIDE
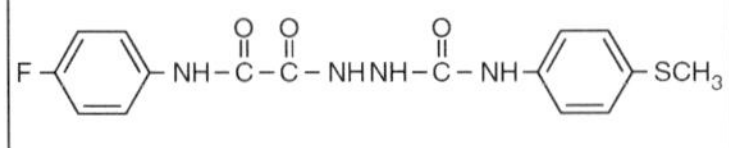
250 mg

S78,068-5
1-(4-CHLORO-2-METHYLPHENYL)-3-[2-(2,4-DINITROPHENYLTHIO)PHENYL]UREA
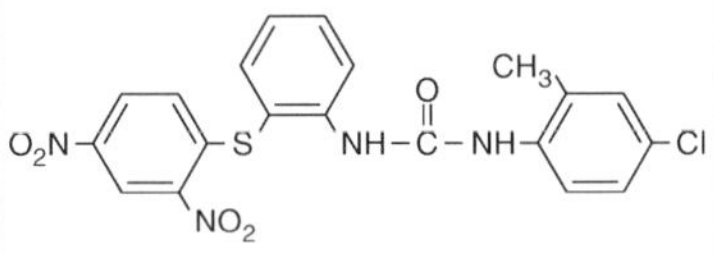
250 mg

S67,634-9
1-(3-CHLOROPHENYL)-3-[3-(METHYLTHIO)PHENYL]UREA
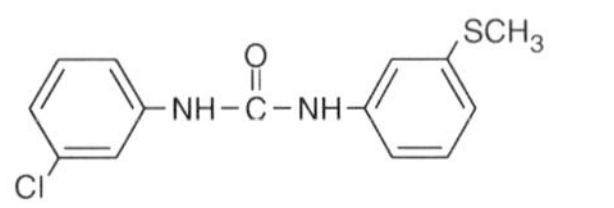
5 gm

S79,790-1
1-(2-BENZOYL-4-CHLOROPHENYL)-3-[4-(METHYLTHIO)PHENYL]UREA
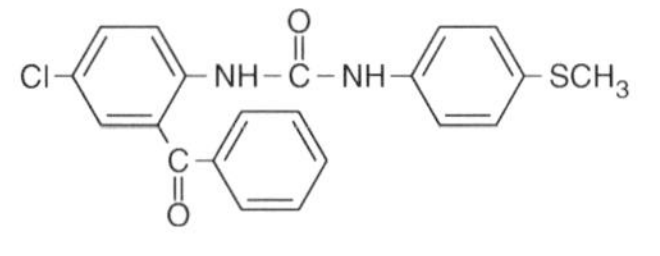
250 mg

S91,085-6
1-(2,4-DICHLOROBENZOYL)-3-[4-(METHYLTHIO)PHENYL]UREA

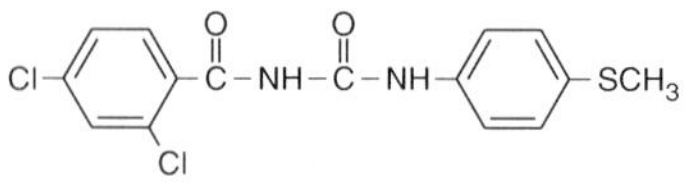

250 mg

S72,077-1
1,1'-(4-BROMO-2-METHYL-1,3-PHENYLENE)BIS(3-[3-(METHYLTHIO)-PHENYL]UREA)

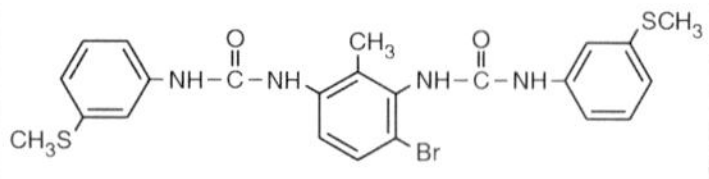

250 mg

S79,702-2
1-(2-FLUOROPHENYL)-3-[4-(METHYLTHIO)PHENYL]UREA

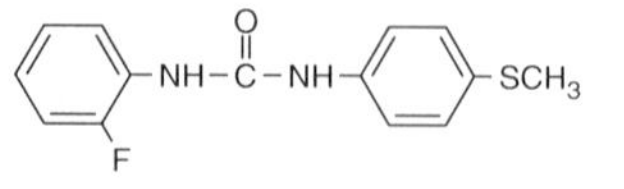

250 mg

S79,703-0
1-(3-FLUOROPHENYL)-3-[4-(METHYLTHIO)PHENYL]UREA

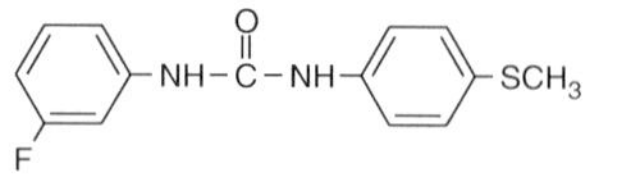

250 mg

S79,704-9
1-(4-FLUOROPHENYL)-3-[4-(METHYLTHIO)PHENYL]UREA

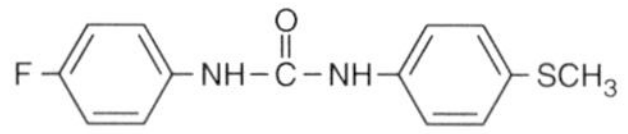

250 mg

S75,237-1
1-(4-CHLOROPHENYL)-3-[3-(METHYLTHIO)PHENYL]UREA

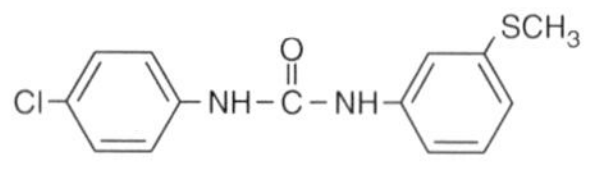

5 gm

S86,791-8
1-(3-CHLORO-2-METHYLPHENYL)-3-[4-(METHYLTHIO)PHENYL]UREA

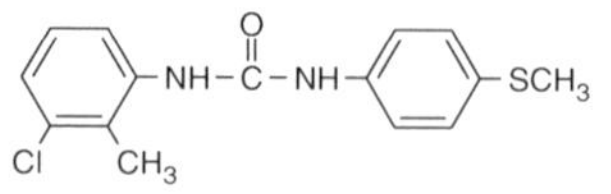

250 mg

S73,669-4
1-(2,5-DIFLUOROPHENYL)-3-[4-(METHYLTHIO)PHENYL]UREA

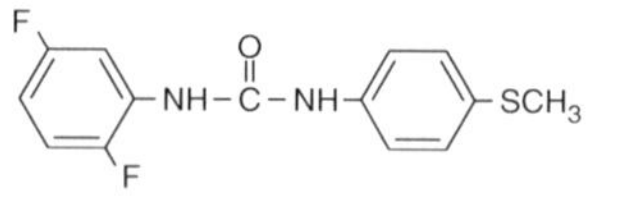

5 gm

S79,700-6
1-(2,4-DICHLOROPHENYL)-3-[4-(METHYLTHIO)PHENYL]UREA

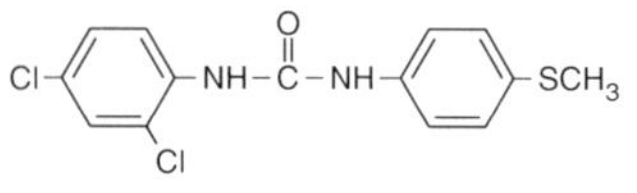

250 mg

S71,573-5
1-(5-CHLORO-2-METHYLPHENYL)-3-[4-(METHYLTHIO)PHENYL]UREA

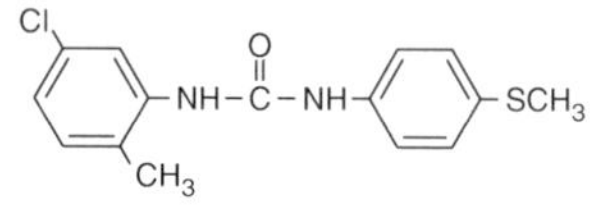

5 gm

S79,728-6
1-(2-CHLORO-5-METHYLPHENYL)-3-[4-(METHYLTHIO)PHENYL]UREA

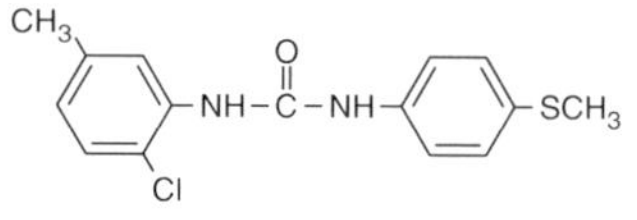

250 mg

S72,297-9
1-(4-BROMO-2,6-DIMETHYLPHENYL)-3-[4-(METHYLTHIO)PHENYL]UREA

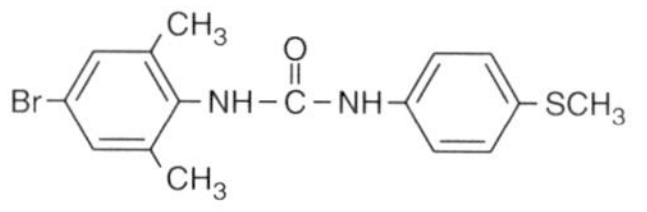

250 mg

S79,718-9
1-(4,5-DICHLORO-2-METHYLPHENYL)-3-[4-(METHYLTHIO)PHENYL]UREA

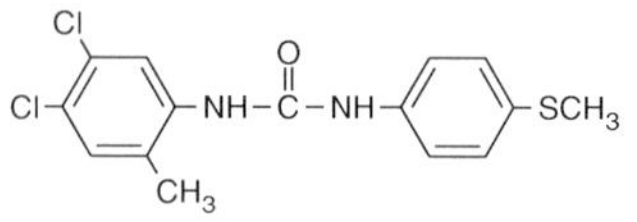

250 mg

S74,781-5
2-[3-(4-BROMOPHENYL)UREIDO]-4-CHLOROPHENYL DISULFIDE

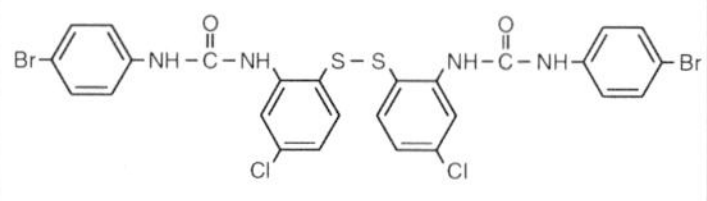

5 gm

S75,105-7
4-CHLORO-2-(3-CHLORO-4-FLUOROPHENYLUREIDO)PHENYL DISULFIDE

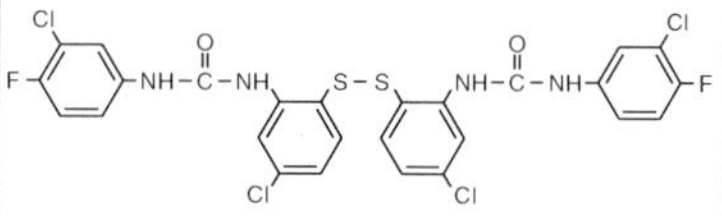

250 mg

S85,052-7
1-(2-AMINO-4-FLUORO-6-NITROPHENYL)-3-(2,5-DICHLOROPHENYL)UREA

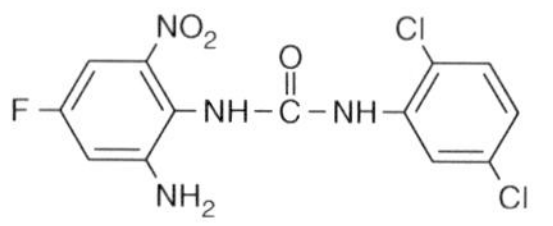

250 mg

S74,733-5
4-(4-BROMOPHENYL)-1-(2,4-DINITROPHENYL)-1-METHYLSEMICARBAZIDE

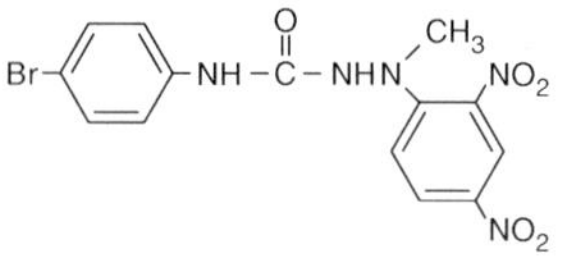

5 gm

S72,394-0
4-(3-CHLOROPHENYL)-1-(2,4-DINITROPHENYL)SEMICARBAZIDE

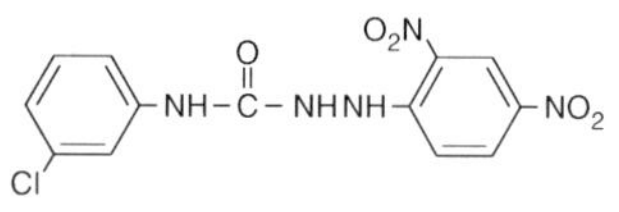

5 gm

S72,376-2
4-(3-CHLOROPHENYL)-1-(2,4-DINITROPHENYL)-1-METHYLSEMICARBAZIDE

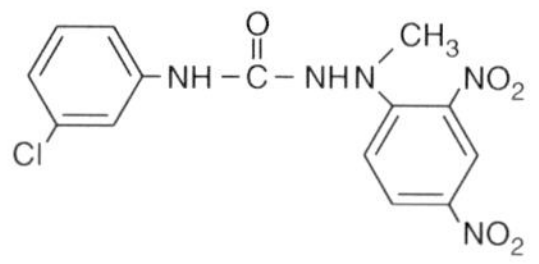

5 gm

S78,069-3
4-(2,4-DIFLUOROPHENYL)-1-(2,4-DINITROPHENYL)-1-METHYLSEMICARBAZIDE

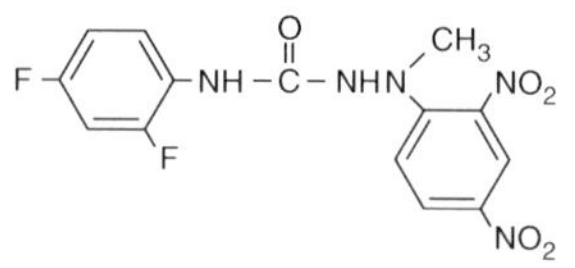

250 mg

S90,425-2
1-(2,4-DINITROPHENYL)-4-(4-FLUOROPHENYL)SEMICARBAZIDE

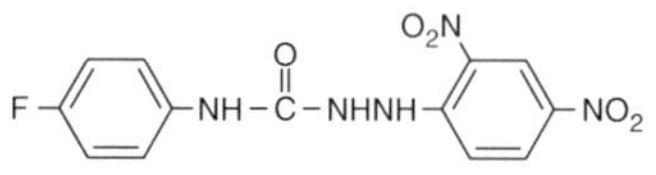

250 mg

S68,779-0
4-(4-CHLOROPHENYL)-1-(2,4-DINITROPHENYL)SEMICARBAZIDE

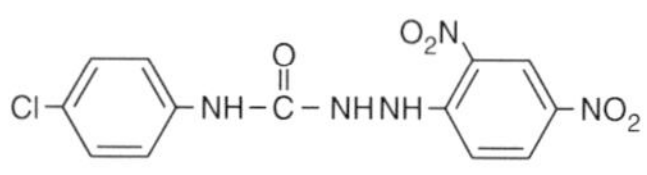

250 mg

S90,312-4
2-[3-(2-CHLORO-4-NITROPHENYL)-UREIDO]ACETIC (2-FLUOROBENZYLIDENE)-HYDRAZIDE

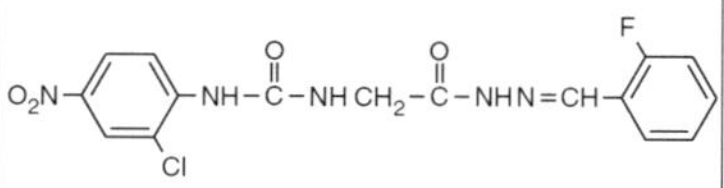

250 mg

S90,210-1
2-[3-(4-FLUOROPHENYL)UREIDO]ACETIC (3-NITROBENZYLIDENE)HYDRAZIDE

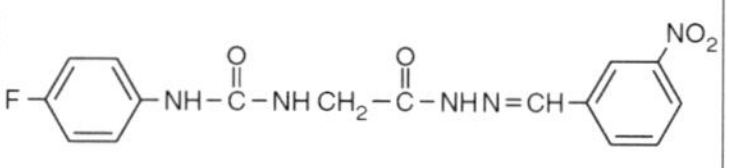

250 mg

S90,311-6
2-[3-(2-CHLORO-4-NITROPHENYL)-UREIDO]ACETIC (4-FLUOROBENZYLIDENE)-HYDRAZIDE

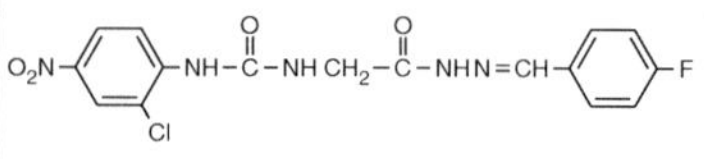

250 mg

S90,316-7
2-[3-(2-CHLORO-4-NITROPHENYL)-UREIDO]ACETIC (4-METHYLBENZYLIDENE)-HYDRAZIDE

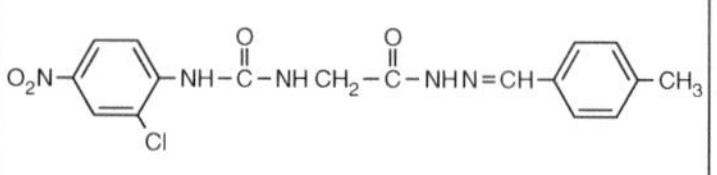

250 mg

S90,314-0
2-[3-(2-CHLORO-4-NITROPHENYL)-UREIDO]ACETIC (3-FLUOROBENZYLIDENE)-HYDRAZIDE

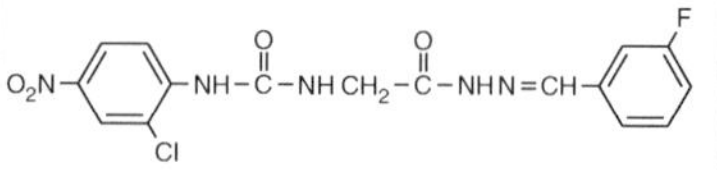

250 mg

S90,317-5
2-[3-(2-CHLORO-4-NITROPHENYL)-UREIDO]ACETIC (2-PHENYLETHYLIDENE)-HYDRAZIDE

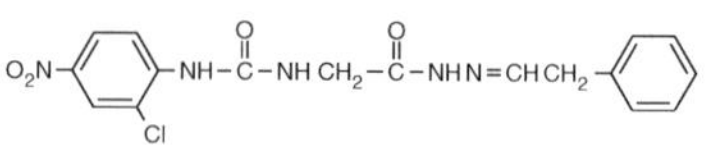

250 mg

S90,241-1
2-[3-(3-CHLOROPHENYL)UREIDO]ACETIC (2-PHENYLETHYLIDENE)HYDRAZIDE

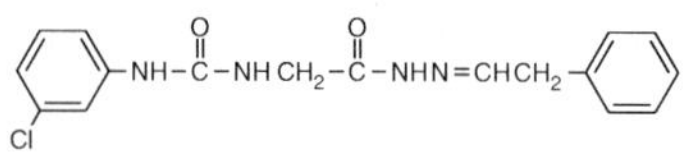

250 mg

S92,328-1
1-(5-CHLORO-2-METHOXYPHENYL)-3-[2-(N-ETHYLANILINO)ETHYL]UREA

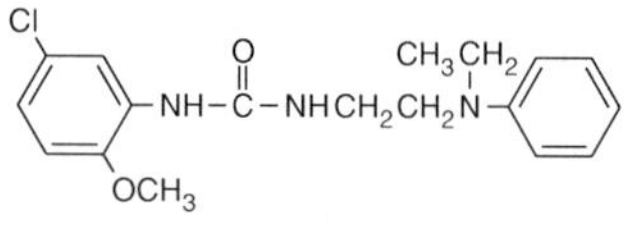

250 mg

S90,303-5
2-(3-BENZYLUREIDO)ACETIC [2-CHLORO-4-(DIMETHYLAMINO)BENZYLIDENE]-HYDRAZIDE

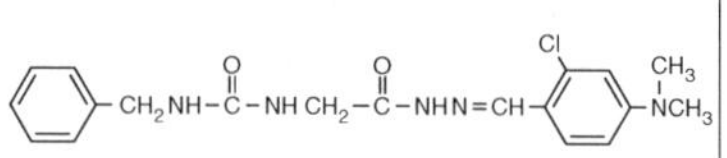

250 mg

S90,240-3
2-[3-(3-CHLOROPHENYL)UREIDO]ACETIC [4-(DIMETHYLAMINO)BENZYLIDENE]-HYDRAZIDE

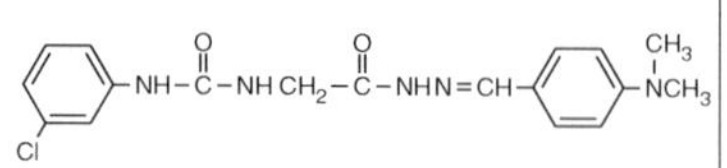

250 mg

S92,422-9
1-(2-CHLOROPHENYL)-3-[2-(1-PYRROLIDINO)ETHYL]UREA
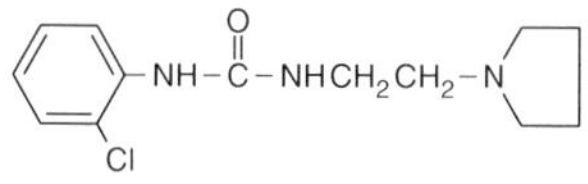
250 mg

S90,597-6
1-(4-BROMOPHENYL)-3-[2-(4-MORPHOLINO)ETHYL]UREA
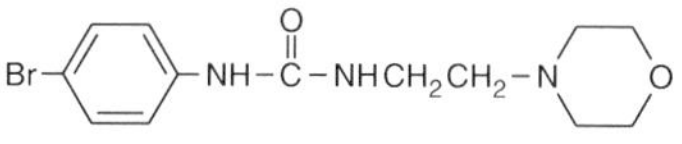
250 mg

S86,781-0
1-BENZYL-3-(4-CHLORO-2-NITROPHENYL)-1-METHYLUREA
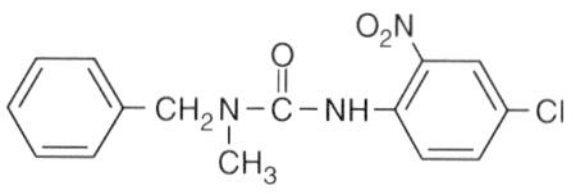
250 mg

S77,870-2
1-(3-CHLOROPHENYL)-4-(3,3-DIMETHYLUREIDO)OXAMIDE
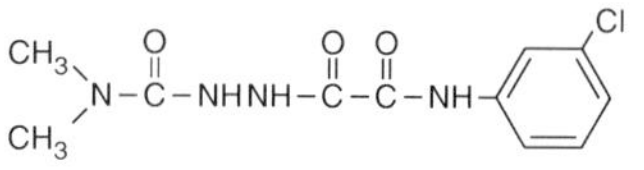
250 mg

S75,131-6
1-(3,4-DICHLOROPHENYL)-4-[3-(3,4-DICHLOROPHENYL)UREIDO]OXAMIDE
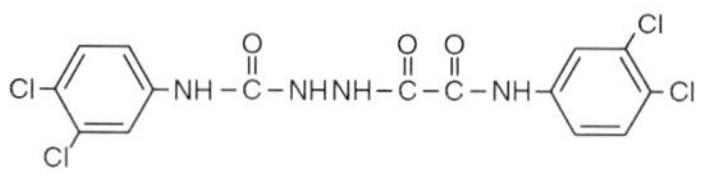
5 gm

S90,302-7
2-(3-BENZYLUREIDO)ACETIC (4-CHLOROBENZYLIDENE)HYDRAZIDE
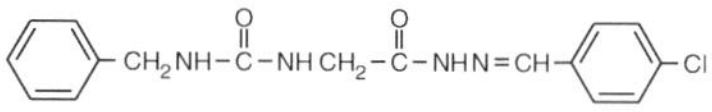
250 mg

S90,212-8
2-[3-(4-FLUOROPHENYL)UREIDO]ACETIC (3-PHENYLALLYLIDENE)HYDRAZIDE
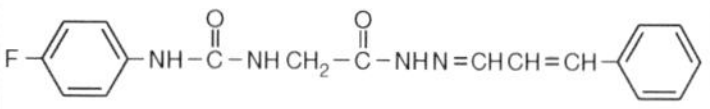
250 mg

S90,207-1
2-[3-(4-FLUOROPHENYL)UREIDO]ACETIC (4-METHYLBENZYLIDENE)HYDRAZIDE
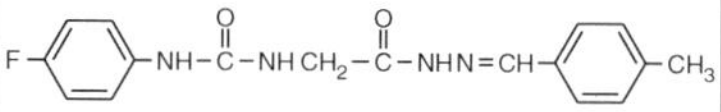
250 mg

S90,238-1
2-[3-(3-CHLOROPHENYL)UREIDO]ACETIC (4-METHYLBENZYLIDENE)HYDRAZIDE
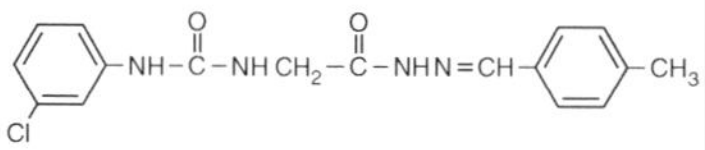
250 mg

S86,800-0
ETHYL 4-[3-(4-CHLORO-2-NITROPHENYL)-UREIDO]BENZOATE
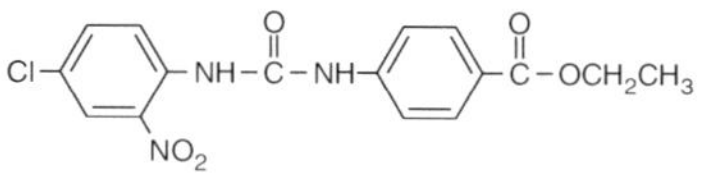
250 mg

S86,764-0
3-(4-CHLORO-2-NITROPHENYL)-1-CYCLOHEXYL-1-METHYLUREA
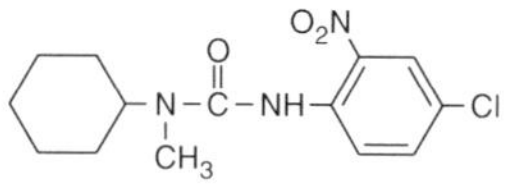
250 mg

S86,730-6
1-(4-CHLORO-2-NITROPHENYL)-3-(4-FLUOROPHENYL)UREA
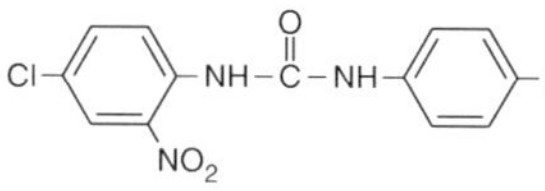
250 mg

S68,748-0
1-(3,4-DICHLOROPHENYL)-3-(2-METHYL-6-NITROPHENYL)UREA
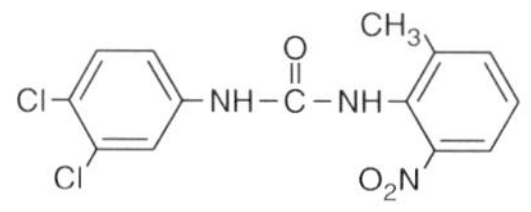
250 mg

S91,308-1
1-(2,5-DIBROMOPHENYL)-3-(4-METHYL-2-NITROPHENYL)UREA
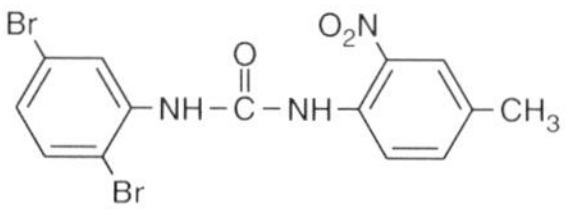
250 mg

S85,155-8
1-(2,6-DINITRO-4-FLUOROPHENYL)-3-OCTADECYLUREA
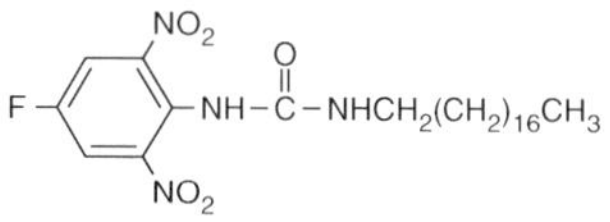
250 mg

S90,105-9
1-(2-[3-(4-FLUOROPHENYL)UREIDO]-ACETYL)-4-METHYLSEMICARBAZIDE
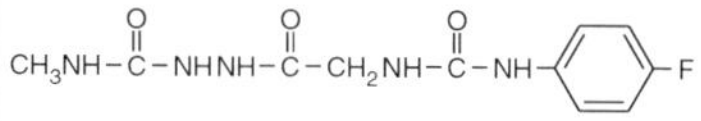
250 mg

S74,340-2
1-(2-CHLOROPHENYL)-3-(2-MORPHOLINOETHYL)UREA
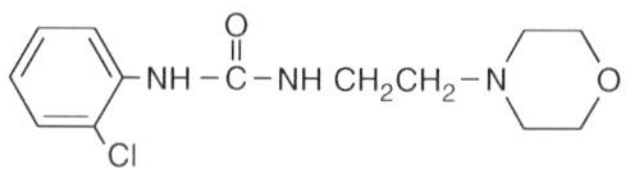
5 gm

S79,647-6
1-[4-(DIMETHYLAMINO)PHENYL]-3-(2-FLUORO-5-NITROPHENYL)UREA
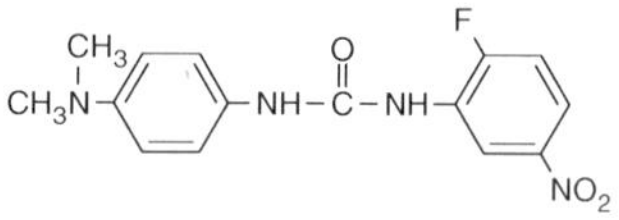
250 mg

S79,445-7
1-[4-(DIMETHYLAMINO)PHENYL]-5-(4-FLUORO-3-NITROPHENYL)BIURET
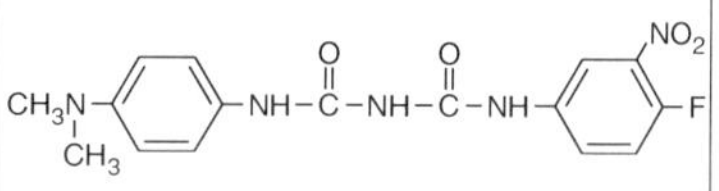
250 mg

S72,096-8
1,1'-(4,6-DIBROMO-1,3-PHENYLENE)-BIS(3-[4-(DIMETHYLAMINO)-PHENYL]UREA)
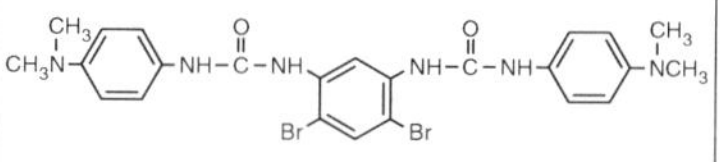
250 mg

S84,553-1
1-(2-FLUORO-5-NITROPHENYL)-3-[4-(N-METHYLACETAMIDO)PHENYL]UREA
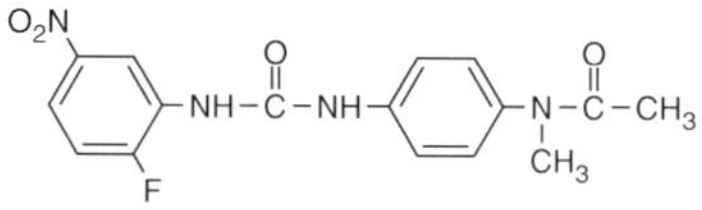
250 mg

S73,720-8
1-(2-CHLORO-5-NITROPHENYL)-3-(N-METHYLACETAMIDO)UREA
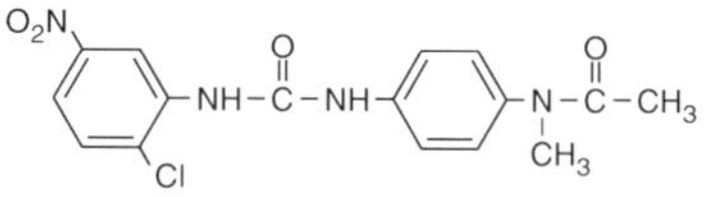
5 gm

S86,729-2
1-(2-CHLORO-4-NITROPHENYL)-3-(4-FLUORO-3-NITROPHENYL)UREA
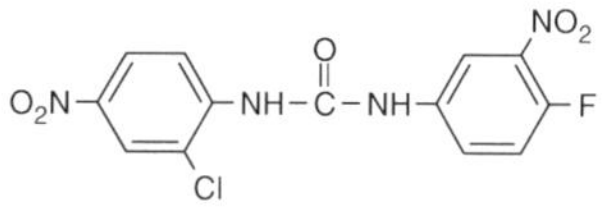
250 mg

S86,887-6
1-(1,2-DIPHENYLETHYL)-3-(4-FLUORO-3-NITROPHENYL)UREA
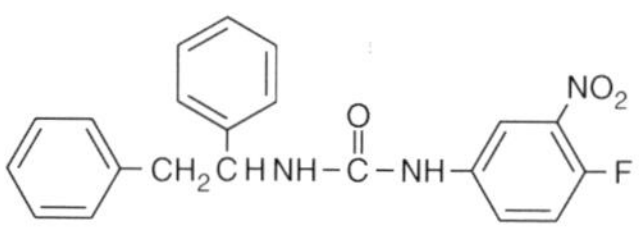
250 mg

S84,529-9
1-(2-FLUORO-5-NITROPHENYL)-3-(PHENETHYL)UREA
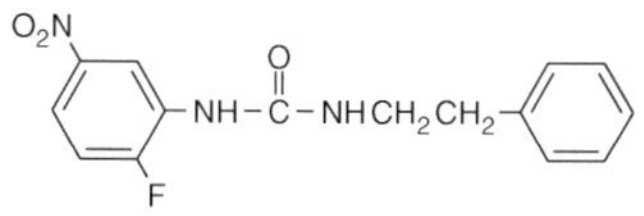
250 mg

S86,788-8
1-BENZYL-3-(4-FLUORO-3-NITROPHENYL)-1-METHYLUREA
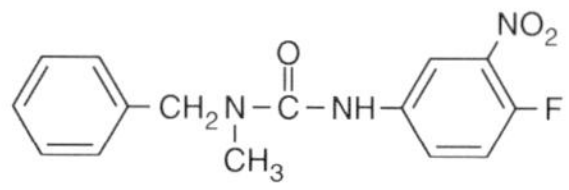
250 mg

S86,870-1
1-BENZYL-3-(4-FLUORO-3-NITROPHENYL)-1-PHENYLUREA
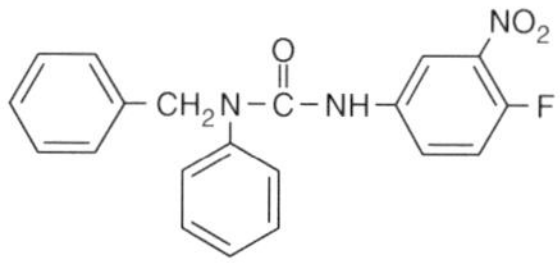
250 mg

S84,794-1
1-(4-CHLOROBENZYL)-3-(2-METHYL-5-NITROPHENYL)UREA
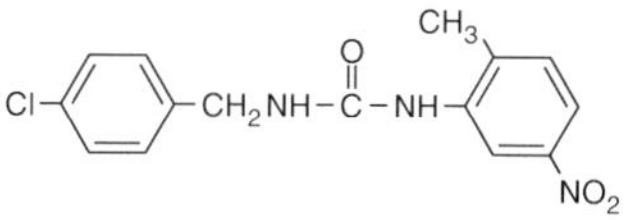
250 mg

S84,796-8
1-(4-CHLOROBENZYL)-1-METHYL-3-(2-METHYL-5-NITROPHENYL)UREA
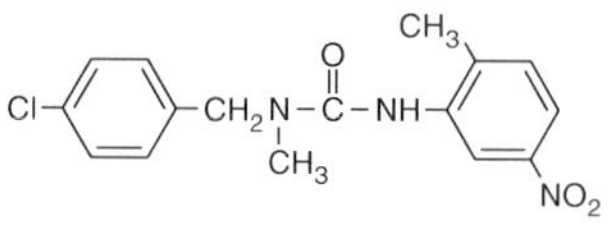
250 mg

S77,937-7
2-[3-(3-NITROPHENYL)UREIDO]ACETIC (2-CHLOROBENZYLIDENE)HYDRAZIDE
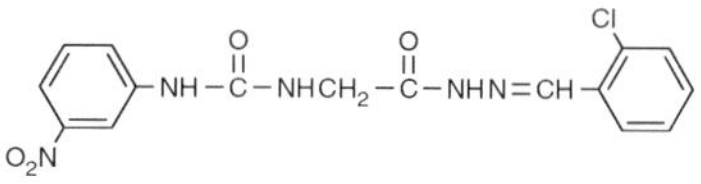
250 mg

S72,276-6
1-(2,4-DIFLUOROPHENYL)-3-[2-(2,4-DINITROPHENYLTHIO)PHENYL]UREA
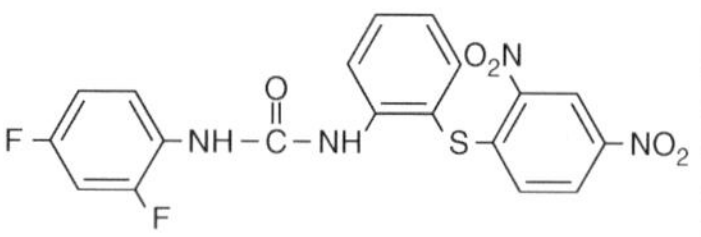
250 mg

S72,334-7
1-(4-ACETYLPHENYL)-3-(4-CHLORO-3-NITROPHENYL)UREA
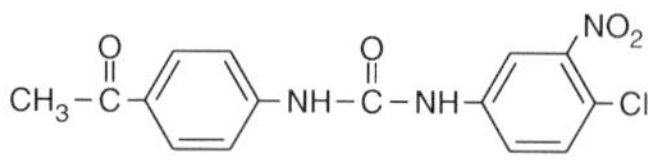
5 gm

S86,771-3
1-(3-ACETYLPHENYL)-3-(4-FLUORO-3-NITROPHENYL)UREA

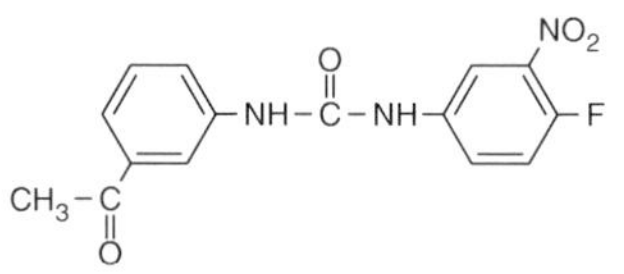

250 mg

S84,460-8
4-[3-(2-FLUORO-5-NITROPHENYL)-UREIDO]BENZOIC ACID

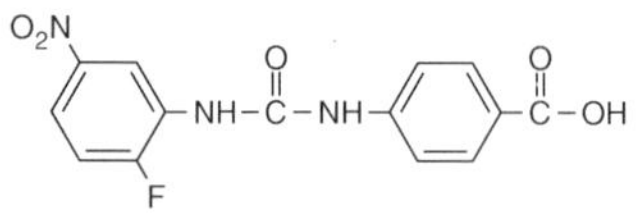

250 mg

S86,772-1
METHYL 2-[3-(4-FLUORO-3-NITROPHENYL)UREIDO]BENZOATE

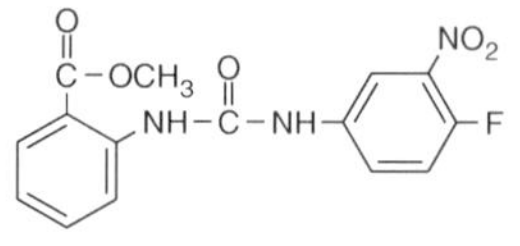

250 mg

S84,548-5
ETHYL 2-[3-(2-FLUORO-5-NITROPHENYL)-UREIDO]BENZOATE

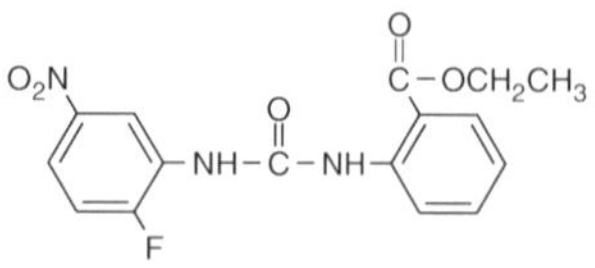

250 mg

S86,801-9
ETHYL 4-[3-(4-FLUORO-3-NITROPHENYL)-UREIDO]BENZOATE

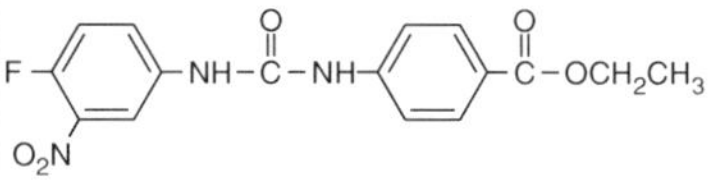

250 mg

S79,682-4
BUTYL 4-[3-(2-FLUORO-5-NITROPHENYL)-UREIDO]BENZOATE

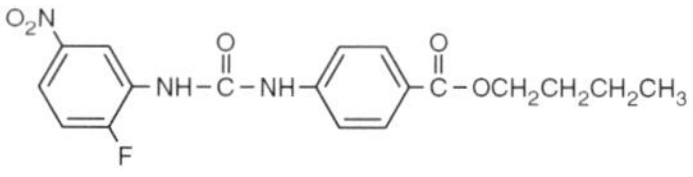

250 mg

S86,850-7
BUTYL 4-[3-(4-FLUORO-3-NITROPHENYL)-UREIDO]BENZOATE

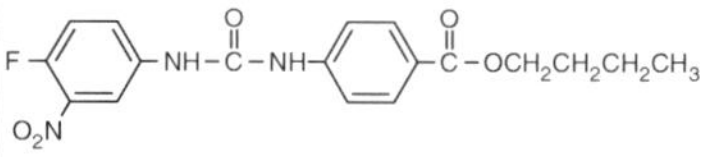

250 mg

S86,755-1
1-(2-CARBAMOYLPHENYL)-3-(4-FLUORO-3-NITROPHENYL)UREA

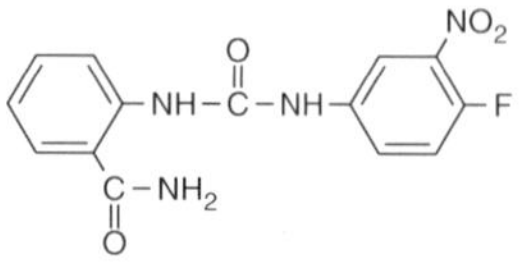

250 mg

S68,713-8
ETHYL 2-[3-(4-CHLORO-3-NITROPHENYL)UREIDO]ACETATE

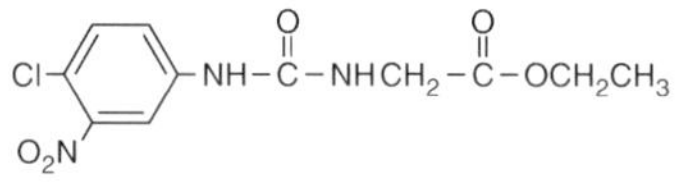

250 mg

S89,827-9
1-BENZYL-3-(2-FLUORO-5-NITROPHENYL)-1-PHENYLUREA

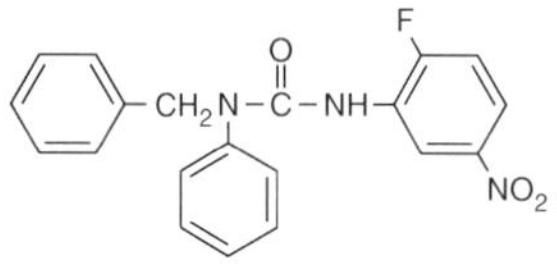

250 mg

S84,444-6
1-(3-FLUOROPHENYL)-3-(3-NITROPHENYL)UREA

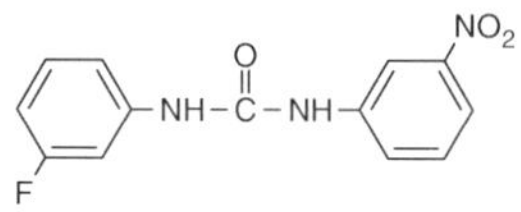

250 mg

S70,081-9
1-(4-CHLORO-3-NITROPHENYL)-3-(3-SULFAMOYLPHENYL)UREA

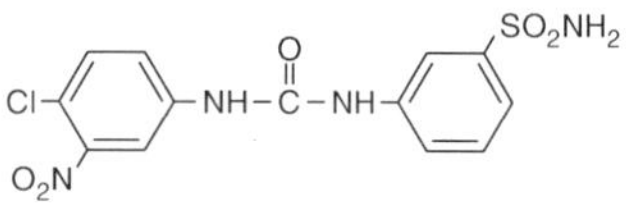

250 mg

S84,765-8
1,1-DICYCLOHEXYL-3-(2-FLUORO-5-NITROPHENYL)UREA

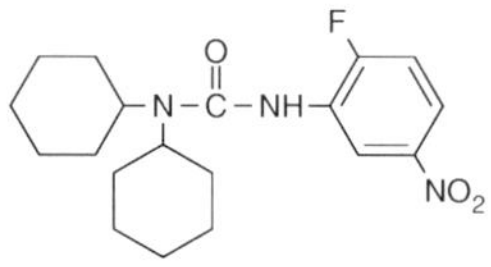

250 mg

S84,560-4
1-(2-FLUORO-5-NITROPHENYL)-3-ISOPROPYL-3-PHENYLUREA

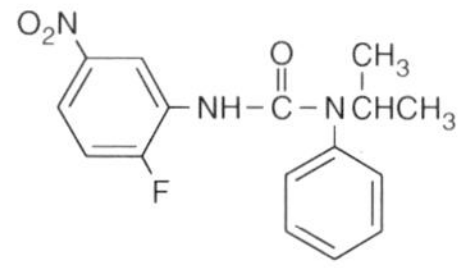

250 mg

S73,637-6
1-(2-FLUOROPHENYL)-3-(3-NITROPHENYL)UREA

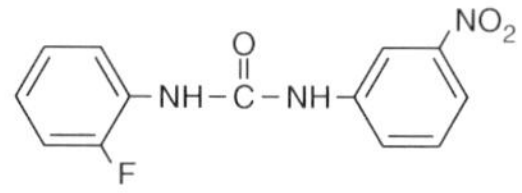

5 gm

S72,488-2
1-(2-CHLOROPHENYL)-3-(3-NITROPHENYL)UREA

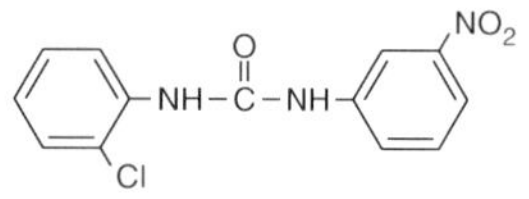

5 gm

S77,309-3
1-(4-FLUORO-3-NITROPHENYL)-3-(2-FLUOROPHENYL)UREA

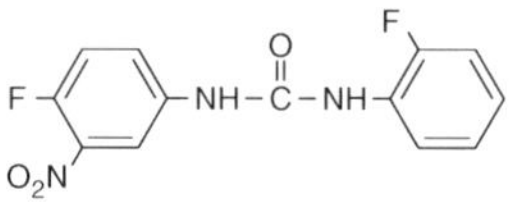

250 mg

S73,668-6
1-(2-FLUOROPHENYL)-3-(2-METHYL-5-NITROPHENYL)UREA

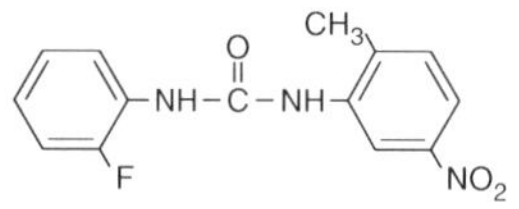

5 gm

S70,866-6
1-(4-CHLOROPHENYL)-3-(3-NITROPHENYL)UREA

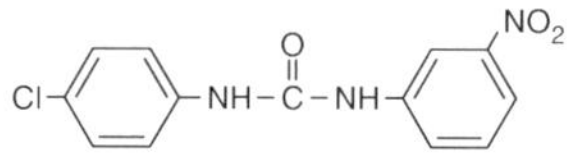

5 gm

S84,437-3
1-(4-FLUOROPHENYL)-3-(2-FLUORO-5-NITROPHENYL)UREA

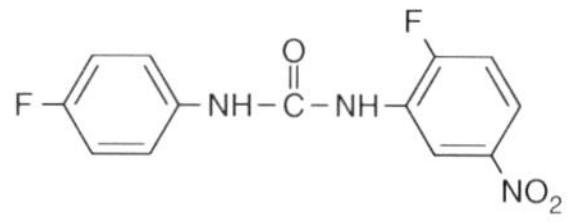

250 mg

S84,429-2
1-(4-BROMOPHENYL)-3-(2-FLUORO-5-NITROPHENYL)UREA

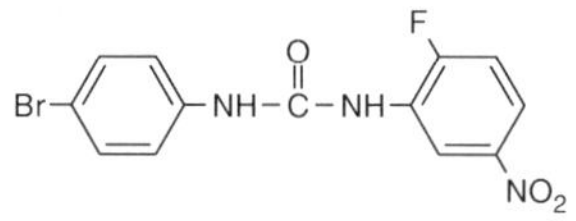

250 mg

S79,635-2
1-(4-CHLORO-3-NITROPHENYL)-3-(4-CHLOROPHENYL)-3-ETHYLUREA

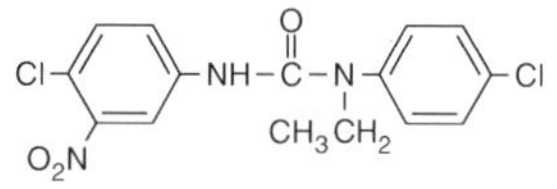

250 mg

S71,203-5
1-(2,6-DICHLOROPHENYL)-3-(3-NITROPHENYL)UREA

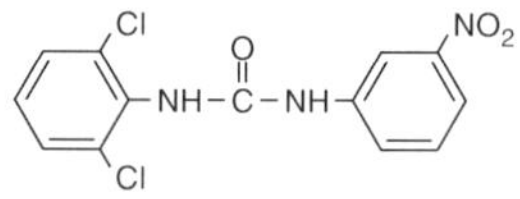

250 mg

S71,384-8
1-(2-CHLORO-6-METHYLPHENYL)-3-(3-NITROPHENYL)UREA

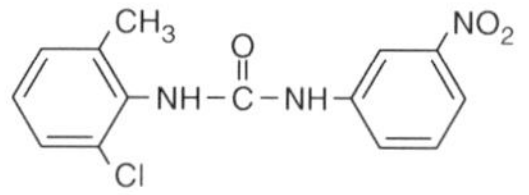

5 gm

S86,751-9
1-(3-CHLORO-2-METHYLPHENYL)-3-(4-FLUORO-3-NITROPHENYL)UREA

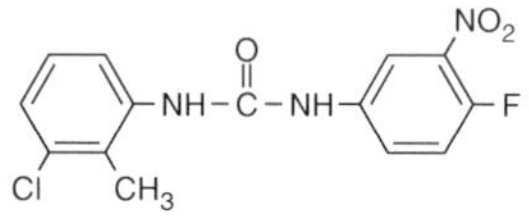

250 mg

S84,817-4
1-(2,3-DICHLOROPHENYL)-3-(2-METHYL-5-NITROPHENYL)UREA

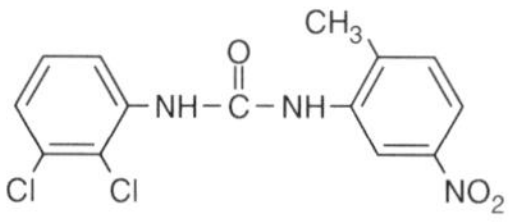

250 mg

S88,963-6
1-(3-CHLORO-2-METHYLPHENYL)-3-(4-METHYL-3-NITROPHENYL)UREA

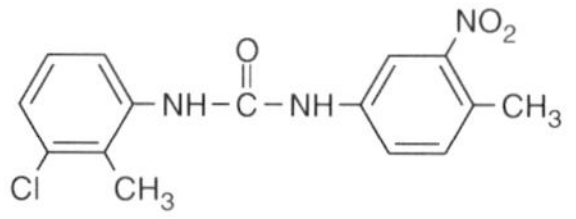

250 mg

S72,155-7
1-(2,5-DIFLUOROPHENYL)-3-(3-NITROPHENYL)UREA

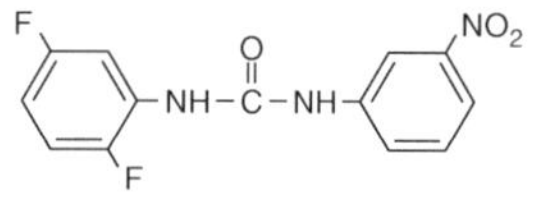

5 gm

S84,436-5
1-(3-FLUOROPHENYL)-3-(2-FLUORO-5-NITROPHENYL)UREA

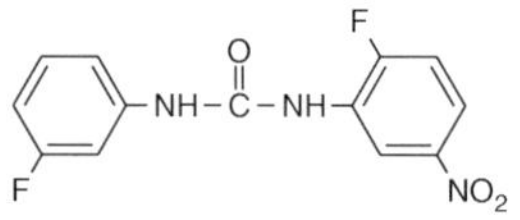

250 mg

S86,732-2
1-(3-FLUOROPHENYL)-3-(4-FLUORO-3-NITROPHENYL)UREA

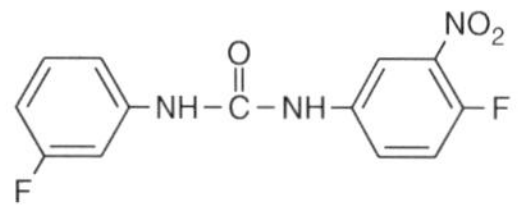

250 mg

S84,431-4
1-(2-CHLORO-5-NITROPHENYL)-3-(3-FLUOROPHENYL)UREA

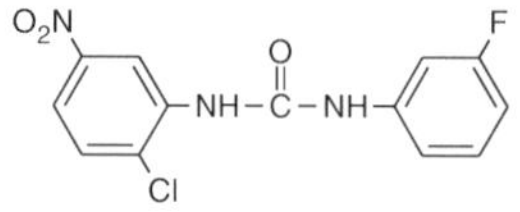

250 mg

S71,555-7
1-(2,5-DIFLUOROPHENYL)-3-(4-FLUORO-3-NITROPHENYL)UREA

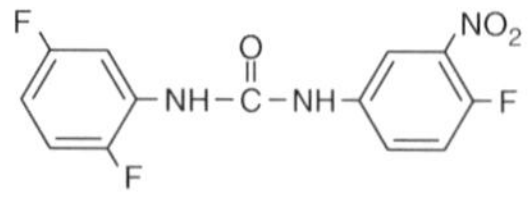

250 mg

S86,731-4
1-(4-FLUORO-3-NITROPHENYL)-3-(3-NITROPHENYL)UREA
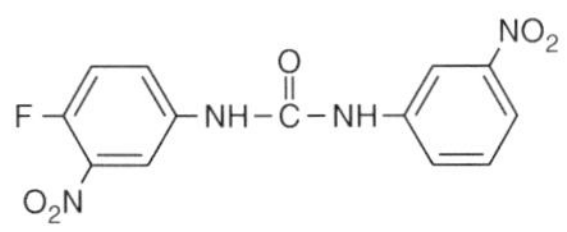
250 mg

S84,428-4
1-(2-FLUORO-5-NITROPHENYL)-3-(4-FLUORO-3-NITROPHENYL)UREA
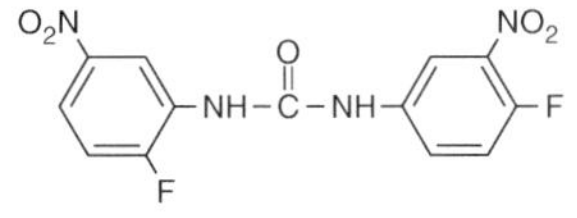
250 mg

S86,728-4
1-(2-CHLORO-5-NITROPHENYL)-3-(4-FLUORO-3-NITROPHENYL)UREA
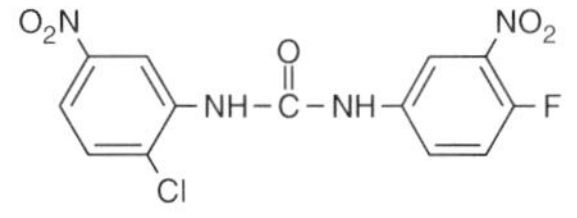
250 mg

S71,783-5
1-(3-FLUOROPHENYL)-3-(2-METHYL-5-NITROPHENYL)UREA
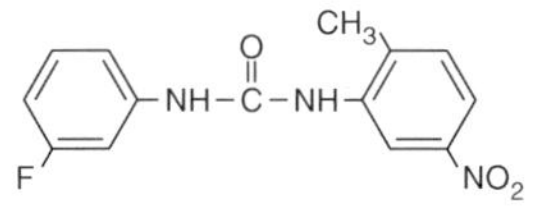
250 mg

S91,615-3
1-(2,5-DIFLUOROPHENYL)-3-(4-METHYL-3-NITROPHENYL)UREA
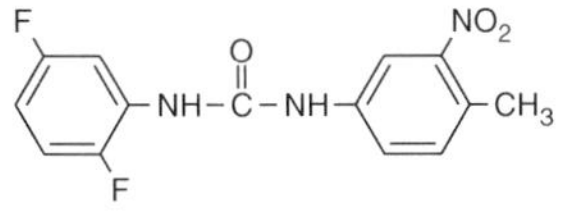
250 mg

S78,070-7
1-(5-CHLORO-2-METHYLPHENYL)-3-(2-FLUORO-5-NITROPHENYL)UREA
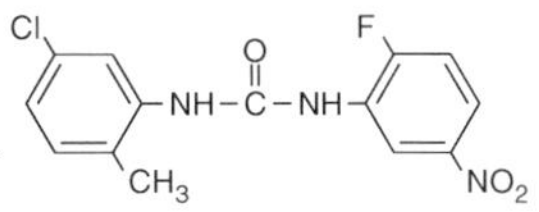
250 mg

S86,757-8
1-(4-FLUORO-3-NITROPHENYL)-3-(4-METHYL-3-NITROPHENYL)UREA
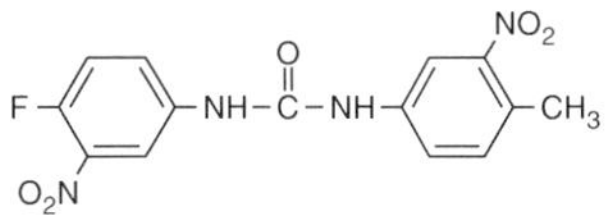
250 mg

S68,751-0
1-(2,5-DICHLOROPHENYL)-3-(2-METHYL-5-NITROPHENYL)UREA
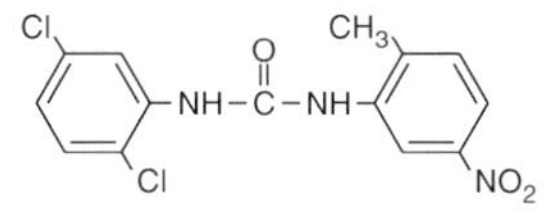
250 mg

S84,530-2
1-(2-FLUORO-5-NITROPHENYL)-3-(3,4-XYLYL)UREA
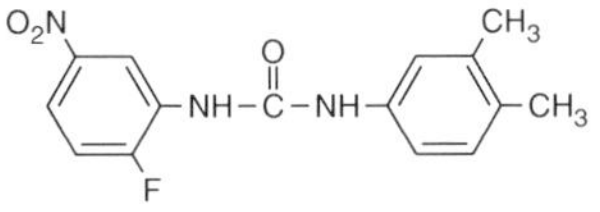
250 mg

S78,038-3
1-(2-CHLORO-5-NITROPHENYL)-3-(2,5-DIFLUOROPHENYL)UREA
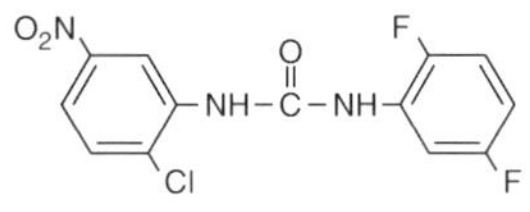
250 mg

S73,092-0
1-(2,5-DIFLUOROPHENYL)-3-(3,5-DINITROPHENYL)UREA
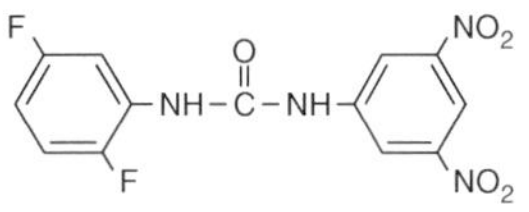
5 gm

S76,804-9
1-(3,5-DICHLOROPHENYL)-3-(3-NITROPHENYL)UREA
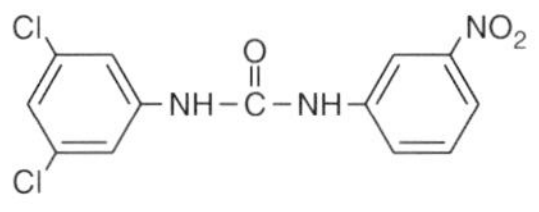
250 mg

S75,038-7
1,1'-(4,6-DIBROMO-1,3-PHENYLENE)-BIS[3-(3-NITROPHENYL)UREA]
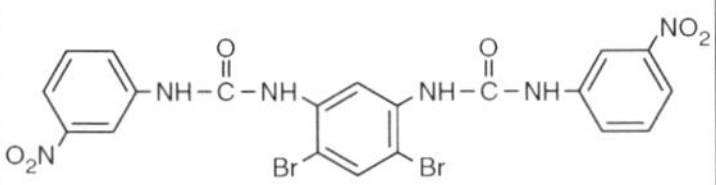
5 gm

S67,058-8
1-CYCLOHEXYL-1-METHYL-3-(4-METHYL-3-NITROPHENYL)UREA
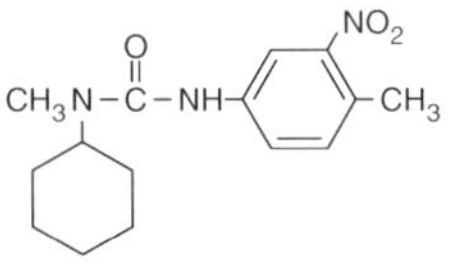
5 gm

S66,007-8
1,1'-(4-METHYL-1,3-PHENYLENE)BIS[3-(2-BENZOYL-4-CHLOROPHENYL)UREA]
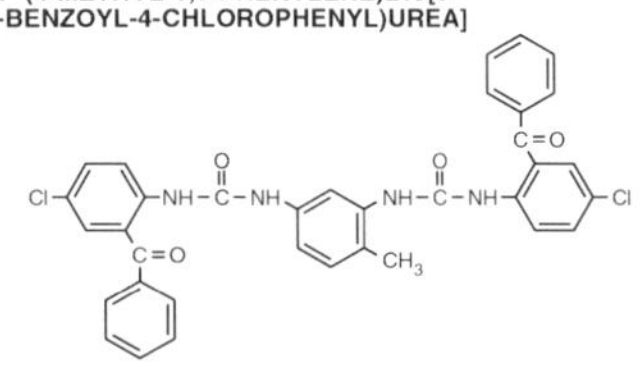
5 gm

S72,079-8
1,1'-(4,6-DIBROMO-1,3-PHENYLENE)-BIS[3-(2-CARBAMOYLPHENYL)UREA]
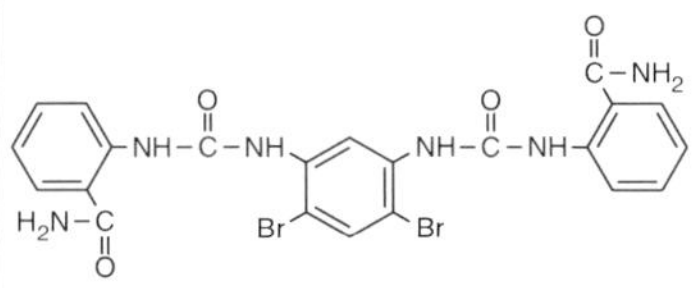
250 mg

S89,125-8
1-(3-ACETAMIDOPHENYL)-3-(2-CHLOROPHENYL)UREA
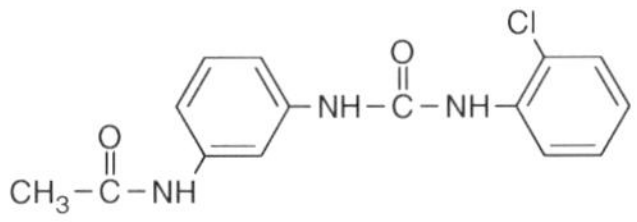
250 mg

S69,523-8
1,1'-(4,6-DIBROMO-1,3-PHENYLENE)-BIS[3-(3-BROMOPHENYL)UREA]
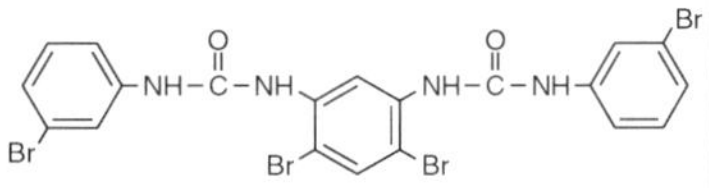
250 mg

S72,395-9
1-(4-CHLORO-2-CYANOPHENYL)-3-(2,5-DIFLUOROPHENYL)UREA
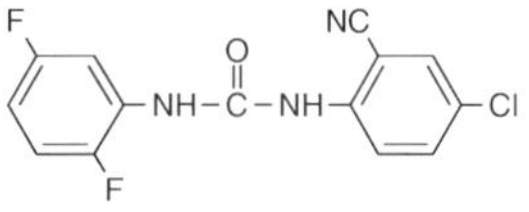
5 gm

S71,295-7
1-(2-CHLORO-6-METHYLPHENYL)-3-[3-(2-METHYLPIPERIDINO)PROPYL]UREA
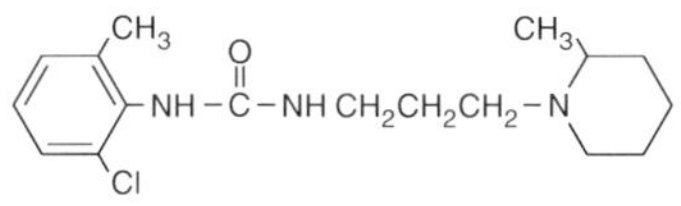
5 gm

S70,925-5
1-(2-CHLORO-6-METHYLPHENYL)-3-(3-MORPHOLINOPROPYL)UREA
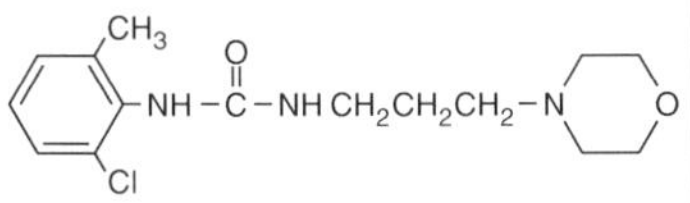
250 mg

S69,583-1
1,1'-(4,6-DIBROMO-1,3-PHENYLENE)-BIS[3-(4-IODOPHENYL)UREA]
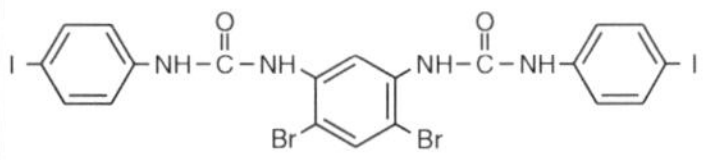
250 mg

S79,166-0
1,1'-(4-METHYL-1,3-PHENYLENE)BIS[3-(4-FLUOROPHENYL)UREA]
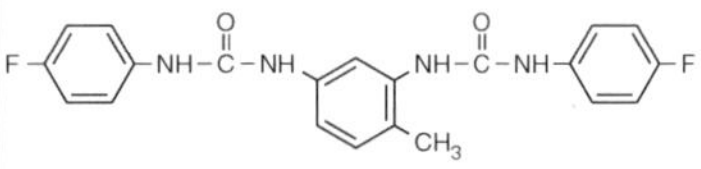
250 mg

S77,980-6
1,1'-(4-METHYL-1,3-PHENYLENE)BIS[3-(3-CHLORO-2-METHYLPHENYL)UREA]
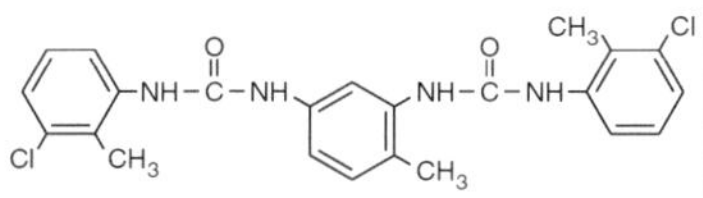
250 mg

S78,058-8
1,1'-(4-METHYL-1,3-PHENYLENE)BIS[3-(3-CHLORO-4-METHYLPHENYL)UREA]
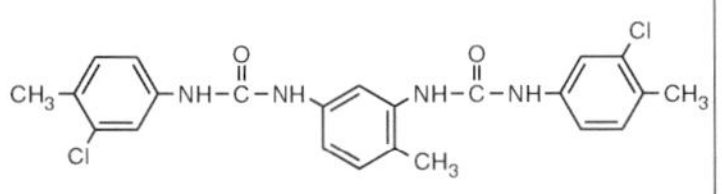
250 mg

S89,092-8
1,1'-(4-METHYL-1,3-PHENYLENE)BIS[3-(5-CHLORO-2-METHYLPHENYL)UREA]
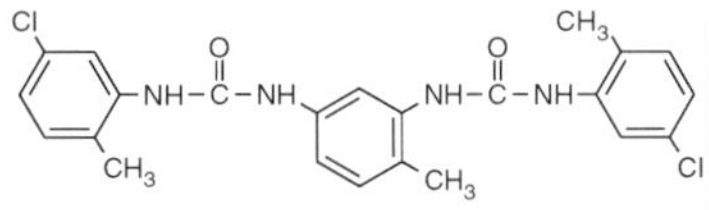
250 mg

S69,959-4
1,1'-(4,6-DIBROMO-1,3-PHENYLENE)-BIS(3-DODECYLUREA)
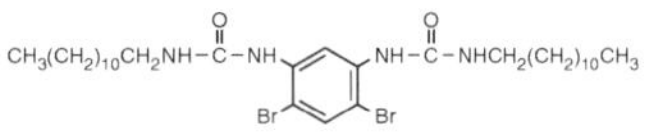
250 mg

S86,872-8
1-(4-ANILINOPHENYL)-3-(3-CHLORO-2-METHYLPHENYL)UREA
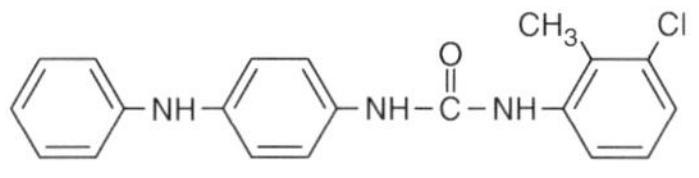
250 mg

S73,680-5
N-(4-[3-(4-FLUORO-3-NITROPHENYL)UREIDO]-PHENYL)GLYCINE
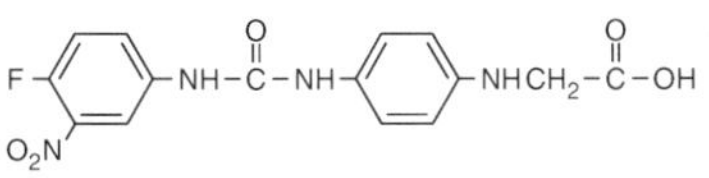
5 gm

S78,101-0
1-(4-ANILINOPHENYL)-3-(3-FLUOROPHENYL)UREA
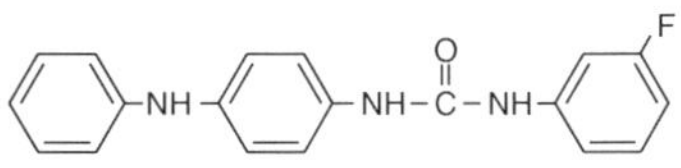
250 mg

S80,075-9
1-(4-ANILINOPHENYL)-3-(4-FLUOROPHENYL)UREA
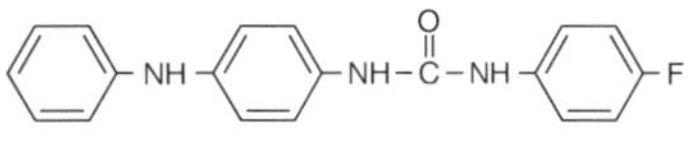
250 mg

S73,908-1
3-(3,4-DICHLOROPHENYL)-1-(4-ANILINOPHENYL)UREA
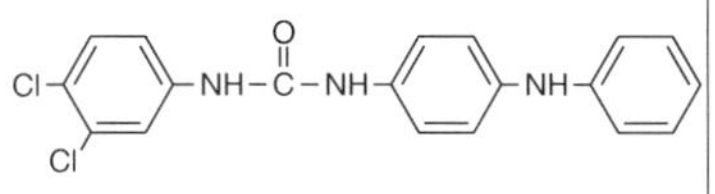
5 gm

S70,207-2
1-(4-ANILINOPHENYL)-3-(3-CHLOROPHENYL)UREA
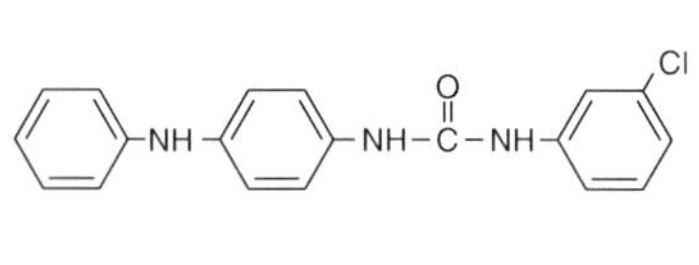
250 mg

S91,630-7
1-[4-(DIETHYLAMINO)PHENYL]-3-(2,5-DIFLUOROPHENYL)UREA
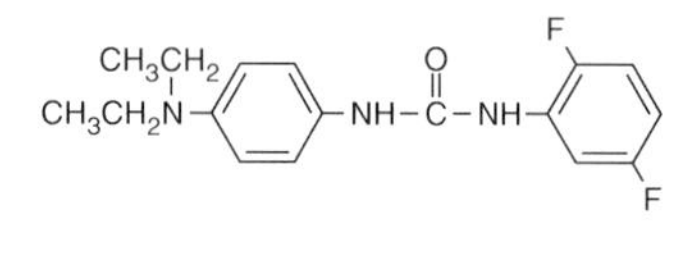
250 mg

S79,650-6
1-[4-(DIMETHYLAMINO)PHENYL]-3-(4-FLUOROPHENYL)UREA
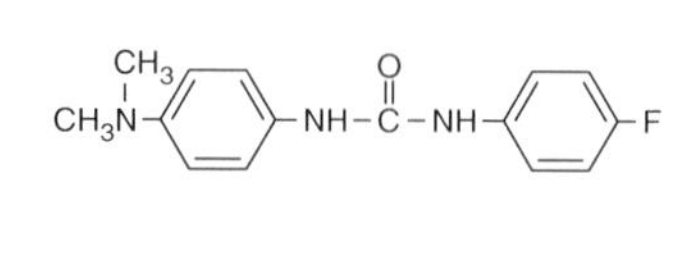
250 mg

S86,812-4
1-(3-CHLORO-2-METHYLPHENYL)-3-[4-(DIMETHYLAMINO)PHENYL]UREA
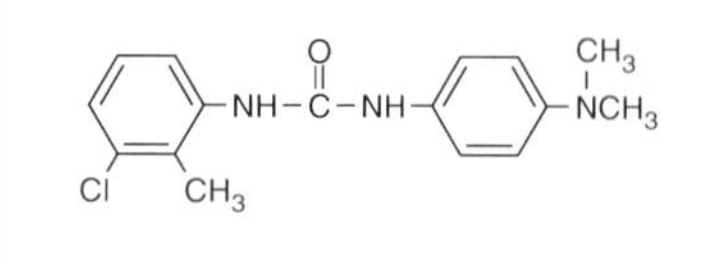
250 mg

S74,298-8
1-(2-CHLOROBENZOYL)-4-(2-CHLORO-4-NITROPHENYL)SEMICARBAZIDE
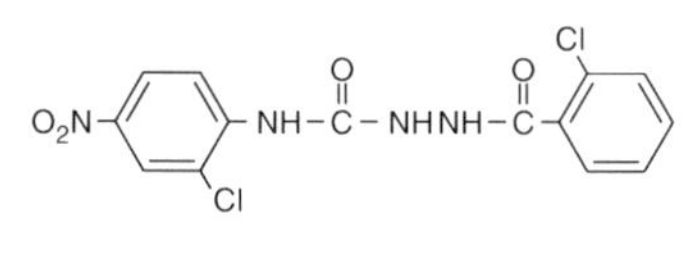
5 gm

S84,468-3
1-(2-CARBAMOYLPHENYL)-3-(2-CHLORO-4-NITROPHENYL)UREA
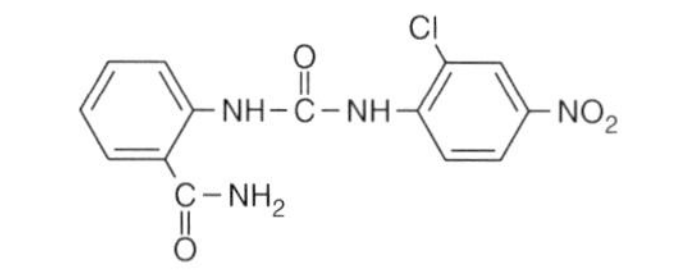
250 mg

S91,647-1
OXAMIC N2-[N-(2-CHLORO-4-NITROPHENYL)CARBAMOYL]HYDRAZIDE
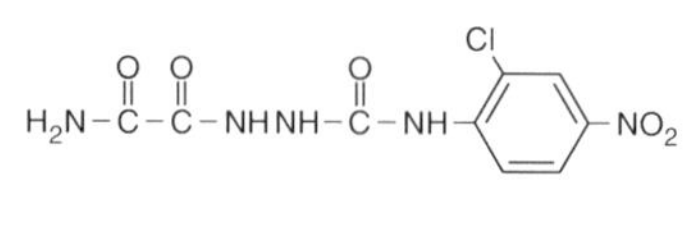
250 mg

S71,538-7
2-[1-METHYL-3-(2-METHYL-4-NITROPHENYL)UREIDO]ETHYL N-(3,4-DICHLOROPHENYL)CARBAMATE
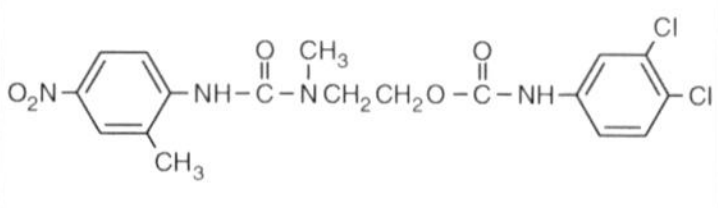
5 gm

S90,427-9
4-(4-FLUOROPHENYL)-1-(4-NITROPHENYL)SEMICARBAZIDE
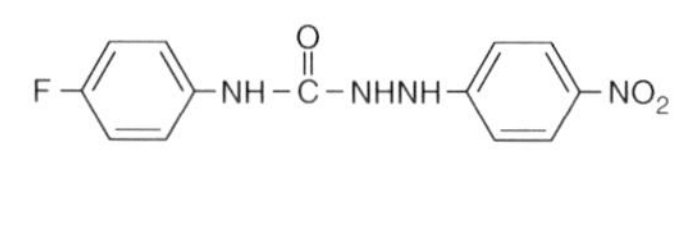
250 mg

S76,868-5
1-(2-CHLORO-4-NITROPHENYL)-3-(1-PIPERIDINO)UREA
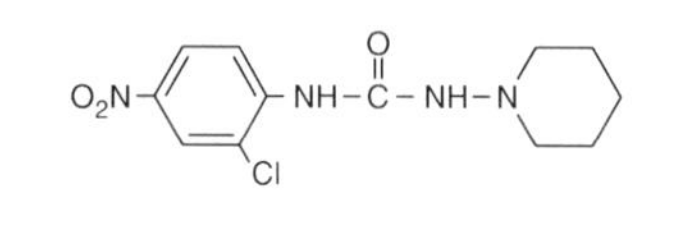
250 mg

S70,043-6
1-(2-CHLORO-4-NITROPHENYL)-3-(DIPHENYLMETHYL)UREA
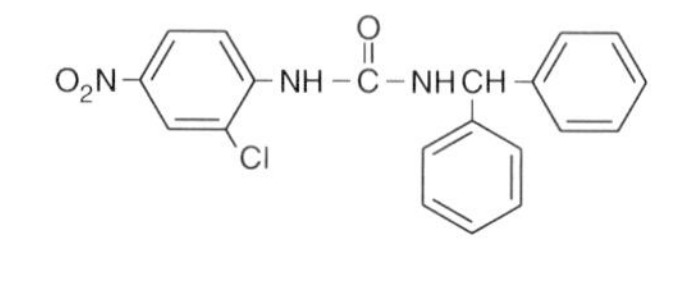
250 mg

S71,218-3
1-(4-CHLOROPHENYL)-3-(4-NITROPHENYL)UREA
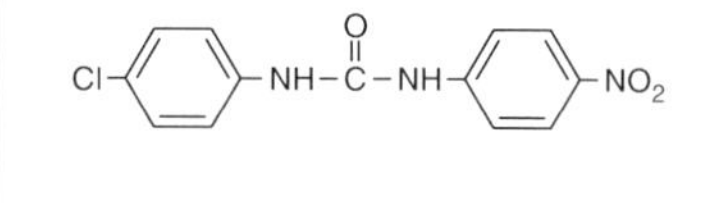
5 gm

S70,915-8
1-(2-CHLORO-6-METHYLPHENYL)-3-(4-NITROPHENYL)UREA
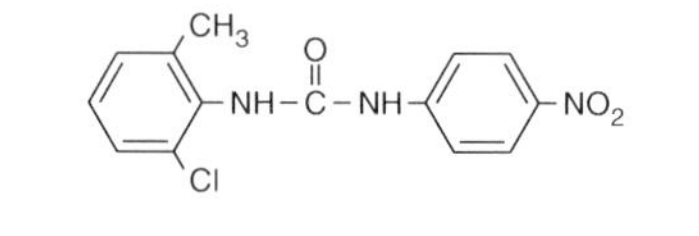
5 gm

S70,861-5
1-(4-CHLOROPHENYL)-3-(2-CHLORO-4-NITROPHENYL)UREA
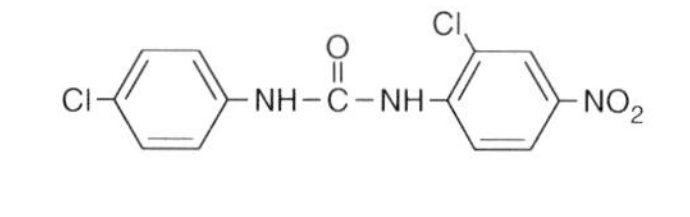
5 gm

S89,128-2
1-(2-BROMO-4-NITROPHENYL)-3-(4-CHLOROPHENYL)UREA
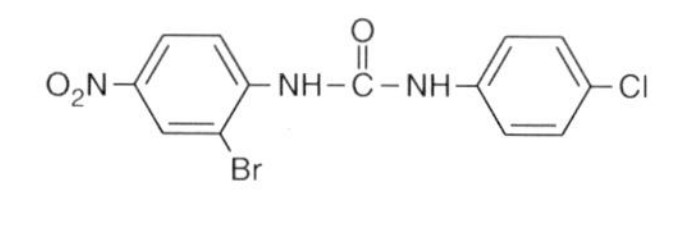
250 mg

S71,252-3
1-(4-CHLOROPHENYL)-3-(2-METHYL-4-NITROPHENYL)UREA
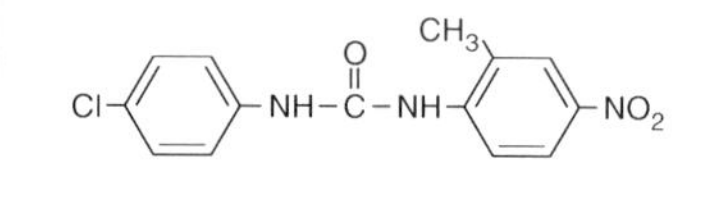
5 gm

S71,238-8
1-(2,6-DICHLOROPHENYL)-3-(2-METHYL-4-NITROPHENYL)UREA
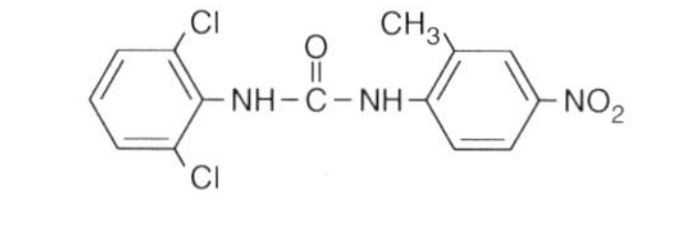
5 gm

S86,782-9
1-(3-CHLORO-2-METHYLPHENYL)-3-(2-METHYL-4-NITROPHENYL)UREA
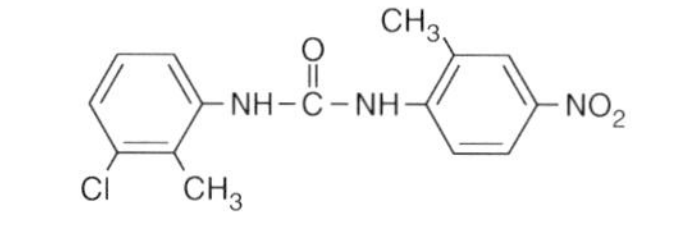
250 mg

S84,430-6
1-(2-CHLORO-4-NITROPHENYL)-3-(3-FLUOROPHENYL)UREA
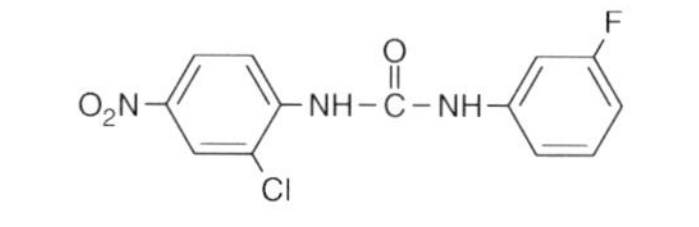
250 mg

S84,482-9
1-(3-FLUOROPHENYL)-3-(2-METHYL-4-NITROPHENYL)UREA
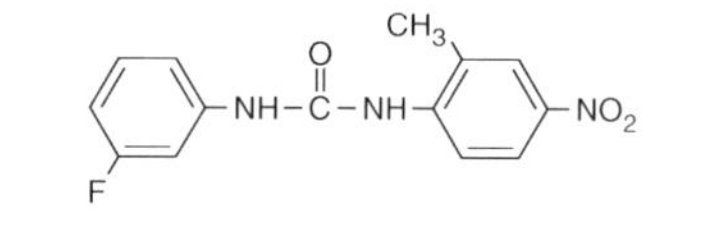
250 mg

S90,990-4
3-(2-CHLORO-4-NITROPHENYL)-1-ETHYL-1-(3-FLUOROPHENYL)UREA
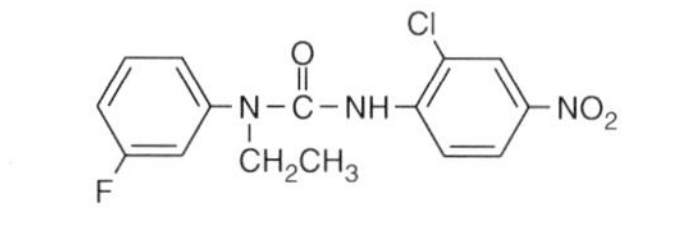
250 mg

S90,992-0
1-(3-CHLORO-4-METHYLPHENYL)-3-(2-CHLORO-4-NITROPHENYL)-1-ETHYLUREA
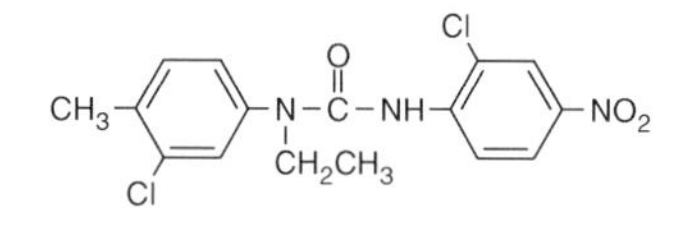
250 mg

S85,173-6
1-(2-CHLORO-4-NITROPHENYL)-3-(2,5-DIMETHYL-4-NITROPHENYL)UREA
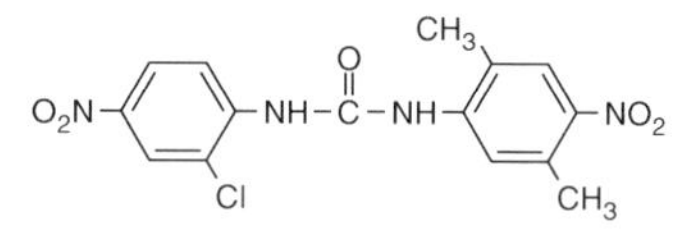
250 mg

S73,687-2
1-(4-ACETAMIDOPHENYL)-3-(2-FLUOROPHENYL)UREA
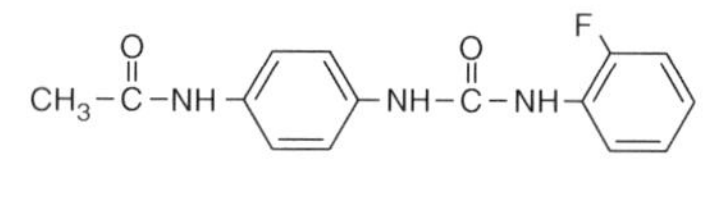
5 gm

S73,686-4
1-(4-ACETAMIDOPHENYL)-3-(3-FLUOROPHENYL)UREA
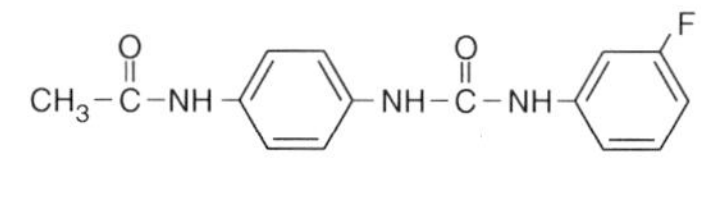
5 gm

S80,067-8
1-(4-ACETAMIDOPHENYL)-3-(4-FLUOROPHENYL)UREA
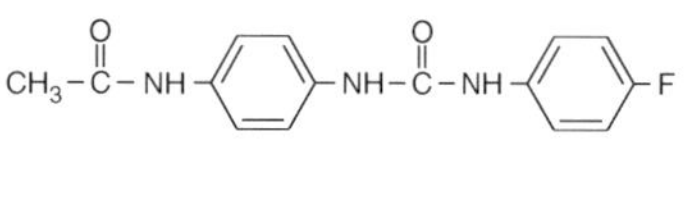
250 mg

S85,521-9
1-(4-ACETAMIDOPHENYL)-3-(4-CHLOROPHENYL)UREA
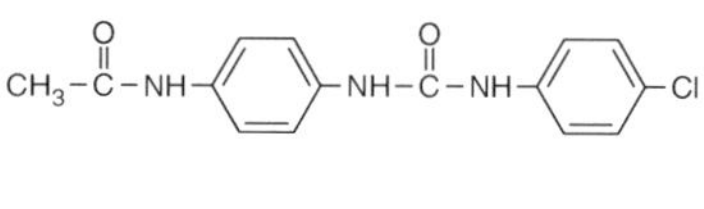
250 mg

S71,751-7
1-(4-ACETAMIDOPHENYL)-3-(2,4,5-TRICHLOROPHENYL)UREA
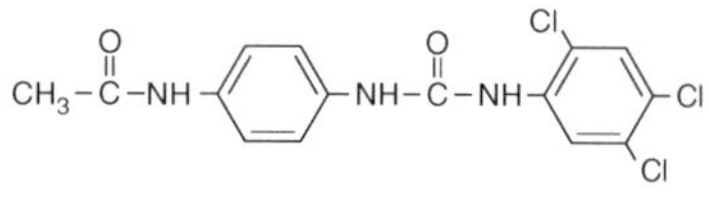
5 gm

S71,753-3
1-(4-ACETAMIDOPHENYL)-3-(2,6-DICHLOROPHENYL)UREA
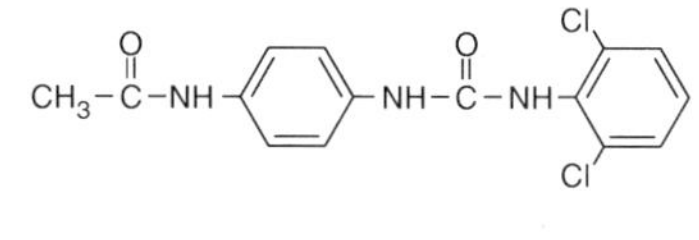
5 gm

S67,670-5
1-(4-ACETAMIDOPHENYL)-3-(3-CHLORO-2-METHYLPHENYL)UREA
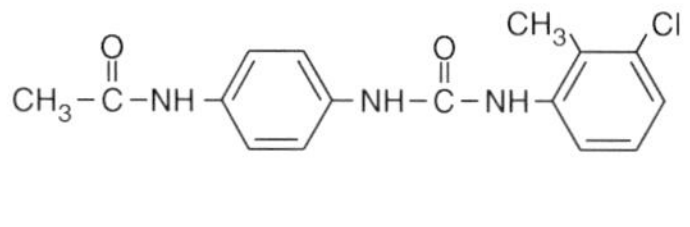
250 mg

S90,580-1
1-(4-ACETAMIDOPHENYL)-3-(3-CHLORO-4-FLUOROPHENYL)UREA
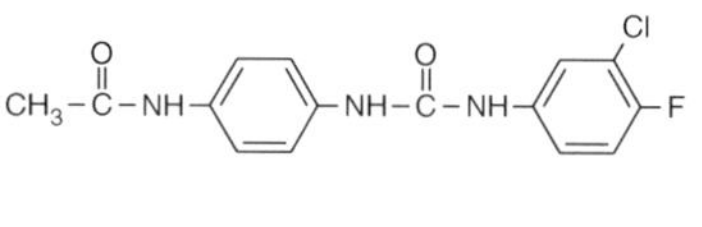
250 mg

S67,499-0
1-(4-ACETAMIDOPHENYL)-3-(5-CHLORO-2-METHYLPHENYL)UREA
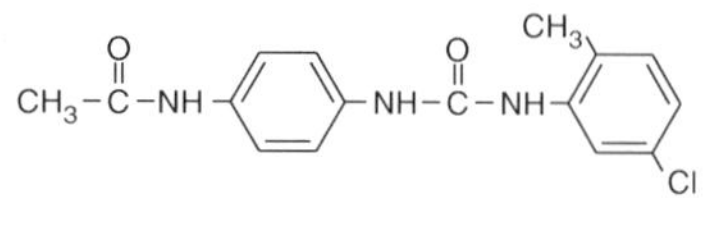
250 mg

S73,705-4
1-(2-FLUOROPHENYL)-3-[4-(N-METHYLACETAMIDO)PHENYL]UREA

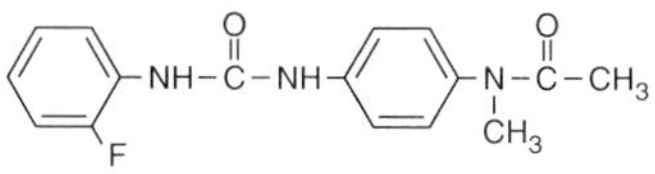

5 gm

S80,071-6
1-(4-FLUOROPHENYL)-3-[4-(N-METHYLACETAMIDO)PHENYL]UREA

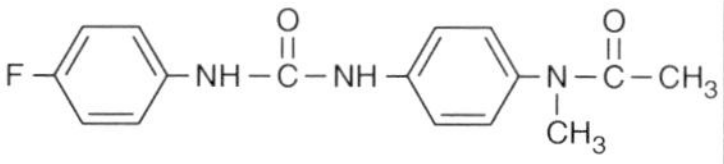

250 mg

S73,199-4
1-(4-ANILINOPHENYL)-3-(2,5-DIFLUOROPHENYL)UREA

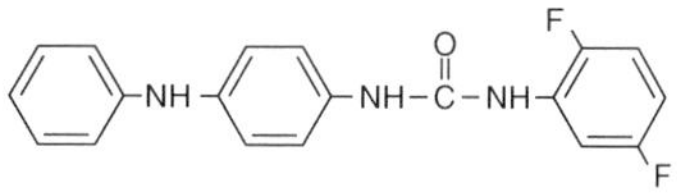

5 gm

S86,810-8
1-(3-CHLORO-2-METHYLPHENYL)-3-PHENETHYLUREA

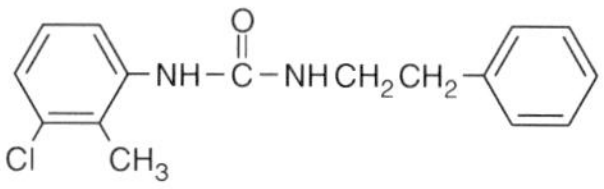

250 mg

S71,069-5
1-PHENETHYL-3-(2,4,5-TRICHLOROPHENYL)UREA

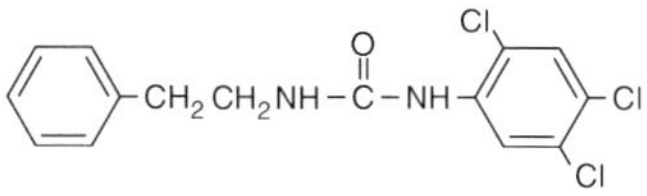

5 gm

S72,660-5
4-(3-CHLOROPHENYL)-1-[2-(METHYLAMINO)BENZOYL]SEMICARBAZIDE

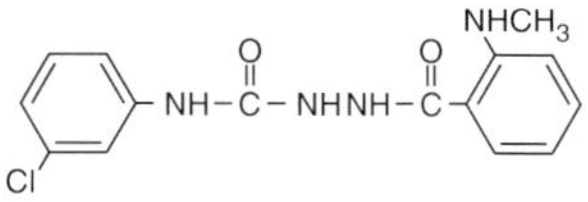

5 gm

S69,372-3
4-(4-FLUOROPHENYL)-1-[2-(METHYLAMINO)BENZOYL]SEMICARBAZIDE

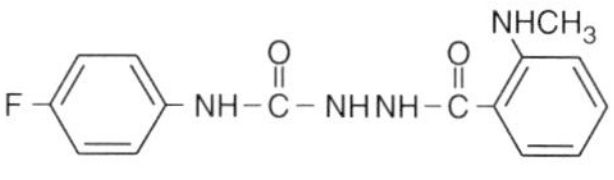

250 mg

S73,766-6
4-(4-CHLORO-2-METHYLPHENYL)-1-[2-(METHYLAMINO)BENZOYL]SEMICARBAZIDE

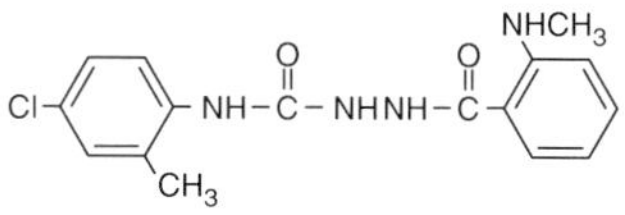

5 gm

S72,313-4
4-(3,4-DICHLOROPHENYL)-1-[2-(METHYLAMINO)BENZOYL]SEMICARBAZIDE

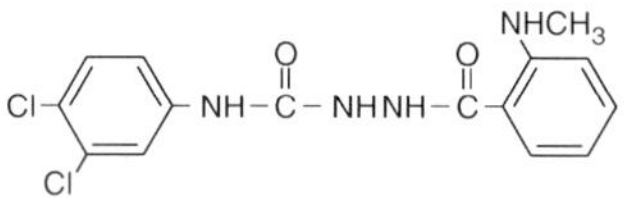

250 mg

S68,819-3
4-(2,4-DIFLUOROPHENYL)-1-(2-METHYLAMINOBENZOYL)SEMICARBAZIDE

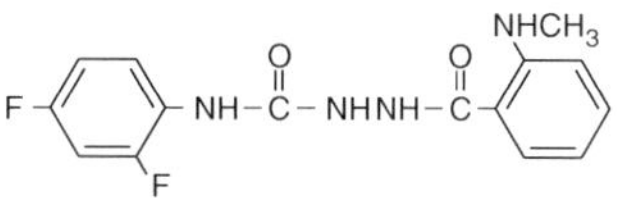

250 mg

S88,039-6
1-(2-BROMOPHENYL)-3-PHENYLUREA

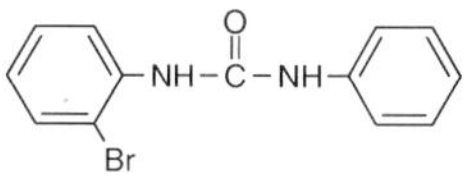

250 mg

S73,600-7
1-(4-ANILINOPHENYL)-3-(2-FLUOROPHENYL)UREA

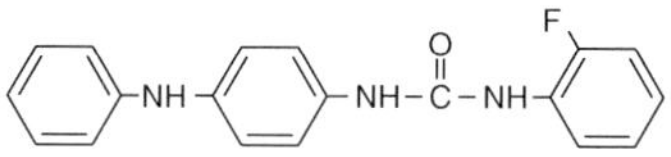

5 gm

S86,903-1
4-[3(3-CHLOROPHENYL)UREIDO]BENZOIC [2-CHLORO-4-(DIMETHYLAMINO)-BENZYLIDENE]HYDRAZIDE

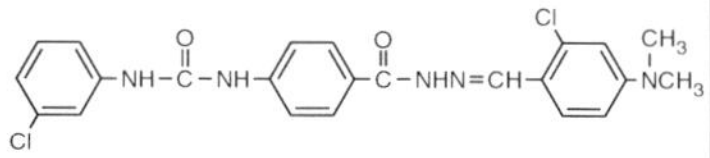

250 mg

S69,158-5
1-[4-(DIMETHYLAMINO)BENZOYL]-4-(4-FLUOROPHENYL)SEMICARBAZIDE

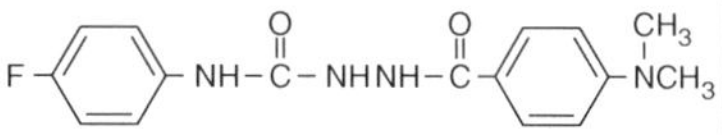

250 mg

S71,782-7
1-[4-(DIMETHYLAMINO)BENZOYL]-4-(2,6-DICHLOROPHENYL)SEMICARBAZIDE

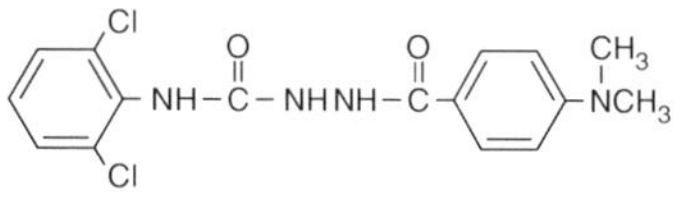

5 gm

S70,967-0
3-(3,5-DICHLOROPHENYL)-1,1-DIISOPROPYLUREA

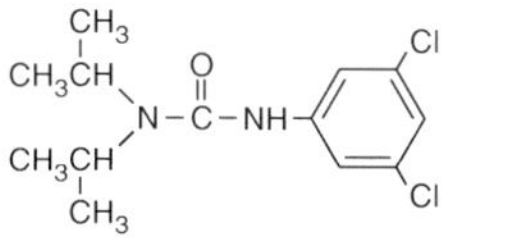

250 mg

S70,963-8
1,1-DIBUTYL-3-(3,5-DICHLOROPHENYL)UREA

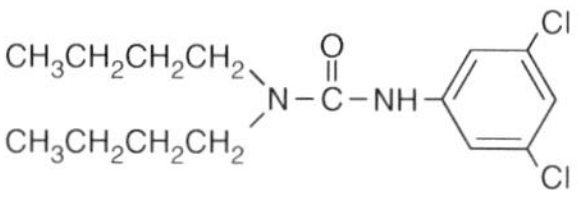

250 mg

S70,932-8
1,1-DIMETHYL-3-(2,4,6-TRICHLOROPHENYL)UREA

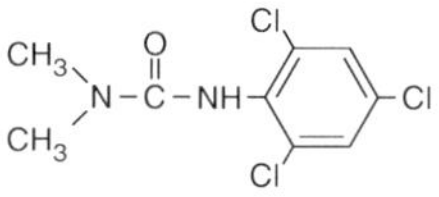

250 mg

S67,635-7
N-(4-CHLORO-3-NITROPHENYL)-N'-3-(3,4-DICHLOROPHENYLUREIDO)OXAMIDE

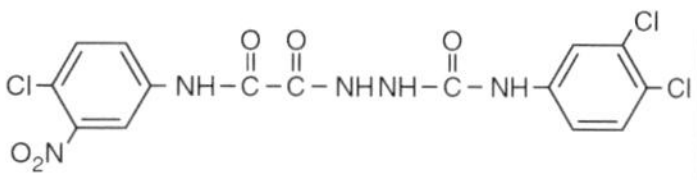

250 mg

S76,950-9
1-TERT-BUTYL-3-(3-CHLOROBENZYL)UREA

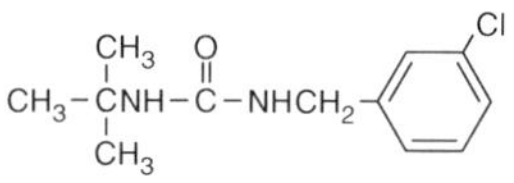

250 mg

S74,332-1
1-(3-CHLOROBENZYL)-3-(3-CHLOROPHENYL)UREA

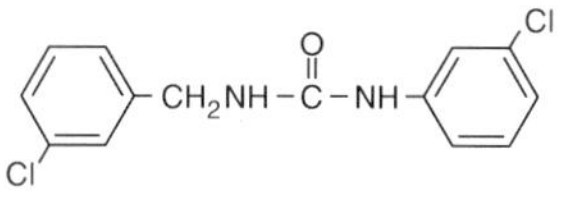

5 gm

S74,702-5
1-(3-CHLOROBENZYL)-3-CYCLOHEXYLUREA

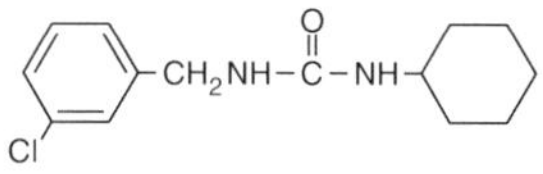

5 gm

S84,492-6
1-BENZYL-3-(3-FLUOROPHENYL)UREA

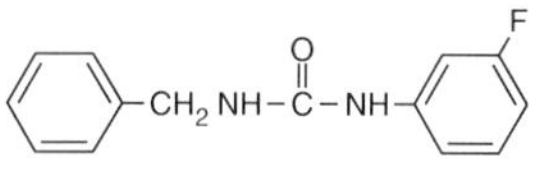

250 mg

S67,459-1
1-(3-CHLOROBENZYL)-3-PHENYLUREA

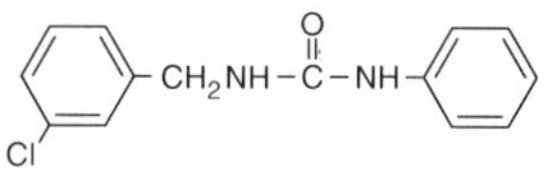

250 mg

S71,051-2
1-BENZYL-3-(3-BROMOPHENYL)UREA

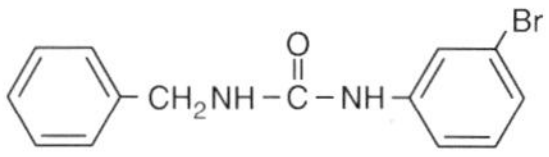

5 gm

S80,190-9
1-BENZYL-3-(3-FLUOROPHENYL)-1-PHENYLUREA

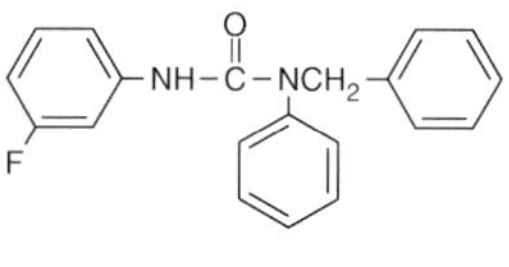

250 mg

S71,790-8
1,1-DIBENZYL-3-(3-FLUOROPHENYL)UREA

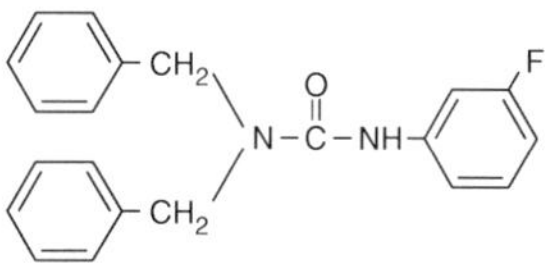

250 mg

S72,981-7
1-BENZYL-3-(2-FLUOROPHENYL)-1-METHYLUREA

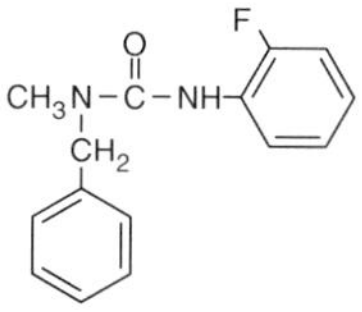

5 gm

S77,374-3
1-BENZYL-3-(2-CHLOROPHENYL)-1-PHENYLUREA

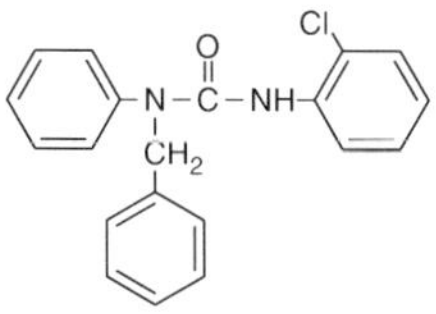

250 mg

S72,779-2
1,1-DIBENZYL-3-(2-FLUOROPHENYL)UREA

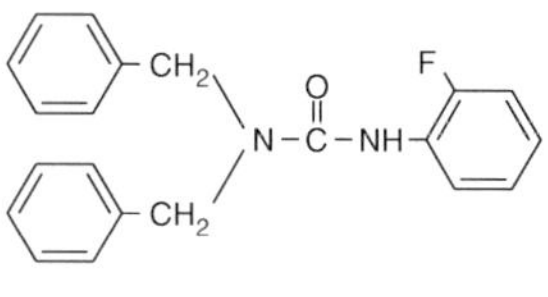

5 gm

S80,054-6
1-BENZYL-3-(4-FLUOROPHENYL)-1-PHENYLUREA

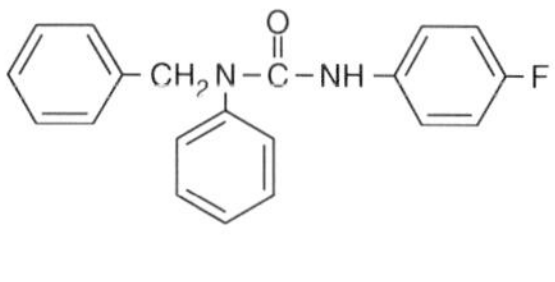

250 mg

S70,808-9
1-BENZYL-3-(2,6-DICHLOROPHENYL)-1-METHYLUREA

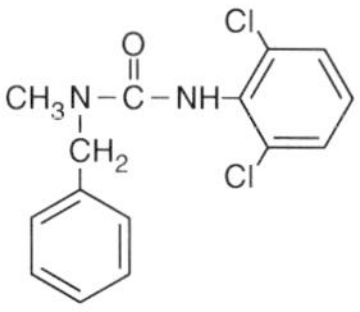

5 gm

S71,502-6
1-BENZYL-3-(2,6-DICHLOROPHENYL)-1-ISOPROPYLUREA
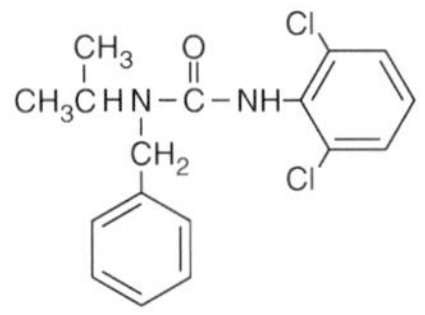
250 mg

S66,577-0
1-BENZYL-3-(2,4,5-TRICHLOROPHENYL)UREA
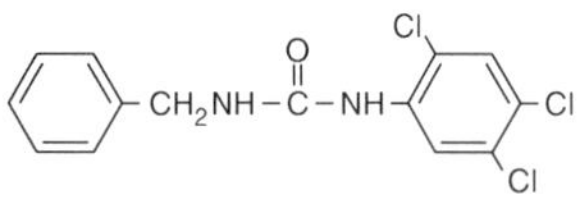
5 gm

S66,552-5
1,1-DIBENZYL-3-(2,4,5-TRICHLOROPHENYL)UREA
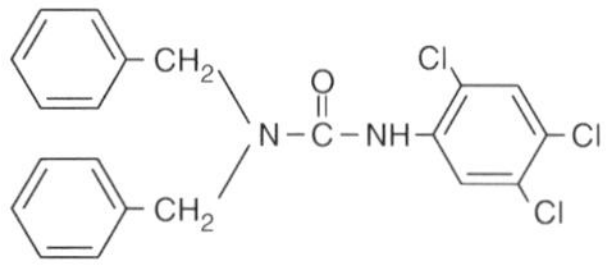
5 gm

S75,032-8
3-TERT-BUTYL-1-(2-CHLOROBENZYL)-1-METHYLUREA
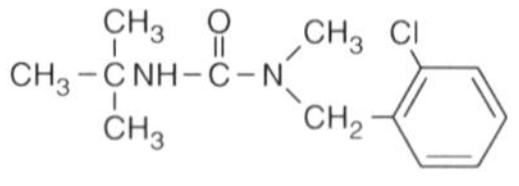
5 gm

S84,475-6
1-(2-CHLOROBENZYL)-3-(3-FLUOROPHENYL)UREA
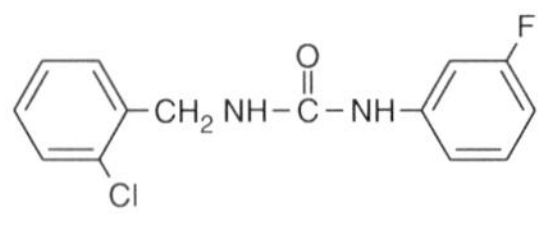
250 mg

S80,062-7
1-(2-CHLOROBENZYL)-3-(4-FLUOROPHENYL)UREA
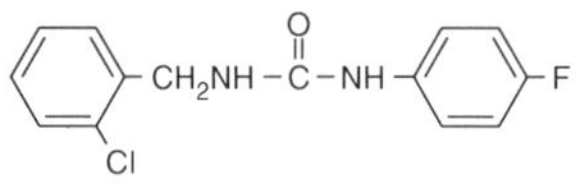
250 mg

S84,563-9
1-(2-CHLOROBENZYL)-3-(2,5-XYLYL)UREA
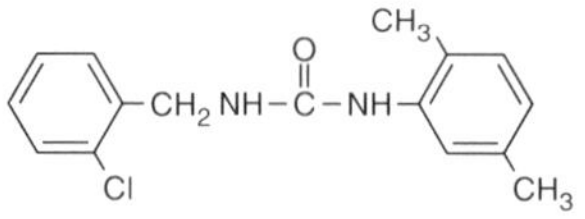
250 mg

S92,419-9
1-(2-CHLOROPHENYL)-3-(2,4-DIMETHYLBENZYL)UREA
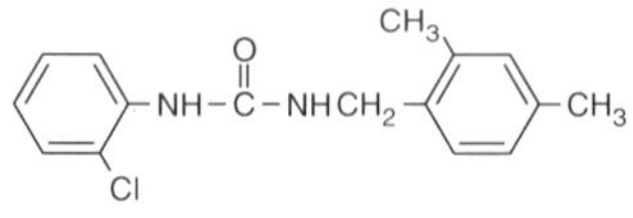
250 mg

S79,561-5
1-(4-CHLOROBENZYL)-3-ETHYL-1-METHYLUREA
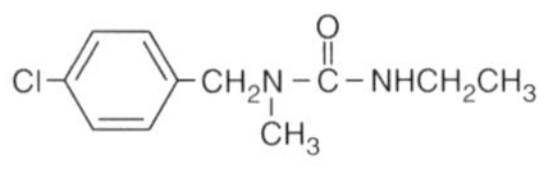
250 mg

S80,198-4
1-(4-CHLOROBENZYL)-3-(3-FLUOROPHENYL)UREA
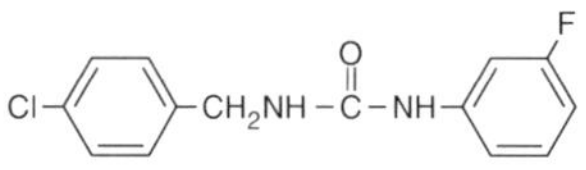
250 mg

S84,589-2
1-(4-CHLOROBENZYL)-1-METHYL-3-(2,5-XYLYL)UREA
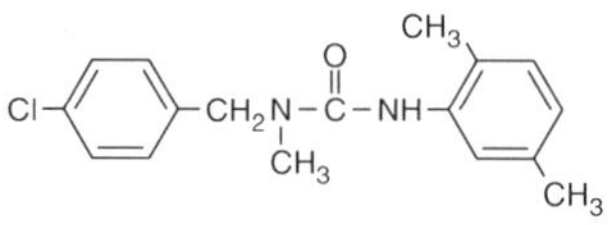
250 mg

S79,673-5
1-[2-(4-CHLOROBENZAMIDO)ACETYL]-4-(2-METHYL-5-NITROPHENYL)-SEMICARBAZIDE
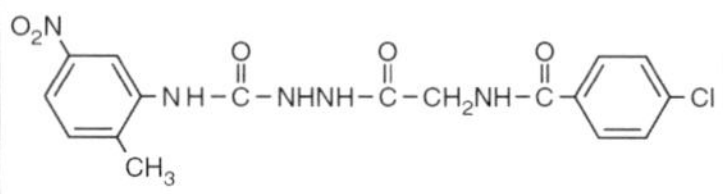
250 mg

S70,005-3
4-(2,4-DIFLUOROPHENYL)-1-(3-NITROBENZOYL)SEMICARBAZIDE
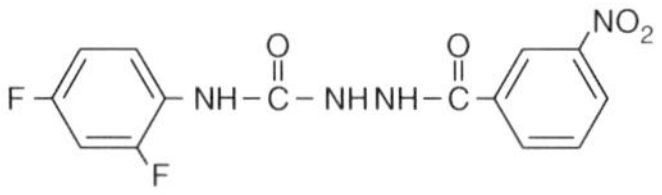
250 mg

S69,435-5
4-[3,5-BIS(METHOXYCARBONYL)PHENYL]-1-(2,4-DICHLOROBENZOYL)-SEMICARBAZIDE
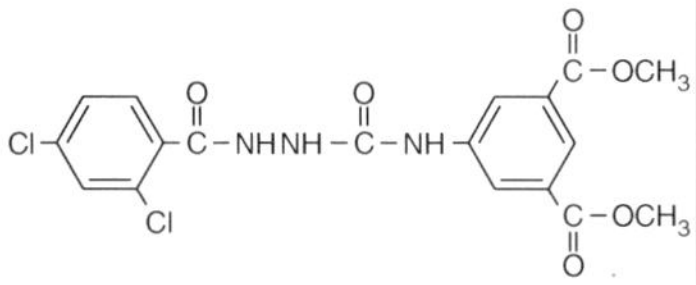
250 mg

S71,855-6
1-(2,4-DICHLOROBENZOYL)-4-[2-(METHOXYCARBONYL)PHENYL]-SEMICARBAZIDE
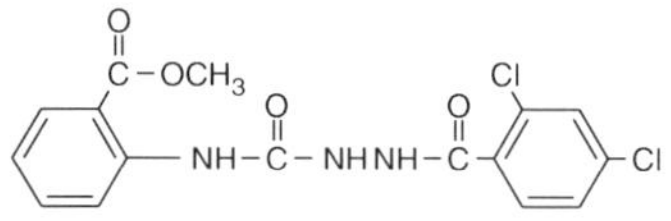
250 mg

S72,636-2
1-(4-CHLOROBENZOYL)-4,4-DIMETHYLSEMICARBAZIDE
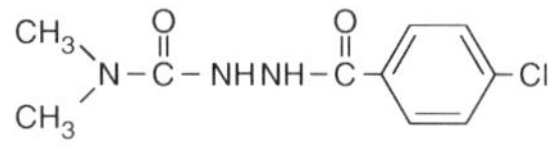
5 gm

S77,215-1
1-(4-CHLOROBENZOYL)-4-PROPYLSEMICARBAZIDE
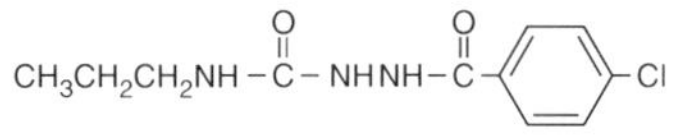
250 mg

S68,646-8
1-(4-CHLOROBENZOYL)-4-(3-ETHYLPHENYL)SEMICARBAZIDE
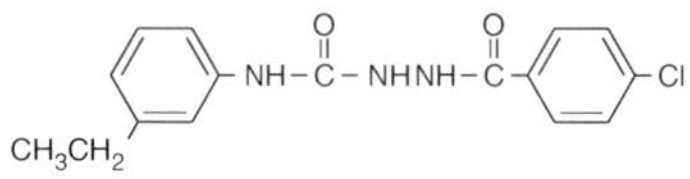
250 mg

S73,919-7
1-(4-CHLOROBENZOYL)-4-(2,6-XYLYL)-SEMICARBAZIDE
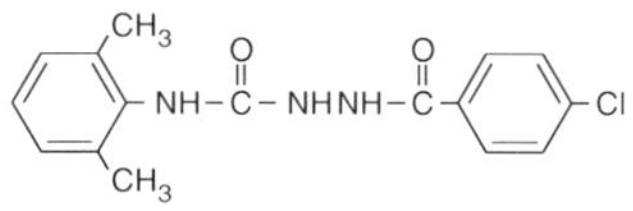
5 gm

S69,365-0
1-(2,4-DICHLOROBENZOYL)-4-(2,6-DIMETHYLPHENYL)SEMICARBAZIDE
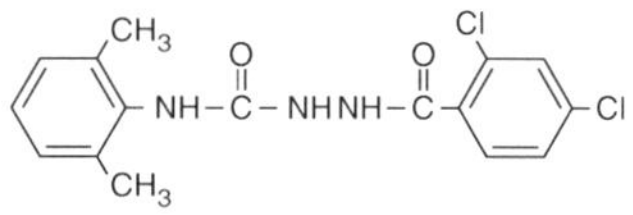
250 mg

S70,022-3
1-(3,4-DICHLOROBENZOYL)-4-PROPYLSEMICARBAZIDE
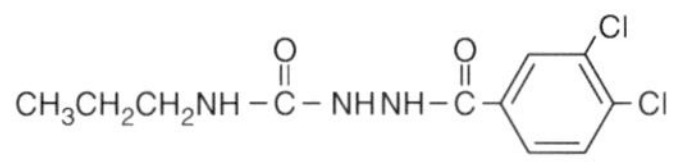
250 mg

S74,772-6
1-(2,4-DIFLUOROPHENYL)-4-(2,4-DIFLUOROBENZOYL)SEMICARBAZIDE
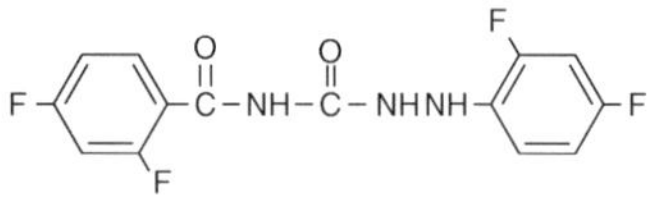
5 gm

S73,752-6
1-(3,4-DICHLOROBENZOYL)-4-(2,4-DIFLUOROPHENYL)SEMICARBAZIDE
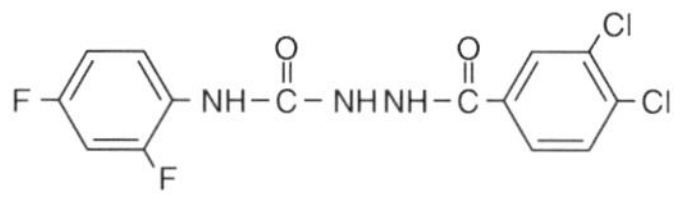
5 gm

S73,923-5
1-(2,4-DICHLOROBENZOYL)-4-(2,4-DIFLUOROPHENYL)SEMICARBAZIDE
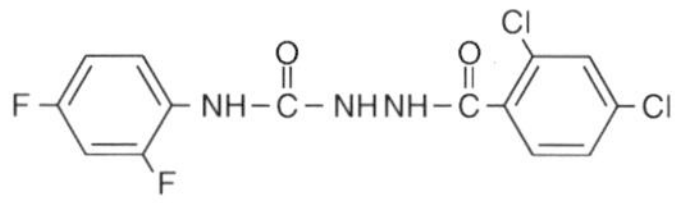
5 gm

S69,356-1
1-(2,4-DICHLOROBENZOYL)-4-(3,4-DICHLOROPHENYL)SEMICARBAZIDE
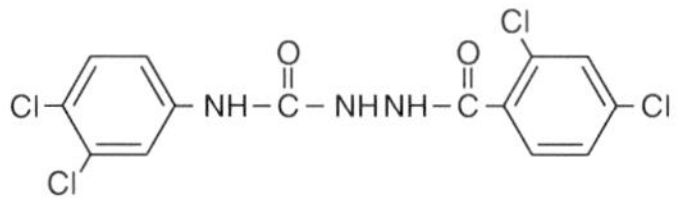
250 mg

S68,426-0
4-(3-CHLOROPHENYL)-1-[3-(4-CHLOROPHENYL)CARBAMOYL]-SEMICARBAZIDE
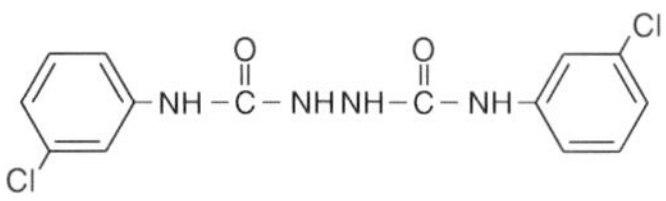
250 mg

S68,528-3
1-ACETYL-4-(3-CHLOROPHENYL)-SEMICARBAZIDE
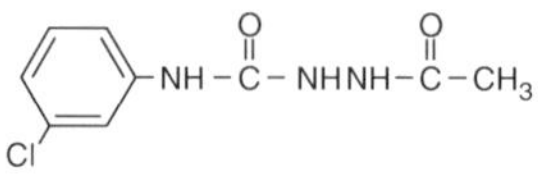
250 mg

S77,351-4
4-(3-CHLOROPHENYL)-1-[3-(4-FLUOROPHENYL)UREIDOACETYL]-SEMICARBAZIDE
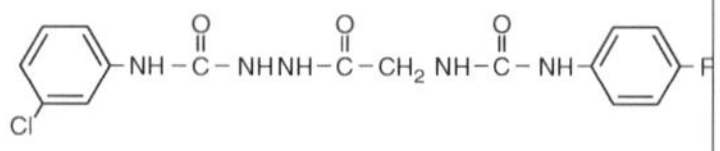
250 mg

S69,255-7
3-(3,4-DICHLOROPHENYLUREIDO)ACETIC ACID N2-(ETHYLCARBAMOYL)HYDRAZIDE
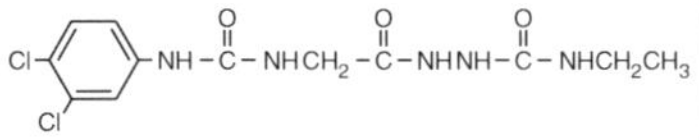
250 mg

S69,333-2
4-(3-CHLOROPHENYL)-1-[3-(3,4-DICHLOROPHENYL)UREIDOACETYL]-SEMICARBAZIDE
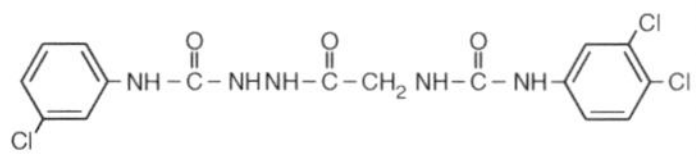
250 mg

S77,205-4
4-(3-CHLOROPHENYL)-1-(PHENYLACETYL)SEMICARBAZIDE
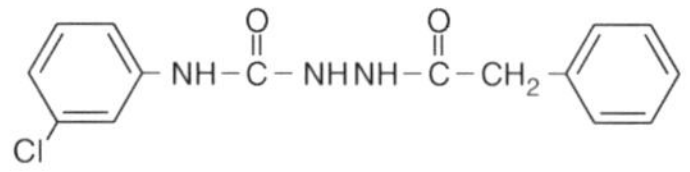
250 mg

S77,208-9
4-(4-FLUOROPHENYL)-1-(PHENYLACETYL)SEMICARBAZIDE
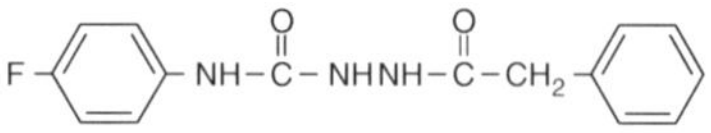
250 mg

S67,373-0
4-(3-CHLOROPHENYL)-1-(3,3-DIETHYLUREIDOACETYL)SEMICARBAZIDE
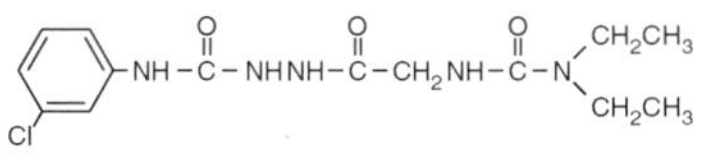
250 mg

S69,264-6
1-[(4-CHLOROPHENYL)CARBAMOYL]-4-ETHYL-1-PHENYLSEMICARBAZIDE
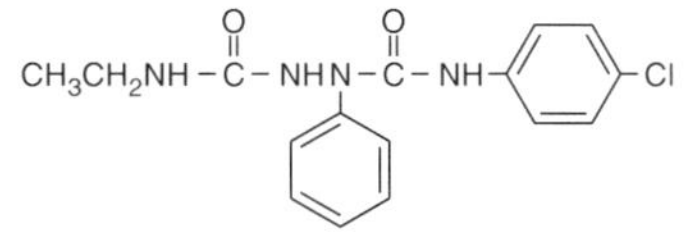
250 mg

S72,855-1
1-(BENZYLCARBONYL)-4-(3,4-DICHLOROPHENYL)SEMICARBAZIDE
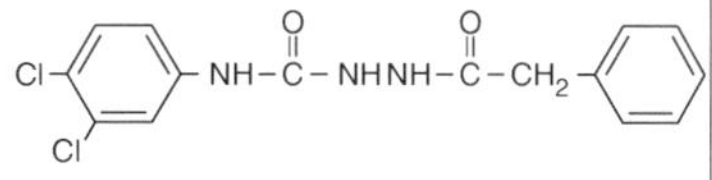
5 gm

S69,265-4
4-(4-CHLORO-3-METHYLPHENYL)-1-(2,4-DIMETHYLPHENYLCARBAMOYL)-1-PHENYLSEMICARBAZIDE
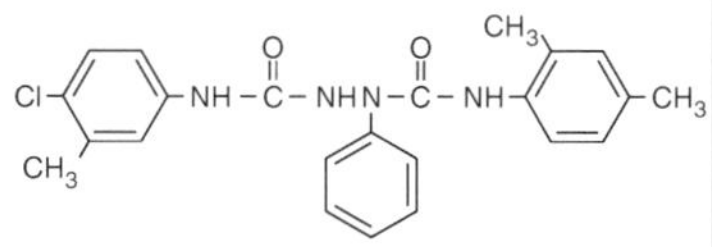
250 mg

S77,222-4
1-(4-BROMO-2,6-DIMETHYLPHENYL)-6-(3-CHLOROPHENYL)-3-PHENYLBIUREA
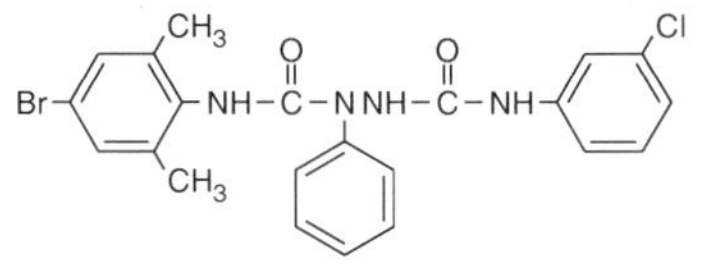
250 mg

S69,422-3
1,6-BIS(4-FLUOROPHENYL)BIUREA
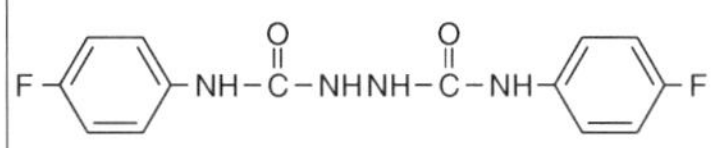
250 mg

S68,955-6
1,6-BIS(3,4-DICHLOROPHENYL)BIUREA
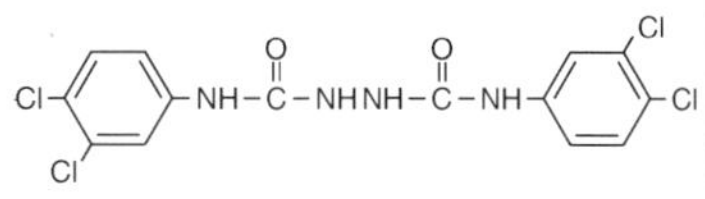
250 mg

S69,062-7
1,6-BIS(2,5-DICHLOROPHENYL)BIUREA
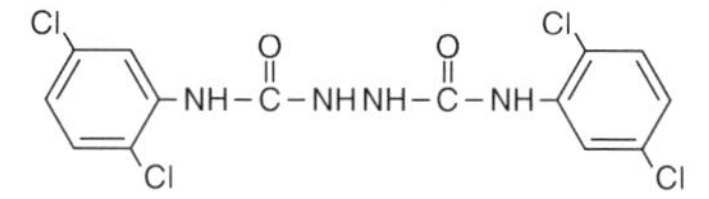
250 mg

S90,313-2
2-[3-(2-CHLORO-4-NITROPHENYL)-UREIDO]ACETIC (3-PHENYLALLYLIDENE)-HYDRAZIDE
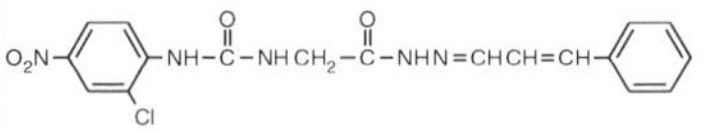
250 mg

S84,787-9
4-[3-(2-FLUOROPHENYL)UREIDO]-BENZOIC (4-ACETAMIDOBENZYLIDENE)-HYDRAZIDE
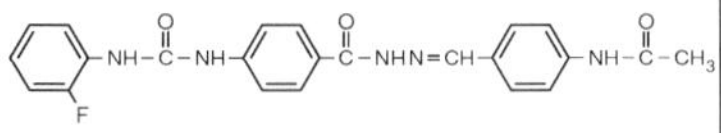
250 mg

S84,784-4
4-[3-(2-FLUOROPHENYL)UREIDO]-BENZOIC (3-PHENYLALLYLIDENE)-HYDRAZIDE
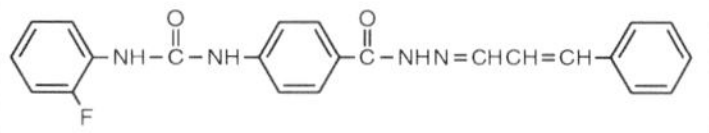
250 mg

S79,692-1
4-[3-(2-FLUOROPHENYL)UREIDO]-BENZOIC (2-CHLOROBENZYLIDENE)-HYDRAZIDE
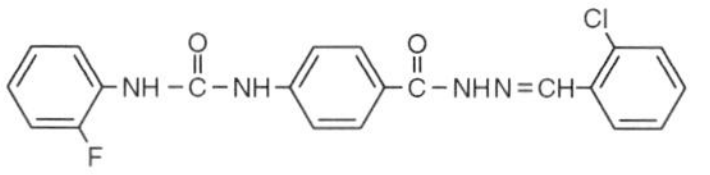
250 mg

S84,775-5
4-[3-(2-FLUOROPHENYL)UREIDO]-BENZOIC (4-METHYLBENZYLIDENE)-HYDRAZIDE
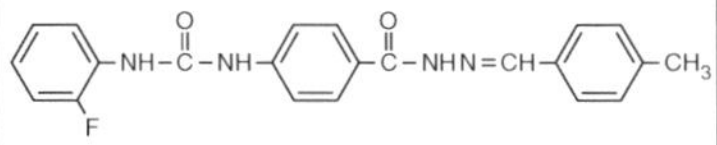
250 mg

S84,771-2
4-[3-(2-FLUOROPHENYL)UREIDO]-BENZOIC (3,4-DICHLOROBENZYLIDENE)-HYDRAZIDE
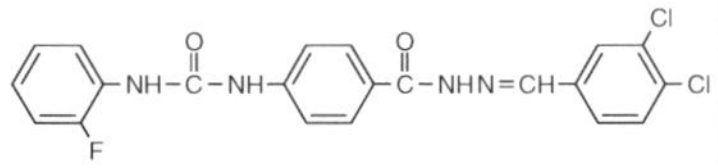
250 mg

S86,845-0
4-(3-ALLYLUREIDO)BENZOIC (3,4-DICHLOROBENZYLIDENE)HYDRAZIDE
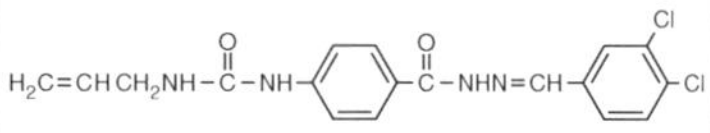
250 mg

S86,849-3
4-(3-PROPYLUREIDO)BENZOIC (2,4-DICHLOROBENZYLIDENE)HYDRAZIDE
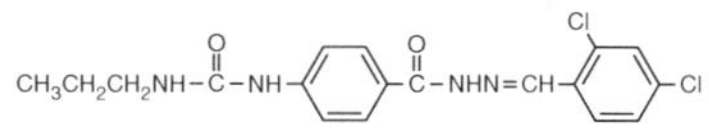
250 mg

S86,879-5
4-[3-(3-CHLOROPHENYL)UREIDO]-BENZOIC (3,4-DICHLOROBENZYLIDENE)-HYDRAZIDE
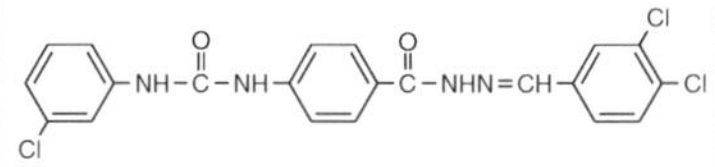
250 mg

S80,080-5
1-(4-CHLOROBENZYLIDENE)-4-PHENYLSEMICARBAZIDE
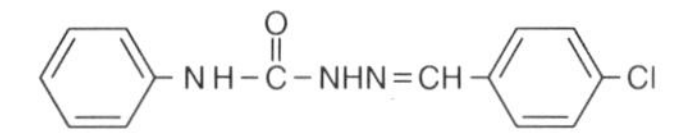
250 mg

S72,342-8
2-[3-(4-TOLYL)UREIDO]ACETIC ACID N2-(2,4-DICHLOROBENZYLIDENE)HYDRAZIDE
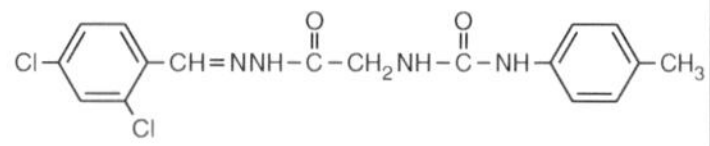
5 gm

S90,204-7
ETHYL [3-(3-CHLOROPHENYL)UREIDO]-ACETATE
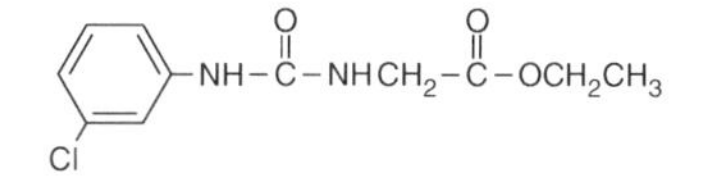
250 mg

S75,640-7
1-(3-PHENYLUREIDO)-2-PROPYL N-(3,4-DICHLOROPHENYL)CARBAMATE
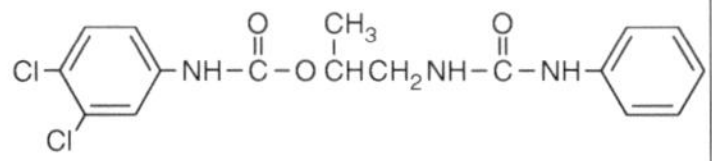
5 gm

S76,993-2
3-(2-FLUOROPHENYL)-1-FURFURYL-1-METHYLUREA
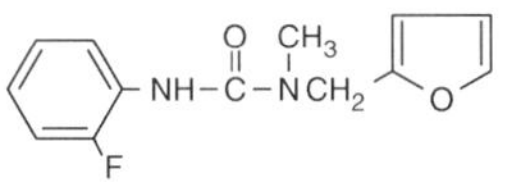
250 mg

S74,324-0
1-(5-CHLORO-2-METHOXYPHENYL)-3-(2-IODOPHENYL)UREA
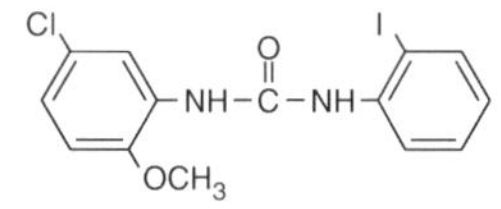
5 gm

S77,253-4
1-(4-CHLORO-2-METHOXYPHENYL)-3-(O-TOLYL)UREA
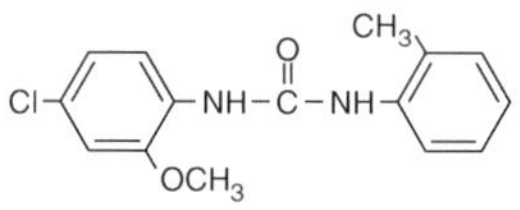
250 mg

S84,480-2
1-(4-CHLORO-2-METHOXYPHENYL)-3-(3-CHLOROPHENYL)UREA
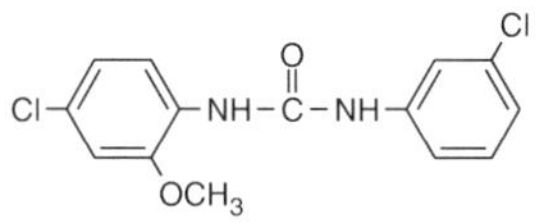
250 mg

S71,098-9
1-(3,4-DICHLOROPHENYL)-3-HYDROXYUREA
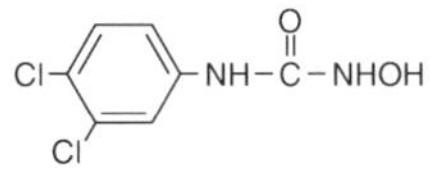
50 mg

S73,549-3
1-(3-CHLORO-4-METHYLPHENYL)-3-HYDROXYUREA
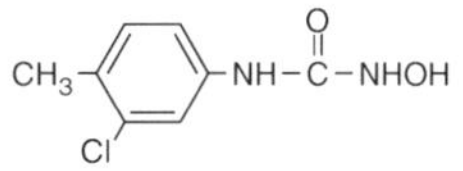
250 mg

S69,906-3
1-(2-BENZOYLPHENYL)-3-(4-CHLORO-3-NITROPHENYL)UREA
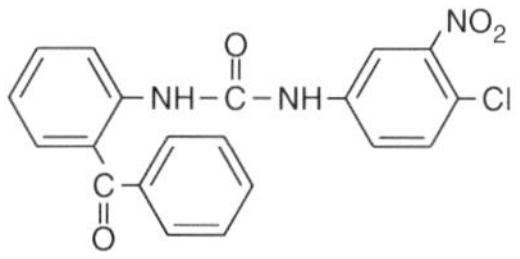
250 mg

S86,820-5
DIMETHYL 5-[3-(4-FLUORO-3-NITROPHENYL)UREIDO]ISOPHTHALATE
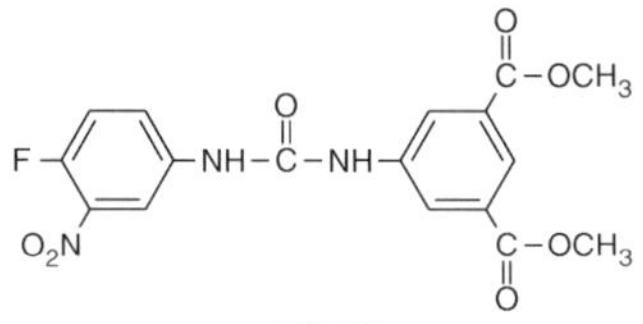
250 mg

S90,568-2
DIMETHYL 5-[3-(4-CHLORO-2-NITROPHENYL)UREIDO]ISOPHTHALATE
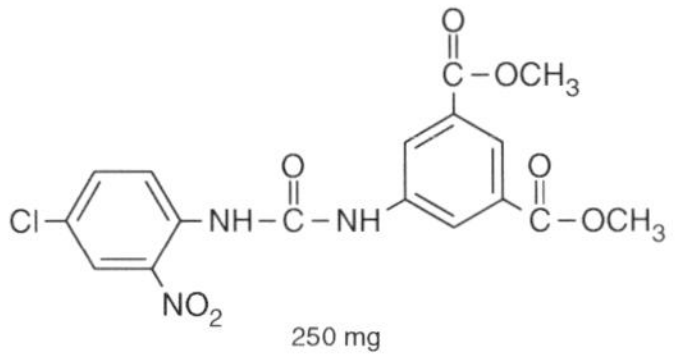
250 mg

S68,736-7
N-(4-CHLORO-3-NITROPHENYL)-N'-[3-(3-METHYLPHENYL)UREIDO]OXAMIDE
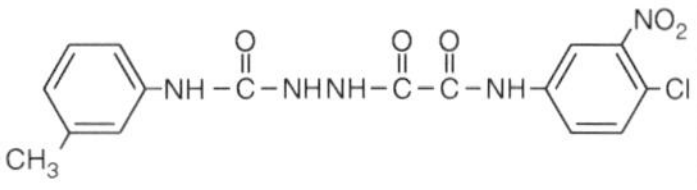
250 mg

S72,528-5
1-(2-CHLOROPHENYL)-3-(2-METHYL-4-NITROPHENYL)UREA
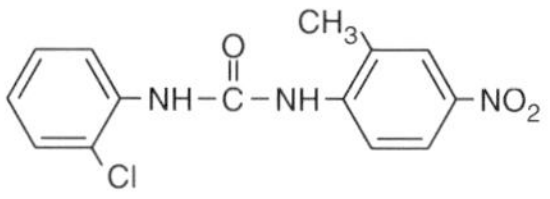
5 gm

S70,859-3
1-(4-NITROPHENYL)-3-(2,4,5-TRICHLOROPHENYL)UREA

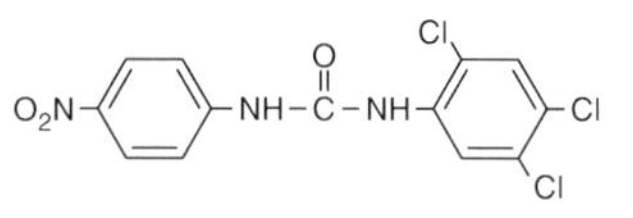

250 mg

S70,860-7
1-(2,6-DICHLOROPHENYL)-3-(4-NITROPHENYL)UREA

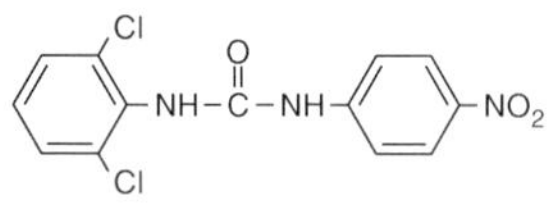

5 gm

S72,143-3
3-(2-CHLORO-4-NITROPHENYL)-1-(4-FLUOROPHENYL)-1-METHYLUREA

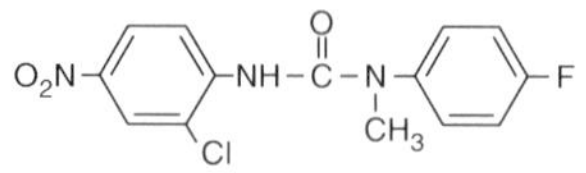

250 mg

S71,191-8
1-(2-CHLORO-4-NITROPHENYL)-3-(2,3-DICHLOROPHENYL)UREA

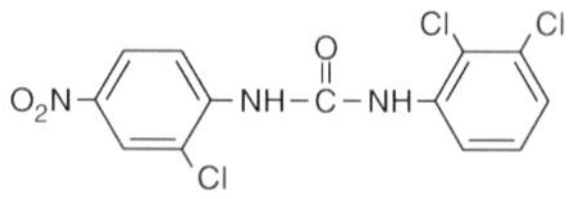

5 gm

S71,373-2
1-(2-CHLORO-4-NITROPHENYL)-3-(2-ETHYL-6-METHYLPHENYL)UREA

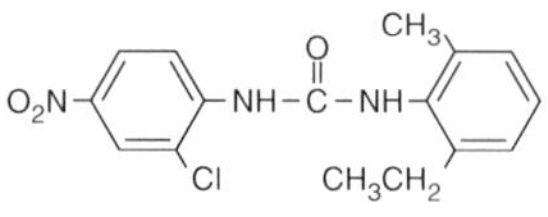

250 mg

S73,677-5
1-(3-ACETYLPHENYL)-3-(3-FLUOROPHENYL)UREA

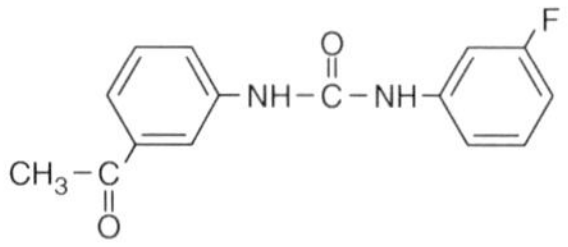

5 gm

S71,487-9
1-(4-BENZOYLPHENYL)-3-(2-CHLOROPHENYL)UREA

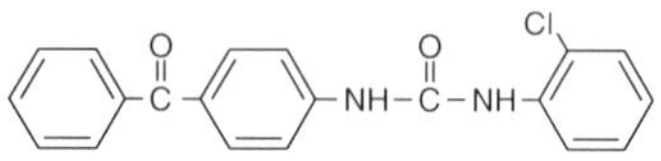

250 mg

S77,368-9
1-(2-BENZOYLPHENYL)-3-(2,5-DIFLUOROPHENYL)UREA

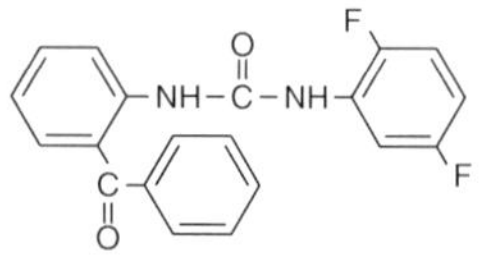

250 mg

S71,486-0
1-(4-BENZOYLPHENYL)-3-(3-CHLOROPHENYL)UREA

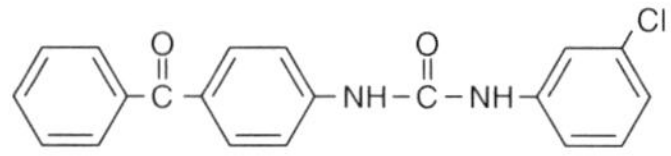

5 gm

S71,378-3
1-(4-BENZOYLPHENYL)-3-(2,6-DICHLOROPHENYL)UREA

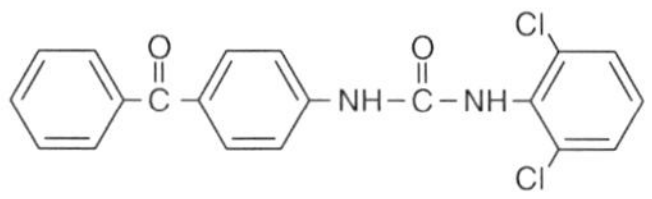

5 gm

S71,662-6
1-(4-BENZOYLPHENYL)-3-(2,5-DICHLOROPHENYL)UREA

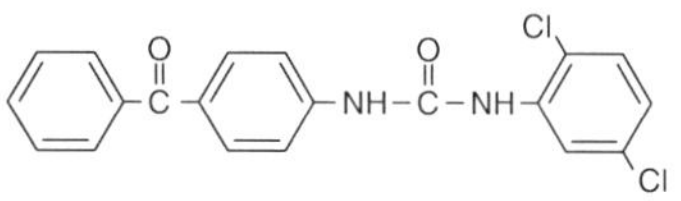

5 gm

S71,757-6
1-(4-BENZOYLPHENYL)-3-(4-CHLORO-2-METHYLPHENYL)UREA

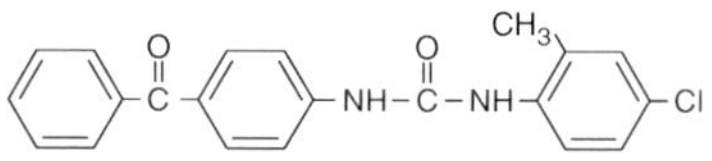

5 gm

S65,979-7
1-(4-BENZOYLPHENYL)-3-(2,4-DICHLOROPHENYL)UREA

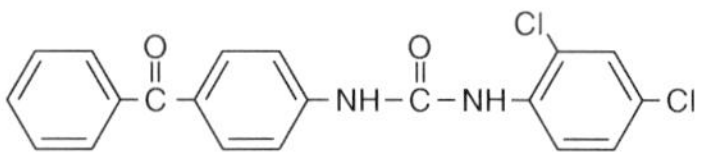

5 gm

S75,639-3
1,1'-(METHYLENEDI-4,4-PHENYLENE)-BIS[3-(2-BENZOYL-4-CHLOROPHENYL)UREA]

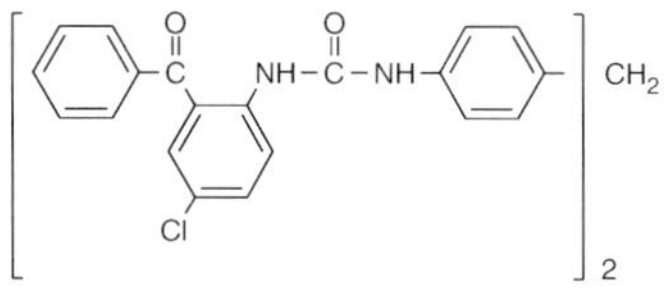

5 gm

S66,730-7
1-(2-BENZOYL-4-CHLOROPHENYL)-3-(2,3-DICHLOROPHENYL)UREA

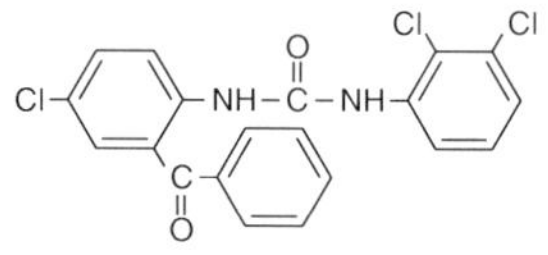

5 gm

S70,743-0
1-(2-BENZOYL-4-CHLOROPHENYL)-3-(2,5-DICHLOROPHENYL)UREA

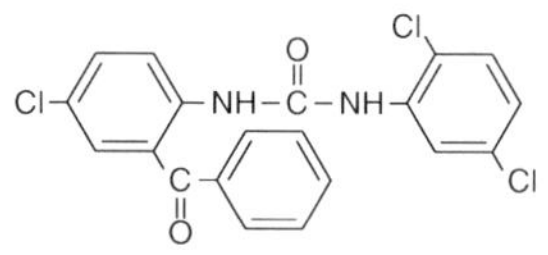

5 gm

S70,048-7
1-(2-BENZOYL-4-CHLOROPHENYL)-3-METHYLUREA

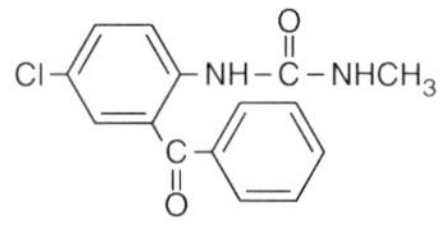

250 mg

S73,676-7
1-(3-ACETYLPHENYL)-3-(2-FLUOROPHENYL)UREA

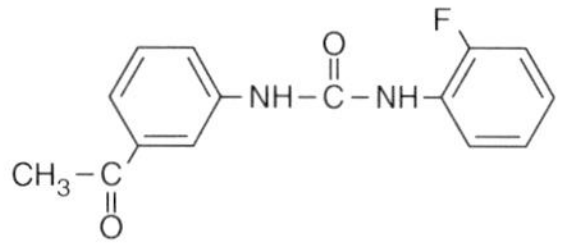

5 gm

S69,604-8
1-(2-BENZOYLPHENYL)-3-(2,4-DIFLUOROPHENYL)UREA

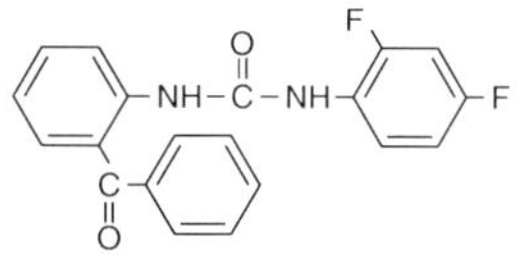

250 mg

S69,860-1
1-(4-ACETYLPHENYL)-3-(3-FLUOROPHENYL)UREA

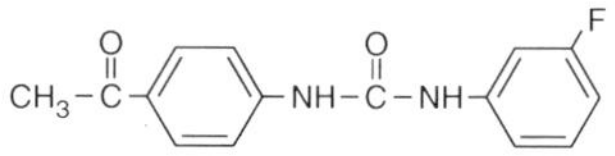

250 mg

S92,410-5
1-(3-ACETYLPHENYL)-3-(4-IODOPHENYL)UREA

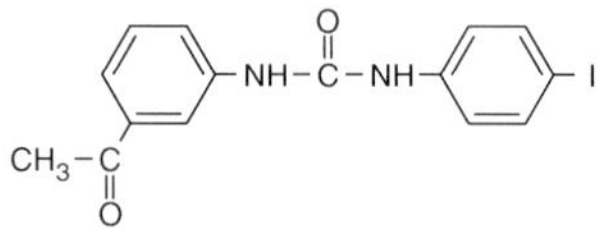

250 mg

S72,328-2
1-(4-ACETYLPHENYL)-3-(2,5-DIFLUOROPHENYL)UREA

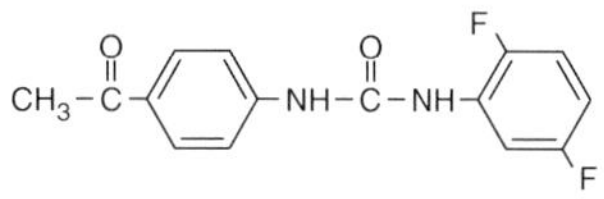

5 gm

S72,035-6
1-(3-ACETYLPHENYL)-3-(2,5-DIFLUOROPHENYL)UREA

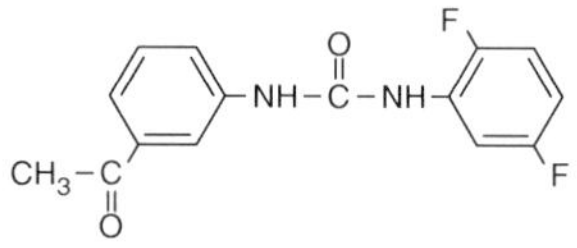

250 mg

S71,140-3
1-(4-BENZOYLPHENYL)-3-(3-CHLORO-4-METHYLPHENYL)UREA

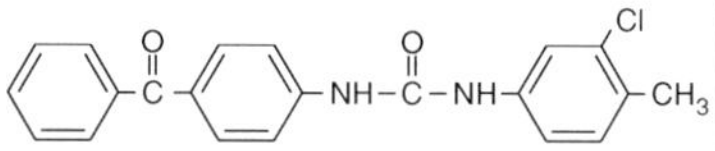

5 gm

S71,144-6
1-(2-BENZOYL-4-CHLOROPHENYL)-3-(2-CHLORO-6-METHYLPHENYL)UREA

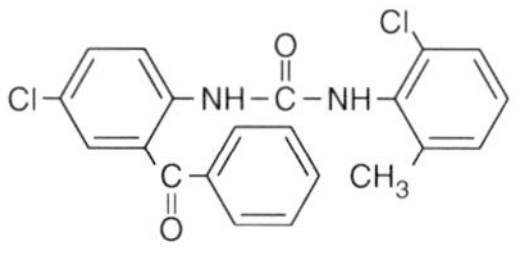

5 gm

S85,057-8
DIMETHYL 5-[3-(3-CHLORO-4-FLUOROPHENYL)UREIDO]ISOPHTHALATE

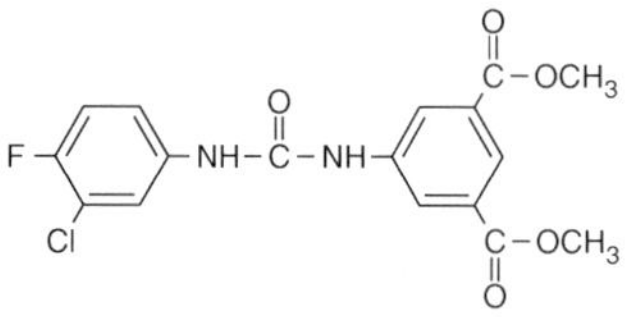

250 mg

S73,163-3
DIMETHYL 5-[3-(2-FLUOROPHENYLUREIDO)] ISOPHTHALATE

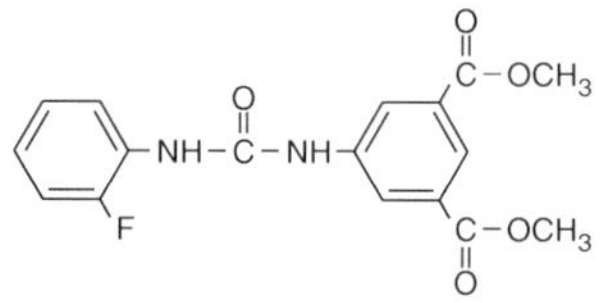

5 gm

S72,549-8
1-(4-CARBOXYPHENYL)-4-(3-CHLORO-4-METHYLPHENYL)SEMICARBAZIDE

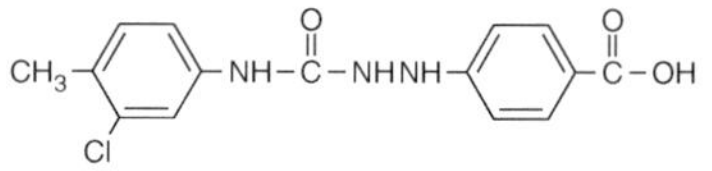

5 gm

S42,283-5
3,3'-HEXAMETHYLENEDIUREIDOBIS (2,4,6-TRIIODOBENZOIC ACID)

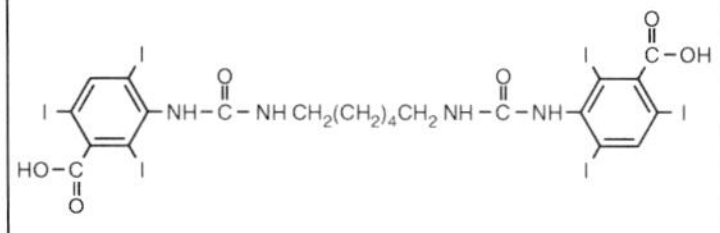

5 gm

S73,553-1
1-(2-CARBOXYPHENYL)-3-(3-FLUOROPHENYL)UREA

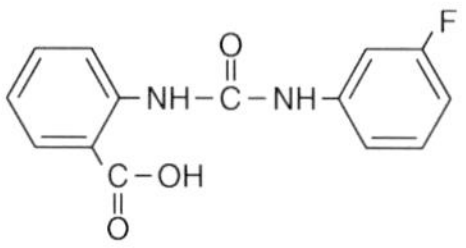

250 mg

S73,644-9
1-(3-CARBOXYPHENYL)-3-(2-FLUOROPHENYL)UREA

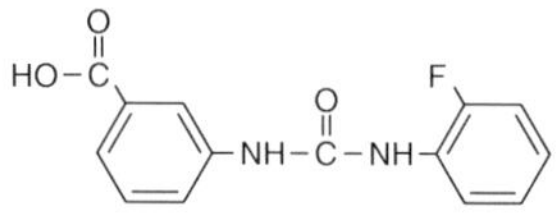

5 gm

S66,900-8
1-(2-CARBOXYPHENYL)-3-(2-CHLOROPHENYL)UREA

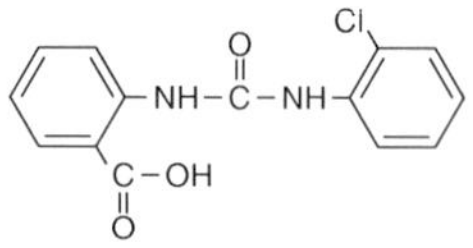

5 gm

S66,959-8
1-(2-CARBOXYPHENYL)-3-(3-CHLOROPHENYL)UREA

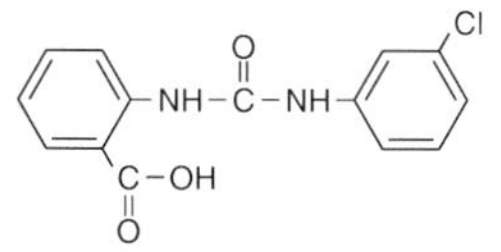

5 gm

S73,643-0
1-(4-CARBOXYPHENYL)-3-(2-FLUOROPHENYL)UREA

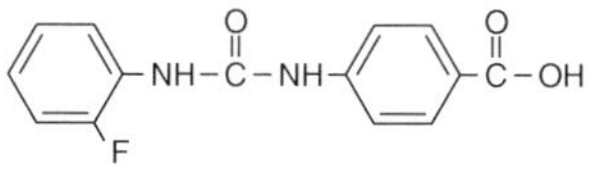

5 gm

S66,921-0
1-(4-CARBOXYPHENYL)-3-(2-CHLOROPHENYL)UREA

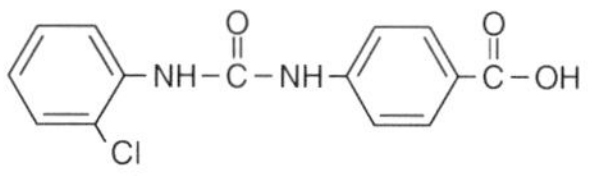

5 gm

S66,957-1
1-(2-CARBOXYPHENYL)-3-(4-CHLOROPHENYL)UREA

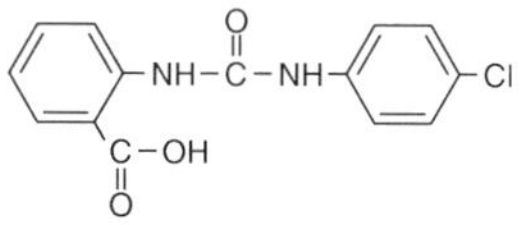

5 gm

S66,886-9
1-(2-CARBOXYPHENYL)-3-(2,3-DICHLOROPHENYL)UREA

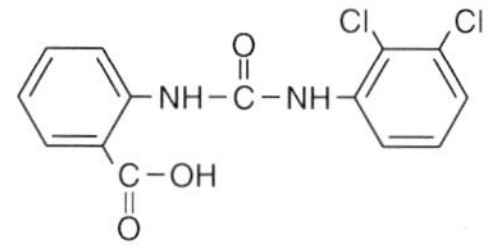

5 gm

S66,887-7
1-(2-CARBOXYPHENYL)-3-(3,4-DICHLOROPHENYL)UREA

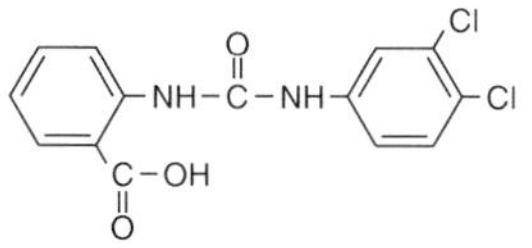

5 gm

S67,007-3
1-(2-CARBOXYPHENYL)-3-(2,4-DICHLOROPHENYL)UREA

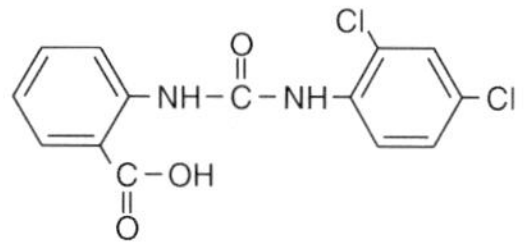

5 gm

S67,010-3
1-(2-CARBOXYPHENYL)-3-(2,5-DICHLOROPHENYL)UREA

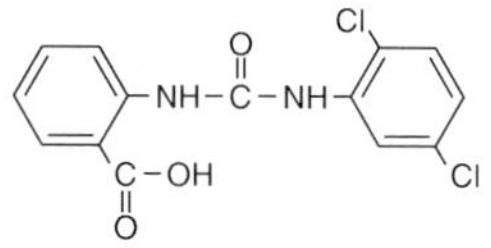

5 gm

S66,881-8
1-(2-CARBOXYPHENYL)-3-(2-CHLORO-6-METHYLPHENYL)UREA

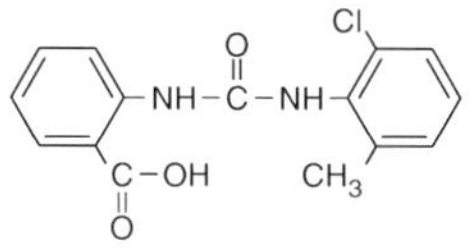

5 gm

S67,016-2
1-(2-CARBOXYPHENYL)-3-(4-CHLORO-2-METHYLPHENYL)UREA

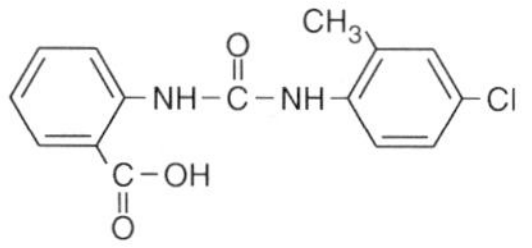

5 gm

S71,872-6
1-(3-CARBOXYPHENYL)-3-(4-FLUOROPHENYL)UREA

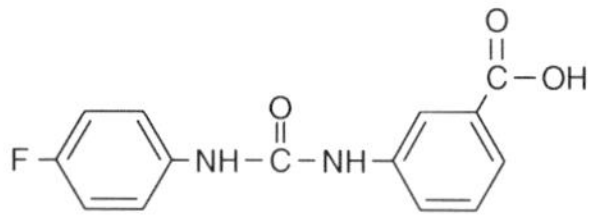

250 mg

S80,197-6
4-[3-(3-FLUOROPHENYL)UREIDO]-BENZOIC ACID

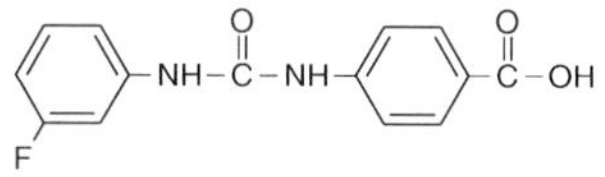

250 mg

S66,917-2
1-(4-CARBOXYPHENYL)-3-(4-CHLOROPHENYL)UREA

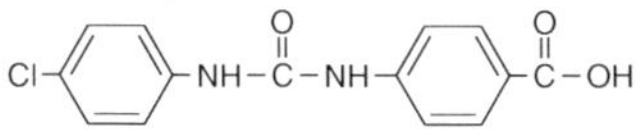

5 gm

S66,984-9
1-(4-CARBOXYPHENYL)-3-(3-CHLOROPHENYL)UREA

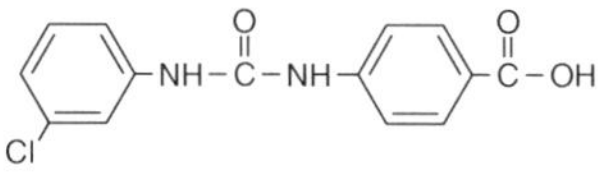

5 gm

S66,970-9
1-(4-CARBOXYPHENYL)-3-(2,3-DICHLOROPHENYL)UREA

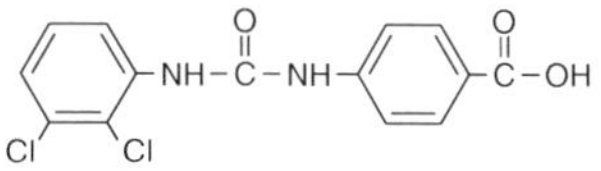

5 gm

S92,411-3
4-[3-(3-CHLORO-4-FLUOROPHENYL)-UREIDO]BENZOIC ACID

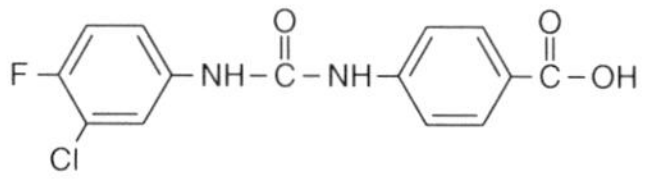

250 mg

S66,981-4
1-(4-CARBOXYPHENYL)-3-(3,4-DICHLOROPHENYL)UREA

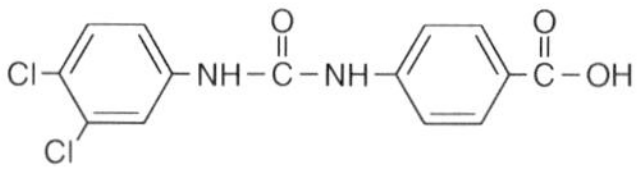

5 gm

S67,011-1
1-(4-CARBOXYPHENYL)-3-(2,4-DICHLOROPHENYL)UREA

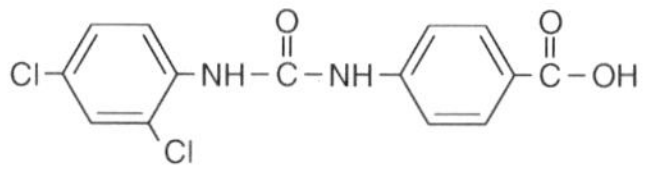

5 gm

S67,045-6
1-(4-CARBOXYPHENYL)-3-(2,5-DICHLOROPHENYL)UREA

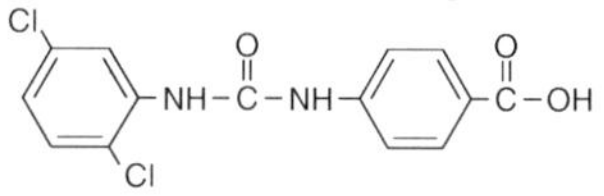

5 gm

S66,888-5
1-(4-CARBOXYPHENYL)-3-(5-CHLORO-2-METHYLPHENYL)UREA

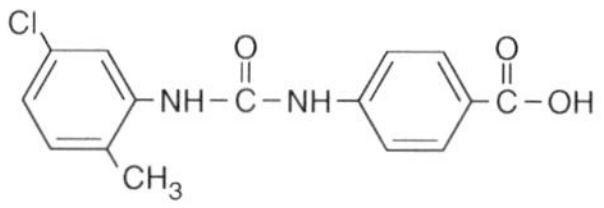

5 gm

S85,148-5
3-[3-(2,3-DICHLOROPHENYL)UREIDO]-BENZOIC ACID

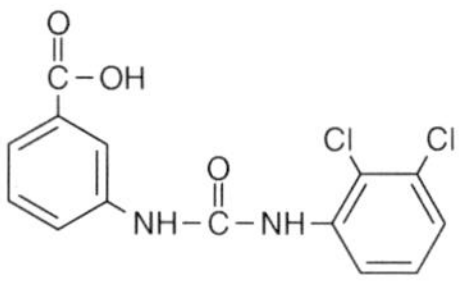

250 mg

S85,144-2
3-[3-(2-CHLORO-6-METHYLPHENYL)-UREIDO]BENZOIC ACID

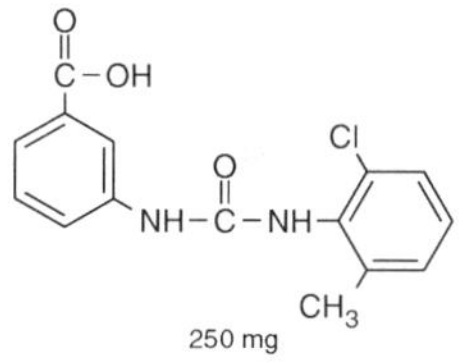

250 mg

S74,392-5
3-(2,5-DIFLUOROPHENYLUREIDO)-BENZOIC ACID

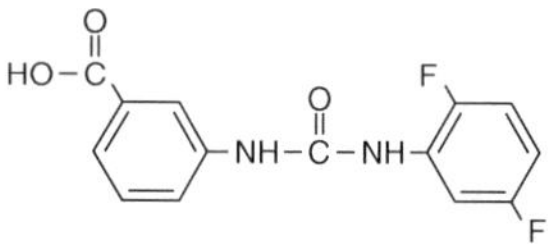

5 gm

S67,502-4
1-(3-CARBOXYPHENYL)-3-(3,4-DICHLOROPHENYL)UREA

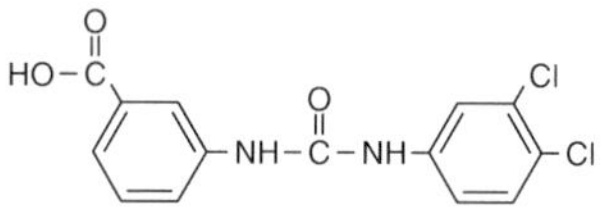

250 mg

S74,856-0
1-(6-CARBOXY-2,4-DICHLOROPHENYL)-3-(2,4,6-TRICHLOROPHENYL)UREA

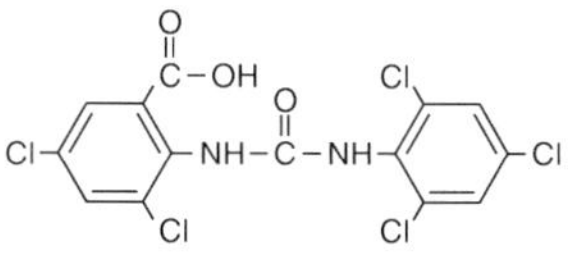

250 mg

S77,953-9
3-[3-(2,4,5-TRICHLOROPHENYL)UREIDO]-BENZOIC ACID

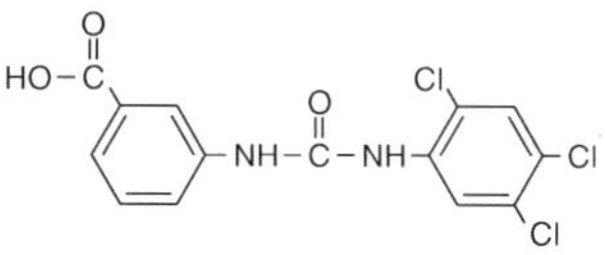

250 mg

S90,484-8
1-(3-CHLOROBENZOYL)-4-(ETHOXYCARBONYLMETHYL)SEMICARBAZIDE

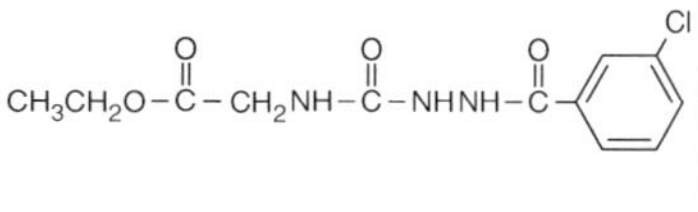

250 mg

S73,679-1
METHYL 2-[3-(2-FLUOROPHENYL)UREIDO]-BENZOATE

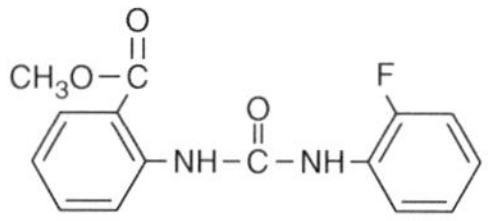

5 gm

S73,718-6
1-[4-(ETHOXYCARBONYL)PHENYL]-3-(2-FLUOROPHENYL)UREA

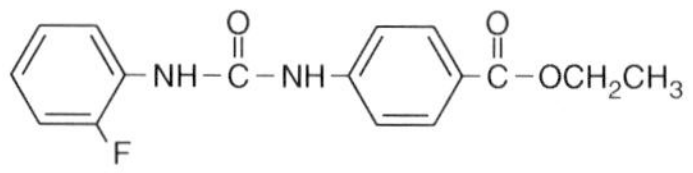

5 gm

S89,068-5
ETHYL 4-[3-(2-CHLOROPHENYL)UREIDO]-BENZOATE

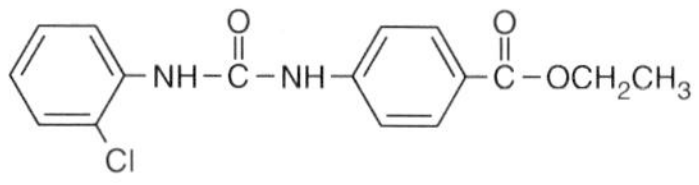

250 mg

S76,931-2
METHYL 2-[3-(3-CHLORO-2-METHYLPHENYL)UREIDO]BENZOATE

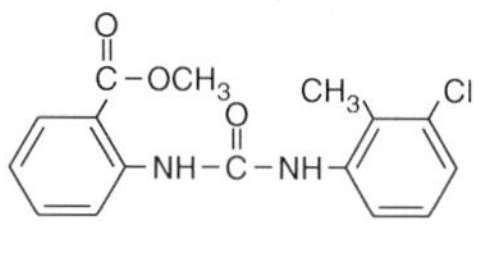

250 mg

S89,955-0
METHYL 2-[3-(2-CHLORO-6-METHYLPHENYL)-1-METHYLUREIDO]-BENZOATE

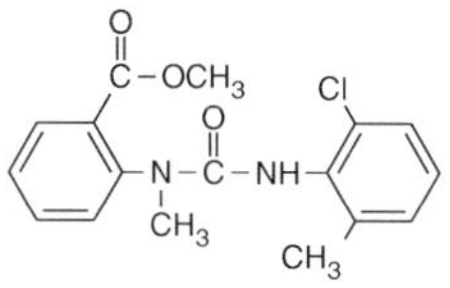

250 mg

S85,062-4
METHYL 2-[3-(3-CHLORO-4-FLUOROPHENYL)UREIDO]BENZOATE

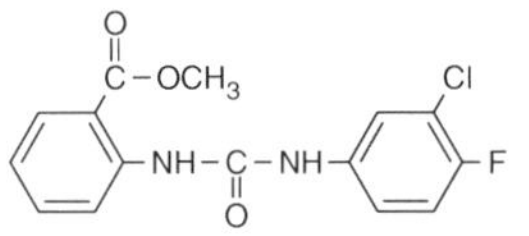

250 mg

S72,591-9
ETHYL 4-[3-(4-FLUOROPHENYL)UREIDO]-BENZOATE

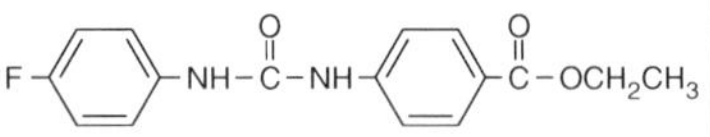

5 gm

S78,092-8
ETHYL 4-[3-(4-CHLOROPHENYL)UREIDO]-BENZOATE

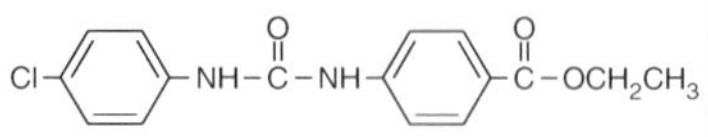

250 mg

S85,592-8
ETHYL 4-[3-(3-CHLOROPHENYL)UREIDO]-BENZOATE

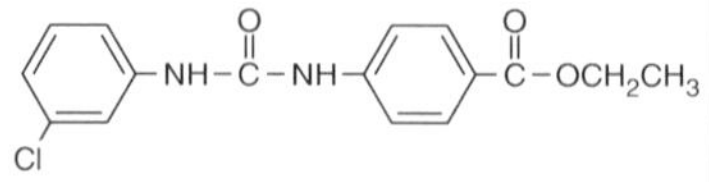

250 mg

S72,590-0
ETHYL 4-[3-(4,5-DICHLORO-2-METHYLPHENYL)UREIDO]BENZOATE

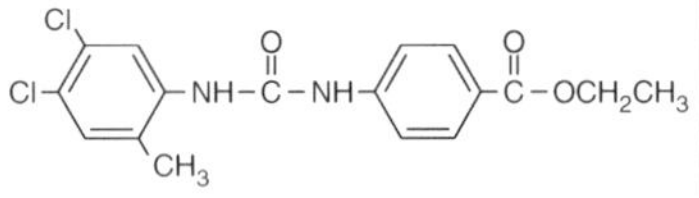

5 gm

S71,834-3
BUTYL 4-[3-(3-FLUOROPHENYL)UREIDO]-BENZOATE

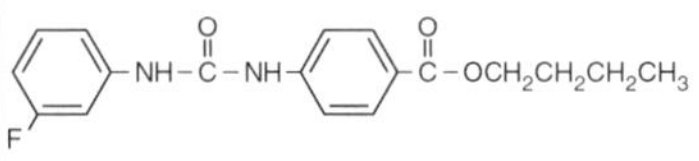

250 mg

S72,080-1
ETHYL 4-[3-(2,5-DIFLUOROPHENYL)-UREIDO]BENZOATE

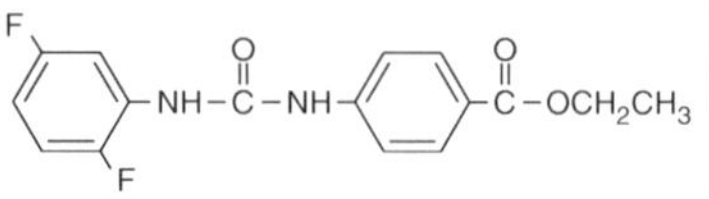

250 mg

S79,731-6
ETHYL 4-[3-(2,4-DICHLOROPHENYL)-UREIDO]BENZOATE

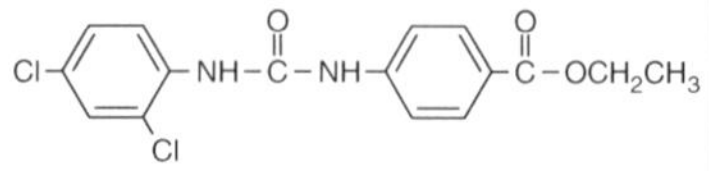

250 mg

S67,500-8
ETHYL 4-[3-(5-CHLORO-2-METHYLPHENYL)UREIDO]BENZOATE

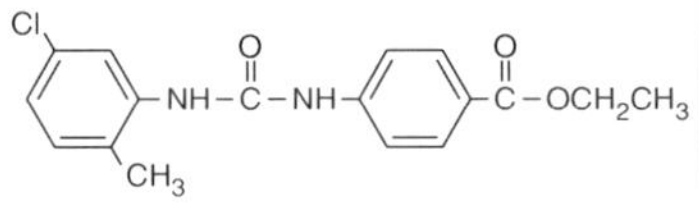

250 mg

S71,887-4
1-(2-CARBAMOYLPHENYL)-3-(2-FLUOROPHENYL)UREA

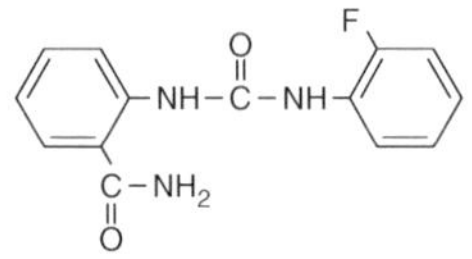

250 mg

S84,481-0
1-(2-CARBAMOYLPHENYL)-3-(3-FLUOROPHENYL)UREA

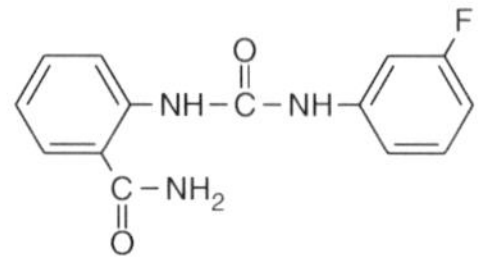

250 mg

S80,028-7
1-(2-CARBAMOYLPHENYL)-3-(4-FLUOROPHENYL)UREA

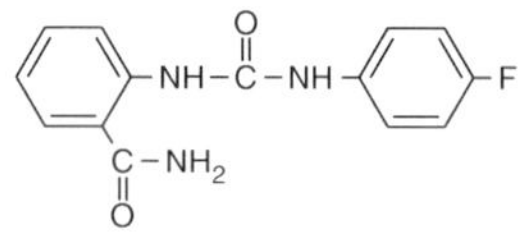

250 mg

S71,415-1
1-BENZOYL-4-(2,6-DICHLOROPHENYL)-SEMICARBAZIDE

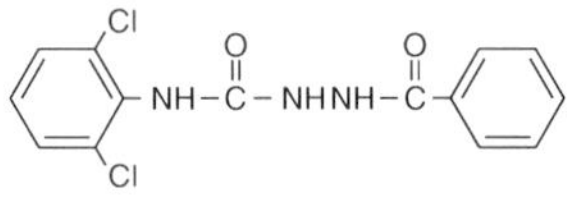

5 gm

S77,220-8
1-(4-CHLOROBENZOYL)-4-(2-CHLOROPHENYL)SEMICARBAZIDE

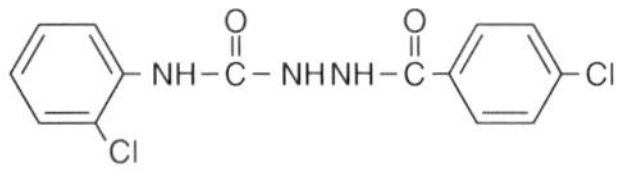

250 mg

S91,025-2
1-(4-CHLOROBENZOYL)-4-ISOPROPYLSEMICARBAZIDE

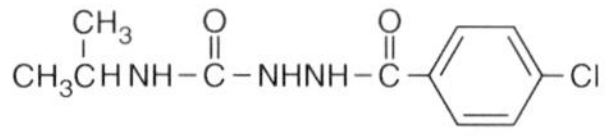

250 mg

S91,031-7
4-BUTYL-1-(4-CHLOROBENZOYL)-SEMICARBAZIDE

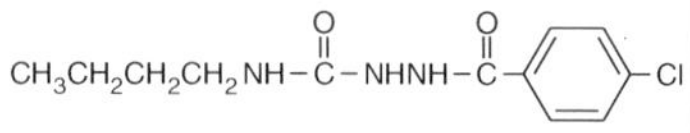

250 mg

S68,619-0
1-(4-CHLOROBENZOYL)-4-(2,4-DIFLUOROPHENYL)SEMICARBAZIDE

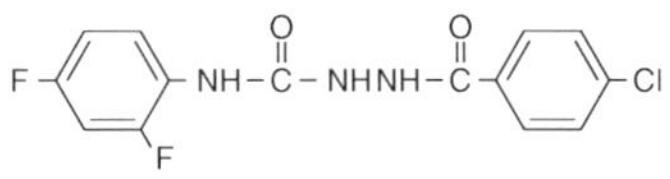

250 mg

S90,415-5
4-(3-CHLOROPHENYL)-1-(2,4-DICHLOROBENZOYL)SEMICARBAZIDE

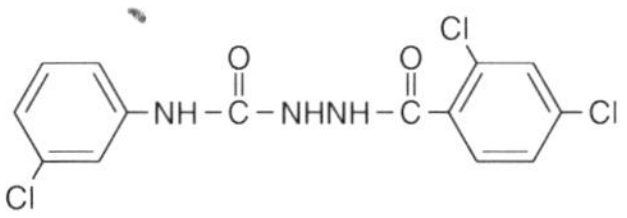

250 mg

S90,416-3
1-(4-CHLOROBENZOYL)-3-(3-CHLOROPHENYL)UREA

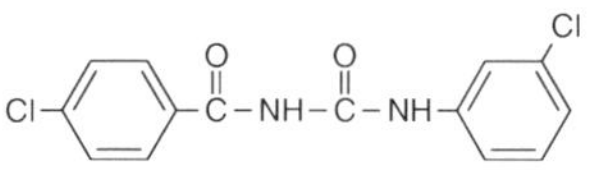

250 mg

S84,493-4
4-[3-(2-FLUOROPHENYL)UREIDO]-BENZOIC HYDRAZIDE

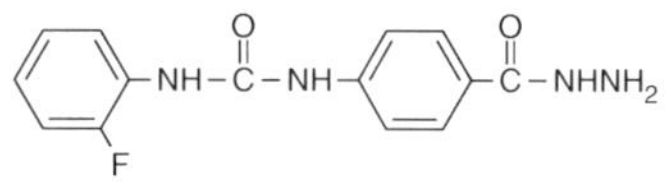

250 mg

S67,764-7
4-[3-(3-CHLOROPHENYL)UREIDO] BENZHYDRAZIDE

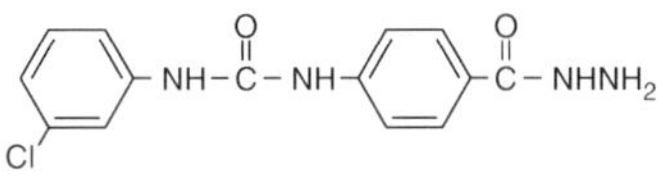

5 gm

S91,635-8
OXAMIC N2-[N-(4-FLUOROPHENYL)-CARBAMOYL]HYDRAZIDE

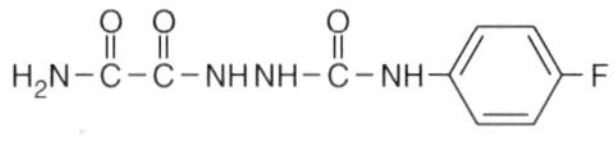

250 mg

S68,636-0
N-[3-(3,4-DICHLOROPHENYL)UREIDO]-N'-(4-FLUOROPHENYL)OXAMIDE

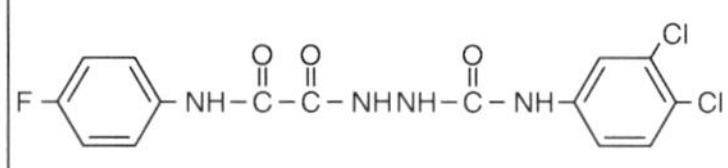

250 mg

S72,413-0
ETHYL [3-(2,4,5-TRICHLOROPHENYL)UREIDO]ACETATE

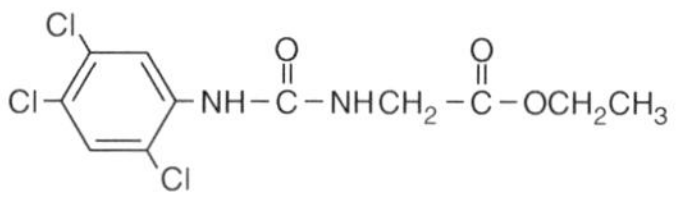

5 gm

S67,337-4
1-(3,5-DICHLOROPHENYL)-3-(ETHOXYCARBONYL)UREA

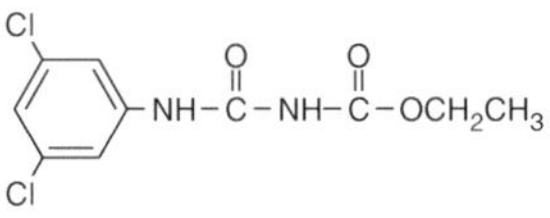

250 mg

S72,666-4
BUTYL (4-[3-(2-FLUOROPHENYL)-UREIDO])BENZOATE

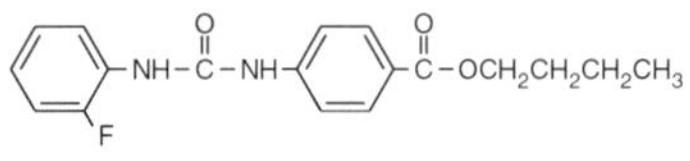

5 gm

S68,500-3
1-(2,4-DIFLUOROPHENYL)-3-(ETHOXYCARBONYLMETHYL)UREA

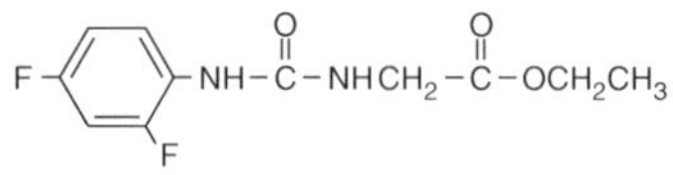

5 gm

S69,905-5
1-(4-BROMO-2,6-DIMETHYLPHENYL)-3-(ETHOXYCARBONYLMETHYL)UREA

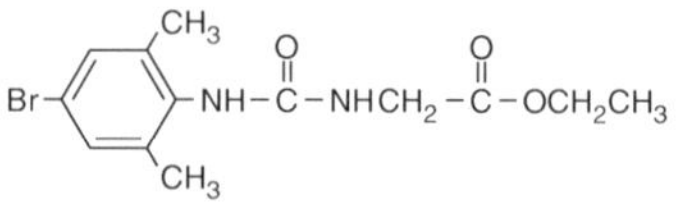

250 mg

S72,340-1
2-[3-(3-METHYL-2H-PYRAN-6-YLUREIDO)]ACETIC N2-[N(3-CHLORO-2-METHYLPHENYL)CARBAMOYL]HYDRAZIDE

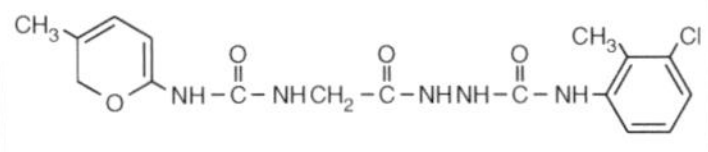

5 gm

S68,694-8
1,6-BIS(2-CHLOROPHENYL)DIUREA

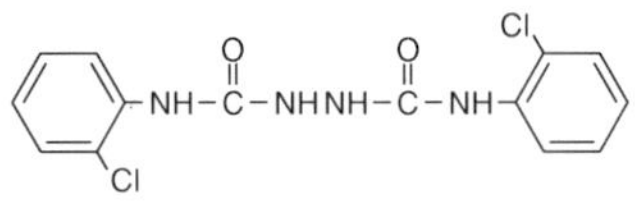

250 mg

S68,670-0
1,6-BIS(4-CHLOROPHENYL)DIUREA

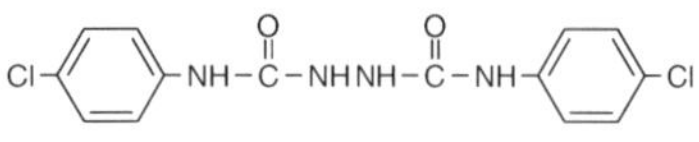

250 mg

S91,311-1
1-ACETYL-4-(3-CHLORO-4-METHYLPHENYL)SEMICARBAZIDE

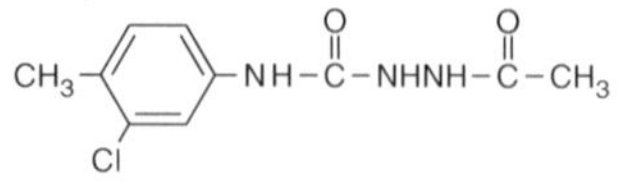

250 mg

S91,637-4
1-(4-FLUOROPHENYL)-3-(PIPERIDINO)-UREA

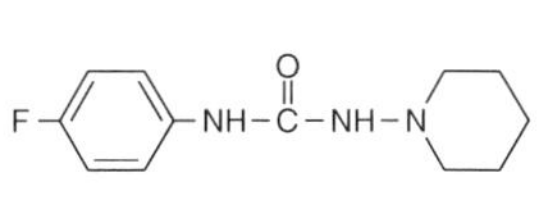

250 mg

S71,036-9
4-(4-BROMOPHENYL)-1-PHENYLSEMICARBAZIDE

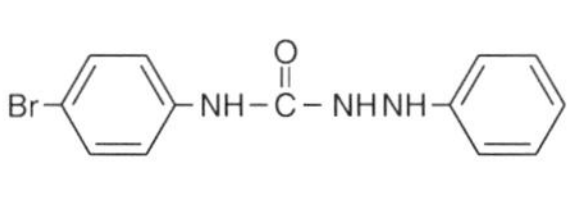

5 gm

S71,053-9
4-(2,6-DICHLOROPHENYL)-1-PHENYLSEMICARBAZIDE

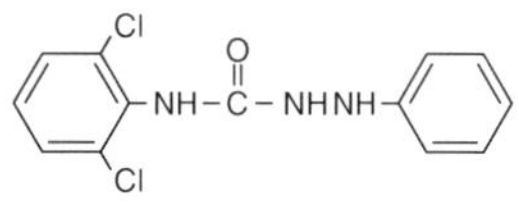

5 gm

S72,124-7
4-(3-CHLORO-2-METHYLPHENYL)-1-METHYL-1-PHENYLSEMICARBAZIDE

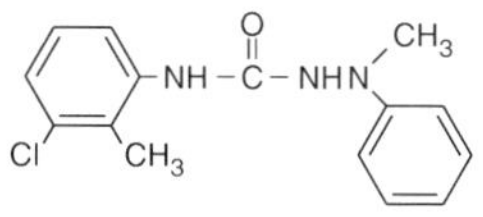

250 mg

S71,048-2
4-(3,4-DICHLOROPHENYL)-1-PHENYLSEMICARBAZIDE

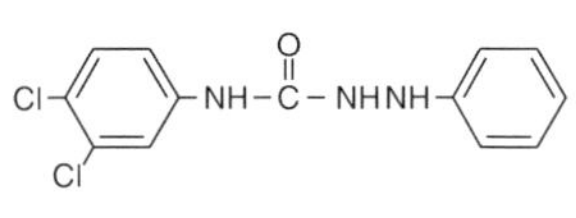

5 gm

S71,054-7
4-(4-BROMOPHENYL)-1,1-DIMETHYLSEMICARBAZIDE

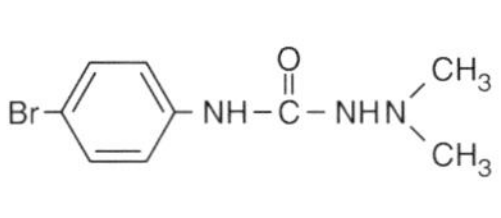

5 gm

S71,169-1
4-(5-CHLORO-2-METHYLPHENYL)-1,1-DIMETHYLSEMICARBAZIDE

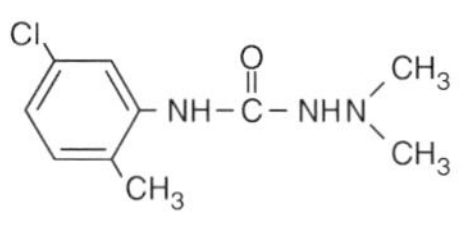

250 mg

S76,980-0
1-(2,5-DICHLOROPHENYL)-4-PROPYLSEMICARBAZIDE

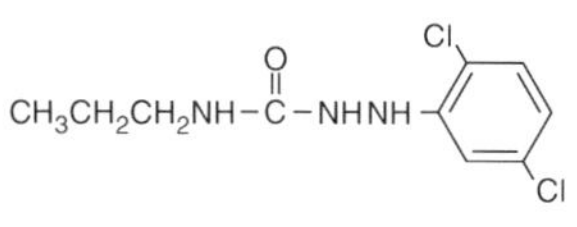

250 mg

S69,243-3
4-(3,4-DICHLOROPHENYL)-1,1-(HEXAMETHYLENE)SEMICARBAZIDE

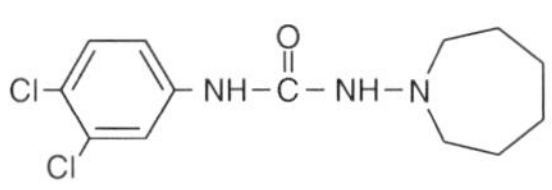

250 mg

S70,892-5
4-(3,5-DICHLOROPHENYL)-1,1-DIMETHYLSEMICARBAZIDE

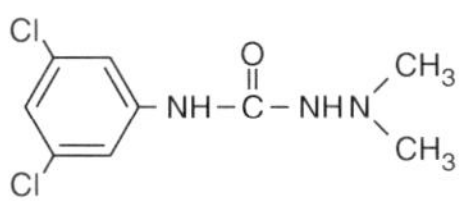

250 mg

S85,644-4
1-(4-CHLOROPHENYL)-3-FORMYLUREA

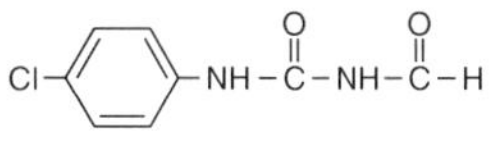

250 mg

S85,609-6
3-(4-CHLOROPHENYL)-1-FORMYL-1-METHYLUREA

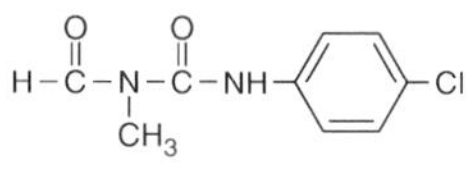

250 mg

S85,631-2
1-ACETYL-3-(4-CHLOROPHENYL)UREA

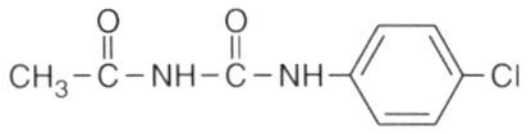

250 mg

S85,613-4
3-(4-CHLOROPHENYL)-1-ETHYL-1-FORMYLUREA

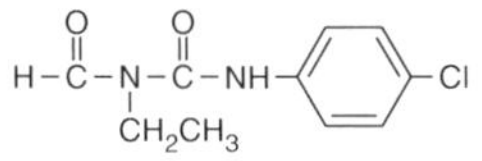

250 mg

S85,632-0
1-(4-CHLOROPHENYL)-3-PROPIONYLUREA

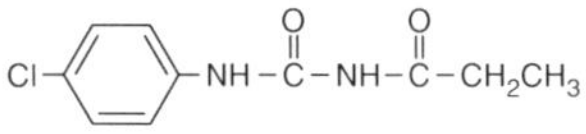

250 mg

S85,605-3
3-(4-CHLOROPHENYL)-1-METHYL-1-PROPIONYLUREA

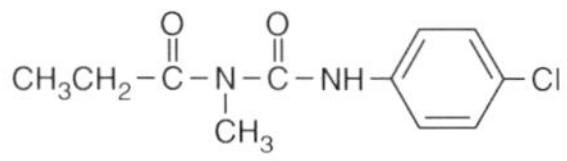

250 mg

S85,612-6
1-ACETYL-3-(4-CHLOROPHENYL)-1-ETHYLUREA

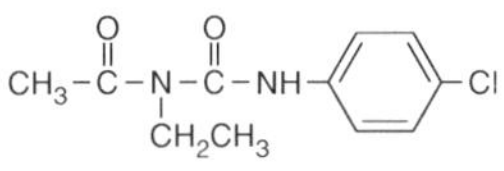

250 mg

S85,633-9
1-BUTYRYL-3-(4-CHLOROPHENYL)UREA

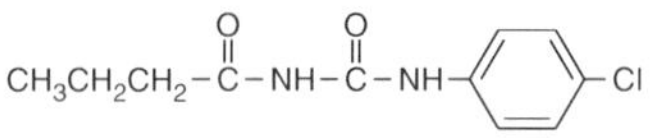

250 mg

S85,651-7
3-(4-CHLOROPHENYL)-1-FORMYL-1-PROPYLUREA

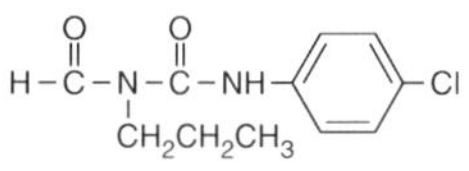

250 mg

S85,606-1
1-BUTYRYL-3-(4-CHLOROPHENYL)-1-METHYLUREA

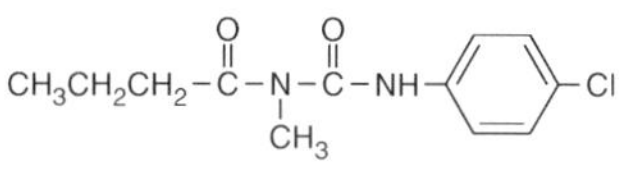

250 mg

S85,616-9
1-BUTYL-3-(4-CHLOROPHENYL)-1-FORMYLUREA

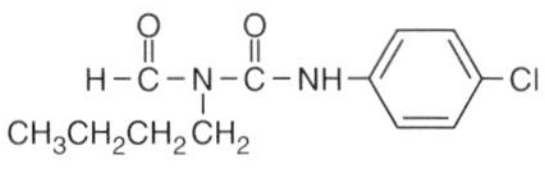

250 mg

S85,634-7
1-ACETYL-3-(4-CHLOROPHENYL)-1-PROPYLUREA

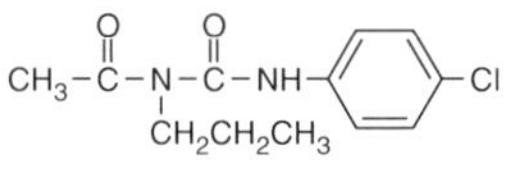

250 mg

S90,430-9
1-(4-FLUOROPHENYL)-3-VALERYLUREA

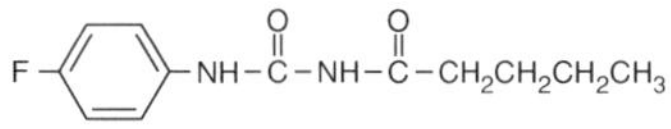

250 mg

S85,642-8
1-ACETYL-1-BUTYL-3-(4-CHLOROPHENYL)UREA

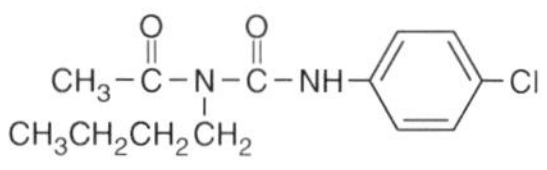

250 mg

S85,648-7
1-(3,4-DICHLOROPHENYL)-3-FORMYLUREA

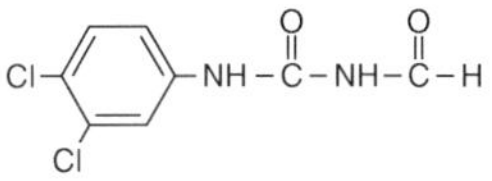

250 mg

S85,608-8
3-(3,4-DICHLOROPHENYL)-1-FORMYL-1-METHYLUREA

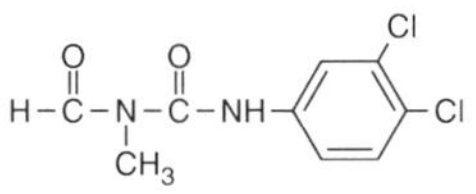

250 mg

S85,646-0
1-ACETYL-3-(3,4-DICHLOROPHENYL)UREA

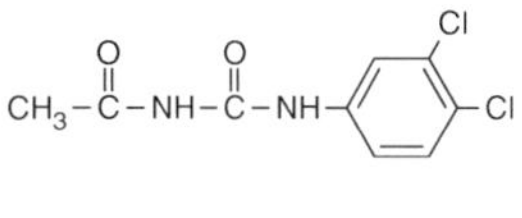

250 mg

S90,163-6
1-(2,5-DICHLOROPHENYL)-5-METHYLBIURET

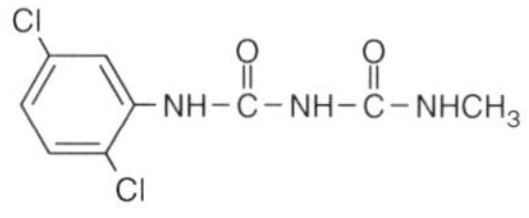

250 mg

S90,164-4
1-(2-CHLORO-5-METHYLPHENYL)-5-METHYLBIURET

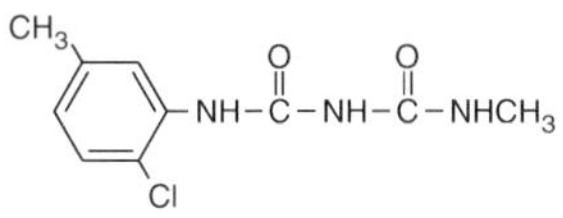

250 mg

S85,602-9
1-ACETYL-3-(3,4-DICHLOROPHENYL)-1-METHYLUREA

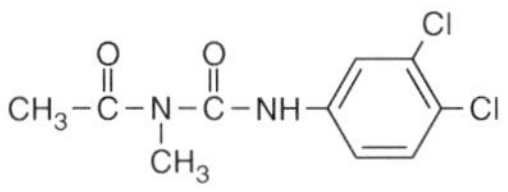

250 mg

S85,611-8
3-(3,4-DICHLOROPHENYL)-1-ETHYL-1-FORMYLUREA

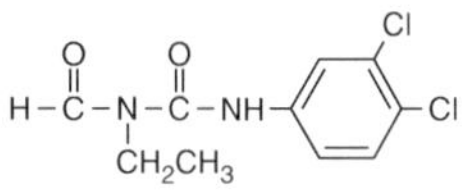

250 mg

S85,645-2
1-(3,4-DICHLOROPHENYL)-3-PROPIONYLUREA

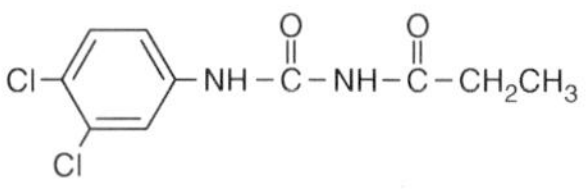

250 mg

S85,601-0
3-(3,4-DICHLOROPHENYL)-1-METHYL-1-PROPIONYLUREA

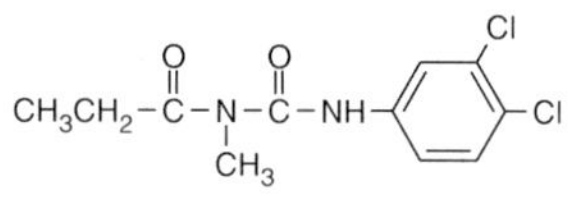

250 mg

S85,614-2
3-(3,4-DICHLOROPHENYL)-1-FORMYL-1-PROPYLUREA

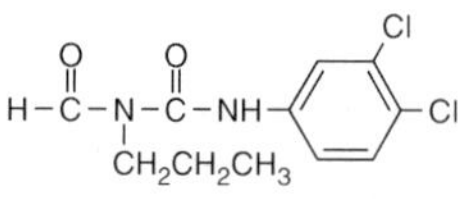

250 mg

S85,643-6
1-BUTYRYL-3-(3,4-DICHLOROPHENYL)UREA

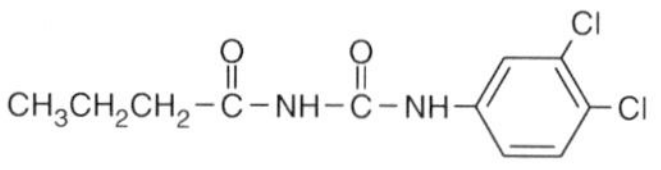

250 mg

S85,604-5
1-BUTYRYL-3-(3,4-DICHLOROPHENYL)-1-METHYLUREA

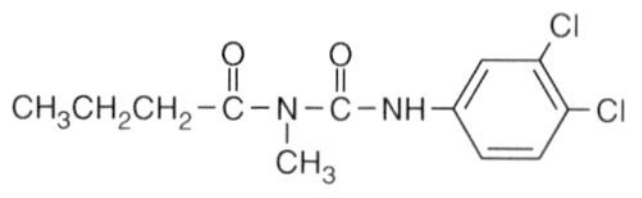

250 mg

S85,615-0
1-BUTYL-3-(3,4-DICHLOROPHENYL)-1-FORMYLUREA

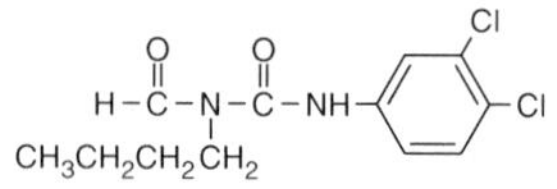

250 mg

S85,640-1
1-ACETYL-1-BUTYL-3-(3,4-DICHLOROPHENYL)UREA

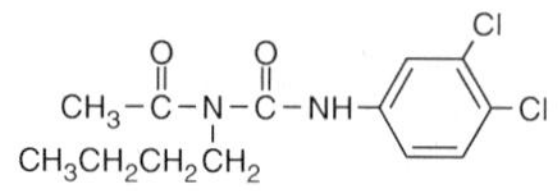

250 mg

S67,418-4
[3-(3-CHLOROPHENYL)UREIDO]ACETIC ACID HYDRAZIDE

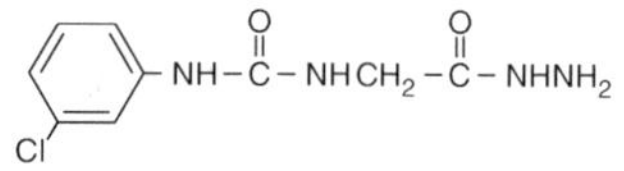

5 gm

S67,414-1
2-[3-(3,4-DICHLOROPHENYL)UREIDO] ACETIC ACID HYDRAZIDE

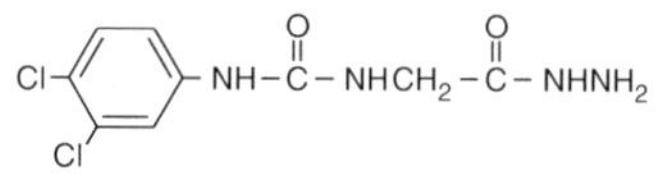

250 mg

S67,767-1
2-[3-(4-FLUOROPHENYL)UREIDO] ACETIC HYDRAZIDE

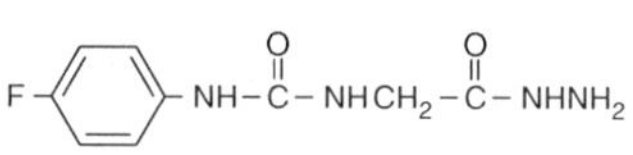

250 mg

S71,129-2
1-(3-BROMOPHENYL)-3-METHYLUREA

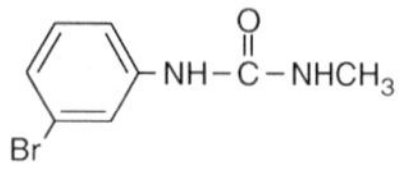

250 mg

S71,109-8
1-(3-BROMOPHENYL)-3-ETHYLUREA

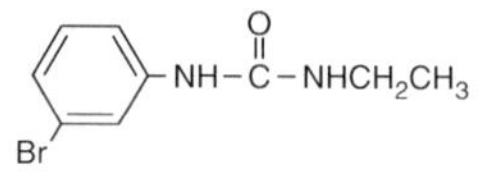

5 gm

S71,787-8
3-(3-BROMOPHENYL)-1,1-DIMETHYLUREA

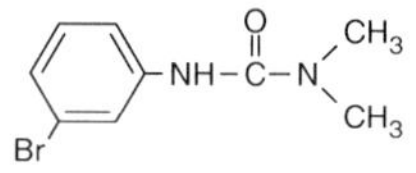

250 mg

S71,124-1
1-ALLYL-3-(3-BROMOPHENYL)UREA

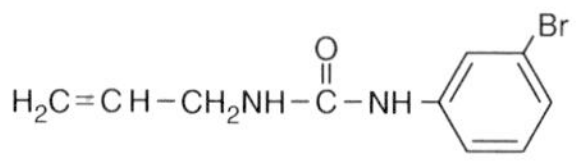

250 mg

S80,196-8
1,1-DIETHYL-3-(3-FLUOROPHENYL)UREA

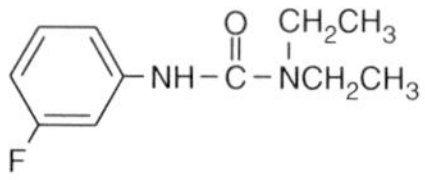

250 mg

S90,545-3
1,3-DIETHYL-1-(3-FLUOROPHENYL)UREA

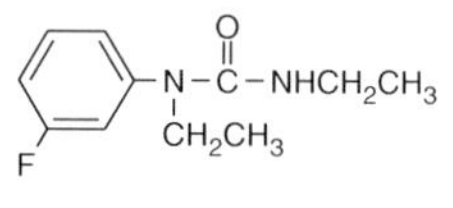

250 mg

S71,792-4
3-(3-BROMOPHENYL)-1-METHYL-1-PROPYLUREA

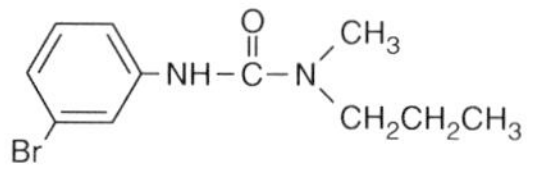

250 mg

S67,528-8
1,3-BIS(3-FLUOROPHENYL)UREA

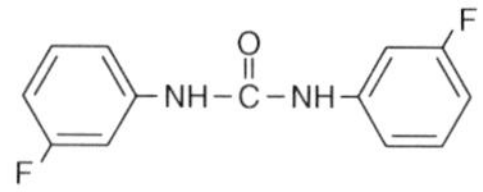

250 mg

S84,440-3
1-(3-CHLOROPHENYL)-3-(3-FLUOROPHENYL)UREA

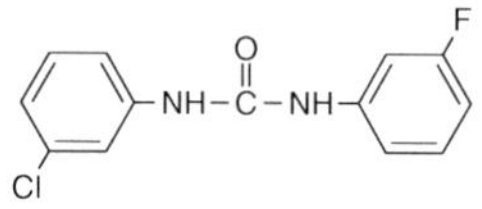

250 mg

S87,532-5
1,3-BIS(3-CHLOROPHENYL)UREA

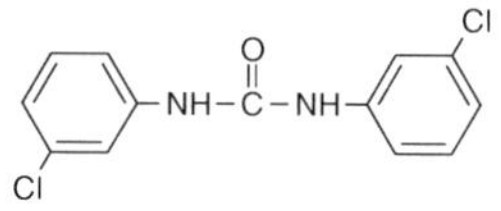

250 mg

S72,034-8
1-(3-BROMOPHENYL)-3-(HEXAMETHYLENEAMINO)UREA

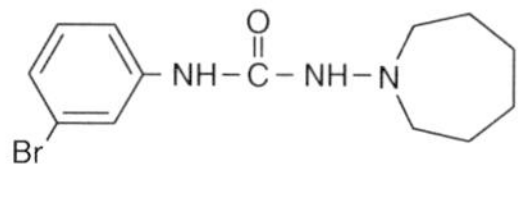

50 mg

S69,329-4
1-(3-BROMOPHENYL)-3-(2,4,5-TRICHLOROPHENYL)UREA

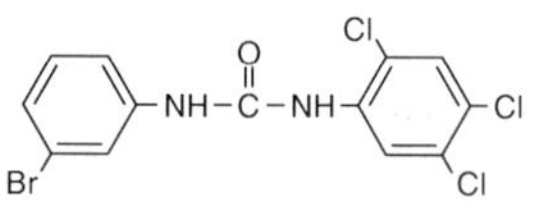

250 mg

S87,531-7
1,3-BIS(3-IODOPHENYL)UREA

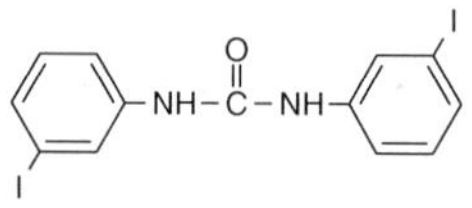

250 mg

S71,255-8
1-(3-CHLOROPHENYL)-3-(3-METHYLPHENYL)UREA

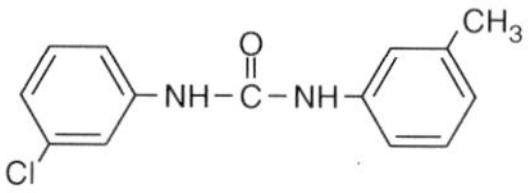

5 gm

S68,820-7
1-(3-CHLOROPHENYL)-1-ETHYL-3-(3-ETHYLPHENYL)UREA

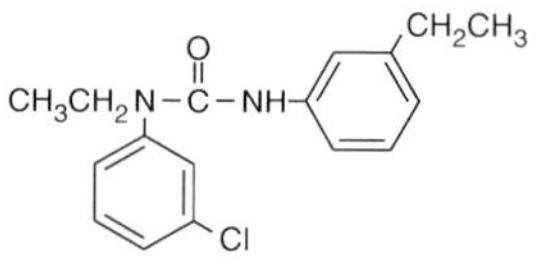

250 mg

S67,075-8
1-(3-CHLOROPHENYL)-3-OCTADECYLUREA

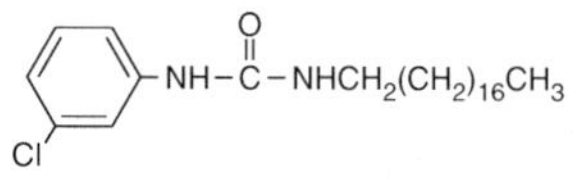

5 gm

S67,040-5
1-(3-BROMOPHENYL)-3-OCTADECYLUREA

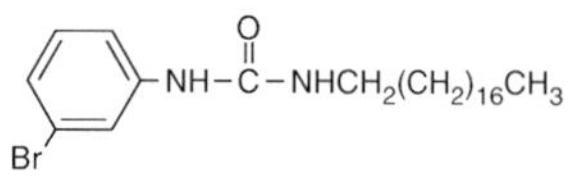

5 gm

S71,558-1
1-(CYCLOPROPYL)-3-(4-FLUOROPHENYL)UREA

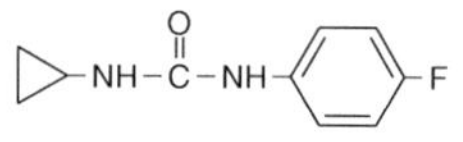

5 gm

S91,690-0
1-CYCLOPROPYL-3-(4-IODOPHENYL)UREA

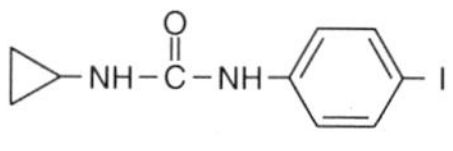

250 mg

S67,437-0
1-CYCLOPROPYL-3-(2,6-DICHLOROPHENYL)UREA

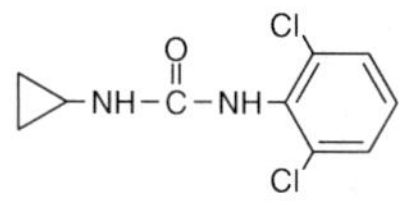

250 mg

S77,267-4
1-(4-CHLORO-2-METHYLPHENYL)-3-CYCLOPROPYLUREA

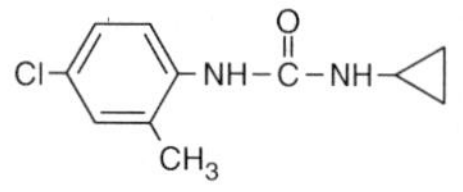

250 mg

S77,299-2
1-(3-CHLOROPHENYL)-3-CYCLOPENTYLUREA

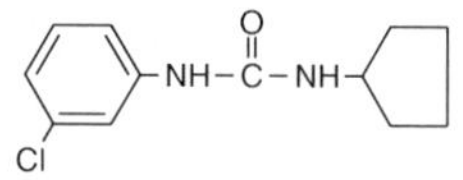

250 mg

S77,298-4
1-(3-BROMOPHENYL)-3-CYCLOPENTYLUREA

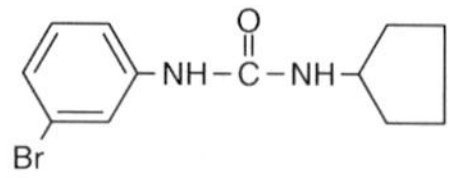

250 mg

S77,301-8
1-CYCLOPENTYL-3-(2-FLUOROPHENYL)UREA

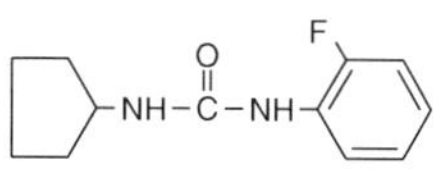

250 mg

S77,303-4
1-(2-CHLOROPHENYL)-3-CYCLOPENTYLUREA

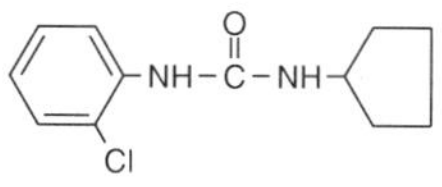

250 mg

S77,293-3
1-CYCLOPENTYL-3-(2,6-DICHLOROPHENYL)UREA

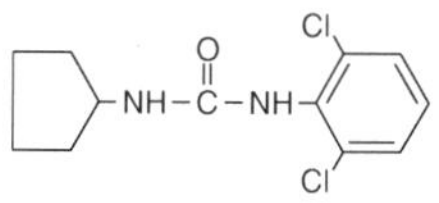

250 mg

S77,263-1
1-(4-FLUOROPHENYL)-3-(4-PHENYL-4-THIAZOLIN-2-YL)UREA

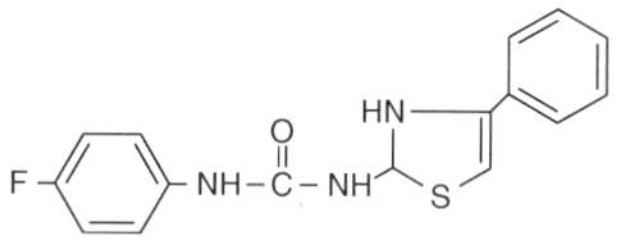

250 mg

S84,570-1
1-CYCLOHEXYL-3-(3-FLUOROPHENYL)-1-ISOPROPYLUREA

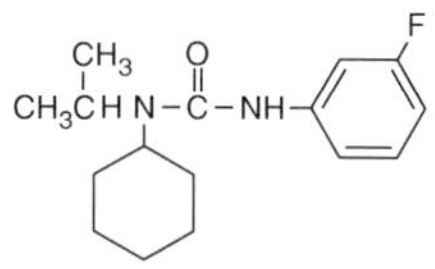

250 mg

S84,728-3
1,1-DICYCLOHEXYL-3-(3-FLUOROPHENYL)UREA

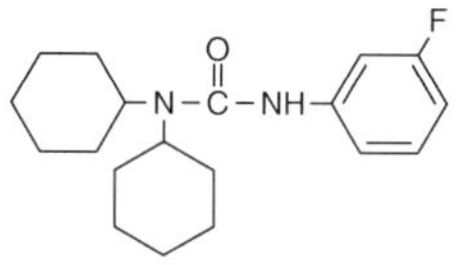

250 mg

S76,919-3
3-(4-CHLOROPHENYL)-1-CYCLOHEXYL-1-PHENYLUREA

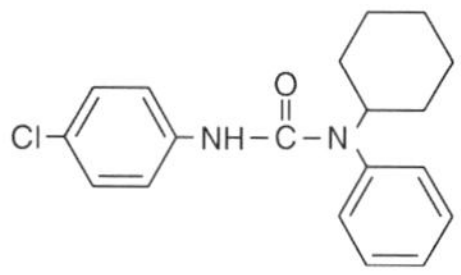

250 mg

S78,017-0
3-(2-CHLOROPHENYL)-1-CYCLOHEXYL-1-ISOPROPYLUREA

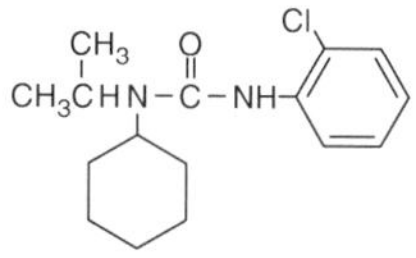

250 mg

S72,823-3
1,1-DICYCLOHEXYL-3-(2-FLUOROPHENYL)UREA

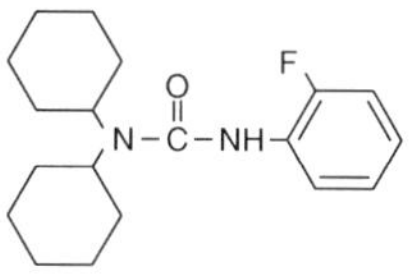

5 gm

S71,503-4
1-CYCLOHEXYL-3-(4-FLUOROPHENYL)UREA

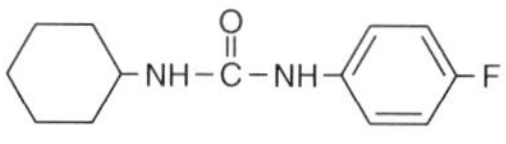

5 gm

S90,044-3
1-(4-CHLOROBENZYL)-3-CYCLOHEXYLUREA

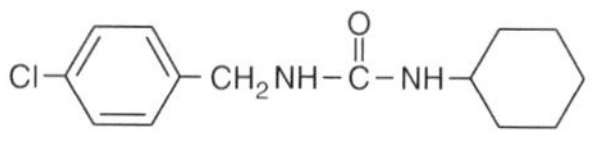

250 mg

S80,076-7
1,1-DICYCLOHEXYL-3-(4-FLUOROPHENYL)UREA

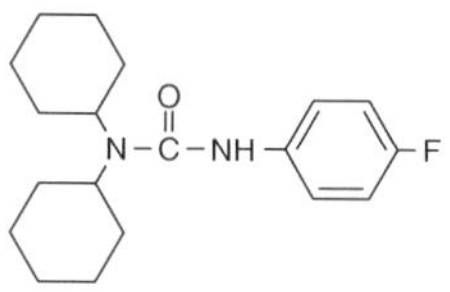

250 mg

S85,138-8
3-(2-CHLORO-6-METHYLPHENYL)-1-CYCLOHEXYL-1-ISOPROPYLUREA

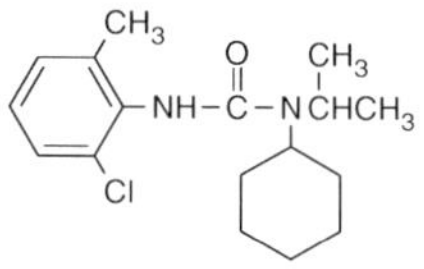

250 mg

S91,344-8
1-CYCLOHEXYL-3-(2,5-DIBROMOPHENYL)UREA

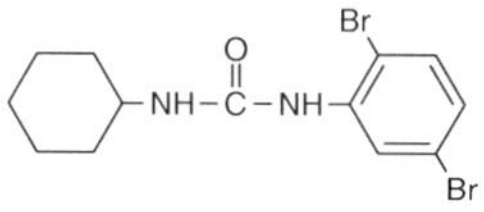

250 mg

S71,088-1
1-(4-CHLORO-2-METHYLPHENYL)-3-CYCLOHEXYLUREA

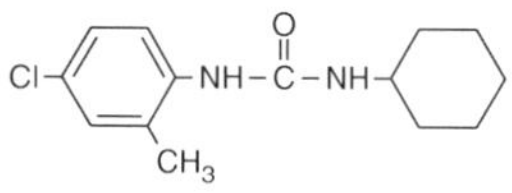

5 gm

S68,609-3
1-CYCLOHEXYL-3-(2,4-DIFLUOROPHENYL)-1-METHYLUREA

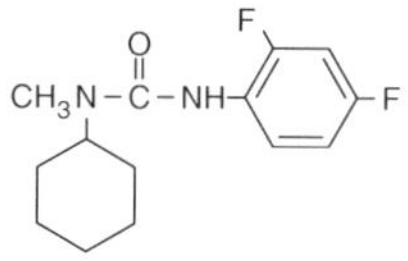

250 mg

S70,891-7
1-CYCLOHEXYL-3-(3,5-DICHLOROPHENYL)-1-METHYLUREA

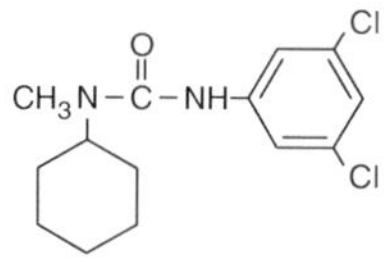

250 mg

S91,626-9
1-[N-(2,5-DIFLUOROPHENYL)CARBAMOYL]-PIPERIDINE

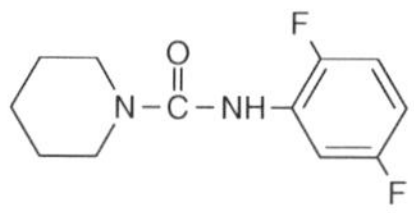

250 mg

S91,628-5
1-[N-(2,5-DIFLUOROPHENYL)CARBAMOYL]-3-METHYLPIPERIDINE

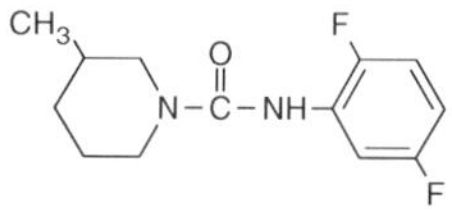

250 mg

S71,219-1
1-(3-CHLOROPHENYL)-3-PHENYLUREA

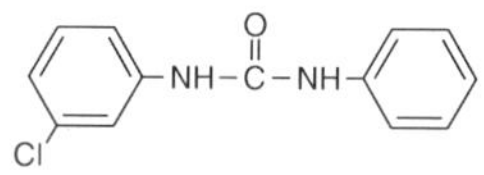

5 gm

S90,070-2
1-(3-CHLOROBENZYL)-1-METHYL-3-PHENYLUREA

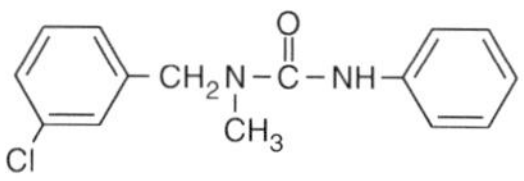

250 mg

S80,191-7
3-(3-FLUOROPHENYL)-1-ISOPROPYL-1-PHENYLUREA

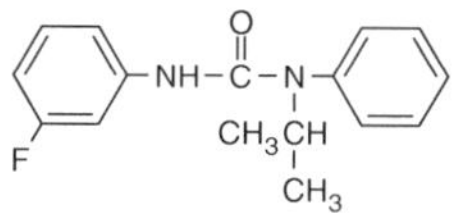

250 mg

S90,034-6
1-BENZYL-1-BUTYL-3-(3-CHLOROPHENYL)UREA

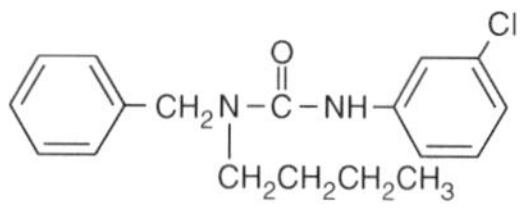

250 mg

S73,493-4
1-(2-CHLOROPHENYL)-3-PHENYLUREA

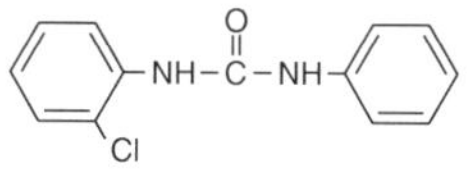

250 mg

S89,281-5
1-(2-CHLOROPHENYL)-3-(1-PHENYLETHYL)UREA

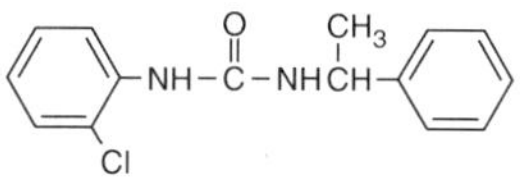

250 mg

S90,060-5
1-(2-CHLOROBENZYL)-1-METHYL-3-PHENYLUREA

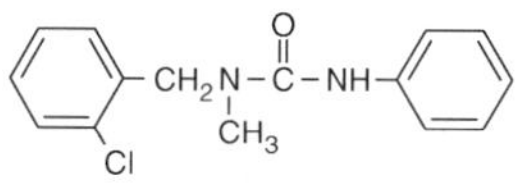

250 mg

S76,937-1
3-(2-CHLOROPHENYL)-1-ISOPROPYL-1-PHENYLUREA

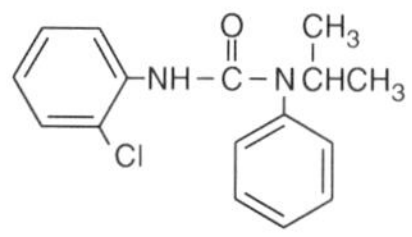

250 mg

S39,746-6
1-(2-FLUOROPHENYL)-3-PHENYLUREA

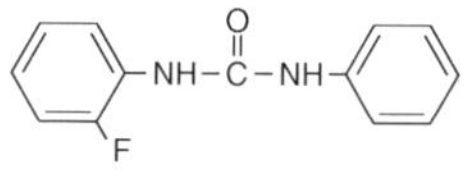

250 mg

S91,296-4
1-(DIPHENYLMETHYL)-3-(2-FLUOROPHENYL)UREA

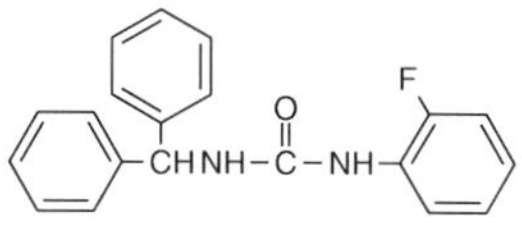

250 mg

S71,220-5
1-(4-CHLOROPHENYL)-3-PHENYLUREA

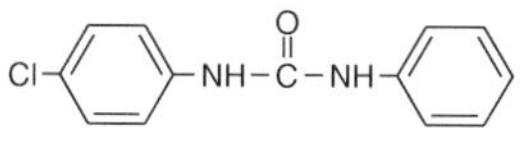

5 gm

S78,080-4
3-(4-CHLOROPHENYL)-1-ETHYL-1-PHENYLUREA

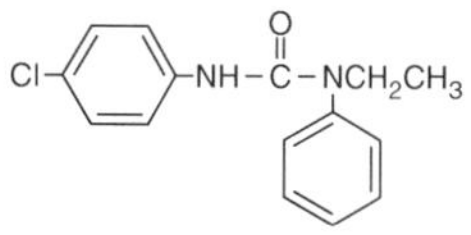

5 gm

S71,666-9
1-(4-BROMOPHENYL)-3-(2-PHENETHYL)UREA

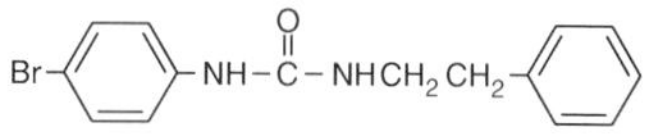

5 gm

S80,041-4
3-(4-FLUOROPHENYL)-1-ISOPROPYL-1-PHENYLUREA

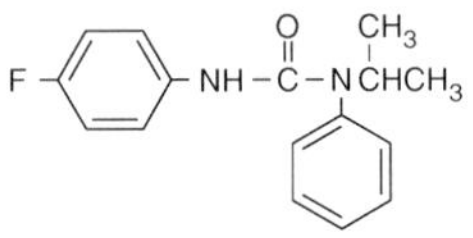

250 mg

S72,449-1
3-(4-BROMOPHENYL)-1,1-DIPHENYLUREA

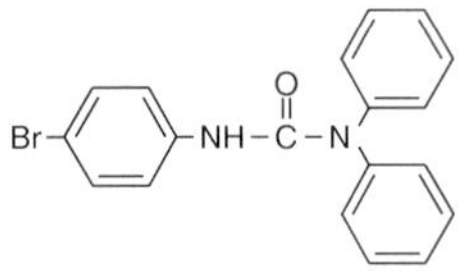

5 gm

S39,719-9
1-(4-FLUOROPHENYL)-3-PHENYLUREA

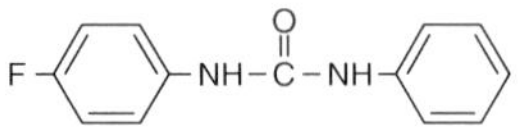

5 gm

S89,019-7
1-(3-CHLORO-2-METHYLPHENYL)-3-PHENYLUREA

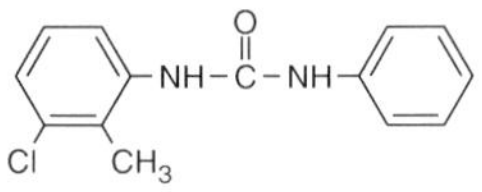

250 mg

S71,696-0
3-(2,6-DICHLOROPHENYL)-1-METHYL-1-PHENYLUREA

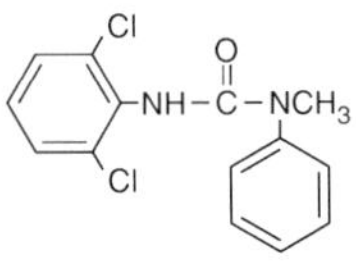

5 gm

S71,494-1
3-(2-CHLORO-6-METHYLPHENYL)-1-METHYL-1-PHENYLUREA

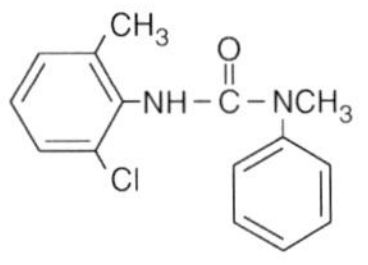

5 gm

S71,143-8
1-(2,6-DICHLOROPHENYL)-3-PHENETHYLUREA

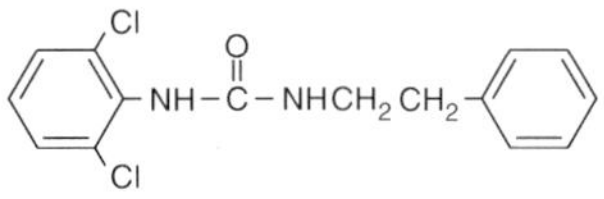

5 gm

S71,754-1
1-BUTYL-3-(2,6-DICHLOROPHENYL)-1-PHENYLUREA

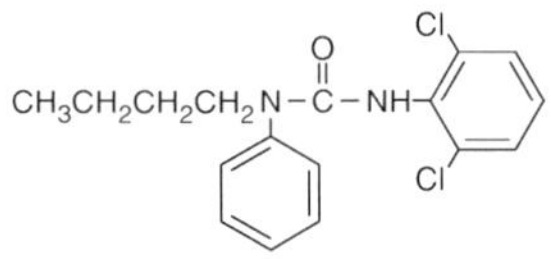

5 gm

S39,760-1
1-(2-CHLORO-3-METHYLPHENYL)-3-PHENYLUREA

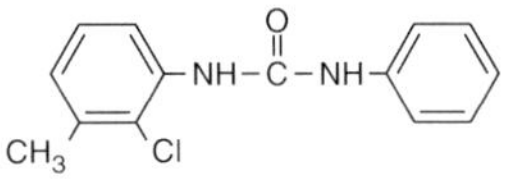

250 mg

S72,493-9
1-BENZYL-3-(3-CHLORO-4-FLUOROPHENYL)-1-METHYLUREA

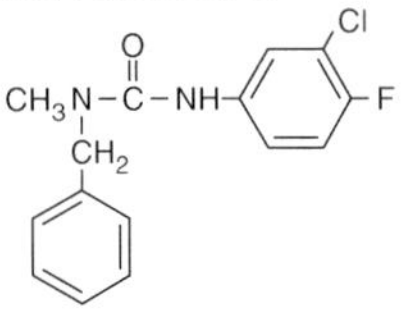

5 gm

S73,562-0
1-BUTYL-3-(2,4-DICHLOROPHENYL)-1-PHENYLUREA

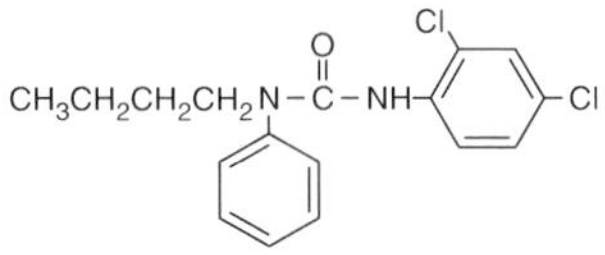

250 mg

S70,026-6
1-(2,4-DIFLUOROPHENYL)-3-(DIPHENYLMETHYL)UREA

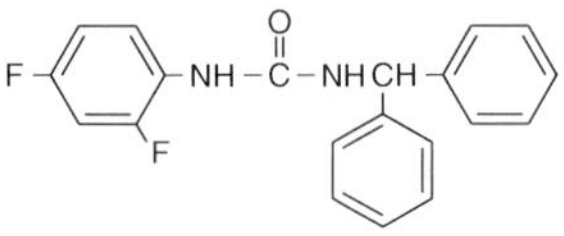

250 mg

S91,316-2
1-(3-CHLORO-4-METHYLPHENYL)-3-(DIPHENYLMETHYL)UREA

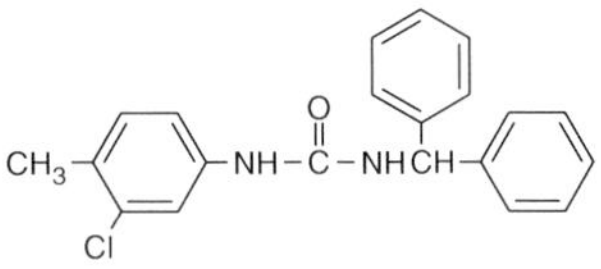

250 mg

S70,867-4
1-PHENYL-3-(2,4,6-TRICHLOROPHENYL)UREA

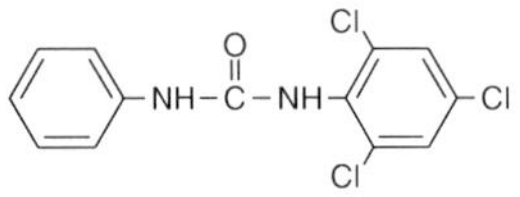

5 gm

S73,057-2
(2-FLUOROPHENYL)UREA

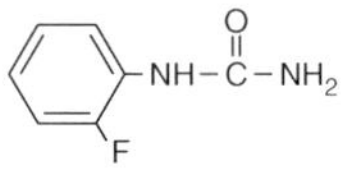

250 mg

S66,798-6
1-(2-FLUOROPHENYL)-3-ISOPROPYLUREA

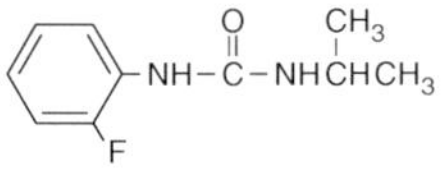

5 gm

S89,816-3
1-BUTYL-3-(2-FLUOROPHENYL)UREA

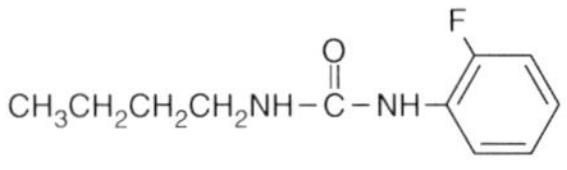

250 mg

S89,287-4
1-SEC-BUTYL-3-(2-CHLOROPHENYL)UREA

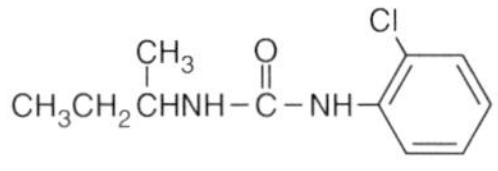

250 mg

S89,982-8
1-TERT-BUTYL-3-(2-CHLOROBENZYL)UREA

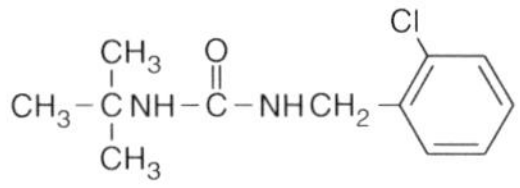

250 mg

S67,476-1
1-(2-CHLOROPHENYL)-3-HEXYLUREA

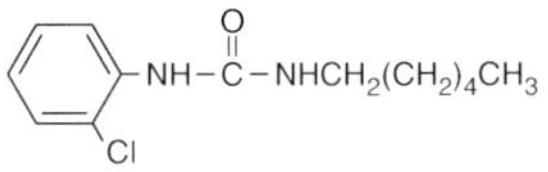

250 mg

S78,096-0
1-(3-CHLOROPHENYL)-3-(2-FLUOROPHENYL)UREA

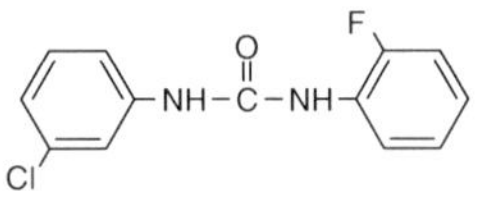

250 mg

S67,535-0
1,3-BIS(2-CHLOROPHENYL)UREA

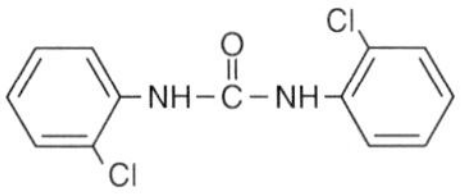

5 gm

S70,868-2
1-(2-CHLOROPHENYL)-3-(2,4,5-TRICHLOROPHENYL)UREA

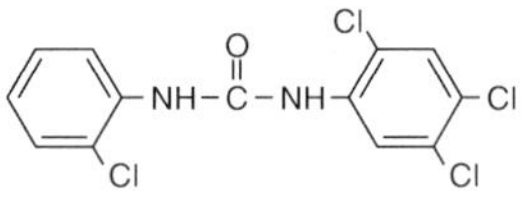

5 gm

S71,253-1
1-(2-CHLOROPHENYL)-3-(2-METHYLPHENYL)UREA

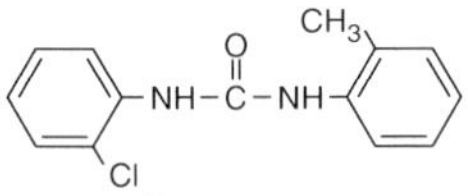

5 gm

S71,852-1
1-(2-CHLOROBENZYL)-3-(2-FLUOROPHENYL)UREA

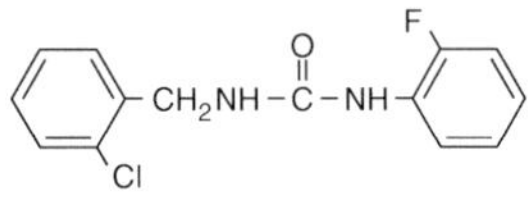

250 mg

S72,815-2
1-(4,5-DICHLORO-2-METHYLPHENYL)-3-(2-FLUOROPHENYL)UREA

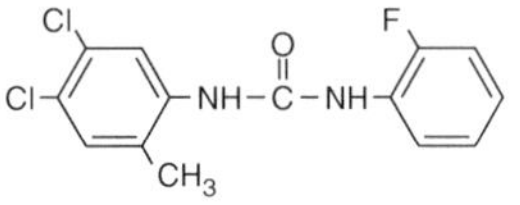

5 gm

S78,078-2
1-(2-FLUOROPHENYL)-3-(2-ETHYLPHENYL)UREA

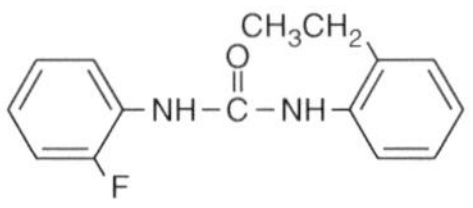

250 mg

S77,327-1
1-(3-CHLOROPHENYL)-3-(2-ETHYLPHENYL)UREA

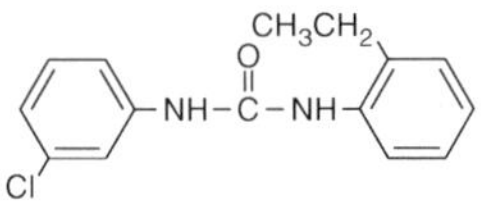

250 mg

S71,878-5
1-(2-FLUOROPHENYL)-3-(2,4,5-TRIMETHYLPHENYL)UREA

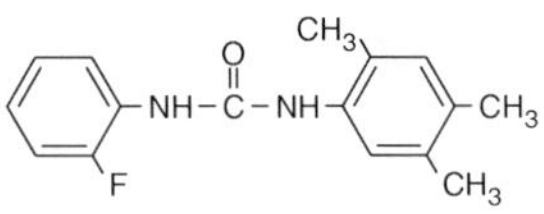

250 mg

S86,876-0
1,1'-HEXAMETHYLENEBIS-[3-(2-FLUOROPHENYL)UREA]

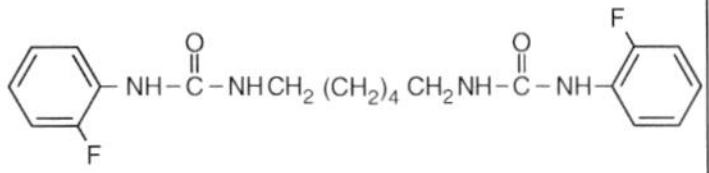

250 mg

S67,019-7
1-(2-FLUOROPHENYL)-3-OCTADECYLUREA

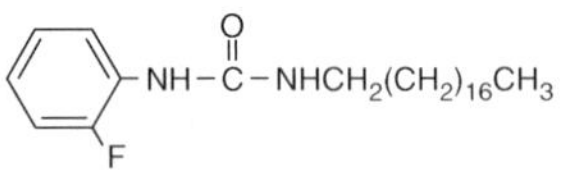

5 gm

S70,266-8
1,1'-[(2-CHLOROBENZYL)-2-METHYLPHENYLENE]BIS[3-(3-CHLOROPHENYL)UREA]

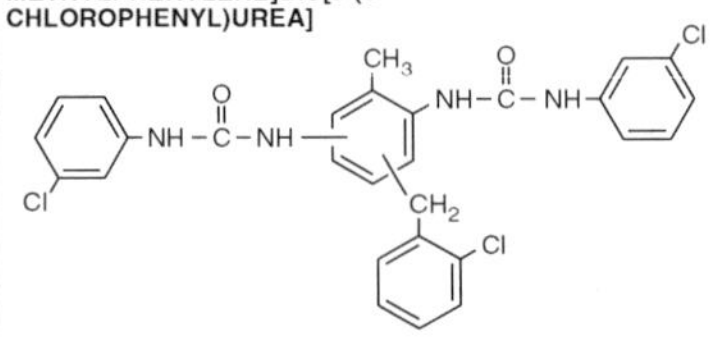

250 mg

S88,983-0
1-(4-CHLOROPHENYL)-3-(2-FLUOROPHENYL)UREA

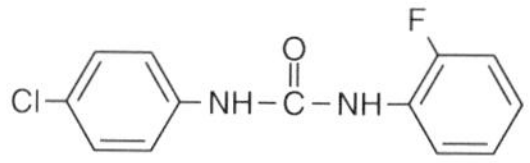

250 mg

S70,865-8
1-(4-BROMOPHENYL)-3-(2-CHLOROPHENYL)UREA

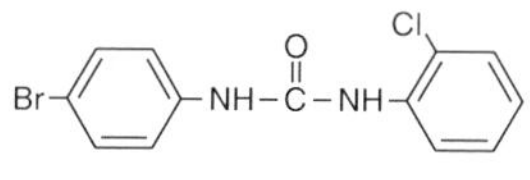

5 gm

S88,997-0
1-(4-FLUOROPHENYL)-3-(O-TOLYL)UREA

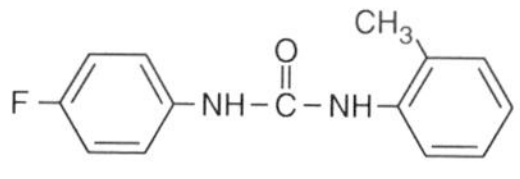

250 mg

S88,964-4
1-(4-ETHYLPHENYL)-3-(2-FLUOROPHENYL)UREA

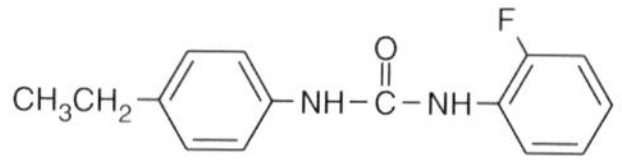

250 mg

S76,841-3
1-(4-CHLOROPHENYL)-3-(2-ETHYLPHENYL)UREA

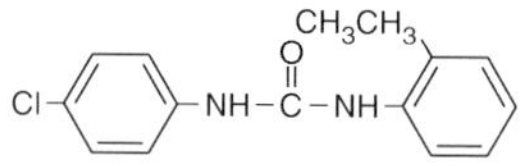

250 mg

S90,019-2
1-(4-CHLOROBENZYL)-3-(2-CHLOROPHENYL)-1-METHYLUREA

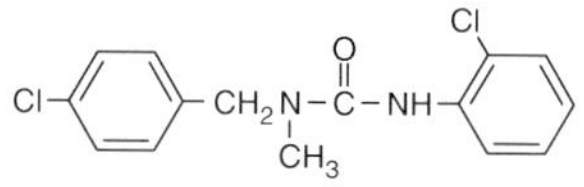

250 mg

S78,087-1
3-(4-CHLOROPHENYL)-1-ETHYL-1-(O-TOLYL)UREA

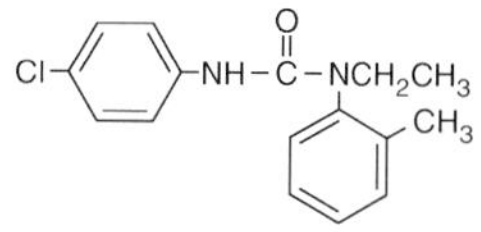

250 mg

S78,016-2
1-(4-BUTYLPHENYL)-3-(2-FLUOROPHENYL)UREA

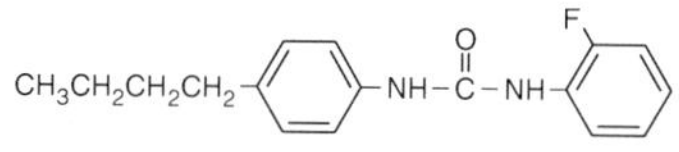

250 mg

S70,267-6
1,1'-[(2-CHLOROBENZYL)-2-METHYLPHENYLENE]BIS[3-(4-BROMOPHENYL)UREA]

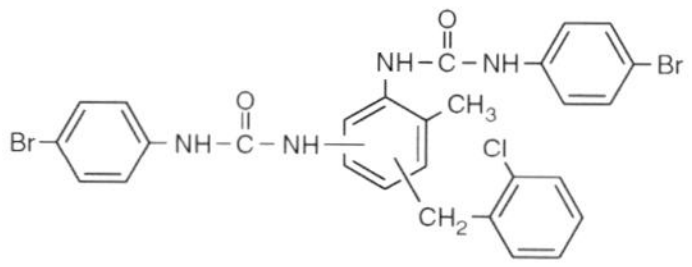

250 mg

S70,268-4
1,1'-[(2-CHLOROBENZYL)-2-METHYLPHENYLENE]BIS[3-(4-IODOPHENYL)UREA]

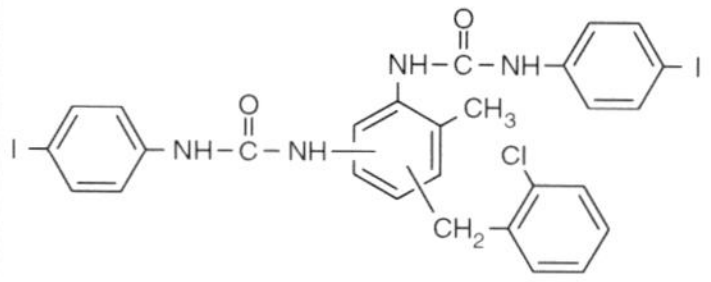

250 mg

S39,725-3
1-(2-CHLOROPHENYL)-3-(4-FLUOROPHENYL)UREA

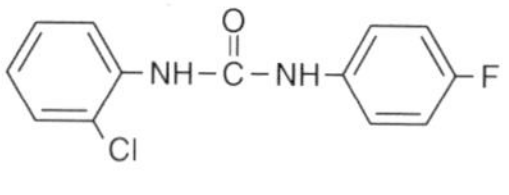

250 mg

S71,210-8
1-(2-CHLOROPHENYL)-3-(2,3-DICHLOROPHENYL)UREA

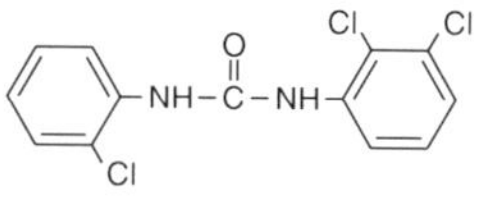

5 gm

S71,212-4
1-(2-CHLOROPHENYL)-3-(2,6-DICHLOROPHENYL)UREA

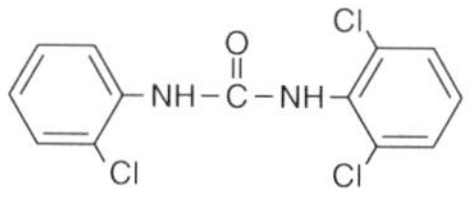

5 gm

S80,124-0
1-(3-CHLORO-2-METHYLPHENYL)-3-(2-FLUOROPHENYL)UREA

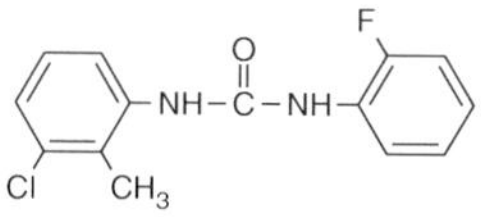

250 mg

S71,248-5
1-(2-CHLOROPHENYL)-3-(2-CHLORO-6-METHYLPHENYL)UREA

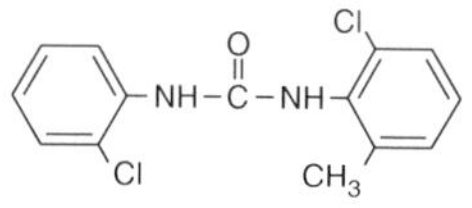

5 gm

S71,497-6
1-(2,6-DICHLOROPHENYL)-3-(O-TOLYL)UREA

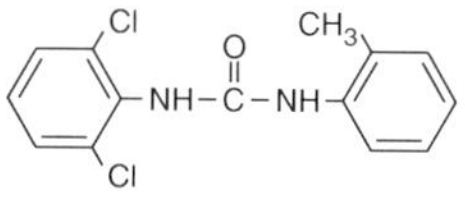

250 mg

S77,378-6
1-(3-CHLORO-2-METHYLPHENYL)-3-(2-CHLOROPHENYL)UREA

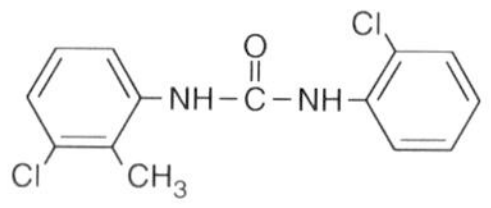

250 mg

S73,693-7
1-(2-FLUOROPHENYL)-3-(2,3-XYLYL)UREA

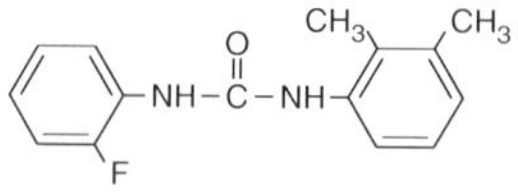

5 gm

S76,880-4
1-(2-FLUOROPHENYL)-3-(2,6-XYLYL)UREA

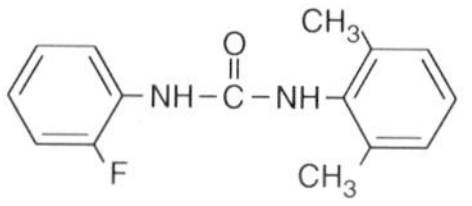

250 mg

S71,493-3
1-(2-CHLOROPHENYL)-3-(2,6-XYLYL)UREA

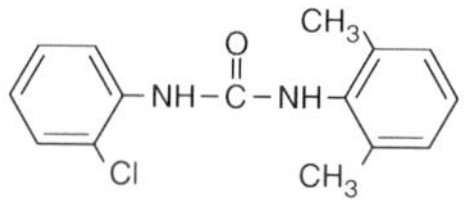

5 gm

S76,877-4
1-(2-CHLOROPHENYL)-3-(2,3-DIMETHYLPHENYL)UREA

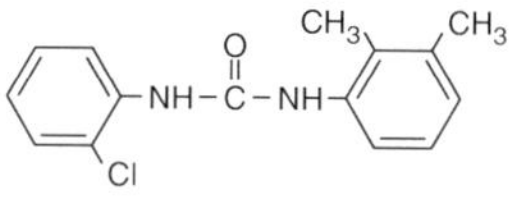

250 mg

S65,968-1
1-(2-CHLOROPHENYL)-3-(2-ETHYL-6-METHYLPHENYL)UREA

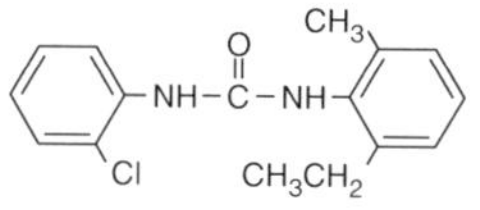

5 gm

S88,969-5
1-(3-CHLORO-2-METHYLPHENYL)-3-(2-ETHYLPHENYL)UREA

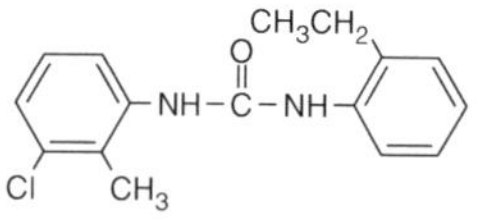

250 mg

S67,627-6
1-(2-CHLOROPHENYL)-3-(2,6-DIETHYLPHENYL)UREA

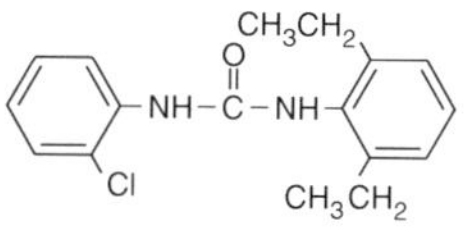

5 gm

S73,633-3
1-(2,4-DICHLOROPHENYL)-3-(2-FLUOROPHENYL)UREA

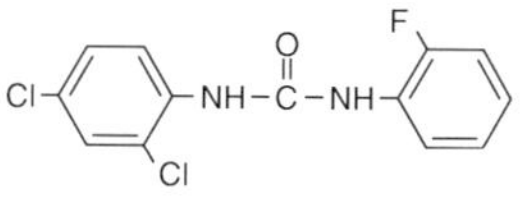

5 gm

S71,211-6
1-(2-CHLOROPHENYL)-3-(2,5-DICHLOROPHENYL)UREA

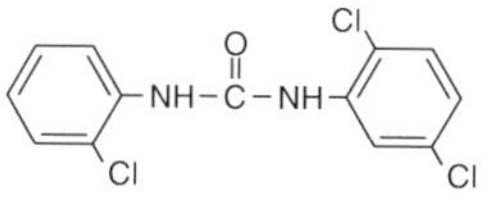

5 gm

S73,646-5
1-(3-CHLORO-4-METHYLPHENYL)-3-(2-FLUOROPHENYL)UREA

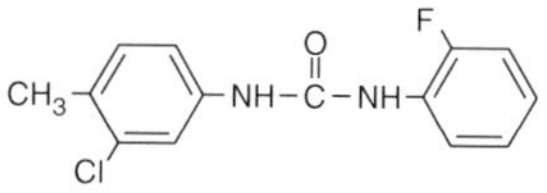

5 gm

S73,647-3
1-(5-CHLORO-2-METHYLPHENYL)-3-(2-FLUOROPHENYL)UREA

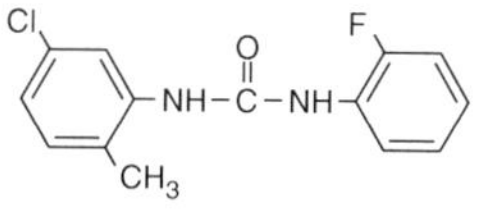

5 gm

S73,648-1
1-(2-CHLORO-5-METHYLPHENYL)-3-(2-FLUOROPHENYL)UREA

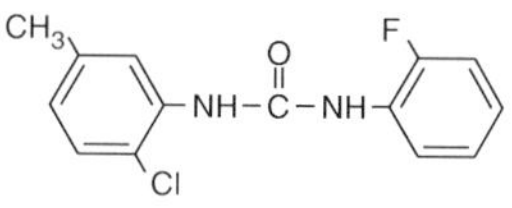

5 gm

S73,650-3
1-(4-CHLORO-2-METHYLPHENYL)-3-(2-FLUOROPHENYL)UREA

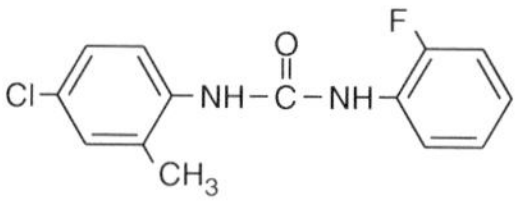

5 gm

S76,832-4
1-(2-CHLOROPHENYL)-3-(3-CHLORO-4-METHYLPHENYL)UREA

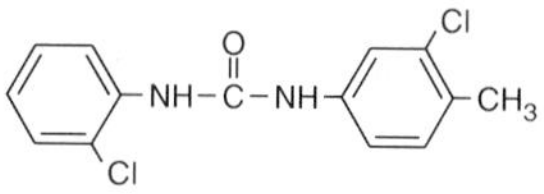

250 mg

S77,332-8
1-(2-CHLOROPHENYL)-3-(4-CHLORO-2-METHYLPHENYL)UREA

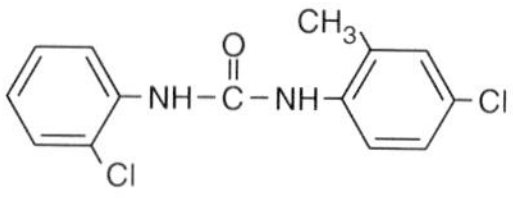

250 mg

S77,373-5
1-(2-CHLORO-5-METHYLPHENYL)-3-(2-CHLOROPHENYL)UREA

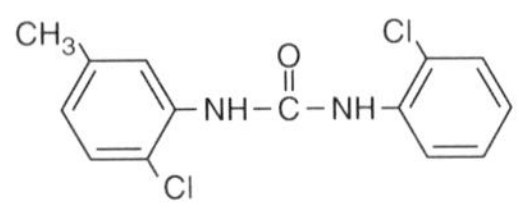

250 mg

S89,035-9
1-(5-CHLORO-2-METHYLPHENYL)-3-(2-CHLOROPHENYL)UREA

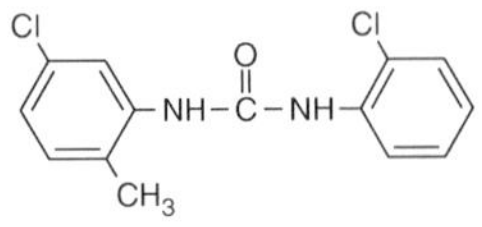

250 mg

S90,018-4
1-(2-CHLOROPHENYL)-3-(3,4-DICHLOROBENZYL)UREA

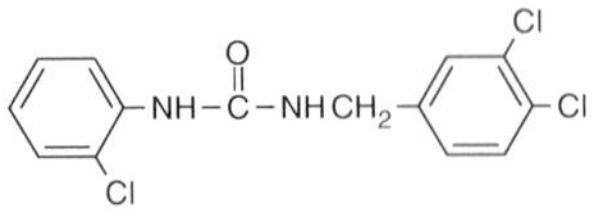

250 mg

S73,692-9
1-(2-FLUOROPHENYL)-3-(2,5-XYLYL)UREA

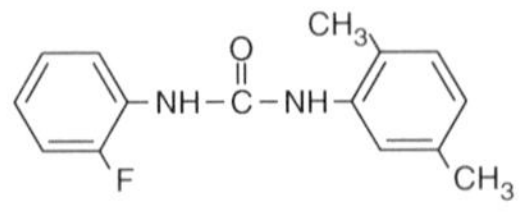

5 gm

S75,364-5
1-(3-CHLORO-4-METHYLPHENYL)-3-(2-ETHYLPHENYL)UREA

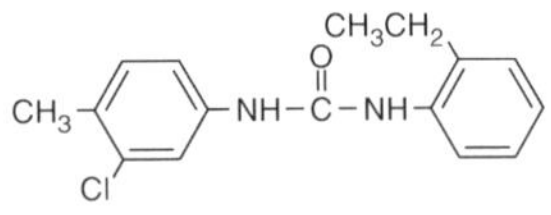

5 gm

S90,026-5
1-(2-CHLOROPHENYL)-3-(2,5-DIMETHYLBENZYL)UREA

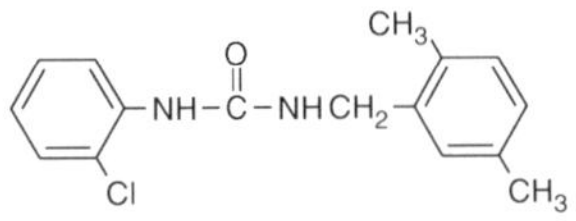

250 mg

S71,876-9
1-(2-FLUOROPHENYL)-3-(5-ISOPROPYL-2-METHYLPHENYL)UREA

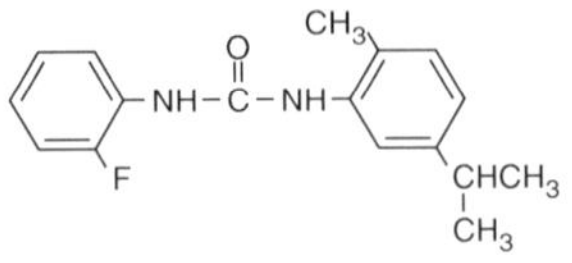

250 mg

S73,563-9
1-(2-CHLOROPHENYL)-3-(3,5-DICHLOROPHENYL)UREA

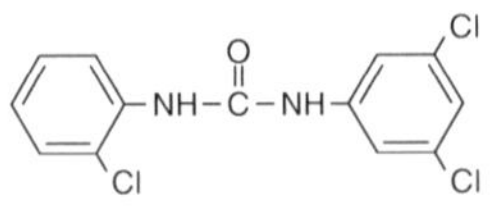

250 mg

S71,194-2
1-(2-CHLOROPHENYL)-3-(2,4,6-TRICHLOROPHENYL)UREA

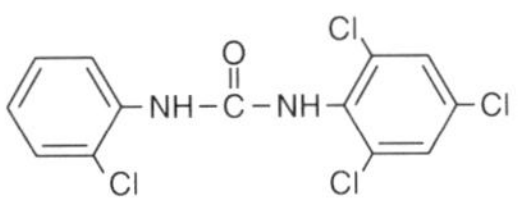

250 mg

S75,363-7
1-(2-CHLOROPHENYL)-3-(2,4,6-TRIMETHYLPHENYL)UREA

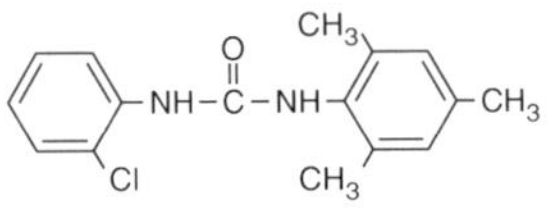

5 gm

S75,339-4
1-(2-CHLOROPHENYL)-3-(4,5-DICHLORO-2-METHYLPHENYL)UREA

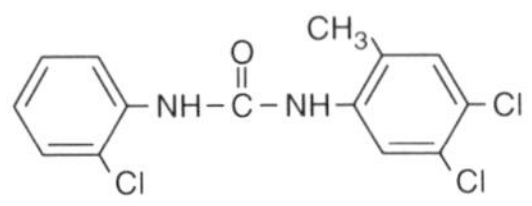

5 gm

S89,122-3
1-(2-CHLOROPHENYL)-3-(2,4,5-TRIMETHYLPHENYL)UREA

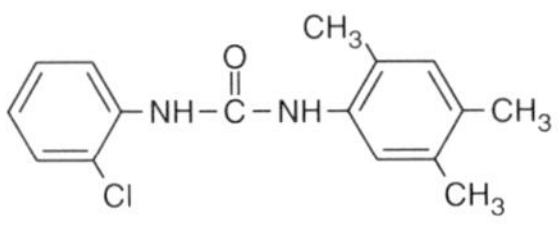

250 mg

S76,965-7
1-(2-FLUOROPHENYL)-3-(1-NAPHTHYL)UREA

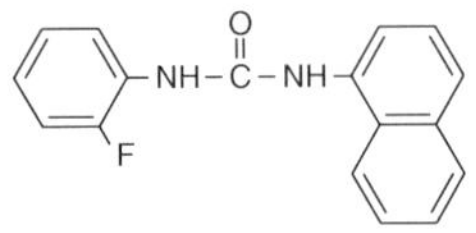

250 mg

S73,160-9
1-(2-FLUOROPHENYL)-3-(2-NAPHTHYL)UREA

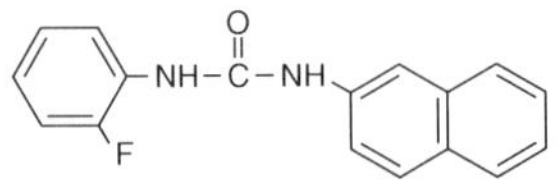

5 gm

S71,432-1
1-(4-FLUOROPHENYL)-3-METHYLUREA

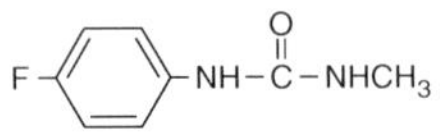

5 gm

S66,508-8
1-(4-BROMOPHENYL)-3-METHYLUREA

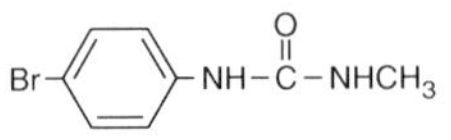

5 gm

S90,535-6
1,1-DIMETHYL-3-(4-IODOPHENYL)UREA

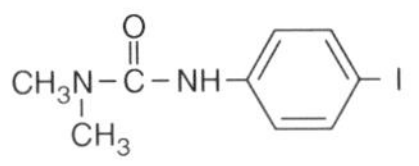

250 mg

S71,388-0
1-(4-FLUOROPHENYL)-3-ISOPROPYLUREA

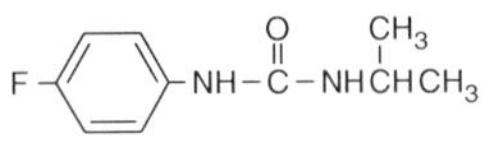

5 gm

S90,359-0
1-(4-CHLOROPHENYL)-1-ETHYL-3-METHYLUREA

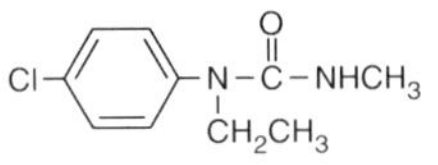

250 mg

S71,042-3
1-ALLYL-3-(4-BROMOPHENYL)UREA

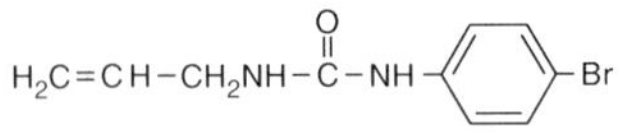

5 gm

S66,504-5
1-(4-BROMOPHENYL)-3-PROPYLUREA

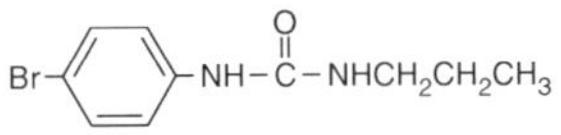

5 gm

S75,329-7
1-TERT-BUTYL-3-(4-FLUOROPHENYL)UREA

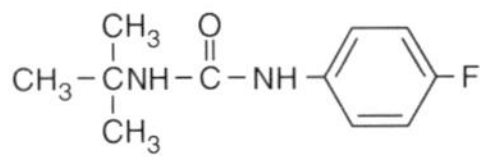

5 gm

S66,497-9
1-(4-BROMOPHENYL)-3-ISOBUTYLUREA

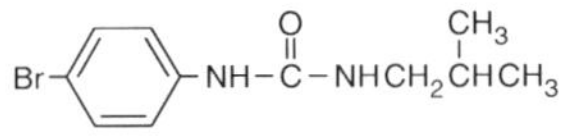

5 gm

S66,568-1
1-(4-BROMOPHENYL)-3-BUTYLUREA

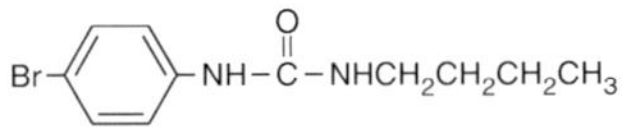

5 gm

S66,589-4
3-(4-BROMOPHENYL)-1,1-DIETHYLUREA

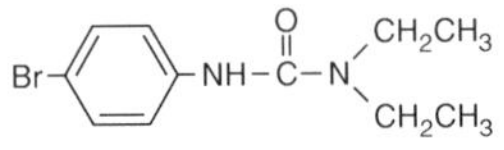

5 gm

S71,496-8
1-(4-CHLOROPHENYL)-3-(1,3-DIMETHYLBUTYL)UREA

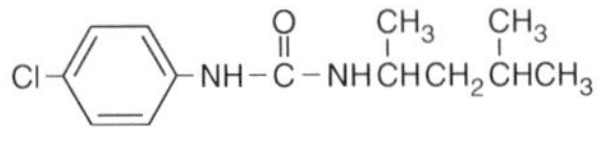

5 gm

S71,216-7
1-(3-CHLOROPHENYL)-3-(4-FLUOROPHENYL)UREA

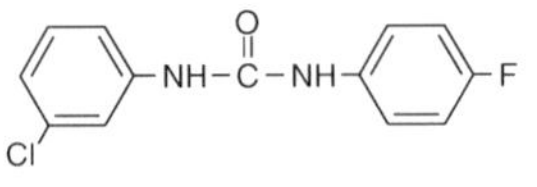

5 gm

S78,095-2
1-(4-CHLOROPHENYL)-3-(3-FLUOROPHENYL)UREA

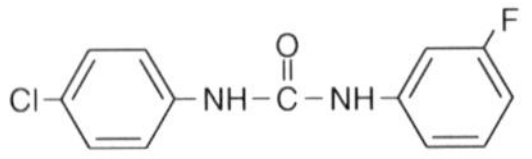

250 mg

S71,217-5
1-(4-BROMOPHENYL)-3-(4-CHLOROPHENYL)UREA

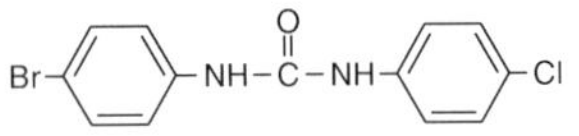

5 gm

S80,020-1
1-(4-FLUOROPHENYL)-3-(4-IODOPHENYL)UREA

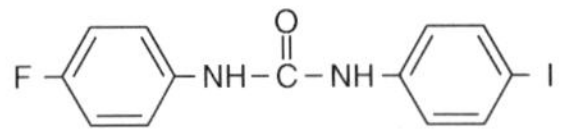

250 mg

S87,530-9
1,3-BIS(4-IODOPHENYL)UREA

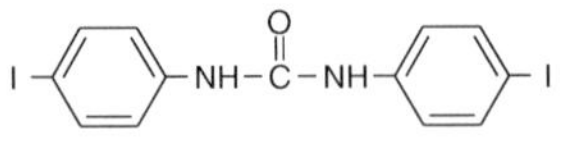

250 mg

S89,112-6
1-(4-FLUOROPHENYL)-3-(M-TOLYL)UREA

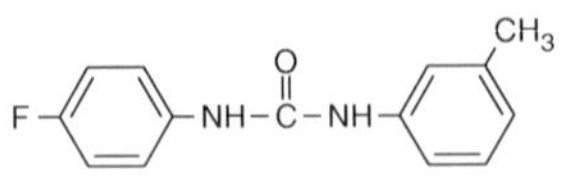

250 mg

S79,135-0
1-(4-CHLOROPHENYL)-3-(4-ETHYLPHENYL)UREA
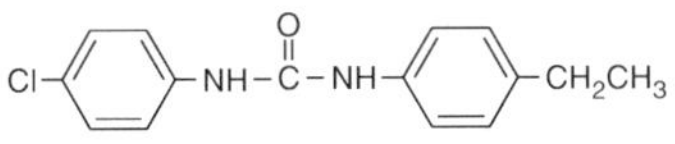
250 mg

S88,999-7
1-(3-CHLOROPHENYL)-3-(4-ETHYLPHENYL)UREA
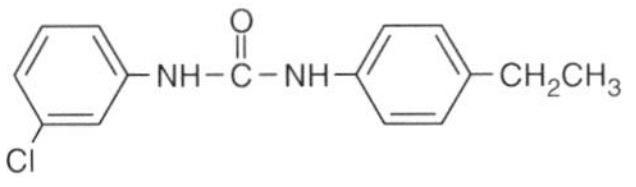
250 mg

S90,022-2
1-(3-CHLOROPHENYL)-3-(4-METHYLBENZYL)UREA
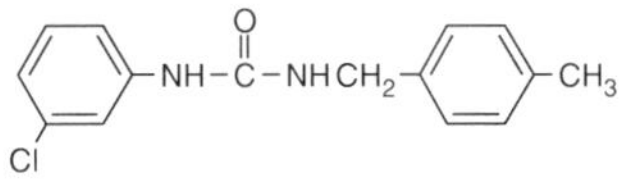
250 mg

S90,429-5
1-(4-CHLOROPHENYL)-1-ETHYL-3-(4-FLUOROPHENYL)UREA
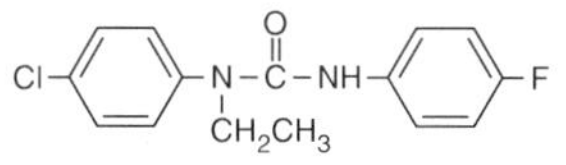
250 mg

S90,419-8
3-(3-CHLOROPHENYL)-1-(4-CHLOROPHENYL)-1-ETHYLUREA
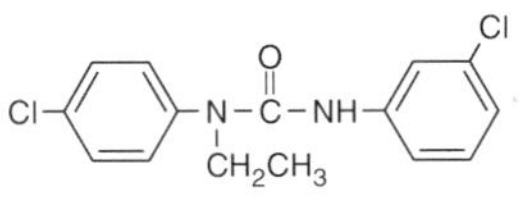
250 mg

S66,564-9
3-(4-BROMOPHENYL)-1,1-DIISOBUTYLUREA
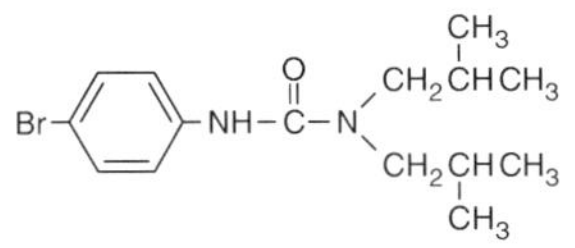
5 gm

S71,040-7
3-(4-BROMOPHENYL)-1,1-DIBUTYLUREA
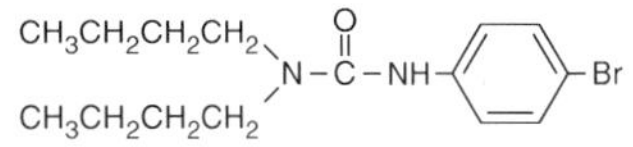
5 gm

S68,815-0
3-(3-ETHYLPHENYL)-1-(4-FLUOROPHENYL)-1-METHYLUREA
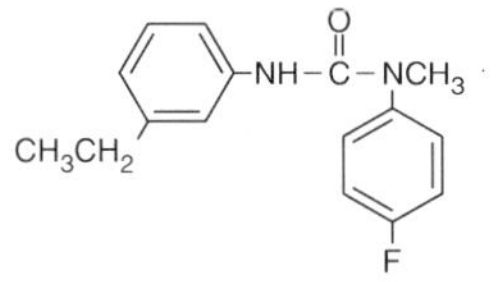
250 mg

S78,015-4
1-(4-BUTYLPHENYL)-3-(4-FLUOROPHENYL)UREA
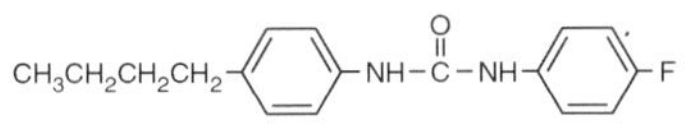
250 mg

S78,014-6
1-(4-BUTYLPHENYL)-3-(4-CHLOROPHENYL)UREA
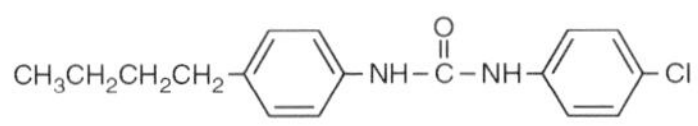
250 mg

S69,940-3
1,1'-(2-BROMO-6-METHYLPHENYLENE)-BIS[3-(4-BROMOPHENYL)UREA], MIXTURE OF ISOMERS
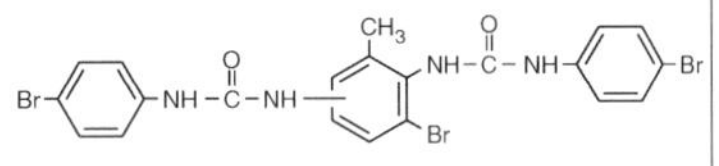
250 mg

S70,008-8
1,1'-(2-BROMO-6-METHYLPHENYLENE)-BIS[3-(4-IODOPHENYL)UREA], MIXTURE OF ISOMERS
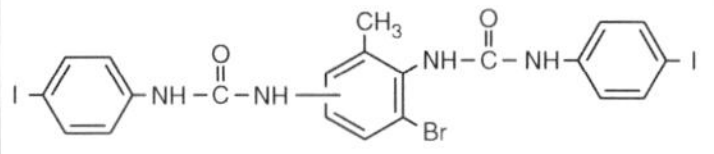
250 mg

S90,040-0
1,1'-HEXAMETHYLENEBIS[3-(4-CHLOROBENZYL)-3-METHYLUREA]
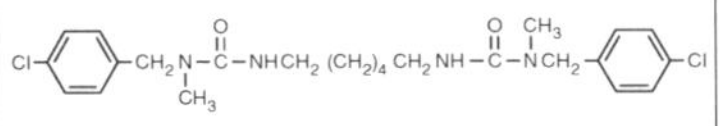
250 mg

S67,038-3
1-(4-FLUOROPHENYL)-3-OCTADECYLUREA
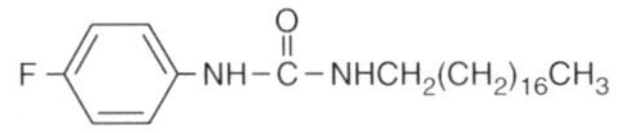
5 gm

S67,017-0
1-(4-CHLOROPHENYL)-3-OCTADECYLUREA
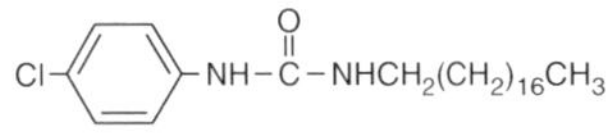
5 gm

S71,118-7
1-(2,3-DICHLOROPHENYL)-3-(4-FLUOROPHENYL)UREA
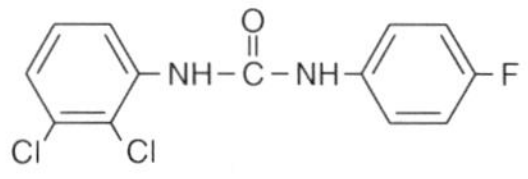
250 mg

S71,204-3
1-(2,6-DICHLOROPHENYL)-3-(4-FLUOROPHENYL)UREA
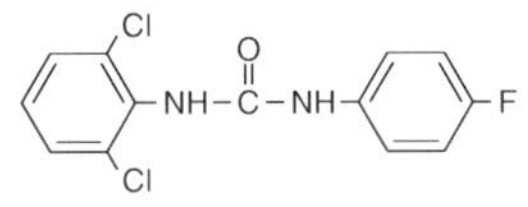
5 gm

S71,209-4
1-(4-CHLOROPHENYL)-3-(2,6-DICHLOROPHENYL)UREA
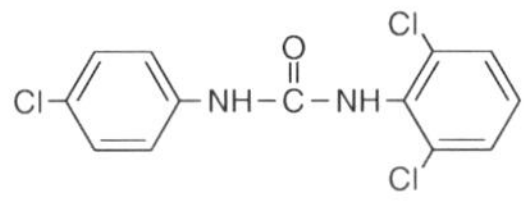
5 gm

S89,129-0
1-(4-CHLOROPHENYL)-3-(2,3-DICHLOROPHENYL)UREA
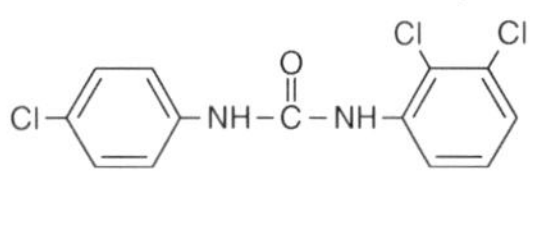
250 mg

S71,247-7
1-(2-CHLORO-6-METHYLPHENYL)-3-(4-FLUOROPHENYL)UREA
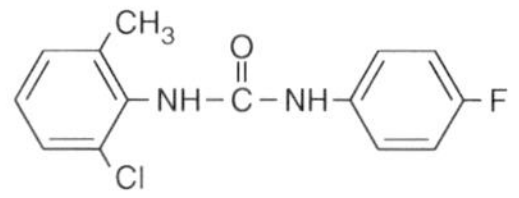
5 gm

S70,822-4
1-(4-CHLOROPHENYL)-3-(2-CHLORO-6-METHYLPHENYL)UREA
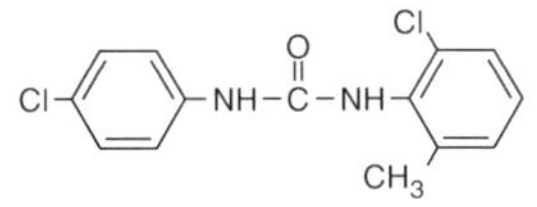
5 gm

S71,148-9
1-(2,6-DICHLOROPHENYL)-3-(4-METHYLPHENYL)UREA
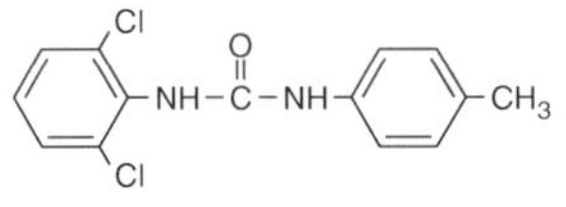
5 gm

S88,984-9
1-(4-CHLOROPHENYL)-3-(2,6-XYLYL)UREA
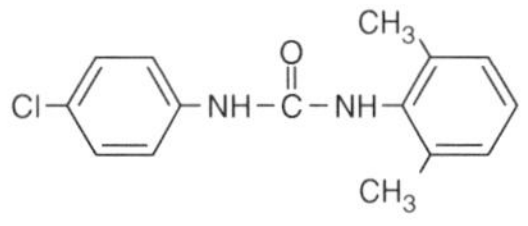
250 mg

S89,952-6
1-(4-CHLOROBENZYL)-3-(2,3-DICHLOROPHENYL)-1-METHYLUREA
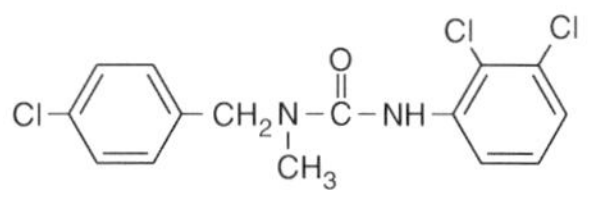
250 mg

S65,974-6
1-(4-CHLOROPHENYL)-3-(2-ETHYL-6-METHYLPHENYL)UREA
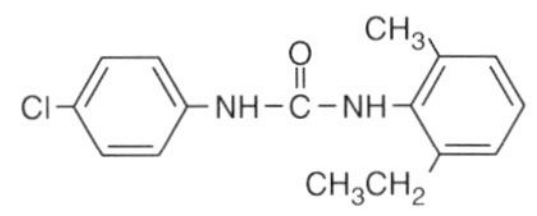
5 gm

S88,970-9
1-(3-CHLORO-2-METHYLPHENYL)-3-(4-ETHYLPHENYL)UREA
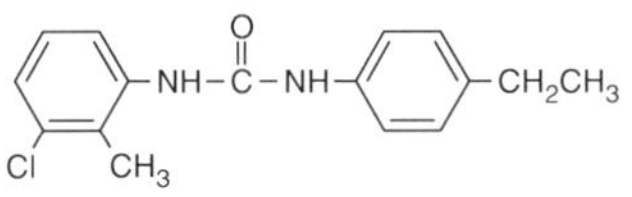
250 mg

S71,672-3
1-(4-CHLOROPHENYL)-3-(2,6-DIETHYLPHENYL)UREA
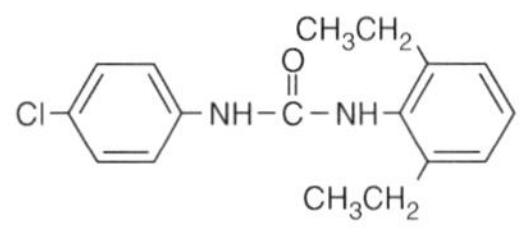
5 gm

S76,837-5
1-(4-BUTYLPHENYL)-3-(3-CHLORO-2-METHYLPHENYL)UREA
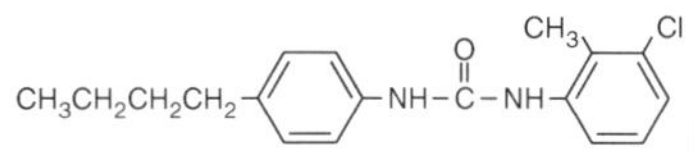
250 mg

S71,385-6
1-(4-CHLOROPHENYL)-3-(2,6-DIISOPROPYLPHENYL)UREA
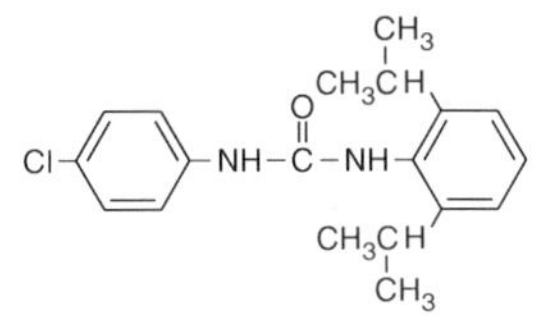
5 gm

S75,039-5
1-(2,5-DIFLUOROPHENYL)-3-(4-FLUOROPHENYL)UREA
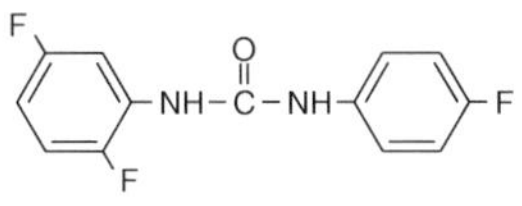
5 gm

S71,205-1
1-(3,4-DICHLOROPHENYL)-3-(4-FLUOROPHENYL)UREA
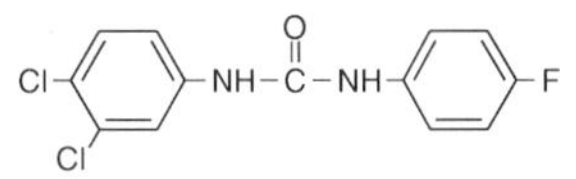
5 gm

S70,970-0
1-(4-CHLOROPHENYL)-3-(3,4-DICHLOROPHENYL)UREA
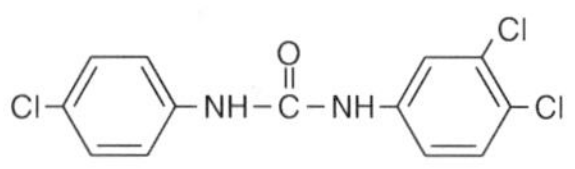
5 gm

S71,207-8
1-(4-CHLOROPHENYL)-3-(2,4-DICHLOROPHENYL)UREA

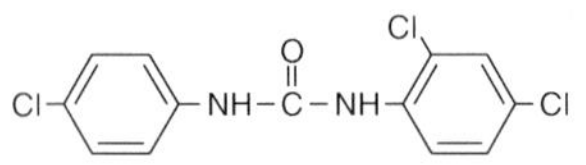

5 gm

S71,208-6
1-(4-CHLOROPHENYL)-3-(2,5-DICHLOROPHENYL)UREA

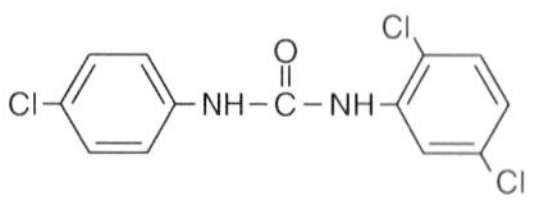

5 gm

S77,960-1
1-(4-BROMOPHENYL)-3-(2,5-DIFLUOROPHENYL)UREA

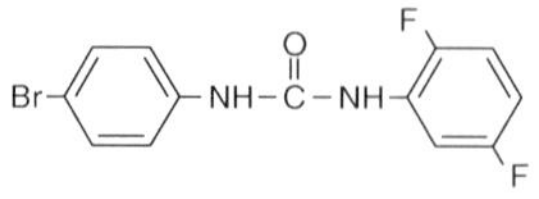

250 mg

S78,037-5
1-(2,5-DIFLUOROPHENYL)-3-(4-IODOPHENYL)UREA

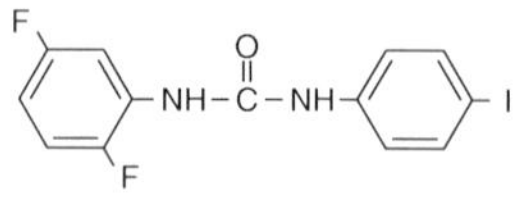

250 mg

S67,493-1
1-(5-CHLORO-2-METHYLPHENYL)-3-(4-FLUOROPHENYL)UREA

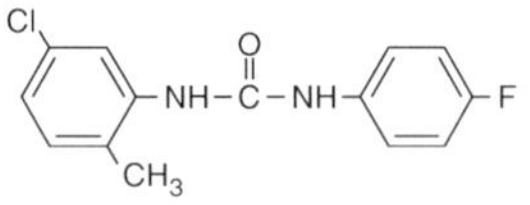

5 gm

S88,985-7
1-(4-CHLOROPHENYL)-3-(2,5-XYLYL)UREA

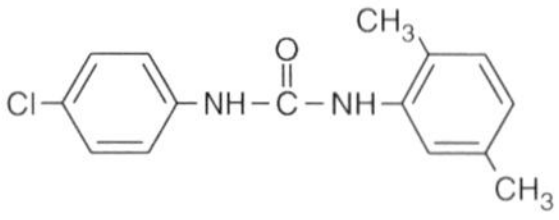

250 mg

S74,330-5
1-(3,4-DICHLOROPHENYL)-3-(4-ETHYLPHENYL)UREA

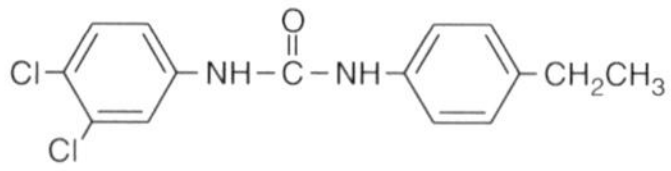

5 gm

S75,365-3
1-(3-CHLORO-4-METHYLPHENYL)-3-(4-ETHYLPHENYL)UREA

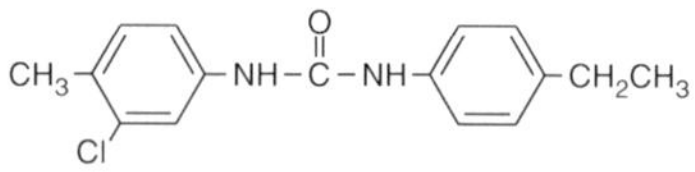

5 gm

S72,299-5
3-(2,5-DICHLOROPHENYL)-1-ETHYL-1-(P-TOLYL)UREA

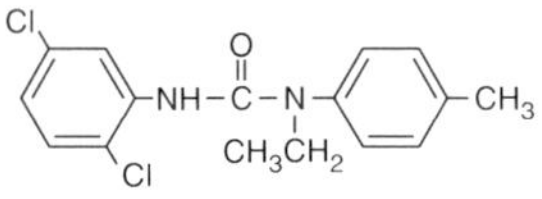

250 mg

S78,010-3
1-(4-BUTYLPHENYL)-3-(3-CHLORO-4-METHYLPHENYL)UREA

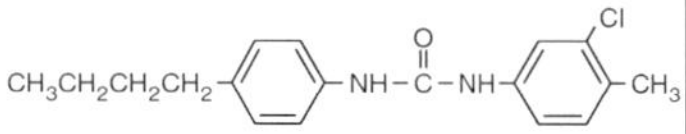

250 mg

S78,013-8
1-(4-BUTYLPHENYL)-3-(4-CHLORO-2-METHYLPHENYL)UREA

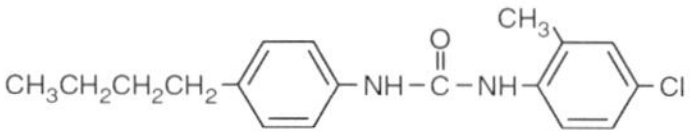

250 mg

S39,742-3
1-(2,5-DICHLOROPHENYL)-3-(4-FLUOROPHENYL)UREA

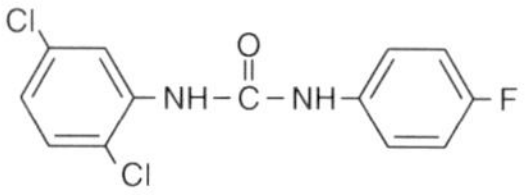

250 mg

S71,196-9
1-(4-CHLOROPHENYL)-3-(2,4,6-TRICHLOROPHENYL)UREA

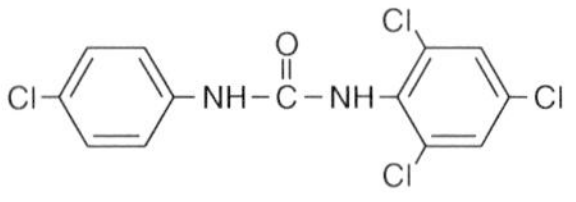

5 gm

S76,953-3
1-(4-CHLOROPHENYL)-3-(2,4,6-TRIMETHYLPHENYL)UREA

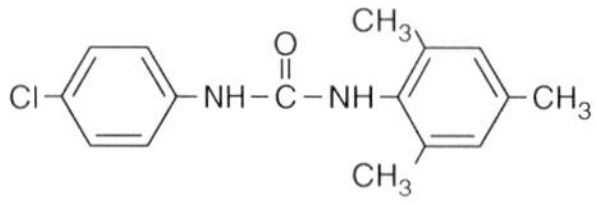

250 mg

S71,193-4
1-(4-FLUOROPHENYL)-3-(2,4,5-TRICHLOROPHENYL)UREA

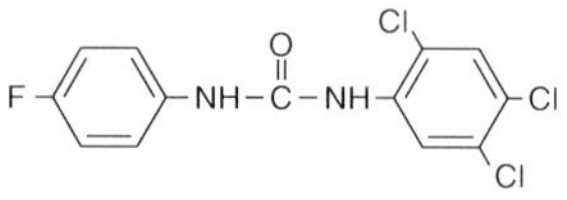

5 gm

S71,198-5
1-(4-CHLOROPHENYL)-3-(2,4,5-TRICHLOROPHENYL)UREA

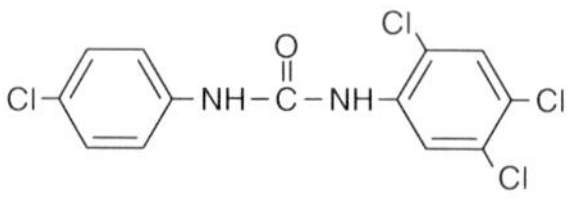

5 gm

S71,192-6
1-(4-BROMOPHENYL)-3-(2,4,5-TRICHLOROPHENYL)UREA

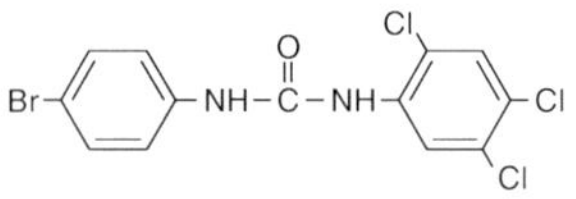

5 gm

S75,340-8
1-(4-CHLOROPHENYL)-3-(4,5-DICHLORO-2-METHYLPHENYL)UREA

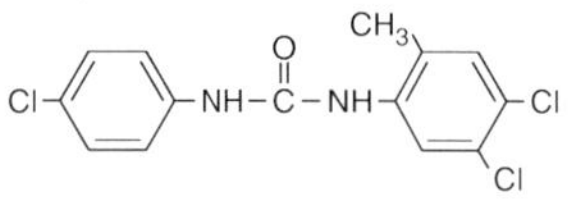

5 gm

S75,368-8
1-(4-BROMOPHENYL)-3-(1-NAPHTHYL)UREA

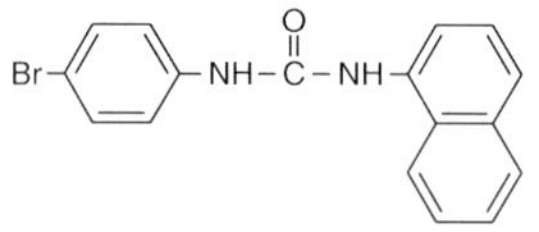

5 gm

S70,885-2
1-(6-CHLORO-O-TOLYL)UREA

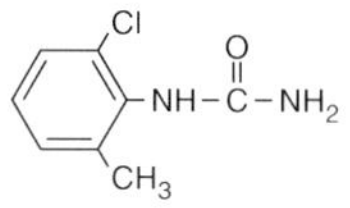

5 gm

S66,566-5
1-(2,6-DICHLOROPHENYL)-3-METHYLUREA

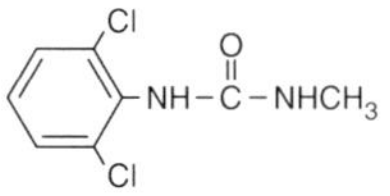

5 gm

S66,562-2
1-(2,6-DICHLOROPHENYL)-3-ETHYLUREA

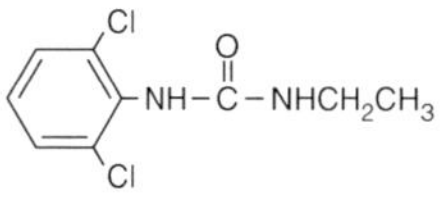

5 gm

S69,387-1
3-(3-CHLORO-2-METHYLPHENYL)-1,1-DIMETHYLUREA

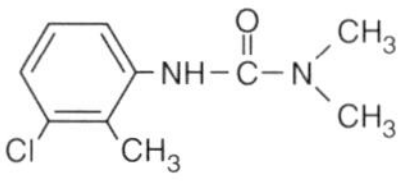

250 mg

S70,911-5
1-(2-CHLORO-6-METHYLPHENYL)-3-ETHYLUREA

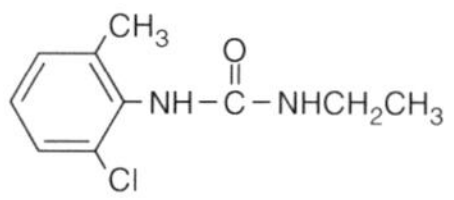

5 gm

S89,028-6
1-(3-CHLORO-2-METHYLPHENYL)-3-ETHYLUREA

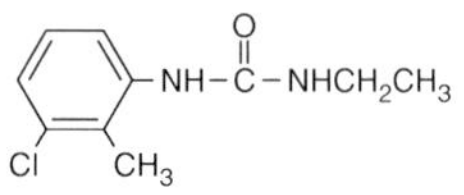

250 mg

S66,505-3
1-(2,6-DICHLOROPHENYL)-3-PROPYLUREA

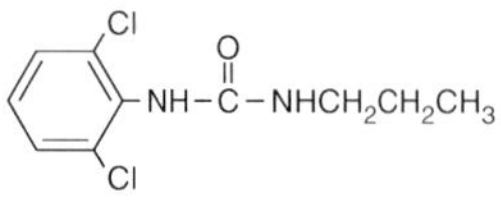

5 gm

S66,756-0
1-(2,6-DICHLOROPHENYL)-3-ISOPROPYLUREA

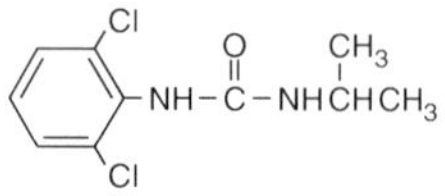

5 gm

S66,035-3
1-ALLYL-3-(2-CHLORO-6-METHYLPHENYL)UREA

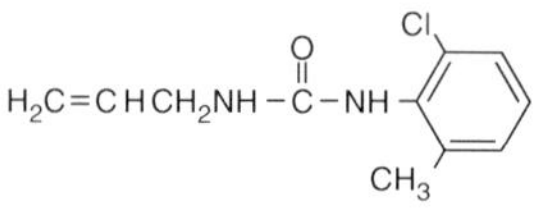

5 gm

S70,876-3
1-(2-CHLORO-6-METHYLPHENYL)-3-PROPYLUREA

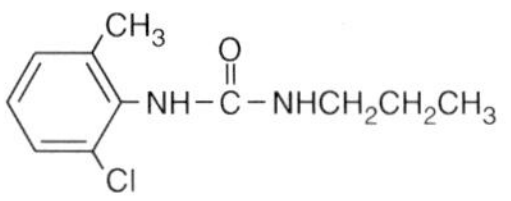

5 gm

S79,102-4
1-(3-CHLORO-2-METHYLPHENYL)-3-PROPYLUREA

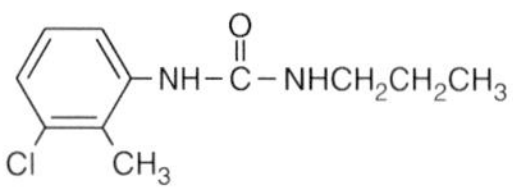

250 mg

S66,506-1
1-BUTYL-3-(2,6-DICHLOROPHENYL)UREA

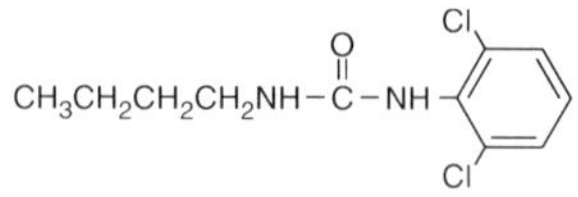

5 gm

S89,116-9
1-TERT-BUTYL-3-(2-CHLORO-6-METHYLPHENYL)UREA

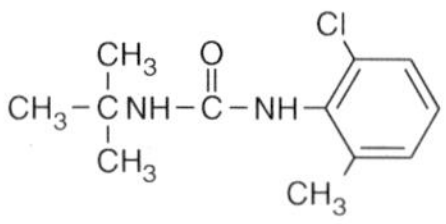

250 mg

S66,745-5
3-(2,6-DICHLOROPHENYL)-1,1-DIISOPROPYLUREA

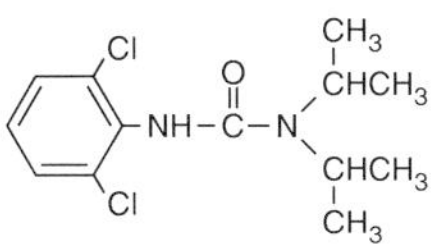

5 gm

S70,862-3
1-(2,6-DICHLOROPHENYL)-3-(1,3-DIMETHYLBUTYL)UREA

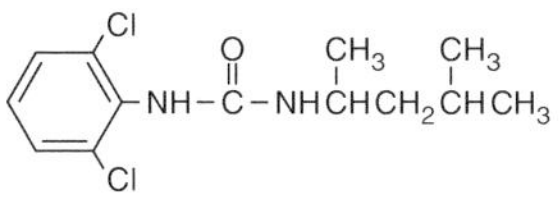

5 gm

S71,379-1
1-BUTYL-3-(2,3-DICHLOROPHENYL)-1-ETHYLUREA

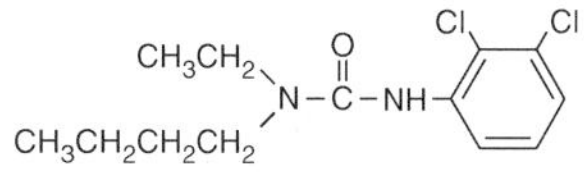

5 gm

S71,213-2
1-(3-CHLOROPHENYL)-3-(2,3-DICHLOROPHENYL)UREA

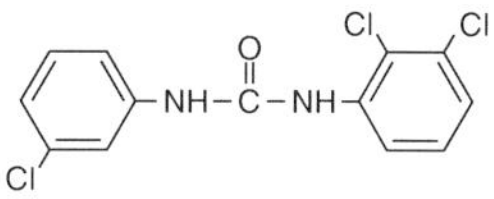

5 gm

S70,870-4
1,3-BIS(2,6-DICHLOROPHENYL)UREA

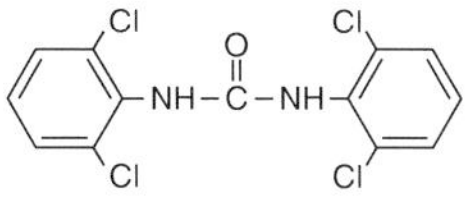

5 gm

S71,491-7
1-(2,3-DICHLOROPHENYL)-3-(2,6-DICHLOROPHENYL)UREA

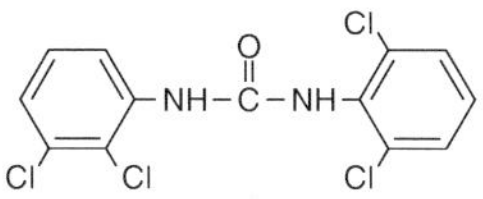

5 gm

S70,849-6
1-(2,3-DICHLOROPHENYL)-3-(2,4,5-TRICHLOROPHENYL)UREA

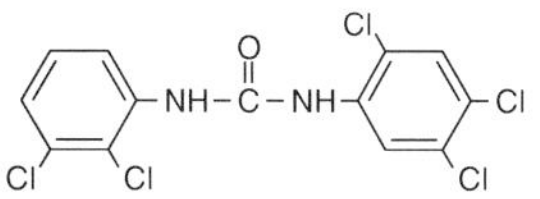

5 gm

S71,492-5
3-(2-CHLORO-6-METHYLPHENYL)-1,1-DIISOPROPYLUREA

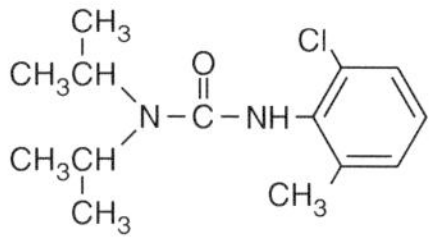

5 gm

S71,249-3
1-(3-CHLOROPHENYL)-3-(2-CHLORO-6-METHYLPHENYL)UREA

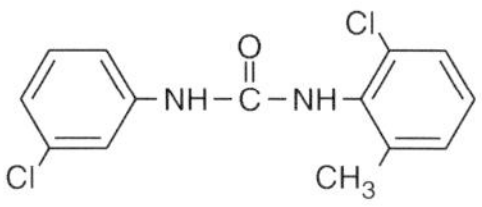

5 gm

S71,250-7
1-(2,6-DICHLOROPHENYL)-3-(M-TOLYL)UREA

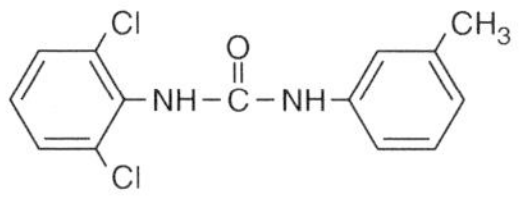

5 gm

S71,242-6
1-(2-CHLORO-6-METHYLPHENYL)-3-(2,3-DICHLOROPHENYL)UREA

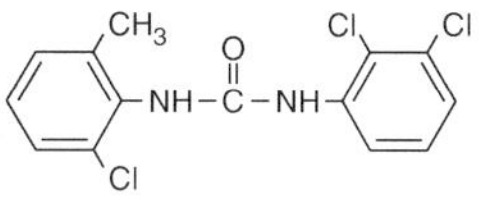

5 gm

S71,246-9
1-(2-CHLORO-6-METHYLPHENYL)-3-(2,6-DICHLOROPHENYL)UREA

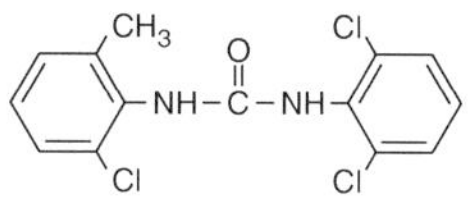

5 gm

S71,377-5
1-(3-CHLOROPHENYL)-3-(2,6-XYLYL)UREA

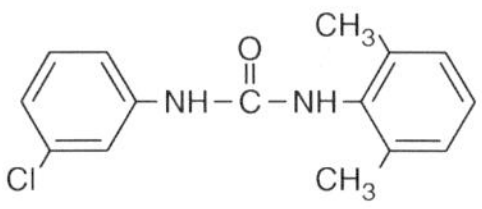

5 gm

S77,334-4
1-(3-CHLOROPHENYL)-3-(2,3-XYLYL)UREA

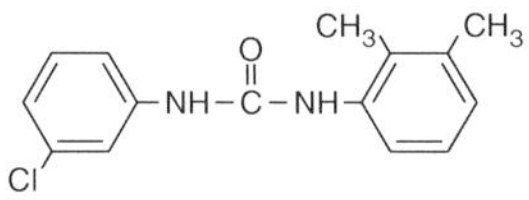

250 mg

S89,108-8
1-(3-CHLORO-2-METHYLPHENYL)-3-(M-TOLYL)UREA

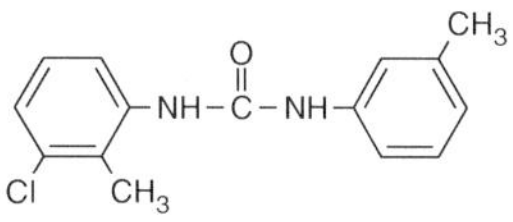

250 mg

S71,264-7
1-(2,6-DICHLOROPHENYL)-3-(2,6-DIMETHYLPHENYL)UREA

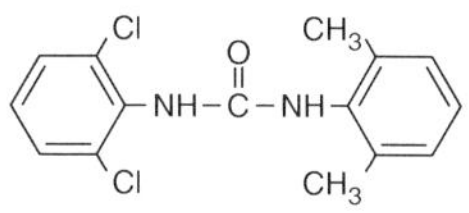

5 gm

S65,971-1
1-(3-CHLOROPHENYL)-3-(2-ETHYL-6-METHYLPHENYL)UREA

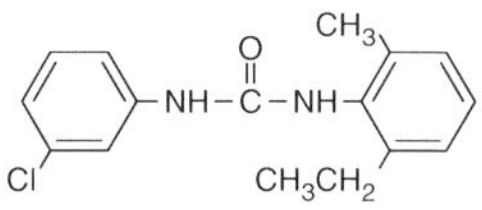

5 gm

S66,503-7
3-(2-CHLORO-6-METHYLPHENYL)-1,1-DIBUTYLUREA

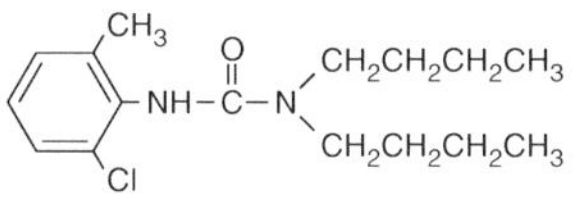

5 gm

S66,000-0
1-(2,3-DICHLOROPHENYL)-3-(2-ETHYL-6-METHYLPHENYL)UREA

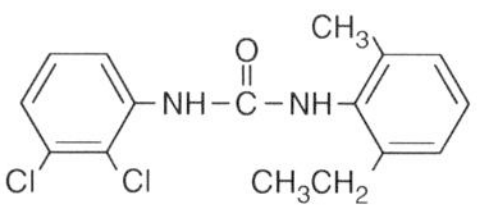

5 gm

S73,575-2
1-(2,6-DICHLOROPHENYL)-3-(2-ETHYL-6-METHYLPHENYL)UREA

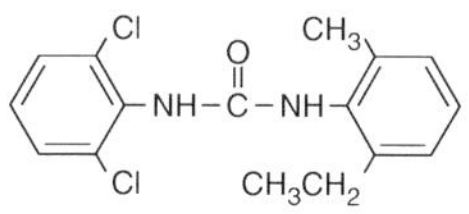

250 mg

S70,770-8
1-(3-CHLOROPHENYL)-3-(2,6-DIETHYLPHENYL)UREA

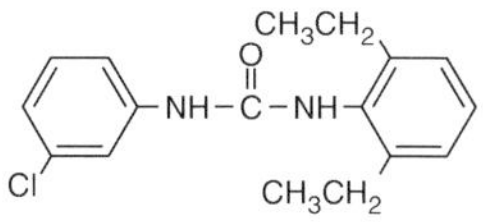

5 gm

S71,286-8
1-(2,3-DICHLOROPHENYL)-3-(2,6-DIETHYLPHENYL)UREA

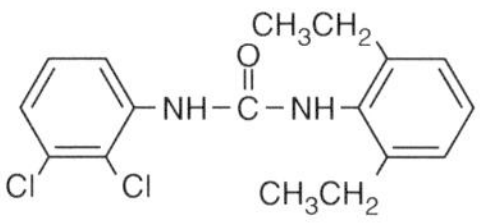

5 gm

S71,387-2
1-(2,6-DICHLOROPHENYL)-3-(2,6-DIETHYLPHENYL)UREA

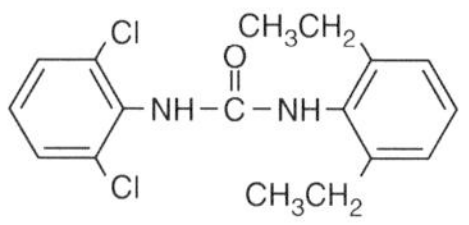

5 gm

S70,747-3
1-(3-CHLOROPHENYL)-3-(2,6-DIISOPROPYLPHENYL)UREA

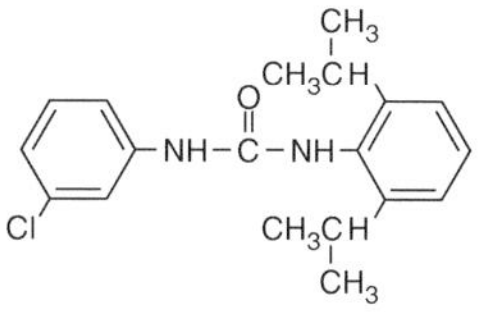

5 gm

S70,750-3
1-(2,3-DICHLOROPHENYL)-3-(2,6-DIISOPROPYLPHENYL)UREA

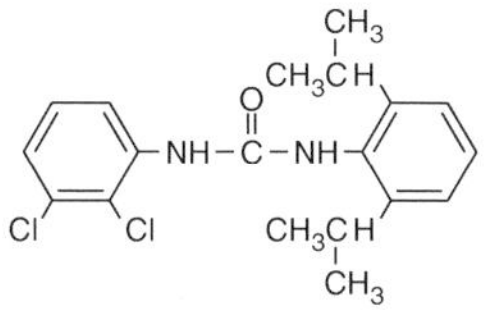

5 gm

S71,540-9
1-(2,6-DICHLOROPHENYL)-3-(2,6-DIISOPROPYLPHENYL)UREA

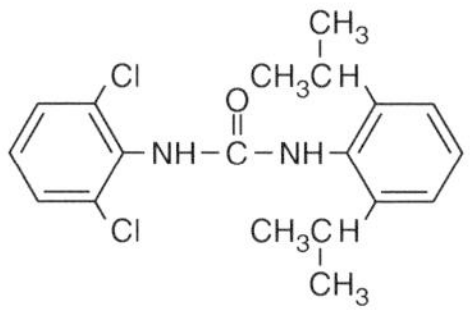

5 gm

S86,896-5
1,1'-HEXAMETHYLENEBIS[3-(3-CHLORO-2-METHYLPHENYL)UREA]

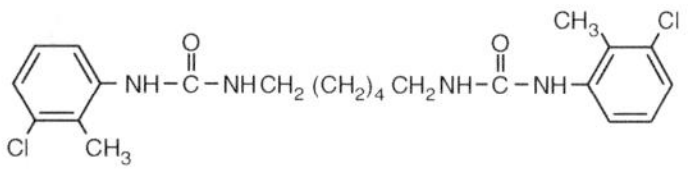

250 mg

S67,047-2
1-(2,3-DICHLOROPHENYL)-3-OCTADECYLUREA

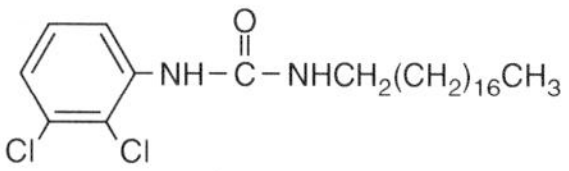

5 gm

S70,869-0
1-(2,4-DICHLOROPHENYL)-3-(2,6-DICHLOROPHENYL)UREA

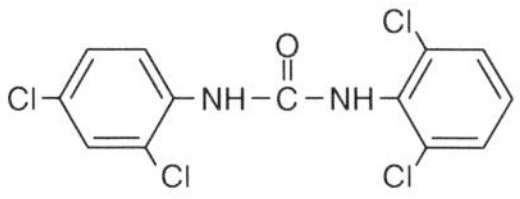

5 gm

S70,871-2
1-(2,3-DICHLOROPHENYL)-3-(2,5-DICHLOROPHENYL)UREA

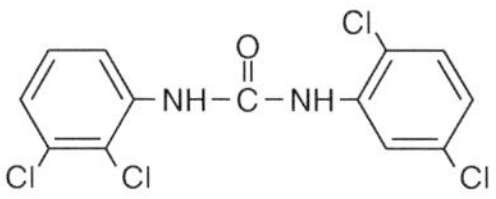

5 gm

S70,873-9
1-(2,3-DICHLOROPHENYL)-3-(3,4-DICHLOROPHENYL)UREA

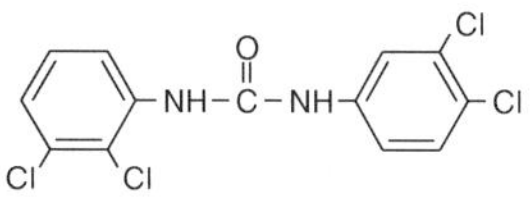

5 gm

S71,199-3
1-(2,3-DICHLOROPHENYL)-3-(2,4-DICHLOROPHENYL)UREA

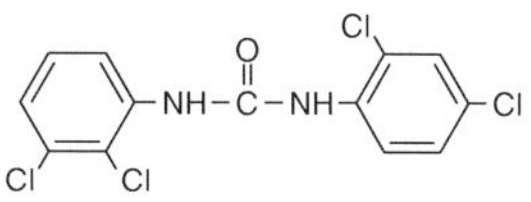

5 gm

S71,200-0
1-(2,5-DICHLOROPHENYL)-3-(2,6-DICHLOROPHENYL)UREA

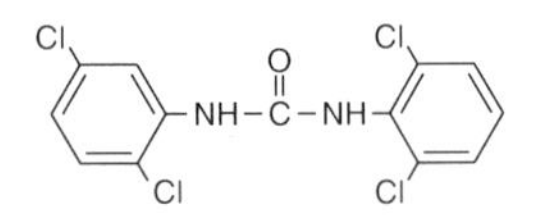

250 mg

S71,201-9
1-(3,4-DICHLOROPHENYL)-3-(2,6-DICHLOROPHENYL)UREA

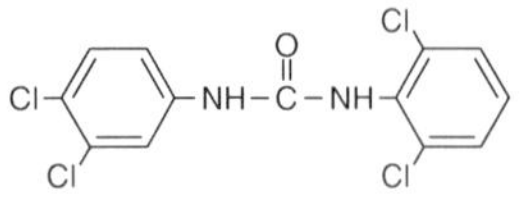

5 gm

S72,951-5
1-(3-CHLORO-2-METHYLPHENYL)-3-(2,5-DIFLUOROPHENYL)UREA

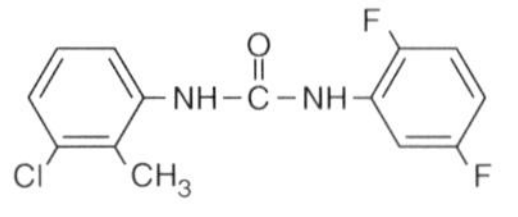

5 gm

S70,825-9
1-(2-CHLORO-6-METHYLPHENYL)-3-(3,4-DICHLOROPHENYL)UREA

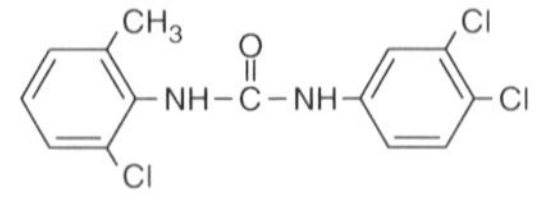

5 gm

S71,243-4
1-(2-CHLORO-6-METHYLPHENYL)-3-(2,4-DICHLOROPHENYL)UREA

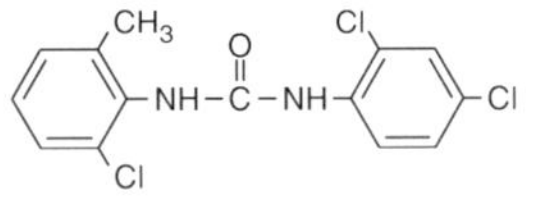

5 gm

S71,244-2
1-(2-CHLORO-6-METHYLPHENYL)-3-(2,5-DICHLOROPHENYL)UREA

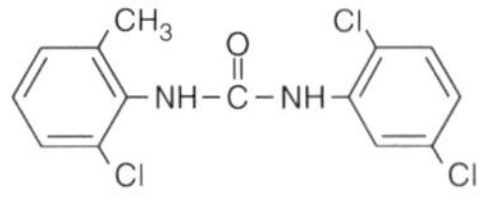

5 gm

S71,245-0
1-(4-CHLORO-2-METHYLPHENYL)-3-(2,6-DICHLOROPHENYL)UREA

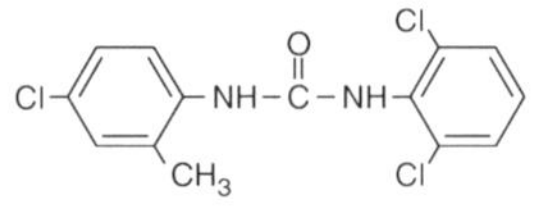

5 gm

S71,681-2
1-(3-CHLORO-4-METHYLPHENYL)-3-(2,6-DICHLOROPHENYL)UREA

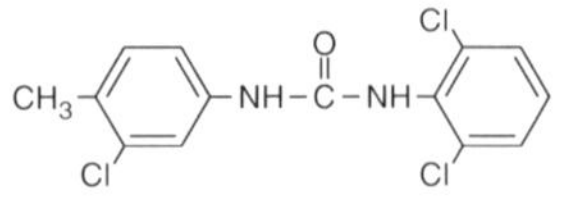

5 gm

S72,613-3
1-(3-CHLORO-2-METHYLPHENYL)-3-(3,4-DICHLOROPHENYL)UREA

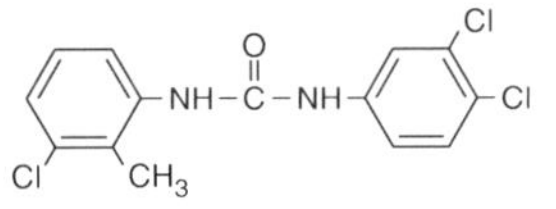

5 gm

S70,908-5
1-(2,5-DICHLOROPHENYL)-3-(2,6-XYLYL)UREA

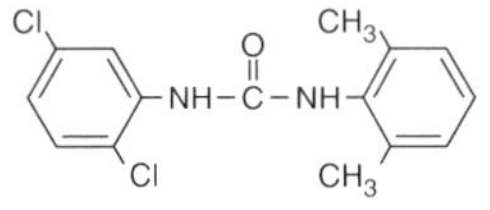

5 gm

S71,263-9
1-(2,6-DICHLOROPHENYL)-3-(2,4-DIMETHYLPHENYL)UREA

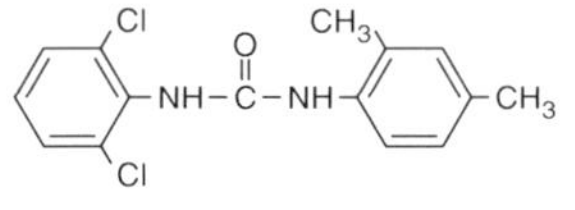

5 gm

S71,265-5
1-(5-CHLORO-2-METHYLPHENYL)-3-(2-CHLORO-6-METHYLPHENYL)UREA

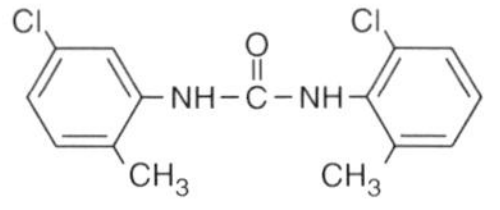

5 gm

S71,266-3
1-(2,4-DICHLOROPHENYL)-3-(2,6-DIMETHYLPHENYL)UREA

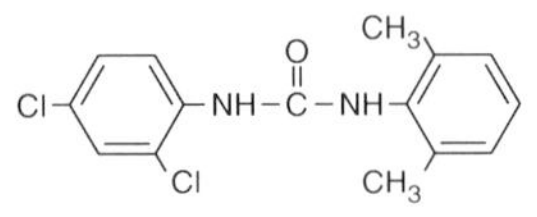

5 gm

S71,749-5
1-(2-CHLORO-6-METHYLPHENYL)-3-(4-CHLORO-2-METHYLPHENYL)UREA

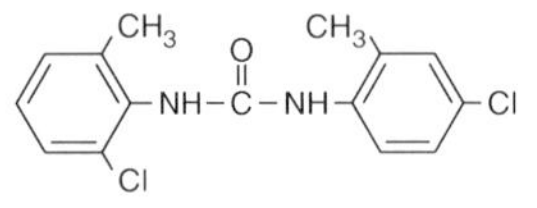

5 gm

S75,350-5
1-(3-CHLORO-2-METHYLPHENYL)-3-(3-CHLORO-4-METHYLPHENYL)UREA

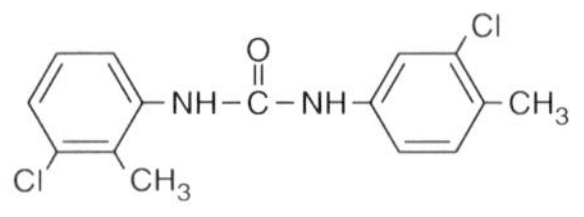

5 gm

S88,942-3
1-(3,4-DICHLOROPHENYL)-3-(2,6-XYLYL)UREA

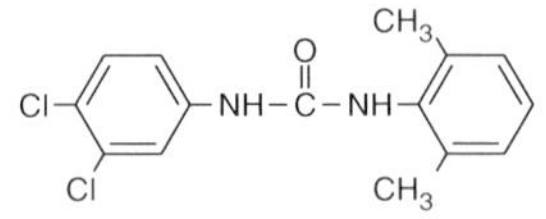

250 mg

S71,277-9
1-(3-CHLORO-4-METHYLPHENYL)-3-(2,6-XYLYL)UREA

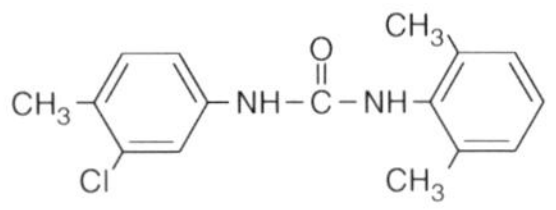

5 gm

S71,686-3
1-(4-CHLORO-2-METHYLPHENYL)-3-(2,6-XYLYL)UREA

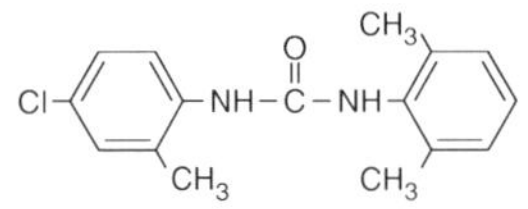

5 gm

S75,366-1
1-(3-CHLORO-4-METHYLPHENYL)-3-(2,3-XYLYL)UREA

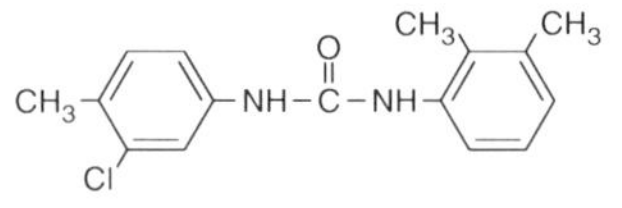

5 gm

S88,947-4
1-(3-CHLORO-2-METHYLPHENYL)-3-(3,4-XYLYL)UREA

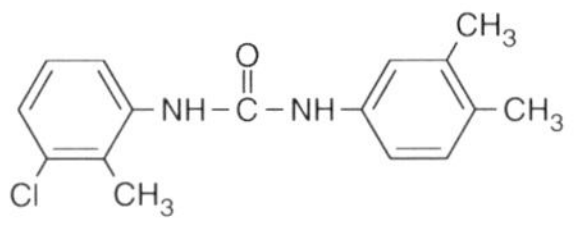

250 mg

S65,970-3
1-(2,4-DICHLOROPHENYL)-3-(2-ETHYL-6-METHYLPHENYL)UREA

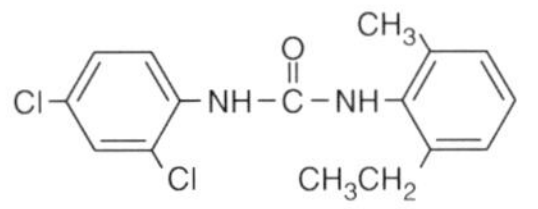

5 gm

S73,565-5
1-(2,5-DICHLOROPHENYL)-3-(2-ETHYL-6-METHYLPHENYL)UREA

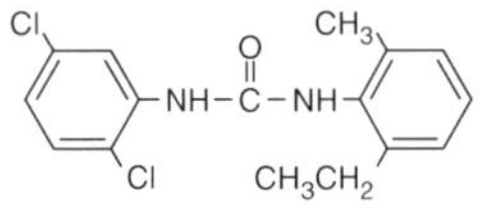

250 mg

S71,289-2
1-(4-CHLORO-2-METHYLPHENYL)-3-(2-ETHYL-6-METHYLPHENYL)UREA

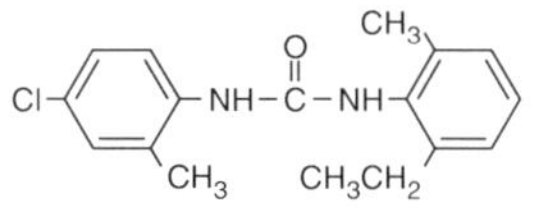

5 gm

S71,290-6
1-(3-CHLORO-4-METHYLPHENYL)-3-(2-ETHYL-6-METHYLPHENYL)UREA

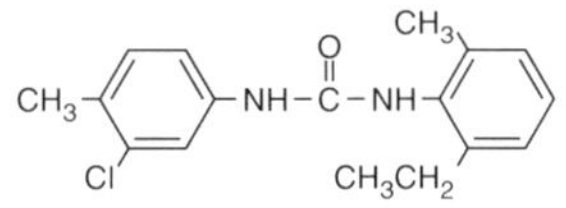

5 gm

S70,910-7
1-(3,4-DICHLOROPHENYL)-3-(2,6-DIETHYLPHENYL)UREA

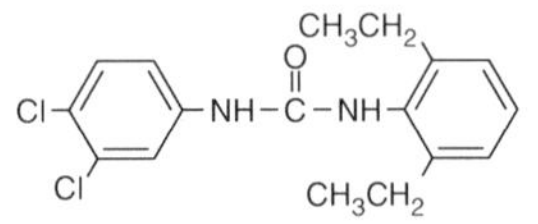

5 gm

S71,287-6
1-(2,4-DICHLOROPHENYL)-3-(2,6-DIETHYLPHENYL)UREA

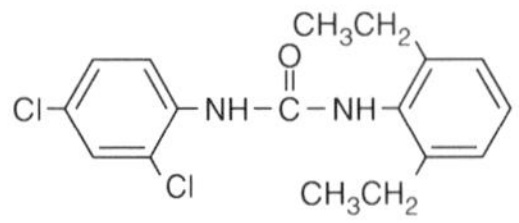

5 gm

S71,288-4
1-(2,5-DICHLOROPHENYL)-3-(2,6-DIETHYLPHENYL)UREA

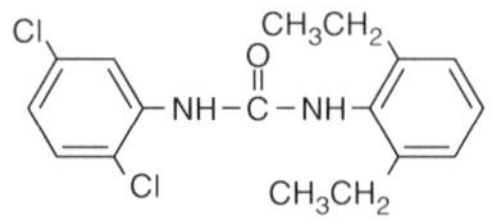

5 gm

S70,762-7
1-(4-CHLORO-2-METHYLPHENYL)-3-(2,6-DIETHYLPHENYL)UREA

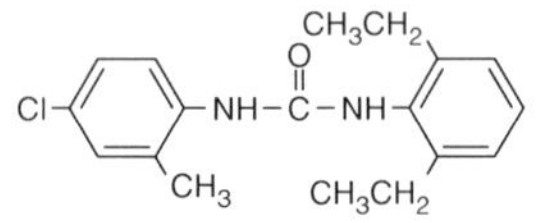

5 gm

S71,297-3
1-(5-CHLORO-2-METHYLPHENYL)-3-(2,6-DIETHYLPHENYL)UREA

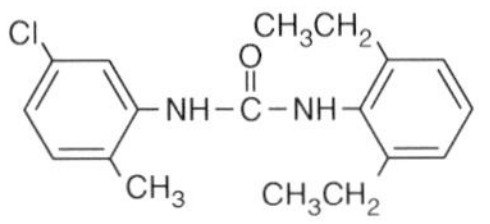

5 gm

S71,301-5
1-(2,4-DICHLOROPHENYL)-3-(2,6-DIISOPROPYLPHENYL)UREA

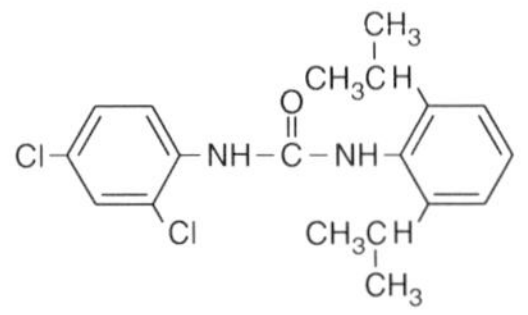

5 gm

S71,307-4
1-(4-CHLORO-2-METHYLPHENYL)-3-(2,6-DIISOPROPYLPHENYL)UREA

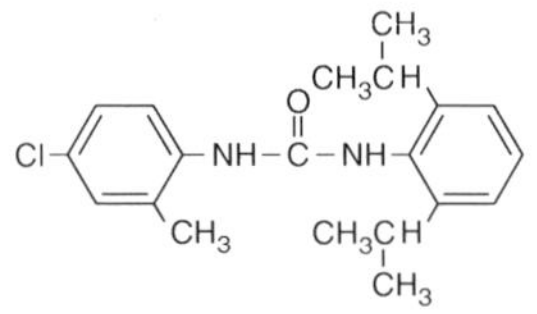

5 gm

S75,195-2
1-(2,3-DICHLOROPHENYL)-3-(3,5-DICHLOROPHENYL)UREA

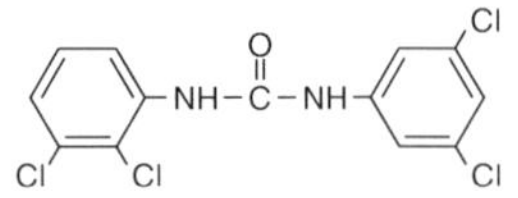

5 gm

S71,671-5
1-(2,6-DICHLOROPHENYL)-3-(2,4,6-TRICHLOROPHENYL)UREA
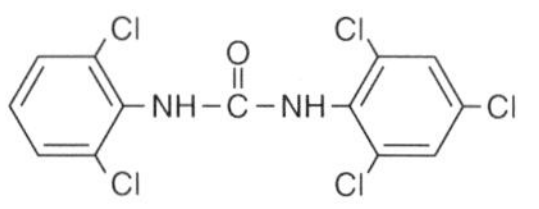
5 gm

S71,522-0
1-(2-CHLORO-6-METHYLPHENYL)-3-(2,4,6-TRICHLOROPHENYL)UREA
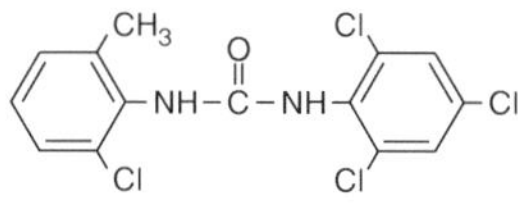
5 gm

S71,184-5
1-(2,6-DICHLOROPHENYL)-3-(2,4,5-TRICHLOROPHENYL)UREA
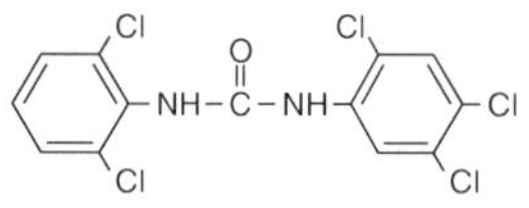
5 gm

S71,261-2
1-(2,6-DIMETHYLPHENYL)-3-(2,4,5-TRICHLOROPHENYL)UREA
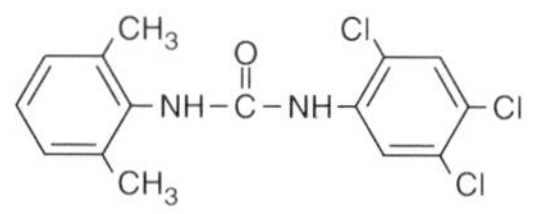
5 gm

S71,283-3
1-(2,6-DIETHYLPHENYL)-3-(2,4,5-TRICHLOROPHENYL)UREA
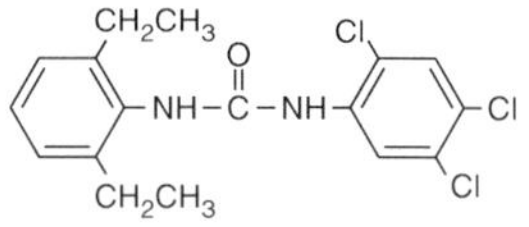
5 gm

S71,282-5
1-(2,6-DICHLOROPHENYL)-3-(1-NAPHTHYL)UREA
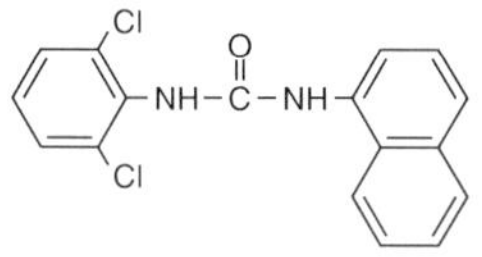
5 gm

S71,155-1
1-(2-CHLORO-6-METHYLPHENYL)-3-(1-NAPHTHYL)UREA
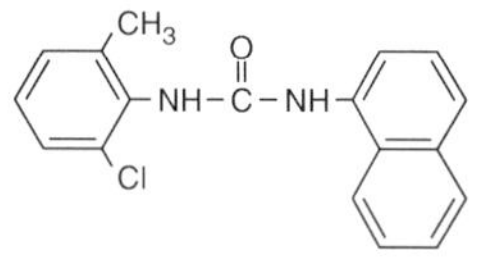
5 gm

S79,160-1
1-(3-CHLORO-2-METHYLPHENYL)-3-(1-NAPHTHYL)UREA
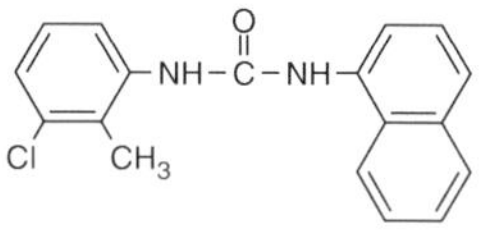
250 mg

S73,056-4
(2,4-DIFLUOROPHENYL)UREA
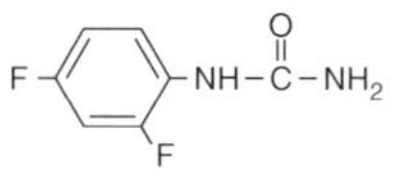
5 gm

S66,633-5
(3,4-DICHLOROPHENYL)UREA
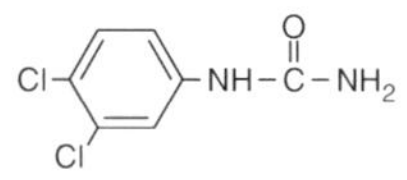
5 gm

S76,411-6
(2,4-DICHLOROPHENYL)UREA
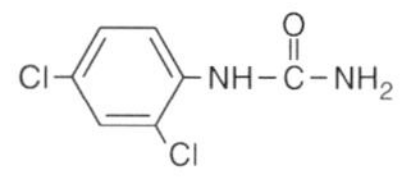
5 gm

S69,554-8
3-(3,4-DICHLOROPHENYL)-1,1-DIMETHYLUREA
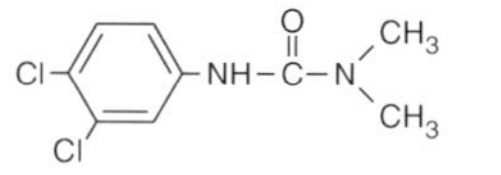
5 gm

S71,425-9
1-(2,5-DICHLOROPHENYL)-3-ETHYLUREA
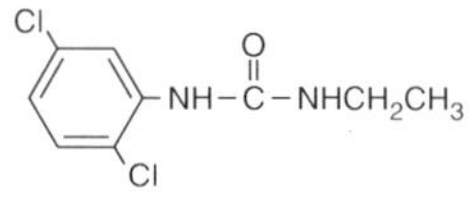
5 gm

S91,188-7
3-(3-CHLORO-4-METHYLPHENYL)-1-METHOXY-1-METHYLUREA
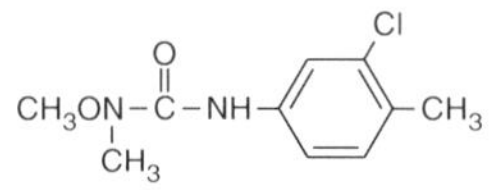
250 mg

S85,647-9
1-(3,4-DICHLOROPHENYL)-1,3,3-TRIMETHYLUREA
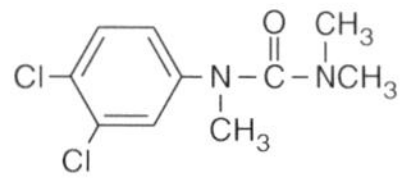
250 mg

S72,165-4
1-(3-FLUORO-4-METHYLPHENYL)-3-PROPYLUREA
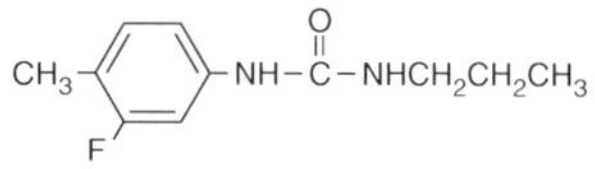
250 mg

S68,754-5
1-BUTYL-3-(2,5-DIFLUOROPHENYL)UREA
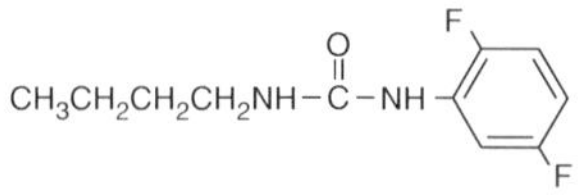
250 mg

S76,818-9
1-(2,4-DIFLUOROPHENYL)-3-ISOBUTYLUREA
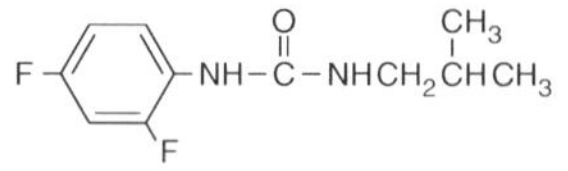
250 mg

S66,742-0
1-(2,5-DICHLOROPHENYL)-3-ISOBUTYLUREA
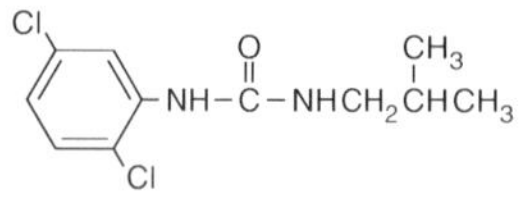
5 gm

S74,334-8
1-SEC-BUTYL-3-(3,4-DICHLOROPHENYL)UREA
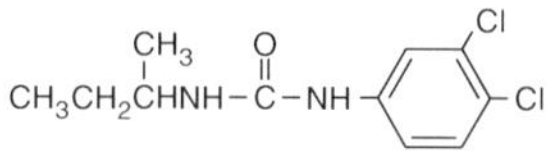
5 gm

S66,005-1
1-(5-CHLORO-2-METHYLPHENYL)-3-ISOBUTYLUREA
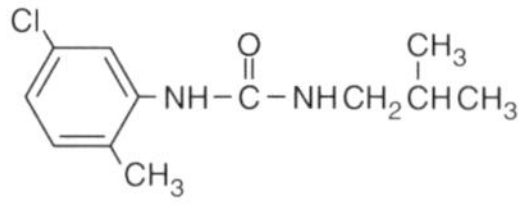
5 gm

S66,578-9
1-(3-CHLORO-4-METHYLPHENYL)-3-ISOBUTYLUREA
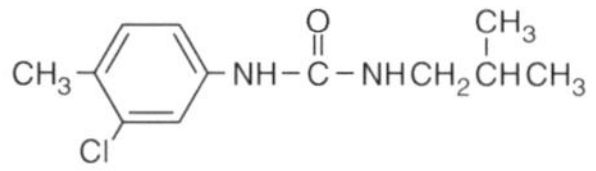
5 gm

S66,585-1
1-(4-CHLORO-2-METHYLPHENYL)-3-ISOBUTYLUREA
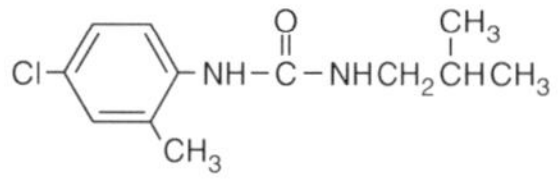
5 gm

S84,419-5
1-TERT-BUTYL-3-(5-CHLORO-2-METHYLPHENYL)UREA
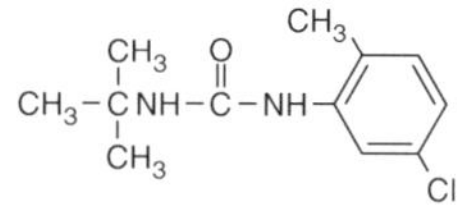
250 mg

S90,548-8
1-(3-CHLORO-4-METHYLPHENYL)-1,3-DIETHYLUREA
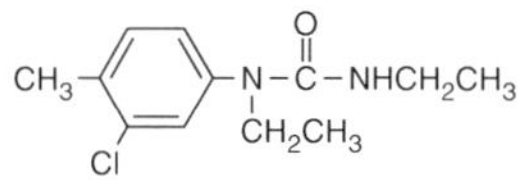
250 mg

S72,156-5
1-(3-CHLOROPHENYL)-3-(2,5-DIFLUOROPHENYL)UREA
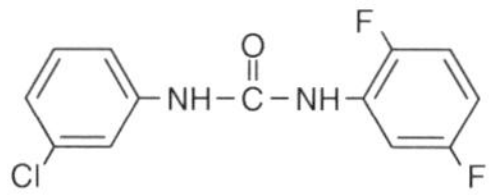
250 mg

S71,050-4
1,1-DIALLYL-3-(2,5-DICHLOROPHENYL)UREA
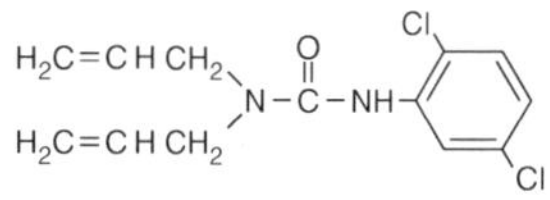
5 gm

S84,433-0
1-(2,4-DICHLOROPHENYL)-3-(3-FLUOROPHENYL)UREA
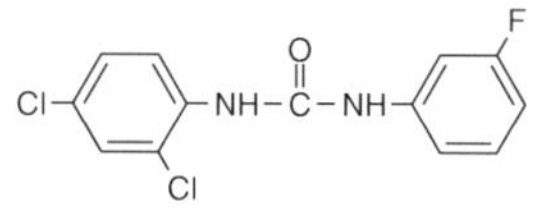
250 mg

S84,434-9
1-(2,5-DICHLOROPHENYL)-3-(3-FLUOROPHENYL)UREA
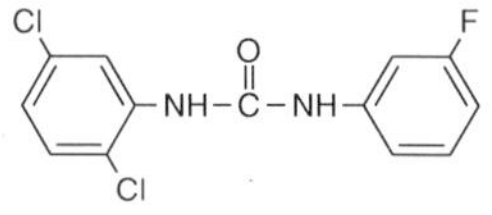
250 mg

S71,214-0
1-(3-CHLOROPHENYL)-3-(2,4-DICHLOROPHENYL)UREA
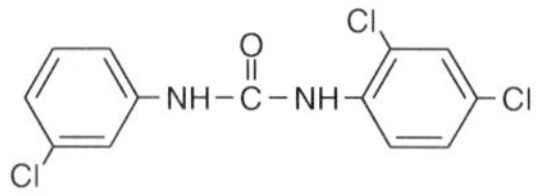
5 gm

S71,215-9
1-(3-CHLOROPHENYL)-3-(2,5-DICHLOROPHENYL)UREA
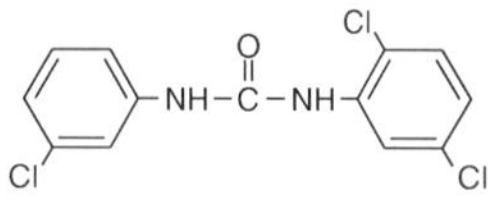
5 gm

S68,753-7
1-(2,5-DICHLOROPHENYL)-3-(2,4-DIFLUOROPHENYL)UREA
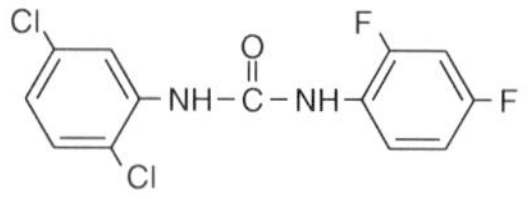
250 mg

S76,984-3
1-(2,5-DICHLOROPHENYL)-3-(2,5-DIFLUOROPHENYL)UREA

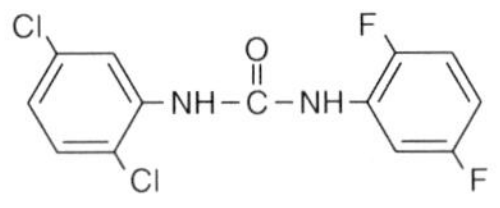

250 mg

S91,619-6
1-(3,4-DICHLOROPHENYL)-3-(2,5-DIFLUOROPHENYL)UREA

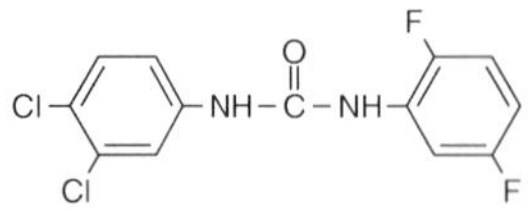

250 mg

S78,036-7
1-(3-BROMOPHENYL)-3-(2,5-DIFLUOROPHENYL)UREA

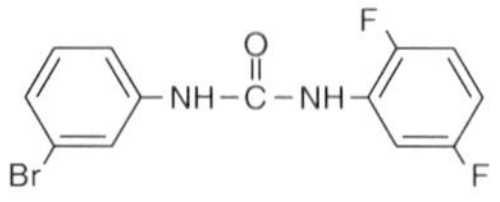

250 mg

S69,161-5
1-(3-BROMOPHENYL)-3-(3-CHLORO-4-FLUOROPHENYL)UREA

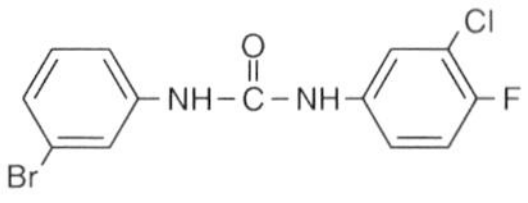

250 mg

S70,872-0
1-(2,4-DICHLOROPHENYL)-3-(3,4-DICHLOROPHENYL)UREA

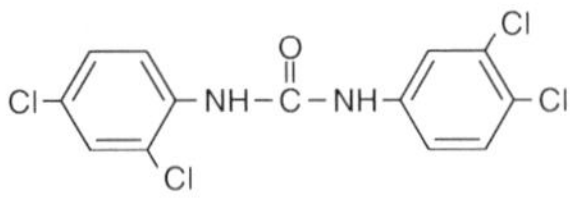

5 gm

S71,202-7
1,3-BIS(2,4-DICHLOROPHENYL)UREA

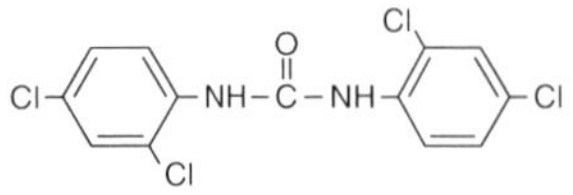

5 gm

S71,693-6
1-(2,5-DICHLOROPHENYL)-3-(3,4-DICHLOROPHENYL)UREA

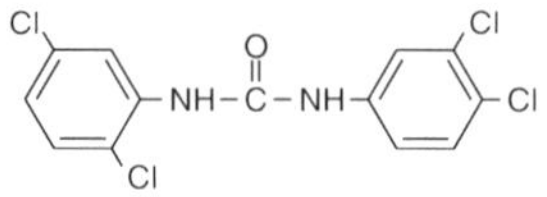

5 gm

S91,613-7
1-(3,4-DIBROMOPHENYL)-3-(2,5-DIFLUOROPHENYL)UREA

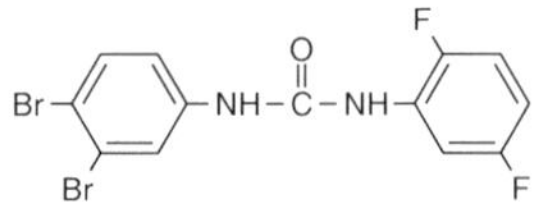

250 mg

S71,785-1
1-(4-CHLORO-2-METHYLPHENYL)-3-(3-FLUOROPHENYL)UREA

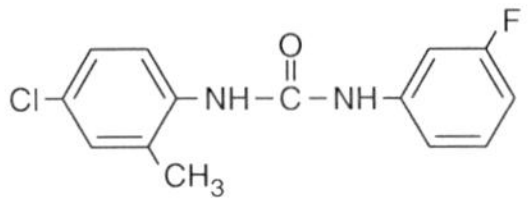

250 mg

S72,762-8
1-(2-CHLORO-5-METHYLPHENYL)-3-(3-FLUOROPHENYL)UREA

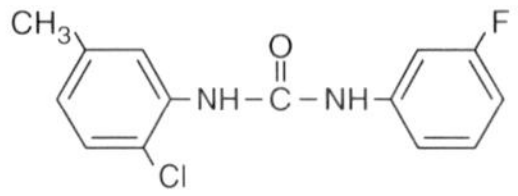

5 gm

S70,057-6
1-(5-CHLORO-2-METHYLPHENYL)-3-(2,4-DIFLUOROPHENYL)UREA

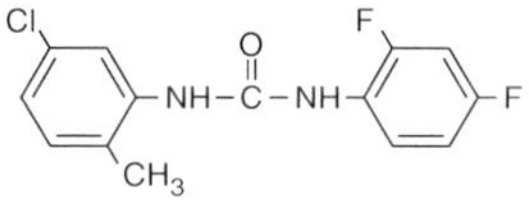

250 mg

S72,142-5
1-(3-CHLORO-4-METHYLPHENYL)-3-(2,5-DIFLUOROPHENYL)UREA

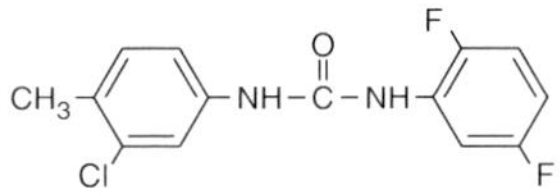

250 mg

S75,232-0
1-(2-CHLORO-4-METHOXYPHENYL)-3-(3-CHLOROPHENYL)UREA

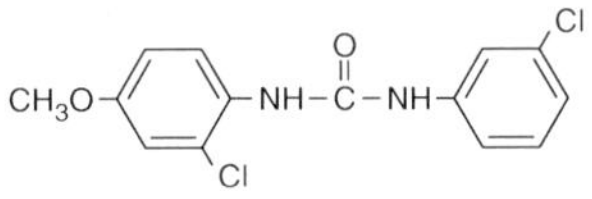

5 gm

S77,315-8
1-(4-CHLORO-2-METHYLPHENYL)-3-(3,4-DICHLOROPHENYL)UREA

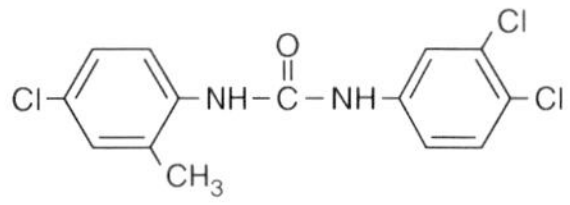

250 mg

S88,939-3
1-(3-CHLORO-4-METHYLPHENYL)-3-(3,4-DICHLOROPHENYL)UREA

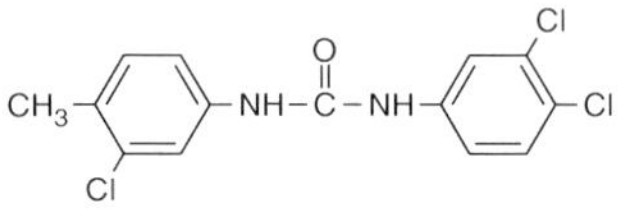

250 mg

S73,642-2
1-(3-CHLORO-4-METHYLPHENYL)-3-(2,5-DIBROMOPHENYL)UREA

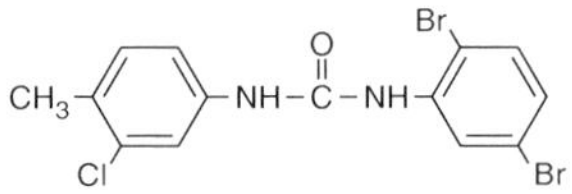

5 gm

S84,533-7
1-(3-FLUOROPHENYL)-3-(3,4-XYLYL)UREA

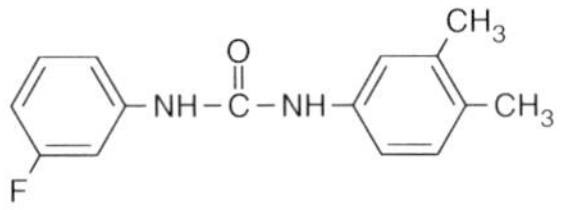

250 mg

S75,351-3
1-(3-CHLORO-4-METHYLPHENYL)-3-(M-TOLYL)UREA

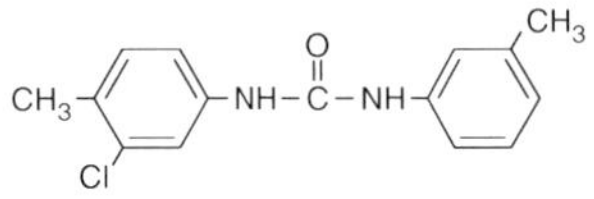

5 gm

S78,022-7
1-(3-CHLOROPHENYL)-3-(2,5-XYLYL)UREA

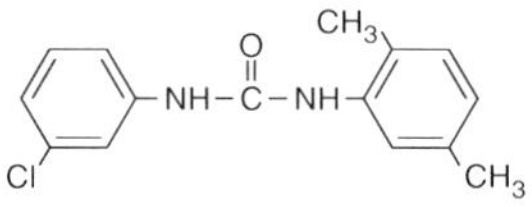

250 mg

S88,998-9
1-(3-CHLOROPHENYL)-3-(3,4-XYLYL)UREA

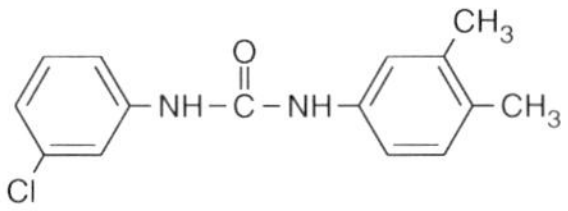

250 mg

S68,415-5
1-(2,5-DIFLUOROPHENYL)-3-(3-ETHYLPHENYL)UREA

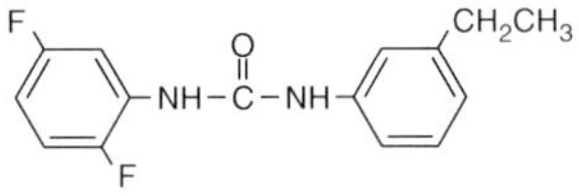

250 mg

S68,718-9
1-(2,4-DIFLUOROPHENYL)-3-(3-ETHYLPHENYL)UREA

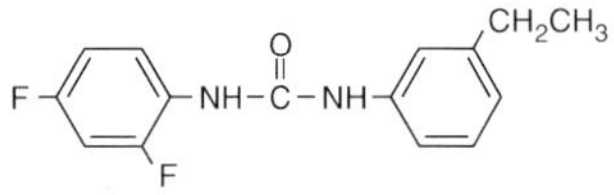

250 mg

S75,349-1
1-(3-CHLORO-4-METHYLPHENYL)-3-(4-CHLORO-2-METHYLPHENYL)UREA

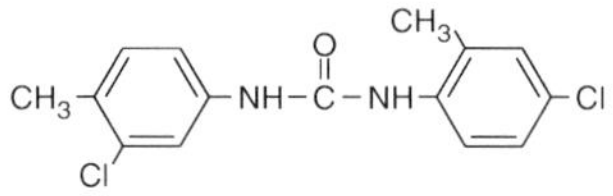

5 gm

S78,024-3
1-(3,4-DICHLOROPHENYL)-3-(2,5-XYLYL)UREA

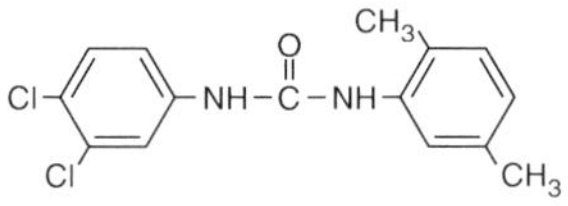

250 mg

S88,941-5
1-(3,4-DICHLOROPHENYL)-3-(2,4-XYLYL)UREA

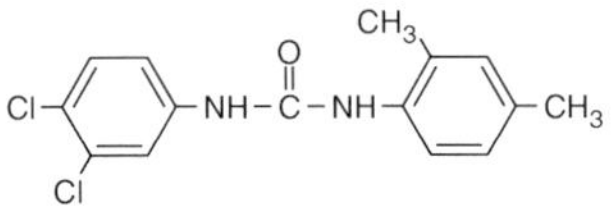

250 mg

S88,957-1
1,3-BIS(3-CHLORO-4-METHYLPHENYL)UREA

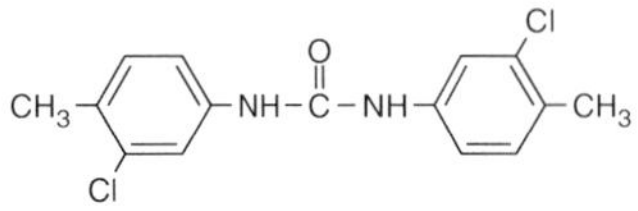

250 mg

S71,046-6
3-(3,4-DICHLOROPHENYL)-1,1-DIISOBUTYLUREA

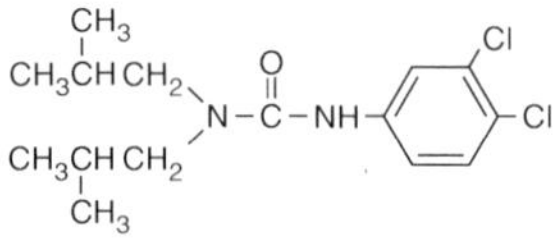

5 gm

S68,405-8
1-(2-BROMO-4,5-DIMETHYLPHENYL)-3-(3-CHLORO-4-FLUOROPHENYL)UREA

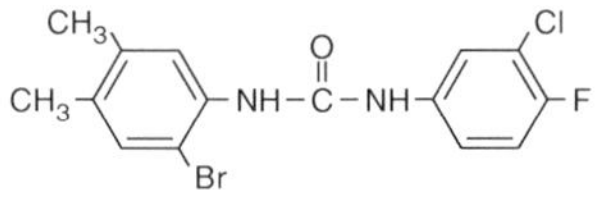

250 mg

S90,025-7
1-(3-CHLOROPHENYL)-3-(3,4-DIMETHYLBENZYL)UREA

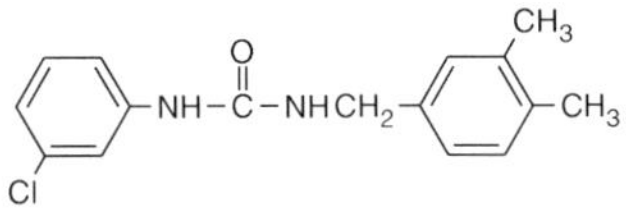

250 mg

S66,036-1
3-(5-CHLORO-2-METHYLPHENYL)-1,1-DIISOBUTYLUREA

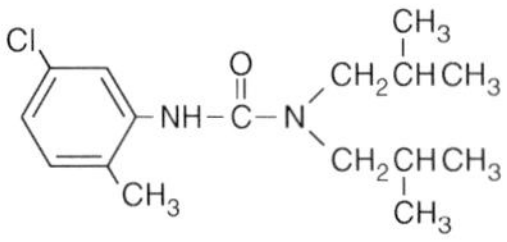

5 gm

S66,411-1
3-(4-CHLORO-2-METHYLPHENYL)-1,1-DIISOBUTYLUREA

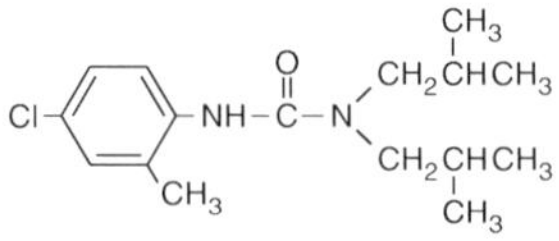

5 gm

S66,499-5
3-(3-CHLORO-4-METHYLPHENYL)-1,1-DIISOBUTYLUREA

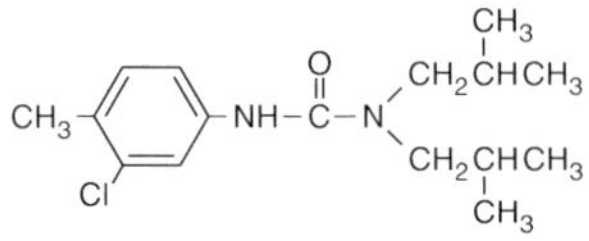

5 gm

S72,559-5
1-(3,4-DICHLOROPHENYL)-3-(2,4,5-TRIMETHYLPHENYL)UREA

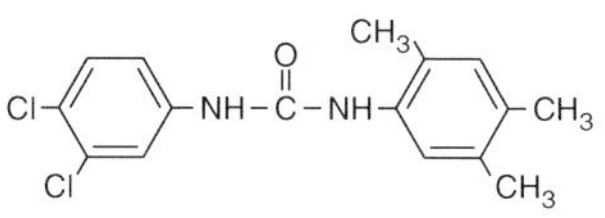

5 gm

S67,495-8
1-(5-CHLORO-2-METHYLPHENYL)-3-(2,4,5-TRIMETHYLPHENYL)UREA

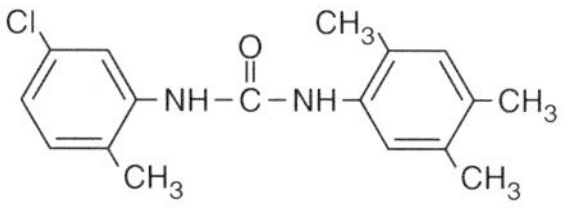

5 gm

S69,380-4
1-(3-CHLORO-4-METHYLPHENYL)-1-ETHYL-3-(3-ETHYLPHENYL)UREA

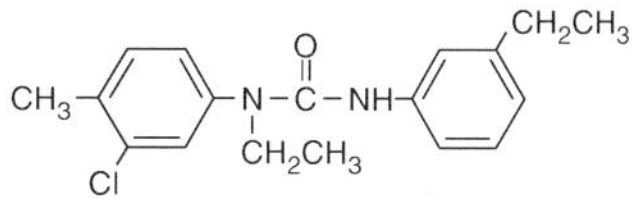

250 mg

S86,895-7
1,1'-HEXAMETHYLENEBIS[3-(3-CHLORO-4-METHYLPHENYL)UREA]

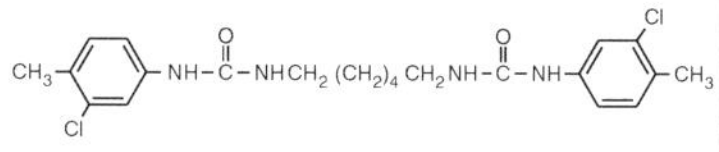

250 mg

S67,027-8
1-(3,4-DICHLOROPHENYL)-3-OCTADECYLUREA

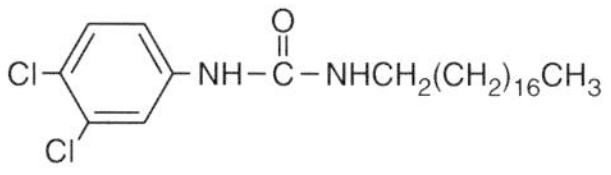

5 gm

S67,018-9
1-(4-CHLORO-2-METHYLPHENYL)-3-OCTADECYLUREA

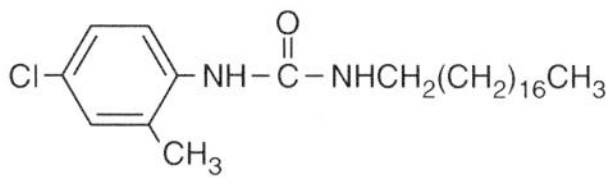

5 gm

S79,538-0
1-(3-CHLORO-4-METHYLPHENYL)-3-OCTADECYLUREA

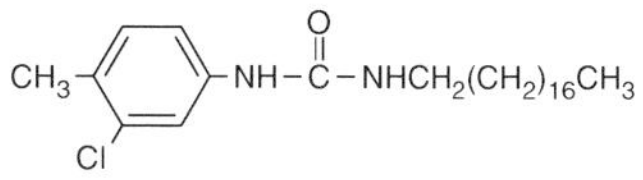

250 mg

S78,011-1
1-(4-BUTYLPHENYL)-3-(5-CHLORO-2-METHYLPHENYL)UREA

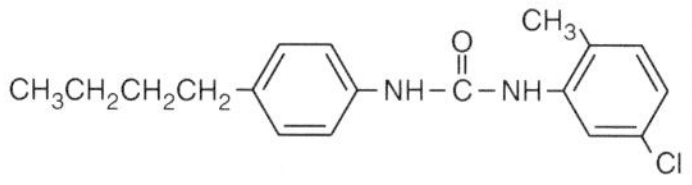

250 mg

S71,490-9
1-(2,5-DICHLOROPHENYL)-3-(3,5-DICHLOROPHENYL)UREA

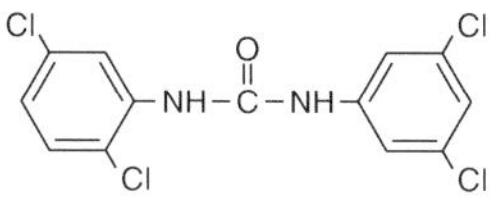

250 mg

S67,665-9
4,4'-BIS[3-(3,5-DICHLOROPHENYL)-UREIDO]-3,3'-DIMETHYLBIPHENYL

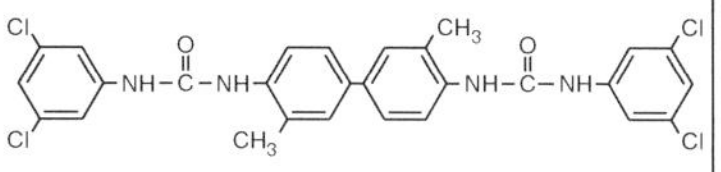

250 mg

S71,185-3
1-(2,4-DICHLOROPHENYL)-3-(2,4,6-TRICHLOROPHENYL)UREA

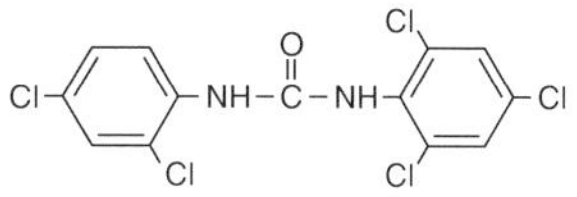

5 gm

S75,372-6
1-(3-CHLORO-4-METHYLPHENYL)-3-(2,4,6-TRIMETHYLPHENYL)UREA

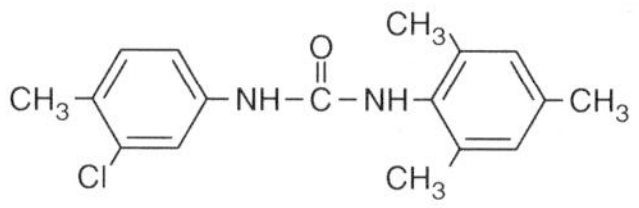

5 gm

S91,623-4
1-(2,5-DIFLUOROPHENYL)-3-(2,4,5-TRICHLOROPHENYL)UREA

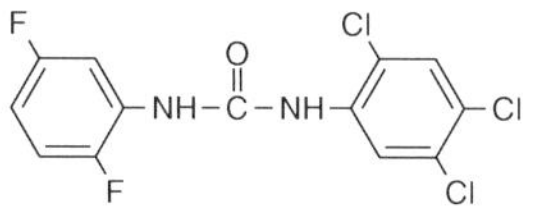

250 mg

S71,182-9
1-(3,4-DICHLOROPHENYL)-3-(2,4,5-TRICHLOROPHENYL)UREA

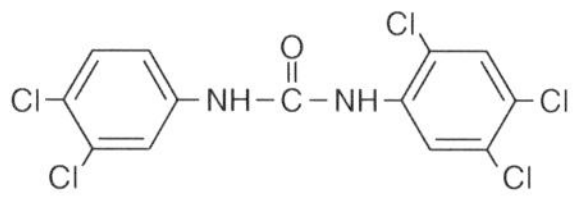

5 gm

S71,183-7
1-(2,5-DICHLOROPHENYL)-3-(2,4,5-TRICHLOROPHENYL)UREA

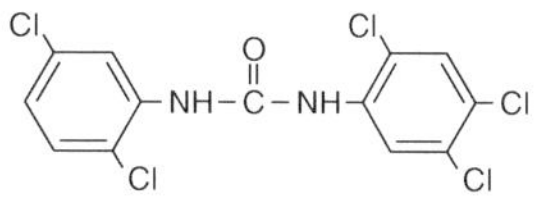

5 gm

S80,796-6
1-(5-CHLORO-2-METHYLPHENYL)-3-(4,5-DICHLORO-2-METHYLPHENYL)UREA

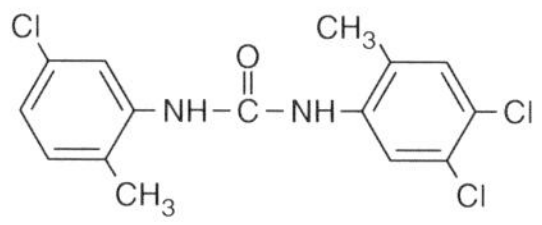

250 mg

S75,373-4
1-(3-CHLORO-4-METHYLPHENYL)-3-(2,4,5-TRIMETHYLPHENYL)UREA

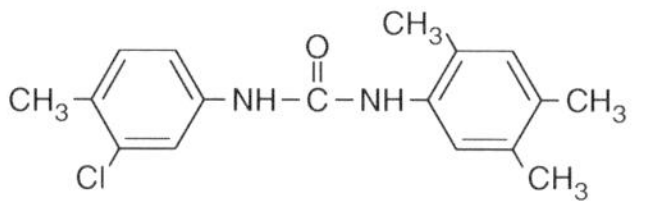

5 gm

S67,336-6
4,4'-BIS[3-(3-CHLOROPHENYL)-UREIDO]-3,3'-DIMETHYLBIPHENYL

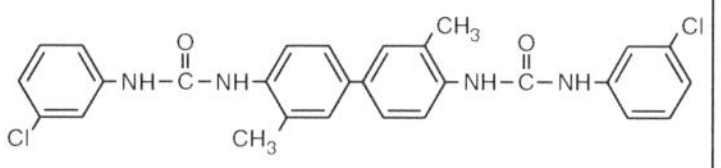

250 mg

S72,113-1
1-(2,5-DIFLUOROPHENYL)-3-(1-NAPHTHYL)UREA

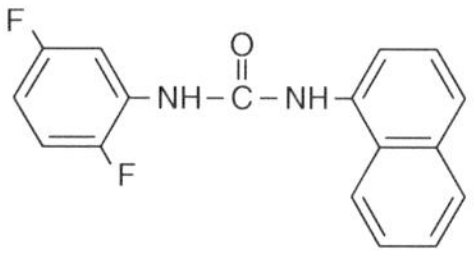

5 gm

S79,159-8
1-(3-CHLORO-4-METHYLPHENYL)-3-(1-NAPHTHYL)UREA

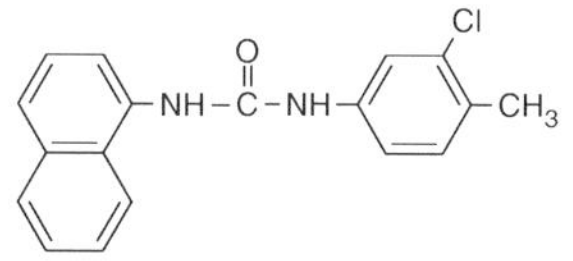

250 mg

S89,071-5
1-(4-CHLORO-2-METHYLPHENYL)-3-(1-NAPHTHYL)UREA

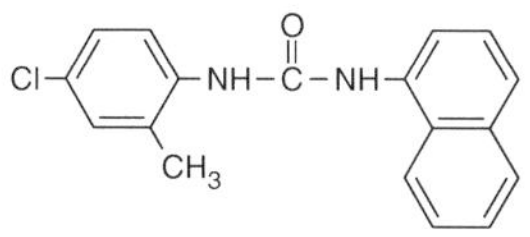

250 mg

S89,072-3
1-(5-CHLORO-2-METHYLPHENYL)-3-(1-NAPHTHYL)UREA

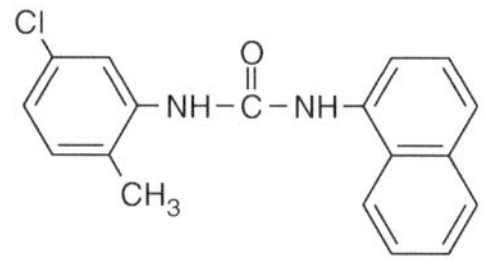

250 mg

S71,499-2
1-(3,5-DICHLOROPHENYL)UREA

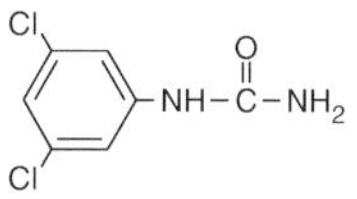

250 mg

S69,867-9
1-(3,5-DICHLOROPHENYL)-3-METHYLUREA

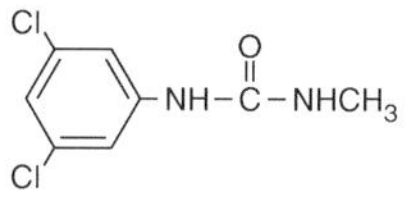

5 gm

S70,884-4
1-ALLYL-3-(3,5-DICHLOROPHENYL)UREA

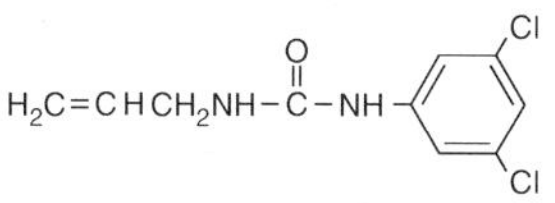

250 mg

S70,880-1
1-(3,5-DICHLOROPHENYL)-3-PROPYLUREA

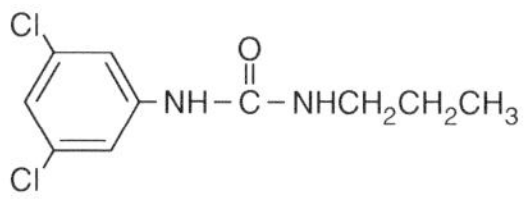

5 gm

S70,882-8
1-(3,5-DICHLOROPHENYL)-3-ISOPROPYLUREA

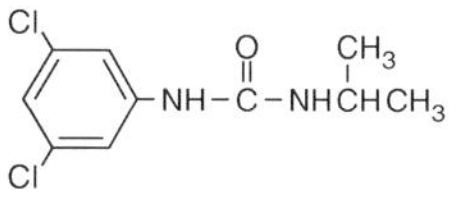

5 gm

S71,127-6
1-BUTYL-3-(3,5-DICHLOROPHENYL)UREA

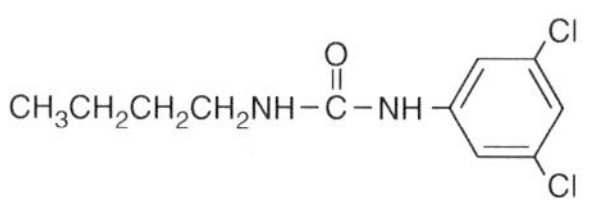

250 mg

S73,577-9
1-(3-CHLOROPHENYL)-3-(3,5-DICHLOROPHENYL)UREA

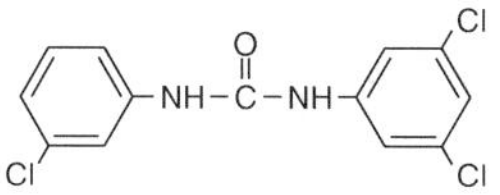

250 mg

S72,631-1
1-(3-BROMOPHENYL)-3-(3,5-DICHLOROPHENYL)UREA

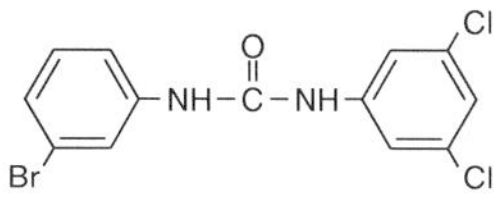

5 gm

S69,450-9
1-(3-BROMOPHENYL)-3-(3,5-DIMETHYLPHENYL)UREA

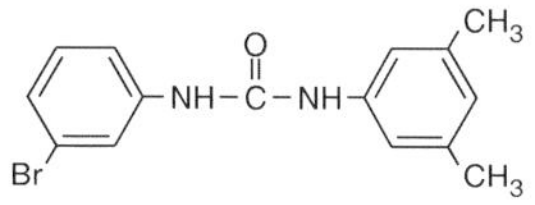

250 mg

S67,079-0
1-(3,5-DICHLOROPHENYL)-3-OCTADECYLUREA

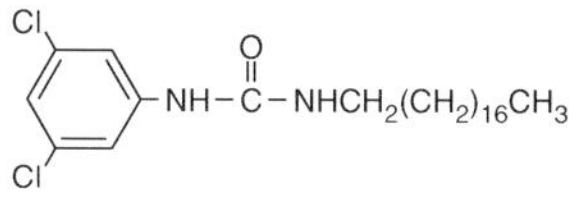

5 gm

S71,789-4
1-(4-BROMO-2,6-DIMETHYLPHENYL)-3-METHYLUREA
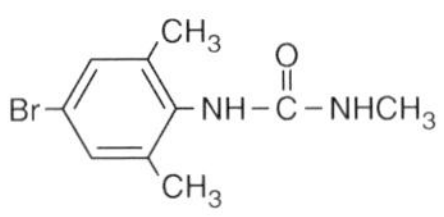
250 mg

S70,883-6
1-ISOPROPYL-3-(2,4,6-TRICHLOROPHENYL)UREA
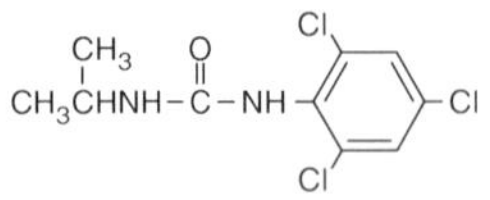
5 gm

S71,195-0
1-(3-CHLOROPHENYL)-3-(2,4,6-TRICHLOROPHENYL)UREA
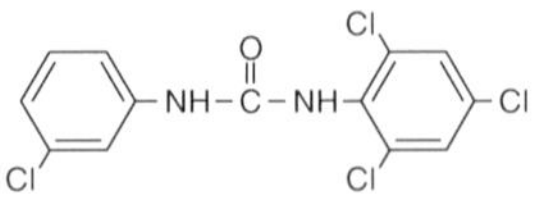
250 mg

S67,524-5
1,3-BIS(2-CHLORO-4,6-DIBROMOPHENYL)UREA
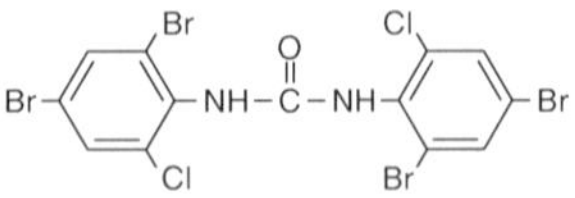
250 mg

S67,529-6
1,3-BIS(4-CHLORO-2,6-DIBROMOPHENYL)UREA
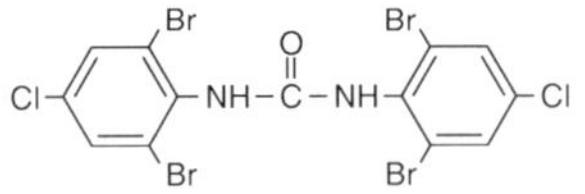
250 mg

S84,564-7
1-(3-FLUOROPHENYL)-3-(2,4,6-TRIMETHYLPHENYL)UREA
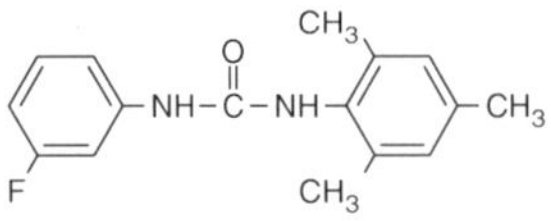
250 mg

S77,325-5
1-(3-CHLOROPHENYL)-3-(2,4,6-TRIMETHYLPHENYL)UREA
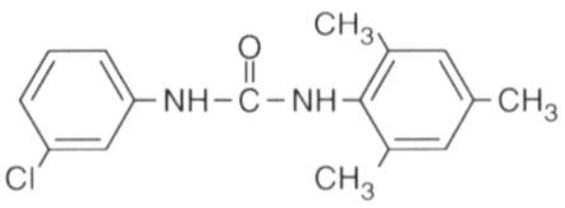
250 mg

S71,796-7
1,1'-(2-BROMO-6-METHYL-1,4-PHENYLENE)BIS[3-(3-BROMOPHENYL)UREA]
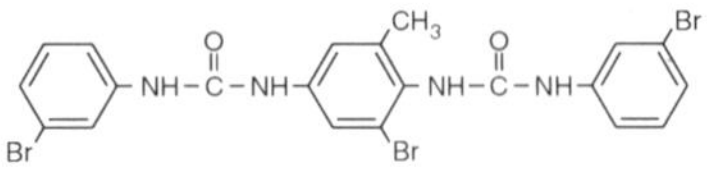
250 mg

S76,412-4
(2,4,5-TRICHLOROPHENYL)UREA
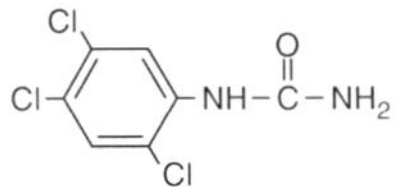
5 gm

S66,696-3
1-METHYL-3-(2,4,5-TRICHLOROPHENYL)UREA
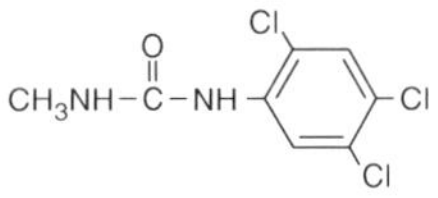
5 gm

S89,289-0
1-(4,5-DICHLORO-2-METHYLPHENYL)-3-METHYLUREA
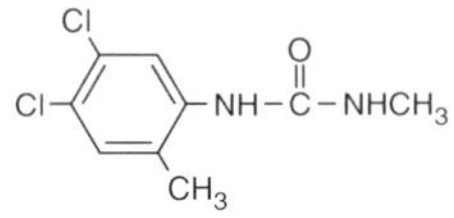
250 mg

S66,744-7
1-ETHYL-3-(2,4,5-TRICHLOROPHENYL)UREA
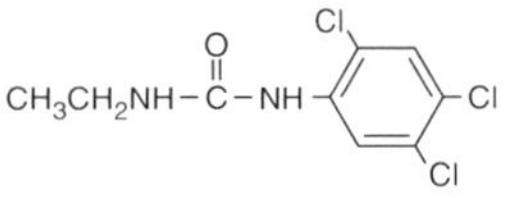
5 gm

S66,565-7
1-ISOBUTYL-3-(2,4,5-TRICHLOROPHENYL)UREA
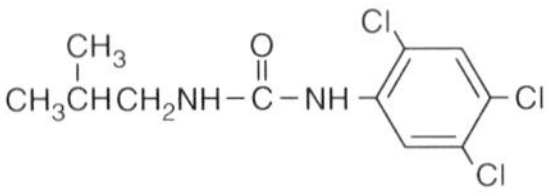
5 gm

S71,197-7
1-(3-CHLOROPHENYL)-3-(2,4,5-TRICHLOROPHENYL)UREA
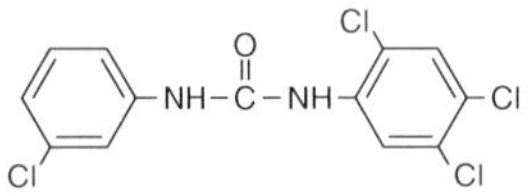
5 gm

S71,189-6
1-(3-NITROPHENYL)-3-(2,4,5-TRICHLOROPHENYL)UREA
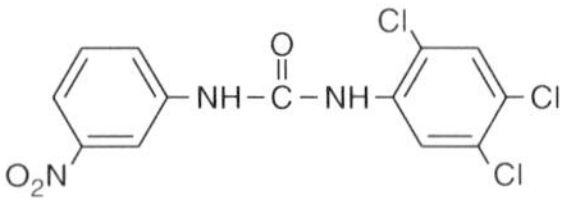
5 gm

S78,071-5
1-(4,5-DICHLORO-2-METHYLPHENYL)-3-(3-FLUOROPHENYL)UREA
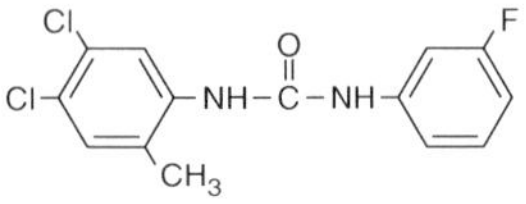
250 mg

S66,657-2
1,1-DIISOBUTYL-3-(2,4,5-TRICHLOROPHENYL)UREA
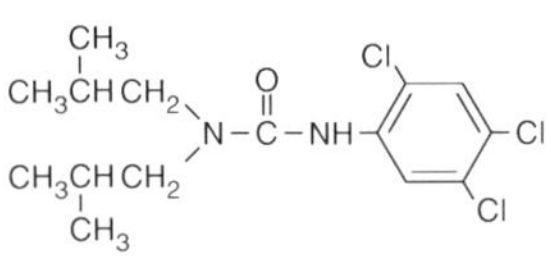
5 gm

S71,146-2
1-(1-NAPHTHYL)-3-(2,4,5-TRICHLOROPHENYL)UREA
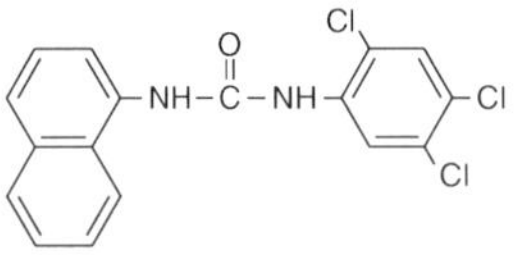
5 gm

S75,374-2
1-(4,5-DICHLORO-2-METHYLPHENYL)-3-(1-NAPHTHYL)UREA
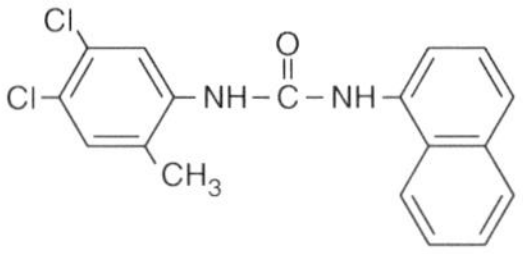
5 gm

S84,576-0
1-(3-FLUOROPHENYL)-3-(2-NAPHTHYL)UREA
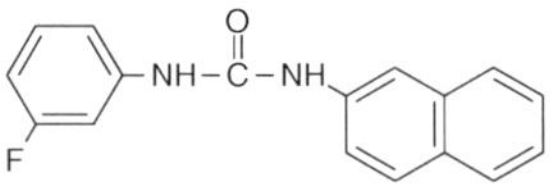
250 mg

S87,652-6
PACTAMYCIN
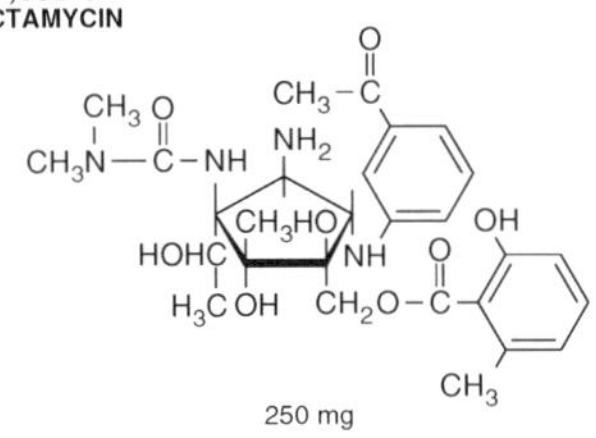
250 mg

S90,561-5
1,1'-HEXAMETHYLENEBIS[3-(3-HYDROXYPHENYL)UREA]
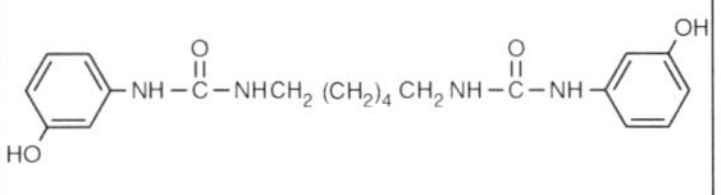
250 mg

S78,063-4
1-CYCLOHEXYL-3-(2,4-DIMETHOXYPHENYL)-1-(2-HYDROXYETHYL)UREA
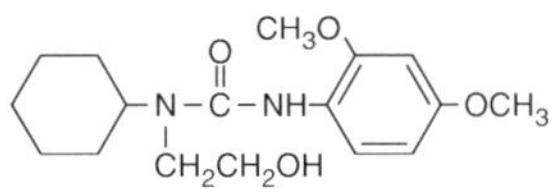
250 mg

S85,112-4
1-(2,4-DIMETHOXYPHENYL)-3-[4-(2-HYDROXYETHYL)PHENYL]UREA
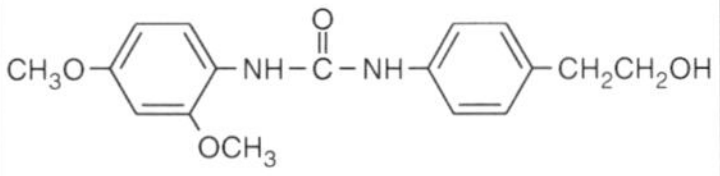
250 mg

S85,124-8
1-(2,4-DIMETHOXYPHENYL)-3-(3-HYDROXYPHENYL)UREA
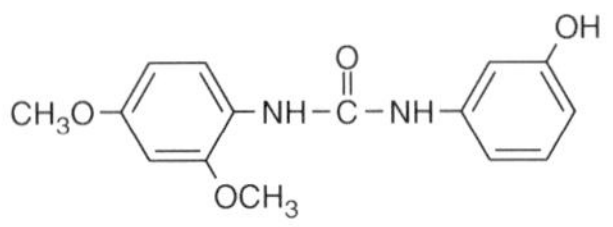
250 mg

S84,869-7
1-(2,5-DIMETHOXYPHENYL)-3-(4-HYDROXYPHENYL)UREA
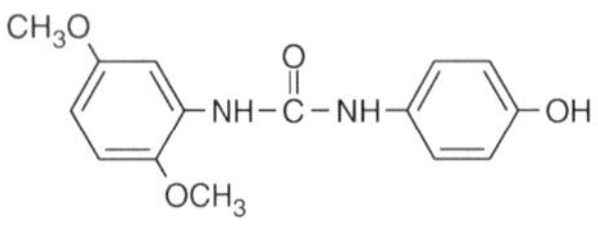
250 mg

S73,779-8
1-(2,5-DIMETHOXYPHENYL)-3-(3-HYDROXYPHENYL)UREA
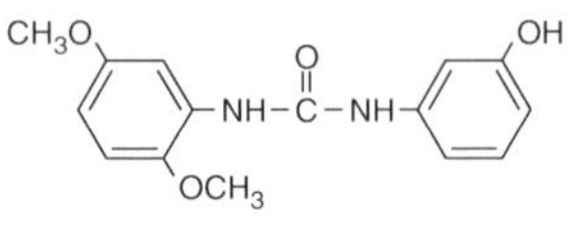
5 gm

S84,868-9
1-(2,5-DIMETHOXYPHENYL)-3-(2-HYDROXY-5-NITROPHENYL)UREA
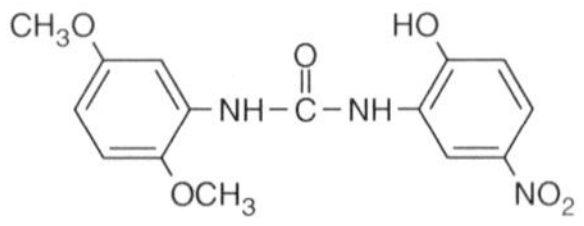
250 mg

S91,730-3
4-(2,5-DIMETHOXYPHENYL)-1-(4-HYDROXYBENZOYL)SEMICARBAZIDE
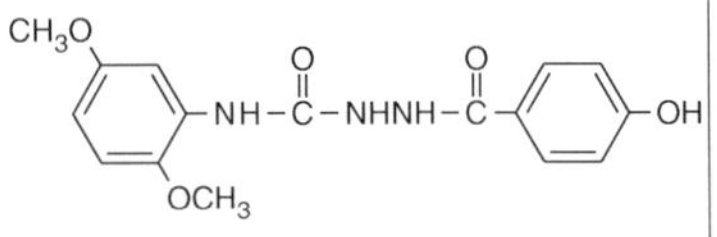
250 mg

S85,434-4
4-(3,3-DIETHYLUREIDO)BENZOIC (4-HYDROXY-3-METHOXYBENZYLIDENE)-HYDRAZIDE
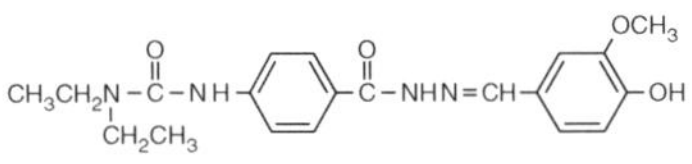
250 mg

S68,951-3
1-(2-HYDROXYBENZOYL)-4-(3,4,5-TRIMETHOXYPHENYL)SEMICARBAZIDE
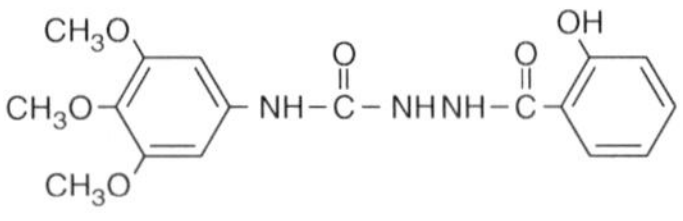
250 mg

S90,294-2
2-[3-(2-ETHOXYPHENYL)UREIDO]ACETIC (2-HYDROXYBENZYLIDENE)HYDRAZIDE
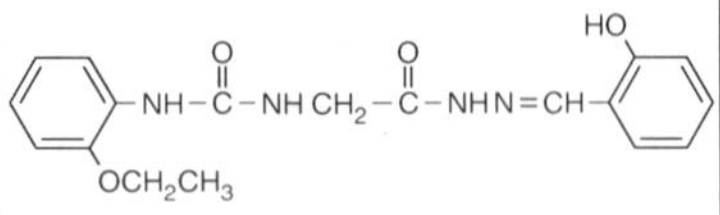
250 mg

S79,488-0
1-[4-(2-HYDROXYETHYL)PHENYL]-3-(2-METHOXY-5-NITROPHENYL)UREA

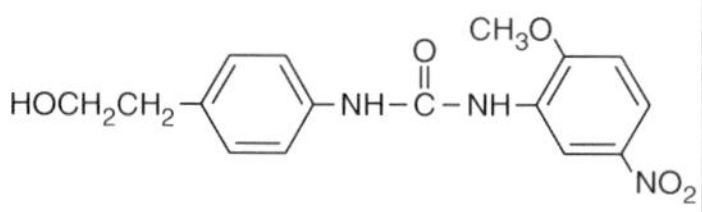

250 mg

S80,029-5
1-(3-HYDROXYPHENYL)-3-(2-METHOXY-5-NITROPHENYL)UREA

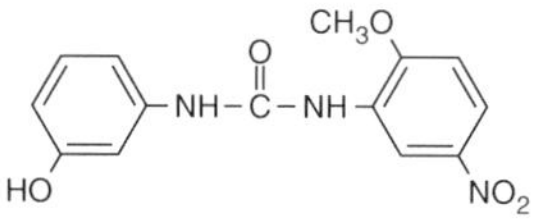

250 mg

S90,231-4
2-[3-(3-NITROPHENYL)UREIDO]ACETIC (4-HYDROXY-3-METHOXYBENZYLIDENE)-HYDRAZIDE

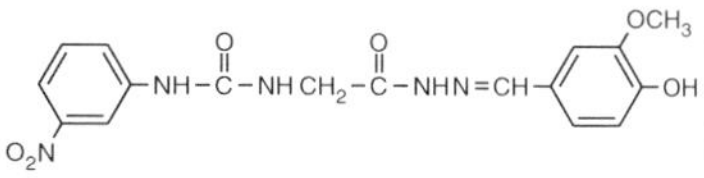

250 mg

S86,875-2
4-(3-ALLYLUREIDO)BENZOIC (3-ETHOXY-4-HYDROXYBENZYLIDENE)HYDRAZIDE

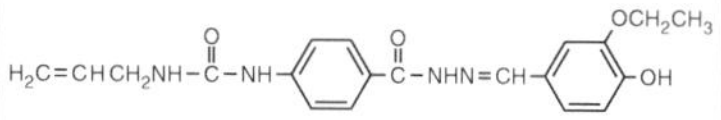

250 mg

S70,070-3
1-(2-HYDROXYETHYL)-3-(2-METHOXY-5-METHYLPHENYL)UREA

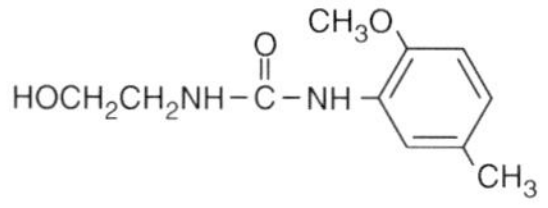

250 mg

S78,052-9
1-CYCLOHEXYL-1-(2-HYDROXY-2-PHENYLETHYL)-3-(2-METHYL-5-NITROPHENYL)UREA

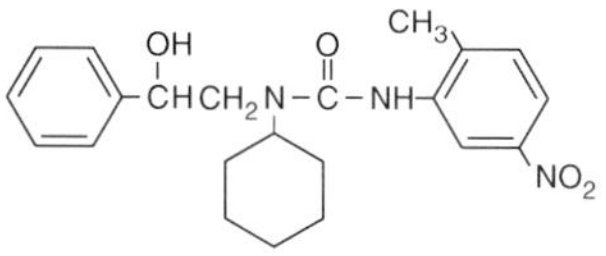

250 mg

S74,728-9
1-(TERT-BUTYL)-3-[3-(1-HYDROXYETHYL)PHENYL]UREA

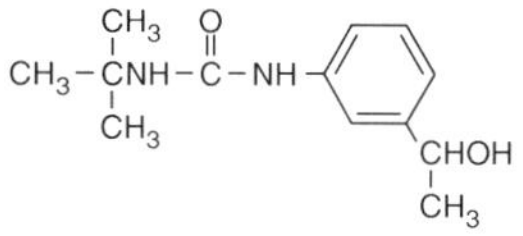

5 gm

S89,948-8
1-TERT-BUTYL-3-[3-(1-HYDROXYETHYL)-PHENYL]UREA

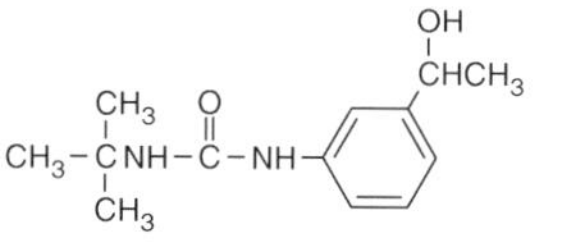

250 mg

S69,175-5
1-(2-HYDROXYBENZOYL)-4-[4-(METHYLTHIO)PHENYL]SEMICARBAZIDE

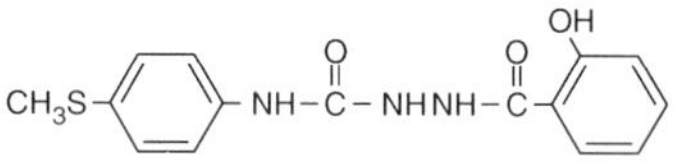

250 mg

S68,996-3
4-(2,6-DIMETHYLPHENYL)-1-(2-HYDROXYBENZOYL)SEMICARBAZIDE

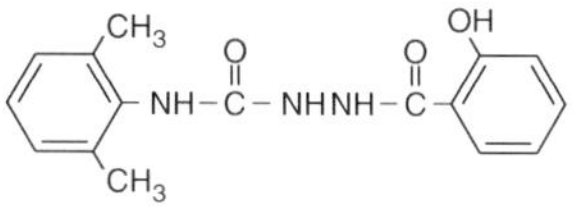

250 mg

S84,561-2
4-[3-(2-ETHYLPHENYL)UREIDO]-SALICYLIC ACID

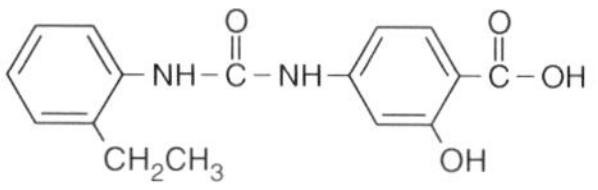

250 mg

S84,562-0
4-[3-(2,6-XYLYL)UREIDO]-SALICYLIC ACID

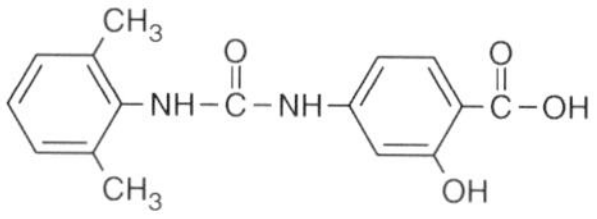

250 mg

S92,425-3
1-CYCLOHEXYL-1-(2-HYDROXY-2-PHENYLETHYL)-3-(2,5-XYLYL)UREA

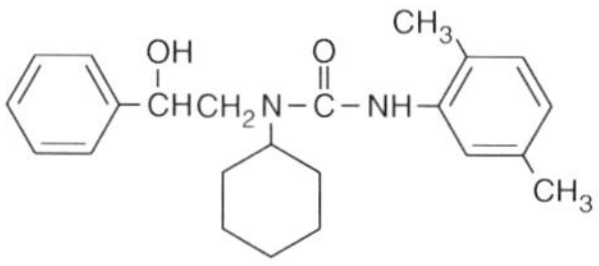

250 mg

S73,748-8
1-(6-HYDROXYHEXANOYL)-4-[4-(METHYLTHIO)PHENYL]SEMICARBAZIDE

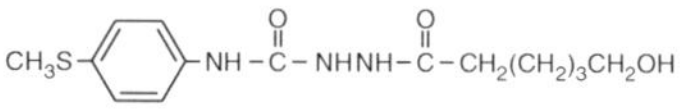

5 gm

S79,705-7
1-(3-HYDROXYPHENYL)-3-[4-(METHYLTHIO)PHENYL]UREA

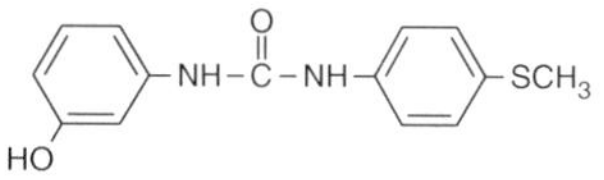

250 mg

S72,283-9
1-(5-HYDROXY-1-NAPHTHYL)-3-[4-(METHYLTHIO)PHENYL]UREA

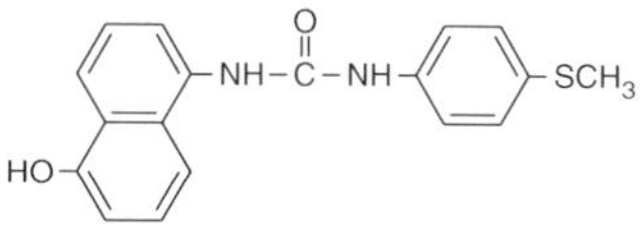

5 gm

S70,247-1
1-(7-HYDROXY-1-NAPHTHYL)-3-[3-(METHYLTHIO)PHENYL]UREA

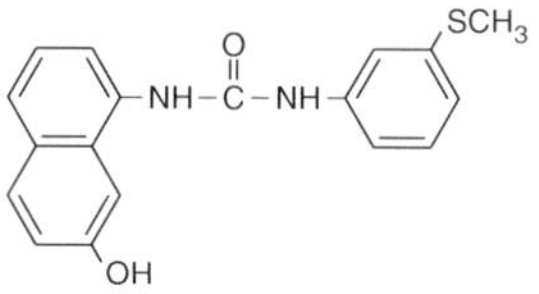

250 mg

S59,182-3
4'-AMINO-2'-HYDROXYCARBANILIDE

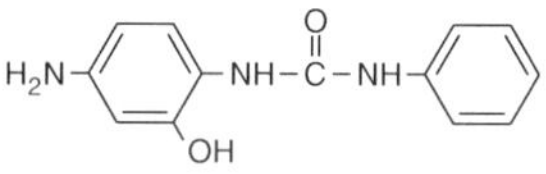

5 gm

S58,728-1
1-(4-AMINO-2-HYDROXYPHENYL)-3-BUTYLUREA

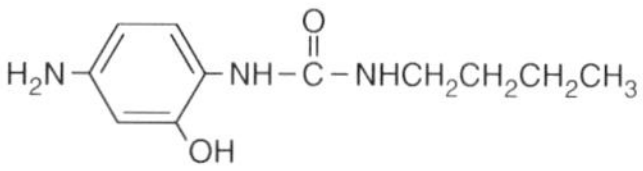

250 mg

S90,299-3
2-(3-BENZYLUREIDO)ACETIC (3-HYDROXYBENZYLIDENE)HYDRAZIDE

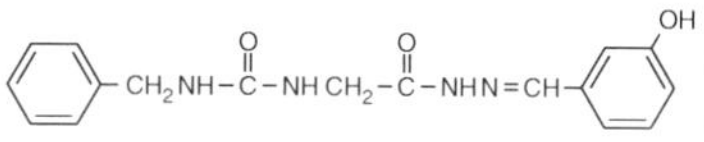

250 mg

S85,334-8
1-CYCLOHEXYL-1-(2-HYDROXYETHYL)-3-(4-METHYL-2-NITROPHENYL)UREA

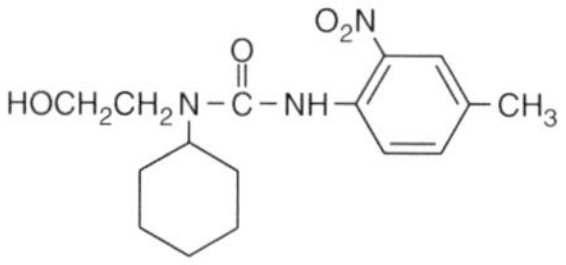

250 mg

S85,325-9
1-(3-HYDROXYPHENYL)-3-(4-METHYL-2-NITROPHENYL)UREA

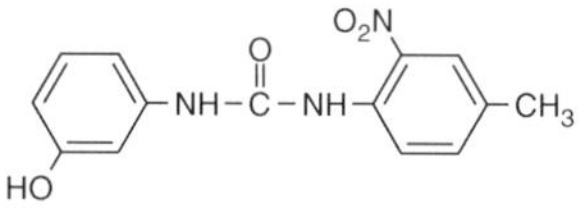

250 mg

S84,840-9
1-CYCLOHEXYL-1-(2-HYDROXYETHYL)-3-(2-METHYL-5-NITROPHENYL)UREA

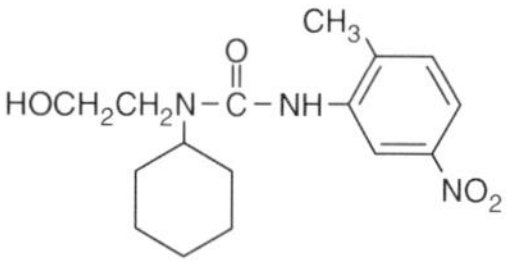

250 mg

S84,791-7
1-(2-HYDROXYPHENYL)-3-(2-METHYL-5-NITROPHENYL)UREA

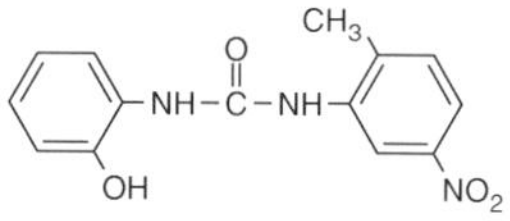

250 mg

S84,790-9
1-(2-HYDROXY-5-NITROPHENYL)-3-(2-METHYL-5-NITROPHENYL)UREA

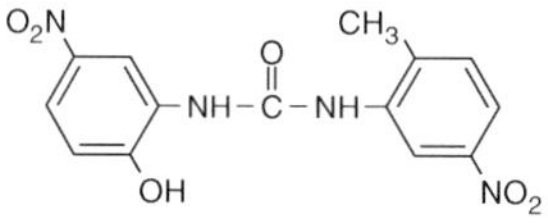

250 mg

S79,667-0
1-[4-(2-HYDROXYETHYL)PHENYL]-3-(2-METHYL-5-NITROPHENYL)UREA

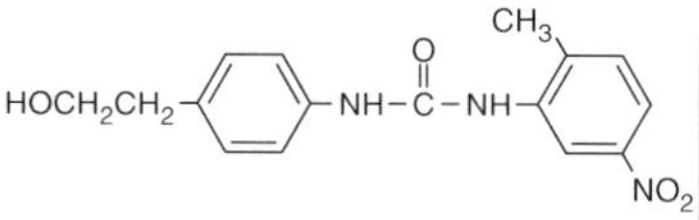

250 mg

S84,792-5
1-(3-HYDROXYPHENYL)-3-(2-METHYL-5-NITROPHENYL)UREA

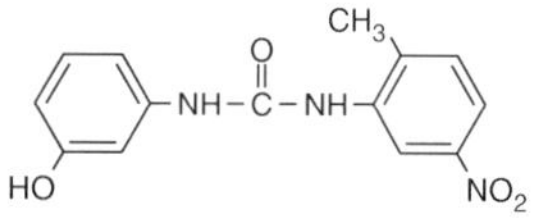

250 mg

S50,245-6
2-HYDROXY-4-NITROCARBANILIDE

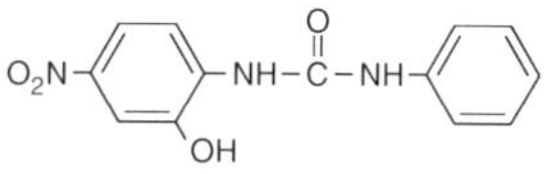

250 mg

S61,130-1
1-(2-HYDROXYETHYL)-1-METHYL-3-(2-METHYL-4-NITROPHENYL)UREA

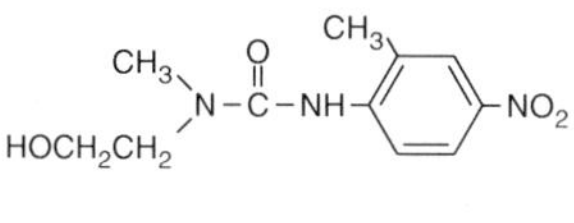

5 gm

S65,507-4
1-(2,5-DICHLOROPHENYL)-3-(2-HYDROXY-4-NITROPHENYL)UREA

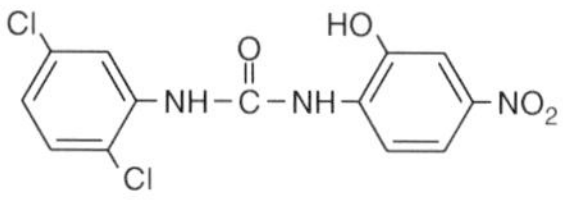

5 gm

S50,463-7
N-CARBAMOYLTYROSINE

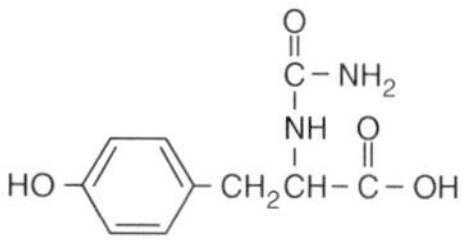

250 mg

S92,518-7
4-HYDROXYBENZOIC (ALPHA,4-DIMETHYLBENZYLIDENE)HYDRAZIDE

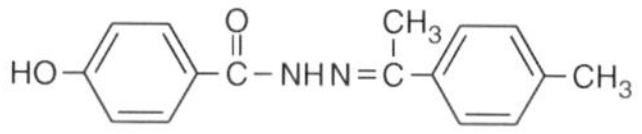

250 mg

S71,892-0
1-(4-HYDROXYBENZOYL)-4-[2-(METHOXYCARBONYL)PHENYL]-SEMICARBAZIDE

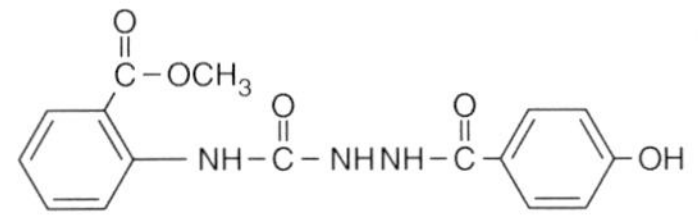

250 mg

S77,390-5
1-(4-HYDROXYBENZOYL)-4-(2,4,5-TRICHLOROPHENYL)SEMICARBAZIDE

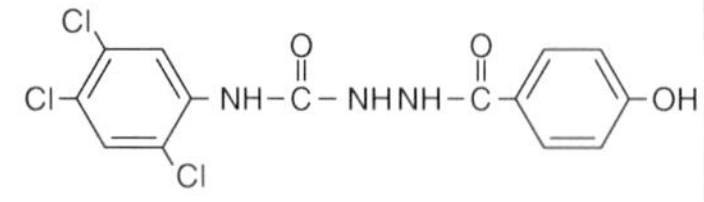

250 mg

S67,510-5
4',4''-METHYLENEBIS[1-(4-HYDROXYBENZOYL)-4-PHENYLSEMICARBAZIDE]

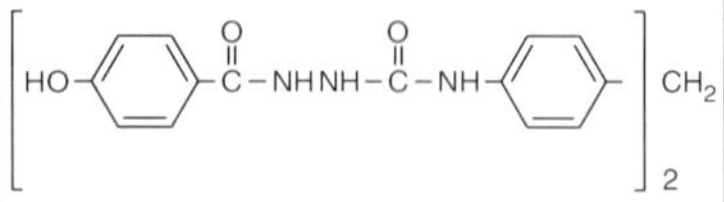

250 mg

S67,385-4
4-(2-ETHYLPHENYL)-1-(2-HYDROXYBENZOYL)SEMICARBAZIDE

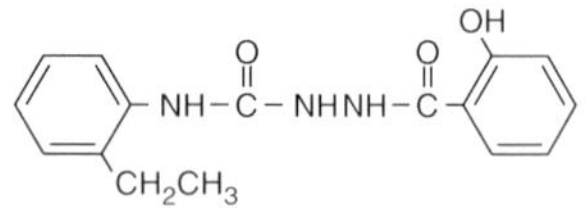

250 mg

S90,235-7
2-[3-(3-NITROPHENYL)UREIDO]ACETIC (3,5-DI-TERT-BUTYL-4-HYDROXYBENZYLIDENE)HYDRAZIDE

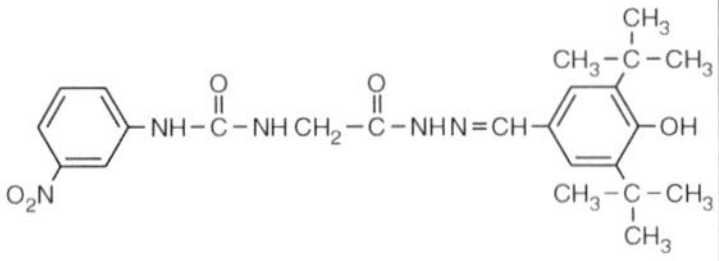

250 mg

S86,851-5
4-(3-ALLYLUREIDO)BENZOIC (2-HYDROXYBENZYLIDENE)HYDRAZIDE

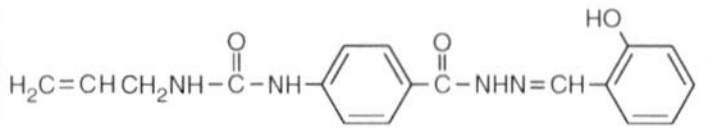

250 mg

S86,858-2
4-(3-PROPYLUREIDO)BENZOIC (3-HYDROXYBENZYLIDENE)HYDRAZIDE

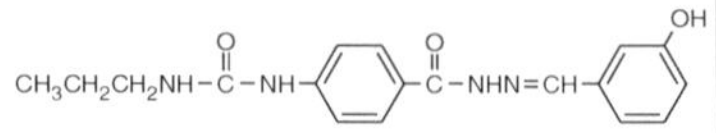

250 mg

S85,435-2
4-(3,3-DIETHYLUREIDO)BENZOIC (4-HYDROXYBENZYLIDENE)HYDRAZIDE

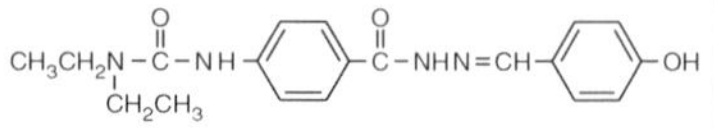

250 mg

S86,927-9
4-[3-(3-CHLOROPHENYL)UREIDO]-BENZOIC (3,5-DI-TERT-BUTYL-4-HYDROXYBENZYLIDENE)HYDRAZIDE

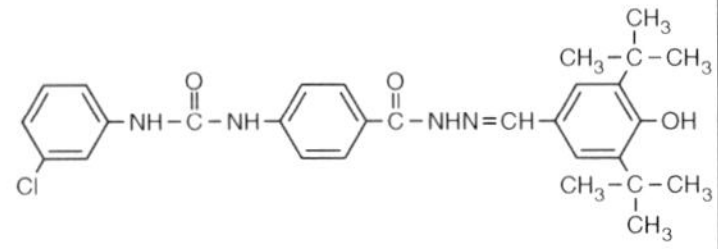

250 mg

S71,885-8
1-CYCLOHEXYL-3-(2-FLUOROPHENYL)-1-(2-HYDROXYETHYL)UREA

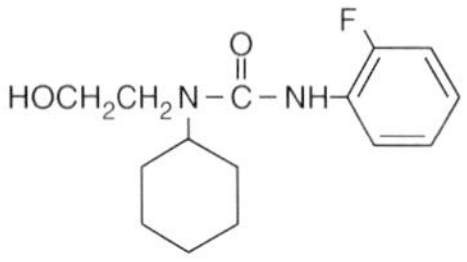

250 mg

S60,937-4
1-(2-HYDROXY-1-PROPYL)-3-PHENYLUREA

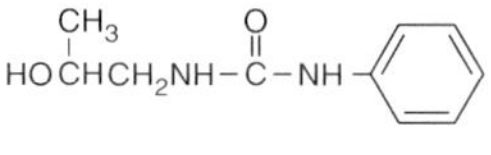

5 gm

S60,797-5
1-ETHYL-1-(2-HYDROXYETHYL)-3-PHENYLUREA

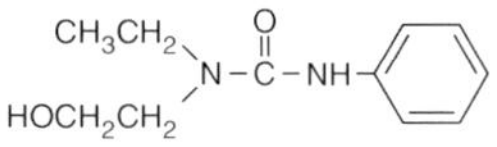

5 gm

S60,988-9
3-(4-CHLOROPHENYL)-1-ETHYL-1-(2-HYDROXYETHYL)UREA

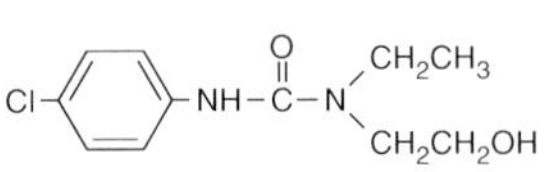

5 gm

S60,969-2
1-(3,4-DICHLOROPHENYL)-3-(2-HYDROXY-1-PROPYL)UREA

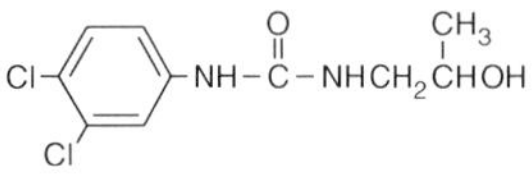

5 gm

S60,945-5
3-(5-CHLORO-2-METHYLPHENYL)-1-(2-HYDROXYETHYL)-1-METHYLUREA

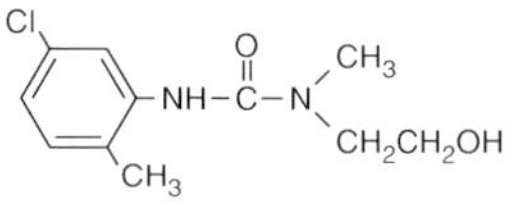

5 gm

S61,023-2
3-(3,4-DICHLOROPHENYL)-1,1-DI(2-HYDROXYETHYL)UREA

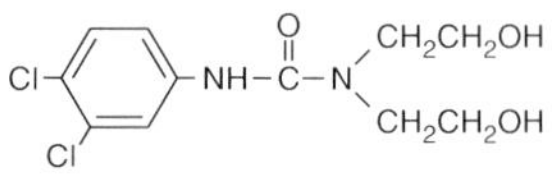

5 gm

S60,968-4
3-(5-CHLORO-2-METHYLPHENYL)-1-ETHYL-1-(2-HYDROXYETHYL)UREA

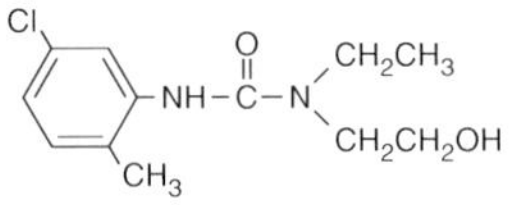

5 gm

S61,083-6
1-(5-CHLORO-2-METHYLPHENYL)-3-(1,1-DIMETHYL-2-HYDROXYETHYL)UREA

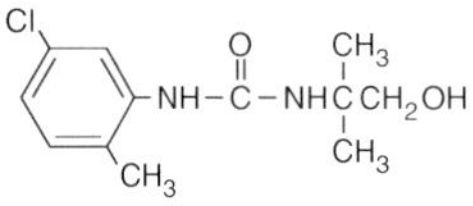

5 gm

S86,861-2
1-(3-ACETYLPHENYL)-3-(5-HYDROXY-1-NAPHTHYL)UREA

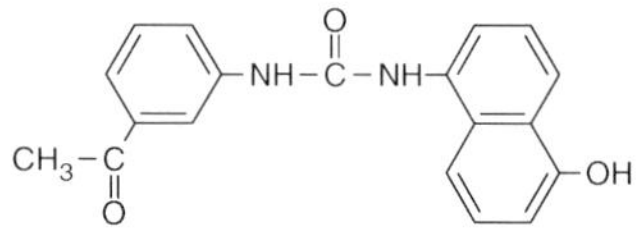

250 mg

S73,706-2
4-[3-(2,5-XYLYL)UREIDO]-SALICYLIC ACID

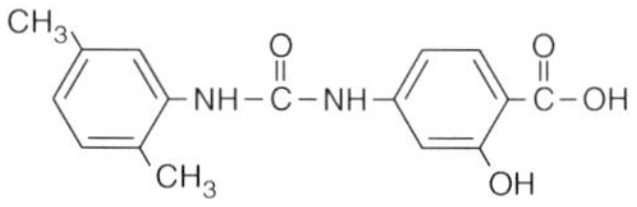

5 gm

S72,770-9
ETHYL 4-[3-(3-HYDROXYPHENYL)-UREIDO]BENZOATE

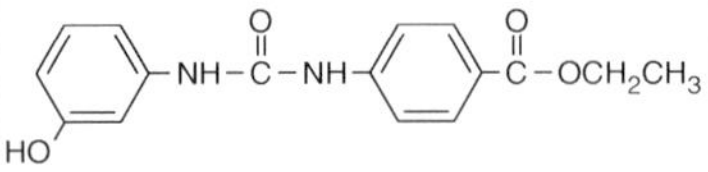

5 gm

S73,195-1
1-[4-(ETHOXYCARBONYL)PHENYL]-3-[4-(2-HYDROXYETHYL)PHENYL]UREA

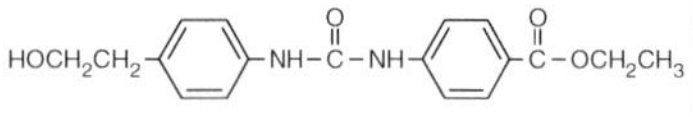

5 gm

S73,187-0
1-[4-(ETHOXYCARBONYL)PHENYL]-3-(5-HYDROXY-1-NAPHTHYL)UREA

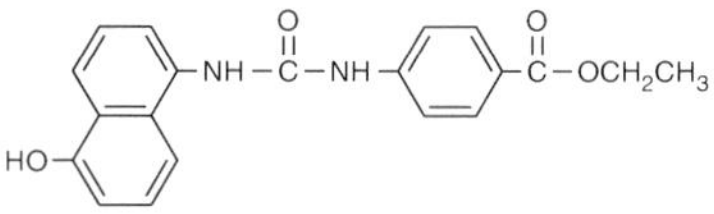

5 gm

S91,716-8
1-(2-HYDROXYBENZOYL)-4-PROPYLSEMICARBAZIDE

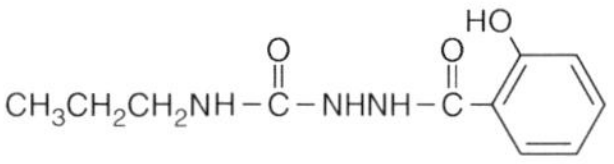

250 mg

S91,721-4
1-(2-HYDROXYBENZOYL)-4-ISOPROPYLSEMICARBAZIDE

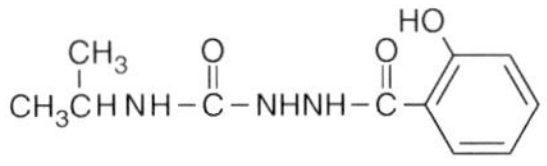

250 mg

S72,330-4
4,4'-(METHYLENEDI-1,1-PHENYLENE)-BIS[1-(2-HYDROXYBENZOYL)-4-SEMICARBAZIDE]

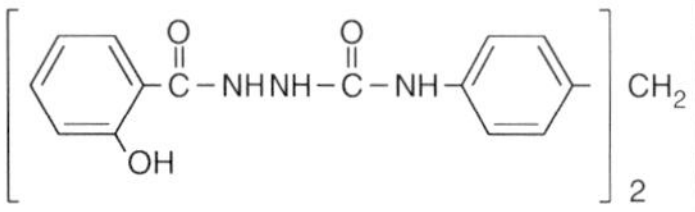

5 gm

S91,717-6
1-(4-HYDROXYBENZOYL)-4-PROPYLSEMICARBAZIDE

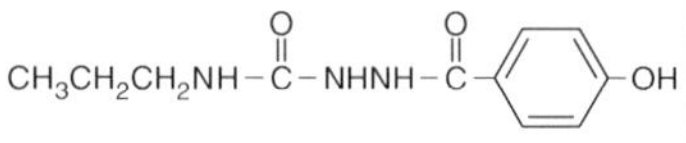

250 mg

S91,722-2
1-(4-HYDROXYBENZOYL)-4-ISOPROPYLSEMICARBAZIDE

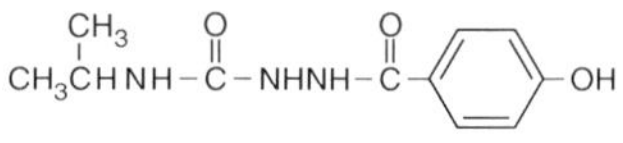

250 mg

S90,482-1
ETHYL 2-[3-(5-HYDROXY-1-NAPHTHYL)-UREIDO]ACETATE

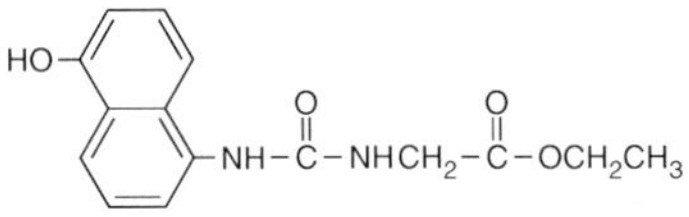

250 mg

S36,269-7
3-HYDROXYPHENYLUREA

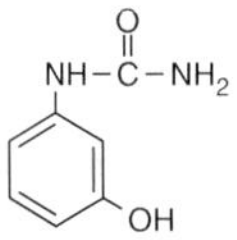

5 gm

S89,944-5
1-TERT-BUTYL-3-(3-HYDROXYPHENYL)UREA

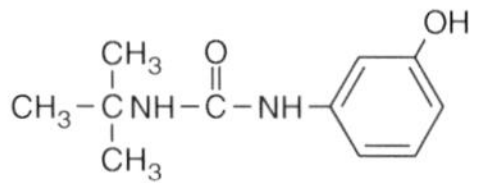

250 mg

S89,293-9
1-(3-HYDROXYPHENYL)-3-OCTADECYLUREA

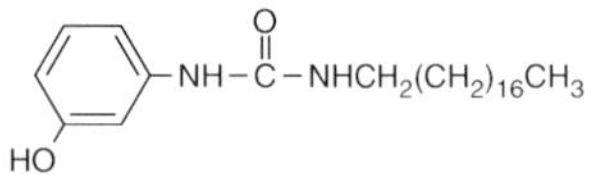

250 mg

S64,720-9
1-(2-CHLOROPHENYL)-3-HYDROXYUREA
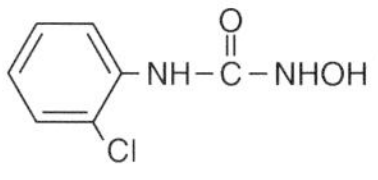
5 gm

S84,591-4
1-(2-ETHYLPHENYL)-3-[4-(2-HYDROXYETHYL)PHENYL]UREA
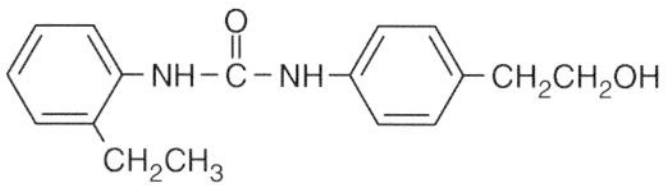
250 mg

S56,241-6
1-BUTYL-3-(4-HYDROXYPHENYL)UREA
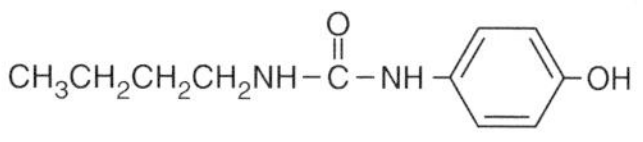
5 gm

S74,329-1
1-TERT-BUTYL-3-(4-HYDROXYPHENYL)UREA
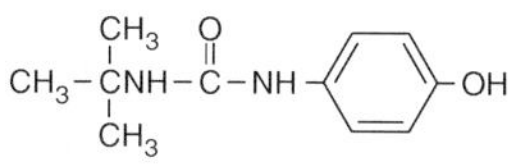
5 gm

S77,974-1
1-[4-(2-HYDROXYETHYL)PHENYL]-3-(2,6-XYLYL)UREA
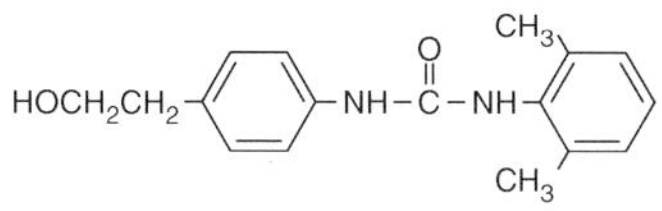
250 mg

S77,976-8
1-[4-(2-HYDROXYETHYL)PHENYL]-3-(2,5-XYLYL)UREA
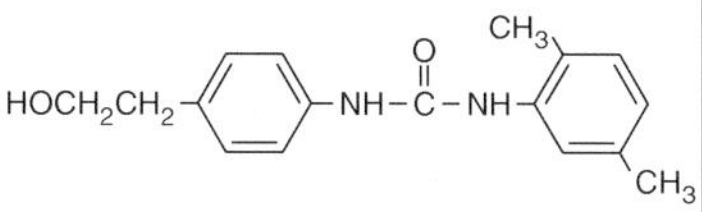
250 mg

S38,171-3
1-(6-CHLORO-ORTHO-TOLYL)-3-(3-HYDROXYPROPYL)UREA
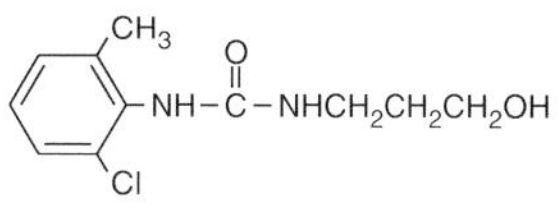
250 mg

S77,387-5
1-(3-HYDROXYPHENYL)-3-(2,6-XYLYL)UREA
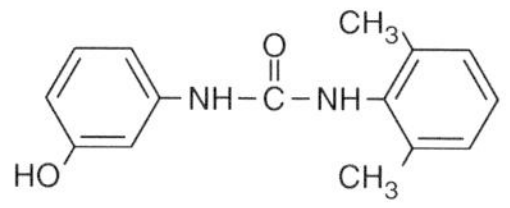
250 mg

S64,726-8
1-(2,4-DICHLOROPHENYL)-3-HYDROXYUREA
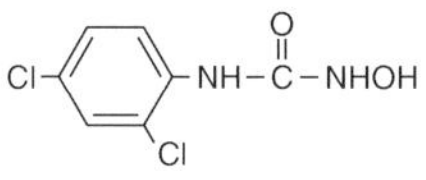
5 gm

S92,417-2
1-(5-HYDROXY-1-NAPHTHYL)-3-METHYLUREA
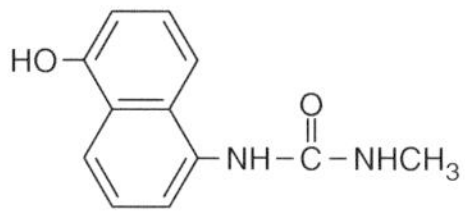
250 mg

S85,458-1
1-TERT-BUTYL-3-(5-HYDROXY-1-NAPHTHYL)UREA
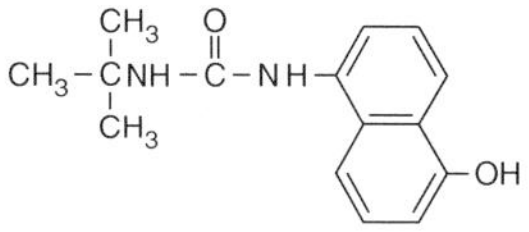
250 mg

S90,301-9
2-(3-BENZYLUREIDO)ACETIC (3,4-DIMETHOXYBENZYLIDENE)HYDRAZIDE
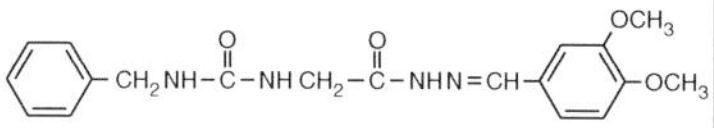
250 mg

S80,005-8
1-(3,4-DIMETHOXYPHENETHYL)-3-(2-METHOXY-4-NITROPHENYL)UREA
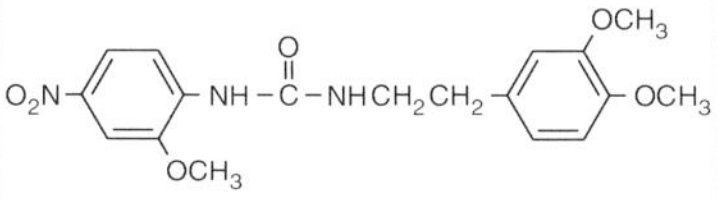
250 mg

S89,268-8
1-BUTYL-3-(3,4-DIMETHOXYPHENETHYL)UREA
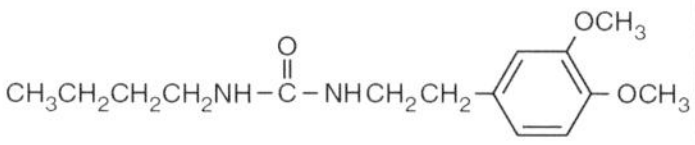
250 mg

S77,979-2
2-[3-(2-ETHOXYPHENYL)UREIDO]ACETIC (3,4-DIMETHOXYBENZYLIDENE)HYDRAZIDE
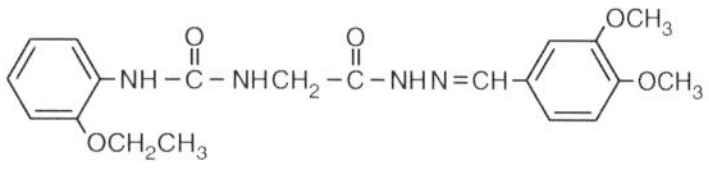
250 mg

S85,433-6
4-(3,3-DIETHYLUREIDO)BENZOIC (3,4-DIMETHOXYBENZYLIDENE)HYDRAZIDE
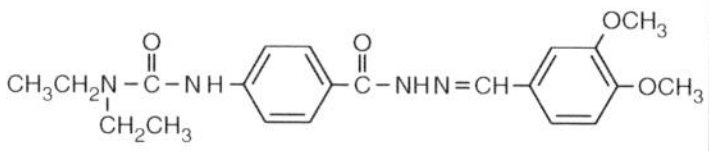
250 mg

S86,889-2
4-(3-PROPYLUREIDO)BENZOIC (4-ETHOXY-3-METHOXYBENZYLIDENE)-HYDRAZIDE
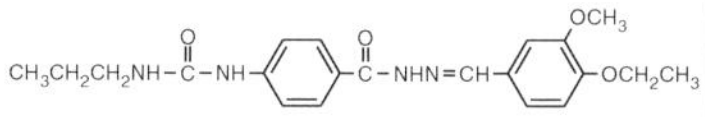
250 mg

S80,094-5
1-(3,4-DIMETHOXYBENZYLIDENE)-4-PHENYLSEMICARBAZIDE
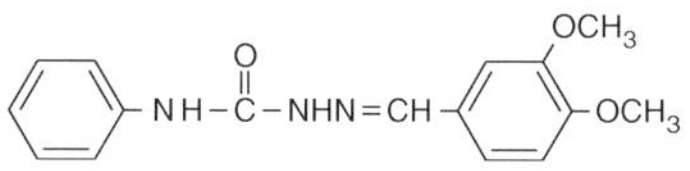
250 mg

S69,000-7
1-(2,3,4-TRIMETHOXYBENZYLIDENE)-SEMICARBAZIDE
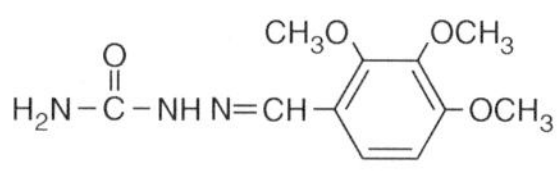
250 mg

S78,061-8
2-[3-(4-METHOXYPHENYL)-UREIDO]ACETIC (2,3,4-TRIMETHOXYBENZYLIDENE)HYDRAZIDE
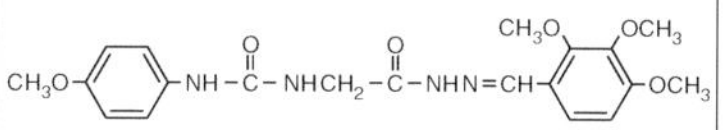
250 mg

S84,838-7
1-(2,4-DIMETHOXYPHENYL)-3-(2-METHYL-5-NITROPHENYL)UREA
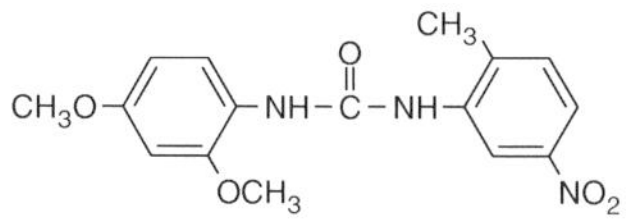
250 mg

S85,365-8
1-(2,4-DIMETHOXYPHENYL)-3-(4-METHYL-3-NITROPHENYL)UREA
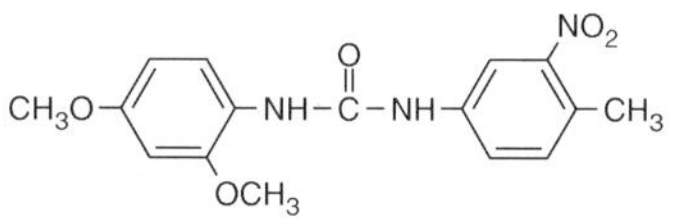
250 mg

S84,833-6
1-(5-CHLORO-2,4-DIMETHOXYPHENYL)-3-(2-METHYL-5-NITROPHENYL)UREA
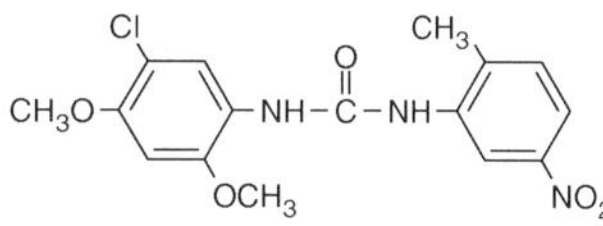
250 mg

S88,864-8
3',5'-DIMETHOXYVALEROPHENONE SEMICARBAZONE
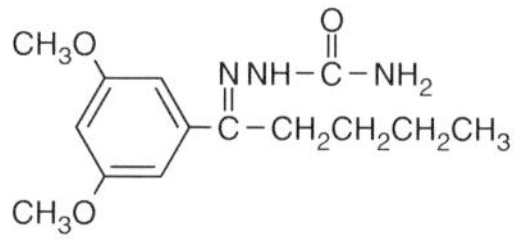
250 mg

S85,118-3
3-[3-(2,4-DIMETHOXYPHENYL)UREIDO]-BENZOIC ACID
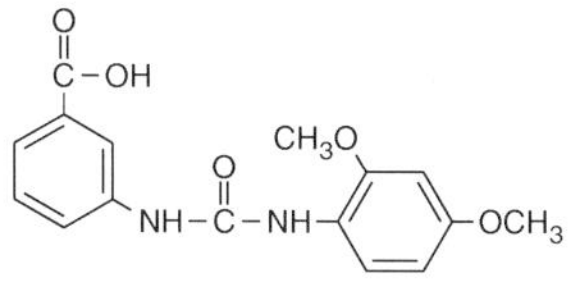
250 mg

S72,826-8
ETHYL 4-[3-(2,4-DIMETHOXYPHENYL)-UREIDO]BENZOATE
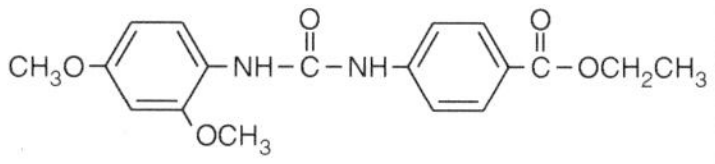
5 gm

S85,120-5
1-BENZOYL-3-(2,4-DIMETHOXYPHENYL)UREA
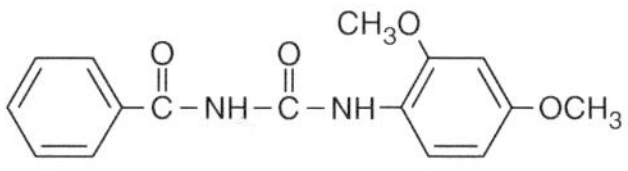
250 mg

S75,252-5
1-(2,4-DIMETHOXYPHENYL)-3-(O-TOLYL)UREA
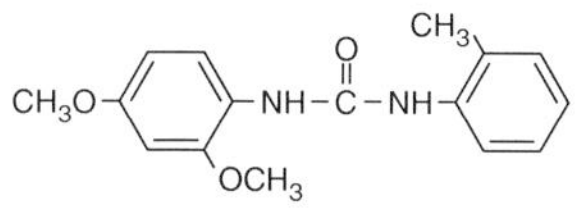
5 gm

S75,253-3
1-(2,4-DIMETHOXYPHENYL)-3-(2-METHOXYPHENYL)UREA
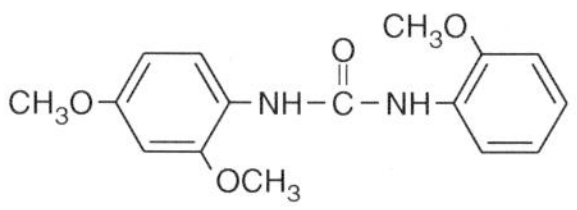
5 gm

S75,254-1
1-(2,4-DIMETHOXYPHENYL)-3-(4-METHOXYPHENYL)UREA
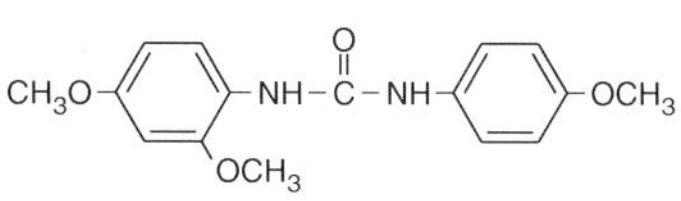
5 gm

S89,274-2
1-BUTYL-3-(2,4-DIMETHOXYPHENYL)UREA
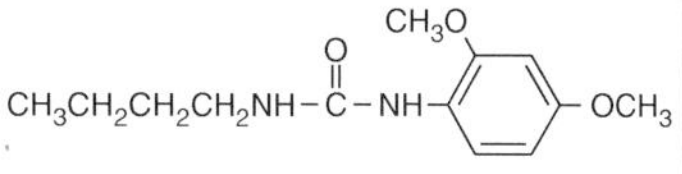
250 mg

S79,543-7
1,3-BIS(2,4-DIMETHOXYPHENYL)UREA
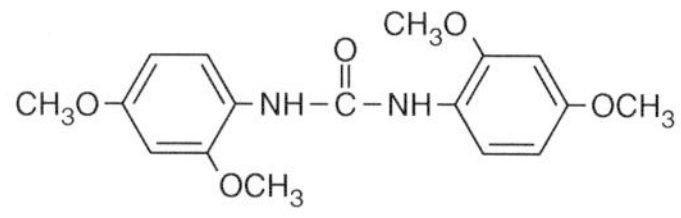
250 mg

S89,305-6
1-(2,4-DIMETHOXYPHENYL)-3-OCTADECYLUREA

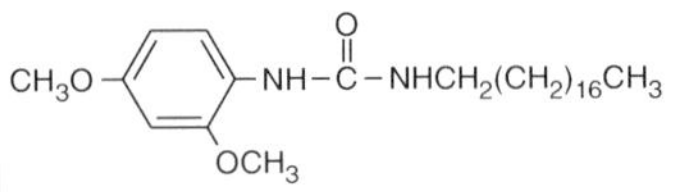

250 mg

S79,741-3
1-(2,5-DIMETHOXYPHENYL)-3-[4-(METHYLTHIO)PHENYL]UREA

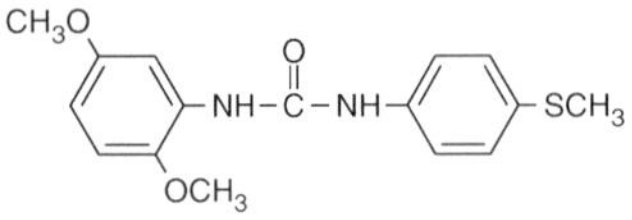

250 mg

S72,774-1
1-(2,5-DIMETHOXYPHENYL)-3-(2-METHYL-5-NITROPHENYL)UREA

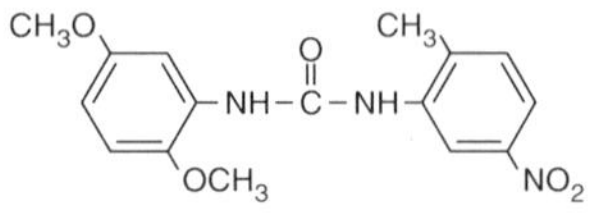

5 gm

S85,379-8
1-(2,5-DIMETHOXYPHENYL)-3-(4-METHYL-3-NITROPHENYL)UREA

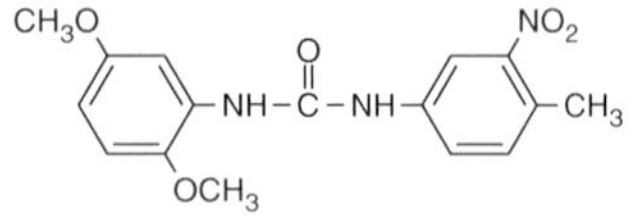

250 mg

S80,043-0
1-(2,5-DIMETHOXYPHENYL)-3-(2-METHOXY-5-NITROPHENYL)UREA

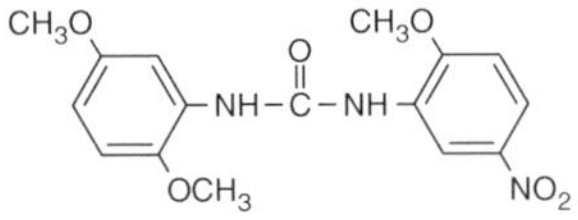

250 mg

S84,871-9
1-(2,5-DIMETHOXYPHENYL)-3-(4-METHOXY-3-NITROPHENYL)UREA

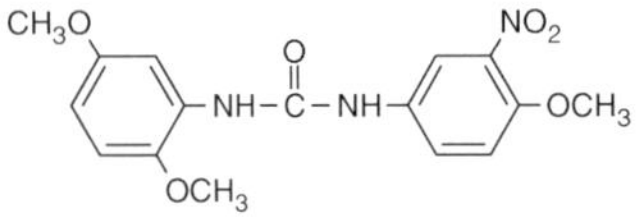

250 mg

S84,863-8
1-(4-ANILINOPHENYL)-3-(2,5-DIMETHOXYPHENYL)UREA

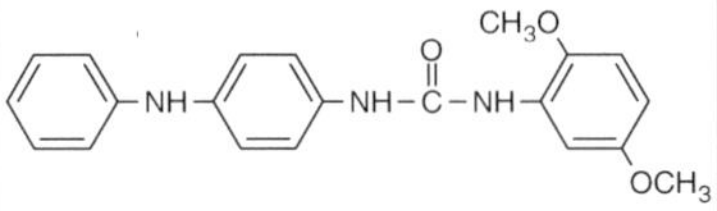

250 mg

S84,852-2
1-(2,5-DIMETHOXYPHENYL)-3-[4-(N-METHYLACETAMIDO)PHENYL]UREA

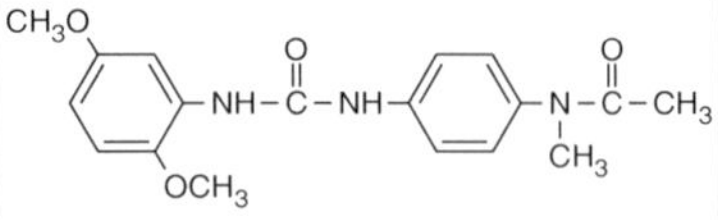

250 mg

S84,877-8
1-BENZYL-3-(2,5-DIMETHOXYPHENYL)-1-METHYLUREA

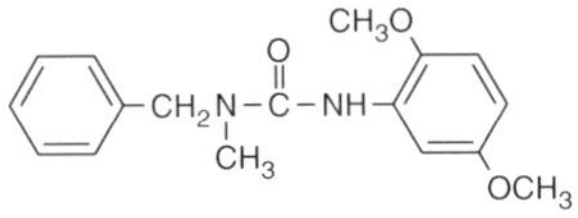

250 mg

S72,767-9
1-(2,5-DIMETHOXYPHENYL)-3-(2-METHOXY-4-NITROPHENYL)UREA

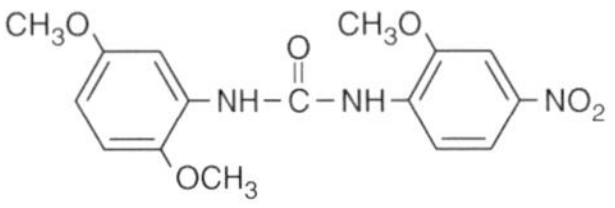

5 gm

S80,783-4
1-(3-ACETYLPHENYL)-3-(2,5-DIMETHOXYPHENYL)UREA

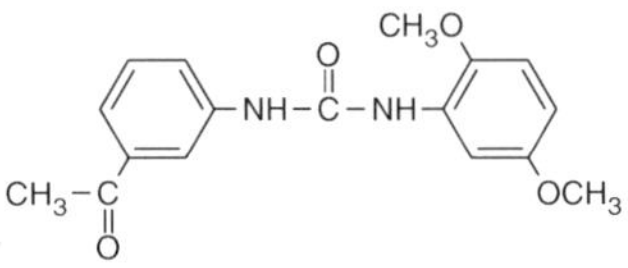

250 mg

S79,773-1
ETHYL 4-[3-(2,5-DIMETHOXYPHENYL)-UREIDO]BENZOATE

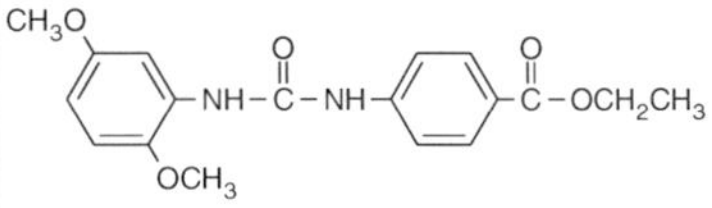

250 mg

S84,837-9
1-(2-CARBAMOYLPHENYL)-3-(2,5-DIMETHOXYPHENYL)UREA

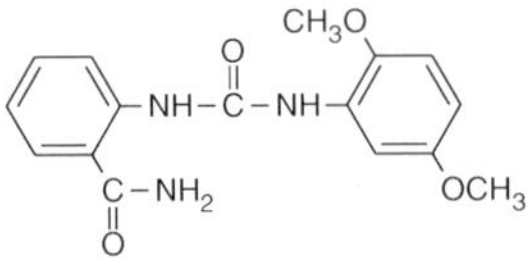

250 mg

S69,613-7
4-(2,5-DIMETHOXYPHENYL)-1-(4-METHOXYBENZOYL)SEMICARBAZIDE

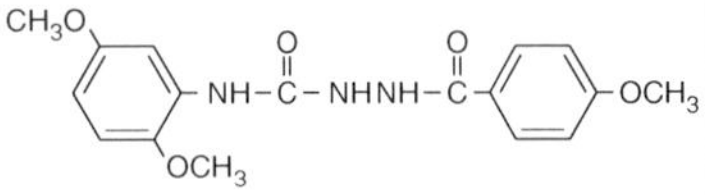

250 mg

S72,499-8
1-BENZOYL-3-(2,5-DIMETHOXYPHENYL)UREA

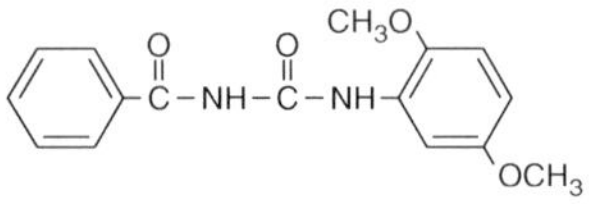

5 gm

S90,252-7
ETHYL 2-[3-(2,5-DIMETHOXYPHENYL)-UREIDO]ACETATE

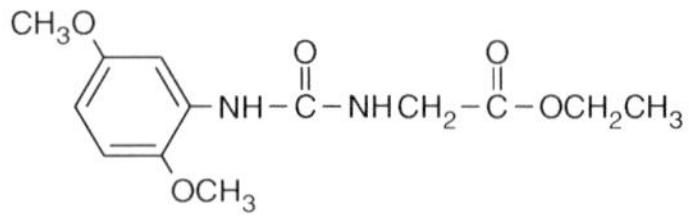

250 mg

S90,367-1
4-(2,5-DIMETHOXYPHENYL)-1-[3-(2-ETHOXYPHENYL)UREIDOACETYL]-SEMI8CARBAZIDE

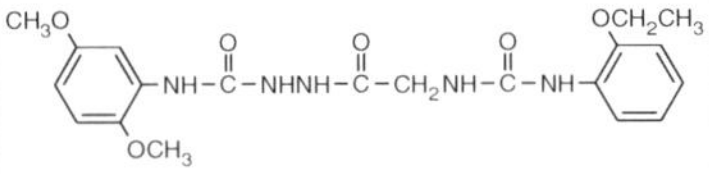

250 mg

S69,239-5
2-[3-(4-METHOXYPHENYL)UREIDO] ACETIC N2-(2,5-DIMETHOXYPHENYL-CARBAMOYL)HYDRAZIDE

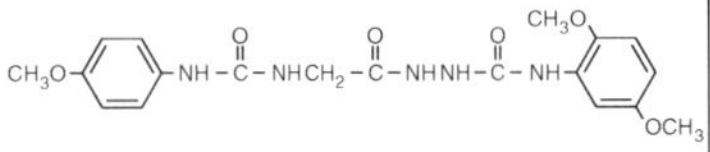

250 mg

S71,391-0
1-(4-ACETAMIDOPHENYL)-3-(2,5-DIMETHOXYPHENYL)UREA

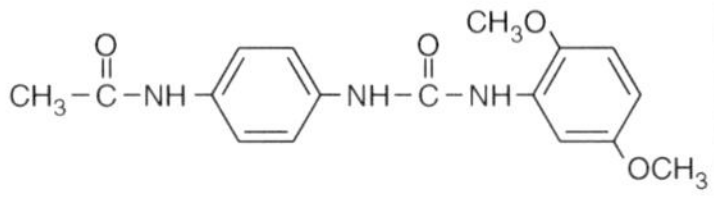

5 gm

S91,650-1
1-CYCLOPROPYL-3-(2,5-DIMETHOXYPHENYL)UREA

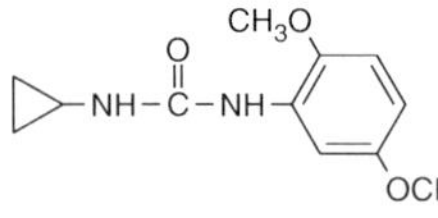

250 mg

S89,164-9
1-(2,5-DIETHOXYPHENYL)-3-(O-TOLYL)UREA

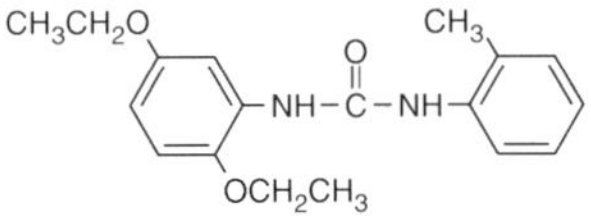

250 mg

S89,168-1
1-(2,5-DIETHOXYPHENYL)-3-(4-METHOXYPHENYL)UREA

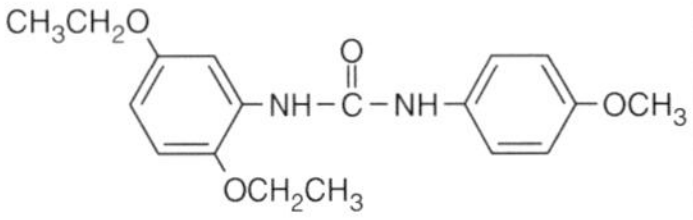

250 mg

S71,527-1
1-(2,6-DICHLOROPHENYL)-3-(2,5-DIMETHOXYPHENYL)UREA

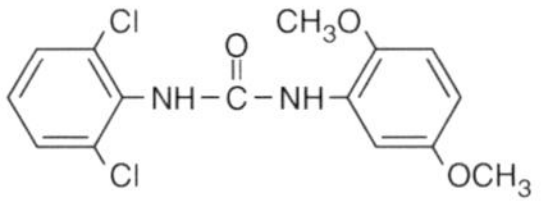

5 gm

S68,660-3
1-(2,5-DIMETHOXYPHENYL)-3-(2,6-DIMETHYLPHENYL)UREA

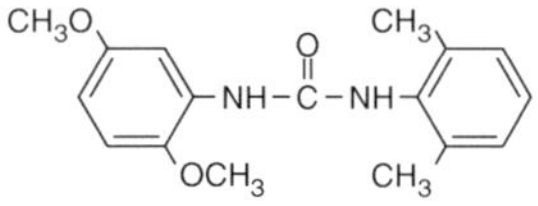

250 mg

S71,302-3
1-(2,6-DIETHYLPHENYL)-3-(2,5-DIMETHOXYPHENYL)UREA

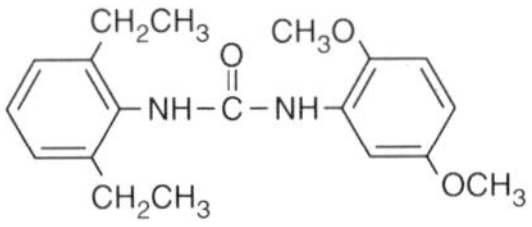

250 mg

S85,517-0
1-TERT-BUTYL-3-(2,5-DIMETHOXYPHENYL)UREA

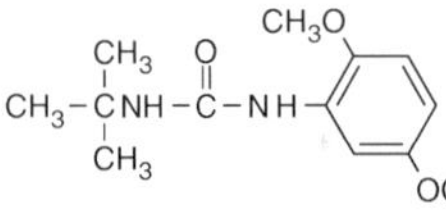

250 mg

S84,878-6
1,3-BIS(2,5-DIMETHOXYPHENYL)UREA

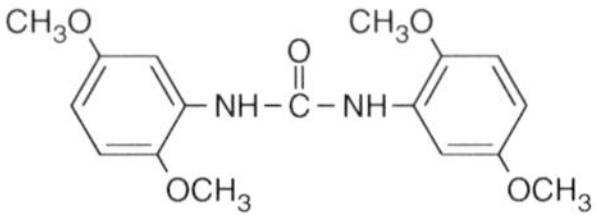

250 mg

S90,372-8
3-(2,5-DIMETHOXYPHENYL)-1-ETHYL-1-(2-METHOXY-5-METHYLPHENYL)UREA

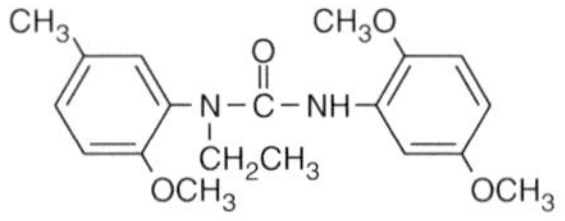

250 mg

S90,232-2
2-[3-(3-NITROPHENYL)UREIDO]ACETIC (3,4-DIMETHOXYBENZYLIDENE)HYDRAZIDE

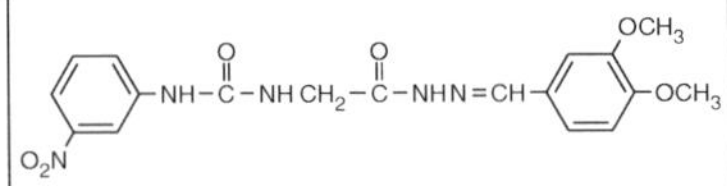

250 mg

S89,173-8
1-(2-METHOXY-3-NITROPHENYL)-3-(4-METHOXYPHENYL)UREA

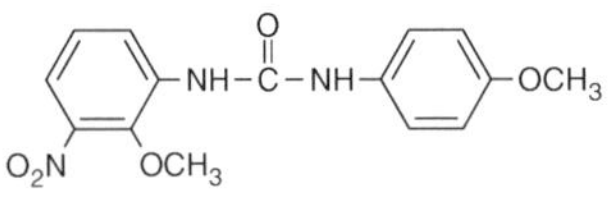

250 mg

S90,363-9
4-(2,5-DIMETHOXYPHENYL)-1-[3-(3-NITROPHENYL)UREIDOACETYL]-SEMICARBAZIDE

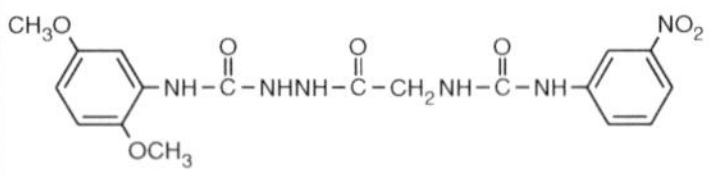

250 mg

S76,872-3
1-(4-METHOXYBENZOYL)-4-(2-METHOXY-5-METHYLPHENYL)SEMICARBAZIDE

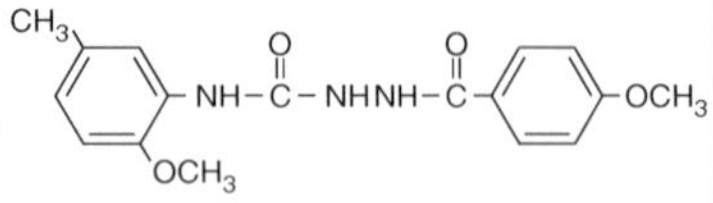

250 mg

S69,095-3
1,6-BIS(2-METHOXYPHENYL)BIUREA
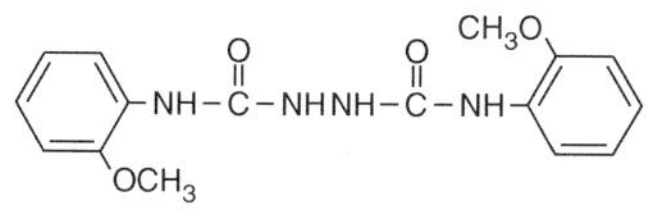
250 mg

S69,897-0
1-(4-ACETYLPHENYL)-3-(2,5-DIMETHOXYPHENYL)UREA
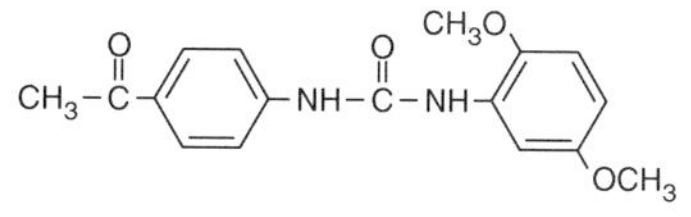
250 mg

S67,461-3
4,4'-BIS(3-CYCLOPROPYLUREIDO)-3,3'-DIMETHOXYBIPHENYL
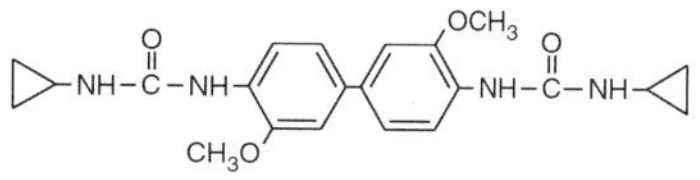
250 mg

S89,066-9
1,3-BIS(2-METHOXYPHENYL)UREA
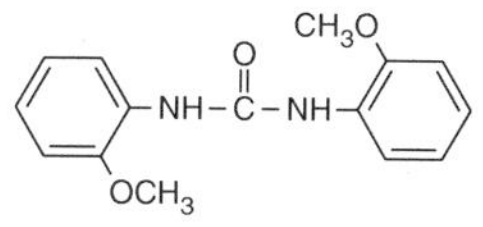
250 mg

S71,569-7
1,1'-HEXAMETHYLENEBIS[3-(2-METHOXYPHENYL)UREA]
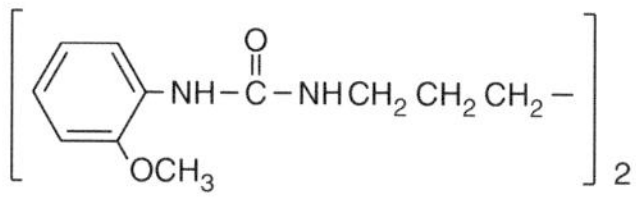
250 mg

S79,152-0
1-(4-ETHOXYPHENYL)-3-(2-METHOXYPHENYL)UREA
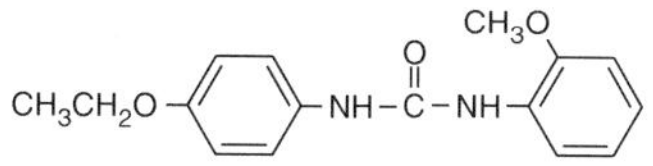
250 mg

S86,912-0
1,1'-HEXAMETHYLENEBIS[3-(2,5-DIMETHOXYPHENYL)UREA]
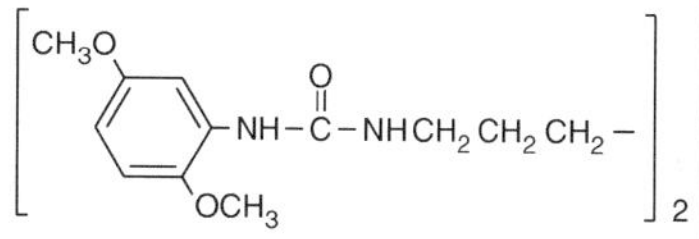
250 mg

S67,340-4
4,4'-BIS(3,3-DIMETHYLUREIDO)-3,3'-DIMETHOXYBIPHENYL
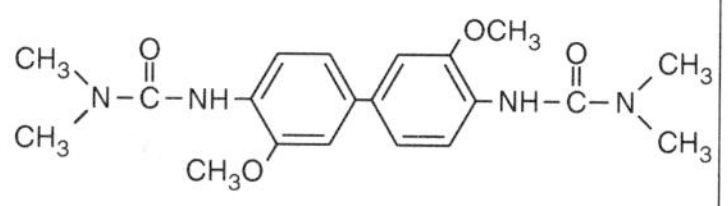
250 mg

S67,664-0
4,4'-BIS[3-(4-METHOXYPHENYL)UREIDO]-3,3'-DIMETHYLBIPHENYL
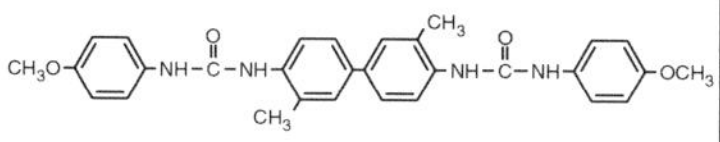
250 mg

S85,182-5
1-[2-(4-ALLYL-2-METHOXYPHENOXY)-ETHYL]-3-(4-METHYL-2-NITROPHENYL)UREA
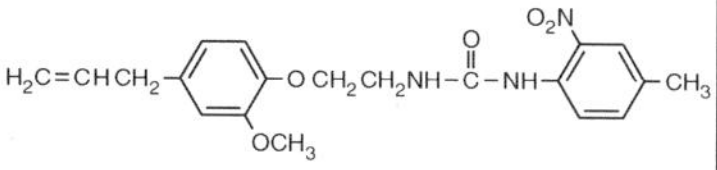
250 mg

S68,960-2
1-(2-METHOXYBENZOYL)-4-[4-(METHYLTHIO)PHENYL]SEMICARBAZIDE
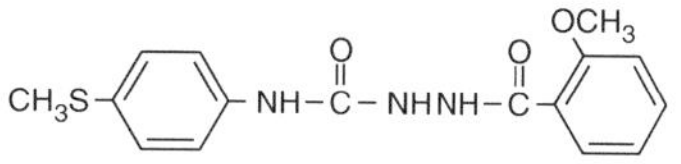
250 mg

S69,151-8
4-(2,6-DIMETHYLPHENYL)-1-(2-METHOXYBENZOYL)SEMICARBAZIDE
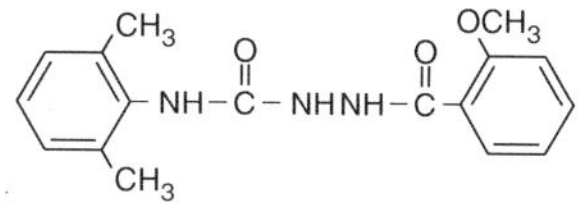
250 mg

S79,721-9
1-(2-METHOXY-5-NITROPHENYL)-3-[4-(METHYLTHIO)PHENYL]UREA
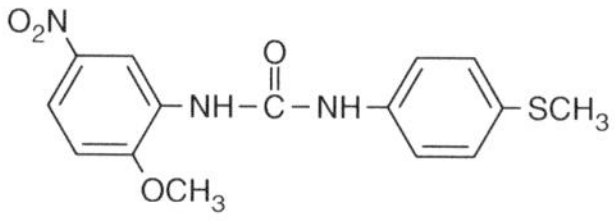
250 mg

S69,253-0
1-[2-(2,4-DINITROPHENYLTHIO)PHENYL]-3-(2-METHOXY-5-METHYLPHENYL)UREA
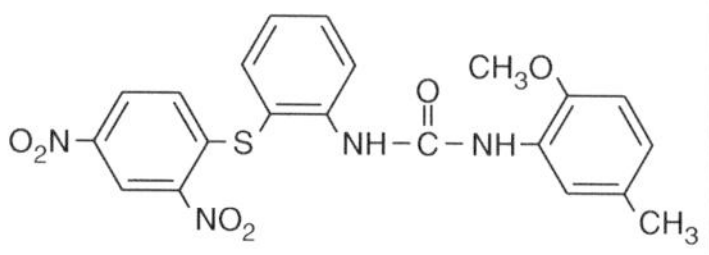
250 mg

S79,720-0
1-(2-METHOXY-3-NITROPHENYL)-3-[4-(METHYLTHIO)PHENYL]UREA
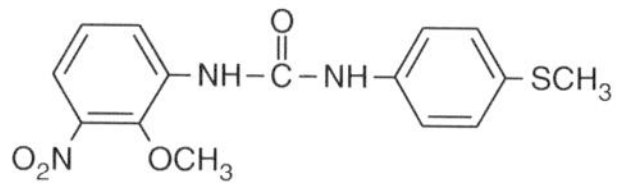
250 mg

S79,742-1
1-(2-METHOXY-5-METHYLPHENYL)-3-[4-(METHYLTHIO)PHENYL]UREA
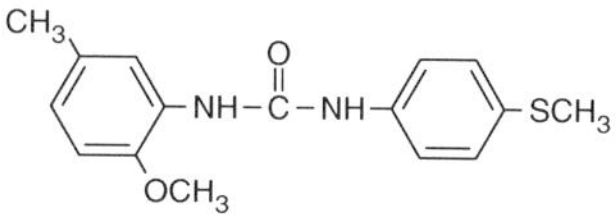
250 mg

S90,551-8
1-(2-METHOXY-5-METHYLPHENYL)-3-[3-(METHYLTHIO)PHENYL]UREA
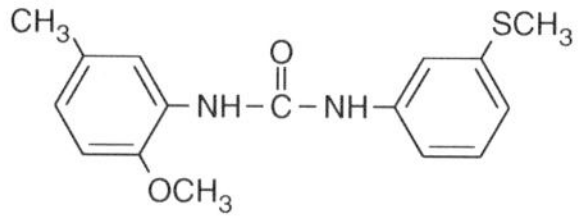
250 mg

S72,294-4
1-(4-METHOXYBENZOYL)-4-[4-(METHYLTHIO)PHENYL]SEMICARBAZIDE
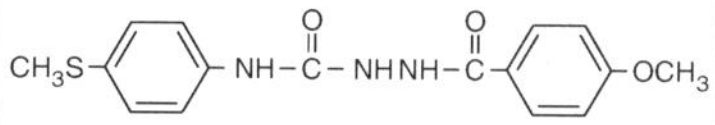
250 mg

S76,847-2
1-(4-METHOXYPHENYL)-3-[3-(METHYLTHIO)PHENYL]UREA
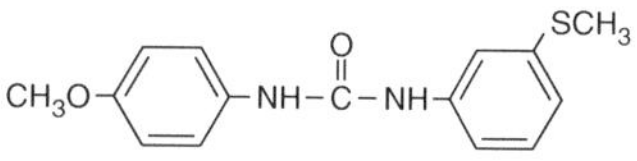
250 mg

S89,156-8
1-(6-AMINOHEXYL)-3-(4-METHOXYPHENYL)UREA
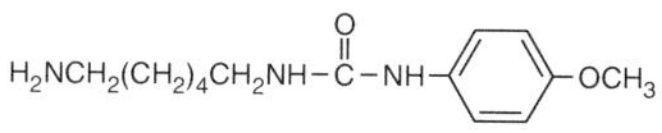
250 mg

S80,047-3
1-[2-(N-ETHYLANILINO)ETHYL]-3-(2-METHOXY-5-NITROPHENYL)UREA
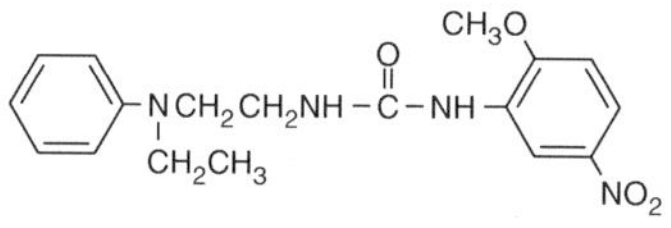
250 mg

S79,436-8
1-(2-METHOXY-4-NITROPHENYL)-3-(1-METHYL-2-PHENYL-4-PYRAZOLIN-4-YL)UREA
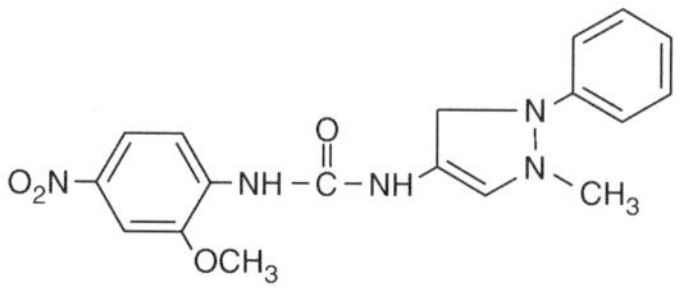
250 mg

S90,005-2
1-(3-METHOXYPHENYL)-3-[2-(N-METHYLANILINO)ETHYL]UREA
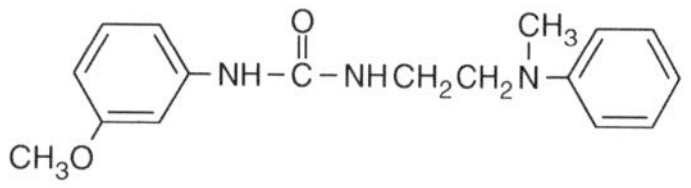
250 mg

S89,998-4
1-(2-METHOXYPHENYL)-3-[2-(4-MORPHOLINO)ETHYL]UREA
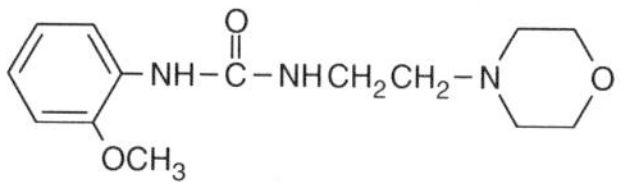
250 mg

S89,997-6
1-(3-METHOXYPHENYL)-3-[2-(1-PYRROLIDINO)ETHYL]UREA
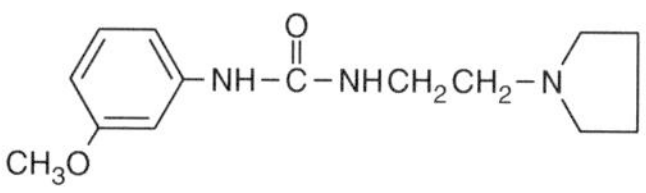
250 mg

S79,530-5
1-[2-(N-ETHYLANILINO)ETHYL]-3-(4-METHOXYPHENYL)UREA
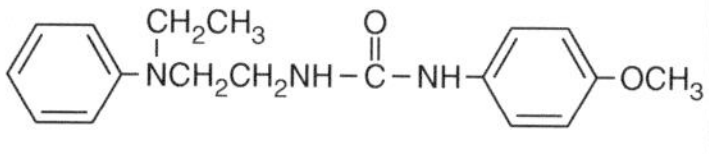
250 mg

S90,036-2
1-[2-(DIETHYLAMINO)ETHYL]-3-(4-METHOXYPHENYL)-1-PHENYLUREA
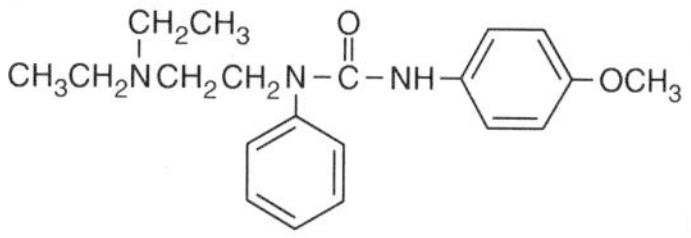
250 mg

S72,542-0
1-[2-(DIETHYLAMINO)ETHYL]-3-(4-METHOXYPHENYL)UREA
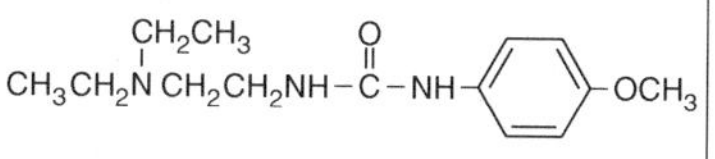
5 gm

S89,189-4
1-(2-METHOXYPHENYL)-3-[2-(N-METHYLANILINO)ETHYL]UREA
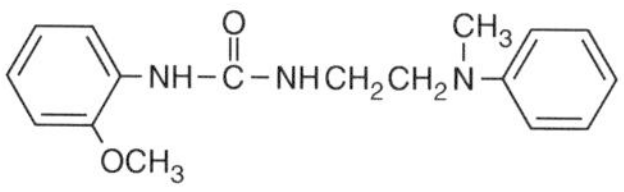
250 mg

S89,190-8
1-(4-METHOXYPHENYL)-3-[2-(N-METHYLANILINO)ETHYL]UREA
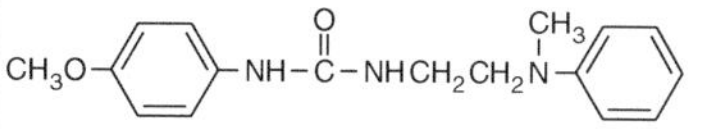
250 mg

S90,297-7
2-[3-(2-ETHOXYPHENYL)UREIDO]ACETIC (4-ACETAMIDOBENZYLIDENE)HYDRAZIDE
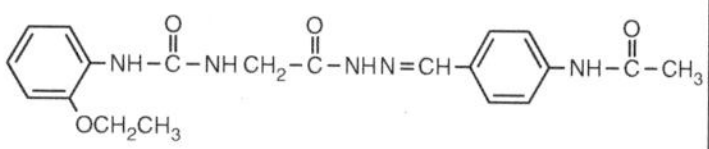
250 mg

S76,831-6
1-[3-(2-METHOXY-5-METHYLPHENYL)-UREIDO]OXAMIDE
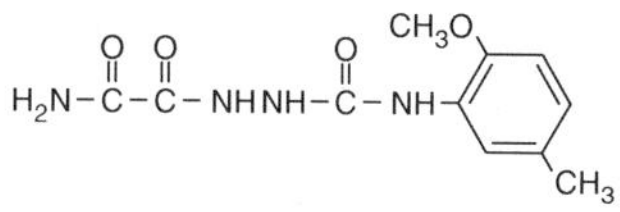
250 mg

S88,994-6
1-MESITYL-3-(2-METHOXYPHENYL)UREA
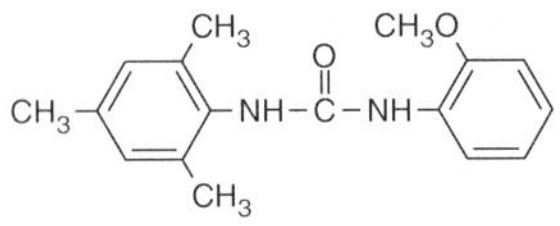
250 mg

S80,042-2
1-(2-METHOXY-5-NITROPHENYL)-3-(PHENETHYL)UREA
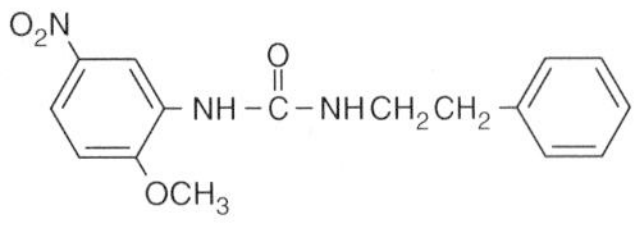
250 mg

S80,055-4
1-(4-ANILINOPHENYL)-3-(2-METHOXY-5-NITROPHENYL)UREA
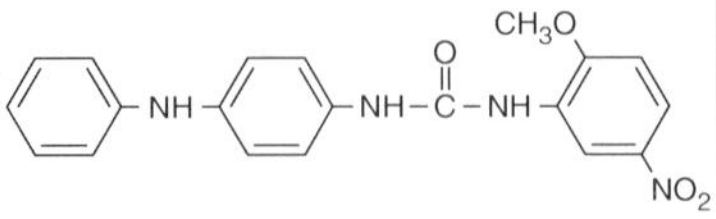
250 mg

S80,056-2
1-BENZYL-3-(2-METHOXY-5-NITROPHENYL)-1-PHENYLUREA
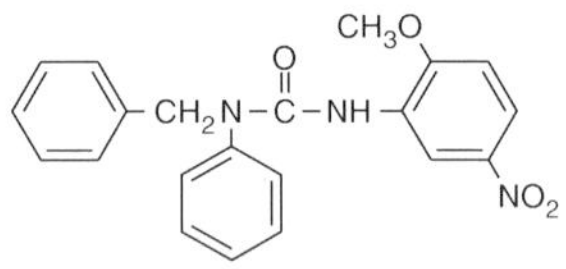
250 mg

S80,031-7
3-[3-(2-METHOXY-5-NITROPHENYL)-UREIDO]BENZOIC ACID
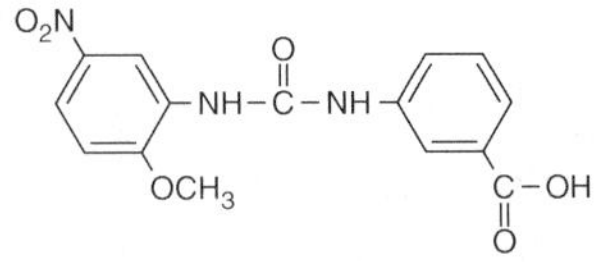
250 mg

S80,040-6
METHYL 2-[3-(2-METHOXY-5-NITROPHENYL)UREIDO]BENZOATE
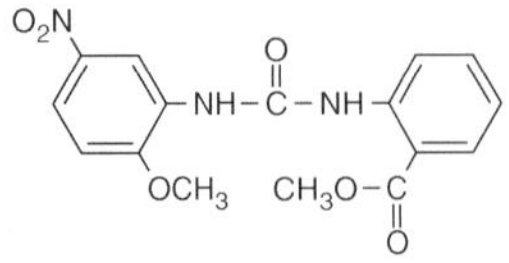
250 mg

S79,450-3
ETHYL 4-[3-(4-METHOXY-3-NITROPHENYL)UREIDO]BENZOATE
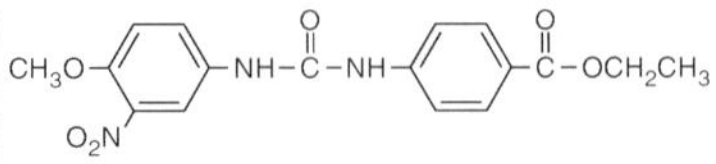
250 mg

S80,036-8
1-(2-CARBAMOYLPHENYL)-3-(2-METHOXY-5-NITROPHENYL)UREA
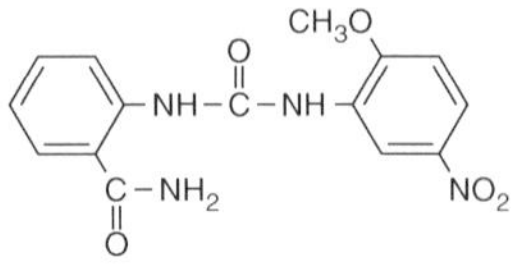
250 mg

S90,356-6
1-[3-(2-ETHOXYPHENYL)UREIDOACETYL]-4-(2-METHYL-5-NITROPHENYL)-SEMICARBAZIDE
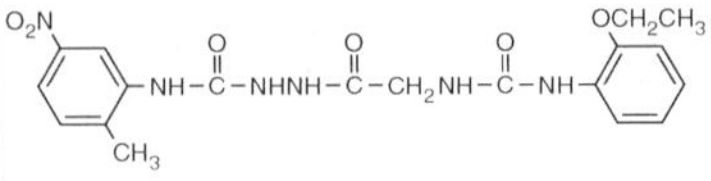
250 mg

S80,057-0
1,1-DICYCLOHEXYL-3-(2-METHOXY-5-NITROPHENYL)UREA
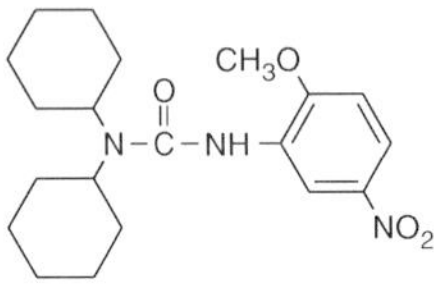
250 mg

S79,493-7
1-ISOPROPYL-3-(2-METHOXY-5-NITROPHENYL)-1-PHENYLUREA
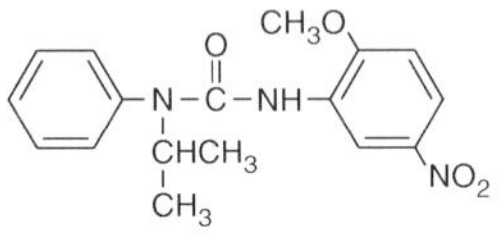
250 mg

S89,165-7
1-(2-METHOXY-3-NITROPHENYL)-3-(O-TOLYL)UREA
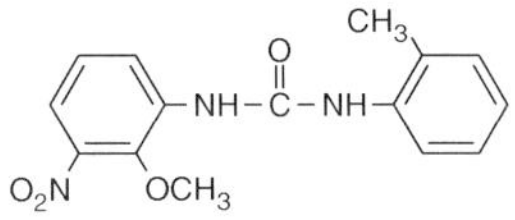
250 mg

S67,015-4
1-(2-METHOXYETHYL)-3-(4-METHYL-3-NITROPHENYL)UREA
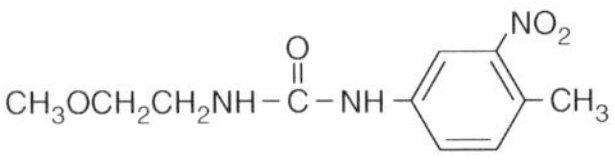
5 gm

S75,236-3
1-(3-CHLOROPHENYL)-3-(2-METHOXY-5-NITROPHENYL)UREA
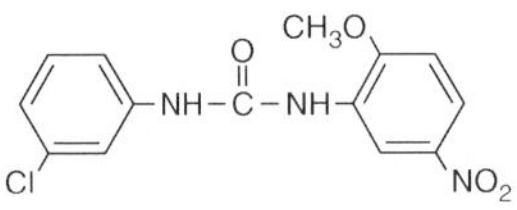
5 gm

S79,641-7
1-(4-METHOXY-3-NITROPHENYL)-3-(2-METHYL-5-NITROPHENYL)UREA
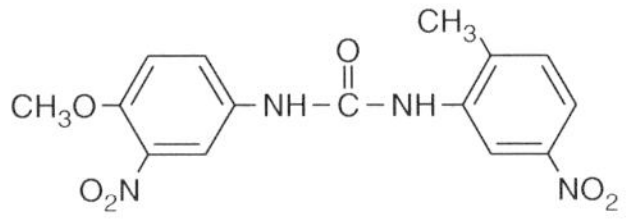
250 mg

S64,963-5
1,1'-(4-METHYL-1,3-PHENYLENE)BIS[3-(2-METHOXYETHYL)UREA]
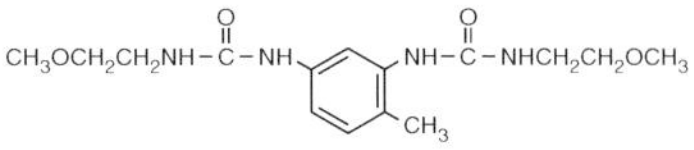
5 gm

S73,543-4
1-(3-METHOXYPHENYL)-3-(3-NITROPHENYL)UREA
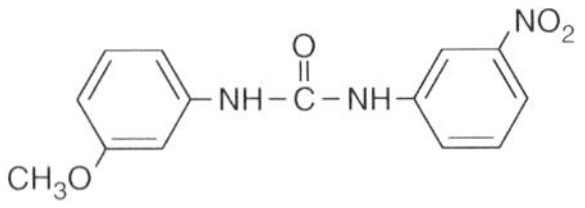
250 mg

S89,159-2
1-(4-METHOXYPHENYL)-3-(3-NITROPHENYL)UREA
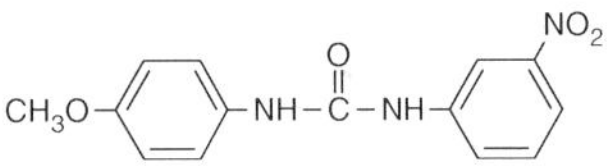
250 mg

S67,618-7
1-(3-ACETAMIDOPHENYL)-3-(4-METHOXYPHENYL)UREA
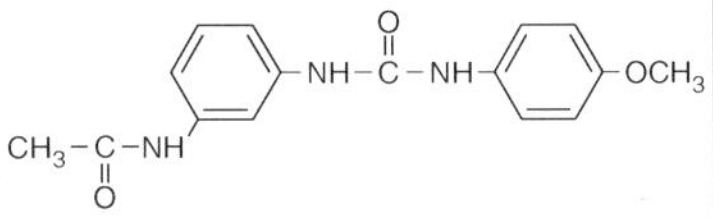
5 gm

S66,446-4
1-(P-ANISYL)-3-(3-[1-(2-METHYLPIPERIDINO)PROPYL]UREA)
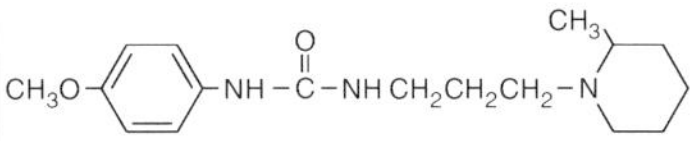
5 gm

S66,522-3
1-(M-ANISYL)-3-[3-(4-MORPHOLINO)PROPYL]UREA
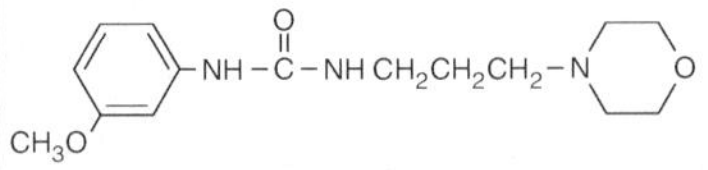
5 gm

S66,615-7
1-(P-ANISYL)-3-[3-(4-MORPHOLINO)PROPYL]UREA
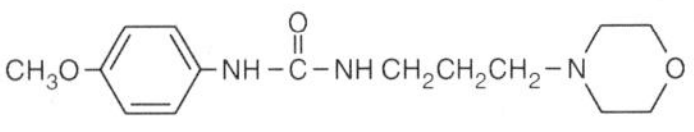
5 gm

S80,013-9
1-(2-METHOXY-4-NITROPHENYL)-3-(PHENETHYL)UREA
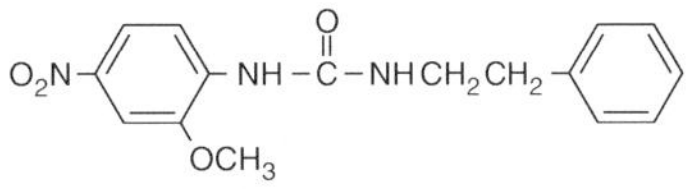
250 mg

S78,049-9
1-[4-(DIPHENYLAMINO)PHENYL]-3-(2-METHOXYPHENYL)UREA
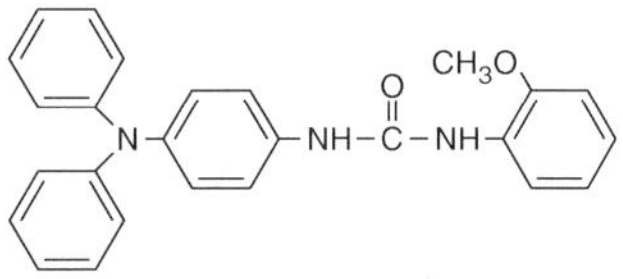
250 mg

S89,089-8
1-[4-(DIETHYLAMINO)PHENYL]-3-(2-METHOXYPHENYL)UREA
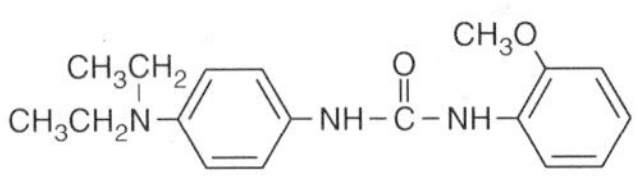
250 mg

S90,433-3
4-(2-METHOXY-5-METHYLPHENYL)-1-(4-NITROPHENYL)SEMICARBAZIDE
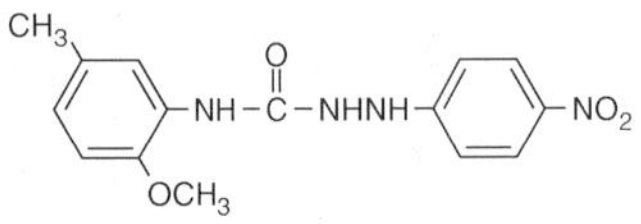
250 mg

S80,010-4
1-CYCLOHEXYL-3-(2-METHOXY-4-NITROPHENYL)-1-METHYLUREA
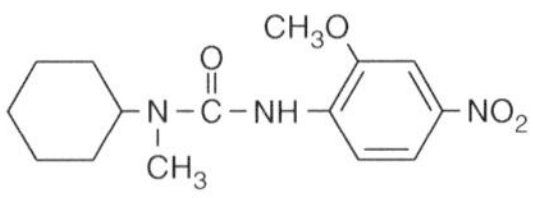
250 mg

S80,016-3
1,1-DICYCLOHEXYL-3-(2-METHOXY-4-NITROPHENYL)UREA
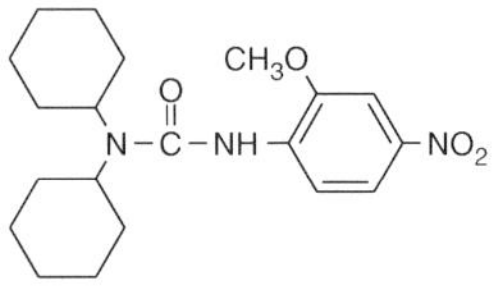
250 mg

S79,453-8
1-ISOPROPYL-3-(2-METHOXY-4-NITROPHENYL)-1-PHENYLUREA
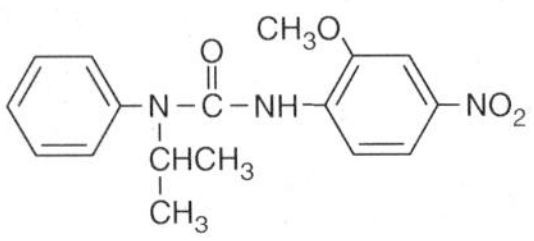
250 mg

S66,915-6
1-(2-METHOXYETHYL)-3-(4-NITROPHENYL)UREA
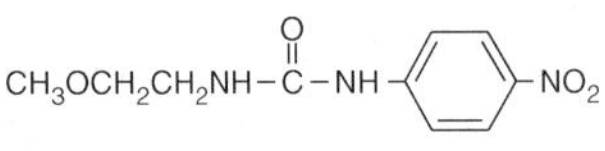
5 gm

S60,789-4
1-(2-CHLORO-4-NITROPHENYL)-3-(2,5-DIMETHOXYPHENYL)UREA
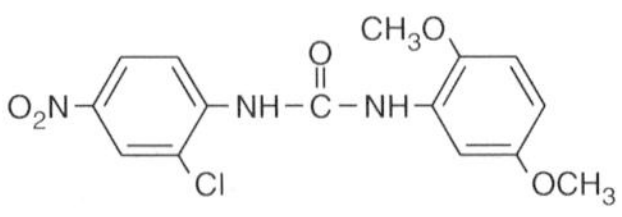
5 gm

S60,183-7
1-(4-ACETAMIDOPHENYL)-3-(2-METHOXYPHENYL)UREA
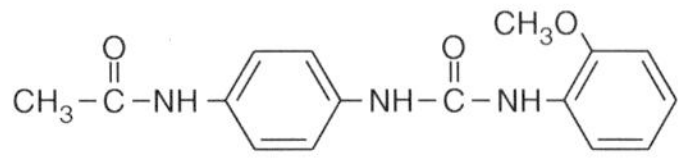
5 gm

S60,174-8
1-(4-ACETAMIDOPHENYL)-3-(4-METHOXYPHENYL)UREA
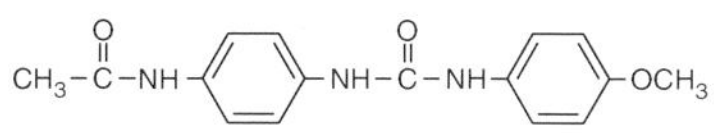
5 gm

S90,436-8
4-(2-METHOXY-5-METHYLPHENYL)-1-(2-PHENYLACETYL)SEMICARBAZIDE
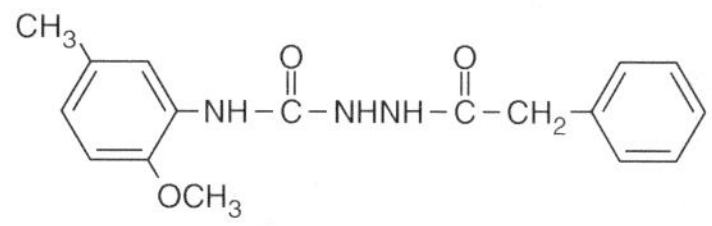
250 mg

S64,513-3
1-(2,5-DIMETHOXYPHENYL)-3-(PHENETHYL)UREA
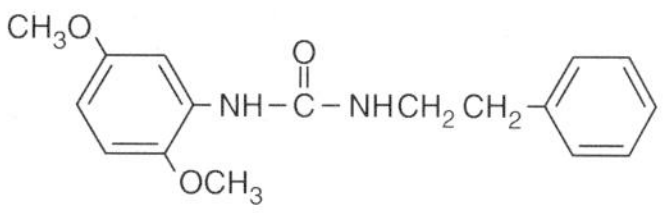
5 gm

S65,019-6
1-(O-ANISYL)-3-PHENETHYLUREA
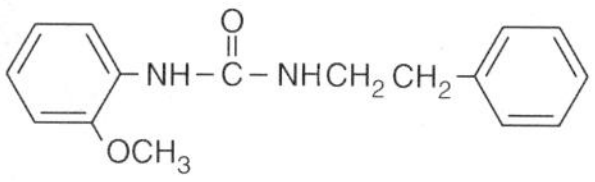
5 gm

S61,155-7
1-(P-ANISYL)-3-PHENETHYLUREA
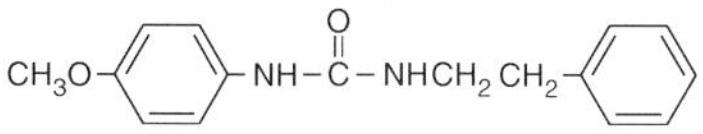
5 gm

S63,382-8
4-(O-ANISYL)-1,1-DIMETHYLSEMICARBAZIDE
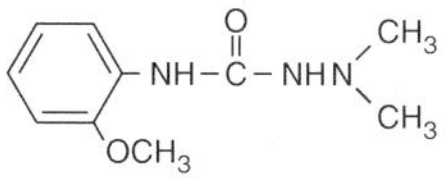
5 gm

S67,420-6
1-(4-ANILINOPHENYL)-3-(2-METHOXYPHENYL)UREA
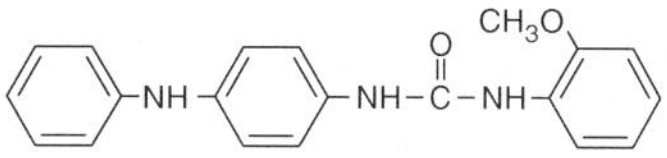
5 gm

S69,586-6
1-(4-ANILINOPHENYL)-3-(4-PHENOXYPHENYL)UREA
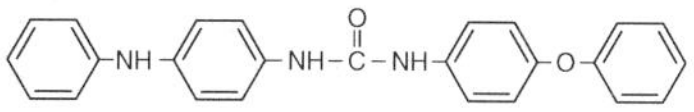
250 mg

S72,781-4
1-BENZYL-3-(2-METHOXY-4-NITROPHENYL)-1-PHENYLUREA
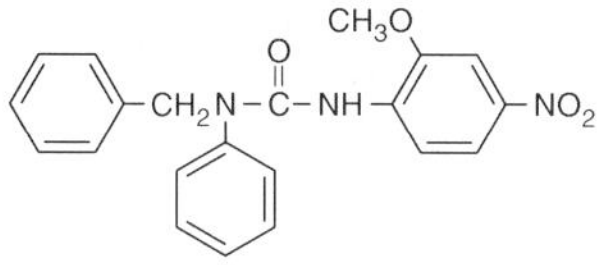
250 mg

S74,704-1
1-(2,5-DIMETHYLBENZYL)-3-(2-METHOXYPHENYL)UREA
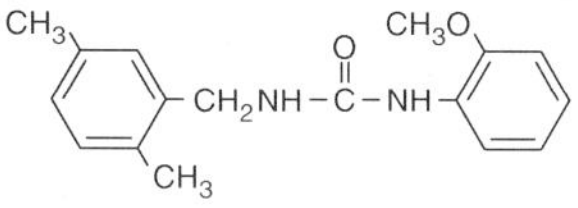
5 gm

S66,581-9
1-BENZYL-3-(2-METHOXY-5-METHYLPHENYL)UREA
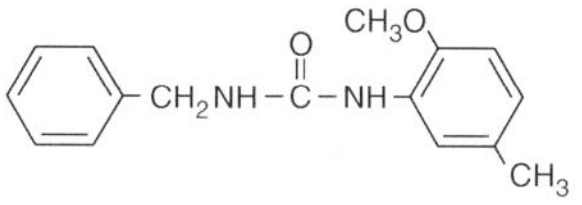
5 gm

S66,592-4
1-BENZYL-3-(2,5-DIMETHOXYPHENYL)UREA
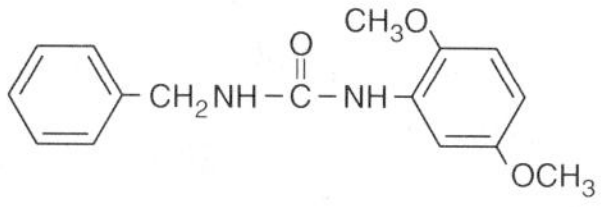
5 gm

S63,806-4
1,1-DIBENZYL-3-(2-METHOXY-5-METHYLPHENYL)UREA
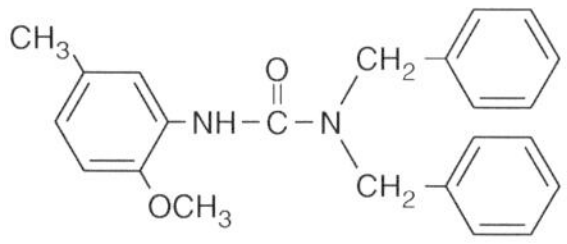
5 gm

S61,053-4
1-(M-ANISYL)-3-BENZYLUREA
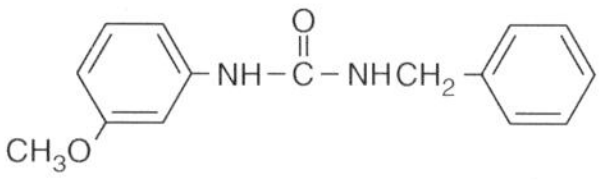
5 gm

S62,161-7
1-(O-ANISYL)-3-BENZYLUREA
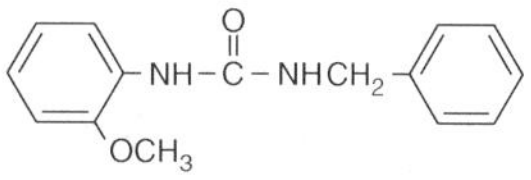
5 gm

S61,136-0
1-(P-ANISYL)-3-BENZYLUREA
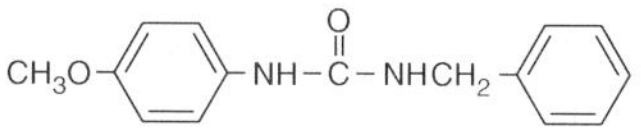
5 gm

S60,154-3
3-(P-ANISYL)-1,1-DIBENZYLUREA
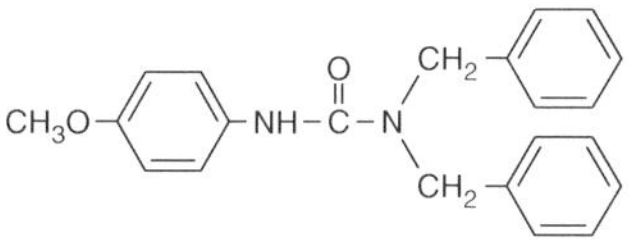
5 gm

S79,612-3
1-ETHOXYCARBONYL-4-(2-METHOXY-5-METHYLPHENYL)SEMICARBAZIDE
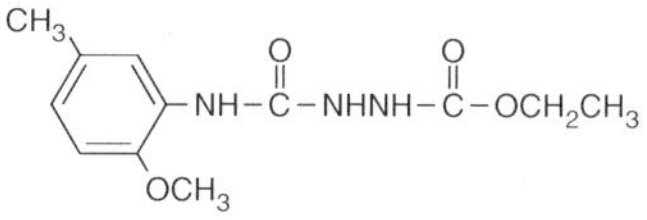
250 mg

S77,231-3
4-(2-METHOXY-5-METHYLPHENYL)-1-(2-PHENYLACETYL)SEMICARBAZIDE
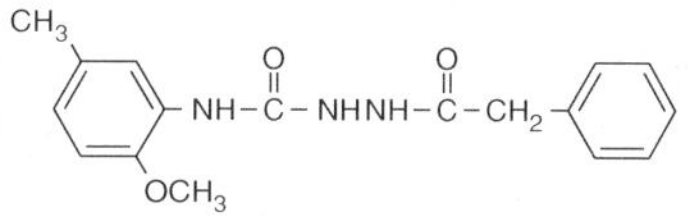
250 mg

S72,368-1
2-[3-(4-METHOXYPHENYL)UREIDO]-ACETIC ACID N2-[N-(2-ETHYLPHENYL)-CARBAMOYL]HYDRAZIDE
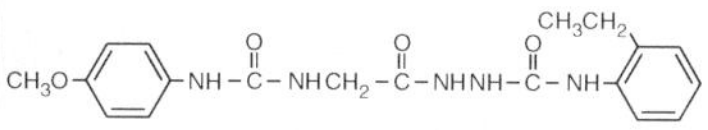
5 gm

S72,367-3
2-[3-(4-METHOXYPHENYL)UREIDO]ACETIC ACID N2-[N-(2,4-XYLYL)CARBAMOYL]-HYDRAZIDE
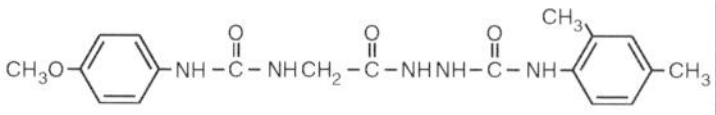
5 gm

S72,364-9
2-[3-(4-METHOXYPHENYL)UREIDO]ACETIC ACID N2-[N-(1-NAPHTHYL)CARBAMOYL]-HYDRAZIDE
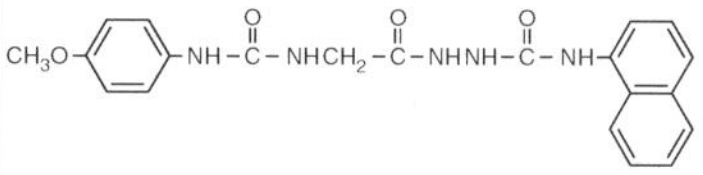
5 gm

S72,357-6
2-[3-(4-METHOXYPHENYL)UREIDO]ACETIC ACID (3-NITROBENZYLIDENE)HYDRAZIDE
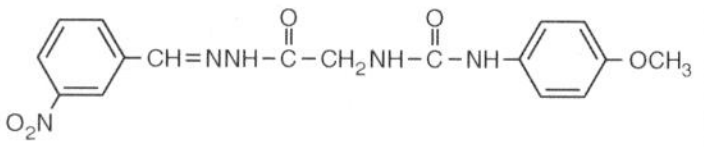
5 gm

S86,060-3
6-METHOXY-ALPHA-TETRALONE SEMICARBAZONE
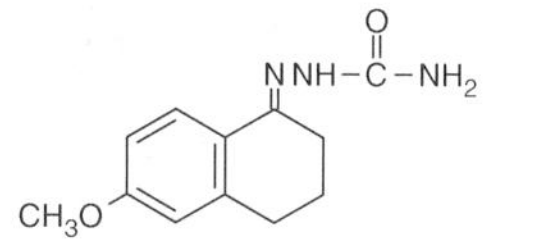
250 mg

S80,100-3
1-(4-BUTOXYBENZYLIDENE)-4-PHENYLSEMICARBAZIDE
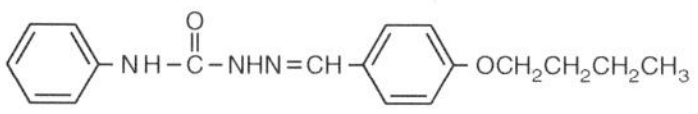
250 mg

S80,884-9
1-[2-(4-METHOXYPHENYL)-1-METHYLBUTYLIDENE]SEMICARBAZIDE
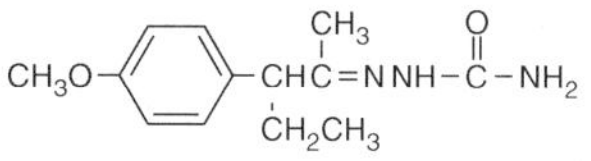
250 mg

S65,027-7
1-(O-ANISYL)-3-HYDROXYUREA
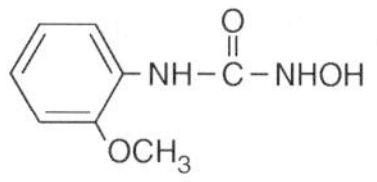
5 gm

S64,969-4
1-(P-ANISYL)-3-HYDROXYUREA
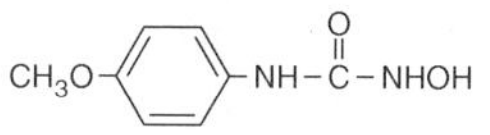
5 gm

S79,426-0
BUTYL 4-[3-(2-METHOXY-4-NITROPHENYL)UREIDO]BENZOATE
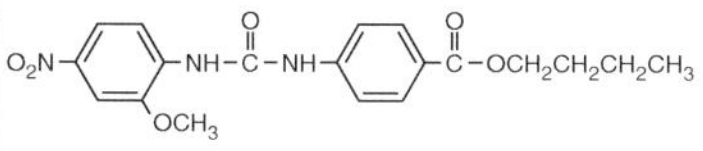
250 mg

S80,050-3
ISOBUTYL 4-[3-(2-METHOXY-5-NITROPHENYL)UREIDO]BENZOATE
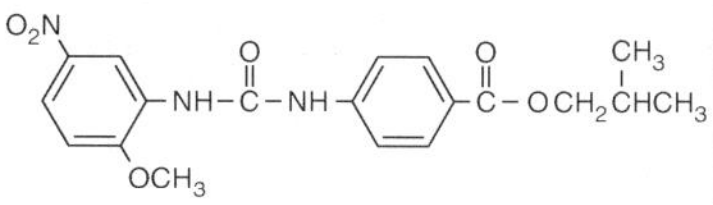
250 mg

S72,793-8
1-(4-ACETYLPHENYL)-3-(2-METHOXY-5-METHYLPHENYL)UREA
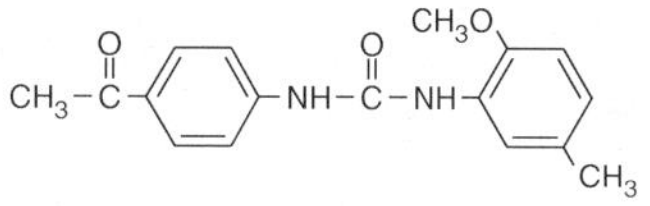
5 gm

S65,417-5
1-(4-BENZOYLPHENYL)-3-(2,5-DIMETHOXYPHENYL)UREA
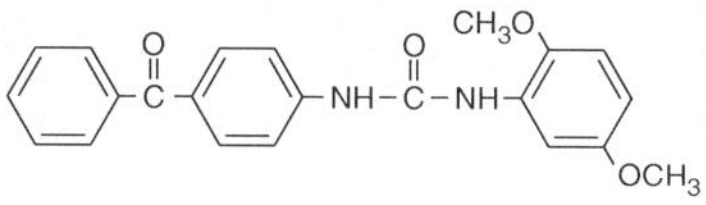
5 gm

S65,392-6
1-(2-BENZOYL-4-CHLOROPHENYL)-3-(2,5-DIMETHOXYPHENYL)UREA

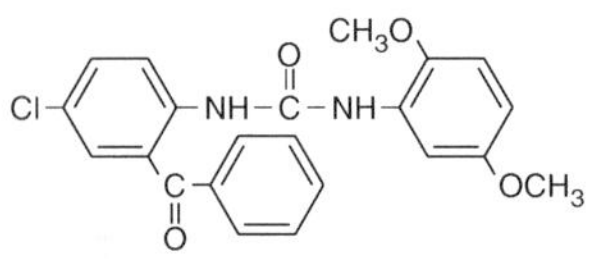

5 gm

S66,923-7
1-(2-CARBOXYPHENYL)-3-(2-METHOXY-5-METHYLPHENYL)UREA

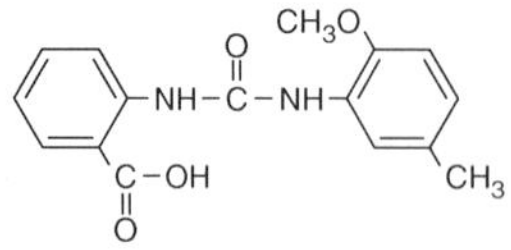

5 gm

S66,910-5
1-(2-CARBOXYPHENYL)-3-(2,5-DIMETHOXYPHENYL)UREA

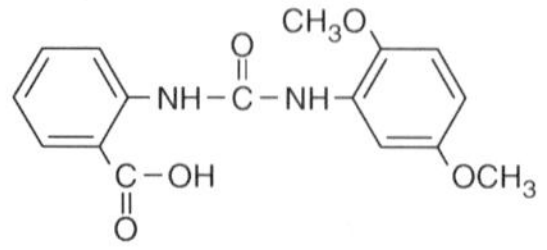

5 gm

S66,973-3
1-(4-CARBOXYPHENYL)-3-(2-METHOXY-5-METHYLPHENYL)UREA

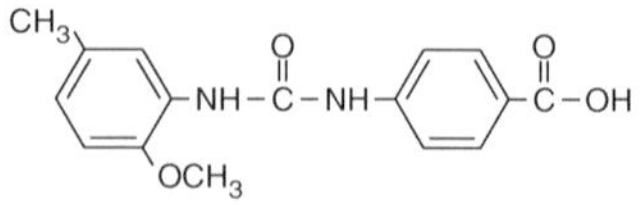

5 gm

S66,912-1
1-(4-CARBOXYPHENYL)-3-(2,5-DIMETHOXYPHENYL)UREA

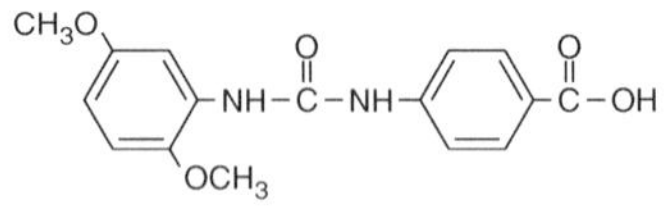

5 gm

S74,615-0
ETHYL 4-(3-[2-METHOXY-5-(N-PHENYLCARBAMOYL)PHENYL]UREIDO)-BENZOATE

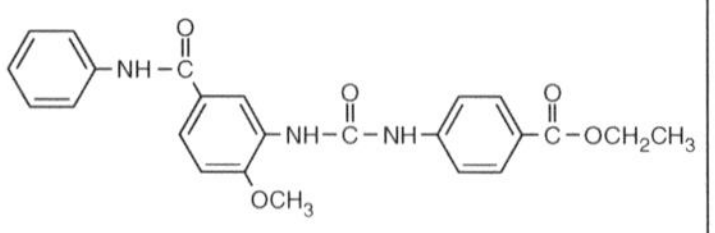

250 mg

S85,456-5
1-ACETYL-3-[2-METHOXY-5-(N-PHENYLCARBAMOYL)PHENYL]UREA

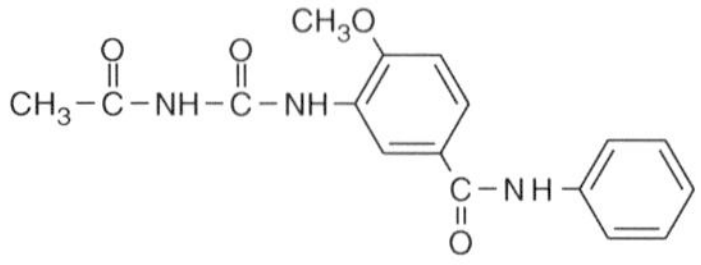

250 mg

S50,872-1
4-[5-CHLORO-4-(4-CHLOROPHENOXY)]-1-(3,4-DICHLOROBENZOYL)SEMICARBAZIDE

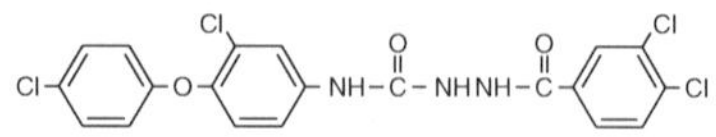

1 gm

S85,457-3
1-TERT-BUTYL-3-[2-METHOXY-5-(N-PHENYLCARBAMOYL)PHENYL]UREA

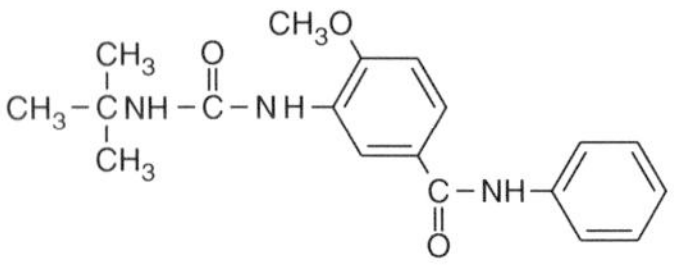

250 mg

S66,546-0
1,1-DIMETHYL-4-(2-METHOXY-5-METHYLPHENYL)SEMICARBAZIDE

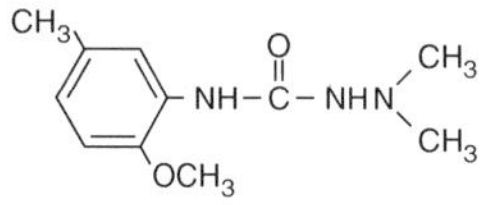

5 gm

S61,076-3
4-(2,5-DIMETHOXYPHENYL)-1,1-DIMETHYLSEMICARBAZIDE

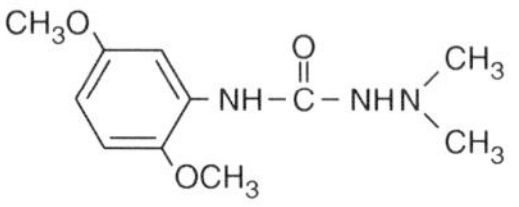

5 gm

S50,851-9
1-[5-CHLORO-2-(PENTACHLOROPHENOXY)-PHENYL]-3-PHENYLUREA

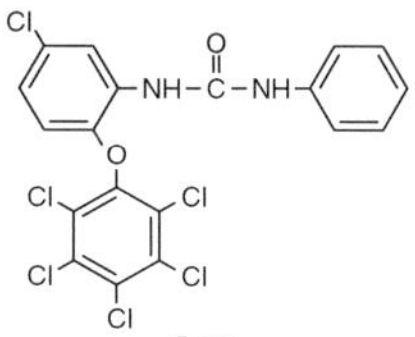

5 gm

S67,435-4
[3-(2-ETHOXYPHENYL)UREIDO]ACETIC ACID HYDRAZIDE

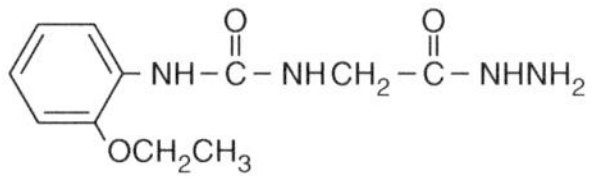

5 gm

S61,126-3
1-(3-CHLOROPHENYL)-3-(2-METHOXYETHYL)UREA

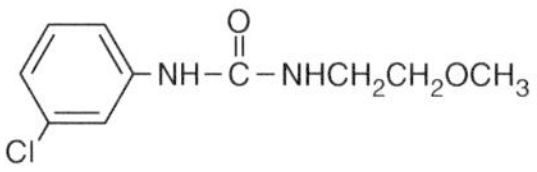

5 gm

S68,789-8
1-CYCLOPROPYL-3-(2-METHOXYPHENYL)UREA

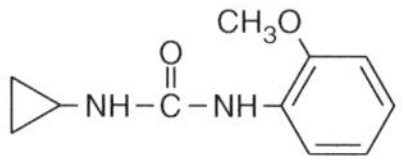

250 mg

S68,780-4
1-CYCLOPROPYL-3-(2-ETHOXYPHENYL)UREA

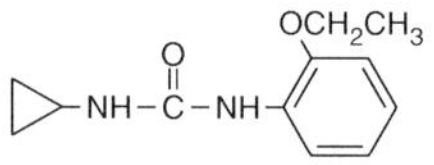

250 mg

S91,643-9
1-CYCLOPROPYL-3-(2-METHOXY-5-METHYLPHENYL)UREA

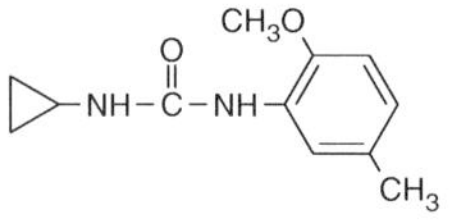

250 mg

S77,294-1
1-CYCLOPENTYL-3-(2-METHOXYPHENYL)UREA

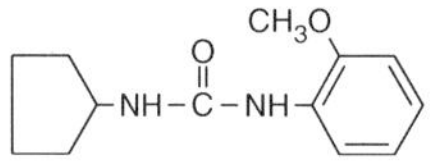

250 mg

S77,322-0
1-CYCLOPENTYL-3-(2-ETHOXYPHENYL)UREA

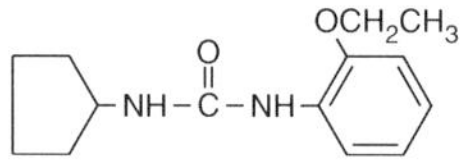

250 mg

S59,754-6
1-CYCLOHEXYL-3-(2-METHOXYPHENYL)UREA

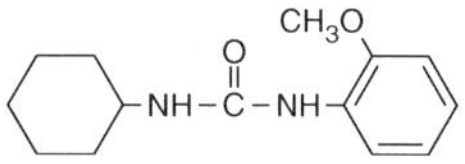

5 gm

S62,180-3
3-(O-ANISYL)-1-CYCLOHEXYL-1-METHYLUREA

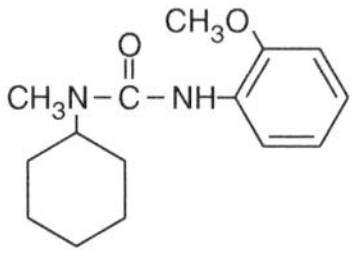

5 gm

S63,816-1
1-CYCLOHEXYL-3-(2-METHOXY-5-METHYLPHENYL)UREA

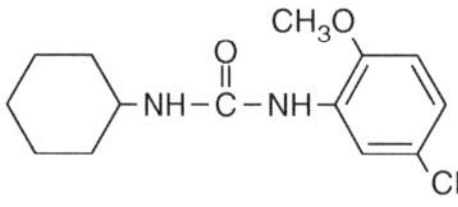

5 gm

S66,548-7
1-CYCLOHEXYL-3-(2,5-DIMETHOXYPHENYL)UREA

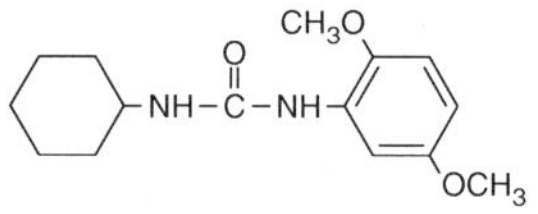

5 gm

S66,575-4
1-CYCLOHEXYL-3-(2-METHOXY-5-METHYLPHENYL)-1-METHYLUREA

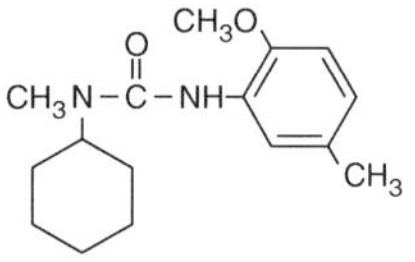

5 gm

S61,063-1
1-CYCLOHEXYL-3-(2,5-DIMETHOXYPHENYL)-1-METHYLUREA

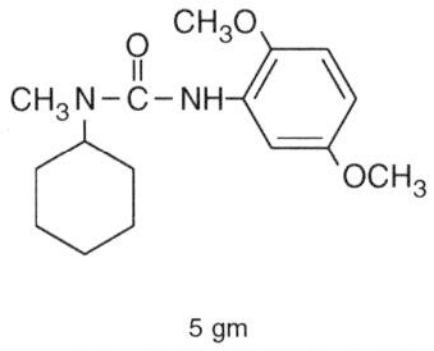

5 gm

S64,527-3
1,1-DICYCLOHEXYL-3-(2,5-DIMETHOXYPHENYL)UREA

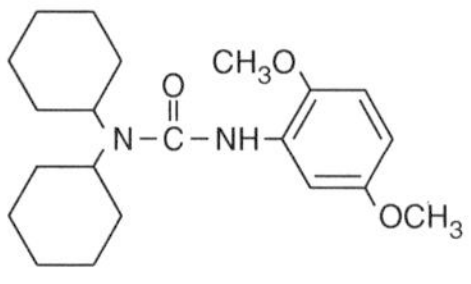

5 gm

S61,070-4
1-(2-METHOXYETHYL)-3-PHENYLUREA

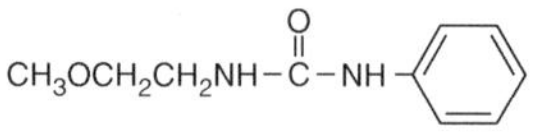

5 gm

S79,138-5
1-(2-ETHOXYPHENYL)-3-PHENYLUREA

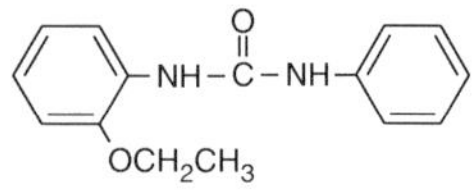

250 mg

S35,253-5
2-METHOXYPHENYLUREA

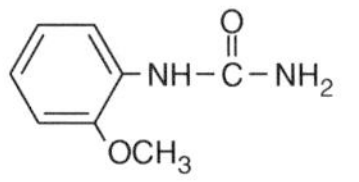

5 gm

S63,392-5
1-(O-ANISYL)-3-METHYLUREA

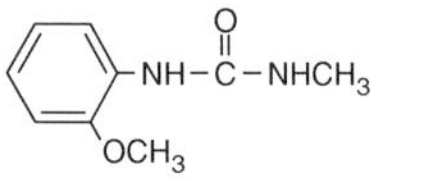

5 gm

S59,761-9
1,1-DIMETHYL-3-(2-METHOXYPHENYL)UREA

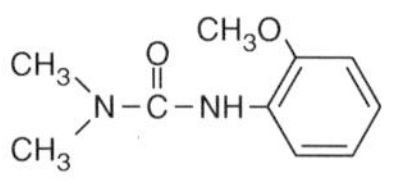

5 gm

S71,041-5
1-ETHYL-3-(2-METHOXYPHENYL)UREA

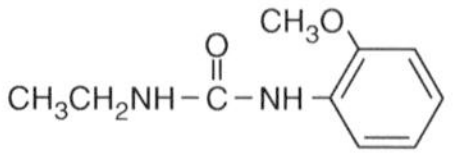

5 gm

S63,408-5
1-(O-ANISYL)-3-ETHYLUREA

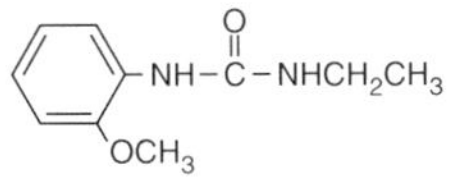

5 gm

S63,666-5
1-(2-CHLOROPHENYL)-3-(2-METHOXYETHYL)UREA

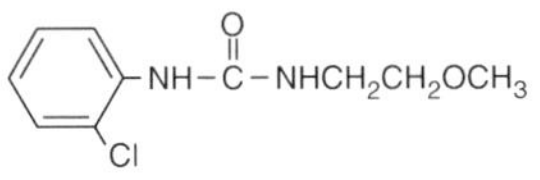

5 gm

S62,178-1
1-ALLYL-3-(O-ANISYL)UREA

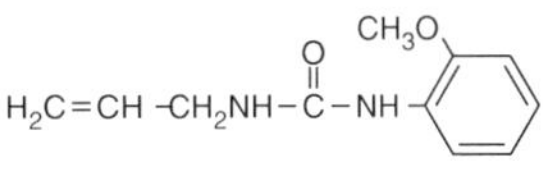

5 gm

S59,744-9
1-ISOPROPYL-3-(2-METHOXYPHENYL)UREA

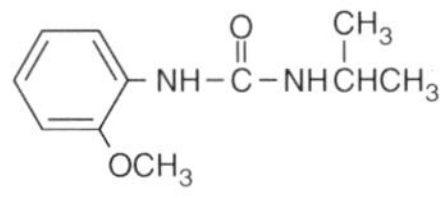

5 gm

S65,041-2
1-(2-METHOXYETHYL)-3-(O-TOLYL)UREA

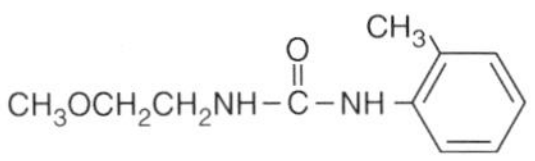

5 gm

S71,052-0
1-(2-METHOXYPHENYL)-3-PROPYLUREA

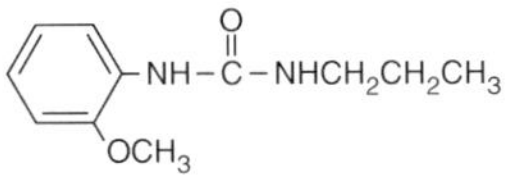

5 gm

S63,830-7
1-(2-METHOXYETHYL)-3-(2-METHOXYPHENYL)UREA

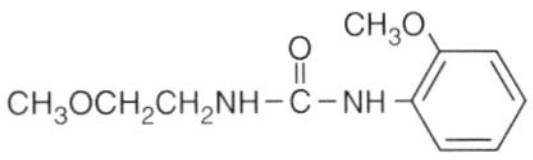

5 gm

S65,368-3
1-(O-ANISYL)-3-ISOBUTYLUREA

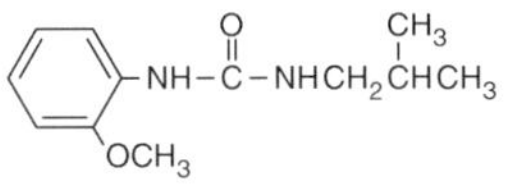

5 gm

S74,714-9
1-(2-ETHOXYPHENYL)-3-PROPYLUREA

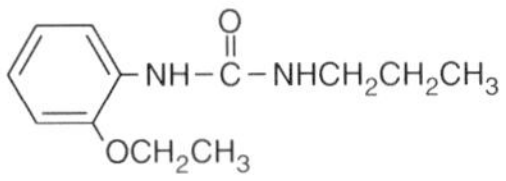

5 gm

S59,750-3
1-(2-METHOXYPHENYL)-3-PHENYLUREA

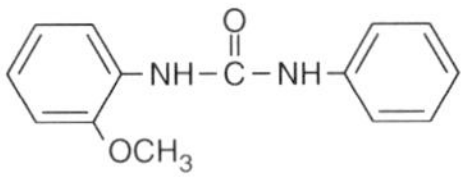

5 gm

S59,747-3
1-(2-CHLOROPHENYL)-3-(2-METHOXYPHENYL)UREA

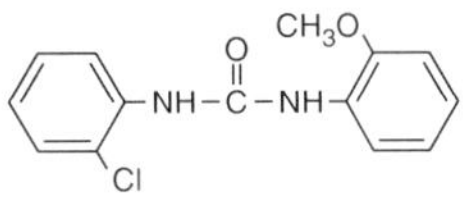

5 gm

S65,025-0
3-(O-ANISYL)-1-METHYL-1-PHENYLUREA

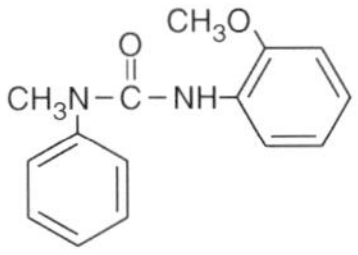

5 gm

S77,943-1
1-(2-METHOXYPHENYL)-3-(O-TOLYL)UREA

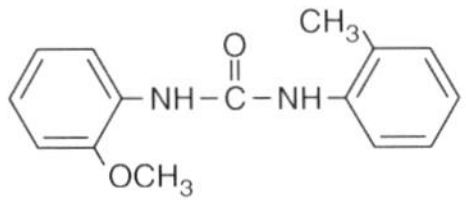

250 mg

S51,346-6
2,2'-DIMETHOXYCARBANILIDE

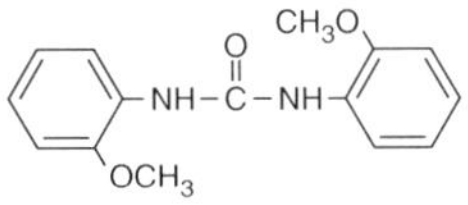

250 mg

S88,993-8
1-(2-ETHYLPHENYL)-3-(2-METHOXYPHENYL)UREA

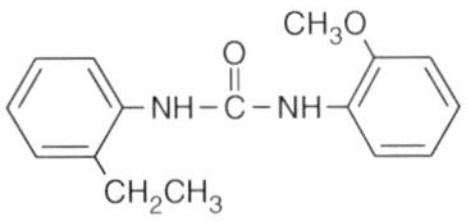

250 mg

S63,506-5
1,1-DIBUTYL-3-(2-METHOXYPHENYL)UREA

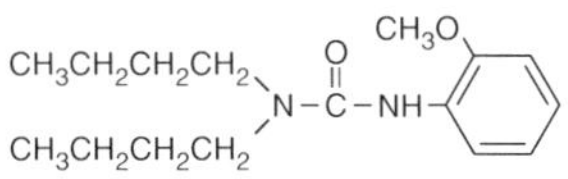

5 gm

S79,153-9
1-(2-ETHOXYPHENYL)-3-(2-METHOXYPHENYL)UREA

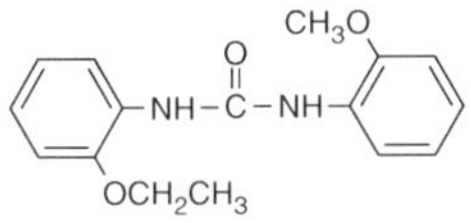

250 mg

S67,084-7
1-(O-ANISYL)-3-OCTADECYLUREA

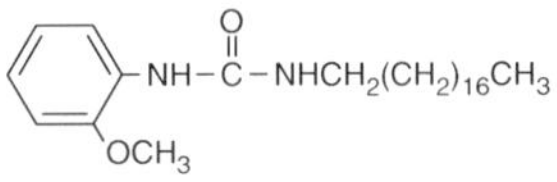

5 gm

S60,475-5
1-(2-METHOXYPHENYL)-3-(4-METHOXYPHENYL)UREA

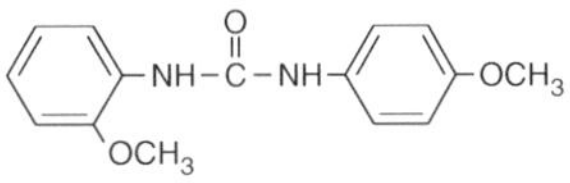

5 gm

S88,991-1
1-(4-ETHYLPHENYL)-3-(2-METHOXYPHENYL)UREA

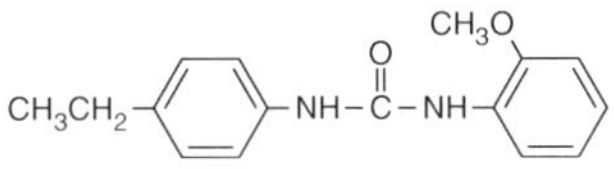

250 mg

S71,149-7
1-(2-METHOXYPHENYL)-3-(2,6-XYLYL)UREA

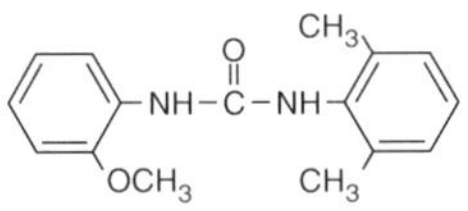

5 gm

S88,986-5
1-(2-METHOXYPHENYL)-3-(2,3-XYLYL)UREA

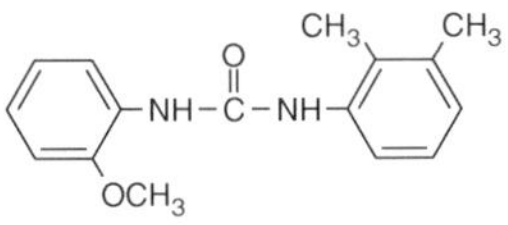

250 mg

S71,293-0
1-(2-ETHYL-6-METHYLPHENYL)-3-(2-METHOXYPHENYL)UREA

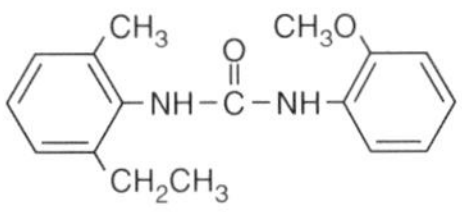

5 gm

S75,271-1
1-(2,6-DIETHYLPHENYL)-3-(2-METHOXYPHENYL)UREA

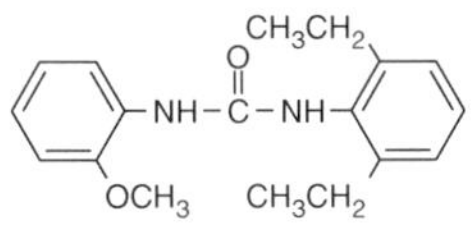

250 mg

S71,748-7
1-(2,6-DIISOPROPYLPHENYL)-3-(2-METHOXYPHENYL)UREA

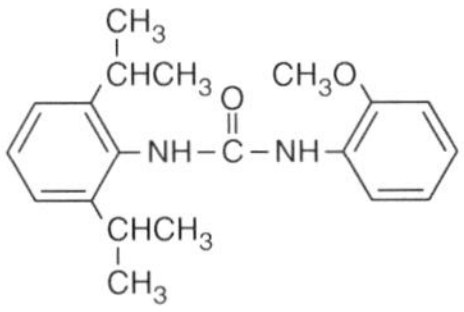

5 gm

S76,936-3
1-(2-METHOXYPHENYL)-3-(2,5-XYLYL)UREA

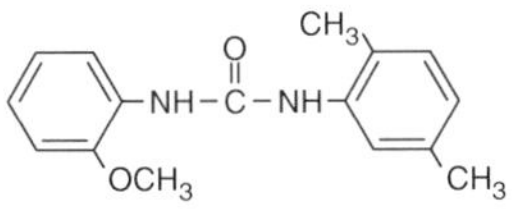

250 mg

S88,990-3
1-(2-METHOXYPHENYL)-3-(3,4-XYLYL)UREA

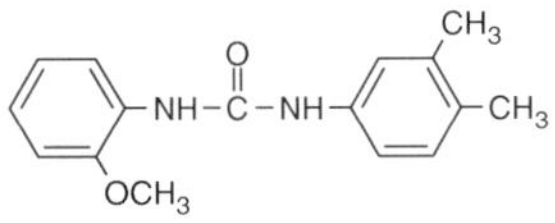

250 mg

S90,031-1
1-(2,4-DIMETHYLBENZYL)-3-(2-METHOXYPHENYL)UREA

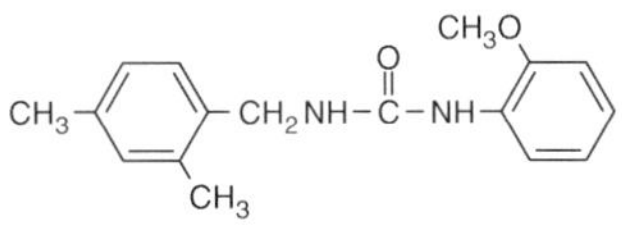

250 mg

S77,323-9
1-(5-ISOPROPYL-2-METHYLPHENYL)-3-(2-METHOXYPHENYL)UREA

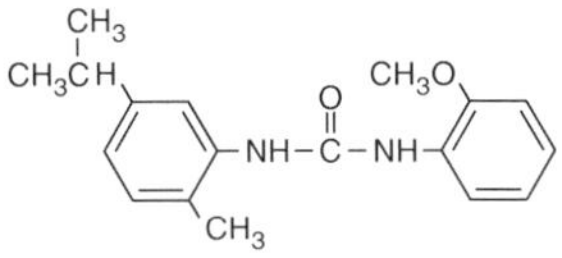

250 mg

S79,163-6
1-(2-ETHOXYPHENYL)-3-(1-NAPHTHYL)UREA

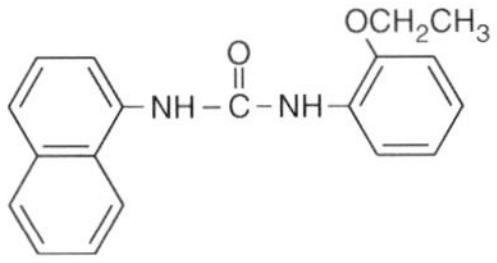

250 mg

S64,475-7
1-(4-BROMOPHENYL)-3-(2-METHOXYETHYL)UREA

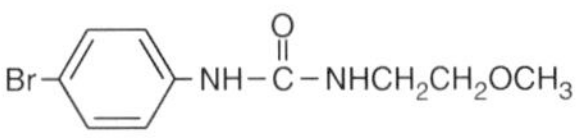

5 gm

S65,112-5
1,1'-[(METHYLENEDI)-4,4'-PHENYLENE]-BIS[3-(2-METHOXYETHYL)UREA]

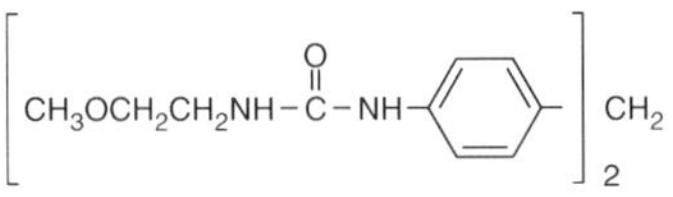

5 gm

S65,074-9
3-(3-CHLORO-2-METHOXYPHENYL)-1,1-DIMETHYLUREA

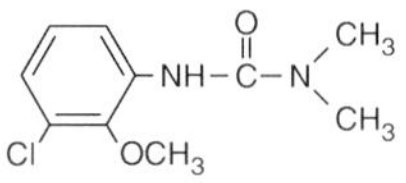

5 gm

S65,494-9
1-(3-CHLORO-2-METHOXYPHENYL)-3-ETHYLUREA

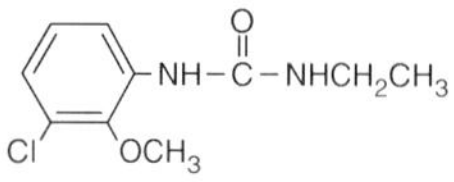

5 gm

S64,440-4
1-(2,3-DICHLOROPHENYL)-3-(2-METHOXYETHYL)UREA

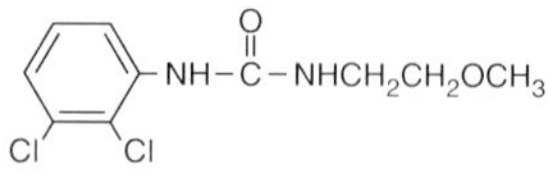

5 gm

S64,048-4
1-(3-CHLORO-2-METHOXYPHENYL)-3-ISOPROPYLUREA

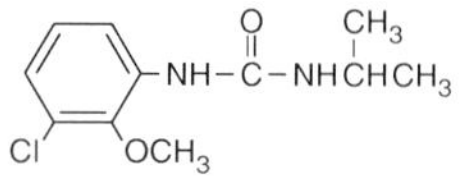

5 gm

S65,336-5
1-(3-CHLORO-2-METHOXYPHENYL)-3-PROPYLUREA

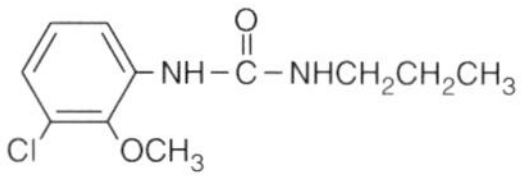

5 gm

S70,889-5
1-(2-METHOXYETHYL)-3-(2,6-XYLYL)UREA

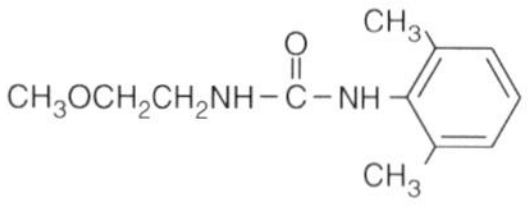

5 gm

S65,338-1
1-BUTYL-3-(3-CHLORO-2-METHOXYPHENYL)UREA

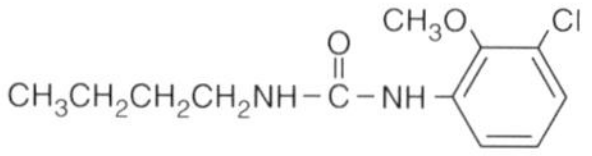

5 gm

S63,819-6
1-(2-METHOXY-5-METHYLPHENYL)-3-METHYLUREA

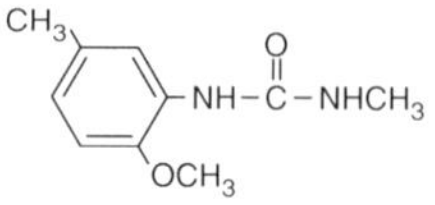

5 gm

S66,753-6
1-(2,5-DIMETHOXYPHENYL)-3-METHYLUREA

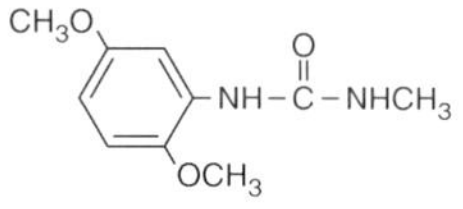

5 gm

S61,084-4
1-(3,4-DICHLOROPHENYL)-3-(2-METHOXYETHYL)UREA

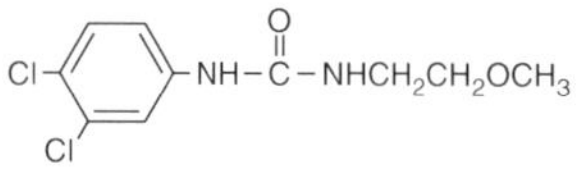

5 gm

S61,631-1
1-(2,4-DICHLOROPHENYL)-3-(2-METHOXYETHYL)UREA

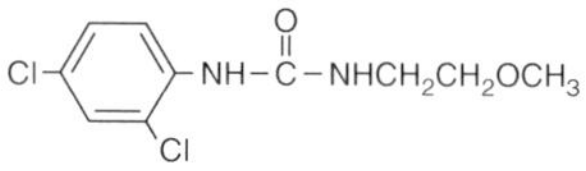

5 gm

S64,419-6
1-(2,5-DICHLOROPHENYL)-3-(2-METHOXYETHYL)UREA

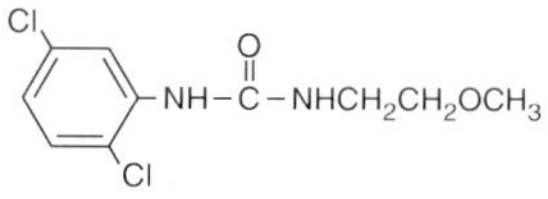

5 gm

S62,481-0
1-ETHYL-3-(2-METHOXY-5-METHYLPHENYL)UREA

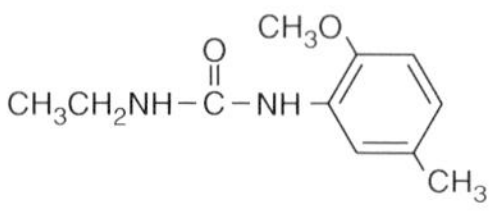

5 gm

S63,797-1
1,1-DIMETHYL-3-(2-METHOXY-5-METHYLPHENYL)UREA

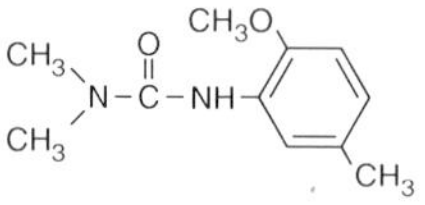

5 gm

S60,983-8
3-(2,5-DIMETHOXYPHENYL)-1,1-DIMETHYLUREA

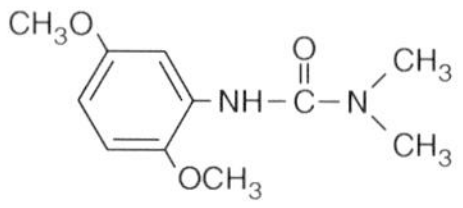

5 gm

S66,567-3
1-(2,5-DIMETHOXYPHENYL)-3-ETHYLUREA

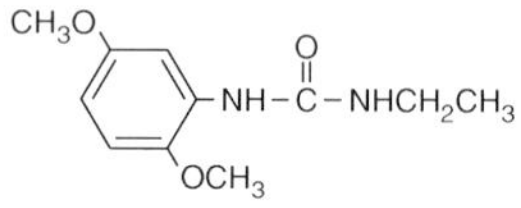

5 gm

S61,022-4
1-(5-CHLORO-2-METHYLPHENYL)-3-(2-METHOXYETHYL)UREA

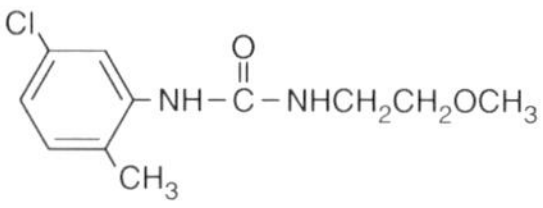

5 gm

S65,224-5
1-ALLYL-3-(2-METHOXY-5-METHYLPHENYL)UREA

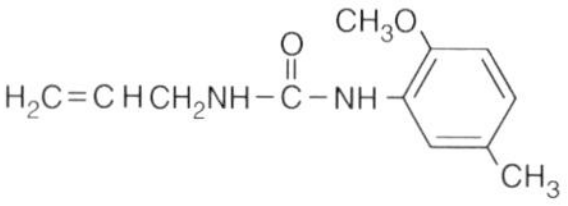

5 gm

S61,069-0
1-(2-METHOXYETHYL)-3-(2,4-XYLYL)UREA

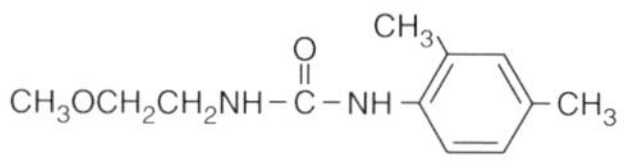

5 gm

S66,087-6
1-(2-METHOXY-5-METHYLPHENYL)-3-PROPYLUREA

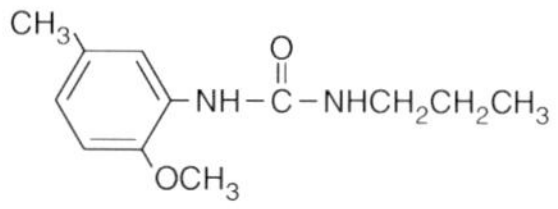

5 gm

S66,406-5
1-ALLYL-3-(2,5-DIMETHOXYPHENYL)UREA

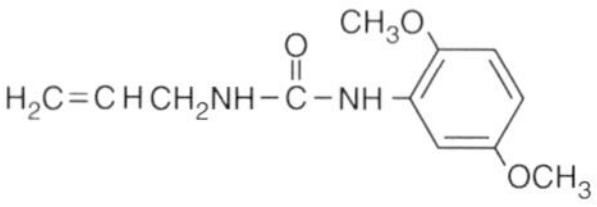

5 gm

S65,643-7
1-(2,5-DIMETHOXYPHENYL)-1-ISOPROPYLUREA

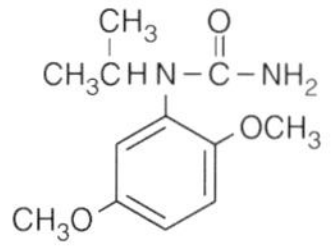

250 mg

S66,400-6
1-(2,5-DIMETHOXYPHENYL)-3-PROPYLUREA

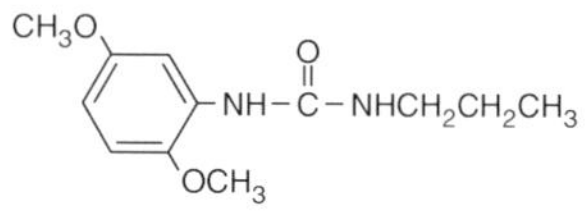

5 gm

S66,463-4
1-(2,5-DIMETHOXYPHENYL)-3-ISOPROPYLUREA

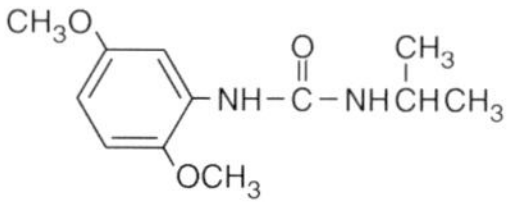

5 gm

S64,525-7
1-(2,5-DIMETHOXYPHENYL)-3-(2-METHOXYETHYL)UREA

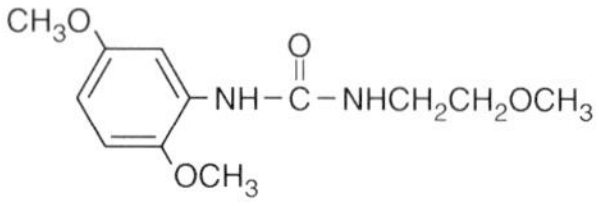

5 gm

S66,093-0
1-BUTYL-3-(2-METHOXY-5-METHYLPHENYL)UREA

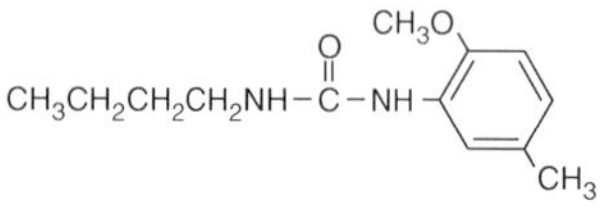

5 gm

S85,516-2
1-TERT-BUTYL-3-(2-METHOXY-5-METHYLPHENYL)UREA

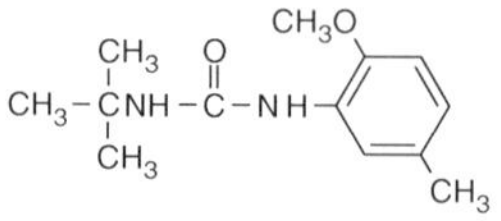

250 mg

S60,991-9
1,1-DIETHYL-3-(2,5-DIMETHOXYPHENYL)UREA

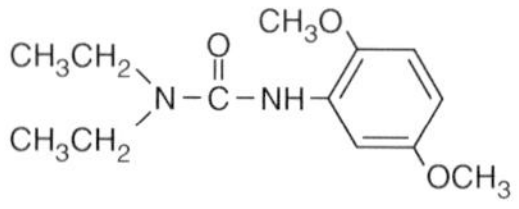

5 gm

S66,544-4
1-BUTYL-3-(2,5-DIMETHOXYPHENYL)UREA

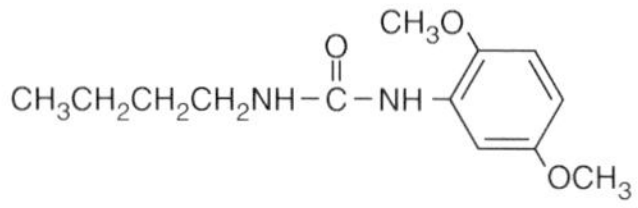

5 gm

S66,582-7
1-(2,5-DIMETHOXYPHENYL)-3-ISOBUTYLUREA

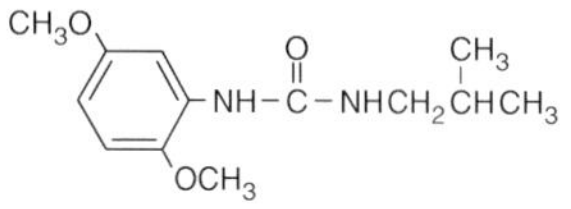

5 gm

S63,802-1
1,1-DIALLYL-3-(2-METHOXY-5-METHYLPHENYL)UREA

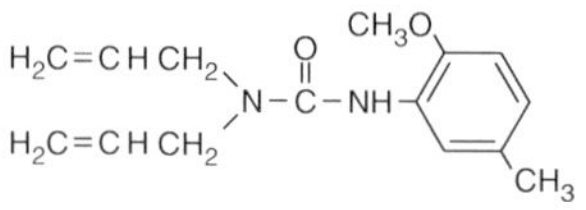

5 gm

S63,796-3
1,1-DIISOPROPYL-3-(2-METHOXY-5-METHYLPHENYL)UREA

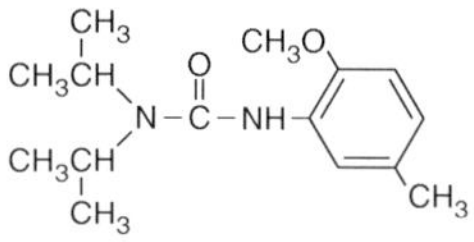

5 gm

S63,410-7
1-(2,5-DIMETHOXYPHENYL)-3-PHENYLUREA
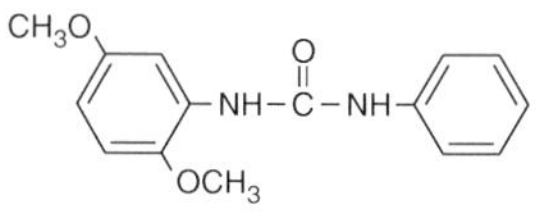
5 gm

S60,981-1
1,1-DIISOPROPYL-3-(2,5-DIMETHOXYPHENYL)UREA
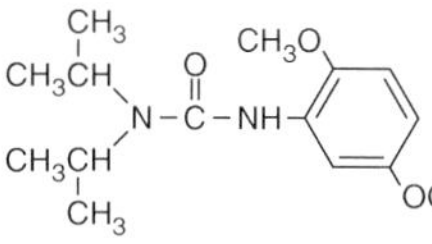
5 gm

S60,992-7
3-(2,5-DIMETHOXYPHENYL)-1,1-DIPROPYLUREA
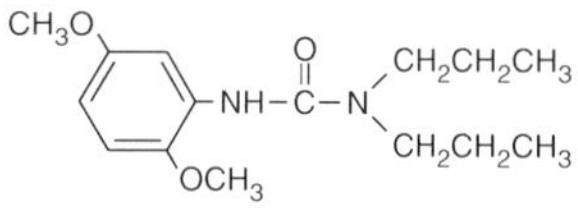
5 gm

S61,009-7
1-(M-ANISYL)-3-(2,5-DIMETHOXYPHENYL)UREA
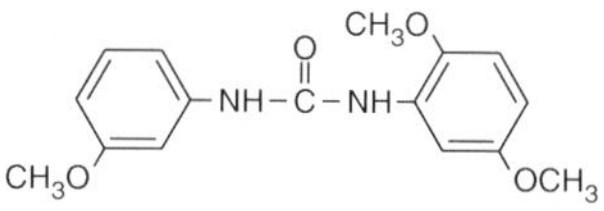
5 gm

S63,647-9
1,1-DIBUTYL-3-(2-METHOXY-5-METHYLPHENYL)UREA
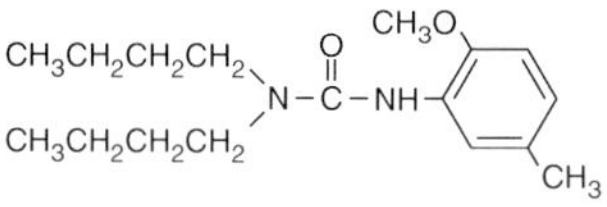
5 gm

S66,047-7
1,1-DIISOBUTYL-3-(2-METHOXY-5-METHYLPHENYL)UREA
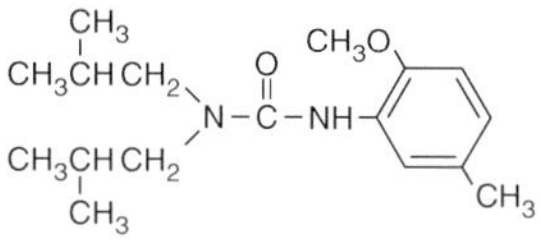
5 gm

S58,787-7
2,2'-DIMETHOXY-5,5'-DIMETHYLCARBANILIDE
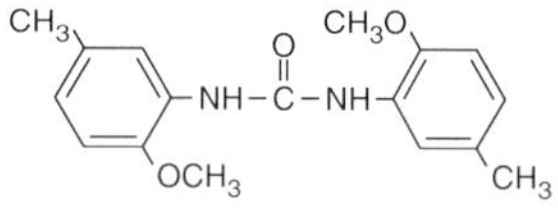
250 mg

S66,523-1
1,1-DIISOBUTYL-3-(2,5-DIMETHOXYPHENYL)UREA
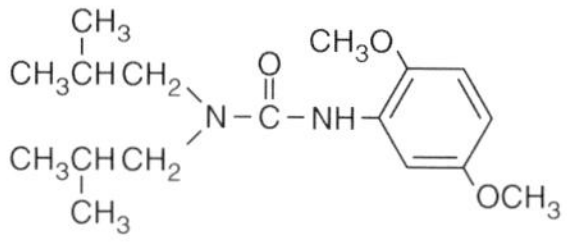
5 gm

S66,529-0
1,1-DIBUTYL-3-(2,5-DIMETHOXYPHENYL)UREA
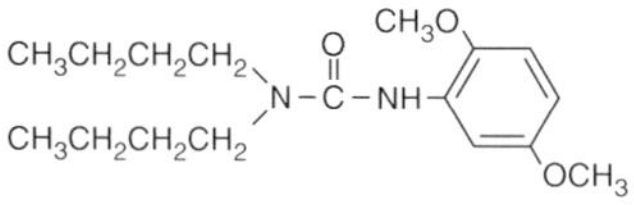
5 gm

S67,008-1
1-(2-METHOXY-5-METHYLPHENYL)-3-OCTADECYLUREA
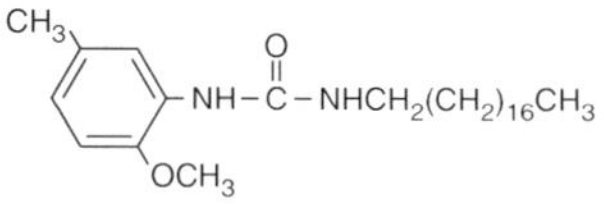
5 gm

S63,693-2
1-(2-METHOXYETHYL)-3-(1-NAPHTHYL)UREA
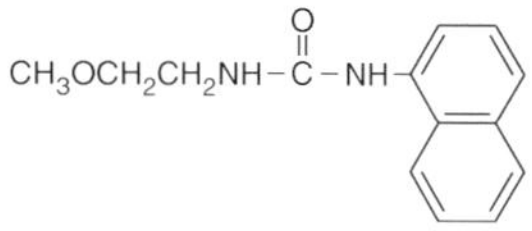
5 gm

S80,795-8
1-(3-ACETYLPHENYL)-3-(3-METHOXYPHENYL)UREA
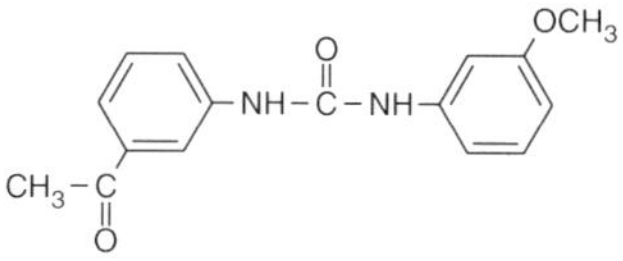
250 mg

S66,882-6
1-(M-ANISYL)-3-(2-CARBOXYPHENYL)UREA
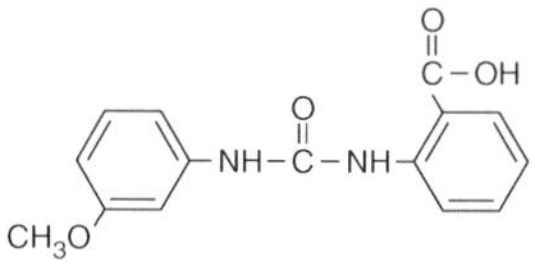
5 gm

S66,974-1
1-(O-ANISYL)-3-(2-CARBOXYPHENYL)UREA
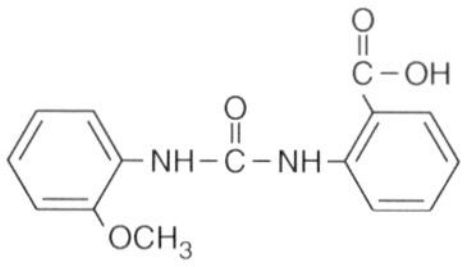
5 gm

S66,904-0
1-(P-ANISYL)-3-(2-CARBOXYPHENYL)UREA
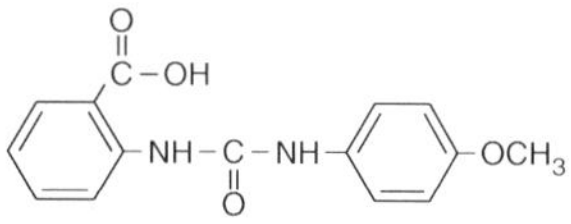
5 gm

S66,972-5
1-(O-ANISYL)-3-(4-CARBOXYPHENYL)UREA
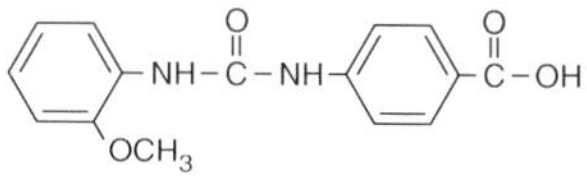
5 gm

S66,977-6
1-(M-ANISYL)-3-(4-CARBOXYPHENYL)UREA
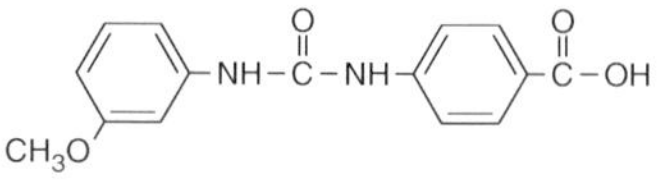
5 gm

S67,025-1
1-(P-ANISYL)-3-(4-CARBOXYPHENYL)UREA
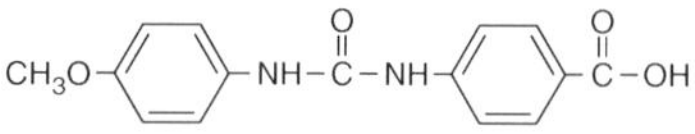
5 gm

S60,472-0
1-(4-METHOXYBENZOYL)-4-PHENYLSEMICARBAZIDE
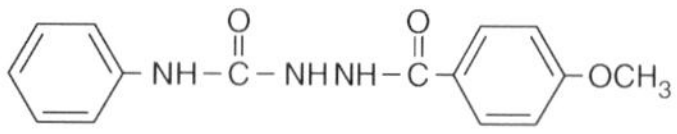
5 gm

S68,715-4
N-(3,3-DIMETHYLUREIDO)-N'-(4-ETHOXYPHENYL)OXAMIDE
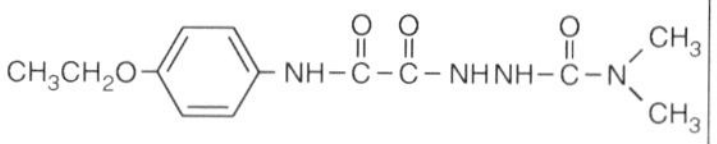
250 mg

S90,251-9
ETHYL 2-[3-(4-METHOXYPHENYL)UREIDO]-ACETATE
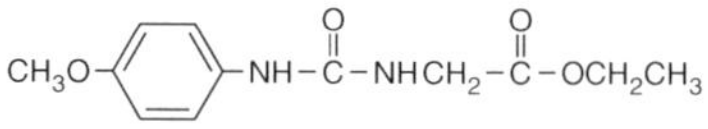
250 mg

S61,054-2
4-(M-ANISYL)-1,1-DIMETHYLSEMICARBAZIDE
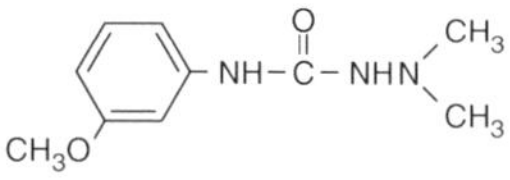
5 gm

S61,295-2
4-(P-ANISYL)-1,1-DIMETHYLSEMICARBAZIDE
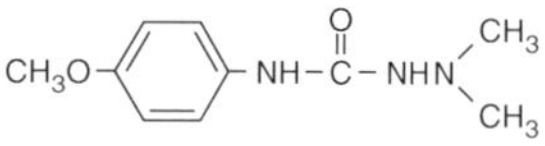
5 gm

S57,674-3
3-METHOXYPHENYLUREA
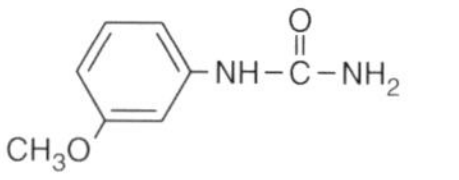
250 mg

S64,721-7
1-(M-ANISYL)-3-METHYLUREA
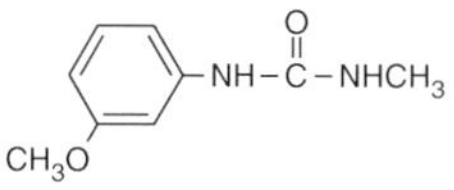
5 gm

S61,048-8
1,1-DIMETHYL-3-(3-METHOXYPHENYL)UREA
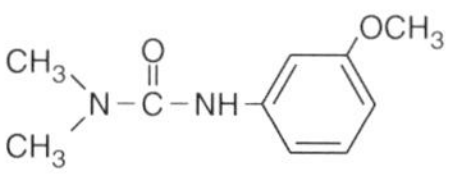
5 gm

S62,355-5
1-(M-ANISYL)-3-ETHYLUREA
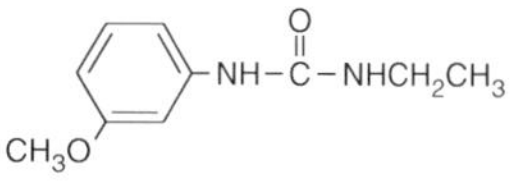
5 gm

S60,999-4
1-ALLYL-3-(M-ANISYL)UREA
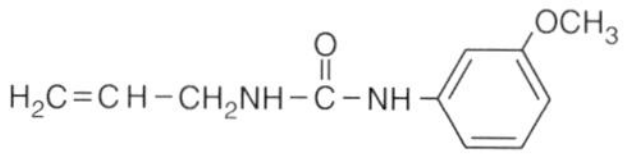
5 gm

S62,360-1
1-(M-ANISYL)-3-PROPYLUREA
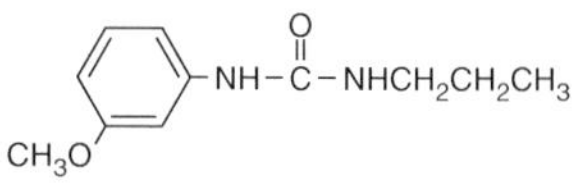
5 gm

S63,453-0
1-ISOPROPYL-3-(3-METHOXYPHENYL)UREA
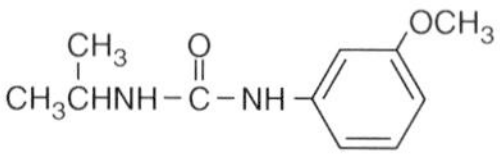
5 gm

S73,253-2
1-ISOPROPYL-1-(3-METHOXYPHENYL)UREA
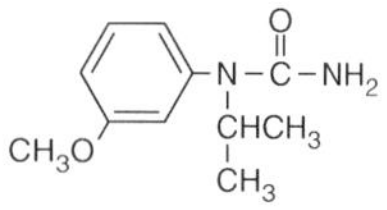
250 mg

S61,085-2
3-(M-ANISYL)-1,1-DIETHYLUREA
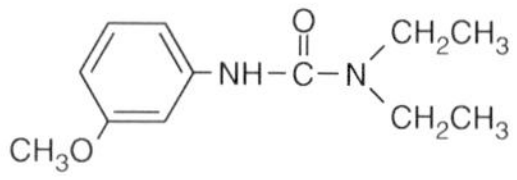
5 gm

S62,372-5
1-(M-ANISYL)-3-BUTYLUREA

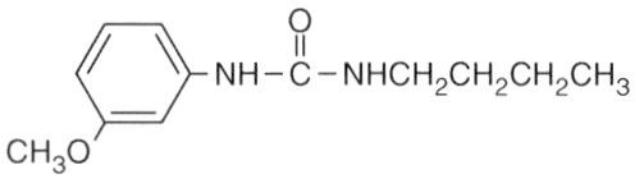

5 gm

S65,382-9
1-(M-ANISYL)-3-ISOBUTYLUREA

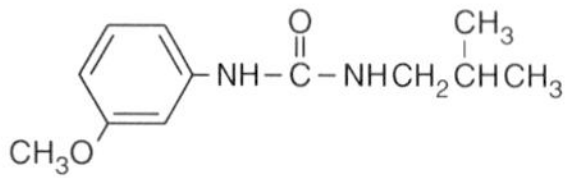

5 gm

S61,132-8
1-(M-ANISYL)-3-PHENYLUREA

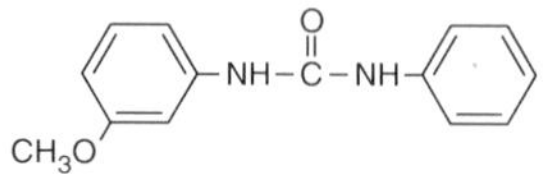

5 gm

S64,463-3
3-(M-ANISYL)-1,1-DIALLYLUREA

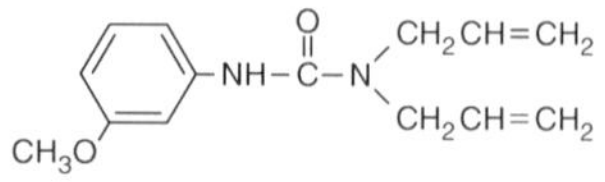

5 gm

S61,051-8
3-(M-ANISYL)-1,1-DIISOPROPYLUREA

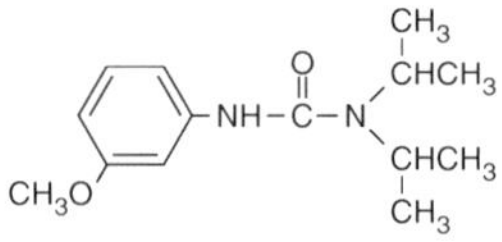

5 gm

S61,055-0
3-(M-ANISYL)-1,1-DIPROPYLUREA

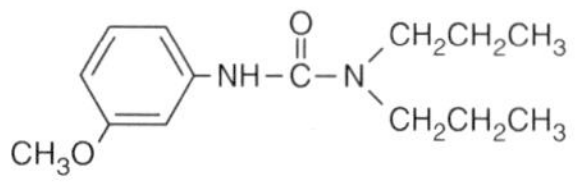

5 gm

S61,112-3
1-(M-ANISYL)-3-(3-CHLOROPHENYL)UREA

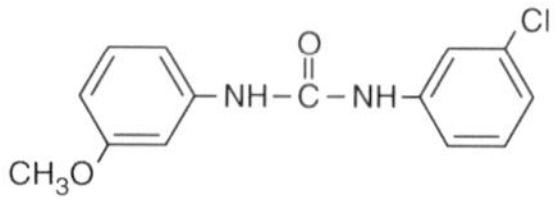

5 gm

S78,021-9
1-(3-BUTOXYPROPYL)-3-(M-TOLYL)UREA

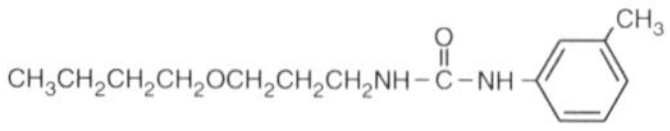

250 mg

S61,011-9
1,3-BIS(M-ANISYL)UREA

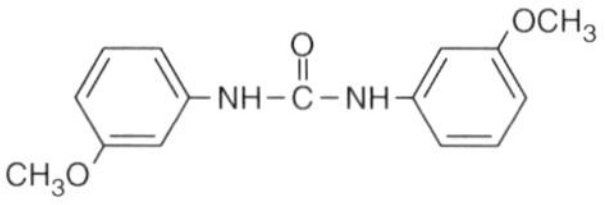

5 gm

S61,032-1
1,1-DIBUTYL-3-(3-METHOXYPHENYL)UREA

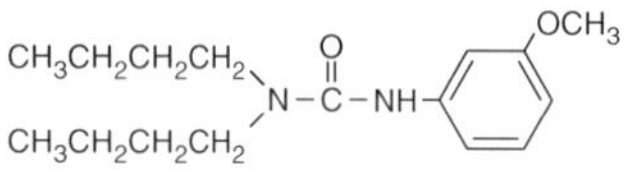

5 gm

S65,400-0
3-(M-ANISYL)-1,1-DIISOBUTYLUREA

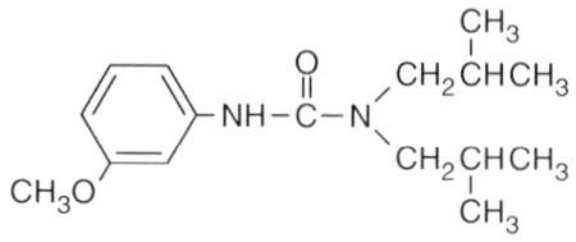

5 gm

S61,004-6
3-(M-ANISYL)-1-BUTYL-1-PHENYLUREA

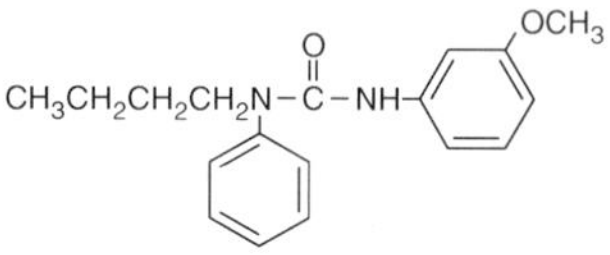

5 gm

S61,024-0
3-(M-ANISYL)-1,1-DIBENZYLUREA

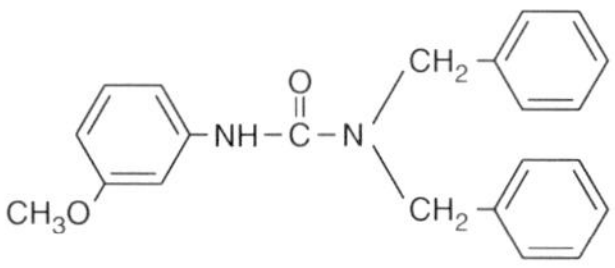

5 gm

S67,073-1
1-(M-ANISYL)-3-OLEYLUREA

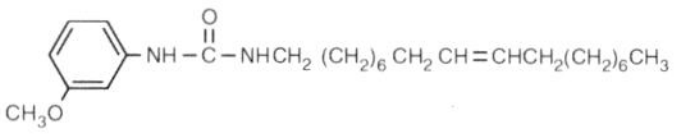

5 gm

S55,503-7
1-(3-METHOXYPHENYL)-3-OCTADECYLUREA

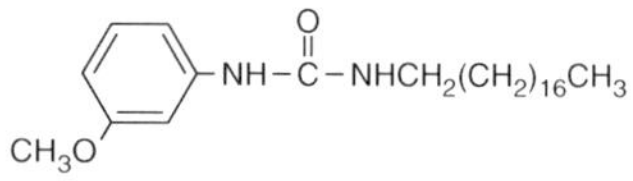

5 gm

S69,030-9
1-CYCLOPROPYL-3-(4-METHOXYPHENYL)UREA

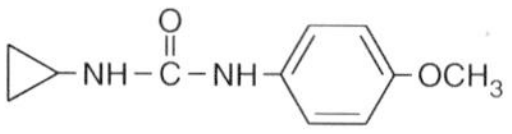

250 mg

S63,389-5
1-CYCLOHEXYL-3-(3-METHOXYPHENYL)UREA

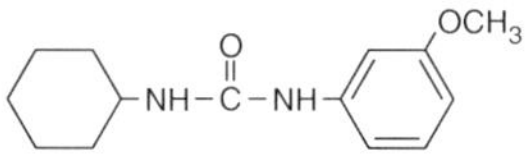

5 gm

S61,089-5
3-(M-ANISYL)-1-CYCLOHEXYL-1-METHYLUREA

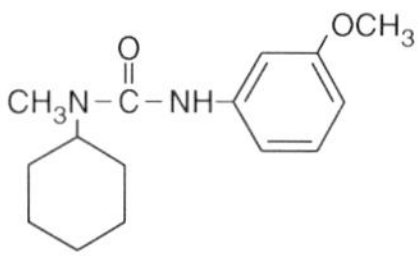

5 gm

S61,285-5
3-(O-ANISYL)-1,1-DICYCLOHEXYLUREA

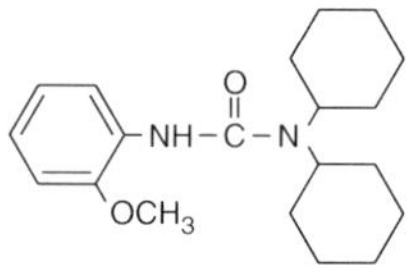

5 gm

S61,154-9
1-(P-ANISYL)-3-CYCLOHEXYLUREA

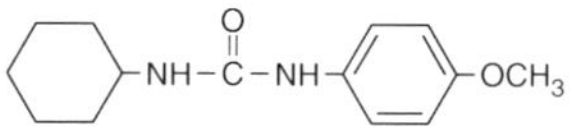

5 gm

S61,146-8
3-(P-ANISYL)-1-CYCLOHEXYL-1-METHYLUREA

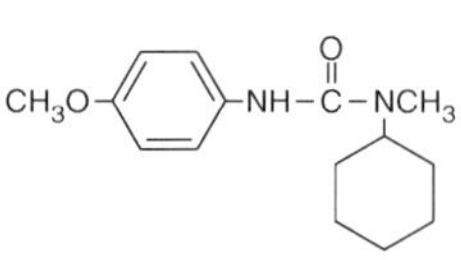

5 gm

S61,360-6
3-(P-ANISYL)-1,1-DICYCLOHEXYLUREA

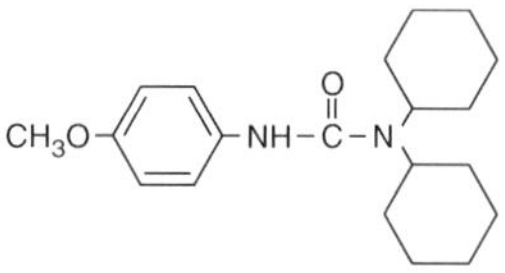

5 gm

S65,990-8
1-(3-CHLORO-2-METHOXYPHENYL)-3-CYCLOHEXYLUREA

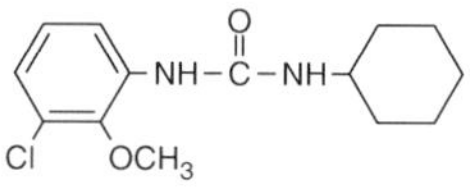

5 gm

S66,053-1
1-(3-CHLORO-2-METHOXYPHENYL)-3-(CYCLOHEXYLMETHYL)UREA

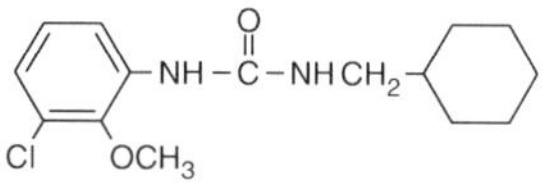

5 gm

S60,159-4
1-[N-(4-METHOXYPHENYL)CARBAMOYL]-PIPERIDINE

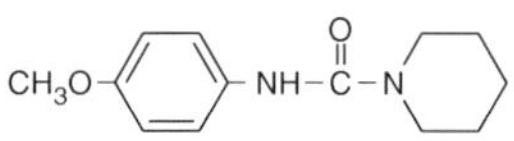

5 gm

S67,490-7
1-[N-(4-METHOXYPHENYL)CARBAMOYL]-4-METHYLPIPERIDINE

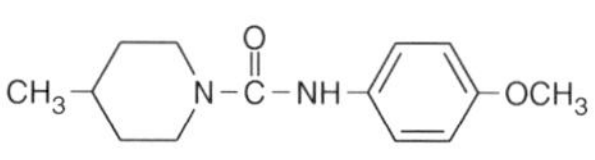

250 mg

S89,064-2
1-(4-ETHOXYPHENYL)-3-PHENYLUREA

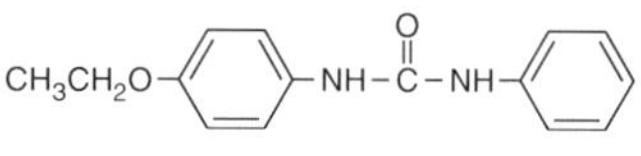

250 mg

S75,033-6
1-BENZYL-1-ETHYL-3-(4-METHOXYPHENYL)UREA

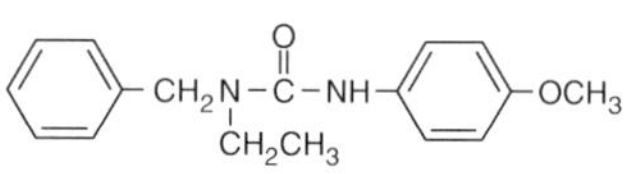

5 gm

S90,035-4
1-BENZYL-1-BUTYL-3-(4-METHOXYPHENYL)UREA

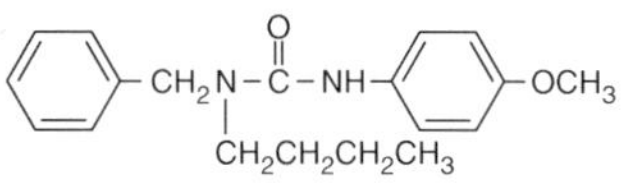

250 mg

S62,167-6
1-(O-ANISYL)-3-PROPYLUREA

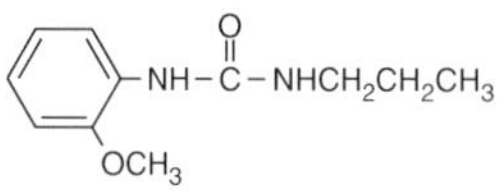

5 gm

S62,171-4
1-(O-ANISYL)-3-BUTYLUREA

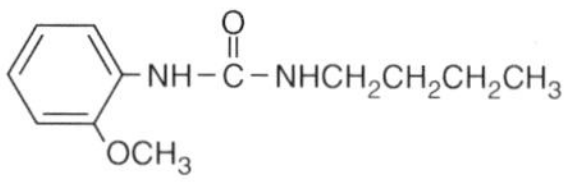

5 gm

S63,444-1
1,1-DIETHYL-3-(2-METHOXYPHENYL)UREA

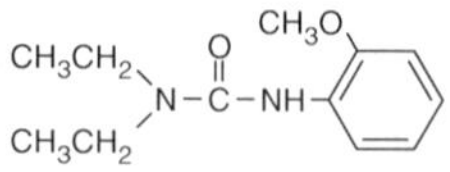

5 gm

S63,031-4
1,1-DIALLYL-3-(2-METHOXYPHENYL)UREA
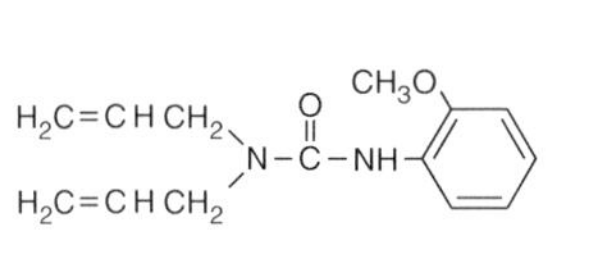
5 gm

S61,312-6
3-(O-ANISYL)-1,1-DIISOPROPYLUREA
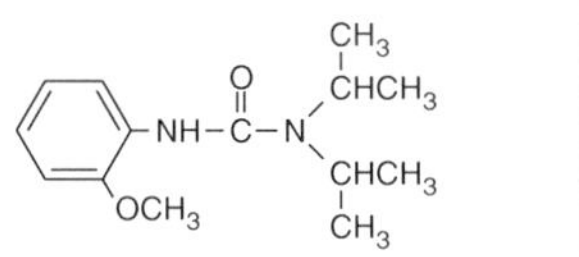
5 gm

S63,117-5
1-(2-CHLOROPHENYL)-3-(3-METHOXYPHENYL)UREA
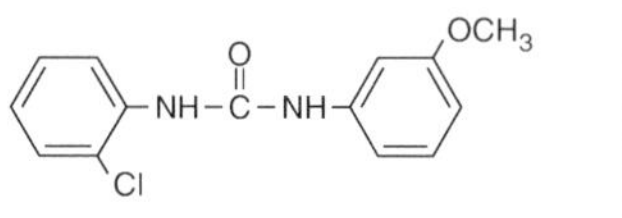
5 gm

S77,942-3
1-(3-METHOXYPHENYL)-3-(O-TOLYL)UREA
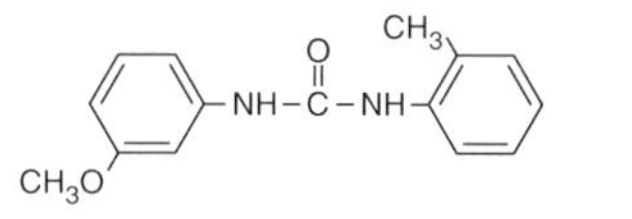
250 mg

S65,419-1
3-(O-ANISYL)-1,1-DIISOBUTYLUREA
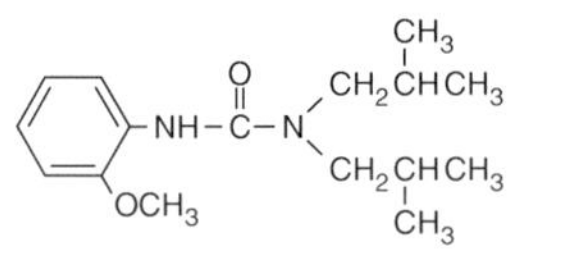
5 gm

S55,774-9
1,3-BIS(2-ETHOXYPHENYL)UREA
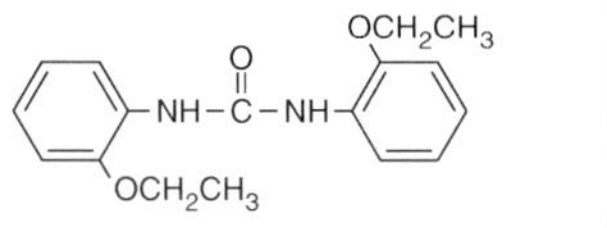
250 mg

S63,149-3
1-(O-CHLOROPHENYL)-3-(P-METHOXYPHENYL)UREA
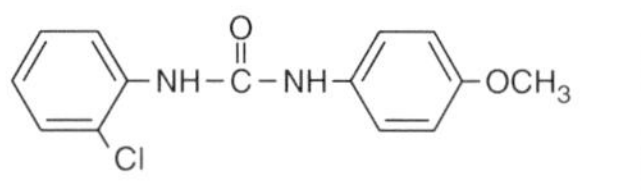
5 gm

S88,988-1
1-(2-ETHYLPHENYL)-3-(4-METHOXYPHENYL)UREA
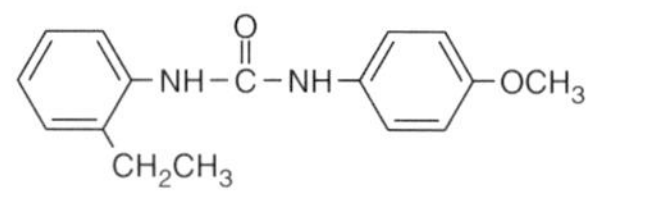
250 mg

S57,675-1
4-METHOXYPHENYLUREA
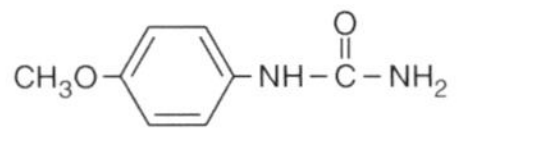
250 mg

S57,658-1
1-(4-ETHOXYPHENYL)UREA
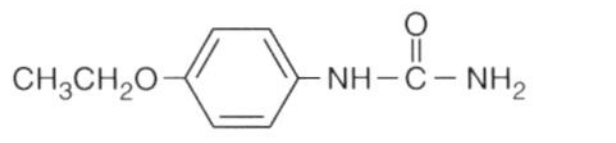
5 gm

S62,364-4
1-(P-ANISYL)-3-METHYLUREA
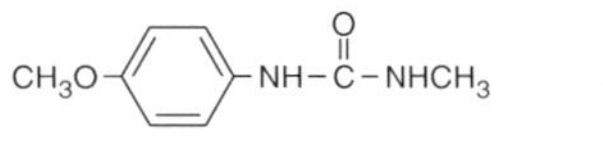
5 gm

S59,762-7
1,1-DIMETHYL-3-(4-METHOXYPHENYL)UREA
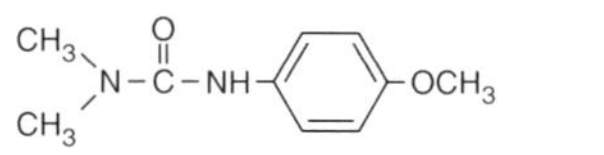
5 gm

S62,373-3
1-(P-ANISYL)-3-ETHYLUREA
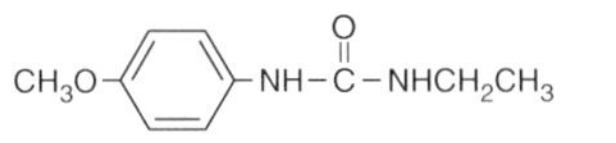
5 gm

S61,164-6
1-ALLYL-3-(P-ANISYL)UREA
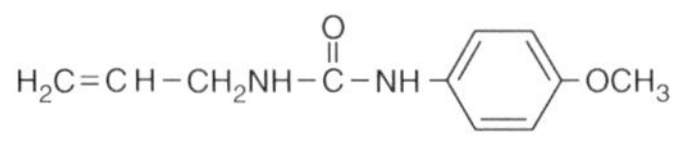
5 gm

S59,755-4
1-ISOPROPYL-3-(4-METHOXYPHENYL)UREA
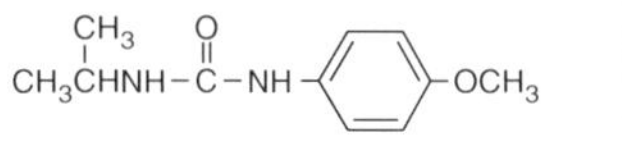
5 gm

S61,150-6
1-(P-ANISYL)-3-PROPYLUREA
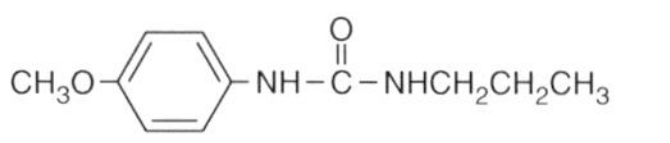
5 gm

S60,157-8
1,1-DIETHYL-3-(4-METHOXYPHENYL)UREA
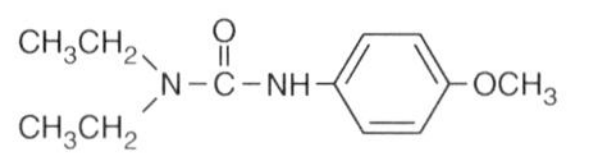
5 gm

S60,166-7
1-BUTYL-3-(4-METHOXYPHENYL)UREA
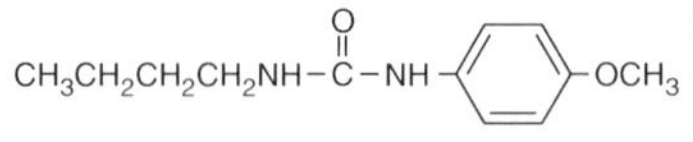
5 gm

S65,486-8
1-(P-ANISYL)-3-ISOBUTYLUREA
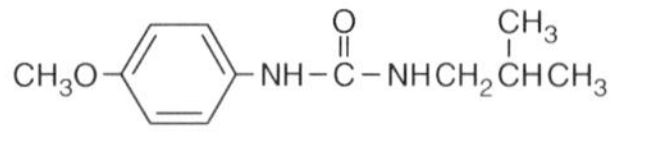
5 gm

S77,963-6
1-BUTYL-3-(4-ETHOXYPHENYL)UREA
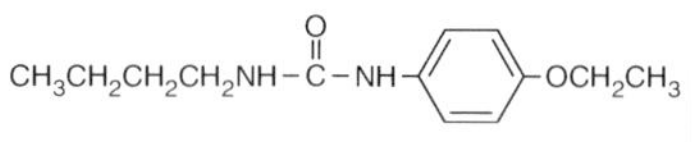
250 mg

S63,058-6
1-(4-METHOXYPHENYL)-3-PHENYLUREA
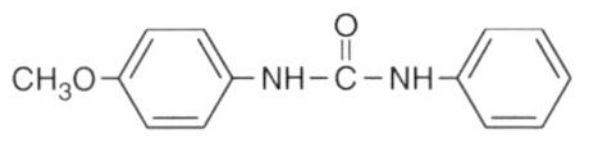
5 gm

S60,448-8
1,1-DIALLYL-3-(4-METHOXYPHENYL)UREA
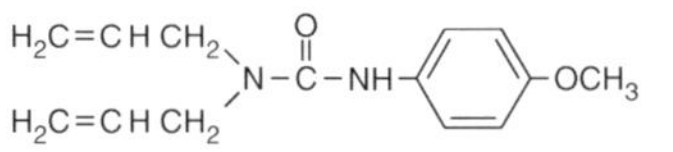
5 gm

S61,308-8
3-(P-ANISYL)-1,1-DIISOPROPYLUREA
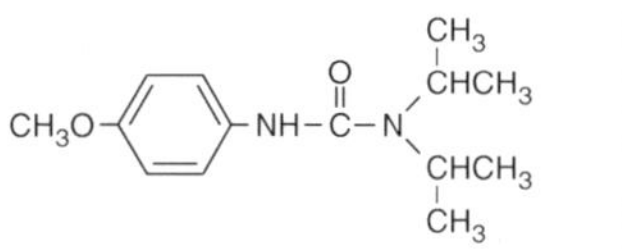
5 gm

S62,367-9
3-(P-ANISYL)-1,1-DIPROPYLUREA
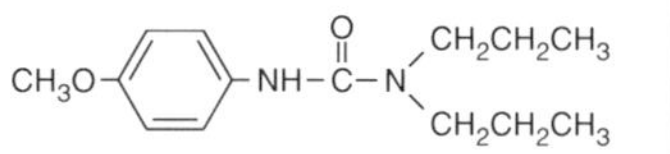
5 gm

S60,170-5
1-(4-FLUOROPHENYL)-3-(4-METHOXYPHENYL)UREA
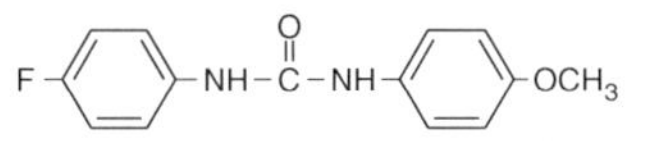
5 gm

S60,467-4
1,3-BIS(4-METHOXYPHENYL)UREA
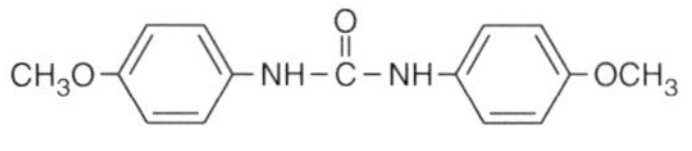
5 gm

S90,027-3
1-(4-METHOXYPHENYL)-3-(4-METHYLBENZYL)UREA
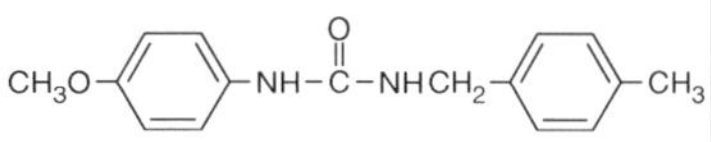
250 mg

S60,167-5
1,1-DIBUTYL-3-(4-METHOXYPHENYL)UREA
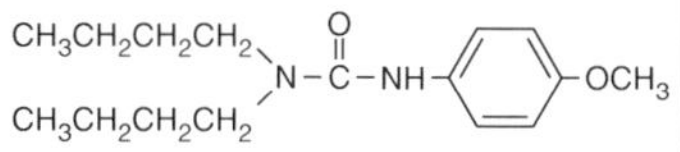
5 gm

S65,411-6
3-(P-ANISYL)-1,1-DIISOBUTYLUREA
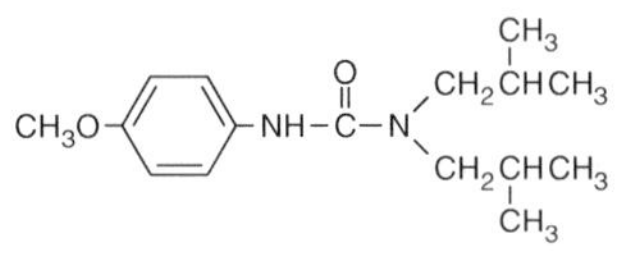
5 gm

S67,078-2
1-(P-ANISYL)-3-OCTADECYLUREA
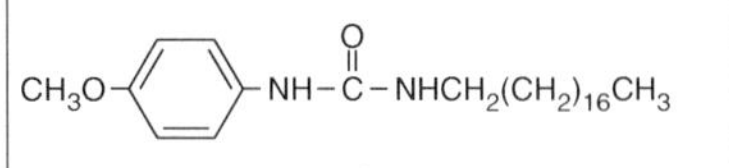
5 gm

S70,948-4
1-(2,6-DIMETHYLPHENYL)-3-(4-METHOXYPHENYL)UREA
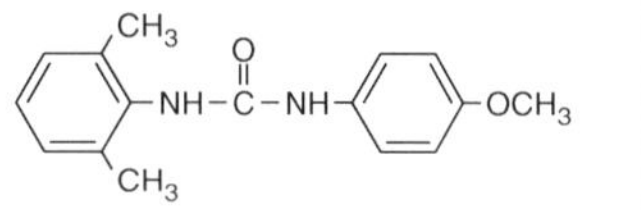
5 gm

S77,329-8
1-(4-METHOXYPHENYL)-3-(2,3-XYLYL)UREA
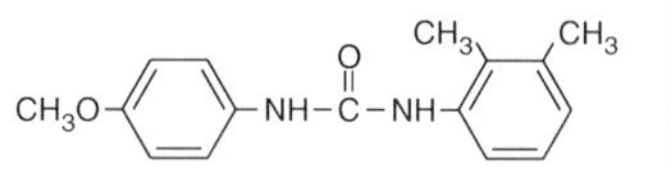
250 mg

S67,644-6
1-(2,6-DIETHYLPHENYL)-3-(4-METHOXYPHENYL)UREA

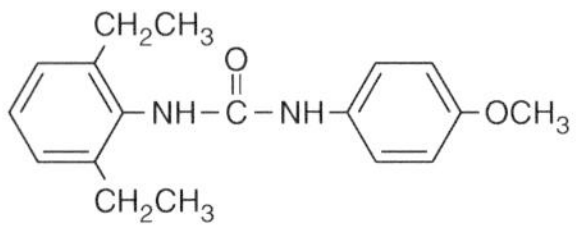

5 gm

S72,140-9
1-(2,5-DIFLUOROPHENYL)-3-(4-METHOXYPHENYL)UREA

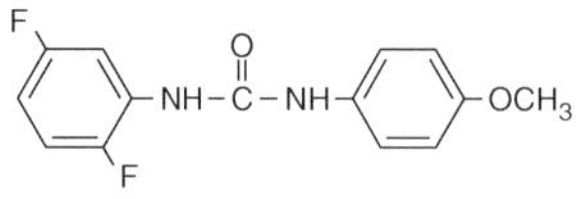

250 mg

S88,987-3
1-(4-METHOXYPHENYL)-3-(3,4-XYLYL)UREA

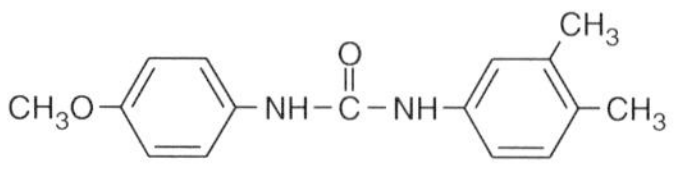

250 mg

S75,270-3
1-(5-ISOPROPYL-2-METHYLPHENYL)-3-(4-METHOXYPHENYL)UREA

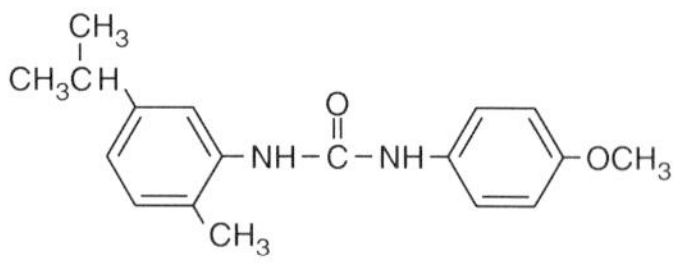

5 gm

S68,871-1
BIS(4-[3-(2,5-DIMETHYLPHENYL)-UREIDO]PHENYL) ETHER

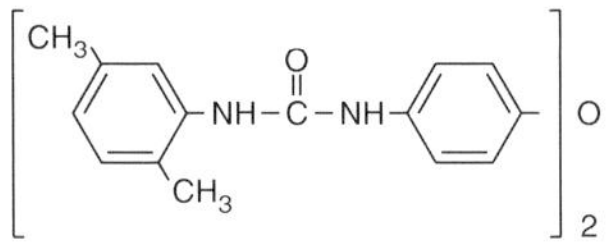

250 mg

S76,879-0
1-(4-METHOXYPHENYL)-3-(2,4,6-TRIMETHYLPHENYL)UREA

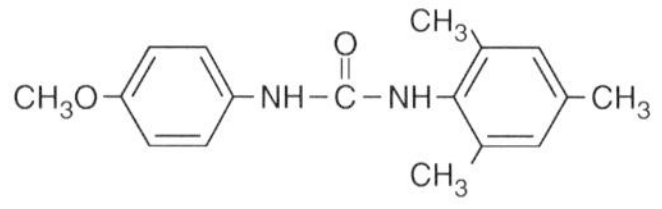

250 mg

S89,073-1
1-(4-ETHOXYPHENYL)-3-(1-NAPHTHYL)UREA

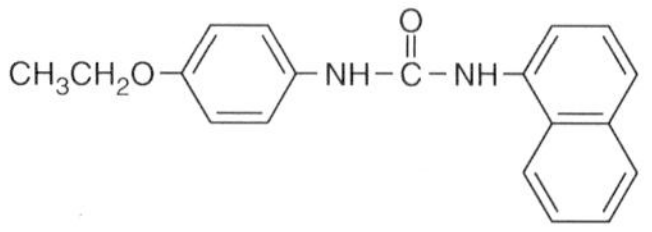

250 mg

S71,483-6
1-(3-METHOXYPHENYL)-3-(2,6-XYLYL)UREA

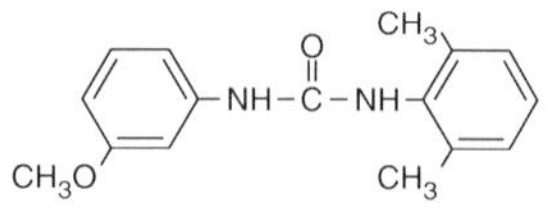

5 gm

S70,874-7
1-ISOPROPYL-3-(2-METHOXY-5-METHYLPHENYL)UREA

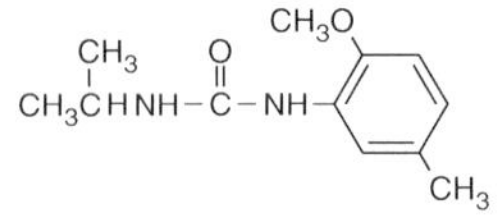

5 gm

S62,500-0
1,1-DIETHYL-3-(2-METHOXY-5-METHYLPHENYL)UREA

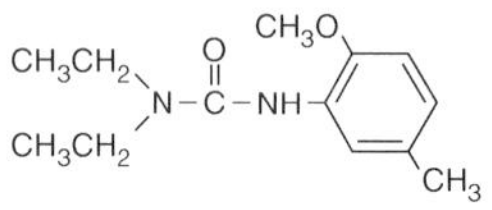

5 gm

S62,499-3
1,1-DIPROPYL-3-(2-METHOXY-5-METHYLPHENYL)UREA

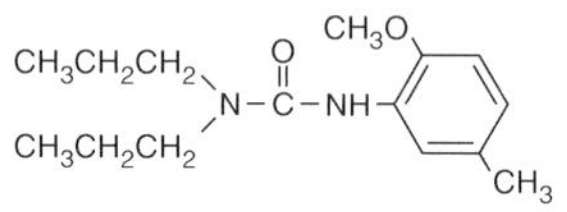

5 gm

S60,995-1
1,1-DIBENZYL-3-(2,5-DIMETHOXYPHENYL)UREA

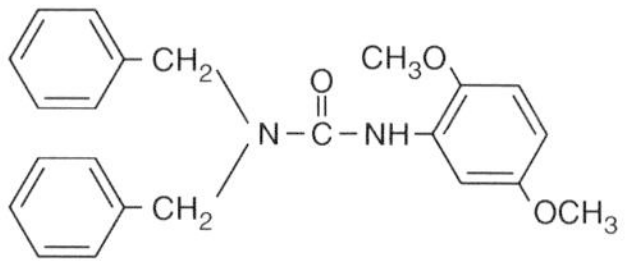

5 gm

S69,918-7
1-(2-MERCAPTOPHENYL)-3-[4-(METHYLTHIO)PHENYL]UREA

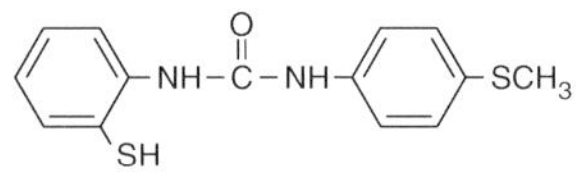

250 mg

S69,611-0
1-(ETHOXYCARBONYLMETHYL)-3-(2-MERCAPTOPHENYL)UREA

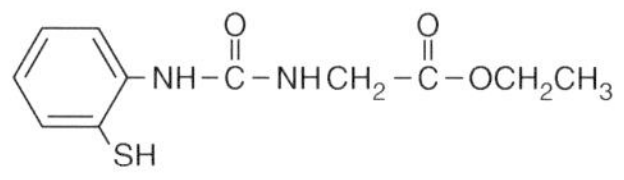

250 mg

S79,793-6
ETHYL 2-(DIETHYLAMINO)-4-(3-[4-(METHYLTHIO)PHENYL]UREIDO)BENZOATE

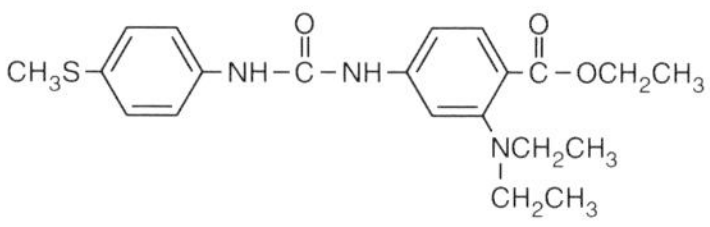

250 mg

S72,273-1
1-[2-(2,4-DINITROPHENYLTHIO)PHENYL]-3-[4-(METHYLTHIO)PHENYL]UREA

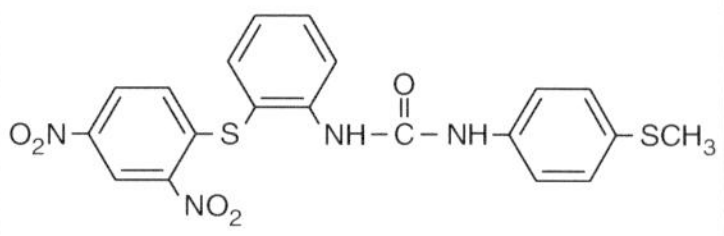

5 gm

S69,334-0
1-[2-(2,4-DINITROPHENYLTHIO)PHENYL]-3-(2-ETHYLPHENYL)UREA

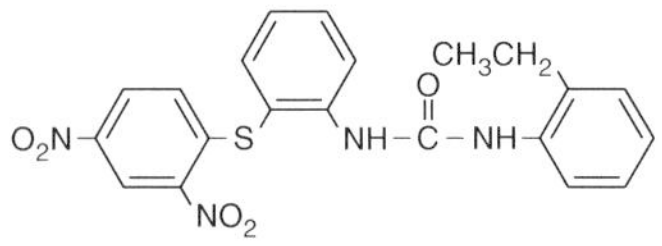

250 mg

S72,349-5
1-[2-(2,4-DINITROPHENYLTHIO)PHENYL]-3-(1-NAPHTHYL)UREA

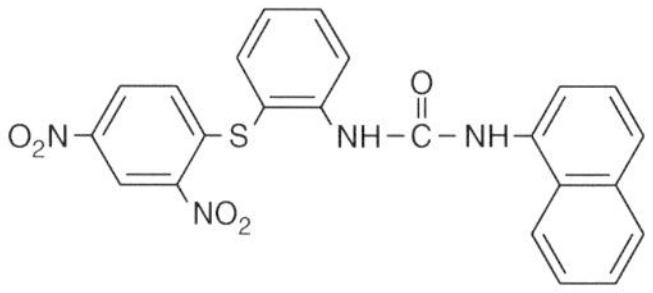

5 gm

S90,573-9
1-(2-CARBAMOYLPHENYL)-3-[4-(METHYLTHIO)PHENYL]UREA

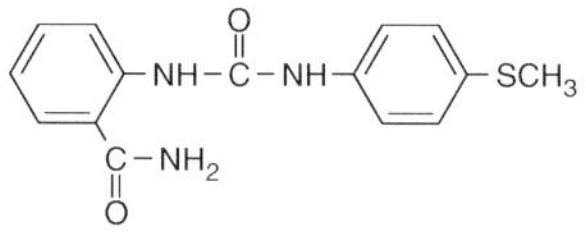

250 mg

S69,919-5
1-[4-(DIMETHYLAMINO)PHENYL]-3-[3-(METHYLTHIO)PHENYL]UREA

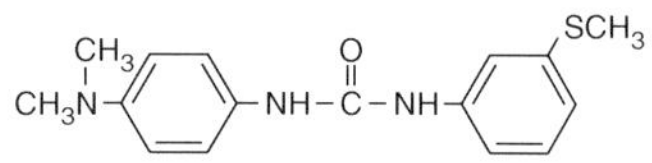

250 mg

S79,746-4
1-[4-(DIMETHYLAMINO)PHENYL]-3-[4-(METHYLTHIO)PHENYL]UREA

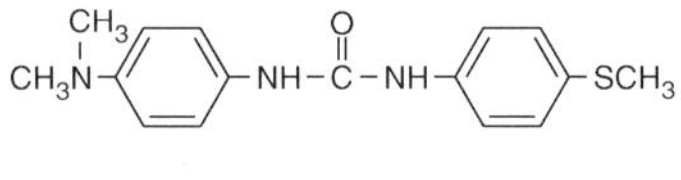

250 mg

S91,705-2
4-[3-(METHYLTHIO)PHENYL]-1-(4-NITROPHENYL)SEMICARBAZIDE

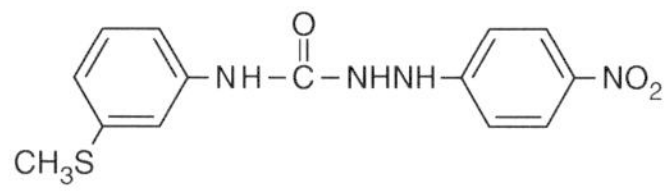

250 mg

S79,722-7
1-(2-METHYL-4-NITROPHENYL)-3-[4-(METHYLTHIO)PHENYL]UREA

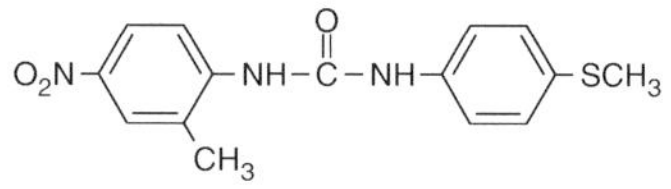

250 mg

S79,740-5
1-(4-ACETAMIDOPHENYL)-3-[4-(METHYLTHIO)PHENYL]UREA

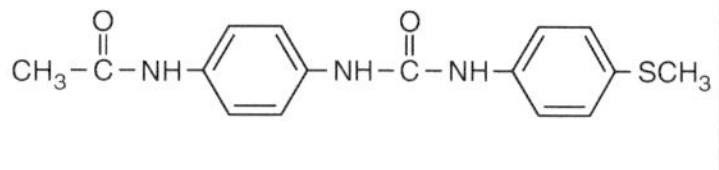

250 mg

S76,946-0
1-(4-ANILINOPHENYL)-3-[4-(METHYLTHIO)PHENYL]UREA

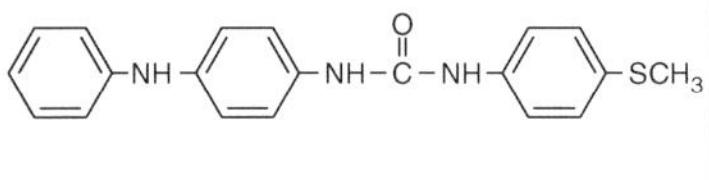

250 mg

S79,792-8
1-BENZYL-3-[4-(METHYLTHIO)PHENYL]-1-PHENYLUREA

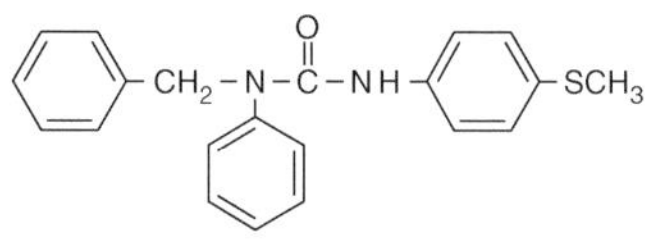

250 mg

S74,778-5
1-(1-NAPHTHYL)-3-[2-(4-NITROPHENYLTHIO)PHENYL]UREA

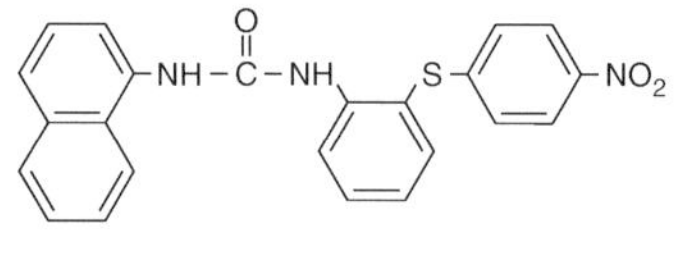

5 gm

S72,771-7
1-(2-ACETYLPHENYL)-3-[4-(METHYLTHIO)PHENYL]UREA

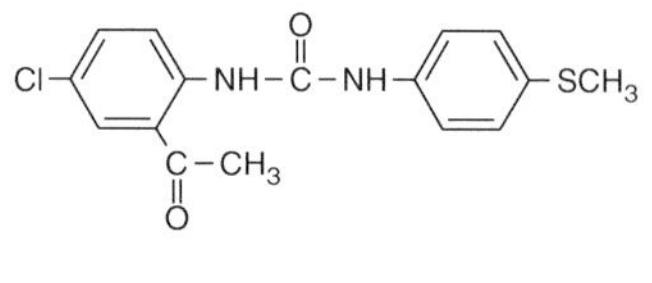

5 gm

S72,298-7
1-(4-ACETYLPHENYL)-3-[4-(METHYLTHIO)PHENYL]UREA

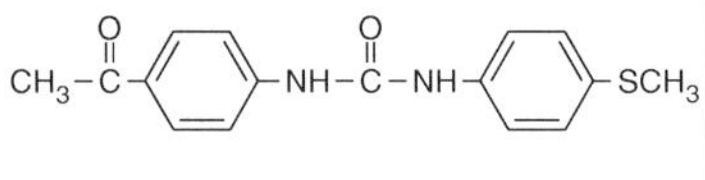

250 mg

S86,809-4
1-(3-ACETYLPHENYL)-3-[4-(METHYLTHIO)PHENYL]UREA

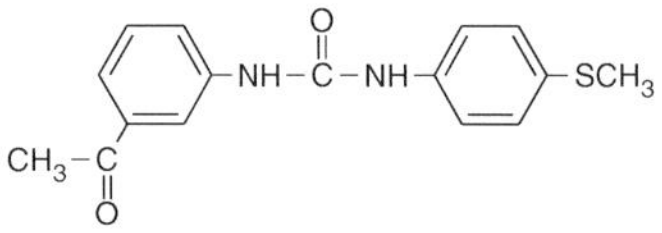

250 mg

S79,736-7
METHYL 2-(3-[4-(METHYLTHIO)PHENYL]-UREIDO)BENZOATE

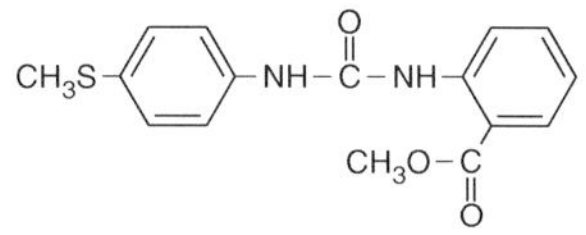

250 mg

S79,752-9
ETHYL 4-(3-[4-(METHYLTHIO)PHENYL]-UREIDO)BENZOATE

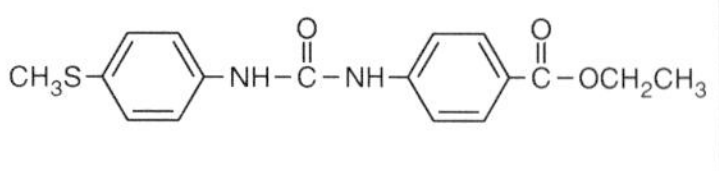

250 mg

S79,779-0
BUTYL 4-(3-[4-(METHYLTHIO)PHENYL]-UREIDO)BENZOATE
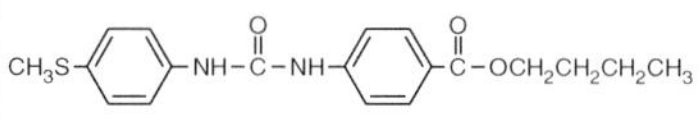
250 mg

S85,352-6
1-(2-CARBAMOYLPHENYL)-3-[3-(METHYLTHIO)PHENYL]UREA
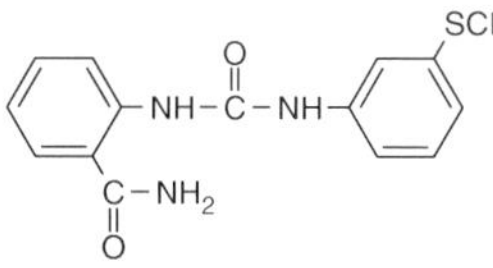
250 mg

S69,310-3
1,1-DIMETHYL-4-(4-[METHYLTHIO)-PHENYL]SEMICARBAZIDE
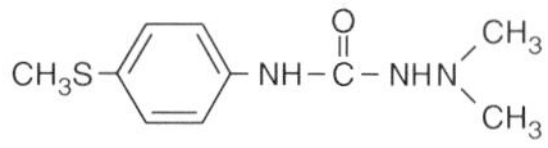
250 mg

S91,736-2
1,1-DIMETHYL-3-[4-(METHYLTHIO)-PHENYL]UREA
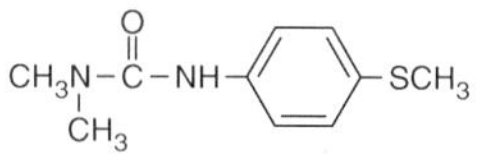
250 mg

S91,704-4
1-METHYL-3-[3-(METHYLTHIO)PHENYL]-1-PROPYLUREA
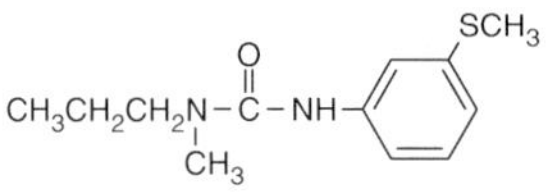
250 mg

S77,356-5
1-CYCLOPROPYL-3-[4-(METHYLTHIO)-PHENYL]UREA
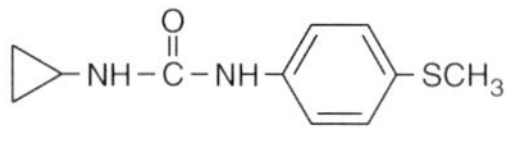
250 mg

S77,288-7
1-CYCLOPENTYL-3-[4-(METHYLTHIO)-PHENYL]UREA
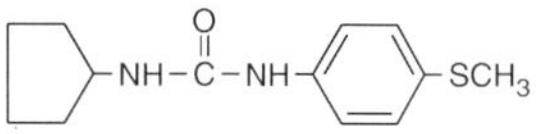
250 mg

S79,761-8
1-CYCLOHEXYL-1-ISOPROPYL-3-[4-(METHYLTHIO)PHENYL]UREA
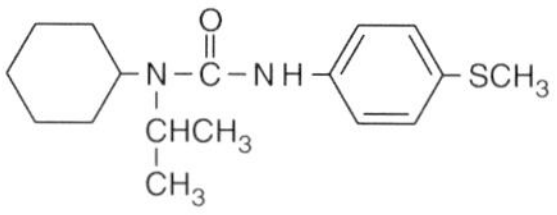
250 mg

S39,339-8
1-[4-(METHYLTHIO)PHENYL-3-PHENYL]UREA
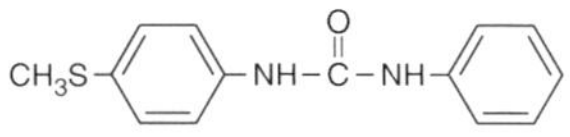
5 gm

S78,082-0
1-[4-(METHYLTHIO)PHENYL]-3-(O-TOLYL)UREA
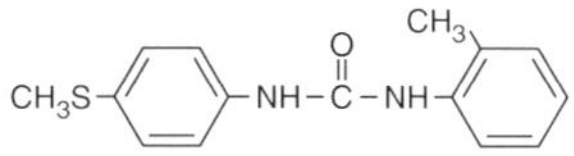
250 mg

S89,288-2
1-[4-(METHYLTHIO)PHENYL]UREA
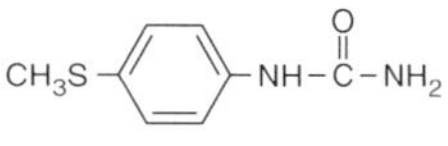
250 mg

S91,190-9
1-METHOXY-1-METHYL-3-[4-(METHYLTHIO)PHENYL]UREA
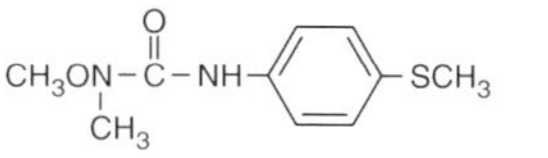
250 mg

S82,905-6
1-[4-(M-TOLYLTHIO)PHENYL]UREA
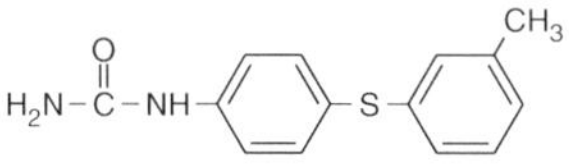
250 mg

S72,557-9
1-[4-(METHYLTHIO)PHENYL]-3-(2,4,5-TRIMETHYLPHENYL)UREA
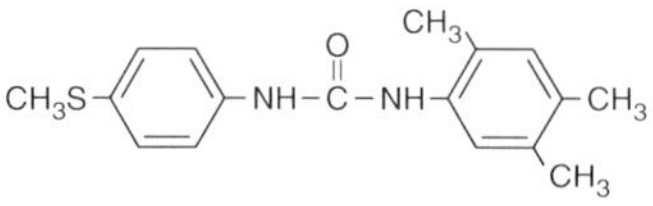
5 gm

S86,897-3
1,1'-HEXAMETHYLENEBIS(3-[4-(METHYLTHIO)PHENYL]UREA)
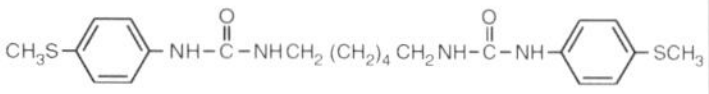
250 mg

S72,667-2
1-[4-(METHYLTHIO)PHENYL]-3-OCTADECYLUREA
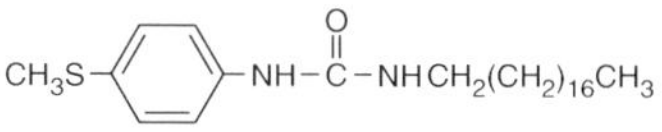
250 mg

S79,744-8
1-[4-(METHYLTHIO)PHENYL]-3-(2,3-XYLYL)UREA
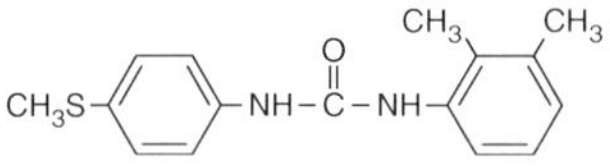
250 mg

S79,777-4
1-(5-ISOPROPYL-2-METHYLPHENYL)-3-[4-(METHYLTHIO)PHENYL]UREA
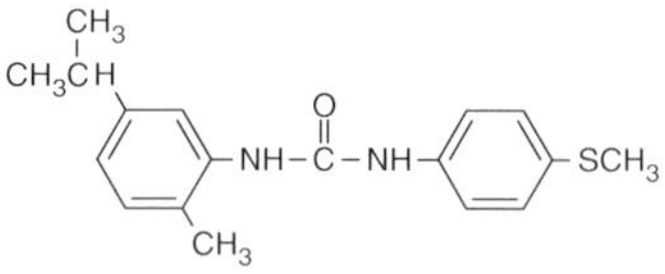
250 mg

S79,755-3
1-[4-(METHYLTHIO)PHENYL]-3-(2,4,6-TRIMETHYLPHENYL)UREA
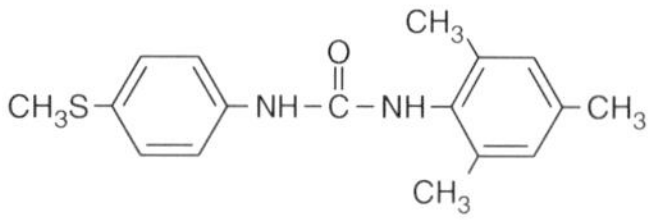
250 mg

S69,400-2
1-(2,4-DINITROPHENYL)-4-(2-ETHYLPHENYL)SEMICARBAZIDE
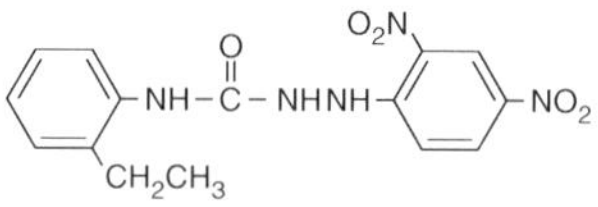
250 mg

S74,732-7
1-(2,4-DINITROPHENYL)-1-METHYL-4-(O-TOLYL)SEMICARBAZIDE
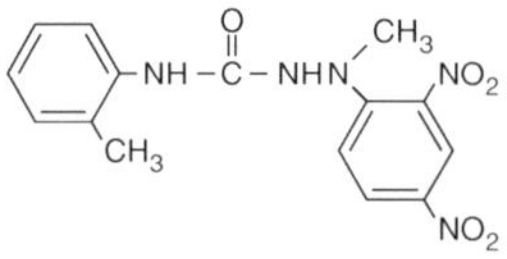
5 gm

S72,849-7
1-(2,4-DINITROPHENYL)-4-METHYLSEMICARBAZIDE
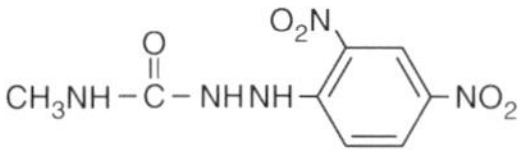
5 gm

S74,720-3
1-(2,4-DINITROPHENYL)-4-(2,4-XYLYL)-SEMICARBAZIDE
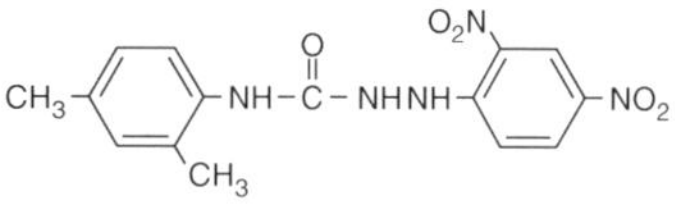
5 gm

S34,751-5
1-(2,4-DINITROPHENYL)SEMICARBAZIDE
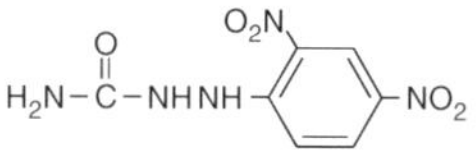
250 mg

S72,836-5
1-(2,4-DINITROPHENYL)-4-PHENYLSEMICARBAZIDE
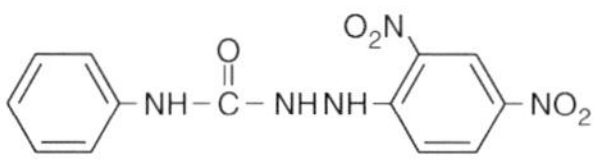
5 gm

S75,040-9
1-(4-ANILINOPHENYL)-3-(4-METHYL-2-NITROPHENYL)UREA
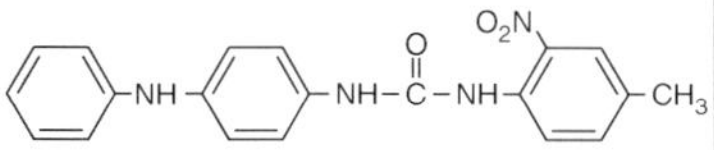
5 gm

S79,666-2
1-[4-(DIMETHYLAMINO)PHENYL]-3-(4-METHYL-2-NITROPHENYL)UREA
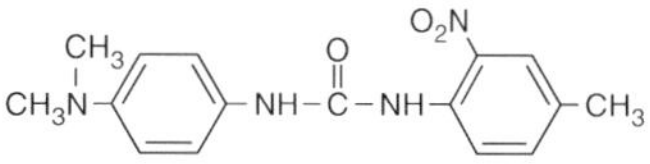
250 mg

S76,994-0
1-[3-(1-ETHYL-3-PIPERIDINYL)]-3-(3,4-XYLYL)UREA
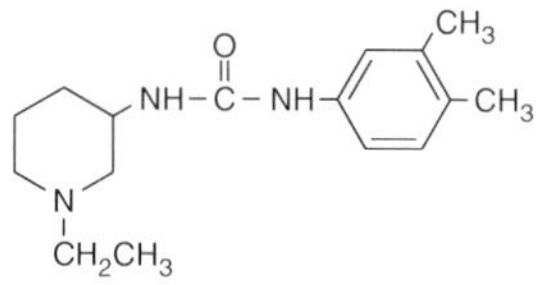
250 mg

S85,331-3
1-(4-ACETAMIDOPHENYL)-3-(4-METHYL-2-NITROPHENYL)UREA
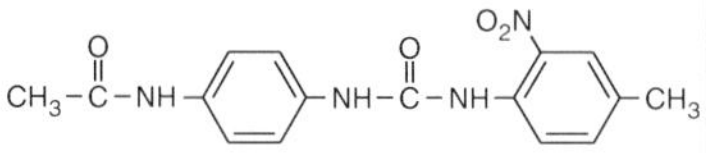
250 mg

S85,335-6
1-[4-(N-METHYLACETAMIDO)PHENYL]-3-(4-METHYL-2-NITROPHENYL)UREA
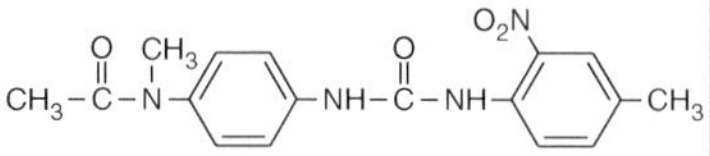
250 mg

S90,233-0
2-[3-(3-NITROPHENYL)UREIDO]ACETIC [4-(DIMETHYLAMINO)BENZYLIDENE]-HYDRAZIDE
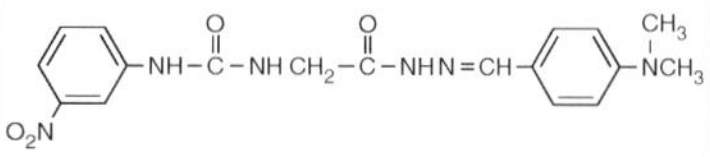
250 mg

S90,300-0
2-(3-BENZYLUREIDO)ACETIC [4-(DIMETHYLAMINO)BENZYLIDENE]-HYDRAZIDE
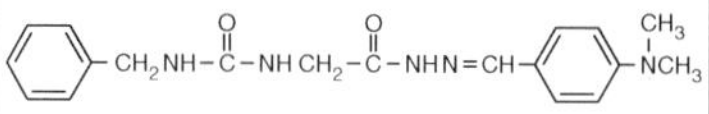
250 mg

S90,064-8
1-[2-(N-METHYLANILINO)ETHYL]-3-PHENYLUREA
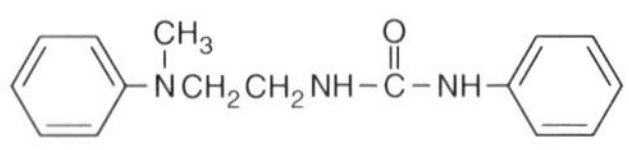
250 mg

S38,224-8
1-(6-CHLORO-ORTHO-TOLYL)-3-[2-(DIETHYLAMINO)ETHYL]UREA
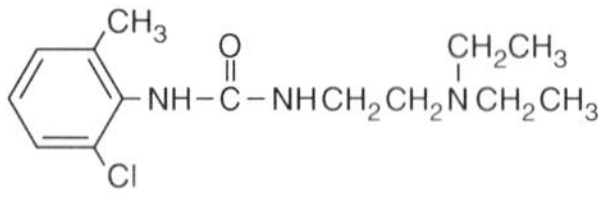
250 mg

S89,217-3
1-[2-(N-METHYLANILINO)ETHYL]-3-OCTADECYLUREA
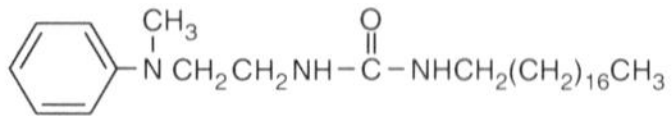
250 mg

S84,732-1
1-[2-(N-METHYLANILINO)ETHYL]-3-(2,5-XYLYL)UREA
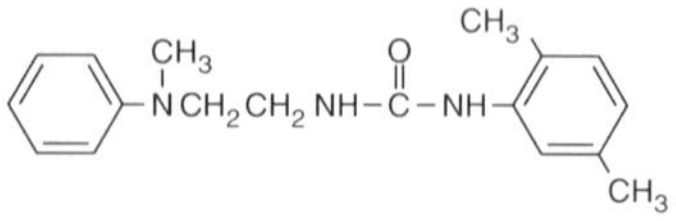
250 mg

S85,332-1
1-BENZYL-1-METHYL-3-(4-METHYL-2-NITROPHENYL)UREA
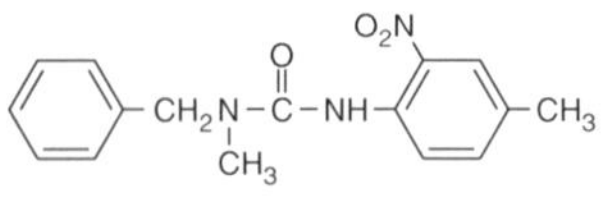
250 mg

S85,337-2
1,1-DIBENZYL-3-(4-METHYL-2-NITROPHENYL)UREA
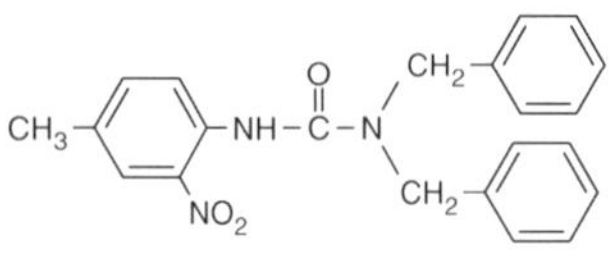
250 mg

S85,336-4
1-(3,4-DIMETHYLBENZYL)-3-(4-METHYL-2-NITROPHENYL)UREA
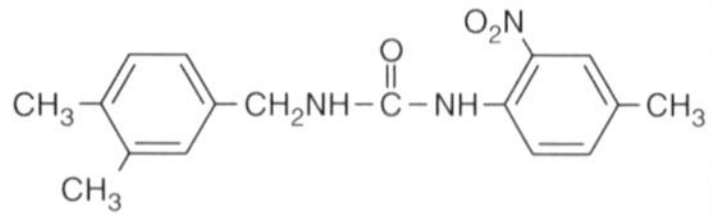
250 mg

S72,831-4
1,6-BIS(4-METHYL-2-NITROPHENYL)-BIUREA
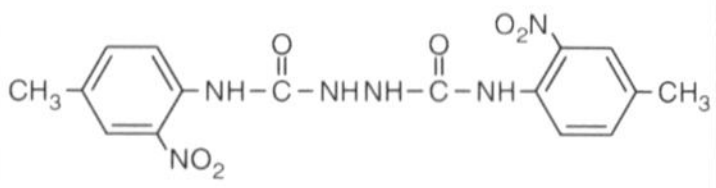
5 gm

S91,304-9
1-(4-CARBOXYPHENYL)-4-(4-METHYL-2-NITROPHENYL)SEMICARBAZIDE
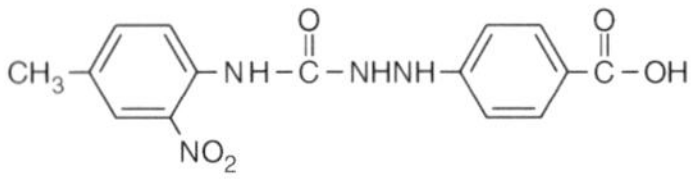
250 mg

S79,644-1
1-(2-CARBAMOYLPHENYL)-3-(4-METHYL-2-NITROPHENYL)UREA
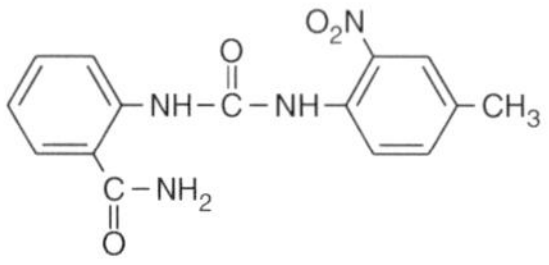
250 mg

S89,780-9
1,1-DIALLYL-3-(2-METHYL-6-NITROPHENYL)UREA
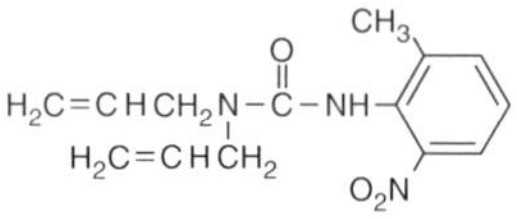
250 mg

S85,327-5
1-[2-(3,4-DIMETHYLTETRAHYDRO-FURYL)]-3-(4-METHYL-2-NITROPHENYL)-UREA
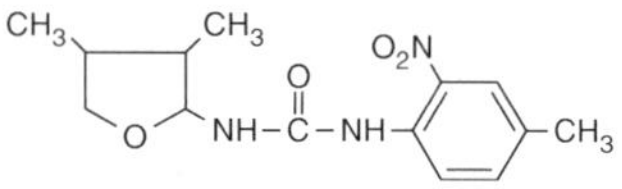
250 mg

S68,430-9
1,1-DIPROPYL-3-(2-METHYL-6-NITROPHENYL)UREA
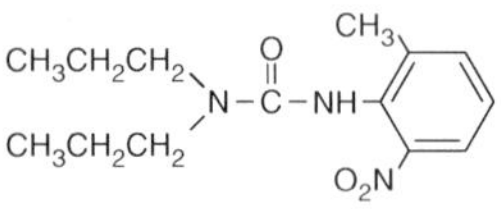
250 mg

S84,857-3
1-(4-ANILINOPHENYL)-3-(2-METHYL-5-NITROPHENYL)UREA
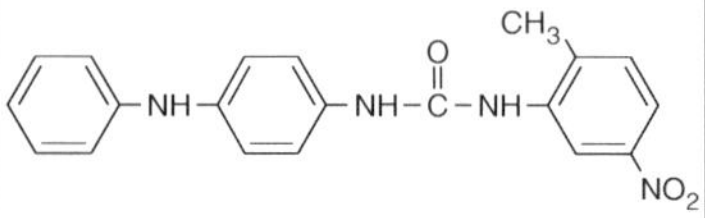
250 mg

S84,839-5
1-[4-(DIMETHYLAMINO)PHENYL]-3-(2-METHYL-5-NITROPHENYL)UREA
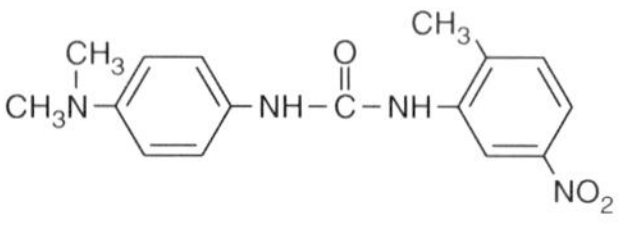
250 mg

S80,046-5
1-(2-METHOXY-5-NITROPHENYL)-3-[4-(N-METHYLACETAMIDO)PHENYL]UREA
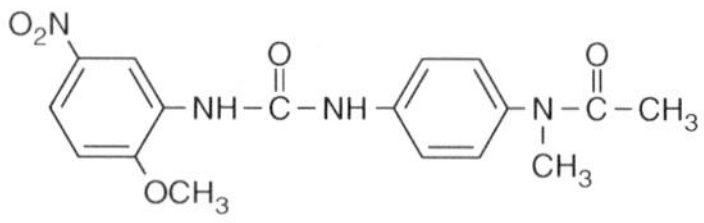
250 mg

S71,565-4
1-(4-ACETAMIDOPHENYL)-3-(4-METHYL-3-NITROPHENYL)UREA
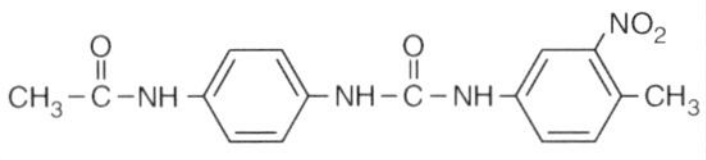
250 mg

S79,663-8
1-(4-ACETAMIDOPHENYL)-3-(2-METHYL-5-NITROPHENYL)UREA
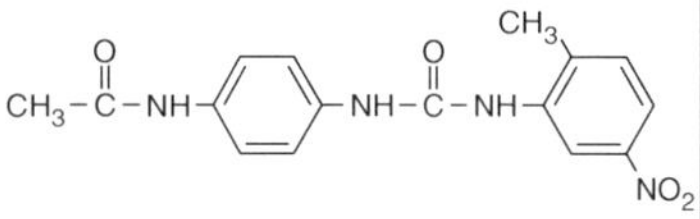
250 mg

S84,846-8
1-[4-(N-METHYLACETAMIDO)PHENYL]-3-(2-METHYL-5-NITROPHENYL)UREA
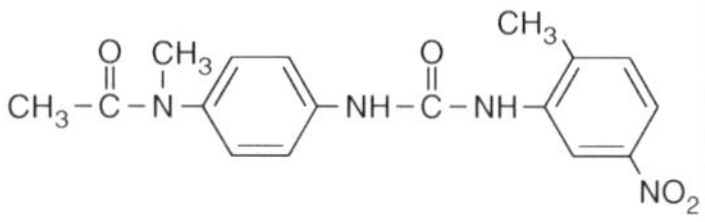
250 mg

S85,367-4
1-[4-(N-METHYLACETAMIDO)PHENYL]-3-(4-METHYL-3-NITROPHENYL)UREA
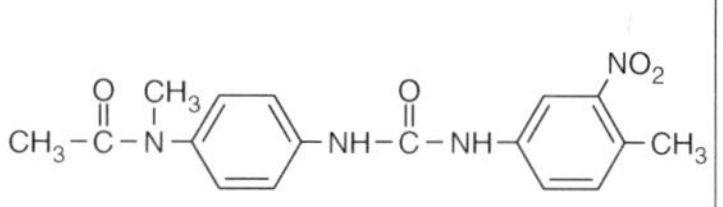
250 mg

S60,177-2
1-(4-ACETAMIDOPHENYL)-3-(3-NITROPHENYL)UREA
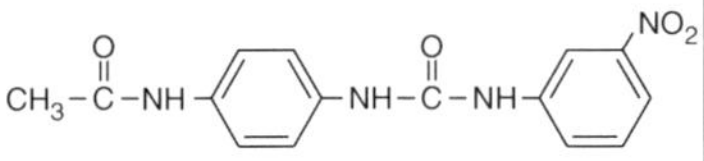
5 gm

S60,800-9
1-(2-CHLORO-4-NITROPHENYL)-3-(3-NITROPHENYL)UREA
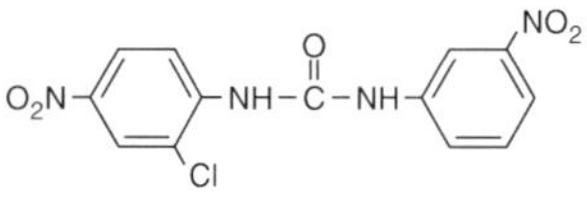
5 gm

S61,304-5
1-(2-METHYL-4-NITROPHENYL)-3-(4-METHYL-3-NITROPHENYL)UREA
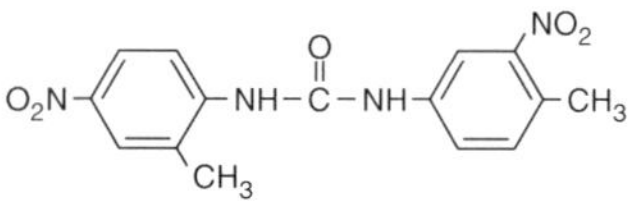
5 gm

S90,234-9
2-[3-(3-NITROPHENYL)UREIDO]ACETIC (2-PHENYLETHYLIDENE)HYDRAZIDE
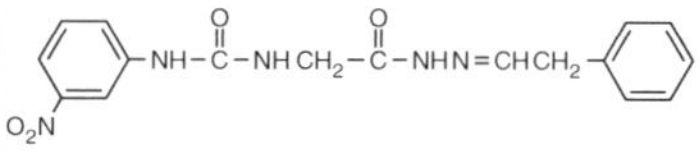
250 mg

S71,393-7
1-(3-NITROPHENYL)-3-(2-PHENETHYL)UREA
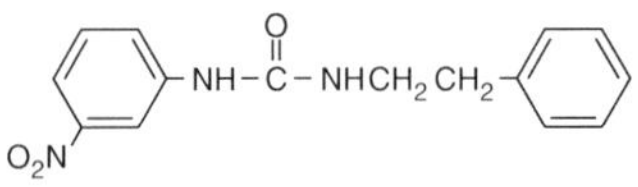
5 gm

S67,076-6
1-(4-METHYL-3-NITROPHENYL)-3-PHENETHYLUREA
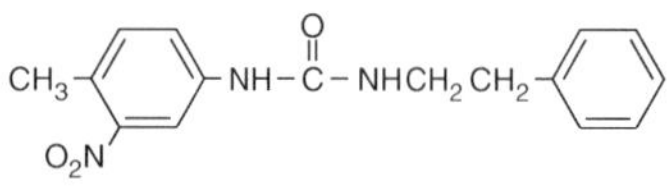
5 gm

S84,797-6
1-(2-METHYL-5-NITROPHENYL)-3-PHENETHYLUREA
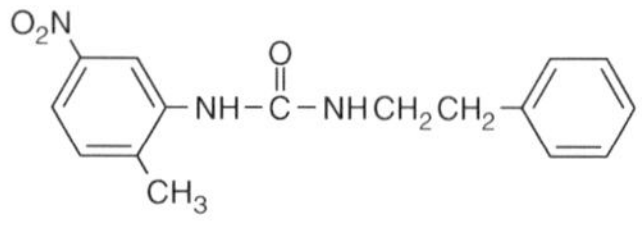
250 mg

S64,948-1
1,1'-(4-METHYL-1,3-PHENYLENE)BIS(3-PHENETHYLUREA)
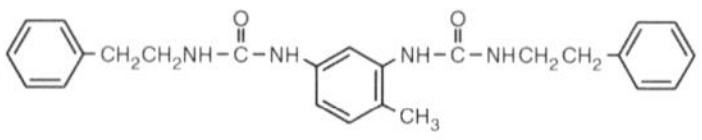
5 gm

S76,326-8
1-METHYL-[3-(2-N-METHYLANILINOMETHYL)PHENYL]-1-PHENYLUREA
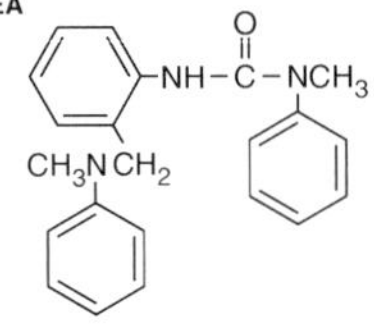
250 mg

S54,375-6
1-[3-(DIMETHYLAMINO)PROPYL]-3-PHENYLUREA
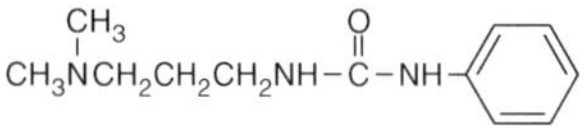
5 gm

S67,080-4
1-BENZYL-1-METHYL-3-(4-METHYL-3-NITROPHENYL)UREA
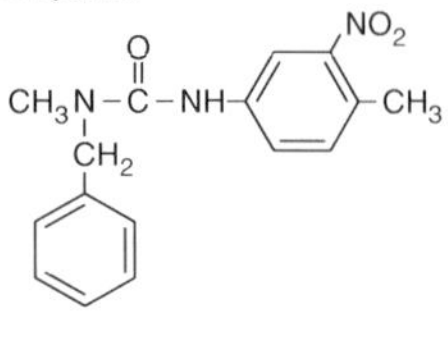
5 gm

S67,081-2
1-BENZYL-1-ISOPROPYL-3-(4-METHYL-3-NITROPHENYL)UREA
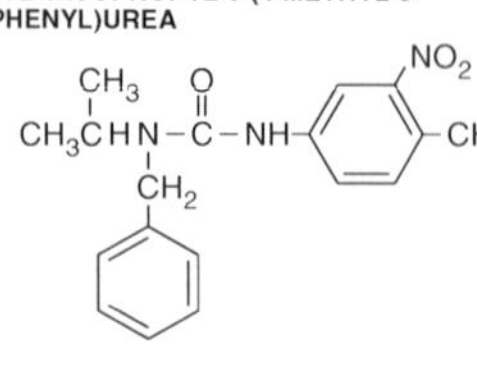
5 gm

S67,020-0
1,1-DIBENZYL-3-(4-METHYL-3-NITROPHENYL)UREA

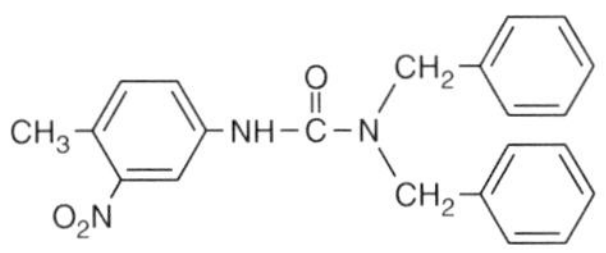

5 gm

S84,810-7
1,1-DIBENZYL-3-(2-METHYL-5-NITROPHENYL)UREA

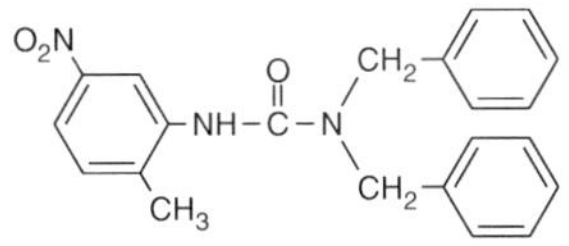

250 mg

S63,398-4
1-BENZYL-3-(3-NITROPHENYL)UREA

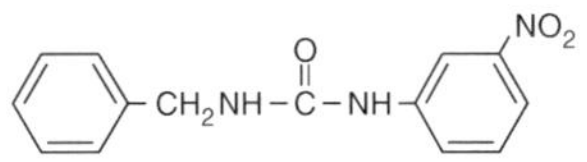

250 mg

S62,526-4
3,3'-(4-METHYL-1,3-PHENYLENE)BIS-(1,1-DIBENZYLUREA)

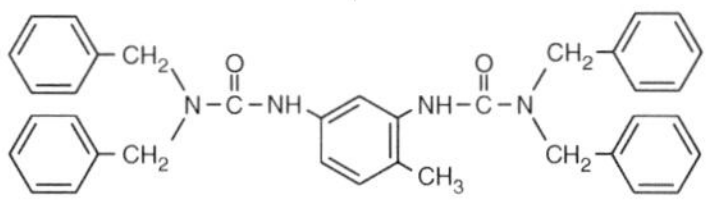

5 gm

S64,472-2
1,1'-(4-METHYL-1,3-PHENYLENE)BIS(3-BENZYLUREA)

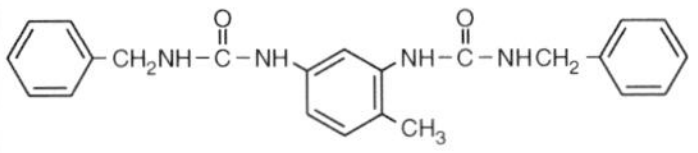

5 gm

S66,002-7
3,3'-(4-METHYL-1,3-PHENYLENE)BIS(1-BENZYL-1-METHYLUREA)

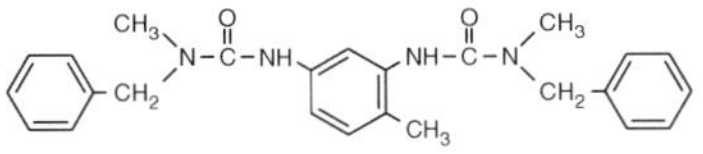

5 gm

S65,966-5
3,3'-(4-METHYL-1,3-PHENYLENE)BIS(1-BENZYL-1-ETHYLUREA)

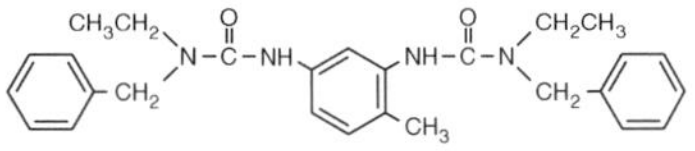

5 gm

S65,998-3
3,3'-(4-METHYL-1,3-PHENYLENE)BIS(1-BENZYL-1-ISOPROPYLUREA)

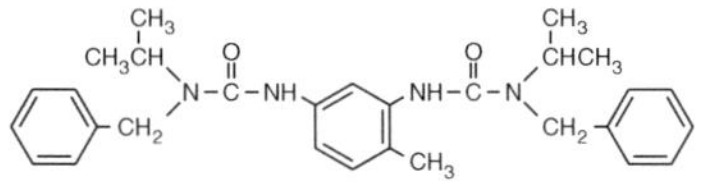

5 gm

S68,982-3
1,6-BIS(2-METHYL-5-NITROPHENYL)-BIUREA

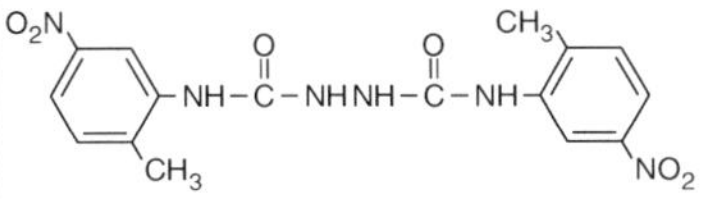

250 mg

S69,358-8
1,6-BIS(4-METHYL-3-NITROPHENYL)-BIUREA

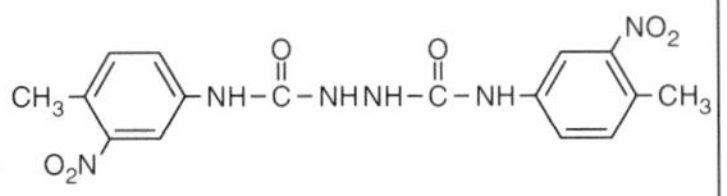

250 mg

S90,236-5
2-[3-(3-NITROPHENYL)UREIDO]ACETIC (2-BUTEN-1-YLIDENE)HYDRAZIDE

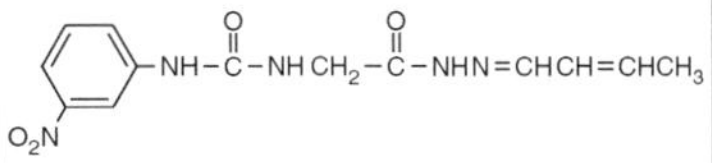

250 mg

S90,203-9
ETHYL 2-[3-(3-NITROPHENYL)UREIDO]-ACETATE

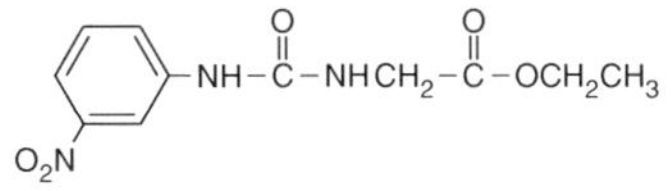

250 mg

S77,240-2
1-(4-CARBOXYPHENYL)-4-(3-NITROPHENYL)SEMICARBAZIDE

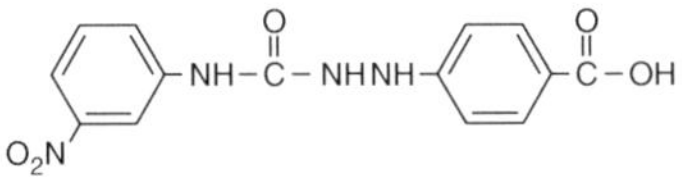

250 mg

S79,639-5
4-[3-(2-METHYL-5-NITROPHENYL)-UREIDO]BENZOIC ACID

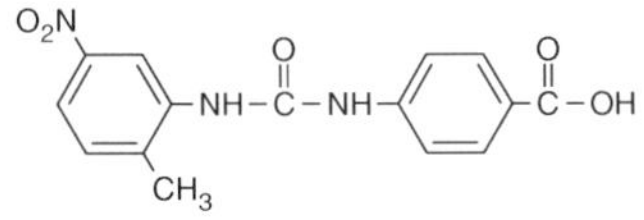

250 mg

S72,568-4
3-[3-(4-METHYL-3-NITROPHENYL)-UREIDO]BENZOIC ACID

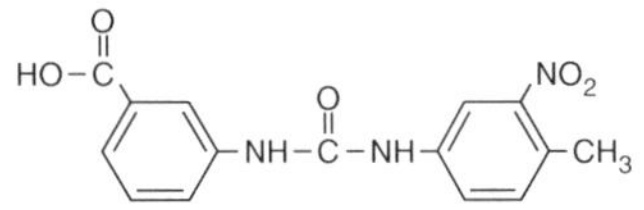

5 gm

S84,848-4
DIMETHYL 5-[3-(2-METHYL-5-NITROPHENYL)UREIDO]ISOPHTHALATE

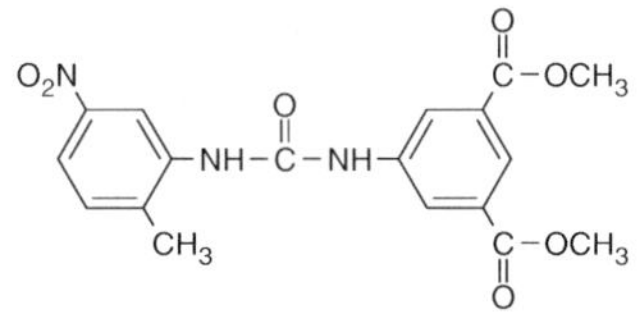

250 mg

S84,582-5
METHYL 2-[1-METHYL-3-(2-METHYL-5-NITROPHENYL)UREIDO]BENZOATE

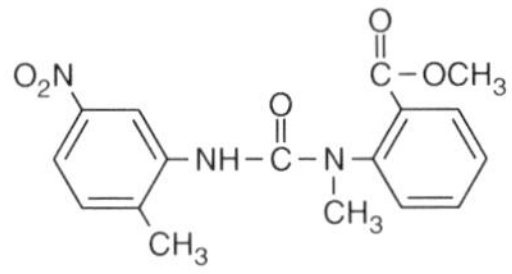

250 mg

S79,687-5
BUTYL 4-[3-(2-METHYL-5-NITROPHENYL)-UREIDO]BENZOATE

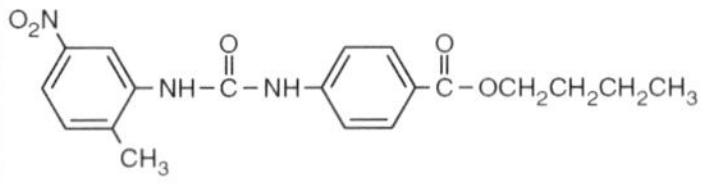

250 mg

S84,829-8
1-(2-CARBAMOYLPHENYL)-3-(2-METHYL-5-NITROPHENYL)UREA

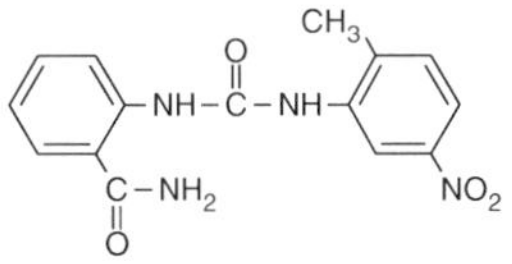

250 mg

S85,370-4
1-(2-CARBAMOYLPHENYL)-3-(4-METHYL-3-NITROPHENYL)UREA

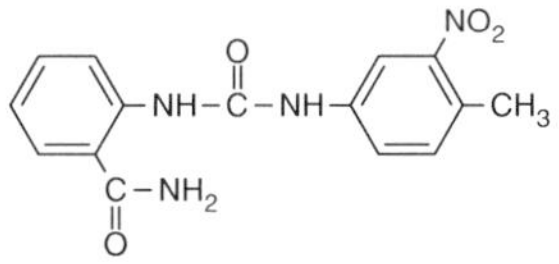

250 mg

S91,225-5
ETHYL 2-[3-(3,5-DINITROPHENYL)-UREIDO]ACETATE

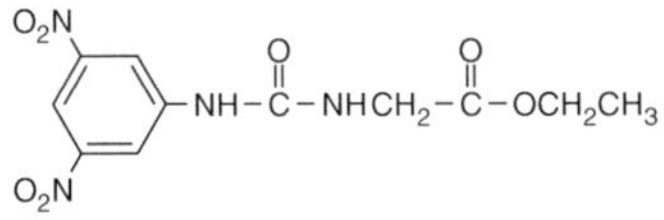

250 mg

S68,756-1
4-ETHYL-1-[3-(3-NITROPHENYL)UREIDOACETYL]-SEMICARBAZIDE

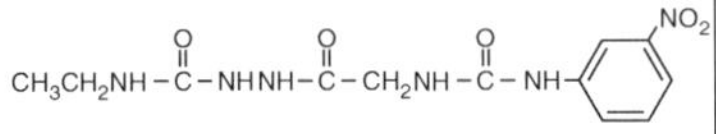

250 mg

S85,382-8
1-ACETYL-3-(4-METHYL-3-NITROPHENYL)UREA

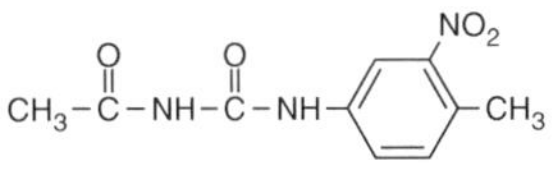

250 mg

S76,419-1
(3-NITROPHENYL)UREA

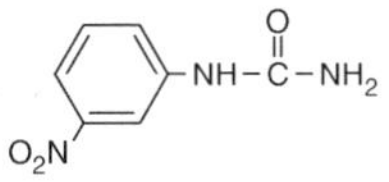

5 gm

S69,946-2
1,1'-(2-BROMO-6-METHYLPHENYLENE)-BIS[3-(3-NITROPHENYL)UREA], MIXTURE OF ISOMERS

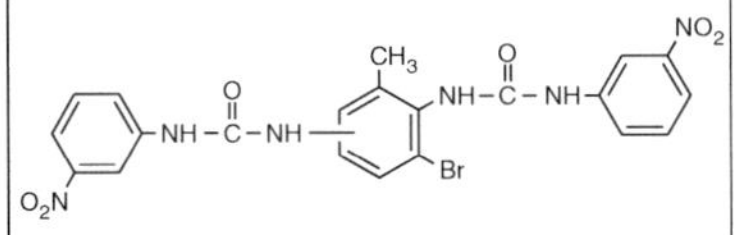

250 mg

S66,919-9
1,1-DICYCLOHEXYL-3-(3-NITROPHENYL)UREA

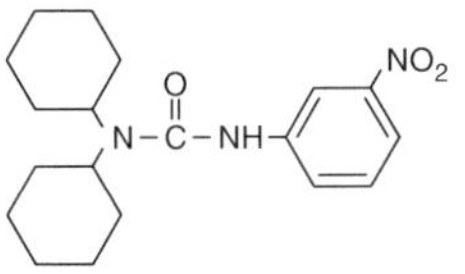

5 gm

S67,021-9
1-CYCLOHEXYL-3-(4-METHYL-3-NITROPHENYL)UREA

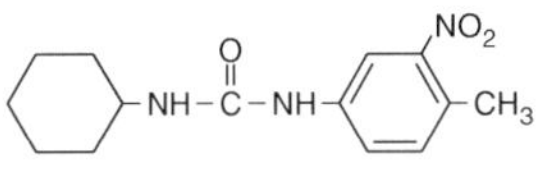

5 gm

S67,082-0
1-CYCLOHEXYL-1-ISOPROPYL-3-(4-METHYL-3-NITROPHENYL)UREA

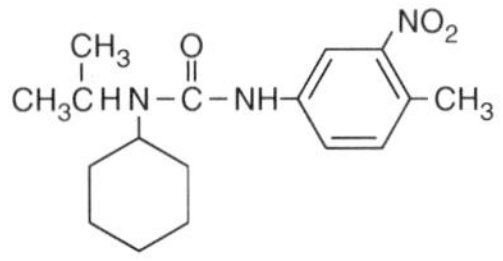

5 gm

S67,053-7
1,1-DICYCLOHEXYL-3-(4-METHYL-3-NITROPHENYL)UREA

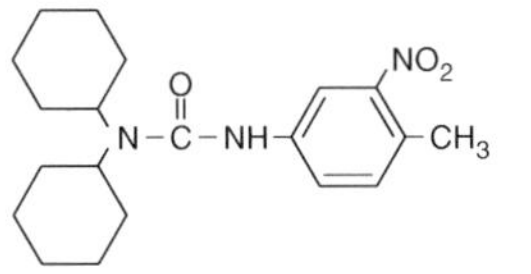

5 gm

S84,806-9
1,1-DICYCLOHEXYL-3-(2-METHYL-5-NITROPHENYL)UREA

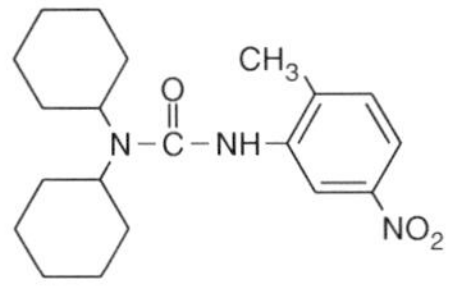

250 mg

S91,352-9
1-(DIPHENYLMETHYL)-3-(3-NITROPHENYL)UREA

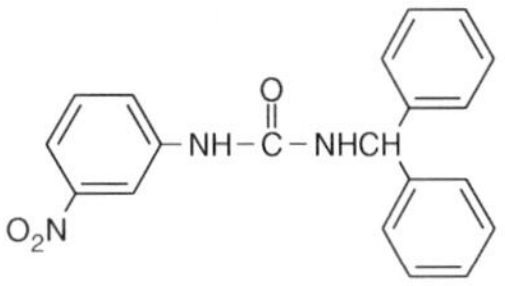

250 mg

S88,972-5
1-(2-ETHYLPHENYL)-3-(4-METHYL-3-NITROPHENYL)UREA

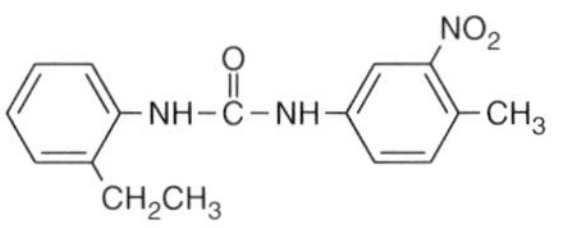

250 mg

S88,971-7
1-(4-ETHYLPHENYL)-3-(4-METHYL-3-NITROPHENYL)UREA
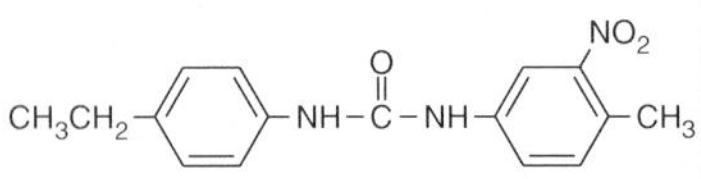
250 mg

S70,817-8
1-(2,6-DIMETHYLPHENYL)-3-(3-NITROPHENYL)UREA
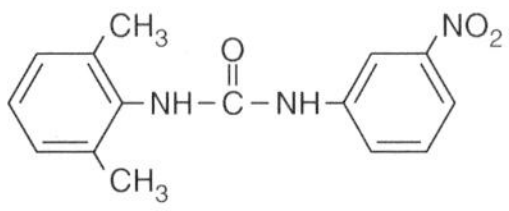
5 gm

S88,974-1
1-(4-METHYL-3-NITROPHENYL)-3-(2,3-XYLYL)UREA
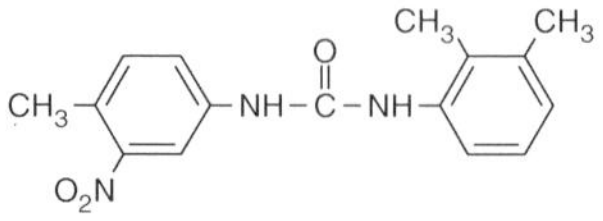
250 mg

S88,976-8
1-(4-METHYL-3-NITROPHENYL)-3-(2,6-XYLYL)UREA
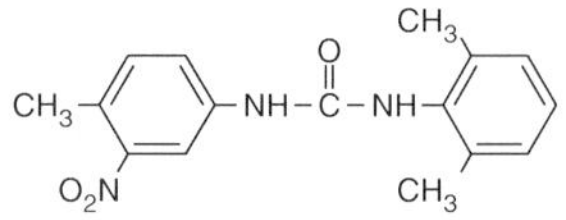
250 mg

S67,071-5
1-METHYL-3-(4-METHYL-3-NITROPHENYL)UREA
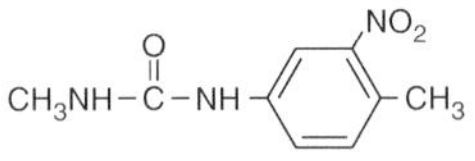
5 gm

S91,189-5
1-METHOXY-1-METHYL-3-(2-METHYL-5-NITROPHENYL)UREA
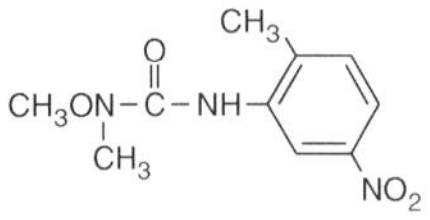
250 mg

S67,049-9
1-ALLYL-3-(4-METHYL-3-NITROPHENYL)UREA
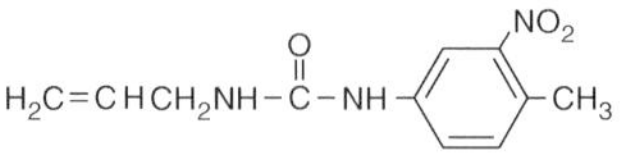
5 gm

S67,023-5
1-(4-METHYL-3-NITROPHENYL)-3-PROPYLUREA
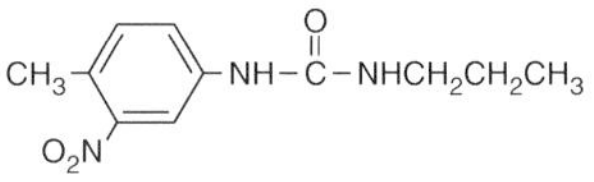
5 gm

S67,051-0
1-ISOPROPYL-3-(4-METHYL-3-NITROPHENYL)UREA
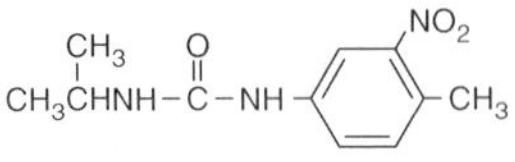
5 gm

S67,024-3
1-BUTYL-3-(4-METHYL-3-NITROPHENYL)UREA
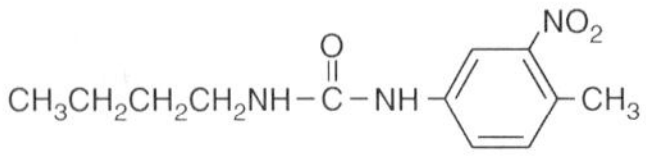
5 gm

S67,055-3
1,1-DIPROPYL-3-(4-METHYL-3-NITROPHENYL)UREA
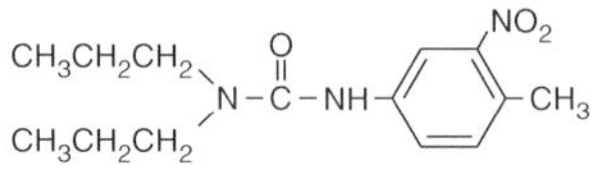
5 gm

S88,955-5
1-HEPTYL-3-(2-METHYL-5-NITROPHENYL)UREA
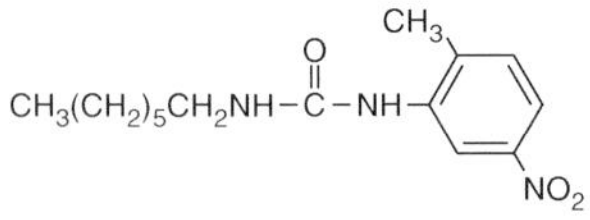
250 mg

S72,600-1
1-(2,4-DIMETHYLPHENYL)-3-(4-METHYL-3-NITROPHENYL)UREA
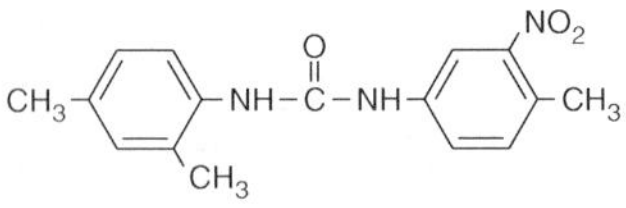
5 gm

S67,022-7
1,1-DIBUTYL-3-(4-METHYL-3-NITROPHENYL)UREA
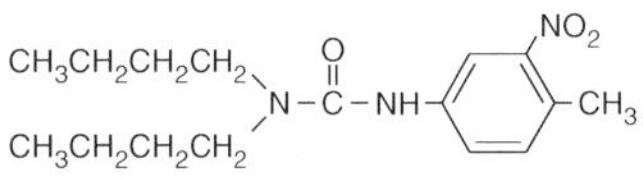
5 gm

S88,982-2
1-(4-METHYL-3-NITROPHENYL)-3-MESITYLUREA
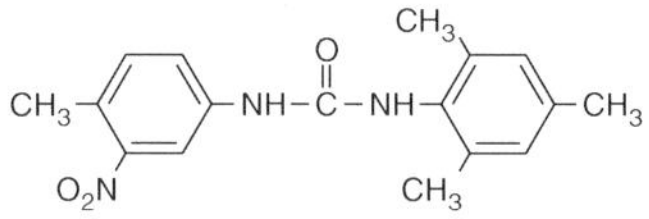
250 mg

S84,799-2
1-(7-HYDROXY-1-NAPHTHYL)-3-(2-METHYL-5-NITROPHENYL)UREA
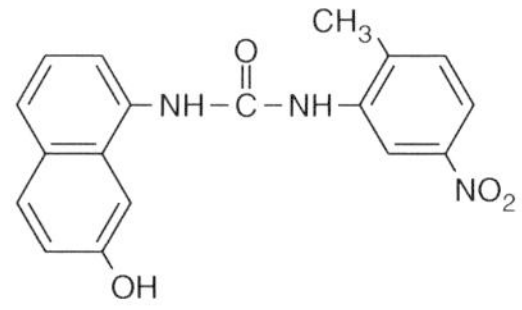
250 mg

S84,798-4
1-(5-HYDROXY-1-NAPHTHYL)-3-(2-METHYL-5-NITROPHENYL)UREA
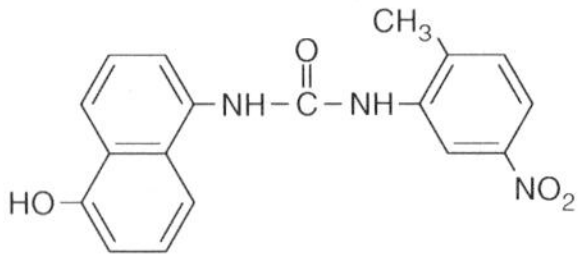
250 mg

S72,669-9
1-(1-NAPHTHYL)-3-(3-NITROPHENYL)UREA
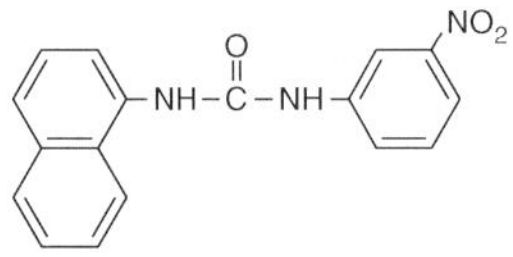
5 gm

S63,457-3
1,1-DIMETHYL-4-(3-NITROPHENYL)-SEMICARBAZIDE
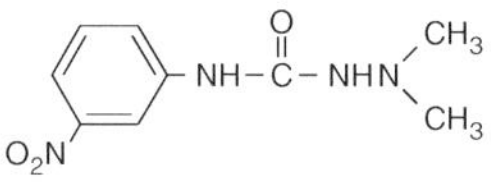
5 gm

S61,003-8
1,1-DIMETHYL-4-(4-METHYL-3-NITROPHENYL)SEMICARBAZIDE
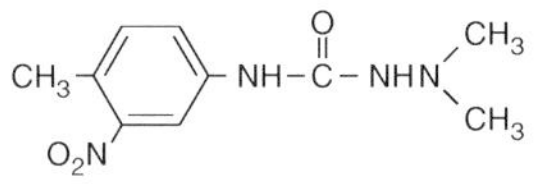
5 gm

S59,927-1
1-METHYL-3-(3-NITROPHENYL)UREA
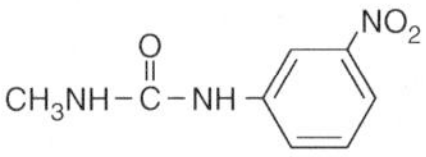
5 gm

S59,773-2
1-ETHYL-3-(3-NITROPHENYL)UREA
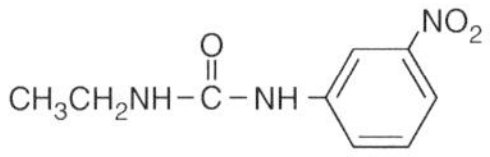
5 gm

S59,917-4
1,1-DIMETHYL-3-(3-NITROPHENYL)UREA
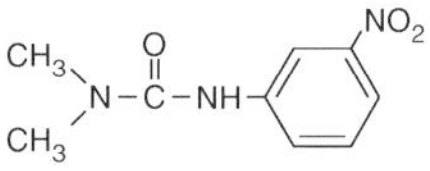
5 gm

S63,465-4
1-ALLYL-3-(3-NITROPHENYL)UREA
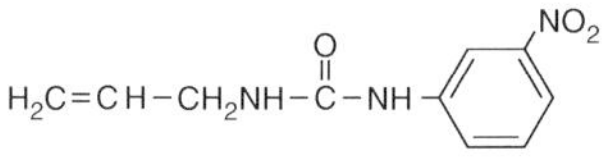
5 gm

S62,166-8
1-(3-NITROPHENYL)-3-PROPYLUREA
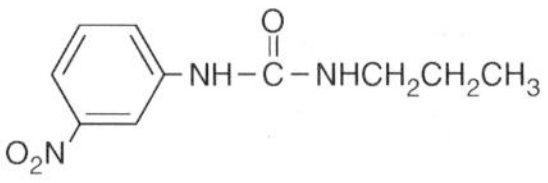
5 gm

S62,169-2
1-ISOPROPYL-3-(3-NITROPHENYL)UREA
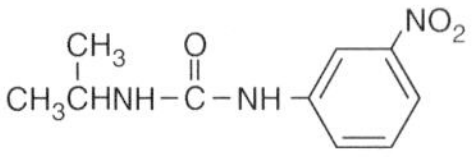
5 gm

S59,915-8
1-BUTYL-3-(3-NITROPHENYL)UREA
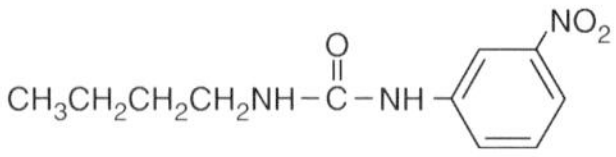
5 gm

S59,920-4
1,1-DIETHYL-3-(3-NITROPHENYL)UREA
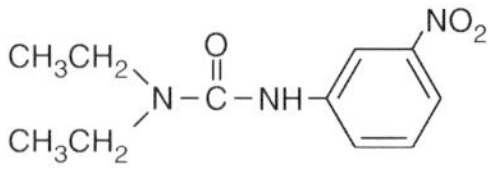
5 gm

S65,337-3
1-ISOBUTYL-3-(3-NITROPHENYL)UREA
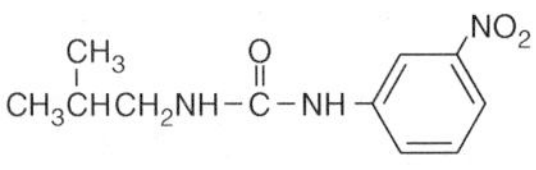
5 gm

S59,919-0
1-(3-NITROPHENYL)-3-PHENYLUREA
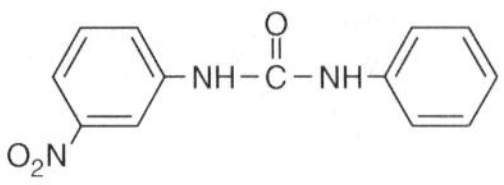
5 gm

S59,765-1
1,1-DIALLYL-3-(3-NITROPHENYL)UREA
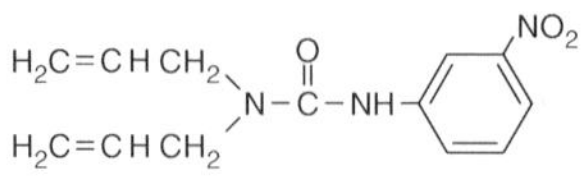
5 gm

S62,172-2
1,1-DIISOPROPYL-3-(3-NITROPHENYL)UREA
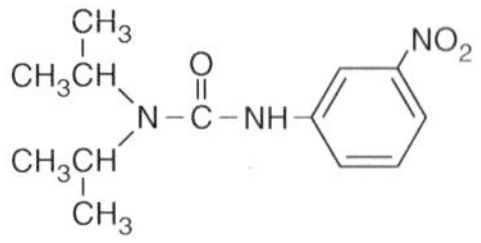
5 gm

S63,399-2
1,1-DIPROPYL-3-(3-NITROPHENYL)UREA

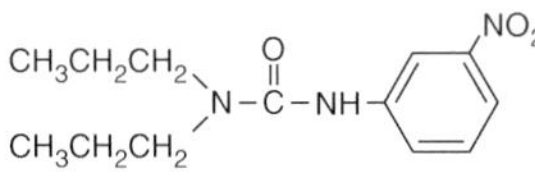

5 gm

S59,751-1
1-(3-CHLOROPHENYL)-3-(3-NITROPHENYL)UREA

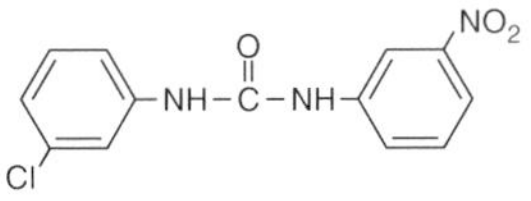

5 gm

S62,164-1
1,1-DIBUTYL-3-(3-NITROPHENYL)UREA

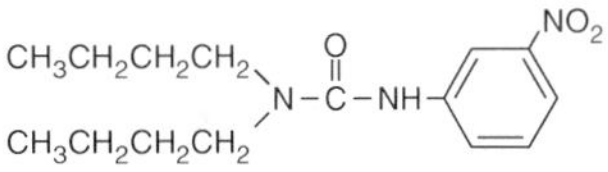

5 gm

S59,764-3
1,1-DIBENZYL-3-(3-NITROPHENYL)UREA

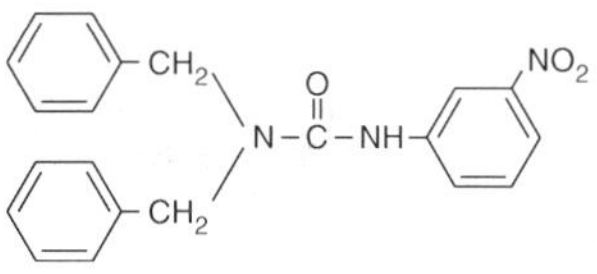

5 gm

S63,406-9
1-CYCLOHEXYL-3-(3-NITROPHENYL)UREA

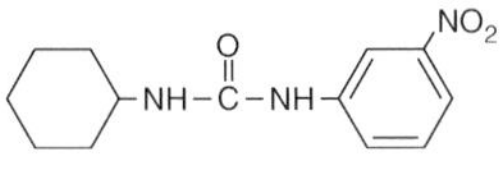

5 gm

S63,467-0
1-CYCLOHEXYL-1-METHYL-3-(3-NITROPHENYL)UREA

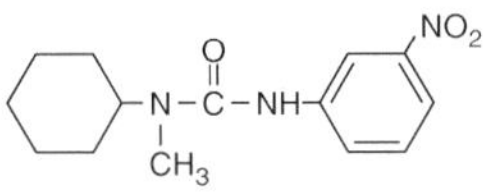

5 gm

S61,134-4
1,1-DIMETHYL-3-(4-METHYL-3-NITROPHENYL)UREA

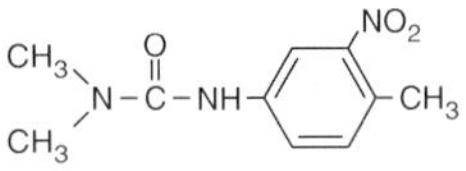

5 gm

S61,115-8
1,1-DIISOPROPYL-3-(4-METHYL-3-NITROPHENYL)UREA

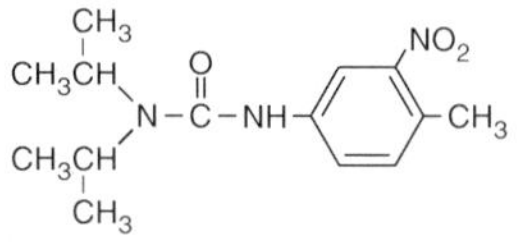

5 gm

S61,013-5
1,1-DIETHYL-3-(4-METHYL-3-NITROPHENYL)UREA

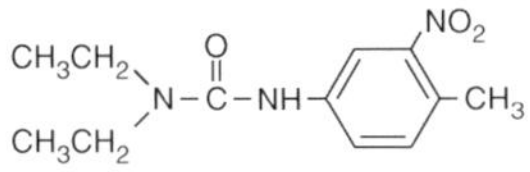

5 gm

S65,992-4
1,1'-(4-METHYL-1,3-PHENYLENE)BIS[3-(4-BENZOYLPHENYL)UREA]

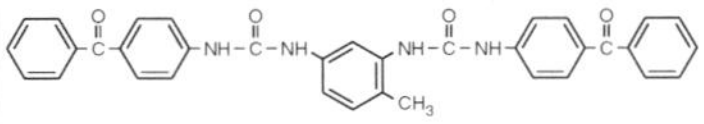

5 gm

S72,891-8
ETHYL 4-[3-(3-ACETAMIDOPHENYL)UREIDO]BENZOATE

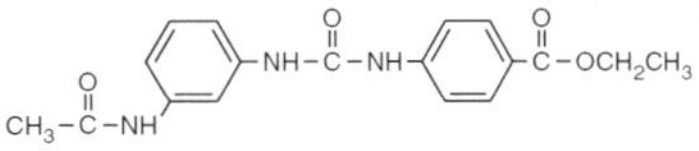

5 gm

S68,641-7
1-(2-CARBAMOYLPHENYL)-3-(3-ETHYLPHENYL)UREA

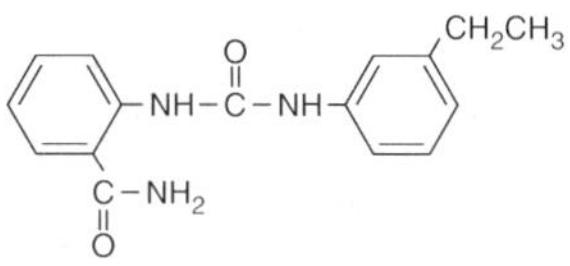

250 mg

S87,106-0
1,3-DIPHENYL-1-(2-MORPHOLINO-1-CYCLOPENTEN-1-YLCARBONYL)UREA

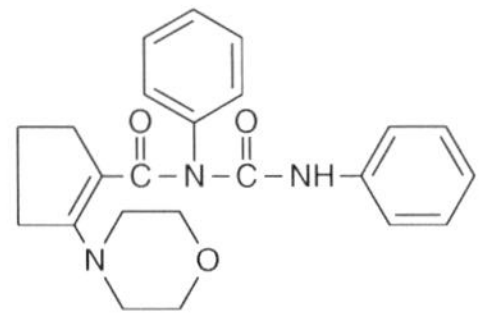

250 mg

S87,110-9
1,3-DIPHENYL-1-(2-MORPHOLINO-1-CYCLOHEXEN-1-YLCARBONYL)UREA

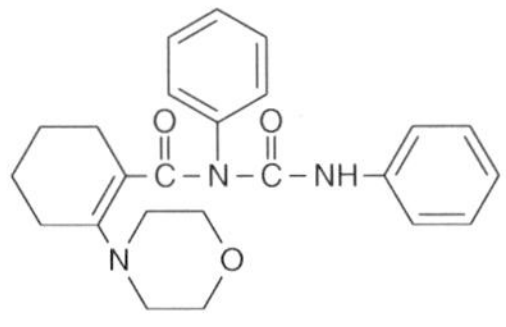

250 mg

S63,436-0
4,4'-(4-METHYL-1,3-PHENYLENE)BIS-(1,1-DIMETHYLSEMICARBAZIDE)

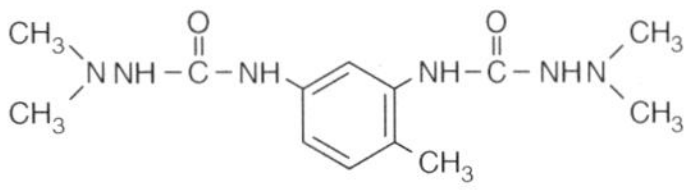

5 gm

S58,625-0
4,4'-(4-METHYL-1,3-PHENYLENE)BIS(1-PHENYLSEMICARBAZIDE)

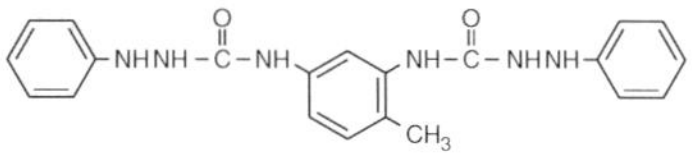

5 gm

S72,489-0
1-(3-ACETAMIDOPHENYL)-3-OCTADECYLUREA

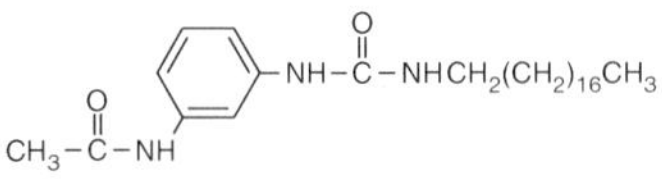

250 mg

S73,541-8
1,1'-(4-METHYL-1,3-PHENYLENE)BIS(3-PROPIONYLUREA)

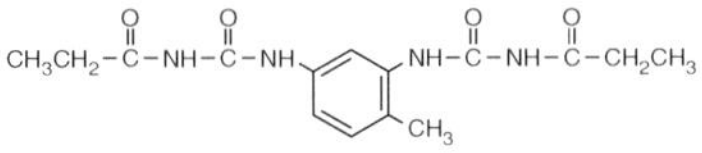

250 mg

S67,481-8
1-(3-ACETAMIDOPHENYL)-3-(1-NAPHTHYL)UREA

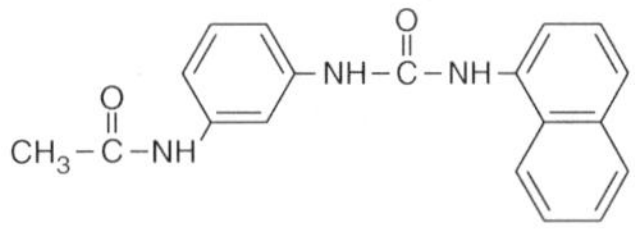

5 gm

S64,411-0
1,1'-(4-METHYL-1,3-PHENYLENE)BIS(3-CYCLOHEXYLUREA)

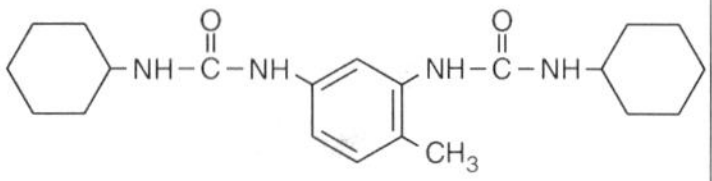

5 gm

S63,463-8
3,3'-(4-METHYL-1,3-PHENYLENE)BIS(1-CYCLOHEXYL-1-METHYLUREA)

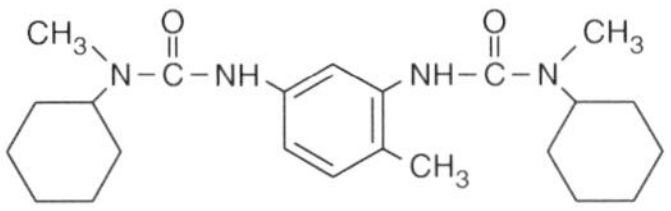

5 gm

S66,032-9
3,3'-(4-METHYL-1,3-PHENYLENE)BIS(1-CYCLOHEXYL-1-ISOPROPYLUREA)

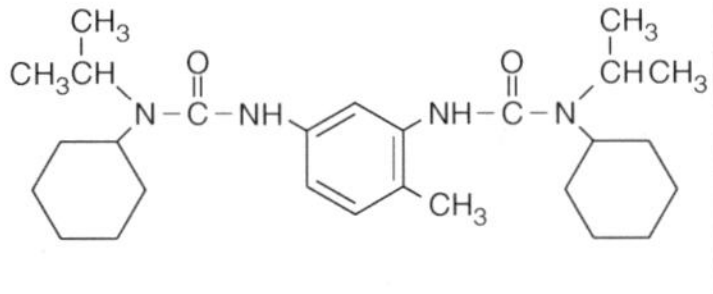

5 gm

S64,968-6
3,3'-(4-METHYL-1,3-PHENYLENE)BIS-(1,1-DICYCLOHEXYLUREA)

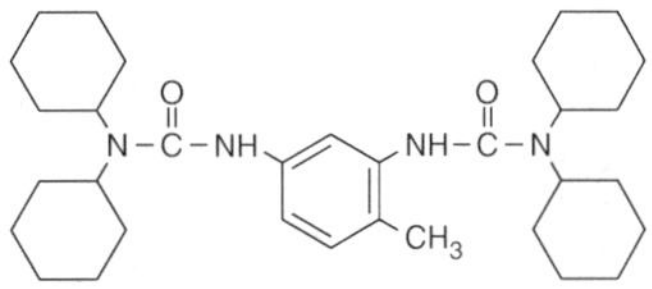

5 gm

S64,957-0
1-[3-(1-[2-METHYLPIPERIDINO])-PROPYL]-3-PHENYLUREA

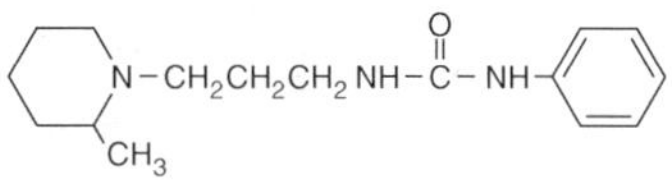

5 gm

S65,111-7
1-(3-CHLOROPHENYL)-3-[3-(2-METHYLPIPERIDINO)PROPYL]UREA

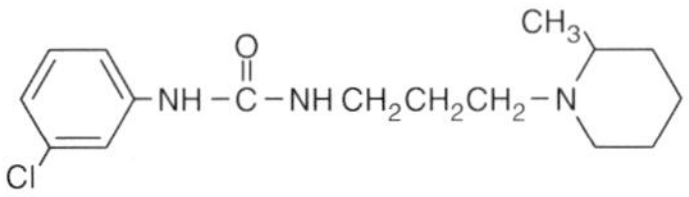

5 gm

S58,613-7
N,N'-(4-METHYL-1,3-PHENYLENE)BIS(1-PIPERIDINECARBOXAMIDE)

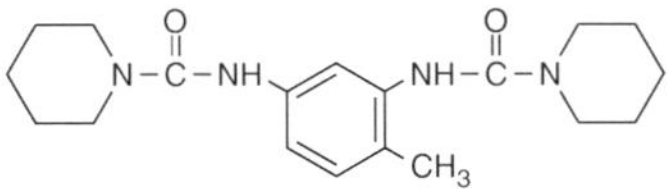

250 mg

S64,956-2
1-(4-CHLOROPHENYL)-3-[3-(2-METHYLPIPERIDINO)PROPYL]UREA

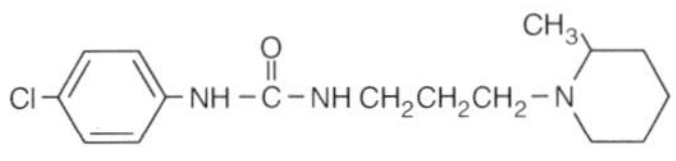

5 gm

S65,849-9
4,4'-BIS(3-[3-(2-METHYLPIPERIDINO)-PROPYL]UREIDO)DIPHENYLMETHANE

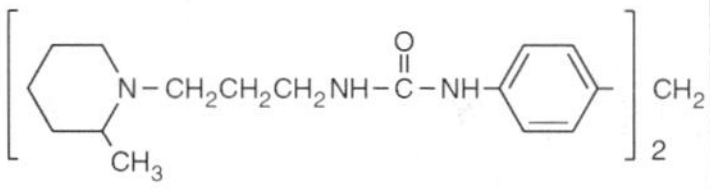

5 gm

S65,219-9
1-(3-[3-(2,6-DICHLOROPHENYL)UREIDO]-PROPYL)-2-METHYLPIPERIDINE

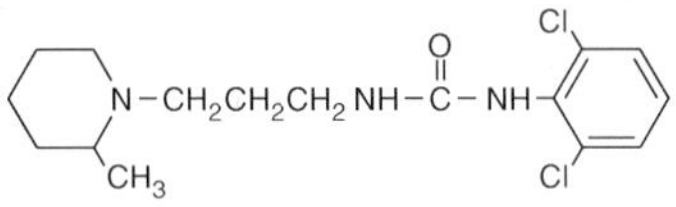

5 gm

S66,062-0
1-(3-[3-(2,3-DICHLOROPHENYL)UREIDO]-PROPYL)-2-METHYLPIPERIDINE

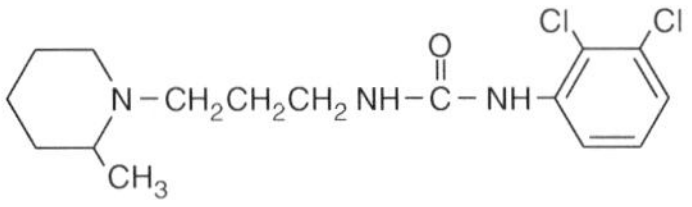

5 gm

S64,678-4
1-[3-(3,4-DICHLOROPHENYLUREIDO)-PROPYL]-2-METHYLPIPERIDINE

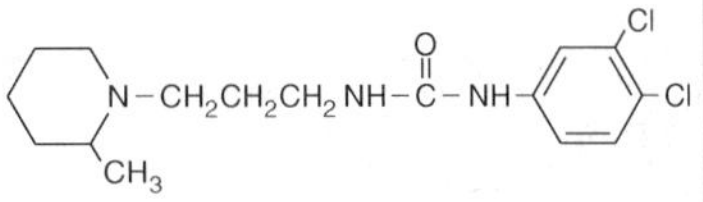

250 mg

S71,759-2
1,1'-(4-METHYL-1,3-PHENYLENE)BIS(3-[3-(2-METHYLPIPERIDINO)PROPYL]UREA)

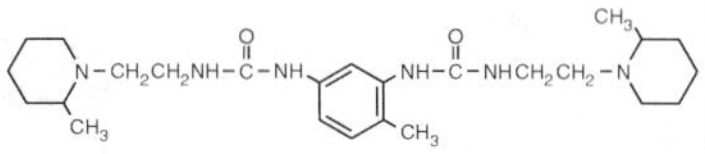

250 mg

S65,022-6
1-(3-MORPHOLINOPROPYL)-3-PHENYLUREA

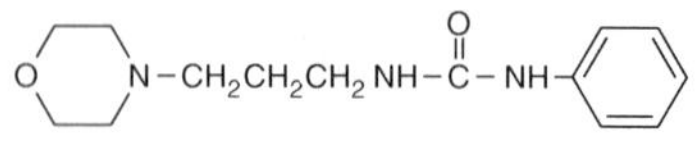

5 gm

S65,221-0
1-(2-CHLOROPHENYL)-3-(3-MORPHOLINOPROPYL)UREA

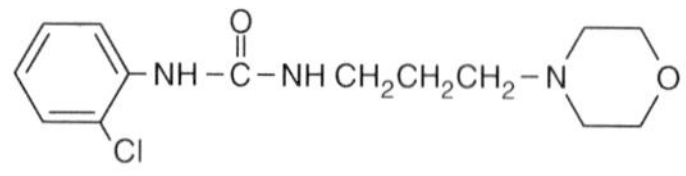

5 gm

S65,217-2
1-(3-CHLOROPHENYL)-3-(3-MORPHOLINOPROPYL)UREA

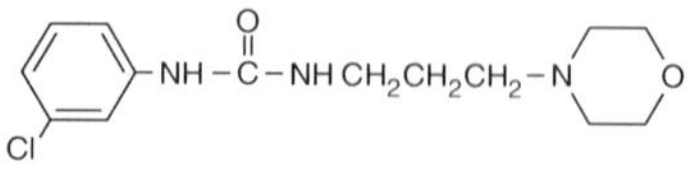

5 gm

S65,279-2
1-(4-CHLOROPHENYL)-3-(3-MORPHOLINOPROPYL)UREA

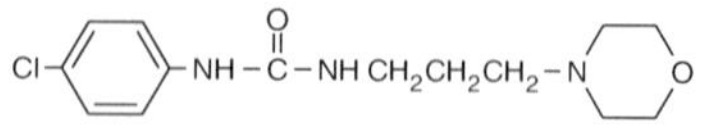

5 gm

S66,708-0
1-(4-BROMOPHENYL)-3-(3-MORPHOLINOPROPYL)UREA

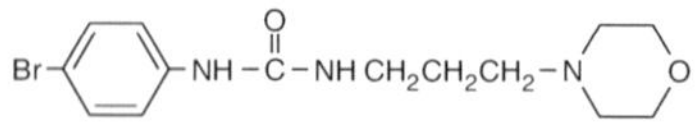

5 gm

S65,996-7
4,4'-METHYLENEBIS[3-(3-MORPHOLINOPROPYL)-1-PHENYLUREA]

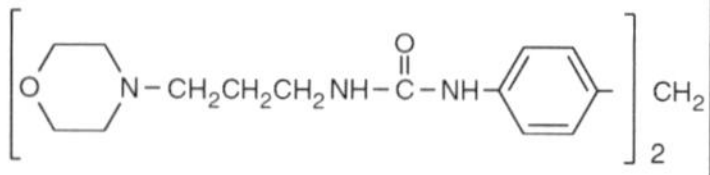

5 gm

S65,039-0
1-(2,3-DICHLOROPHENYL)-3-(3-MORPHOLINOPROPYL)UREA

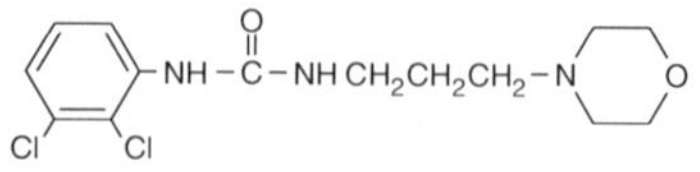

5 gm

S65,266-0
1-(2,6-DICHLOROPHENYL)-3-(3-MORPHOLINOPROPYL)UREA

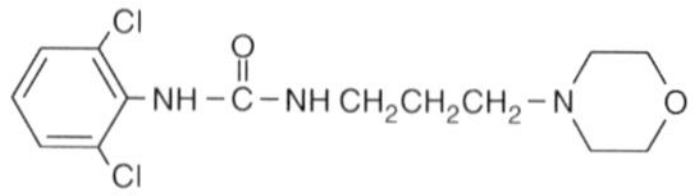

5 gm

S65,089-7
1-(3,4-DICHLOROPHENYL)-3-(3-MORPHOLINOPROPYL)UREA

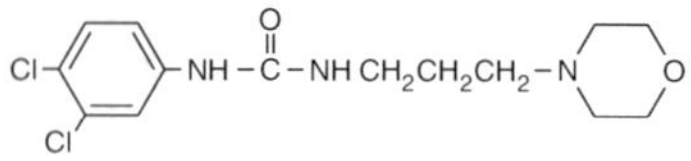

5 gm

S65,964-9
1,1'-(4-METHYL-1,3-PHENYLENE)BIS[3-(3-MORPHOLINOPROPYL)UREA]

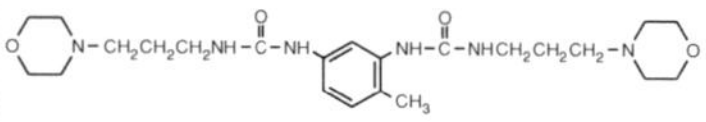

5 gm

S58,998-5
1,1'-(4-METHYL-1,3-PHENYLENE)-BISUREA

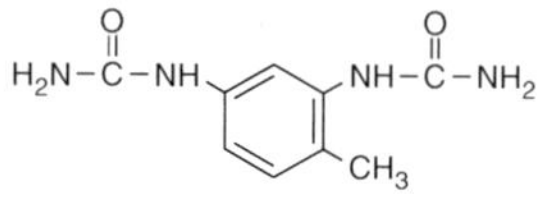

5 gm

S62,522-1
1,1'-(4-METHYL-1,3-PHENYLENE)BIS(3-METHYLUREA)

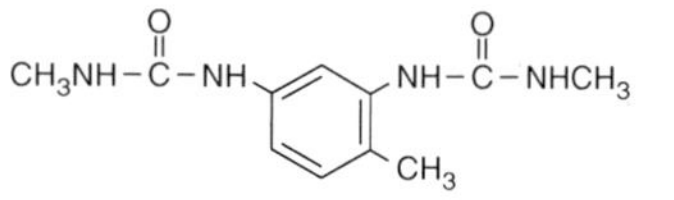

5 gm

S59,042-8
3,3'-(4-METHYL-1,3-PHENYLENE)BIS-(1,1-DIMETHYLUREA)

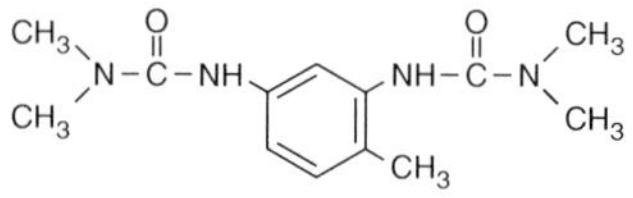

5 gm

S64,474-9
1,1'-(4-METHYL-1,3-PHENYLENE)BIS(3-ETHYLUREA)

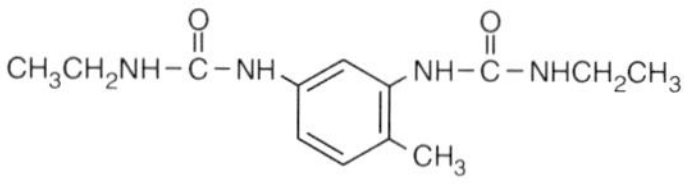

5 gm

S64,693-8
1,1'-(4-METHYL-1,3-PHENYLENE)BIS(3-ALLYLUREA)

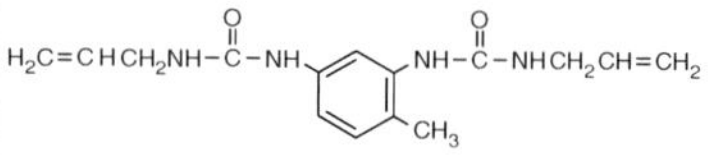

5 gm

S62,525-6
1,1'-(4-METHYL-1,3-PHENYLENE)BIS(3-ISOPROPYLUREA)

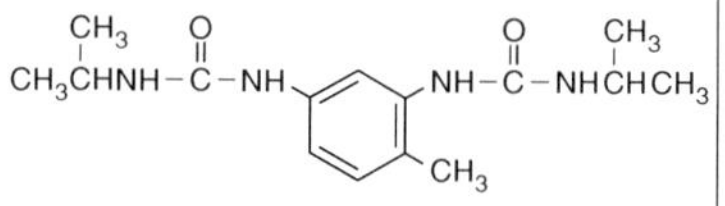

5 gm

S62,531-0
1,1'-(4-METHYL-1,3-PHENYLENE)BIS(3-PROPYLUREA)

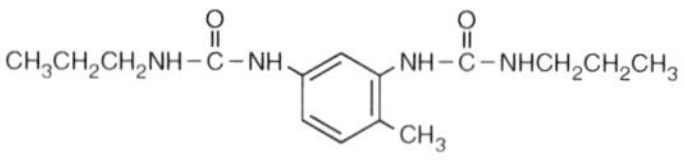

5 gm

S57,955-6
3,3'-(4-METHYL-1,3-PHENYLENE)BIS-(1,1-DIETHYLUREA)

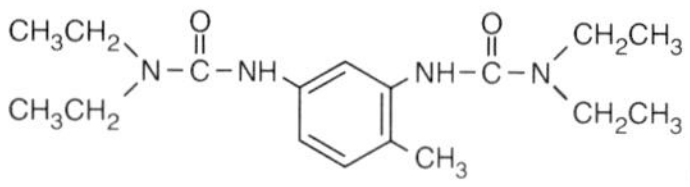

5 gm

S62,507-8
1,1'-(4-METHYL-1,3-PHENYLENE)BIS(3-BUTYLUREA)

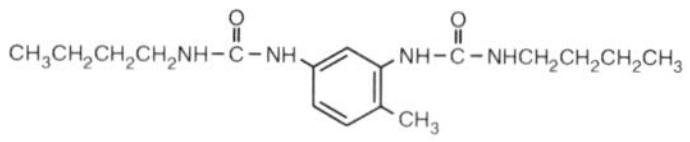

5 gm

S65,975-4
1,1'-(4-METHYL-1,3-PHENYLENE)BIS(3-ISOBUTYLUREA)

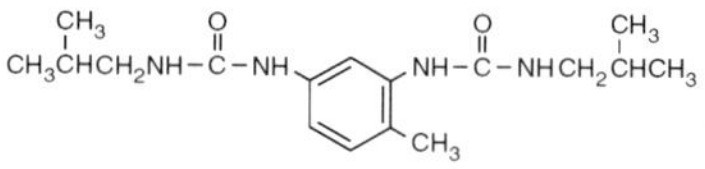

5 gm

S65,110-9
3,3'-(4-METHYL-1,3-PHENYLENE)BIS(1-BUTYL-1-METHYLUREA)

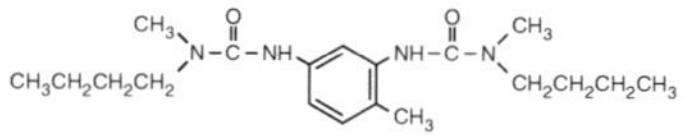

5 gm

S64,421-8
3,3'-(4-METHYL-1,3-PHENYLENE)BIS-(1,1-DIALLYLUREA)

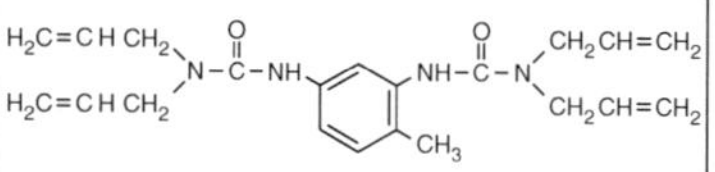

5 gm

S62,518-3
3,3'-(4-METHYL-1,3-PHENYLENE)BIS-(1,1-DIISOPROPYLUREA)

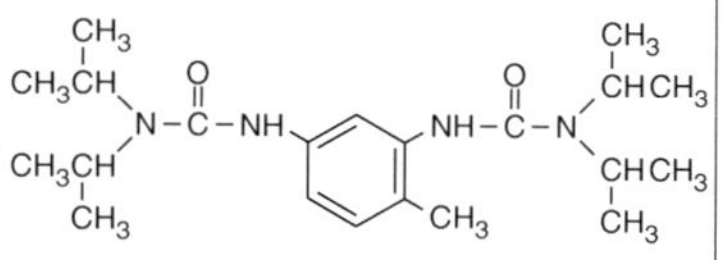

5 gm

S62,528-0
3,3'-(4-METHYL-1,3-PHENYLENE)BIS-(1,1-DIPROPYLUREA)

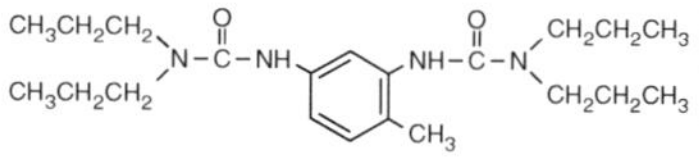

5 gm

S66,015-9
3,3'-(4-METHYL-1,3-PHENYLENE)BIS(1-BUTYL-1-ETHYLUREA)

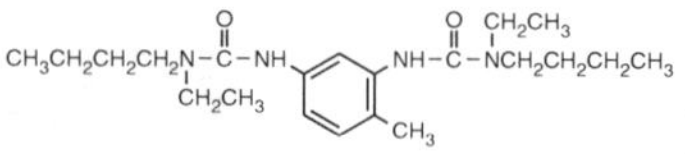

5 gm

S70,733-3
1,1'-(4-METHYL-1,3-PHENYLENE)BIS[3-(1,3-DIMETHYLBUTYL)UREA]

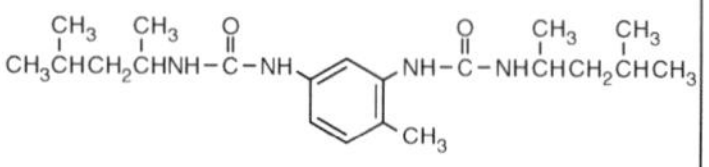

5 gm

S66,001-9
3,3'-(4-METHYL-1,3-PHENYLENE)BIS(1-ETHYL-1-PHENYLUREA)

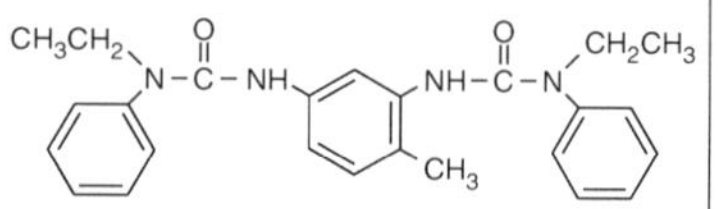

5 gm

S77,319-0
1,1'-(4-METHYL-1,3-PHENYLENE)BIS[3-(2,4-XYLYL)UREA]

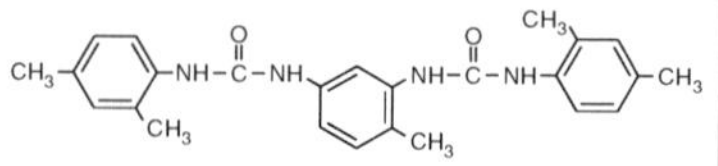

250 mg

S63,446-8
3,3'-(4-METHYL-1,3-PHENYLENE)BIS-(1,1-DIBUTYLUREA)

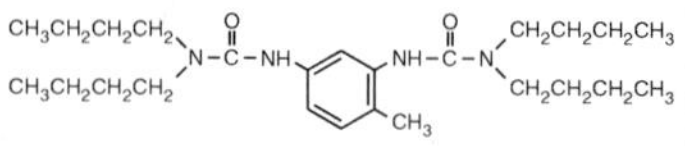

250 mg

S66,066-3
3,3'-(4-METHYL-1,3-PHENYLENE)BIS-(1,1-DIISOBUTYLUREA)

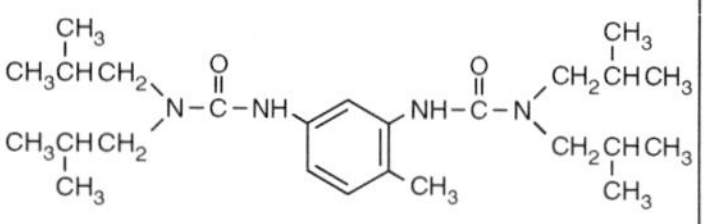

5 gm

S66,940-7
1-(4-NITROPHENYL)-3-(2-PHENETHYL)UREA

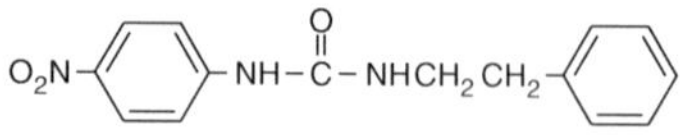

5 gm

S61,129-8
1-(2-METHYL-4-NITROPHENYL)-3-PHENETHYLUREA

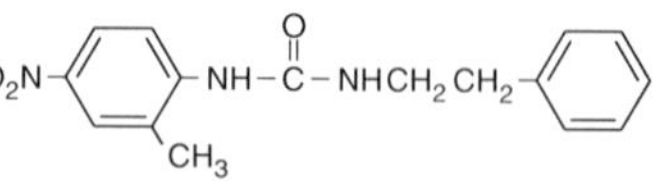

5 gm

S86,888-4
1-(3-ACETYLPHENYL)-3-(4-ANILINOPHENYL)UREA

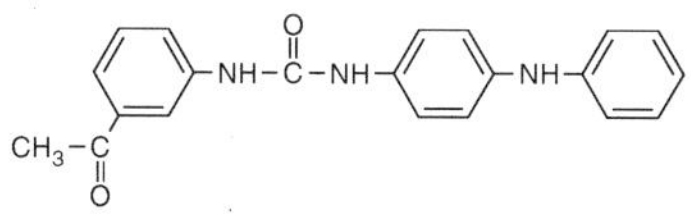

250 mg

S86,873-6
1-(4-ANILINOPHENYL)-3-(M-TOLYL)UREA

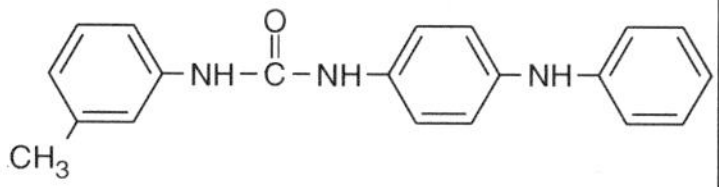

250 mg

S72,929-9
1-(4-ANILINOPHENYL)-3-(1-NAPHTHYL)UREA

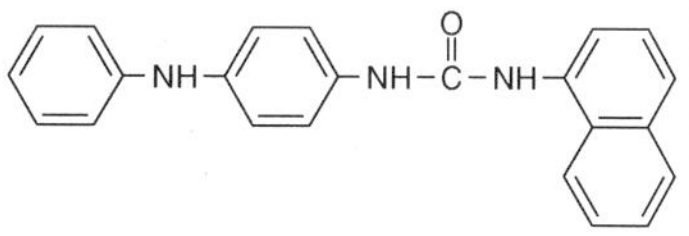

5 gm

S70,205-6
1-(4-ANILINOPHENYL)-3-(4-ETHOXYCARBONYLPHENYL)UREA

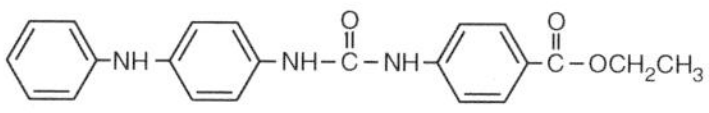

250 mg

S71,348-1
3,3'-(1,5-NAPHTHALENEDIYL)BIS(1-CYCLOHEXYL-1-METHYLUREA)

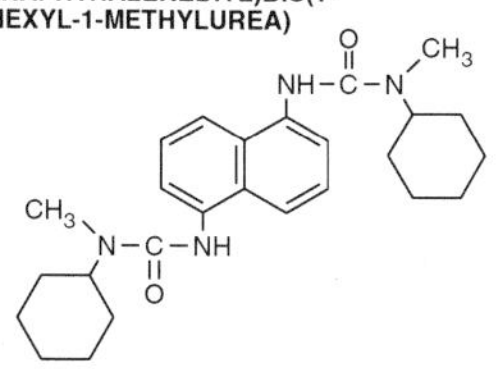

250 mg

S71,572-7
1-(4-ANILINOPHENYL)-3-METHYLUREA

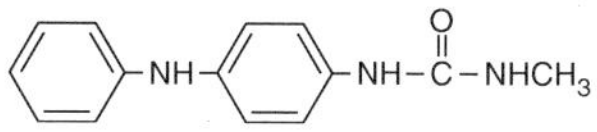

5 gm

S73,189-7
1-(4-ANILINOPHENYL)-3-(M-TOLYL)UREA

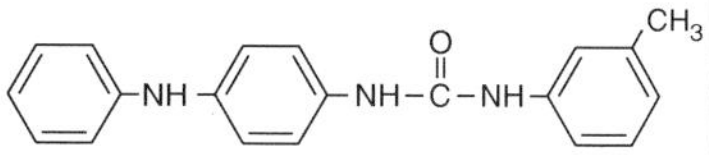

5 gm

S91,237-9
1-[N-[4-(DIMETHYLAMINO)PHENYL]-CARBAMOYL)-2-METHYLPIPERIDINE

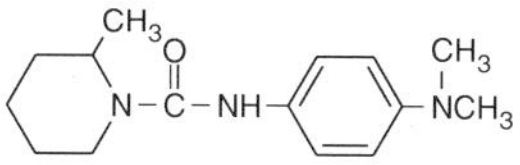

250 mg

S85,467-0
1-TERT-BUTYL-3-[4-(DIMETHYLAMINO)-PHENYL]UREA

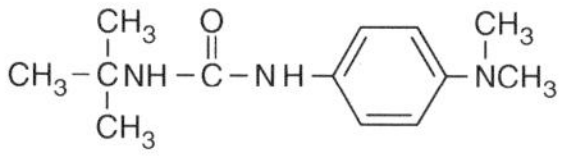

250 mg

S86,834-5
1-(3-ACETYLPHENYL)-3-[4-(DIMETHYLAMINO)PHENYL]UREA

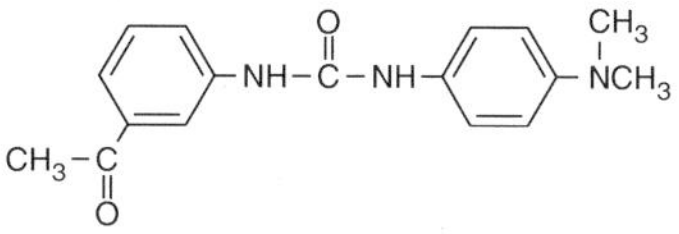

250 mg

S84,592-2
1-[4-(DIMETHYLAMINO)PHENYL]-3-(2-ETHYLPHENYL)UREA

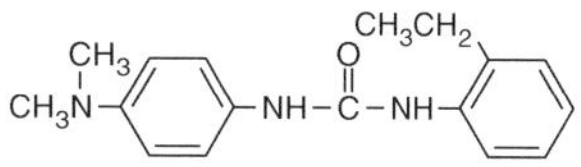

250 mg

S77,347-6
1-BUTYL-3-[4-(DIMETHYLAMINO)-PHENYL]UREA

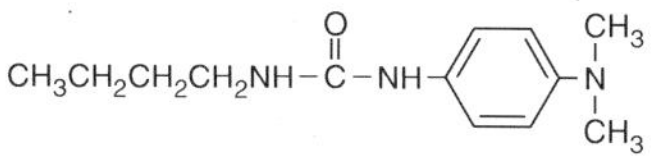

250 mg

S89,081-2
1-[4-(DIETHYLAMINO)PHENYL]-3-ISOPROPYLUREA

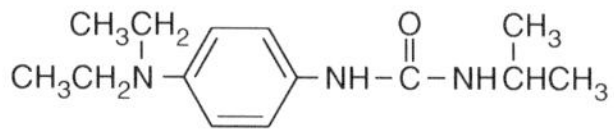

250 mg

S89,084-7
1-BUTYL-3-[4-(DIETHYLAMINO)-PHENYL]UREA

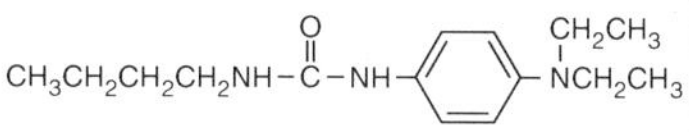

250 mg

S84,594-9
1-[4-(DIMETHYLAMINO)PHENYL]-3-(2,6-XYLYL)UREA

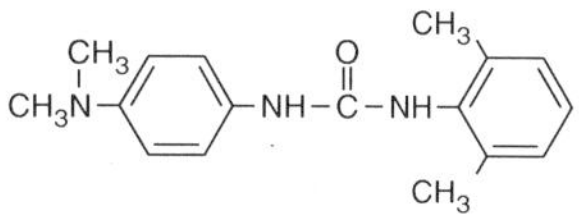

250 mg

S84,593-0
1-[4-(DIMETHYLAMINO)PHENYL]-3-(2,5-XYLYL)UREA

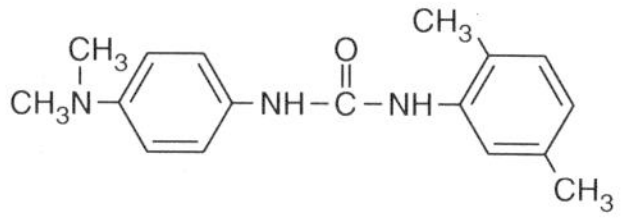

250 mg

S90,187-3
1-[4-(DIMETHYLAMINO)PHENYL]-3-(3,4-XYLYL)UREA

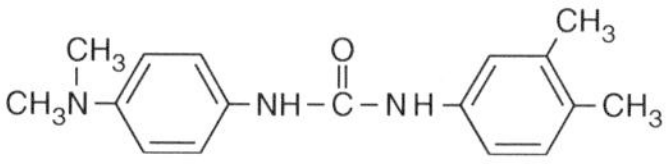

250 mg

S66,932-6
1-BENZYL-3-(4-NITROPHENYL)UREA

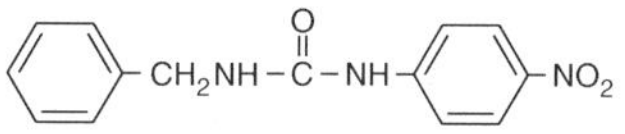

5 gm

S66,951-2
1-BENZYL-1-METHYL-3-(4-NITROPHENYL)UREA

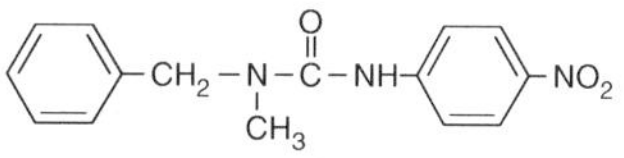

5 gm

S66,906-7
1-BENZYL-1-ETHYL-3-(4-NITROPHENYL)UREA

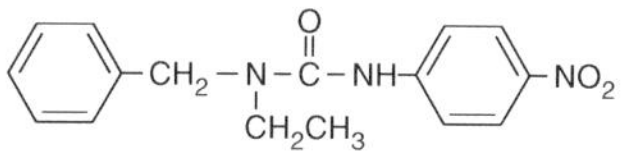

5 gm

S66,780-3
1,1-DIBENZYL-3-(4-NITROPHENYL)UREA

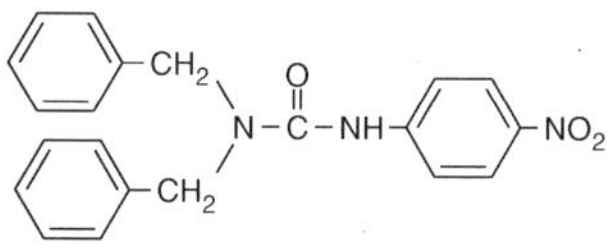

5 gm

S60,780-0
3-(2-CHLORO-4-NITROPHENYL)-1,1-DIBENZYLUREA

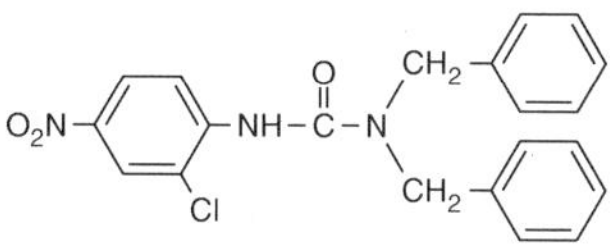

5 gm

S71,033-4
3,3'-(1,5-NAPHTHALENEDIYL)BIS(1,1-DIBENZYLUREA)-

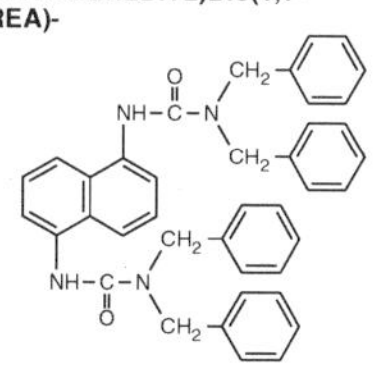

5 gm

S90,554-2
4-(ETHOXYCARBONYLMETHYL)-1-(4-NITROPHENYL)SEMICARBAZIDE

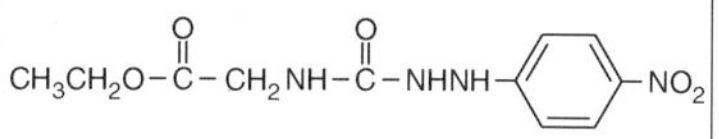

250 mg

S35,382-5
1-(3-NITROBENZOYL)UREA

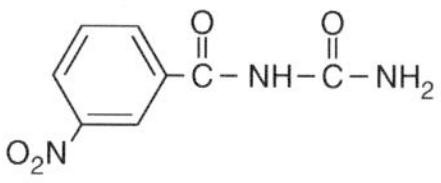

5 gm

S90,417-1
4-(3-CHLOROPHENYL)-1-(4-NITROPHENYL)SEMICARBAZIDE

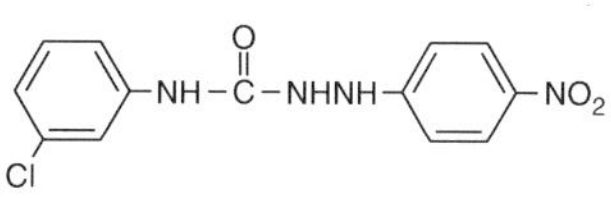

250 mg

S69,972-1
4-ETHYL-1-(4-NITROPHENYL)-SEMICARBAZIDE

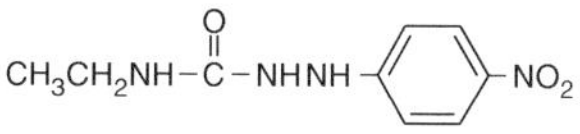

250 mg

S66,787-0
1-CYCLOHEXYL-3-(4-NITROPHENYL)UREA

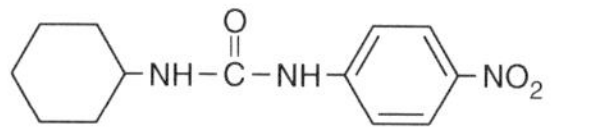

5 gm

S67,166-5
1-CYCLOHEXYLMETHYL-3-(4-NITROPHENYL)UREA

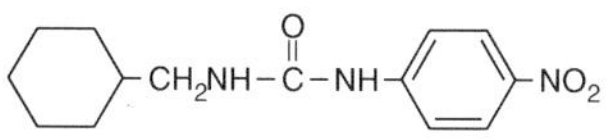

5 gm

S66,960-1
1-CYCLOHEXYL-1-ISOPROPYL-3-(4-NITROPHENYL)UREA

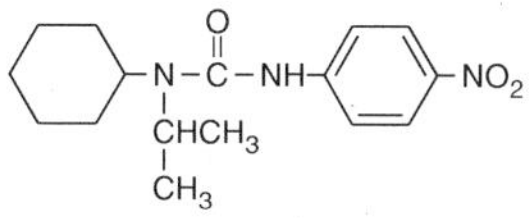

5 gm

S66,939-3
1,1-DICYCLOHEXYL-3-(4-NITROPHENYL)UREA

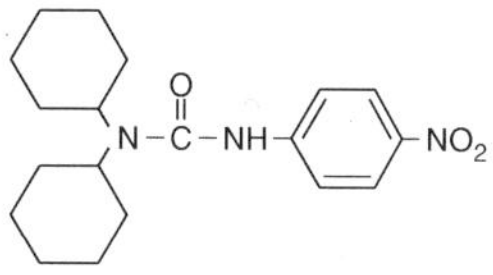

5 gm

S66,905-9
1-ETHYL-3-(4-NITROPHENYL)UREA

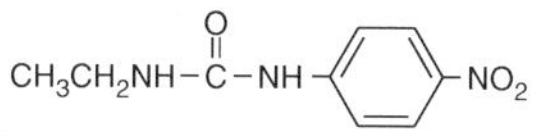

5 gm

S66,807-9
1-ALLYL-3-(4-NITROPHENYL)UREA

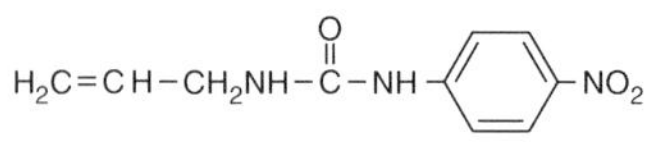

5 gm

S66,916-4
1-ISOPROPYL-3-(4-NITROPHENYL)UREA

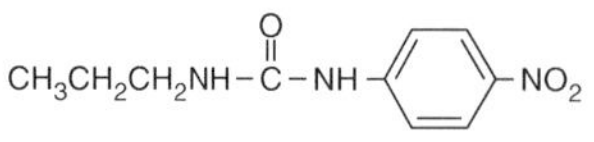

5 gm

S66,924-5
1-(4-NITROPHENYL)-3-PROPYLUREA

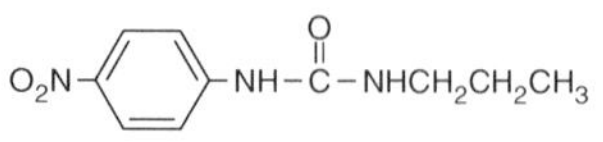

5 gm

S66,956-3
1-BUTYL-3-(4-NITROPHENYL)UREA

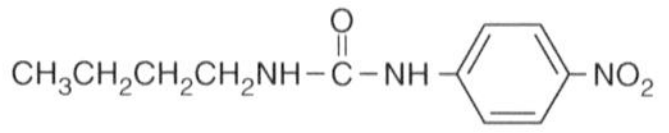

5 gm

S66,983-0
1,1-DIETHYL-3-(4-NITROPHENYL)UREA

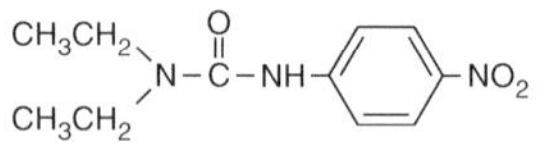

5 gm

S66,911-3
1,1-DIALLYL-3-(4-NITROPHENYL)UREA

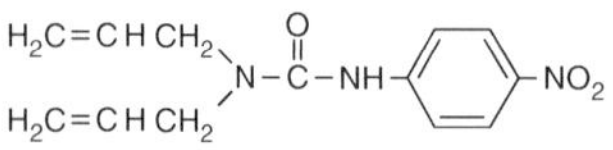

5 gm

S66,967-9
1-BUTYL-1-ETHYL-3-(4-NITROPHENYL)UREA

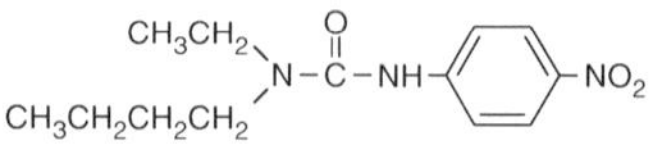

5 gm

S67,151-7
1,1-DIISOPROPYL-3-(4-NITROPHENYL)UREA

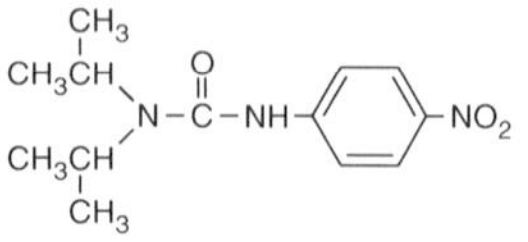

5 gm

S67,152-5
1,1-DIPROPYL-3-(4-NITROPHENYL)UREA

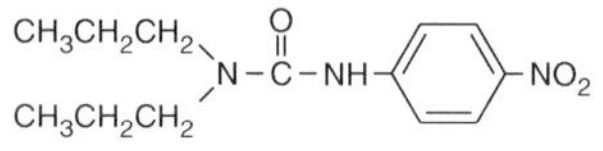

5 gm

S66,936-9
1,1-DIISOBUTYL-3-(4-NITROPHENYL)UREA

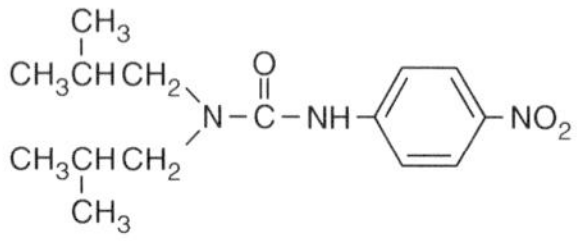

5 gm

S66,948-2
1,1-DIBUTYL-3-(4-NITROPHENYL)UREA

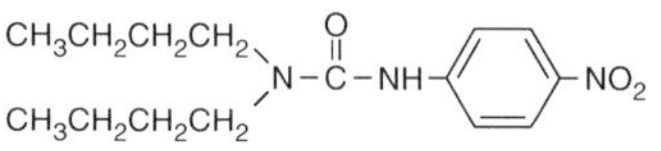

5 gm

S71,380-5
1-(4-NITROPHENYL)-3-(2,6-XYLYL)UREA

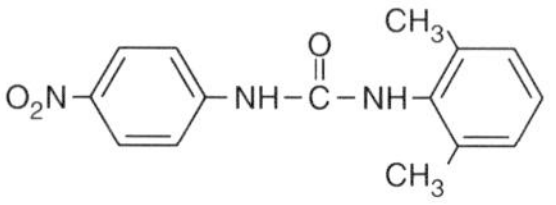

5 gm

S71,267-1
1-(2,4-DIMETHYLPHENYL)-3-(4-NITROPHENYL)UREA

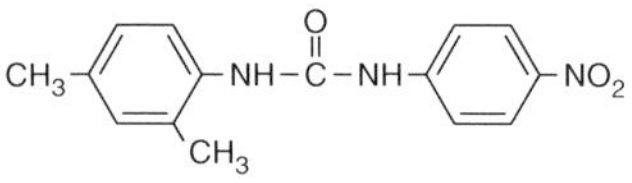

5 gm

S89,119-3
1-(2-METHYL-4-NITROPHENYL)-3-(1-NAPHTHYL)UREA

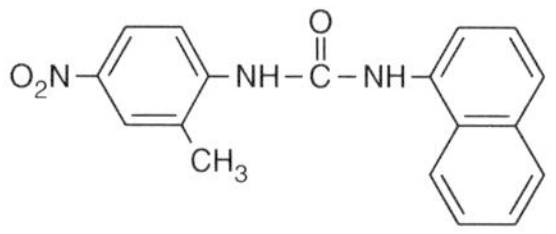

250 mg

S60,765-7
4-(2-CHLORO-4-NITROPHENYL)-1,1-DIMETHYLSEMICARBAZIDE

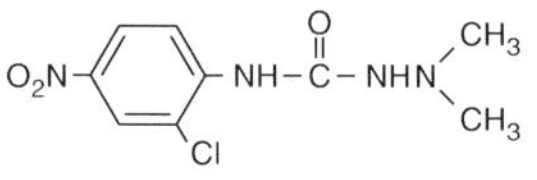

5 gm

S60,745-2
4-(2-CHLORO-4-NITROPHENYL)-1-PHENYLSEMICARBAZIDE

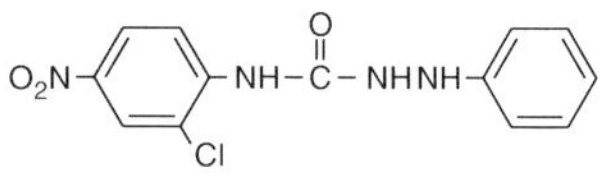

5 gm

S34,748-5
1,5-BIS(4-NITROPHENYL)CARBAZIDE

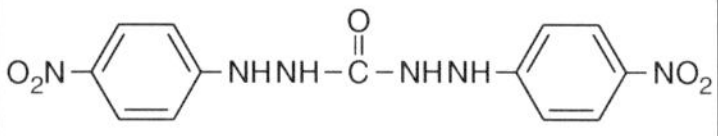

250 mg

S60,721-5
3-(2-CHLORO-4-NITROPHENYL)-1-CYCLOHEXYL-1-METHYLUREA

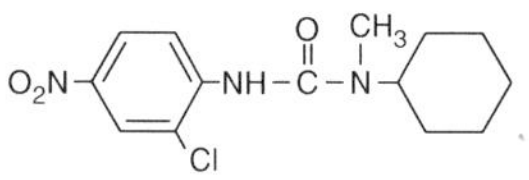

5 gm

S60,997-8
1,1-DICYCLOHEXYL-3-(2-METHYL-4-NITROPHENYL)UREA

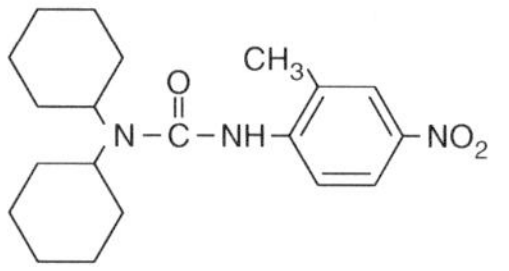

5 gm

S59,549-7
1-(2-CHLOROPHENYL)-3-(4-NITROPHENYL)UREA

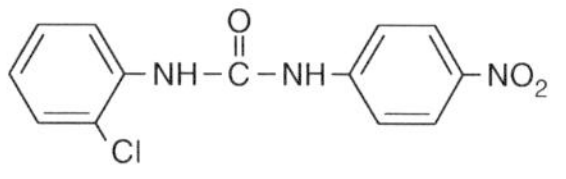

5 gm

S60,729-0
1-(2-CHLOROPHENYL)-3-(2-CHLORO-4-NITROPHENYL)UREA

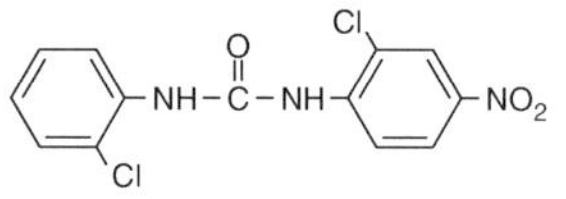

5 gm

S59,482-2
1-(4-NITROPHENYL)-3-PHENYLUREA

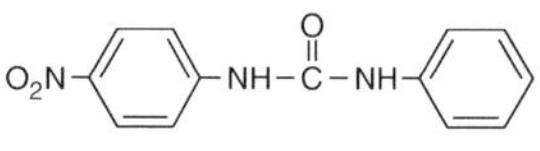

5 gm

S60,744-4
1-(2-CHLORO-4-NITROPHENYL)-3-METHYLUREA

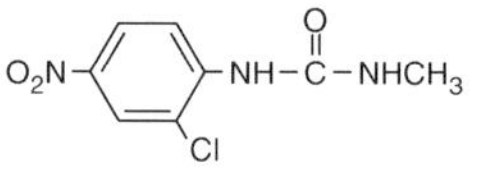

5 gm

S61,343-6
1-METHYL-3-(2-METHYL-4-NITROPHENYL)UREA

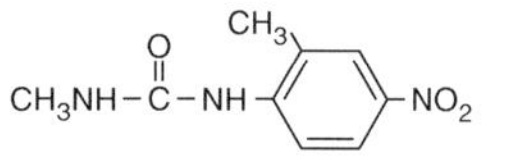

5 gm

S60,778-9
3-(2-CHLORO-4-NITROPHENYL)-1,1-DIMETHYLUREA

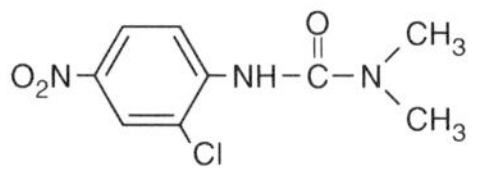

5 gm

S61,088-7
1,1-DIMETHYL-3-(2-METHYL-4-NITROPHENYL)UREA

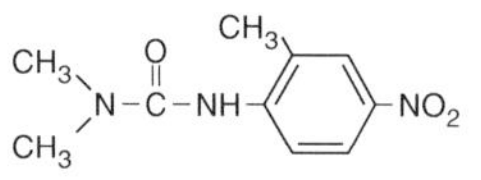

5 gm

S61,292-8
1-ETHYL-3-(2-METHYL-4-NITROPHENYL)UREA

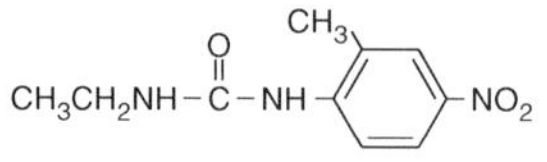

5 gm

S64,698-9
1-ALLYL-3-(2-CHLORO-4-NITROPHENYL)UREA

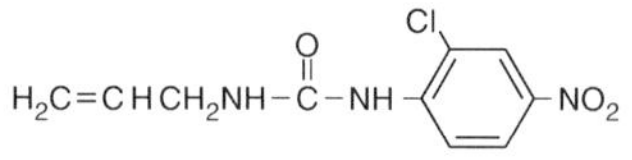

5 gm

S64,041-7
1-(2-CHLORO-4-NITROPHENYL)-3-ISOPROPYLUREA

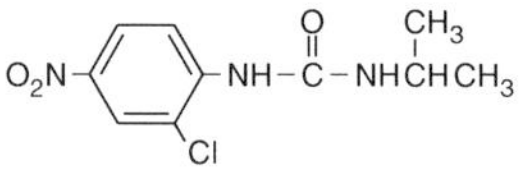

5 gm

S60,782-7
3-(2-CHLORO-4-NITROPHENYL)-1,1-DIETHYLUREA

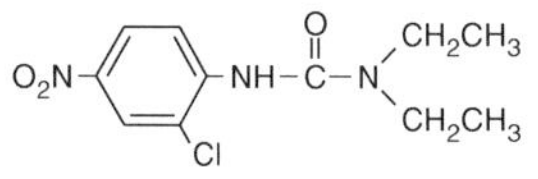

5 gm

S61,044-5
1,1-DIETHYL-3-(2-METHYL-4-NITROPHENYL)UREA

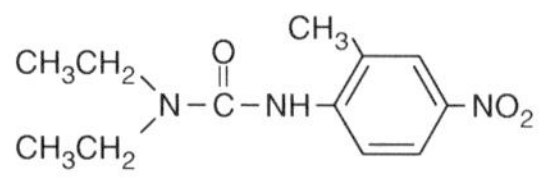

5 gm

S60,758-4
1-(2-CHLORO-4-NITROPHENYL)-3-PHENYLUREA

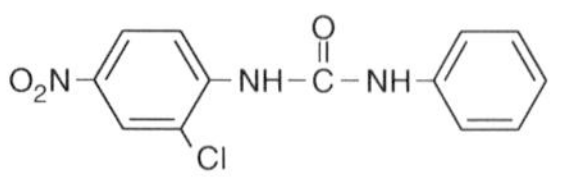

5 gm

S60,783-5
3-(2-CHLORO-4-NITROPHENYL)-1,1-DIALLYLUREA
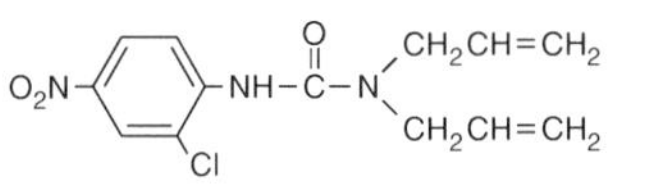
5 gm

S60,770-3
3-(2-CHLORO-4-NITROPHENYL)-1,1-DIPROPYLUREA
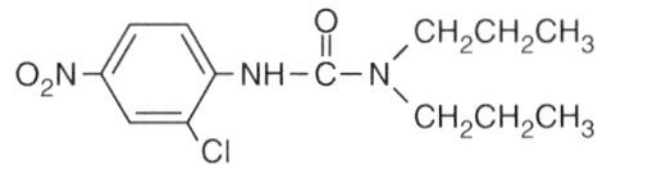
5 gm

S60,725-8
1-(3-CHLOROPHENYL)-3-(2-CHLORO-4-NITROPHENYL)UREA
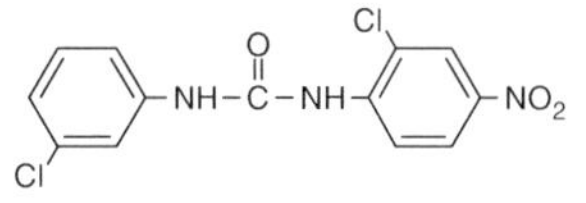
5 gm

S60,719-3
1,3-BIS(2-CHLORO-4-NITROPHENYL)UREA
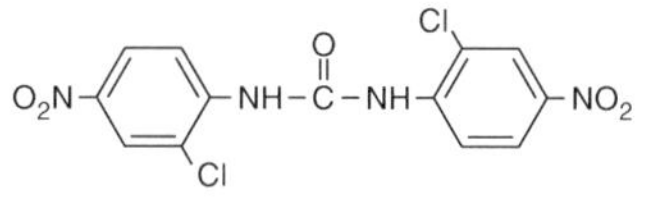
5 gm

S60,726-6
1-(2-CHLORO-4-NITROPHENYL)-3-(2,4,5-TRICHLOROPHENYL)UREA
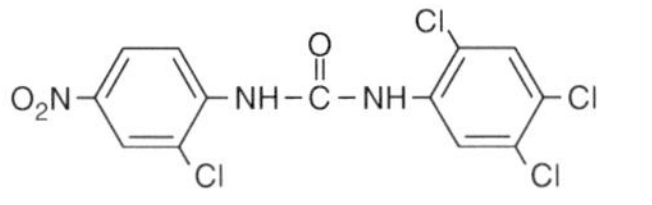
5 gm

S61,046-1
1-(2-METHYL-4-NITROPHENYL)-3-PHENYLUREA
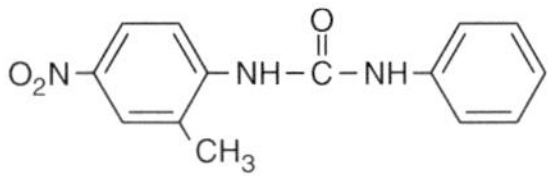
5 gm

S61,031-3
1-BENZYL-3-(2-METHYL-4-NITROPHENYL)UREA
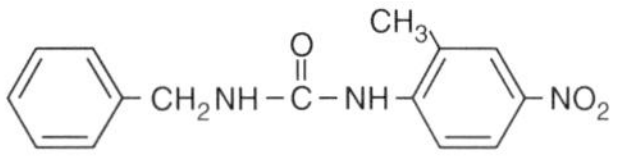
5 gm

S60,779-7
3-(2-CHLORO-4-NITROPHENYL)-1,1-DIBUTYLUREA
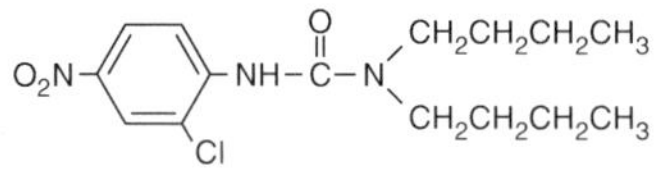
5 gm

S61,060-7
1-BUTYL-3-(2-METHYL-4-NITROPHENYL)-1-PHENYLUREA
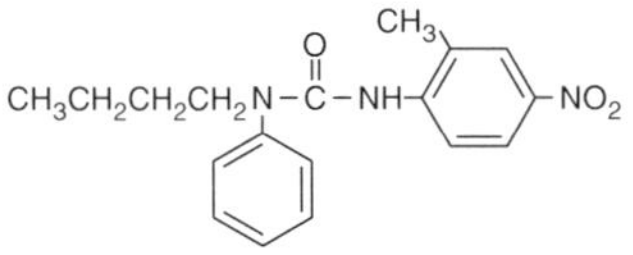
5 gm

S61,059-3
1,1-DIPROPYL-3-(2-METHYL-4-NITROPHENYL)UREA
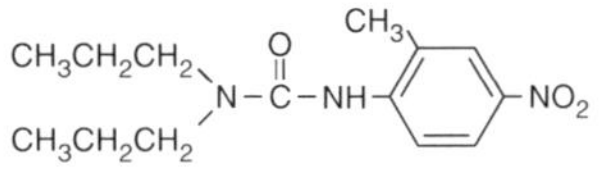
5 gm

S90,570-4
1-(4-ACETAMIDOPHENYL)-3-(3-ACETYLPHENYL)UREA
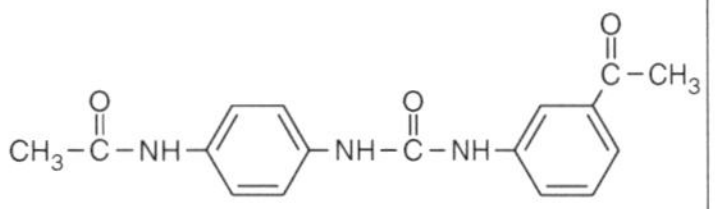
250 mg

S79,770-7
ETHYL 4-[3-(4-ACETAMIDOPHENYL)-UREIDO]BENZOATE
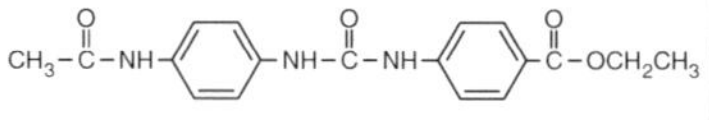
250 mg

S79,427-9
ETHYL 4-(3-[4-(N-METHYLACETAMIDO)-PHENYL]UREIDO)BENZOATE
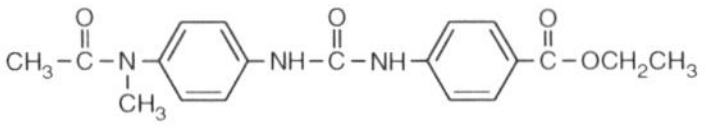
250 mg

S85,514-6
1-(4-ACETAMIDOPHENYL)-3-TERT-BUTYLUREA
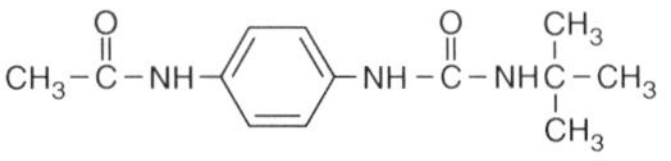
250 mg

S60,346-5
1-(4-ACETAMIDOPHENYL)-3-PHENYLUREA
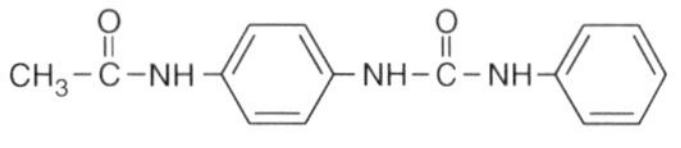
5 gm

S60,186-1
1-(4-ACETAMIDOPHENYL)-3-(2,3-DICHLOROPHENYL)UREA
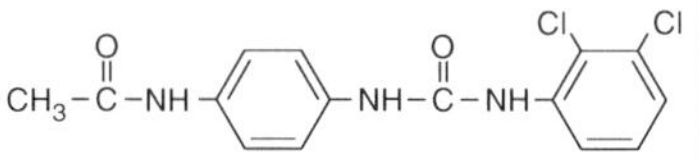
5 gm

S60,172-1
1-(4-ACETAMIDOPHENYL)-3-(4-BROMOPHENYL)UREA
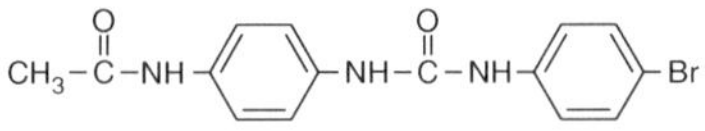
5 gm

S73,894-8
1-(4-ACETAMIDOPHENYL)-3-(P-TOLYL)UREA
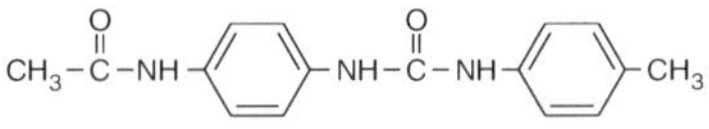
5 gm

S77,258-5
1,1'-HEXAMETHYLENEBIS[3-(4-ACETAMIDOPHENYL)UREA]
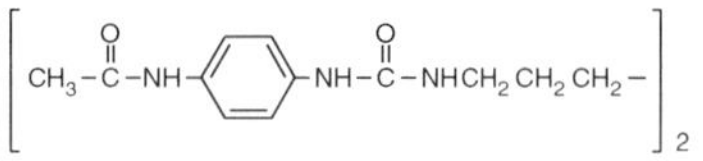
250 mg

S77,344-1
1-[(4-ACETAMIDO)PHENYL]-3-(2,5-XYLYL)UREA
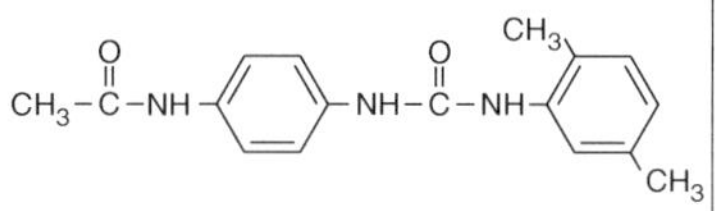
250 mg

S60,188-8
1-(4-ACETAMIDOPHENYL)-3-(2,5-DICHLOROPHENYL)UREA
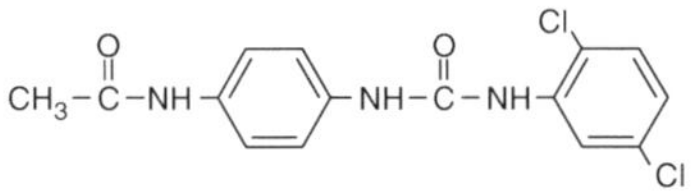
5 gm

S79,584-4
N-METHYL-4'-(3-PHENYLUREIDO)-ACETANILIDE
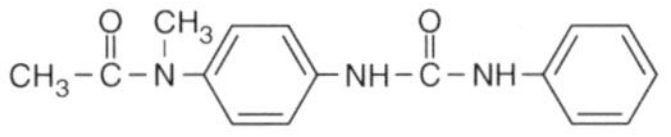
250 mg

S76,917-7
4'-(3-ETHYLUREIDO)-N-METHYLACETANILIDE
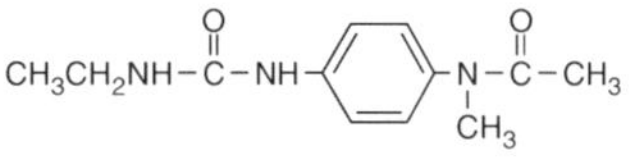
250 mg

S90,014-1
1-ETHYL-3-[4-(N-METHYLACETAMIDO)-PHENYL]UREA
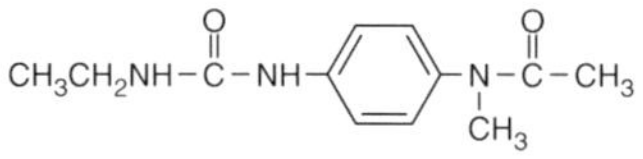
250 mg

S84,730-5
1-[4-(N-METHYLACETAMIDO)PHENYL]-3-(2,6-XYLYL)UREA
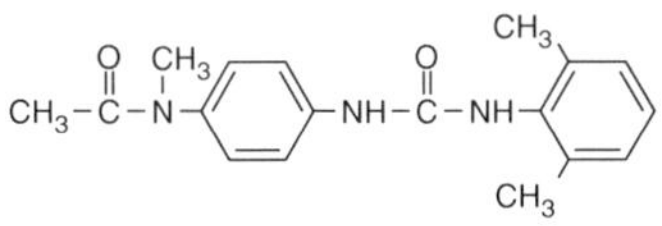
250 mg

S58,624-2
N,N'-(1,5-NAPHTHALENEDIYL)BIS(1-PIPERIDINECARBOXAMIDE)
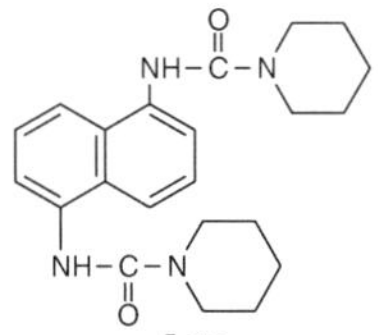
5 gm

S77,370-0
1-(4-ANILINOPHENYL)-3-BUTYLUREA
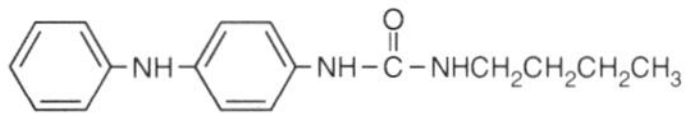
250 mg

S59,045-2
3,3'-(1,5-NAPHTHALENEDIYL)BIS(1,1-DIMETHYLUREA)
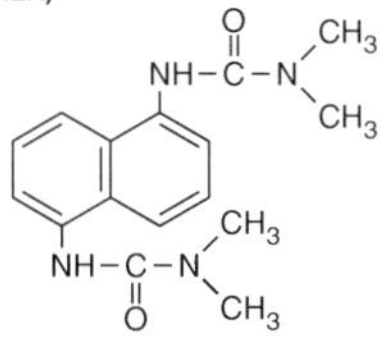
5 gm

S71,057-1
3,3'-(1,5-NAPHTHALENEDIYL)BIS(1,1-DIETHYLUREA)
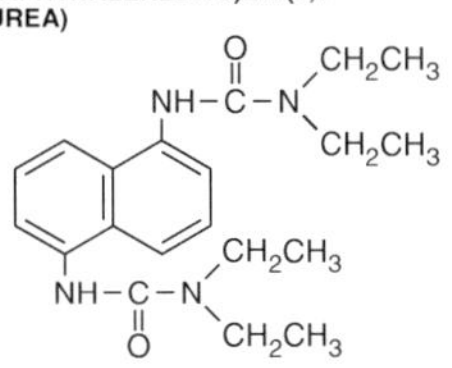
5 gm

S71,085-7
3,3'-(1,5-NAPHTHALENEDIYL)BIS(1,1-DIALLYLUREA)
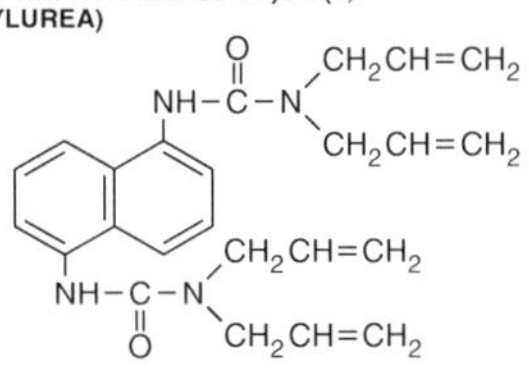
5 gm

S71,086-5
3,3'-(1,5-NAPHTHALENEDIYL)BIS(1,1-DIISOPROPYLUREA)
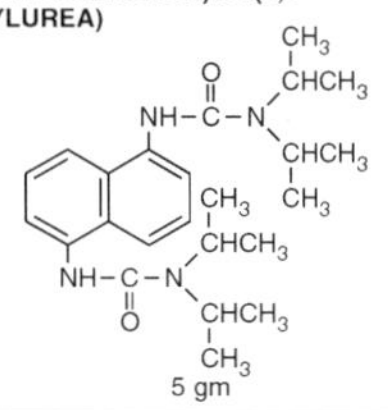
5 gm

S71,158-6
3,3'-(1,5-NAPHTHALENEDIYL)BIS(1,1-DIPROPYLUREA)
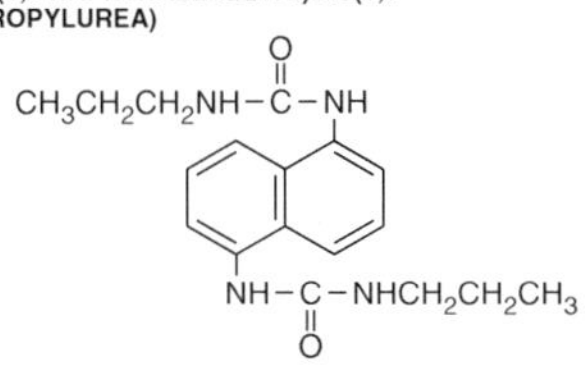
5 gm

S39,793-8
1-ACETYL-3-(1,2-DIPHENYLETHYL)UREA

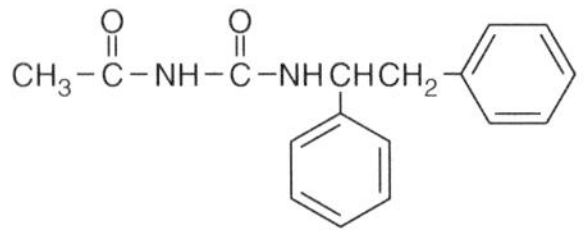

250 mg

S89,337-4
2-[3-(2-PHENYLCYCLOPROPYL)-UREIDO]ACETIC ACID

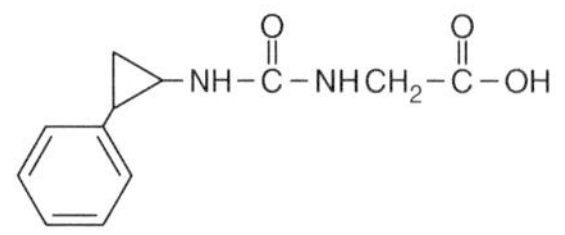

250 mg

S86,832-9
1-(3-ACETYLPHENYL)-3-PHENETHYLUREA

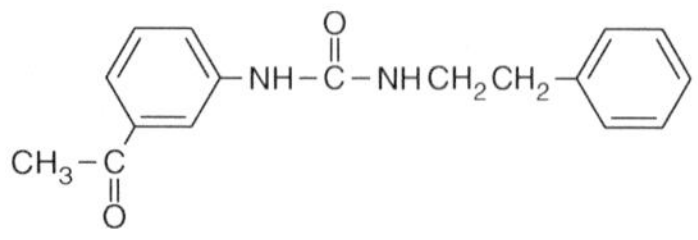

250 mg

S63,631-2
1-(3-CHLOROPHENYL)-3-PHENETHYLUREA

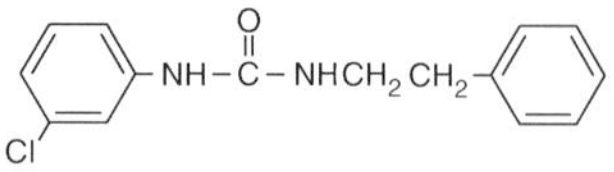

5 gm

S61,030-5
1-PHENETHYL-3-PHENYLUREA

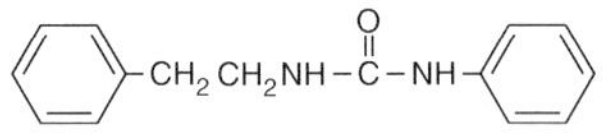

5 gm

S75,296-7
1-OCTADECYL-3-(2-PHENYLETHYL)UREA

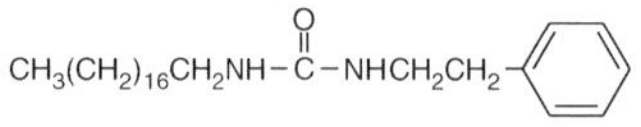

5 gm

S77,456-1
1-METHYL-1-(1-METHYL-2-PHENYLETHYL)-3-(O-TOLYL)UREA

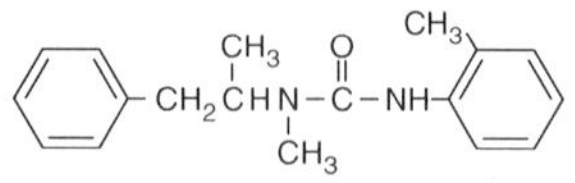

250 mg

S63,923-0
1-(1-NAPHTHYL)-3-(2-PHENYLETHYL)UREA

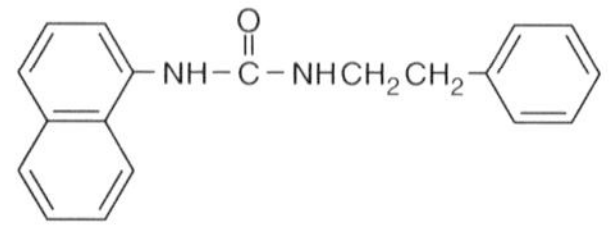

5 gm

S63,470-0
1-(2-CHLOROPHENYL)-3-PHENETHYLUREA

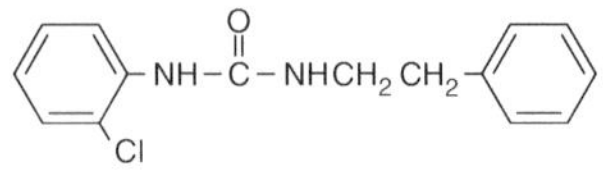

5 gm

S64,523-0
1-PHENETHYL-3-(O-TOLYL)UREA

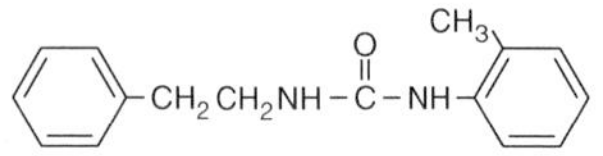

5 gm

S61,337-1
1-(4-CHLOROPHENYL)-3-PHENETHYLUREA

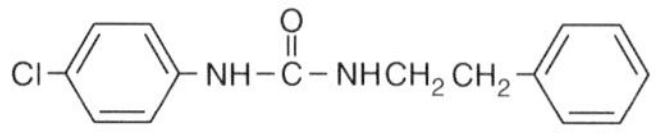

5 gm

S65,136-2
4,4'-METHYLENEBIS[3-(2-PHENETHYL)-1-PHENYLUREA]

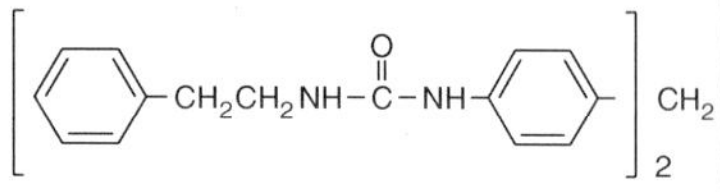

5 gm

S64,691-1
1-(2,3-DICHLOROPHENYL)-3-PHENETHYLUREA

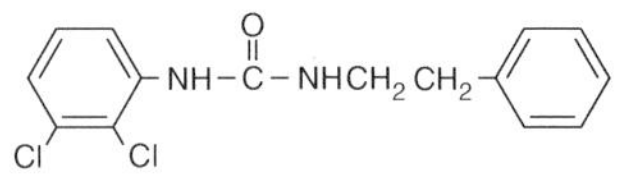

5 gm

S65,238-5
1-(2-CHLORO-6-METHYLPHENYL)-3-PHENETHYLUREA

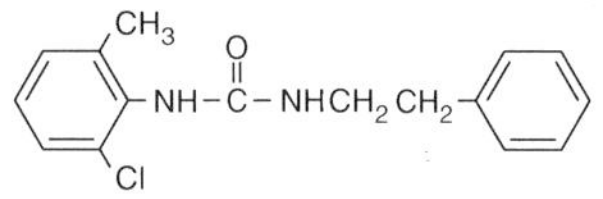

5 gm

S60,961-7
1-(3,4-DICHLOROPHENYL)-3-PHENETHYLUREA

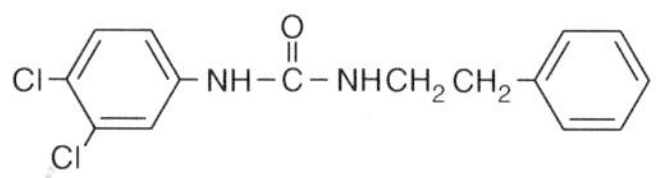

5 gm

S61,314-2
1-(2,4-DICHLOROPHENYL)-3-PHENETHYLUREA

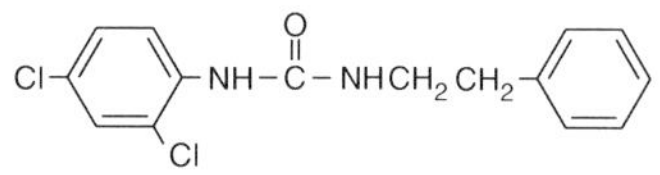

5 gm

S64,715-2
1-(2,5-DICHLOROPHENYL)-3-PHENETHYLUREA

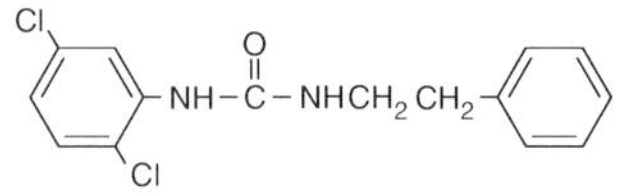

5 gm

S61,181-6
1-(5-CHLORO-2-METHYLPHENYL)-3-PHENETHYLUREA

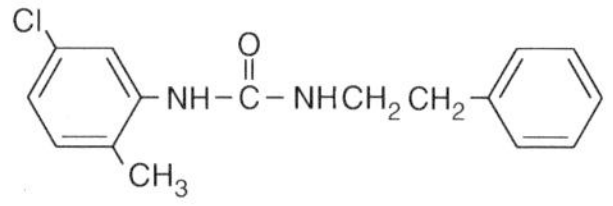

5 gm

S61,111-5
1-PHENETHYL-3-(2,4-XYLYL)UREA

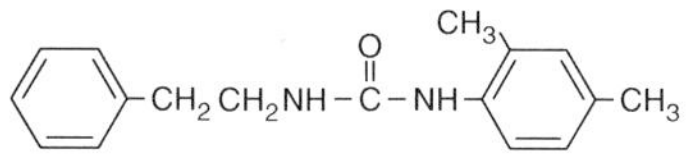

5 gm

S69,266-2
4-(2,6-DIMETHYLPHENYL)-1-[2-(METHYLAMINO)BENZOYL]SEMICARBAZIDE

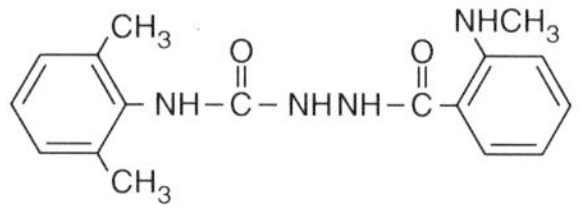

250 mg

S94,354-1
3-(UREIDOMETHYLENE)-2,4-CHROMANEDIONE

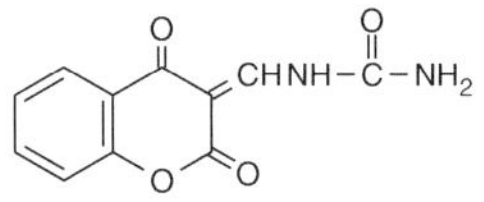

5 gm

S69,113-5
4-(3,5-DIMETHYLPHENYL)-1-[2-(METHYLAMINO)BENZOYL]SEMICARBAZIDE

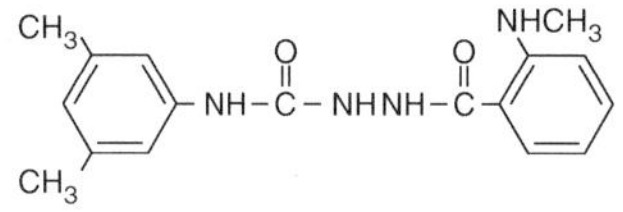

250 mg

S63,384-4
4-(2-CHLOROPHENYL)-1,1-DIMETHYLSEMICARBAZIDE

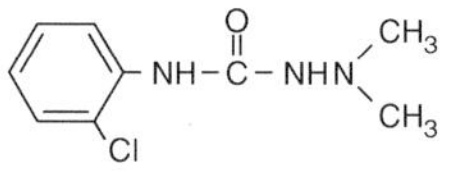

5 gm

S86,267-3
1-PHENYLBIURET

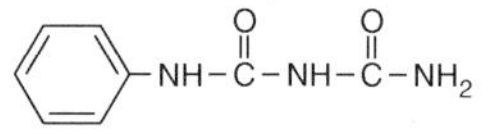

250 mg

S69,489-4
1-(4-ANILINOPHENYL)-3-(P-TOLYL)UREA

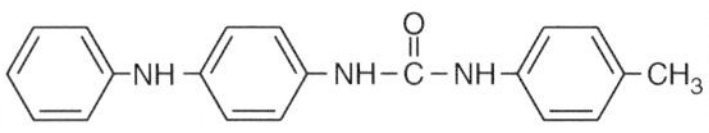

250 mg

S79,479-1
1-(4-ANILINOPHENYL)-3-(2,5-XYLYL)UREA

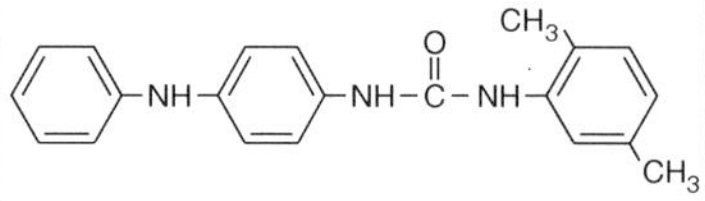

250 mg

S86,877-9
4-(3-PROPYLUREIDO)BENZOIC [4-(DIMETHYLAMINO)BENZYLIDENE]-HYDRAZIDE

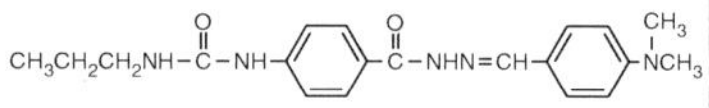

250 mg

S90,009-5
4-(3,3-DIETHYLUREIDO)BENZOIC [4-(DIMETHYLAMINO)BENZYLIDENE]-HYDRAZIDE

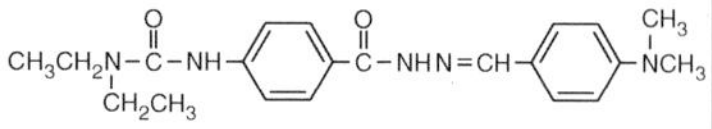

250 mg

S71,577-8
1-[2-(N-ETHYLANILINO)ETHYL]-3-PROPYLUREA

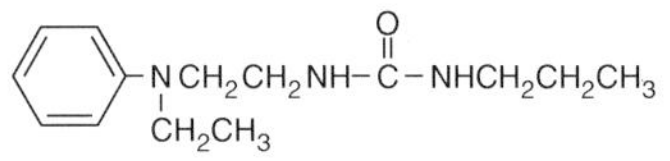

250 mg

S80,095-3
1-[4-(DIMETHYLAMINO)BENZYLIDENE]-4-PHENYLSEMICARBAZIDE

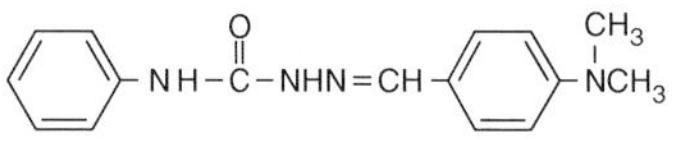

250 mg

S60,466-6
1-[4-(DIMETHYLAMINO)BENZOYL]-4-PHENYLSEMICARBAZIDE

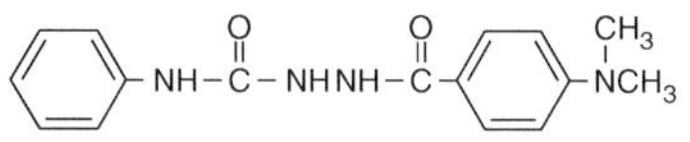

5 gm

S63,925-7
1,1-DIMETHYL-4-(1-NAPHTHYL)-SEMICARBAZIDE

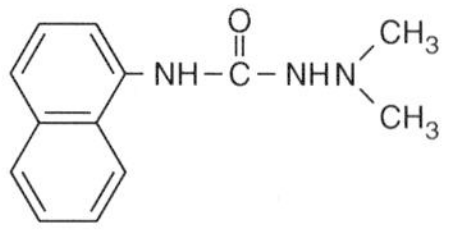

5 gm

S58,177-1
1-METHYL-4-PENTADECYL-2(1H)-QUINOLONE SEMICARBAZONE

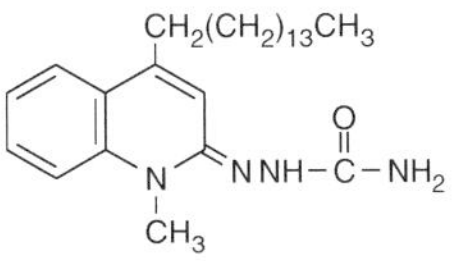

250 mg

S94,676-1
3-PHENYL-2-THIO-2,4-THIAZOLIDINEDIONE

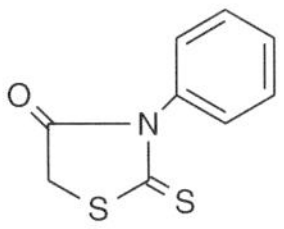

250 mg

S50,286-3
3-(4-CHLOROPHENYL)-1,1-DIMETHYLUREA

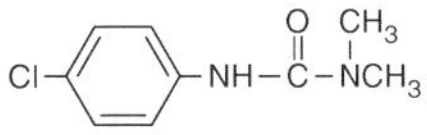

250 mg

S70,815-1
3-(2,6-DIISOPROPYLPHENYL)-1,1-DIMETHYLUREA

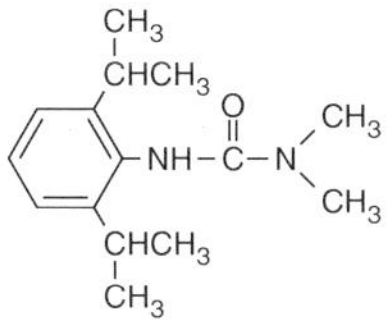

5 gm

S86,831-0
3-(3-ACETYLPHENYL)-1-BENZYL-1-METHYLUREA

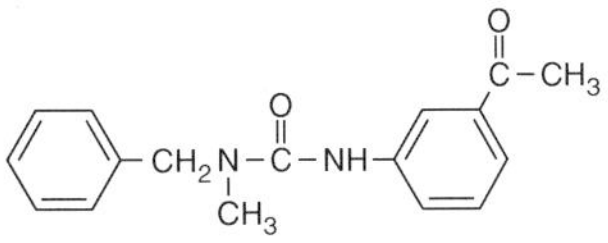

250 mg

S67,765-5
(2,4,6-TRIMETHYLPHENYL)UREA

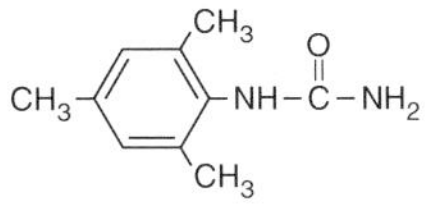

5 gm

S83,468-8
PENTANOPHENONE SEMICARBAZONE

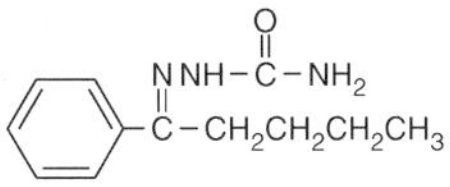

250 mg

S83,802-0
2-BENZYL-4,4-DIPHENYL-1-(DIPHENYLMETHYLENE)SEMICARBAZIDE

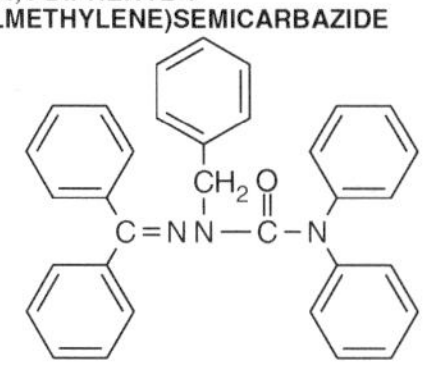

250 mg

S68,377-9
1-METHYL-1-[ALPHA-(4'-METHYLPHENACYL)BENZYL]UREA

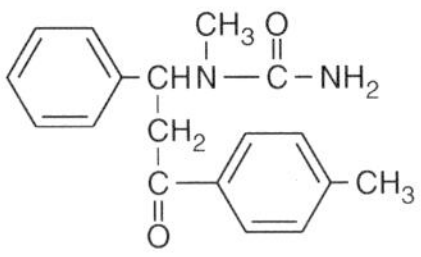

5 gm

S66,941-5
1-BENZYL-1-ISOPROPYL-3-(4-NITROPHENYL)UREA

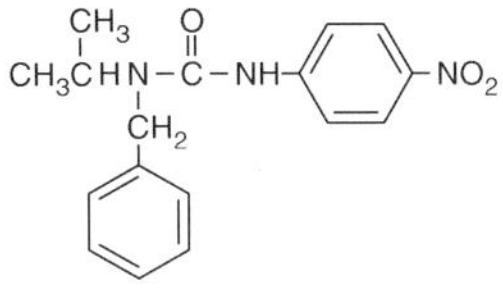

5 gm

S59,426-1
1-BENZYL-3-(3-CHLOROPHENYL)UREA

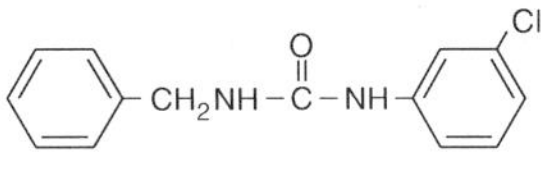

5 gm

S60,635-9
1-BENZYL-3-(M-TOLYL)UREA

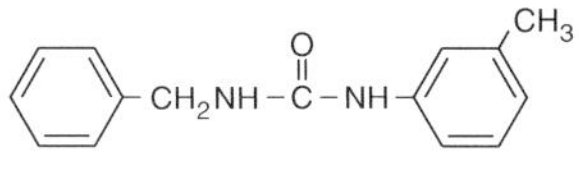

5 gm

S63,832-3
1-BENZYL-3-(3-CHLOROPHENYL)-1-METHYLUREA

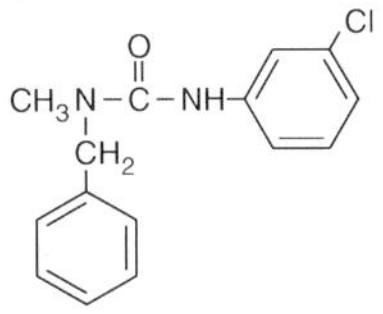

5 gm

S62,485-3
1-BENZYL-3-(3-CHLOROPHENYL)-1-ETHYLUREA

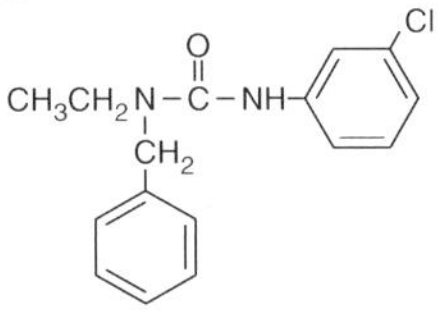

5 gm

S62,478-0
1-BENZYL-3-(3-CHLOROPHENYL)-1-ISOPROPYLUREA

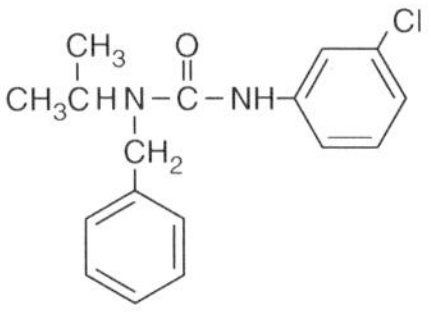

5 gm

S59,431-8
3-(3-CHLOROPHENYL)-1,1-DIBENZYLUREA

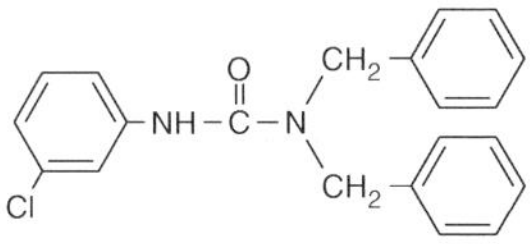

5 gm

S60,712-6
1,1-DIBENZYL-3-(M-TOLYL)UREA

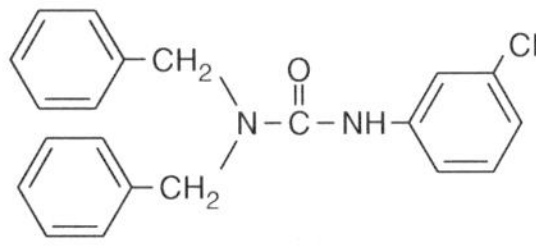

5 gm

S60,348-1
1-BENZYL-3-CYCLOHEXYLUREA

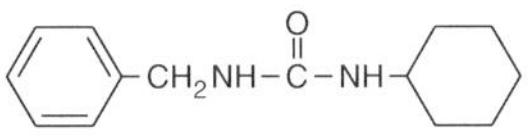

5 gm

S63,468-9
3-CYCLOHEXYL-1,1-DIBENZYLUREA

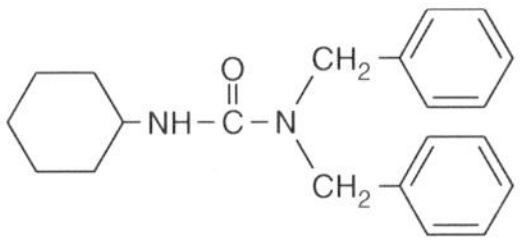

5 gm

S66,950-4
1-BENZYL-3-CYCLOHEXYL-3-ISOPROPYLUREA

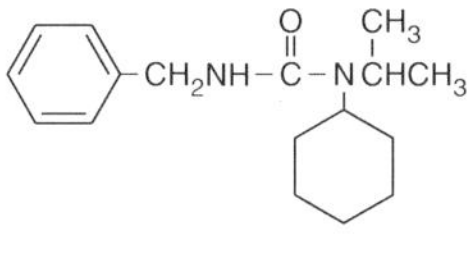

5 gm

S66,908-3
3-BENZYL-1,1-DICYCLOHEXYLUREA

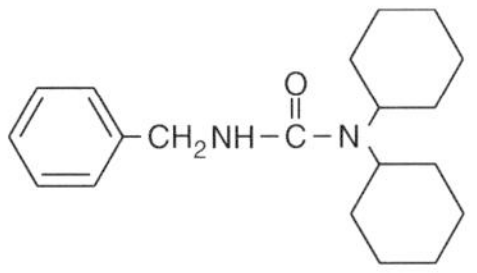

5 gm

S69,848-2
1-BENZYL-3-ISOPROPYLUREA

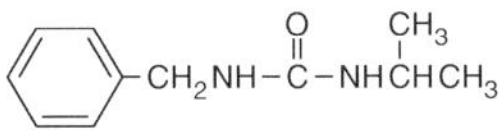

250 mg

S77,476-6
1-BENZYL-3-ISOPROPYL-1-METHYLUREA

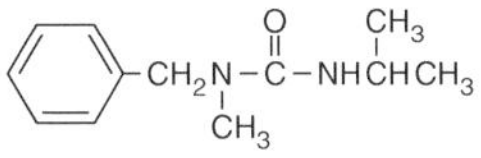

250 mg

S74,326-7
1-TERT-BUTYL-3-(ALPHA-METHYLBENZYL)UREA

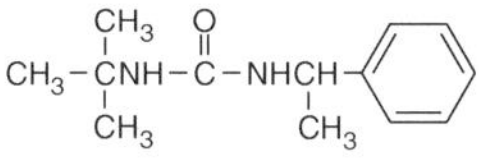

5 gm

S75,036-0
1-BENZYL-3-TERT-BUTYL-1-METHYLUREA

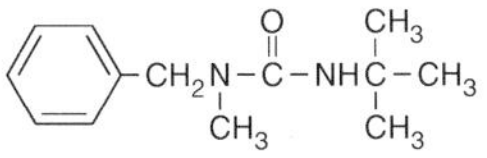

5 gm

S51,777-1
1-BENZYL-3-PHENYLUREA

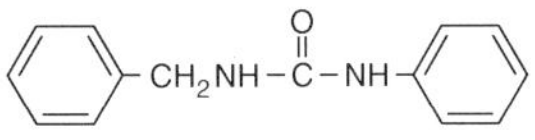

5 gm

S62,436-5
1-BENZYL-1-METHYL-3-PHENYLUREA

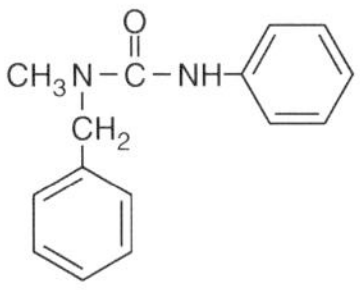

5 gm

S62,435-7
1-BENZYL-1-ETHYL-3-PHENYLUREA

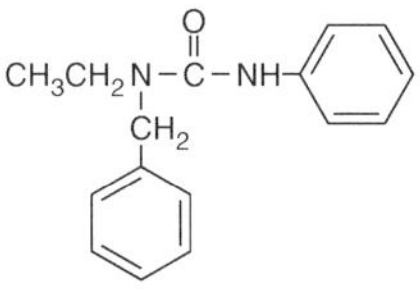

5 gm

S62,421-7
1-BENZYL-1-ISOPROPYL-3-PHENYLUREA

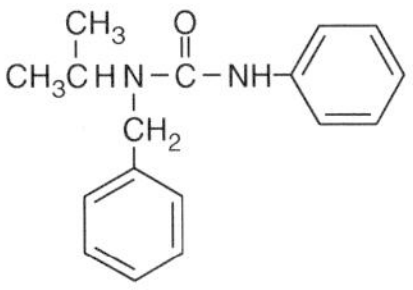

5 gm

S70,639-6
1,3-BIS(ALPHA-METHYLBENZYL)UREA

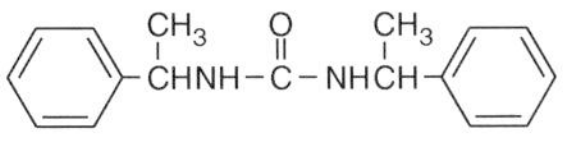

5 gm

S74,328-3
1,1-DIBENZYL-3-ETHYLUREA

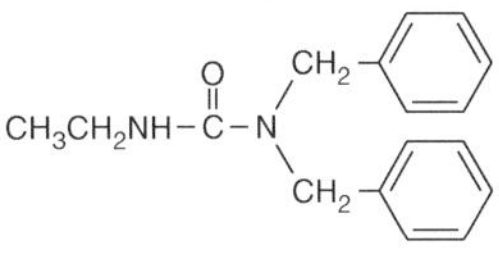

250 mg

S72,431-9
3-ALLYL-1,1-DIBENZYLUREA

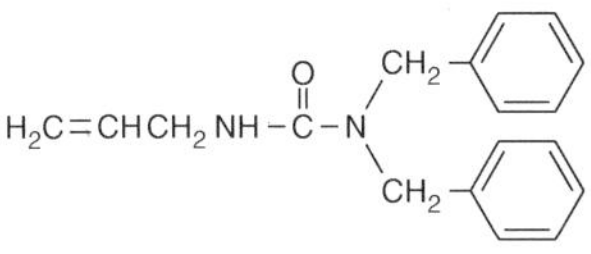

5 gm

S60,790-8
3-BUTYL-1,1-DIBENZYLUREA

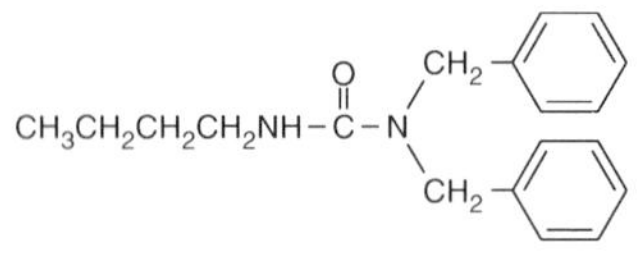

5 gm

S77,519-3
1-BENZYL-3-ETHYL-1-METHYL-3-(ALPHA-METHYLBENZYL)UREA

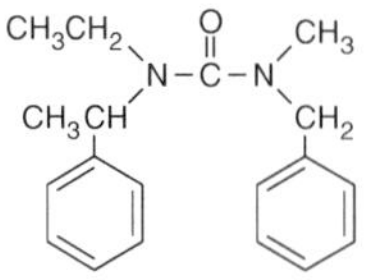

250 mg

S66,907-5
1,1,3-TRIBENZYLUREA

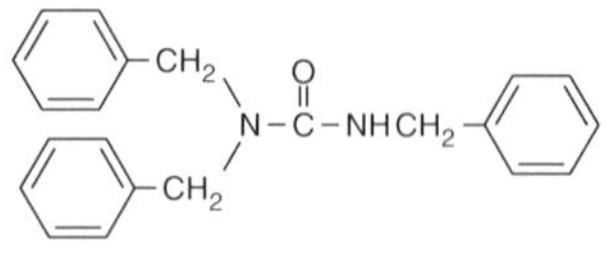

5 gm

S63,206-6
1,1'-HEXAMETHYLENEBIS(3-BENZYLUREA)

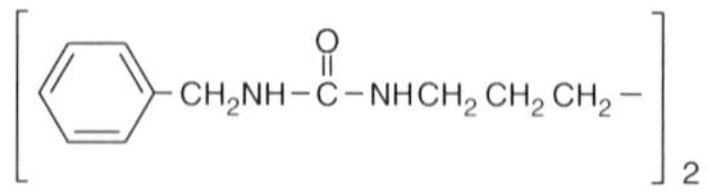

5 gm

S75,297-5
1-BENZYL-1-METHYL-3-OCTADECYLUREA

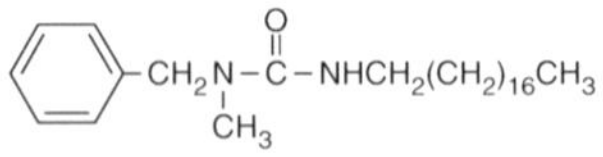

250 mg

S64,699-7
3,3'-HEXAMETHYLENEBIS(1,1-DIBENZYLUREA)

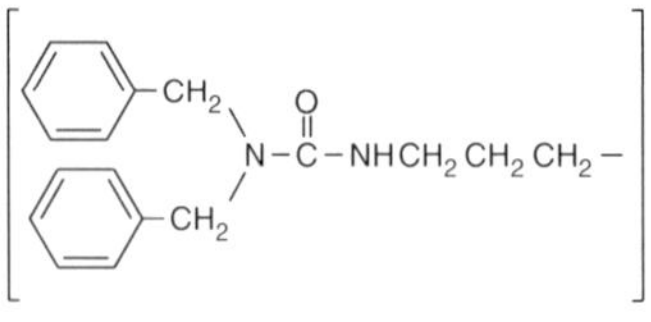

5 gm

S63,477-8
4,4'-METHYLENEBIS(1-BENZYL-3-PHENYLUREA)

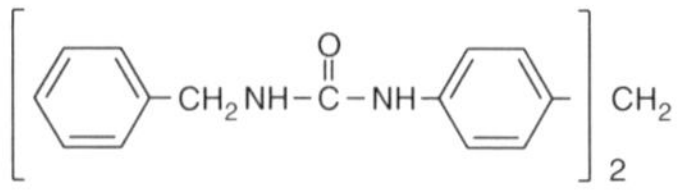

250 mg

S63,489-1
4,4'-METHYLENEBIS(1,1-DIBENZYL-3-PHENYLUREA)

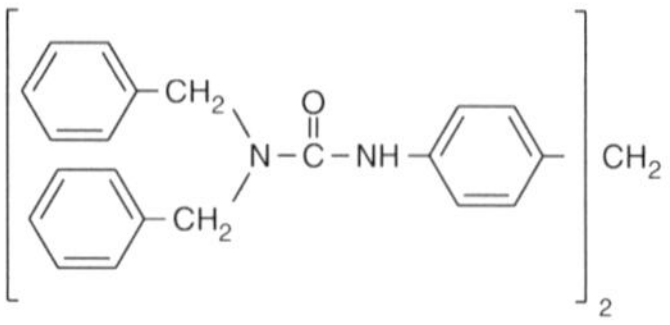

5 gm

S71,374-0
1-BENZYL-3-(2,6-DIISOPROPYLPHENYL)UREA

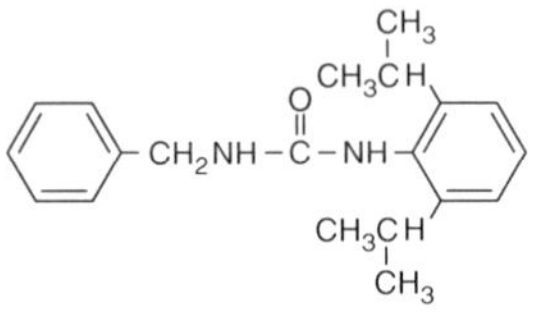

250 mg

S59,003-7
1-BENZYL-3-(1-NAPHTHYL)UREA

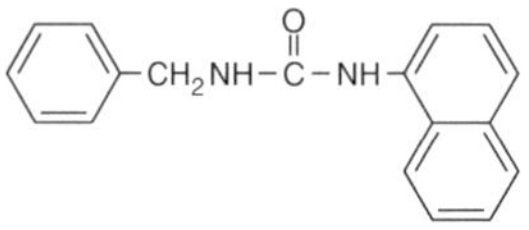

5 gm

S58,995-0
1,1-DIBENZYL-3-(1-NAPHTHYL)UREA

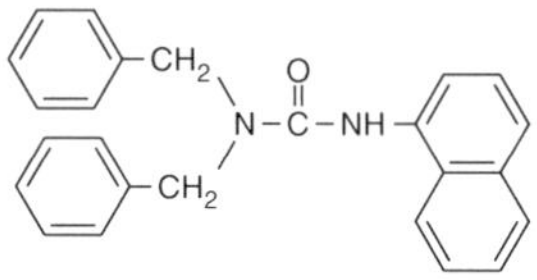

5 gm

S71,145-4
1,1'-(1,5-NAPHTHALENEDIYL)BIS(3-BENZYLUREA)

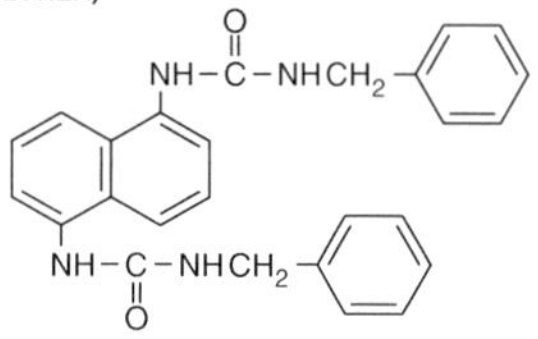

250 mg

S64,025-5
1-CYCLOHEXYL-3-(O-TOLYL)UREA

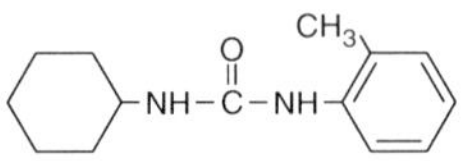

5 gm

S63,405-0
1-BENZYL-3-(2-CHLOROPHENYL)UREA

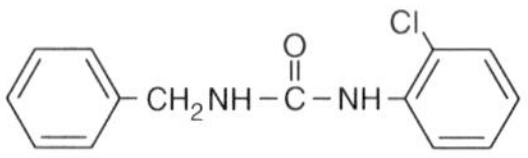

5 gm

S64,032-8
1-BENZYL-3-(O-TOLYL)UREA

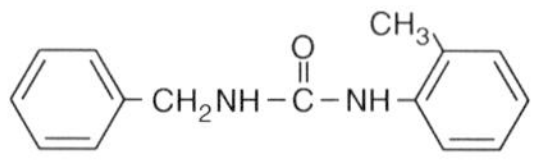

5 gm

S65,123-0
1-BENZYL-1-METHYL-3-(O-TOLYL)UREA

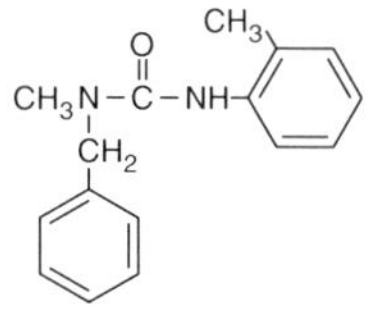

5 gm

S65,148-6
1-BENZYL-3-(2-ETHYLPHENYL)UREA

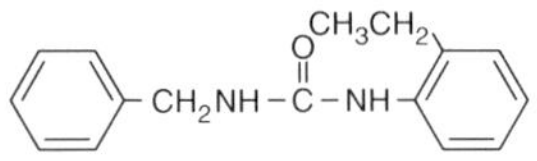

5 gm

S63,801-3
1-BENZYL-3-(2-CHLOROPHENYL)-1-ETHYLUREA

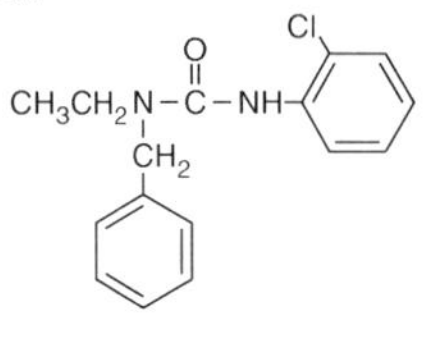

5 gm

S64,981-3
1-BENZYL-1-ETHYL-3-(O-TOLYL)UREA

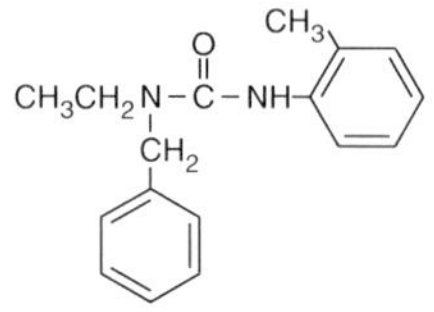

5 gm

S62,177-3
3-(2-CHLOROPHENYL)-1,1-DIBENZYLUREA

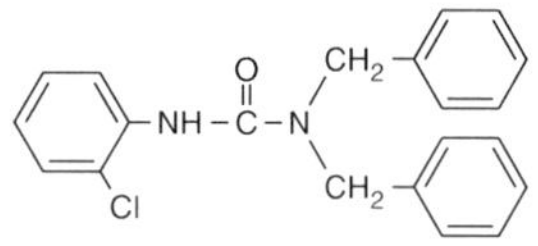

5 gm

S60,665-3
1,1-DIBENZYL-3-(O-TOLYL)UREA

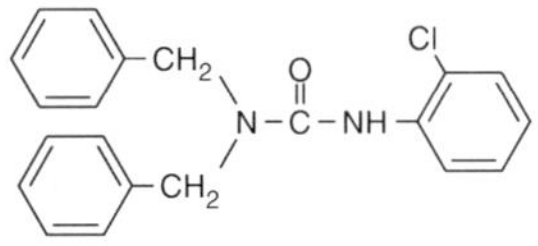

5 gm

S66,608-4
1,1-DIBENZYL-3-(2-ETHYLPHENYL)UREA

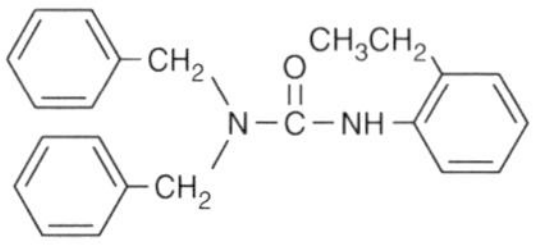

5 gm

S84,727-5
1-(3,4-DIMETHYLBENZYL)-3-(2-ETHYLPHENYL)UREA

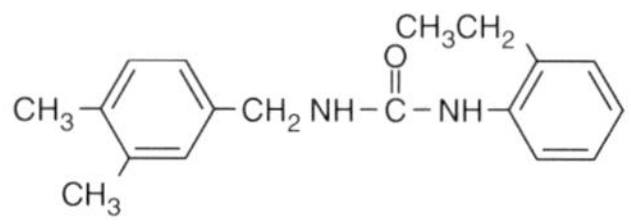

250 mg

S59,149-1
1-BENZYL-3-(4-CHLOROPHENYL)UREA

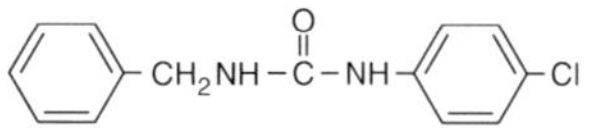

5 gm

S64,024-7
1-BENZYL-3-(4-BROMOPHENYL)UREA

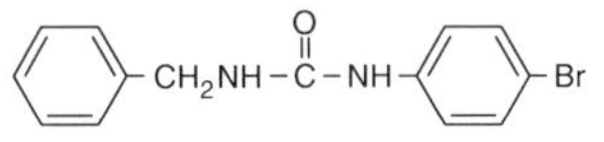

5 gm

S64,034-4
1-BENZYL-3-(P-TOLYL)UREA

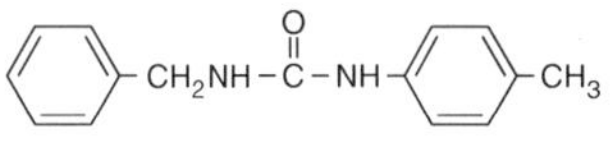

5 gm

S59,138-6
3-(4-CHLOROPHENYL)-1,1-DIBENZYLUREA

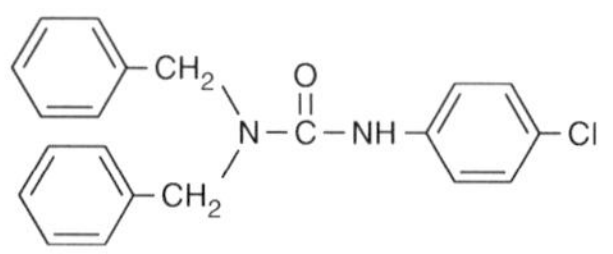

5 gm

S59,544-6
3-(4-BROMOPHENYL)-1,1-DIBENZYLUREA

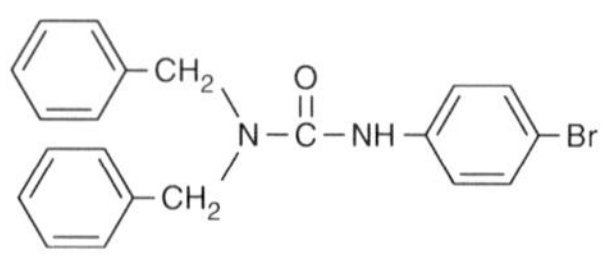

5 gm

S61,333-9
1,1-DIBENZYL-3-(P-TOLYL)UREA

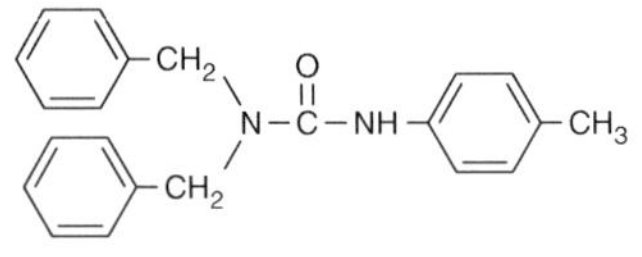

5 gm

S74,339-9
1,1'-HEXAMETHYLENEBIS[3-(4-METHYLBENZYL)UREA]

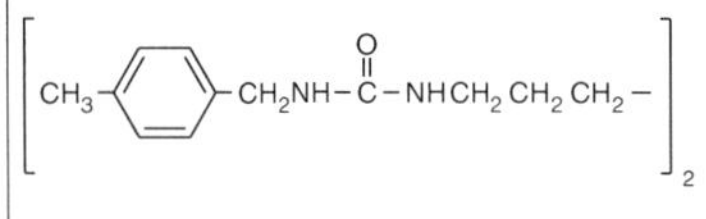

5 gm

S66,031-0
4,4'-METHYLENEBIS(1-BENZYL-1-METHYL-3-PHENYLUREA)

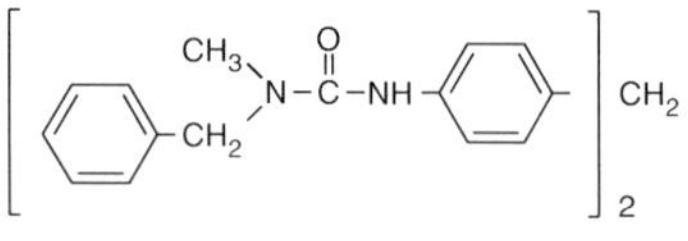

5 gm

S66,030-2
4,4'-METHYLENEBIS(1-BENZYL-1-ETHYL-3-PHENYLUREA)

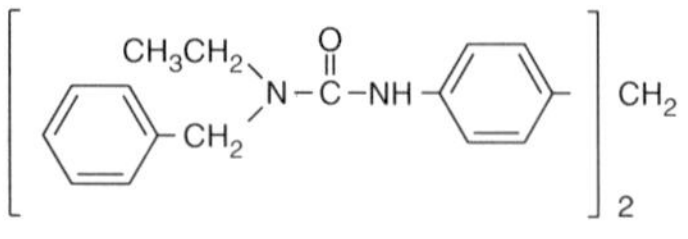

5 gm

S66,014-0
4,4'-METHYLENEBIS(1-BENZYL-1-ISOPROPYL-3-PHENYLUREA)
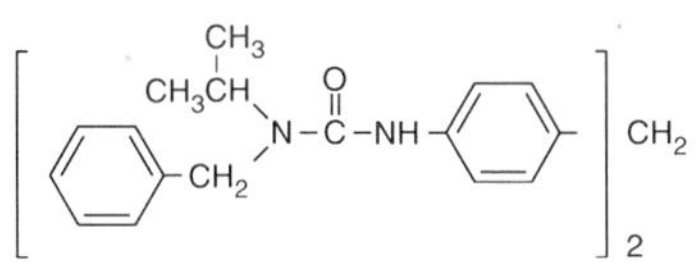
5 gm

S62,378-4
1-BENZYL-3-(2,3-DICHLOROPHENYL)UREA
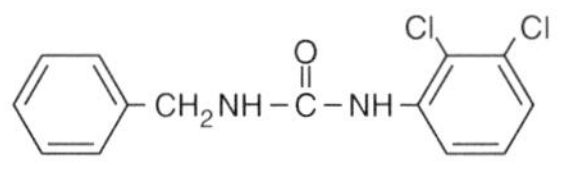
5 gm

S64,464-1
1-BENZYL-3-(2,6-DICHLOROPHENYL)UREA
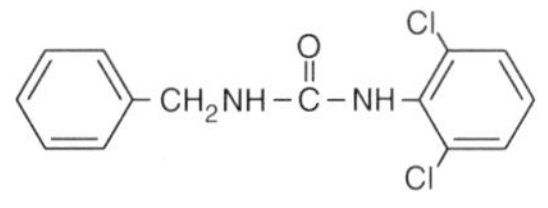
5 gm

S65,240-7
1-BENZYL-3-(2-CHLORO-6-METHYLPHENYL)UREA
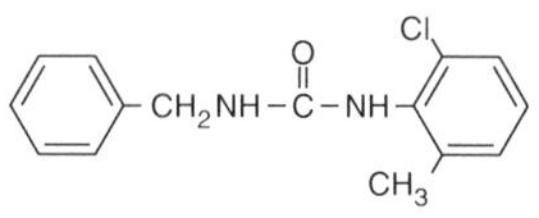
5 gm

S64,405-6
1-BENZYL-3-(2,3-XYLYL)UREA
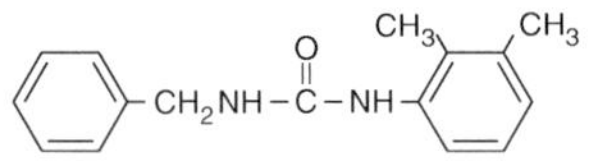
5 gm

S65,386-1
1-BENZYL-3-(2,6-XYLYL)UREA
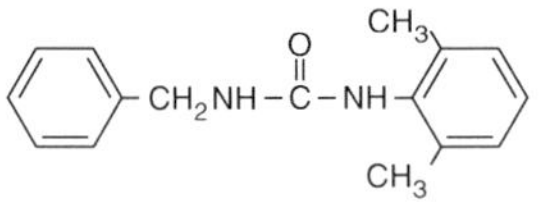
5 gm

S66,074-4
1-BENZYL-3-(2-ETHYL-6-METHYLPHENYL)UREA
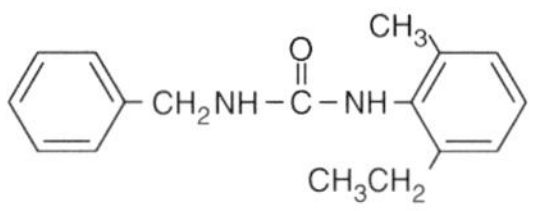
5 gm

S66,600-9
1-BENZYL-3-(2,6-DIETHYLPHENYL)UREA
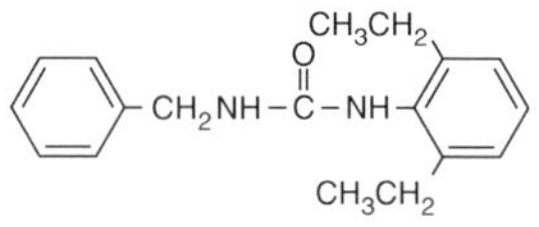
5 gm

S61,196-4
1,1-DIBENZYL-3-(2,6-DICHLOROPHENYL)UREA
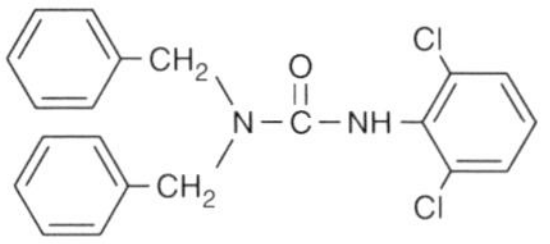
5 gm

S61,205-7
1,1-DIBENZYL-3-(2,3-DICHLOROPHENYL)UREA
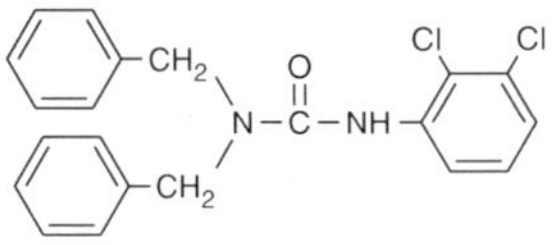
5 gm

S64,664-4
3-(2-CHLORO-6-METHYLPHENYL)-1,1-DIBENZYLUREA
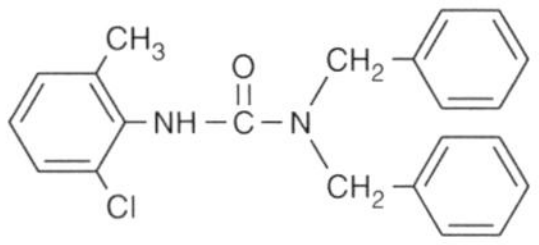
5 gm

S64,999-6
1,1-DIBENZYL-3-(2,3-XYLYL)UREA
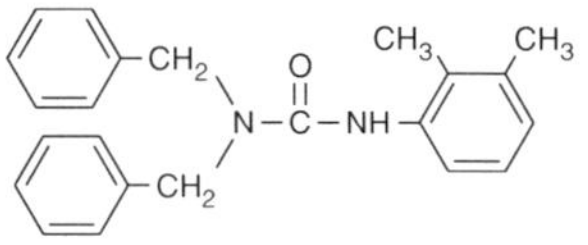
5 gm

S65,046-3
1,1-DIBENZYL-3-(2,6-XYLYL)UREA
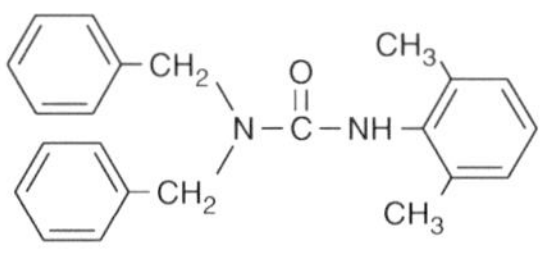
5 gm

S66,175-9
1,1-DIBENZYL-3-(2-ETHYL-6-METHYLPHENYL)UREA
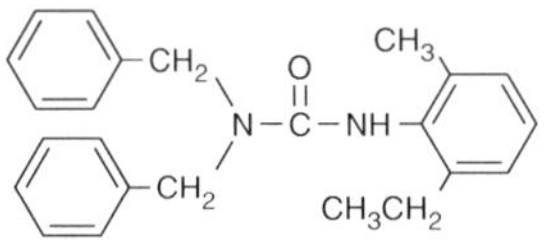
5 gm

S66,579-7
1,1-DIBENZYL-3-(2,6-DIETHYLPHENYL)UREA
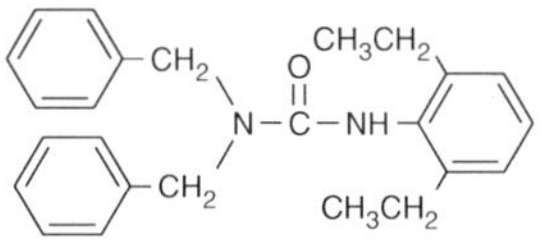
5 gm

S66,594-0
1,1-DIBENZYL-3-(2,6-DIISOPROPYLPHENYL)UREA
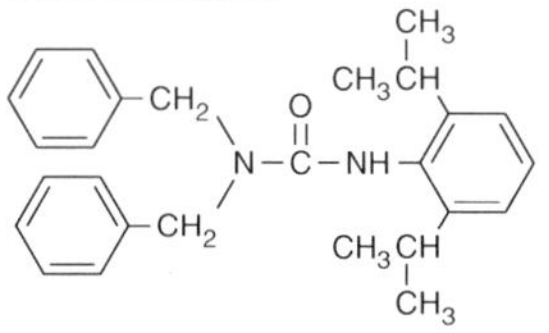
5 gm

S84,731-3
1-(3,4-DIMETHYLBENZYL)-3-(2,6-XYLYL)UREA
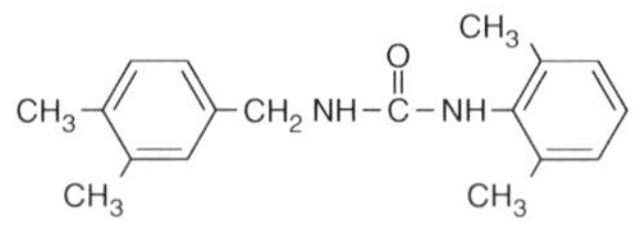
250 mg

S74,707-6
1-(TERT-BUTYL)-3-(2,5-DIMETHYLBENZYL)UREA
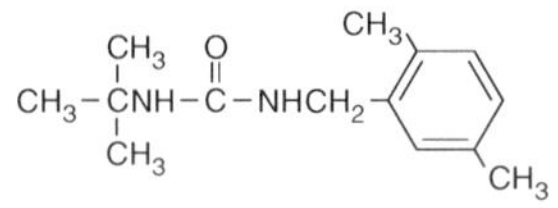
5 gm

S61,273-1
1-BENZYL-3-(2,4-DICHLOROPHENYL)UREA
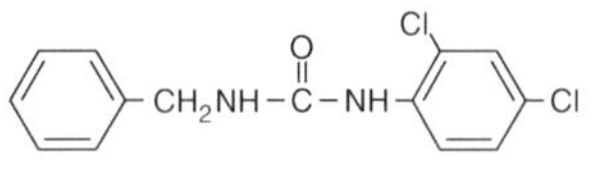
5 gm

S61,353-3
1-BENZYL-3-(3,4-DICHLOROPHENYL)UREA
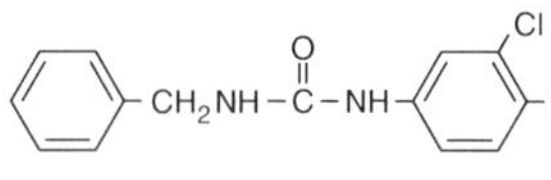
5 gm

S60,339-2
1-BENZYL-3-(4-CHLORO-2-METHYLPHENYL)UREA
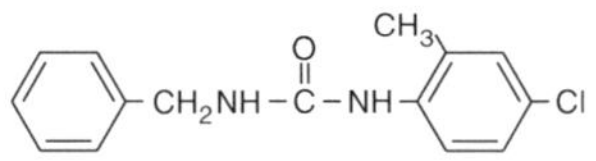
5 gm

S60,954-4
1-BENZYL-3-(5-CHLORO-2-METHYLPHENYL)UREA
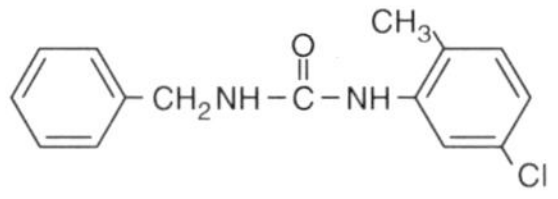
5 gm

S63,809-9
1-BENZYL-3-(3-CHLORO-4-METHYLPHENYL)UREA
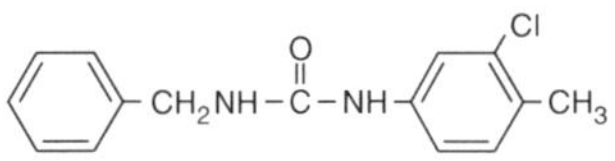
5 gm

S64,413-7
1-BENZYL-3-(3,4-DICHLOROPHENYL)-1-METHYLUREA
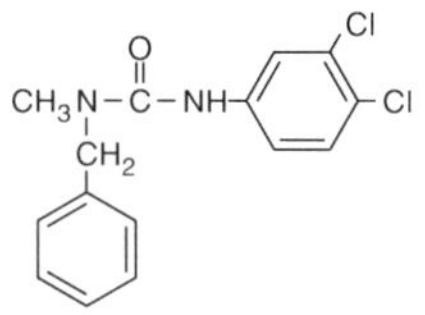
5 gm

S60,629-4
1-BENZYL-3-(2,4-XYLYL)UREA
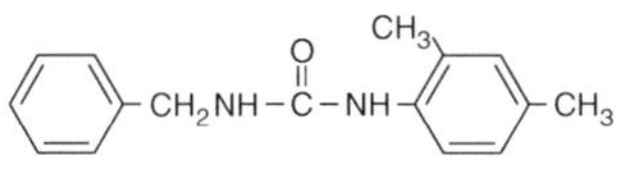
5 gm

S65,047-1
1-BENZYL-3-(2,5-XYLYL)UREA
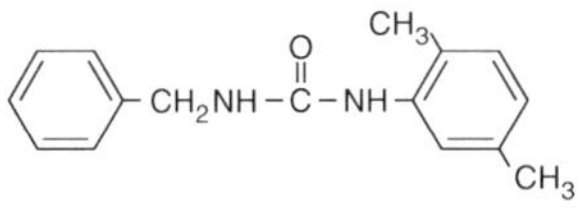
5 gm

S66,372-7
1-BENZYL-3-(3,4-XYLYL)UREA
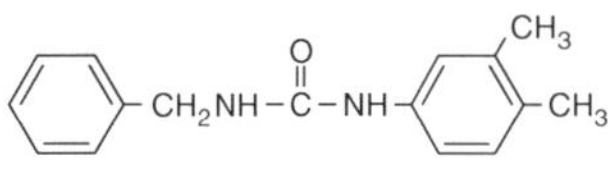
5 gm

S62,816-6
1-BENZYL-3-(3,4-DICHLOROPHENYL)-1-ETHYLUREA
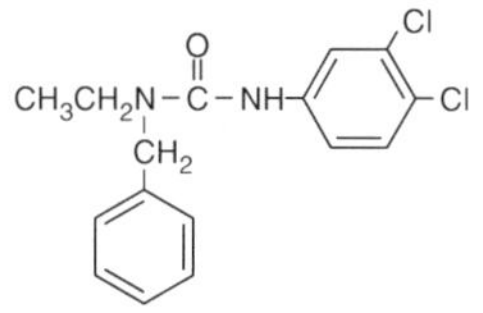
5 gm

S60,740-1
1,1-DIBENZYL-3-(3,4-DICHLOROPHENYL)UREA
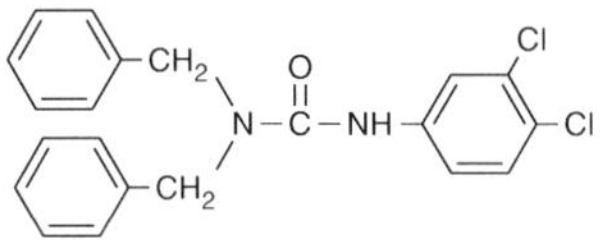
5 gm

S61,194-8
1,1-DIBENZYL-3-(2,5-DICHLOROPHENYL)UREA
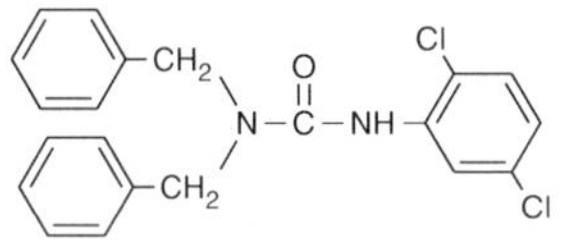
5 gm

S61,266-9
1,1-DIBENZYL-3-(2,4-DICHLOROPHENYL)UREA
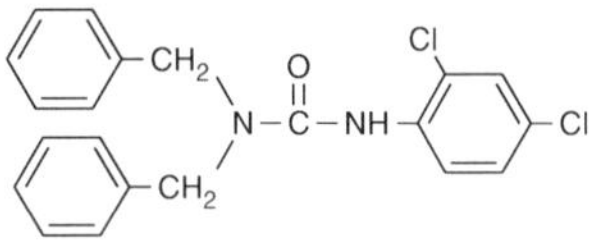
5 gm

S60,455-0
3-(4-CHLORO-2-METHYLPHENYL)-1,1-DIBENZYLUREA
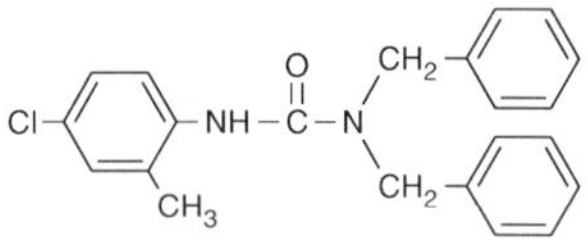
5 gm

S61,062-3
3-(3-CHLORO-4-METHYLPHENYL)-1,1-DIBENZYLUREA

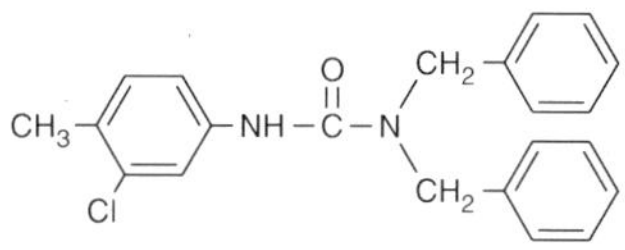

5 gm

S60,633-2
1,1-DIBENZYL-3-(2,4-XYLYL)UREA

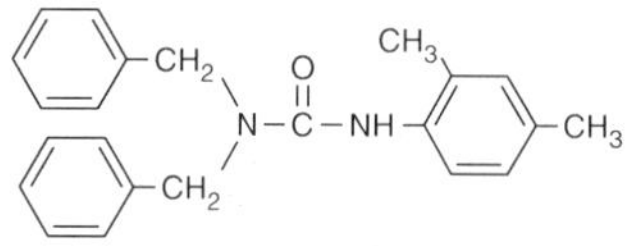

250 mg

S64,382-3
1,1-DIBENZYL-3-(2,5-XYLYL)UREA

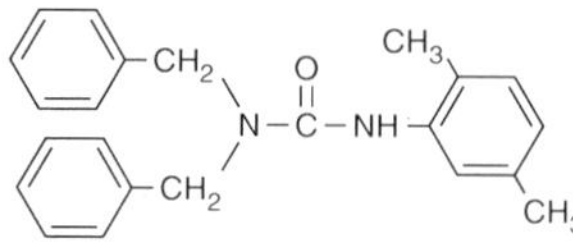

5 gm

S66,380-8
1,1-DIBENZYL-3-(3,4-XYLYL)UREA

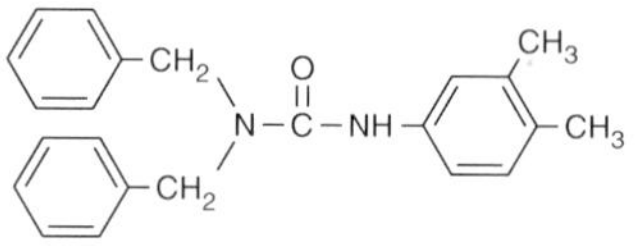

5 gm

S66,133-3
1-BENZYL-3-(3,5-XYLYL)UREA

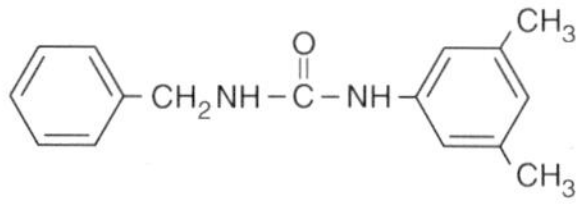

5 gm

S71,163-2
1,1-DIBENZYL-3-PHENYLUREA

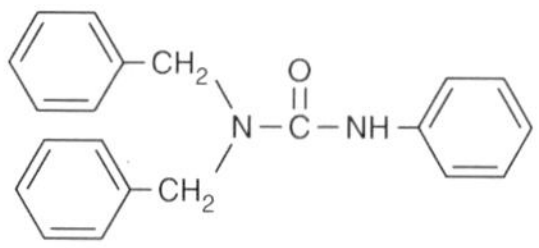

5 gm

S66,146-5
1,1-DIBENZYL-3-(3,5-XYLYL)UREA

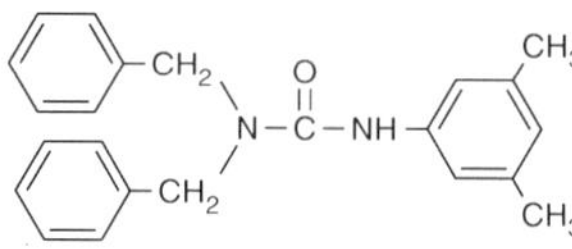

5 gm

S69,491-6
4-PROPYL-1-[2-(3-PROPYLUREIDO)-BENZOYL]SEMICARBAZIDE

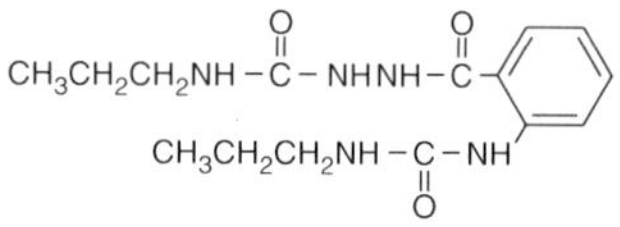

250 mg

S70,063-0
4-METHYL-1-[4-(3-METHYLUREIDO)-BENZOYL]SEMICARBAZIDE

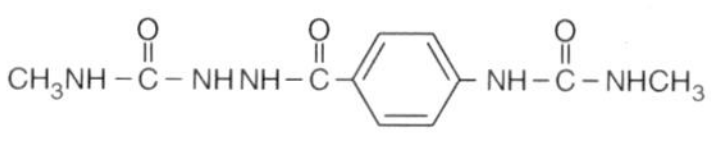

250 mg

S70,071-1
4-METHYL-1-(4-METHYLBENZOYL)-SEMICARBAZIDE

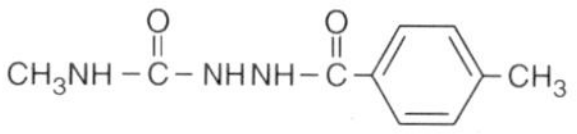

250 mg

S69,204-2
N-[3-(3,5-DIMETHYLPHENYL)UREIDO] OXAMIDE

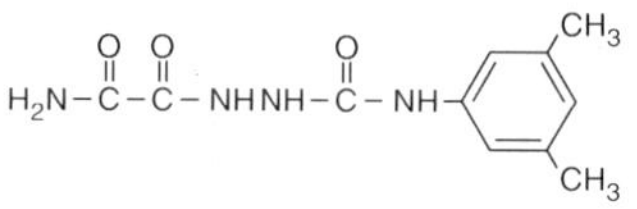

250 mg

S68,548-8
4-(ETHOXYCARBONYLMETHYL)-1-(PHENYLACETYL)SEMICARBAZIDE

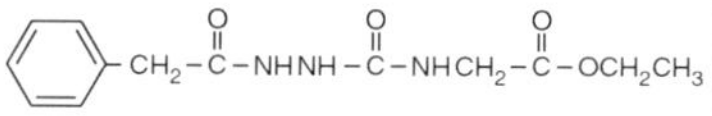

250 mg

S72,354-1
2-(3,3-DIETHYLUREIDO)ACETIC ACID N2-(N-PHENYLCARBAMOYL)HYDRAZIDE

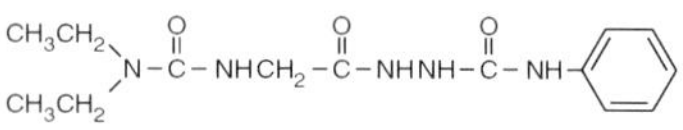

5 gm

S69,548-3
1-(DIETHYLUREIDOACETYL)-4-(2-ETHYLPHENYL)SEMICARBAZIDE

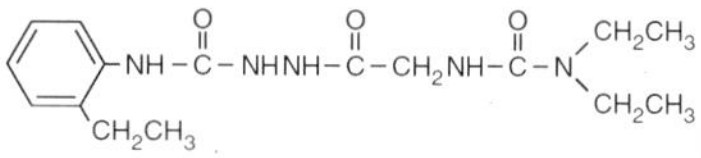

250 mg

S69,466-5
1-(DIETHYLUREIDOACETYL)-4-(2,4-DIMETHYLPHENYL)SEMICARBAZIDE

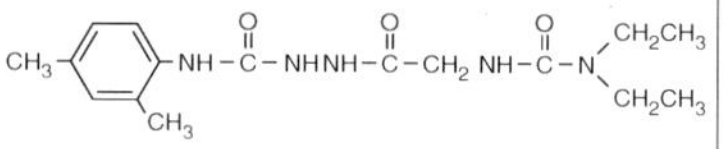

250 mg

S69,186-0
1-(DIETHYLUREIDOACETYL)-4-(1-NAPHTHYL)SEMICARBAZIDE

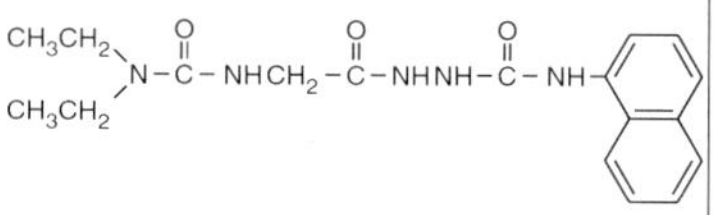

250 mg

S68,406-6
4,4-DIPHENYL-1-[N-(HYDRAZINOCARBONYLMETHYL)CARBAMOYL]-SEMICARBAZIDE

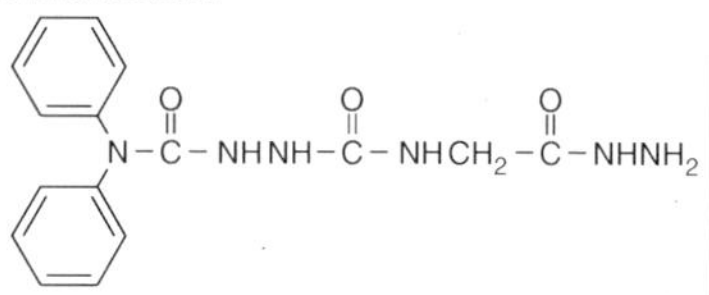

250 mg

S69,020-1
1-ACETYL-4-(2,6-DIMETHYLPHENYL)-SEMICARBAZIDE

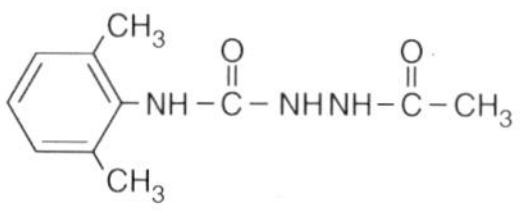

250 mg

S68,349-3
1-(2,6-DIMETHYLPHENYL)-6-(2-METHYLPHENYL)BIUREA

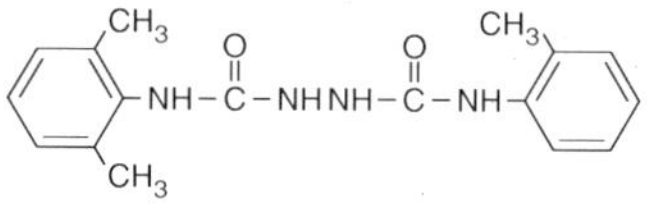

250 mg

S68,073-7
1,1,5,5-TETRAPHENYLCARBAZIDE

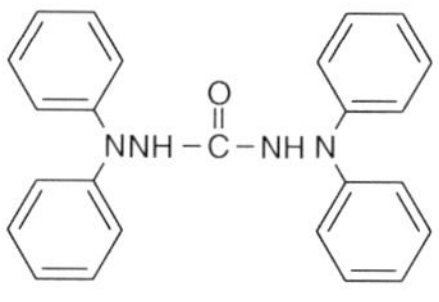

5 gm

S91,488-6
3-METHYL-1,2,4-CYCLOPENTANETRIONE 1-SEMICARBAZONE

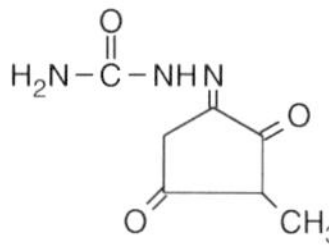

250 mg

S86,874-4
4-(3-ALLYLUREIDO)BENZOIC (3-PHENYLPROPYLIDENE)HYDRAZIDE

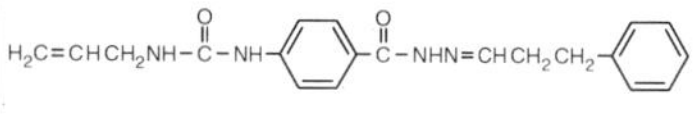

250 mg

S80,092-9
1-(4-ACETAMIDOBENZYLIDENE)-4-PHENYLSEMICARBAZIDE

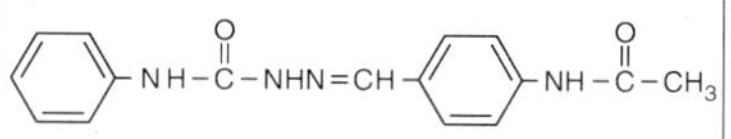

250 mg

S87,547-3
1-(ALPHA-METHYLBENZYLIDENE)-SEMICARBAZIDE

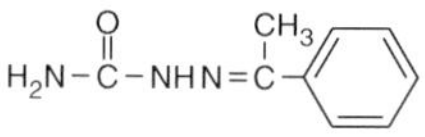

250 mg

S83,773-3
1-(DIPHENYLMETHYLENE)-2,4,4-TRIPHENYLSEMICARBAZIDE

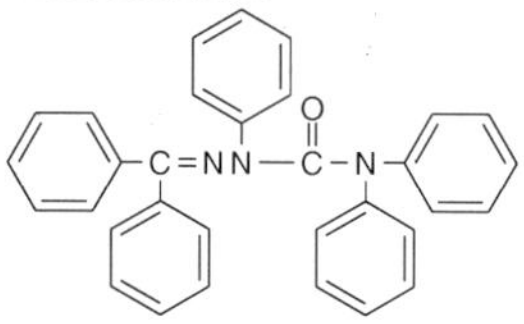

250 mg

S89,573-3
1,2-PHENYLENEDIACETALDEHYDE DISEMICARBAZONE

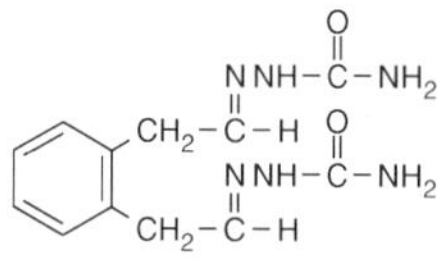

250 mg

S84,007-6
4'-METHYLACETOPHENONE SEMICARBAZONE

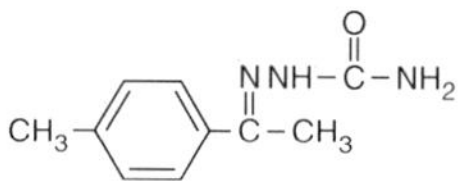

250 mg

S60,946-3
2-[1-METHYL-3-(2,4-XYLYL)UREIDO]-ETHYL N-(2,4-XYLYL)CARBAMATE

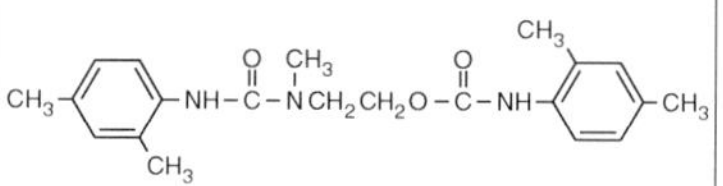

5 gm

S90,254-3
ETHYL 2-[3-(3,4-XYLYL)UREIDO]-ACETATE

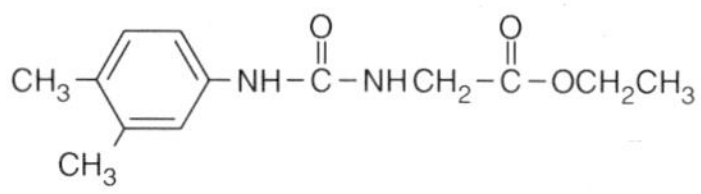

250 mg

S90,253-5
ETHYL 2-[3-(1-NAPHTHYL)UREIDO]-ACETATE

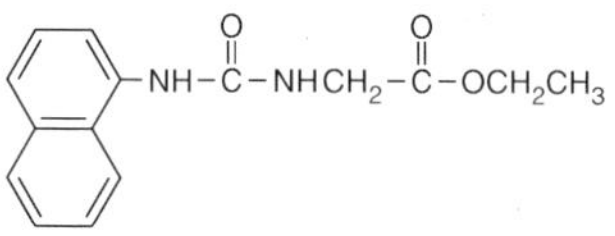

250 mg

S59,756-2
1,1-DIBENZYL-3-(2-METHOXYPHENYL)UREA

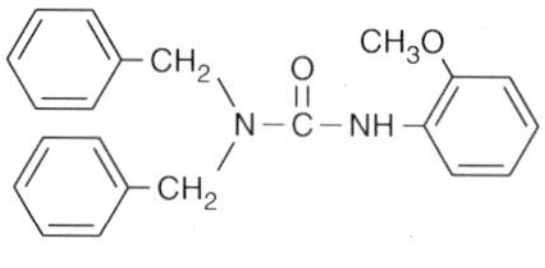

5 gm

S65,226-1
1-(3-CHLOROPHENYL)-3-HYDROXYUREA

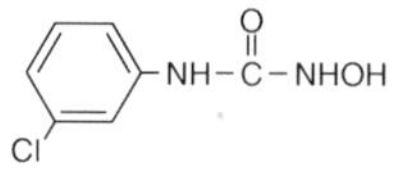

5 gm

S71,679-0
1-HYDROXY-3-PHENYLUREA

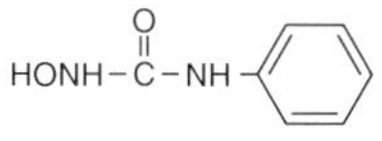

5 gm

S58,626-9
1-METHOXY-1-METHYL-3-PHENYLUREA

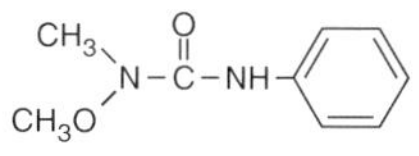

5 gm

S58,619-6
3-(4-BROMOPHENYL)-1-METHOXY-1-METHYLUREA

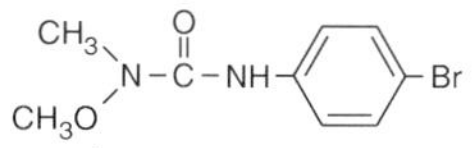

5 gm

S73,544-2
1-(2,6-DIISOPROPYLPHENYL)-3-HYDROXYUREA

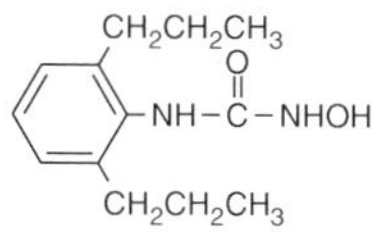

250 mg

S73,555-8
1-HYDROXY-3-(2,4-XYLYL)UREA

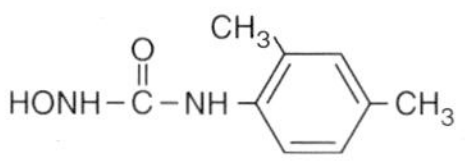

250 mg

S71,035-0
1-HYDROXY-3-(1-NAPHTHYL)UREA

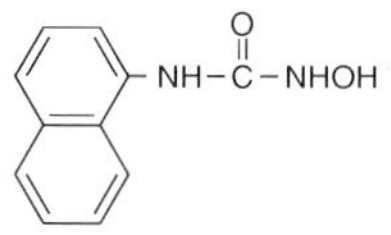

50 mg

S65,431-0
1-(4-NITROBENZOYL)SEMICARBAZIDE

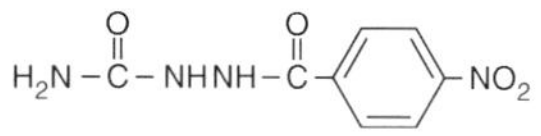

250 mg

S67,417-6
2-[3-(3-NITROPHENYL)UREIDO]ACETIC ACID HYDRAZIDE

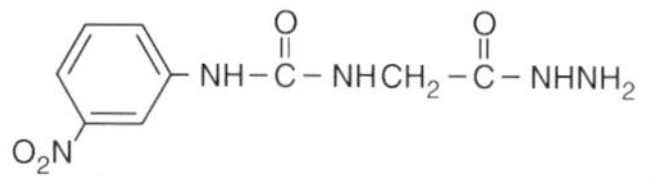

5 gm

S72,280-4
ETHYL 2-[3-(2-BENZOYLPHENYL)UREIDO]-ACETATE

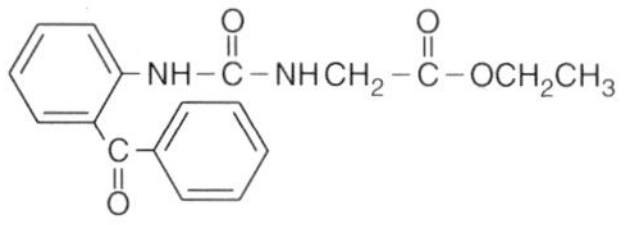

250 mg

S79,415-5
ETHYL 4-[3-(3-ACETYLPHENYL)UREIDO]-BENZOATE

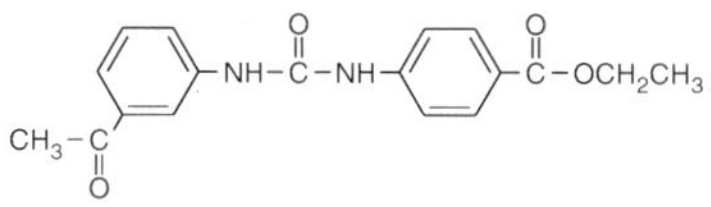

250 mg

S86,814-0
3-(3-ACETYLPHENYL)-1-CYCLOHEXYL-1-METHYLUREA

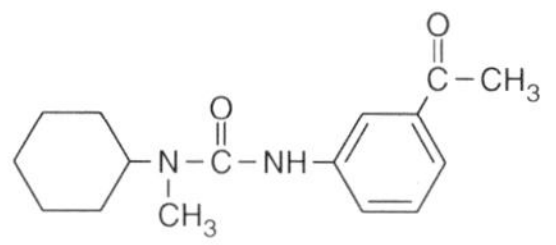

250 mg

S69,593-9
1-(2-BENZOYLPHENYL)-3-(2-METHYLPHENYL)UREA

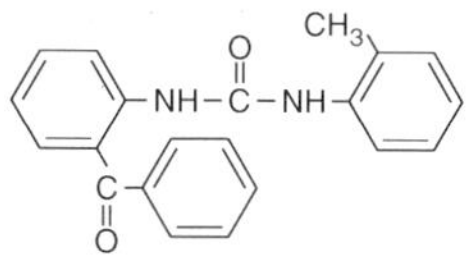

250 mg

S71,659-6
1-(4-BENZOYLPHENYL)-3-(O-TOLYL)UREA

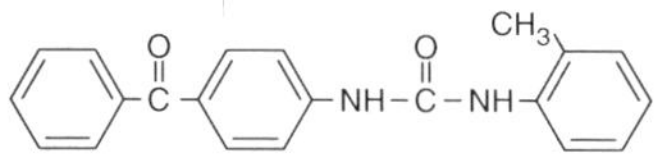

250 mg

S69,974-8
1-(2-BENZOYLPHENYL)-3-METHYLUREA

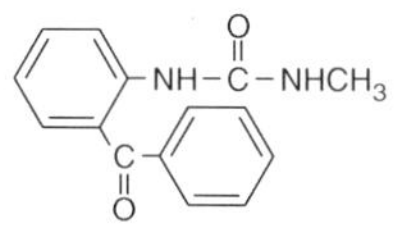

250 mg

S71,049-0
1-(4-BENZOYLPHENYL)-3-(3-METHYLPHENYL)UREA

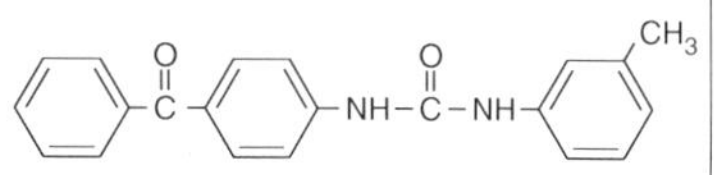

250 mg

S71,392-9
1-(4-BENZOYLPHENYL)-3-(2,4-XYLYL)UREA

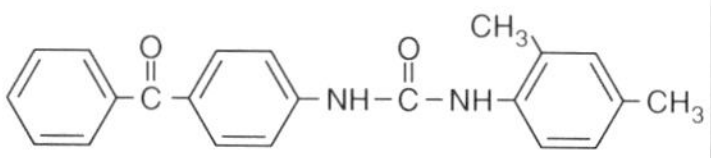

5 gm

S64,531-1
1-(2-BENZOYL-4-CHLOROPHENYL)-3-(2-CHLOROPHENYL)UREA

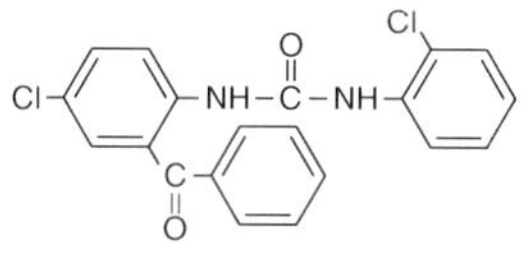

5 gm

S73,068-8
(4-ACETYLPHENYL)UREA

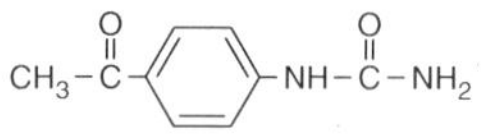

5 gm

S70,187-4
1-(4-ACETYLPHENYL)-3-ETHYLUREA

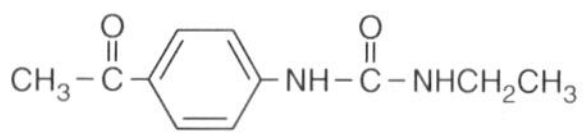

250 mg

S64,660-1
1-(2-BENZOYL-4-CHLOROPHENYL)-3-(4-BROMOPHENYL)UREA

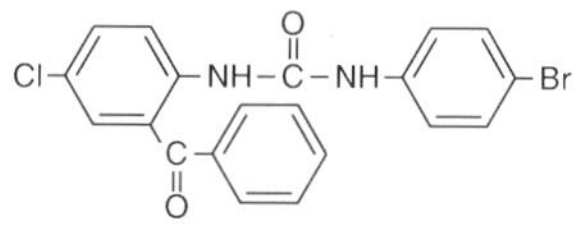

5 gm

S65,481-7
1-(4-BENZOYLPHENYL)-3-(3,5-DICHLOROPHENYL)UREA

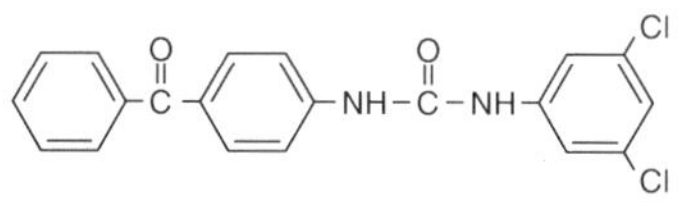

5 gm

S59,420-2
1-(4-BENZOYLPHENYL)-3-PHENYLUREA

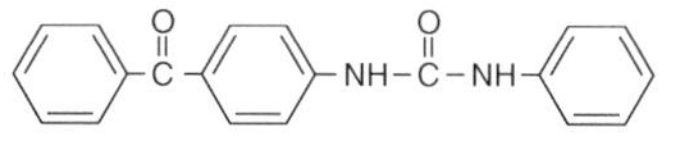

5 gm

S59,403-2
1-(4-BENZOYLPHENYL)-3-(4-CHLOROPHENYL)UREA

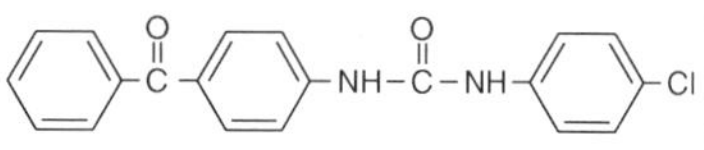

5 gm

S65,774-3
4,4'-METHYLENEBIS(N-[N-(4-BENZOYLPHENYL)CARBAMOYL]ANILINE)

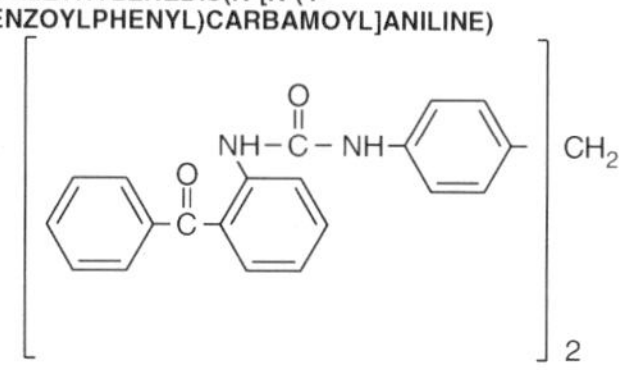

5 gm

S65,223-7
1-(2-BENZOYL-4-CHLOROPHENYL)-3-(2,6-DICHLOROPHENYL)UREA

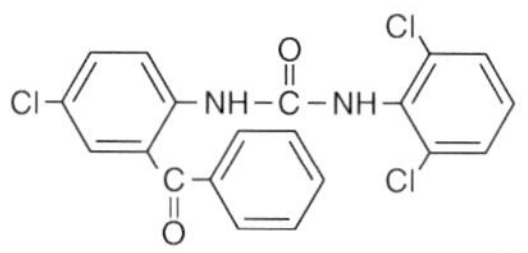

5 gm

S64,514-1
1-(2-BENZOYL-4-CHLOROPHENYL)-3-PHENYLUREA

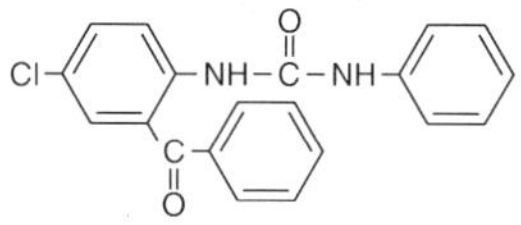

5 gm

S64,510-9
1-(2-BENZOYL-4-CHLOROPHENYL)-3-(3-CHLOROPHENYL)UREA

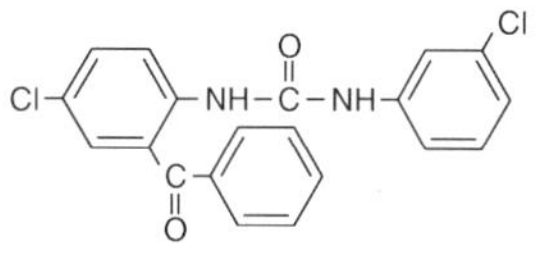

5 gm

S64,679-2
1-(2-BENZOYL-4-CHLOROPHENYL)-3-(2,4-DICHLOROPHENYL)UREA

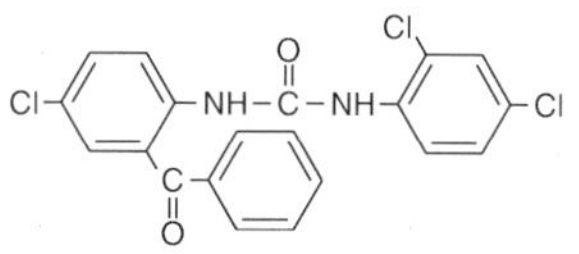

5 gm

S64,530-3
1-(2-BENZOYL-4-CHLOROPHENYL)-3-(2,4,5-TRICHLOROPHENYL)UREA

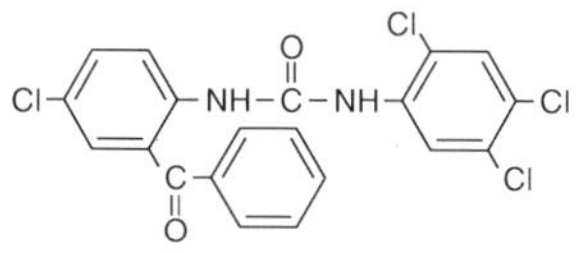

5 gm

S72,787-3
1-(3-ACETYLPHENYL)-3-(1-NAPHTHYL)UREA

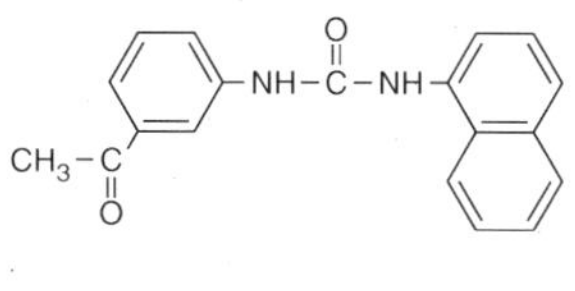

250 mg

S87,078-1
1,3-DIPHENYL-1-(2-OXO-1-CYCLOPENTYLCARBONYL)UREA

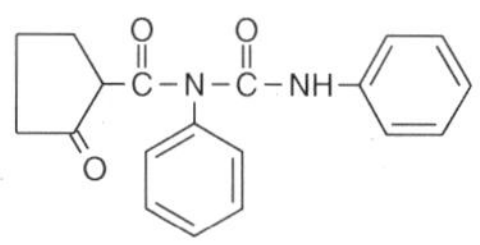

250 mg

S87,077-3
1,3-DIPHENYL-1-(2-OXO-1-CYCLOHEXYLCARBONYL)UREA

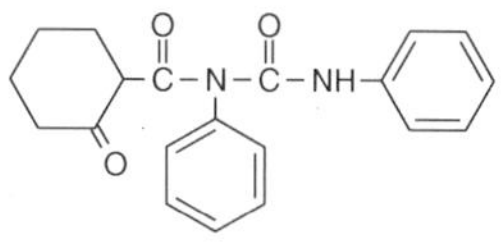

250 mg

S85,510-3
DIMETHYL 5-(3-METHYLUREIDO)-ISOPHTHALATE

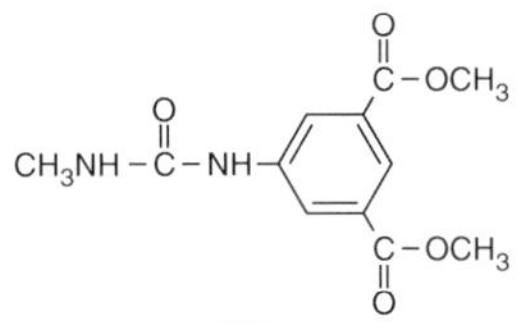

250 mg

S69,487-8
3-[3,5-BIS(METHOXYCARBONYL)PHENYL]-1,1-DIETHYLUREA

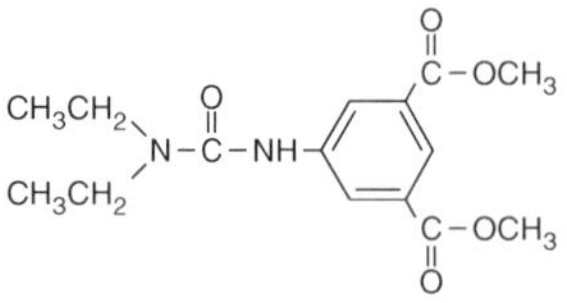

250 mg

S62,783-6
4-UREIDOBENZOIC ACID

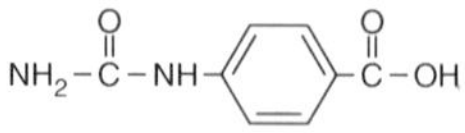

250 mg

S76,846-4
1-(4-CARBOXYPHENYL)-4-CYCLOHEXYLSEMICARBAZIDE

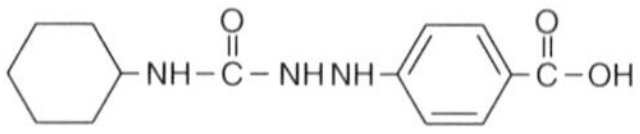

250 mg

S91,256-5
1-(4-CARBOXYPHENYL)-4-PROPYLSEMICARBAZIDE

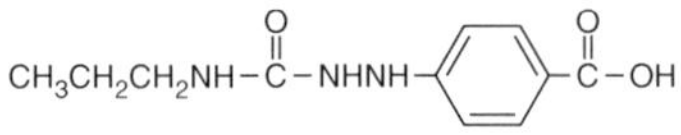

250 mg

S72,295-2
1-(4-CARBOXYPHENYL)-4-(3,4-XYLYL)-SEMICARBAZIDE

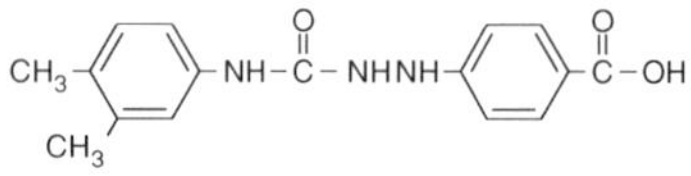

250 mg

S66,979-2
1-(2-CARBOXYPHENYL)-3-CYCLOHEXYLUREA

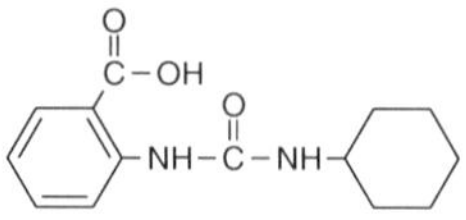

5 gm

S65,598-8
1-(4-CARBOXYPHENYL)-3-CYCLOHEXYLUREA

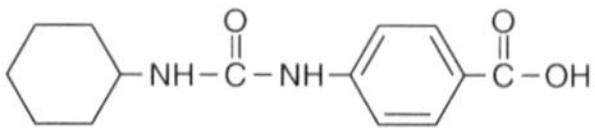

5 gm

S66,980-6
2-(3-PHENYLUREIDO)BENZOIC ACID

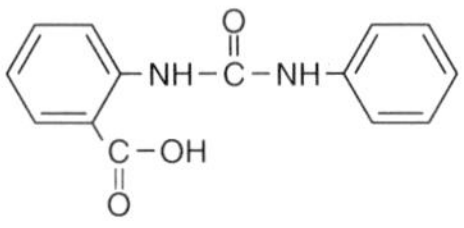

5 gm

S65,578-3
1-(2-CARBOXYPHENYL)-3-ETHYLUREA

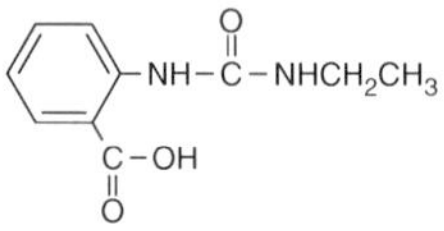

5 gm

S65,564-3
1-BUTYL-3-(2-CARBOXYPHENYL)UREA

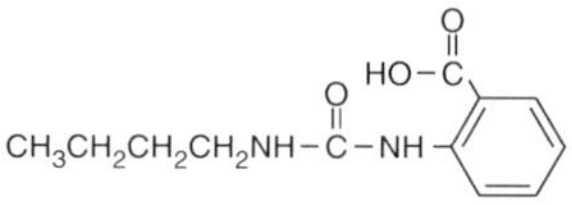

250 mg

S66,922-9
1-(2-CARBOXYPHENYL)-3-(O-TOLYL)UREA

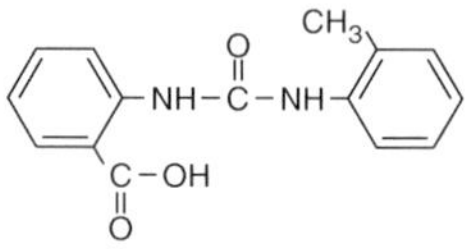

5 gm

S66,971-7
1-(2-CARBOXYPHENYL)-3-(M-TOLYL)UREA

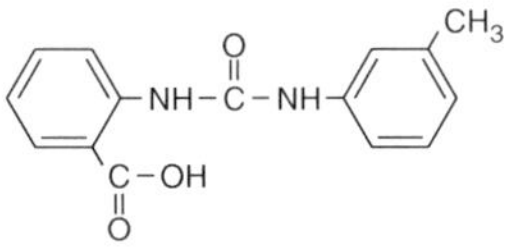

5 gm

S66,952-0
1-(2-CARBOXYPHENYL)-3-(P-TOLYL)UREA

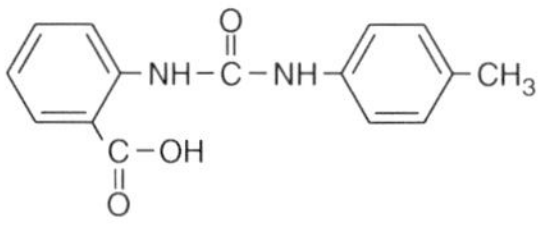

5 gm

S66,964-4
1-(4-CARBOXYPHENYL)-3-(O-TOLYL)UREA

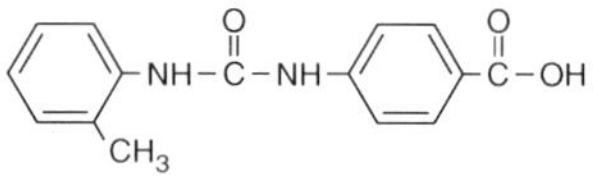

5 gm

S67,052-9
1-(2-CARBOXYPHENYL)-3-(2,6-XYLYL)UREA

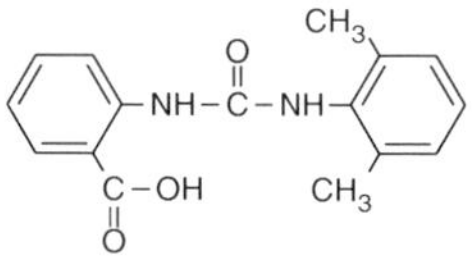

5 gm

S66,934-2
1-(2-CARBOXYPHENYL)-3-(2-ETHYL-6-METHYLPHENYL)UREA

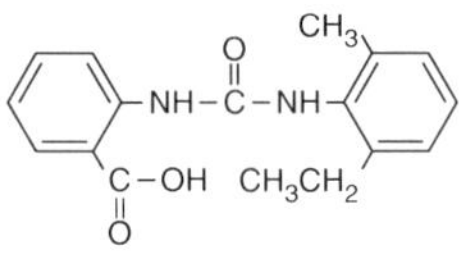

5 gm

S66,935-0
1-(2-CARBOXYPHENYL)-3-(2,6-DIETHYLPHENYL)UREA

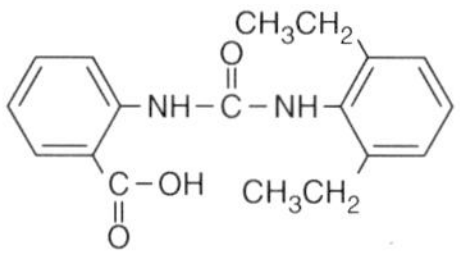

5 gm

S66,942-3
2-[3-(2,4-XYLYL)UREIDO]BENZOIC ACID

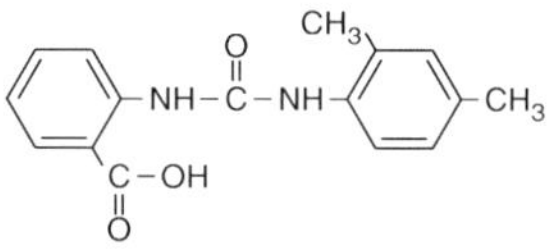

5 gm

S79,589-5
4-(3-METHYLUREIDO)BENZOIC ACID

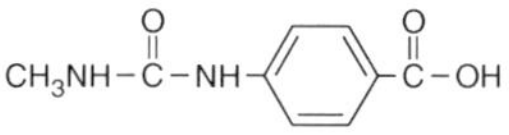

250 mg

S65,360-8
4-(3-PHENYLUREIDO)BENZOIC ACID

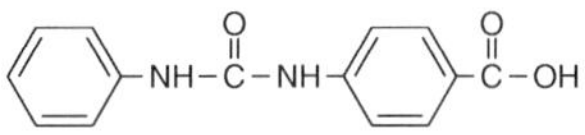

5 gm

S66,883-4
1-(4-CARBOXYPHENYL)-3-(M-TOLYL)UREA

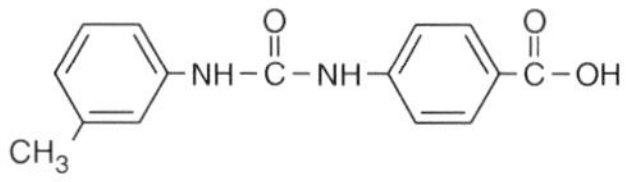

5 gm

S66,985-7
1-(4-CARBOXYPHENYL)-3-(P-TOLYL)UREA

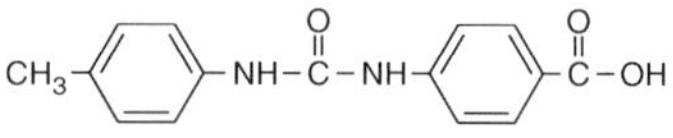

5 gm

S66,938-5
4-[3-(2,3-XYLYL)UREIDO]BENZOIC ACID

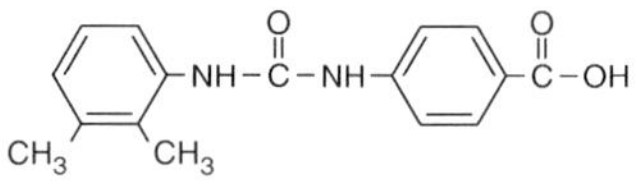

5 gm

S66,963-6
4-[3-(2,6-XYLYL)UREIDO]BENZOIC ACID

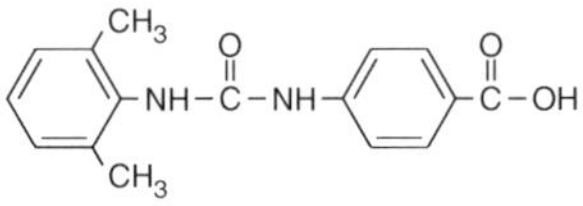

5 gm

S66,954-7
1-(4-CARBOXYPHENYL)-3-(2-ETHYL-6-METHYLPHENYL)UREA

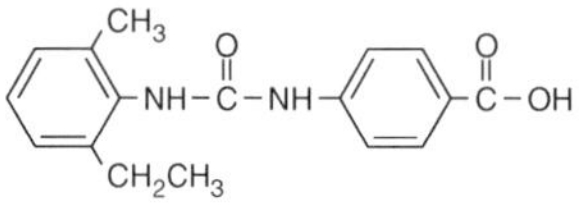

5 gm

S66,933-4
1-(4-CARBOXYPHENYL)-3-(2,6-DIETHYLPHENYL)UREA

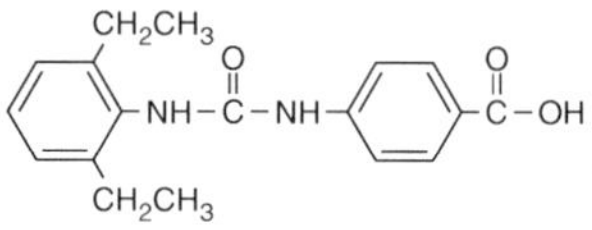

5 gm

S66,953-9
4-[3-(2,4-XYLYL)UREIDO]BENZOIC ACID

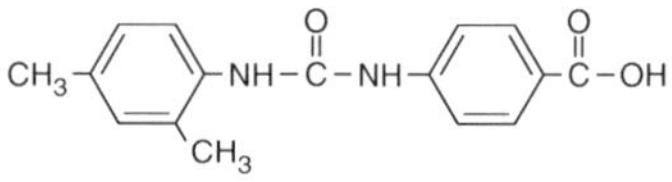

5 gm

S77,397-2
4-[3-(2,5-XYLYL)UREIDO]BENZOIC ACID

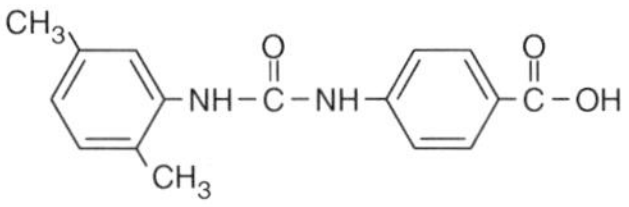

250 mg

S95,279-6
DIMETHYL [3-(BENZAMIDOMETHYL)-CARBAMOYL]-5,5-DIMETHYL-2,4-THIAZOLIDINEDICARBOXYLATE

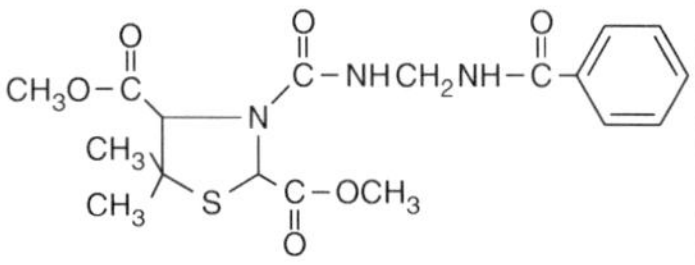

50 mg

S67,743-4
1,3-BIS[2-(METHOXYCARBONYL)-PHENYL]UREA

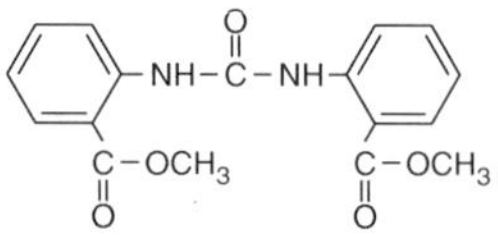

5 gm

S72,942-6
1-[4-(ETHOXYCARBONYL)PHENYL]-3-[2-(METHOXYCARBONYL)PHENYL]UREA

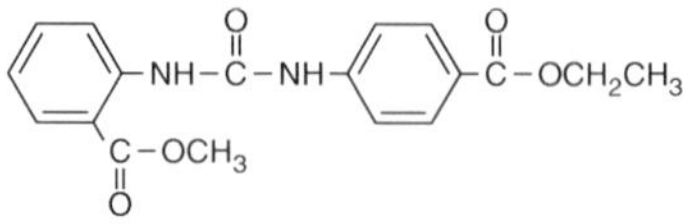

5 gm

S86,919-8
1,1'-HEXAMETHYLENEBIS(3-[4-(ETHOXYCARBONYL)PHENYL]UREA)

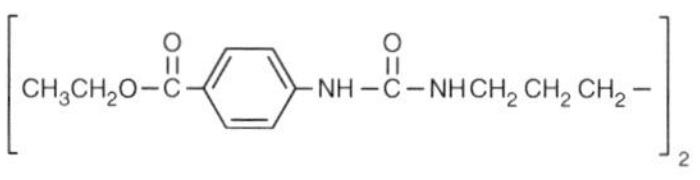

250 mg

S67,330-7
4,4'-BIS[4-(CARBETHOXY)-PHENYLUREIDO]-3,3'-DIMETHYLBIPHENYL

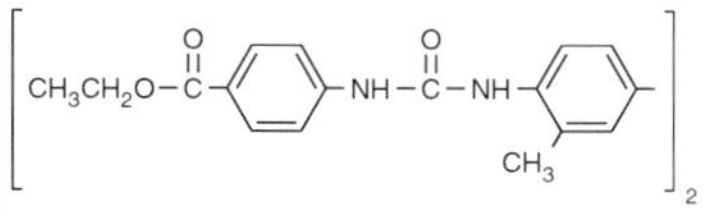

5 gm

S76,861-8
1-(HEXAHYDRO-1H-AZEPIN-1-YL)-3-(2-METHOXYCARBONYL)PHENYLUREA

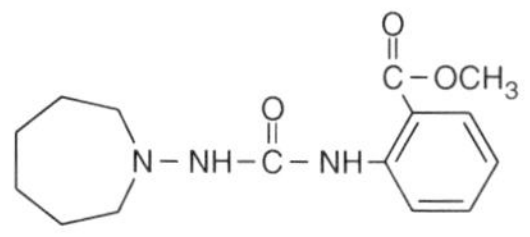

250 mg

S79,758-8
ETHYL 4-(3-CYCLOHEXYL-3-METHYLUREIDO)BENZOATE

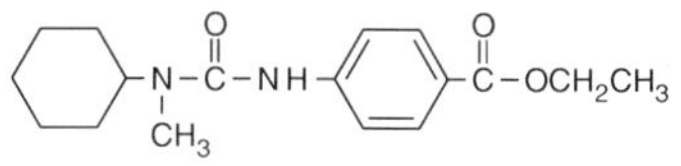

250 mg

S72,812-8
ETHYL 4-(3,3-DICYCLOHEXYLUREIDO)-BENZOATE

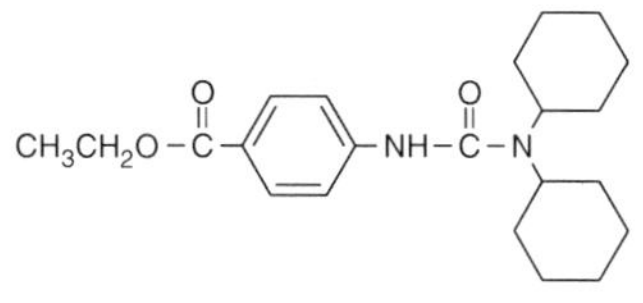

250 mg

S79,780-4
ETHYL 4-(3-ISOPROPYL-3-PHENYLUREIDO)BENZOATE

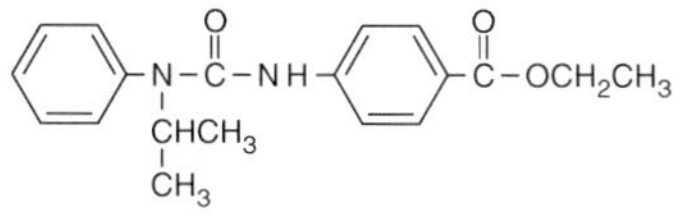

250 mg

S70,224-2
1-(4-ETHOXYCARBONYLPHENYL)-3-(1,2-DIPHENYLETHYL)UREA

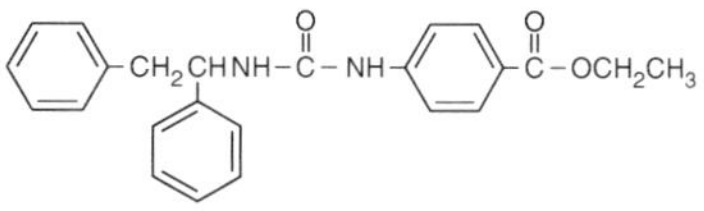

250 mg

S84,729-1
1-[4-(BUTOXYCARBONYL)PHENYL]-3-(2-ETHYLPHENYL)UREA

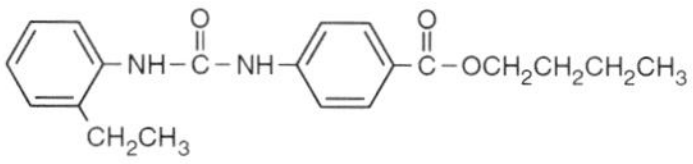

250 mg

S85,455-7
BUTYL 4-(3-METHYLUREIDO)BENZOATE

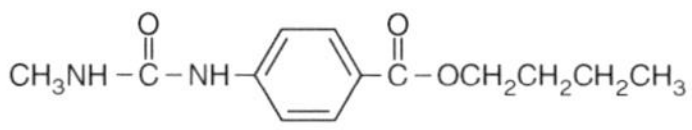

250 mg

S79,141-5
BUTYL 4-(3-PROPYLUREIDO)BENZOATE

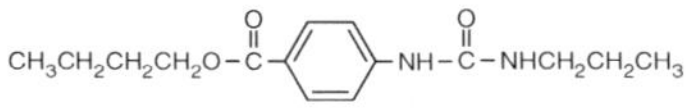

250 mg

S79,154-7
BUTYL 4-(3-BUTYLUREIDO)BENZOATE

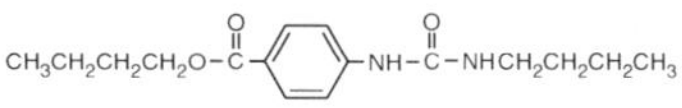

250 mg

S79,425-2
1-[4-(ETHOXYCARBONYL)PHENYL]-3-ETHYL-3-(M-TOLYL)UREA

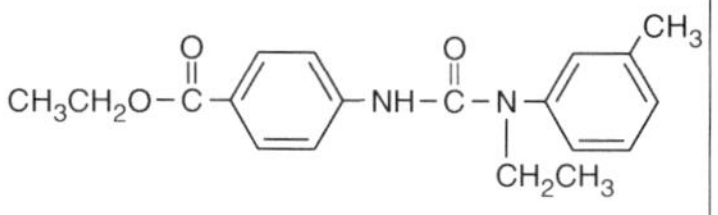

250 mg

S67,486-9
1-(4-BUTOXYCARBONYLPHENYL)-3-(2,6-DIMETHYLPHENYL)UREA

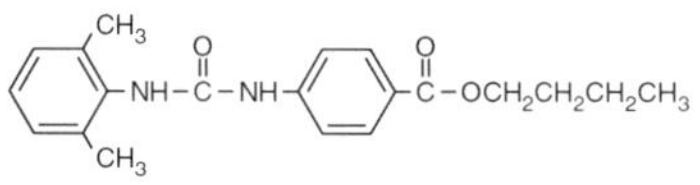

250 mg

S77,379-4
1-(4-BUTOXYCARBONYLPHENYL)-3-(2,5-XYLYL)UREA

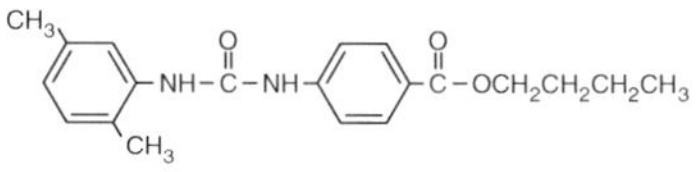

250 mg

S78,073-1
1-(2-CARBAMOYLPHENYL)-3-PHENYLUREA

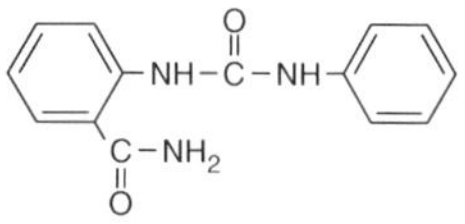

250 mg

S84,569-8
1-(2-CARBAMOYLPHENYL)-3-(2-ETHYLPHENYL)UREA

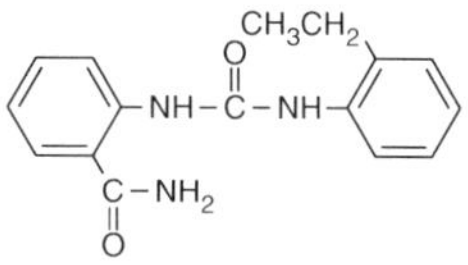

250 mg

S72,502-1
1-(2-CARBAMOYLPHENYL)-3-(2,6-XYLYL)UREA

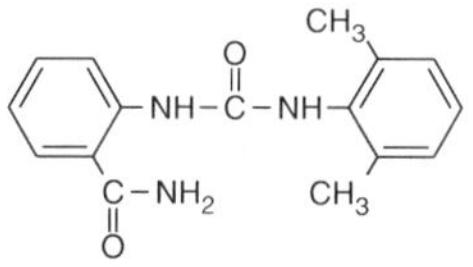

250 mg

S77,386-7
1-(2-CARBAMOYLPHENYL)-3-(2,5-XYLYL)UREA

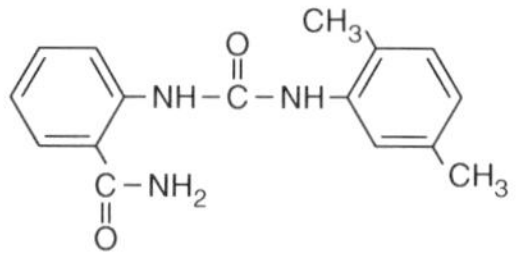

250 mg

S60,178-0
1-(BENZOYL)-4-PHENYLSEMICARBAZIDE

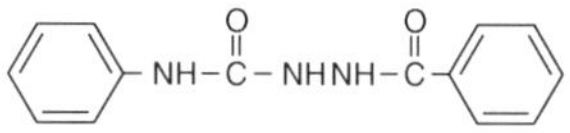

5 gm

S43,306-3
1-BENZOYLSEMICARBAZIDE

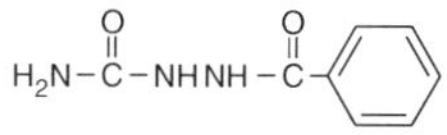

5 gm

S45,054-5
BENZOYLUREA

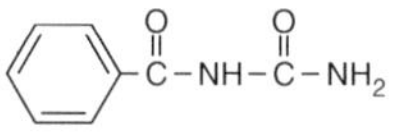

5 gm

S67,599-7
4-(CYCLOHEXYLUREIDO)BENZOIC ACID HYDRAZIDE

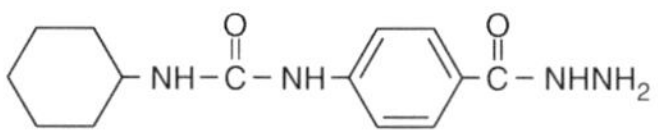

5 gm

S67,597-0
2-(METHYLUREIDO)BENZOIC ACID HYDRAZIDE

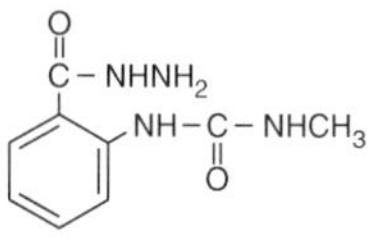

5 gm

S67,438-9
4-(TERT-BUTYLUREIDO)BENZOIC ACID HYDRAZIDE

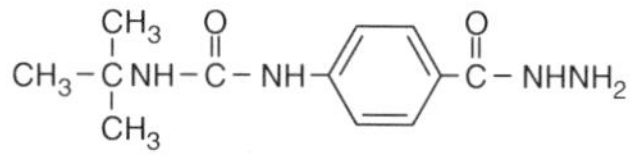

5 gm

S80,862-8
4-(3-PHENYLUREIDO)BUTYRIC ACID

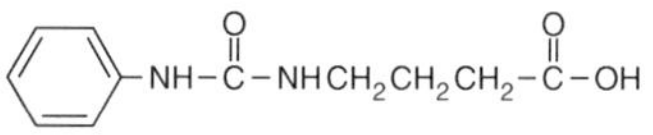

250 mg

S71,568-9
4,4'-BIS-(ETHOXYCARBONYLMETHYLUREIDO)-3,3'-DIMETHYLBIPHENYL

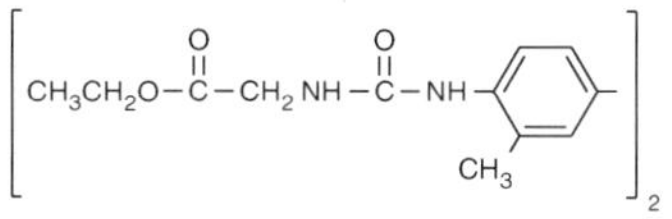

250 mg

S76,383-7
1-(ETHOXYCARBONYL)-4-PHENYLSEMICARBAZIDE

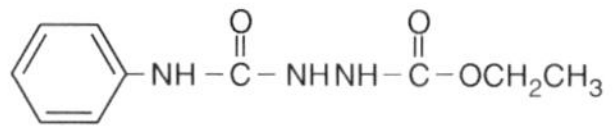

250 mg

S60,730-4
3-(3,4-DICHLOROPHENYL)-1,1-DIISOPROPYLUREA

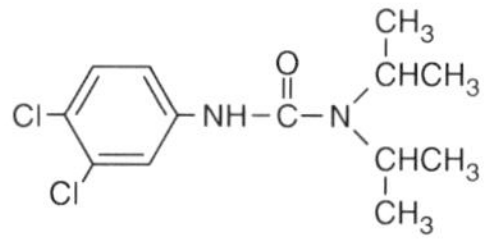

5 gm

S53,444-7
1-(3-CHLORO-PARA-TOLYL)UREA

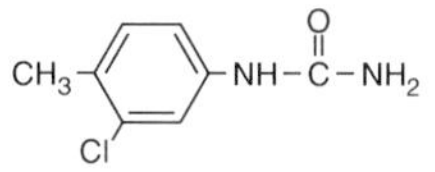

5 gm

S61,101-8
4-(3-CHLOROPHENYL)-1,1-DIMETHYLSEMICARBAZIDE

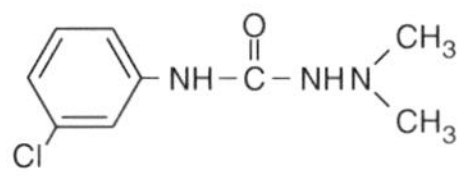

5 gm

S60,625-1
1,1-DIMETHYL-4-(M-TOLYL)-SEMICARBAZIDE

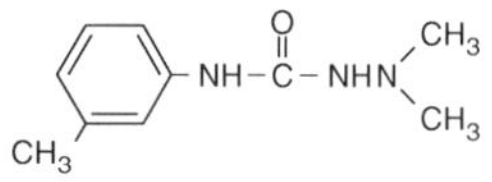

5 gm

S64,697-0
4,4'-HEXAMETHYLENEBIS(1,1-DIMETHYLSEMICARBAZIDE)

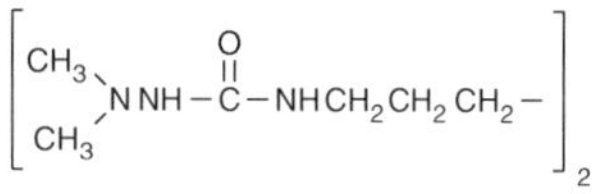

5 gm

S60,646-4
1-PHENYL-4-(M-TOLYL)SEMICARBAZIDE

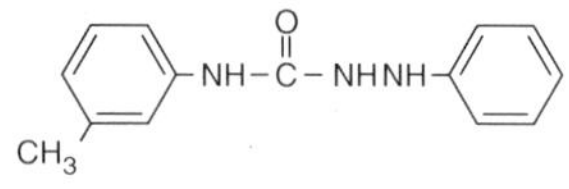

5 gm

S68,644-1
4-(2-METHYLPHENYL)-1-(2-METHYLPHENYLCARBAMOYL)SEMICARBAZIDE

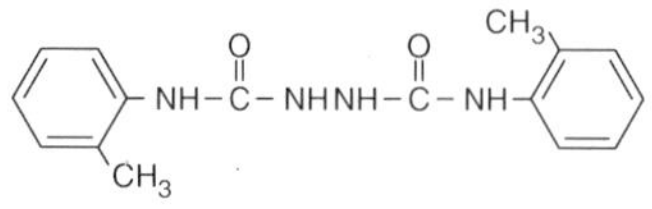

250 mg

S91,319-7
1-ACETYL-4-(3,4-XYLYL)SEMICARBAZIDE

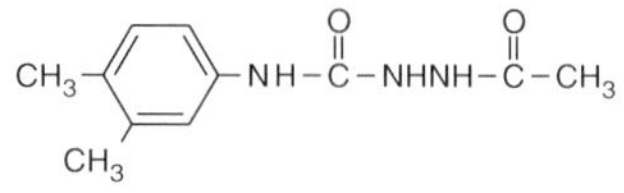

250 mg

S76,858-8
3-PHENYL-6-(M-TOLYL)-1-(O-TOLYL)-BIUREA

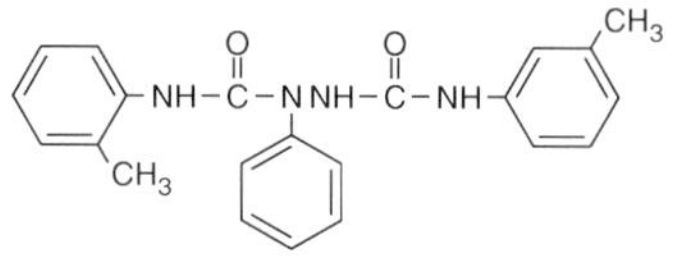

250 mg

S55,234-8
1-(4-METHYL-4-PHENYLCYCLOHEXYL)-SEMICARBAZIDE

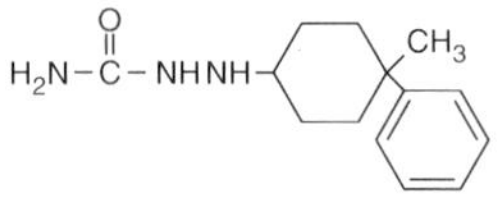

250 mg

S76,388-8
4-METHYL-1-PHENYLSEMICARBAZIDE

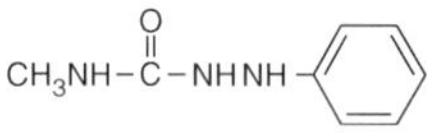

250 mg

S60,628-6
1,1-DIMETHYL-4-PHENYLSEMICARBAZIDE

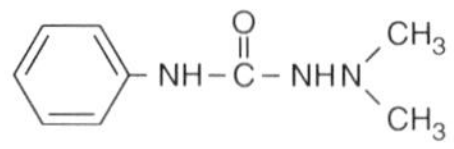

5 gm

S82,962-5
PROPIOPHENONE SEMICARBAZONE

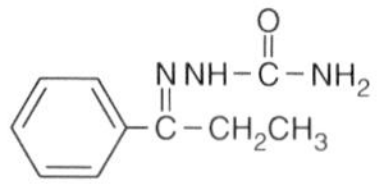

250 mg

S72,948-5
1-METHYL-1-PHENYL-4-PROPYLSEMICARBAZIDE

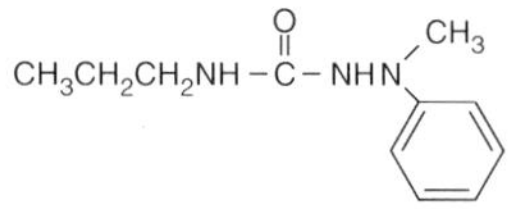

5 gm

S59,185-8
1,4-DIPHENYLSEMICARBAZIDE

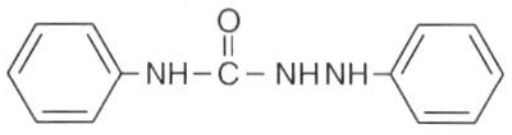

250 mg

S91,961-6
4-PHENYL-1-(M-TOLYL)SEMICARBAZIDE

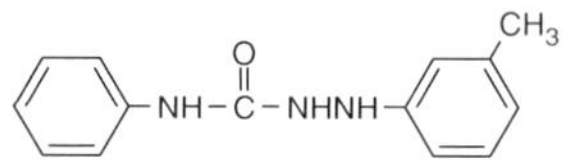

250 mg

S68,732-4
1-(DIPHENYLCARBAMOYL)-4-ETHYLSEMICARBAZIDE

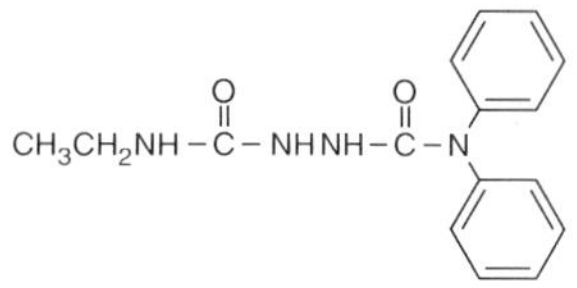

250 mg

S64,705-5
4,4'-HEXAMETHYLENEBIS(1-PHENYLSEMICARBAZIDE)

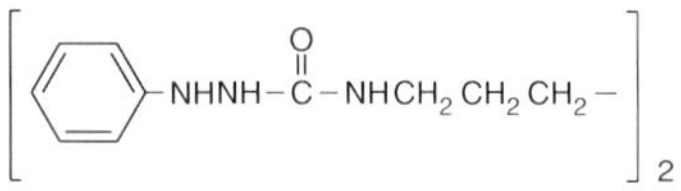

5 gm

S76,807-3
6-ETHYL-3-PHENYL-1-(O-TOLYL)BIUREA

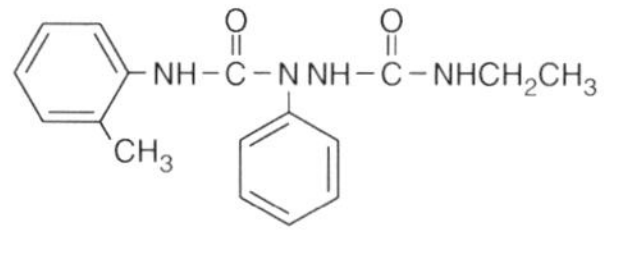

250 mg

S91,962-4
4-PHENYL-1-(P-TOLYL)SEMICARBAZIDE

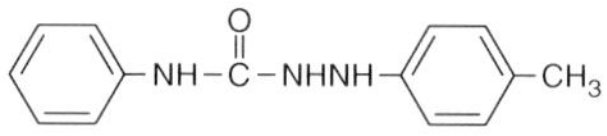

250 mg

S71,294-9
4-(2,6-DIETHYLPHENYL)-1-PHENYLSEMICARBAZIDE

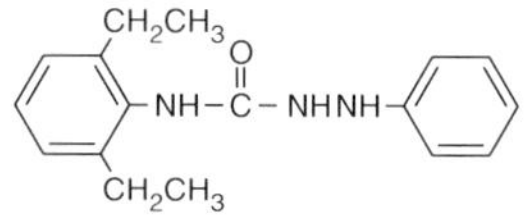

5 gm

S72,118-2
4-(2,4-DIMETHYLBENZYL)-1-PHENYLSEMICARBAZIDE

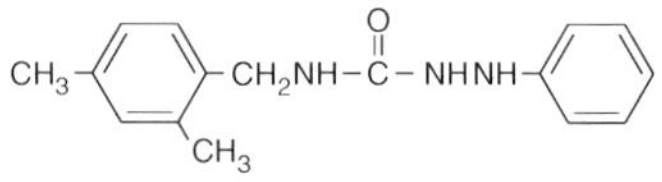

250 mg

S71,097-0
4-(1-NAPHTHYL)-1-PHENYLSEMICARBAZIDE

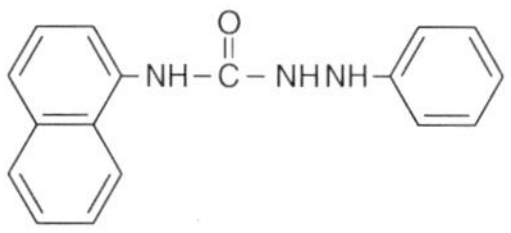

5 gm

S60,649-9
1,1-DIMETHYL-4-(O-TOLYL)-SEMICARBAZIDE

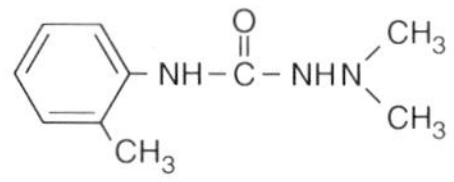

5 gm

S66,430-8
1,1-DIMETHYL-4-(2-ETHYLPHENYL)-SEMICARBAZIDE

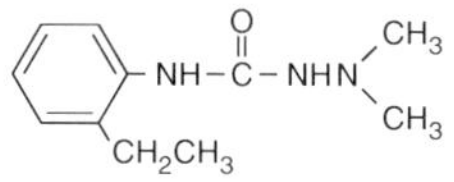

5 gm

S60,661-8
1-PHENYL-4-(O-TOLYL)SEMICARBAZIDE

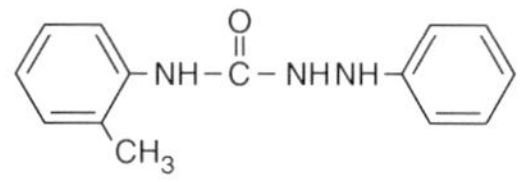

5 gm

S60,697-9
4-(4-CHLOROPHENYL)-1,1-DIMETHYLSEMICARBAZIDE

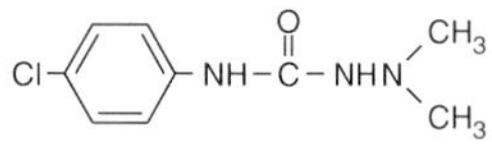

5 gm

S61,300-2
1,1-DIMETHYL-4-(P-TOLYL)-SEMICARBAZIDE

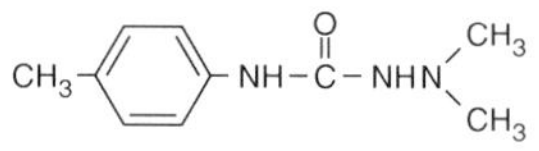

5 gm

S60,198-5
4-(4-FLUOROPHENYL)-1-PHENYLSEMICARBAZIDE

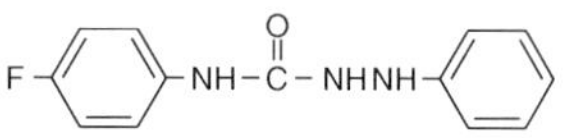

5 gm

S59,143-2
4-(4-CHLOROPHENYL)-1-PHENYLSEMICARBAZIDE

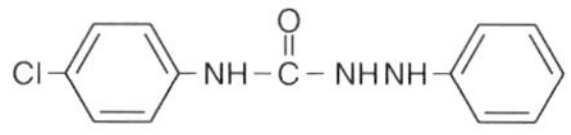

5 gm

S61,180-8
4-(2,6-DICHLOROPHENYL)-1,1-DIMETHYLSEMICARBAZIDE

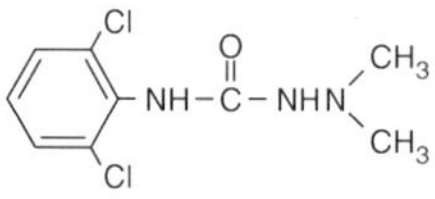

5 gm

S61,204-9
4-(2,3-DICHLOROPHENYL)-1,1-DIMETHYLSEMICARBAZIDE

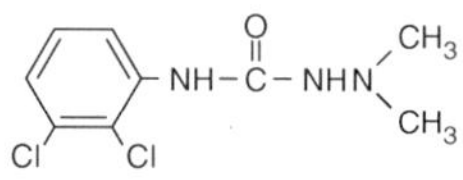

5 gm

S66,072-8
4-(2-CHLORO-6-METHYLPHENYL)-1,1-DIMETHYLSEMICARBAZIDE

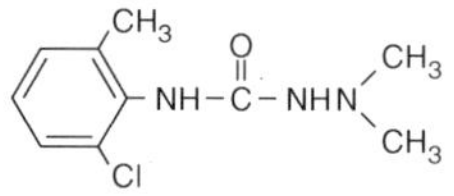

5 gm

S66,444-8
1,1-DIMETHYL-4-(2,3-XYLYL)-SEMICARBAZIDE

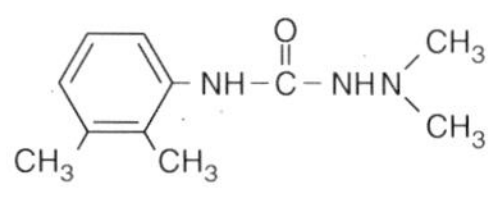

5 gm

S76,722-0
1-(2,6-XYLYL)-4-ETHYLSEMICARBAZIDE

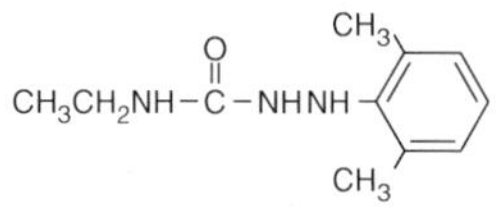

250 mg

S66,414-6
1,1-DIMETHYL-4-(2-ETHYL-6-METHYLPHENYL)SEMICARBAZIDE

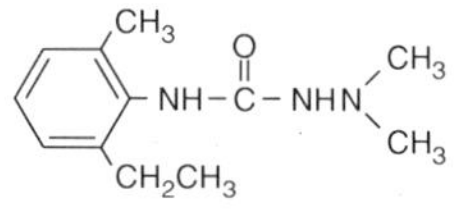

5 gm

S66,607-6
4-(2,6-DIISOPROPYLPHENYL)-1,1-DIMETHYLSEMICARBAZIDE

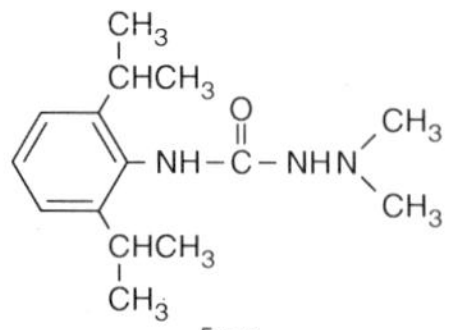

5 gm

S92,451-2
4-(2,4-XYLYL)SEMICARBAZIDE

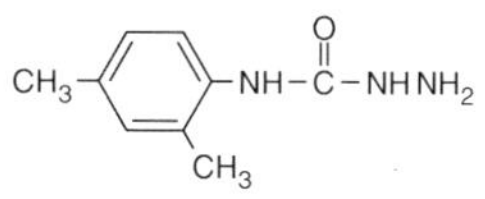

250 mg

S60,759-2
4-(3,4-DICHLOROPHENYL)-1,1-DIMETHYLSEMICARBAZIDE

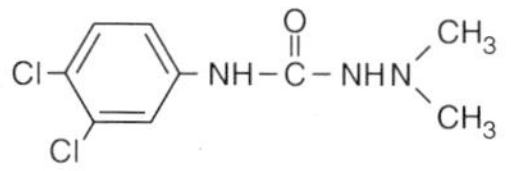

5 gm

S61,167-0
4-(2,5-DICHLOROPHENYL)-1,1-DIMETHYLSEMICARBAZIDE

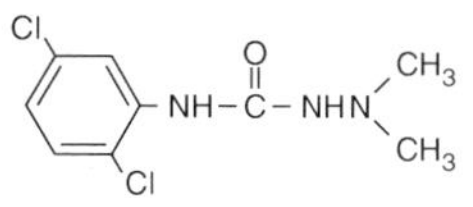

5 gm

S61,269-3
4-(2,4-DICHLOROPHENYL)-1,1-DIMETHYLSEMICARBAZIDE

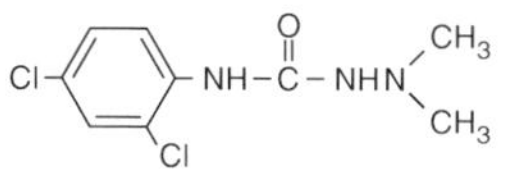

5 gm

S60,785-1
4-(4-CHLORO-2-METHYLPHENYL)-1,1-DIMETHYLSEMICARBAZIDE

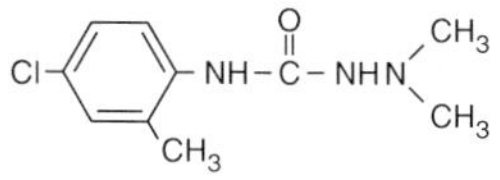

5 gm

S61,040-2
4-(3-CHLORO-4-METHYLPHENYL)-1,1-DIMETHYLSEMICARBAZIDE

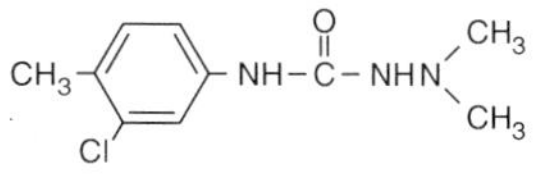

5 gm

S61,077-1
1,1-DIMETHYL-4-(2,4-XYLYL)-SEMICARBAZIDE

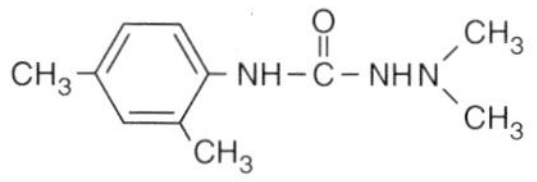

5 gm

S64,528-1
1,1-DIMETHYL-4-(2,5-XYLYL)-SEMICARBAZIDE

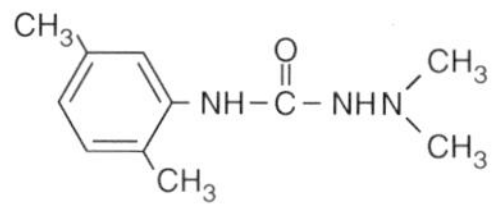

5 gm

S60,451-8
N-(4-CHLORO-2-METHYLPHENYL)-CARBAMIC ACID PHENYLHYDRAZIDE

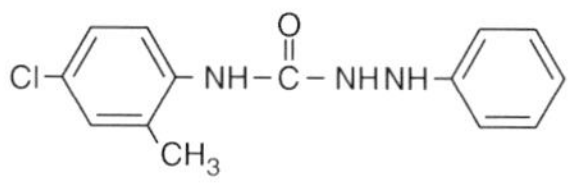

5 gm

S68,846-0
1,6-BIS(2,5-DIMETHYLPHENYL)BIUREA

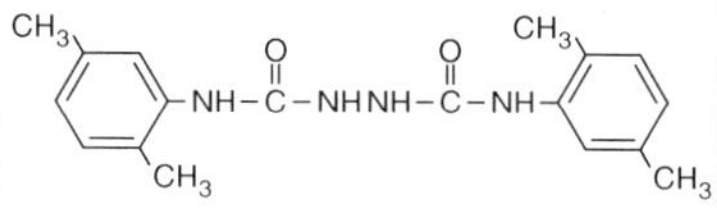

250 mg

S66,739-0
1,1-DIMETHYL-4-(3,5-XYLYL)-SEMICARBAZIDE

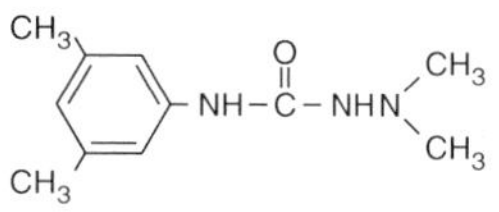

5 gm

S64,415-3
1,1-DIMETHYL-4-(2,4,5-TRICHLOROPHENYL)SEMICARBAZIDE

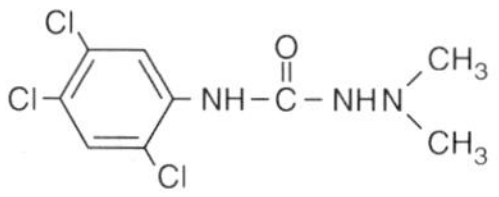

5 gm

S85,649-5
1-FORMYL-1-PHENYLUREA

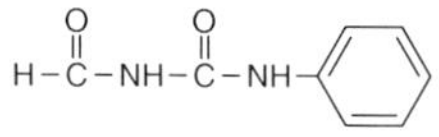

250 mg

S45,003-0
1-(2-CHLORO-2-PHENYLACETYL)UREA

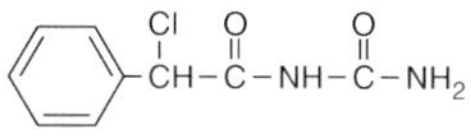

5 gm

S85,629-0
1-PHENYL-3-PROPIONYLUREA

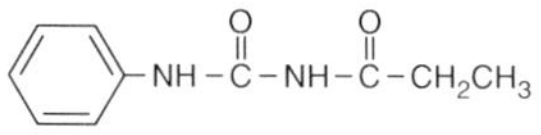

250 mg

S85,650-9
1-ACETYL-1-METHYL-3-PHENYLUREA

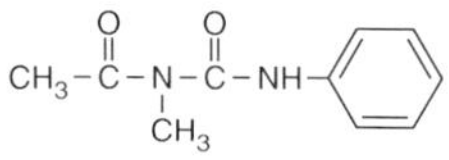

250 mg

S85,630-4
1-BUTYRYL-3-PHENYLUREA

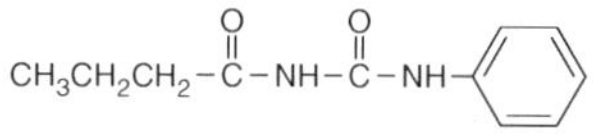

250 mg

S90,160-1
1-PHENYL-5-(2,4-XYLYL)BIURET

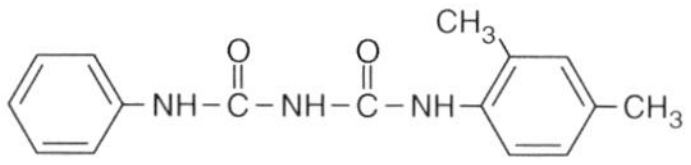

250 mg

S77,500-2
1-ACETYL-1,3-DI(O-TOLYL)UREA

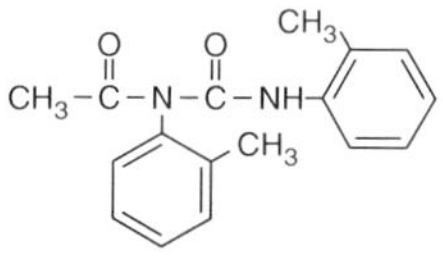

250 mg

S90,152-0
1-METHYL-5-(2,4-XYLYL)BIURET

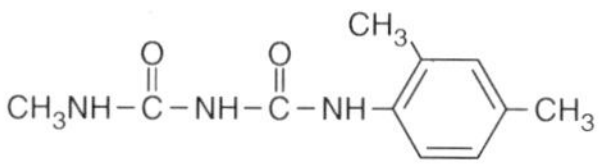

250 mg

S90,166-0
1-(5-ISOPROPYL-2-METHYLPHENYL)-5-METHYLBIURET

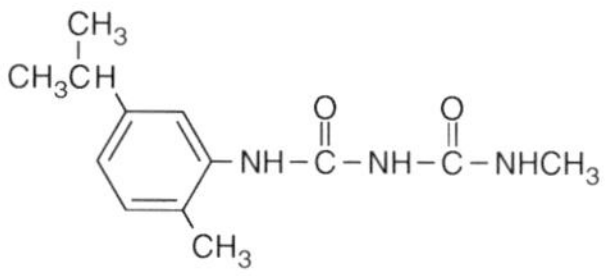

250 mg

S34,056-1
1-ACETYL-3-PHENYLUREA

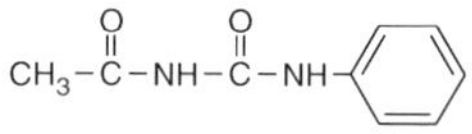

5 gm

S76,830-8
1-PHENYL-3-VALERYLUREA

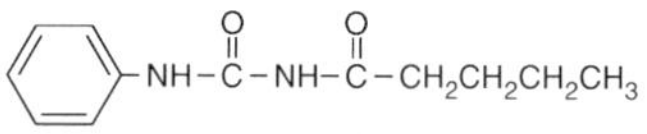

250 mg

S79,591-7
4'-(3-METHYLUREIDO)ACETANILIDE

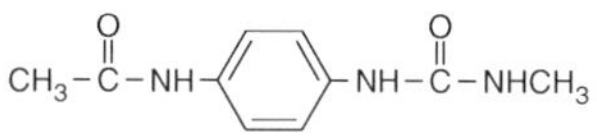

250 mg

S53,614-8
1-(2,3-XYLYL)UREA

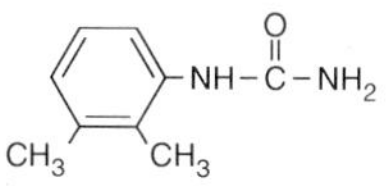

5 gm

S52,775-0
1-(2-NAPHTHYL)UREA

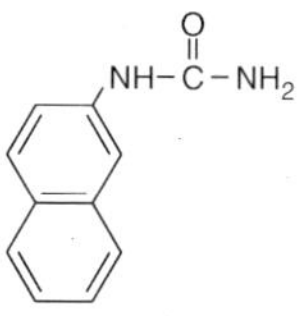

5 gm

S59,040-1
4-[N-(1-NAPHTHYL)CARBAMOYL]-MORPHOLINE

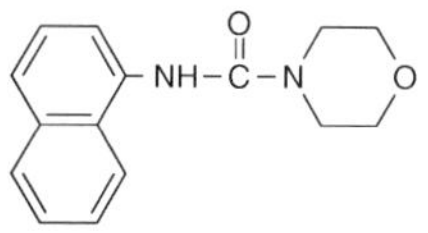

5 gm

S88,995-4
1-BENZYL-1,3-DIPHENYLUREA

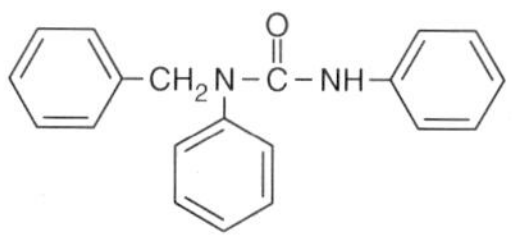

250 mg

S89,004-9
1-ETHYL-3-OCTADECYL-1-PHENYLUREA

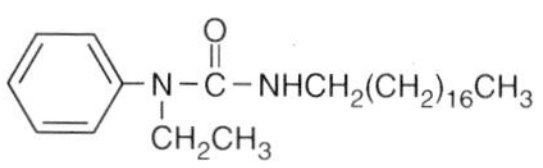

250 mg

S89,003-0
1-ETHYL-3-OCTADECYL-1-(O-TOLYL)UREA

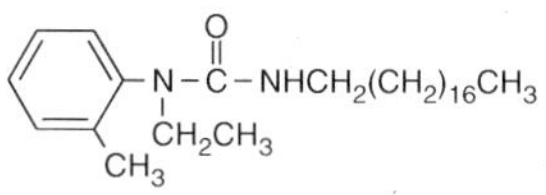

250 mg

S66,387-5
4-(2,6-DIETHYLPHENYL)-1,1-DIMETHYLSEMICARBAZIDE

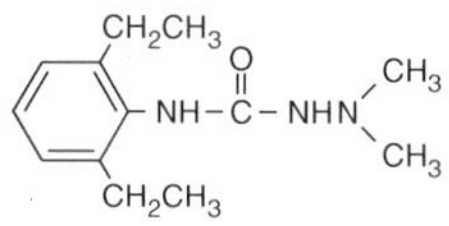

5 gm

S44,122-8
4-CHLOROPHENYLUREA

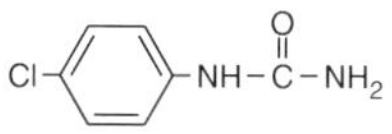

5 gm

S57,647-6
3-CHLOROPHENYLUREA
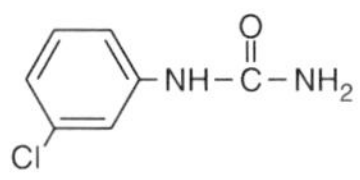
5 gm

S41,023-3
META-TOLYLUREA
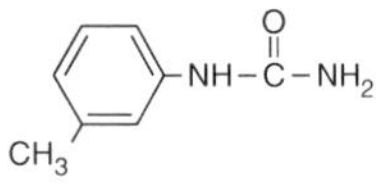
5 gm

S61,109-3
1-(3-CHLOROPHENYL)-3-METHYLUREA
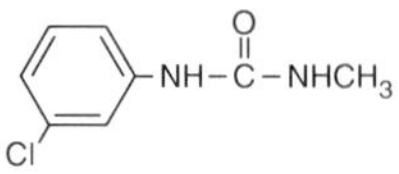
5 gm

S65,274-1
1-METHYL-3-(M-TOLYL)UREA
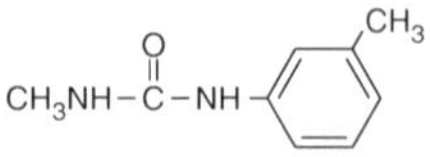
5 gm

S59,037-1
3-(3-CHLOROPHENYL)-1,1-DIMETHYLUREA
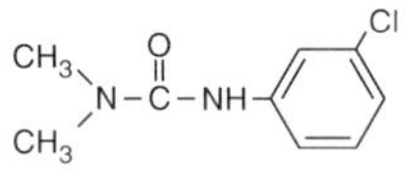
5 gm

S62,371-7
1-(3-CHLOROPHENYL)-3-ETHYLUREA
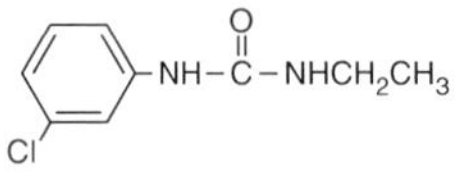
5 gm

S60,463-1
1,1-DIMETHYL-3-(3-METHYLPHENYL)UREA
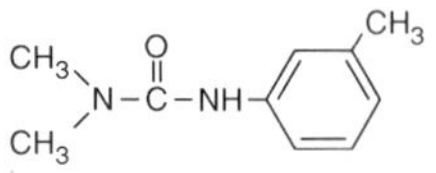
5 gm

S64,662-8
1-ETHYL-3-(M-TOLYL)UREA
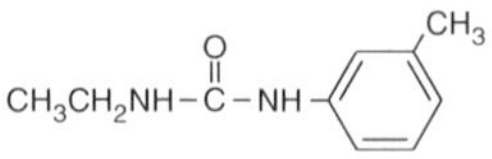
5 gm

S62,365-2
1-ALLYL-3-(3-CHLOROPHENYL)UREA
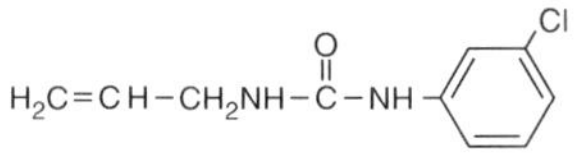
5 gm

S61,029-1
1-(3-CHLOROPHENYL)-3-PROPYLUREA
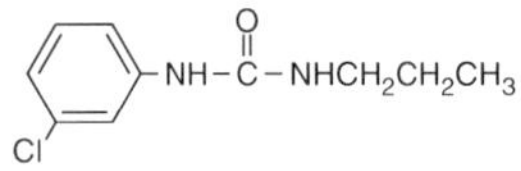
5 gm

S61,094-1
1-(3-CHLOROPHENYL)-3-ISOPROPYLUREA
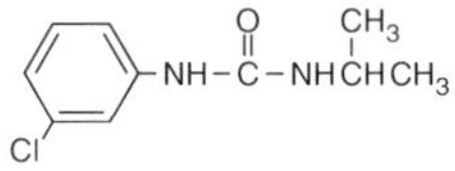
5 gm

S64,037-9
1-ALLYL-3-(M-TOLYL)UREA
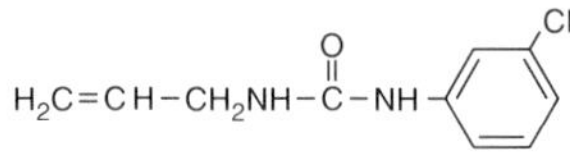
5 gm

S63,636-3
1-ISOPROPYL-3-(M-TOLYL)UREA
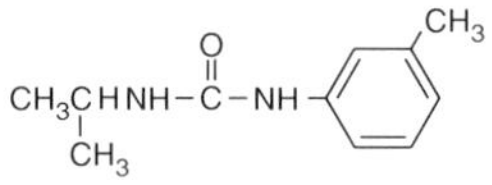
5 gm

S64,649-0
1-PROPYL-3-(M-TOLYL)UREA
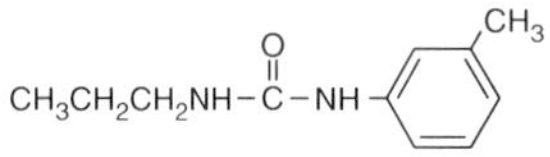
5 gm

S73,762-3
1,1-DIMETHYL-3-(3-ETHYLPHENYL)UREA
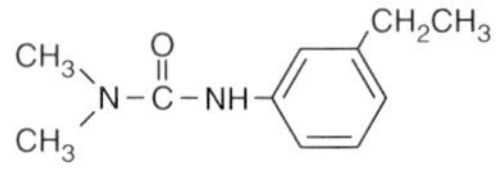
5 gm

S59,428-8
3-(3-CHLOROPHENYL)-1,1-DIETHYLUREA
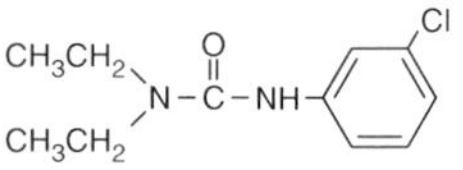
5 gm

S62,370-9
1-BUTYL-3-(3-CHLOROPHENYL)UREA
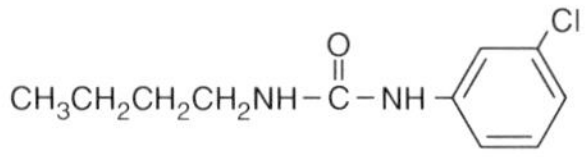
5 gm

S65,444-2
1-(3-CHLOROPHENYL)-3-ISOBUTYLUREA
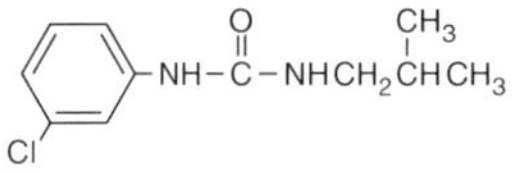
5 gm

S60,708-8
1,1-DIETHYL-3-(M-TOLYL)UREA
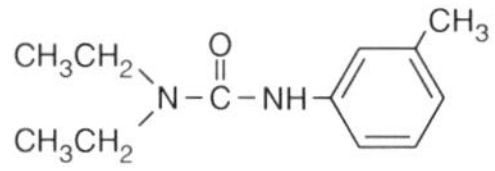
5 gm

S64,653-9
1-BUTYL-3-(M-TOLYL)UREA
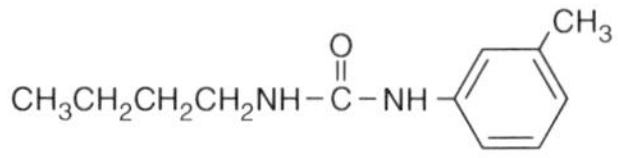
5 gm

S65,376-4
1-ISOBUTYL-3-(M-TOLYL)UREA
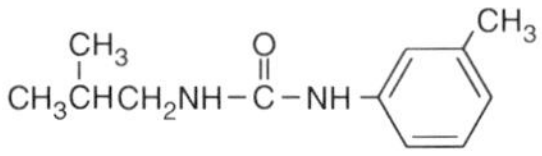
5 gm

S89,117-7
1-TERT-BUTYL-3-(M-TOLYL)UREA
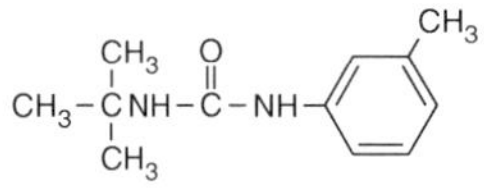
250 mg

S62,376-8
1-BUTYL-3-(3-CHLOROPHENYL)-1-METHYLUREA
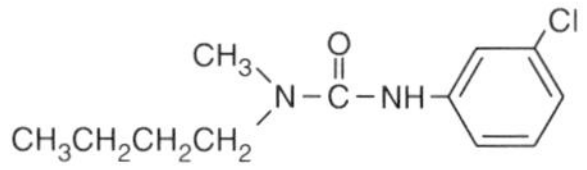
5 gm

S61,315-0
1-BUTYL-1-METHYL-3-(M-TOLYL)UREA
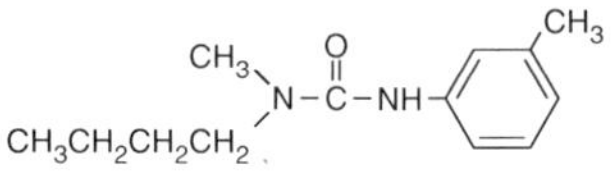
5 gm

S89,795-7
1-TERT-BUTYL-3-(3-ETHYLPHENYL)UREA
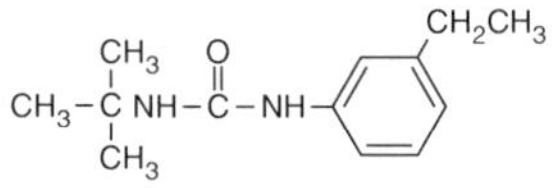
250 mg

S61,356-8
3-(3-CHLOROPHENYL)-1,1-DIALLYLUREA
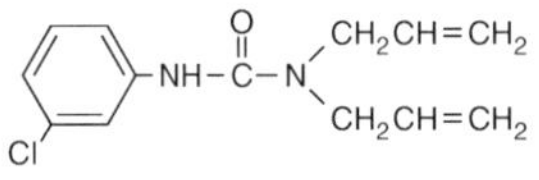
5 gm

S59,424-5
3-(3-CHLOROPHENYL)-1,1-DIPROPYLUREA
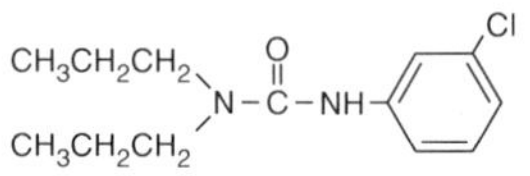
5 gm

S61,035-6
3-(3-CHLOROPHENYL)-1,1-DIISOPROPYLUREA
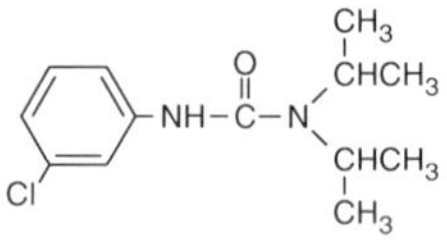
5 gm

S62,441-1
1-BUTYL-3-(3-CHLOROPHENYL)-1-ETHYLUREA
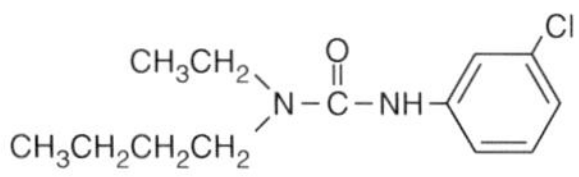
5 gm

S59,015-0
3'-METHYLCARBANILIDE
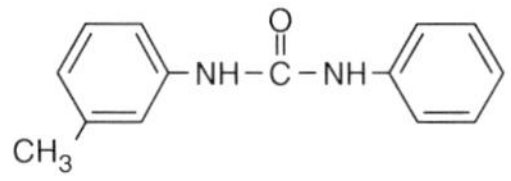
5 gm

S60,711-8
1,1-DIALLYL-3-(M-TOLYL)UREA
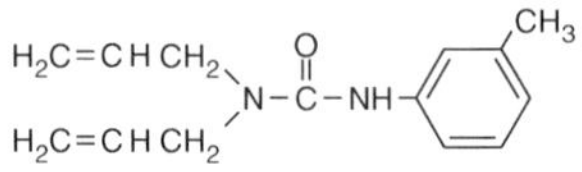
5 gm

S60,670-7
1,1-DIPROPYL-3-(M-TOLYL)UREA
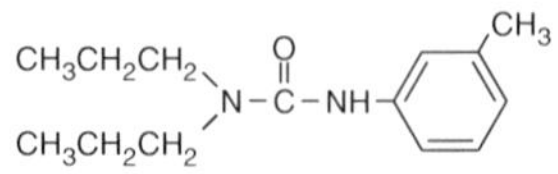
5 gm

S60,671-5
1,1-DIISOPROPYL-3-(M-TOLYL)UREA
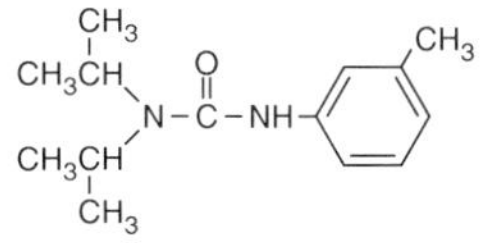
5 gm

S68,419-8
1-BUTYL-3-(3-ETHYLPHENYL)-1-METHYLUREA
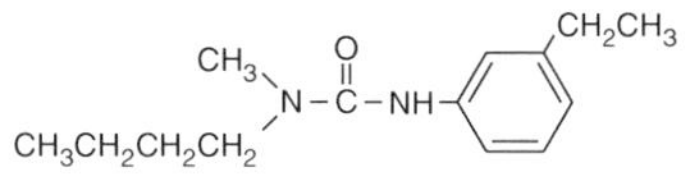
250 mg

S63,847-1
3-(3-CHLOROPHENYL)-1-METHYL-1-PHENYLUREA
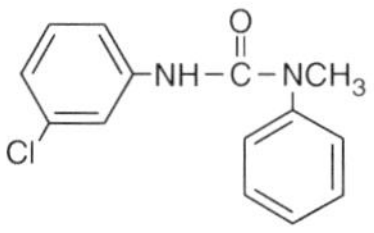
5 gm

S76,908-8
1,1-DIPROPYL-3-(3-ETHYLPHENYL)UREA
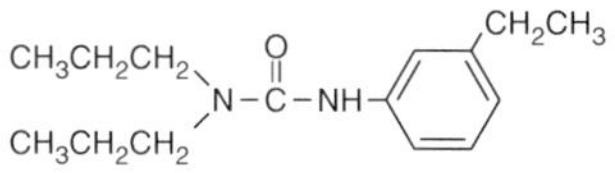
250 mg

S62,482-9
3-(3-CHLOROPHENYL)-1-ETHYL-1-PHENYLUREA
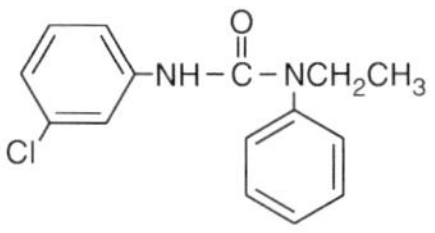
5 gm

S59,423-7
3-(3-CHLOROPHENYL)-1,1-DIBUTYLUREA
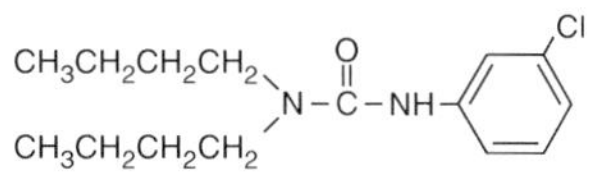
5 gm

S65,422-1
3-(3-CHLOROPHENYL)-1,1-DIISOBUTYLUREA
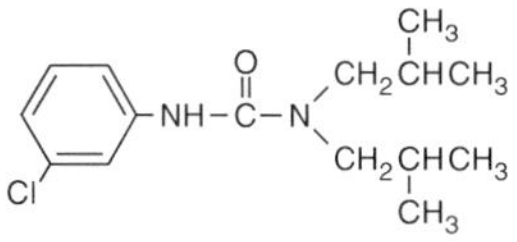
5 gm

S60,693-6
1,1-DIBUTYL-3-(M-TOLYL)UREA
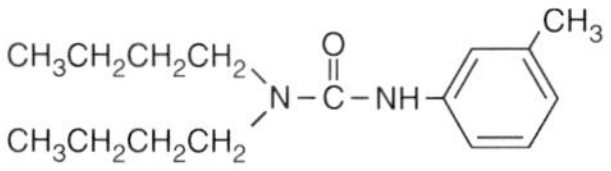
5 gm

S65,395-0
1,1-DIISOBUTYL-3-(M-TOLYL)UREA
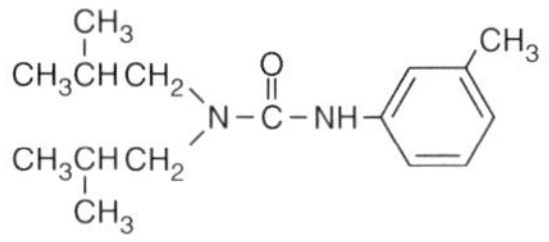
5 gm

S64,674-1
1,1-DIISOBUTYL-3-(3-TRIFLUOROMETHYLPHENYL)UREA
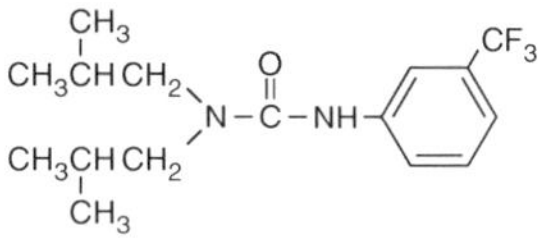
5 gm

S81,068-1
1-HEXADECYLUREA
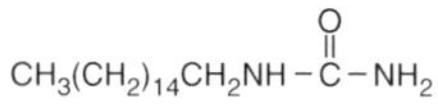
250 mg

S62,491-8
1-BUTYL-3-(3-CHLOROPHENYL)-1-PHENYLUREA
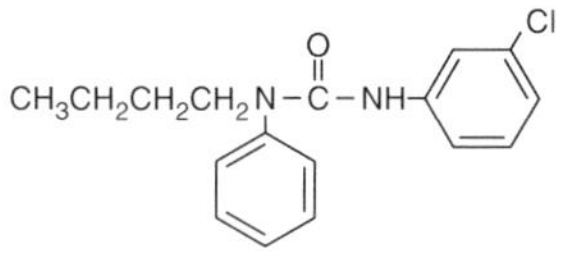
5 gm

S64,694-6
3,3'-HEXAMETHYLENEBIS(1,1-DIPROPYLUREA)
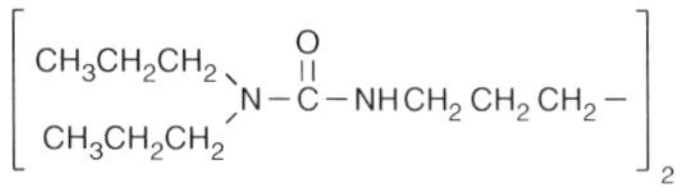
5 gm

S67,006-5
1-OCTADECYL-3-(M-TOLYL)UREA
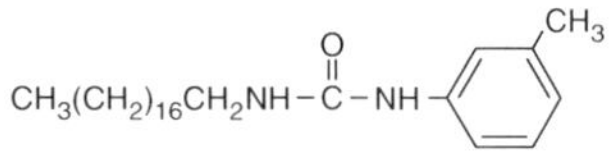
5 gm

S92,421-0
1-PHENYL-3-(2-THIAZOLIN-2-YL)UREA
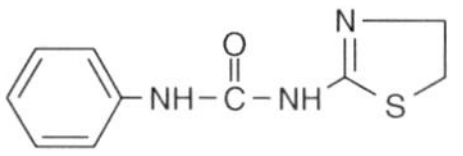
250 mg

S71,275-2
1-(4-CYANOPHENYL)-3-(2,6-DIMETHYLPHENYL)UREA
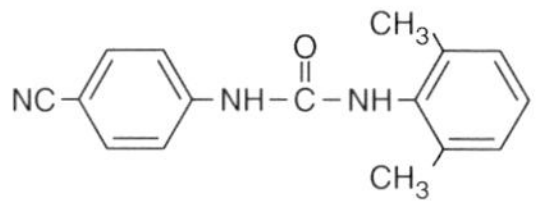
5 gm

S73,579-5
1-(4-CYANOPHENYL)-3-(2,4-XYLYL)UREA
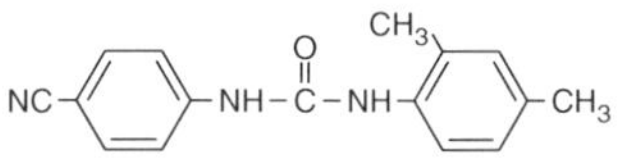
250 mg

S70,736-8
1,1'-(METHYLENEDI-4,1-PHENYLENE)-BIS[3-(4-CYANOPHENYL)UREA]
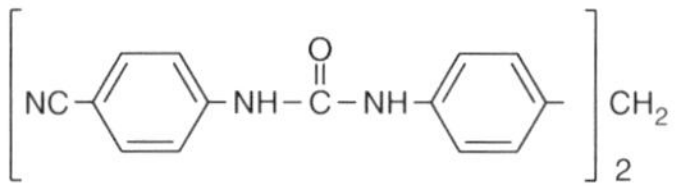
5 gm

S94,232-4
1-PHENYL-3-(3-SULFOLANYL)UREA
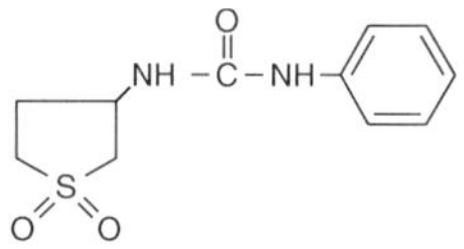
250 mg

S70,269-2
1,1'-[(2-CHLOROBENZYL)-2-METHYLPHENYLENE]BIS[3-(4-SULFAMOYLPHENYL)UREA]
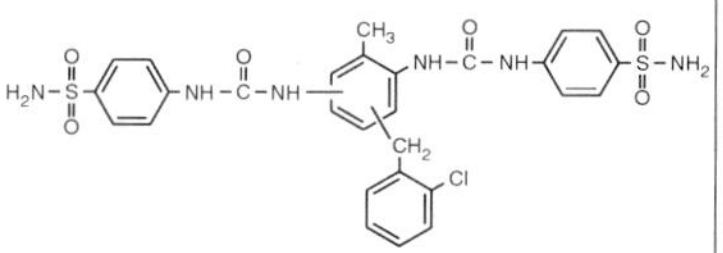
250 mg

S76,816-2
1-(2-CHLOROPHENYL)-3-CYCLOPROPYLUREA
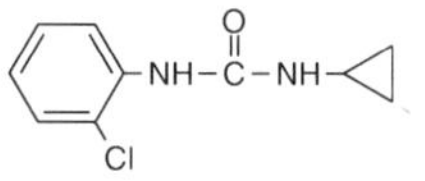
250 mg

S69,199-2
1-CYCLOPROPYL-3-(2-ETHYLPHENYL)UREA
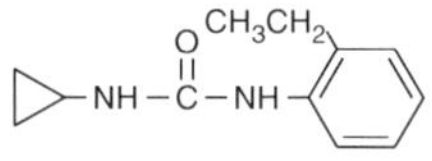
250 mg

S69,429-0
1-CYCLOPROPYL-3-(2,6-DIMETHYLPHENYL)UREA
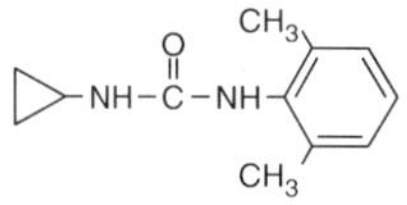
250 mg

S69,325-1
1-CYCLOPROPYL-3-(2,4-XYLYL)UREA
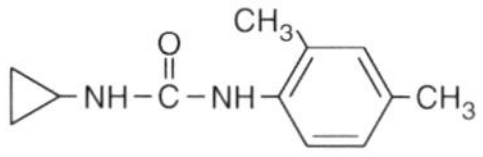
250 mg

S77,291-7
1-CYCLOPENTYL-3-(M-TOLYL)UREA
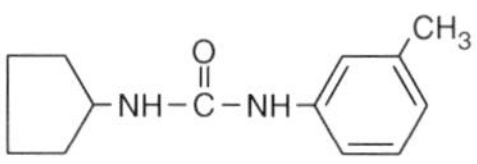
250 mg

S77,302-6
1-CYCLOPENTYL-3-(O-TOLYL)UREA
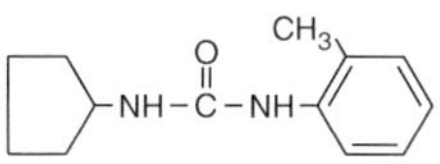
250 mg

S77,290-9
1-CYCLOPENTYL-3-(P-TOLYL)UREA
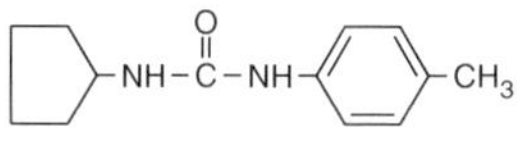
250 mg

S77,255-0
1-CYCLOPENTYL-3-(4-ISOPROPYLPHENYL)UREA
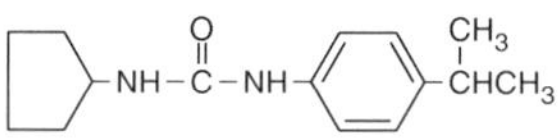
250 mg

S77,312-3
1-CYCLOPENTYL-3-(2-ISOPROPYL-6-METHYLPHENYL)UREA
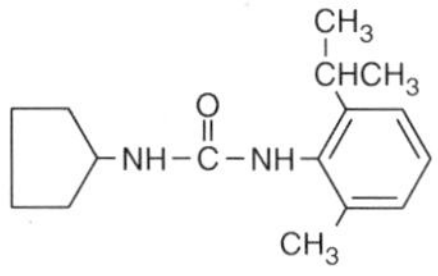
250 mg

S77,313-1
1-CYCLOPENTYL-3-(3,4-XYLYL)UREA
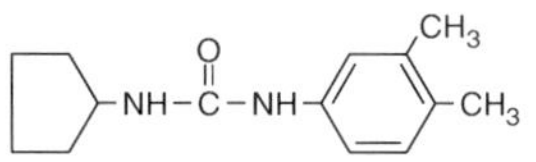
250 mg

S77,316-6
1-CYCLOPENTYL-3-(2,4,5-TRIMETHYLPHENYL)UREA
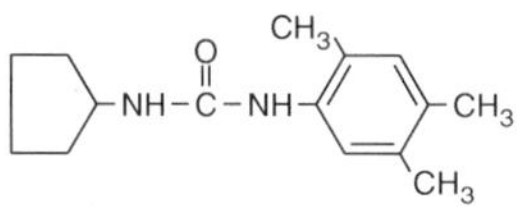
250 mg

S73,214-1
1,3-BENZODIOXOL-5-YLUREA
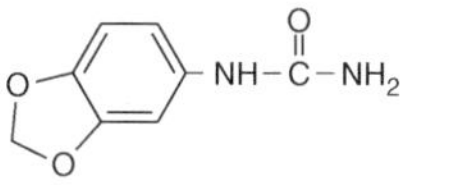
250 mg

S34,765-5
4,5-DIPHENYL-2(3H)-IMIDAZOLONE
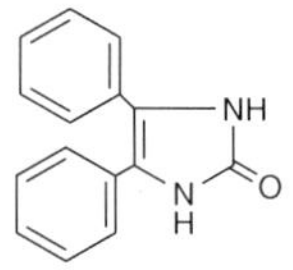
5 gm

S59,132-7
1-CYCLOHEXYL-3-PHENYLUREA
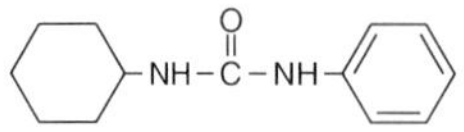
5 gm

S61,107-7
1-(3-CHLOROPHENYL)-3-CYCLOHEXYLUREA
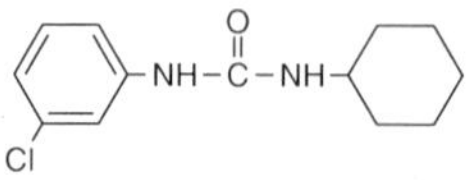
5 gm

S64,438-2
1-(2-CHLOROPHENYL)-3-CYCLOHEXYLUREA
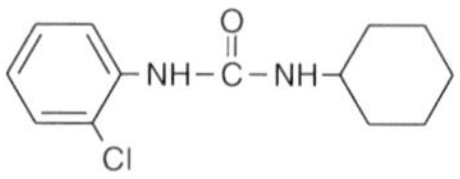
5 gm

S62,159-5
1-CYCLOHEXYL-3-(2,4,5-TRICHLOROPHENYL)UREA
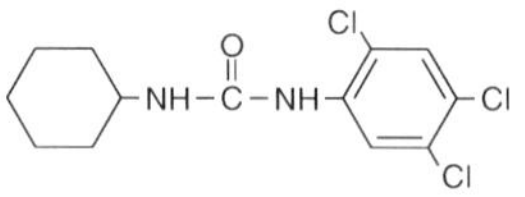
5 gm

S60,666-9
1-CYCLOHEXYL-1-METHYL-3-PHENYLUREA
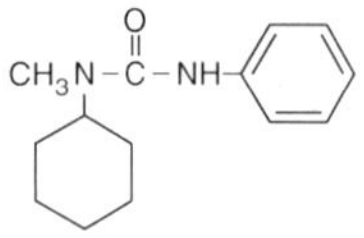
5 gm

S60,689-8
1-CYCLOHEXYL-3-(M-TOLYL)UREA
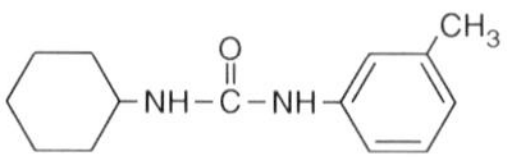
5 gm

S61,068-2
3-(3-CHLOROPHENYL)-1-CYCLOHEXYL-1-METHYLUREA
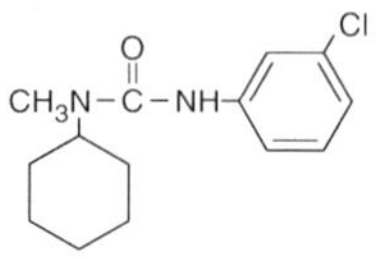
5 gm

S66,729-3
1-CYCLOHEXYL-1-METHYL-3-(2,4,5-TRICHLOROPHENYL)UREA
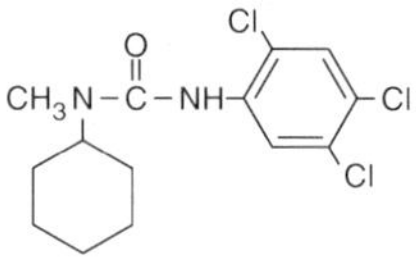
5 gm

S60,713-4
1-CYCLOHEXYL-1-METHYL-3-(M-TOLYL)UREA
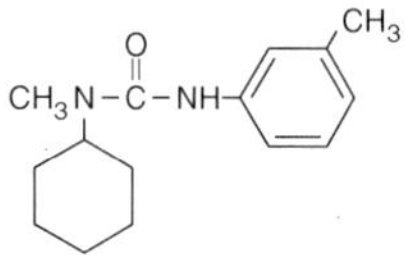
5 gm

S62,437-3
1-CYCLOHEXYL-1-ISOPROPYL-3-PHENYLUREA
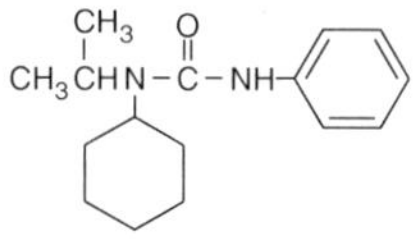
5 gm

S74,331-3
1-CYCLOHEXYL-3-(2,4,5-TRIMETHYLPHENYL)UREA
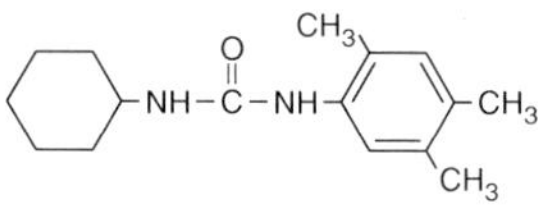
5 gm

S62,483-7
3-(3-CHLOROPHENYL)-1-CYCLOHEXYL-1-ISOPROPYLUREA
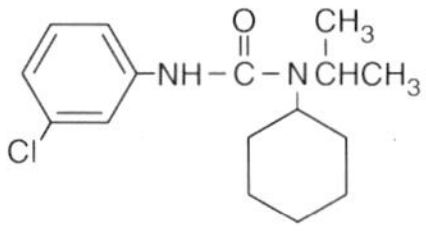
5 gm

S61,113-1
1,1-DICYCLOHEXYL-3-PHENYLUREA
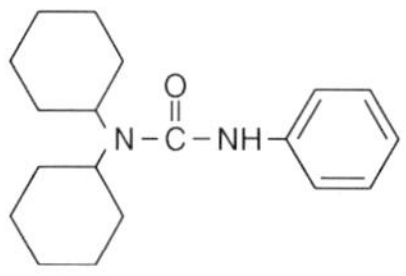
5 gm

S62,494-2
3-(3-CHLOROPHENYL)-1,1-DICYCLOHEXYLUREA
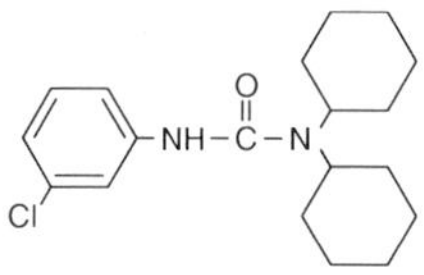
5 gm

S61,158-1
3-(2,6-DICHLOROPHENYL)-1,1-DICYCLOHEXYLUREA
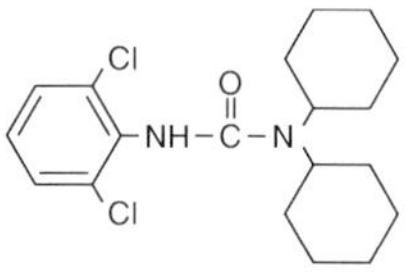
5 gm

S63,630-4
1,1-DICYCLOHEXYL-3-(2,4,5-TRICHLOROPHENYL)CARBAMATE
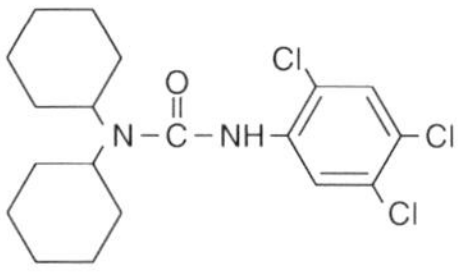
5 gm

S61,363-0
1,1-DICYCLOHEXYL-3-(M-TOLYL)UREA
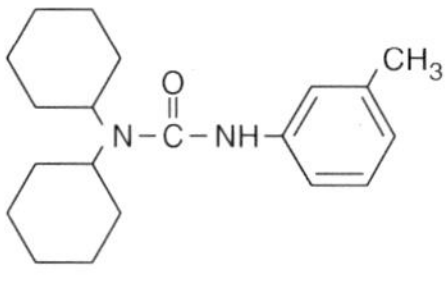
5 gm

S63,225-2
1,1'-HEXAMETHYLENEBIS(3-CYCLOHEXYLUREA)
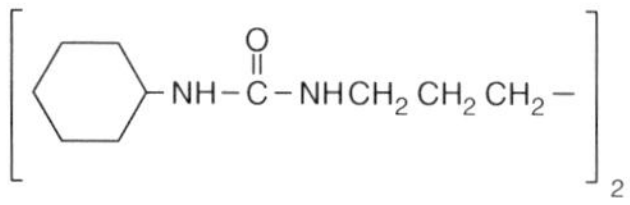
5 gm

S89,941-0
1-BENZYL-3-CYCLOHEXYL-1-METHYLUREA
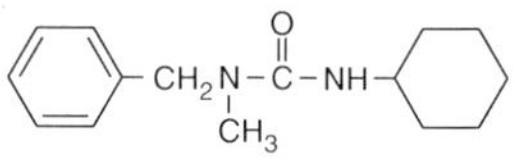
250 mg

S89,291-2
1-CYCLOHEXYL-3-OCTADECYL-1-PHENYLUREA
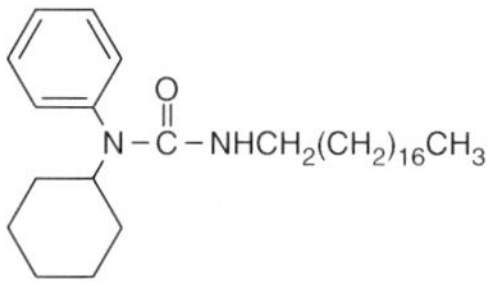
250 mg

S75,289-4
1-CYCLOHEXYL-3-(1-NAPHTHYL)-1-PHENYLUREA
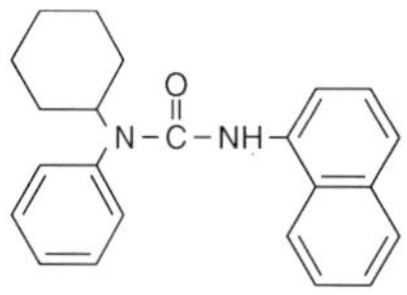
250 mg

S63,404-2
3-(2-CHLOROPHENYL)-1-CYCLOHEXYL-1-METHYLUREA
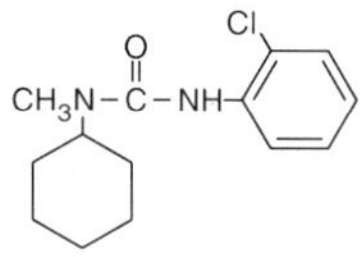
5 gm

S64,666-0
1-CYCLOHEXYL-1-METHYL-3-(O-TOLYL)UREA
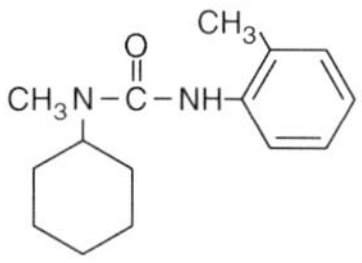
5 gm

S66,044-2
1-CYCLOHEXYL-3-(2-ETHYLPHENYL)UREA
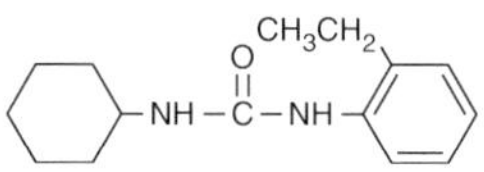
5 gm

S66,379-4
1-CYCLOHEXYL-3-(2-ETHYLPHENYL)-1-METHYLUREA
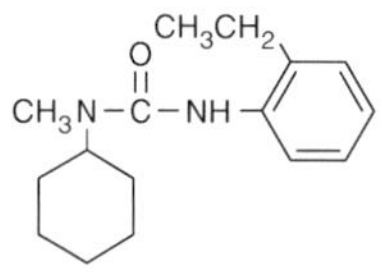
5 gm

S61,282-0
1,1-DICYCLOHEXYL-3-(O-TOLYL)UREA
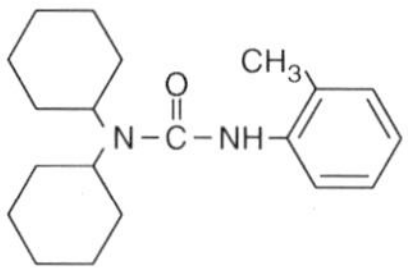
5 gm

S59,139-4
1-(4-CHLOROPHENYL)-3-CYCLOHEXYLUREA
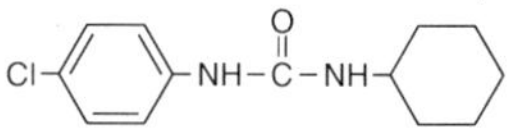
5 gm

S64,030-1
1-(4-BROMOPHENYL)-3-CYCLOHEXYLUREA
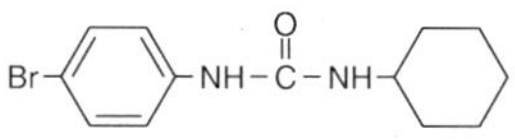
5 gm

S61,358-4
1-CYCLOHEXYL-3-(P-TOLYL)UREA
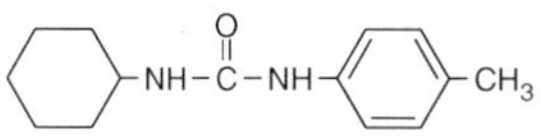
5 gm

S60,663-4
3-(4-CHLOROPHENYL)-1-CYCLOHEXYL-1-METHYLUREA
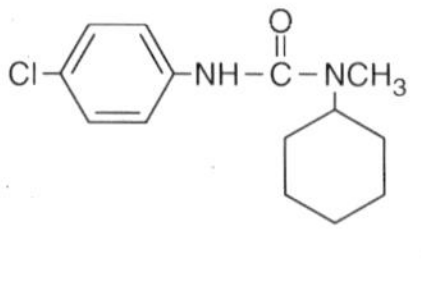
5 gm

S62,832-8
3-(4-BROMOPHENYL)-1-CYCLOHEXYL-1-METHYLUREA
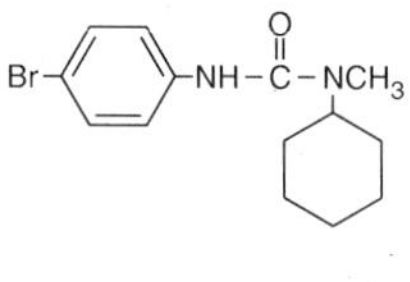
5 gm

S61,287-1
1-CYCLOHEXYL-1-METHYL-3-(P-TOLYL)UREA

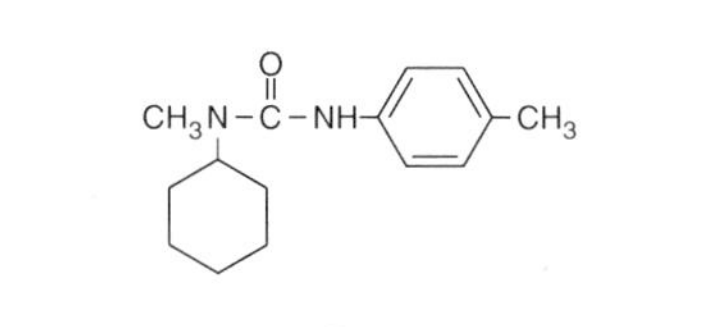

5 gm

S76,864-2
1-CYCLOHEXYL-3-(4-ISOPROPYLPHENYL)UREA

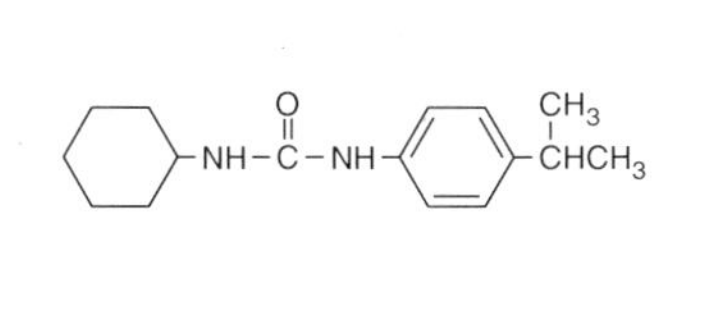

250 mg

S62,440-3
3-(4-CHLOROPHENYL)-1-CYCLOHEXYL-1-ISOPROPYLUREA

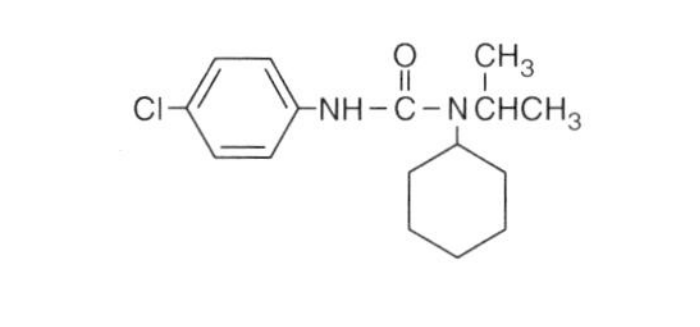

5 gm

S61,124-7
1-(4-CHLOROPHENYL)-3,3-DICYCLOHEXYLUREA

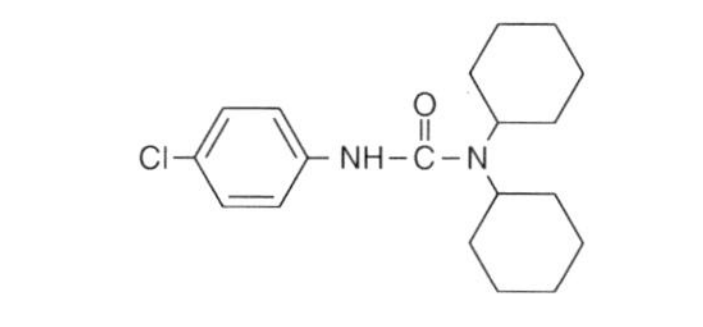

5 gm

S61,330-4
1,1-DICYCLOHEXYL-3-(P-TOLYL)UREA

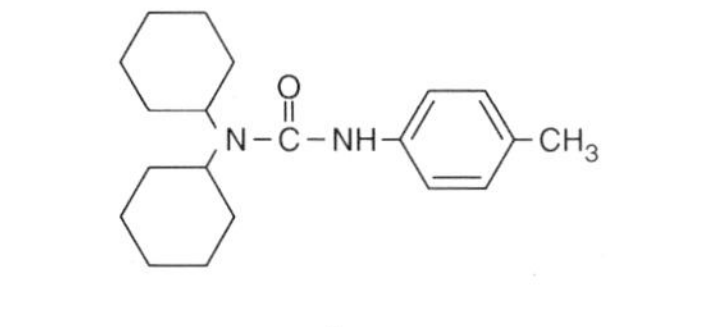

5 gm

S66,019-1
4,4'-METHYLENEBIS(1-CYCLOHEXYL-1-ISOPROPYL-3-PHENYLUREA)

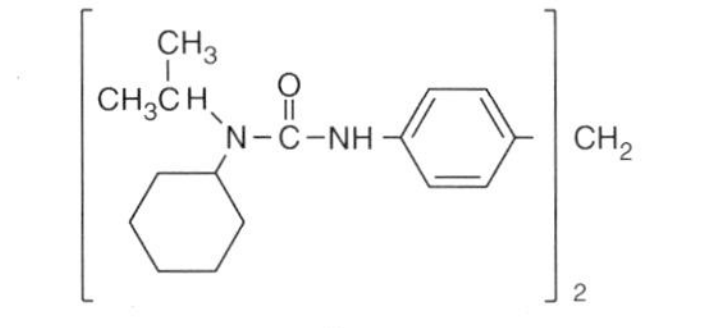

5 gm

S65,160-5
4,4'-METHYLENEBIS(1,1-DICYCLOHEXYL-3-PHENYLUREA)

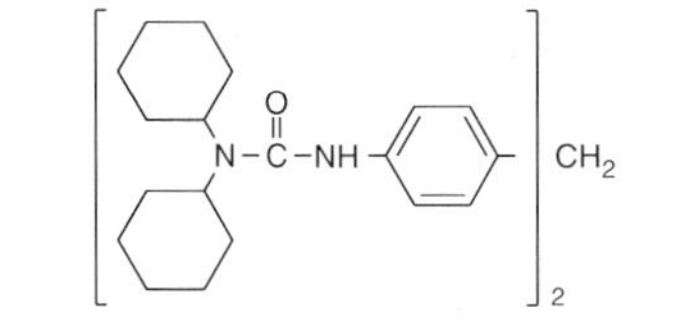

5 gm

S61,143-3
1-CYCLOHEXYL-3-(2,6-DICHLOROPHENYL)UREA

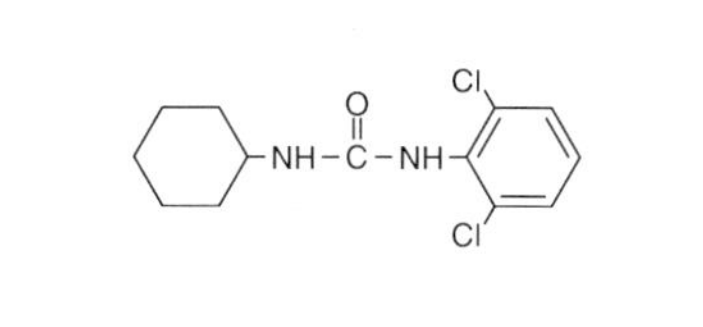

5 gm

S61,186-7
1-CYCLOHEXYL-3-(2,3-DICHLOROPHENYL)UREA

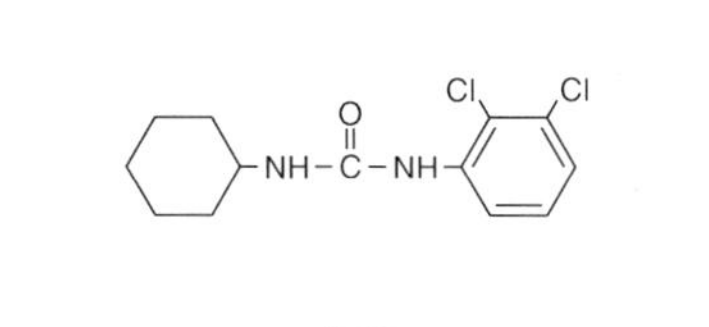

5 gm

S65,237-7
1-(2-CHLORO-6-METHYLPHENYL)-3-CYCLOHEXYLUREA

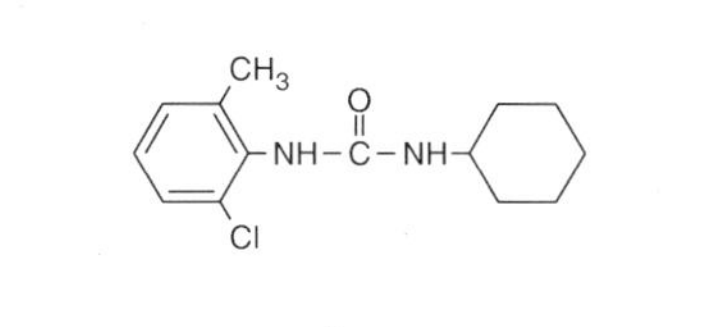

5 gm

S61,161-1
1-CYCLOHEXYL-3-(2,6-DICHLOROPHENYL)-1-METHYLUREA

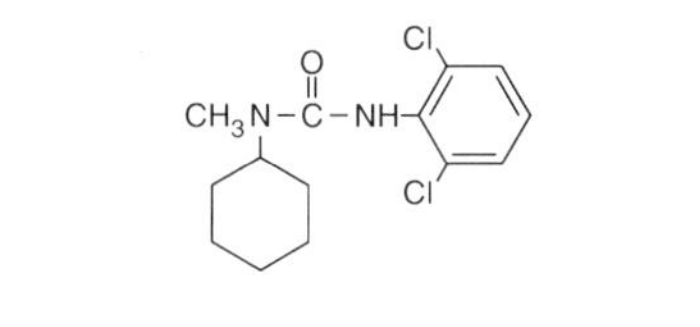

5 gm

S61,177-8
1-CYCLOHEXYL-3-(2,3-DICHLOROPHENYL)-1-METHYLUREA

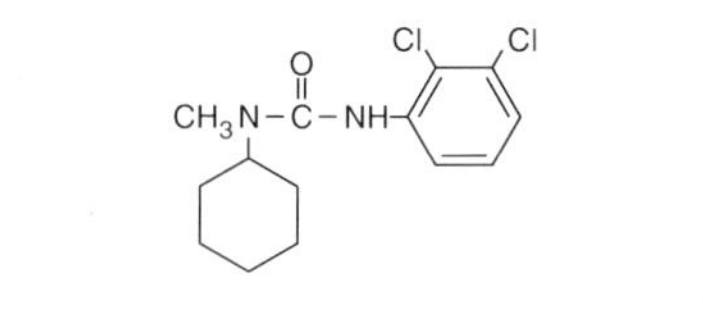

5 gm

S64,406-4
1-CYCLOHEXYL-3-(2,3-XYLYL)UREA

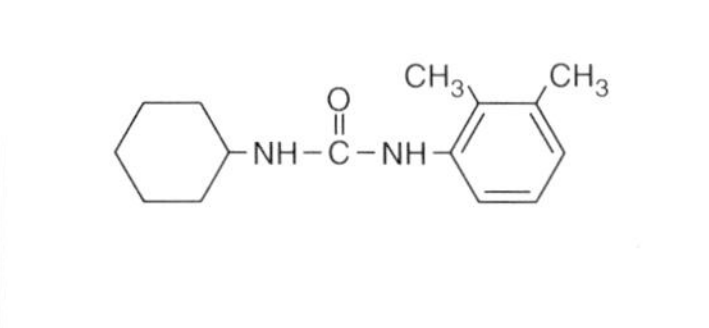

5 gm

S65,329-2
1-CYCLOHEXYL-3-(2,6-XYLYL)UREA

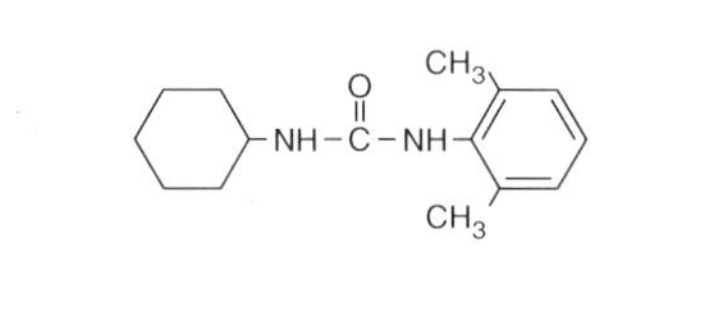

5 gm

S66,616-5
3-(2-CHLORO-6-METHYLPHENYL)-1-CYCLOHEXYL-1-METHYLUREA

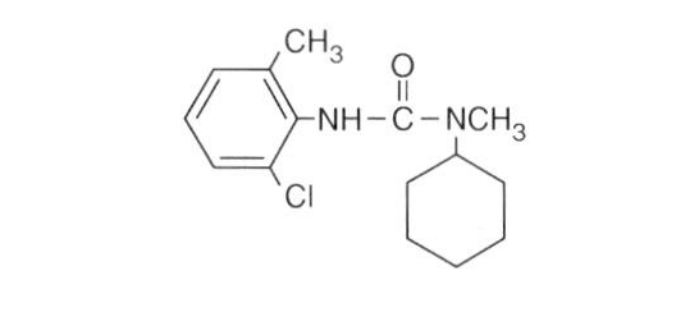

5 gm

S66,064-7
1-CYCLOHEXYL-3-(2-ETHYL-6-METHYLPHENYL)UREA

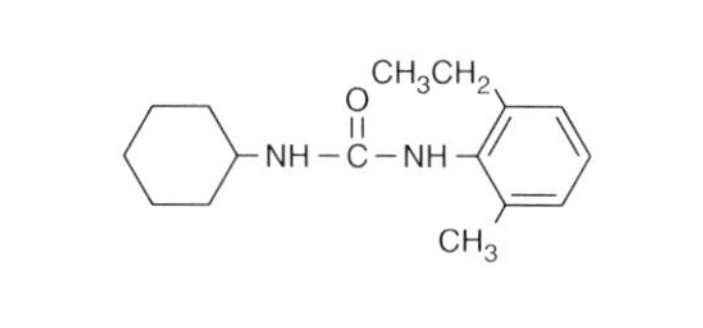

5 gm

S66,076-0
1-CYCLOHEXYL-1-METHYL-3-(2,3-XYLYL)UREA

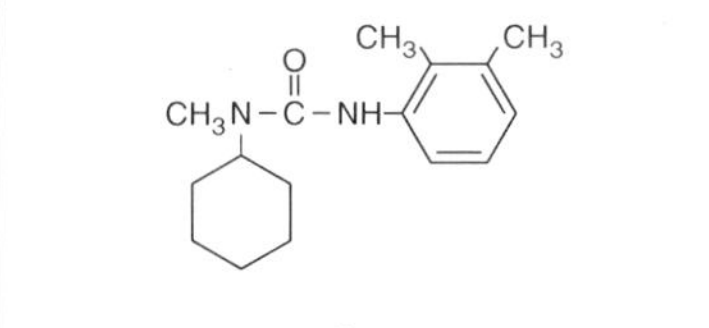

5 gm

S66,382-4
1-CYCLOHEXYL-1-METHYL-3-(2,6-XYLYL)UREA

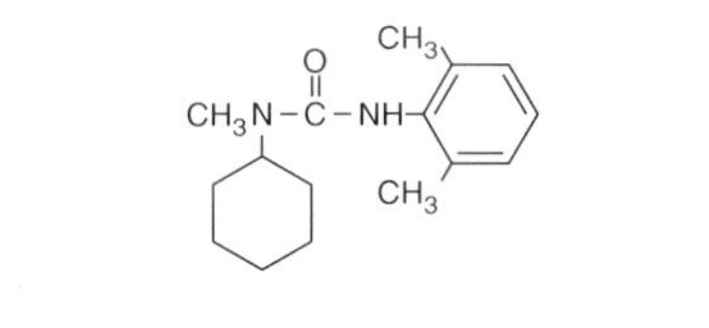

5 gm

S66,364-6
1-CYCLOHEXYL-3-(2-ETHYL-6-METHYLPHENYL)-1-METHYLUREA

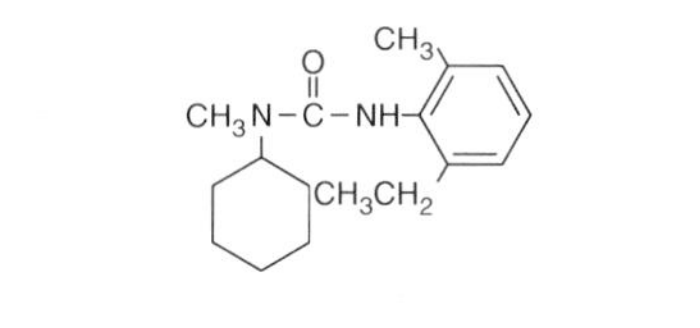

5 gm

S71,296-5
1-CYCLOHEXYL-3-(2,6-DIETHYLPHENYL)UREA

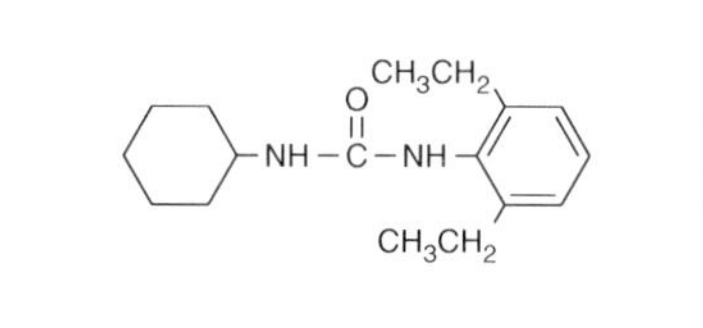

5 gm

S66,845-1
1-CYCLOHEXYL-3-(2,6-DIETHYLPHENYL)-1-METHYLUREA

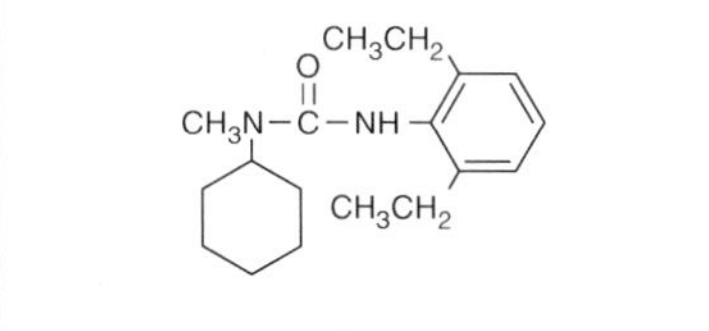

5 gm

S64,442-0
1-CYCLOHEXYL-3-(2,6-DIISOPROPYLPHENYL)UREA

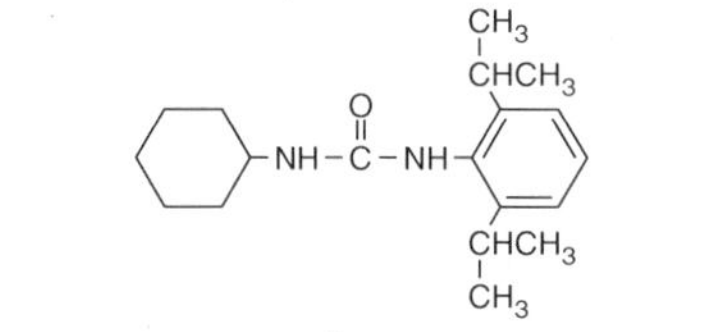

5 gm

S61,144-1
3-(2,3-DICHLOROPHENYL)-1,1-DICYCLOHEXYLUREA

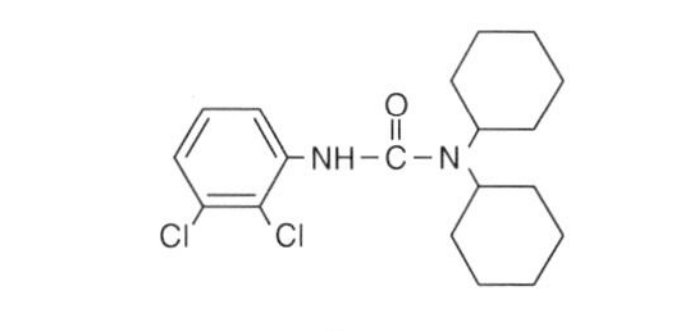

5 gm

S66,389-1
1-CYCLOHEXYL-3-(2,6-DIISOPROPYLPHENYL)-1-METHYLUREA

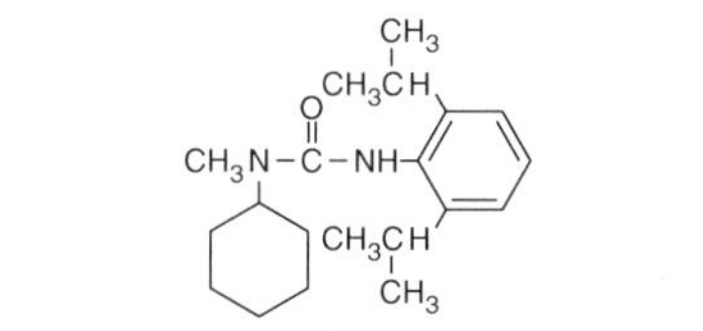

5 gm

S61,168-9
1-CYCLOHEXYL-3-(2,5-DICHLOROPHENYL)UREA

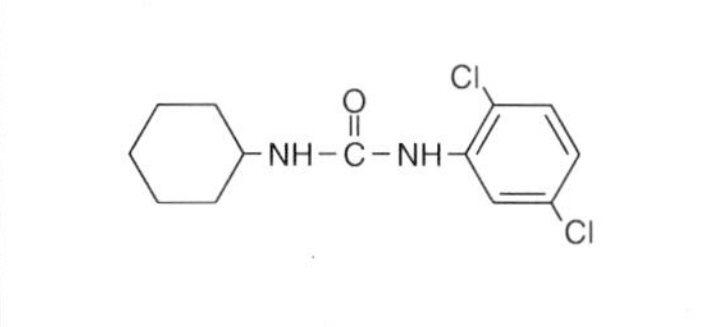

5 gm

S61,275-8
1-CYCLOHEXYL-3-(2,4-DICHLOROPHENYL)UREA

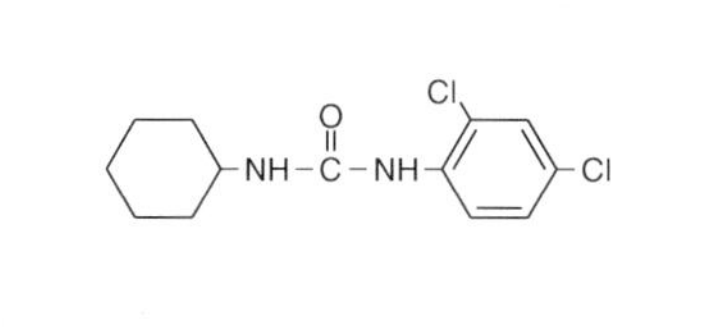

5 gm

S61,297-9
1-CYCLOHEXYL-3-(3,4-DICHLOROPHENYL)UREA

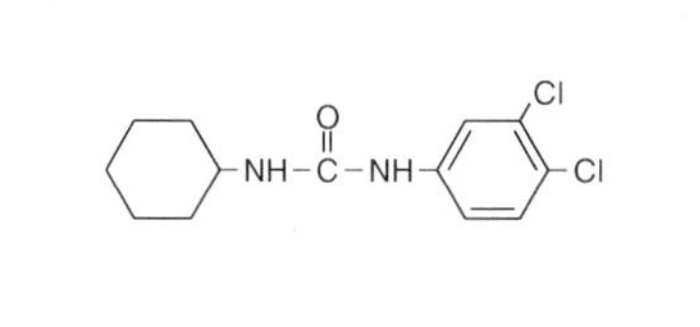

5 gm

S61,103-4
1-(5-CHLORO-2-METHYLPHENYL)-3-CYCLOHEXYLUREA

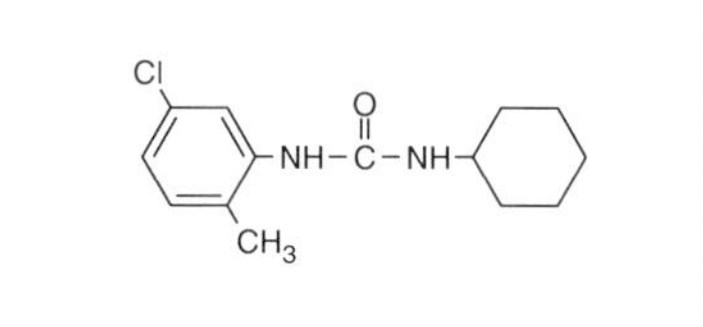

5 gm

S63,814-5
1-(3-CHLORO-4-METHYLPHENYL)-3-CYCLOHEXYLUREA

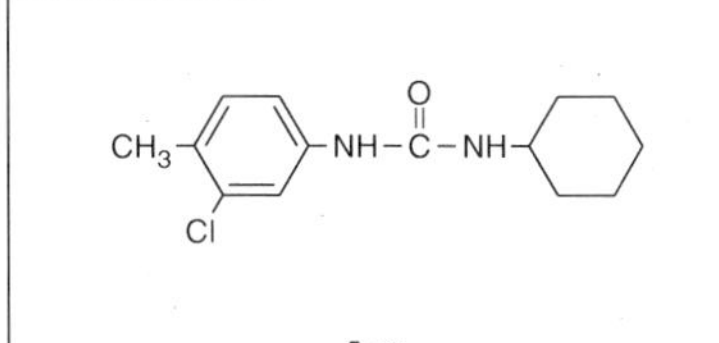

5 gm

S61,151-4
1-CYCLOHEXYL-3-(2,5-DICHLOROPHENYL)-1-METHYLUREA

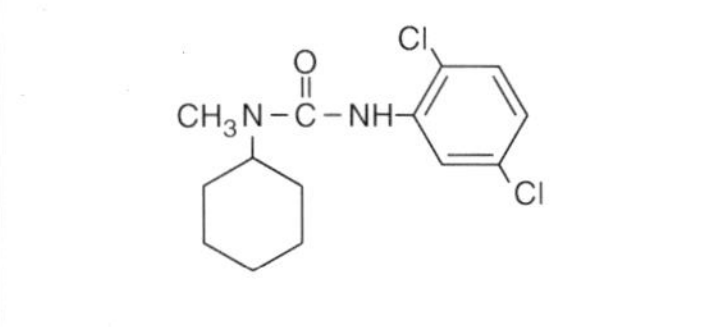

5 gm

S61,280-4
1-CYCLOHEXYL-3-(2,4-DICHLOROPHENYL)-1-METHYLUREA

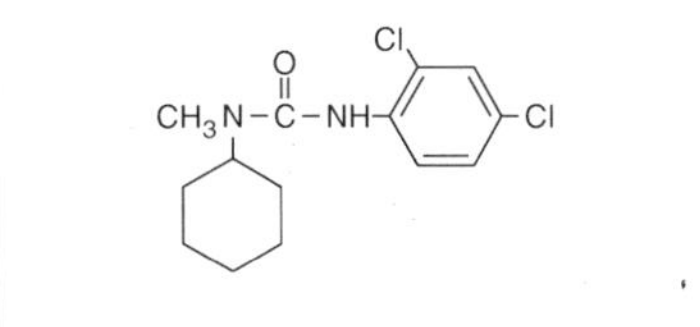

5 gm

S61,362-2
1-CYCLOHEXYL-3-(3,4-DICHLOROPHENYL)-1-METHYLUREA

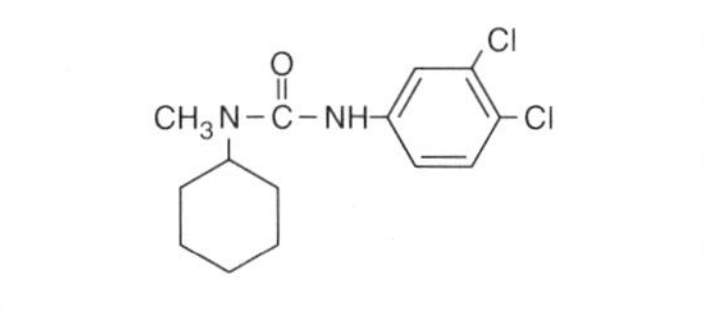

5 gm

S60,664-2
1-CYCLOHEXYL-3-(3,4-XYLYL)UREA

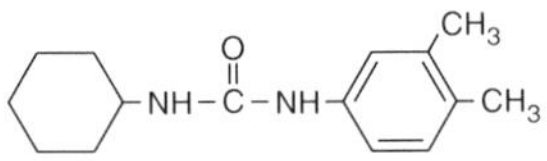

5 gm

S63,658-4
1-CYCLOHEXYL-3-(2,5-XYLYL)UREA

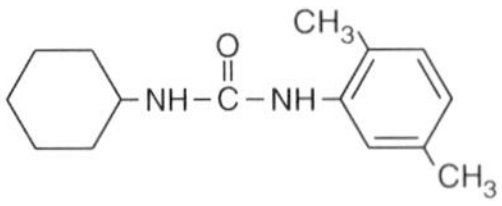

5 gm

S60,798-3
3-(4-CHLORO-2-METHYLPHENYL)-1-CYCLOHEXYL-1-METHYLUREA

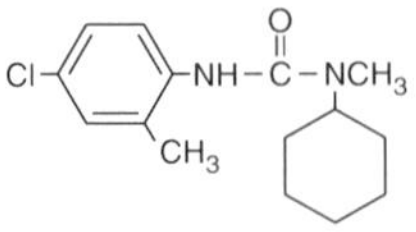

5 gm

S61,012-7
3-(5-CHLORO-2-METHYLPHENYL)-1-CYCLOHEXYL-1-METHYLUREA

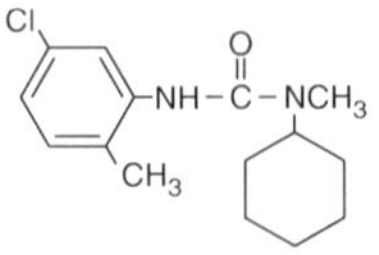

5 gm

S63,839-0
3-(3-CHLORO-4-METHYLPHENYL)-1-CYCLOHEXYL-1-METHYLUREA

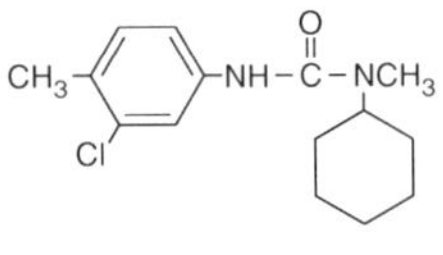

5 gm

S61,052-6
1-CYCLOHEXYL-1-METHYL-3-(2,4-XYLYL)UREA

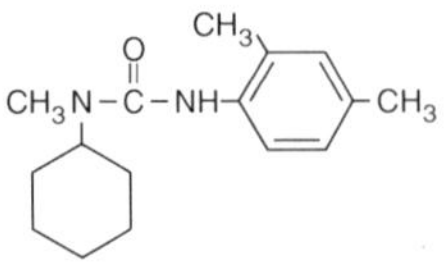

5 gm

S66,119-8
1-CYCLOHEXYL-1-METHYL-3-(3,4-XYLYL)UREA

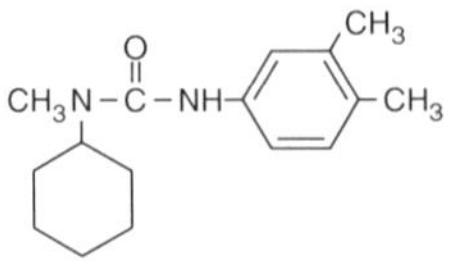

5 gm

S66,450-2
1-CYCLOHEXYL-1-METHYL-3-(2,5-XYLYL)UREA

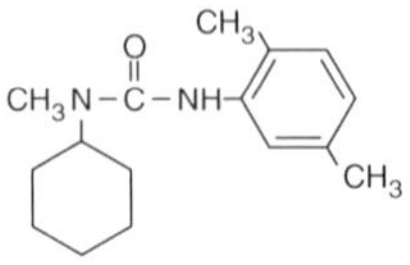

5 gm

S90,052-4
1-CYCLOHEXYL-3-(3,4-DIMETHYLBENZYL)UREA

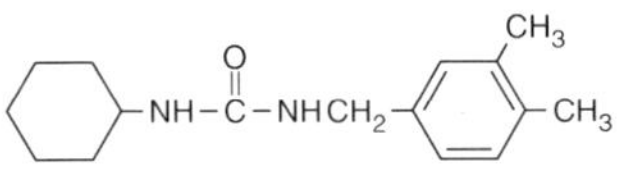

250 mg

S61,125-5
3-(3,4-DICHLOROPHENYL)-1,1-DICYCLOHEXYLUREA

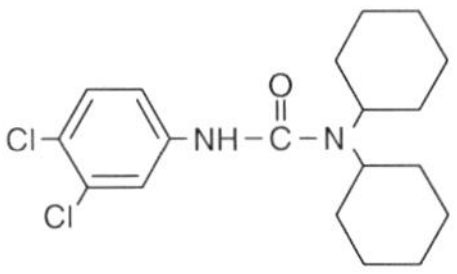

5 gm

S61,208-1
3-(2,5-DICHLOROPHENYL)-1,1-DICYCLOHEXYLUREA

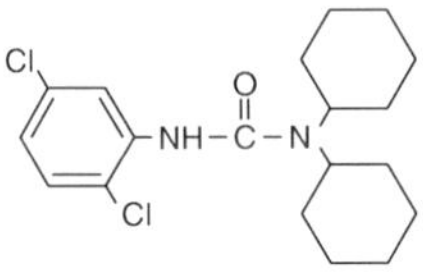

5 gm

S63,645-2
3-(2,4-DICHLOROPHENYL)-1,1-DICYCLOHEXYLUREA

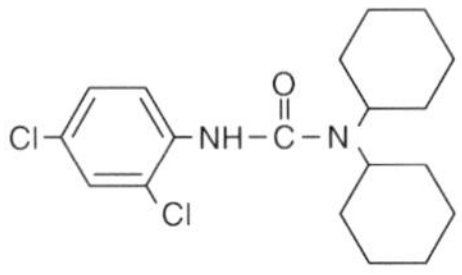

5 gm

S61,108-5
3-(5-CHLORO-2-METHYLPHENYL)-1,1-DICYCLOHEXYLUREA

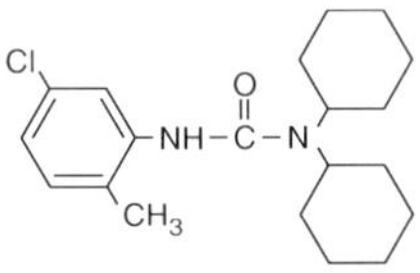

5 gm

S61,121-2
3-(3-CHLORO-4-METHYLPHENYL)-1,1-DICYCLOHEXYLUREA

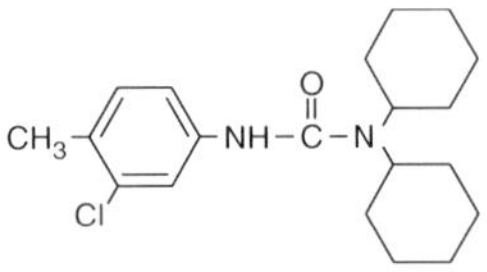

5 gm

S61,597-8
1,1-DICYCLOHEXYL-3-(2,4-XYLYL)UREA

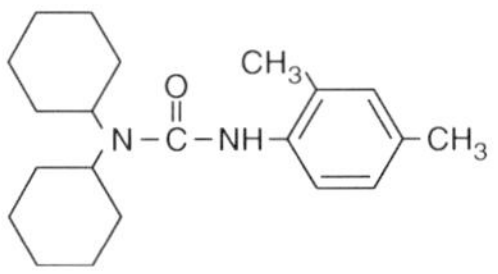

5 gm

S77,348-4
1,1-DICYCLOHEXYL-3-(2,5-XYLYL)UREA

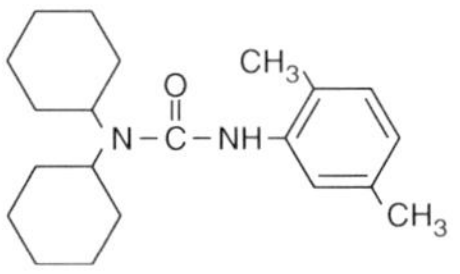

250 mg

S66,371-9
1-CYCLOHEXYL-3-(3,5-XYLYL)UREA

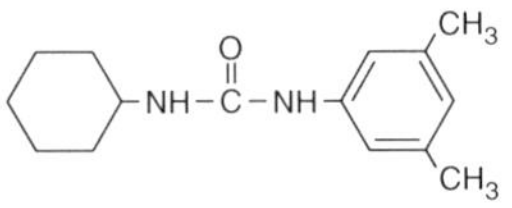

5 gm

S66,378-6
1-CYCLOHEXYL-1-METHYL-3-(3,5-XYLYL)UREA

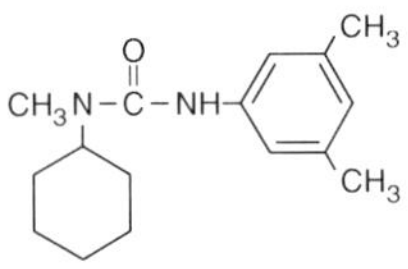

5 gm

S74,319-4
1-CYCLOHEXYL-3-(2,4,6-TRIMETHYLPHENYL)UREA

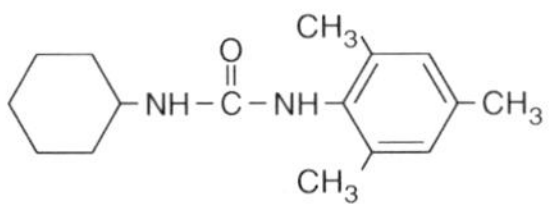

5 gm

S63,490-5
4,4'-METHYLENEBIS(1-CYCLOHEXYL-3-PHENYLUREA)

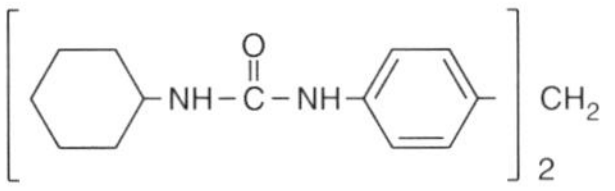

5 gm

S59,028-2
1-CYCLOHEXYL-3-(1-NAPHTHYL)UREA

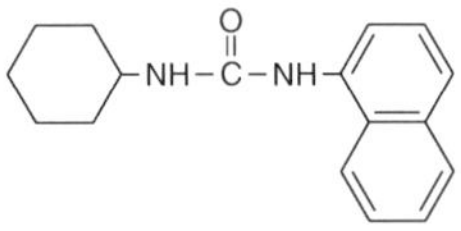

5 gm

S62,520-5
1-CYCLOHEXYL-1-METHYL-3-(1-NAPHTHYL)UREA

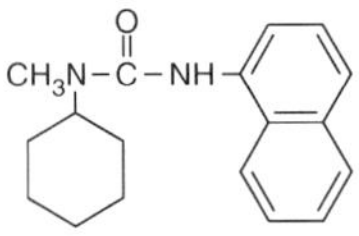

5 gm

S72,665-6
1-CYCLOHEXYL-1-ETHYL-3-(1-NAPHTHYL)UREA

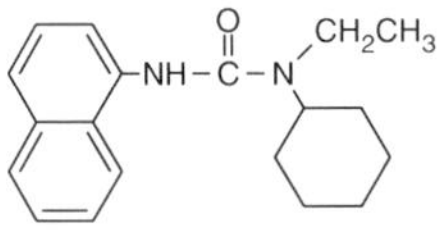

5 gm

S75,281-9
1-CYCLOHEXYL-1-ISOPROPYL-3-(1-NAPHTHYL)UREA

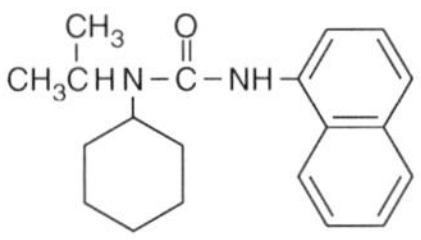

50 mg

S63,924-9
1,1-DICYCLOHEXYL-3-(1-NAPHTHYL)UREA

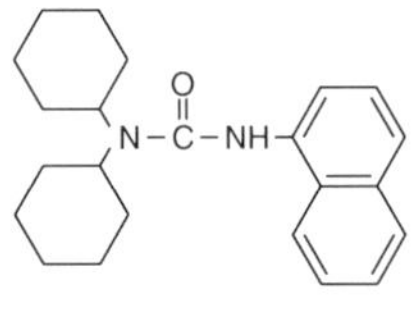

5 gm

S63,837-4
1,1'-(1,5-NAPHTHALENEDIYL)BIS(3-CYCLOHEXYLUREA)

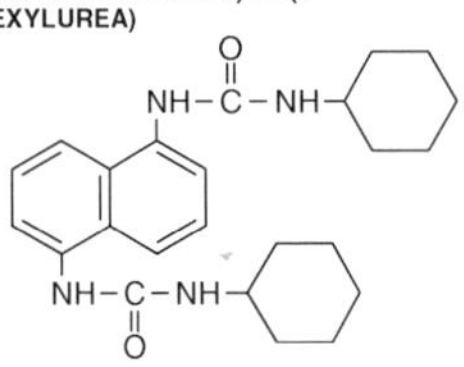

5 gm

S63,850-1
1,1'-(1,5-NAPHTHALENEDIYL)BIS(3-CYCLOHEXYL-3-METHYLUREA)

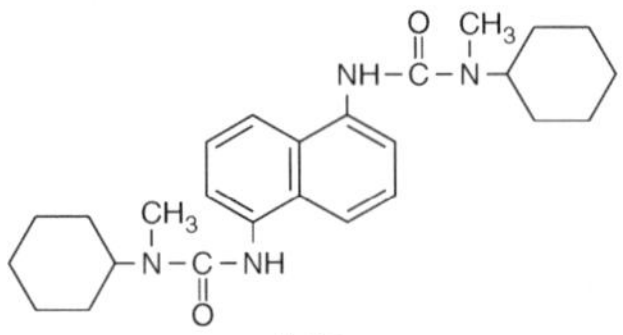

5 gm

S59,140-8
1-(4-CHLOROPHENYLCARBAMOYL)-PIPERIDINE

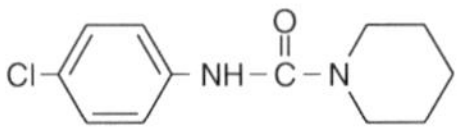

5 gm

S59,422-9
1-(3-CHLOROPHENYLCARBAMOYL)-PIPERIDINE

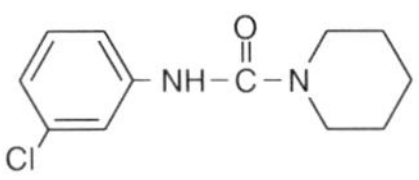

5 gm

S59,543-8
1-(2-CHLOROPHENYLCARBAMOYL)-PIPERIDINE

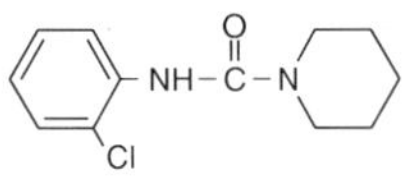

5 gm

S62,821-2
1-[(4-BROMOPHENYL)CARBAMOYL]-PIPERIDINE

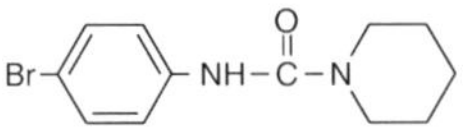

5 gm

S91,236-0
2-METHYL-1-[N-(M-TOLYL)CARBAMOYL]-PIPERIDINE

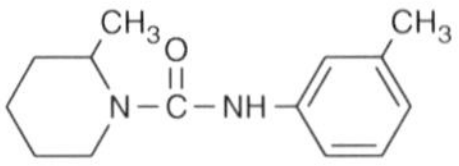

250 mg

S91,238-7
4-METHYL-1-[N-(M-TOLYL)CARBAMOYL]-PIPERIDINE

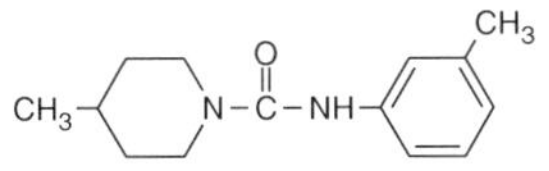

250 mg

S65,099-4
1-(2-CHLOROPHENYL)-3-[3-(2-METHYLPIPERIDINO)PROPYL]UREA

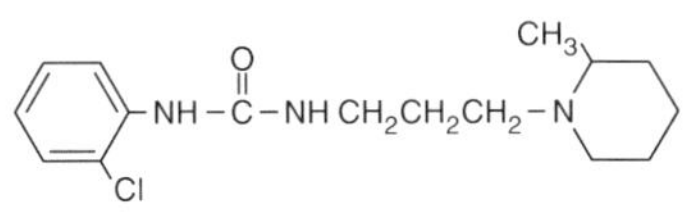

5 gm

S91,703-6
1-[N-(4-BUTYLPHENYL)CARBAMOYL]-2,6-DIMETHYLPIPERIDINE

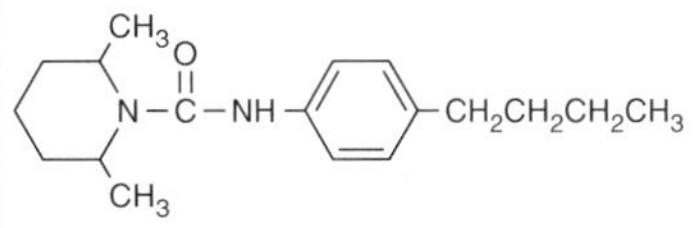

250 mg

S91,217-4
1-[N-(2,6-DIETHYLPHENYL)CARBAMOYL]-2-METHYLPIPERIDINE

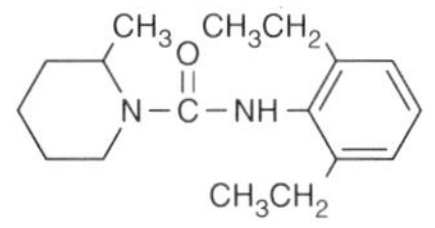

250 mg

S58,615-3
3-(1-NAPHTHYL)-1,1-PENTAMETHYLENEUREA

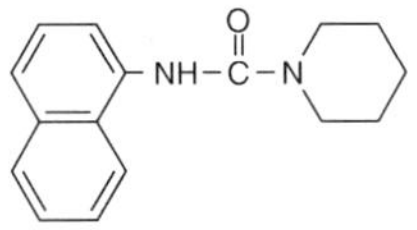

5 gm

S59,151-3
4-(N-PHENYLCARBAMOYL)MORPHOLINE

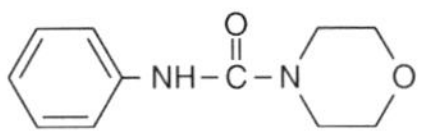

5 gm

S59,145-9
4-(4-CHLOROPHENYLCARBAMOYL)-MORPHOLINE

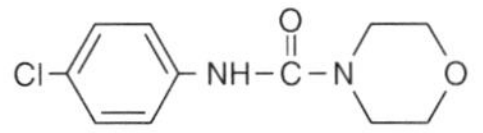

5 gm

S92,560-8
1-(2,6-DIETHYLPHENYL)-3-(2-MORPHOLINOETHYL)UREA

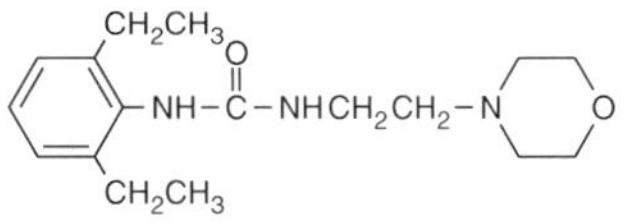

250 mg

S70,764-3
1-(2-ETHYL-6-METHYLPHENYL)-3-(3-MORPHOLINOPROPYL)UREA

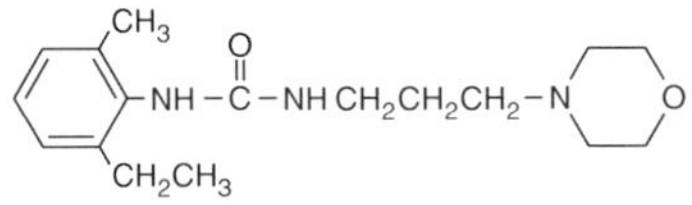

5 gm

S58,040-6
N,N'-(METHYLENEDI-4,1-PHENYLENE)-BIS(4-MORPHOLINECARBOXAMIDE)

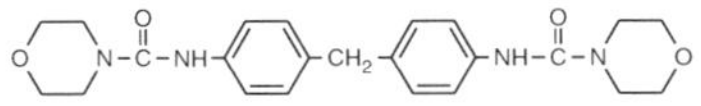

250 mg

S57,973-4
1-METHYL-3-PHENYLUREA

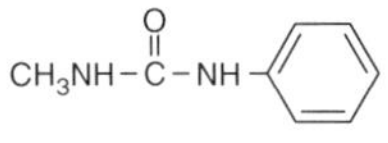

5 gm

S59,163-7
1-ETHYL-3-PHENYLUREA

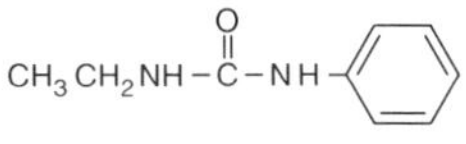

5 gm

S85,622-3
1,3-DIMETHYL-1-PHENYLUREA

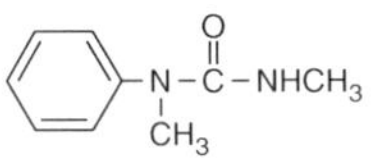

250 mg

S85,603-7
3-PHENYL-1,1,3-TRIMETHYLUREA

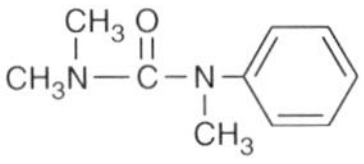

250 mg

S59,142-4
1-ALLYL-3-PHENYLUREA

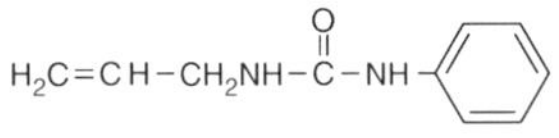

5 gm

S59,137-8
1-PHENYL-3-PROPYLUREA

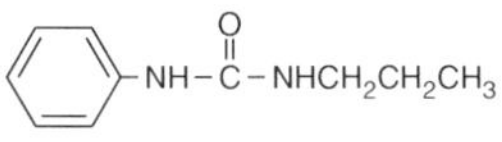

5 gm

S61,319-3
1-ISOPROPYL-3-PHENYLUREA

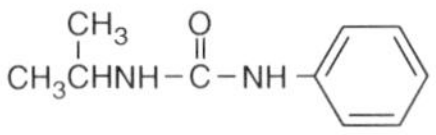

5 gm

S59,124-6
1-BUTYL-3-PHENYLUREA

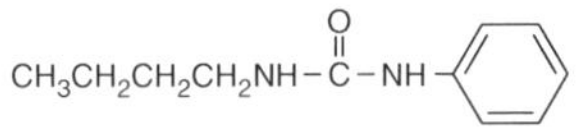

5 gm

S64,380-7
1-ISOBUTYL-3-PHENYLUREA

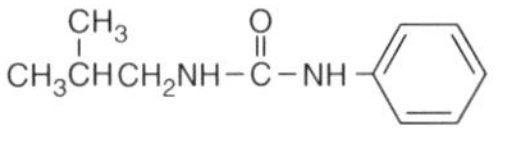

5 gm

S71,122-5
1,1-DIETHYL-3-PHENYLUREA

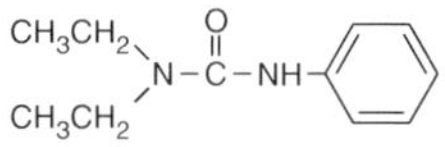

5 gm

S77,540-1
1-BENZYL-3-ETHYL-1-METHYLUREA

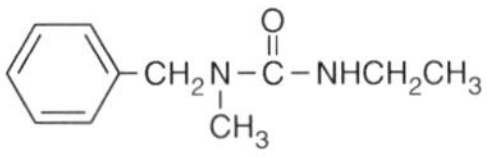

250 mg

S60,344-9
3-BUTYL-1-METHYL-1-PHENYLUREA

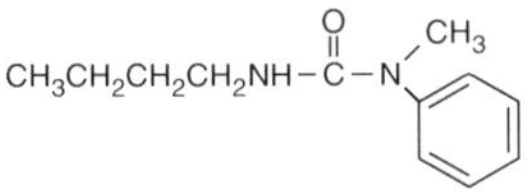

5 gm

S61,073-9
1-BUTYL-1-METHYL-3-PHENYLUREA

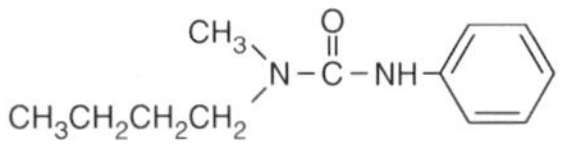

5 gm

S71,554-9
1-(BENZYL)-3-(TERT-BUTYL)UREA

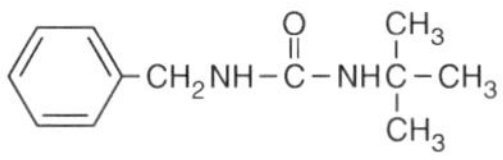

250 mg

S85,617-7
1,1-DIETHYL-3-METHYL-3-PHENYLUREA

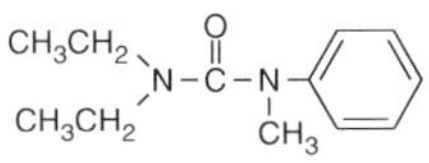

250 mg

S89,983-6
1-BENZYL-3-TERT-BUTYLUREA

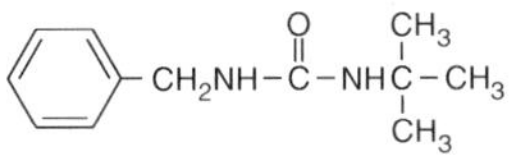

250 mg

S63,680-0
1,1-DIPHENYLUREA

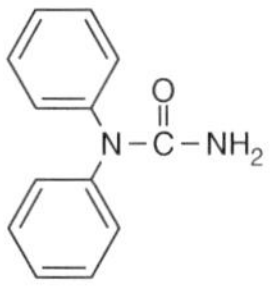

250 mg

S59,129-7
1,1-DIALLYL-3-PHENYLUREA

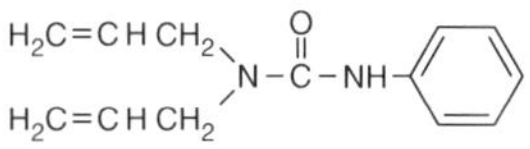

5 gm

S59,135-1
1,1-DIPROPYL-3-PHENYLUREA

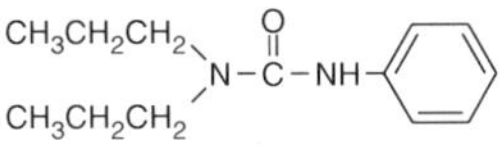

5 gm

S62,366-0
1,1-DIISOPROPYL-3-PHENYLUREA

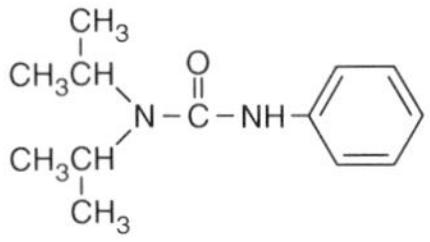

5 gm

S71,517-4
1-BUTYL-1-ETHYL-3-PHENYLUREA

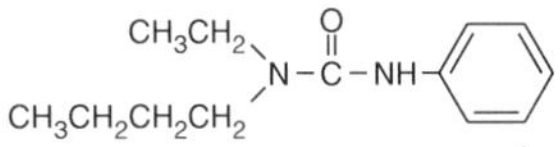

5 gm

S89,946-1
1-BENZYL-3-TERT-BUTYL-1-METHYLUREA

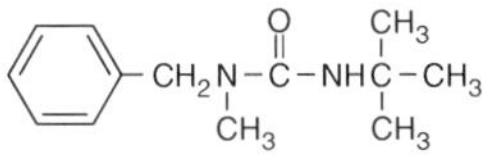

250 mg

S87,128-1
1-PHENYL-3-(PHENYLIMINO)UREA

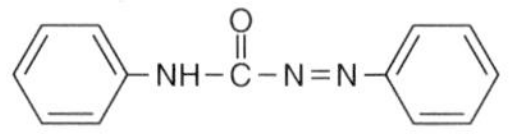

250 mg

S63,466-2
1,3-DIPHENYL-1-METHYLUREA

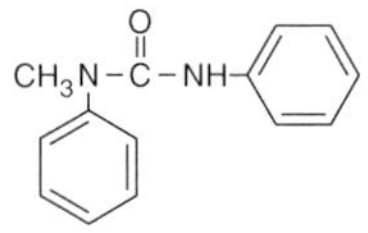

5 gm

S71,665-0
1-PHENYL-3-(M-TOLYL)UREA

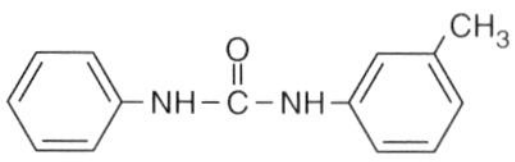

5 gm

S63,926-5
1-ETHYL-1,3-DIPHENYLUREA

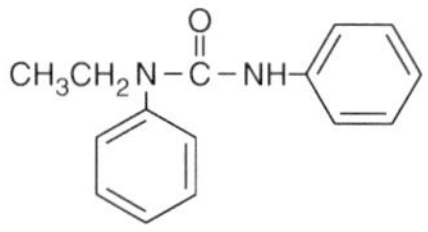

5 gm

S58,623-4
1,1-DIBUTYL-3-PHENYLUREA

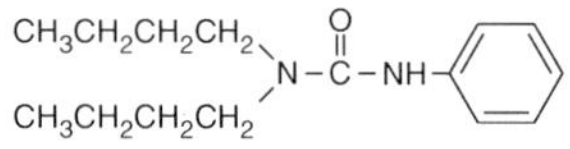

5 gm

S64,385-8
1,1-DIISOBUTYL-3-PHENYLUREA

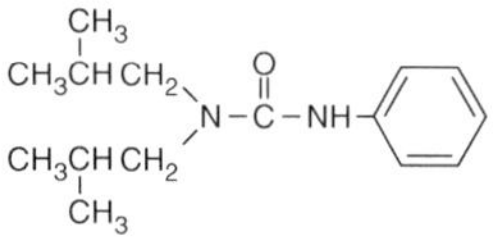

5 gm

S89,989-5
1-BENZYL-1-BUTYL-3-TERT-BUTYLUREA

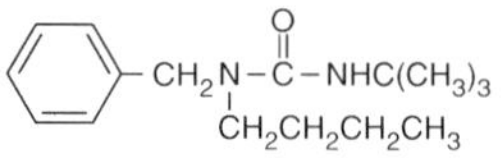

250 mg

S59,147-5
1-BUTYL-1,3-DIPHENYLUREA

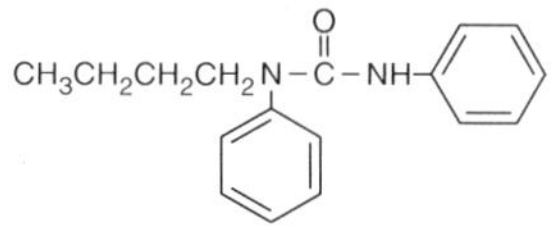

5 gm

S81,336-2
1,1,3-TRIPHENYLUREA

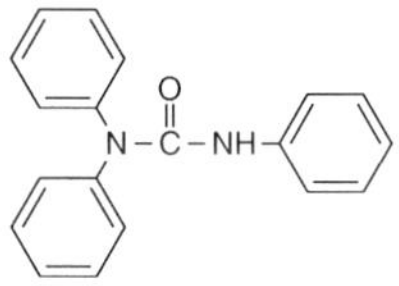

250 mg

S64,690-3
3,3'-HEXAMETHYLENEBIS(1-METHYL-1-PHENYLUREA)

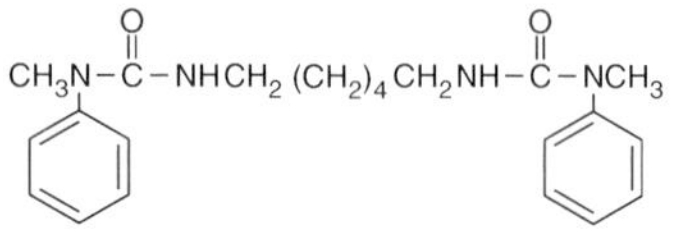

5 gm

S81,990-5
1,1,3,3-TETRAPHENYLUREA

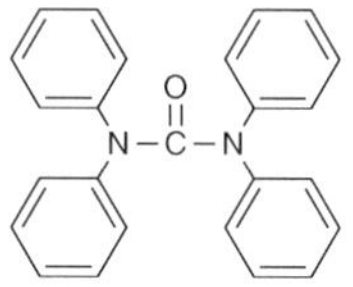

250 mg

S67,069-3
1-OCTADECYL-3-PHENYLUREA

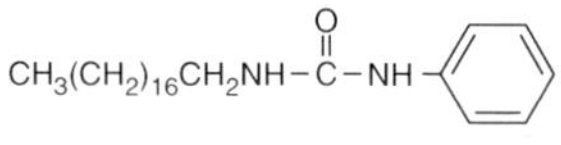

5 gm

S71,669-3
1-PHENYL-3-(O-TOLYL)UREA

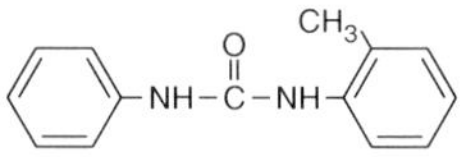

5 gm

S91,960-8
1-PHENYL-3-(O-TOLYLIMINO)UREA

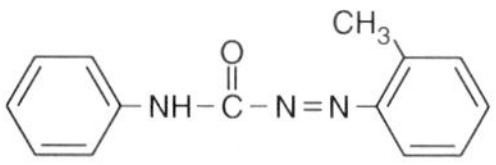

250 mg

S76,853-7
1-(2-ETHYLPHENYL)-3-PHENYLUREA

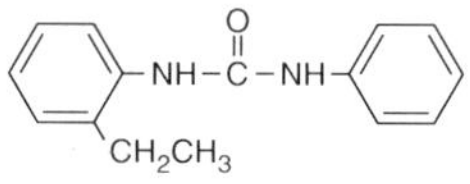

250 mg

S76,904-5
1-ETHYL-3-PHENYL-1-(O-TOLYL)UREA

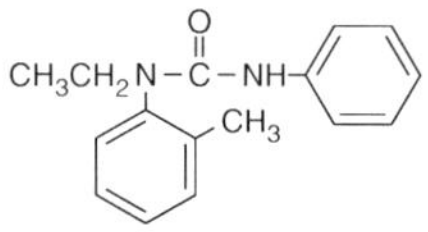

250 mg

S71,670-7
1-PHENYL-3-(P-TOLYL)UREA

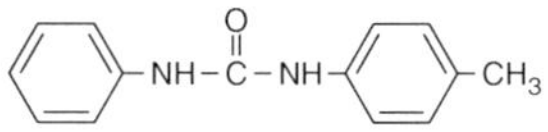

5 gm

S89,000-6
1-(4-ETHYLPHENYL)-3-PHENYLUREA

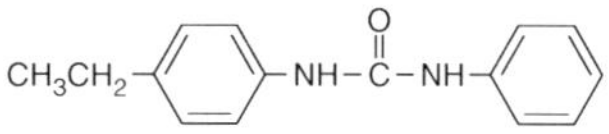

250 mg

S71,482-8
1-PHENYL-3-(2,6-XYLYL)UREA

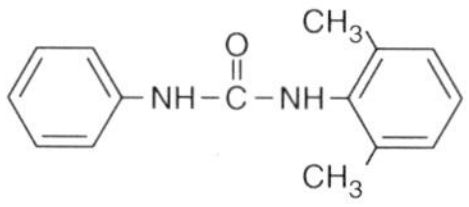

5 gm

S89,001-4
1-PHENYL-3-(2,3-XYLYL)UREA

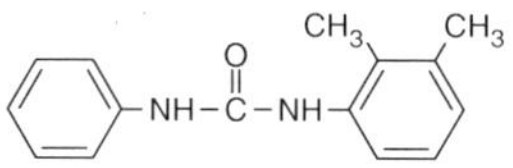

250 mg

S70,949-2
3-(2,6-DIMETHYLPHENYL)-1-METHYL-1-PHENYLUREA

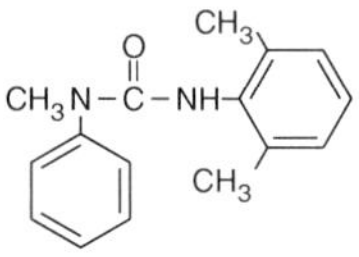

250 mg

S71,292-2
1-(2,6-DIETHYLPHENYL)-3-PHENYLUREA

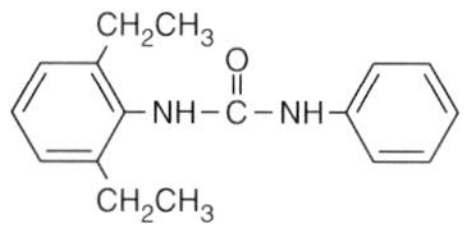

5 gm

S71,694-4
1-PHENETHYL-3-(2,6-XYLYL)UREA

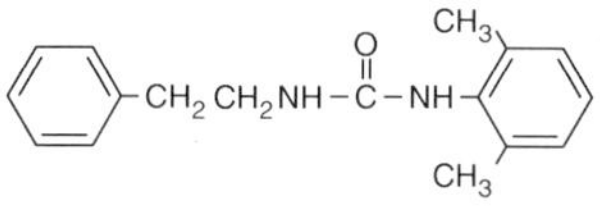

5 gm

S77,248-8
1-PHENYL-3-(2,5-XYLYL)UREA

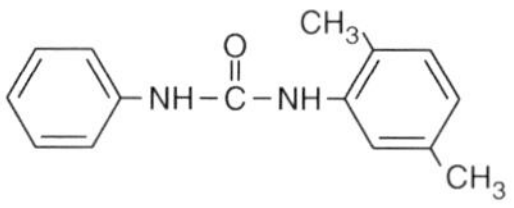

250 mg

S71,484-4
1-METHYL-1-PHENYL-3-(3,4-XYLYL)UREA

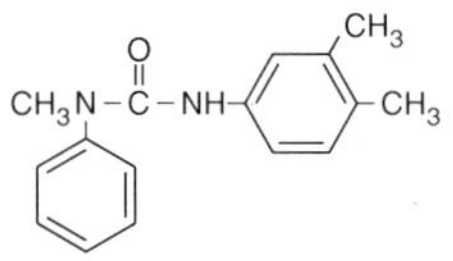

5 gm

S90,051-6
1-(2,4-DIMETHYLBENZYL)-3-PHENYLUREA

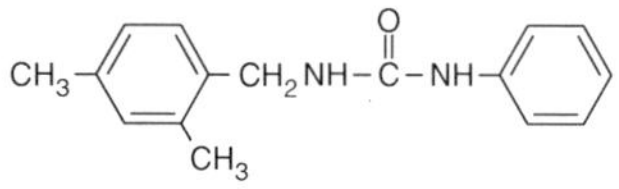

250 mg

S76,992-4
1-(DIPHENYLMETHYL)-3-(3,4-XYLYL)UREA

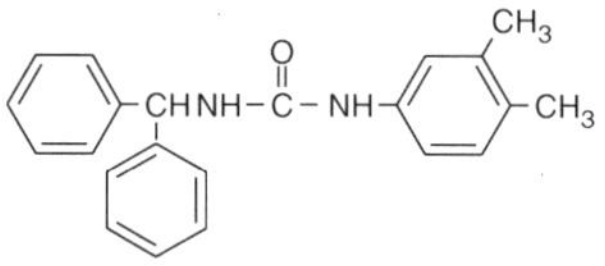

250 mg

S76,925-8
1-PHENYL-3-(2,4,6-TRIMETHYLPHENYL)UREA

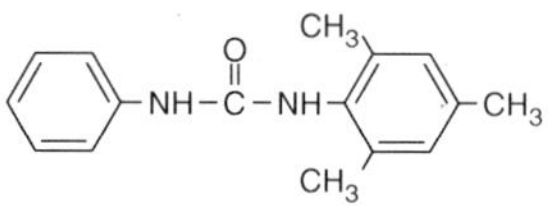

250 mg

S64,446-3
1-PHENYL-3-(2,4,5-TRICHLOROPHENYL)UREA

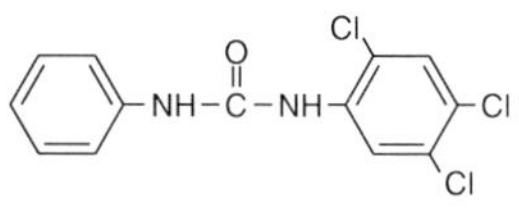

5 gm

S75,029-8
1-PHENYL-3-(2,4,5-TRIMETHYLPHENYL)UREA

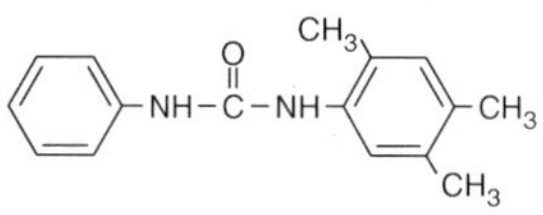

5 gm

S59,004-5
1-(1-NAPHTHYL)-3-PHENYLUREA

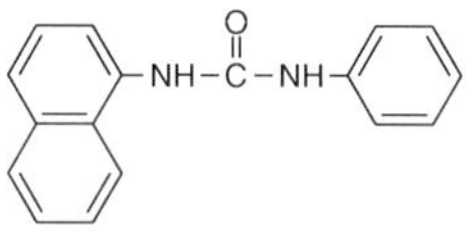

5 gm

S62,811-5
1-METHYL-3-(1-NAPHTHYL)-1-PHENYLUREA

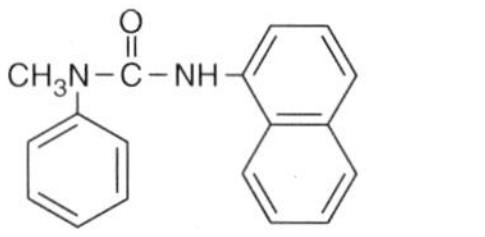

5 gm

S77,982-2
1-ISOPROPYL-3-(1-NAPHTHYL)-1-PHENYLUREA

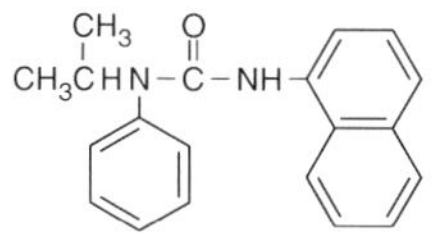

250 mg

S59,056-8
1-BUTYL-3-(1-NAPHTHYL)-1-PHENYLUREA

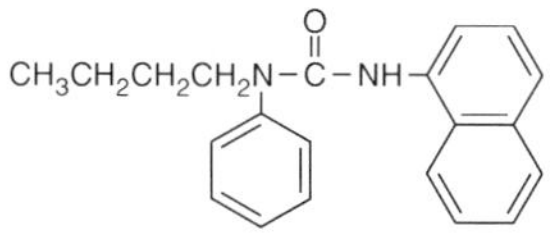

5 gm

S43,564-3
ORTHO-TOLYLUREA

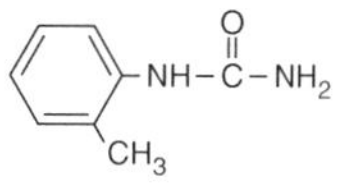

5 gm

S59,538-1
1-(2-CHLOROPHENYL)-3-METHYLUREA

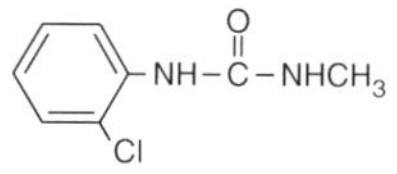

5 gm

S60,652-9
1-METHYL-3-(O-TOLYL)UREA

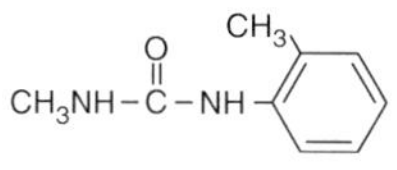

5 gm

S59,568-3
1-(2-CHLOROPHENYL)-3-ETHYLUREA

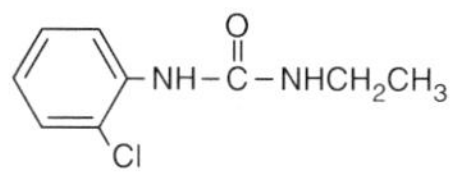

5 gm

S60,627-8
1,1-DIMETHYL-3-(O-TOLYL)UREA

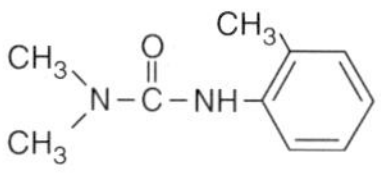

5 gm

S64,033-6
1-ETHYL-3-(O-TOLYL)UREA

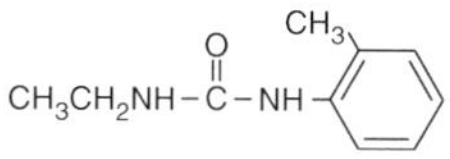

5 gm

S65,270-9
1-(2-ETHYLPHENYL)-3-METHYLUREA

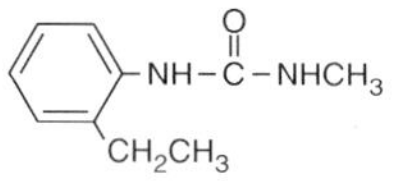

5 gm

S77,529-0
1-ETHYL-1-(O-TOLYL)UREA

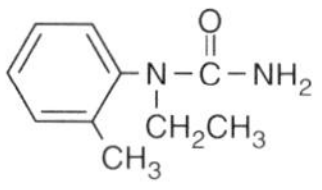

250 mg

S63,178-7
1-ALLYL-3-(2-CHLOROPHENYL)UREA

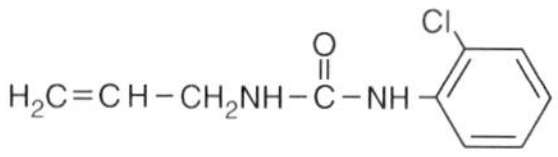

5 gm

S62,162-5
1-(2-CHLOROPHENYL)-3-PROPYLUREA

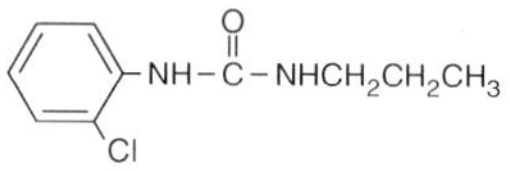

5 gm

S62,168-4
1-(2-CHLOROPHENYL)-3-ISOPROPYLUREA

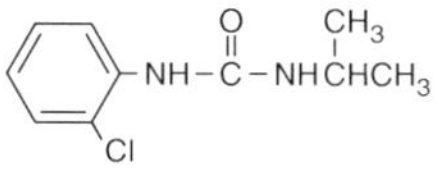

5 gm

S64,023-9
1-ALLYL-3-(O-TOLYL)UREA

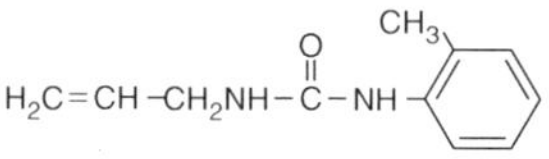

5 gm

S64,021-2
1-ISOPROPYL-3-(O-TOLYL)UREA

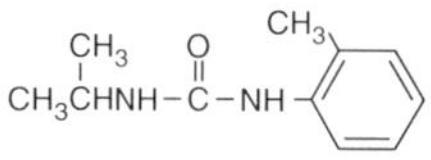

5 gm

S64,656-3
1-PROPYL-3-(O-TOLYL)UREA

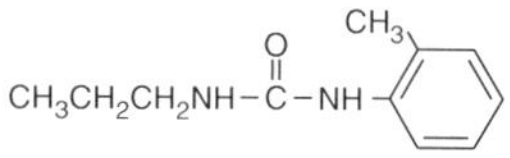

5 gm

S65,053-6
1-ETHYL-3-(2-ETHYLPHENYL)UREA

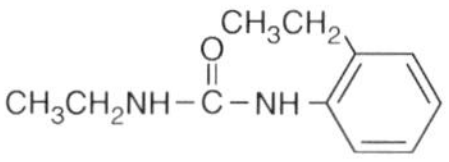

5 gm

S65,239-3
1,1-DIMETHYL-3-(2-ETHYLPHENYL)UREA

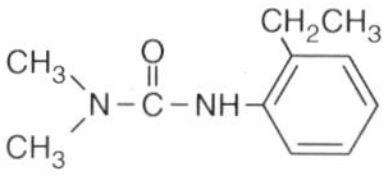

5 gm

S77,473-1
1-ETHYL-3-METHYL-1-(O-TOLYL)UREA

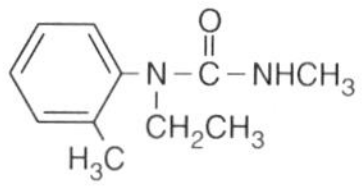

250 mg

S62,158-7
1-BUTYL-3-(2-CHLOROPHENYL)UREA

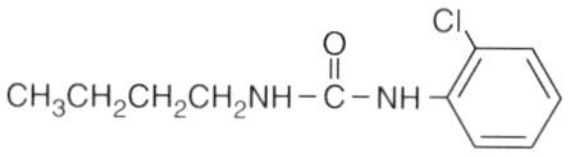

5 gm

S62,173-0
3-(2-CHLOROPHENYL)-1,1-DIETHYLUREA

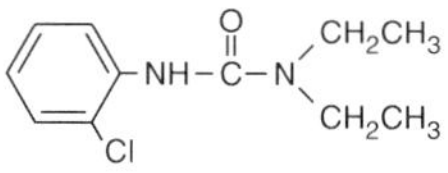

5 gm

S64,507-9
1-(2-CHLOROPHENYL)-3-ISOBUTYLUREA

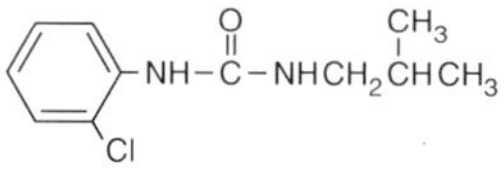

5 gm

S65,770-0
1-ALLYL-3-(2-ETHYLPHENYL)UREA

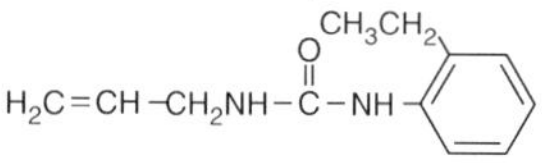

5 gm

S62,820-4
1,1-DIETHYL-3-(O-TOLYL)UREA

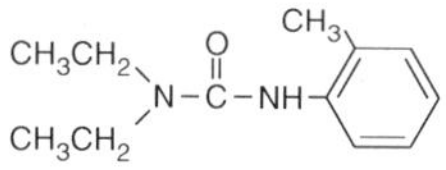

5 gm

S64,659-8
1-BUTYL-3-(O-TOLYL)UREA

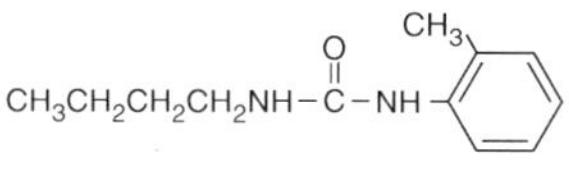

5 gm

S65,159-1
1-(2-ETHYLPHENYL)-3-PROPYLUREA

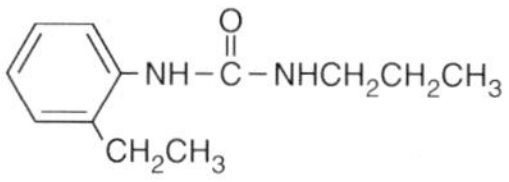

5 gm

S65,235-0
1-(2-ETHYLPHENYL)-3-ISOPROPYLUREA

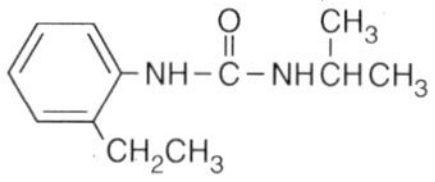

5 gm

S65,362-4
1-ISOBUTYL-3-(O-TOLYL)UREA

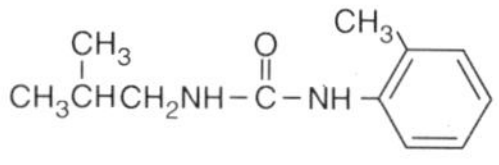

5 gm

S62,417-9
1-BUTYL-3-(2-CHLOROPHENYL)-1-METHYLUREA

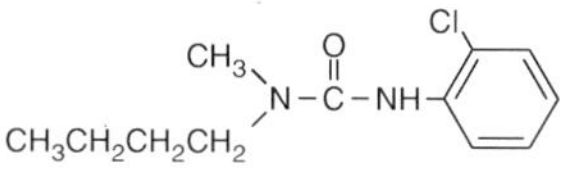

5 gm

S61,286-3
1-BUTYL-1-METHYL-3-(O-TOLYL)UREA

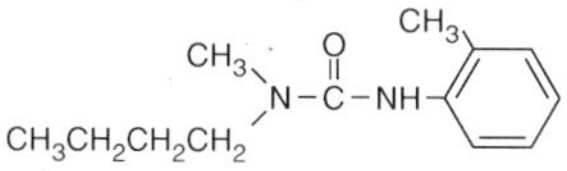

5 gm

S64,980-5
1-BUTYL-3-(2-ETHYLPHENYL)UREA

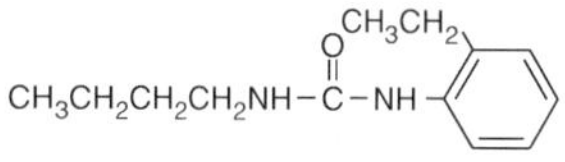

5 gm

S65,772-7
1,1-DIETHYL-3-(2-ETHYLPHENYL)UREA

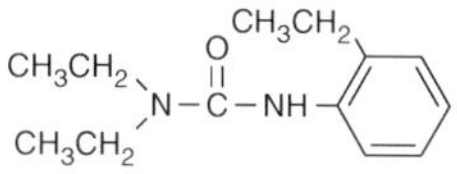

5 gm

S65,961-4
1-(2-ETHYLPHENYL)-3-ISOBUTYLUREA
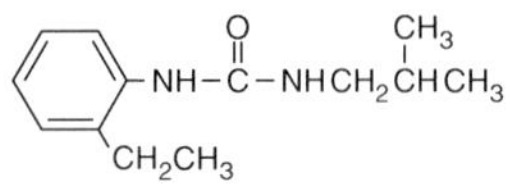
5 gm

S79,109-1
1-TERT-BUTYL-3-(2-ETHYLPHENYL)UREA
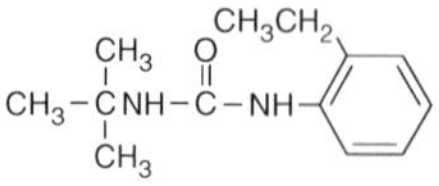
250 mg

S63,437-9
3-(2-CHLOROPHENYL)-1,1-DIALLYLUREA
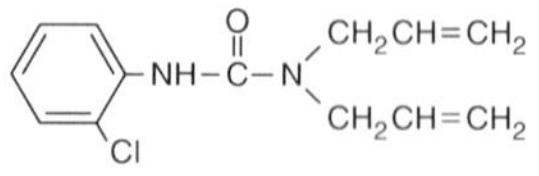
5 gm

S62,163-3
3-(2-CHLOROPHENYL)-1,1-DIISOPROPYLUREA
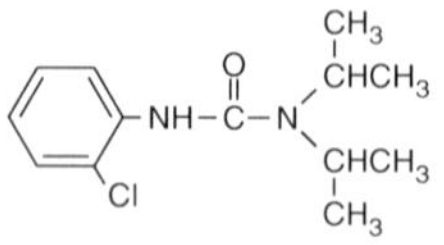
5 gm

S62,175-7
3-(2-CHLOROPHENYL)-1,1-DIPROPYLUREA
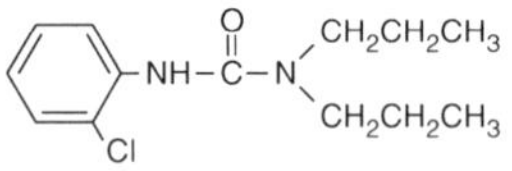
5 gm

S62,422-5
1-BUTYL-3-(2-CHLOROPHENYL)-1-ETHYLUREA
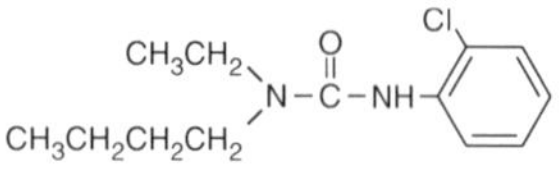
5 gm

S84,438-1
1-(2-CHLOROPHENYL)-3-(3-FLUOROPHENYL)UREA
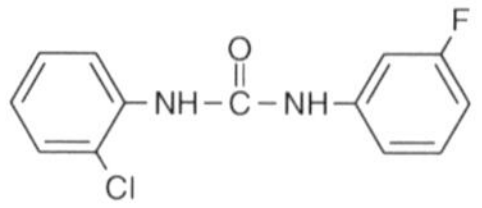
250 mg

S59,566-7
1-(2-CHLOROPHENYL)-3-(3-CHLOROPHENYL)UREA
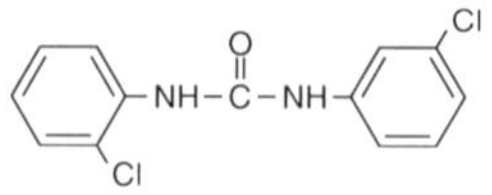
5 gm

S60,632-4
1,1-DIALLYL-3-(O-TOLYL)UREA
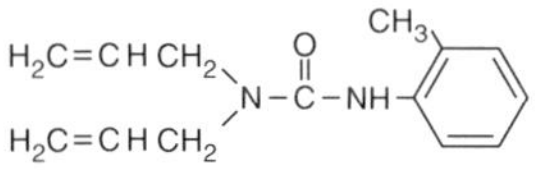
5 gm

S64,035-2
1,1-DIISOPROPYL-3-(O-TOLYL)UREA
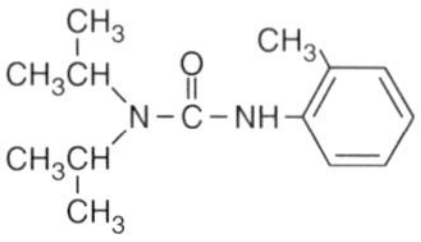
5 gm

S64,647-4
1,1-DIPROPYL-3-(O-TOLYL)UREA
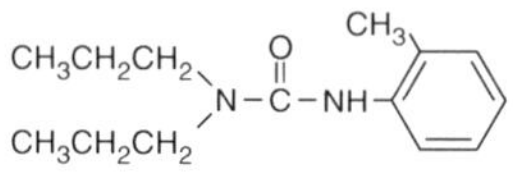
5 gm

S63,075-6
1-(3-CHLOROPHENYL)-3-(O-TOLYL)UREA
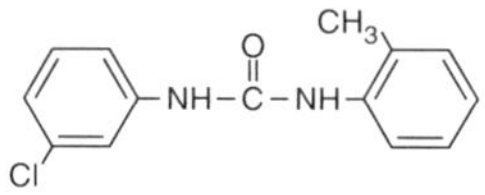
5 gm

S63,175-2
1-(2-CHLOROPHENYL)-3-(M-TOLYL)UREA
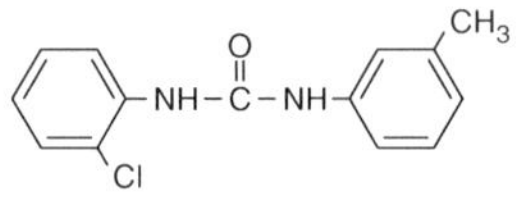
5 gm

S63,813-7
3-(2-CHLOROPHENYL)-1-METHYL-1-PHENYLUREA
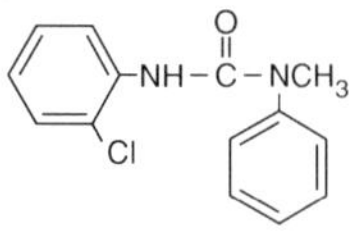
5 gm

S60,673-1
1,3-BIS(O-TOLYL)UREA
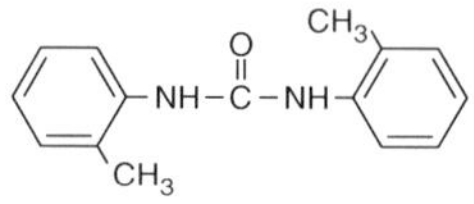
5 gm

S64,506-0
1-METHYL-1-PHENYL-3-(O-TOLYL)UREA
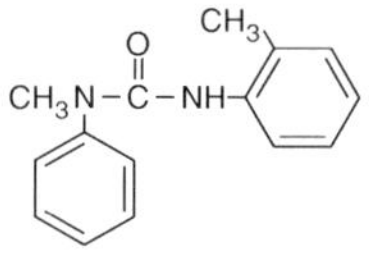
5 gm

S65,786-7
1,1-DIALLYL-3-(2-ETHYLPHENYL)UREA
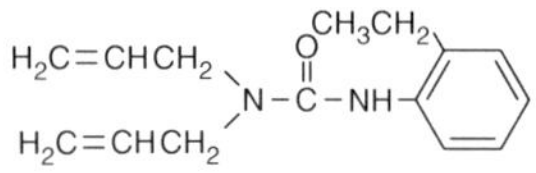
5 gm

S66,011-6
1,1-DIISOPROPYL-3-(2-ETHYLPHENYL)UREA
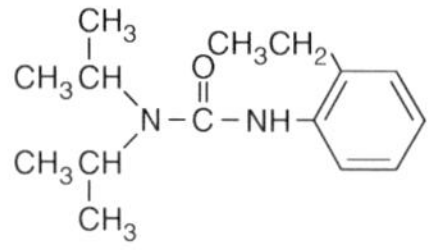
5 gm

S66,428-6
1,1-DIPROPYL-3-(2-ETHYLPHENYL)UREA
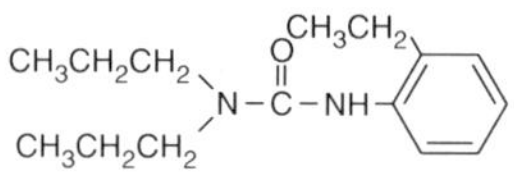
5 gm

S63,412-3
3-(2-CHLOROPHENYL)-1,1-DIBUTYLUREA
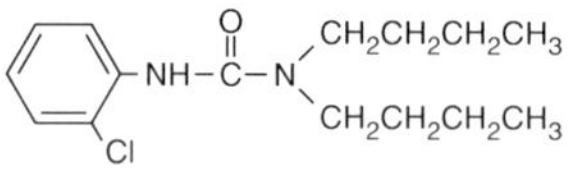
5 gm

S64,508-7
3-(2-CHLOROPHENYL)-1,1-DIISOBUTYLUREA
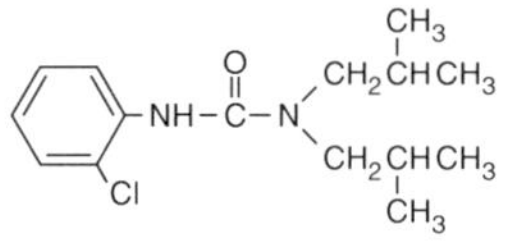
5 gm

S65,147-8
1-ETHYL-1-PHENYL-3-(O-TOLYL)UREA
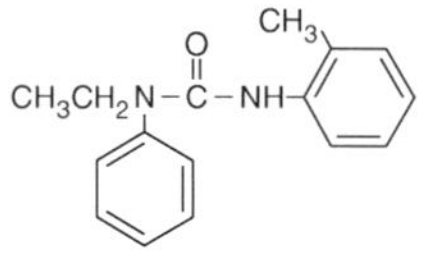
5 gm

S88,977-6
1-(2-ETHYLPHENYL)-3-(M-TOLYL)UREA
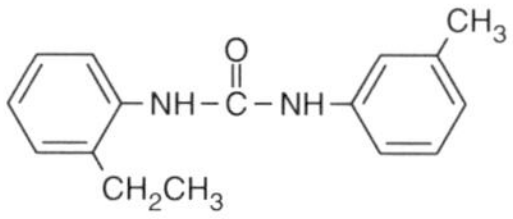
250 mg

S60,742-8
1,1-DIBUTYL-3-(O-TOLYL)UREA
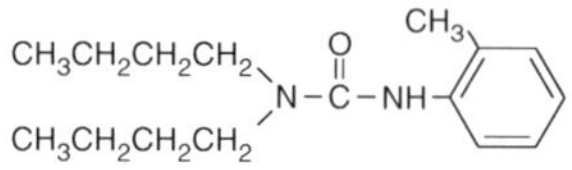
5 gm

S65,480-9
1,1-DIISOBUTYL-3-(O-TOLYL)UREA
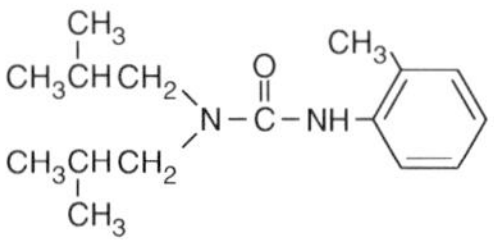
5 gm

S63,456-5
1-BUTYL-3-(2-CHLOROPHENYL)-1-PHENYLUREA
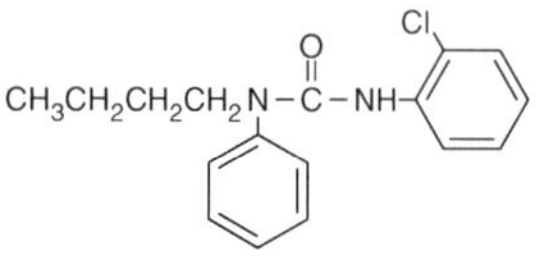
5 gm

S67,044-8
1-OCTADECYL-3-(O-TOLYL)UREA
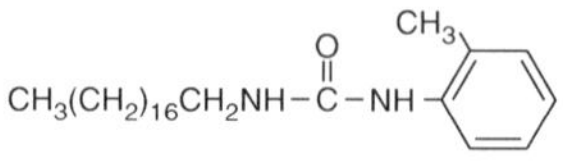
5 gm

S64,178-2
1-(2-ETHYLPHENYL)UREA
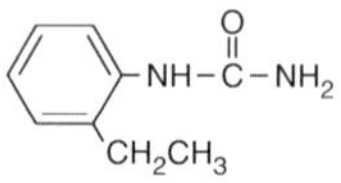
5 gm

S59,537-3
1-(2-CHLOROPHENYL)-3-(4-CHLOROPHENYL)UREA
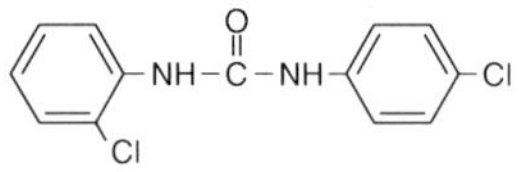
5 gm

S59,022-3
1-(4-CHLOROPHENYL)-3-(O-TOLYL)UREA
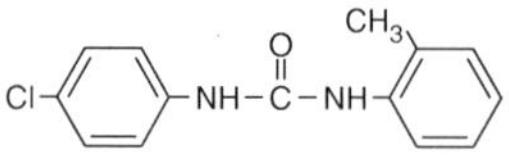
5 gm

S62,995-2
1-(2-CHLOROPHENYL)-3-(P-TOLYL)UREA
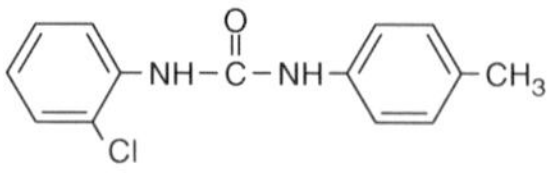
5 gm

S79,149-0
1-(4-ETHYLPHENYL)-3-(O-TOLYL)UREA
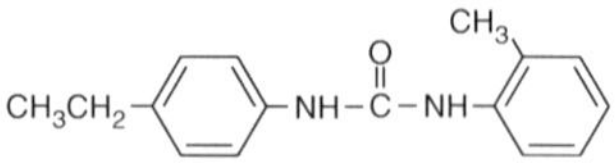
250 mg

S75,386-6
1-(4-BUTYLPHENYL)-3-(2-ETHYLPHENYL)UREA
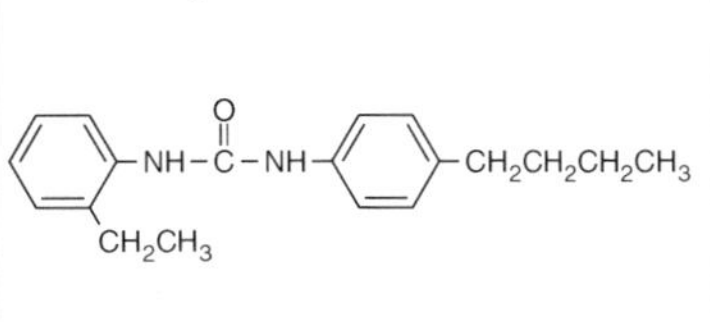
250 mg

S89,002-2
1-(O-TOLYL)-3-(2,3-XYLYL)UREA
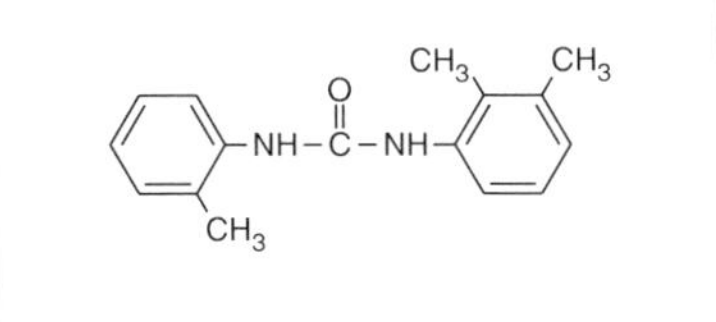
250 mg

S71,308-2
1-(2,6-DIISOPROPYLPHENYL)-3-(2-METHYLPHENYL)UREA
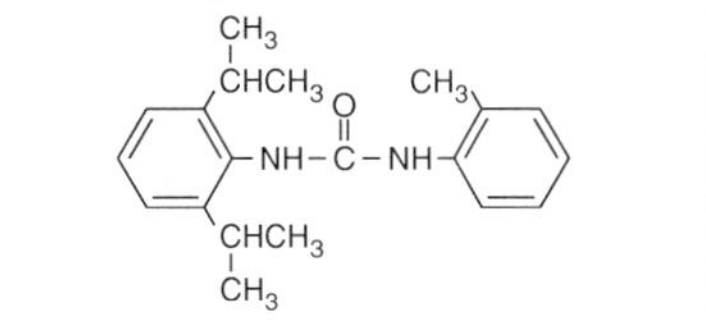
5 gm

S59,562-4
1-(2-CHLOROPHENYL)-3-(2,4-DICHLOROPHENYL)UREA
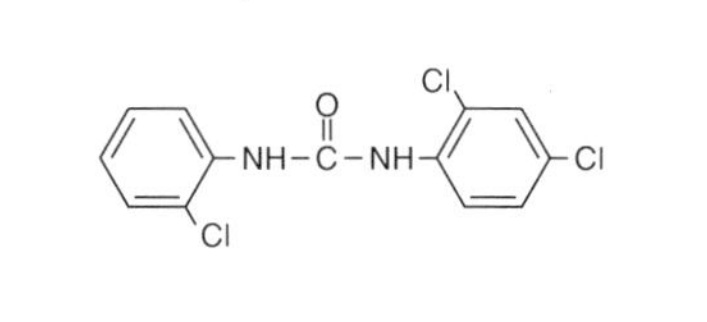
5 gm

S64,445-5
1-(2-CHLOROPHENYL)-3-(3,4-DICHLOROPHENYL)UREA
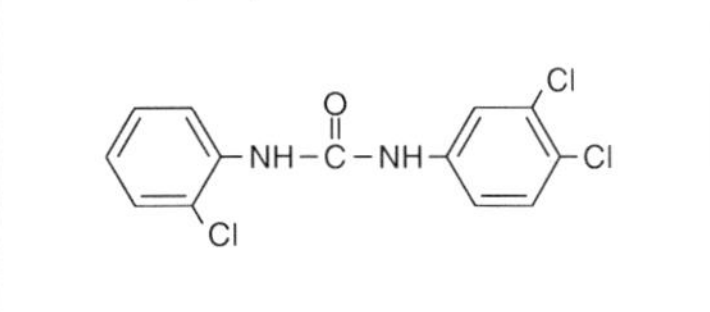
5 gm

S60,630-8
1-(2-CHLOROPHENYL)-3-(2,4-XYLYL)UREA
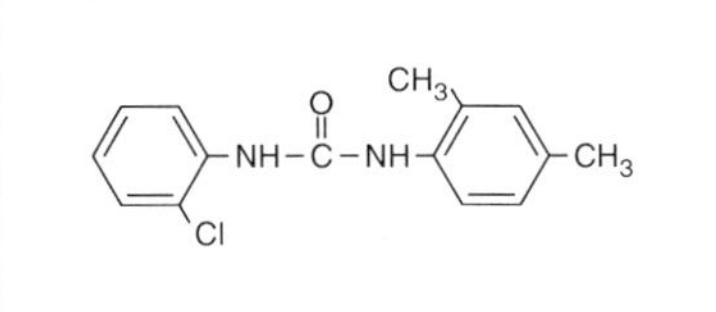
5 gm

S60,681-2
1-(O-TOLYL)-3-(2,4-XYLYL)UREA
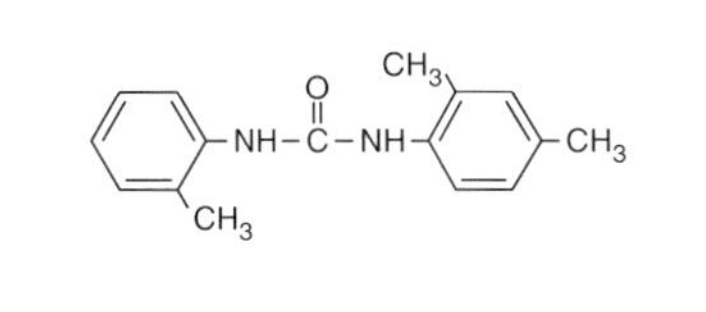
5 gm

S79,151-2
1-(O-TOLYL)-3-(2,5-XYLYL)UREA
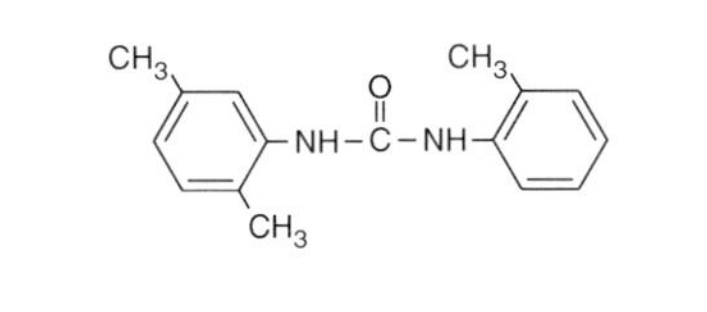
250 mg

S84,590-6
1-(2-ETHYLPHENYL)-3-(3,4-XYLYL)UREA
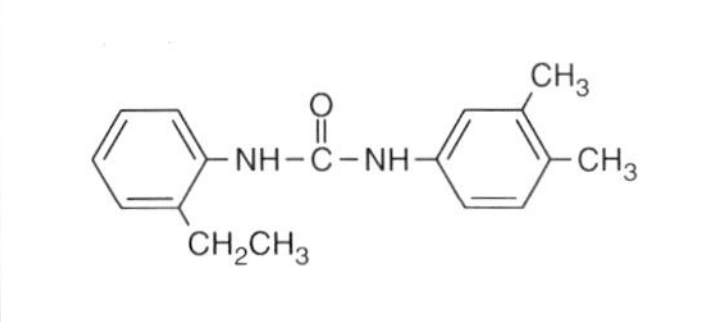
250 mg

S74,338-0
1,1'-(3,3'-DIMETHYL-1,1'-BIPHENYL-4,4'-DIYL)BIS[3-(O-TOLYL)UREA]
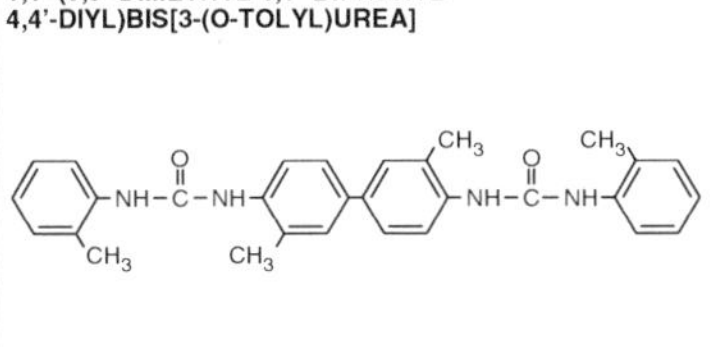
5 gm

S59,030-4
1-(2-CHLOROPHENYL)-3-(1-NAPHTHYL)UREA
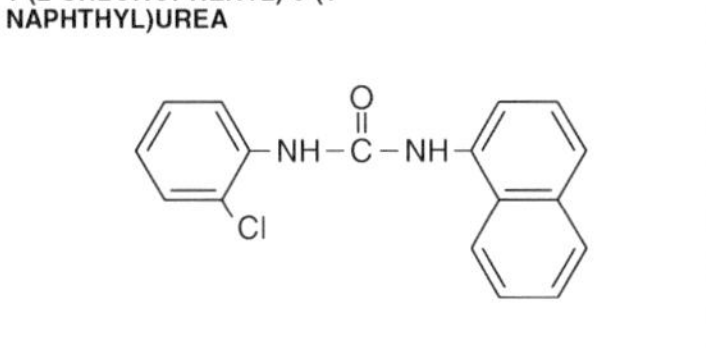
5 gm

S59,031-2
1-(1-NAPHTHYL)-3-(O-TOLYL)UREA
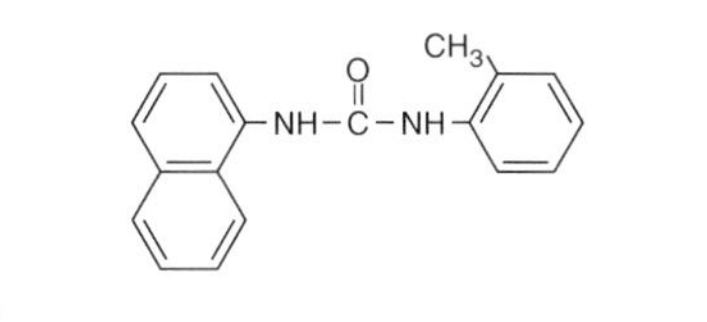
5 gm

S78,003-0
1-(2-ETHYLPHENYL)-3-(1-NAPHTHYL)UREA
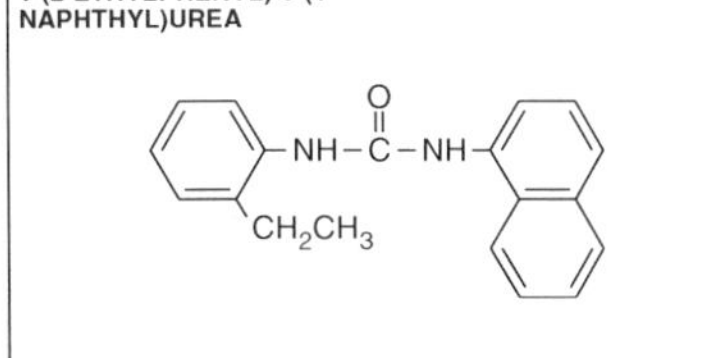
250 mg

S79,164-4
1-ETHYL-3-(1-NAPHTHYL)-1-(O-TOLYL)UREA
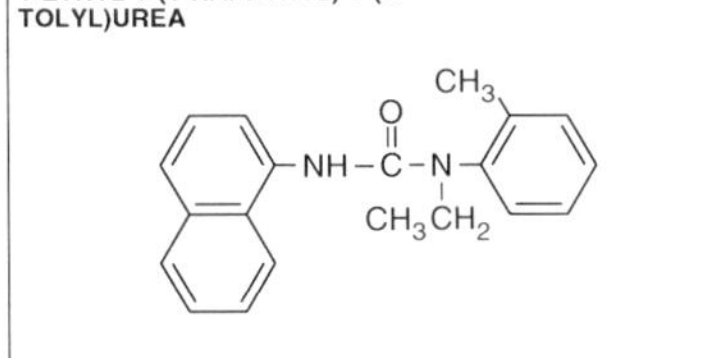
250 mg

S60,152-7
4-FLUOROPHENYLUREA
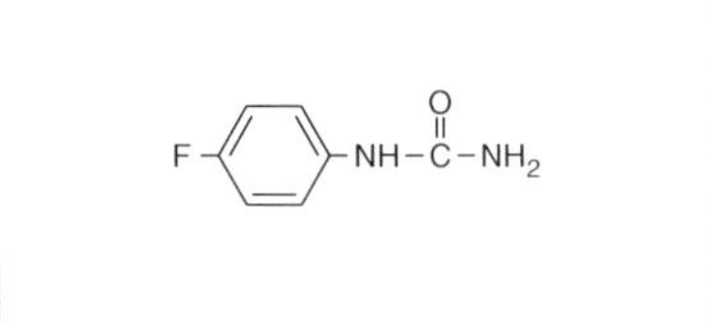
5 gm

S59,567-5
1-(4-CHLOROPHENYL)-3-HYDROXYUREA
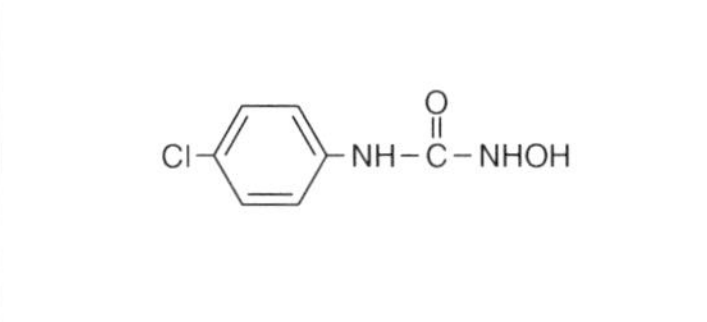
5 gm

S41,322-4
4-BROMOPHENYLUREA
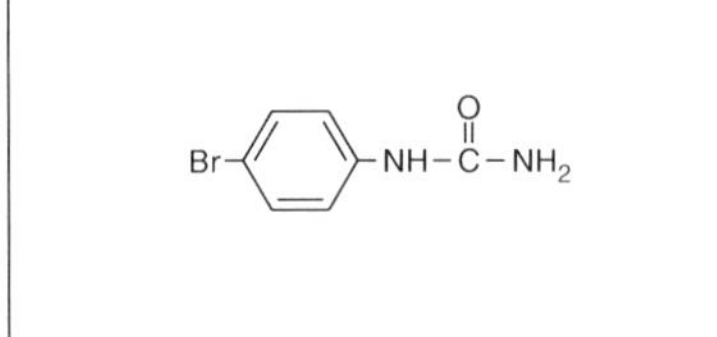
5 gm

S36,514-9
PARA-TOLYLUREA
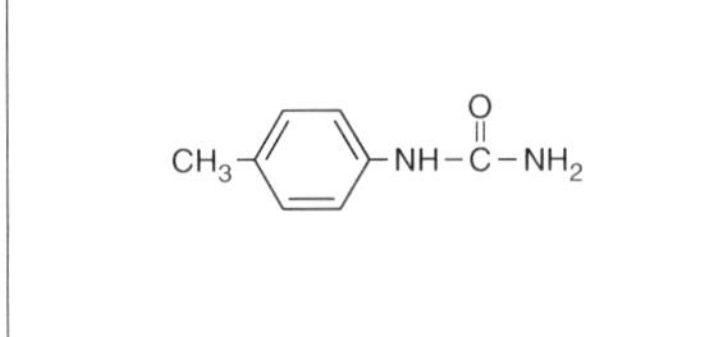
5 gm

S59,173-4
1-(4-CHLOROPHENYL)-3-METHYLUREA
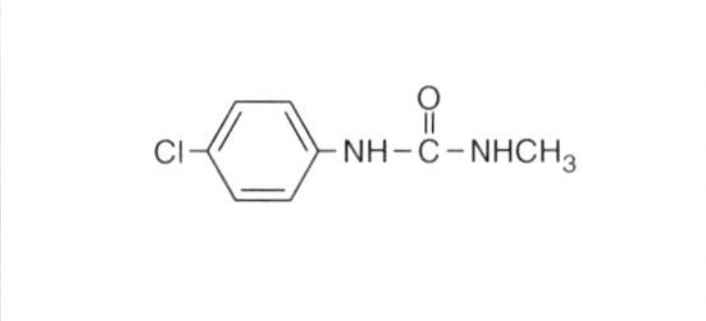
5 gm

S63,978-8
1-(4-ETHYLPHENYL)UREA
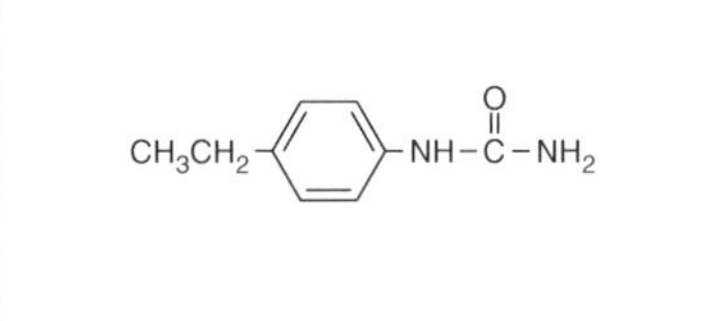
5 gm

S64,431-5
1-METHYL-3-(P-TOLYL)UREA
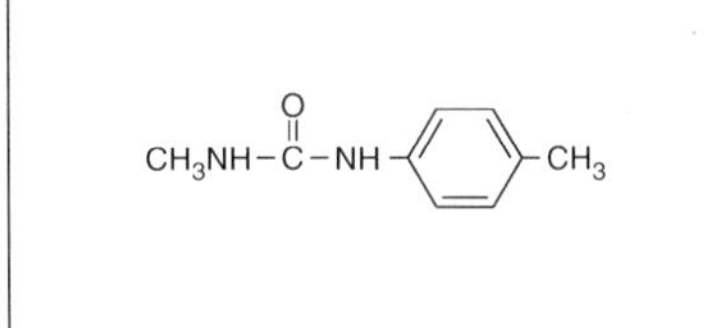
5 gm

S60,164-0
1,1-DIMETHYL-3-(4-FLUOROPHENYL)UREA
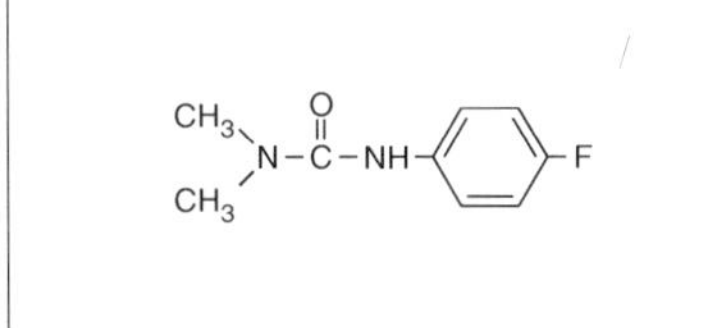
5 gm

S59,152-1
1-(4-CHLOROPHENYL)-3-ETHYLUREA
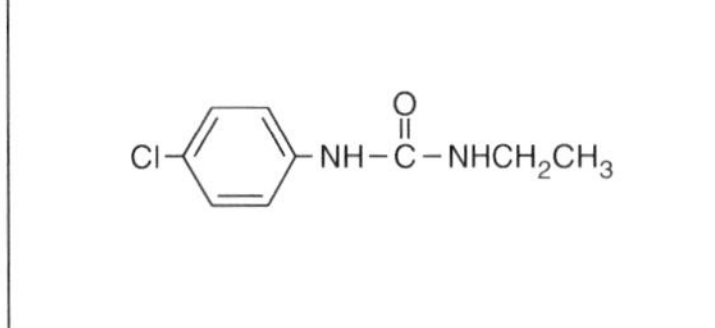
5 gm

S59,770-8
3-(4-BROMOPHENYL)-1,1-DIMETHYLUREA
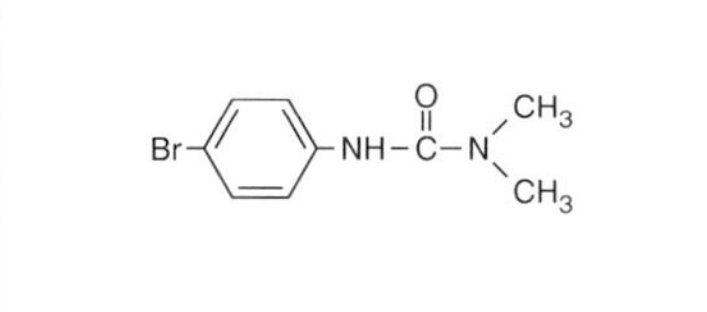
5 gm

S62,825-5
1-(4-BROMOPHENYL)-3-ETHYLUREA
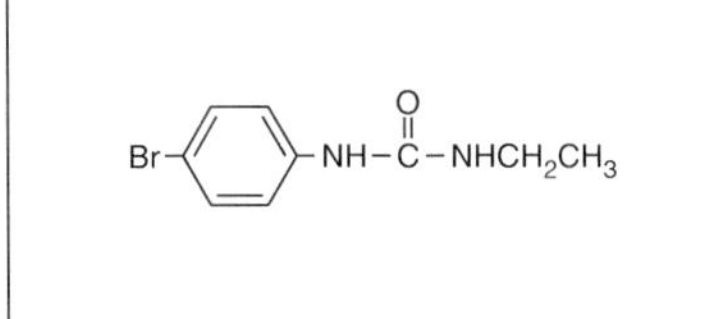
5 gm

S61,321-5
1,1-DIMETHYL-3-(P-TOLYL)UREA
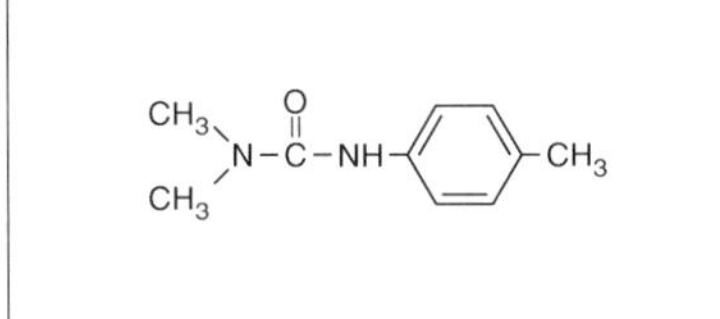
5 gm

S64,428-5
1-ETHYL-3-(P-TOLYL)UREA
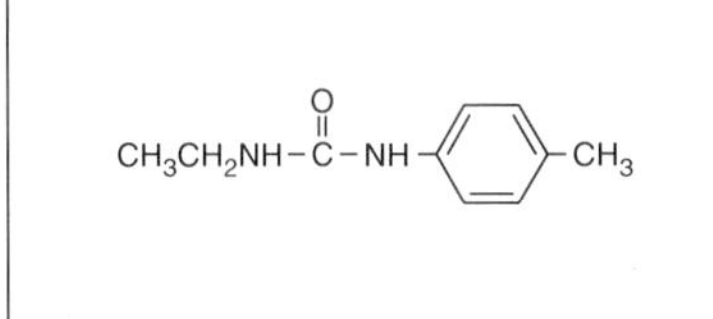
5 gm

S63,699-1
1-ALLYL-3-(4-FLUOROPHENYL)UREA
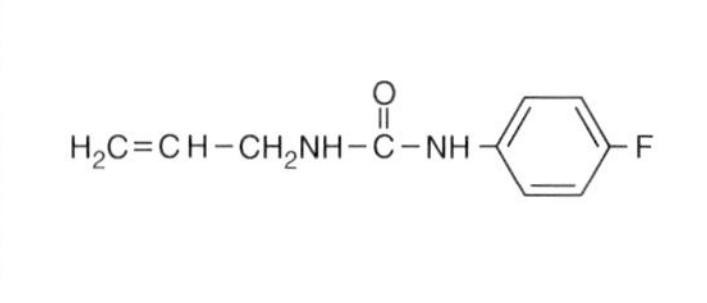
5 gm

S59,148-3
1-ALLYL-3-(4-CHLOROPHENYL)UREA
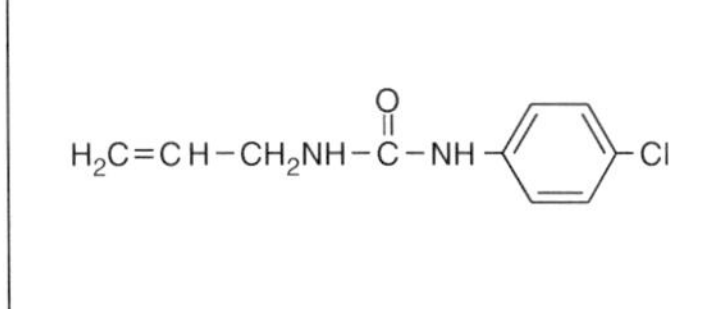
5 gm

S59,134-3
1-(4-CHLOROPHENYL)-3-PROPYLUREA
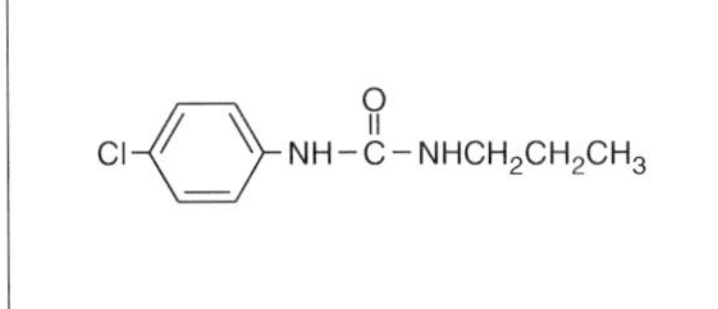
5 gm

S61,348-7
1-(4-CHLOROPHENYL)-3-ISOPROPYLUREA
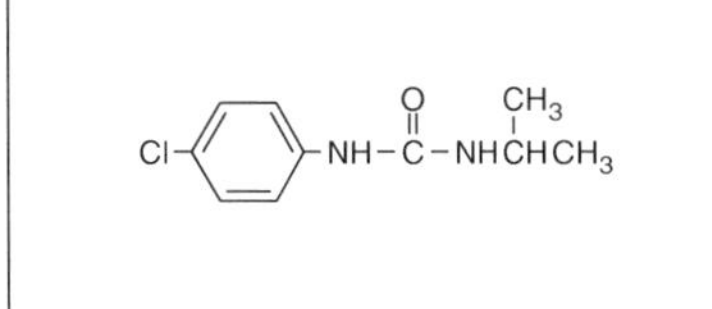
5 gm

S59,406-7
1-(4-BROMOPHENYL)-3-ISOPROPYLUREA
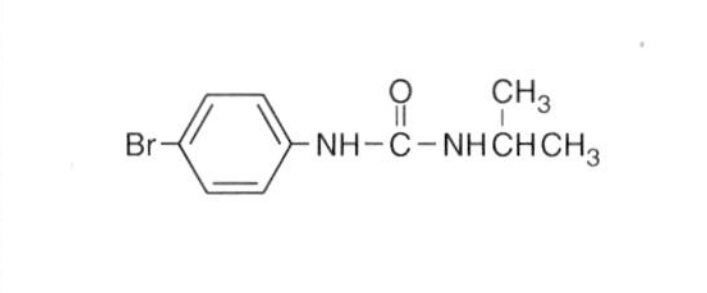
5 gm

S61,317-7
1-ALLYL-3-(P-TOLYL)UREA
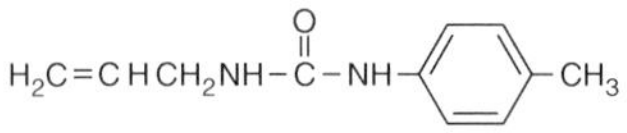
5 gm

S64,437-4
1-PROPYL-3-(P-TOLYL)UREA
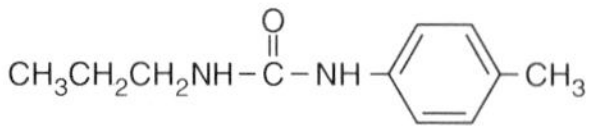
5 gm

S64,466-8
1-ISOPROPYL-3-(P-TOLYL)UREA
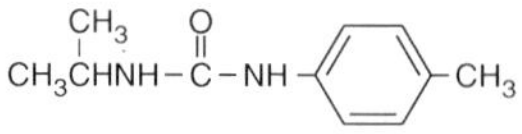
5 gm

S90,012-5
1-ETHYL-3-(4-METHYLBENZYL)UREA
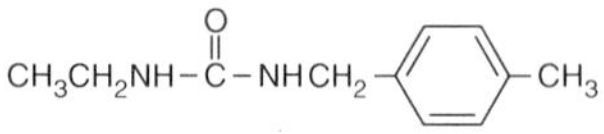
250 mg

S60,165-9
1,1-DIETHYL-3-(4-FLUOROPHENYL)UREA
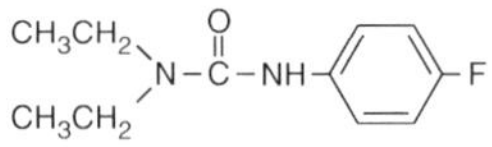
5 gm

S60,169-1
1-BUTYL-3-(4-FLUOROPHENYL)UREA
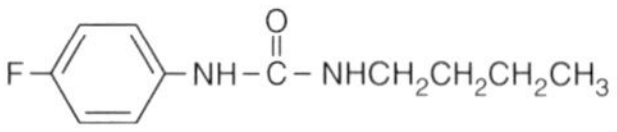
5 gm

S41,156-6
1-(4-CHLORO-ALPHA,ALPHA-DIMETHYLPHENETHYL)UREA
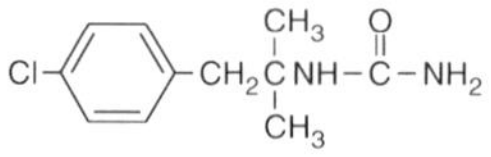
5 gm

S58,627-7
3-(4-CHLOROPHENYL)-1,1-DIETHYLUREA
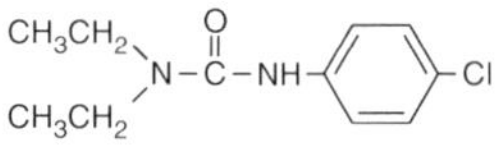
5 gm

S59,141-6
1-BUTYL-3-(4-CHLOROPHENYL)UREA
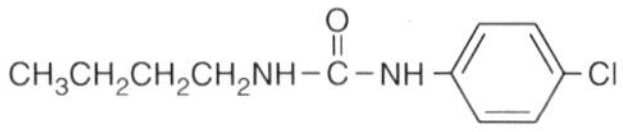
5 gm

S64,518-4
1-(4-CHLOROPHENYL)-3-ISOBUTYLUREA
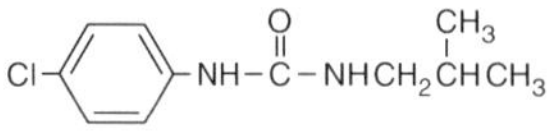
5 gm

S61,309-6
1-BUTYL-3-(P-TOLYL)UREA
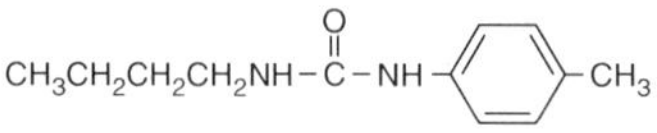
5 gm

S61,326-6
1,1-DIETHYL-3-(P-TOLYL)UREA
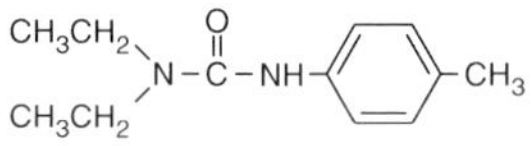
5 gm

S65,366-7
1-ISOBUTYL-3-(P-TOLYL)UREA
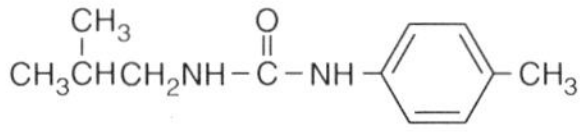
5 gm

S61,082-8
1-BUTYL-3-(4-CHLOROPHENYL)-1-METHYLUREA
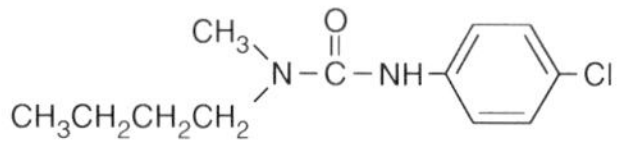
5 gm

S61,318-5
1-BUTYL-1-METHYL-3-(P-TOLYL)UREA
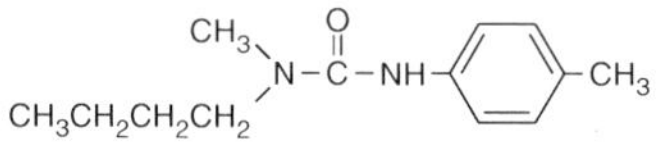
5 gm

S71,667-7
3-(4-TERT-BUTYLPHENYL)-1,1-DIMETHYLUREA
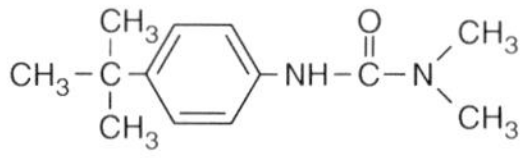
5 gm

S89,113-4
1-TERT-BUTYL-3-(4-ETHYLPHENYL)UREA
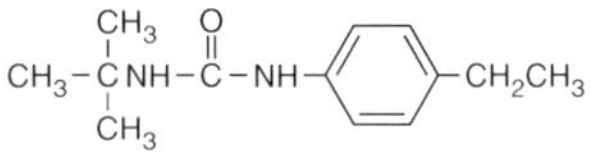
250 mg

S60,199-3
1,1-DIALLYL-3-(4-FLUOROPHENYL)UREA
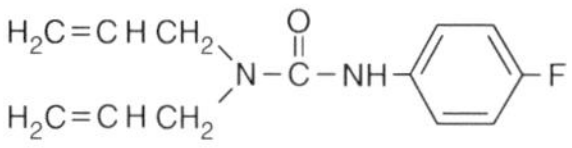
5 gm

S60,160-8
1,1-DIISOPROPYL-3-(4-FLUOROPHENYL)UREA
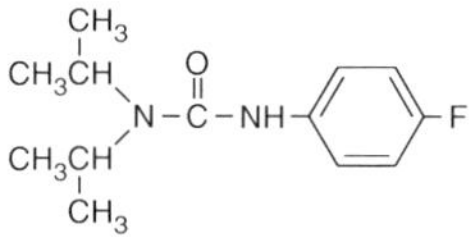
5 gm

S60,168-3
1,1-DIPROPYL-3-(4-FLUOROPHENYL)UREA
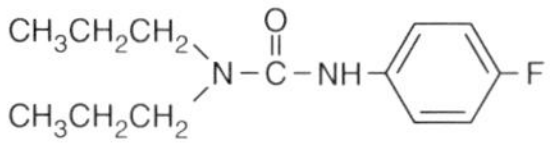
5 gm

S60,161-6
1,3-BIS(4-FLUOROPHENYL)UREA
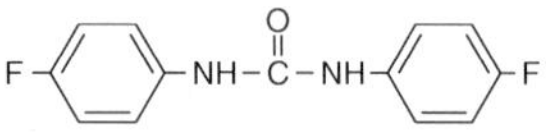
5 gm

S59,130-0
3-(4-CHLOROPHENYL)-1,1-DIALLYLUREA
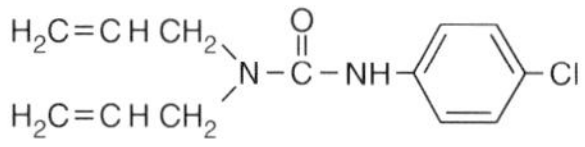
5 gm

S59,128-9
3-(4-CHLOROPHENYL)-1,1-DIPROPYLUREA
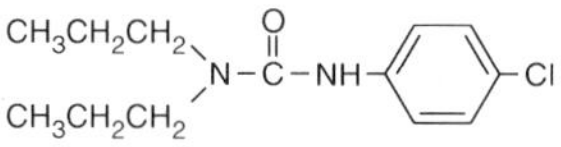
5 gm

S62,359-8
1-BUTYL-3-(4-CHLOROPHENYL)-1-ETHYLUREA
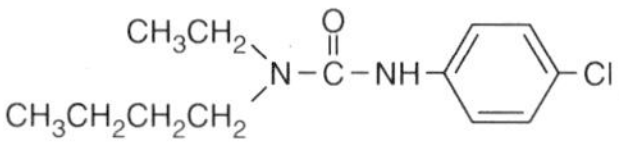
5 gm

S63,458-1
3-(4-CHLOROPHENYL)-1,1-DIISOPROPYLUREA
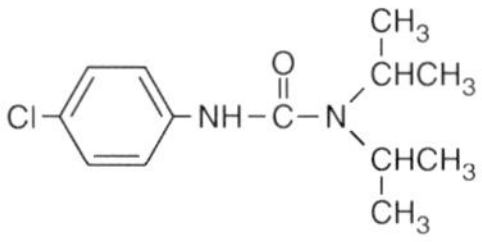
5 gm

S60,153-5
1-(4-CHLOROPHENYL)-3-(4-FLUOROPHENYL)UREA
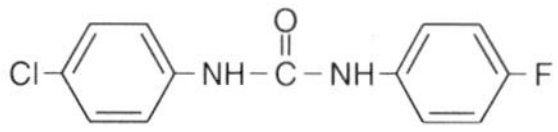
5 gm

S58,027-9
1,3-BIS(4-CHLOROPHENYL)UREA
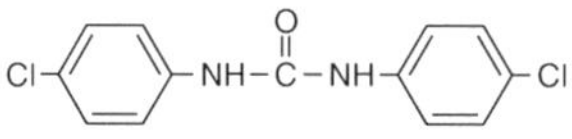
250 mg

S59,569-1
1-(3-CHLOROPHENYL)-3-(4-CHLOROPHENYL)UREA
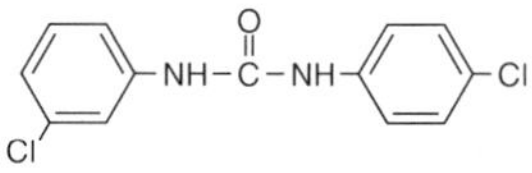
5 gm

S59,410-5
1-(4-BROMOPHENYL)-3-PHENYLUREA
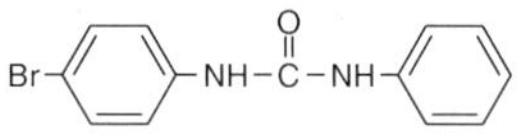
5 gm

S64,038-7
3-(4-BROMOPHENYL)-1,1-DIALLYLUREA
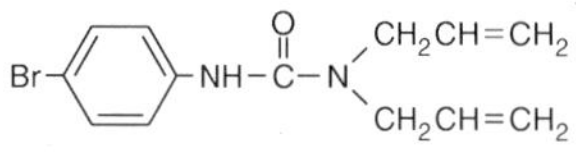
5 gm

S59,565-9
3-(4-BROMOPHENYL)-1,1-DIPROPYLUREA
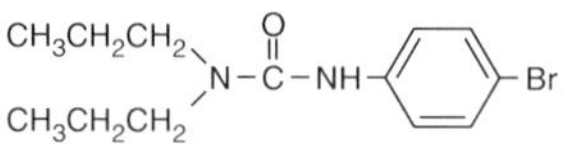
5 gm

S64,026-3
3-(4-BROMOPHENYL)-1,1-DIISOPROPYLUREA
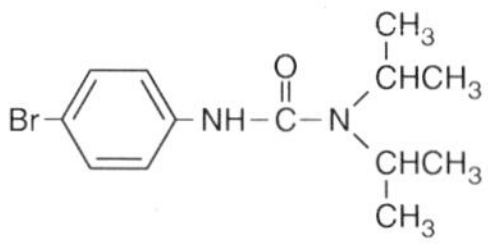
5 gm

S59,019-3
4'-METHYLCARBANILIDE

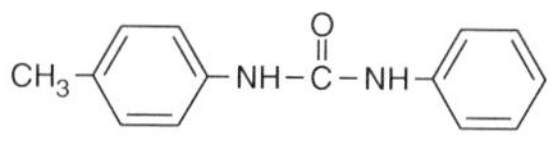

5 gm

S61,336-3
1,1-DIALLYL-3-(P-TOLYL)UREA

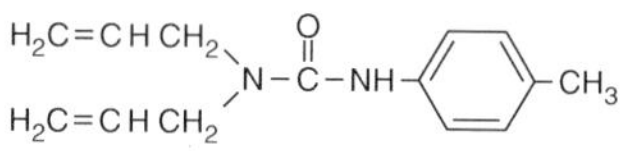

5 gm

S61,320-7
1,1-DIISOPROPYL-3-(P-TOLYL)UREA

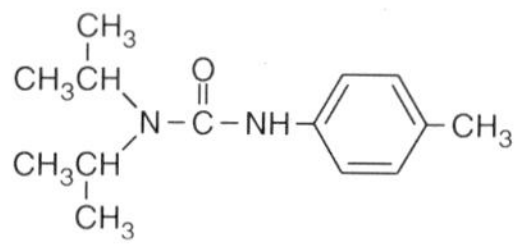

5 gm

S61,329-0
1,1-DIPROPYL-3-(P-TOLYL)UREA

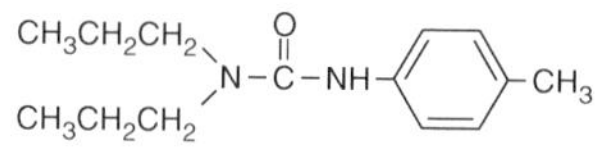

5 gm

S59,020-7
1-(4-CHLOROPHENYL)-3-(P-TOLYL)UREA

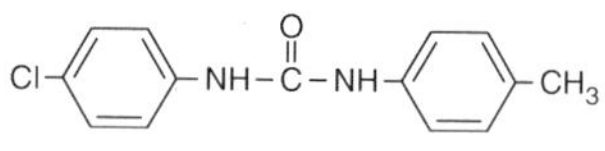

5 gm

S59,021-5
1-(4-CHLOROPHENYL)-3-(M-TOLYL)UREA

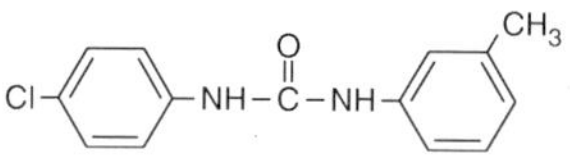

5 gm

S63,637-1
3-(4-CHLOROPHENYL)-1-METHYL-1-PHENYLUREA

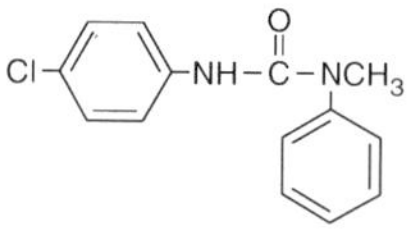

5 gm

S55,602-5
1,3-DI-P-TOLYLUREA

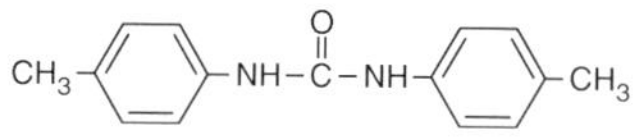

250 mg

S60,155-1
1,1-DIBUTYL-3-(4-FLUOROPHENYL)UREA

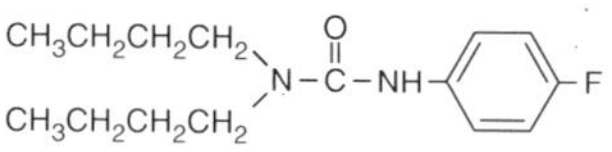

5 gm

S59,127-0
3-(4-CHLOROPHENYL)-1,1-DIBUTYLUREA

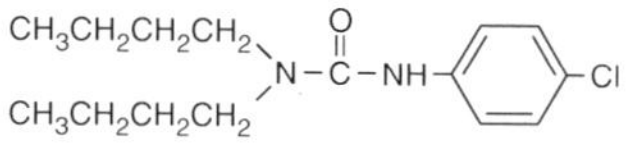

5 gm

S64,677-6
3-(4-CHLOROPHENYL)-1,1-DIISOBUTYLUREA

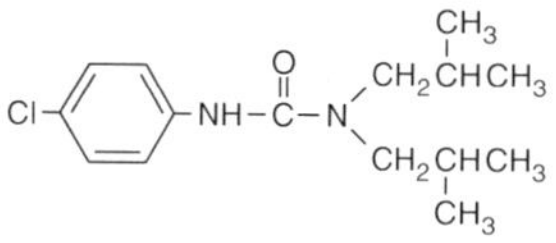

5 gm

S88,978-4
1-(4-ETHYLPHENYL)-3-(M-TOLYL)UREA

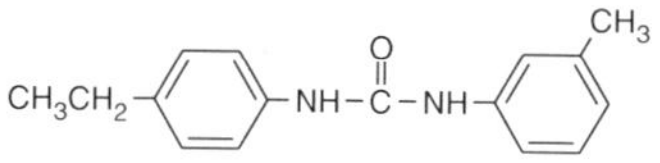

250 mg

S61,334-7
1,1-DIBUTYL-3-(P-TOLYL)UREA

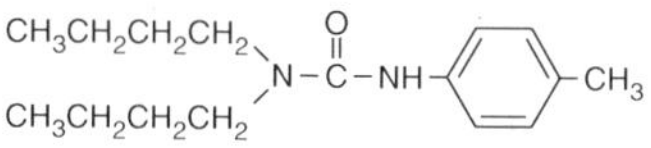

5 gm

S65,394-2
1,1-DIISOBUTYL-3-(P-TOLYL)UREA

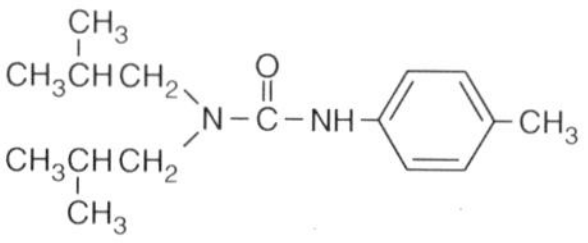

5 gm

S59,131-9
1-BUTYL-3-(4-CHLOROPHENYL)-1-PHENYLUREA

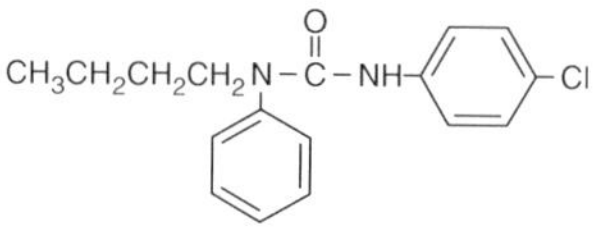

5 gm

S75,377-7
1-(4-BUTYLPHENYL)-3-(M-TOLYL)UREA

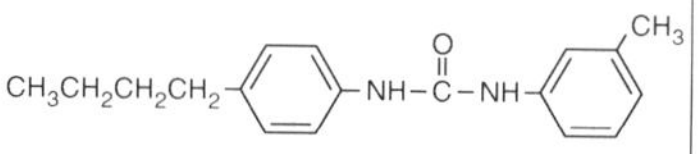

5 gm

S75,378-5
1-(4-BUTYLPHENYL)-3-(P-TOLYL)UREA

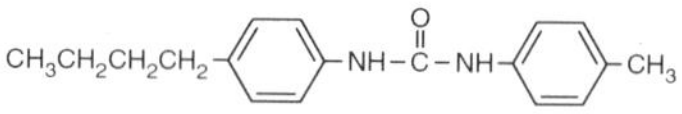

5 gm

S75,387-4
1-(4-BUTYLPHENYL)-3-(4-ETHYLPHENYL)UREA

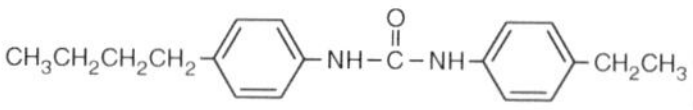

5 gm

S75,388-2
1-DODECYL-3-(P-TOLYL)UREA

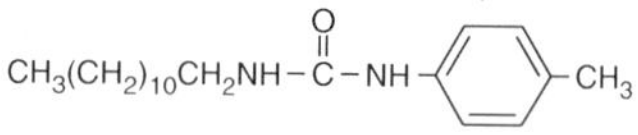

5 gm

S67,041-3
1-OCTADECYL-3-(P-TOLYL)UREA

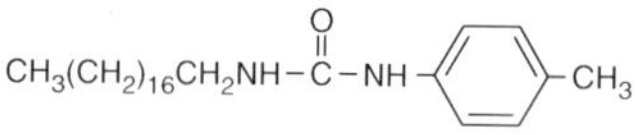

5 gm

S79,171-7
1-(4-ETHYLPHENYL)-3-OCTADECYLUREA

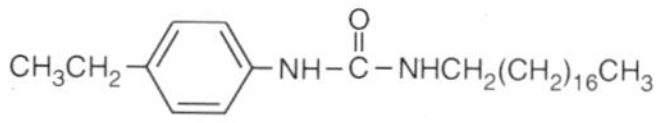

250 mg

S77,311-5
1-(4-ISOPROPYLPHENYL)-3-OCTADECYLUREA

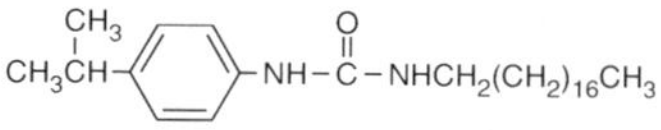

250 mg

S59,049-5
2,3,4'-TRICHLOROCARBANILIDE

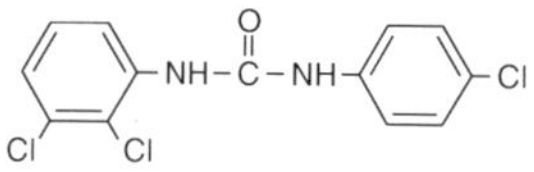

5 gm

S71,485-2
1-(2,6-DIMETHYLPHENYL)-3-(4-METHYLPHENYL)UREA

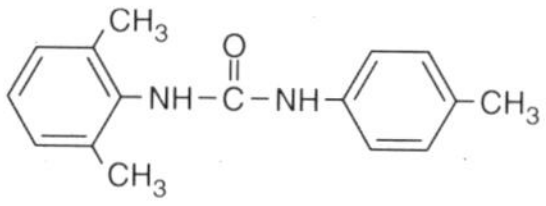

5 gm

S39,429-7
4'-BROMO-2,6-DIETHYLCARBANILIDE

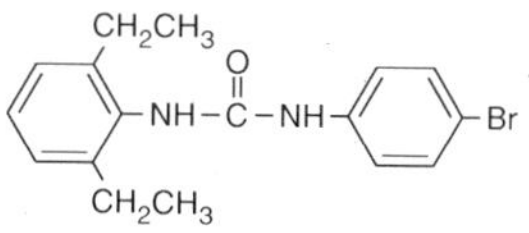

5 gm

S75,382-3
1-(4-BUTYLPHENYL)-3-(2,3-XYLYL)UREA

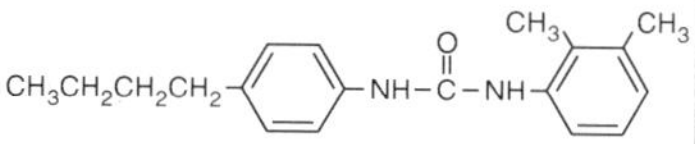

5 gm

S75,385-8
1-(4-BUTYLPHENYL)-3-(2,6-XYLYL)UREA

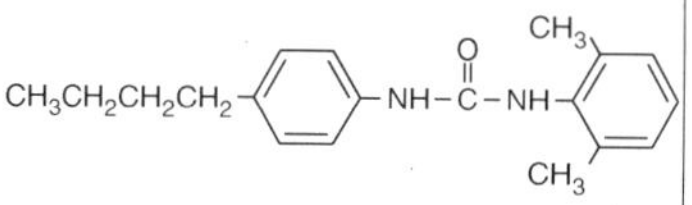

5 gm

S59,026-6
2,4,4'-TRICHLOROCARBANILIDE

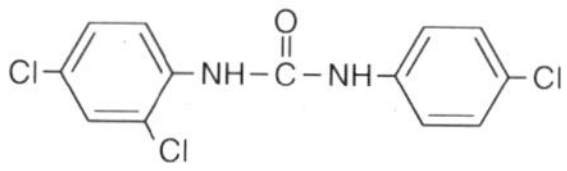

5 gm

S59,483-0
1-(4-BROMOPHENYL)-3-(3,4-DICHLOROPHENYL)UREA

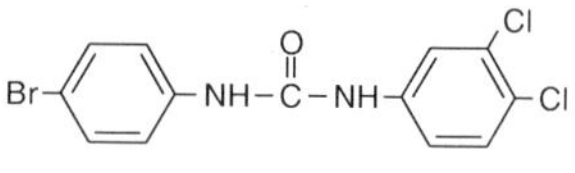

5 gm

S60,180-2
1-(4-CHLOROPHENYL)-3-(4-CHLORO-2-METHYLPHENYL)UREA

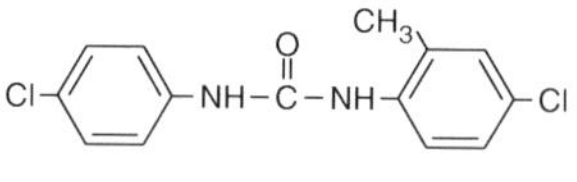

5 gm

S60,342-2
1-(3-CHLORO-4-METHYLPHENYL)-3-P-TOLYLUREA

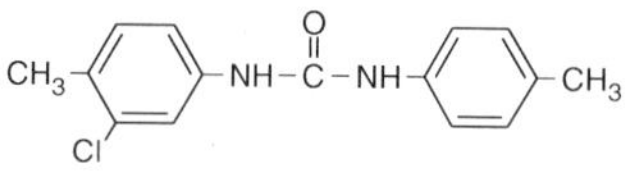

5 gm

S60,659-6
1-(P-TOLYL)-3-(2,4-XYLYL)UREA

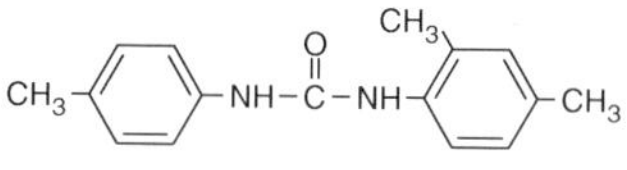

5 gm

S70,202-1
1-(2,5-DIMETHYLPHENYL)-3-(P-TOLYL)UREA

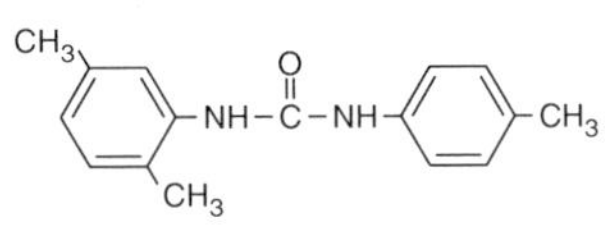

250 mg

S76,903-7
1-(4-METHOXYPHENYL)-3-(2,5-XYLYL)UREA

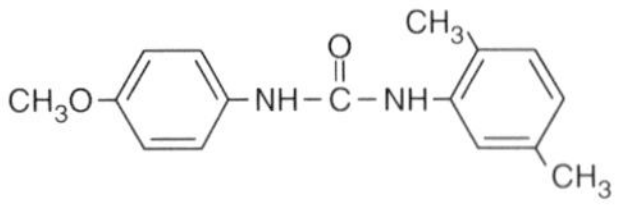

250 mg

S75,383-1
1-(4-BUTYLPHENYL)-3-(2,4-XYLYL)UREA

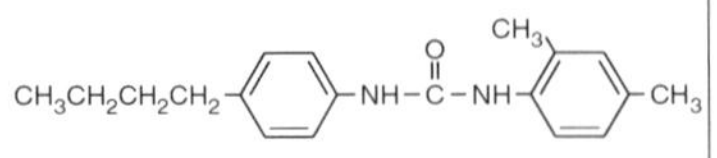

5 gm

S71,797-5
4,4'-BIS(P-TOLYLUREIDO)-3,3'-DIMETHYLBIPHENYL

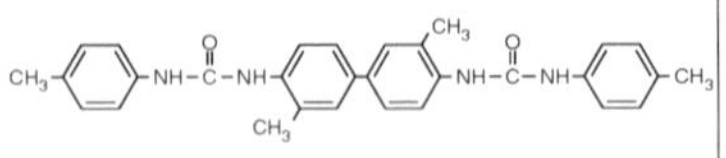

5 gm

S75,390-4
1-(4-BUTYLPHENYL)-3-(2,4,6-TRIMETHYLPHENYL)UREA

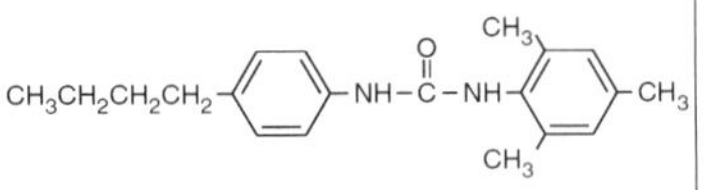

5 gm

S75,389-0
1-(4-BUTYLPHENYL)-3-(2,4,5-TRIMETHYLPHENYL)UREA

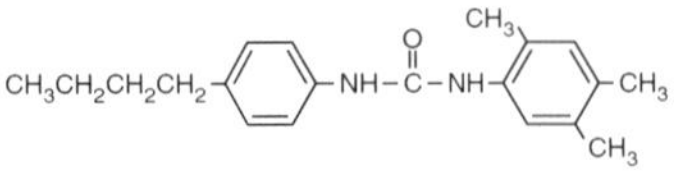

5 gm

S59,009-6
4,4'-METHYLENEBIS(PHENYLUREA)

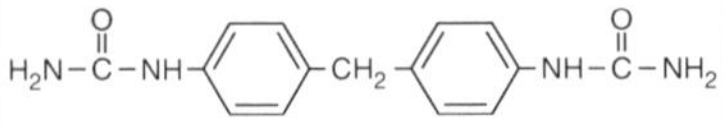

5 gm

S64,520-6
4,4'-METHYLENEBIS(1-METHYL-3-PHENYLUREA)

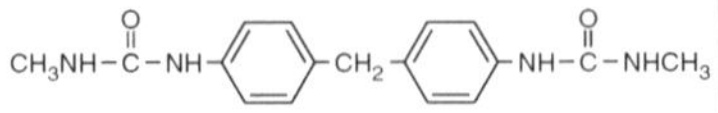

5 gm

S59,039-8
4,4'-METHYLENEBIS(1,1-DIMETHYL-3-PHENYLUREA)

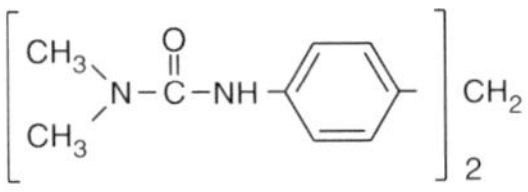

5 gm

S64,517-6
4,4'-METHYLENEBIS(1-ETHYL-3-PHENYLUREA)

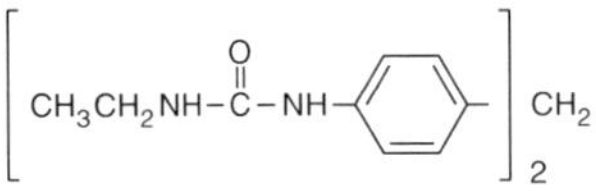

5 gm

S66,757-9
4,4'-METHYLENEBIS(1-ALLYL-3-PHENYLUREA)

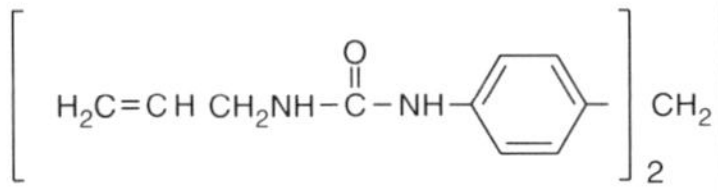

5 gm

S63,491-3
4,4'-METHYLENEBIS(1-ISOPROPYL-3-PHENYLUREA)

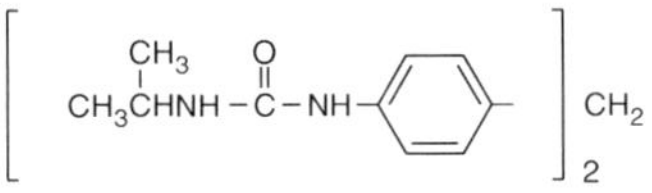

5 gm

S64,505-2
4,4'-METHYLENEBIS(3-PHENYL-1-PROPYLUREA)

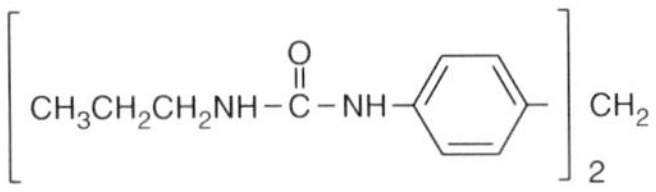

5 gm

S57,944-0
4,4'-METHYLENEBIS(1,1-DIETHYL-3-PHENYLUREA)

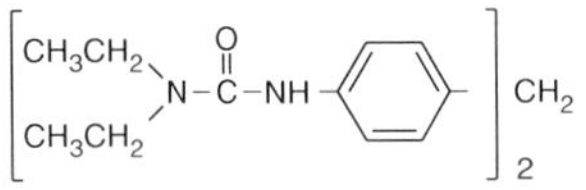

5 gm

S64,516-8
4,4'-METHYLENEBIS(1-BUTYL-3-PHENYLUREA)

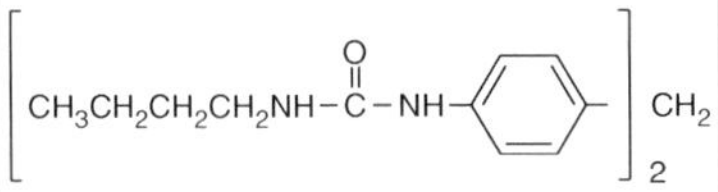

5 gm

S66,553-3
4,4'-METHYLENEBIS(1-ISOBUTYL-3-PHENYLUREA)

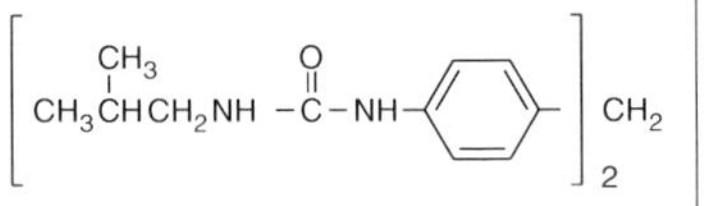

5 gm

S64,993-7
4,4'-METHYLENEBIS(1-BUTYL-1-METHYL-3-PHENYLUREA)

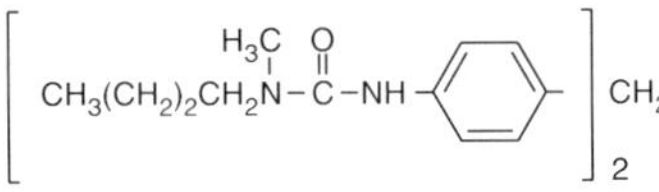

5 gm

S63,486-7
4,4'-METHYLENEBIS(1,1-DIALLYL-3-PHENYLUREA)

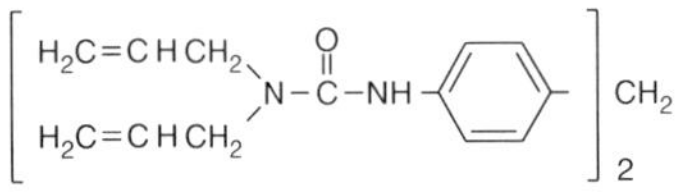

5 gm

S63,480-8
4,4'-METHYLENEBIS(1,1-DIPROPYL-3-PHENYLUREA)

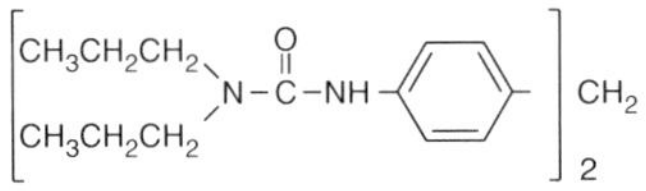

5 gm

S63,494-8
4,4'-METHYLENEBIS(1,1-DIISOPROPYL-3-PHENYLUREA)

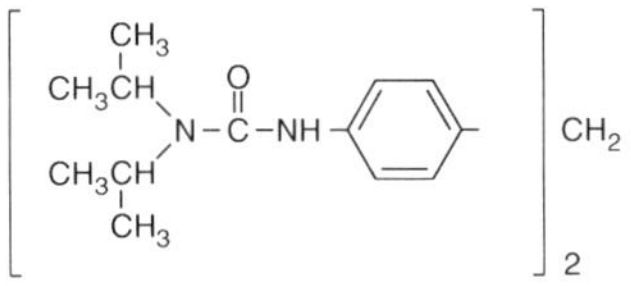

5 gm

S66,017-5
4,4'-METHYLENEBIS[3-(1,3-DIMETHYLBUTYL)-1-PHENYLUREA]

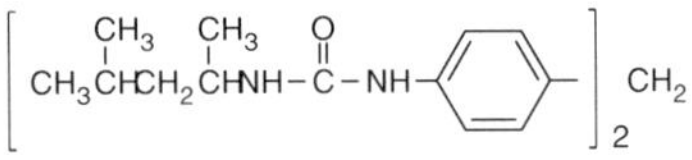

5 gm

S65,098-6
4,4'-BIS(3-METHYL-3-PHENYLUREIDO)-DIPHENYLMETHANE

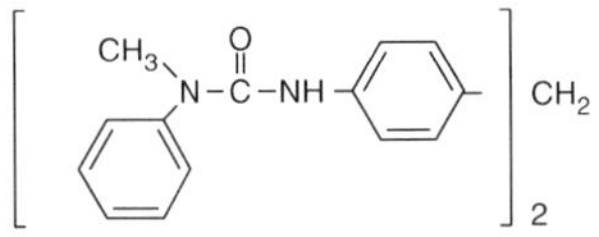

5 gm

S66,022-1
4,4'-METHYLENEBIS(1,3-DIPHENYL-1-ETHYLUREA)

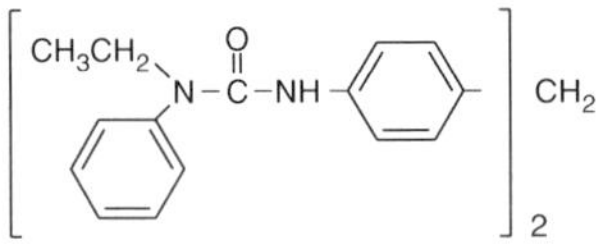

5 gm

S57,968-8
4,4'-METHYLENEBIS(1,1-DIBUTYL-3-PHENYLUREA)

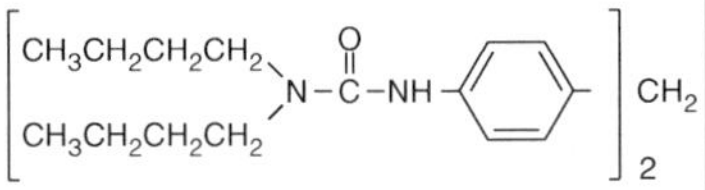

5 gm

S65,785-9
4,4'-METHYLENEBIS[N-(N-BUTYL-N-PHENYLCARBAMOYL)ANILINE]

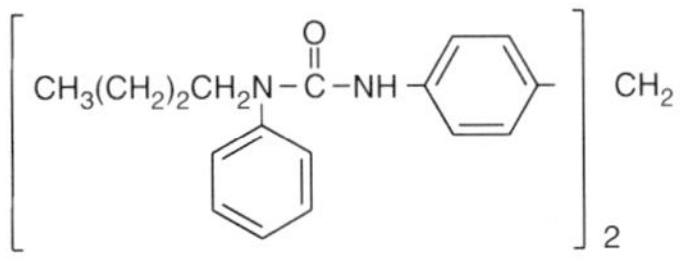

5 gm

S59,034-7
1-(4-CHLOROPHENYL)-3-(1-NAPHTHYL)UREA

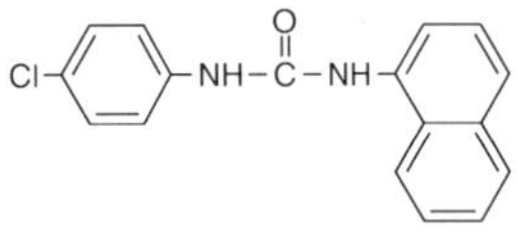

5 gm

S59,057-6
1-(1-NAPHTHYL)-3-(P-TOLYL)UREA

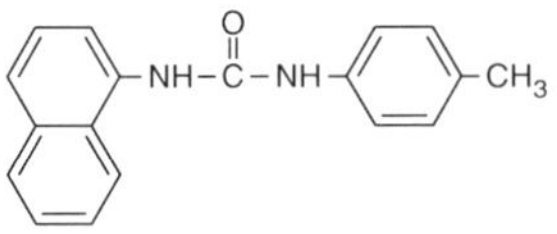

5 gm

S78,004-9
1-(4-ETHYLPHENYL)-3-(1-NAPHTHYL)UREA

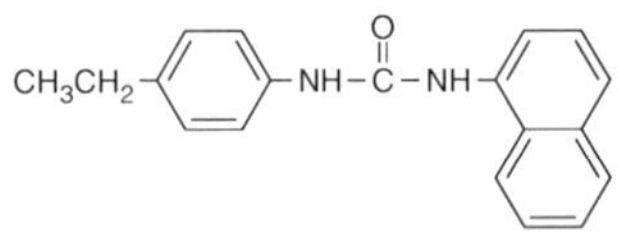

250 mg

S76,933-9
1-(4-ISOPROPYLPHENYL)-3-(1-NAPHTHYL)UREA

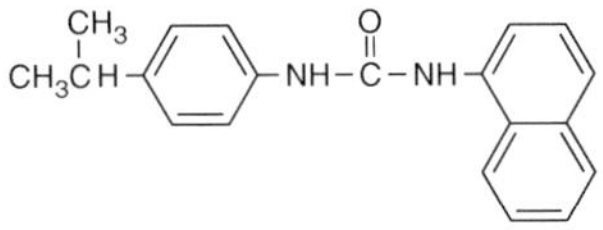

250 mg

S62,368-7
1-(2,3-DICHLOROPHENYL)-3-METHYLUREA

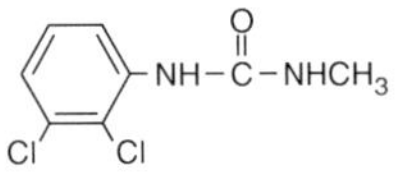

5 gm

S41,160-4
1-(2,6-XYLYL)UREA

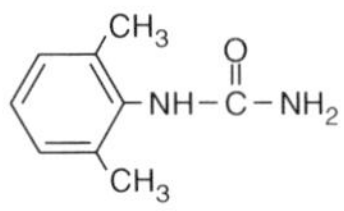

5 gm

S64,717-9
1-(2-CHLORO-6-METHYLPHENYL)-3-METHYLUREA

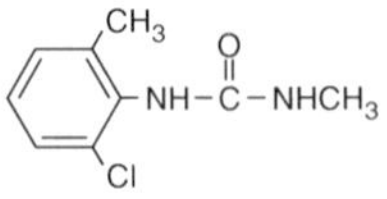

5 gm

S59,772-4
3-(2,3-DICHLOROPHENYL)-1,1-DIMETHYLUREA

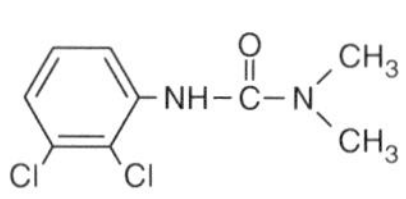

5 gm

S61,613-3
3-(2,6-DICHLOROPHENYL)-1,1-DIMETHYLUREA

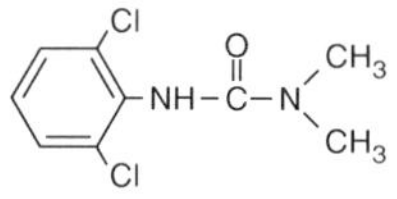

5 gm

S62,357-1
1-(2,3-DICHLOROPHENYL)-3-ETHYLUREA

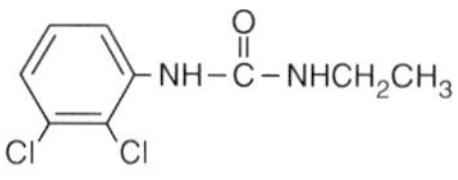

5 gm

S64,402-1
1-METHYL-3-(2,3-XYLYL)UREA

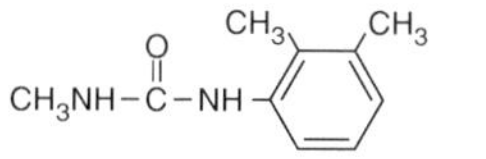

5 gm

S66,452-9
1-METHYL-3-(2,6-XYLYL)UREA

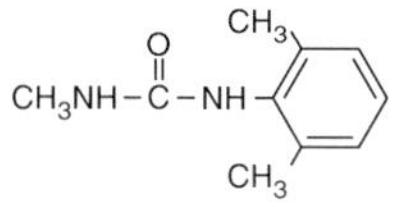

5 gm

S65,233-4
3-(2-CHLORO-6-METHYLPHENYL)-1,1-DIMETHYLUREA

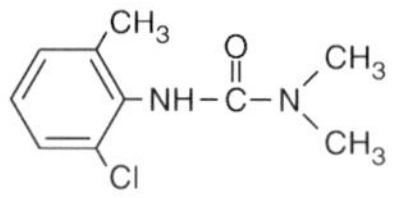

5 gm

S61,152-2
1-ALLYL-3-(2,6-DICHLOROPHENYL)UREA

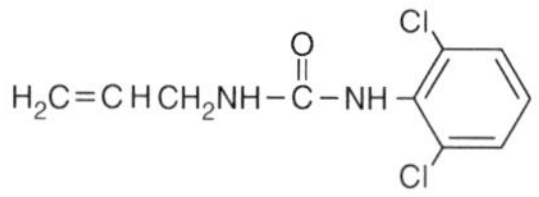

5 gm

S63,448-4
1-ALLYL-3-(2,3-DICHLOROPHENYL)UREA

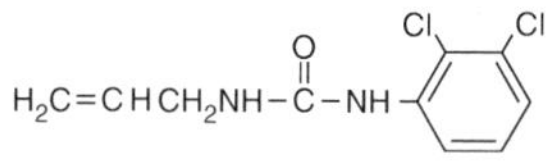

5 gm

S59,742-2
1-(2,3-DICHLOROPHENYL)-3-ISOPROPYLUREA

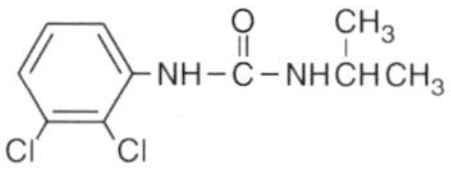

5 gm

S62,377-6
1-(2,3-DICHLOROPHENYL)-3-PROPYLUREA

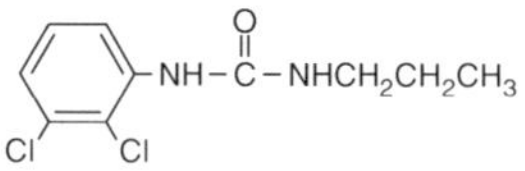

5 gm

S63,652-5
1-ETHYL-3-(2,3-XYLYL)UREA

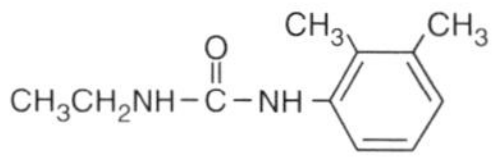

5 gm

S64,671-7
1,1-DIMETHYL-3-(2,3-XYLYL)UREA

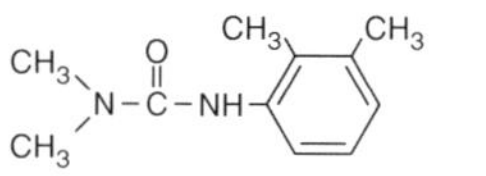

5 gm

S65,028-5
1-ETHYL-3-(2,6-XYLYL)UREA

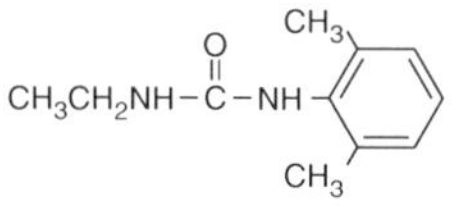

5 gm

S66,596-7
1-(2-ETHYL-6-METHYLPHENYL)-3-METHYLUREA

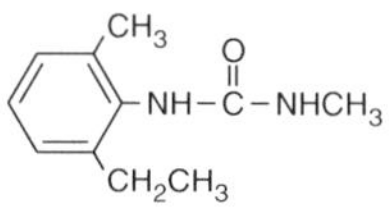

5 gm

S66,606-8
1,1-DIMETHYL-3-(2,6-XYLYL)UREA

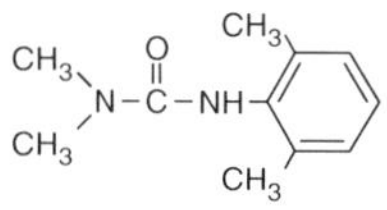

5 gm

S71,172-1
1-(2,6-DIETHYLPHENYL)UREA

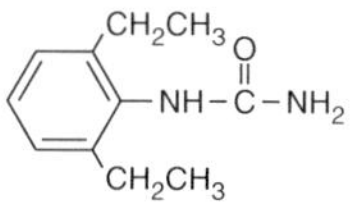

5 gm

S64,724-1
1-(2-CHLORO-6-METHYLPHENYL)-3-ISOPROPYLUREA

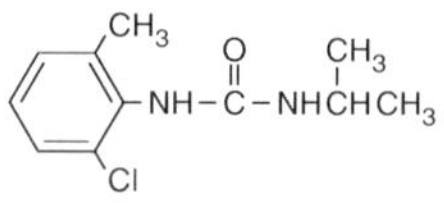

5 gm

S61,141-7
3-(2,6-DICHLOROPHENYL)-1,1-DIETHYLUREA

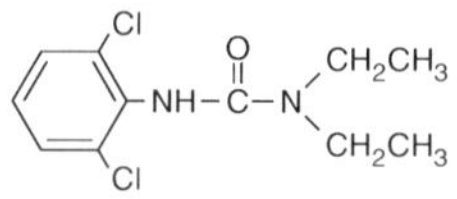

5 gm

S61,201-4
3-(2,3-DICHLOROPHENYL)-1,1-DIETHYLUREA

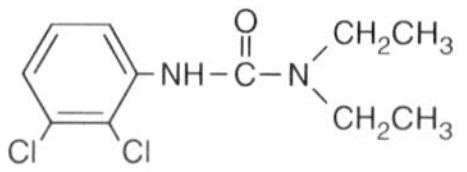

5 gm

S62,356-3
1-BUTYL-3-(2,3-DICHLOROPHENYL)UREA

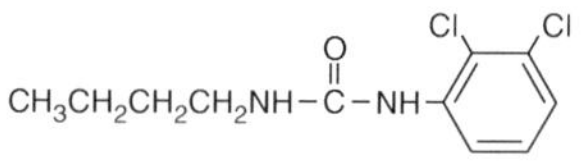

5 gm

S65,229-6
1-(2,6-DICHLOROPHENYL)-3-ISOBUTYLUREA

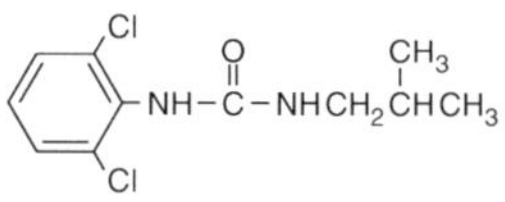

5 gm

S65,377-2
1-(2,3-DICHLOROPHENYL)-3-ISOBUTYLUREA

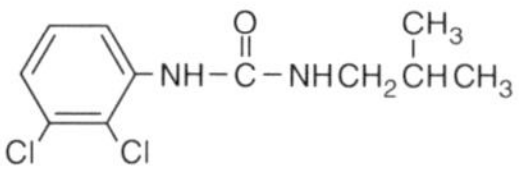

5 gm

S65,381-0
1-ALLYL-3-(2,6-XYLYL)UREA

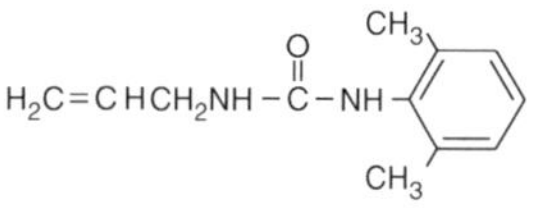

5 gm

S65,778-6
1-ALLYL-3-(2,3-XYLYL)UREA

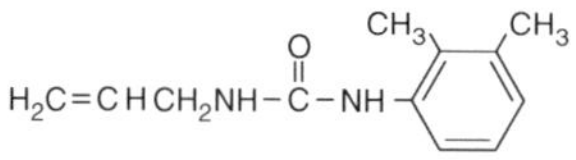

5 gm

S63,653-3
1-PROPYL-3-(2,3-XYLYL)UREA

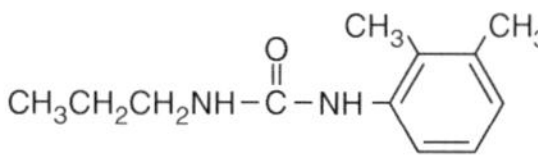

5 gm

S64,401-3
1-ISOPROPYL-3-(2,3-XYLYL)UREA

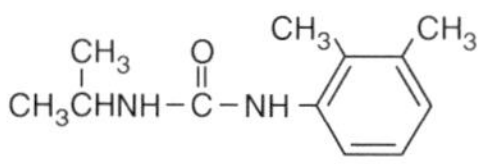

5 gm

S65,228-8
1-PROPYL-3-(2,6-XYLYL)UREA

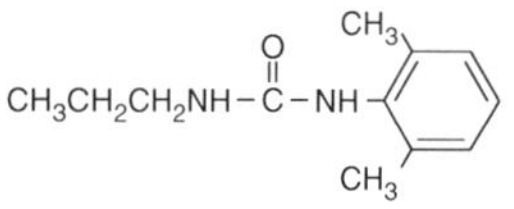

5 gm

S65,433-7
1-ISOPROPYL-3-(2,6-XYLYL)UREA

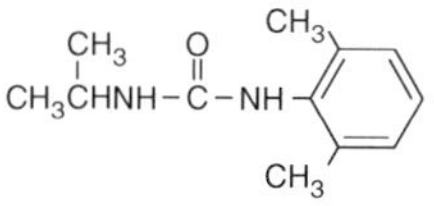

5 gm

S65,988-6
1-ETHYL-3-(2-ETHYL-6-METHYLPHENYL)UREA

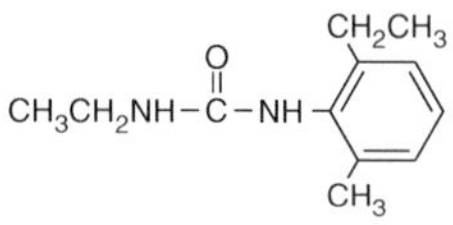

5 gm

S66,369-7
1-(2,6-DIETHYLPHENYL)-3-METHYLUREA

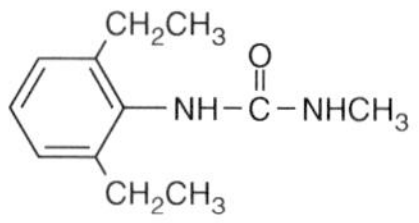

5 gm

S66,587-8
1,1-DIMETHYL-3-(2-ETHYL-6-METHYLPHENYL)UREA

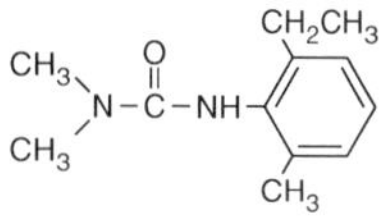

5 gm

S64,668-7
3-(2-CHLORO-6-METHYLPHENYL)-1,1-DIETHYLUREA

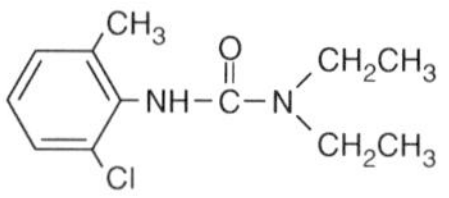

5 gm

S65,268-7
1-(2-CHLORO-6-METHYLPHENYL)-3-ISOBUTYLUREA

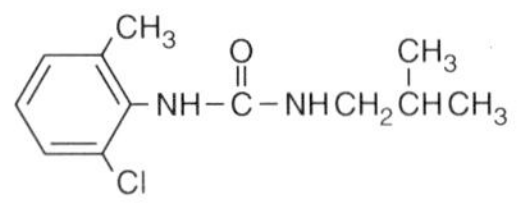

5 gm

S65,331-4
1-BUTYL-3-(2-CHLORO-6-METHYLPHENYL)UREA

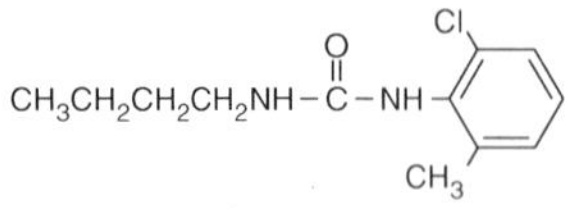

5 gm

S61,179-4
1-BUTYL-3-(2,6-DICHLOROPHENYL)-1-METHYLUREA

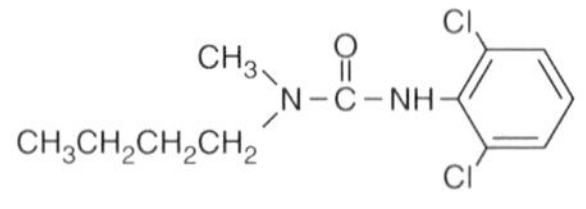

5 gm

S64,408-0
1-BUTYL-3-(2,3-DICHLOROPHENYL)-1-METHYLUREA

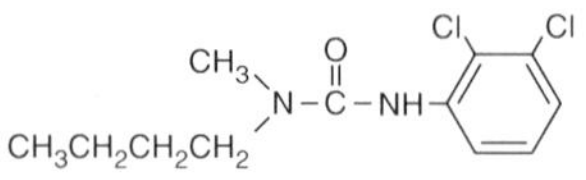

5 gm

S66,346-8
1-ALLYL-3-(2-ETHYL-6-METHYLPHENYL)UREA

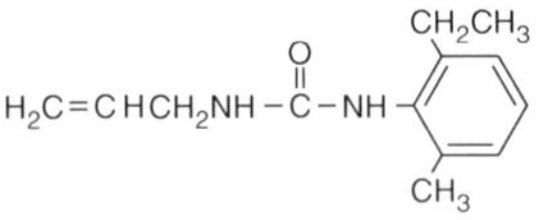

5 gm

S63,654-1
1-BUTYL-3-(2,3-XYLYL)UREA

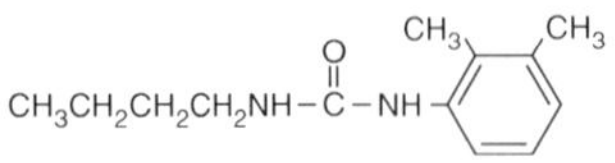

5 gm

S65,030-7
1-BUTYL-3-(2,6-XYLYL)UREA

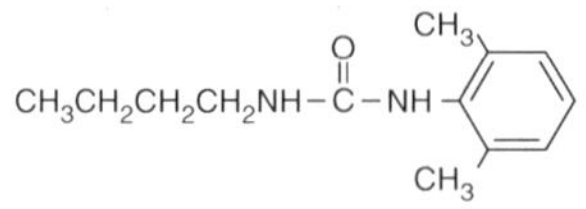

5 gm

S65,049-8
1-ISOBUTYL-3-(2,6-XYLYL)UREA

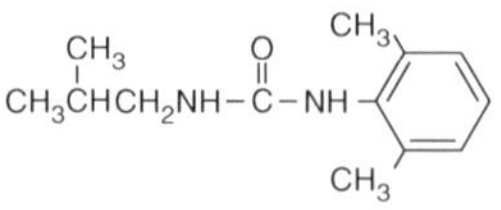

5 gm

S65,113-3
1,1-DIETHYL-3-(2,3-XYLYL)UREA

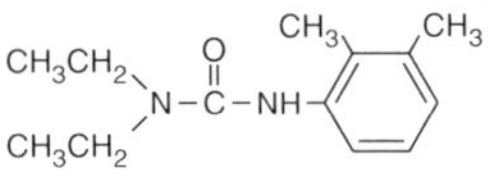

5 gm

S65,383-7
3-(2,6-DIETHYLPHENYL)-1,1-DIMETHYLUREA

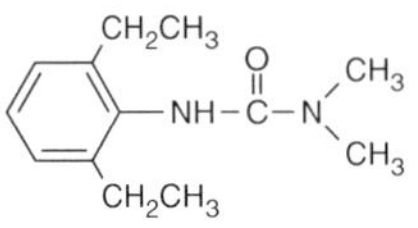

5 gm

S65,446-9
1-(2,6-DIETHYLPHENYL)-3-ETHYLUREA

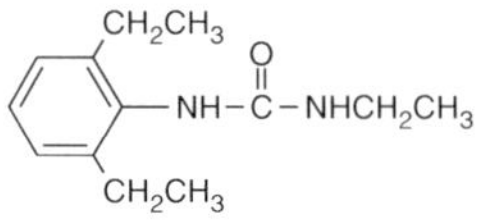

5 gm

S65,989-4
1-(2-ETHYL-6-METHYLPHENYL)-3-PROPYLUREA

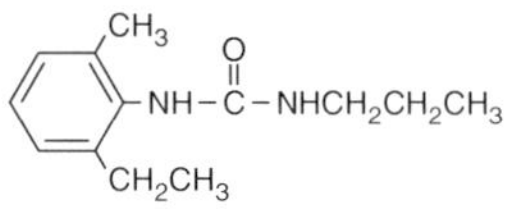

5 gm

S66,080-9
1-ISOBUTYL-3-(2,3-XYLYL)UREA

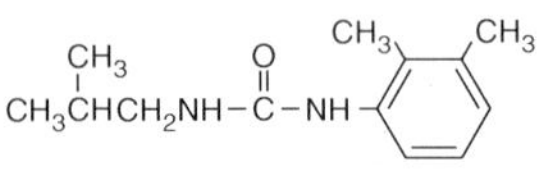

5 gm

S66,432-4
1-(2-ETHYL-6-METHYLPHENYL)-3-ISOPROPYLUREA

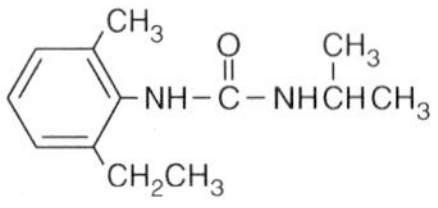

5 gm

S66,604-1
1,1-DIETHYL-3-(2,6-XYLYL)UREA

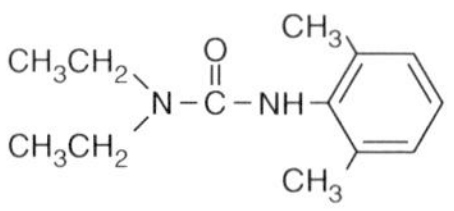

5 gm

S71,382-1
1-(2,6-DIISOPROPYLPHENYL)UREA

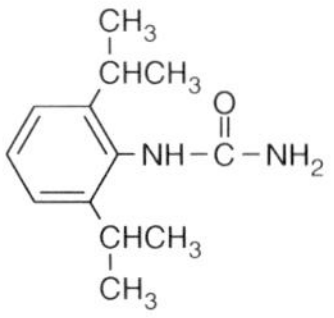

250 mg

S79,110-5
1-TERT-BUTYL-3-(2,6-XYLYL)UREA

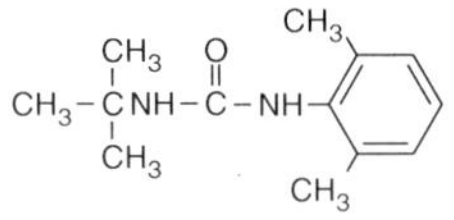

250 mg

S79,111-3
1-TERT-BUTYL-3-(2,3-XYLYL)UREA

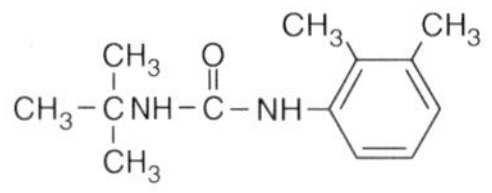

250 mg

S65,220-2
1-BUTYL-3-(2-CHLORO-6-METHYLPHENYL)-1-METHYLUREA

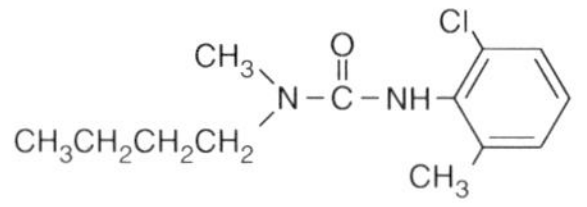

5 gm

S59,048-7
1-(2,3-DICHLOROPHENYL)-3-PHENYLUREA

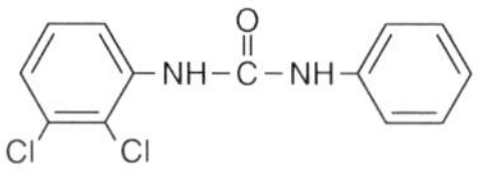

5 gm

S64,704-7
1-(2,6-DICHLOROPHENYL)-3-PHENYLUREA

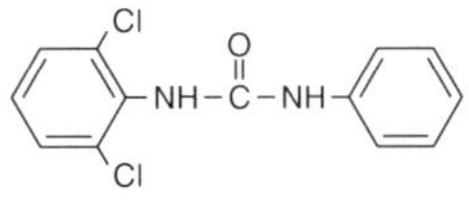

5 gm

S61,199-9
1,1-DIALLYL-3-(2,3-DICHLOROPHENYL)UREA

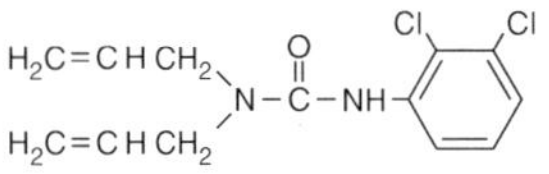

5 gm

S61,207-3
1,1-DIALLYL-3-(2,6-DICHLOROPHENYL)UREA

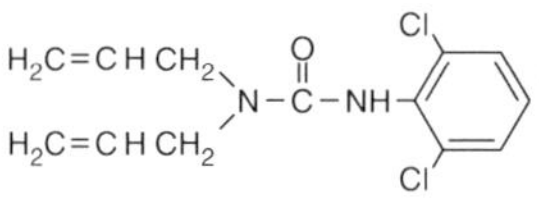

5 gm

S61,600-1
3-(2,3-DICHLOROPHENYL)-1,1-DIPROPYLUREA

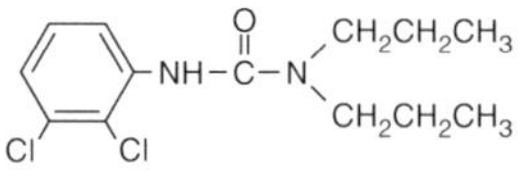

5 gm

S61,605-2
3-(2,3-DICHLOROPHENYL)-1,1-DIISOPROPYLUREA

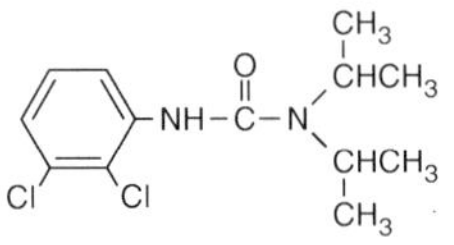

5 gm

S64,684-9
1-BUTYL-3-(2,6-DICHLOROPHENYL)-1-ETHYLUREA

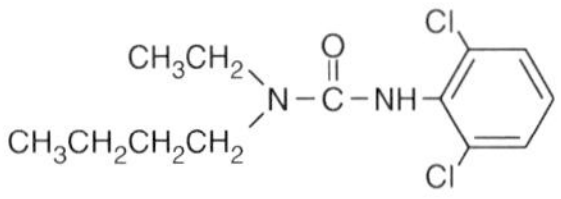

5 gm

S60,353-8
1,3-BIS(2,3-DICHLOROPHENYL)UREA

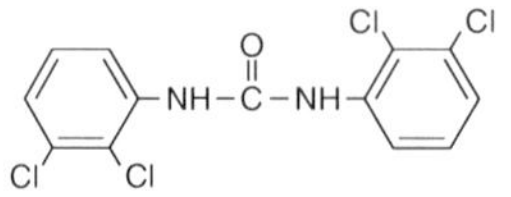

5 gm

S65,784-0
1-ALLYL-3-(2,6-DIETHYLPHENYL)UREA

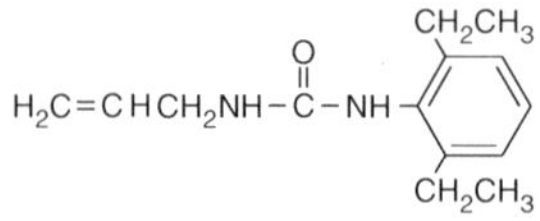

5 gm

S66,056-6
1-(2-ETHYL-6-METHYLPHENYL)-3-ISOBUTYLUREA

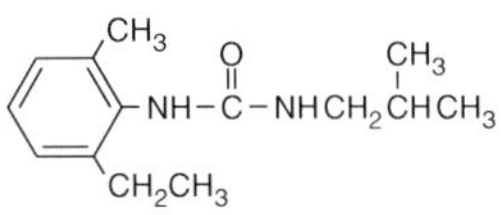

5 gm

S66,156-2
1-(2,6-DIETHYLPHENYL)-3-PROPYLUREA

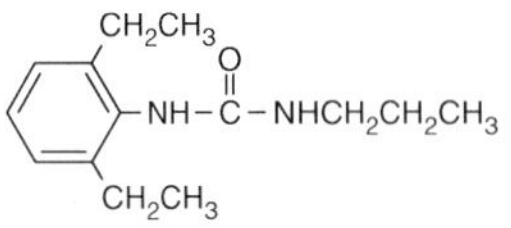

5 gm

S66,162-7
1,1-DIETHYL-3-(2-ETHYL-6-METHYLPHENYL)UREA

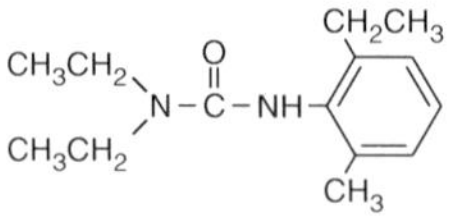

5 gm

S66,448-0
1-(2,6-DIISOPROPYLPHENYL)-3-METHYLUREA

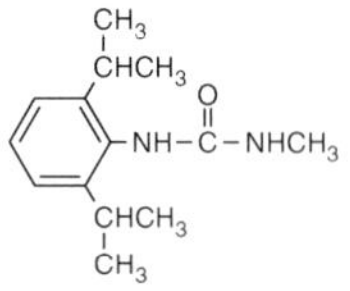

5 gm

S66,473-1
1-BUTYL-3-(2-ETHYL-6-METHYLPHENYL)UREA

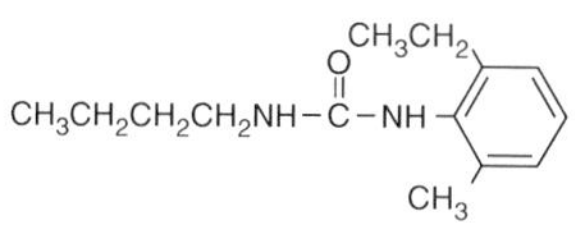

5 gm

S70,820-8
1-(2,6-DIETHYLPHENYL)-3-ISOPROPYLUREA

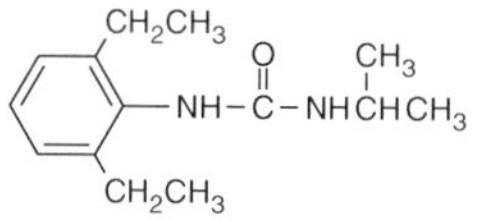

5 gm

S64,657-1
3-(2-CHLORO-6-METHYLPHENYL)-1,1-DIALLYLUREA

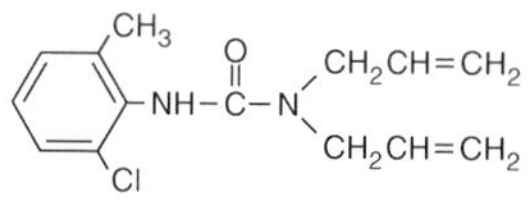

5 gm

S64,683-0
3-(2-CHLORO-6-METHYLPHENYL)-1,1-DIPROPYLUREA

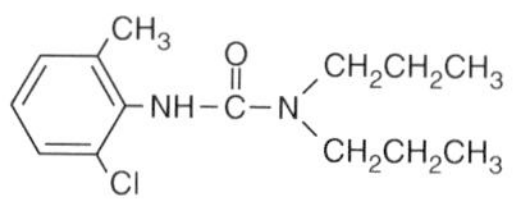

5 gm

S64,714-4
3-(2,3-DICHLOROPHENYL)-1-METHYL-1-PHENYLUREA

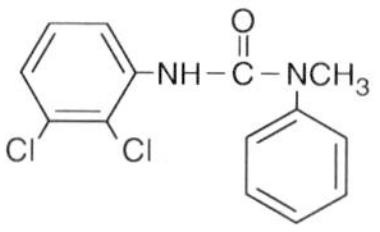

5 gm

S64,995-3
1,1-DIALLYL-3-(2,3-XYLYL)UREA

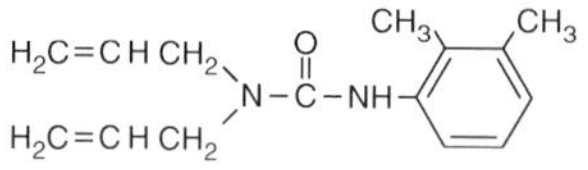

5 gm

S65,044-7
1,1-DIALLYL-3-(2,6-XYLYL)UREA

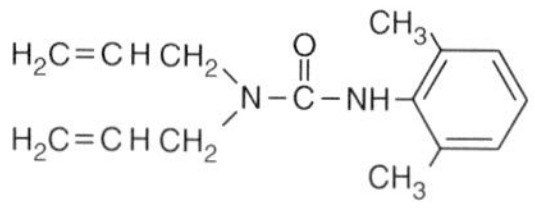

5 gm

S64,994-5
1,1-DIPROPYL-3-(2,6-XYLYL)UREA

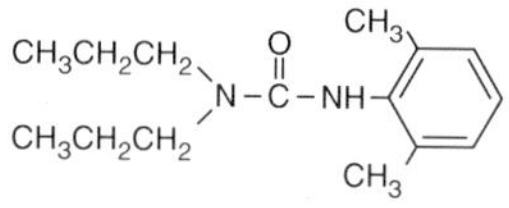

5 gm

S65,001-3
1,1-DIISOPROPYL-3-(2,6-XYLYL)UREA

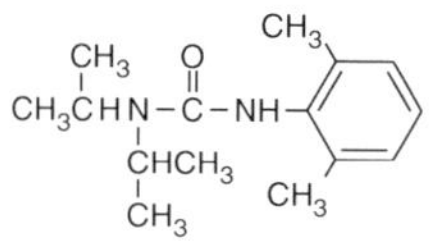

5 gm

S65,004-8
1,1-DIPROPYL-3-(2,3-XYLYL)UREA

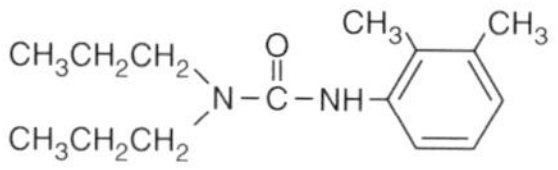

5 gm

S65,427-2
1-BUTYL-3-(2,6-DIETHYLPHENYL)UREA

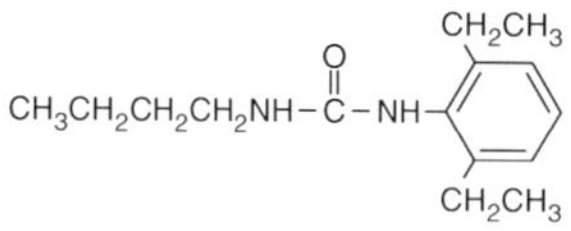

5 gm

S65,962-2
1-(2,6-DIISOPROPYLPHENYL)-3-ETHYLUREA

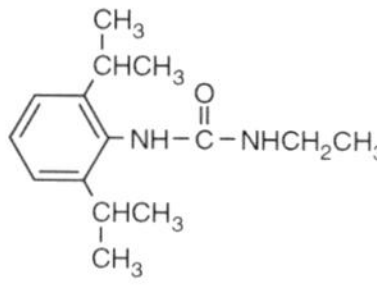

250 mg

S66,082-5
1,1-DIISOPROPYL-3-(2,3-XYLYL)UREA

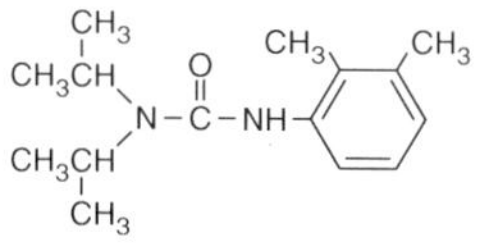

5 gm

S66,350-6
1-(2,6-DIETHYLPHENYL)-3-ISOBUTYLUREA

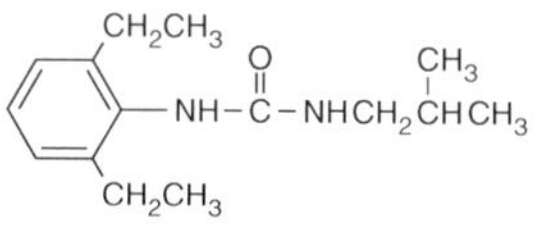

5 gm

S66,354-9
1,1-DIETHYL-3-(2,6-DIETHYLPHENYL)UREA

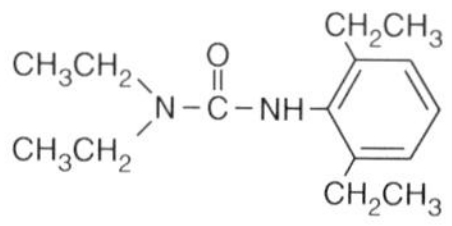

5 gm

S89,109-6
1-TERT-BUTYL-3-(2,6-DIETHYLPHENYL)UREA

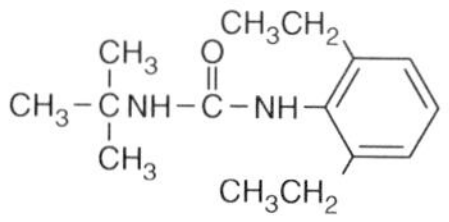

250 mg

S65,277-6
3-(2,6-DICHLOROPHENYL)-1-ETHYL-1-PHENYLUREA

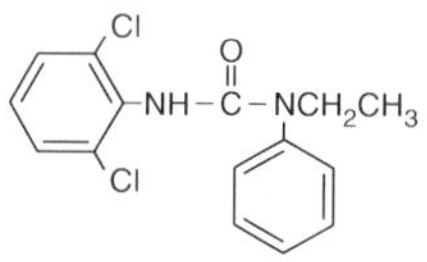

5 gm

S61,149-2
1,1-DIBUTYL-3-(2,6-DICHLOROPHENYL)UREA

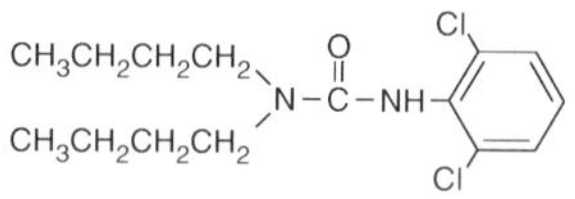

5 gm

S63,411-5
1,1-DIBUTYL-3-(2,3-DICHLOROPHENYL)UREA

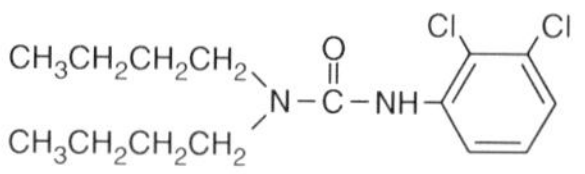

5 gm

S65,216-4
3-(2,6-DICHLOROPHENYL)-1,1-DIISOBUTYLUREA

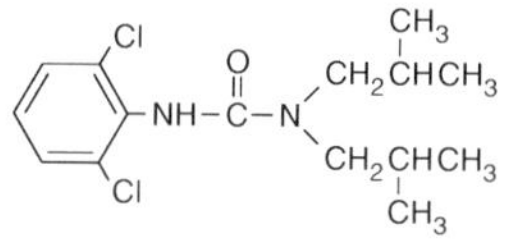

5 gm

S65,349-7
3-(2,3-DICHLOROPHENYL)-1,1-DIISOBUTYLUREA

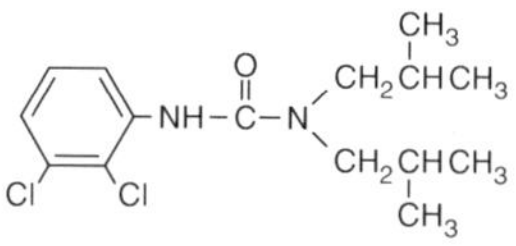

5 gm

S66,009-4
1-(2-ETHYL-6-METHYLPHENYL)-3-PHENYLUREA

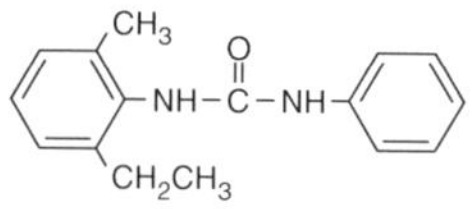

5 gm

S71,279-5
1-(2,6-DIMETHYLPHENYL)-3-(3-METHYLPHENYL)UREA

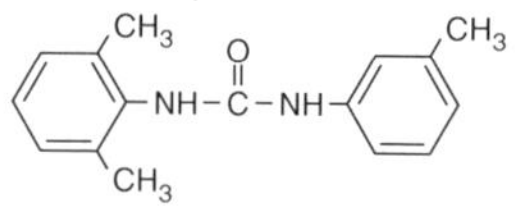

5 gm

S88,979-2
1-(M-TOLYL)-3-(2,6-XYLYL)UREA

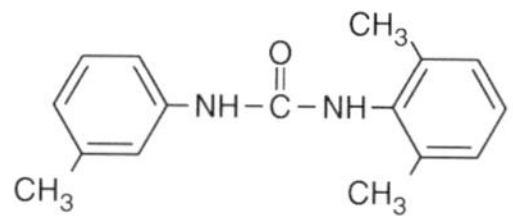

250 mg

S66,344-1
1,1-DIALLYL-3-(2-ETHYL-6-METHYLPHENYL)UREA

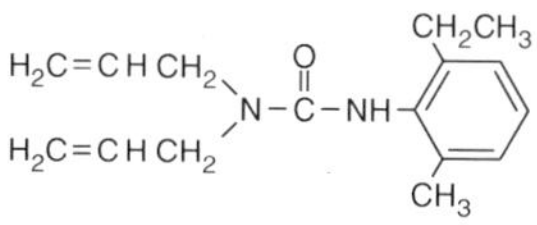

5 gm

S65,451-5
1-ALLYL-3-(2,6-DIISOPROPYLPHENYL)UREA

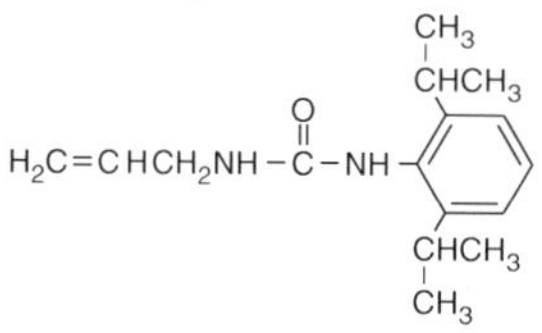

5 gm

S66,356-5
1,1-DIISOPROPYL-3-(2-ETHYL-6-METHYLPHENYL)UREA

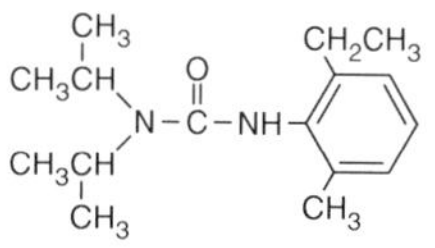

5 gm

S66,416-2
1-(2,6-DIISOPROPYLPHENYL)-3-PROPYLUREA

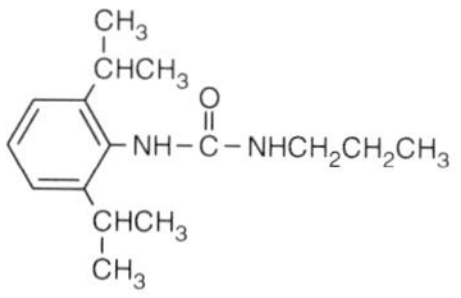

5 gm

S66,560-6
3-(2-ETHYL-6-METHYLPHENYL)-1,1-DIPROPYLUREA

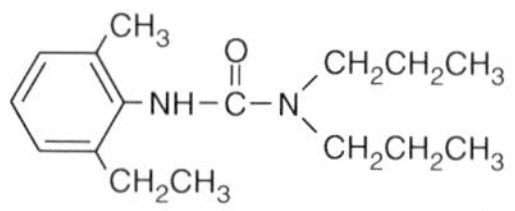

5 gm

S66,598-3
1-(2,6-DIISOPROPYLPHENYL)-3-ISOPROPYLUREA

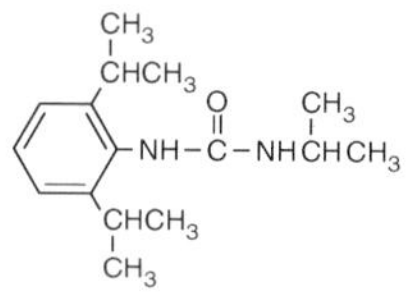

5 gm

S65,273-3
3-(2-CHLORO-6-METHYLPHENYL)-1,1-DIISOBUTYLUREA

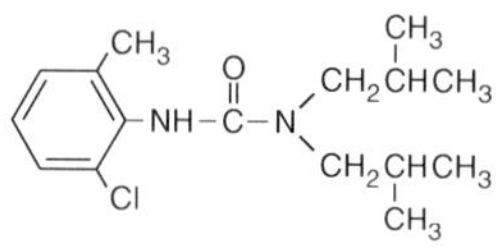

5 gm

S66,130-9
1,1-DIALLYL-3-(2,6-DIETHYLPHENYL)UREA

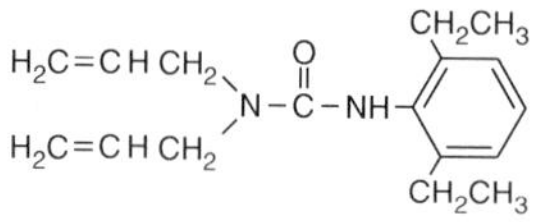

5 gm

S65,781-6
1,1-DIETHYL-3-(2,6-DIISOPROPYLPHENYL)UREA

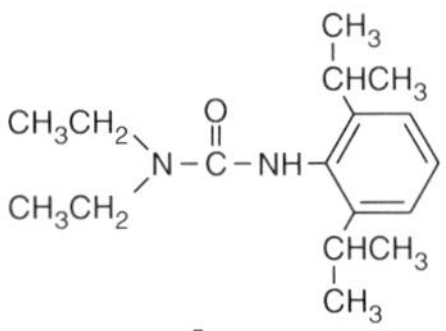

5 gm

S65,983-5
1-BUTYL-3-(2,6-DIISOPROPYLPHENYL)UREA

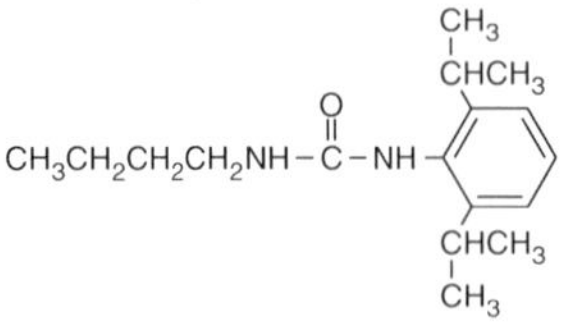

5 gm

S66,045-0
1,1-DIISOBUTYL-3-(2,3-XYLYL)UREA

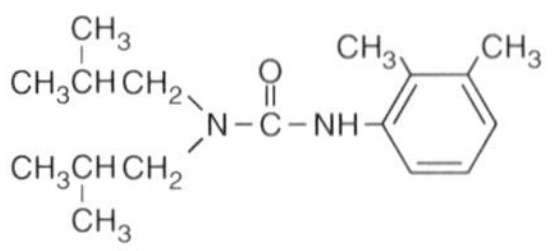

5 gm

S66,084-1
1,1-DIBUTYL-3-(2,3-XYLYL)UREA

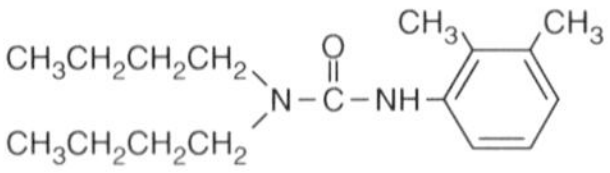

5 gm

S66,377-8
3-(2,6-DIETHYLPHENYL)-1,1-DIISOPROPYLUREA

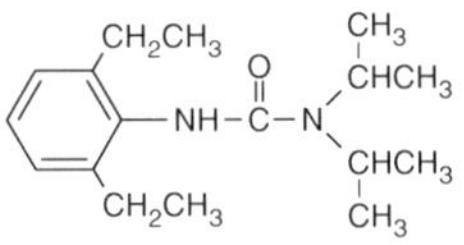

5 gm

S66,398-0
1,1-DIISOBUTYL-3-(2,6-XYLYL)UREA

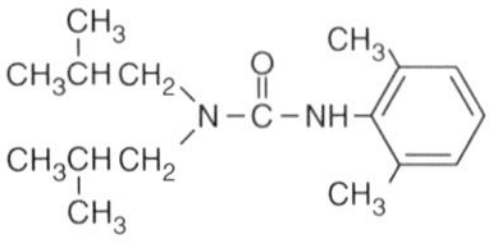

5 gm

S66,402-2
1,1-DIBUTYL-3-(2,6-XYLYL)UREA

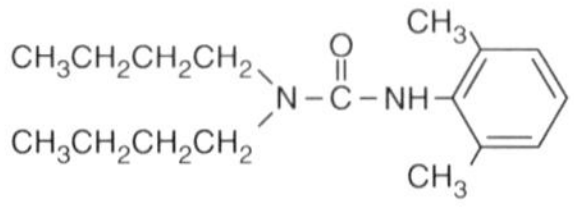

5 gm

S66,404-9
3-(2,6-DIETHYLPHENYL)-1,1-DIPROPYLUREA

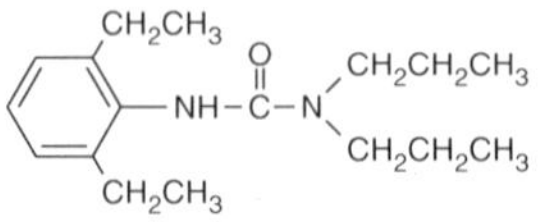

250 mg

S70,763-5
1-(2,6-DIISOPROPYLPHENYL)-3-ISOBUTYLUREA

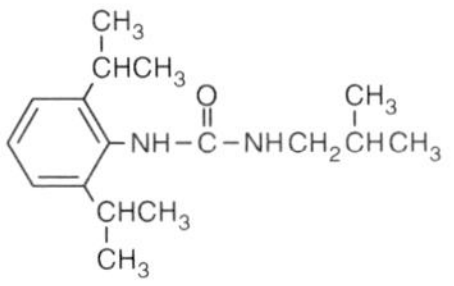

5 gm

S64,688-1
1-BUTYL-3-(2,3-DICHLOROPHENYL)-1-PHENYLUREA

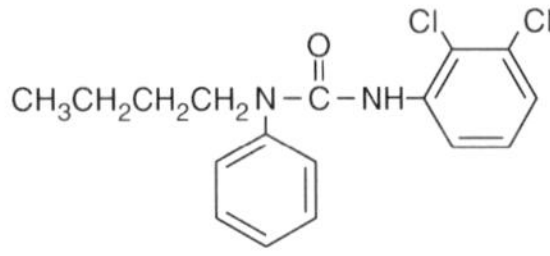

5 gm

S66,418-9
1,1-DIISOBUTYL-3-(2-ETHYL-6-METHYLPHENYL)UREA

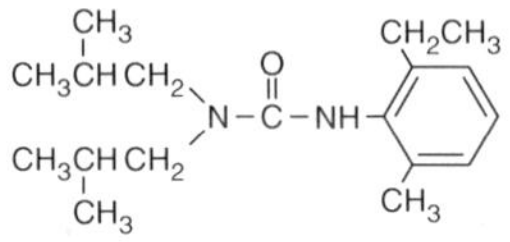

5 gm

S66,620-3
1,1-DIBUTYL-3-(2-ETHYL-6-METHYLPHENYL)UREA

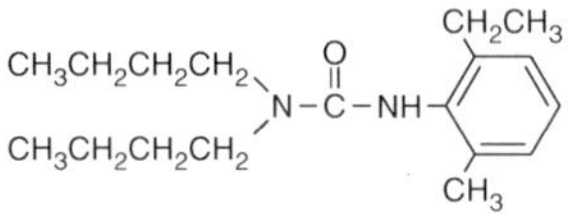

5 gm

S67,367-6
1,3-BIS(2-ETHYL-6-METHYLPHENYL)UREA

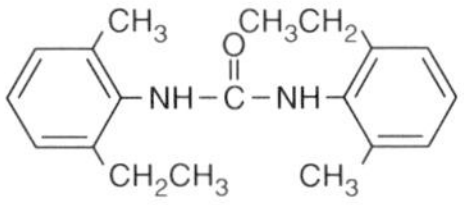

5 gm

S65,960-6
1,1-DIALLYL-3-(2,6-DIISOPROPYLPHENYL)UREA

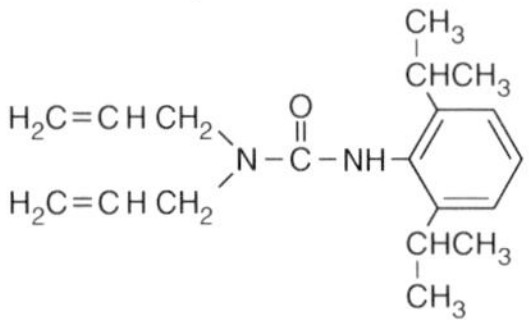

5 gm

S66,051-5
1,1-DIISOPROPYL-3-(2,6-DIISOPROPYLPHENYL)UREA

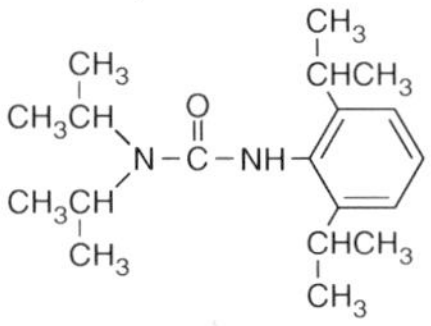

5 gm

S66,496-0
3-(2,6-DIISOPROPYLPHENYL)-1,1-DIPROPYLUREA

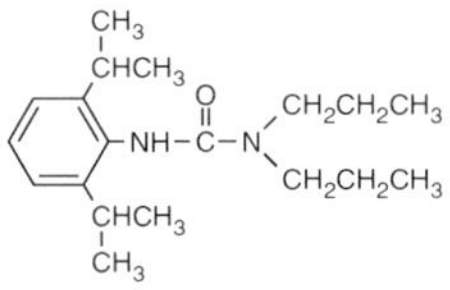

5 gm

S66,857-5
3-(2,6-DIETHYLPHENYL)-1,1-DIISOBUTYLUREA

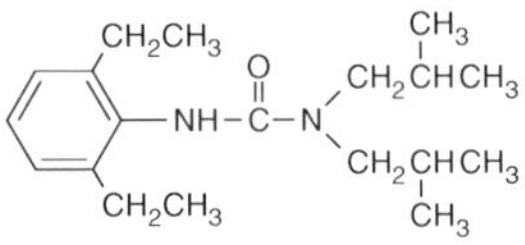

5 gm

S66,454-5
3-(2,6-DIISOPROPYLPHENYL)-1-METHYL-1-PHENYLUREA

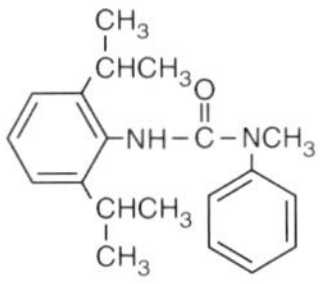

5 gm

S66,475-8
1,1-DIISOBUTYL-3-(2,6-DIISOPROPYLPHENYL)UREA

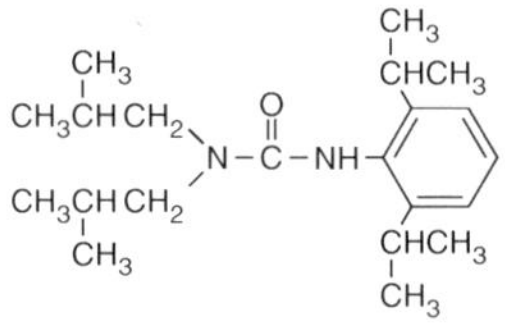

5 gm

S66,621-1
1,1-DIBUTYL-3-(2,6-DIISOPROPYLPHENYL)UREA

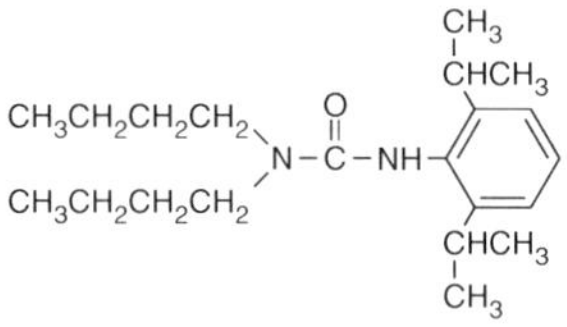

5 gm

S67,013-8
1-OCTADECYL-3-(2,6-XYLYL)UREA

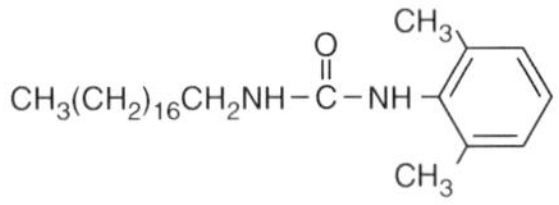

5 gm

S67,061-8
1-OCTADECYL-3-(2,3-XYLYL)UREA

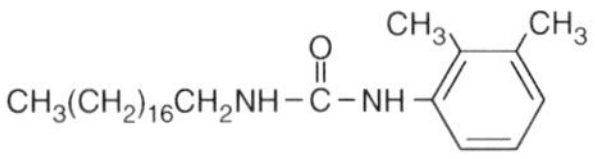

5 gm

S67,083-9
1-(2-ETHYL-6-METHYLPHENYL)-3-OCTADECYLUREA

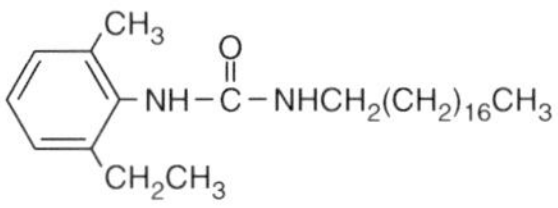

5 gm

S67,072-3
1-(2,6-DIETHYLPHENYL)-3-OCTADECYLUREA

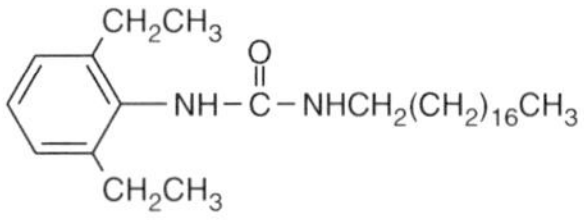

5 gm

S71,291-4
1-(2,4-DIMETHYLPHENYL)-3-(2,6-DIMETHYLPHENYL)UREA

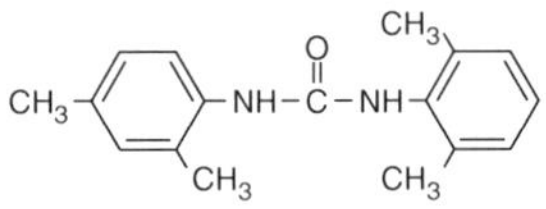

5 gm

S70,899-2
1-(2,4-DIMETHYLPHENYL)-3-(2-ETHYL-6-METHYLPHENYL)UREA

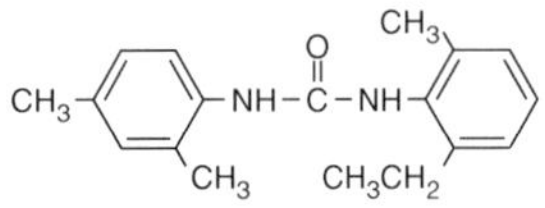

5 gm

S70,745-7
1-(2,6-DIETHYLPHENYL)-3-(2,4-DIMETHYLPHENYL)UREA

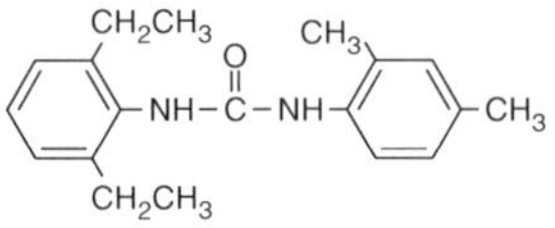

5 gm

S71,697-9
1-(2,6-DIISOPROPYLPHENYL)-3-(2,4-XYLYL)UREA

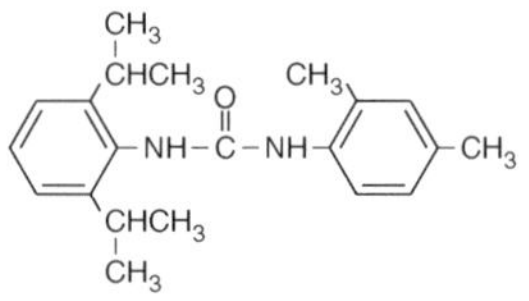

5 gm

S67,663-2
4,4'-BIS(2,6-DIMETHYLPHENYLUREIDO)-3,3'-DIMETHYLBIPHENYL

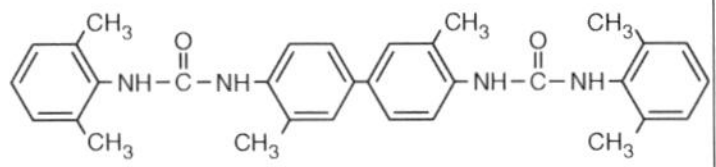

5 gm

S59,053-3
1-(2,3-DICHLOROPHENYL)-3-(1-NAPHTHYL)UREA

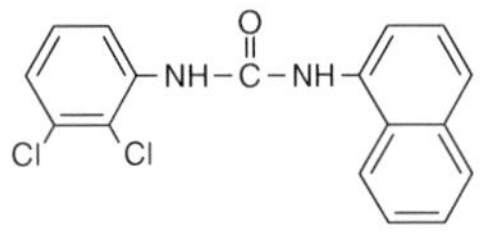

5 gm

S71,298-1
1-(1-NAPHTHYL)-3-(2,6-XYLYL)UREA

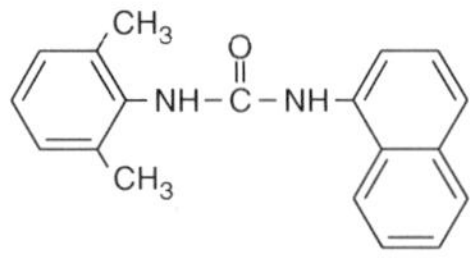

5 gm

S78,006-5
1-(1-NAPHTHYL)-3-(2,3-XYLYL)UREA

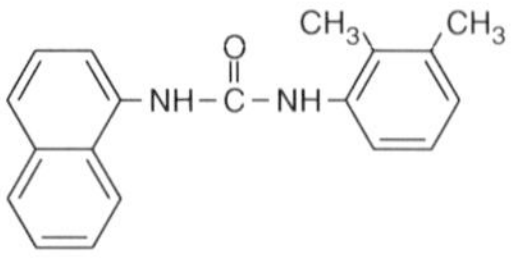

250 mg

S72,861-6
1-(2,6-DIETHYLPHENYL)-3-(1-NAPHTHYL)UREA
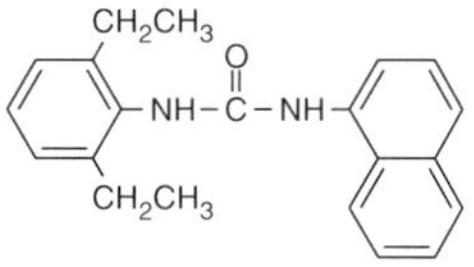
5 gm

S43,135-4
1-(5-CHLORO-ORTHO-TOLYL)UREA
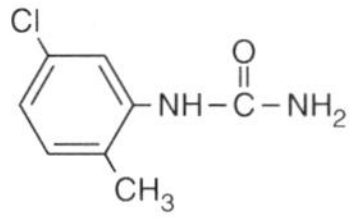
5 gm

S60,151-9
4-CHLORO-2-METHYLPHENYLUREA
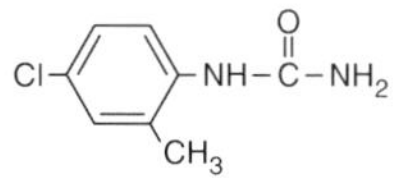
5 gm

S61,305-3
1-(3,4-DICHLOROPHENYL)-3-METHYLUREA
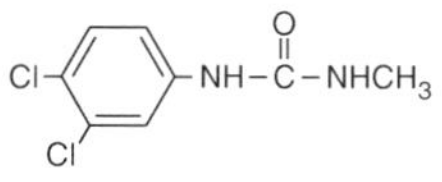
5 gm

S61,351-7
1-(2,4-DICHLOROPHENYL)-3-METHYLUREA
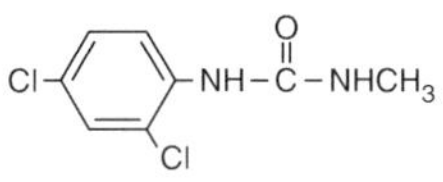
5 gm

S63,649-5
1-(2,5-DICHLOROPHENYL)-3-METHYLUREA
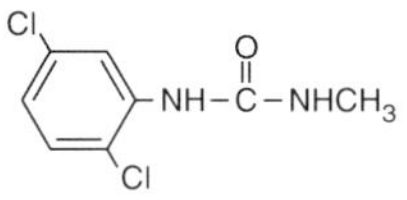
5 gm

S41,159-0
1-(2,4-XYLYL)UREA
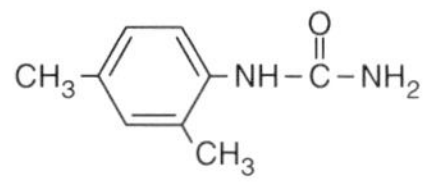
5 gm

S57,605-0
1-(3,4-XYLYL)UREA
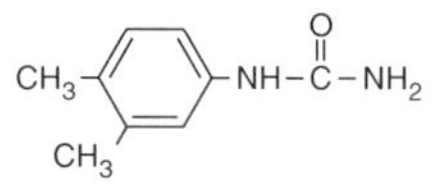
5 gm

S57,648-4
1-(2,5-XYLYL)UREA
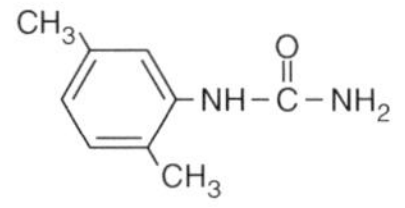
250 mg

S61,102-6
1-(5-CHLORO-2-METHYLPHENYL)-3-METHYLUREA
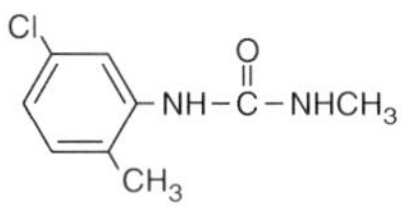
5 gm

S63,812-9
1-(4-CHLORO-2-METHYLPHENYL)-3-METHYLUREA
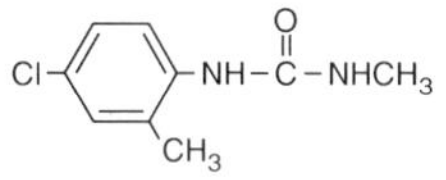
5 gm

S64,672-5
1-(3-CHLORO-4-METHYLPHENYL)-3-METHYLUREA
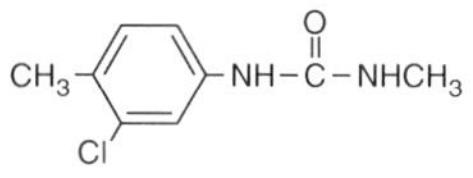
5 gm

S59,757-0
3-(2,5-DICHLOROPHENYL)-1,1-DIMETHYLUREA
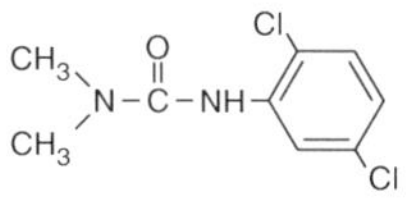
5 gm

S61,332-0
3-(2,4-DICHLOROPHENYL)-1,1-DIMETHYLUREA
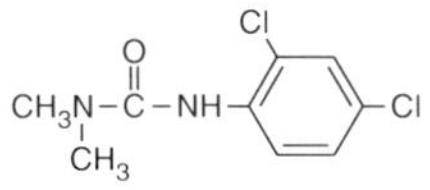
5 gm

S61,344-4
1-(3,4-DICHLOROPHENYL)-3-ETHYLUREA
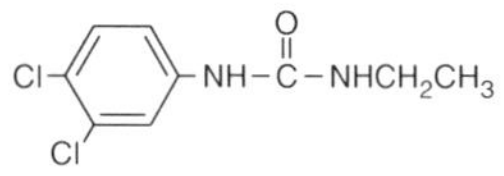
5 gm

S62,374-1
1-(2,4-DICHLOROPHENYL)-3-ETHYLUREA
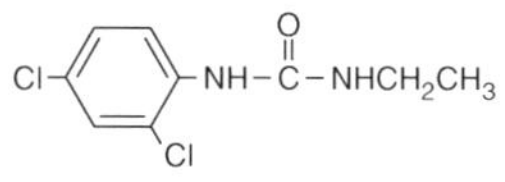
5 gm

S60,678-2
1-METHYL-3-(2,4-XYLYL)UREA
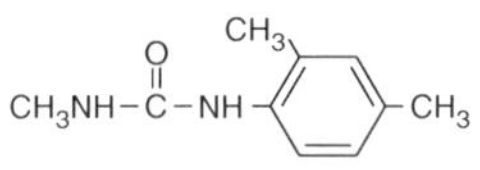
5 gm

S63,665-7
1-METHYL-3-(2,5-XYLYL)UREA
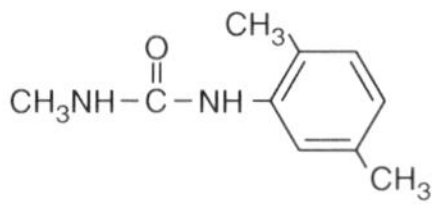
5 gm

S65,086-2
1-METHYL-3-(3,4-XYLYL)UREA
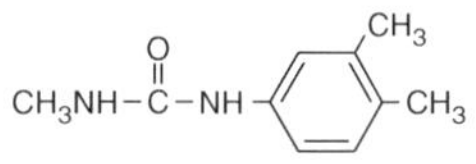
5 gm

S60,203-5
3-(4-CHLORO-2-METHYLPHENYL)-1,1-DIMETHYLUREA
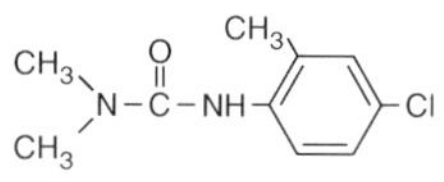
5 gm

S60,357-0
3-(3-CHLORO-4-METHYLPHENYL)-1,1-DIMETHYLUREA
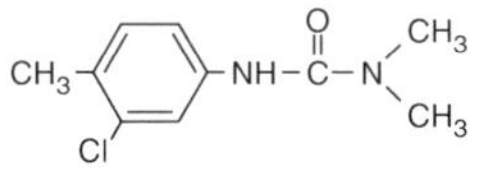
5 gm

S60,462-3
1-(4-CHLORO-2-METHYLPHENYL)-3-ETHYLUREA
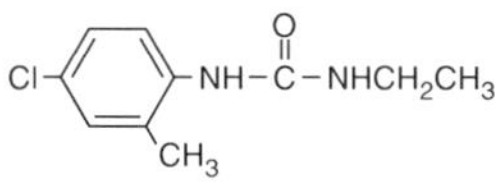
5 gm

S61,072-0
3-(5-CHLORO-2-METHYLPHENYL)-1,1-DIMETHYLUREA
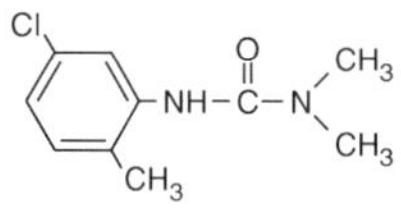
5 gm

S61,352-5
1-(5-CHLORO-2-METHYLPHENYL)-3-ETHYLUREA
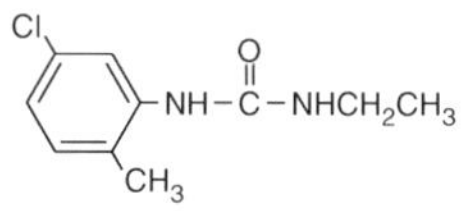
5 gm

S64,663-6
1-(3-CHLORO-4-METHYLPHENYL)-3-ETHYLUREA
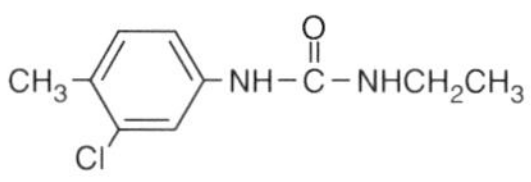
5 gm

S61,160-3
1-ALLYL-3-(2,5-DICHLOROPHENYL)UREA
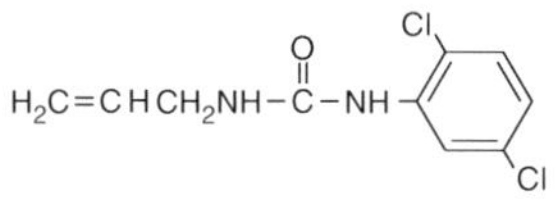
5 gm

S61,296-0
1-ALLYL-3-(2,4-DICHLOROPHENYL)UREA
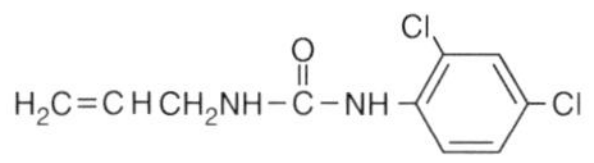
5 gm

S61,327-4
1-ALLYL-3-(3,4-DICHLOROPHENYL)UREA
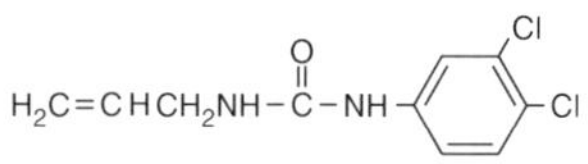
5 gm

S61,148-4
1-(2,5-DICHLOROPHENYL)-3-ISOPROPYLUREA
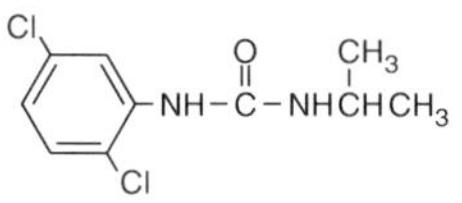
5 gm

S61,175-1
1-(2,5-DICHLOROPHENYL)-3-PROPYLUREA
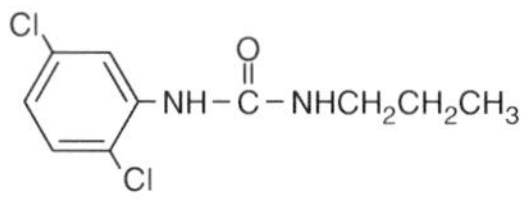
5 gm

S61,339-8
1-(3,4-DICHLOROPHENYL)-3-ISOPROPYLUREA
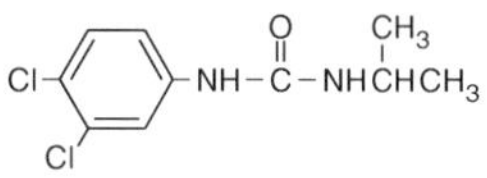
5 gm

S61,354-1
1-(3,4-DICHLOROPHENYL)-3-PROPYLUREA
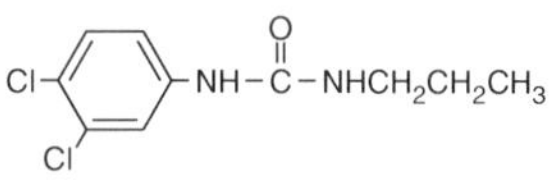
5 gm

S62,369-5
1-(2,4-DICHLOROPHENYL)-3-PROPYLUREA

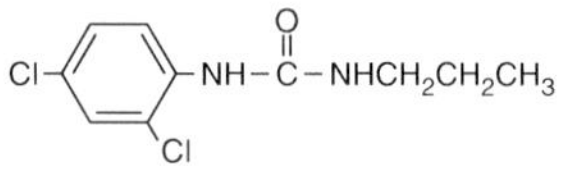

5 gm

S63,454-9
1-(2,4-DICHLOROPHENYL)-3-ISOPROPYLUREA

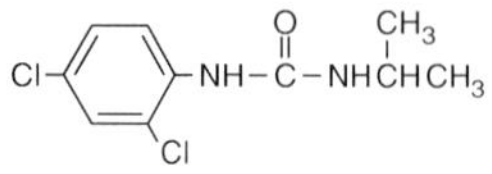

5 gm

S60,631-6
1-ETHYL-3-(2,4-XYLYL)UREA

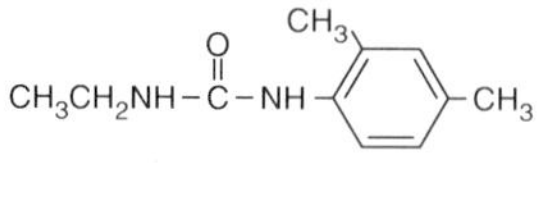

5 gm

S60,690-1
1,1-DIMETHYL-3-(2,4-XYLYL)UREA

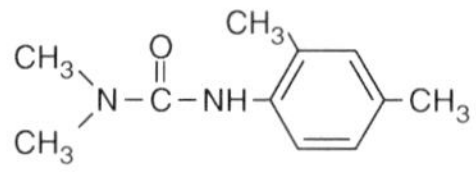

5 gm

S63,659-2
1,1-DIMETHYL-3-(2,5-XYLYL)UREA

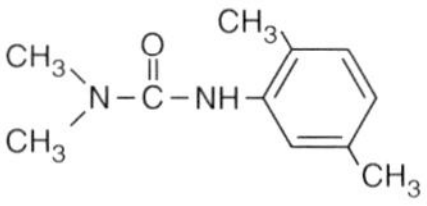

5 gm

S65,062-5
1-ETHYL-3-(3,4-XYLYL)UREA

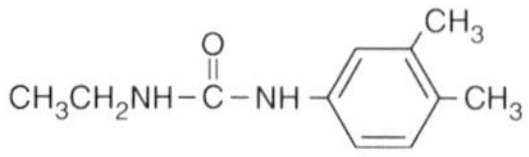

5 gm

S65,135-4
1,1-DIMETHYL-3-(3,4-XYLYL)UREA

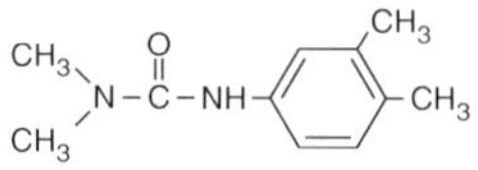

5 gm

S61,096-8
1-ALLYL-3-(5-CHLORO-2-METHYLPHENYL)UREA

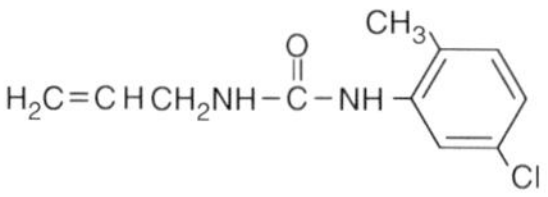

5 gm

S63,822-6
1-ALLYL-3-(4-CHLORO-2-METHYLPHENYL)UREA

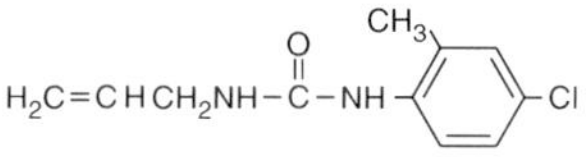

5 gm

S64,681-4
1-ALLYL-3-(3-CHLORO-4-METHYLPHENYL)UREA

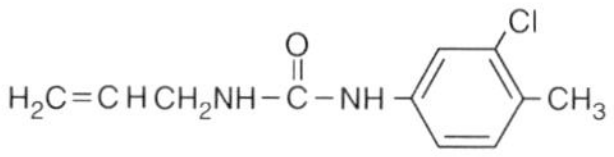

5 gm

S61,061-5
1-(5-CHLORO-2-METHYLPHENYL)-3-ISOPROPYLUREA

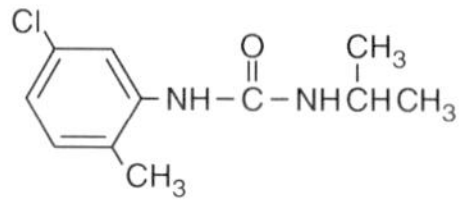

5 gm

S61,299-5
1-(5-CHLORO-2-METHYLPHENYL)-3-PROPYLUREA

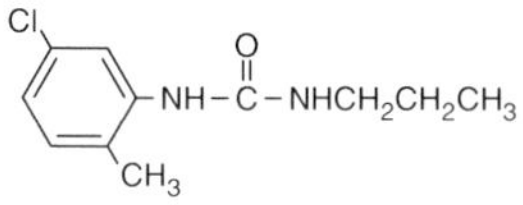

5 gm

S63,811-0
1-(4-CHLORO-2-METHYLPHENYL)-3-ISOPROPYLUREA

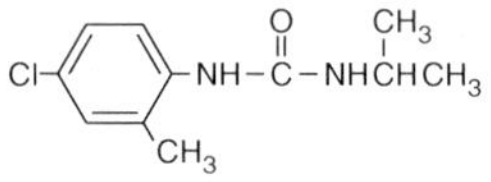

5 gm

S64,670-9
1-(4-CHLORO-2-METHYLPHENYL)-3-PROPYLUREA

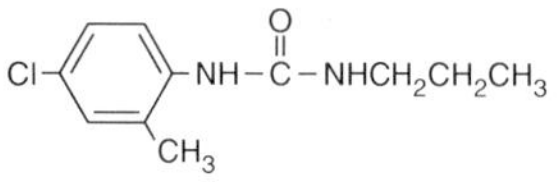

5 gm

S64,680-6
1-(3-CHLORO-4-METHYLPHENYL)-3-PROPYLUREA

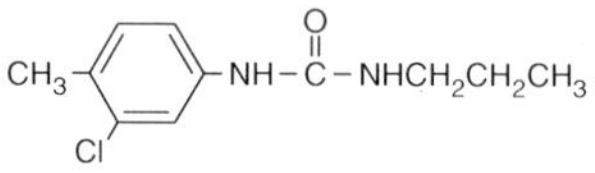

5 gm

S64,682-2
1-(3-CHLORO-4-METHYLPHENYL)-3-ISOPROPYLUREA

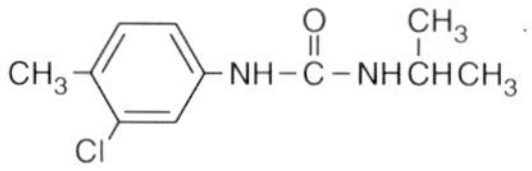

5 gm

S60,672-3
3-(3,4-DICHLOROPHENYL)-1,1-DIETHYLUREA

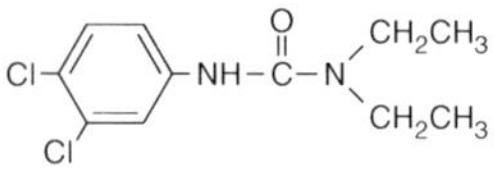

5 gm

S61,192-1
3-(2,5-DICHLOROPHENYL)-1,1-DIETHYLUREA

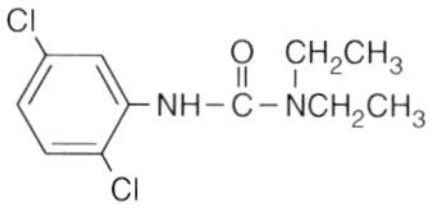

5 gm

S61,316-9
1-BUTYL-3-(2,4-DICHLOROPHENYL)UREA

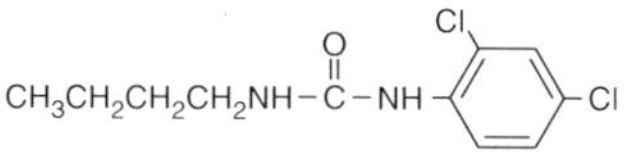

5 gm

S61,342-8
1-BUTYL-3-(3,4-DICHLOROPHENYL)UREA

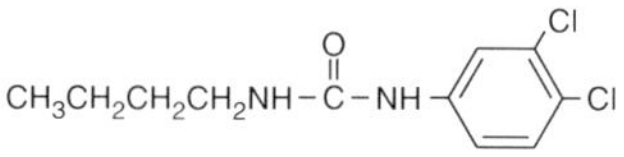

5 gm

S61,361-4
3-(2,4-DICHLOROPHENYL)-1,1-DIETHYLUREA

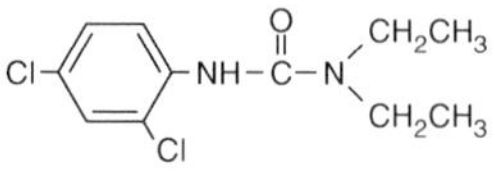

5 gm

S63,650-9
1-BUTYL-3-(2,5-DICHLOROPHENYL)UREA

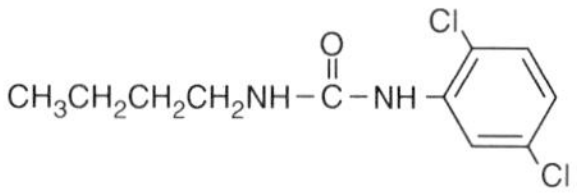

5 gm

S64,509-5
1-(3,4-DICHLOROPHENYL)-3-ISOBUTYLUREA

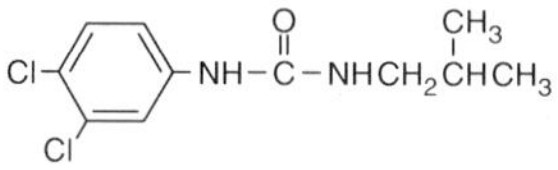

5 gm

S65,344-6
1-(2,4-DICHLOROPHENYL)-3-ISOBUTYLUREA

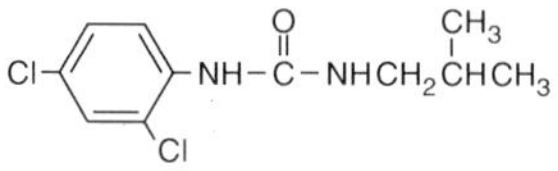

5 gm

S60,620-0
1-ALLYL-3-(2,4-XYLYL)UREA

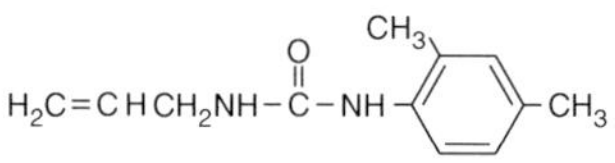

5 gm

S66,127-9
1-ALLYL-3-(3,4-XYLYL)UREA

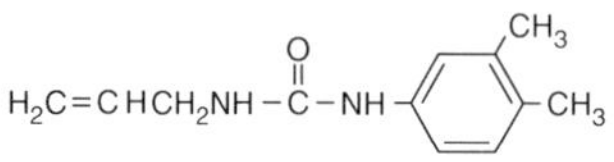

5 gm

S66,597-5
1-ALLYL-3-(2,5-XYLYL)UREA

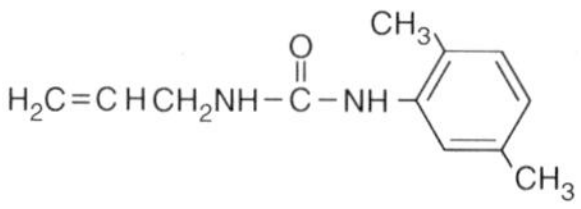

5 gm

S61,027-5
1-PROPYL-3-(2,4-XYLYL)UREA

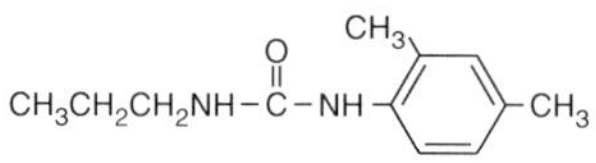

5 gm

S61,602-8
1-ISOPROPYL-3-(2,4-XYLYL)UREA

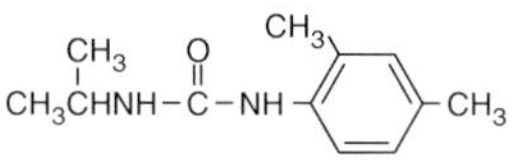

5 gm

S63,657-6
1-ISOPROPYL-3-(2,5-XYLYL)UREA

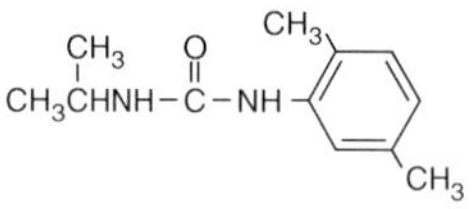

5 gm

S64,951-1
1-ISOPROPYL-3-(3,4-XYLYL)UREA

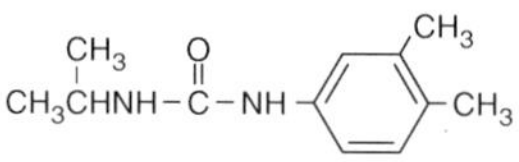

5 gm

S64,953-8
1-PROPYL-3-(3,4-XYLYL)UREA

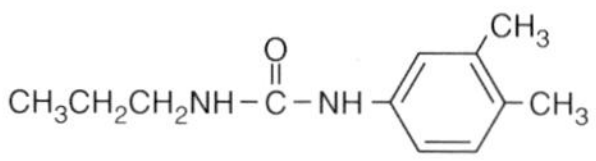

5 gm

S66,624-6
1-PROPYL-3-(2,5-XYLYL)UREA
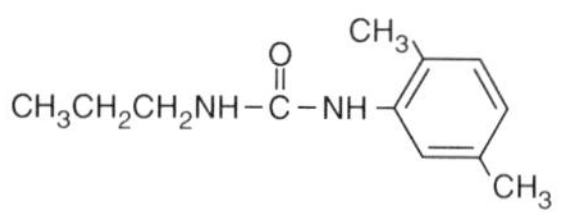
5 gm

S72,536-6
1-TERT-BUTYL-3-(3,4-DIMETHYLBENZYL)UREA
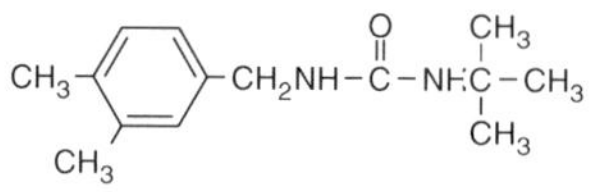
5 gm

S60,347-3
1-BUTYL-3-(4-CHLORO-2-METHYLPHENYL)UREA
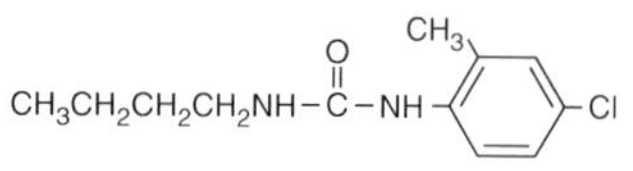
5 gm

S60,355-4
3-(3-CHLORO-4-METHYLPHENYL)-1,1-DIETHYLUREA
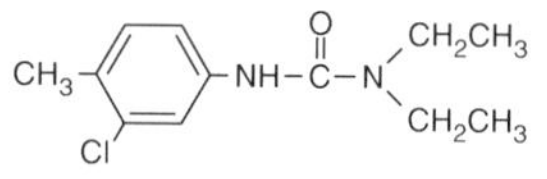
5 gm

S60,358-9
1-BUTYL-3-(3-CHLORO-4-METHYLPHENYL)UREA
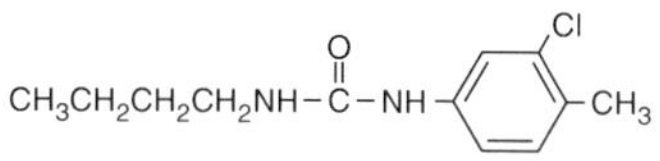
5 gm

S60,794-0
1,1-DIETHYL-3-(4-CHLORO-2-METHYLPHENYL)UREA
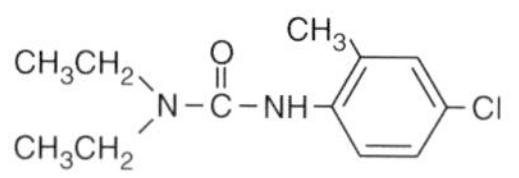
5 gm

S61,014-3
3-(5-CHLORO-2-METHYLPHENYL)-1,1-DIETHYLUREA
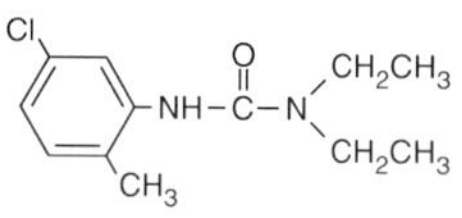
5 gm

S61,323-1
1-BUTYL-3-(5-CHLORO-2-METHYLPHENYL)UREA
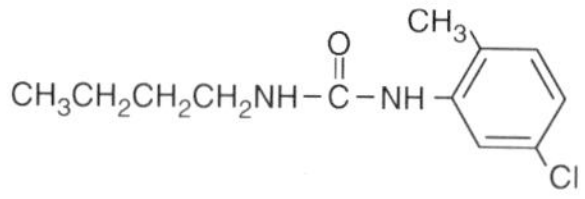
5 gm

S61,037-2
1-BUTYL-3-(3,4-DICHLOROPHENYL)-1-METHYLUREA
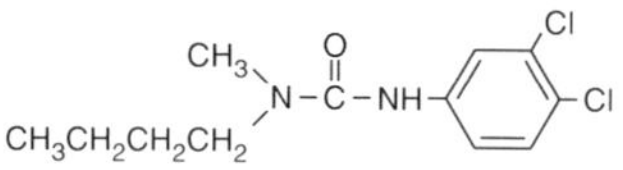
5 gm

S64,416-1
1-BUTYL-3-(2,4-DICHLOROPHENYL)-1-METHYLUREA
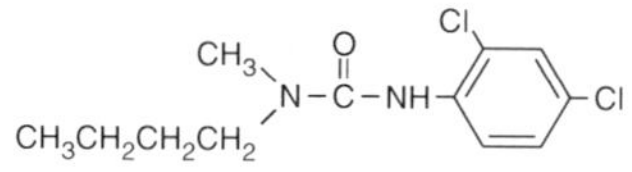
250 mg

S65,215-6
1-BUTYL-3-(2,5-DICHLOROPHENYL)-1-METHYLUREA
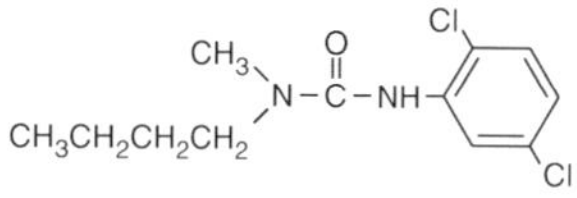
5 gm

S60,465-8
1,1-DIETHYL-3-(2,4-XYLYL)UREA
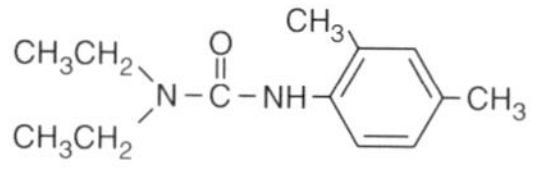
5 gm

S60,680-4
1-BUTYL-3-(2,4-XYLYL)UREA
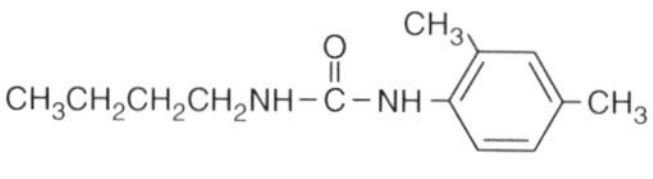
5 gm

S65,430-2
1-ISOBUTYL-3-(2,4-XYLYL)UREA
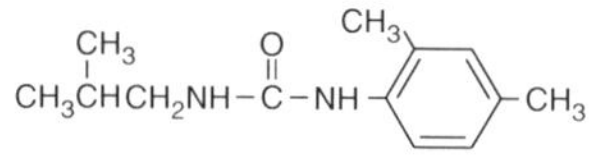
5 gm

S66,150-3
1,1-DIETHYL-3-(3,4-XYLYL)UREA
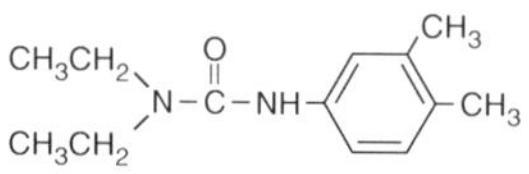
5 gm

S66,368-9
1-ISOBUTYL-3-(3,4-XYLYL)UREA
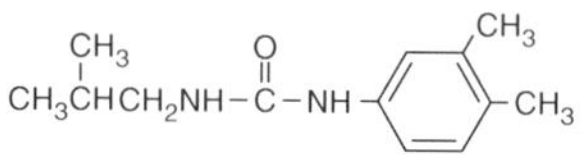
5 gm

S66,388-3
1-BUTYL-3-(3,4-XYLYL)UREA
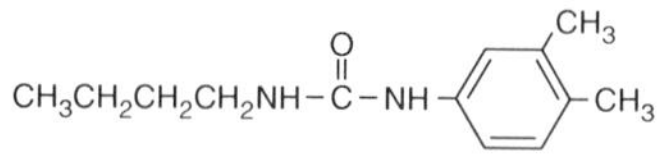
5 gm

S66,459-6
1-ISOBUTYL-3-(2,5-XYLYL)UREA
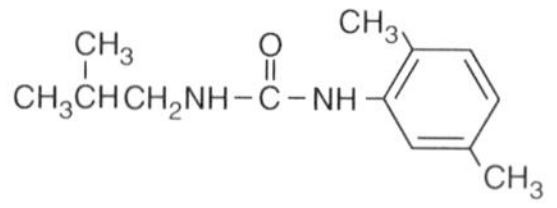
5 gm

S66,471-5
1-BUTYL-3-(2,5-XYLYL)UREA
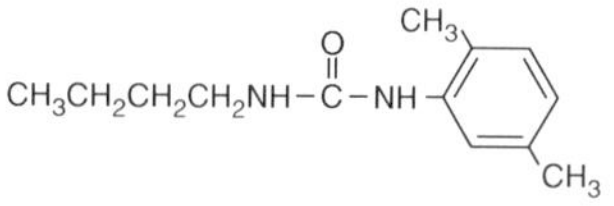
5 gm

S66,601-7
1,1-DIETHYL-3-(2,5-XYLYL)UREA
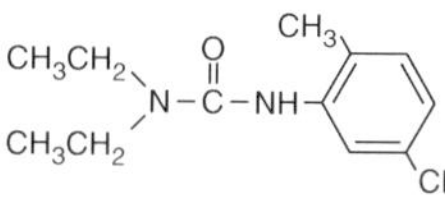
5 gm

S72,150-6
1-(TERT-BUTYL)-3-(2,5-XYLYL)UREA
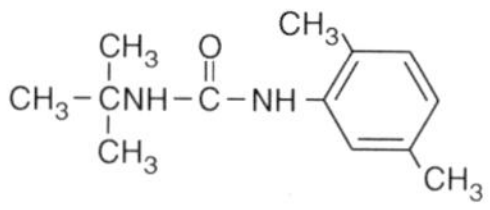
250 mg

S89,114-2
1-TERT-BUTYL-3-(2,4-XYLYL)UREA
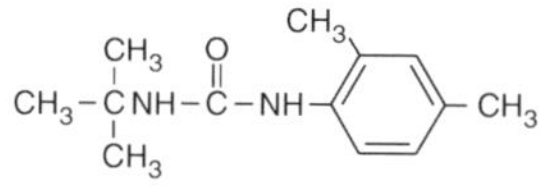
250 mg

S61,028-3
1-BUTYL-3-(3-CHLORO-4-METHYLPHENYL)-1-METHYLUREA
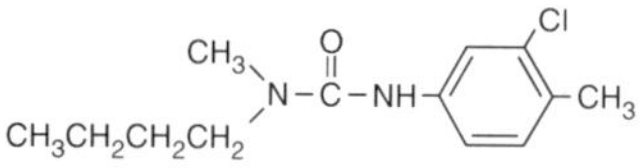
5 gm

S61,086-0
1-BUTYL-3-(5-CHLORO-2-METHYLPHENYL)-1-METHYLUREA
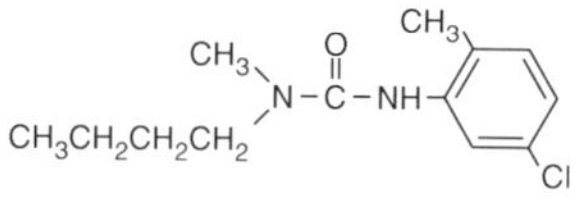
5 gm

S59,047-9
1-(2,4-DICHLOROPHENYL)-3-PHENYLUREA
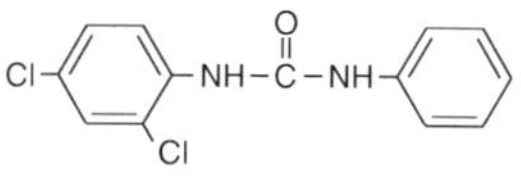
5 gm

S59,564-0
1-(3,4-DICHLOROPHENYL)-3-PHENYLUREA
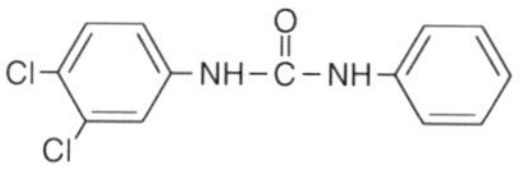
5 gm

S64,728-4
1-(2,5-DICHLOROPHENYL)-3-PHENYLUREA
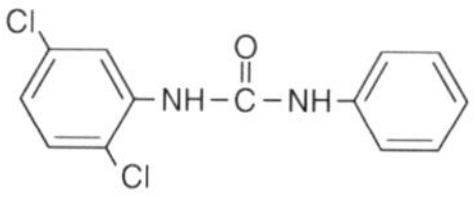
5 gm

S61,263-4
1,1-DIALLYL-3-(3,4-DICHLOROPHENYL)UREA
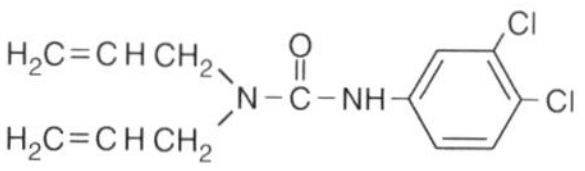
5 gm

S61,340-1
1,1-DIALLYL-3-(2,4-DICHLOROPHENYL)UREA
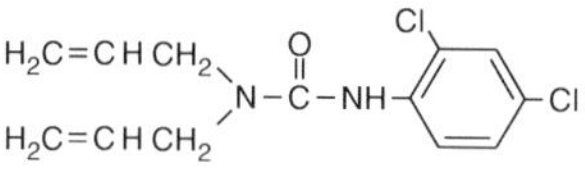
5 gm

S60,749-5
3-(3,4-DICHLOROPHENYL)-1,1-DIPROPYLUREA
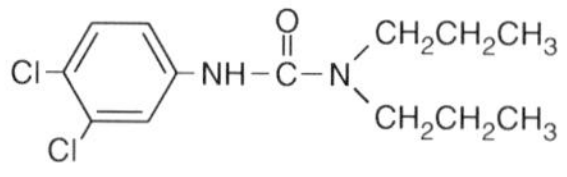
5 gm

S61,302-9
3-(2,4-DICHLOROPHENYL)-1,1-DIISOPROPYLUREA
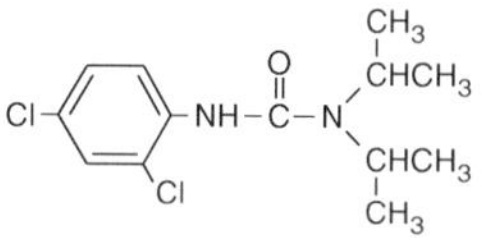
5 gm

S61,599-4
3-(2,5-DICHLOROPHENYL)-1,1-DIISOPROPYLUREA
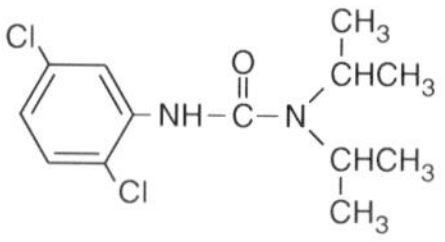
5 gm

S61,606-0
3-(2,5-DICHLOROPHENYL)-1,1-DIPROPYLUREA

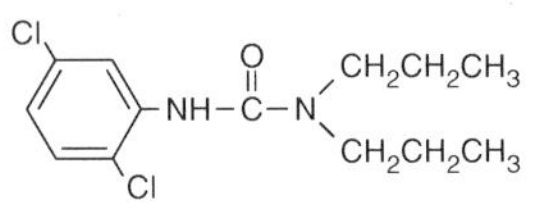

5 gm

S62,363-6
3-(2,4-DICHLOROPHENYL)-1,1-DIPROPYLUREA

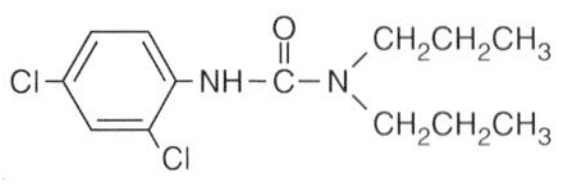

5 gm

S63,831-5
1-BUTYL-3-(3,4-DICHLOROPHENYL)-1-ETHYLUREA

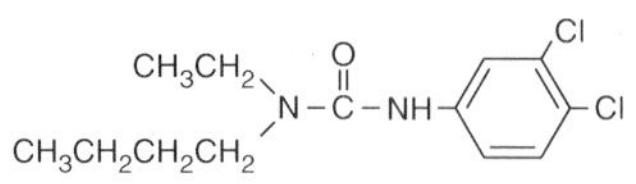

5 gm

S64,444-7
1-(3-CHLOROPHENYL)-3-(3,4-DICHLOROPHENYL)UREA

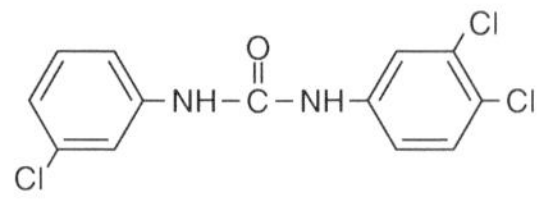

5 gm

S62,981-2
1,3-BIS(3,4-DICHLOROPHENYL)UREA

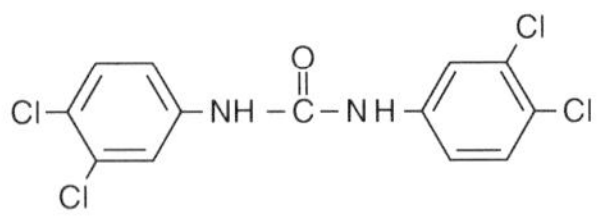

250 mg

S61,311-8
1-BUTYL-1-METHYL-3-(2,4-XYLYL)UREA

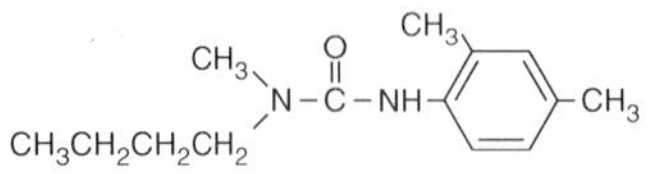

5 gm

S89,985-2
1-TERT-BUTYL-3-(2,4-DIMETHYLBENZYL)UREA

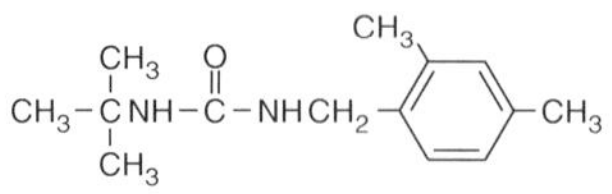

250 mg

S60,723-1
1-(4-CHLORO-2-METHYLPHENYL)-3-PHENYLUREA

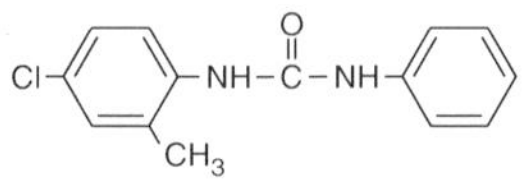

5 gm

S61,005-4
1-(5-CHLORO-2-METHYLPHENYL)-3-PHENYLUREA

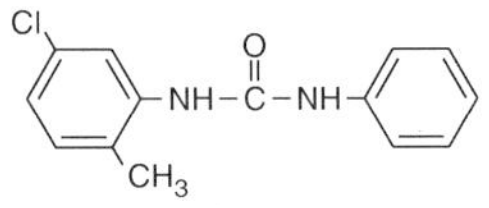

5 gm

S63,065-9
1-(3-CHLORO-4-METHYLPHENYL)-3-PHENYLUREA

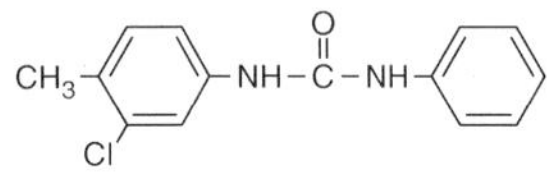

5 gm

S60,360-0
3-(4-CHLORO-2-METHYLPHENYL)-1,1-DIALLYLUREA

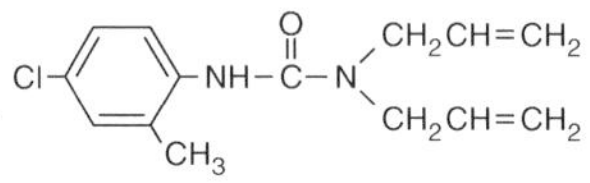

5 gm

S60,458-5
3-(3-CHLORO-4-METHYLPHENYL)-1,1-DIALLYLUREA

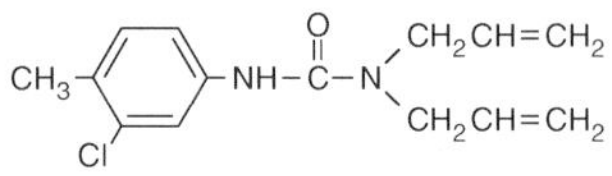

5 gm

S61,133-6
3-(5-CHLORO-2-METHYLPHENYL)-1,1-DIALLYLUREA

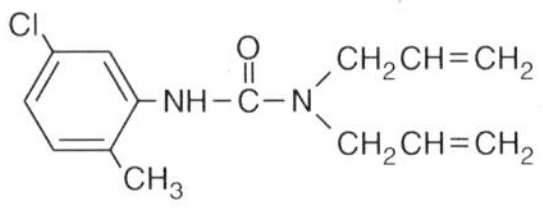

5 gm

S60,341-4
3-(3-CHLORO-4-METHYLPHENYL)-1,1-DIPROPYLUREA

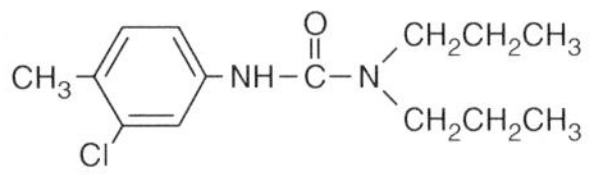

5 gm

S60,350-3
3-(3-CHLORO-4-METHYLPHENYL)-1,1-DIISOPROPYLUREA

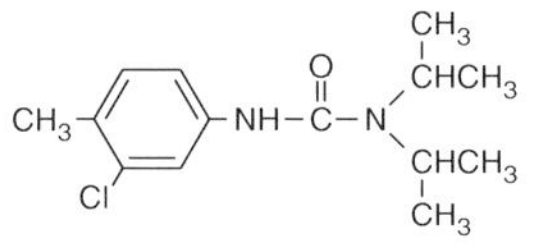

5 gm

S61,056-9
3-(5-CHLORO-2-METHYLPHENYL)-1,1-DIPROPYLUREA

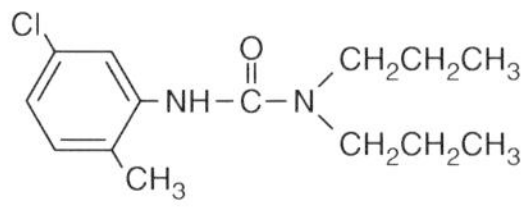

5 gm

S61,099-2
3-(5-CHLORO-2-METHYLPHENYL)-1,1-DIISOPROPYLUREA

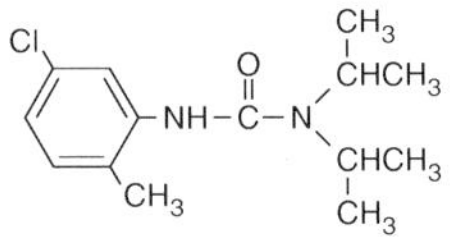

5 gm

S63,829-3
3-(4-CHLORO-2-METHYLPHENYL)-1,1-DIISOPROPYLUREA

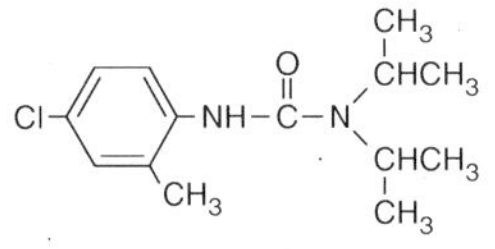

5 gm

S64,650-4
3-(4-CHLORO-2-METHYLPHENYL)-1,1-DIPROPYLUREA

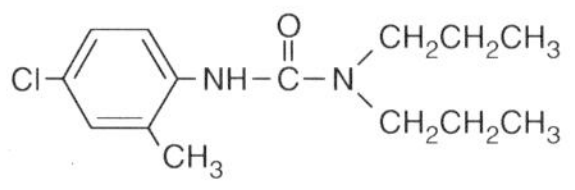

5 gm

S59,759-7
1-BENZYL-3-(2,5-DICHLOROPHENYL)UREA

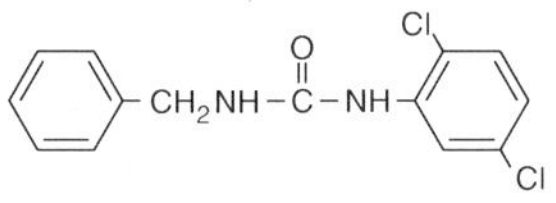

5 gm

S60,191-8
1-(3-CHLOROPHENYL)-3-(4-CHLORO-2-METHYLPHENYL)UREA

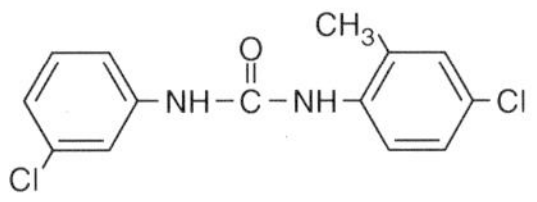

5 gm

S60,354-6
1-(3-CHLOROPHENYL)-3-(3-CHLORO-4-METHYLPHENYL)UREA

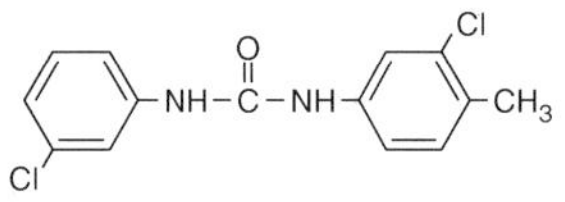

5 gm

S61,034-8
1-(3-CHLOROPHENYL)-3-(5-CHLORO-2-METHYLPHENYL)UREA

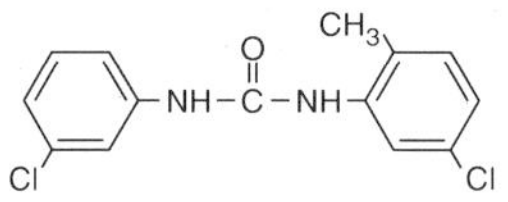

5 gm

S63,846-3
3-(3,4-DICHLOROPHENYL)-1-METHYL-1-PHENYLUREA

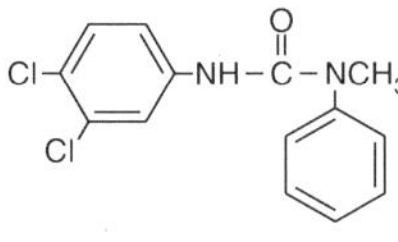

5 gm

S64,455-2
3-(2,4-DICHLOROPHENYL)-1-METHYL-1-PHENYLUREA

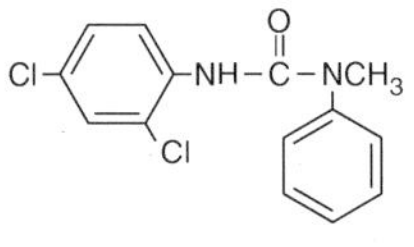

5 gm

S64,456-0
3-(2,5-DICHLOROPHENYL)-1-METHYL-1-PHENYLUREA

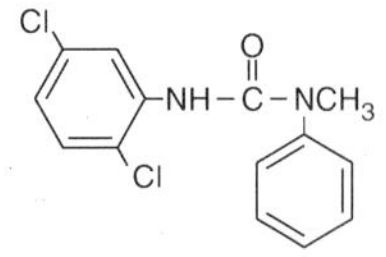

5 gm

S60,647-2
1-PHENYL-3-(2,4-XYLYL)UREA

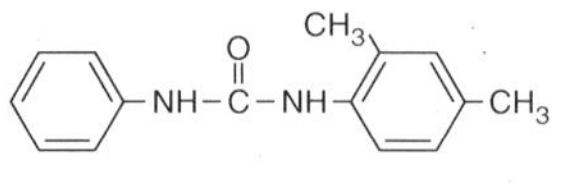

5 gm

S60,650-2
1,1-DIALLYL-3-(2,4-XYLYL)UREA

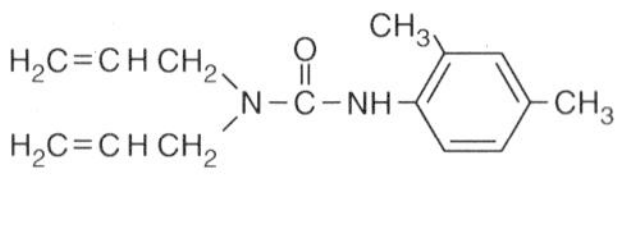

5 gm

S64,386-6
1,1-DIALLYL-3-(2,5-XYLYL)UREA

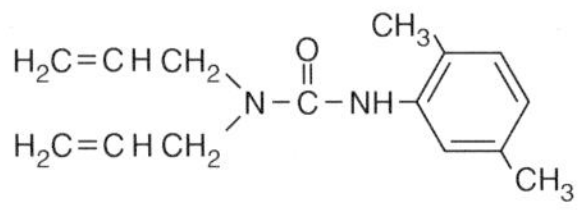

5 gm

S66,425-1
1,1-DIALLYL-3-(3,4-XYLYL)UREA

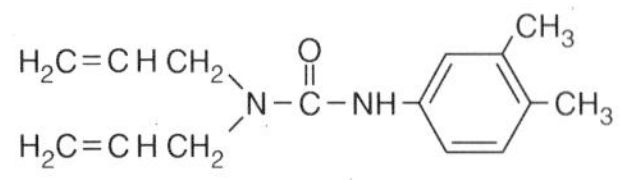

5 gm

S60,622-7
1,1-DIPROPYL-3-(2,4-XYLYL)UREA

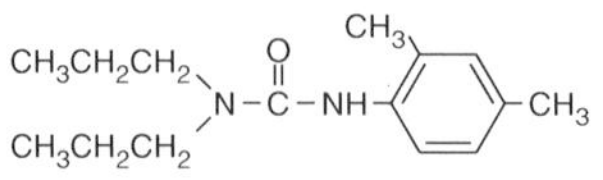

5 gm

S60,667-7
1,1-DIISOPROPYL-3-(2,4-XYLYL)UREA

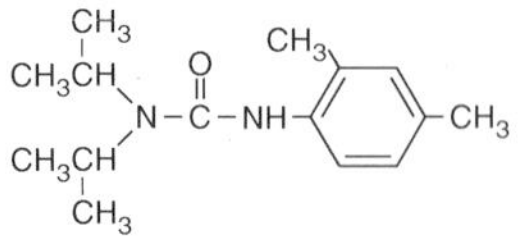

5 gm

S64,399-8
1,1-DIISOPROPYL-3-(2,5-XYLYL)UREA
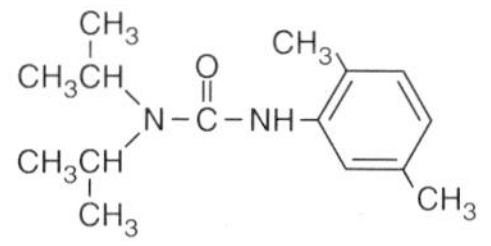
5 gm

S66,384-0
1,1-DIISOPROPYL-3-(3,4-XYLYL)UREA
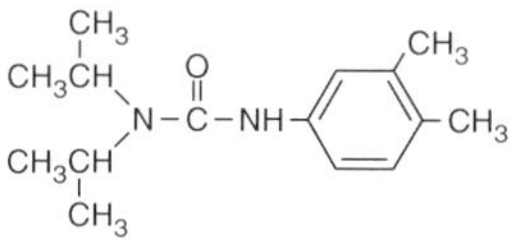
5 gm

S66,396-4
1,1-DIPROPYL-3-(2,5-XYLYL)UREA
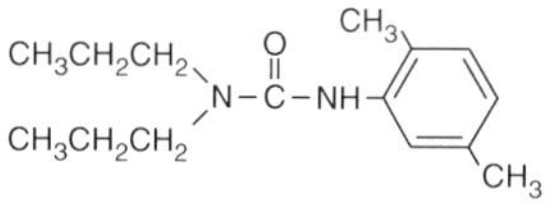
5 gm

S66,692-0
1,1-DIPROPYL-3-(3,4-XYLYL)UREA
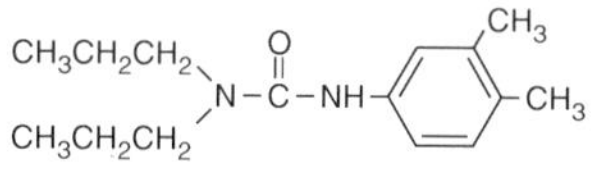
5 gm

S60,634-0
1-(3-CHLOROPHENYL)-3-(2,4-XYLYL)UREA
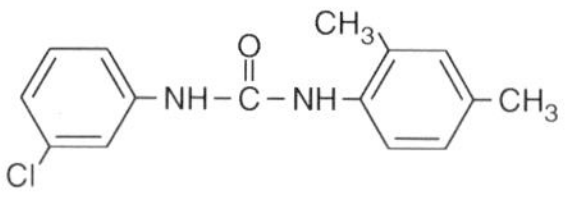
5 gm

S60,194-2
1,3-BIS(4-CHLORO-2-METHYLPHENYL)UREA
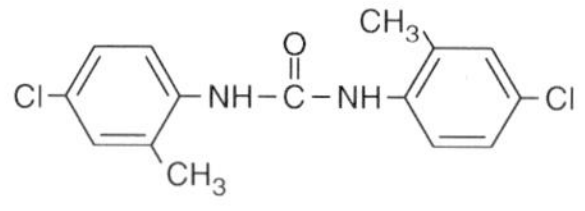
5 gm

S61,018-6
1-(3-CHLORO-4-METHYLPHENYL)-3-(5-CHLORO-2-METHYLPHENYL)UREA
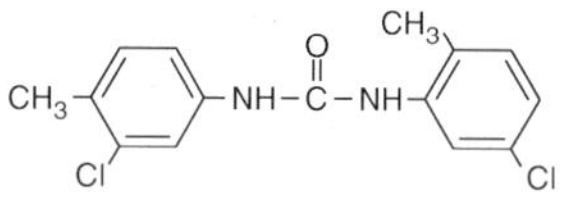
5 gm

S61,058-5
1,3-BIS(5-CHLORO-2-METHYLPHENYL)UREA
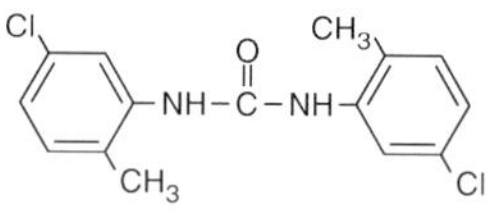
5 gm

S61,074-7
1-(2,4-DICHLOROPHENYL)-3-(2,4-XYLYL)UREA
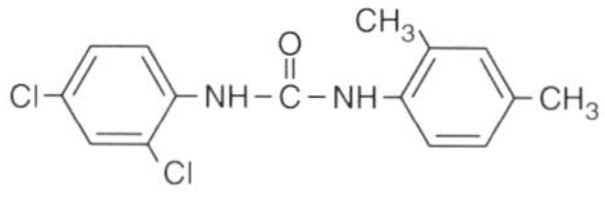
5 gm

S62,812-3
3-(3,4-DICHLOROPHENYL)-1-ETHYL-1-PHENYLUREA
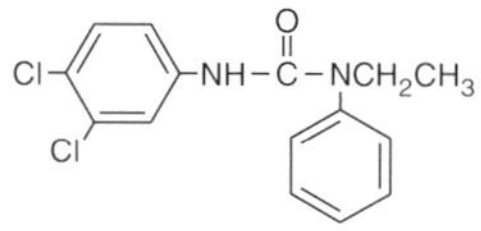
5 gm

S60,781-9
1,1-DIBUTYL-3-(3,4-DICHLOROPHENYL)UREA
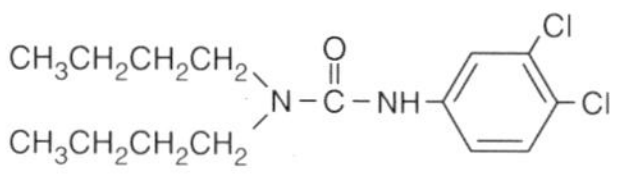
5 gm

S61,170-0
1,1-DIBUTYL-3-(2,5-DICHLOROPHENYL)UREA
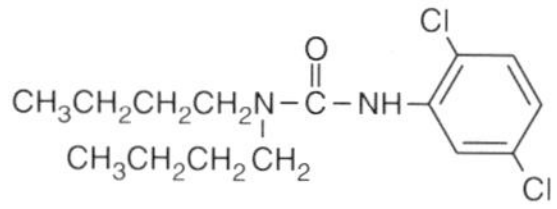
5 gm

S64,478-1
1,1-DIBUTYL-3-(2,4-DICHLOROPHENYL)UREA
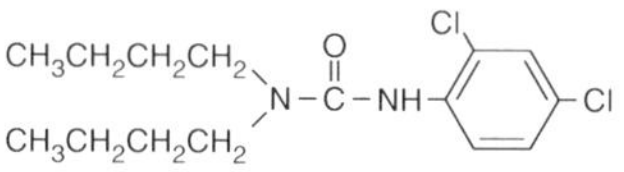
5 gm

S65,353-5
3-(2,4-DICHLOROPHENYL)-1,1-DIISOBUTYLUREA
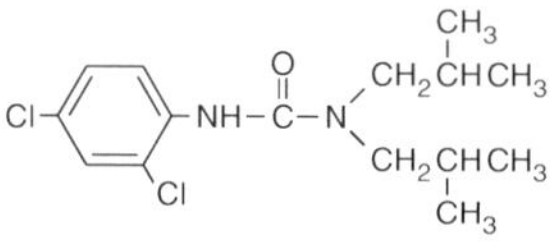
5 gm

S60,669-3
1-(M-TOLYL)-3-(2,4-XYLYL)UREA
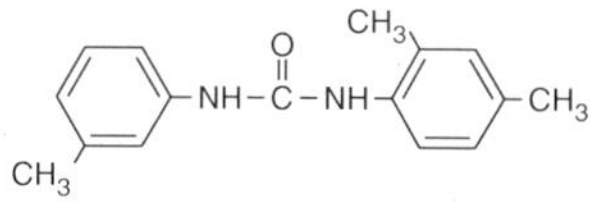
5 gm

S60,975-7
1-(5-CHLORO-2-METHYLPHENYL)-3-(2,4-XYLYL)UREA
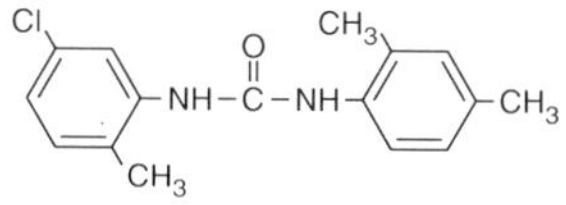
5 gm

S61,071-2
1-(3-CHLORO-4-METHYLPHENYL)-3-(2,4-XYLYL)UREA
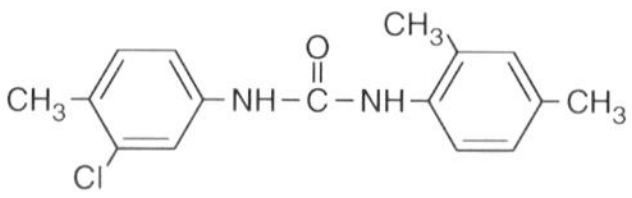
5 gm

S61,090-9
1-(4-CHLORO-2-METHYLPHENYL)-3-(2,4-XYLYL)UREA
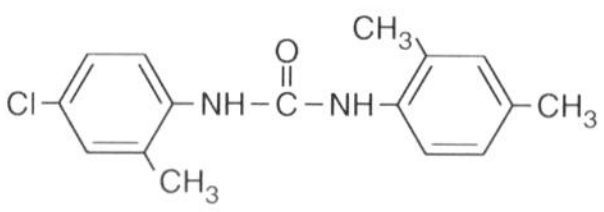
5 gm

S60,338-4
3-(4-CHLORO-2-METHYLPHENYL)-1,1-DIBUTYLUREA
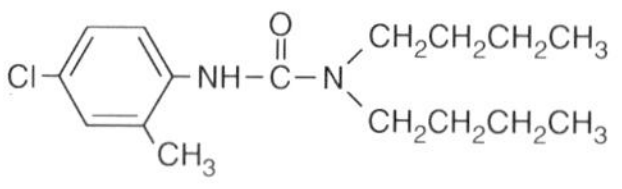
5 gm

S60,960-9
3-(5-CHLORO-2-METHYLPHENYL)-1,1-DIBUTYLUREA
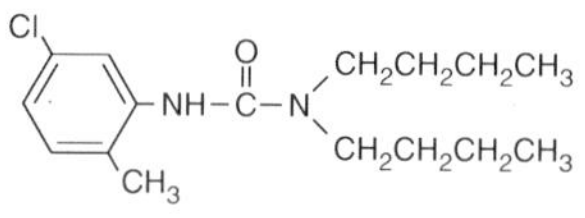
5 gm

S64,055-7
3-(3-CHLORO-4-METHYLPHENYL)-1,1-DIBUTYLUREA
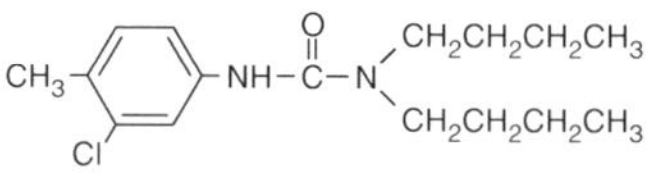
5 gm

S60,654-5
1,3-DI(2,4-XYLYL)UREA
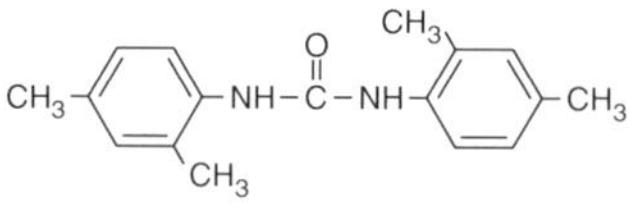
5 gm

S60,676-6
1,1-DIBUTYL-3-(2,4-XYLYL)UREA
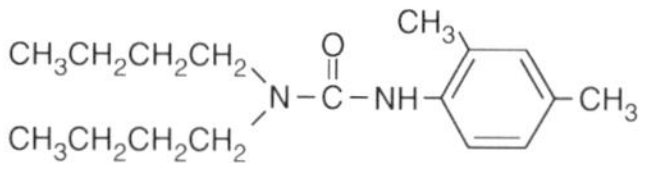
5 gm

S65,461-2
1,1-DIISOBUTYL-3-(2,4-XYLYL)UREA
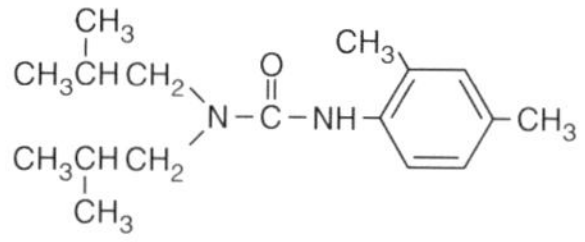
5 gm

S66,545-2
1,1-DIISOBUTYL-3-(3,4-XYLYL)UREA
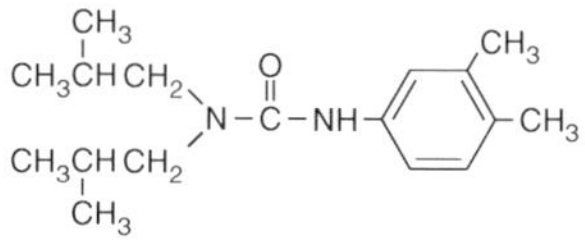
5 gm

S66,590-8
1,1-DIISOBUTYL-3-(2,5-XYLYL)UREA
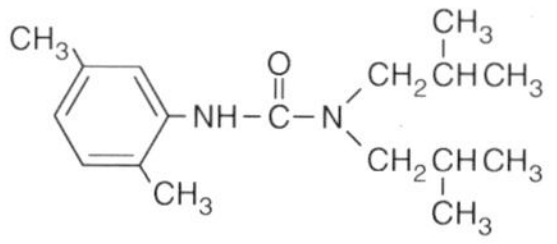
5 gm

S66,629-7
1,1-DIBUTYL-3-(2,5-XYLYL)UREA
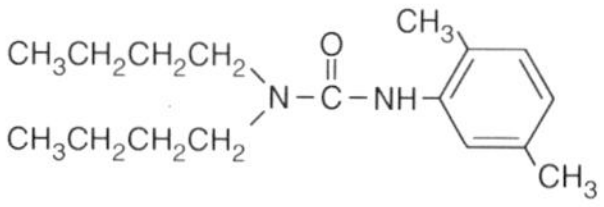
5 gm

S71,047-4
1,1-DIBUTYL-3-(3,4-DIMETHYLPHENYL)UREA
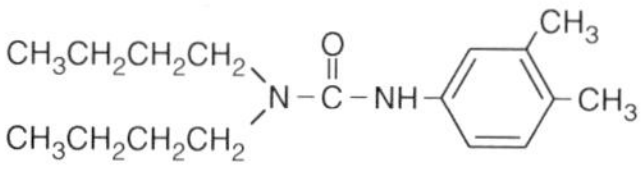
5 gm

S65,024-2
1-BUTYL-3-(2,5-DICHLOROPHENYL)-1-PHENYLUREA
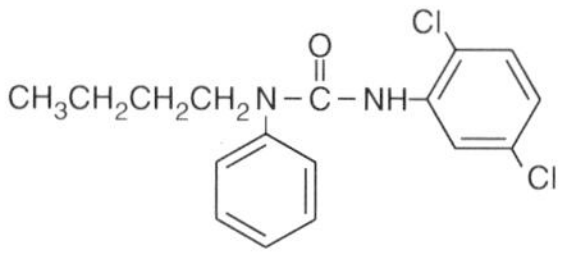
5 gm

S84,734-8
1-NONYL-3-(2,5-XYLYL)UREA
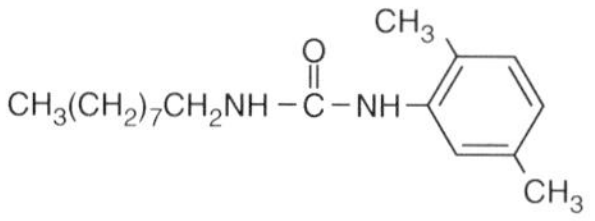
250 mg

S61,119-0
1-BUTYL-3-(5-CHLORO-2-METHYLPHENYL)-1-PHENYLUREA
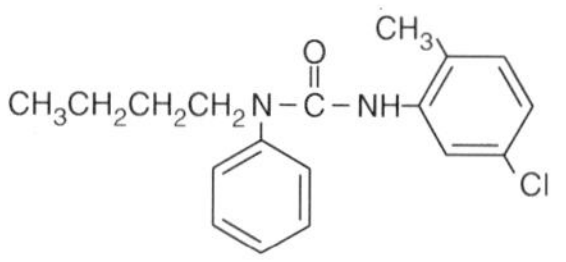
5 gm

S60,679-0
1-BUTYL-1-PHENYL-3-(2,4-XYLYL)UREA
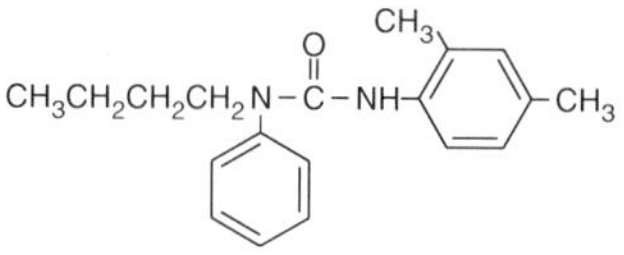
5 gm

S65,641-0
N,N'-BIS[N-(3,4-DICHLOROPHENYL)-CARBAMOYL]HEXAMETHYLENEDIAMINE

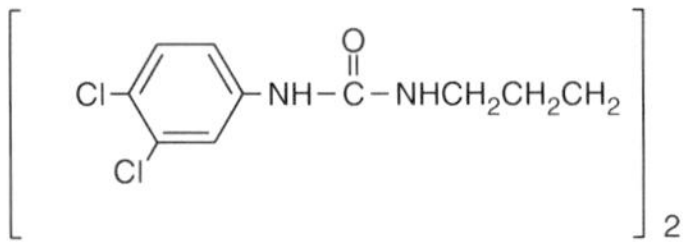

250 mg

S61,001-1
3-(5-CHLORO-2-METHYLPHENYL)-1,1-DIBENZYLUREA

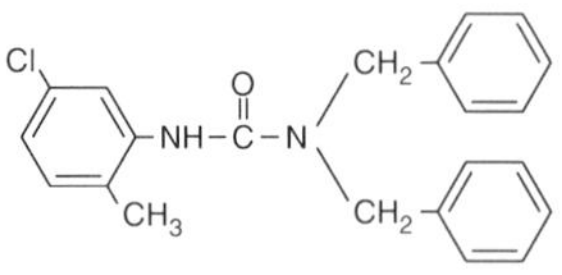

5 gm

S67,042-1
1-OCTADECYL-3-(2,4-XYLYL)UREA

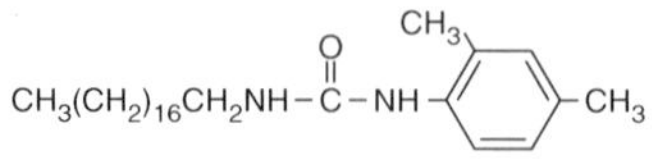

5 gm

S79,170-9
1-OCTADECYL-3-(3,4-XYLYL)UREA

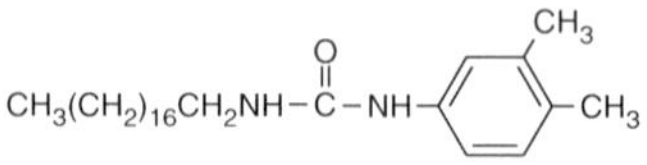

250 mg

S89,005-7
1-OCTADECYL-3-(2,5-XYLYL)UREA

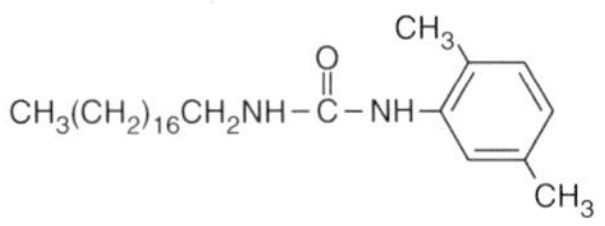

250 mg

S89,292-0
1-(5-ISOPROPYL-2-METHYLPHENYL)-3-OCTADECYLUREA

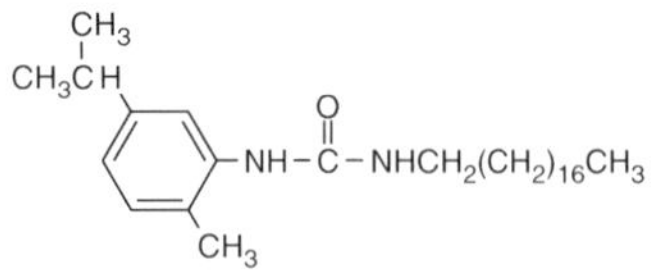

250 mg

S71,798-3
4,4'-BIS(3-PROPYLUREIDO)-3,3'-DIMETHYLBIPHENYL

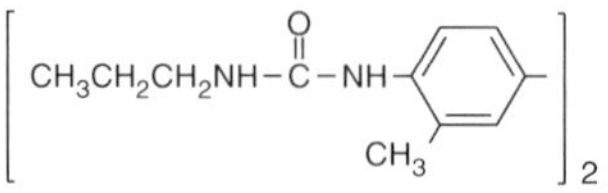

250 mg

S74,337-2
1,1'-(3,3'-DIMETHYL-1,1'-BIPHENYL-4,4'-DIYL)BIS[3-(2,5-XYLYL)UREA]

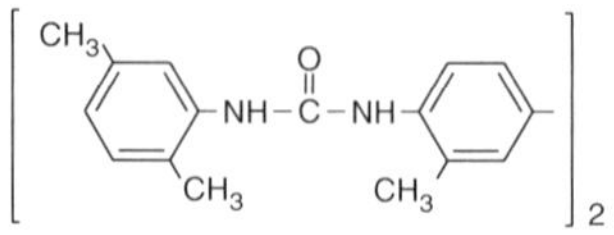

5 gm

S59,032-0
1-(3,4-DICHLOROPHENYL)-3-(1-NAPHTHYL)UREA

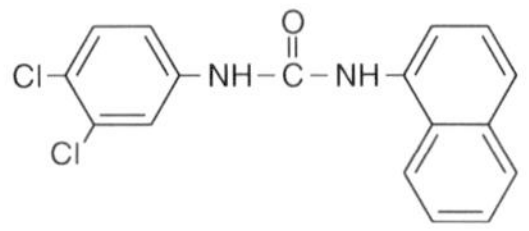

5 gm

S59,044-4
1-(2,4-DICHLOROPHENYL)-3-(1-NAPHTHYL)UREA

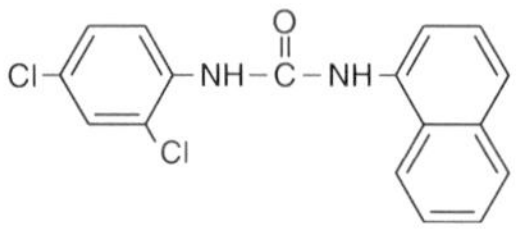

5 gm

S77,326-3
1-(1-NAPHTHYL)-3-(2,5-XYLYL)UREA

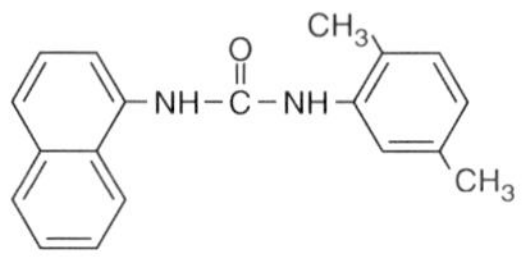

250 mg

S78,005-7
1-(1-NAPHTHYL)-3-(2,4-XYLYL)UREA

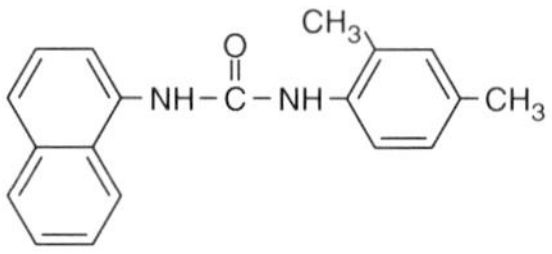

250 mg

S75,284-3
1-(5-ISOPROPYL-2-METHYLPHENYL)-3-(1-NAPHTHYL)UREA

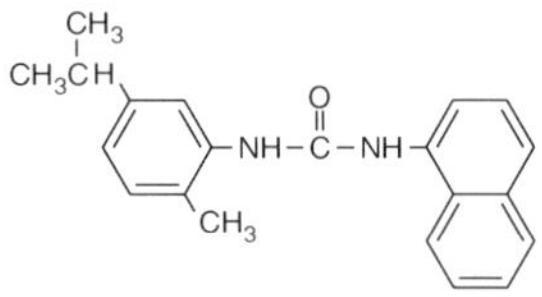

5 gm

S65,768-9
3-(3,5-DICHLOROPHENYL)-1,1-DIMETHYLUREA

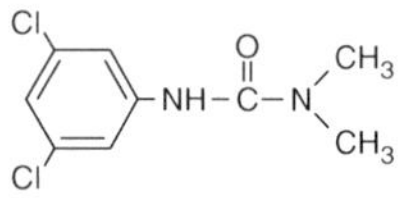

5 gm

S66,108-2
1-METHYL-3-(3,5-XYLYL)UREA

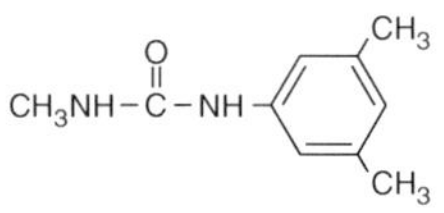

5 gm

S66,355-7
1,1-DIMETHYL-3-(3,5-XYLYL)UREA

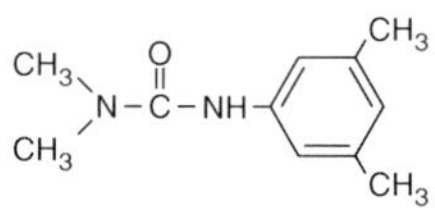

5 gm

S66,375-1
1-ETHYL-3-(3,5-XYLYL)UREA

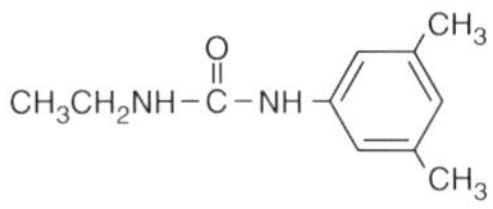

5 gm

S66,348-4
1-ALLYL-3-(3,5-XYLYL)UREA

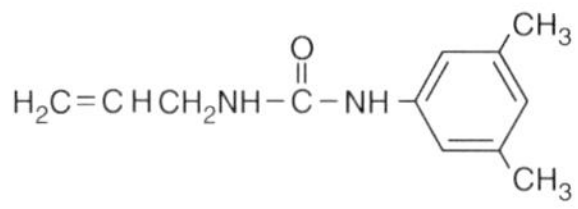

5 gm

S66,111-2
1-PROPYL-3-(3,5-XYLYL)UREA

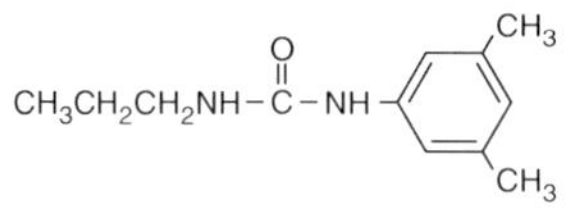

5 gm

S66,352-2
1-ISOPROPYL-3-(3,5-XYLYL)UREA

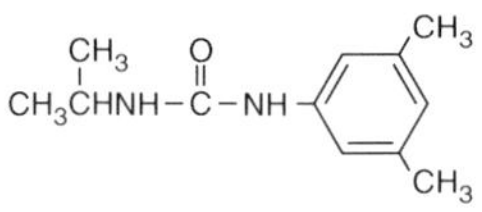

5 gm

S66,360-3
1,1-DIETHYL-3-(3,5-XYLYL)UREA

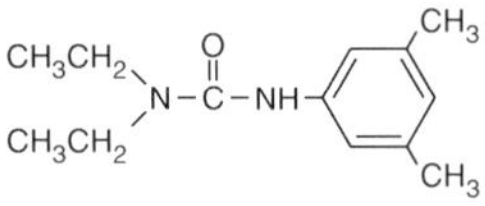

5 gm

S66,373-5
1-ISOBUTYL-3-(3,5-XYLYL)UREA

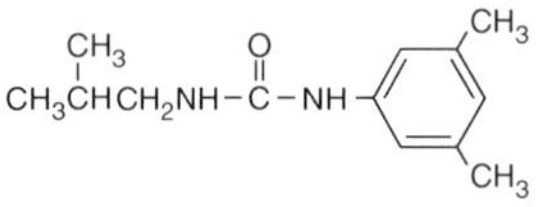

5 gm

S66,381-6
1-BUTYL-3-(3,5-XYLYL)UREA

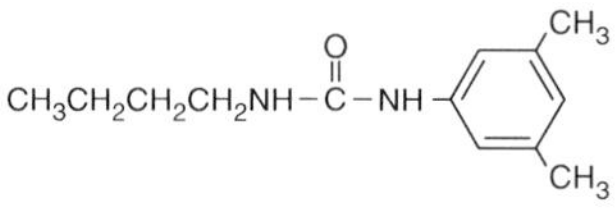

5 gm

S66,370-0
1,1-DIALLYL-3-(3,5-XYLYL)UREA

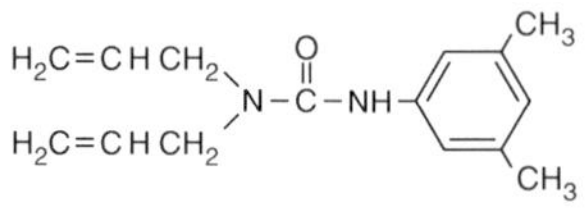

5 gm

S66,169-4
1,1-DIISOPROPYL-3-(3,5-XYLYL)UREA

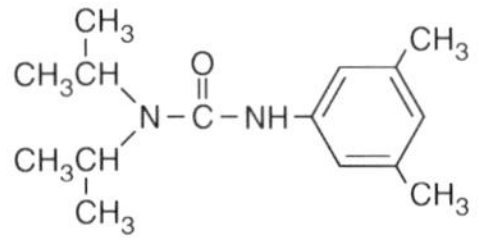

5 gm

S66,140-6
1,1-DIBUTYL-3-(3,5-XYLYL)UREA

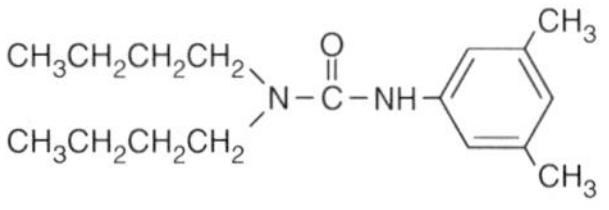

5 gm

S50,891-8
1-(2,4,6-TRICHLOROPHENYL)UREA

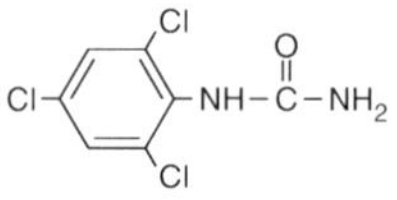

5 gm

S88,996-2
1-ISOPROPYL-3-MESITYLUREA

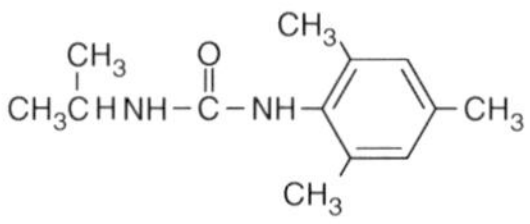

250 mg

S79,118-0
1-TERT-BUTYL-3-MESITYLUREA

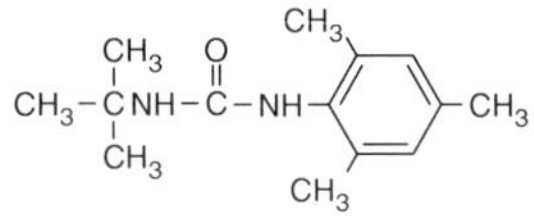

250 mg

S79,157-1
1-MESITYL-3-(M-TOLYL)UREA

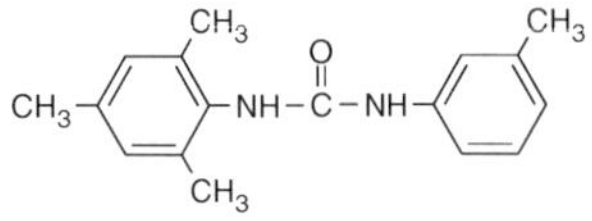

250 mg

S79,172-5
1-MESITYL-3-OCTADECYLUREA

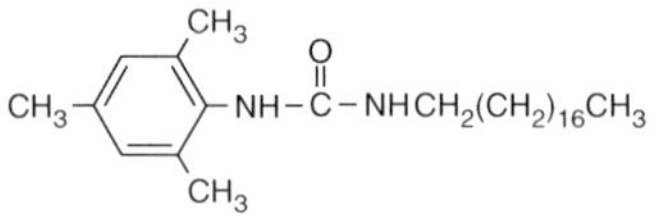

250 mg

S76,966-5
1-(1-NAPHTHYL)-3-(2,4,6-TRIMETHYLPHENYL)UREA

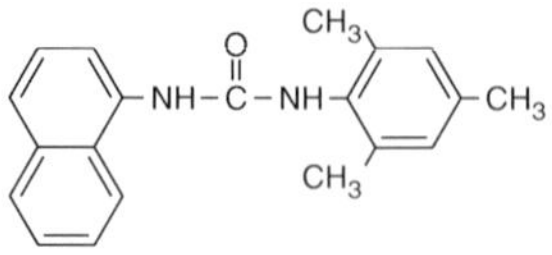

250 mg

S60,751-7
1,1-DIMETHYL-3-(2,4,5-TRICHLOROPHENYL)UREA

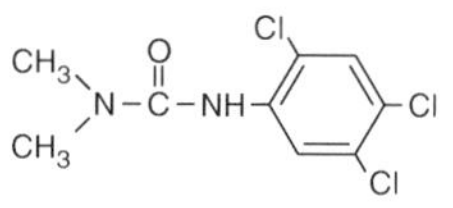

5 gm

S64,439-0
1-ALLYL-3-(2,4,5-TRICHLOROPHENYL)UREA

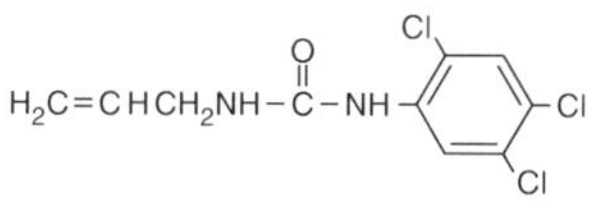

5 gm

S62,160-9
1-PROPYL-3-(2,4,5-TRICHLOROPHENYL)UREA

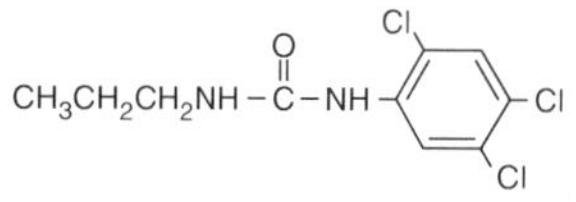

5 gm

S63,394-1
1-ISOPROPYL-3-(2,4,5-TRICHLOROPHENYL)UREA

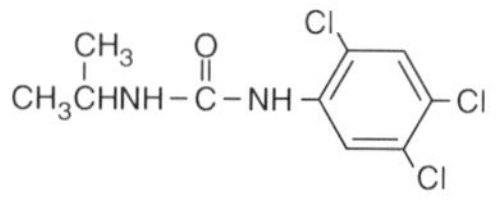

5 gm

S60,727-4
1-BUTYL-3-(2,4,5-TRICHLOROPHENYL)UREA

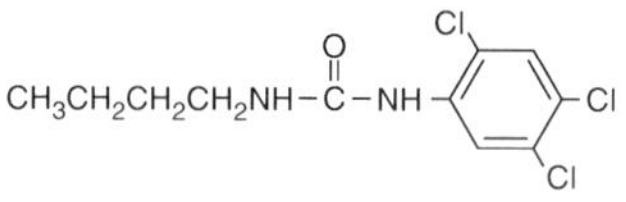

5 gm

S60,777-0
1,1-DIETHYL-3-(2,4,5-TRICHLOROPHENYL)UREA

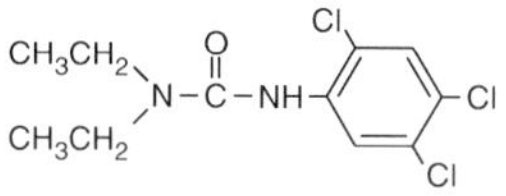

5 gm

S64,432-3
1,1-DIALLYL-3-(2,4,5-TRICHLOROPHENYL)UREA

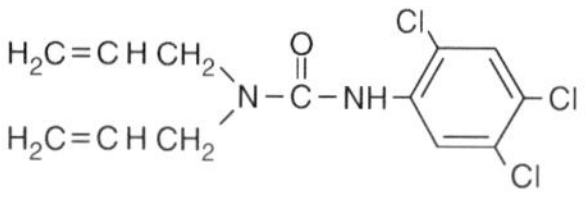

5 gm

S64,453-6
1,1-DIPROPYL-3-(2,4,5-TRICHLOROPHENYL)UREA

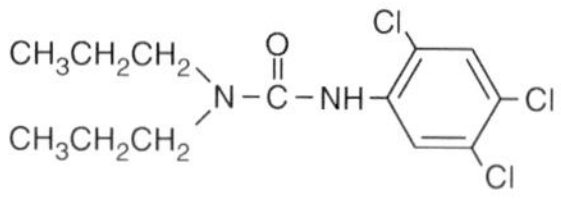

5 gm

S64,476-5
1,1-DIISOPROPYL-3-(2,4,5-TRICHLOROPHENYL)UREA

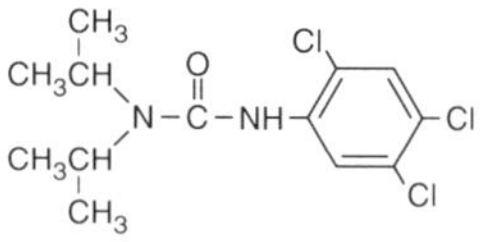

5 gm

S71,181-0
1,3-BIS(2,4,5-TRICHLOROPHENYL)UREA

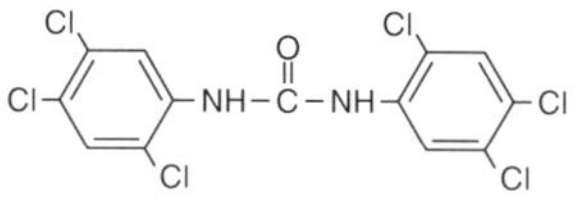

5 gm

S89,111-8
1-TERT-BUTYL-3-(2,4,5-TRIMETHYLPHENYL)UREA

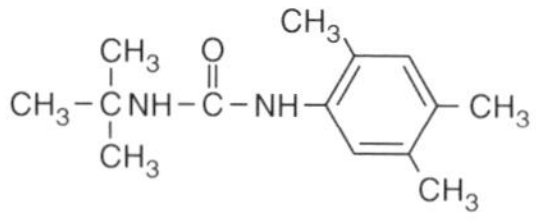

250 mg

S64,418-8
1,1-DIBUTYL-3-(2,4,5-TRICHLOROPHENYL)UREA

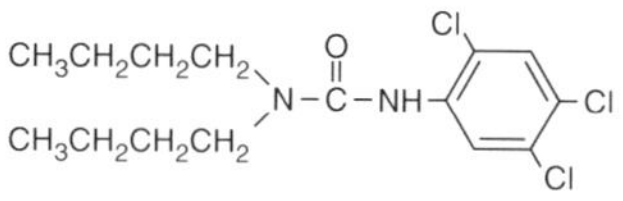

5 gm

S35,378-7
1-(1-NAPHTHYL)UREA

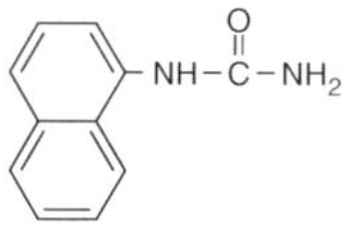

5 gm

S58,997-7
1-METHYL-3-(1-NAPHTHYL)UREA

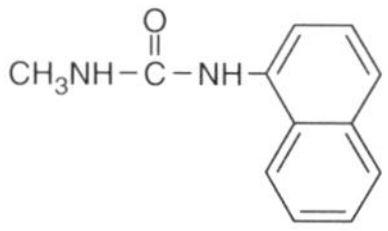

5 gm

S58,611-0
1,1-DIMETHYL-3-(1-NAPHTHYL)UREA

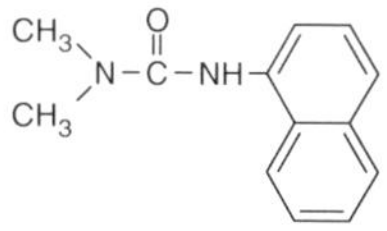

5 gm

S59,023-1
1-ETHYL-3-(1-NAPHTHYL)UREA

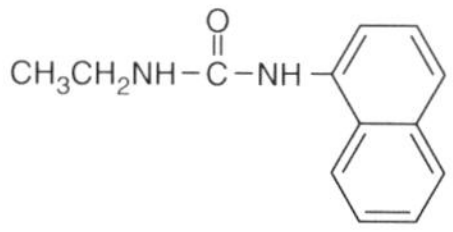

5 gm

S59,059-2
1-ALLYL-3-(1-NAPHTHYL)UREA

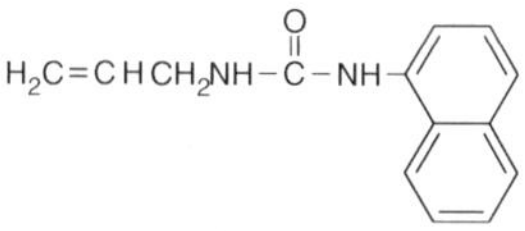

5 gm

S59,013-4
1-(1-NAPHTHYL)-3-PROPYLUREA

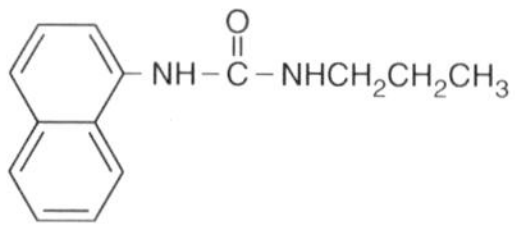

5 gm

S62,519-1
1-ISOPROPYL-3-(1-NAPHTHYL)UREA

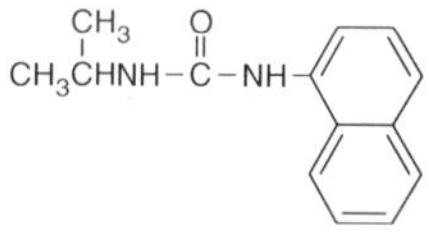

5 gm

S74,703-3
1-ETHYL-3-(1-NAPHTHYLMETHYL)UREA

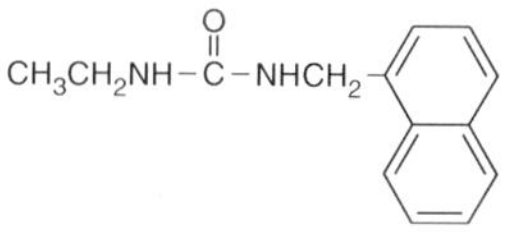

5 gm

S58,638-2
1,1-DIETHYL-3-(1-NAPHTHYL)UREA

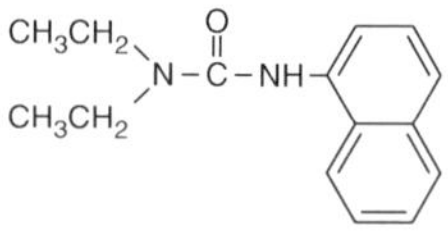

5 gm

S59,006-1
1-BUTYL-3-(1-NAPHTHYL)UREA

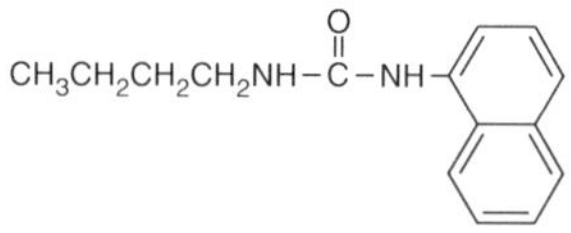

5 gm

S62,830-1
1-BUTYL-1-METHYL-3-(1-NAPHTHYL)UREA

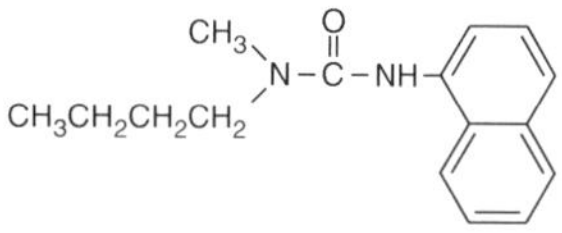

5 gm

S74,342-9
1-TERT-BUTYL-3-(1-NAPHTHYLMETHYL)UREA

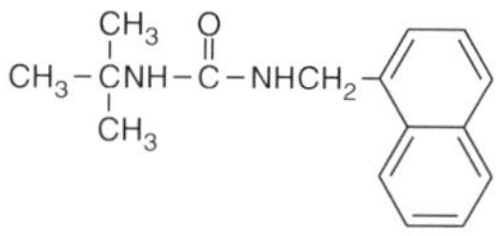

5 gm

S59,033-9
1,1-DIALLYL-3-(1-NAPHTHYL)UREA

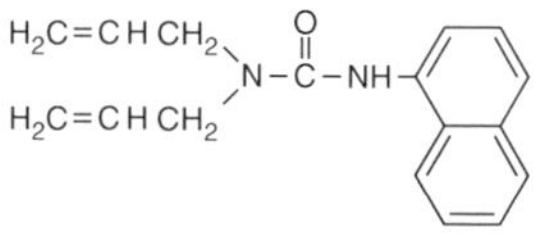

5 gm

S59,027-4
1,1-DIPROPYL-3-(1-NAPHTHYL)UREA

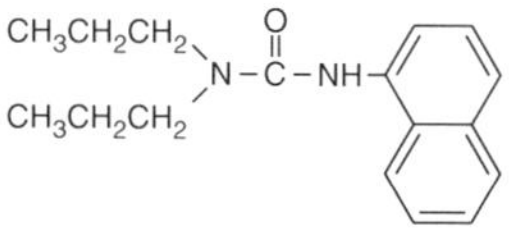

5 gm

S62,829-8
1,1-DIISOPROPYL-3-(1-NAPHTHYL)UREA

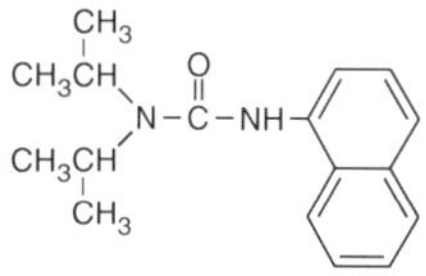

5 gm

S59,052-5
1-(3-CHLOROPHENYL)-3-(1-NAPHTHYL)UREA

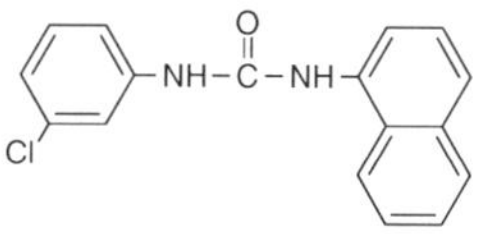

5 gm

S59,051-7
1-(1-NAPHTHYL)-3-(M-TOLYL)UREA

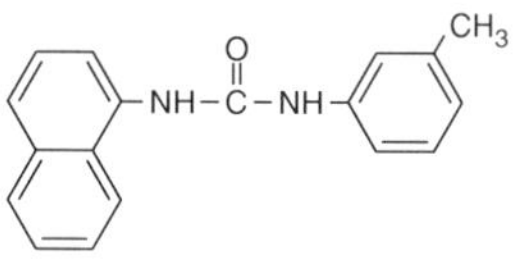

5 gm

S71,303-1
1,1-DIBUTYL-3-(1-NAPHTHYL)UREA

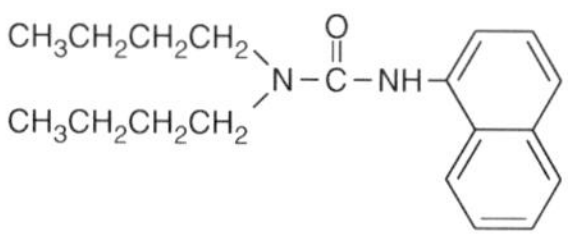

5 gm

S79,539-9
1-(1-NAPHTHYL)-3-OCTADECYLUREA

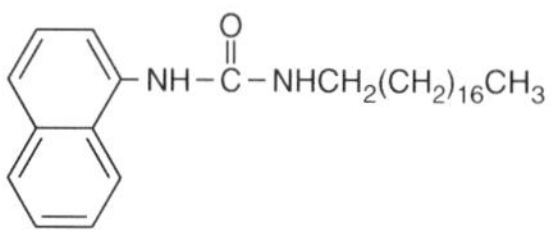

250 mg

S67,587-3
1,1'-HEXAMETHYLENEBIS[3-(1-NAPHTHYLMETHYL)]UREA

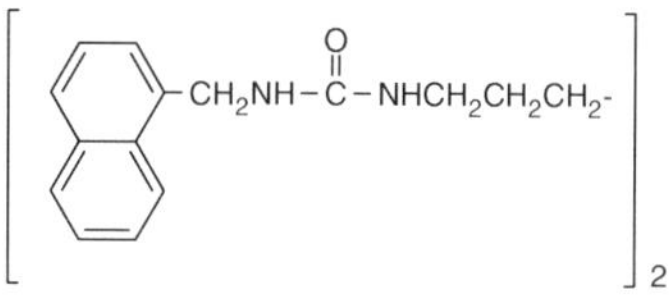

5 gm

S57,975-0
1,1'-(1,5-NAPHTHALENEDIYL)DIUREA
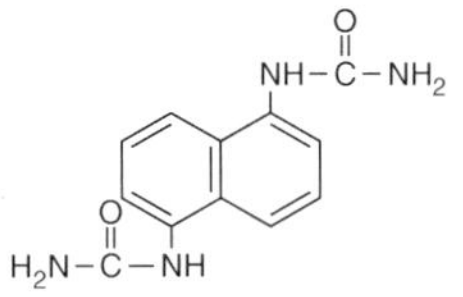
250 mg

S63,676-2
1,1'-(1,5-NAPHTHALENEDIYL)BIS(3-METHYLUREA)
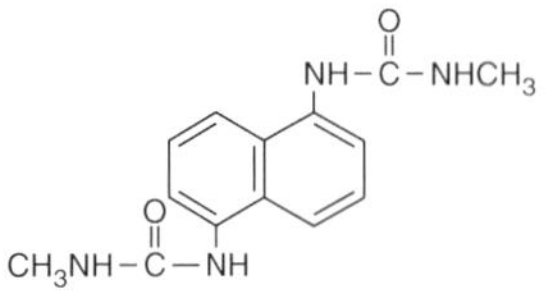
5 gm

S63,439-5
1,1'-(1,5-NAPHTHALENEDIYL)BIS(3-PROPYLUREA)
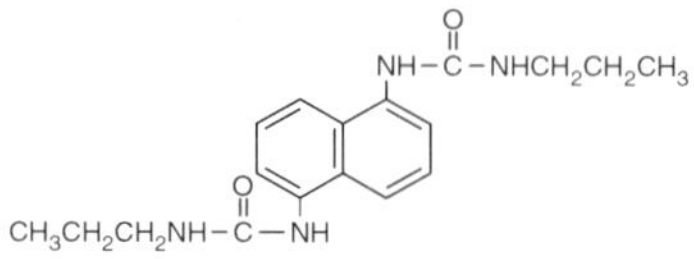
5 gm

S63,451-4
1,1'-(1,5-NAPHTHALENEDIYL)BIS(3-ISOPROPYLUREA)
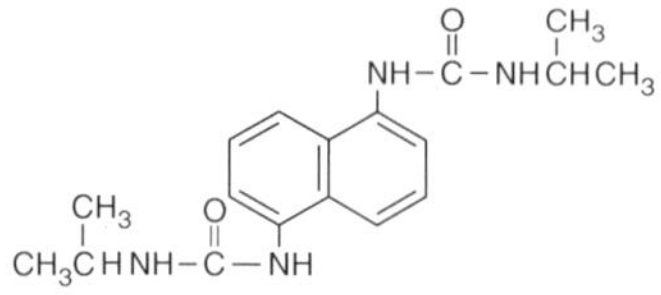
5 gm

S63,464-6
1,1'-(1,5-NAPHTHALENEDIYL)BIS(3-BUTYLUREA)
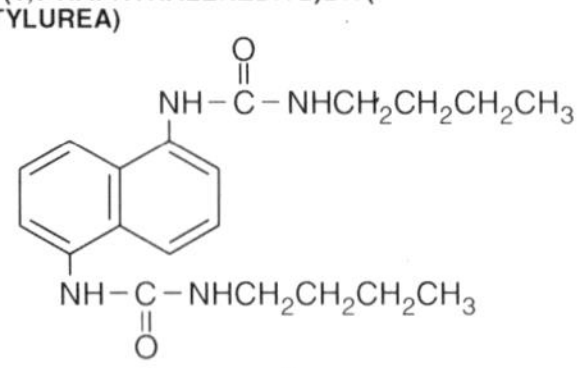
5 gm

S71,865-3
1-METHYL-3-(2-NAPHTHYL)UREA
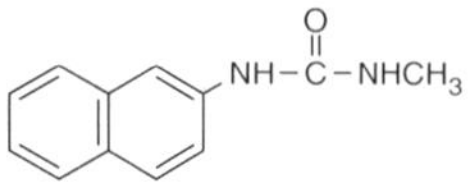
5 gm

S85,471-9
1-TERT-BUTYL-3-(2-NAPHTHYL)UREA
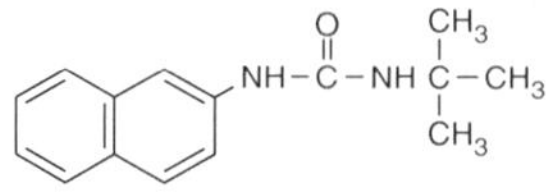
250 mg

S66,690-4
2-CHLOROETHYL N-(2-CHLORO-5-TRIFLUOROMETHYLPHENYL)CARBAMATE
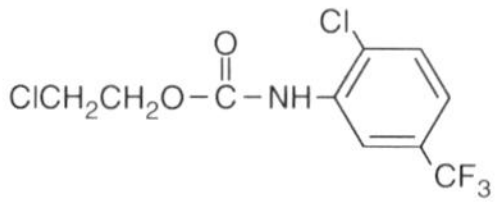
5 gm

S71,168-3
2-BROMOETHYL N-(3-TRIFLUOROMETHYLPHENYL)-CARBAMATE
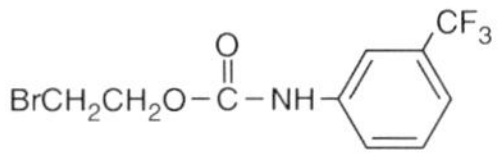
250 mg

S66,736-6
2-CHLOROETHYL N-(2-TRIFLUOROMETHYLPHENYL)-CARBAMATE
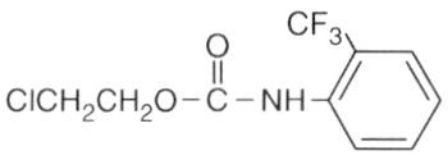
5 gm

S71,166-7
2-CHLOROETHYL N-(2-FLUOROPHENYL)-CARBAMATE
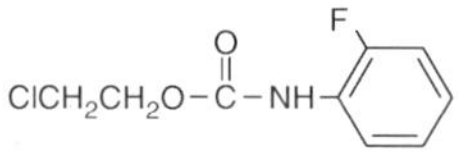
5 gm

S71,429-1
2-CHLOROETHYL N-(4-FLUOROPHENYL)-CARBAMATE
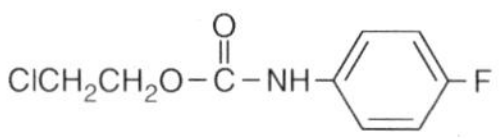
5 gm

S66,658-0
2-CHLOROETHYL N-(2-CHLORO-6-METHYLPHENYL)CARBAMATE
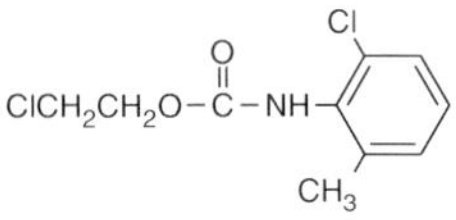
5 gm

S71,775-4
2-CHLORO-1-METHYLETHYL N-(2,6-DICHLOROPHENYL)CARBAMATE
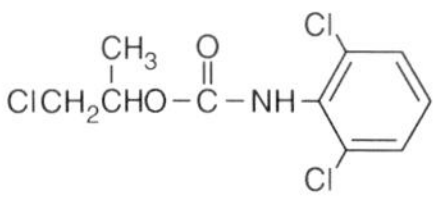
5 gm

S71,165-9
2-BROMOETHYL N-(3,4-DICHLOROPHENYL)-ARBAMATE
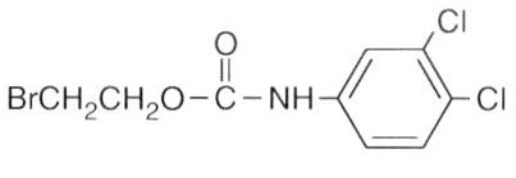
5 gm

S70,912-3
2-CHLOROETHYL N-(3,5-DICHLOROPHENYL)CARBAMATE
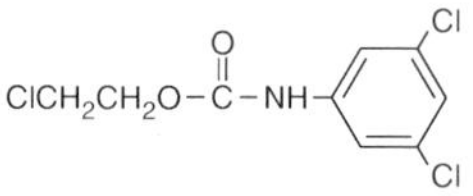
5 gm

S70,919-0
2-CHLOROETHYL N-(2,5-DIMETHOXYPHENYL)CARBAMATE
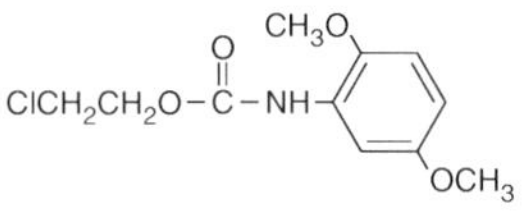
5 gm

S63,440-9
2-CHLOROETHYL N-(O-ANISYL)CARBAMATE
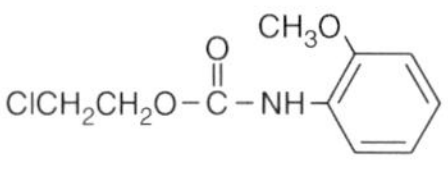
5 gm

S65,435-3
2-CHLOROETHYL N-(3-CHLORO-2-METHOXYPHENYL)CARBAMATE
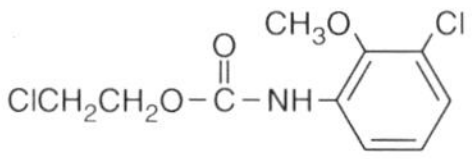
5 gm

S63,930-3
2-CHLOROETHYL N-(2-METHOXY-5-METHYLPHENYL)CARBAMATE
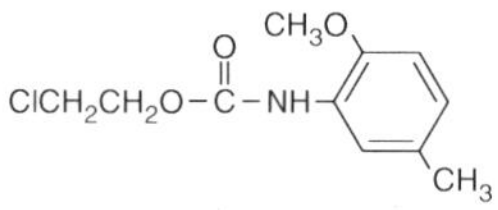
5 gm

S66,534-7
2-CHLOROETHYL N-(M-ANISYL)CARBAMATE
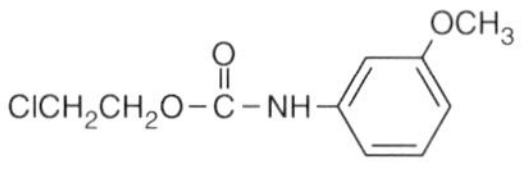
5 gm

S61,203-0
2-CHLOROETHYL N-(4-METHOXYPHENYL)-CARBAMATE
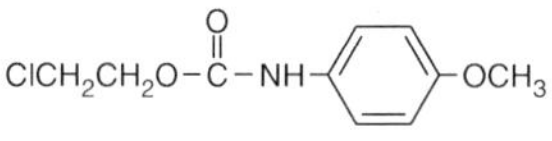
5 gm

S87,642-9
BENZYL N-(2-CHLORO-1-PHENYLETHYL)-CARBAMATE
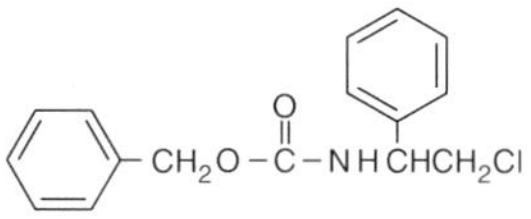
250 mg

S36,726-5
2-CHLOROETHYL LACTATE CARBANILATE
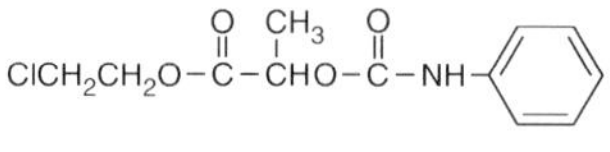
5 gm

S62,514-0
N,N'-(4-METHYL-1,3-PHENYLENE)BIS(2-CHLOROETHYL CARBAMATE)
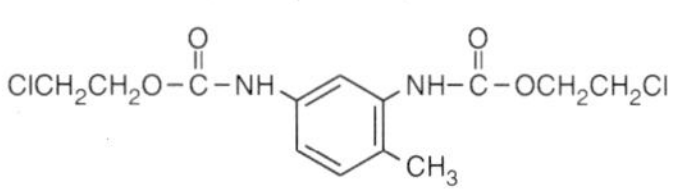
5 gm

S84,708-9
2-CHLOROETHYL N-(2-METHYL-5-NITROPHENYL)CARBAMATE
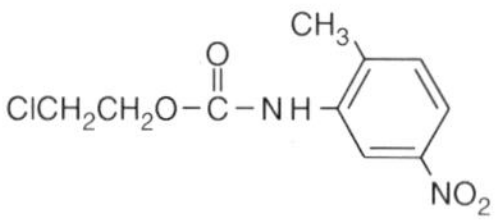
250 mg

S63,383-6
2-CHLOROETHYL N-(3-NITROPHENYL)-CARBAMATE
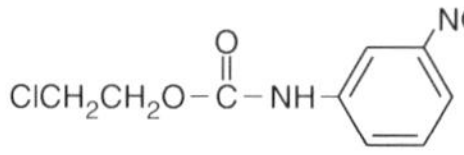
5 gm

S63,842-0
2-CHLOROETHYL N-(2-CHLORO-4-NITROPHENYL)CARBAMATE
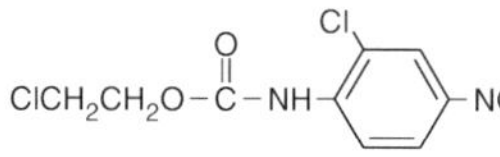
5 gm

S74,986-9
7-[N,N-BIS(CHLOROMETHYL)AMINO]-2-(ETHOXYCARBOXAMIDO)FLUORENE
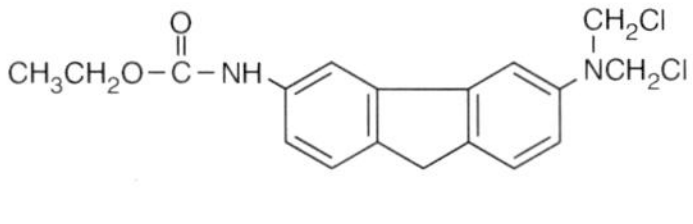
250 mg

S64,462-5
BIS(2-CHLOROETHYL)-N,N'-(METHYLENEDI-4,1-PHENYLENE)-BISCARBAMATE
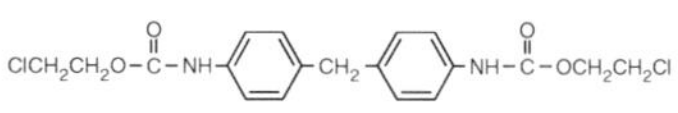
5 gm

S67,878-3
CARBOBENZYLOXYGLYCINE 2-CHLOROMETHYL-4-NITROPHENYL ESTER
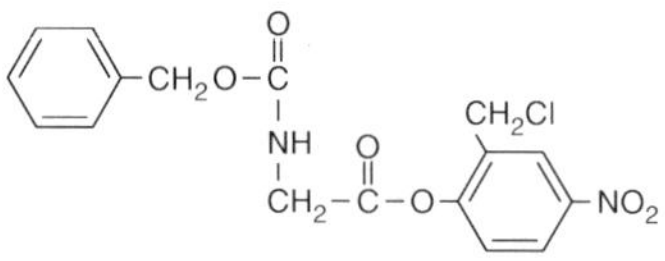
5 gm

S59,433-4
2-CHLOROETHYL N-(3-CHLOROPHENYL)-CARBAMATE
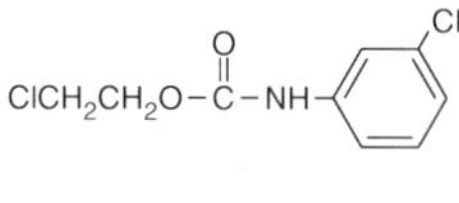
5 gm

S63,478-6
2-CHLOROETHYL N-(M-TOLYL)CARBAMATE
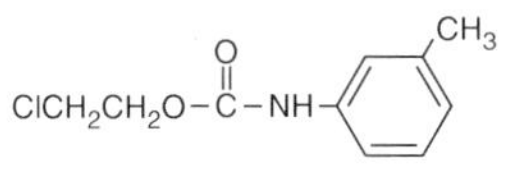
5 gm

S64,481-1
2-CHLORO-1-METHYLETHYL-N-(3-CHLOROPHENYL)CARBAMATE
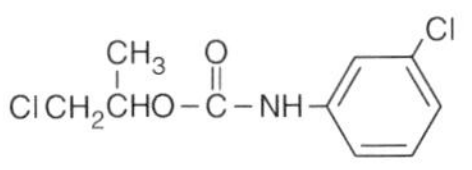
250 mg

S34,567-9
1,3-DICHLORO-2-PROPYL N-(3-CHLOROPHENYL)CARBAMATE
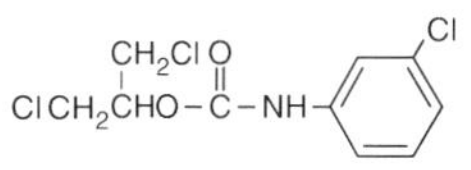
250 mg

S59,417-2
2-CHLOROETHYL N-PHENYLCARBAMATE
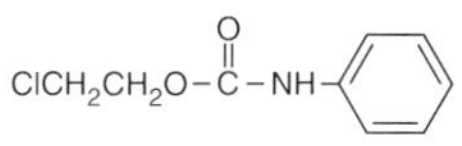
5 gm

S71,167-5
2-BROMOETHYL N-PHENYLCARBAMATE
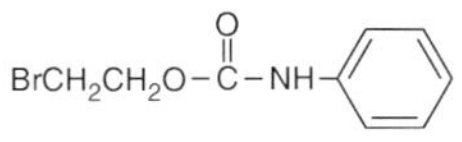
5 gm

S64,468-4
2-CHLORO-1-METHYLETHYL-N-PHENYLCARBAMATE
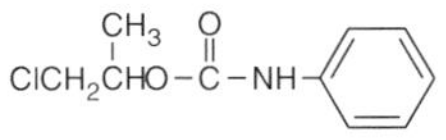
5 gm

S63,379-8
2-CHLOROETHYL N-(2-CHLOROPHENYL)-CARBAMATE
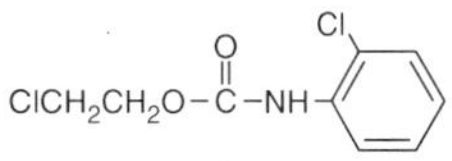
5 gm

S60,750-9
2-CHLOROETHYL N-(O-TOLYL)CARBAMATE
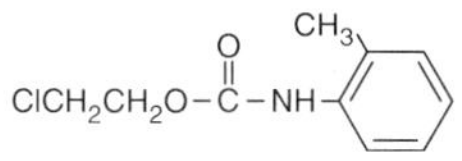
5 gm

S64,991-0
2-CHLOROETHYL N-(2-ETHYLPHENYL)-CARBAMATE
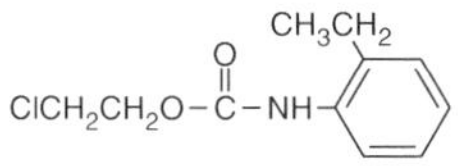
5 gm

S61,272-3
2-CHLOROETHYL N-(4-CHLOROPHENYL)-CARBAMATE
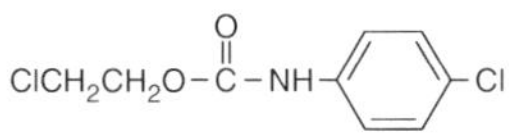
5 gm

S62,815-8
2-CHLOROETHYL N-(4-BROMOPHENYL)-CARBAMATE
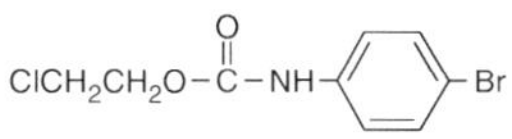
5 gm

S61,276-6
2-CHLOROETHYL N-(P-TOLYL)CARBAMATE
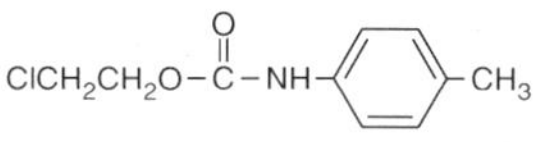
5 gm

S65,271-7
2-CHLORO-1-METHYLETHYL-N-(4-CHLOROPHENYL)CARBAMATE
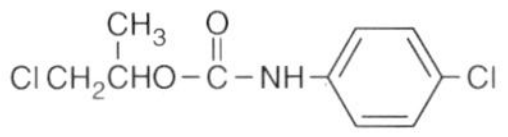
5 gm

S59,746-5
2-CHLOROETHYL N-(2,3-DICHLOROPHENYL)CARBAMATE
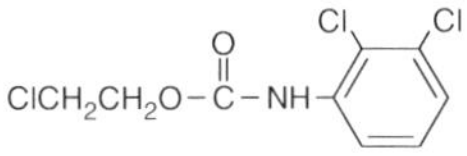
5 gm

S61,200-6
2-CHLOROETHYL N-(2,6-DICHLOROPHENYL)CARBAMATE
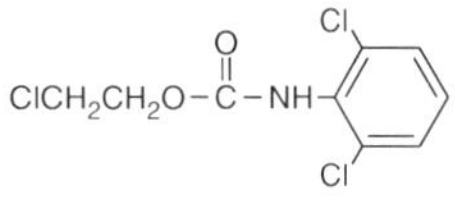
5 gm

S65,100-1
2-BROMOETHYL N-(2,3-DICHLOROPHENYL)-ARBAMATE
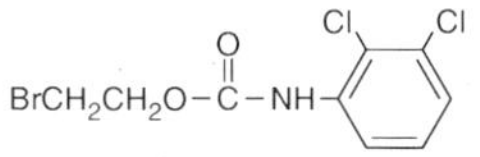
5 gm

S65,222-9
2-BROMOETHYL N-(2,6-DICHLOROPHENYL)-CARBAMATE
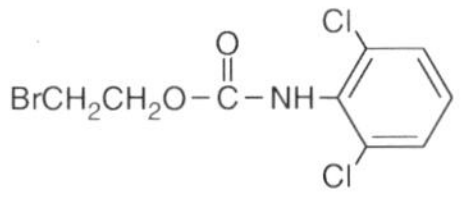
5 gm

S63,662-2
2-CHLOROETHYL N-(2,3-XYLYL)-CARBAMATE
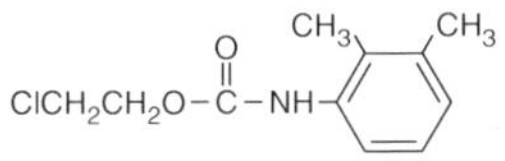
5 gm

S66,637-8
2-CHLOROETHYL N-(2,6-XYLYL)-CARBAMATE
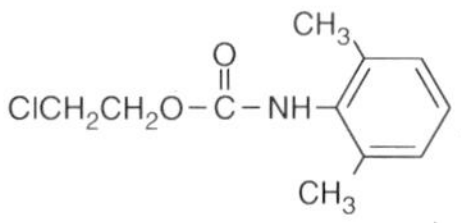
5 gm

S65,035-8
2-CHLOROETHYL N-(2-ETHYL-6-METHYLPHENYL)CARBAMATE
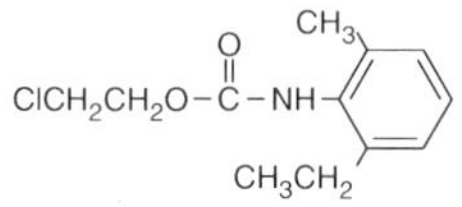
5 gm

S71,471-2
2-CHLOROETHYL N-(2,6-DIETHYLPHENYL)-CARBAMATE
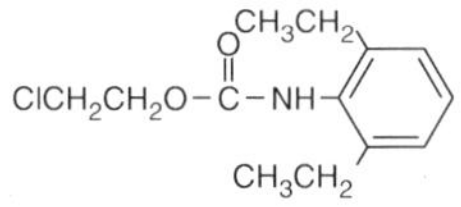
5 gm

S71,712-6
2-CHLOROETHYL N-(2,6-DIISOPROPYLPHENYL)CARBAMATE
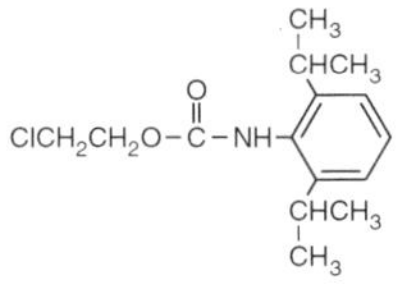
5 gm

S60,651-0
2-CHLOROETHYL N-(3,4-DICHLOROPHENYL)CARBAMATE
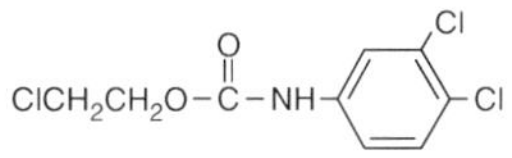
5 gm

S61,202-2
2-CHLOROETHYL N-(2,5-DICHLOROPHENYL)CARBAMATE
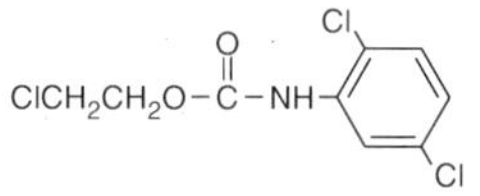
5 gm

S64,499-4
2-CHLOROETHYL N-(2,4-DICHLOROPHENYL)CARBAMATE
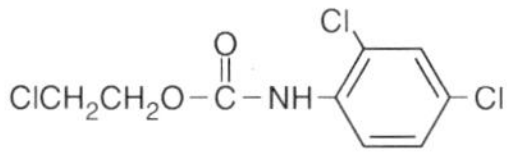
5 gm

S64,954-6
2-BROMOETHYL N-(2,4-DICHLOROPHENYL)-CARBAMATE
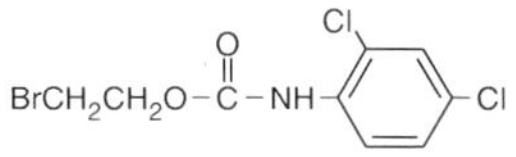
5 gm

S60,185-3
2-CHLOROETHYL N-(4-CHLORO-2-METHYLPHENYL)CARBAMATE
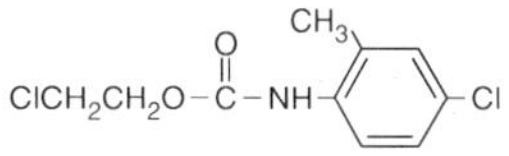
5 gm

S63,487-5
2-CHLOROETHYL N-(3-CHLORO-4-METHYLPHENYL)CARBAMATE
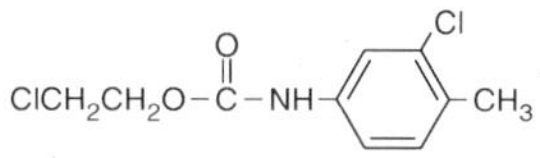
5 gm

S60,658-8
2-CHLOROETHYL N-(2,4-XYLYL)-CARBAMATE
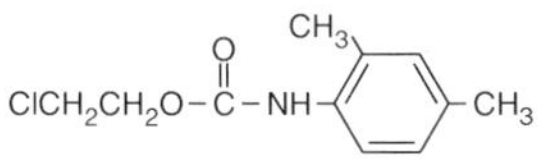
5 gm

S64,049-2
2-CHLOROETHYL N-(2,5-XYLYL)-CARBAMATE
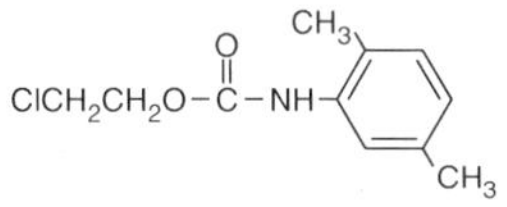
5 gm

S71,518-2
2-CHLOROETHYL N-(3,4-XYLYL)-CARBAMATE
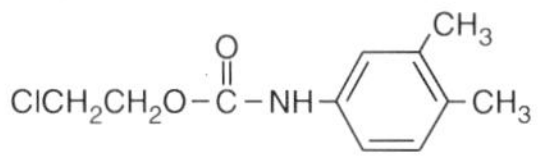
5 gm

S77,413-8
2-CHLOROETHYL N-(3,5-XYLYL)-CARBAMATE
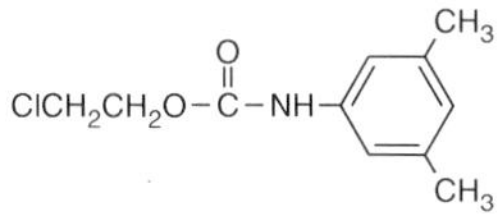
250 mg

S60,771-1
2-CHLOROETHYL N-(2,4,5-TRICHLOROPHENYL)CARBAMATE

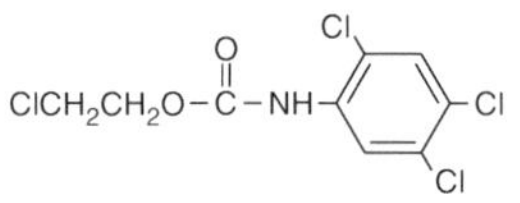

5 gm

S63,940-0
2-CHLOROETHYL N-(1-NAPHTHYL)-CARBAMATE

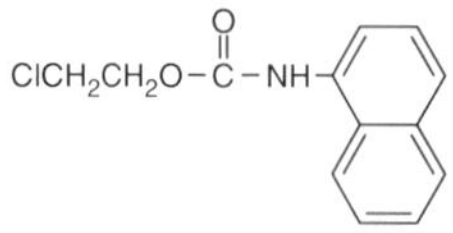

5 gm

S34,421-4
2-CHLOROETHYL N-(5-CHLORO-2-METHYLPHENYL)CARBAMATE

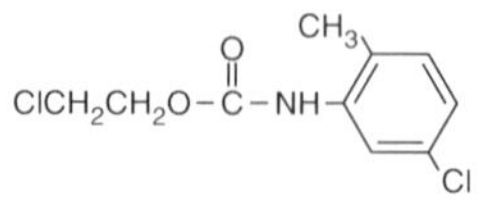

5 gm

S71,096-2
S-(2-CHLOROETHYL)-N-PHENYLTHIOCARBAMATE

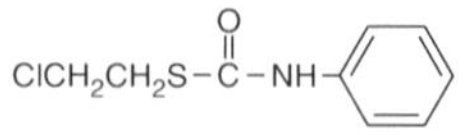

5 gm

S90,259-4
S-(2-HYDROXYPHENYL) N-[4-CHLORO-2-(TRIFLUOROMETHYL)PHENYL]-THIOCARBAMATE

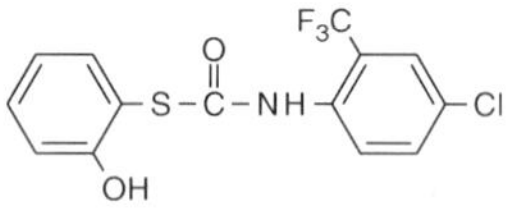

250 mg

S80,101-1
4-BENZYLOXYPHENYL N-[2-CHLORO-5-(TRIFLUOROMETHYL)PHENYL]CARBAMATE

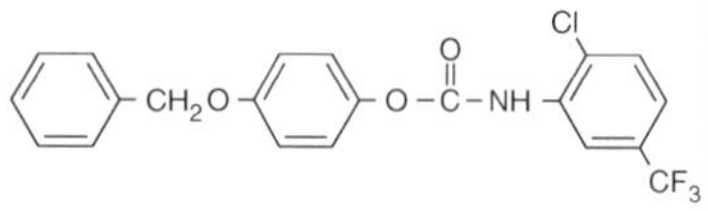

250 mg

S79,688-3
2-[4-(2-CHLOROVINYL)-2-METHOXYPHENOXY]ETHYL N-[4-CHLORO-2-(TRIFLUOROMETHYL)PHENYL]CARBAMATE

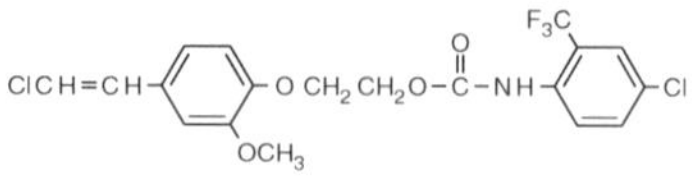

250 mg

S79,651-4
4-FORMYL-2-METHOXYPHENYL N-[4-CHLORO-2-(TRIFLUOROMETHYL)-PHENYL]CARBAMATE

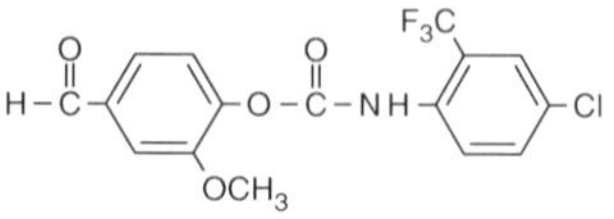

250 mg

S90,255-1
2-(4-ACETAMIDOPHENOXY)ETHYL N-[4-CHLORO-2-(TRIFLUOROMETHYL)-PHENYL]CARBAMATE

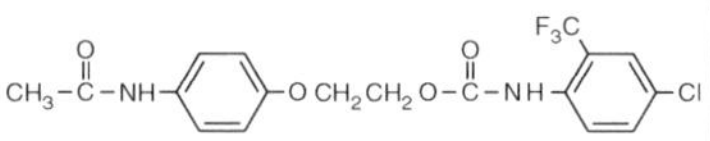

250 mg

S85,330-5
2-(4-BROMOPHENOXY)ETHYL N-[2-(TRIFLUOROMETHYL)PHENYL]-CARBAMATE

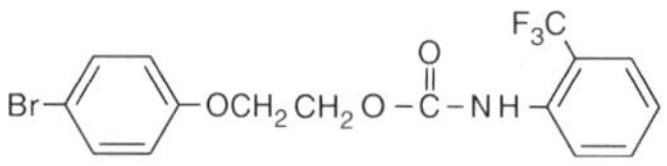

250 mg

S73,717-8
2-(4-CHLOROPHENOXY)ETHYL N-[4-CHLORO-3-(TRIFLUOROMETHYL)-PHENYL]CARBAMATE

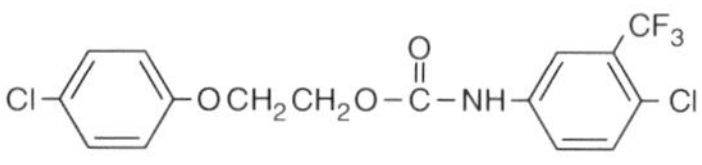

5 gm

S90,257-8
2-(4-CHLORO-3-METHYLPHENOXY)ETHYL N-[4-CHLORO-2(TRIFLUOROMETHYL)-PHENYL]CARBAMATE

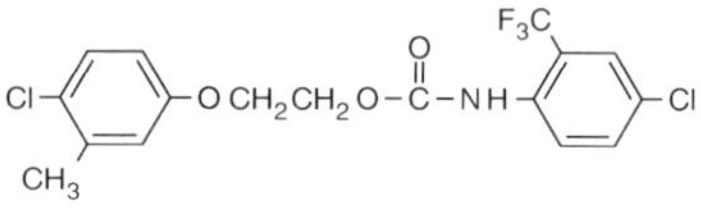

250 mg

S76,897-9
2-(2,4-DICHLOROPHENOXY)ACETYL N-[2-CHLORO-4-(TRIFLUOROMETHYL)-PHENYL]CARBAMATE

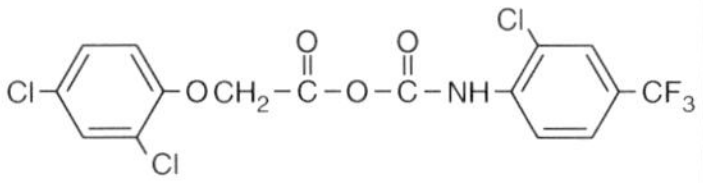

250 mg

S85,676-2
METHYL N-[1-(4-CHLOROPHENOXY)-2,2,2-TRICHLOROETHYL]CARBAMATE

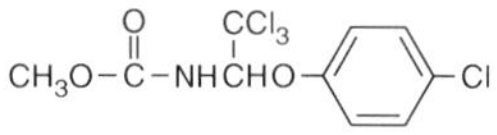

250 mg

S85,708-4
ETHYL N-[1-(4-CHLOROPHENOXY)-2,2,2-TRICHLOROETHYL]CARBAMATE

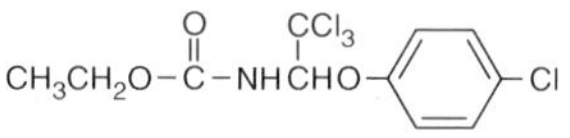

250 mg

S85,714-9
ETHYL N-[2,2,2-TRICHLORO-1-(2,4,5-TRICHLOROPHENOXY)ETHYL]CARBAMATE

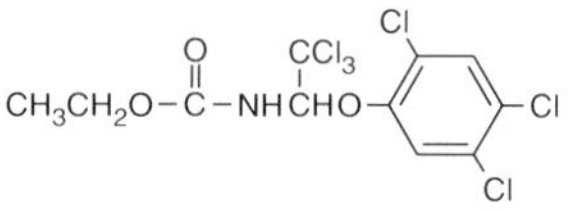

250 mg

S76,889-8
1-(4-CHLOROPHENYLTHIO)-2,2,2-TRICHLOROETHYL N-(3-TRIFLUOROMETHYLPHENYL)CARBAMATE

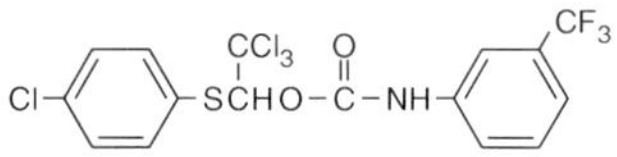

250 mg

S69,471-1
2,2,2-TRICHLOROETHYL N-(4-FLUORO-2-NITROPHENYL)CARBAMATE

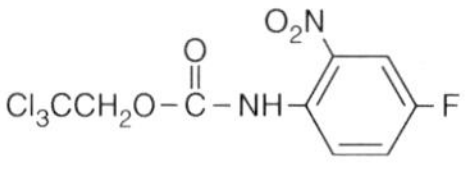

250 mg

S79,478-3
2,2,2-TRICHLOROETHYL N-(4-FLUORO-3-NITROPHENYL)CARBAMATE

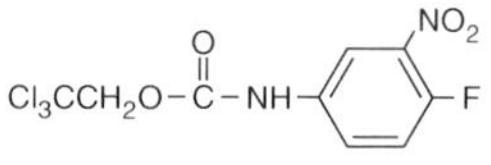

250 mg

S66,699-8
2-DIMETHYLAMINOETHYL N-(2-CHLORO-5-TRIFLUOROMETHYLPHENYL)CARBAMATE

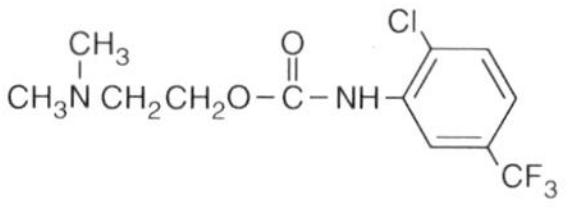

5 gm

S70,918-2
2-(DIETHYLAMINO)ETHYL N-(2-CHLORO-5-TRIFLUOROMETHYLPHENYL)CARBAMATE

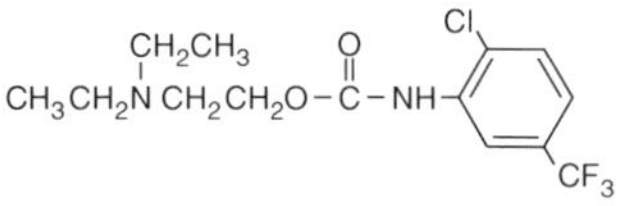

250 mg

S91,303-0
3-(DIMETHYLAMINO)PROPYL N-[2-CHLORO-5-(TRIFLUOROMETHYL)-PHENYL]CARBAMATE

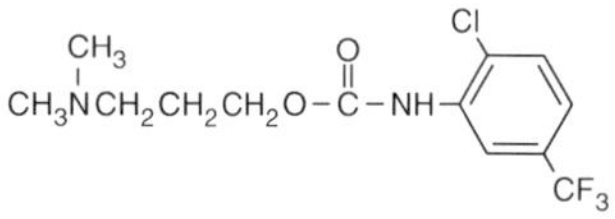

250 mg

S69,416-9
4-[4-CHLORO-2-(TRIFLUOROMETHYL)-PHENYL]-1-(ETHOXYCARBONYL)-SEMICARBAZIDE

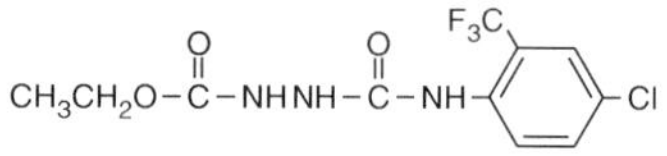

250 mg

S69,206-9
2,6-DIBROMO-4-NITROPHENYL N-[4-CHLORO-2-(TRIFLUOROMETHYL)-PHENYL]CARBAMATE

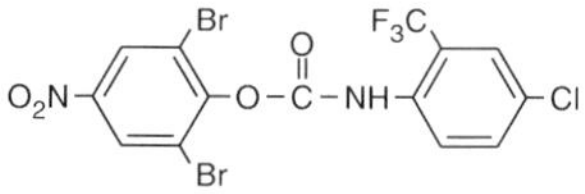

250 mg

S85,668-1
METHYL N-[1-(4-CHLOROBENZOYLOXY)-2,2,2-TRICHLOROETHYL]CARBAMATE

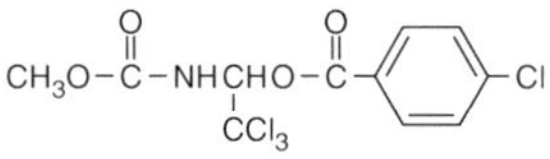

250 mg

S85,694-0
ETHYL N-[1-(4-CHLOROBENZOYLOXY)-2,2,2-TRICHLOROETHYL]CARBAMATE

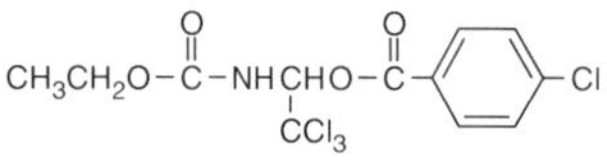

250 mg

S85,691-6
ISOPROPYL N-[1-(4-CHLOROBENZOYLOXY)-2,2,2-TRICHLOROETHYL]CARBAMATE

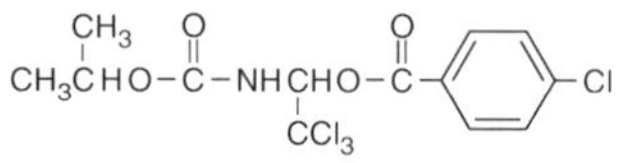

250 mg

S85,692-4
PROPYL N-[1-(4-CHLOROBENZOYLOXY)-2,2,2-TRICHLOROETHYL]CARBAMATE

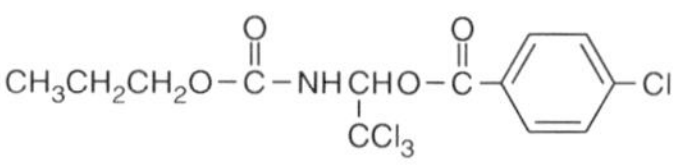

250 mg

S85,696-7
BUTYL N-[4-(4-CHLOROBENZOYLOXY)-2,2,2-TRICHLOROETHYL]CARBAMATE

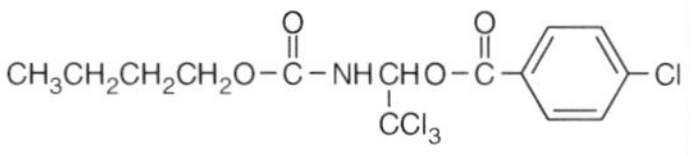

250 mg

S78,118-5
METHYL 2-(N-[2-CHLORO-5-(TRIFLUOROMETHYL)PHENYL]-CARBAMOYLOXY)-2-METHYLPROPIONATE

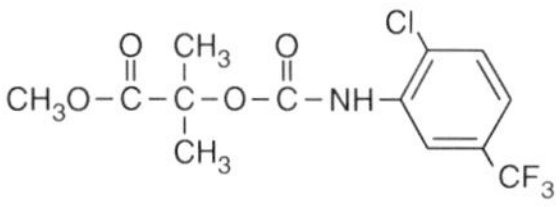

250 mg

S72,566-8
1-[2-CHLORO-5-(TRIFLUOROMETHYL)-PHENYL]-3-(2-OXO-3-OXAZOLIDINYL)UREA

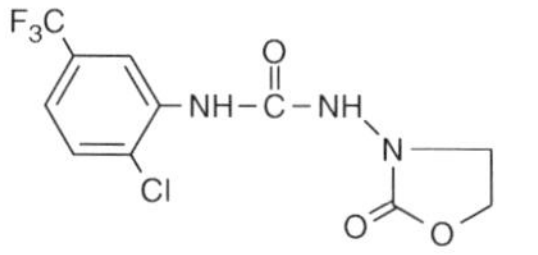

250 mg

S79,652-2
4-ACETAMIDOPHENYL N-[4-CHLORO-2-(TRIFLUOROMETHYL)PHENYL]CARBAMATE

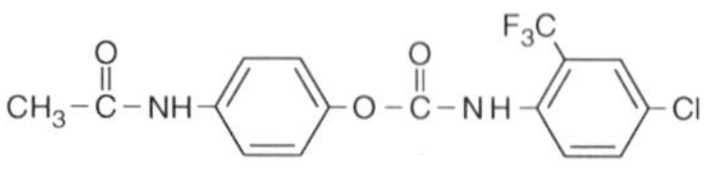

250 mg

S80,089-9
4-ACETAMIDOPHENYL N-[2-CHLORO-5-(TRIFLUOROMETHYL)PHENYL]CARBAMATE

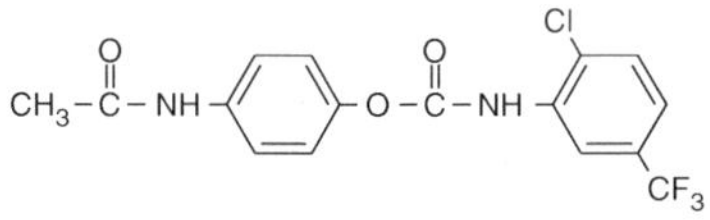

250 mg

S79,510-0
2,2,2-TRIFLUOROETHYL N-(3-FLUOROPHENYL)CARBAMATE

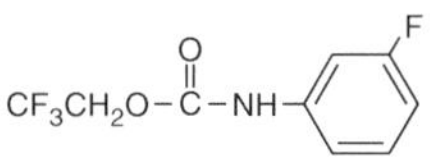

250 mg

S71,723-1
2,4,5-TRICHLOROPHENYL N-[3-(TRIFLUOROMETHYL)PHENYL]-CARBAMATE

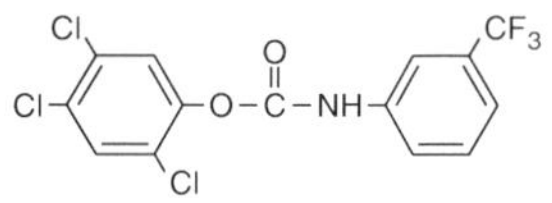

5 gm

S68,695-6
2,4-DIBROMO-5,6-DIMETHYLPHENYL N-[3-(TRIFLUOROMETHYL)PHENYL]-CARBAMATE

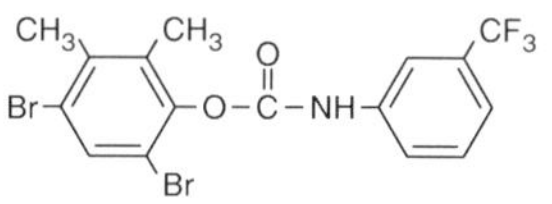

250 mg

S68,590-9
PIPERIDINO N-[2-CHLORO-5-(TRIFLUOROMETHYL)PHENYL]CARBAMATE

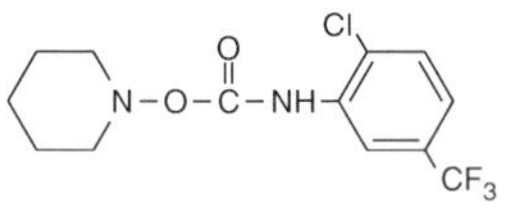

250 mg

S80,179-8
2-BENZYL-4-CHLOROPHENYL N-[2-(TRIFLUOROMETHYL)PHENYL]-CARBAMATE

250 mg

S73,730-5
(2-BROMO-4-PHENYL)PHENYL N-[2-(TRIFLUOROMETHYL)PHENYL]-CARBAMATE

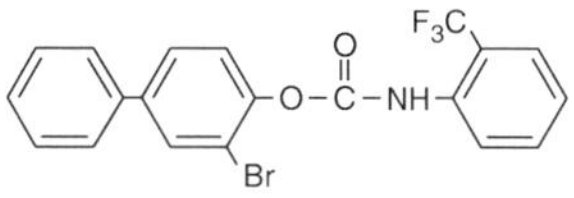

5 gm

S80,077-5
2-BENZYL-4-CHLOROPHENYL N-[2-CHLORO-5-(TRIFLUOROMETHYL)-PHENYL]CARBAMATE

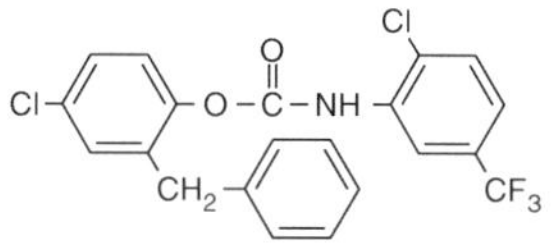

250 mg

S89,805-8
2,2,2-TRIFLUOROETHYL N-(2-FLUOROPHENYL)CARBAMATE

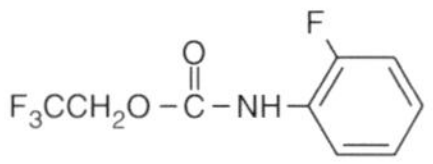

250 mg

S79,092-3
2,2,2-TRICHLOROETHYL N-(2-CHLOROPHENYL)CARBAMATE

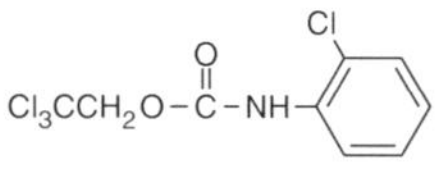

250 mg

S72,505-6
3-(TRIFLUOROMETHYL)PHENYL N-(2-FLUOROPHENYL)CARBAMATE

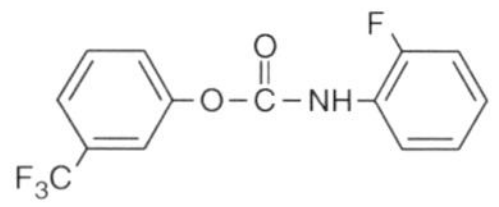

5 gm

S69,063-5
2-BROMO-4,5-DIMETHYLPHENYL N-[2-(TRIFLUOROMETHYL)PHENYL]-CARBAMATE

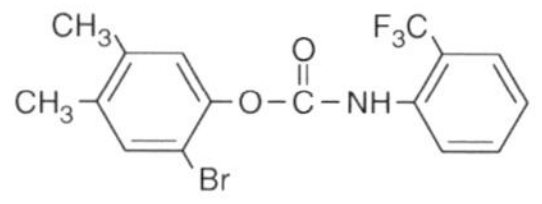

250 mg

S90,258-6
2-FLUOROPHENYL N-[4-CHLORO-2-(TRIFLUOROMETHYL)PHENYL]CARBAMATE

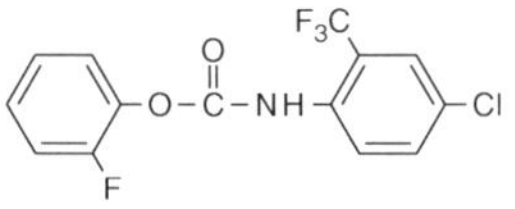

250 mg

S77,959-8
2,4-DICHLOROPHENYL N-[2-(TRIFLUOROMETHYL)PHENYL]-CARBAMATE

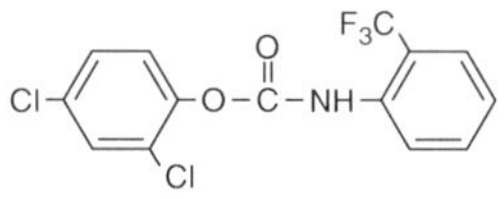

250 mg

S69,446-0
2-BROMO-4-CHLORO-6-METHYLPHENYL N-[2-(TRIFLUOROMETHYL)PHENYL]-CARBAMATE

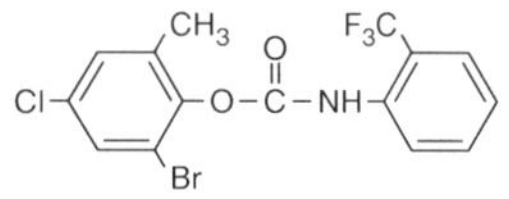

250 mg

S77,213-5
(2-BROMO-4-CHLORO-5-METHYL)PHENYL N-[2-(TRIFLUOROMETHYL)PHENYL]-CARBAMATE

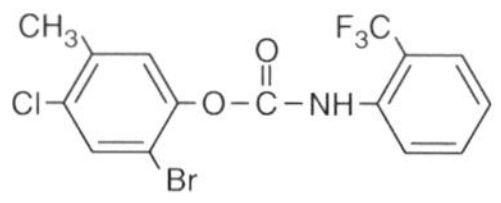

250 mg

S69,182-8
4-BROMOPHENYL N-[4-BROMO-2 (TRIFLUOROMETHYL)PHENYL]CARBAMATE

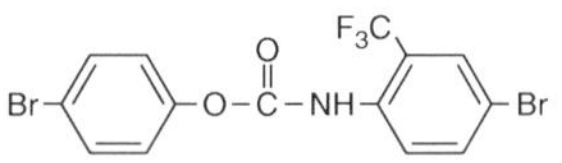

250 mg

S85,071-3
3-(TRIFLUOROMETHYL)PHENYL N-(2,3-DICHLOROPHENYL)CARBAMATE

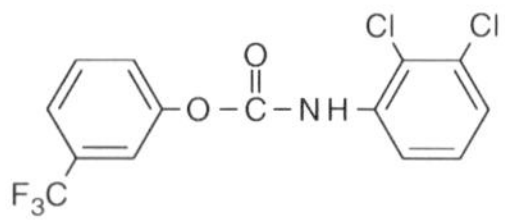

250 mg

S65,532-0
METHYL N-(2-CHLORO-5-TRIFLUOROMETHYLPHENYL)CARBAMATE

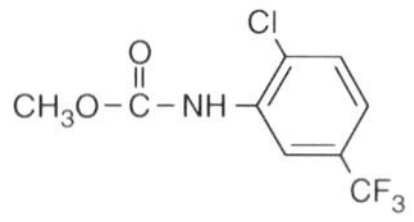

5 gm

S90,322-1
2,2,2-TRICHLOROETHYL N-(2,4-DICHLOROPHENYL)CARBAMATE

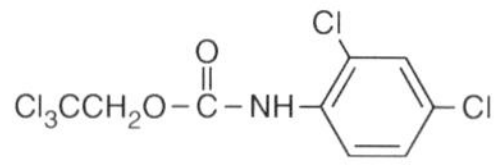

250 mg

S66,674-2
ETHYL N-(2-CHLORO-5-TRIFLUOROMETHYLPHENYL)CARBAMATE

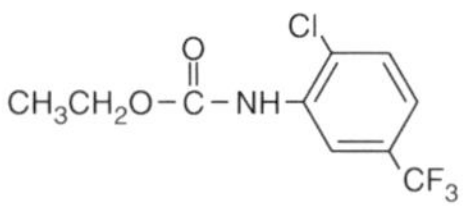

5 gm

S79,098-2
2,2,2-TRICHLOROETHYL N-(3-CHLORO-4-METHYLPHENYL)CARBAMATE

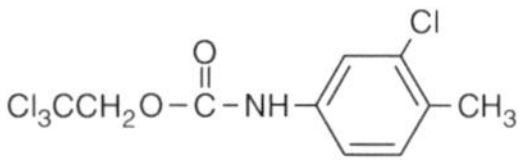

250 mg

S66,530-4
PROPYL N-[2-CHLORO-5-(TRIFLUOROMETHYL)PHENYL]CARBAMATE

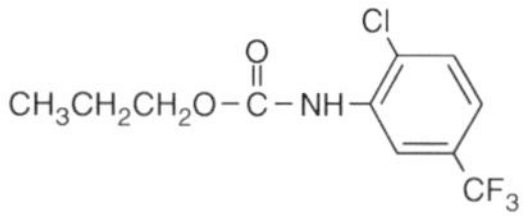

5 gm

S66,540-1
ISOPROPYL N-[2-CHLORO-5-(TRIFLUOROMETHYL)PHENYL]CARBAMATE

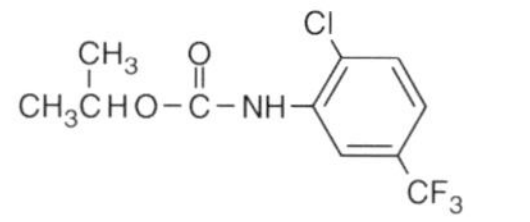

5 gm

S90,200-4
ISOPROPYL N-[4-CHLORO-3-(TRIFLUOROMETHYL)PHENYL]CARBAMATE

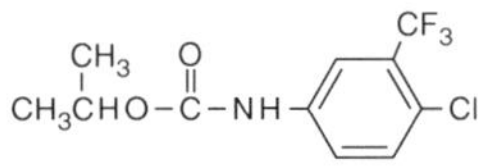

250 mg

S79,505-4
2-CHLORO-4-FLUOROPHENYL N-[3-(TRIFLUOROMETHYL)PHENYL]-CARBAMATE

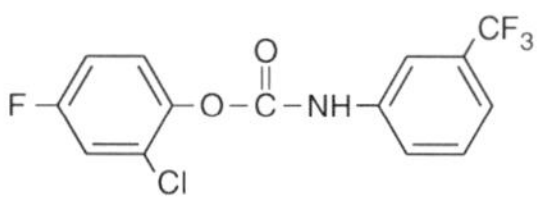

250 mg

S68,878-9
3,4-DICHLOROPHENYL N-[3-(TRIFLUOROMETHYL)PHENYL]-CARBAMATE

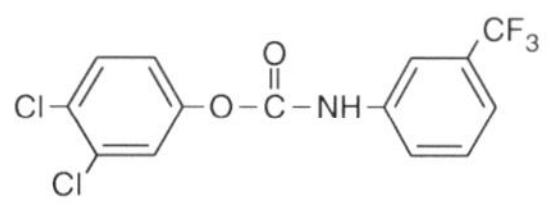

250 mg

S88,960-1
2,4-DICHLOROPHENYL N-[3-(TRIFLUOROMETHYL)PHENYL]-CARBAMATE

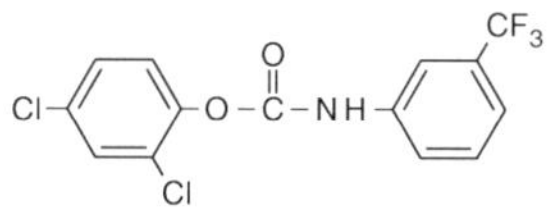

250 mg

S76,898-7
2,5-DICHLOROPHENYL N-[4-CHLORO-2-(TRIFLUOROMETHYL)PHENYL]CARBAMATE

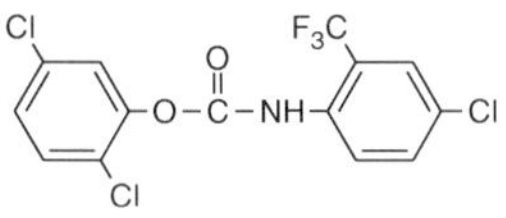

250 mg

S77,945-8
4-CHLORO-3-METHYLPHENYL N-[3-(TRIFLUOROMETHYL)PHENYL]-CARBAMATE

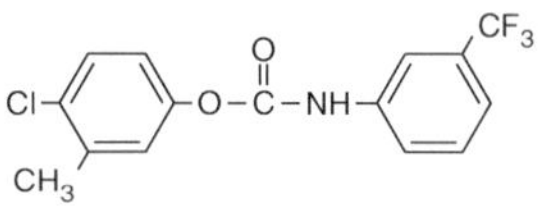

250 mg

S77,282-8
3-FLUOROBENZYL N-[4-CHLORO-2-(TRIFLUOROMETHYL)PHENYL]CARBAMATE

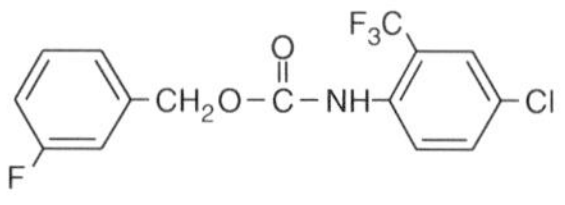

250 mg

S68,877-0
2-BROMO-4,5-DIMETHYLPHENYL N-[4-CHLORO-2-(TRIFLUOROMETHYL)-PHENYL]CARBAMATE

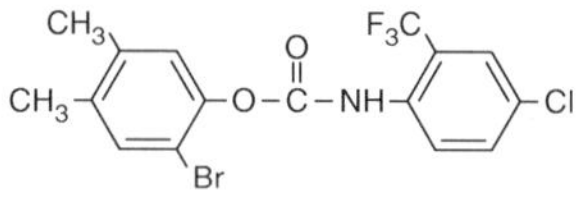

250 mg

S69,051-1
2-BROMO-4,5-DIMETHYLPHENYL N-[2-CHLORO-5-(TRIFLUOROMETHYL)-PHENYL]CARBAMATE
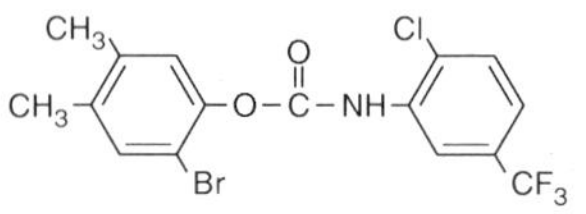
250 mg

S68,868-1
2,4-DIBROMO-5,6-DIMETHYLPHENYL N-[4-CHLORO-2-(TRIFLUOROMETHYL)-PHENYL]CARBAMATE
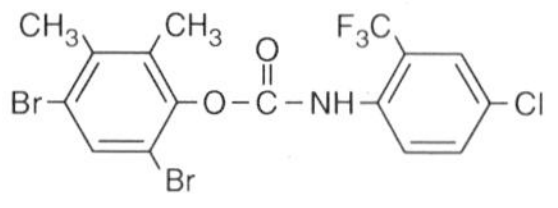
250 mg

S68,647-6
2-BROMO-4,6-DICHLOROPHENYL N-[4-CHLORO-2-(TRIFLUOROMETHYL)-PHENYL]CARBAMATE
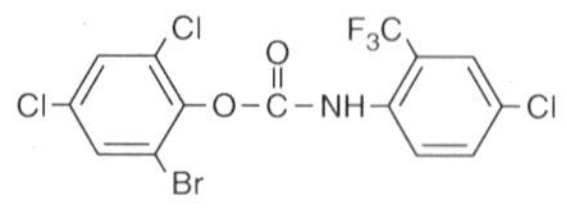
250 mg

S76,876-6
(2-CHLORO-4,6-DIBROMOPHENYL) N-[4-CHLORO-2-(TRIFLUOROMETHYL)-PHENYL]CARBAMATE
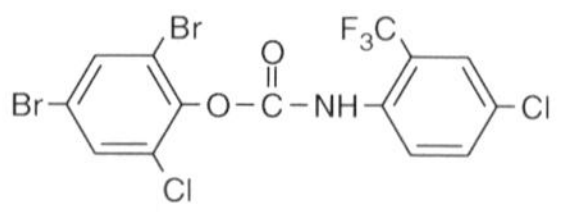
250 mg

S68,735-9
2-BROMO-4-CHLORO-6-METHYLPHENYL N-[4-CHLORO-2-(TRIFLUOROMETHYL)-PHENYL]CARBAMATE
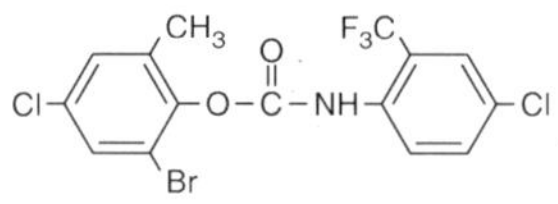
250 mg

S80,069-4
4-CHLORO-3,5-DIMETHYLPHENYL N-[2-CHLORO-5-(TRIFLUOROMETHYL)-PHENYL]CARBAMATE
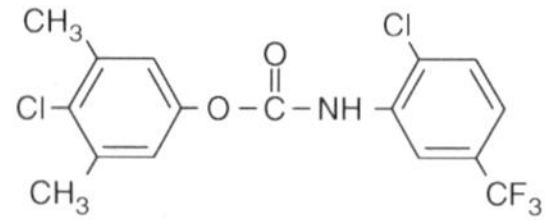
250 mg

S68,852-5
4-BROMO-3,5-DIMETHYLPHENYL N-[4-CHLORO-2-(TRIFLUOROMETHYL)-PHENYL]CARBAMATE
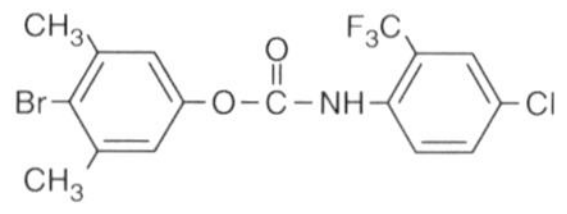
250 mg

S79,669-7
2,3,5-TRIMETHYLPHENYL N-[4-CHLORO-2-(TRIFLUOROMETHYL)PHENYL]CARBAMATE
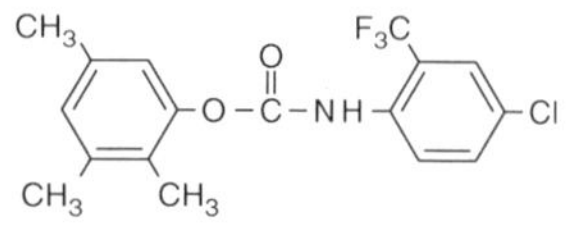
250 mg

S90,201-2
2,4,5-TRICHLOROPHENYL N-[4-CHLORO-3-(TRIFLUOROMETHYL)PHENYL]CARBAMATE
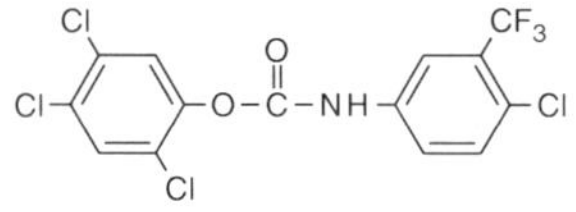
250 mg

S74,448-4
1,2-DIHYDRO-1-METHYL-4-PYRIDYL N-[4-CHLORO-2-(TRIFLUOROMETHYL)-PHENYL]CARBAMATE
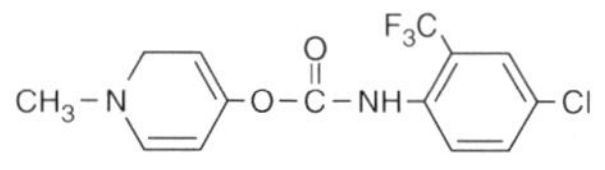
250 mg

S71,720-7
2,4,6-TRICHLOROPHENYL N-[3-(TRIFLUOROMETHYL)PHENYL]-CARBAMATE
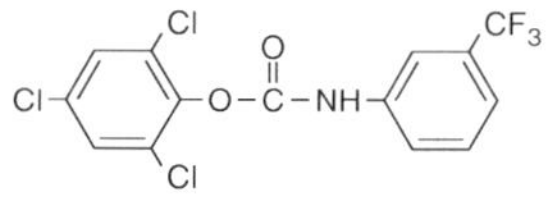
250 mg

S68,706-5
2-BROMO-4,6-DICHLOROPHENYL N-[3-(TRIFLUOROMETHYL)PHENYL]-CARBAMATE
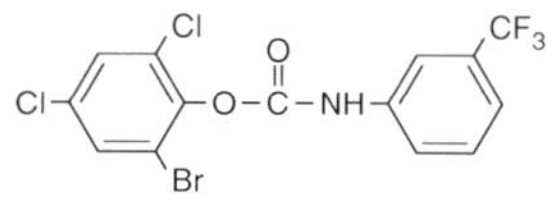
250 mg

S71,766-5
2,4,6-TRIBROMOPHENYL N-[3-(TRIFLUOROMETHYL)PHENYL]-CARBAMATE
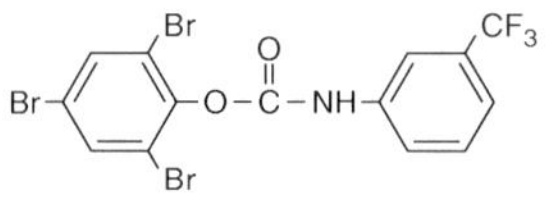
5 gm

S68,690-5
2-BROMO-4-CHLORO-6-METHYLPHENYL N-[3-(TRIFLUOROMETHYL)PHENYL]-CARBAMATE
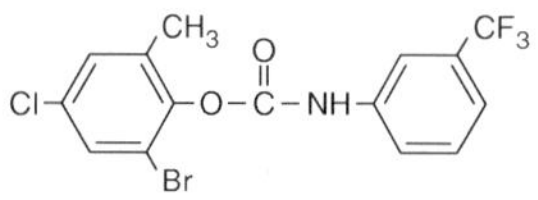
250 mg

S68,355-8
4-BROMO-2-TERT-BUTYL-6-METHYLPHENYL N-[3-(TRIFLUOROMETHYL)-PHENYL]CARBAMATE
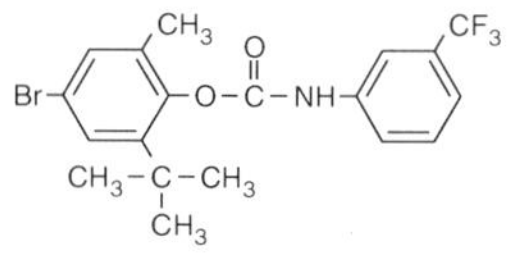
250 mg

S80,108-9
4-CHLORO-2-ISOPROPYL-5-METHYLPHENYL N-[3-(TRIFLUOROMETHYL)-PHENYL]CARBAMATE
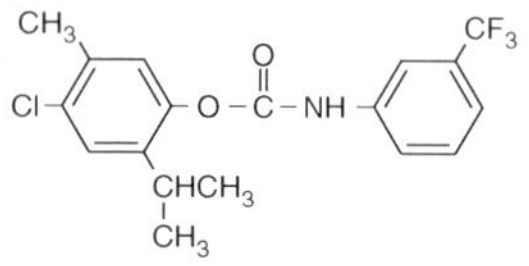
250 mg

S84,507-8
S-(O-TOLYL) N-[2-CHLORO-5-(TRIFLUOROMETHYL)PHENYL]-THIOCARBAMATE
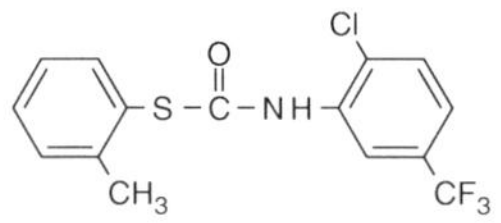
250 mg

S84,508-6
S-(M-TOLYL) N-[2-CHLORO-6-(TRIFLUOROMETHYL)PHENYL]-THIOCARBAMATE
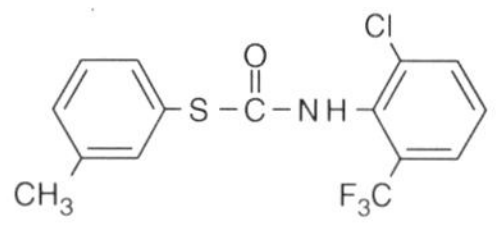
250 mg

S85,675-4
METHYL N-[1-(BENZYLOXY)-2,2,2-TRICHLOROETHYL]CARBAMATE
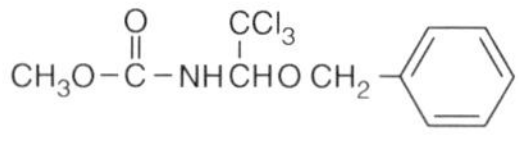
250 mg

S85,699-1
ETHYL N-(1-BENZYLOXY-2,2,2-TRICHLOROETHYL)CARBAMATE
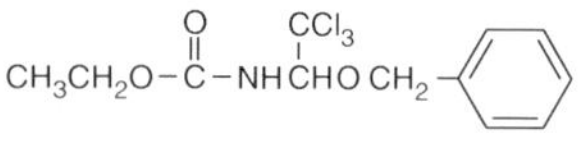
250 mg

S69,478-9
2,2,2-TRIFLUOROETHYL N-(2,5-DIMETHOXYPHENYL)CARBAMATE
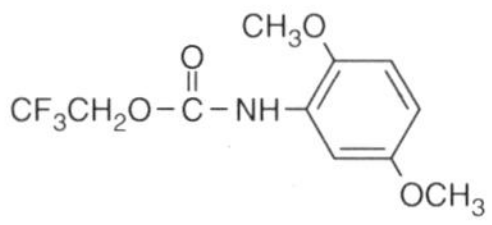
250 mg

S85,713-0
METHYL N-[1-(4-METHOXYPHENOXY)-2,2,2-TRICHLOROETHYL]CARBAMATE
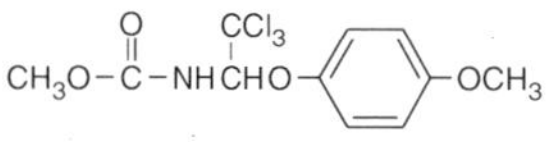
250 mg

S79,549-6
2-(4-ACETAMIDOPHENOXY)ETHYL N-[3-(TRIFLUOROMETHYL)PHENYL]-CARBAMATE
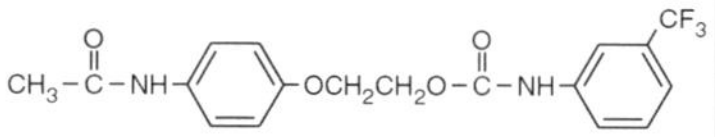
250 mg

S71,722-3
2-METHOXYETHYL N-[3-(TRIFLUOROMETHYL)PHENYL]-CARBAMATE
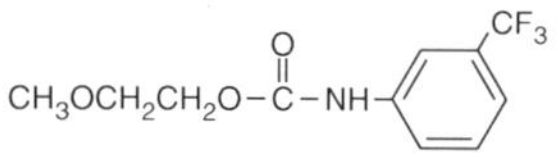
250 mg

S71,177-2
2-ETHOXYETHYL N-[3-(TRIFLUOROMETHYL)PHENYL]-CARBAMATE
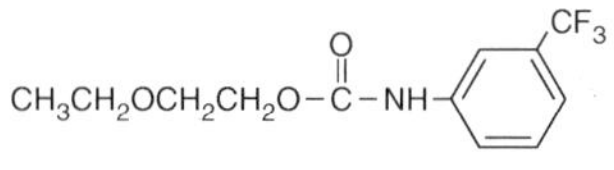
5 gm

S79,551-8
2-(4-BIPHENYLOXY)ETHYL N-[3-(TRIFLUOROMETHYL)PHENYL]-CARBAMATE
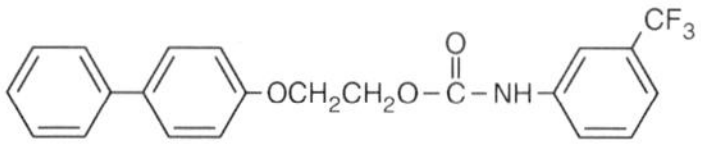
250 mg

S80,180-1
2-BENZOYL-5-METHOXYPHENYL N-[2-(TRIFLUOROMETHYL)PHENYL]-CARBAMATE
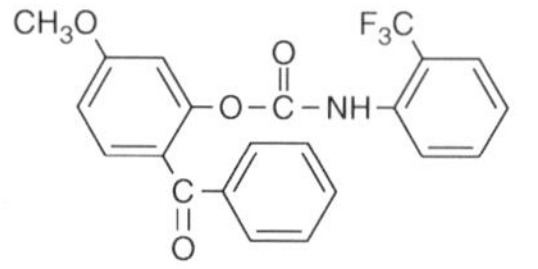
250 mg

S80,174-7
4-BUTOXYPHENYL N-[2-(TRIFLUOROMETHYL)PHENYL]-CARBAMATE
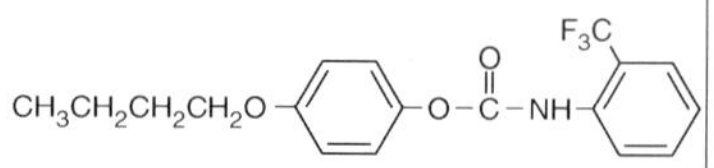
250 mg

S79,196-2
2,2,2-TRICHLOROETHYL N-(2-METHOXYPHENYL)CARBAMATE
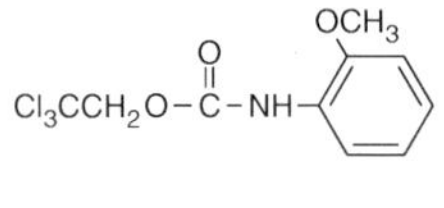
250 mg

S85,715-7
ETHYL N-[1-(4-NITROPHENOXY)-2,2,2-TRICHLOROETHYL]CARBAMATE
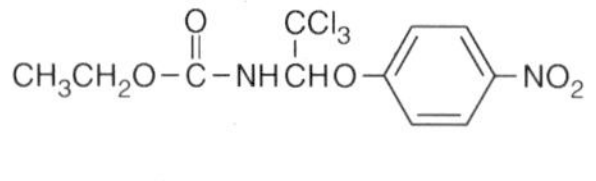
250 mg

S85,662-2
METHYL N-[1-(4-METHOXYBENZOYLOXY)-2,2,2-TRICHLOROETHYL]CARBAMATE
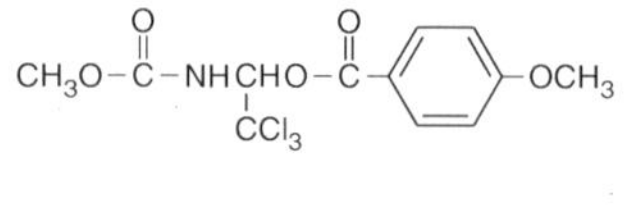
250 mg

S85,687-8
ETHYL N-[1-(4-METHOXYBENZOYLOXY)-2,2,2-TRICHLOROETHYL]CARBAMATE
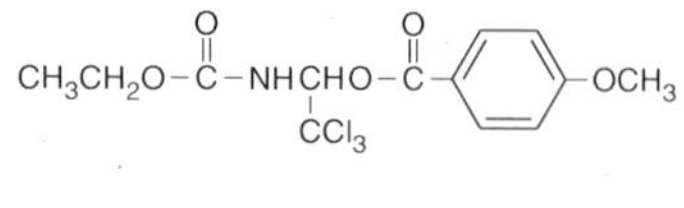
250 mg

S85,688-6
ISOPROPYL N-[1-(4-METHOXYBENZOYLOXY)-2,2,2-TRICHLOROETHYL]CARBAMATE

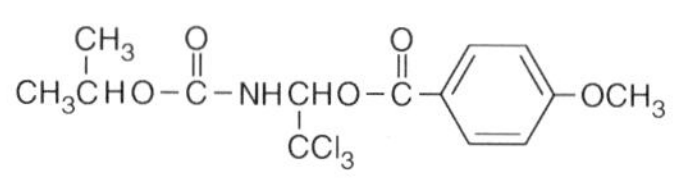

250 mg

S85,693-2
PROPYL N-[1-(4-METHOXYBENZOYLOXY)-2,2,2-TRICHLOROETHYL]CARBAMATE

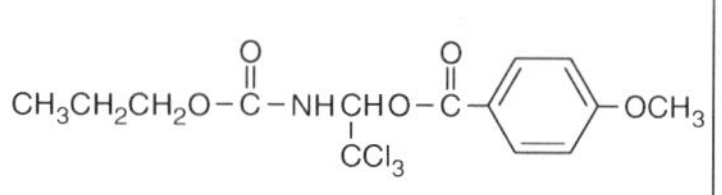

250 mg

S85,712-2
METHYL N-(1-PHENOXY-2,2,2-TRICHLOROETHYL)CARBAMATE

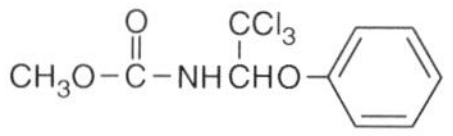

250 mg

S85,709-2
ETHYL N-(1-PHENOXY-2,2,2-TRICHLOROETHYL)CARBAMATE

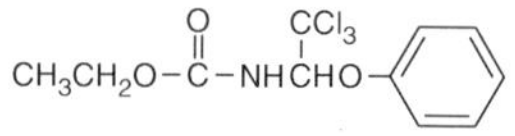

250 mg

S91,712-5
2,2,2-TRIFLUOROETHYL N-(4-BUTOXYPHENYL)CARBAMATE

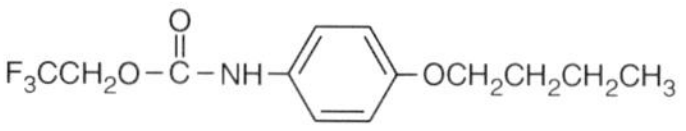

250 mg

S79,122-9
4-METHOXYPHENYL N-[3-(TRIFLUOROMETHYL)PHENYL]-CARBAMATE

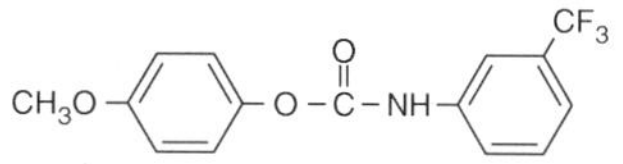

250 mg

S80,182-8
4-(BENZYLOXYCARBONYL)PHENYL N-[2-(TRIFLUOROMETHYL)PHENYL]-CARBAMATE

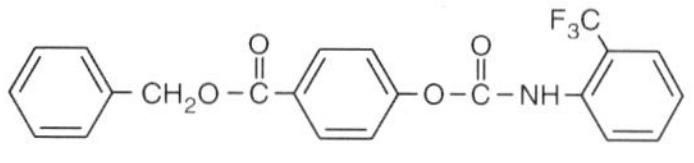

250 mg

S65,278-4
BENZYL N-(3-TRIFLUOROMETHYLPHENYL)-CARBAMATE

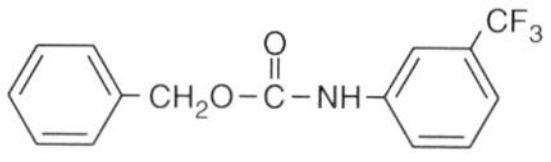

5 gm

S65,343-8
2-(2,4-DICHLOROPHENOXY)ETHYL N-(3-TRIFLUOROMETHYLPHENYL)-CARBAMATE

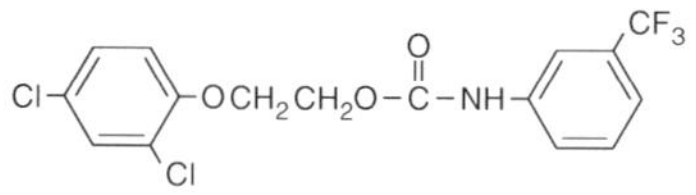

5 gm

S76,859-6
4-METHYL-2-OXO-2H-1-BENZOPYRAN-7-YL N-[3-(TRIFLUOROMETHYL)PHENYL]-CARBAMATE

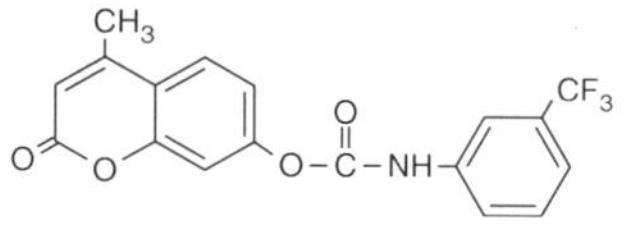

250 mg

S76,819-7
2-(ISOPROPENYLOXYCARBONYL)ETHYL N-(3-TRIFLUOROMETHYLPHENYL)-CARBAMATE

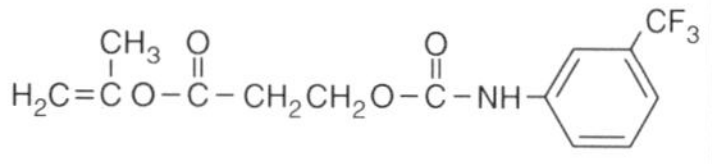

250 mg

S80,171-2
3-ACETYLPHENYL N-[2-(TRIFLUOROMETHYL)PHENYL]-CARBAMATE

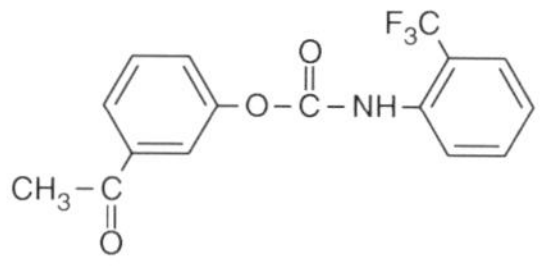

250 mg

S85,732-7
METHYL N-[1-(PHENYLTHIO)-2,2,2-TRICHLOROETHYL]CARBAMATE

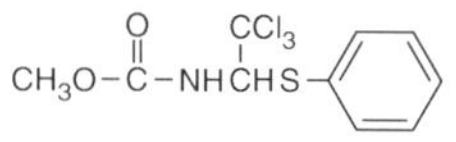

250 mg

S88,962-8
4-(METHYLTHIO)PHENYL N-[3-(TRIFLUOROMETHYL)PHENYL]-CARBAMATE

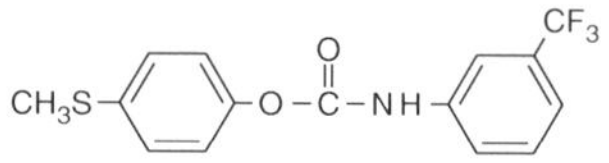

250 mg

S79,143-1
3-METHYL-4-(METHYLTHIO)PHENYL N-[3-(TRIFLUOROMETHYL)PHENYL]-CARBAMATE

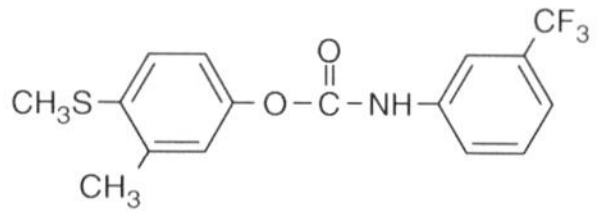

250 mg

S85,795-5
METHYL N-[2,6-DINITRO-4-(TRIFLUOROMETHYL)PHENYL]CARBAMATE

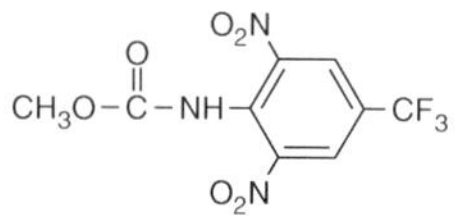

250 mg

S85,796-3
ETHYL N-[2,6-DINITRO-4-(TRIFLUOROMETHYL)PHENYL]CARBAMATE

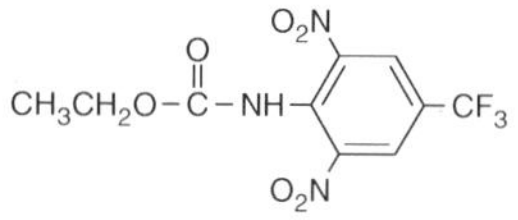

250 mg

S85,824-2
METHYL N-[2,6-DINITRO-4-(TRIFLUOROMETHYL)PHENYL]-N-METHYLCARBAMATE

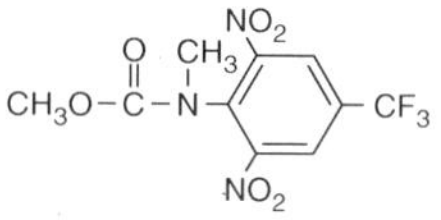

250 mg

S85,797-1
PROPYL N-[2,6-DINITRO-4-(TRIFLUOROMETHYL)PHENYL]CARBAMATE

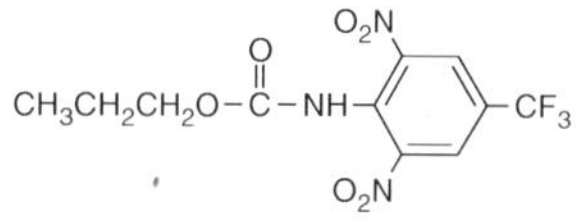

250 mg

S85,799-8
ISOPROPYL N-[2,6-DINITRO-4-(TRIFLUOROMETHYL)PHENYL]CARBAMATE

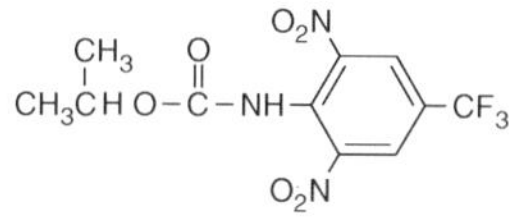

250 mg

S85,825-0
METHYL N-[2,6-DINITRO-4-(TRIFLUOROMETHYL)PHENYL]-N-ETHYLCARBAMATE

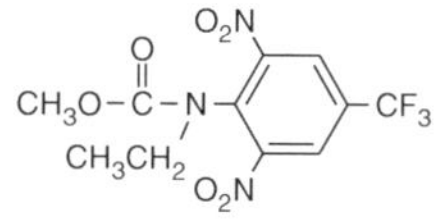

250 mg

S85,826-9
ETHYL N-[2,6-DINITRO-4-(TRIFLUOROMETHYL)PHENYL]-N-METHYLCARBAMATE

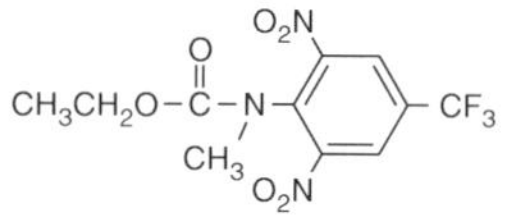

250 mg

S85,833-1
ISOPROPYL N-[2,6-DINITRO-4-(TRIFLUOROMETHYL)PHENYL]-N-METHYLCARBAMATE

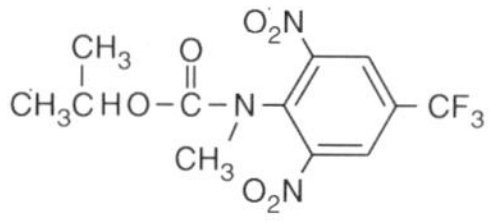

250 mg

S85,827-7
ETHYL N-[2,6-DINITRO-4-(TRIFLUOROMETHYL)PHENYL]-N-ETHYLCARBAMATE

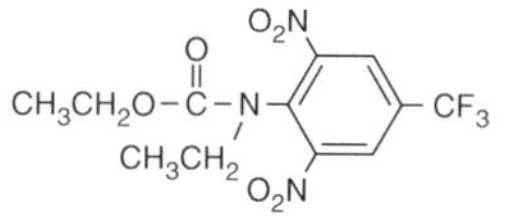

250 mg

S85,829-3
PROPYL N-[2,6-DINITRO-4-(TRIFLUOROMETHYL)PHENYL]-N-METHYLCARBAMATE

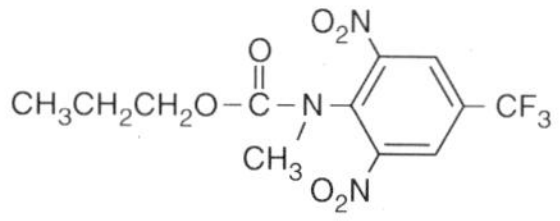

250 mg

S85,828-5
ETHYL N-[2,6-DINITRO-4-(TRIFLUOROMETHYL)PHENYL]-N-PROPYLCARBAMATE

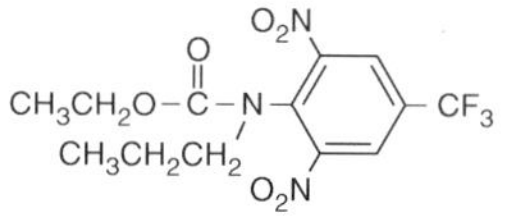

250 mg

S85,830-7
PROPYL N-[2,6-DINITRO-4-(TRIFLUOROMETHYL)PHENYL]-N-ETHYLCARBAMATE

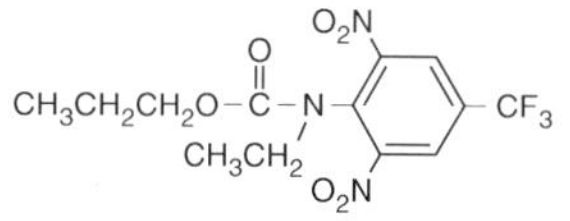

250 mg

S85,835-8
ISOPROPYL N-[2,6-DINITRO-4-(TRIFLUOROMETHYL)PHENYL]-N-ETHYLCARBAMATE

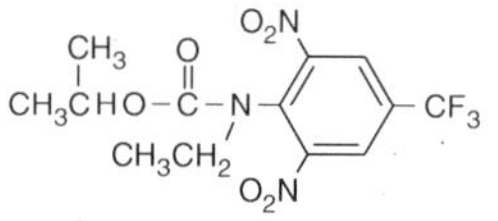

250 mg

S85,836-6
METHYL N-BUTYL-N-[2,6-DINITRO-4-(TRIFLUOROMETHYL)PHENYL]CARBAMATE

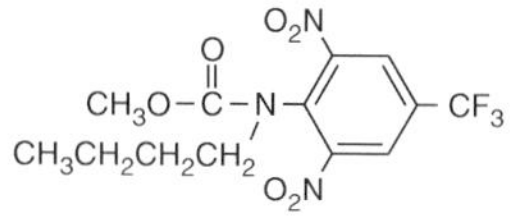

250 mg

S85,831-5
PROPYL N-[2,6-DINITRO-4-(TRIFLUOROMETHYL)PHENYL]-N-PROPYLCARBAMATE

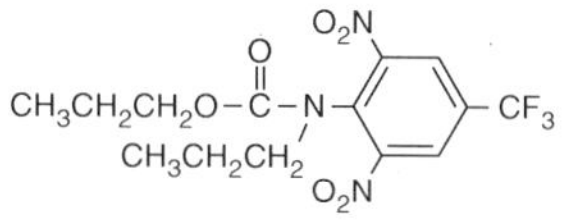

250 mg

S85,832-3
PROPYL N-BUTYL-N-[2,6-DINITRO-4-(TRIFLUOROMETHYL)PHENYL]CARBAMATE

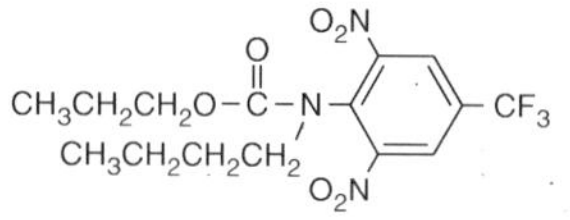

250 mg

S85,724-6
METHYL N-(1-ANILINO-2,2,2-TRICHLOROETHYL)CARBAMATE

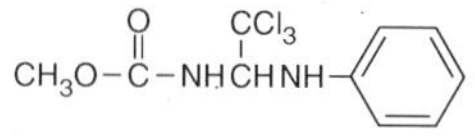

250 mg

S85,741-6
METHYL N-[1-(BENZYLAMINO)-2,2,2-TRICHLOROETHYL]CARBAMATE

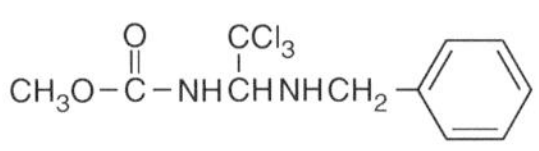

250 mg

S85,725-4
ETHYL N-[1-(BENZYLAMINO)-2,2,2-TRICHLOROETHYL]CARBAMATE

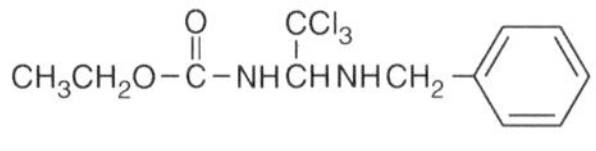

250 mg

S85,742-4
METHYL N-[1-(N-METHYLANILINO)-2,2,2-TRICHLOROETHYL]CARBAMATE

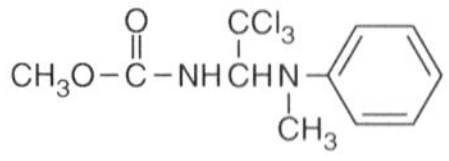

250 mg

S66,810-9
2-DIMETHYLAMINOETHYL N-(3-TRIFLUOROMETHYLPHENYL)-CARBAMATE

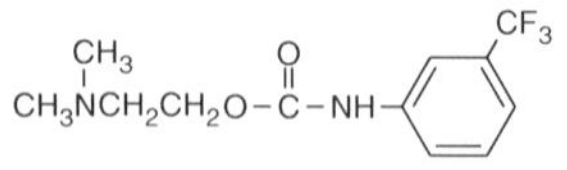

5 gm

S71,362-7
3-DIETHYLAMINOPHENYL N-(3-TRIFLUOROMETHYLPHENYL)-CARBAMATE

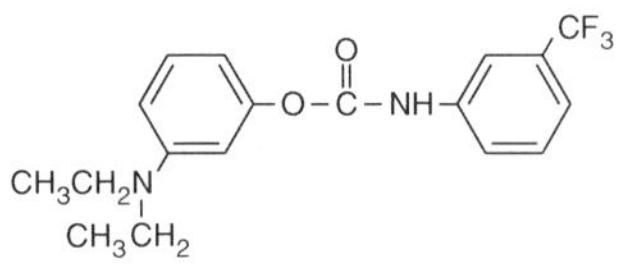

5 gm

S71,480-1
2-DIMETHYLAMINOETHYL N-(2-TRIFLUOROMETHYLPHENYL)-CARBAMATE

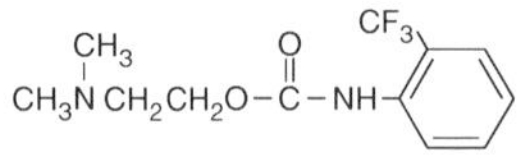

250 mg

S80,175-5
3-(DIETHYLAMINO)PHENYL N-[2-(TRIFLUOROMETHYL)PHENYL]-CARBAMATE

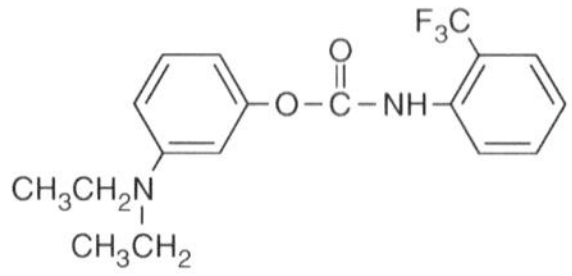

250 mg

S85,711-4
ETHYL N-[1-(4-NITROBENZOYLOXY)-2,2,2-TRICHLOROETHYL]CARBAMATE

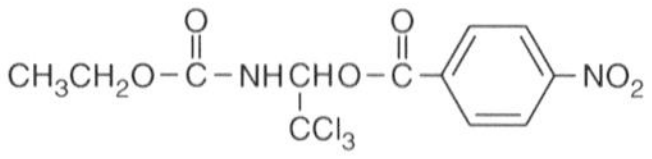

250 mg

S71,760-6
2,6-DIMETHYL-4-NITROPHENYL N-(3-TRIFLUOROMETHYLPHENYL)-CARBAMATE

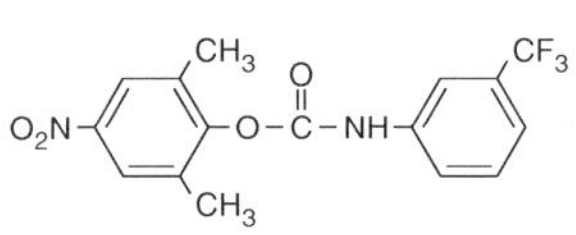

5 gm

S60,452-6
2-BENZOYL-4-CHLOROPHENYL N-(3-TRIFLUOROMETHYLPHENYL)-CARBAMATE

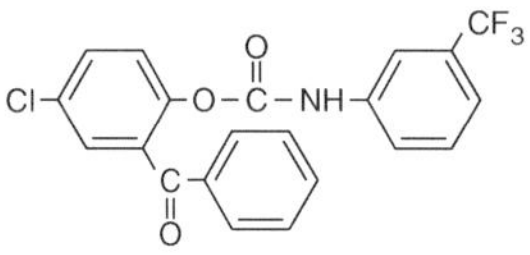

5 gm

S85,660-6
METHYL N-[1-(BENZOYLOXY)-2,2,2-TRICHLOROETHYL]CARBAMATE

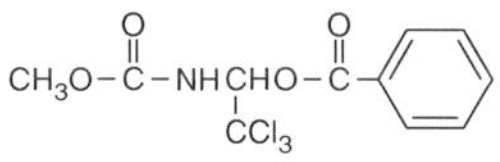

250 mg

S85,657-6
ETHYL N-[1-(BENZOYLOXY)-2,2,2-TRICHLOROETHYL]CARBAMATE

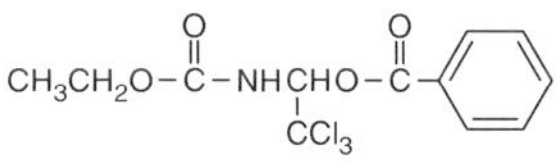

250 mg

S85,665-7
PROPYL N-[1-(BENZOYLOXY)-2,2,2-TRICHLOROETHYL]CARBAMATE

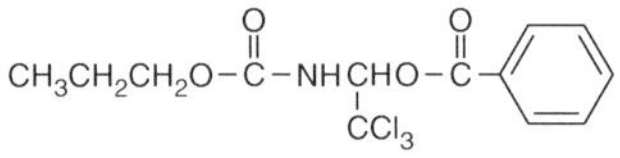

250 mg

S85,678-9
ISOPROPYL N-[1-(BENZOYLOXY)-2,2,2-TRICHLOROETHYL]CARBAMATE

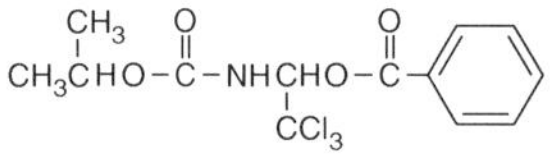

250 mg

S85,658-4
BUTYL N-[1-(BENZOYLOXY)-2,2,2-TRICHLOROETHYL]CARBAMATE

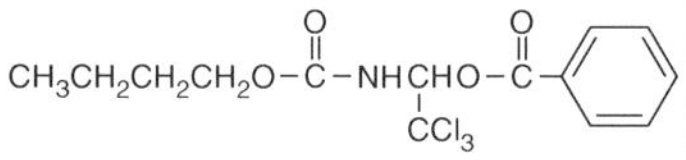

250 mg

S78,117-7
METHYL 2-(N-[2-(TRIFLUOROMETHYL)-PHENYL]CARBAMOYLOXY)-2-METHYLPROPIONATE

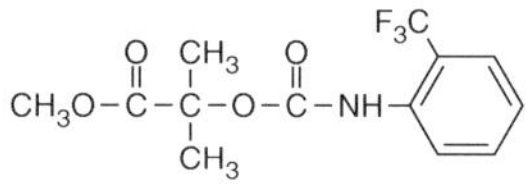

250 mg

S80,176-3
4-(BUTOXYCARBONYL)PHENYL N-[2-(TRIFLUOROMETHYL)PHENYL]-CARBAMATE

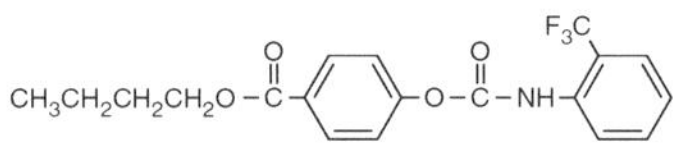

250 mg

S54,538-4
ETHYL 3-(TRIFLUOROMETHYL)-CARBANILATE

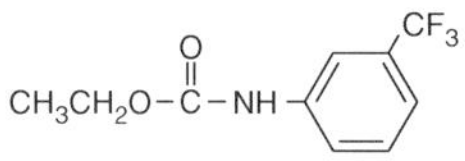

5 gm

S89,879-1
1,2,2,2-TETRACHLOROETHYL N-ETHYL-N-PHENYLCARBAMATE

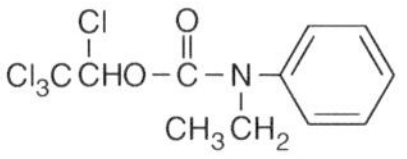

250 mg

S59,769-4
METHYL N-(3-TRIFLUOROMETHYLPHENYL)-CARBAMATE

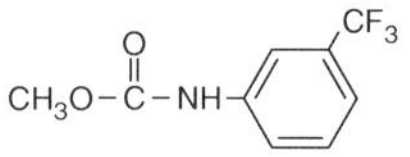

5 gm

S66,901-6
ALLYL N-(3-TRIFLUOROMETHYLPHENYL)-CARBAMATE

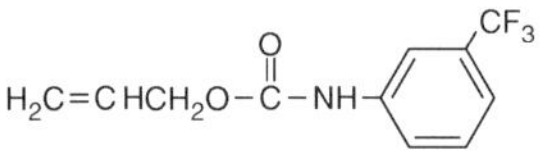

5 gm

S60,087-3
ISOPROPYL N-(3-TRIFLUOROMETHYLPHENYL)-CARBAMATE

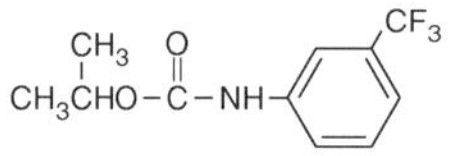

5 gm

S66,794-3
PROPYL N-[3-(TRIFLUOROMETHYL)-PHENYL]CARBAMATE

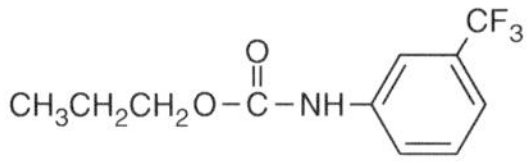

5 gm

S68,684-0
2,2,2-TRICHLOROETHYL N-(3-ETHYLPHENYL)CARBAMATE

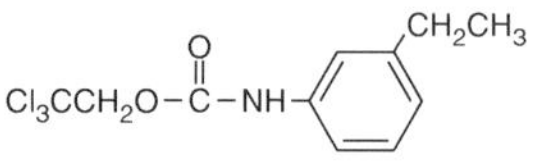

250 mg

S71,176-4
BUTYL N-(3-TRIFLUOROMETHYLPHENYL)-CARBAMATE

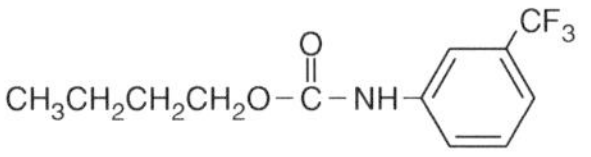

250 mg

S65,367-5
PHENYL N-[3-(TRIFLUOROMETHYL)-PHENYL]CARBAMATE

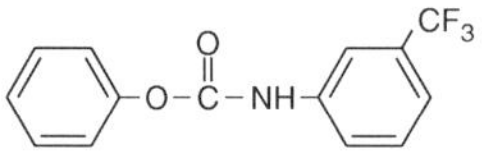

5 gm

S85,573-1
3-(TRIFLUOROMETHYL)PHENYL N-(M-TOLYL)CARBAMATE

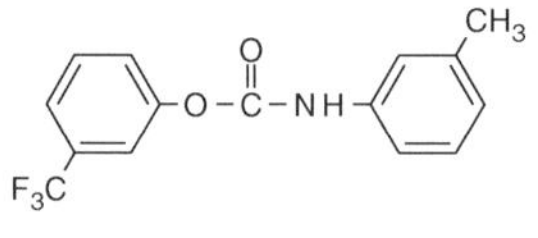

250 mg

S78,059-6
1,2-EPOXY-HEXAHYDRO-4,7-METHANO-5-INDANYL N-3-(TRIFLUOROMETHYL)-PHENYLCARBAMATE

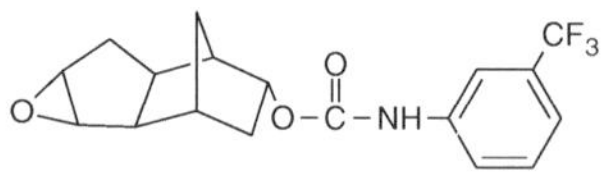

250 mg

S71,761-4
PIPERONYL N-(3-TRIFLUOROMETHYLPHENYL)-CARBAMATE

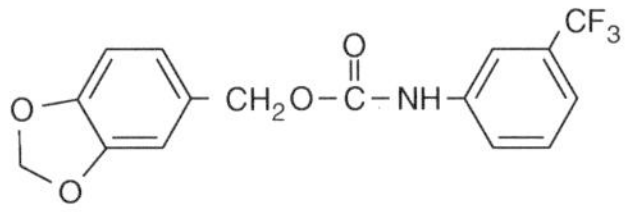

250 mg

S61,210-3
CYCLOHEXYL N-(3-TRIFLUOROMETHYLPHENYL)-CARBAMATE

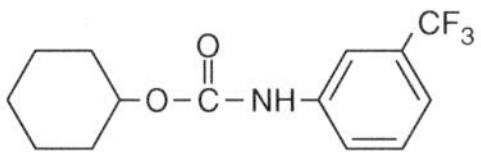

5 gm

S61,187-5
2-METHYLCYCLOHEXYL N-(3-TRIFLUOROMETHYLPHENYL)-CARBAMATE

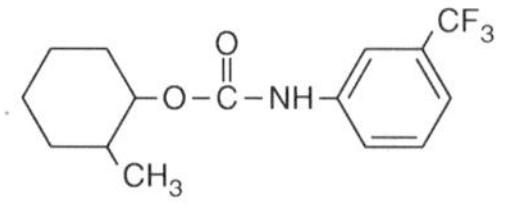

5 gm

S80,177-1
4-CYCLOHEXYLPHENYL N-[2-(TRIFLUOROMETHYL)PHENYL]-CARBAMATE

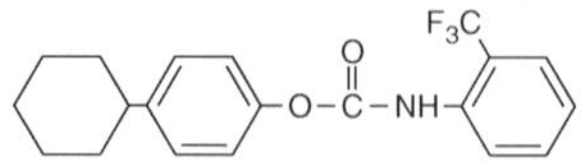

250 mg

S66,542-8
CYCLOHEXYL N-(2-CHLORO-5-TRIFLUOROMETHYLPHENYL)CARBAMATE
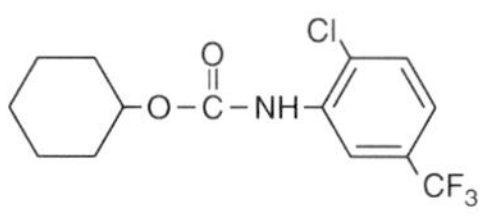
5 gm

S71,544-1
METHYL N-(2-TRIFLUOROMETHYLPHENYL)-CARBAMATE
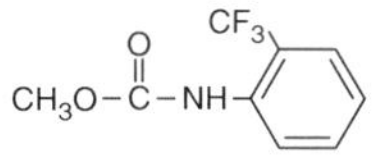
5 gm

S66,732-3
ISOPROPYL N-(2-TRIFLUOROMETHYLPHENYL)-CARBAMATE
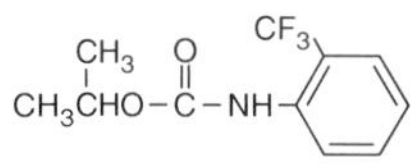
5 gm

S80,167-4
2-BUTENYL N-[2-(TRIFLUOROMETHYL)-PHENYL]CARBAMATE
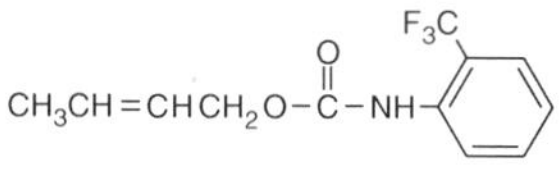
250 mg

S84,549-3
3-(TRIFLUOROMETHYL)PHENYL N-(2-ETHYLPHENYL)CARBAMATE
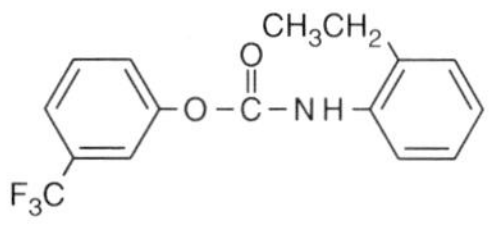
250 mg

S65,348-9
4-CHLOROPHENYL N-(3-TRIFLUOROMETHYLPHENYL)-CARBAMATE
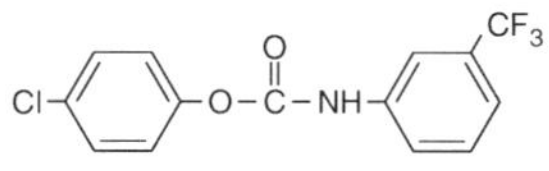
5 gm

S80,119-4
4-TERT-BUTYLPHENYL N-[3-(TRIFLUOROMETHYL)PHENYL]-CARBAMATE
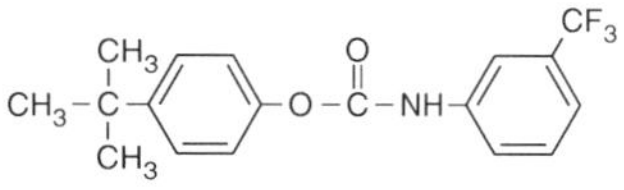
250 mg

S76,999-1
2,2,2-TRIFLUOROETHYL N-(2,5-XYLYL)-CARBAMATE
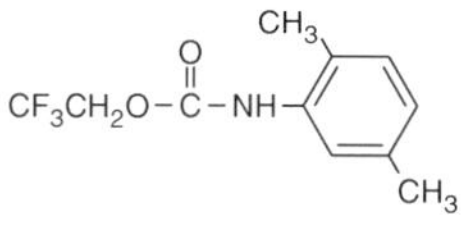
250 mg

S84,550-7
3-(TRIFLUOROMETHYL)PHENYL N-(2,5-XYLYL)CARBAMATE
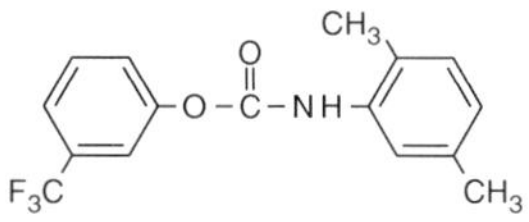
250 mg

S84,443-8
2,2,2-TRICHLOROETHYL N-(1-NAPHTHYL)-CARBAMATE
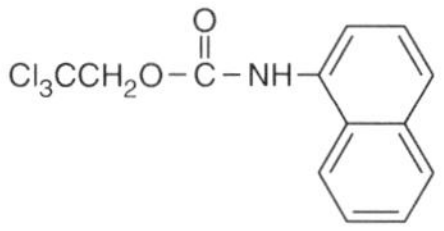
250 mg

S89,837-6
S-(O-TOLYL) N-[3-(TRIFLUOROMETHYL)-PHENYL]THIOCARBAMATE
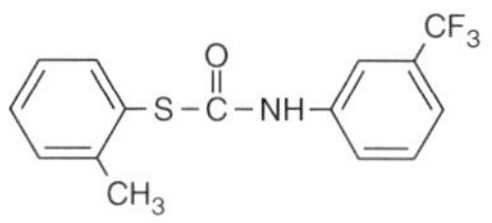
250 mg

S71,703-7
S-ISOPROPYL N-[3-(TRIFLUOROMETHYL)-PHENYL]THIOLCARBAMATE
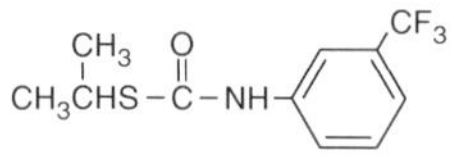
5 gm

S71,704-5
S-PROPYL N-[3-(TRIFLUOROMETHYL)-PHENYL]THIOLCARBAMATE
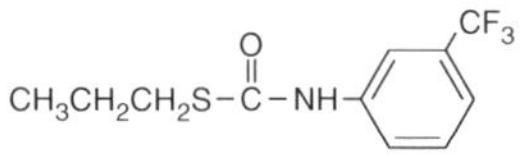
5 gm

S80,183-6
S-BENZYL N-[2-(TRIFLUOROMETHYL)-PHENYL]THIOCARBAMATE
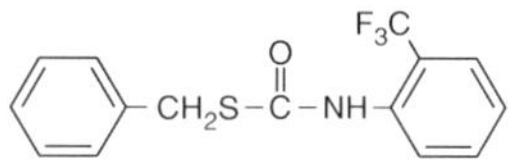
250 mg

S65,363-2
S-PROPYL N-[2-(TRIFLUOROMETHYL)-PHENYL]THIOLCARBAMATE
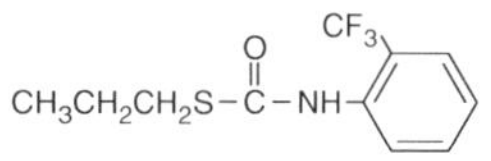
5 gm

S73,183-8
S-DODECYL N-[2-(TRIFLUOROMETHYL)-PHENYL]THIOCARBAMATE
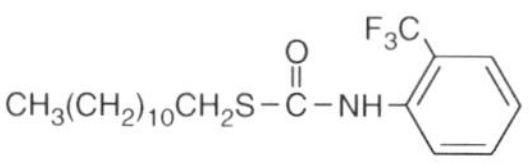
5 gm

S80,187-9
S-(4-TERT-BUTYL-2-METHYLPHENYL) N-[2-(TRIFLUOROMETHYL)PHENYL]-THIOCARBAMATE
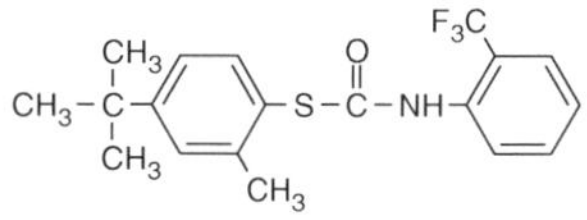
250 mg

S65,012-9
S-(P-TOLYL) N-[3-(TRIFLUOROMETHYL)-PHENYL]THIOLCARBAMATE
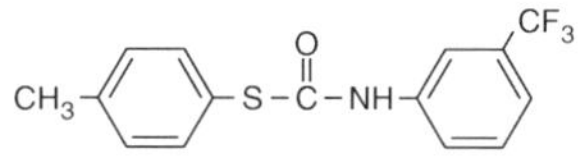
5 gm

S65,345-4
S-PROPYL N-[2-CHLORO-5-(TRIFLUOROMETHYL)PHENYL]-THIOLCARBAMATE
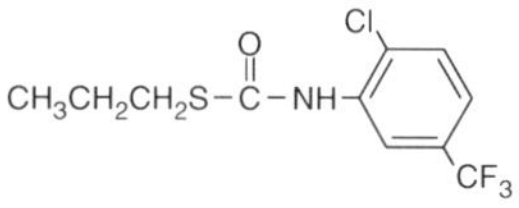
5 gm

S70,240-4
S-(4-HYDROXYPHENYL) N-(5-CHLORO-2,4-DIMETHOXYPHENYL)THIOCARBAMATE
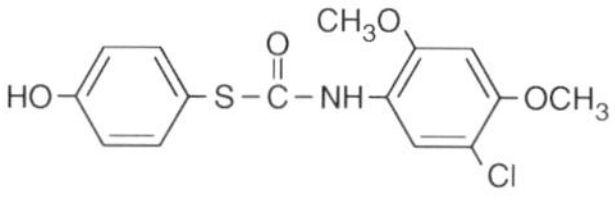
250 mg

S85,439-5
S-(2-HYDROXYPHENYL) N-(4-CHLORO-2-METHOXYPHENYL)THIOCARBAMATE
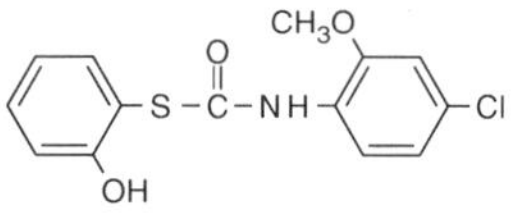
250 mg

S73,704-6
[2-(HYDROXYMETHYL)-6-METHYL-4-METHYLTHIO]PHENYL N-(2-FLUOROPHENYL)CARBAMATE
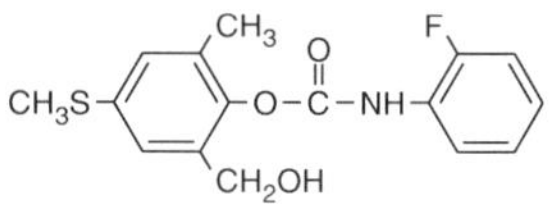
5 gm

S86,776-4
2-[N-(HYDROXYMETHYL)-CARBAMOYL]PHENYL N-(4-CHLOROPHENYL)-CARBAMATE
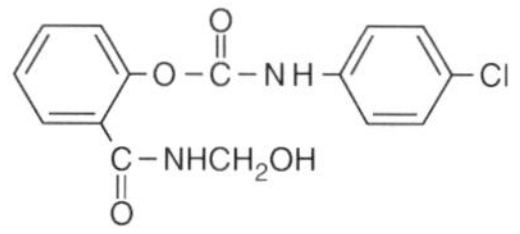
250 mg

S70,065-7
S-(4-HYDROXYPHENYL) N-(2-CHLORO-4-NITROPHENYL)THIOCARBAMATE
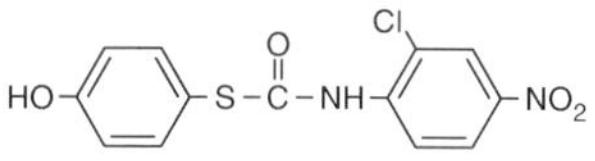
250 mg

S71,881-5
S-(2-HYDROXYPHENYL)-N-(2-FLUOROPHENYL)THIOCARBAMATE
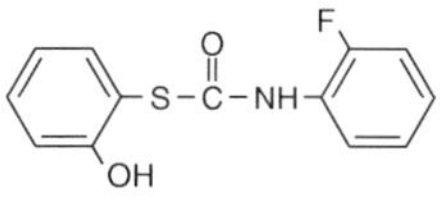
250 mg

S84,445-4
S-(2-HYDROXYPHENYL)-N-(4-FLUOROPHENYL)THIOCARBAMATE
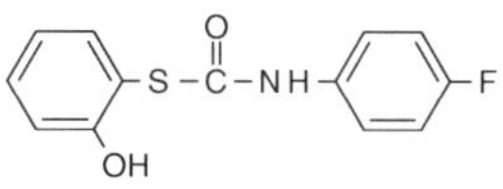
250 mg

S84,477-2
S-(2-HYDROXYPHENYL) N-(3-CHLORO-2-METHYLPHENYL)THIOCARBAMATE
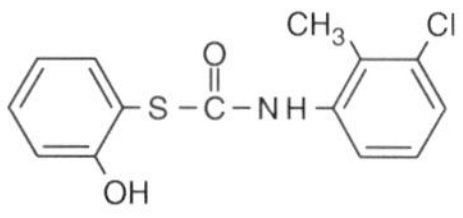
250 mg

S85,130-2
S-(2-HYDROXYPHENYL) N-(2-CHLORO-6-METHYL)THIOCARBAMATE
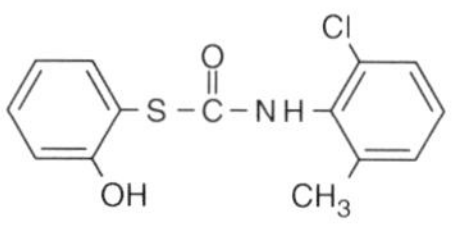
250 mg

S73,634-1
S-(4-HYDROXYPHENYL)-N-(3-FLUOROPHENYL)THIOCARBAMATE
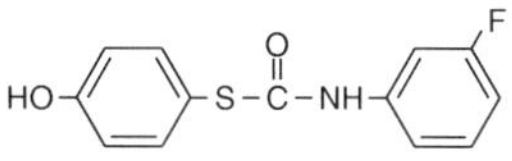
5 gm

S69,086-4
S-(4-HYDROXYPHENYL) N-(2,4,5-TRICHLOROPHENYL)THIOCARBAMATE
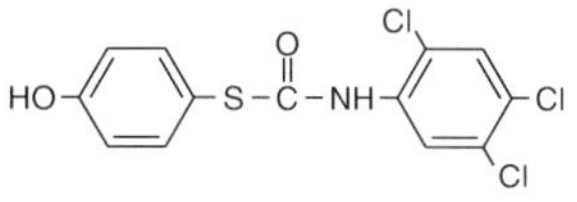
250 mg

S70,264-1
N,N'-(4-BROMO-2-METHYLPHENYLENE)-BIS[S-(4-HYDROXYPHENYL)]-THIOCARBAMATE, ISOMERS
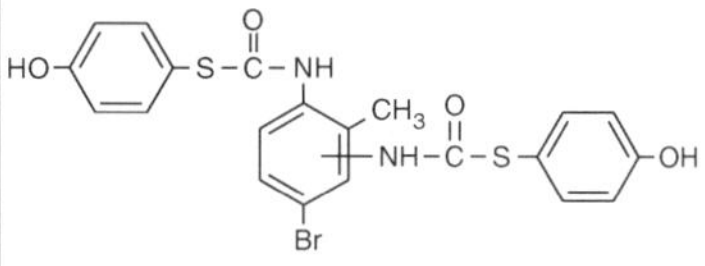
250 mg

S70,011-8
S-(4-HYDROXYPHENYL) N-(2-CHLORO-6-METHYLPHENYL)THIOCARBAMATE
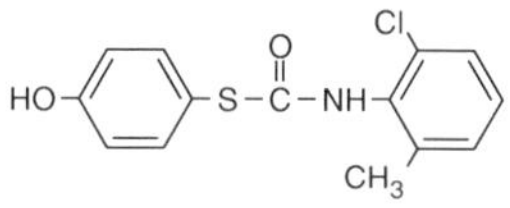
250 mg

S72,797-0
S-(4-HYDROXYPHENYL) N-(2,5-DIFLUOROPHENYL)THIOCARBAMATE
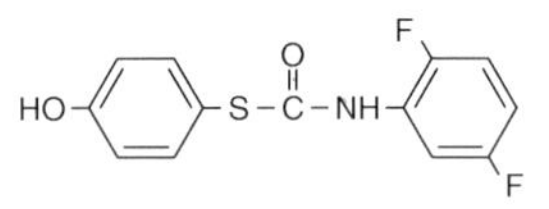
5 gm

S73,902-2
S-(2-HYDROXYETHYL) N-(3-CHLORO-2-METHYLPHENYL)THIOCARBAMATE
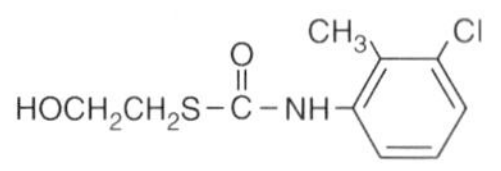
5 gm

S77,360-3
2,3-DIMETHOXYBENZYL N-(2,4,5-TRICHLOROPHENYL)CARBAMATE
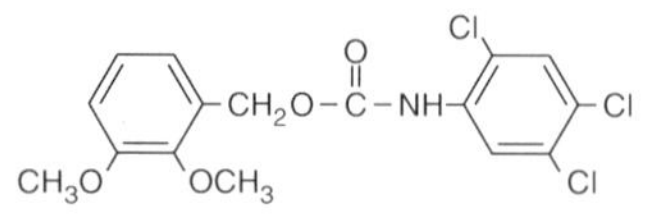
250 mg

S74,751-3
3-FLUOROBENZYL N-(3,4,5-TRIMETHOXYPHENYL)CARBAMATE
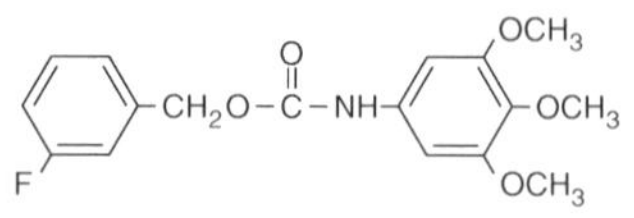
5 gm

S76,944-4
2,4-DIMETHOXYBENZYL N-(4-BROMOPHENYL)CARBAMATE
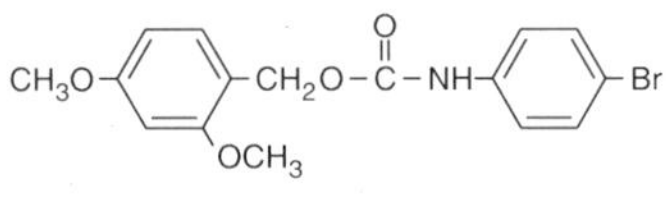
250 mg

S85,084-5
4-(BENZYLOXY)PHENYL N-(2-CHLORO-4-METHOXYPHENYL)CARBAMATE
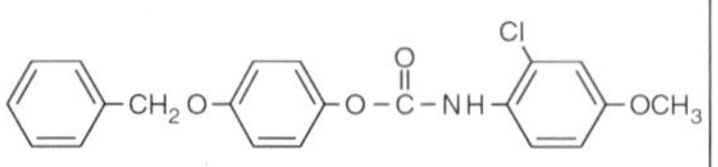
250 mg

S67,391-9
3,4,5-TRIMETHOXYBENZYL N-(2-FLUOROPHENYL)CARBAMATE
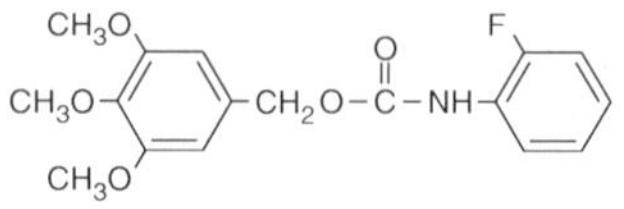
250 mg

S89,966-6
3,4-DIMETHOXYBENZYL N-(3,4-DICHLOROPHENYL)CARBAMATE
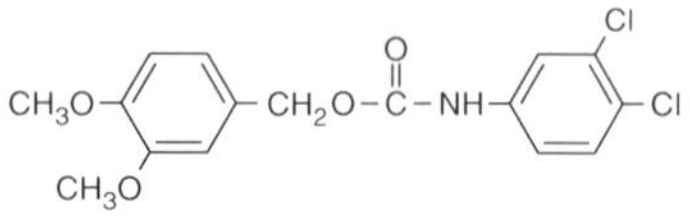
250 mg

S73,178-1
4-(BENZYLOXY)PHENYL N-(2-FLUOROPHENYL)CARBAMATE
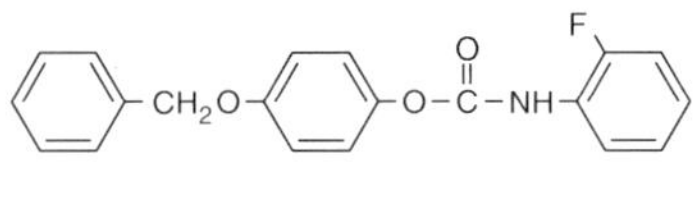
5 gm

S80,053-8
4-(BENZYLOXY)PHENYL N-(4-FLUOROPHENYL)CARBAMATE
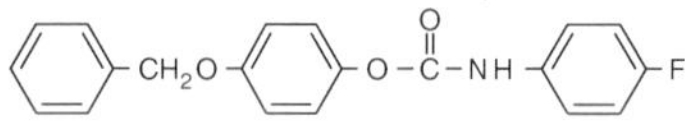
250 mg

S72,962-0
4-(BENZYLOXY)PHENYL N-(3-CHLOROPHENYL)CARBAMATE
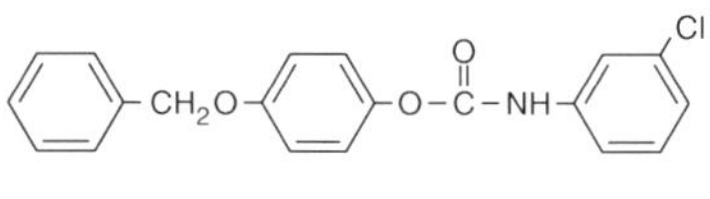
100 mg

S72,818-7
4-METHOXYBENZYL N-(2-CHLORO-5-NITROPHENYL)CARBAMATE
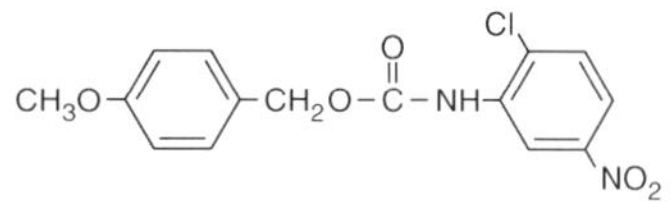
5 gm

S69,403-7
4-BENZYLOXYPHENYL N-(5-CHLORO-2-METHOXYPHENYL)CARBAMATE
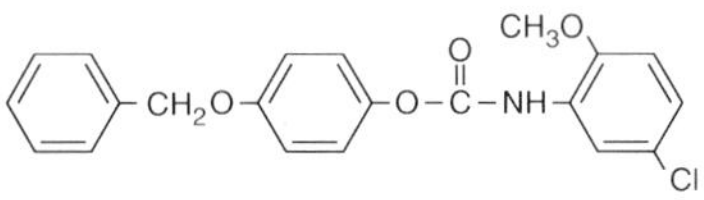
250 mg

S72,679-6
2-FLUOROBENZYL N-(5-CHLORO-2-METHOXYPHENYL)CARBAMATE
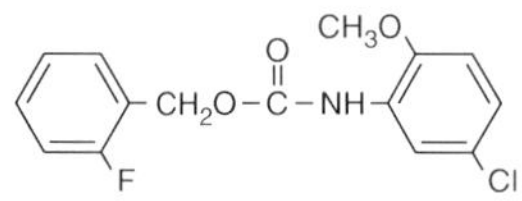
5 gm

S69,440-1
4-FLUOROBENZYL N-(5-CHLORO-2-METHOXYPHENYL)CARBAMATE
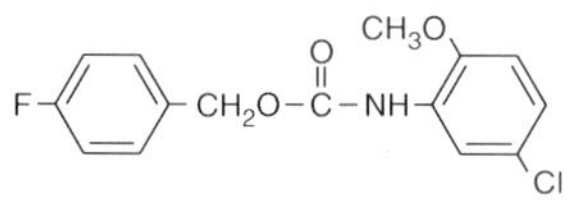
250 mg

S84,802-6
4-(BENZYLOXY)PHENYL N-(2-FLUORO-5-NITROPHENYL)CARBAMATE
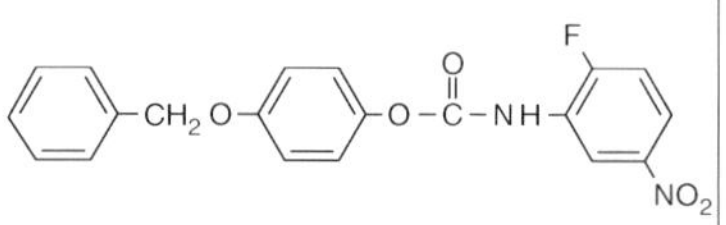
250 mg

S84,748-8
4-(BENZYLOXY)PHENYL N-(2-CHLORO-4-NITROPHENYL)CARBAMATE
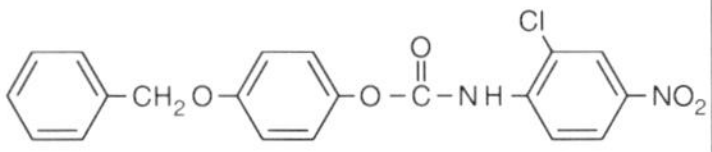
250 mg

S80,032-5
4-METHOXYBENZYL N-(4-FLUOROPHENYL)-CARBAMATE
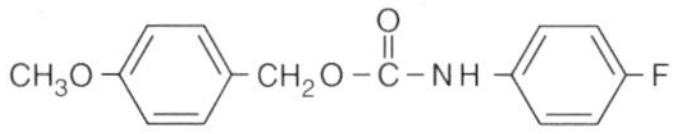
250 mg

S89,999-2
4-METHOXYBENZYL N-(4-BROMOPHENYL)-CARBAMATE
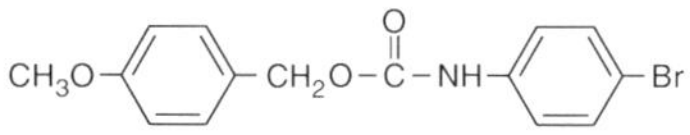
250 mg

S86,808-6
4-METHOXYBENZYL N-(3-CHLORO-2-METHYLPHENYL)CARBAMATE
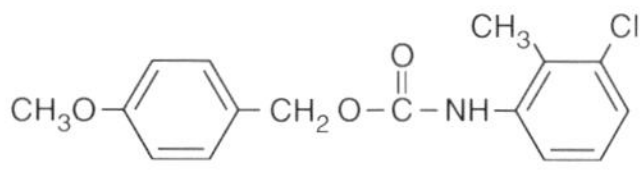
250 mg

S80,789-3
4-METHOXYBENZYL N-(5-CHLORO-2-METHYLPHENYL)CARBAMATE
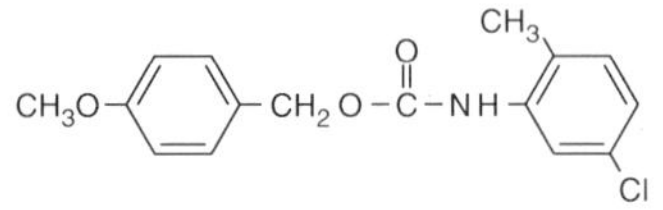
250 mg

S85,405-0
TETRAHYDROFURFURYL N-(5-CHLORO-2,4-DIMETHOXYPHENYL)CARBAMATE
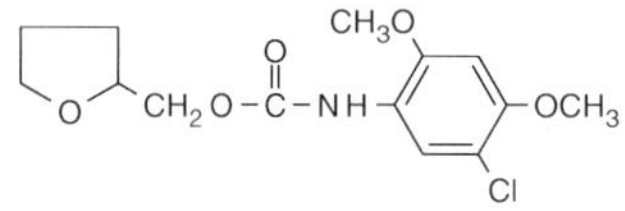
250 mg

S85,394-1
2-(4-BROMOPHENOXY)ETHYL N-(5-CHLORO-2,4-DIMETHOXYPHENYL)-CARBAMATE
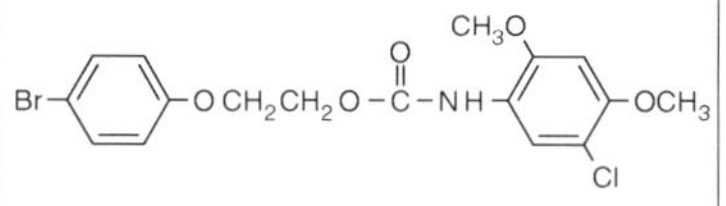
250 mg

S73,170-6
2,6-DIMETHOXYPHENYL N-(5-CHLORO-2,4-DIMETHOXYPHENYL)CARBAMATE
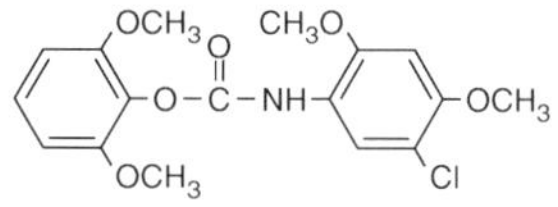
5 gm

S85,387-9
2-(2-CHLORO-4-METHYLPHENOXY)ETHYL N-(5-CHLORO-2,4-DIMETHOXYPHENYL)-CARBAMATE
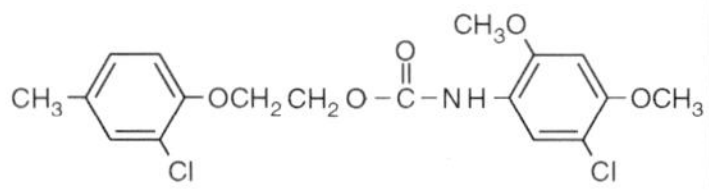
250 mg

S85,385-2
2-(3,4,5-TRIMETHYLPHENOXY)ETHYL N-(5-CHLORO-2,4-DIMETHOXYPHENYL)-CARBAMATE
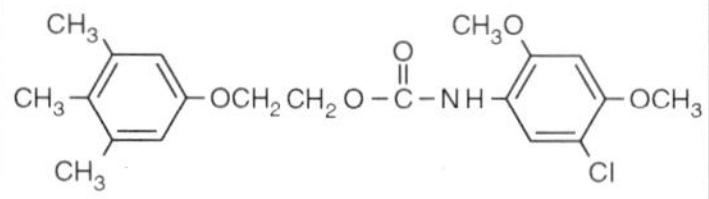
250 mg

S76,942-8
2-(4-METHOXYPHENOXY)ETHYL N-(2-FLUOROPHENYL)CARBAMATE
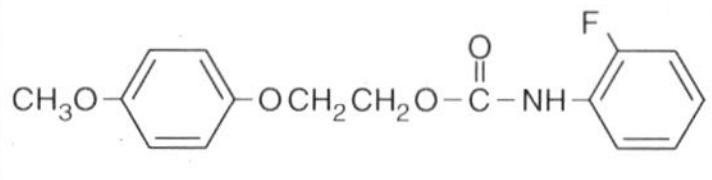
250 mg

S85,393-3
2-(4-NITROPHENOXY)ETHYL N-(5-CHLORO-2,4-DIMETHOXYPHENYL)-CARBAMATE
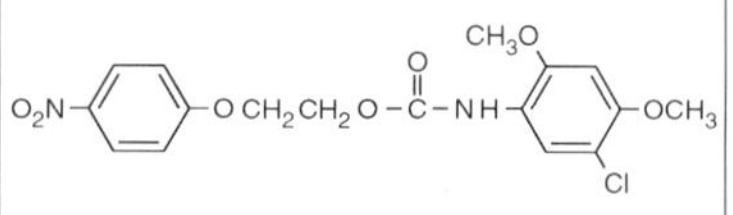
250 mg

S85,095-0
4-FORMYL-2-METHOXYPHENYL N-(4-CHLORO-2-METHOXYPHENYL)-CARBAMATE
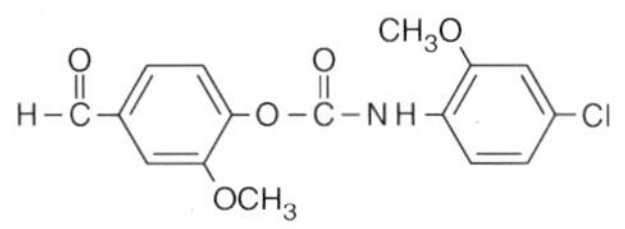
250 mg

S91,180-1
2,6-DIMETHOXYPHENYL N-(5-CHLORO-2-METHOXYPHENYL)CARBAMATE
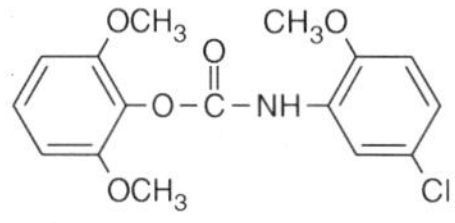
250 mg

S79,439-2
4-ALLYL-2-METHOXYPHENYL N-(5-CHLORO-2-METHOXYPHENYL)-CARBAMATE
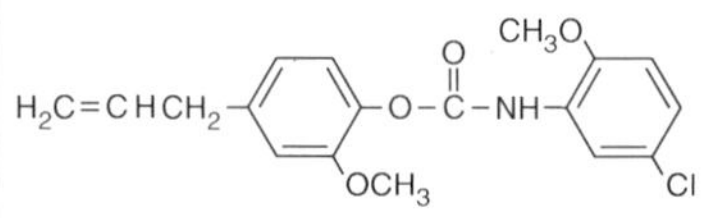
250 mg

S70,001-0
2,6-DIMETHOXYPHENYL N-(4-CHLORO-3-NITROPHENYL)CARBAMATE
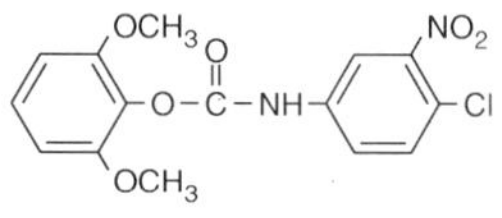
250 mg

S79,672-7
4-FORMYL-2-METHOXYPHENYL N-(5-CHLORO-2,4-DIMETHOXYPHENYL)-CARBAMATE

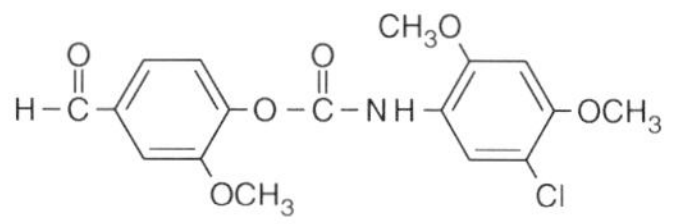

250 mg

S90,246-2
2-(2-CHLOROPHENOXY)ETHYL N-(2-METHOXY-5-METHYLPHENYL)-CARBAMATE

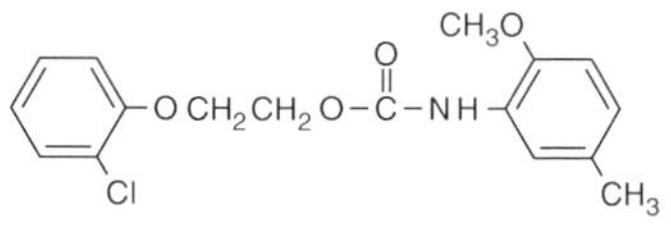

250 mg

S85,343-7
2,6-DIMETHOXYPHENYL N-(4-FLUOROPHENYL)CARBAMATE

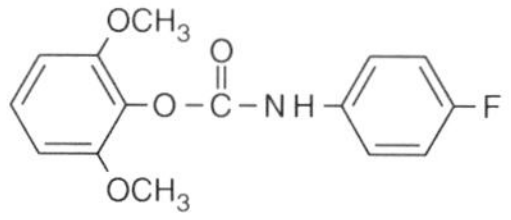

250 mg

S85,530-8
2-METHOXY-4-(4-METHOXYPHENYLIMINOMETHYL)PHENYL N-(3-CHLOROPHENYL)CARBAMATE

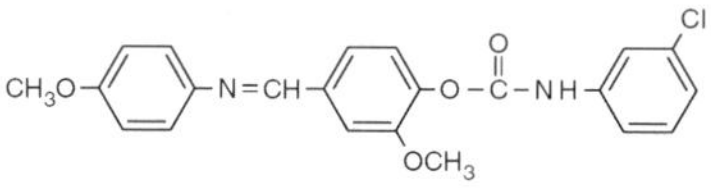

250 mg

S91,108-9
2,6-DIMETHOXYPHENYL N-(2,3-DICHLOROPHENYL)CARBAMATE

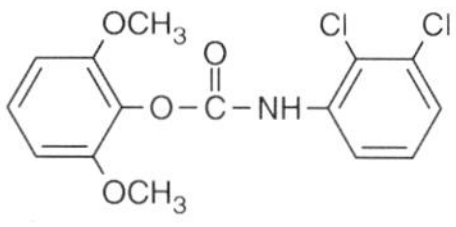

250 mg

S72,777-6
2,6-DIMETHOXYPHENYL N-(2,5-DIFLUOROPHENYL)CARBAMATE

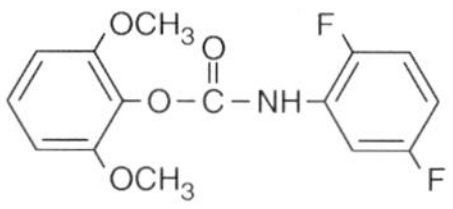

5 gm

S85,426-3
2-[2-METHOXY-4-(1-PROPENYL)PHENOXY]-ETHYL N-(5-CHLORO-2,4-DIMETHOXYPHENYL)CARBAMATE

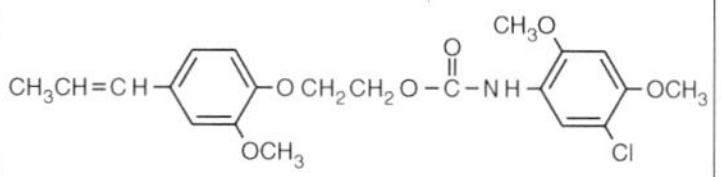

250 mg

S85,390-9
(4-METHOXYPHENOXY)METHYL N-(5-CHLORO-2,4-DIMETHOXYPHENYL)-CARBAMATE

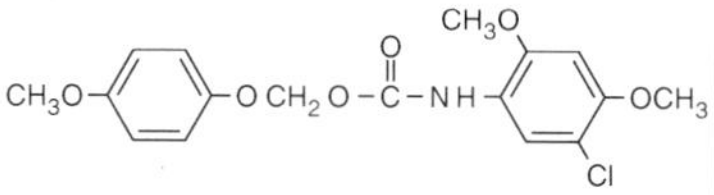

250 mg

S91,160-7
4-BUTOXYPHENYL N-(5-CHLORO-2,4-DIMETHOXYPHENYL)CARBAMATE

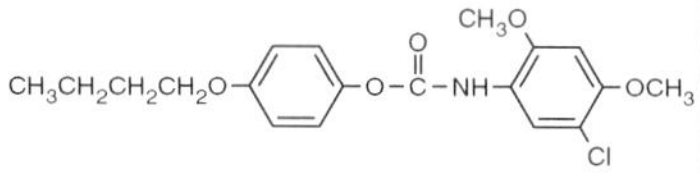

250 mg

S69,123-2
2-CHLORO-4-METHOXYPHENYL N-(5-CHLORO-2,4-DIMETHOXYPHENYL)-CARBAMATE

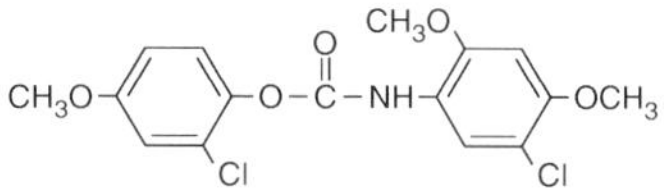

250 mg

S85,366-6
3-(4-METHOXYPHENOXY)PROPYL N-(2,6-DICHLOROPHENYL)CARBAMATE

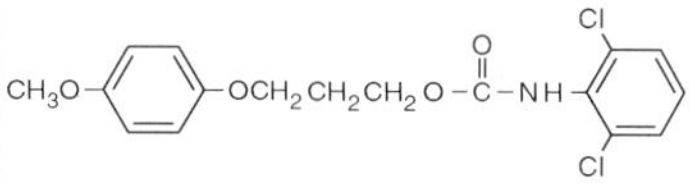

250 mg

S85,398-4
4-(METHYLTHIO)PHENYL N-(5-CHLORO-2,4-DIMETHOXYPHENYL)CARBAMATE

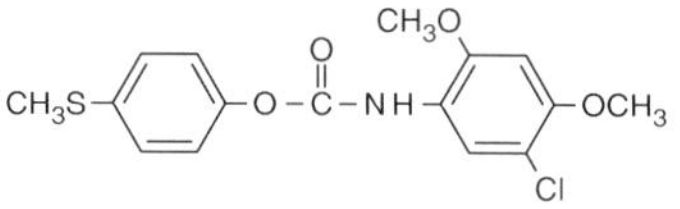

250 mg

S85,391-7
2-(4-CHLOROPHENYLTHIO)ETHYL N-(5-CHLORO-2,4-DIMETHOXYPHENYL)-CARBAMATE

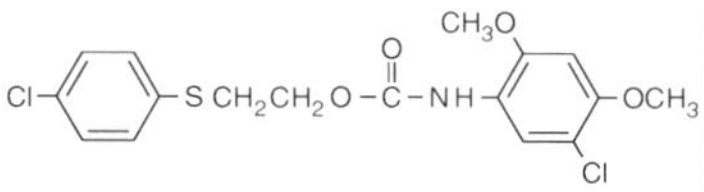

250 mg

S85,392-5
4-ACETAMIDOPHENYL N-(5-CHLORO-2,4-DIMETHOXYPHENYL)CARBAMATE

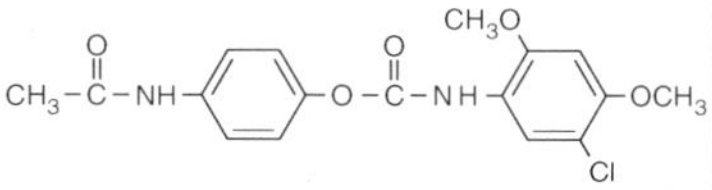

250 mg

S75,129-4
4,5-DIHYDRO-2-IMIDAZOLYL N-(5-CHLORO-2,4-DIMETHOXYPHENYL)-CARBAMATE

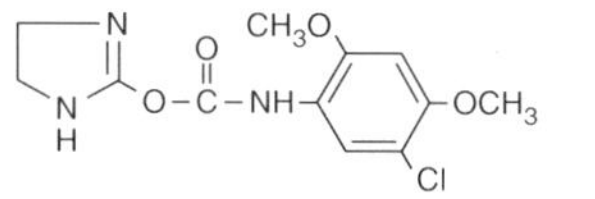

5 gm

S85,402-6
2-FLUOROPHENYL N-(5-CHLORO-2,4-DIMETHOXYPHENYL)CARBAMATE

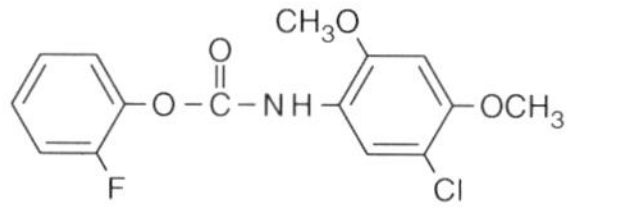

250 mg

S85,413-1
4-FLUOROPHENYL N-(5-CHLORO-2,4-DIMETHOXYPHENYL)CARBAMATE

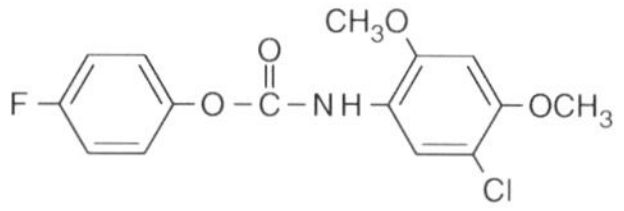

250 mg

S85,408-5
2,6-DICHLOROPHENYL N-(5-CHLORO-2,4-DIMETHOXYPHENYL)CARBAMATE

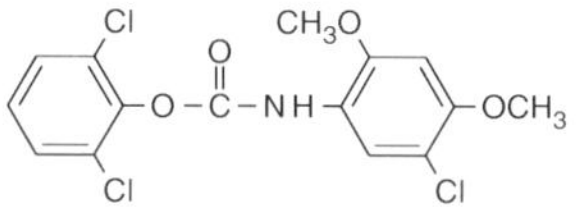

250 mg

S79,638-7
2-CHLORO-4-FLUOROPHENYL N-(2,4-DIMETHOXYPHENYL)CARBAMATE

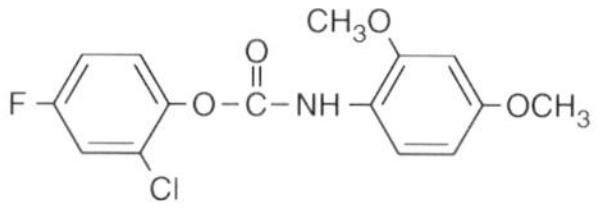

250 mg

S79,636-0
2,4-DIBROMOPHENYL N-(2,4-DIMETHOXYPHENYL)CARBAMATE

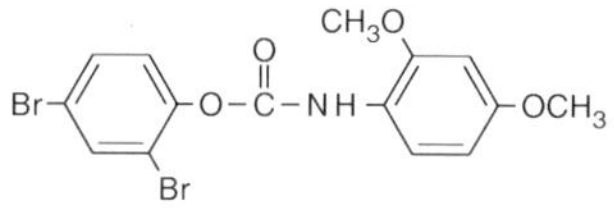

250 mg

S85,407-7
2-CHLORO-4-FLUOROPHENYL N-(5-CHLORO-2,4-DIMETHOXYPHENYL)-CARBAMATE

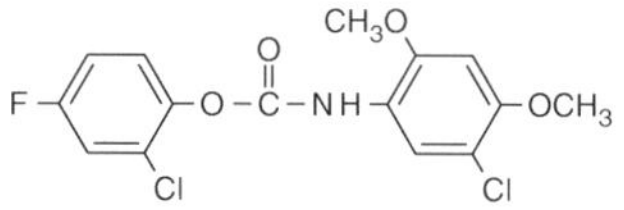

250 mg

S85,411-5
2,4-DIBROMOPHENYL N-(5-CHLORO-2,4-DIMETHOXYPHENYL)CARBAMATE

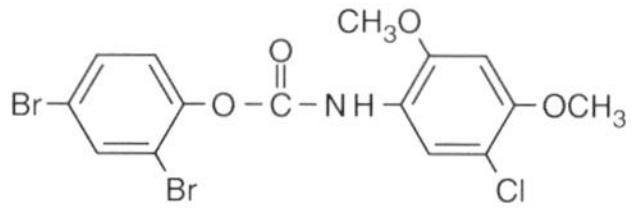

250 mg

S85,424-7
2-BROMO-4-TERT-BUTYLPHENYL N-(5-CHLORO-2,4-DIMETHOXYPHENYL)-CARBAMATE

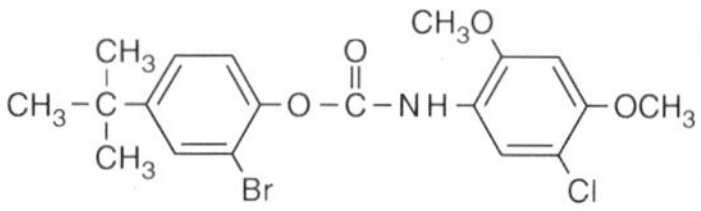

250 mg

S85,410-7
2,4,5-TRICHLOROPHENYL N-(5-CHLORO-2,4-DIMETHOXYPHENYL)CARBAMATE

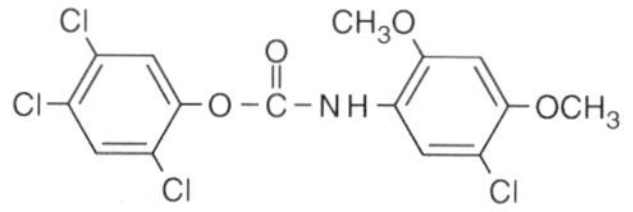

250 mg

S69,031-7
2-BROMO-4-CHLORO-5-METHYLPHENYL N-(5-CHLORO-2,4-DIMETHOXYPHENYL)-CARBAMATE

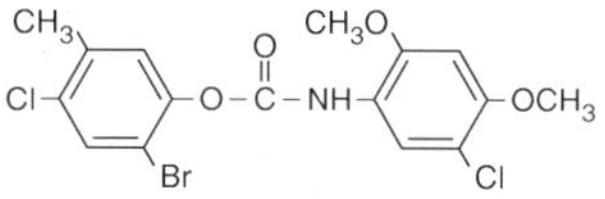

250 mg

S85,396-8
2,4-DICHLORO-3,5-DIMETHYLPHENYL N-(5-CHLORO-2,4-DIMETHOXYPHENYL)-CARBAMATE

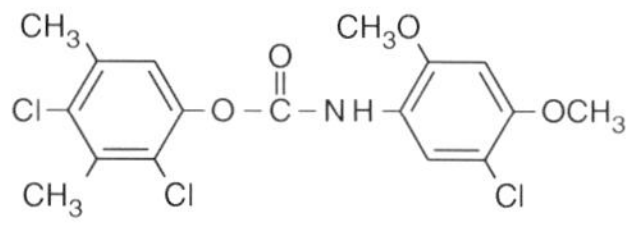

250 mg

S69,368-5
2-BROMO-4,5-DIMETHYLPHENYL N-(5-CHLORO-2,4-DIMETHOXYPHENYL)-CARBAMATE

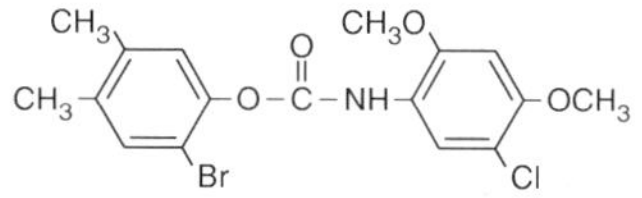

250 mg

S69,428-2
4-BROMO-2,5-DIMETHYLPHENYL N-(5-CHLORO-2,4-DIMETHOXYPHENYL)-CARBAMATE

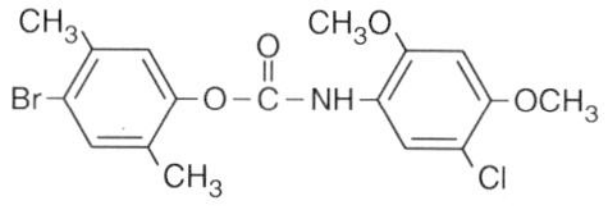

250 mg

S85,427-1
CYCLOTRIDECYL N-(5-CHLORO-2,4-DIMETHOXYPHENYL)CARBAMATE

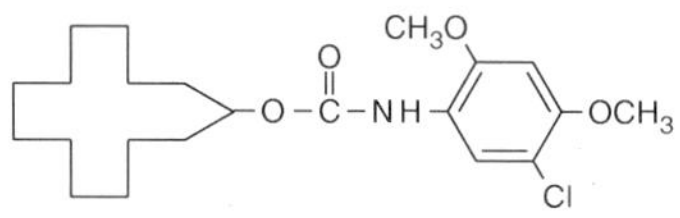

250 mg

S91,159-3
5-INDANYL N-(5-CHLORO-2,4-DIMETHOXYPHENYL)CARBAMATE

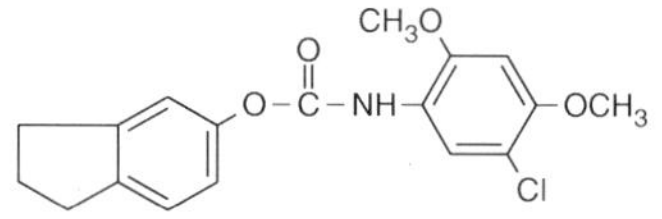

250 mg

S69,375-8
PIPERIDINO N-(5-CHLORO-2,4-DIMETHOXYPHENYL)CARBAMATE

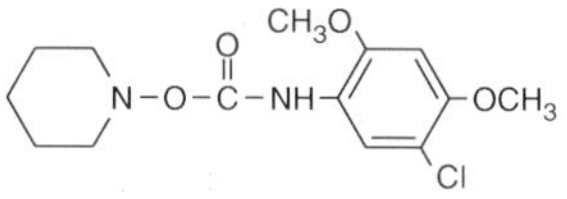

250 mg

S85,418-2
S-BENZYL N-(5-CHLORO-2,4-DIMETHOXYPHENYL)THIOCARBAMATE

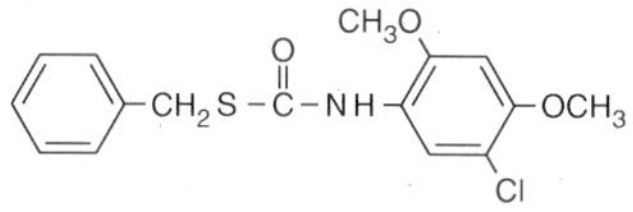

250 mg

S85,096-9
2-FLUOROPHENYL N-(2,4-DIMETHOXYPHENYL)CARBAMATE
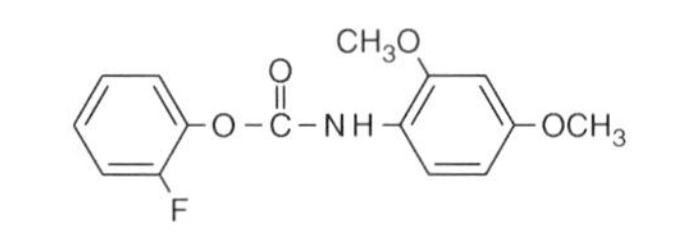
250 mg

S68,618-2
2-CHLORO-4-FLUOROPHENYL N-(3,5-DIMETHOXYPHENYL)CARBAMATE
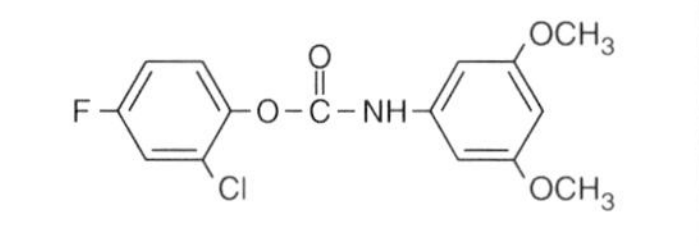
250 mg

S69,355-3
2-BROMO-4-CHLORO-6-METHYLPHENYL N-(5-CHLORO-2,4-DIMETHOXYPHENYL)-CARBAMATE
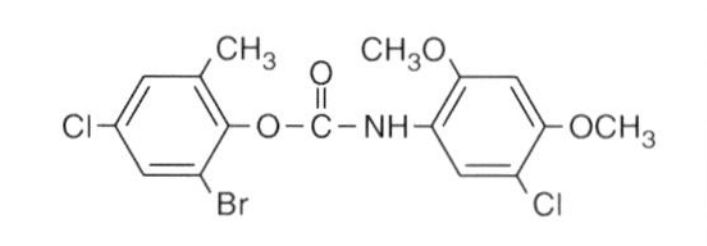
250 mg

S73,905-7
2-CHLORO-4-NITROPHENYL N-(2,5-DIMETHOXYPHENYL)CARBAMATE
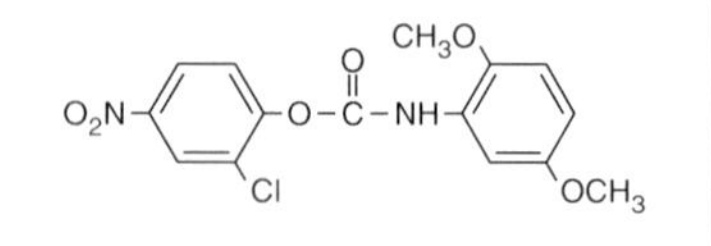
5 gm

S84,861-1
3-CHLORO-4-BIPHENYLYL N-(2,5-DIMETHOXYPHENYL)CARBAMATE
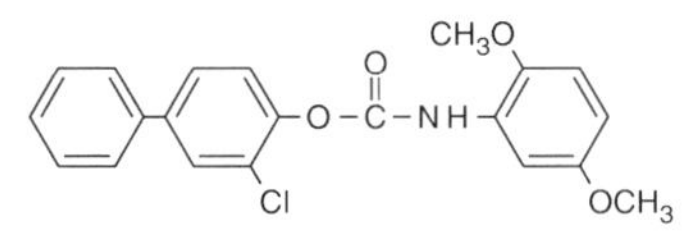
250 mg

S84,824-7
2-FLUOROPHENYL N-(2,5-DIMETHOXYPHENYL)CARBAMATE
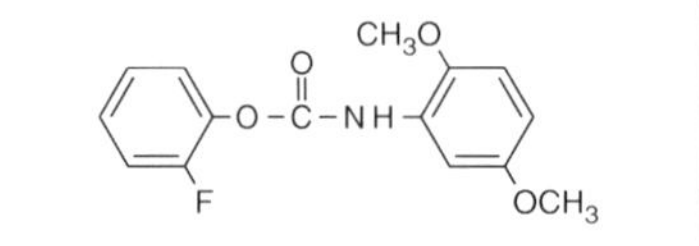
250 mg

S84,825-5
4-FLUOROPHENYL N-(2,5-DIMETHOXYPHENYL)CARBAMATE
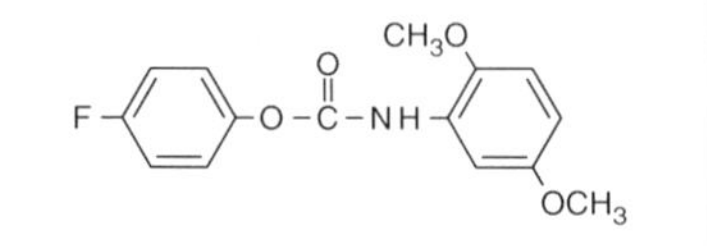
250 mg

S84,822-0
2-CHLORO-4-FLUOROPHENYL N-(2,5-DIMETHOXYPHENYL)CARBAMATE
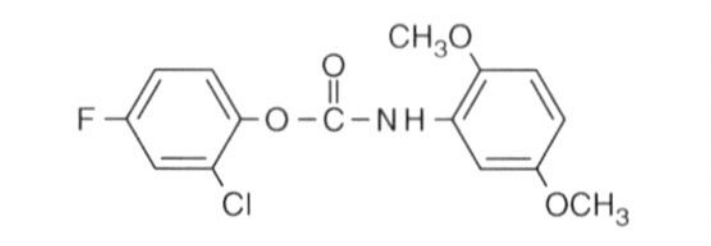
250 mg

S69,346-4
2-CHLORO-4-METHOXYPHENYL N-(4-CHLORO-2-METHOXYPHENYL)-CARBAMATE
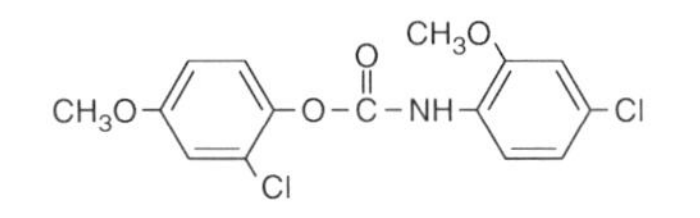
250 mg

S69,351-0
2-CHLORO-4-METHOXYPHENYL N-(5-CHLORO-2-METHOXYPHENYL)-CARBAMATE
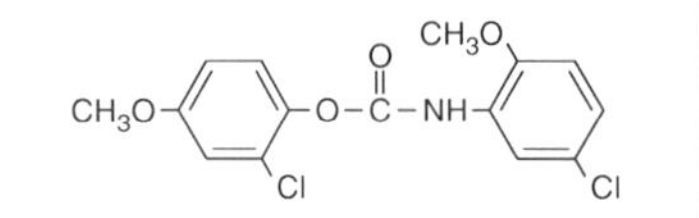
250 mg

S72,320-7
3,4-DICHLOROPHENYL N-(2,5-DIMETHOXYPHENYL)CARBAMATE
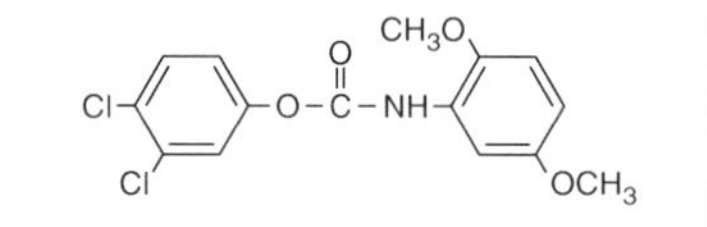
250 mg

S84,821-2
2,4-DIBROMOPHENYL N-(2,5-DIMETHOXYPHENYL)CARBAMATE
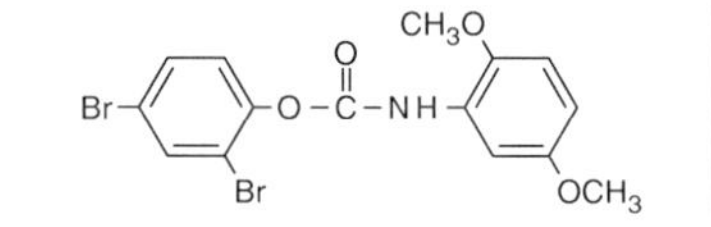
250 mg

S79,665-4
4-CHLORO-2-METHYLPHENYL N-(2,5-DIMETHOXYPHENYL)CARBAMATE
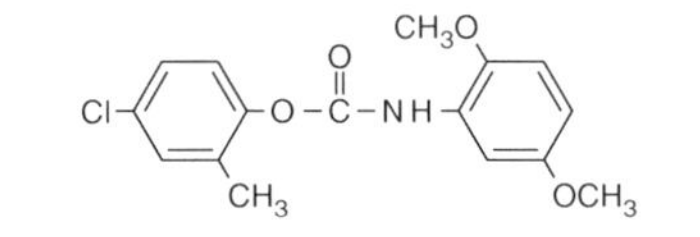
250 mg

S69,124-0
2-CHLORO-4-METHOXYPHENYL N-(2,5-DIMETHOXYPHENYL)CARBAMATE
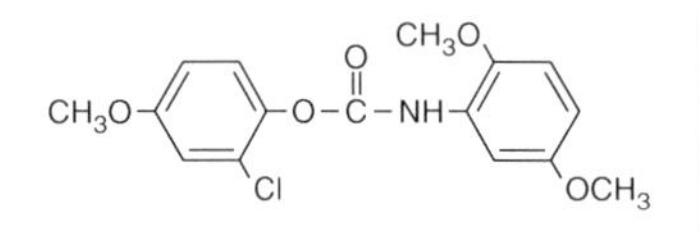
250 mg

S69,171-2
4-BROMO-2,5-DIMETHYLPHENYL N-(2,5-DIMETHOXYPHENYL)CARBAMATE
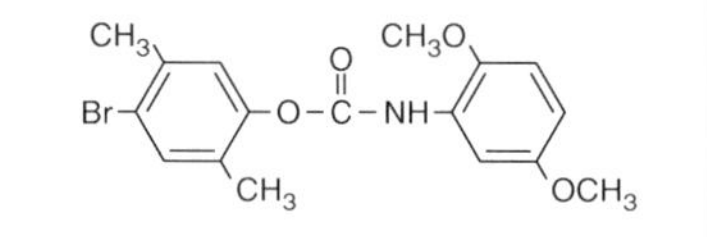
5 gm

S68,941-6
2,5-DIMETHYL-4-IODOPHENYL N-(2,5-DIMETHOXYPHENYL)CARBAMATE
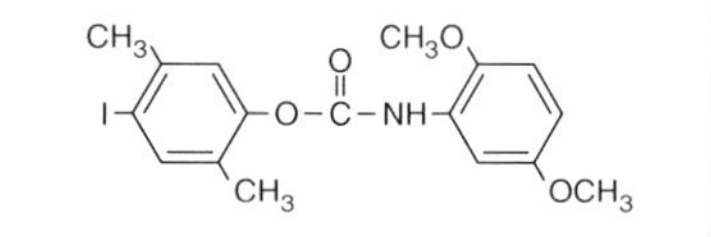
250 mg

S72,763-6
3,5-DICHLOROPHENYL N-(2,5-DIMETHOXYPHENYL)CARBAMATE
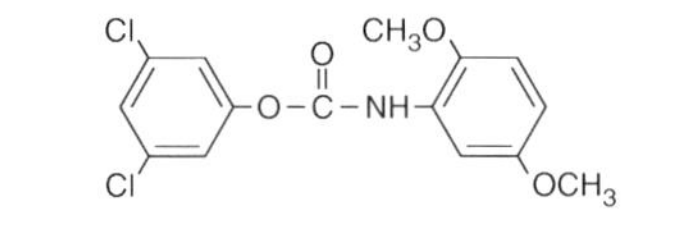
5 gm

S69,445-2
4-BROMO-3,5-DIMETHYLPHENYL N-(2,5-DIMETHOXYPHENYL)CARBAMATE
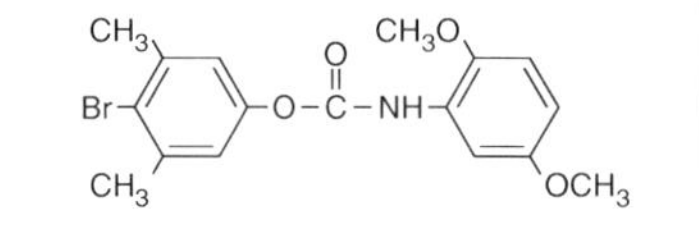
250 mg

S69,137-2
2-CHLORO-4-METHOXYPHENYL N-(3-METHOXYPHENYL)CARBAMATE
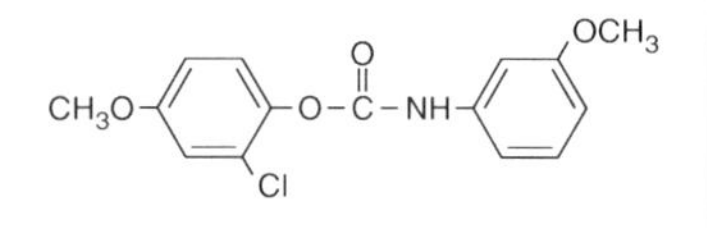
250 mg

S85,412-3
2-CHLORO-4-NITROPHENYL N-(5-CHLORO-2,4-DIMETHOXYPHENYL)CARBAMATE
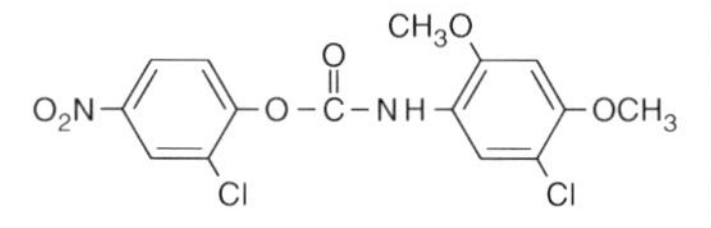
250 mg

S91,156-9
2-ACETYLPHENYL N-(5-CHLORO-2,4-DIMETHOXYPHENYL)CARBAMATE
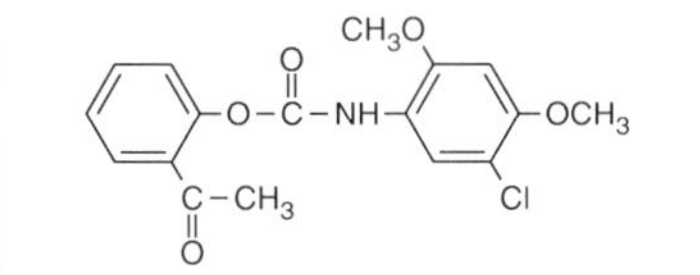
250 mg

S85,423-9
4-(ETHOXYCARBONYL)PHENYL N-(5-CHLORO-2,4-DIMETHOXYPHENYL)-CARBAMATE
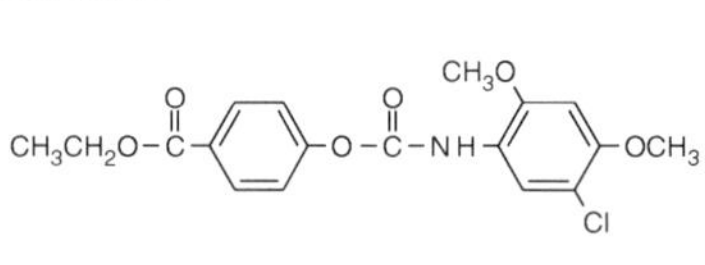
250 mg

S73,722-4
2,4-DICHLORO-6-METHYLPHENYL N-(2,5-DIMETHOXYPHENYL)CARBAMATE
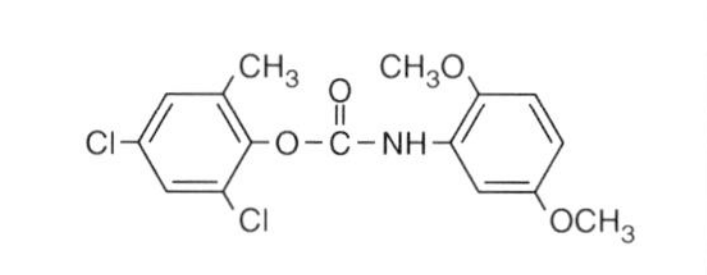
5 gm

S85,444-1
S,S'-ETHYLENE BIS[N-(4-CHLORO-2-METHOXYPHENYL)THIOCARBAMATE]
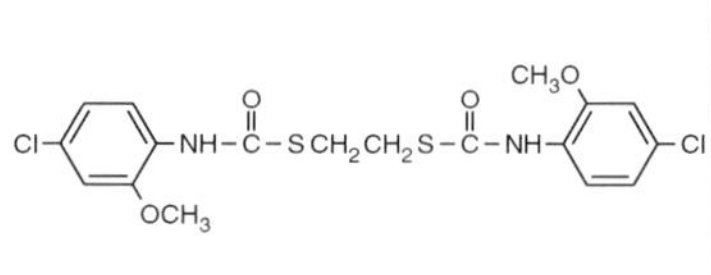
250 mg

S77,440-5
BIS[2-(2,4-DICHLOROPHENOXY)ETHYL] N,N'-(METHYLENEDI-4,1-PHENYLENE)-BISCARBAMATE
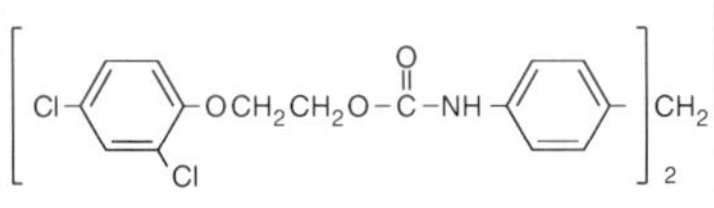
250 mg

S92,433-4
2-(2,4-DICHLOROPHENOXY)ETHYL N-(ETHOXYCARBONYLMETHYL)CARBAMATE
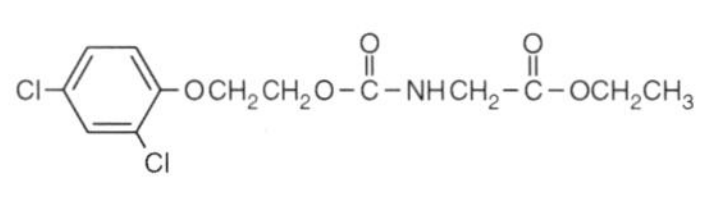
250 mg

S92,434-2
2-(4-CHLORO-3-METHYLPHENOXY)ETHYL N-(ETHOXYCARBONYLMETHYL)CARBAMATE
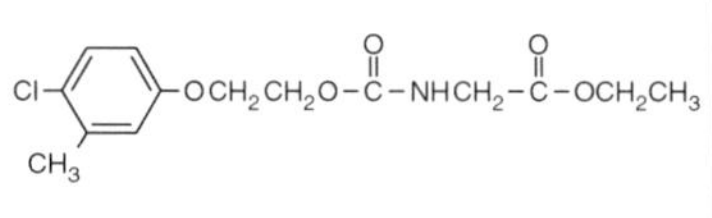
250 mg

S85,386-0
2-NAPHTHYL N-(5-CHLORO-2,4-DIMETHOXYPHENYL)CARBAMATE
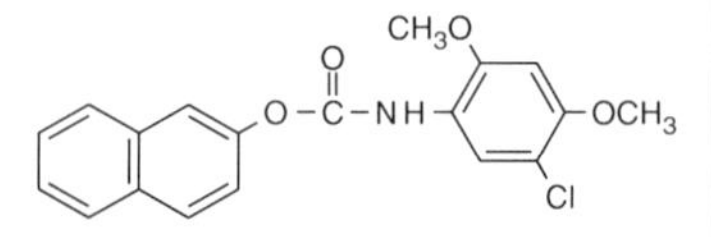
250 mg

S91,157-7
2-(2-NAPHTHOXY)ETHYL N-(5-CHLORO-2,4-DIMETHOXYPHENYL)CARBAMATE
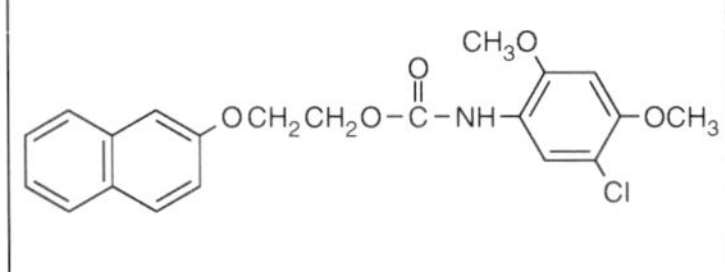
250 mg

S80,787-7
4-FORMYL-2-METHOXYPHENYL N-(5-CHLORO-2-METHYLPHENYL)-CARBAMATE
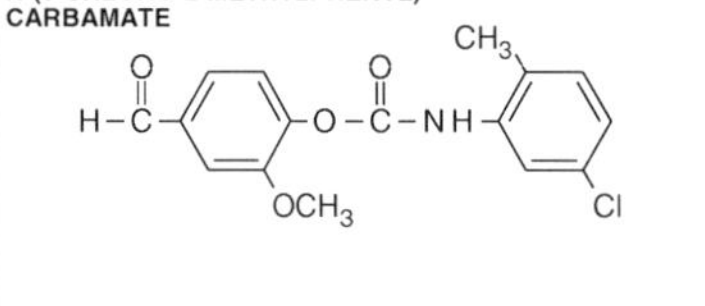
250 mg

S86,797-7
4-FORMYL-2-METHOXYPHENYL N-(2-BROMO-4-METHYLPHENYL)CARBAMATE
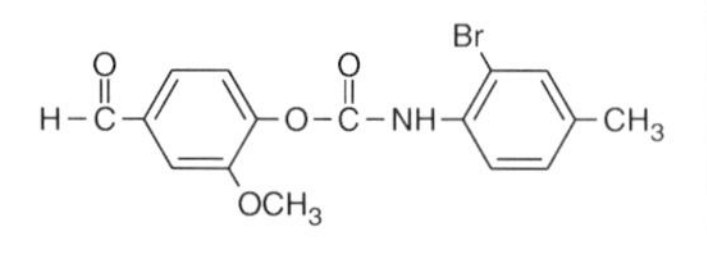
250 mg

S85,355-0
2-(2-CHLOROPHENOXY)ETHYL N-[3-(METHYLTHIO)PHENYL]CARBAMATE
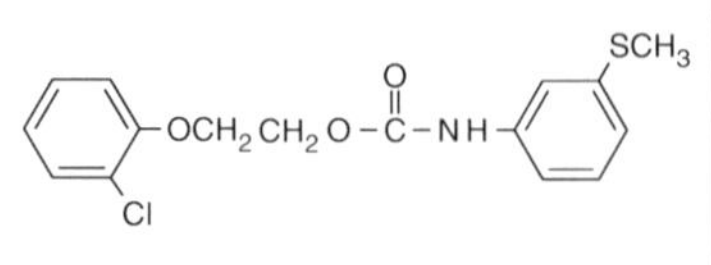
250 mg

S85,354-2
2-(4-BROMOPHENOXY)ETHYL N-[3-(METHYLTHIO)PHENYL]CARBAMATE

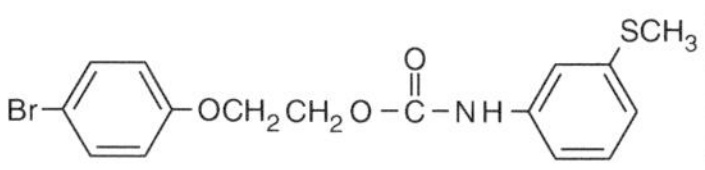

250 mg

S85,359-3
2-(4-CHLORO-3,5-DIMETHYLPHENOXY)-ETHYL N-[3-(METHYLTHIO)PHENYL]-CARBAMATE

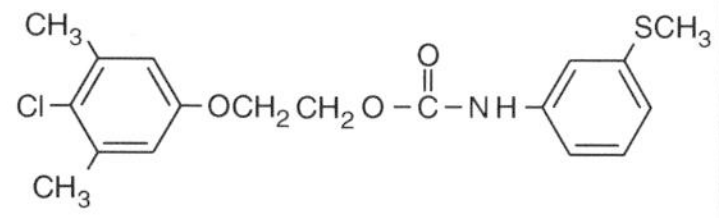

250 mg

S79,655-7
2-(2,4-DICHLOROPHENOXY)ETHYL N-(4-METHYL-2-NITROPHENYL)CARBAMATE

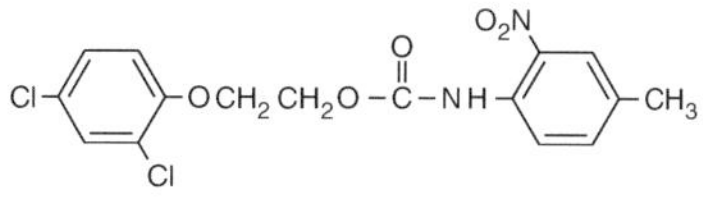

250 mg

S72,820-9
2-PHENOXYETHYL N-(2-CHLORO-5-NITROPHENYL)CARBAMATE

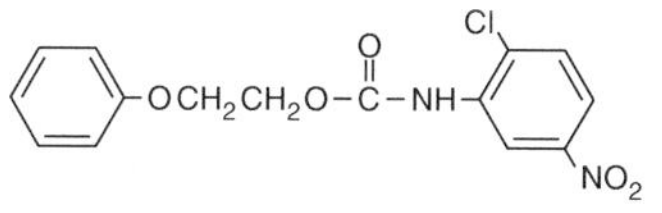

5 gm

S84,520-5
2-PHENOXYETHYL N-(4-CHLORO-3-NITROPHENYL)CARBAMATE

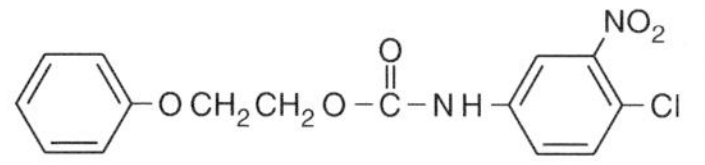

250 mg

S85,375-5
2-(2-CHLOROPHENOXY)ETHYL N-(4-METHYL-3-NITROPHENYL)CARBAMATE

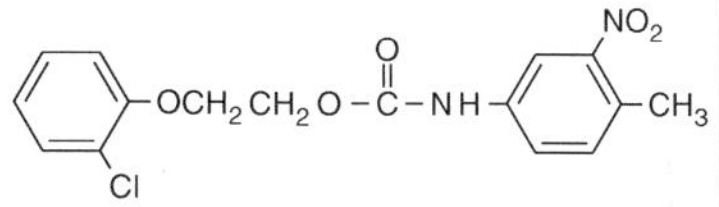

250 mg

S85,374-7
2-(4-CHLOROPHENOXY)ETHYL N-(4-METHYL-3-NITROPHENYL)CARBAMATE

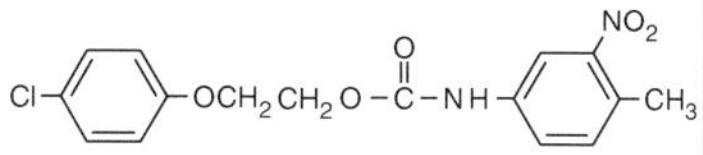

250 mg

S72,821-7
4-ALLYL-2-METHOXYPHENYL N-(2-CHLORO-5-NITROPHENYL)CARBAMATE

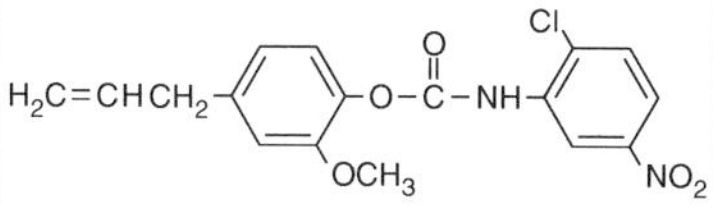

5 gm

S62,691-0
2-(2,4-DICHLOROPHENOXY)ETHYL O-(3-CHLOROPHENYL)CARBAMOYLLACTATE

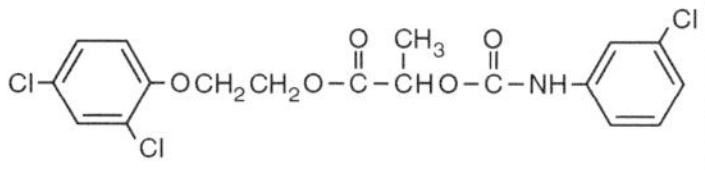

5 gm

S73,734-8
4-FORMYL-2-METHOXYPHENYL N-(5-CHLORO-2-METHOXYPHENYL)-CARBAMATE

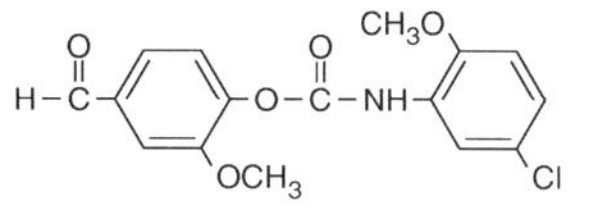

5 gm

S89,104-5
N,N-BIS[2-(N-DODECYLCARBAMOYLOXY)-ETHYL]-2-(4-CHLOROPHENOXY)ACETAMIDE

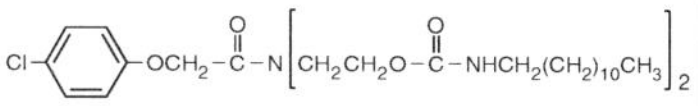

250 mg

S84,509-4
4-FORMYL-2-METHOXYPHENYL N-(2-CHLORO-4-NITROPHENYL)CARBAMATE

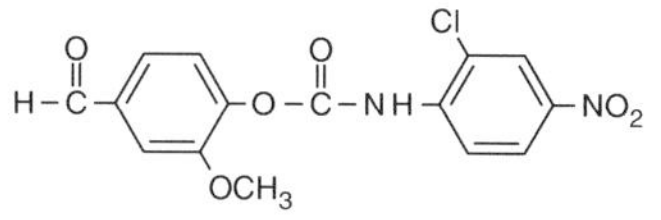

250 mg

S85,342-9
2-(4-NITROPHENOXY)ETHYL N-(4-FLUOROPHENYL)CARBAMATE

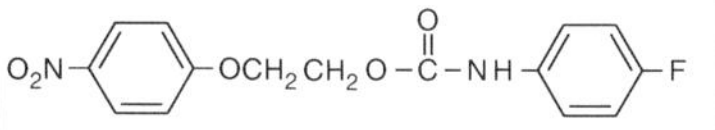

250 mg

S71,821-1
2-(4-NITROPHENOXY)ETHYL N-(2,4-DIFLUOROPHENYL)CARBAMATE

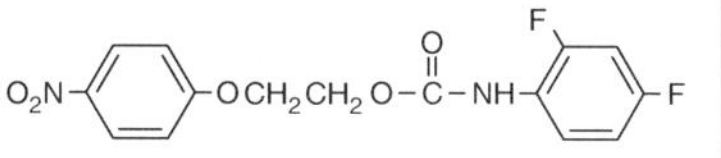

250 mg

S85,369-0
2-(4-NITROPHENOXY)ETHYL N-(2,5-DIFLUOROPHENYL)CARBAMATE

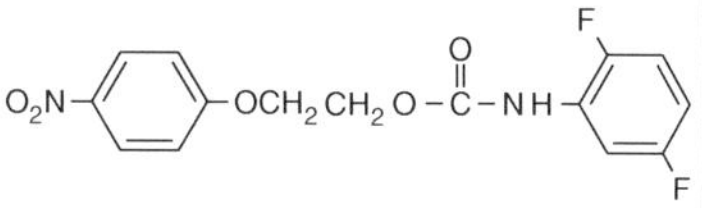

250 mg

S85,121-3
4-FORMYL-2-METHOXYPHENYL N-(2-CHLORO-6-METHYLPHENYL)-CARBAMATE

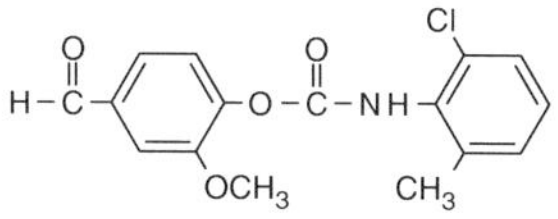

250 mg

S90,565-8
4-FORMYL-2-METHOXYPHENYL N-(3-CHLORO-2-METHYLPHENYL)-CARBAMATE

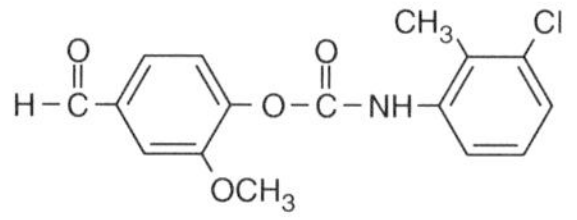

250 mg

S85,064-0
4-FORMYL-2-METHOXYPHENYL N-(3-CHLORO-4-FLUOROPHENYL)-CARBAMATE

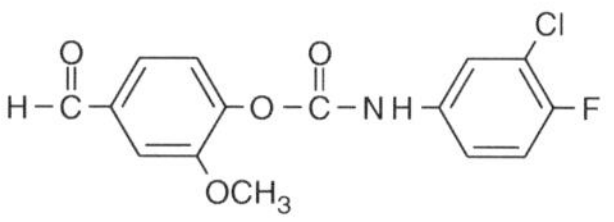

250 mg

S84,528-0
2-PHENOXYETHYL N-(3-FLUOROPHENYL)-CARBAMATE

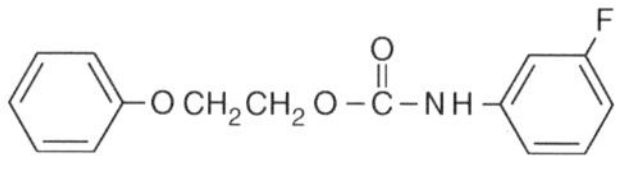

250 mg

S80,128-3
2-PHENOXYETHYL N-(2-FLUOROPHENYL)-CARBAMATE

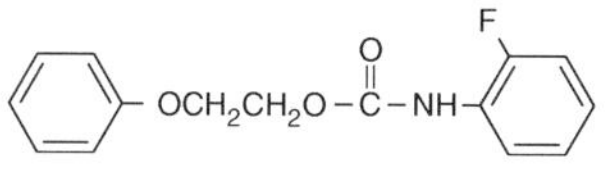

250 mg

S76,902-9
2-PHENOXYETHYL N-(2-CHLOROPHENYL)-CARBAMATE

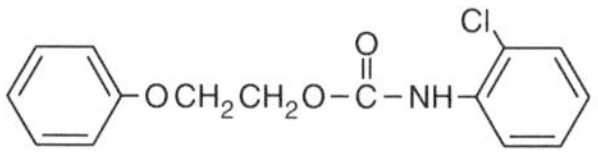

250 mg

S85,537-5
2-PHENOXYETHYL N-(4-BROMOPHENYL)-CARBAMATE

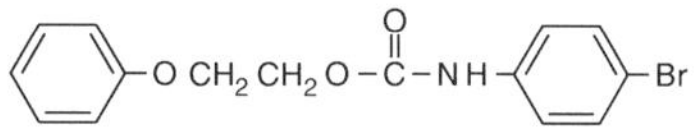

250 mg

S80,790-7
2-PHENOXYETHYL N-(5-CHLORO-2-METHYLPHENYL)CARBAMATE

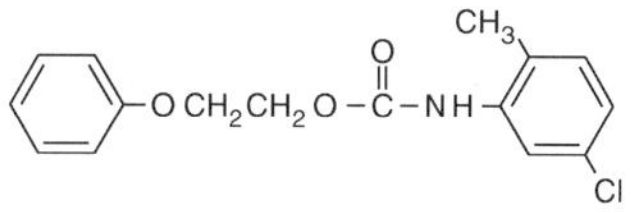

250 mg

S73,685-6
2-ETHOXYPHENYL N-(2-FLUOROPHENYL)-CARBAMATE

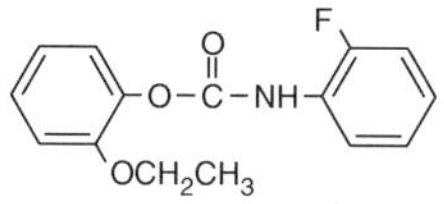

5 gm

S85,363-1
2-(2-CHLOROPHENOXY)ETHYL N-(2,6-DICHLOROPHENYL)CARBAMATE

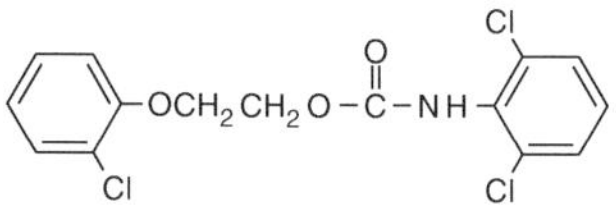

250 mg

S68,682-4
2-(2-CHLOROPHENOXY)ETHYL N-(2,4-DIFLUOROPHENYL)CARBAMATE

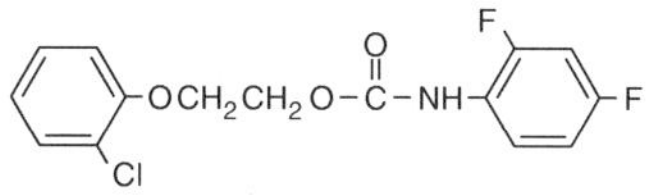

250 mg

S79,147-4
2-ETHOXYPHENYL N-(3-CHLORO-4-METHYLPHENYL)CARBAMATE

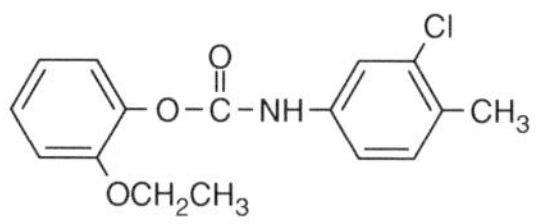

250 mg

S66,593-2
2-ETHOXYETHYL N-(4-BROMOPHENYL)-CARBAMATE

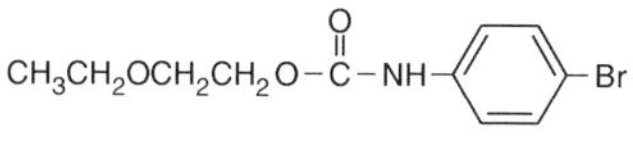

5 gm

S90,587-9
2-(4-BROMOPHENOXY)ETHYL N-(4-FLUOROPHENYL)CARBAMATE

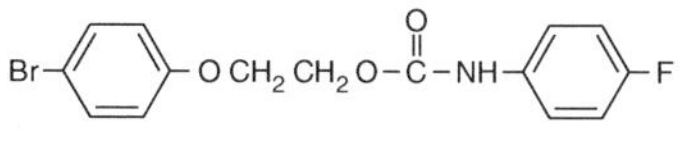

250 mg

S89,831-7
2-(4-TERT-BUTYLPHENOXY)ETHYL N-(4-FLUOROPHENYL)CARBAMATE

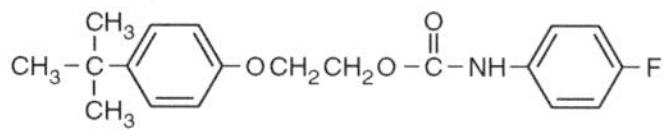

250 mg

S85,351-8
2-(4-BROMOPHENOXY)ETHYL N-(2,6-DICHLOROPHENYL)CARBAMATE

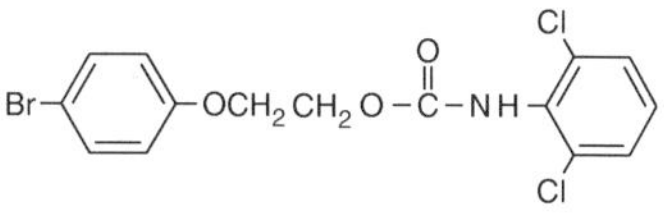

250 mg

S74,308-9
2-(4-CHLOROPHENOXY)ETHYL N-(2,4-DIFLUOROPHENYL)CARBAMATE

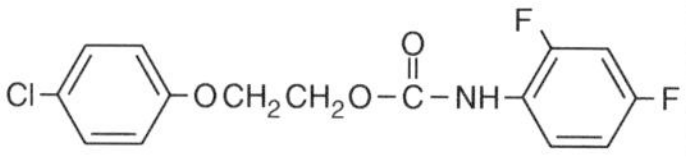

5 gm

S90,585-2
2-(2,4-DICHLOROPHENOXY)ETHYL N-(4-FLUOROPHENYL)CARBAMATE
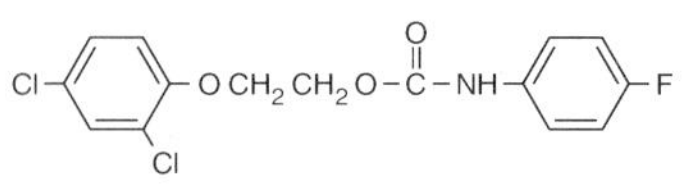
250 mg

S90,586-0
2-(4-CHLORO-3-METHYLPHENOXY)ETHYL N-(4-FLUOROPHENYL)CARBAMATE
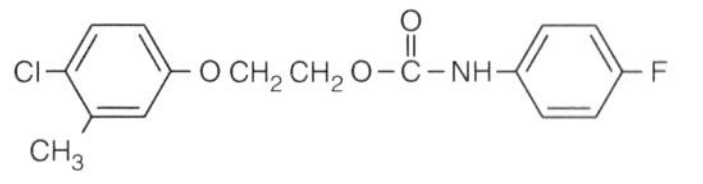
250 mg

S72,481-5
4-ALLYL-2-METHOXYPHENYL N-(4-CHLOROPHENYL)CARBAMATE
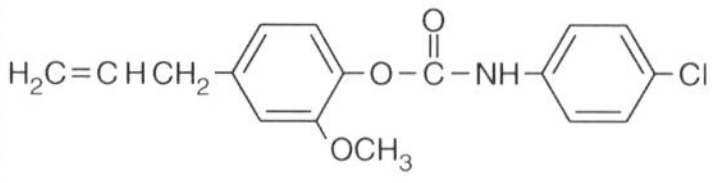
5 gm

S79,525-9
2-METHOXY-4-(1-PROPENYL)PHENYL N-(4-CHLOROPHENYL)CARBAMATE
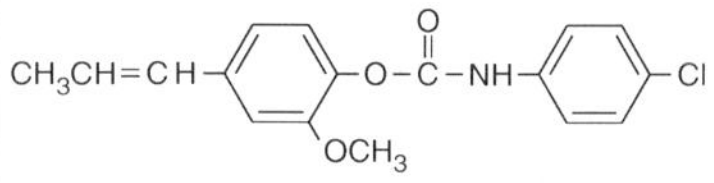
250 mg

S66,688-2
2-METHOXYETHYL N-(2,3-DICHLOROPHENYL)CARBAMATE
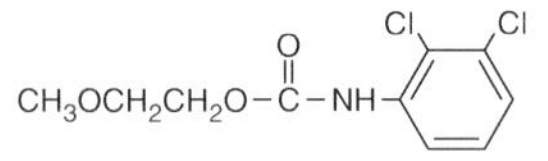
5 gm

S70,903-4
2-METHOXYETHYL N-(2,6-DICHLOROPHENYL)CARBAMATE
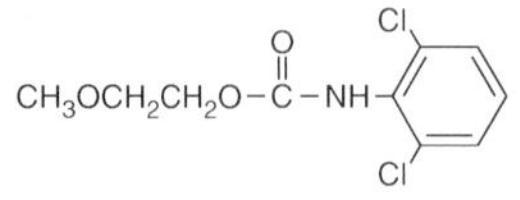
250 mg

S66,771-4
2-ETHOXYETHYL N-(2,3-DICHLOROPHENYL)CARBAMATE
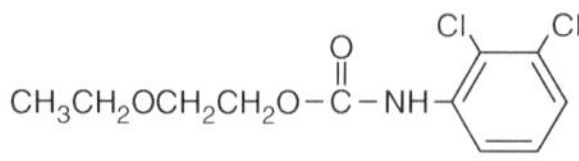
5 gm

S71,123-3
2-ETHOXYETHYL N-(2,6-DICHLOROPHENYL)CARBAMATE
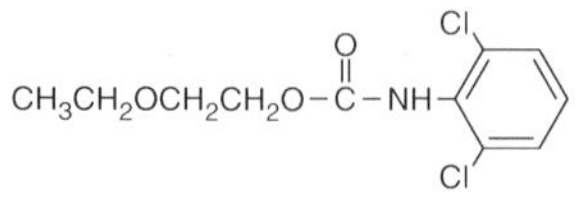
250 mg

S86,847-7
4-ALLYL-2-METHOXYPHENYL N-(3-CHLORO-2-METHYLPHENYL)-CARBAMATE
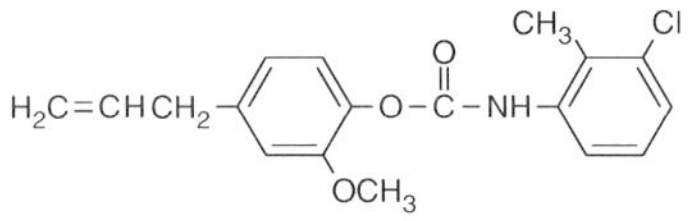
250 mg

S90,248-9
2-(4-CHLORO-3-METHYLPHENOXY)ETHYL N-(2,6-XYLYL)CARBAMATE
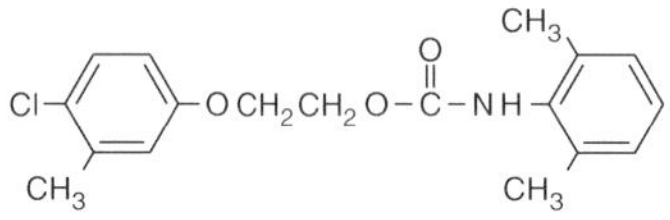
250 mg

S72,541-2
2-[2-METHOXY-4-(1-PROPENYL)PHENOXY]-ETHYL N-(2,6-DICHLOROPHENYL)-CARBAMATE
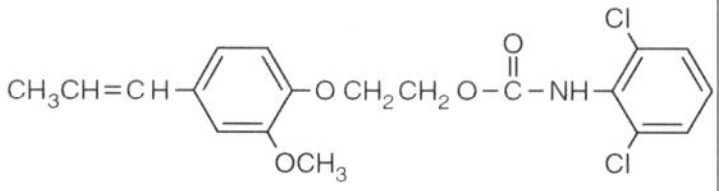
250 mg

S72,110-7
2-(4-CHLORO-3,5-DIMETHYLPHENOXY)ETHYL N-(2,6-DICHLOROPHENYL)CARBAMATE
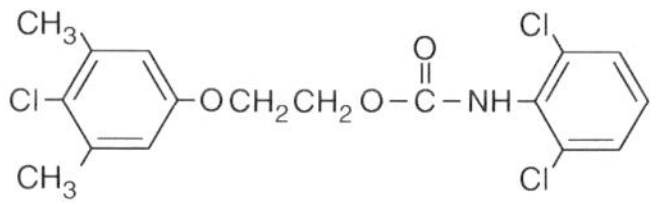
250 mg

S66,679-3
2-METHOXYETHYL N-(3,4-DICHLOROPHENYL)CARBAMATE
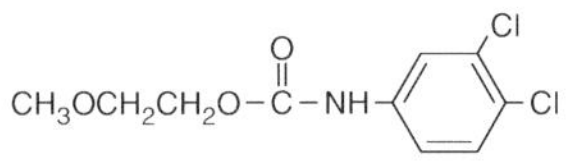
5 gm

S66,500-2
2-METHOXYETHYL N-(4-CHLORO-2-METHYLPHENYL)CARBAMATE
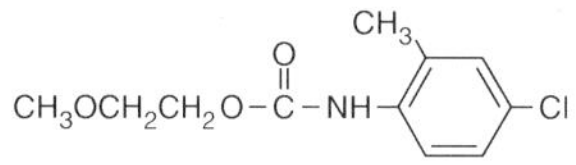
5 gm

S66,686-6
2-ETHOXYETHYL N-(3,4-DICHLOROPHENYL)CARBAMATE
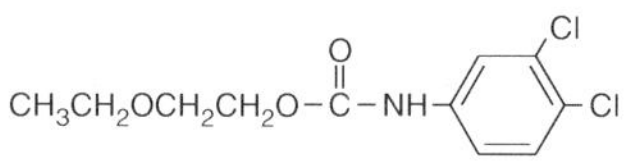
5 gm

S66,710-2
2-ETHOXYETHYL N-(5-CHLORO-2-METHYLPHENYL)CARBAMATE
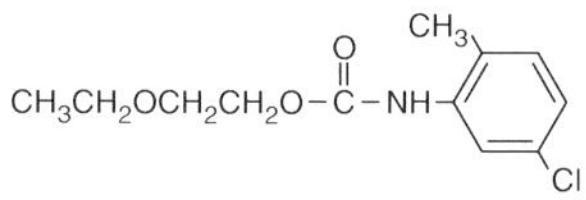
5 gm

S66,752-8
2-ETHOXYETHYL N-(4-CHLORO-2-METHYLPHENYL)CARBAMATE
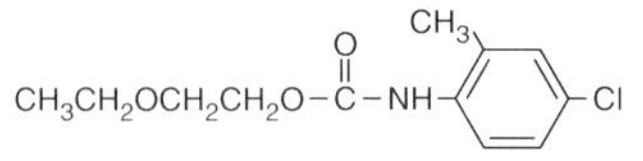
5 gm

S76,899-5
2-(3-METHYLPHENOXY)ETHYL N-(2,4-DIFLUOROPHENYL)CARBAMATE
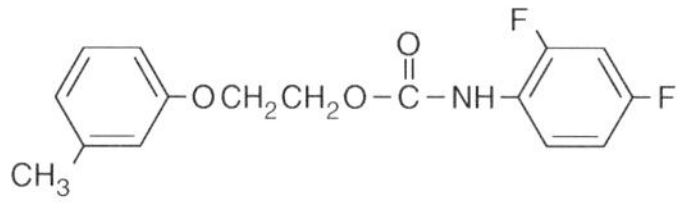
250 mg

S75,259-2
2-METHOXY-4-(1-PROPENYL)PHENYL N-(3-CHLOROPHENYL)CARBAMATE
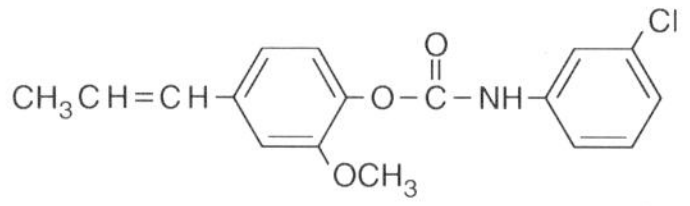
5 gm

S89,206-8
4-ALLYL-2-METHOXYPHENYL N-(3-CHLOROPHENYL)CARBAMATE
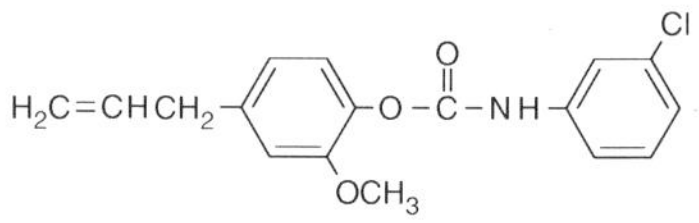
250 mg

S68,643-3
2-(2,4-DICHLOROPHENOXY)ETHYL N-(3-ETHYLPHENYL)CARBAMATE
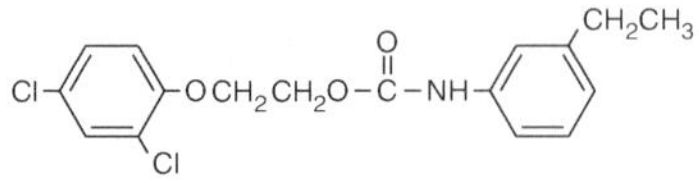
250 mg

S86,846-9
4-ALLYL-2-METHOXYPHENYL N-(2-CHLORO-4-METHYLPHENYL)-CARBAMATE
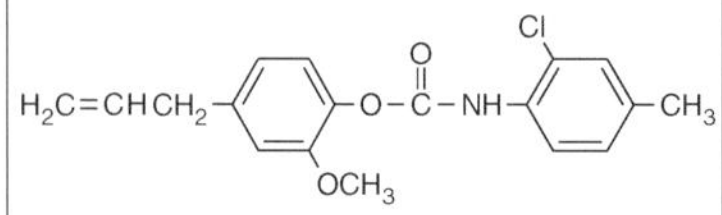
250 mg

S85,529-4
4-(3,4-DICHLOROPHENYLIMINOMETHYL)-2-METHOXYPHENYL N-(3-CHLOROPHENYL)-CARBAMATE
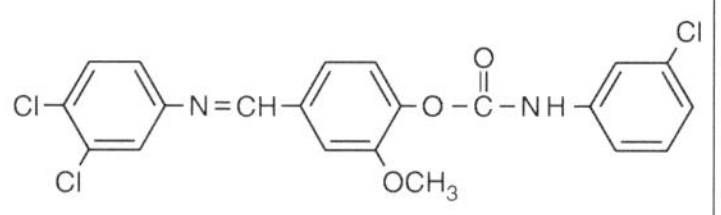
250 mg

S90,191-1
2-[2-METHOXY-4-(1-PROPENYL)PHENOXY]-ETHYL N-(4-CHLORO-2-METHYLPHENYL)-CARBAMATE
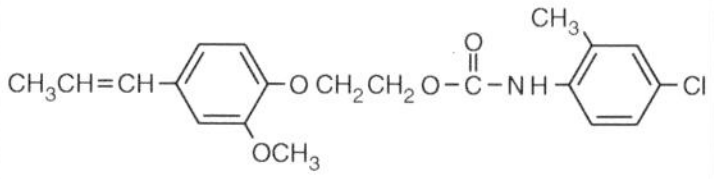
250 mg

S70,894-1
2-METHOXYETHYL N-(3,5-DICHLOROPHENYL)CARBAMATE
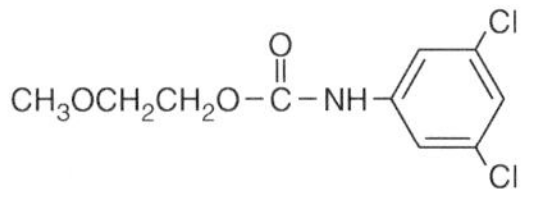
5 gm

S70,939-5
2-METHOXYETHYL N-(2,4,5-TRICHLOROPHENYL)CARBAMATE
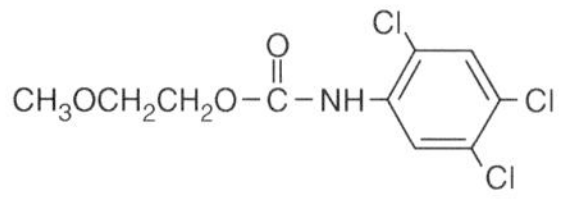
5 gm

S66,707-2
2-ETHOXYETHYL N-(2,4,5-TRICHLOROPHENYL)CARBAMATE
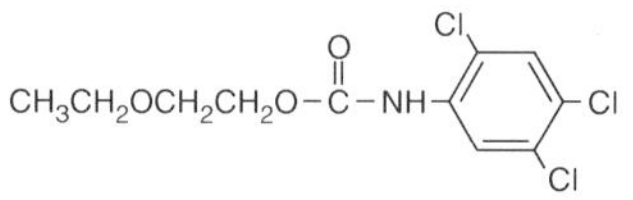
5 gm

S70,064-9
3-METHOXYPHENYL N-(4-FLUORO-2-NITROPHENYL)CARBAMATE
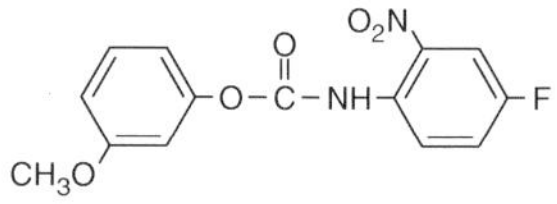
5 gm

S72,383-5
3-METHOXYPHENYL N-(4-CHLORO-3-NITROPHENYL)CARBAMATE
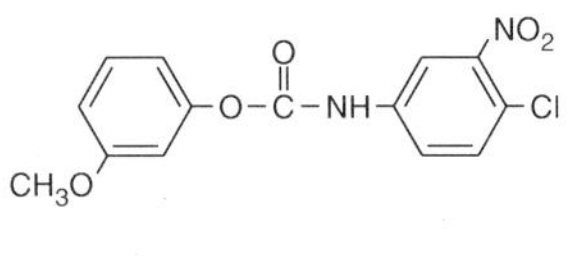
5 gm

S73,712-7
2-(METHOXYCARBONYL)PHENYL N-(5-CHLORO-2-METHOXYPHENYL)-CARBAMATE
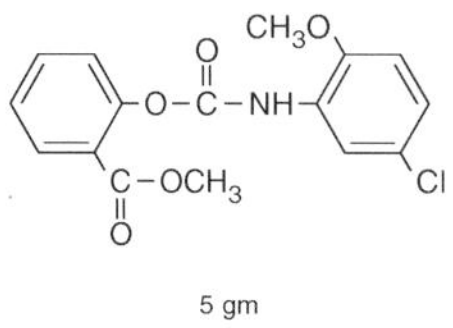
5 gm

S92,438-5
4-FORMYL-2-METHOXYPHENYL N-(4-FLUORO-3-NITROPHENYL)CARBAMATE
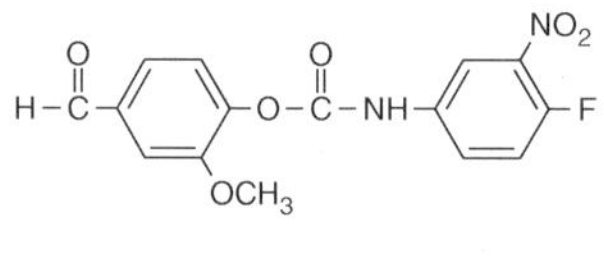
250 mg

S68,617-4
2,6-DIBROMO-4-NITROPHENYL N-(3,5-DIMETHOXYPHENYL)CARBAMATE
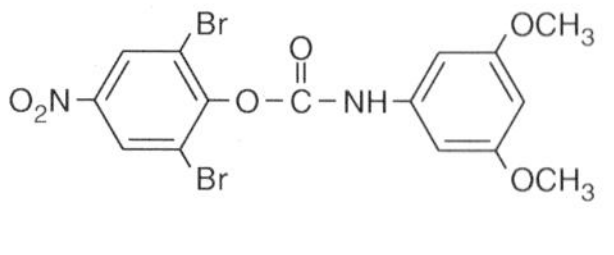
250 mg

S90,583-6
2-(METHOXYCARBONYL)PHENYL N-(2-CHLORO-4-METHOXYPHENYL)-CARBAMATE

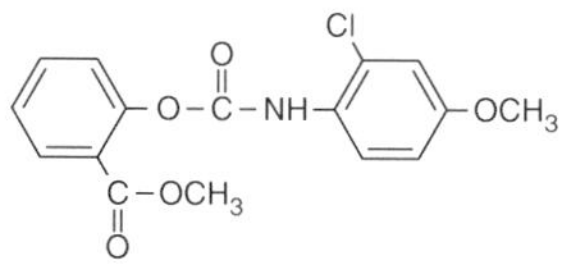

250 mg

S79,784-7
2-(ISOPENTYLOXYCARBONYL)PHENYL N-(5-CHLORO-2-METHOXYPHENYL)-CARBAMATE

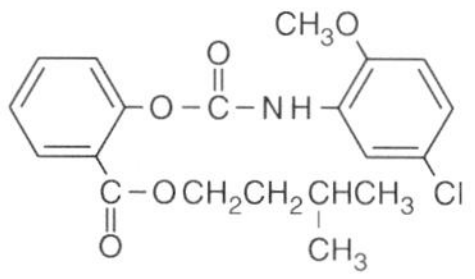

250 mg

S85,078-0
2-CARBAMOYLPHENYL N-(2-CHLORO-4-METHOXYPHENYL)CARBAMATE

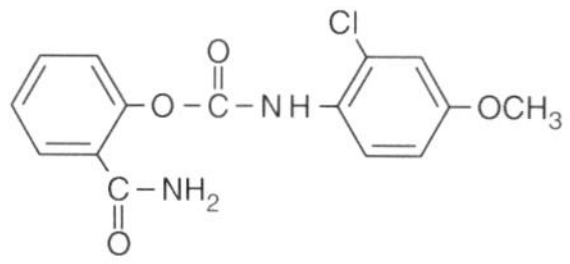

250 mg

S85,085-3
4-COUMARINYL N-(2-CHLORO-4-METHOXYPHENYL)CARBAMATE

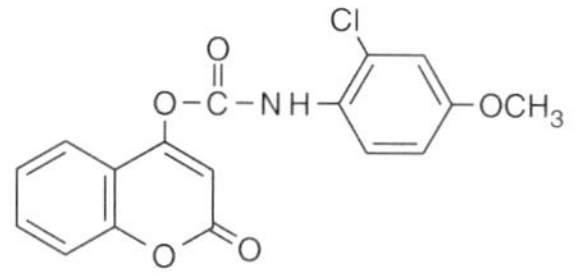

250 mg

S69,619-6
3-METHOXYPHENYL N-(3-FLUOROPHENYL)-CARBAMATE

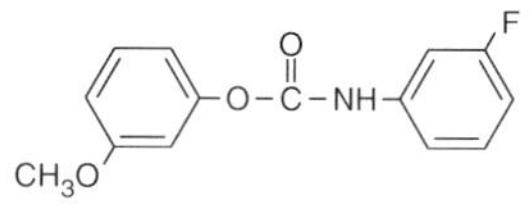

250 mg

S73,883-2
3-METHOXYPHENYL N-(2,4-DIFLUOROPHENYL)CARBAMATE

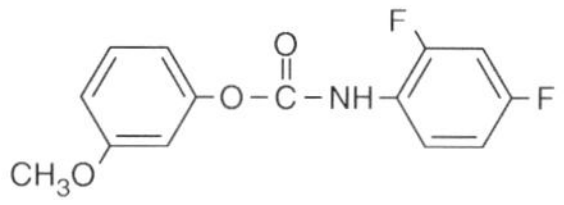

5 gm

S72,380-0
3-METHOXYPHENYL N-(3,4-DICHLOROPHENYL)CARBAMATE

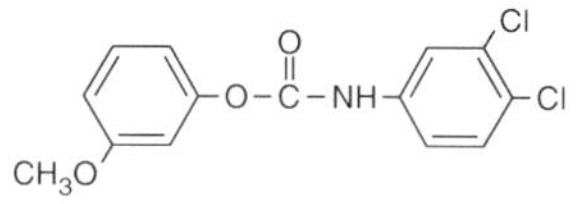

5 gm

S77,364-6
2-CHLORO-4-METHOXYPHENYL N-(4-FLUORO-3-NITROPHENYL)CARBAMATE

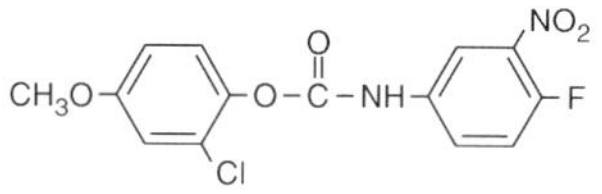

250 mg

S69,348-0
2-CHLORO-4-METHOXYPHENYL N-(2-METHYL-5-NITROPHENYL)CARBAMATE

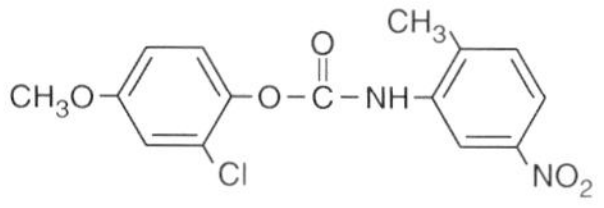

250 mg

S84,472-1
4-METHOXYPHENYL N-(2-FLUORO-5-NITROPHENYL)CARBAMATE

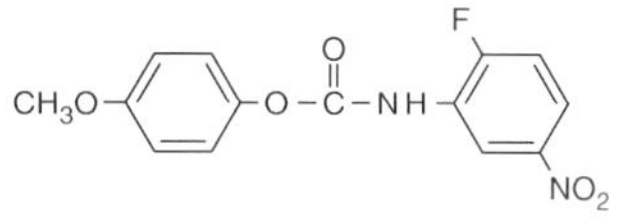

250 mg

S79,461-9
4-METHOXYPHENYL N-(5-CHLORO-2-METHOXYPHENYL)CARBAMATE

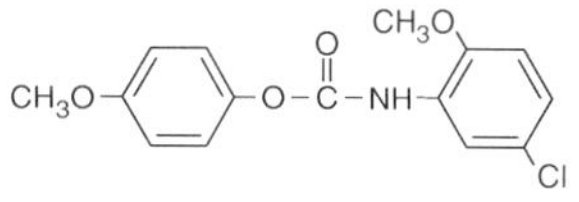

250 mg

S73,776-3
2,4,5-TRICHLOROPHENYL N-(2,5-DIMETHOXYPHENYL)CARBAMATE

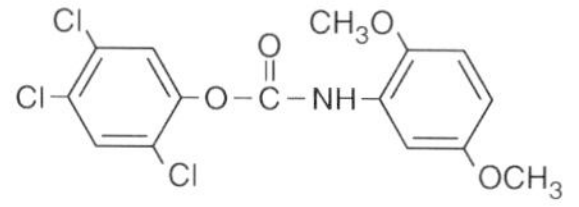

5 gm

S69,398-7
2-CHLORO-4-METHOXYPHENYL N-PHENYLCARBAMATE

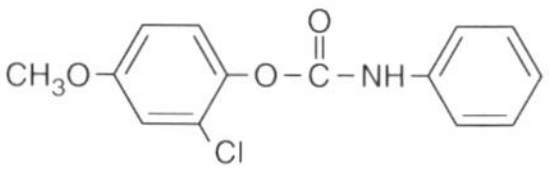

250 mg

S73,667-8
4-METHOXYPHENYL N-(2-FLUOROPHENYL)-CARBAMATE

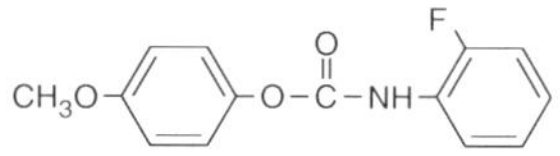

5 gm

S89,034-0
4-METHOXYPHENYL N-(2-CHLOROPHENYL)-CARBAMATE

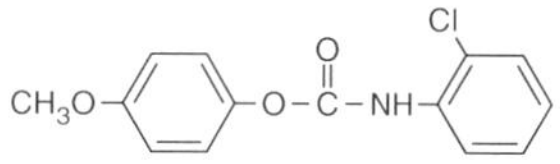

250 mg

S80,027-9
4-METHOXYPHENYL N-(4-FLUOROPHENYL)-CARBAMATE

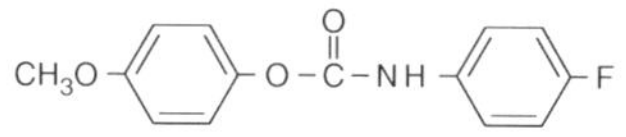

250 mg

S85,077-2
4-METHOXYPHENYL N-(2,3-DICHLOROPHENYL)CARBAMATE

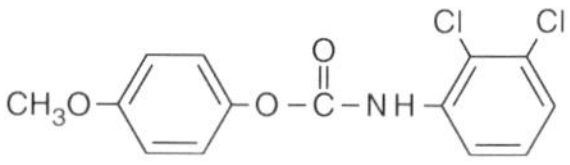

250 mg

S91,109-7
4-BUTOXYPHENYL N-(2,3-DICHLOROPHENYL)CARBAMATE

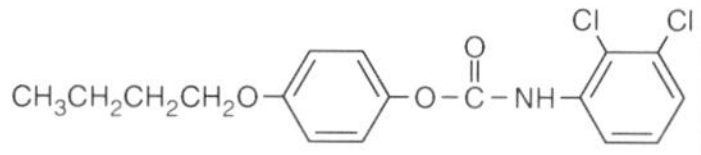

250 mg

S89,961-5
4-METHOXYPHENYL N-(3,4-DICHLOROPHENYL)CARBAMATE

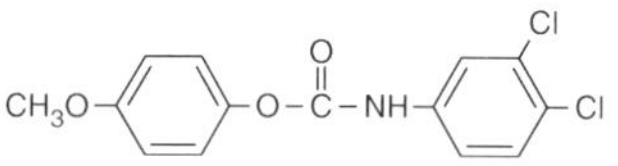

250 mg

S79,134-2
4-METHOXYPHENYL N-(3-CHLORO-4-METHYLPHENYL)CARBAMATE

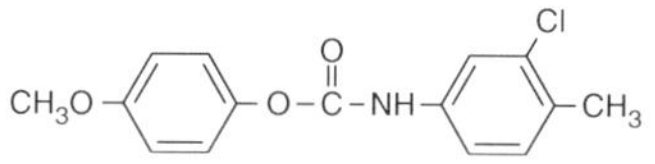

250 mg

S86,779-9
4-METHOXYPHENYL N-(5-CHLORO-2-METHYLPHENYL)CARBAMATE

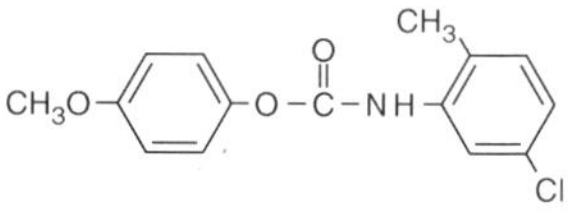

250 mg

S86,780-2
4-METHOXYPHENYL N-(2-CHLORO-4-METHYLPHENYL)CARBAMATE

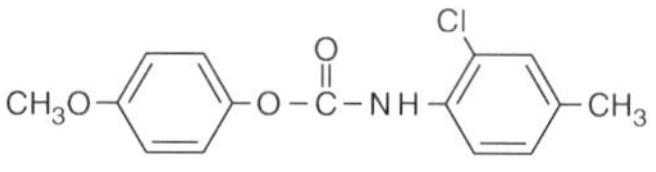

250 mg

S77,270-4
4-BUTOXYPHENYL N-(5-CHLORO-2-METHYLPHENYL)CARBAMATE

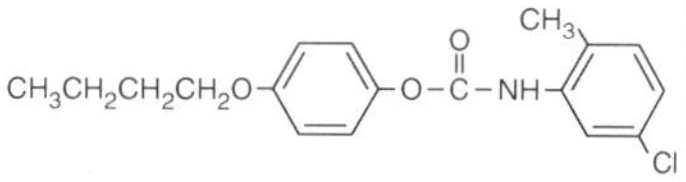

250 mg

S92,440-7
4-BUTOXYPHENYL N-(5-CHLORO-2-METHOXYPHENYL)CARBAMATE

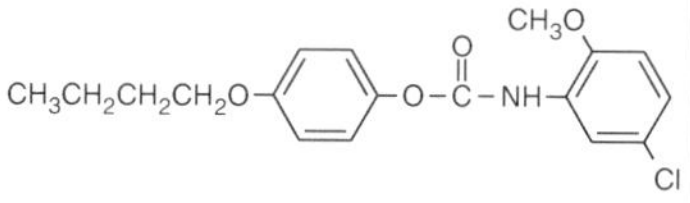

250 mg

S77,352-2
2-CHLORO-4-METHOXYPHENYL N-(2,4-DIFLUOROPHENYL)CARBAMATE

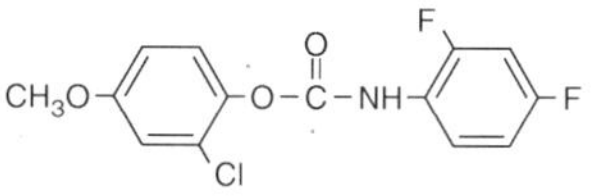

250 mg

S69,166-6
2-CHLORO-4-METHOXYPHENYL N-(5-CHLORO-2-METHYLPHENYL)-CARBAMATE

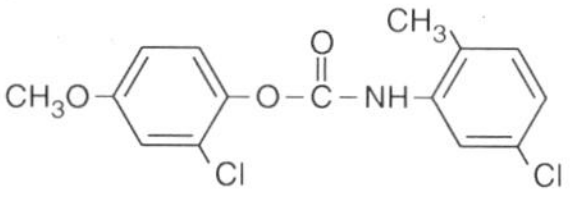

250 mg

S79,719-7
4-(METHYLTHIO)PHENYL N-(5-CHLORO-2-METHOXYPHENYL)CARBAMATE

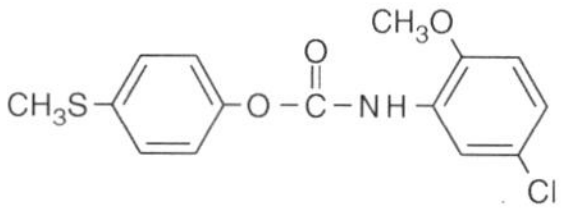

250 mg

S85,443-3
2-(4-CHLOROPHENYLTHIO)ETHYL N-(4-CHLORO-2-METHOXYPHENYL)-CARBAMATE

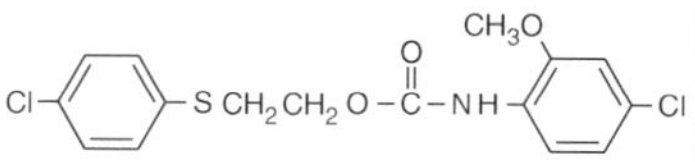

250 mg

S79,737-5
3-METHYL-4-(METHYLTHIO)PHENYL N-(5-CHLORO-2-METHOXYPHENYL)-CARBAMATE

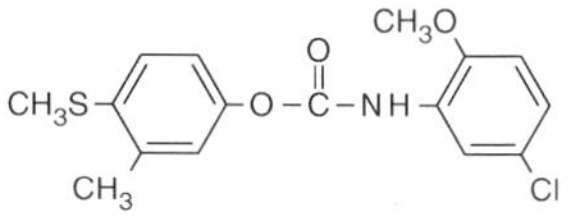

250 mg

S70,021-5
N,N'-(4-BROMO-2-METHYLPHENYLENE)-BIS(2-[4-(METHYLTHIO)PHENOXY]ETHYL)-CARBAMATE, ISOMERS

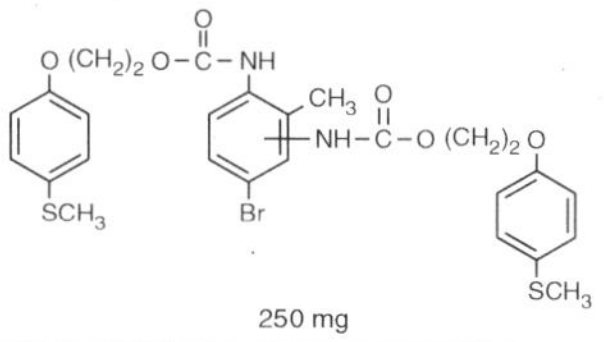

250 mg

S85,081-0
3-METHYL-4-(METHYLTHIO)PHENYL N-(2-CHLORO-4-METHOXYPHENYL)-CARBAMATE

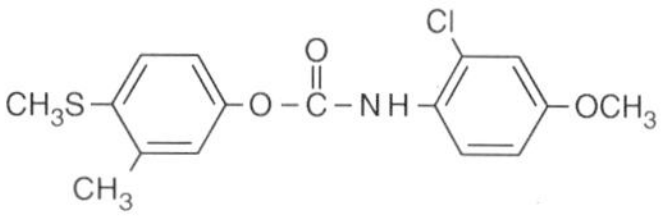

250 mg

S80,026-0
4-FLUOROPHENYL N-(2-METHOXY-5-NITROPHENYL)CARBAMATE

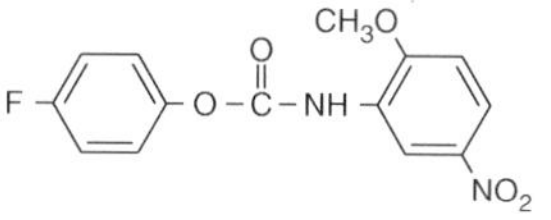

250 mg

S80,023-6
2-CHLORO-4-FLUOROPHENYL N-(2-METHOXY-5-NITROPHENYL)-CARBAMATE

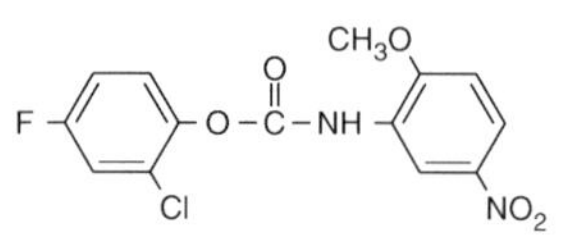

250 mg

S70,066-5
2-METHOXYETHYL N-(2-CHLORO-5-NITROPHENYL)CARBAMATE

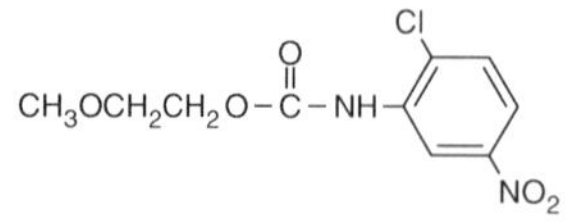

250 mg

S84,462-4
S-(4-METHOXYPHENYL) N-(2-CHLORO-5-NITROPHENYL)THIOCARBAMATE

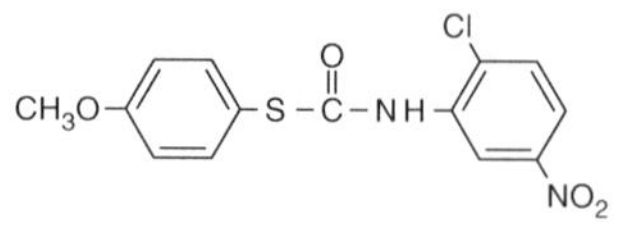

250 mg

S79,471-6
2-FLUOROPHENYL N-(2-METHOXY-4-NITROPHENYL)CARBAMATE

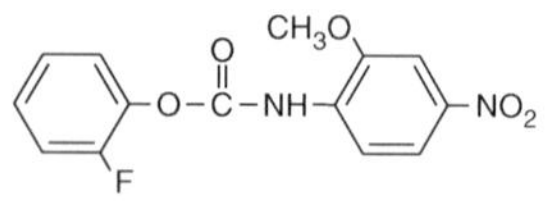

250 mg

S80,007-4
2-CHLORO-4-FLUOROPHENYL N-(2-METHOXY-4-NITROPHENYL)-CARBAMATE

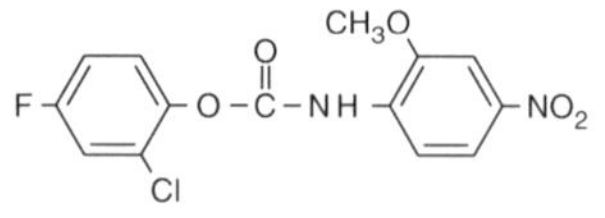

250 mg

S80,006-6
2,4-DIBROMOPHENYL N-(2-METHOXY-4-NITROPHENYL)CARBAMATE

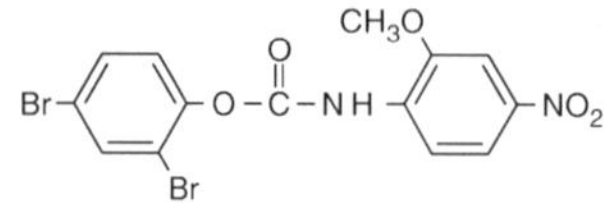

250 mg

S80,009-0
4-CHLORO-2-METHYLPHENYL N-(2-METHOXY-4-NITROPHENYL)-CARBAMATE

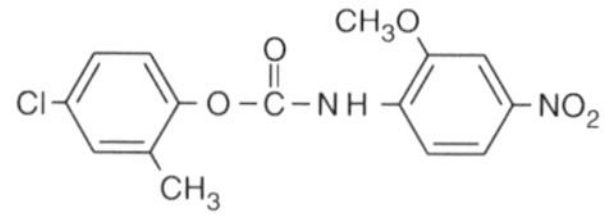

250 mg

S73,721-6
4-CHLORO-3,5-DIMETHYLPHENYL N-(2-METHOXY-4-NITROPHENYL)-CARBAMATE

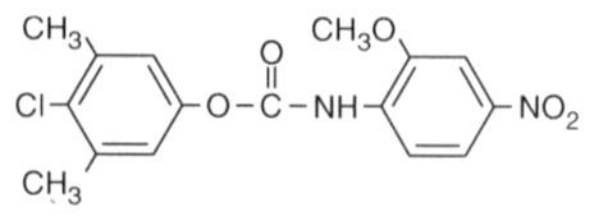

5 gm

S77,230-5
4-(5-CHLORO-2-METHOXYPHENYL)-1-(ETHOXYCARBONYL)SEMICARBAZIDE

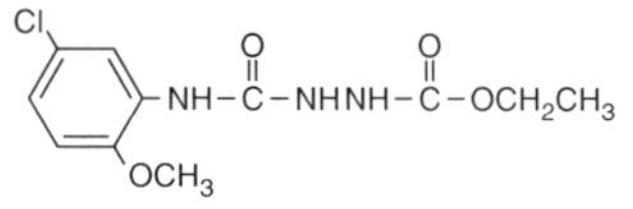

250 mg

S79,424-4
3-CHLORO-1,1'-BIPHENYL-4-YL N-(2-METHOXY-4-NITROPHENYL)CARBAMATE

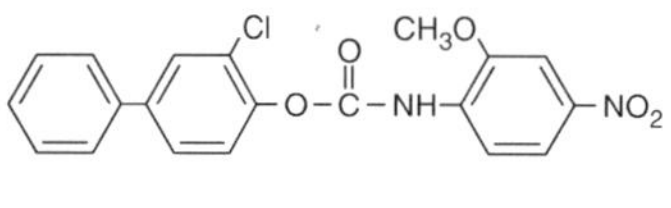

250 mg

S91,130-5
2-ACETYLPHENYL N-(4-CHLORO-2-METHOXYPHENYL)CARBAMATE

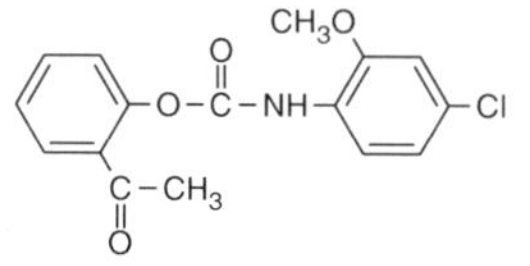

250 mg

S79,716-2
2-CARBAMOYLPHENYL N-(5-CHLORO-2-METHOXYPHENYL)CARBAMATE

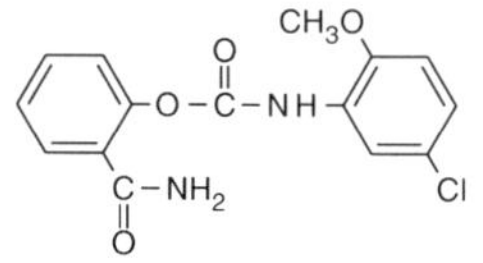

250 mg

S85,086-1
4-METHYL-7-COUMARINYL N-(4-CHLORO-2-METHOXYPHENYL)CARBAMATE

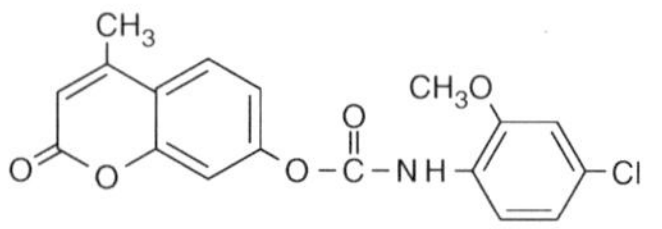

250 mg

S79,735-9
4-ACETAMIDOPHENYL N-(5-CHLORO-2-METHOXYPHENYL)CARBAMATE

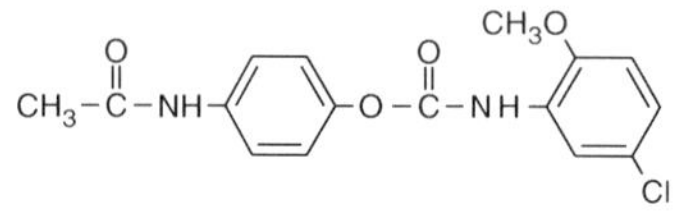

250 mg

S85,094-2
4-ACETAMIDOPHENYL N-(4-CHLORO-2-METHOXYPHENYL)CARBAMATE

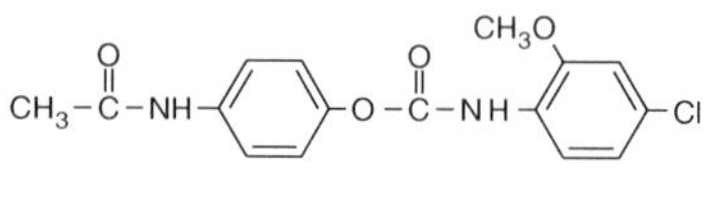

250 mg

S79,658-1
3-ACETAMIDOPHENYL N-(4-CHLORO-2-METHOXYPHENYL)CARBAMATE

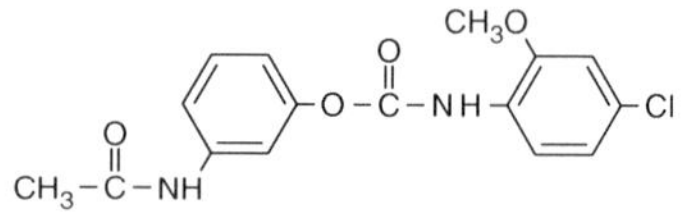

250 mg

S90,282-9
4-[3-(5-CHLORO-2-METHOXYPHENYL)-UREIDO]PHENYL N-METHYLCARBAMATE

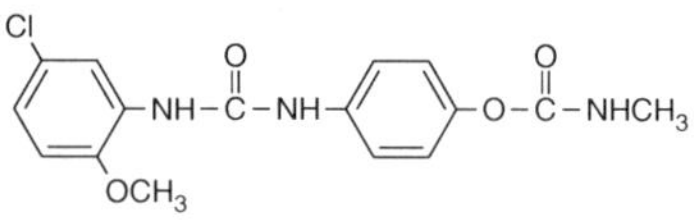

250 mg

S89,834-1
4-METHYLCYCLOHEXYL N-(4-CHLORO-2-METHOXYPHENYL)CARBAMATE

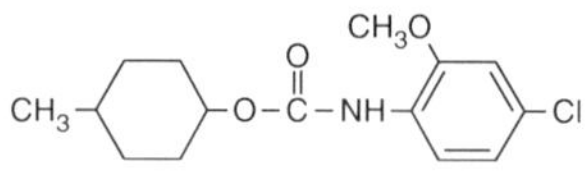

250 mg

S76,947-9
(2-BROMO-4-PHENYL)PHENYL N-(2-METHOXYPHENYL)CARBAMATE

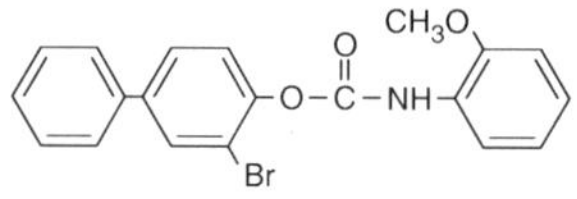

250 mg

S79,750-2
3-PHENYLALLYL N-(5-CHLORO-2-METHOXYPHENYL)CARBAMATE

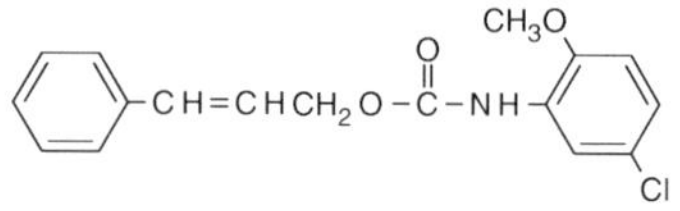

250 mg

S72,544-7
3-PHENYLPROPYL N-(5-CHLORO-2-METHOXYPHENYL)CARBAMATE

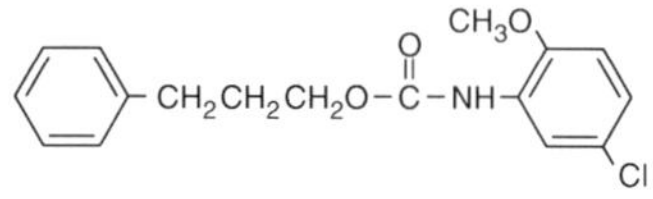

5 gm

S70,192-0
3-CHLORO-1,1'-BIPHENYL-4-YL N-(2-METHOXY-5-METHYLPHENYL)CARBAMATE

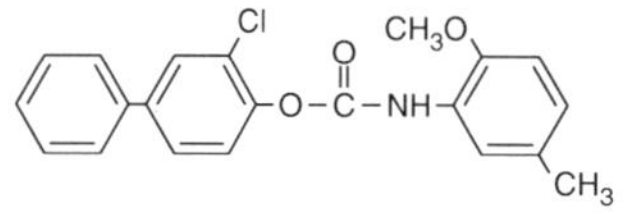

250 mg

S69,419-3
4-BROMO-2,5-DIMETHYLPHENYL N-(2-METHOXYPHENYL)CARBAMATE

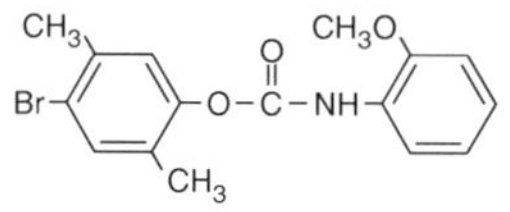

250 mg

S69,436-3
2-BROMO-4,5-DIMETHYLPHENYL N-(2-METHOXYPHENYL)CARBAMATE

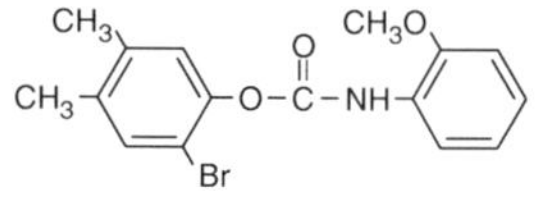

250 mg

S85,099-3
2-FLUOROPHENYL N-(4-CHLORO-2-METHOXYPHENYL)CARBAMATE

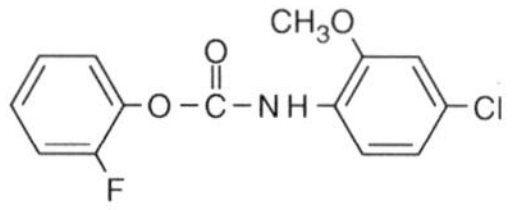

250 mg

S89,185-1
2,4-DICHLOROPHENYL N-(2-METHOXYPHENYL)CARBAMATE

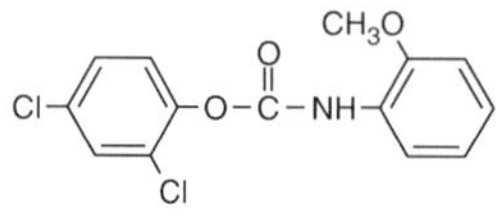

250 mg

S72,126-3
4-CHLORO-3-METHYLPHENYL N-(2-METHOXYPHENYL)CARBAMATE

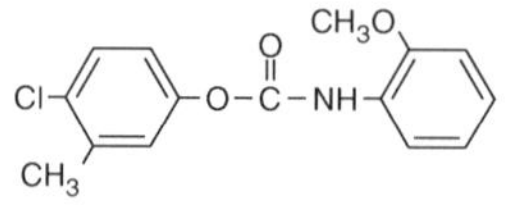

250 mg

S79,753-7
2-ISOPROPYLPHENYL N-(5-CHLORO-2-METHOXYPHENYL)CARBAMATE

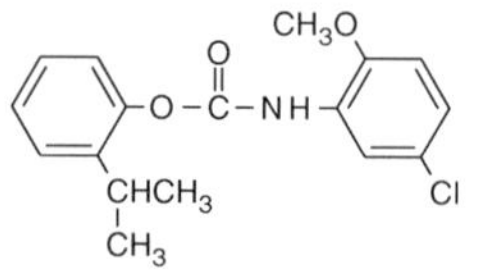

250 mg

S72,529-3
3-CHLORO-1,1'-BIPHENYL-4-YL N-(5-CHLORO-2-METHOXYPHENYL)CARBAMATE

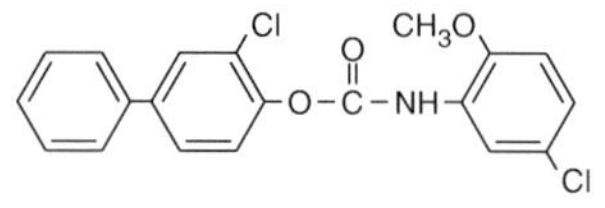

5 gm

S69,180-1
2-BROMO-4-CHLORO-6-METHYLPHENYL N-(2-METHOXYPHENYL)CARBAMATE

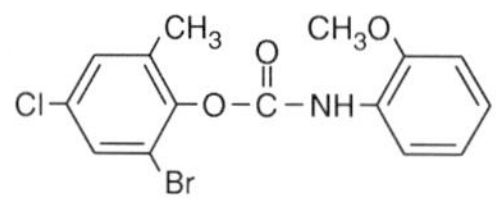

250 mg

S85,545-6
4-CHLORO-3,5-DIMETHYLPHENYL N-(2-ETHOXYPHENYL)CARBAMATE

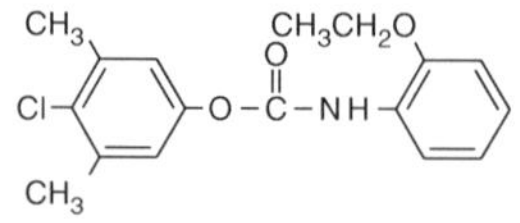

250 mg

S79,472-4
4-FLUOROPHENYL N-(5-CHLORO-2-METHOXYPHENYL)CARBAMATE

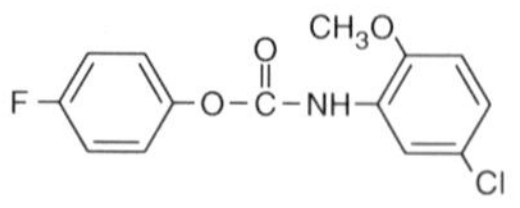

250 mg

S66,501-0
4-CHLOROPHENYL N-(2-METHOXY-5-METHYLPHENYL)CARBAMATE

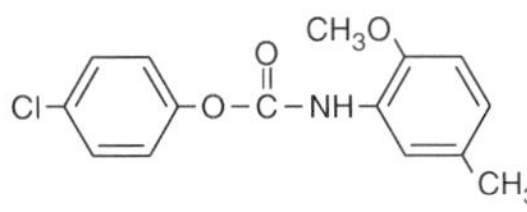

5 gm

S73,727-5
4-(METHOXYCARBONYLAMINO)PHENYL N-(5-CHLORO-2-METHOXYPHENYL)-CARBAMATE

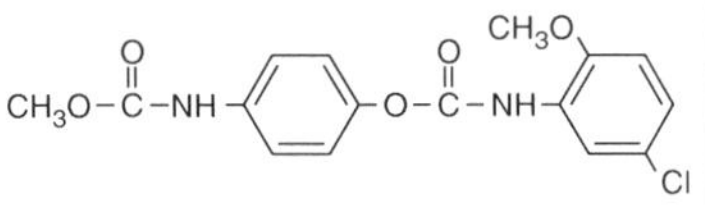

5 gm

S70,980-8
METHYL N-(3-CHLORO-2-METHOXYPHENYL)-CARBAMATE

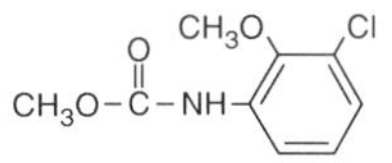

5 gm

S70,002-9
2,3-DICHLOROPHENYL N-(2-METHOXY-5-METHYLPHENYL)CARBAMATE

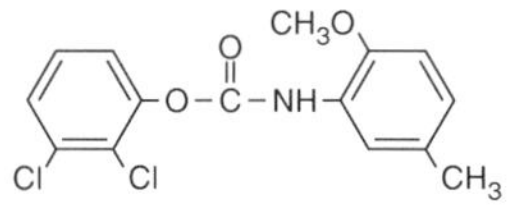

250 mg

S79,446-5
2,6-XYLYL N-(5-CHLORO-2-METHOXYPHENYL)CARBAMATE

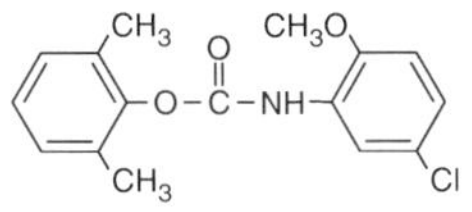

250 mg

S85,584-7
ETHYL N-(5-CHLORO-2-METHOXYPHENYL)-CARBAMATE

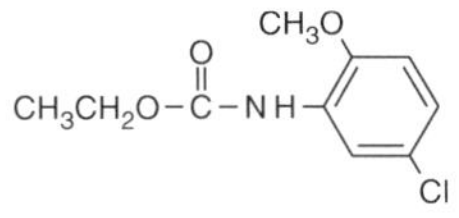

250 mg

S69,038-4
3,4-DICHLOROPHENYL N-(5-CHLORO-2-METHOXYPHENYL)CARBAMATE

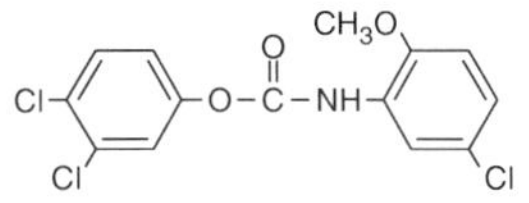

250 mg

S69,432-0
2,5-DICHLOROPHENYL N-(5-CHLORO-2-METHOXYPHENYL)CARBAMATE

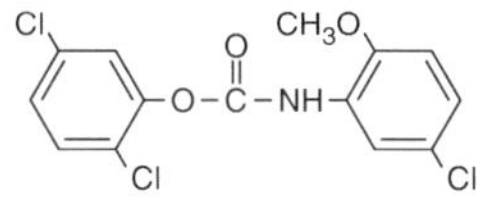

250 mg

S85,437-9
2-BROMO-4-CHLOROPHENYL N-(4-CHLORO-2-METHOXYPHENYL)CARBAMATE

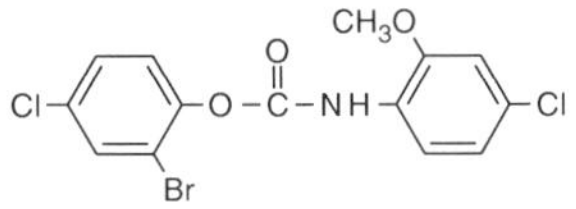

250 mg

S72,537-4
2,5-DICHLOROPHENYL N-(2-METHOXY-5-METHYLPHENYL)CARBAMATE

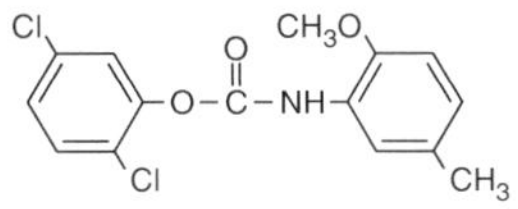

5 gm

S68,992-0
2-BROMO-4,5-DIMETHYLPHENYL N-(4-CHLORO-2-METHOXYPHENYL)-CARBAMATE

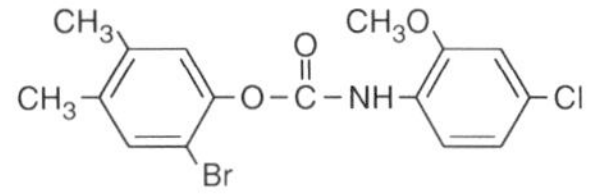

250 mg

S69,023-6
4-BROMO-2,5-DIMETHYLPHENYL N-(4-CHLORO-2-METHOXYPHENYL)-CARBAMATE

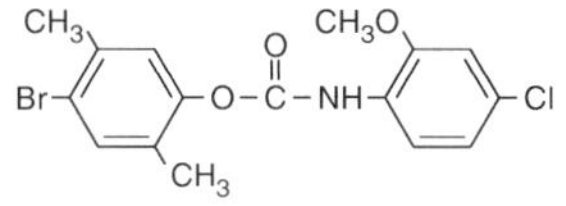

250 mg

S69,318-9
2-BROMO-4-CHLORO-5-METHYLPHENYL N-(2-METHOXY-5-METHYLPHENYL)-CARBAMATE

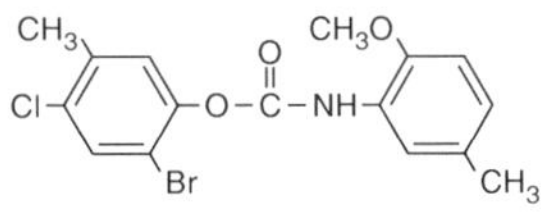

250 mg

S69,323-5
4-BROMO-2,5-DIMETHYLPHENYL N-(5-CHLORO-2-METHOXYPHENYL)-CARBAMATE

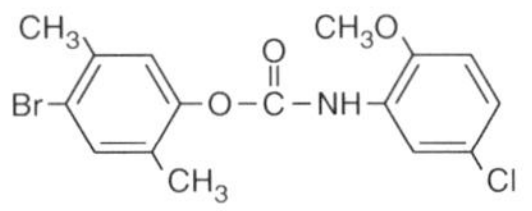

250 mg

S72,759-8
2-CHLORO-4,5-DIMETHYLPHENYL N-(2-METHOXY-5-METHYLPHENYL)-CARBAMATE

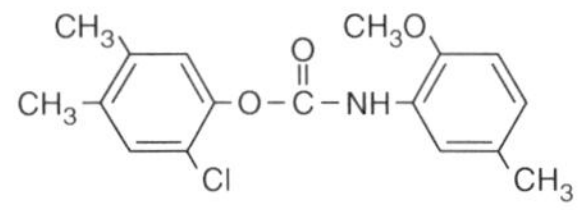

5 gm

S68,990-4
2-BROMO-4,5-DIMETHYLPHENYL N-(2-METHOXY-5-METHYLPHENYL)-CARBAMATE

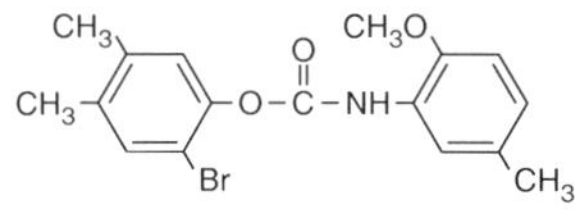

250 mg

S69,366-9
4-BROMO-2,5-DIMETHYLPHENYL N-(2-METHOXY-5-METHYLPHENYL)-CARBAMATE

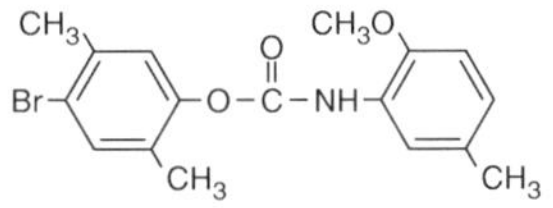

250 mg

S79,451-1
3-ETHYL-5-METHYLPHENYL N-(5-CHLORO-2-METHOXYPHENYL)CARBAMATE

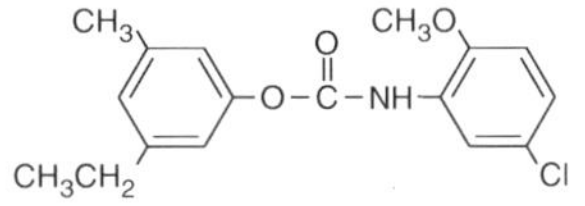

250 mg

S68,945-9
2-BROMO-4-CHLORO-6-METHYLPHENYL N-(5-CHLORO-2-METHOXYPHENYL)-CARBAMATE

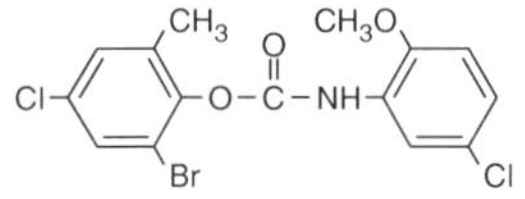

250 mg

S68,974-2
2-BROMO-4-CHLORO-6-METHYLPHENYL N-(4-CHLORO-2-METHOXYPHENYL)-CARBAMATE

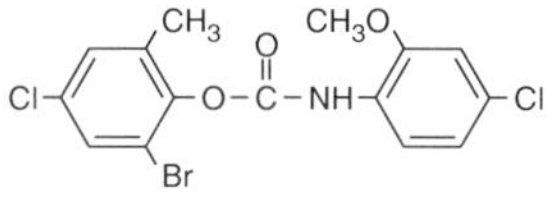

5 gm

S71,850-5
4-CHLORO-3,5-DIMETHYLPHENYL N-(5-CHLORO-2-METHOXYPHENYL)-CARBAMATE

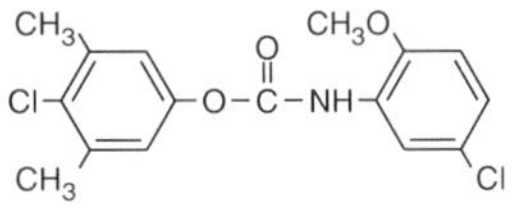

5 gm

S68,991-2
4-BROMO-3,5-DIMETHYLPHENYL N-(4-CHLORO-2-METHOXYPHENYL)-CARBAMATE

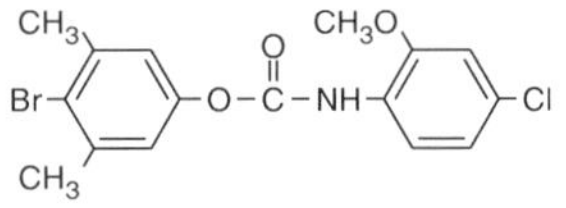

5 gm

S68,997-1
2-BROMO-4-CHLORO-6-METHYLPHENYL N-(2-METHOXY-5-METHYLPHENYL)-CARBAMATE

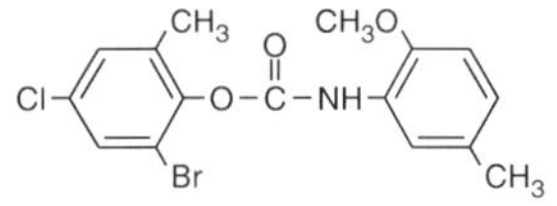

5 gm

S69,579-3
4-CHLORO-2,6-DIMETHYLPHENYL N-(2-METHOXY-5-METHYLPHENYL)-CARBAMATE

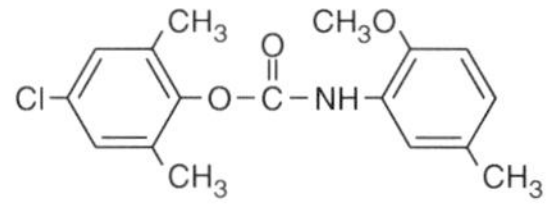

5 gm

S68,940-8
4-BROMO-3,5-DIMETHYLPHENYL N-(2-METHOXY-5-METHYLPHENYL)-CARBAMATE

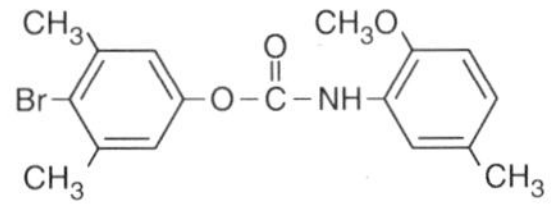

250 mg

S91,182-8
5-INDANYL N-(5-CHLORO-2-METHOXYPHENYL)CARBAMATE

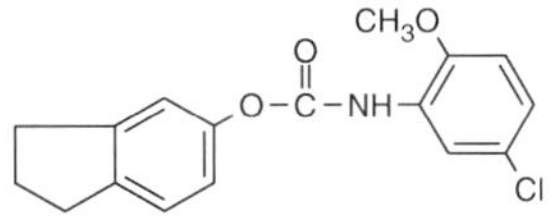

250 mg

S89,784-1
1-NAPHTHYL N-(4-CHLORO-2-METHOXYPHENYL)CARBAMATE

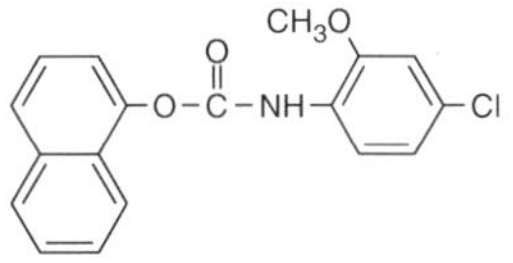

250 mg

S79,765-0
2-NAPHTHYL N-(5-CHLORO-2-METHOXYPHENYL)CARBAMATE

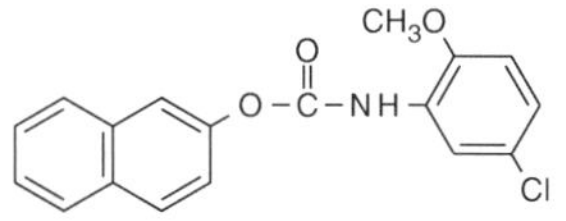

250 mg

S85,092-6
2-NAPHTHYL N-(4-CHLORO-2-METHOXYPHENYL)CARBAMATE

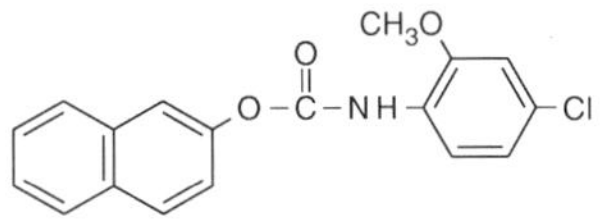

250 mg

S68,794-4
O-[N-(5-CHLORO-2-METHOXYPHENYL)-CARBAMOYL]CYCLOHEXANONE OXIME

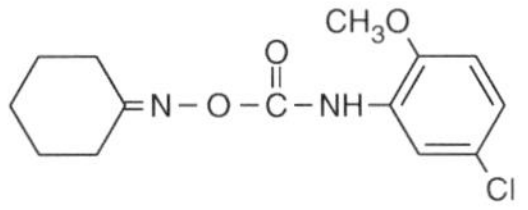

250 mg

S69,139-9
PIPERIDINO N-(5-CHLORO-2-PHENOXYPHENYL)CARBAMATE

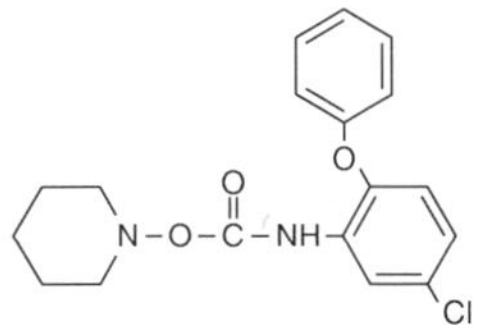

250 mg

S69,145-3
PIPERIDINO N-(4-CHLORO-2-METHOXYPHENYL)CARBAMATE

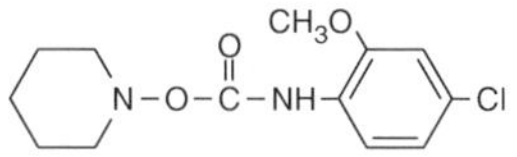

250 mg

S69,378-2
PIPERIDINO N-(5-CHLORO-2-METHOXYPHENYL)CARBAMATE
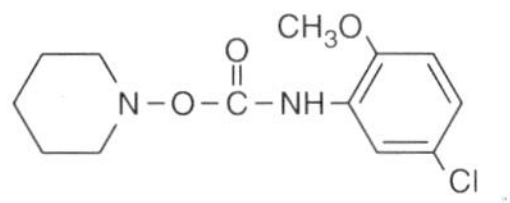
250 mg

S85,440-9
S-BENZYL N-(4-CHLORO-2-METHOXYPHENYL)THIOCARBAMATE
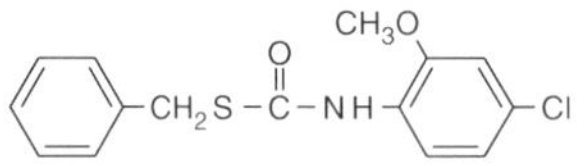
250 mg

S85,441-7
S-(O-TOLYL) N-(4-CHLORO-2-METHOXYPHENYL)THIOCARBAMATE
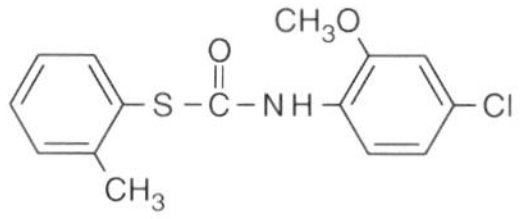
250 mg

S85,438-7
S-(4-CHLOROPHENYL) N-(4-CHLORO-2-METHOXYPHENYL)THIOCARBAMATE
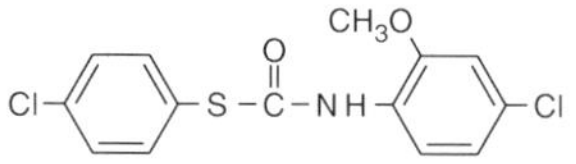
250 mg

S85,446-8
S-(4-TERT-BUTYLPHENYL) N-(4-CHLORO-2-METHOXYPHENYL)THIOCARBAMATE
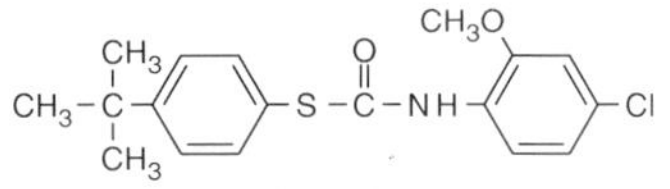
250 mg

S84,807-7
2-BENZOYL-5-METHOXYPHENYL N-(2-FLUORO-5-NITROPHENYL)CARBAMATE
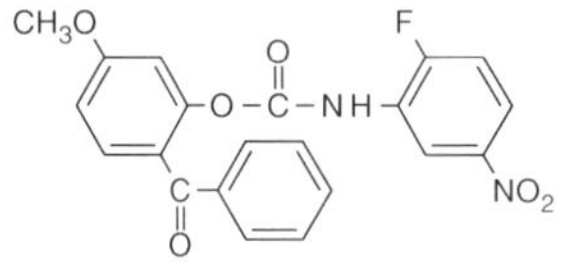
250 mg

S91,172-0
2-(2-NAPHTHOXY)ETHYL N-(2-FLUORO-5-NITROPHENYL)CARBAMATE
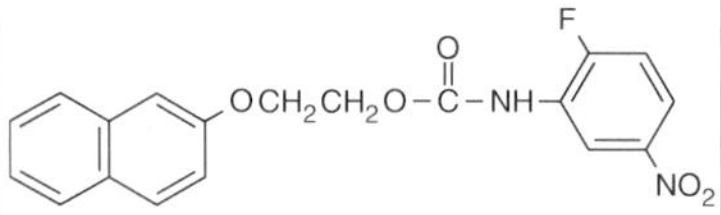
250 mg

S71,360-0
4-CHLORO-2-METHYL-6-NITROPHENYL N-(4-METHOXYPHENYL)CARBAMATE
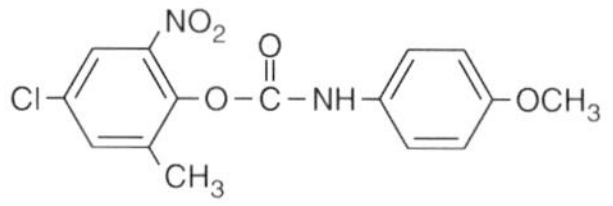
5 gm

S85,083-7
4-(BUTOXYCARBONYL)PHENYL N-(2-CHLORO-4-METHOXYPHENYL)-CARBAMATE
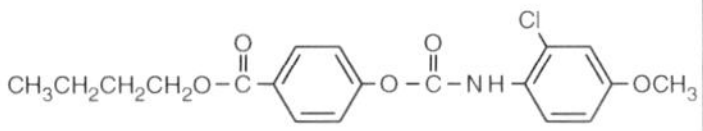
250 mg

S85,531-6
4-[4-(ETHOXYCARBONYL)-PHENYLIMINOMETHYL]-2-METHOXYPHENYL N-(3-CHLOROPHENYL)CARBAMATE
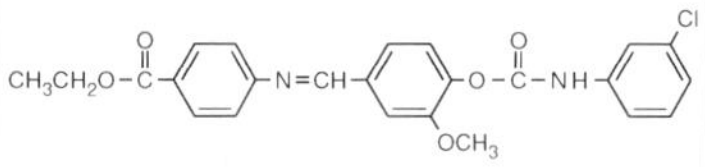
250 mg

S73,609-0
3-(N-PHENYLCARBAMOYL)-2-NAPHTHYL N-(5-CHLORO-2-METHOXYPHENYL)-CARBAMATE
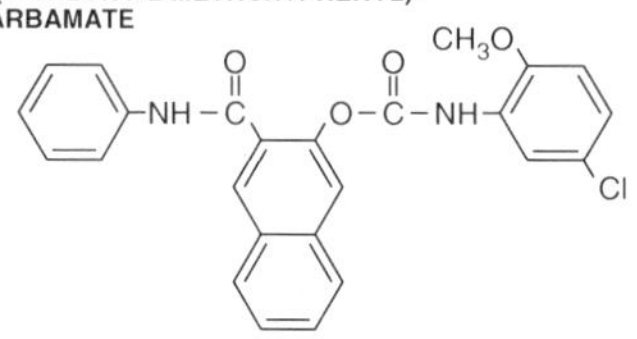
5 gm

S78,112-6
2-(2,4-DICHLOROPHENOXY)ACETYL N-(2,4-XYLYL)CARBAMATE
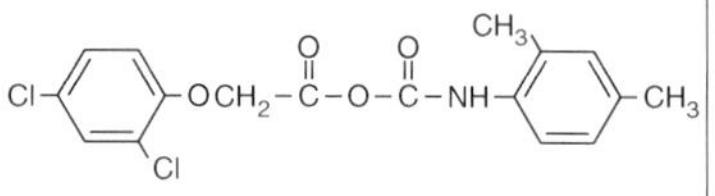
250 mg

S76,813-8
2-(2,4-DICHLOROPHENOXY)ACETYL N-(2,4,6-TRIMETHYLPHENYL)CARBAMATE
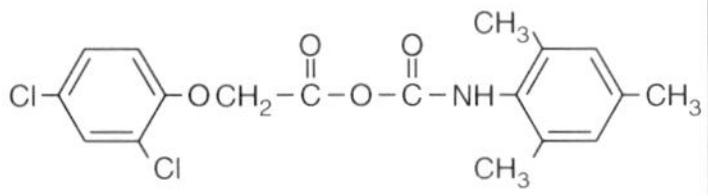
250 mg

S80,769-9
3-[N-(2-METHOXYPHENYL)CARBAMOYL]-2-NAPHTHYL N-(5-CHLORO-2-METHOXYPHENYL)CARBAMATE
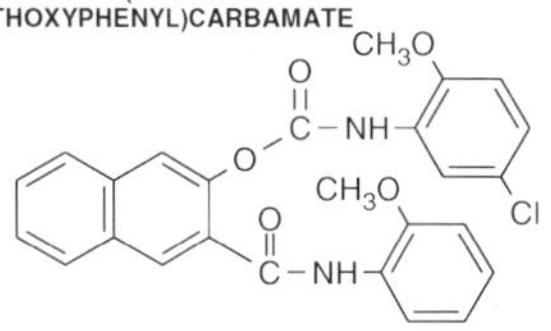
250 mg

S85,080-2
4-ACETAMIDOPHENYL N-(2-CHLORO-4-METHOXYPHENYL)CARBAMATE
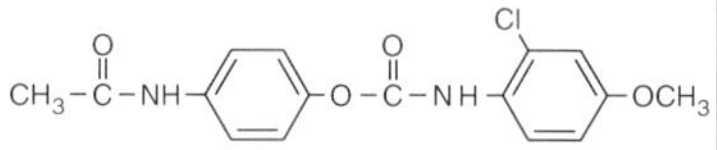
250 mg

S76,918-5
3-[N-(2,5-DIMETHOXYPHENYL)-CARBAMOYL]-2-NAPHTHYL N-(5-CHLORO-2-METHOXYPHENYL)CARBAMATE
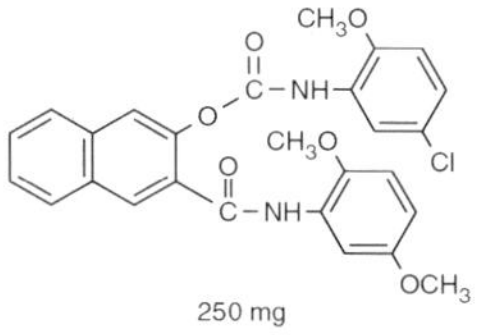
250 mg

S69,067-8
2-BROMO-4,5-DIMETHYLPHENYL N-(3-METHOXYPHENYL)CARBAMATE
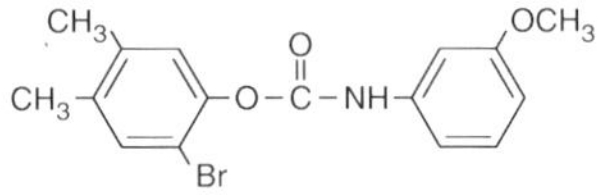
250 mg

S85,070-5
CYCLOPENTYL N-(2-CHLORO-4-METHOXYPHENYL)CARBAMATE
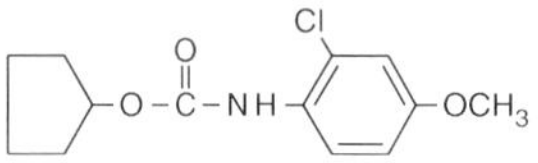
250 mg

S71,897-1
3-CHLORO-1,1'-BIPHENYL-4-YL N-(4-METHOXYPHENYL)CARBAMATE
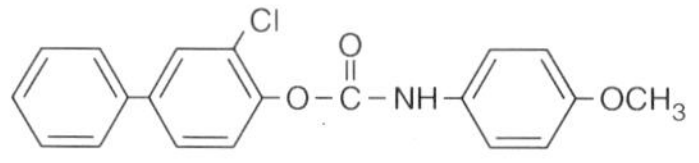
250 mg

S76,957-6
3-BROMO-1,1'-BIPHENYL-4-YL N-(4-METHOXYPHENYL)CARBAMATE
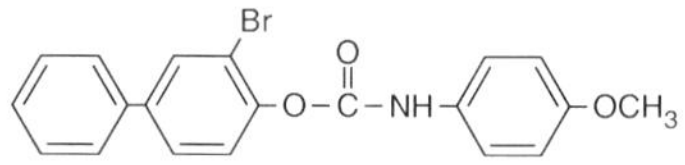
250 mg

S89,230-0
4-METHOXYBENZYL N-(2-CHLOROPHENYL)-CARBAMATE
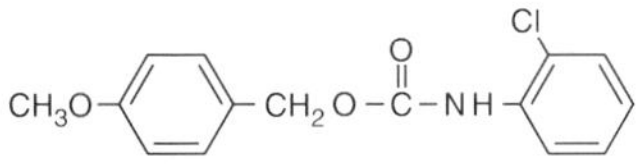
250 mg

S85,075-6
2-FLUOROPHENYL N-(2-CHLORO-4-METHOXYPHENYL)CARBAMATE
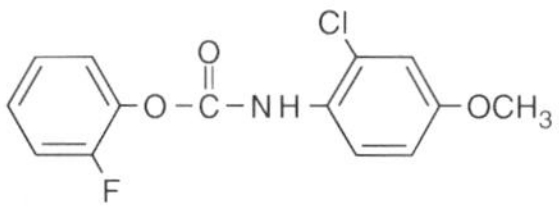
250 mg

S69,547-5
2-CHLORO-4-METHOXYPHENYL N-(2-ETHYLPHENYL)CARBAMATE
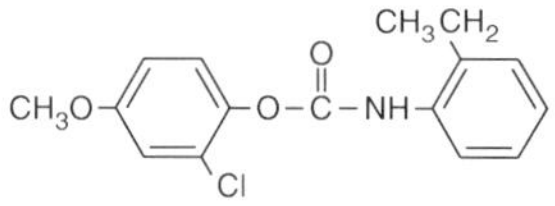
250 mg

S66,711-0
2-METHOXYETHYL N-(4-BROMOPHENYL)-CARBAMATE
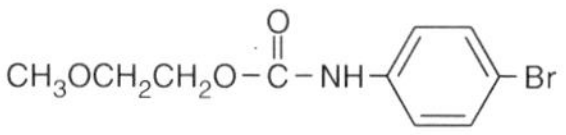
5 gm

S69,219-0
2-BROMO-4,5-DIMETHYLPHENYL N-(4-ETHOXYPHENYL)CARBAMATE
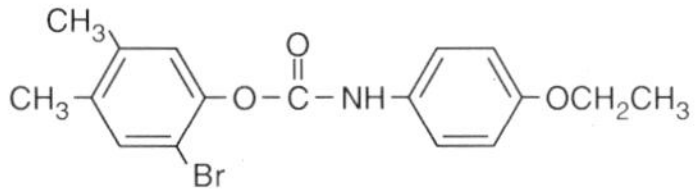
250 mg

S85,076-4
4-FLUOROPHENYL N-(2-CHLORO-4-METHOXYPHENYL)CARBAMATE
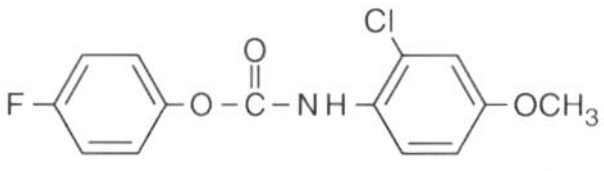
250 mg

S72,519-6
2,4-DICHLOROPHENYL N-(4-METHOXYPHENYL)CARBAMATE
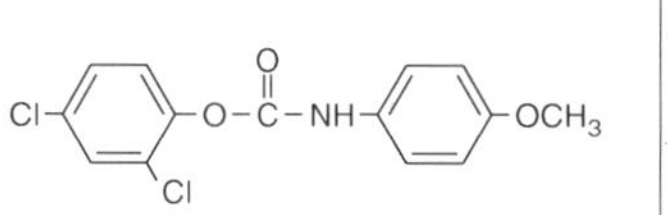
5 gm

S77,324-7
4-CHLORO-3-METHYLPHENYL N-(4-METHOXYPHENYL)CARBAMATE
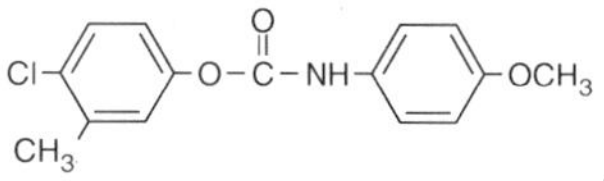
250 mg

S92,249-8
4-ACETYLPHENYL N-(5-CHLORO-2-METHOXYPHENYL)CARBAMATE
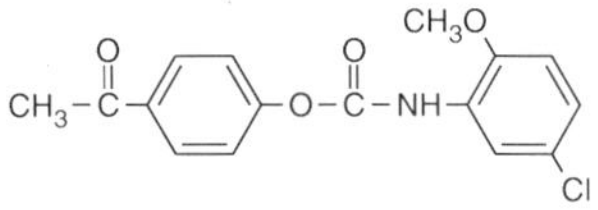
250 mg

S78,056-1
2,6-DI-TERT-BUTYL-4-(METHOXYMETHYL)PHENYL N-(4-CHLOROPHENYL)CARBAMATE
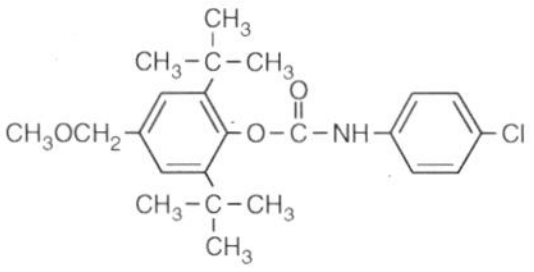
250 mg

S85,554-5
4-CHLORO-2-ISOPROPYL-5-METHYLPHENYL N-(4-ETHOXYPHENYL)-CARBAMATE
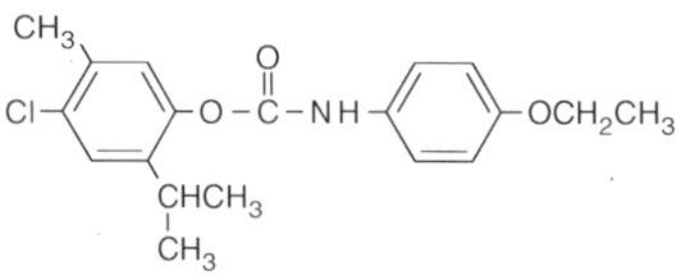
250 mg

S66,778-1
2-METHOXYETHYL N-(2,4-DICHLOROPHENYL)CARBAMATE
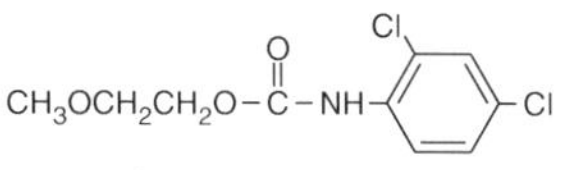
5 gm

S66,713-7
2-METHOXYETHYL N-(5-CHLORO-2-METHYLPHENYL)CARBAMATE
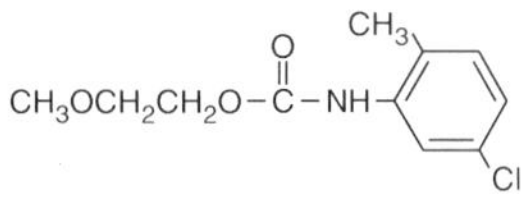
5 gm

S66,731-5
2-ETHOXYETHYL N-(2,4-DICHLOROPHENYL)CARBAMATE
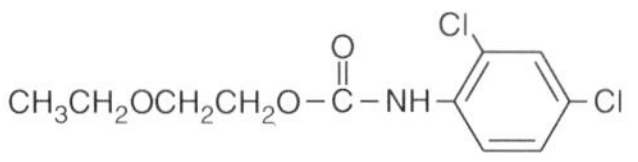
5 gm

S89,994-1
2,4-DICHLOROPHENYL N-(3-METHOXYPHENYL)CARBAMATE
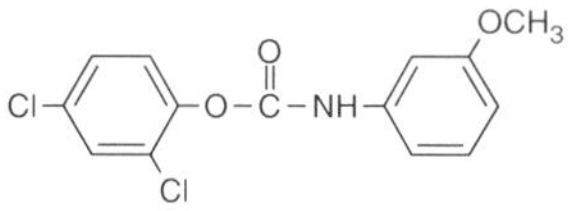
250 mg

S85,074-8
2-CHLORO-4-FLUOROPHENYL N-(2-CHLORO-4-METHOXYPHENYL)-CARBAMATE
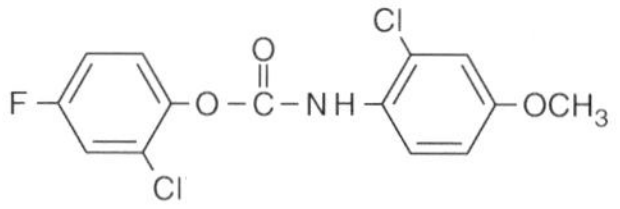
250 mg

S85,066-7
2,4-DICHLOROPHENYL N-(2-CHLORO-4-METHOXYPHENYL)CARBAMATE
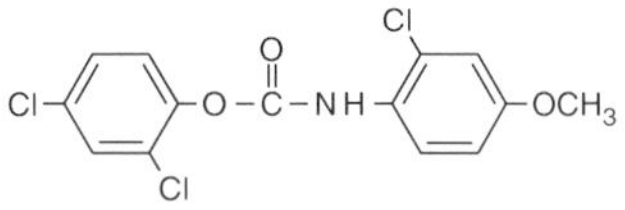
250 mg

S85,072-1
2,4-DIBROMOPHENYL N-(2-CHLORO-4-METHOXYPHENYL)CARBAMATE
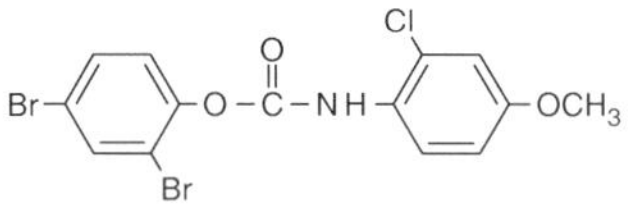
250 mg

S69,070-8
2-BROMO-4-CHLORO-5-METHYLPHENYL N-(4-CHLORO-2-METHOXYPHENYL)-CARBAMATE
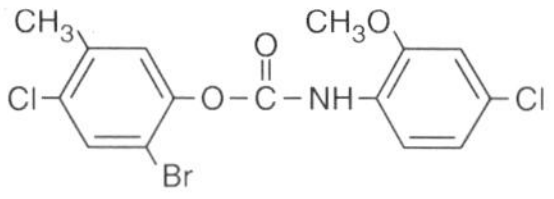
250 mg

S69,369-3
2-CHLORO-4-METHOXYPHENYL N-(2,4-DIMETHYLPHENYL)CARBAMATE
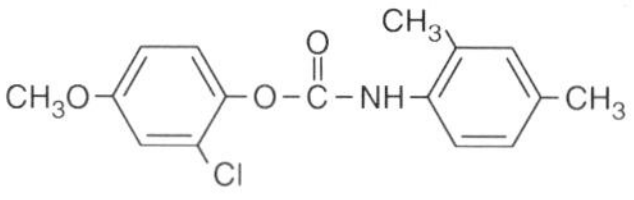
250 mg

S69,077-5
2-BROMO-4,5-DIMETHYLPHENYL N-(5-CHLORO-2-METHOXYPHENYL)-CARBAMATE
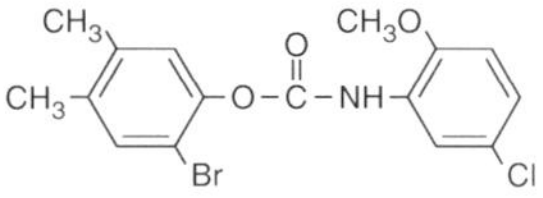
250 mg

S85,054-3
4-CHLORO-2-ISOPROPYL-5-METHYLPHENYL N-(2-CHLORO-4-METHOXYPHENYL)CARBAMATE
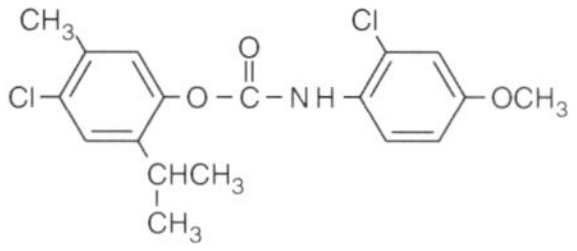
250 mg

S85,082-9
1-NAPHTHYL N-(2-CHLORO-4-METHOXYPHENYL)CARBAMATE
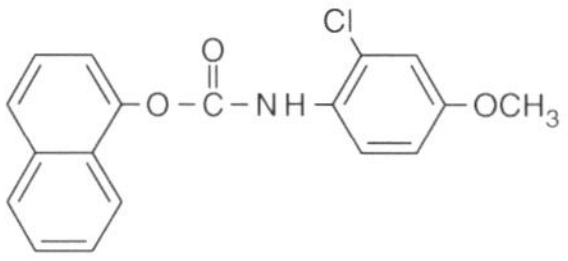
250 mg

S77,336-0
4-CHLOROPHENOXYACETYL N-(2-CHLOROPHENYL)CARBAMATE
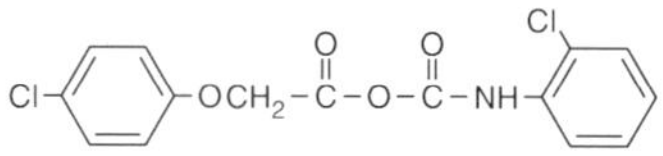
250 mg

S79,626-3
S-(4-METHOXYPHENYL) N-(2,3-DICHLOROPHENYL)THIOCARBAMATE
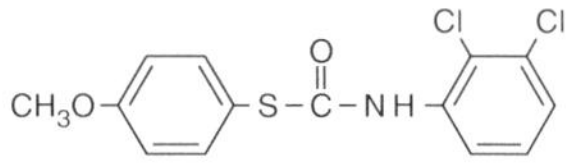
250 mg

S85,126-4
S-(4-METHOXYPHENYL) N-(2-CHLORO-6-METHYLPHENYL)THIOCARBAMATE
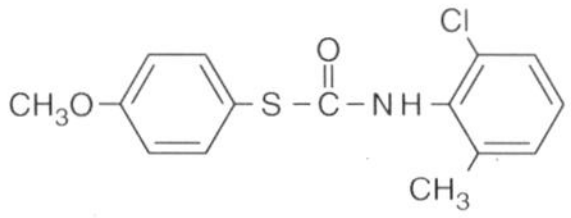
250 mg

S86,881-7
2-(BENZYLOXYCARBONYL)PHENYL N-(3-CHLOROPHENYL)CARBAMATE
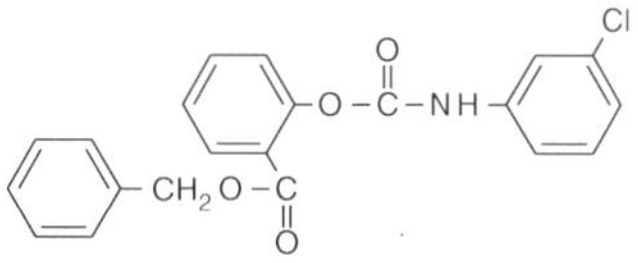
250 mg

S86,890-6
2-(BENZYLOXYCARBONYL)PHENYL N-(5-CHLORO-2-METHYLPHENYL)-CARBAMATE
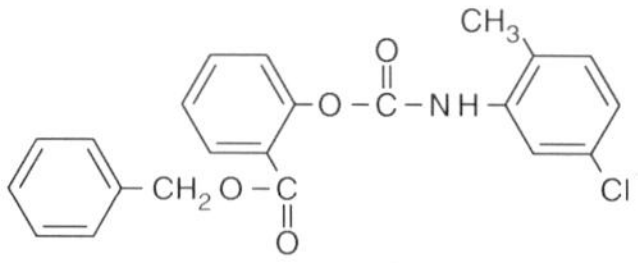
250 mg

S69,476-2
2-NITROBENZYL N-(4-FLUORO-2-NITROPHENYL)CARBAMATE
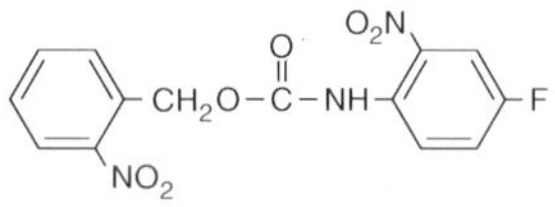
250 mg

S68,739-1
2-FLUOROBENZYL N-(2-CHLORO-4-NITROPHENYL)CARBAMATE
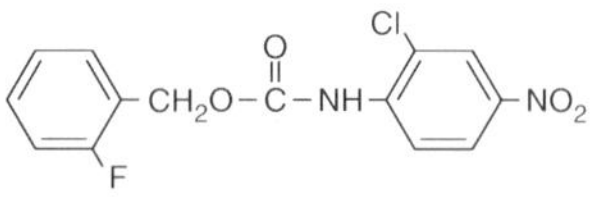
250 mg

S69,996-9
2-NITROBENZYL N-(2-CHLORO-4-NITROPHENYL)CARBAMATE
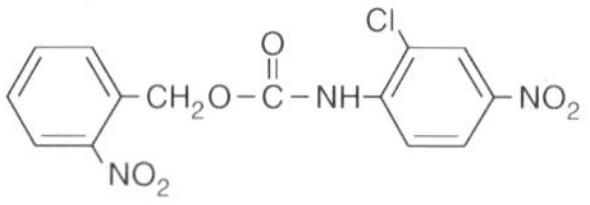
5 gm

S86,880-9
2-(BENZYLOXYCARBONYL)PHENYL N-(4-FLUORO-3-NITROPHENYL)CARBAMATE
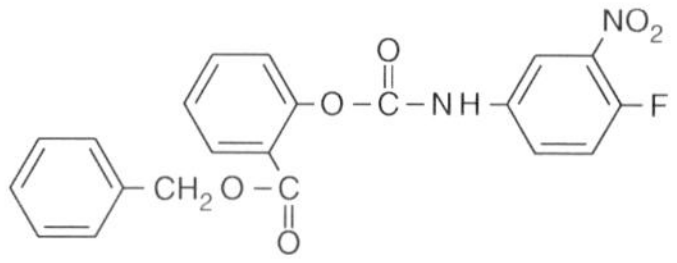
250 mg

S86,878-7
4-(BENZYLOXYCARBONYL)PHENYL N-(4-CHLORO-2-NITROPHENYL)CARBAMATE
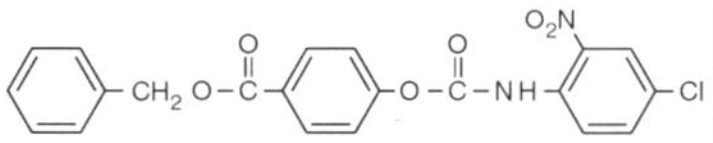
250 mg

S72,388-6
2-NITROBENZYL N-(2,5-DIFLUOROPHENYL)CARBAMATE
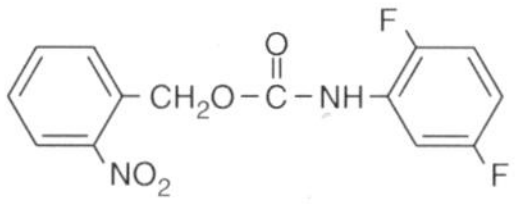
5 gm

S72,389-4
2-NITROBENZYL N-(3,4-DICHLOROPHENYL)CARBAMATE
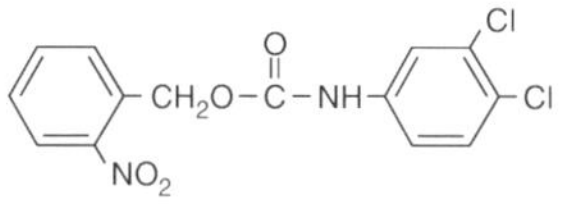
5 gm

S68,680-8
4-NITROBENZYL N-(2,4-DIFLUOROPHENYL)CARBAMATE
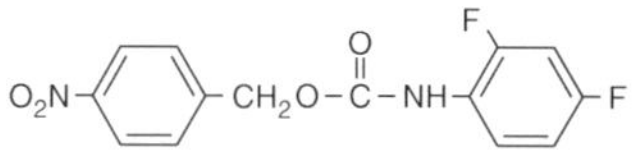
250 mg

S74,306-2
4-NITROBENZYL N-(2,5-DIFLUOROPHENYL)CARBAMATE
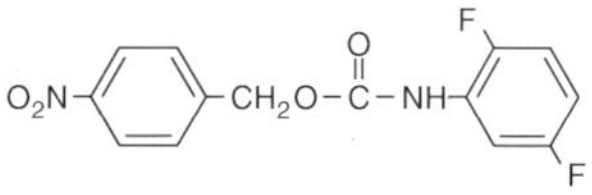
5 gm

S68,681-6
3-FLUOROBENZYL N-(2-CHLORO-4-NITROPHENYL)CARBAMATE
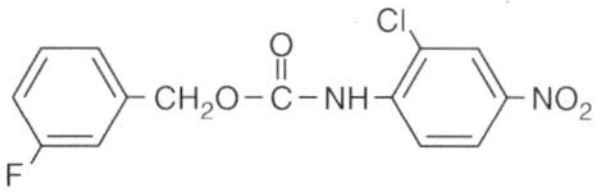
250 mg

S79,694-8
4-(BENZYLOXYCARBONYL)PHENYL N-(2,3-DICHLOROPHENYL)CARBAMATE
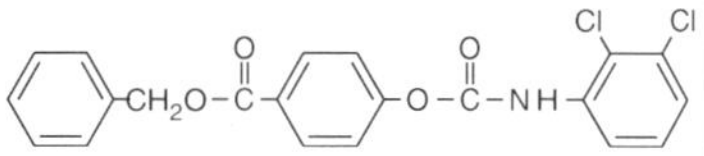
250 mg

S86,891-4
4-(BENZYLOXYCARBONYL)PHENYL N-(5-CHLORO-2-METHYLPHENYL)-CARBAMATE
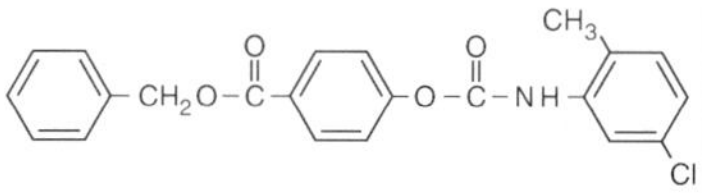
250 mg

S72,144-1
2-FLUOROBENZYL N-(3-CHLOROPHENYL)-CARBAMATE
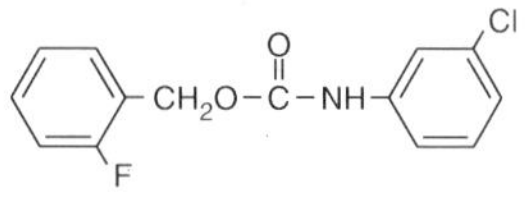
250 mg

S72,391-6
2-FLUOROBENZYL N-(2,4,5-TRICHLOROPHENYL)CARBAMATE
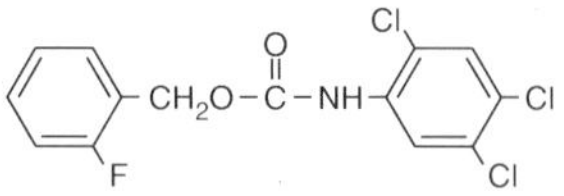
5 gm

S72,649-4
4-FLUOROBENZYL N-(2-FLUOROPHENYL)-CARBAMATE
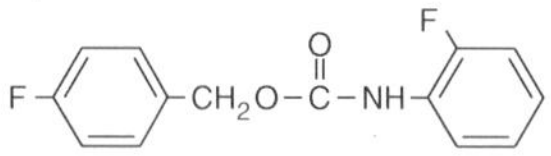
5 gm

S68,965-3
4-FLUOROBENZYL N-(2-CHLOROPHENYL)-CARBAMATE
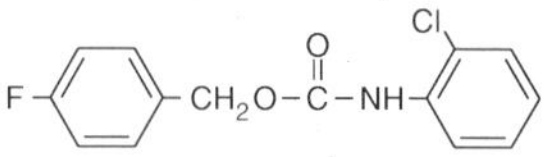
250 mg

S74,304-6
2-FLUOROBENZYL N-(2,5-DIFLUOROPHENYL)CARBAMATE

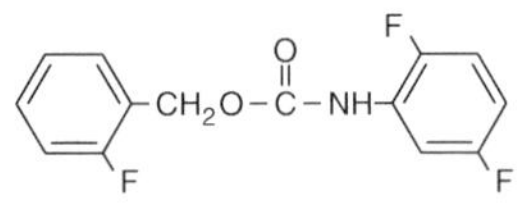

5 gm

S72,387-8
2-FLUOROBENZYL N-(3,4-DICHLOROPHENYL)CARBAMATE

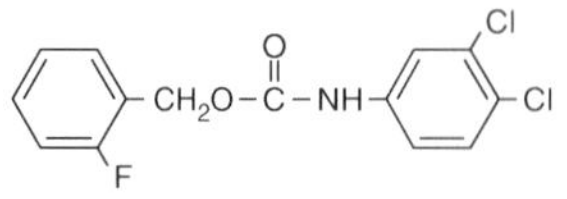

5 gm

S68,773-1
4-FLUOROBENZYL N-(3-CHLOROPHENYL)-CARBAMATE

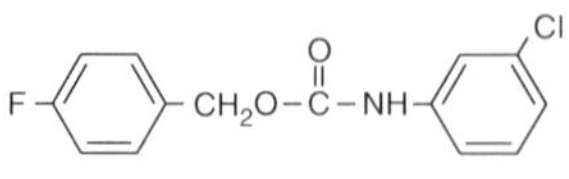

250 mg

S69,084-8
4-FLUOROBENZYL N-(2,4,5-TRICHLOROPHENYL)CARBAMATE

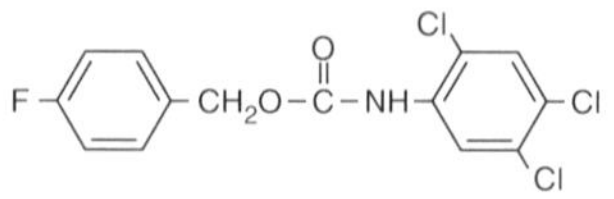

250 mg

S69,037-6
4-FLUOROBENZYL N-(2,3-DICHLOROPHENYL)CARBAMATE

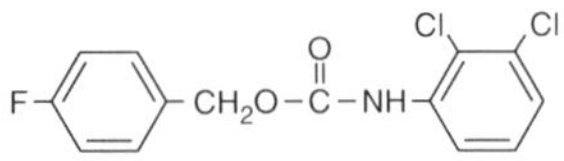

250 mg

S68,675-1
4-FLUOROBENZYL N-(2,5-DICHLOROPHENYL)CARBAMATE

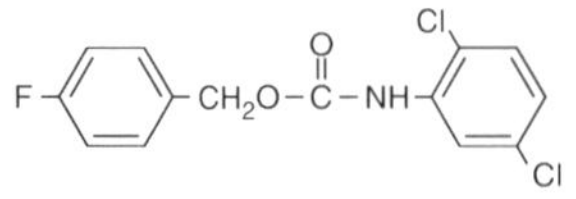

250 mg

S72,607-9
4-FLUOROBENZYL N-(3,4-DICHLOROPHENYL)CARBAMATE

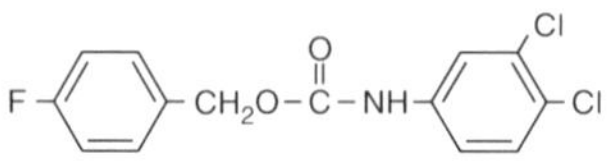

5 gm

S74,695-9
3-FLUOROBENZYL N-(2,5-DIFLUOROPHENYL)CARBAMATE

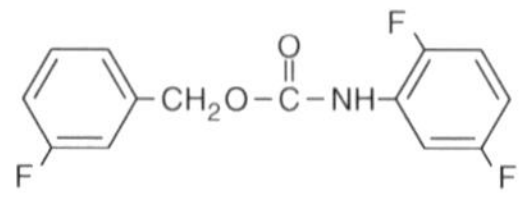

5 gm

S72,603-6
3-FLUOROBENZYL N-(3,4-DICHLOROPHENYL)CARBAMATE

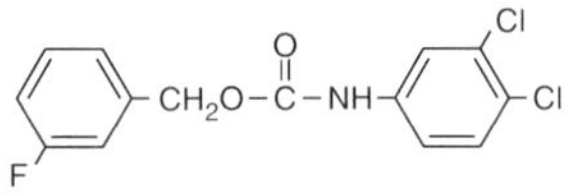

250 mg

S68,691-3
3-FLUOROBENZYL N-(2,4-DIMETHYLPHENYL)CARBAMATE

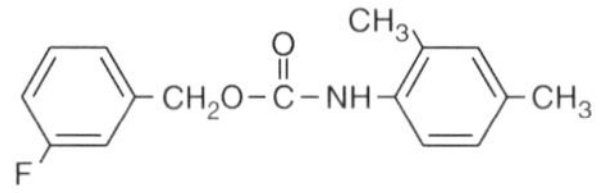

250 mg

S84,512-4
4-FORMYL-2-METHOXYPHENYL N-(2-FLUORO-5-NITROPHENYL)CARBAMATE

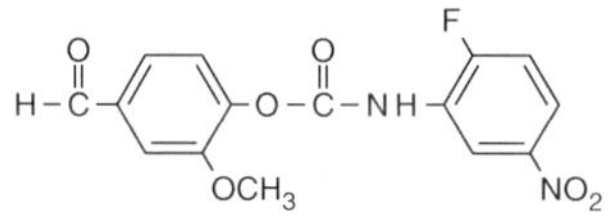

250 mg

S85,373-9
2-(4-BROMOPHENOXY)ETHYL N-(4-METHYL-3-NITROPHENYL)CARBAMATE

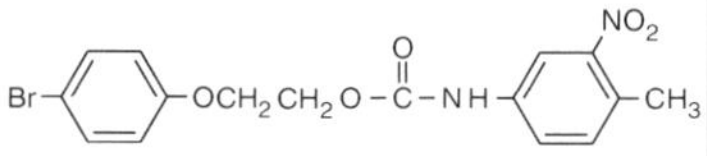

250 mg

S85,371-2
2-(2,4-DICHLOROPHENOXY)ETHYL N-(4-METHYL-3-NITROPHENYL)CARBAMATE

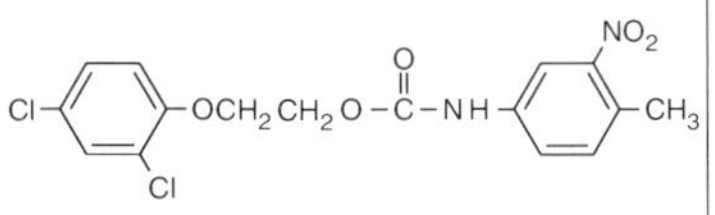

250 mg

S77,991-1
2-(METHACRYLOYLOXY)ETHYL N-(3-BROMOPHENYL)CARBAMATE

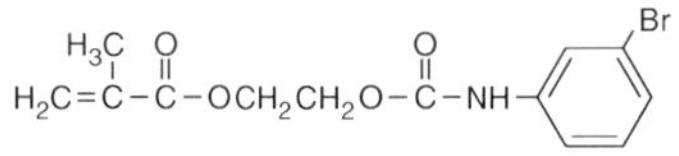

250 mg

S34,148-7
1-(CARBOBUTOXY)ETHYL N-(3-CHLOROPHENYL)CARBAMATE

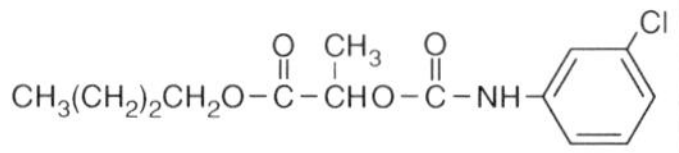

5 gm

S71,889-0
TETRAHYDROFURFURYL N-(3-CHLOROPHENYL)CARBAMATE

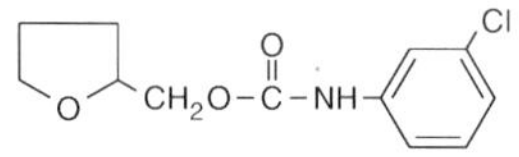

250 mg

S77,439-1
(4H-PYRAN-2-YL)METHYL N-(4-BROMOPHENYL)CARBAMATE

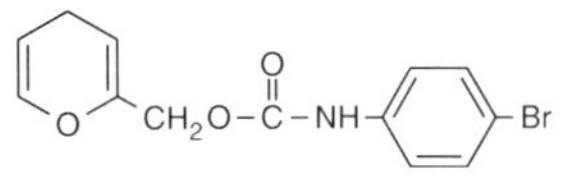

250 mg

S70,999-9
2-(2,4-DICHLOROPHENOXY)ETHYL N-(2,6-DICHLOROPHENYL)CARBAMATE

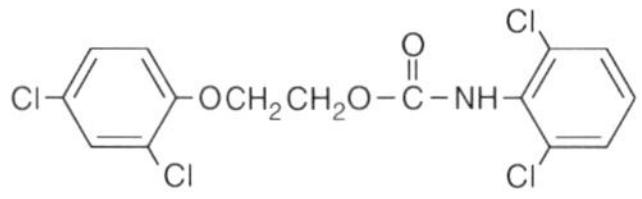

250 mg

S72,753-9
4-METHYL-2-OXO-2H-1-BENZOPYRAN-7-YL N-(4-FLUORO-3-NITROPHENYL)-CARBAMATE

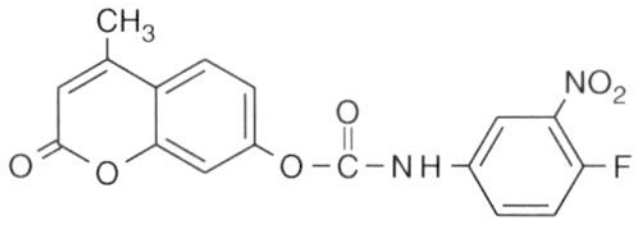

5 gm

S86,748-9
2-CARBAMOYLPHENYL N-(4-FLUORO-3-NITROPHENYL)CARBAMATE

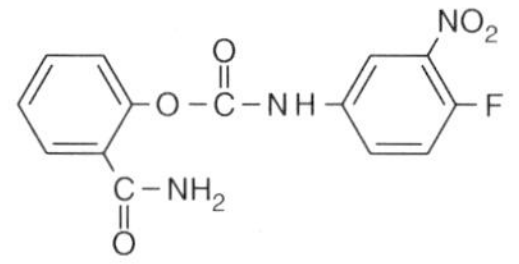

250 mg

S71,356-2
2-BENZOYL-4-CHLOROPHENYL N-(4-CHLOROPHENYL)CARBAMATE

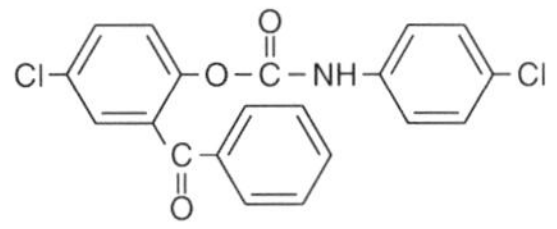

250 mg

S71,076-8
2-BENZOYL-4-CHLOROPHENYL N-(4-CHLORO-2-METHYLPHENYL)-CARBAMATE

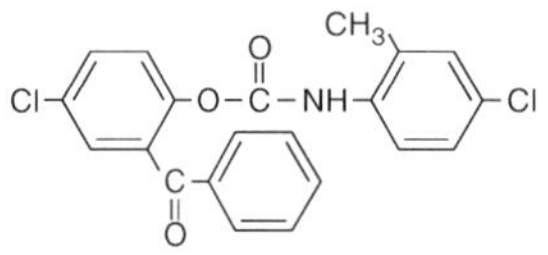

250 mg

S85,089-6
2-(ISOPENTYLOXYCARBONYL)PHENYL N-(2,3-DICHLOROPHENYL)CARBAMATE

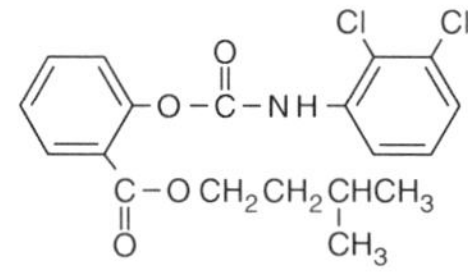

250 mg

S71,848-3
2-CARBAMOYLPHENYL N-(2-FLUOROPHENYL)CARBAMATE

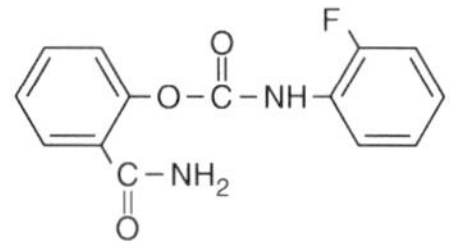

5 gm

S84,818-2
2-CARBAMOYLPHENYL N-(3-FLUOROPHENYL)CARBAMATE

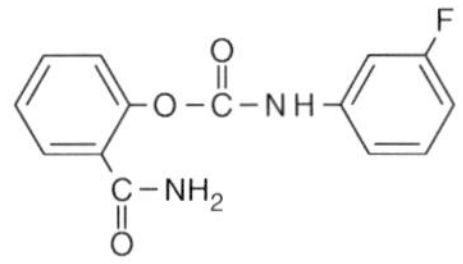

250 mg

S80,025-2
2-CARBAMOYLPHENYL N-(4-FLUOROPHENYL)CARBAMATE

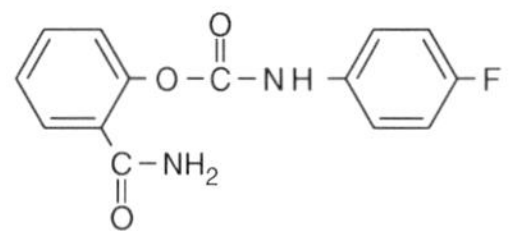

250 mg

S79,637-9
2-CARBAMOYLPHENYL N-(2-CHLORO-6-METHYLPHENYL)CARBAMATE

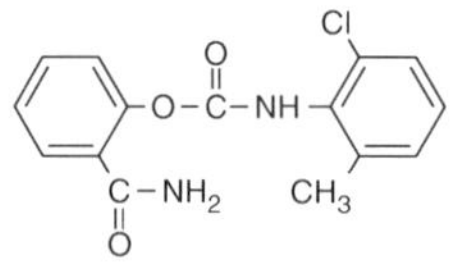

250 mg

S80,064-3
2-(METHOXYCARBONYL)PHENYL N-(4-FLUOROPHENYL)CARBAMATE

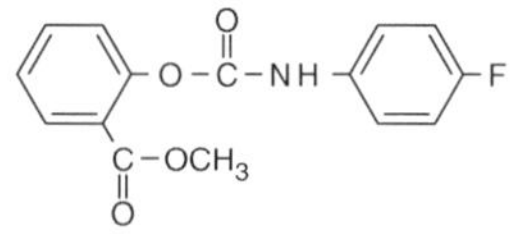

250 mg

S85,098-5
2-(METHOXYCARBONYL)PHENYL N-(2,3-DICHLOROPHENYL)CARBAMATE

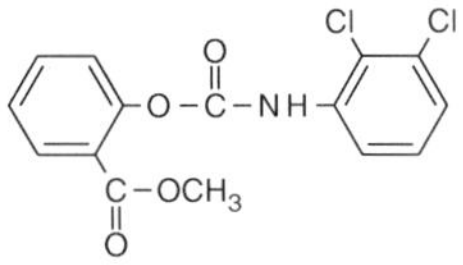

250 mg

S86,799-3
3-ACETOXYPHENYL N-(3-CHLORO-2-METHYLPHENYL)CARBAMATE

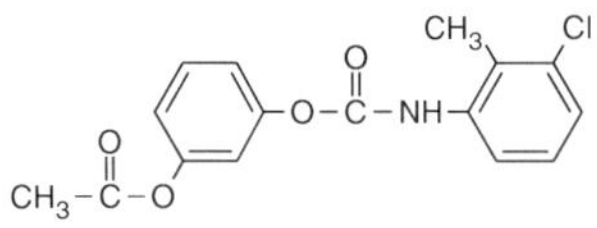

250 mg

S77,973-3
3,4-DIHYDRO-2H-1-BENZOPYRAN-3-YLMETHYL N-(4-FLUOROPHENYL)-CARBAMATE

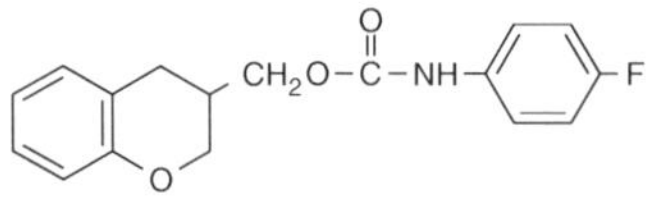

250 mg

S72,809-8
4-(METHOXYCARBONYL)PHENYL N-(2-CHLORO-5-NITROPHENYL)CARBAMATE

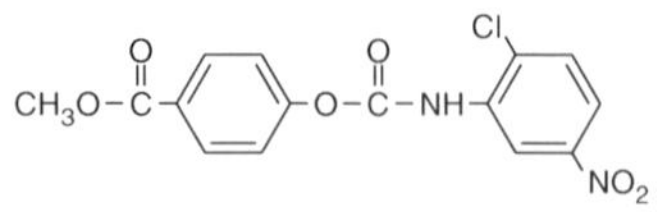

5 gm

S72,321-5
2,3-DICHLOROPHENYL N-(2,5-DIMETHOXYPHENYL)CARBAMATE
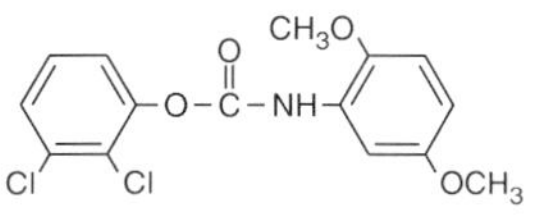
250 mg

S85,519-7
3-ACETYLPHENYL N-(4-CHLOROPHENYL)-CARBAMATE
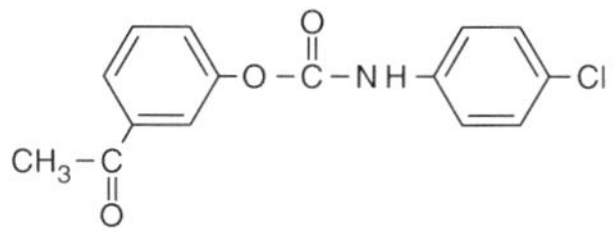
250 mg

S75,018-2
3-ACETYLPHENYL N-(2,3-DICHLOROPHENYL)CARBAMATE
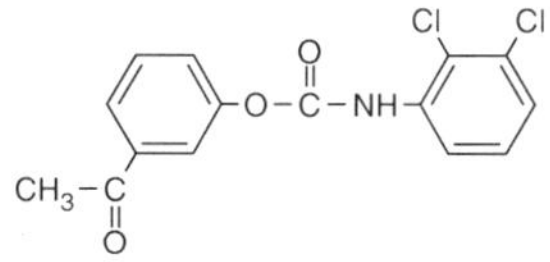
5 gm

S85,088-8
O,O'-(1,4-CYCLOHEXYLENE) BIS[N-(2,3-DICHLOROPHENYL)CARBAMATE]
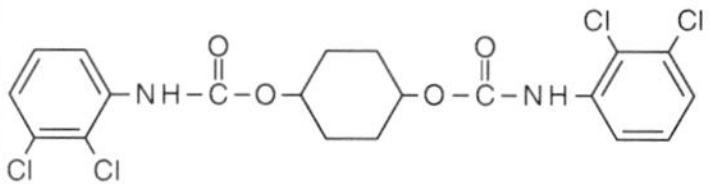
250 mg

S70,079-7
4-(METHYLTHIO)PHENYL N-(4-FLUORO-2-NITROPHENYL)CARBAMATE
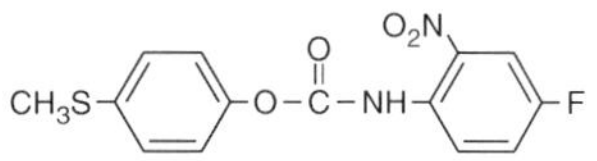
250 mg

S69,564-5
N,N'-(4,6-DIBROMO-1,3-PHENYLENE)-BIS[2-(4-CHLOROPHENYLTHIO)ETHYL-CARBAMATE]
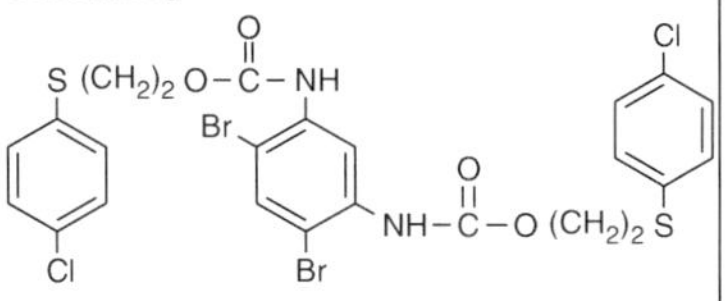
250 mg

S69,982-9
2,6-DIBROMO-4-NITROPHENYL N-[4-(METHYLTHIO)PHENYL]CARBAMATE
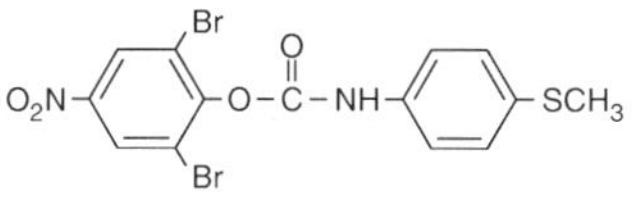
250 mg

S80,768-0
3-(4-CHLOROPHENYLCARBAMOYL)-2-NAPHTHYL N-[4-(METHYLTHIO)PHENYL]-CARBAMATE
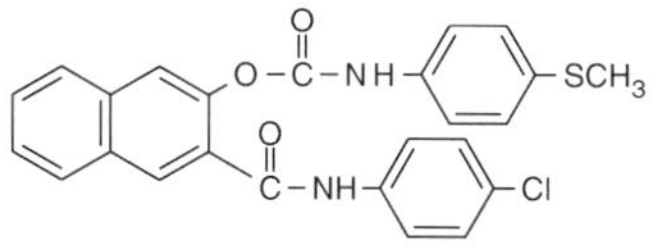
250 mg

S85,353-4
2,4-DICHLORO-3,5-DIMETHYLPHENYL N-[3-(METHYLTHIO)PHENYL]CARBAMATE
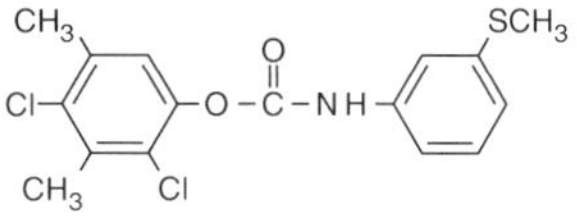
250 mg

S72,516-1
4-FLUOROPHENYL N-[3-(METHYLTHIO)-PHENYL]CARBAMATE
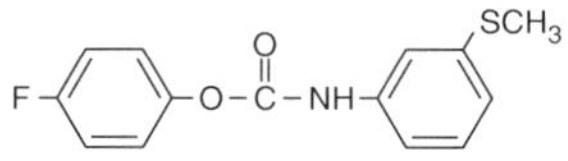
5 gm

S79,701-4
4-FLUOROPHENYL N-[4-(METHYLTHIO)-PHENYL]CARBAMATE
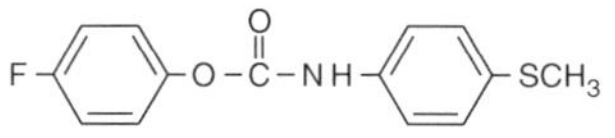
250 mg

S72,465-3
4-(METHYLTHIO)PHENYL N-(4-CHLOROPHENYL)CARBAMATE
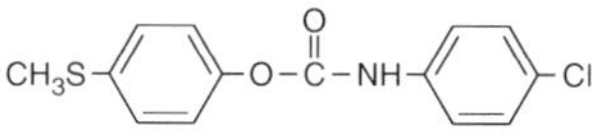
5 gm

S75,230-4
4-(METHYLTHIO)PHENYL N-(3-CHLOROPHENYL)CARBAMATE
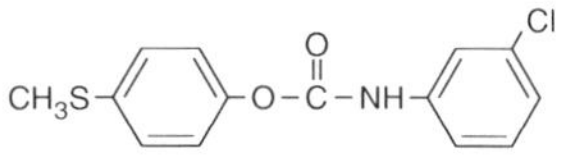
5 gm

S69,594-7
2,4,5-TRICHLOROPHENYL N-[4-(METHYLTHIO)PHENYL]CARBAMATE
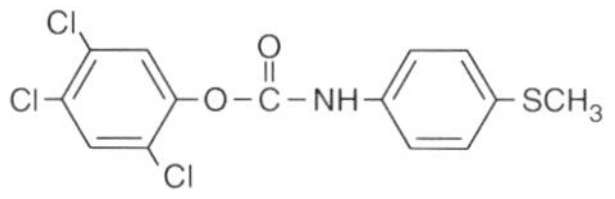
250 mg

S69,936-5
N,N'(4-BROMO-2-METHYL-M-PHENYLENE)-BIS[2-(4-CHLOROPHENYLTHIO)ETHYL CARBAMATE]
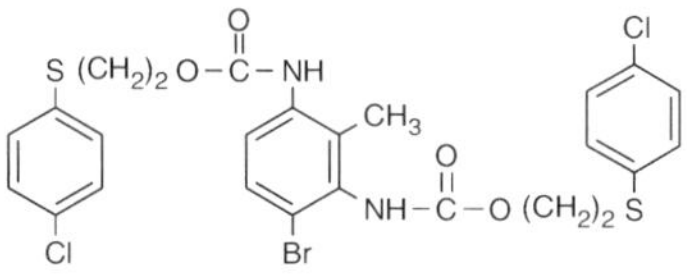
250 mg

S77,380-8
2,3-DICHLOROPHENYL N-[4-(METHYLTHIO)PHENYL]CARBAMATE
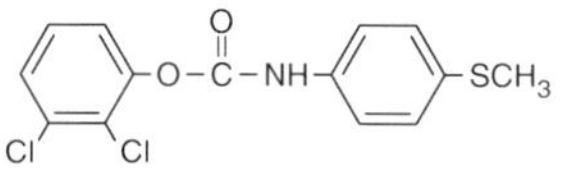
250 mg

S72,303-7
2-(4-CHLOROPHENYLTHIO)ETHYL N-(3-CHLORO-2-METHYLPHENYL)-CARBAMATE
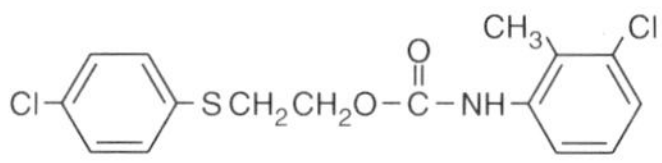
250 mg

S72,377-0
4-(METHYLTHIO)PHENYL N-(2,5-DIFLUOROPHENYL)CARBAMATE
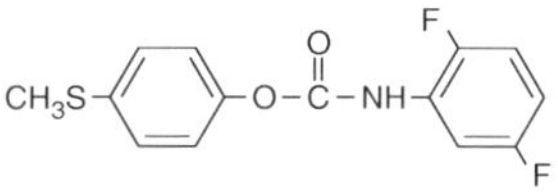
5 gm

S70,237-4
2,4-DICHLOROPHENYL N-[4-(METHYLTHIO)PHENYL]CARBAMATE
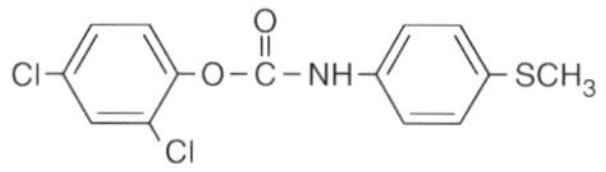
250 mg

S72,381-9
2,5-DICHLOROPHENYL N-[4-(METHYLTHIO)PHENYL]CARBAMATE
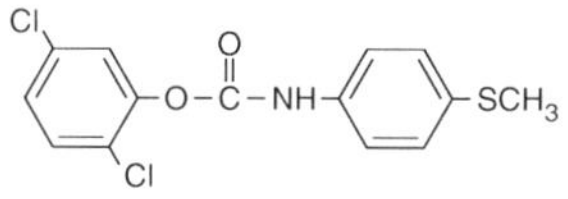
5 gm

S72,382-7
3,5-DICHLOROPHENYL N-[4-(METHYLTHIO)PHENYL]CARBAMATE
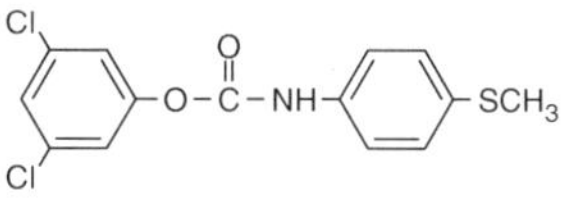
5 gm

S77,385-9
3,4-DICHLOROPHENYL N-[4-(METHYLTHIO)PHENYL]CARBAMATE
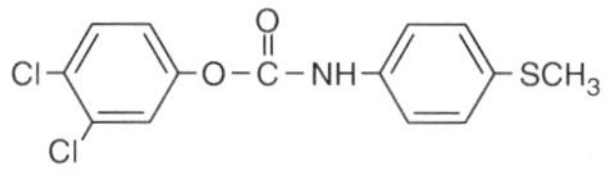
250 mg

S79,133-4
4-(METHYLTHIO)PHENYL N-(3-CHLORO-4-METHYLPHENYL)CARBAMATE
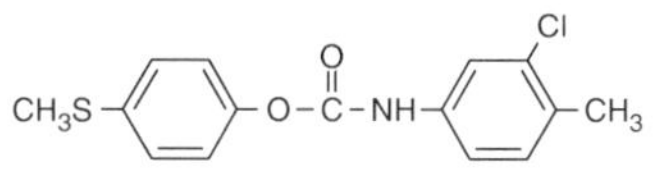
250 mg

S89,208-4
3-METHYL-4-(METHYLTHIO)PHENYL N-(4-CHLOROPHENYL)CARBAMATE
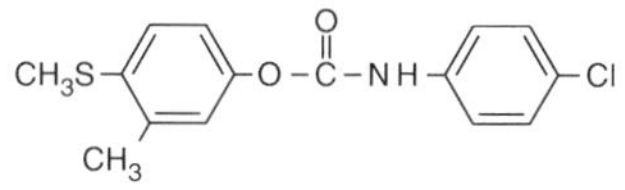
250 mg

S73,911-1
2-(4-CHLOROPHENYLTHIO)ETHYL N-(3,4-DICHLOROPHENYL)CARBAMATE
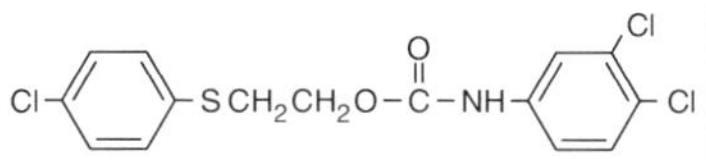
5 gm

S79,434-1
2-BROMO-4-TERT-BUTYLPHENYL N-[4-(METHYLTHIO)PHENYL]CARBAMATE
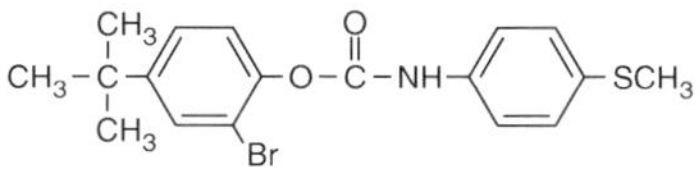
250 mg

S91,080-5
4-CHLORO-2,6-DIMETHYLPHENYL N-[4-(METHYLTHIO)PHENYL]CARBAMATE
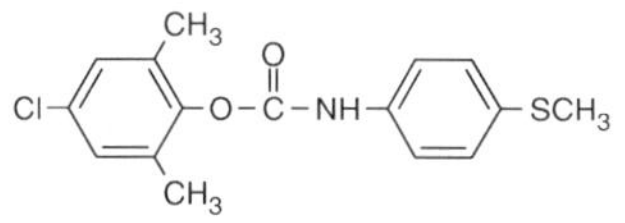
250 mg

S79,775-8
4-CHLORO-2-ISOPROPYL-5-METHYLPHENYL N-[4-(METHYLTHIO)-PHENYL]CARBAMATE
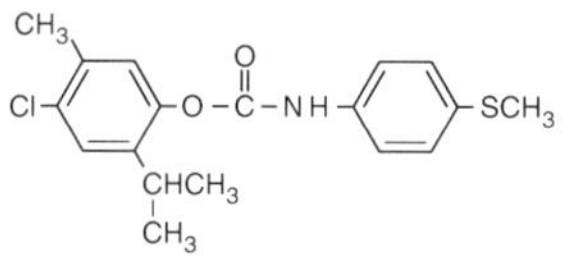
250 mg

S85,346-1
2-CHLORO-4-NITROPHENYL N-[3-(METHYLTHIO)PHENYL]CARBAMATE
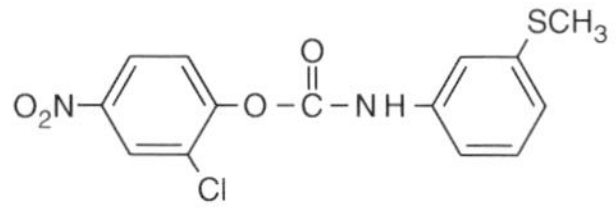
250 mg

S89,951-8
3-METHYL-4-(METHYLTHIO)PHENYL N-(3,4-DICHLOROPHENYL)CARBAMATE
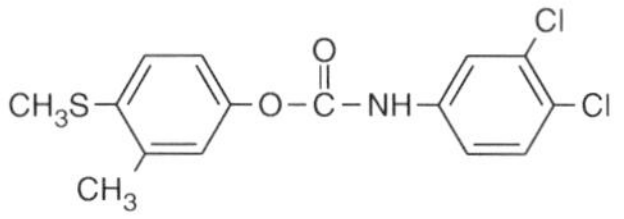
250 mg

S86,807-8
3-METHYL-4-(METHYLTHIO)PHENYL N-(2-BROMO-4-METHYLPHENYL)CARBAMATE
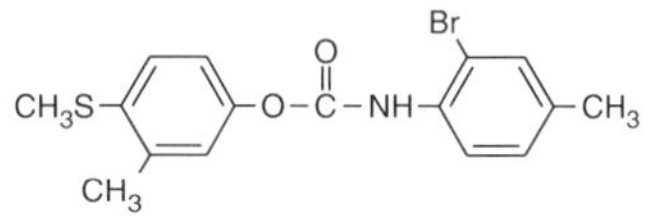
250 mg

S84,834-4
2,6-DIMETHYL-4-(METHYLTHIO)PHENYL N-(3-FLUOROPHENYL)CARBAMATE
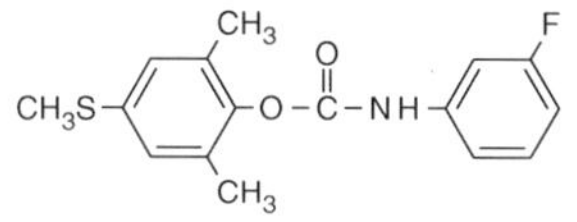
250 mg

S69,490-8
4-CHLORO-2,6-DIMETHYLPHENYL N-[3-(METHYLTHIO)PHENYL]CARBAMATE
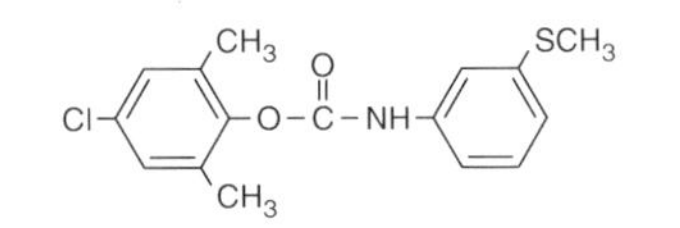
250 mg

S74,613-4
4-CHLORO-3,5-DIMETHYLPHENYL N-[3-(METHYLTHIO)PHENYL]CARBAMATE
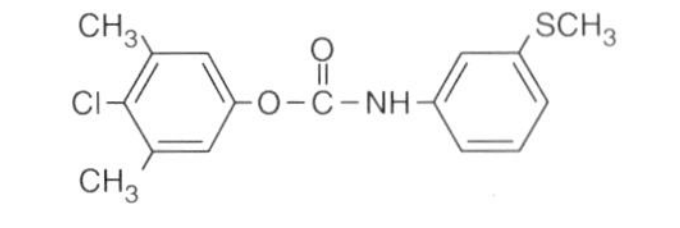
5 gm

S69,993-4
S-(4-CHLOROPHENYL) N-[4-(METHYLTHIO)PHENYL]-THIOCARBAMATE
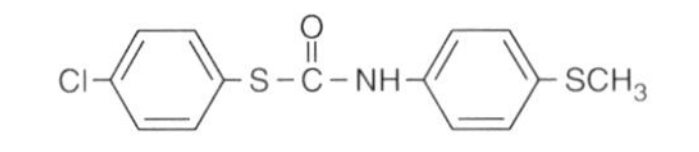
250 mg

S80,543-2
ETHYL N-(2-AMINO-4-CHLOROPHENYL)-CARBAMATE
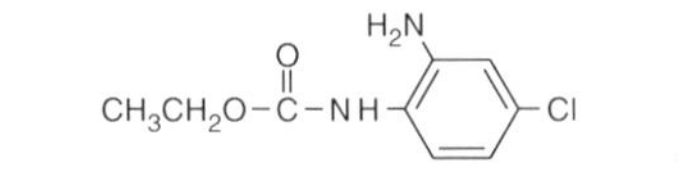
250 mg

S90,567-4
4-ACETAMIDOPHENYL N-(4-CHLORO-2-NITROPHENYL)CARBAMATE
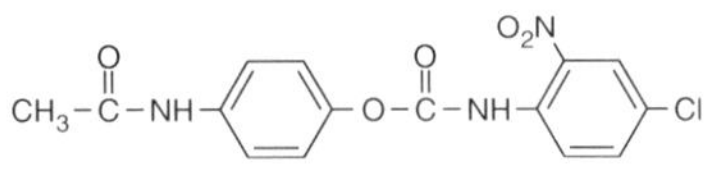
250 mg

S67,522-9
4-CYCLOHEXYLPHENYL N-(4-CHLORO-2-NITROPHENYL)CARBAMATE
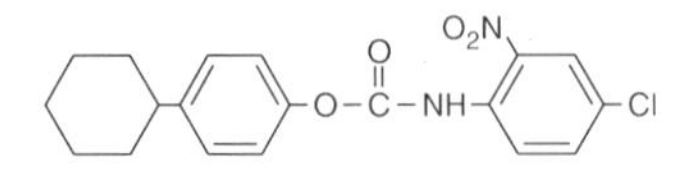
250 mg

S86,869-8
2-BENZYLPHENYL N-(4-CHLORO-2-NITROPHENYL)CARBAMATE
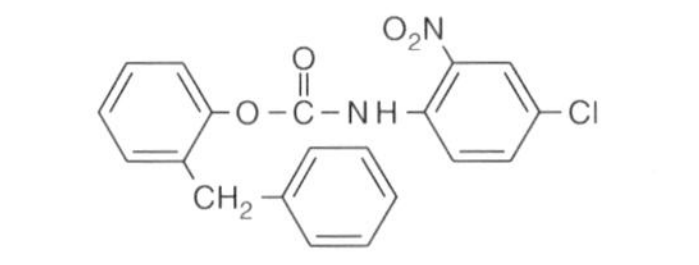
250 mg

S86,727-6
2-FLUOROPHENYL N-(4-CHLORO-2-NITROPHENYL)CARBAMATE
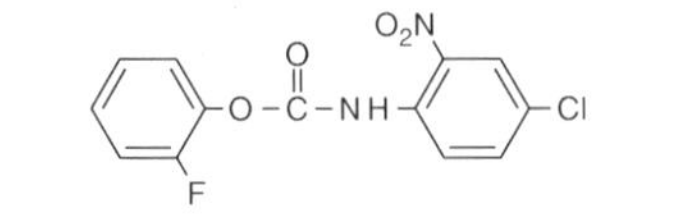
250 mg

S85,176-0
2-FLUOROPHENYL N-(4-METHYL-2-NITROPHENYL)CARBAMATE
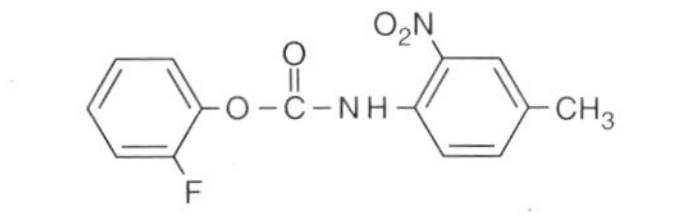
250 mg

S85,536-7
4-CHLORO-3,5-DIMETHYL N-(2-NITROPHENYL)CARBAMATE
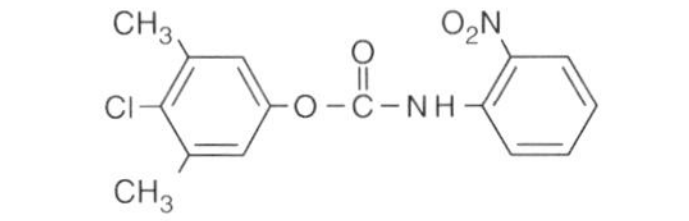
250 mg

S70,183-1
4-BROMOPHENYL N-(4-FLUORO-2-NITROPHENYL)CARBAMATE
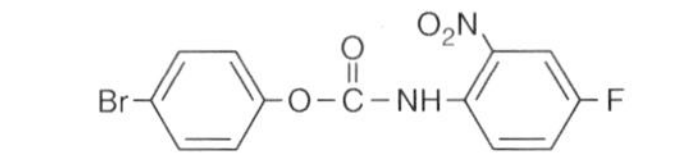
250 mg

S85,177-9
4-FLUOROPHENYL N-(4-METHYL-2-NITROPHENYL)CARBAMATE
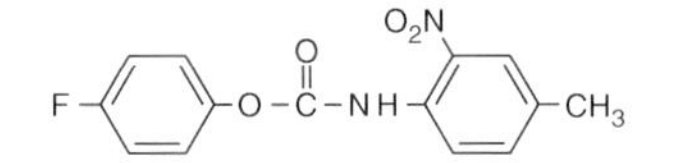
250 mg

S85,178-7
4-CHLOROPHENYL N-(4-METHYL-2-NITROPHENYL)CARBAMATE
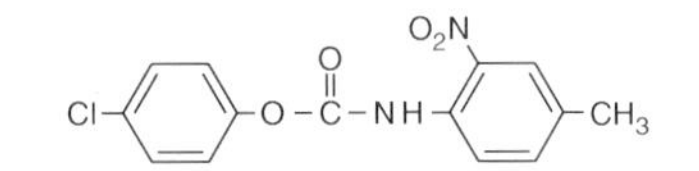
250 mg

S85,179-5
4-BROMOPHENYL N-(4-METHYL-2-NITROPHENYL)CARBAMATE
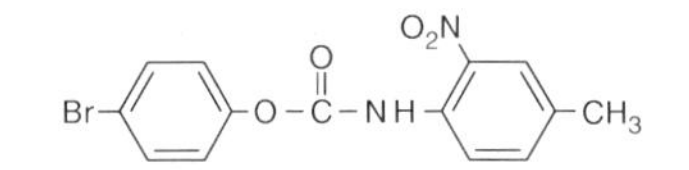
250 mg

S76,978-9
3,4-DICHLOROPHENYL N-(2-METHYL-6-NITROPHENYL)CARBAMATE
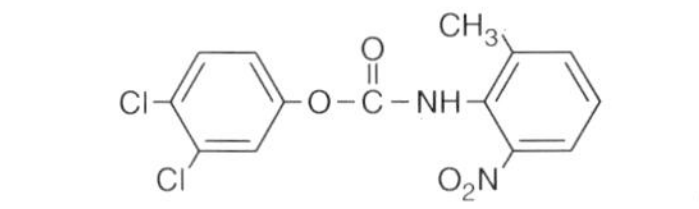
250 mg

S80,211-5
ETHYL N-(4-CHLORO-2-NITROPHENYL)-CARBAMATE
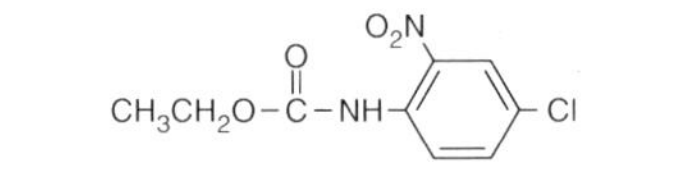
250 mg

S75,135-9
2,5-DICHLOROPHENYL N-(4-FLUORO-2-NITROPHENYL)CARBAMATE
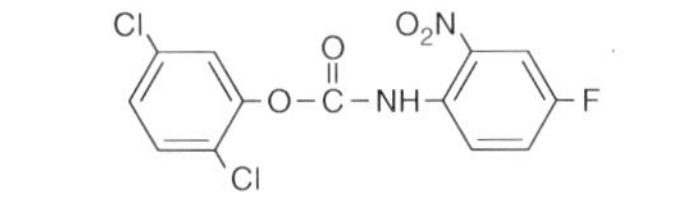
5 gm

S86,726-8
2,4-DIBROMOPHENYL N-(4-CHLORO-2-NITROPHENYL)CARBAMATE
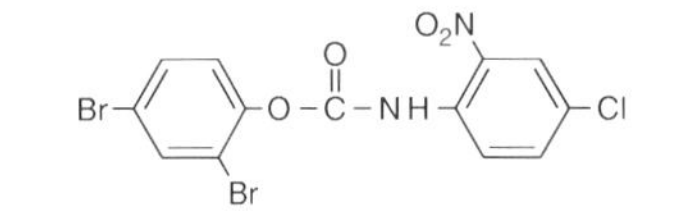
250 mg

S85,180-9
2,4-DICHLOROPHENYL N-(4-METHYL-2-NITROPHENYL)CARBAMATE
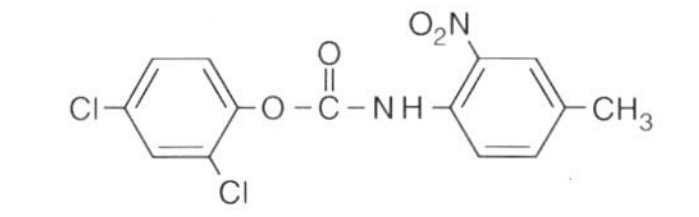
250 mg

S86,747-0
4-CHLORO-2-METHYLPHENYL N-(4-CHLORO-2-NITROPHENYL)CARBAMATE
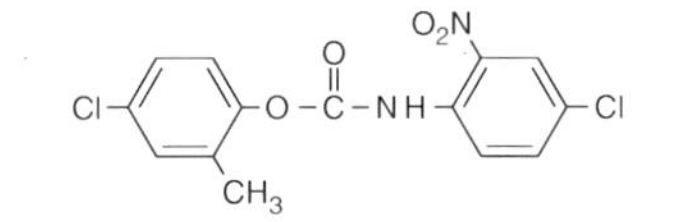
250 mg

S72,447-5
3,4-XYLYL N-(4-FLUORO-2-NITROPHENYL)CARBAMATE
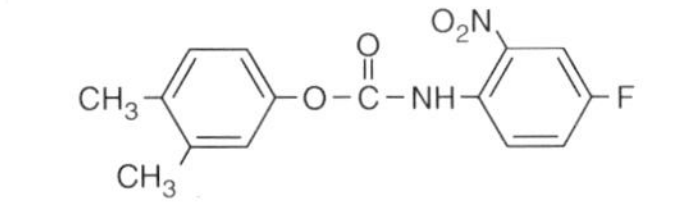
5 gm

S85,174-4
4-CHLORO-3-METHYLPHENYL N-(4-METHYL-2-NITROPHENYL)CARBAMATE
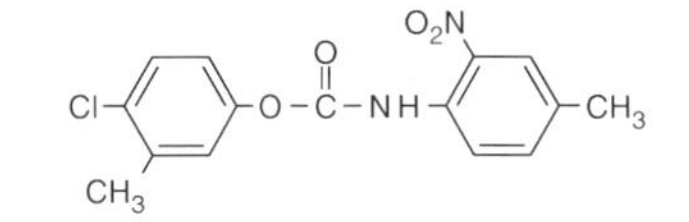
250 mg

S68,805-3
4-BROMO-2,5-DIMETHYLPHENYL N-(4-CHLORO-2-NITROPHENYL)CARBAMATE
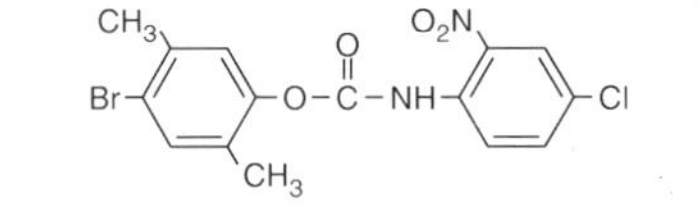
250 mg

S85,170-1
2,4-DICHLORO-3,5-DIMETHYLPHENYL N-(4-METHYL-2-NITROPHENYL)CARBAMATE
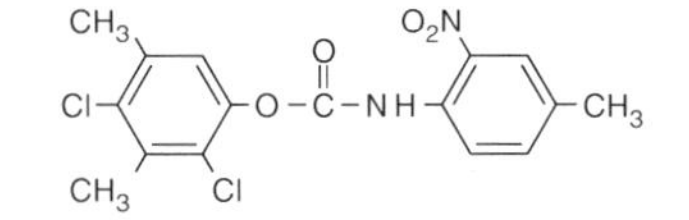
250 mg

S86,822-1
2-BROMO-4-TERT-BUTYLPHENYL N-(4-CHLORO-2-NITROPHENYL)CARBAMATE
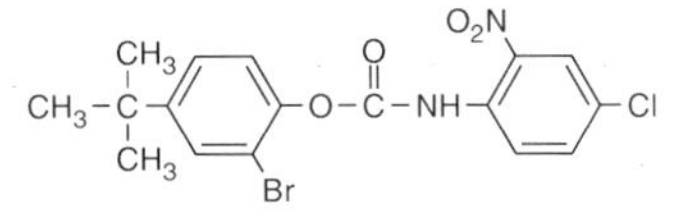
250 mg

S85,161-2
2-BROMO-4-TERT-BUTYLPHENYL N-(4-METHYL-2-NITROPHENYL)CARBAMATE
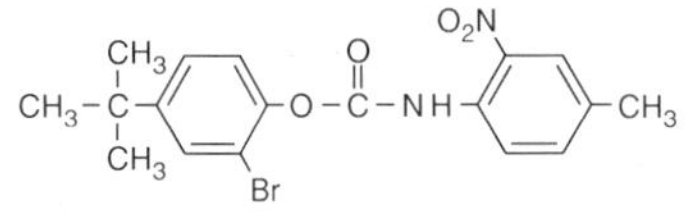
250 mg

S69,618-8
2,4-DICHLORO-6-METHYLPHENYL N-(4-FLUORO-2-NITROPHENYL)CARBAMATE
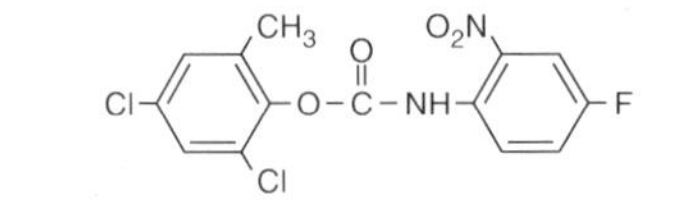
5 gm

S70,054-1
4-CHLORO-3,5-DIMETHYLPHENYL N-(4-FLUORO-2-NITROPHENYL)CARBAMATE
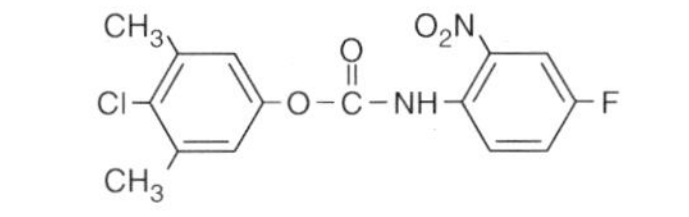
5 gm

S72,337-1
4-CHLORO-2,6-DIMETHYLPHENYL N-(4-FLUORO-2-NITROPHENYL)CARBAMATE
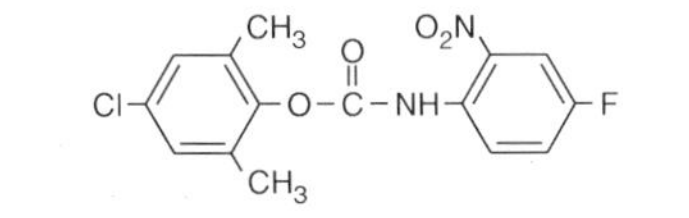
5 gm

S85,165-5
4-CHLORO-2,6-DIMETHYLPHENYL N-(4-METHYL-2-NITROPHENYL)CARBAMATE
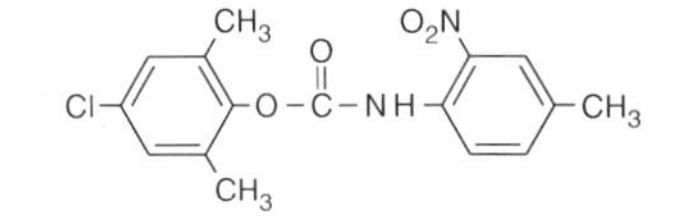
250 mg

S85,166-3
4-CHLORO-2,5-DIMETHYLPHENYL N-(4-METHYL-2-NITROPHENYL)CARBAMATE
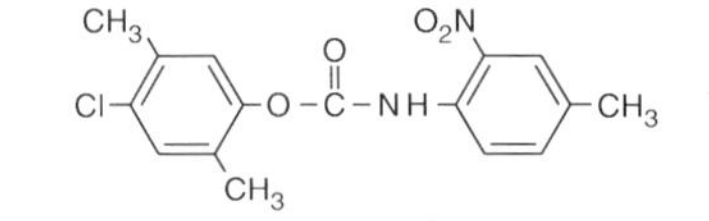
250 mg

S85,167-1
2-CHLORO-4,5-DIMETHYLPHENYL N-(4-METHYL-2-NITROPHENYL)CARBAMATE
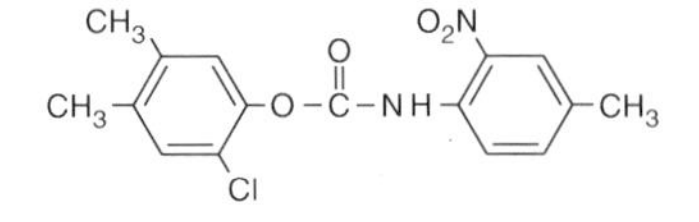
250 mg

S85,190-6
S-(4-CHLOROPHENYL) N-(4-METHYL-2-NITROPHENYL)THIOCARBAMATE

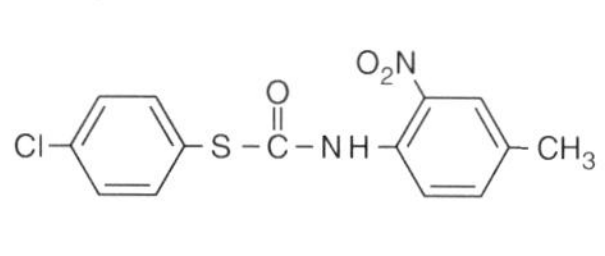

250 mg

S69,920-9
6-BROMO-2-NAPHTHYL N-(4-FLUORO-2-NITROPHENYL)CARBAMATE

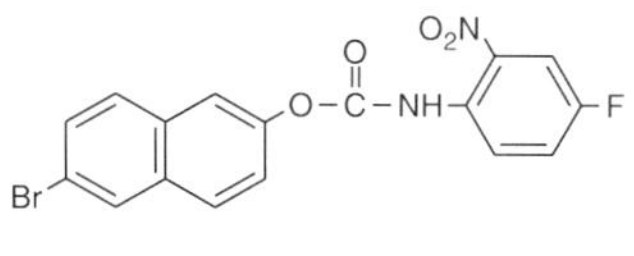

250 mg

S84,511-6
4-ACETYLPHENYL N-(2-FLUORO-5-NITROPHENYL)CARBAMATE

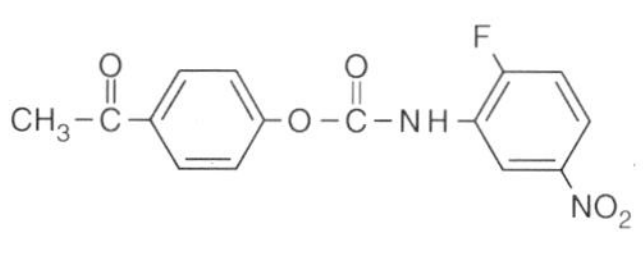

250 mg

S78,113-4
3,5-DINITROBENZOYL N-(2-CHLOROPHENYL)CARBAMATE

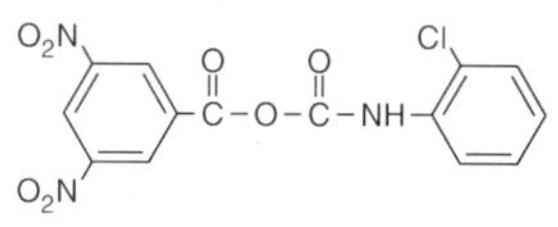

250 mg

S84,461-6
2-CARBAMOYLPHENYL N-(2-FLUORO-5-NITROPHENYL)CARBAMATE

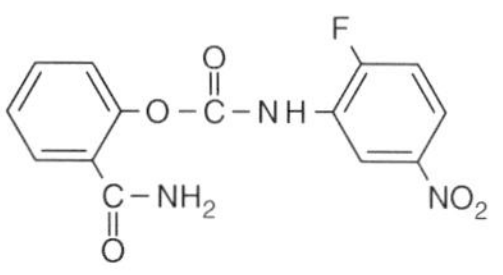

250 mg

S84,517-5
4-ACETAMIDOPHENYL N-(4-CHLORO-3-NITROPHENYL)CARBAMATE

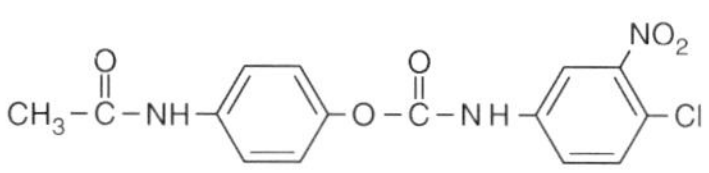

250 mg

S84,513-2
2-(METHOXYCARBONYL)PHENYL N-(2-FLUORO-5-NITROPHENYL)CARBAMATE

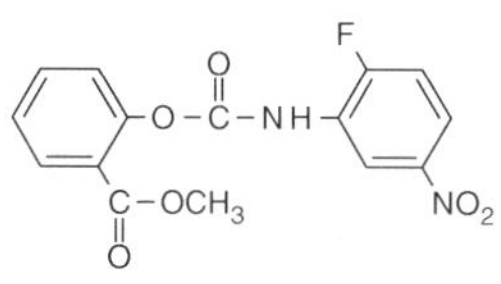

250 mg

S72,810-1
4-(METHOXYCARBONYL)PHENYL N-(4-FLUORO-3-NITROPHENYL)CARBAMATE

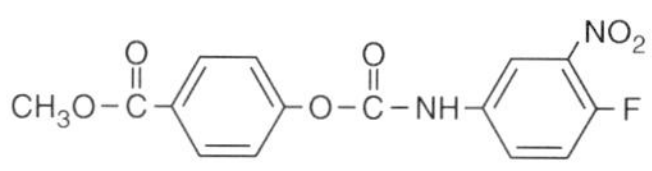

5 gm

S84,546-9
4-(ETHOXYCARBONYL)PHENYL N-(4-CHLORO-3-NITROPHENYL)CARBAMATE

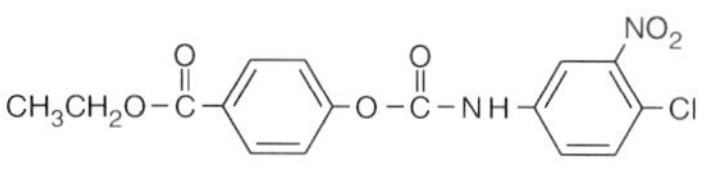

250 mg

S72,828-4
4-ACETAMIDOPHENYL N-(4-FLUORO-3-NITROPHENYL)CARBAMATE

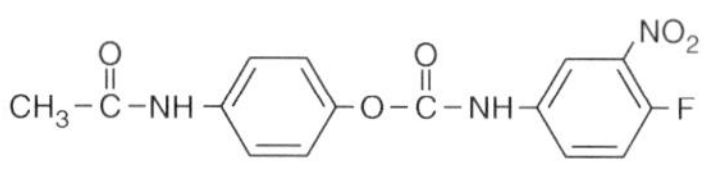

5 gm

S69,994-2
TETRAHYDROFURFURYL N-(4-CHLORO-3-NITROPHENYL)CARBAMATE

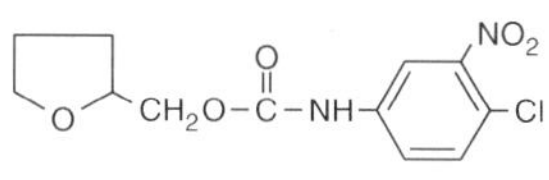

250 mg

S79,431-7
4-CYCLOHEXYLPHENYL N-(4-FLUORO-3-NITROPHENYL)CARBAMATE

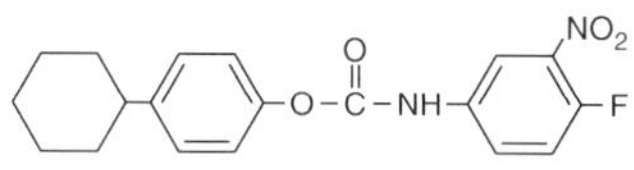

250 mg

S79,456-2
3-PHENYLALLYL N-(4-FLUORO-3-NITROPHENYL)CARBAMATE

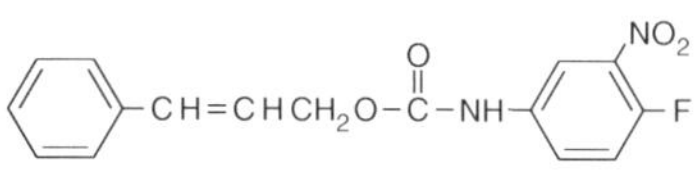

250 mg

S79,476-7
2-FLUOROPHENYL N-(4-FLUORO-3-NITROPHENYL)CARBAMATE

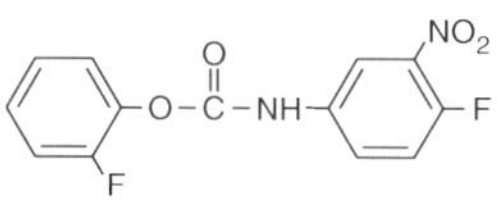

250 mg

S84,427-6
2-FLUOROPHENYL N-(2-FLUORO-5-NITROPHENYL)CARBAMATE

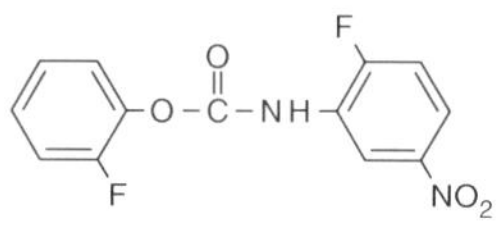

250 mg

S84,424-1
2-FLUOROPHENYL N-(4-CHLORO-3-NITROPHENYL)CARBAMATE

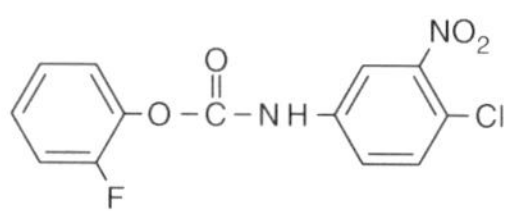

250 mg

S84,426-8
4-FLUOROPHENYL N-(4-CHLORO-3-NITROPHENYL)CARBAMATE

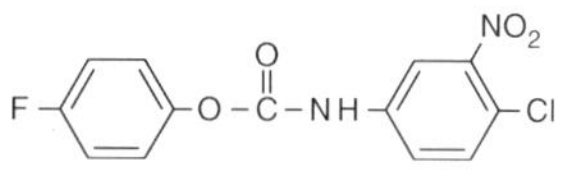

250 mg

S79,116-4
4-CHLOROPHENYL N-(2-METHYL-5-NITROPHENYL)CARBAMATE

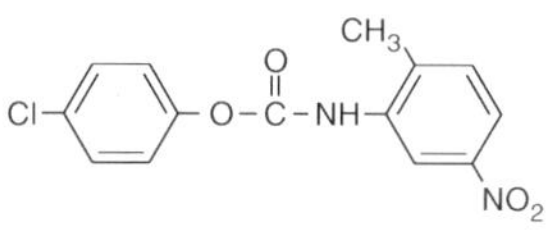

250 mg

S84,519-1
4-METHYLBENZYL N-(4-CHLORO-3-NITROPHENYL)CARBAMATE

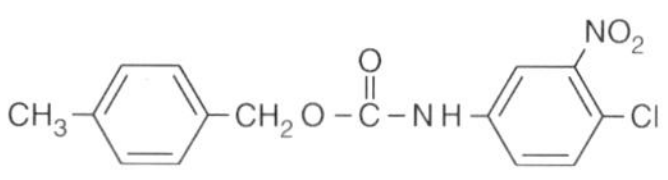

250 mg

S69,910-1
2,3-DICHLOROPHENYL N-(4-CHLORO-3-NITROPHENYL)CARBAMATE

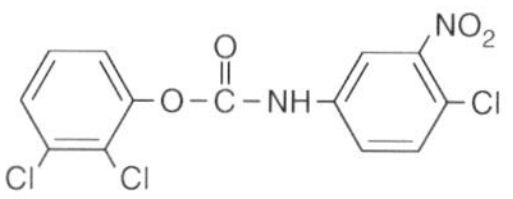

250 mg

S86,741-1
HEXYL N-(4-FLUORO-3-NITROPHENYL)-CARBAMATE

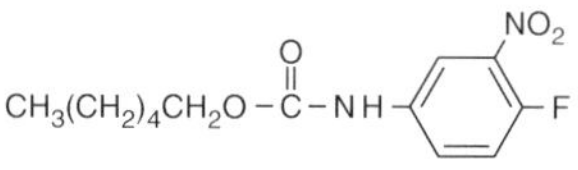

250 mg

S84,421-7
2-CHLORO-4-FLUOROPHENYL N-(2-FLUORO-5-NITROPHENYL)CARBAMATE

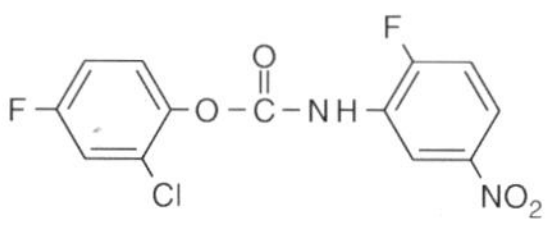

250 mg

S72,816-0
2-CHLORO-4-FLUOROPHENYL N-(2-CHLORO-5-NITROPHENYL)CARBAMATE

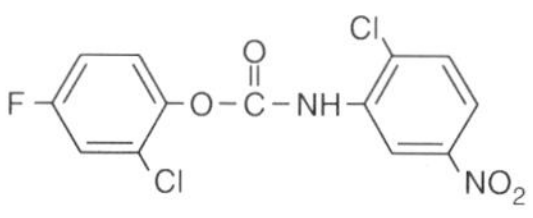

5 gm

S69,981-0
3,4-DICHLOROPHENYL N-(4-CHLORO-3-NITROPHENYL)CARBAMATE

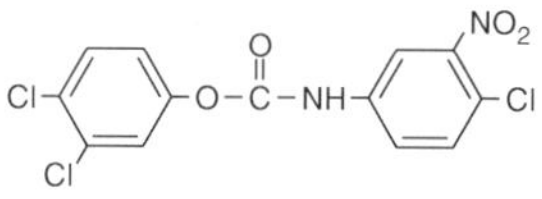

5 gm

S69,986-1
2,5-DICHLOROPHENYL N-(4-CHLORO-3-NITROPHENYL)CARBAMATE

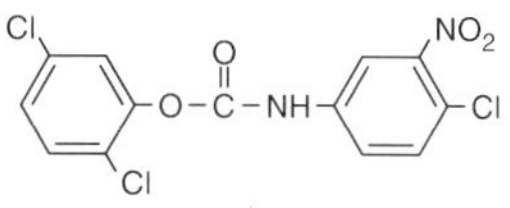

250 mg

S86,746-2
4-CHLORO-2-METHYLPHENYL N-(4-FLUORO-3-NITROPHENYL)CARBAMATE

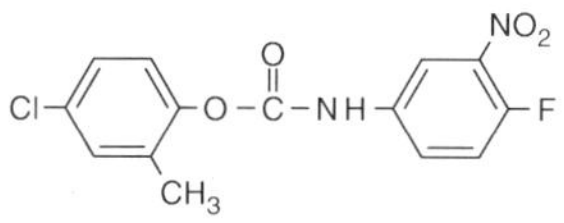

250 mg

S69,935-7
4-CHLORO-2,5-DIMETHYLPHENYL N-(4-CHLORO-3-NITROPHENYL)CARBAMATE

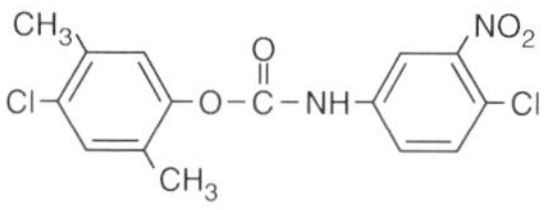

250 mg

S69,118-6
2-BROMO-4-CHLORO-5-METHYLPHENYL N-(4-METHYL-3-NITROPHENYL)CARBAMATE

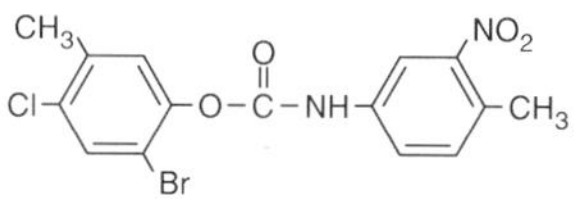

250 mg

S68,938-6
4-BROMO-2,5-DIMETHYLPHENYL N-(4-METHYL-3-NITROPHENYL)CARBAMATE

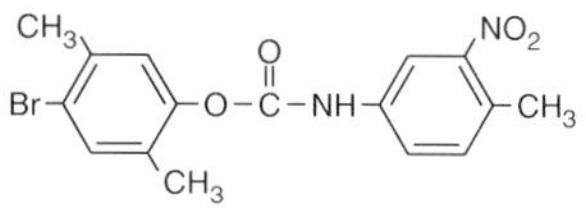

250 mg

S68,939-4
4-BROMO-2,5-DIMETHYLPHENYL N-(2-METHYL-5-NITROPHENYL)CARBAMATE

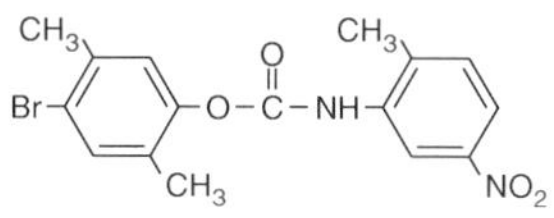

250 mg

S69,074-0
2-BROMO-4,5-DIMETHYLPHENYL N-(4-METHYL-3-NITROPHENYL)CARBAMATE

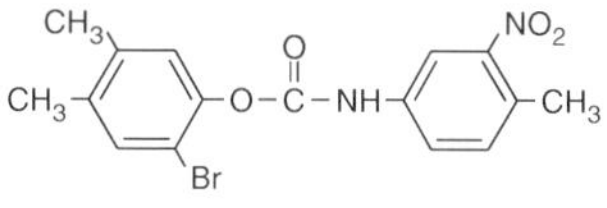

250 mg

S69,339-1
2-BROMO-4,5-DIMETHYLPHENYL N-(2-METHYL-5-NITROPHENYL)CARBAMATE

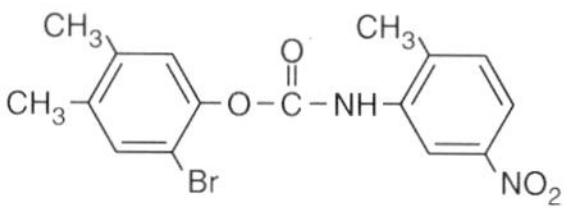

250 mg

S68,994-7
2,5-DIMETHYL-4-IODOPHENYL N-(2-METHYL-5-NITROPHENYL)CARBAMATE

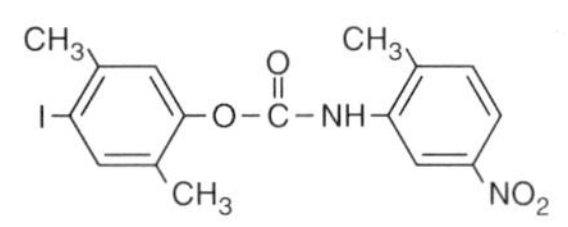

250 mg

S73,877-8
2-BROMO-6-CHLORO-4-NITROPHENYL N-(4-CHLORO-3-NITROPHENYL)CARBAMATE

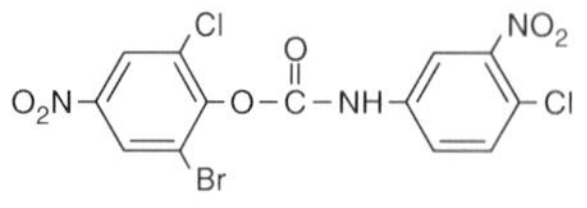

250 mg

S68,958-0
2,6-DIBROMO-4-FLUOROPHENYL N-(4-CHLORO-3-NITROPHENYL)CARBAMATE

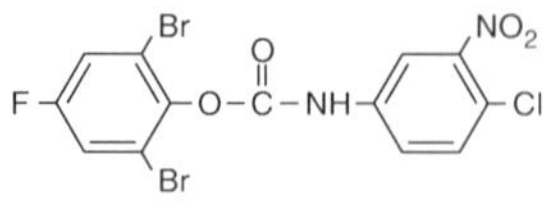

250 mg

S69,354-5
2-BROMO-4-CHLORO-6-METHYLPHENYL N-(4-FLUORO-3-NITROPHENYL)CARBAMATE

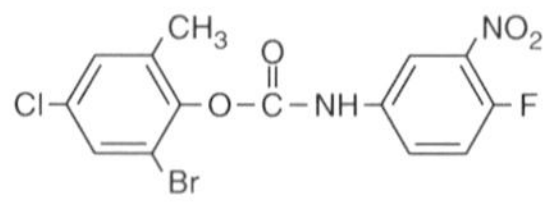

250 mg

S69,399-5
2-BROMO-4-CHLORO-6-METHYLPHENYL N-(4-CHLORO-3-NITROPHENYL)CARBAMATE

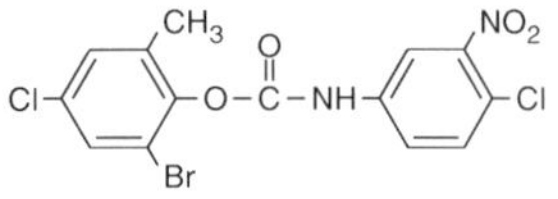

250 mg

S73,672-4
4-CHLORO-3,5-DIMETHYLPHENYL N-(4-FLUORO-3-NITROPHENYL)CARBAMATE

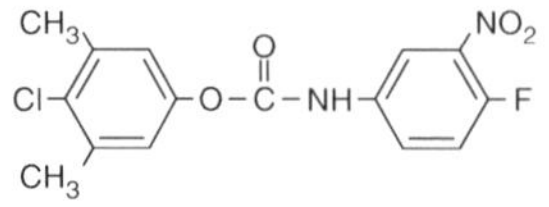

5 gm

S84,515-9
4-CHLORO-3,5-DIMETHYL N-(2-FLUORO-5-NITROPHENYL)CARBAMATE

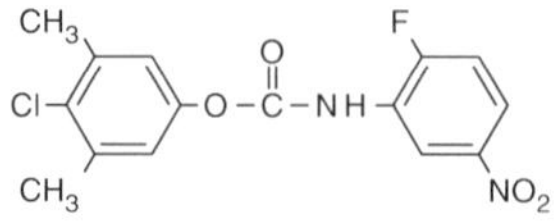

250 mg

S69,327-8
2-BROMO-4-CHLORO-6-METHYLPHENYL N-(2-METHYL-5-NITROPHENYL)CARBAMATE

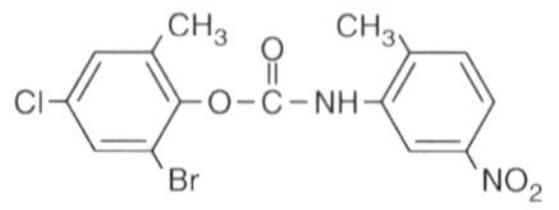

250 mg

S75,228-2
2,4,5-TRICHLOROPHENYL N-(4-METHYL-3-NITROPHENYL)CARBAMATE

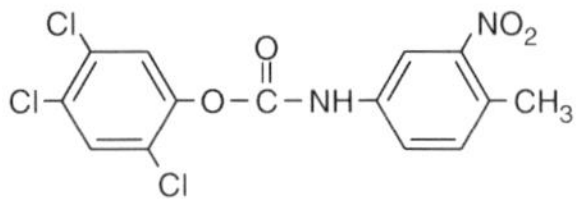

5 gm

S91,203-4
5-INDANYL N-(4-FLUORO-3-NITROPHENYL)CARBAMATE

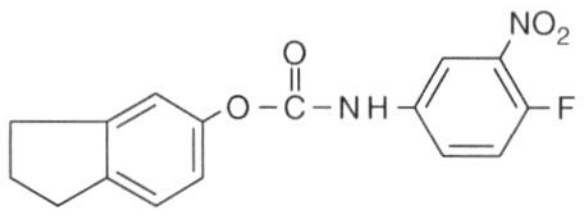

250 mg

S84,467-5
S-BENZYL N-(2-CHLORO-5-NITROPHENYL)-THIOCARBAMATE

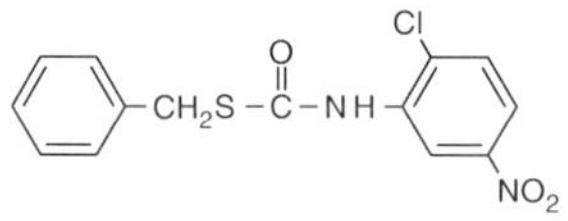

250 mg

S84,464-0
S-(O-TOLYL) N-(2-CHLORO-5-NITROPHENYL)THIOCARBAMATE

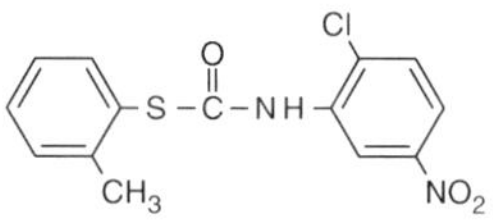

250 mg

S69,977-2
S-(4-CHLOROPHENYL) N-(4-CHLORO-3-NITROPHENYL)THIOCARBAMATE

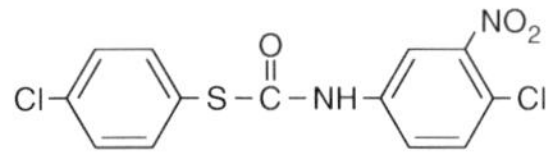

250 mg

S84,522-1
S-(4-TERT-BUTYL-2-METHYLPHENYL)-N-(2-CHLORO-5-NITROPHENYL)-THIOCARBAMATE

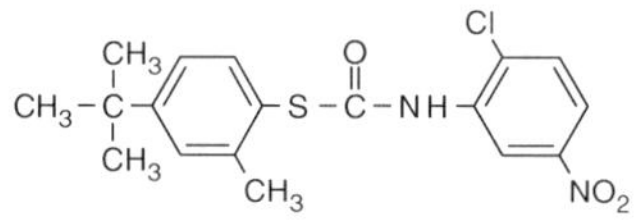

250 mg

S86,882-5
BIS(4-FLUOROPHENYL) N,N'-(4-METHYL-1,3-PHENYLENE)BISCARBAMATE

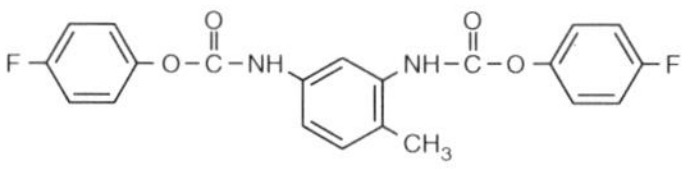

250 mg

S79,411-2
BIS(2-CHLORO-4-FLUOROPHENYL)-N,N'-(4-METHYL-1,3-PHENYLENE)-BISCARBAMATE

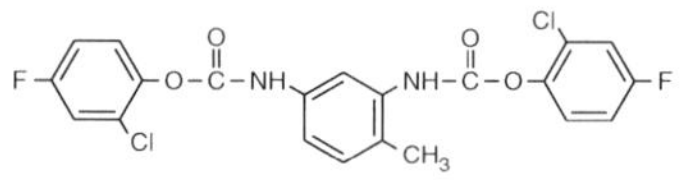

250 mg

S73,177-3
BIS(2,4-DIBROMOPHENYL) N,N'-(4-METHYL-1,3-PHENYLENE)BISCARBAMATE

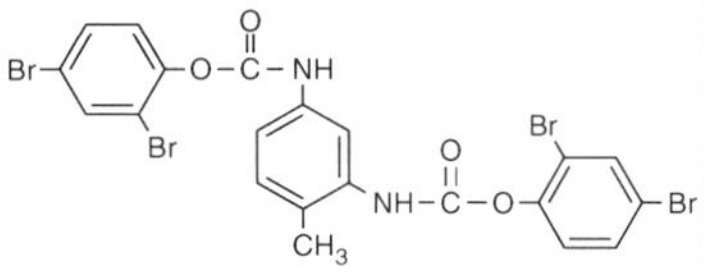

5 gm

S71,868-8
BIS(4-CHLORO-2-METHYLPHENYL)-N,N'-(4-METHYL-1,3-PHENYLENE)-BISCARBAMATE

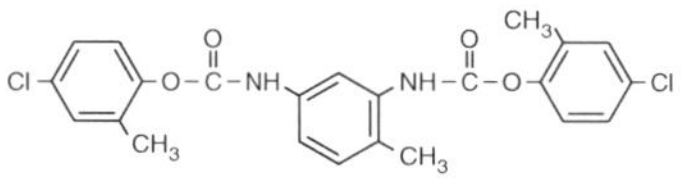

5 gm

S73,610-4
BIS(4-CHLORO-3,5-DIMETHYLPHENYL) N,N'-(4-METHYL-1,3-PHENYLENE)-BISCARBAMATE

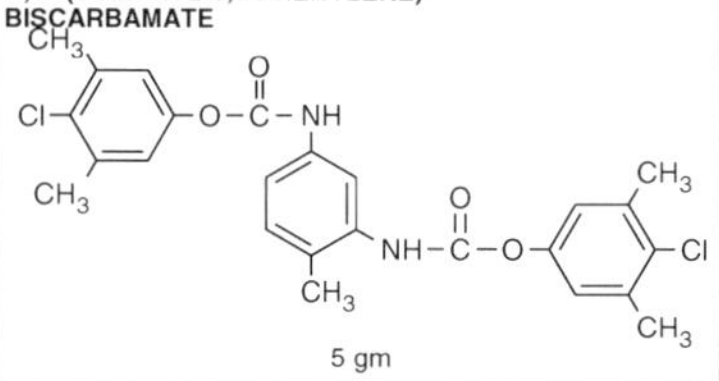

5 gm

S72,076-3
BIS[S-(4-CHLOROPHENYL)] N,N'-(4,6-DIBROMO-1,3-PHENYLENE)BIS-(THIOCARBAMATE)

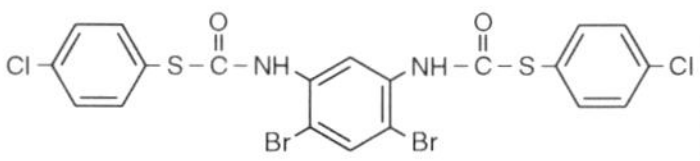

250 mg

S69,498-3
1-METHYL-4-PIPERIDINYL N-(2-CHLORO-4-NITROPHENYL)CARBAMATE

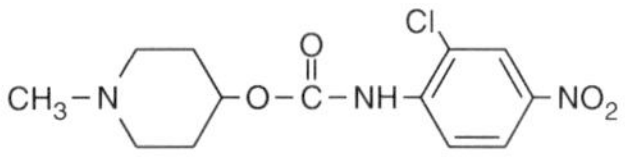

250 mg

S84,518-3
4-ACETAMIDOPHENYL N-(2-CHLORO-4-NITROPHENYL)CARBAMATE

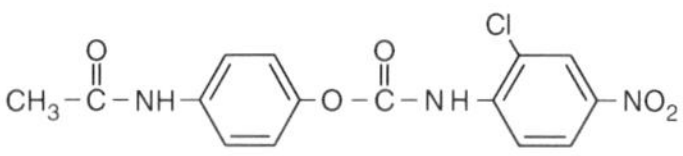

250 mg

S84,516-7
3-ACETAMIDOPHENYL N-(2-CHLORO-4-NITROPHENYL)CARBAMATE

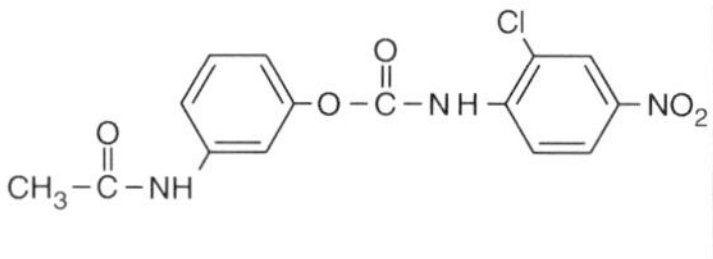

250 mg

S69,915-2
1-METHYL-3-PIPERIDINYL N-(2-CHLORO-4-NITROPHENYL)CARBAMATE

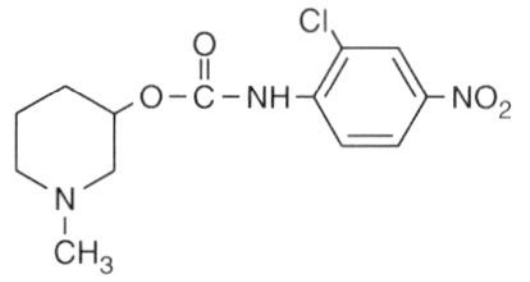

250 mg

S84,746-1
2-BENZYL-4-CHLOROPHENYL N-(2-CHLORO-4-NITROPHENYL)CARBAMATE

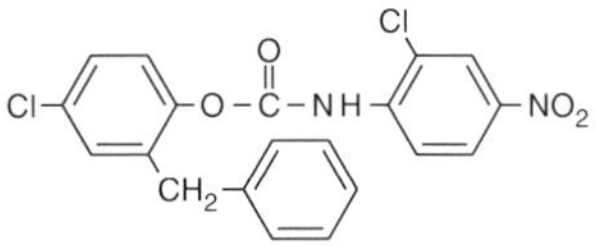

250 mg

S84,423-3
4-FLUOROPHENYL N-(2-CHLORO-4-NITROPHENYL)CARBAMATE

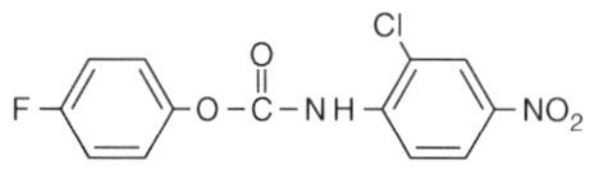

250 mg

S68,620-4
4-FLUOROBENZYL N-(2-CHLORO-4-NITROPHENYL)CARBAMATE

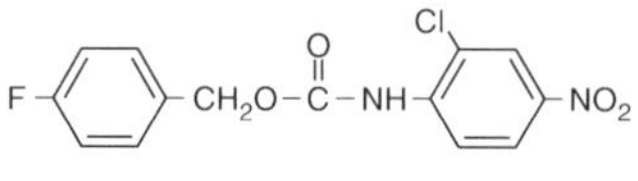

250 mg

S72,765-2
2-CHLORO-4-FLUOROPHENYL N-(2-CHLORO-4-NITROPHENYL)CARBAMATE

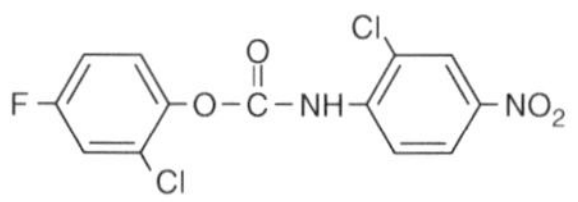

5 gm

S70,188-2
2,4-DICHLOROPHENYL N-(2-CHLORO-4-NITROPHENYL)CARBAMATE

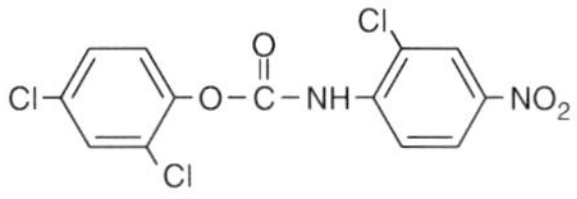

250 mg

S73,631-7
2,4-DIBROMOPHENYL N-(2-CHLORO-4-NITROPHENYL)CARBAMATE

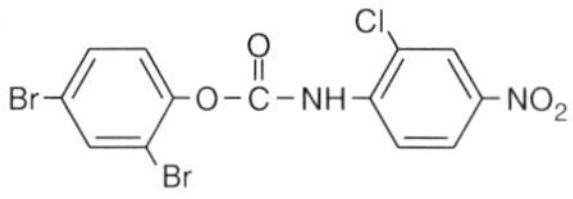

5 gm

S72,838-1
2,4-DICHLOROPHENYL N-(2-METHYL-4-NITROPHENYL)CARBAMATE

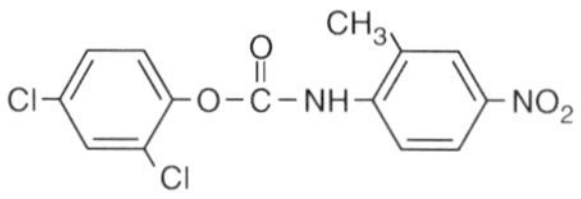

5 gm

S72,784-9
2,5-DIMETHYLPHENYL N-(2-CHLORO-4-NITROPHENYL)CARBAMATE

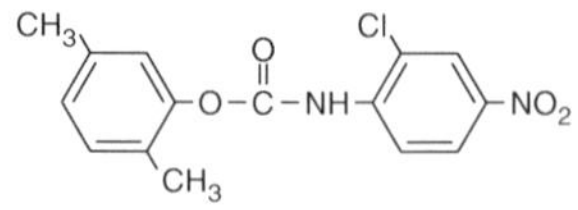

5 gm

S72,161-1
2,6-DIBROMO-4-NITROPHENYL N-(2-CHLORO-4-NITROPHENYL)CARBAMATE

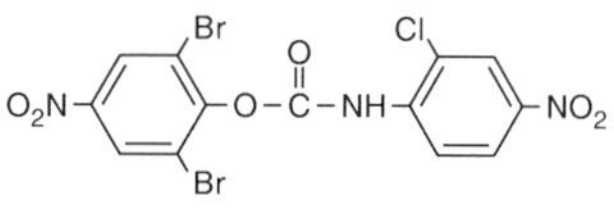

250 mg

S84,466-7
S-BENZYL N-(2-CHLORO-4-NITROPHENYL)-THIOCARBAMATE

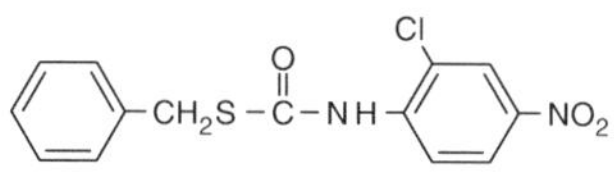

250 mg

S84,463-2
S-(O-TOLYL) N-(2-CHLORO-4-NITROPHENYL)THIOCARBAMATE

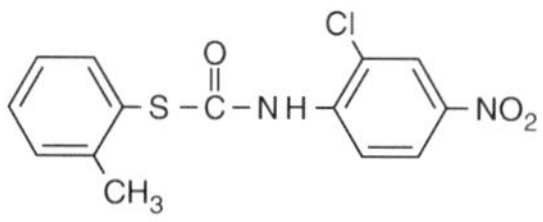

250 mg

S79,674-3
S-(4-TERT-BUTYLPHENYL) N-(2-CHLORO-4-NITROPHENYL)THIOCARBAMATE

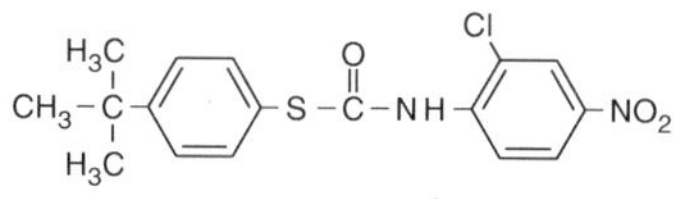

250 mg

S84,465-9
S-(M-TOLYL) N-(2-CHLORO-4-NITROPHENYL)THIOCARBAMATE

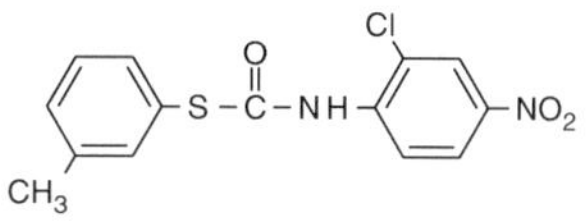

250 mg

S79,653-0
S,S'-ETHYLENE BIS[N-(2-CHLORO-4-NITROPHENYL)THIOCARBAMATE]

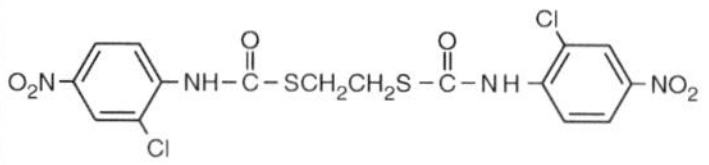

250 mg

S85,553-7
4-ANILINOPHENYL N-(4-BROMOPHENYL)-CARBAMATE

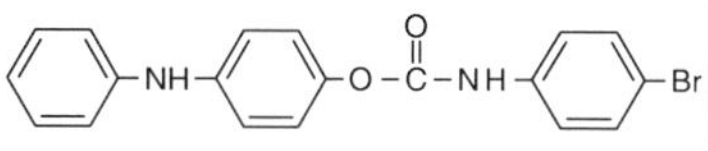

250 mg

S80,074-0
2-(N-ETHYLANILINO)ETHYL N-(4-FLUOROPHENYL)CARBAMATE

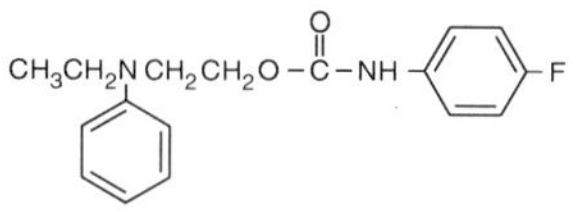

250 mg

S89,127-4
2-MORPHOLINOETHYL N-(4-CHLOROPHENYL)CARBAMATE

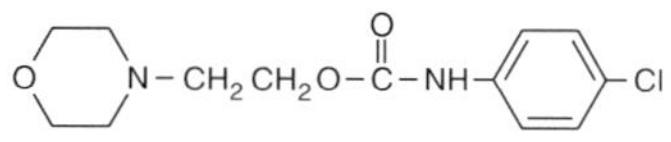

250 mg

S71,424-0
2-(DIMETHYLAMINO)ETHYL N-(4-FLUOROPHENYL)CARBAMATE

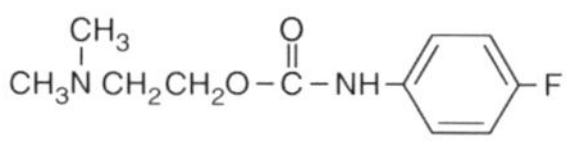

5 gm

S71,171-3
2-(DIMETHYLAMINO)ETHYL N-(4-BROMOPHENYL)CARBAMATE

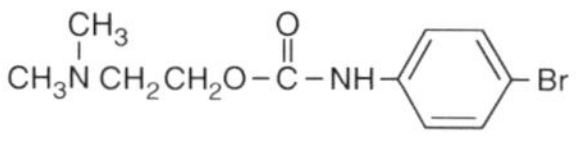

5 gm

S70,988-3
2-(DIMETHYLAMINO)ETHYL N-(2,6-DICHLOROPHENYL)CARBAMATE

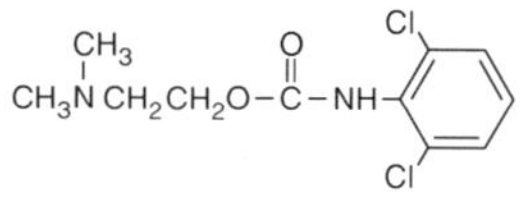

250 mg

S71,004-0
2-(DIETHYLAMINO)ETHYL N-(2,6-DICHLOROPHENYL)CARBAMATE

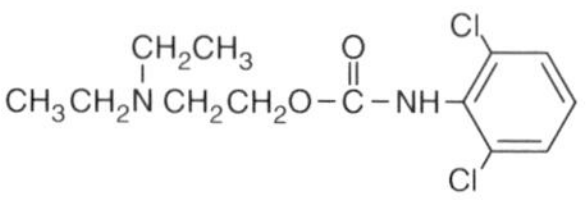

250 mg

S71,745-2
2-(DIMETHYLAMINO)ETHYL N-(2,4-DICHLOROPHENYL)CARBAMATE

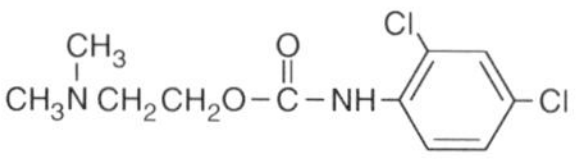

5 gm

S70,936-0
2-(DIMETHYLAMINO)ETHYL N-(3,5-DICHLOROPHENYL)CARBAMATE

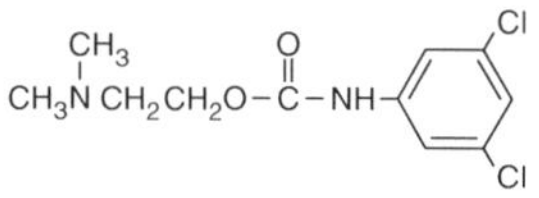

250 mg

S70,969-7
2-(DIETHYLAMINO)ETHYL N-(3,5-DICHLOROPHENYL)CARBAMATE

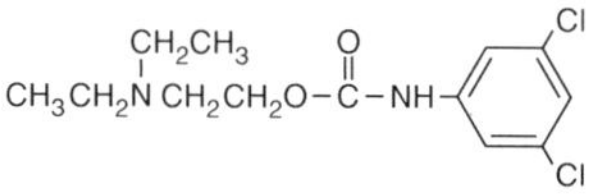

250 mg

S71,170-5
2-(DIMETHYLAMINO)ETHYL N-(2,4,6-TRICHLOROPHENYL)CARBAMATE

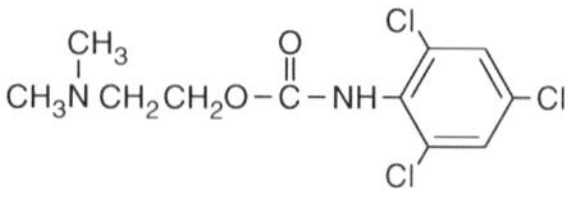

250 mg

S70,983-2
2-(DIMETHYLAMINO)ETHYL N-(2,4,6-TRIBROMOPHENYL)CARBAMATE

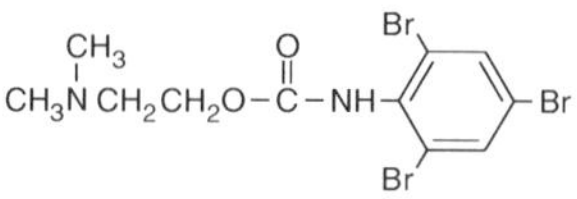

5 gm

S72,961-2
1-METHYL-4-PIPERIDINYL N-(3-CHLOROPHENYL)CARBAMATE

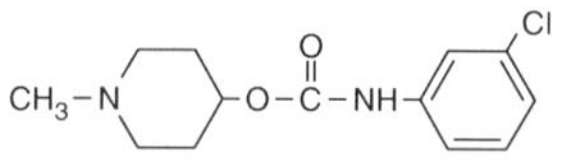

5 gm

S71,823-8
1-METHYL-4-PIPERIDINYL N-(3-BROMOPHENYL)CARBAMATE

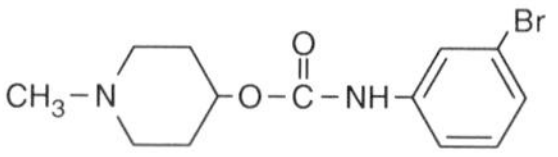

250 mg

S91,638-2
1-METHYL-4-PIPERIDINYL N-(4-FLUOROPHENYL)CARBAMATE

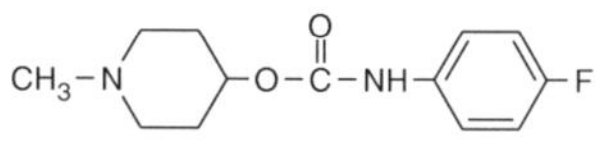

250 mg

S73,640-6
1-METHYL-4-PIPERIDINYL N-(2,4-DIFLUOROPHENYL)CARBAMATE

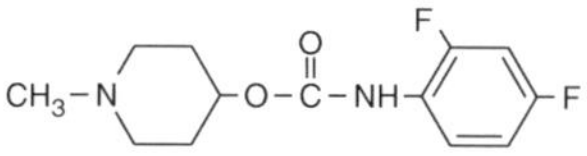

5 gm

S91,625-0
1-METHYL-4-PIPERIDINYL N-(2,5-DIFLUOROPHENYL)CARBAMATE

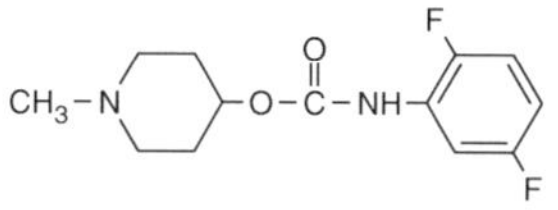

250 mg

S73,578-7
3-(DIMETHYLAMINO)PHENYL N-(2,6-DICHLOROPHENYL)CARBAMATE

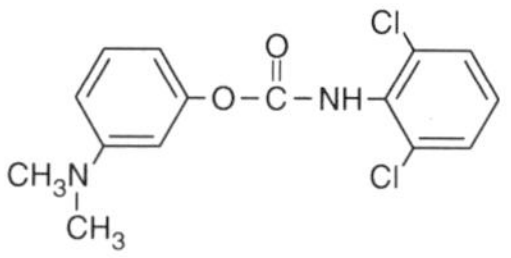

250 mg

S71,354-6
3-(DIETHYLAMINO)PHENYL N-(2,3-DICHLOROPHENYL)CARBAMATE

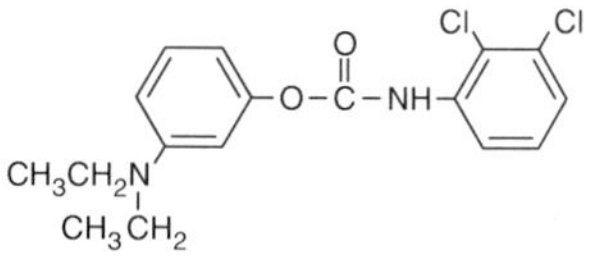

5 gm

S73,167-6
2-(N-ETHYLANILINO)ETHYL N-(2-FLUOROPHENYL)CARBAMATE

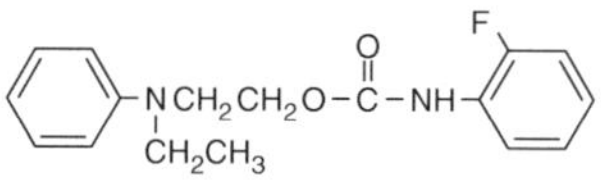

5 gm

S73,794-1
4-(4-CHLORO-2-METHYLPHENYL)-1-(ETHOXYCARBONYL)SEMICARBAZIDE

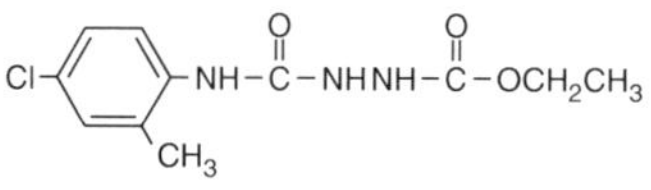

5 gm

S74,949-4
4-CHLORO-2-METHYL-6-NITROPHENYL N-PHENYLCARBAMATE

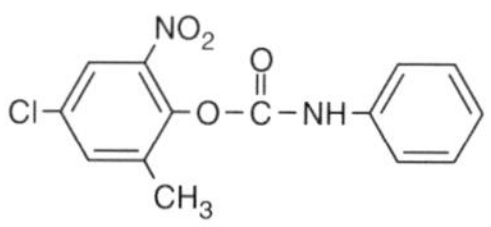

250 mg

S72,957-4
2,4-DIBROMOPHENYL N-(2-METHOXYPHENYL)CARBAMATE

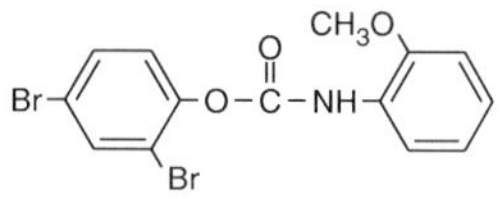

5 gm

S73,197-8
4-TERT-BUTYLPHENYL N-(5-CHLORO-2-METHOXYPHENYL)CARBAMATE

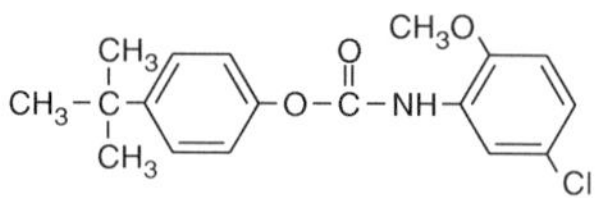

5 gm

S85,436-0
2,4,5-TRICHLOROPHENYL N-(4-CHLORO-2-METHOXYPHENYL)CARBAMATE

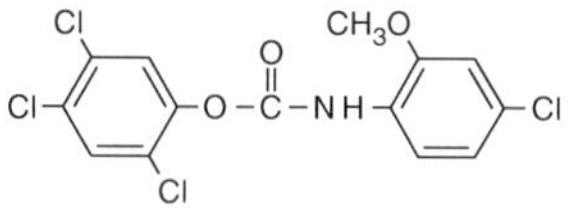

250 mg

S85,785-8
METHYL N-(3-CHLOROPHENYL)-N-HYDROXYCARBAMATE

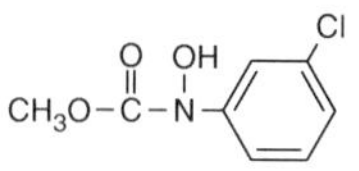

250 mg

S85,782-3
ETHYL N-(3-CHLOROPHENYL)-N-HYDROXYCARBAMATE

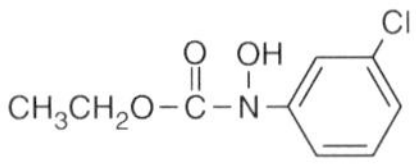

250 mg

S85,786-6
METHYL N-(3-CHLOROPHENYL)-N-METHOXYCARBAMATE

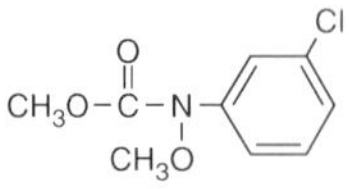

250 mg

S85,783-1
ETHYL N-(3-CHLOROPHENYL)-N-METHOXYCARBAMATE

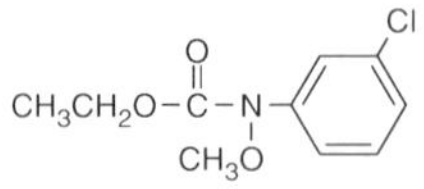

250 mg

S85,754-8
ETHYL N-(4-CHLOROPHENYL)-N-METHOXYCARBAMATE

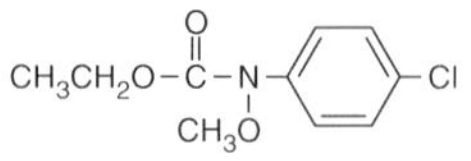

250 mg

S85,778-5
ISOPROPYL N-(4-CHLOROPHENYL)-N-HDYROXYCARBAMATE

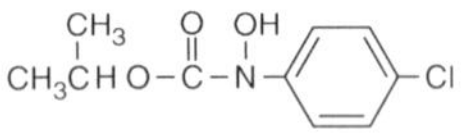

250 mg

S85,775-0
BUTYL N-(4-CHLOROPHENYL)-N-HYDROXYCARBAMATE

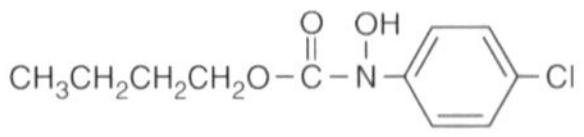

250 mg

S85,779-3
ISOPROPYL N-(4-CHLOROPHENYL)-N-METHOXYCARBAMATE

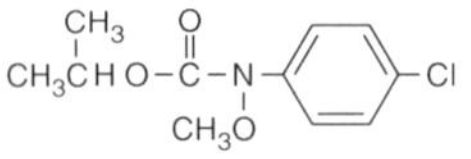

250 mg

S85,787-4
METHYL N-(3,4-DICHLOROPHENYL)-N-HYDROXYCARBAMATE

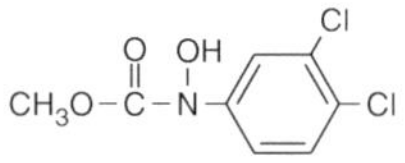

250 mg

S85,788-2
ETHYL N-(3,4-DICHLOROPHENYL)-N-HYDROXYCARBAMATE

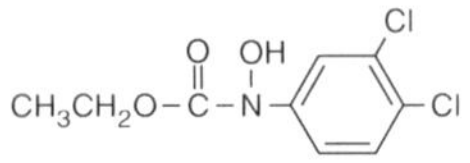

250 mg

S85,806-4
METHYL N-(3,4-DICHLOROPHENYL)-N-METHOXYCARBAMATE

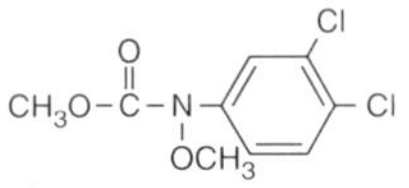

250 mg

S85,789-0
ETHYL N-(3,4-DICHLOROPHENYL)-N-METHOXYCARBAMATE

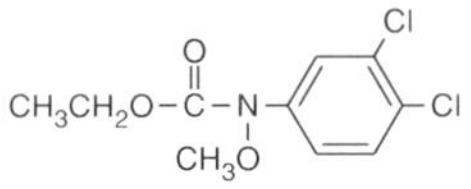

250 mg

S85,821-8
ISOPROPYL N-(3,4-DICHLOROPHENYL)-N-HYDROXYCARBAMATE

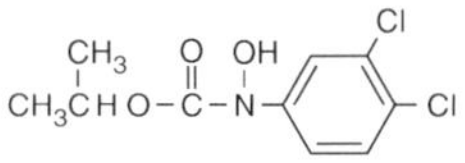

250 mg

S86,794-2
4-COUMARINYL N-(4-FLUORO-3-NITROPHENYL)CARBAMATE

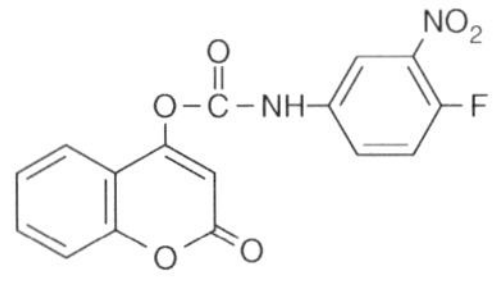

250 mg

S86,817-5
4-METHYL-7-COUMARINYL N-(4-CHLORO-2-NITROPHENYL)CARBAMATE

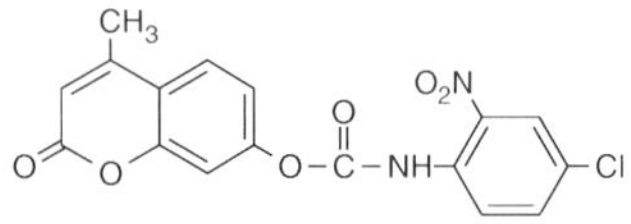

250 mg

S86,928-7
2-(BENZYLOXYCARBONYL)PHENYL N-(4-CHLORO-2-NITROPHENYL)CARBAMATE

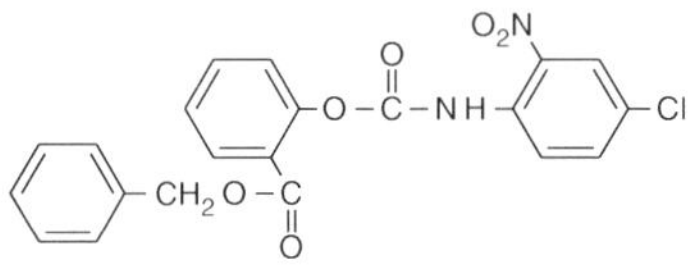

250 mg

S90,566-6
2-(N-PHENYLCARBAMOYL)PHENYL N-(4-CHLORO-2-NITROPHENYL)CARBAMATE

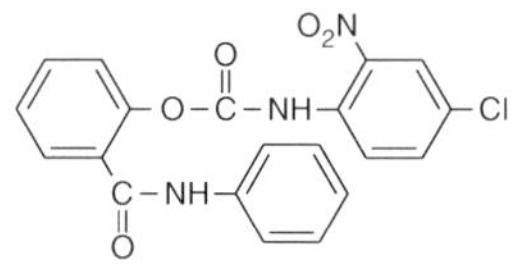

250 mg

S86,792-6
5-NORBORNEN-2-YLMETHYL N-(4-CHLORO-2-NITROPHENYL)CARBAMATE

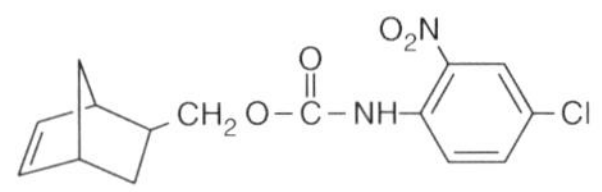

250 mg

S85,340-2
2-CHLORO-4-NITROPHENYL N-(2-FLUOROPHENYL)CARBAMATE

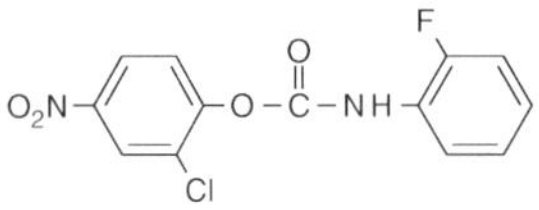

250 mg

S89,033-2
3-METHYL-4-NITROPHENYL N-(2-CHLOROPHENYL)CARBAMATE

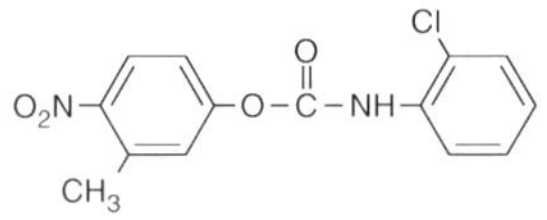

250 mg

S85,344-5
4-NITROPHENYL N-(4-FLUOROPHENYL)-CARBAMATE

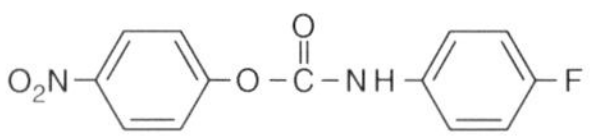

250 mg

S85,532-4
4-NITROPHENYL N-(4-BROMOPHENYL)-CARBAMATE

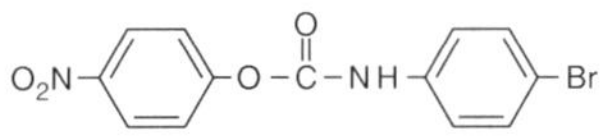

250 mg

S71,673-1
2,6-DIMETHYL-4-NITROPHENYL N-(4-CHLOROPHENYL)CARBAMATE

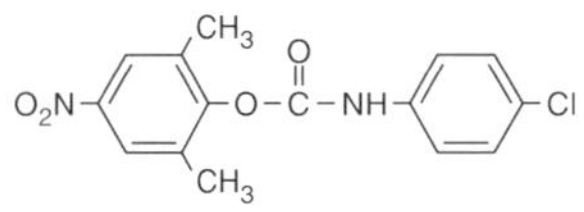

5 gm

S72,690-7
2-CHLORO-4-NITROPHENYL N-(3-CHLORO-2-METHYLPHENYL)CARBAMATE

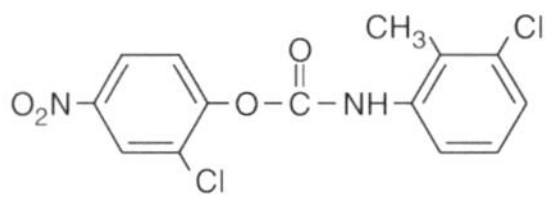

5 gm

S86,775-6
3-METHYL-4-NITROPHENYL N-(3-CHLORO-2-METHYLPHENYL)CARBAMATE

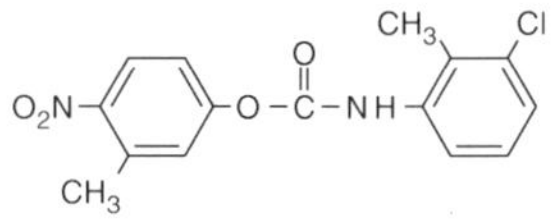

250 mg

S70,933-6
ETHYL N-(2-CHLORO-4-NITROPHENYL)-CARBAMATE

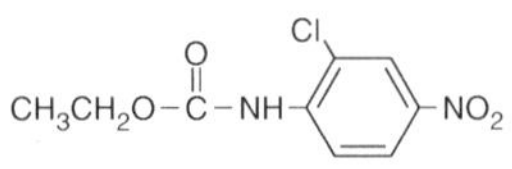

5 gm

S68,935-1
2-BROMO-4-CHLORO-5-METHYLPHENYL N-(4-CHLORO-2-NITROPHENYL)CARBAMATE

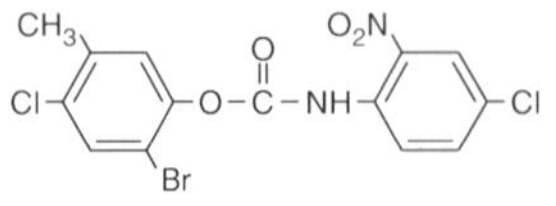

250 mg

S79,123-7
3-METHYL-4-NITROPHENYL N-(3-CHLORO-4-METHYLPHENYL)CARBAMATE

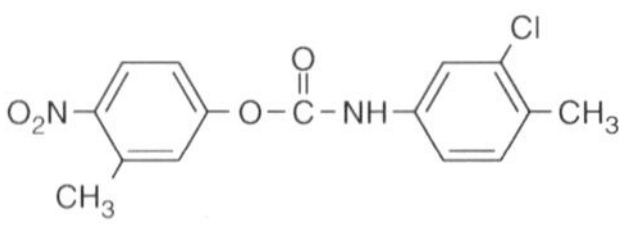

250 mg

S67,335-8
2-BROMO-4-CHLORO-5-METHYLPHENYL N-(2-METHYL-5-NITROPHENYL)CARBAMATE

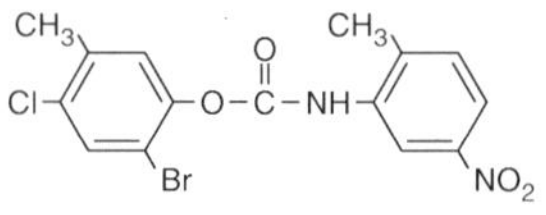

250 mg

S68,778-2
2-BROMO-4,5-DIMETHYLPHENYL N-(4-CHLORO-2-NITROPHENYL)CARBAMATE

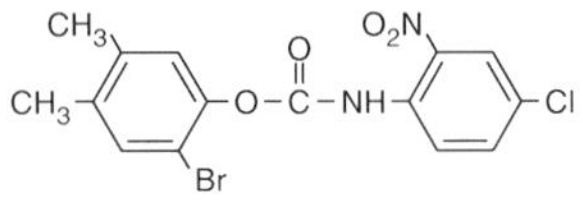

250 mg

S69,173-9
2-BROMO-4,5-DIMETHYLPHENYL N-(2-CHLORO-4-NITROPHENYL)CARBAMATE

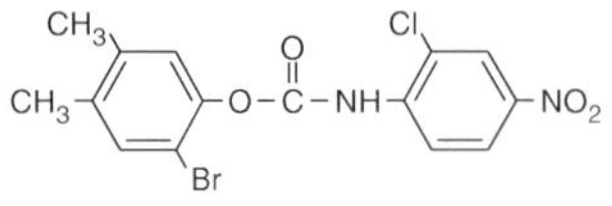

250 mg

S68,949-1
2,5-DIMETHYL-4-IODOPHENYL N-(4-CHLORO-2-NITROPHENYL)CARBAMATE

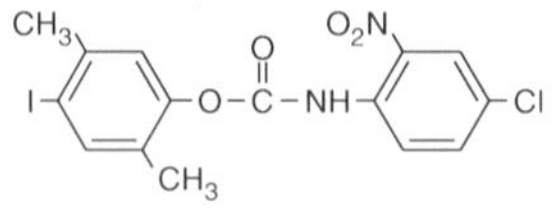

250 mg

S72,403-3
2,6-DIBROMO-4-NITROPHENYL N-(2,5-DIFLUOROPHENYL)CARBAMATE

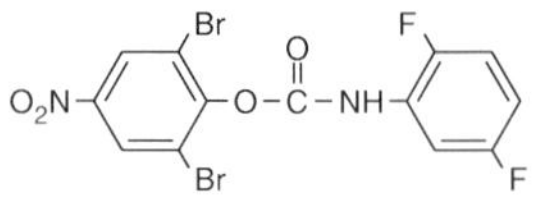

5 gm

S69,144-5
2-BROMO-4-CHLORO-6-METHYLPHENYL N-(4-CHLORO-2-NITROPHENYL)CARBAMATE

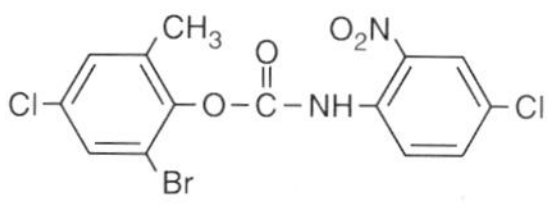

5 gm

S69,027-9
4-BROMO-3,5-DIMETHYLPHENYL N-(2-METHYL-5-NITROPHENYL)CARBAMATE

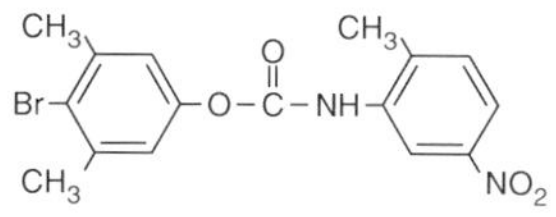

250 mg

S69,577-7
2,4-DICHLORO-1-NAPHTHYL N-(2-CHLORO-4-NITROPHENYL)CARBAMATE

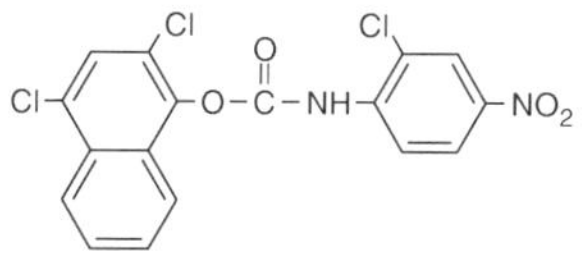

250 mg

S84,572-8
1-NAPHTHYL N-(2-CHLORO-4-NITROPHENYL)CARBAMATE

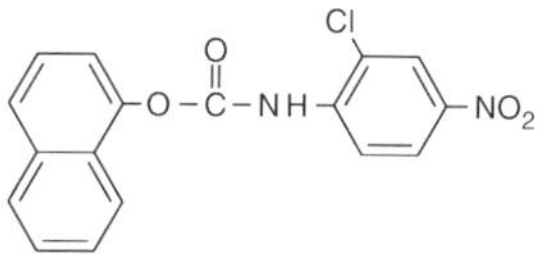

250 mg

S84,575-2
2-NAPHTHYL N-(2-FLUORO-5-NITROPHENYL)CARBAMATE

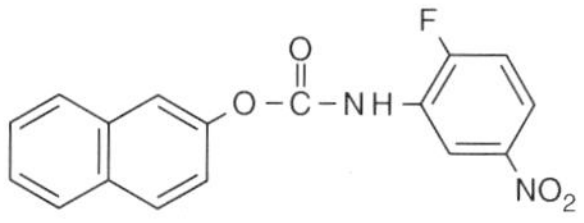

250 mg

S86,818-3
2-NAPHTHYL N-(4-FLUORO-3-NITROPHENYL)CARBAMATE

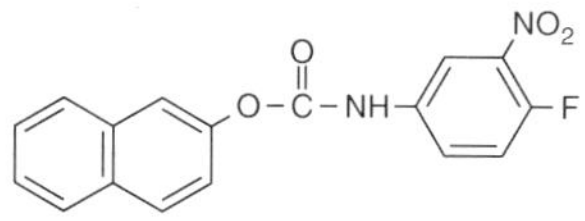

250 mg

S73,161-7
2-NAPHTHYL N-(2-CHLORO-5-NITROPHENYL)CARBAMATE

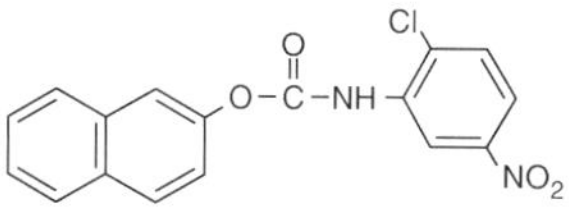

5 gm

S84,573-6
2-NAPHTHYL N-(2-CHLORO-4-NITROPHENYL)CARBAMATE

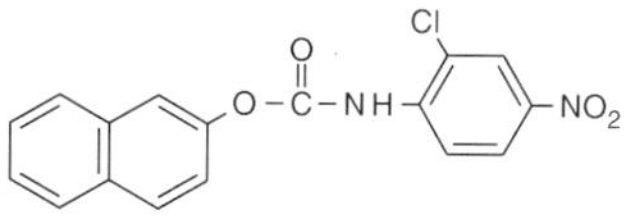

250 mg

S84,574-4
2-NAPHTHYL N-(4-CHLORO-3-NITROPHENYL)CARBAMATE

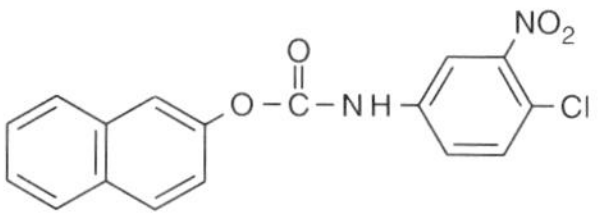

250 mg

S86,816-7
2-NAPHTHYL N-(4-CHLORO-2-NITROPHENYL)CARBAMATE

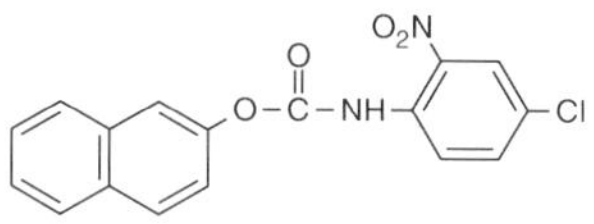

250 mg

S68,532-1
2,6-DIBROMO-4-NITROPHENYL N-(ETHOXYCARBONYLMETHYL)CARBAMATE

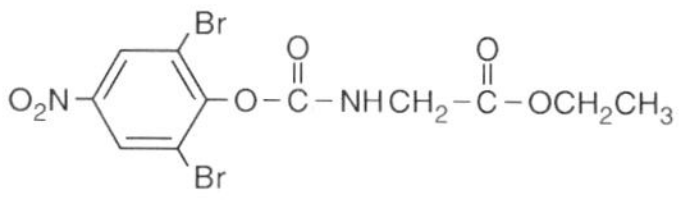

50 mg

S79,408-2
3-(N-PHENYLCARBAMOYL)-2-NAPHTHYL N-(4-CHLORO-2-NITROPHENYL)CARBAMATE

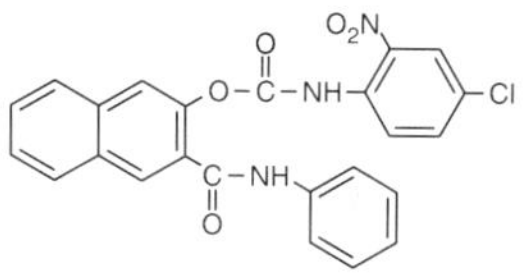

250 mg

S71,713-4
4-BENZOYLPHENYL N-(4-CHLOROPHENYL)-CARBAMATE

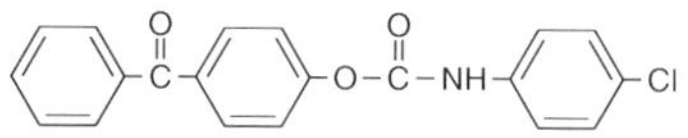

5 gm

S71,716-9
4-BENZOYLPHENYL N-(3-CHLOROPHENYL)-CARBAMATE

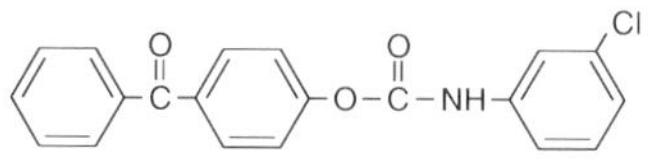

5 gm

S71,427-5
2-BENZOYL-4-CHLOROPHENYL N-(M-TOLYL)CARBAMATE

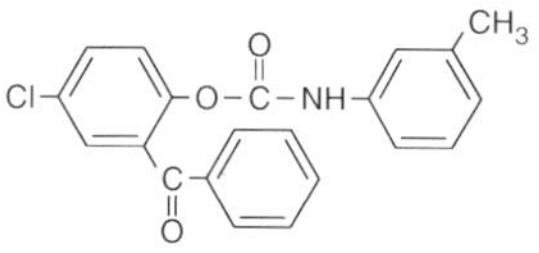

5 gm

S86,770-5
2-FLUOROPHENYL N-(3-ACETYLPHENYL)-CARBAMATE

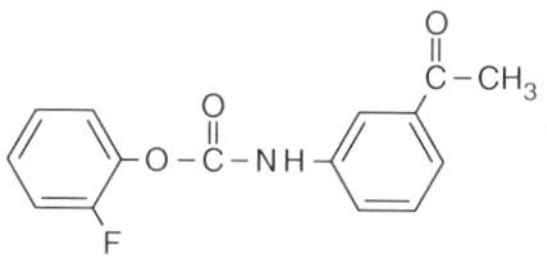

250 mg

S91,113-5
2-ACETYLPHENYL N-(2,3-DICHLOROPHENYL)CARBAMATE

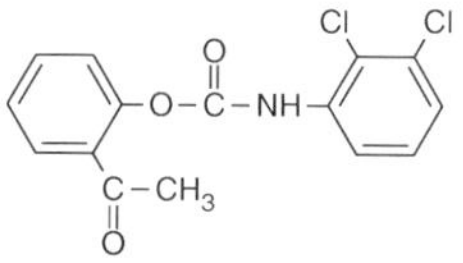

250 mg

S71,368-6
4-BENZOYLPHENYL N-(2-CHLOROPHENYL)-CARBAMATE

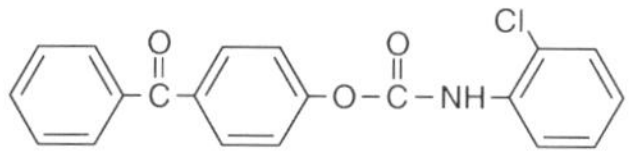

5 gm

S84,820-4
4-ACETYLPHENYL N-(3-FLUOROPHENYL)-CARBAMATE

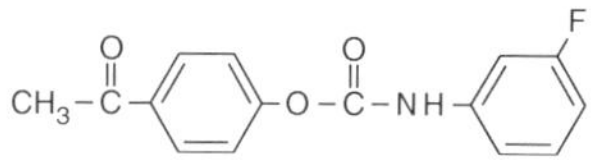

250 mg

S85,520-0
4-ACETYLPHENYL N-(4-CHLOROPHENYL)-CARBAMATE

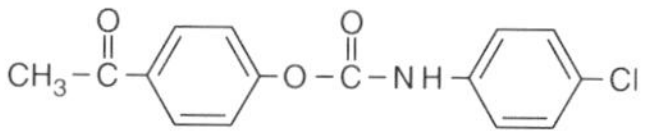

250 mg

S91,112-7
4-ACETYLPHENYL N-(2,3-DICHLOROPHENYL)CARBAMATE

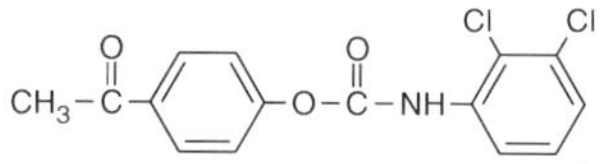

250 mg

S74,315-1
4-ACETYLPHENYL N-(2,5-DIFLUOROPHENYL)CARBAMATE

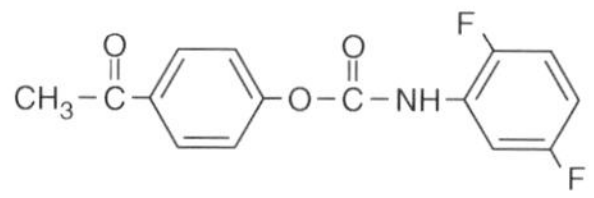

5 gm

S76,680-1
METHYL N-(2-[2-(4-CHLOROBENZYL)-1-OXO-2-INDANYL]ETHYL)CARBAMATE

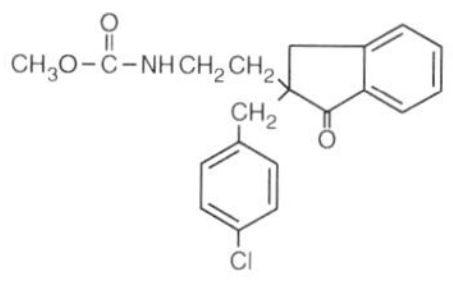

250 mg

S86,766-7
2-CHLORO-4-FLUOROPHENYL N-(3-ACETYLPHENYL)CARBAMATE

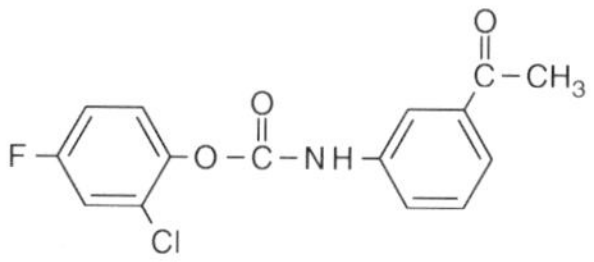

250 mg

S86,863-9
2-BROMO-4-TERT-BUTYLPHENYL N-(3-ACETYLPHENYL)CARBAMATE

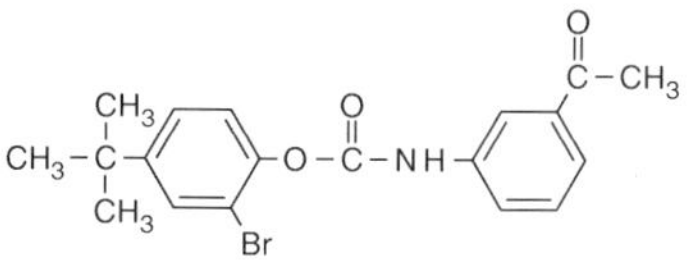

250 mg

S70,040-1
4-ACETYLPHENYL N-(2,4-DIFLUOROPHENYL)CARBAMATE

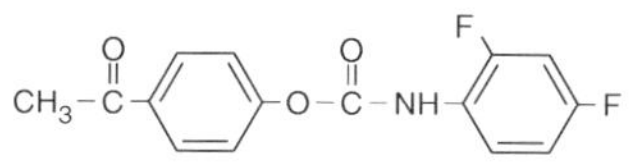

250 mg

S69,415-0
2-CHLORO-4-FLUOROPHENYL N-[3,5-BIS-(METHOXYCARBONYL)PHENYL]CARBAMATE

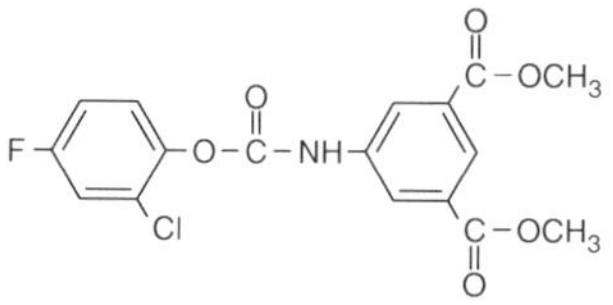

250 mg

S69,130-5
3,4-DICHLOROPHENYL N-[3,5-BIS-(METHOXYCARBONYL)PHENYL]CARBAMATE

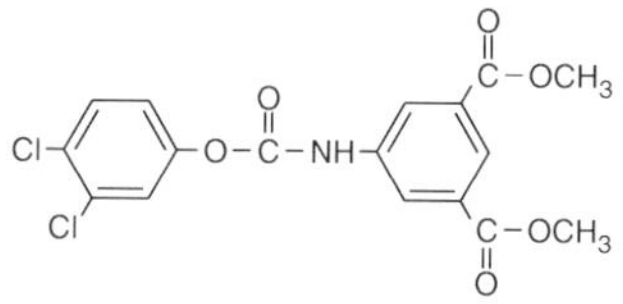

250 mg

S79,548-8
2-(PHENOXYCARBONYL)PHENYL N-(3,4-DICHLOROPHENYL)CARBAMATE

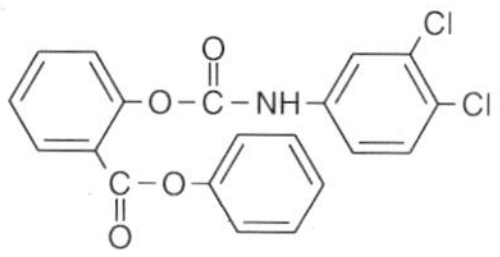

250 mg

S77,947-4
4-(METHOXYCARBONYL)PHENYL N-(2-FLUOROPHENYL)CARBAMATE

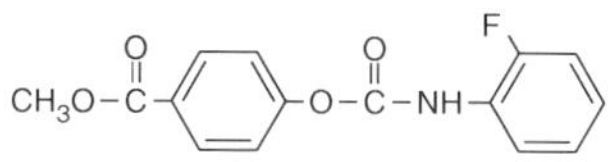

250 mg

S86,768-3
4-(METHOXYCARBONYL)PHENYL N-(3-CHLOROPHENYL)CARBAMATE

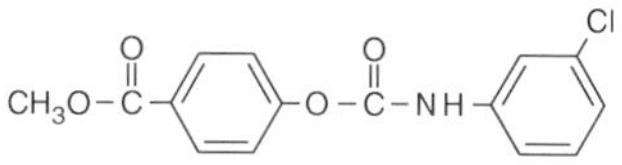

250 mg

S80,770-2
4-(METHOXYCARBONYL)PHENYL N-(4-BROMOPHENYL)CARBAMATE

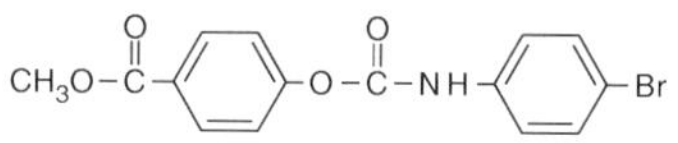

250 mg

S80,039-2
4-(ETHOXYCARBONYL)PHENYL N-(4-FLUOROPHENYL)CARBAMATE

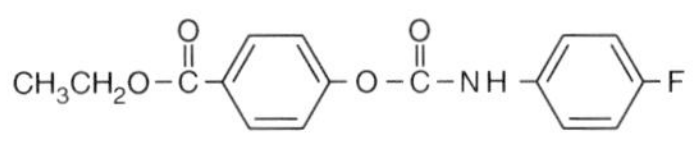

250 mg

S79,142-3
4-ETHOXYCARBONYLPHENYL N-(4-CHLOROPHENYL)CARBAMATE

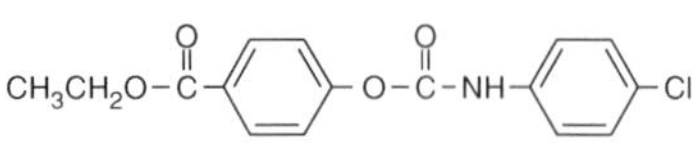

250 mg

S86,848-5
4-(BUTOXYCARBONYL)PHENYL N-(3-CHLOROPHENYL)CARBAMATE

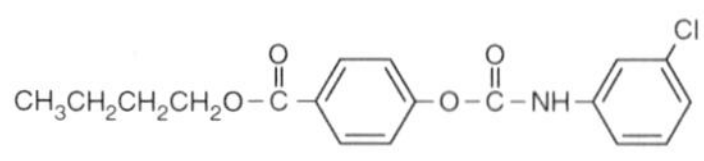

250 mg

S79,671-9
4-(ETHOXYCARBONYL)PHENYL N-(2-CHLORO-6-METHYLPHENYL)-CARBAMATE

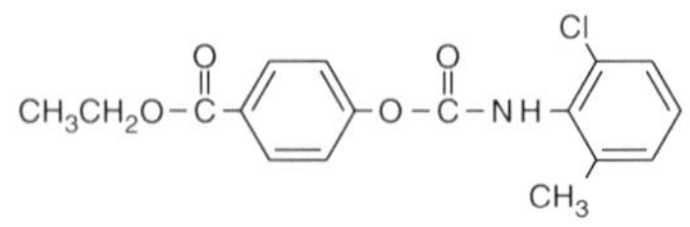

250 mg

S85,108-6
4-(BUTOXYCARBONYL)PHENYL N-(2-CHLORO-6-METHYLPHENYL)-CARBAMATE

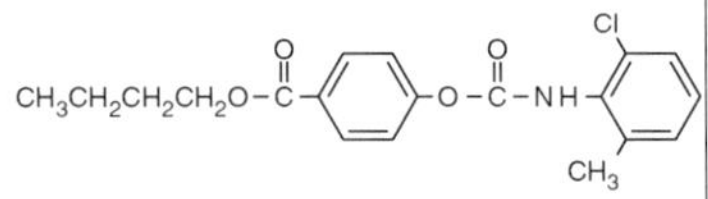

250 mg

S86,798-5
4-(METHOXYCARBONYL)PHENYL N-(5-CHLORO-2-METHYLPHENYL)-CARBAMATE

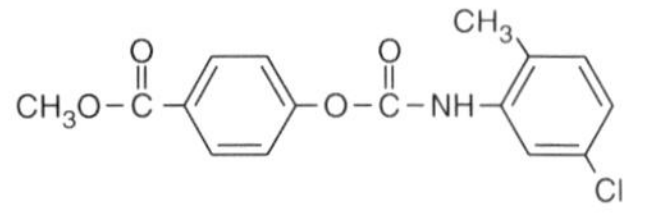

250 mg

S79,457-0
2-CHLORO-4-FLUOROPHENYL N-[4-(ETHOXYCARBONYL)PHENYL]-CARBAMATE

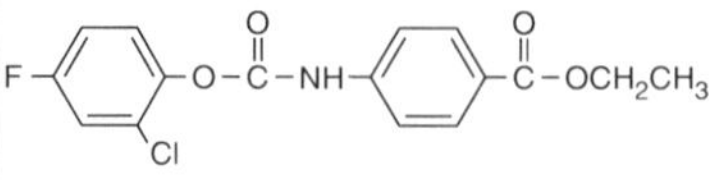

250 mg

S79,455-4
2,4-DICHLOROPHENYL N-[4-(ETHOXYCARBONYL)PHENYL]-CARBAMATE

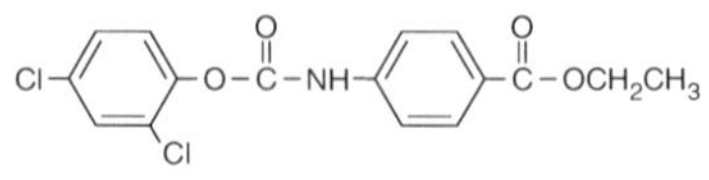

250 mg

S86,796-9
4-(METHOXYCARBONYL)PHENYL N-(2-BROMO-4-METHYLPHENYL)CARBAMATE

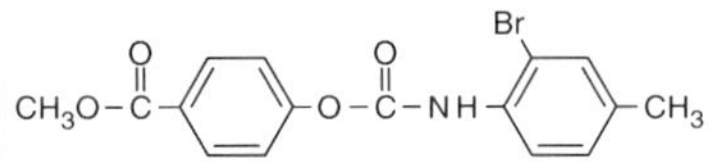

250 mg

S79,751-0
4-CHLORO-2-METHYLPHENYL N-[4-(ETHOXYCARBONYL)PHENYL]-CARBAMATE

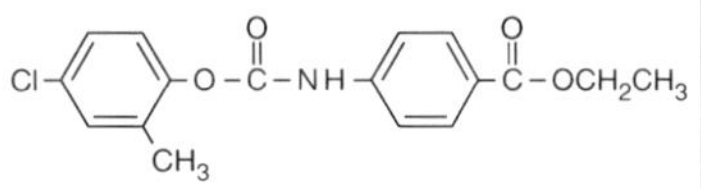

250 mg

S91,144-5
4-CHLORO-2,6-DIMETHYLPHENYL N-[4-(ETHOXYCARBONYL)PHENYL]-CARBAMATE

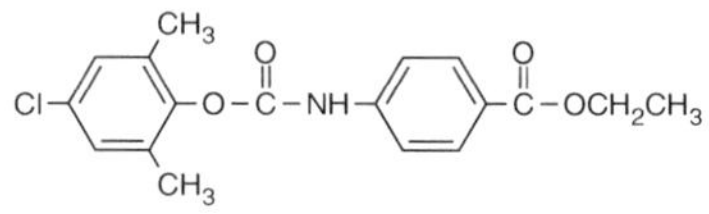

250 mg

S77,941-5
4-(4-CHLOROBENZAMIDO)PHENYL N-METHYLCARBAMATE

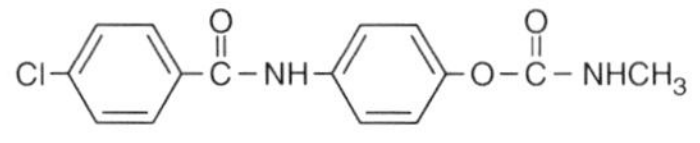

250 mg

S73,774-7
2-CHLORO-4-FLUOROPHENYL N-[2-(METHOXYCARBONYL)PHENYL]-CARBAMATE

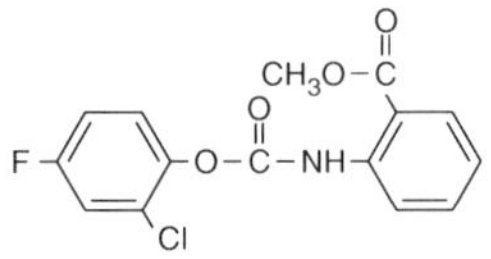

5 gm

S71,860-2
3,4-DICHLOROPHENYL N-[2-(METHOXYCARBONYL)PHENYL]-CARBAMATE

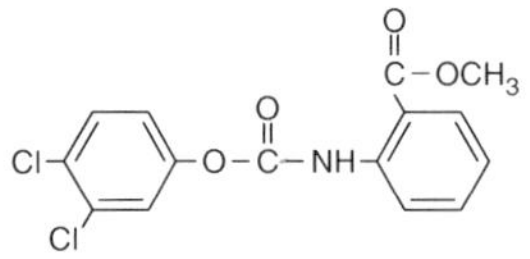

250 mg

S72,943-4
2-CHLORO-4,5-DIMETHYLPHENYL N-[4-(ETHOXYCARBONYL)PHENYL]-CARBAMATE

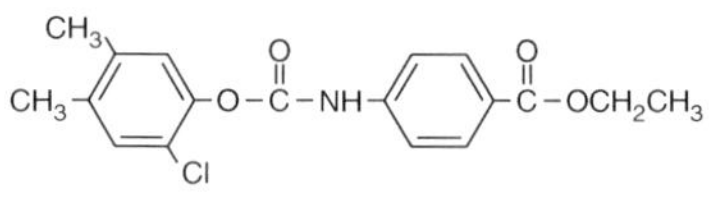

5 gm

S85,091-8
4-(BUTOXYCARBONYL)PHENYL N-(2,3-DICHLOROPHENYL)CARBAMATE

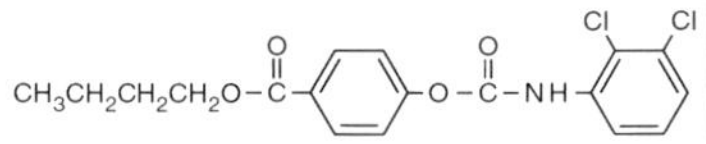

250 mg

S85,058-6
4-(ETHOXYCARBONYL)PHENYL N-(3-CHLORO-4-FLUOROPHENYL)-CARBAMATE

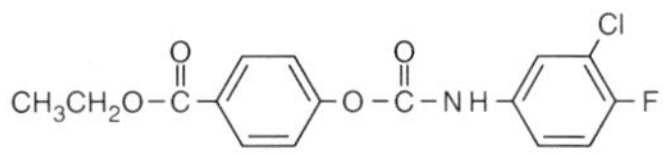

250 mg

S85,055-1
4-(BUTOXYCARBONYL)PHENYL N-(3-CHLORO-4-FLUOROPHENYL)-CARBAMATE

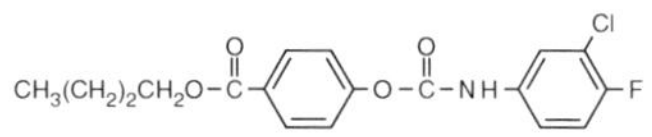

250 mg

S73,192-7
4-CHLORO-3,5-DIMETHYLPHENYL N-[4-(ETHOXYCARBONYL)PHENYL]-CARBAMATE

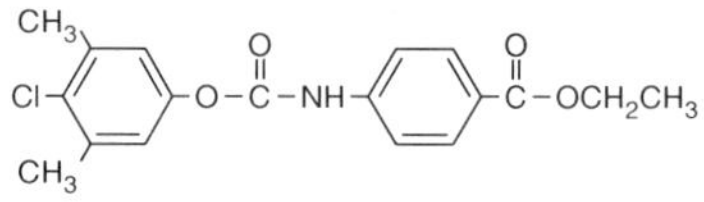

5 gm

S68,797-9
S-(4-CHLOROPHENYL) N-[3,5-BIS-(METHOXYCARBONYL)PHENYL]-THIOCARBAMATE

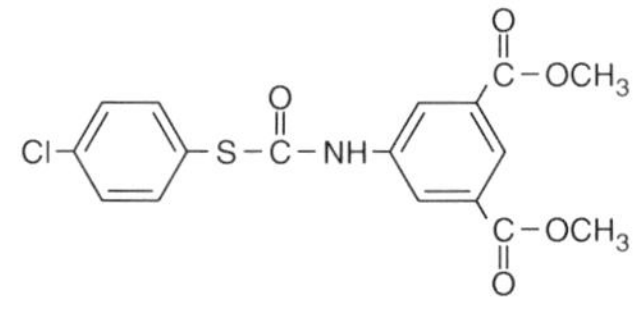

250 mg

S69,563-7
N,N'-(4-BROMO-2-METHYLPHENYLENE)-BIS(4-METHYLCOUMARIN-7-YL CARBAMATE), ISOMERS

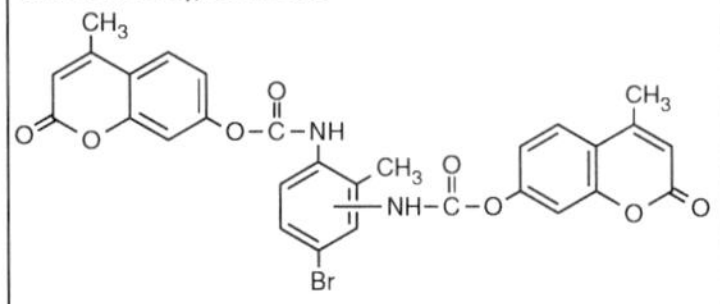

250 mg

S76,911-8
1-(METHOXYCARBONYL)-1-METHYLETHYL N-(2-CHLOROPHENYL)CARBAMATE

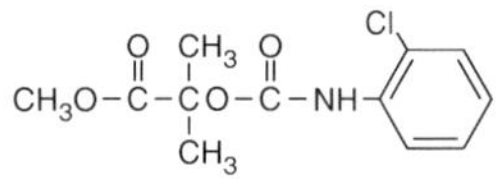

250 mg

S68,986-6
N,N'-BIS(2,4,5-TRICHLOROPHENOXYCARBONYL)-2-METHYL-1,3-PHENYLENEDIAMINE

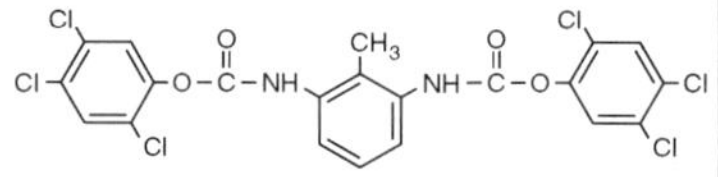

250 mg

S79,699-9
2,4-DIBROMOPHENYL N-(BUTOXYCARBONYLOXYMETHYL)-CARBAMATE

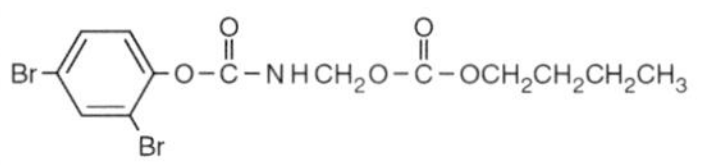

250 mg

S79,708-1
4-CHLORO-2-METHYLPHENYL N-(BUTOXYCARBONYLOXYMETHYL)-CARBAMATE

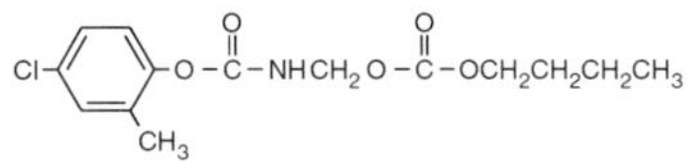

250 mg

S92,558-6
1-(3-CHLORO-4-METHYLPHENYL)-3-(2-OXO-3-OXAZOLIDINYL)UREA

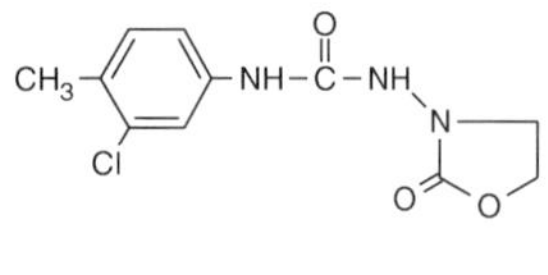

250 mg

S85,621-5
ETHYL N-[N-(3,4-DICHLOROPHENYL)-CARBAMOYL]CARBAMATE

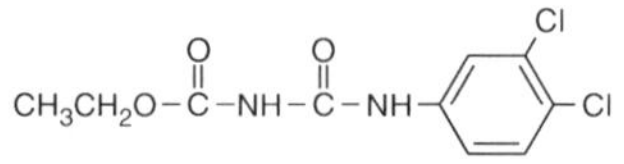

250 mg

S85,623-1
ETHYL N-(3,4-DICHLOROPHENYL)-N-(N,N-DIMETHYLCARBAMOYL)CARBAMATE

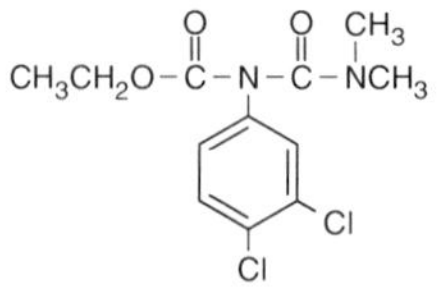

250 mg

S85,635-5
PROPYL N-[N-(3,4-DICHLOROPHENYL)-CARBAMOYL]-N-METHYLCARBAMATE

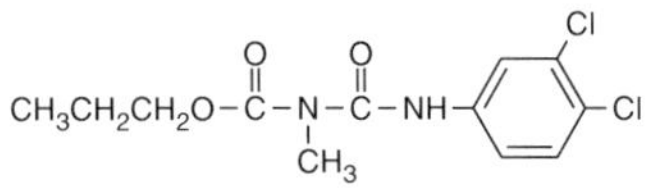

250 mg

S85,636-3
BUTYL N-[N-(3,4-DICHLOROPHENYL)-CARBAMOYL]CARBAMATE

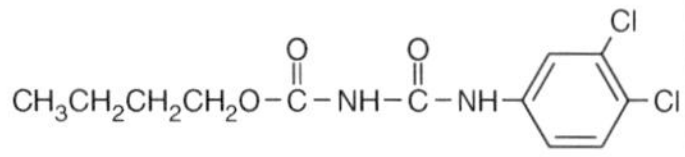

250 mg

S85,654-1
METHYL N-[N-(3,4-DICHLOROPHENYL)-CARBAMOYL]-N-PROPYLCARBAMATE

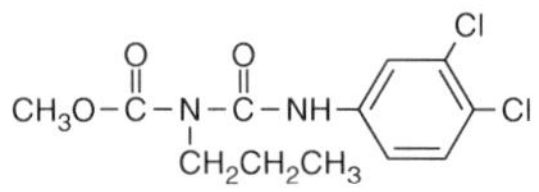

250 mg

S85,639-8
BUTYL N-[N-(3,4-DICHLOROPHENYL)-CARBAMOYL]-N-METHYLCARBAMATE

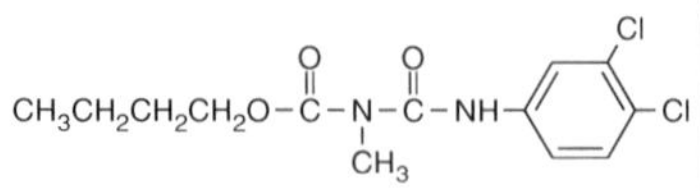

250 mg

S73,678-3
4-ACETAMIDOPHENYL N-(2-FLUOROPHENYL)CARBAMATE

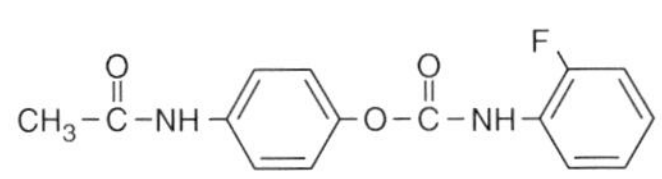

5 gm

S89,038-3
4-ACETAMIDOPHENYL N-(2-CHLOROPHENYL)CARBAMATE

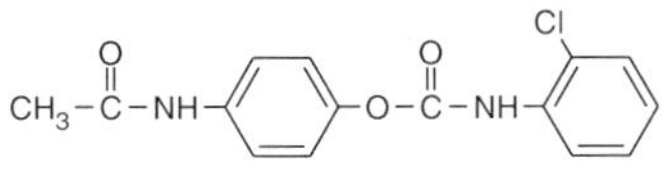

250 mg

S84,823-9
4-ACETAMIDOPHENYL N-(3-FLUOROPHENYL)CARBAMATE

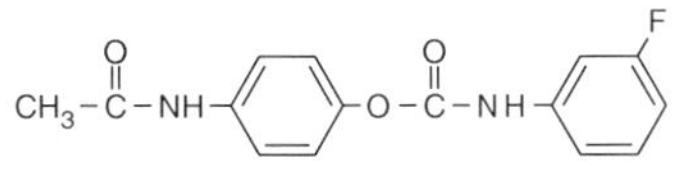

250 mg

S85,097-7
4-ACETAMIDOPHENYL N-(2,3-DICHLOROPHENYL)CARBAMATE

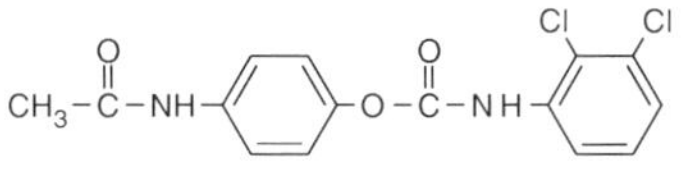

250 mg

S86,802-7
4-ACETAMIDOPHENYL N-(3-CHLORO-2-METHYLPHENYL)CARBAMATE

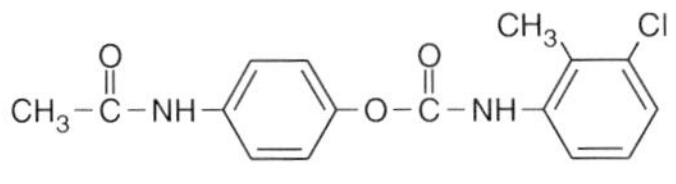

250 mg

S85,063-2
4-ACETAMIDOPHENYL N-(3-CHLORO-4-FLUOROPHENYL)CARBAMATE

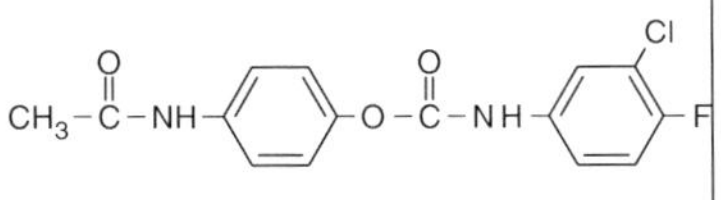

250 mg

S85,597-9
ETHYL N-[N-(3,4-DICHLOROPHENYL)-CARBAMOYL]-N-METHYLCARBAMATE

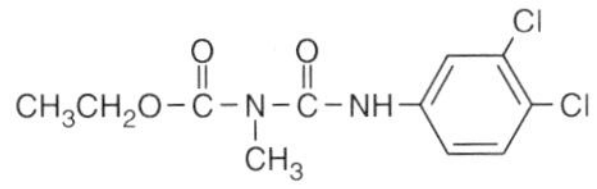

250 mg

S85,812-9
BUTYL N-(3-CHLOROPHENYL)CARBAMATE

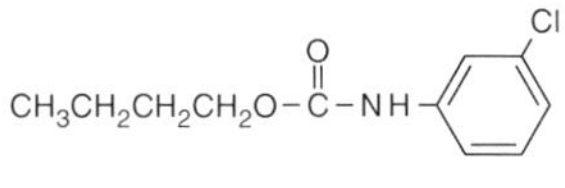

250 mg

S71,891-2
2,4,5-TRICHLOROPHENYL N-(3-CHLOROPHENYL)CARBAMATE

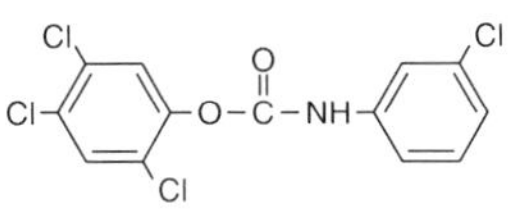

250 mg

S89,210-6
M-TOLYL N-(3-CHLOROPHENYL)CARBAMATE

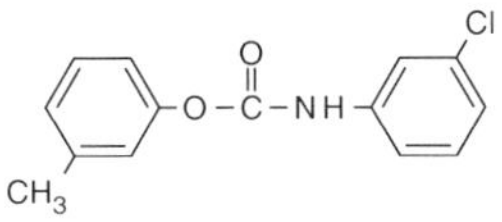

250 mg

S67,557-1
2-BROMO-4,5-DIMETHYLPHENYL N-(3-CHLOROPHENYL)CARBAMATE

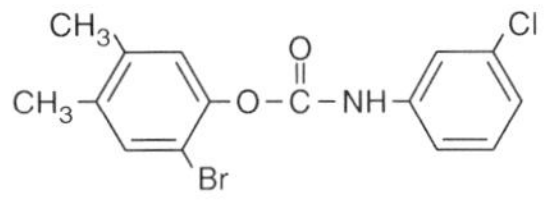

250 mg

S71,861-0
4-BROMO-2,5-DIMETHYLPHENYL N-(3-CHLOROPHENYL)CARBAMATE

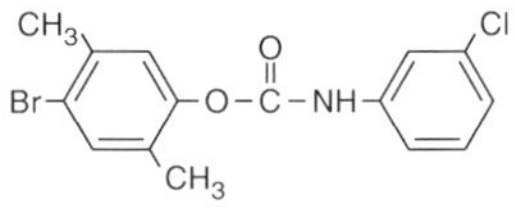

250 mg

S72,685-0
4-IODO-2,5-DIMETHYLPHENYL N-(3-CHLOROPHENYL)CARBAMATE

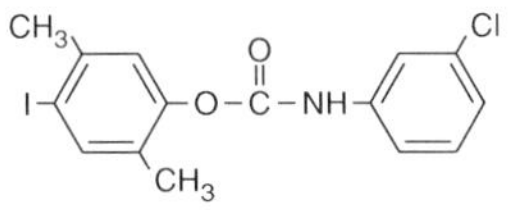

5 gm

S72,550-1
4-CHLORO-2-ISOPROPYL-5-METHYLPHENYL N-(M-TOLYL)CARBAMATE

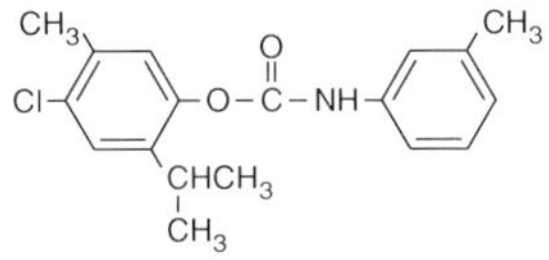

5 gm

S68,728-6
4-CHLORO-2-ISOPROPYL-5-METHYLPHENYL N-(3-ETHYLPHENYL)-CARBAMATE

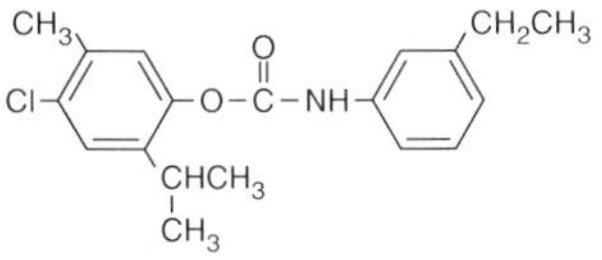

250 mg

S89,900-3
1,2-EPOXY-HEXAHYDRO-4,7-METHANO-5-INDANYL N-(4-CHLOROPHENYL)-CARBAMATE

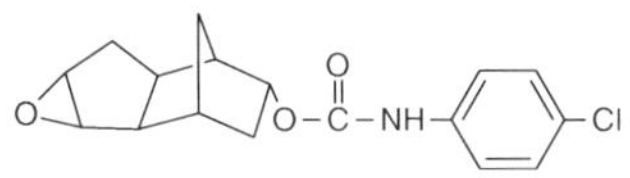

250 mg

S89,905-4
1,2-EPOXY-HEXAHYDRO-4,7-ETHANO-5-INDANYL N-(3,5-DICHLOROPHENYL)-CARBAMATE

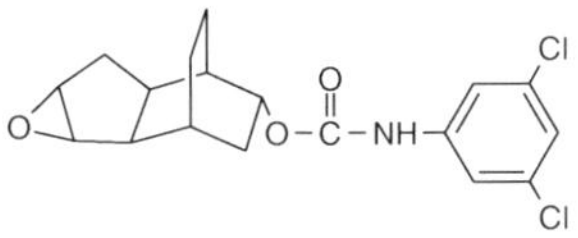

250 mg

S77,286-0
1,2-EPOXY-HEXAHYDRO-4,7-METHANO-5-INDANYL N-(3-BROMOPHENYL)CARBAMATE

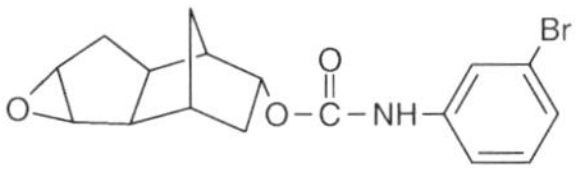

250 mg

S80,068-6
5-NORBORNEN-2-YLMETHYL N-(4-FLUOROPHENYL)CARBAMATE

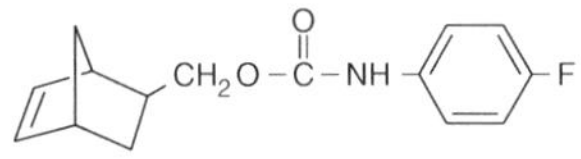

250 mg

S85,101-9
CYCLOPENTYL N-(2,3-DICHLOROPHENYL)-CARBAMATE

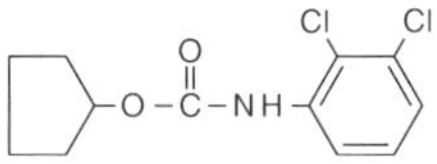

250 mg

S85,132-9
CYCLOPENTYL N-(2-CHLORO-6-METHYLPHENYL)CARBAMATE

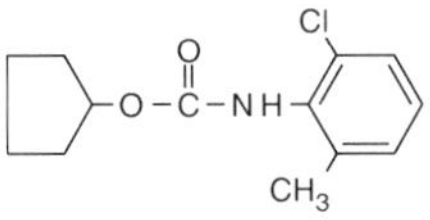

250 mg

S86,737-3
CYCLOPENTYL N-(3-CHLORO-2-METHYLPHENYL)CARBAMATE

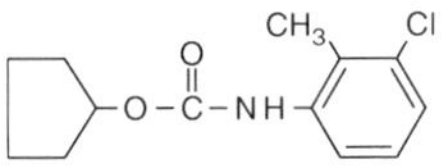

250 mg

S85,069-1
CYCLOPENTYL N-(3-CHLORO-4-FLUOROPHENYL)CARBAMATE

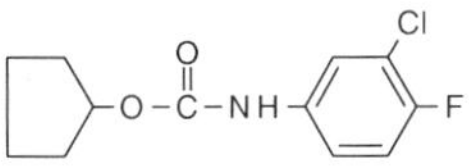

250 mg

S70,957-3
PIPERONYL N-(2,6-DICHLOROPHENYL)-CARBAMATE

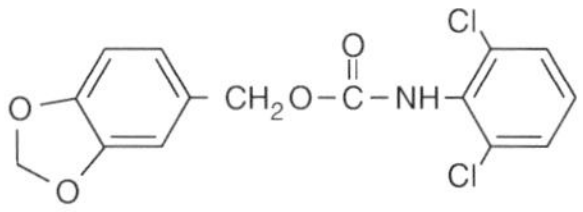

250 mg

S71,475-5
2-METHYLCYCLOHEXYL N-(3-BROMOPHENYL)CARBAMATE

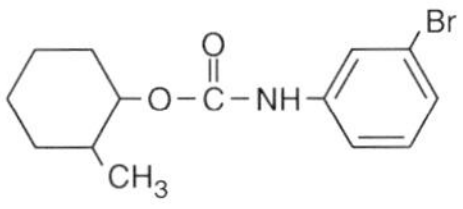

5 gm

S71,698-7
2-METHYLCYCLOHEXYL N-(4-FLUOROPHENYL)CARBAMATE

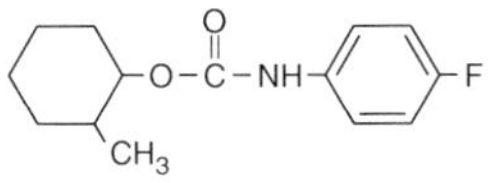

5 gm

S71,658-8
2-METHYLCYCLOHEXYL N-(4-CHLOROPHENYL)CARBAMATE

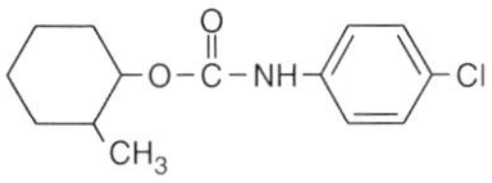

5 gm

S80,779-6
4-(TERT-BUTYL)CYCLOHEXYL N-(5-CHLORO-2-METHYLPHENYL)-CARBAMATE

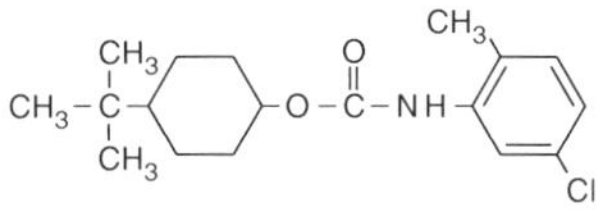

250 mg

S73,607-4
4-CYCLOHEXYLPHENYL N-(2-FLUOROPHENYL)CARBAMATE

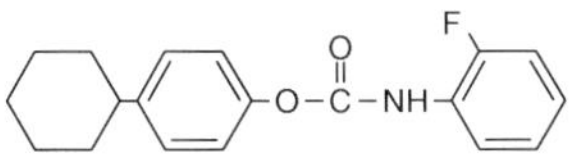

5 gm

S75,380-7
4-CYCLOHEXYLPHENYL N-(4-CHLOROPHENYL)CARBAMATE

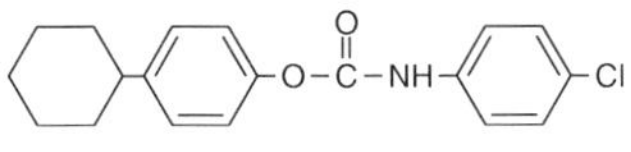

5 gm

S89,969-0
4-CYCLOHEXYLPHENYL N-(3,4-DICHLOROPHENYL)CARBAMATE

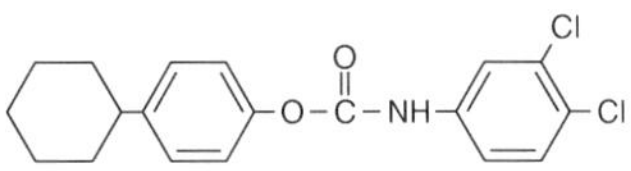

250 mg

S70,997-2
2-METHYLCYCLOHEXYL N-(2,6-DICHLOROPHENYL)CARBAMATE

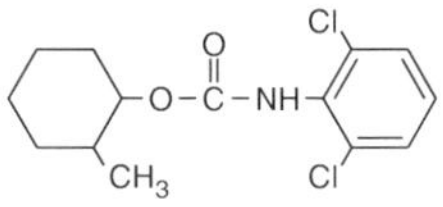

5 gm

S70,960-3
CYCLOHEXYL N-(3-CHLORO-4-METHYLPHENYL)CARBAMATE

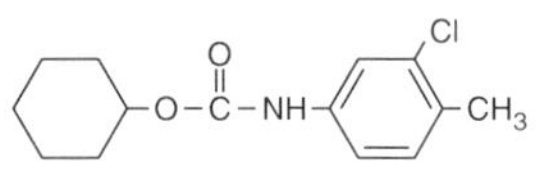

5 gm

S89,962-3
4-CHLORO-2-METHYLPHENYL N-CYCLOHEXYLCARBAMATE

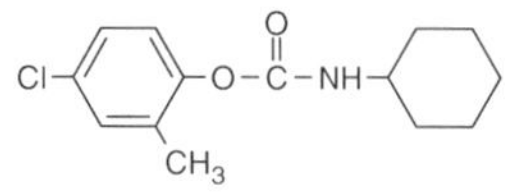

250 mg

S70,998-0
CYCLOHEXYL N-(3,5-DICHLOROPHENYL)-CARBAMATE

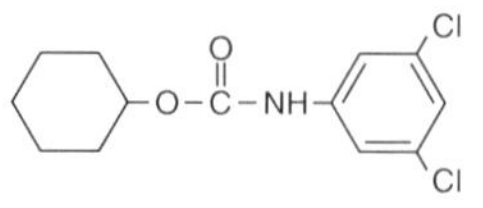

5 gm

S85,133-7
3-METHYLCYCLOHEXYL N-(2-CHLORO-6-METHYLPHENYL)CARBAMATE

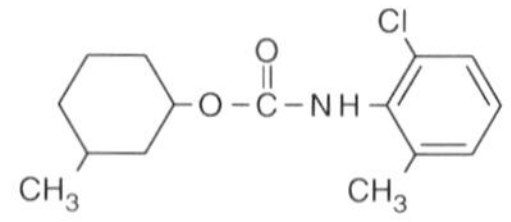

250 mg

S70,734-1
2,4,5-TRICHLOROPHENYL N-PHENYLCARBAMATE

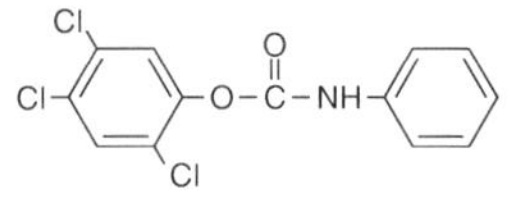

5 gm

S39,296-0
2-CHLOROPHENYL CARBANILATE

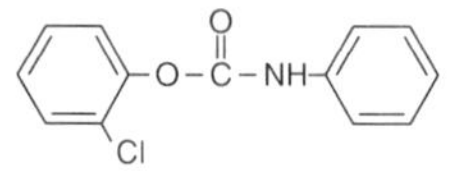

5 gm

S89,227-0
3-PHENYLALLYL N-(2-CHLOROPHENYL)-CARBAMATE

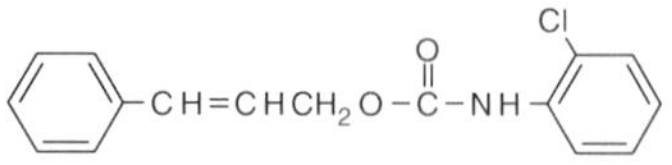

250 mg

S89,048-0
2-BENZYLPHENYL N-(2-CHLOROPHENYL)-CARBAMATE

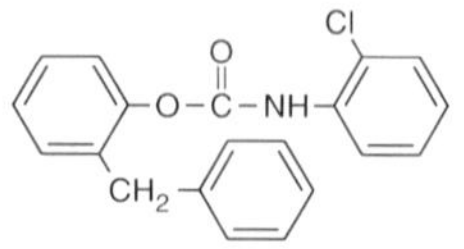

250 mg

S86,886-8
2-BENZYLPHENYL N-(3-CHLORO-2-METHYLPHENYL)CARBAMATE

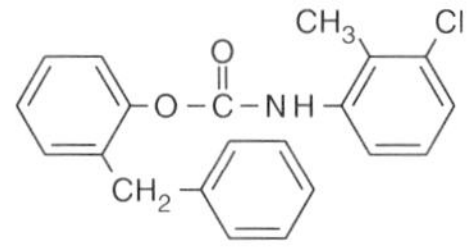

250 mg

S72,485-8
3-CHLORO-1,1'-BIPHENYL-4-YL N-(O-TOLYL)CARBAMATE

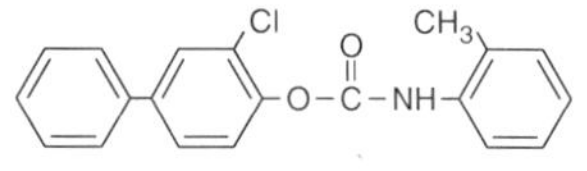

5 gm

S73,186-2
2-BENZYL-4-CHLOROPHENYL N-(2-FLUOROPHENYL)CARBAMATE

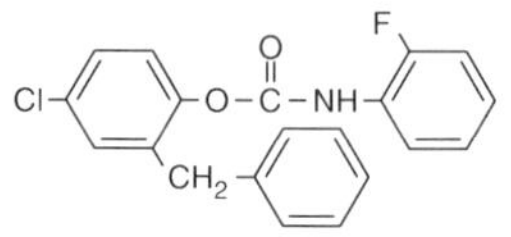

5 gm

S89,226-2
2-BENZYL-4-CHLOROPHENYL N-(2-CHLOROPHENYL)CARBAMATE

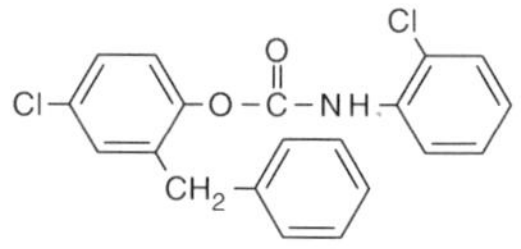

250 mg

S90,003-6
3-PHENYLALLYL N-(4-BROMOPHENYL)-CARBAMATE

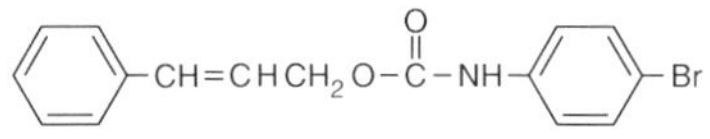

250 mg

S80,051-1
2-BENZYL-4-CHLOROPHENYL N-(4-FLUOROPHENYL)CARBAMATE

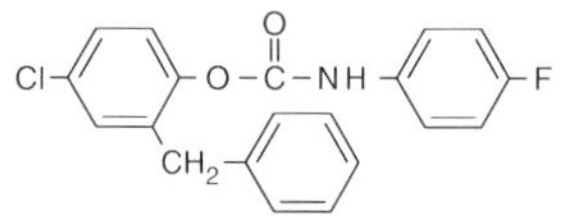

250 mg

S89,957-7
4-BIPHENYLYL N-(3,4-DICHLOROPHENYL)-CARBAMATE

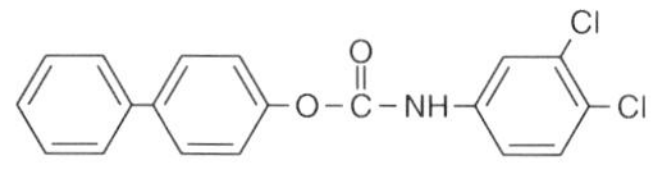

250 mg

S89,202-5
2-BENZYL-4-CHLOROPHENYL N-(4-CHLOROPHENYL)CARBAMATE

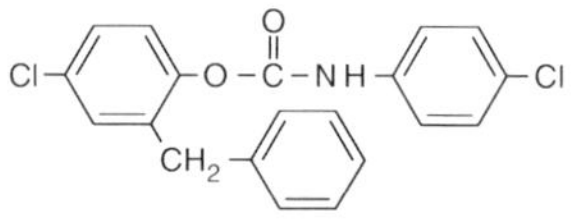

250 mg

S85,551-0
4-BIPHENYLYL N-(4-BROMOPHENYL)-CARBAMATE

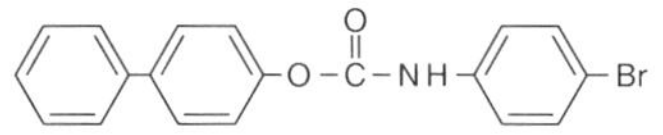

250 mg

S66,777-3
PHENYL N-(2-CHLORO-6-METHYLPHENYL)-CARBAMATE

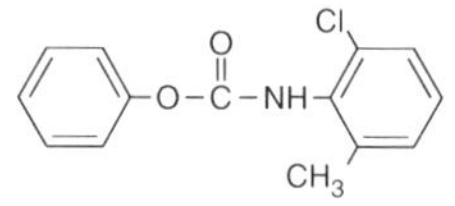

5 gm

S89,979-8
PHENETHYL N-(2,6-DICHLOROPHENYL)-CARBAMATE

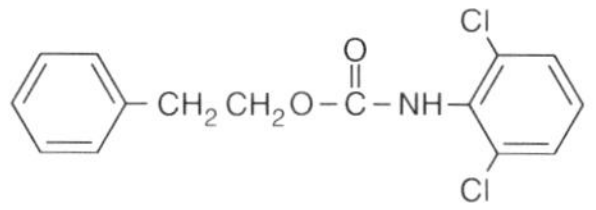

250 mg

S86,835-3
CINNAMYL N-(3-CHLORO-2-METHYLPHENYL)CARBAMATE

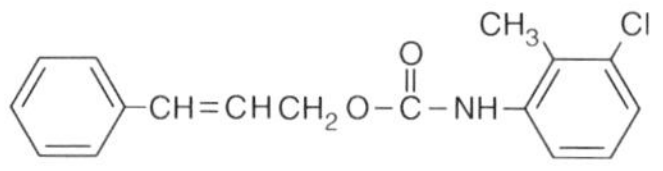

250 mg

S86,829-9
3-PHENYLPROPYL N-(3-CHLORO-2-METHYLPHENYL)CARBAMATE

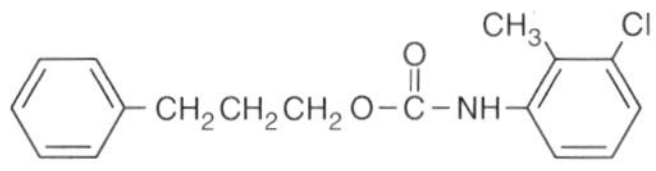

250 mg

S84,740-2
2-BENZYL-4-CHLOROPHENYL N-(2,6-DIMETHYLPHENYL)CARBAMATE

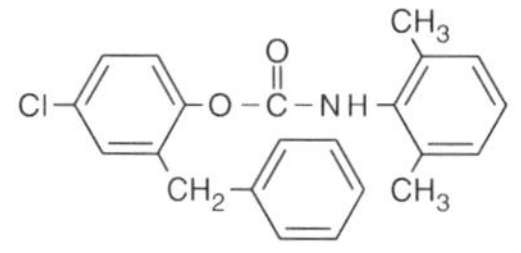

250 mg

S86,836-1
CINNAMYL N-(2-CHLORO-4-METHYLPHENYL)CARBAMATE

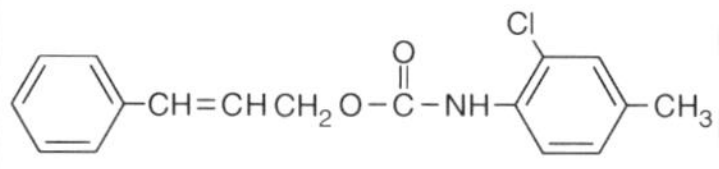

250 mg

S72,104-2
2-BENZYL-4-CHLOROPHENYL N-PHENYLCARBAMATE

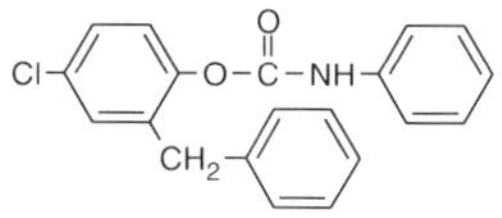

250 mg

S75,280-0
2-BENZYL-4-CHLOROPHENYL N-(3-CHLOROPHENYL)CARBAMATE

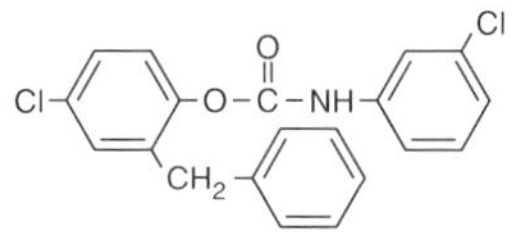

5 gm

S68,860-6
2-BENZYL-4-CHLOROPHENYL N-(2,5-DIMETHYLPHENYL)CARBAMATE

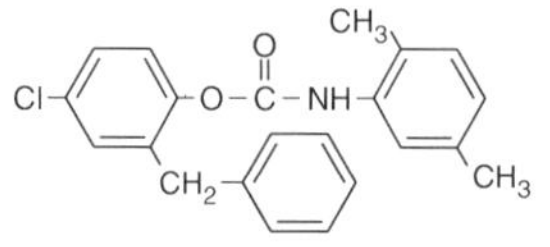

250 mg

S84,856-5
2-BENZYL-4-CHLOROPHENYL N-(3-FLUOROPHENYL)CARBAMATE

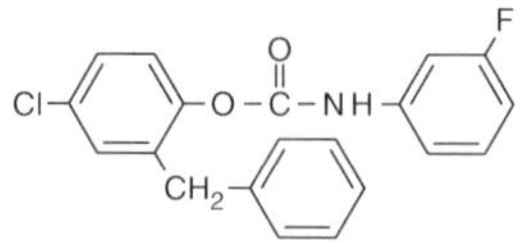

250 mg

S89,956-9
3-CHLORO-1,1'-BIPHENYL-4-YL N-TERT-BUTYLCARBAMATE

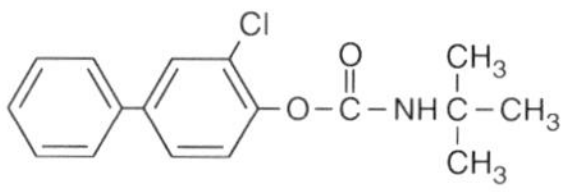

250 mg

S75,104-9
3-CHLORO-1,1'-BIPHENYL-4-YL N-(2,4-DICHLOROPHENYL)CARBAMATE

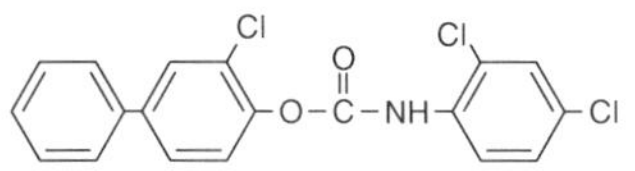

5 gm

S86,867-1
3-BROMO-1,1'-BIPHENYL-4-YL N-(2-BROMO-4-METHYLPHENYL)CARBAMATE

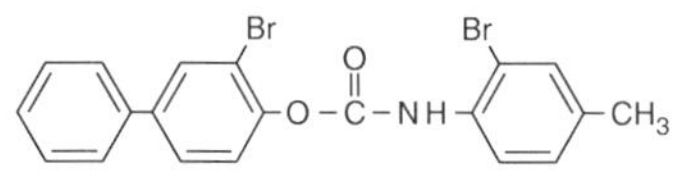

250 mg

S68,746-4
3-CHLORO-1,1'-BIPHENYL-4-YL N-(2,5-XYLYL)CARBAMATE

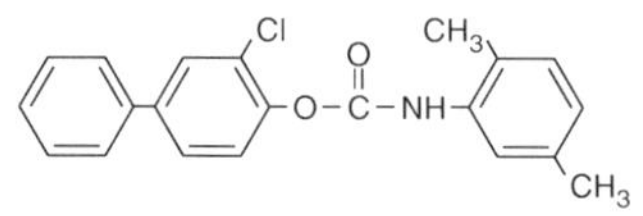

250 mg

S79,167-9
3-BROMO-1,1'-BIPHENYL-4-YL N-(1-NAPHTHYL)CARBAMATE

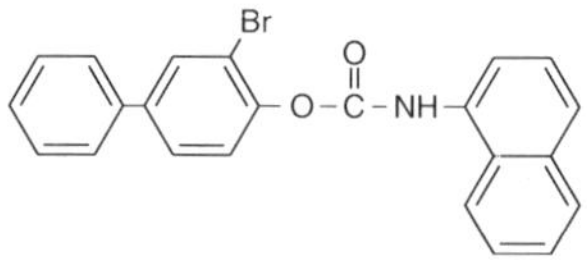

250 mg

S75,290-8
2-BENZYL-4-CHLOROPHENYL N-(1-NAPHTHYL)CARBAMATE

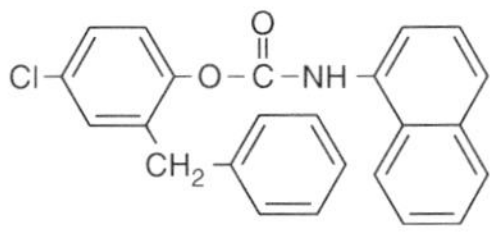

5 gm

S70,946-8
PHENYL N-(3,5-DICHLOROPHENYL)-CARBAMATE

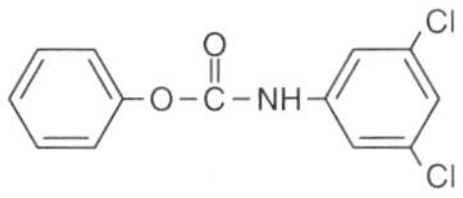

5 gm

S71,059-8
2,4,6-TRICHLOROPHENYL N-PHENYLCARBAMATE

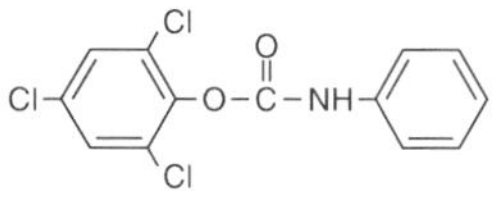

5 gm

S71,764-9
2,4,6-TRIBROMOPHENYL N-PHENYLCARBAMATE

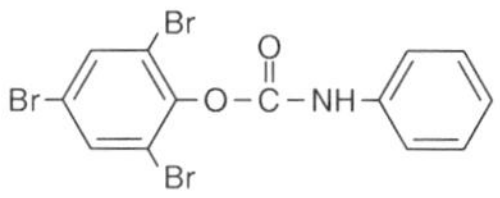

5 gm

S66,715-3
PHENYL N-(2,4,5-TRICHLOROPHENYL)-CARBAMATE

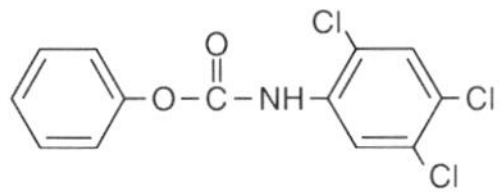

5 gm

S76,998-3
2-FLUOROPHENYL N-ISOPROPYLCARBAMATE

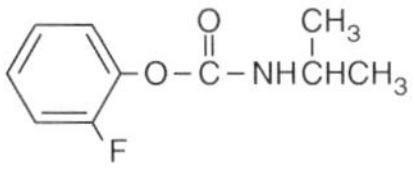

250 mg

S89,316-1
ETHYL N-(2-CHLOROBENZYLOXY)-CARBAMATE

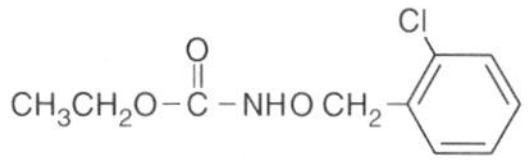

250 mg

S79,507-0
2-BUTENYL N-(2-FLUOROPHENYL)-CARBAMATE

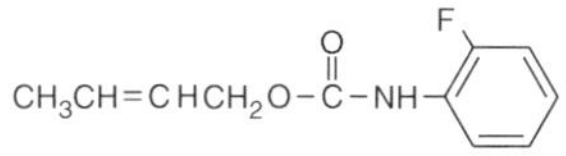

250 mg

S72,692-3
2-FLUOROBENZYL N-(2-FLUOROPHENYL)-CARBAMATE

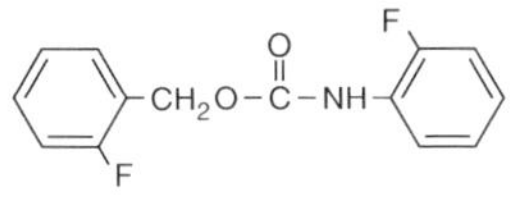

5 gm

S69,065-1
4-BROMO-2,5-DIMETHYLPHENYL N-(2-FLUOROPHENYL)CARBAMATE

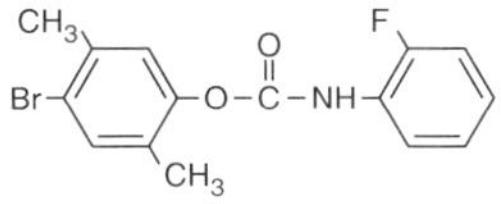

250 mg

S72,289-8
4-BROMO-2,5-DIMETHYLPHENYL N-(2-ETHYLPHENYL)CARBAMATE

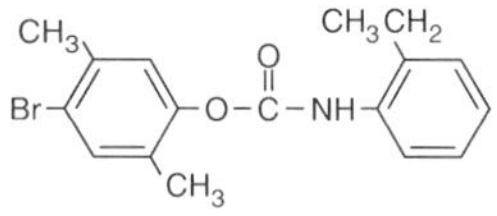

5 gm

S80,019-8
2-FLUOROPHENYL N-(4-FLUOROPHENYL)-CARBAMATE

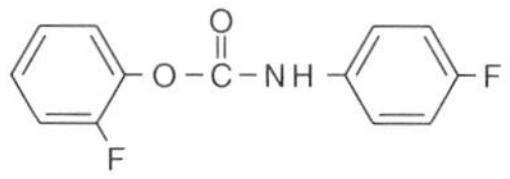

250 mg

S80,125-9
P-TOLYL N-(2-FLUOROPHENYL)CARBAMATE

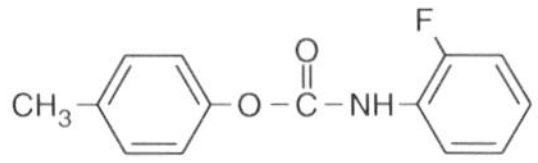

250 mg

S66,947-4
4-CHLOROPHENYL N-(2-ETHYLPHENYL)-CARBAMATE

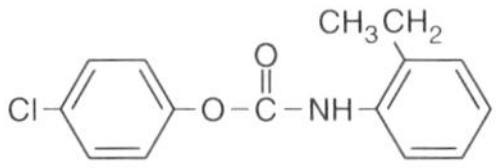

5 gm

S80,073-2
2-TERT-BUTYLPHENYL N-(4-FLUOROPHENYL)CARBAMATE

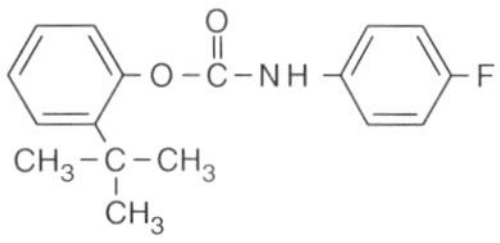

250 mg

S71,186-1
2-CHLOROPHENYL N-(2,6-DICHLOROPHENYL)CARBAMATE

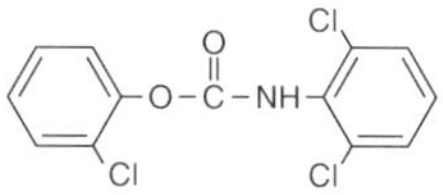

5 gm

S72,504-8
3,4-DICHLOROPHENYL N-(2-FLUOROPHENYL)CARBAMATE

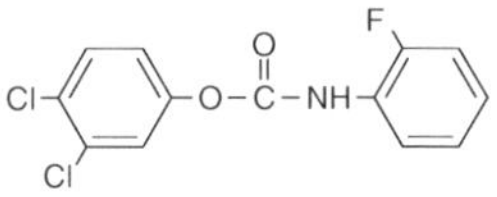

5 gm

S73,819-0
2,5-DICHLOROPHENYL N-(2-FLUOROPHENYL)CARBAMATE

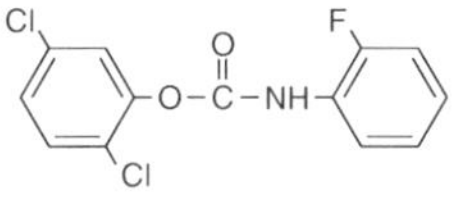

5 gm

S76,974-6
2-CHLORO-4-FLUOROPHENYL N-(2-CHLOROPHENYL)CARBAMATE

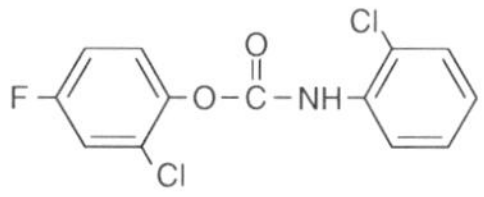

250 mg

S71,353-8
2-CHLOROPHENYL N-(3,4-DICHLOROPHENYL)CARBAMATE

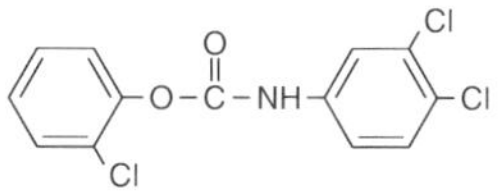

5 gm

S72,654-0
2-IODOPHENYL N-(2,5-DIFLUOROPHENYL)-CARBAMATE

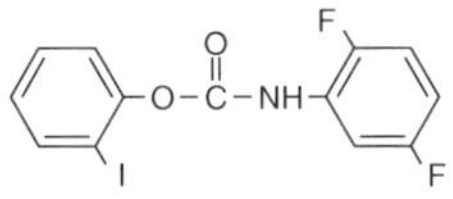

5 gm

S80,122-4
2,4-DIBROMOPHENYL N-(2-FLUOROPHENYL)CARBAMATE

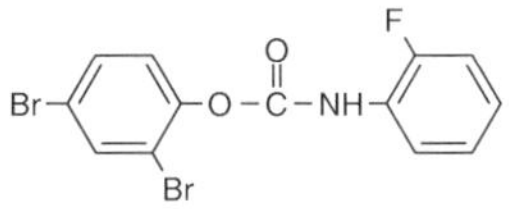

250 mg

S73,792-5
4-CHLORO-3-METHYLPHENYL N-(2-FLUOROPHENYL)CARBAMATE

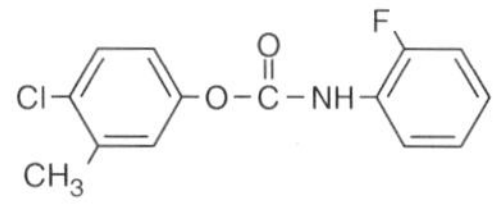

5 gm

S86,749-7
2-FLUOROPHENYL N-(2-CHLORO-4-METHYLPHENYL)CARBAMATE

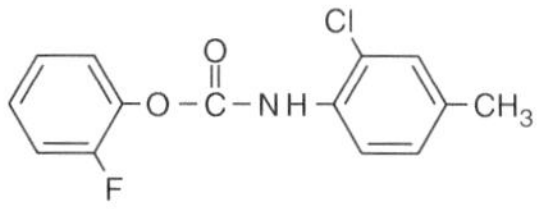

250 mg

S73,682-1
2,5-XYLYL N-(2-FLUOROPHENYL)-CARBAMATE

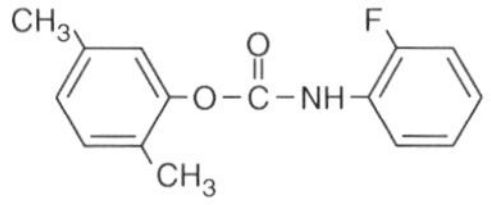

5 gm

S73,684-8
2-FLUOROPHENYL N-(2,5-XYLYL)-CARBAMATE

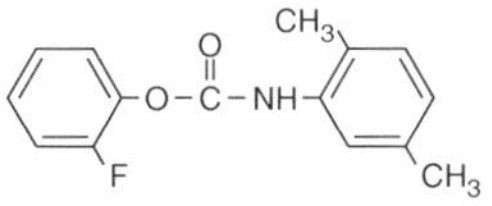

5 gm

S79,515-1
4-CHLORO-3-METHYLPHENYL N-(O-TOLYL)-CARBAMATE

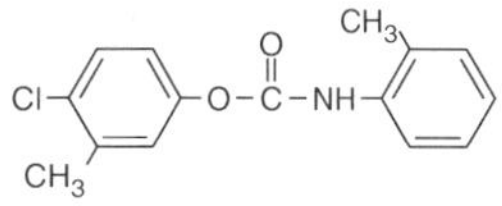

250 mg

S89,183-5
4-CHLORO-2-METHYLPHENYL N-(O-TOLYL)-CARBAMATE

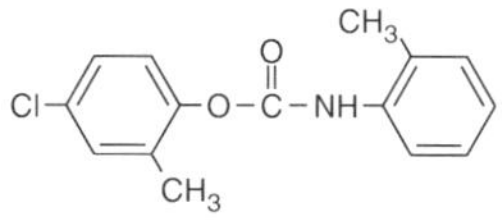

250 mg

S89,042-1
3,5-XYLYL N-(2-CHLOROPHENYL)-CARBAMATE

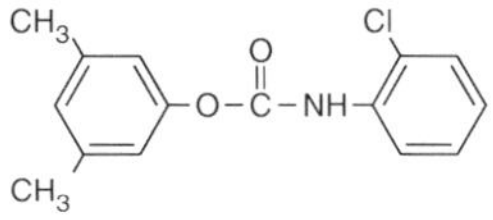

250 mg

S76,973-8
2-CHLORO-4,6-DIBROMOPHENYL N-(2-CHLOROPHENYL)CARBAMATE

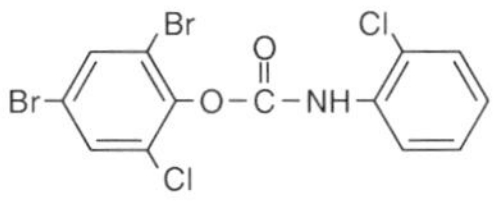

250 mg

S71,702-9
2,4,6-TRIBROMOPHENYL N-(2-CHLOROPHENYL)CARBAMATE

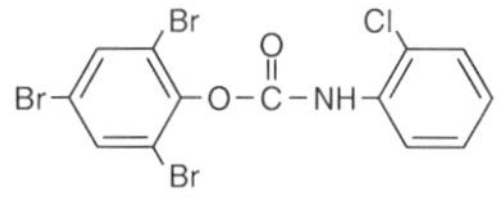

5 gm

S69,135-6
2-BROMO-4-CHLORO-6-METHYLPHENYL N-(2-CHLOROPHENYL)CARBAMATE

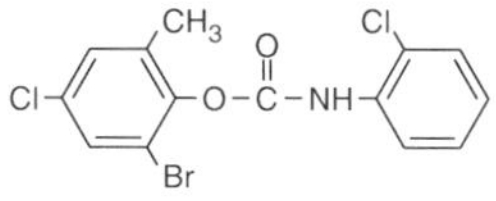

250 mg

S69,439-8
4-BROMO-3,5-DIMETHYLPHENYL N-(2-CHLOROPHENYL)CARBAMATE

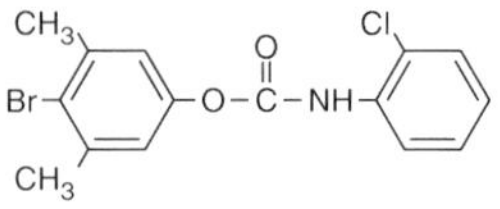

250 mg

S75,257-6
2,4-DICHLORO-1-NAPHTHYL N-(2-CHLOROPHENYL)CARBAMATE

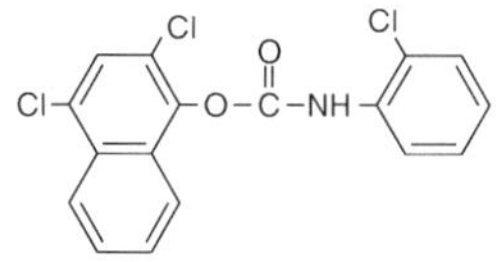

5 gm

S78,064-2
1-NAPHTHYL N-(2-FLUOROPHENYL)-CARBAMATE

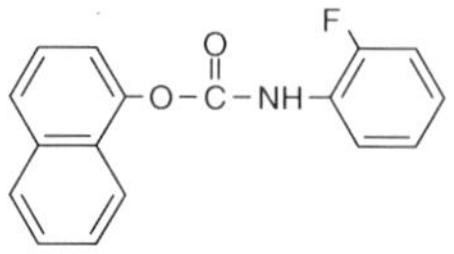

250 mg

S72,466-1
2-NAPHTHYL N-(2-CHLOROPHENYL)-CARBAMATE

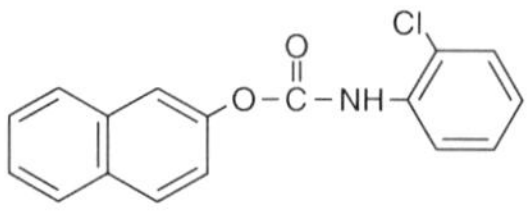

5 gm

S77,280-1
4-FLUOROPHENYL N-METHYLCARBAMATE

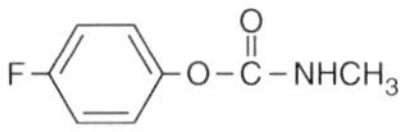

250 mg

S66,769-2
METHYL N-(4-BROMOPHENYL)CARBAMATE

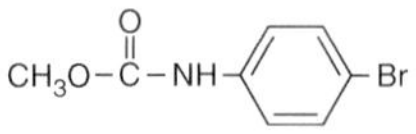

5 gm

S76,318-7
4-BROMOPHENYL N-METHYLCARBAMATE

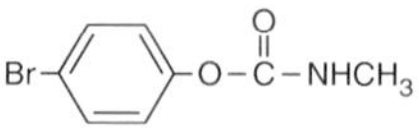

250 mg

S76,313-6
4-CHLOROBENZYL N-METHYLCARBAMATE

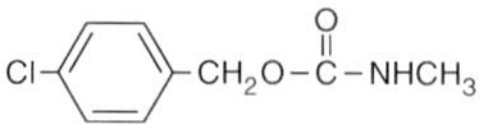

250 mg

S66,659-9
ETHYL N-(4-BROMOPHENYL)CARBAMATE

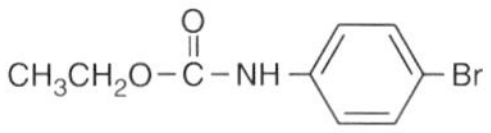

5 gm

S91,020-1
4-FLUOROPHENYL N-ISOPROPYLCARBAMATE

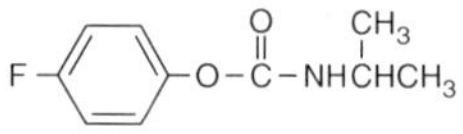

250 mg

S66,661-0
ALLYL 4'-BROMOCARBANILATE

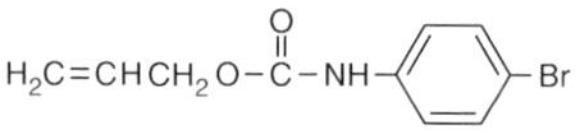

5 gm

S91,028-7
4-FLUOROPHENYL N-BUTYLCARBAMATE

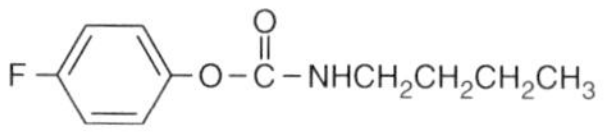

250 mg

S84,815-8
4-FLUOROPHENYL N-(3-FLUOROPHENYL)-CARBAMATE

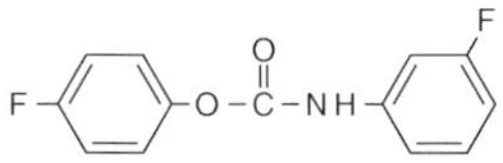

250 mg

S71,893-9
4-FLUOROPHENYL N-(3-CHLOROPHENYL)-CARBAMATE

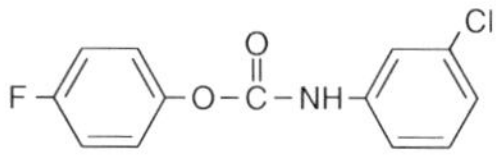

250 mg

S71,727-4
3-CHLOROPHENYL N-(4-CHLOROPHENYL)-CARBAMATE

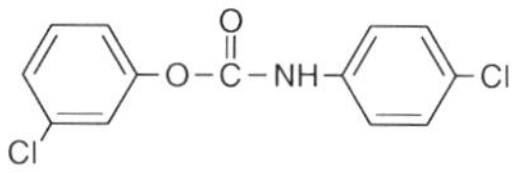

5 gm

S71,708-8
2,4,5-TRICHLOROPHENYL N-(4-CHLOROPHENYL)CARBAMATE

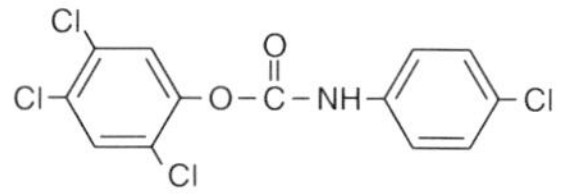

5 gm

S70,179-3
4-BROMOPHENYL N-(3-BROMOPHENYL)-CARBAMATE

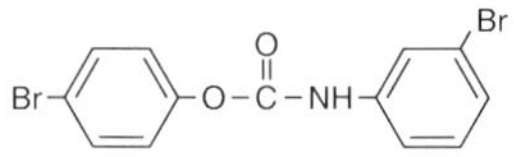

250 mg

S72,445-9
M-TOLYL N-(4-CHLOROPHENYL)CARBAMATE

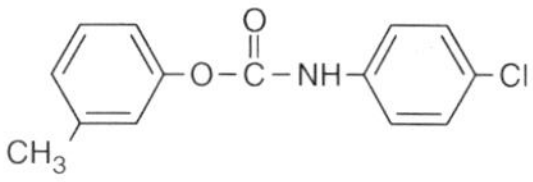

5 gm

S72,327-4
2-BROMO-4,5-DIMETHYLPHENYL N-(4-FLUOROPHENYL)CARBAMATE

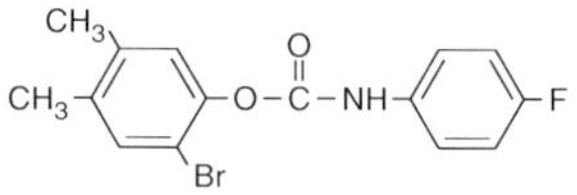

5 gm

S79,156-3
4-ISOBUTYLPHENYL N-(4-CHLOROPHENYL)-CARBAMATE

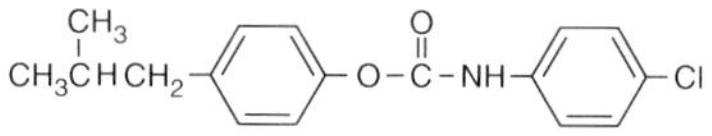

250 mg

S69,136-4
4-BROMO-2,5-DIMETHYLPHENYL N-(4-ISOPROPYLPHENYL)CARBAMATE

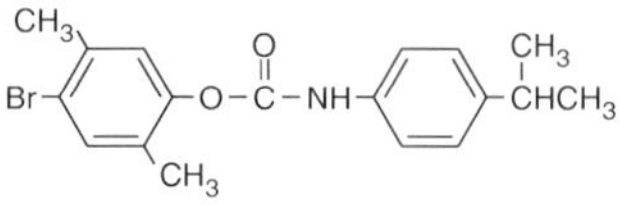

250 mg

S70,943-3
4-CHLOROPHENYL N-(2,6-DICHLOROPHENYL)CARBAMATE

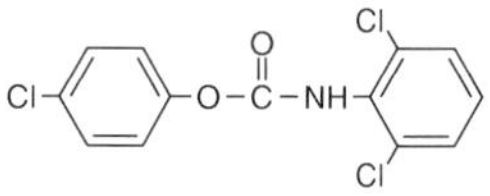

5 gm

S79,613-1
4-BROMOPHENYL N-(2,6-DICHLOROPHENYL)CARBAMATE

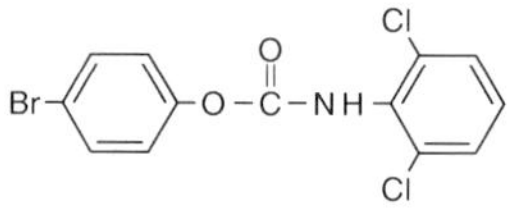

250 mg

S85,131-0
4-FLUOROPHENYL N-(2-CHLORO-6-METHYLPHENYL)CARBAMATE

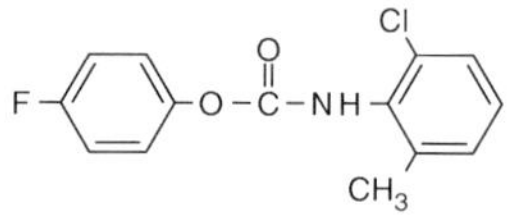

250 mg

S69,999-3
4-FLUOROPHENYL N-(2,3-DIMETHYLPHENYL)CARBAMATE

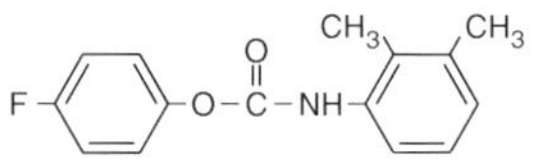

250 mg

S66,464-2
4-CHLOROPHENYL N-(2,6-XYLYL)-CARBAMATE

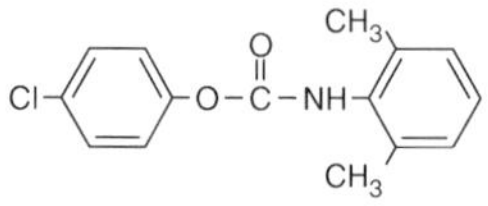

5 gm

S89,207-6
2,6-XYLYL N-(4-CHLOROPHENYL)-CARBAMATE

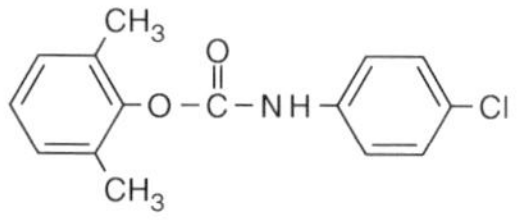

250 mg

S66,617-3
4-CHLOROPHENYL N-(2-ETHYL-6-METHYLPHENYL)CARBAMATE

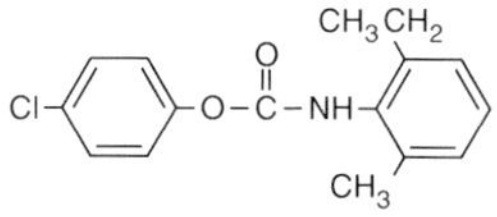

5 gm

S71,351-1
4-ISOPROPYLPHENYL N-(2,6-DICHLOROPHENYL)CARBAMATE

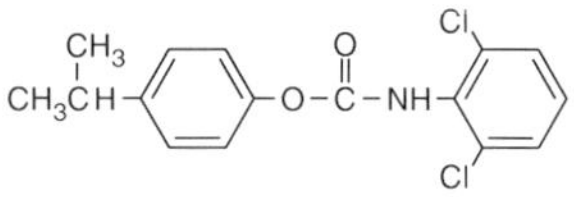

5 gm

S80,017-1
2-CHLORO-4-FLUOROPHENYL N-(4-FLUOROPHENYL)CARBAMATE

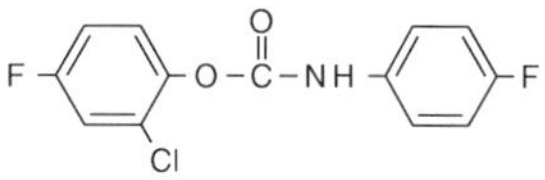

250 mg

S72,158-1
4-BROMOPHENYL N-(2,5-DIFLUOROPHENYL)CARBAMATE

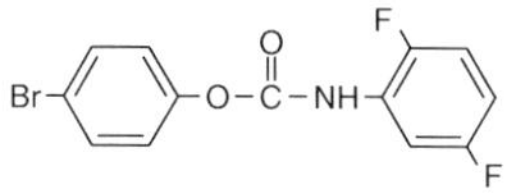

250 mg

S72,398-3
4-BROMOPHENYL N-(3,4-DICHLOROPHENYL)CARBAMATE

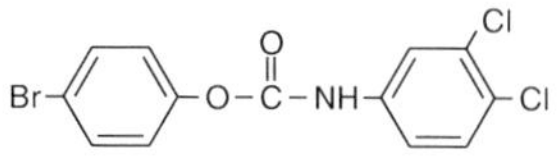

5 gm

S69,073-2
4-FLUOROPHENYL N-(4-CHLORO-2-METHYLPHENYL)CARBAMATE

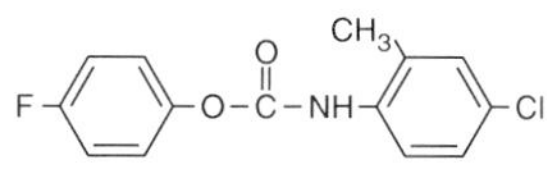

250 mg

S86,750-0
4-FLUOROPHENYL N-(5-CHLORO-2-METHYLPHENYL)CARBAMATE

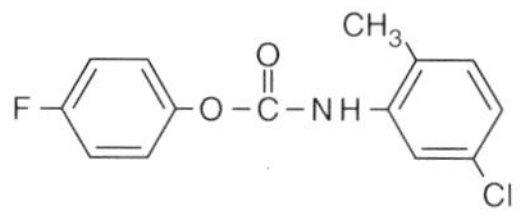

250 mg

S66,551-7
4-CHLOROPHENYL N-(4-CHLORO-2-METHYLPHENYL)CARBAMATE

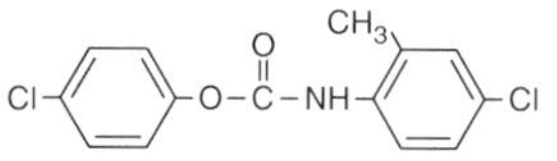

5 gm

S66,588-6
4-CHLOROPHENYL N-(3-CHLORO-4-METHYLPHENYL)CARBAMATE

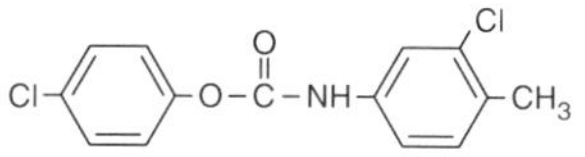

5 gm

S66,613-0
4-CHLOROPHENYL N-(5-CHLORO-2-METHYLPHENYL)CARBAMATE

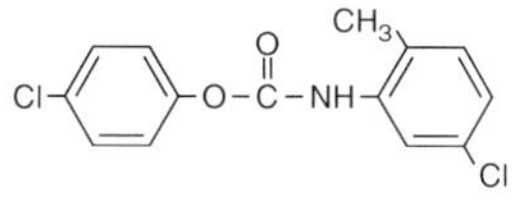

5 gm

S74,312-7
P-TOLYL N-(3,4-DICHLOROPHENYL)-CARBAMATE

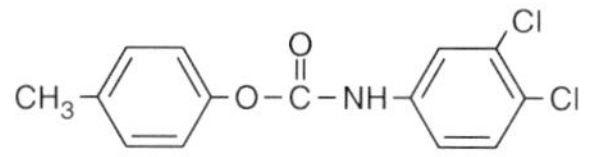

5 gm

S84,470-5
4-CHLORO-2-METHYLPHENYL N-(4-CHLOROPHENYL)CARBAMATE

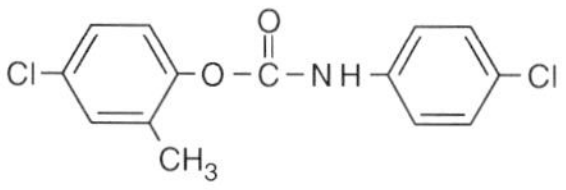

250 mg

S84,471-3
4-CHLORO-3-METHYLPHENYL N-(4-CHLOROPHENYL)CARBAMATE

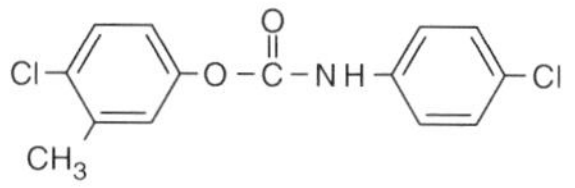

250 mg

S80,065-1
2,5-XYLYL N-(4-FLUOROPHENYL)-CARBAMATE

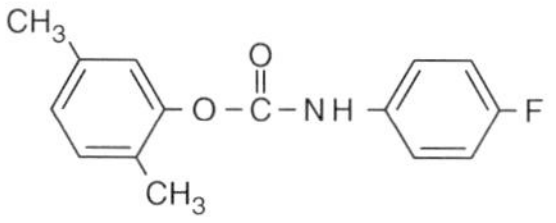

250 mg

S84,527-2
4-FLUOROPHENYL N-(2,5-XYLYL)-CARBAMATE

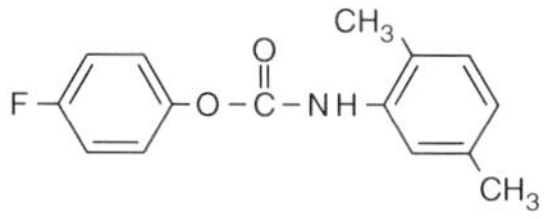

250 mg

S66,518-5
4-CHLOROPHENYL N-(2,5-XYLYL)-CARBAMATE

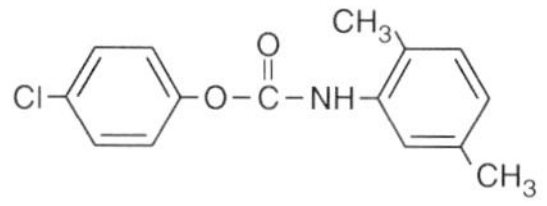

5 gm

S66,611-4
4-CHLOROPHENYL N-(3,4-XYLYL)-CARBAMATE

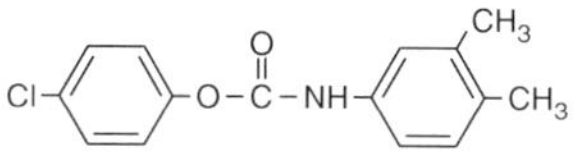

5 gm

S86,826-4
2-ISOPROPYL-5-METHYLPHENYL N-(4-CHLOROPHENYL)CARBAMATE

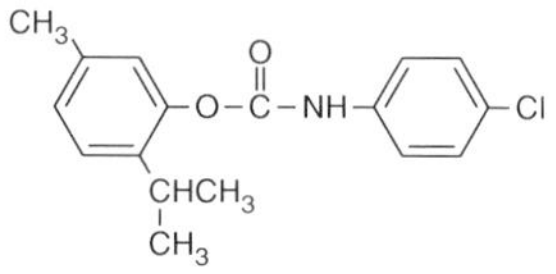

250 mg

S90,180-6
4-(1-NAPHTHYLIMINOMETHYL)PHENYL N-(3,4-DICHLOROPHENYL)CARBAMATE

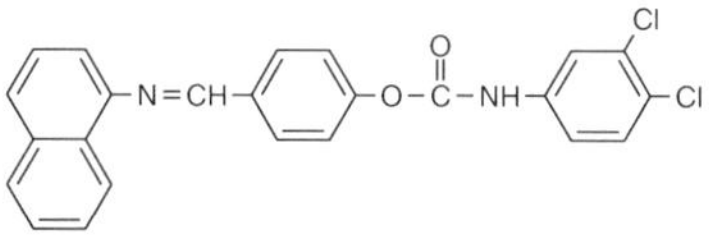

250 mg

S66,800-1
4-CHLOROPHENYL N-(3,5-XYLYL)-CARBAMATE

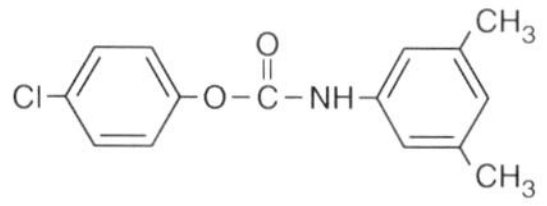

5 gm

S76,952-5
3-ETHYL-5-METHYLPHENYL N-(4-CHLOROPHENYL)CARBAMATE

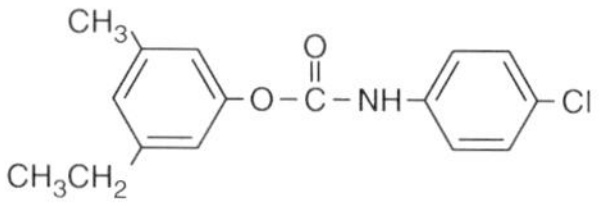

250 mg

S71,711-8
2,4,6-TRICHLOROPHENYL N-(4-CHLOROPHENYL)CARBAMATE

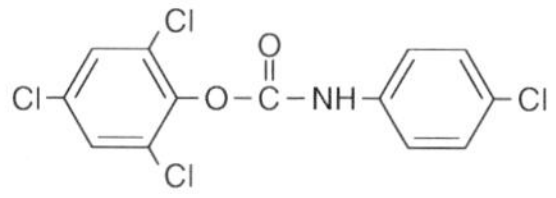

5 gm

S72,390-8
2-BROMO-4-CHLORO-6-METHYLPHENYL N-(4-FLUOROPHENYL)CARBAMATE

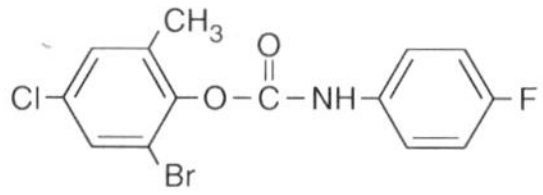

5 gm

S85,341-0
4-CHLORO-2,6-DIMETHYLPHENYL N-(4-FLUOROPHENYL)CARBAMATE

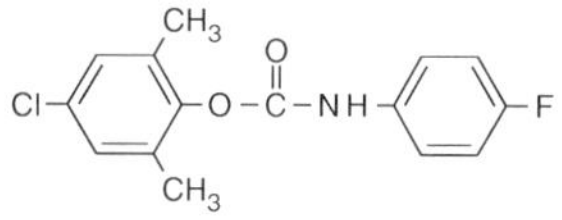

250 mg

S69,364-2
2-BROMO-4-CHLORO-6-METHYLPHENYL N-(4-ISOPROPYLPHENYL)CARBAMATE

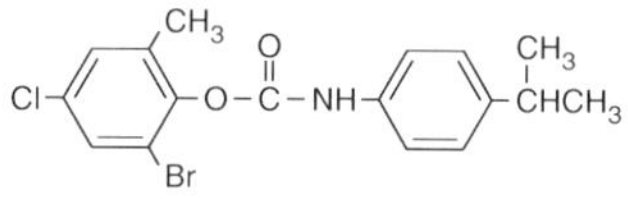

250 mg

S66,704-8
4-CHLOROPHENYL N-(2,4,5-TRICHLOROPHENYL)CARBAMATE

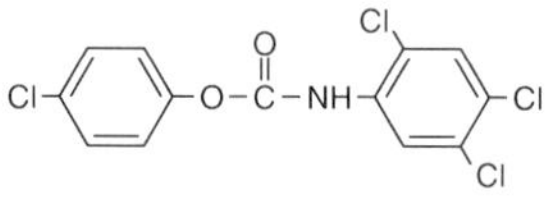

5 gm

S77,940-7
4-BROMO-2,5-DIMETHYLPHENYL N-(4-FLUOROPHENYL)CARBAMATE

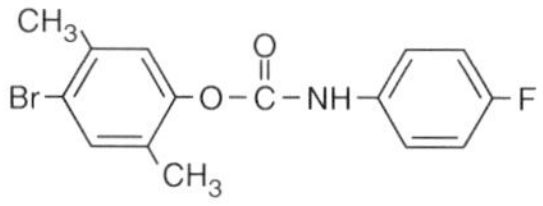

250 mg

S80,045-7
4-CHLORO-2-ISOPROPYL-5-METHYLPHENYL N-(4-FLUOROPHENYL)-CARBAMATE

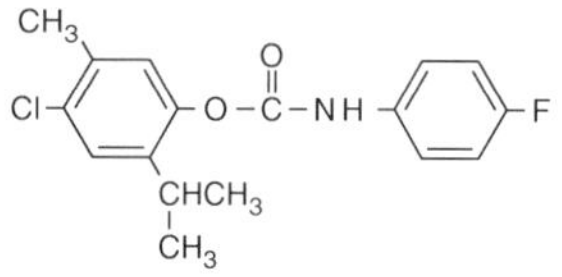

250 mg

S85,583-9
4-CHLORO-2-ISOPROPYL-5-METHYLPHENYL N-(P-TOLYL)CARBAMATE

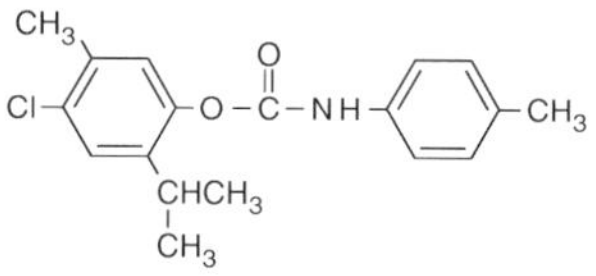

250 mg

S77,388-3
2-BROMO-4,5-DIMETHYLPHENYL N-(4-ISOPROPYLPHENYL)CARBAMATE

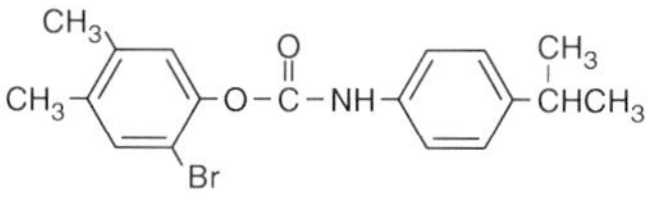

250 mg

S75,258-4
2-NAPHTHYL N-(4-CHLOROPHENYL)-CARBAMATE

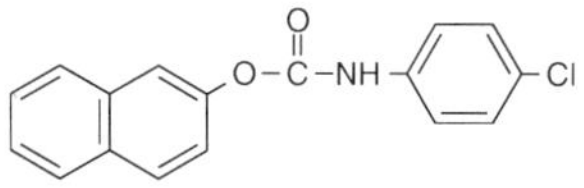

250 mg

S72,553-6
2,3-DICHLOROPHENYL N-METHYLCARBAMATE

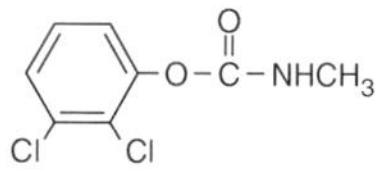

5 gm

S66,762-5
METHYL N-(2-CHLORO-6-METHYLPHENYL)-CARBAMATE

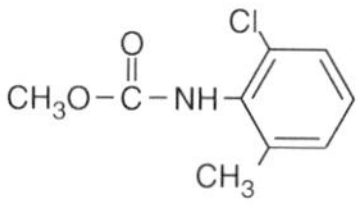

5 gm

S70,934-4
ETHYL N-(2,6-DICHLOROPHENYL)-CARBAMATE

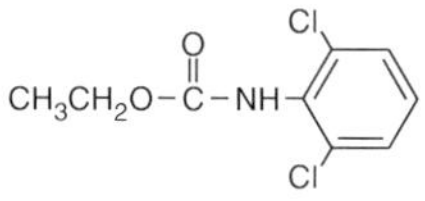

5 gm

S70,987-5
ALLYL N-(2,6-DICHLOROPHENYL)-CARBAMATE

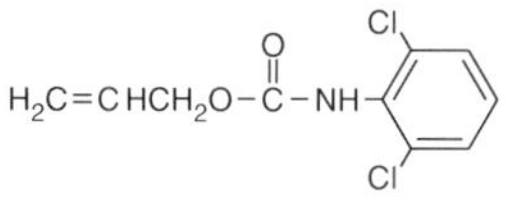

5 gm

S91,720-6
2,3-DICHLOROPHENYL N-ISOPROPYLCARBAMATE

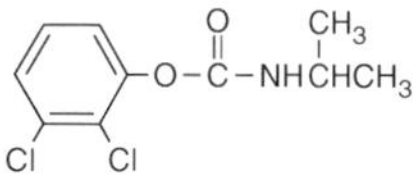

250 mg

S66,770-6
ALLYL N-(2-CHLORO-6-METHYLPHENYL)-CARBAMATE

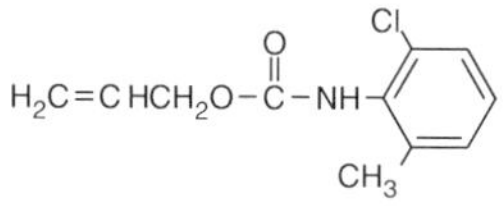

5 gm

S66,570-3
ISOPROPYL N-(2-CHLORO-6-METHYLPHENYL)CARBAMATE

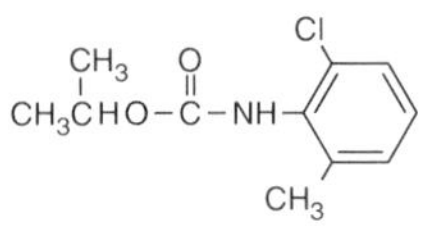

5 gm

S66,761-7
PROPYL N-(2-CHLORO-6-METHYLPHENYL)-CARBAMATE

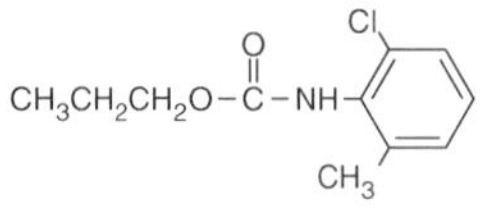

5 gm

S91,726-5
2,3-DICHLOROPHENYL N-BUTYLCARBAMATE

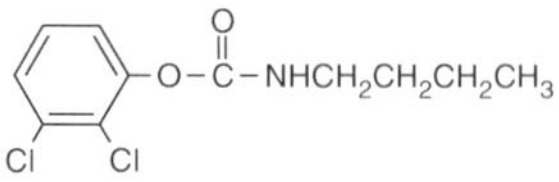

250 mg

S71,188-8
3-CHLOROPHENYL N-(2,6-DICHLOROPHENYL)CARBAMATE

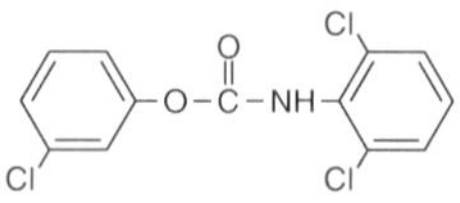

5 gm

S72,697-4
2,3-DICHLOROPHENYL N-(3-CHLOROPHENYL)CARBAMATE

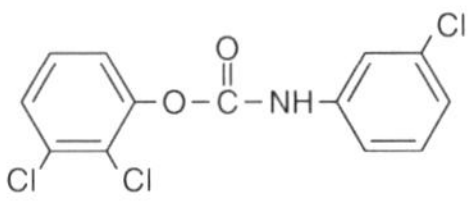

5 gm

S72,331-2
2-BROMO-4-CHLORO-5-METHYL N-(2,3-DICHLOROPHENYL)CARBAMATE

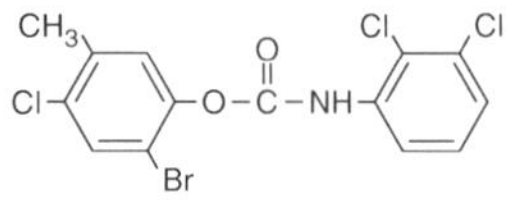

5 gm

S75,242-8
2,6-XYLYL N-(3-CHLOROPHENYL)-CARBAMATE

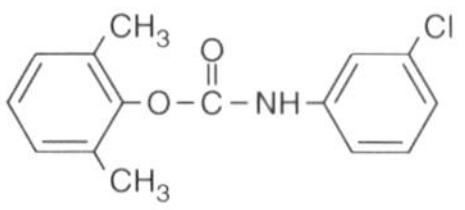

5 gm

S85,349-6
2,4-DICHLORO-3,5-DIMETHYLPHENYL N-(2,6-DICHLOROPHENYL)CARBAMATE

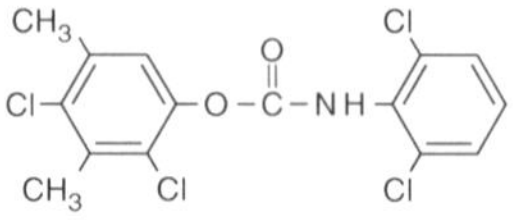

250 mg

S69,068-6
4-BROMO-2,5-DIMETHYLPHENYL N-(2,3-DICHLOROPHENYL)CARBAMATE

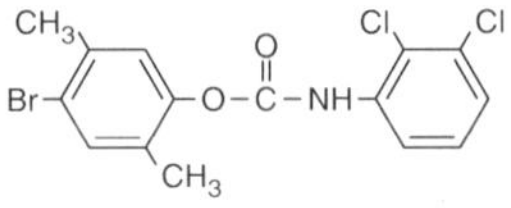

250 mg

S72,339-8
2-BROMO-4,5-DIMETHYLPHENYL N-(2,3-DICHLOROPHENYL)CARBAMATE

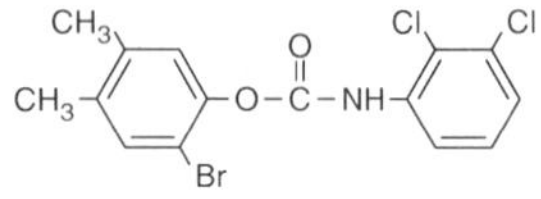

5 gm

S70,211-0
4-CHLORO-2,5-DIMETHYLPHENYL N-(3-CHLORO-2-METHYLPHENYL)-CARBAMATE

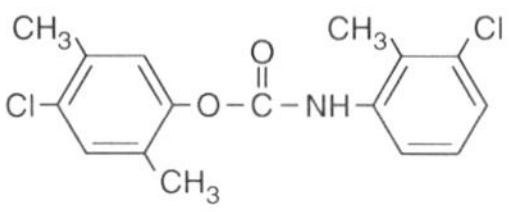

250 mg

S73,608-2
2,6-DIISOPROPYLPHENYL N-(3-CHLOROPHENYL)CARBAMATE

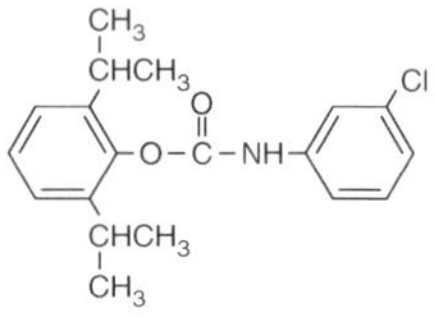

5 gm

S85,362-3
CYCLODODECYL N-(2,6-DICHLOROPHENYL)-CARBAMATE

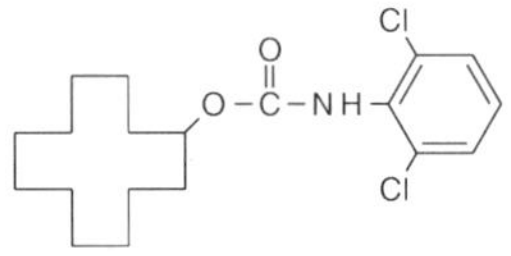

250 mg

S67,426-5
2,5-DICHLOROPHENYL N-(2,6-DICHLOROPHENYL)CARBAMATE

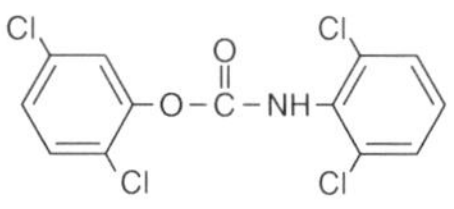

5 gm

S69,096-1
2,5-DICHLOROPHENYL N-(2,3-DICHLOROPHENYL)CARBAMATE

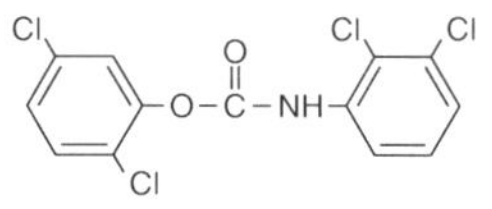

250 mg

S69,359-6
3,4-DICHLOROPHENYL N-(2,3-DICHLOROPHENYL)CARBAMATE

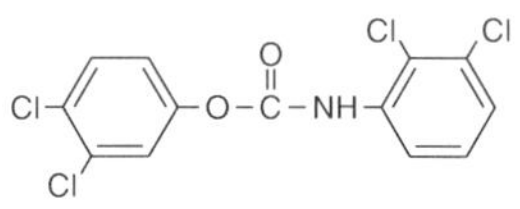

250 mg

S72,648-6
2,5-DICHLOROPHENYL N-(3-CHLORO-2-METHYLPHENYL)CARBAMATE

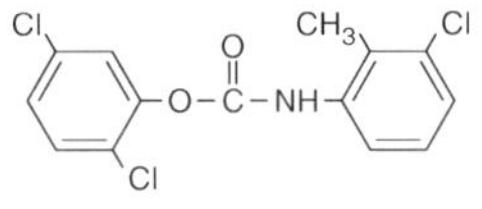

5 gm

S70,230-7
2-CHLORO-4-FLUOROPHENYL N-(2,3-DIMETHYLPHENYL)CARBAMATE

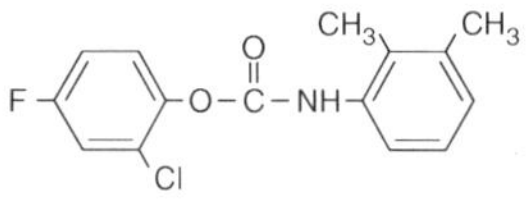

250 mg

S79,500-3
2-CHLORO-4-FLUOROPHENYL N-(2,6-XYLYL)CARBAMATE

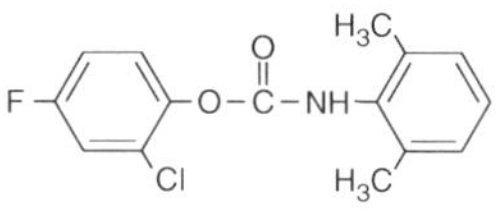

250 mg

S68,683-2
2,5-DICHLOROPHENYL N-(2,6-DIMETHYLPHENYL)CARBAMATE

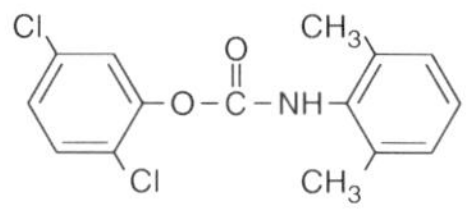

250 mg

S70,031-2
2,5-DICHLOROPHENYL N-(2,3-DIMETHYLPHENYL)CARBAMATE

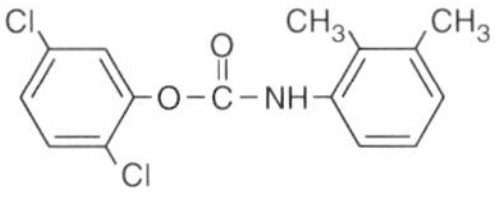

250 mg

S72,614-1
3,4-DICHLOROPHENYL N-(2,3-DIMETHYLPHENYL)CARBAMATE

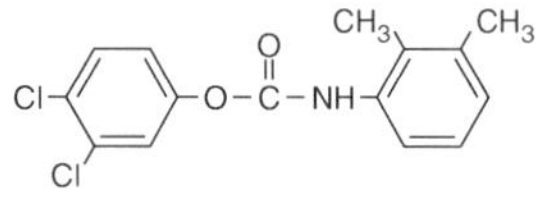

5 gm

S84,558-2
4-CHLORO-2-METHYLPHENYL N-(2,6-XYLYL)CARBAMATE

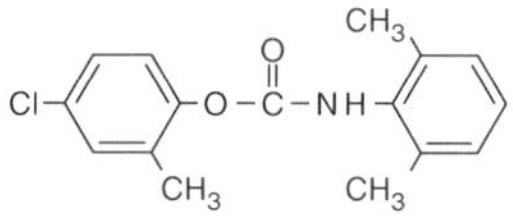

250 mg

S79,414-7
2-BROMO-4-(TERT-BUTYL)PHENYL N-(3-CHLORO-2-METHYLPHENYL)-CARBAMATE

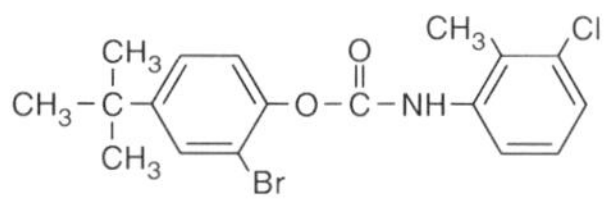

250 mg

S73,914-6
3,5-DICHLOROPHENYL N-(3-CHLORO-2-METHYLPHENYL)CARBAMATE

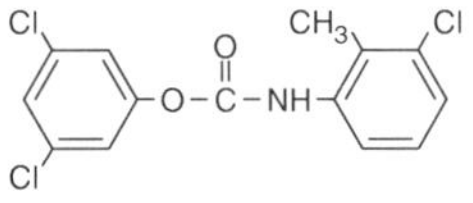

5 gm

S72,129-8
3,5-DICHLOROPHENYL N-(2,3-DIMETHYLPHENYL)CARBAMATE

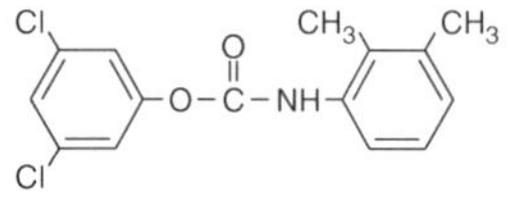

250 mg

S71,180-2
2,4,6-TRICHLOROPHENYL N-(2,6-DICHLOROPHENYL)CARBAMATE

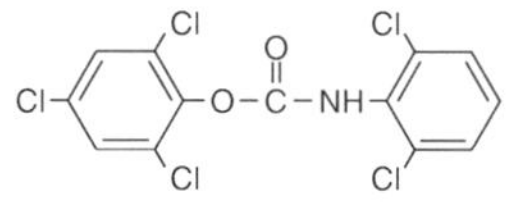

250 mg

S70,947-6
2,4,6-TRIBROMOPHENYL N-(2,6-DICHLOROPHENYL)CARBAMATE

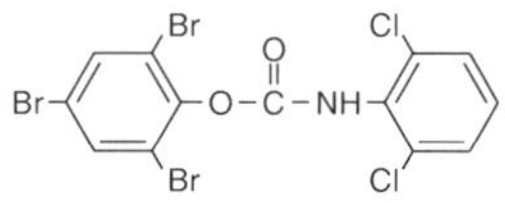

250 mg

S91,111-9
2,4-DICHLORO-6-METHYLPHENYL N-(2,3-DICHLOROPHENYL)CARBAMATE

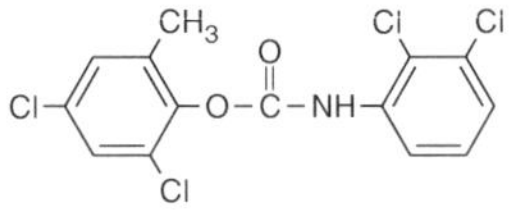

250 mg

S72,392-4
2-BROMO-4-CHLORO-6-METHYLPHENYL N-(2,3-DICHLOROPHENYL)CARBAMATE

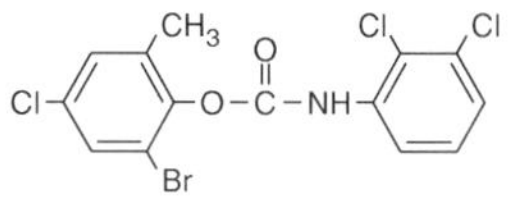

5 gm

S91,110-0
4-CHLORO-2,6-DIMETHYLPHENYL N-(2,3-DICHLOROPHENYL)CARBAMATE

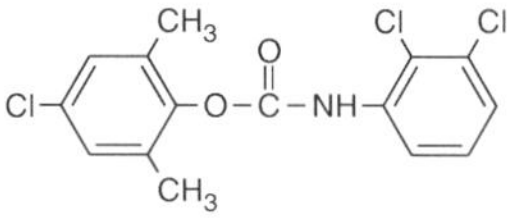

250 mg

S91,114-3
4-CHLORO-3,5-DIMETHYLPHENYL N-(2,3-DICHLOROPHENYL)CARBAMATE

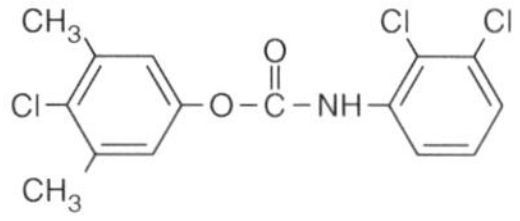

250 mg

S69,418-5
4-BROMO-3,5-DIMETHYLPHENYL N-(2,3-DICHLOROPHENYL)CARBAMATE

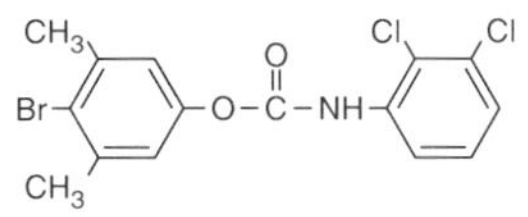

5 gm

S69,474-6
4-CHLORO-2,6-DIMETHYLPHENYL N-(3-CHLORO-2-METHYLPHENYL)-CARBAMATE

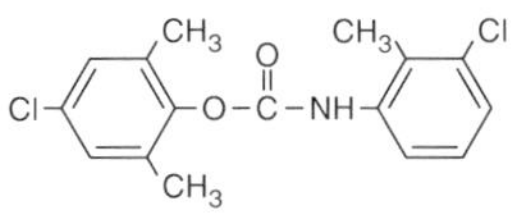

250 mg

S84,584-1
4-CHLORO-3,5-DIMETHYLPHENYL N-(2,6-XYLYL)CARBAMATE

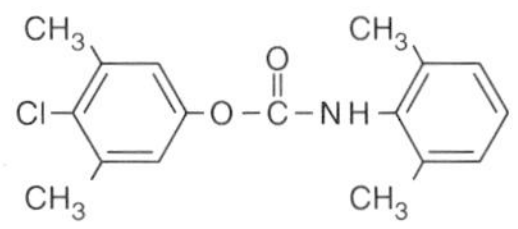

250 mg

S69,071-6
4-BROMO-3,5-DIMETHYLPHENYL N-(2,6-DIMETHYLPHENYL)CARBAMATE

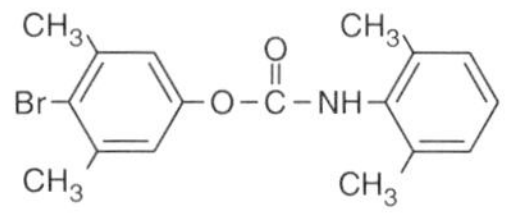

250 mg

S91,107-0
2-CHLORO-4,5-DIMETHYLPHENYL N-(2,3-DICHLOROPHENYL)CARBAMATE

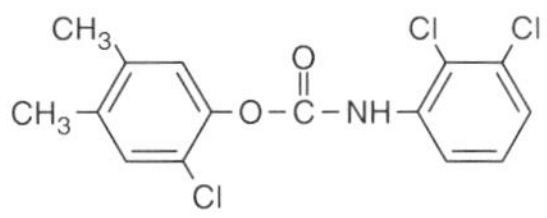

250 mg

S85,109-4
3-CHLORO-1,1'-BIPHENYL-4-YL N-(2-CHLORO-6-METHYLPHENYL)CARBAMATE

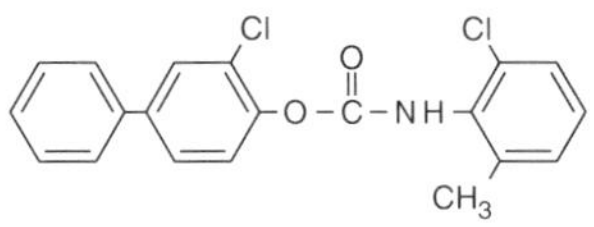

250 mg

S75,017-4
2,4-DICHLORO-1-NAPHTHYL N-(3-CHLORO-2-METHYLPHENYL)-CARBAMATE

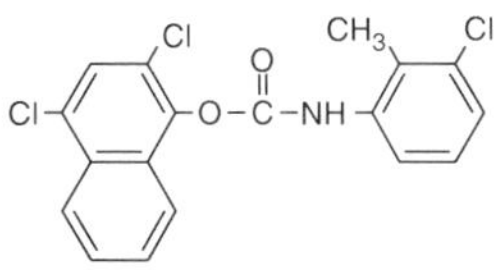

5 gm

S69,998-5
2,4-DIBROMO-8-QUINOLYL N-(3-CHLORO-2-METHYLPHENYL)CARBAMATE

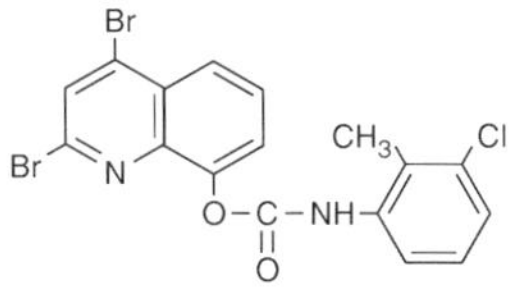

250 mg

S72,437-8
2,4-DICHLOROPHENYL N-METHYLCARBAMATE

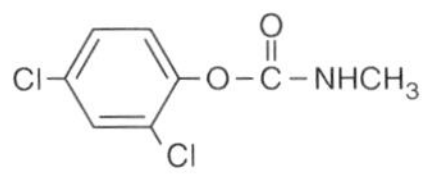

5 gm

S72,532-3
3,4-DICHLOROPHENYL N-METHYLCARBAMATE

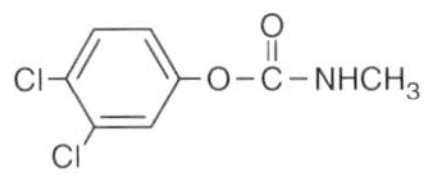

5 gm

S86,720-9
2,4-DIBROMOPHENYL N-METHYLCARBAMATE

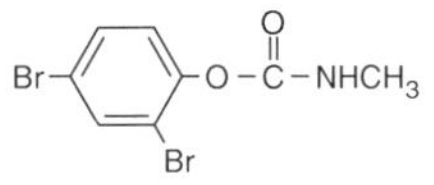

250 mg

S89,197-5
4-CHLORO-3-METHYLPHENYL N-METHYLCARBAMATE

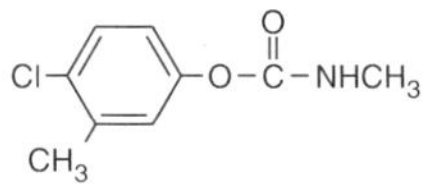

250 mg

S66,709-9
ETHYL N-(2,4-DICHLOROPHENYL)-CARBAMATE

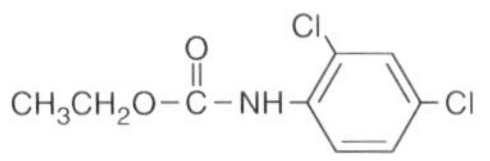

5 gm

S91,023-6
2-CHLORO-4-FLUOROPHENYL N-ISOPROPYLCARBAMATE

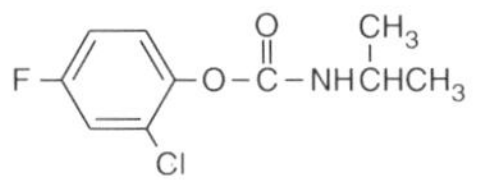

250 mg

S91,713-3
3,4-DICHLOROPHENYL N-PROPYLCARBAMATE

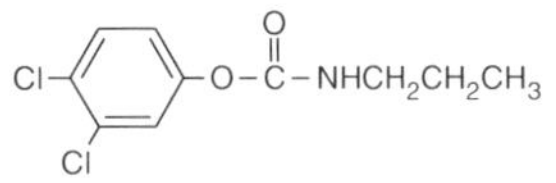

250 mg

S91,719-2
3,4-DICHLOROPHENYL N-ISOPROPYLCARBAMATE

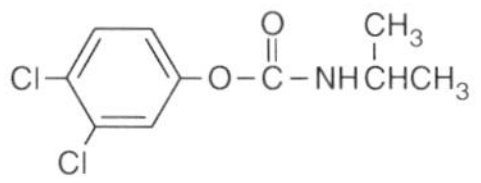

250 mg

S79,590-9
2,4-DIBROMOPHENYL N-ISOPROPYLCARBAMATE

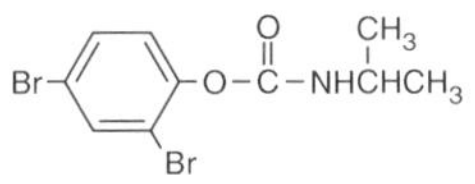

250 mg

S71,700-2
BUTYL N-(2,5-DICHLOROPHENYL)-CARBAMATE

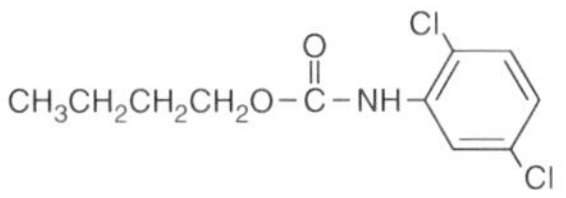

5 gm

S92,436-9
2,4-DICHLOROPHENYL N-TERT-BUTYLCARBAMATE

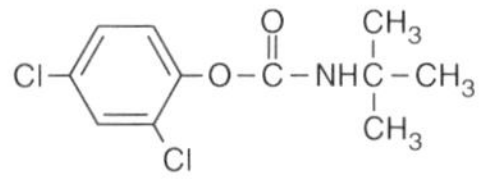

250 mg

S91,032-5
2-BROMO-4-CHLOROPHENYL N-BUTYLCARBAMATE

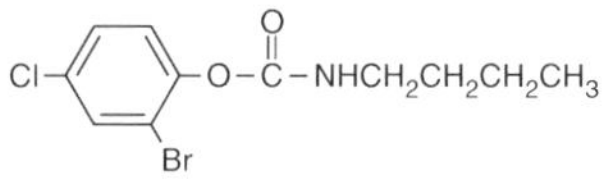

250 mg

S84,813-1
2-CHLORO-4-FLUOROPHENYL N-(3-FLUOROPHENYL)CARBAMATE

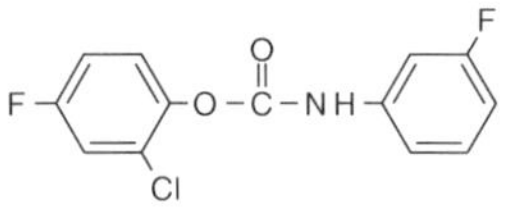

250 mg

S70,901-8
3-CHLOROPHENYL N-(3,4-DICHLOROPHENYL)CARBAMATE

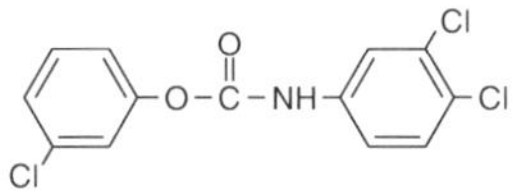

5 gm

S72,974-4
3,4-DICHLOROPHENYL N-(3-CHLOROPHENYL)CARBAMATE

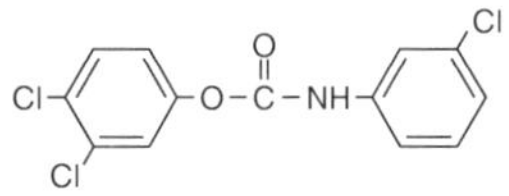

5 gm

S73,844-1
2,5-DICHLOROPHENYL N-(3-CHLOROPHENYL)CARBAMATE

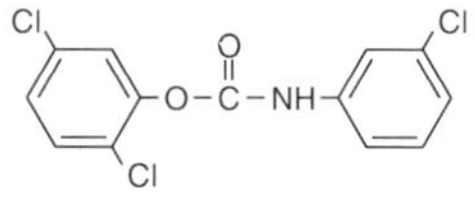

5 gm

S89,212-2
2,4-DICHLOROPHENYL N-(3-CHLOROPHENYL)CARBAMATE

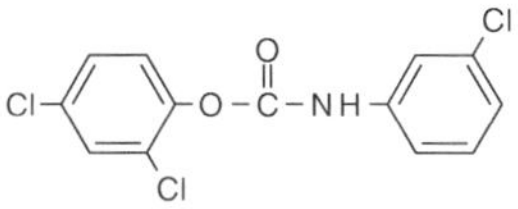

250 mg

S74,305-4
2,4-DICHLOROPHENYL N-(2,5-DIFLUOROPHENYL)CARBAMATE

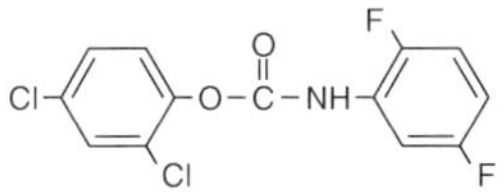

5 gm

S90,582-8
2-CHLORO-4-FLUOROPHENYL N-(3-CHLORO-4-FLUOROPHENYL)-CARBAMATE

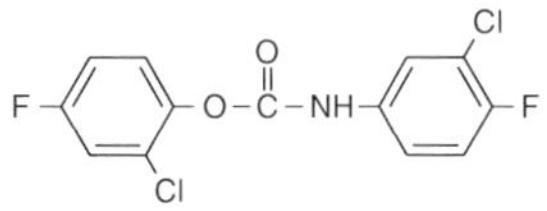

250 mg

S91,733-8
3,4-DICHLOROPHENYL N-(2,5-DIFLUOROPHENYL)CARBAMATE

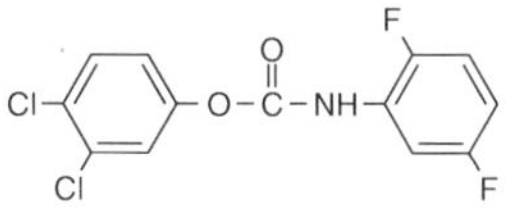

250 mg

S72,401-7
2-CHLORO-4-FLUOROPHENYL N-(3,4-DICHLOROPHENYL)CARBAMATE

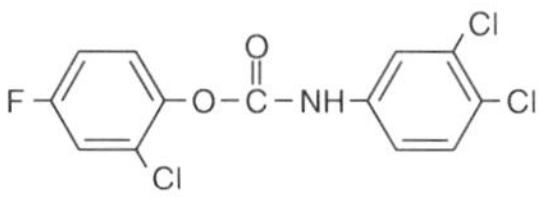

5 gm

S85,067-5
2,4-DICHLOROPHENYL N-(3-CHLORO-4-FLUOROPHENYL)CARBAMATE

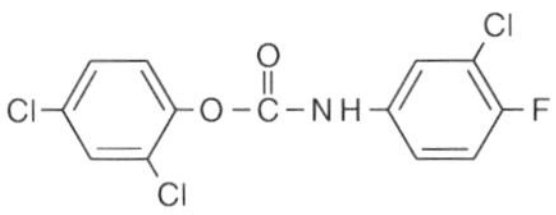

250 mg

S72,400-9
2,5-DICHLOROPHENYL N-(3,4-DICHLOROPHENYL)CARBAMATE

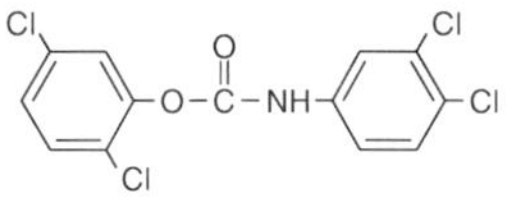

5 gm

S72,766-0
3,4-DICHLOROPHENYL N-(3,4-DICHLOROPHENYL)CARBAMATE

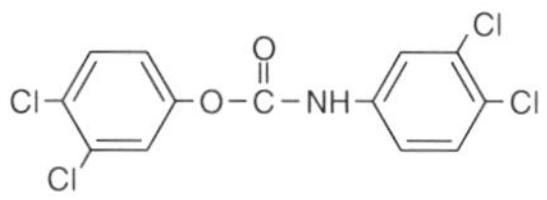

5 gm

S67,505-9
2,4,5-TRICHLOROPHENYL N-(3,4-DICHLOROPHENYL)CARBAMATE
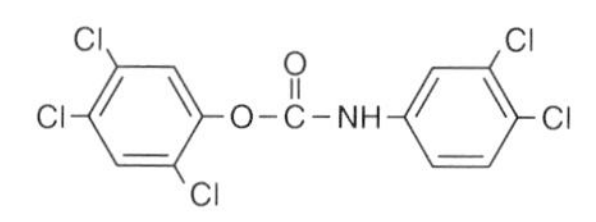
5 gm

S68,783-9
3,4-DICHLOROPHENYL N-(2,4,5-TRICHLOROPHENYL)CARBAMATE
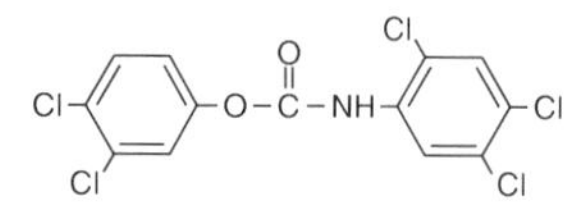
250 mg

S84,709-7
2,4-DIBROMOPHENYL N-(3-FLUOROPHENYL)CARBAMATE
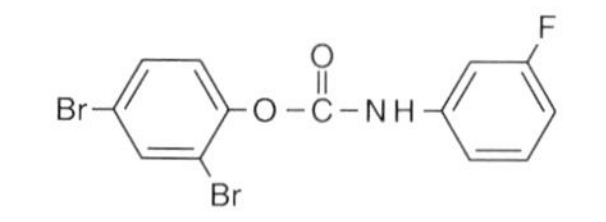
250 mg

S90,321-3
2-BROMO-4-CHLOROPHENYL N-(2,4-DICHLOROPHENYL)CARBAMATE
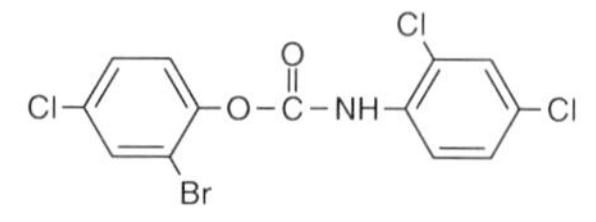
250 mg

S67,408-7
2,4-DIBROMOPHENYL N-(3-CHLOROPHENYL)CARBAMATE
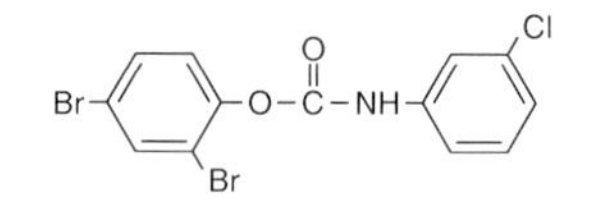
250 mg

S85,068-3
2,4-DIBROMOPHENYL N-(3-CHLORO-4-FLUOROPHENYL)CARBAMATE
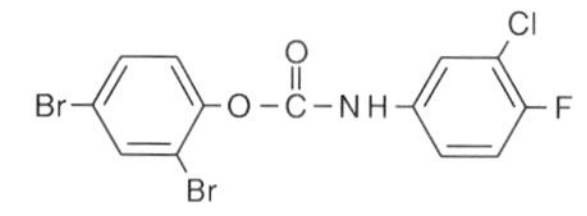
250 mg

S71,579-4
2,4-DIBROMOPHENYL N-(3,4-DICHLOROPHENYL)CARBAMATE
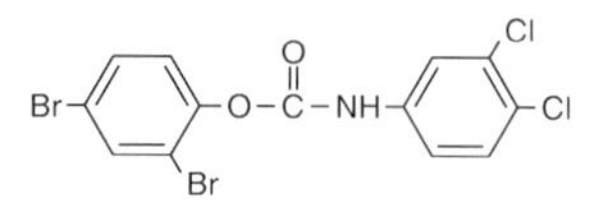
250 mg

S75,335-1
4-CHLORO-3-METHYLPHENYL N-(3-CHLOROPHENYL)CARBAMATE
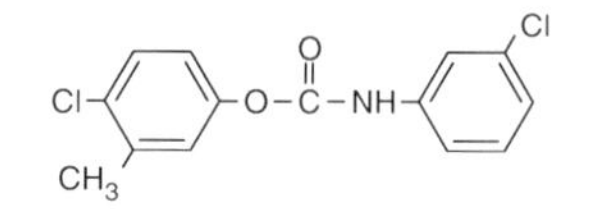
5 gm

S69,079-1
2-CHLORO-4-FLUOROPHENYL N-(4-CHLORO-2-METHYLPHENYL)-CARBAMATE
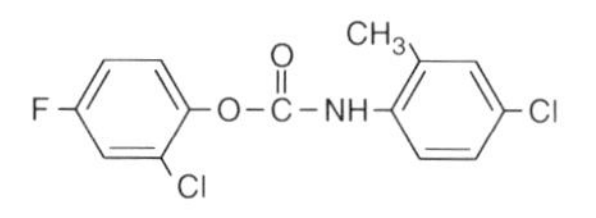
250 mg

S69,078-3
3,4-DICHLOROPHENYL N-(4-CHLORO-2-METHYLPHENYL)CARBAMATE
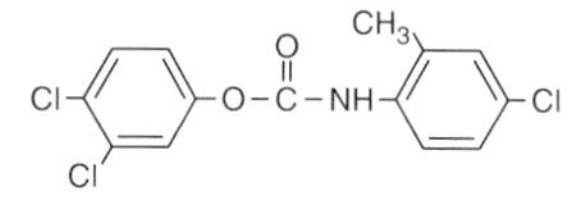
250 mg

S72,670-2
2,4-DICHLOROPHENYL N-(3-CHLORO-4-METHYLPHENYL)CARBAMATE
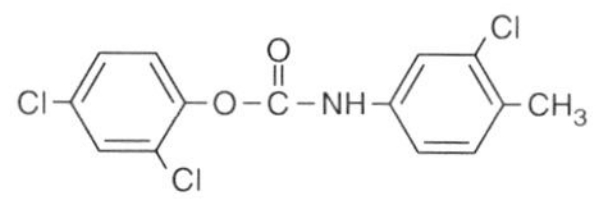
5 gm

S72,695-8
4-BROMO-2-CHLORO-5-METHYLPHENYL N-(2,4-DIFLUOROPHENYL)CARBAMATE
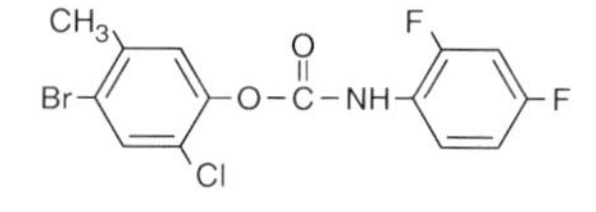
5 gm

S86,745-4
2,4-DIBROMOPHENYL N-(2-CHLORO-4-METHYLPHENYL)CARBAMATE
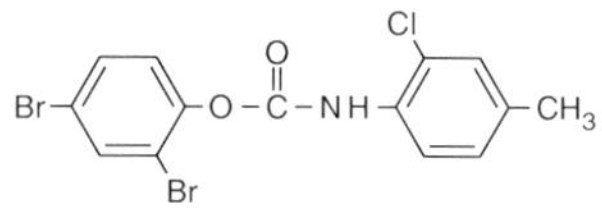
250 mg

S68,685-9
2,5-DICHLOROPHENYL N-(3-ETHYLPHENYL)CARBAMATE
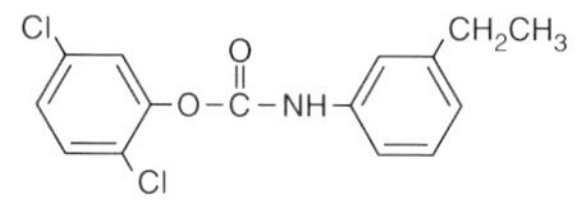
250 mg

S68,750-2
3,4-DIMETHYLPHENYL N-(2,5-DICHLOROPHENYL)CARBAMATE
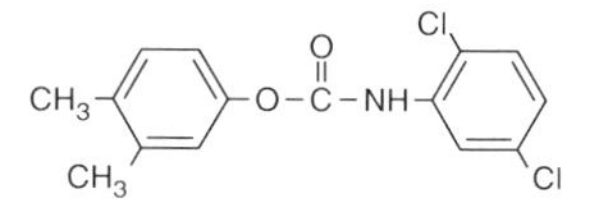
250 mg

S77,339-5
4-CHLORO-2-METHYLPHENYL N-(5-CHLORO-2-METHYLPHENYL)-CARBAMATE
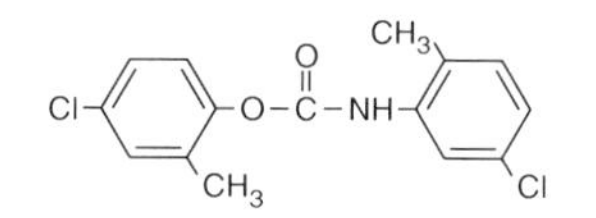
250 mg

S79,125-3
4-CHLORO-2-METHYLPHENYL N-(3-CHLORO-4-METHYLPHENYL)-CARBAMATE
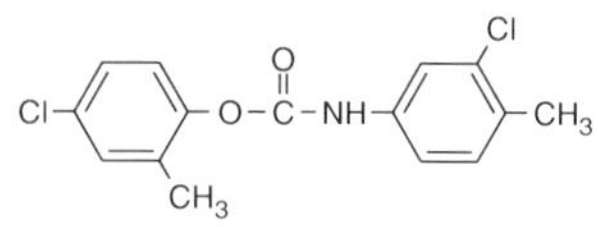
250 mg

S79,128-8
4-CHLORO-3-METHYLPHENYL N-(3-CHLORO-4-METHYLPHENYL)CARBAMATE
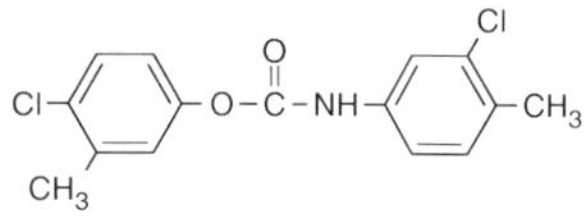
250 mg

S84,521-3
2,4-DICHLOROPHENYL N-(2,5-XYLYL)-CARBAMATE
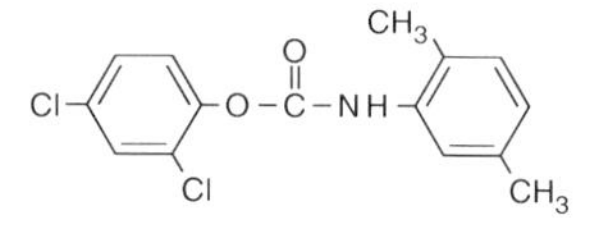
250 mg

S72,843-8
2-BROMO-4,5-DIMETHYLPHENYL N-(2,4-DIFLUOROPHENYL)CARBAMATE
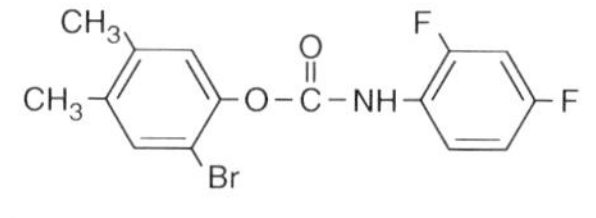
5 gm

S69,069-4
2-BROMO-4,5-DIMETHYLPHENYL N-(2,5-DICHLOROPHENYL)CARBAMATE
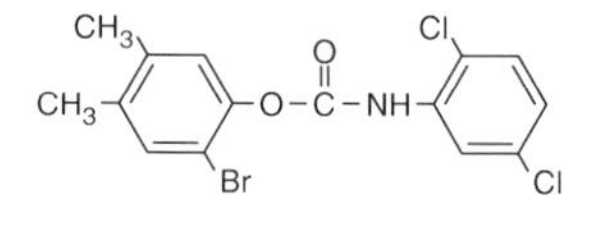
250 mg

S69,177-1
2-BROMO-4-CHLORO-5-METHYLPHENYL N-(3-CHLORO-4-METHYLPHENYL)-CARBAMATE
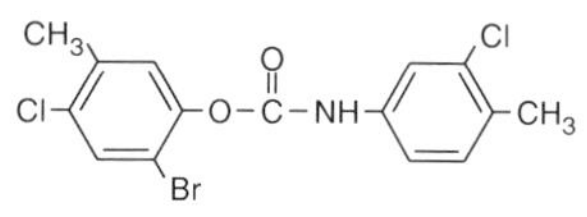
250 mg

S69,417-7
4-BROMO-2,5-DIMETHYLPHENYL N-(2,5-DICHLOROPHENYL)CARBAMATE
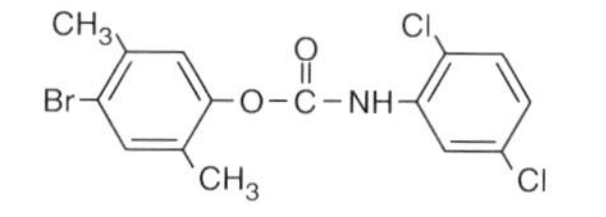
250 mg

S72,442-4
4-BROMO-2,5-DIMETHYLPHENYL N-(3,4-DICHLOROPHENYL)CARBAMATE
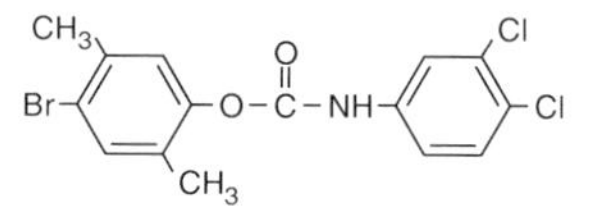
5 gm

S73,742-9
2-BROMO-4-CHLORO-5-METHYLPHENYL N-(4-CHLORO-2-METHYLPHENYL)-CARBAMATE
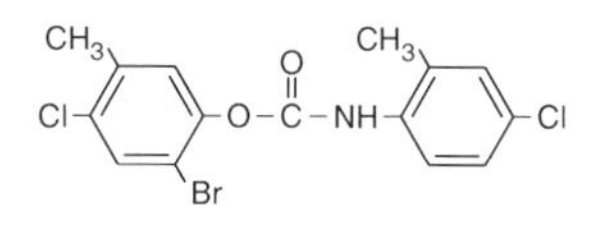
5 gm

S73,903-0
2-BROMO-4,5-DIMETHYLPHENYL N-(3,4-DICHLOROPHENYL)CARBAMATE
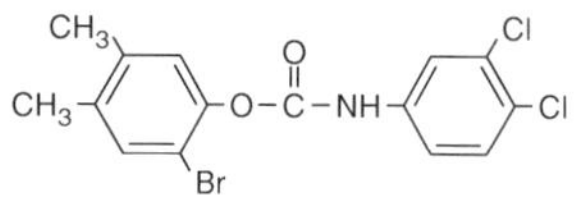
5 gm

S68,830-4
2,4-DIBROMOPHENYL N-(2,5-DIMETHYLPHENYL)CARBAMATE
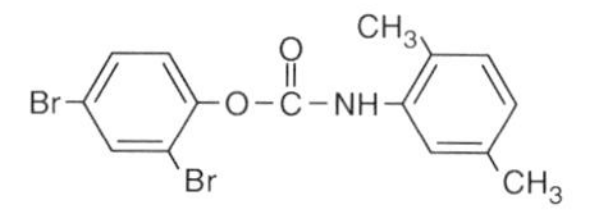
250 mg

S72,432-7
4-CHLORO-2-METHYLPHENYL N-(2,5-DIMETHYLPHENYL)CARBAMATE
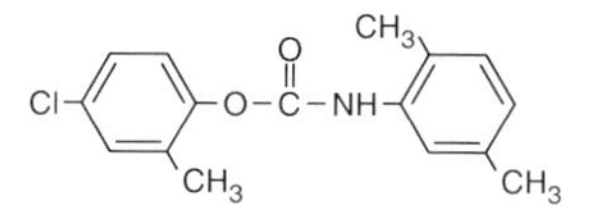
5 gm

S84,559-0
4-CHLORO-3-METHYLPHENYL N-(2,5-XYLYL)CARBAMATE
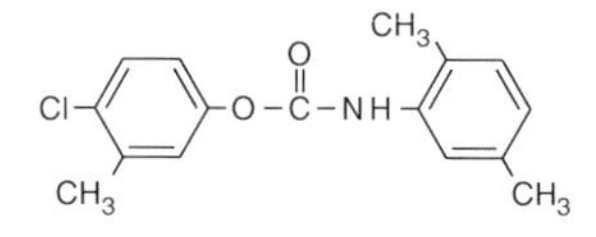
250 mg

S90,193-8
2,4-DICHLORO-3,5-DIMETHYLPHENYL N-(4-CHLORO-2-METHYLPHENYL)-CARBAMATE
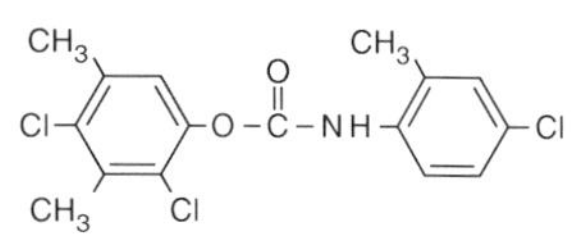
250 mg

S68,988-2
2-BROMO-4,5-DIMETHYLPHENYL N-(3-CHLORO-4-METHYLPHENYL)-CARBAMATE
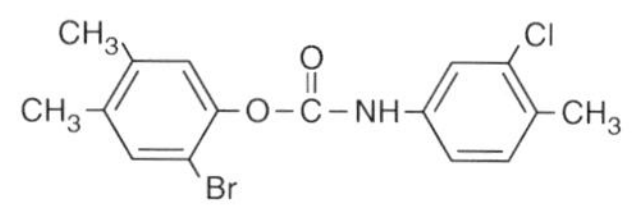
250 mg

S69,036-8
4-BROMO-2,5-DIMETHYLPHENYL N-(3-CHLORO-4-METHYLPHENYL)-CARBAMATE
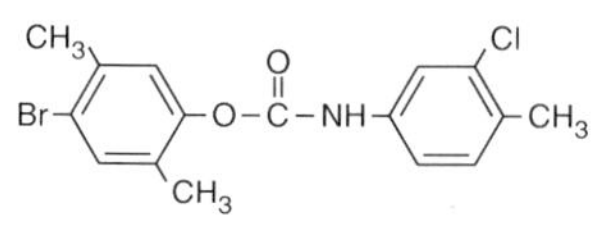
250 mg

S69,076-7
2-BROMO-4,5-DIMETHYLPHENYL N-(4-CHLORO-2-METHYLPHENYL)-CARBAMATE

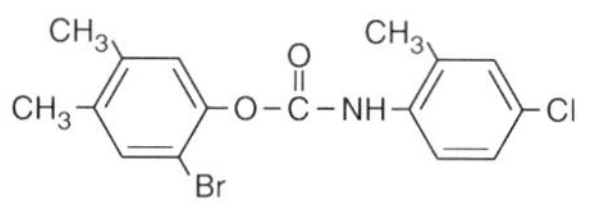

250 mg

S69,337-5
4-BROMO-2,5-DIMETHYLPHENYL N-(4-CHLORO-2-METHYLPHENYL)-CARBAMATE

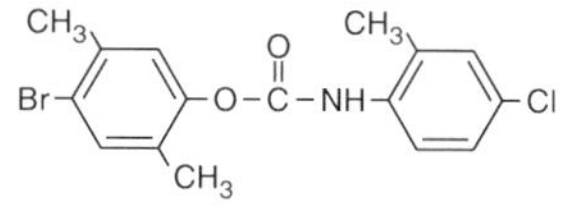

250 mg

S73,850-6
4-BROMO-2-CHLORO-5-METHYLPHENYL N-(2,4-XYLYL)CARBAMATE

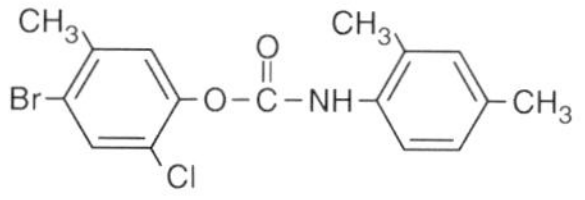

5 gm

S69,053-8
4-IODO-2,5-DIMETHYLPHENYL N-(3-CHLORO-4-METHYLPHENYL)-CARBAMATE

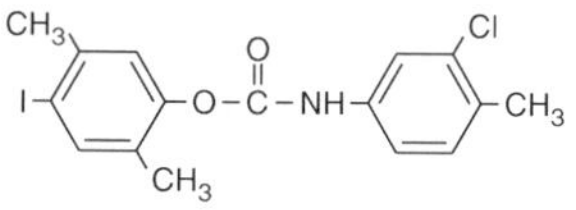

250 mg

S84,843-3
2-ISOPROPYL-5-METHYLPHENYL N-(3-FLUOROPHENYL)CARBAMATE

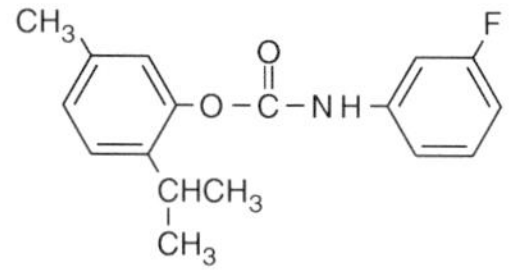

250 mg

S69,578-5
2-ISOPROPYL-5-METHYLPHENYL N-(2,4-DIFLUOROPHENYL)CARBAMATE

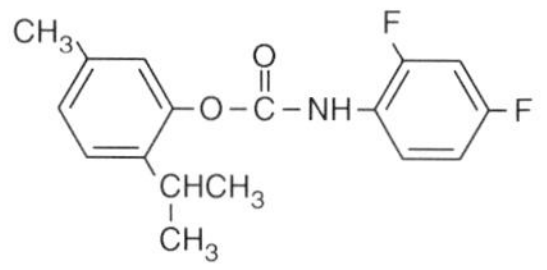

250 mg

S90,186-5
2,4-DICHLORO-3,5-DIMETHYLPHENYL N-(3,4-XYLYL)CARBAMATE

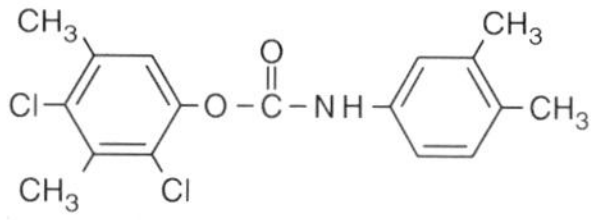

250 mg

S69,322-7
4-BROMO-2,5-DIMETHYLPHENYL N-(2,4-DIMETHYLPHENYL)CARBAMATE

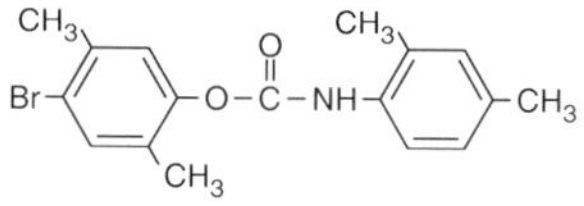

250 mg

S86,857-4
2-ISOPROPYL-5-METHYLPHENYL N-(5-CHLORO-2-METHYLPHENYL)-CARBAMATE

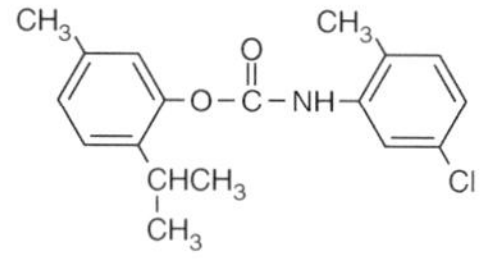

250 mg

S86,852-3
2-BROMO-4-TERT-BUTYLPHENYL N-(5-CHLORO-2-METHYLPHENYL)-CARBAMATE

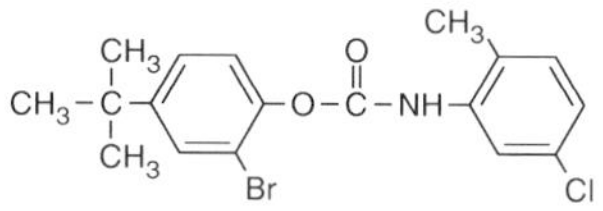

250 mg

S71,584-0
BIS(2,4-DIBROMOPHENYL) N,N'-(4,6-DIBROMO-1,3-PHENYLENE)BISCARBAMATE

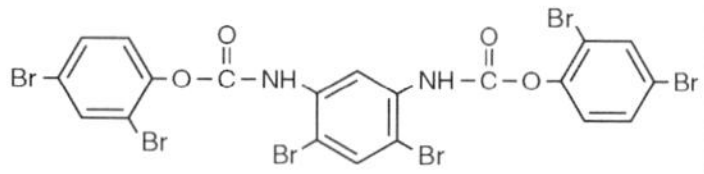

250 mg

S72,402-5
3,5-DICHLOROPHENYL N-(2,5-DIFLUOROPHENYL)CARBAMATE

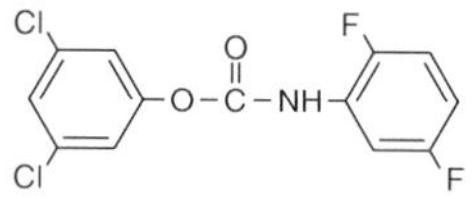

5 gm

S72,399-1
3,5-DICHLOROPHENYL N-(3,4-DICHLOROPHENYL)CARBAMATE

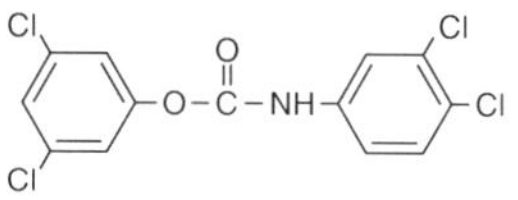

5 gm

S68,347-7
2,5-DICHLOROPHENYL N-(3,5-DIMETHYLPHENYL)CARBAMATE

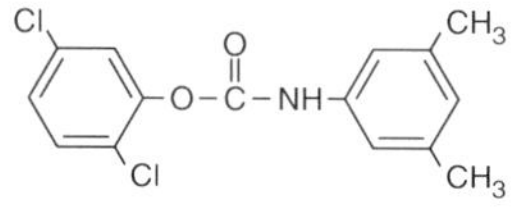

250 mg

S88,946-6
3-ETHYL-5-METHYLPHENYL N-(3,4-DICHLOROPHENYL)CARBAMATE

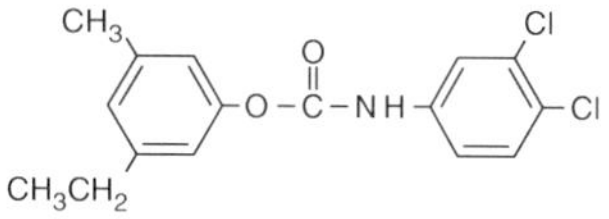

250 mg

S86,828-0
3-ETHYL-5-METHYLPHENYL N-(5-CHLORO-2-METHYLPHENYL)CARBAMATE

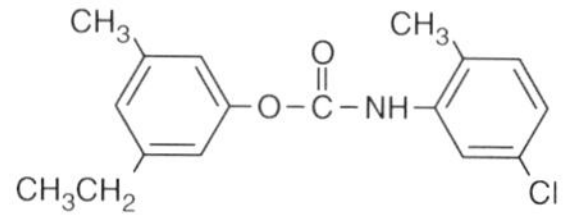

250 mg

S69,421-5
2-BROMO-4-TERT-BUTYLPHENYL N-(3,5-DIMETHYLPHENYL)CARBAMATE

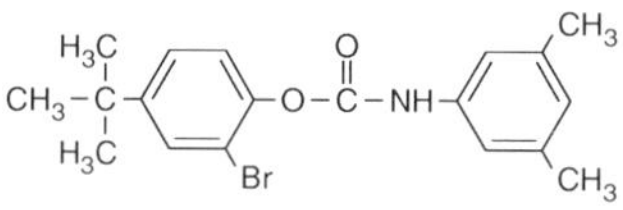

250 mg

S71,179-9
2,4,6-TRICHLOROPHENYL N-(3,4-DICHLOROPHENYL)CARBAMATE

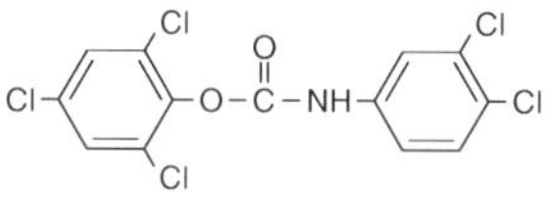

5 gm

S74,738-6
2-BROMO-4-CHLORO-6-METHYLPHENYL N-(2,4-DIFLUOROPHENYL)CARBAMATE

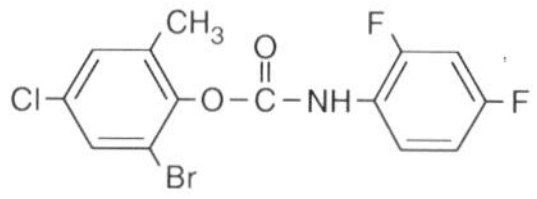

5 gm

S69,448-7
2-BROMO-4-CHLORO-6-METHYLPHENYL N-(2,5-DICHLOROPHENYL)CARBAMATE

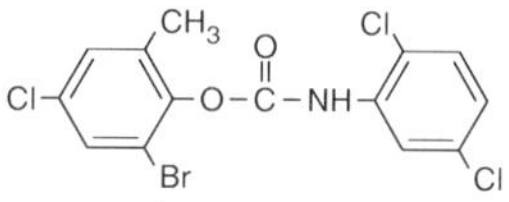

5 gm

S73,906-5
2-BROMO-4-CHLORO-6-METHYLPHENYL N-(3,4-DICHLOROPHENYL)CARBAMATE

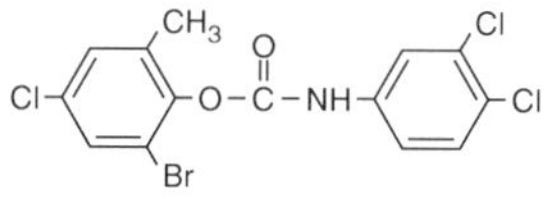

5 gm

S72,130-1
4-CHLORO-3,5-DIMETHYLPHENYL N-(2,5-DIFLUOROPHENYL)CARBAMATE

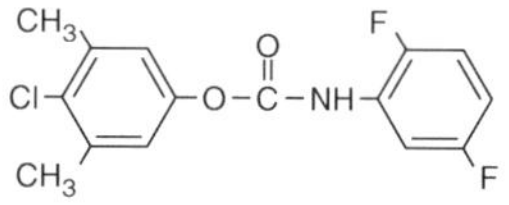

250 mg

S72,791-1
4-CHLORO-2,6-DIMETHYLPHENYL N-(2,4-DIFLUOROPHENYL)CARBAMATE

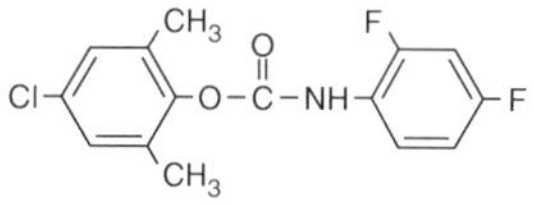

5 gm

S69,121-6
2-BROMO-4-CHLORO-6-METHYLPHENYL N-(3-CHLORO-4-METHYLPHENYL)-CARBAMATE

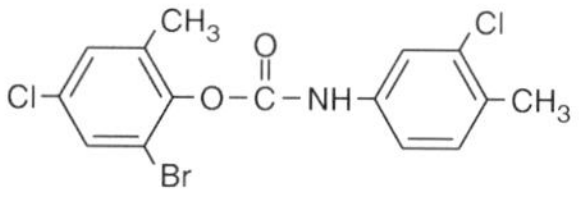

250 mg

S69,453-3
2-BROMO-4-CHLORO-6-METHYLPHENYL N-(4-CHLORO-2-METHYLPHENYL)-CARBAMATE

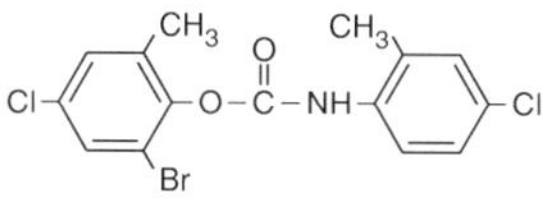

250 mg

S73,759-3
4-BROMO-3,5-DIMETHYLPHENYL N-(3,4-DICHLOROPHENYL)CARBAMATE

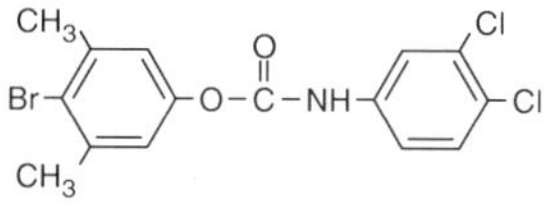

5 gm

S77,283-6
4-CHLORO-2,6-DIMETHYLPHENYL N-(5-CHLORO-2-METHYLPHENYL)-CARBAMATE

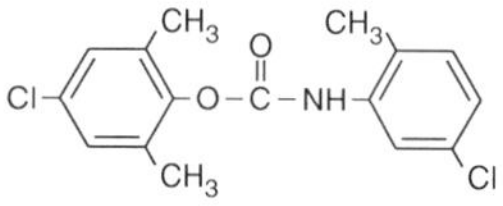

250 mg

S86,803-5
4-CHLORO-3,5-DIMETHYLPHENYL N-(5-CHLORO-2-METHYLPHENYL)-CARBAMATE

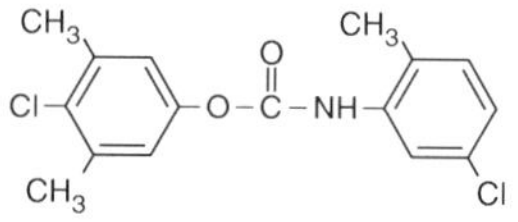

250 mg

S67,538-5
4-BROMO-3,5-DIMETHYLPHENYL N-(4-CHLORO-2-METHYLPHENYL)-CARBAMATE

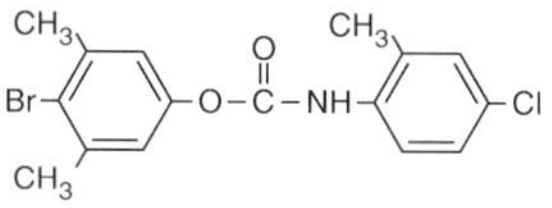

5 gm

S68,989-0
2-BROMO-4-CHLORO-6-METHYLPHENYL N-(2,4-DIMETHYLPHENYL)CARBAMATE

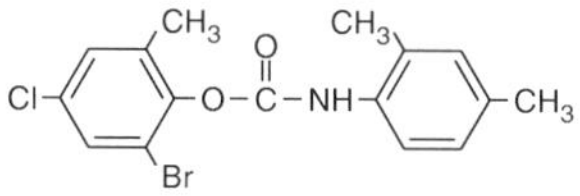

250 mg

S84,583-3
4-CHLORO-3,5-DIMETHYLPHENYL N-(2,5-XYLYL)CARBAMATE

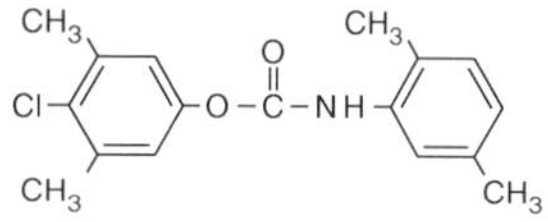

250 mg

S86,827-2
2,3,5-TRIMETHYLPHENYL N-(5-CHLORO-2-METHYLPHENYL)CARBAMATE

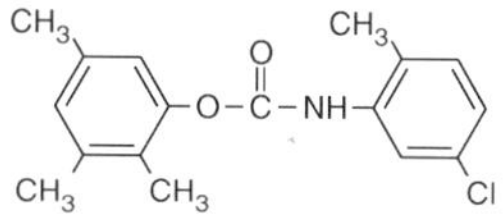

250 mg

S93,392-9
2-BROMO-4-CHLORO-5-METHYLPHENYL N-(3,4-DICHLOROPHENYL)CARBAMATE
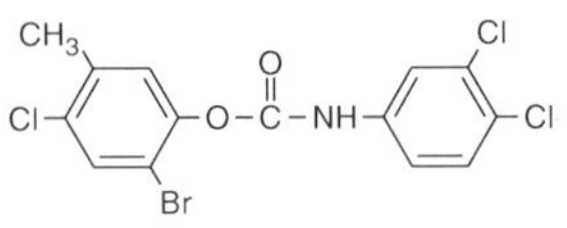
250 mg

S77,237-2
4-CHLORO-2,5-DIMETHYLPHENYL N-(2,5-DIFLUOROPHENYL)CARBAMATE
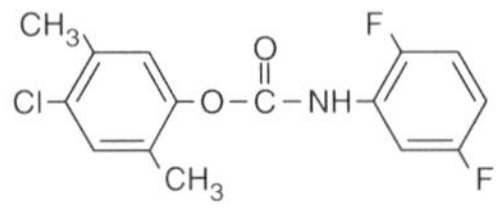
250 mg

S90,196-2
2,4,5-TRICHLOROPHENYL N-(2,4-XYLYL)-CARBAMATE
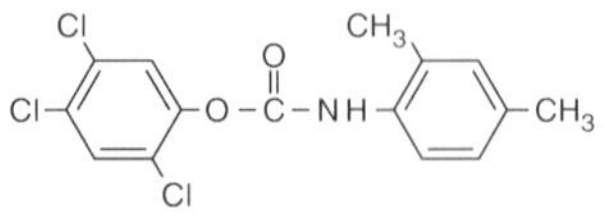
250 mg

S90,195-4
4-CHLORO-2,5-DIMETHYLPHENYL N-(2,4-XYLYL)CARBAMATE
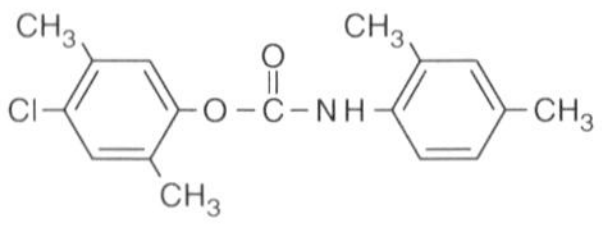
250 mg

S77,395-6
4-CHLORO-2-ISOPROPYL-5-METHYLPHENYL N-(2,5-XYLYL)CARBAMATE
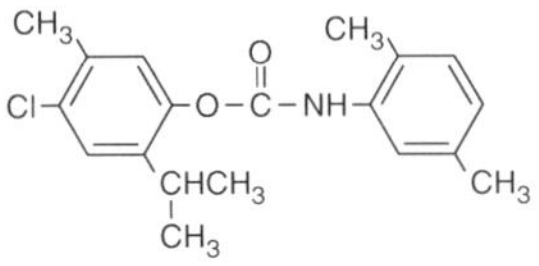
250 mg

S91,624-2
1-NAPHTHYL N-(2,5-DIFLUOROPHENYL)-CARBAMATE
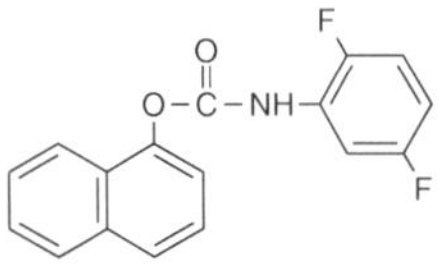
250 mg

S72,459-9
2,4-DICHLOROPHENYL N-(1-NAPHTHYL)-CARBAMATE
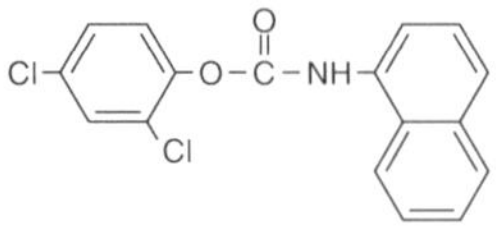
5 gm

S86,815-9
2,4-DIBROMOPHENYL N-(1-NAPHTHYL)-CARBAMATE
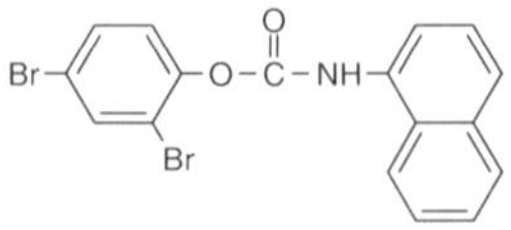
250 mg

S72,454-8
4-CHLORO-3-METHYLPHENYL N-(1-NAPHTHYL)CARBAMATE
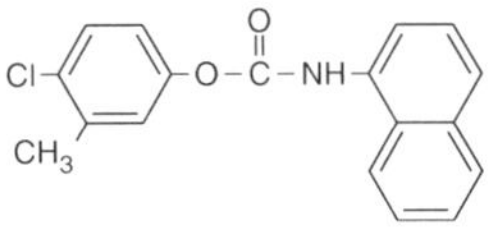
5 gm

S75,261-4
4-CHLORO-2-METHYLPHENYL N-(1-NAPHTHYL)CARBAMATE
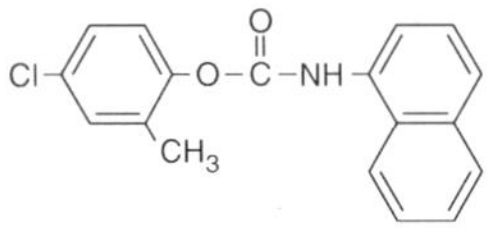
5 gm

S89,876-7
2-NAPHTHYL N-(3-CHLORO-4-FLUOROPHENYL)CARBAMATE
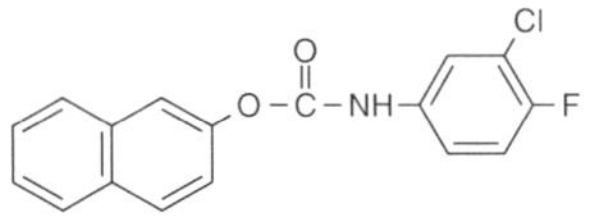
250 mg

S68,637-9
6-BROMO-2-NAPHTHYL N-(2,4-DIFLUOROPHENYL)CARBAMATE
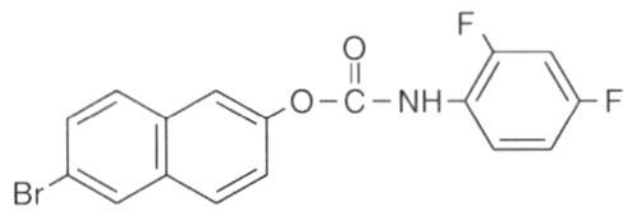
250 mg

S72,286-3
6-BROMO-2-NAPHTHYL N-(2,5-DIFLUOROPHENYL)CARBAMATE
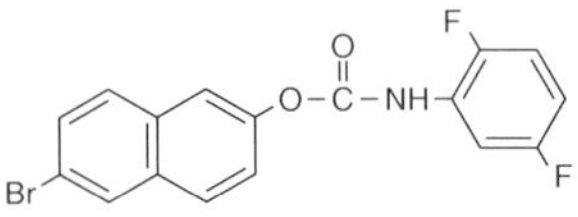
250 mg

S67,559-8
6-BROMO-2-NAPHTHYL N-(3,4-DICHLOROPHENYL)CARBAMATE
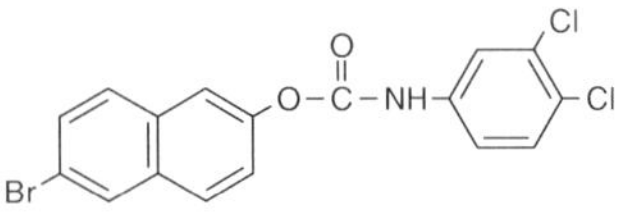
5 gm

S70,025-8
2-NAPHTHYL N-(2,4-DIFLUOROPHENYL)-CARBAMATE
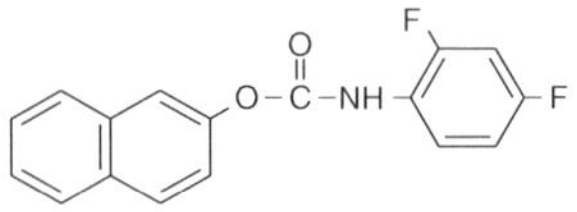
250 mg

S70,985-9
PROPYL N-(3,5-DICHLOROPHENYL)-CARBAMATE
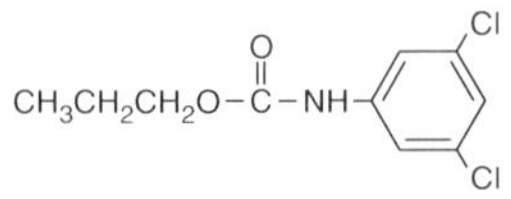
250 mg

S72,168-9
3,5-DICHLOROPHENYL N-ISOPROPYLCARBAMATE
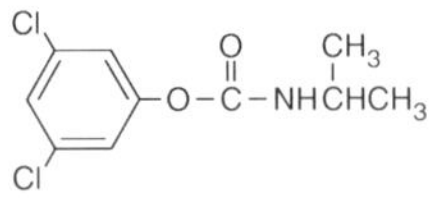
250 mg

S70,937-9
BUTYL N-(3,5-DICHLOROPHENYL)-CARBAMATE
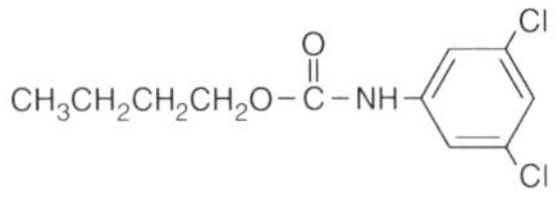
250 mg

S91,725-7
3,5-DICHLOROPHENYL N-BUTYLCARBAMATE
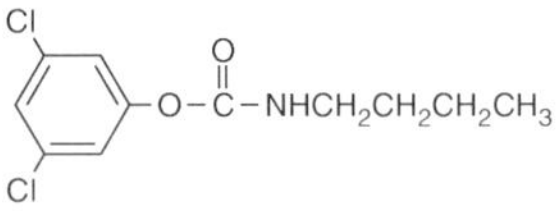
250 mg

S72,796-2
3,5-DICHLOROPHENYL N-(ETHOXYCARBONYLMETHYL)CARBAMATE
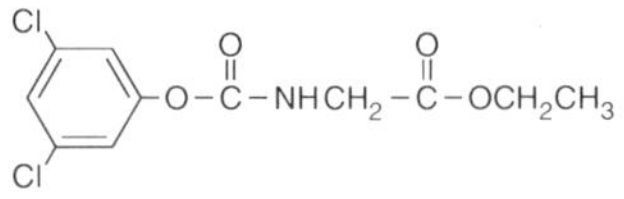
5 gm

S68,621-2
2,4,5-TRICHLOROPHENYL N-(3,5-DIMETHYLPHENYL)CARBAMATE
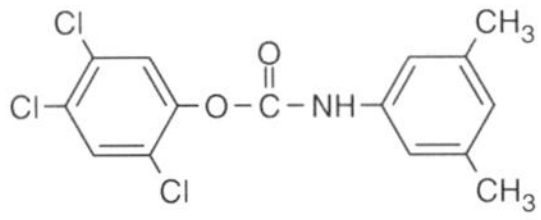
250 mg

S79,145-8
3-ETHYL-5-METHYLPHENYL N-(3-CHLOROPHENYL)CARBAMATE
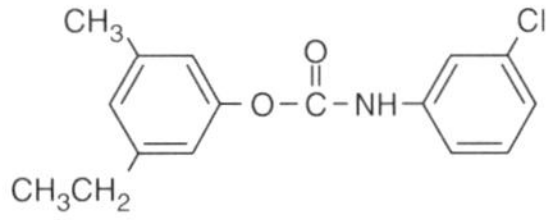
250 mg

S68,688-3
2-BROMO-4-CHLORO-6-METHYLPHENYL N-(3,5-DIMETHYLPHENYL)CARBAMATE
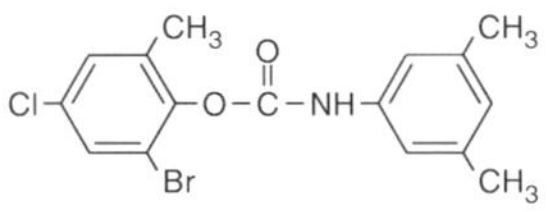
250 mg

S78,077-4
2-BROMO-4-CHLORO-5-METHYLPHENYL N-(3,5-DICHLOROPHENYL)CARBAMATE
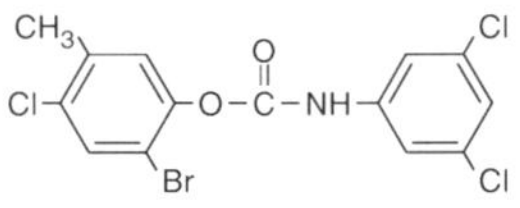
250 mg

S77,246-1
2-CHLORO-6-ISOPROPYL-3-METHYLPHENYL N-PROPYLCARBAMATE
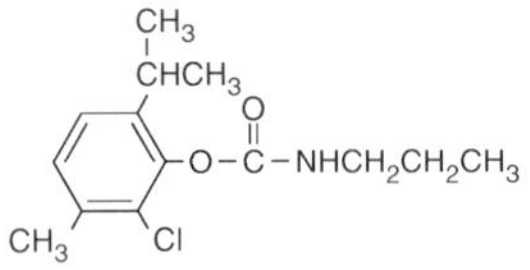
250 mg

S72,452-1
2-BROMO-4,6-DICHLOROPHENYL N-METHYLCARBAMATE
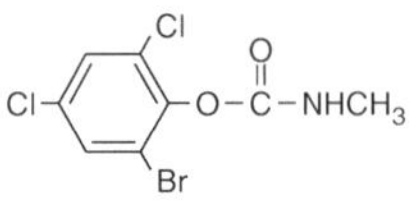
5 gm

S68,811-8
2-BROMO-4-CHLORO-6-METHYLPHENYL N-METHYLCARBAMATE
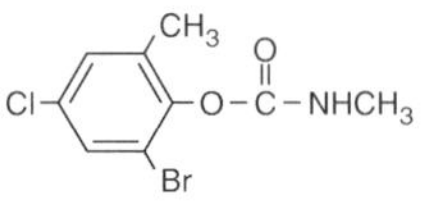
250 mg

S69,395-2
2-BROMO-4-CHLORO-6-METHYLPHENYL N-ETHYLCARBAMATE
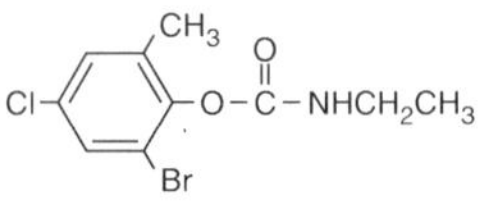
250 mg

S68,944-0
2-BROMO-4-CHLORO-6-METHYLPHENYL N-(3-METHYLPHENYL)CARBAMATE
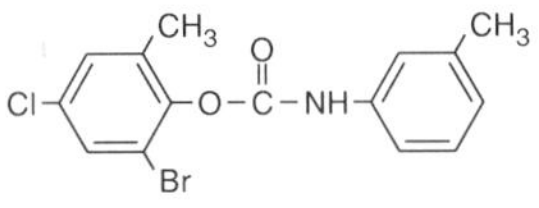
250 mg

S73,675-9
4-CHLORO-3,5-DIMETHYLPHENYL N-(3-BROMOPHENYL)CARBAMATE
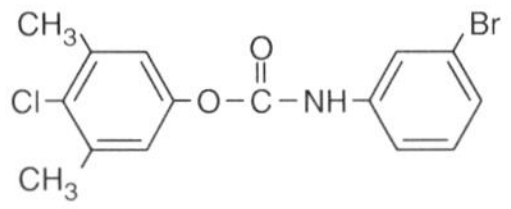
5 gm

S72,371-1
4-BROMO-3,5-DIMETHYLPHENYL N-(2,4,5-TRICHLOROPHENYL)CARBAMATE
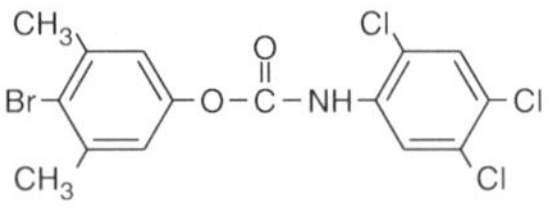
5 gm

S80,786-9
4-CHLORO-3,5-DIMETHYLPHENYL N-(M-TOLYL)CARBAMATE
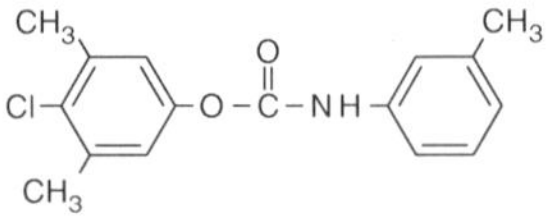
250 mg

S68,769-3
2-BROMO-4,5-DIMETHYLPHENYL N-ETHYLCARBAMATE
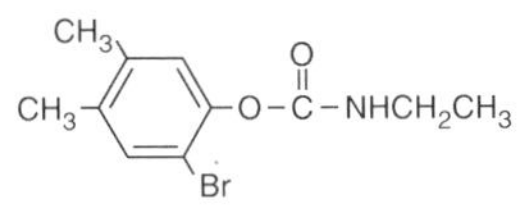
250 mg

S68,804-5
4-BROMO-2,5-DIMETHYLPHENYL N-PROPYLCARBAMATE
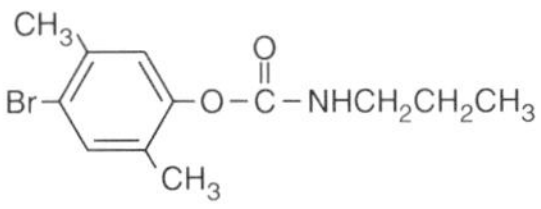
250 mg

S71,718-5
2,4,5-TRICHLOROPHENYL N-(2,4,5-TRICHLOROPHENYL)CARBAMATE
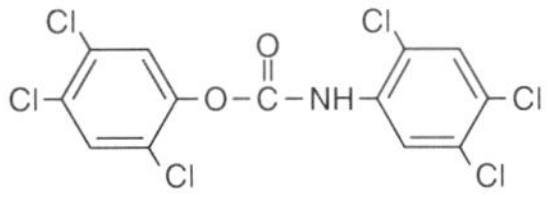
5 gm

S72,370-3
2-BROMO-4,5-DIMETHYLPHENYL N-(2,4,5-TRICHLOROPHENYL)CARBAMATE
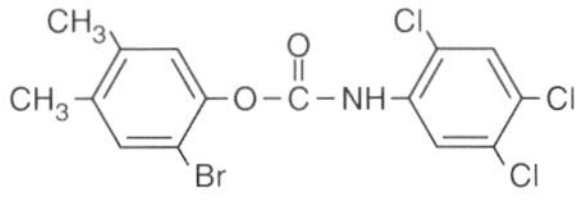
5 gm

S77,389-1
4-BROMO-2,5-DIMETHYLPHENYL N-(M-TOLYL)CARBAMATE
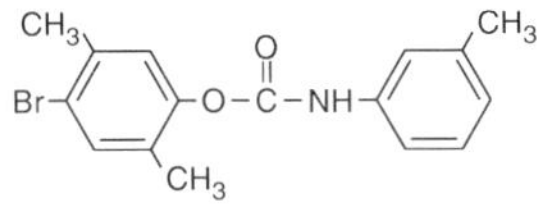
250 mg

S79,527-5
2-NAPHTHYL N-(3-CHLOROPHENYL)-CARBAMATE
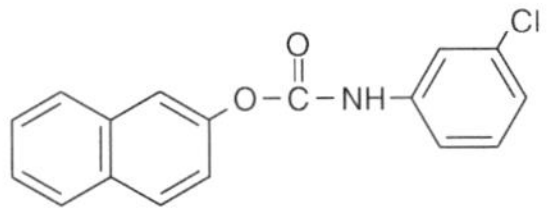
250 mg

S69,947-0
6-BROMO-2-NAPHTHYL N-(3-FLUOROPHENYL)CARBAMATE
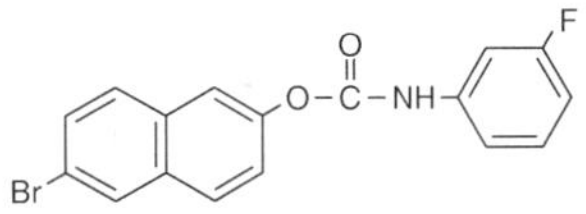
5 gm

S71,005-9
O-[N-(4-CHLORO-2-METHYLPHENYL)-CARBAMOYL]CYCLOHEXANONE OXIME
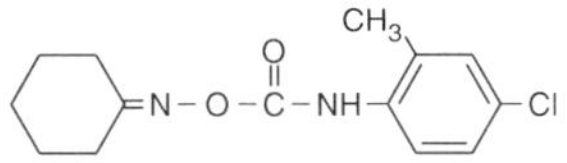
5 gm

S67,363-3
PIPERIDINO N-(4-BROMOPHENYL)-CARBAMATE
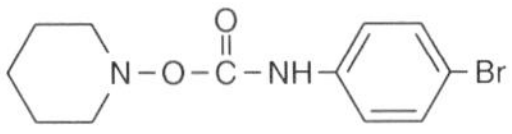
5 gm

S72,671-0
S-2-THIAZOLIN-2-YL N-(3-CHLOROPHENYL)THIOCARBAMATE
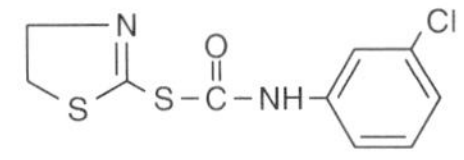
5 gm

S80,199-2
S-BENZYL N-(3-FLUOROPHENYL)-THIOCARBAMATE
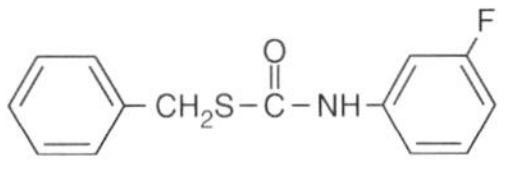
250 mg

S73,790-9
S-BENZYL N-(2-FLUOROPHENYL)-THIOCARBAMATE
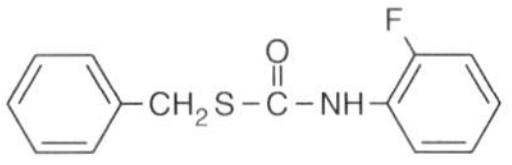
5 gm

S85,128-0
S-BENZYL N-(2-CHLORO-6-METHYLPHENYL)THIOCARBAMATE
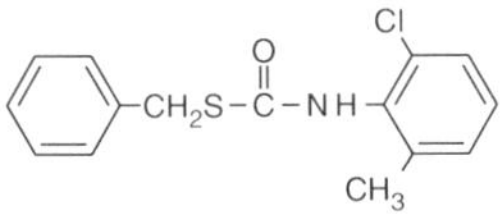
250 mg

S85,147-7
S-(O-TOLYL) N-(2,3-DICHLOROPHENYL)-THIOCARBAMATE
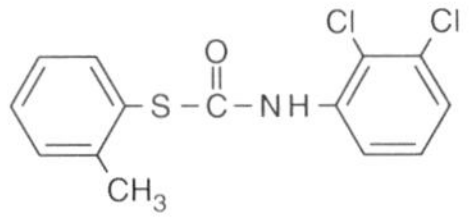
250 mg

S79,643-3
S-O-TOLYL N-(2-CHLORO-6-METHYLPHENYL)THIOCARBAMATE
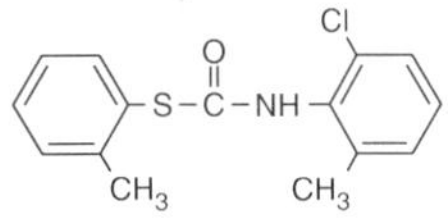
250 mg

S90,361-2
S-(4-CHLOROPHENYL) N-METHYLTHIOCARBAMATE
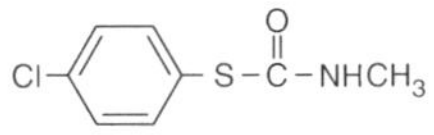
250 mg

S84,406-3
S-(4-CHLOROPHENYL) N-(PROPYL)-THIOCARBAMATE
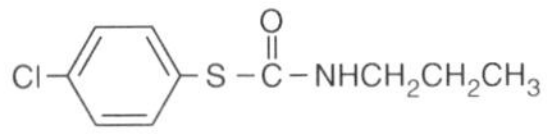
250 mg

S74,310-0
S-(4-CHLOROPHENYL) N-TERT-BUTYLTHIOCARBAMATE
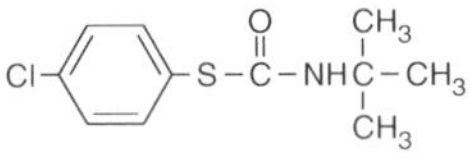
5 gm

S71,854-8
S-(4-CHLOROPHENYL) N-(3-CHLOROPHENYL)THIOCARBAMATE
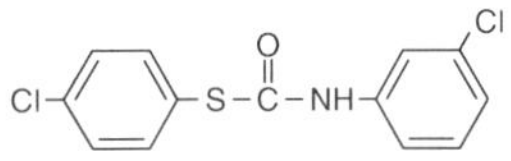
250 mg

S84,586-8
S-(4-TERT-BUTYLPHENYL) N-(4-FLUOROPHENYL)THIOCARBAMATE
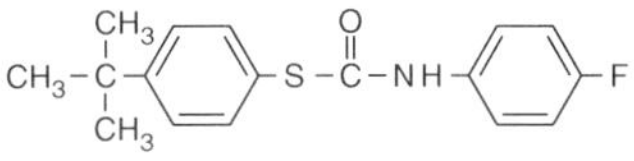
250 mg

S84,753-4
S-(4-TERT-BUTYLPHENYL) N-(3-CHLORO-2-METHYLPHENYL)THIOCARBAMATE
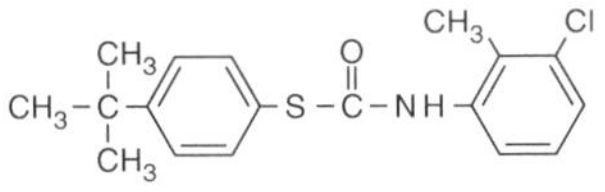
250 mg

S72,397-5
S-(4-CHLOROPHENYL) N-(2,5-DIFLUOROPHENYL)THIOCARBAMATE
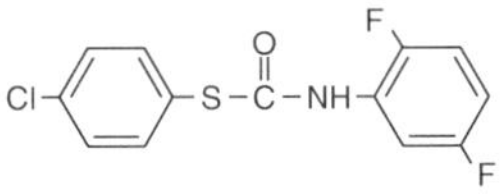
5 gm

S73,875-1
S-(4-CHLOROPHENYL) N-(3,4-DICHLOROPHENYL)THIOCARBAMATE
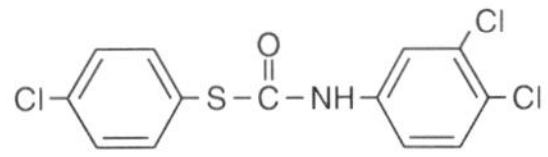
5 gm

S90,326-4
S-(4-CHLOROPHENYL) N-(2,4-DICHLOROPHENYL)THIOCARBAMATE
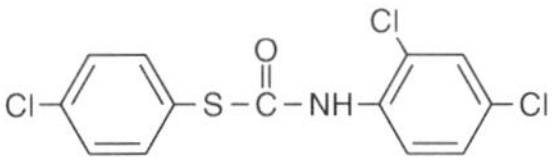
250 mg

S69,353-7
S-(4-CHLOROPHENYL) N-(4-CHLORO-2-METHYLPHENYL)THIOCARBAMATE
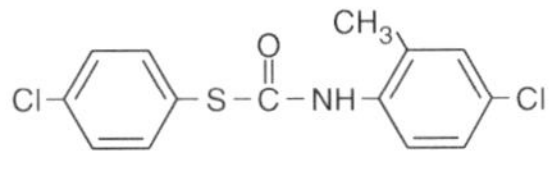
250 mg

S77,269-0
S-(4-TERT-BUTYLPHENYL) N-(3,4-DICHLOROPHENYL)THIOCARBAMATE
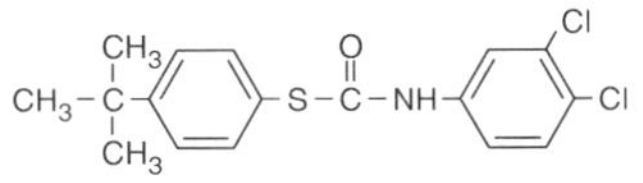
250 mg

S70,989-1
S-PROPYL N-(2,6-DICHLOROPHENYL)-THIOLCARBAMATE
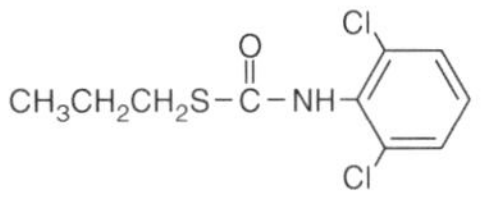
5 gm

S71,765-7
S-PROPYL N-(2,3-DICHLOROPHENYL)-THIOLCARBAMATE
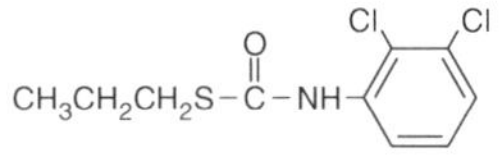
5 gm

S71,117-9
S-ISOPROPYL N-(2-CHLORO-6-METHYLPHENYL)THIOLCARBAMATE
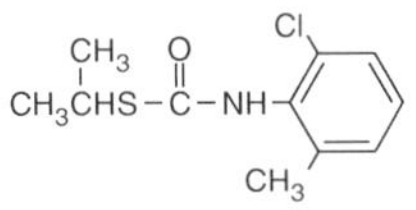
5 gm

S71,120-9
S-PROPYL N-(2-CHLORO-6-METHYLPHENYL)THIOLCARBAMATE
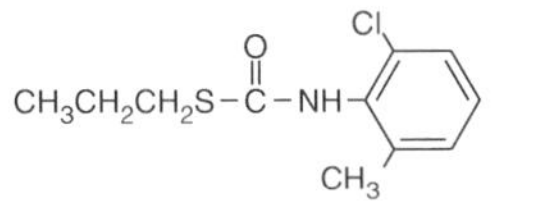
5 gm

S71,003-2
S-BUTYL N-(2,6-DICHLOROPHENYL)-THIOCARBAMATE
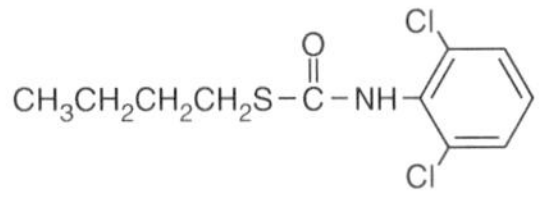
5 gm

S84,526-4
S-(M-TOLYL) N-(3-CHLORO-2-METHYLPHENYL)THIOCARBAMATE
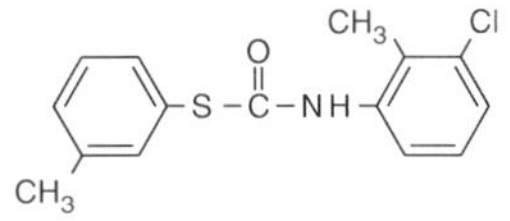
250 mg

S85,127-2
S-(M-TOLYL) N-(2-CHLORO-6-METHYLPHENYL)THIOCARBAMATE

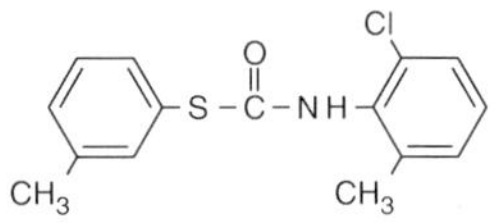

250 mg

S85,141-8
S,S'-ETHYLENE BIS[N-(2,3-DICHLOROPHENYL)THIOCARBAMATE]

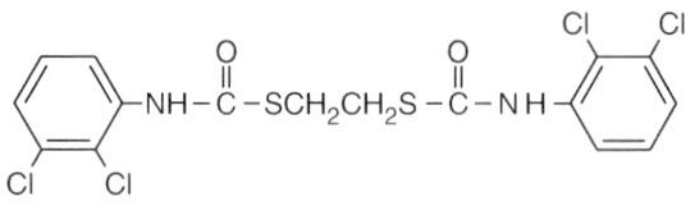

250 mg

S70,935-2
S-PROPYL N-(2,4,5-TRICHLOROPHENYL)-THIOLCARBAMATE

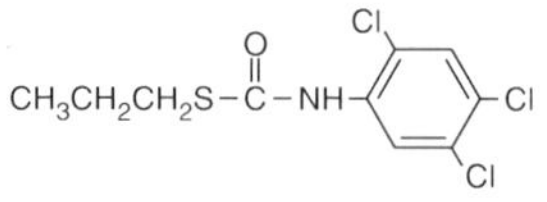

5 gm

S70,992-1
S-ISOPROPYL N-(2,4,5-TRICHLOROPHENYL)THIOLCARBAMATE

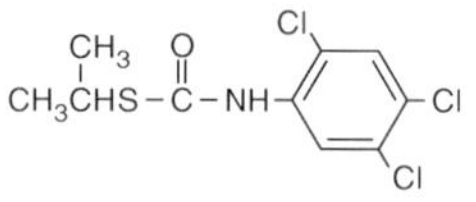

5 gm

S70,938-7
S-BUTYL N-(2,4,5-TRICHLOROPHENYL)-THIOCARBAMATE

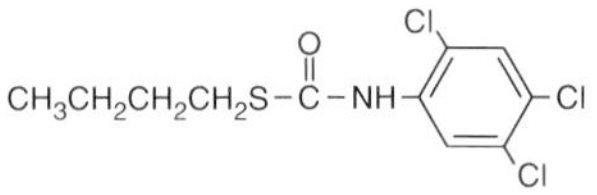

5 gm

S77,757-9
CARBOBENZYLOXY-L-SERYL-L-SERINE METHYL ESTER

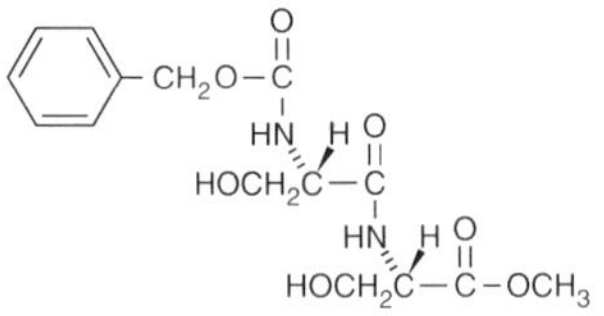

250 mg

S77,694-7
CARBOBENZYLOXY-L-THREONYL-L-SERINE METHYL ESTER

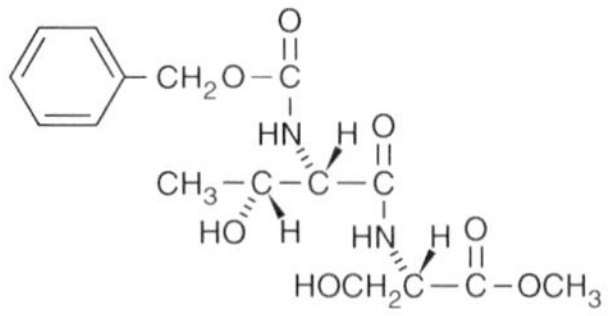

250 mg

S78,985-2
CARBOBENZYLOXY-L-THREONYL-L-TYROSINE METHYL ESTER

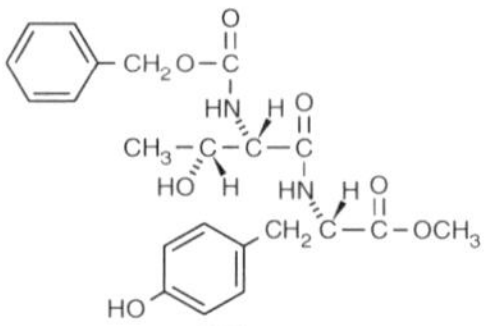

250 mg

S78,827-9
CARBOBENZYLOXY-L-SERYL-L-TYROSINE METHYL ESTER

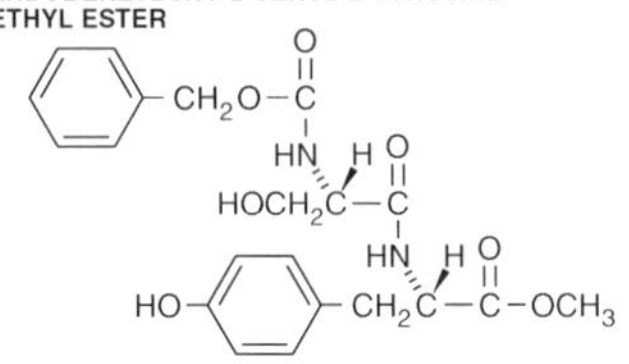

250 mg

S78,478-8
CARBOBENZYLOXY-L-TYROSYL-L-THREONINAMIDE

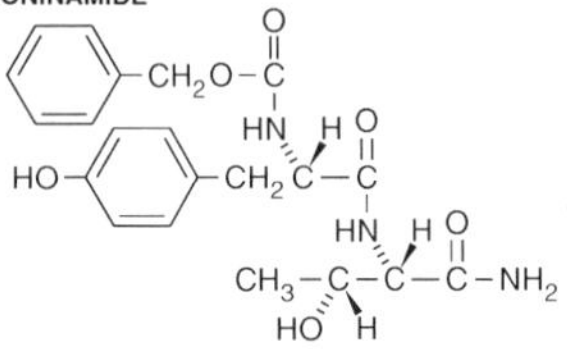

250 mg

S78,651-9
CARBOBENZYLOXY-L-SERYL-L-TYROSINE

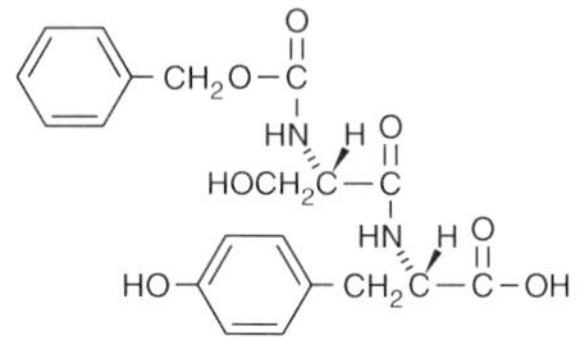

250 mg

S70,161-0
S-(4-HYDROXYPHENYL) N-(2,5-DIMETHOXYPHENYL)THIOCARBAMATE

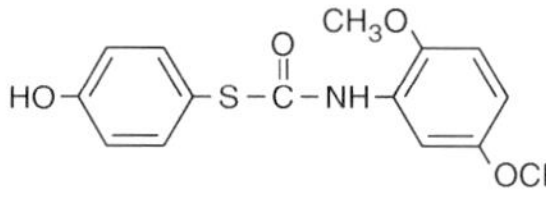

250 mg

S79,648-4
S-(2-HYDROXYPHENYL) N-(2-METHOXY-5-METHYLPHENYL)THIOCARBAMATE

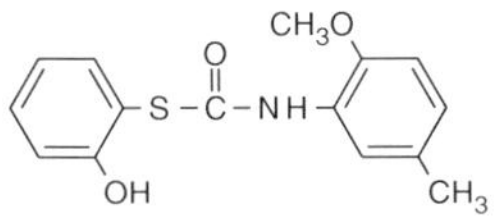

250 mg

S76,921-5
S-(4-HYDROXYPHENYL) N-(2-METHOXY-5-METHYLPHENYL)THIOCARBAMATE

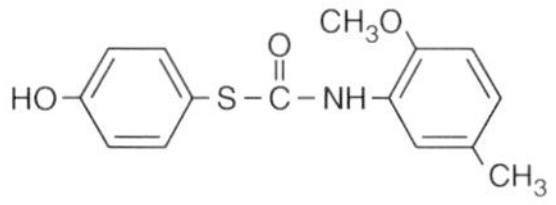

250 mg

S84,743-7
2-HYDROXYMETHYL-6-METHYL-4-(METHYLTHIO)PHENYL N-(2,6-XYLYL)-CARBAMATE

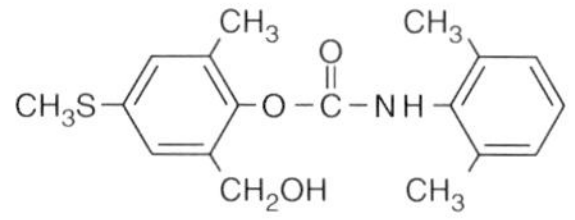

250 mg

S78,802-3
CARBOBENZYLOXY-L-GLYCYL-L-THREONINE METHYL ESTER

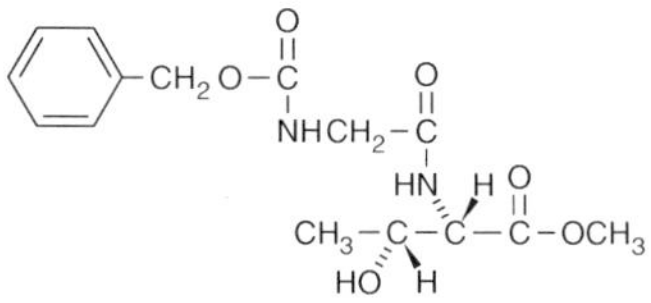

250 mg

S78,806-6
CARBOBENZYLOXY-L-ALANYL-L-THREONINE METHYL ESTER

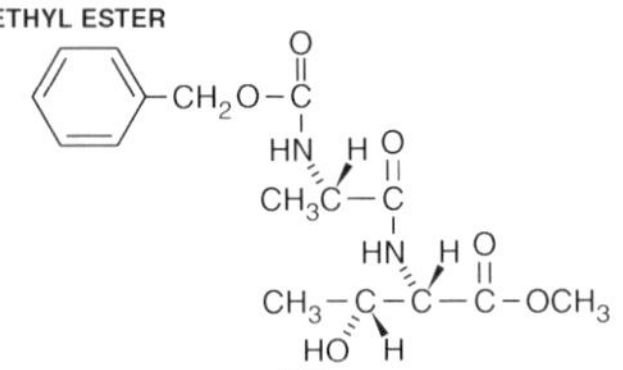

250 mg

S78,817-1
CARBOBENZYLOXY-L-ISOLEUCYL-L-SERINE METHYL ESTER

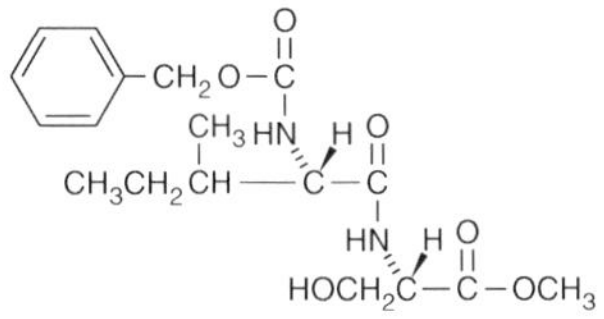

250 mg

S78,496-6
CARBOBENZYLOXY-S-BENZYL-L-CYSTEINYL-L-TYROSINAMIDE

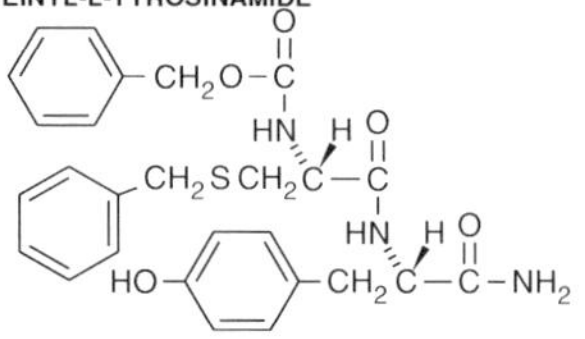

250 mg

S78,654-3
CARBOBENZYLOXY-L-SERYL-L-METHIONINE METHYL ESTER

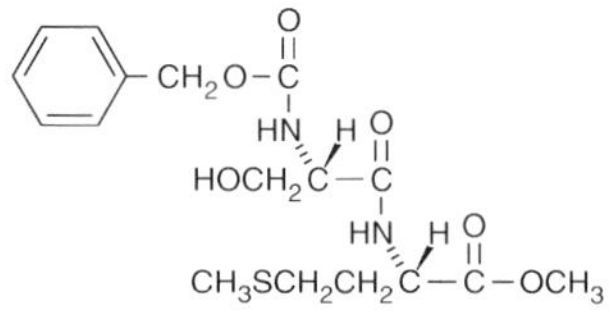

250 mg

S77,785-4
CARBOBENZYLOXY-L-METHIONYL-L-THREONINAMIDE

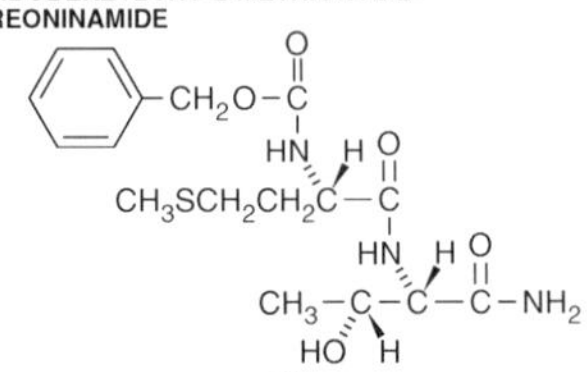

250 mg

S77,678-5
CARBOBENZYLOXY-L-METHIONYL-L-THREONINE

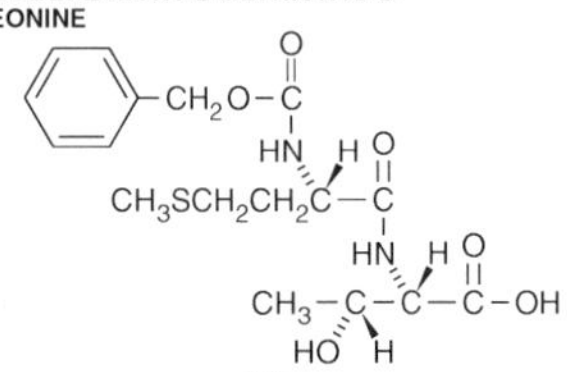

250 mg

S79,020-6
CARBOBENZYLOXY-L-METHIONYL-L-THREONINE METHYL ESTER

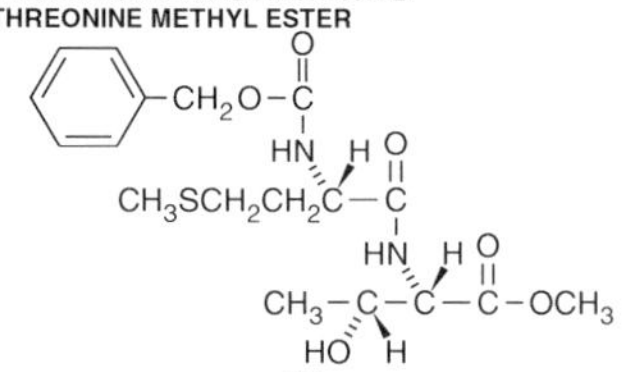

250 mg

S78,481-8
CARBOBENZYLOXY-L-THREONYL-L-PHENYLALANINE

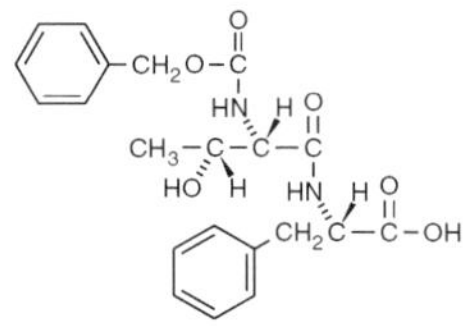

250 mg

S78,986-0
CARBOBENZYLOXY-L-NORLEUCYL-L-TYROSINE

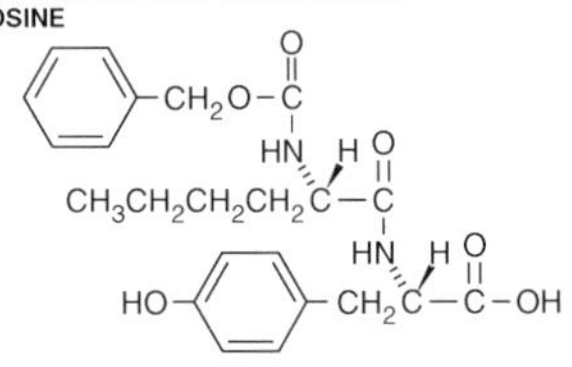

250 mg

S78,610-1
CARBOBENZYLOXY-L-SARCOSYL-L-TYROSINE BENZYL ESTER

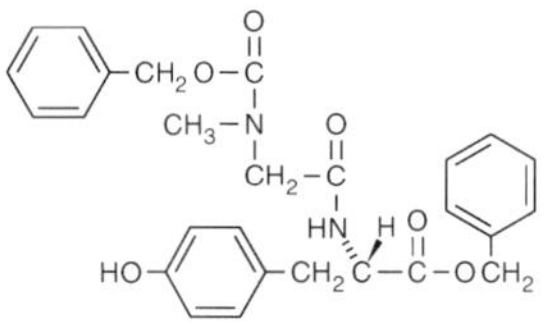

250 mg

S79,011-7
CARBOBENZYLOXY-L-TYROSYL-L-TYROSINE METHYL ESTER

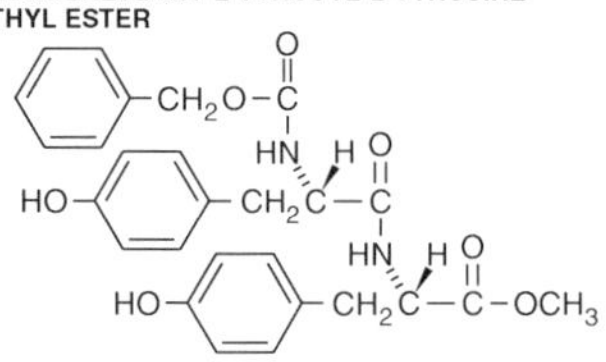

250 mg

S78,466-4
CARBOBENZYLOXY-L-TYROSINAMIDE

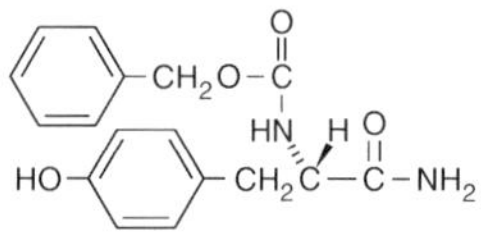

250 mg

S77,759-5
CARBOBENZYLOXY-L-NORVALYL-L-SERINE

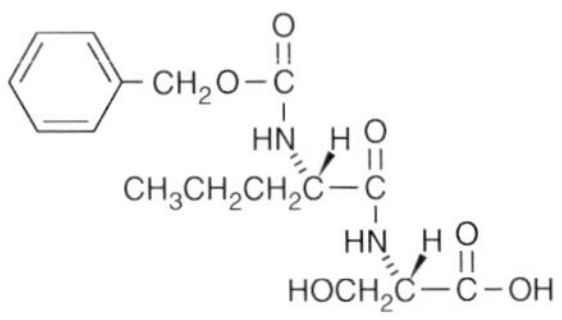

250 mg

S78,586-5
CARBOBENZYLOXY-L-SERYL-L-GLYCINE METHYL ESTER

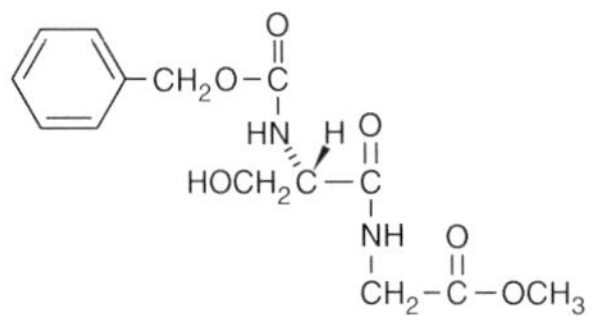

250 mg

S78,588-1
CARBOBENZYLOXY-L-SERYL-L-ALANINE METHYL ESTER

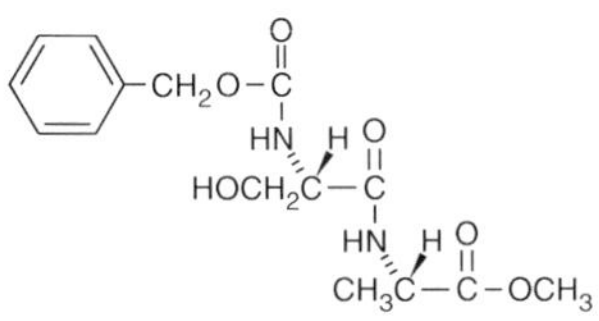

250 mg

S78,590-3
CARBOBENZYLOXY-L-SARCOSYL-L-SERINE METHYL ESTER

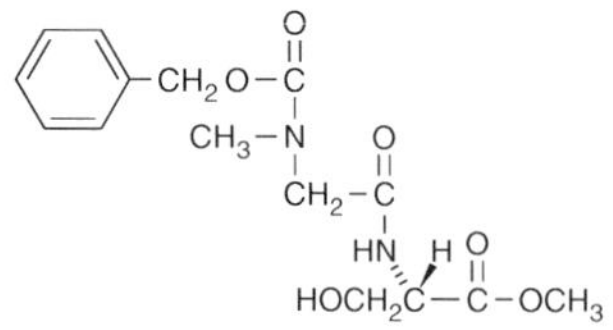

250 mg

S78,653-5
CARBOBENZYLOXY-L-SERYL-L-GLYCINE BENZYL ESTER
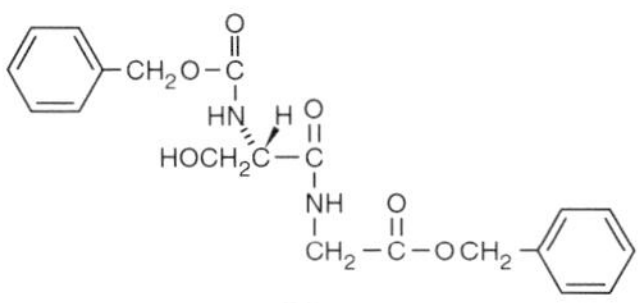
250 mg

S78,997-6
CARBOBENZYLOXY-L-NORVALYL-L-SERINE METHYL ESTER
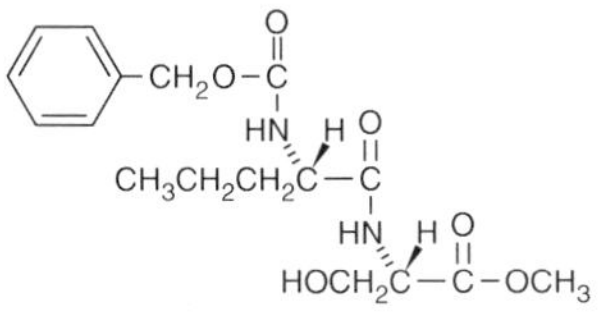
250 mg

S78,984-4
CARBOBENZYLOXYGLYCYL-L-SERINE BENZYL ESTER
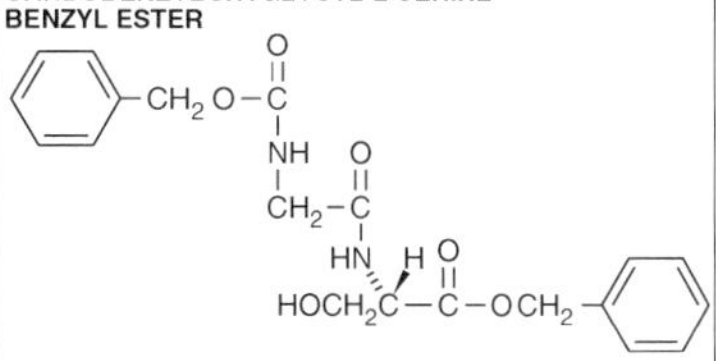
250 mg

S77,665-3
CARBOBENZYLOXY-L-THREONYLGLYCYLGLYCINE BENZYL ESTER
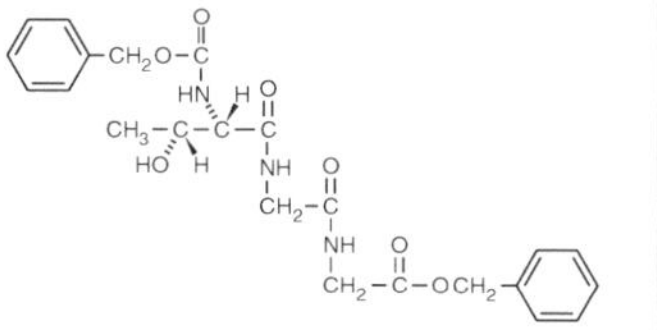
250 mg

S78,578-4
CARBOBENZYLOXY-DL-SERINE METHYL ESTER
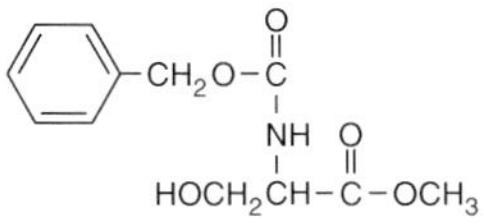
250 mg

S78,957-7
CARBOBENZYLOXY-L-SERINE METHYL ESTER
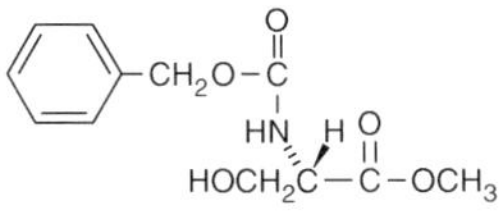
250 mg

S78,937-2
CARBOBENZYLOXY-L-THREONINE METHYL ESTER
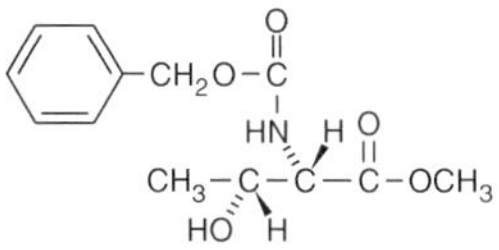
250 mg

S77,630-0
CARBOBENZYLOXY-L-THREONYL-L-ALANINE BENZYL ESTER
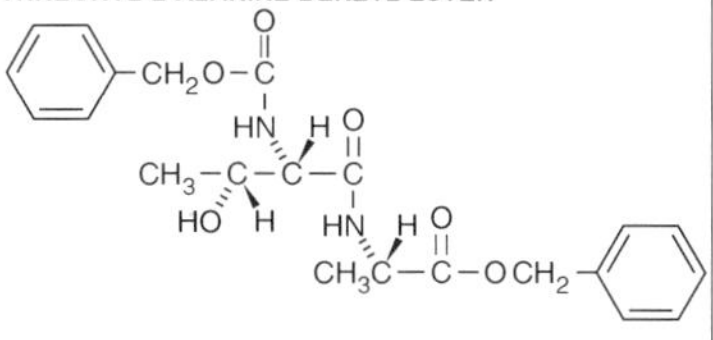
250 mg

S78,526-1
CARBOBENZYLOXY-L-ALANYL-L-SERINAMIDE
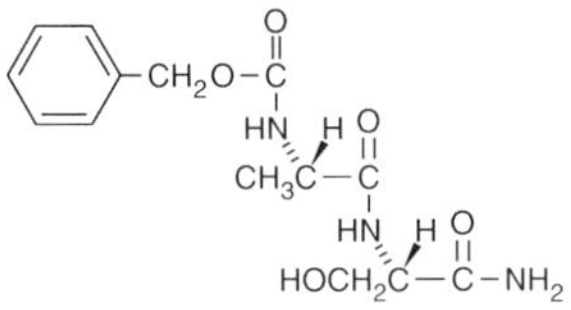
250 mg

S78,648-9
CARBOBENZYLOXY-L-SERYL-L-LEUCINAMIDE
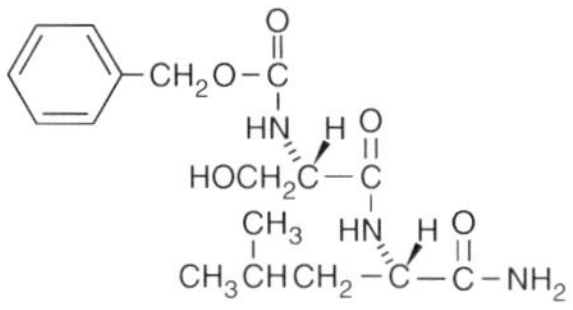
250 mg

S78,576-8
CARBOBENZYLOXY-DL-SERINAMIDE
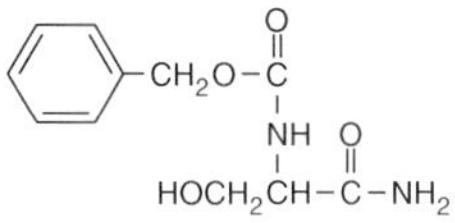
250 mg

S86,437-4
MENTHYL N-(2-HYDROXY-1-METHYL-2-PHENYLETHYL)CARBAMATE
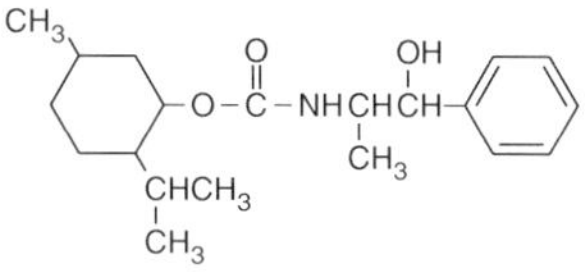
250 mg

S87,643-7
BENZYL N-(2-HYDROXY-1-PHENYLETHYL)-CARBAMATE
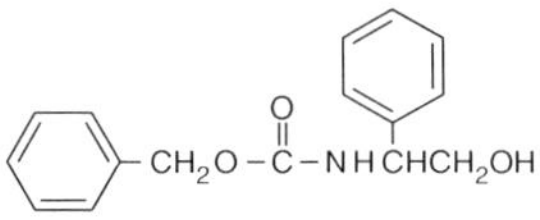
250 mg

S95,863-8
N-CARBOBENZYLOXY-S-BENZOYL-L-CYSTEINE
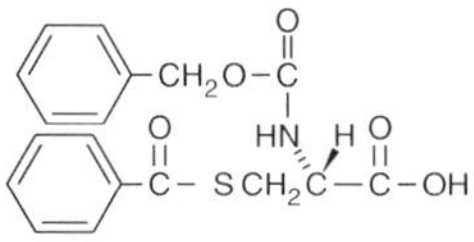
50 mg

S78,462-1
CARBOBENZYLOXY-L-VALYL-L-SERINE
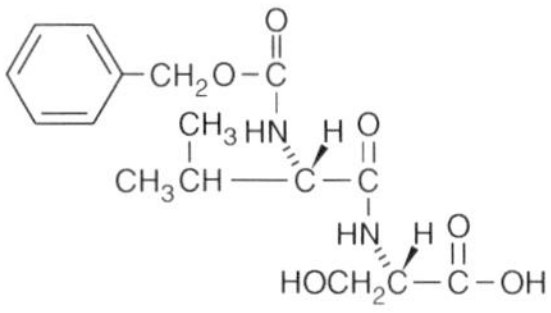
250 mg

S78,940-2
CARBOBENZYLOXYGLYCYL-L-THREONINE DICYCLOHEXYLAMINE SALT HYDRATE
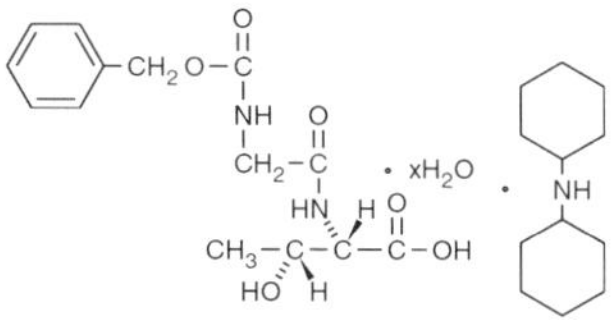
250 mg

S78,319-6
CARBOBENZYLOXYGLYCYLGLYCYL-L-THREONINE
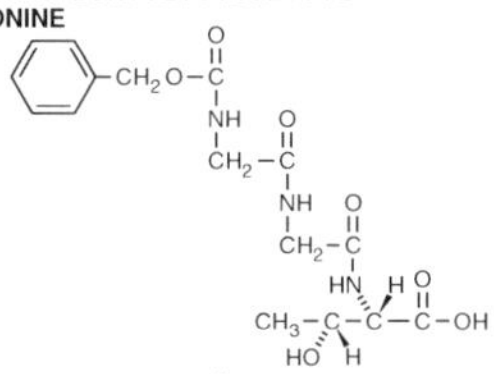
5 gm

S78,736-1
CARBOBENZYLOXY-DL-THREONINE
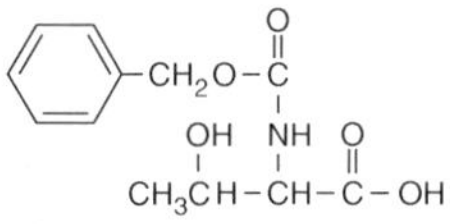
250 mg

S78,744-2
CARBOBENZYLOXY-L-ALANYL-L-SERINE METHYL ESTER
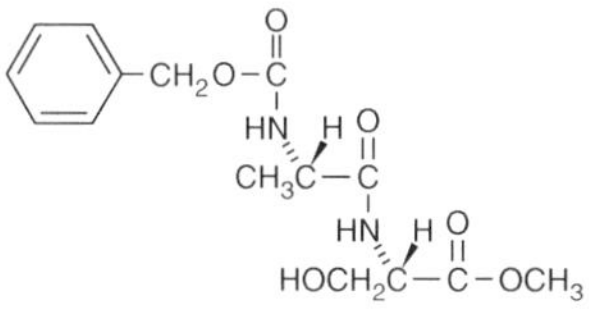
250 mg

S78,538-5
CARBOBENZYLOXY-L-NORLEUCYL-L-SERINE METHYL ESTER
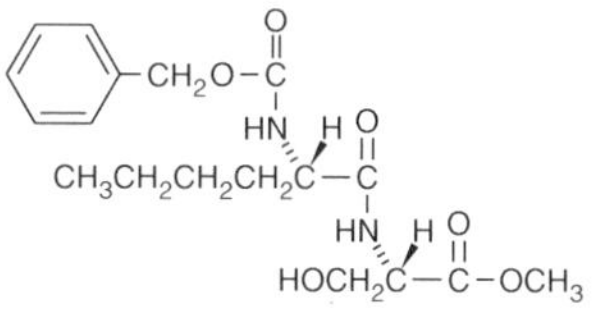
250 mg

S93,316-3
CARBOBENZYLOXY-L-THREONYLGLYCINE BENZYL ESTER
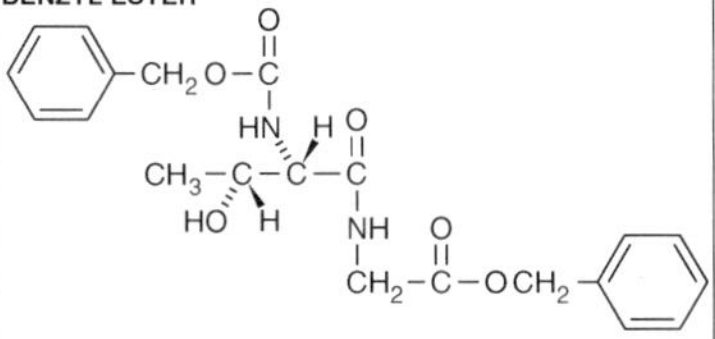
250 mg

S78,359-5
N,N'-BIS(CARBOBENZYLOXY)-L-LYSYL-L-SERINE METHYL ESTER
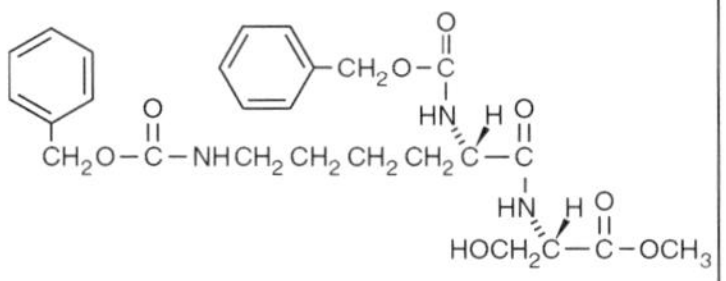
5 gm

S78,330-7
CARBOBENZYLOXY-L-LEUCYL-L-SERINE METHYL ESTER
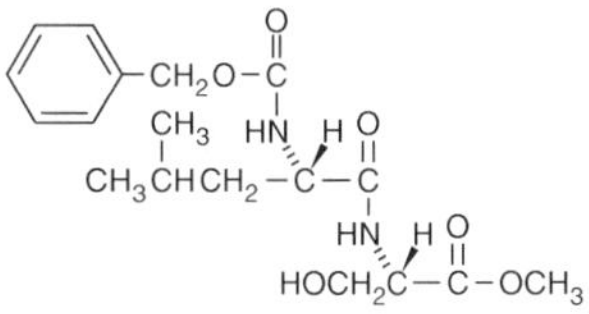
5 gm

S78,761-2
CARBOBENZYLOXYGLYCYL-L-THREONINE BENZYL ESTER
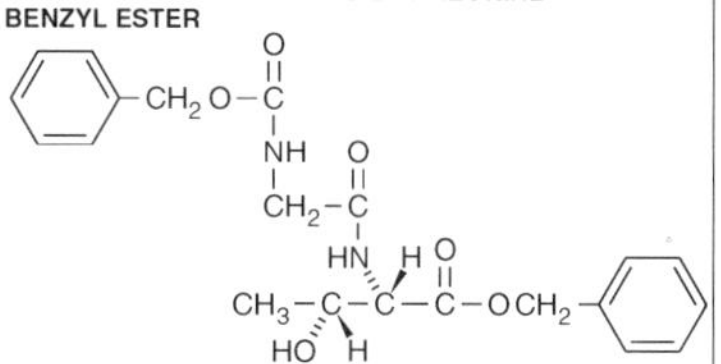
250 mg

S78,959-3
CARBOBENZYLOXY-L-SERYL-L-VALINE BENZYL ESTER
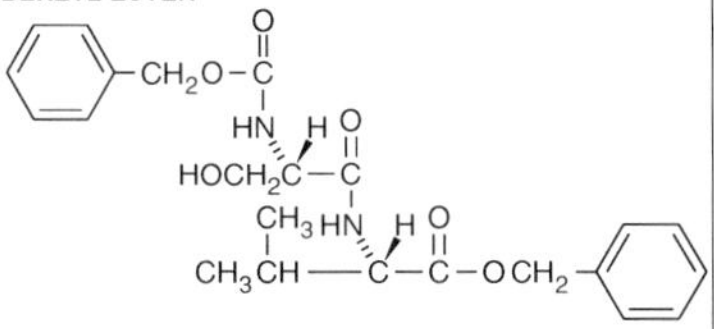
250 mg

S78,960-7
CARBOBENZYLOXY-L-THREONYL-L-VALINE BENZYL ESTER
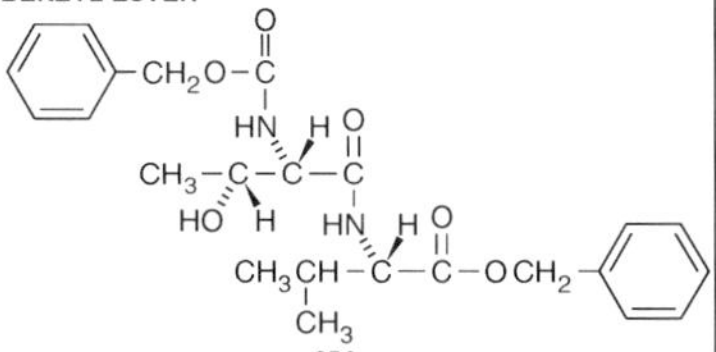
250 mg

S78,969-0
CARBOBENZYLOXY-L-THREONYL-L-LEUCINE BENZYL ESTER
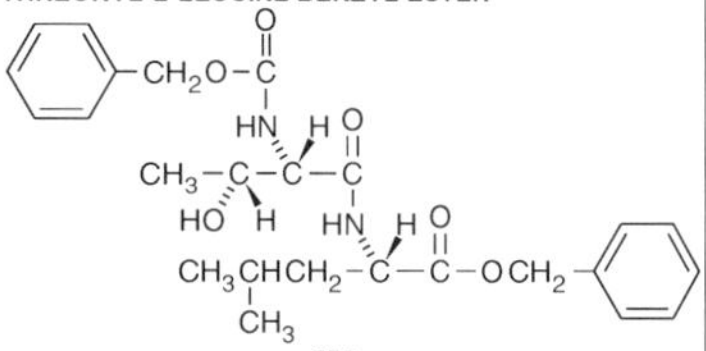
250 mg

S78,333-1
CARBOBENZYLOXY-L-ISOLEUCYL-L-TYROSINE BENZYL ESTER
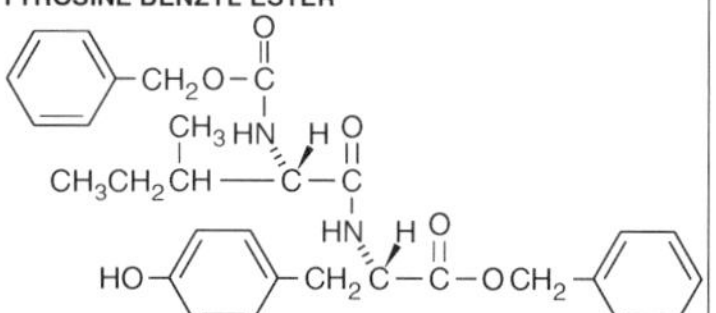
5 gm

S77,794-3
CARBOBENZYLOXYSARCOSYL-L-TYROSINE METHYL ESTER
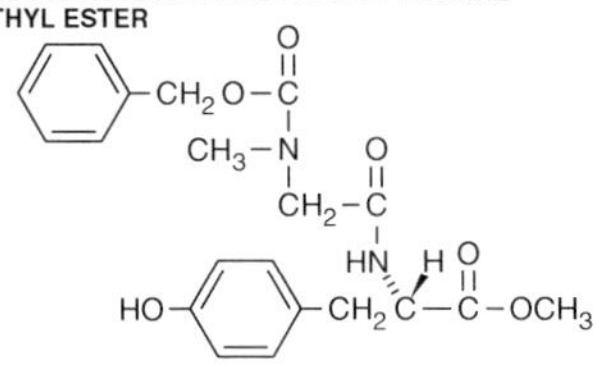
250 mg

S77,690-4
CARBOBENZYLOXY-L-LEUCYL-L-SERINAMIDE
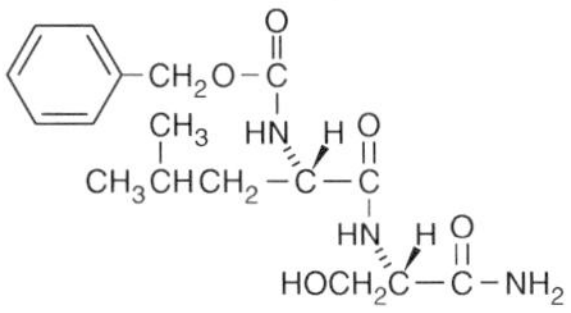
250 mg

S79,072-9
CARBOBENZYLOXY-L-THREONYL-L-LEUCINAMIDE
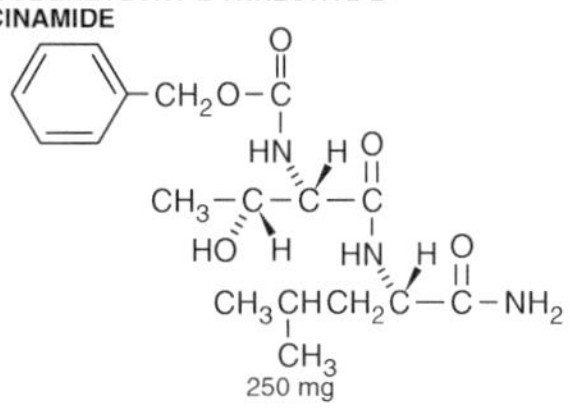
250 mg

S77,668-8
CARBOBENZYLOXY-L-VALYL-L-TYROSINAMIDE
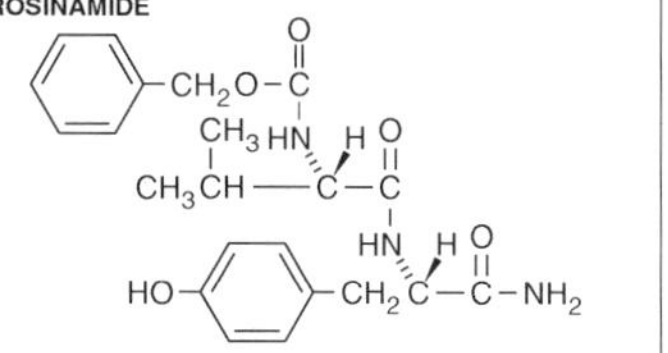
250 mg

S86,443-9
BENZYL CIS-N-(4-HYDROXYCYCLOHEXYL)-CARBAMATE

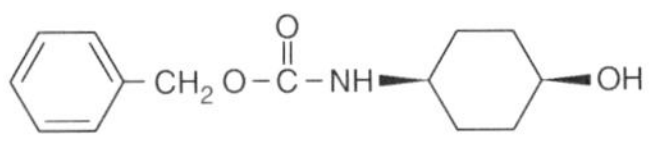

250 mg

S86,444-7
BENZYL TRANS-N-(4-HYDROXYCYCLOHEXYL)CARBAMATE

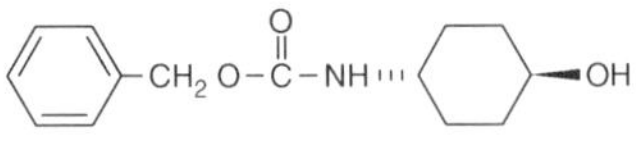

250 mg

S74,968-0
2-HYDROXYETHYL CARBAZATE

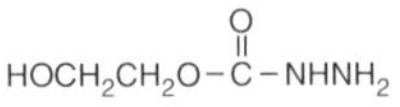

250 mg

S86,724-1
2-[N-(HYDROXYMETHYL)-CARBAMOYL]PHENYL N-PROPYLCARBAMATE

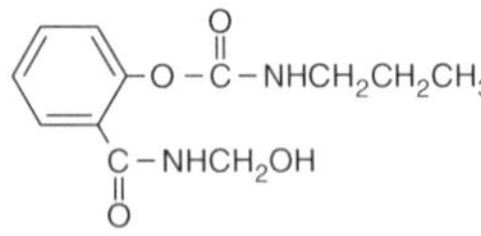

250 mg

S34,176-2
S-BENZYL-N-CARBOBENZYLOXY-L-CYSTEINYL-L-TYROSINE

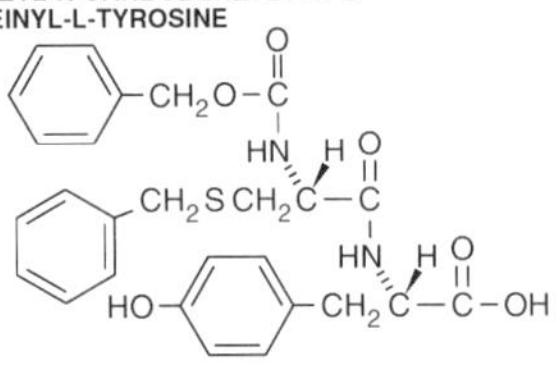

5 gm

S78,551-2
CARBOBENZYLOXY-S-BENZYL-L-CYSTEINYL-L-TYROSINE METHYL ESTER

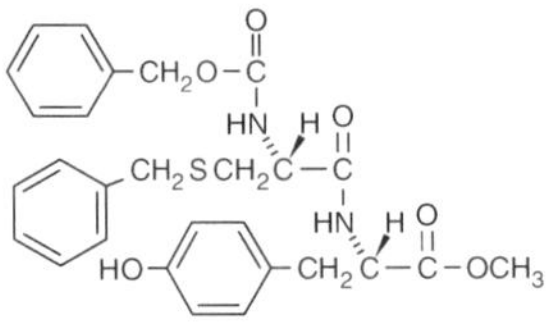

250 mg

S78,385-4
CARBOBENZYLOXY-S-BENZYL-L-CYSTEINYL-L-TYROSINE BENZYL ESTER

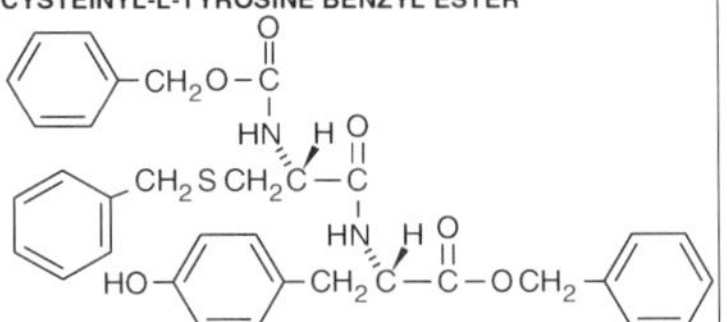

5 gm

S77,669-6
CARBOBENZYLOXY-L-METHIONYL-L-TYROSINAMIDE

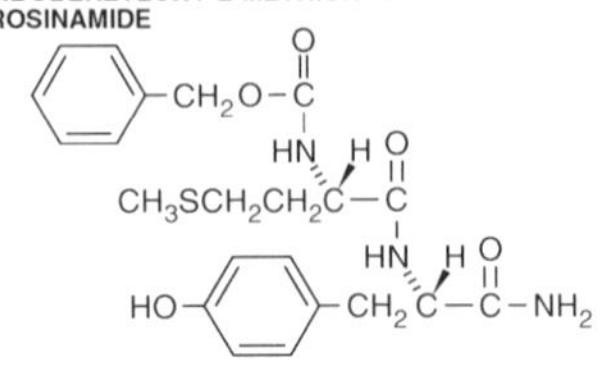

250 mg

S72,986-8
S-(4-HYDROXYPHENYL) N-[3-(METHYLTHIO)PHENYL]-THIOCARBAMATE

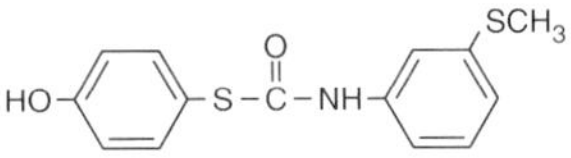

5 gm

S60,555-7
ETHYL N-(4-AMINO-2-HYDROXYPHENYL)-CARBAMATE

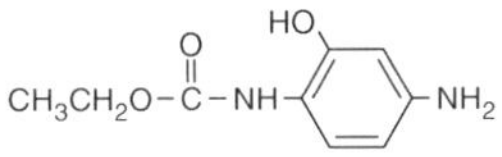

250 mg

S69,306-5
S-(4-HYDROXYPHENYL) N-(2-METHYL-6-NITROPHENYL)THIOCARBAMATE

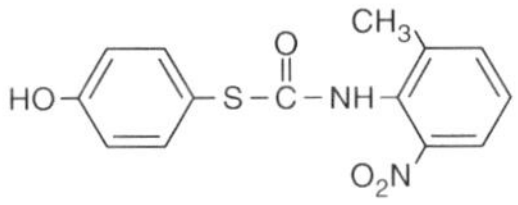

250 mg

S84,819-0
S-(2-HYDROXYPHENYL) N-(2-METHYL-5-NITROPHENYL)THIOCARBAMATE

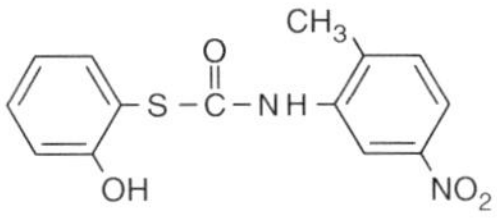

250 mg

S69,901-2
S-(4-HYDROXYPHENYL) N-(2-METHYL-5-NITROPHENYL)THIOCARBAMATE

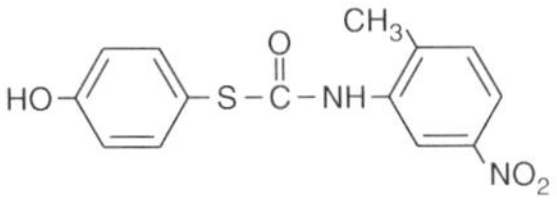

250 mg

S78,826-0
CARBOBENZYLOXY-L-PHENYLALANYL-L-SERINE METHYL ESTER

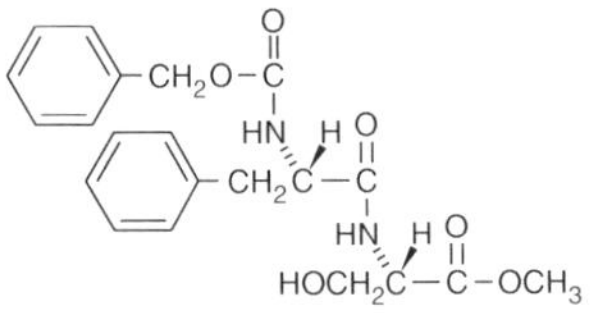

250 mg

S78,609-8
CARBOBENZYLOXY-L-SERYL-L-TYROSINE BENZYL ESTER

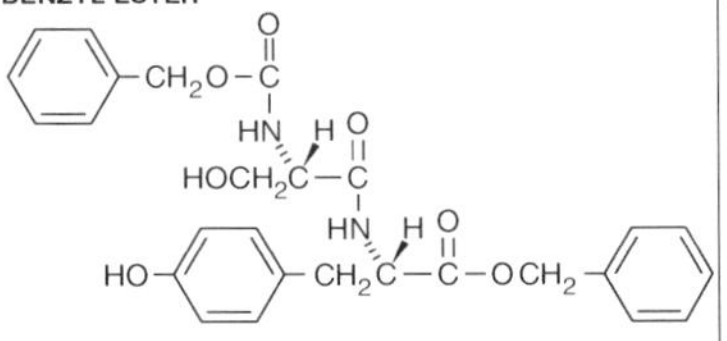

250 mg

S78,183-5
CARBOBENZYLOXY-L-THREONYL-L-PHENYLALANINAMIDE

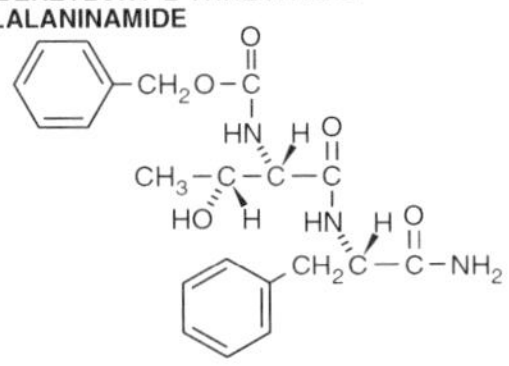

250 mg

S77,638-6
CARBOBENZYLOXY-L-TYROSINE 4-NITROPHENYL ESTER

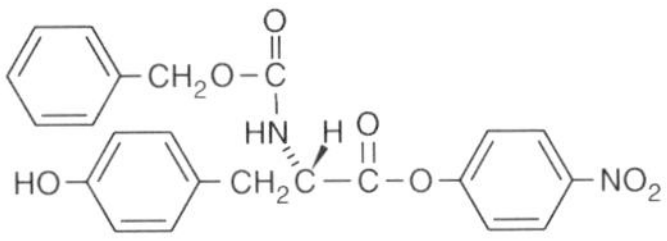

250 mg

S77,762-5
CARBOBENZYLOXY-L-PHENYLALANYL-L-SERINE

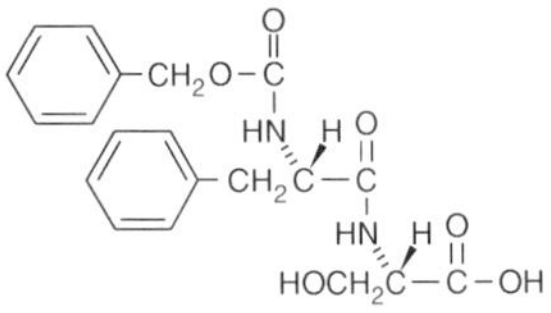

250 mg

S78,275-0
CARBOBENZYLOXY-L-SERYL-L-PHENYLALANINE

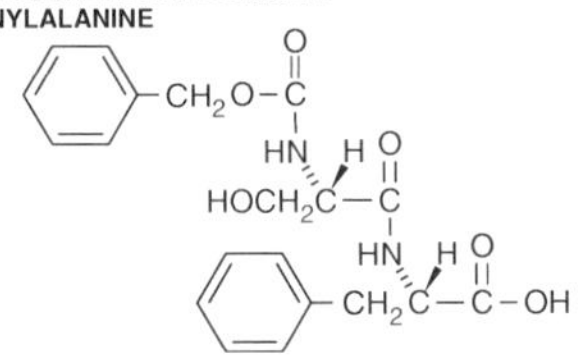

5 gm

S78,492-3
CARBOBENZYLOXY-L-TYROSYL-L-PHENYLALANINE

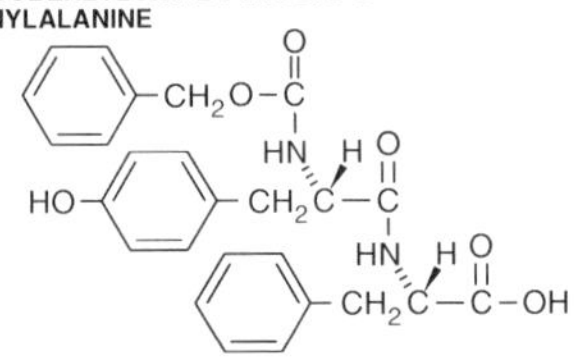

250 mg

S78,187-8
CARBOBENZYLOXY-L-TYROSYL-L-ALANINE

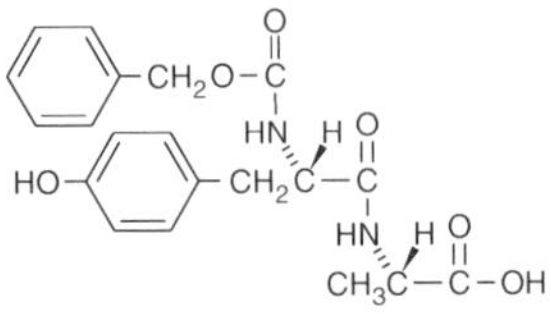

250 mg

S78,159-2
CARBOBENZYLOXY-L-TYROSYL-L-THREONINE

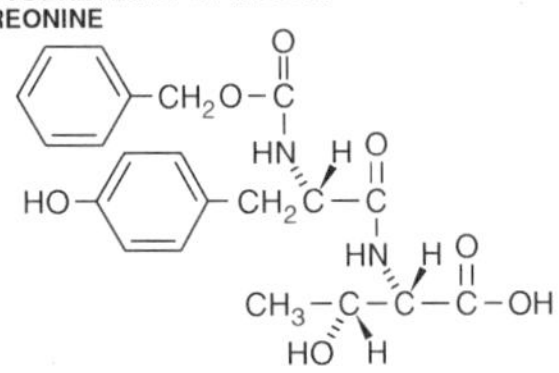

250 mg

S78,128-2
CARBOBENZYLOXY-L-THREONYL-L-PHENYLALANINE BENZYL ESTER

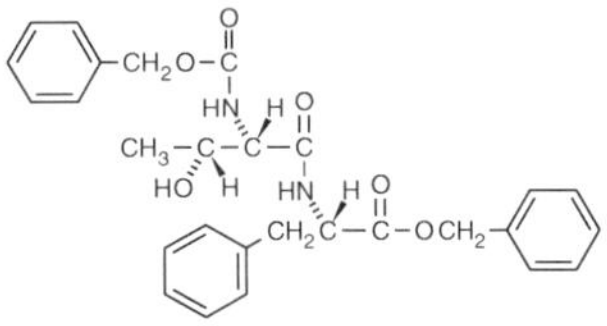

250 mg

S78,764-7
CARBOBENZYLOXY-L-GLYCYLGLYCYL-L-TYROSINE METHYL ESTER

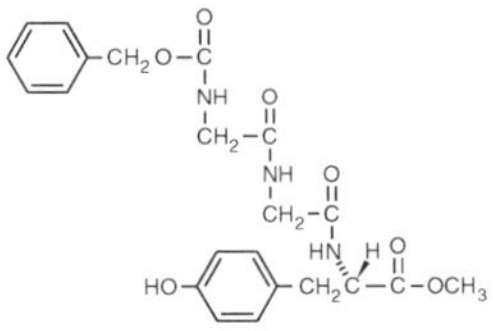

250 mg

S78,848-1
CARBOBENZYLOXY-L-ISOLEUCYL-L-TYROSINE METHYL ESTER

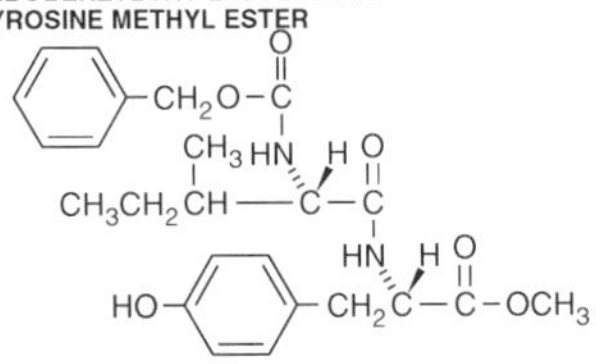

250 mg

S78,845-7
CARBOBENZYLOXY-L-NORVALYL-L-TYROSINE METHYL ESTER

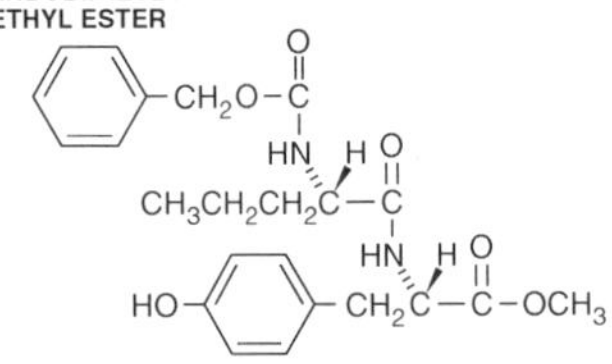

250 mg

S78,944-5
CARBOBENZYLOXY-L-METHIONYL-L-TYROSINE METHYL ESTER

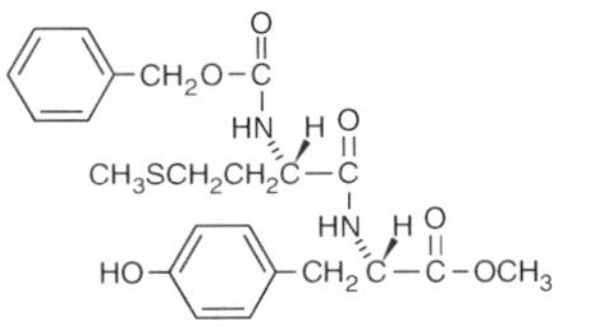

250 mg

S78,615-2
CARBOBENZYLOXYGLYCYL-L-TYROSINE BENZYL ESTER

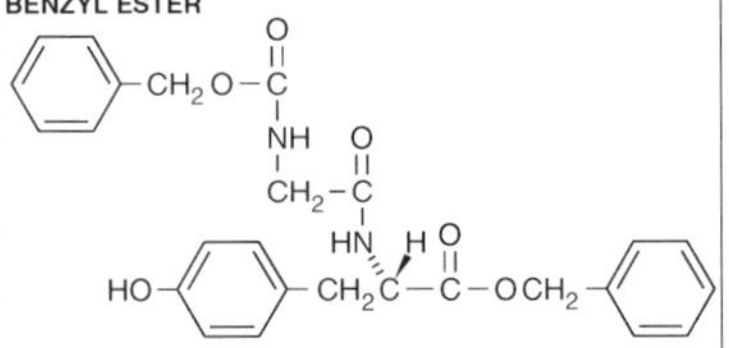

250 mg

S78,776-0
CARBOBENZYLOXY-L-TYROSYL-L-PHENYLALANINE METHYL ESTER

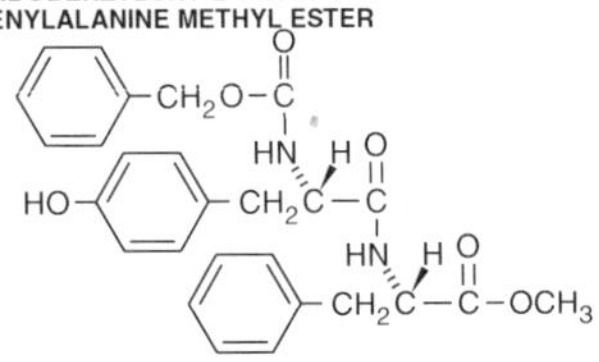

250 mg

S78,856-2
CARBOBENZYLOXY-L-ALANYL-L-TYROSINE BENZYL ESTER

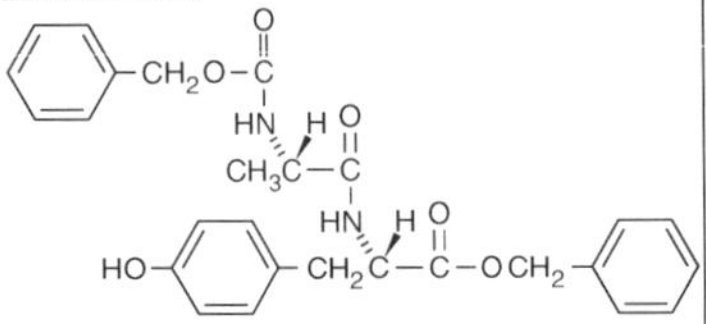

250 mg

S79,068-0
CARBOBENZYLOXY-L-PHENYLALANYL-L-TYROSINE METHYL ESTER

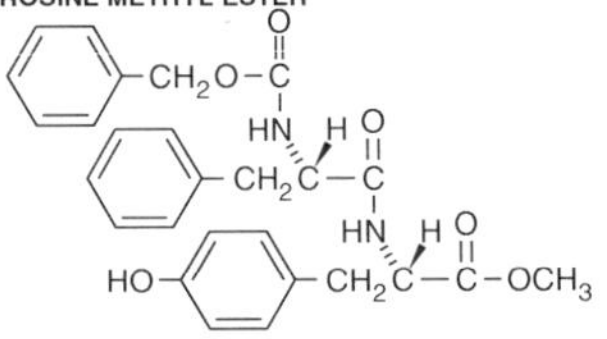

250 mg

S78,859-7
CARBOBENZYLOXY-L-THREONYL-L-TYROSINE BENZYL ESTER

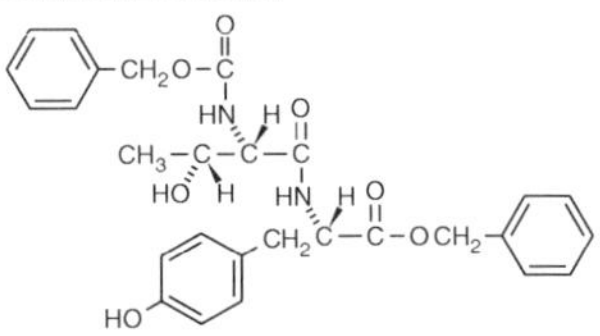

250 mg

S79,039-7
CARBOBENZYLOXYGLYCYLGLYCYL-L-TYROSINE BENZYL ESTER
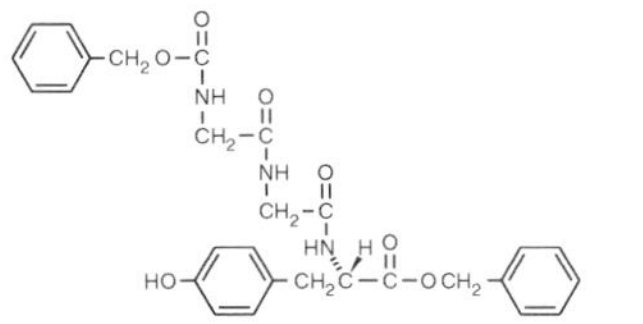
250 mg

S78,862-7
N,N'-BIS(CARBOBENZYLOXY)-L-LYSYL-L-TYROSINE METHYL ESTER
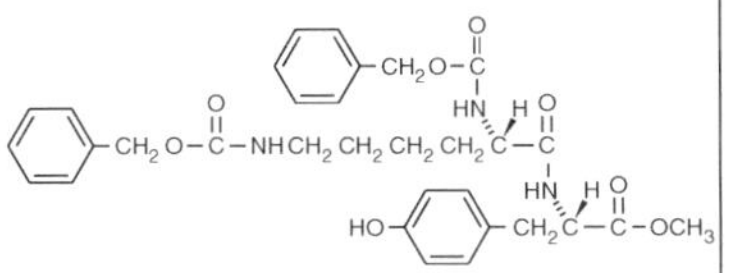
250 mg

S79,040-0
CARBOBENZYLOXY-L-PHENYLALANYL-L-TYROSINE BENZYL ESTER
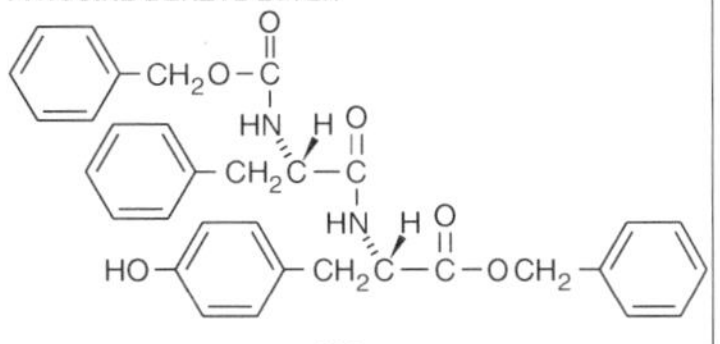
250 mg

S78,185-1
CARBOBENZYLOXY-L-TYROSYL-L-ALANINAMIDE
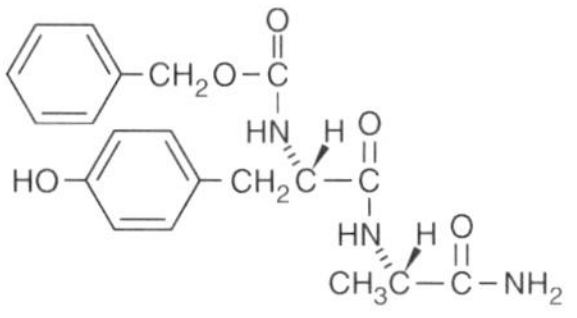
250 mg

S78,635-7
CARBOBENZYLOXY-L-SARCOSYL-L-TYROSINAMIDE
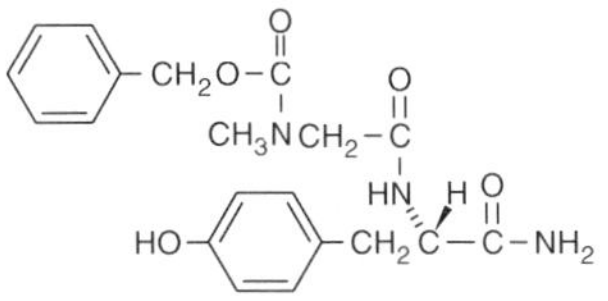
250 mg

S78,184-3
CARBOBENZYLOXY-L-TYROSYLGLYCYLGLYCINAMIDE
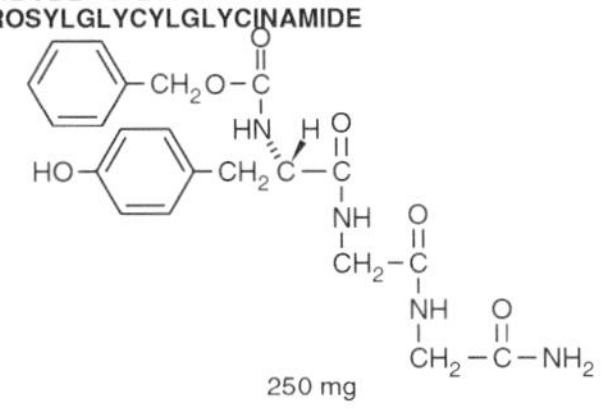
250 mg

S78,547-4
CARBOBENZYLOXY-L-ISOLEUCYL-L-TYROSINAMIDE
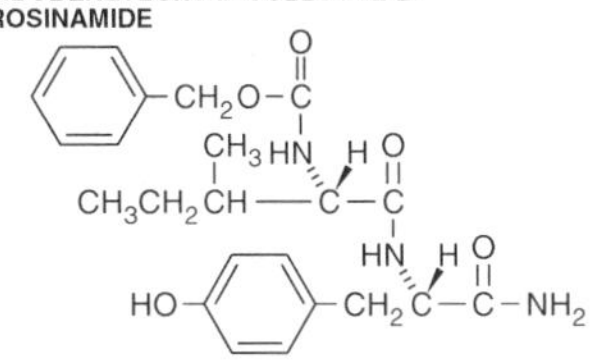
250 mg

S79,086-9
CARBOBENZYLOXY-L-NORLEUCYL-L-TYROSINAMIDE
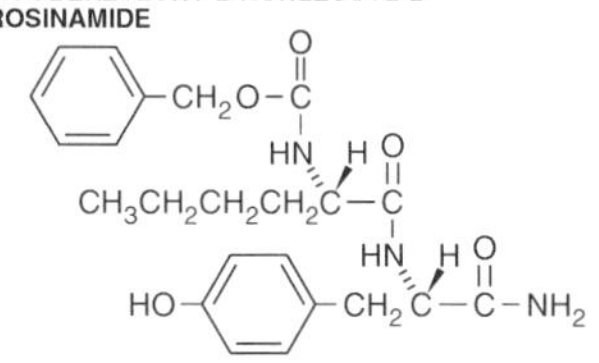
250 mg

S77,791-9
CARBOBENZYLOXY-L-SERYL-L-TYROSINAMIDE
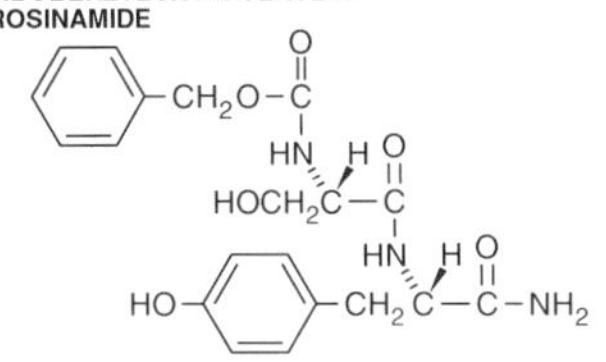
250 mg

S77,709-9
CARBOBENZYLOXY-L-THREONYL-L-TYROSINAMIDE
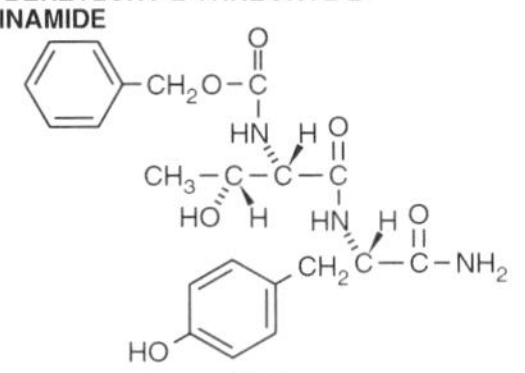
250 mg

S78,634-9
CARBOBENZYLOXY-L-SERYL-L-ALANINE BENZYL ESTER
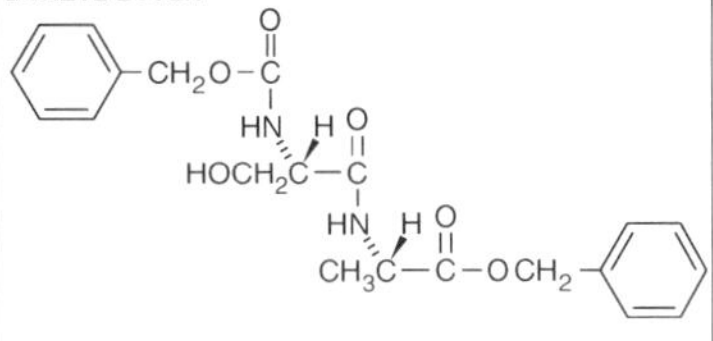
250 mg

S78,612-8
CARBOBENZYLOXY-L-SERYL-L-PHENYLALANINE BENZYL ESTER
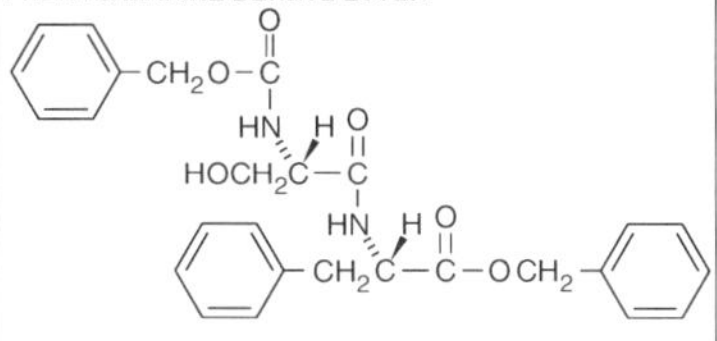
250 mg

S78,374-9
CARBOBENZYLOXY-L-THREONYL-L-VALINE METHYL ESTER
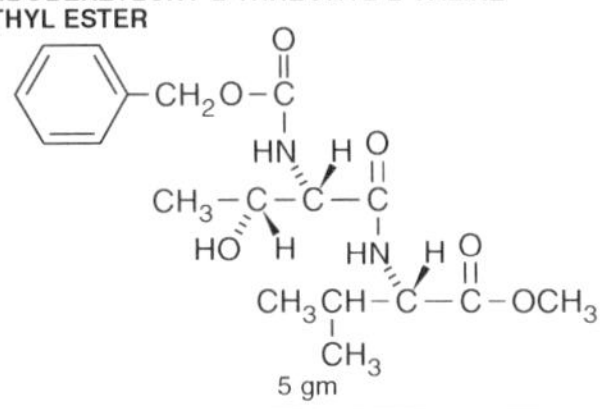
5 gm

S78,489-3
N,N'-BIS(CARBOBENZYLOXY)-L-LYSYL-L-SERINE
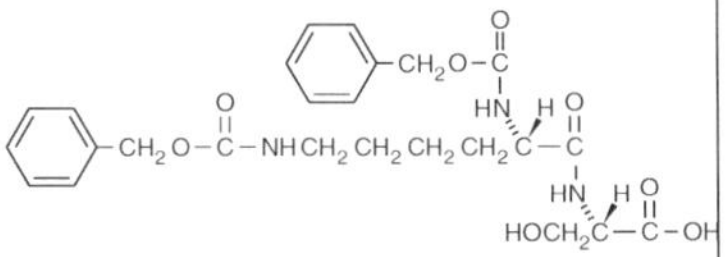
250 mg

S78,541-5
CARBOBENZYLOXYGLYCYL-L-TYROSINE
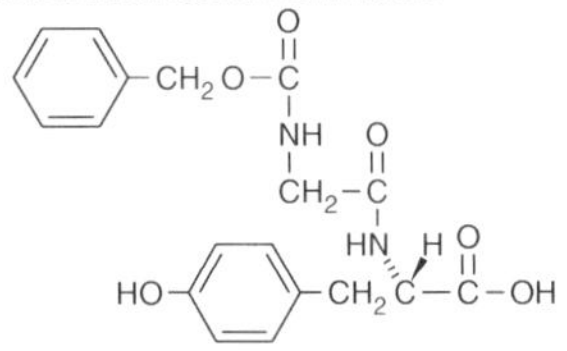
250 mg

S78,756-6
CARBOBENZYLOXY-L-TYROSYLGLYCINE
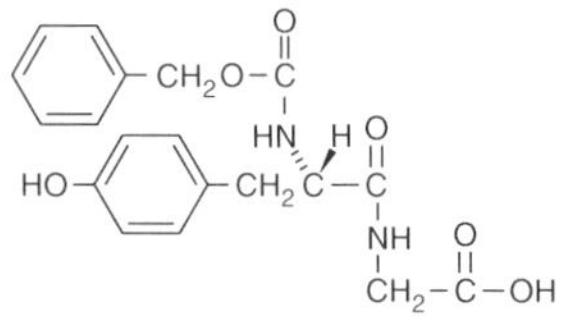
250 mg

S78,483-4
CARBOBENZYLOXY-L-TYROSYL-L-VALINE
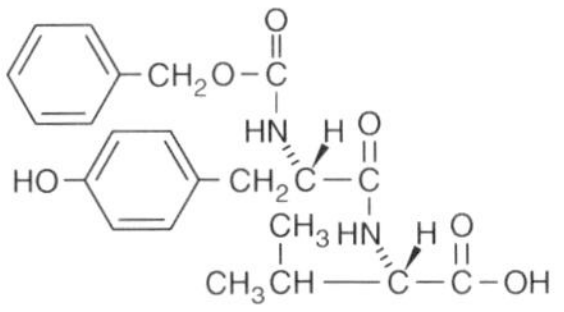
250 mg

S78,837-6
CARBOBENZYLOXY-L-VALYL-L-TYROSINE
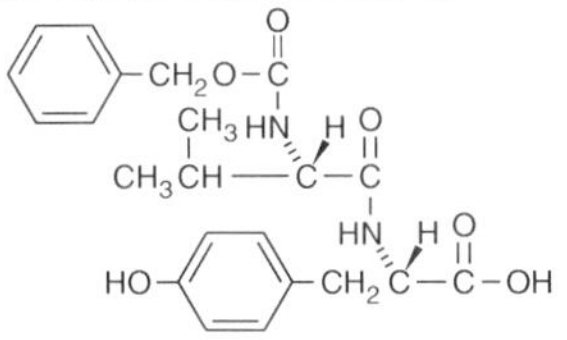
250 mg

S78,632-2
CARBOBENZYLOXY-L-ALANYL-L-TYROSINE METHYL ESTER
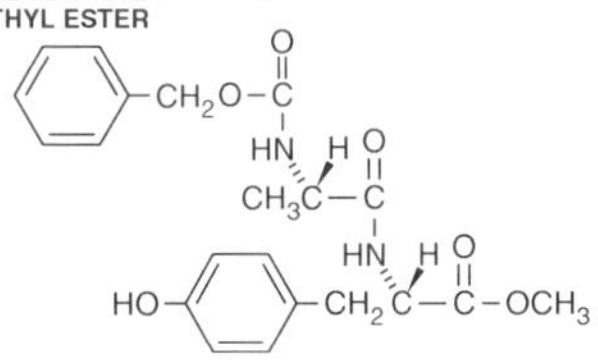
250 mg

S78,629-2
CARBOBENZYLOXY-L-TYROSYL-L-GLYCYLGLYCINE METHYL ESTER
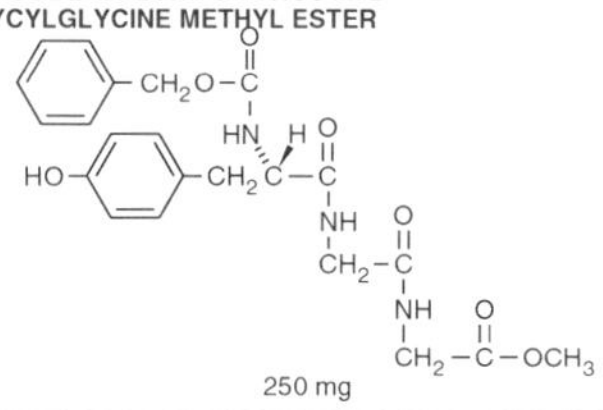
250 mg

S78,675-6
CARBOBENZYLOXY-DL-NORVALYLGLYCYL-DL-TYROSINE METHYL ESTER
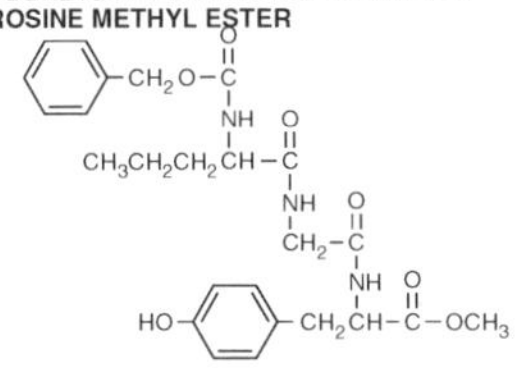
250 mg

S78,623-3
CARBOBENZYLOXY-L-TYROSINE BENZYL ESTER
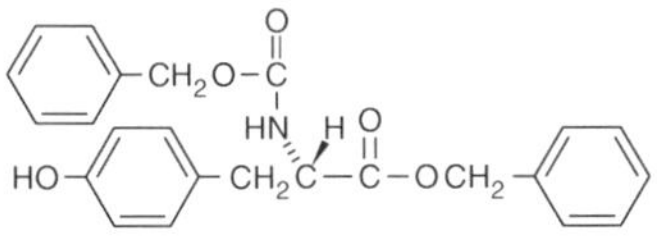
250 mg

S78,619-5
CARBOBENZYLOXY-L-TYROSYL-L-LEUCINE METHYL ESTER
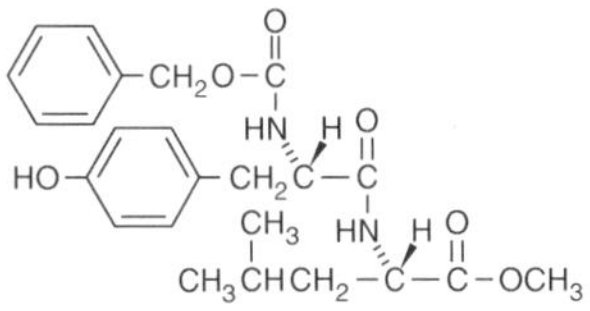
250 mg

S78,990-9
CARBOBENZYLOXY-L-TYROSYL-L-TYROSINE BENZYL ESTER
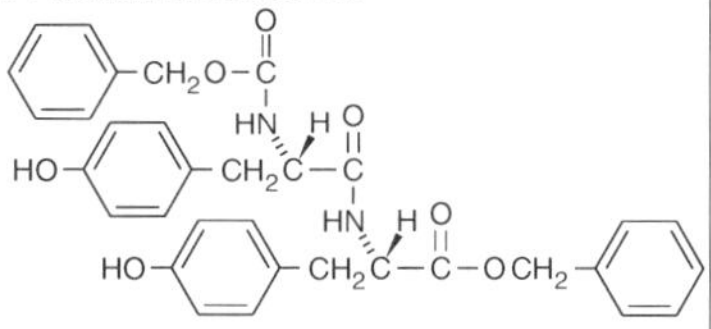
250 mg

S67,909-7
CARBOBENZYLOXY-L-VALYL-L-TYROSINE METHYL ESTER
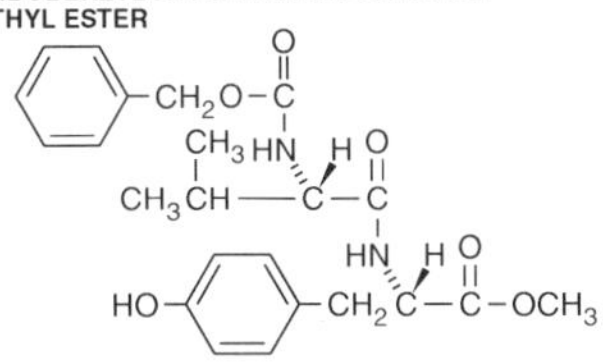
250 mg

S78,126-6
CARBOBENZYLOXY-L-NORLEUCYL-L-TYROSINE METHYL ESTER
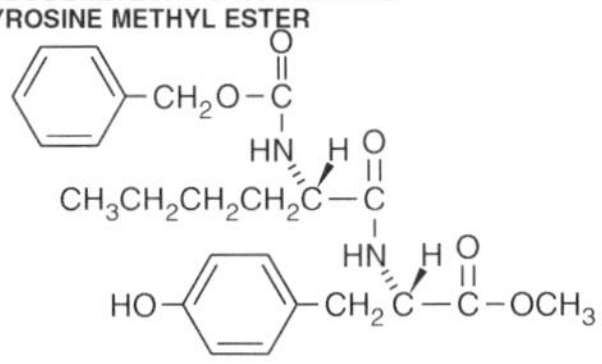
250 mg

S78,487-7
CARBOBENZYLOXY-L-TYROSYL-L-ISOLEUCINE METHYL ESTER
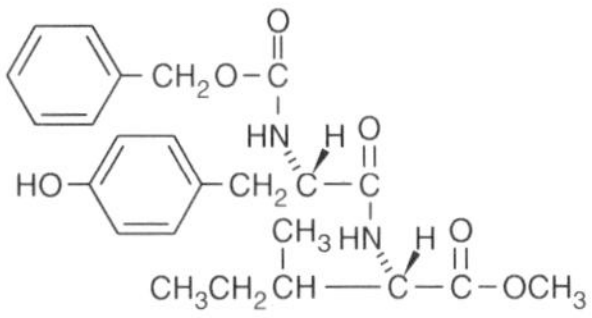
250 mg

S78,132-0
N,N'-BIS(CARBOBENZYLOXY)-L-LYSYL-L-TYROSINAMIDE
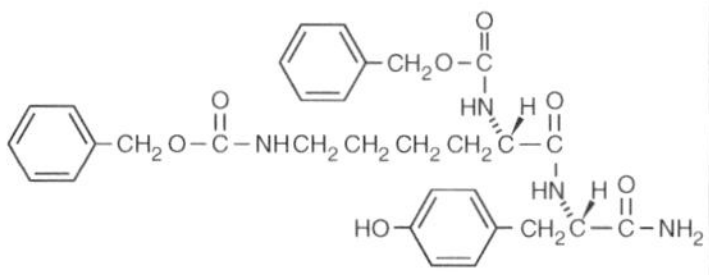
250 mg

S78,143-6
CARBOBENZYLOXY-L-LEUCYL-L-TYROSINAMIDE
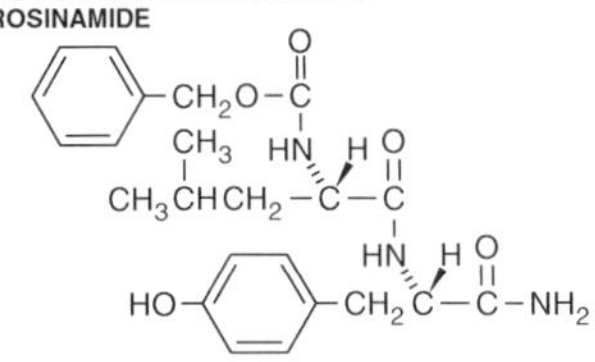
250 mg

S77,666-1
CARBOBENZYLOXY-L-PROLYL-L-TYROSINAMIDE
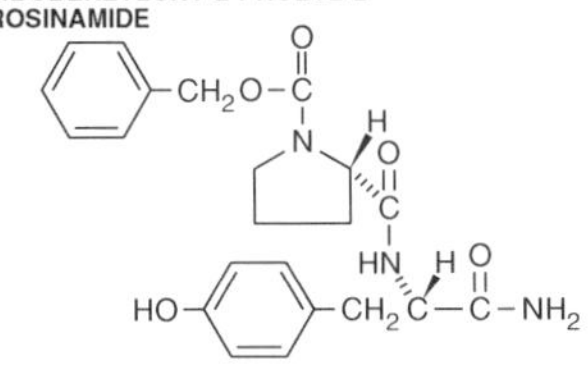
250 mg

S76,546-5
CARBOBENZYLOXY-L-VALYL-L-TYROSINE HYDRAZIDE
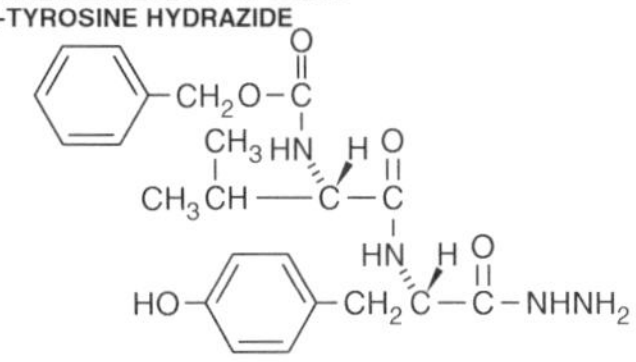
5 gm

S78,828-7
CARBOBENZYLOXY-O-(TERT-BUTYL)-L-TYROSINE DICYCLOHEXYLAMINE SALT
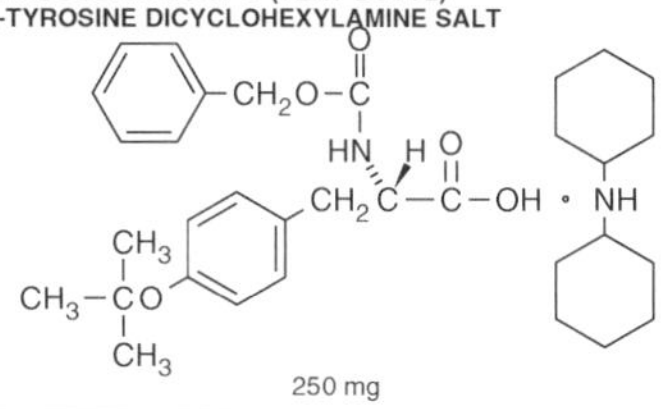
250 mg

S76,512-0
O-BENZYL-N-BENZYLOXYCARBONYL-L-TYROSINE 4-NITROPHENYL ESTER

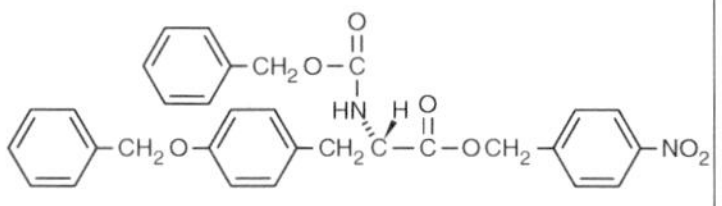

5 gm

S78,633-0
N-CARBOBENZYLOXYGLYCYL-O-BENZYL-L-THREONINE

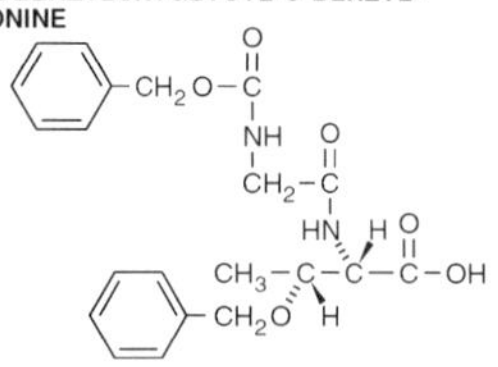

250 mg

S78,941-0
CARBOBENZYLOXY-L-LEUCINE HYDRAZIDE

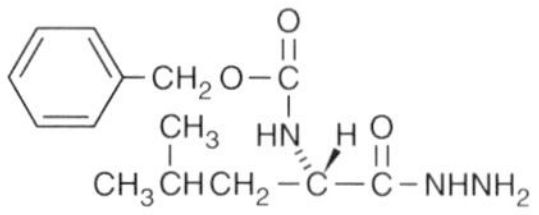

250 mg

S72,794-6
S-(4-HYDROXYPHENYL) N-(ETHOXYCARBONYLMETHYL)-THIOCARBAMATE

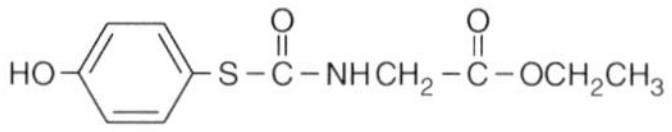

5 gm

S85,442-5
S-(2-HYDROXYPHENYL) N-(3,4-XYLYL)-THIOCARBAMATE

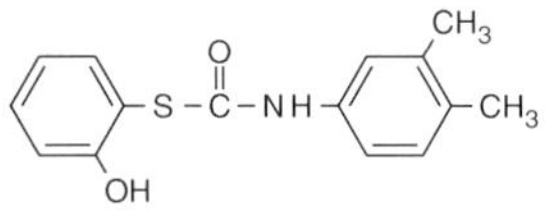

250 mg

S72,795-4
S-(4-HYDROXYPHENYL) N-BUTYLTHIOCARBAMATE

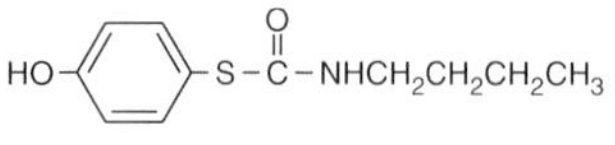

5 gm

S77,211-9
3,4,5-TRIMETHOXYBENZYL N-ETHYLCARBAMATE

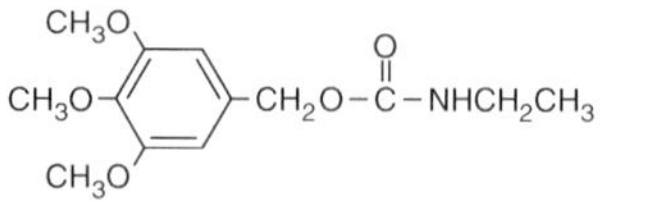

250 mg

S76,916-9
4-(BENZYLOXY)PHENYL N-(2-METHOXYPHENYL)CARBAMATE

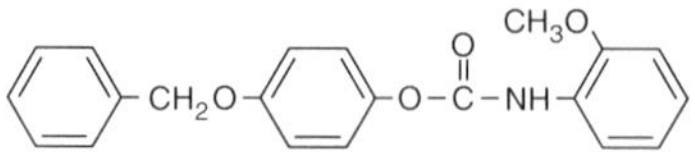

250 mg

S84,864-6
4-(BENZYLOXY)PHENYL N-(2,5-DIMETHOXYPHENYL)CARBAMATE

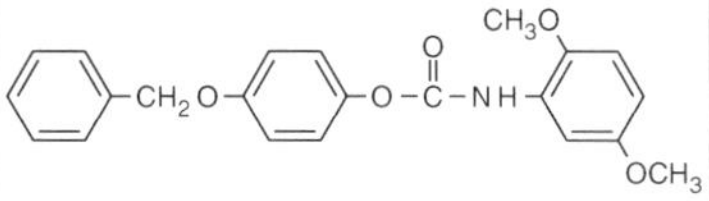

250 mg

S77,381-6
3,4,5-TRIMETHOXYBENZYL N-(2,6-DIISOPROPYLPHENYL)CARBAMATE

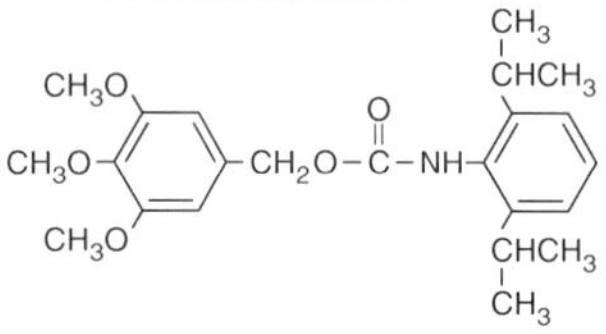

250 mg

S72,801-2
4-METHOXYBENZYL N-(2-METHOXY-4-NITROPHENYL)CARBAMATE

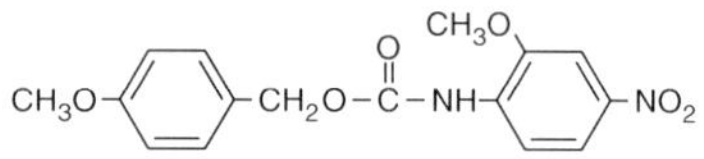

5 gm

S84,865-4
4-(BENZYLOXYCARBONYL)PHENYL N-(2,5-DIMETHOXYPHENYL)CARBAMATE

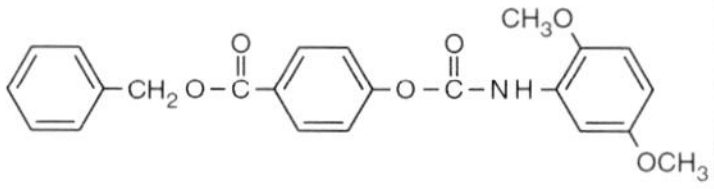

250 mg

S85,102-7
4-(BENZYLOXY)PHENYL N-(2,4-DIMETHOXYPHENYL)CARBAMATE

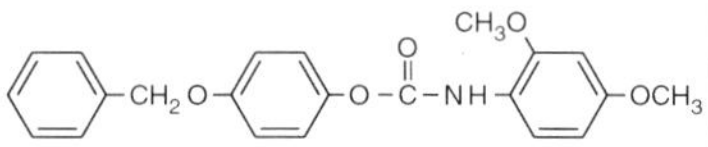

250 mg

S69,094-5
3,4,5-TRIMETHOXYBENZYL N-PROPYLCARBAMATE

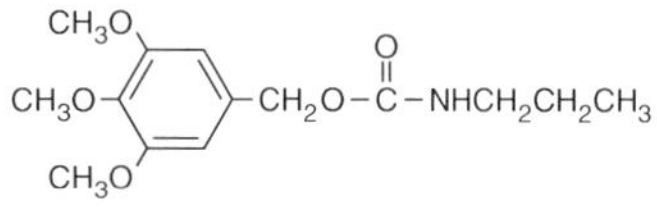

250 mg

S60,796-7
PIPERONYL N-(4-CHLOROPHENYL)-CARBAMATE

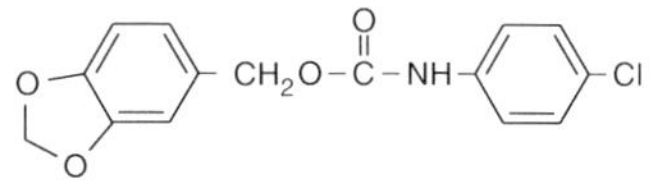

5 gm

S63,409-3
PIPERONYL N-(4-BROMOPHENYL)-CARBAMATE

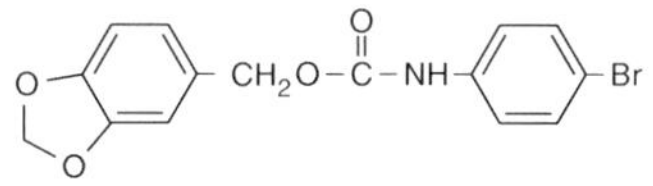

5 gm

S60,192-6
PIPERONYL N-(4-METHOXYPHENYL)-CARBAMATE

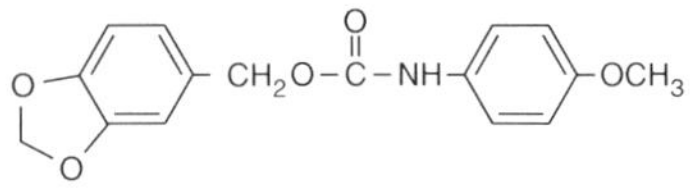

5 gm

S60,459-3
PIPERONYL N-(4-CHLORO-2-METHYLPHENYL)CARBAMATE

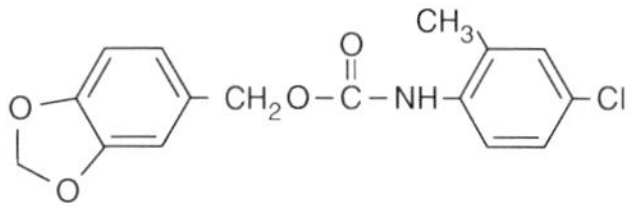

5 gm

S60,621-9
PIPERONYL N-(2,4-XYLYL)CARBAMATE

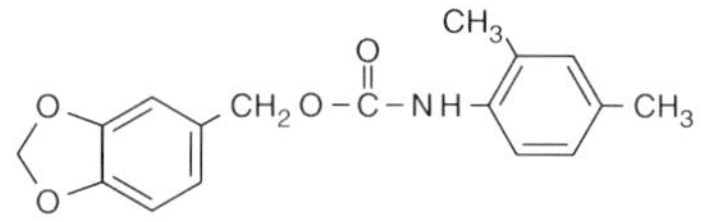

5 gm

S60,476-3
PIPERONYL N-(3-CHLORO-4-METHYLPHENYL)CARBAMATE

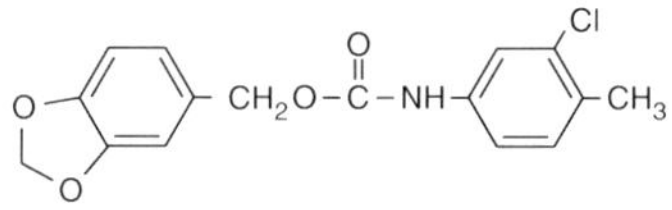

5 gm

S85,157-4
4-(BENZYLOXY)PHENYL N-(4-METHYL-2-NITROPHENYL)CARBAMATE

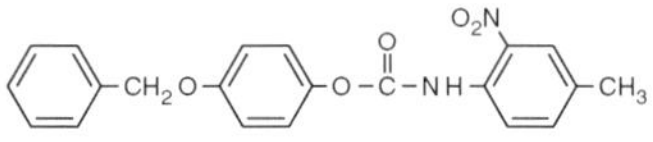

250 mg

S79,162-8
4-(BENZYLOXY)PHENYL N-BUTYLCARBAMATE

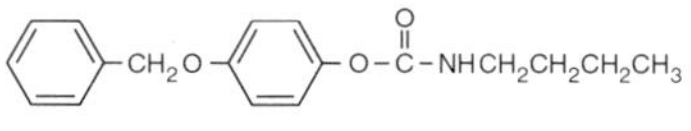

250 mg

S79,739-1
4-METHOXYBENZYL N-[4-(METHYLTHIO)-PHENYL]CARBAMATE

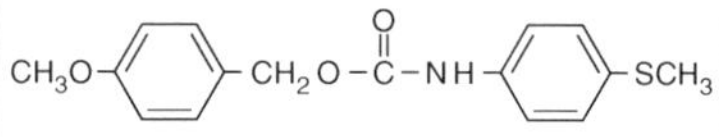

250 mg

S86,219-3
BENZYL 3-(4-METHOXYBENZYLIDENE)-CARBAZATE

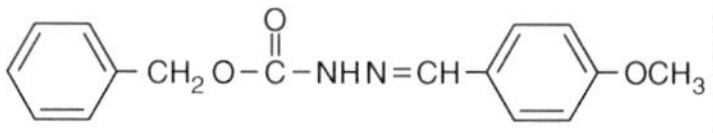

250 mg

S66,672-6
PIPERONYL N-(O-ANISYL)CARBAMATE

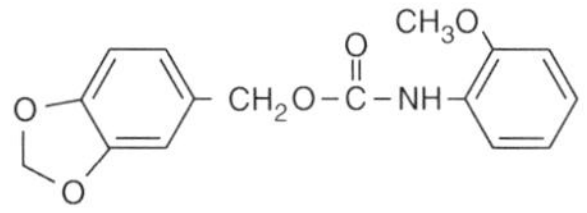

5 gm

S79,448-1
4-METHOXYBENZYL N-(3-ACETYLPHENYL)-CARBAMATE

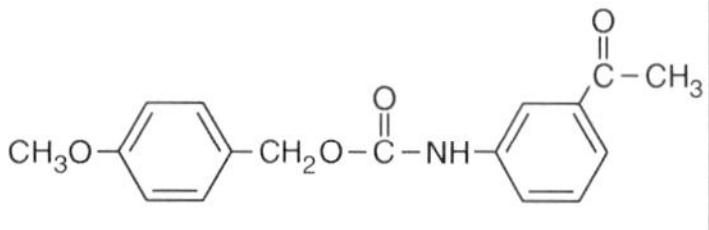

250 mg

S95,377-6
BENZYL N-(CARBOXYMETHYL)-N-METHYLCARBAMATE

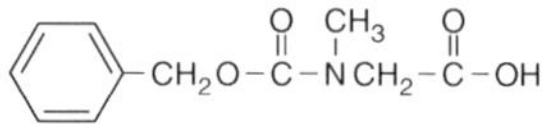

250 mg

S63,455-7
PIPERONYL N-(2,3-DICHLOROPHENYL)-CARBAMATE

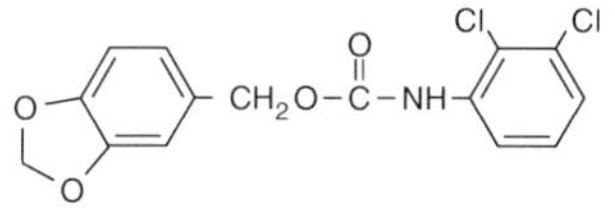

5 gm

S63,415-8
PIPERONYL N-(3-METHOXYPHENYL)-CARBAMATE

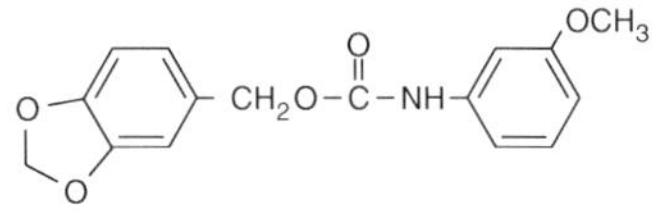

5 gm

S85,113-2
2-(PHENOXY)ETHYL N-(2,4-DIMETHOXYPHENYL)CARBAMATE

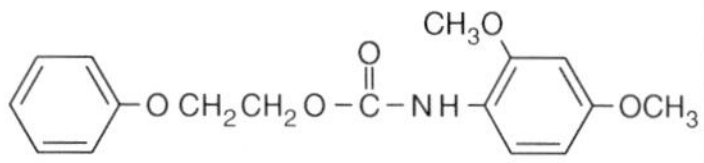

250 mg

S85,107-8
4-ALLYL-2-METHOXYPHENYL N-(2,4-DIMETHOXYPHENYL)CARBAMATE

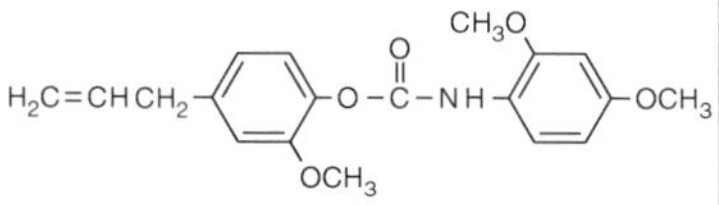

250 mg

S76,884-7
2,3,4-TRIMETHOXYBENZYL N-PROPYLCARBAMATE

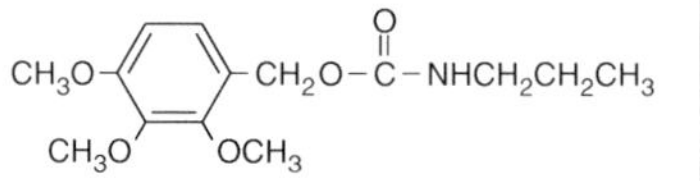

250 mg

S84,842-5
4-FORMYL-2-METHOXYPHENYL N-(2,5-DIMETHOXYPHENYL)CARBAMATE

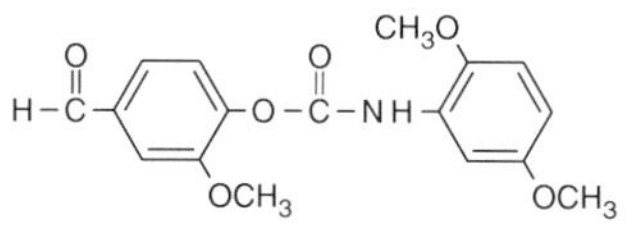

250 mg

S84,875-1
2-PHENOXYETHYL N-(2,5-DIMETHOXYPHENYL)CARBAMATE

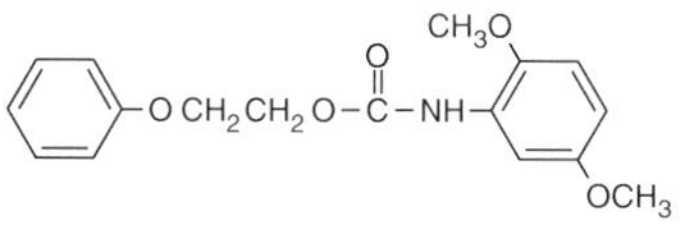

250 mg

S85,447-6
2-(4-METHOXYPHENOXY)ETHYL N-(3,4-XYLYL)CARBAMATE

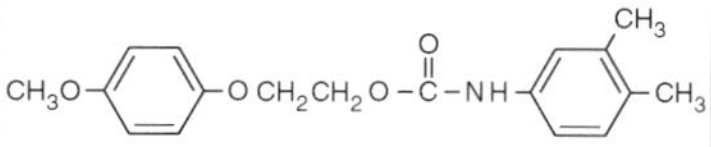

250 mg

S73,773-9
3,4,5-TRIMETHOXYBENZYL N-BUTYLCARBAMATE

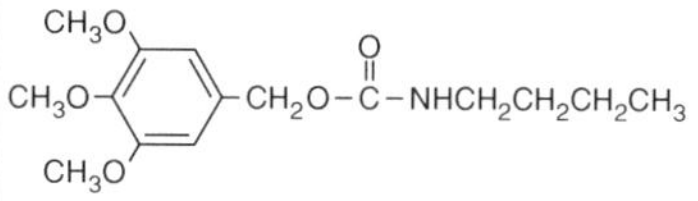

5 gm

S85,358-5
2-(4-METHOXYPHENOXY)ETHYL N-[3-(METHYLTHIO)PHENYL]CARBAMATE

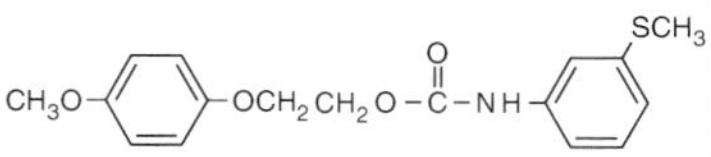

250 mg

S91,078-3
2,6-DIMETHOXYPHENYL N-[4-(METHYLTHIO)PHENYL]CARBAMATE

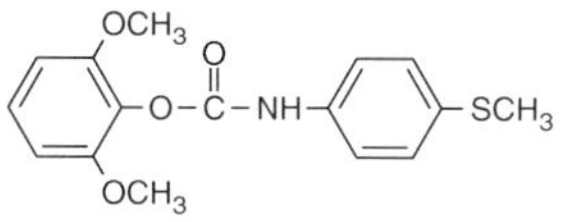

250 mg

S85,357-7
2,6-DIMETHOXYPHENYL N-[3-(METHYLTHIO)PHENYL]CARBAMATE

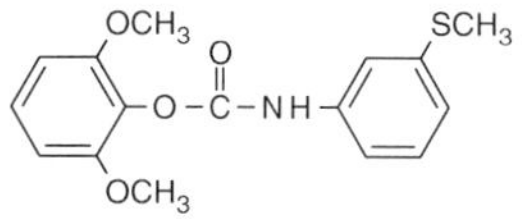

250 mg

S71,580-8
2-(4-ALLYL-2-METHOXYPHENOXY)ETHYL N-[3-(METHYLTHIO)PHENYL]CARBAMATE

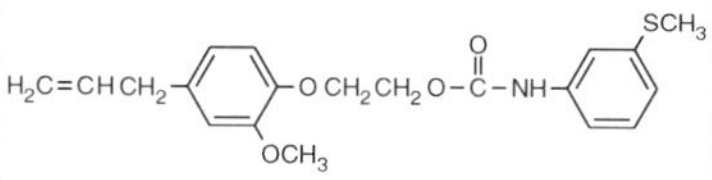

250 mg

S66,685-8
N,N'-BIS(2-METHOXYETHOXYCARBONYL)-4-METHYL-1,3-PHENYLENEDIAMINE

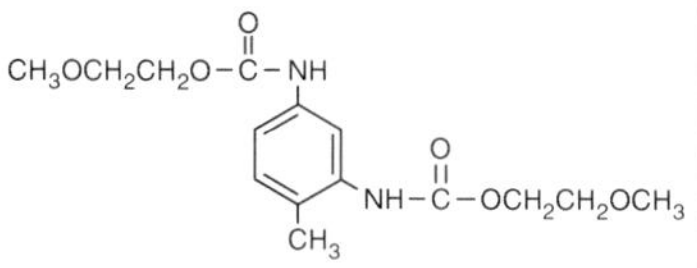

5 gm

S65,787-5
N,N'-BIS(2-ETHOXYETHOXYCARBONYL)-4-METHYL-1,3-PHENYLENEDIAMINE

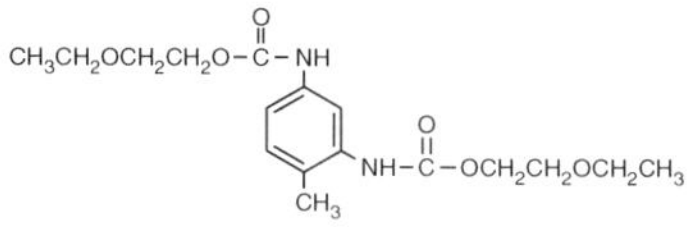

5 gm

S85,380-1
2-(4-METHOXYPHENOXY)ETHYL N-(4-METHYL-3-NITROPHENYL)CARBAMATE

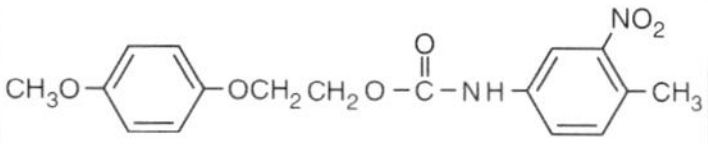

250 mg

S85,548-0
2-PHENOXYETHYL N-(2-ETHOXYPHENYL)-CARBAMATE

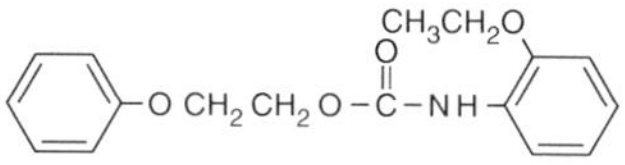

250 mg

S79,520-8
2-ETHOXYPHENYL N-(2-METHOXYPHENYL)-CARBAMATE

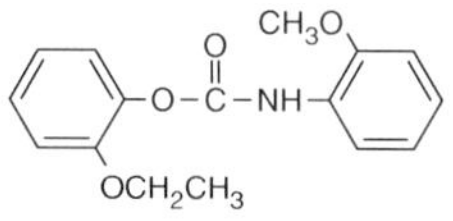

250 mg

S72,287-1
2,6-DIMETHOXYPHENYL N-(2,5-DIMETHOXYPHENYL)CARBAMATE

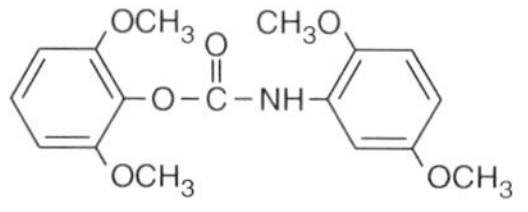

250 mg

S90,247-0
2-[2-METHOXY-4-(1-PROPENYL)PHENOXY]-ETHYL N-(2-METHOXY-5-METHYLPHENYL)-CARBAMATE

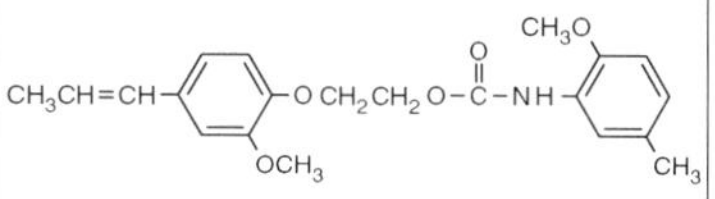

250 mg

S85,383-6
2-[2-METHOXY-4-(1-PROPENYL)PHENOXY]-ETHYL N-(4-METHYL-3-NITROPHENYL)-CARBAMATE

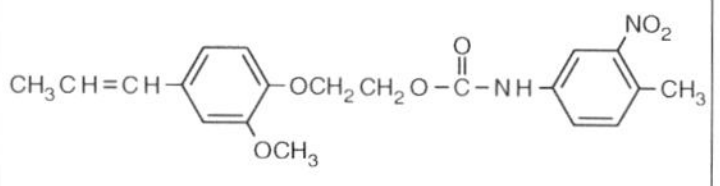

250 mg

S91,145-3
2,6-DIMETHOXYPHENYL N-[4-(ETHOXYCARBONYL)PHENYL]-CARBAMATE

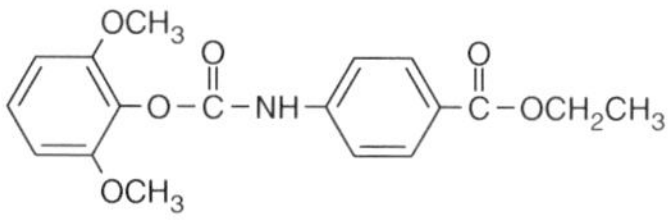

250 mg

S85,549-9
2-PHENOXYETHYL N-(4-ETHOXYPHENYL)-CARBAMATE

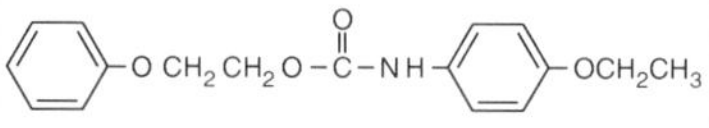

250 mg

S89,192-4
2-ETHOXYPHENYL N-(4-METHOXYPHENYL)-CARBAMATE

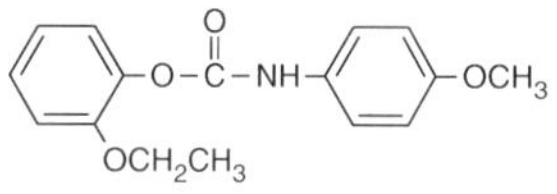

250 mg

S91,041-4
2,6-DIMETHOXYPHENYL N-ISOPROPYLCARBAMATE

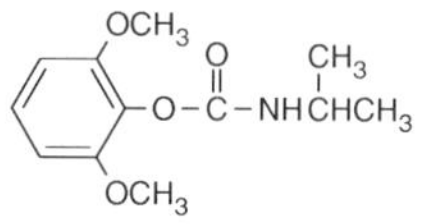

250 mg

S73,107-2
2,6-DIMETHOXYPHENYL N-BUTYLCARBAMATE

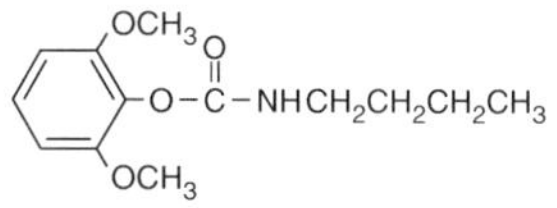

5 gm

S91,220-4
2,6-DIMETHOXYPHENYL N-(2,6-DIETHYLPHENYL)CARBAMATE

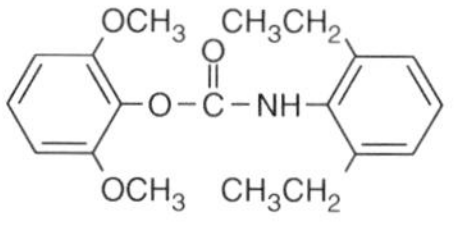

250 mg

S74,313-5
4-ALLYL-2-METHOXYPHENYL N-(3-METHOXYPHENYL)CARBAMATE

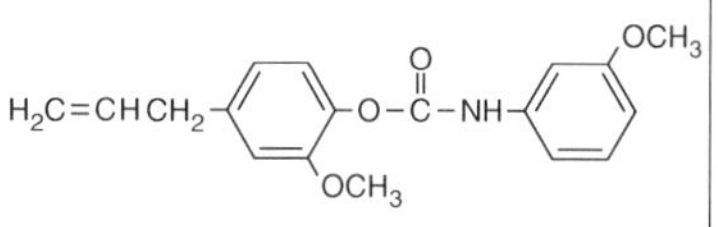

5 gm

S68,845-2
2-[2-METHOXY-4-(1-PROPENYL)PHENOXY]-ETHYL N-(3-ETHYLPHENYL)CARBAMATE

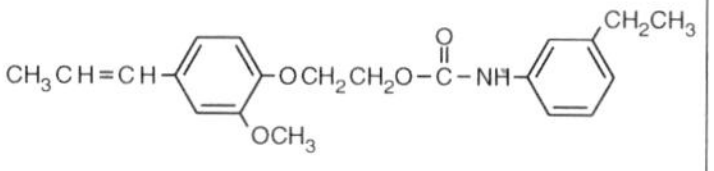

250 mg

S85,448-4
2-[2-METHOXY-4-(1-PROPENYL)PHENOXY]-ETHYL N-(3,4-XYLYL)CARBAMATE

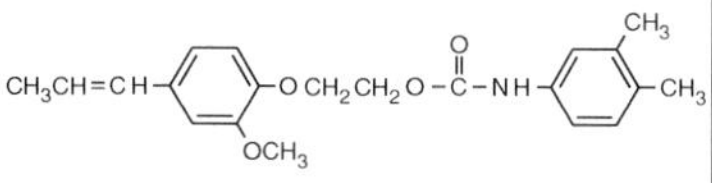

250 mg

S84,841-7
2-(METHOXYCARBONYL)PHENYL N-(2,5-DIMETHOXYPHENYL)CARBAMATE

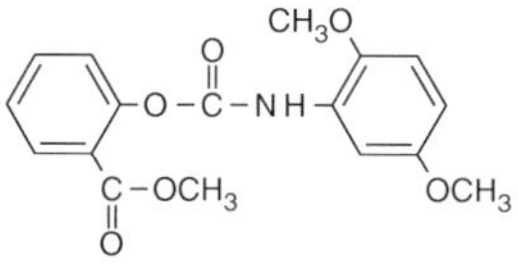

250 mg

S85,105-1
4-BUTOXYPHENYL N-(2,4-DIMETHOXYPHENYL)CARBAMATE

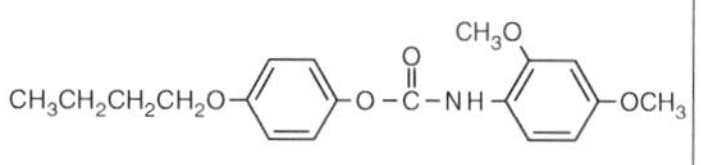

250 mg

S85,116-7
3-ACETYLPHENYL N-(2,4-DIMETHOXYPHENYL)CARBAMATE

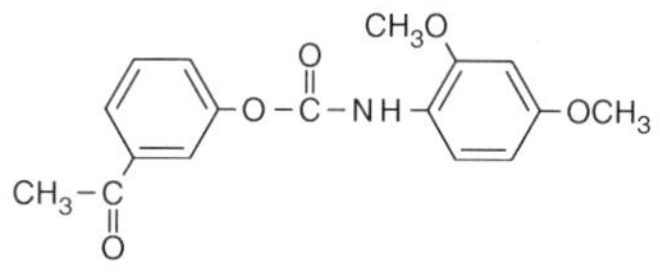

250 mg

S79,677-8
2-(METHOXYCARBONYL)PHENYL N-(2,4-DIMETHOXYPHENYL)CARBAMATE

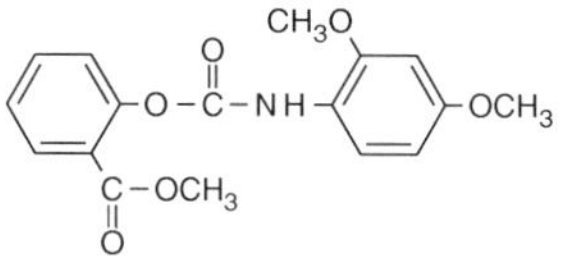

250 mg

S85,103-5
4-(BUTOXYCARBONYLPHENYL) N-(2,4-DIMETHOXYPHENYL)CARBAMATE

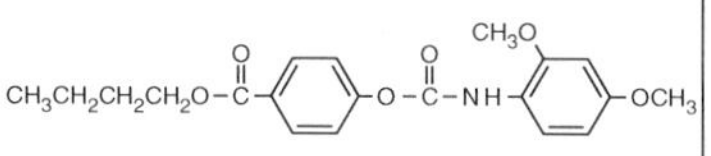

250 mg

S85,119-1
2-CARBAMOYLPHENYL N-(2,4-DIMETHOXYPHENYL)CARBAMATE

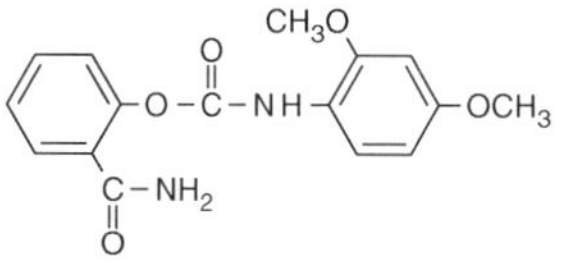

250 mg

S84,880-8
4-METHYL-7-COUMARINYL N-(2,5-DIMETHOXYPHENYL)CARBAMATE

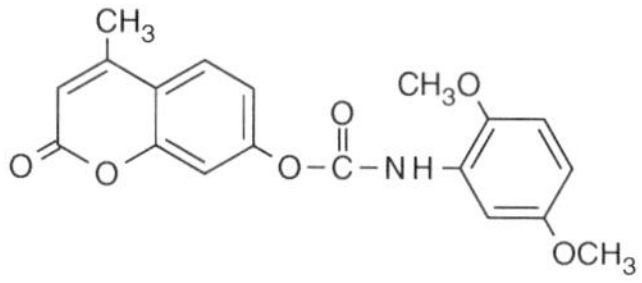

250 mg

S85,110-8
4-(ETHOXYCARBONYLPHENYL) N-(2,4-DIMETHOXYPHENYL)CARBAMATE

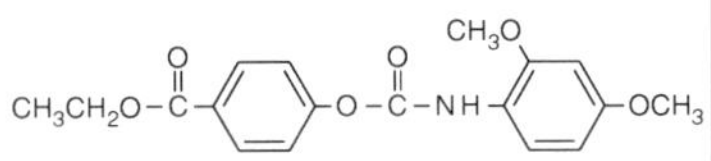

250 mg

S85,114-0
4-ACETAMIDOPHENYL N-(2,4-DIMETHOXYPHENYL)CARBAMATE

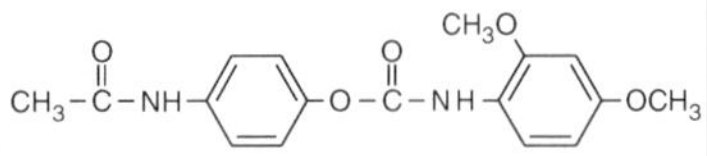

250 mg

S85,115-9
3-ACETAMIDOPHENYL N-(2,4-DIMETHOXYPHENYL)CARBAMATE

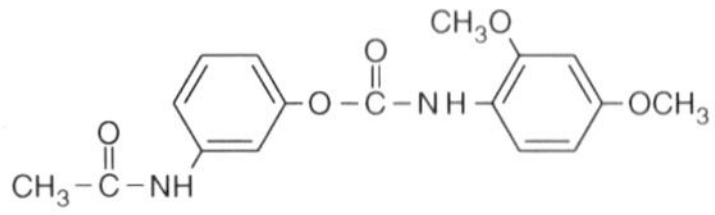

250 mg

S85,129-9
CYCLOPENTYL N-(2,4-DIMETHOXYPHENYL)-CARBAMATE

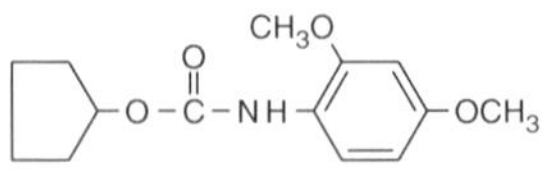

250 mg

S85,111-6
3-PHENYLALLYL N-(2,4-DIMETHOXYPHENYL)CARBAMATE

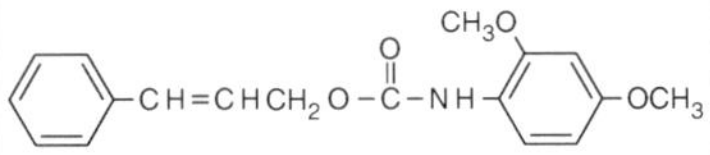

250 mg

S84,836-0
4-(METHYLTHIO)PHENYL N-(2,5-DIMETHOXYPHENYL)CARBAMATE

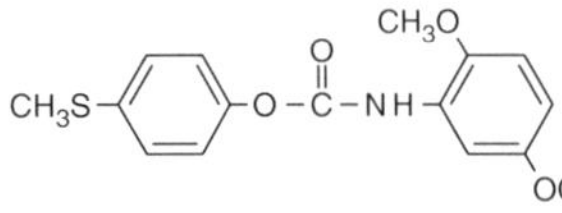

250 mg

S72,778-4
4-METHOXYPHENYL N-(2-METHOXY-4-NITROPHENYL)CARBAMATE

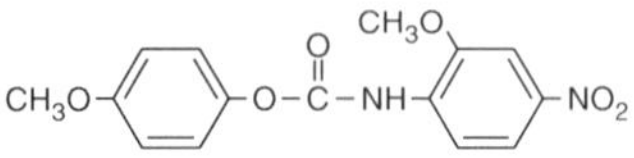

5 gm

S72,984-1
4-NITROBENZYL N-(2,5-DIMETHOXYPHENYL)CARBAMATE

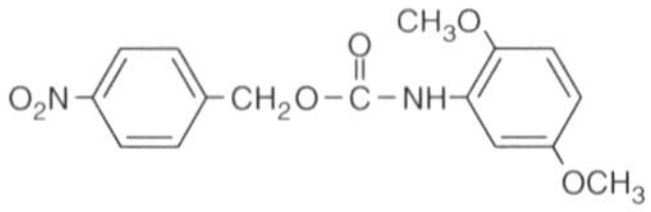

5 gm

S84,858-1
4-(BUTOXYCARBONYL)PHENYL N-(2,5-DIMETHOXYPHENYL)CARBAMATE

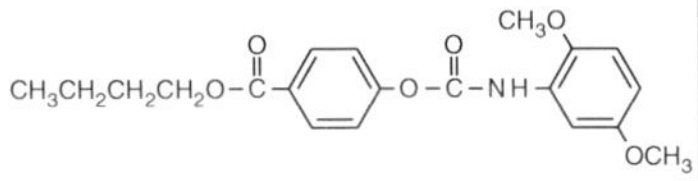

250 mg

S84,879-4
4-(ETHOXYCARBONYL)PHENYL N-(2,5-DIMETHOXYPHENYL)CARBAMATE

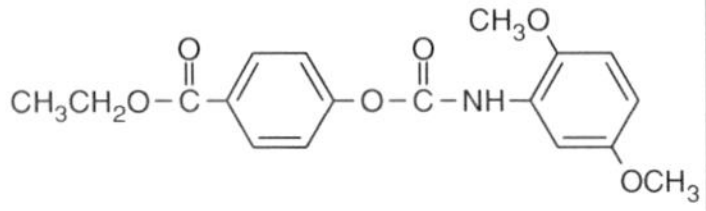

250 mg

S90,373-6
4-(2,5-DIMETHOXYPHENYL)-1-(ETHOXYCARBONYL)SEMICARBAZIDE

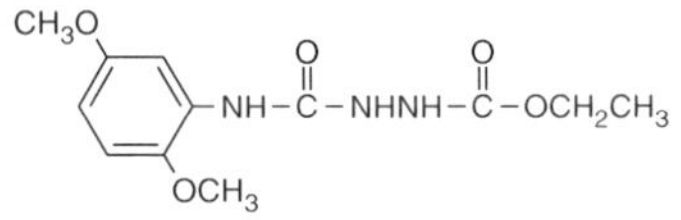

250 mg

S84,844-1
4-ACETAMIDOPHENYL N-(2,5-DIMETHOXYPHENYL)CARBAMATE

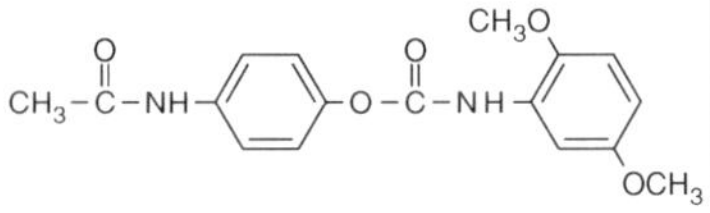

250 mg

S84,872-7
3-ACETAMIDOPHENYL N-(2,5-DIMETHOXYPHENYL)CARBAMATE

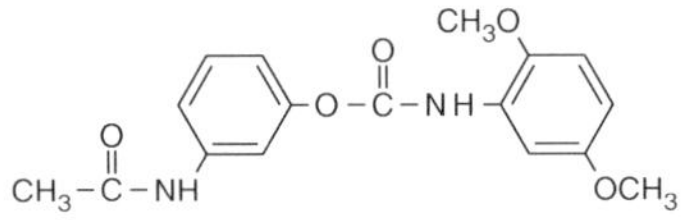

250 mg

S71,269-8
CYCLOHEXYL N-(2,5-DIMETHOXYPHENYL)-CARBAMATE

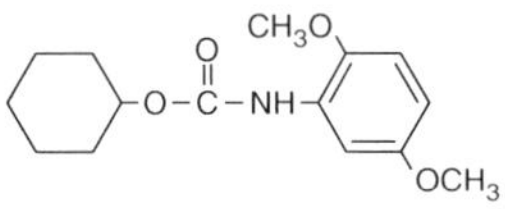

5 gm

S79,137-7
4-METHOXYPHENYL N-(2-METHOXPHENYL)-CARBAMATE

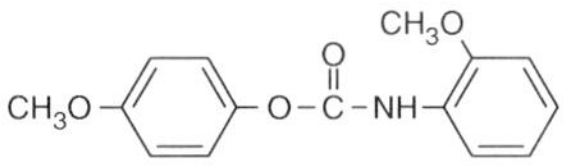

250 mg

S84,855-7
2-ISOPROPYL-5-METHYLPHENYL N-(2,5-DIMETHOXYPHENYL)CARBAMATE

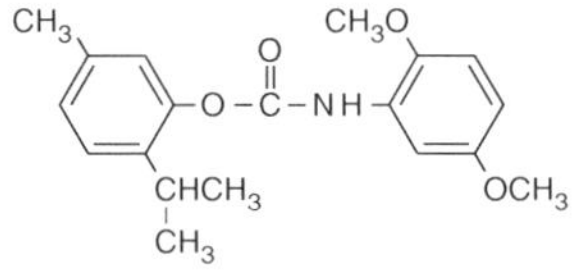

250 mg

S79,684-0
3-ETHYL-5-METHYLPHENYL N-(2,5-DIMETHOXYPHENYL)CARBAMATE

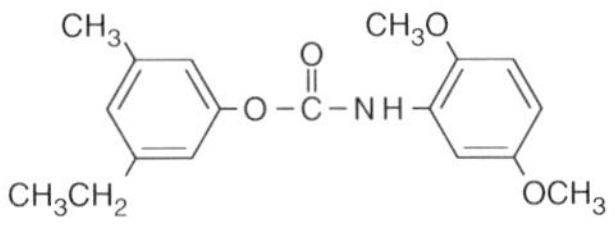

250 mg

S85,049-7
S-(4-METHOXYPHENYL) N-(2,5-DIMETHOXYPHENYL)THIOCARBAMATE

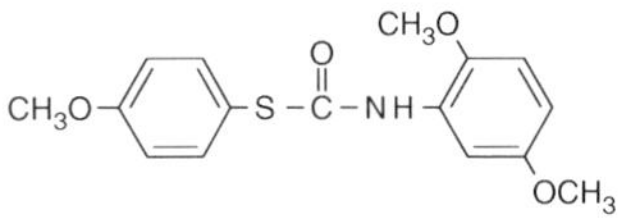

250 mg

S70,991-3
S-ISOPROPYL N-(2,5-DIMETHOXYPHENYL)-THIOLCARBAMATE

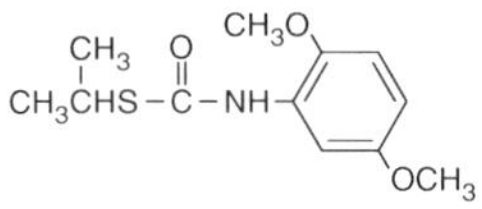

5 gm

S71,113-6
S-PROPYL N-(2,5-DIMETHOXYPHENYL)-THIOLCARBAMATE

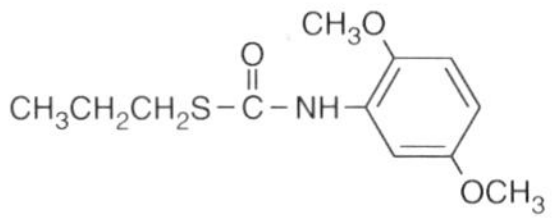

5 gm

S70,968-9
S-BUTYL N-(2,5-DIMETHOXYPHENYL)-THIOCARBAMATE

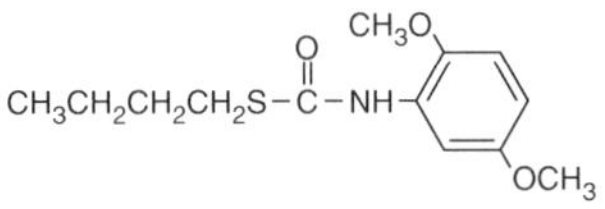

5 gm

S85,047-0
S,S'-ETHYLENE BIS[N-(2,5-DIMETHOXYPHENYL)THIOCARBAMATE]

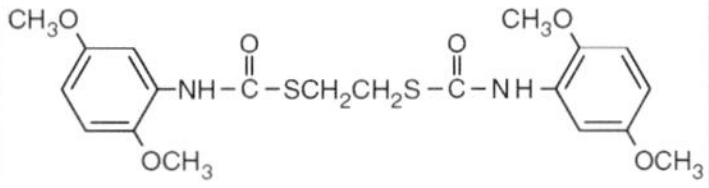

250 mg

S89,025-1
4-METHOXYPHENYL N-(4-METHOXYPHENYL)-CARBAMATE

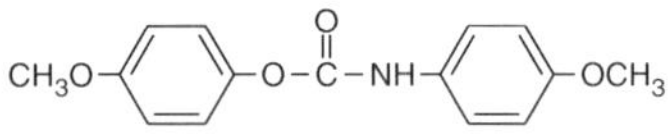

250 mg

S80,034-1
4-METHOXYPHENYL N-(2-METHOXY-5-NITROPHENYL)CARBAMATE

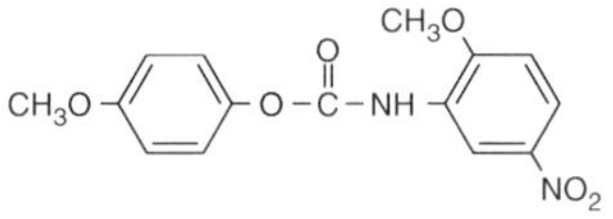

250 mg

S90,245-4
2-(4-BIPHENYLYLOXY)ETHYL N-(2-METHOXY-5-METHYLPHENYL)-CARBAMATE

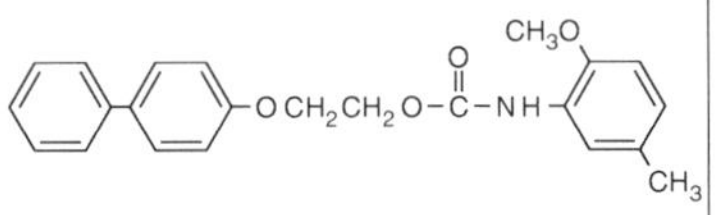

250 mg

S84,853-0
2-NAPHTHYL N-(2,5-DIMETHOXYPHENYL)-CARBAMATE

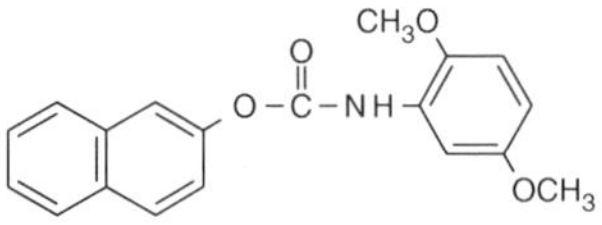

250 mg

S72,279-0
6-BROMO-2-NAPHTHYL N-(2,5-DIMETHOXYPHENYL)CARBAMATE

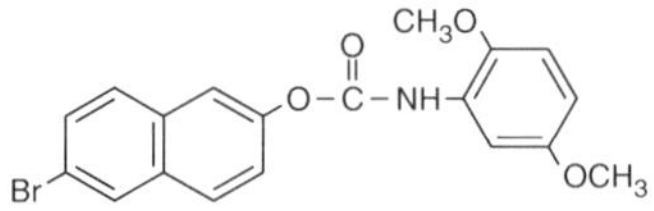

5 gm

S92,432-6
2-(4-NITROPHENOXY)ETHYL N-(ETHOXYCARBONYLMETHYL)CARBAMATE

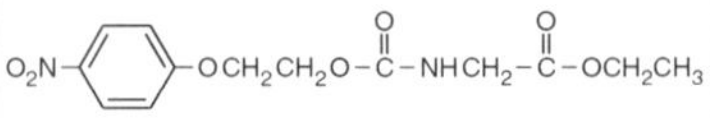

250 mg

S71,344-9
DIMETHYL 4',4''-METHYLENEDICARBANILATE

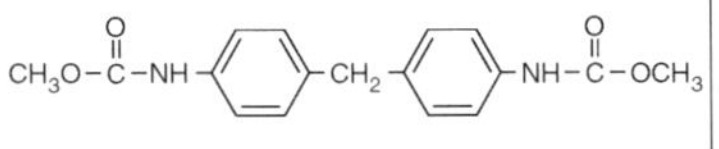

5 gm

S85,184-1
2-(4-METHOXYPHENOXY)ETHYL N-(4-METHYL-2-NITROPHENYL)CARBAMATE

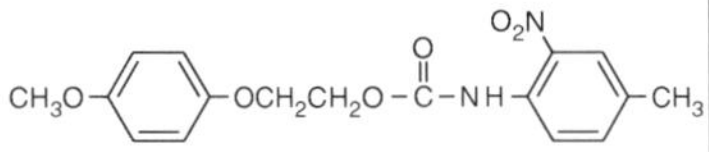

250 mg

S79,485-6
4-FORMYL-2-METHOXYPHENYL N-(2,5-XYLYL)CARBAMATE

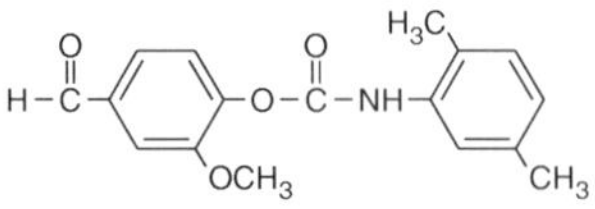

250 mg

S85,356-9
2-(4-NITROPHENOXY)ETHYL N-[3-(METHYLTHIO)PHENYL]CARBAMATE

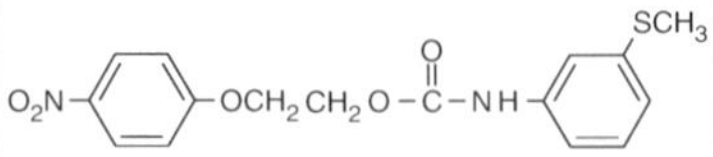

250 mg

S79,732-4
4-FORMYL-2-METHOXYPHENYL N-[4-(METHYLTHIO)PHENYL]CARBAMATE

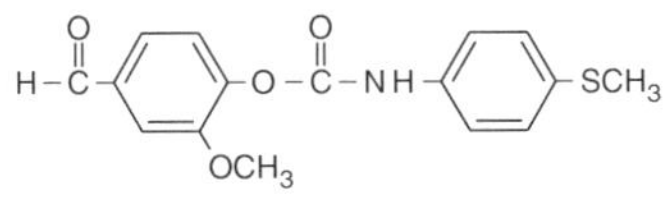

250 mg

S79,738-3
2-(PHENOXY)ETHYL N-[4-(METHYLTHIO)-PHENYL]CARBAMATE

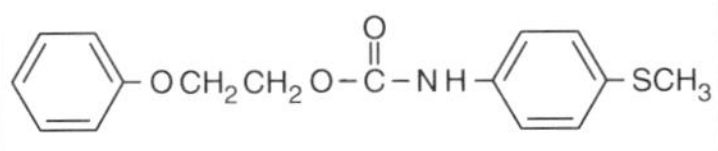

250 mg

S79,545-3
2-(4-TERT-BUTYLPHENOXY)ETHYL N-[3-(METHYLTHIO)PHENYL]CARBAMATE

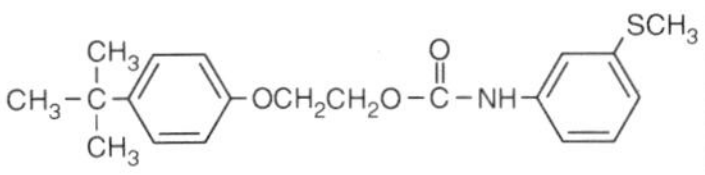

250 mg

S79,772-3
4-ALLYL-2-METHOXYPHENYL N-[4-(METHYLTHIO)PHENYL]CARBAMATE

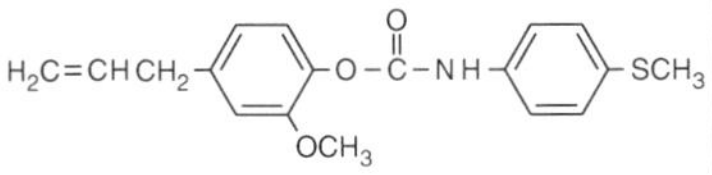

250 mg

S85,538-3
2-PHENOXYETHYL N-(2-NITROPHENYL)-CARBAMATE

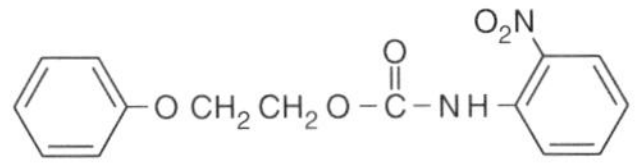

250 mg

S85,181-7
2-(2-TERT-BUTYL-6-METHYLPHENOXY)-ETHYL N-(4-METHYL-2-NITROPHENYL)-CARBAMATE

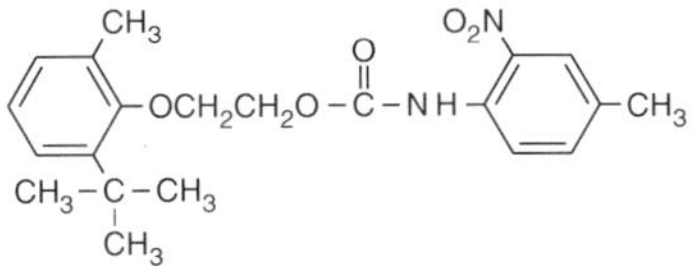

250 mg

S85,162-0
4-ALLYL-2-METHOXYPHENYL N-(4-METHYL-2-NITROPHENYL)CARBAMATE

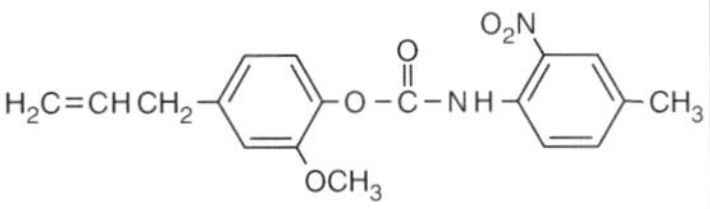

250 mg

S84,835-2
2-PHENOXYETHYL N-(2-METHYL-5-NITROPHENYL)CARBAMATE

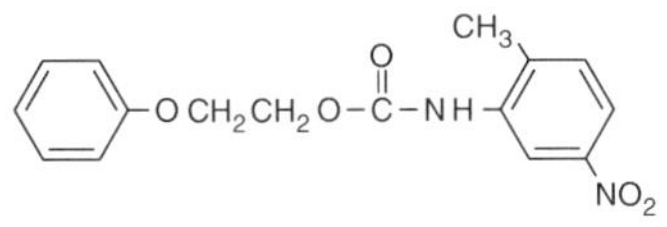

250 mg

S85,368-2
4-ALLYL-2-METHOXYPHENYL N-(4-METHYL-3-NITROPHENYL)CARBAMATE

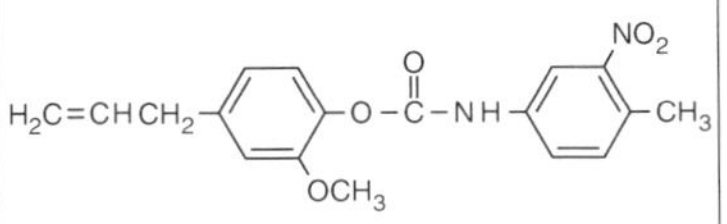

250 mg

S90,562-3
BIS(2-PHENOXYETHYL) N,N'-(4-METHYL-1,3-PHENYLENE)BISCARBAMATE

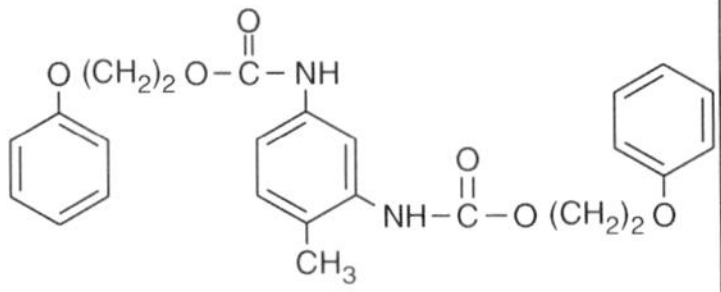

250 mg

S80,014-7
4-ALLYL-2-METHOXYPHENYL N-(2-METHOXY-4-NITROPHENYL)-CARBAMATE

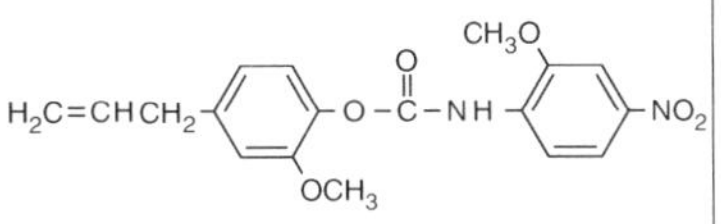

250 mg

S80,001-5
2-PHENOXYETHYL N-(2-METHOXY-4-NITROPHENYL)CARBAMATE

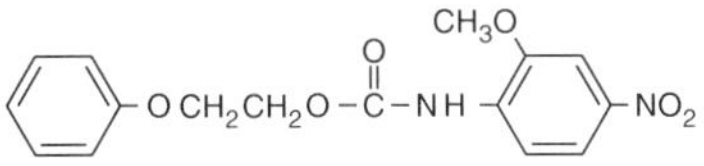

250 mg

S66,174-0
2-ETHOXYETHYL N-[3-(DIETHYLAMINO)-PHENYL]CARBAMATE

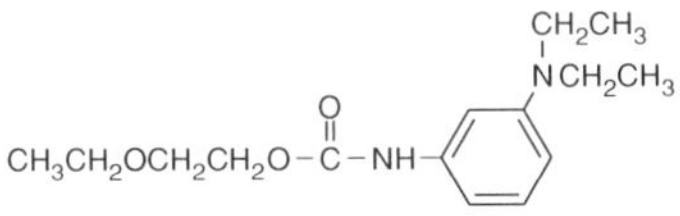

5 gm

S90,068-0
TETRAHYDROFURFURYL N-(2-METHOXY-5-METHYLPHENYL)CARBAMATE

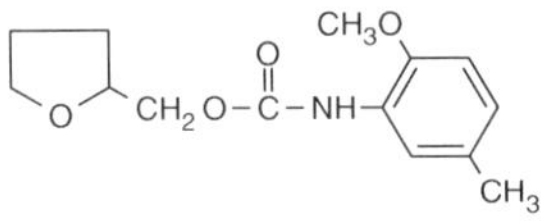

250 mg

S90,571-2
4-FORMYL-2-METHOXYPHENYL N-(3-ACETYLPHENYL)CARBAMATE

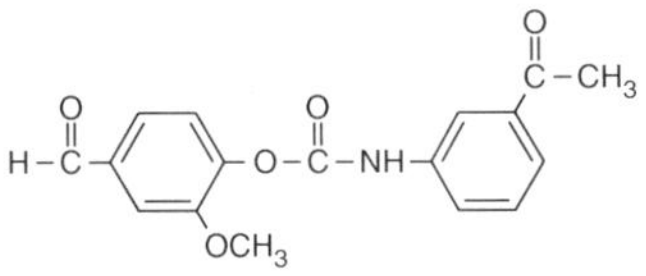

250 mg

S67,566-0
4-FORMYL-2-METHOXYPHENYL N-[4-(ETHOXYCARBONYL)PHENYL]CARBAMATE

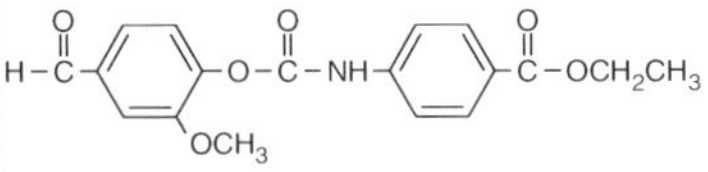

5 gm

S60,474-7
2-BENZOYL-4-CHLOROPHENYL N-(4-METHOXYPHENYL)CARBAMATE

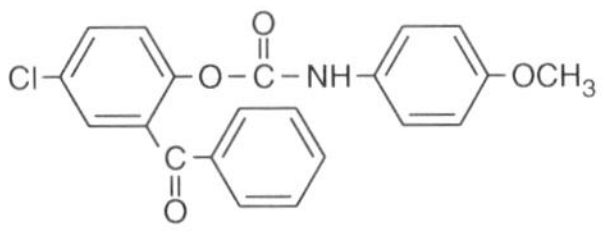

5 gm

S79,771-5
2-ETHOXYPHENYL N-[4-(ETHOXYCARBONYL)PHENYL]-CARBAMATE

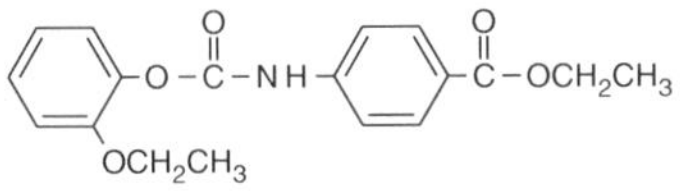

250 mg

S69,500-9
2-METHOXYETHYL N-[4-(ETHOXYCARBONYL)PHENYL]CARBAMATE

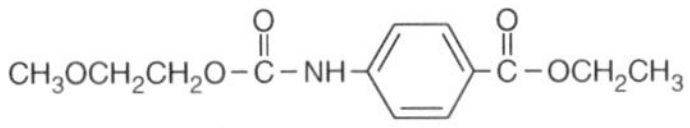

250 mg

S79,783-9
4-ALLYL-2-METHOXYPHENYL N-[4-(ETHOXYCARBONYL)PHENYL]-CARBAMATE

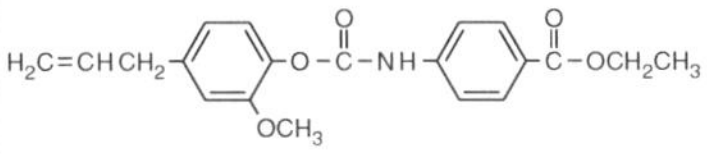

250 mg

S91,229-8
2-(2-NAPHTHOXY)ETHYL N-(ETHOXYCARBONYLMETHYL)CARBAMATE

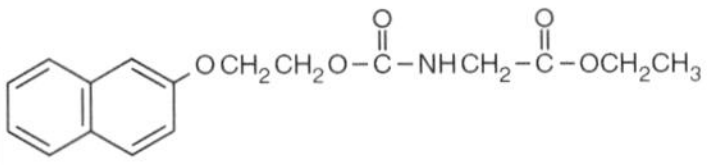

250 mg

S88,290-9
5-(TRITYLOXYMETHYL)-TETRAHYDROFURFURYL N-OCTADECYLCARBAMATE

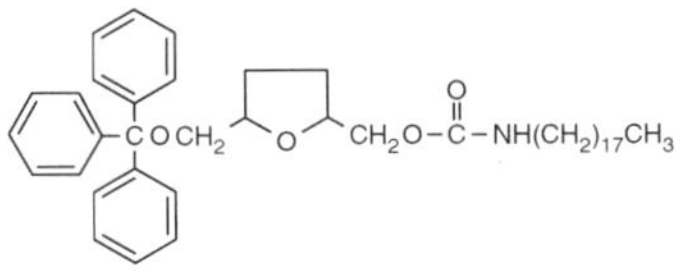

250 mg

S63,807-2
2-METHOXYETHYL N-(3-CHLOROPHENYL)-CARBAMATE

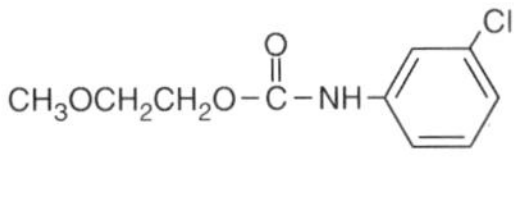

5 gm

S66,662-9
2-METHOXYETHYL N-(M-TOLYL)CARBAMATE

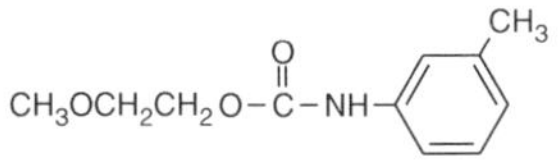

5 gm

S63,163-9
2-ETHOXYETHYL N-(3-CHLOROPHENYL)-CARBAMATE

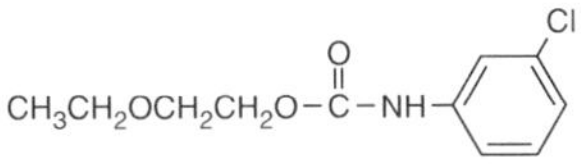

5 gm

S66,668-8
2-ETHOXYETHYL N-(M-TOLYL)CARBAMATE

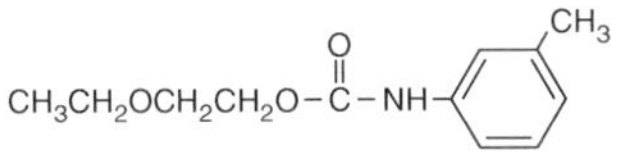

5 gm

S68,615-8
2-PHENOXYETHYL N-(3-ETHYLPHENYL)-CARBAMATE

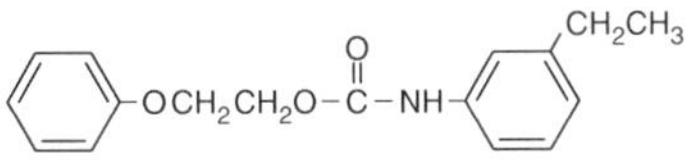

250 mg

S68,857-6
2-(3-METHYLPHENOXY)ETHYL N-(3-ETHYLPHENYL)CARBAMATE

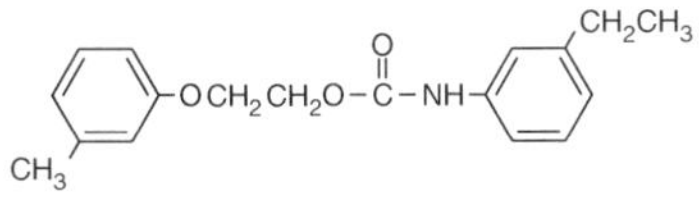

250 mg

S60,757-6
PIPERONYL N-PHENYLCARBAMATE

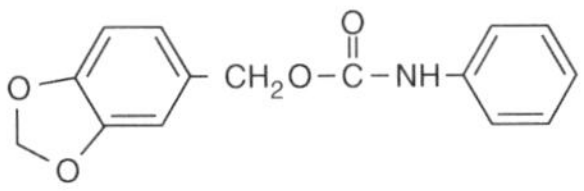

5 gm

S74,309-7
2-PHENOXYETHYL N-CYCLOHEXYLCARBAMATE

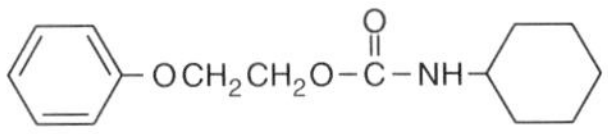

5 gm

S64,407-2
2-METHOXYETHYL N-PHENYLCARBAMATE

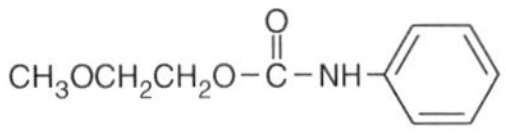

250 mg

S62,434-9
2-ETHOXYETHYL N-PHENYLCARBAMATE

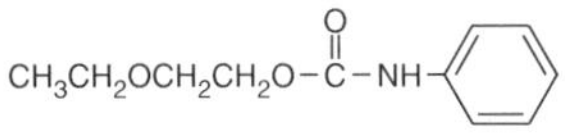

5 gm

S85,576-6
2-PHENOXYETHYL N-(P-TOLYL)CARBAMATE
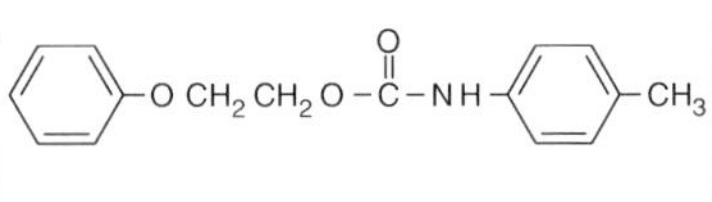
250 mg

S34,551-2
2-(2,4-DICHLOROPHENOXY)ETHYL N-PHENYLCARBAMATE
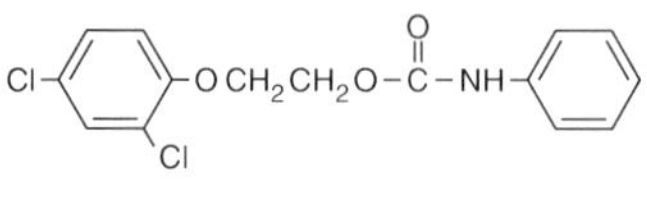
5 gm

S67,407-9
2-PHENOXYETHYL N-(2,5-DIMETHYLPHENYL)CARBAMATE
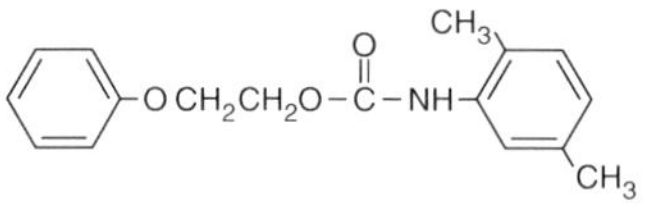
250 mg

S63,854-4
2-METHOXYETHYL N-(2-CHLOROPHENYL)-CARBAMATE
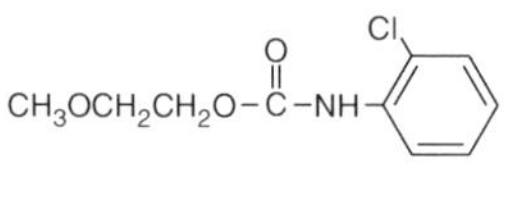
5 gm

S66,665-3
2-METHOXYETHYL N-(O-TOLYL)CARBAMATE
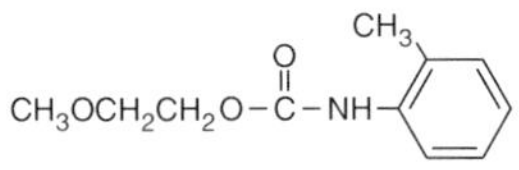
5 gm

S63,012-8
2-ETHOXYETHYL N-(2-CHLOROPHENYL)-CARBAMATE
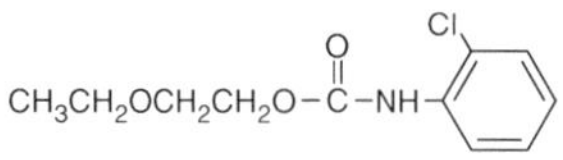
5 gm

S66,666-1
2-ETHOXYETHYL N-(O-TOLYL)CARBAMATE
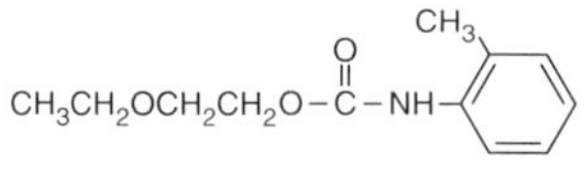
5 gm

S66,795-1
2-METHOXYETHYL N-(2-ETHYLPHENYL)-CARBAMATE
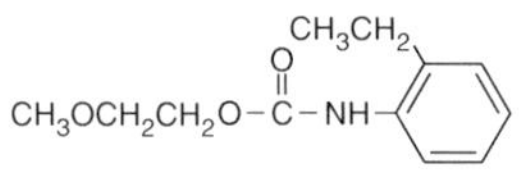
5 gm

S66,781-1
2-ETHOXYETHYL N-(2-ETHYLPHENYL)-CARBAMATE
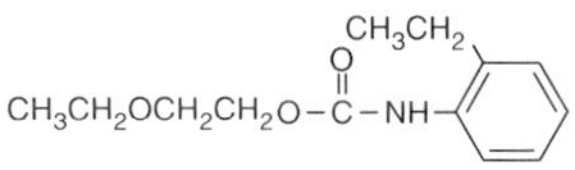
5 gm

S63,815-3
2-(2,4-DICHLOROPHENOXY)ETHYL N-(2-CHLOROPHENYL)CARBAMATE
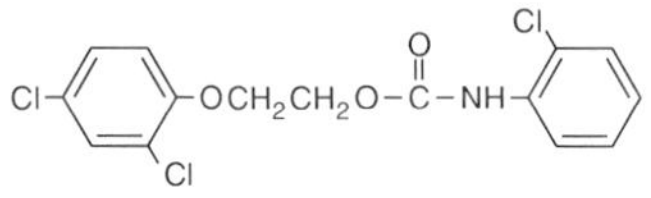
5 gm

S86,853-1
4-ALLYL-2-METHOXYPHENYL N-(O-TOLYL)-CARBAMATE
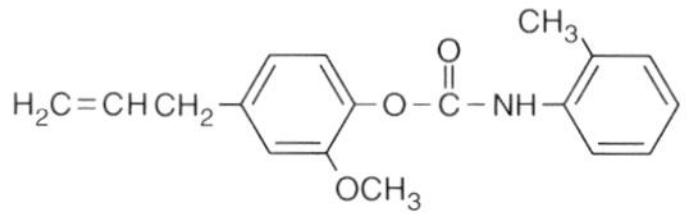
250 mg

S64,414-5
2-METHOXYETHYL N-(4-CHLOROPHENYL)-CARBAMATE
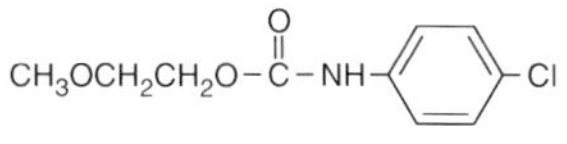
5 gm

S66,663-7
2-METHOXYETHYL N-(P-TOLYL)CARBAMATE
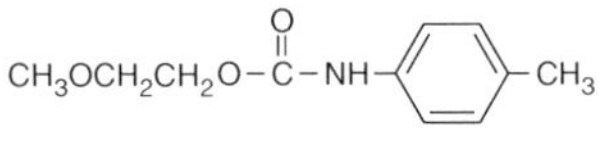
5 gm

S64,467-6
2-ETHOXYETHYL N-(4-CHLOROPHENYL)-CARBAMATE
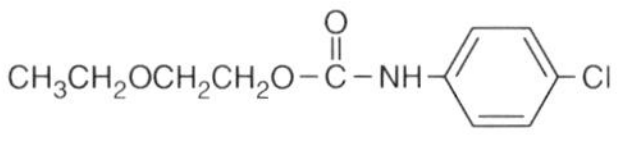
5 gm

S66,664-5
2-ETHOXYETHYL N-(P-TOLYL)CARBAMATE
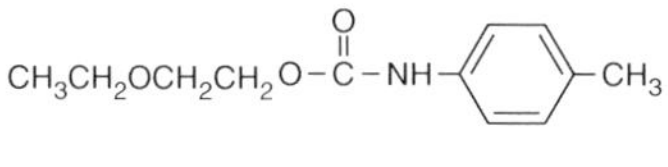
5 gm

S77,377-8
2-(4-TERT-BUTYLPHENOXY)ETHYL N-(3-ETHYLPHENYL)CARBAMATE
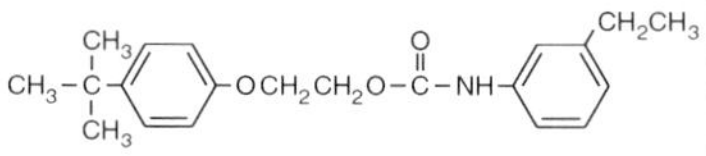
250 mg

S64,450-1
2-(2,4-DICHLOROPHENOXY)ETHYL N-(4-BROMOPHENYL)CARBAMATE
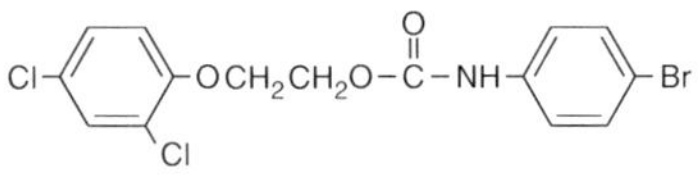
5 gm

S65,042-0
2-METHOXYETHYL N-(2-CHLORO-6-METHYLPHENYL)CARBAMATE
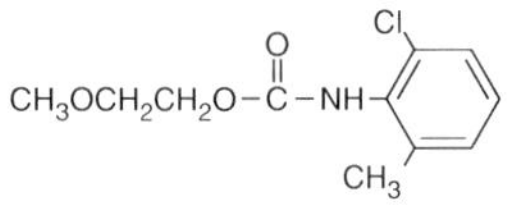
5 gm

S66,520-7
2-METHOXYETHYL N-(2,3-XYLYL)-CARBAMATE
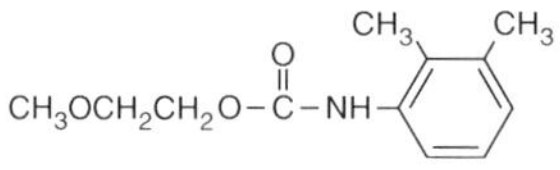
5 gm

S66,656-4
2-METHOXYETHYL N-(2,6-XYLYL)-CARBAMATE
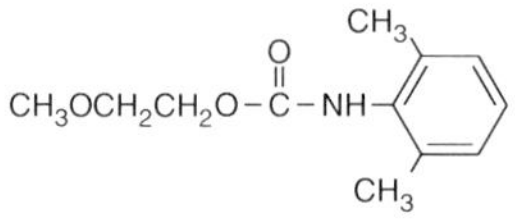
5 gm

S65,033-1
2-ETHOXYETHYL N-(2-CHLORO-6-METHYLPHENYL)CARBAMATE
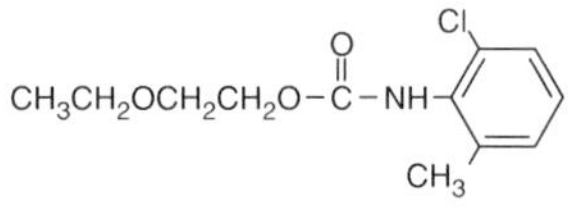
5 gm

S66,509-6
2-ETHOXYETHYL N-(2,3-XYLYL)-CARBAMATE
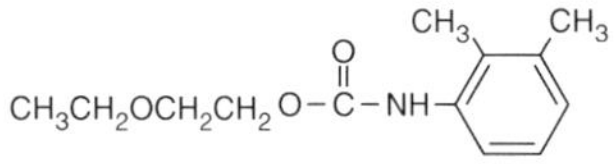
5 gm

S66,539-8
2-METHOXYETHYL N-(2-ETHYL-6-METHYLPHENYL)CARBAMATE
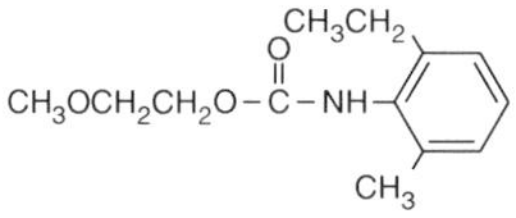
5 gm

S66,643-2
2-ETHOXYETHYL N-(2,6-XYLYL)-CARBAMATE
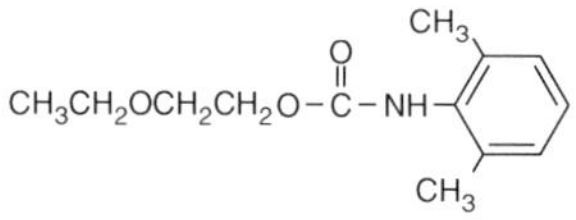
5 gm

S66,543-6
2-ETHOXYETHYL N-(2-ETHYL-6-METHYLPHENYL)CARBAMATE
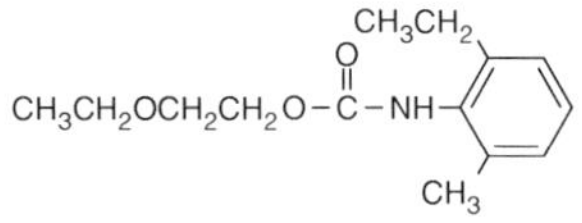
5 gm

S90,249-7
2-(M-TOLYLOXY)ETHYL N-(2,6-XYLYL)-CARBAMATE
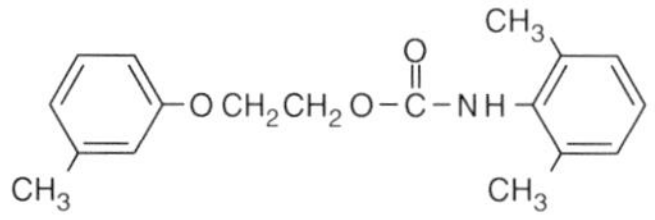
250 mg

S64,470-6
2-(2,4-DICHLOROPHENOXY)ETHYL N-(2,3-DICHLOROPHENYL)CARBAMATE
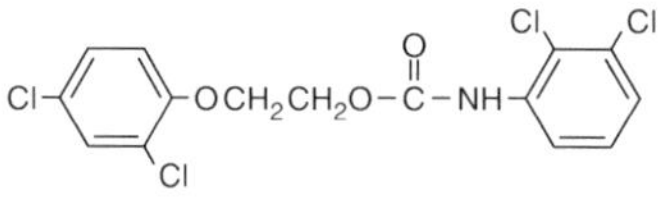
5 gm

S64,955-4
2-(2,4-DICHLOROPHENOXY)ETHYL N-(2,3-XYLYL)CARBAMATE
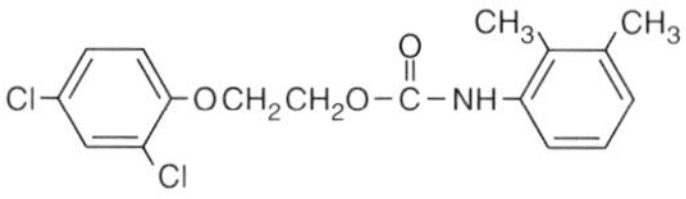
5 gm

S66,527-4
2-METHOXYETHYL N-(2,5-XYLYL)-CARBAMATE
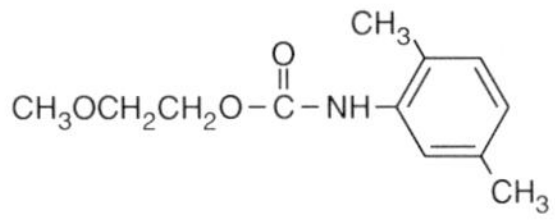
5 gm

S66,583-5
2-METHOXYETHYL N-(2,4-XYLYL)-CARBAMATE
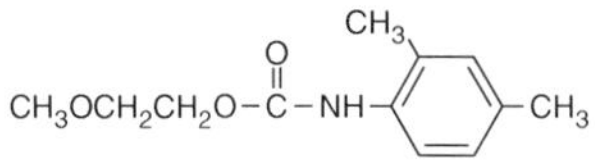
5 gm

S66,536-3
2-ETHOXYETHYL N-(2,5-XYLYL)-CARBAMATE
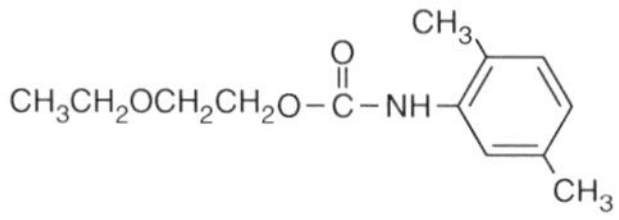
5 gm

S66,591-6
2-ETHOXYETHYL N-(2,4-XYLYL)-CARBAMATE
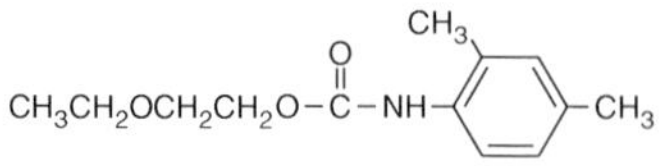
5 gm

S63,805-6
2-(2,4-DICHLOROPHENOXY)ETHYL N-(3-CHLOROPHENYL)CARBAMATE

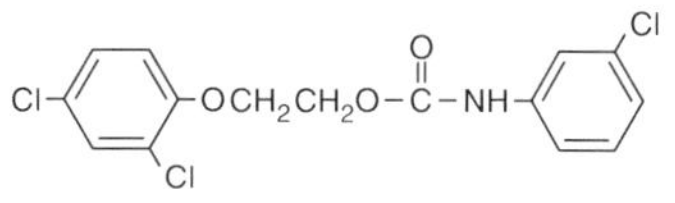

5 gm

S61,245-6
2-(2,4-DICHLOROPHENOXY)ETHYL N-(3,4-DICHLOROPHENYL)CARBAMATE

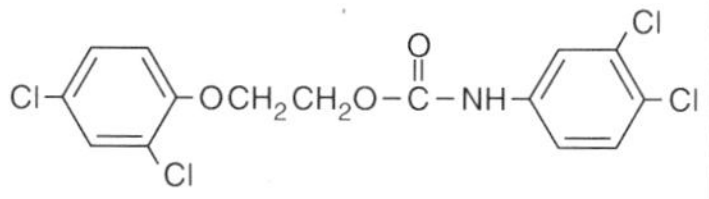

5 gm

S61,092-5
2-(2,4-DICHLOROPHENOXY)ETHYL N-(2,4-XYLYL)CARBAMATE

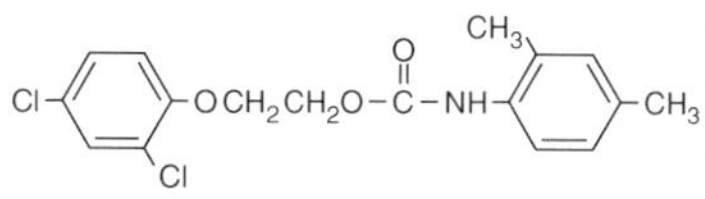

5 gm

S86,855-8
4-ALLYL-2-METHOXYPHENYL N-(M-TOLYL)-CARBAMATE

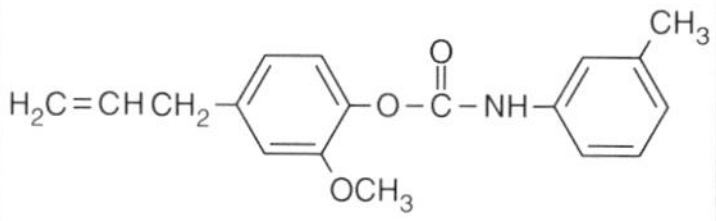

250 mg

S66,783-8
2-METHOXYETHYL N-(3,5-XYLYL)-CARBAMATE

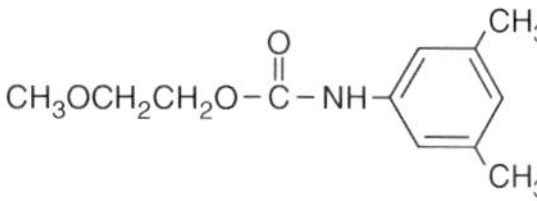

5 gm

S72,311-8
3-METHOXYPHENYL N-[4-(METHYLTHIO)-PHENYL]CARBAMATE

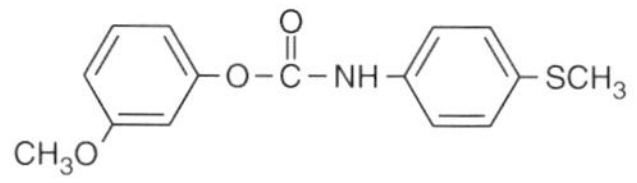

250 mg

S72,813-6
(4-FORMYL-3-METHOXYPHENYL) N-(2-METHOXY-4-NITROPHENYL)-CARBAMATE

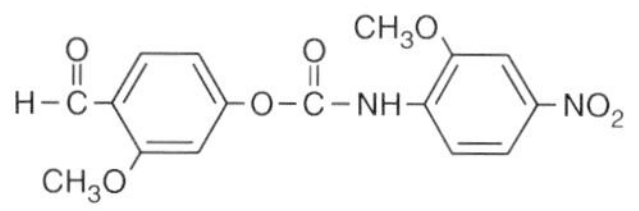

5 gm

S77,241-0
2-(ISOPROPENYLOXYCARBONYL)ETHYL N-(2-METHOXYPHENYL)CARBAMATE

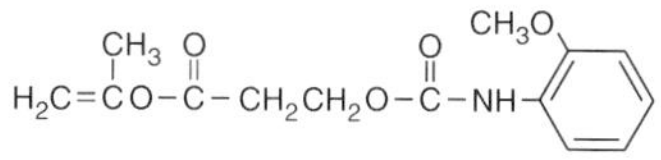

250 mg

S72,103-4
3-(BENZOYLOXY)PHENYL N-(2-METHOXYPHENYL)CARBAMATE

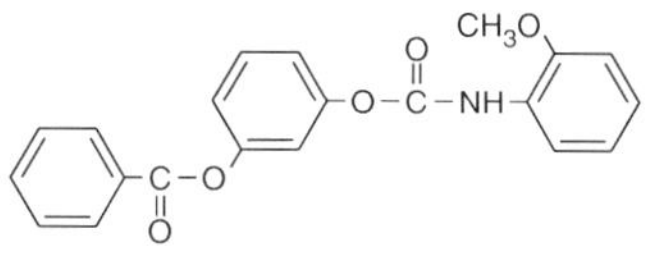

250 mg

S79,401-5
2-(BENZYLOXYCARBONYL)PHENYL N-(3-METHOXYPHENYL)CARBAMATE

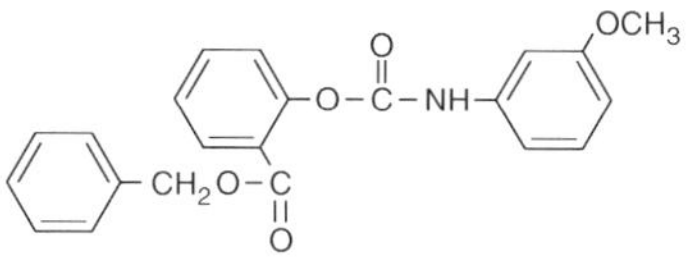

250 mg

S75,283-5
3-(BENZOYLOXY)PHENYL N-(4-METHOXYPHENYL)CARBAMATE

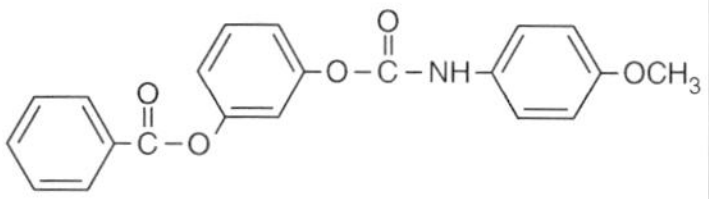

5 gm

S79,723-5
4-METHOXYPHENYL N-[4-(METHYLTHIO)-PHENYL]CARBAMATE

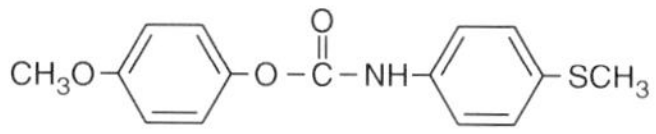

250 mg

S91,079-1
4-BUTOXYPHENYL N-[4-(METHYLTHIO)-PHENYL]CARBAMATE

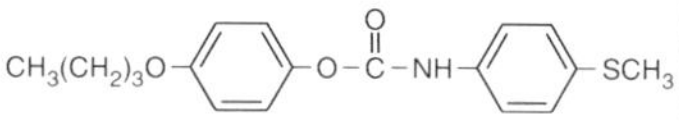

250 mg

S85,189-2
4-METHOXYPHENYL N-(4-METHYL-2-NITROPHENYL)CARBAMATE

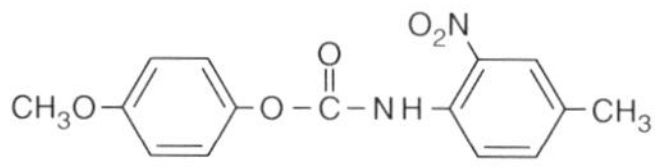

250 mg

S80,003-1
4-BUTOXYPHENYL N-(2-METHOXY-4-NITROPHENYL)CARBAMATE

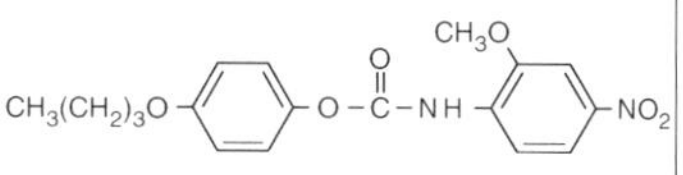

250 mg

S86,805-1
4-METHOXYPHENYL N-(3-ACETYLPHENYL)-CARBAMATE

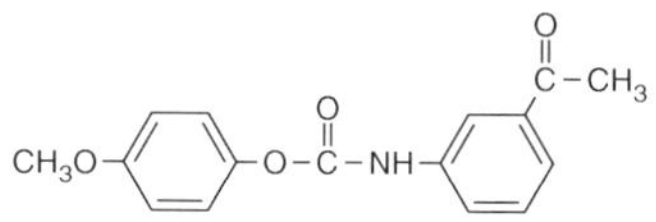

250 mg

S79,428-7
4-BUTOXYPHENYL N-(3-ACETYLPHENYL)-CARBAMATE

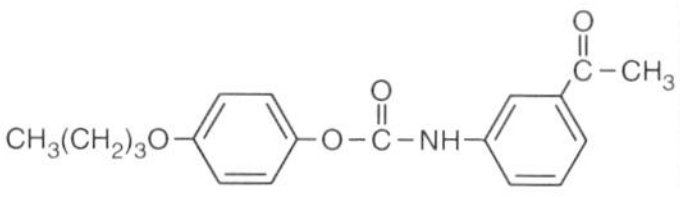

250 mg

S79,749-9
4-METHOXYPHENYL N-[4-(ETHOXYCARBONYL)PHENYL]-CARBAMATE

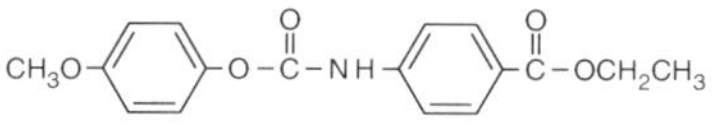

250 mg

S79,787-1
4-BUTOXYPHENYL N-[4-(ETHOXYCARBONYL)PHENYL]-CARBAMATE

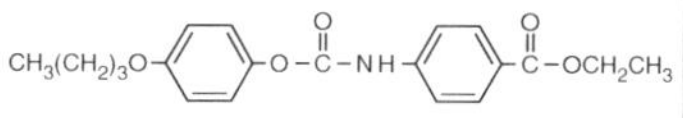

250 mg

S79,759-6
4-BUTOXYPHENYL N-(BUTOXYCARBONYLMETHYL)CARBAMATE

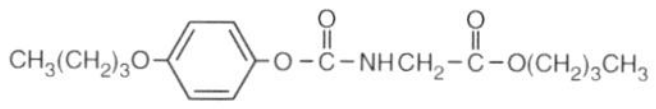

250 mg

S79,463-5
4-METHOXYPHENYL N-(BUTOXYCARBONYLOXYMETHYL)-CARBAMATE

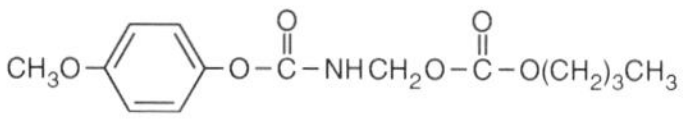

250 mg

S84,567-1
4-METHOXYPHENYL N-(2,6-XYLYL)-CARBAMATE

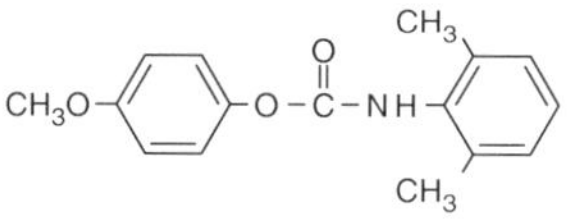

250 mg

S91,221-2
4-BUTOXYPHENYL N-(2,6-DIETHYLPHENYL)CARBAMATE

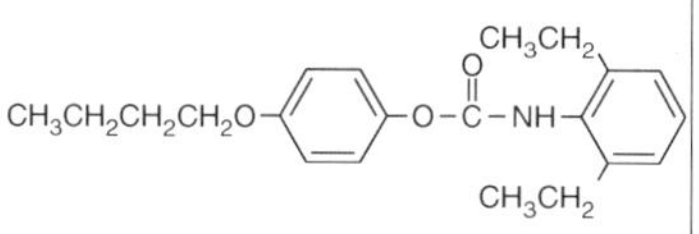

250 mg

S77,363-8
4-METHOXYPHENYL N-(2,5-XYLYL)-CARBAMATE

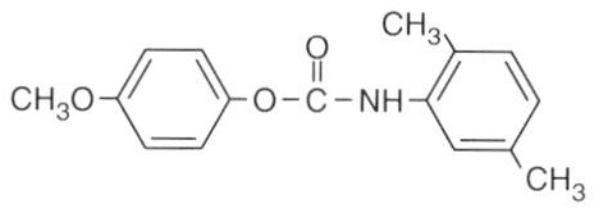

250 mg

S84,737-2
4-BUTOXYPHENYL N-(2,5-XYLYL)-CARBAMATE

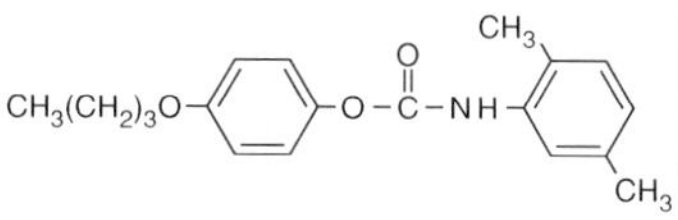

250 mg

S80,033-3
4-(METHYLTHIO)PHENYL N-(2-METHOXY-5-NITROPHENYL)CARBAMATE

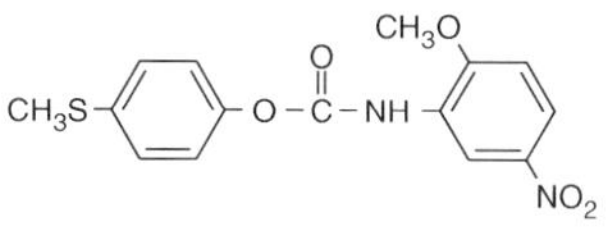

250 mg

S79,518-6
3-METHYL-4-(METHYLTHIO)PHENYL N-(2-METHOXYPHENYL)CARBAMATE

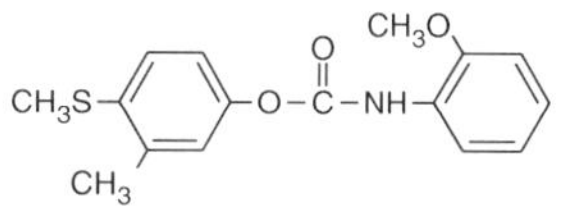

250 mg

S85,546-4
3-METHYL-4-(METHYLTHIO)PHENYL N-(2-ETHOXYPHENYL)CARBAMATE

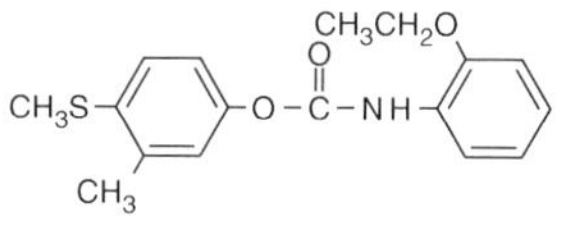

250 mg

S89,196-7
4-(METHYLTHIO)PHENYL N-(4-METHOXYPHENYL)CARBAMATE

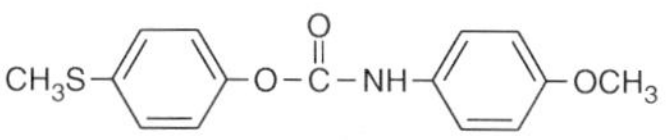

250 mg

S89,193-2
3-METHYL-4-(METHYLTHIO)PHENYL N-(4-METHOXYPHENYL)CARBAMATE

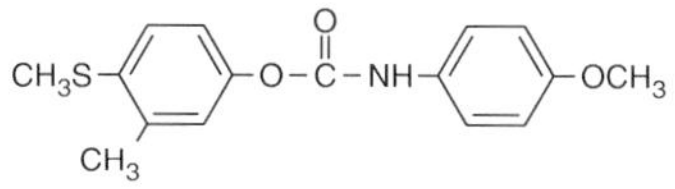

250 mg

S85,547-2
3-METHYL-4-(METHYLTHIO)PHENYL N-(4-ETHOXYPHENYL)CARBAMATE

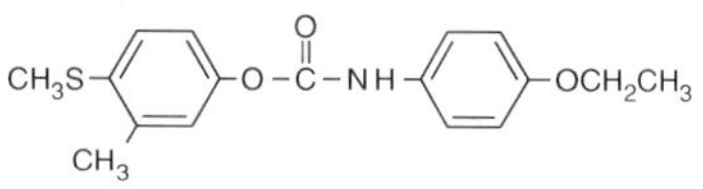

250 mg

S85,156-6
2-(4-BIPHENYLYLOXY)ETHYL N-(4-METHYL-2-NITROPHENYL)CARBAMATE

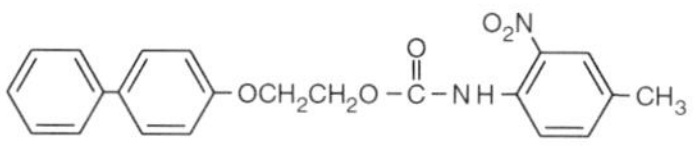

250 mg

S80,497-5
ETHYL N-(4-METHOXY-2-NITROPHENYL)-CARBAMATE

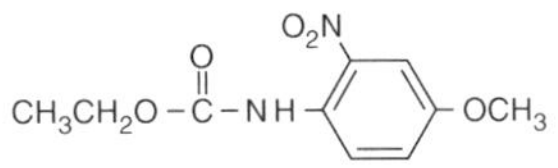

250 mg

S80,506-8
ETHYL N-(4-ETHOXY-2-NITROPHENYL)-CARBAMATE

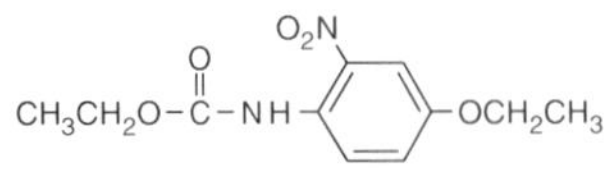

250 mg

S85,585-5
ETHYL N-(2-METHOXY-5-NITROPHENYL)-CARBAMATE

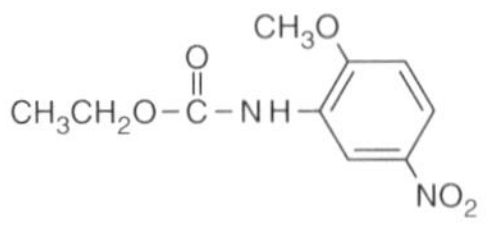

250 mg

S69,307-3
PIPERIDINO N-(2-METHOXY-5-NITROPHENYL)CARBAMATE

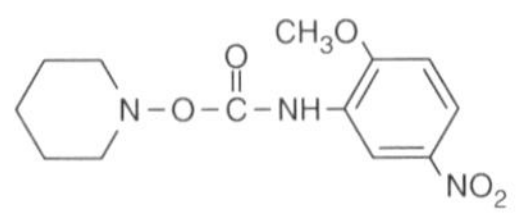

250 mg

S72,552-8
2-(4-BIPHENYLYLOXY)ETHYL N-(4-METHYL-3-NITROPHENYL)CARBAMATE

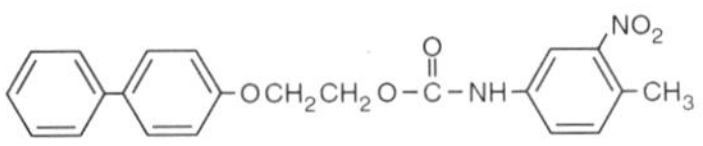

250 mg

S84,827-1
S-(4-METHOXYPHENYL) N-(2-METHYL-5-NITROPHENYL)THIOCARBAMATE

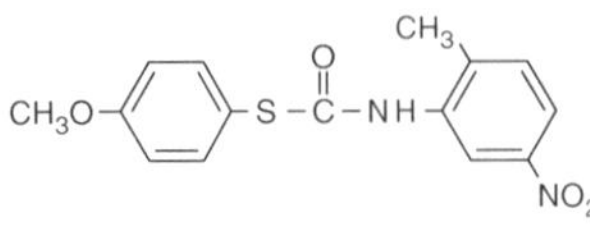

250 mg

S79,734-0
4-ACETAMIDOPHENYL N-(2-METHOXY-4-NITROPHENYL)CARBAMATE

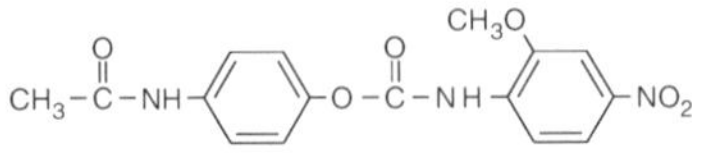

250 mg

S80,000-7
HEXYL N-(2-METHOXY-4-NITROPHENYL)-CARBAMATE

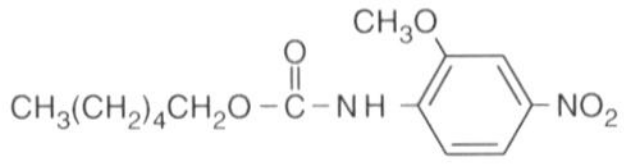

250 mg

S80,015-5
2-ISOPROPYL-5-METHYLPHENYL N-(2-METHOXY-4-NITROPHENYL)-CARBAMATE

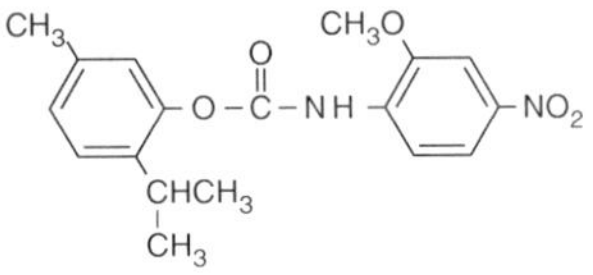

250 mg

S79,443-0
5-NORBORNEN-2-YLMETHYL N-(2-METHOXY-4-NITROPHENYL)-CARBAMATE

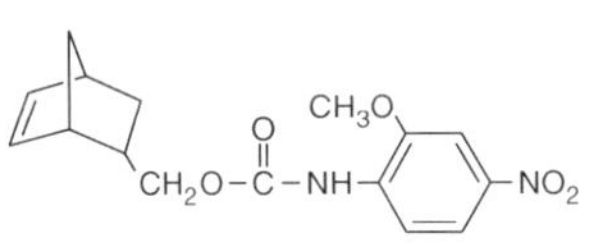

250 mg

S76,533-3
O-BENZYL-N-TERT-BUTOXYCARBONYL-L-TYROSINE-4-NITROPHENYL ESTER

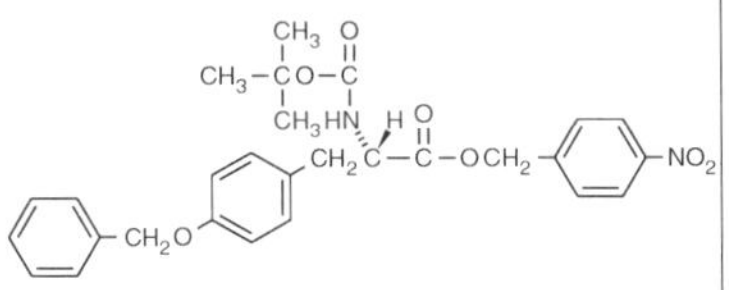

5 gm

S63,646-0
2-(DIMETHYLAMINO)ETHYL N-(3-METHOXYPHENYL)CARBAMATE

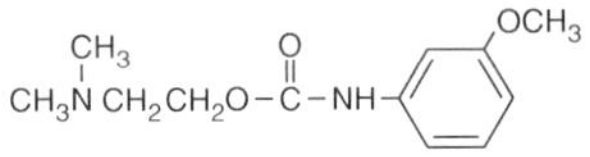

5 gm

S65,305-5
2-(DIETHYLAMINO)ETHYL N-(M-ANISYL)-CARBAMATE

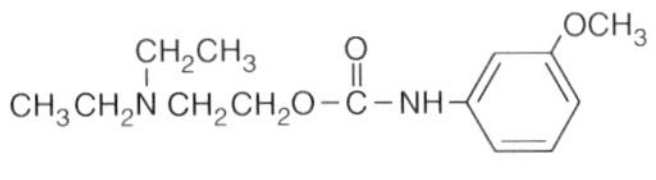

5 gm

S63,015-2
2-(DIMETHYLAMINO)ETHYL N-(2-METHOXYPHENYL)CARBAMATE

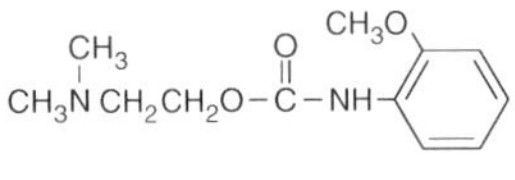

250 mg

S65,306-3
2-(DIETHYLAMINO)ETHYL N-(O-ANISYL)-CARBAMATE

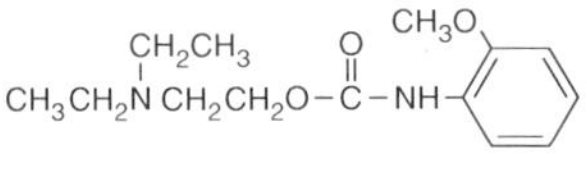

5 gm

S71,070-9
2-(DIMETHYLAMINO)ETHYL N-(4-METHOXYPHENYL)CARBAMATE

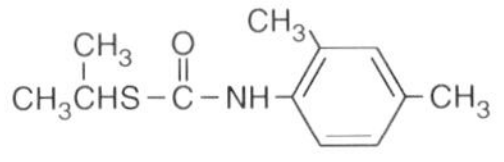

5 gm

S65,211-3
2-(DIETHYLAMINO)ETHYL N-(P-ANISYL)-CARBAMATE

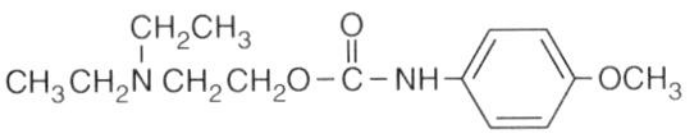

5 gm

S66,353-0
2-(N-ETHYL-N-PHENYLAMINO)ETHYL N-(P-ANISYL)CARBAMATE

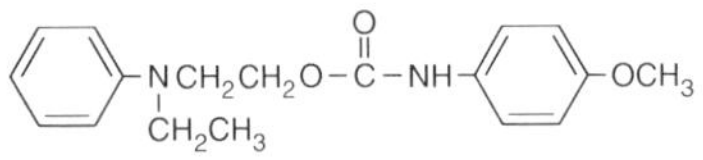

5 gm

S65,061-7
2-DIMETHYLAMINOETHYL N-(3-CHLORO-2-METHOXYPHENYL)CARBAMATE

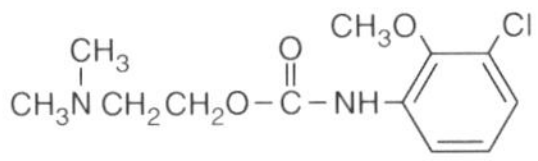

5 gm

S65,069-2
2-DIETHYLAMINOETHYL N-(3-CHLORO-2-METHOXYPHENYL)CARBAMATE

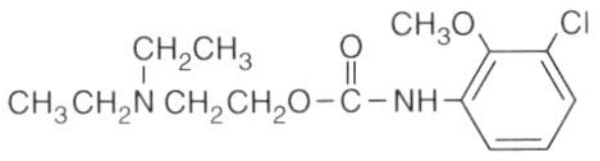

250 mg

S66,678-5
2-DIMETHYLAMINOETHYL N-(2-METHOXY-5-METHYLPHENYL)CARBAMATE

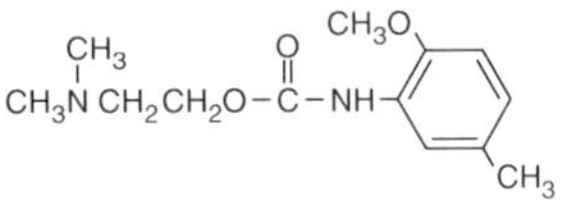

5 gm

S63,479-4
2-(DIMETHYLAMINO)ETHYL N-(2,5-DIMETHOXYPHENYL)CARBAMATE

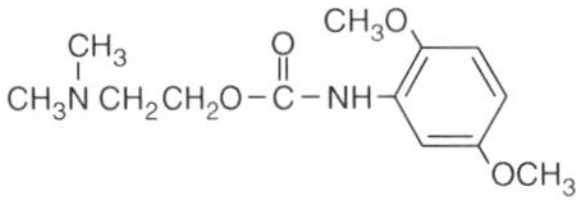

5 gm

S65,060-9
2-DIETHYLAMINOETHYL N-(2-METHOXY-5-METHYLPHENYL)CARBAMATE

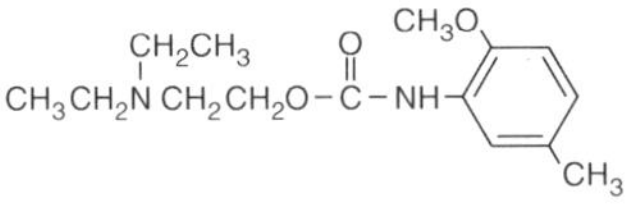

5 gm

S65,068-4
2-DIETHYLAMINOETHYL N-(2,5-DIMETHOXYPHENYL)CARBAMATE

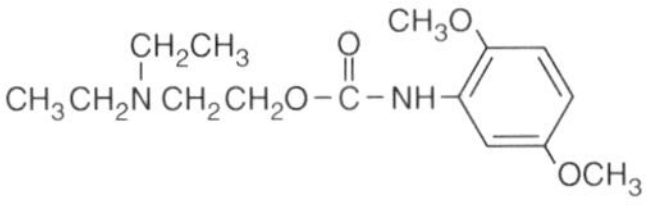

5 gm

S92,435-0
1-METHYL-4-PIPERIDINYL N-(2,5-DIMETHOXYPHENYL)CARBAMATE

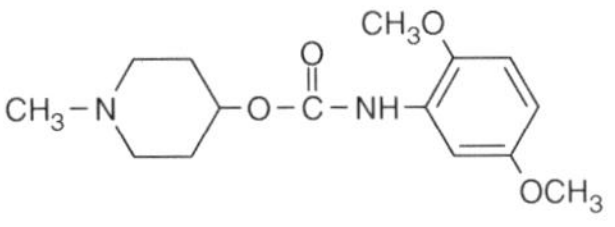

250 mg

S90,283-7
2-METHOXYBENZOIC [4-(N-METHYLCARBAMOYLOXY)BENZYLIDENE]-HYDRAZIDE

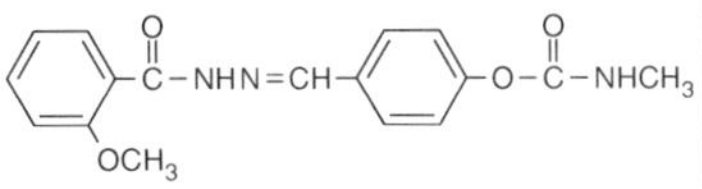

250 mg

S86,213-4
TERT-BUTYL 3-(4-METHOXYBENZYLIDENE)-CARBAZATE

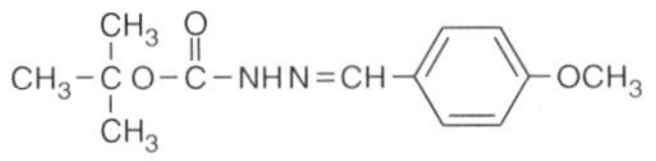

250 mg

S80,011-2
4-(METHOXYCARBONYL)PHENYL N-(2-METHOXY-4-NITROPHENYL)-CARBAMATE

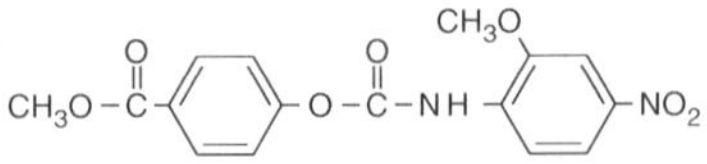

250 mg

S80,048-1
4-(BUTOXYCARBONYL)PHENYL N-(2-METHOXY-4-NITROPHENYL)-CARBAMATE

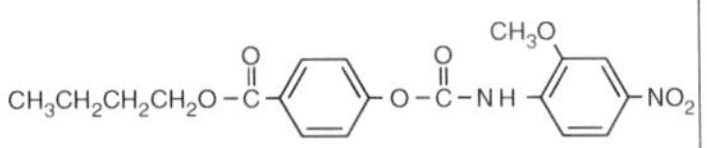

250 mg

S79,717-0
2-CARBAMOYLPHENYL N-(2-METHOXY-4-NITROPHENYL)CARBAMATE

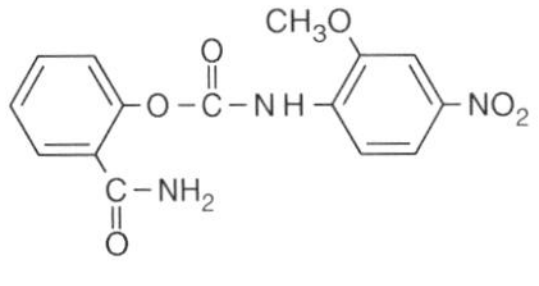

250 mg

S80,002-3
3-PHENYLALLYL N-(2-METHOXY-4-NITROPHENYL)CARBAMATE

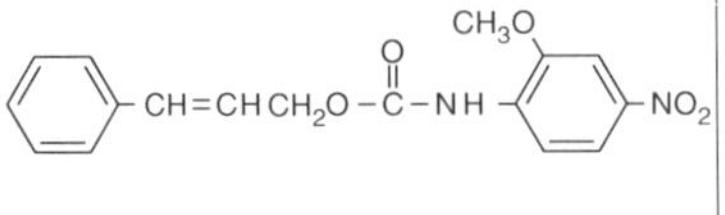

250 mg

S60,364-3
2-BENZOYL-4-CHLOROPHENYL N-(O-ANISYL)CARBAMATE

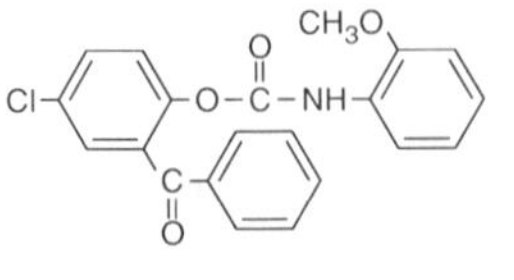

5 gm

S85,542-1
4-(METHOXYCARBONYL)PHENYL N-(2-ETHOXYPHENYL)CARBAMATE

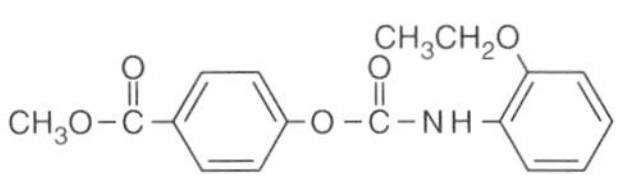

250 mg

S89,087-1
4-(ETHOXYCARBONYL)PHENYL N-(2-METHOXYPHENYL)CARBAMATE

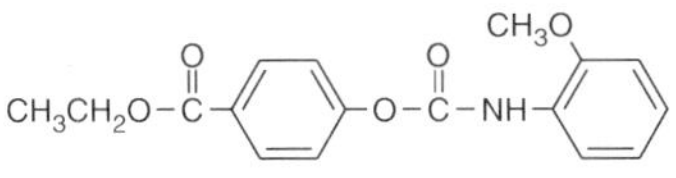

250 mg

S63,642-8
2-METHYLCYCLOHEXYL N-(2-METHOXYPHENYL)CARBAMATE

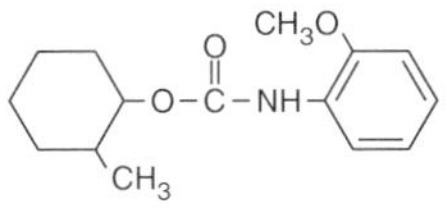

5 gm

S66,681-5
2-METHYLCYCLOHEXYL N-(2-METHOXY-5-METHYLPHENYL)CARBAMATE

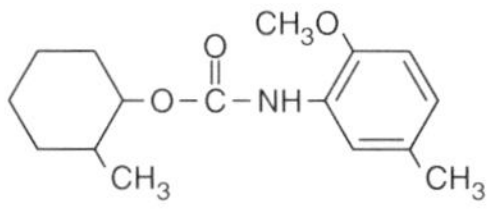

5 gm

S62,427-6
CYCLOHEXYL N-(O-ANISYL)CARBAMATE

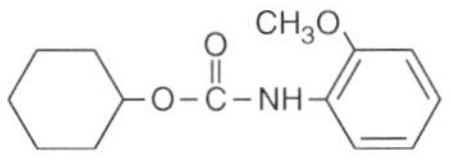

5 gm

S72,461-0
4-CYCLOHEXYLPHENYL N-(2-METHOXYPHENYL)CARBAMATE

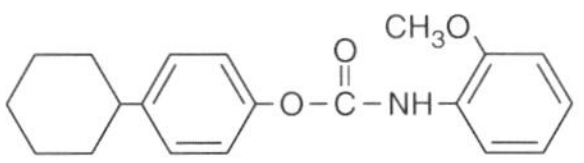

5 gm

S74,870-6
CYCLOHEXYL N-(2-METHOXY-5-METHYLPHENYL)CARBAMATE

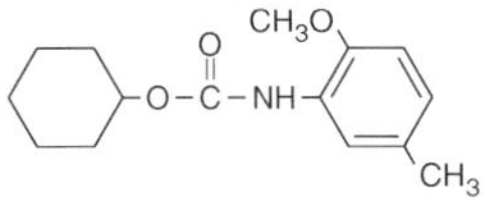

250 mg

S84,502-7
2-MORPHOLINOETHYL N-(4-METHOXYPHENYL)CARBAMATE

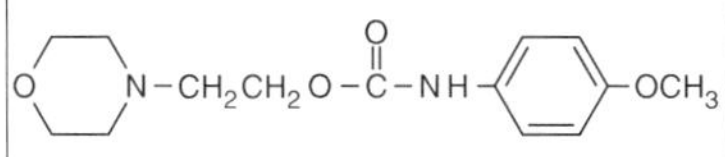

250 mg

S85,555-3
4-BIPHENYLYL N-(2-ETHOXYPHENYL)-CARBAMATE

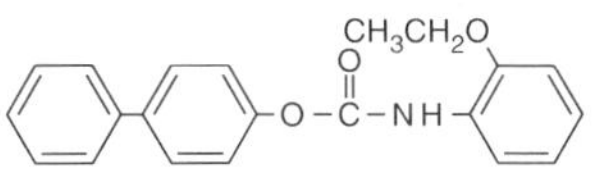

250 mg

S71,899-8
2-BIPHENYLYL N-(2-METHOXYPHENYL)-CARBAMATE

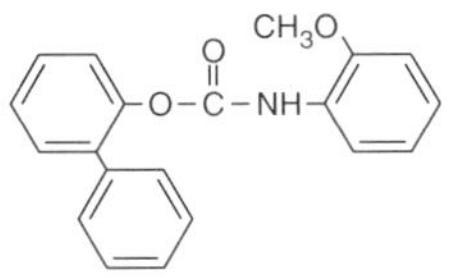

250 mg

S63,853-6
METHYL N-(2-METHOXYPHENYL)CARBAMATE

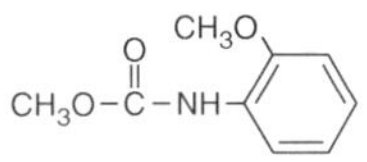

5 gm

S62,185-4
ETHYL N-(O-ANISYL)CARBAMATE

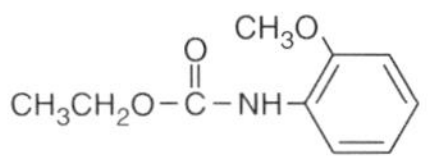

5 gm

S64,710-1
ALLYL N-(2-METHOXYPHENYL)CARBAMATE

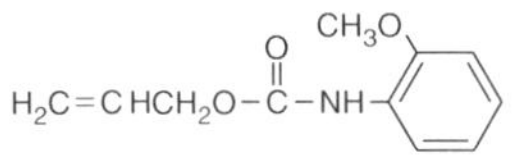

5 gm

S62,517-5
PROPYL N-(2-METHOXYPHENYL)CARBAMATE

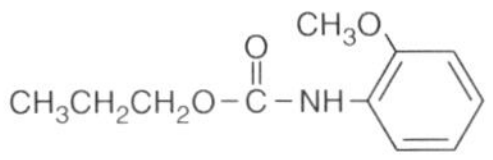

5 gm

S85,589-8
ETHYL N-(2-ETHOXYPHENYL)CARBAMATE

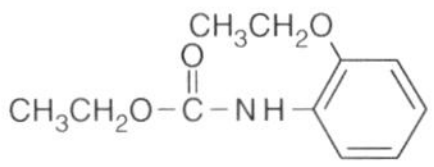

250 mg

S65,359-4
PHENYL N-(O-ANISYL)CARBAMATE

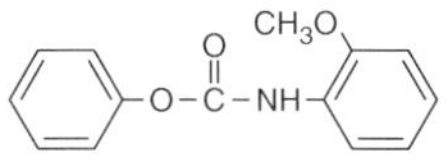

5 gm

S89,182-7
M-TOLYL N-(2-METHOXYPHENYL)-CARBAMATE

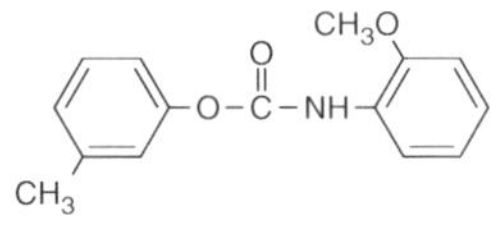

250 mg

S72,106-9
4-TERT-BUTYLPHENYL N-(2-METHOXYPHENYL)CARBAMATE

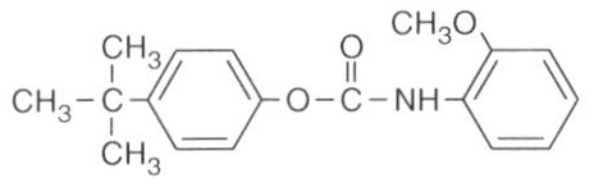

5 gm

S79,165-2
2,6-DIISOPROPYLPHENYL N-(2-METHOXYPHENYL)CARBAMATE

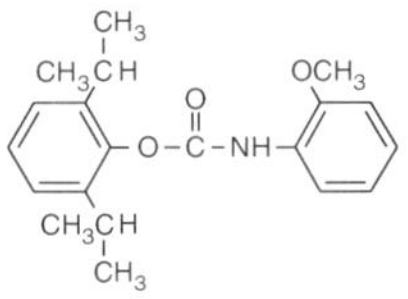

250 mg

S89,178-9
2-NAPHTHYL N-(2-METHOXYPHENYL)-CARBAMATE

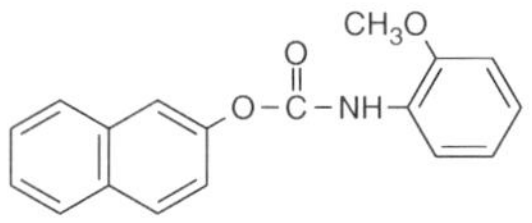

250 mg

S66,533-9
PHENYL N-(2-METHOXY-5-METHYLPHENYL)-CARBAMATE

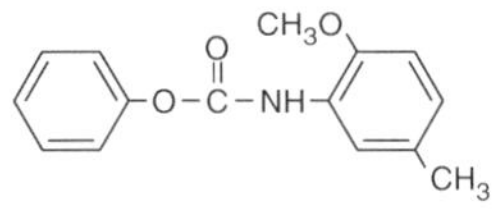

5 gm

S63,475-1
METHYL N-(2,5-DIMETHOXYPHENYL)-CARBAMATE

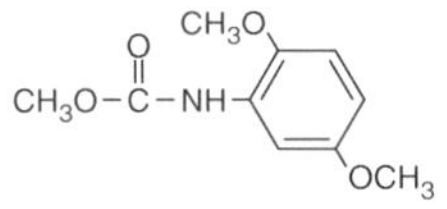

5 gm

S63,947-8
ETHYL N-(2-METHOXY-5-METHYLPHENYL)-CARBAMATE

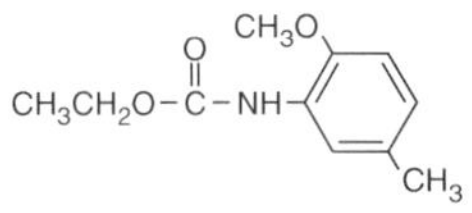

5 gm

S66,511-8
ALLYL N-(2-METHOXY-5-METHYLPHENYL)-CARBAMATE

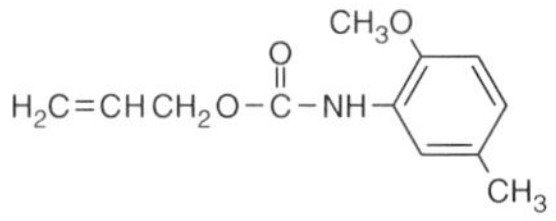

5 gm

S62,498-5
PROPYL N-(2-METHOXY-5-METHYLPHENYL)-CARBAMATE

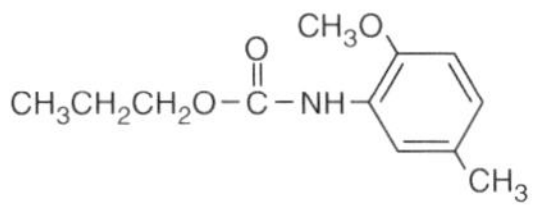

5 gm

S63,952-4
ISOPROPYL N-(2-METHOXY-5-METHYLPHENYL)CARBAMATE

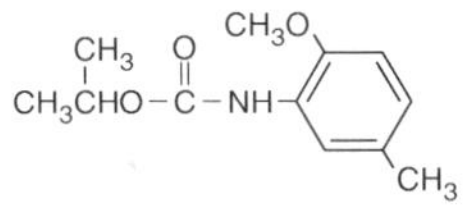

5 gm

S65,006-4
ALLYL N-(2,5-DIMETHOXYPHENYL)-CARBAMATE

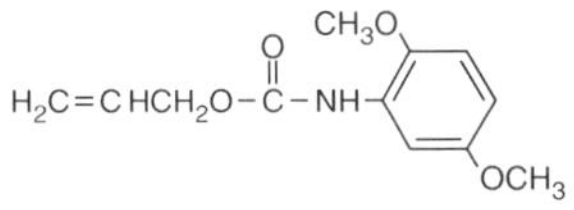

5 gm

S63,472-7
ISOPROPYL N-(2,5-DIMETHOXYPHENYL)-CARBAMATE

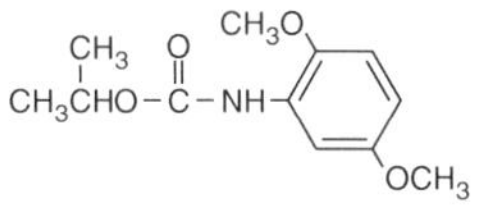

5 gm

S66,531-2
BUTYL N-(2-METHOXY-5-METHYLPHENYL)-CARBAMATE

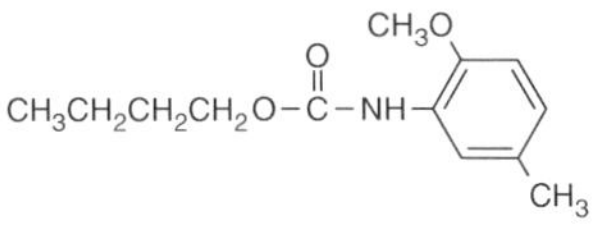

5 gm

S69,165-8
PIPERIDINO N-(2-METHOXY-5-METHYLPHENYL)CARBAMATE

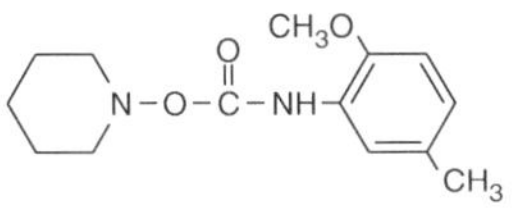

250 mg

S71,110-1
S-ISOPROPYL N-(2-METHOXY-5-METHYLPHENYL)THIOLCARBAMATE

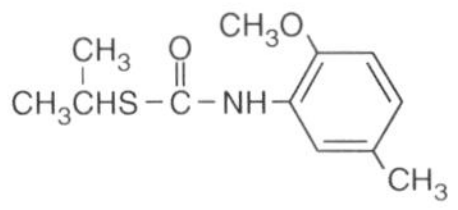

5 gm

S65,354-3
S-BUTYL N-(2-METHOXY-5-METHYLPHENYL)THIOLCARBAMATE

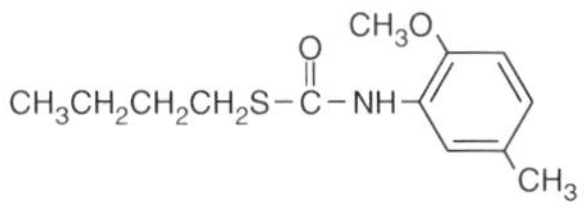

5 gm

S73,181-1
S,S'-ETHYLENE BIS[N-(2-METHOXY-5-METHYLPHENYL)THIOCARBAMATE]

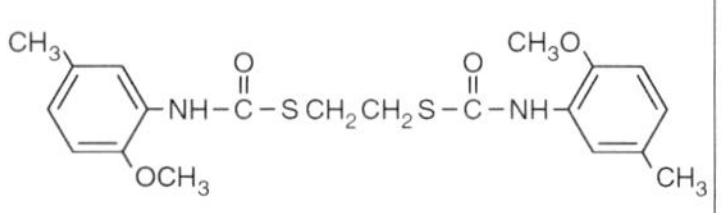

5 gm

S72,785-7
2-NAPHTHYL N-(2-METHOXY-4-NITROPHENYL)CARBAMATE

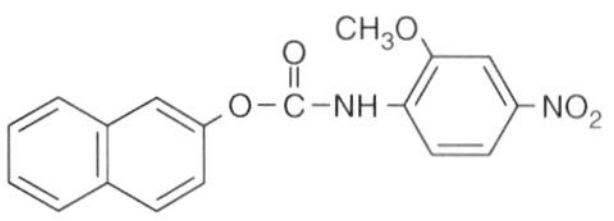

5 gm

S95,650-3
1,4-DIHYDRO-1,4-DIOXO-3-METHYL-2-NAPHTHYLMETHYL N-BENZYLCARBAMATE

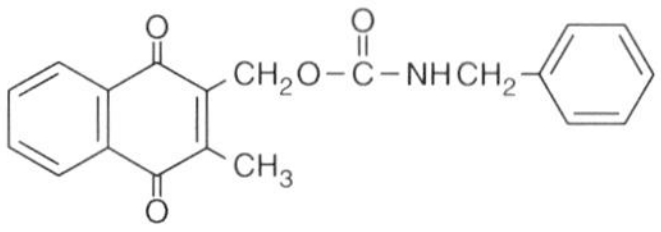

50 mg

S85,524-3
4-ACETYLPHENYL N-(4-METHOXYPHENYL)-CARBAMATE

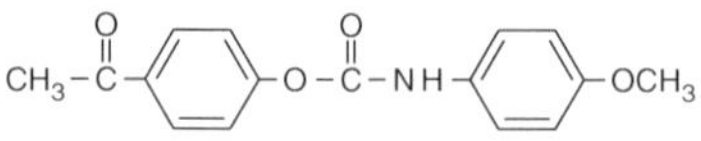

250 mg

S85,525-1
4-PROPIONYLPHENYL N-(4-METHOXYPHENYL)CARBAMATE

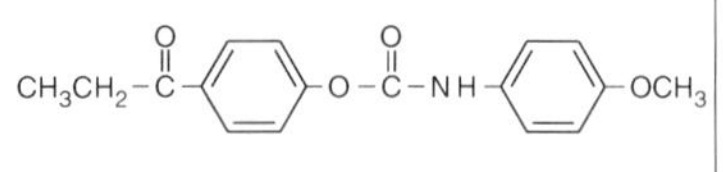

250 mg

S65,410-8
2-BENZOYL-5-METHOXYPHENYL N-(4-CHLOROPHENYL)CARBAMATE

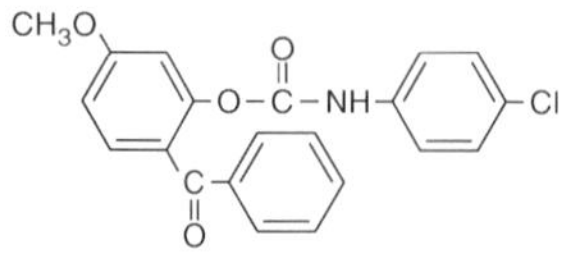

5 gm

S85,544-8
4-(METHOXYCARBONYL)PHENYL N-(4-ETHOXYPHENYL)CARBAMATE

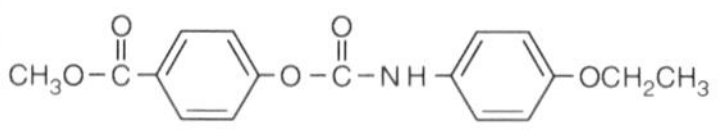

250 mg

S73,835-2
3-(4-METHOXYPHENYLCARBAMOYL)-2-NAPHTHYL N-METHYLCARBAMATE

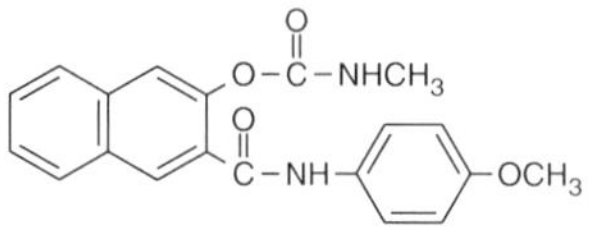

5 gm

S95,850-6
TERT-BUTYL N-MESITOYLOXYCARBAMATE

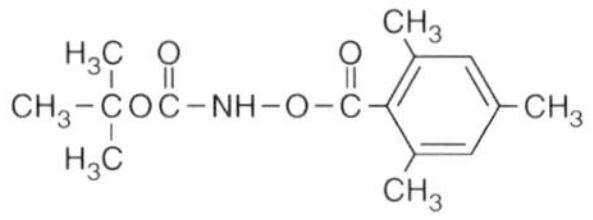

50 mg

S84,580-9
S-(4-METHOXYPHENYL) N-[4-(ETHOXYCARBONYL)PHENYL]-THIOCARBAMATE

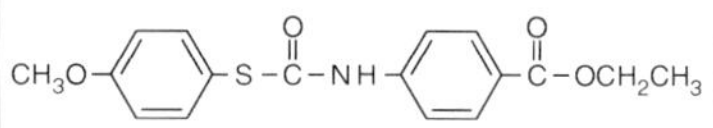

250 mg

S95,449-7
N-(BENZYLOXYCARBONYL)-N-(BENZYLOXYCARBOXAMIDO)GLYCINE ETHYL ESTER

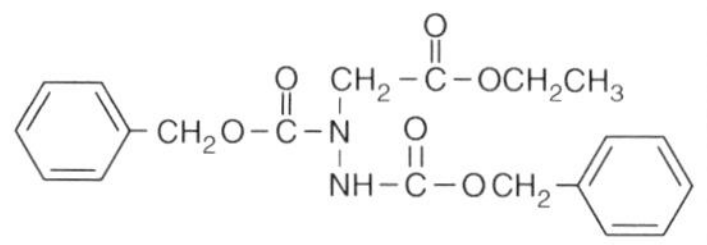

50 mg

S45,030-8
4-METHOXYBENZYL HYDRAZINOCARBOXYLATE

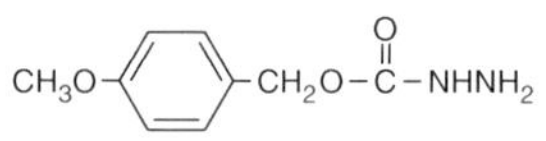

5 gm

S60,936-6
METHYL N-(3-METHOXYPHENYL)CARBAMATE

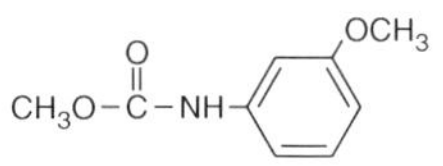

5 gm

S64,712-8
ALLYL N-(3-METHOXYPHENYL)CARBAMATE

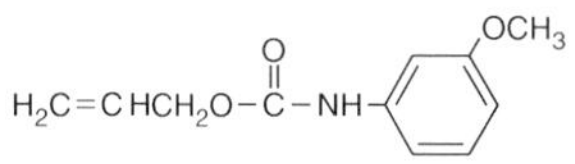

5 gm

S60,959-5
ISOPROPYL N-(3-METHOXYPHENYL)-CARBAMATE

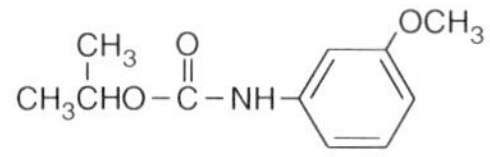

5 gm

S60,980-3
BUTYL N-(M-ANISYL)CARBAMATE

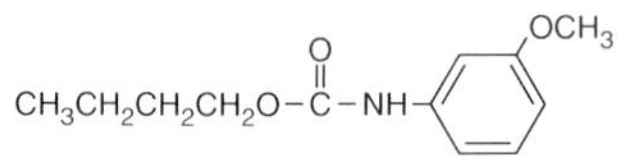

5 gm

S60,964-1
PHENYL N-(M-ANISYL)CARBAMATE

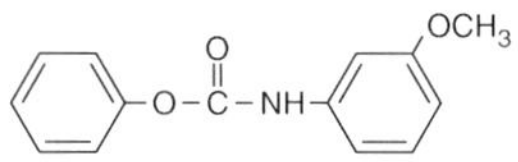

5 gm

S77,292-5
1,2-EPOXY-HEXAHYDRO-4,7-METHANO-5-INDANYL N-(4-METHOXYPHENYL)-CARBAMATE

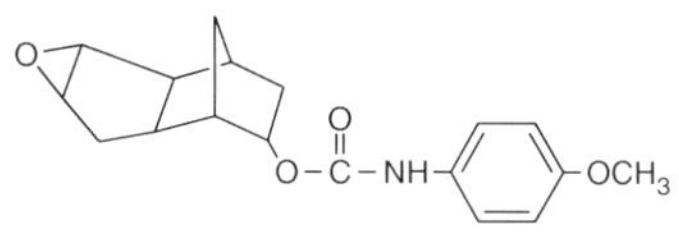

250 mg

S77,349-2
1,2-EPOXY-HEXAHYDRO-4,7-METHANO-5-INDANYL N-(3-ETHOXYPHENYL)-CARBAMATE

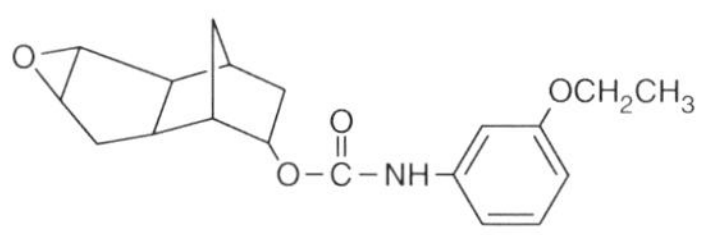

250 mg

S61,025-9
CYCLOHEXYL N-(M-ANISYL)CARBAMATE

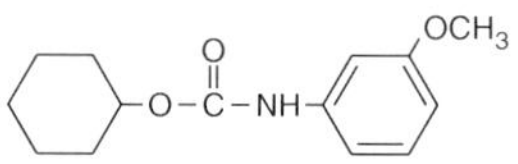

5 gm

S60,984-6
2-METHYLCYCLOHEXYL N-(M-ANISYL)-CARBAMATE

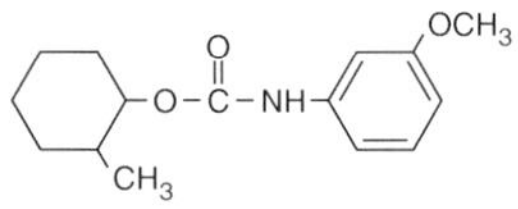

5 gm

S61,156-5
2-METHYLCYCLOHEXYL N-(P-ANISYL)-CARBAMATE

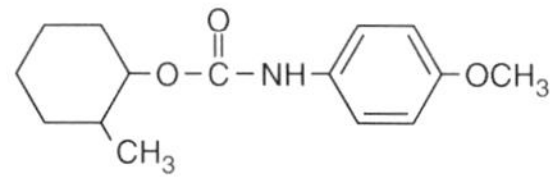

5 gm

S60,471-2
CYCLOHEXYL N-(4-METHOXYPHENYL)-CARBAMATE

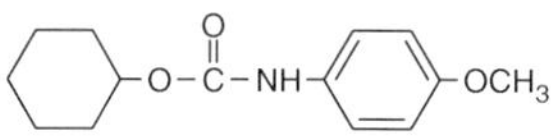

5 gm

S89,187-8
4-CYCLOHEXYLPHENYL N-(4-METHOXYPHENYL)CARBAMATE

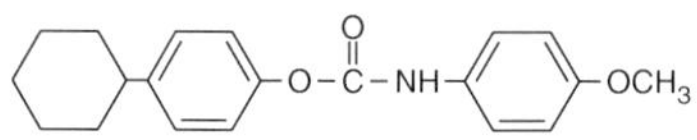

250 mg

S76,820-0
2-BENZYLPHENYL N-(4-METHOXYPHENYL)-CARBAMATE

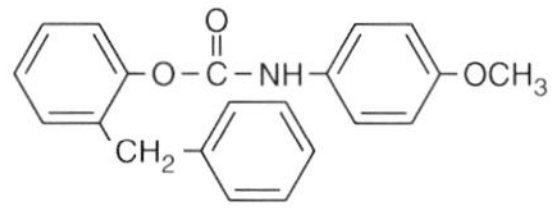

250 mg

S85,550-2
3-PHENYLALLYL N-(4-ETHOXYPHENYL)-CARBAMATE

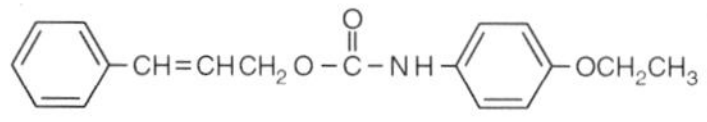

250 mg

S63,648-7
ISOPROPYL N-(2-METHOXYPHENYL)-CARBAMATE

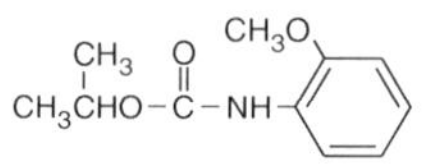

5 gm

S64,947-3
2-CHLOROPHENYL N-(O-ANISYL)-CARBAMATE

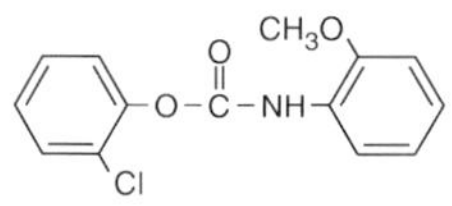

5 gm

S89,191-6
2-ISOPROPYLPHENYL N-(4-METHOXYPHENYL)CARBAMATE

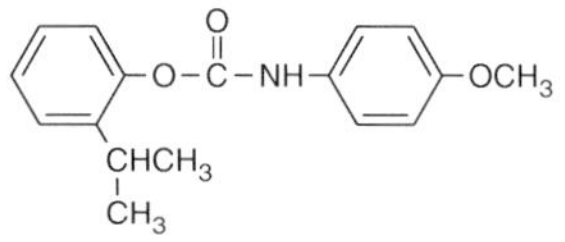

250 mg

S89,023-5
2-SEC-BUTYLPHENYL N-(4-METHOXYPHENYL)CARBAMATE

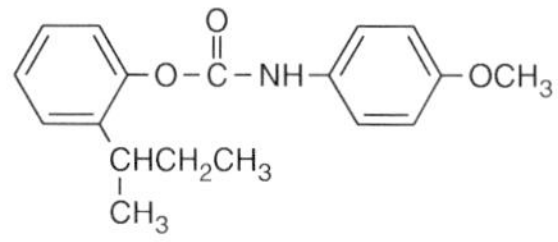

250 mg

S64,047-6
METHYL N-(4-METHOXYPHENYL)CARBAMATE

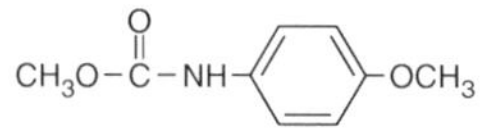

5 gm

S64,051-4
ETHYL N-(4-METHOXYPHENYL)CARBAMATE

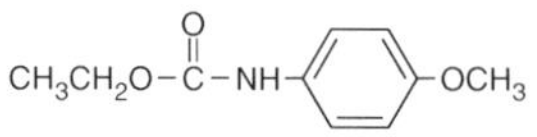

5 gm

S60,362-7
ALLYL N-(4-METHOXYPHENYL)CARBAMATE
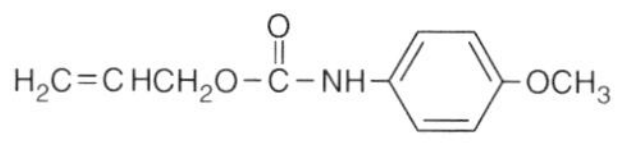
5 gm

S60,158-6
ISOPROPYL N-(4-METHOXYPHENYL)-CARBAMATE
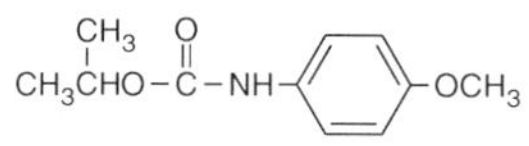
5 gm

S64,052-2
PROPYL N-(4-METHOXYPHENYL)CARBAMATE
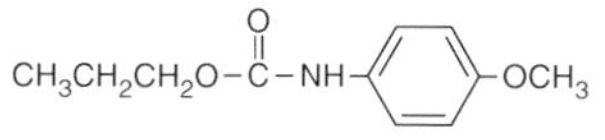
5 gm

S85,587-1
ETHYL N-(4-ETHOXYPHENYL)CARBAMATE
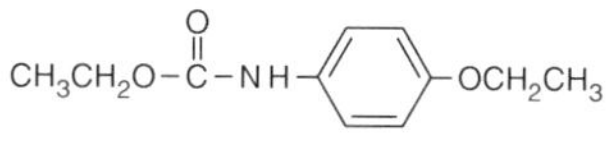
250 mg

S65,361-6
PHENYL N-(P-ANISYL)CARBAMATE
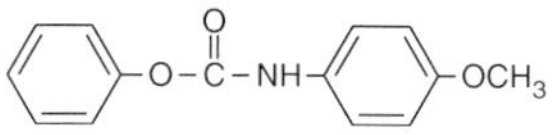
5 gm

S89,195-9
M-TOLYL N-(4-METHOXYPHENYL)-CARBAMATE
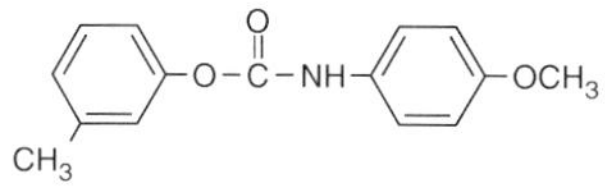
250 mg

S93,333-3
DIMETHYL 4,4'-DITHIOBIS(N-PHENYLCARBAMATE)
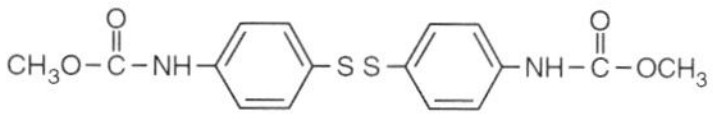
250 mg

S89,188-6
4-TERT-BUTYLPHENYL N-(4-METHOXYPHENYL)CARBAMATE
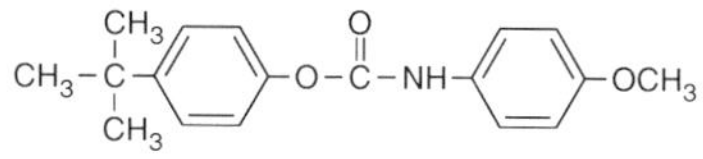
250 mg

S89,194-0
2,6-XYLYL N-(4-METHOXYPHENYL)-CARBAMATE
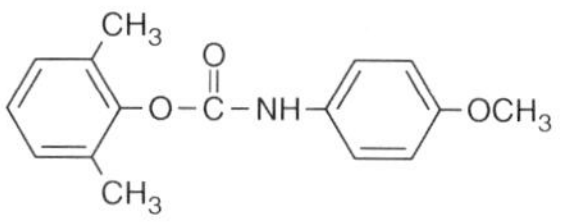
250 mg

S79,528-3
5-ISOPROPYL-2-METHYLPHENYL N-(4-METHOXYPHENYL)CARBAMATE
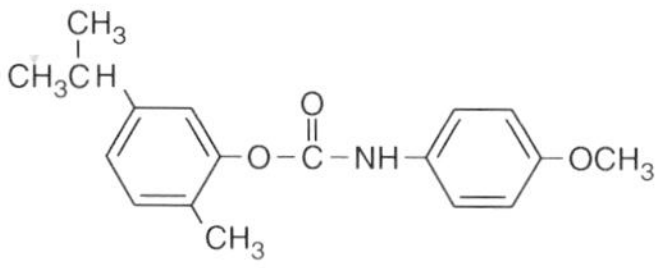
250 mg

S89,024-3
3,5-XYLYL N-(4-METHOXYPHENYL)-CARBAMATE
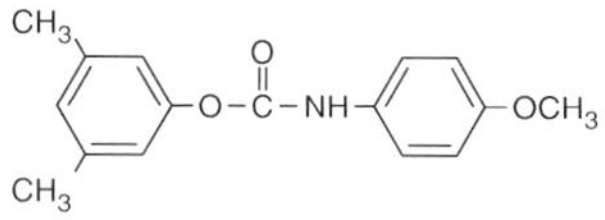
250 mg

S72,108-5
2,4-DICHLORO-1-NAPHTHYL N-(4-METHOXYPHENYL)CARBAMATE
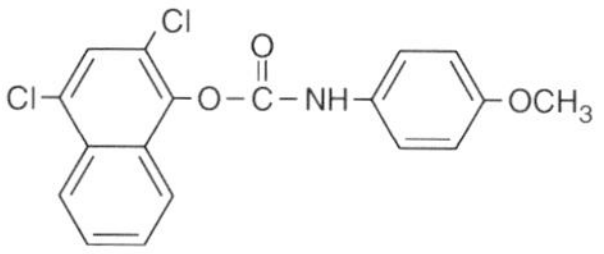
5 gm

S72,107-7
2-NAPHTHYL N-(4-METHOXYPHENYL)-CARBAMATE
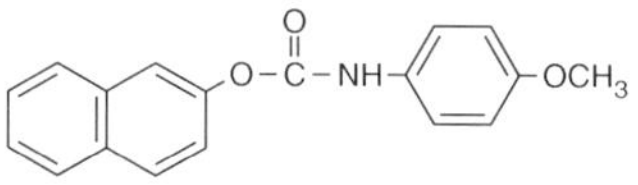
250 mg

S62,426-8
METHYL N-(2-METHOXY-5-METHYLPHENYL)-CARBAMATE
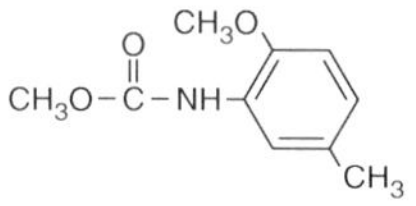
5 gm

S89,198-3
4-ALLYL-2-METHOXYPHENYL N-(1-NAPHTHYL)CARBAMATE
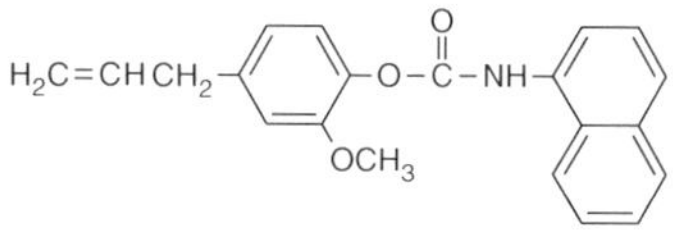
250 mg

S72,265-0
2,6-DI-TERT-BUTYL-4-METHOXYMETHYLPHENYL N-(2-NAPHTHYL)-CARBAMATE
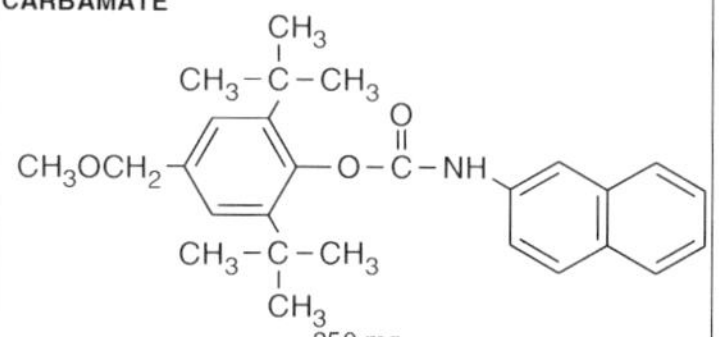
250 mg

S65,008-0
S-ISOPROPYL N-(P-ANISYL)-THIOLCARBAMATE
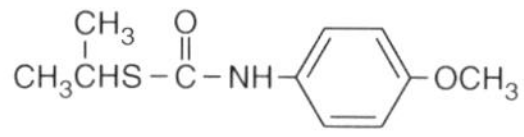
5 gm

S61,349-5
PIPERONYL N-(2,4-DICHLOROPHENYL)-CARBAMATE
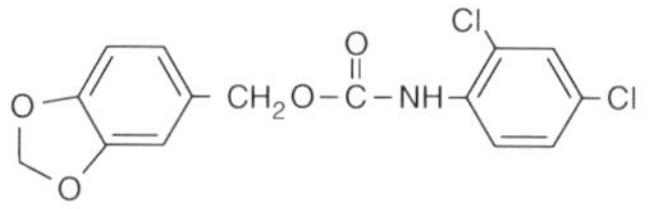
5 gm

S79,795-2
2-(BENZYLOXYCARBONYL)PHENYL N-[4-(METHYLTHIO)PHENYL]CARBAMATE
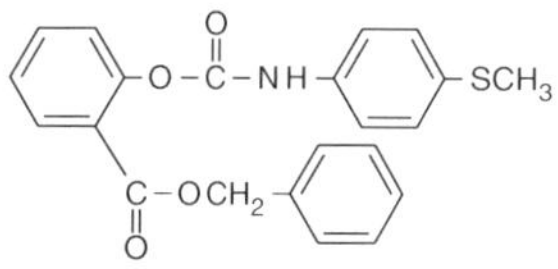
250 mg

S86,893-0
2-(BENZYLOXYCARBONYL)PHENYL N-(O-TOLYL)CARBAMATE
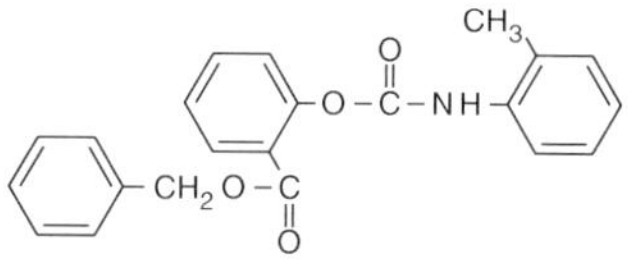
250 mg

S79,799-5
2-(BENZYLOXYCARBONYL)PHENYL N-[4-(ETHOXYCARBONYL)PHENYL]-CARBAMATE
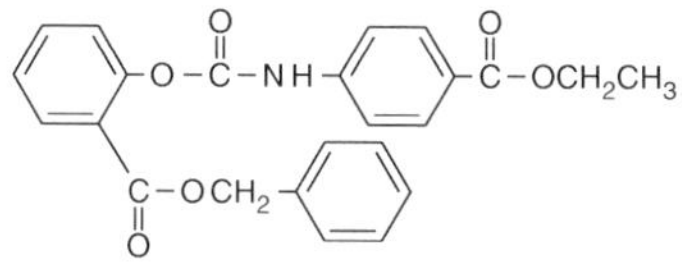
250 mg

S77,603-3
CARBOBENZYLOXY-L-PHENYLALANYL-L-METHIONINE METHYL ESTER
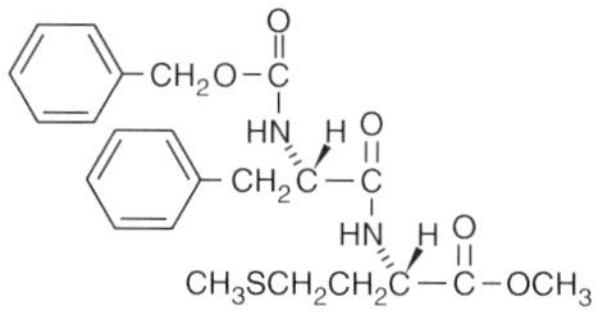
250 mg

S78,858-9
CARBOBENZYLOXY-L-ALANYL-S-BENZYLCYSTEINE BENZYL ESTER
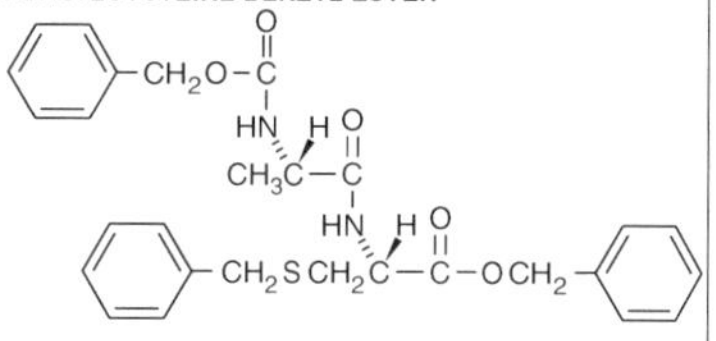
250 mg

S76,940-1
4-(BENZYLOXYCARBONYL)PHENYL N-[4-(METHYLTHIO)PHENYL]CARBAMATE
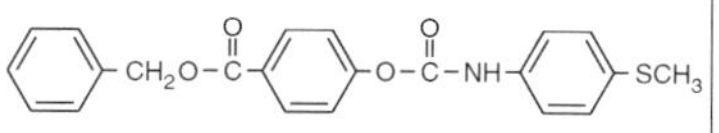
250 mg

S78,467-2
CARBOBENZYLOXYGLYCYLGLYCYL-L-METHIONINE
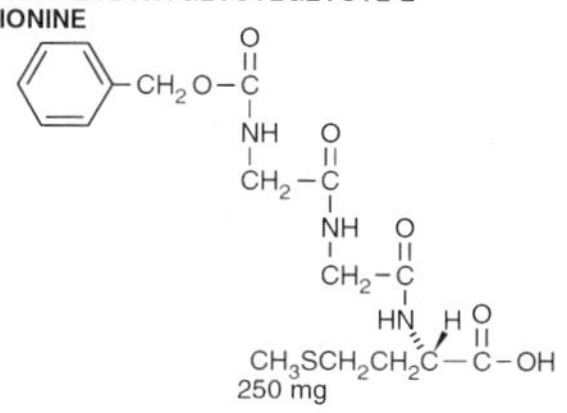
250 mg

S76,537-6
CARBOBENZYLOXY-GLYCYL-L-METHIONINE
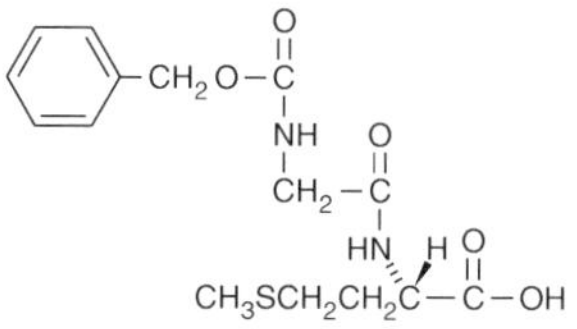
5 gm

S78,266-1
CARBOBENZYLOXYSARCOSYL-L-METHIONINE
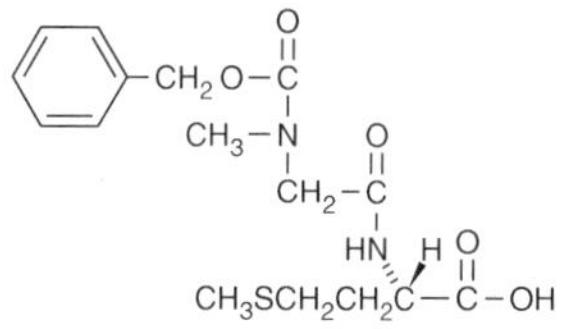
5 gm

S78,461-3
CARBOBENZYLOXY-L-ALANYL-L-METHIONINE
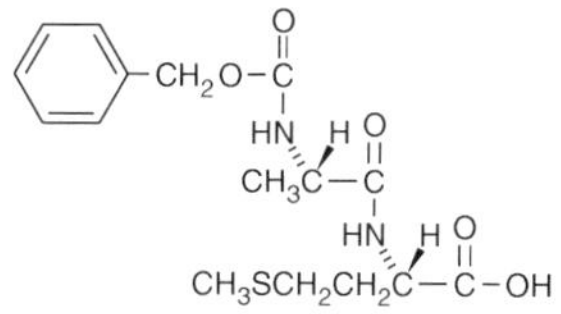
250 mg

S77,659-9
CARBOBENZYLOXY-L-NORVALYL-L-METHIONINE
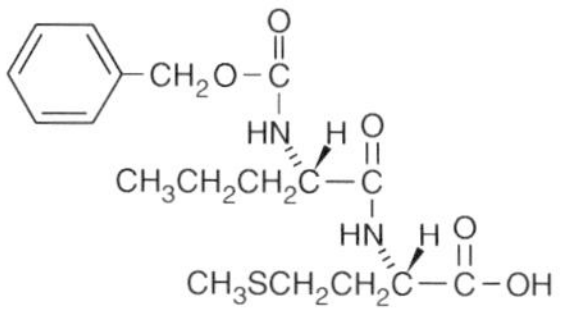
250 mg

S77,700-5
CARBOBENZYLOXY-L-METHIONYL-L-LEUCINE
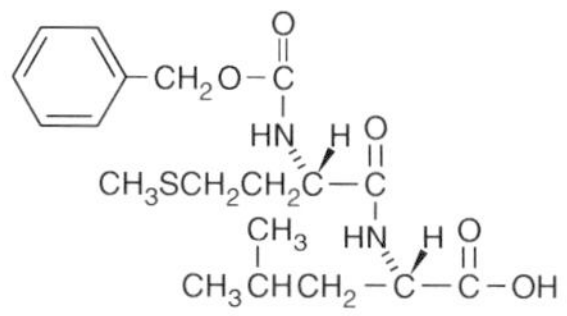
250 mg

S77,790-0
CARBOBENZYLOXY-L-METHIONYL-L-ISOLEUCINE
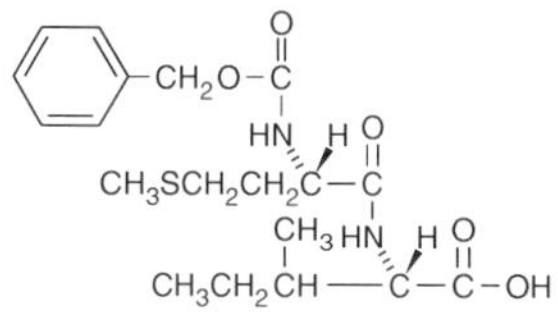
250 mg

S78,757-4
CARBOBENZYLOXY-L-ISOLEUCYL-METHIONINE
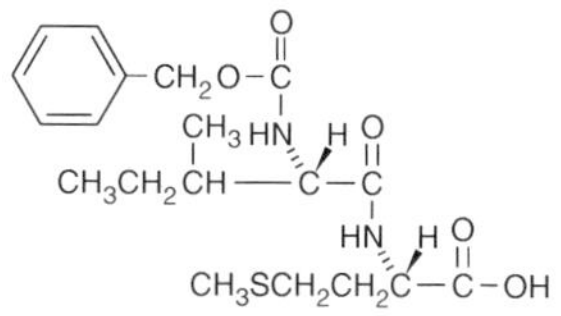
250 mg

S70,512-8
CARBOBENZYLOXY-DL-METHIONYL-GLYCINE
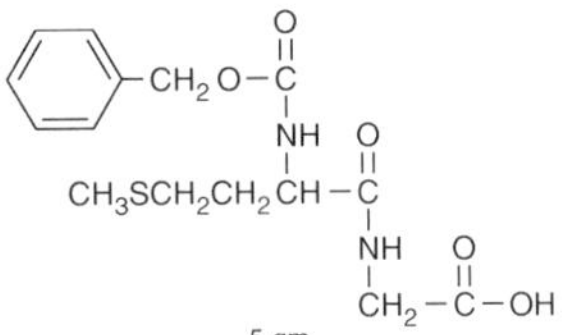
5 gm

S70,516-0
CARBOBENZYLOXY-L-METHIONYL-GLYCINE
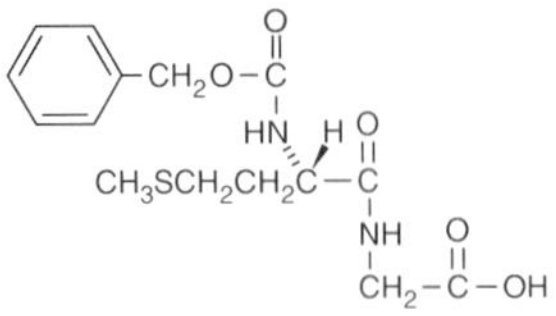
5 gm

S70,522-5
CARBOBENZYLOXY-L-METHIONYL-L-ALANINE
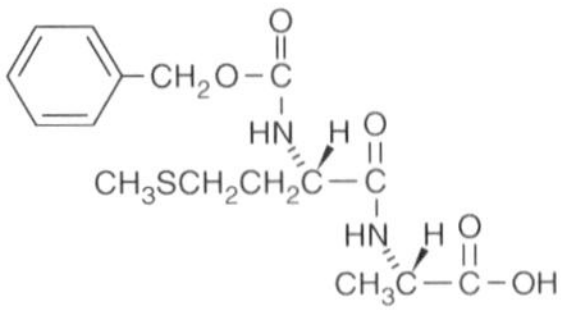
5 gm

S70,520-9
CARBOBENZYLOXY-L-METHIONYL-GLYCYL-GLYCINE
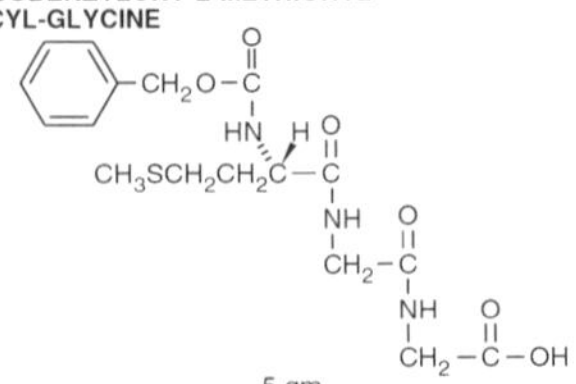
5 gm

S70,510-1
CARBOBENZYLOXY-L-LEUCYL-L-METHIONINE
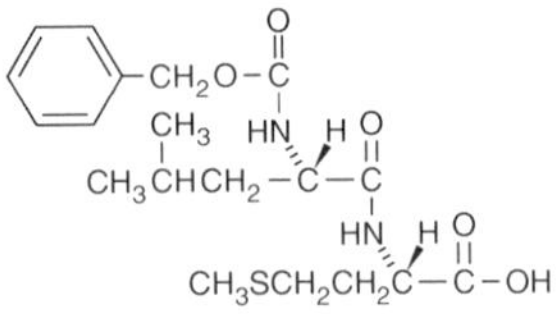
5 gm

S79,037-0
CARBOBENZYLOXY-S-BENZYL-L-CYSTEINYL-L-LEUCINE DICYCLOHEXYLAMINE SALT
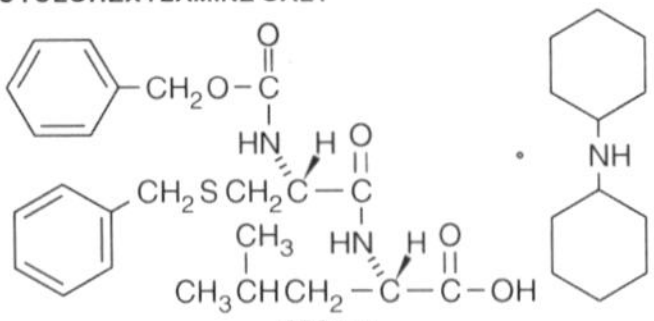
250 mg

S45,132-0
CARBOBENZYLOXY-DL-METHIONINE
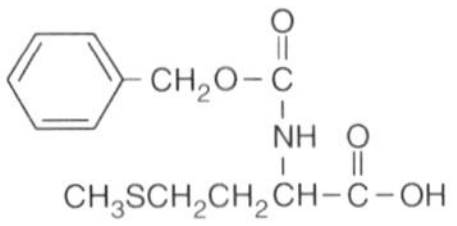
5 gm

S57,939-4
CARBOBENZYLOXY-L-METHIONINE
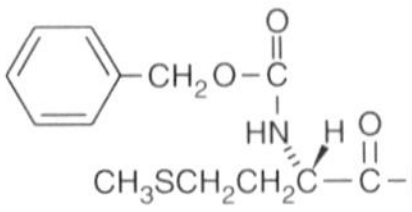
5 gm

S77,712-9
CARBOBENZYLOXY-D-METHIONINE
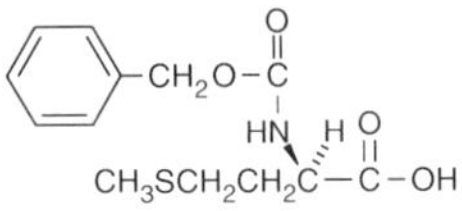
250 mg

S87,086-2
N-CARBOBENZYLOXY-S-BENZYL-DL-HOMOCYSTEINE
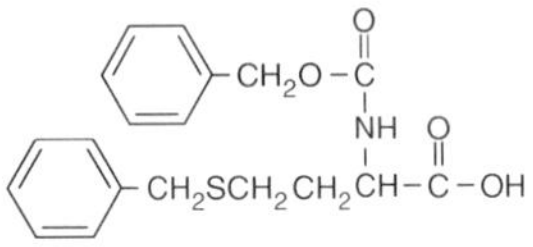
250 mg

S78,468-0
CARBOBENZYLOXY-L-ALANYL-L-METHIONINE METHYL ESTER
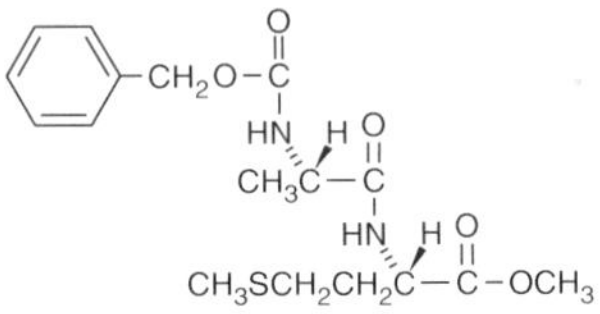
250 mg

S78,331-5
CARBOBENZYLOXY-L-METHIONYL-L-ALANINE METHYL ESTER
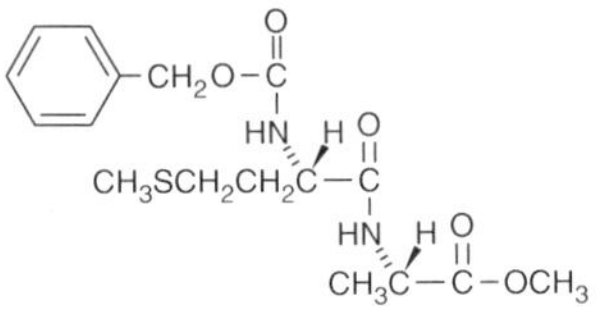
5 gm

S79,062-1
CARBOBENZYLOXYGLYCYLGLYCYL-L-METHIONINE METHYL ESTER
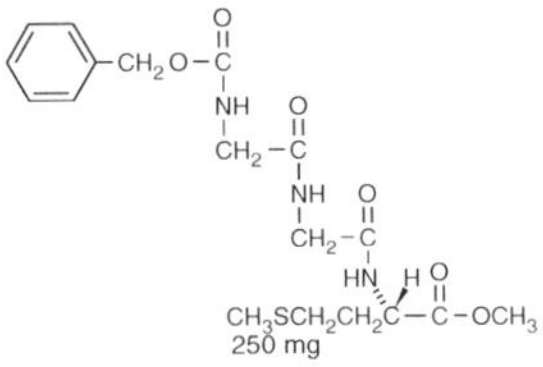
250 mg

S76,582-1
CARBOBENZYLOXY-L-VALYL-L-METHIONINE METHYL ESTER
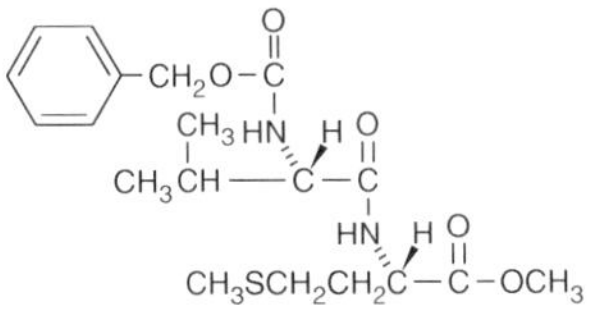
5 gm

S78,822-8
CARBOBENZYLOXY-L-METHIONYL-L-ISOLEUCINE METHYL ESTER
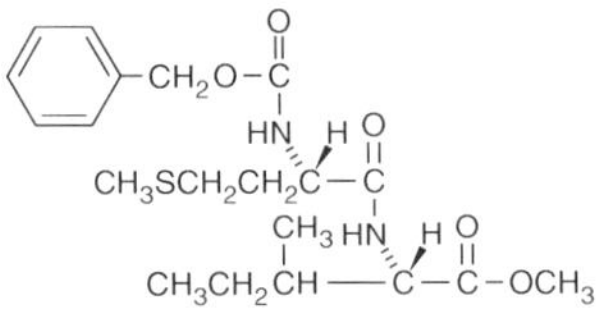
250 mg

S78,965-8
CARBOBENZYLOXY-L-ISOLEUCYL-L-METHIONINE METHYL ESTER
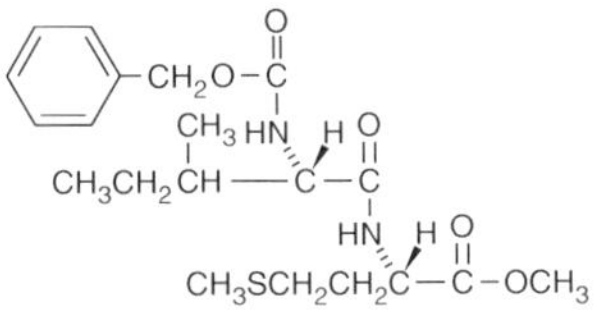
250 mg

S78,482-6
CARBOBENZYLOXY-L-ALANYL-S-BENZYL-L-CYSTEINE METHYL ESTER
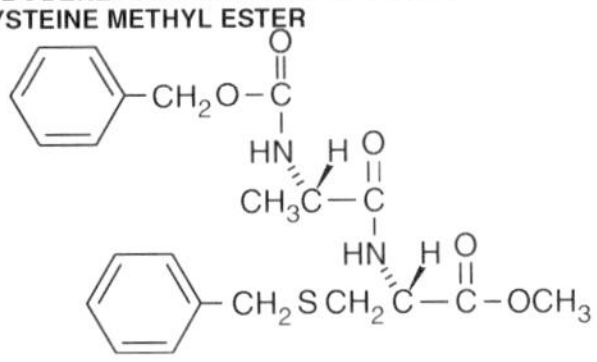
250 mg

S78,140-1
CARBOBENZYLOXY-S-BENZYL-L-CYSTEINYL-L-LEUCINE METHYL ESTER
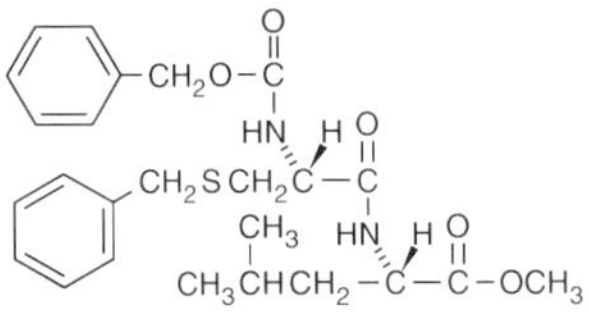
250 mg

S79,055-9
CARBOBENZYLOXY-S-BENZYL-L-CYSTEINYL-L-ALANINE BENZYL ESTER
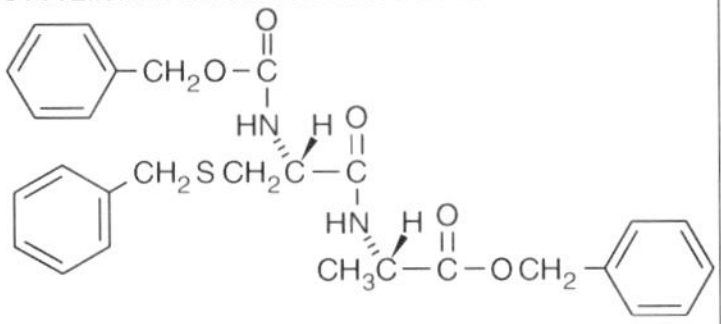
250 mg

S78,307-2
CARBOBENZYLOXY-S-BENZYL-L-CYSTEINE METHYL ESTER
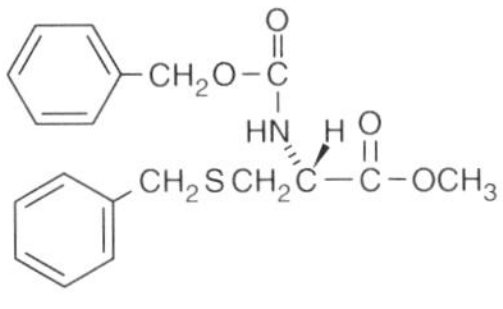
5 gm

S77,677-7
CARBOBENZYLOXY-L-METHIONYL-L-ALANINAMIDE
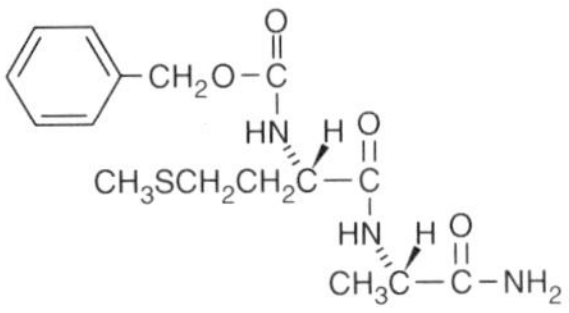
250 mg

S77,758-7
CARBOBENZYLOXYSARCOSYL-L-METHIONINAMIDE
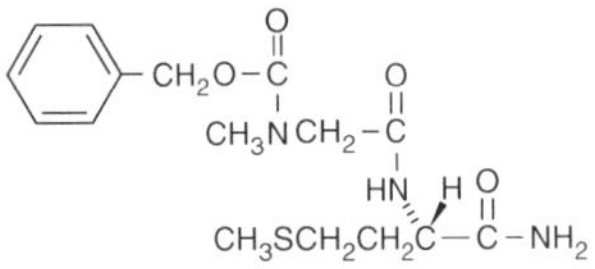
250 mg

S78,188-6
CARBOBENZYLOXY-L-ALANYL-L-METHIONINAMIDE
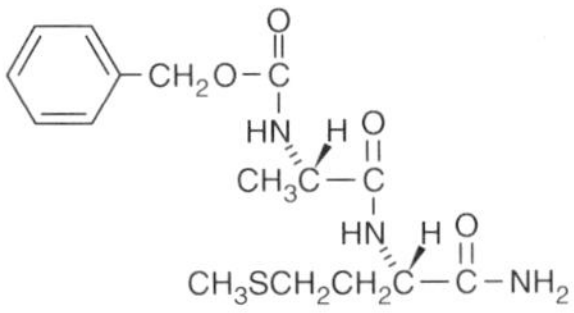
250 mg

S77,760-9
CARBOBENZYLOXY-L-METHIONYL-GLYCYLGLYCINAMIDE
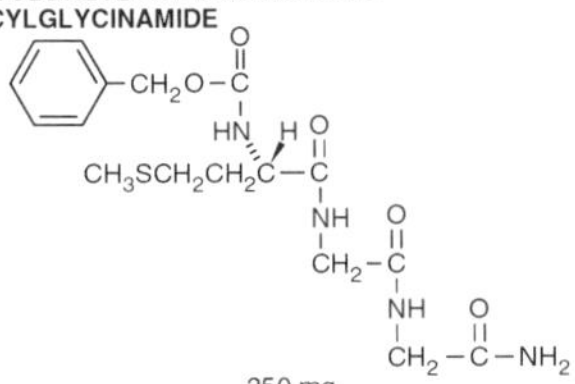
250 mg

S78,161-4
CARBOBENZYLOXYGLYCYLGLYCYL-L-METHIONINAMIDE
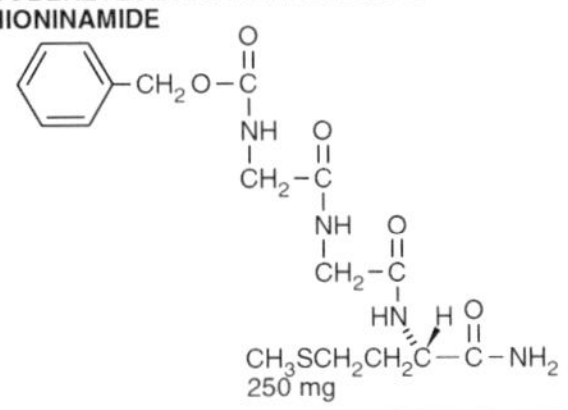
250 mg

S77,697-1
CARBOBENZYLOXY-L-METHIONYL-L-VALINAMIDE
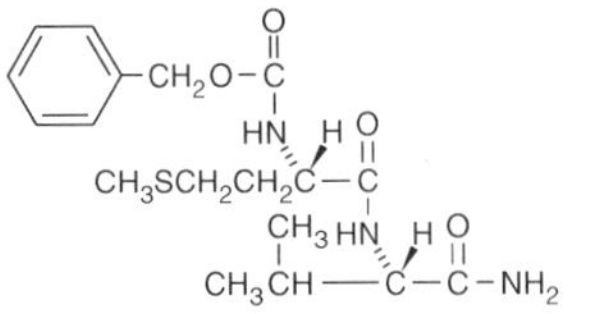
250 mg

S78,471-0
CARBOBENZYLOXY-L-NORVALYL-L-METHIONINAMIDE
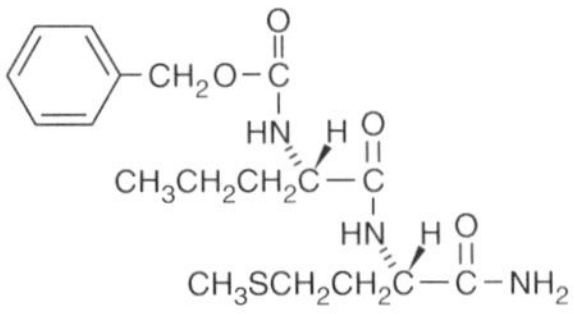
250 mg

S77,702-1
CARBOBENZYLOXY-L-METHIONYL-L-ISOLEUCINAMIDE
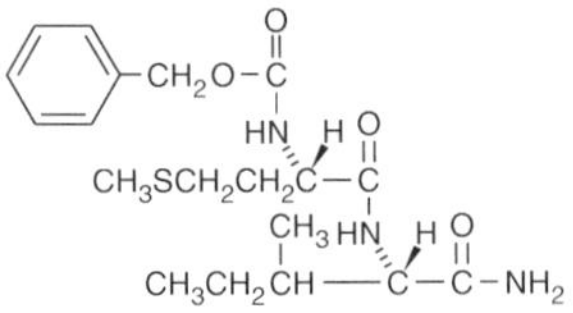
250 mg

S77,761-7
CARBOBENZYLOXY-L-NORLEUCYL-L-METHIONINAMIDE
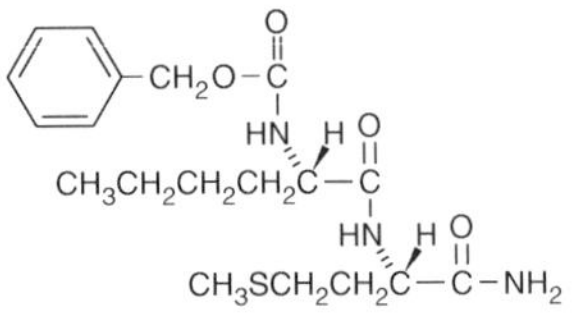
250 mg

S77,789-7
CARBOBENZYLOXY-L-METHIONYL-L-LEUCINAMIDE
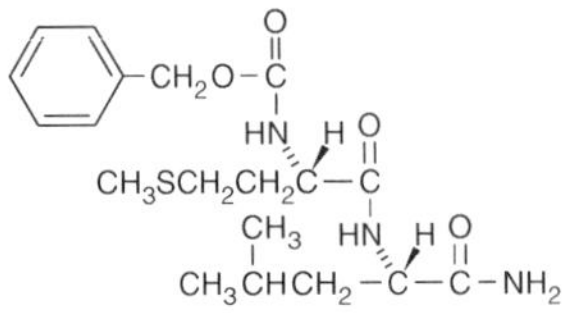
250 mg

S78,542-3
CARBOBENZYLOXY-L-ISOLEUCYL-L-METHIONINAMIDE
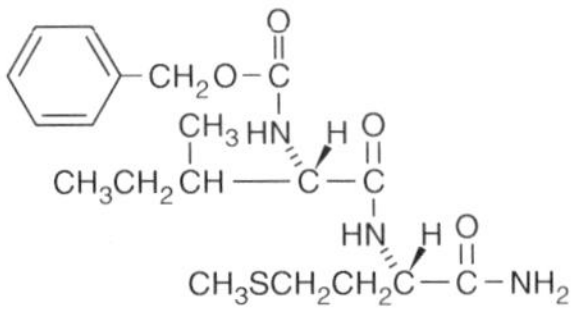
250 mg

S78,548-2
CARBOBENZYLOXY-S-BENZYL-L-CYSTEINYL-L-VALINAMIDE
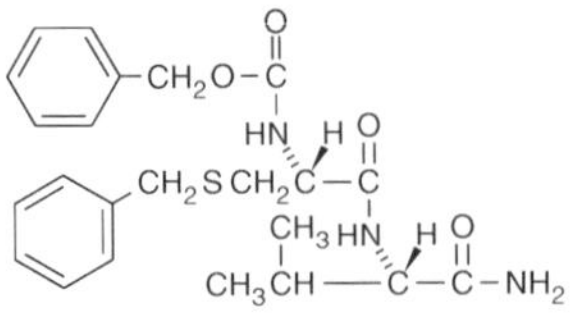
250 mg

S78,120-7
CARBOBENZYLOXY-L-METHIONINAMIDE

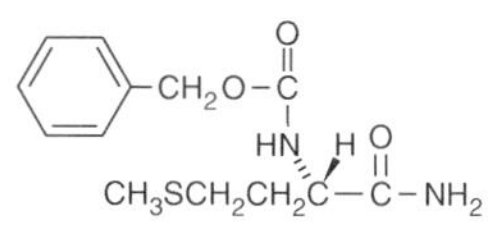

250 mg

S34,899-6
N,N'-DICARBOBENZYLOXY-L-CYSTINE

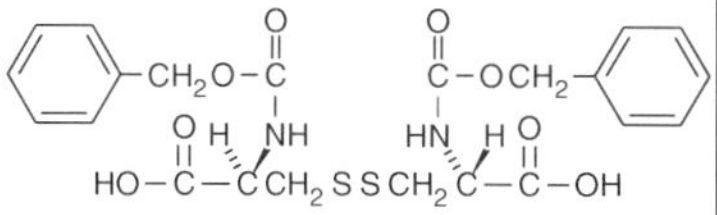

5 gm

S60,283-3
N-6-CARBOBENZYLOXY-L-LYSINE METHYL ESTER HYDROCHLORIDE

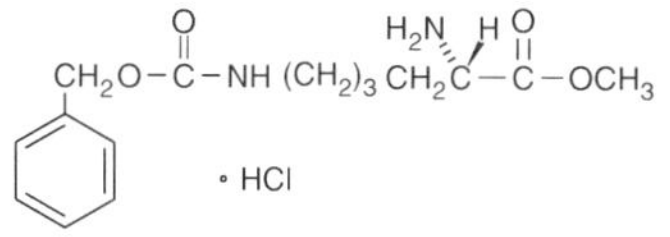

250 mg

S45,058-8
N-6-CARBOBENZYLOXY-L-LYSINE BENZYL ESTER HYDROCHLORIDE

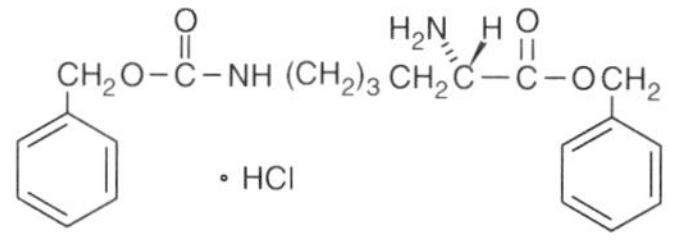

5 gm

S67,682-9
N,N'-(4-METHYL-1,3-PHENYLENE)BIS[4-(BENZYLOXYCARBONYL)PHENYL CARBAMATE]

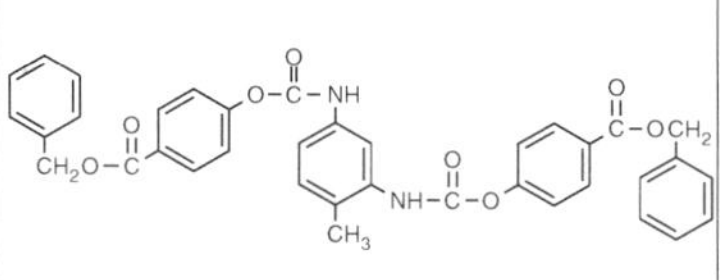

250 mg

S57,597-6
CARBOBENZYLOXY-DL-ASPARAGINE

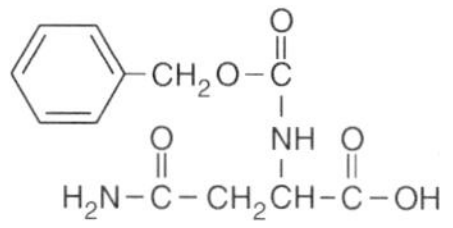

5 gm

S73,094-7
BIS(4-METHYLBENZYL) N,N'-(4-METHYL-1,3-PHENYLENE)BISCARBAMATE

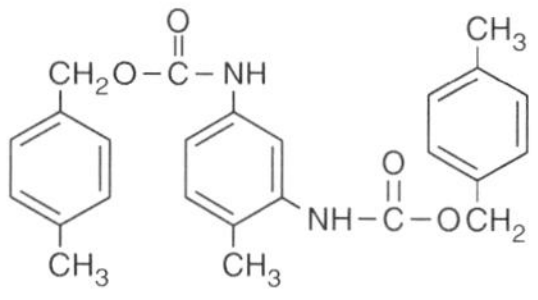

5 gm

S60,265-5
CARBOBENZYLOXY-DL-GLUTAMINE

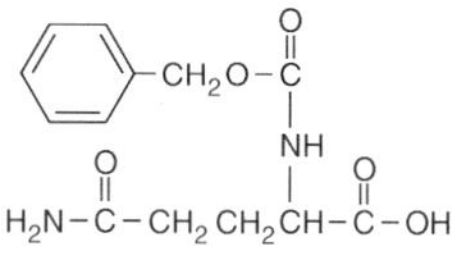

5 gm

S60,728-2
PIPERONYL N-(2-CHLORO-4-NITROPHENYL)CARBAMATE

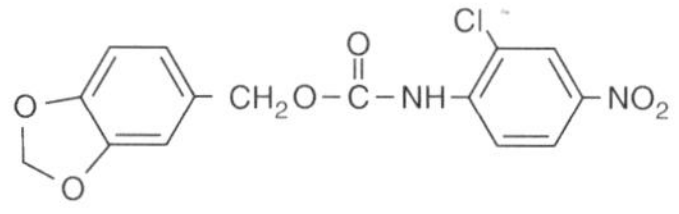

5 gm

S34,861-9
ALPHA, ALPHA'-DIBENZYL CARBOBENZYLOXY-GAMMA-L-GLUTAMYL-L-GLUTAMATE

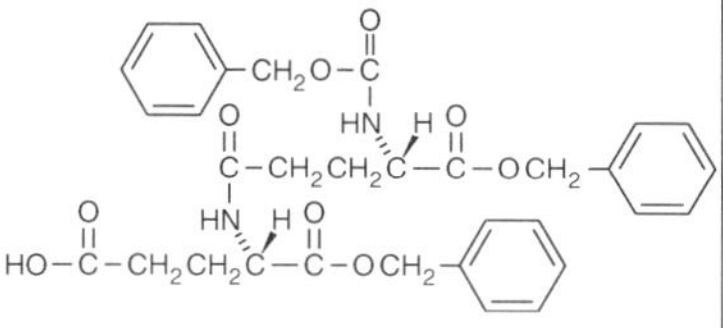

250 mg

S79,019-2
N-DELTA-(TERT-BUTOXYCARBONYL)-N-ALPHA-CARBOBENZYLOXY-L-ORNITHINE

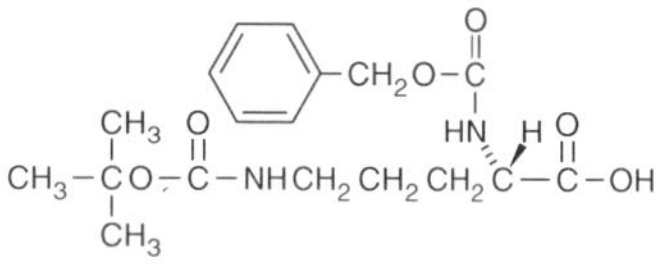

250 mg

S44,711-0
N,N'-DICARBOBENZYLOXY-L-ORNITHINE

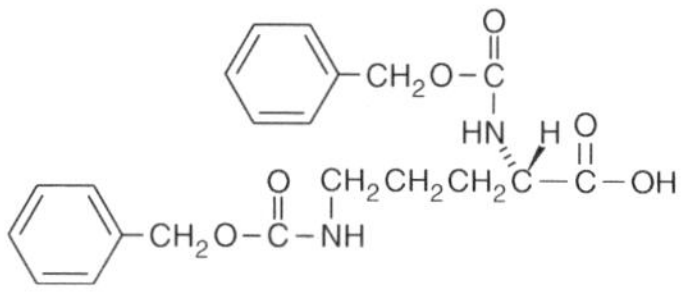

5 gm

S77,681-5
N2-CARBOBENZYLOXY-L-ARGININE HYDROBROMIDE

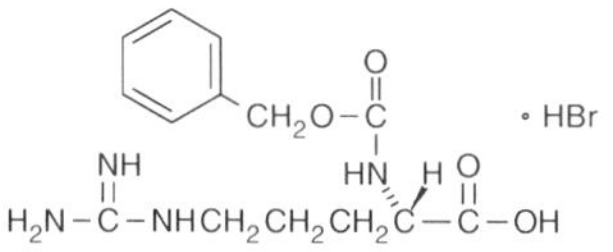

250 mg

S92,823-2
4-METHYL-5-PHENYL-2-OXAZOLIDINONE

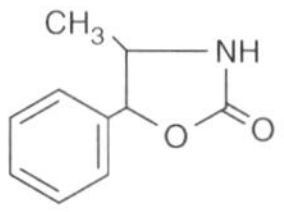

250 mg

S66,251-8
5-(3,5-DI-TERT-BUTYLPHENYL)-3-NITROSO-2-OXAZOLIDINONE

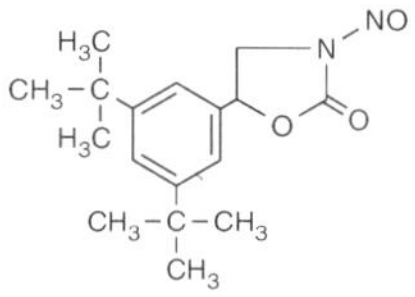

250 mg

S76,523-6
CARBOBENZYLOXY-L-ALANYL-L-PHENYLALANINE

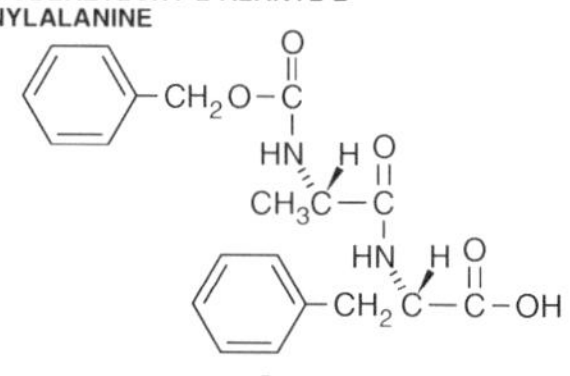

5 gm

S78,545-8
CARBOBENZYLOXYGLYCYLGLYCYL-L-PHENYLALANINE

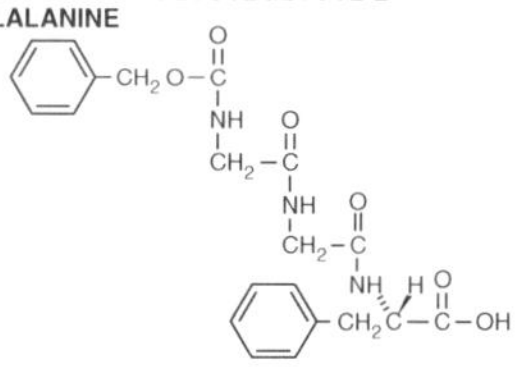

250 mg

S77,746-3
CARBOBENZYLOXY-L-PHENYLALANYL-L-NORVALINE

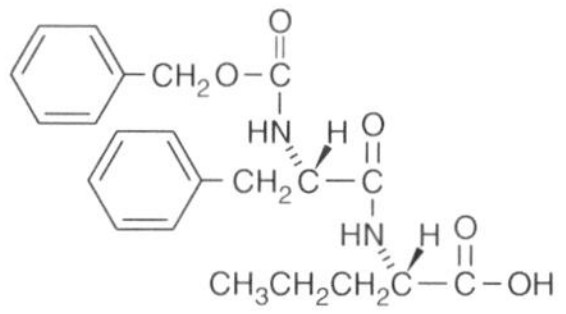

250 mg

S78,387-0
CARBOBENZYLOXYGLYCYL-D-PHENYLALANINE

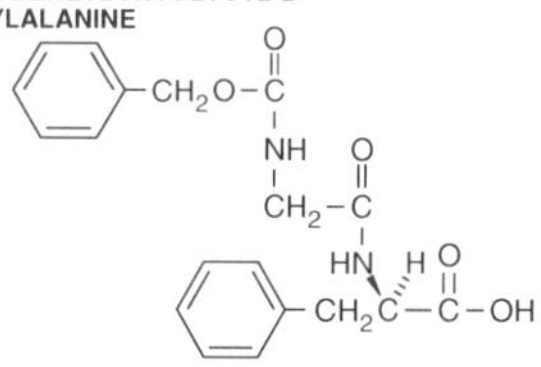

5 gm

S78,655-1
CARBOBENZYLOXY-BETA-ALANYL-L-PHENYLALANINE

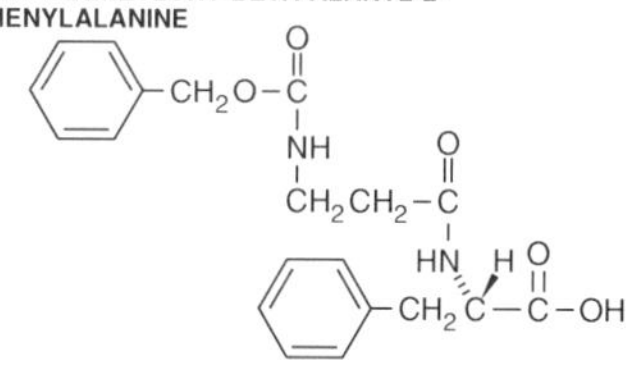

250 mg

S34,358-7
N-CARBOBENZYLOXY-DL-PHENYLALANINE

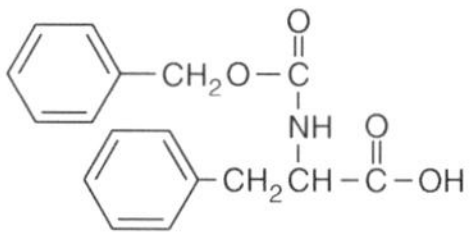

5 gm

S79,009-5
CARBOBENZYLOXYSARCOSYL-L-PHENYLALANINE METHYL ESTER

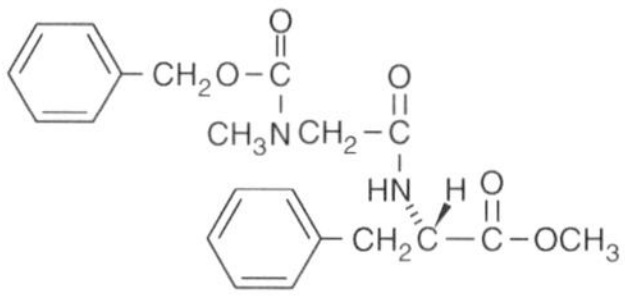

250 mg

S78,838-4
CARBOBENZYLOXY-L-THREONYL-L-PHENYLALANINE METHYL ESTER

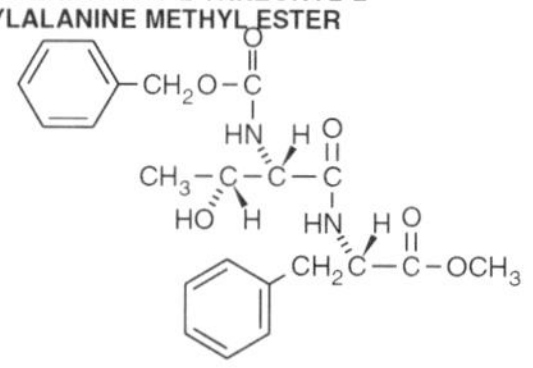

250 mg

S79,090-7
CARBOBENZYLOXY-L-PHENYLALANYL-L-GLYCYLGLYCINE METHYL ESTER

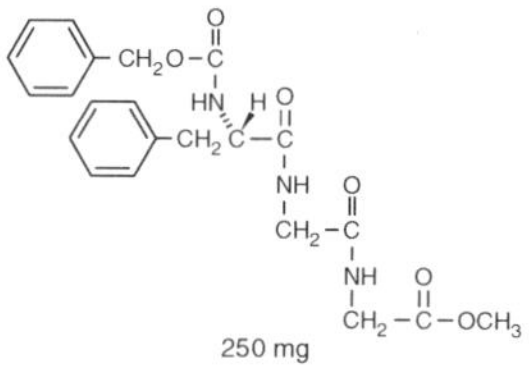

250 mg

S78,842-2
CARBOBENZYLOXY-L-NORVALYL-L-PHENYLALANINE METHYL ESTER

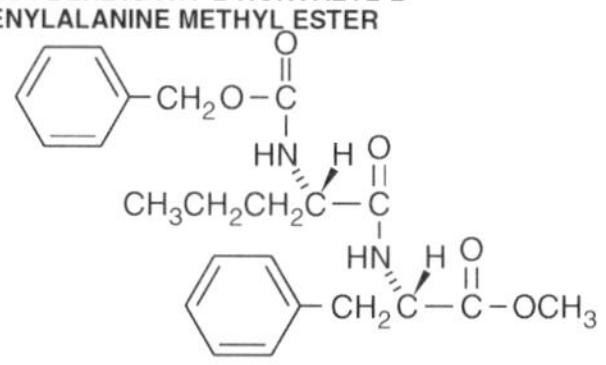

250 mg

S79,088-5
CARBOBENZYLOXY-L-PHENYLALANYL-L-NORVALINE METHYL ESTER

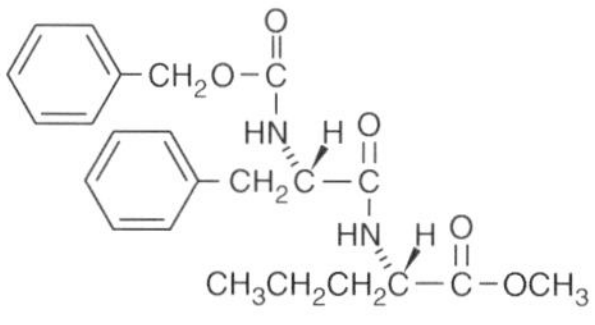

250 mg

S78,620-9
CARBOBENZYLOXY-L-NORLEUCYL-L-PHENYLALANINE METHYL ESTER

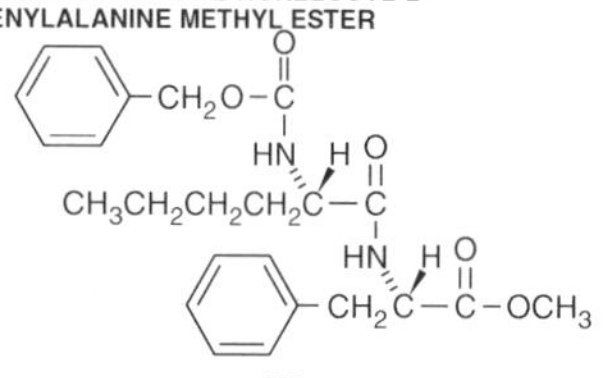

250 mg

S78,847-3
CARBOBENZYLOXY-L-PHENYLALANYL-L-ISOLEUCINE METHYL ESTER

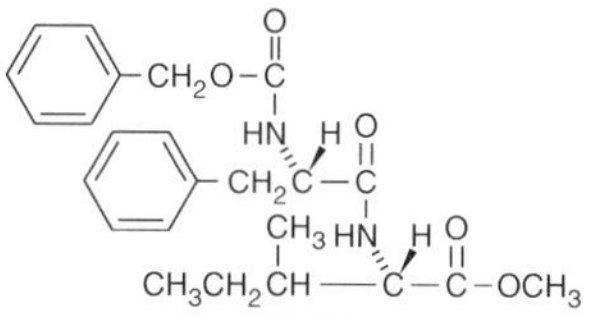

250 mg

S78,860-0
CARBOBENZYLOXY-L-PHENYLALANYL-L-NORLEUCINE METHYL ESTER

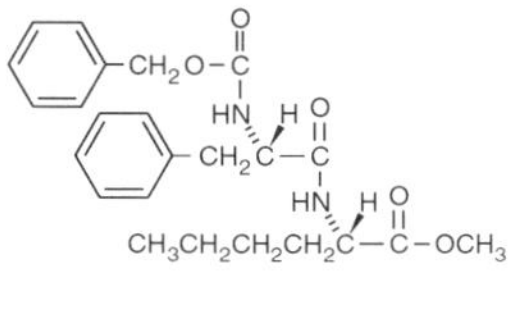

250 mg

S78,968-2
CARBOBENZYLOXY-L-ISOLEUCYL-L-PHENYLALANINE METHYL ESTER

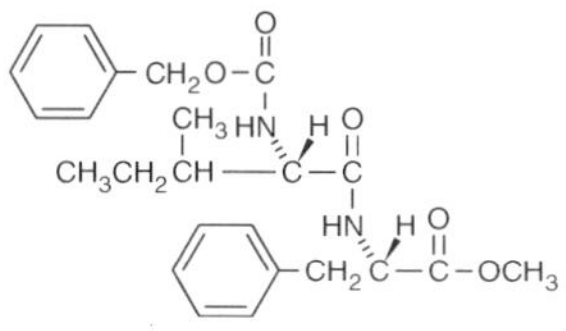

250 mg

S78,772-8
CARBOBENZYLOXYGLYCYL-L-PHENYLALANINE BENZYL ESTER

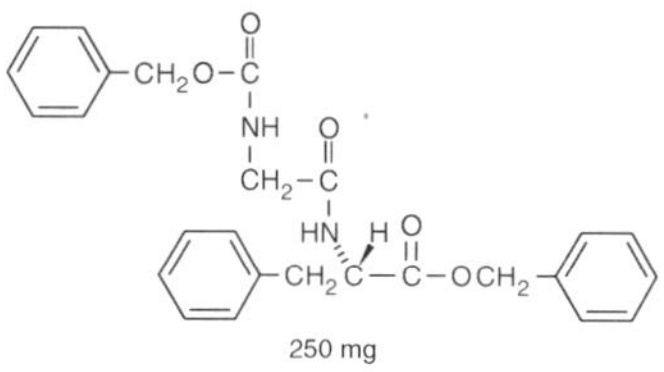

250 mg

S78,853-8
CARBOBENZYLOXY-L-PHENYLALANYL-GLYCINE BENZYL ESTER

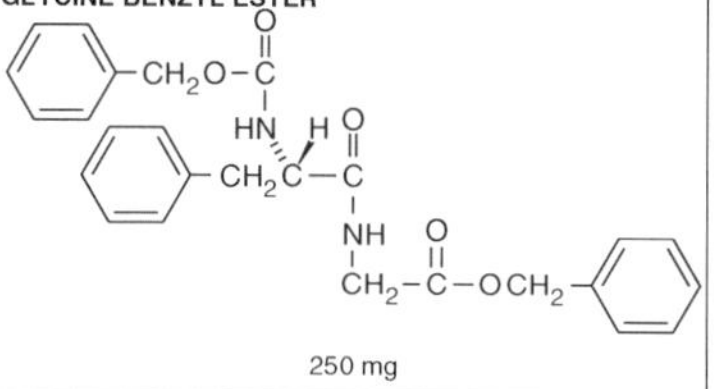

250 mg

S78,775-2
CARBOBENZYLOXY-L-ALANYL-L-PHENYLALANINE BENZYL ESTER
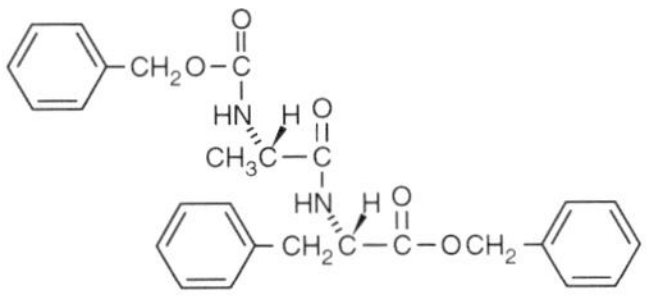
250 mg

S78,855-4
CARBOBENZYLOXY-L-PHENYL-ALANYL-L-ALANINE BENZYL ESTER
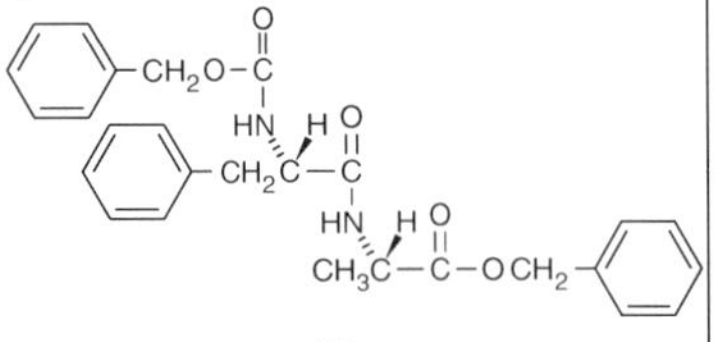
250 mg

S79,079-6
CARBOBENZYLOXY-L-PHENYLALANYL-L-PHENYLALANINE METHYL ESTER
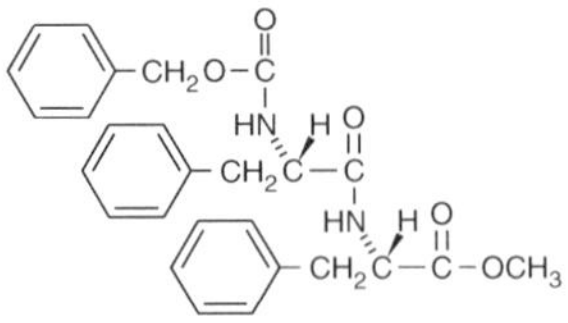
250 mg

S79,081-8
CARBOBENZYLOXYSARCOSYL-L-PHENYL-ALANINE BENZYL ESTER
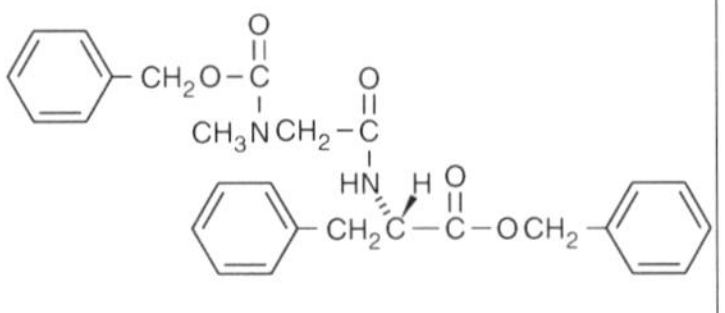
250 mg

S78,777-9
CARBOBENZYLOXYGLYCYLGLYCYL-L-PHENYLALANINE BENZYL ESTER
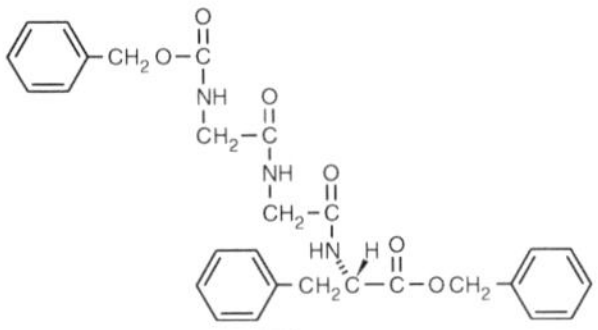
250 mg

S79,076-1
CARBOBENZYLOXY-L-PHENYLALANYL-L-NORVALINE BENZYL ESTER
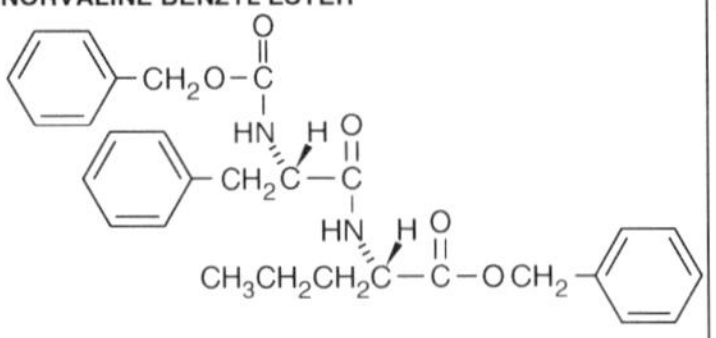
250 mg

S78,989-5
CARBOBENZYLOXY-L-ISOLEUCYL-L-PHENYLALANINE BENZYL ESTER
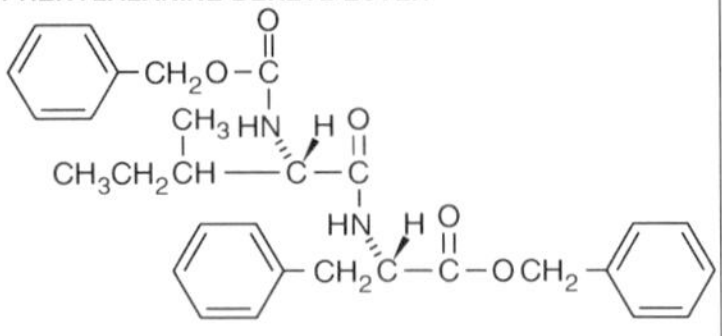
250 mg

S79,074-5
CARBOBENZYLOXY-L-PHENYLALANYL-L-ISOLEUCINE BENZYL ESTER
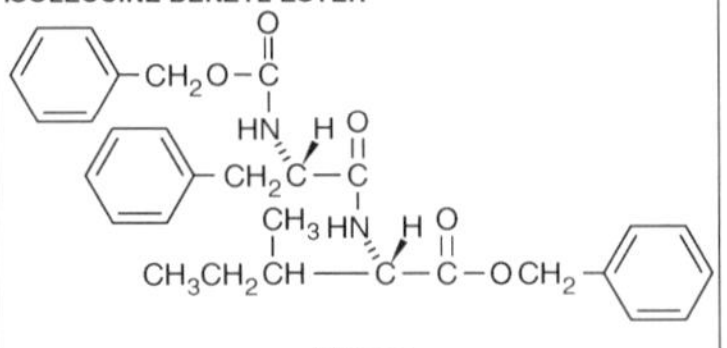
250 mg

S78,552-0
N,N'-BIS(CARBOBENZYLOXY)-L-LYSYL-L-PHENYLALANINE METHYL ESTER
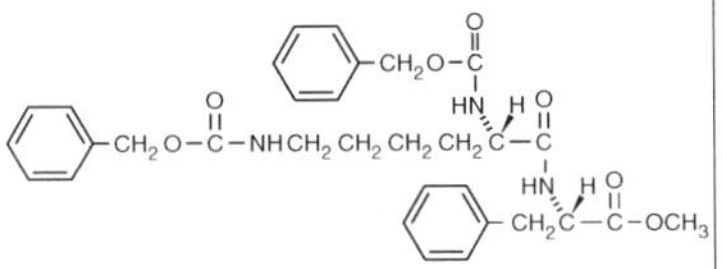
250 mg

S77,792-7
CARBOBENZYLOXY-BETA-ALANYL-L-PHENYLALANINE METHYL ESTER
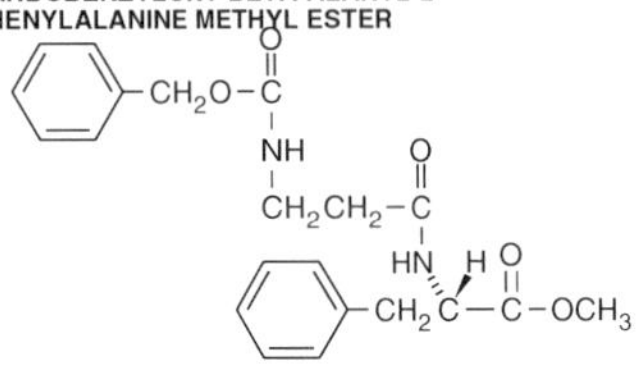
250 mg

S78,613-6
CARBOBENZYLOXY-BETA-ALANYL-L-PHENYLALANINE BENZYL ESTER
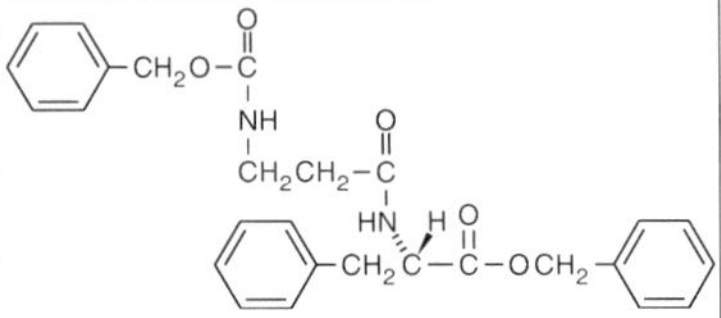
250 mg

S78,820-1
CARBOBENZYLOXY-L-PHENYLALANYL-GLYCINE METHYL ESTER
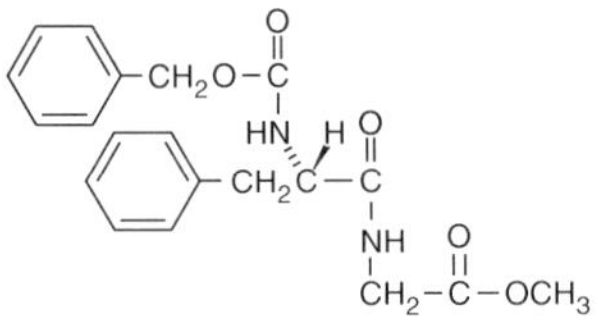
250 mg

S78,825-2
CARBOBENZYLOXY-D-ALANYL-L-PHENYLALANINE METHYL ESTER
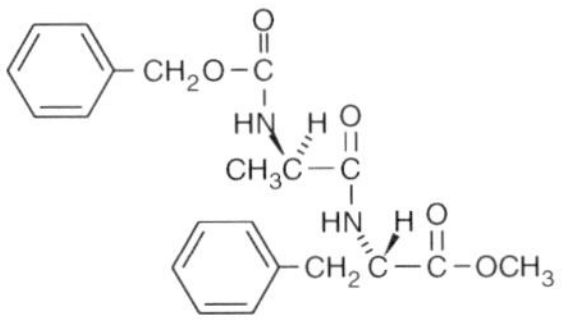
250 mg

S78,263-7
CARBOBENZYLOXY-L-VALYL-L-PHENYLALANINE METHYL ESTER
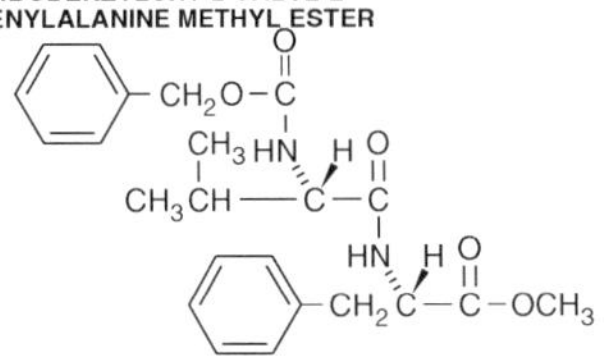
5 gm

S78,137-1
CARBOBENZYLOXY-L-PHENYLALANYL-L-VALINE BENZYL ESTER
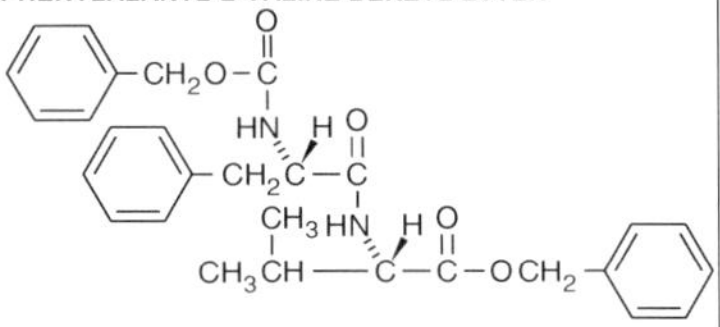
250 mg

S78,863-5
CARBOBENZYLOXY-L-PHENYLALANYL-L-PHENYLALANINE BENZYL ESTER
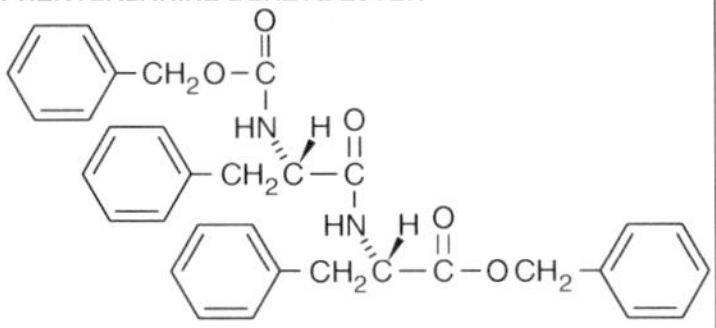
250 mg

S77,705-6
CARBOBENZYLOXY-L-NORVALYL-L-PHENYLALANINAMIDE
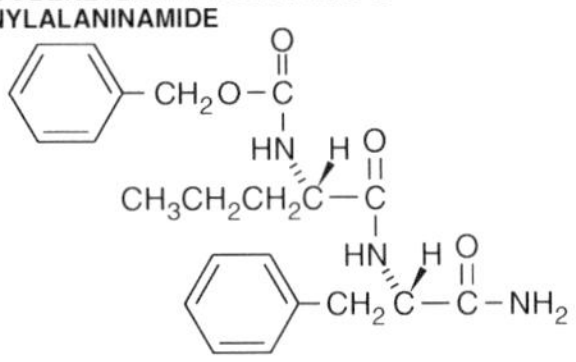
250 mg

S78,372-2
CARBOBENZYLOXY-L-PHENYLALANYL-L-NORLEUCINAMIDE
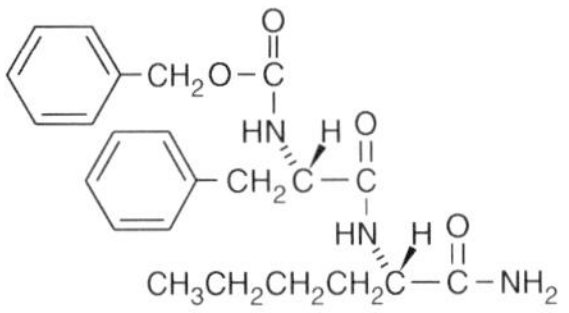
5 gm

S79,087-7
CARBOBENZYLOXY-L-PHENYLALANYL-L-ISOLEUCIN AMIDE
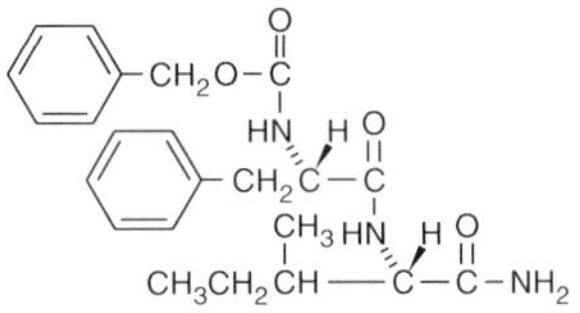
250 mg

S77,664-5
CARBOBENZYLOXY-L-ISOLEUCYL-L-PHENYLALANINAMIDE
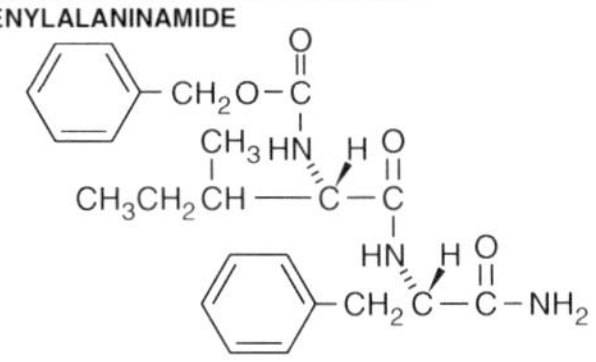
250 mg

S78,475-3
CARBOBENZYLOXY-BETA-ALANYL-L-PHENYLALANINAMIDE
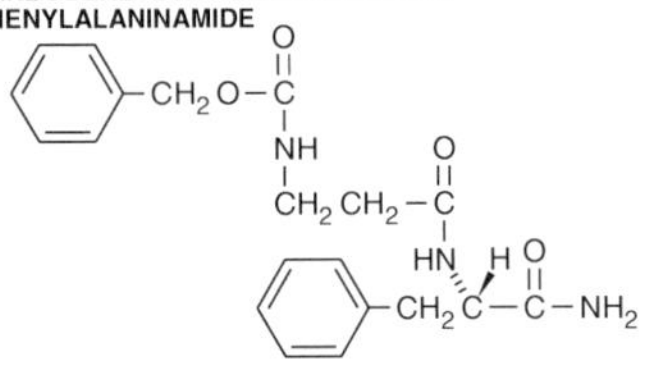
250 mg

S77,710-2
CARBOBENZYLOXYSARCOSYL-L-PHENYLALANINAMIDE
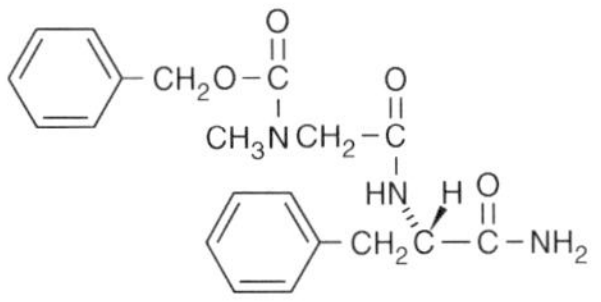
250 mg

S78,287-4
CARBOBENZYLOXY-L-PHENYLALANYLGLYCYLGLYCINAMIDE
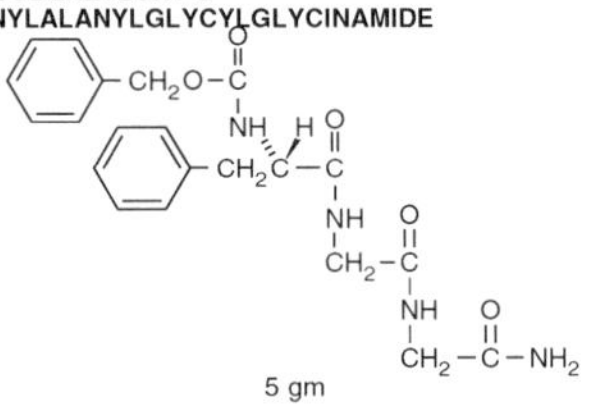
5 gm

S77,713-7
CARBOBENZYLOXY-L-PHENYLALANYL-L-LEUCINAMIDE
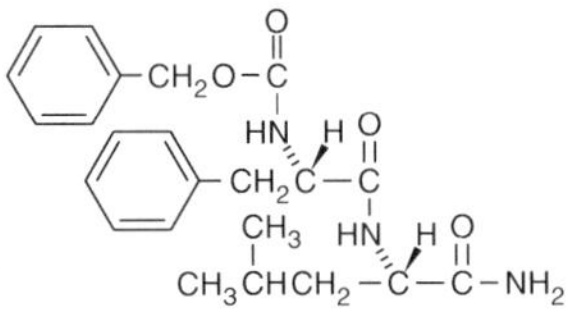
250 mg

S34,359-5
CARBOBENZYLOXY-D-(-)-PHENYLGLYCINE
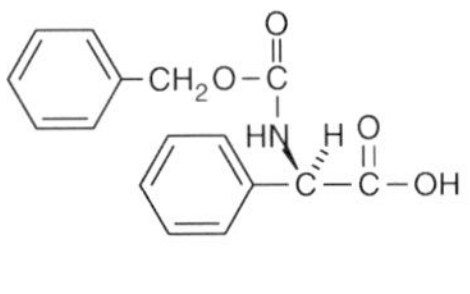
5 gm

S70,493-8
CARBOBENZYLOXY-L-ISOLEUCYL-L-ISOLEUCINE
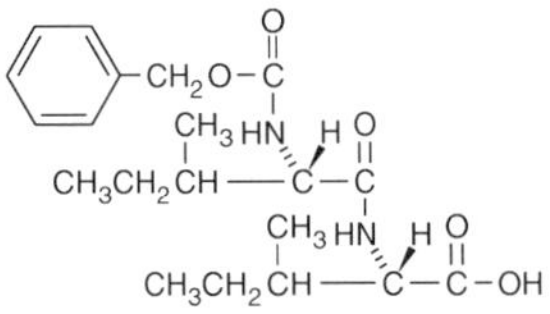
5 gm

S77,696-3
CARBOBENZYLOXY-O-BENZYL-L-SERINE
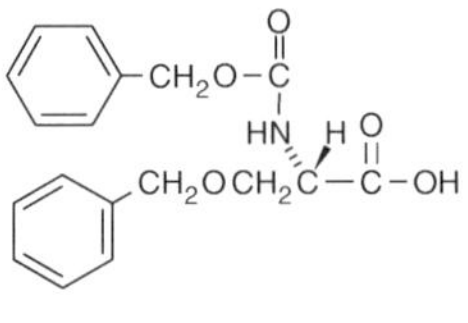
250 mg

S79,047-8
CARBOBENZYLOXY-L-ASPARTIC ANHYDRIDE
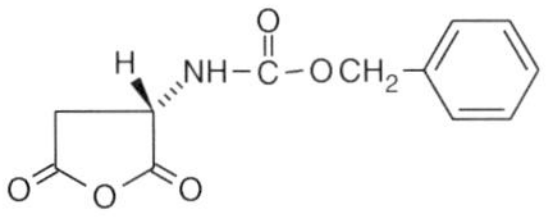
250 mg

S78,537-7
CARBOBENZYLOXY-L-ISOLEUCYL-L-ALANINAMIDE
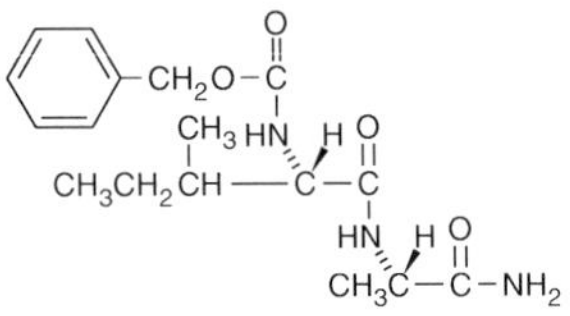
250 mg

S78,810-4
CARBOBENZYLOXY-BETA-ALANYL-L-VALINAMIDE
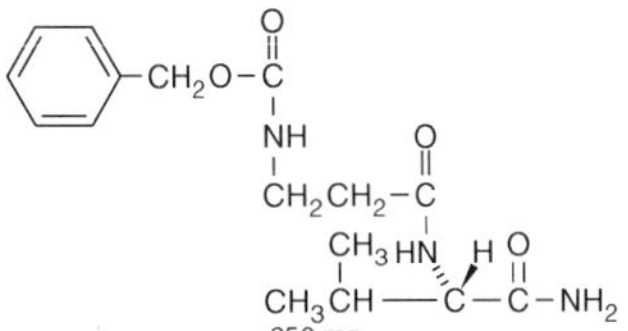
250 mg

S78,581-4
CARBOBENZYLOXYGLYCYLSARCOSINAMIDE
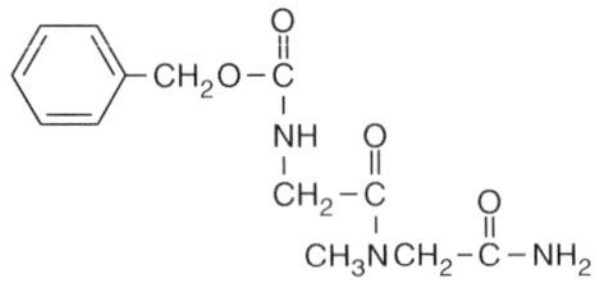
250 mg

S60,110-1
CARBOBENZYLOXYGLYCINE 4-NITROPHENYL ESTER
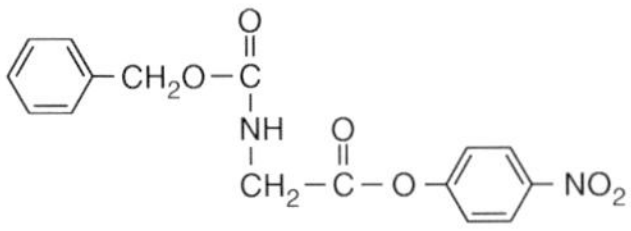
5 gm

S60,113-6
CARBOBENZYLOXY-L-LEUCINE 4-NITROPHENYL ESTER

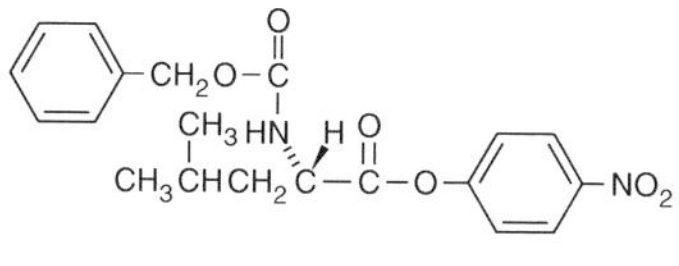

5 gm

S70,521-7
N,N'-BIS(CARBOBENZYLOXY)-L-LYSINE 4-NITROPHENYL ESTER

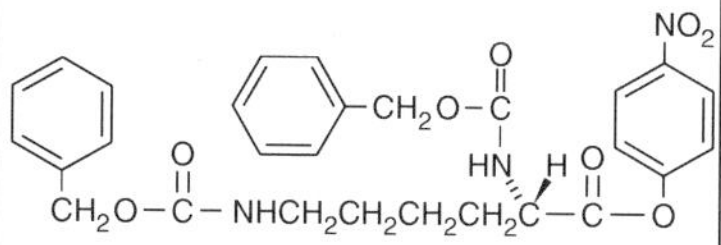

5 gm

S69,569-6
N-(2-NITROBENZYLOXYCARBONYL)GLYCINE ETHYL ESTER

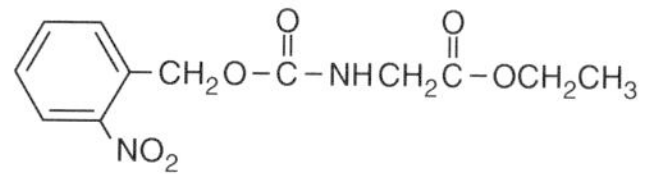

250 mg

S79,031-1
N-ALPHA-CARBOBENZYLOXY-N-EPSILON-FORMYL-L-LYSINE

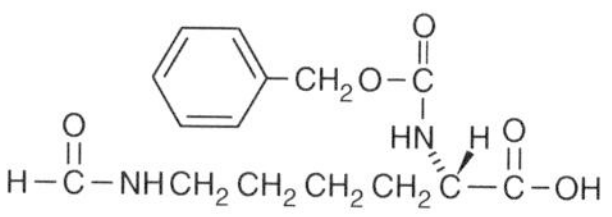

250 mg

S79,022-2
N,N'-BIS(CARBOBENZYLOXY)-L-LYSINE

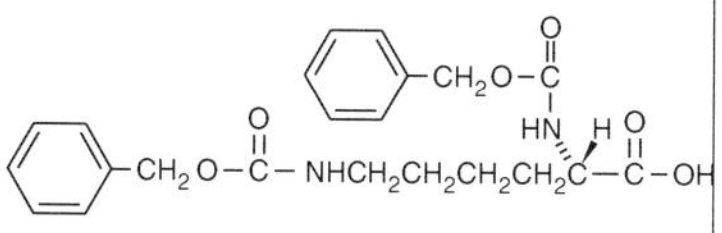

250 mg

S84,741-0
4-(CARBOBENZYLOXY)PHENYL N-(2,5-XYLYL)CARBAMATE

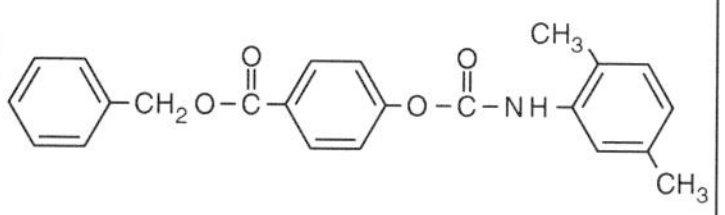

250 mg

S78,448-6
CARBOBENZYLOXY-DL-GLUTAMIC ACID

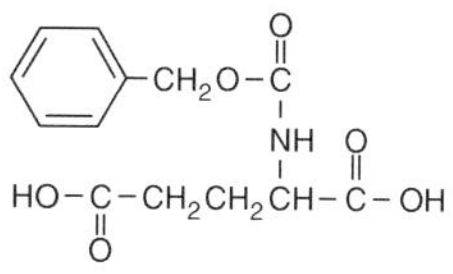

250 mg

S80,509-2
N-CARBOBENZYLOXY-2-METHYL-L-GLUTAMIC ACID DICYCLOHEXYLAMINE SALT

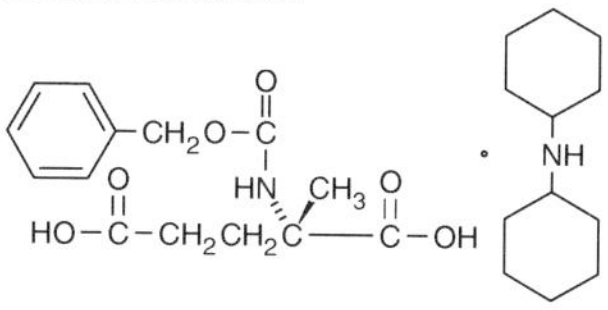

250 mg

S79,028-1
CARBOBENZYLOXY-L-GLUTAMIC ACID ALPHA-ETHYL ESTER DICYCLOHEXYLAMINE SALT

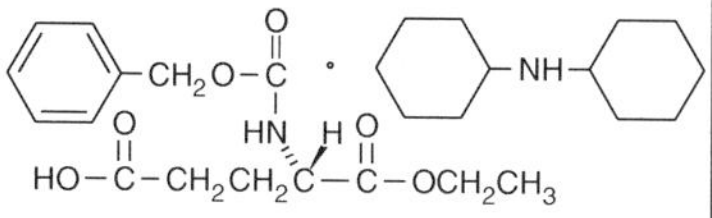

250 mg

S78,972-0
CARBOBENZYLOXY-L-GLUTAMIC ACID 1-TERT-BUTYL ESTER

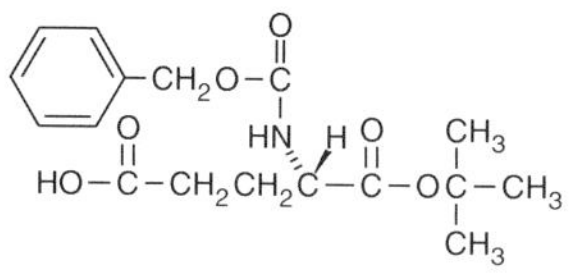

250 mg

S79,048-6
CARBOBENZYLOXY-L-ASPARTIC ACID 1-TERT-BUTYL ESTER DICYCLOHEXYLAMINE SALT

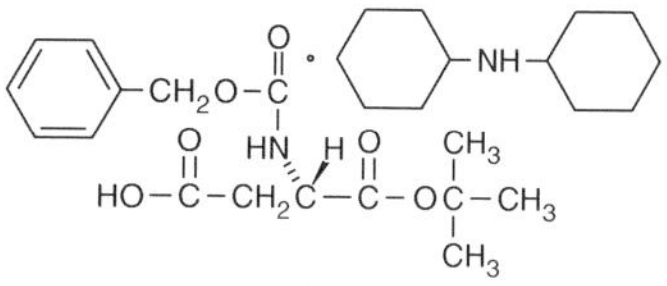

250 mg

S78,603-9
CARBOBENZYLOXY-L-ALANYL-L-ASPARTIC ACID DIBENZYL ESTER

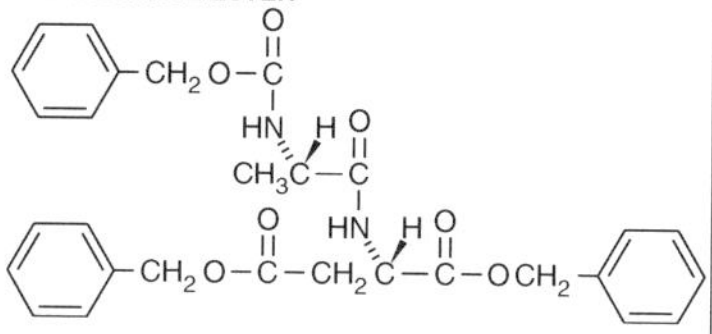

250 mg

S78,709-4
CARBOBENZYLOXY-L-ALANYL-L-GLUTAMIC ACID DIBENZYL ESTER

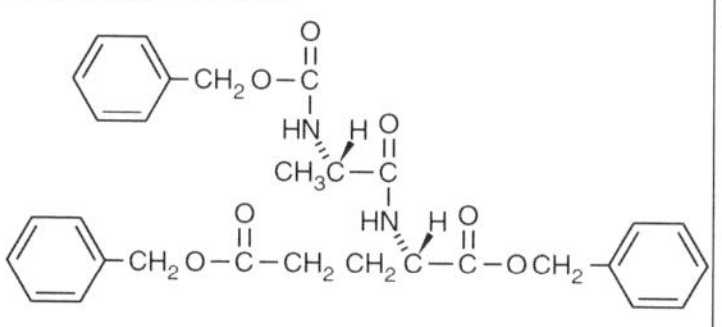

250 mg

S78,276-9
N(T-BUTOXYCARBONYL)L-GLUTAMIC ACID GAMMA-BENZYL ESTER ALPHA(4NITROPHENYL) ESTER

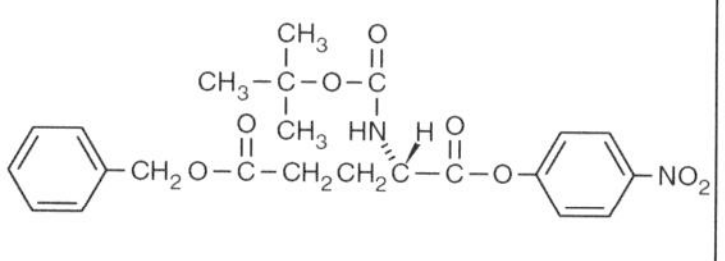

5 gm

S78,484-2
CARBOBENZYLOXY-ALPHA-L-GLUTAMYL-L-PHENYLALANINE

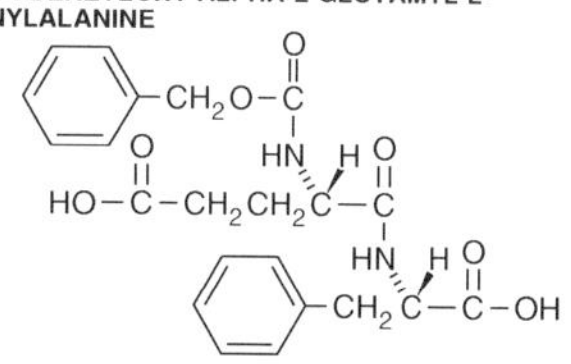

250 mg

S57,634-4
CARBOBENZYLOXY-DL-ASPARTIC ACID

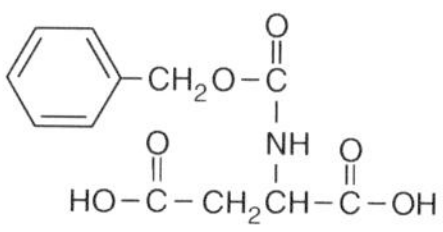

5 gm

S78,500-8
N,N'-BIS(CARBOBENZYLOXY)-L-LYSYL-L-ISOLEUCINE

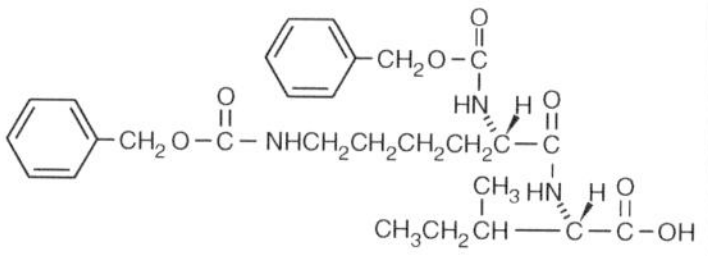

250 mg

S78,134-7
N,N'-BIS(CARBOBENZYLOXY)-L-LYSYL-L-PHENYLALANINE

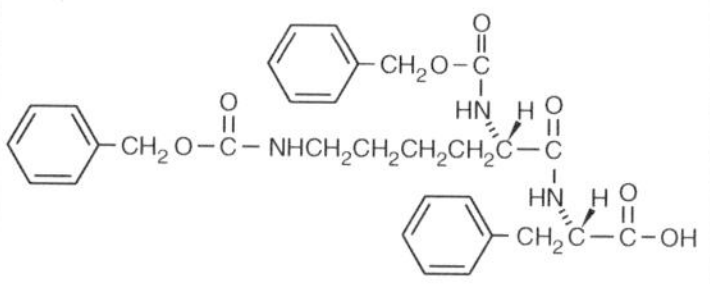

250 mg

S78,474-5
N-EPSILON-TERT-BUTOXYCARBONYL-N-ALPHA-CARBOBENZYLOXY-L-LYSINE

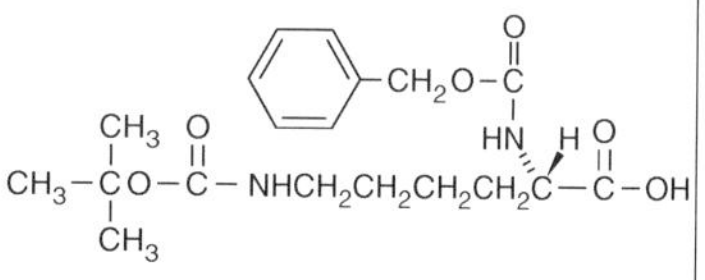

250 mg

S76,536-8
CARBOBENZYLOXY-L-GLUTAMIC ACID 5-TERT-BUTYL ESTER DICYCLOHEXYLAMINE SALT

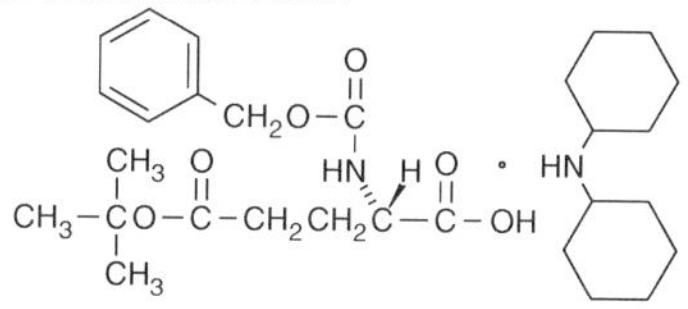

5 gm

S44,718-8
N,N'-BIS(CARBOBENZYLOXY)-L-LYSINE

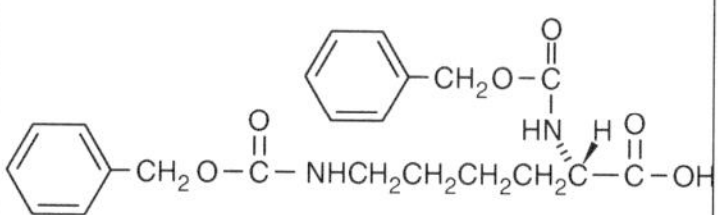

5 gm

S45,183-5
N,N'-BIS(CARBOBENZYLOXY)-DL-LYSINE

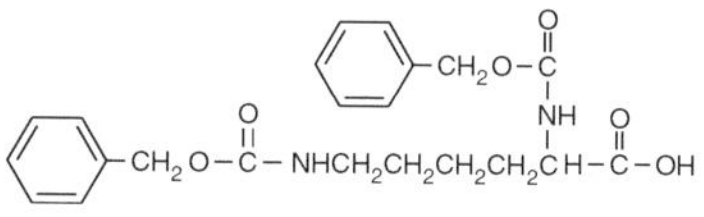

5 gm

S70,514-4
N,O-BIS(CARBOBENZYLOXY)-L-TYROSINE

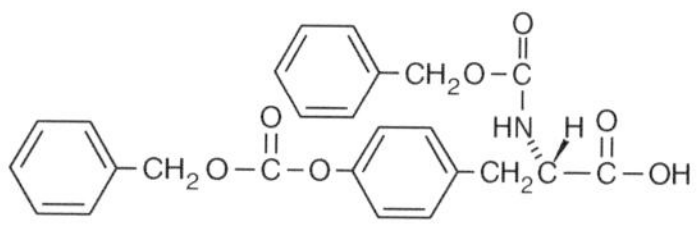

5 gm

S76,524-4
CARBOBENZYLOXY-L-ALANYLGLYCINE

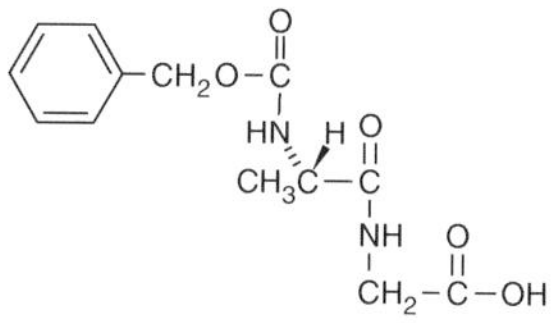

5 gm

S77,756-0
CARBOBENZYLOXYSARCOSYLGLYCINE

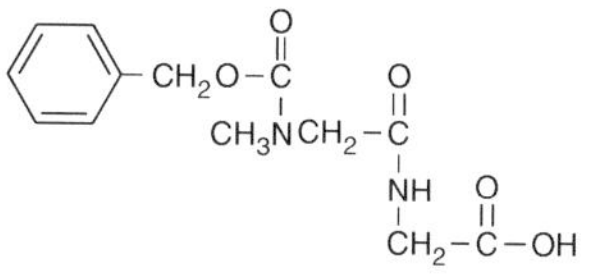

250 mg

S77,671-8
CARBOBENZYLOXY-L-NORVALYLGLYCINE

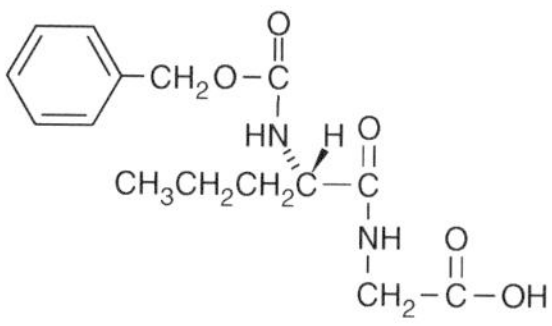

250 mg

S57,129-6
N-CARBOBENZYLOXY-DL-NORVALYLGLYCINE

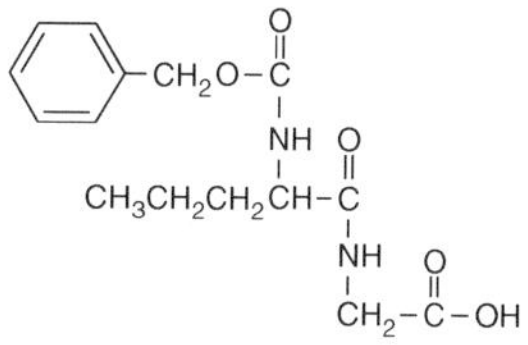

5 gm

S78,150-9
CARBOBENZYLOXYGLYCYL-L-NORVALINE

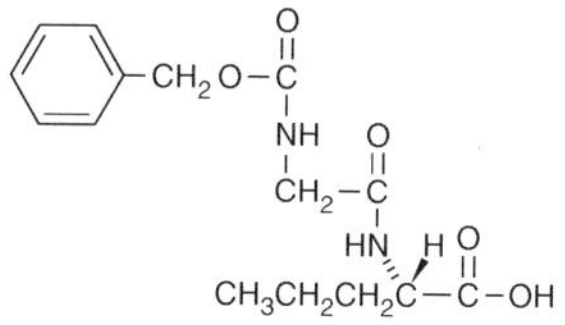

250 mg

S78,151-7
CARBOBENZYLOXY-L-NORVALYLGLYCINE

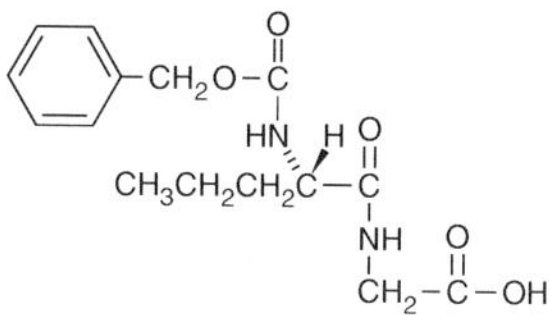

250 mg

S78,347-1
CARBOBENZYLOXY-L-VALYLGLYCINE

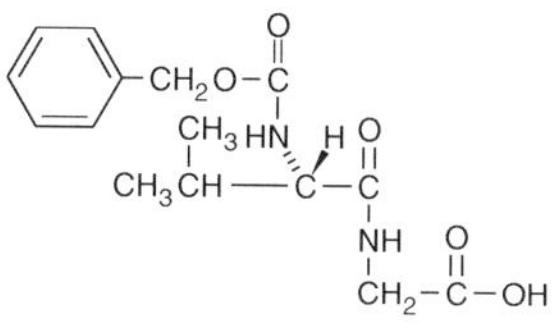

5 gm

S78,534-2
CARBOBENZYLOXY-L-ALANYL-2-AMINOBUTYRIC ACID

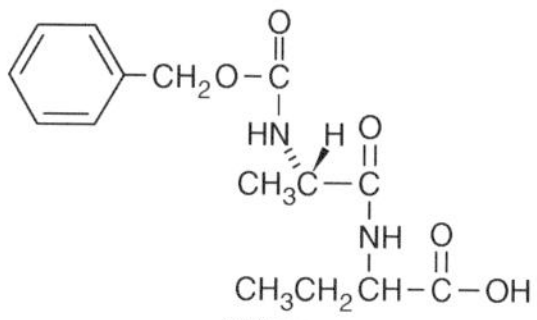

250 mg

S78,531-8
CARBOBENZYLOXY-BETA-ALANYLGLYCYLGLYCINE

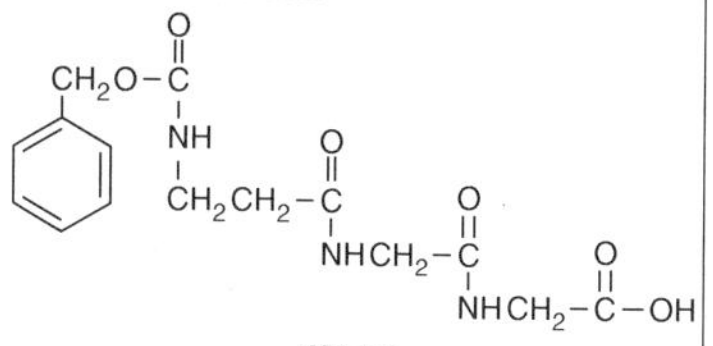

250 mg

S76,532-5
CARBOBENZYLOXY-GLYCYL-L-ISOLEUCINE
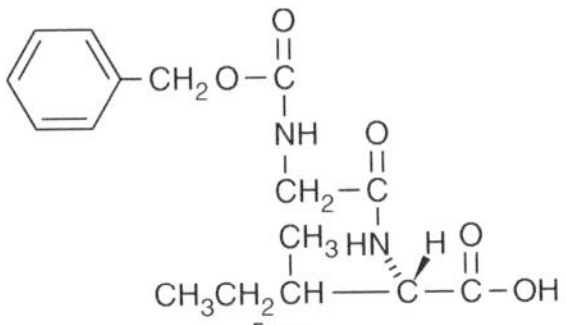
5 gm

S76,545-7
CARBOBENZYLOXY-L-LEUCYL-GLYCINE
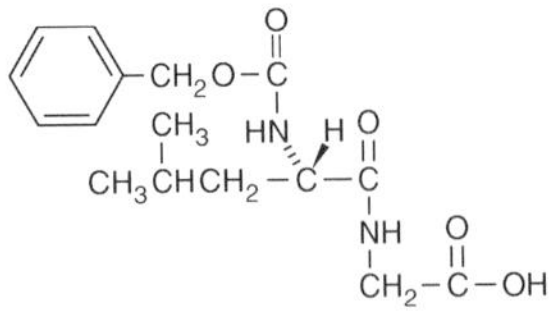
5 gm

S77,611-4
CARBOBENZYLOXY-L-ALANYL-L-NORVALINE
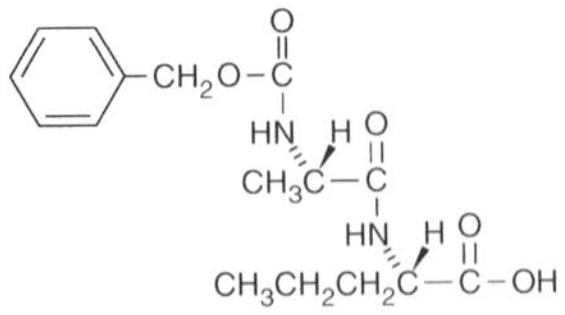
250 mg

S77,693-9
CARBOBENZYLOXY-L-NORVALYL-L-ALANINE
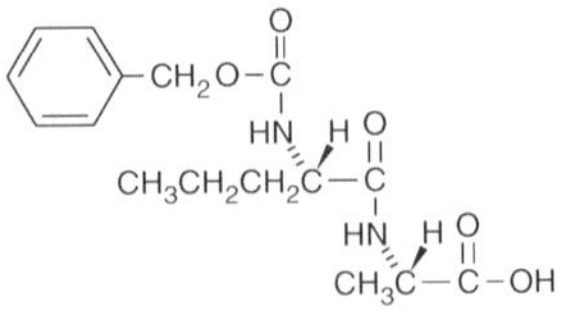
250 mg

S78,748-5
CARBOBENZYLOXY-L-ALANYL-L-LEUCINE
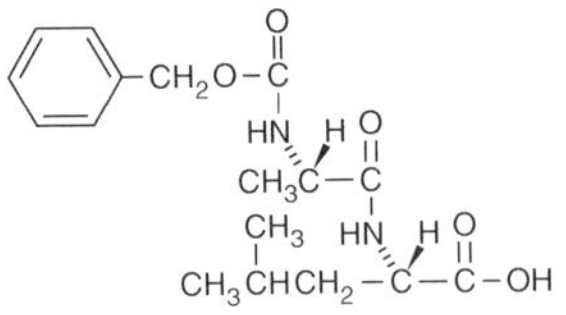
250 mg

S78,749-3
CARBOBENZYLOXY-L-ALANYL-L-NORLEUCINE
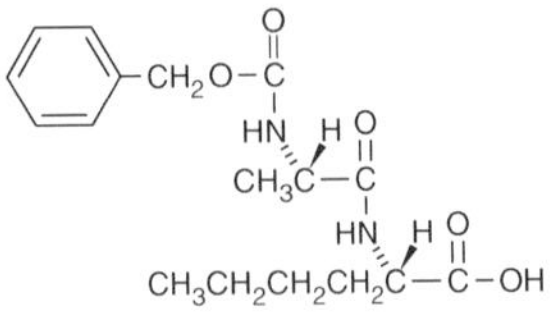
250 mg

S77,787-0
CARBOBENZYLOXY-L-NORVALYL-L-NORVALINE
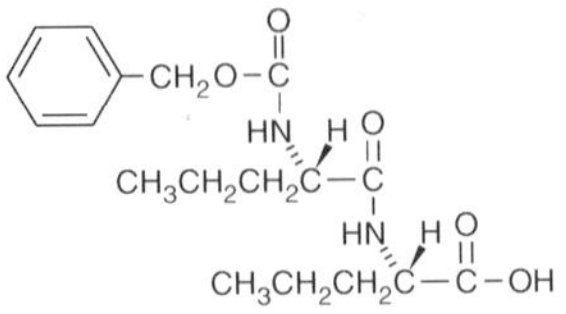
250 mg

S78,945-3
CARBOBENZYLOXYGLYCYLGLYCYL-L-NORLEUCINE
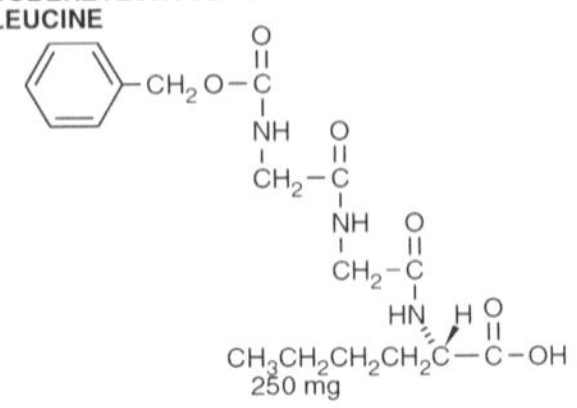
250 mg

S76,550-3
CARBOBENZYLOXY-L-PHENYLALANYLGLYCINE
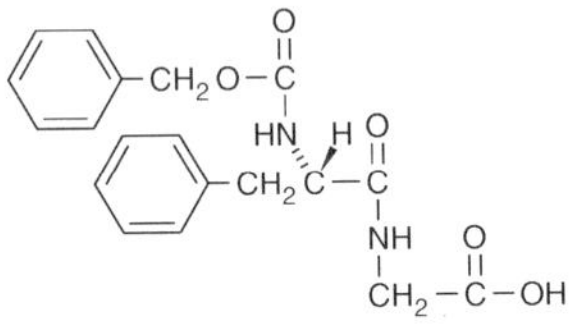
5 gm

S76,553-8
CARBOBENZYLOXY-L-ISOLEUCYL-L-VALINE
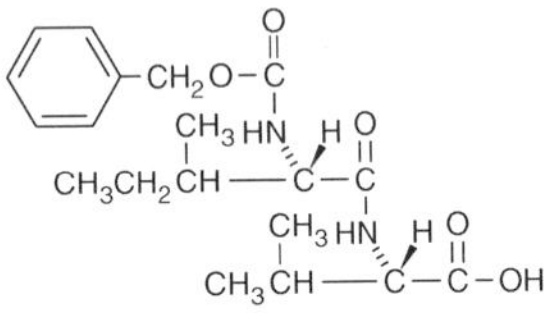
5 gm

S77,658-0
CARBOBENZYLOXY-L-NORVALYL-L-NORLEUCINE
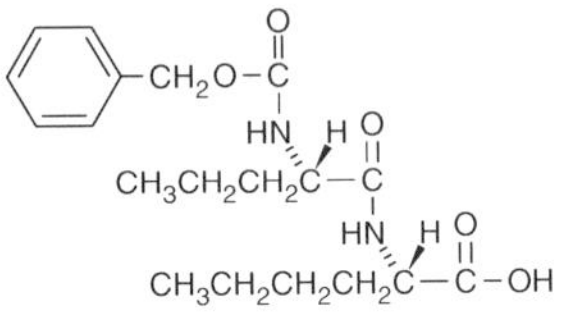
250 mg

S77,673-4
CARBOBENZYLOXY-L-NORLEUCYL-L-VALINE
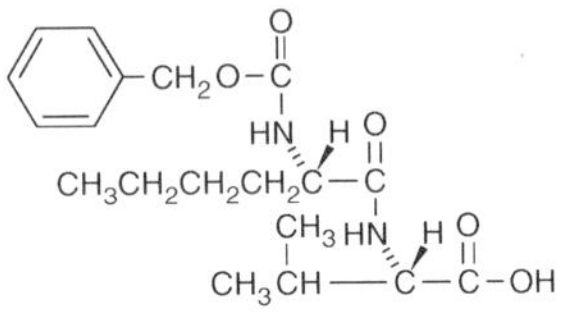
250 mg

S77,674-2
CARBOBENZYLOXY-L-NORVALYL-L-LEUCINE
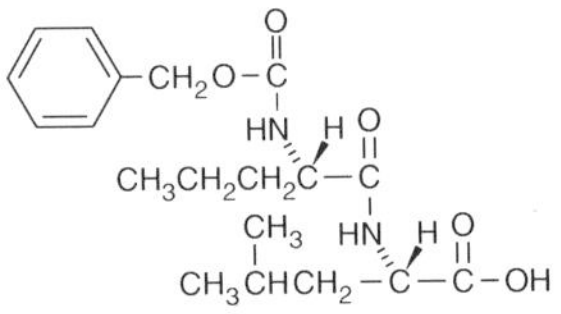
250 mg

S77,675-0
CARBOBENZYLOXY-L-NORVALYL-L-ISOLEUCINE
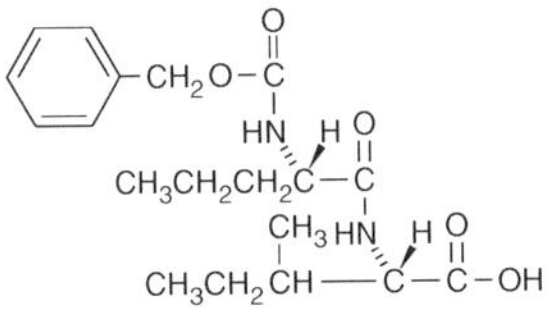
250 mg

S78,145-2
CARBOBENZYLOXY-L-ISOLEUCYL-L-NORVALINE
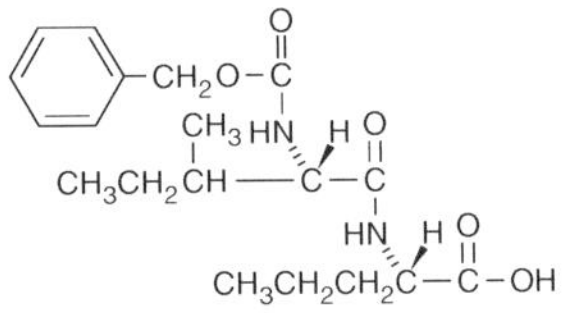
250 mg

S76,556-2
CARBOBENZYLOXY-L-LEUCYL-L-LEUCINE
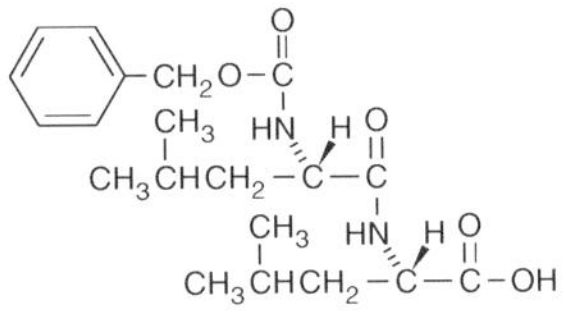
5 gm

S78,928-3
CARBOBENZYLOXY-D-2-AMINOBUTYRIC ACID
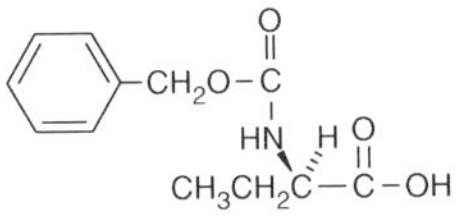
250 mg

S78,192-4
CARBOBENZYLOXY-BETA-ALANYLGLYCINE
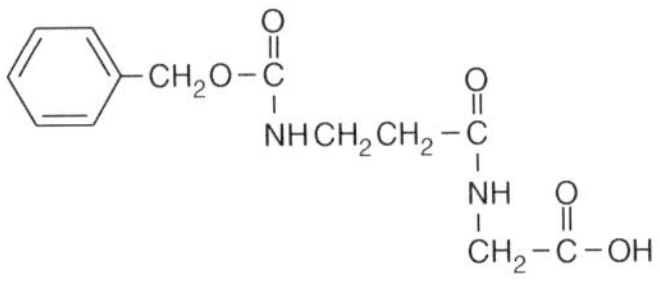
250 mg

S78,529-6
CARBOBENZYLOXY-BETA-ALANYL-L-ALANINE
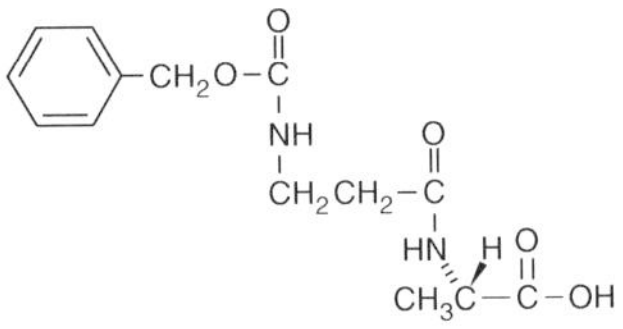
250 mg

S78,293-9
N-EPSILON-CARBOBENZYLOXY-N-ALPHA-ACETYL-DL-LYSINE
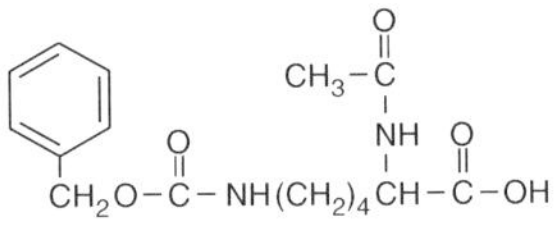
5 gm

S78,982-8
N-EPSILON-CARBOBENZYLOXY-N-ALPHA-ACETYL-L-LYSINE
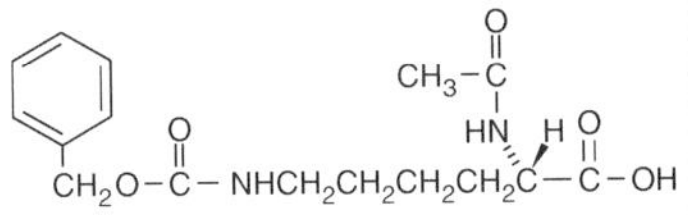
250 mg

S78,162-2
CARBOBENZYLOXY-BETA-ALANYL-L-LEUCINE
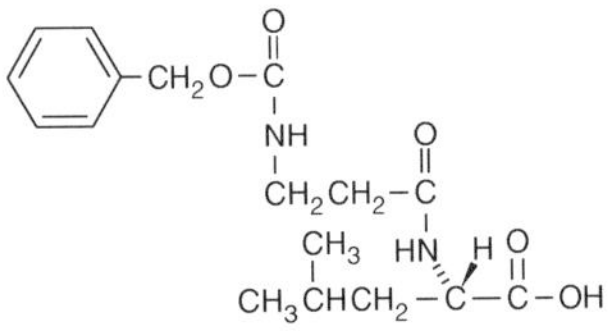
250 mg

S78,532-6
CARBOBENZYLOXYGLYCYLGLYCYLSARCOSINE
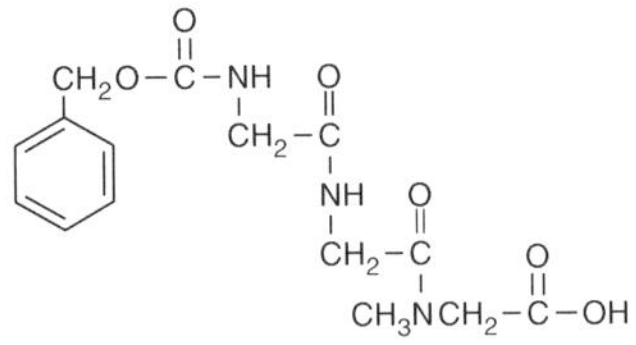
250 mg

S77,729-3
CARBOBENZYLOXYSARCOSYL-L-ALANINE
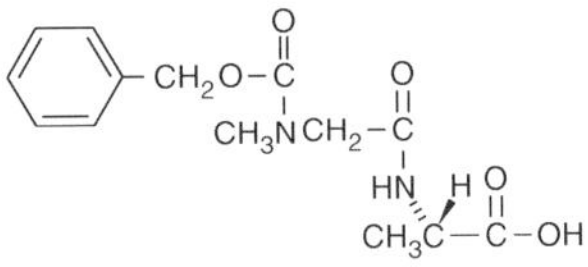
250 mg

S70,517-9
N-CARBOBENZYLOXYLGLYCYL-L-VALINE
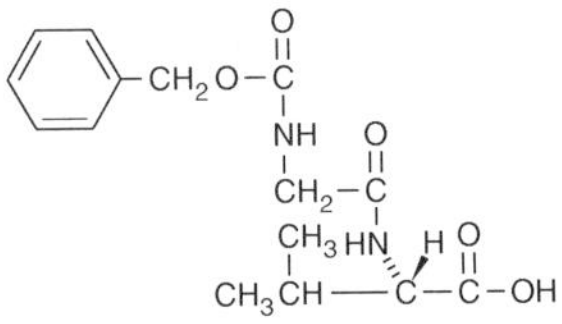
5 gm

S70,524-1
CARBOBENZYLOXY-L-ISOLEUCYL-GLYCINE
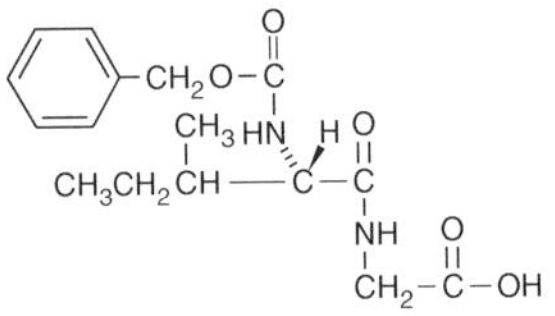
5 gm

S77,651-3
CARBOBENZYLOXY-L-ALANYL-L-VALINE
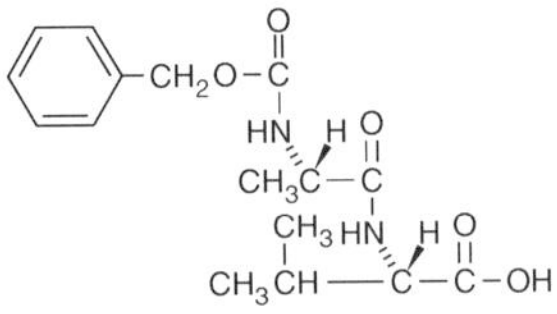
250 mg

S77,652-1
CARBOBENZYLOXY-SARCOSYL-L-VALINE
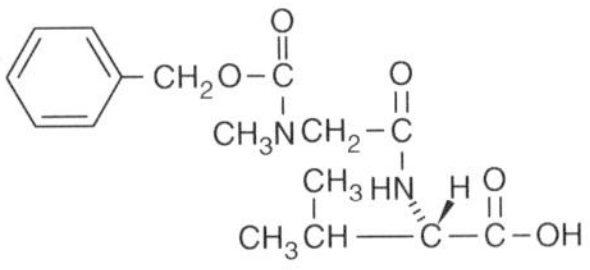
250 mg

S70,494-6
CARBOBENZYLOXY-L-ISOLEUCYL-L-ALANINE
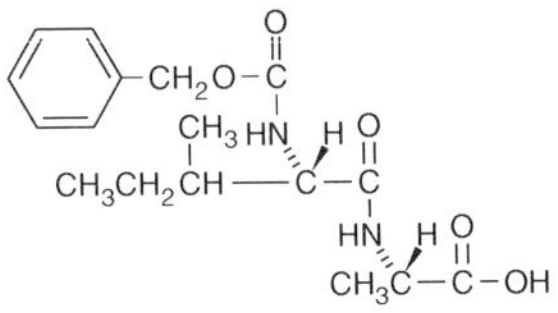
5 gm

S70,515-2
CARBOBENZYLOXY-L-LEUCYL-L-ISOLEUCINE
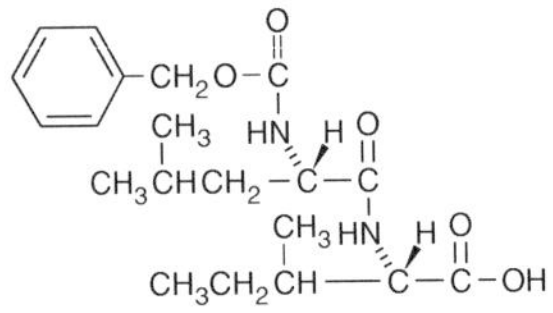
5 gm

S70,506-3
CARBOBENZYLOXY-L-PHENYLALANYL-L-ISOLEUCINE
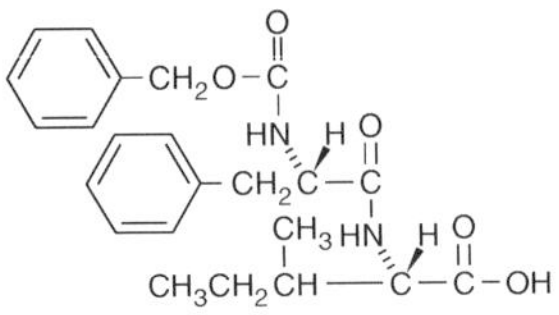
250 mg

S70,523-3
CARBOBENZYLOXY-L-ISOLEUCYL-L-PHENYLALANINE
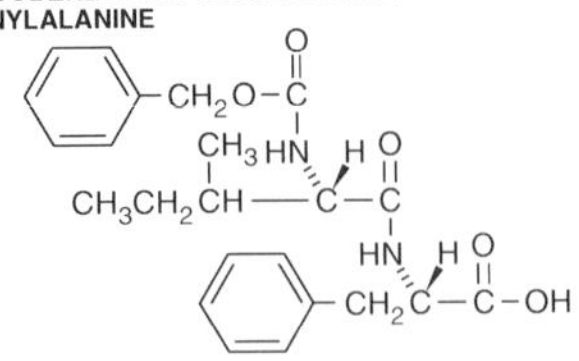
5 gm

S78,283-1
CARBOBENZYLOXY-L-2-AMINOBUTYRIC ACID

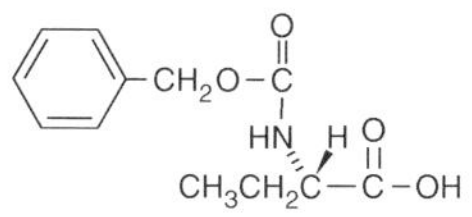

5 gm

S45,133-9
CARBOBENZYLOXY-DL-NORVALINE

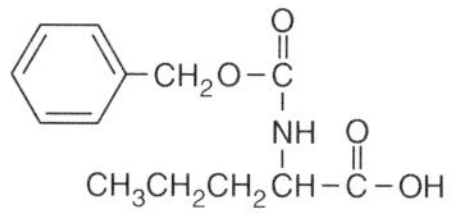

5 gm

S45,134-7
CARBOBENZYLOXY-DL-VALINE

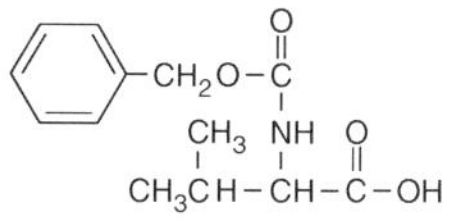

5 gm

S78,363-3
CARBOBENZYLOXY-L-NORVALINE

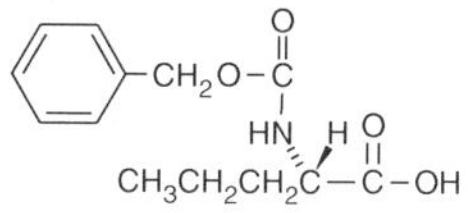

5 gm

S76,577-5
CARBOBENZYLOXY-DL-NORLEUCINE

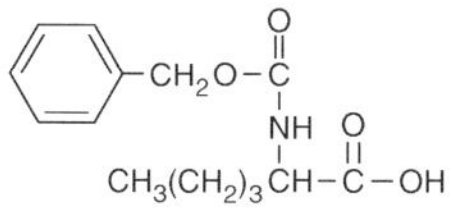

5 gm

S71,067-9
N-CARBOBENZYLOXY-DL-2-PHENYLGLYCINE

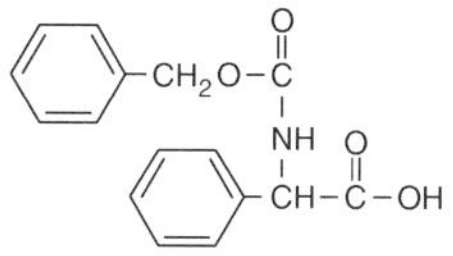

5 gm

S79,060-5
CARBOBENZYLOXY-D-NORVALINE

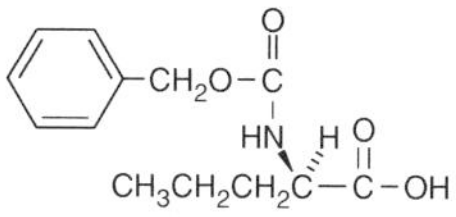

250 mg

S34,354-4
CARBOBENZYLOXY-L-LEUCINE

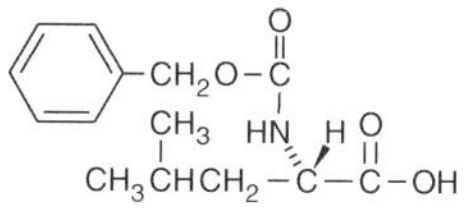

5 gm

S52,077-2
CARBOBENZYLOXY-L-ISOLEUCINE

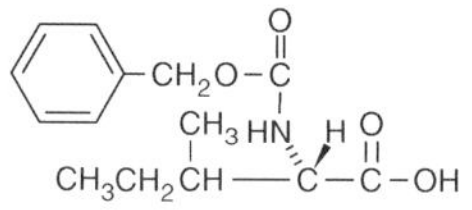

250 mg

S85,959-1
CARBOBENZYLOXY-BETA-ALANINE

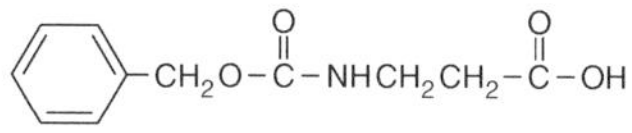

250 mg

S79,004-4
N,O-BIS(CARBOBENZYLOXY)-L-TYROSYL-L-LEUCINE METHYL ESTER

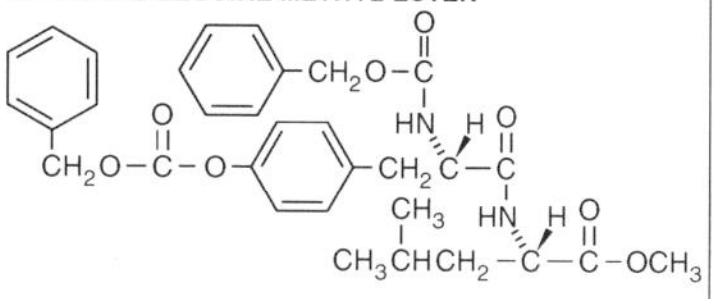

250 mg

S78,763-9
CARBOBENZYLOXYGLYCYLGLYCYLSARCOSINE BENZYL ESTER

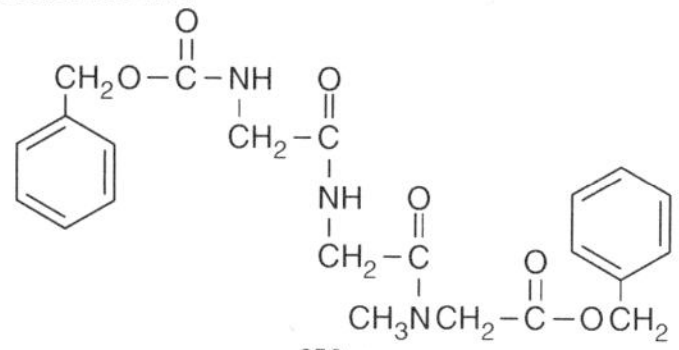

250 mg

S79,033-8
CARBOBENZYLOXY-L-GLYCYL-L-NORVALINE METHYL ESTER

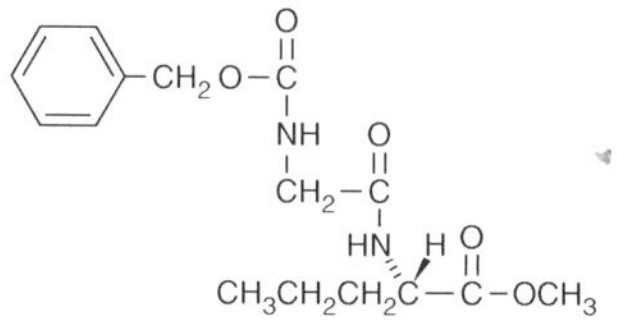

250 mg

S78,469-9
CARBOBENZYLOXY-L-ALANYL-L-NORVALINE METHYL ESTER

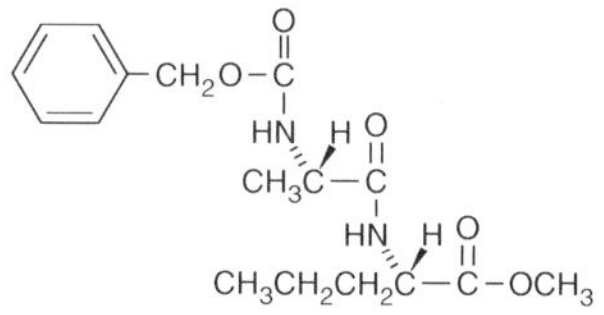

250 mg

S79,064-8
CARBOBENZYLOXY-L-NORLEUCYL-L-VALINE METHYL ESTER

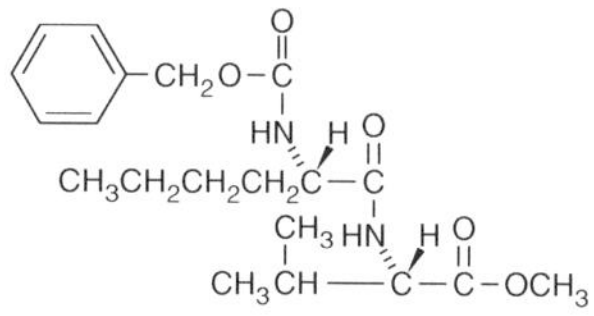

250 mg

S78,987-9
CARBOBENZYLOXY-L-VALYLGLYCYLGLYCINE BENZYL ESTER

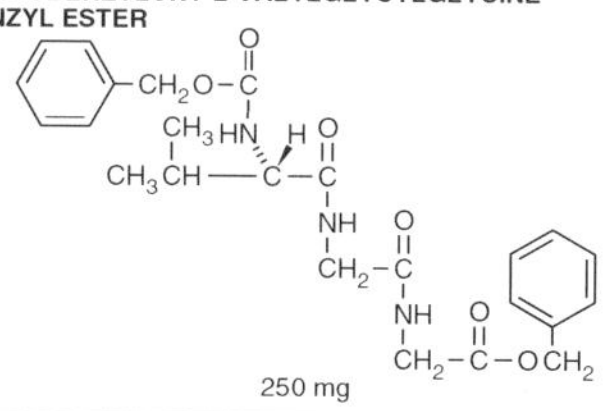

250 mg

S79,027-3
CARBOBENZYLOXYGLYCYLGLYCINE METHYL ESTER

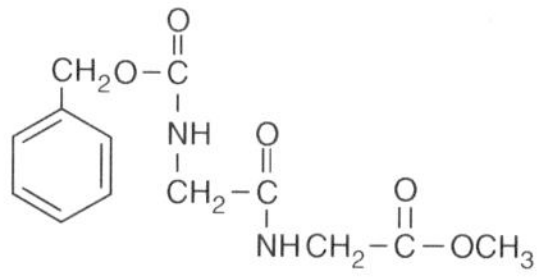

250 mg

S78,992-5
CARBOBENZYLOXYSARCOSYLGLYCINE METHYL ESTER

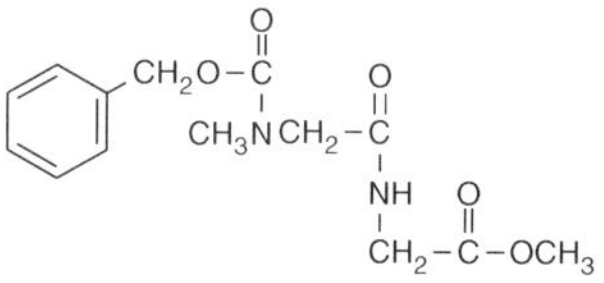

250 mg

S78,943-7
CARBOBENZYLOXYSARCOSYL-L-ALANINE METHYL ESTER

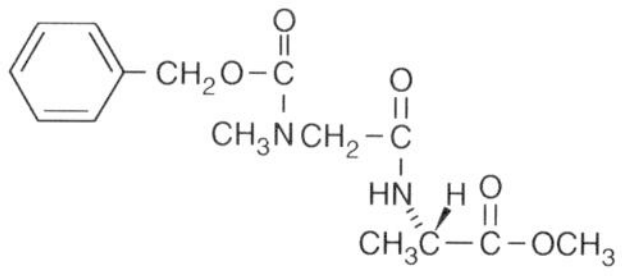

250 mg

S79,030-3
CARBOBENZYLOXYGLYCYLGLYCYLGLYCINE METHYL ESTER

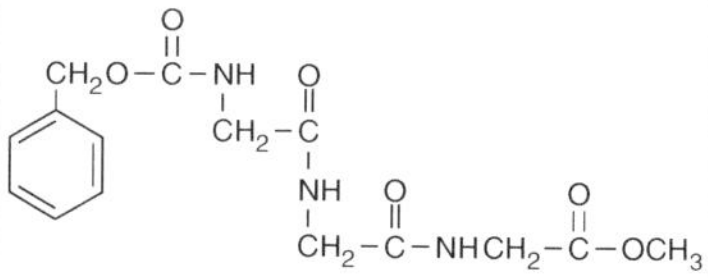

250 mg

S34,352-8
CARBOBENZYLOXY-L-ALANYL-L-ALANINE ETHYL ESTER

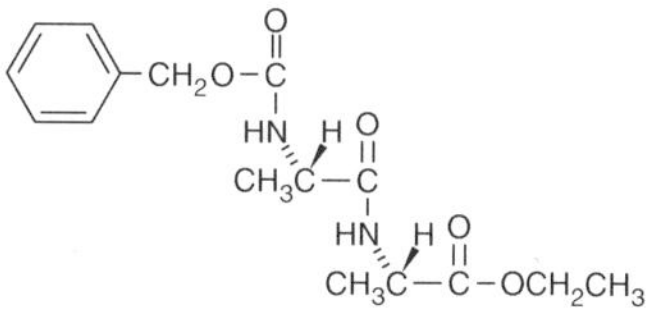

5 gm

S57,017-6
CARBOBENZYLOXY-DL-NORVALYLGLYCINE METHYL ESTER

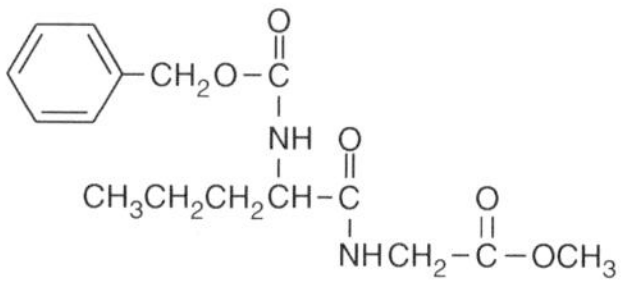

5 gm

S78,596-2
CARBOBENZYLOXY-L-ALANYL-2-AMINOBUTYRIC ACID METHYL ESTER

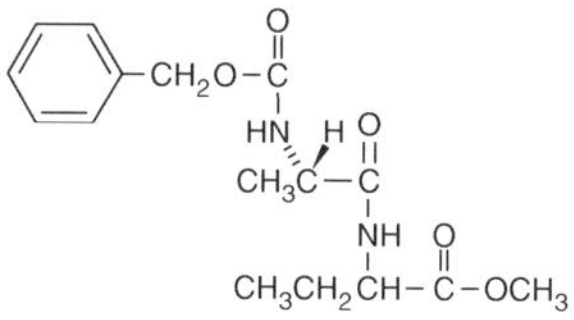

250 mg

S78,597-0
CARBOBENZYLOXYGLYCYL-L-VALINE METHYL ESTER

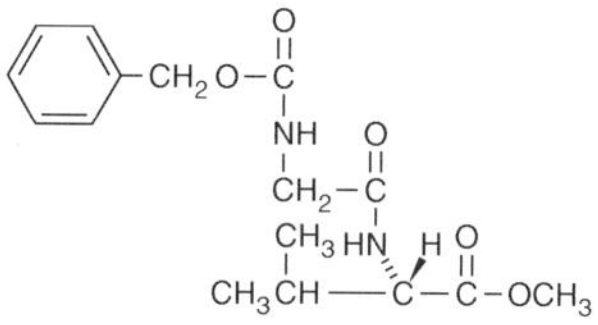

250 mg

S78,807-4
CARBOBENZYLOXY-BETA-ALANYLGLYCYLGLYCINE METHYL ESTER

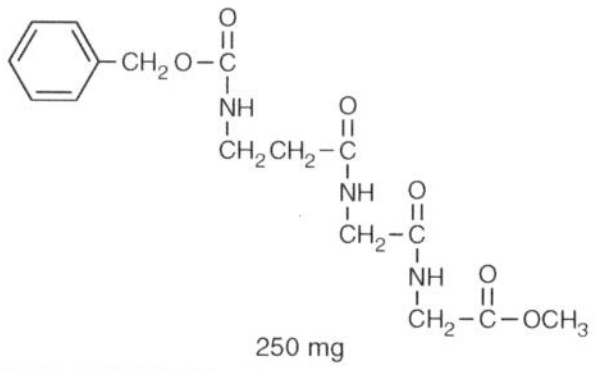

250 mg

S78,808-2
CARBOBENZYLOXY-L-ALANYLGLYCYLGLYCINE METHYL ESTER

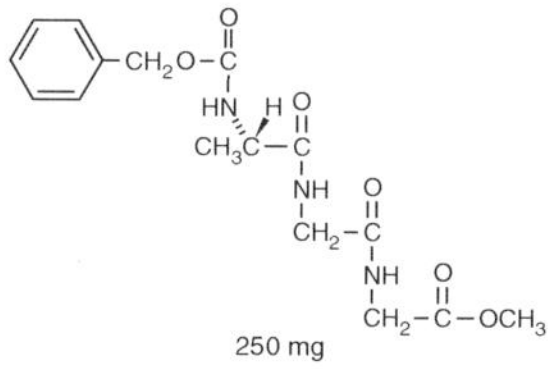

250 mg

S79,049-4
CARBOBENZYLOXYGLYCYLGLYCYL-L-ALANINE METHYL ESTER

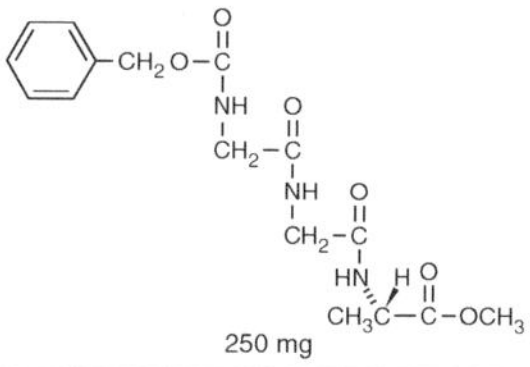

250 mg

S78,640-3
CARBOBENZYLOXY-L-ALANYL-L-VALINE METHYL ESTER

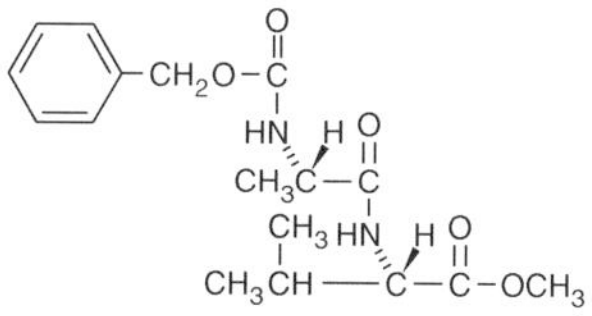

250 mg

S78,747-7
CARBOBENZYLOXYGLYCYL-L-LEUCINE METHYL ESTER

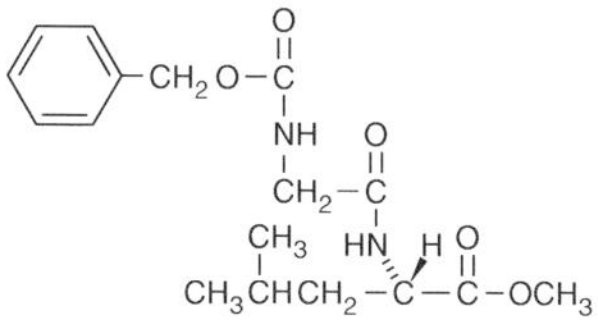

250 mg

S78,811-2
CARBOBENZYLOXY-L-LEUCYLGLYCINE METHYL ESTER

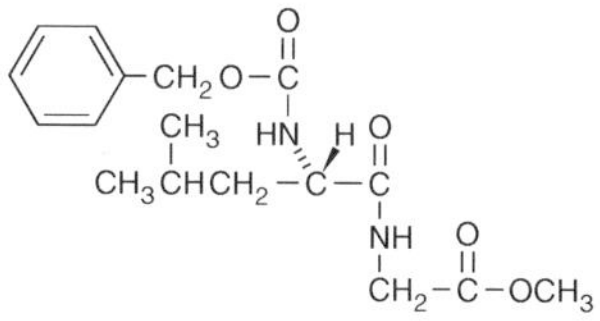

250 mg

S78,996-8
CARBOBENZYLOXYSARCOSYL-L-VALINE METHYL ESTER

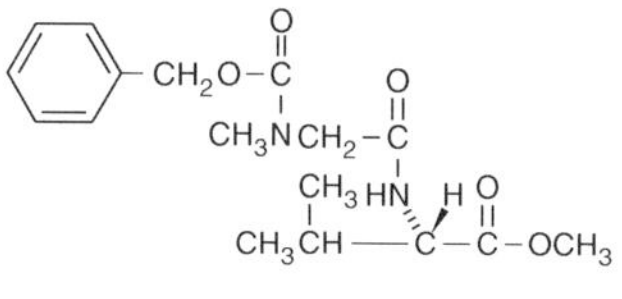

250 mg

S34,355-2
CARBOBENZYLOXY-L-LEUCYLGLYCINE ETHYL ESTER

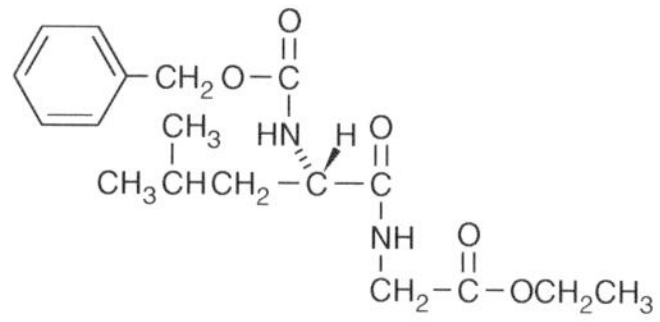

250 mg

S78,314-5
CARBOBENZYLOXY-L-LEUCYL-L-ALANINE METHYL ESTER
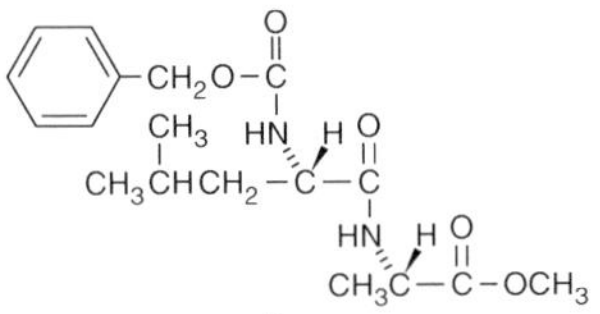
5 gm

S78,326-9
CARBOBENZYLOXY-L-ALANYL-L-NORLEUCINE METHYL ESTER
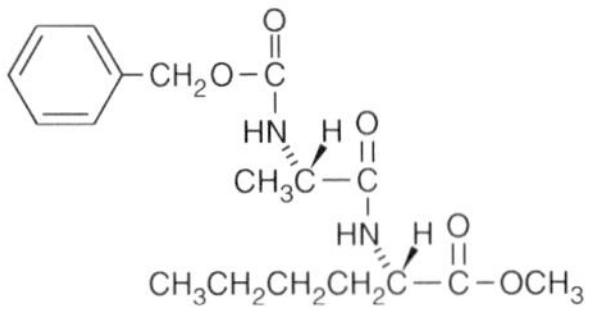
5 gm

S78,755-8
CARBOBENZYLOXY-L-ISOLEUCYL-L-ALANINE METHYL ESTER
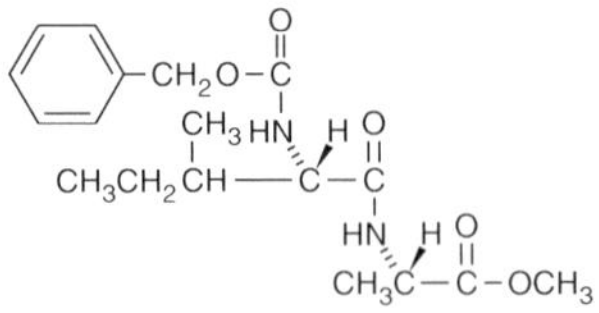
250 mg

S78,998-4
CARBOBENZYLOXYSARCOSYL-L-ISOLEUCINE METHYL ESTER
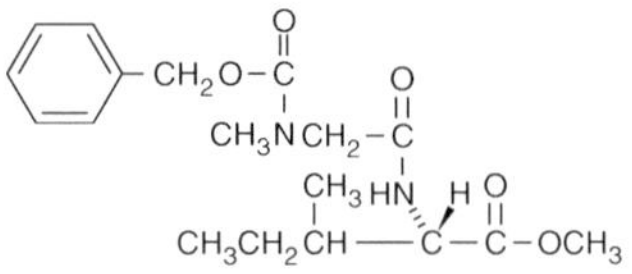
250 mg

S77,660-2
CARBOBENZYLOXY-L-NORVALYLGLYCYLGLYCINE METHYL ESTER
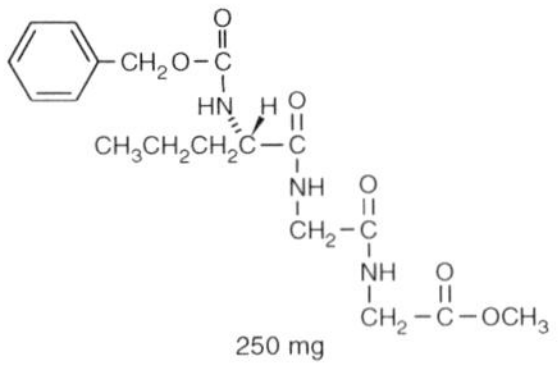
250 mg

S78,752-3
CARBOBENZYLOXYGLYCYLGLYCYL-L-VALINE METHYL ESTER
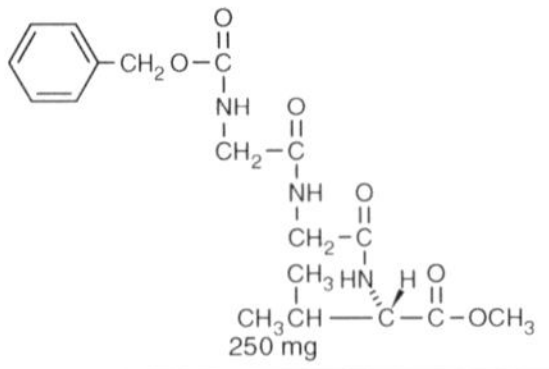
250 mg

S78,753-1
CARBOBENZYLOXYGLYCYLGLYCYL-L-NORVALINE METHYL ESTER
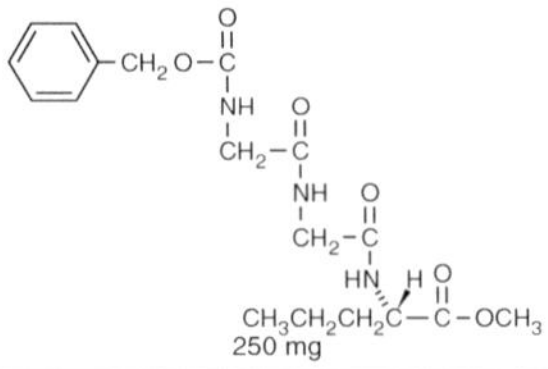
250 mg

S78,819-8
CARBOBENZYLOXY-L-NORVALYL-L-VALINE METHYL ESTER
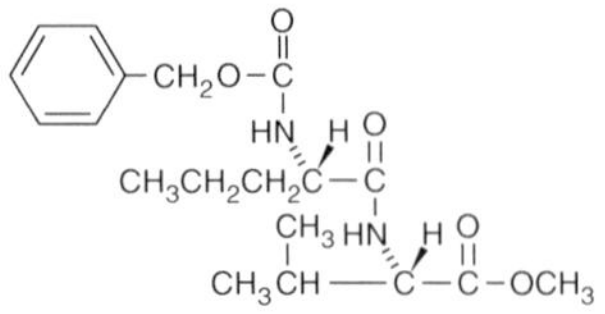
250 mg

S78,638-1
CARBOBENZYLOXYGLYCYLGLYCYL-L-NORLEUCINE METHYL ESTER
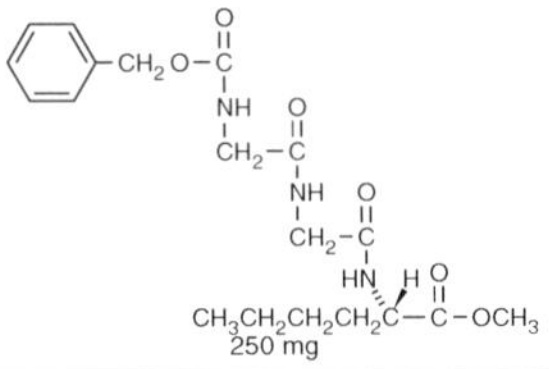
250 mg

S78,963-1
CARBOBENZYLOXY-L-NORVALYL-L-LEUCINE METHYL ESTER
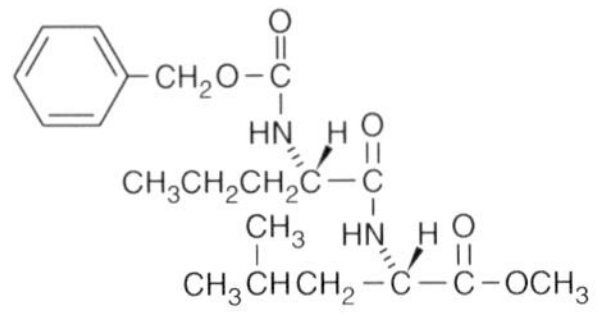
250 mg

S78,973-9
CARBOBENZYLOXY-L-ISOLEUCYL-L-NORVALINE METHYL ESTER
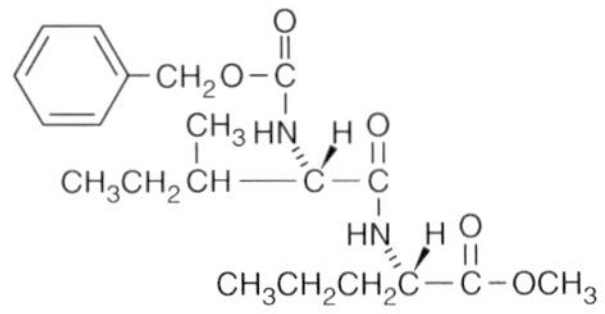
250 mg

S78,999-2
CARBOBENZYLOXY-L-NORVALYL-L-NORLEUCINE METHYL ESTER
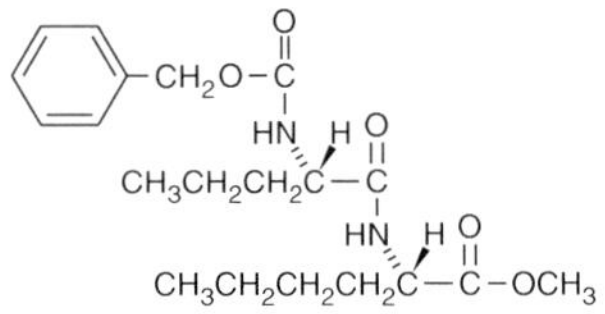
250 mg

S77,604-1
CARBOBENZYLOXY-L-ALANYL-L-ALANINE BENZYL ESTER
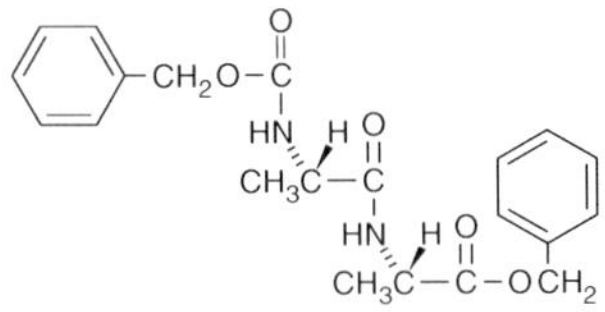
250 mg

S77,763-3
CARBOBENZYLOXY-L-LEUCYL-L-LEUCINE METHYL ESTER
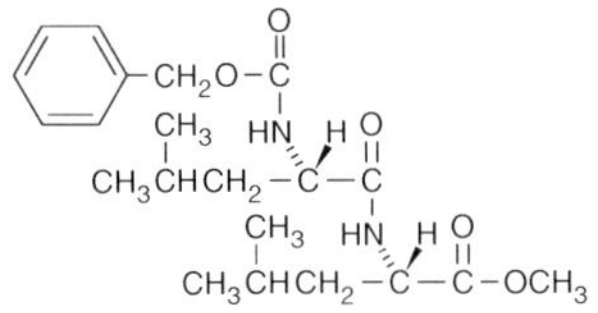
250 mg

S78,958-5
CARBOBENZYLOXY-L-NORLEUCYL-L-LEUCINE METHYL ESTER
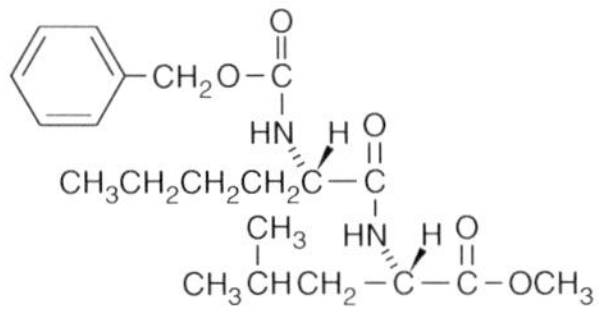
250 mg

S79,010-9
CARBOBENZYLOXY-L-NORLEUCYL-L-NORLEUCINE METHYL ESTER
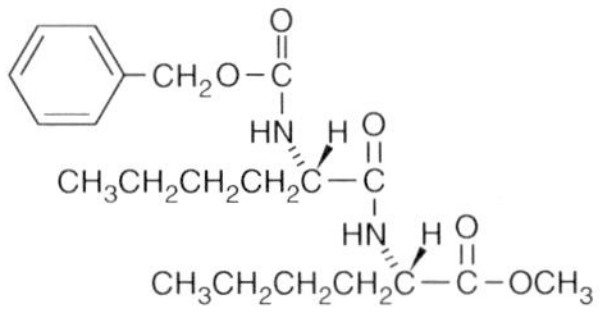
250 mg

S78,762-0
CARBOBENZYLOXYGLYCYLGLYCYLGLYCINE BENZYL ESTER
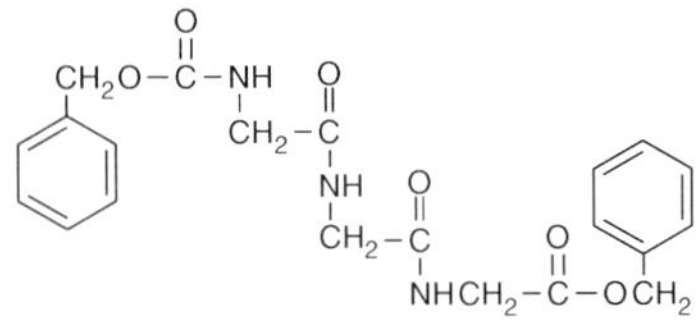
250 mg

S78,836-8
CARBOBENZYLOXY-L-ALANYL-L-2-AMINOBUTYRIC ACID BENZYL ESTER
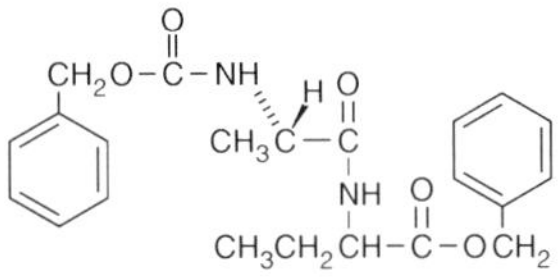
250 mg

S78,631-4
CARBOBENZYLOXYGLYCYLGLYCYL-L-ALANINE BENZYL ESTER
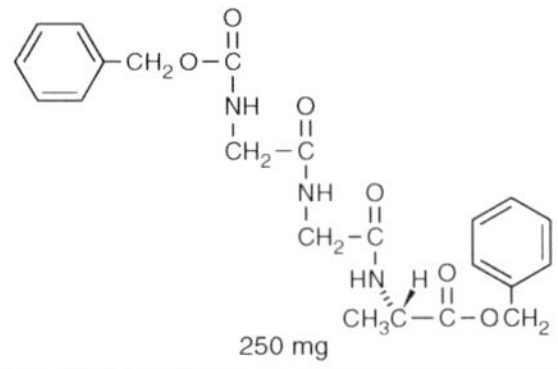
250 mg

S78,832-5
CARBOBENZYLOXY-L-ALANYLGLYCYL-GLYCINE BENZYL ESTER
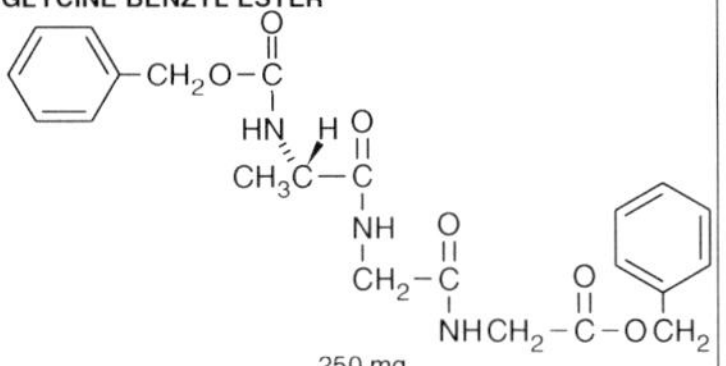
250 mg

S78,834-1
CARBOBENZYLOXY-BETA-ALANYLGLYCYLGLYCINE BENZYL ESTER
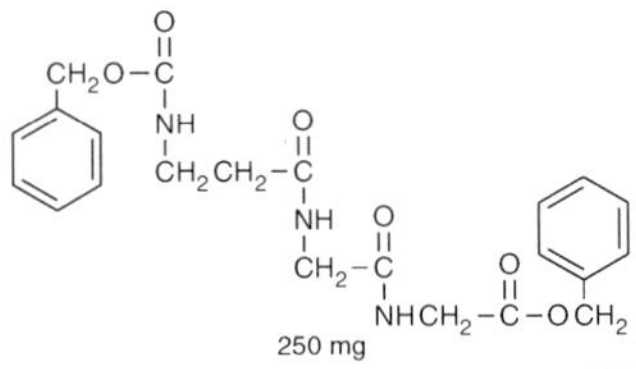
250 mg

S77,798-6
CARBOBENZYLOXYSARCOSYL-L-VALINE BENZYL ESTER
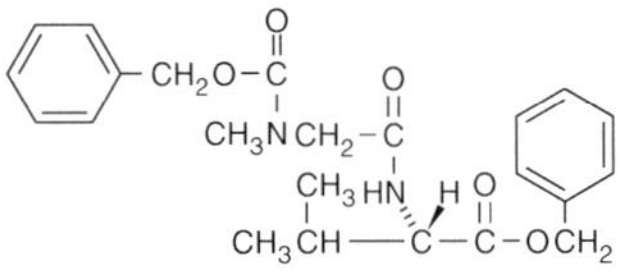
250 mg

S78,393-5
CARBOBENZYLOXY-L-ISOLEUCYLGLYCINE BENZYL ESTER
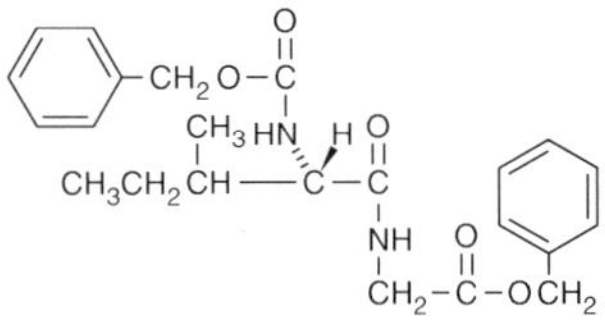
5 gm

S78,624-1
CARBOBENZYLOXY-L-VALYL-L-ALANINE BENZYL ESTER
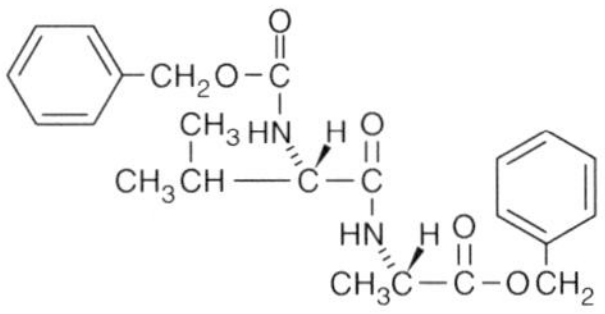
250 mg

S78,841-4
CARBOBENZYLOXY-L-PHENYLALANYL-L-VALINE METHYL ESTER
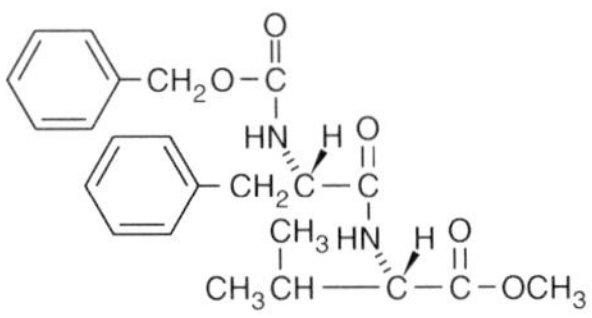
250 mg

S78,844-9
CARBOBENZYLOXY-L-ALANYL-L-NORVALINE BENZYL ESTER
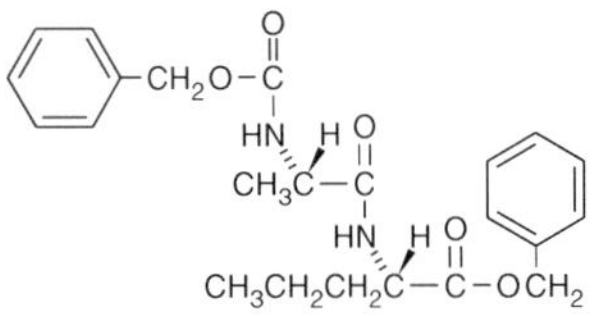
250 mg

S78,766-3
CARBOBENZYLOXY-L-ISOLEUCYL-L-LEUCYL GLYCINE METHYL ESTER
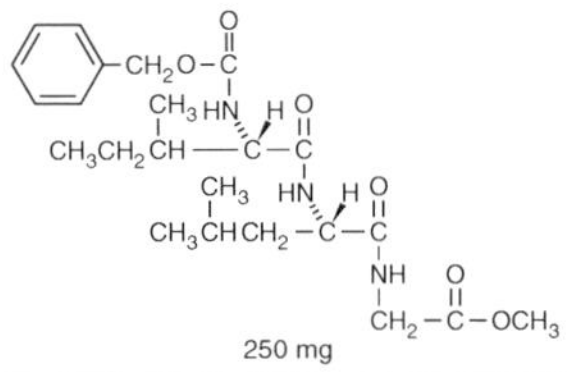
250 mg

S77,799-4
CARBOBENZYLOXY-L-ALANYL-L-ISOLEUCINE BENZYL ESTER
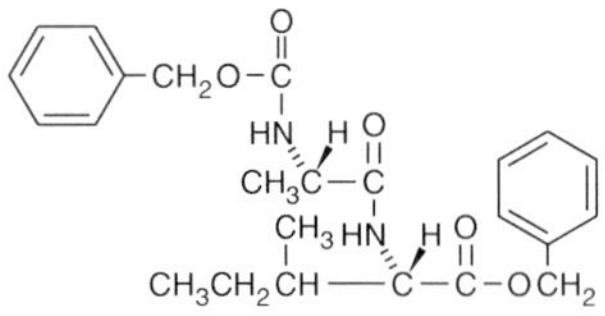
250 mg

S78,121-5
CARBOBENZYLOXY-L-ISOLEUCYL-L-ALANINE BENZYL ESTER
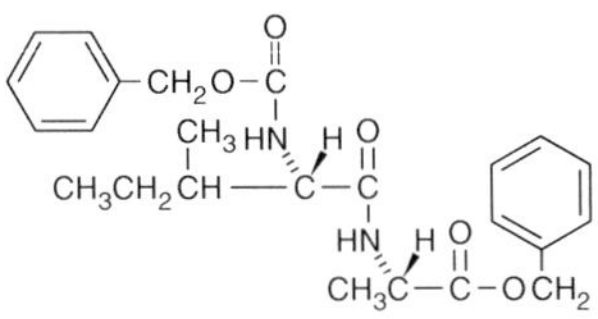
250 mg

S78,846-5
CARBOBENZYLOXY-L-ALANYL-L-LEUCINE BENZYL ESTER
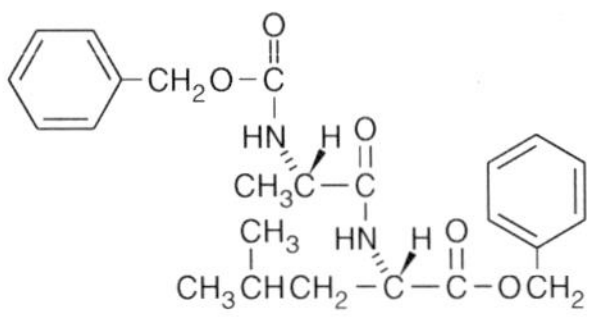
250 mg

S79,066-4
CARBOBENZYLOXY-L-LEUCYL-L-ALANINE BENZYL ESTER
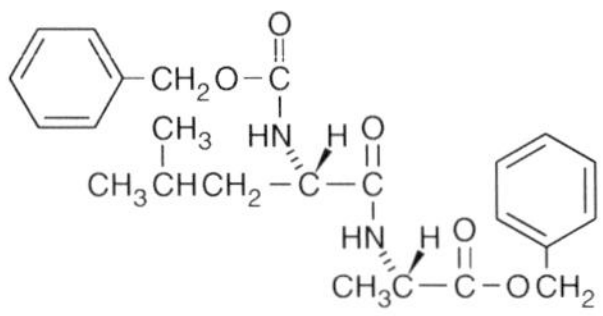
250 mg

S79,054-0
CARBOBENZYLOXYGLYCYLGLYCYL-L-VALINE BENZYL ESTER
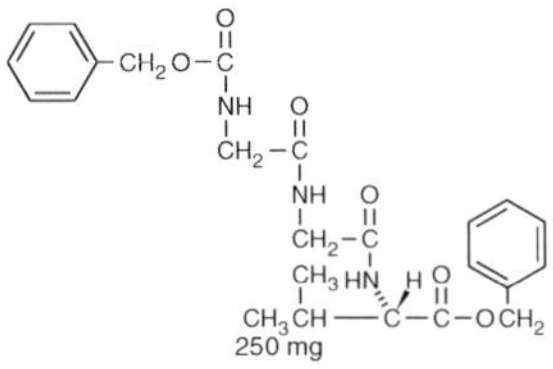
250 mg

S78,770-1
CARBOBENZYLOXYGLYCYL-L-LEUCYL-GLYCINE BENZYL ESTER
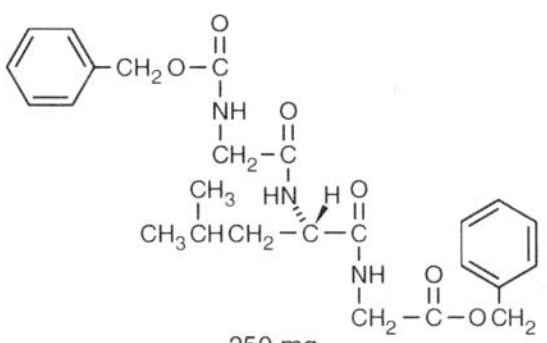
250 mg

S79,038-9
CARBOBENZYLOXYGLYCYLGLYCYL-L-ISOLEUCINE BENZYL ESTER
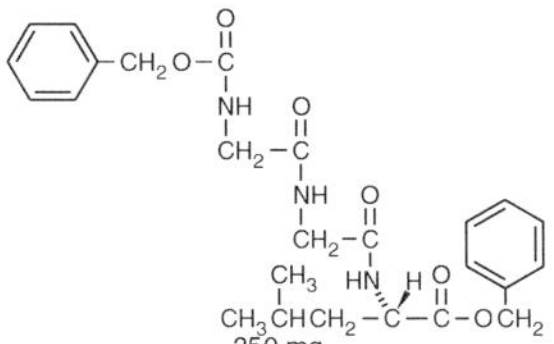
250 mg

S78,316-1
N,N'-BIS(CARBOBENZYLOXY)-L-LYSYLGLYCINE METHYL ESTER
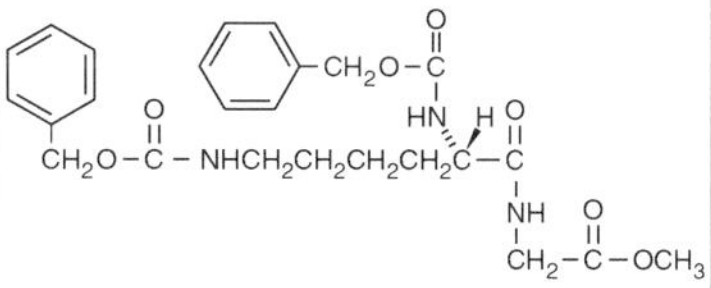
5 gm

S78,977-1
CARBOBENZYLOXY-L-ISOLEUCYL-L-VALINE BENZYL ESTER
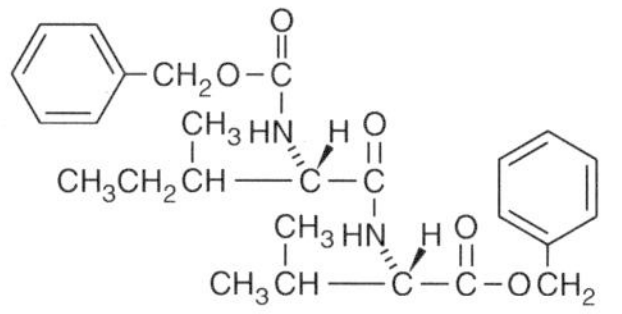
250 mg

S78,607-1
CARBOBENZYLOXY-L-ISOLEUCYL-L-ISOLEUCINE BENZYL ESTER
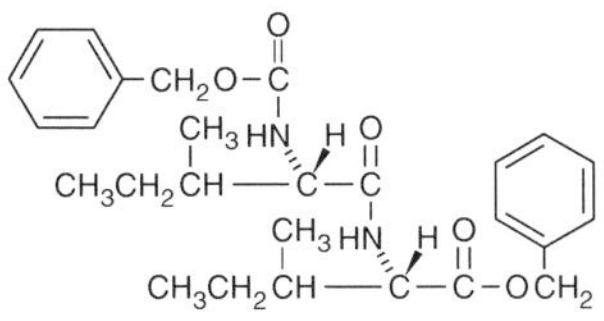
250 mg

S78,857-0
CARBOBENZYLOXY-L-ISOLEUCYL-L-LEUCINE BENZYL ESTER
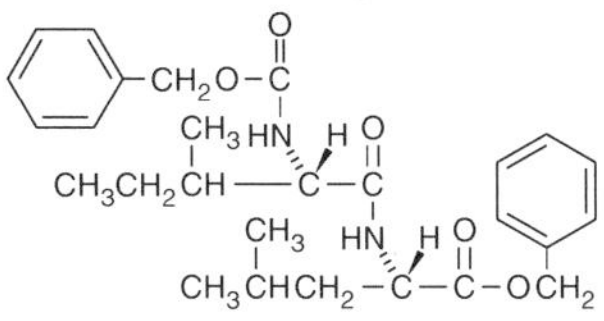
250 mg

S78,988-7
CARBOBENZYLOXY-L-LEUCYL-L-LEUCINE BENZYL ESTER
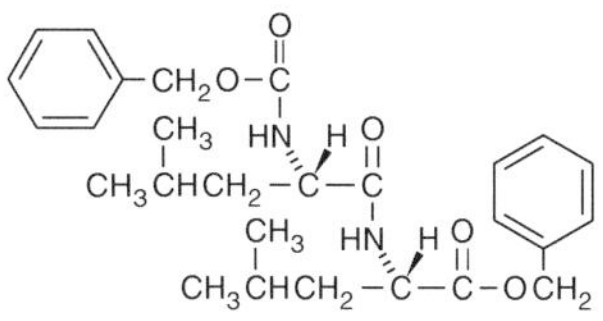
250 mg

S78,495-8
N,N'-BIS(CARBOBENZYLOXY)-L-LYSYLGLYCYLGLYCINE METHYL ESTER
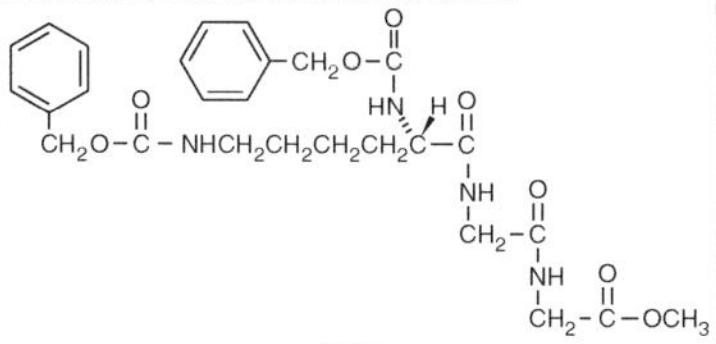
250 mg

S78,399-4
N,N'-BIS(CARBOBENZYLOXY)-L-LYSYLVALINE METHYL ESTER
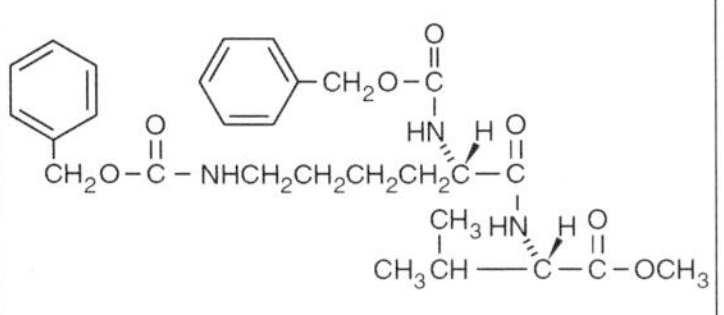
5 gm

S78,956-9
N,N'-BIS(CARBOBENZYLOXY)-L-LYSYL-L-ISOLEUCINE METHYL ESTER
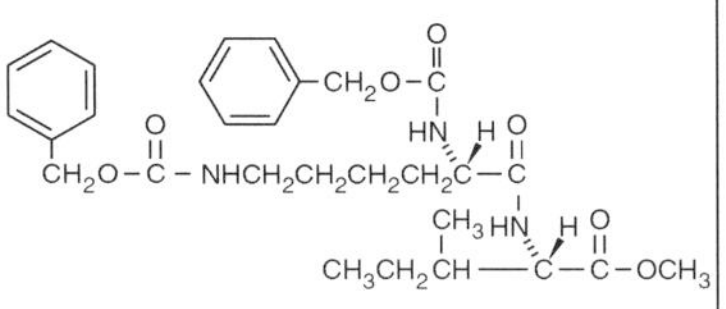
250 mg

S78,157-6
CARBOBENZYLOXYGLYCYLGLYCYL-L-GLUTAMIC ACID DIBENZYL ESTER
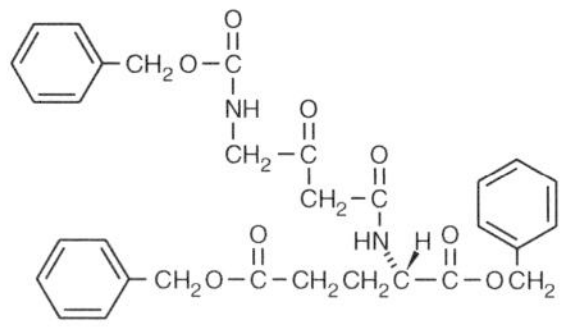
250 mg

S78,929-1
CARBOBENZYLOXY-DL-ALANINE METHYL ESTER
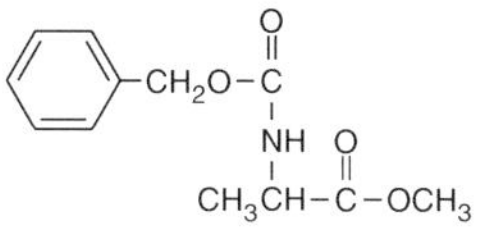
250 mg

S78,942-9
CARBOBENZYLOXY-BETA-ALANYL-L-ALANINE METHYL ESTER
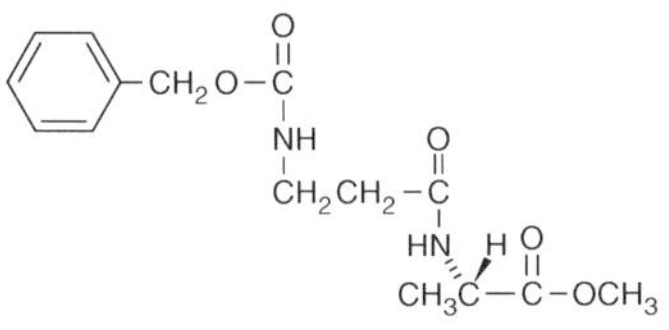
250 mg

S78,652-7
CARBOBENZYLOXY-BETA-ALANYLGLYCINE BENZYL ESTER
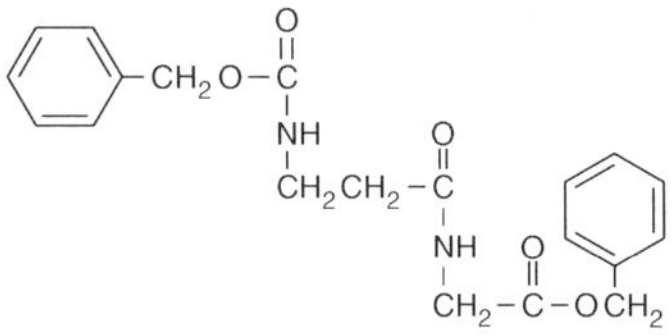
250 mg

S78,843-0
CARBOBENZYLOXY-BETA-ALANYL-L-VALINE BENZYL ESTER
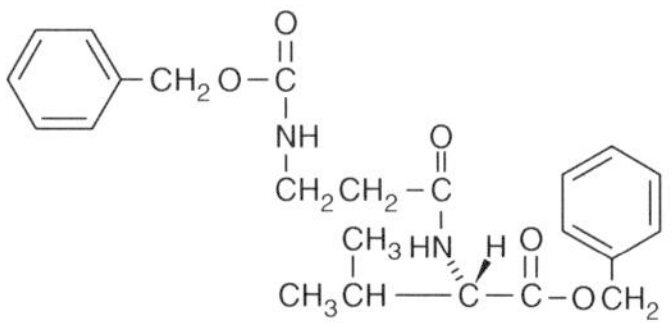
250 mg

S78,758-2
CARBOBENZYLOXYGLYCYLSARCOSINE BENZYL ESTER
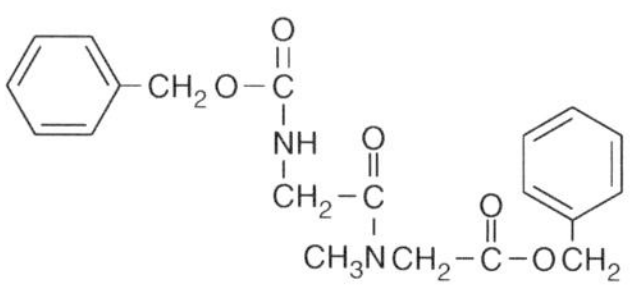
250 mg

S78,598-9
CARBOBENZYLOXYGLYCYL-L-NORLEUCINE METHYL ESTER
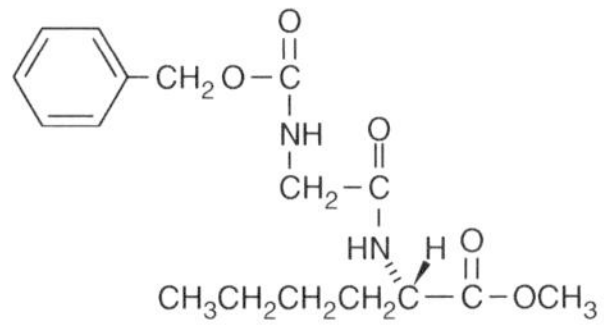
250 mg

S78,599-7
CARBOBENZYLOXY-L-VALYL-L-ALANINE METHYL ESTER
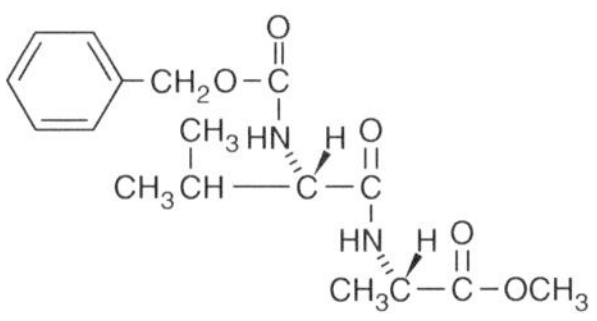
250 mg

S78,641-1
CARBOBENZYLOXYGLYCYLGLYCYL-L-ISOLEUCINE METHYL ESTER
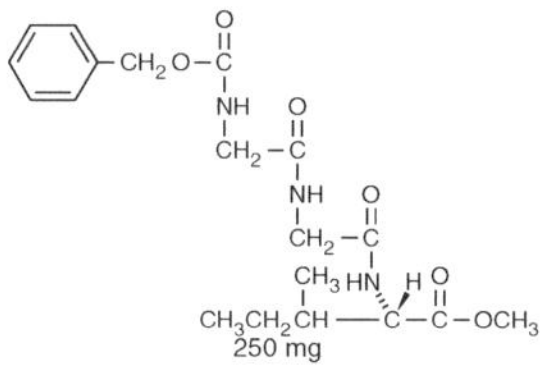
250 mg

S78,643-8
CARBOBENZYLOXYGLYCYLGLYCYL-L-LEUCINE METHYL ESTER
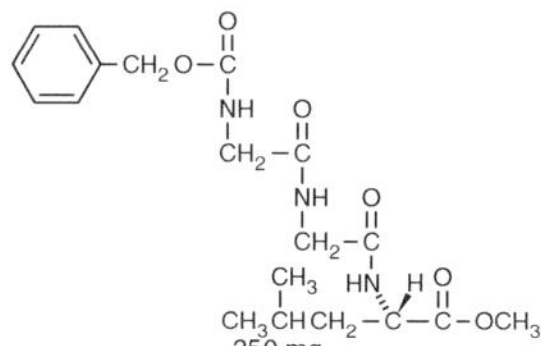
250 mg

S78,831-7
CARBOBENZYLOXY-L-ISOLEUCYL-L-LEUCINE METHYL ESTER
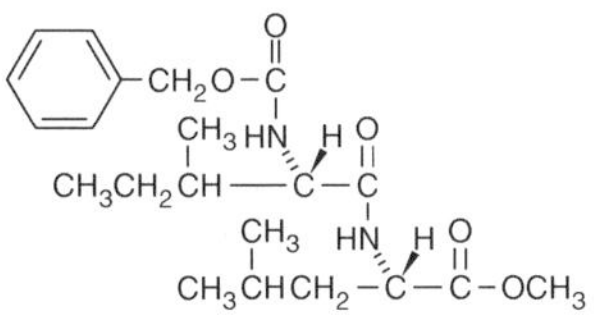
250 mg

S78,628-4
CARBOBENZYLOXY-L-VALYLGLYCINE BENZYL ESTER
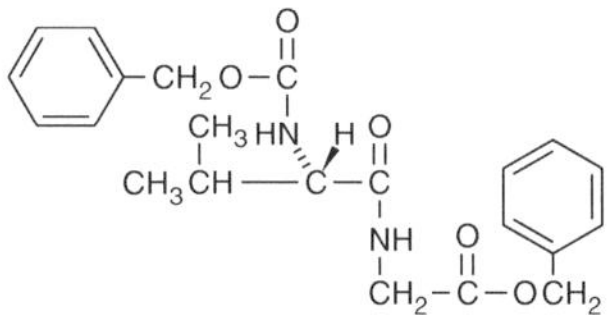
250 mg

S78,617-9
CARBOBENZYLOXYGLYCYLGLYCYL-L-LEUCINE BENZYL ESTER
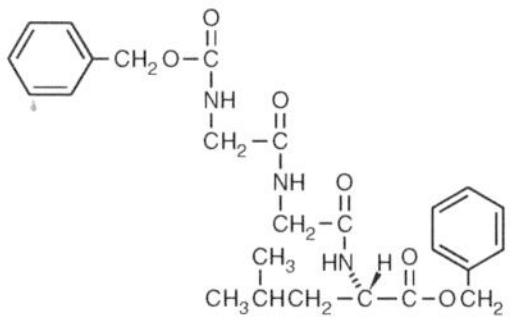
250 mg

S78,443-5
CARBOBENZYLOXY-D-ALANINE METHYL ESTER
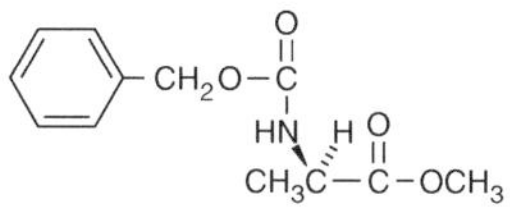
250 mg

S78,824-4
CARBOBENZYLOXY-BETA-ALANYL-L-ALANINE BENZYL ESTER
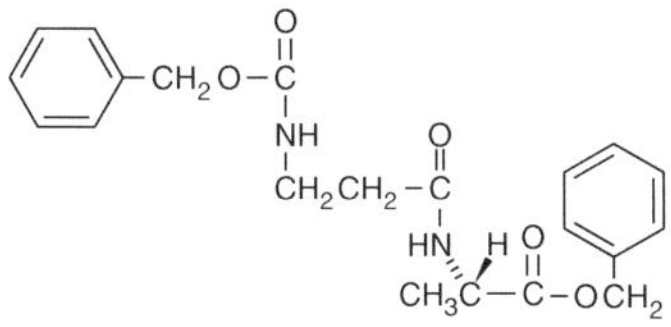
250 mg

S57,941-6
CARBOBENZYLOXY-DL-NORLEUCYLGLYCINE ETHYL ESTER
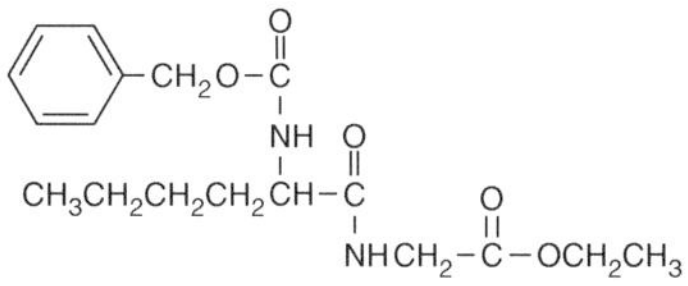
5 gm

S46,020-6
CARBOBENZYLOXY-DL-LEUCINE
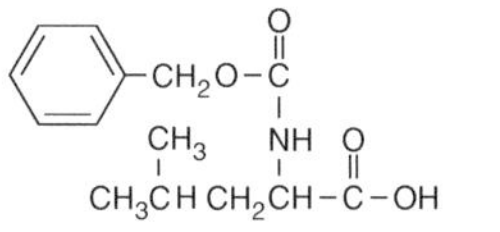
5 gm

S78,976-3
CARBOBENZYLOXYGLYCYL-L-LEUCINE BENZYL ESTER
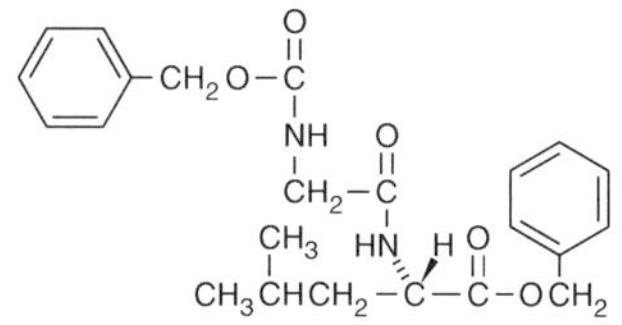
250 mg

S78,488-5
N,N'-BIS(CARBOBENZYLOXY)-L-LYSYLGLYCINAMIDE
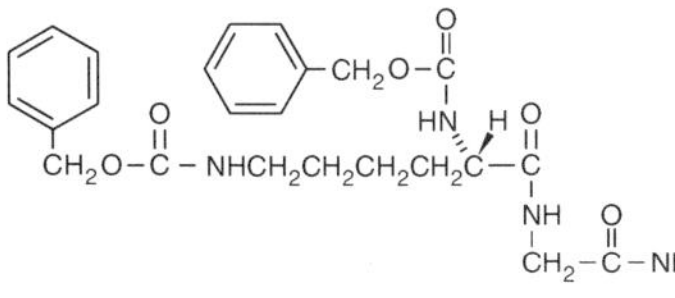
250 mg

S78,493-1
N,N'-BIS(CARBOBENZYLOXY)-L-LYSYL-L-VALINAMIDE
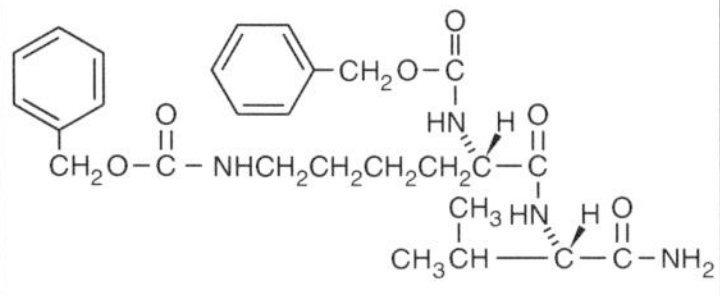
250 mg

S78,497-4
N,N'-BIS(CARBOBENZYLOXY)-L-LYSYL-L-ISOLEUCINAMIDE
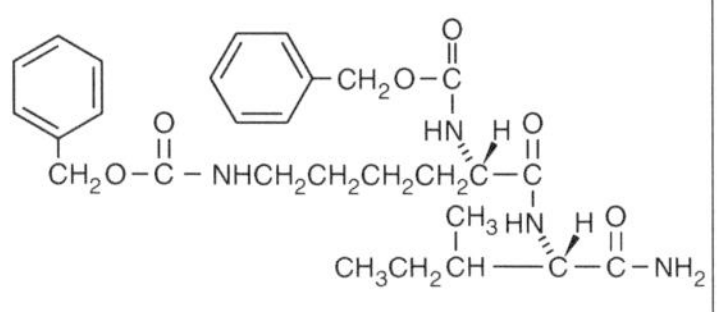
250 mg

S78,498-2
N,N'-BIS(CARBOBENZYLOXY)-L-LYSYL-L-LEUCINAMIDE
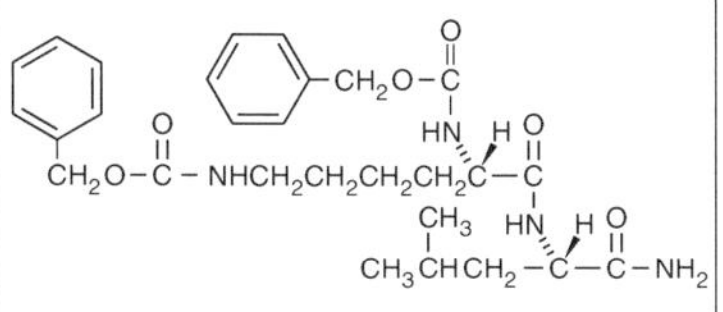
250 mg

S78,392-7
N,N'-BIS(CARBOBENZYLOXY)-L-LYSINAMIDE

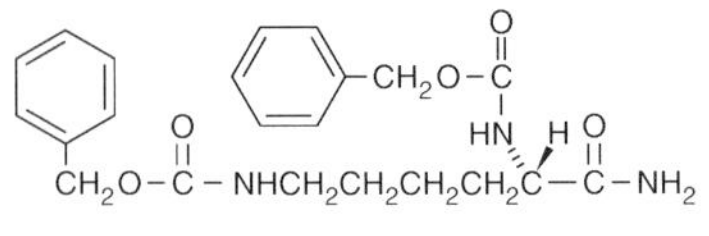

5 gm

S78,801-5
CARBOBENZYLOXY-L-ALANYL-BETA-ALANINE METHYL ESTER

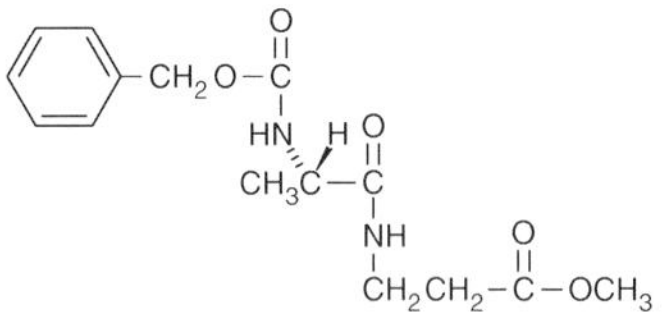

250 mg

S77,601-7
CARBOBENZYLOXY-L-ALANYL-L-ISOLEUCINE METHYL ESTER

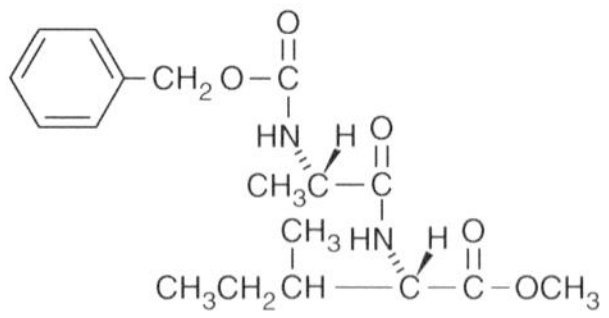

250 mg

S78,833-3
CARBOBENZYLOXYGLYCYLGLYCYL-BETA-ALANINE BENZYL ESTER

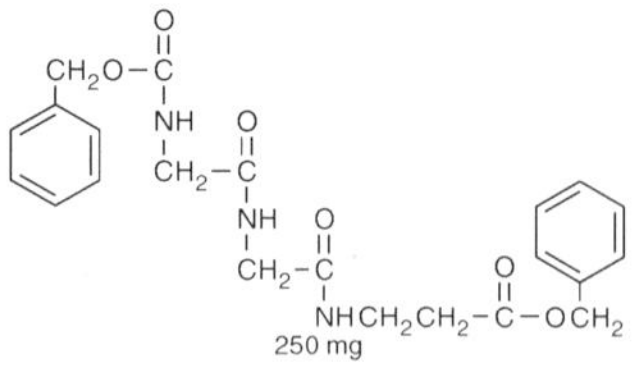

250 mg

S78,661-6
CARBOBENZYLOXY-BETA-ALANYLGLYCINE METHYL ESTER

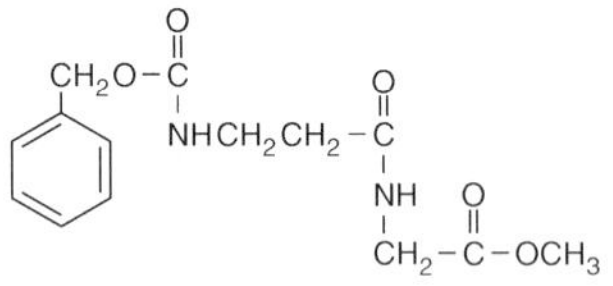

250 mg

S78,760-4
CARBOBENZYLOXY-BETA-ALANYL-BETA-ALANINE BENZYL ESTER

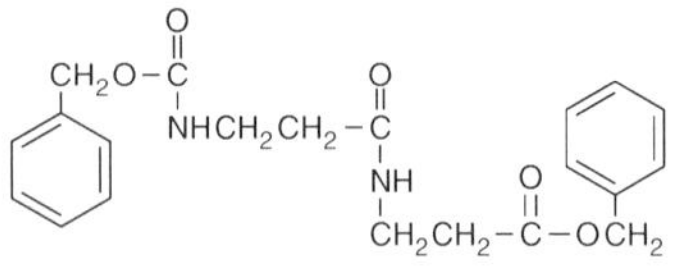

250 mg

S78,454-0
CARBOBENZYLOXYSARCOSYL-L-ALANINAMIDE

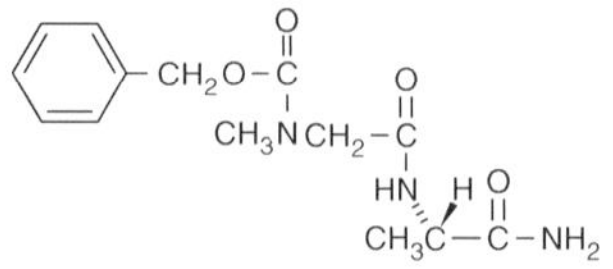

250 mg

S78,457-5
CARBOBENZYLOXYGLYCYL-L-VALINAMIDE

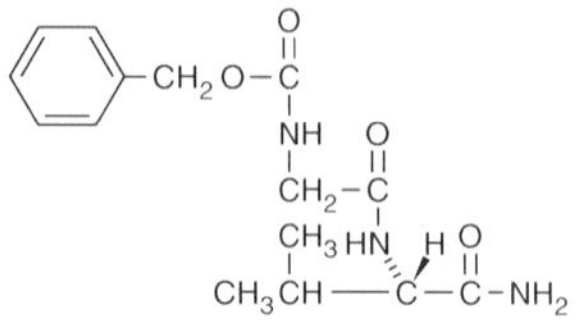

250 mg

S78,533-4
CARBOBENZYLOXYGLYCYL-L-NORVALINAMIDE

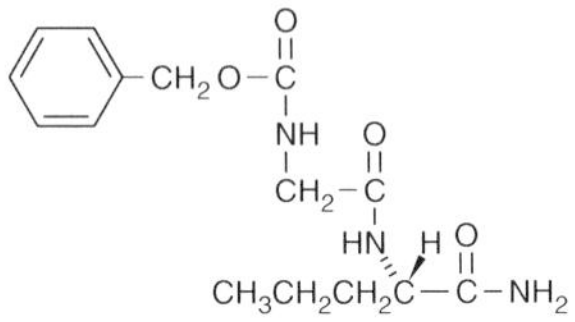

250 mg

S77,711-0
CARBOBENZYLOXYGLYCYLGLYCYL-L-ALANINAMIDE

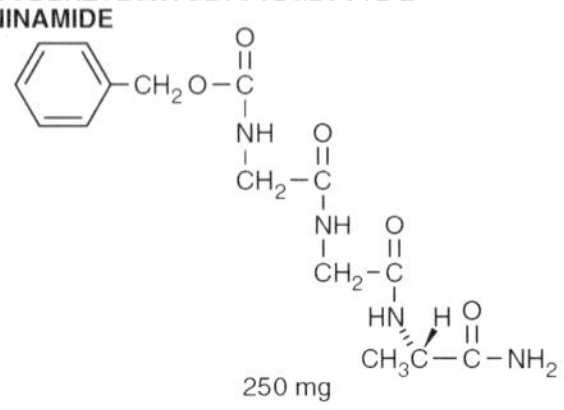

250 mg

S77,613-0
CARBOBENZYLOXY-L-ALANYL-L-NORVALINAMIDE

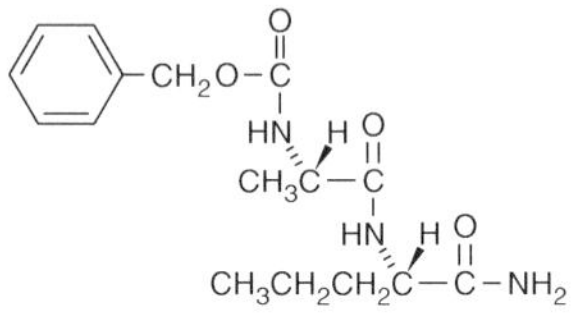

250 mg

S77,786-2
CARBOBENZYLOXY-L-ALANYL-L-LEUCINAMIDE

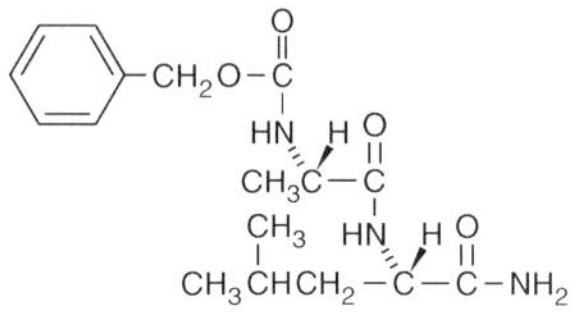

250 mg

S78,163-0
CARBOBENZYLOXY-BETA-ALANYL-L-LEUCINAMIDE

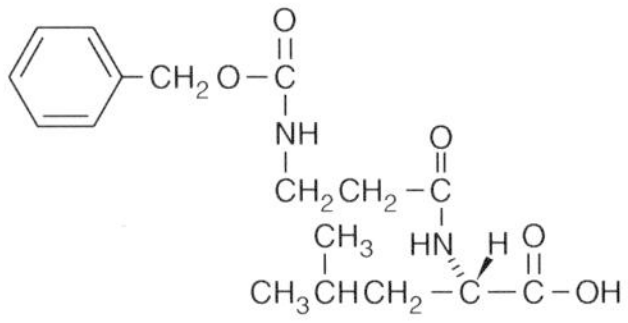

250 mg

S78,649-7
CARBOBENZYLOXY-L-ALANYL-L-NORLEUCINAMIDE

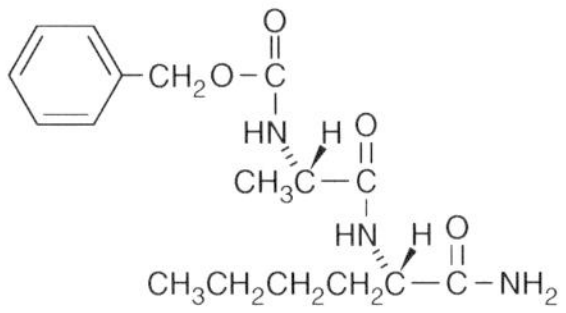

250 mg

S77,699-8
CARBOBENZYLOXY-L-VALYLGLYCYLGLYCINAMIDE

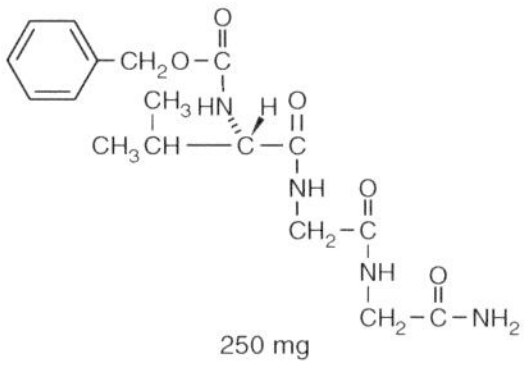

250 mg

S77,701-3
CARBOBENZYLOXY-L-NORVALYL-L-LEUCINAMIDE

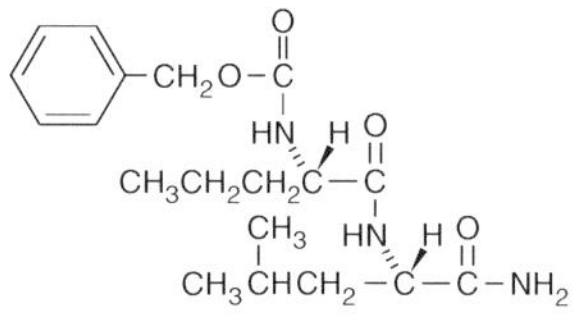

250 mg

S78,955-0
CARBOBENZYLOXY-L-LEUCYL-L-LEUCINAMIDE

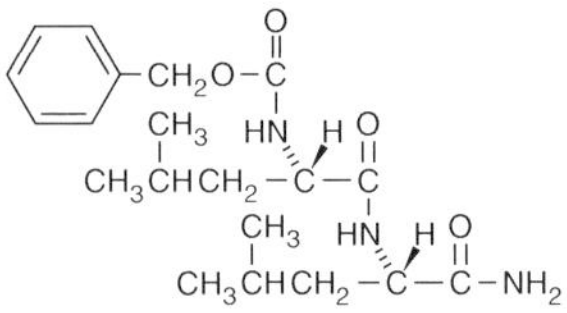

250 mg

S79,034-6
CARBOBENZYLOXY-L-LEUCYL-L-ISOLEUCINAMIDE

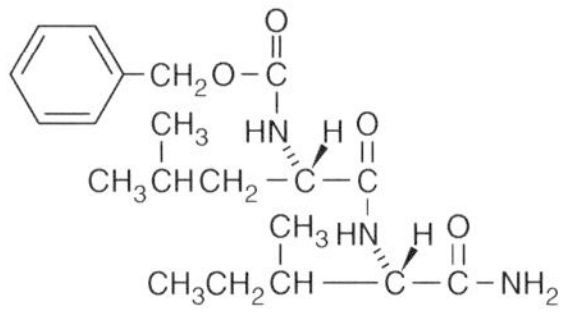

250 mg

S77,616-5
CARBOBENZYLOXY-BETA-ALANYLGLYCINAMIDE

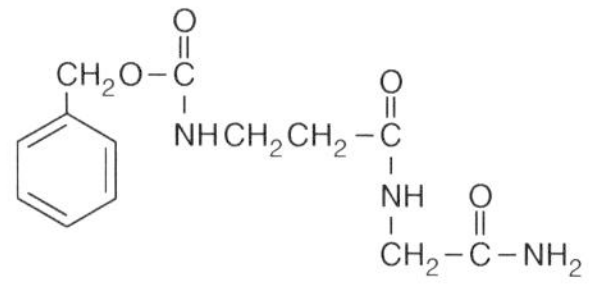

250 mg

S78,191-6
CARBOBENZYLOXY-BETA-ALANYL-BETA-ALANINAMIDE

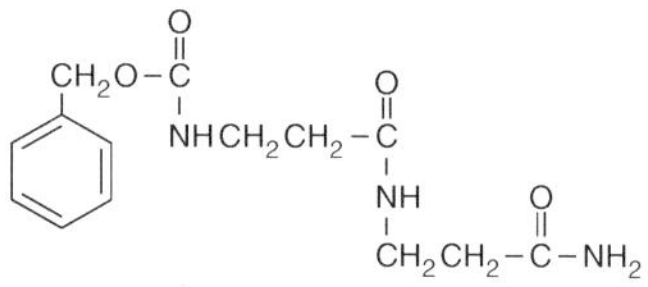

250 mg

S77,612-2
CARBOBENZYLOXY-L-VALYL-L-ALANINAMIDE

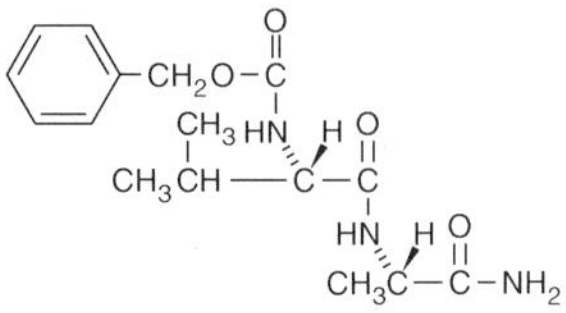

250 mg

S78,327-7
CARBOBENZYLOXY-L-ISOLEUCYL-L-LEUCINAMIDE

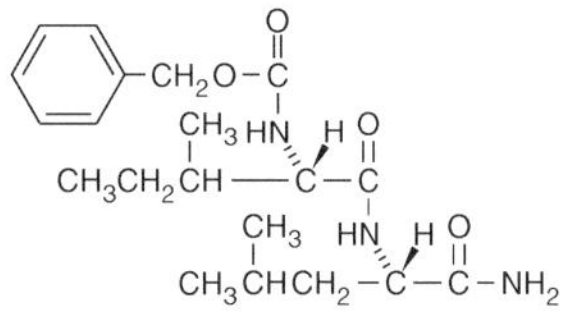

5 gm

S78,731-0
CARBOBENZYLOXY-DL-ALANINAMIDE

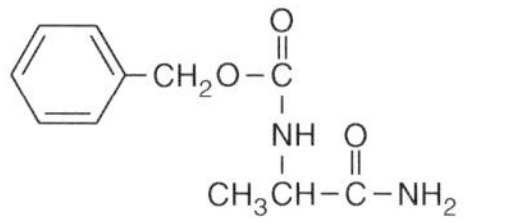

250 mg

S78,953-4
CARBOBENZYLOXY-DL-LEUCINAMIDE

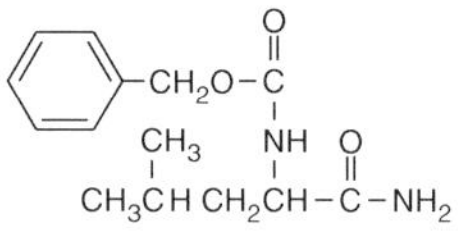

250 mg

S78,954-2
CARBOBENZYLOXY-L-LEUCINAMIDE

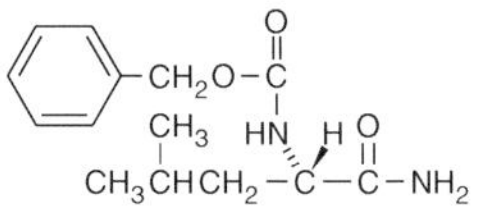

250 mg

S54,104-4
BENZYL N-(2,2-DIETHOXYETHYL)-CARBAMATE

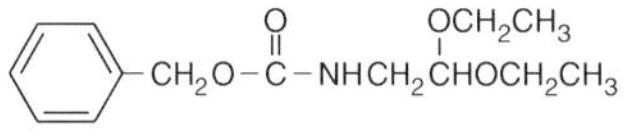

250 mg

S78,708-6
CARBOBENZYLOXY-D-ALANINE BETA-NAPHTHYLAMIDE

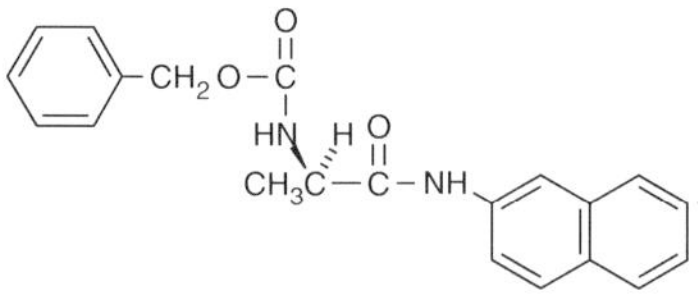

250 mg

S76,517-1
N-(BENZYLOXYCARBONYL)-GLYCINE HYDRAZIDE

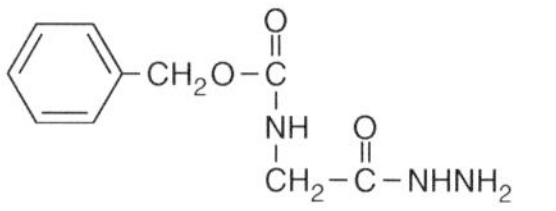

5 gm

S54,598-8
BENZYL 3-PHENYLCARBAZATE

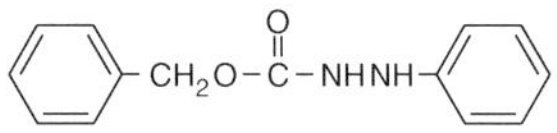

250 mg

S95,032-7
DIBENZYL 1-METHYL-1,2-HYDRAZINEDICARBOXYLATE

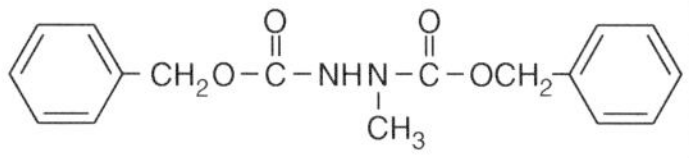

50 mg

S66,635-1
PIPERONYL N-(2-CHLOROPHENYL)-CARBAMATE

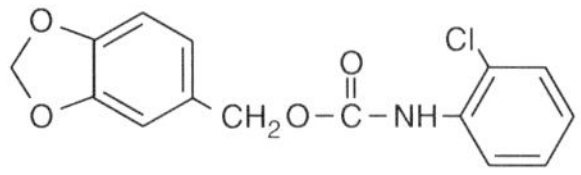

5 gm

S66,642-4
PIPERONYL N-(O-TOLYL)CARBAMATE

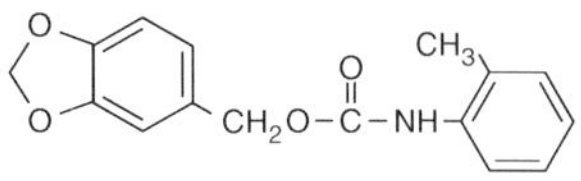

5 gm

S66,646-7
PIPERONYL N-(P-TOLYL)CARBAMATE

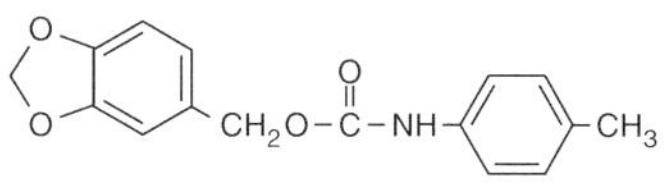

5 gm

S66,655-6
PIPERONYL N-(2-CHLORO-6-METHYLPHENYL)CARBAMATE

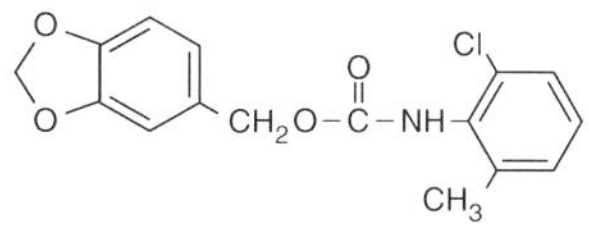

5 gm

S66,639-4
PIPERONYL N-(2,6-XYLYL)CARBAMATE

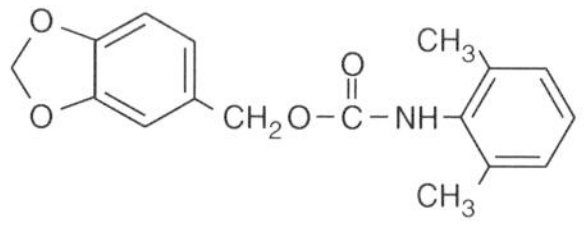

5 gm

S66,640-8
PIPERONYL N-(2,3-XYLYL)CARBAMATE

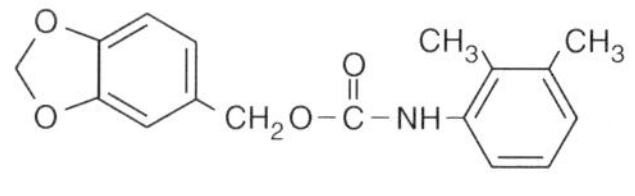

5 gm

S66,766-8
PIPERONYL N-(3-CHLOROPHENYL)-CARBAMATE

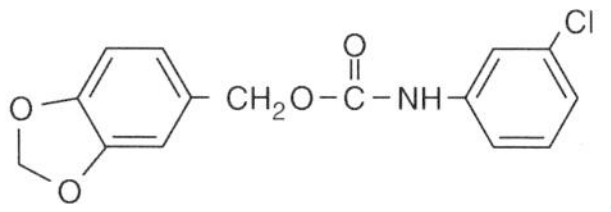

5 gm

S66,701-3
PIPERONYL N-(3,4-DICHLOROPHENYL)-CARBAMATE

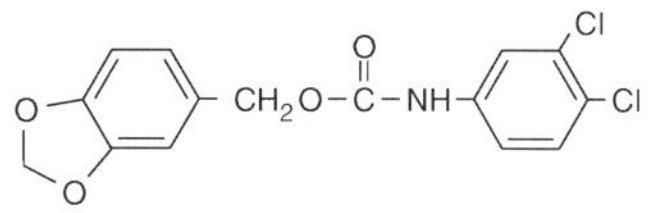

5 gm

S66,651-3
PIPERONYL N-(M-TOLYL)CARBAMATE

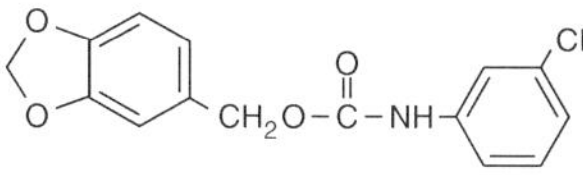

5 gm

S66,716-1
PIPERONYL N-(5-CHLORO-2-METHYLPHENYL)CARBAMATE

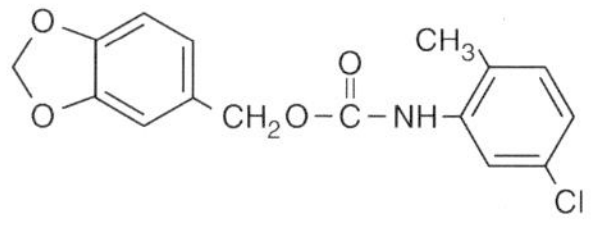

5 gm

S76,319-5
BENZYL N-METHYLCARBAMATE

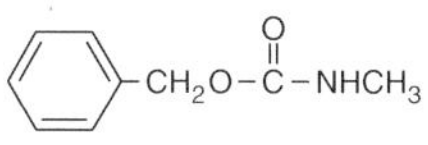

250 mg

S89,801-5
BENZYL N-BUTYLCARBAMATE

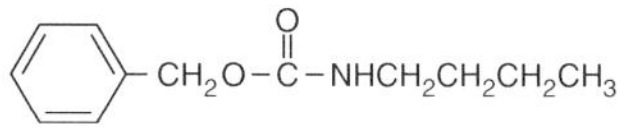

250 mg

S85,361-5
2-(4-BIPHENYLYLOXY)ETHYL N-[3-(METHYLTHIO)PHENYL]CARBAMATE

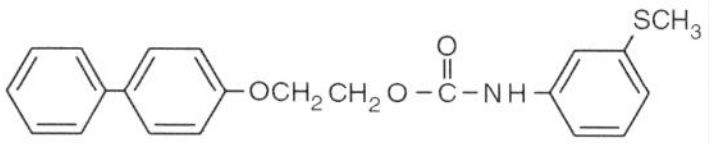

250 mg

S85,160-4
2-(4-ACETAMIDOPHENOXY)ETHYL N-(4-METHYL-2-NITROPHENYL)CARBAMATE

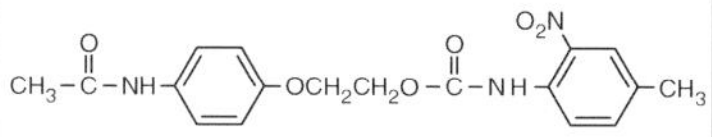

250 mg

S85,186-8
2-(PHENOXY)ETHYL N-(4-METHYL-2-NITROPHENYL)CARBAMATE

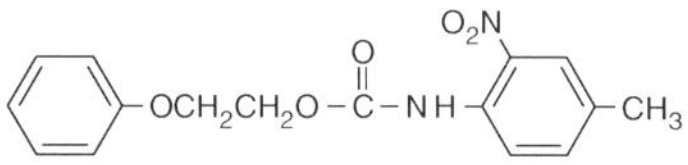

250 mg

S85,188-4
2-(4-NITROPHENOXY)ETHYL N-(4-METHYL-2-NITROPHENYL)CARBAMATE

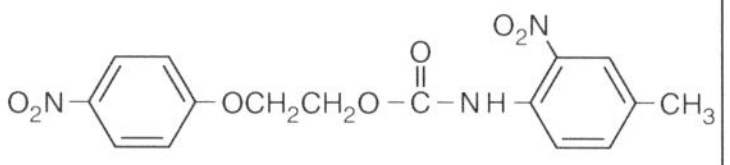

250 mg

S85,183-3
2-(1-NAPHTHOXY)ETHYL N-(4-METHYL-2-NITROPHENYL)CARBAMATE

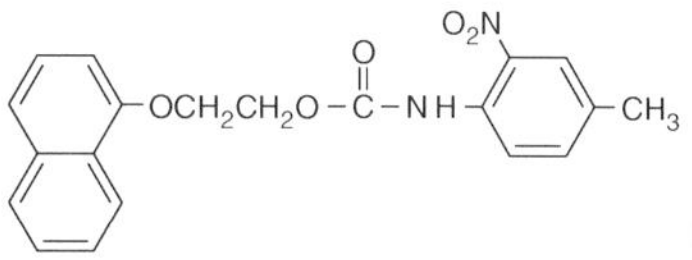

250 mg

S60,470-4
2-BENZOYL-4-CHLOROPHENYL N-(3-NITROPHENYL)CARBAMATE

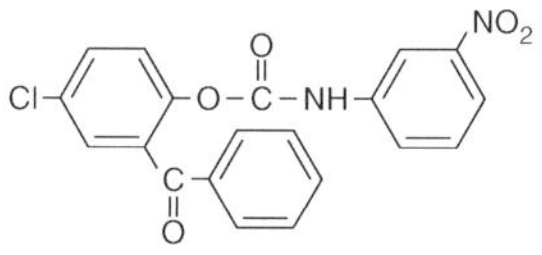

5 gm

S65,045-5
2-DIETHYLAMINOETHYL N-(2-CHLORO-4-NITROPHENYL)CARBAMATE

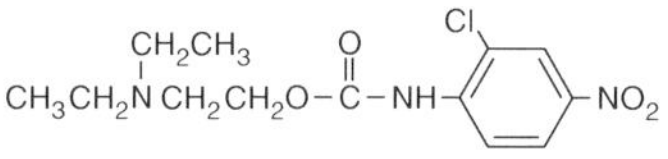

250 mg

S36,727-3
CYCLOHEXYL LACTATE 3'-CHLOROCARBANILATE

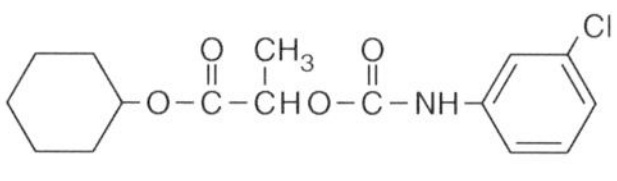

5 gm

S36,728-1
CYCLOHEXYL LACTATE CARBANILATE

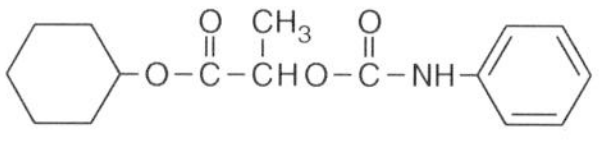

5 gm

S60,468-2
2-BENZOYL-4-CHLOROPHENYL N-(2,5-DICHLOROPHENYL)CARBAMATE

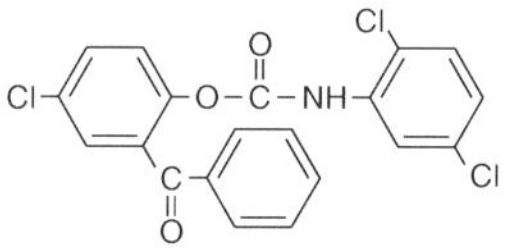

5 gm

S60,473-9
2-BENZOYL-4-CHLOROPHENYL N-(2,3-DICHLOROPHENYL)CARBAMATE

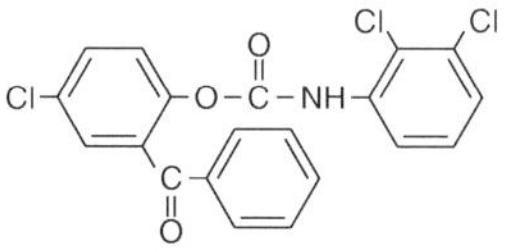

5 gm

S60,469-0
2-BENZOYL-4-CHLOROPHENYL N-PHENYLCARBAMATE

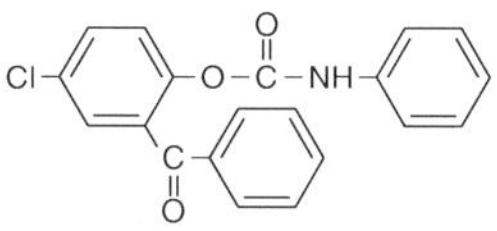

5 gm

S79,782-0
2-METHOXY-4-(1-PROPENYL)PHENYL N-[4-(ETHOXYCARBONYL)PHENYL]-CARBAMATE

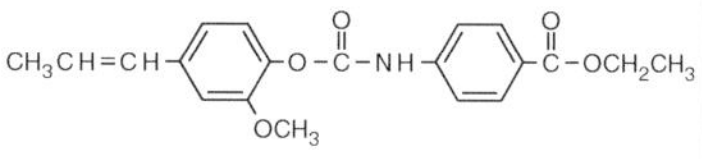

250 mg

S34,366-8
2-(3-CHLOROPHENYLCARBAMOYLOXY)-PROPIONIC ACID

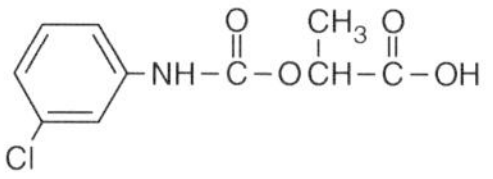

250 mg

S34,367-6
2-(PHENYLCARBAMOYLOXY)-PROPIONIC ACID

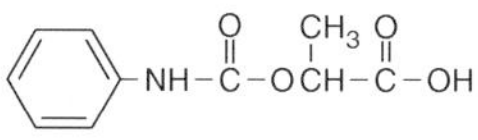

5 gm

S94,738-5
R-(+)-2-(PHENYLCARBAMOYLOXY)-PROPIONIC ACID

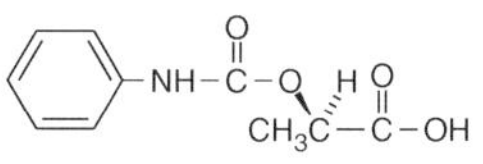

250 mg

S34,362-5
BUTYL 2-(PHENYLCARBAMOYLOXY)-PROPIONATE

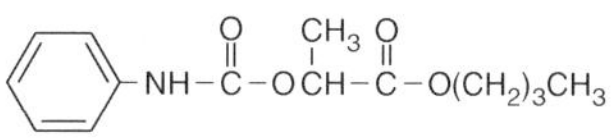

5 gm

S77,994-6
2-(METHACRYLOYLOXY)ETHYL N-(4-ISOPROPYLPHENYL)CARBAMATE

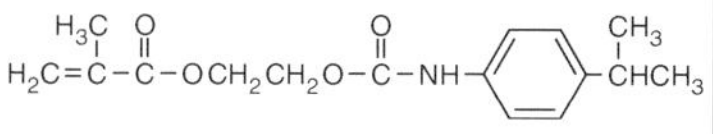

250 mg

S63,954-0
PIPERONYL N-(2,5-DICHLOROPHENYL)-CARBAMATE

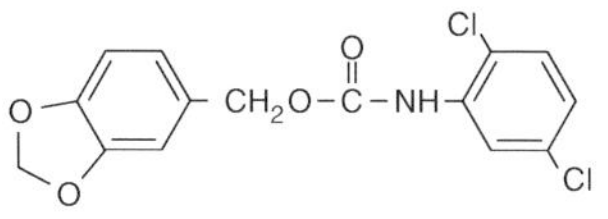

5 gm

S64,667-9
2-(2,4-DICHLOROPHENOXY)ETHYL N-(O-TOLYL)CARBAMATE

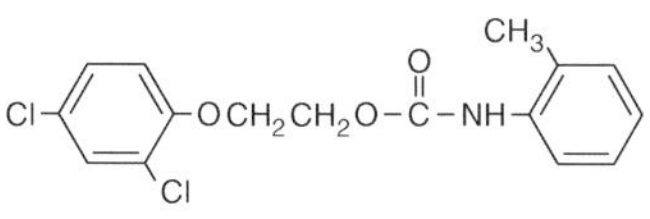

5 gm

S64,412-9
2-(2,4-DICHLOROPHENOXY)ETHYL N-(4-CHLOROPHENYL)CARBAMATE

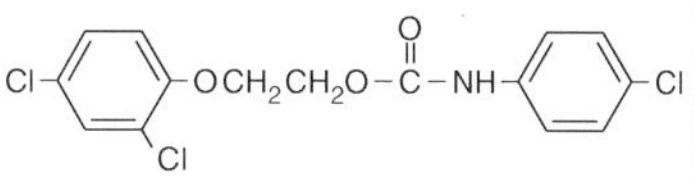

5 gm

S65,023-4
2-(2,4-DICHLOROPHENOXY)ETHYL N-(P-TOLYL)CARBAMATE

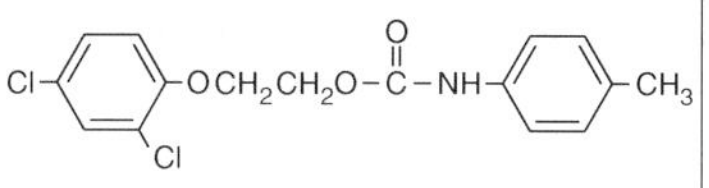

5 gm

S65,040-4
2-(2,4-DICHLOROPHENOXY)ETHYL N-(2-CHLORO-6-METHYLPHENYL)-CARBAMATE
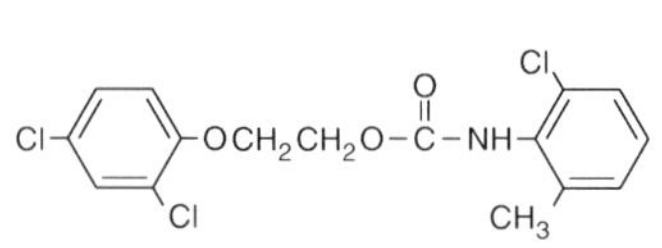
5 gm

S65,457-4
2-(2,4-DICHLOROPHENOXY)ETHYL N-(M-TOLYL)CARBAMATE
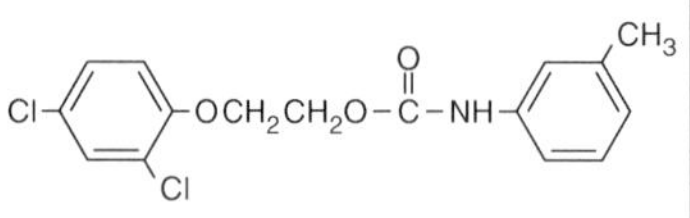
5 gm

S73,690-2
2-CARBAMOYLPHENYL N-[4-(METHYLTHIO)-PHENYL]CARBAMATE
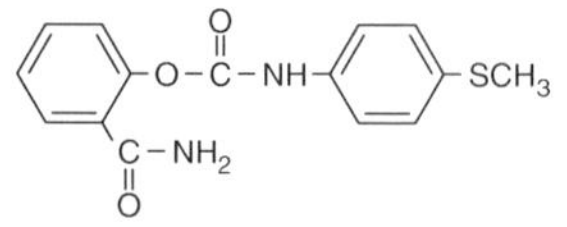
5 gm

S79,733-2
3-(ACETOXY)PHENYL N-[4-(METHYLTHIO)-PHENYL]CARBAMATE
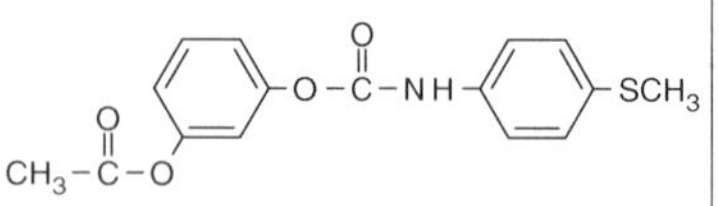
250 mg

S85,169-8
3-ACETOXYPHENYL N-(4-METHYL-2-NITROPHENYL)CARBAMATE
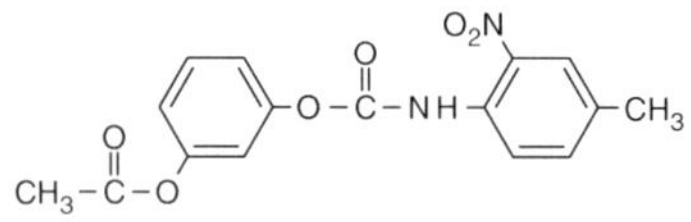
250 mg

S84,847-6
4-METHYL-7-COUMARINYL N-(2-METHYL-5-NITROPHENYL)CARBAMATE
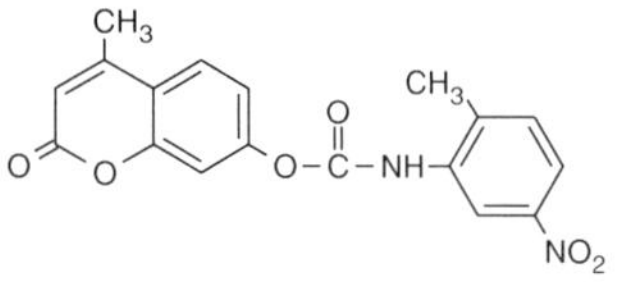
250 mg

S85,372-0
3-ACETOXYPHENYL N-(4-METHYL-3-NITROPHENYL)CARBAMATE
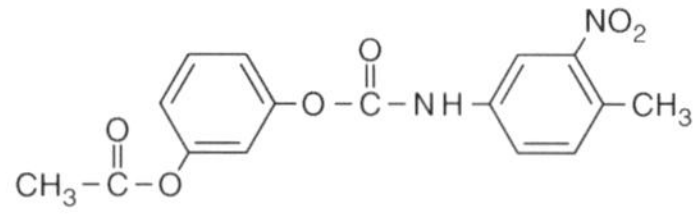
250 mg

S86,913-9
BIS(2-ACETYLPHENYL) N,N'-(4-METHYL-1,3-PHENYLENE)BISCARBAMATE
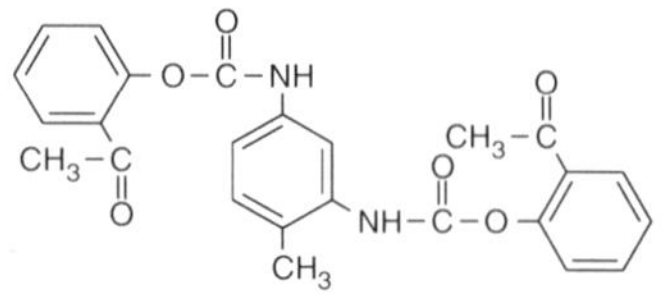
250 mg

S86,914-7
BIS(3-ACETOXYPHENYL) N,N'-(4-METHYL-1,3-PHENYLENE)BISCARBAMATE
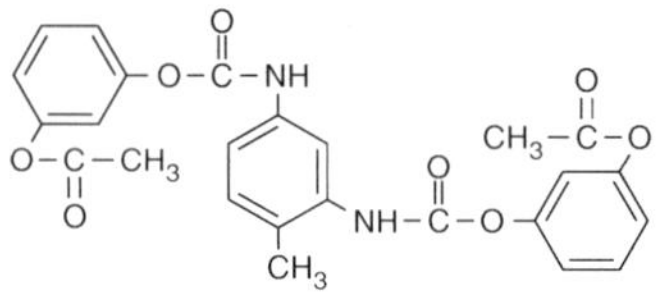
250 mg

S71,871-8
2-FORMYLPHENYL N-PHENYLCARBAMATE
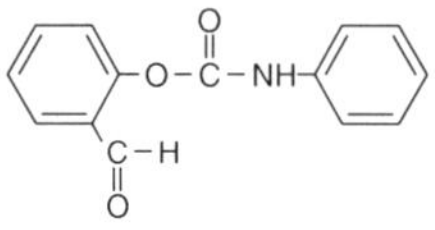
250 mg

S86,841-8
4-COUMARINYL N-(3-ACETYLPHENYL)-CARBAMATE
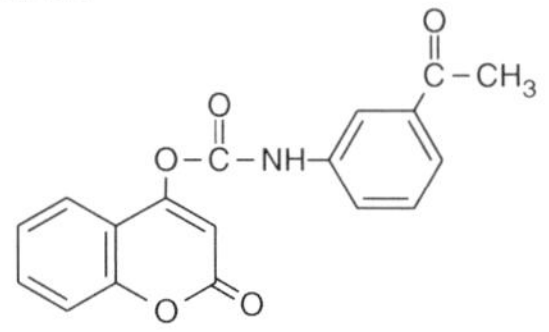
250 mg

S86,860-4
4-METHYL-7-COUMARINYL N-(3-ACETYLPHENYL)CARBAMATE
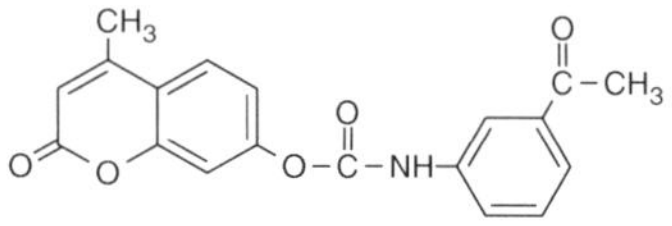
250 mg

S79,422-8
2-(ISOBUTOXYCARBONYL)PHENYL N-(3-ACETYLPHENYL)CARBAMATE
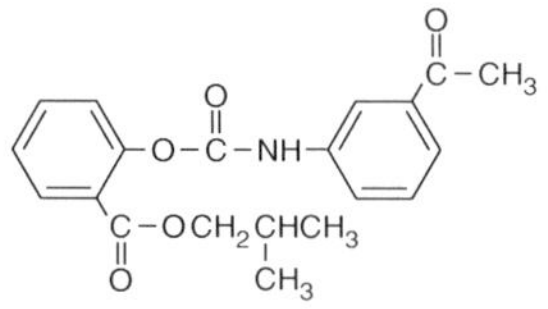
250 mg

S79,794-4
2-(ISOBUTOXYCARBONYL)PHENYL N-[4-(ETHOXYCARBONYL)PHENYL]-CARBAMATE
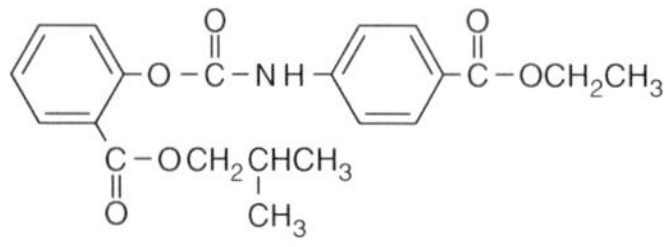
250 mg

S74,725-4
2-CARBAMOYLPHENYL N-CYCLOHEXYLCARBAMATE
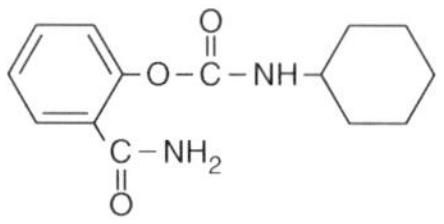
5 gm

S86,759-4
2-CARBAMOYLPHENYL N-PHENYLCARBAMATE
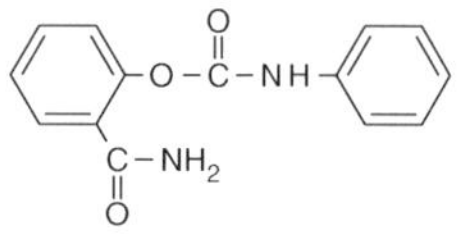
250 mg

S86,789-6
2-CARBAMOYLPHENYL N-(M-TOLYL)-CARBAMATE
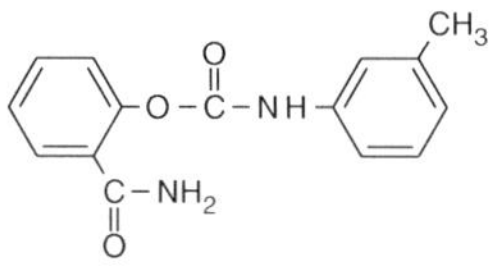
250 mg

S73,190-0
2-CARBAMOYLPHENYL N-(1-NAPHTHYL)-CARBAMATE
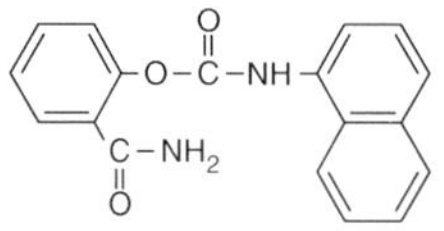
5 gm

S85,497-2
2-(N-PHENYLCARBAMOYL)PHENYL N-METHYLCARBAMATE
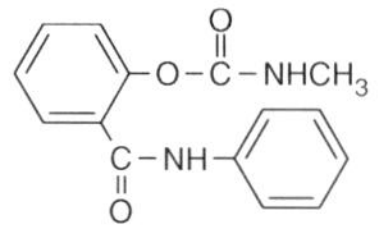
250 mg

S85,528-6
2-(PHENYLCARBAMOYL)PHENYL N-TERT-BUTYLCARBAMATE
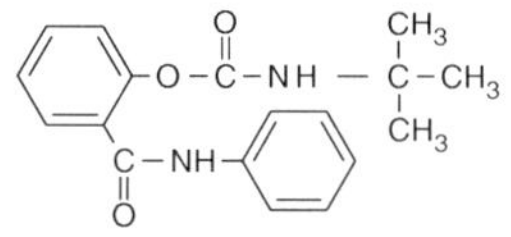
250 mg

S72,458-0
3-(BENZOYLOXY)PHENYL N-(O-TOLYL)-CARBAMATE
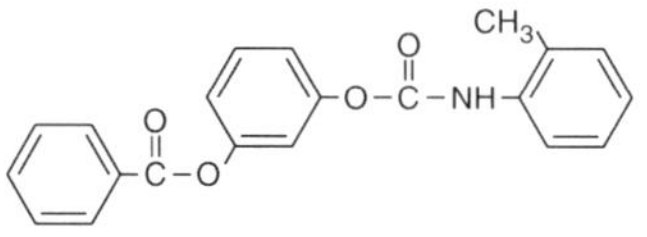
5 gm

S77,108-2
3-ACETOXY-1-PHENYLPROPYL CARBAMATE
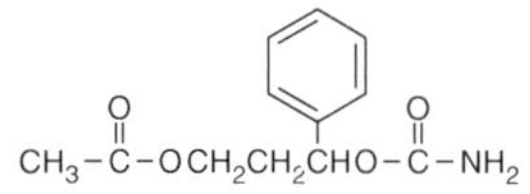
250 mg

S89,179-7
3-(PROPIONYLOXY)PHENYL N-(O-TOLYL)-CARBAMATE
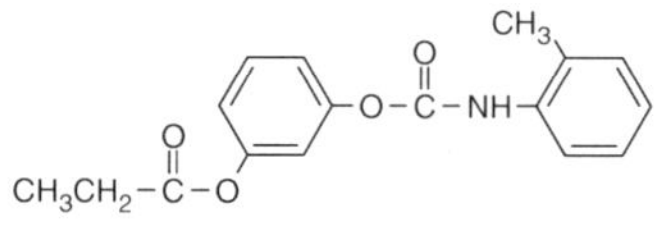
250 mg

S85,523-5
3-ACETYLPHENYL N-(P-TOLYL)CARBAMATE
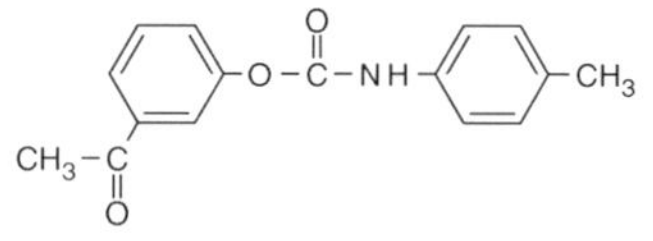
250 mg

S72,967-1
3-ACETYLPHENYL N-[4-(ETHOXYCARBONYL)PHENYL]-CARBAMATE
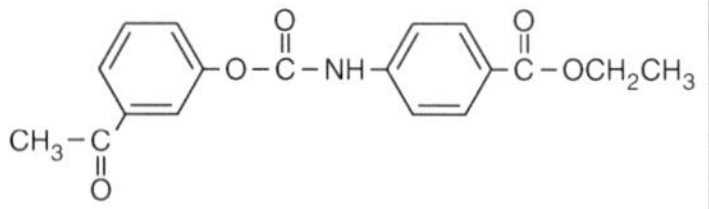
5 gm

S93,332-5
METHYL N-(4-MERCAPTOPHENYL)-CARBAMATE
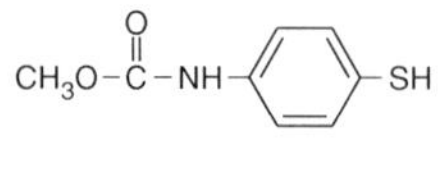
250 mg

S50,578-1
2-(CARBAMOYLTHIO)ACETIC ACID 2-PHENYLHYDRAZIDE
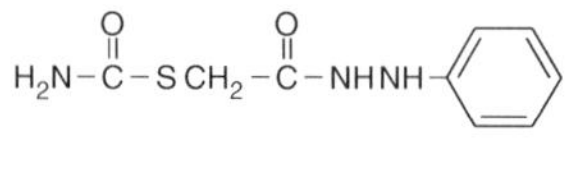
250 mg

S85,172-8
4-(METHYLTHIO)PHENYL N-(4-METHYL-2-NITROPHENYL)CARBAMATE
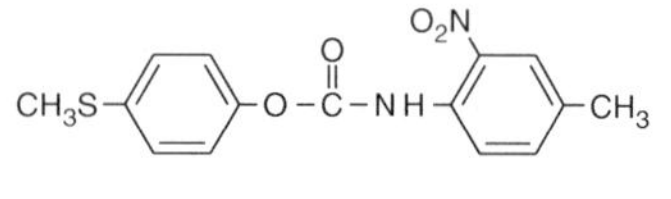
250 mg

S86,905-8
BIS[4-(METHYLTHIO)PHENYL] N,N'-(4-METHYL-1,3-PHENYLENE)BISCARBAMATE
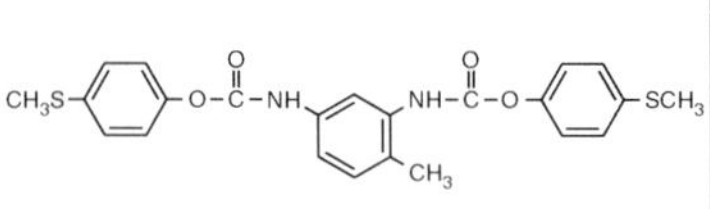
250 mg

S77,704-8
CARBOBENZYLOXY-L-METHIONYL-L-PHENYLALANINE
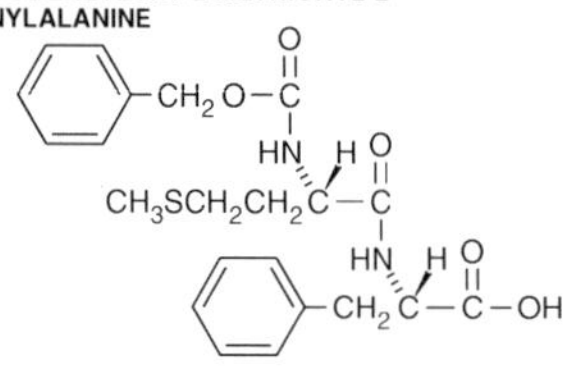
250 mg

S78,354-4
CARBOBENZYLOXY-S-BENZYL-L-CYSTEINYL-L-PHENYLALANINE BENZYL ESTER
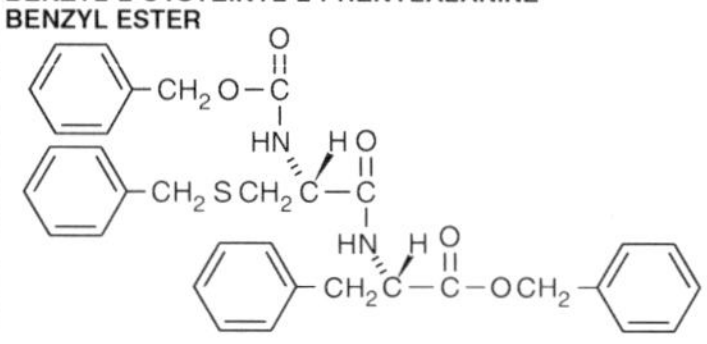
5 gm

S77,739-0
CARBOBENZYLOXY-L-METHIONYL-L-PHENYLALANINAMIDE
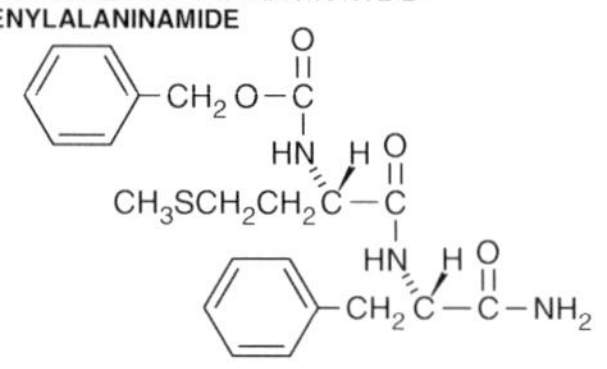
250 mg

S79,778-2
2-(N-ETHYLANILINO)ETHYL N-[4-(METHYLTHIO)PHENYL]CARBAMATE

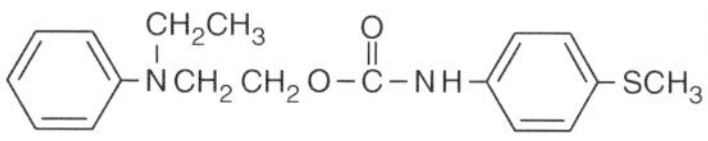

250 mg

S72,450-5
2-(1-AZIRIDINYL)ETHYL N-[4-(METHYLTHIO)PHENYL]CARBAMATE

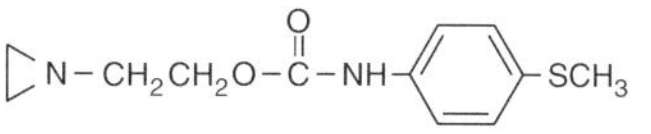

5 gm

S34,175-4
S-BENZYL-N-CARBOBENZYLOXY-L-CYSTEINE 4-NITROPHENYL ESTER

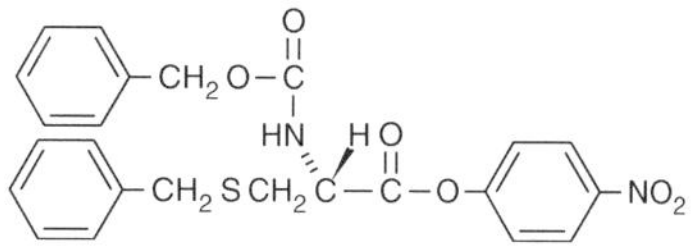

5 gm

S78,146-0
CARBOBENZYLOXY-L-METHIONINE 4-NITROPHENYL ESTER

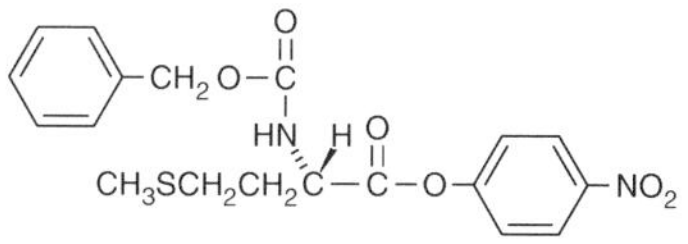

250 mg

S70,246-3
2-NITROBENZYL N-[4-(METHYLTHIO)-PHENYL]CARBAMATE

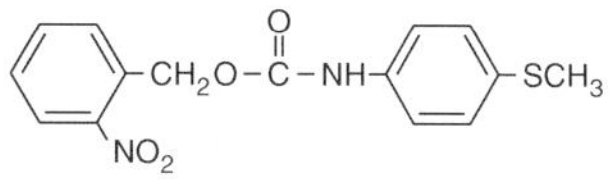

250 mg

S69,621-8
3-METHYL-4-NITROPHENYL N-[4-(METHYLTHIO)PHENYL]CARBAMATE

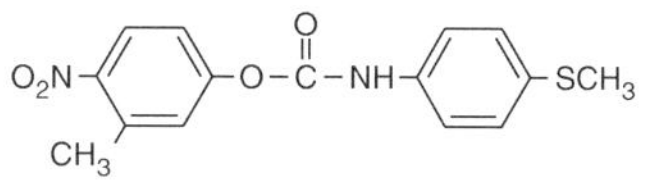

250 mg

S78,301-3
N-(TERT-BUTOXYCARBONYL)-S-BENZYL-L-CYSTEINE 4-NITROPHENYL ESTER

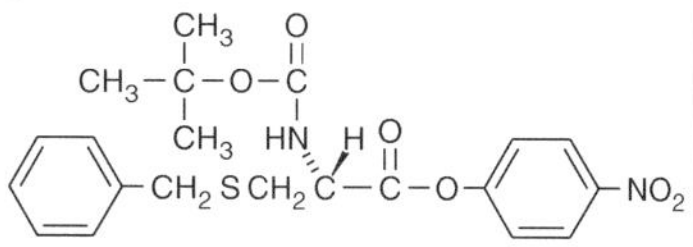

5 gm

S69,595-5
BENZOYLMETHYL N-[4-(METHYLTHIO)-PHENYL]CARBAMATE

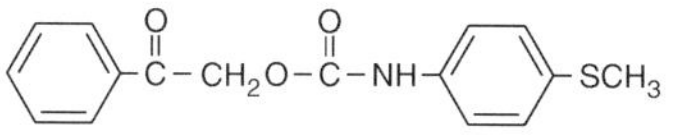

250 mg

S72,995-7
4-ACETYLPHENYL N-[3-(METHYLTHIO)-PHENYL]CARBAMATE

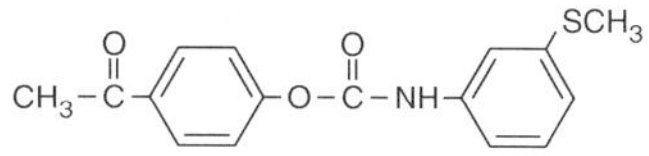

5 gm

S91,081-3
4-ACETYLPHENYL N-[4-(METHYLTHIO)-PHENYL]CARBAMATE

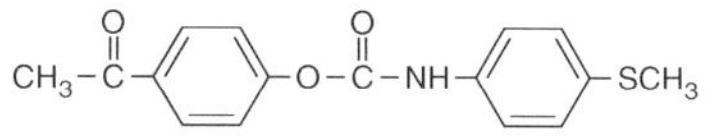

250 mg

S92,439-3
2-(METHOXYCARBONYL)PHENYL N-[4-(METHYLTHIO)PHENYL]CARBAMATE

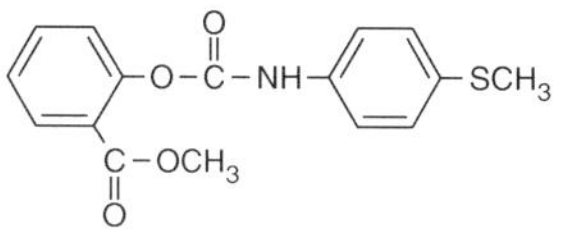

250 mg

S79,447-3
4-(METHOXYCARBONYL)PHENYL N-[4-(METHYLTHIO)PHENYL]CARBAMATE

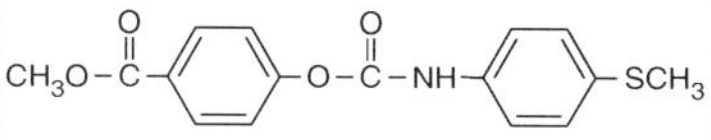

250 mg

S79,748-0
4-(METHYLTHIO)PHENYL N-[4-(ETHOXYCARBONYL)PHENYL]-CARBAMATE

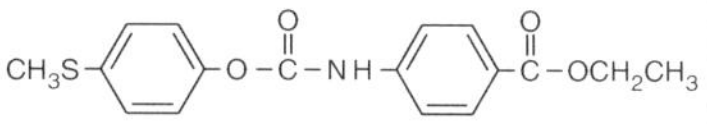

250 mg

S67,885-6
CARBOBENZYLOXY-L-VALYL-L-METHIONINE

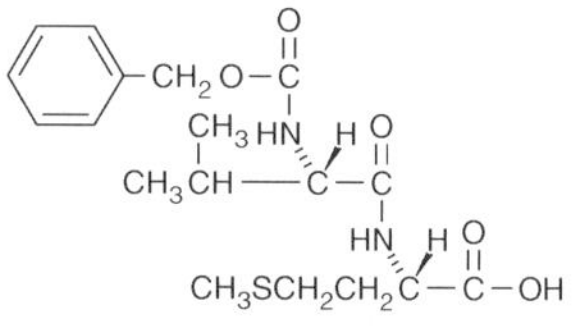

5 gm

S67,881-3
CARBOBENZYLOXY-L-PHENYLALANYL-L-METHIONINE

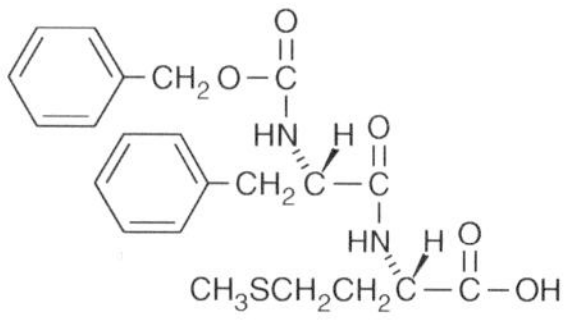

5 gm

S93,328-7
CARBOBENZYLOXY-L-METHIONYLGLYCINE METHYL ESTER

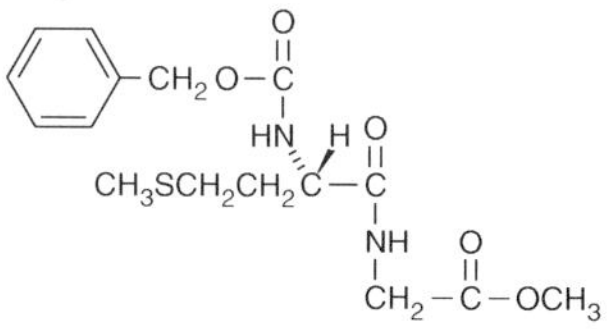

250 mg

S78,983-6
CARBOBENZYLOXY-L-METHIONYLGLYCYLGLYCINE METHYL ESTER

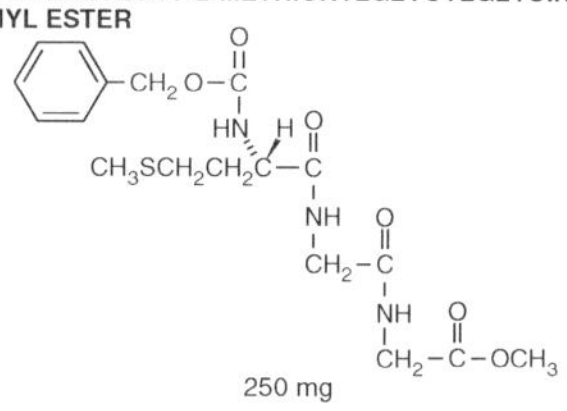

250 mg

S76,575-9
CARBOBENZYLOXY-S-BENZYL-L-CYSTEINYL-L-VALINE METHYL ESTER

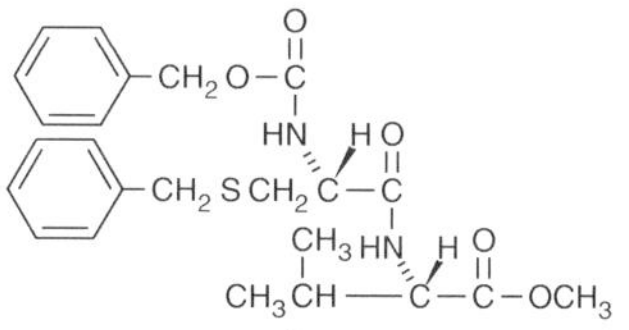

5 gm

S78,501-6
CARBOBENZYLOXY-S-BENZYL-L-CYSTEINYL-L-LEUCINE BENZYL ESTER

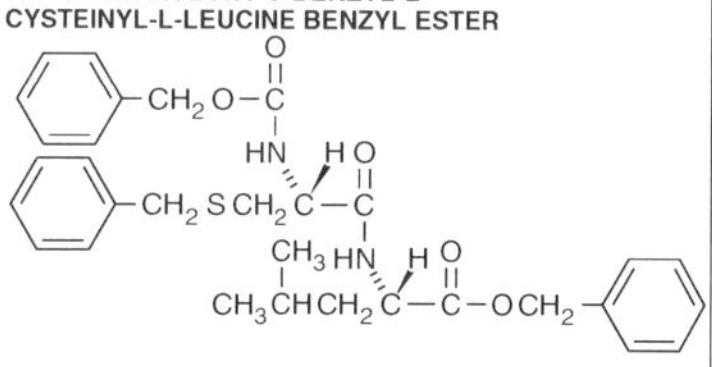

250 mg

S78,486-9
CARBOBENZYLOXY-S-BENZYL-L-CYSTEINYL-L-LEUCINAMIDE

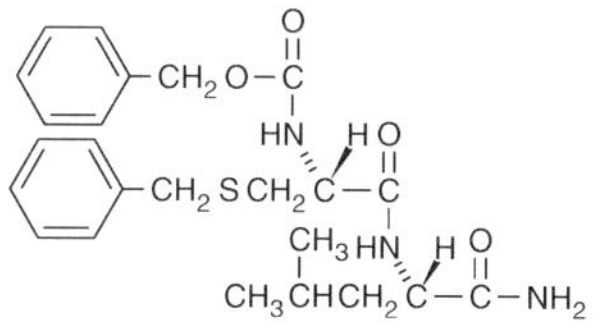

250 mg

S71,824-6
3-(N-PHENYLCARBAMOYL)-2-NAPHTHYL N-[4-(METHYLTHIO)PHENYL]CARBAMATE

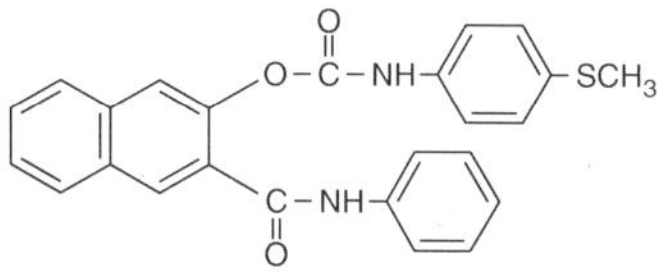

250 mg

S53,155-3
4-(METHYLTHIO)PHENYL CARBANILATE

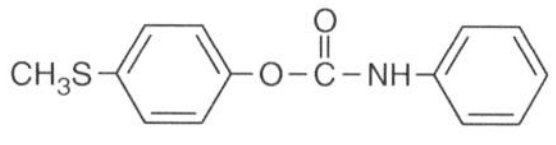

5 gm

S71,858-0
4-ACETAMIDOPHENYL N-[4-(METHYLTHIO)-PHENYL]CARBAMATE

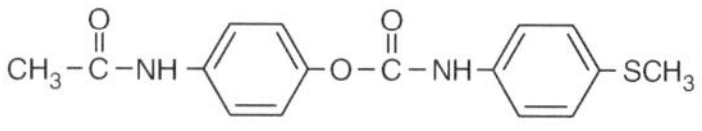

250 mg

S85,360-7
CYCLODODECYL N-[3-(METHYLTHIO)-PHENYL]CARBAMATE

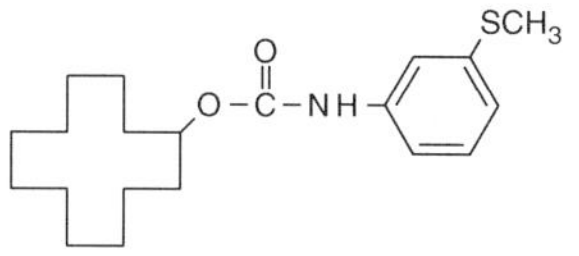

250 mg

S79,698-0
CYCLOPENTYL N-[4-(METHYLTHIO)-PHENYL]CARBAMATE

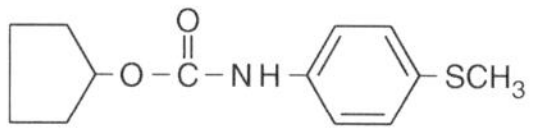

250 mg

S89,965-8
3-METHYL-4-(METHYLTHIO)PHENYL N-CYCLOHEXYLCARBAMATE

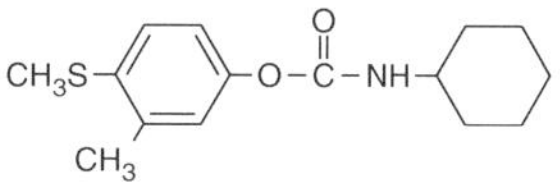

250 mg

S75,130-8
PIPERIDINO N-[4-(METHYLTHIO)PHENYL]-CARBAMATE

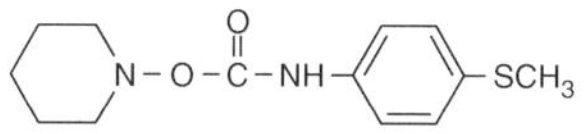

5 gm

S75,306-8
3-METHYL-4-(METHYLTHIO)PHENYL N-(2-METHYLPHENYL)CARBAMATE

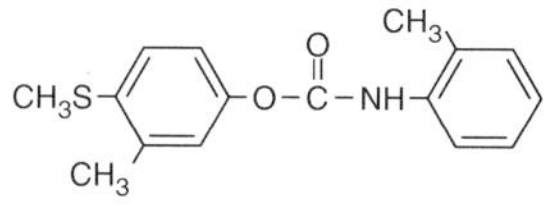

5 gm

S79,195-4
4-(METHYLTHIO)PHENYL N-METHYLCARBAMATE

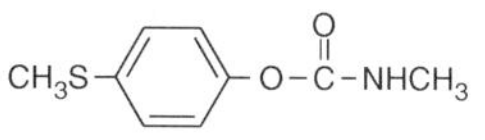

250 mg

S75,226-6
4-(METHYLTHIO)PHENYL N-PROPYLCARBAMATE

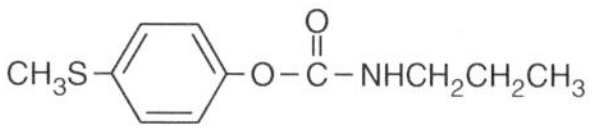

5 gm

S79,711-1
HEXYL N-[4-(METHYLTHIO)PHENYL]-CARBAMATE

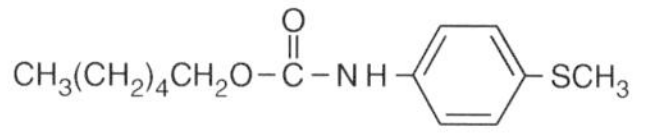

250 mg

S68,692-1
4-(METHYLTHIO)PHENYL N-(3-ETHYLPHENYL)CARBAMATE

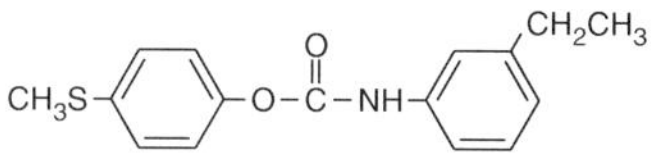

250 mg

S69,868-7
4-(METHYLTHIO)PHENYL N-(2,3-DIMETHYLPHENYL)CARBAMATE
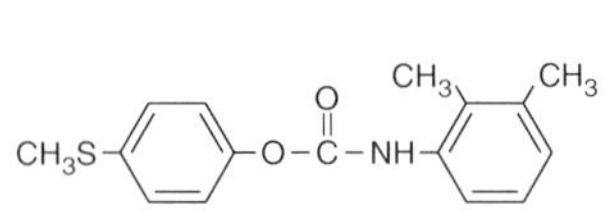
250 mg

S70,076-2
3,4-DIMETHYLPHENYL N-[4-(METHYLTHIO)PHENYL]CARBAMATE
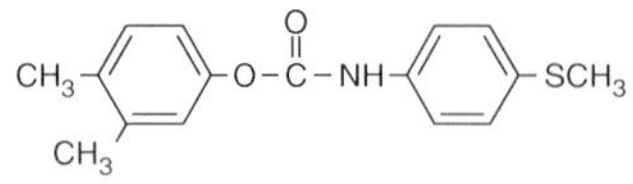
250 mg

S84,566-3
4-(METHYLTHIO)PHENYL N-(2,5-XYLYL)-CARBAMATE
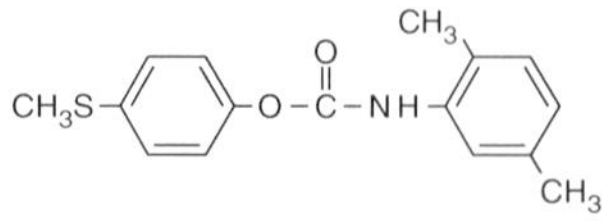
250 mg

S79,776-6
2-ISOPROPYL-5-METHYLPHENYL N-[4-(METHYLTHIO)PHENYL]CARBAMATE
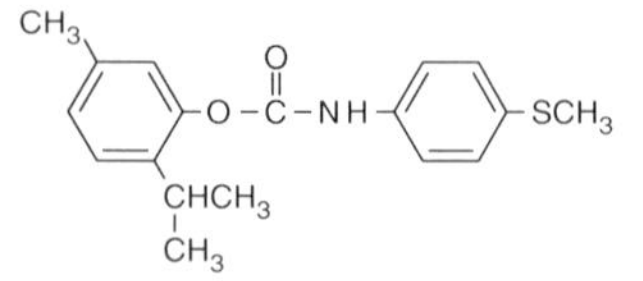
250 mg

S79,754-5
2,3,5-TRIMETHYLPHENYL N-[4-(METHYLTHIO)PHENYL]CARBAMATE
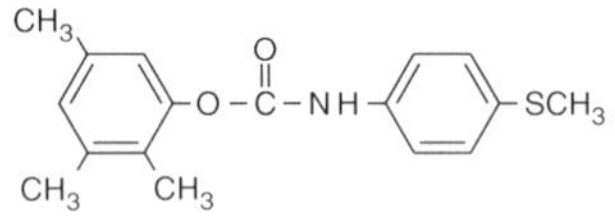
250 mg

S77,929-6
2-CHLORO-4,5-DIMETHYLPHENYL N-[4-(METHYLTHIO)PHENYL]CARBAMATE
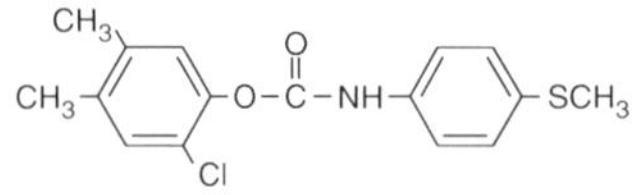
250 mg

S79,745-6
5-NORBORNEN-2-YLMETHYL N-[4-(METHYLTHIO)PHENYL]CARBAMATE
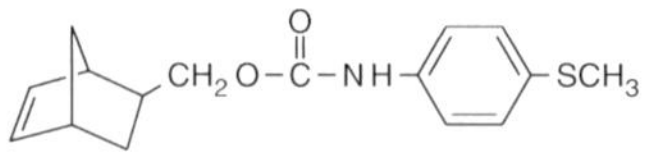
250 mg

S91,082-1
5-INDANYL N-[4-(METHYLTHIO)PHENYL]-CARBAMATE
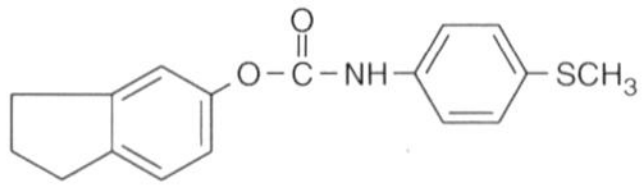
250 mg

S89,203-3
4-(METHYLTHIO)PHENYL N-(1-NAPHTHYL)-CARBAMATE
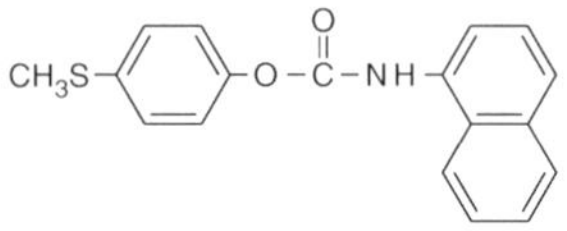
250 mg

S79,764-2
2-NAPHTHYL N-[4-(METHYLTHIO)PHENYL]-CARBAMATE
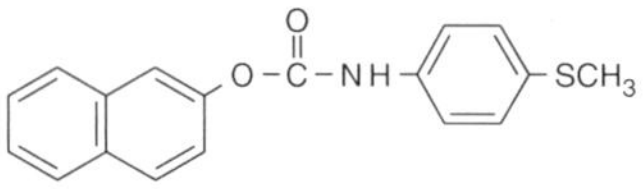
250 mg

S72,285-5
6-BROMO-2-NAPHTHYL N-[4-(METHYLTHIO)PHENYL]CARBAMATE
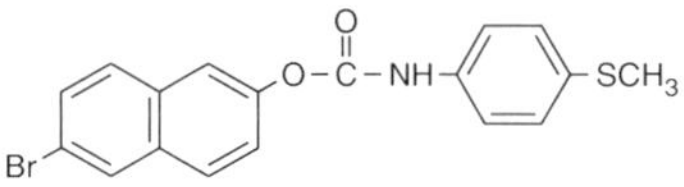
250 mg

S72,518-8
3-METHYL-4-(METHYLTHIO)PHENYL N-METHYLCARBAMATE
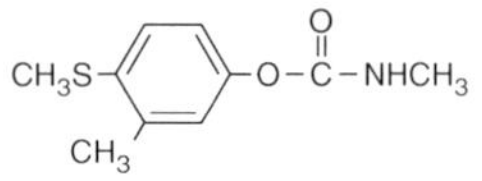
5 gm

S91,029-5
3-METHYL-4-(METHYLTHIO)PHENYL N-BUTYLCARBAMATE
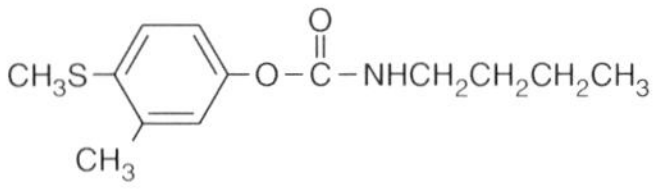
250 mg

S77,384-0
2,6-DIMETHYL-4-(METHYLTHIO)PHENYL N-(2,5-XYLYL)CARBAMATE
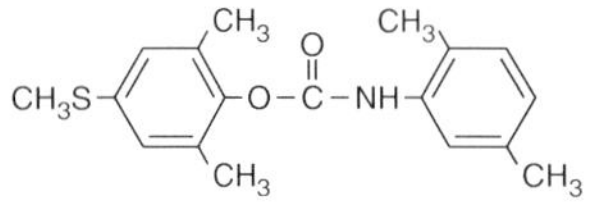
250 mg

S89,201-7
3-METHYL-4-(METHYLTHIO)PHENYL N-(1-NAPHTHYL)CARBAMATE
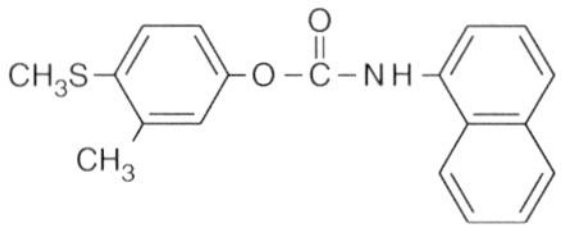
250 mg

S91,981-0
DIETHYL N,N'-(2,2'-DITHIODIPHENYL)BISCARBAMATE
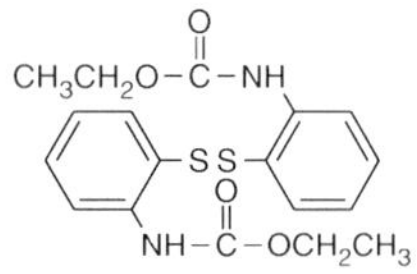
250 mg

S38,814-9
4-(ETHOXYCARBONYLAMINO)-ANTHRANILIC ACID
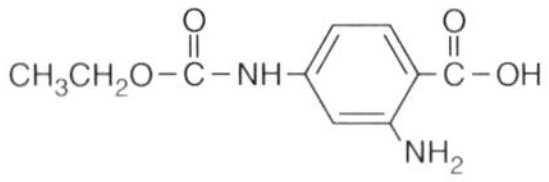
250 mg

S38,220-5
4-NITROPHENYL N-[2-(DIETHYLAMINO)-ETHYL]CARBAMATE
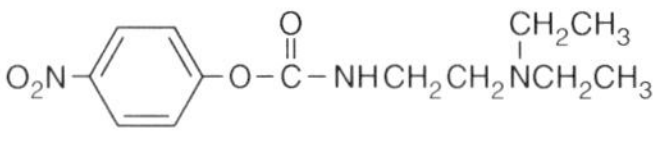
250 mg

S42,162-6
2,6-XYLYL N-[2-(DIETHYLAMINO)-ETHYL]CARBAMATE HYDROCHLORIDE
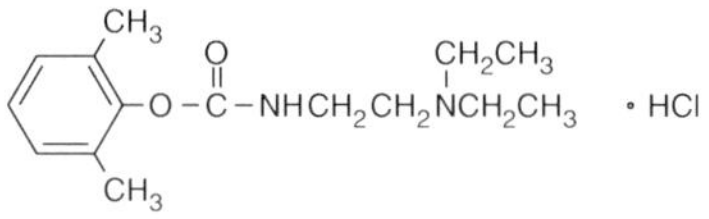
5 gm

S85,535-9
4-(METHOXYCARBONYL)PHENYL N-(2-NITROPHENYL)CARBAMATE
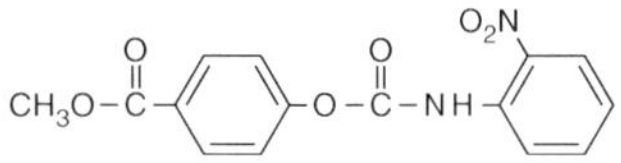
250 mg

S85,158-2
4-(BUTOXYCARBONYL)PHENYL N-(4-METHYL-2-NITROPHENYL)CARBAMATE
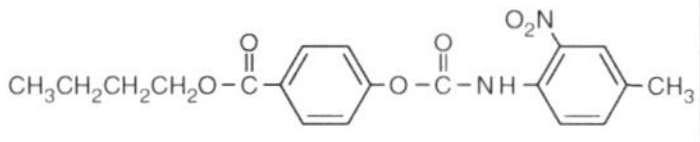
250 mg

S72,116-6
4-ACETAMIDOPHENYL N-(4-METHYL-2-NITROPHENYL)CARBAMATE
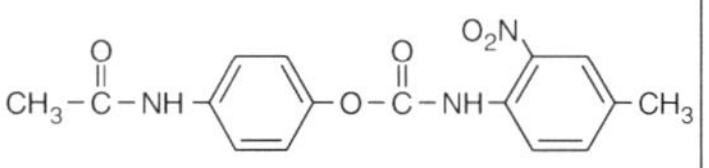
5 gm

S85,324-0
CYCLOPENTYL N-(4-METHYL-2-NITROPHENYL)CARBAMATE
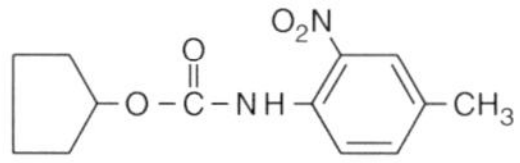
250 mg

S85,159-0
4-TERT-BUTYLCYCLOHEXYL N-(4-METHYL-2-NITROPHENYL)CARBAMATE
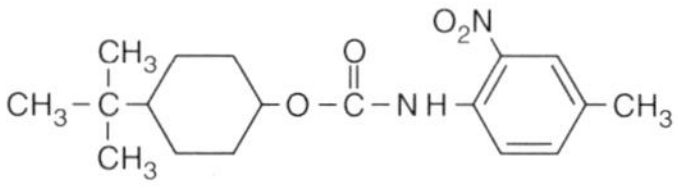
250 mg

S85,539-1
3-PHENYLALLYL N-(2-NITROPHENYL)-CARBAMATE
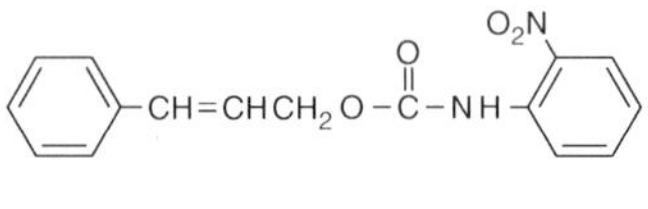
250 mg

S85,552-9
4-BIPHENYLYL N-(2-NITROPHENYL)-CARBAMATE
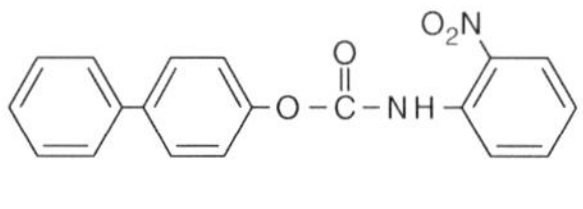
250 mg

S80,483-5
ETHYL N-(2-NITROPHENYL)CARBAMATE
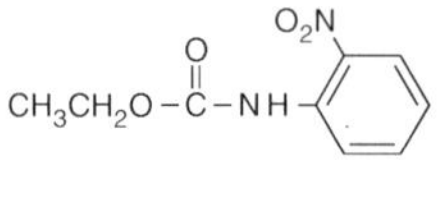
250 mg

S85,164-7
2,6-XYLYL N-(4-METHYL-2-NITROPHENYL)CARBAMATE
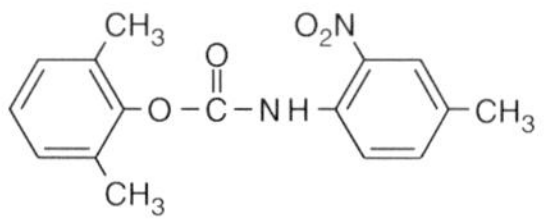
250 mg

S80,485-1
ETHYL N-(4-METHYL-2-NITROPHENYL)-CARBAMATE
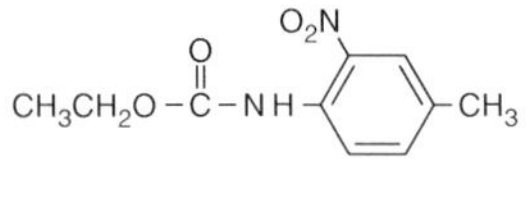
250 mg

S74,307-0
2-NAPHTHYL N-(4-METHYL-2-NITROPHENYL)CARBAMATE
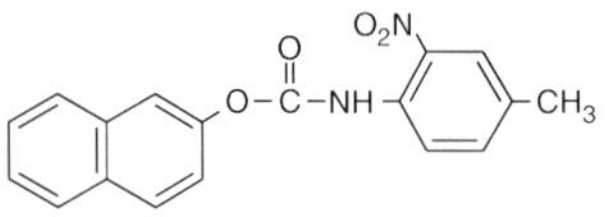
5 gm

S80,507-6
ETHYL N-(4,5-DIMETHYL-2-NITROPHENYL)CARBAMATE
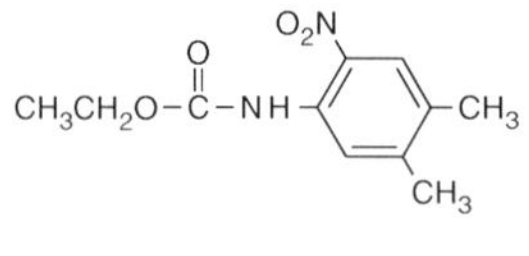
250 mg

S85,329-1
S-BENZYL N-(4-METHYL-2-NITROPHENYL)-THIOCARBAMATE
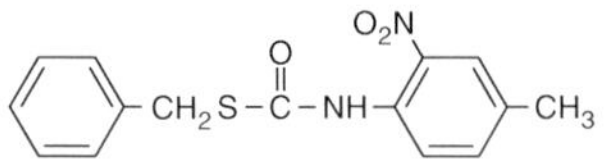
250 mg

S85,328-3
S-(O-TOLYL) N-(4-METHYL-2-NITROPHENYL)THIOCARBAMATE
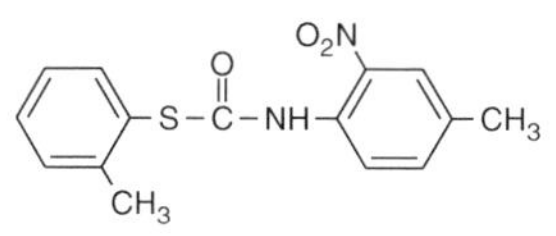
250 mg

S75,831-0
ETHYL N-(3-NITRO-9-OXO-2-FLUORENYL)-CARBAMATE
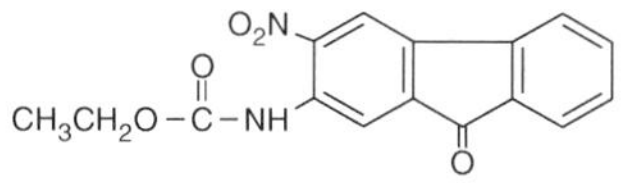
250 mg

S34,896-1
N',N2-DICARBETHOXY-2-METHYL-N'-PHENYL-1,2-PROPANEDIAMINE
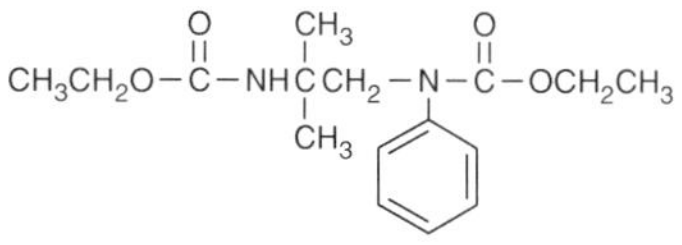
5 gm

S84,851-4
2-(N-ETHYLANILINO)ETHYL N-(2-METHYL-5-NITROPHENYL)CARBAMATE
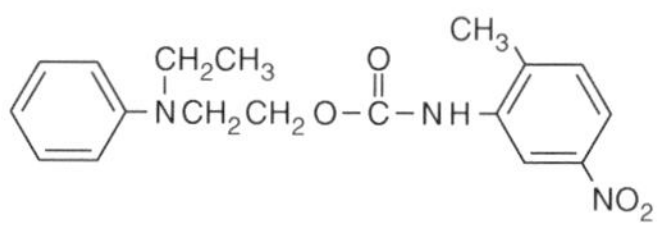
250 mg

S65,057-9
N,N'-BIS(2-DIETHYLAMINOETHOXYCARBONYL)-4-METHYL-1,3-PHENYLENEDIAMINE
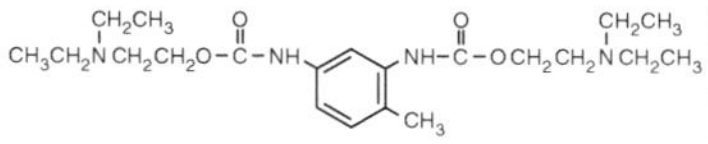
5 gm

S63,933-8
N,N'-BIS(2-DIMETHYLAMINOETHOXYCARBONYL)-4-METHYL-1,3-PHENYLENEDIAMINE
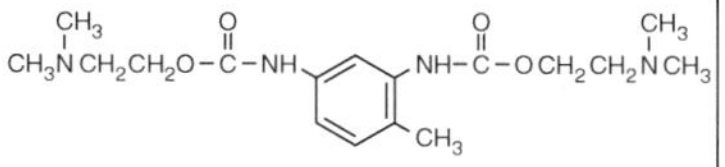
5 gm

S62,184-6
2-DIMETHYLAMINOETHYL N-(3-NITROPHENYL)CARBAMATE
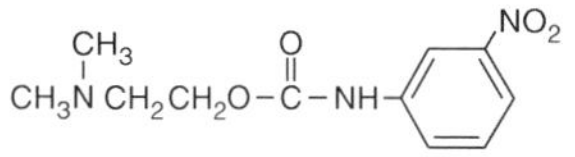
5 gm

S67,491-5
N,N'-BIS(4-ACETAMIDOPHENOXYCARBONYL)-2,4-TOLUENEDIAMINE
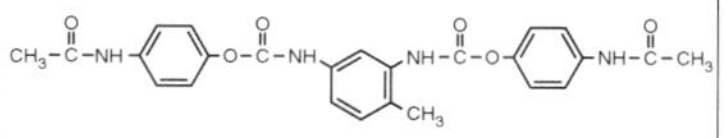
250 mg

S63,474-3
N,N'-BIS(ALLYLOXYCARBONYL)-4-METHYL-1,3-PHENYLENEDIAMINE
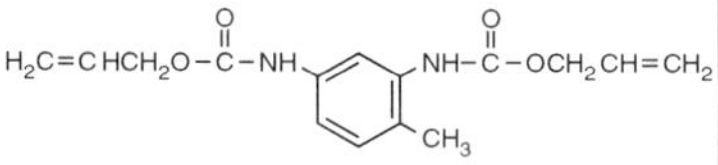
5 gm

S62,813-1
N,N'-BIS(CYCLOHEXYLOXYCARBONYL)-4-METHYL-1,3-PHENYLENEDIAMINE
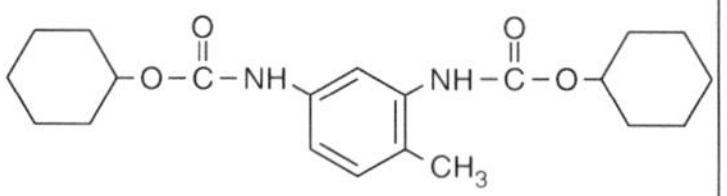
5 gm

S63,471-9
N,N'-BIS(PHENOXYCARBONYL)-4-METHYL-1,3-PHENYLENEDIAMINE
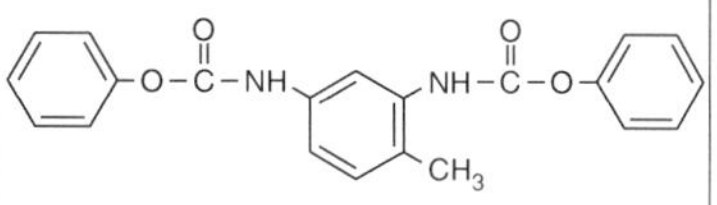
5 gm

S66,609-2
N,N'-BIS(4-CHLOROPHENOXYCARBONYL)-4-METHYL-1,3-PHENYLENEDIAMINE
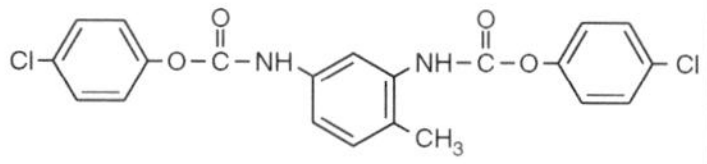
5 gm

S63,207-4
N,N'-BIS(2-METHYLCYCLOHEXYLOXYCARBONYL)-4-METHYL-1,3-PHENYLENEDIAMINE
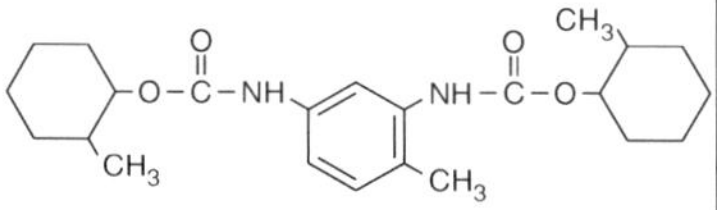
5 gm

S63,945-1
N,N'-BIS(ETHOXYCARBONYL)-4-METHYL-1,3-PHENYLENEDIAMINE
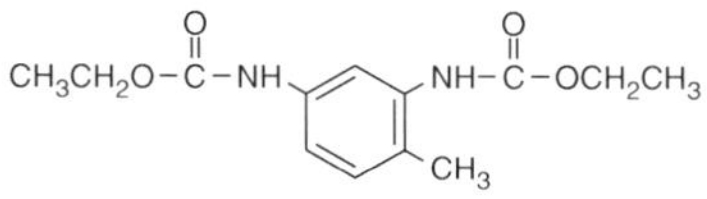
5 gm

S62,508-6
N,N'-(4-METHYL-1,3-PHENYLENE)BIS (PROPYL CARBAMATE)
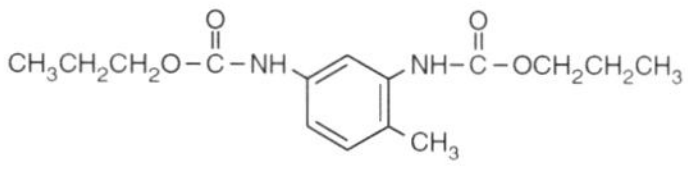
5 gm

S63,920-6
N,N'-BIS(ISOPROPOXYCARBONYL)-4-METHYL-1,3-PHENYLENEDIAMINE
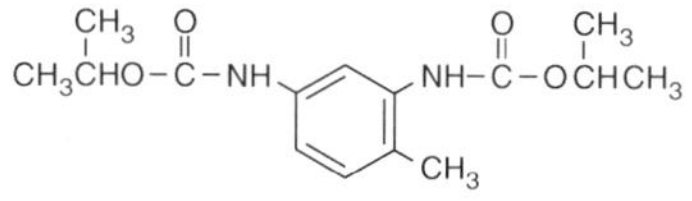
5 gm

S62,827-1
N,N'-BIS(BUTOXYCARBONYL)-4-METHYL-1,3-PHENYLENEDIAMINE
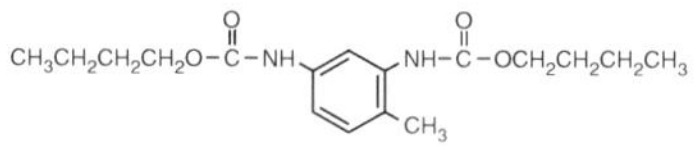
5 gm

S91,137-2
2-ACETYLPHENYL N-(4-METHYL-3-NITROPHENYL)CARBAMATE
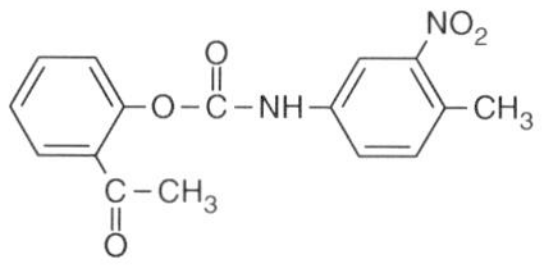
250 mg

S80,364-2
ETHYL N-(2-CARBOXY-5-NITROPHENYL)-CARBAMATE
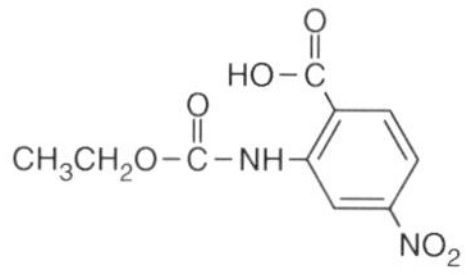
250 mg

S73,900-6
4-(METHOXYCARBONYL)PHENYL N-(3-NITROPHENYL)CARBAMATE
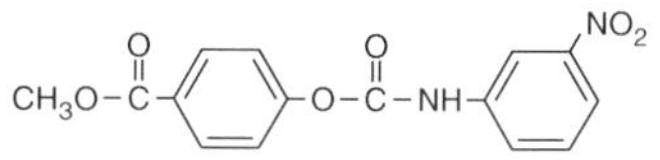
5 gm

S90,357-4
1-(ETHOXYCARBONYL)-4-(2-METHYL-5-NITROPHENYL)SEMICARBAZIDE
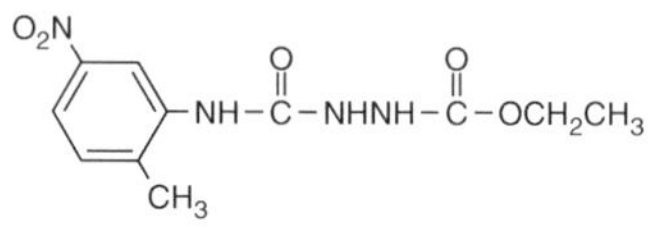
250 mg

S84,830-1
3-METHYLCYCLOHEXYL N-(2-METHYL-5-NITROPHENYL)CARBAMATE
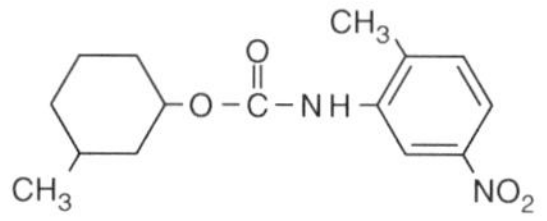
250 mg

S69,997-7
2-METHYL-4-NITROPHENYL N-(4-METHYL-3-NITROPHENYL)CARBAMATE
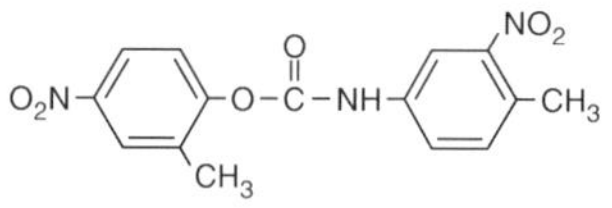
5 gm

S69,376-6
PIPERIDINO N-(3-NITROPHENYL)-CARBAMATE
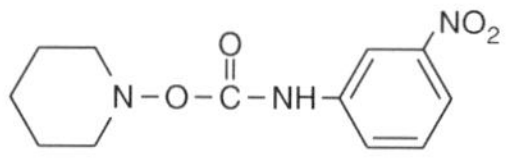
250 mg

S84,850-6
S-(4-TERT-BUTYLPHENYL) N-(2-METHYL-5-NITROPHENYL)THIOCARBAMATE
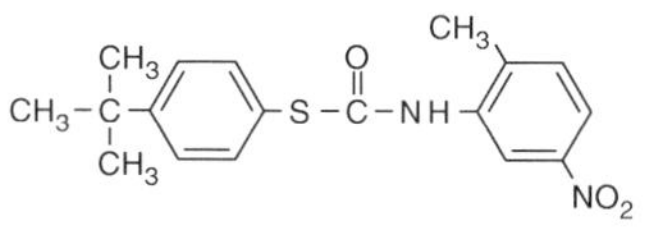
250 mg

S84,826-3
S-(M-TOLYL) N-(2-METHYL-5-NITROPHENYL)THIOCARBAMATE
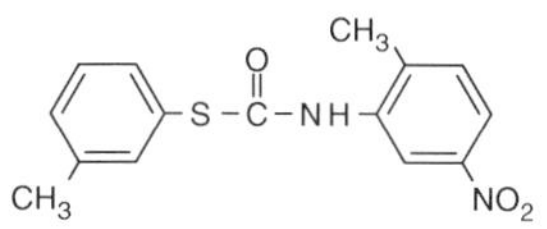
250 mg

S84,849-2
S,S'-ETHYLENE BIS[N-(2-METHYL-5-NITROPHENYL)THIOCARBAMATE]
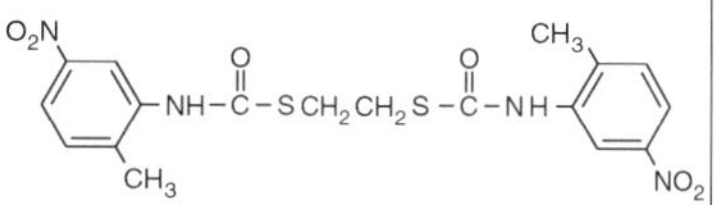
250 mg

S84,854-9
S-(4-TERT-BUTYL-2-METHYLPHENYL)-N-(2-METHYL-5-NITROPHENYL)-THIOCARBAMATE
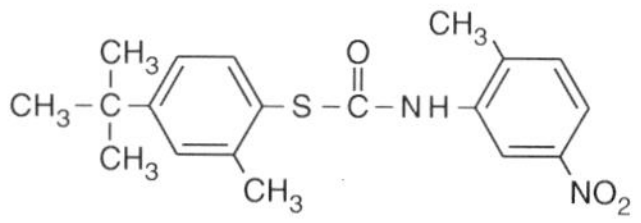
250 mg

S84,860-3
S-DODECYL N-(2-METHYL-5-NITROPHENYL)THIOCARBAMATE
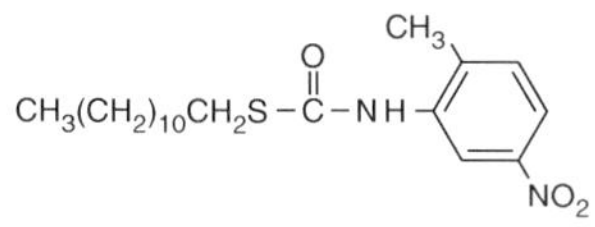
250 mg

S63,625-8
ALLYL N-(3-NITROPHENYL)CARBAMATE
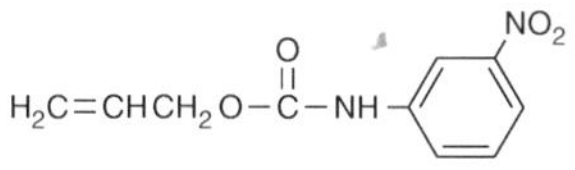
5 gm

S59,760-0
ISOPROPYL N-(3-NITROPHENYL)-CARBAMATE
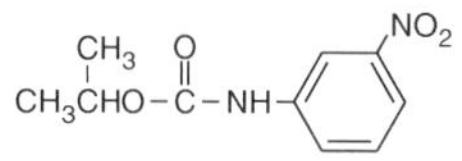
5 gm

S63,434-4
PROPYL N-(3-NITROPHENYL)CARBAMATE
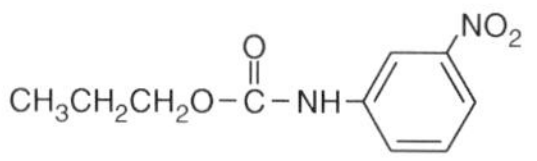
5 gm

S62,486-1
BUTYL N-(3-NITROPHENYL)CARBAMATE
670

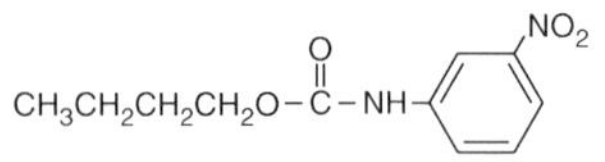

5 gm

S62,512-4
PHENYL N-(3-NITROPHENYL)CARBAMATE

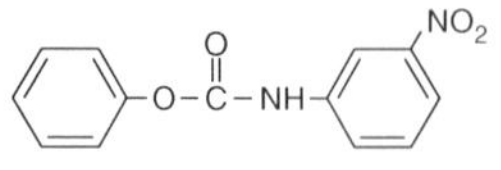

5 gm

S59,763-5
CYCLOHEXYL N-(3-NITROPHENYL)-CARBAMATE

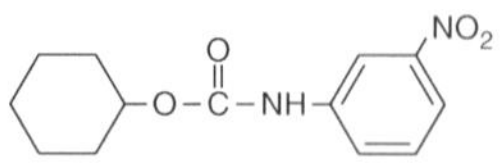

5 gm

S63,387-9
2-METHYLCYCLOHEXYL N-(3-NITROPHENYL)CARBAMATE

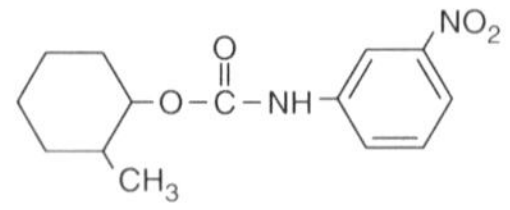

5 gm

S60,987-0
CYCLOHEXYL N-(4-METHYL-3-NITROPHENYL)CARBAMATE

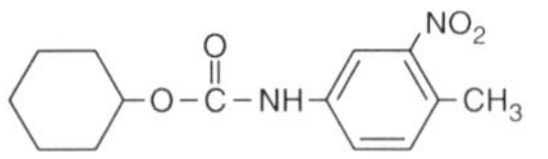

5 gm

S60,977-3
ISOPROPYL N-(4-METHYL-3-NITROPHENYL)CARBAMATE

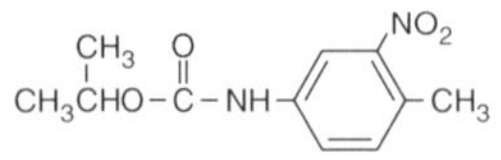

5 gm

S34,945-3
ETHYL 3-NITROPHENYLCARBAMATE

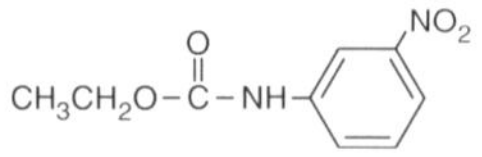

5 gm

S63,113-2
S-ISOPROPYL N-(3-NITROPHENYL)-THIOLCARBAMATE

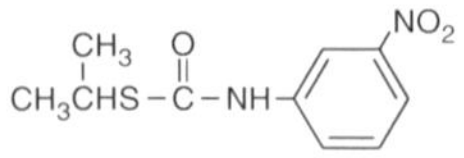

5 gm

S86,929-5
BIS[2-(N-PHENYLCARBAMOYL)PHENYL]-N,N'-(4-METHYL-1,3-PHENYLENE)-BISCARBAMATE

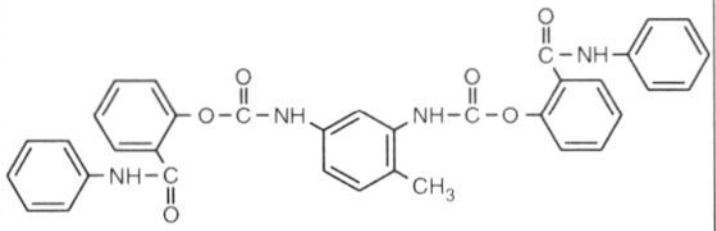

250 mg

S73,806-9
BIS[4-(METHOXYCARBONYL)PHENYL]-N,N'-(4-METHYL-1,3-PHENYLENE)-BISCARBAMATE

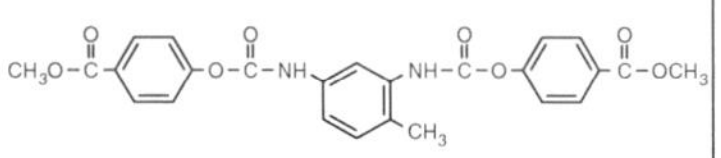

5 gm

S73,807-7
BIS(4-PROPIONYLPHENYL) N,N'-(4-METHYL-1,3-PHENYLENE)BISCARBAMATE

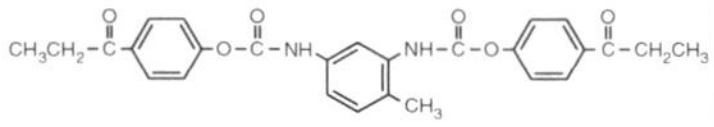

5 gm

S58,642-0
DIMETHYL N,N'-(4-METHYL-1,3-PHENYLENE)BISCARBAMATE

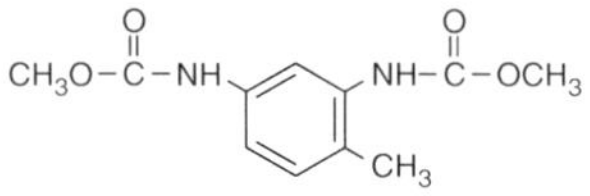

5 gm

S89,921-6
DIETHYL N,N'-(4-METHYL-1,3-PHENYLENE)BISCARBAMATE

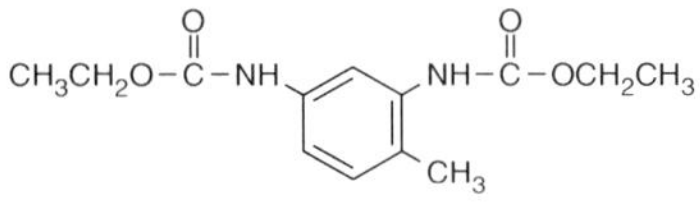

250 mg

S86,924-4
BIS(2-NAPHTHYL) N,N'-(4-METHYL-1,3-PHENYLENE)BISCARBAMATE

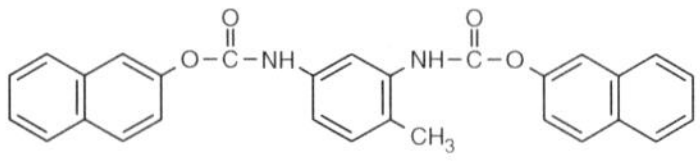

250 mg

S79,403-1
BIS(5-NORBORNEN-2-YLMETHYL)-N,N'-(4-METHYL-1,3-PHENYLENE)-BISCARBAMATE

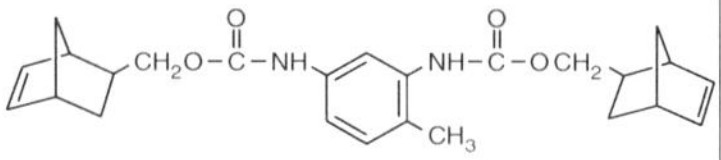

250 mg

S73,093-9
BIS(4-TERT-BUTYLCYCLOHEXYL)-N,N'-(4-METHYL-1,3-PHENYLENE)-BISCARBAMATE

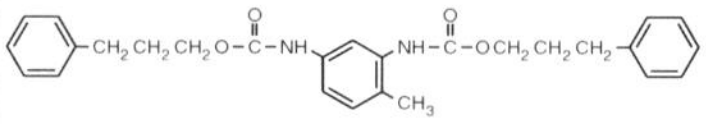

5 gm

S73,808-5
BIS(3-PHENYLPROPYL) N,N'-(4-METHYL-1,3-PHENYLENE)BISCARBAMATE

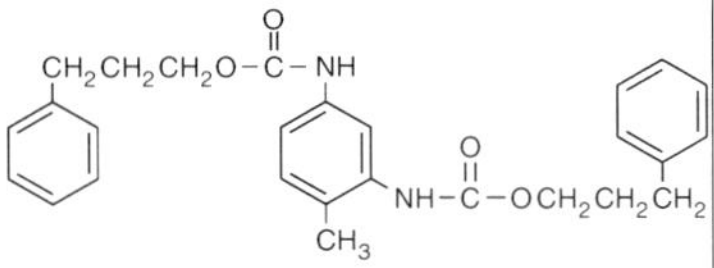

5 gm

S76,960-6
BIS(2,3-XYLYL) N,N'-(4-METHYL-1,3-PHENYLENE)BISCARBAMATE

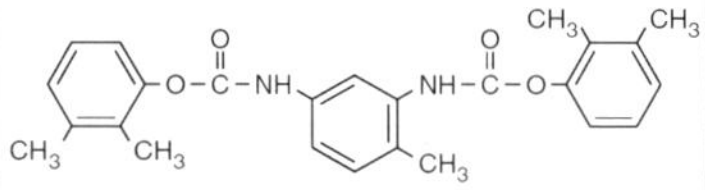

250 mg

S60,802-5
2-(DIMETHYLAMINO)ETHYL N-(2-CHLORO-4-NITROPHENYL)CARBAMATE

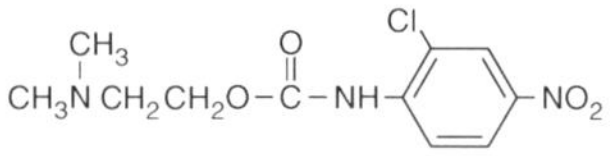

5 gm

S50,560-9
ETHYL 4-(DIETHYLAMINO)CARBANILATE

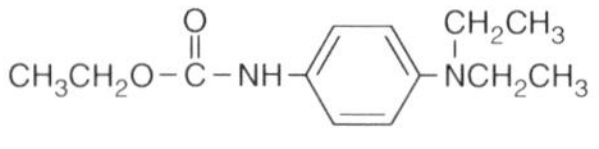

250 mg

S71,090-3
2-[1-METHYL-3-(2-METHYL-4-NITROPHENYL)UREIDO]ETHYL N-PHENYLCARBAMATE

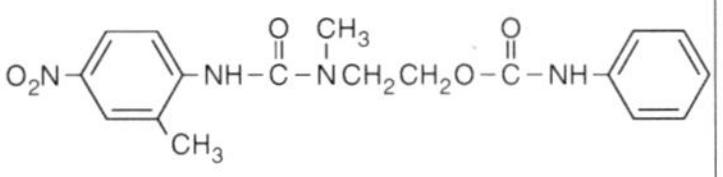

5 gm

S74,314-3
4-FORMYLPHENYL N-(4-NITROPHENYL)-CARBAMATE

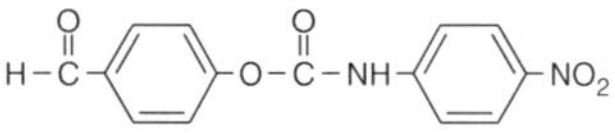

5 gm

S80,217-4
ETHYL N-(2-CARBOXY-4-NITROPHENYL)-CARBAMATE

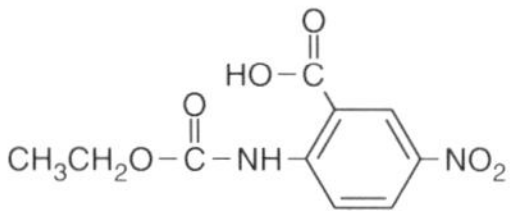

250 mg

S78,002-2
4-BIPHENYLYL N-(4-NITROPHENYL)-CARBAMATE

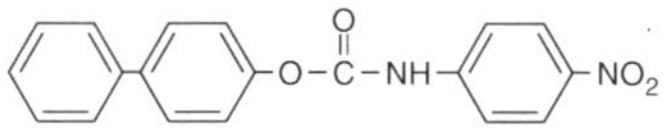

250 mg

S59,408-3
METHYL N-(4-NITROPHENYL)CARBAMATE

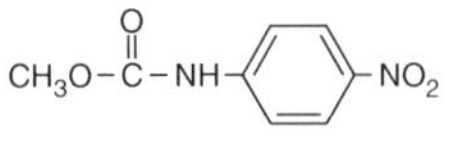

5 gm

S71,331-7
ISOPROPYL N-(2-METHYL-4-NITROPHENYL)CARBAMATE

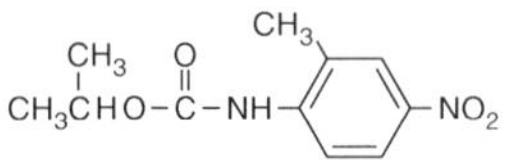

250 mg

S59,536-5
4-NITROBENZYL CARBANILATE

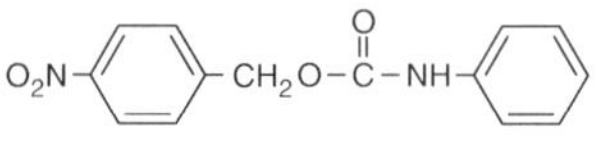

5 gm

S60,743-6
2-METHYLCYCLOHEXYL N-(2-CHLORO-4-NITROPHENYL)CARBAMATE

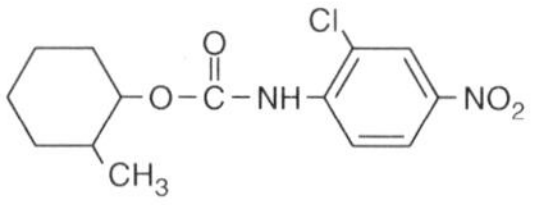

5 gm

S60,774-6
CYCLOHEXYL N-(2-CHLORO-4-NITROPHENYL)CARBAMATE

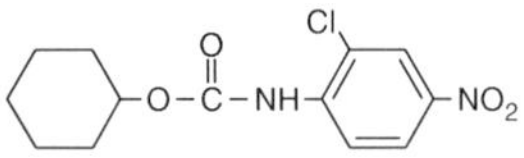

5 gm

S61,075-5
METHYL N-(2-METHYL-4-NITROPHENYL)-CARBAMATE

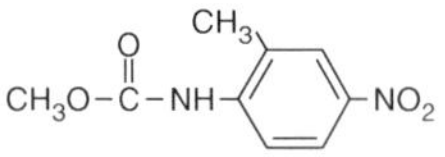

5 gm

S64,503-6
ALLYL N-(2-CHLORO-4-NITROPHENYL)-CARBAMATE

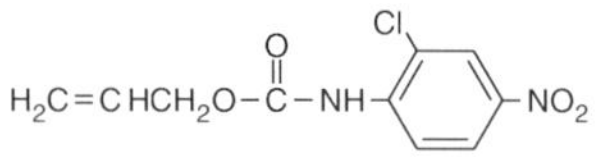

5 gm

S60,756-8
ISOPROPYL N-(2-CHLORO-4-NITROPHENYL)CARBAMATE

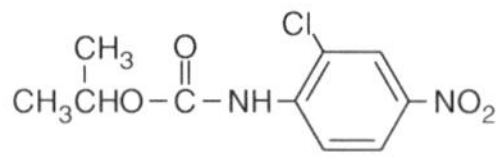

5 gm

S64,498-6
PHENYL N-(2-CHLORO-4-NITROPHENYL)-CARBAMATE

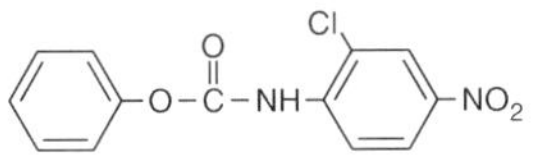

5 gm

S34,946-1
ETHYL 4-NITROPHENYLCARBAMATE

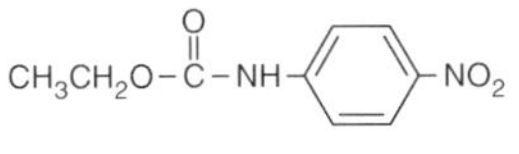

5 gm

S71,079-2
N,N'-(1,5-NAPHTHALENEDIYL)BIS (ETHYL CARBAMATE)

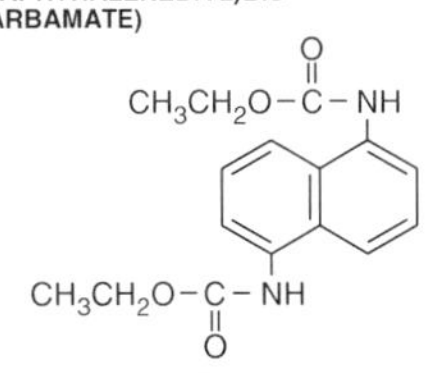

5 gm

S70,994-8
N,N'-(1,5-NAPHTHALENEDIYL)BIS (ISOPROPYL CARBAMATE)

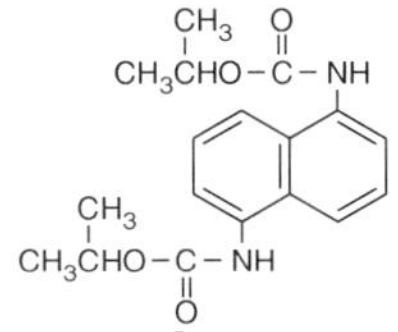

5 gm

S78,358-7
N,O-BIS(CARBOBENZYLOXY)DL-TYROSINE

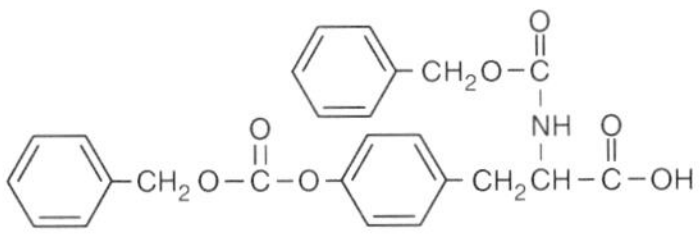

5 gm

S77,764-1
CARBOBENZYLOXY-L-VALYL-L-PHENYLALANINE

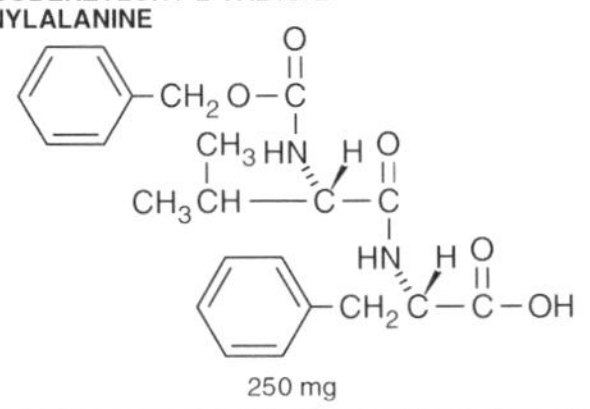

250 mg

S77,642-4
CARBOBENZYLOXY-L-TYROSYL-L-TYROSINE

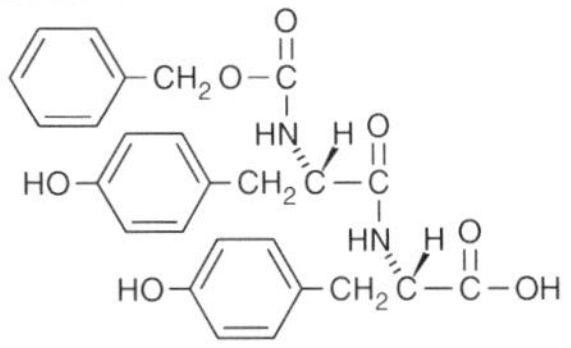

250 mg

S59,944-1
CARBOBENZYLOXY-L-ASPARAGINYL-L-PHENYLALANINE METHYL ESTER

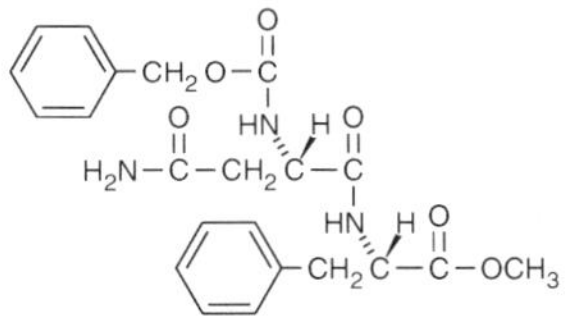

5 gm

S78,630-6
CARBOBENZYLOXYGLYCYLGLYCYL-L-PHENYLALANINE METHYL ESTER

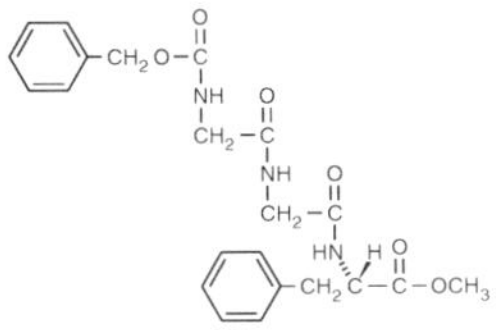

250 mg

S78,861-9
CARBOBENZYLOXY-L-LEUCYL-L-PHENYLALANINE BENZYL ESTER

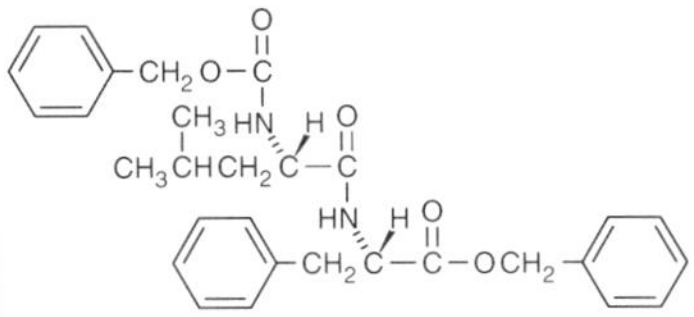

250 mg

S36,612-9
ETHYL N-PHENETHYLCARBAMATE

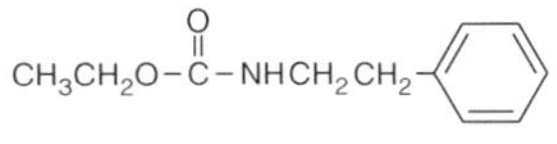

5 gm

S39,227-8
ETHYL N-(ALPHA,ALPHA-DIMETHYLPHENETHYL)CARBAMATE

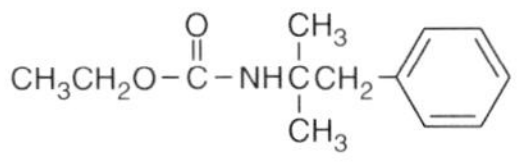

5 gm

S59,937-9
ETHYL N-(ALPHA-ALLYLPHENETHYL)-CARBAMATE

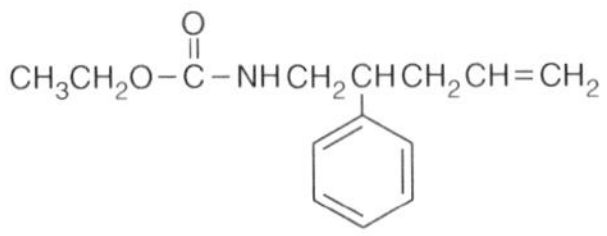

5 gm

S51,987-1
ETHYL N-ACETYL-N-PHENETHYLCARBAMATE

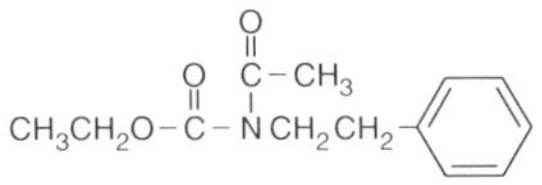

250 mg

S69,304-9
2-(ANILINO)ETHYL N-ETHYL-N-PHENYLCARBAMATE

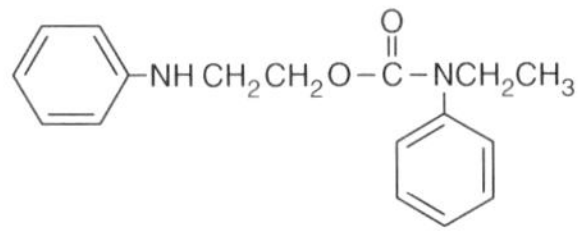

5 gm

S86,393-9
ETHYL N-(9,10-DIHYDRO-9,10-ETHANOANTHRACEN-11-YL)CARBAMATE

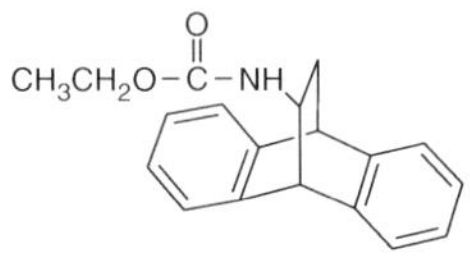

250 mg

S76,945-2
4-(ANILINO)PHENYL N-METHYLCARBAMATE

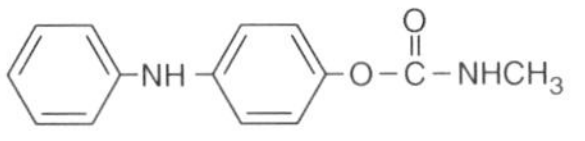

250 mg

S73,892-1
4-ANILINOPHENYL N-(P-TOLYL)-CARBAMATE

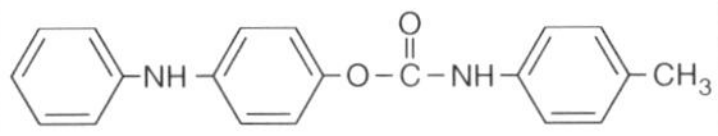

5 gm

S86,865-5
2-(N-ETHYLANILINO)ETHYL N-(3-ACETYLPHENYL)CARBAMATE

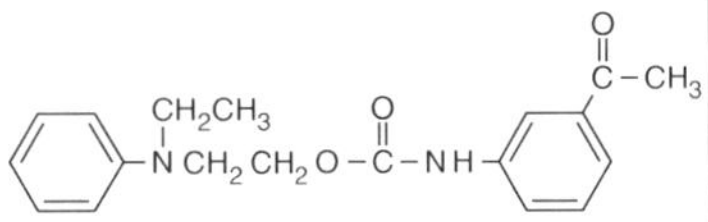

250 mg

S52,863-3
3-DIETHYLAMINOPROPYL N-(4-CHLOROBENZYL)-N-CYCLOPROPYLCARBAMATE

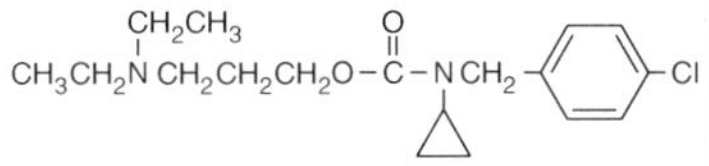

5 gm

S79,419-8
2-(N-ETHYLANILINO)ETHYL N-[4-(ETHOXYCARBONYL)PHENYL]-CARBAMATE

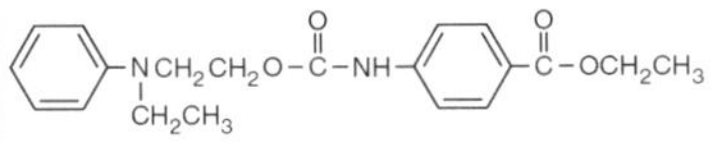

250 mg

S60,966-8
2-DIMETHYLAMINOETHYL N-(3-CHLOROPHENYL)CARBAMATE

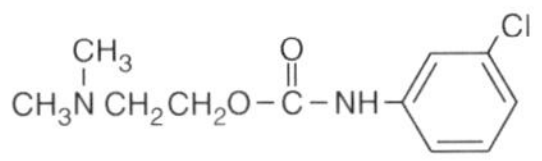

5 gm

S66,683-1
2-DIMETHYLAMINOETHYL N-(M-TOLYL)-CARBAMATE

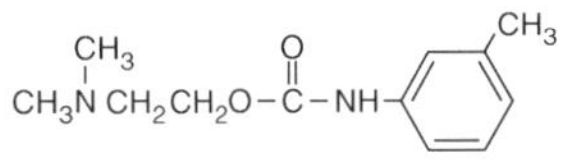

5 gm

S63,438-7
2-(DIETHYLAMINO)ETHYL N-(3-CHLOROPHENYL)CARBAMATE

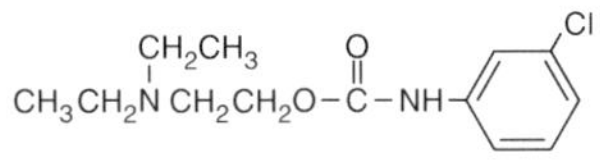

5 gm

S65,059-5
2-DIETHYLAMINOETHYL N-(M-TOLYL)-CARBAMATE

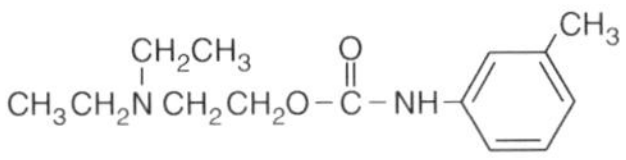

5 gm

S66,065-5
2-[N-ETHYL-N-(M-TOLYL)AMINO]ETHYL N-PHENYLCARBAMATE

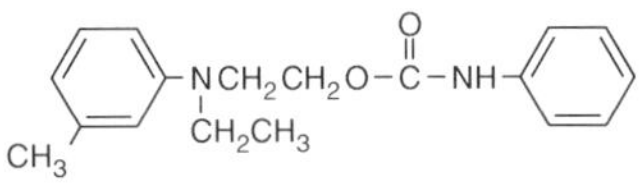

5 gm

S63,835-8
2-DIMETHYLAMINOETHYL N-PHENYLCARBAMATE

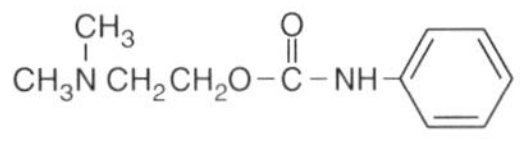

5 gm

S62,418-7
2-(DIETHYLAMINO)ETHYL N-PHENYLCARBAMATE

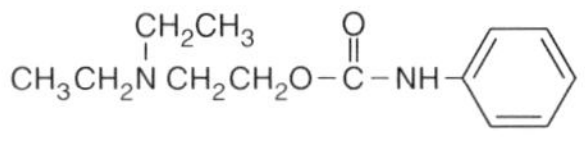

5 gm

S83,979-5
2-(DIETHYLAMINO)ETHYL N-PHENYLCARBAMATE HYDROCHLORIDE

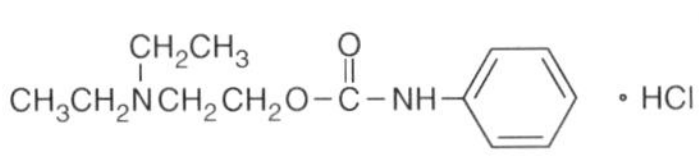

250 mg

S66,172-4
2-(N-ETHYLANILINO)ETHYL N-PHENYLCARBAMATE

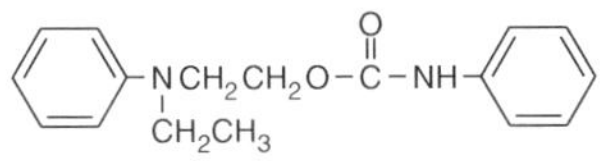

5 gm

S63,833-1
2-DIMETHYLAMINOETHYL N-(2-CHLOROPHENYL)CARBAMATE

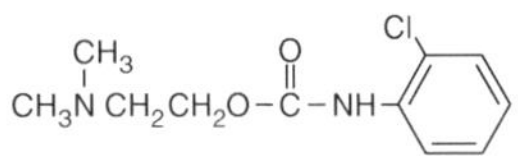

250 mg

S66,675-0
2-DIMETHYLAMINOETHYL N-(O-TOLYL)-CARBAMATE

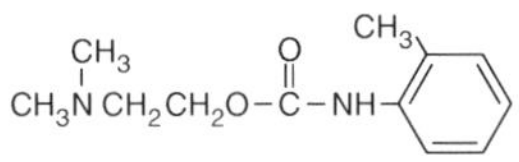

5 gm

S65,442-6
2-(DIMETHYLAMINO)ETHYL N-(2-ETHYLPHENYL)CARBAMATE

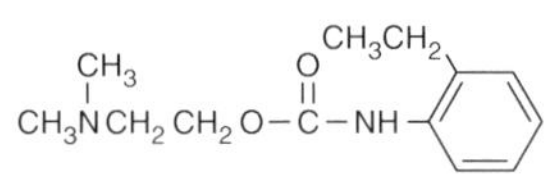

5 gm

S63,111-6
2-(DIETHYLAMINO)ETHYL N-(2-CHLOROPHENYL)CARBAMATE

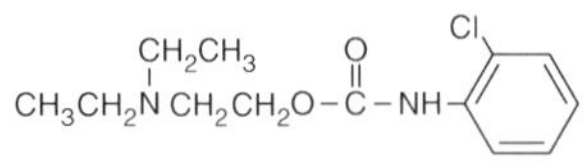

5 gm

S66,669-6
2-DIETHYLAMINOETHYL N-(O-TOLYL)-CARBAMATE

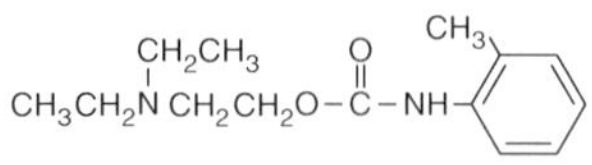

5 gm

S66,811-7
2-DIETHYLAMINOETHYL N-(2-ETHYLPHENYL)CARBAMATE

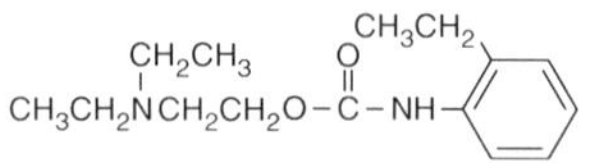

5 gm

S60,762-2
2-(DIMETHYLAMINO)ETHYL N-(4-CHLOROPHENYL)CARBAMATE

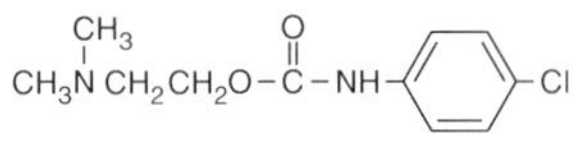

5 gm

S66,652-1
2-DIMETHYLAMINOETHYL N-(P-TOLYL)-CARBAMATE

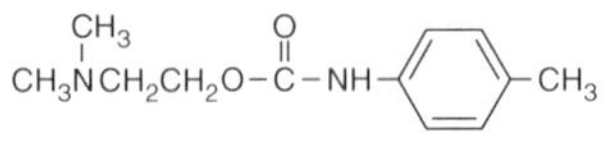

5 gm

S65,058-7
2-DIETHYLAMINOETHYL N-(P-TOLYL)-CARBAMATE

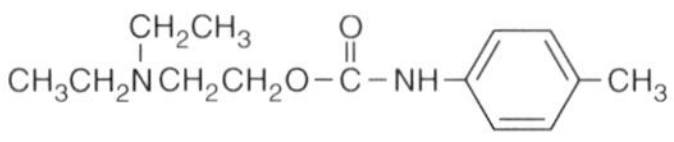

5 gm

S66,143-0
2-(N-ETHYLANILINO)ETHYL N-(P-TOLYL)-CARBAMATE

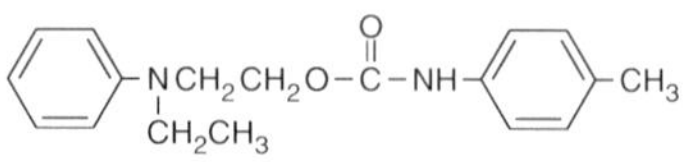

5 gm

S71,105-5
BIS(2-DIMETHYLAMINOETHYL)-4,4'-METHYLENEBIS(N-PHENYLCARBAMATE)

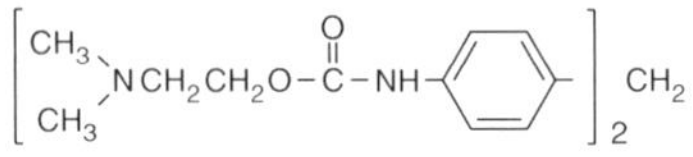

5 gm

S63,442-5
2-(DIMETHYLAMINO)ETHYL N-(2,3-DICHLOROPHENYL)CARBAMATE

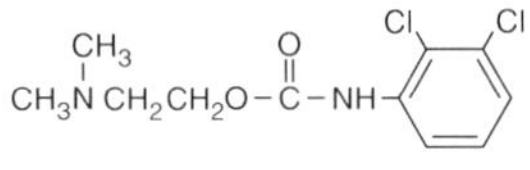

250 mg

S64,982-1
2-DIMETHYLAMINOETHYL N-(2-CHLORO-6-METHYLPHENYL)CARBAMATE

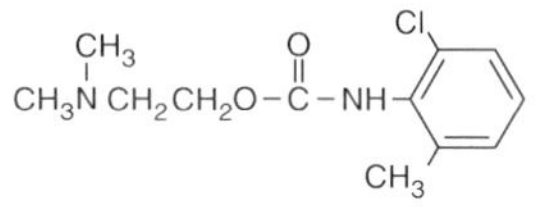

250 mg

S65,054-4
2-DIMETHYLAMINOETHYL N-(2,3-XYLYL)-CARBAMATE

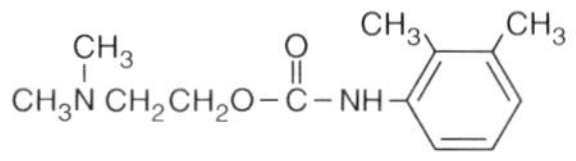

5 gm

S66,647-5
2-DIMETHYLAMINOETHYL N-(2,6-XYLYL)-CARBAMATE

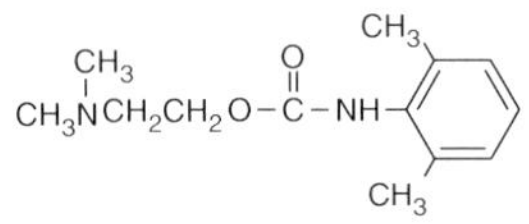

5 gm

S65,307-1
2-(DIETHYLAMINO)ETHYL N-(2,3-DICHLOROPHENYL)CARBAMATE

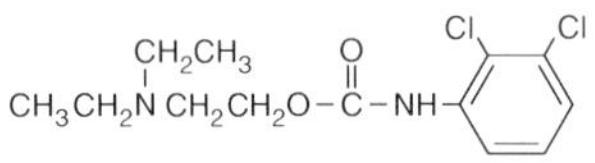

5 gm

S73,592-2
2-DIMETHYLAMINOETHYL N-(2-ETHYL-6-METHYLPHENYL)CARBAMATE

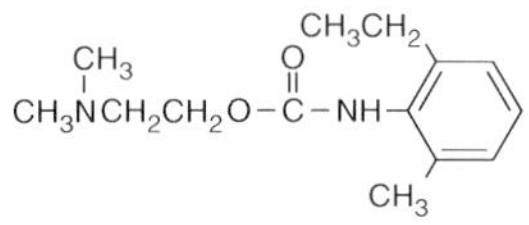

250 mg

S65,043-9
2-DIETHYLAMINOETHYL N-(2-CHLORO-6-METHYLPHENYL)CARBAMATE

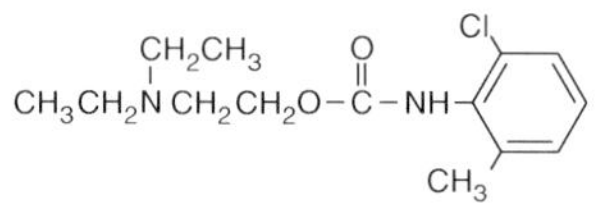

5 gm

S66,645-9
2-DIETHYLAMINOETHYL N-(2,3-XYLYL)-CARBAMATE

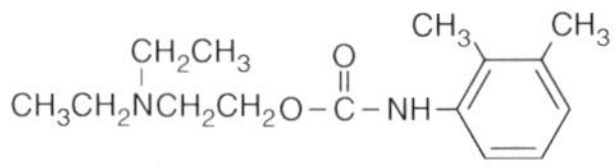

5 gm

S66,676-9
2-DIMETHYLAMINOETHYL N-(2,6-DIETHYLPHENYL)CARBAMATE

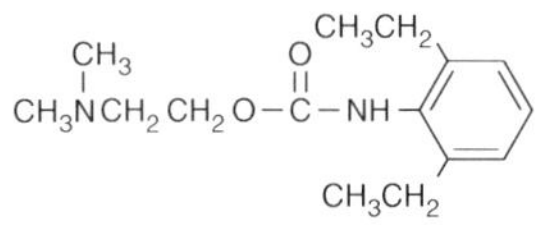

5 gm

S66,682-3
2-DIETHYLAMINOETHYL N-(2,6-XYLYL)-CARBAMATE

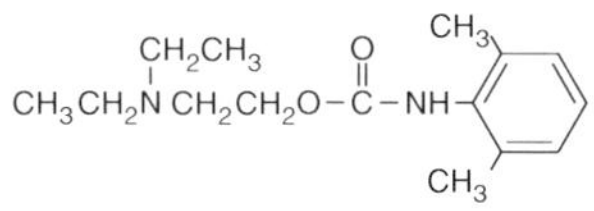

5 gm

S73,542-6
2-(DIETHYLAMINO)ETHYL N-(2-ETHYL-6-METHYLPHENYL)CARBAMATE

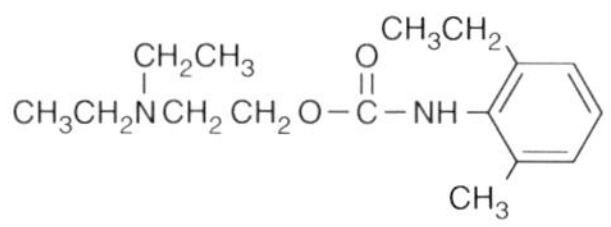

250 mg

S63,376-3
2-DIMETHYLAMINOETHYL N-(3,4-DICHLOROPHENYL)CARBAMATE

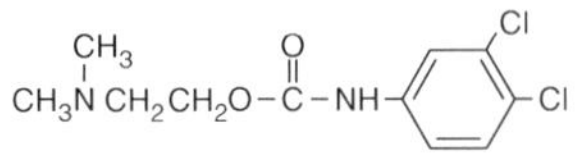

5 gm

S60,792-4
2-(DIMETHYLAMINO)ETHYL N-(4-CHLORO-2-METHYLPHENYL)CARBAMATE

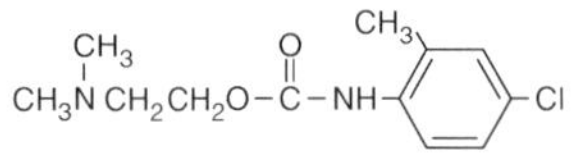

5 gm

S60,973-0
2-DIMETHYLAMINOETHYL N-(5-CHLORO-2-METHYLPHENYL)CARBAMATE

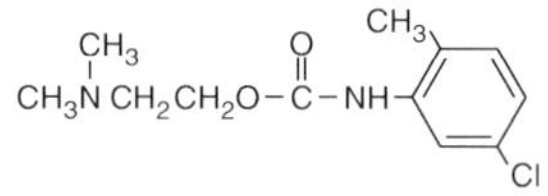

5 gm

S60,974-9
2-DIMETHYLAMINOETHYL N-(3-CHLORO-4-METHYLPHENYL)CARBAMATE

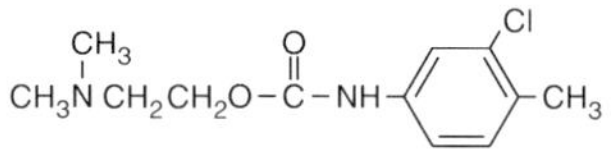

5 gm

S61,594-3
2-(DIMETHYLAMINO)ETHYL N-(2,4-XYLYL)CARBAMATE

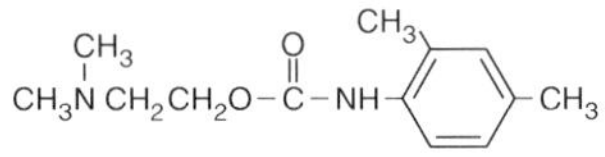

5 gm

S65,280-6
2-(DIMETHYLAMINO)ETHYL N-(2,5-XYLYL)CARBAMATE

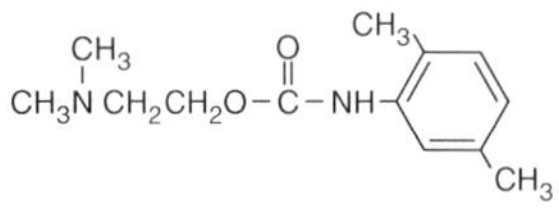

5 gm

S65,281-4
2-(DIMETHYLAMINO)ETHYL N-(3,4-XYLYL)CARBAMATE

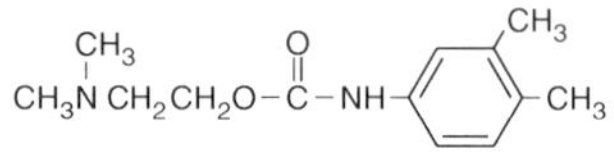

5 gm

S65,210-5
2-(DIETHYLAMINO)ETHYL N-(2,4-DICHLOROPHENYL)CARBAMATE

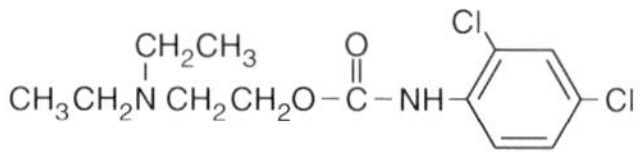

5 gm

S65,212-1
2-(DIETHYLAMINO)ETHYL N-(3,4-DICHLOROPHENYL)CARBAMATE

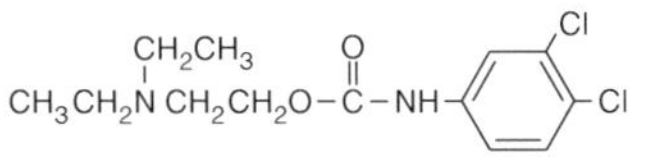

5 gm

S65,013-7
2-DIETHYLAMINOETHYL N-(4-CHLORO-2-METHYLPHENYL)CARBAMATE

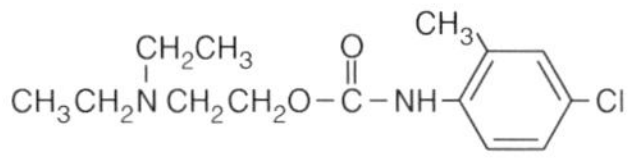

5 gm

S65,055-2
2-DIETHYLAMINOETHYL N-(3-CHLORO-4-METHYLPHENYL)CARBAMATE

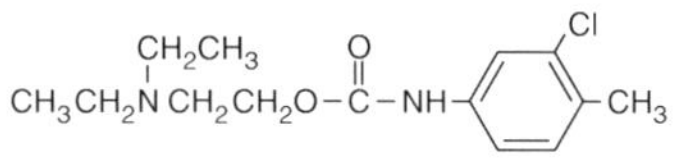

5 gm

S65,056-0
2-DIETHYLAMINOETHYL N-(2,4-XYLYL)-CARBAMATE

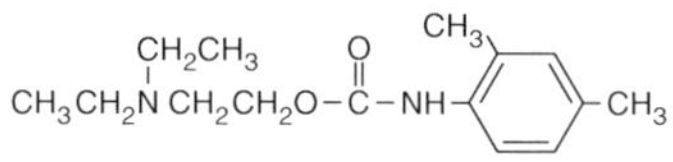

5 gm

S65,304-7
2-(DIMETHYLAMINO)ETHYL N-(3,5-XYLYL)CARBAMATE

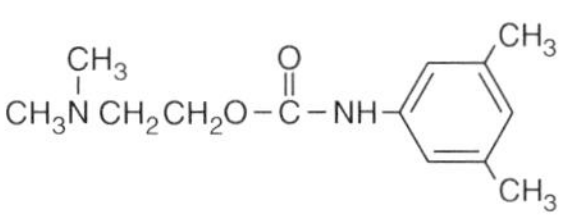

5 gm

S71,270-1
2-DIETHYLAMINOETHYL N-(3,5-DIMETHYLPHENYL)CARBAMATE

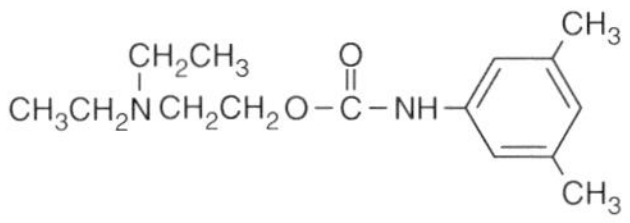

5 gm

S62,174-9
2-DIMETHYLAMINOETHYL N-(2,4,5-TRICHLOROPHENYL)CARBAMATE

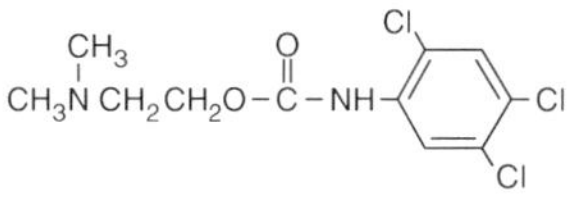

5 gm

S63,844-7
2-DIETHYLAMINOETHYL N-(2,4,5-TRICHLOROPHENYL)CARBAMATE

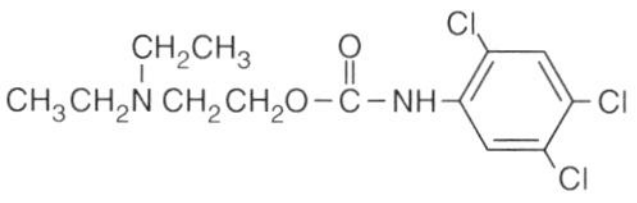

5 gm

S63,948-6
2-DIMETHYLAMINOETHYL N-(1-NAPHTHYL)-CARBAMATE

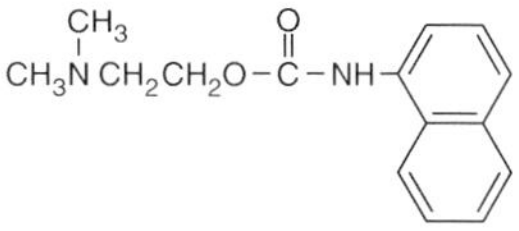

5 gm

S79,789-8
3-(DIETHYLAMINO)PHENYL N-[4-(ETHOXYCARBONYL)PHENYL]-CARBAMATE

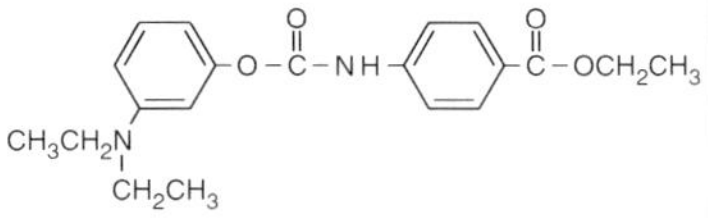

250 mg

S71,583-2
3-(DIETHYLAMINO)PHENYL N-METHYLCARBAMATE

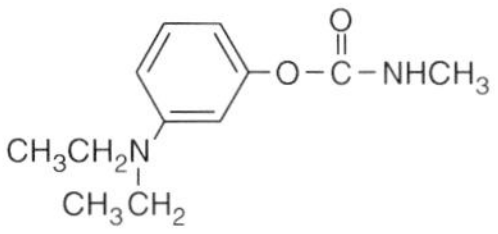

5 gm

S64,703-9
3-DIETHYLAMINOPHENYL N-(3-CHLOROPHENYL)CARBAMATE

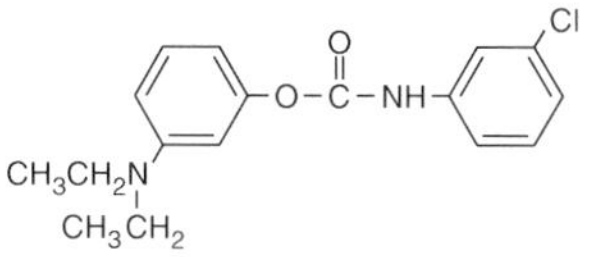

5 gm

S64,702-0
3-DIETHYLAMINOPHENYL N-(2-CHLOROPHENYL)CARBAMATE

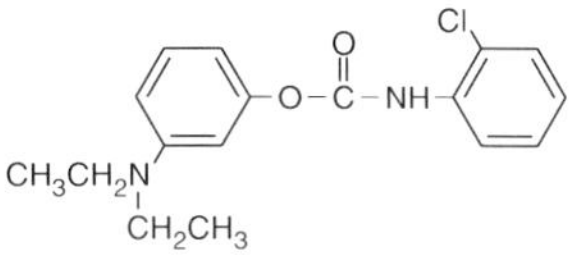

5 gm

S64,709-8
2-(DIETHYLAMINO)ETHYL N-(4-CHLOROPHENYL)CARBAMATE

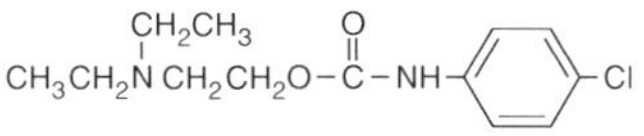

5 gm

S71,358-9
2-DIMETHYLAMINOETHYL N-(2,6-DIISOPROPYLPHENYL)CARBAMATE

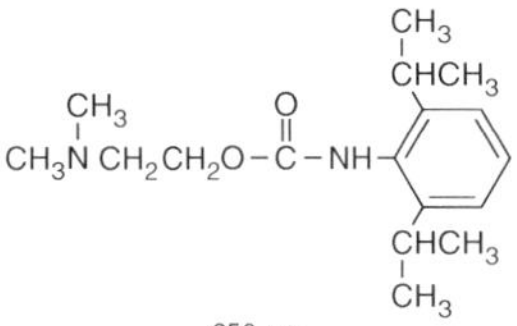

250 mg

S71,365-1
2-(DIETHYLAMINO)ETHYL N-(2,6-DIETHYLPHENYL)CARBAMATE

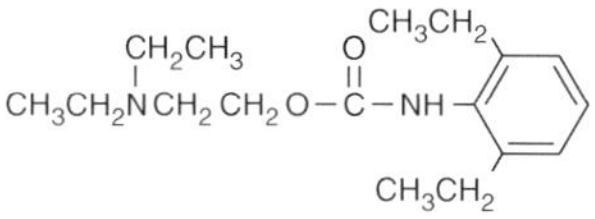

5 gm

S65,783-2
3-DIETHYLAMINOPHENYL N-(2,6-DICHLOROPHENYL)CARBAMATE

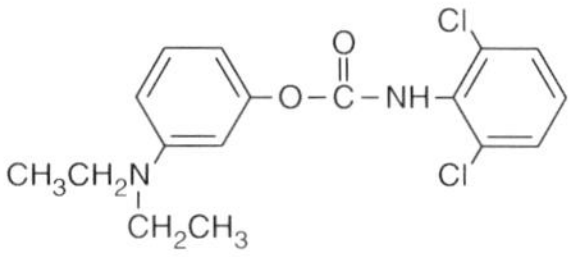

5 gm

S71,367-8
2-(DIETHYLAMINO)ETHYL N-(2,6-DIISOPROPYLPHENYL)CARBAMATE

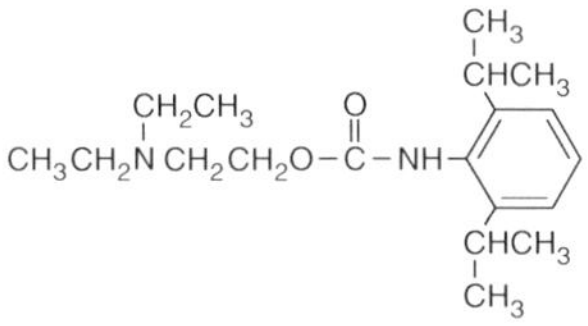

5 gm

S64,718-7
2-DIMETHYLAMINOETHYL N-(2,5-DICHLOROPHENYL)CARBAMATE

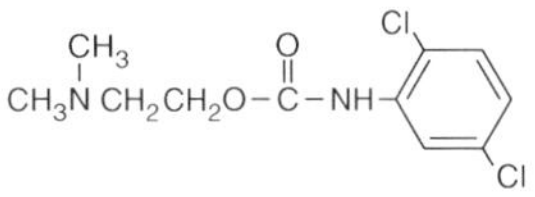

5 gm

S81,903-4
3-(DIMETHYLAMINO)-4-ISOPROPYLPHENYL N-METHYLCARBAMATE

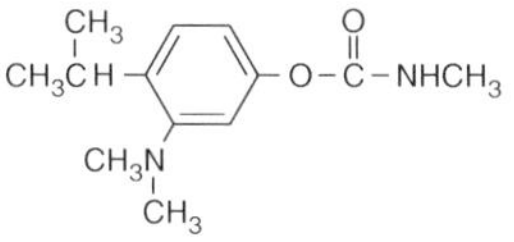

50 mg

S83,747-4
4-(DIMETHYLAMINO)-3-ISOPROPYLPHENYL N-METHYLCARBAMATE

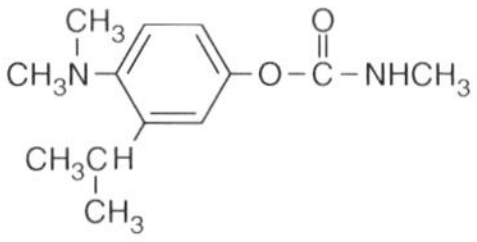

250 mg

S83,189-1
4-(DIMETHYLAMINO)-3-ISOPROPYLPHENYL N,N-DIMETHYLCARBAMATE

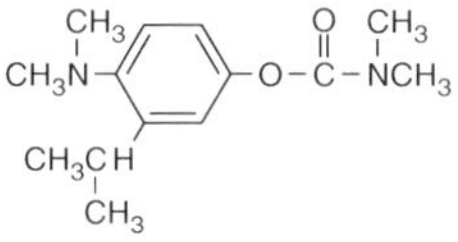

250 mg

S53,091-3
4-DIMETHYLAMINOMETHYL-3,5-XYLYL N-METHYLCARBAMATE

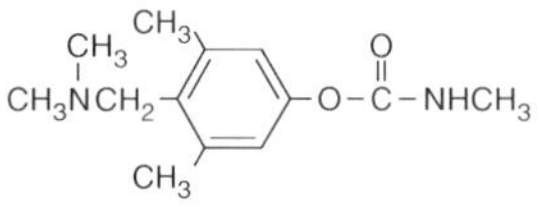

5 gm

S64,389-0
4,4'-METHYLENEBIS(2-METHYLCYCLOHEXYL N-PHENYLCARBAMATE)

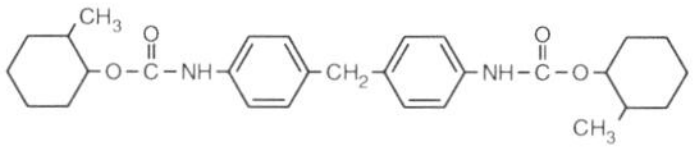

5 gm

S64,383-1
4,4'-METHYLENEBIS(CYCLOHEXYL N-PHENYLCARBAMATE)

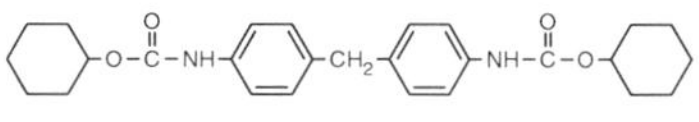

5 gm

S65,124-9
DIPHENYL 4,4'-METHYLENEDICARBANILATE

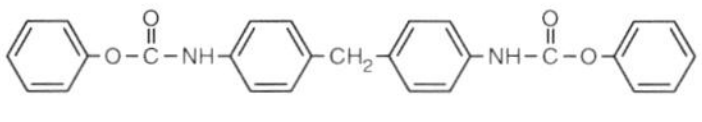

5 gm

S63,626-6
DIETHYL 4,4'-METHYLENEBIS(N-PHENYLCARBAMATE)

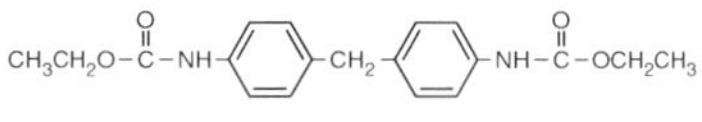

5 gm

S65,029-3
DIALLYL 4,4'-METHYLENEBIS(N-PHENYLCARBAMATE)

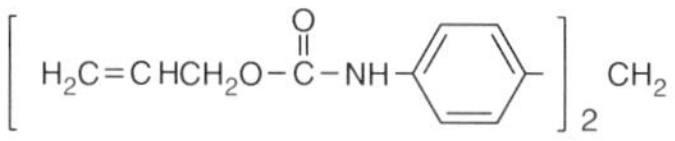

5 gm

S58,029-5
DIPROPYL 4,4'-METHYLENEBIS(N-PHENYLCARBAMATE)

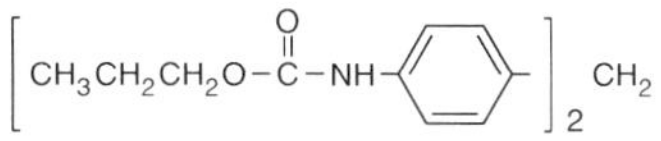

250 mg

S59,429-6
DIISOPROPYL 4,4'-METHYLENEBIS(N-PHENYLCARBAMATE)

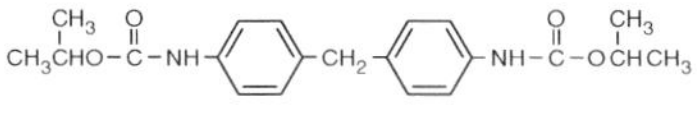

5 gm

S63,838-2
DIBUTYL 4,4'-METHYLENEBIS(N-PHENYLCARBAMATE)

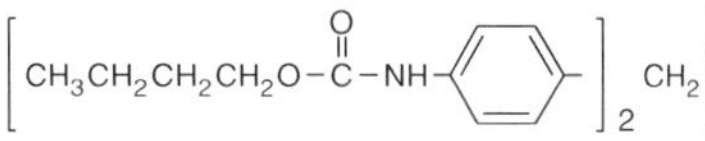

5 gm

S59,414-8
METHYL N-BENZYLCARBAMATE

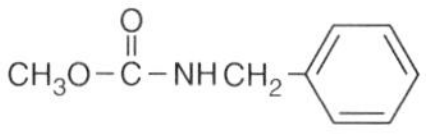

5 gm

S34,811-2
ETHYL N-BENZYLCARBAMATE

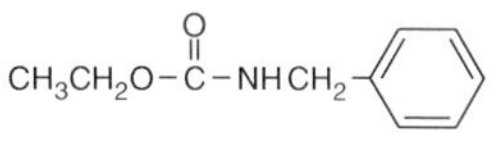

5 gm

S68,440-6
4-[3,5-BIS(METHOXYCARBONYL)PHENYL]-1-(ETHOXYCARBONYL)SEMICARBAZIDE

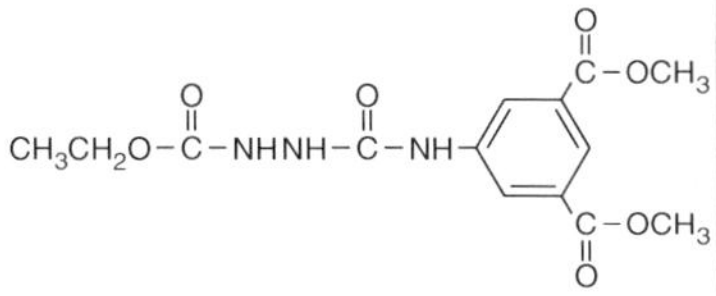

250 mg

S74,530-8
DIETHYL 9,10-DIHYDROANTHRACENE-9,10-BIIMINE-11,12-DICARBOXYLATE

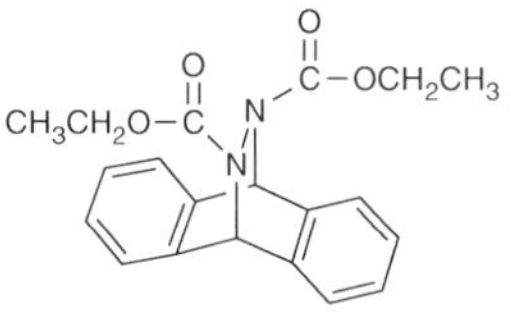

250 mg

S76,823-5
1-(ETHOXYCARBONYL)-4-(2,6-XYLYL)-SEMICARBAZIDE

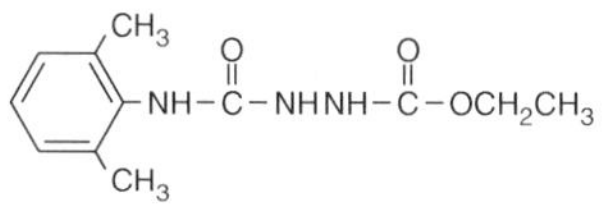

250 mg

S86,222-3
BENZALDEHYDE (TERT-BUTOXYCARBONYL)-HYDRAZONE
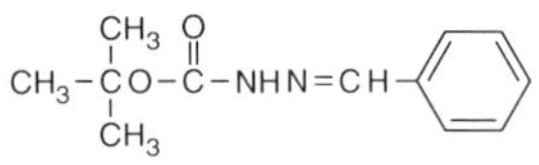
250 mg

S90,281-0
2-(3-PHENYLUREIDO)ETHYL N-METHYLCARBAMATE
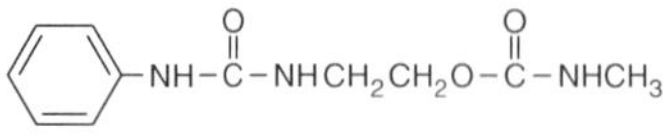
250 mg

S61,122-0
2-(3-PHENYLUREIDO)ETHYL N-PHENYLCARBAMATE
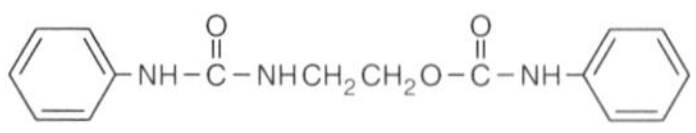
5 gm

S61,043-7
1-(3-PHENYLUREIDO)-2-PROPYL N-PHENYLCARBAMATE
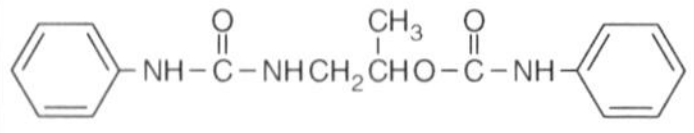
5 gm

S60,787-8
2-(1-ETHYL-3-PHENYLUREIDO)ETHYL N-PHENYLCARBAMATE
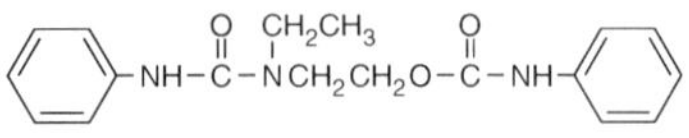
5 gm

S60,994-3
2-[3-(4-CHLOROPHENYL)-1-ETHYLUREIDO]ETHYL N-(4-CHLOROPHENYL)CARBAMATE
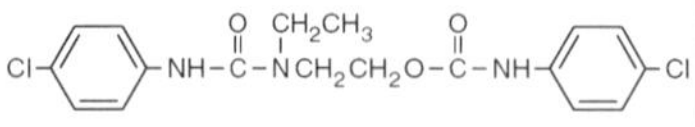
5 gm

S61,066-6
2-[3-PHENYLUREIDO)ETHYL N-(3,4-DICHLOROPHENYL]CARBAMATE
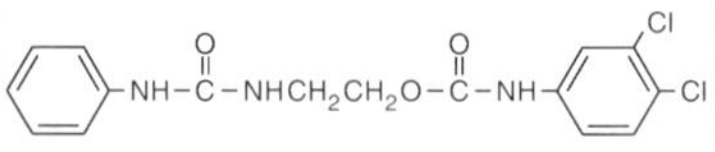
5 gm

S60,972-2
2-[3-(5-CHLORO-2-METHYLPHENYL)-1-METHYLUREIDO]ETHYL N-(3-CHLOROPHENYL)-CARBAMATE
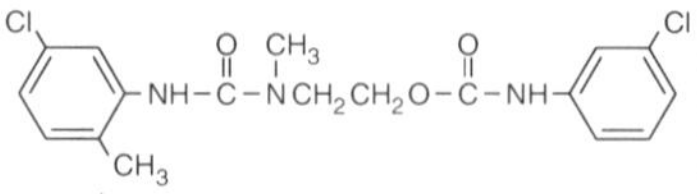
5 gm

S61,008-9
2-[3-(5-CHLORO-2-METHYLPHENYL)-1-METHYLUREIDO]ETHYL N-(5-CHLORO-2-METHYLPHENYL)CARBAMATE
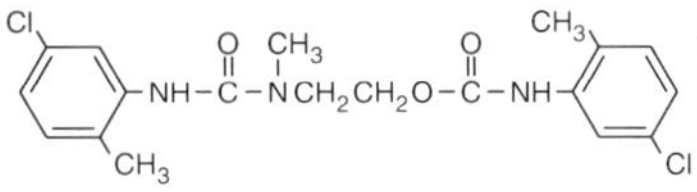
5 gm

S61,098-4
2-[3-(5-CHLORO-2-METHYLPHENYL)-1-ETHYLUREIDO]ETHYL N-(5-CHLORO-2-METHYL PHENYL)CARBAMATE
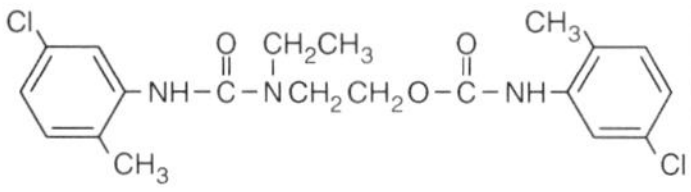
5 gm

S61,110-7
2-[3-(5-CHLORO-2-METHYLPHENYL]-UREIDO)ISOBUTYL N-(5-CHLORO-2-METHYLPHENYL)CARBAMATE
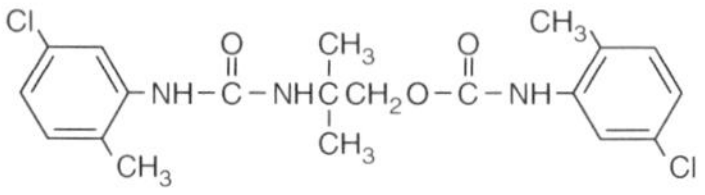
5 gm

S60,967-6
2-[1-ETHYL-3-(2,4-XYLYL)UREIDO]-ETHYL N-(2,4-XYLYL)CARBAMATE
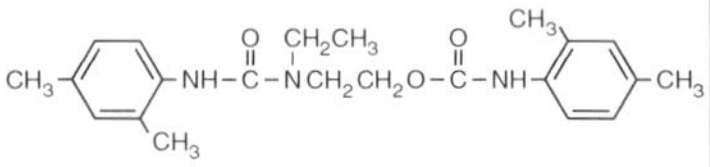
5 gm

S85,750-5
ETHYL N-HYDROXY-N-PHENYLCARBAMATE
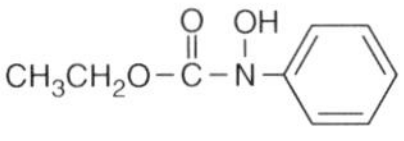
250 mg

S85,751-3
ETHYL N-METHOXY-N-PHENYLCARBAMATE
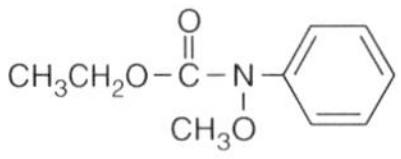
250 mg

S85,755-6
ISOPROPYL N-HYDROXY-N-PHENYLCARBAMATE
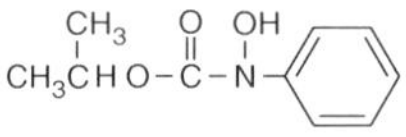
250 mg

S85,764-5
PROPYL N-HYDROXY-N-PHENYLCARBAMATE
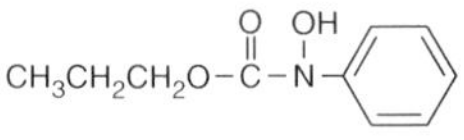
250 mg

S85,756-4
ISOPROPYL N-METHOXY-N-PHENYLCARBAMATE
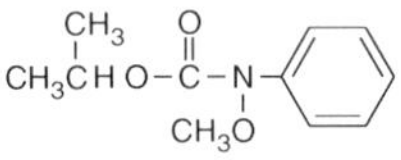
250 mg

S85,765-3
PROPYL N-METHOXY-N-PHENYLCARBAMATE
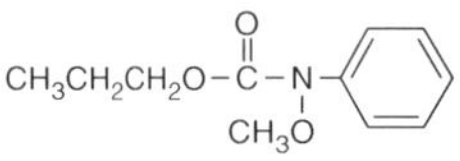
250 mg

S78,746-9
T-BOC-L-GLUTAMIC ACID GAMMA-(4-NITROBENZYL) ESTER DICYCLOHEXYLAMINE SALT
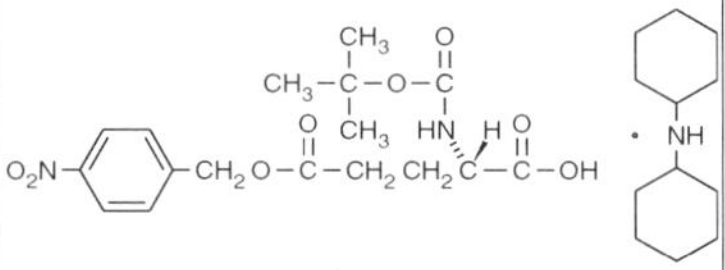
250 mg

S78,751-5
CARBOBENZYLOXY-L-ALANINE 4-NITROBENZYL ESTER
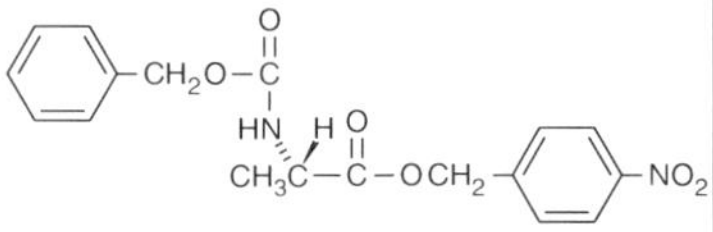
250 mg

S78,765-5
N-ALPHA-NEPSILON-BIS(TERT-BUTOXYCARBONYL)-L-LYSINE 4-NITROPHENYL ESTER
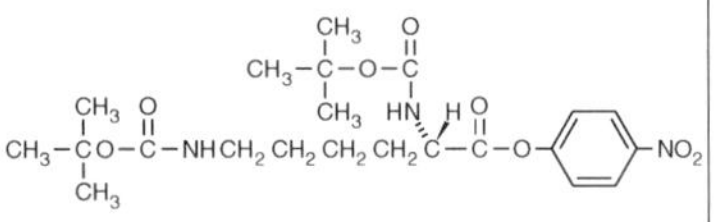
250 mg

S78,265-3
CARBOBENZYLOXY-L-VALINE 4-NITROPHENYL ESTER
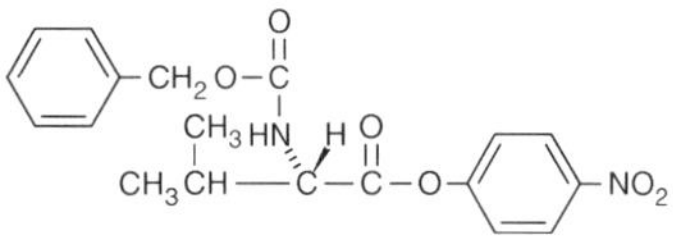
5 gm

S78,616-0
N-ALPHA-TERT-BUTOXYCARBONYL-NEPSILON-CARBOBENZYLOXY-L-LYSINE 4-NITROPHENYL ESTER
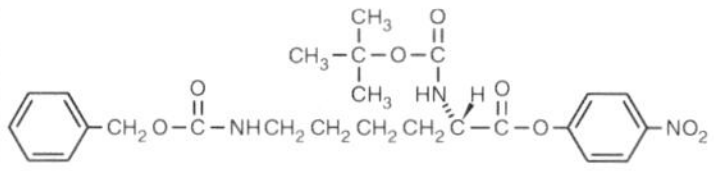
250 mg

S67,884-8
CARBOBENZYLOXY-L-ASPARAGINE 4-NITROPHENYL ESTER
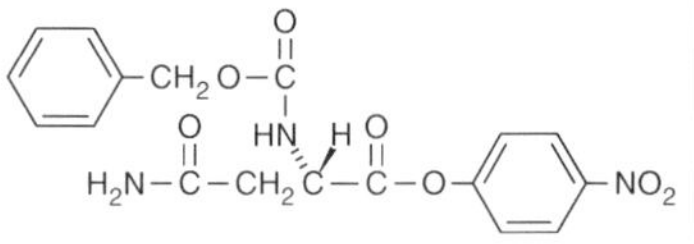
5 gm

S68,742-1
3-NITROPHENYL N-BUTYLCARBAMATE
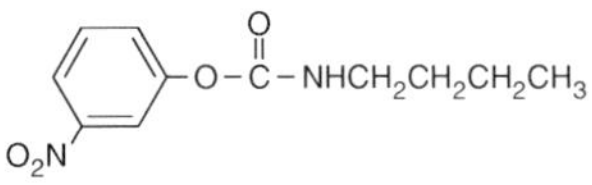
5 gm

S89,045-6
3-METHYL-4-NITROPHENYL N-(1-NAPHTHYL)CARBAMATE
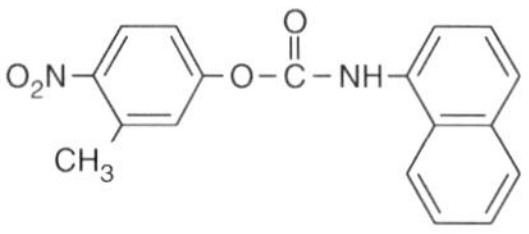
250 mg

S71,363-5
2,6-DIMETHYL-4-NITROPHENYL N-(2,6-DICHLOROPHENYL)CARBAMATE
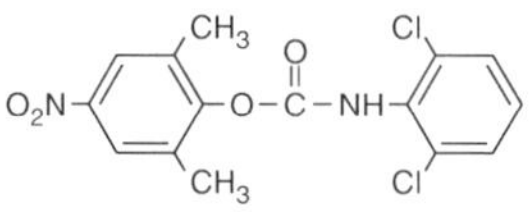
5 gm

S67,462-1
2-FORMYLPHENYL N-METHYLCARBAMATE
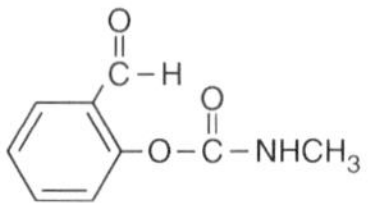
250 mg

S88,956-3
4-FORMYLPHENYL N-PHENYLCARBAMATE
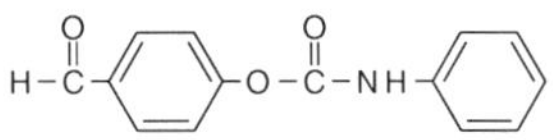
250 mg

S90,165-2
2-FORMYLPHENYL N-BUTYLCARBAMATE
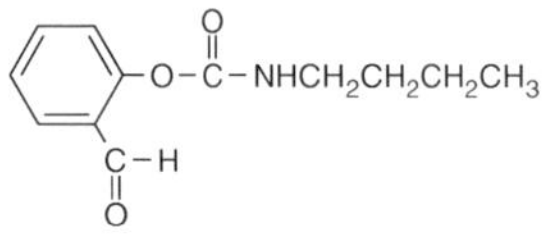
250 mg

S86,821-3
4-(METHOXYCARBONYL)PHENYL N-(3-ACETYLPHENYL)CARBAMATE
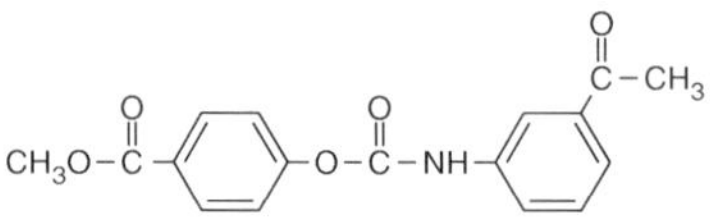
250 mg

S77,273-9
4-ACETYLPHENYL N-[4-(ETHOXYCARBONYL)PHENYL]-CARBAMATE
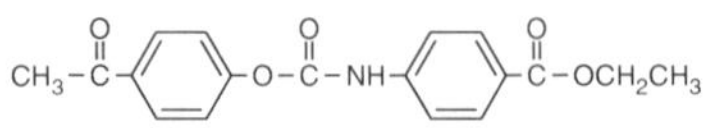
250 mg

S86,824-8
4-ACETAMIDOPHENYL N-(3-ACETYLPHENYL)CARBAMATE

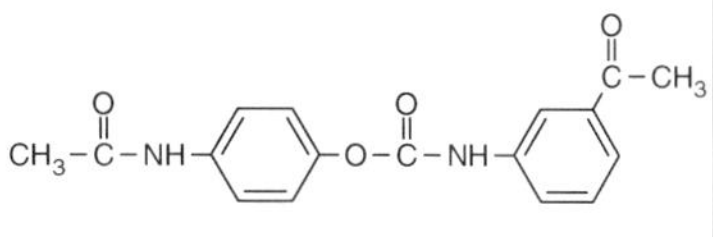

250 mg

S79,467-8
CYCLOPENTYL N-(3-ACETYLPHENYL)-CARBAMATE

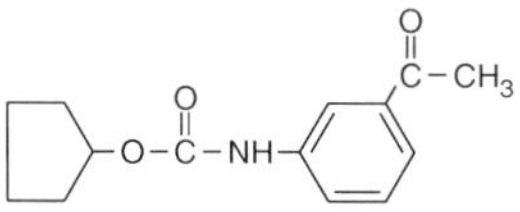

250 mg

S86,833-7
5-NORBORNEN-2-YLMETHYL N-(3-ACETYLPHENYL)CARBAMATE

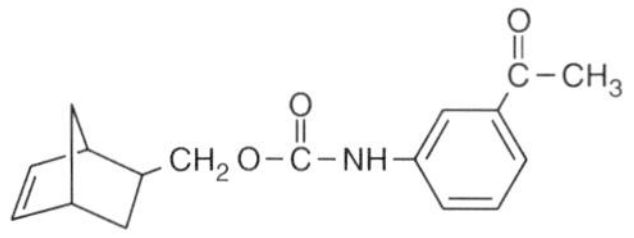

250 mg

S86,813-2
3-METHYLCYCLOHEXYL N-(3-ACETYLPHENYL)CARBAMATE

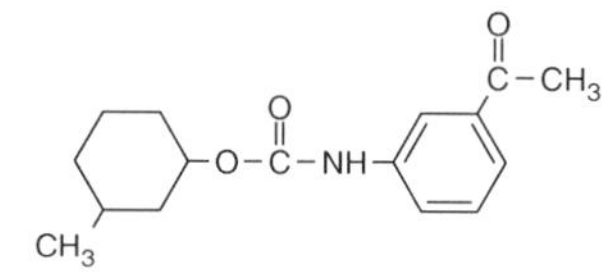

250 mg

S86,866-3
4-TERT-BUTYLCYCLOHEXYL N-(3-ACETYLPHENYL)CARBAMATE

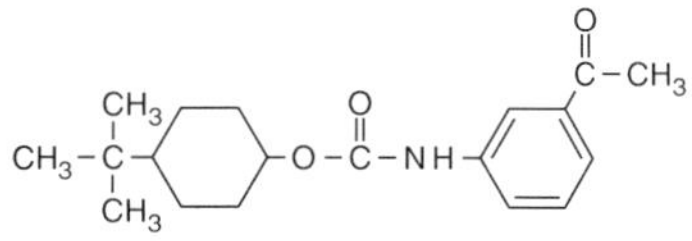

250 mg

S85,527-8
4-PROPIONYLPHENYL N-(4-METHYLCYCLOHEXYL)CARBAMATE

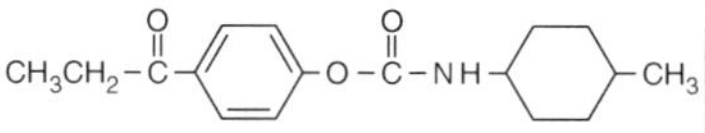

250 mg

S79,437-6
3-PHENYLPROPYL N-(3-ACETYLPHENYL)-CARBAMATE

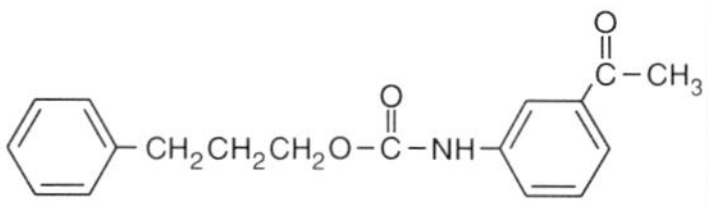

250 mg

S63,629-0
2-BENZOYL-4-CHLOROPHENYL N-(2-CHLOROPHENYL)CARBAMATE

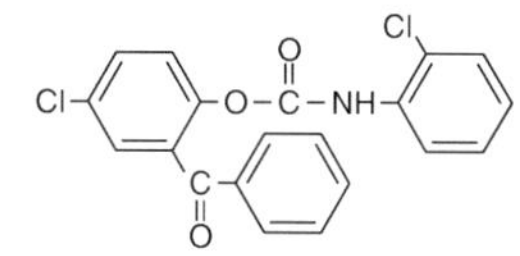

5 gm

S85,507-3
4-ACETYLPHENYL N-METHYLCARBAMATE

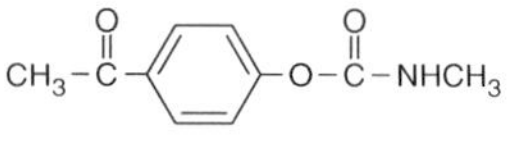

250 mg

S85,505-7
4-PROPIONYLPHENYL N-METHYLCARBAMATE

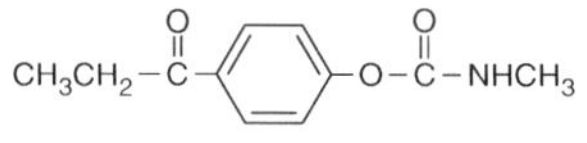

250 mg

S73,890-5
4-PROPIONYLPHENYL N-(P-TOLYL)-CARBAMATE

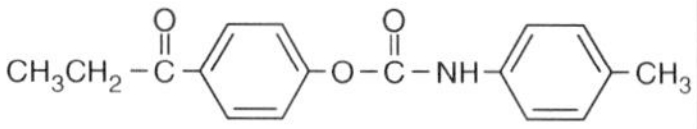

5 gm

S64,427-7
4-BENZOYLPHENYL N-(3,4-DICHLOROPHENYL)CARBAMATE

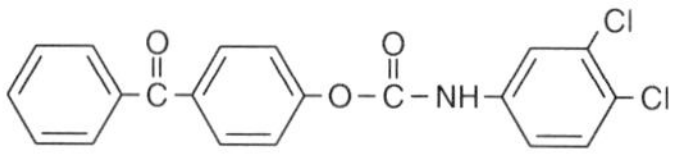

250 mg

S59,405-9
METHYL N-(4-BENZOYLPHENYL)CARBAMATE

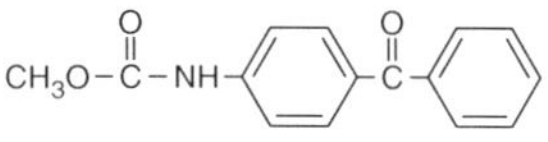

5 gm

S63,622-3
4-BENZOYLPHENYL N-PHENYLCARBAMATE

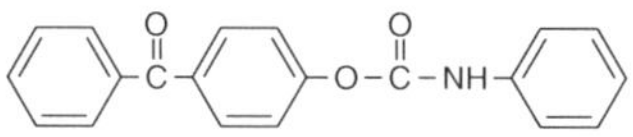

5 gm

S79,441-4
2,6-XYLYL N-(3-ACETYLPHENYL)-CARBAMATE

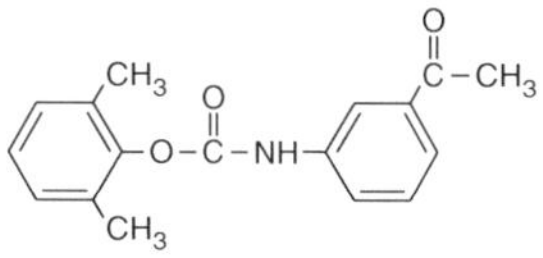

250 mg

S63,828-5
2-BENZOYL-4-CHLOROPHENYL N-(3-CHLOROPHENYL)CARBAMATE

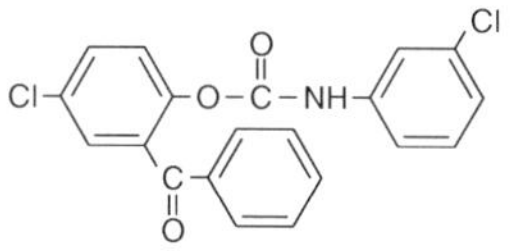

5 gm

S61,364-9
2-BENZOYL-4-CHLOROPHENYL N-(2,4-DICHLOROPHENYL)CARBAMATE

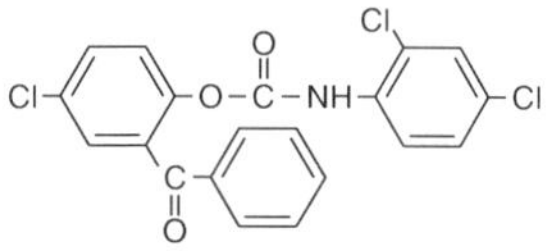

5 gm

S63,800-5
2-BENZOYL-4-CHLOROPHENYL N-(2,4,5-TRICHLOROPHENYL)CARBAMATE

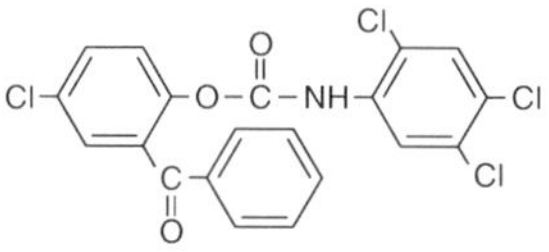

5 gm

S86,859-0
2-NAPHTHYL N-(3-ACETYLPHENYL)-CARBAMATE

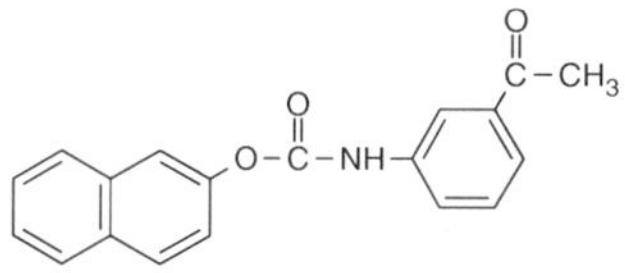

250 mg

S79,730-8
4-(METHOXYCARBONYL)PHENYL N-(BUTOXYCARBONYLOXYMETHYL)-CARBAMATE

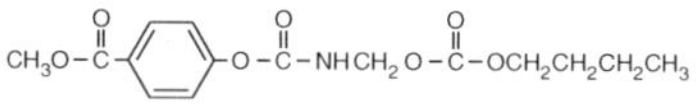

250 mg

S79,766-9
4-ACETAMIDOPHENYL N-[4-(ETHOXYCARBONYL)PHENYL]-CARBAMATE

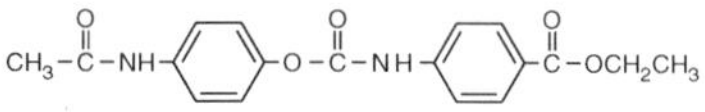

250 mg

S79,767-7
3-ACETAMIDOPHENYL N-[4-(ETHOXYCARBONYL)PHENYL]-CARBAMATE

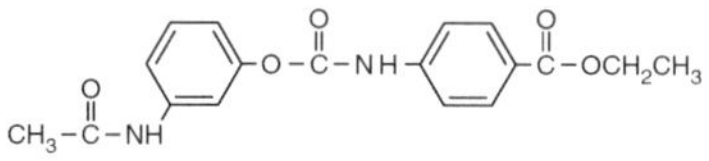

250 mg

S79,458-9
CYCLOPENTYL N-[4-(ETHOXYCARBONYL)-PHENYL]CARBAMATE

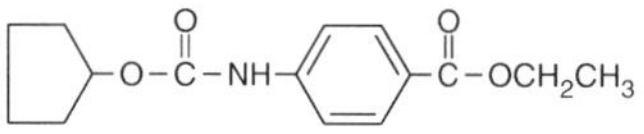

250 mg

S79,756-1
3-METHYLCYCLOHEXYL N-[4-(ETHOXYCARBONYL)PHENYL]-CARBAMATE

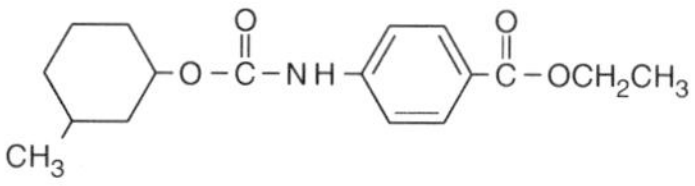

250 mg

S86,856-6
4-(BUTOXYCARBONYL)PHENYL N-PHENYLCARBAMATE

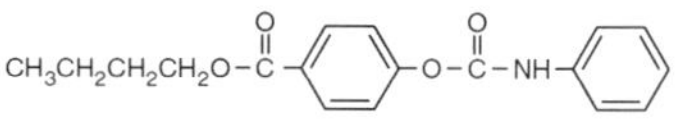

250 mg

S79,417-1
4-BIPHENYLYL N-[4-(ETHOXYCARBONYL)-PHENYL]CARBAMATE

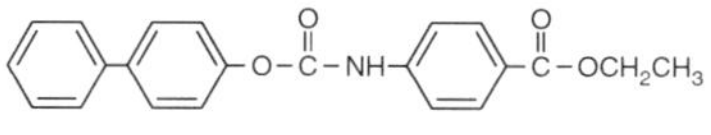

250 mg

S84,597-3
4-(ETHOXYCARBONYL)PHENYL N-(2-ETHYLPHENYL)CARBAMATE

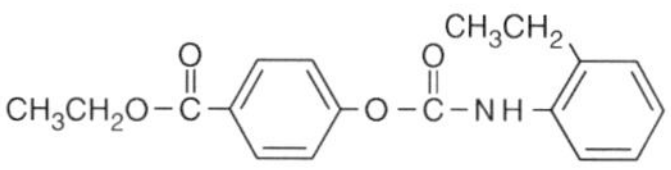

250 mg

S79,440-6
2-(METHOXYCARBONYL)PHENYL N-[4-(ETHOXYCARBONYL)PHENYL]-CARBAMATE

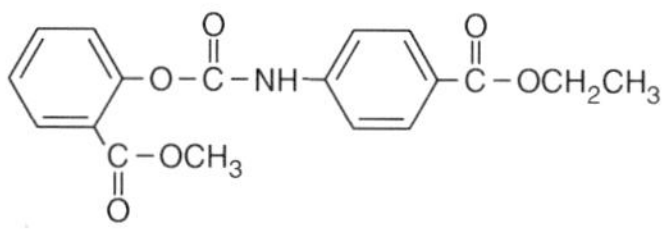

250 mg

S79,429-5
2-ISOPROPYLPHENYL N-[4-(ETHOXYCARBONYL)PHENYL]-CARBAMATE

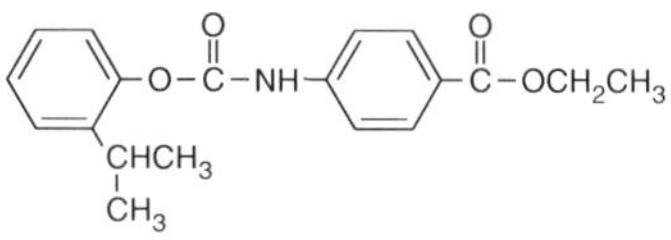

250 mg

S79,786-3
2-TERT-BUTYLPHENYL N-[4-(ETHOXYCARBONYL)PHENYL]-CARBAMATE

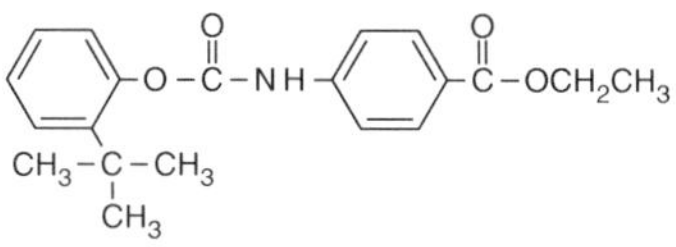

250 mg

S85,513-8
4-(BUTOXYCARBONYL)PHENYL N-METHYLCARBAMATE

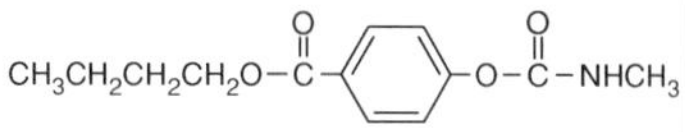

250 mg

S79,707-3
2-BUTENYL N-[4-(ETHOXYCARBONYL)-PHENYL]CARBAMATE

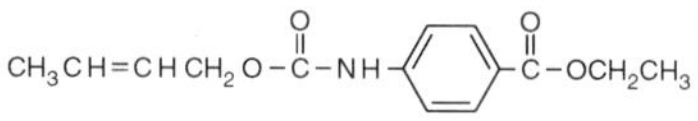

250 mg

S90,560-7
4-(BUTOXYCARBONYL)PHENYL N-PROPYLCARBAMATE

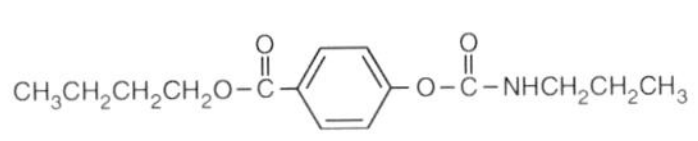

250 mg

S68,353-1
4-(METHOXYCARBONYL)PHENYL N-(3-ETHYLPHENYL)CARBAMATE

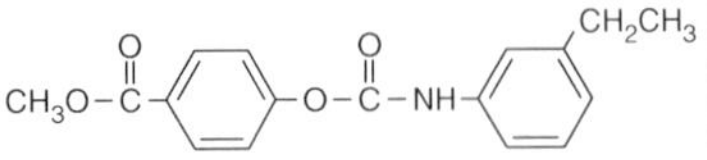

250 mg

S79,762-6
4-(METHOXYCARBONYL)PHENYL N-[4-(ETHOXYCARBONYL)PHENYL] CARBAMATE

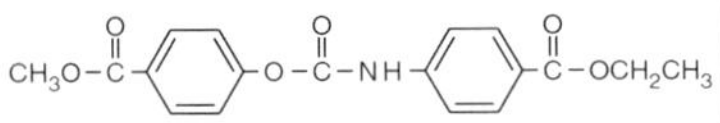

250 mg

S73,663-5
N,N'-HEXAMETHYLENEBIS[4-(METHOXYCARBONYL)PHENYL CARBAMATE]

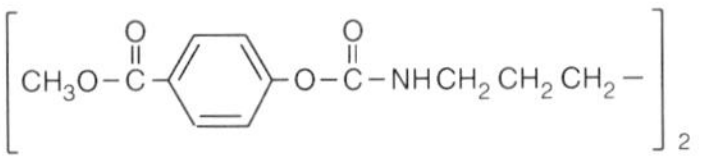

5 gm

S84,739-9
4-(BUTOXYCARBONYL)PHENYL N-(2,6-XYLYL)CARBAMATE

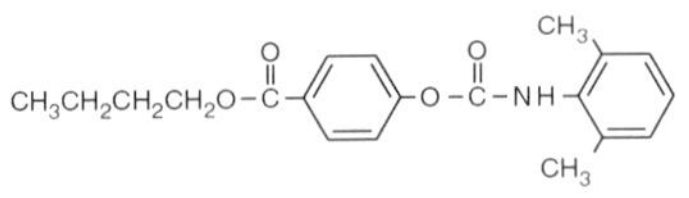

250 mg

S79,769-3
2,5-XYLYL N-[4-(ETHOXYCARBONYL)-PHENYL]CARBAMATE

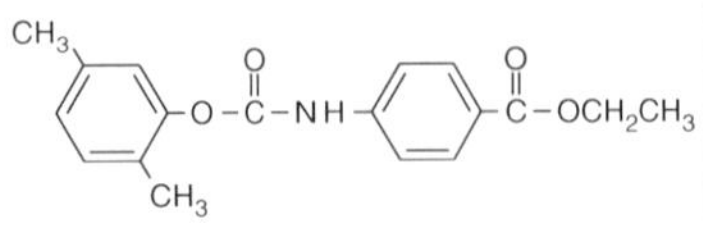

250 mg

S84,588-4
4-(ETHOXYCARBONYL)PHENYL N-(2,5-XYLYL)CARBAMATE

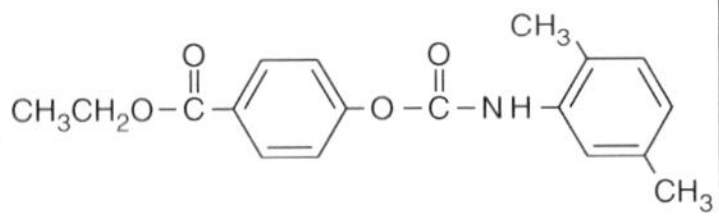

250 mg

S79,420-1
5-ISOPROPYL-2-METHYLPHENYL N-[4-(ETHOXYCARBONYL)PHENYL]-CARBAMATE

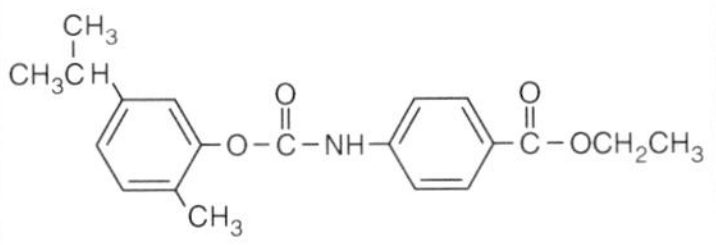

250 mg

S79,785-5
2-ISOPROPYL-5-METHYLPHENYL N-[4-(ETHOXYCARBONYL)PHENYL]-CARBAMATE

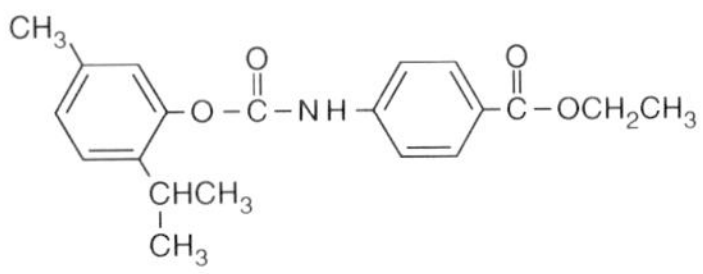

250 mg

S84,738-0
4-(BUTOXYCARBONYL)PHENYL N-(2,5-XYLYL)CARBAMATE

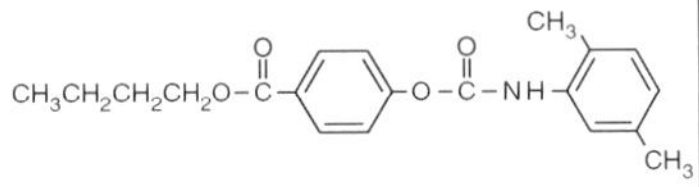

250 mg

S79,781-2
2-NAPHTHYL N-[4-(ETHOXYCARBONYL)-PHENYL]CARBAMATE

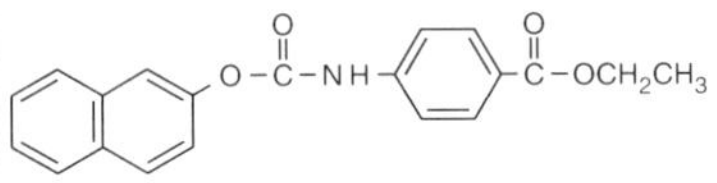

250 mg

S76,985-1
5-INDANYL N-[3-(ETHOXYCARBONYL)-PHENYL]CARBAMATE

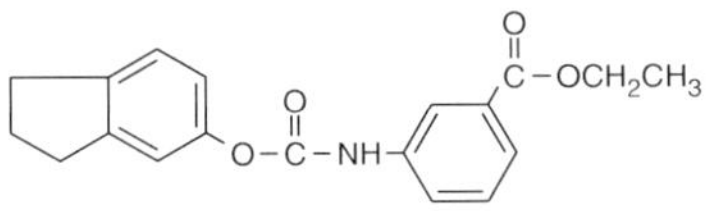

250 mg

S84,811-5
S,S'-ETHYLENE BIS(N-[4-(ETHOXYCARBONYL)PHENYL]-THIOCARBAMATE)

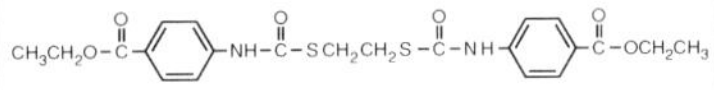

250 mg

S92,667-1
METHYL N-(N-BENZOYLCARBAMOYL)-CARBAMATE

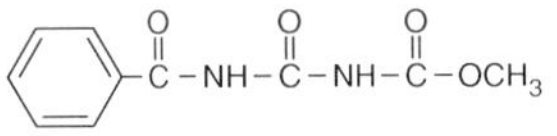

250 mg

S71,585-9
4-METHYL-2-OXO-2H-1-BENZOPYRAN-7-YL N-METHYLCARBAMATE

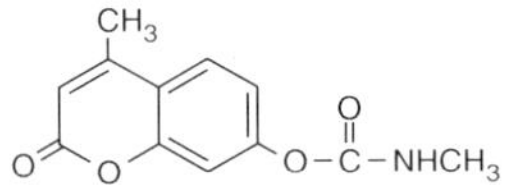

250 mg

S69,184-4
PIPERIDINO N-[4-(ETHOXYCARBONYL)-PHENYL]CARBAMATE

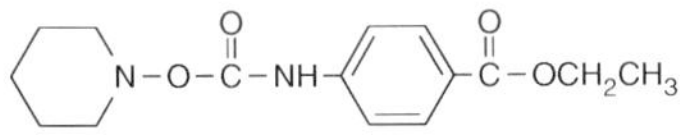

250 mg

S71,645-6
4-(BUTOXYCARBONYL)PHENYL N-PHENYLCARBAMATE

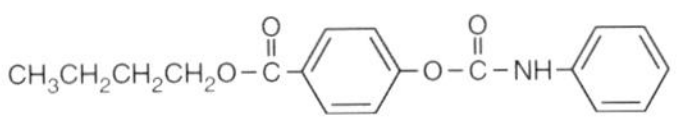

250 mg

S75,100-6
4-(PROPOXYCARBONYL)PHENYL N-METHYLCARBAMATE

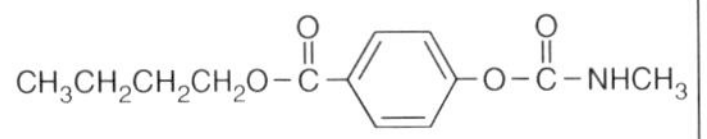

250 mg

S79,675-1
S-(2-TOLYL) N-[4-(ETHOXYCARBONYL)-PHENYL]THIOCARBAMATE

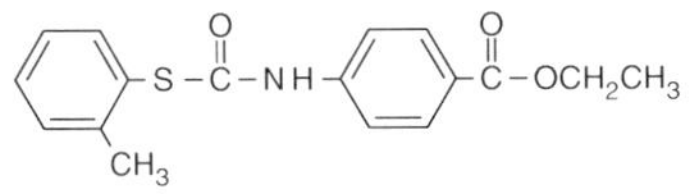

250 mg

S84,803-4
S-(4-TERT-BUTYLPHENYL) N-[4-(ETHOXYCARBONYL)PHENYL]-THIOCARBAMATE

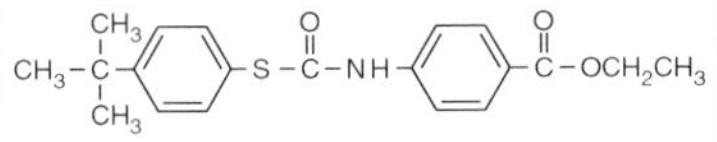

250 mg

S84,812-3
S-DODECYL N-[4-(ETHOXYCARBONYL)-PHENYL]THIOCARBAMATE

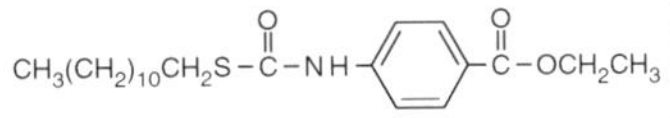

250 mg

S84,809-3
S-(4-TERT-BUTYL-2-METHYLPHENYL) N-[4-(ETHOXYCARBONYL)PHENYL]-THIOCARBAMATE

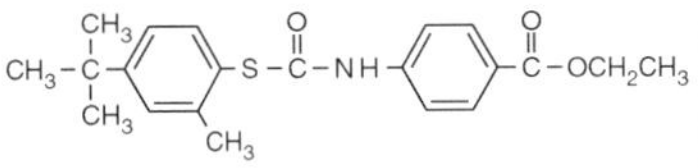

250 mg

S78,386-2
N,N'-BIS(CARBOBENZYLOXY)-L-LYSYL-L-ASPARTIC ACID DIBENZYL ESTER

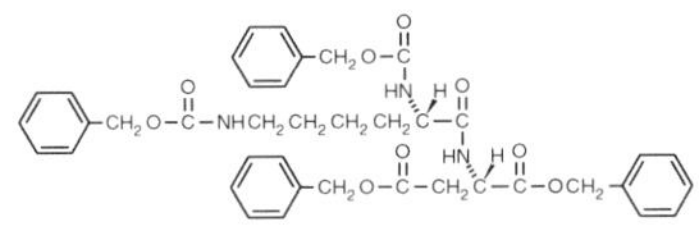

5 gm

S78,864-3
CARBOBENZYLOXY-L-GLUTAMIC ACID GAMMA-BENZYL ESTER 2-NAPHTHYLAMIDE

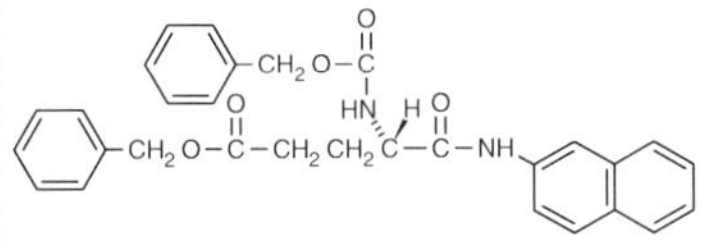

250 mg

S56,750-7
CARBOBENZYLOXY-DL-ALANYLGLYCINE

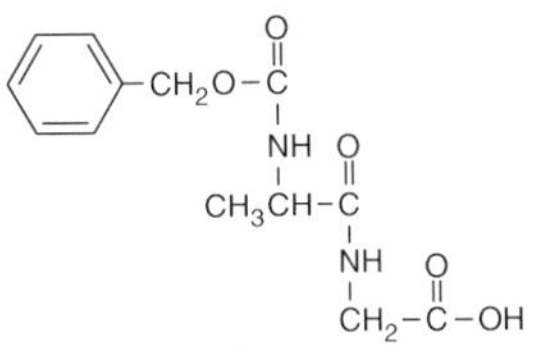

5 gm

S67,889-9
CARBOBENZYLOXY-L-ALANYLGLYCYLGLYCINE

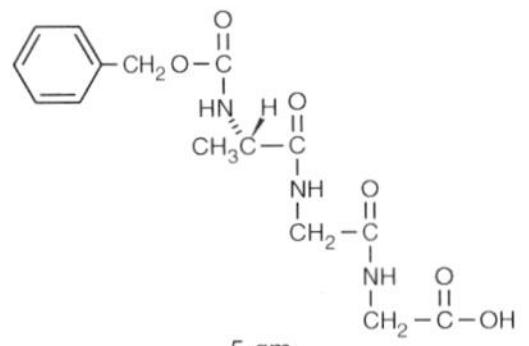

5 gm

S67,877-5
CARBOBENZYLOXY-L-VALYL-L-ALANINE

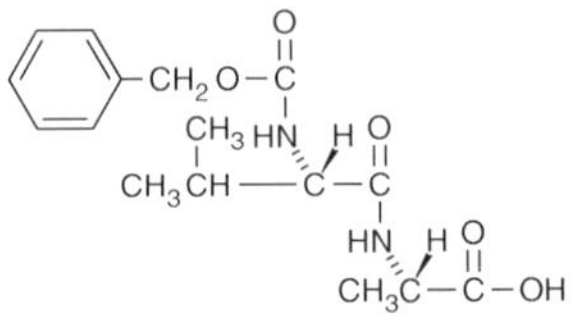

5 gm

S67,886-4
CARBOBENZYLOXY-L-VALYLGLYCYLGLYCINE

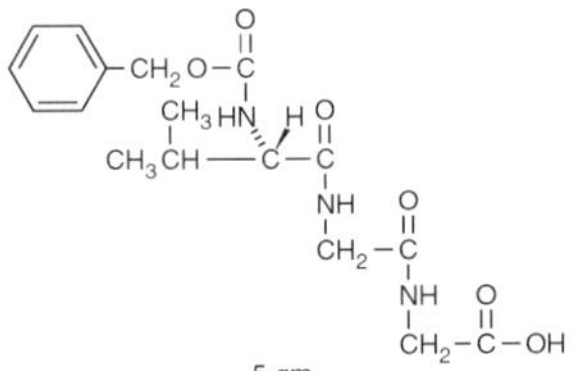

5 gm

S78,646-2
CARBOBENZYLOXYGLYCYLGLYCYL-L-VALINE

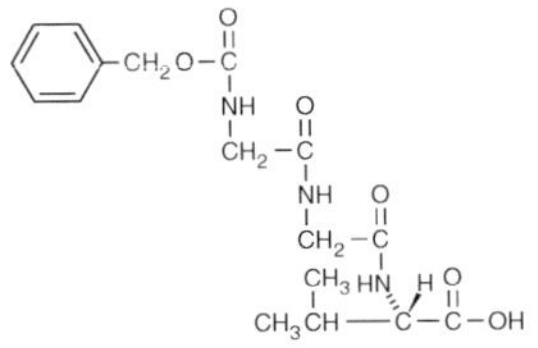

250 mg

S67,888-0
CARBOBENZYLOXY-L-LEUCYLGLYCYLGLYCINE

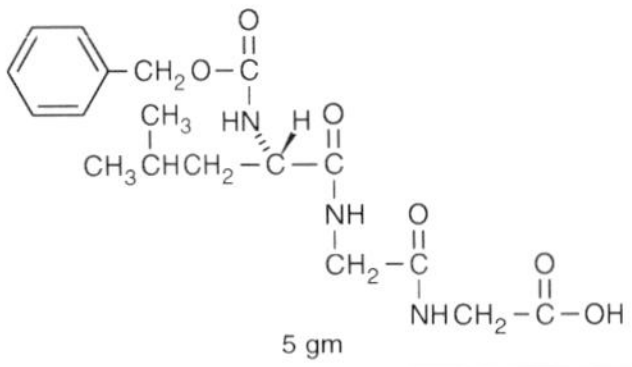

5 gm

S67,880-5
CARBOBENZYLOXY-L-VALYL-L-ISOLEUCINE

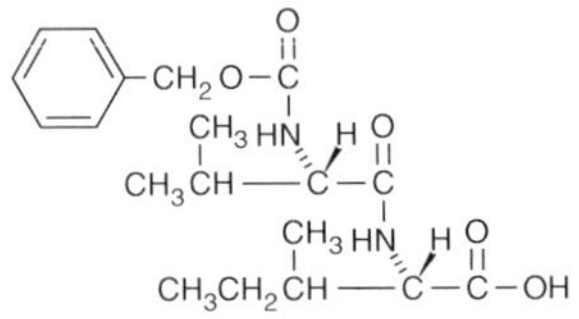

5 gm

S67,906-2
CARBOBENZYLOXY-L-LEUCINE DICYCLOHEXYLAMINE SALT

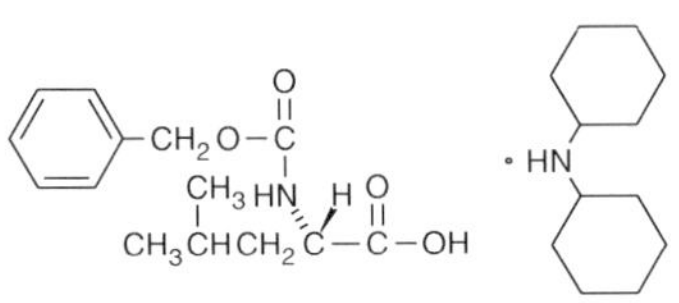

5 gm

S67,912-7
CARBOBENZYLOXY-GLYCYL-GLYCINE ETHYL ESTER
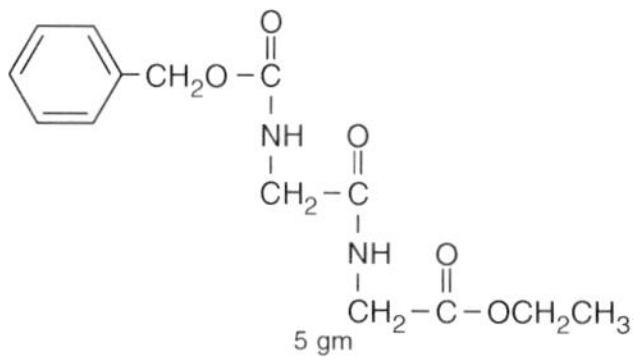
5 gm

S78,778-7
N-(TERT-BUTOXYCARBONYL)-L-VALYL-L-ASPARTIC ACID DIBENZYL ESTER
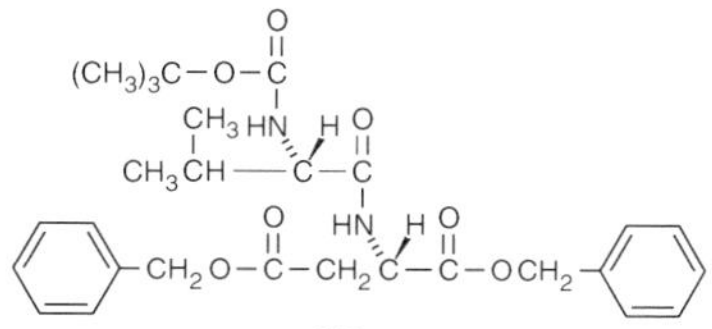
250 mg

S76,596-1
CARBOBENZYLOXY-BETA-ALANYL-L-VALINE METHYL ESTER
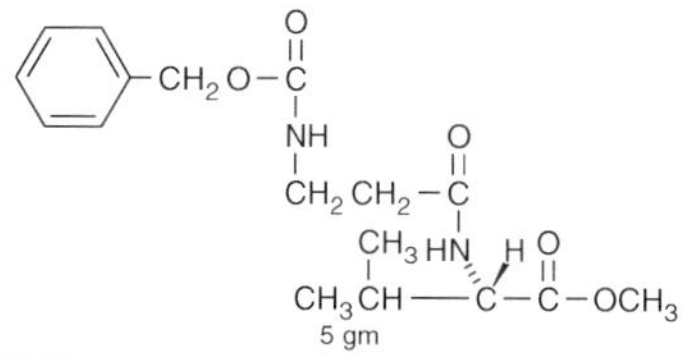
5 gm

S78,585-7
CARBOBENZYLOXYGLYCYLSARCOSINE METHYL ESTER
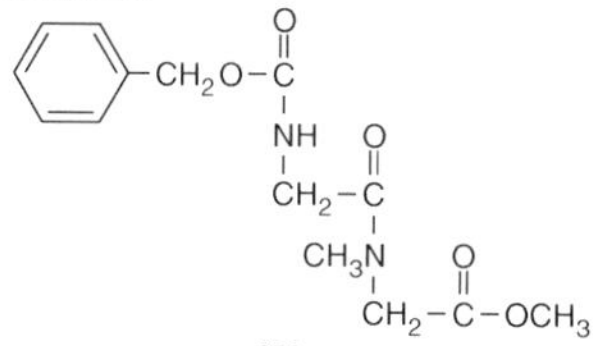
250 mg

S78,277-7
CARBOBENZYLOXY-L-ALANYL-L-VALINE BENZYL ESTER
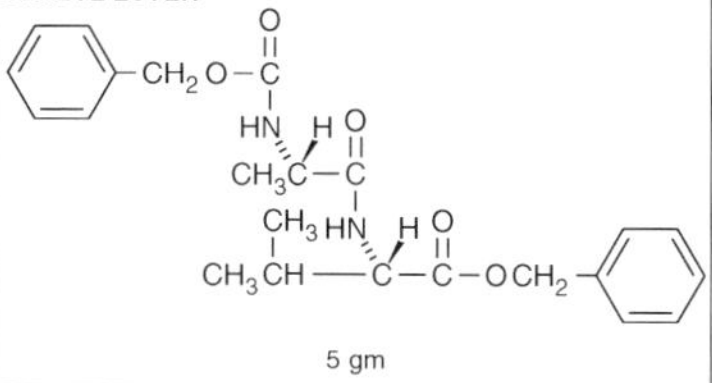
5 gm

S89,487-7
ETHYL N-[2,4-DIMETHYL-ALPHA-(ETHOXYCARBONYL)BENZYL]CARBAMATE
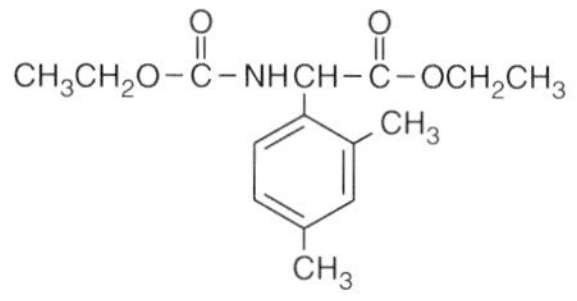
250 mg

S63,818-8
N,N'-HEXAMETHYLENEBIS(2-METHYLCYCLOHEXYL CARBAMATE)
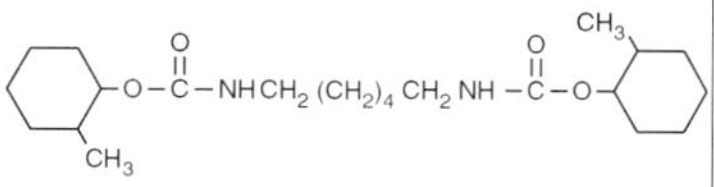
5 gm

S44,137-6
DIPHENYL AZODICARBOXYLATE
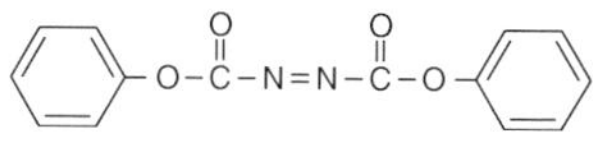
5 gm

S67,632-2
N,N'-HEXAMETHYLENEBIS(4-TERT-BUTYLPHENYL CARBAMATE)
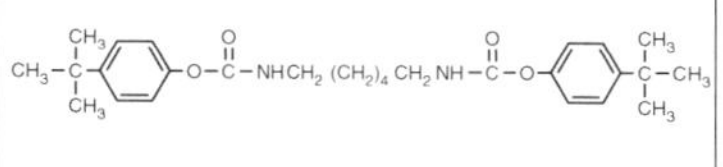
250 mg

S41,616-9
CARBOBENZYLOXY-L-GLUTAMIC ANHYDRIDE
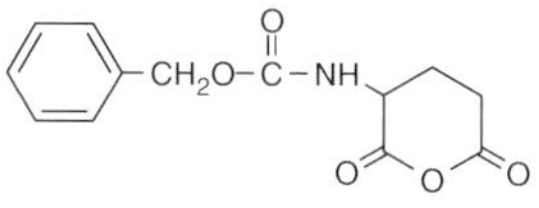
250 mg

S85,598-7
METHYL N-METHYL-N-(PHENYLCARBAMOYL)-CARBAMATE
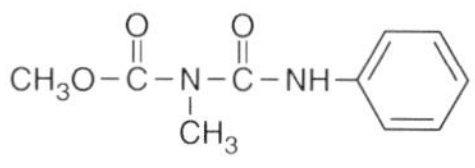
250 mg

S85,637-1
ETHYL N-METHYL-N-(N-PHENYLCARBAMOYL)CARBAMATE
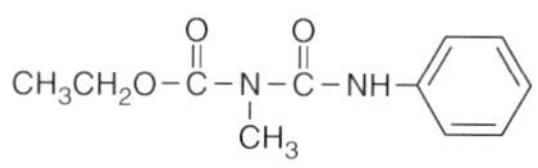
250 mg

S71,676-6
3-ACETAMIDOPHENYL N-(TERT-BUTYL)-CARBAMATE
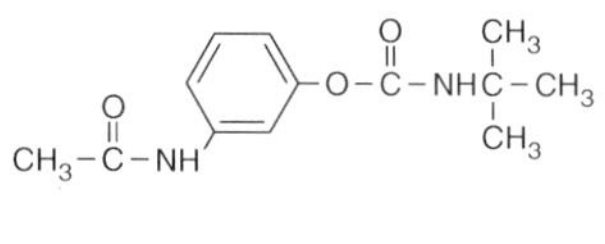
5 gm

S55,601-7
SEC-BUTYL CARBANILATE
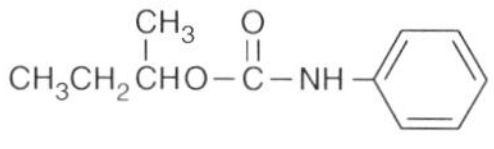
5 gm

S55,551-7
4-(TERT-BUTYL)PHENYL 2-METHYLCARBANILATE
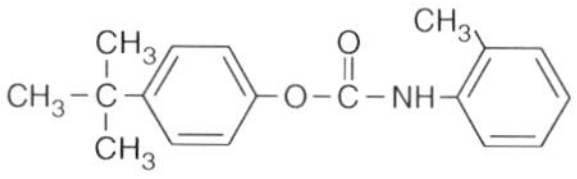
5 gm

S55,520-7
2,6-XYLYL 2-METHYLCARBANILATE
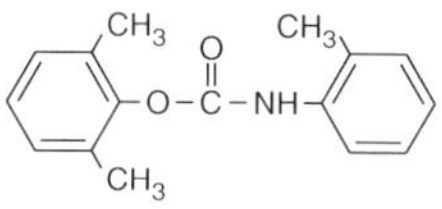
5 gm

S55,540-1
2-NAPHTHYL 2'-METHYLCARBANILATE
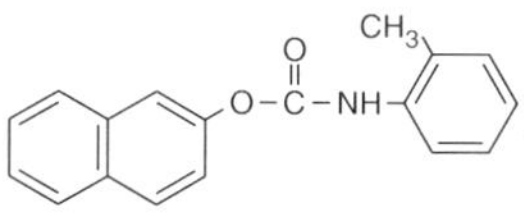
5 gm

S55,519-3
4-(TERT-BUTYL)PHENYL N-METHYLCARBAMATE
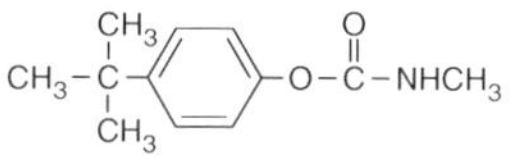
250 mg

S91,022-8
4-ACETAMIDOPHENYL N-ISOPROPYLCARBAMATE
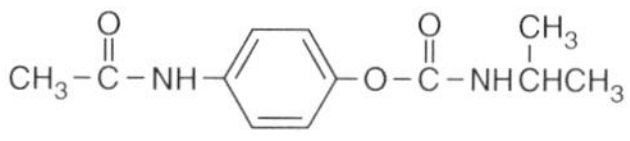
250 mg

S54,371-3
4-CHLOROPHENYL CARBANILATE
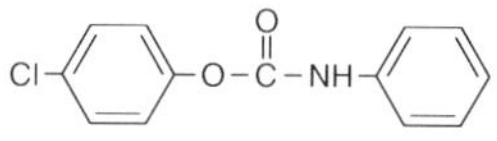
5 gm

S68,613-1
4-ACETAMIDOPHENYL N-(3-ETHYLPHENYL)-CARBAMATE
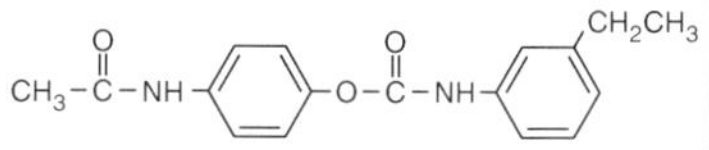
5 gm

S73,813-1
4-STEAROYLAMINOPHENYL N-METHYLCARBAMATE
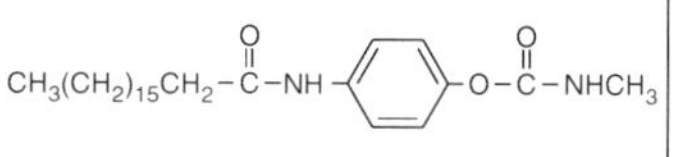
5 gm

S84,587-6
4-ACETAMIDOPHENYL N-(2,6-XYLYL)-CARBAMATE
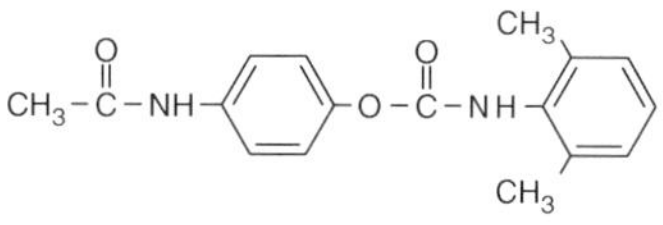
250 mg

S77,382-4
4-ACETAMIDOPHENYL N-(2,5-XYLYL)-CARBAMATE
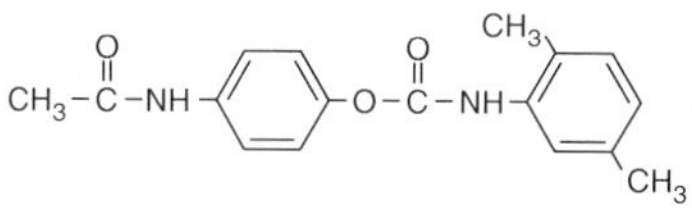
250 mg

S89,046-4
4-ACETAMIDOPHENYL N-(1-NAPHTHYL)-CARBAMATE
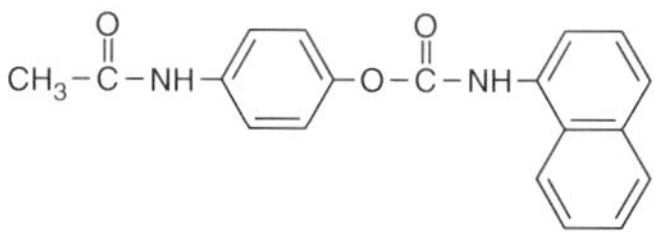
250 mg

S89,886-4
1-CHLOROETHYL N-METHYL-N-PHENYLCARBAMATE
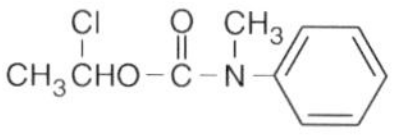
250 mg

S89,880-5
1-CHLOROETHYL N-ETHYL-N-PHENYLCARBAMATE
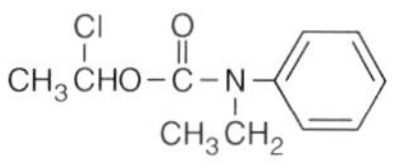
250 mg

S53,893-0
ISOBUTYL 3-PHENYLCARBAZATE
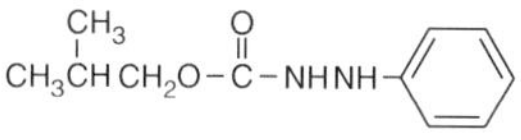
250 mg

S59,432-6
METHYL N-(3-CHLOROPHENYL)CARBAMATE
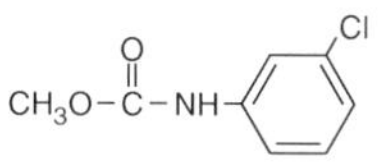
5 gm

S60,687-1
METHYL N-(M-TOLYL)CARBAMATE
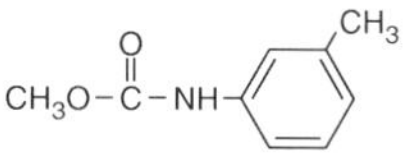
5 gm

S76,317-9
3-ETHYLPHENYL N-METHYLCARBAMATE
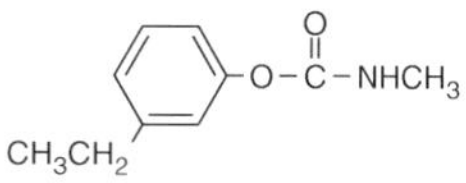
250 mg

S62,888-3
ALLYL 3'-CHLOROCARBANILATE
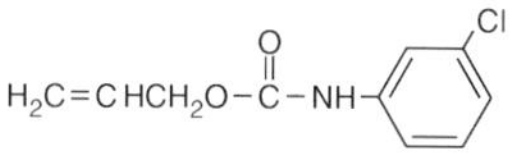
5 gm

S59,425-3
PROPYL N-(3-CHLOROPHENYL)CARBAMATE

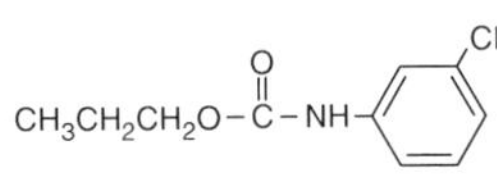

5 gm

S66,689-0
ALLYL N-(M-TOLYL)CARBAMATE

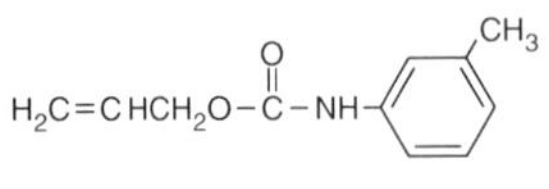

5 gm

S60,640-5
ISOPROPYL N-(M-TOLYL)CARBAMATE

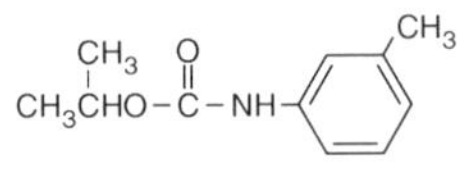

5 gm

S60,699-5
PROPYL N-(M-TOLYL)CARBAMATE

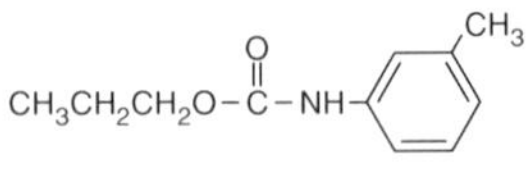

5 gm

S70,629-9
3-ISOPROPYLPHENYL N-METHYLCARBAMATE

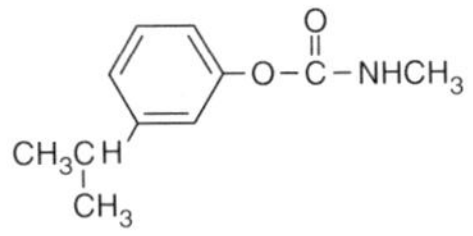

250 mg

S89,214-9
M-TOLYL N-PROPYLCARBAMATE

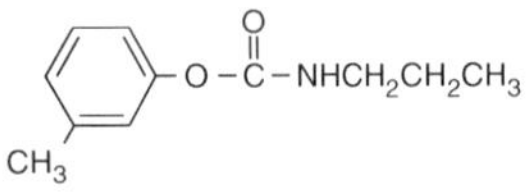

250 mg

S62,419-5
PHENYL N-(3-CHLOROPHENYL)CARBAMATE

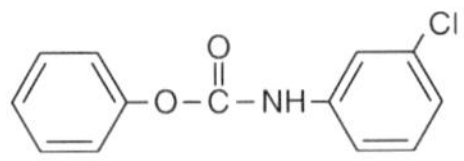

5 gm

S62,524-8
PHENYL N-(M-TOLYL)CARBAMATE

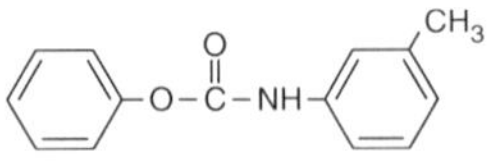

5 gm

S80,793-1
M-TOLYL N-(M-TOLYL)CARBAMATE

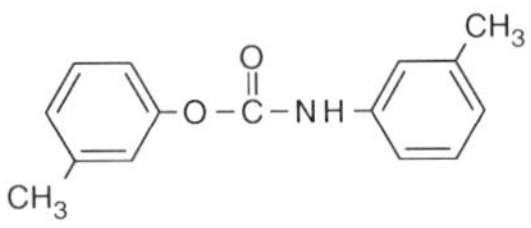

250 mg

S77,289-5
OCTAHYDRO-2,5-METHANO-2H-INDENO-(1,2-B)OXIREN-3-YL N-(P-TOLYL)CARBAMATE

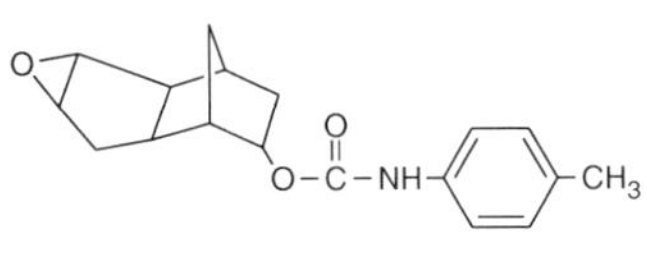

250 mg

S77,350-6
OCTAHYDRO-2,5-METHANO-2H-INDENO-(1,2-B)OXIREN-3-YL N-(2,4,6-TRIMETHYLPHENYL)CARBAMATE

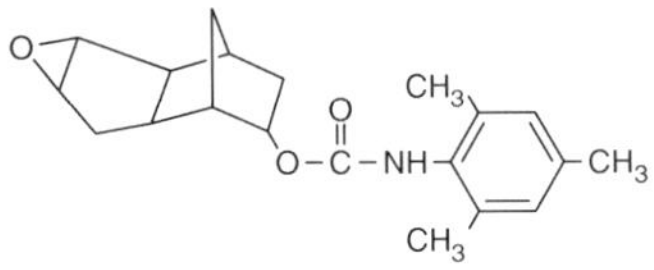

250 mg

S73,763-1
5-NORBORNEN-2-YLMETHYL N-(P-TOLYL)-CARBAMATE

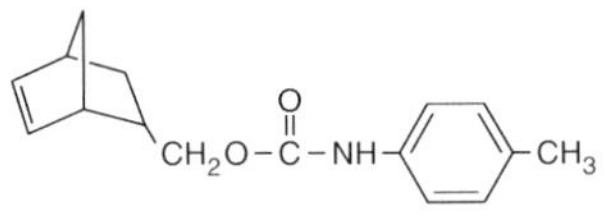

5 gm

S85,580-4
5-NORBORNEN-2-YLMETHYL N-(M-TOLYL)-CARBAMATE

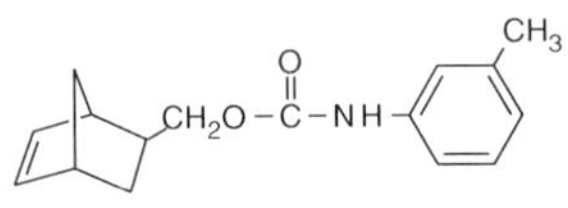

250 mg

S86,723-3
CYCLOPENTYL N-PHENYLCARBAMATE

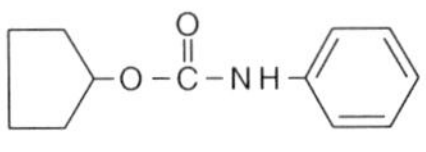

250 mg

S84,499-3
CYCLOPENTYL N-(2,5-XYLYL)CARBAMATE

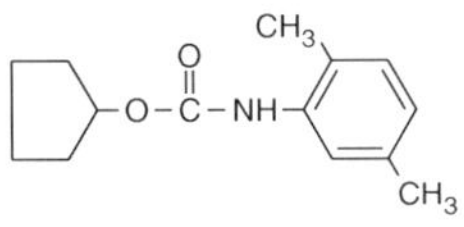

250 mg

S65,455-8
4,4'-METHYLENEBIS[3,4-(METHYLENEDIOXY)PHENYL CARBANILATE]

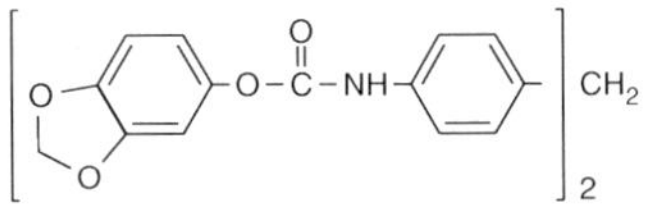

5 gm

S58,616-1
CYCLOHEXYL CARBANILATE

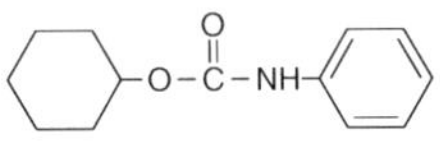

5 gm

S60,953-6
CYCLOHEXYL 3'-CHLOROCARBANILATE

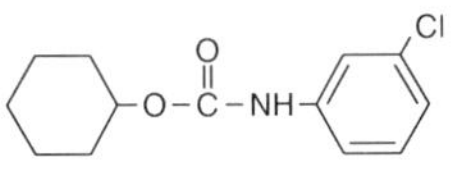

5 gm

S60,760-6
CYCLOHEXYL N-(2,4,5-TRICHLOROPHENYL)CARBAMATE

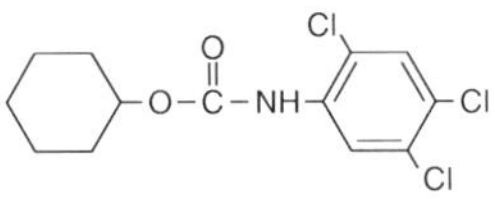

5 gm

S60,677-4
CYCLOHEXYL N-(M-TOLYL)CARBAMATE

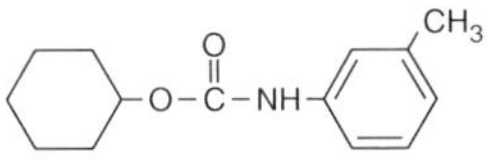

5 gm

S61,039-9
2-METHYLCYCLOHEXYL N-(3-CHLOROPHENYL)CARBAMATE

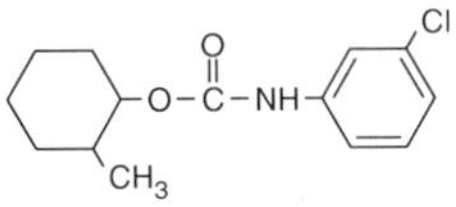

5 gm

S63,390-9
2-METHYLCYCLOHEXYL N-(2,4,5-TRICHLOROPHENYL)CARBAMATE

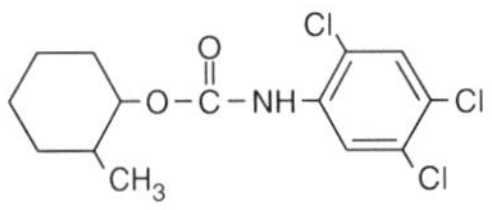

5 gm

S66,634-3
2-METHYLCYCLOHEXYL N-(M-TOLYL)-CARBAMATE

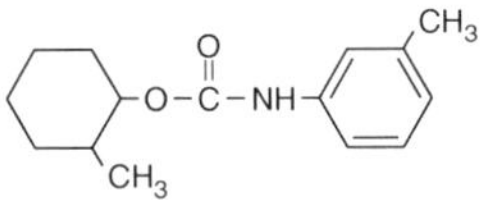

5 gm

S71,474-7
2-METHYLCYCLOHEXYL N-PHENYLCARBAMATE

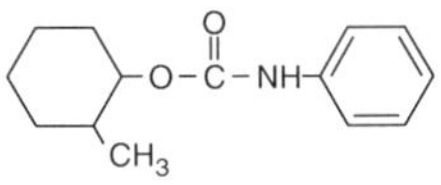

5 gm

S62,183-8
2-METHYLCYCLOHEXYL N-(2-CHLOROPHENYL)CARBAMATE

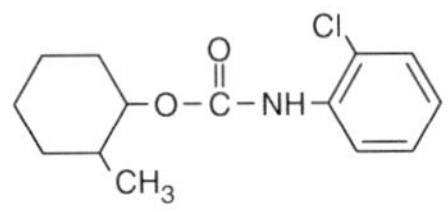

5 gm

S60,735-5
2-METHYLCYCLOHEXYL N-(O-TOLYL)-CARBAMATE

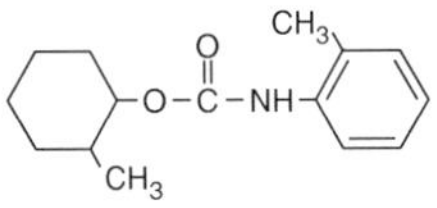

5 gm

S66,705-6
2-METHYLCYCLOHEXYL N-(4-BROMOPHENYL)CARBAMATE

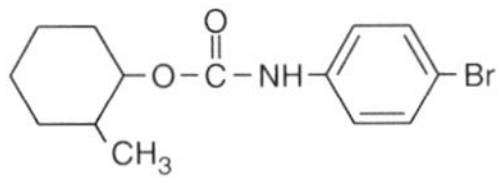

5 gm

S74,813-7
2-METHYLCYCLOHEXYL N-(P-TOLYL)-CARBAMATE

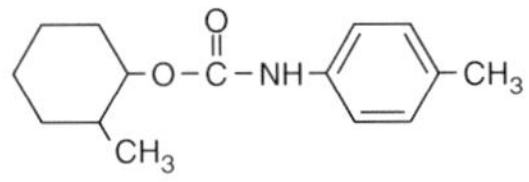

250 mg

S66,738-2
2-METHYLCYCLOHEXYL N-(P-TOLYL)-CARBAMATE

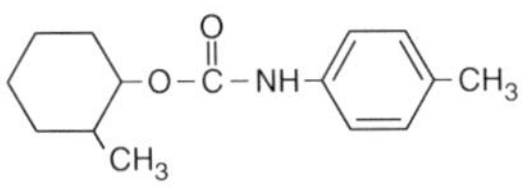

5 gm

S63,059-4
2-METHYLCYCLOHEXYL N-(2,3-DICHLOROPHENYL)CARBAMATE

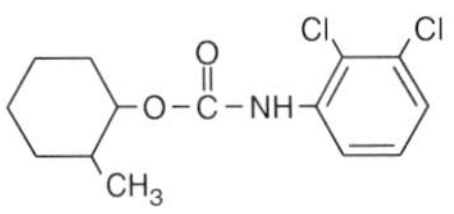

5 gm

S66,764-1
2-METHYLCYCLOHEXYL N-(2-CHLORO-6-METHYLPHENYL)CARBAMATE

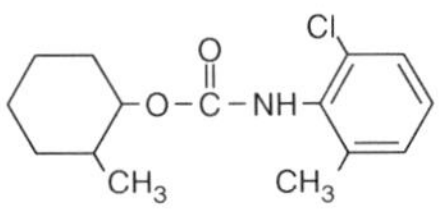

5 gm

S66,525-8
2-METHYLCYCLOHEXYL N-(2,3-XYLYL)-CARBAMATE

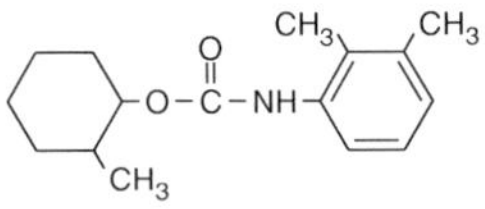

5 gm

S66,746-3
2-METHYLCYCLOHEXYL N-(2,6-XYLYL)-CARBAMATE

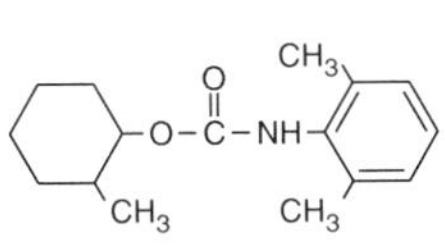

5 gm

S66,772-2
2-METHYLCYCLOHEXYL N-(2,6-DIETHYLPHENYL)CARBAMATE

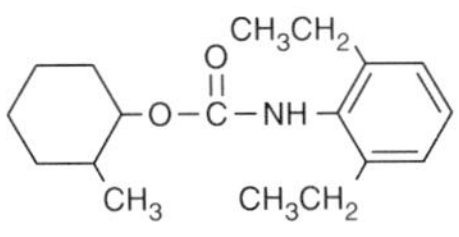

5 gm

S61,267-7
2-METHYLCYCLOHEXYL N-(2,4-DICHLOROPHENYL)CARBAMATE

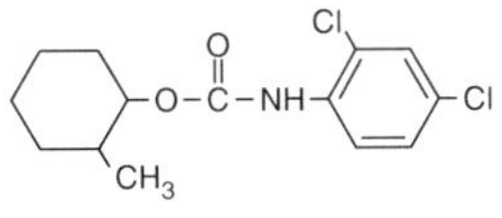

5 gm

S61,345-2
2-METHYLCYCLOHEXYL N-(3,4-DICHLOROPHENYL)CARBAMATE

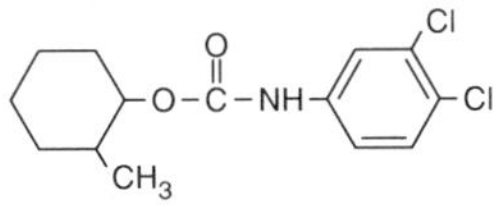

5 gm

S64,397-1
2-METHYLCYCLOHEXYL N-(2,5-DICHLOROPHENYL)CARBAMATE

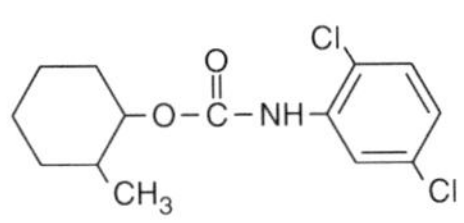

5 gm

S60,731-2
2-METHYLCYCLOHEXYL N-(4-CHLORO-2-METHYLPHENYL)CARBAMATE

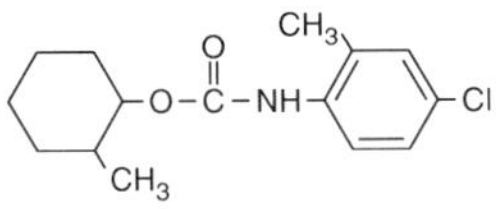

5 gm

S61,021-6
2-METHYLCYCLOHEXYL N-(5-CHLORO-2-METHYLPHENYL)CARBAMATE

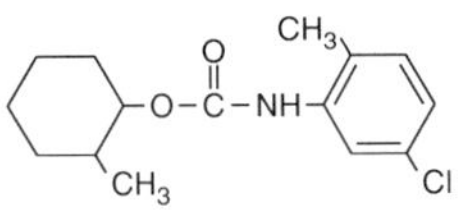

5 gm

S61,614-1
2-METHYLCYCLOHEXYL N-(2,4-XYLYL)-CARBAMATE

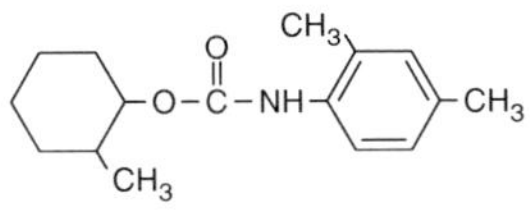

5 gm

S85,962-1
PHENYL N-CYCLOHEXYLCARBAMATE

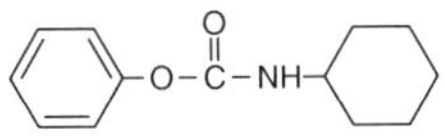

250 mg

S72,482-3
4-CYCLOHEXYLPHENYL N-PHENYLCARBAMATE

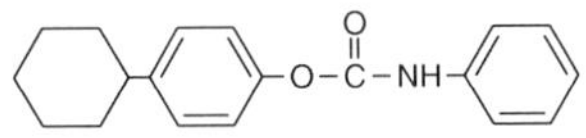

5 gm

S74,303-8
4-BIPHENYLYL N-CYCLOHEXYLCARBAMATE

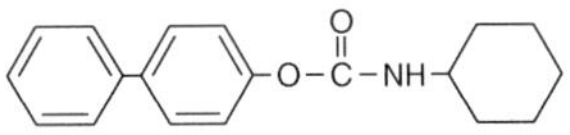

5 gm

S63,799-8
CYCLOHEXYL 2'-CHLOROCARBANILATE

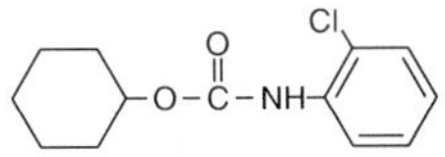

5 gm

S60,656-1
CYCLOHEXYL N-(O-TOLYL)CARBAMATE

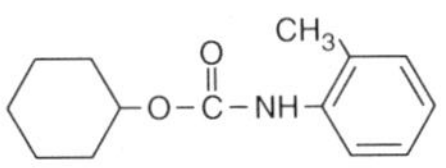

5 gm

S89,967-4
2-ISOPROPYLPHENYL N-CYCLOHEXYLCARBAMATE

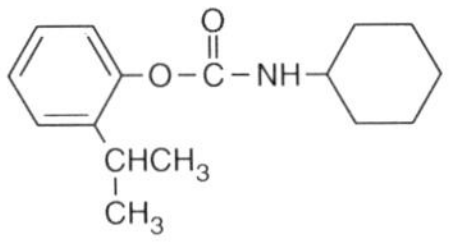

250 mg

S89,176-2
4-CYCLOHEXYLPHENYL N-(O-TOLYL)-CARBAMATE

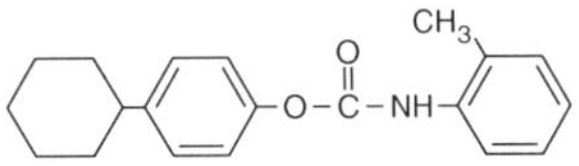

250 mg

S71,357-0
2-METHYLCYCLOHEXYL N-(2,6-DIISOPROPYLPHENYL)CARBAMATE

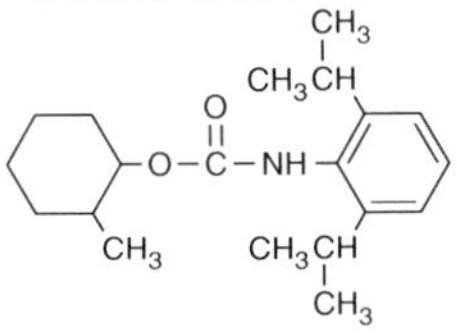

250 mg

S60,184-5
CYCLOHEXYL N-(4-FLUOROPHENYL)-CARBAMATE

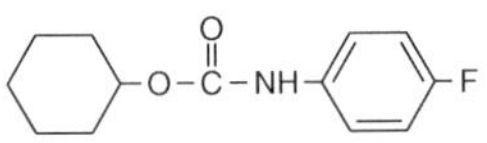

5 gm

S59,144-0
CYCLOHEXYL 4'-CHLOROCARBANILATE

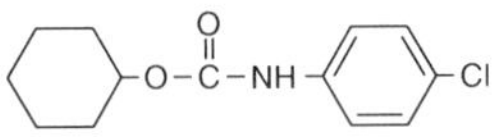

5 gm

S66,726-9
CYCLOHEXYL N-(4-BROMOPHENYL)-CARBAMATE

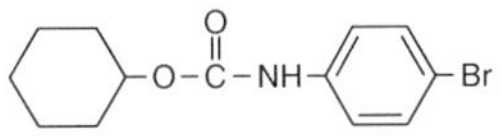

5 gm

S61,328-2
CYCLOHEXYL N-(P-TOLYL)CARBAMATE

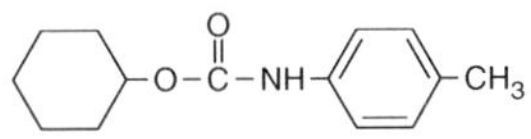

5 gm

S74,311-9
P-TOLYL N-CYCLOHEXYLCARBAMATE

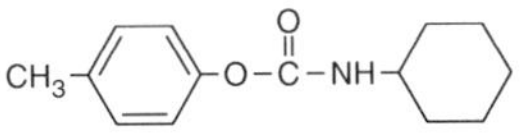

5 gm

S84,749-6
4-CYCLOHEXYLPHENYL N-(2,6-XYLYL)-CARBAMATE

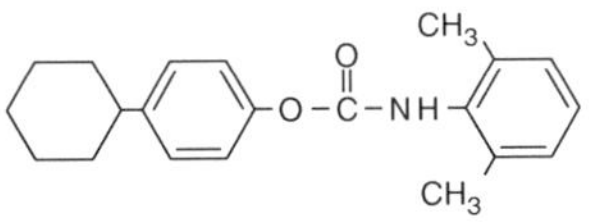

250 mg

S79,537-2
4-CYCLOHEXYLPHENYL N-(1-NAPHTHYL)-CARBAMATE

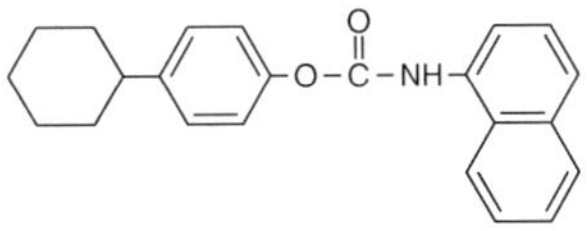

250 mg

S62,424-1
CYCLOHEXYL N-(2,3-DICHLOROPHENYL)-CARBAMATE

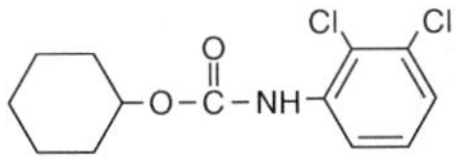

5 gm

S66,763-3
CYCLOHEXYL N-(2-CHLORO-6-METHYLPHENYL)CARBAMATE

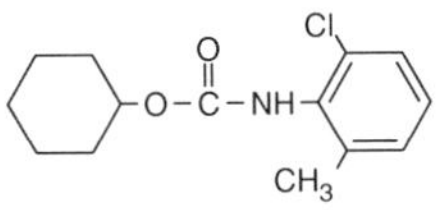

5 gm

S64,430-7
CYCLOHEXYL N-(2,3-XYLYL)CARBAMATE

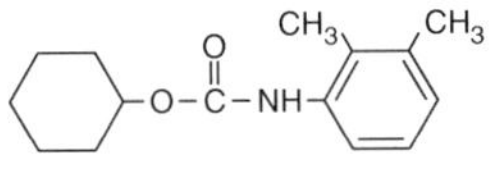

5 gm

S66,630-0
CYCLOHEXYL N-(2,6-XYLYL)CARBAMATE

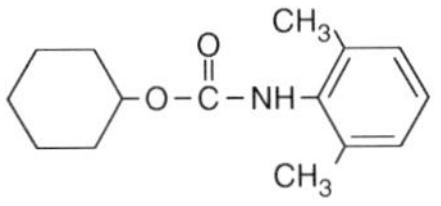

5 gm

S74,301-1
2,3-XYLYL N-CYCLOHEXYLCARBAMATE

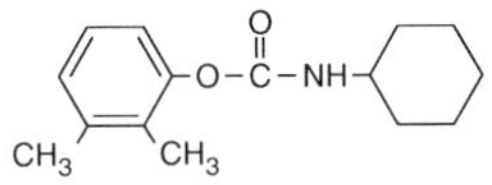

5 gm

S79,541-0
2,6-XYLYL N-CYCLOHEXYLCARBAMATE

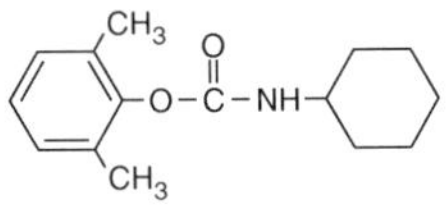

250 mg

S65,276-8
CYCLOHEXYL N-(2-ETHYL-6-METHYLPHENYL)CARBAMATE

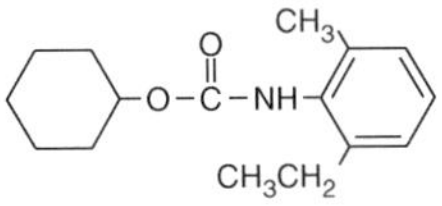

5 gm

S70,929-8
2-METHYLCYCLOHEXYL N-(2-ETHYL-6-METHYLPHENYL)CARBAMATE

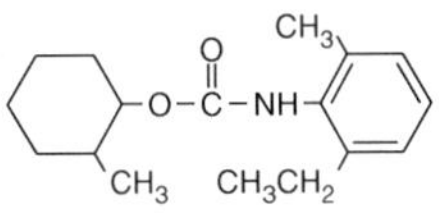

5 gm

S71,305-8
CYCLOHEXYL N-(2,6-DIISOPROPYLPHENYL)CARBAMATE

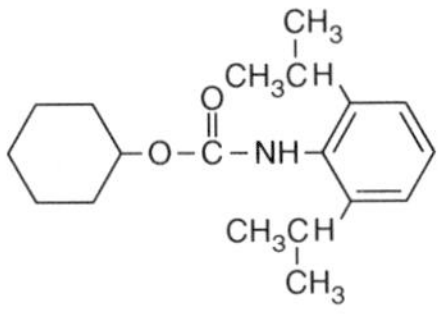

250 mg

S60,700-2
CYCLOHEXYL N-(3,4-DICHLOROPHENYL)-CARBAMATE
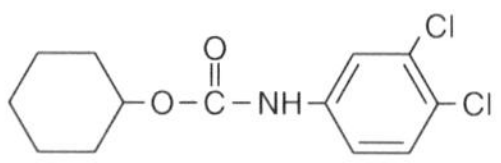
5 gm

S61,301-0
CYCLOHEXYL N-(2,4-DICHLOROPHENYL)-CARBAMATE
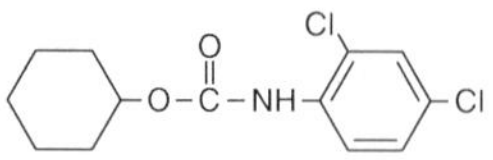
5 gm

S64,388-2
CYCLOHEXYL N-(2,5-DICHLOROPHENYL)-CARBAMATE
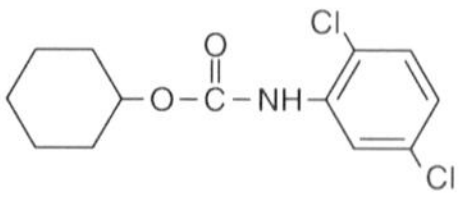
5 gm

S89,960-7
2,4-DICHLOROPHENYL N-CYCLOHEXYLCARBAMATE
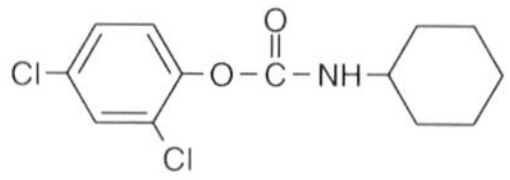
250 mg

S60,181-0
CYCLOHEXYL N-(4-CHLORO-2-METHYLPHENYL)CARBAMATE
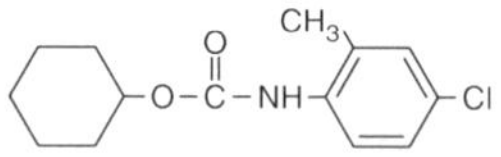
5 gm

S61,049-6
CYCLOHEXYL N-(5-CHLORO-2-METHYLPHENYL)CARBAMATE
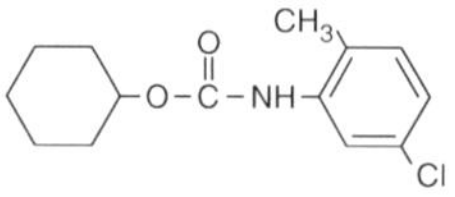
5 gm

S60,668-5
CYCLOHEXYL N-(2,4-XYLYL)CARBAMATE
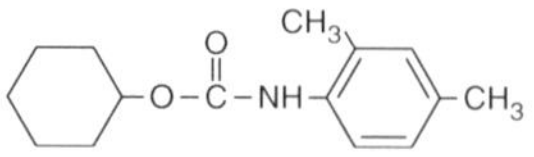
5 gm

S64,044-1
CYCLOHEXYL N-(2,5-XYLYL)CARBAMATE
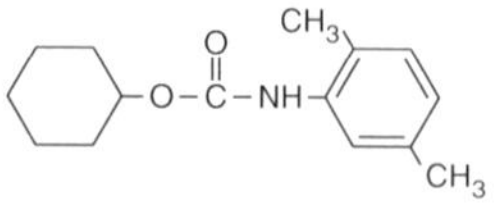
5 gm

S79,544-5
5-ISOPROPYL-2-METHYLPHENYL N-CYCLOHEXYLCARBAMATE
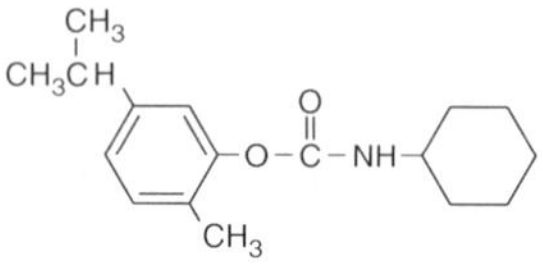
250 mg

S58,996-9
CYCLOHEXYL N-(1-NAPHTHYL)CARBAMATE
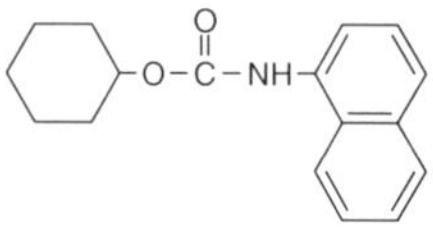
5 gm

S89,968-2
2-NAPHTHYL N-CYCLOHEXYLCARBAMATE
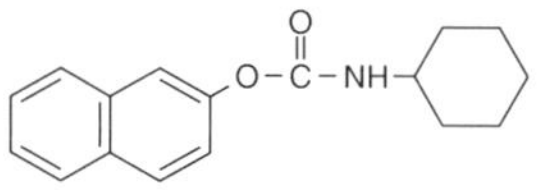
250 mg

S70,241-2
N,N'-(2-BROMO-6-METHYLPHENYLENE)-BIS(1,4-BENZODIOXAN-2-YLMETHYL CARBAMATE)
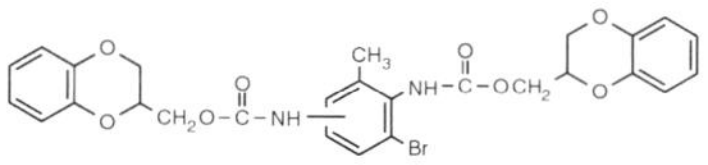
250 mg

S71,729-0
METHYL N-PHENYLCARBAMATE
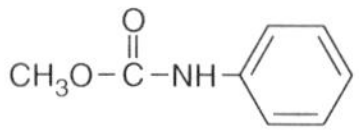
5 gm

S36,443-6
PHENYL N,N-DIMETHYLCARBAMATE
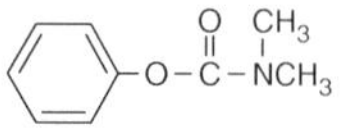
5 gm

S71,461-5
ETHYL N-PHENYLCARBAMATE
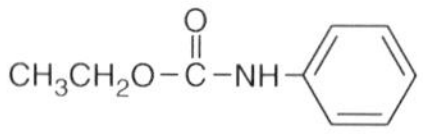
5 gm

S58,641-2
ALLYL N-PHENYLCARBAMATE
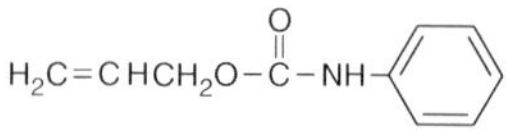
5 gm

S35,339-6
ETHYL N-METHYLCARBANILATE
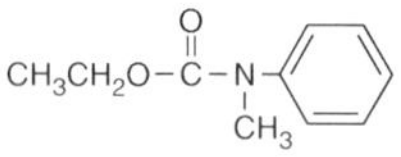
5 gm

S58,033-3
ISOPROPYL N-PHENYLCARBAMATE
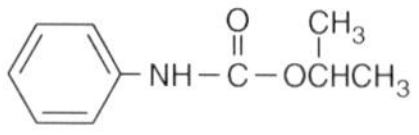
5 gm

S58,644-7
PROPYL N-PHENYLCARBAMATE
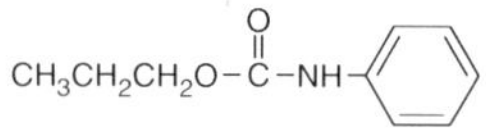
5 gm

S72,440-8
2-BUTENYL N-PHENYLCARBAMATE
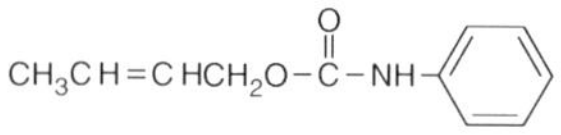
5 gm

S58,636-6
BUTYL N-PHENYLCARBAMATE
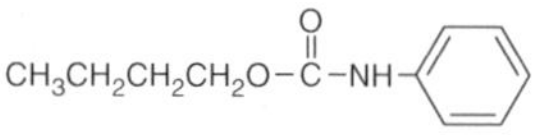
5 gm

S58,999-3
PENTYL N-PHENYLCARBAMATE
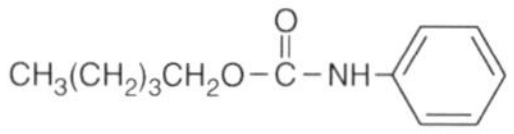
5 gm

S58,031-7
PHENYL N-PHENYLCARBAMATE
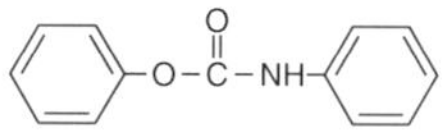
5 gm

S36,155-0
ETHYL N-BUTYL-N-PHENYLCARBAMATE
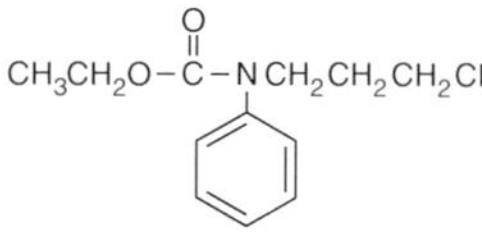
5 gm

S59,011-8
HEPTYL N-PHENYLCARBAMATE
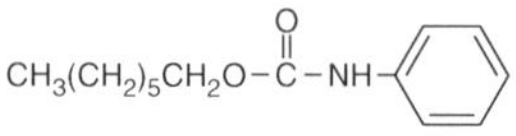
5 gm

S59,018-5
OCTYL N-PHENYLCARBAMATE
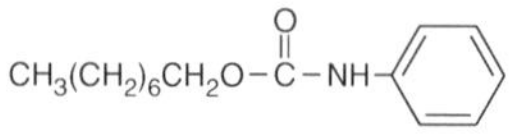
5 gm

S62,251-6
2-OCTYL N-PHENYLCARBAMATE
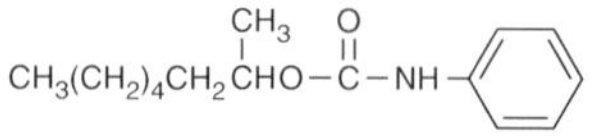
5 gm

S89,228-9
3-PHENYLALLYL N-PHENYLCARBAMATE
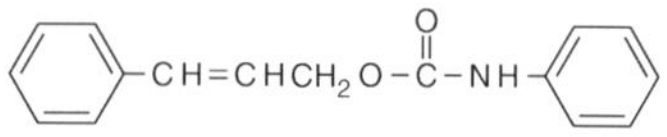
250 mg

S85,581-2
3-PHENYLALLYL N-(M-TOLYL)CARBAMATE
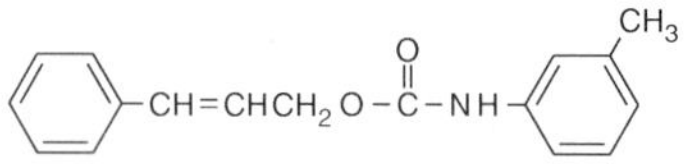
250 mg

S66,799-4
PHENYL N-(2-ETHYLPHENYL)CARBAMATE
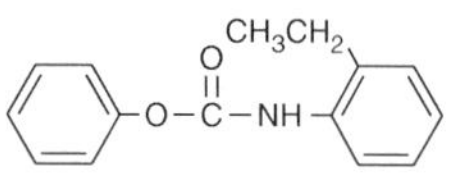
5 gm

S79,168-7
2-BENZYLPHENYL N-(1-NAPHTHYL)-CARBAMATE
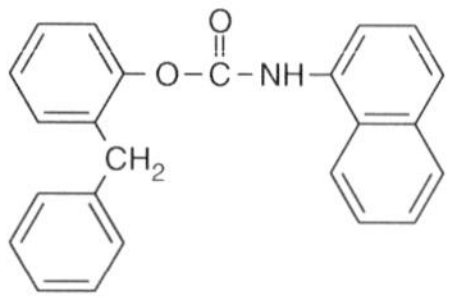
250 mg

S89,177-0
2-BIPHENYLYL N-(O-TOLYL)CARBAMATE
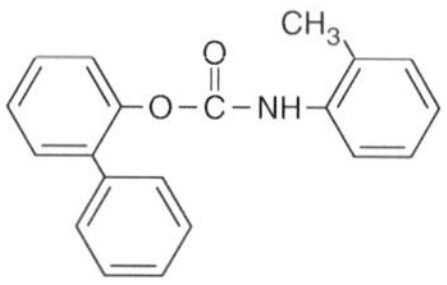
250 mg

S73,090-4
CINNAMYL N-(P-TOLYL)CARBAMATE

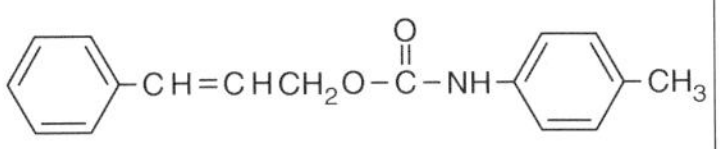

5 gm

S34,197-5
4-BENZYLPHENYL CARBAMATE

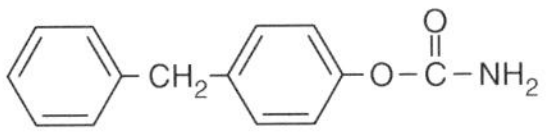

5 gm

S73,897-2
4-BIPHENYLYL N-(M-TOLYL)CARBAMATE

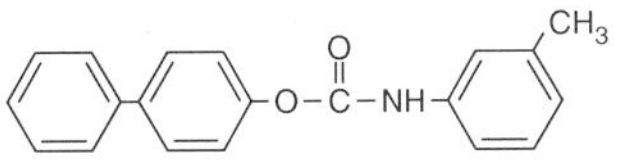

5 gm

S70,990-5
PHENYL N-(2,6-DICHLOROPHENYL)-CARBAMATE

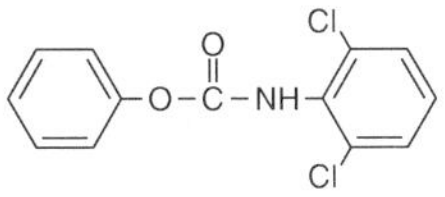

5 gm

S89,229-7
2,6-XYLYL N-PHENYLCARBAMATE

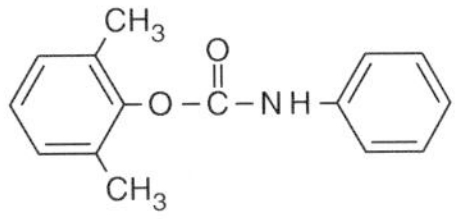

250 mg

S72,998-1
2,6-DIISOPROPYLPHENYL N-PHENYLCARBAMATE

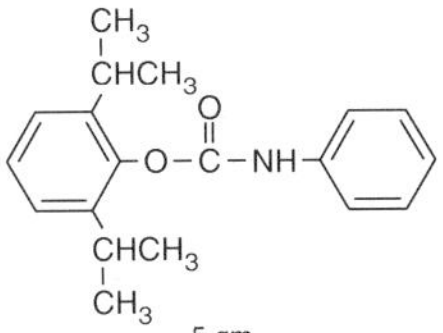

5 gm

S84,525-6
3-PHENYLALLYL N-(2,5-XYLYL)-CARBAMATE

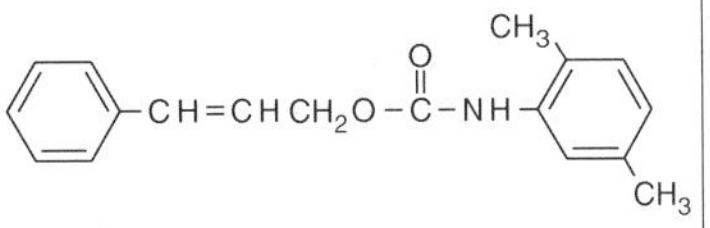

250 mg

S77,383-2
4-BIPHENYLYL N-(2,5-XYLYL)CARBAMATE

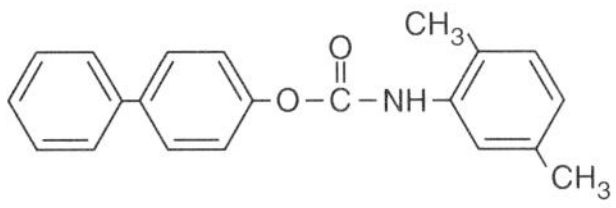

250 mg

S79,148-2
3-ETHYL-5-METHYLPHENYL N-PHENYLCARBAMATE

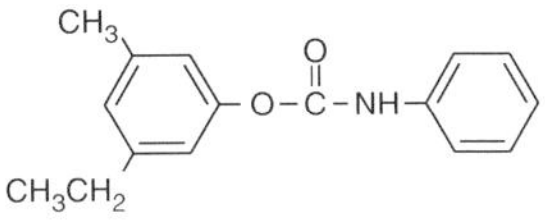

250 mg

S64,825-6
2,3,4-TRICHLOROPHENYL N-PHENYLCARBAMATE

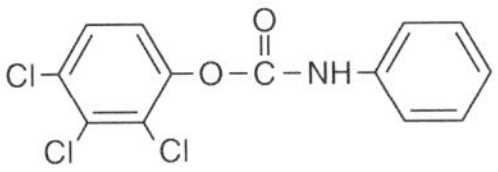

5 gm

S72,170-0
2,4-DICHLORO-1-NAPHTHYL N-PHENYLCARBAMATE

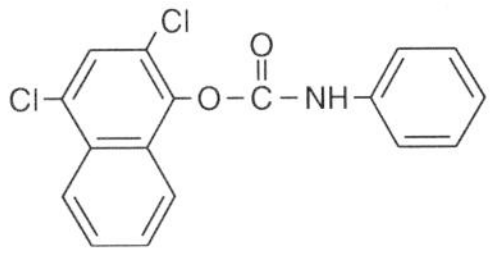

5 gm

S72,673-7
CINNAMYL N-(1-NAPHTHYL)CARBAMATE

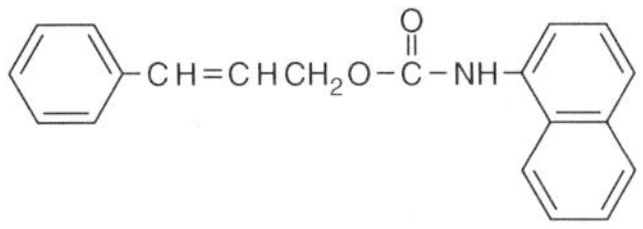

5 gm

S87,509-0
2-NAPHTHYL N-PHENYLCARBAMATE

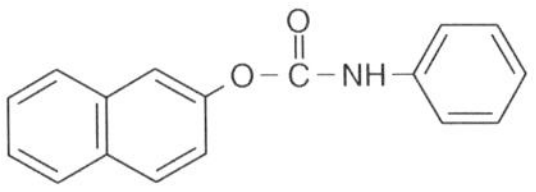

250 mg

S63,435-2
METHYL N-(2-CHLOROPHENYL)CARBAMATE

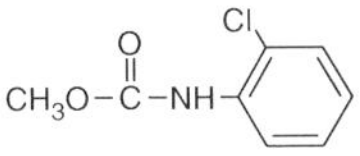

5 gm

S62,833-6
METHYL N-(O-TOLYL)CARBAMATE

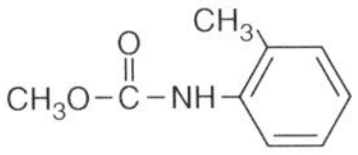

5 gm

S59,535-7
ETHYL N-(2-CHLOROPHENYL)-CARBAMATE

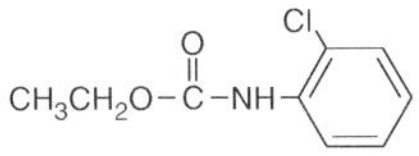

5 gm

S65,063-3
METHYL N-(2-ETHYLPHENYL)CARBAMATE

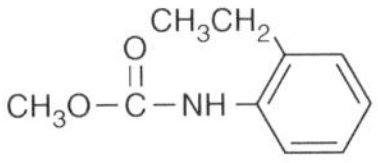

250 mg

S63,013-6
ALLYL N-(2-CHLOROPHENYL)-CARBAMATE

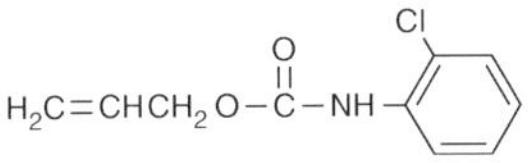

5 gm

S63,852-8
PROPYL N-(2-CHLOROPHENYL)CARBAMATE

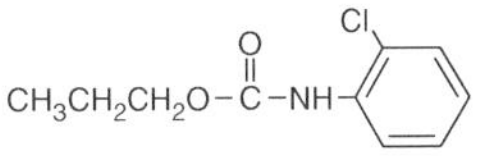

5 gm

S63,855-2
ISOPROPYL N-(2-CHLOROPHENYL)-CARBAMATE

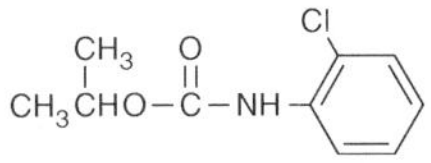

5 gm

S66,563-0
ALLYL N-(O-TOLYL)CARBAMATE

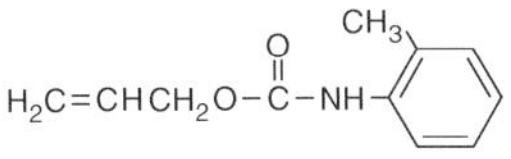

5 gm

S60,653-7
ISOPROPYL N-(O-TOLYL)CARBAMATE

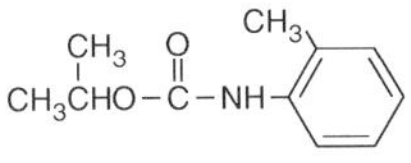

5 gm

S64,409-9
PROPYL N-(O-TOLYL)CARBAMATE

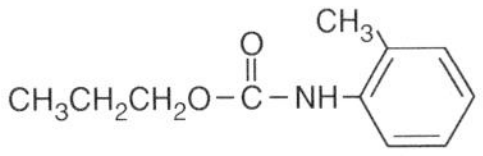

5 gm

S65,014-5
ETHYL N-(2-ETHYLPHENYL)CARBAMATE

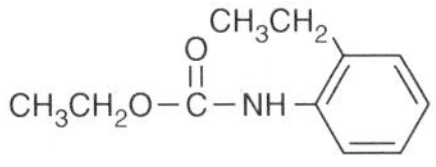

5 gm

S84,412-8
2-ISOPROPYLPHENYL N-METHYLCARBAMATE

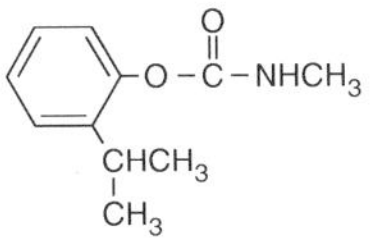

250 mg

S60,694-4
BUTYL N-(O-TOLYL)CARBAMATE

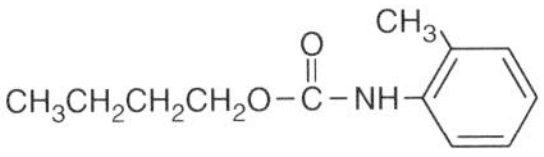

5 gm

S64,958-9
PROPYL N-(2-ETHYLPHENYL)CARBAMATE

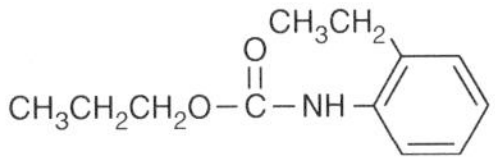

5 gm

S65,134-6
ISOPROPYL N-(2-ETHYLPHENYL)-CARBAMATE

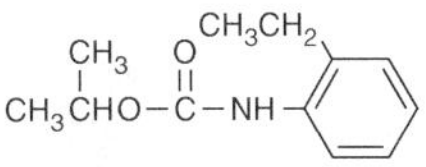

5 gm

S66,809-5
BUTYL N-(2-ETHYLPHENYL)CARBAMATE

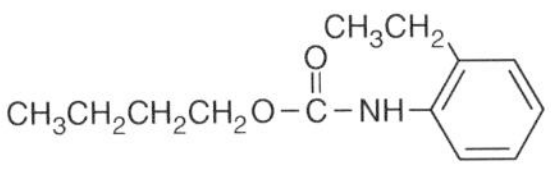

5 gm

S65,428-0
PHENYL N-(2-CHLOROPHENYL)CARBAMATE

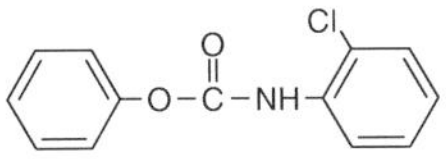

5 gm

S62,513-2
PHENYL N-(O-TOLYL)CARBAMATE

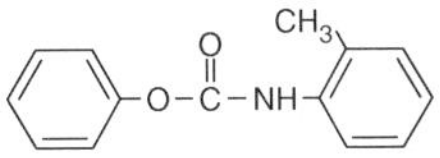

5 gm

S89,180-0
2-ETHYLPHENYL N-(O-TOLYL)CARBAMATE

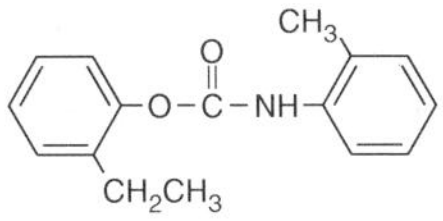

250 mg

S59,000-2
2-CHLOROPHENYL N-(4-CHLOROPHENYL)-CARBAMATE

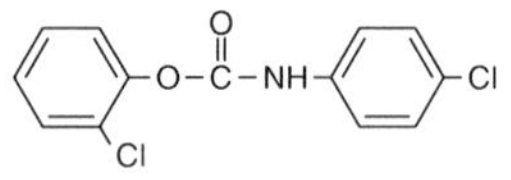

5 gm

S63,452-2
4-CHLOROPHENYL N-(2-CHLOROPHENYL)-CARBAMATE

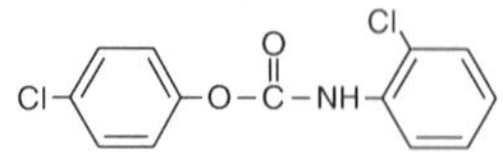

5 gm

S64,673-3
4-CHLOROPHENYL N-(O-TOLYL)CARBAMATE

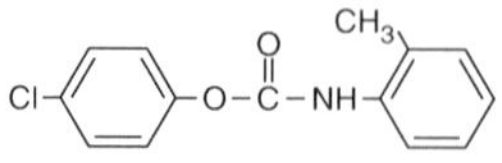

5 gm

S80,785-0
2,3-XYLYL N-(O-TOLYL)CARBAMATE

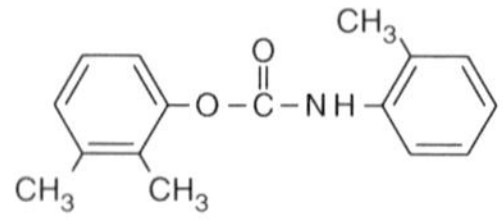

250 mg

S65,272-5
O-TOLYL N-(2-ETHYL-6-METHYLPHENYL)-CARBAMATE

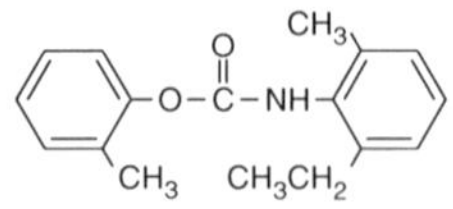

5 gm

S64,978-3
2-CHLOROPHENYL N-(2,4-DICHLOROPHENYL)CARBAMATE

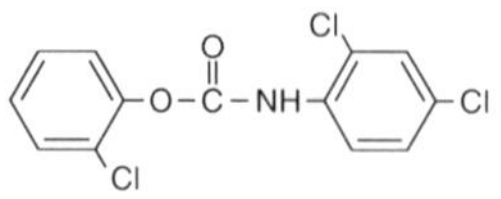

5 gm

S65,015-3
2-CHLOROPHENYL N-(2,5-DICHLOROPHENYL)CARBAMATE

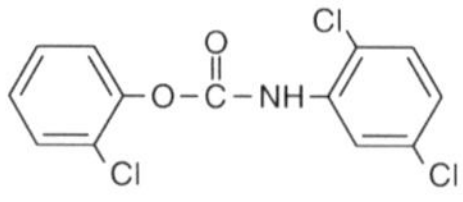

5 gm

S66,803-6
PHENYL N-(3,5-XYLYL)CARBAMATE

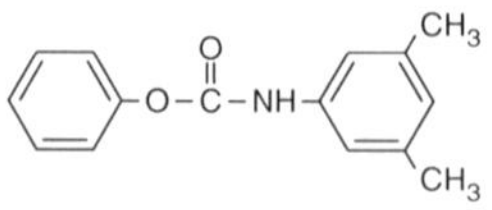

5 gm

S89,181-9
3,5-XYLYL N-(O-TOLYL)CARBAMATE

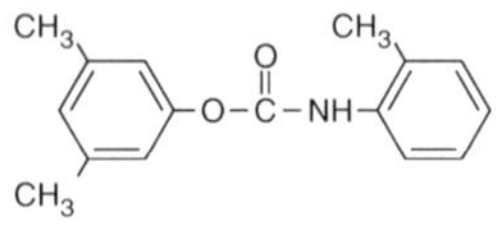

250 mg

S72,848-9
2-BIPHENYLYL N-METHYLCARBAMATE

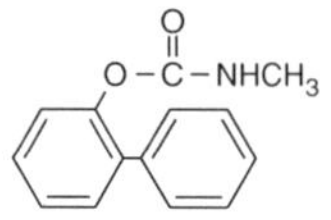

5 gm

S75,260-6
2,4-DICHLORO-1-NAPHTHYL N-(O-TOLYL)-CARBAMATE

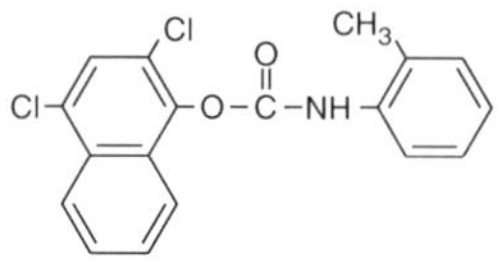

5 gm

S89,051-0
2-SEC-BUTYLPHENYL N-(1-NAPHTHYL)-CARBAMATE

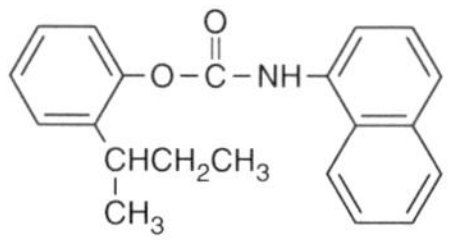

250 mg

S58,631-5
METHYL N-(4-CHLOROPHENYL)CARBAMATE

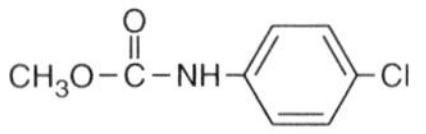

5 gm

S61,270-7
METHYL N-(P-TOLYL)CARBAMATE

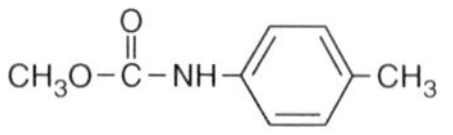

5 gm

S64,396-3
ETHYL N-(P-TOLYL)CARBAMATE

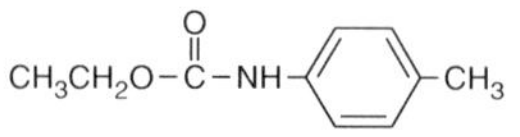

5 gm

S64,410-2
ETHYL N-(O-TOLYL)CARBAMATE

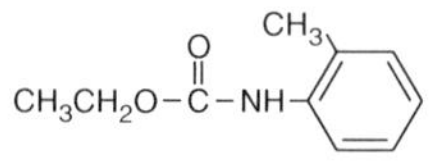

5 gm

S69,058-9
P-TOLYL N-ETHYLCARBAMATE

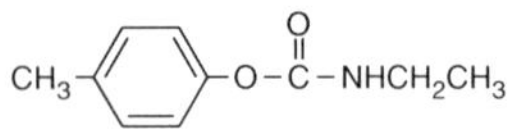

5 gm

S60,182-9
ALLYL N-(4-FLUOROPHENYL)CARBAMATE

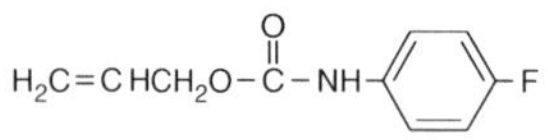

5 gm

S60,205-1
ISOPROPYL N-(4-FLUOROPHENYL)-CARBAMATE

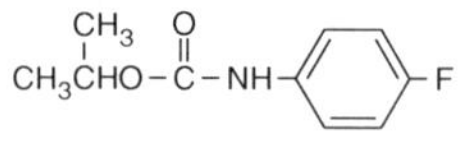

5 gm

S58,629-3
ALLYL N-(4-CHLOROPHENYL)-CARBAMATE

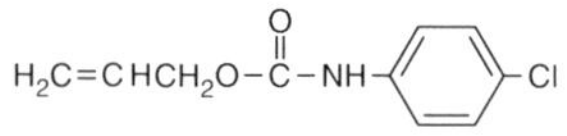

5 gm

S58,628-5
PROPYL N-(4-CHLOROPHENYL)CARBAMATE

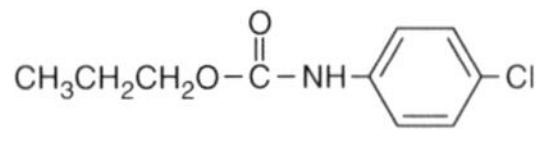

5 gm

S58,635-8
ISOPROPYL N-(4-CHLOROPHENYL)-CARBAMATE

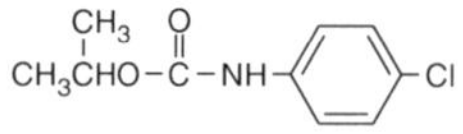

5 gm

S59,419-9
ISOPROPYL N-(4-BROMOPHENYL)-CARBAMATE

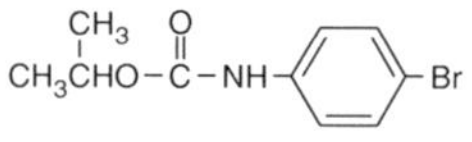

5 gm

S66,660-2
ALLYL N-(P-TOLYL)CARBAMATE

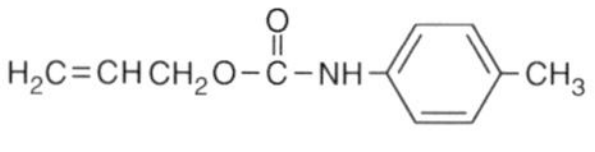

5 gm

S61,268-5
ISOPROPYL N-(P-TOLYL)CARBAMATE

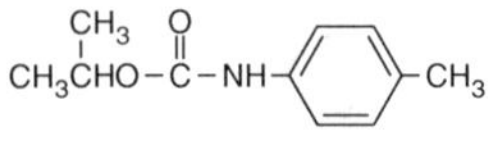

5 gm

S64,387-4
PROPYL N-(P-TOLYL)CARBAMATE

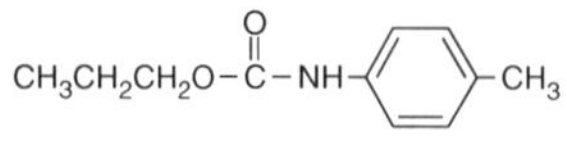

5 gm

S58,640-4
BUTYL N-(4-CHLOROPHENYL)-CARBAMATE

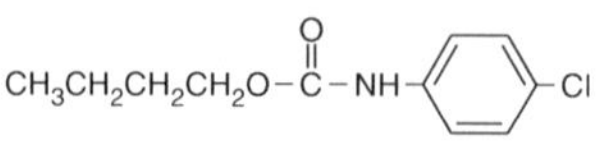

5 gm

S59,407-5
BUTYL N-(4-BROMOPHENYL)CARBAMATE

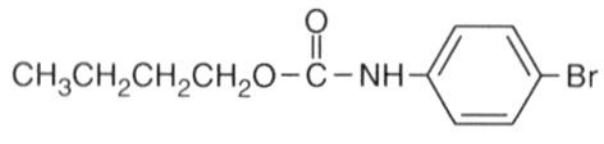

5 gm

S64,381-5
BUTYL N-(P-TOLYL)CARBAMATE

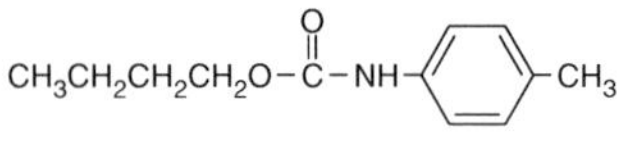

5 gm

S59,540-3
PENTYL N-(4-CHLOROPHENYL)-CARBAMATE

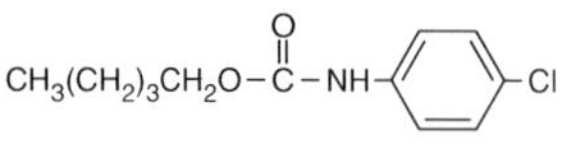

5 gm

S59,025-8
PHENYL N-(4-CHLOROPHENYL)CARBAMATE

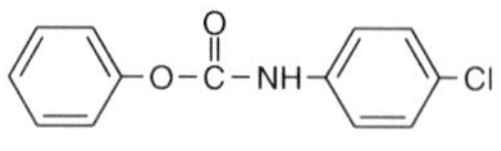

5 gm

S63,823-4
4-CHLOROPHENYL N-(3-CHLOROPHENYL)-CARBAMATE

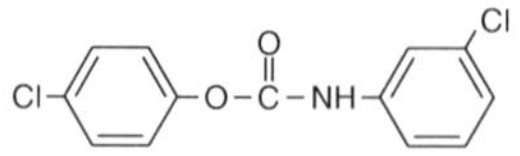

5 gm

S64,424-2
4-CHLOROPHENYL N-(4-CHLOROPHENYL)-CARBAMATE

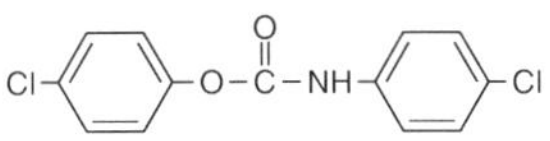

5 gm

S64,435-8
PHENYL N-(4-BROMOPHENYL)CARBAMATE

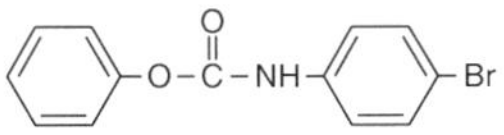

5 gm

S64,423-4
4-CHLOROPHENYL N-(4-BROMOPHENYL)-CARBAMATE

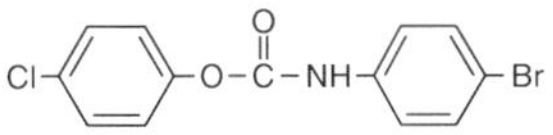

5 gm

S66,468-5
PHENYL N-(P-TOLYL)CARBAMATE

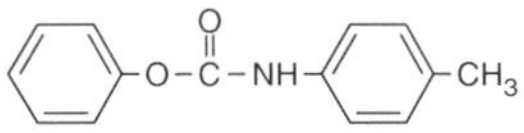

5 gm

S64,960-0
4-CHLOROPHENYL N-(M-TOLYL)CARBAMATE

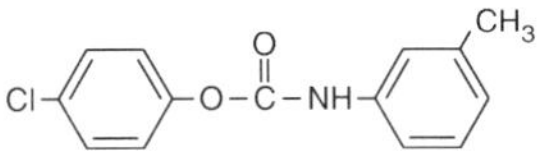

5 gm

S65,002-1
4-CHLOROPHENYL N-(P-TOLYL)CARBAMATE

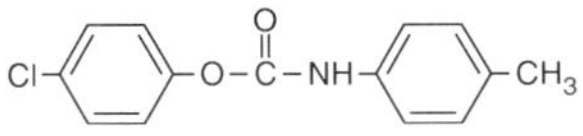

5 gm

S59,012-6
HEPTYL N-(4-CHLOROPHENYL)CARBAMATE

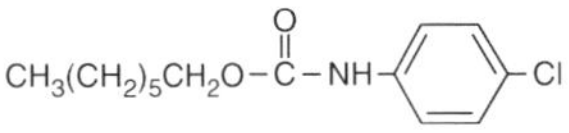

5 gm

S59,551-9
OCTYL N-(4-CHLOROPHENYL)CARBAMATE

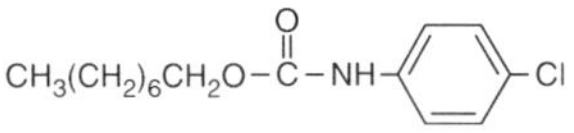

5 gm

S64,425-0
4-CHLOROPHENYL N-(2,3-DICHLOROPHENYL)CARBAMATE

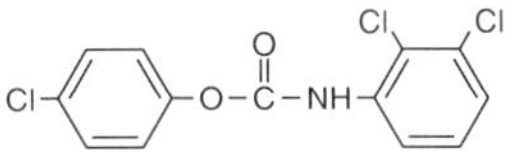

5 gm

S65,067-6
4-CHLOROPHENYL N-(2-CHLORO-6-METHYLPHENYL)CARBAMATE

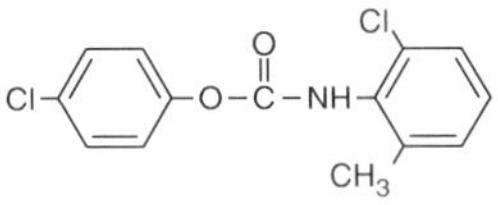

5 gm

S64,992-9
4-CHLOROPHENYL N-(2,3-XYLYL)-CARBAMATE

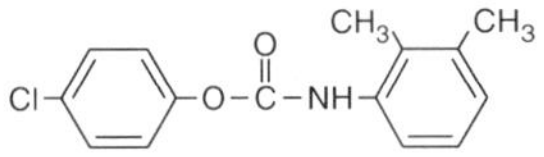

5 gm

S63,198-1
4-CHLOROPHENYL N-(3,4-DICHLOROPHENYL)CARBAMATE

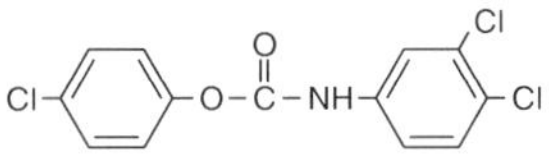

5 gm

S65,371-3
4-CHLOROPHENYL N-(2,4-DICHLOROPHENYL)CARBAMATE

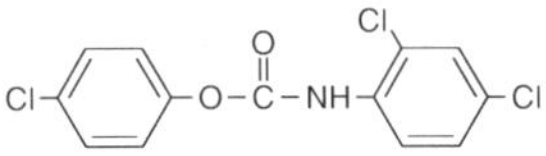

5 gm

S62,818-2
PHENYL N-(3-CHLORO-4-METHYLPHENYL)-CARBAMATE

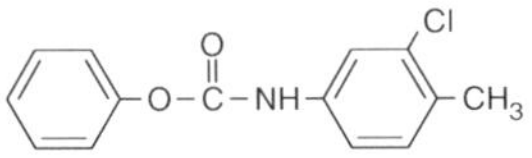

5 gm

S65,370-5
4-CHLOROPHENYL N-(2,4-XYLYL)-CARBAMATE

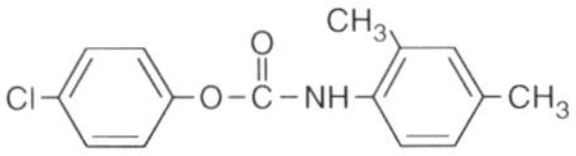

5 gm

S59,017-7
1-NAPHTHYL N-(4-CHLOROPHENYL)-CARBAMATE

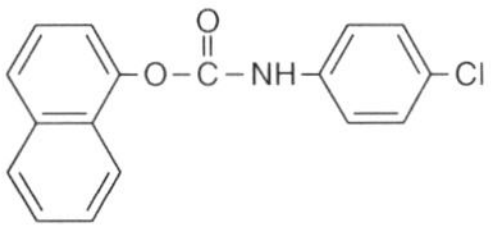

5 gm

S80,044-9
2-NAPHTHYL N-(4-FLUOROPHENYL)-CARBAMATE

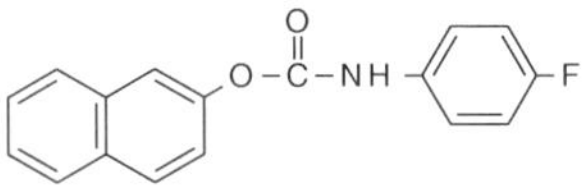

250 mg

S59,728-7
METHYL N-(2,6-DICHLOROPHENYL)-CARBAMATE

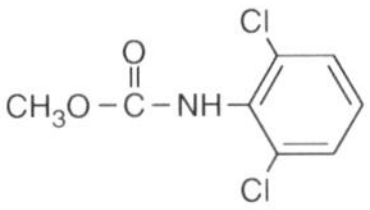

5 gm

S63,810-2
METHYL N-(2,3-DICHLOROPHENYL)-CARBAMATE

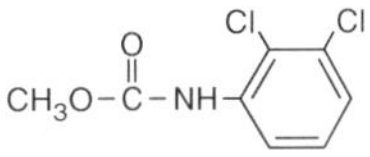

5 gm

S34,829-5
ETHYL 2,3-DICHLOROPHENYLCARBAMATE

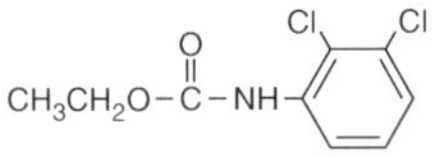

5 gm

S39,142-5
2,6-DICHLOROPHENYL N,N-DIMETHYLCARBAMATE

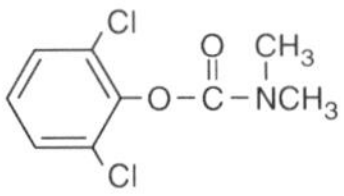

5 gm

S64,046-8
METHYL N-(2,3-XYLYL)CARBAMATE

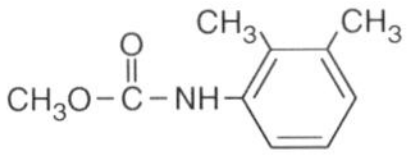

5 gm

S66,776-5
METHYL N-(2,6-XYLYL)CARBAMATE

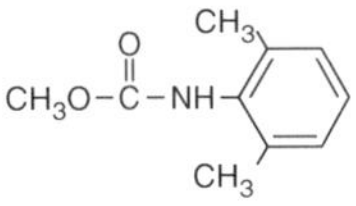

5 gm

S66,571-1
ETHYL N-(2-CHLORO-6-METHYLPHENYL)-CARBAMATE

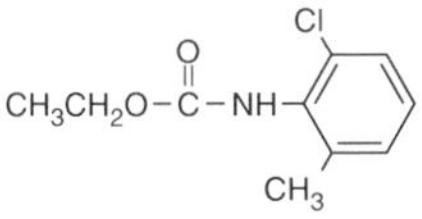

5 gm

S63,903-6
ALLYL N-(2,3-DICHLOROPHENYL)-CARBAMATE

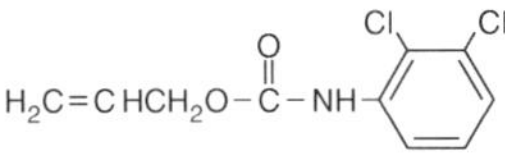

5 gm

S59,758-9
ISOPROPYL N-(2,3-DICHLOROPHENYL)-CARBAMATE

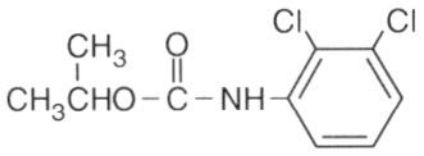

5 gm

S61,195-6
ISOPROPYL N-(2,6-DICHLOROPHENYL)-CARBAMATE

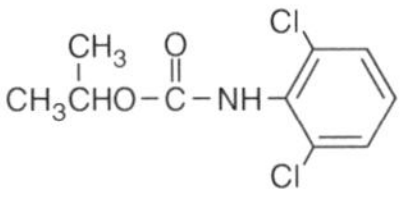

5 gm

S62,423-3
PROPYL N-(2,3-DICHLOROPHENYL)-CARBAMATE

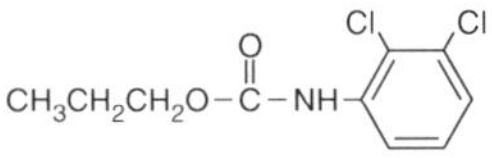

5 gm

S65,018-8
PROPYL N-(2,6-DICHLOROPHENYL)-CARBAMATE

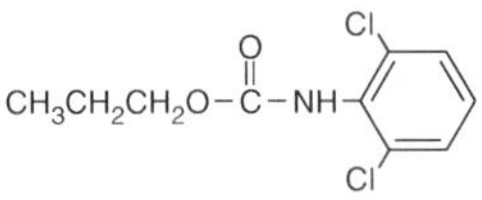

5 gm

S65,009-9
ETHYL N-(2,3-XYLYL)CARBAMATE

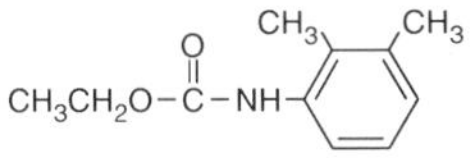

5 gm

S66,706-4
ETHYL N-(2,6-XYLYL)CARBAMATE

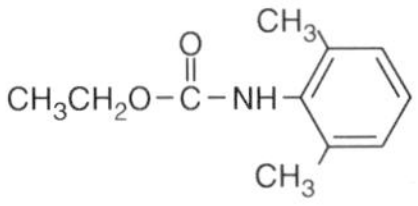

5 gm

S66,775-7
METHYL N-(2-ETHYL-6-METHYLPHENYL)-CARBAMATE

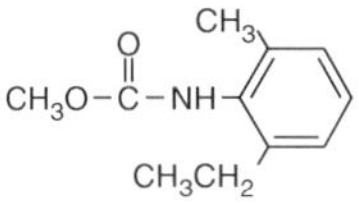

5 gm

S63,644-4
BUTYL N-(2,3-DICHLOROPHENYL)-CARBAMATE
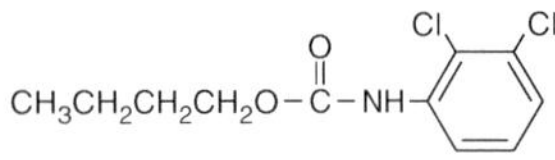
5 gm

S66,537-1
ALLYL N-(2,3-XYLYL)CARBAMATE
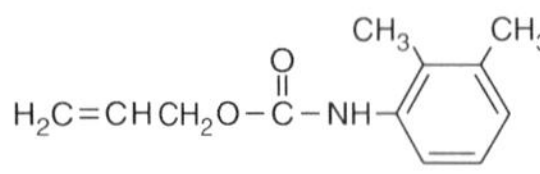
5 gm

S66,747-1
ALLYL N-(2,6-XYLYL)CARBAMATE
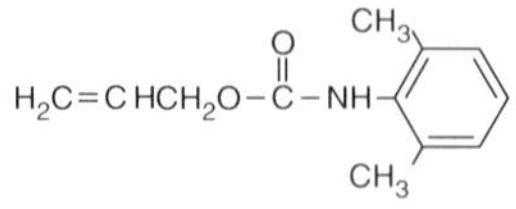
5 gm

S64,053-0
ISOPROPYL N-(2,3-XYLYL)CARBAMATE
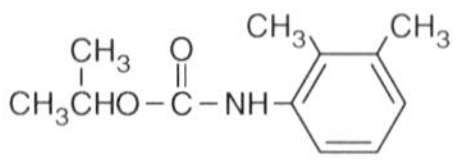
5 gm

S66,561-4
PROPYL N-(2,3-XYLYL)CARBAMATE
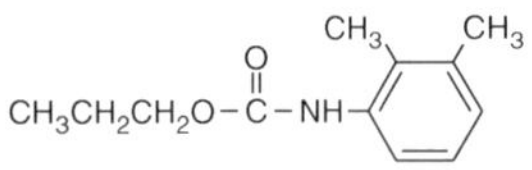
5 gm

S66,631-9
METHYL N-(2,6-DIETHYLPHENYL)-CARBAMATE
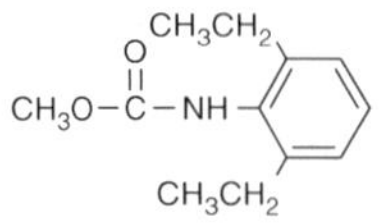
5 gm

S66,641-6
ETHYL N-(2-ETHYL-6-METHYLPHENYL)-CARBAMATE
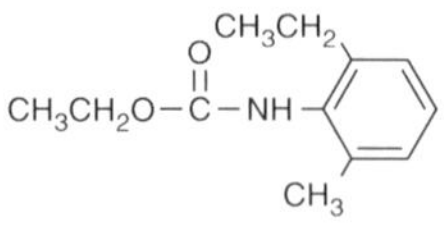
5 gm

S66,671-8
PROPYL N-(2,6-XYLYL)CARBAMATE
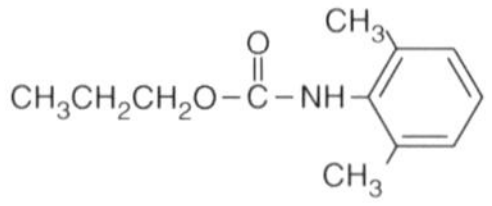
250 mg

S66,741-2
ISOPROPYL N-(2,6-XYLYL)CARBAMATE
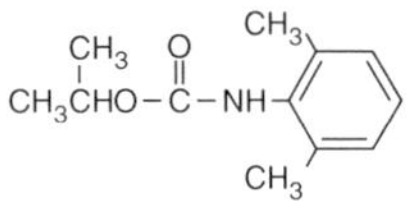
5 gm

S66,521-5
ALLYL N-(2-ETHYL-6-METHYLPHENYL)-CARBAMATE
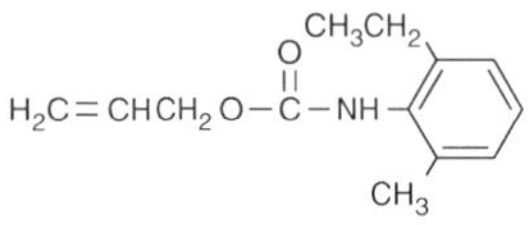
5 gm

S66,559-2
BUTYL N-(2,3-XYLYL)CARBAMATE
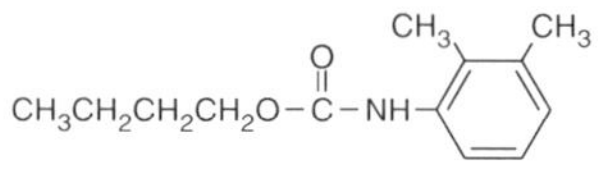
5 gm

S66,649-1
ISOPROPYL N-(2-ETHYL-6-METHYLPHENYL)CARBAMATE
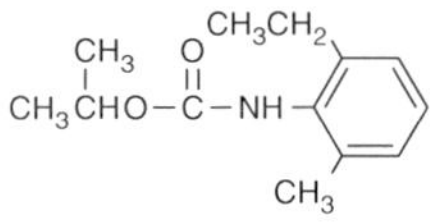
5 gm

S66,673-4
PROPYL N-(2-ETHYL-6-METHYLPHENYL)-CARBAMATE
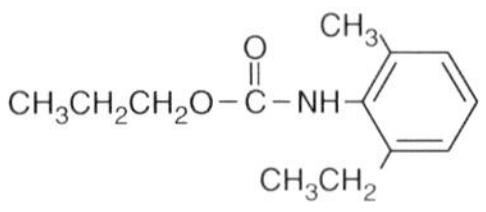
5 gm

S66,687-4
BUTYL N-(2,6-XYLYL)CARBAMATE
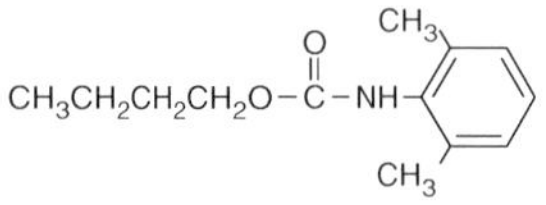
5 gm

S66,749-8
ETHYL N-(2,6-DIETHYLPHENYL)-CARBAMATE
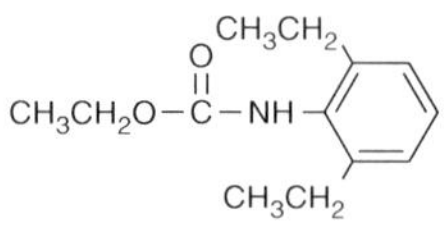
5 gm

S64,447-1
PHENYL N-(2,3-DICHLOROPHENYL)-CARBAMATE
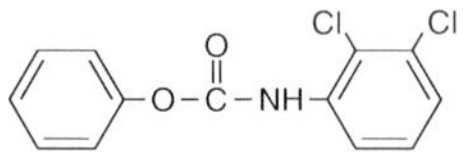
5 gm

S65,088-9
BUTYL N-(2-ETHYL-6-METHYLPHENYL)-CARBAMATE
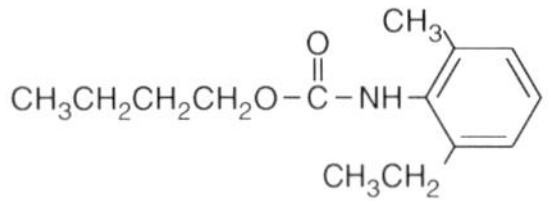
5 gm

S66,644-0
ISOPROPYL N-(2,6-DIETHYLPHENYL)-CARBAMATE
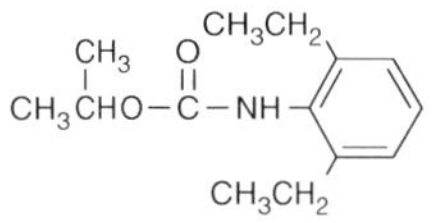
5 gm

S66,719-6
METHYL N-(2,6-DIISOPROPYLPHENYL)-CARBAMATE
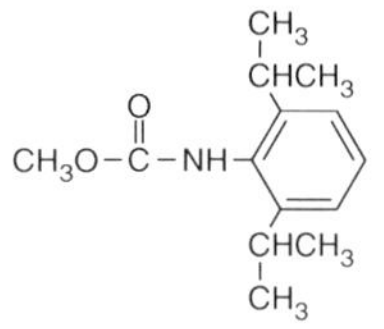
5 gm

S66,773-0
PROPYL N-(2,6-DIETHYLPHENYL)-CARBAMATE
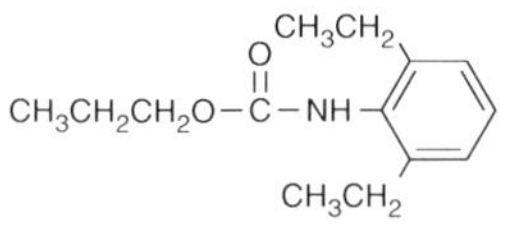
5 gm

S66,623-8
PHENYL N-(2,3-XYLYL)CARBAMATE
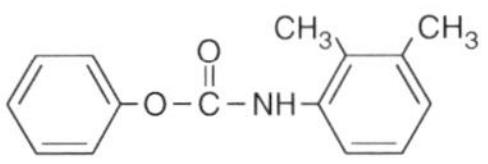
5 gm

S66,628-9
PHENYL N-(2,6-XYLYL)CARBAMATE
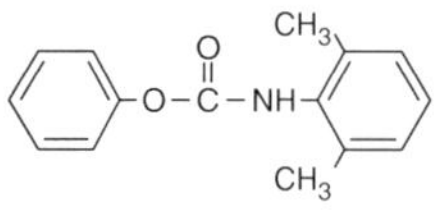
5 gm

S66,677-7
BUTYL N-(2,6-DIETHYLPHENYL)-CARBAMATE
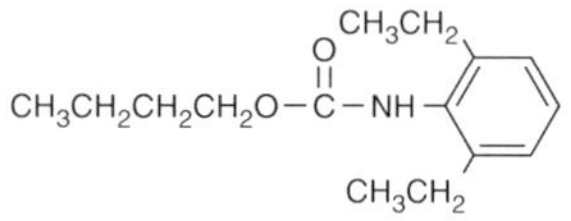
5 gm

S66,735-8
ETHYL N-(2,6-DIISOPROPYLPHENYL)-CARBAMATE
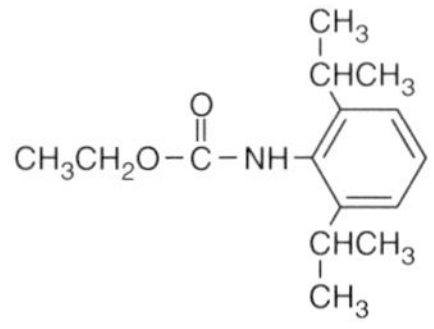
5 gm

S66,462-6
PHENYL N-(2-ETHYL-6-METHYLPHENYL)-CARBAMATE
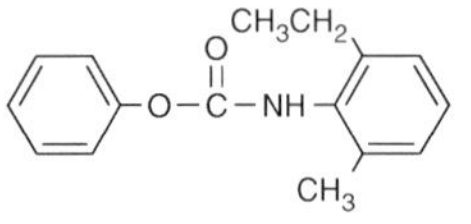
5 gm

S86,811-6
2,3-XYLYL N-(M-TOLYL)CARBAMATE
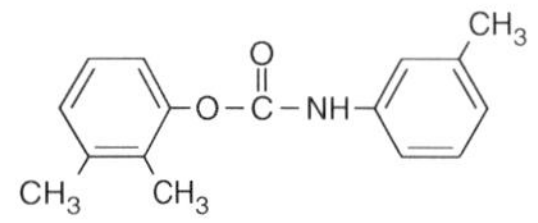
250 mg

S71,717-7
ALLYL N-(2,6-DIISOPROPYLPHENYL)-CARBAMATE
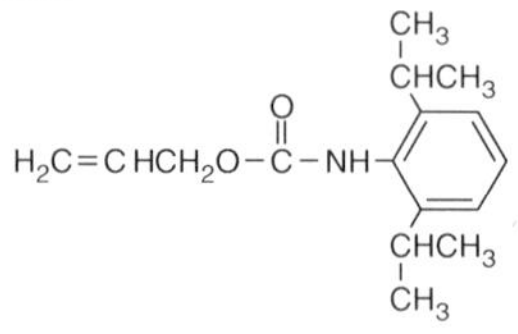
5 gm

S71,280-9
PROPYL N-(2,6-DIISOPROPYLPHENYL)-CARBAMATE
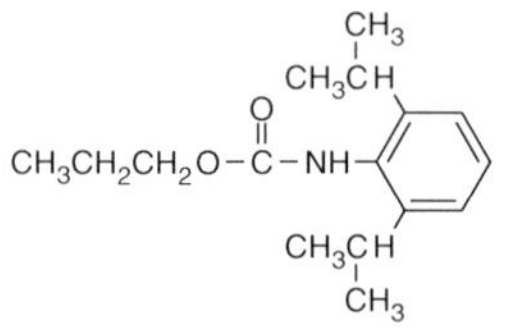
5 gm

S71,361-9
ISOPROPYL N-(2,6-DIISOPROPYL-PHENYL)CARBAMATE
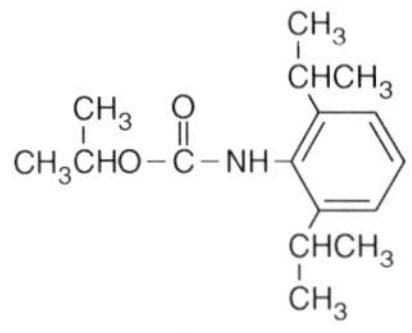
5 gm

S66,627-0
PHENYL N-(2,6-DIETHYLPHENYL)-CARBAMATE
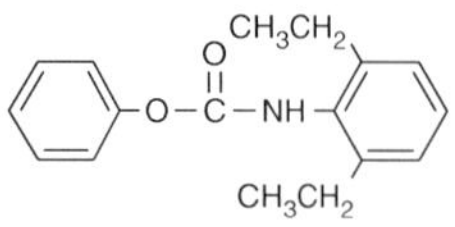
5 gm

S79,679-4
2,6-XYLYL N-(2,6-XYLYL)CARBAMATE
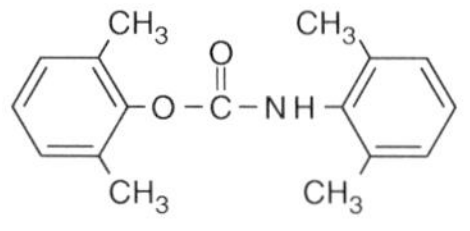
250 mg

S66,519-3
BUTYL N-(2,6-DIISOPROPYLPHENYL)-CARBAMATE
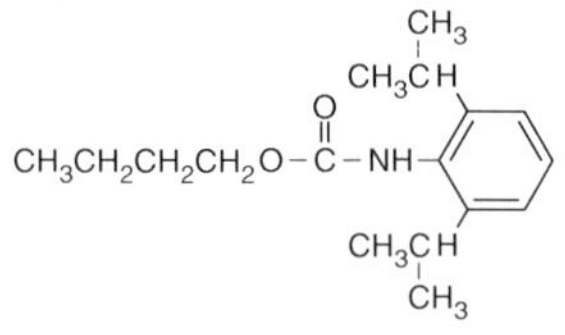
5 gm

S66,557-6
PHENYL N-(2,6-DIISOPROPYLPHENYL)-CARBAMATE
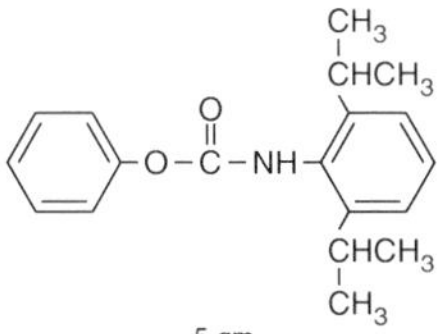
5 gm

S65,218-0
BENZYL N-(2,6-DICHLOROPHENYL)-CARBAMATE
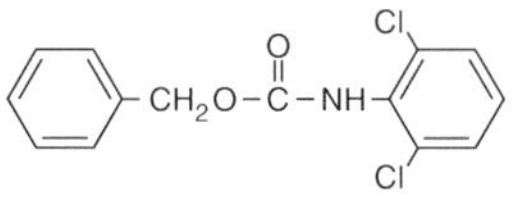
5 gm

S84,736-4
2-ISOPROPYL-5-METHYLPHENYL N-(2,6-XYLYL)CARBAMATE
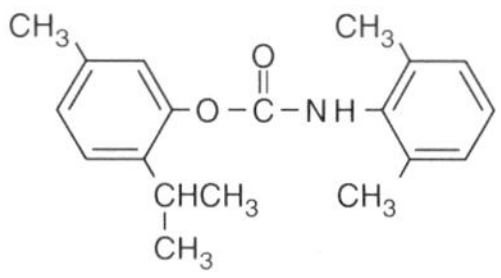
250 mg

S89,200-9
2,6-XYLYL N-(1-NAPHTHYL)CARBAMATE
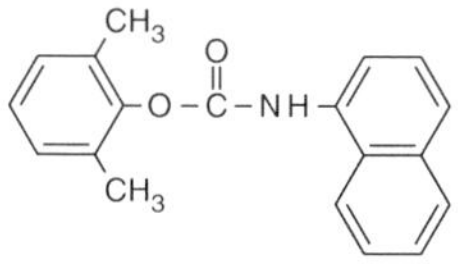
250 mg

S84,745-3
1-NAPHTHYL N-(2,6-XYLYL)CARBAMATE
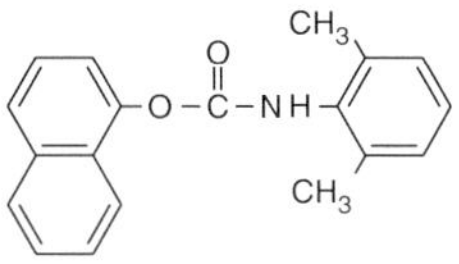
250 mg

S84,744-5
2-NAPHTHYL N-(2,6-XYLYL)CARBAMATE
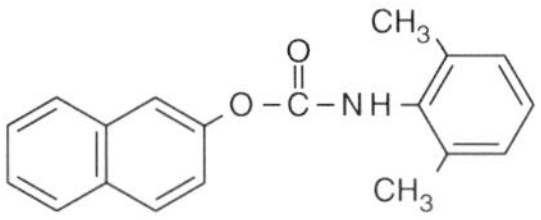
250 mg

S59,411-3
METHYL N-(2,5-DICHLOROPHENYL)-CARBAMATE
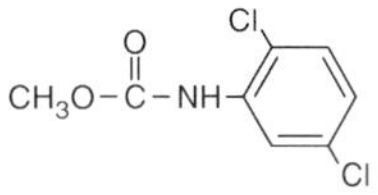
5 gm

S60,644-8
METHYL N-(3,4-DICHLOROPHENYL)-CARBAMATE
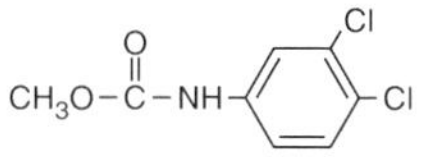
5 gm

S61,283-9
METHYL N-(2,4-DICHLOROPHENYL)-CARBAMATE
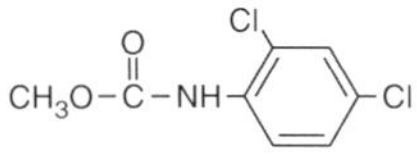
5 gm

S61,019-4
METHYL N-(5-CHLORO-2-METHYLPHENYL)-CARBAMATE
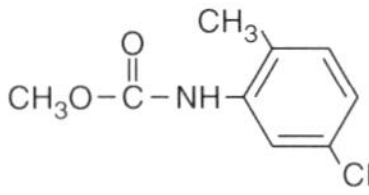
5 gm

S62,515-9
METHYL N-(3-CHLORO-4-METHYLPHENYL)-CARBAMATE
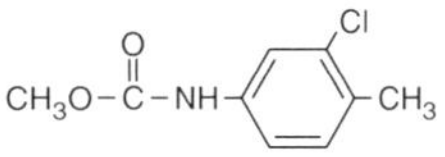
5 gm

S62,527-2
METHYL N-(4-CHLORO-2-METHYLPHENYL)-CARBAMATE
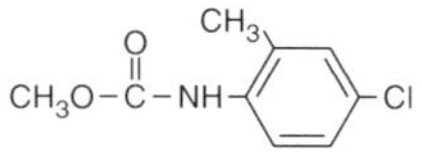
5 gm

S63,443-3
ETHYL N-(3,4-DICHLOROPHENYL)-CARBAMATE
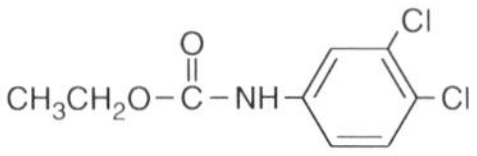
5 gm

S60,683-9
METHYL N-(2,4-XYLYL)CARBAMATE
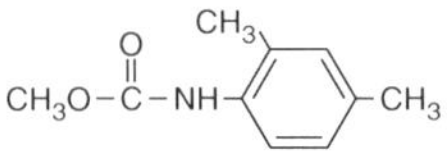
5 gm

S63,623-1
METHYL N-(2,5-XYLYL)CARBAMATE
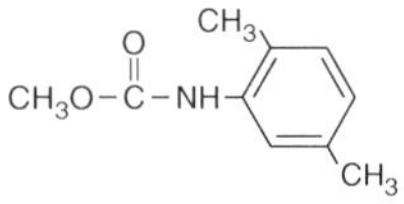
5 gm

S60,179-9
ETHYL N-(4-CHLORO-2-METHYLPHENYL)-CARBAMATE
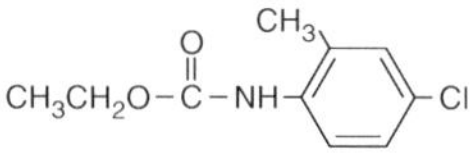
5 gm

S63,488-3
ETHYL N-(5-CHLORO-2-METHYLPHENYL)-CARBAMATE
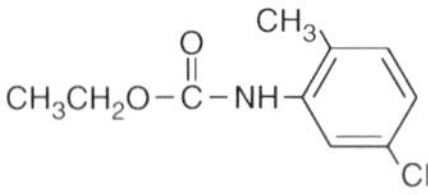
5 gm

S61,264-2
ALLYL N-(3,4-DICHLOROPHENYL)-CARBAMATE
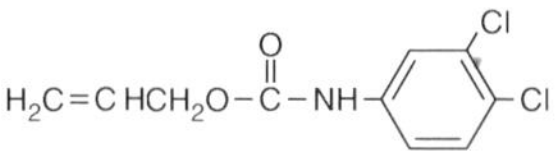
5 gm

S61,265-0
ALLYL N-(2,4-DICHLOROPHENYL)-CARBAMATE
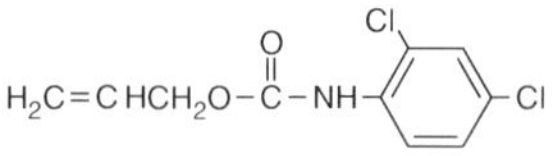
5 gm

S64,706-3
ALLYL N-(2,5-DICHLOROPHENYL)-CARBAMATE
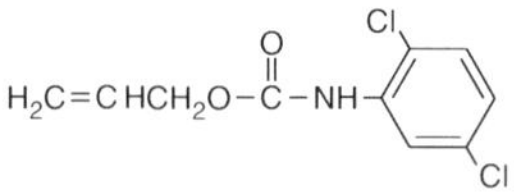
5 gm

S59,741-4
ISOPROPYL N-(2,5-DICHLOROPHENYL)-CARBAMATE
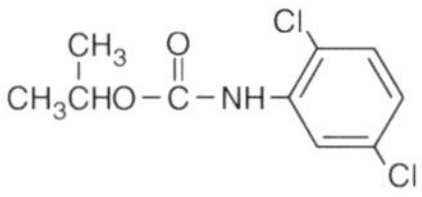
5 gm

S60,642-1
PROPYL N-(3,4-DICHLOROPHENYL)-CARBAMATE
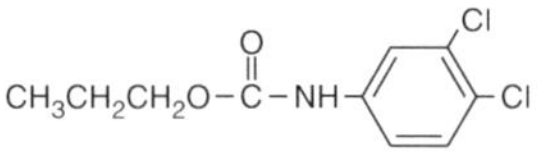
5 gm

S60,692-8
ISOPROPYL N-(3,4-DICHLOROPHENYL)-CARBAMATE
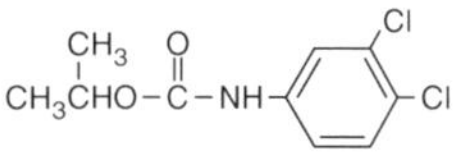
5 gm

S61,284-7
ISOPROPYL N-(2,4-DICHLOROPHENYL)-CARBAMATE
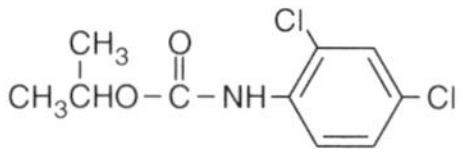
5 gm

S62,438-1
PROPYL N-(2,4-DICHLOROPHENYL)-CARBAMATE
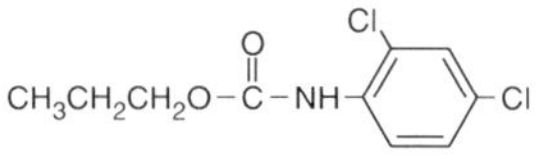
5 gm

S64,390-4
PROPYL N-(2,5-DICHLOROPHENYL)-CARBAMATE
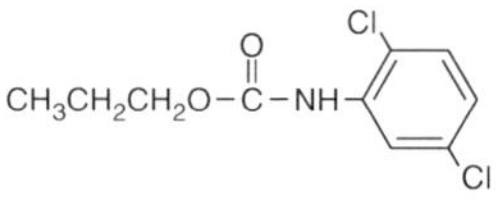
5 gm

S61,598-6
ETHYL N-(2,4-XYLYL)CARBAMATE
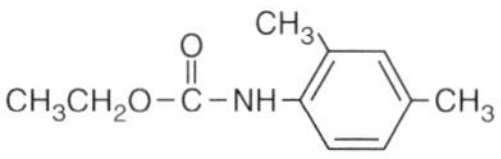
250 mg

S64,676-8
ETHYL N-(3,4-XYLYL)CARBAMATE
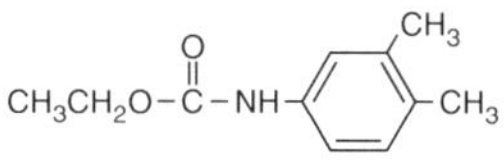
5 gm

S66,512-6
ETHYL N-(2,5-XYLYL)CARBAMATE
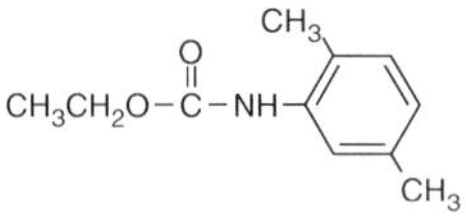
5 gm

S60,351-1
ALLYL N-(4-CHLORO-2-METHYLPHENYL)-CARBAMATE
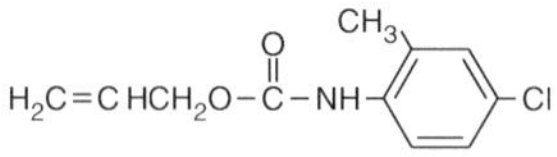
5 gm

S60,460-7
ALLYL N-(3-CHLORO-4-METHYLPHENYL)-CARBAMATE
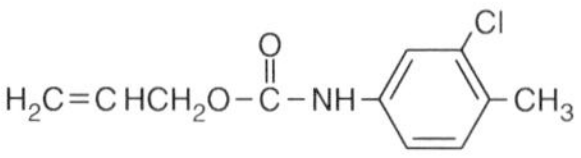
5 gm

S60,189-6
ISOPROPYL N-(4-CHLORO-2-METHYLPHENYL)CARBAMATE
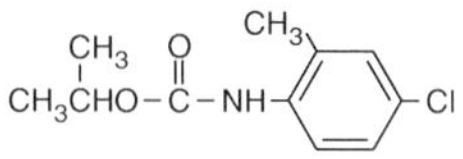
5 gm

S60,941-2
ISOPROPYL N-(3-CHLORO-4-METHYLPHENYL)CARBAMATE
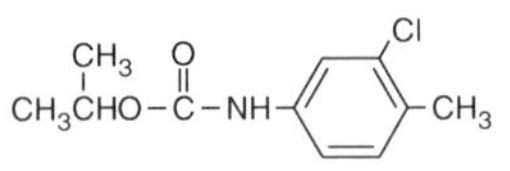
5 gm

S61,065-8
ISOPROPYL N-(5-CHLORO-2-METHYLPHENYL)CARBAMATE
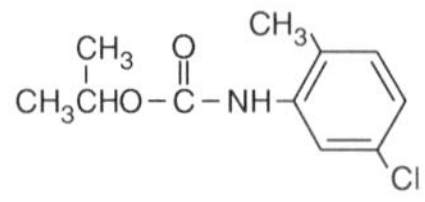
5 gm

S62,529-9
PROPYL N-(4-CHLORO-2-METHYLPHENYL)-CARBAMATE
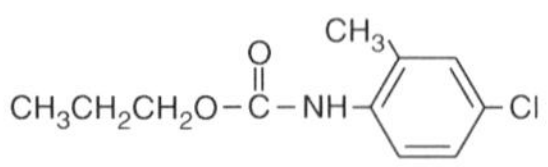
5 gm

S62,817-4
PROPYL N-(3-CHLORO-4-METHYLPHENYL)-CARBAMATE
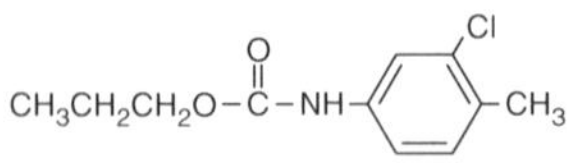
5 gm

S63,391-7
BUTYL N-(3,4-DICHLOROPHENYL)-CARBAMATE
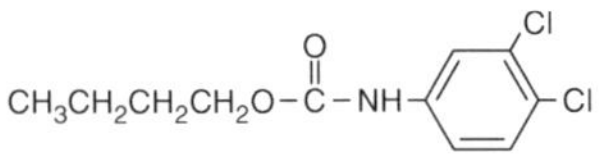
5 gm

S60,685-5
ALLYL N-(2,4-XYLYL)CARBAMATE
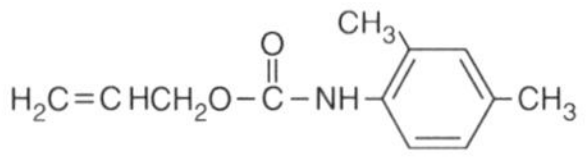
5 gm

S64,990-2
ALLYL N-(2,5-XYLYL)CARBAMATE
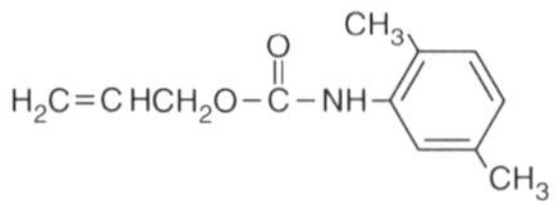
5 gm

S65,005-6
ALLYL N-(3,4-XYLYL)CARBAMATE
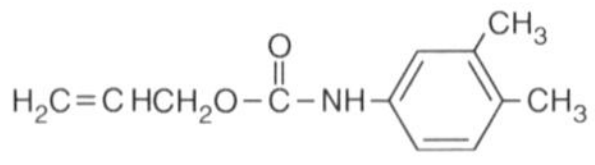
5 gm

S60,623-5
ISOPROPYL N-(2,4-XYLYL)CARBAMATE
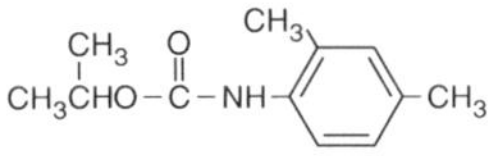
5 gm

S63,216-3
PROPYL N-(2,4-XYLYL)CARBAMATE
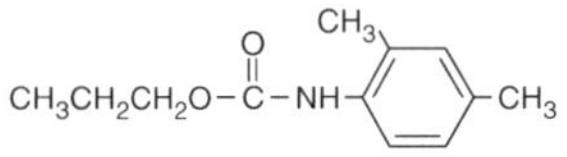
5 gm

S64,027-1
ISOPROPYL N-(2,5-XYLYL)CARBAMATE
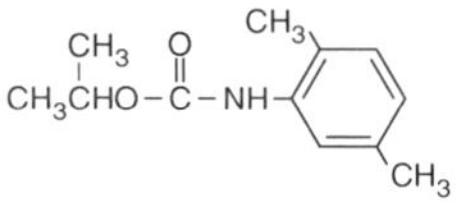
5 gm

S66,516-9
PROPYL N-(2,5-XYLYL)CARBAMATE
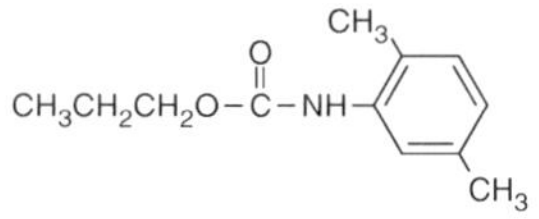
5 gm

S66,712-9
PROPYL N-(3,4-XYLYL)CARBAMATE
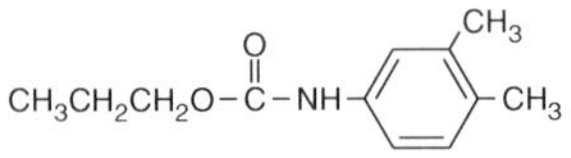
5 gm

S61,117-4
BUTYL N-(2,4-XYLYL)CARBAMATE
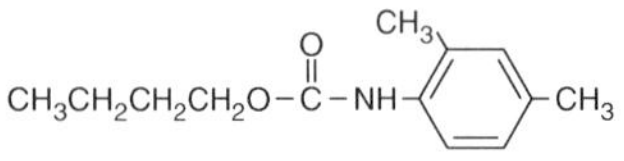
5 gm

S66,526-6
BUTYL N-(2,5-XYLYL)CARBAMATE
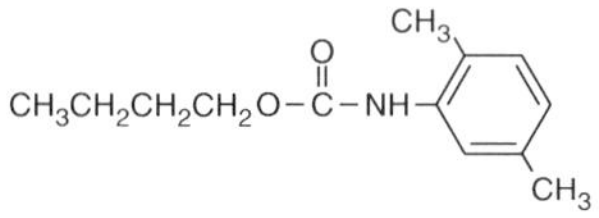
5 gm

S61,322-3
PHENYL N-(2,4-DICHLOROPHENYL)-CARBAMATE
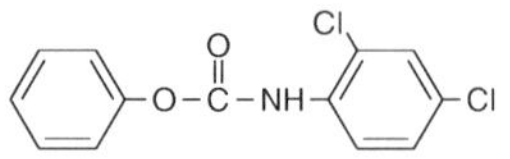
5 gm

S63,222-8
PHENYL N-(3,4-DICHLOROPHENYL)-CARBAMATE
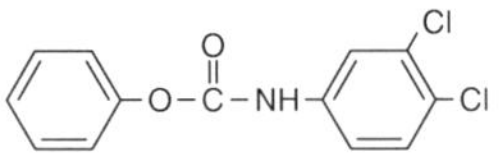
5 gm

S64,449-8
PHENYL N-(2,5-DICHLOROPHENYL)-CARBAMATE
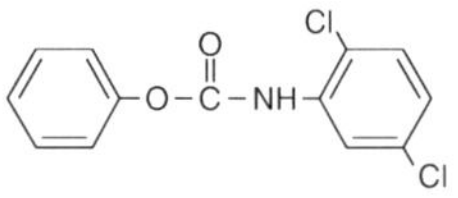
5 gm

S65,075-7
3-CHLOROPHENYL N-(2,4-DICHLOROPHENYL)CARBAMATE
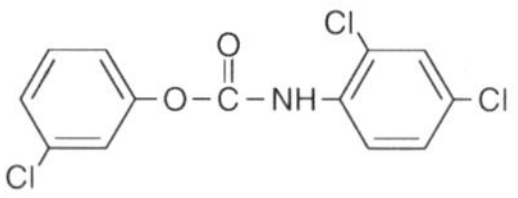
5 gm

S64,533-8
2,4,5-TRICHLOROPHENYL N-(2,4-DICHLOROPHENYL)CARBAMATE
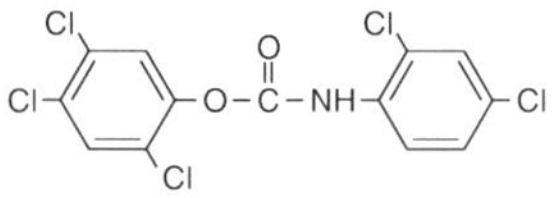
5 gm

S61,006-2
PHENYL N-(5-CHLORO-2-METHYLPHENYL)-CARBAMATE
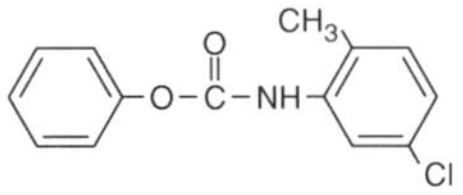
5 gm

S64,422-6
PHENYL N-(4-CHLORO-2-METHYLPHENYL)-CARBAMATE
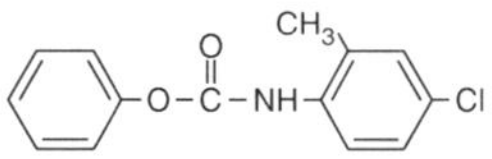
5 gm

S61,026-7
PHENYL N-(2,4-XYLYL)CARBAMATE
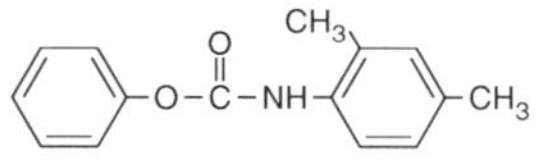
5 gm

S64,950-3
PHENYL N-(2,5-XYLYL)CARBAMATE
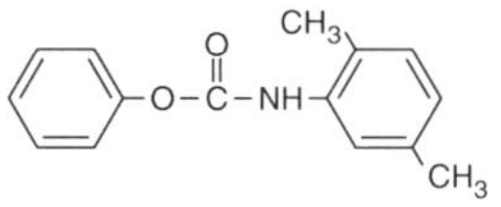
5 gm

S66,422-7
PHENYL N-(3,4-XYLYL)CARBAMATE
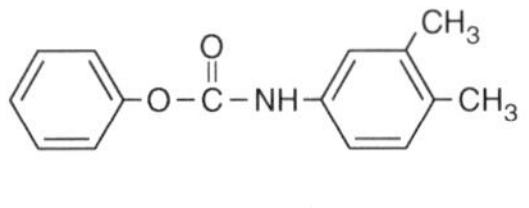
5 gm

S68,727-8
2-ISOPROPYL-5-METHYLPHENYL-N-(3-ETHYLPHENYL)CARBAMATE
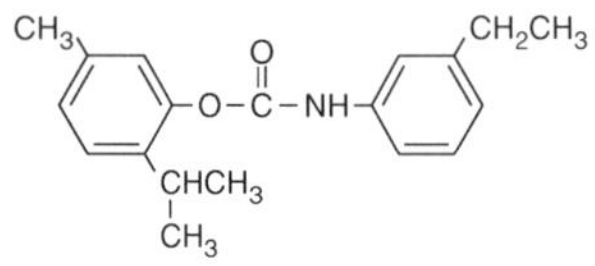
250 mg

S73,612-0
N,N'-HEXAMETHYLENEBIS(2-ISOPROPYL-5-METHYLPHENYL CARBAMATE)
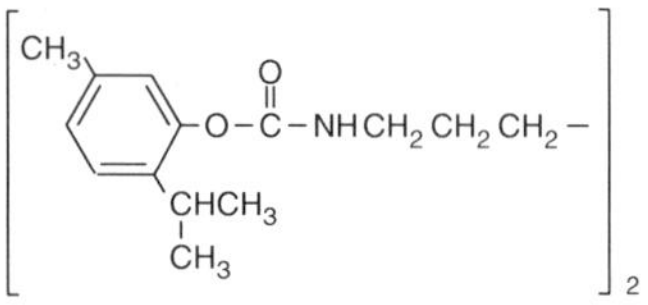
5 gm

S73,603-1
1-NAPHTHYL N-(2,5-DIMETHYLPHENYL)-CARBAMATE
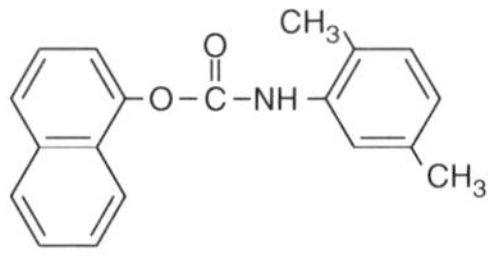
5 gm

S84,735-6
2-NAPHTHYL N-(2,5-XYLYL)CARBAMATE
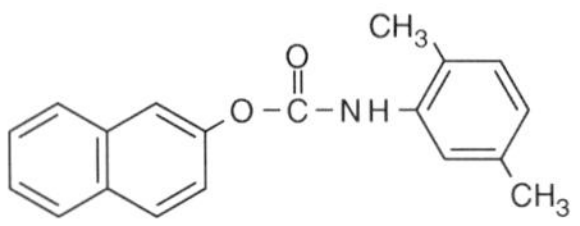
250 mg

S66,784-6
ALLYL N-(3,5-XYLYL)CARBAMATE
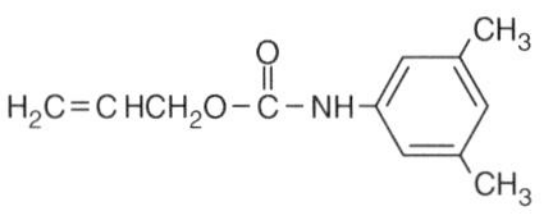
5 gm

S66,786-2
PROPYL N-(3,5-XYLYL)CARBAMATE
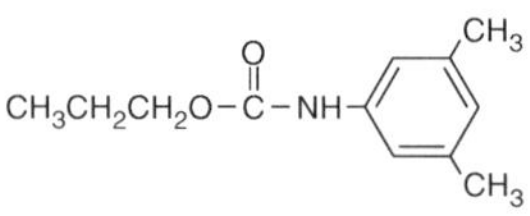
5 gm

S66,789-7
ISOPROPYL N-(3,5-XYLYL)CARBAMATE
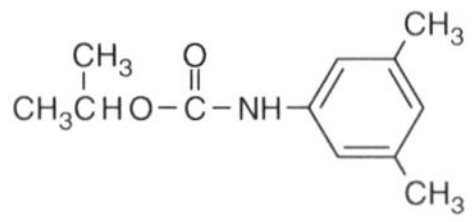
5 gm

S66,806-0
BUTYL N-(3,5-XYLYL)CARBAMATE
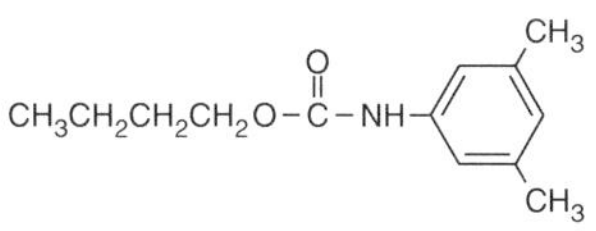
5 gm

S89,047-2
3,5-XYLYL N-(1-NAPHTHYL)CARBAMATE
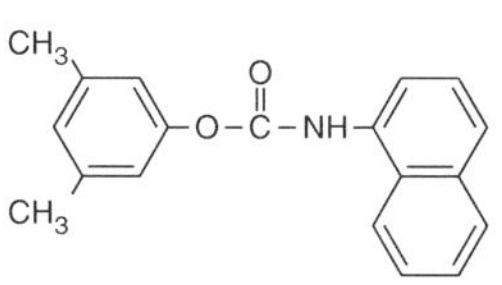
250 mg

S85,590-1
ETHYL N-(MESITYL)CARBAMATE
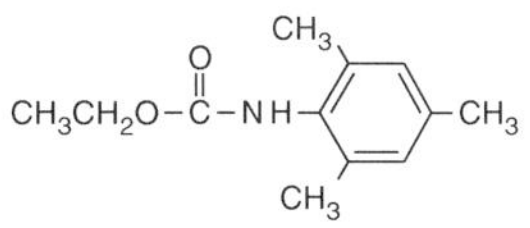
250 mg

S60,768-1
METHYL N-(2,4,5-TRICHLOROPHENYL)-CARBAMATE
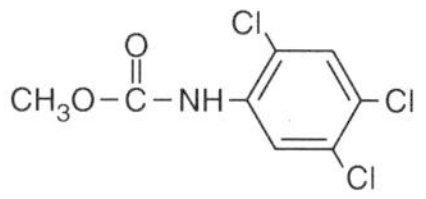
5 gm

S60,752-5
ETHYL N-(2,4,5-TRICHLOROPHENYL)-CARBAMATE
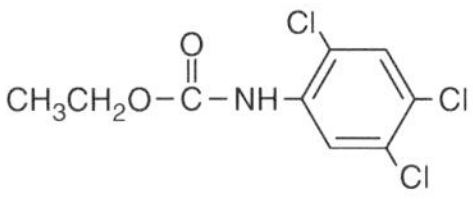
5 gm

S60,776-2
ALLYL N-(2,4,5-TRICHLOROPHENYL)-CARBAMATE
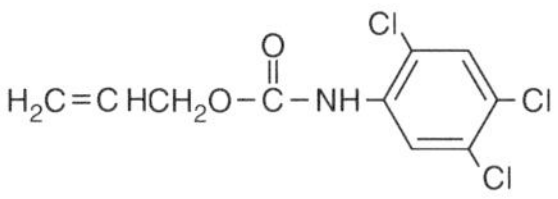
5 gm

S60,775-4
ISOPROPYL N-(2,4,5-TRICHLOROPHENYL)CARBAMATE
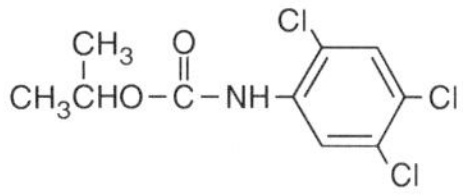
5 gm

S63,388-7
PROPYL N-(2,4,5-TRICHLOROPHENYL)-CARBAMATE
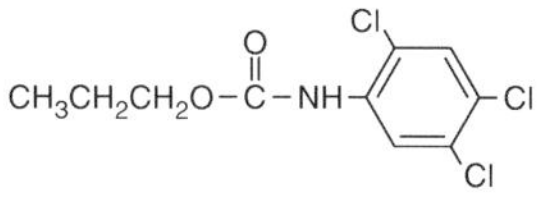
5 gm

S63,380-1
BUTYL N-(2,4,5-TRICHLOROPHENYL)-CARBAMATE
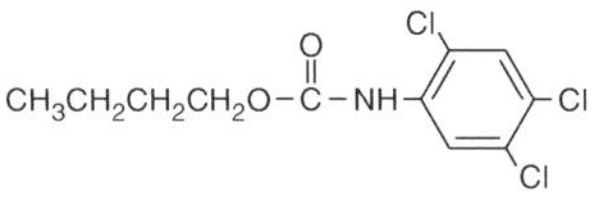
5 gm

S75,282-7
2-NAPHTHYL N-(1-NAPHTHYL)CARBAMATE
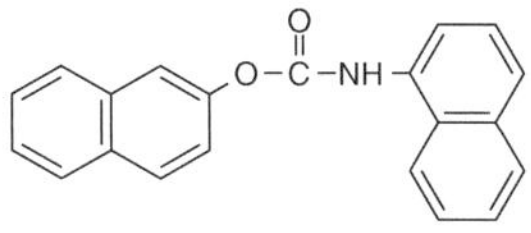
5 gm

S58,639-0
1-NAPHTHYL N-(METHYL)CARBAMATE
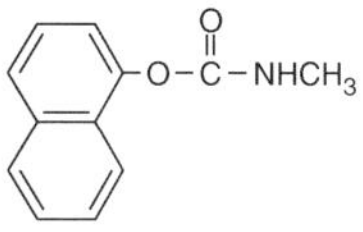
250 mg

S59,029-0
METHYL N-(1-NAPHTHYL)CARBAMATE
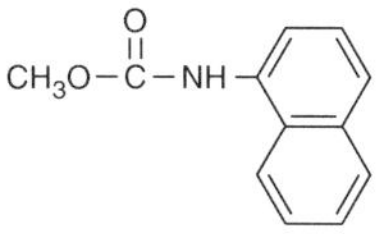
5 gm

S59,001-0
ETHYL N-(1-NAPHTHYL)CARBAMATE
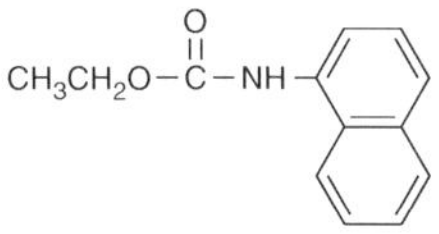
5 gm

S59,050-9
ALLYL N-(1-NAPHTHYL)CARBAMATE
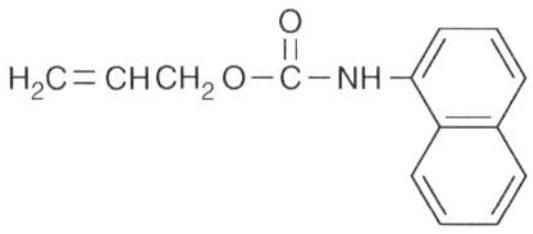
5 gm

S59,058-4
ISOPROPYL N-(1-NAPHTHYL)CARBAMATE
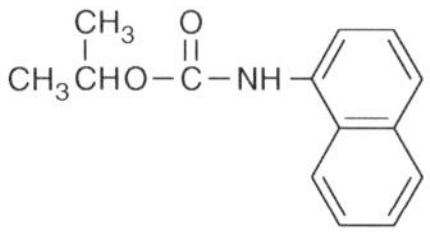
5 gm

S59,008-8
BUTYL N-(1-NAPHTHYL)CARBAMATE
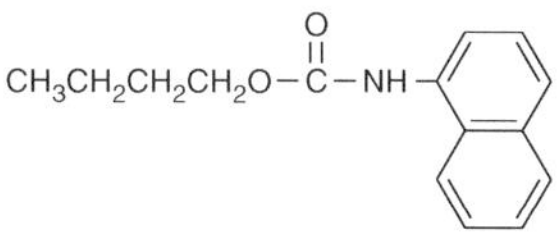
5 gm

S59,005-3
PENTYL N-(1-NAPHTHYL)CARBAMATE
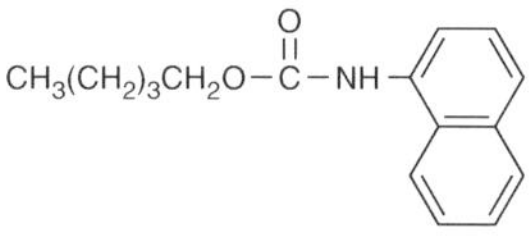
5 gm

S89,204-1
M-TOLYL N-(1-NAPHTHYL)CARBAMATE
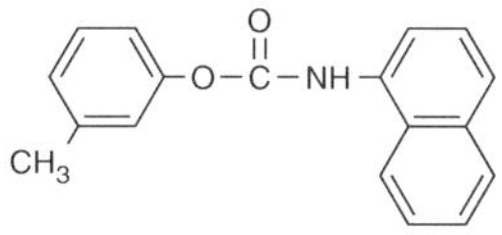
250 mg

S70,955-7
N,N'-(1,5-NAPHTHALENEDIYL)BIS (METHYL CARBAMATE)
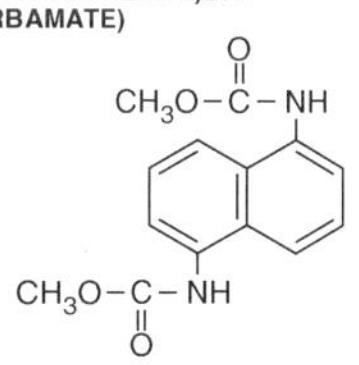
250 mg

S72,473-4
2-NAPHTHYL N-(METHYL)CARBAMATE
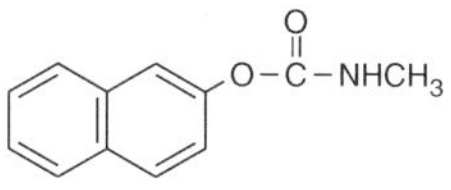
5 gm

S71,582-4
2-NAPHTHYL N-(TERT-BUTYL)CARBAMATE
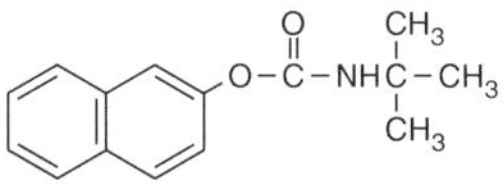
5 gm

S72,138-7
6-BROMO-2-NAPHTHYL N-(ISOPROPYL)CARBAMATE
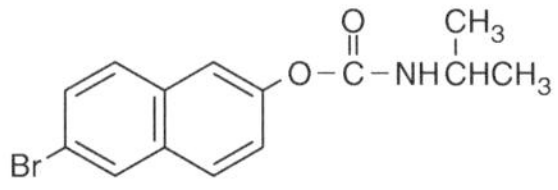
250 mg

S72,531-5
6-BROMO-2-NAPHTHYL N-(PROPYL)CARBAMATE
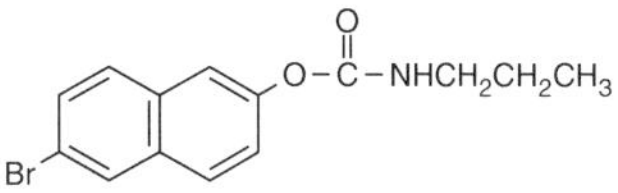
5 gm

S72,312-6
6-BROMO-2-NAPHTHYL N-(BUTYL)-CARBAMATE
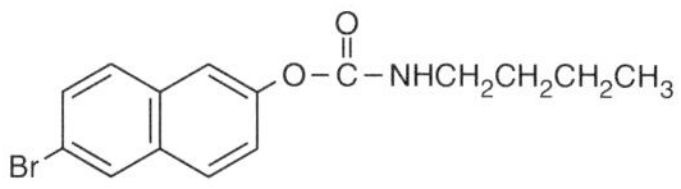
250 mg

S59,014-2
PROPYL N-(1-NAPHTHYL)CARBAMATE
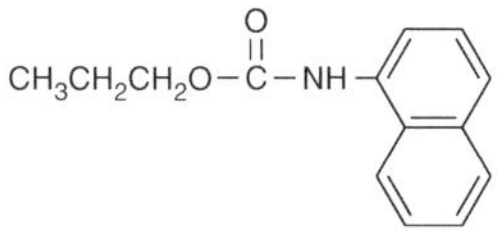
5 gm

S61,289-8
ALLYL N-(5-CHLORO-2-METHYLPHENYL)-CARBAMATE
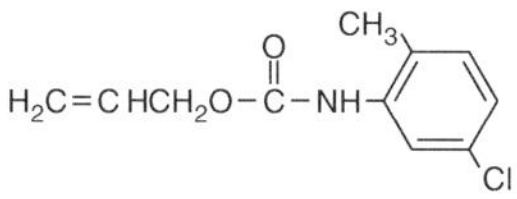
5 gm

S36,156-9
ETHYL 3-CHLOROPHENYLCARBAMATE
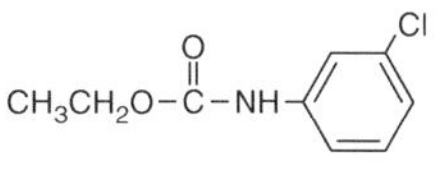
5 gm

S67,371-4
PIPERIDINO N-(4-BUTYLPHENYL)-CARBAMATE
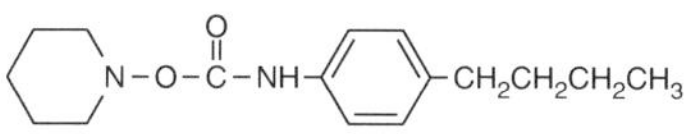
250 mg

S51,827-1
2-CHLOROPHENYL N-(3-CHLOROPHENYL)-CARBAMATE
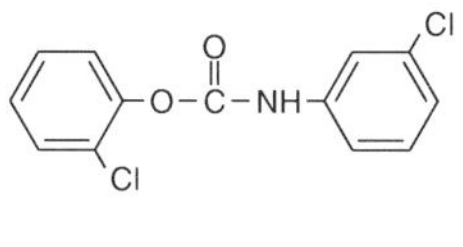
5 gm

S36,157-7
ETHYL N-(4-CHLOROPHENYL)-CARBAMATE
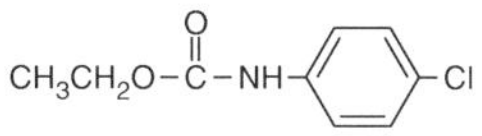
5 gm

S36,166-6
ETHYL N-(2,5-DICHLOROPHENYL)-CARBAMATE
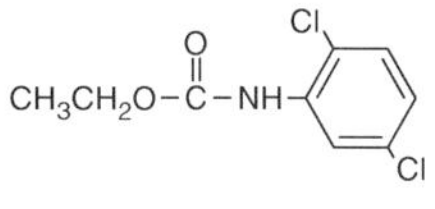
5 gm

S69,409-6
9-FLUORENYLIDENEAMINO-N-(1-NAPHTHYL)CARBAMATE
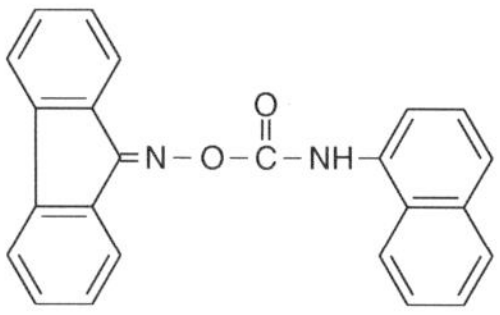
250 mg

S59,582-9
S-PROPYL N-(3-CHLOROPHENYL)-THIOLCARBAMATE
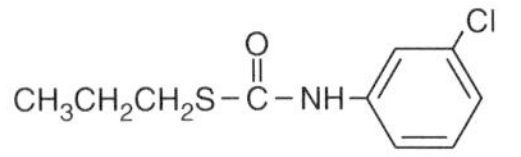
5 gm

S60,162-4
S-ISOPROPYL N-(3-CHLOROPHENYL)-THIOLCARBAMATE
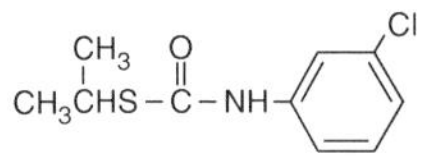
5 gm

S64,695-4
S-PROPYL N-(M-TOLYL)THIOLCARBAMATE
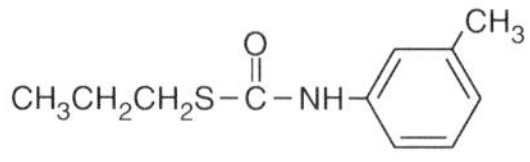
5 gm

S64,716-0
S-ISOPROPYL N-(M-TOLYL)-THIOLCARBAMATE
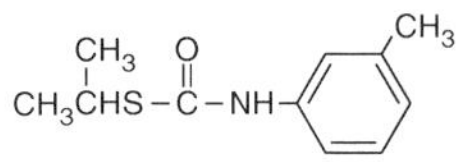
5 gm

S59,581-0
S-BUTYL N-(3-CHLOROPHENYL)-THIOCARBAMATE
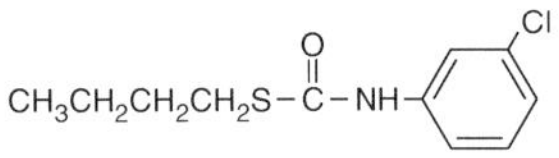
5 gm

S64,707-1
S-BUTYL N-(M-TOLYL)THIOCARBAMATE
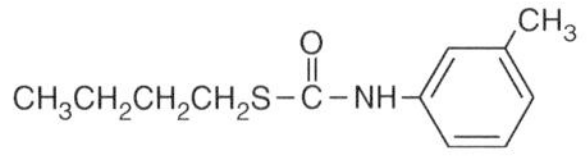
250 mg

S76,906-1
S-(2-THIAZOLIN-2-YL)-N-METHYLTHIOCARBAMATE
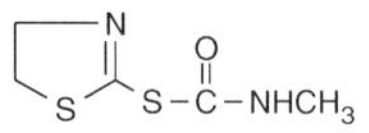
250 mg

S95,459-4
S-PHENYL N-(4-CYCLOHEXYLCYCLO-HEXYL)THIOCARBAMATE
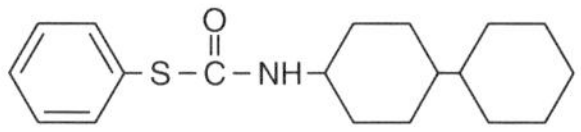
50 mg

S64,393-9
S-ISOPROPYL N-(PHENYL)THIOL-CARBAMATE
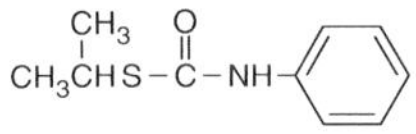
5 gm

S64,394-7
S-PROPYL N-(PHENYL)THIOLCARBAMATE
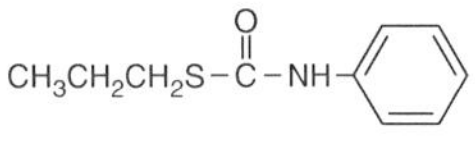
5 gm

S77,437-5
S-BENZYL N,N-(DIMETHYL)THIO-CARBAMATE
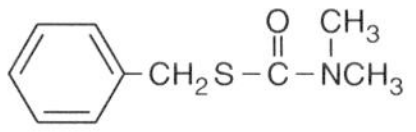
250 mg

S64,400-5
S-BUTYL N-(PHENYL)THIOCARBAMATE
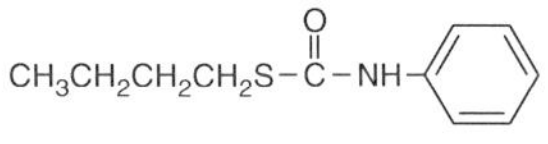
5 gm

S82,272-8
S-(P-TOLYL) N-(PHENYL)THIOCARBAMATE
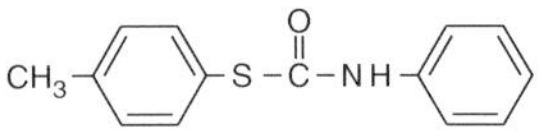
250 mg

S63,450-6
S-ISOPROPYL N-(2-CHLOROPHENYL)-THIOLCARBAMATE
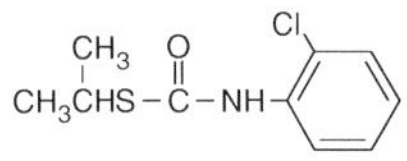
5 gm

S65,282-2
S-ISOPROPYL N-(O-TOLYL)-THIOLCARBAMATE
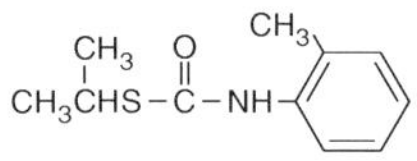
5 gm

S84,565-5
S-(M-TOLYL) N-(2-ETHYLPHENYL)-THIOCARBAMATE
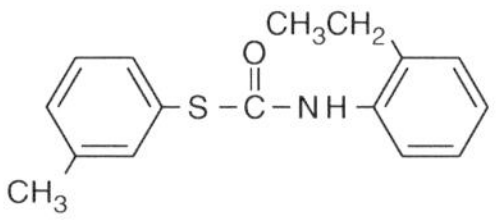
250 mg

S84,805-0
S,S'-ETHYLENE BIS[N-(2-ETHYLPHENYL)THIOCARBAMATE]
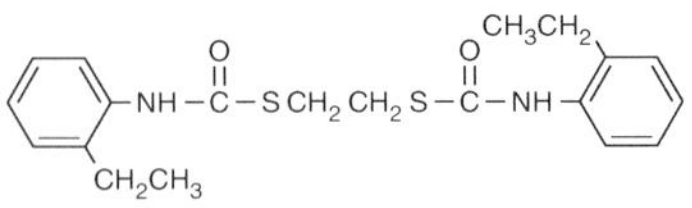
250 mg

S64,441-2
S-ISOPROPYL N-(4-CHLOROPHENYL)-THIOLCARBAMATE
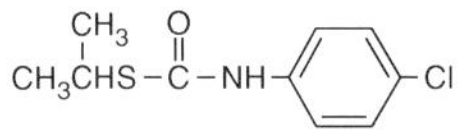
5 gm

S60,171-3
S-ISOPROPYL N-(4-BROMOPHENYL)-THIOLCARBAMATE
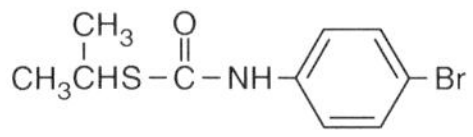
5 gm

S63,165-5
S-PROPYL N-(4-BROMOPHENYL)-THIOLCARBAMATE
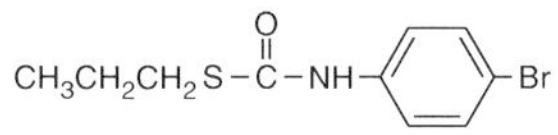
5 gm

S65,231-8
S-ISOPROPYL N-(P-TOLYL)-THIOLCARBAMATE
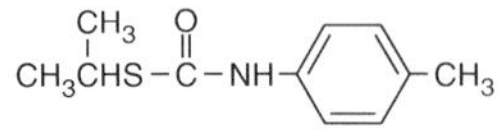
5 gm

S65,232-6
S-PROPYL N-(P-TOLYL)THIOLCARBAMATE
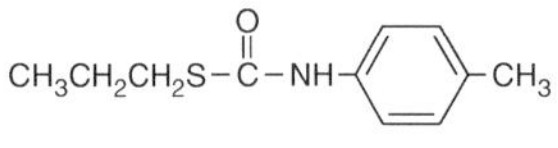
5 gm

S59,580-2
S-BUTYL N-(4-BROMOPHENYL)-THIOCARBAMATE
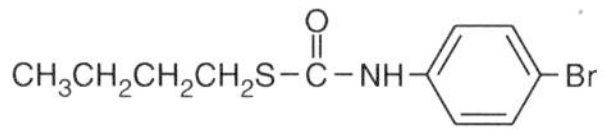
5 gm

S65,374-8
S-(P-TOLYL) N-(2,6-DICHLOROPHENYL)-THIOLCARBAMATE
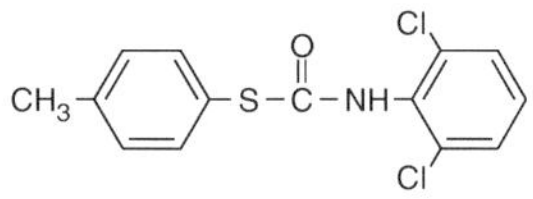
5 gm

S65,459-0
4,4'-METHYLENEBIS[S-ISOPROPYL N-(PHENYL)THIOLCARBAMATE]
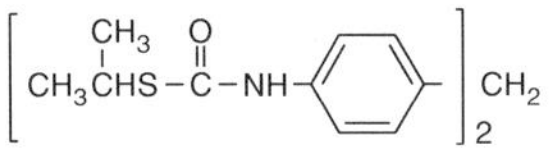
5 gm

S71,350-3
BIS(S-PROPYL) N,N'-(METHYLENEDI-4,1-PHENYLENE)BISTHIOCARBAMATE
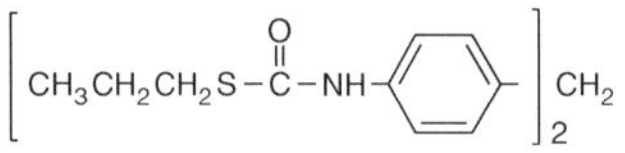
250 mg

S65,408-6
4,4'-METHYLENEBIS(S-BUTYL-N-PHENYLTHIOLCARBAMATE)
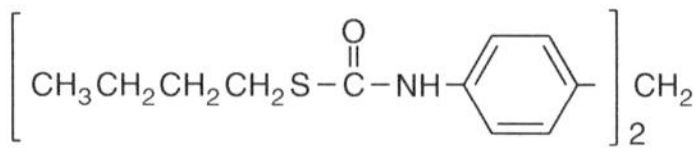
250 mg

S60,196-9
S-ISOPROPYL N-(2,6-DICHLOROPHENYL)-THIOLCARBAMATE
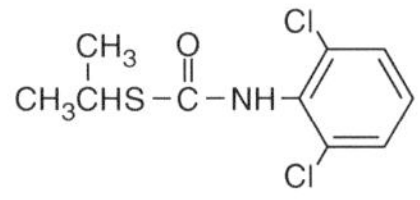
5 gm

S63,937-0
S-ISOPROPYL N-(2,3-DICHLOROPHENYL)-THIOLCARBAMATE
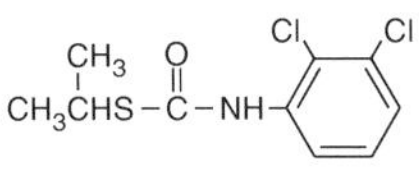
5 gm

S65,379-9
S-BUTYL N-(2-ETHYL-6-METHYLPHENYL)-THIOCARBAMATE
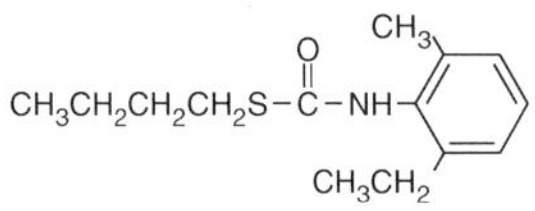
5 gm

S64,696-2
S-ISOPROPYL N-(3,4-DICHLOROPHENYL)-THIOLCARBAMATE
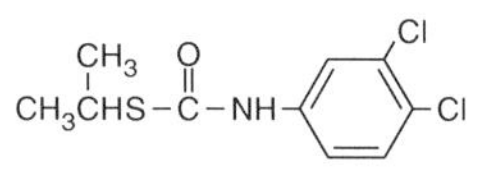
5 gm

S64,727-6
S-ISOPROPYL N-(2,4-DICHLOROPHENYL)-THIOLCARBAMATE
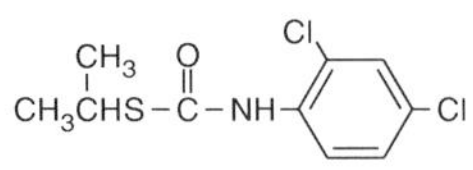
5 gm

S65,010-2
S-ISOPROPYL N-(4-CHLORO-2-METHYLPHENYL)THIOLCARBAMATE

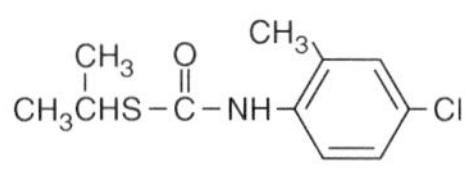

5 gm

S65,031-5
S-PROPYL N-(4-CHLORO-2-METHYLPHENYL)THIOLCARBAMATE

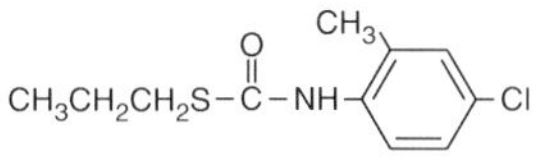

5 gm

S65,032-3
S-ISOPROPYL N-(3-CHLORO-4-METHYLPHENYL)THIOLCARBAMATE

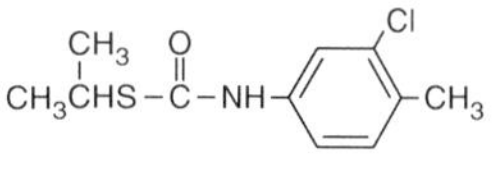

5 gm

S71,070-9
S-ISOPROPYL N-(2,4-XYLYL)-THIOLCARBAMATE

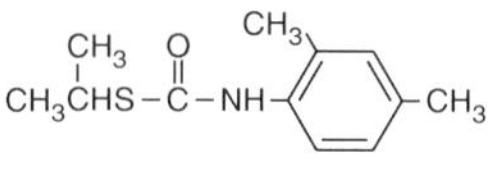

5 gm

S71,705-3
S-PROPYL N-(2,4-XYLYL)-THIOLCARBAMATE

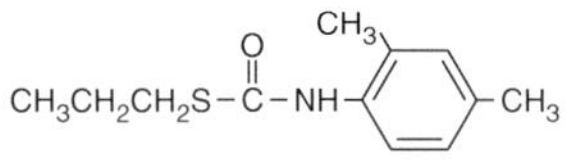

5 gm

S65,066-8
S-BUTYL N-(4-CHLORO-2-METHYL-PHENYL)THIOCARBAMATE

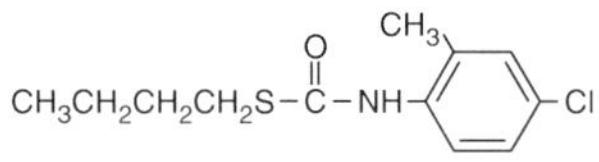

5 gm

S71,071-7
S-BUTYL N-(2,4-XYLYL)THIOCARBAMATE

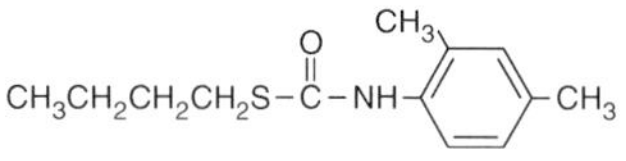

5 gm

S63,660-6
S-ISOPROPYL N-(1-NAPHTHYL)-THIOLCARBAMATE

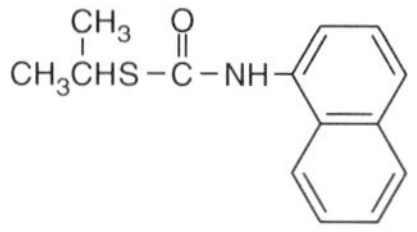

5 gm

S77,015-9
3-(2,4-DICHLOROBENZYLIDENE)-6-(TRIFLUOROMETHYL)-2-INDOLINONE

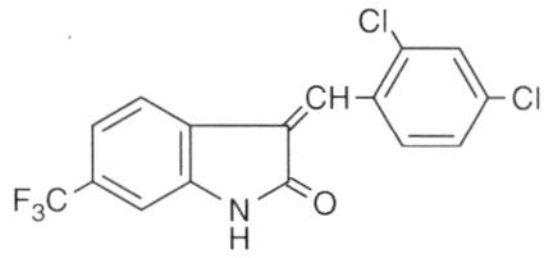

250 mg

S76,920-7
3,4-DIHYDRO-7-METHOXY-4-(TRIFLUOROMETHYL)-2(1H)-QUINOLINONE

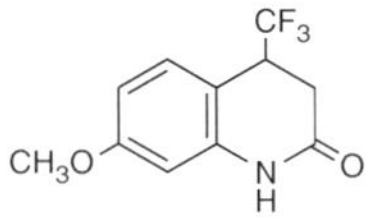

250 mg

S85,910-9
1-[2,6-DINITRO-4-(TRIFLUOROMETHYL)-PHENYL]-2-PYRROLIDINONE

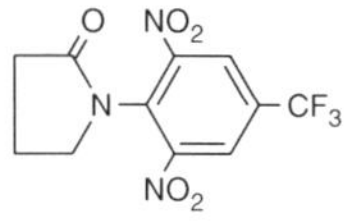

250 mg

S66,187-2
2-(2-AMINOETHYL)-1-(4-CHLORO-PHENYL)-1-HYDROXY-3-OXOISOINDOLINE

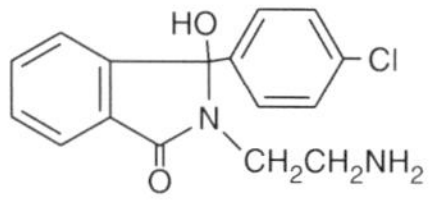

5 gm

S76,085-4
3-(4-CHLOROPHENYL)-3-HYDROXY-2-(TETRAHYDROPYRAN-2-YLOXY)-1-ISOINDOLINONE

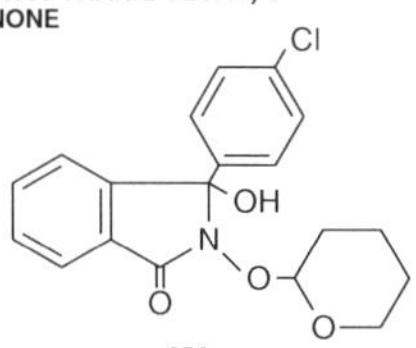

250 mg

S73,525-6
ETHYL 2-[3-(4-CHLOROPHENYL)-3-HYDROXY-1-OXO-2-ISOINDOLINYL]-ACETATE

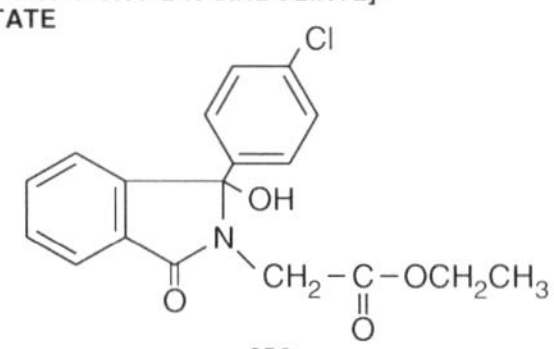

250 mg

S73,355-5
3-(4-CHLOROPHENYL)-2-CYCLOHEXYL-3-HYDROXY-1-ISOINDOLINONE

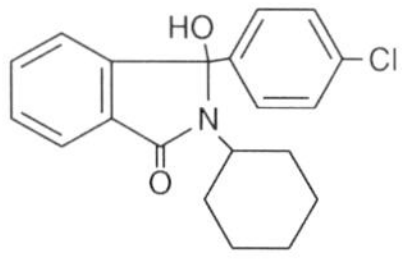

250 mg

S76,056-0
2-(4-CHLOROPHENYL)-1-[2-(3,4-DIMETHOXYPHENYL)ETHYL]-6-OXO-3-PIPERIDINECARBOXYLIC ACID

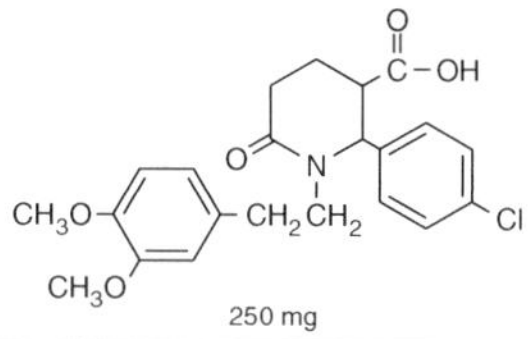

250 mg

S81,114-9
1-(4-CHLOROPHENYL)-4-(4-METHOXYPHENYL)-2-AZETIDINONE

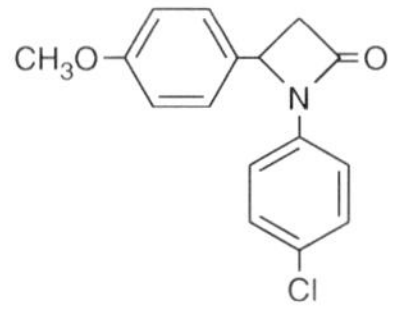

250 mg

S76,394-2
3-[5-CHLORO-2-(METHYLAMINO)PHENYL]-PHTHALIDE

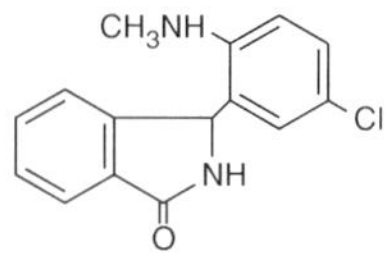

250 mg

S76,382-9
1,1'-TRIMETHYLENEBIS[2-(4-CHLOROPHENYL)-6-OXO-3-PIPERIDINECARBOXYLIC ACID]

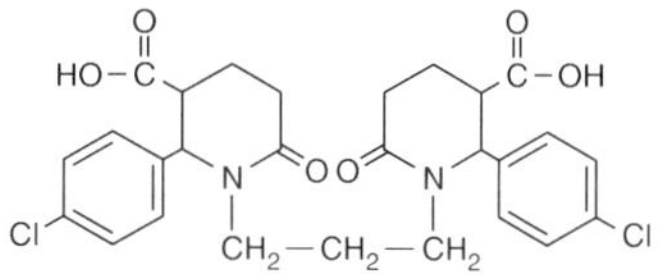

250 mg

S71,653-7
1-(4-CHLOROBENZYL)ISATIN

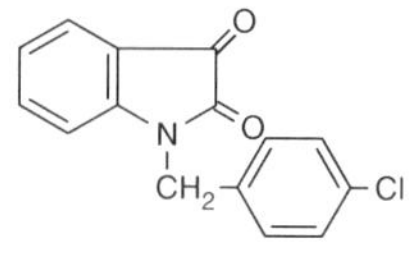

5 gm

S76,334-9
N,2-DIMETHYL-3-(2-FLUOROPHENYL)-1-OXO-1,2,3,4-TETRAHYDRO-4-ISOQUINOLINECARBOXAMIDE

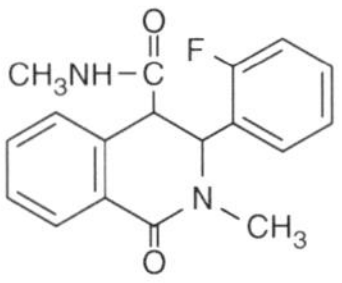

250 mg

S73,284-2
2-(4-FLUOROPHENYL)-1-METHYL-6-OXO-3-PIPERIDINECARBOXYLIC ACID

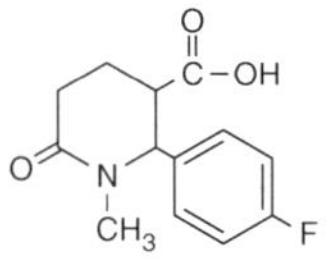

250 mg

S76,661-5
2-(4-CHLOROPHENYL)-N-METHYL-6-OXO-3-PIPERIDINECARBOXAMIDE

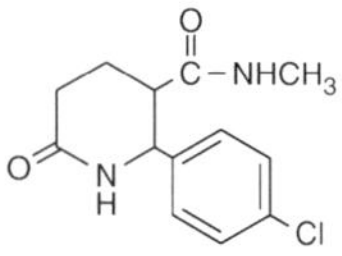

250 mg

S83,965-5
4-(2-CHLOROPHENYL)-1-PHENYL-2-AZETIDINONE

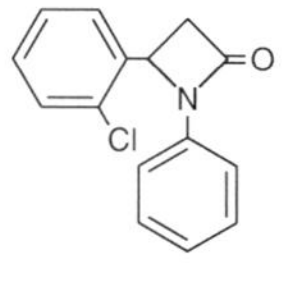

250 mg

S81,113-0
1-(4-CHLOROPHENYL)-4-PHENYL-2-AZETIDINONE

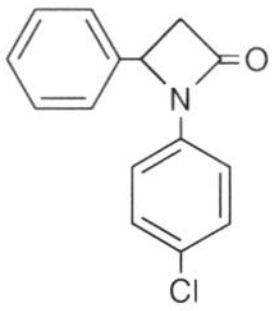

250 mg

S67,540-7
3,3-BIS(4-HYDROXYPHENYL)OXINDOLE

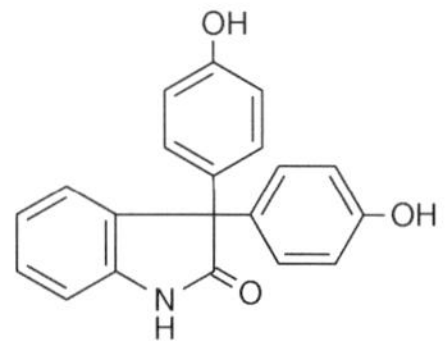

5 gm

S95,567-1
2-(2,3-DIMETHOXYPHENYL)-N-[2-HYDROXY-2-(2-OXOINDOLIN-3-YLIDENE)-ETHYL]ACETAMIDE

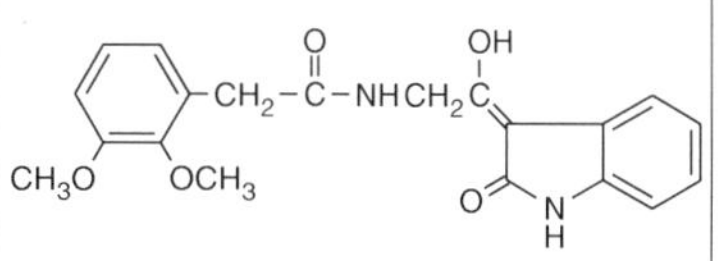

50 mg

S76,669-0
3-TERT-BUTYL-3-HYDROXY-5-METHOXY-2-METHYL-1-ISOINDOLINONE

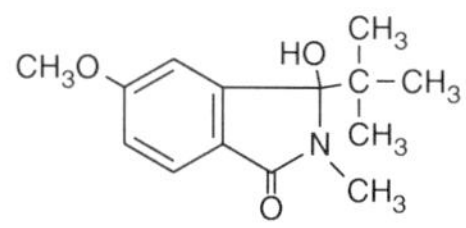

250 mg

S60,330-9
BENZYL 1,2-DIPHENYL-4-HYDROXY-5-OXO-3-PYRROLINE-3-CARBOXYLATE

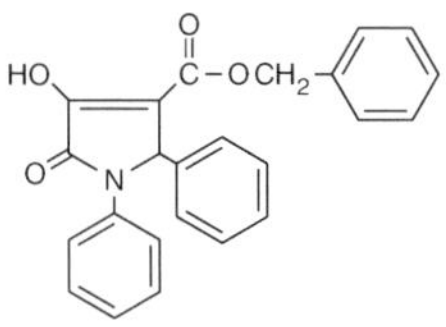

250 mg

S65,683-6
1-TERT-BUTYL-1-HYDROXY-2-PHENYL-3-OXOISOINDOLINE

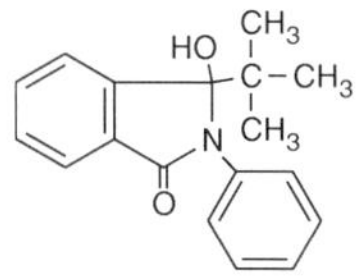

5 gm

S60,888-2
3-HYDROXY-3-NITROMETHYLOXINDOLE

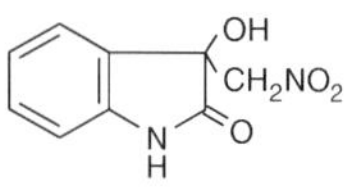

250 mg

S89,917-8
ETHYL 1-HYDROXY-2-OXO-1,2,3,10-TETRAHYDROINDENO(2,1-E)INDOLE-1-CARBOXYLATE

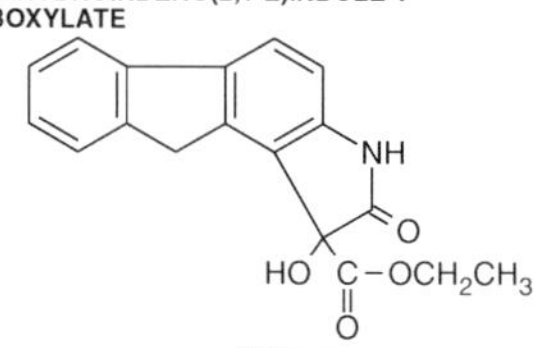

250 mg

S76,162-1
ETHYL 3-HYDROXY-3-(2-OXO-3-INDOLINYLIDENE)PROPIONATE

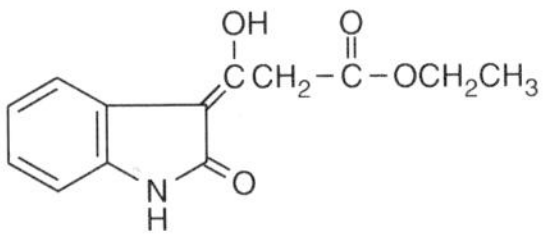

250 mg

S88,018-3
3-HYDROXY-2-PHENYL-1-ISOINDOLINONE

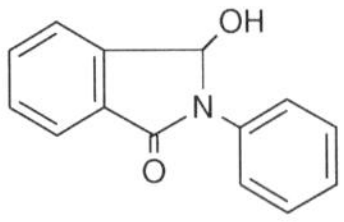

250 mg

S73,343-1
1-TERT-BUTYL-1-HYDROXY-2-METHYL-6-PHENYL-3-ISOINDOLINONE

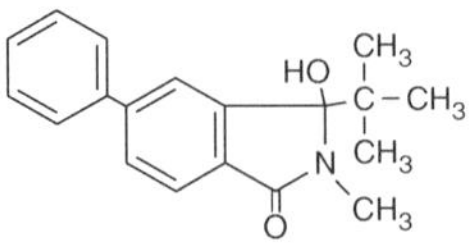

250 mg

S75,842-6
3-HYDROXY-2-(2-PHENYLETHYL)-1-ISOINDOLINONE

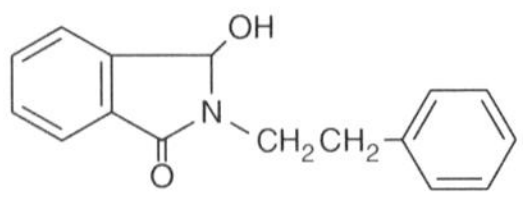

250 mg

S76,720-4
3-(2-HYDROXYETHYLAMINO)-3-PHENYL-2-INDOLINONE

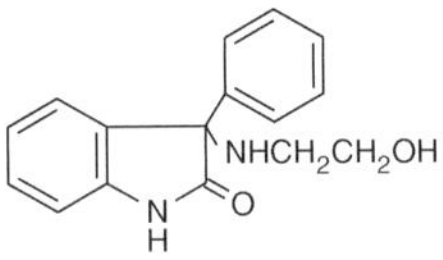

250 mg

S65,688-7
1-(4-CHLOROPHENYL)-1-HYDROXY-3-OXOISOINDOLINE

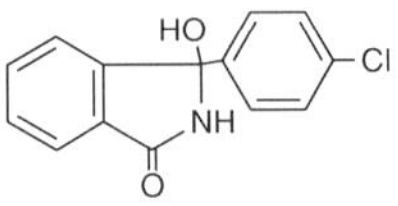

5 gm

S65,690-9
1-(TERT-BUTYL)-1-HYDROXY-2-PHENYLHEXAHYDRO-3-OXOISOINDOLINE

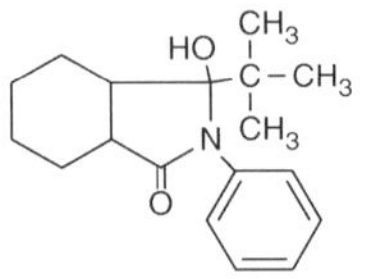

5 gm

S73,531-0
3-(1,1-DIPHENYLETHYL)-3-HYDROXY-2-METHYL-1-ISOINDOLINONE

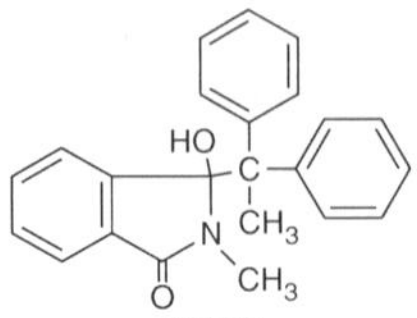

250 mg

S74,853-6
3-TERT-BUTYL-2-ETHYL-3-HYDROXY-1-ISOINDOLINONE

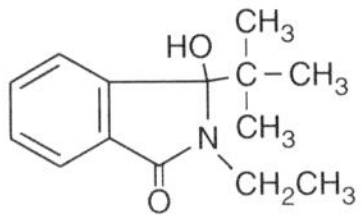

250 mg

S68,117-2
2-(3,4-DIMETHOXYPHENYL)-1-METHYL-5-OXOPYRROLIDINE-3-CARBOXYLIC ACID

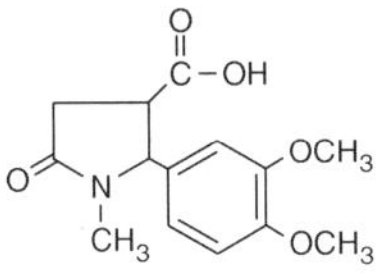

5 gm

S66,227-5
3',4'-DIHYDRO-6',7'-DIMETHOXYSPIRO [3H-INDOLE-3,1'(2H)ISOQUINOLIN]-2 (1H)ONE

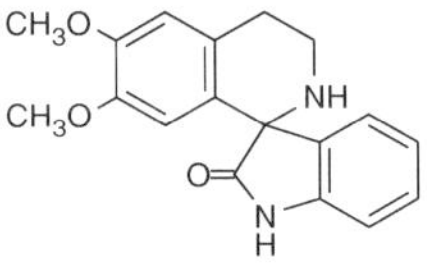

250 mg

S52,626-6
6,7-DIMETHOXY-1(2H)ISOQUINOLONE

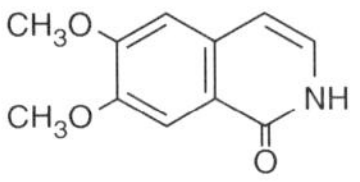

250 mg

S76,775-1
6,11-DIHYDRO-1-METHOXY-5H-DIBENZ (B,E)AZEPIN-6,11-DIONE

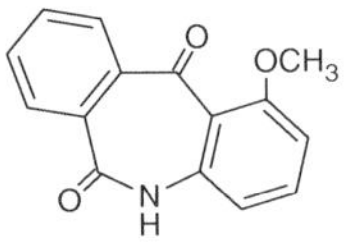

250 mg

S95,550-7
ETHYL 2-OXO-3-INDOLINEGLYOXYLATE

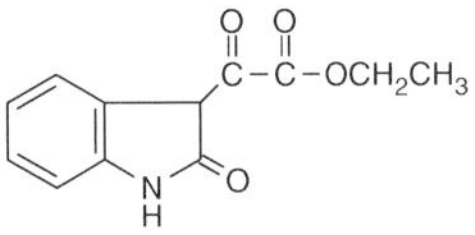

50 mg

S87,699-2
ETHYL 1,2-DIHYDRO-6-METHOXY-1-METHYL-2-OXO-4-QUINOLINECARBOXYLATE

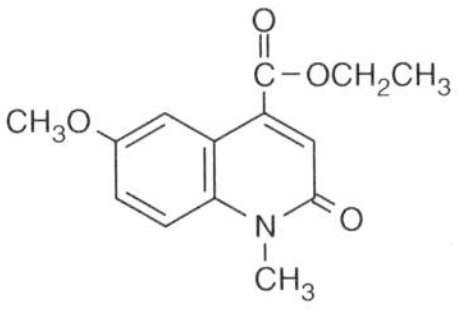

250 mg

S94,007-0
3-AMINO-4-(3,4-DIMETHOXYPHENYL)-2-HEXYL-7-NITRO-1(2H)ISOQUINOLINONE

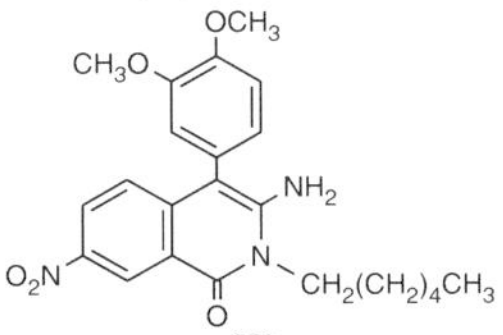

250 mg

S65,670-4
1-(4-CHLOROPHENYL)-1-METHOXY-3-OXOISOINDOLINE

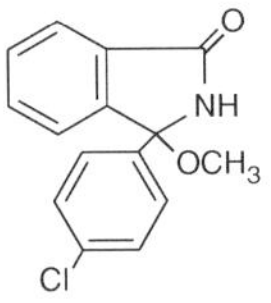

5 gm

S60,813-0
1,2-DIHYDRO-6-METHOXY-1-METHYL-2-OXO-4-QUINOLINECARBOXYLIC ACID

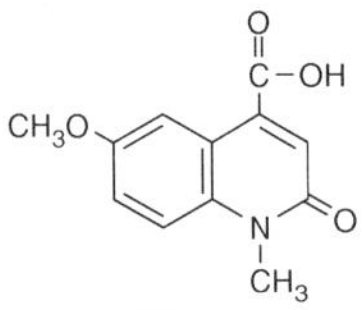

250 mg

S80,658-7
2-(4-METHOXY-1-NAPHTHYL)-1-METHYL-5-OXO-3-PYRROLIDINECARBOXYLIC ACID

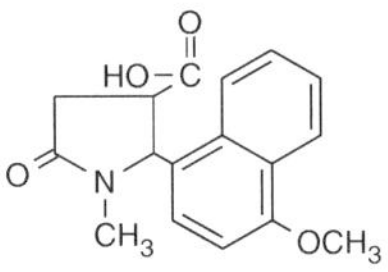

250 mg

S76,071-4
2-(2-METHOXYPHENYL)-1-METHYL-6-OXO-3-PIPERIDINECARBOXYLIC ACID

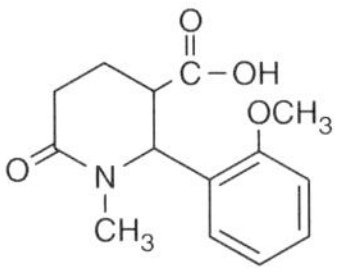

250 mg

S78,873-2
PYROGLUTAMIC ACID 4-METHOXY-2-NAPHTHYLAMIDE

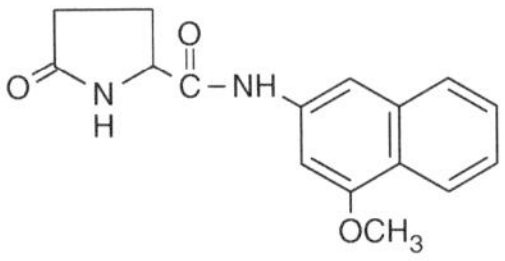

250 mg

S81,116-5
1-(4-METHOXYPHENYL)-4-PHENYL-2-AZETIDINONE

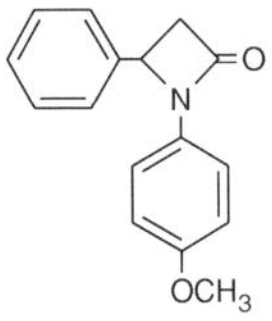

250 mg

S76,109-5
7-CHLORO-5-ETHOXY-1-METHYL-1H-1,4-BENZODIAZEPIN-2(3H)ONE

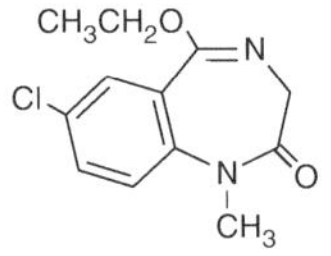

250 mg

S76,021-8
BENZYL N-[3-(5-CHLORO-2-OXO-3-PHENYL-3-INDOLINYL)PROPYL]CARBAMATE

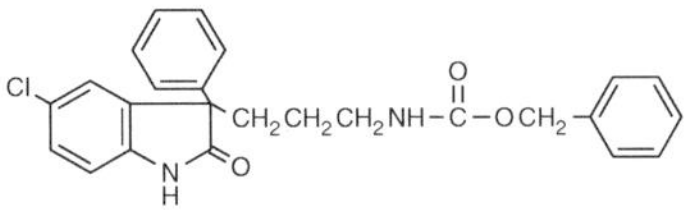

250 mg

S93,798-3
2H-1,4-BENZOXAZIN-3(4H)ONE

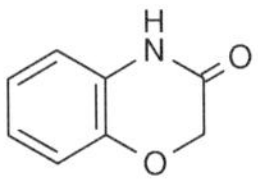

250 mg

S38,429-1
1,5-DIPHENYL-3-[2-(METHYLTHIO)-ETHYL]-2,4-PYRROLIDINEDIONE

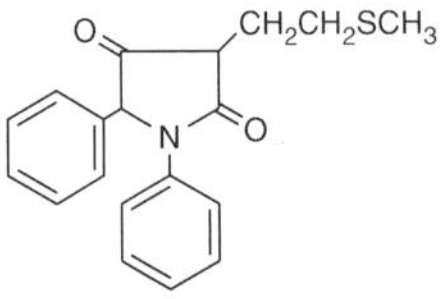

250 mg

S75,413-7
4-AMINO-2-BROMO-6(5H)-PHENANTHRIDINONE

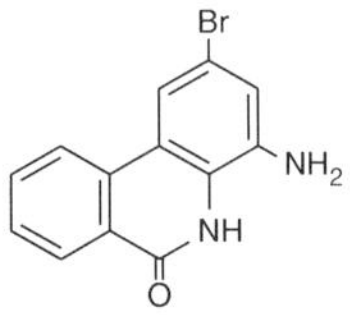

250 mg

S42,754-3
2-AMINOBENZOYL HYDRAZIDE

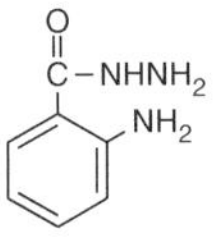

5 gm

S90,870-3
1-(4-AMINOPHENYL)-2-PYRROLIDINONE

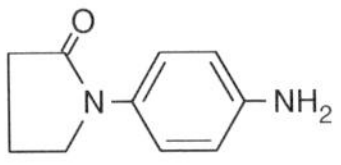

250 mg

S94,006-2
3-AMINO-2-METHYL-7-NITRO-4-PHENYL-1(2H)ISOQUINOLINONE

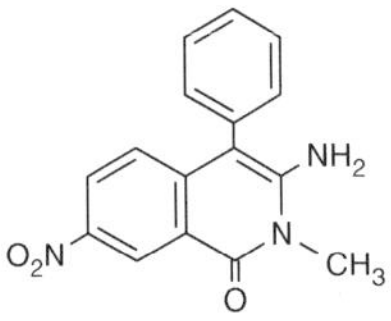

250 mg

S94,008-9
3-AMINO-2-HEXYL-7-NITRO-4-PHENYL-1 (2H)ISOQUINOLINONE

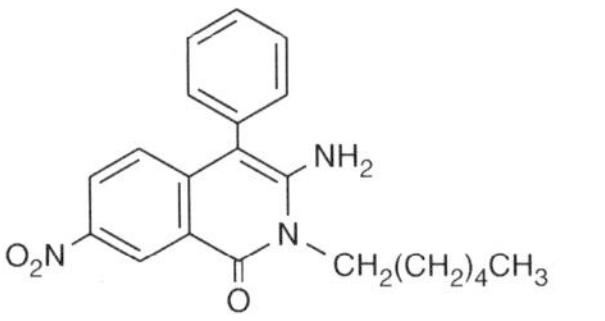

250 mg

S75,699-7
2-BROMO-4-NITRO-6(5H)-PHENANTHRIDINONE

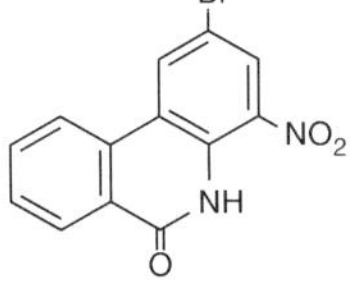

5 gm

S45,470-2
N-PHTHALOYLGLYCYL-DL-PHENYLALANINE
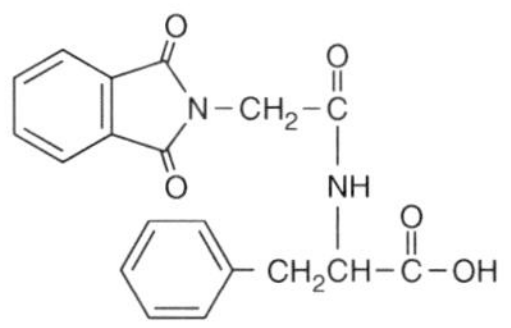
5 gm

S95,584-1
(1,1'-BIISOINDOLINE)-3,3'-DIONE
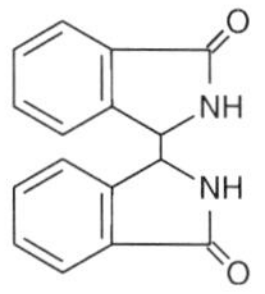
50 mg

S61,759-8
1-ETHYL-7-(METHACRYLAMIDO)-3-PHENYL-2(1H)QUINOLINONE
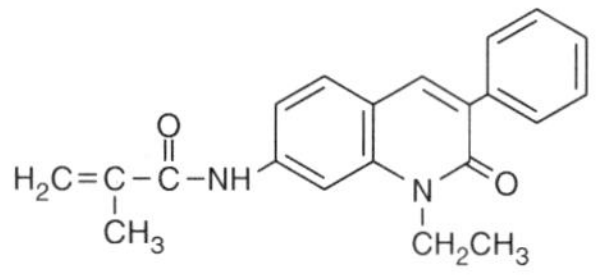
250 mg

S76,037-4
5-METHYL-1-(METHYLAMINO)-5H-DIBENZ (B,E)AZEPINE-6,11-DIONE
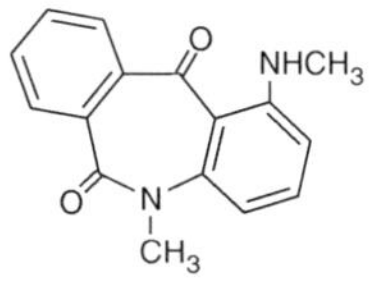
250 mg

S75,161-8
3-NITRO-6(5H)PHENANTHRIDINONE
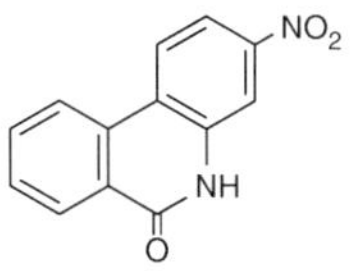
5 gm

S89,597-0
3-[2-(BENZYLIDENEAMINO)ETHYL]-4-METHYL-3-PYRROLIN-2-ONE
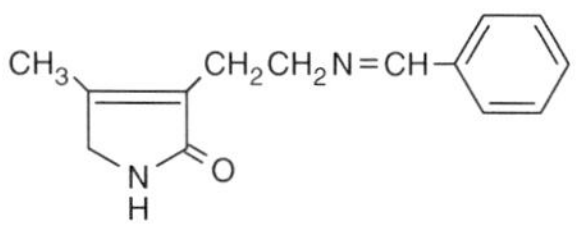
250 mg

S90,675-1
3-PHENYL-2-AZETIDINONE
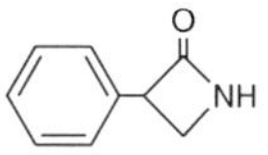
250 mg

S65,689-5
1,4-DIHYDRO-1-PHENYL-2,4,4-TRIMETHYL-3(2H)ISOQUINOLINONE
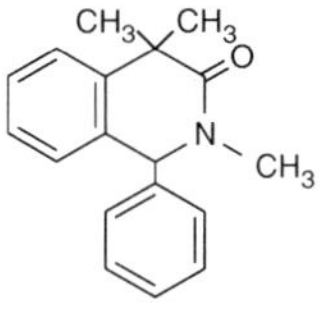
5 gm

S51,755-0
1-(MORPHOLINOMETHYL)ISATIN
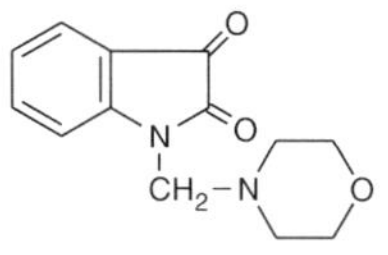
5 gm

S51,724-0
1-(DIETHYLAMINOMETHYL)ISATIN
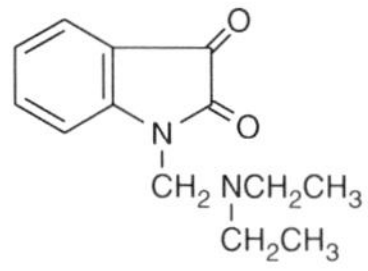
250 mg

S38,437-2
3-[3-(DIMETHYLAMINO)PROPYL]-1,5-DIPHENYL-2,4-PYRROLIDINEDIONE
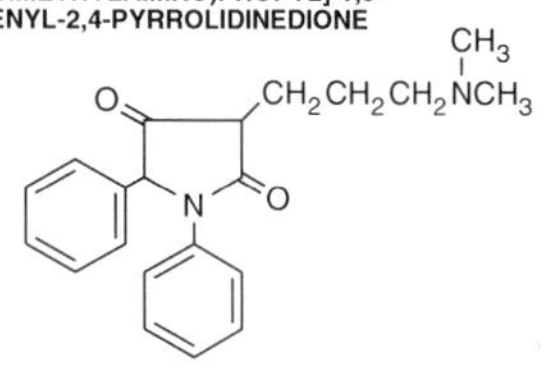
250 mg

S76,708-5
1-[4-(2,2-DIMETHYLPROPIONYL)BENZYL]-ISATIN
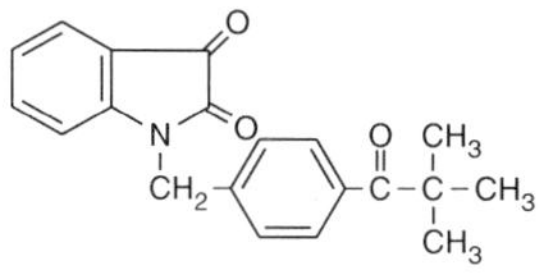
250 mg

S65,648-8
1-(4-CHLOROPHENYL)-3-OXOISOINDOLINE
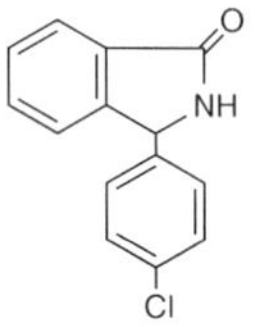
5 gm

S65,692-5
1,4-DIHYDRO-1-PHENYL-3(2H)-ISOQUINOLINONE
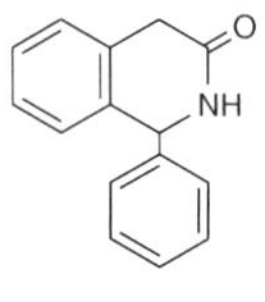
5 gm

S65,710-7
4-(TERT-BUTYL)-1,4-DIHYDRO-1-PHENYL-3(2H)ISOQUINOLINONE
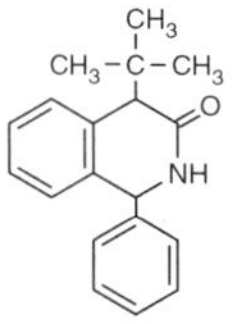
5 gm

S81,758-9
1,4-DIPHENYL-3-PROPYL-2-AZETIDINONE
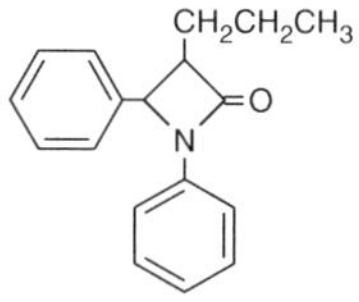
250 mg

S81,759-7
1-BENZYL-4-PHENYL-3-PROPYL-2-AZETIDINONE
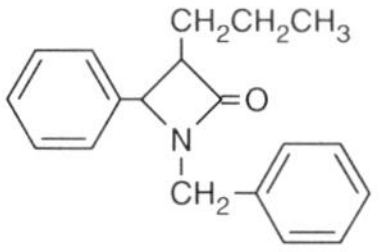
250 mg

S81,760-0
3-METHYL-1-(2-NAPHTHYL)-4-PHENYL-2-AZETIDINONE
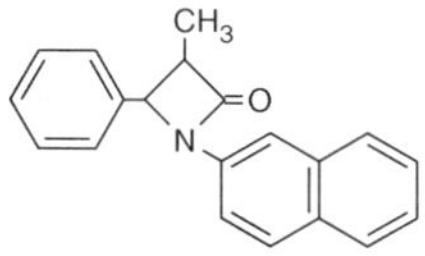
250 mg

S92,774-0
1-METHYL-6-NITRO-2(1H)QUINOLONE
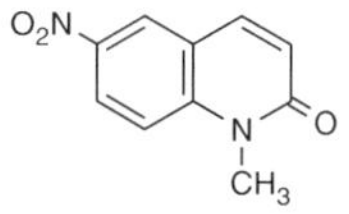
250 mg

S75,562-1
10-CHLORO-3-NITRO-6(5H)-PHENANTHRIDINONE
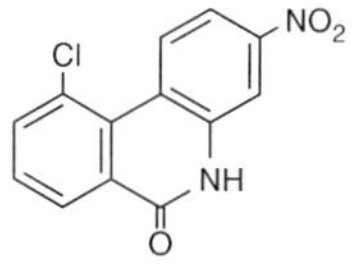
250 mg

S56,112-6
1-ACETYLISATIN
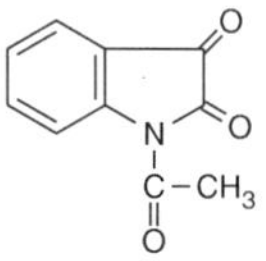
5 gm

S76,074-9
2-CHLORO-5-METHYL-5H-DIBENZ(B,E)-AZEPINE-6,11-DIONE
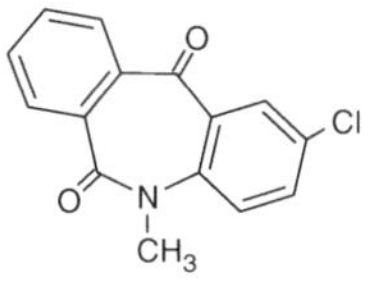
250 mg

S65,534-1
2,3-DIHYDRO-3-OXO-(1H)-ISOINDOL-1-YLIDENEACETIC ACID
@
gm "
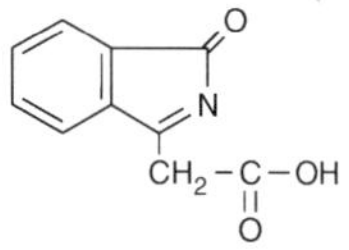

S70,647-7
2,3-DIHYDRO-5-METHYLINDOLE-2,3-DIONE
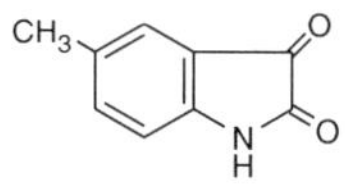
5 gm

S95,854-9
3-PHENYLACETYL-2-OXOINDOLINE
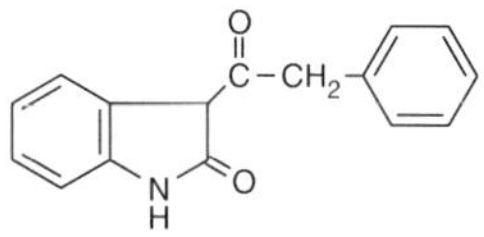
50 mg

S76,691-7
ETHYL ALPHA-(2-IMIDAZOLIDINYLIDENE)-3-OXO-2-(2-PHENYLETHYL)-1-ISOINDOLINEACETATE
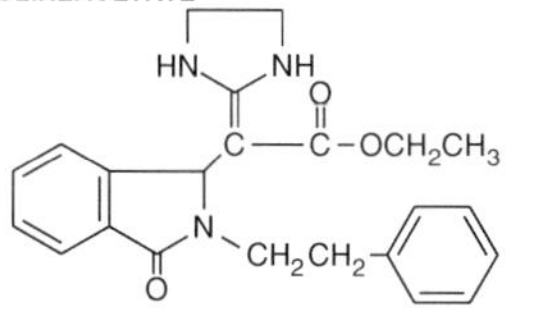
250 mg

S89,595-4
5-BENZYLIDENE-4-ETHYL-3-METHYL-3-PYRROLIN-2-ONE
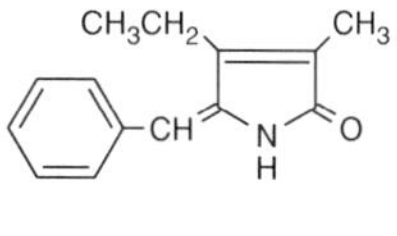
250 mg

S89,596-2
5-BENZYLIDENE-3-ETHYL-4-METHYL-3-PYRROLIN-2-ONE
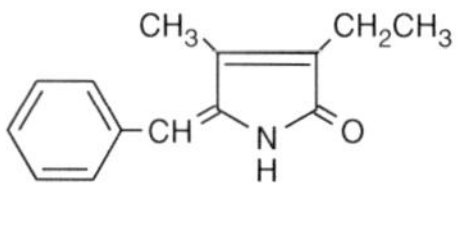
250 mg

S92,023-1
2-ISOPROPYL-2H-1,3-BENZOTHIAZIN-4(3H)-ONE
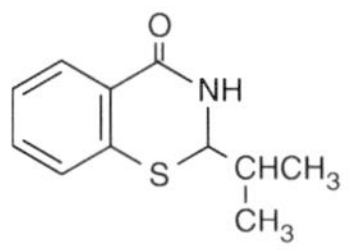
250 mg

S76,043-9
3-CINNAMYLIDENE-2(3H)INDOLONE
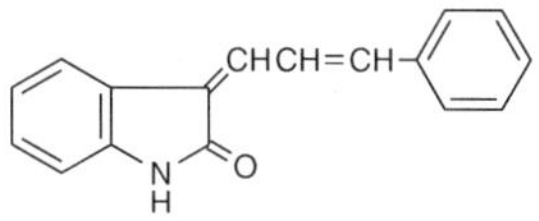
250 mg

S95,559-0
3-(2-BENZAMIDOETHYL)-2-INDOLINONE
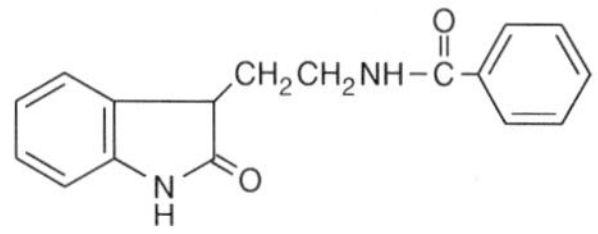
50 mg

S74,922-2
2-CHLORO-6(5H)PHENANTHRIDINONE
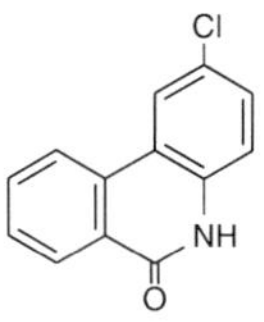
250 mg

S75,159-6
2-BROMO-6(5H)PHENANTHRIDINONE

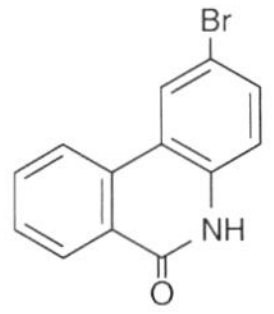

5 gm

S68,315-9
1-ETHYL-2-PYRIDONE

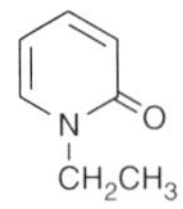

5 gm

S75,849-3
2'-METHYLSPIRO(BENZOSUBERAN-5,1'-ISOINDOLIN)-3'-ONE

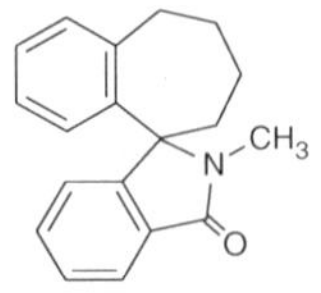

250 mg

S70,155-6
2-METHYL-1H-BENZ(E)ISOINDOL-3-ONE

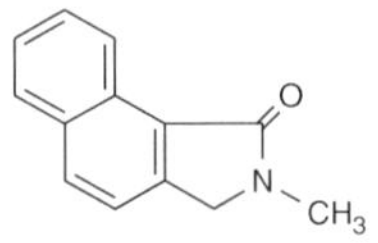

250 mg

S84,607-4
2-METHYL-1H-BENZ(F)ISOINDOL-3(2H)-ONE

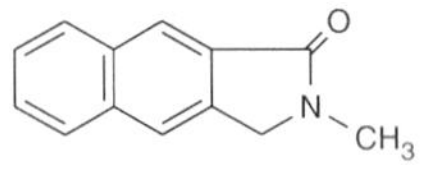

250 mg

S50,965-5
4-METHYL-1-PHENYL-2(1H)QUINOLONE

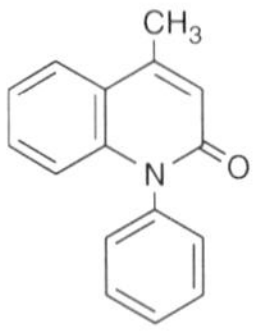

250 mg

S36,386-3
1-METHYL-2(1H)QUINOLONE

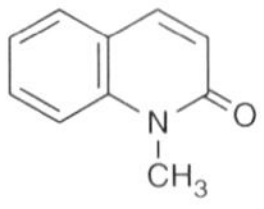

5 gm

S76,121-4
2,11-DICHLORO-5,11-DIHYDRO-5-METHYL-6H-DIBENZ(B,E)AZEPIN-6-ONE

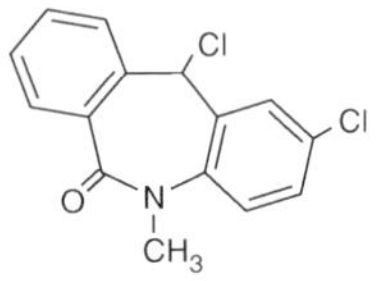

250 mg

S92,750-3
ETHYL 1-BENZYL-4,5-DIOXO-3-PYRROLIDINECARBOXYLATE

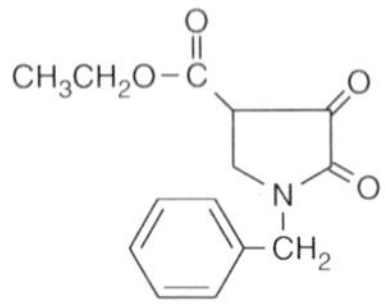

250 mg

S94,951-5
5,7-DIBROMOISATIN

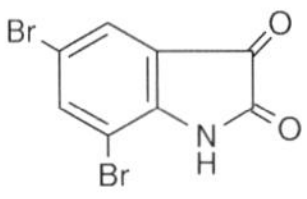

250 mg

S76,764-6
2-OXO-4-PHENYL-3-PYRROLIDINECARBOXYLIC ACID

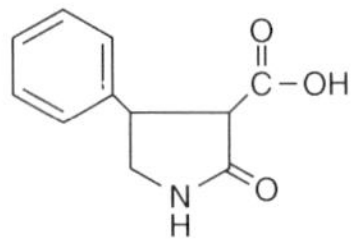

250 mg

S76,179-6
ETHYL (2-OXO-4-PHENYL)-3-PYRROLIDINECARBOXYLATE

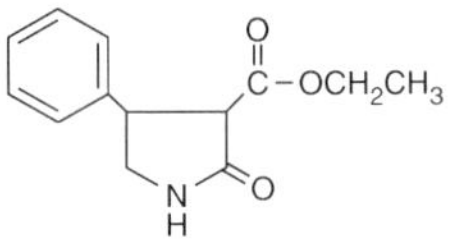

250 mg

S68,601-8
2-(1,3-BENZODIOXOL-5-YL)-5-OXOPYRROLIDINE-3-CARBOXYLIC ACID

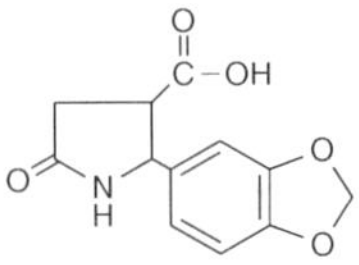

250 mg

S68,279-9
2-(1,3-BENZODIOXOL-5-YL)-1-METHYL-5-OXO-3-PYRROLIDINECARBOXYLIC ACID

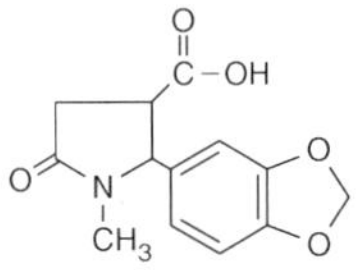

250 mg

S76,038-2
1,3,3A,4,7,7A-HEXAHYDRO-2-METHYL-3-PHENYL-1-OXO-3A-ISOINDOLECARBOXYLIC ACID

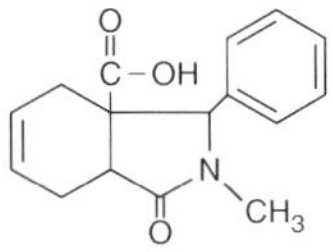

250 mg

S76,108-7
ETHYL 2-(2-OXO-3-INDOLINYLIDENE)-ACETATE

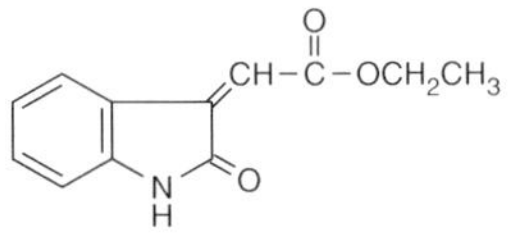

250 mg

S73,462-4
METHYL 1-METHYL-6-OXO-2-PHENYL-3-PIPERIDINECARBOXYLATE

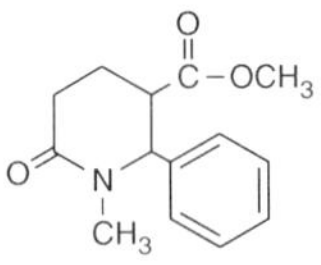

250 mg

S84,610-4
2-OXO-2,3,4,5-TETRAHYDRO-1H-BENZAZEPINE-4-PROPIONIC ACID

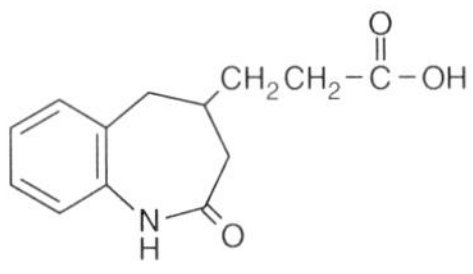

250 mg

S38,134-9
1-BENZYL-5-OXO-3-PYRROLIDINECARBOXYLIC ACID

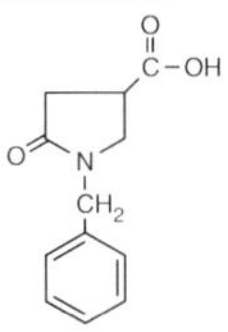

5 gm

S68,271-3
2-(1,3-BENZODIOXOL-5-YL)-1-METHYL-6-OXO-3-PIPERIDINECARBOXYLIC ACID

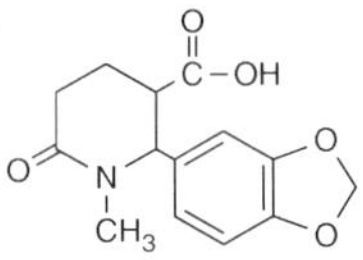

250 mg

S60,094-6
3,3-BIS(4-ACETOXYPHENYL)OXINDOLE

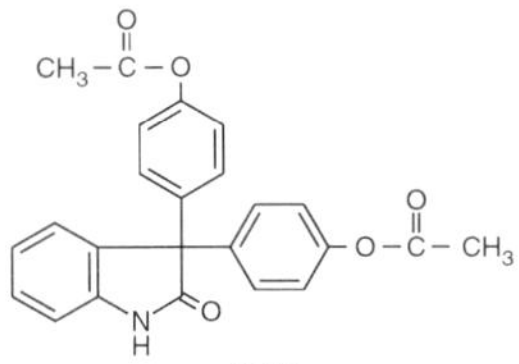

5 gm

S35,335-3
5-METHYL-5-PHENYL-2-PIPERIDONE

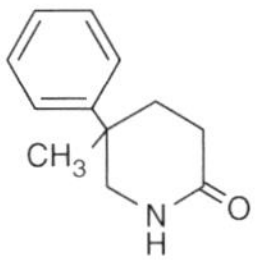

1 gm

S93,199-3
3-PHENYLOXINDOLE

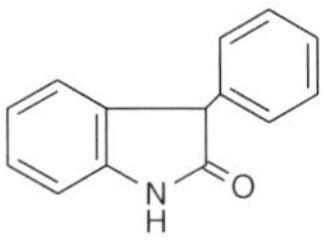

250 mg

S76,164-8
5,6-DICHLORO-3-PHENYL-2-INDOLINONE

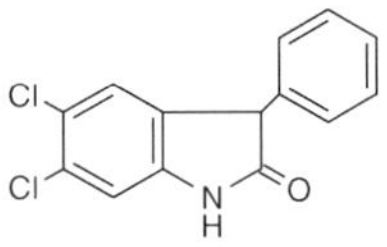

250 mg

S76,117-6
3-PHENYL-4,5,6-TRICHLORO-2-INDOLINONE

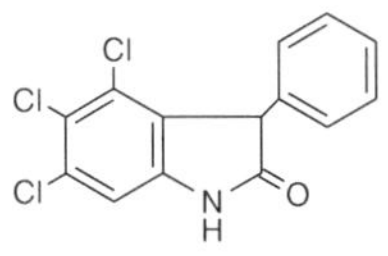

250 mg

S86,080-8
3,3-DIMETHYL-2-INDOLINONE

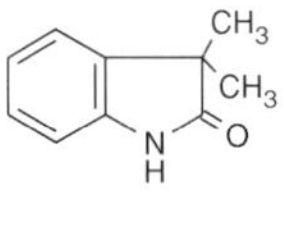

250 mg

S87,787-5
3,4-DIHYDRO-2(1H)-QUINOLINONE

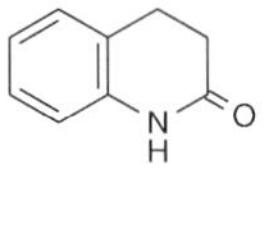

250 mg

S83,836-5
1-ETHYL-3-METHYL-4-PHENYL-2-AZETIDINONE

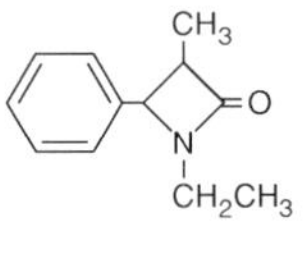

250 mg

S81,754-6
1,4-DIPHENYL-2-AZETIDINONE

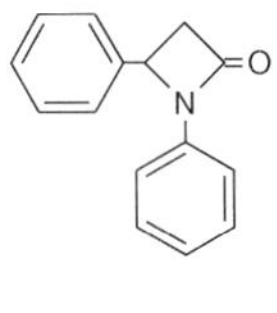

250 mg

S81,115-7
1,4-DIPHENYL-3-METHYL-2-AZETIDINONE

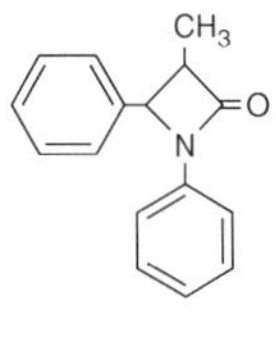

250 mg

S81,756-2
1,4-DIPHENYL-3-ETHYL-2-AZETIDINONE

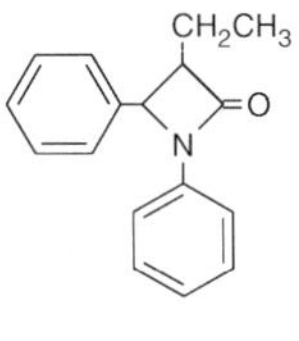

250 mg

S83,840-3
1,3,4-TRIPHENYL-2-AZETIDINONE

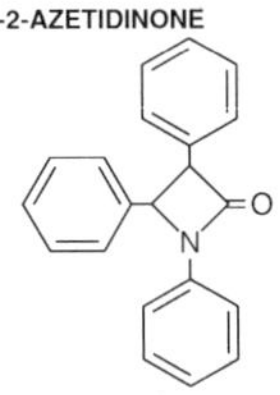

250 mg

S95,137-4
1-BENZOYL-2-PYRROLIDINONE
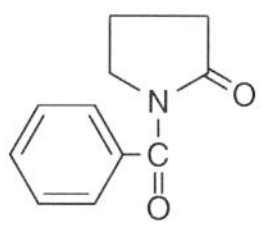
50 mg

S70,156-4
2-METHYLISOINDOLIN-1-ONE
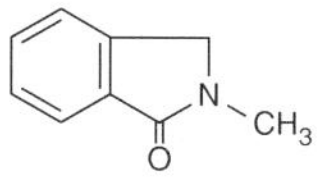
5 gm

S83,837-3
3-ETHYL-1-METHYL-4-PHENYL-2-AZETIDINONE
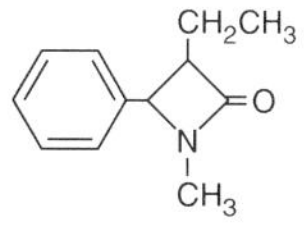
250 mg

S79,865-7
3,4-DIHYDRO-1-PHENYL-2(1H)-QUINOLINONE
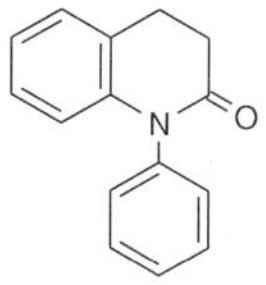
250 mg

S75,925-2
3-TERT-BUTYL-2-METHYL-4-(O-TOLYL)-1(2H)ISOQUINOLINONE
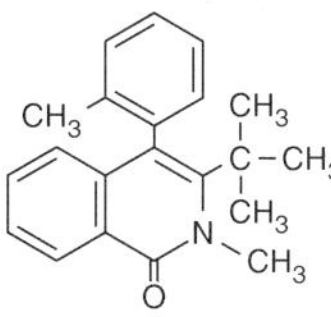
250 mg

S95,543-4
1-METHYL-2-INDOLINONE
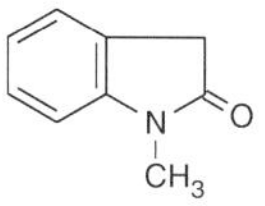
50 mg

S38,767-3
1-METHYL-4-PENTADECYL-2(1H)-QUINOLONE
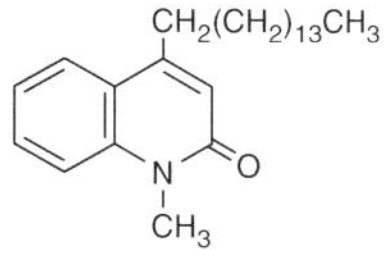
250 mg

S70,716-3
1-(3-CHLOROPROPIONYL)-4-(2-METHOXYPHENYL)PIPERAZINE
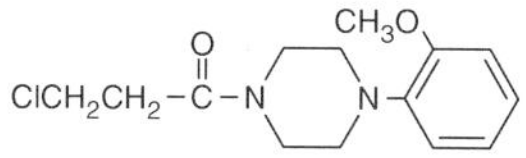
5 gm

S34,389-7
4-(2-CHLOROACETYL)ANTIPYRINE
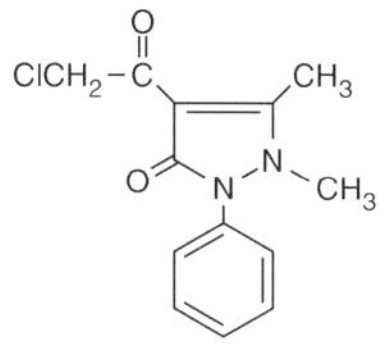
5 gm

S88,369-7
2-(2-BROMOETHYL)-1,4(2H,3H)-ISOQUINOLINEDIONE
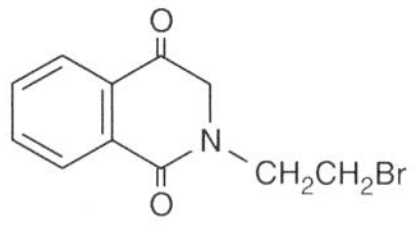
250 mg

S38,245-0
2-CHLOROETHYL ANTIPYRATE
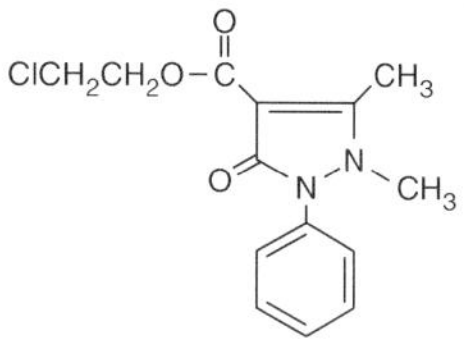
5 gm

S38,380-5
2-CHLORO-N-(3,5-DIOXO-1,2-DIPHENYL-4-PYRAZOLIDINYL)ACETAMIDE
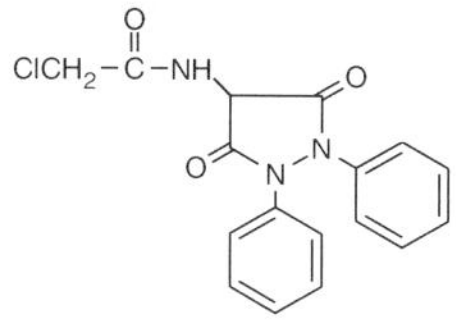
250 mg

S90,512-7
2-(CHLOROMETHYL)(1,2,4)TRIAZOLO(4,3-A)QUINOLIN-1(2H)-ONE
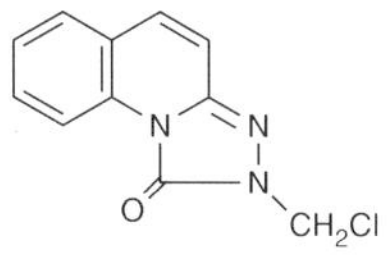
250 mg

S90,795-2
1-(2-CHLOROACETYL)INDOLINE
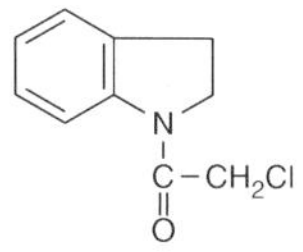
250 mg

S90,796-0
1-(2-BROMOACETYL)INDOLINE
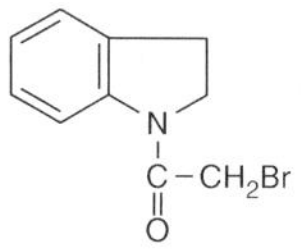
250 mg

S76,059-5
2-CHLOROACETYL-1,2,3,4-TETRAHYDROISOQUINOLINE
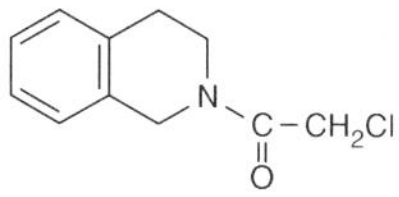
250 mg

S51,274-5
1-[4-(DIFLUOROMETHOXY)BENZOYL]-PIPERIDINE
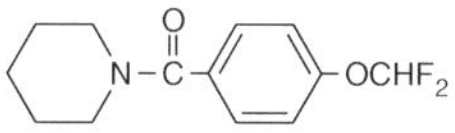
250 mg

S79,632-8
N-[4-CHLORO-3-(TRIFLUOROMETHYL)-PHENYL]-3-METHYL-1-PIPERIDINECARBOXAMIDE
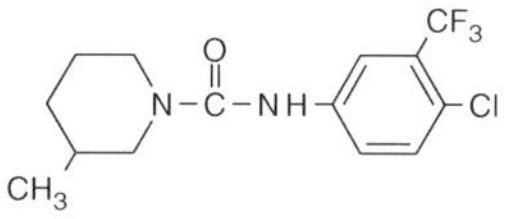
250 mg

S68,536-4
4-[4-CHLORO-2-(TRIFLUOROMETHYL)-PHENYLCARBAMOYL]-2,6-DIMETHYLMORPHOLINE
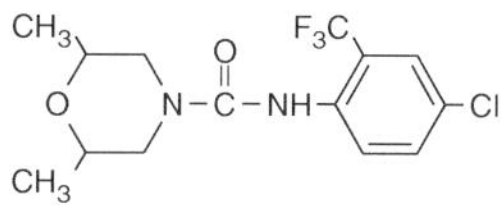
250 mg

S38,331-7
2,3-DIMETHYL-4-(1-HYDROXY-2,2,2-TRICHLOROETHYL)-1-PHENYL-3-PYRAZOLIN-5-ONE
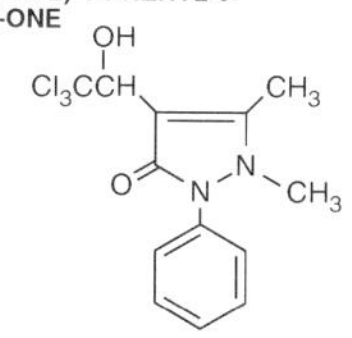
250 mg

S55,849-4
3-(3-AMINOBENZAMIDO)-1-(2,2,2-TRIFLUOROETHYL)-2-PYRAZOLIN-5-ONE
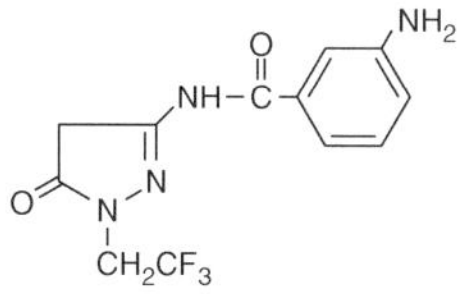
5 gm

S64,229-0
3-(METHACRYLAMIDO)-1-(2,2,2-TRIFLUOROETHYL)-2-PYRAZOLIN-5-ONE
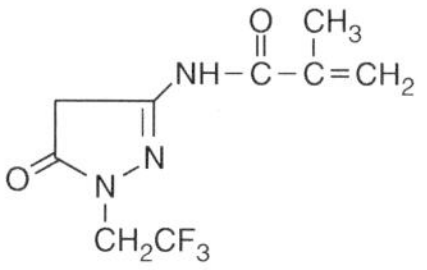
250 mg

S66,213-5
4,5-DICHLORO-2-(3-TRIFLUOROMETHYLPHENYL)PYRIDAZIN-3(2H)ONE
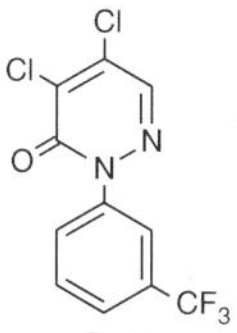
5 gm

S75,863-9
5-[2,5-BIS(TRIFLUOROMETHYL)PHENYL]-3,5-DIHYDRO-4,1-BENZOTHIAZEPIN-2(1H)ONE
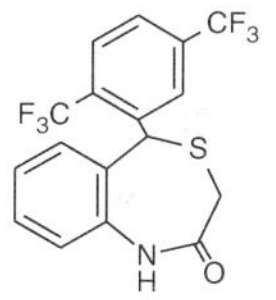
250 mg

S64,457-9
3-METHYL-1-[N-(3-TRIFLUOROMETHYLPHENYL)CARBAMOYL]-PIPERIDINE
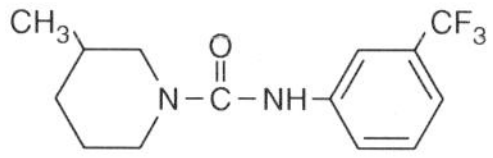
5 gm

S64,967-8
2-METHYL-1-[N-(3-TRIFLUOROMETHYLPHENYL)CARBAMOYL]-PIPERIDINE
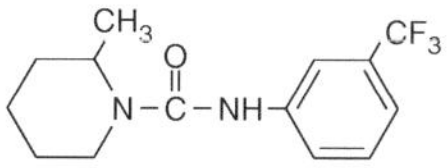
5 gm

S73,665-1
4-METHYL-1-[N-(3-TRIFLUOROMETHYLPHENYL)CARBAMOYL]-PIPERIDINE
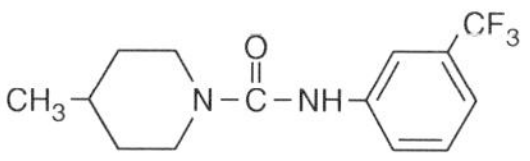
5 gm

S61,153-0
4-(N-[3-(TRIFLUOROMETHYL)PHENYL]-CARBAMOYL)MORPHOLINE
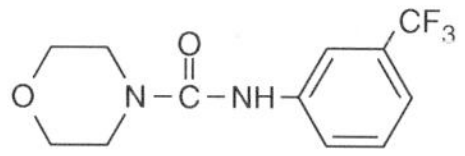
5 gm

S65,332-2
4-(N-[3-(TRIFLUOROMETHYL)PHENYL]-CARBAMOYL)-2,6-DIMETHYLMORPHOLINE
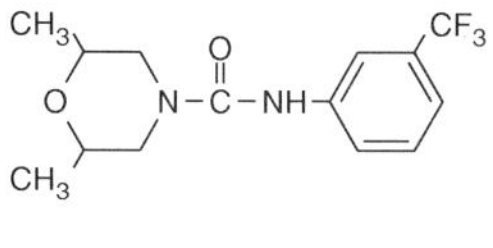
5 gm

S76,300-4
1-(4-CHLORO-ALPHA-HYDROXYBENZYL)-1,2,3,4-TETRAHYDRO-2-ISOQUINOLINECARBOXAMIDE
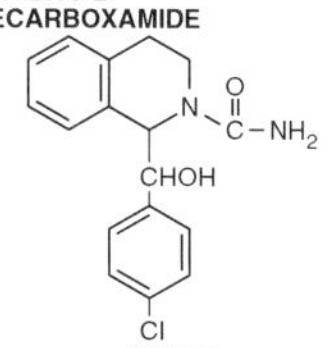
250 mg

S87,927-4
5-ACETYL-3,4-BIS(4-CHLOROPHENYL)-1-(2-HYDROXYETHYL)-6(1H)PYRIDAZINONE
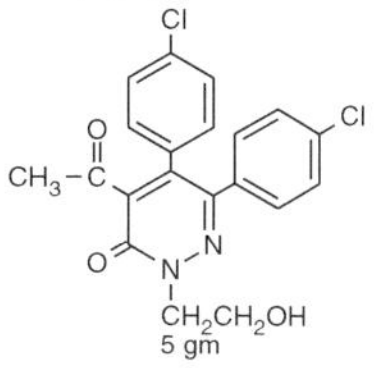
5 gm

S73,472-1
6-(4-CHLORO-3-METHYLPHENYL)-2-(2-HYDROXYBUTYL)-2,3,4,5-TETRAHYDRO-3-PYRIDAZINONE
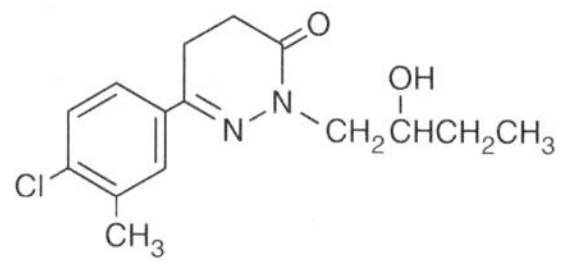
250 mg

S75,455-2
2-CARBAMOYL-3-(4-CHLOROPHENYL)-3-HYDROXY-1-ISOINDOLINONE
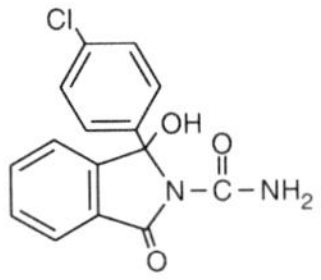
250 mg

S73,291-5
6-(4,5-DICHLORO-2-METHYLPHENYL)-DIHYDRO-2-(3-HYDROXYPROPYL)-3(2H)-PYRIDAZINONE
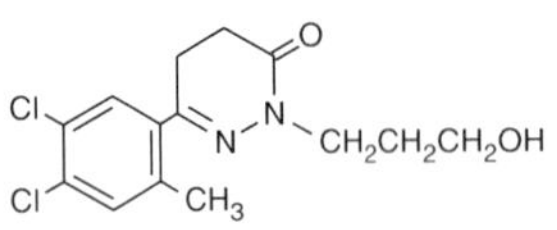
250 mg

S76,067-6
6-(3,4-DICHLOROPHENYL)-4,5-DIHYDRO-2-(3-HYDROXYPROPYL)-3(2H)-PYRIDAZINONE
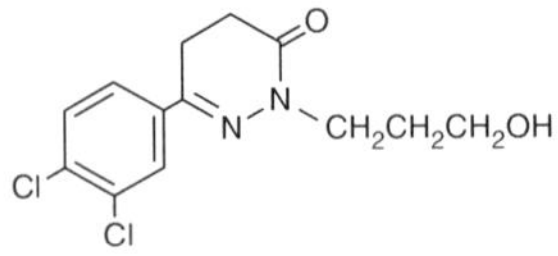
250 mg

S76,141-9
1-(4-CHLOROPHENYL)-1,5,6,7,8,8A-HEXAHYDRO-1-HYDROXY-2-PHENYLIMIDAZO(1,5-A)PYRIDIN-3-(2H)ONE
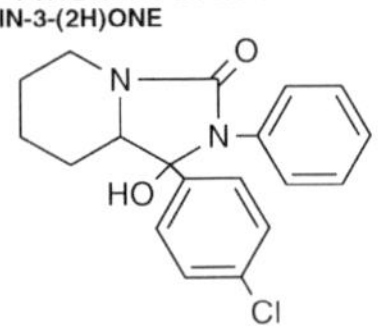
250 mg

S85,406-9
N-(5-CHLORO-2,4-DIMETHOXYPHENYL)-4-METHYL-1-PIPERAZINECARBOXAMIDE
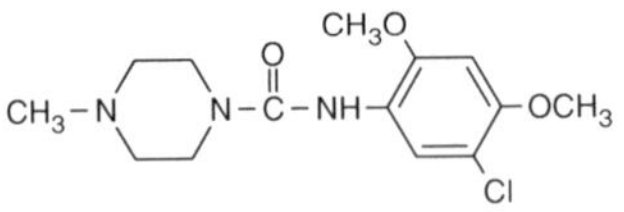
250 mg

S85,429-8
1-(5-CHLORO-2,4-DIMETHOXYPHENYL)-3-ETHYL-2,4,5-IMIDAZOLIDINETRIONE
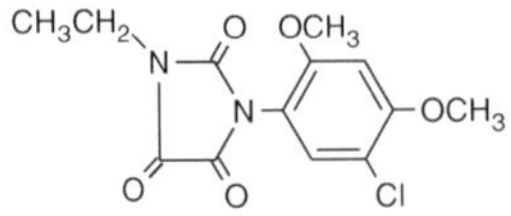
250 mg

S85,415-8
1-BUTYL-3-(5-CHLORO-2,4-DIMETHOXYPHENYL)-2,4,5-IMIDAZOLIDINETRIONE
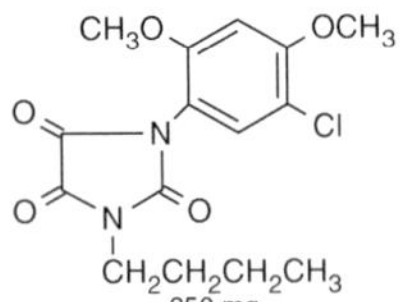
250 mg

S76,615-1
12B-(4-CHLOROBENZYL)-2,3-DIMETHOXYHEXAHYDROISOINDOLO(12A)-ISOQUINOLIN-8(8AH)ONE
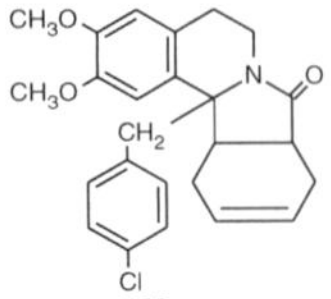
250 mg

S85,409-3
2-(2-OXO-1-IMIDAZOLIDINYLOXY)ETHYL-N-(5-CHLORO-2,4-DIMETHOXYPHENYL)-CARBAMATE
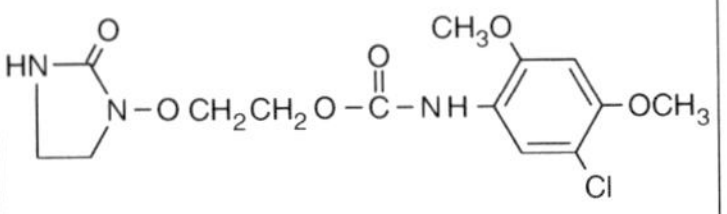
250 mg

S79,474-0
1-[N-(5-CHLORO-2-METHOXYPHENYL)-CARBAMOYL]-4-METHYLPIPERAZINE
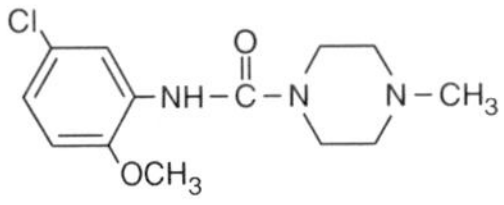
250 mg

S72,687-7
1-[N-(5-CHLORO-2-METHOXYPHENYL)-CARBAMOYL]PIPERIDINE
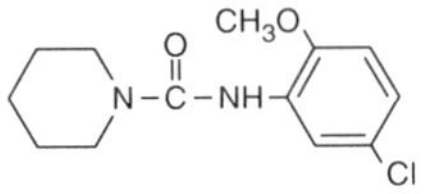
5 gm

S79,464-3
1-[N-(5-CHLORO-2-METHOXYPHENYL)-CARBOMOYL]-3-METHYLPIPERIDINE
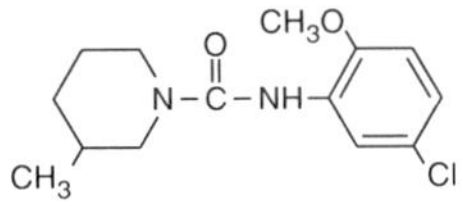
250 mg

S79,465-1
1-[N-(5-CHLORO-2-METHOXYPHENYL)-CARBAMOYL]-2-METHYLPIPERIDINE
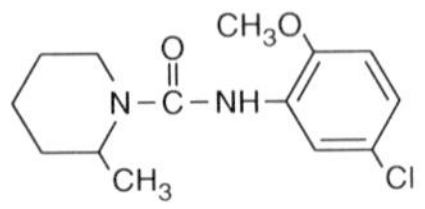
250 mg

S79,432-5
1-[N-(5-CHLORO-2-METHOXYPHENYL)-CARBAMOYL]-4-PHENYLPIPERAZINE
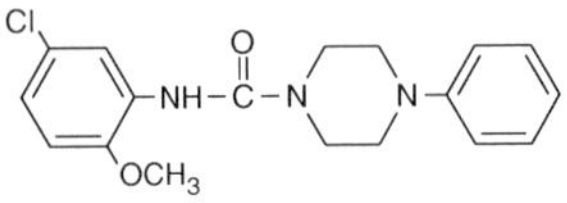
250 mg

S76,848-0
5-(2-CHLOROBENZYLIDENE)-3-(3-METHOXYPROPYL)-2,4-THIAZOLIDINEDIONE
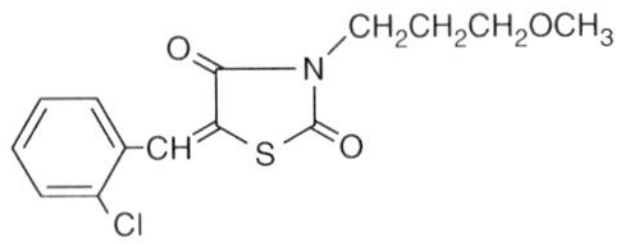
250 mg

S75,927-9
4-(4-CHLOROPHENYL)-6-(4-METHOXYPHENYL)-1-METHYL-2(1H)-PYRIMIDINONE
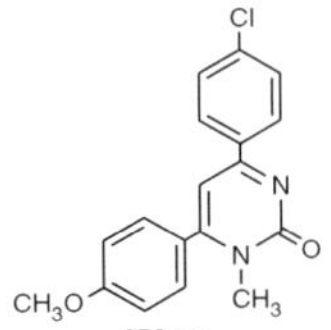
250 mg

S93,636-7
3-AMINO-2-BENZYL-4-(4-FLUOROPHENYL)-7-NITRO-1(2H)-ISOQUINOLINONE
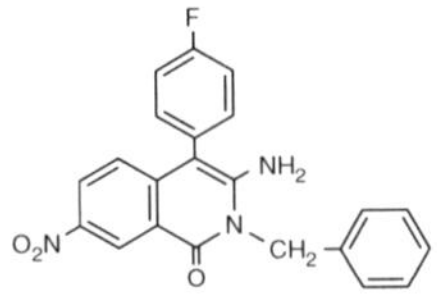
250 mg

S94,711-3
DIETHYL 6-AMINO-1-BENZAMIDO-4-(4-CHLOROPHENYL)-2-OXO-TETRAHYDRO-3,5-PYRIDINEDICARBOXYLATE
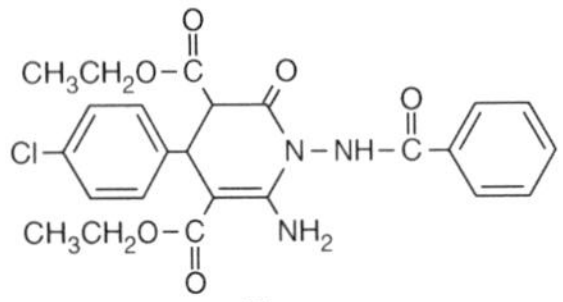
50 mg

S90,569-0
4'-CHLORO-2'-NITRO-4-PHENYL-1-PIPERAZINECARBOXANILIDE
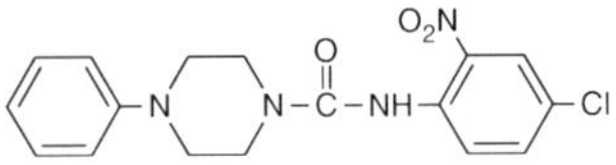
250 mg

S86,738-1
N-(4-CHLORO-2-NITROPHENYL)-2-METHYL-1-PIPERIDINECARBOXAMIDE
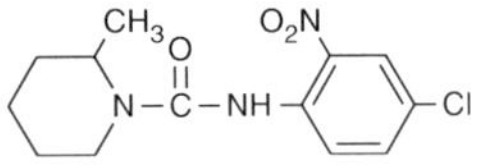
250 mg

S86,740-3
N-(4-CHLORO-2-NITROPHENYL)-4-METHYL-1-PIPERIDINECARBOXAMIDE
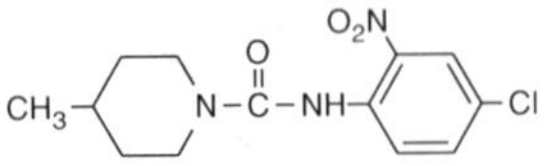
250 mg

S86,825-6
N-(4-FLUORO-3-NITROPHENYL)-4-PHENYL-1-PIPERAZINECARBOXAMIDE
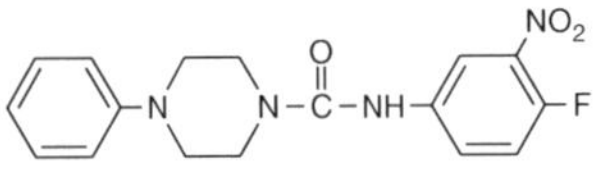
250 mg

S91,173-9
1-[N-(2-FLUORO-5-NITROPHENYL)-CARBAMOYL]-4-METHYLPIPERAZINE
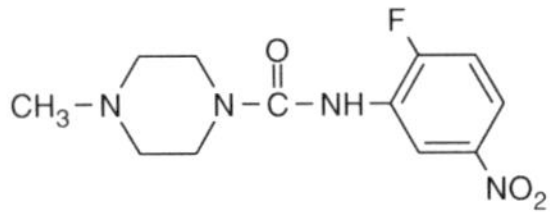
250 mg

S52,969-9
1-(5-CHLORO-2-NITROBENZOYL)-4-PHENYL-1,2,3,6-TETRAHYDROPYRIDINE
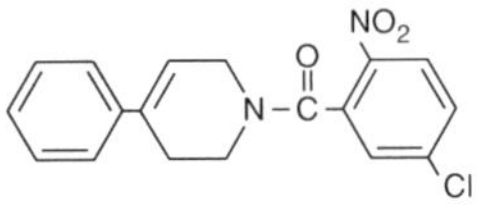
250 mg

S67,059-6
2,6-DIMETHYL-4-[N-(4-METHYL-3-NITROPHENYL)CARBAMOYL]MORPHOLINE
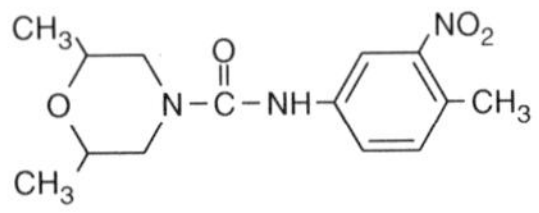
5 gm

S90,008-7
4-(4-METHYL-1-PIPERAZINAMIDO)-BENZOIC (4-FLUOROBENZYLIDENE)-HYDRAZIDE
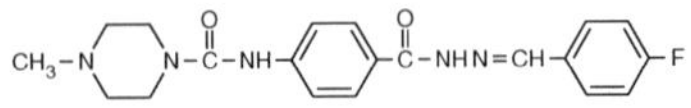
250 mg

S75,008-5
2-(4-METHYLPIPERAZINOCARBONYL)-3,4,5,6-TETRABROMOBENZOIC ACID
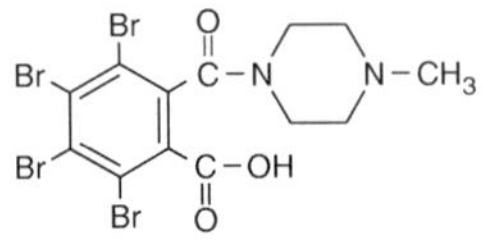
5 gm

S88,294-1
4B-(4-CHLOROPHENYL)-5-METHYLISOINDOLO(1,2-B)QUINAZOLINE-10,12(4BH,5H)DIONE
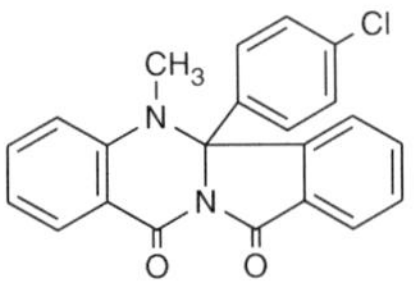
250 mg

S84,418-7
1-[N-(3-FLUOROPHENYL)CARBAMOYL]-4-METHYLPIPERAZINE
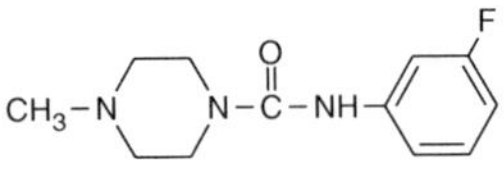
250 mg

S67,342-0
1-METHYL-4-(2,3,4,5,6-PENTACHLOROPHENYLCARBAMOYL)-PIPERAZINE
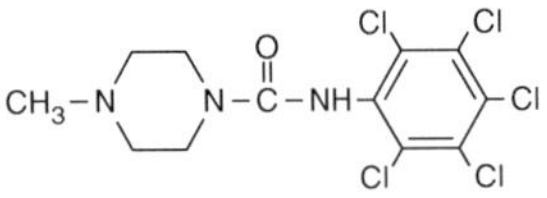
250 mg

S55,139-2
2-(4-CHLOROPHENYL)-1,2-DIHYDRO-4-(3H)QUINAZOLINONE
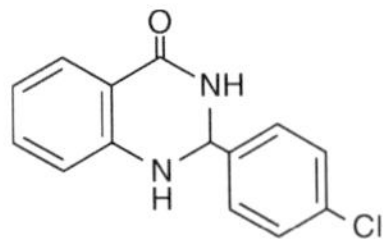
250 mg

S54,229-6
2-(2,6-DICHLOROPHENYL)-1,2-DIHYDRO-4(3H)QUINAZOLINONE
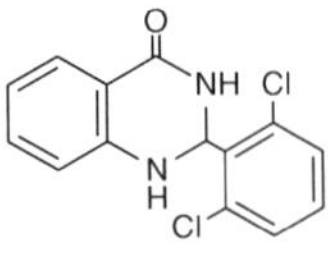
250 mg

S55,148-1
2-(3,4-DICHLOROPHENYL)-1,2-DIHYDRO-4(3H)QUINAZOLINONE
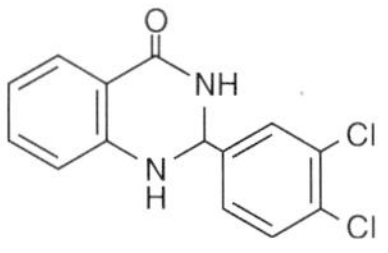
250 mg

S69,465-7
4-(4-CHLOROPHENYL)-1-(PYRROLIDINOCARBOXAMIDOACETYL)-SEMICARBAZIDE
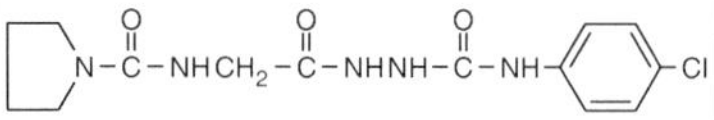
250 mg

S69,317-0
4-(4-CHLORO-2-METHYLPHENYL)-1-(PYRROLIDINOCARBOXAMIDOACETYL)-SEMICARBAZIDE
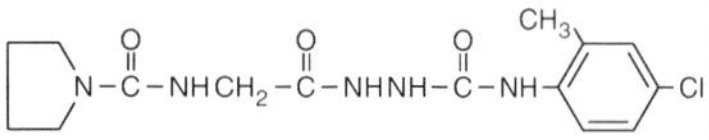
250 mg

S78,102-9
1-(2,3-DIHYDRO-1,5-DIMETHYL-3-OXO-2-PHENYL-4-PYRAZOLYL)-3-(3-CHLORO-2-METHYLPHENYL)UREA
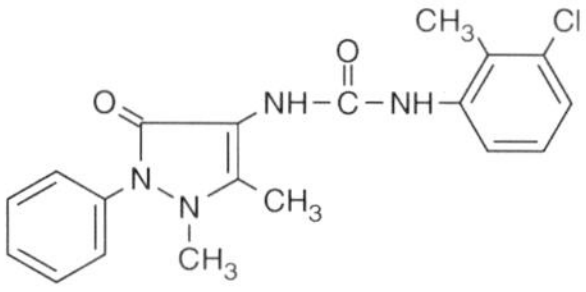
250 mg

S76,875-8
(5,5-DIMETHYL-2,4-DIOXO-1-IMIDAZOLIDINYLMETHYL)-N-(2CHLORO4NITROPHENYL)CARBAMATE
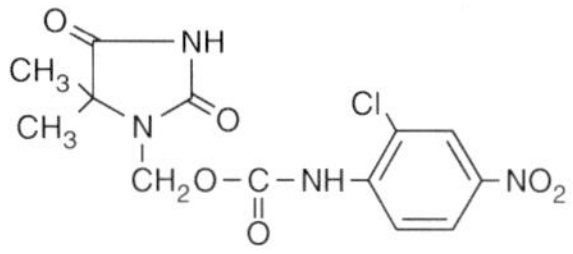
250 mg

S67,093-6
2-(5-CARBOXY-2-ETHYL-3-OXOPENTYL)-6-(4-CHLOROPHENYL)-4,5-DIHYDRO-3-(2H)-PYRIDAZINONE
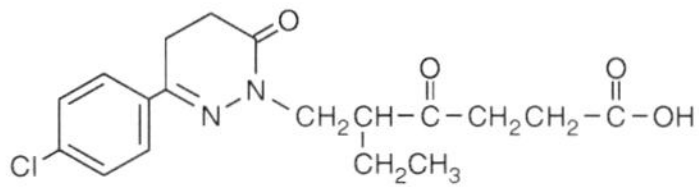
5 gm

S73,317-2
1,2-DIHYDRO-2,N-DIMETHYL-3-(2-FLUOROPHENYL)-1-OXO-4-ISOQUINOLINECARBOXAMIDE
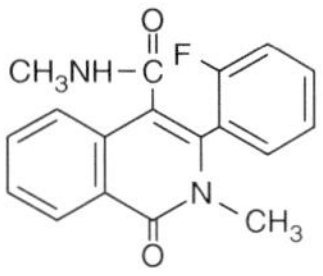
250 mg

S90,132-6
3-CHLOROPHENYL-4-METHYLPIPERIDINO KETONE
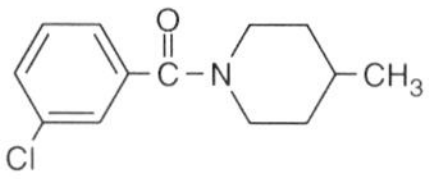
250 mg

S89,235-1
4-CHLOROPHENYL-4-PHENETHYLPIPERIDINO KETONE
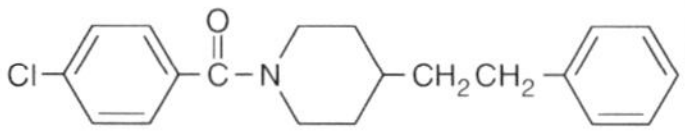
250 mg

S89,234-3
1-(4-CHLOROBENZOYL)-3-METHYLPIPERIDINE
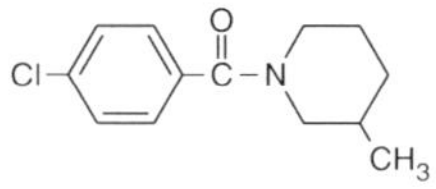
250 mg

S70,263-3
1,1'-[(2-CHLOROBENZYL)-2-METHYLPHENYLENE]BIS(3-ANTIPYRINYLUREA) MIXT. OF ISOMERS
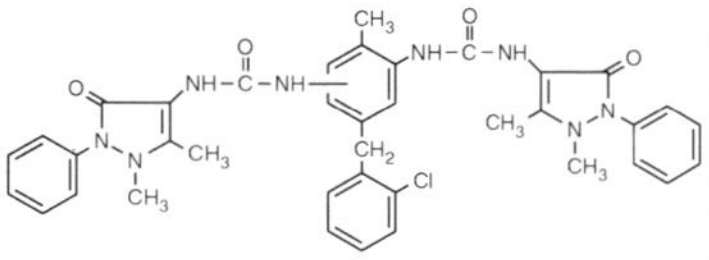
250 mg

S68,219-5
2-(4-CHLOROPHENYL)SPIRO(ISOINDOLINE-1,1'-PHTHALAN)-3,3'-DIONE
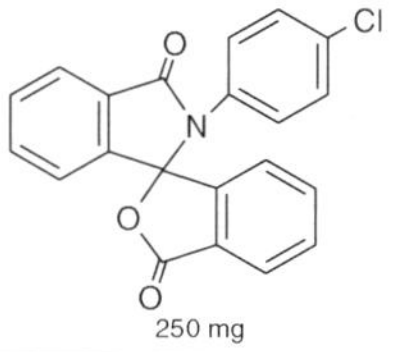
250 mg

S67,562-8
5-(2,4-DICHLOROBENZYLIDENE)-1-METHYLHYDANTOIN
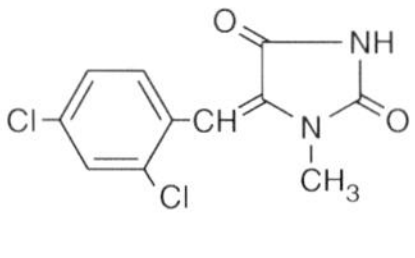
250 mg

S67,355-2
3-(4-CHLOROPHENYL)-2,4-(1H,3H)-QUINAZOLINEDIONE
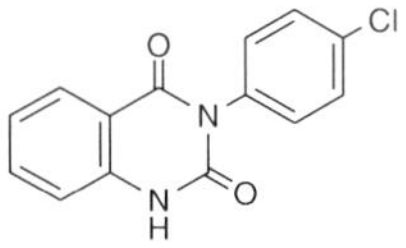
250 mg

S69,557-2
3-(4-BROMOPHENYL)-2,4-(1H,3H)-QUINAZOLINEDIONE
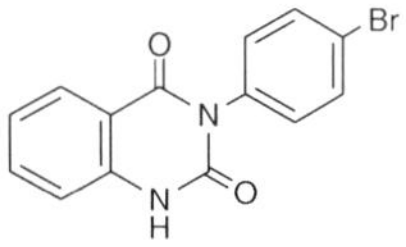
250 mg

S68,177-6
3-(4-FLUORO-2-METHYLPHENYL)-2,4-(1H,3H)QUINAZOLINEDIONE
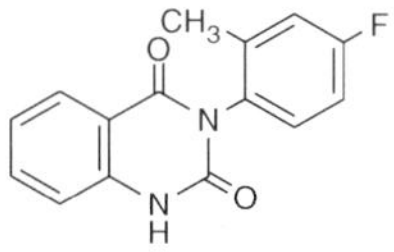
5 gm

S73,852-2
3-(4-FLUOROPHENYL)-2,4-THIAZOLIDINEDIONE
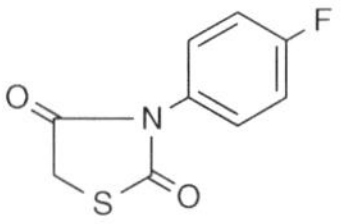
5 gm

S77,219-4
3-BUTYL-5-(4-FLUOROBENZYLIDENE)-2,4-THIAZOLIDINEDIONE
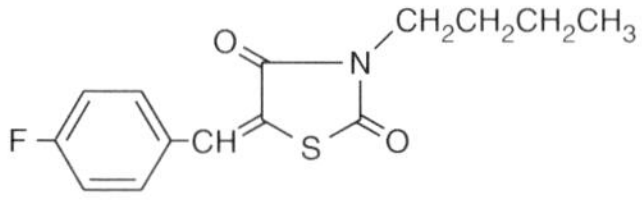
250 mg

S73,299-0
9B-(3-FLUOROPHENYL)-1,2,3,9B-TETRAHYDRO-5H-IMIDAZO(2,1-A)-ISOINDOL-5-ONE
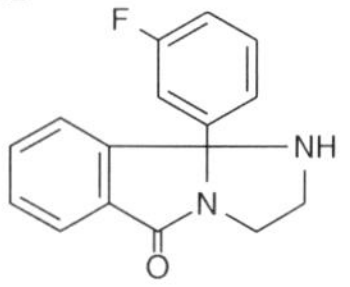
250 mg

S76,170-2
9B-(3-CHLOROPHENYL)-2,3,5,9B-TETRAHYDRO-1H-IMIDAZO(2,1-A)-ISOINDOL-5-ONE
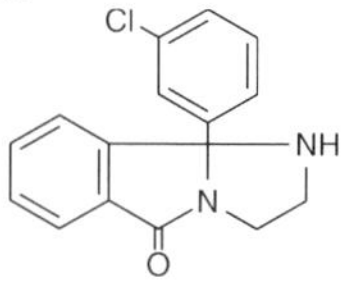
250 mg

S68,162-8
10B-(4-FLUOROPHENYL)-1,2,3,4,6,10B-HEXAHYDROPYRIMIDO(2,1-A)-ISOINDOL-6-ONE
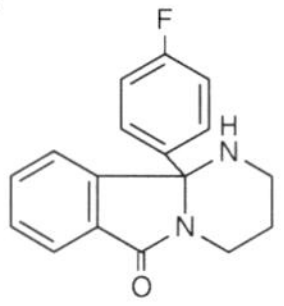
250 mg

S93,958-7
3-(4-CHLOROBENZYL)-4-PHENYL-2,5-OXAZOLIDINEDIONE
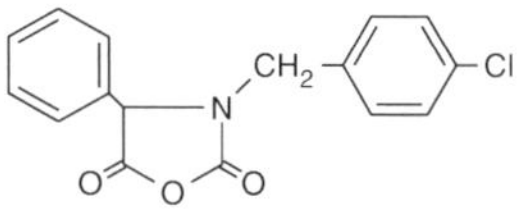
250 mg

S76,007-2
11B-(4-CHLOROPHENYL)-2,3,4,5,7,11B-HEXAHYDRO(1,3)OXAZEPINO(2,3-A)-ISOINDOL-7-ONE
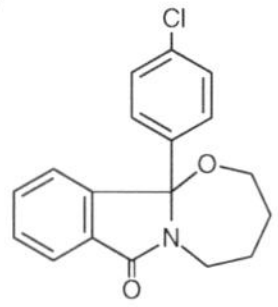
250 mg

S74,117-5
1-(4-FLUOROPHENYL)-5-IMINO-3-PHENYL-4-THIOHYDANTOIN
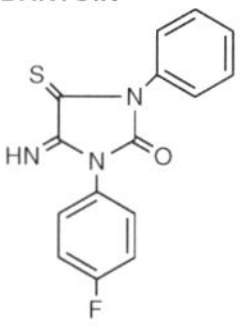
250 mg

S74,112-4
1-(4-BROMOPHENYL)-5-IMINO-3-PHENYL-4-THIOHYDANTOIN
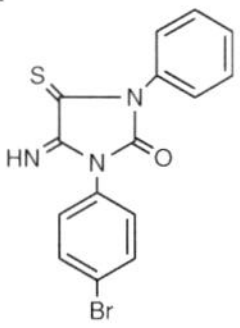
250 mg

S77,296-8
1-[N-(4-FLUOROPHENYL)CARBAMOYL]-AZIRIDINE
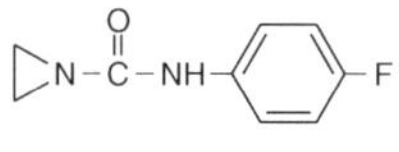
250 mg

S67,431-1
1-(2-CHLOROPHENYLCARBAMOYL)-PYRROLIDINE
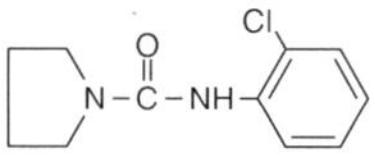
250 mg

S91,103-8
1-[N-(2,3-DICHLOROPHENYL)-CARBAMOYL]-2-PYRROLINE
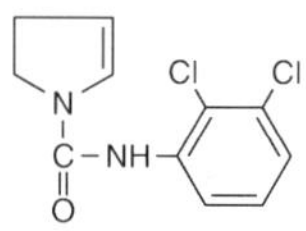
250 mg

S68,808-8
1-[N-(4-CHLORO-2-METHYLPHENYL)-CARBAMOYL]PYRROLIDINE
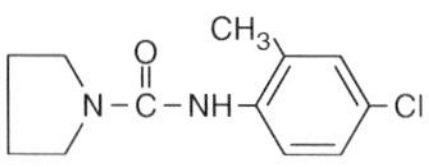
250 mg

S84,451-9
1-[N-(3-FLUOROPHENYL)CARBAMOYL]-2-METHYLPIPERIDINE
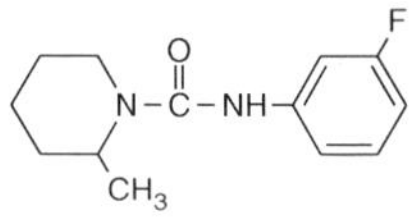
250 mg

S71,826-2
1-[N-(3-BROMOPHENYL)CARBAMOYL]-4-METHYLPIPERIDINE
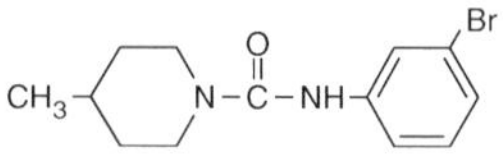
250 mg

S84,498-5
2,6-DIMETHYL-1-[N-(3-FLUOROPHENYL)-CARBAMOYL]PIPERIDINE
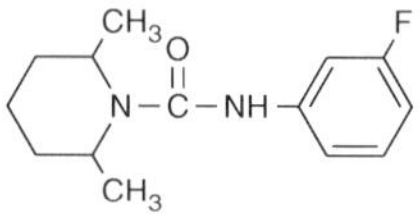
250 mg

S71,825-4
1-[N-(3-BROMOPHENYL)CARBAMOYL]-2,6-DIMETHYLPIPERIDINE
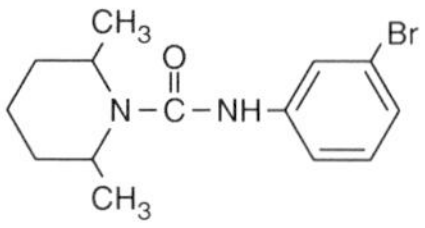
250 mg

S74,618-5
1-[N-(2-FLUOROPHENYL)CARBAMOYL]-3-METHYLPIPERIDINE
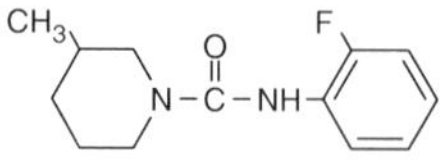
5 gm

S84,496-9
1-[N-(2-CHLOROPHENYL)CARBAMOYL]-2,6-DIMETHYLPIPERIDINE
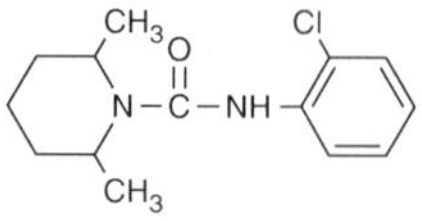
250 mg

S84,497-7
1-[N-(4-CHLOROPHENYL)CARBAMOYL]-2,6-DIMETHYLPIPERIDINE
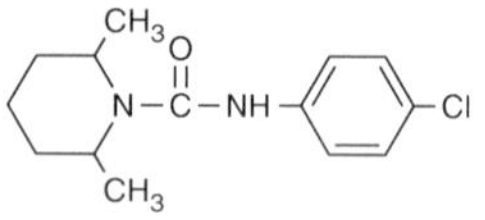
250 mg

S70,952-2
1-[N-(2,6-DICHLOROPHENYL)-CARBAMOYL]-4-METHYLPIPERIDINE
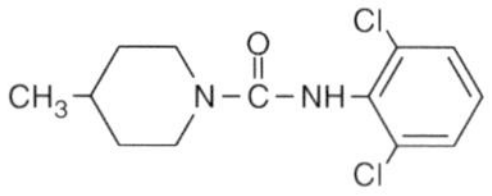
5 gm

S71,126-8
1-[N-(2,6-DICHLOROPHENYL)-CARBAMOYL]-2-METHYLPIPERIDINE
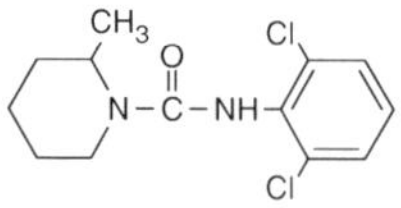
250 mg

S72,543-9
1-[N-(3-CHLORO-2-METHYLPHENYL)-CARBAMOYL]-4-METHYLPIPERIDINE
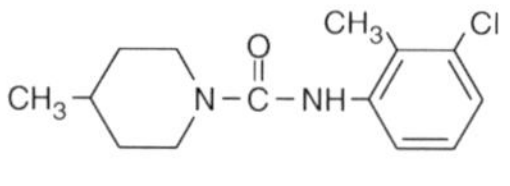
5 gm

S86,765-9
N-(3-CHLORO-2-METHYLPHENYL)-3-METHYL-1-PIPERIDINECARBOXAMIDE
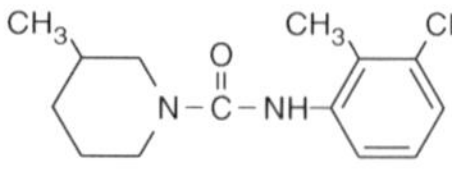
250 mg

S90,451-1
3'-CHLORO-3,4'-DIMETHYL-1-PIPERIDINECARBOXANILIDE
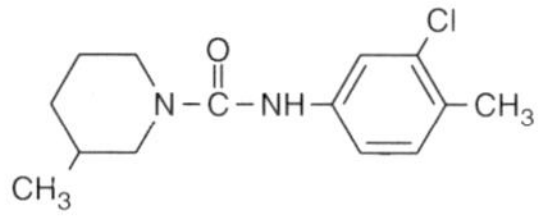
250 mg

S90,453-8
3'-CHLORO-4,4'-DIMETHYL-1-PIPERIDINECARBOXANILIDE
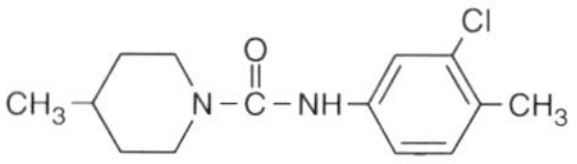
250 mg

S65,352-7
4-[N-(2-CHLOROPHENYL)CARBAMOYL]-2,6-DIMETHYLMORPHOLINE
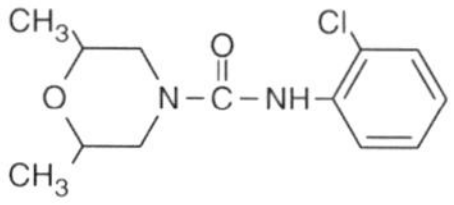
5 gm

S73,823-9
2,6-DIMETHYL-4-[N-(4-FLUOROPHENYL)-CARBAMOYL]MORPHOLINE
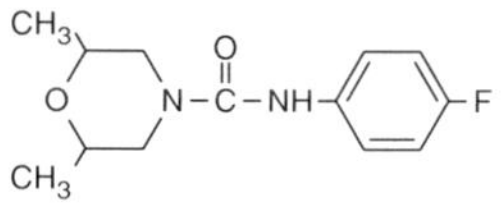
5 gm

S85,989-3
4-HYDROXY-2-(3-HYDROXYBUTYL)-1(2H)-PHTHALAZINONE
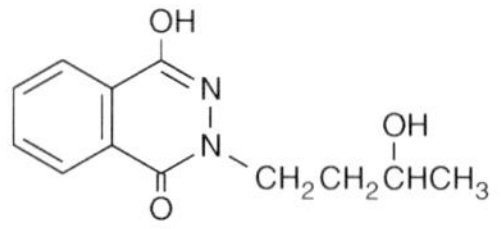
250 mg

S80,496-7
3,6-BIS(3-HYDROXYBENZYLIDENE)-2,5-PIPERAZINEDIONE
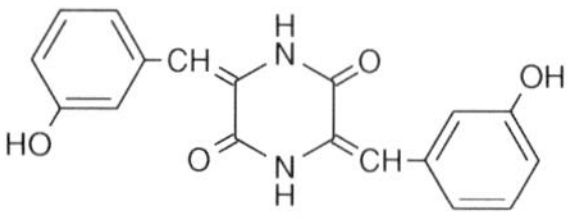
250 mg

S36,560-2
5-CINNAMYLIDENEBARBITURIC ACID
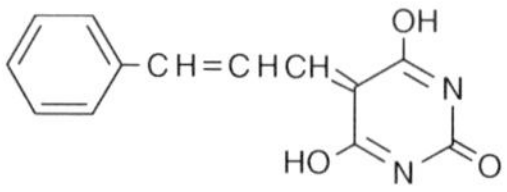
5 gm

S95,675-9
2-ACETYL-6-(DIMETHYLAMINO)-1-(3-HYDROXY-4-METHOXYBENZYL)-1,2,3,4-TETRAHYDRO-ISOQUINOLINE
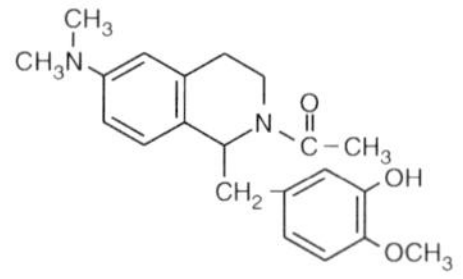

S76,138-9
10B-(2-HYDROXY-4-METHOXYPHENYL)-3,4,6,10B-TETRAHYDRO-2H-(1,3)OXAZINO (2,3-A)ISOINDOL-6-ONE
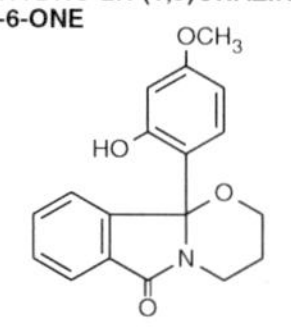
250 mg

S44,936-9
5'-O-TRITYLTHYMIDINEBENZENE SOLVATE
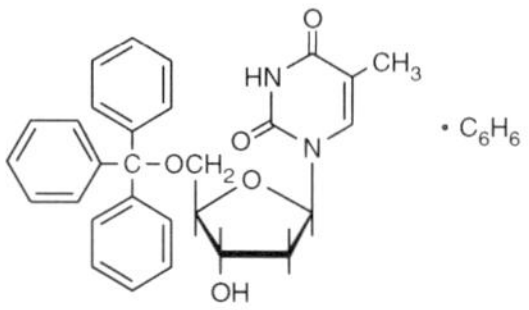
250 mg

S76,724-7
10B-(2-HYDROXY-4-METHOXYPHENYL)-1,2,3,4-TETRAHYDROPYRIMIDO(2,1-A)-ISOINDOL-6-(10-BH)ONE
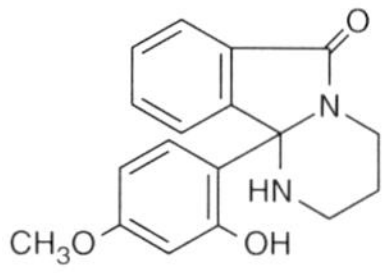
250 mg

S76,072-2
4,5-DIHYDRO-2-(3-HYDROXYPROPYL)-6-(4-METHOXYPHENYL)-3(2H)-PYRIDAZINONE
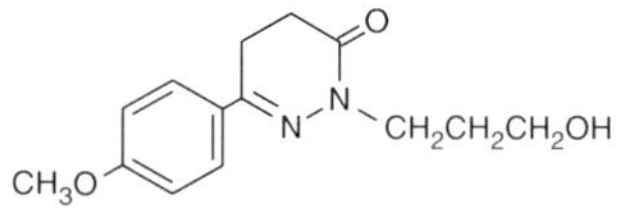
250 mg

S78,647-0
CARBOBENZYLOXYSERYLPROLINE METHYL ESTER
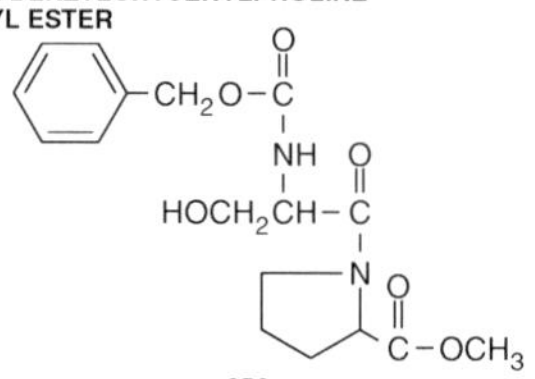
250 mg

S76,088-9
1,2,3,4,12,12A-HEXAHYDRO-12-HYDROXY-12-PHENYLPYRIDO(1,2-B)(2)-BENZAZEPIN-6(11H)ONE
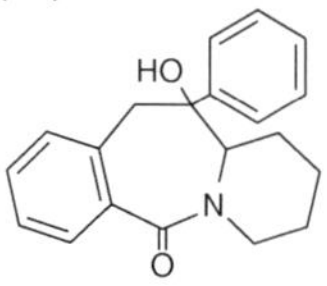
250 mg

S79,051-6
CARBOBENZYLOXYTHREONYLPROLINE METHYL STER
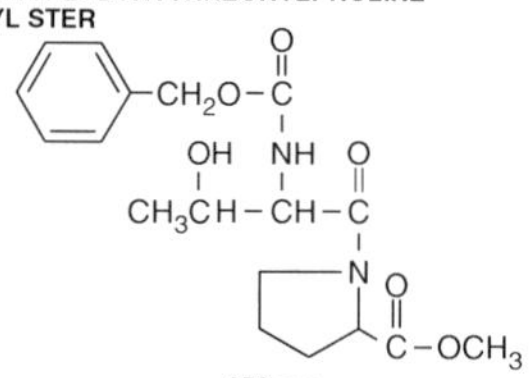
250 mg

S45,292-0
2',3'-DIBENZOYLURIDINE
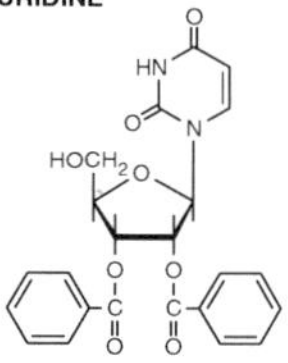
250 mg

S58,365-0
RIBOFLAVIN-5'-PHOSPHATE
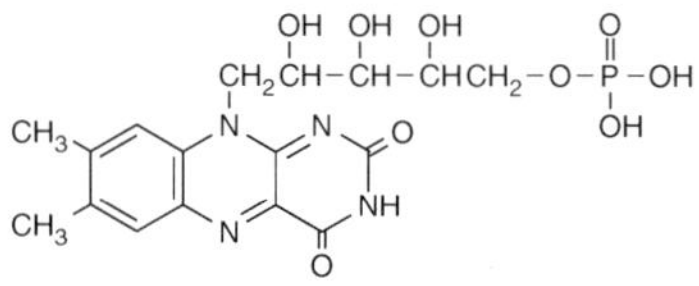
5 gm

S57,276-4
5-CHLORO-3-HYDROXY-1-(3-NITROPHENYL)-2(1H)PYRIDONE
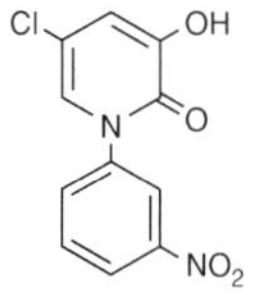
250 mg

S51,923-5
ANTIPYRINE SALICYLATE
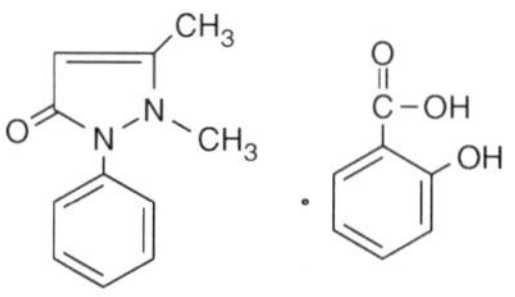
5 gm

S89,327-7
1-BENZOYL-3-HYDROXY-2-PIPERIDINYLMETHYL BENZOATE
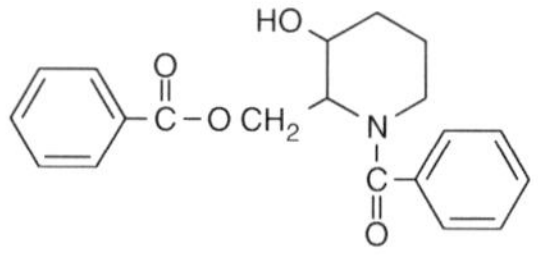
250 mg

S77,742-0
L-PHENYLALANYL-L-SERYLGLYCINE ANHYDRIDE HEMIHYDRATE
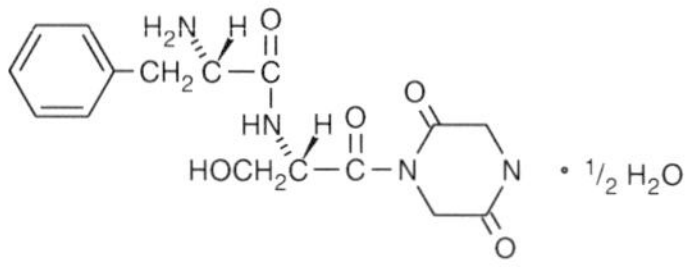
250 mg

S78,865-1
L-PHENYLALANYL-D-TYROSYLGLYCINE ANHYDRIDE
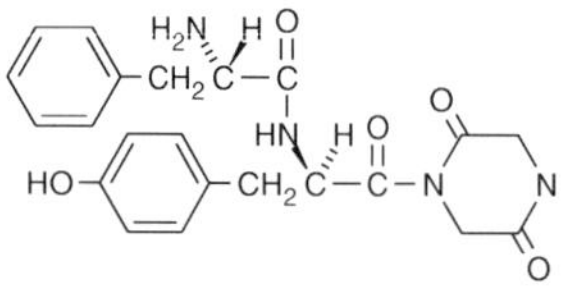
250 mg

S52,481-6
ETHYL 4-(2-HYDROXYANILINO)-1-PIPERIDINECARBOXYLATE
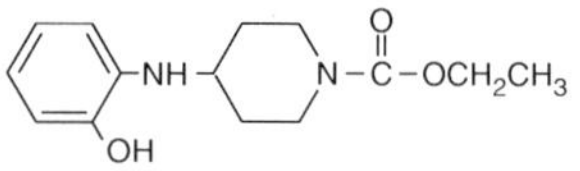
250 mg

S64,200-2
1-(4-AMINOPHENYL)-4-(2-HYDROXYETHYL)-3-METHYL-2-PYRAZOLIN-5-ONE
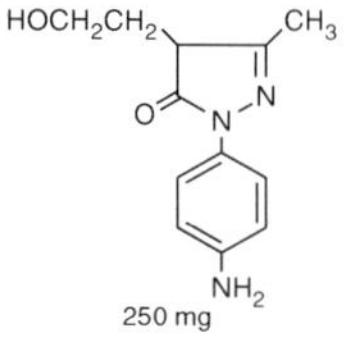
250 mg

S74,549-9
VOMICINE

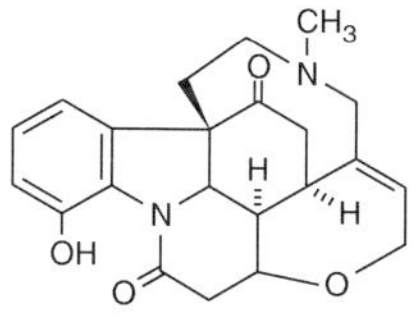

250 mg

S66,482-0
6-(3,4-DICHLOROPHENYL)-4,5-DIHYDRO-2-(2-HYDROXYBUTYL)-3(2H)-PYRIDAZINONE

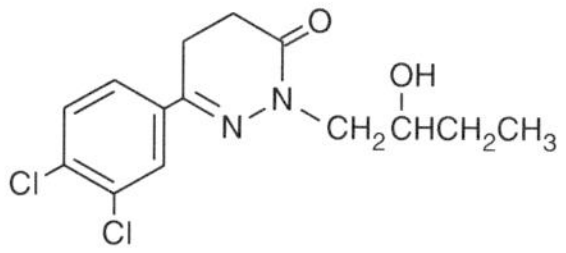

5 gm

S68,265-9
10B-(4-HYDROXYPHENYL)-1,2,3,4,6,10B-HEXAHYDROPYRIMIDO(2,1-A)ISOINDOL-6-ONE

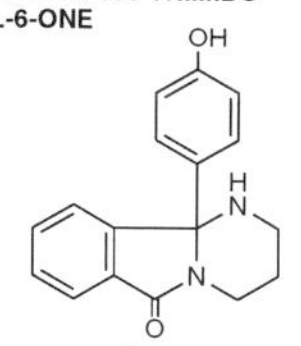

5 gm

S75,446-3
1-(2-BENZOYLBENZOYL)-2-PIPERIDINEMETHANOL

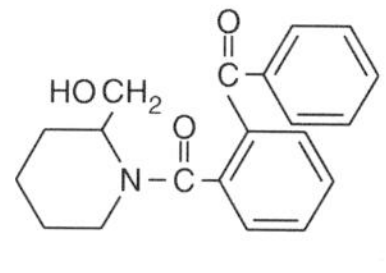

250 mg

S88,370-0
2-(2-HYDROXYETHYL)-1,4-(2H,3H)-ISOQUINOLINEDIONE

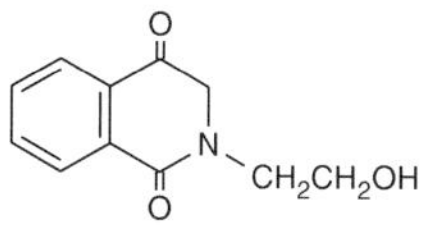

250 mg

S92,741-4
4-HYDROXY-6-METHYL-2-PHENYL-3(2H)-PYRIDAZINONE

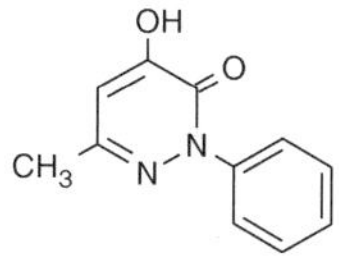

250 mg

S76,727-1
2-(2-HYDROXYETHYL)-1(2H)-PHTHALAZINONE

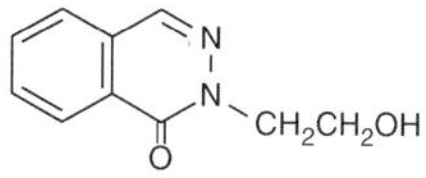

250 mg

S80,210-7
1-(HYDROXYMETHYL)ISATIN

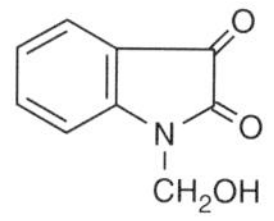

250 mg

S50,657-5
2-HYDROXY-5-(INDANYLMETHYLENE)-1-PHENYL-4,6-(1H,5H)-PYRIMIDINEDIONE

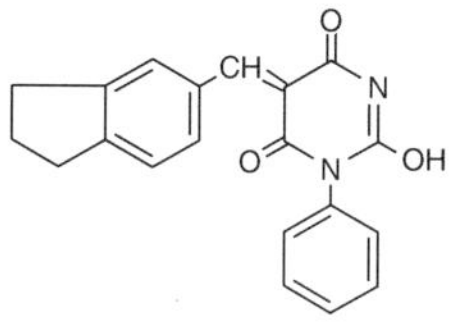

250 mg

S57,037-0
1-SALICYLOYLPYRROLIDINE

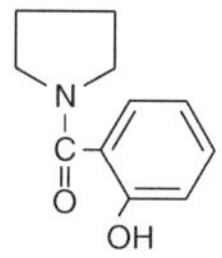

5 gm

S60,296-5
4-(2-HYDROXY-1-NAPHTHOYL)MORPHOLINE

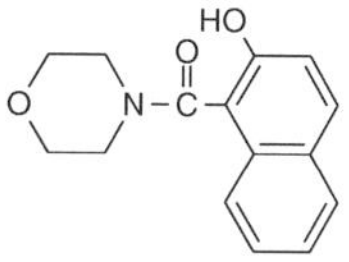

5 gm

S76,777-8
4-(4-HYDROXYPHENYL)-2-PHENYL-1(2H)-PHTHALAZINONE

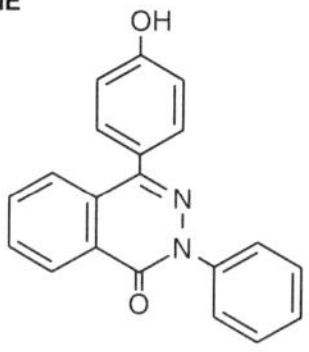

250 mg

S76,168-0
4-(4-HYDROXYPHENYL)-2-METHYL-1(2H)-PHTHALAZINONE

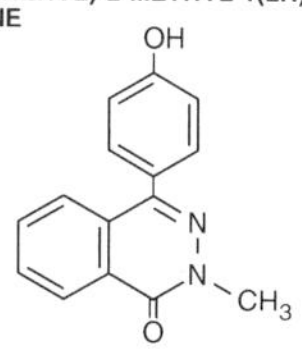

250 mg

S77,637-8
CARBOBENZYLOXY-L-PROLYL-L-TYROSINE

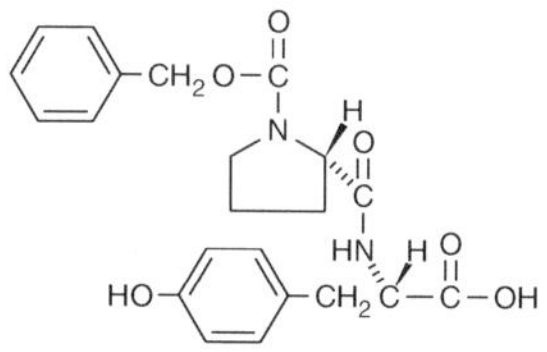

250 mg

S77,767-6
CARBOBENZYLOXY-L-SERYL-L-PROLINE BENZYL ESTER

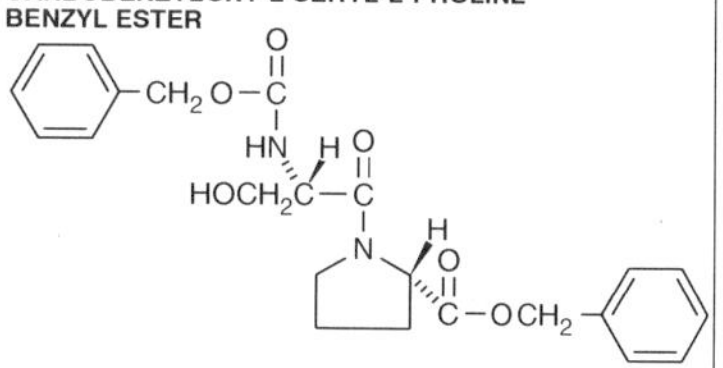

250 mg

S79,053-2
CARBOBENZYLOXY-L-PROLYL-L-TYROSINE METHYL ESTER

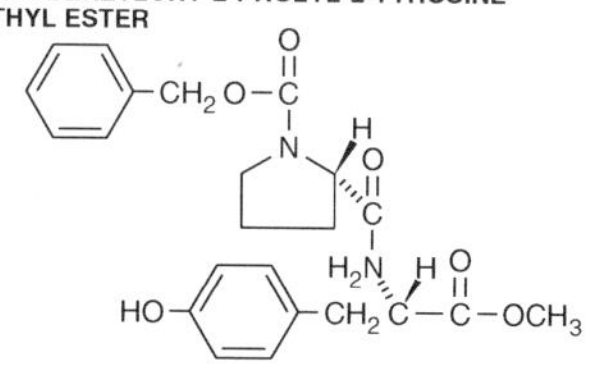

250 mg

S79,069-9
CARBOBENZYLOXY-L-PROLYL-L-TYROSINE-BENZYL ESTER

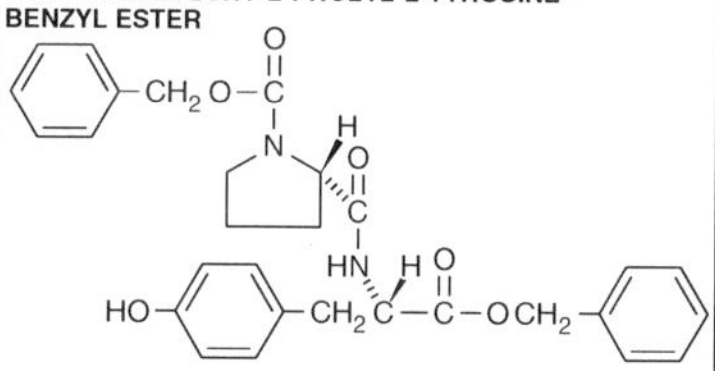

250 mg

S54,790-5
5-(4-HYDROXYBENZYL)HYDANTOIN

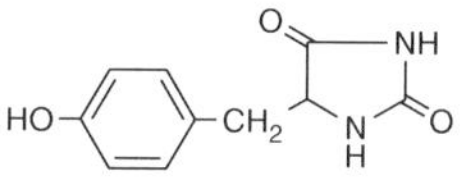

250 mg

S38,758-4
N-(7-HYDROXY-1-NAPHTHYL)ACETAMIDE

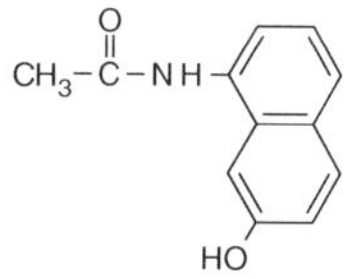

250 mg

S76,656-9
2,3-DIHYDRO-9B-(4-HYDROXYPHENYL)-1H-IMIDAZO(2,1-A)ISOINDOL-5(9BH)ONE

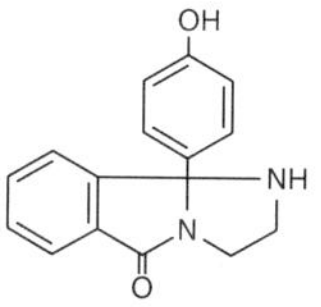

250 mg

S73,416-0
HYDROXY-METHYL-TETRAHYDRO-SPIRO[5H-DIBENZO(A,D)CYCLOHEPTENE-(1H)-ISOINDOL]-3'-ONE

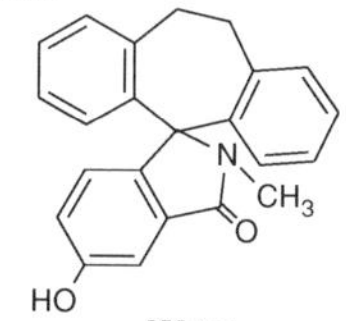

250 mg

S75,938-4
9B-(4-HYDROXYPHENYL)-1,2,5,9B-TETRAHYDRO-3H-IMIDAZO(2,1A)-ISOINDOL-5-ONE

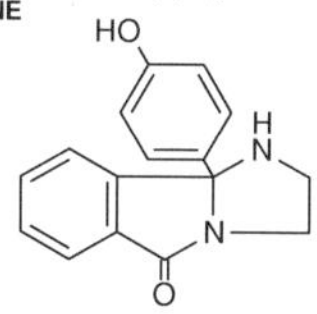

250 mg

S41,976-1
4-HYDROXY-1-METHYL-2(1H)QUINOLONE

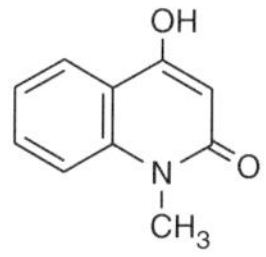

250 mg

S34,427-3
4-CHLORO-3-HYDROXY-1-PHENYL-6-PYRIDAZONE

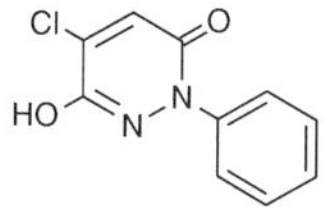

5 gm

S86,048-4
2'-DEOXYCYTIDINE PICRATE

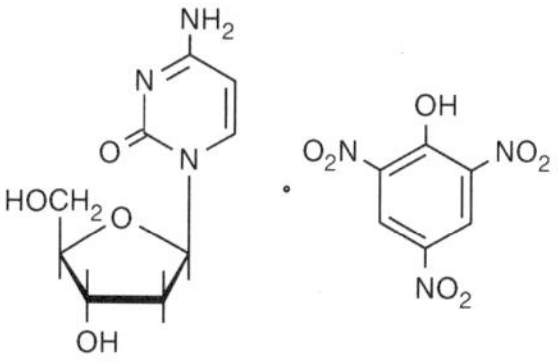

250 mg

S76,511-2
1-CARBOBENZYLOXY-4-TERT-BUTOXYPROLINE-DICYCLOHEXYLAMINE SALT

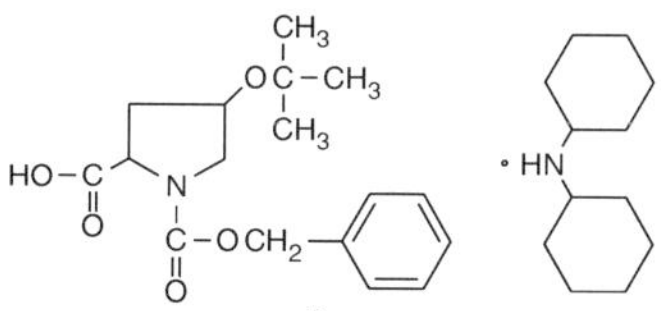

5 gm

S77,682-3
N,2',3',5'-TETRABENZYLCYTIDINE

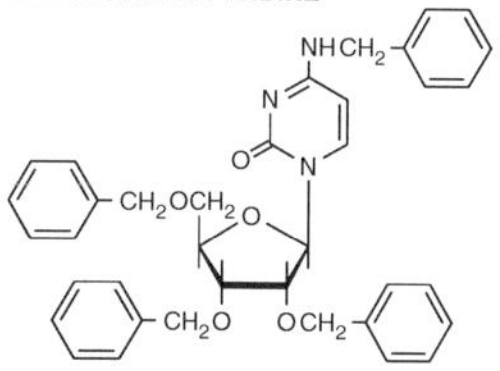

250 mg

S84,303-2
6,7-DIMETHOXY-1,4(2H,3H)-PHTHALAZINEDIONE

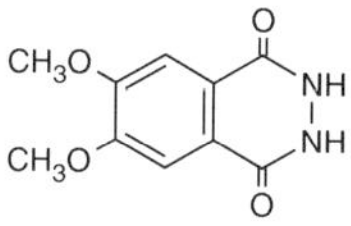

250 mg

S79,861-4
1-(3,4-DIMETHOXYPHENETHYL)-3,4,5,6,7,8-HEXAHYDRO-2(1H)-QUINOLINONE

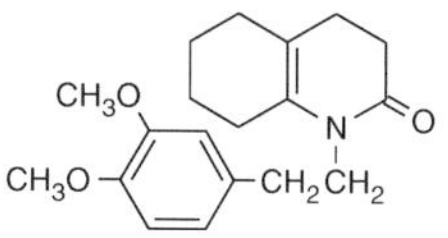

250 mg

S95,676-7
2-ACETYL-1-(3,4-DIMETHOXYBENZYL)-6-(DIMETHYLAMINO)1,2,3,4-TETRAHYDROISOQUINOLINE

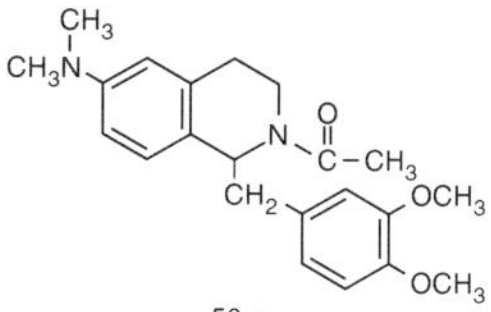

50 mg

S86,199-5
1-(4-METHOXYBENZYLIDENEAMINO)-4-(4-METHOXYPHENYL)-3-PHENOXY-2AZETIDINONE N-OXIDE

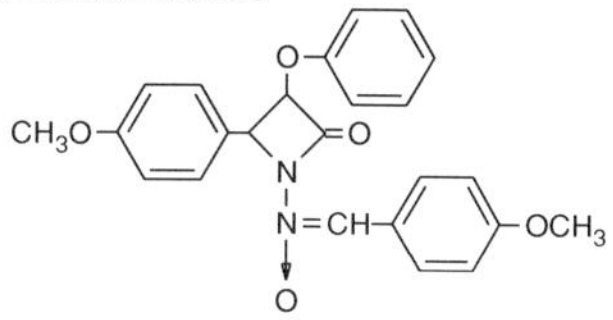

250 mg

S76,008-0
2,3-DIMETHOXY-5,6,8,12B-TETRAHYDROISOINDOLO(1,2-A)-ISOQUINOLIN-8-ONE

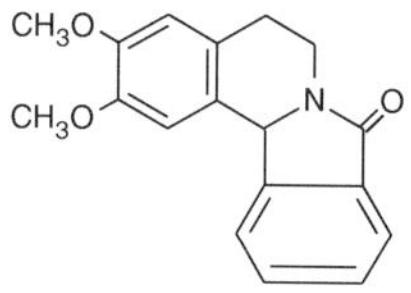

250 mg

S77,372-7
5-(2,3-DIMETHOXYPHENYL)HYDANTOIN

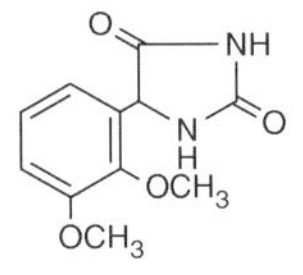

250 mg

S76,851-0
5-(3,4-DIMETHOXYBENZYLIDENE)-3-ETHYL-2,4-THIAZOLIDINEDIONE

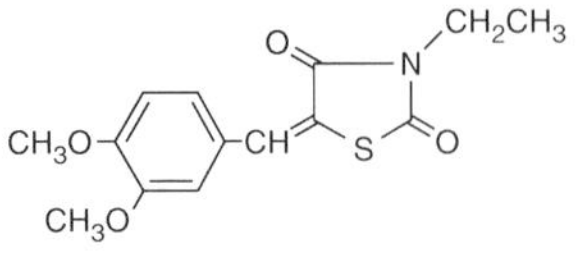

250 mg

S68,302-7
5,6-DIHYDRO-2,3-DIMETHOXY-12B-PHENYL-ISOINDOLO(1,2-A)ISOQUINOLIN-8(12BH)ONE

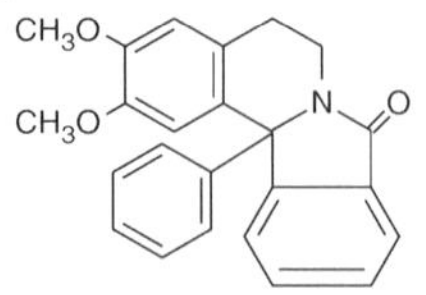

250 mg

S76,175-3
8,9-DIMETHOXY-10B-PHENYL-1,5,6,10B-TETRAHYDROPYRROLO(2,1-A)-ISOQUINOLIN-3(2H)ONE

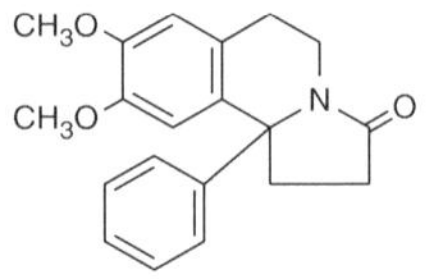

250 mg

S73,332-6
2,3-DIMETHOXY-5,6,8,8A,9,12,12A,12B-OCTAHYDRO-ISOINDOLO(1,2-B)ISOQUINOLIN-8-ONE

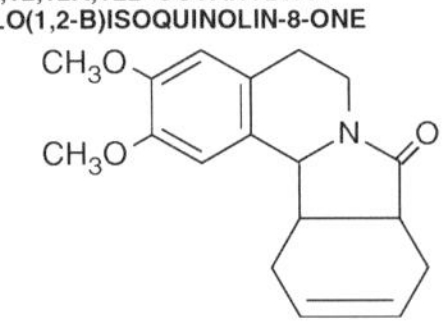

250 mg

S78,099-5
2',5'-DIMETHOXY-1-PYRROLIDINECARBOXANILIDE

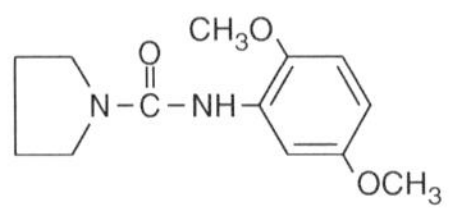

250 mg

S90,513-5
2',5'-DIMETHOXY-2-METHYL-2-IMIDAZOLINE-1-CARBOXANILIDE

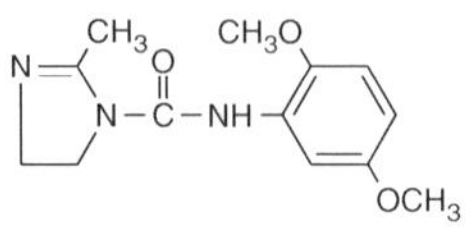

250 mg

S80,838-5
2-(1-BENZOYL-2-PIPERIDYL)-2',5'-DIMETHOXYACETOPHENONE

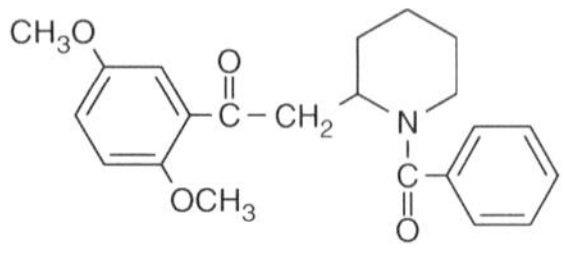

250 mg

S55,143-0
1,2-DIHYDRO-2-(2,3-DIMETHOXYPHENYL)--4(3H)QUINAZOLINONE

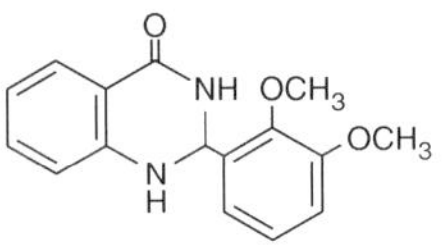

250 mg

S52,586-3
1,2-DIHYDRO-2-(4-ETHOXY-3-METHOXYPHENYL)-4(3H)QUINAZOLINONE

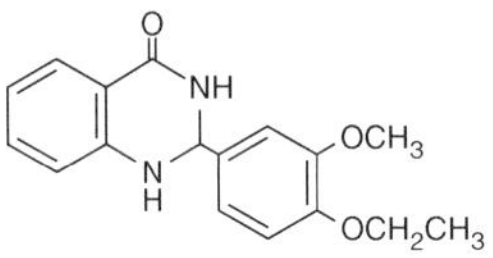

250 mg

S45,330-7
1,2-DIHYDRO-2-(2-ETHOXY-3-METHOXYPHENYL)-4-(3H)QUINAZOLINONE

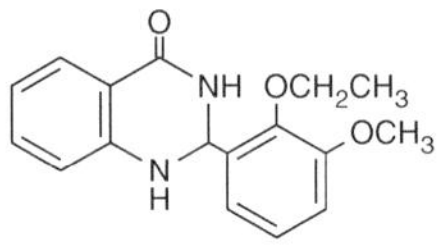

5 gm

S76,128-1
2-[2-(CARBOXYMETHOXY)PHENYL]-1,2-DIHYDRO-4(3H)QUINAZOLINONE

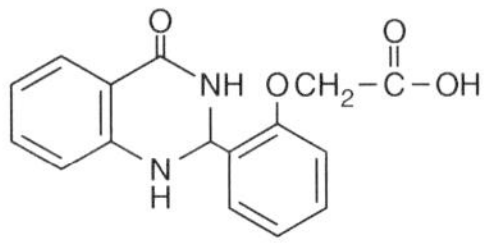

250 mg

S52,584-7
2-(3,4-DIBENZYLOXYPHENYL)-1,2-DIHYDRO-4(3H)QUINAZOLINONE

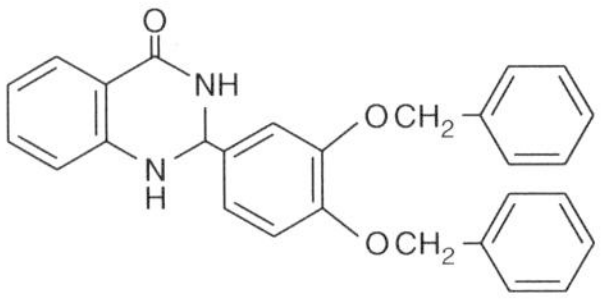

250 mg

S85,976-1
5'-O-TRITYLTHYMIDINE

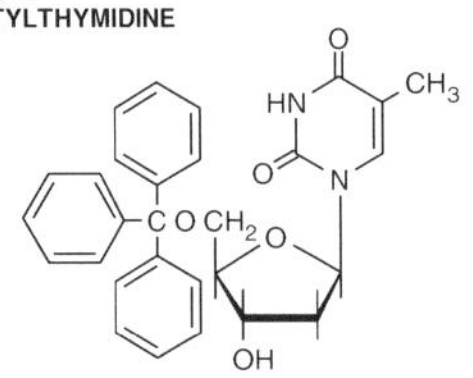

250 mg

S64,245-2
3-AMINO-1-(4-METHOXYBENZYL)-2-PYRAZOLIN-5-ONE

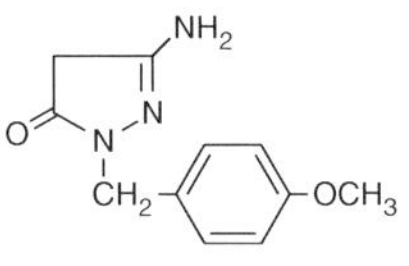

5 gm

S94,710-5
DIETHYL 6-AMINO-1-BENZAMIDO-4-(4-METHOXYPHENYL)-2-OXO-TETRAHYDRO-3,5-PYRIDINEDICARBOXYLATE

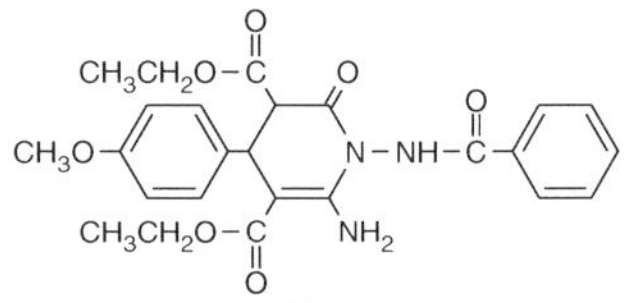

50 mg

S65,424-8
1,2-DIHYDRO-2-(4-METHOXYPHENYL)-4-(3H)QUINAZOLINONE

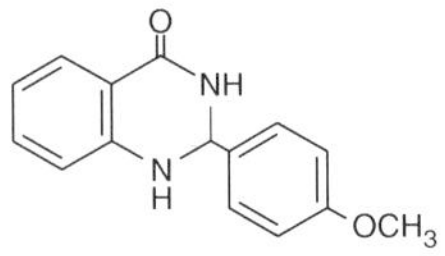

250 mg

S54,225-3
1,2-DIHYDRO-2-(2-ETHOXYPHENYL)-4-(3H)QUINAZOLINONE

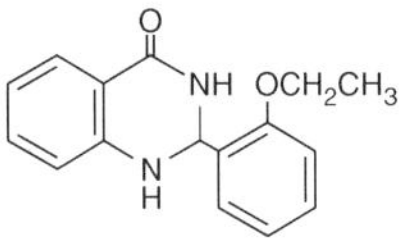

250 mg

S54,226-1
1,2-DIHYDRO-2-(4-ETHOXYPHENYL)-4-(3H)QUINAZOLINONE

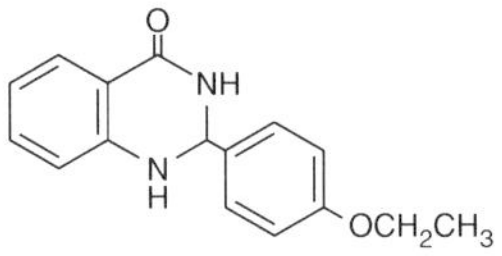

250 mg

S55,140-6
1,2-DIHYDRO-2-(2,5-DIMETHOXY-PHENYL)4(3H)QUINAZOLINONE

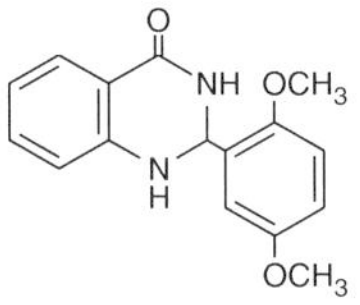

250 mg

S55,146-5
1,2-DIHYDRO-2-(3-METHOXYPHENYL)-4-(3H)QUINAZOLINONE

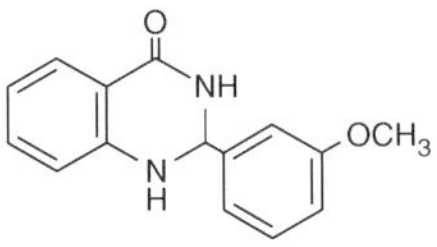

250 mg

S52,595-2
2-(3-BENZYLOXYPHENYL)-1,2-DIHYDRO-4(3H)QUINAZOLINONE

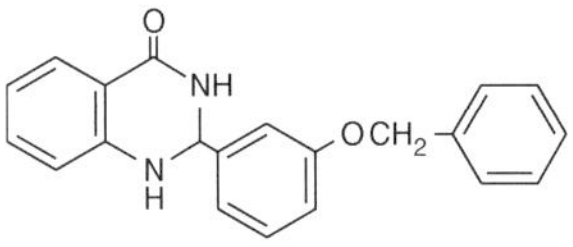

5 gm

S55,142-2
1,2-DIHYDRO-2-(3,4-DIMETHOXY-PHENYL)4(3H)QUINAZOLINONE

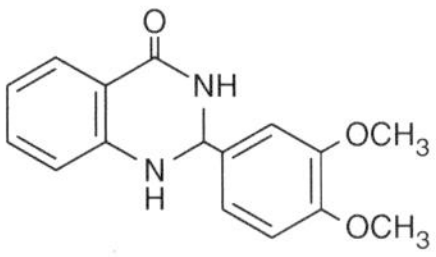

250 mg

S89,154-1
2'-METHOXY-2-METHYL-1-PIPERIDINECARBOXANILIDE

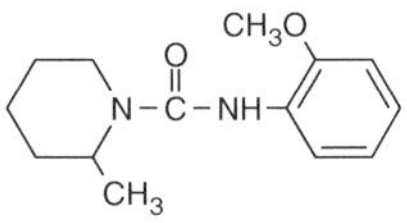

250 mg

S43,456-6
2-NITROSTRYCHNINE

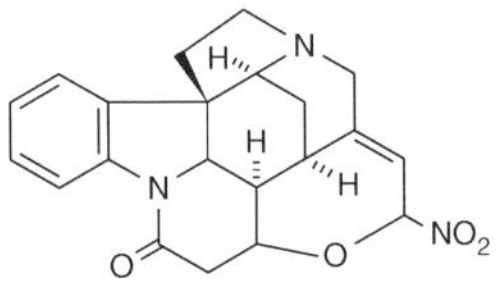

5 gm

S87,171-0
2-(4-METHOXYANILINO)AZACYCLOHEPT-1-ENE HYDROCHLORIDE

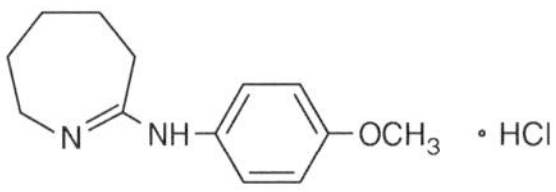

250 mg

S63,308-9
3-(4-PENTYLOXYPHENYL)-1-PHENYL-2-PYRAZOLIN-5-ONE

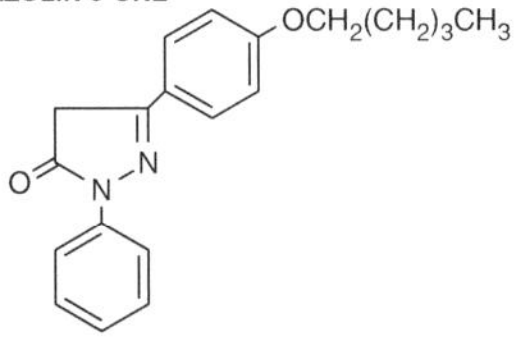

250 mg

S54,631-3
3-ETHOXY-1-PHENYL-2-PYRAZOLIN-5-ONE

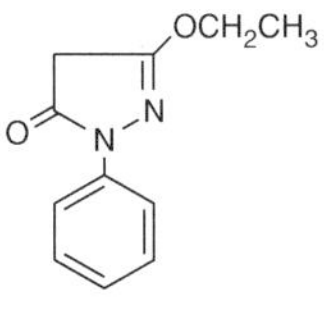

5 gm

S87,237-7
2-(2-ETHOXYETHYL)-1(2H)-ISOQUINOLINONE

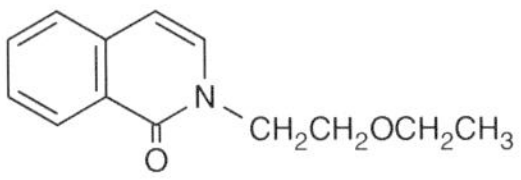

250 mg

S84,278-8
3-(2-METHOXYPHENYL)-1,2,3-BENZOTRIAZIN-4(3H)-ONE

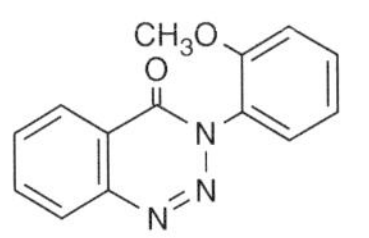

250 mg

S85,416-6
1-BUTYL-3-(2-METHOXY-5-METHYLPHENYL)-2,4,5-IMIDAZOLIDINETRIONE

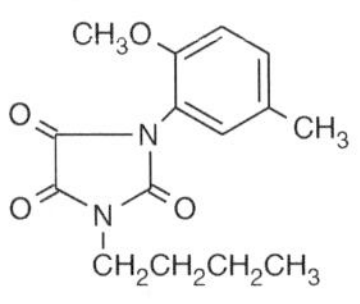

250 mg

S63,824-2
1-[N-(2-METHOXY-5-METHYLPHENYL)-CARBAMOYL]PIPERIDINE

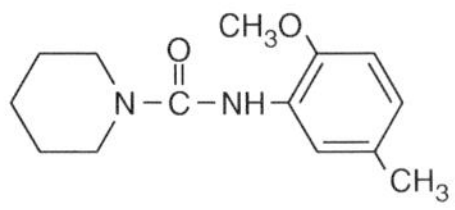

5 gm

S84,500-0
1-[N-(2-METHOXYPHENYL)CARBAMOYL]-3-METHYLPIPERIDINE

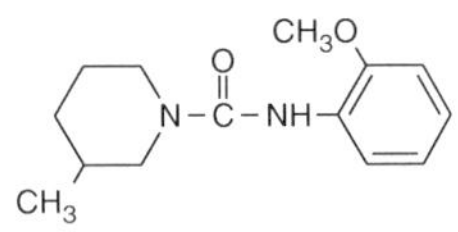

250 mg

S89,152-5
2,6-DIMETHYL-2'-METHOXY-1-PIPERIDINECARBOXANILIDE

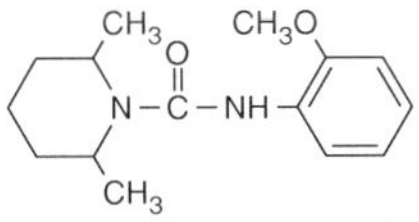

250 mg

S66,535-5
1-[N-(2,5-DIMETHOXYPHENYL)-CARBAMOYL]PIPERIDINE

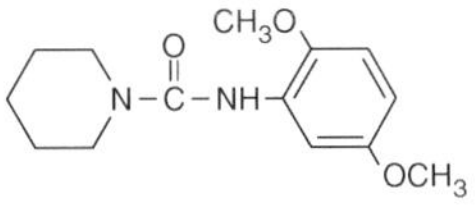

5 gm

S60,449-6
4-(P-ANISYLCARBAMOYL)MORPHOLINE

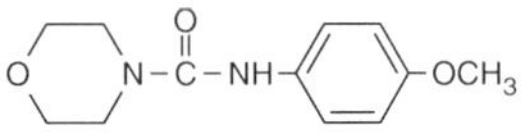

5 gm

S63,385-2
4-(O-ANISYLCARBAMOYL)MORPHOLINE

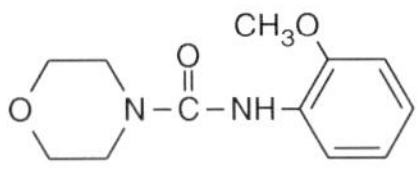

5 gm

S63,821-8
4-[N-(2-METHOXY-5-METHYLPHENYL)-CARBAMOYL]MORPHOLINE

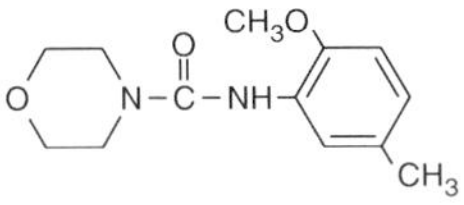

5 gm

S65,347-0
4-[N-(O-ANISYL)CARBAMOYL]-2,6-DIMETHYLMORPHOLINE

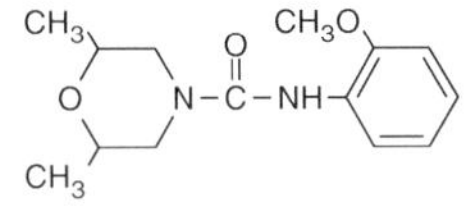

5 gm

S65,425-6
4-[N-(3-CHLORO-2-METHOXYPHENYL)-CARBAMOYL]MORPHOLINE

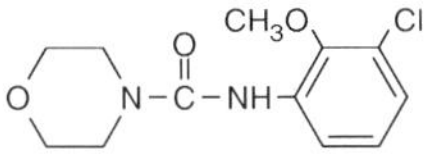

5 gm

S66,580-0
4-[N-(3-CHLORO-2-METHOXYPHENYL)-CARBAMOYL]-2,6-DIMETHYLMORPHOLINE

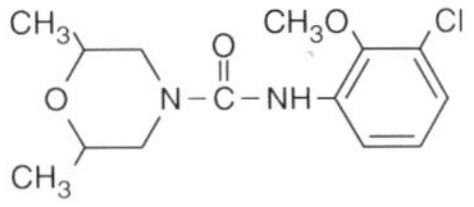

5 gm

S66,517-7
4-[N-(2,5-DIMETHOXYPHENYL)-CARBAMOYL]MORPHOLINE

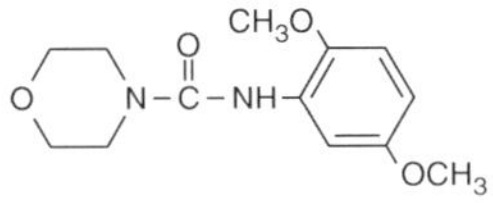

5 gm

S66,165-1
2,6-DIMETHYL-4-[N-(2-METHOXY-5-METHYLPHENYL)CARBAMOYL]MORPHOLINE

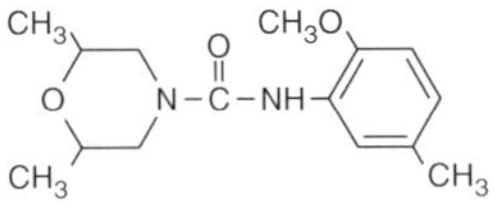

5 gm

S66,513-4
4-[N-(2,5-DIMETHOXYPHENYL)-CARBAMOYL]-2,6-DIMETHYLMORPHOLINE

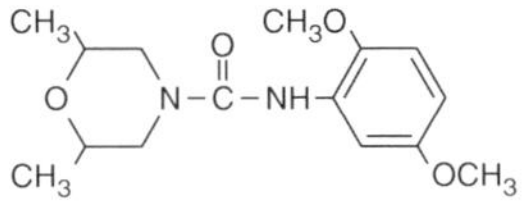

5 gm

S95,689-9
CIS-4-(3-CHLOROACRYLOYL)-3A,8B-DIMETHYL-1,2,3,3A,4,8B-HEXAHYDRO-6-METHOXYCYCLOPENT(B)-INDOL-2-ONE

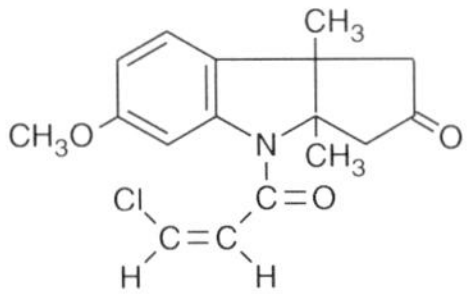

50 mg

S57,062-1
4-(3,4,5-TRIMETHOXYBENZOYL)-MORPHOLINE

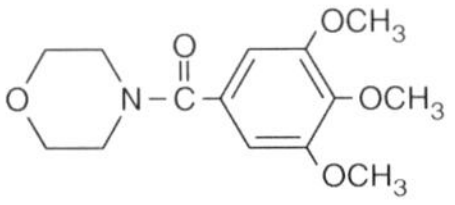

5 gm

S95,730-5
10A,10B-DIMETHYL-9,9-ETHYLENEDIOXY-HEXAHYDRO-3-METHOXYBENZ(B)CYCLOPENT(HI)-INDOLIZIN-6-ONE

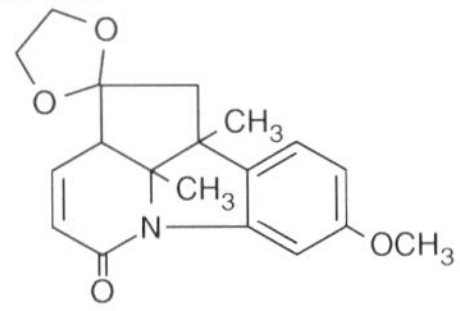

50 mg

S51,014-9
1,2-DIMETHYL-3-[4-(HEXADECYLOXY)-PHENYL]-3-PYRAZOLIN-5-ONE

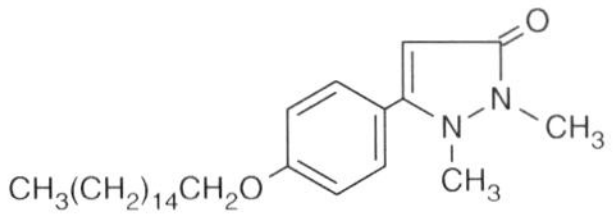

250 mg

S89,988-7
1-ETHYL-3-(4-METHOXYPHENYL)-PARABANIC ACID

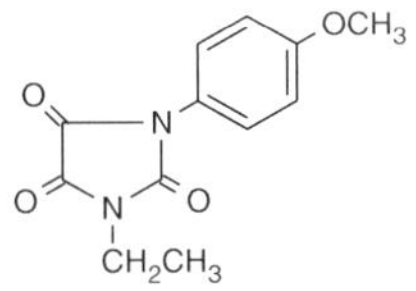

250 mg

S91,885-7
3-METHYL-8-OXO-7-(2-PHENOXYACETAMIDO)-5-THIA-1-AZABICYCLO[4.2.0]-OCT-2-ENE-2-CARBOXYLIC ACID, LITHIUM SALT

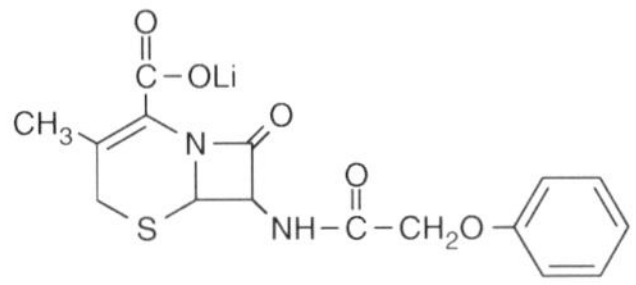

250 mg

S75,890-6
2-[4-(3,4-DIHYDRO-4-OXO-3-PHENYLPHTHALAZIN-1-YL)PHENOXY]-PROPIONIC ACID

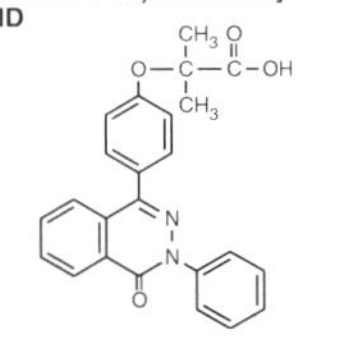

250 mg

S76,347-0
ETHYL METHYL-OXO-4H-SPIRO[5H-DIBENZO(A,D)CYCLOHEPTENE(1H)-ISOINDOLYLOXY]ACETATE

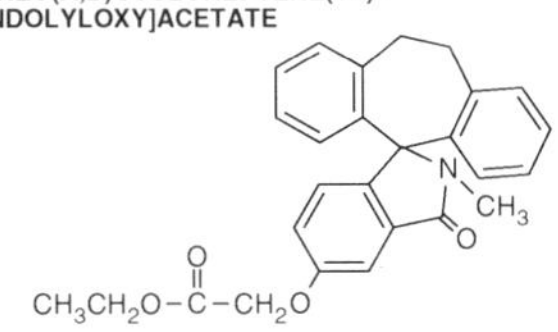

250 mg

S66,891-5
5-(4-HYDROXY-3-METHOXYBENZYLIDENE)-BARBITURIC ACID

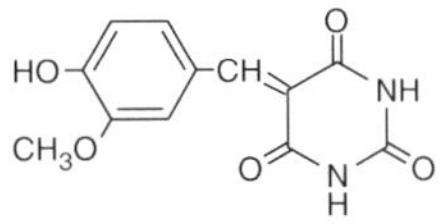

5 gm

S95,735-6
9,9-(ETHYLENEDIOXY)-3-METHOXYOCTAHYDRO-8,10A,10B,TRIMETHYLBENZ[B]-CYCLOPENT[HI]INDOLIZIN-6-ONE

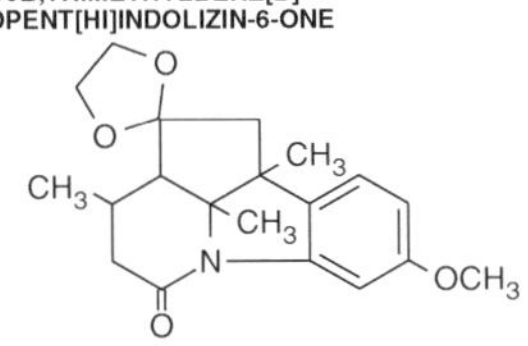

50 mg

S95,737-2
DECAHYDRO-11,11-(ETHYLENEDIOXY)-3-METHOXY-8,12A,12B-TRIMETHYLBENZO[B]CYCLOPENTA[HI]-PYRIDO[3,4-F]INDOLIZIN-6,9-ONE

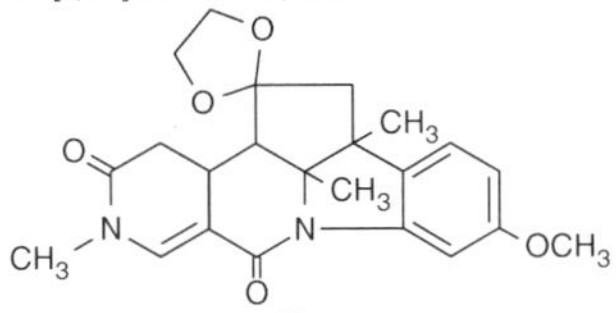

50 mg

S62,181-1
1-[N-(O-ANISYL)CARBAMOYL]PIPERIDINE

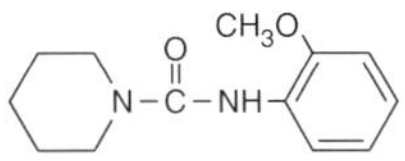

5 gm

S63,402-6
1-[N-(M-ANISYL)CARBAMOYL]PIPERIDINE

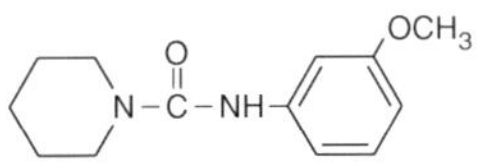

5 gm

S72,650-8
1-[N-(4-METHOXYPHENYL)CARBAMOYL]-3-METHYLPIPERIDINE

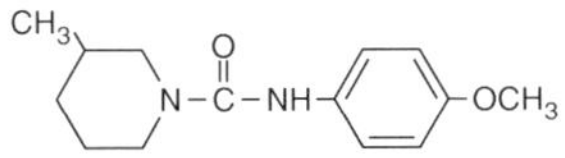

5 gm

S84,501-9
1-[N-(4-METHOXYPHENYL)CARBAMOYL]-2-METHYLPIPERIDINE

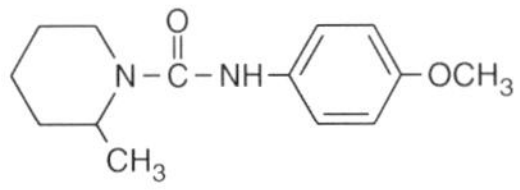

250 mg

S89,153-3
2,6-DIMETHYL-4'-METHOXY-1-PIPERIDINECARBOXANILIDE

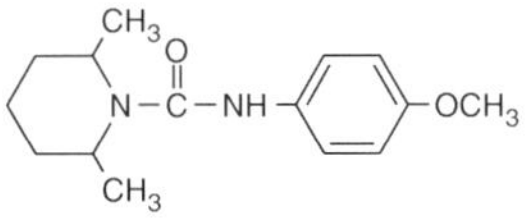

250 mg

S66,351-4
1-[N-(3-CHLORO-2-METHOXYPHENYL)-CARBAMOYL]PIPERIDINE

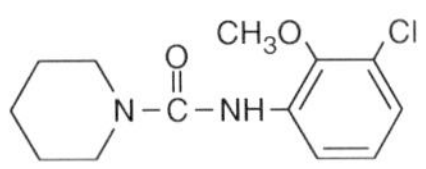

5 gm

S63,393-3
4-(M-ANISYLCARBAMOYL)MORPHOLINE

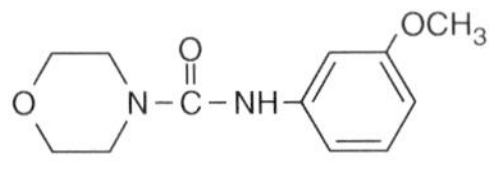

5 gm

S65,335-7
4-[N-(M-ANISYL)CARBAMOYL]-2,6-DIMETHYLMORPHOLINE

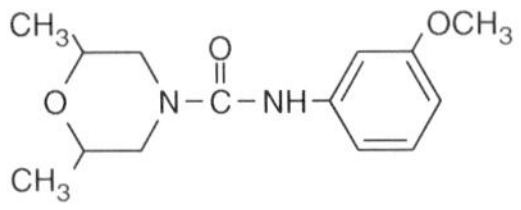

5 gm

S65,365-9
4-[N-(P-ANISYL)CARBAMOYL]-2,6-DIMETHYLMORPHOLINE

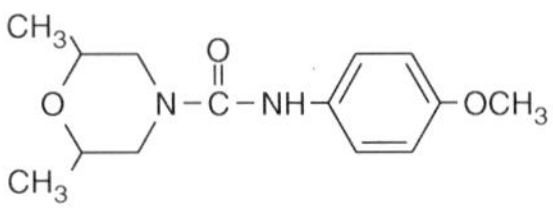

5 gm

S68,285-3
1-CARBAMOYL-5-METHOXY-2-METHYLINDOLINE

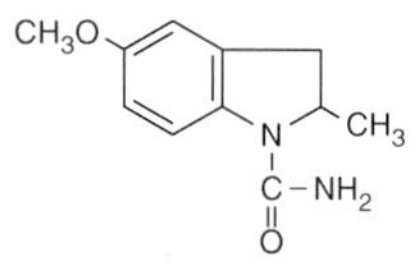

250 mg

S77,708-0
CARBOBENZYLOXY-L-PHENYLALANYL-L-PROLINAMIDE

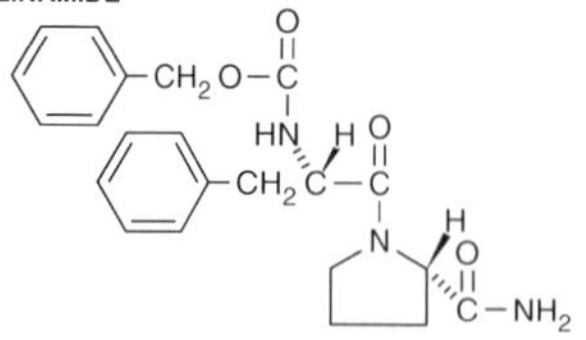

250 mg

S92,133-5
2-PHENYL-4H-1,3-BENZOXAZIN-4-ONE

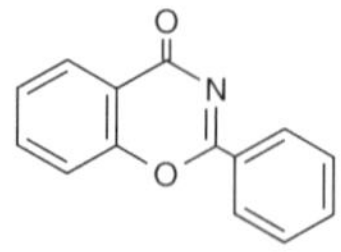

250 mg

S79,050-8
CARBOBENZYLOXYPROLYLGLYCYLGLYCINE-METHYL ESTER

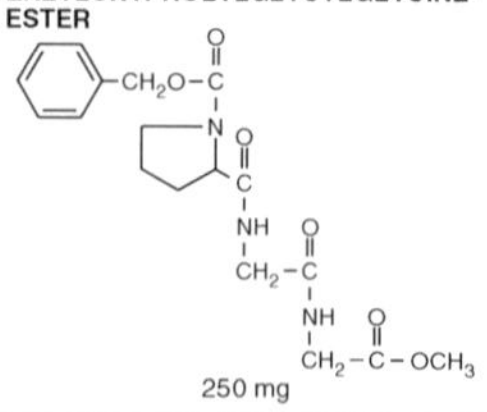

250 mg

S78,961-5
CARBOBENZYLOXYPROLYLVALINE-METHYL ESTER

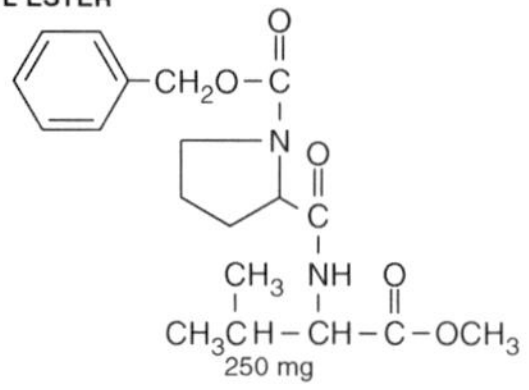

250 mg

S78,974-7
CARBOBENZYLOXYPROLYLGLYCINE-BENZYL ESTER

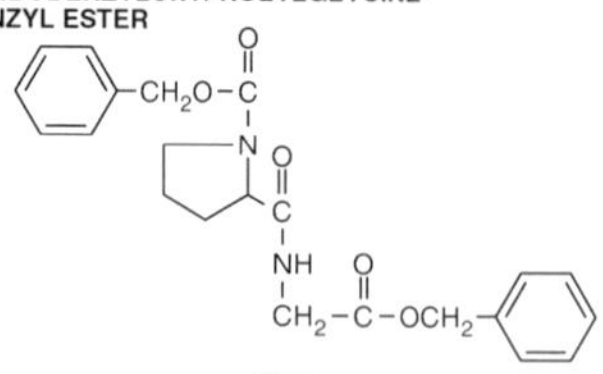

250 mg

S79,083-4
CARBOBENZYLOXYGLYCYLPROLYLALANINE-BENZYL ESTER

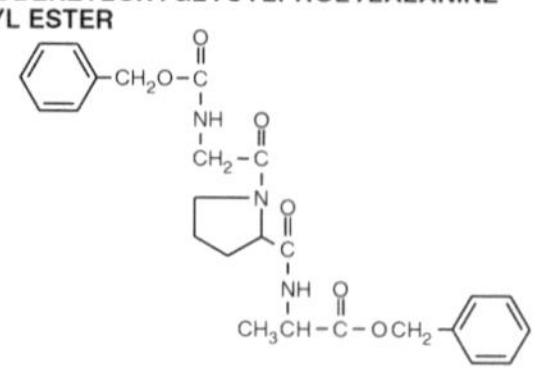

250 mg

S78,774-4
CARBOBENZYLOXYPROLYLNORLEUCINE-BENZYL ESTER

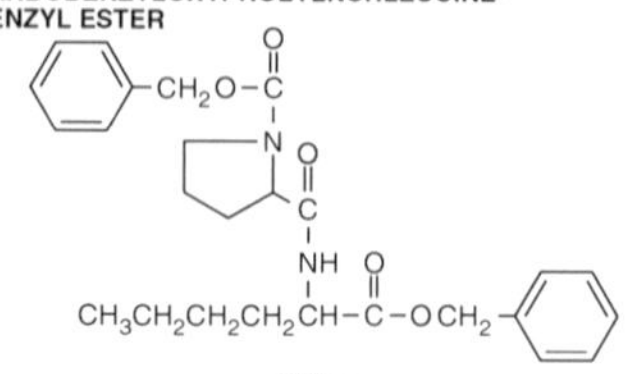

250 mg

S78,993-3
CARBOBENZYLOXYPROLYLGLYCINE-METHYL ESTER

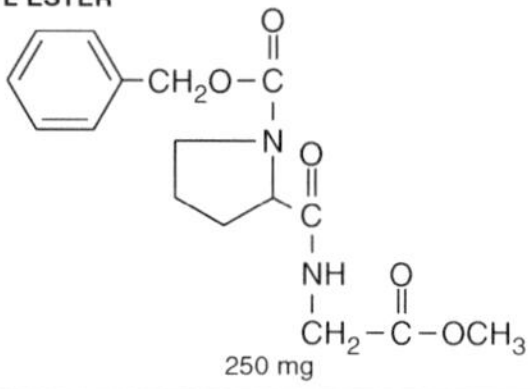

250 mg

S78,618-7
CARBOBENZYLOXYGLYCYLGLYCYL-L-PROLINE BENZYL ESTER

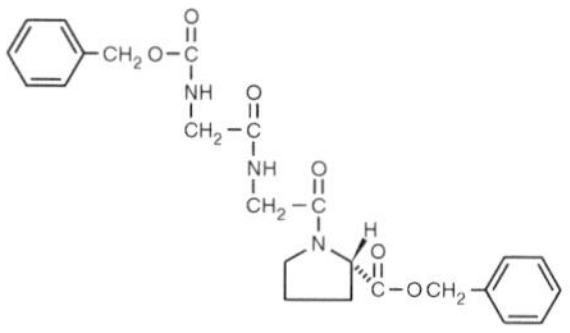

250 mg

S77,741-2
CARBOBENZYLOXY-L-ALANYL-L-PROLYLGLYCINE BENZYL ESTER

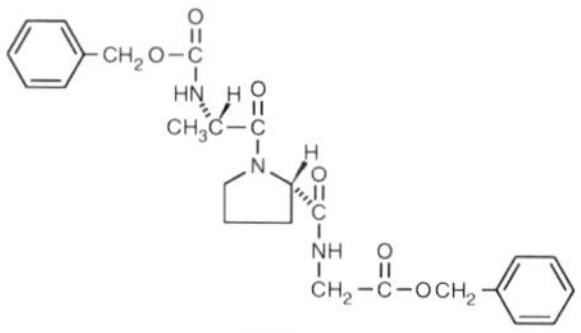

250 mg

S79,067-2
CARBOBENZYLOXYPROLYLLEUCINE-BENZYL ESTER

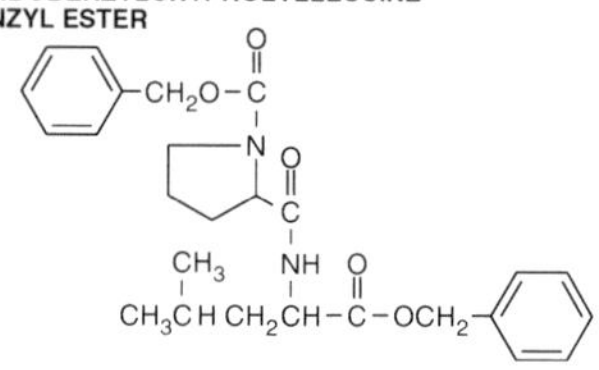

250 mg

S77,772-2
CARBOBENZYLOXY-L-PROLINAMIDE

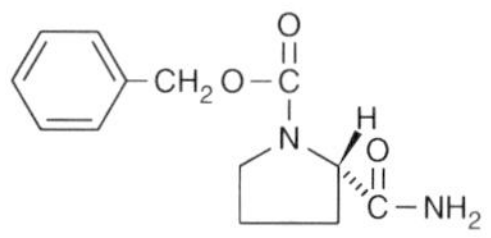

250 mg

S74,002-0
DIBENZYL 2,3-DIAZABICYCLO[2.2.1]-HEPT-5-ENE-2,3-DICARBOXYLATE

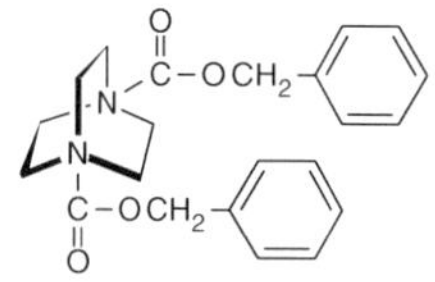

250 mg

S55,144-9
1,2-DIHYDRO-2-[3,4-(METHYLENE-DIOXY)PHENYL]-4(3H)QUINAZOLINONE

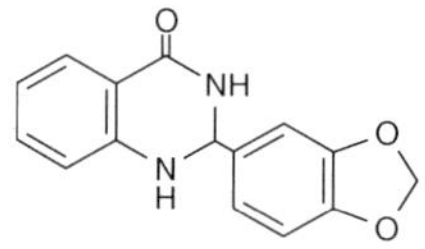

250 mg

S95,164-1
METHYL 1-BENZYL-4,5-DIOXO-3-PYRROLIDINECARBOXYLATE

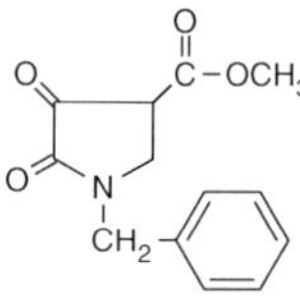

250 mg

S51,770-4
1,2-DIHYDRO-2-(3,4,5-TRIMETHOXYPHENYL)-4(3H)-QUINAZOLINONE

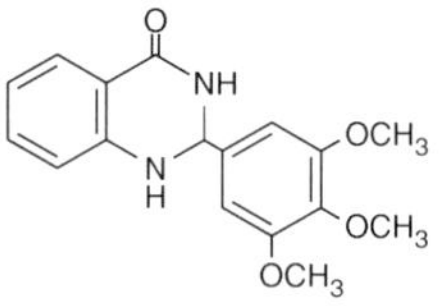

250 mg

S94,348-7
5-(ANILINOMETHYLENE)BARBITURIC ACID

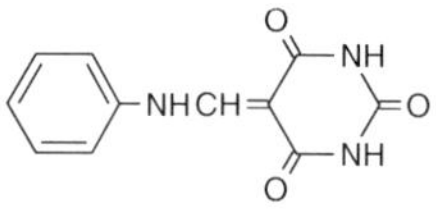

250 mg

S94,347-9
5-(ANILINOMETHYLENE)-1,3-DIMETHYLBARBITURIC ACID

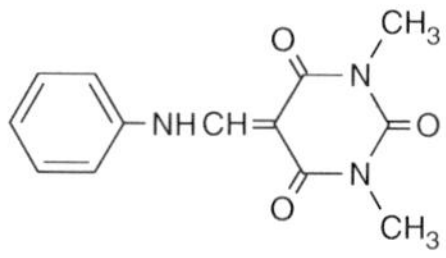

5 gm

S42,247-9
4-[2-(DIMETHYLAMINO)ACETAMIDO]-1,2-DIPHENYL-3,5-PYRAZOLIDINEDIONE

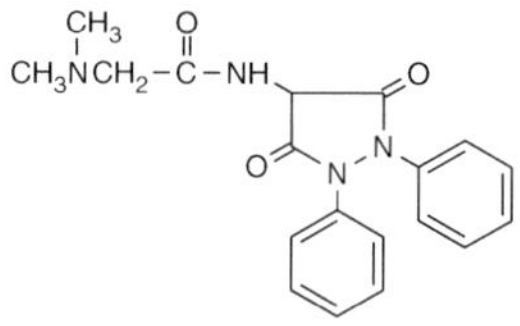

5 gm

S94,903-5
2H-PYRIDO(1,2-A)PYRIMIDINE-2,4-(3H)-DIONE

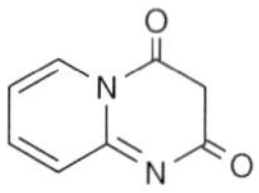

250 mg

S90,104-0
2-(2-ACETOXYPHENYL)-1-ACETYL-1,2-DIHYDRO-4H-3,1-BENZOXAZIN-4-ONE

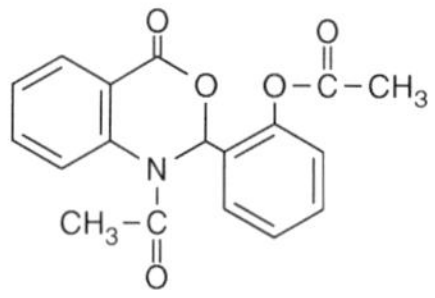

250 mg

S94,356-8
3,4-DIHYDRO-3-PHENYL-2H-PYRIDO(1,2-A)-PYRIMIDINE-2,4-DIONE

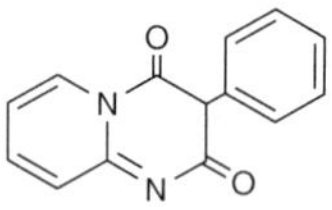

250 mg

S94,357-6
3-BUTYL-3,4-DIHYDRO-2H-PYRIDO-(1,2-A)-PYRIMIDINE-2,4-DIONE

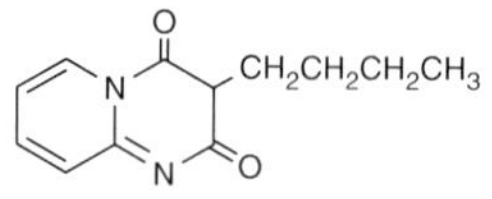

250 mg

S95,545-0
3-(2,4-DIOXOIMIDAZOLIDIN-5-YLIDENE)-2-INDOLINONE

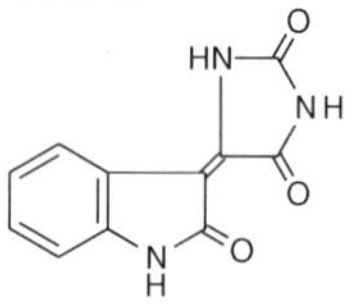

50 mg

S44,771-4
5'-TRITYLURIDINE

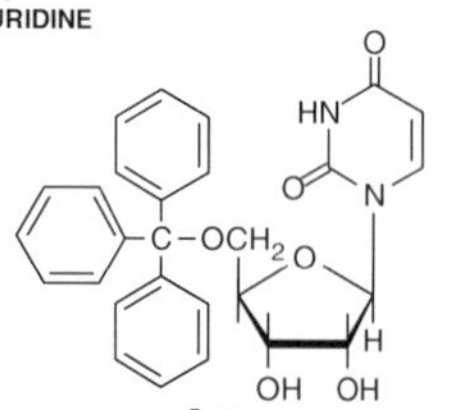

5 gm

S86,047-6
2',3'-O-BENZYLIDENE-N4,5'-O-DIBENZOYLCYTIDINE

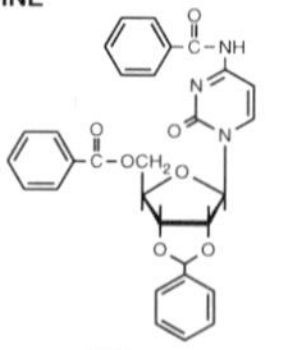

250 mg

S35,177-6
2',3'-ISOPROPYLIDENE-5'-TRITYLURIDINE

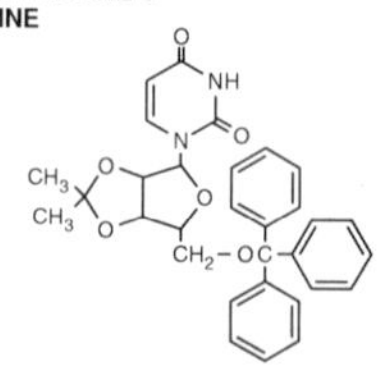

250 mg

S66,199-6
2,5-DIHYDRO-1,4-BENZOTHIAZEPIN-3-(4H)-ONE

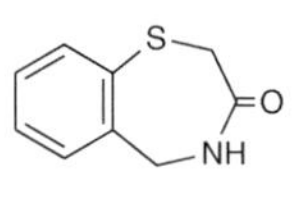

250 mg

S40,226-5
10-[2-(2-PIPERIDINOETHYLTHIO)-ACETYL]PHENOTHIAZINE HYDROCHLORIDE

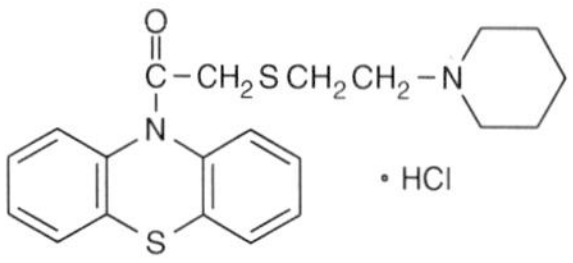

250 mg

S78,962-3
CARBOBENZYLOXY-L-PROLYL-L-METHIONINE METHYL ESTER

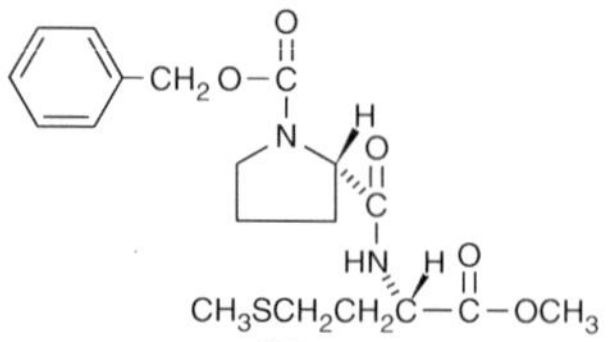

250 mg

S78,946-1
CARBOBENZYLOXY-L-PROLYL-L-METHIONINE-2-NAPHTHYLAMIDE

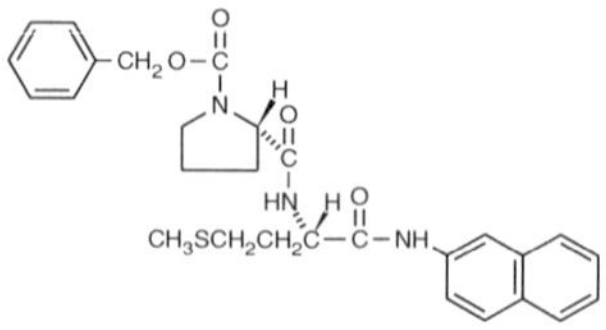

250 mg

S76,160-5
1-(METHYLTHIO)ISOQUINOLIN-3(4H)-ONE
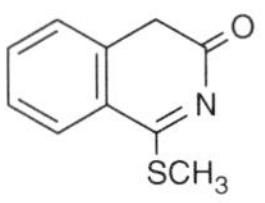
250 mg

S92,400-8
1-(2,3-DIMETHYL-5-OXO-1-PHENYL-3-PYRAZOLIN-4-YL)-3-[(3-METHYLTHIO)-PHENYL]UREA
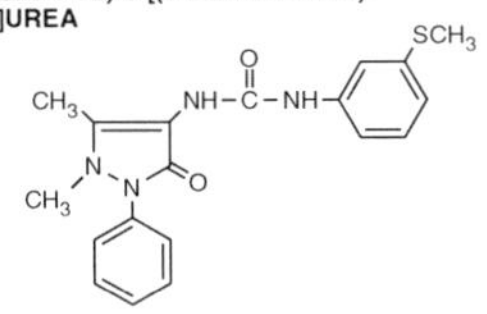
250 mg

S85,348-8
2-METHYL-N-[3-(METHYLTHIO)PHENYL]-1-PIPERIDINECARBOXAMIDE
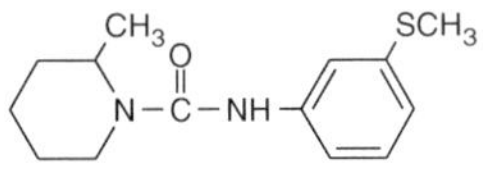
250 mg

S79,798-7
1-(N-[4-(METHYLTHIO)PHENYL]-CARBAMOYL)-4-(3-PHENYLPROPYL)-PIPERIDINE
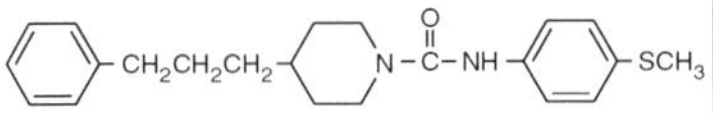
250 mg

S79,710-3
4-METHYL-1-(N-[4-(METHYLTHIO)-PHENYL]CARBAMOYL)PIPERIDINE
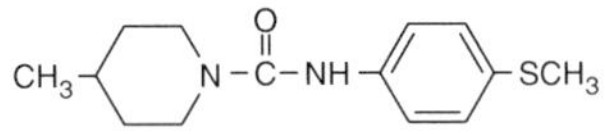
250 mg

S80,272-7
7-AMINO-1,3-DIETHYL-6-NITRO-2,4-(1H,3H)QUINAZOLINEDIONE
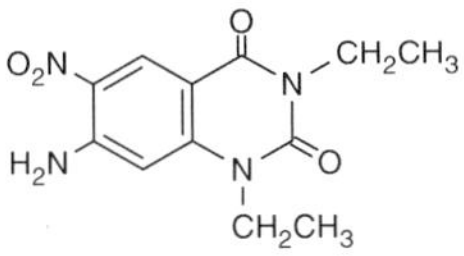
250 mg

S60,561-1
1-(4-AMINOPHENYL)-3,4-DIMETHYL-2-PYRAZOLIN-5-ONE
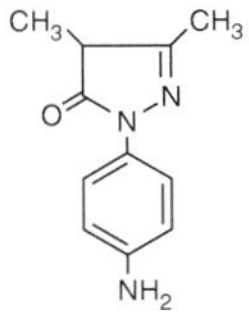
5 gm

S64,232-0
3-(3-AMINOPHENYLAMINO)-1-PHENYL-2-PYRAZOLIN-5-ONE
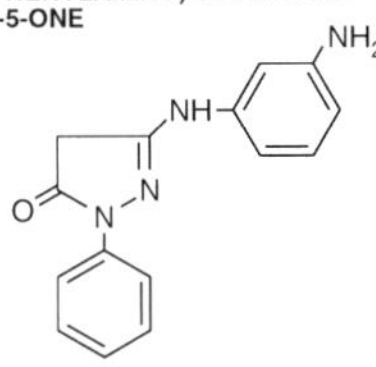
5 gm

S76,046-3
10B-(2-AMINOPHENYL)-1,2,3,4,6,10B-HEXAHYDROPYRIMIDO(2,1-A)-ISOINDOL-6-ONE
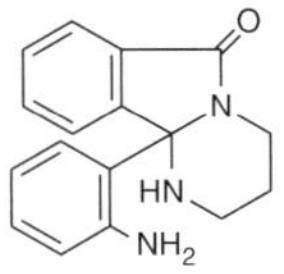
250 mg

S62,213-3
3-AMINO-1-(1-NAPHTHYLMETHYL)-2-PYRAZOLIN-5-ONE
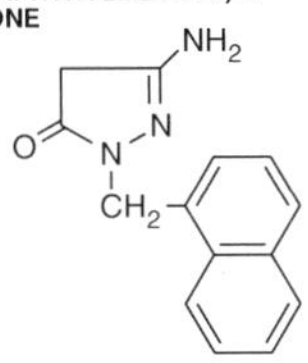
5 gm

S93,627-8
3-AMINO-2-BENZYL-7-NITRO-4-PHENYL-1(2H)-ISOQUINOLINONE
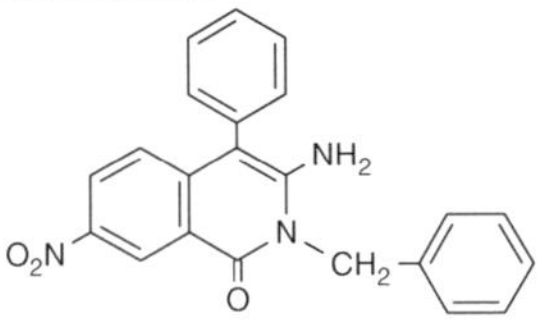
250 mg

S50,087-9
1-(4-AMINOPHENYL)-2-ETHYL-3-METHYL-3-PYRAZOLIN-5-ONE
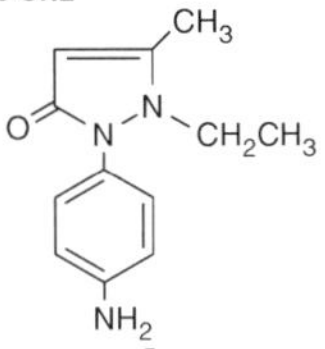
5 gm

S50,210-3
3-(3-AMINOPHENYL)-2-ETHYL-1-PHENYL-3-PYRAZOLIN-5-ONE
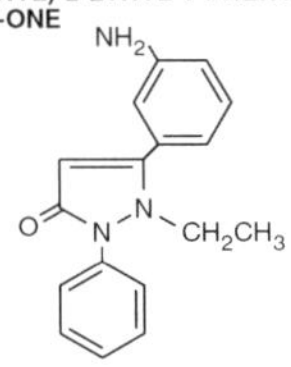
5 gm

S50,216-2
1-(4-AMINOPHENYL)-3-HEPTADECYL-2-METHYL-3-PYRAZOLIN-5-ONE
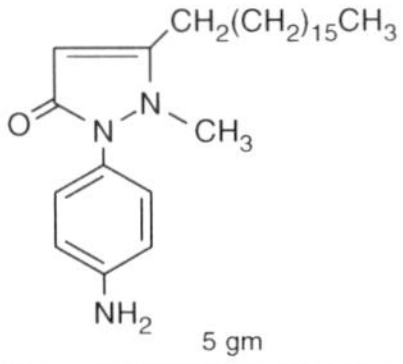
5 gm

S92,738-4
4-AMINO-6-METHYL-2-(4-NITROPHENYL)-3(2H)-PYRIDAZINONE
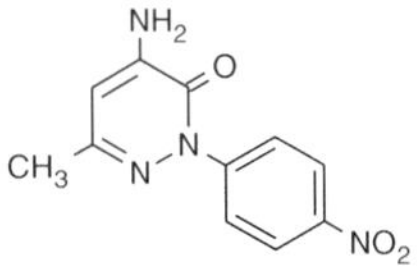
250 mg

S80,481-9
5-AMINOISATIN
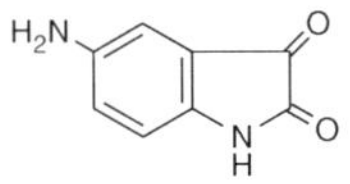
250 mg

S57,004-4
METHYL-4'-(4-AMINO-2-METHYL-5-OXO-3-PENTADECYL-3-PYRAZOLIN-1-YL)-SUCCINANILATE
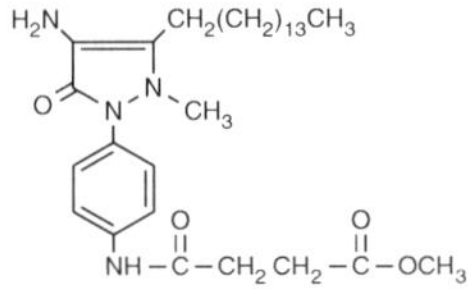
5 gm

S54,052-8
1-ACETYL-5-AMINOINDOLINE
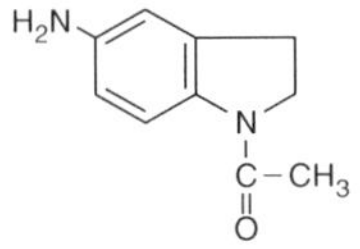
5 gm

S76,011-0
6-AMINO-1-ISOPROPYL-7-METHYL-4-PHENYLQUINAZOLIN-2(1H)ONE
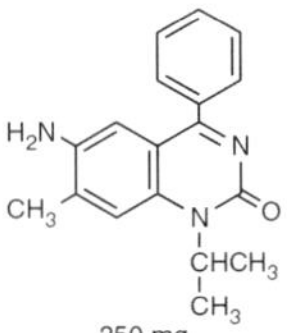
250 mg

S45,278-5
L-PHENYLALANYL-L-PROLINE
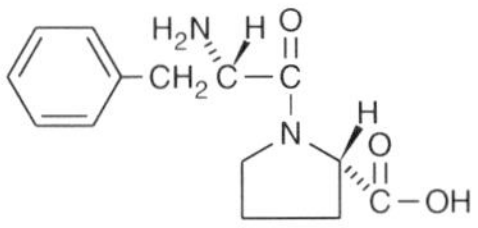
250 mg

S77,607-6
L-PHENYLALANYL-D-ALANYLGLYCINE ANHYDRIDE
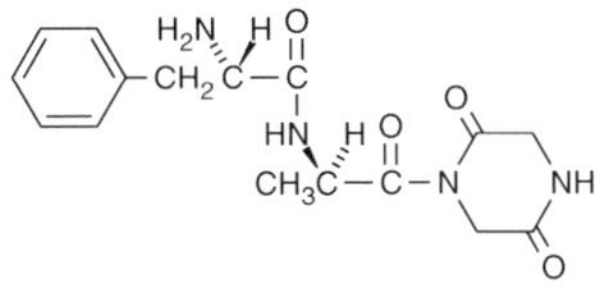
250 mg

S77,680-7
PHENYLALANYL-L-ALANYLGLYCINE ANHYDRIDE
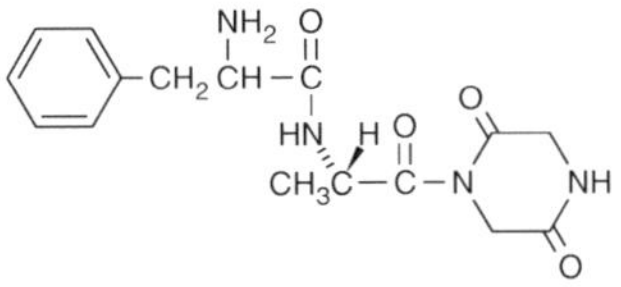
250 mg

S76,070-6
3-(4-AMINO-1,3-DIMETHYL-6-OXO-1,2,3,6-TETRAHYDROPYRIMIDIN-5-YL)-PHTHALIDE
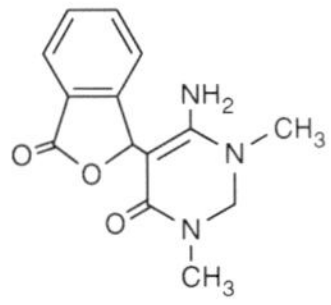
250 mg

S76,417-5
1-(2-AMINOBENZOYL)PYRROLIDINE
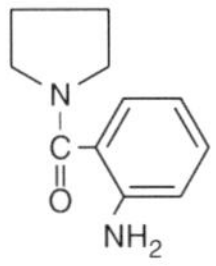
250 mg

S76,341-1
4-(2-AMINOBENZOYL)MORPHOLINE HYDROCHLORIDE
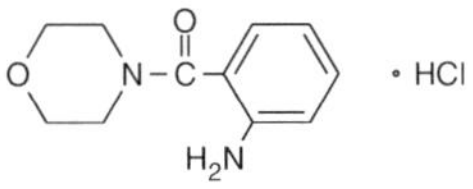
250 mg

S76,672-0
1-(2-AMINOBENZOYL)-1,2,3,6-TETRAHYDROPYRIDAZINE
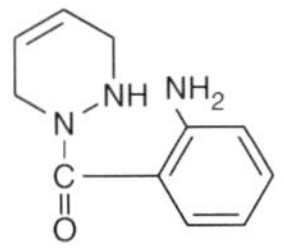
250 mg

S51,743-7
3-AMINO-2-METHYL-1-PHENYL-3-PYRAZOLIN-5-ONE
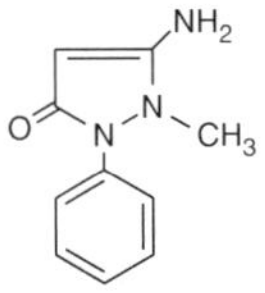
250 mg

S94,712-1
DIETHYL 6-AMINO-1-BENZAMIDO-4-(4-NITROPHENYL)-2-OXO-TETRAHYDRO-3,5-PYRIDINEDICARBOXYLATE
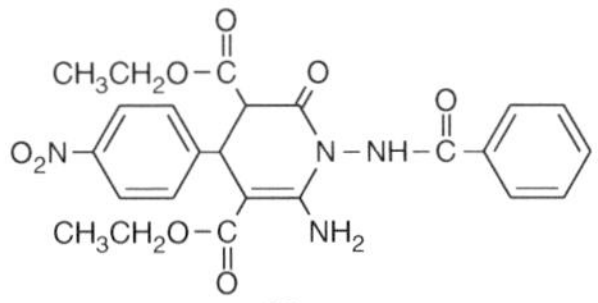
50 mg

S80,449-5
5,5'-BENZYLIDENEBIS(6-AMINO-1,3-DIMETHYLURACIL)-
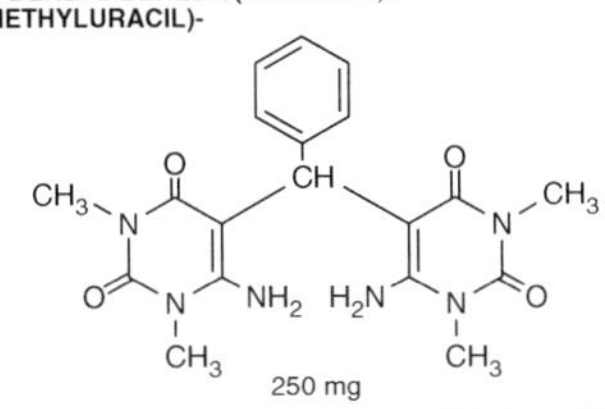
250 mg

S76,224-5
10-AMINO-5H-DIBENZ(B,E)AZEPINE-6,11-DIONE
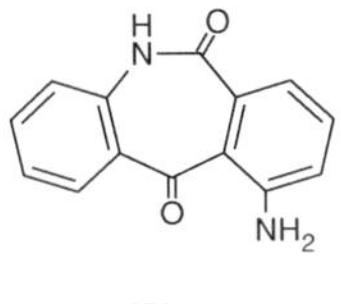
250 mg

S76,403-5
2,3-DIMETHYL-1-PHENYL-4-(2,4,6-TRINITROPHENYLAMINO)-3-PYRAZOLIN-5-ONE
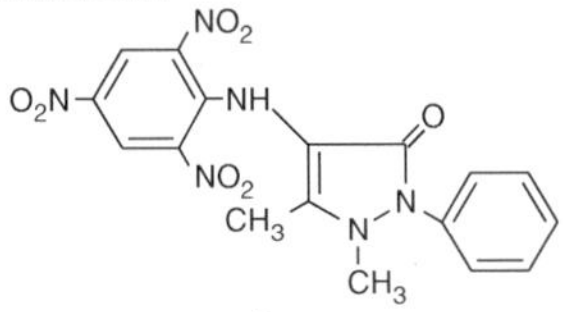
5 gm

S66,179-1
7-CHLORO-3H-1,4-BENZODIAZEPIN-2,5(1H,4H)-DIONE
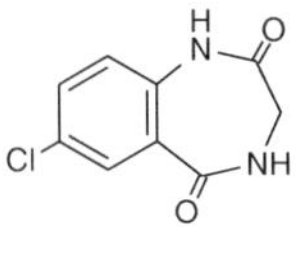
5 gm

S34,627-6
4,5-DIHYDRO-7,8-DIMETHYL-1H-1,5-BENZODIAZEPIN-2(3H)-ONE

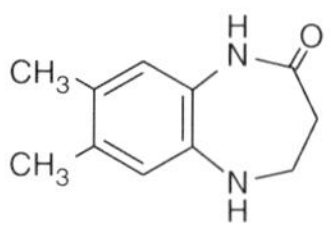

250 mg

S34,628-4
4,5-DIHYDRO-4,7,8-TRIMETHYL-1H-1,5-BENZODIAZEPIN-2(3H)-ONE

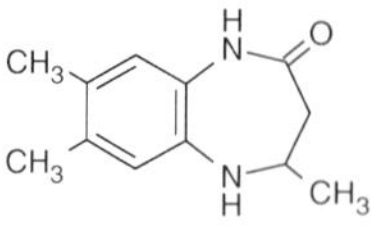

250 mg

S41,706-8
5-CYCLOBUTYL-5-PHENYLHYDANTOIN

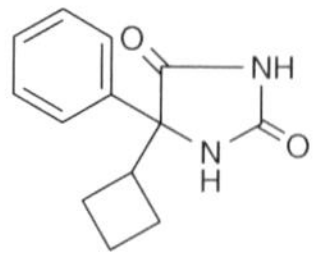

250 mg

S61,500-5
4-METHYL-2,3,4,5-TETRAHYDRO-1H-1,5-BENZODIAZEPIN-2-ONE

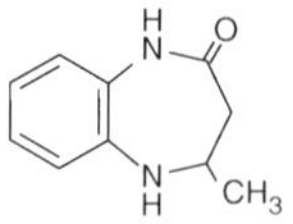

250 mg

S76,894-4
5-IMINO-1-(4-METHYL-2-NITROPHENYL)-3-PHENYLHYDANTOIN

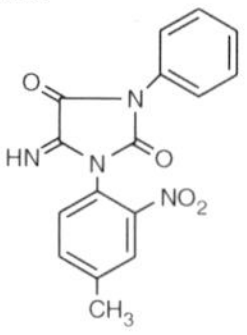

250 mg

S95,541-8
5-METHYL-7-NITROISATIN

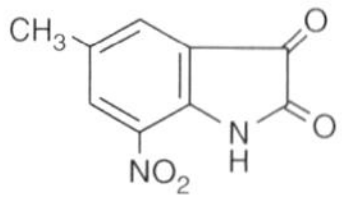

50 mg

S90,574-7
ETHYL 4-[3-(1,5-DIMETHYL-3-OXO-2-PHENYL-4-PYRAZOLIN-4-YL)UREIDO]-BENZOATE

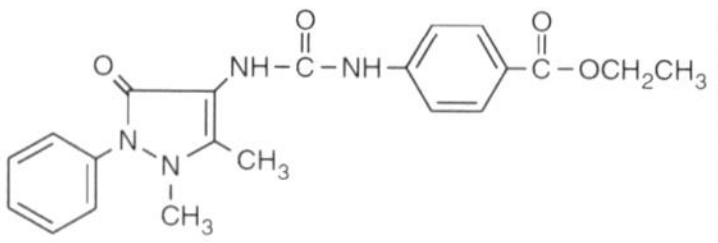

250 mg

S34,680-2
7,8-DIMETHYL-1H-1,5-BENZODIAZEPINE-2,4(3H,5H)-DIONE

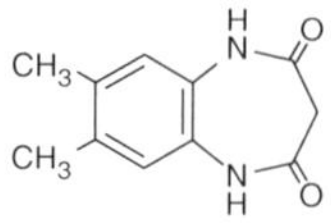

5 gm

S94,973-6
1,1'-THIOBIS(5,5-DIMETHYLHYDANTOIN)

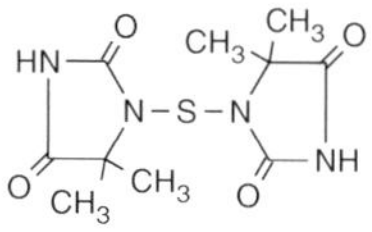

250 mg

S50,090-9
N-(5-OXO-3-PENTADECYL-1-PHENYL-2-PYRAZOLIN-4-YL)FORMAMIDE

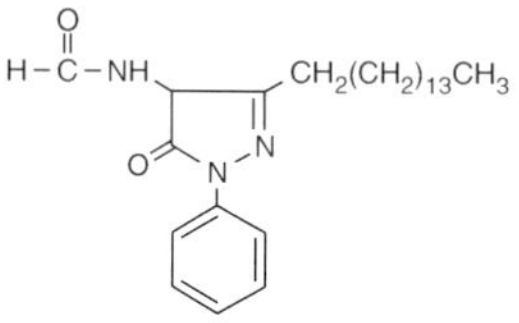

250 mg

S37,597-7
7,8-DIMETHYL-4-PHENYL-1H-1,5-BENZODIAZEPIN-2(3H)-ONE

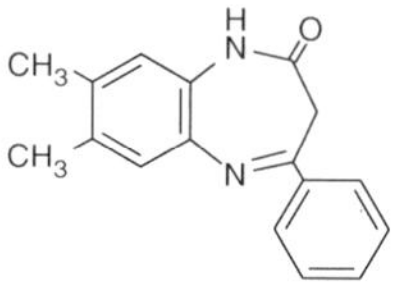

5 gm

S37,596-9
4,7-DIMETHYL-1H-1,5-BENZODIAZEPIN-2(3H)-ONE

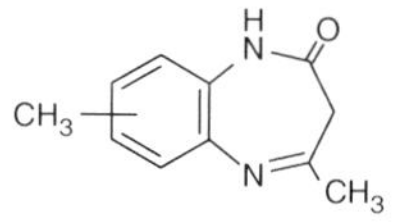

250 mg

S35,664-6
4,7,8-TRIMETHYL-1H-1,5-BENZODIAZEPIN-2-(3H)-ONE

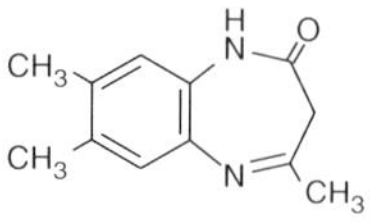

5 gm

S35,576-3
1,3,4,5-TETRAHYDRO-3,7,8-TRIMETHYL-2H-1,5-BENZODIAZEPIN-2-ONE

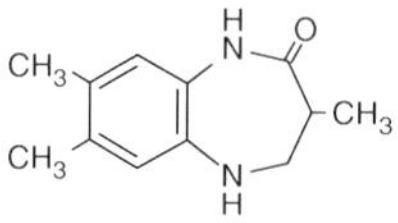

5 gm

S37,604-3
4-HEPTADECYL-1H-1,5-BENZODIAZEPIN-2(3H)-ONE

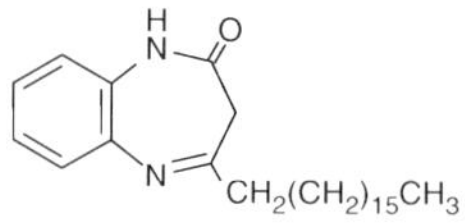

5 gm

S37,607-8
7,8-DIMETHYL-4-HEPTADECYL-1H-1,5-BENZODIAZEPIN-2(3H)-ONE

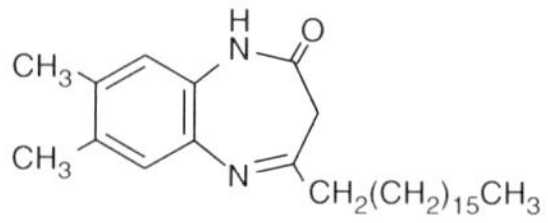

5 gm

S60,533-6
1-BENZYL-3-DIMETHYLAMINO-2-PYRAZOLIN-5-ONE

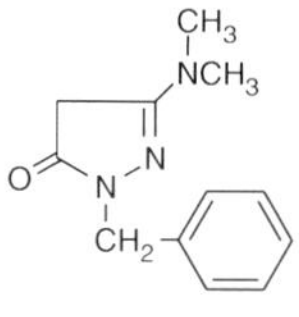

250 mg

S62,501-9
(2,3-DIMETHYL-5-OXO-1-PHENYL-3-PYRAZOLIN-4-YLMETHYL)-TRIMETHYLAMMONIUM IODIDE

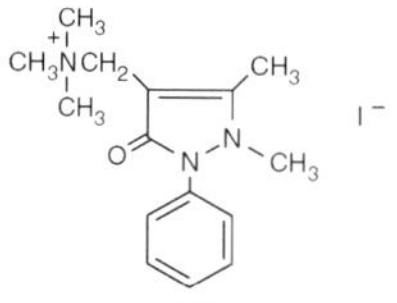

5 gm

S76,080-3
N-[2-(MORPHOLINOMETHYL)PHENYL]-4-MORPHOLINECARBOXAMIDE

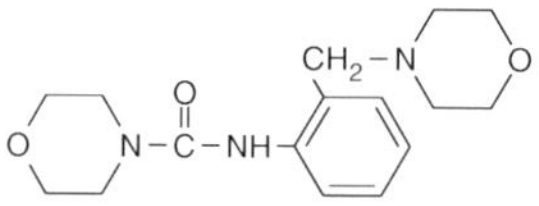

250 mg

S67,077-4
1-[N-(4-METHYL-3-NITROPHENYL)-CARBAMOYL]PIPERIDINE

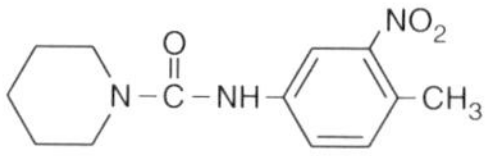

5 gm

S67,014-6
2-METHYL-1-[N-(4-METHYL-3-NITROPHENYL)CARBAMOYL]PIPERIDINE

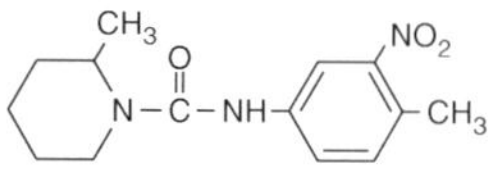

5 gm

S67,028-6
3-METHYL-1-[N-(4-METHYL-3-NITROPHENYL)CARBAMOYL]PIPERIDINE

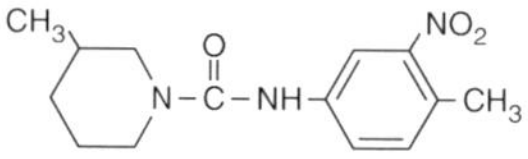

5 gm

S67,085-5
4-METHYL-1-[N-(4-METHYL-3-NITROPHENYL)CARBAMOYL]PIPERIDINE

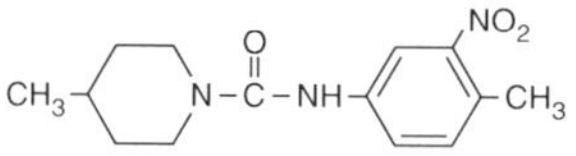

5 gm

S67,070-7
4-[N-(4-METHYL-3-NITROPHENYL)-CARBAMOYL]MORPHOLINE

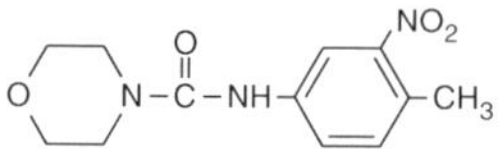

5 gm

S50,088-7
2-METHYL-1-(3-NITROPHENYL)-3-PENTADECYL-3-PYRAZOLIN-5-ONE

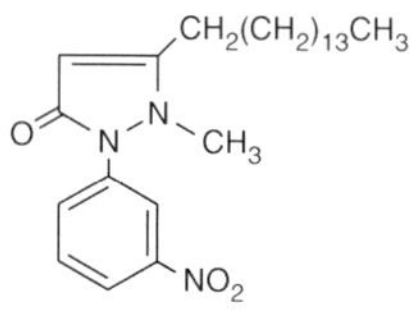

5 gm

S43,109-5
6-NITRO-1-INDOLINECARBOXALDEHYDE

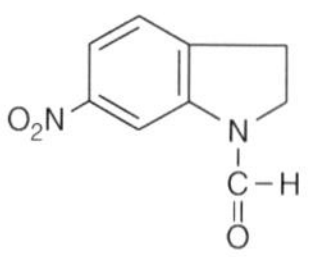

250 mg

S44,560-6
1-ACETYL-6-NITROINDOLINE

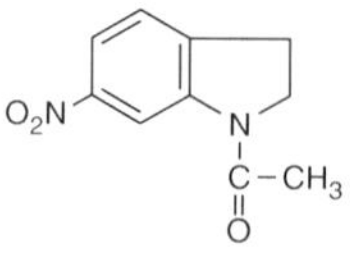

250 mg

S63,403-4
1-[N-(3-NITROPHENYL)CARBAMOYL]-PIPERIDINE

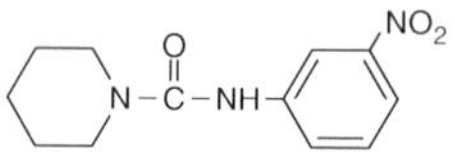

5 gm

S63,460-3
4-[N-(3-NITROPHENYL)CARBAMOYL]-MORPHOLINE

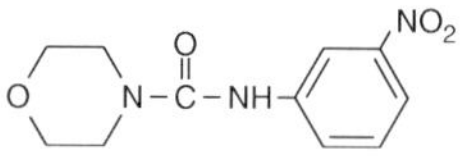

5 gm

S64,338-6
ETHYL N-(1-PHENYL-5-OXO-2-PYRAZOLIN-3-YL)CARBAMATE

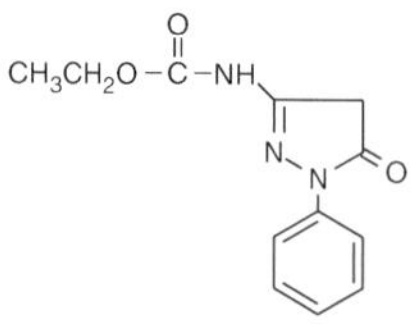

5 gm

S65,508-2
1-(5-OXO-1-PHENYL-3-PYRAZOLIN-3-YL)-3-PHENYLUREA

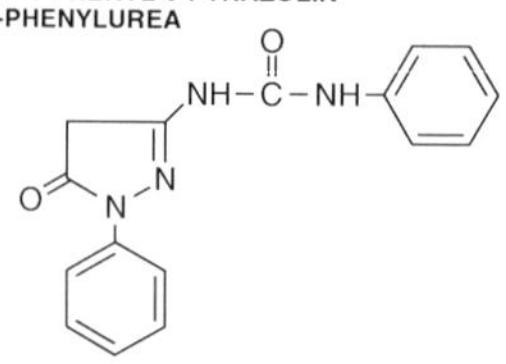

5 gm

S50,735-0
N-HEXADECYL-3'-(2-METHYL-5-OXO-1-PHENYL-3-PYRAZOLIN-3-YL)ACETANILIDE

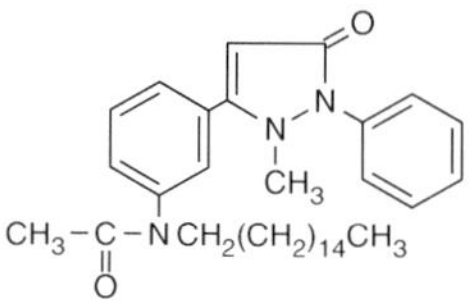

250 mg

S76,740-9
1,3-BIS[2-(1,2-DIHYDRO-2-METHYL-1-OXO-3-ISOQUINOLINYL)ETHYL]UREA

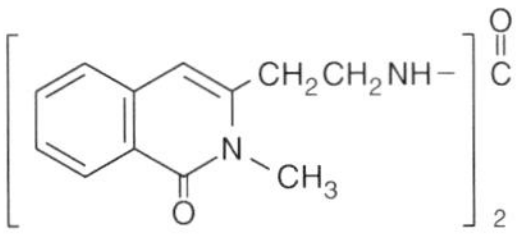

250 mg

S66,026-4
N,N'-(4-METHYL-1,3-PHENYLENE)BIS(1-CARBAMOYL-2-METHYLPIPERIDINE)

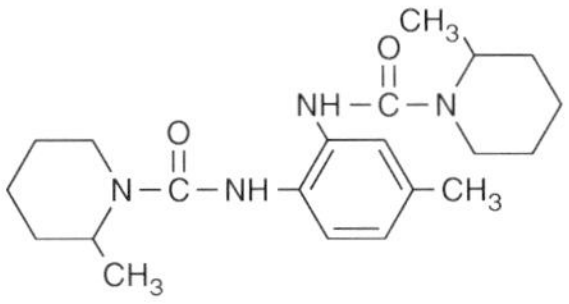

5 gm

S62,521-3
2,4-BIS(MORPHOLINOCARBOXAMIDO)-TOLUENE

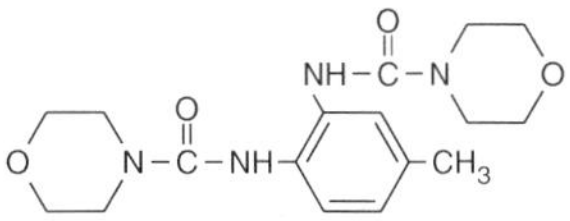

5 gm

S65,416-7
2,4-BIS(2,6-DIMETHYLMORPHOLINOCARBOXAMIDO)-TOLUENE

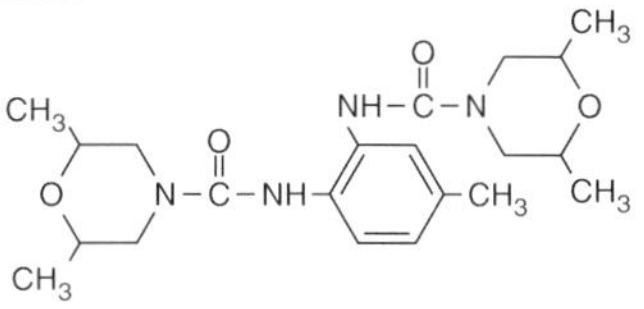

5 gm

S36,451-7
4-PHENYL-1,4-(PIPERIDINOCARBONYL)-PIPERIDINE HYDROCHLORIDE

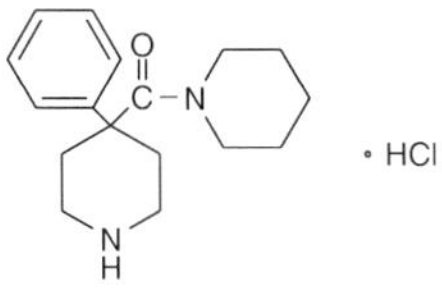

250 mg

S76,384-5
2-BENZOYL-1-[2-(METHYLAMINOMETHYL)-PHENYL]-1,2,3,4-TETRAHYDROISOQUINOLINE HYDROCHLORIDE

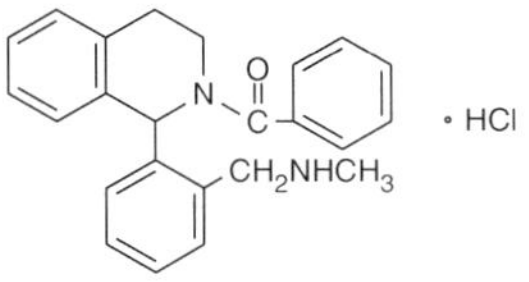

250 mg

S41,599-5
4-PHENYL-4-(PYRROLIDINOCARBONYL)-PIPERIDINE HYDROCHLORIDE

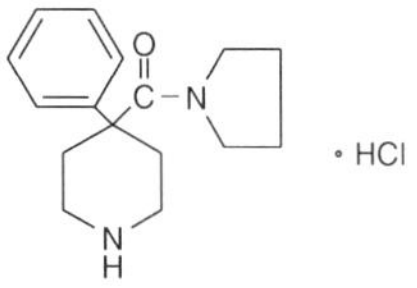

250 mg

S76,648-8
2-CHLOROACETYL-1-[2-(DIMETHYLAMINOMETHYL)PHENYL]-1,2,3,4-TETRAHYDROISOQUINOLINE HYDROCHLORIDE

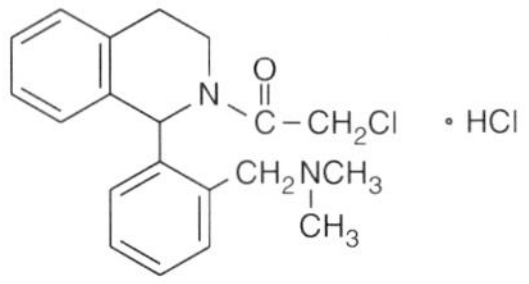

250 mg

S63,010-1
3-(3-NITROPHENYL)-1-PHENYL-2-PYRAZOLIN-5-ONE

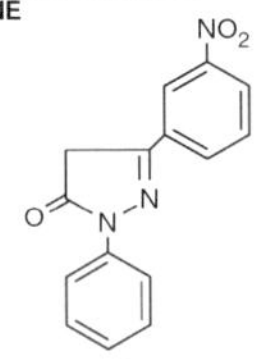

5 gm

S50,215-4
3-[4-(2-ETHYL-3-METHYL-5-OXO-3-PYRAZOLIN-1-YL)PHENYLCARBAMOYL]-5-NONADECENOIC ACID

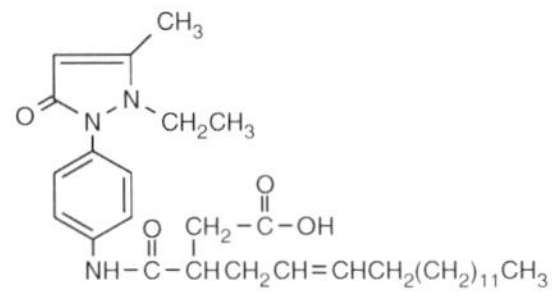

250 mg

S50,218-9
3-[3-(2-METHYL-5-OXO-1-PHENYL-3-PYRAZOLIN-3-YL)PHENYLCARBAMOYL]-5-NONADECENOIC ACID

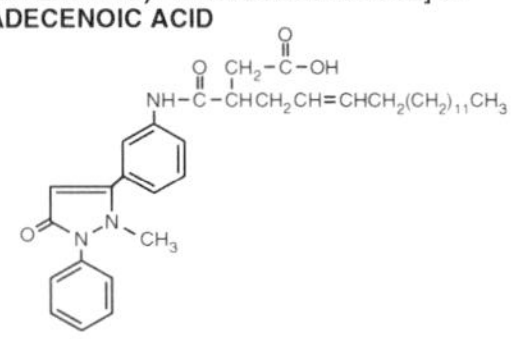

250 mg

S61,045-3
4-[N-(2-METHYL-4-NITROPHENYL)-CARBAMOYL]MORPHOLINE

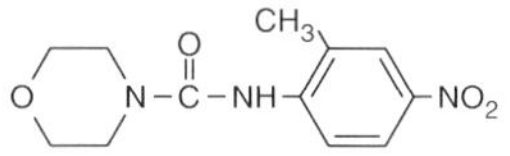

5 gm

S90,099-0
1-ACETYL-1,2-DIHYDRO-2-(3-NITROPHENYL)-4H-3,1-BENZOXAZIN-4-ONE

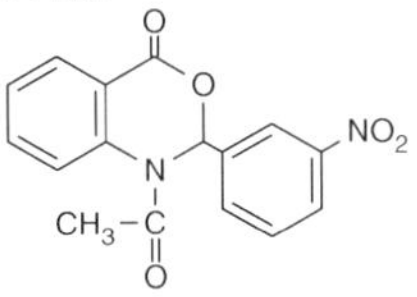

250 mg

S92,739-2
4-CHLORO-6-METHYL-2-(4-NITROPHENYL)-3(2H)PYRIDAZINONE

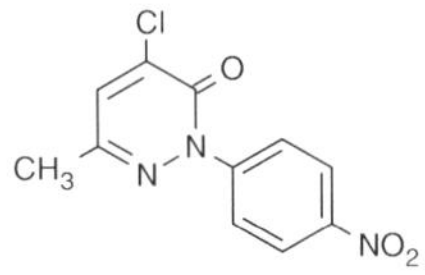

250 mg

S92,737-6
6-METHYL-2-(4-NITROPHENYL)-3(2H)-PYRIDAZINONE

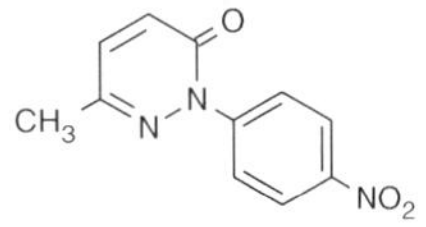

250 mg

S66,968-7
1-[N-(4-NITROPHENYL)CARBAMOYL]-PIPERIDINE

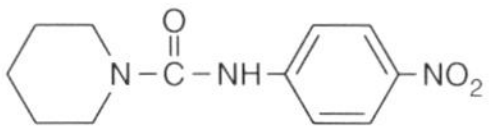

5 gm

S66,918-0
3-METHYL-1-[N-(4-NITROPHENYL)-CARBAMOYL]PIPERIDINE

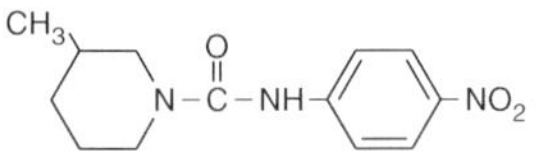

5 gm

S66,966-0
2-METHYL-1-[N-(4-NITROPHENYL)-CARBAMOYL]PIPERIDINE

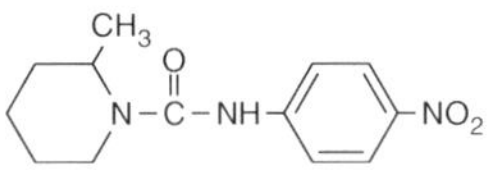

5 gm

S66,899-0
4-[N-(4-NITROPHENYL)CARBAMOYL]-MORPHOLINE

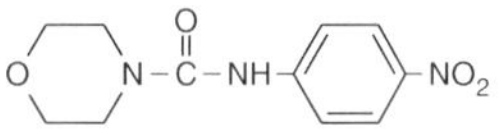

5 gm

S66,945-8
2,6-DIMETHYL-4-[N-(4-NITROPHENYL)-CARBAMOYL]MORPHOLINE

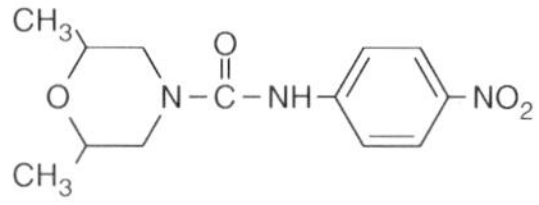

5 gm

S80,347-2
1-METHYL-5-NITROISATIN

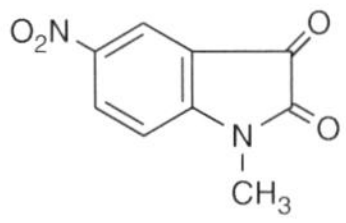

250 mg

S88,807-9
1-METHYL-6-NITROISATOIC ANHYDRIDE

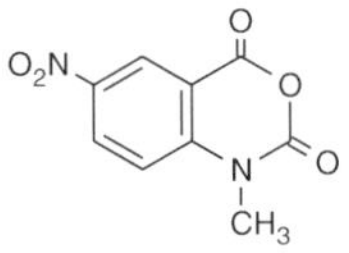

250 mg

S88,907-5
6-NITROISATOIC ANHYDRIDE

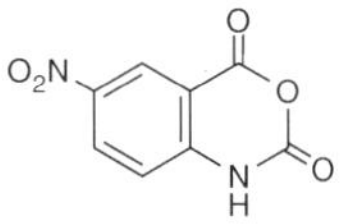

250 mg

S76,184-2
3,4-DIHYDRO-7-NITRO-2H-1,4-BENZOXAZIN-3-ONE

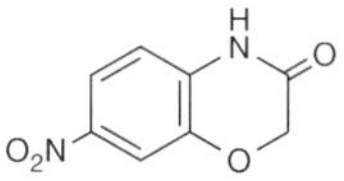

250 mg

S37,286-2
3-METHYL-1-(4-NITROPHENYL)-2-PYRAZOLIN-5-ONE

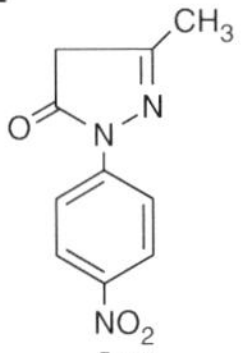

5 gm

S61,106-9
1-[N-(2-METHYL-4-NITROPHENYL)-CARBAMOYL]PIPERIDINE

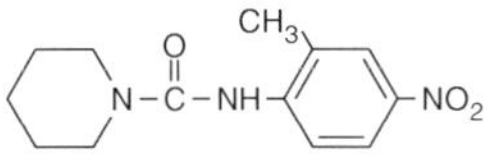

5 gm

S60,739-8
4-[N-(2-CHLORO-4-NITROPHENYL)-CARBAMOYL]MORPHOLINE

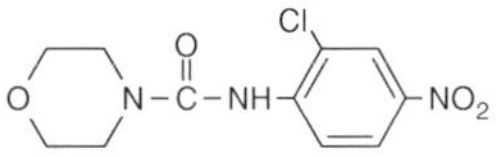

5 gm

S42,839-6
5-NITROISATOIC ANHYDRIDE

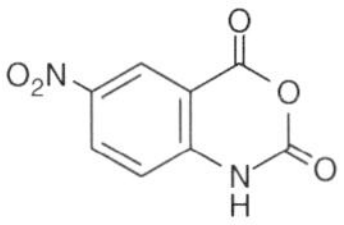

5 gm

S35,361-2
4-MORPHOLINOCARBONYL-4-PHENYLPIPERIDINE HYDROCHLORIDE

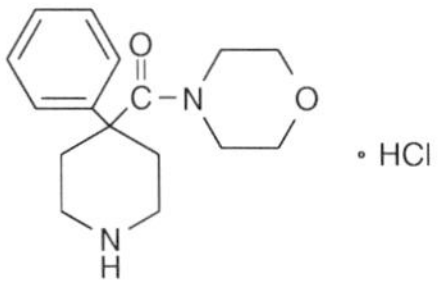

250 mg

S80,812-1
2-BENZOYL-1,2-DIHYDRO-2-PHENYL-4-(3H)-QUINAZOLINONE ZINC CHLORIDE SALT

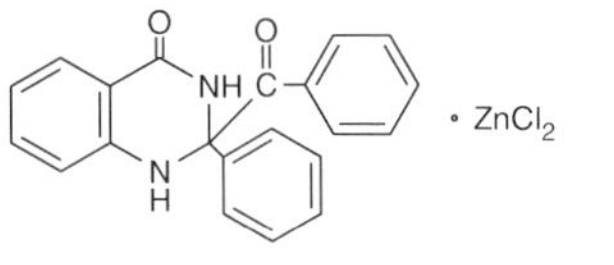

250 mg

S80,817-2
2-BENZOYL-1,2-DIHYDRO-2-PHENYL-4-(3H)-QUINAZOLINONE

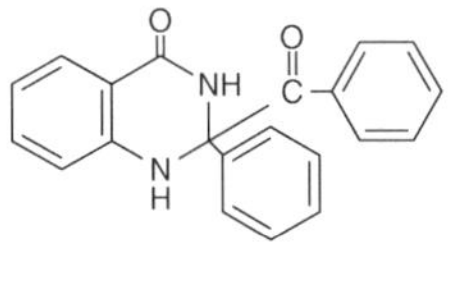

250 mg

S60,879-3
2-(4-ACETAMIDOPHENYL)-2,3-DIHYDRO-4H-1,3-BENZOXAZIN-4-ONE

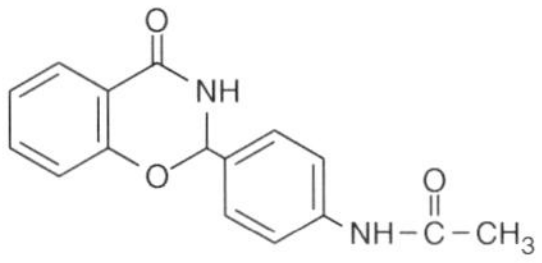

5 gm

S78,840-6
CARBOBENZYLOXYPHENYLALANYLPROLINE METHYL ESTER

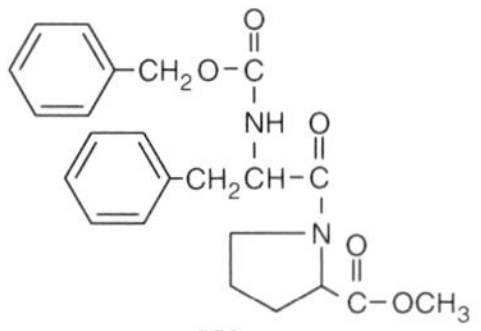

250 mg

S79,056-7
CARBOBENZYLOXY-L-PROLYL-L-PHENYLALANINE BENZYL ESTER

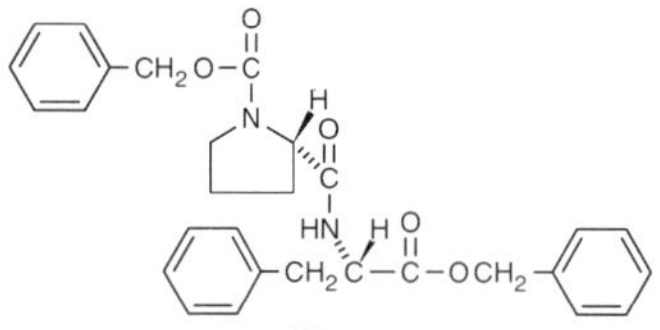

250 mg

S77,766-8
CARBOBENZYLOXY-L-PROLYL-L-PHENYLALANINAMIDE

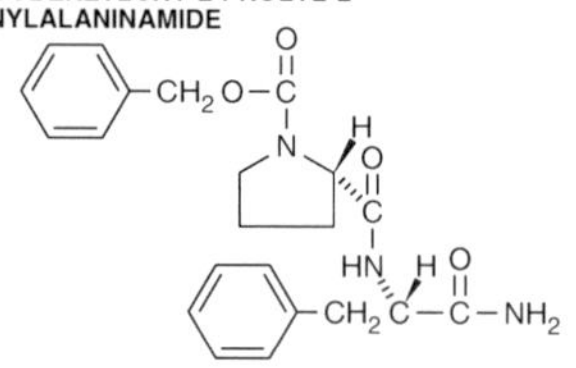

250 mg

S92,288-9
3,5-DIPHENYLHYDANTOIN

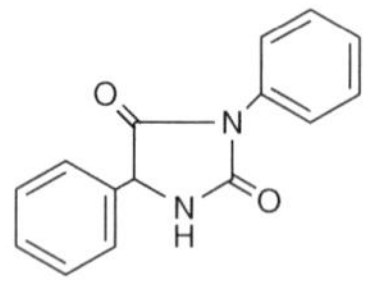

250 mg

S79,863-0
3,4,5,6,7,8-HEXAHYDRO-1-PHENETHYL-2(1H)-QUINOLINONE

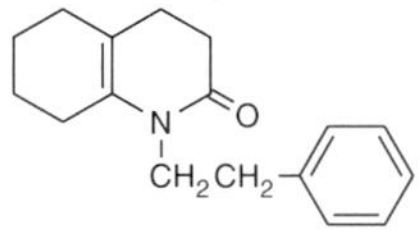

250 mg

S73,510-8
1-(2,2-DIPHENYLVINYL)-2-PHENYL-3-PYRAZOLIDINONE

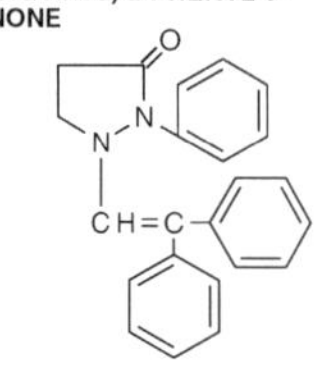

250 mg

S65,418-3
1,2-DIHYDRO-2-(4-DIMETHYLAMINOPHENYL)-4(3H)-QUINAZOLINONE

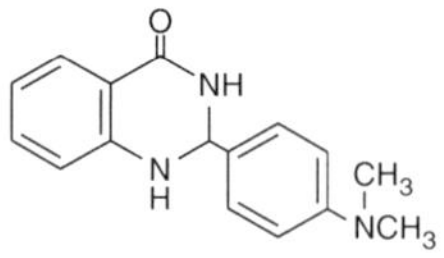

250 mg

S38,192-6
2,3-DIMETHYL-5-OXO-1-PHENYL-3-PYRAZOLINE-4-ACETIC ACID

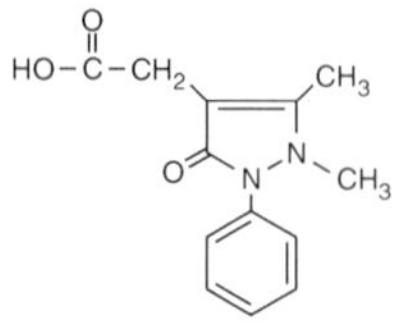

5 gm

S64,593-1
1-(4-BROMOANILINOMETHYL)ISATIN

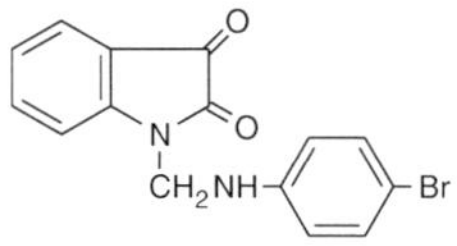

250 mg

S53,900-7
1-(4-FLUOROANILINOMETHYL)ISATIN

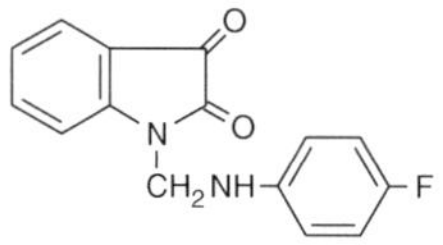

250 mg

S55,137-6
4'-(4-OXO-1,2,3,4-TETRAHYDROQUINAZOLIN-2-YL)-ACETANILIDE

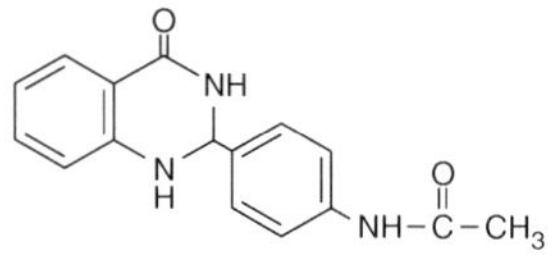

250 mg

S55,141-4
1,2-DIHYDRO-2-(4-ISOPROPYLPHENYL)-4(3H)-QUINAZOLINONE

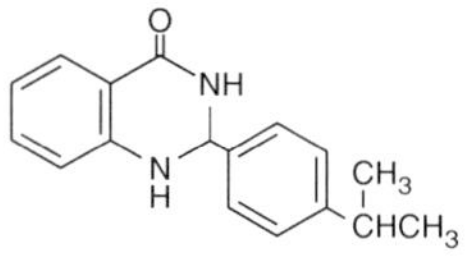

250 mg

S54,207-5
1,2-DIHYDRO-2-(2,4-XYLYL)-4(3H)-QUINAZOLINONE

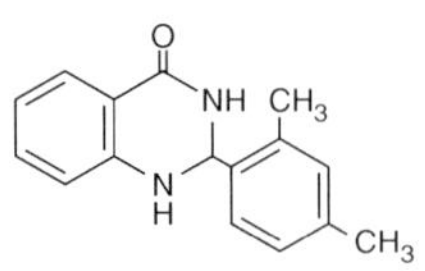

250 mg

S54,208-3
1,2-DIHYDRO-2-(2,5-XYLYL)-4(3H)-QUINAZOLINONE

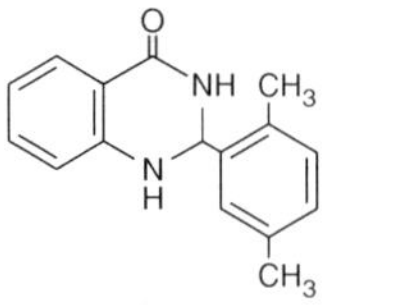

250 mg

S55,138-4
1,2-DIHYDRO-2-DIPHENYLMETHYL-4(3H)-QUINAZOLINONE

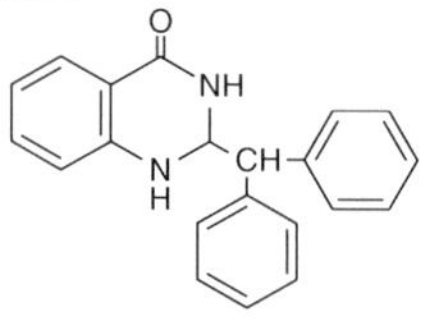

250 mg

S55,147-3
2-(3-CHLOROPHENYL)-1,2-DIHYDRO-4(3H)-QUINAZOLINONE

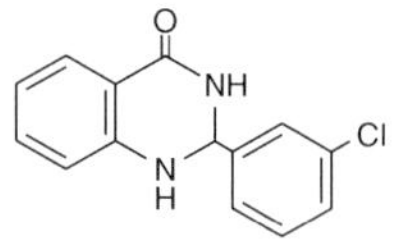

250 mg

S54,224-5
1,2-DIHYDRO-2-(META-TOLYL)-4(3H)-QUINAZOLINONE

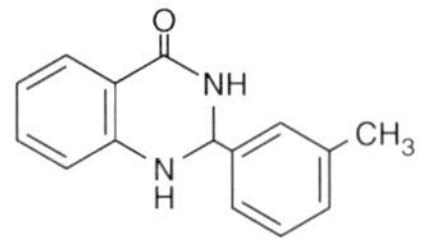

250 mg

S43,299-7
3A,6A-DIPHENYLGLYCOLURIL

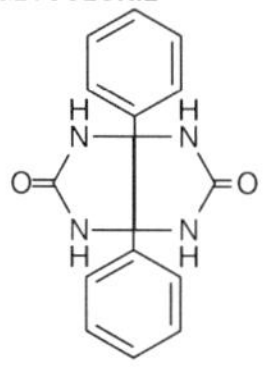

5 gm

S65,716-6
1,4-DIHYDRO-1-(P-FLUOROBENZYL)-2-METHYLAMINO-4(4H)-QUINAZOLINONE

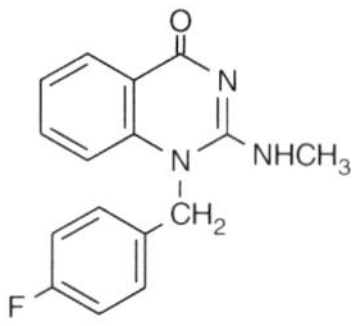

250 mg

S67,099-5
1,4-DIHYDRO-1-(4-FLUOROBENZYL)-2-METHYLAMINO-4-QUINAZOLINONE

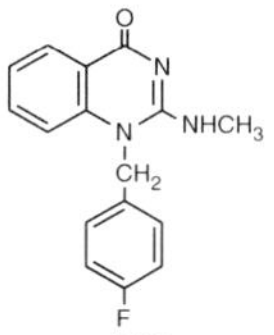

5 gm

S53,609-1
5,7-DIMETHYLISATIN

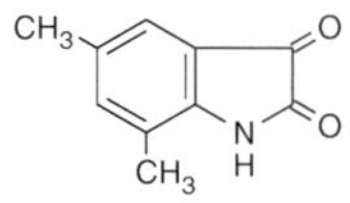

250 mg

S92,753-8
6,7-DIMETHYLISATIN

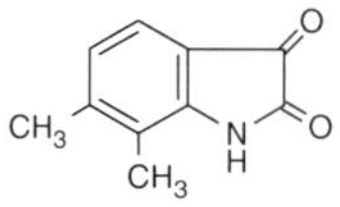

250 mg

S65,735-2
3,4,5,6-TETRAHYDRO-1-BENZAZOCIN-2(1H)-ONE

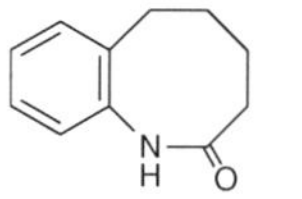

250 mg

S54,228-8
1,2-DIHYDRO-2-(2-NAPHTHYL)-4(3H)-QUINAZOLINONE

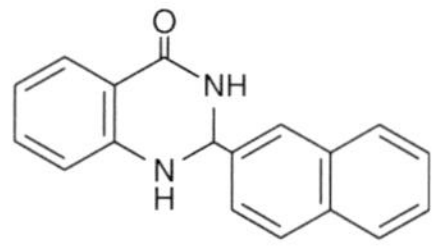

250 mg

S54,059-5
1,2-DIHYDRO-2-(5-NORBORNEN-2-YL)-4-(3H)-QUINAZOLINONE

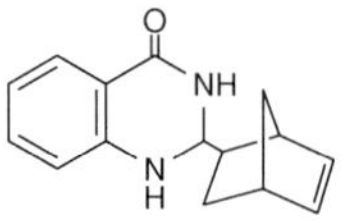

5 gm

S54,215-6
1,2-DIHYDRO-2-(2,3,4-TRIMETHOXYPHENYL)-4(3H)-QUINAZOLINONE

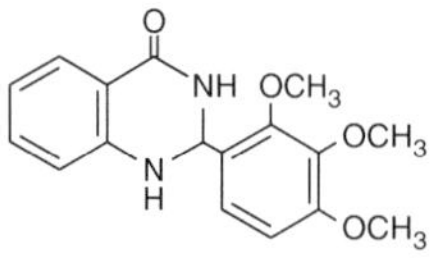

250 mg

S52,597-9
1,2-DIHYDRO-2-(5-INDANYL)-4(3H)-QUINAZOLINONE

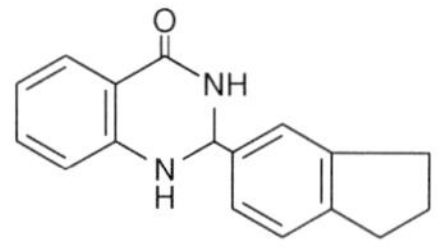

250 mg

S87,316-0
FORMIC (3,3-DIMETHYL-1-ETHYL-2-INDOLINYLIDENE)HYDRAZIDE

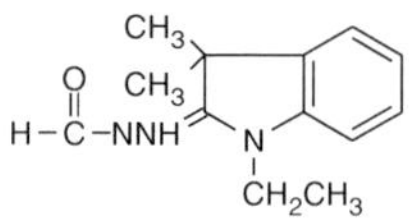

250 mg

S42,622-9
1-(PIPERIDINOMETHYL)ISATIN

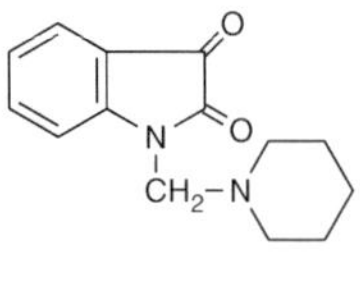

5 gm

S42,526-5
1-ISATINACETIC ACID

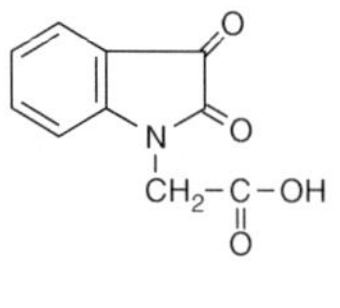

250 mg

S86,864-7
N-(3-ACETYLPHENYL)-4-PHENYL-1-PIPERAZINECARBOXAMIDE

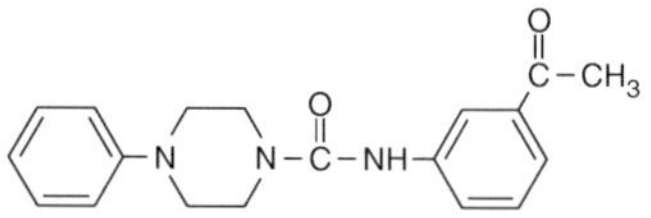

250 mg

S91,039-2
1-METHYL-4-(3-METHYLBENZOYL)-PIPERAZINE HYDROCHLORIDE

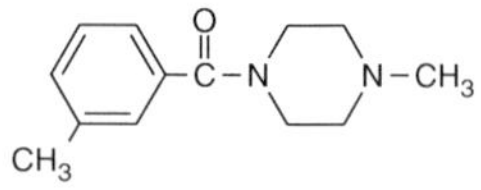

250 mg

S75,636-9
3,5-DIMETHYL-5-PHENYLHYDANTOIN

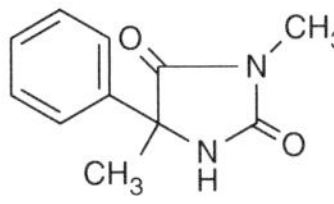

5 gm

S73,502-7
1,2,5,6,7,8-HEXAHYDRO-6-ISOPROPYL-4-PHENYLPYRIDO(4,3-D)-PYRIMIDIN-2-ONE

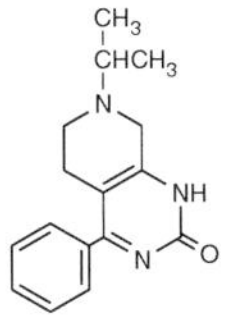

250 mg

S75,116-2
1-[N-(4-BUTYLPHENYL)CARBAMOYL]-4-METHYLPIPERAZINE

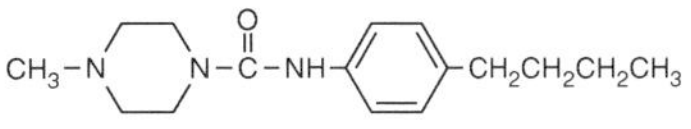

5 gm

S84,504-3
4-METHYL-1-[N-(2,6-XYLYL)-CARBAMOYL]PIPERAZINE

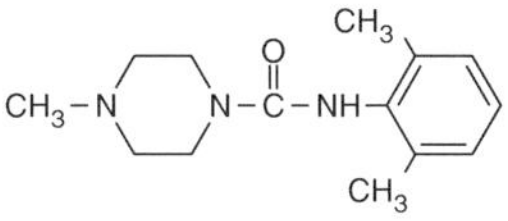

250 mg

S84,503-5
4-METHYL-1-[N-(2,5-XYLYL)-CARBAMOYL]PIPERAZINE

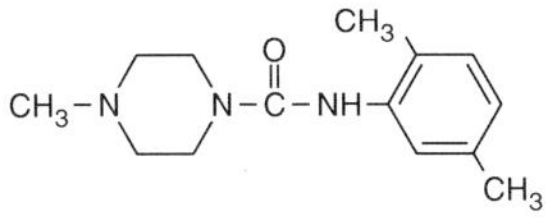

250 mg

S90,188-1
3',4,4'-TRIMETHYL-1-PIPERAZINECARBOXANILIDE

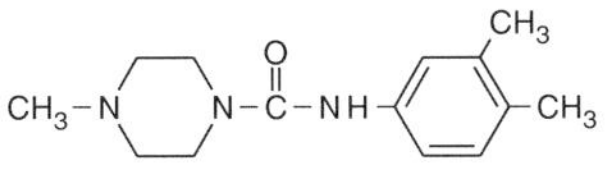

250 mg

S65,539-2
1-ACETYL-2,2,4-TRIMETHYL-1,2-DIHYDROQUINOLINE

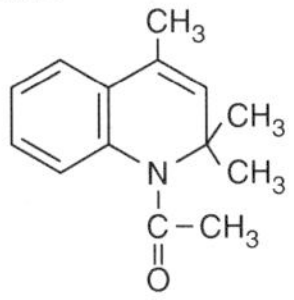

250 mg

S76,086-2
1-ISOPROPYL-3-METHYL-6,7-(METHYLENEDIOXY)-2-OXO-4-PHENYL-3-QUINAZOLINIUM IODIDE

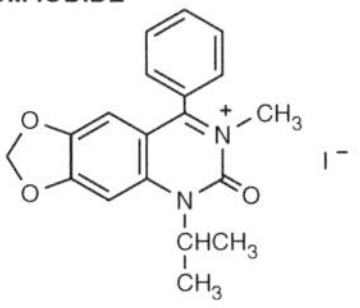

250 mg

S66,423-5
5-HYDRAZONO-3-PHENYL-2-PYRAZOLINE-1-CARBOXAMIDE

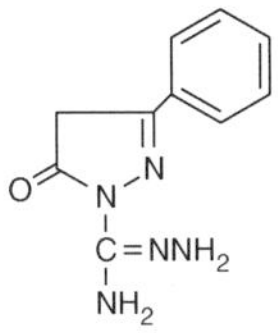

250 mg

S55,145-7
1,2-DIHYDRO-2-(2-NITROPHENYL)-4(3H)-QUINAZOLINONE

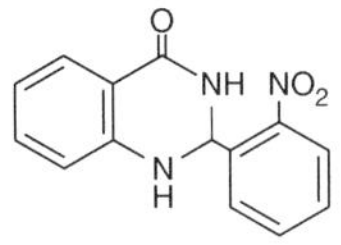

250 mg

S95,203-6
3-BENZAMIDO-1-BENZOYL-2,4-PYRROLIDINEDIONE

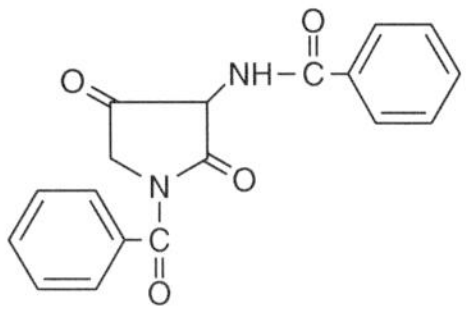

50 mg

S66,189-9
1-ISOPROPYL-6,7-METHYLENEDIOXY-4-PHENYL-2(1H)-QUINAZOLINONE

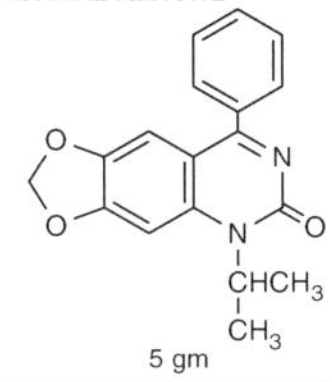

5 gm

S78,922-4
1-BENZYLURACIL

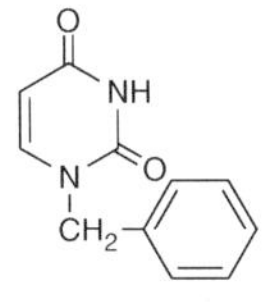

250 mg

S42,898-1
5-CYCLOPROPYL-5-PHENYLHYDANTOIN

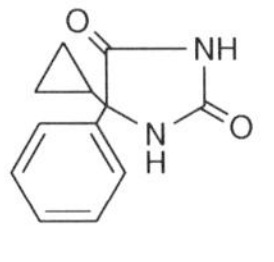

250 mg

S94,970-1
1-MERCAPTO-5-METHYL-5-PHENYLHYDANTOIN

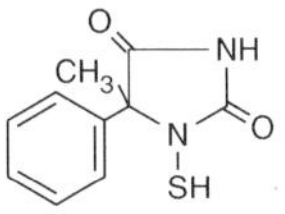

250 mg

S94,969-8
1-BROMO-5-METHYL-5-PHENYLHYDANTOIN

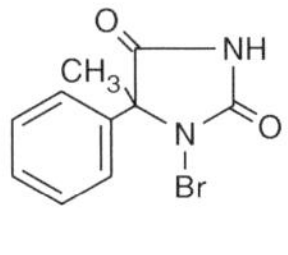

250 mg

S76,145-1
3-CHLORO-6A-PHENYLISOINDOLO(2,1-A)-QUINAZOLINE-5,11(6H,6AH)-DIONE

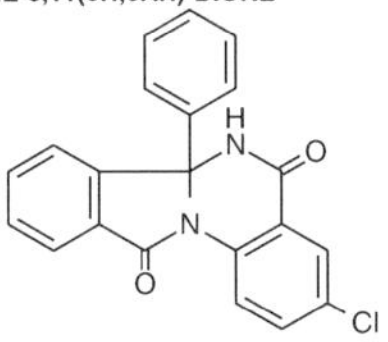

250 mg

S95,285-0
2,3,6,7-TETRAHYDRO-1,2,5,6-TETRA-PHENYL-1H,5H-S-TRIAZOLO(1,2-A)-S-TRIAZOLE-3,7-DIONE

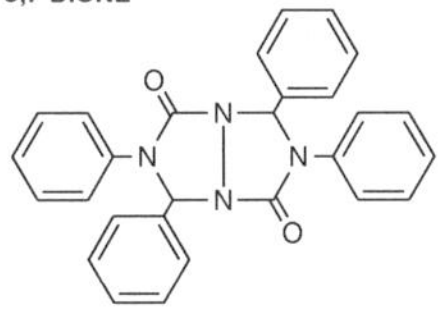

50 mg

S95,283-4
1,5-DIPHENYL-2,3,6,7-TETRAHYDRO-1H,5H-S-TRIAZOLO(1,2-A)-S-TRIAZOLE-3,7-DIONE

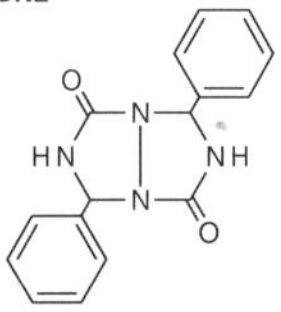

50 mg

S66,893-1
4-(N-BENZYLCARBAMOYL)-2,6-DIMETHYLMORPHOLINE

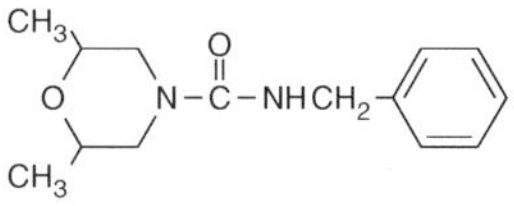

5 gm

S44,693-9
4-METHYL-1-PHENYL-3-PYRAZOLIDINONE

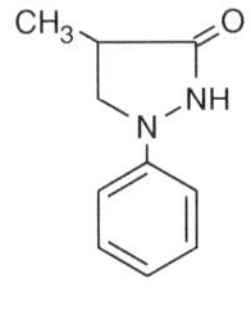

5 gm

S76,069-2
3-HYDRAZINO-4-PHENYL-1H-PYRIDO(1,2-C)PYRIMIDIN-1-ONE

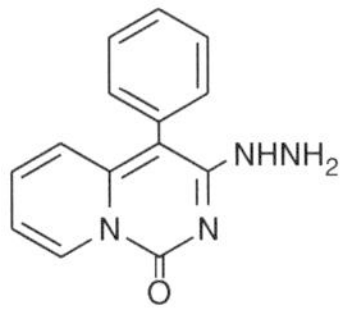

250 mg

S34,956-9
ETHYL 5-OXO-1-PHENYL-2-PYRAZOLINE-3-CARBOXYLATE

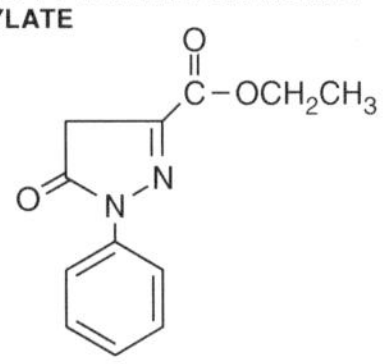

5 gm

S53,107-3
4,4-DIBROMO-3-METHYL-1-PHENYL-2-PYRAZOLIN-5-ONE

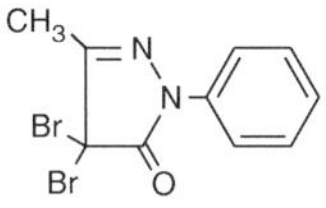

250 mg

S44,727-7
4,5-DIHYDRO-6-METHYL-2-PHENYL-3(2H)-PYRIDAZINONE

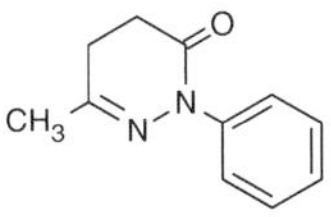

5 gm

S94,879-9
N-(4,5-DIHYDRO-5-METHYL-5-PHENYL-1,3,4-THIADIAZOL-2-YL)ACETAMIDE

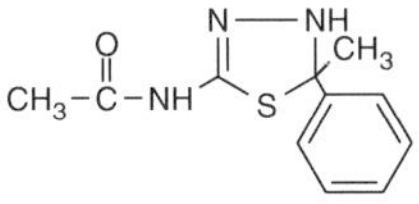

250 mg

S60,938-2
4-(3-CHLOROPHENYLCARBAMOYL)-MORPHOLINE

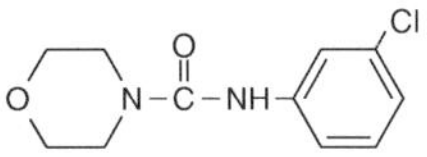

5 gm

S61,172-7
4-[N-(2,6-DICHLOROPHENYL)-CARBAMOYL]MORPHOLINE

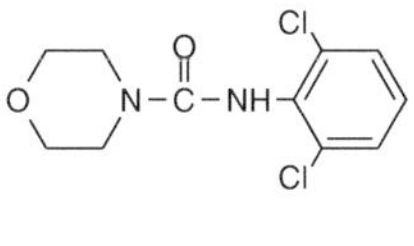

5 gm

S61,191-3
4-[N-(2,5-DICHLOROPHENYL)-CARBAMOYL]MORPHOLINE

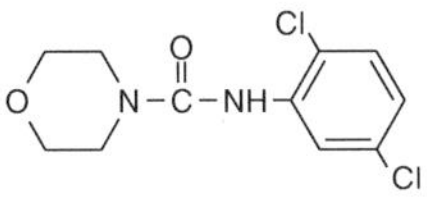

5 gm

S61,294-4
4-[N-(2,4-DICHLOROPHENYL)-CARBAMOYL]MORPHOLINE

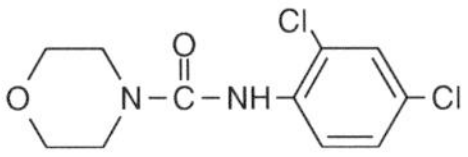

5 gm

S61,347-9
4-[N-(3,4-DICHLOROPHENYL)-CARBAMOYL]MORPHOLINE

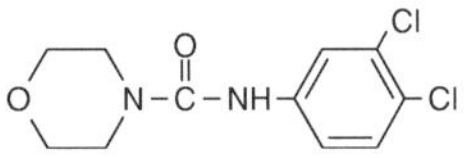

5 gm

S60,665-0
4-[N-(O-TOLYL)CARBAMOYL]MORPHOLINE

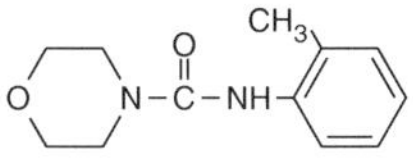

5 gm

S61,293-6
4-[N-(P-TOLYL)CARBAMOYL]MORPHOLINE
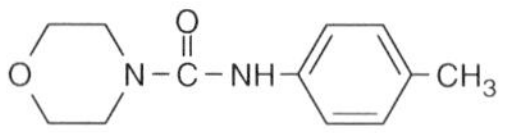
5 gm

S61,104-2
4-[N-(5-CHLORO-2-METHYLPHENYL)-CARBAMOYL]MORPHOLINE
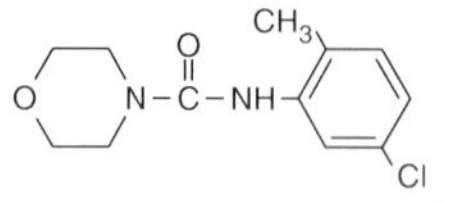
5 gm

S66,477-4
2,3-DIHYDRO-4-METHYL-1,4-BENZOTHIAZEPIN-5(4H)-ONE
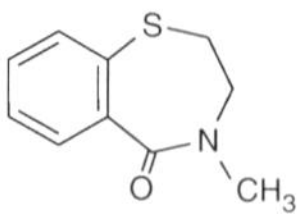
5 gm

S68,146-6
2,3,4,5-TETRAHYDRO-1,5-BENZOTHIAZEPIN-4-ONE
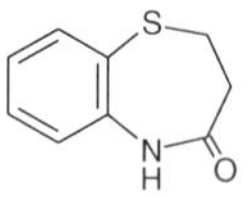
250 mg

S68,164-4
2,3-DIHYDRO-2-METHYL-1,5-BENZOTHIAZEPIN-4(5H)-ONE
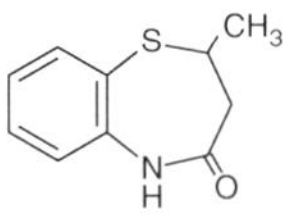
250 mg

S36,421-5
4-NITROSOANTIPYRINE
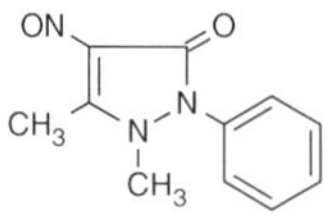
5 gm

S67,356-0
3-(4-METHYL-3-NITROPHENYL)-2,4-(1H,3H)-QUINAZOLINEDIONE
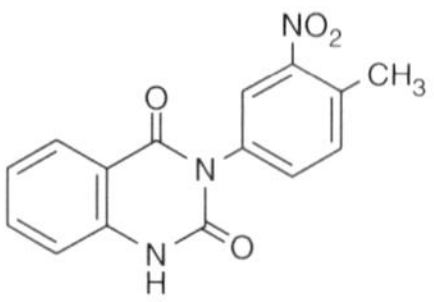
250 mg

S89,922-4
1,3-DIETHYL-5-NITRO-6-STYRYLURACIL
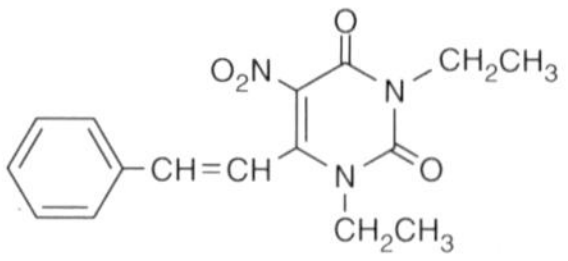
250 mg

S40,701-1
CARBOBENZYLOXY-L-PROLINE 4-NITROPHENYL ESTER
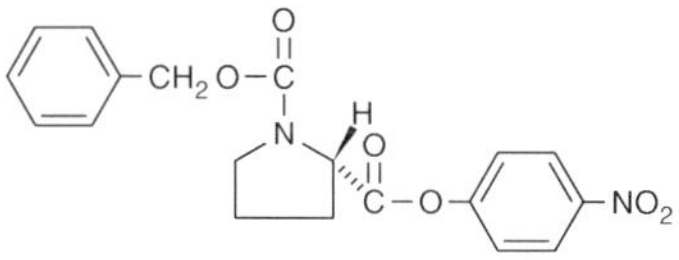
5 gm

S95,741-0
1-BENZOYL-1,2,2A,3-TETRAHYDROBENZ (CD)INDOLE-5-CARBOXALDEHYDE
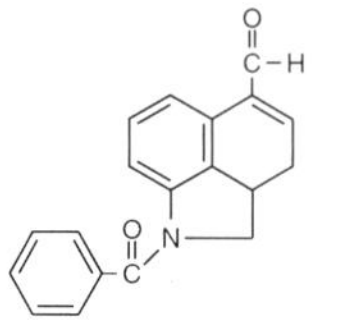
50 mg

S45,240-8
6-CHLOROISATIN
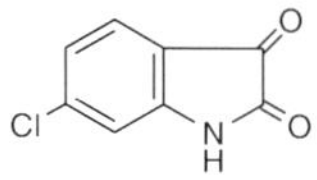
5 gm

S56,459-1
5-CHLOROISATIN
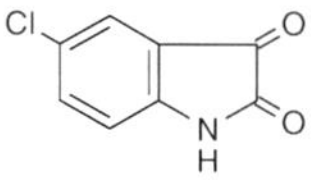
250 mg

S80,667-6
ETHYL 1-BENZOYL-4-OXO-3-PIPERIDINECARBOXYLATE
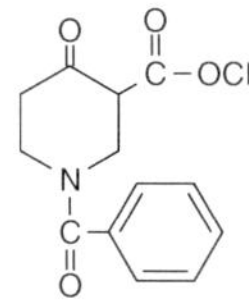
250 mg

S72,058-5
1-[2-(BENZYLCARBONYL)BENZOYL]-PYRROLIDINE
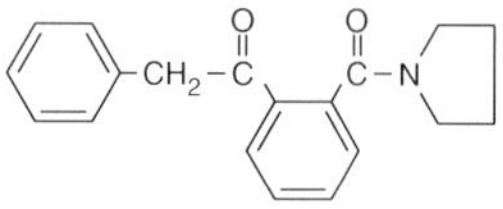
5 gm

S73,008-4
4-[2-(PHENYLACETYL)BENZOYL]-PIPERIDINE
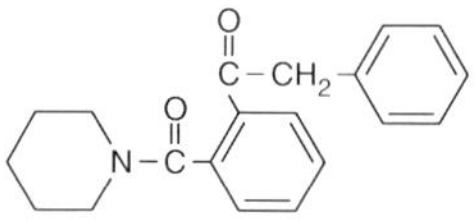
5 gm

S73,007-6
4-[2-(PHENYLACETYL)BENZOYL]-MORPHOLINE
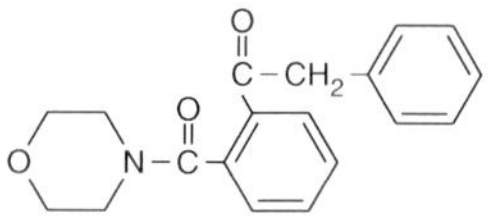
5 gm

S64,201-0
1-(3-CHLOROPHENYL)-3-METHACRYLAMIDO-2-PYRAZOLIN-5-ONE
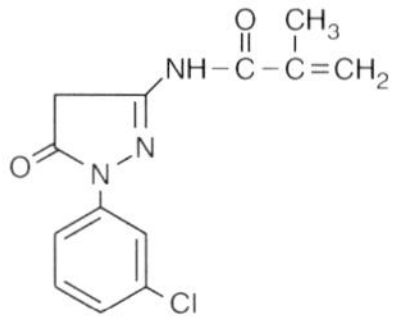
5 gm

S76,063-3
2,3,4,5,6,7-HEXAHYDRO-1H-1-BENZAZONINE-2,7-DIONE
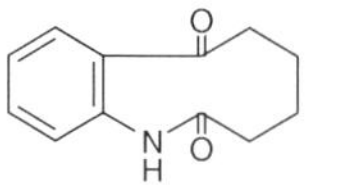
250 mg

S65,702-6
N-ETHYLISATOIC ANHYDRIDE
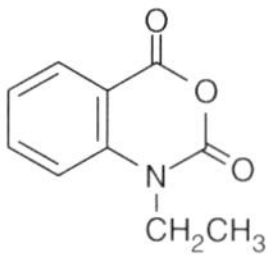
5 gm

S76,469-8
2-METHYL-3-(2-OXOPROPYL)-1(2H)-ISOQUINOLINONE
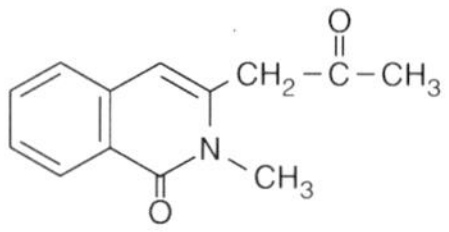
5 gm

S86,793-4
N-(3-ACETYLPHENYL)-2-METHYL-1-PIPERIDINECARBOXAMIDE
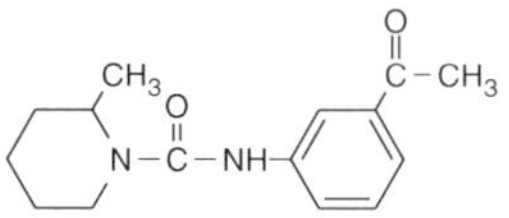
250 mg

S64,609-1
2-(4-CHLOROBENZYLIDENE)-3-OXO-4H-1,4-BENZOTHIAZINE
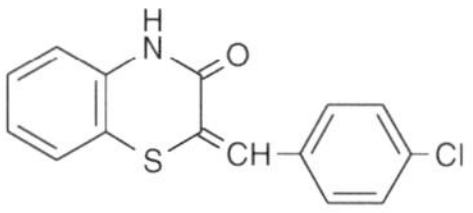
250 mg

S64,607-5
2,3-DIHYDRO-3-OXO-2-(3,4,5-TRIMETHOXYBENZYLIDENE)-4H-1,4-BENZOTHIAZINE
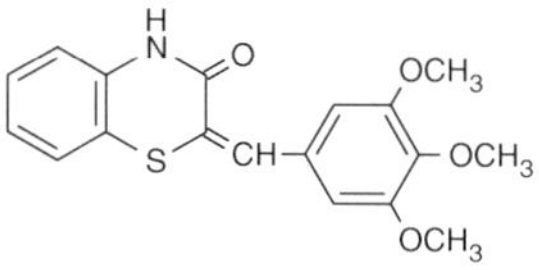
250 mg

S80,413-4
N,N'-METHYLENEDIISATIN
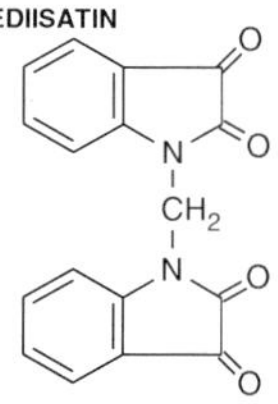
250 mg

S63,912-5
7-METHYLISATIN
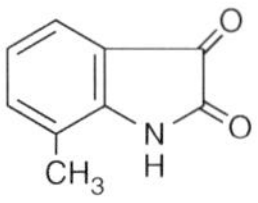
250 mg

S37,593-4
1-(ACETOACETYL)INDOLINE
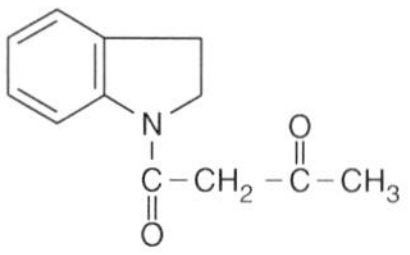
5 gm

S92,381-8
4-HEXADECANOYL-3-METHYL-1-PHENYL-2-PYRAZOLIN-5-ONE
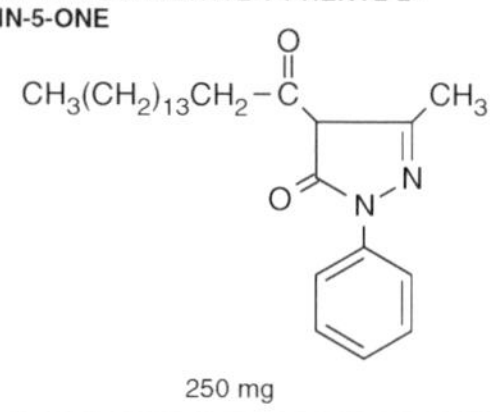
250 mg

S95,731-3
12-ACETYL-DECAHYDRO-9H-PYRIDO (1,2,3-LM)PYRROLO(2,3-D)-CARBAZOL-9-ONE
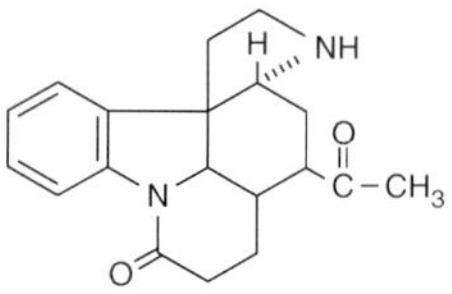
50 mg

S45,869-4
1,3-DIPHENYL-5-PYRAZOLONE
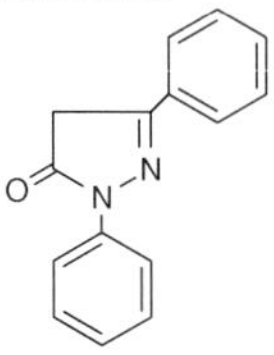
5 gm

S63,771-8
4,4-DICHLORO-3-METHYL-1-PHENYL-3-PYRAZOLIN-5-ONE
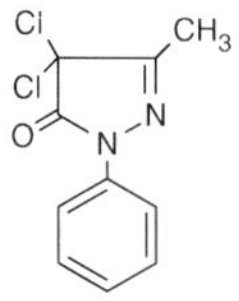
5 gm

S67,103-7
4,6-DIPHENYL-1-ETHYL-2(1H)-PYRIMIDINONE
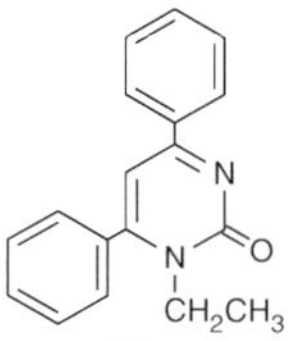
250 mg

S64,603-2
2-(4-ISOPROPYLBENZYLIDENE)-3-OXO-4H-1,4-BENZOTHIAZINE
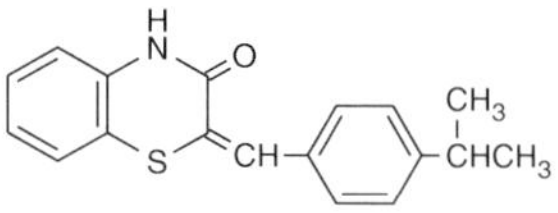
250 mg

S67,250-5
2-CHLORO-5,9,10,14B-TETRAHYDRO-5-METHYLISOQUINO(2,1-D)(1,4)-BENZODIAZEPIN-6(7H)-ONE
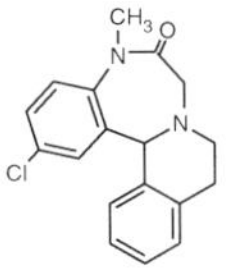
5 gm

S91,473-8
1-PROPIONYLISATIN
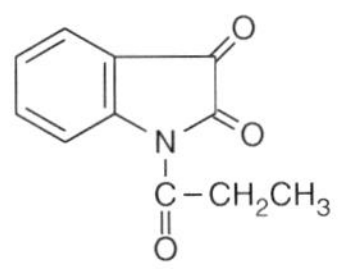
250 mg

S36,669-2
ANTIPYRIC ACID
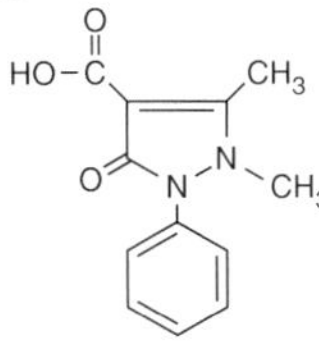
5 gm

S89,903-8
1,4-BIS(2-CARBOXYBENZOYL)PIPERAZINE
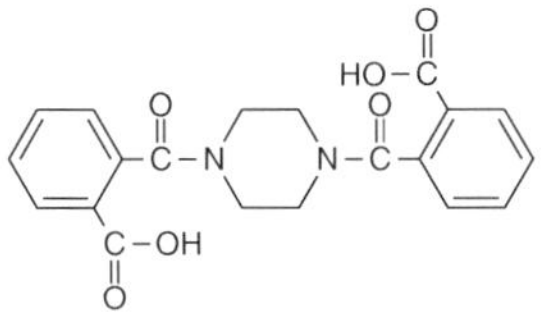
250 mg

S69,447-9
4-[3,5-BIS(METHOXYCARBONYL)-PHENYLCARBAMOYL]-2,6-DIMETHYLMORPHOLINE
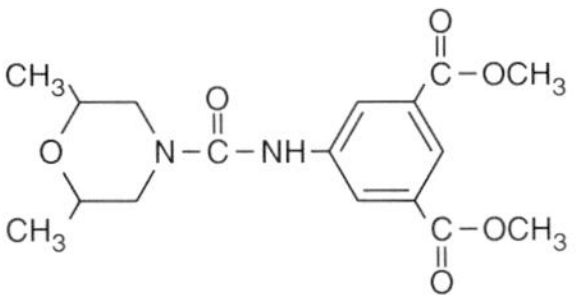
250 mg

S76,473-6
1,2-DIHYDRO-2-METHYL-1-OXO-3-ISOQUINOLINEBUTYRIC ACID
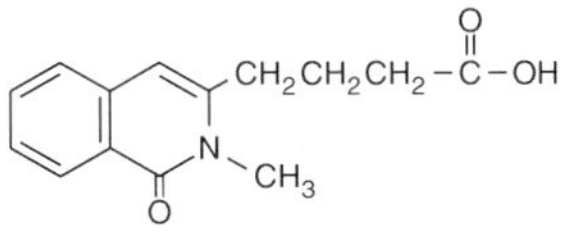
5 gm

S38,232-9
ANTIPYRINE-4-PROPIONIC ACID
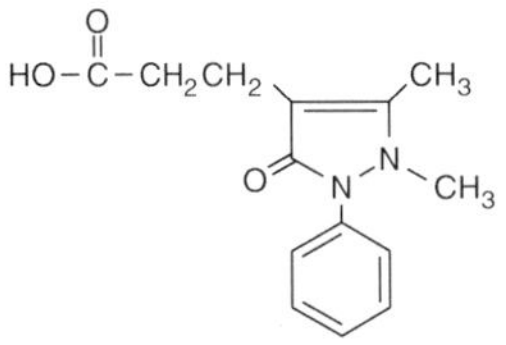
250 mg

S75,454-4
4-(3,4-DIHYDRO-4-OXO-1,2,3-BENZOTRIAZIN-3-YL)PHENYLACETIC ACID
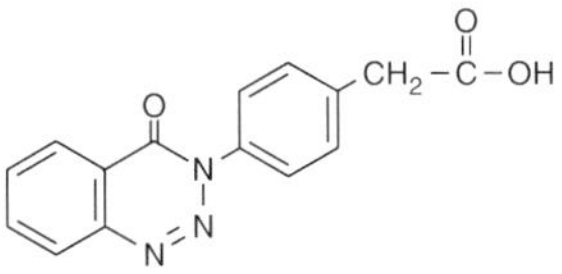
250 mg

S76,324-1
4-[8-CHLORO-5,6-DIHYDRO-3-OXO-3H-PYRROLO(3,2,1-IJ)QUINAZOLIN-1-YL]-BENZOIC ACID
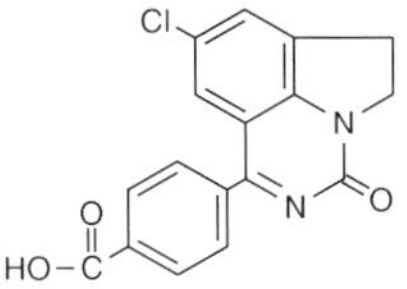
250 mg

S51,741-0
2-ETHYL-5-OXO-1-PHENYL-3-PYRAZOLINE-3-CARBOXAMIDE
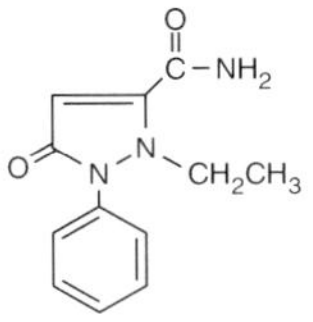
5 gm

S76,001-3
ETHYL 7-ACETOXYMETHYL-6-BENZAMIDOHEXAHYDRO-7-METHYL-5-OXOIMIDAZO(1,2-A)PYRIDINE-8-CARBOBOXYLATE
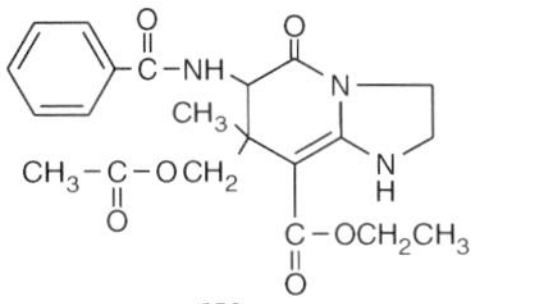
250 mg

S76,161-3
METHYL 2-(3,4-DIHYDRO-3-OXO-2H-1,4-BENZOTHIAZIN-2-YL)ACRYLATE
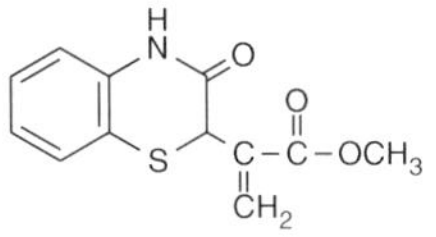
250 mg

S50,983-3
1-ACETYL-1,2-DIHYDRO-2-(5-INDANYL)-3,1,4-BENZOXAZIN-4-ONE
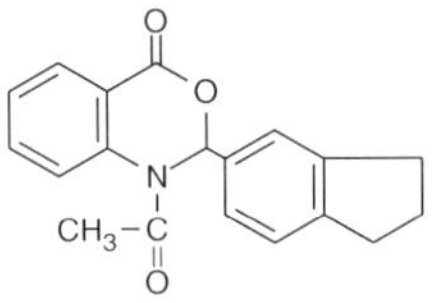
250 mg

S89,779-5
ETHYL 4-[3-(1-METHYL-3-OXO-2-PHENYL-4-PYRAZOLIN-4-YL)UREIDO]-BENZOATE
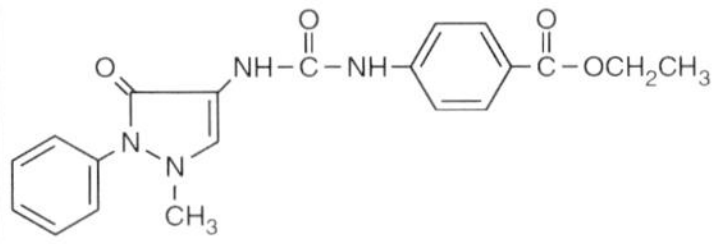
250 mg

S79,416-3
ETHYL 4-[4-(3-PHENYLPROPYL)-PIPERIDINOCARBOXAMIDO]BENZOATE
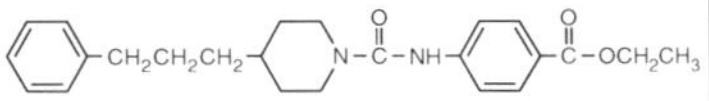
250 mg

S79,442-2
ETHYL 4-(2-METHYLPIPERIDINO)-CARBOXAMIDOBENZOATE
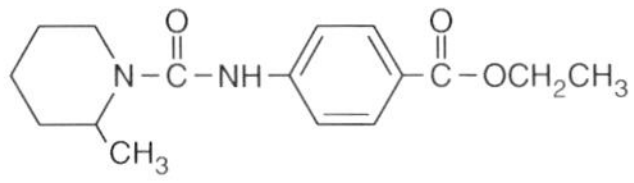
250 mg

S73,466-7
4-(1,2-DIHYDRO-2-METHYL-1-OXO-3-ISOQUINOLYL)BUTYRYL CHLORIDE
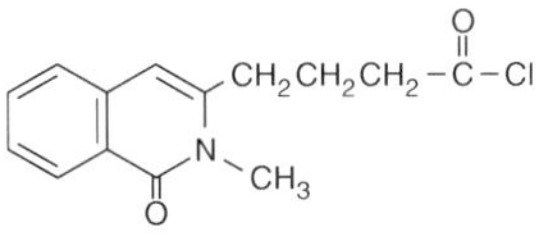
250 mg

S51,270-2
1,3-DIMETHYL-2,4(1H,3H)-QUINAZOLINEDIONE
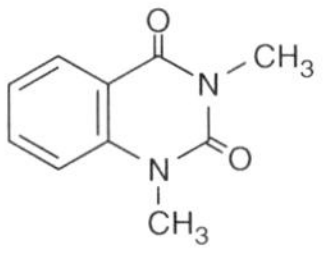
250 mg

S36,581-5
3-PHENYL-2,4(1H,3H)-QUINAZOLINEDIONE
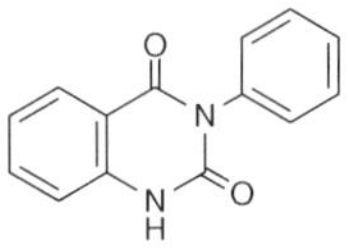
250 mg

S80,602-1
1-ISOPROPYL-7-METHYL-2,4(1H,3H)-QUINAZOLINEDIONE
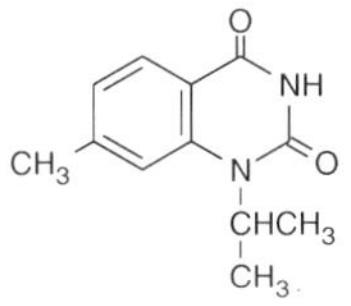
250 mg

S76,016-1
7A-BENZAMIDO-2,3,3A,4,5,7A-HEXAHYDRO-2-METHYL-5-PHENYL-1H-ISOINDOL-1-ONE
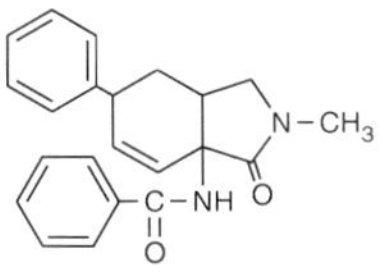
250 mg

S36,448-7
2-PHENYL-4-PHENYLACETAMIDO-3-PYRAZOLIN-5-ONE
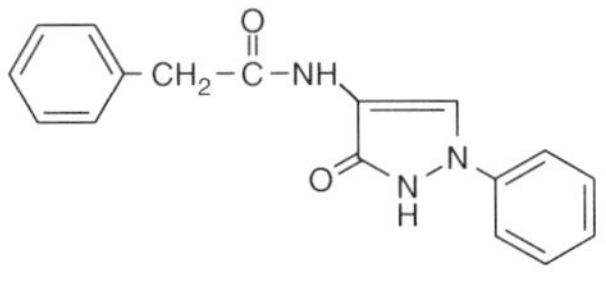
5 gm

S91,658-7
N-(2,3-DIMETHYL-5-OXO-1-PHENYL-3-PYRAZOLIN-4-YL)-2-BUTENAMIDE
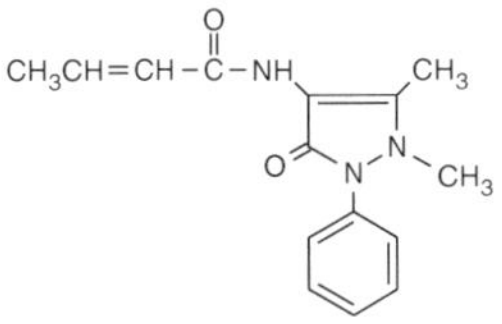
250 mg

S34,994-1
4-FORMYLAMINOANTIPYRINE
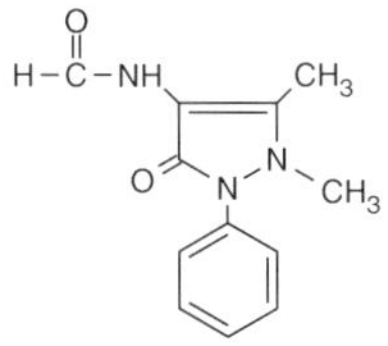
5 gm

S50,301-0
ALPHA,3-BIS(PHENYLCARBAMOYL)-2-METHYLENEOXAZOLIDINE
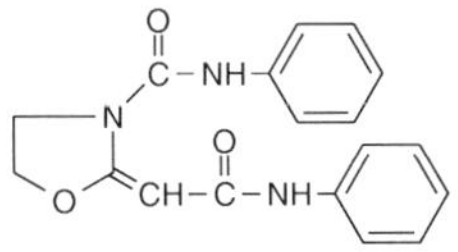
5 gm

S76,012-9
1,2,3,3A-TETRAHYDRO-3A-(2,4-XYLYL)-PYRROLO(1,2-A)QUINAZOLIN-5(4H)-ONE
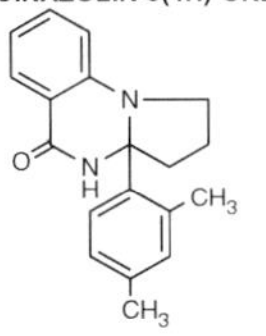
250 mg

S51,383-0
2-METHYL-3-PHENYL-4(3H)-QUINAZOLINONE
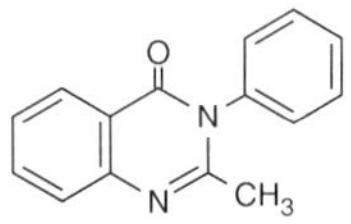
250 mg

S88,020-5
3-BENZYLIDENE-2-INDOLINONE
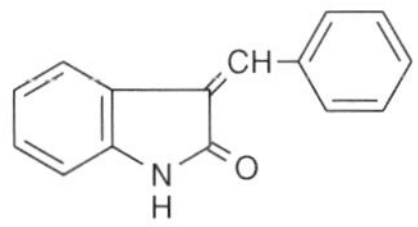
250 mg

S79,875-4
4B,5-DIHYDRO-4B-PHENYL-11H-ISOINDOLO(2,1-A)BENZIMIDAZOL-11-ONE
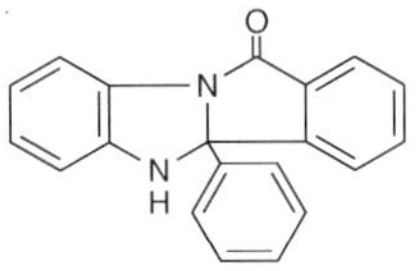
250 mg

S76,178-8
2'TERT-BUTYL-10,11-DIHYDROSPIRO-[5H-DIBENZO[A,D]CYCLOHEPTENE-5,1'-[1H]ISOINDOL]-3'(2H)-ONE
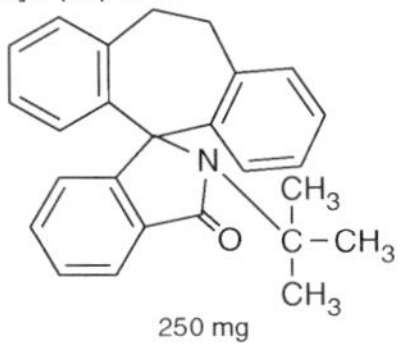
250 mg

S34,157-6
1-BENZOYLPIPERIDINE
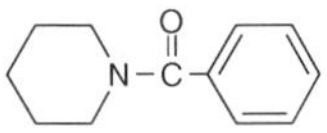
5 gm

S56,831-7
4-(2,4-DICHLOROBENZOYL)MORPHOLINE
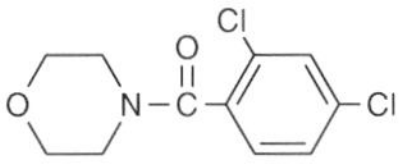
250 mg

S89,634-9
1-BENZOYLINDOLINE
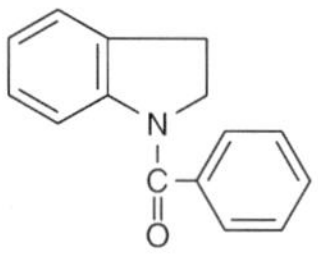
250 mg

S34,810-4
ETHYL N-(4-ANTIPYRYL)CARBAMATE
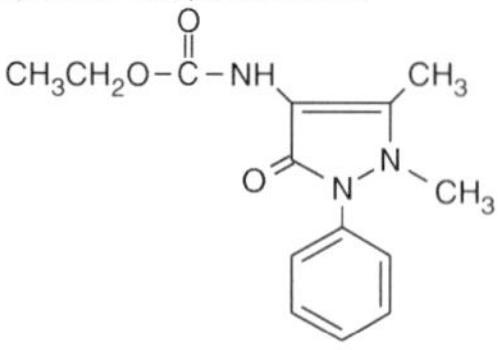
5 gm

S51,761-5
N,N'-METHYLENEBIS[4-(METHYLAMINO)-ANTIPYRINE]-
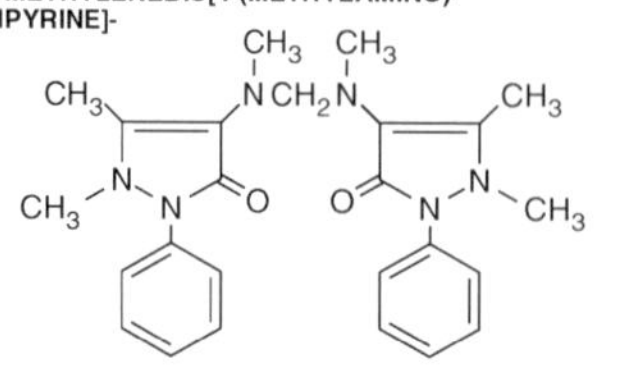
5 gm

S76,482-5
2,3-DIMETHYL-1-PHENYL-3-PYRAZOLIN-5-ONE
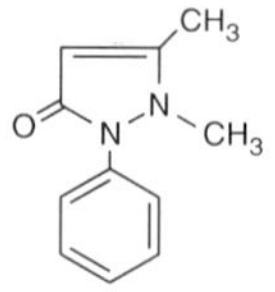
5 gm

S92,740-6
4-CHLORO-6-METHYL-2-PHENYL-3(2H)-PYRIDAZINONE
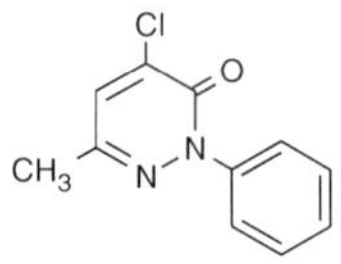
250 mg

S84,710-0
5,7,12,14-TETRAHYDRO-5,14(1',4')-BENZENOPHTHALAZINO(2,3-B)-PHTHALAZINE-7,12-DIONE
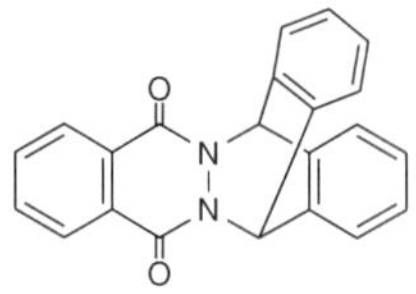
250 mg

S94,249-9
1,3-DIMETHYL-5-(1-NAPHTHYLMETHYLENE)BARBITURIC ACID
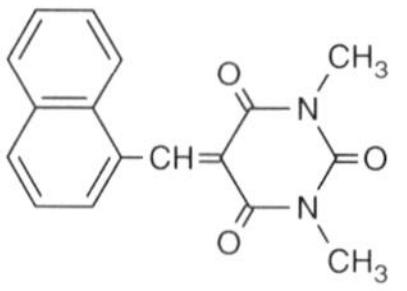
250 mg

S34,769-8
1,3-DIPHENYLPARABANIC ACID
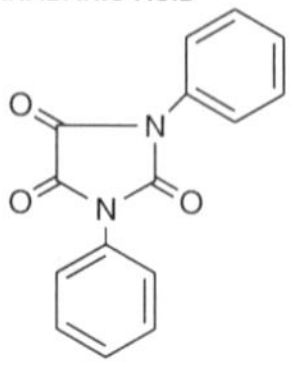
5 gm

S74,050-0
3-METHYL-1-PHENYL-2,4,5-IMIDAZOLIDINETRIONE
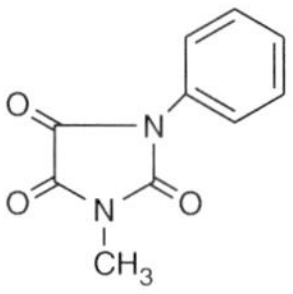
250 mg

S76,142-7
4,7-DIHYDRO-4-ETHYL-4-PHENYL-1H-2,6-BENZODIAZONINE-3,5(2H,6H)-DIONE
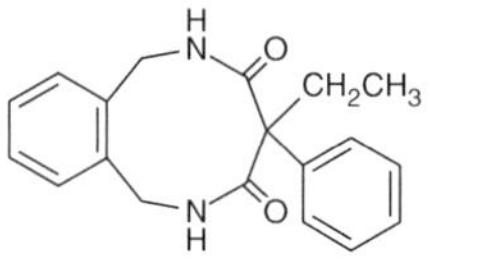
250 mg

S68,885-1
1,3-DIACETYL-5-ETHYL-5-PHENYLHEXAHYDROPYRIMIDINE-4,6-DIONE
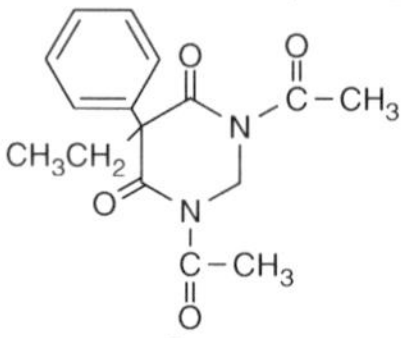
5 gm

S35,914-9
4-BUTYL-3,5-DIKETO-1-PHENYLPYRAZOLIDINE
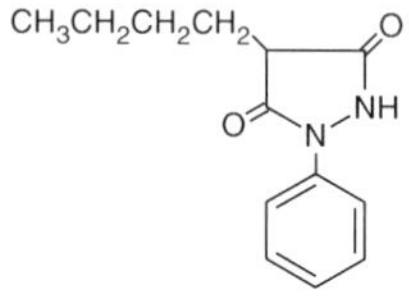
5 gm

S38,561-1
4-ALLYL-1,2-DIPHENYL-3,5-PYRAZOLIDINEDIONE
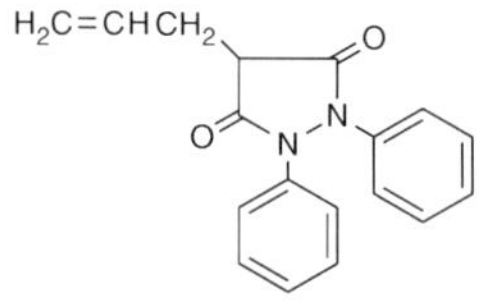
250 mg

S85,202-3
1,2-DIPHENYL-4-[3-(1-NAPHTHYL)-PROPYL]-3,5-PYRAZOLIDINEDIONE
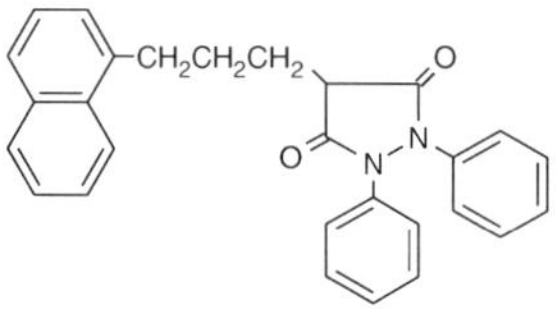
250 mg

S42,082-4
4,4-DIETHYL-1-PHENYL-3,5-PYRAZOLIDINEDIONE
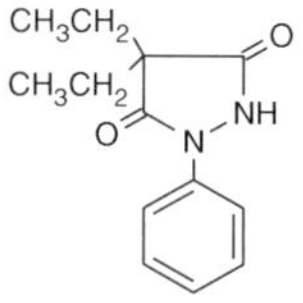
250 mg

S72,053-4
CIS-HEXAHYDRO-N-(2-METHYL-5-OXO-1-PHENYL-3-PYRAZOLIN-4-YL)-PHTHALAMIC ACID
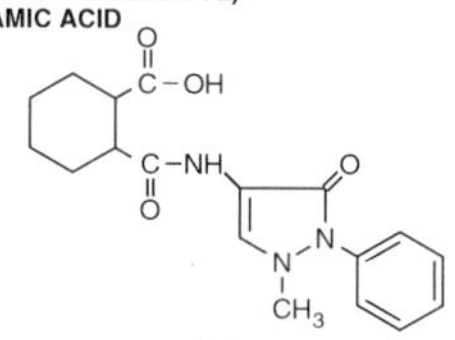
250 mg

S73,282-6
1,2,3,5,6,10B-HEXAHYDRO-3-OXOPYRROLO(2,1-A)ISOQUINOLINE-1-CARBOXYLIC ACID
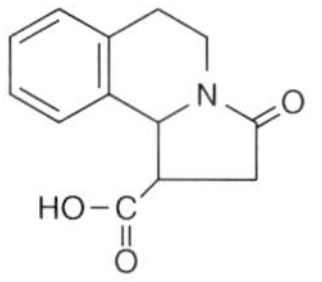
250 mg

S67,879-1
CARBOBENZYLOXY-L-PROLYL-L-ISOLEUCINE
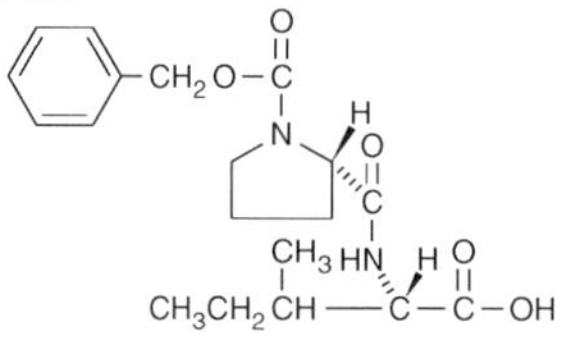
5 gm

S77,606-8
CARBOBENZYLOXY-L-PROLYL-L-LEUCINE
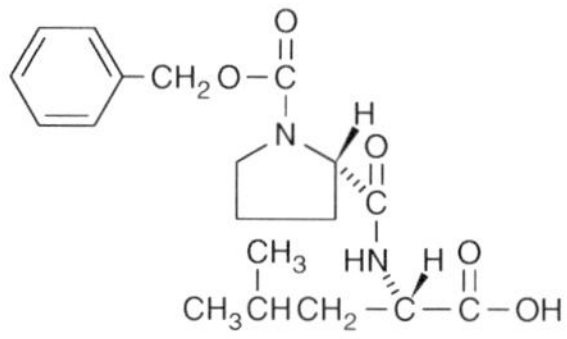
250 mg

S76,534-1
CARBOBENZYLOXY-L-PROLYL-L-VALINE
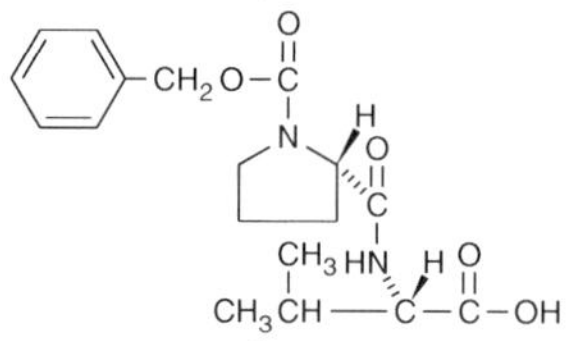
5 gm

S76,547-3
N-BENZYLOXYCARBONYL-L-PHENYLALANYL-L-PROLINE
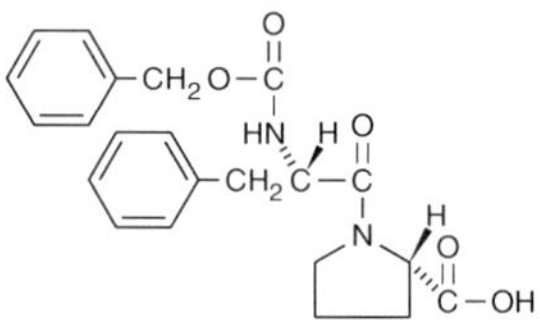
5 gm

S67,847-3
2-[1-(3-CARBOXYPROPIONYLOXY)BUTYL]-4,5-DIHYDRO-6-PHENYLPYRIDAZIN-3(2H)-ONE
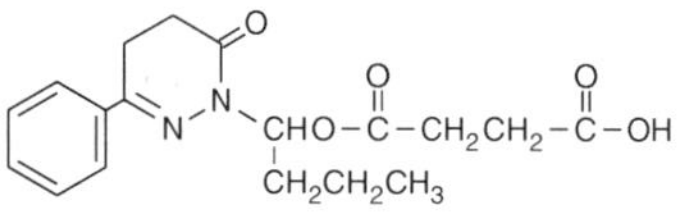
5 gm

S56,027-8
2,3-DIHYDRO-3-OXO-4H-1,4-BENZOTHIAZINE-2-ACETIC ACID
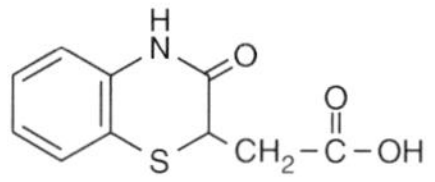
250 mg

S95,729-1
1,2,3,3A,4,5,11B,11C-OCTAHYDRO-6-OXO-6H-PYRIDO[3,2,1-JK]CARBAZOLE-11B-PROPIONIC ACID
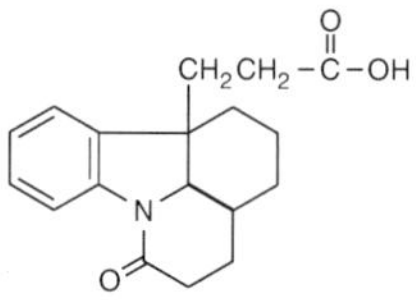
50 mg

S95,728-3
OCTAHYDRO-9-OXO-9H-PYRIDO(1,2,3-LM)-PYRROLO(2,3-D)CARBAZOLE-12-CARBOXYLIC ACID
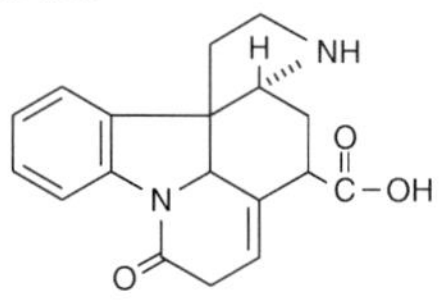
50 mg

S79,089-3
CARBOBENZYLOXYPROLYL-L-ALANINE BENZYL ESTER
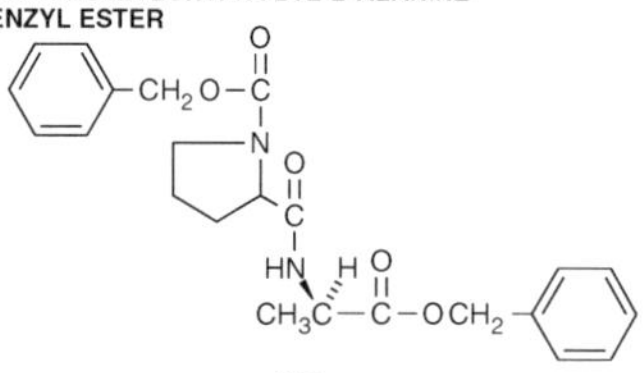
250 mg

S79,082-6
CARBOBENZYLOXYPROLYLVALINE BENZYL ESTER
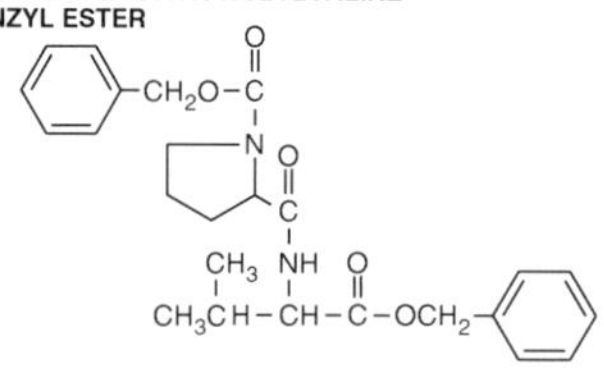
250 mg

S78,994-1
CARBOBENZYLOXYPROLYLALANINE METHYL ESTER
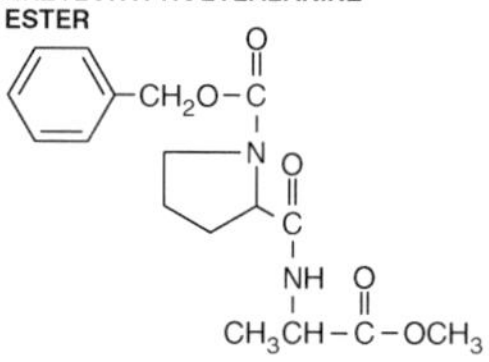
250 mg

S78,546-6
CARBOBENZYLOXYGLYCYL-L-PROLINE BENZYL ESTER
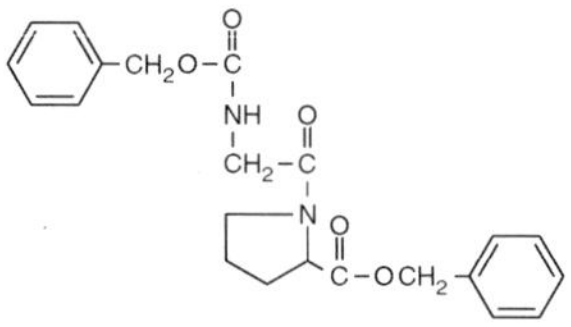
250 mg

S81,543-8
ETHYL 2,3-DIHYDRO-2-OXO-3-BENZOXAZOLECARBOXYLATE
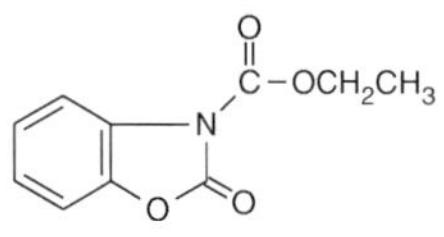
250 mg

S77,661-0
CARBOBENZYLOXY-L-PROLYL-L-VALINAMIDE HYDRATE
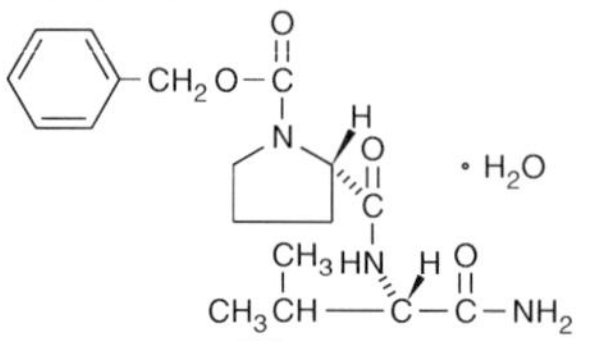
250 mg

S72,045-3
3-(ETHOXYCARBONYL)-5-PHENYLURACIL
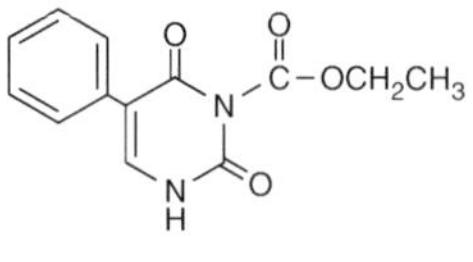
5 gm

S76,047-1
8-(1,3-DIHYDRO-3-OXO-1-ISOBENZOFURANYL)-1,2,3,5,6,7-HEXAHYDRO-6-METHYLIMIDAZO[1,2-C]-PYRIMIDINE-5,7-DIONE

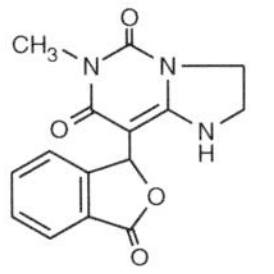

250 mg

S52,482-4
ETHYL 3,4-BENZO-2-OXO-1,5,9-TRIAZASPIRO(5.5)UNDEC-3-ENE-9-CARBOXYLATE

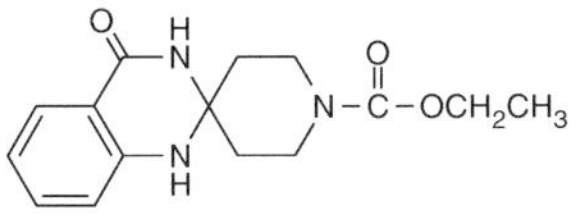

250 mg

S50,093-3
METHYL 4'-(3-HEPTADECYL-2-METHYL-5-OXO-3-PYRAZOLIN-1-YL)SUCCINANILATE

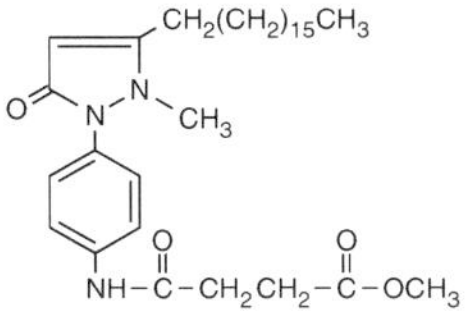

250 mg

S78,874-0
CARBOBENZYLOXYPROLINE-2-NAPHTHYLAMIDE

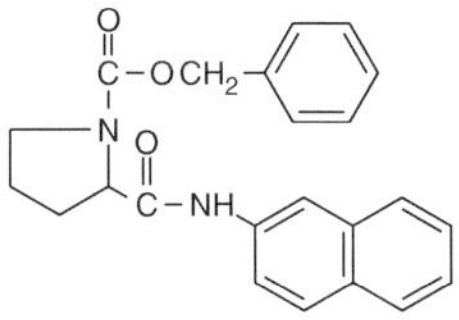

250 mg

S61,749-0
ETHYL 2-(2,3-DIHYDRO-3-OXO-4H-1,4-BENZOTHIAZIN-2-YL)ACETATE

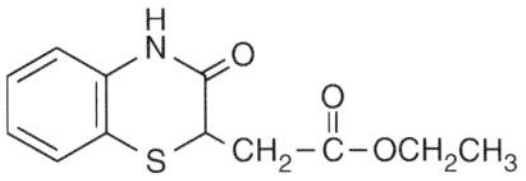

250 mg

S53,732-2
METHYL 2,3-DIHYDRO-3-OXO-4H-1,4-BENZOTHIAZINE-2-ACETATE

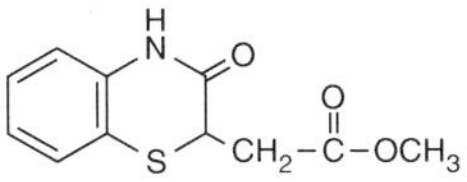

5 gm

S81,541-1
3-ACETYL-2(3H)-BENZOXAZOLONE

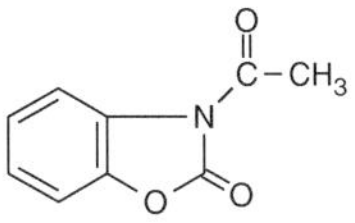

250 mg

S76,073-0
2-CHLORO-6A,11-DIHYDRO-5H-ISOINDOLO(2,1-A)(3,1)BENZOXAZINE-5,11-DIONE

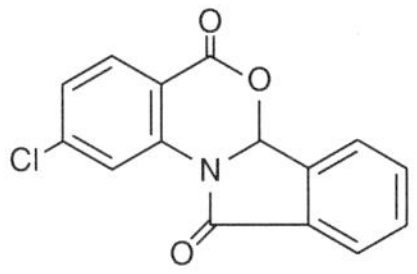

250 mg

S73,318-0
4A-PHENYL-2,3,4,4A-TETRAHYDRO-1H,6H-PYRIDO(1,2-A)(3,1)-BENZOXAZINE-1,6-DIONE

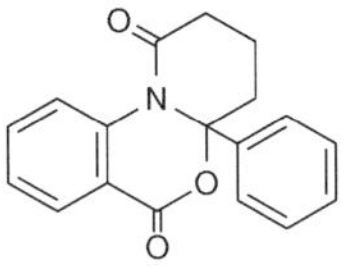

250 mg

S81,618-3
3-(PHENYLCARBAMOYL)-2(3H)-BENZOXAZOLONE

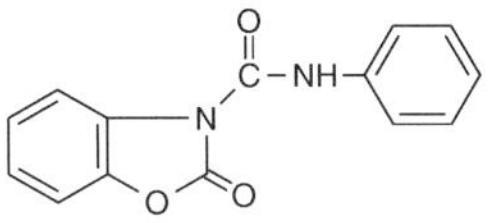

250 mg

S64,230-4
PHENYL 1-PIPERIDINECARBOXYLATE

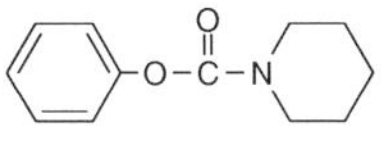

5 gm

S45,776-0
PHENOTHIAZINE-10-CARBONYL CHLORIDE

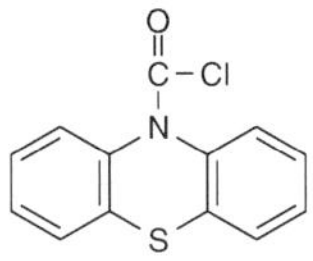

5 gm

S76,355-1
6-CHLOROISATOIC ANHYDRIDE

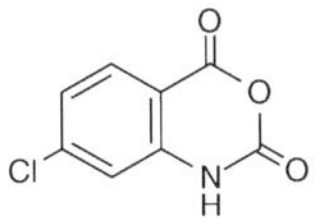

250 mg

S69,246-8
2-(PYRROLIDINOCARBOXAMIDO)ACETIC ACID N2-[N-(1-NAPHTHYL)CARBAMOYL)-HYDRAZIDE

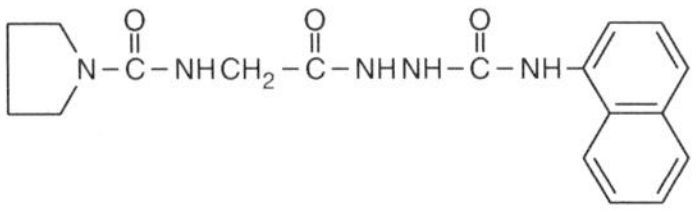

250 mg

S74,001-2
2-PHENYL-2,3,5,8,-TETRAHYDRO-5,8-METHANO-1H-1,2,4-TRIAZOLO(1,2-A)-PYRIDAZINE-1,3-DIONE

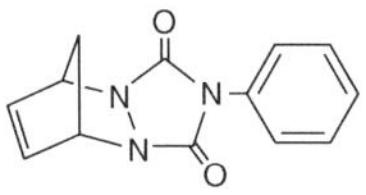

250 mg

S73,257-5
1,2,3,5,6,7-HEXAHYDRO-6-PHENYLIMIDAZO(1,2-C)-PYRIMIDINE-5,7-DIONE

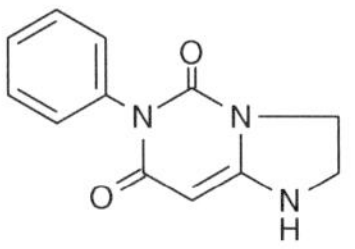

250 mg

S92,302-8
5-(4-METHOXYPHENYL)-5-PHENYLHYDANTOIN

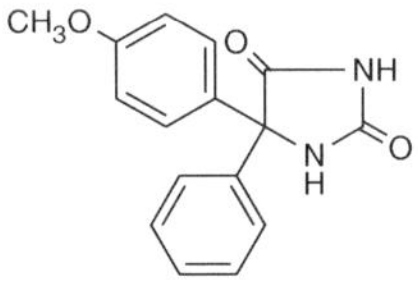

250 mg

S74,884-6
SPIRO(FLUOREN-9,4'-IMIDAZOLIDINE)-2',5'-DIONE

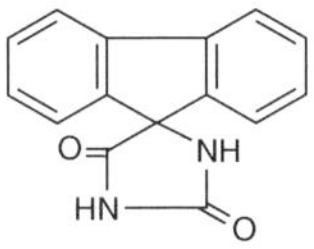

250 mg

S75,405-6
1,3-DICHLORO-5-(2-FLUORENYL)-5-METHYLHYDANTOIN

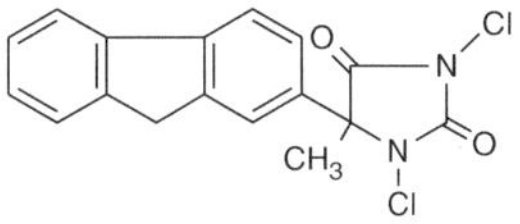

250 mg

S76,022-6
4,8-DIPHENYL-1,2,4,5,6,7-HEXAHYDRO-6-MORPHOLINO-3H,8H-2A-4,8A-TRIAZAACENAPHTHLENE-3,5-DIONE

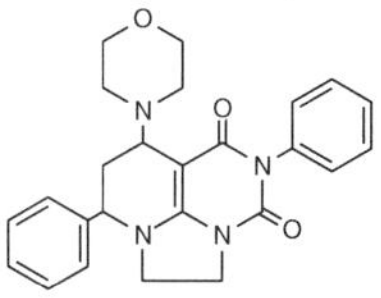

250 mg

S71,345-7
1,3-DIPHENYLURETIDINE-2,4-DIONE

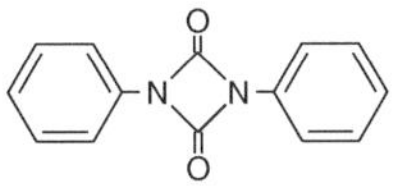

5 gm

S55,136-8
1,2-DIHYDRO-2-(P-TOLYL)-4(3H)-QUINAZOLINONE

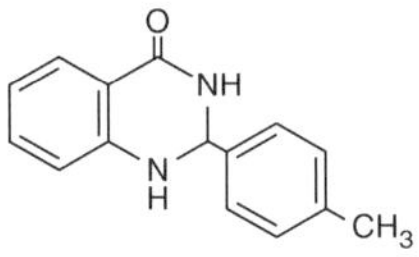

250 mg

S84,202-8
1-ACETYL-5-PHENYL-2-THIOHYDANTOIN

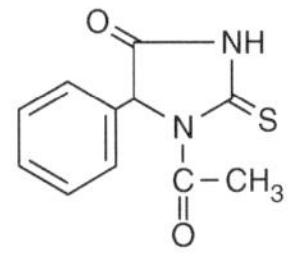

250 mg

S40,771-2
1,2-DIHYDRO-1-METHYL-2-PHENYL-3,6-PYRIDAZINEDIONE

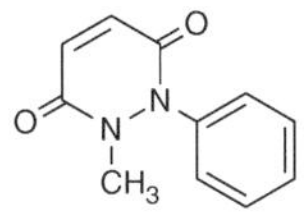

5 gm

S94,248-0
5-(9-ANTHRACENYLMETHYLENE)-BARBITURIC ACID

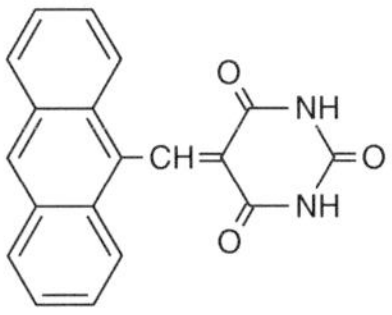

250 mg

S89,336-6
3,4-DIHYDRONAPHTHO(1,8-EF)-1,4-DIAZEPIN-2(1H)-ONE

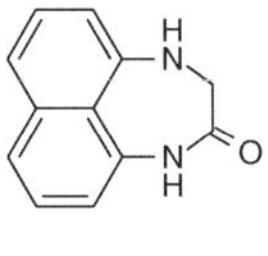

250 mg

S80,864-4
4-METHYL-3-PHENYL-2-PYRROLIN-5-ONE

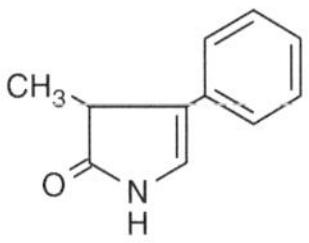

250 mg

S94,543-9
5-METHYL-1-PHENYL-1H-PYRAZOL-3-(2H)-ONE

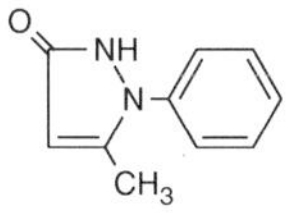

250 mg

S93,515-8
10-BENZOYL-10H-PHENOTHIAZINE

250 mg

S87,689-5
(3S,6S)-3,6-DIPHENYL-2,5-PIPERAZINEDIONE

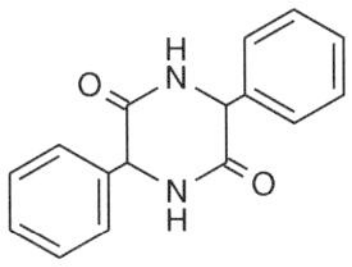

250 mg

S89,340-4
2,3-DIHYDRO-2-PHENYL-1,5-BENZOTHIAZEPIN-4(5H)-ONE

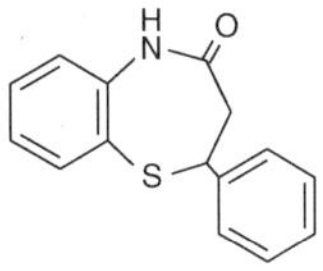

250 mg

S57,690-5
1,4-DIBENZYL-2,5-PIPERAZINEDIONE

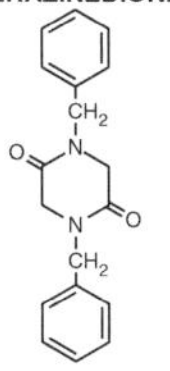

5 gm

S89,757-4
1-(2-IMIDAZOLIN-2-YL)-2-PYRROLIDINONE PICRATE

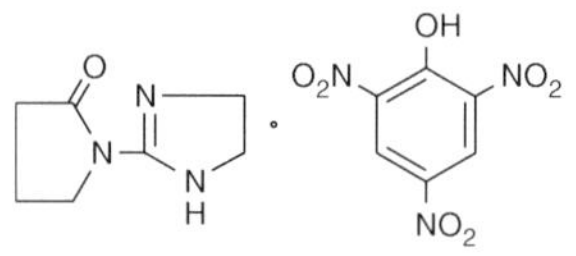

250 mg

S76,127-3
9B-PHENYL-2,3,5,9B-TETRAHYDRO-1H-IMIDAZO(2,1-A)ISOINDOL-5-ONE

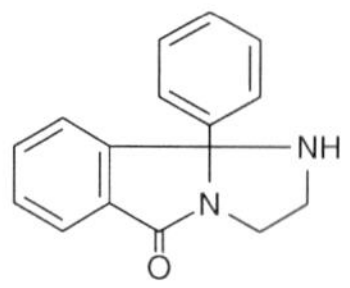

250 mg

S73,350-4
4B,11-DIHYDRO-4B-PHENYLISOINDOLO-(1,2-B)BENZOTHIAZOL-11-ONE

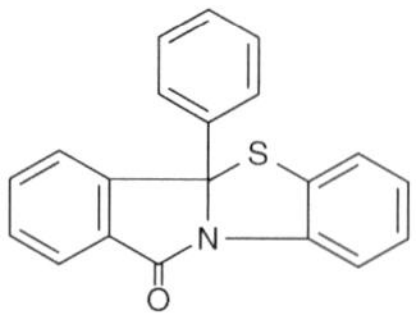

250 mg

S73,307-5
1,2,3,4,6,10B-HEXAHYDRO-10B-PHENYLPYRIMIDO(2,1-A)ISOINDOL-6-ONE

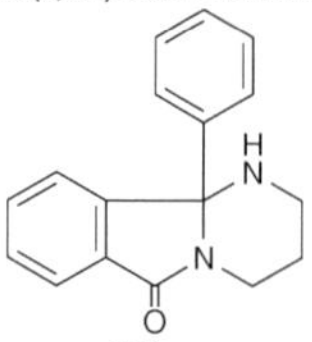

250 mg

S68,216-0
1,2,3,4,6,10B-HEXAHYDROPYRIMIDO-(2,1-A)ISOINDOL-6-ONE

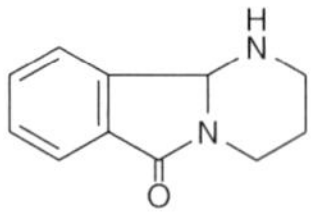

250 mg

S76,693-3
1,2,3,3A,4,10A-HEXAHYDRO-3-PHENYL-9A(H)PYRIDO(21B)PYRROLO(32E)(13)THIAZIN4ONE

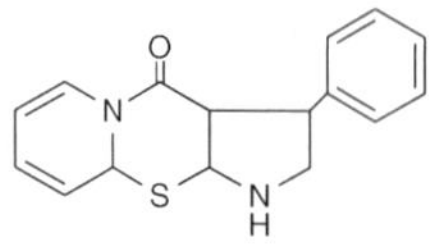

250 mg

S67,793-0
1',2'-DIHYDRO-2,2'-DIMETHYLSPIRO-[ISOINDOLINE-1,3'(4'H)ISOQUINOLINE]-1',3-DIONE

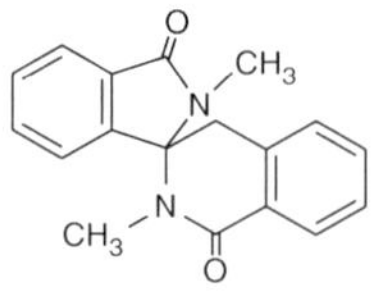

250 mg

S38,357-0
2,3-DIPHENYL-4-THIAZOLIDINONE

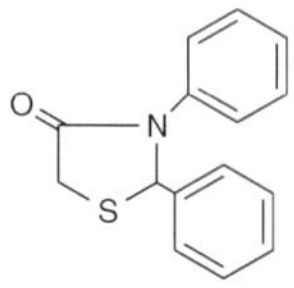

5 gm

S74,544-8
HEXAHYDRO-SPIRO(3-CYCLOHEXENE-1,1'-[2'H-INDOLO(2,3-A)QUINOLIZIN])-4'-ONE

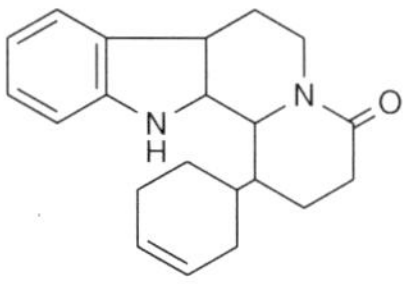

250 mg

S74,040-3
1-ETHYL-1,2,3,4,5,6,7,12B-OCTAHYDROINDOLO(2,3-A)-QUINOLIZIN-4-ONE

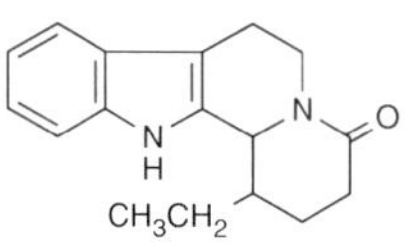

250 mg

S79,534-8
1-(2,2-DIPHENYLACETYL)-4-METHYLPIPERIDINE

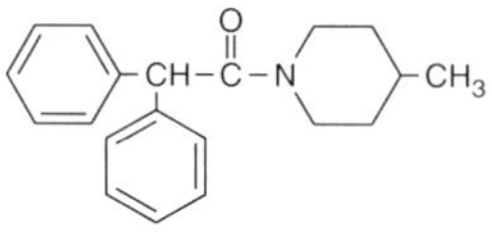

250 mg

S35,338-8
3,4-DIHYDRO-8-METHYL-3-PHENYL-2H-PYRIDO(1,2-A)PYRIMIDINE-2,4-DIONE

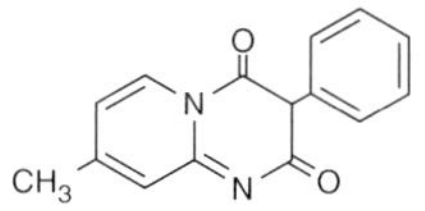

1 gm

S35,474-0
3-PHENYL-4(3H)-QUINAZOLINONE

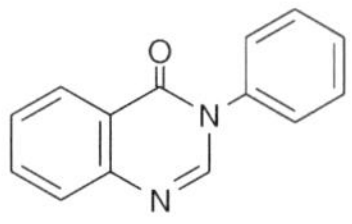

250 mg

S50,443-2
3-(3-CHLORO-O-TOLYL)-2-METHYL-4(3H)-QUINAZOLONE

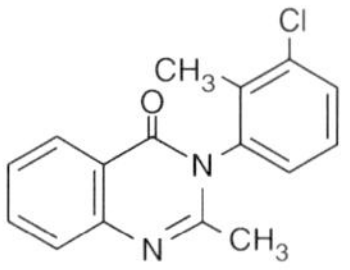

5 gm

S76,042-0
6-CHLORO-6,7,8,9,10,12-HEXAHYDROAZEPINO(2,1-B)-QUINAZOLIN-12-ONE

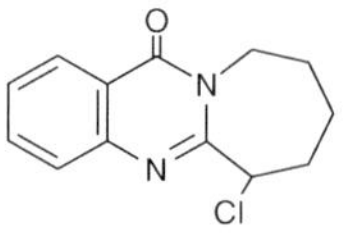

250 mg

S95,514-0
3,6-DIMETHYLENE-1-METHYL-4-PHENYL-2,5-PIPERAZINEDIONE

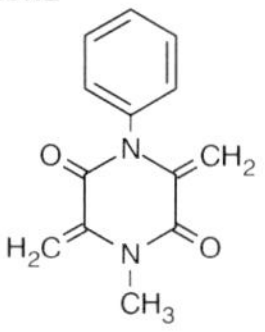

50 mg

S50,224-3
4-(2-CHLOROBENZOYL)-MORPHOLINE

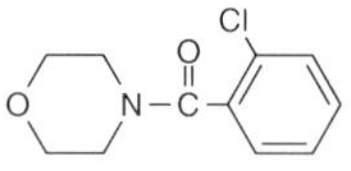

5 gm

S83,297-9
10-ACETYLPHENOXAZINE

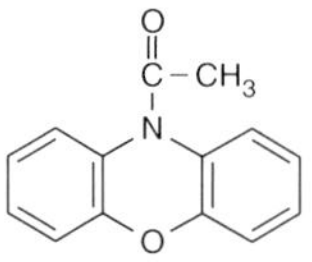

250 mg

S45,571-7
10-ACETYLPHENOTHIAZINE

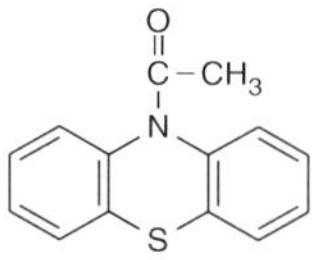

5 gm

S82,150-0
10-OCTADECANOYLPHENOTHIAZINE

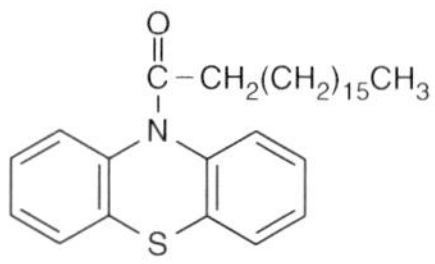

250 mg

S41,046-9
1-ACETYL-5-IODOINDOLINE

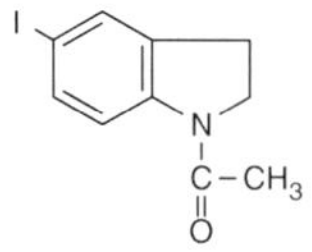

5 gm

S43,460-4
1,2,3,4-TETRAHYDROQUINOLINE-1-CARBOXALDEHYDE

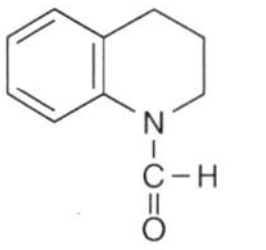

5 gm

S69,568-8
1,1'-(2-BROMO-6-METHYLPHENYLENE)-BIS(3-ANTIPYRINYLUREA) MIXTURE OF ISOMERS

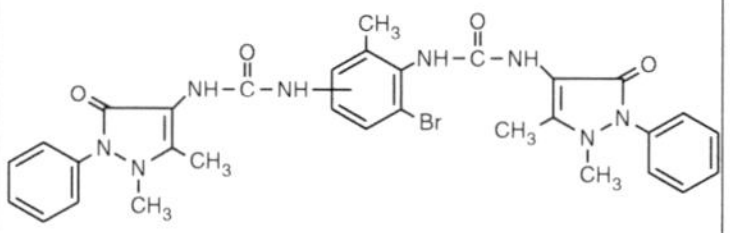

5 gm

S67,249-1
1-PHENYL-3-THIOUREIDO-2-PYRAZOLIN-5-ONE

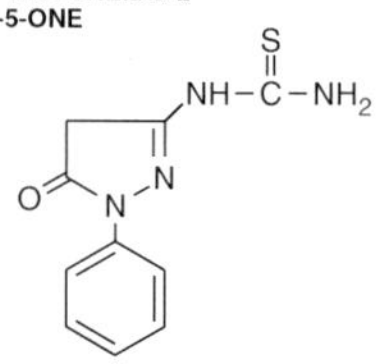

5 gm

S38,277-9
5-OXO-1-PHENYL-2-PYRAZOLINE-3-CARBOXYLIC ACID HYDRAZIDE

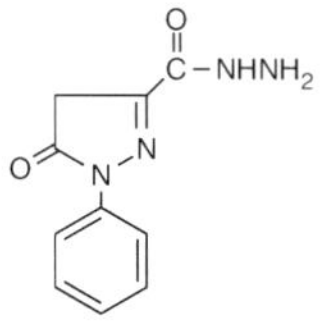

250 mg

S58,283-2
5-METHYL-1-PHENYL-3-PYRAZOLIDINONE

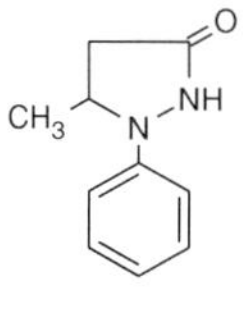

5 gm

S39,023-2
1-(3-CHLOROPHENYL)-3-METHYL-2-PYRAZOLIN-5-ONE

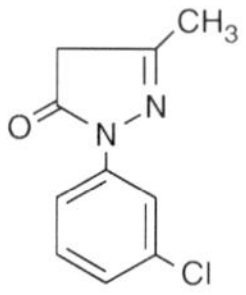

5 gm

S39,060-7
4-CHLORO-3-METHYL-1-PHENYL-2-PYRAZOLIN-5-ONE

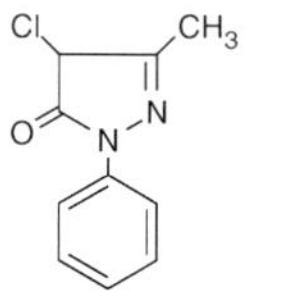

250 mg

S34,247-5
4-BROMOANTIPYRINE

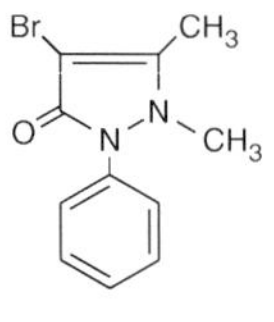

5 gm

S92,323-0
4-HEXANOYL-3-METHYL-1-PHENYL-2-PYRAZOLIN-2-ONE

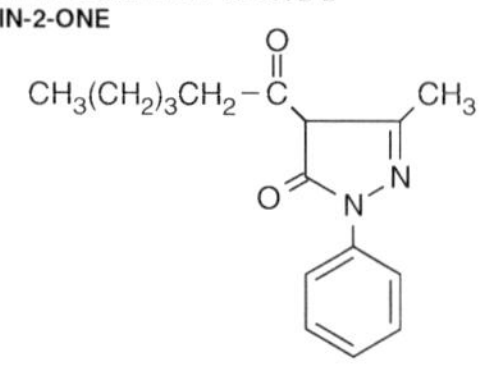

250 mg

S73,250-8
5,5-DIMETHYL-2-PHENYL-3-PYRAZOLIDINONE

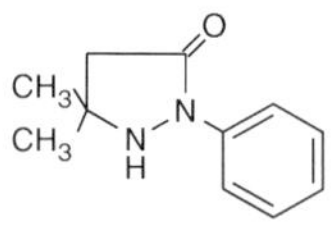

250 mg

S92,285-4
4-BUTYRYL-3-METHYL-1-PHENYL-2-PYRAZOLIN-5-ONE

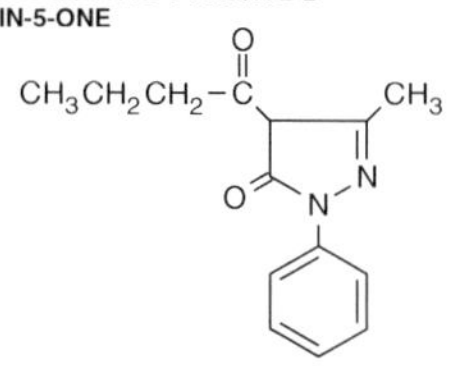

250 mg

S92,297-8
3-METHYL-1-PHENYL-4-VALEROYL-2-PYRAZOLIN-5-ONE

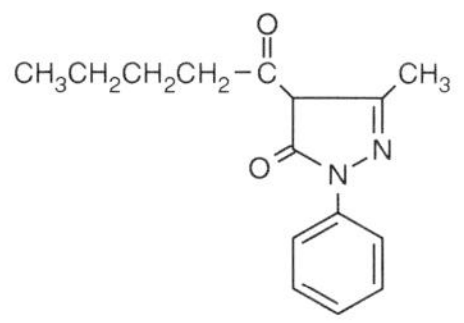

250 mg

S95,257-5
4-BENZYLIDENE-3-METHYL-1-PHENYL-2-PYRAZOLIN-5-ONE

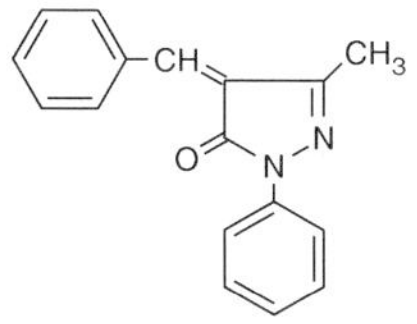

50 mg

S76,756-6
4A,5-DIHYDRO-2H-INDENO(1,2-C)-PYRIDAZIN-3(4H)-ONE

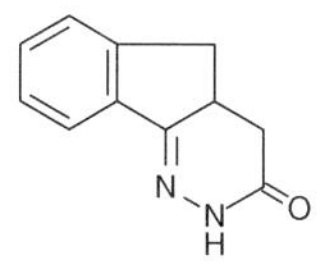

250 mg

S73,262-1
2-METHYL-3,4,4A,5-TETRAHYDROINDENO(1,2-C)PYRIDAZIN-3(2H)-ONE

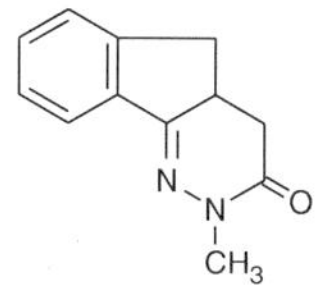

250 mg

S75,471-4
3-AMINO-2,6-DIMETHYL-4(3H)-QUINAZOLINONE

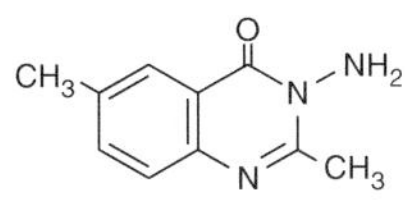

250 mg

S54,564-3
3-(3,5-XYLYL)-1,2,3-BENZOTRIAZIN-4-(3H)-ONE

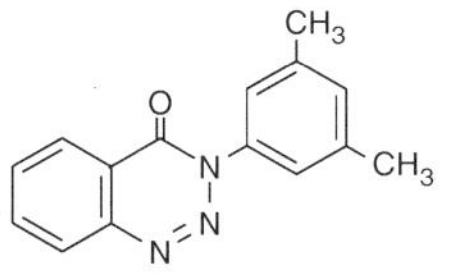

5 gm

S69,254-9
1-[N-(2,4-XYLYL)CARBAMOYL]-PYRROLIDINE

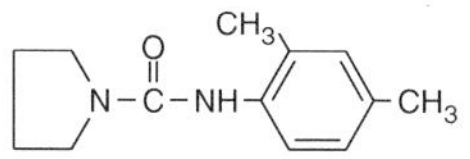

5 gm

S72,309-6
N-(1-NAPHTHYL)-1-PYRROLIDINECARBOXAMIDE

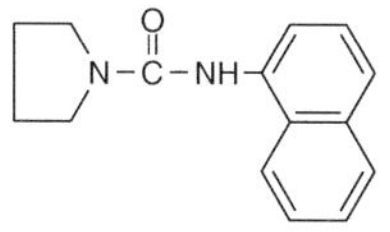

250 mg

S68,231-4
7,8-DIHYDRO-1,3-DIOXOLO(4,5-G)-QUINAZOLIN-6(5H)-ONE

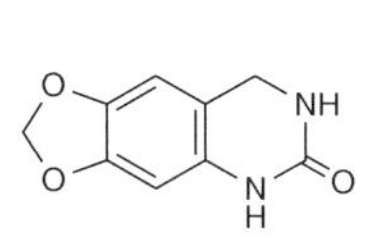

5 gm

S61,990-6
1-(3-CHLOROPHENYL)-2-IMIDAZOLIDINONE

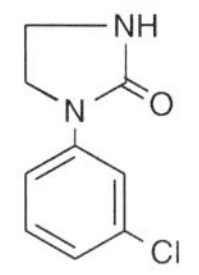

5 gm

S52,536-7
1-(META-TOLYL)-2-IMIDAZOLIDINONE

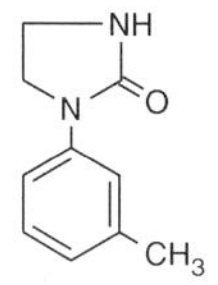

250 mg

S70,636-1
3A,6A-DIPHENYL-1,3,4,6-TETRACHLOROGLYCOLURIL

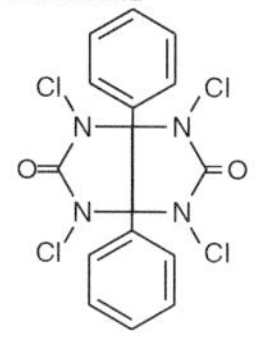

250 mg

S61,135-2
1-[N-(2,5-DICHLOROPHENYL)-CARBAMOYL]PIPERIDINE

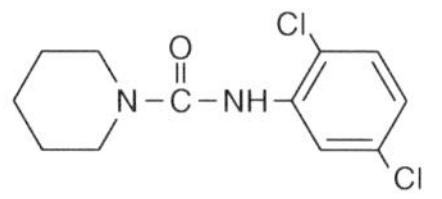

5 gm

S61,147-6
1-[N-(2,3-DICHLOROPHENYL)-CARBAMOYL]PIPERIDINE

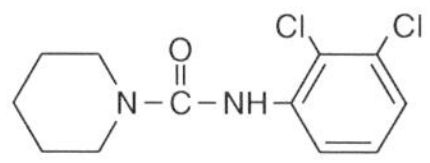

5 gm

S61,184-0
1-[N-(2,6-DICHLOROPHENYL)-CARBAMOYL]PIPERIDINE

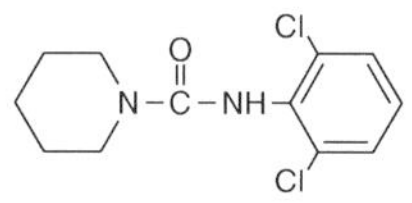

5 gm

S61,313-4
1-[N-(2,4-DICHLOROPHENYL)-CARBAMOYL]PIPERIDINE

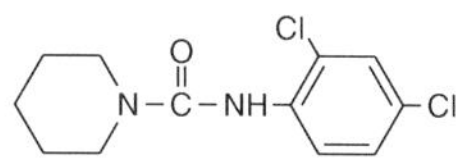

5 gm

S61,325-8
1-[N-(3,4-DICHLOROPHENYL)-CARBAMOYL]PIPERIDINE

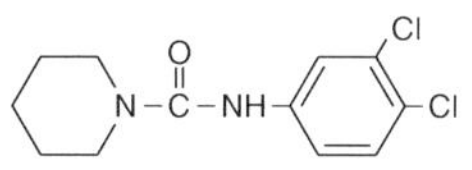

5 gm

S64,471-4
1-[N-(2,4,5-TRICHLOROPHENYL)-CARBAMOYL]PIPERIDINE

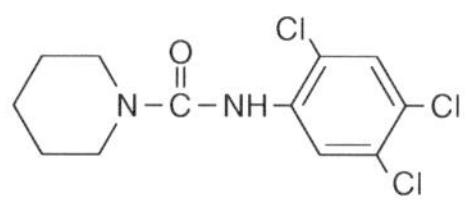

5 gm

S60,637-5
1-[N-(M-TOLYL)CARBAMOYL]PIPERIDINE

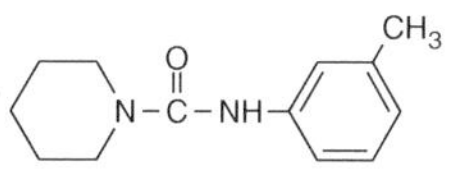

5 gm

S61,350-9
1-[N-(P-TOLYL)CARBAMOYL]PIPERIDINE

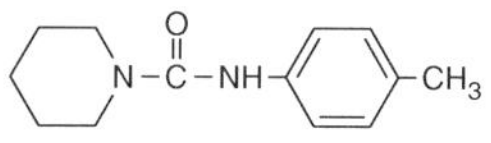

5 gm

S63,203-1
1-(N-PHENYLCARBAMOYL)-3-METHYLPIPERIDINE

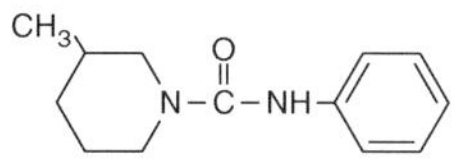

5 gm

S63,211-1
1-(N-PHENYLCARBAMOYL)-4-METHYLPIPERIDINE

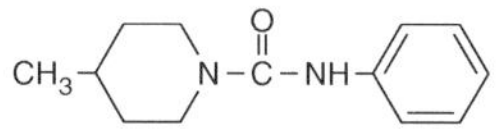

5 gm

S63,224-4
1-(N-PHENYLCARBAMOYL)-2-METHYLPIPERIDINE

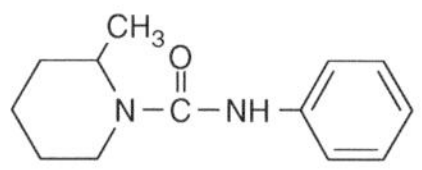

5 gm

S60,503-4
1-[N-(4-CHLORO-3-METHYLPHENYL)-CARBAMOYL]PIPERIDINE

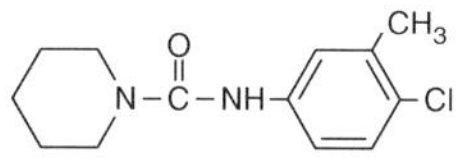

5 gm

S61,036-4
1-[N-(5-CHLORO-2-METHYLPHENYL)-CARBAMOYL]PIPERIDINE

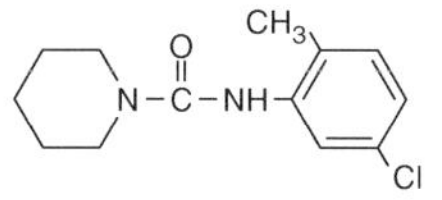

5 gm

S62,823-9
1-[N-(3-CHLORO-4-METHYLPHENYL)-CARBAMOYL]PIPERIDINE

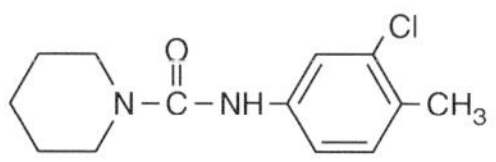

5 gm

S63,202-3
1-(4-CHLOROPHENYLCARBAMOYL)-2-METHYLPIPERIDINE

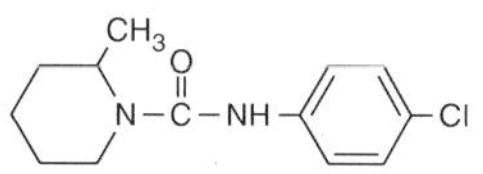

5 gm

S63,217-1
1-(4-CHLOROPHENYLCARBAMOYL)-3-METHYLPIPERIDINE

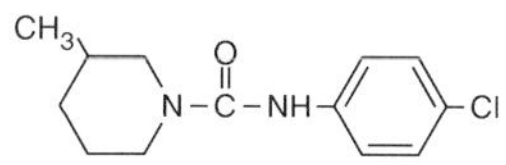

5 gm

S63,220-1
1-(4-CHLOROPHENYLCARBAMOYL)-4-METHYLPIPERIDINE

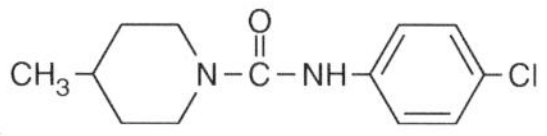

5 gm

S63,481-6
1-(2-CHLOROPHENYLCARBAMOYL)-2-METHYLPIPERIDINE

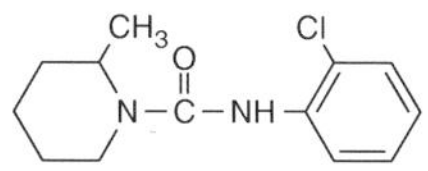

5 gm

S63,482-4
1-(2-CHLOROPHENYLCARBAMOYL)-4-METHYLPIPERIDINE

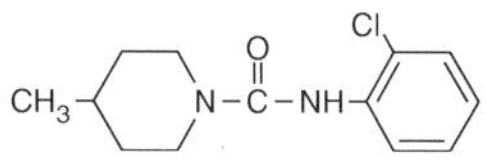

5 gm

S63,483-2
1-(2-CHLOROPHENYLCARBAMOYL)-3-METHYLPIPERIDINE

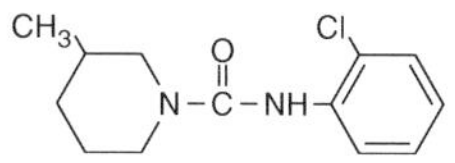

5 gm

S63,484-0
1-(3-CHLOROPHENYLCARBAMOYL)-3-METHYLPIPERIDINE
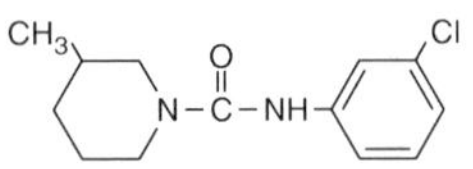
5 gm

S63,485-9
1-(3-CHLOROPHENYLCARBAMOYL)-2-METHYLPIPERIDINE
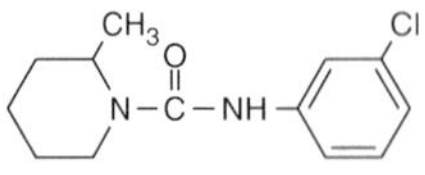
5 gm

S63,492-1
1-(3-CHLOROPHENYLCARBAMOYL)-4-METHYLPIPERIDINE
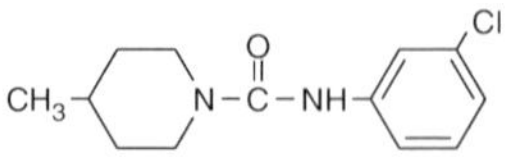
5 gm

S63,931-1
1-[N-(3,4-DICHLOROPHENYL)CARBAMOYL]-4-METHYLPIPERIDINE
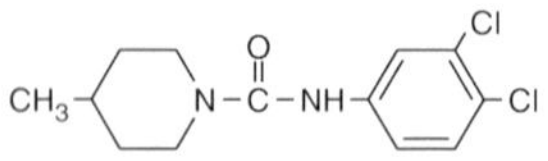
5 gm

S63,934-6
1-[N-(3,4-DICHLOROPHENYL)CARBAMOYL]-2-METHYLPIPERIDINE
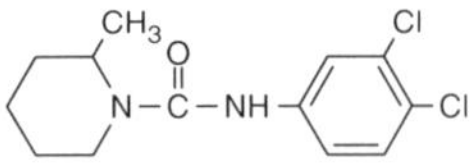
5 gm

S63,943-5
1-[N-(3,4-DICHLOROPHENYL)CARBAMOYL]-3-METHYLPIPERIDINE
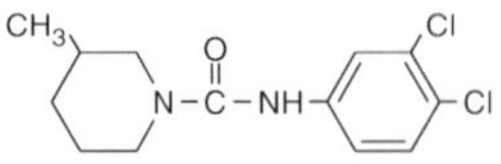
5 gm

S63,664-9
1-[N-(2,3-XYLYL)CARBAMOYL]-PIPERIDINE
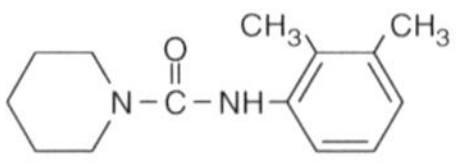
5 gm

S68,424-4
1-[N-(3-ETHYLPHENYL)CARBAMOYL]-PIPERIDINE
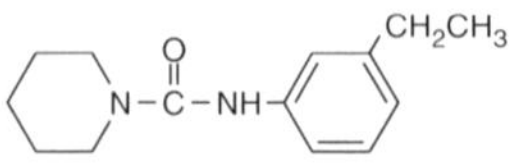
250 mg

S64,648-2
1-[N-(O-TOLYL)CARBAMOYL]PIPERIDINE
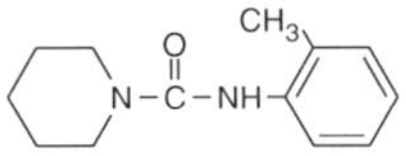
5 gm

S65,757-3
1-[N-(2-ETHYLPHENYL)CARBAMOYL]-PIPERIDINE
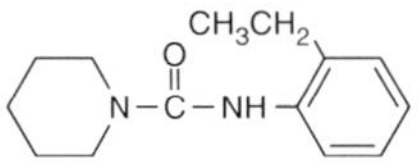
5 gm

S84,543-4
1-[N-(2-ETHYLPHENYL)CARBAMOYL]-2-METHYLPIPERIDINE
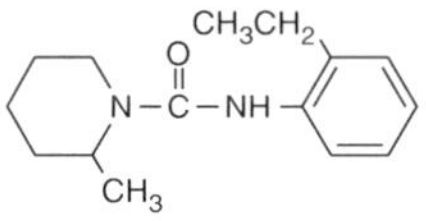
250 mg

S64,511-7
4,4'-METHYLENEBIS(1,1-PENTAMETHYLENE-3-PHENYLUREA)
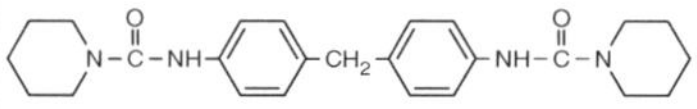
5 gm

S64,962-7
N,N'-(METHYLENEDI-4,1-PHENYLENE)-BIS(2-METHYL-1-PIPERIDINECARBOXAMIDE)
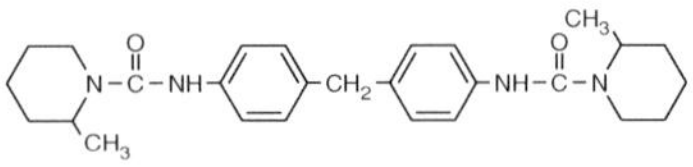
5 gm

S65,050-1
N,N'-(METHYLENEDI-4,1-PHENYLENE)-BIS(3-METHYL-1-PIPERIDINECARBOXAMIDE)
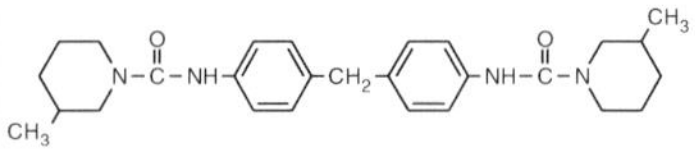
5 gm

S65,122-2
N,N'-(METHYLENEDI-4,1-PHENYLENE)-BIS(4-METHYL-1-PIPERIDINECARBOXAMIDE)
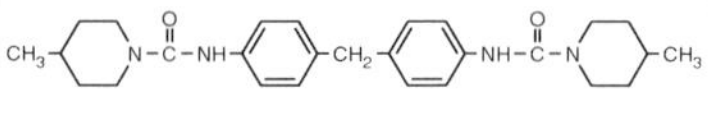
5 gm

S64,685-7
1-[N-(2-CHLORO-6-METHYLPHENYL)-CARBAMOYL]PIPERIDINE
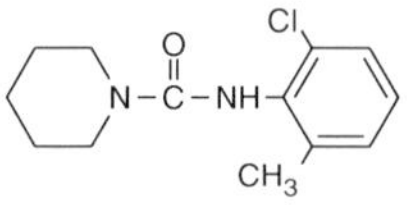
5 gm

S64,686-5
1-[N-(2,6-DICHLOROPHENYL)CARBAMOYL]-3-METHYLPIPERIDINE
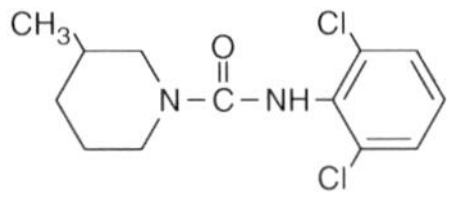
5 gm

S66,367-0
1-[N-(2,6-XYLYL)CARBAMOYL]-PIPERIDINE
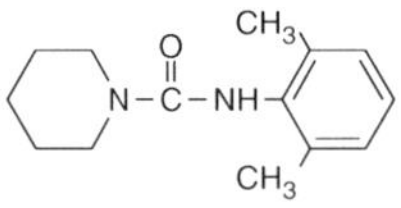
5 gm

S66,115-5
1-[N-(2-ETHYL-6-METHYLPHENYL)-CARBAMOYL]PIPERIDINE
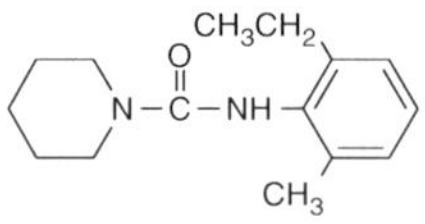
5 gm

S84,545-0
2-METHYL-1-[N-(2,6-XYLYL)-CARBAMOYL]PIPERIDINE
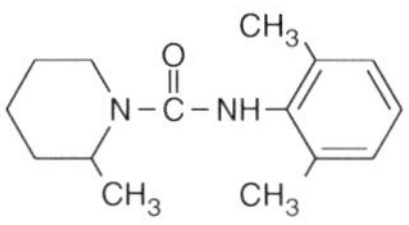
250 mg

S66,550-9
1-[N-(2,6-DIETHYLPHENYL)CARBAMOYL]-PIPERIDINE
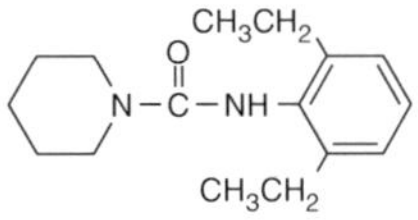
5 gm

S64,512-5
1-[N-(2,5-XYLYL)CARBAMOYL]-PIPERIDINE
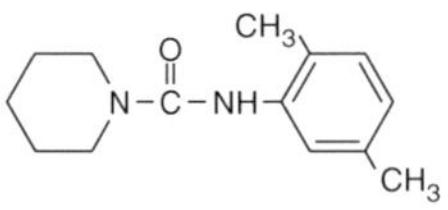
5 gm

S66,366-2
1-[N-(3,4-XYLYL)CARBAMOYL]-PIPERIDINE
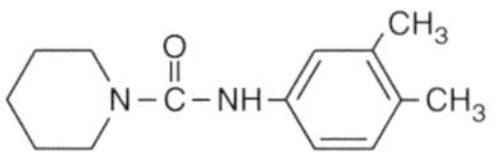
5 gm

S71,164-0
1-[N-(2,4-XYLYL)CARBAMOYL]-PIPERIDINE
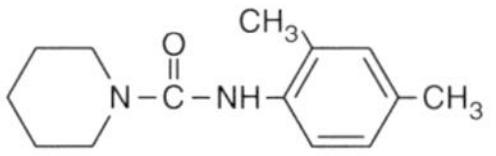
5 gm

S84,544-2
2-METHYL-1-[N-(2,5-XYLYL)-CARBAMOYL]PIPERIDINE
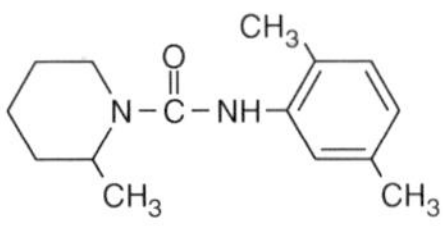
250 mg

S66,374-3
1-[N-(3,5-XYLYL)CARBAMOYL]-PIPERIDINE
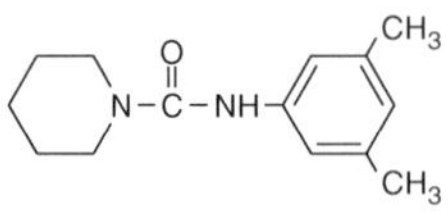
5 gm

S77,410-3
1-[N-(1-NAPHTHYL)CARBAMOYL]-PIPERIDINE
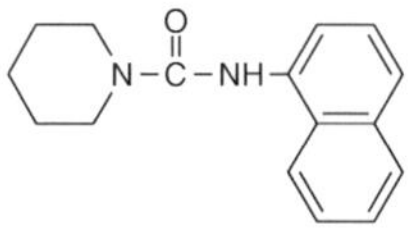
250 mg

S79,524-0
2-METHYL-N-(1-NAPHTHYL)-1-PIPERIDINECARBOXAMIDE
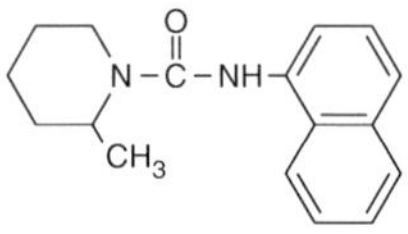
250 mg

S89,120-7
3-METHYL-N-(1-NAPHTHYL)-1-PIPERIDINECARBOXAMIDE
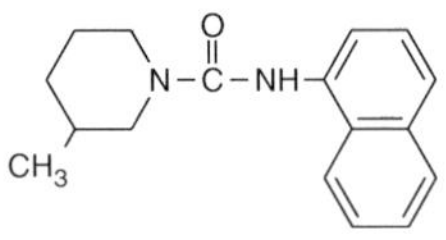
250 mg

S89,121-5
4-METHYL-N-(1-NAPHTHYL)-1-PIPERIDINECARBOXAMIDE
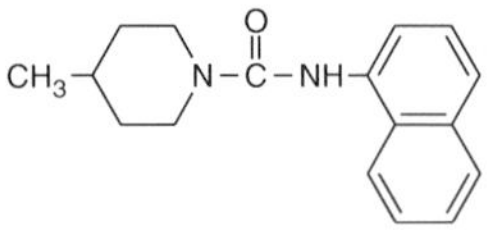
250 mg

S79,529-1
2,6-DIMETHYL-N-(1-NAPHTHYL)-1-PIPERIDINECARBOXAMIDE
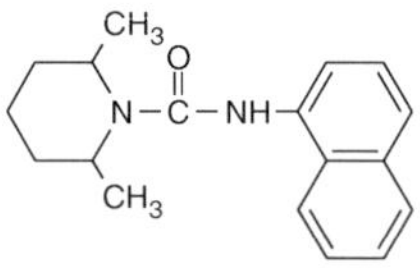
250 mg

S76,172-9
5,6-DIPHENYL-1-METHYL-2(1H)-PYRIMIDINONE
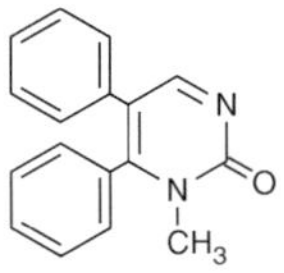
250 mg

S76,150-8
5,6-DIPHENYL-1-ETHYL-2(1H)-PYRIMIDINONE
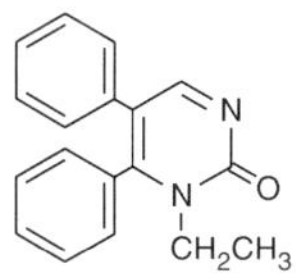
250 mg

S73,470-5
3,4-DIHYDRO-1-ISOPROPYL-7-METHYL-4-PHENYL-2(1H)-QUINAZOLINONE
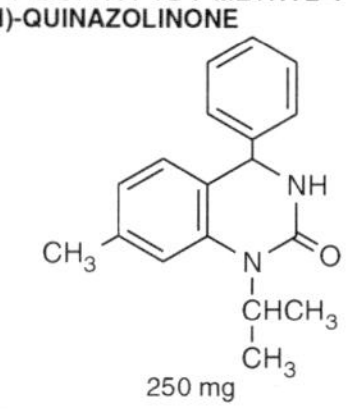
250 mg

S75,928-7
4,6-DI(P-TOLYL)-1-METHYL-2(1H)-PYRIMIDINONE
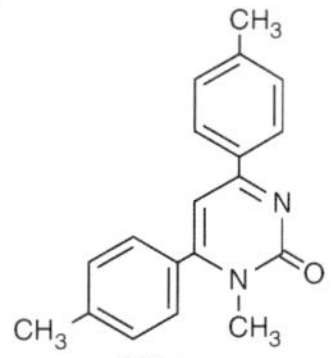
250 mg

S63,401-8
4-(2-CHLOROPHENYLCARBAMOYL)-MORPHOLINE
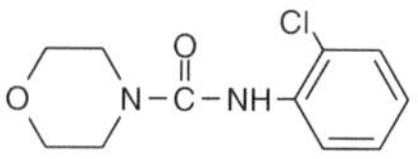
5 gm

S61,139-5
4-[N-(2,3-DICHLOROPHENYL)-CARBAMOYL]MORPHOLINE
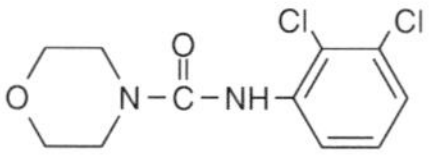
5 gm

S62,826-3
4-(4-BROMOPHENYLCARBAMOYL)-MORPHOLINE
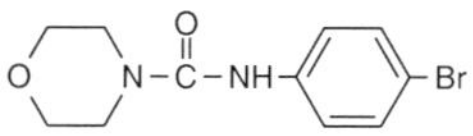
5 gm

S63,635-5
4-[N-(M-TOLYL)CARBAMOYL]MORPHOLINE
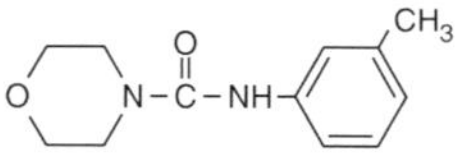
5 gm

S61,630-3
4-[N-(4-CHLORO-2-METHYLPHENYL)-CARBAMOYL]MORPHOLINE
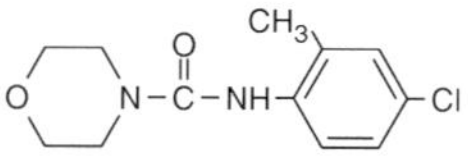
5 gm

S62,824-7
4-[N-(3-CHLORO-4-METHYLPHENYL)-CARBAMOYL]MORPHOLINE
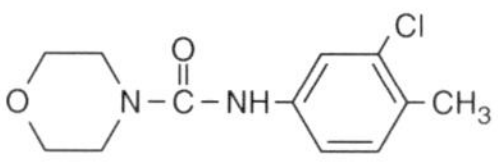
5 gm

S63,656-8
4-[N-(2,3-XYLYL)CARBAMOYL]-MORPHOLINE
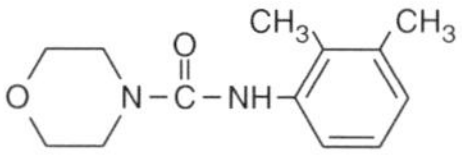
5 gm

S65,448-5
4-(N-PHENYLCARBAMOYL)-2,6-DIMETHYLMORPHOLINE
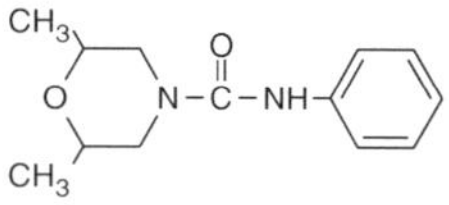
5 gm

S65,340-3
4-[N-(3-CHLOROPHENYL)CARBAMOYL]-2,6-DIMETHYLMORPHOLINE
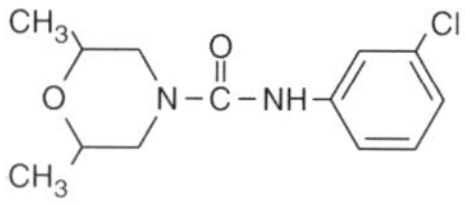
5 gm

S66,558-4
2,6-DIMETHYL-4-[N-(2,4,5-TRICHLOROPHENYL)CARBAMOYL]-MORPHOLINE
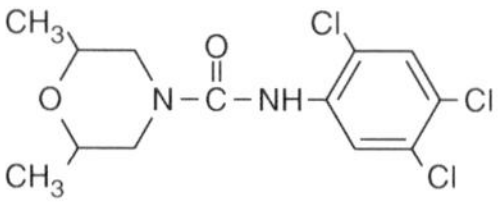
5 gm

S65,372-1
4-[N-(M-TOLYL)CARBAMOYL]-2,6-DIMETHYLMORPHOLINE
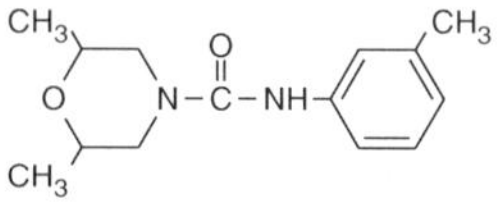
5 gm

S63,843-9
N,N'-HEXAMETHYLENEBIS(4-CARBAMOYLMORPHOLINE)
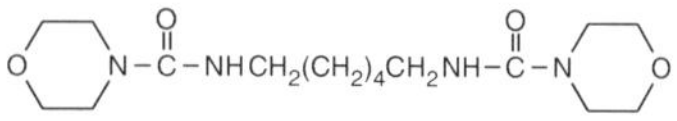
5 gm

S65,765-4
4-[N-(2-ETHYLPHENYL)CARBAMOYL]-MORPHOLINE
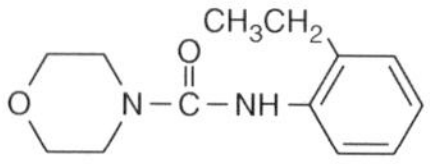
5 gm

S65,351-9
4-[N-(O-TOLYL)CARBAMOYL]-2,6-DIMETHYLMORPHOLINE
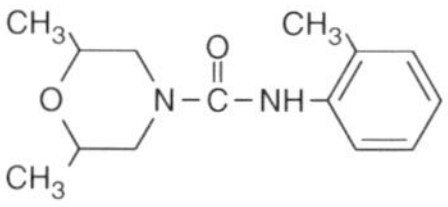
5 gm

S65,848-0
2,6-DIMETHYL-4-[N-(2-ETHYLPHENYL)-CARBAMOYL]MORPHOLINE
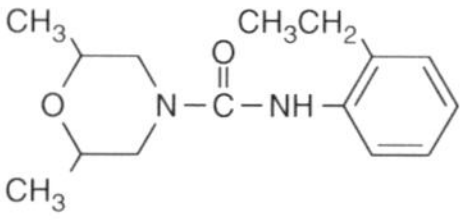
5 gm

S65,328-4
4-[N-(4-CHLOROPHENYL)CARBAMOYL]-2,6-DIMETHYLMORPHOLINE
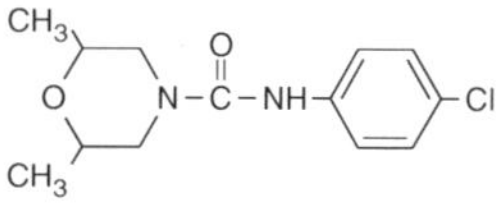
5 gm

S66,556-8
4-(4-BROMOPHENYLCARBAMOYL)-2,6-DIMETHYLMORPHOLINE
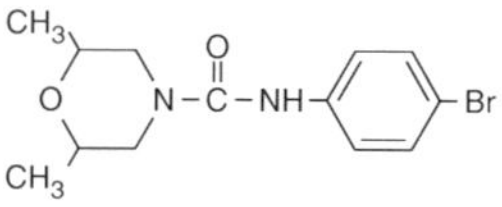
5 gm

S65,346-2
4-[N-(P-TOLYL)CARBAMOYL]-2,6-DIMETHYLMORPHOLINE
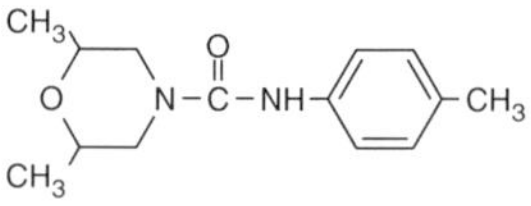
5 gm

S65,463-9
4,4'-METHYLENEBIS[N-(2,6-DIMETHYLMORPHOLINOCARBOXAMIDO)ANILINE)
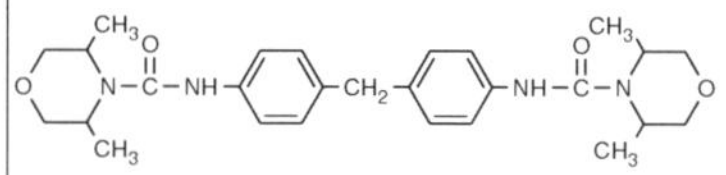
5 gm

S65,498-1
4-[N-(6-CHLORO-2-METHYLPHENYL)-CARBAMOYL]MORPHOLINE
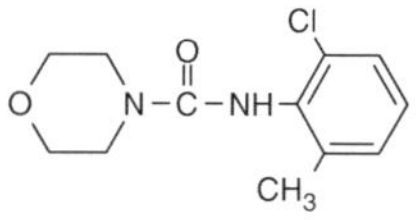
5 gm

S66,137-6
4-[N-(2,6-XYLYL)CARBAMOYL]-MORPHOLINE
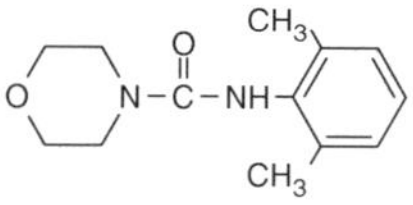
5 gm

S65,333-0
4-[N-(2,6-DICHLOROPHENYL)-CARBAMOYL]-2,6-DIMETHYLMORPHOLINE
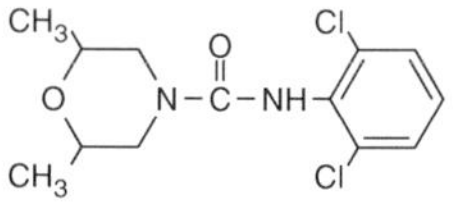
5 gm

S65,438-8
4-[N-(2,3-DICHLOROPHENYL)CARBAMOYL]-2,6-DIMETHYLMORPHOLINE
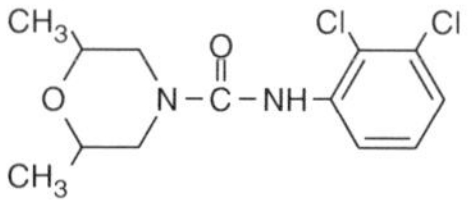
5 gm

S66,365-4
4-[N-(2-ETHYL-6-METHYLPHENYL)-CARBAMOYL]MORPHOLINE
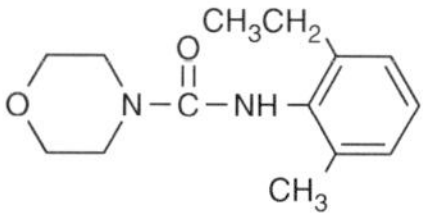
5 gm

S65,330-6
4-[N-(2-CHLORO-6-METHYLPHENYL)-CARBAMOYL]-2,6-DIMETHYLMORPHOLINE
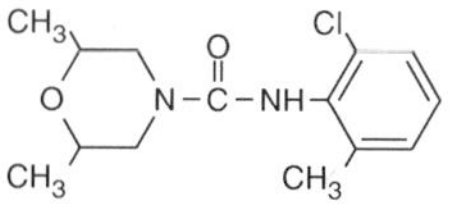
5 gm

S65,432-9
4-[N-(2,6-DIETHYLPHENYL)CARBAMOYL]-MORPHOLINE
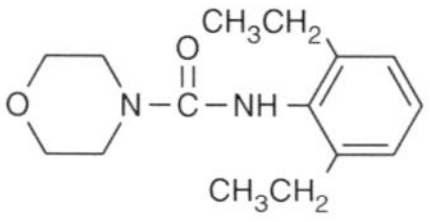
5 gm

S66,078-7
2,6-DIMETHYL-4-[N-(2,3-XYLYL)-CARBAMOYL]MORPHOLINE
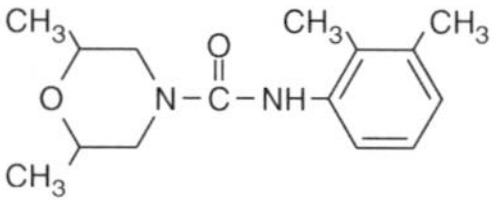
5 gm

S66,358-1
2,6-DIMETHYL-4-[N-(2,6-XYLYL)-CARBAMOYL]MORPHOLINE
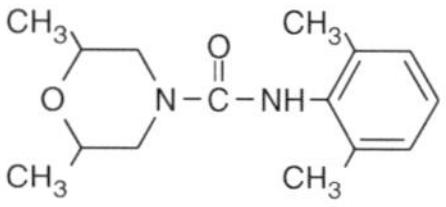
5 gm

S66,390-5
2,6-DIMETHYL-4-[N-(2-ETHYL-6-METHYLPHENYL)CARBAMOYL]MORPHOLINE
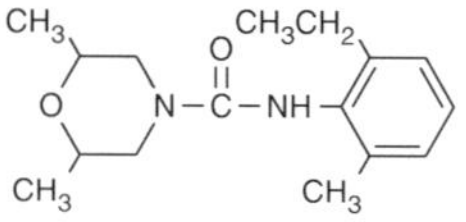
5 gm

S65,398-5
4-[N-(2,6-DIETHYLPHENYL)CARBAMOYL]-2,6-DIMETHYLMORPHOLINE

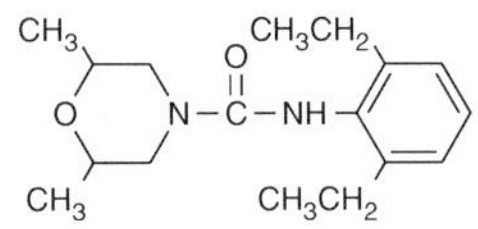

5 gm

S65,779-4
4-[N-(2,6-DIISOPROPYLPHENYL)-CARBAMOYL]MORPHOLINE

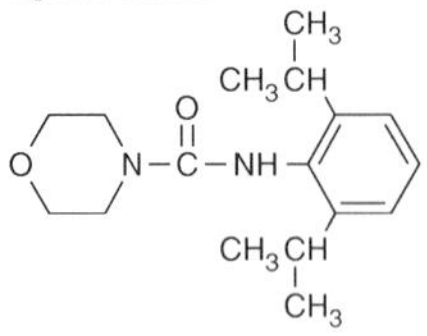

5 gm

S65,760-3
4-[N-(2,6-DIISOPROPYLPHENYL)-CARBAMOYL]-2,6-DIMETHYLMORPHOLINE

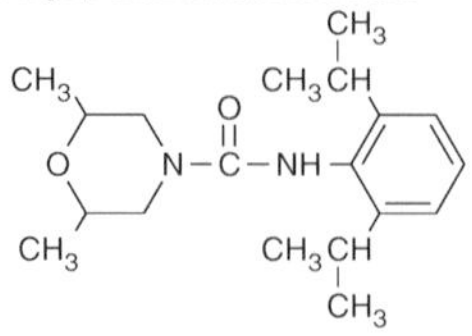

5 gm

S64,536-2
4-[N-(2,5-XYLYL)CARBAMOYL]-MORPHOLINE

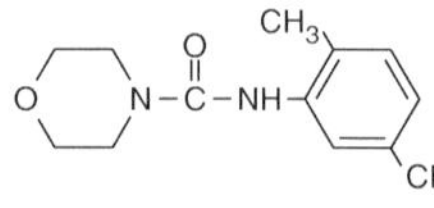

5 gm

S66,349-2
4-[N-(3,4-XYLYL)CARBAMOYL]-MORPHOLINE

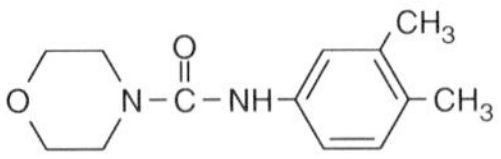

5 gm

S65,355-1
4-[N-(2,4-DICHLOROPHENYL)CARBAMOYL]-2,6-DIMETHYLMORPHOLINE

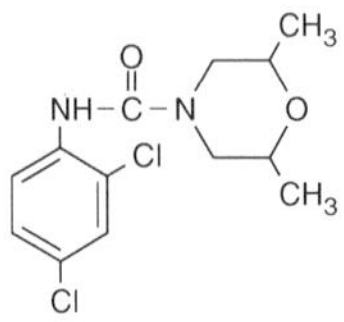

5 gm

S66,528-2
2,6-DIMETHYL-4-[N-(3,4-DICHLOROPHENYL)CARBAMOYL]MORPHOLINE

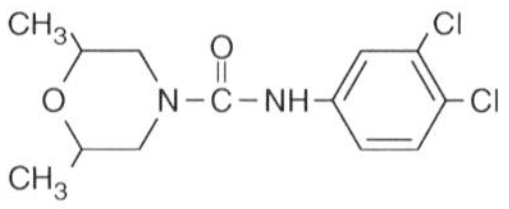

5 gm

S66,573-8
2,6-DIMETHYL-4-[N-(2,5-DICHLOROPHENYL)CARBAMOYL]MORPHOLINE

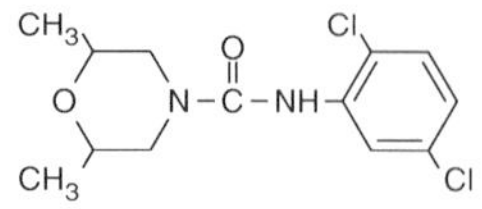

5 gm

S65,421-3
4-[N-(4-CHLORO-2-METHYLPHENYL)-CARBAMOYL]-2,6-DIMETHYLMORPHOLINE

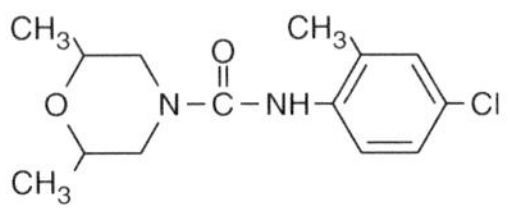

5 gm

S66,058-2
4-[N-(5-CHLORO-2-METHYLPHENYL)-CARBAMOYL]-2,6-DIMETHYLMORPHOLINE

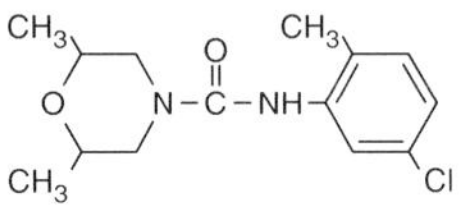

5 gm

S66,549-5
4-[N-(3-CHLORO-4-METHYLPHENYL)-CARBAMOYL]-2,6-DIMETHYLMORPHOLINE

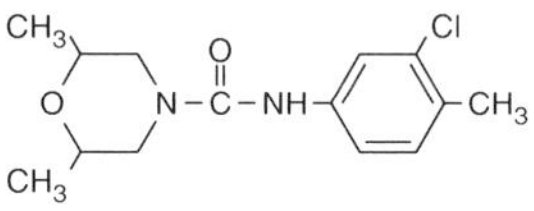

5 gm

S65,357-8
2,6-DIMETHYL-4-[N-(2,4-XYLYL)-CARBAMOYL]MORPHOLINE

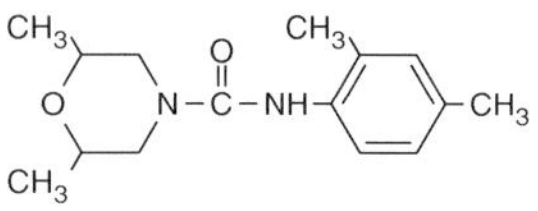

5 gm

S66,105-8
2,6-DIMETHYL-4-[N-(3,4-XYLYL)-CARBAMOYL]MORPHOLINE

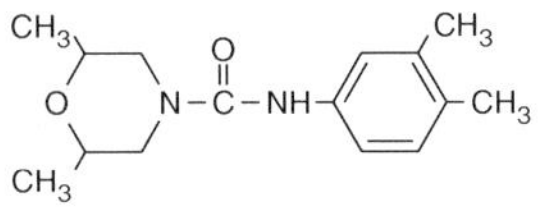

5 gm

S66,603-3
2,6-DIMETHYL-4-[N-(2,5-XYLYL)-CARBAMOYL]MORPHOLINE

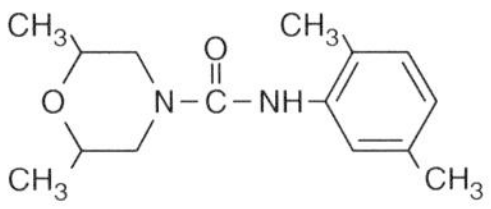

5 gm

S66,123-6
4-[N-(3,5-XYLYL)CARBAMOYL]-MORPHOLINE

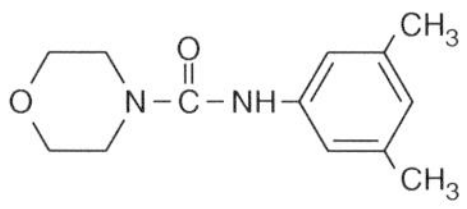

5 gm

S66,361-1
2,6-DIMETHYL-4-[N-(3,5-XYLYL)-CARBAMOYL]MORPHOLINE

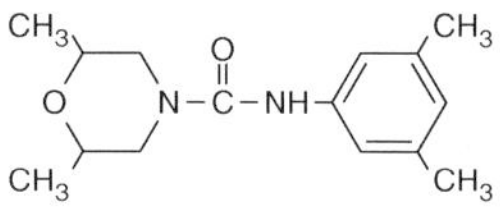

5 gm

S71,038-5
4-[N-(2,4,5-TRICHLOROPHENYL)-CARBAMOYL]MORPHOLINE

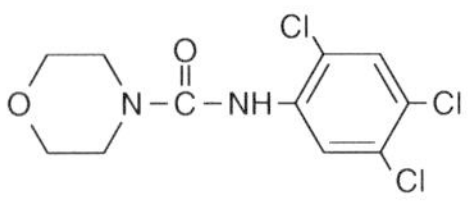

5 gm

S65,699-2
1-CARBAMOYL-5-CHLOROINDOLINE

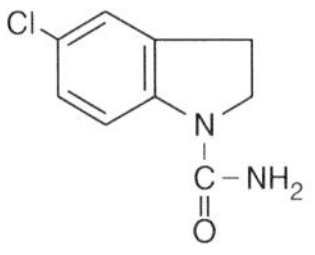

5 gm

S68,236-5
1-CARBAMOYL-1,2,3,4-TETRAHYDROQUINOLINE

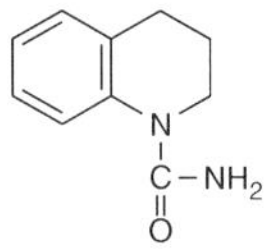

5 gm

S41,305-4
N-(3-CHLORO-2-HYDROXYPROPYL)-PHTHALIMIDE

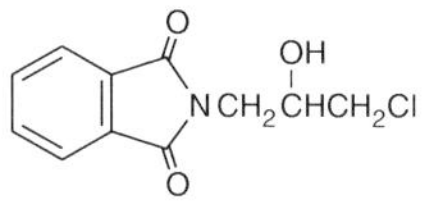

5 gm

S95,135-8
N-[2-(BROMOMETHYL)PHENYL]-SUCCINIMIDE

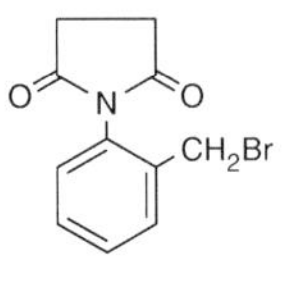

50 mg

S87,235-0
N-(2-CHLOROETHYL)PHTHALIMIDE

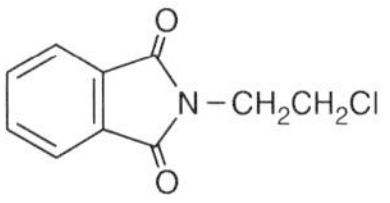

250 mg

S52,540-5
N-(3-CHLOROPROPOXY)PHTHALIMIDE

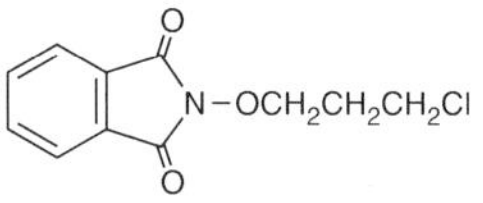

250 mg

S85,928-1
N-(2-BROMOETHOXY)PHTHALIMIDE

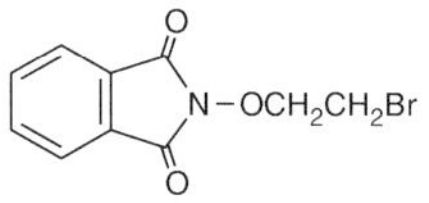

250 mg

S38,082-2
N-(3-CHLOROPROPYL)PHTHALIMIDE

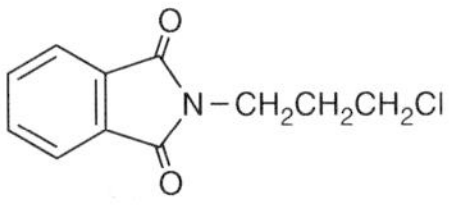

250 mg

S85,921-4
N-(3-BROMOPROPOXY)PHTHALIMIDE

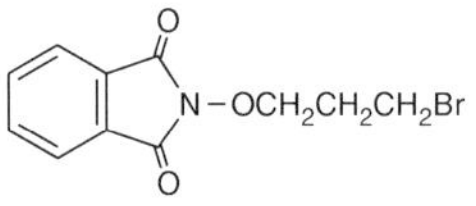

250 mg

S83,978-7
N-(5-BROMOPENTYL)PHTHALIMIDE

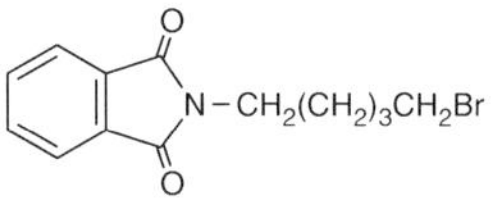

250 mg

S70,717-1
2-(6-BROMOHEXYL)PHTHALIMIDE

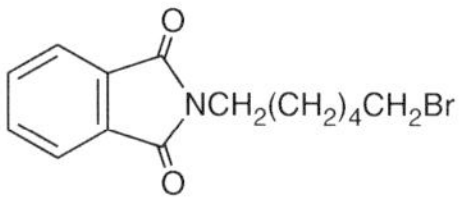

5 gm

S87,827-8
N-[2-CHLORO-5-(TRIFLUOROMETHYL)-PHENYL]GLUTARIMIDE

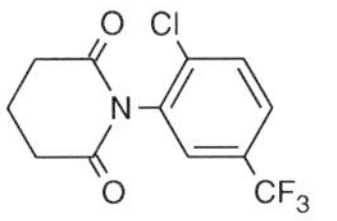

5 gm

S85,909-5
N-[2,6-DINITRO-4-(TRIFLUOROMETHYL)-PHENYL]SUCCINIMIDE

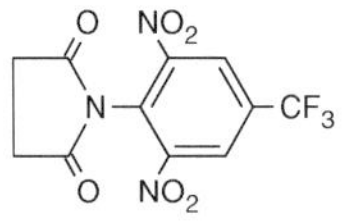

250 mg

S85,908-7
N-[2,6-DINITRO-4-(TRIFLUOROMETHYL)-PHENYL]PHTHALIMIDE

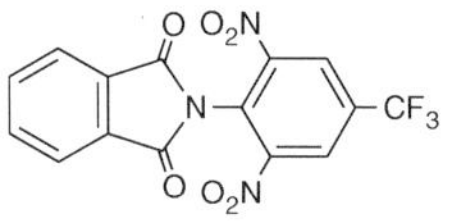

250 mg

S74,640-1
N-(3-TRIFLUOROMETHYLPHENYL)-SUCCINIMIDE

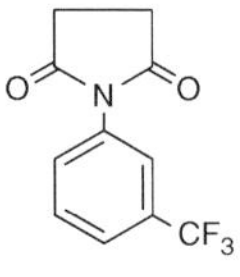

5 gm

S69,336-7
PHTHALIMIDO N-[3-(TRIFLUOROMETHYL)-PHENYL]CARBAMATE

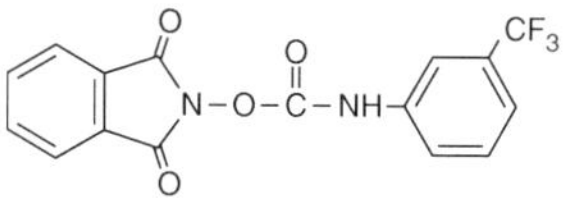

250 mg

S94,991-4
N-(TRICHLOROMETHYLTHIO)PHTHALIMIDE

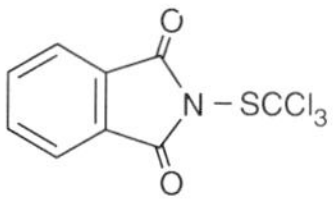

1 gm

S68,499-6
PHTHALIMIDO N-(5-CHLORO-2-METHOXYPHENYL)CARBAMATE

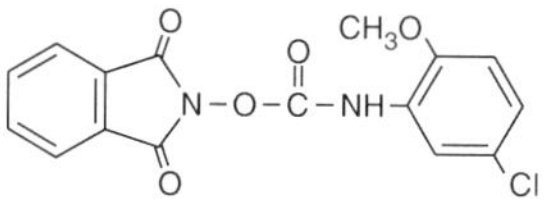

5 gm

S89,271-8
N-(5-CHLORO-2-METHOXYPHENYL)-1,4,5,6,7,7-HEXACHLORO-5-NORBORNENE-2,3-DICARBOXIMIDE

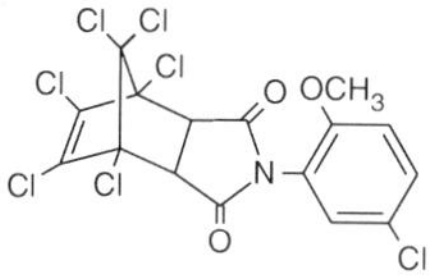

250 mg

S87,847-2
N-(4-CHLOROPHENYL)SUCCINIMIDE

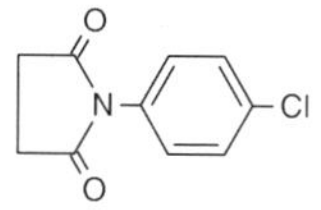

5 gm

S54,747-6
2-(2-AMINOPHENYLTHIO)-N-(3,4-DICHLOROPHENYL)SUCCINIMIDE

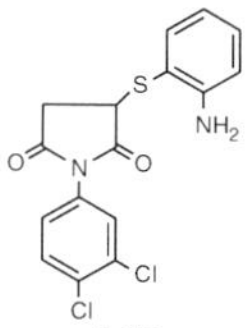

5 gm

S54,751-4
2-(2-AMINOPHENYLTHIO)-N-PHENYLSUCCINIMIDE

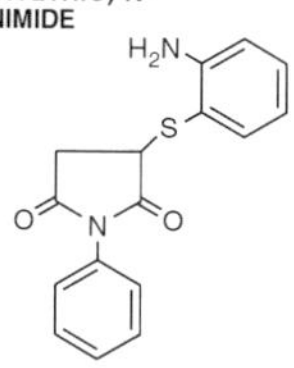

5 gm

S54,266-0
2-(2-AMINOPHENYLTHIO)-N-(4-IODOPHENYL)-SUCCINIMIDE

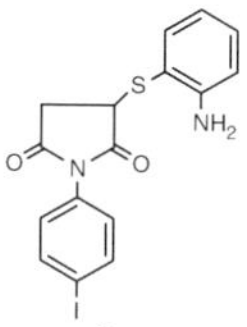

5 gm

S54,754-9
2-(2-AMINOPHENYLTHIO)-N-(P-TOLYL)-SUCCINIMIDE

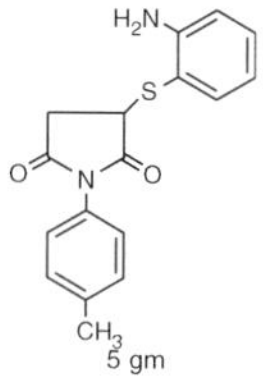

5 gm

S54,741-7
2-(2-AMINOPHENYLTHIO)-N-(3-CHLORO-O-TOLYL)SUCCINIMIDE

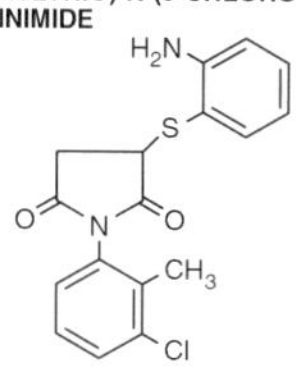

250 mg

S92,404-0
1,4,5,6,7,7-HEXACHLORO-N-(4-METHYLTHIOPHENYL)-5-NORBORNENE-2,3-DICARBOXIMIDE

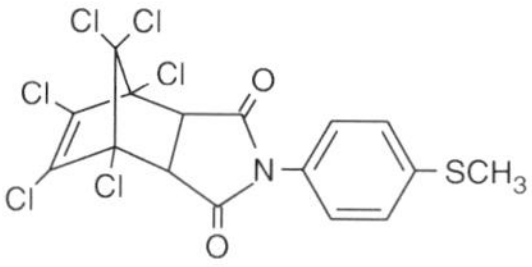

250 mg

S71,941-2
1-(4-BROMOPHENYL)-3-(P-TOLYLTHIO)-SUCCINIMIDE

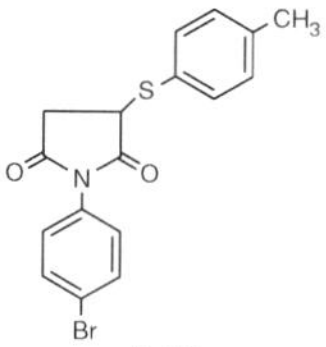

5 gm

S71,940-4
1-(4-BROMOPHENYL)-3-(PHENETHYLAMINO)SUCCINIMIDE

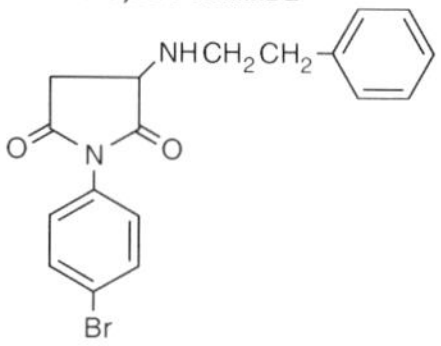

5 gm

S71,939-0
1-(4-BROMOPHENYL)-3-(CYCLOHEXYLAMINO)SUCCINIMIDE

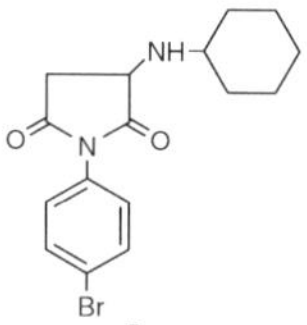

5 gm

S71,938-2
N-(4-BROMOPHENYL)-2-PIPERIDINOSUCCINIMIDE

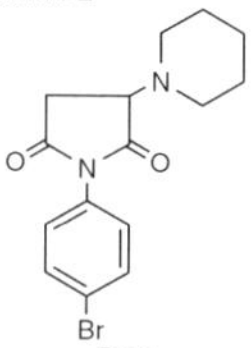

5 gm

S72,907-8
2-(2-CHLORO-5-NITROPHENYL)HEXACHLORO-1,3,3A,4,7,7A-HEXAHYDRO-4,7-METHANOISOINDOLE-1,3-DIONE

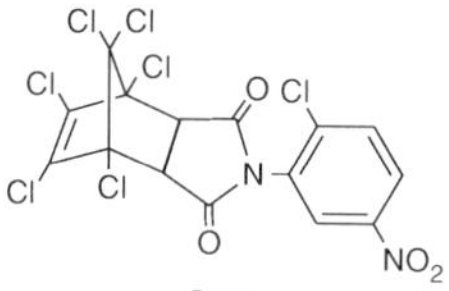

5 gm

S87,850-2
N-(4-CHLOROPHENYL)GLUTARIMIDE

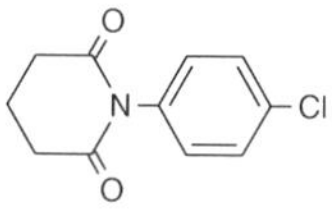

5 gm

S87,824-3
N-(4-IODOPHENYL)GLUTARIMIDE

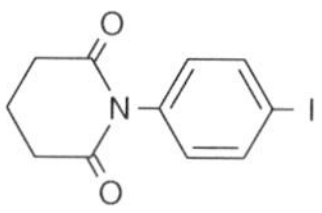

5 gm

S87,829-4
N-(3-CHLORO-4-METHYLPHENYL)-GLUTARIMIDE

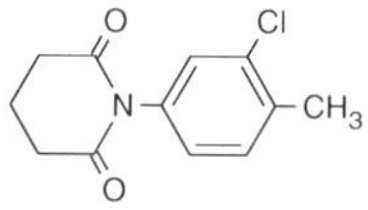

5 gm

S77,106-6
2-(2-BROMOBENZYL)PHTHALIMIDE

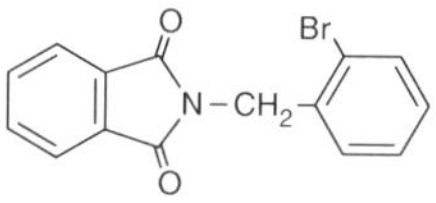

250 mg

S90,538-0
N-(2-CHLORO-5-NITROPHENYL)-1,4,5,6,7,7-HEXACHLORO-5-NORBORNENE-2,3-DICARBOXIMIDE

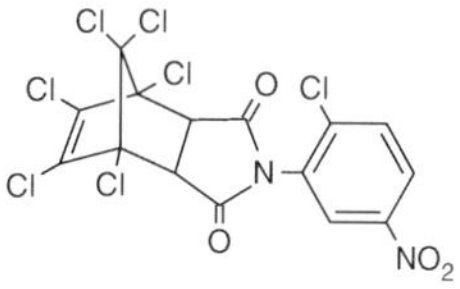

250 mg

S85,556-1
2,3-DIBROMO-N-(2-FLUOROPHENYL)-SUCCINIMIDE

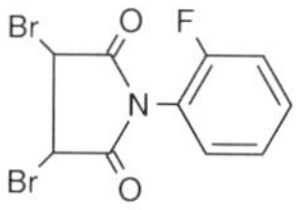

250 mg

S74,248-1
4-CHLORO-2-PHTHALIMIDOBENZOIC ACID

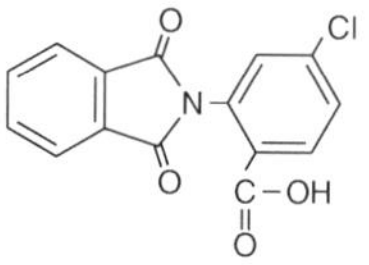

5 gm

S76,839-1
1,3-DIOXOISOINDOLIN-2-YL N-(3-CHLOROPHENYL)CARBAMATE

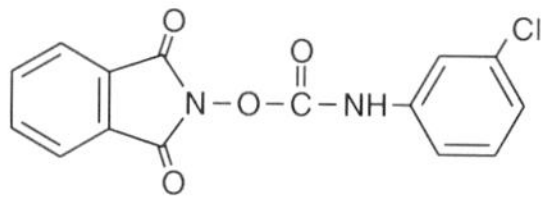

250 mg

S72,373-8
PHTHALIMIDO N-(2,4-DICHLOROPHENYL)-CARBAMATE

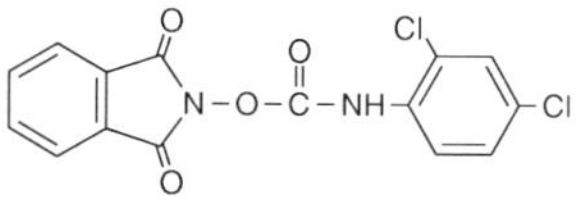

5 gm

S79,512-7
N-(4-BROMOPHENYL)-1,4,5,6,7,7-HEXACHLORO-5-NORBORNENE-2,3-DICARBOXIMIDE

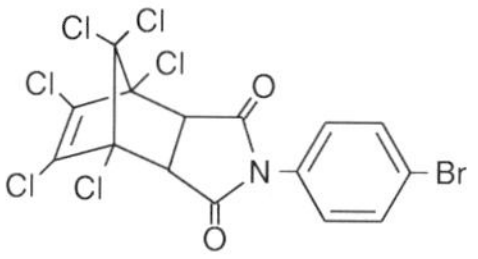

250 mg

S90,543-7
N-(2,3-DICHLOROPHENYL)-1,4,5,6,7,7-HEXACHLORO-5-NORBORNENE-2,3-DICARBOXIMIDE

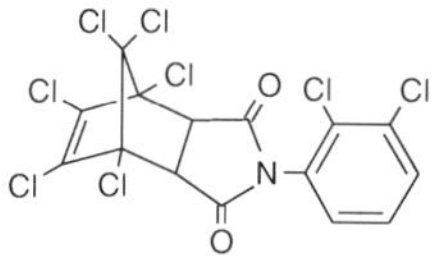

250 mg

S79,511-9
N-(3,4-DICHLOROPHENYL)-1,4,5,6,7,7-HEXACHLORO-5-NORBORNENE-2,3-DICARBOXIMIDE

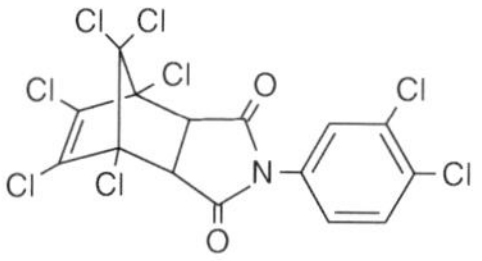

250 mg

S90,542-9
N-(3-CHLORO-4-METHYLPHENYL)-1,4,5,6,7,7-HEXACHLORO-5-NORBORNENE-2,3DICARBOXIMIDE

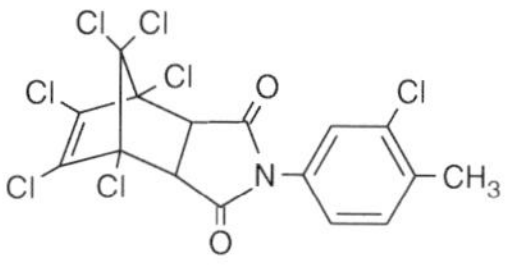

250 mg

S74,645-2
N-(2-FLUOROPHENYL)MALEIMIDE

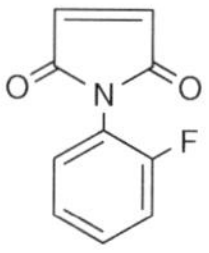

5 gm

S76,201-6
2-(4-CHLOROPHENYL)SUCCINIMIDE

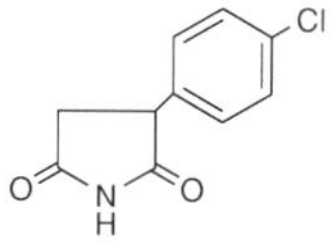

250 mg

S71,944-7
N-(4-BROMOPHENYL)MALEIMIDE

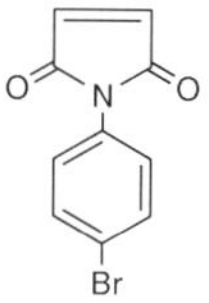

5 gm

S74,643-6
N-(4-BROMOPHENYL)SUCCINIMIDE

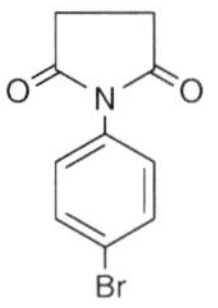

5 gm

S74,642-8
N-(2,3-DICHLOROPHENYL)SUCCINIMIDE

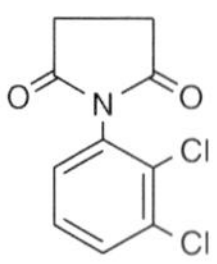

5 gm

S74,646-0
N-(3-CHLORO-2-METHYLPHENYL)-SUCCINIMIDE

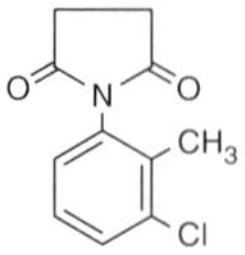

5 gm

S73,223-0
N-(2,4-DIFLUOROPHENYL)MALEIMIDE

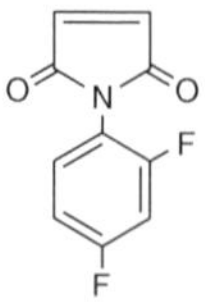

250 mg

S74,647-9
N-(3-CHLORO-2-METHYLPHENYL)-GLUTARIMIDE

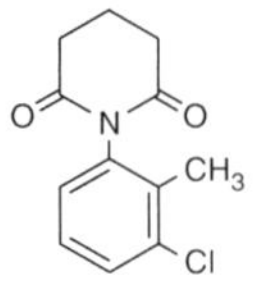

5 gm

S74,639-8
N-(4-IODOPHENYL)SUCCINIMIDE

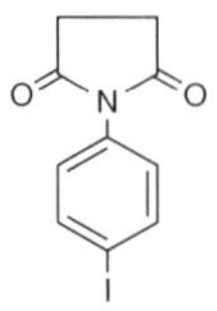

5 gm

S84,622-8
N-[2-(4-CHLOROPHENYL)ETHYL]-PHTHALIMIDE

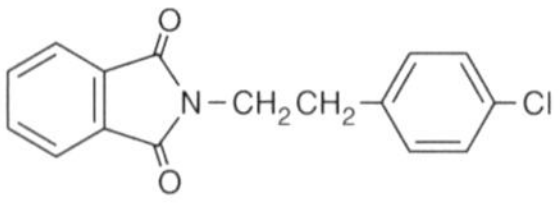

250 mg

S69,163-1
PHTHALIMIDO N-(3-CHLORO-4-METHYLPHENYL)CARBAMATE

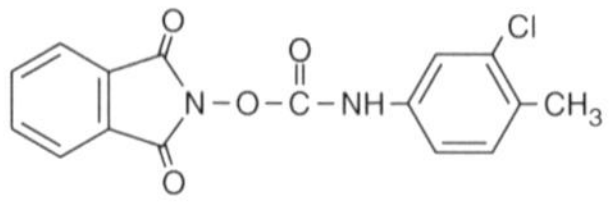

250 mg

S69,708-7
N-(2-BROMOPHENYL)PHTHALIMIDE

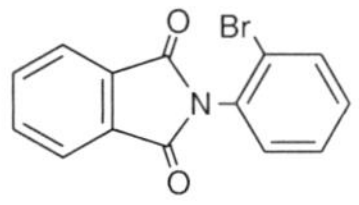

250 mg

S95,587-6
2-BENZYL-5-BROMO-3A-(2,3-DIMETHOXYPHENYL)-HEXAHYDRO-HYDROXY-1,3-ISOINDOLINEDIONE

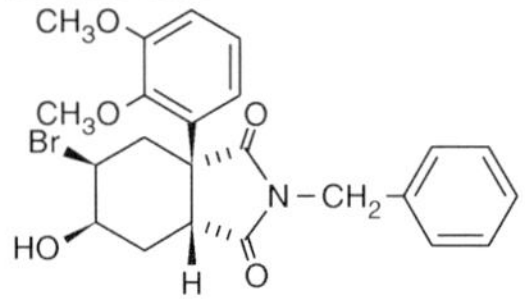

50 mg

S78,035-9
1-[2-HYDROXY-3-(PHENOXY)PROPYL]-SUCCINIMIDE

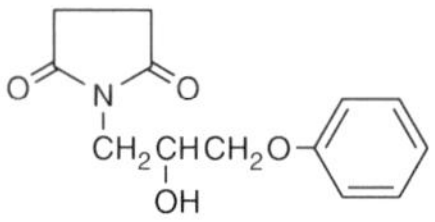

250 mg

S51,951-0
2-(4-BROMOPHENYL)-3-HYDROXYMALEIMIDE

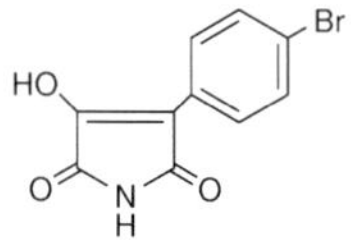

5 gm

S65,516-3
N-(2-HYDROXY-4-NITROPHENYL)-2-OCTADECYLSUCCINIMIDE

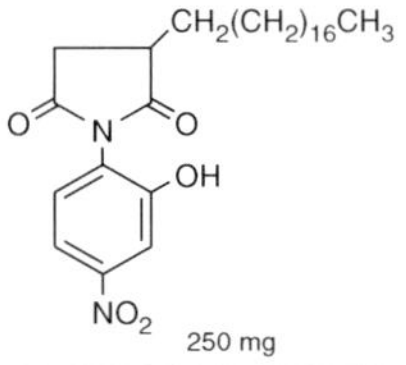

250 mg

S64,891-4
N-(HYDROXYMETHYL)-4-NITROPHTHALIMIDE

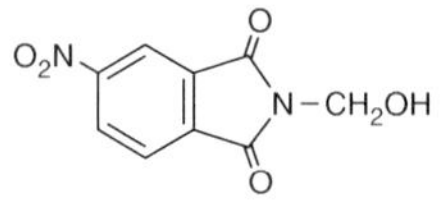

5 gm

S50,595-1
4-MALEIMIDOSALICYLIC ACID

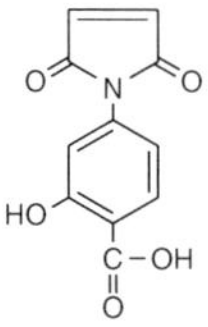

250 mg

S55,175-9
N-(4-HYDROXYPHENYL)PHTHALIMIDE

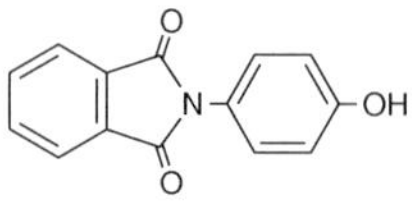

250 mg

S73,999-5
N-(2-HYDROXYETHYL)-1,8-NAPHTHALIMIDE

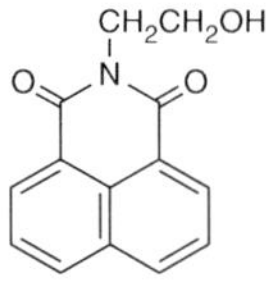

5 gm

S76,329-2
5-CHLORO-N-(2-HYDROXYETHYL)-PHTHALIMIDE

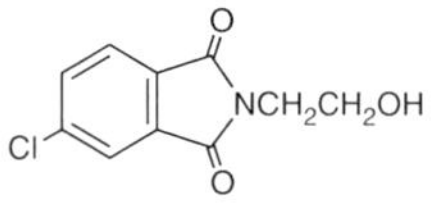

250 mg

S89,662-4
N-(3-HYDROXYCYCLOHEXYL)PHTHALIMIDE

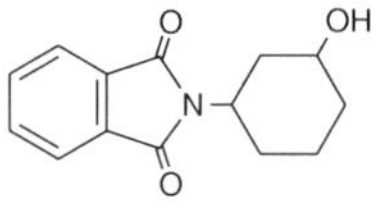

250 mg

S74,700-9
N-(2-HYDROXYPHENYL)-3,4,5,6-TETRACHLOROPHTHALIMIDE

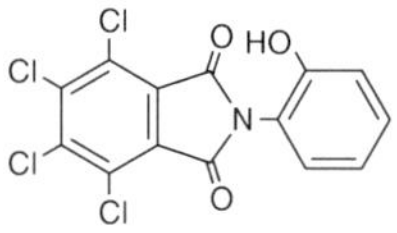

5 gm

S69,929-2
2-(3-BENZYLOXYPHENYL)-N-METHYLSUCCINIMIDE

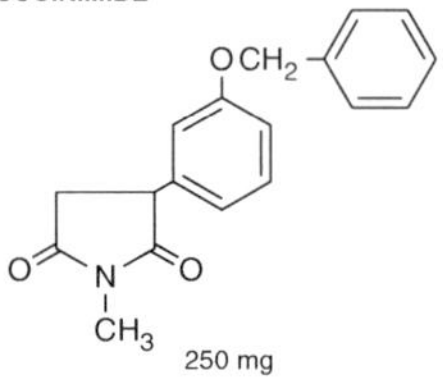

250 mg

S76,087-0
2-[2-(3,4-DIMETHOXYPHENYL)ETHYL]-ISOQUINOLINE-1,3(2H,4H)-DIONE

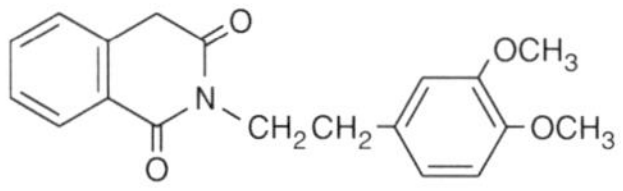

250 mg

S74,860-9
2-[2-(3,4-DIMETHOXYPHENYL)ETHYL]-1,3,3A,4,7,7A-HEXAHYDRO-4,7-METHANOISOINDOLE-1,3-DIONE

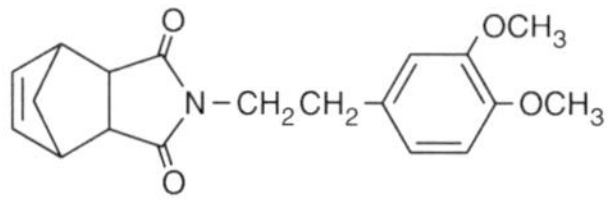

250 mg

S75,450-1
N-[2-(3,4-DIMETHOXYPHENYL)ETHYL]-PHTHALIMIDE

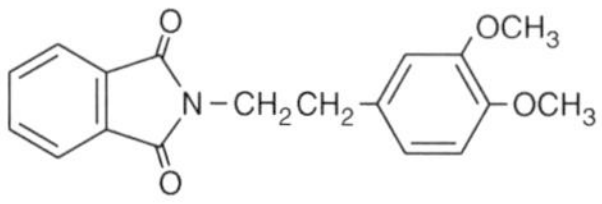

250 mg

S72,293-6
PHTHALIMIDO N-(2,5-DIMETHOXYPHENYL)CARBAMATE

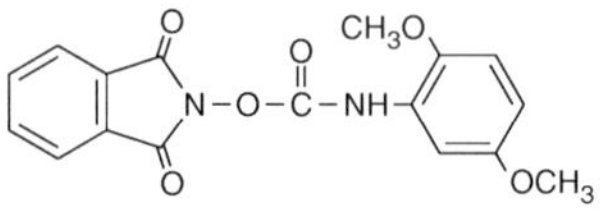

250 mg

S87,584-8
1,1'-(3,3'-DIMETHYL-1,1'-BIPHENYL-4,4'-DIYL)BISMALEIMIDE

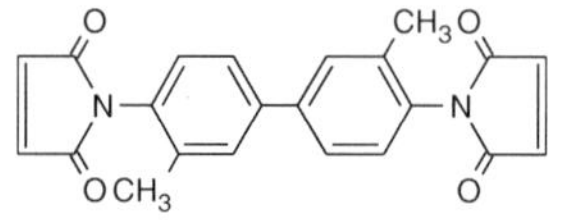

250 mg

S95,047-5
3A-(2,3-DIMETHOXYPHENYL)-5,6-EPOXYHEXAHYDROISOINDOLINE-1,3-DIONE

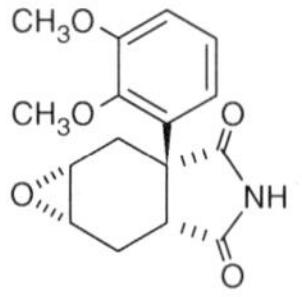

50 mg

S95,210-9
N-BENZYL-2-(2,3-DIMETHOXYPHENYL)-MALEIMIDE

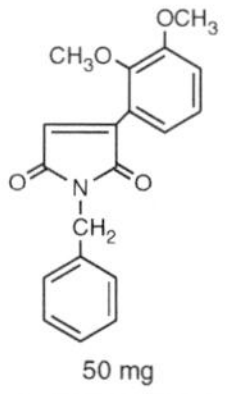

50 mg

S55,009-4
2-(2-AMINOPHENYLTHIO)-N-(4-ETHOXYPHENYL)SUCCINIMIDE

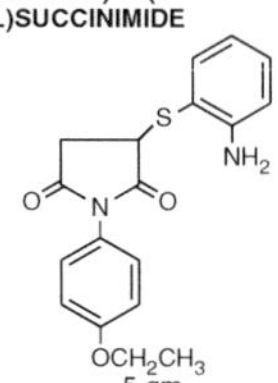

5 gm

S55,388-3
2-(2-AMINOPHENYLTHIO)-N-(2-METHOXYPHENYL)SUCCINIMIDE

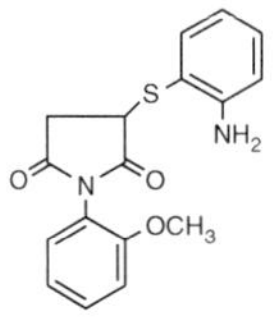

250 mg

S50,598-6
N-(4-METHOXY-3-NITROPHENYL)-MALEIMIDE

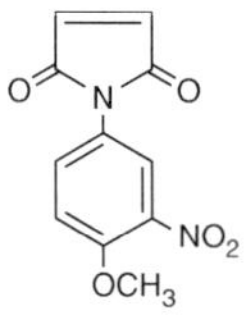

250 mg

S53,665-2
N-(PARA-PHENETIDINOMETHYL)-PHTHALIMIDE

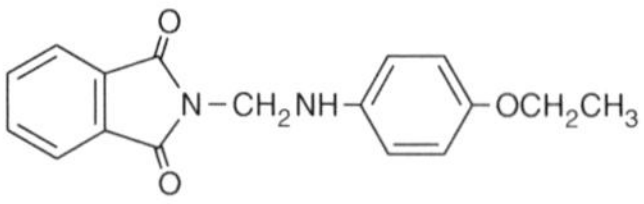

5 gm

S53,671-7
N-[4-(4-CHLOROPHENOXY)-ANILINOMETHYL]PHTHALIMIDE

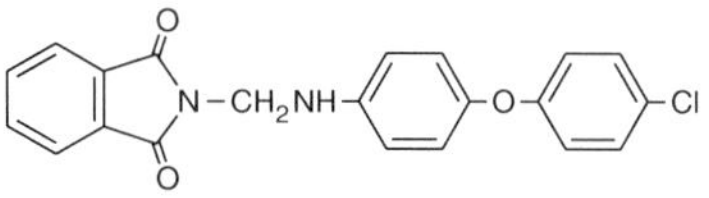

5 gm

S52,714-9
N-(ORTHO-ANISIDINOMETHYL)-PHTHALIMIDE

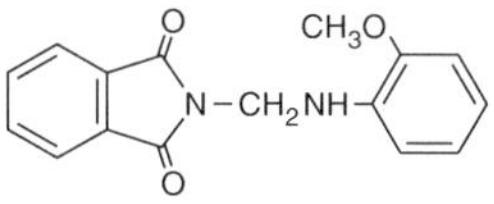

250 mg

S87,840-5
N-(4-PHENOXYPHENYL)GLUTARIMIDE

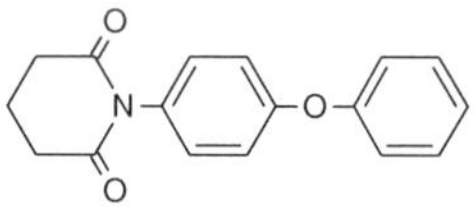

5 gm

S88,900-8
N-(3-BENZYLOXYPHENYL)PHTHALIMIDE

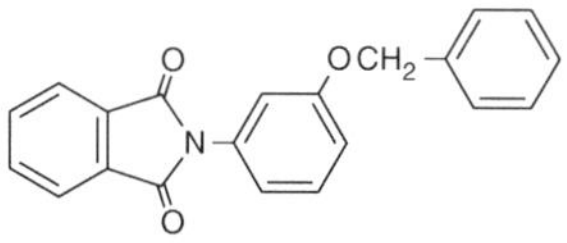

250 mg

S55,775-7
N-(ORTHO-PHENETIDINOMETHYL)-PHTHALIMIDE

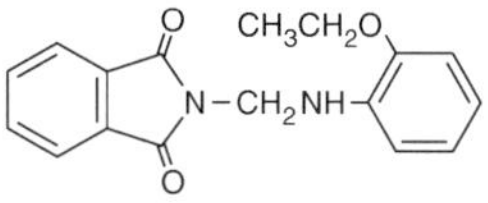

250 mg

S70,090-8
N-(2-ETHOXYPHENYL)MALEIMIDE

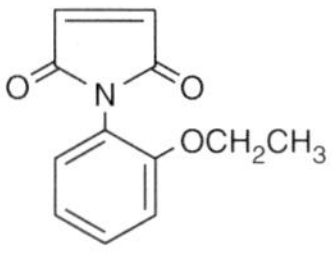

5 gm

S70,404-0
1-(2-METHOXY-5-METHYLPHENYL)-MALEIMIDE

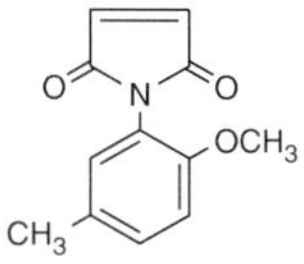

5 gm

S74,644-4
N-(2-METHOXY-5-METHYLPHENYL)-SUCCINIMIDE

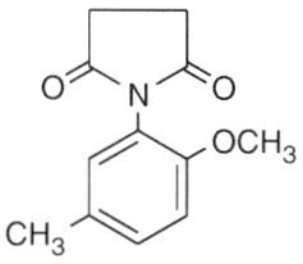

5 gm

S67,121-5
N-(P-ANISYL)-SUCCINIMIDE

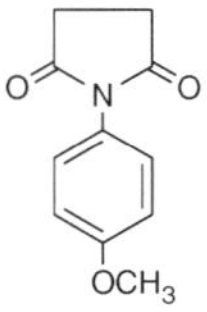

5 gm

S54,863-4
N-(2-METHOXYPHENYL)MALEIMIDE

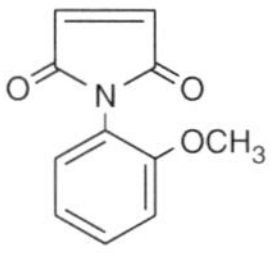

250 mg

S53,794-2
N-(4-PHENOXYPHENYL)MALEIMIDE

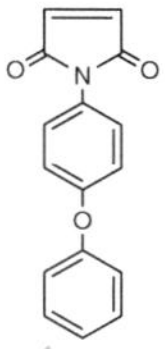

5 gm

S79,025-7
CARBOBENZYLOXY-O-BENZYL-L-TYROSINE N-HYDROXYSUCCINIMIDE ESTER

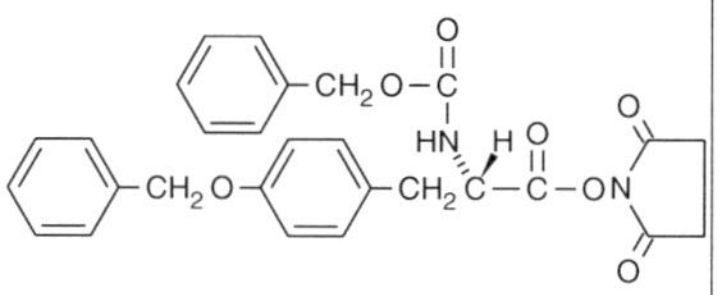

250 mg

S78,160-6
O-BENZYL-N-(TERT-BUTOXYCARBONYL)-L-SERINE N-HYDROXYSUCCINIMIDE ESTER

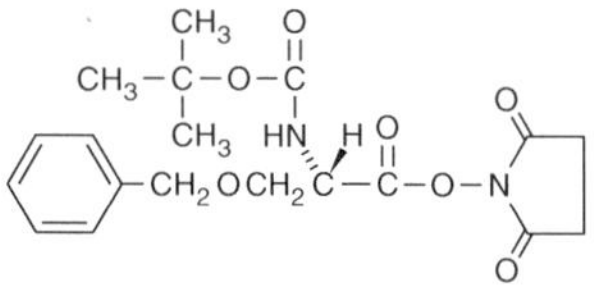

250 mg

S89,272-6
1,4,5,6,7,7-HEXACHLORO-N-(3-METHOXYPHENYL)-5-NORBORNENE-2,3-DICARBOXIMIDE

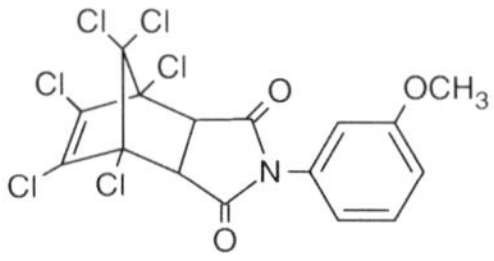

250 mg

S52,359-3
N-(4-METHOXYPHENYL)MALEIMIDE

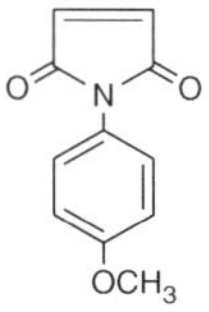

5 gm

S76,385-3
N-(3-METHOXYPHENYL)MALEIMIDE

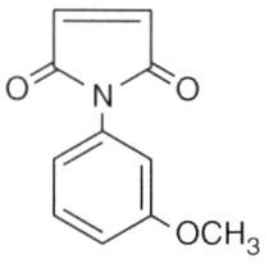

250 mg

S74,648-7
N-(4-METHOXYPHENYL)GLUTARIMIDE

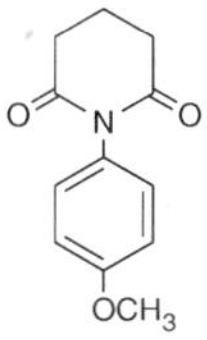

5 gm

S84,634-1
N-[2-(4-METHOXYPHENYL)ETHYL]-PHTHALIMIDE

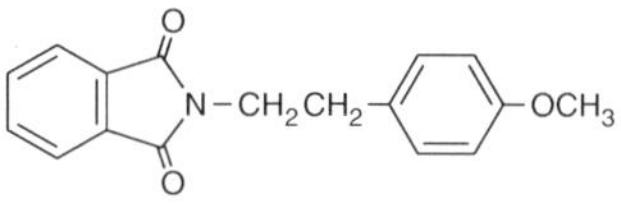

250 mg

S45,027-8
N-(BENZYLOXYCARBONYLOXY)-PHTHALIMIDE

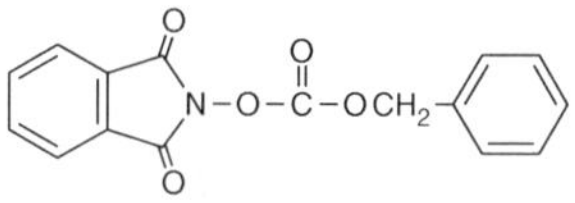

5 gm

S79,061-3
CARBOBENZYLOXY-L-VALINE N-HYDROXYSUCCINIMIDE ESTER

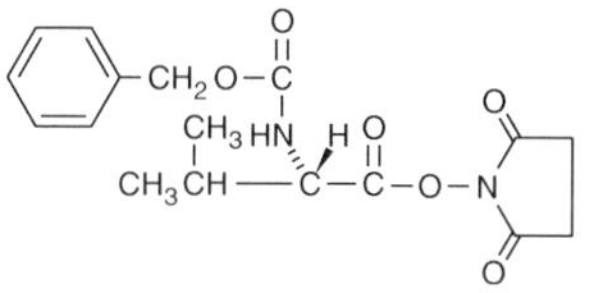

250 mg

S78,815-5
CARBOBENZYLOXY-L-ISOLEUCINE N-HYDROXYSUCCINIMIDE ESTER

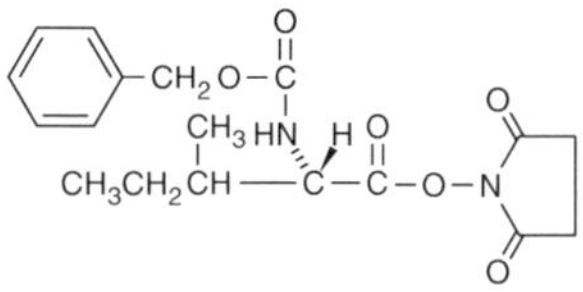

250 mg

S35,853-3
N-BENZYLOXYPHTHALIMIDE

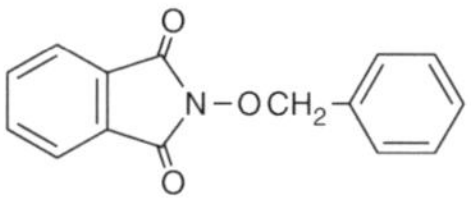

5 gm

S45,448-6
PHTHALIMIDOACETALDEHYDE DIMETHYL ACETAL

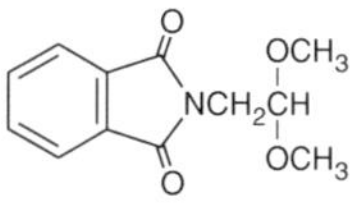

5 gm

S67,217-3
1-ACETOXY-3-PHTHALIMIDO-2-PROPANONE

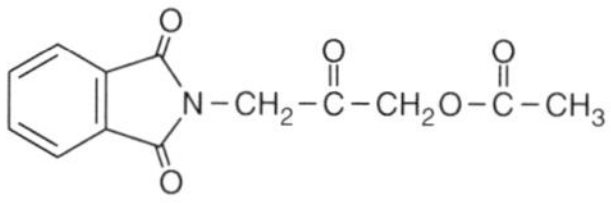

5 gm

S54,264-4
N-(3,4-DIMETHOXYPHENETHYL)-MALEIMIDE

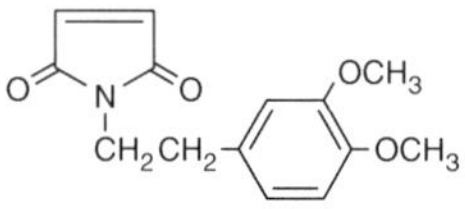

250 mg

S85,926-5
N,N'-(ETHYLENEDIOXY)DIPHTHALIMIDE

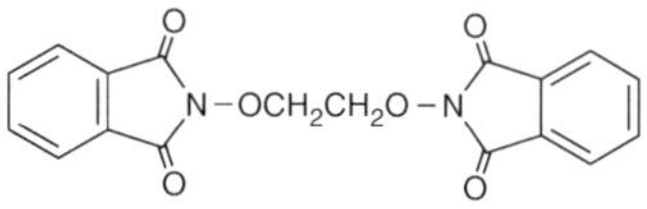

250 mg

S55,122-8
2,2-BIS-(PHTHALIMIDOMETHYL)-1,3-PROPANEDIOL DIACETATE

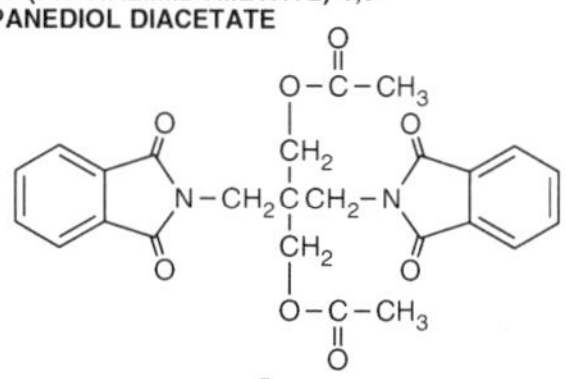

5 gm

S74,275-9
N-(2-MERCAPTOETHYL)PHTHALIMIDE

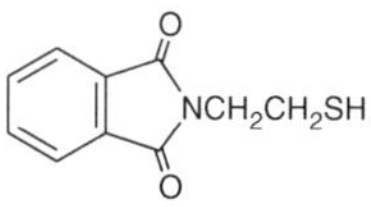

5 gm

S54,200-8
2-(ORTHO-AMINOPHENYLTHIO)-N-(PARA-CHLOROPHENYL)-SUCCINIMIDE

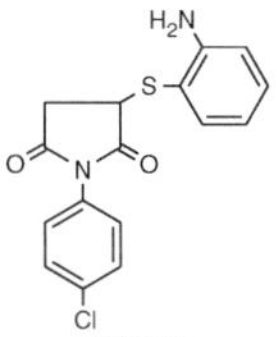

250 mg

S54,851-0
2-(2-AMINOPHENYLTHIO)-N-(2-CHLOROPHENYL)-SUCCINIMIDE

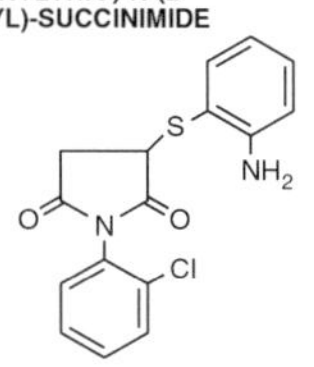

5 gm

S55,339-5
2-(2-AMINOPHENYLTHIO)-N-(3-BROMOPHENYL)-SUCCINIMIDE

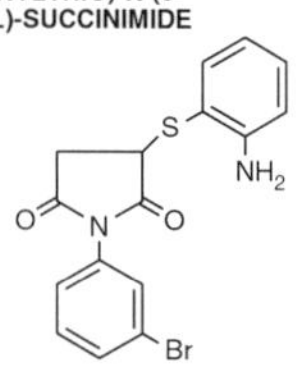

5 gm

S64,937-6
2-(2-AMINOPHENYLTHIO)-N-(3-METHOXYPHENYL)-SUCCINIMIDE

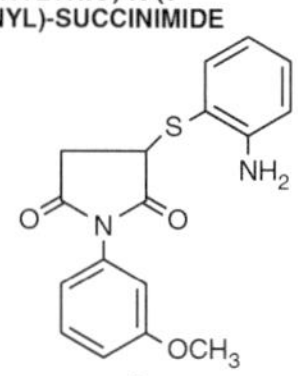

5 gm

S55,477-4
ETHYL 4-[2-(2-AMINOPHENYLTHIO)-SUCCINIMIDO]BENZOATE
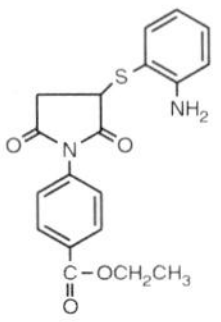
5 gm

S55,385-9
2-(2-AMINOPHENYLTHIO)-N-(4-METHOXYPHENYL)-SUCCINIMIDE
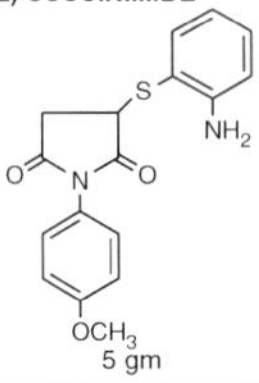
5 gm

S54,746-8
2-(2-AMINOPHENYLTHIO)-N-(3-CHLORO-P-TOLYL)SUCCINIMIDE
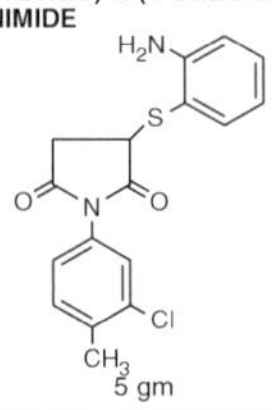
5 gm

S54,854-5
2-(2-AMINOPHENYLTHIO)-N-(1-NAPHTHYL)SUCCINIMIDE
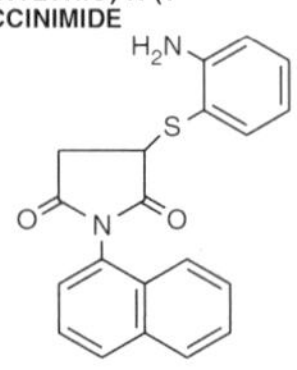
5 gm

S52,790-4
N-PHTHALOYL-L-METHIONINE
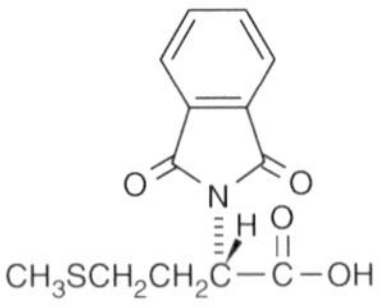
5 gm

S45,086-3
N-PHTHALOYL-DL-METHIONINE
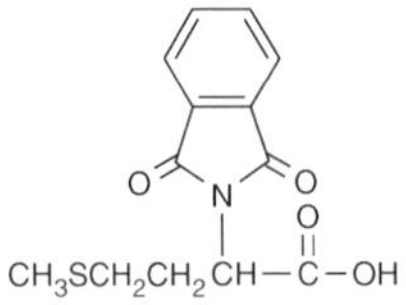
5 gm

S79,035-4
CARBOBENZYLOXY-S-BENZYL-L-CYSTEINE N-HYDROXY SUCCINIMIDE
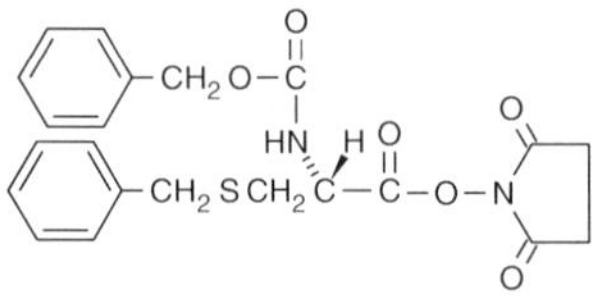
250 mg

S78,868-6
N-(TERT-BUTOXYCARBONYL)-S-BENZYLCYSTEINE N-HYDROXYSUCCINIMIDE ESTER
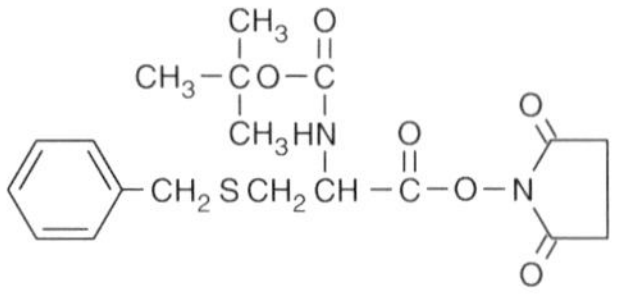
250 mg

S90,544-5
1,4,5,6,7,7-HEXACHLORO-N-[3-(METHYLTHIO)PHENYL]-5-NORBORNENE-2,3-DICARBOXIMIDE
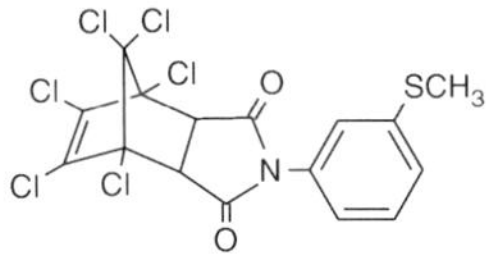
250 mg

S74,990-7
N-(2-FLUORENYL)-2-(HEXADECYLTHIO)-SUCCINIMIDE
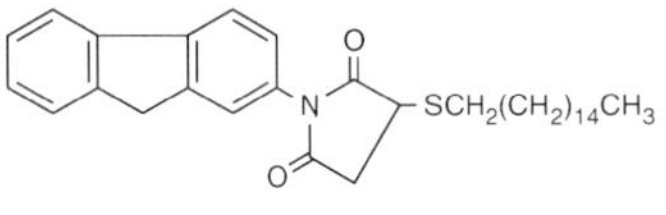
250 mg

S82,011-3
N-[3-(CYCLOHEXYLTHIO)PROPYL]-PHTHALIMIDE
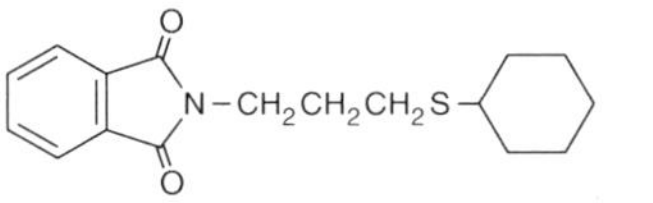
250 mg

S70,497-0
N,N'-DIPHTHALYL-L-CYSTINE
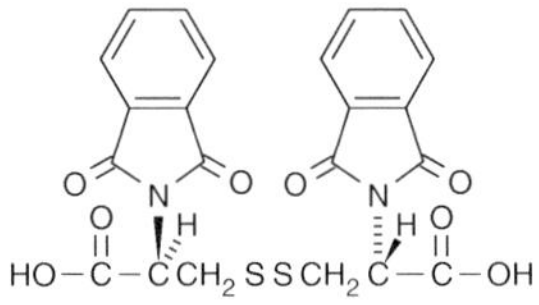
5 gm

S58,185-2
N-(2-AMINOPHENYL)SUCCINIMIDE
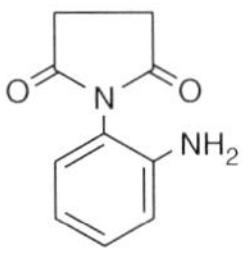
5 gm

S90,877-0
N-[2-(3-ACETAMIDO-N-ETHYLANILINO)-ETHYL]SUCCINIMIDE
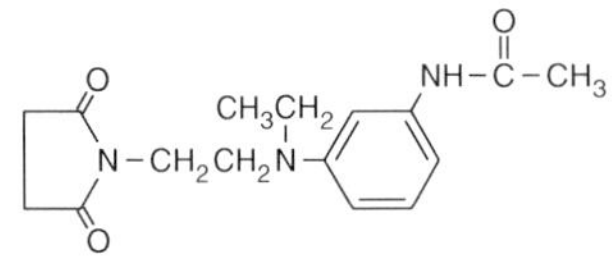
250 mg

S73,152-8
2'-NITRO-4'-PHTHALIMIDOACETANILIDE
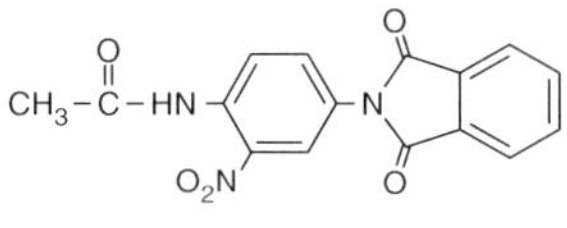
5 gm

S50,019-4
N-(2-ANILINOETHYL)-PHTHALIMIDE
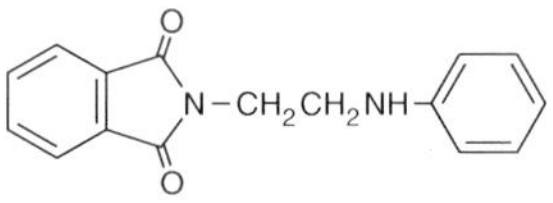
250 mg

S90,896-7
N-[2-(N-ETHYL-3-METHYLANILINO)-ETHYL]SUCCINIMIDE
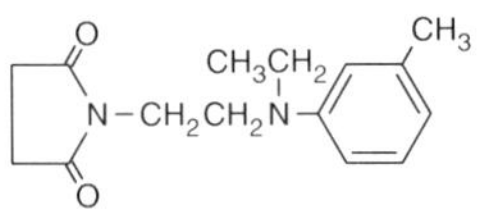
250 mg

S90,887-8
N-[2-(N-ETHYL-3-METHYLANILINO)-ETHYL]PHTHALIMIDE
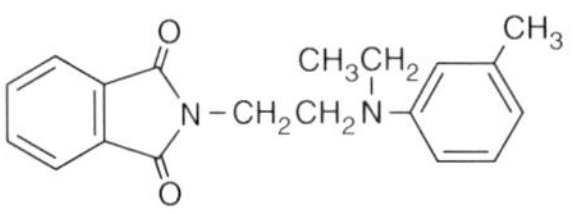
250 mg

S39,279-0
N-(2-NITROPHENYL)MALEIMIDE
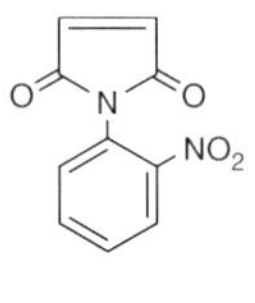
5 gm

S52,786-6
N-PHTHALOYLGLYCYLGLYCINE ETHYL ESTER
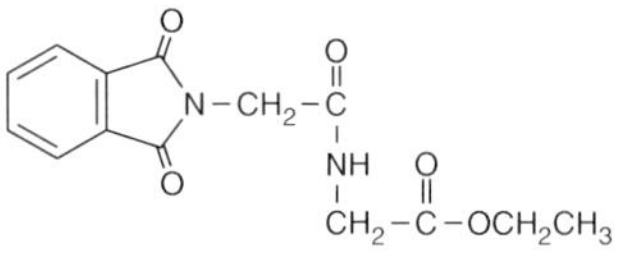
5 gm

S45,090-1
N-PHTHALOYLGLYCYLGLYCINE
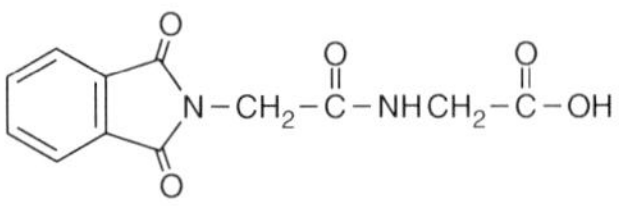
250 mg

S54,703-4
2-PHTHALIMIDOACETAMIDE
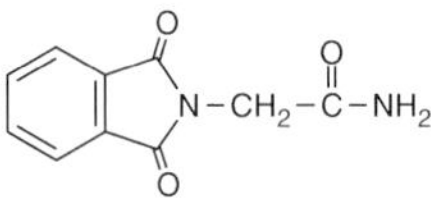
5 gm

S38,276-0
N,N'-ETHYLENEDIPHTHALIMIDE
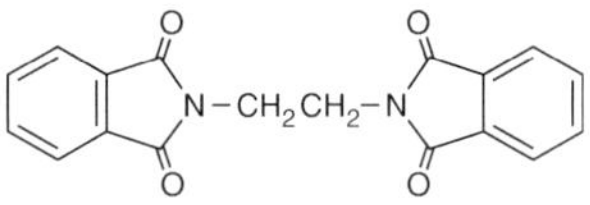
250 mg

S87,188-5
N-(3-ANILINOPROPYL)PHTHALIMIDE
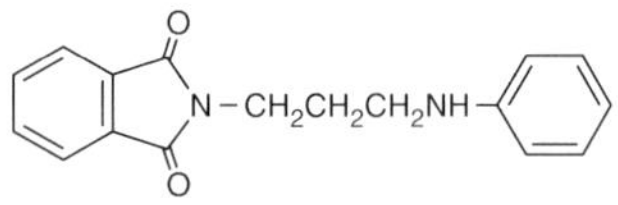
250 mg

S87,825-1
N-(3-NITROPHENYL)GLUTARIMIDE
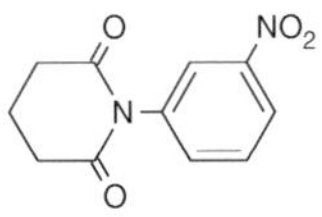
5 gm

S87,830-8
N-(4-METHYL-3-NITROPHENYL)-GLUTARIMIDE
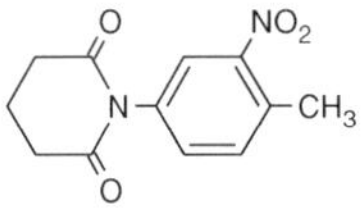
5 gm

S53,922-8
N,N'-(1,3-PHENYLENE)BIS[N-(AMINOMETHYL)PHTHALIMIDE]
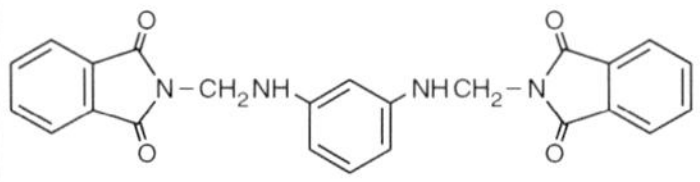
250 mg

S74,692-4
4-NITRO-2-PHTHALIMIDOBENZOIC ACID
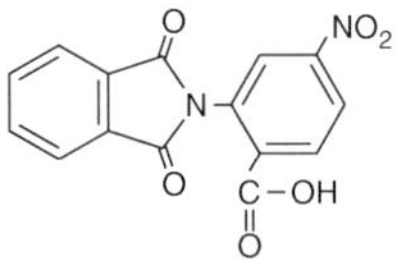
5 gm

S86,122-7
N-(3-NITROPHENYL)SUCCINIMIDE
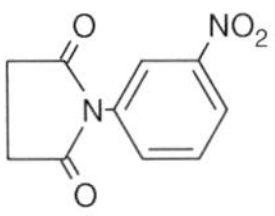
250 mg

S68,813-4
PHTHALIMIDO N-(2-METHYL-5-NITROPHENYL)CARBAMATE
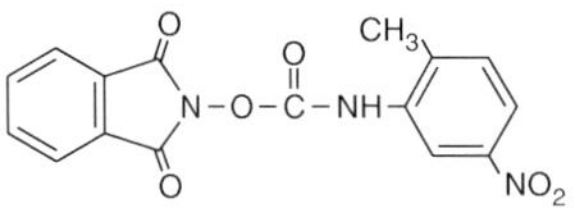
250 mg

S92,290-0
2,2-DIPHENYLSUCCINIMIDE
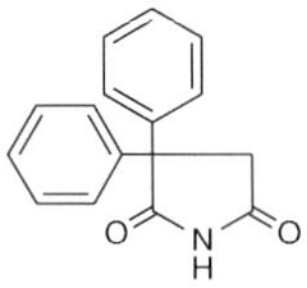
250 mg

S87,557-0
1,4-BIS(3-PHTHALIMIDOPROPYL)-PIPERAZINE
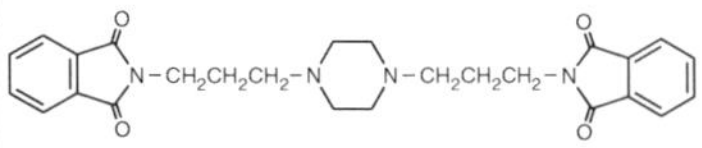
250 mg

S74,236-8
N,N'-TRIMETHYLENEDIPHTHALIMIDE
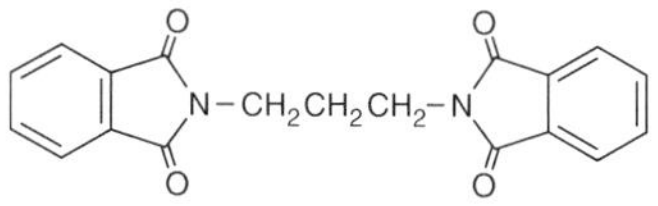
5 gm

S87,841-3
N-(4-ANILINOPHENYL)GLUTARIMIDE
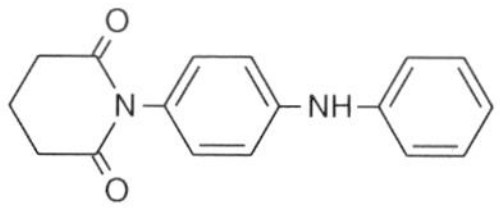
5 gm

S39,277-4
N-(4-NITROPHENYL)MALEIMIDE
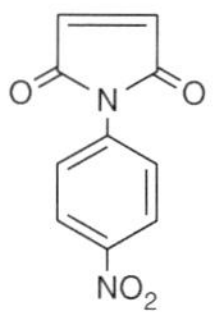
250 mg

S64,914-7
N-(4-MORPHOLINOANILINOMETHYL)-PHTHALIMIDE
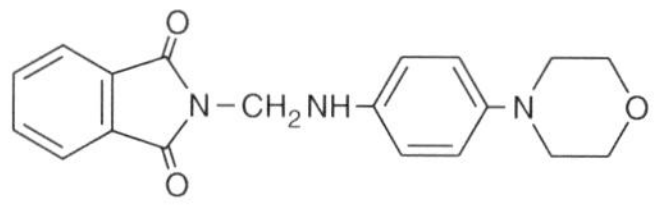
250 mg

S80,572-6
DIETHYL 2-ACETAMIDO-2-(3-PHTHALIMIDOPROPYL)MALONATE
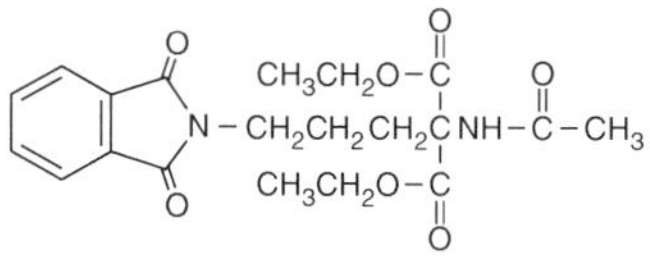
250 mg

S76,410-8
N-(2-METHYL-4-NITROPHENYL)-PHTHALIMIDE
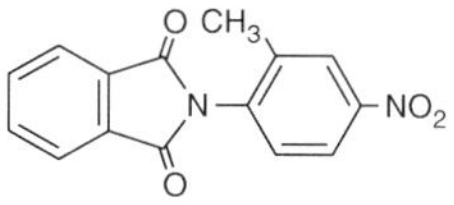
5 gm

S64,604-0
N-ACETYL-4-NITROPHTHALIMIDE
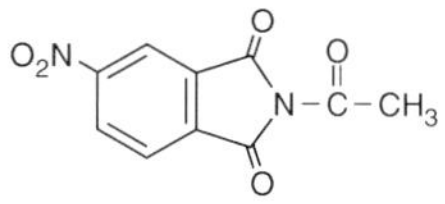
5 gm

S55,171-6
N-(4-NITROPHENYL)PHTHALIMIDE
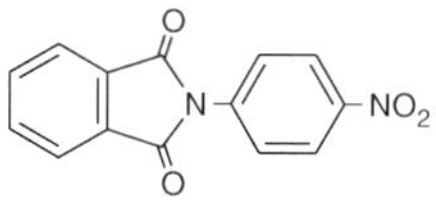
5 gm

S63,360-7
N-(2,6-DICHLORO-4-NITROPHENYL)-PHTHALIMIDE
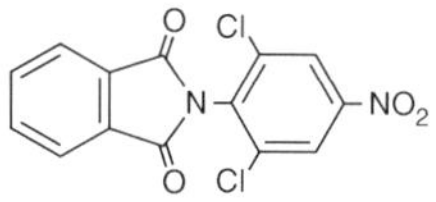
250 mg

S53,628-8
N-(1,2-DIPHENYLETHYLAMINOMETHYL)-PHTHALIMIDE
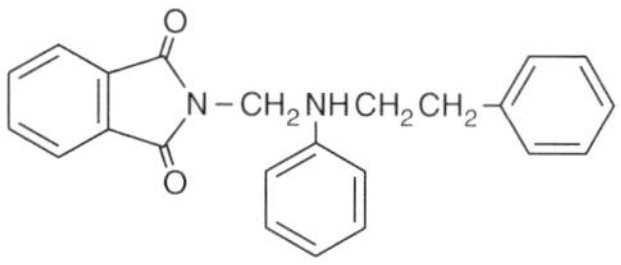
250 mg

S60,106-3
N-PHTHALOYL-DL-PHENYLALANYLGLYCINE
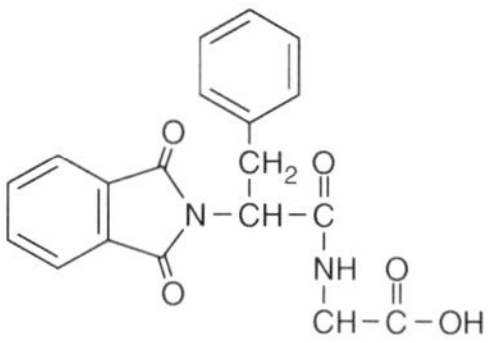
250 mg

S43,980-0
N-(BENZOYLMETHYL)-PHTHALIMIDE
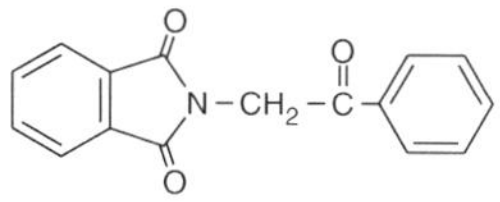
5 gm

S52,792-0
N-PHTHALOYL-DL-PHENYLALANINE
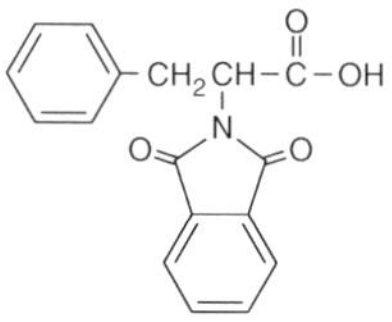
5 gm

S86,131-6
3-PHENYL-2-PHTHALIMIDOPROPIONIC ACID
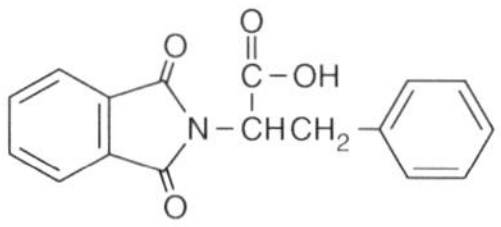
250 mg

S86,132-4
METHYL 3-PHENYL-2-PHTHALIMIDOPROPIONATE
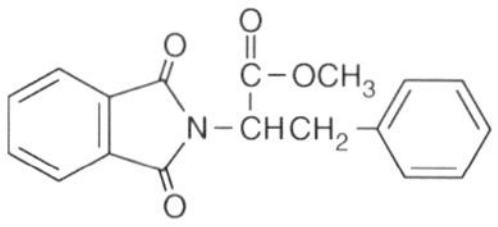
250 mg

S84,142-0
2-PHENYLSUCCINIMIDE
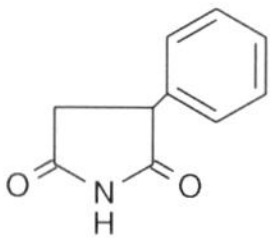
250 mg

S78,251-3
1-(PHENYLETHYL)SUCCINIMIDE
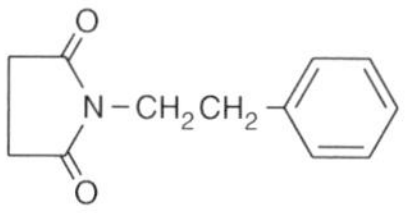
5 gm

S53,963-5
N-(3,4-XYLIDINOMETHYL)PHTHALIMIDE
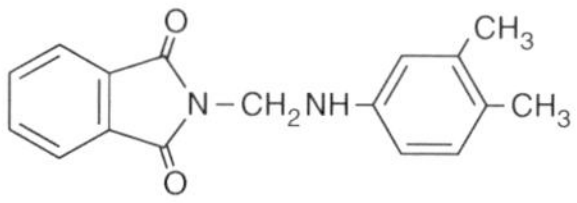
250 mg

S53,243-6
N-(5-NITRO-ORTHO-TOLUIDINOMETHYL)-PHTHALIMIDE
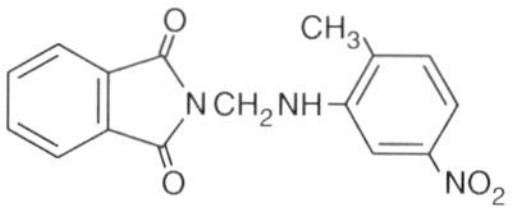
5 gm

S53,244-4
N-(3-NITRO-PARA-TOLUIDINOMETHYL)-PHTHALIMIDE
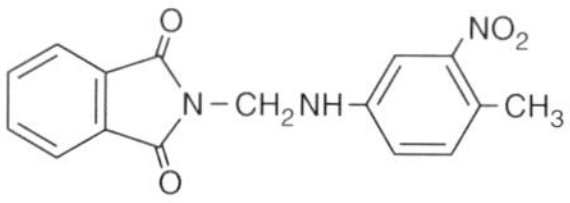
5 gm

S53,783-7
N-(4-ACETYLANILINOMETHYL)-PHTHALIMIDE
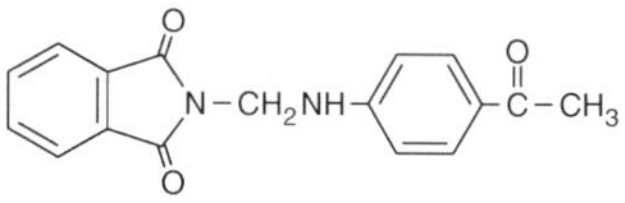
5 gm

S55,776-5
N-[3,4-(METHYLENEDIOXY)-ANILINOMETHYL]PHTHALIMIDE
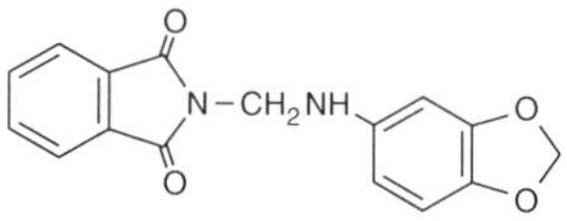
250 mg

S53,821-3
ETHYL 4-(PHTHALIMIDOMETHYLAMINO)-BENZOATE
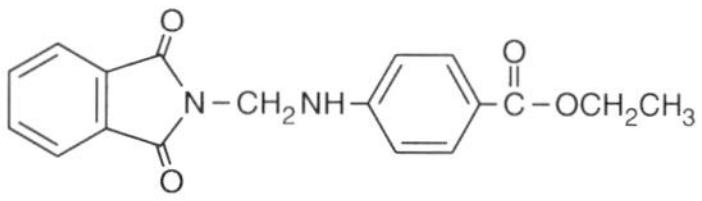
5 gm

S52,368-2
N-(4-CHLORO-ORTHO-TOLUIDINOMETHYL)-PHTHALIMIDE
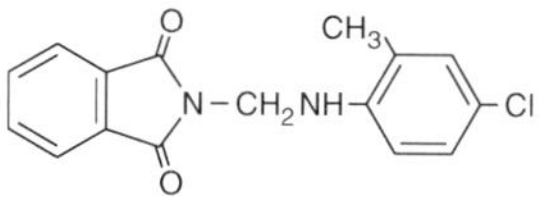
5 gm

S52,718-1
N-(5-CHLORO-ORTHO-TOLUIDINOMETHYL)-PHTHALIMIDE
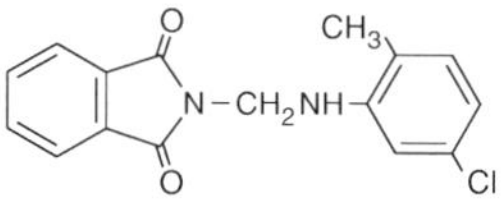
250 mg

S64,608-3
N-(4-CHLOROANILINOMETHYL)-PHTHALIMIDE
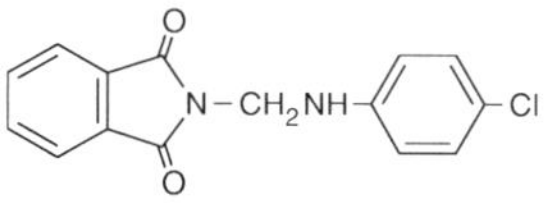
5 gm

S53,175-8
N-(4-BROMOANILINOMETHYL)-PHTHALIMIDE
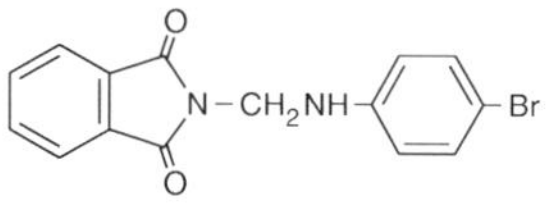
5 gm

S64,919-8
N-(4-ISOPROPYLANILINOMETHYL)-PHTHALIMIDE
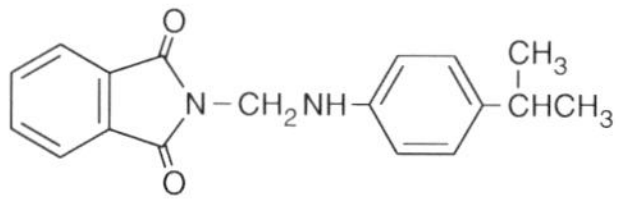
250 mg

S64,596-6
N-(2,6-DICHLOROPHENYLAMINOMETHYL)-PHTHALIMIDE
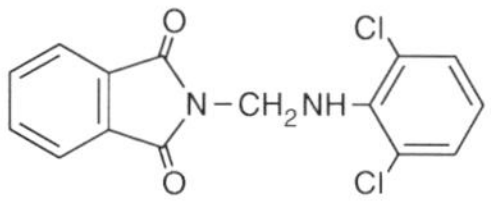
250 mg

S52,731-9
N-(3-CHLORO-ORTHO-TOLUIDINOMETHYL)-PHTHALIMIDE
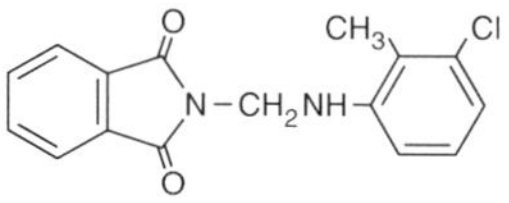
250 mg

S52,352-6
N-(3-CHLORO-PARA-TOLUIDINOMETHYL)-PHTHALIMIDE
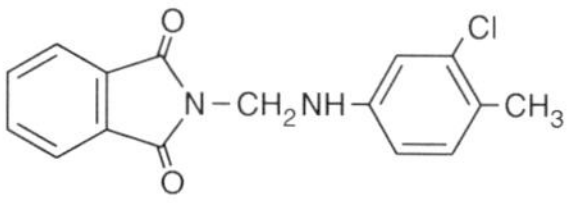
5 gm

S53,902-3
N-(4-BROMO-ORTHO-TOLUIDINOMETHYL)-PHTHALIMIDE
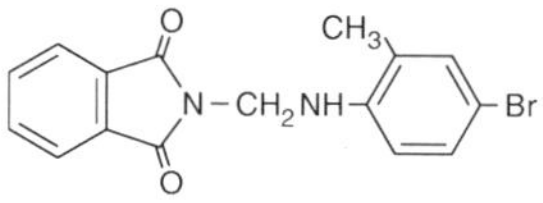
5 gm

S54,943-6
N-(DIMETHYLAMINOMETHYL)PHTHALIMIDE
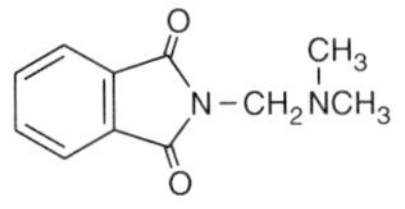
250 mg

S88,521-5
N-(4-OXOCYCLOHEXYL)PHTHALIMIDE
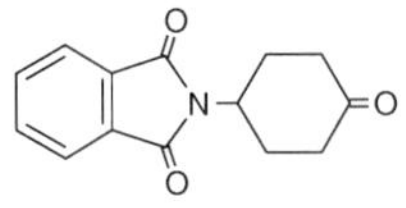
250 mg

S93,518-2
(L)-2-PHTHALIMIDOGLUTARIC ACID
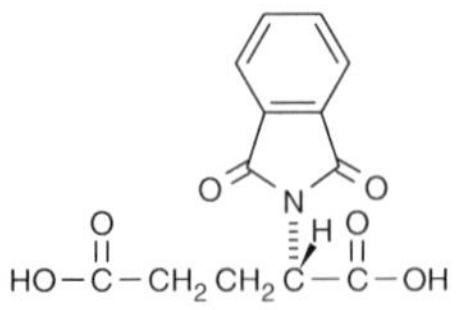
250 mg

S50,597-8
2-CHLORO-5-MALEIMIDOBENZOIC ACID
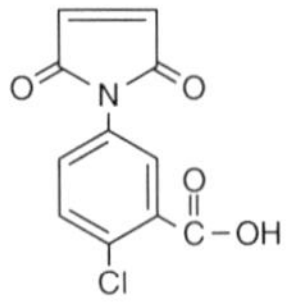
250 mg

S50,605-2
4-CHLORO-3-MALEIMIDOBENZOIC ACID
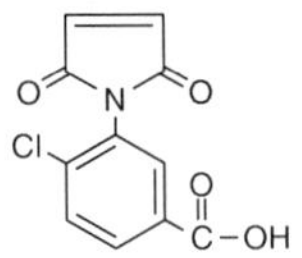
5 gm

S45,007-3
2-PHTHALIMIDOBENZOIC ACID
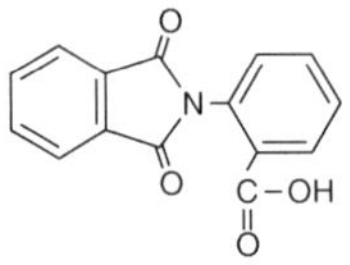
5 gm

S55,472-3
ETHYL 4-MALEIMIDOBENZOATE
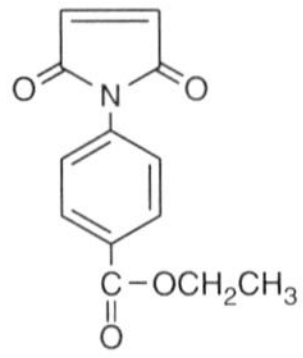
250 mg

S39,276-6
N-(4-CHLOROPHENYL)MALEIMIDE
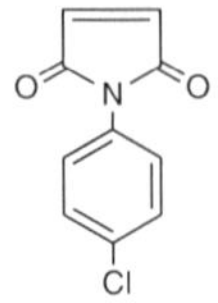
5 gm

S50,607-9
N-(2-CHLOROPHENYL)MALEIMIDE
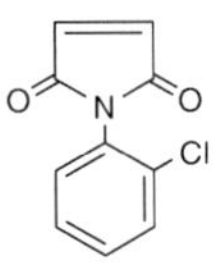
5 gm

S36,283-2
N-(4-IODOPHENYL)MALEIMIDE
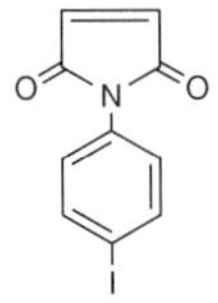
250 mg

S39,275-8
N-BENZYLMALEIMIDE
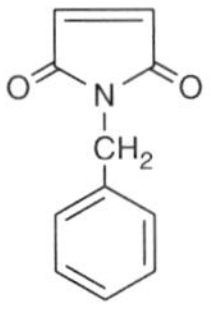
5 gm

S50,608-7
N-(ORTHO-TOLYL)MALEIMIDE
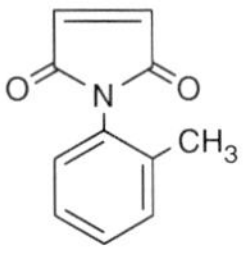
5 gm

S36,549-1
N-(2,6-XYLYL)MALEIMIDE
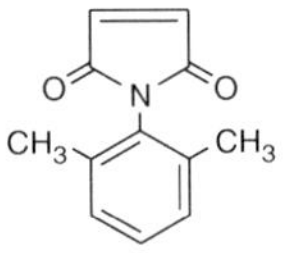
5 gm

S87,828-6
N-(2-ETHYLPHENYL)MALEIMIDE
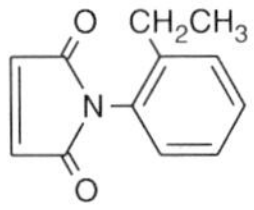
5 gm

S87,854-5
N-(4-ETHYLPHENYL)MALEIMIDE
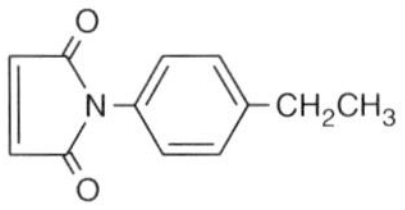
5 gm

S88,009-4
N-(3-PHENYLALLYL)PHTHALIMIDE
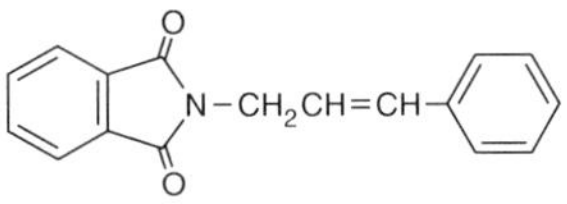
250 mg

S53,962-7
N-(3,4-XYLYL)MALEIMIDE
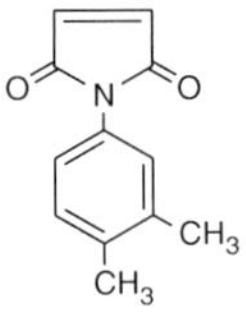
5 gm

S50,535-8
3-PHTHALIMIDOPROPIONIC ACID
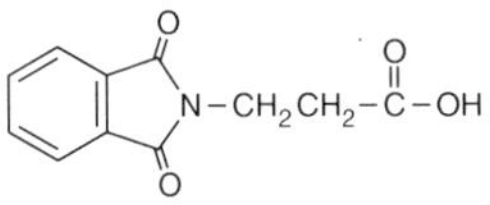
250 mg

S38,161-6
N-(2,6-XYLYL)SUCCINIMIDE
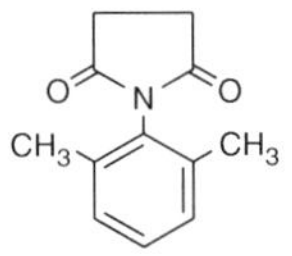
5 gm

S95,136-6
N-(O-TOLYL)SUCCINIMIDE
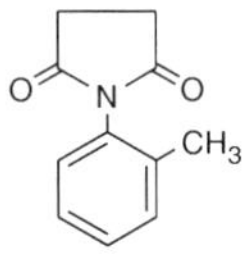
50 mg

S86,499-4
N-(P-TOLYL)SUCCINIMIDE
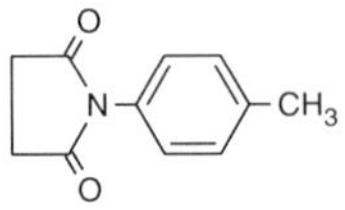
250 mg

S87,831-6
N-(P-TOLYL)GLUTARIMIDE
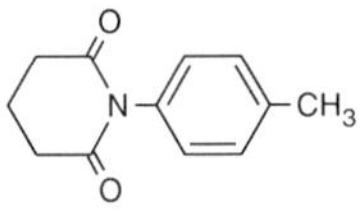
5 gm

S87,833-2
N-(2,5-XYLYL)SUCCINIMIDE
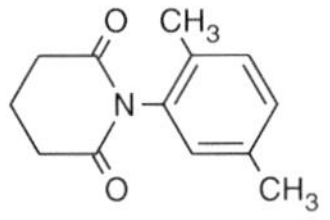
5 gm

S87,834-0
N-(3,5-DIMETHYLPHENYL)GLUTARIMIDE
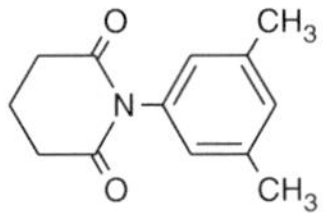
5 gm

S87,836-7
N-(2,4,6-TRIMETHYLPHENYL)-GLUTARIMIDE
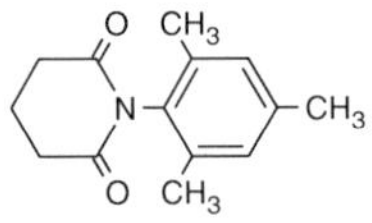
5 gm

S55,157-0
N-CYCLOHEXYLPHTHALIMIDE
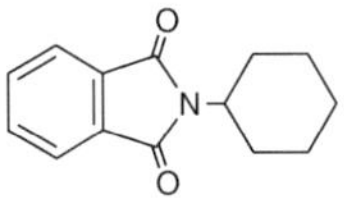
5 gm

S55,176-7
N-PHENYLPHTHALIMIDE
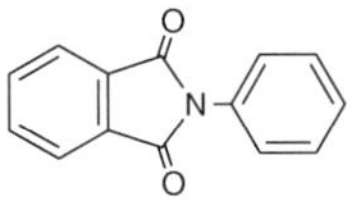
250 mg

S56,807-4
N-BENZOYLPHTHALIMIDE
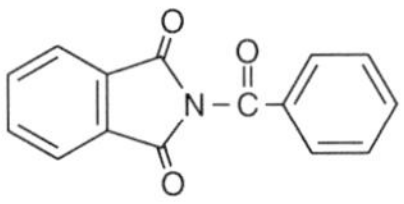
250 mg

S56,851-1
N-ACETYLPHTHALIMIDE
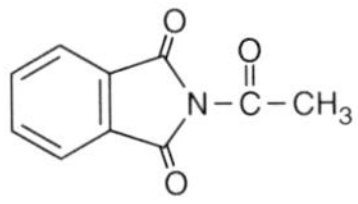
250 mg

S88,207-0
N-ALLYLPHTHALIMIDE
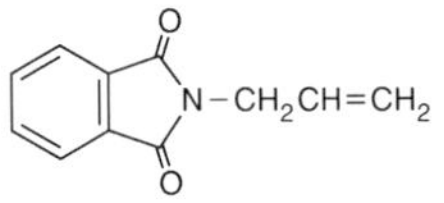
250 mg

S88,206-2
N-(3-BROMOALLYL)PHTHALIMIDE
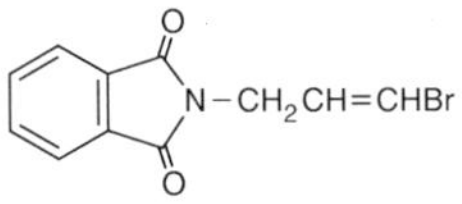
250 mg

S70,292-7
3-ANILINO-1-PHENYLSUCCINIMIDE
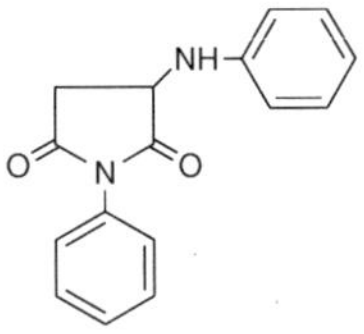
5 gm

S74,839-0
DIETHYL [2-(PHTHALIMIDOMETHYL)-BENZOYL]MALONATE

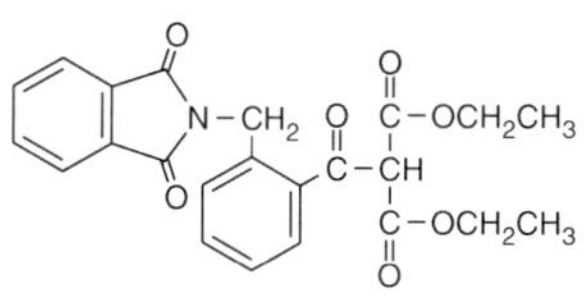

250 mg

S95,583-3
N-(2-CARBOXYBENZYL)PHTHALIMIDE

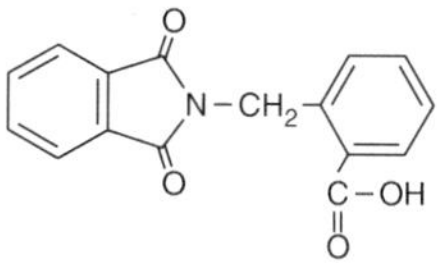

50 mg

S60,096-2
N-(ETHOXYCARBONYLOXY)PHTHALIMIDE

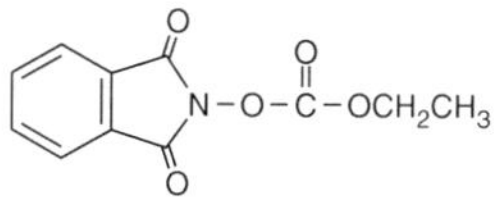

5 gm

S94,598-6
3-BENZYL-3-AZABICYCLO[3.1.0]-HEXANE-2,4-DIONE

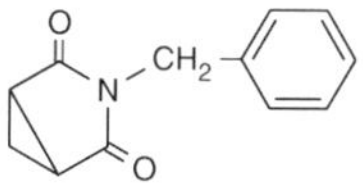

250 mg

S95,586-8
N-(BENZYLCARBAMOYLMETHYL)-PHTHALIMIDE

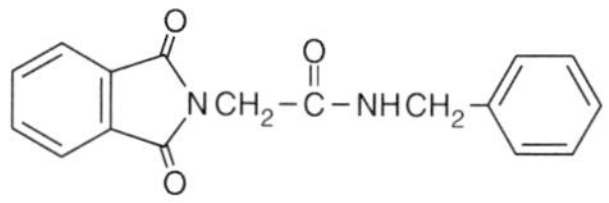

50 mg

S61,626-5
N-(3-IODOBENZYL)PHTHALIMIDE

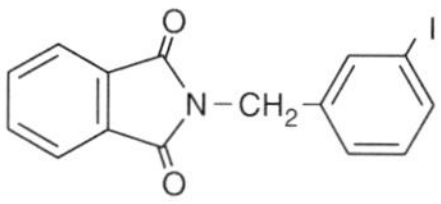

5 gm

S54,774-3
N-ANILINOMALEIMIDE

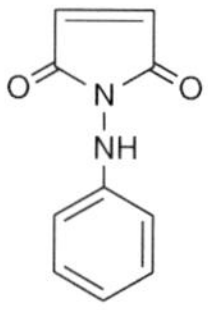

250 mg

S60,507-7
N-ANILINOPHTHALIMIDE

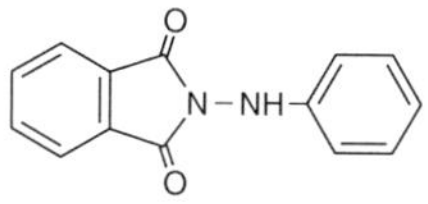

250 mg

S67,905-4
2-PHENYL-1-PHTHALIMIDOAZIRIDINE

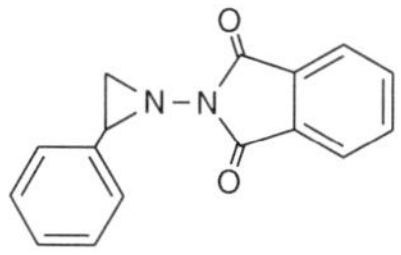

5 gm

S52,442-5
4,4'-TRIMETHYLENEBIS[N-(PIPERIDINOMETHYL)PHTHALIMIDE]

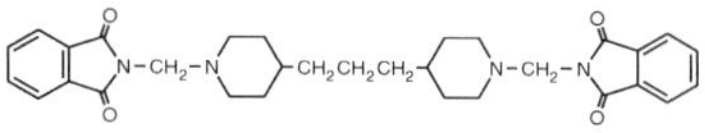

250 mg

S65,948-7
2-PHTHALIMIDOACETALDEHYDE HYDRATE

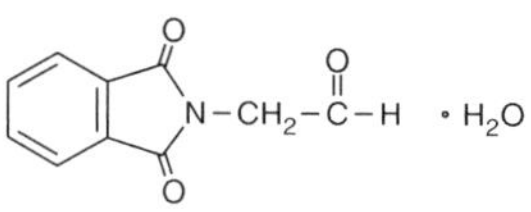

250 mg

S66,487-1
2-PHTHALIMIDOSUCCINIC ANHYDRIDE

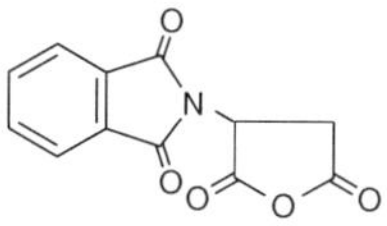

5 gm

S35,482-1
PHTHALOYL-DL-ALANINE

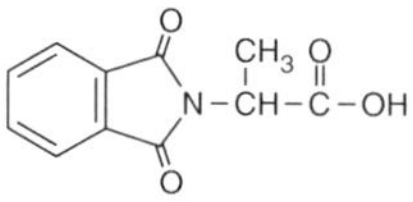

5 gm

S89,315-3
2-PHTHALIMIDOACETIC ACID

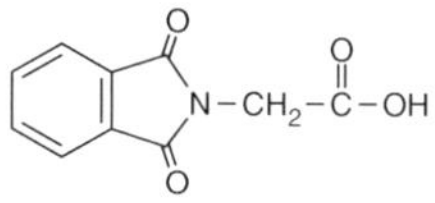

1 gm

S89,347-1
PHTHALOYL GLYCINE METHYL ESTER

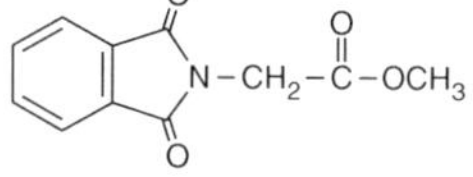

250 mg

S70,692-2
2-PHTHALIMIDOETHYL ACETATE

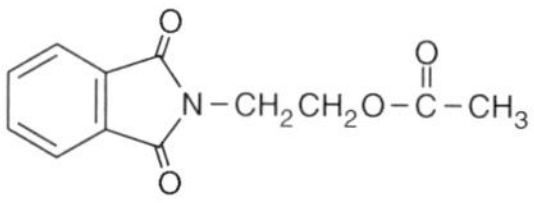

5 gm

S67,165-7
N-PHTHALOYLGLYCINE TERT-BUTYL ESTER

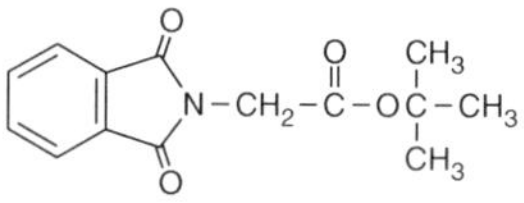

250 mg

S61,531-5
N-PHTHALOYLALANINE METHYL ESTER

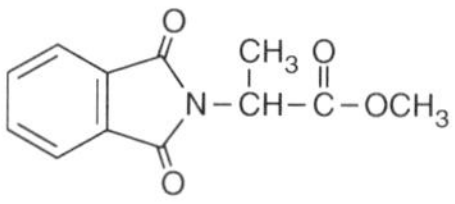

250 mg

S51,899-9
N-HYDROXY-2-PHENYLMALEIMIDE

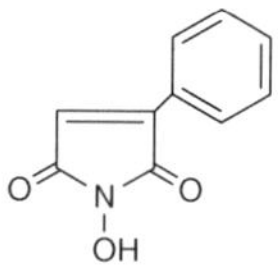

5 gm

S51,900-6
2-(4-BROMOPHENYL)-N-HYDROXYMALEIMIDE, SODIUM SALT

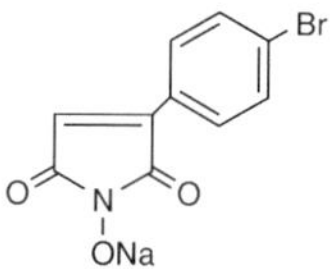

5 gm

S71,648-0
2-(4-FLUORO-3-NITROPHENYL)-HEXACHLORO-HEXAHYDRO-4,7-METHANOISOINDOLE-1,3-DIONE

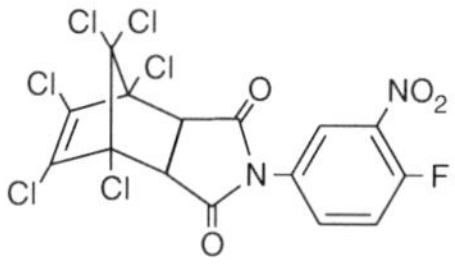

250 mg

S53,245-2
N-(5-NITRO-ORTHO-TOLYL)MALEIMIDE

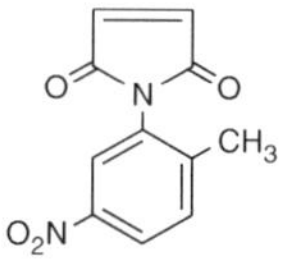

5 gm

S66,258-5
1,4-DIMETHYL-2,3-NAPHTHALENEDICARBOXIMIDE

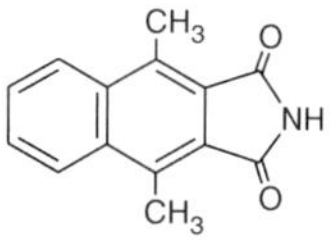

250 mg

S69,392-8
PHTHALIMIDO N-(4-ACETYLPHENYL)-CARBAMATE

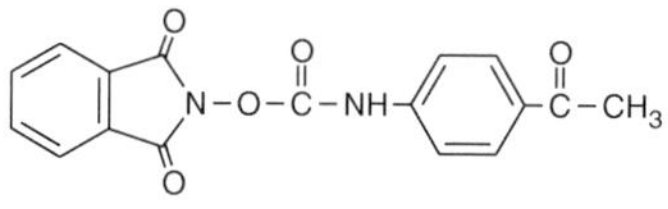

250 mg

S52,709-2
N-(4'-BROMOPHENACYL)PHTHALIMIDE

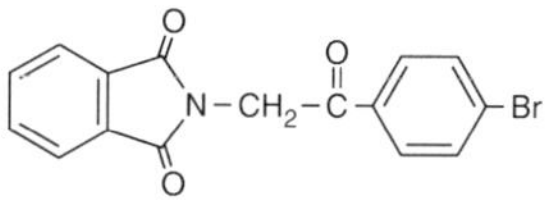

5 gm

S74,795-5
2-CHLORO-3-(3,4,5,6-TETRACHLOROPHTHALIMIDO)-1,4-NAPHTHOQUINONE

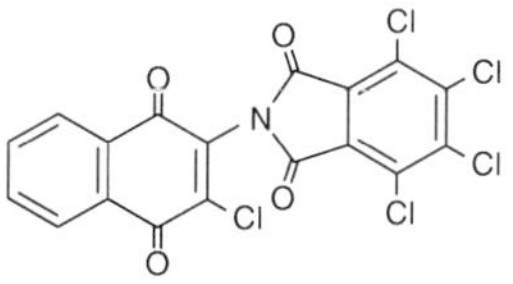

5 gm

S39,282-0
3-MALEIMIDOBENZOIC ACID

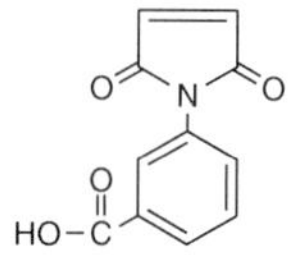

250 mg

S80,672-2
5-PHTHALIMIDO-2-PENTENOIC ACID

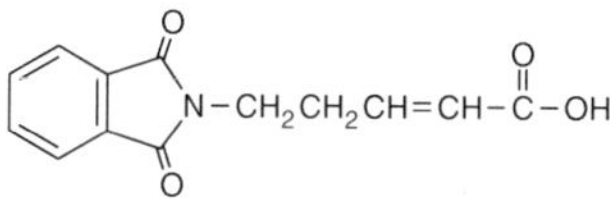

250 mg

S93,182-9
9-(PHTHALIMIDO)NONANOIC ACID

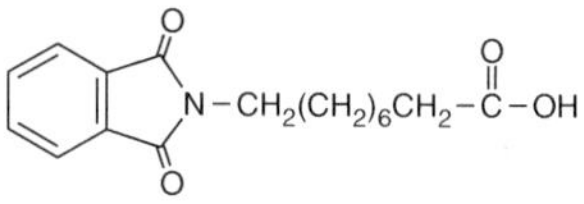

250 mg

S52,789-0
N-PHTHALOYL-DL-LEUCINE

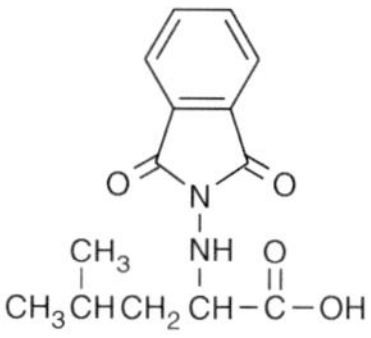

5 gm

S64,586-9
4-PHTHALIMIDOBUTYL BENZOATE

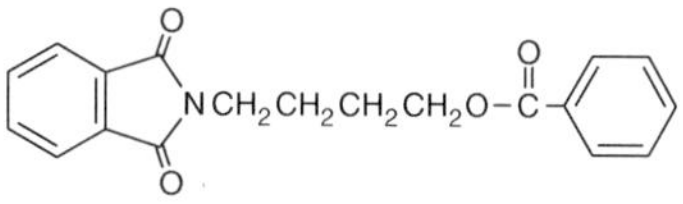

250 mg

S54,862-6
N-(3-CHLOROPHENYL)MALEIMIDE

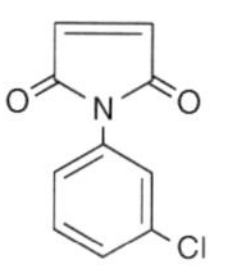

250 mg

S39,280-4
N-(2,5-DICHLOROPHENYL)MALEIMIDE

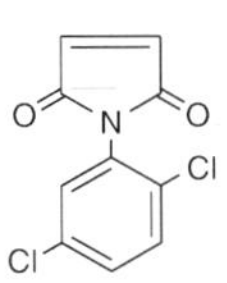

250 mg

S54,525-2
N-(4-BROMO-2-CHLOROPHENYL)-MALEIMIDE

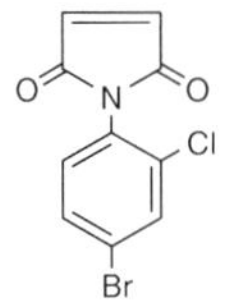

250 mg

S39,284-7
N-(PARA-TOLYL)MALEIMIDE

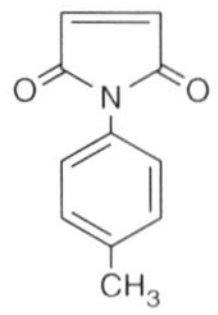

250 mg

S86,713-6
N,N'-(1,1'-BIPHENYL-4,4'-DIYL)-BISMALEIMIDE

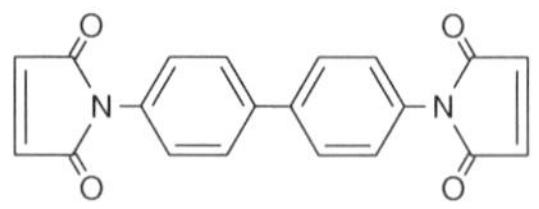

250 mg

S39,285-5
N-(1-NAPHTHYL)MALEIMIDE

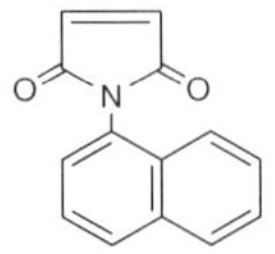

250 mg

S57,115-6
N-HYDROXYNAPHTHALIMIDE

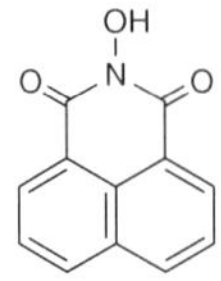

5 gm

S57,082-6
N-PHTHALOYL-DL-GLUTAMIC ACID

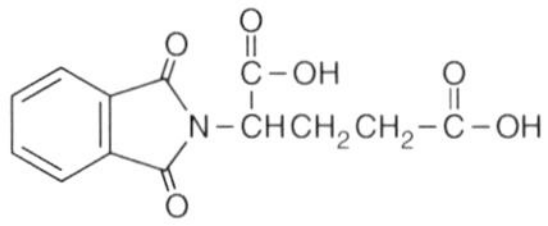

5 gm

S61,641-6
N-PHTHALOYL-L-LEUCINE

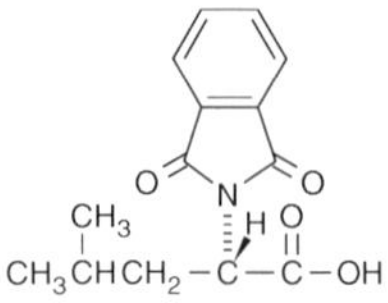

250 mg

S55,162-7
4-SUCCINIMIDOBENZOIC ACID

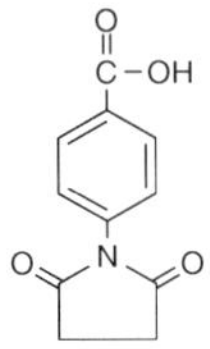

5 gm

S55,163-5
2-SUCCINIMIDOBENZOIC ACID

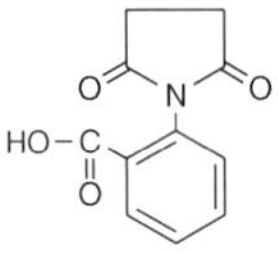

5 gm

S84,627-9
4-(2-PHTHALIMIDOPHENYL)BUTYRIC ACID

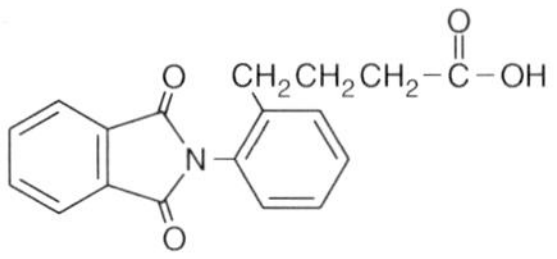

250 mg

S55,159-7
4-PHTHALIMIDOBENZOIC ACID

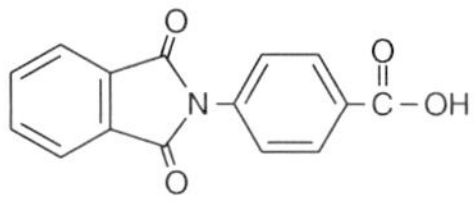

5 gm

S70,159-9
4-(2-PHTHALIMIDOPHENYL)BUTYRIC ACID

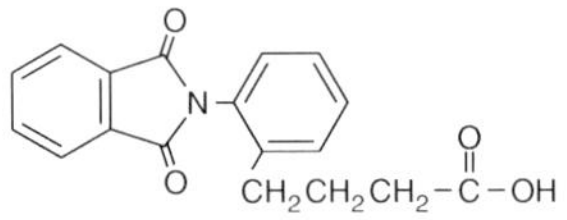

5 gm

S78,966-6
CARBOBENZYLOXY-L-ASPARTIC ACID 4-BENZYL 1-(N-HYDROXYSUCCINIMIDE)-ESTERS

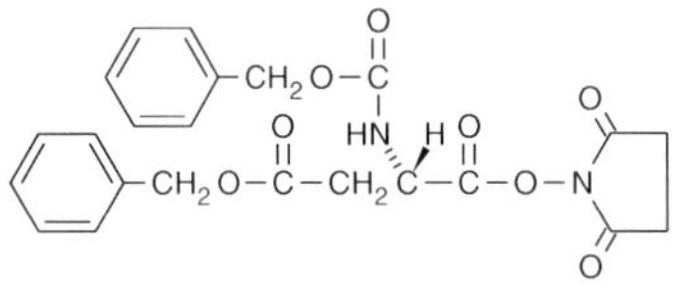

250 mg

S93,957-9
N-PHTHALOYL-L-GLUTAMIC ANHYDRIDE

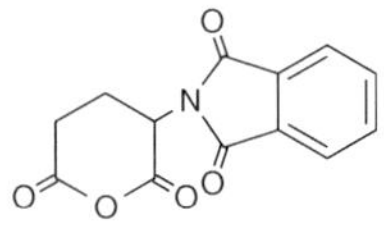

250 mg

S76,050-1
ETHYL 1-(2-ACETAMIDOETHYL)-5-(1,3-DIACETYL-2-IMIDAZOLIDINYLIDENE)-2,6-DIOXO-4-PHENYL-3-PIPERIDINECARBOXYLATE

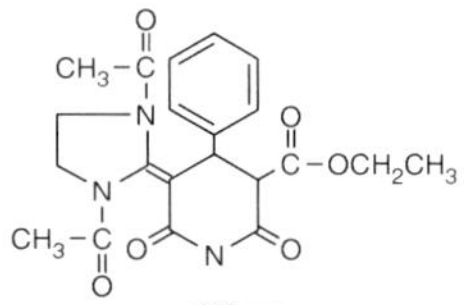

250 mg

S75,396-3
2-MALEIMIDO-9-FLUORENYL ACETATE

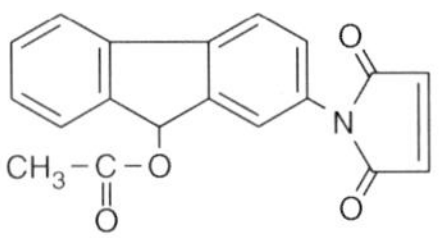

250 mg

S52,454-9
PHTHALOYL GLYCINE ETHYL ESTER

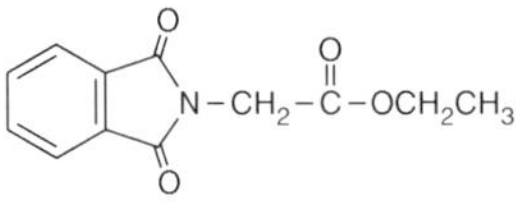

5 gm

S64,744-6
ETHYL 3-PHTHALIMIDOPROPIONATE

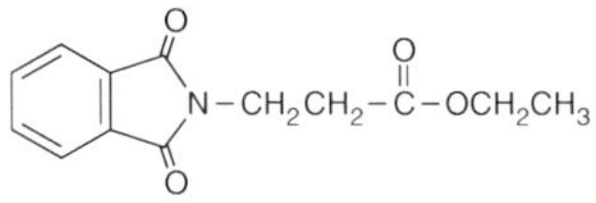

250 mg

S63,299-6
3-PHTHALIMIDOPROPIONYL CHLORIDE

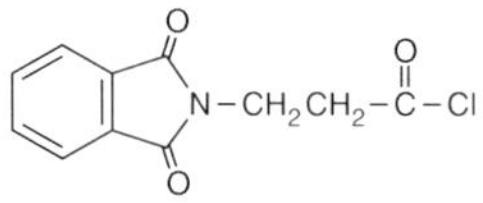

5 gm

S53,461-7
N-(4-METHYLBENZYL)PHTHALIMIDE

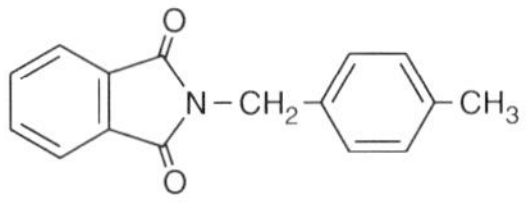

250 mg

S67,442-7
PHTHALIMIDO N-(1-NAPHTHYL)CARBAMATE

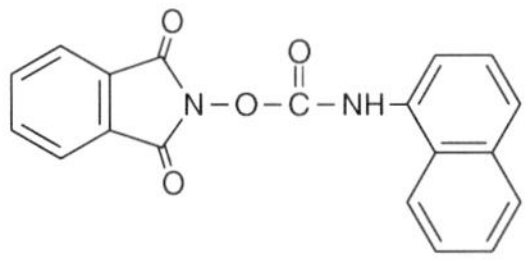

250 mg

S89,265-3
1,4,5,6,7,7-HEXACHLORO-N-(2,3-XYLYL)-5-NORBORNENE-2,3-DICARBOXIMIDE

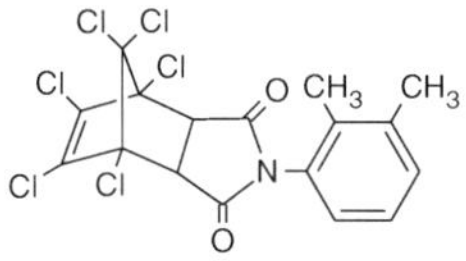

250 mg

S79,532-1
1,4,5,6,7,7-HEXACHLORO-N-(5-ISOPROPYL-2-METHYLPHENYL)-5-NORBORNENE-2,3-DICARBOXIMIDE

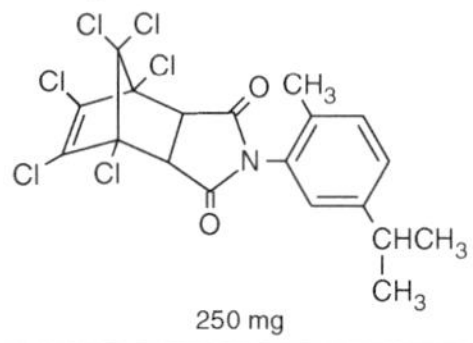

250 mg

S89,262-9
1,4,5,6,7,7-HEXACHLORO-N-(1-NAPHTHYL)-5-NORBORNENE-2,3-DICARBOXIMIDE

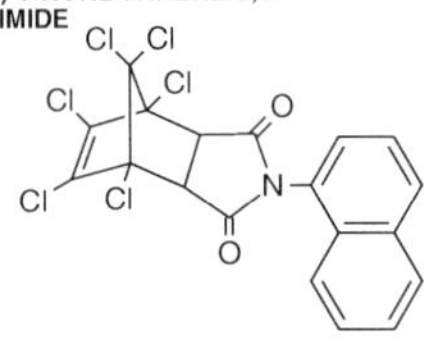

250 mg

S55,158-9
N-(3-CHLOROPHENYL)SUCCINIMIDE

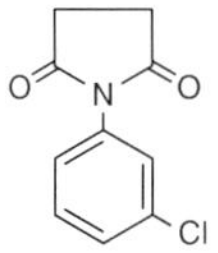

5 gm

S52,338-0
N-(6-CHLORO-ORTHO-TOLYL)MALEIMIDE

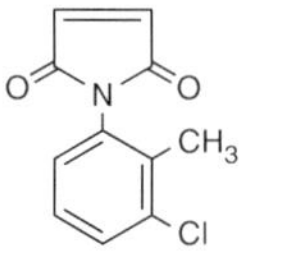

5 gm

S52,387-9
N-(3-CHLORO-PARA-TOLYL)MALEIMIDE

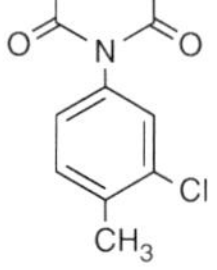

5 gm

S85,222-8
9,10-DIHYDRO-9,10-ETHANOANTHRACENE-11,12-DICARBOXIMIDE

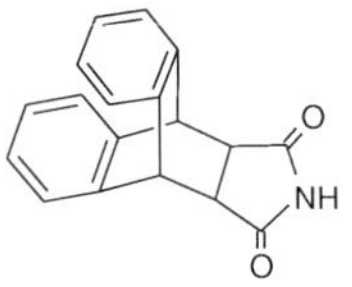

250 mg

S70,293-5
N-PHENYL-2-PIPERIDINOSUCCINIMIDE

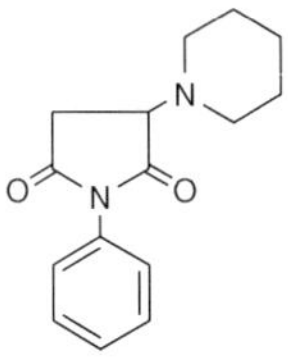

5 gm

S70,320-6
ALPHA-MORPHOLINO-N-PHENYLSUCCINIMIDE

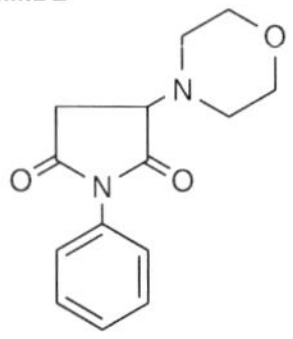

5 gm

S95,155-2
N-[2-(1-BROMOETHYL)PHENYL]-SUCCINIMIDE

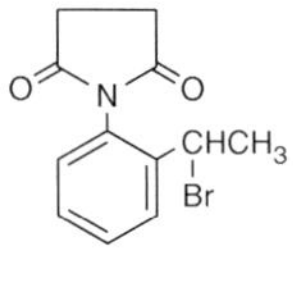

50 mg

S73,226-5
N-(4-FLUOROPHENYL)MALEIMIDE

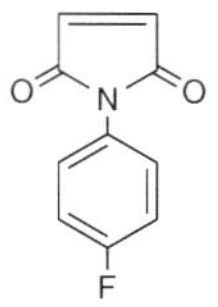

250 mg

S71,919-6
N-(2,3-XYLYL)MALEIMIDE

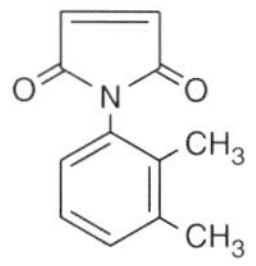

5 gm

S71,943-9
N-(2,4-XYLYL)MALEIMIDE

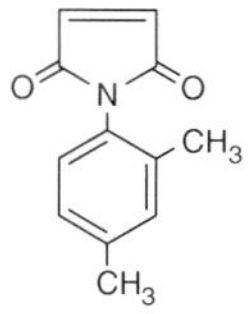

5 gm

S75,498-6
2-MALEIMIDOFLUORENE

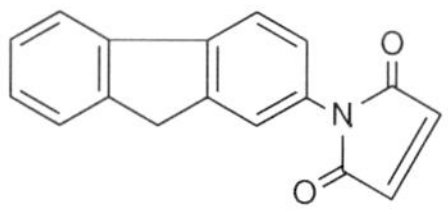

250 mg

S93,549-2
N-PHENYLGLUTARIMIDE

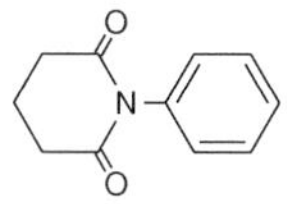

250 mg

S94,585-4
2-(4-ISOPROPYL-3-METHYLPHENYL)-1,3-(2H,4H)-ISOQUINOLINEDIONE

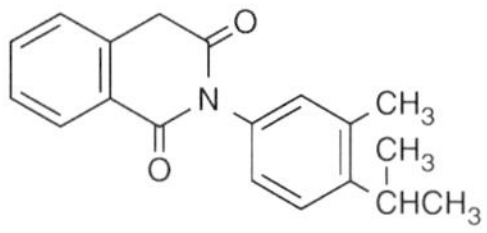

250 mg

S65,872-3
1,4,5,8-NAPHTHALENETETRACARBOXYLIC DIIMIDE

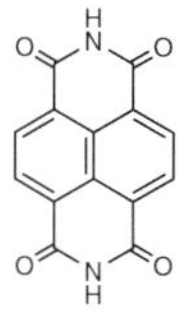

250 mg

S40,971-5
3,4,9,10-PERYLENETETRACARBOXYLIC DIIMIDE

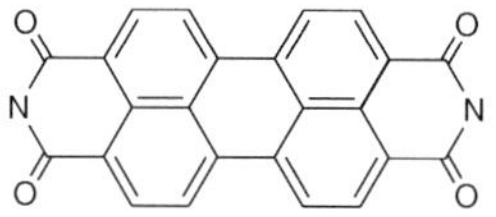

5 gm

S75,829-9
N-(2-FLUORENYL)PHTHALIMIDE

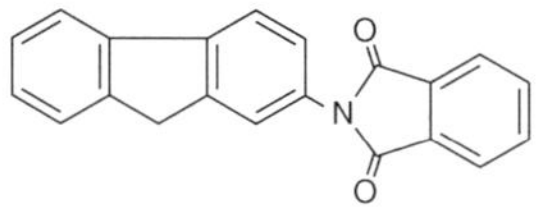

250 mg

S52,511-1
N-(2-METHYLALLYL)PHTHALIMIDE

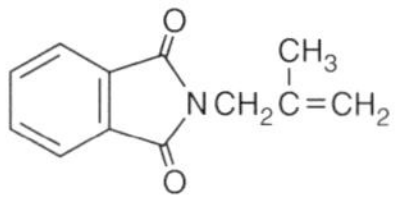

5 gm

S69,345-6
N-(4-PHTHALIMIDOPHENYL)ACETAMIDE

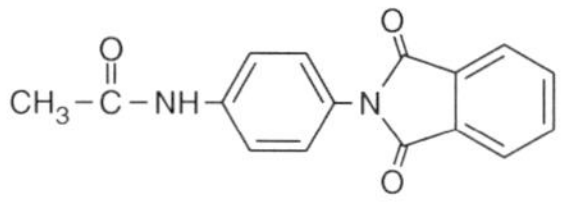

5 gm

S55,160-0
N-(4-CHLOROPHENYL)PHTHALIMIDE

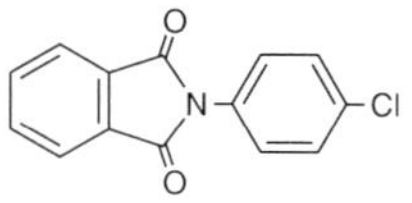

5 gm

S55,164-3
N-(3-CHLOROPHENYL)PHTHALIMIDE

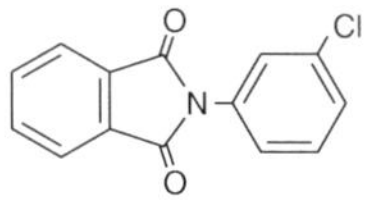

5 gm

S54,523-6
PHTHALIC ANHYDRIDE AZINE

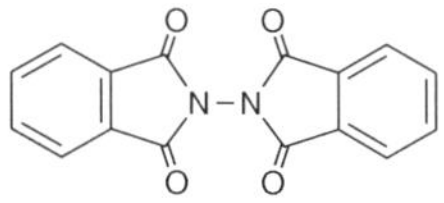

250 mg

S55,150-3
N-(3,4-DICHLOROPHENYL)PHTHALIMIDE

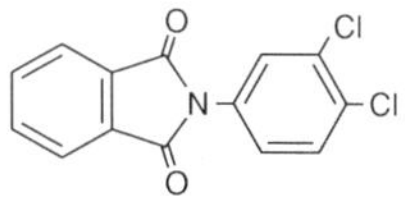

5 gm

S79,825-8
TETRAFLUOROPHTHALIMIDE

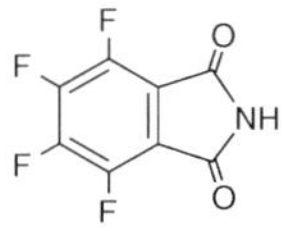

250 mg

S74,265-1
N-METHYLPHTHALIMIDE

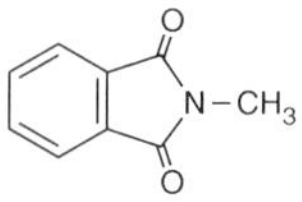

250 mg

S70,254-4
N-ISOPROPYLPHTHALIMIDE

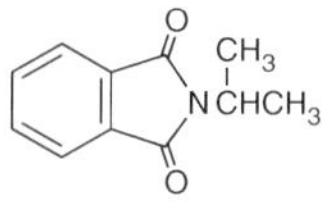

5 gm

S85,305-4
N-(2-BUTENYL)PHTHALIMIDE

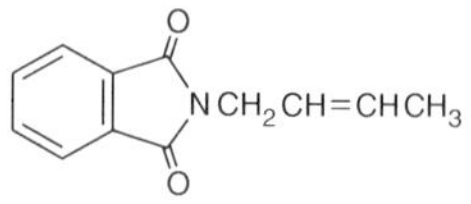

250 mg

S92,746-5
N-(3-METHYL-2-BUTEN-1-YL)-PHTHALIMIDE

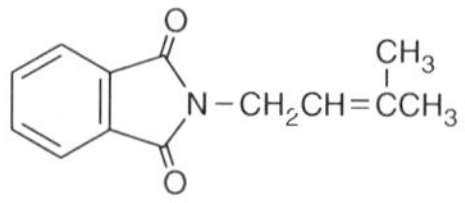

250 mg

S93,272-8
N-(2,3-DIMETHYL-2-BUTEN-1-YL)-PHTHALIMIDE

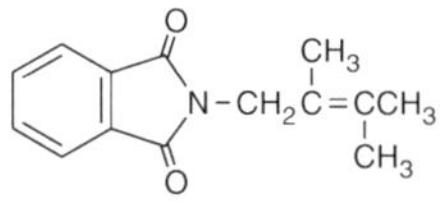

250 mg

S61,678-8
N,N'-THIODIPHTHALIMIDE

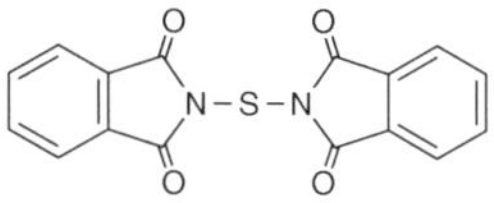

5 gm

S68,890-8
1-AMINO-3-(3-METHYLPHENYL)GUANIDINE

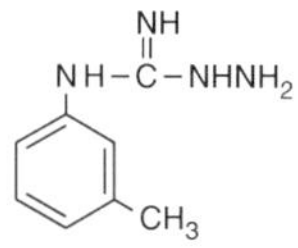

5 gm

S61,476-9
1-[3-(TRIFLUOROMETHYL)PHENYL]-GUANIDINE CARBONATE

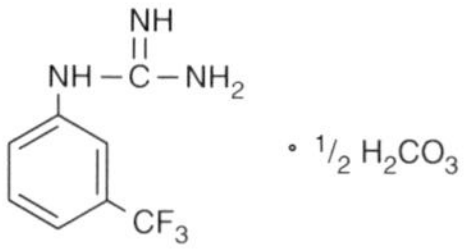

5 gm

S87,861-8
N,N'-BIS[3-(TRIFLUOROMETHYL)PHENYL]-FORMAMIDINE

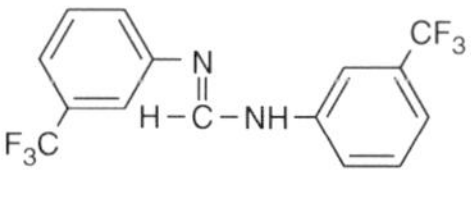

5 gm

S90,989-0
1-(2-CHLORO-4-NITROPHENYL)-3-(2,2,2-TRICHLOROACETIMIDOYL)UREA

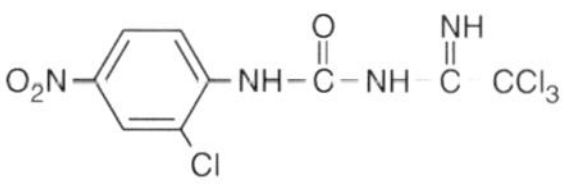

250 mg

S90,995-5
1-(2,5-DIMETHOXYPHENYL)-3-(2,2,2-TRICHLOROACETIMIDOYL)UREA

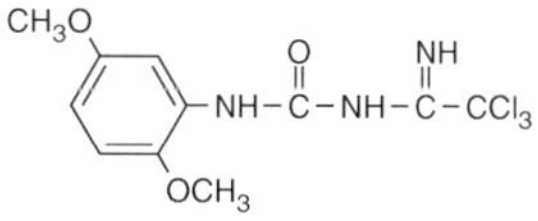

250 mg

S74,166-3
5-IMINO-3-(P-TOLYL)-1-(3-TRIFLUOROMETHYLPHENYL)HYDANTOIN

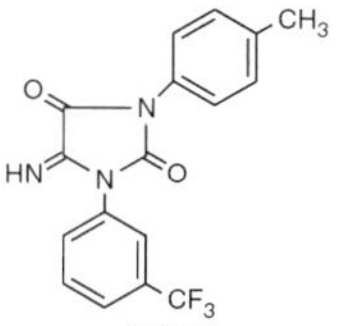

250 mg

S73,258-3
2-METHYL-N-[3-(TRIFLUOROMETHYL)-PHENYL]-2-IMIDAZOLINE-1-CARBOXAMIDE

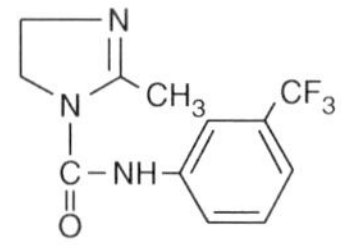

250 mg

S76,738-7
1-(2,6-DICHLORO-ALPHA-HYDROXYBENZYLAMINO)GUANIDINE HYDROCHLORIDE

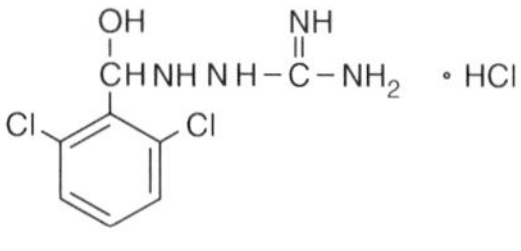

250 mg

S74,208-2
N-[1-(4-CHLOROPHENYL)-1-METHOXY-1H-ISOINDOL-3-YL]GLYCINE METHYL ESTER HCL

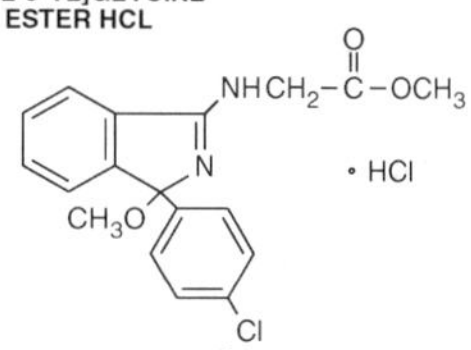

5 gm

S70,336-2
1-[1-AMINO-2-(3,4-DICHLOROPHENYL)-ETHYLIDENE]UREA HYDROCHLORIDE

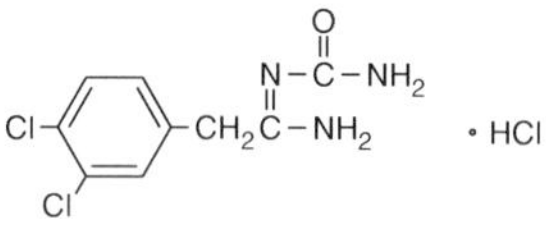

5 gm

S87,919-3
N,N'-BIS(4-CHLORO-3-NITROPHENYL)-FORMAMIDINE

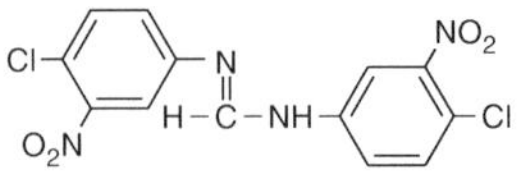

5 gm

S67,097-9
10-(3-FLUOROBENZYL)-2,3,4,4A-TETRAHYDROBENZO(B)NAPHTHYRIDIN-5-(10H)-ONE

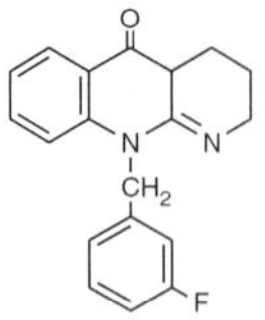

5 gm

S50,104-2
N,N'-BIS(4-CHLOROPHENYL)-BENZAMIDINE

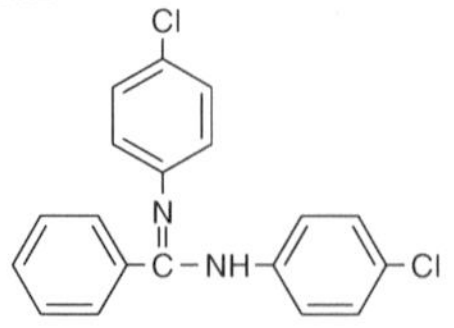

5 gm

S74,401-8
1-(3-CHLOROPHENYL)-5-IMINO-3-(P-TOLYL)HYDANTOIN

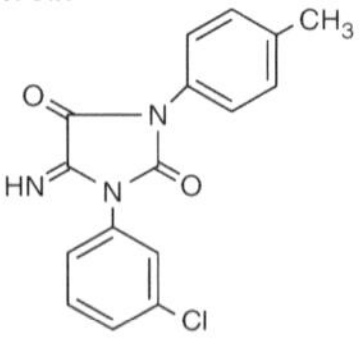

250 mg

S77,162-7
1-[(3,4-DICHLOROBENZOYL)AMINO]-GUANIDINE ACETATE

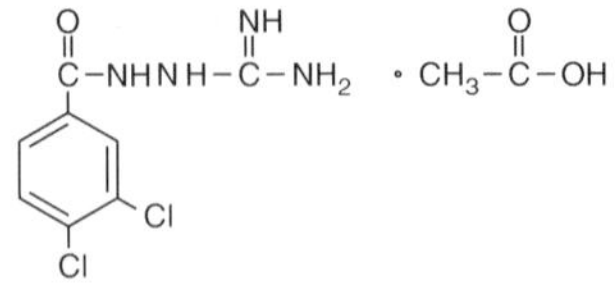

250 mg

S78,108-8
1-(3-CHLORO-2-METHYLPHENYL)-5-IMINO-3-PHENYLHYDANTOIN

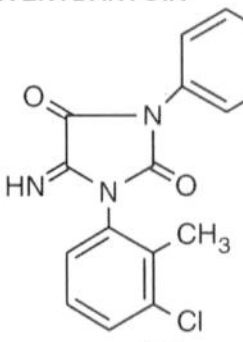

250 mg

S78,109-6
1-(4-CHLORO-2-METHYLPHENYL)-5-IMINO-3-PHENYLHYDANTOIN

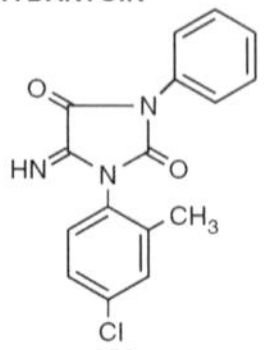

250 mg

S74,066-7
1-(3,4-DICHLOROPHENYL)-5-IMINO-3-(P-TOLYL)HYDANTOIN

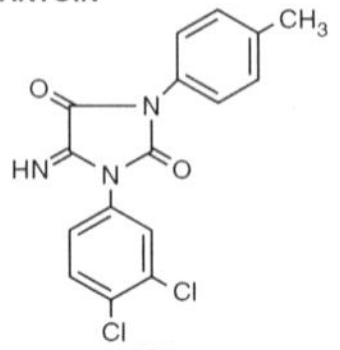

250 mg

S74,067-5
1-(3,5-DICHLOROPHENYL)-5-IMINO-3-(P-TOLYL)HYDANTOIN

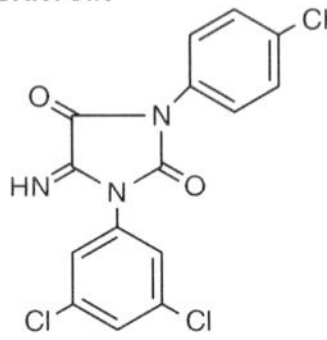

250 mg

S44,736-6
1-(4-CHLOROPHENYL)BIGUANIDE HYDRATE

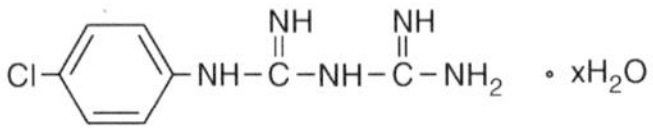

5 gm

S87,908-8
N,N'-BIS(4-FLUOROPHENYL)FORMAMIDINE

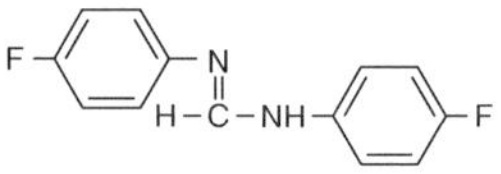

5 gm

S87,856-1
N,N'-BIS(4-BROMOPHENYL)FORMAMIDINE

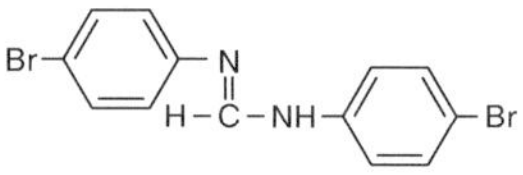

5 gm

S68,889-4
1,5-BIS(4-CHLOROPHENYL)BIGUANIDE HYDROCHLORIDE

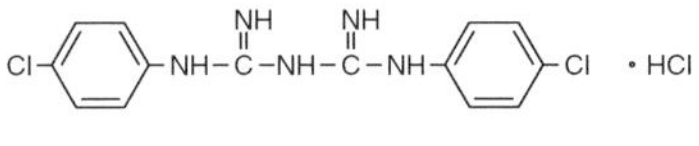

5 gm

S77,148-1
GUANIDINO 2-(2,6-DICHLOROPHENYL)-ACETATE

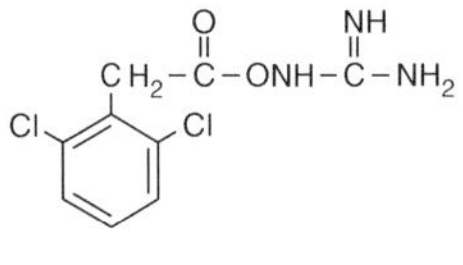

250 mg

S87,920-7
N,N'-BIS(3,4-DICHLOROPHENYL)-FORMAMIDINE

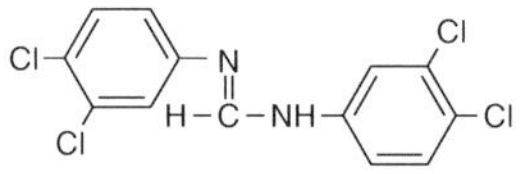

5 gm

S87,921-5
N,N'-BIS(3-CHLORO-4-METHYLPHENYL)-FORMAMIDINE

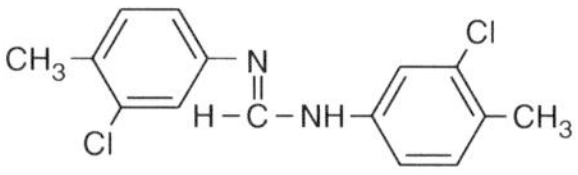

5 gm

S87,922-3
N,N'-BIS(4-CHLORO-2-METHYLPHENYL)-FORMAMIDINE

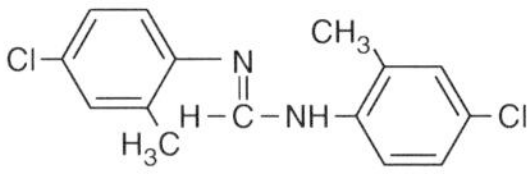

5 gm

S94,864-0
2-AMINO-2-OXAZOLINE PICRATE

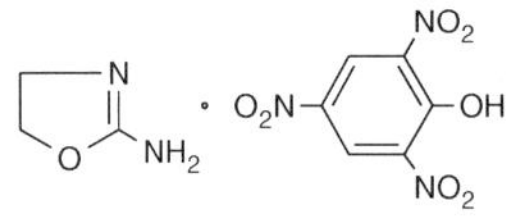

250 mg

S55,897-4
1-BENZYL-5-IMINO-2-PYRAZOLIN-3-OL

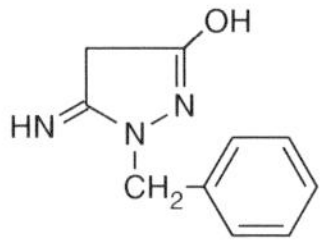

250 mg

S76,155-9
4-HYDROXYBENZAMIDINE HYDROCHLORIDE

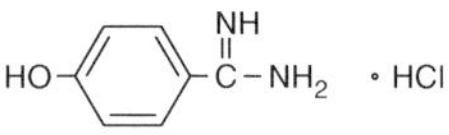

5 gm

S83,935-3
1,3-BIS(4-METHOXYPHENYL)-2-PHENYLGUANIDINE

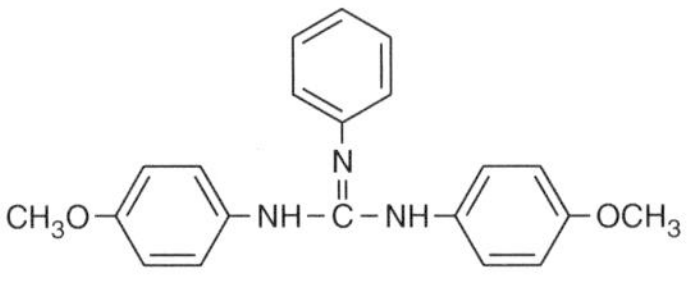

250 mg

S87,915-0
N,N'-BIS(4-METHOXYPHENYL)-FORMAMIDINE

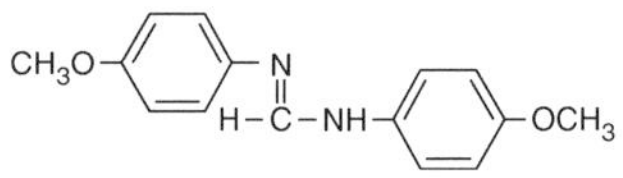

5 gm

S42,055-7
MESITYLOXYACETAMIDINE HYDROCHLORIDE

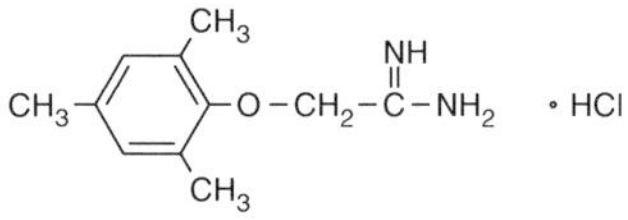

250 mg

S86,242-8
1-(4-ETHOXYPHENYL)GUANIDINE P-TOLUENESULFONATE

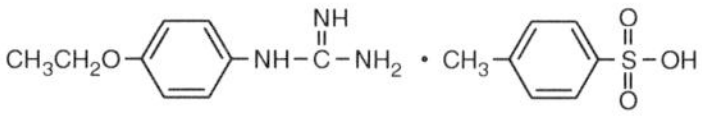

250 mg

S54,093-5
1-(2-BENZOXAZOLYL)GUANIDINE

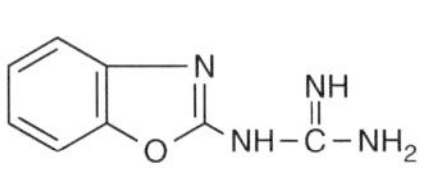

5 gm

S73,268-0
2-(4-METHOXYSTYRYL)-2-IMIDAZOLINE

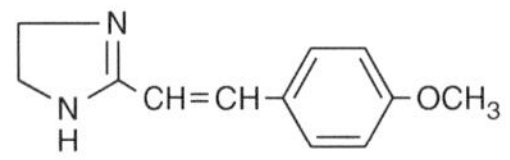

250 mg

S93,563-8
N-(ALPHA-PHENOXYBENZYLIDENE)-2-NAPHTHYLAMINE

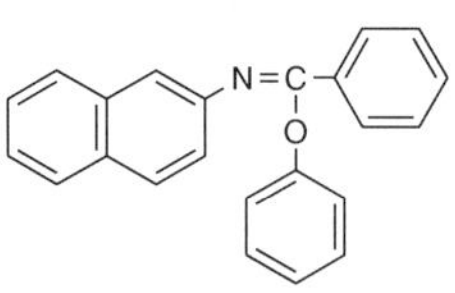

250 mg

S61,449-1
1-(4-METHOXYPHENYL)-GUANIDINE CARBONATE

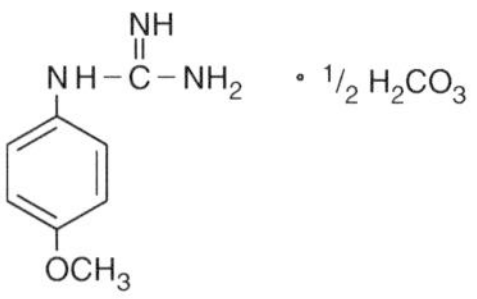

5 gm

S36,141-0
1-4-(4-ETHOXYPHENYL)BIGUANIDE HYDROCHLORIDE

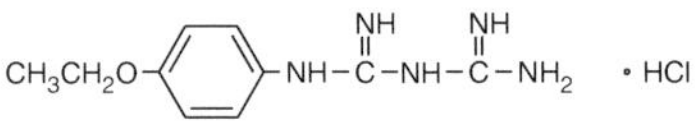

5 gm

S78,562-8
CARBOBENZYLOXYNITROARGINYLALANINE BENZYL ESTER

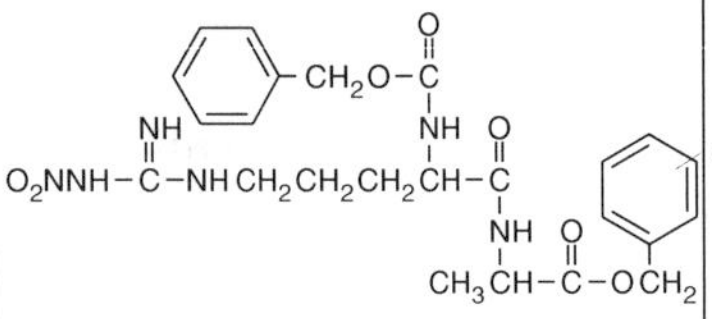

250 mg

S78,879-1
L-NITROARGININE 4-NITROANILIDEHYDROBROMIDE

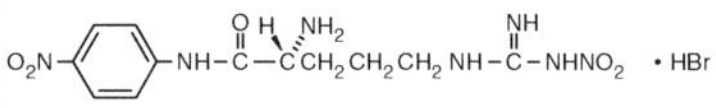

250 mg

S95,346-6
ETHYL 4-AMIDINOBENZOATE

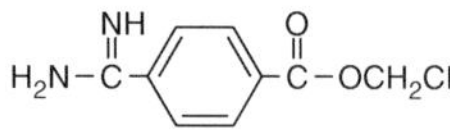

50 mg

S61,468-8
1-(2-NITROPHENYL)GUANIDINE NITRATE

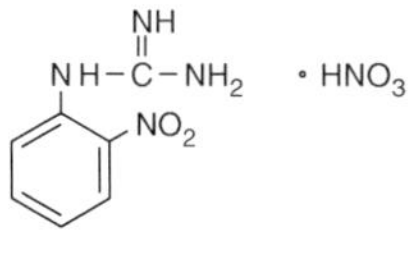

5 gm

S57,594-1
2-GUANYLBENZIMIDAZOLE

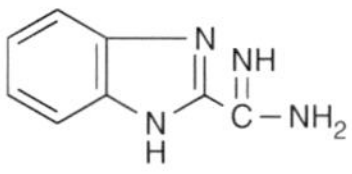

5 gm

S76,110-9
7-IMINO-3,4,6,7-TETRAHYDRO-2H-PYRIMIDO(2,1-A)ISOQUINOLINE

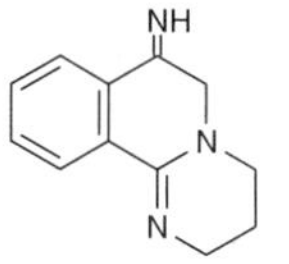

250 mg

S50,102-6
N,N'-BIS(3-NITROPHENYL)-BENZAMIDINE

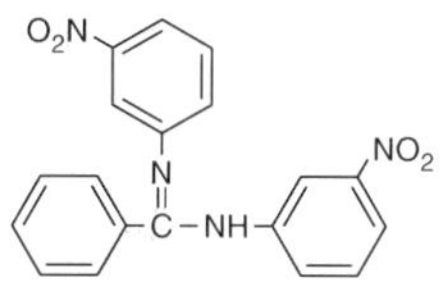

5 gm

S78,090-1
5-IMINO-1-(2-METHYL-5-NITROPHENYL)-3-PHENYLHYDANTOIN

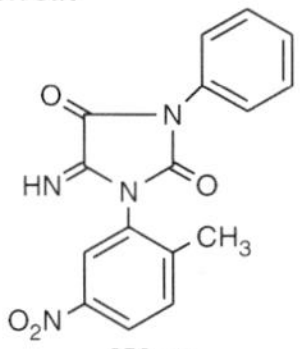

250 mg

S87,863-4
N,N'-BIS(2-METHYL-5-NITROPHENYL)-FORMAMIDINE

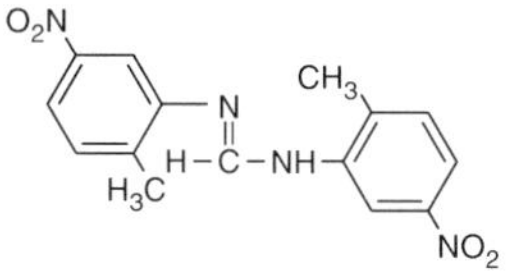

5 gm

S69,745-1
3-IMINO-2-PHENYLPYRAZOLIDINE

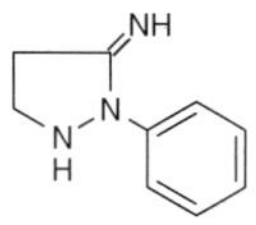

250 mg

S70,491-1
N-ALPHA-TRITYL-L-ARGININE

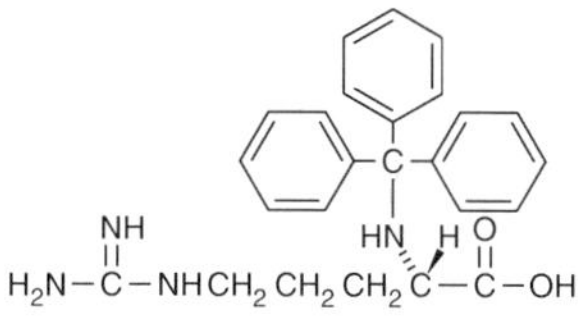

5 gm

S70,498-9
N-ALPHA-TRITYL-L-ARGININE MONOHYDRATE

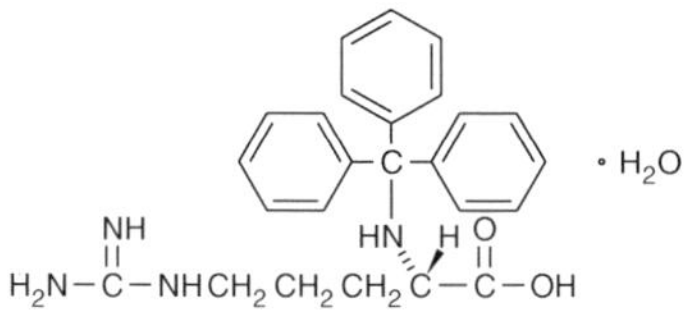

5 gm

S87,449-3
4,5-DIMETHYL-2-[4-(DIMETHYLAMINO)-PHENYLIMINO]-3-ETHYL-4-THIAZOLINE

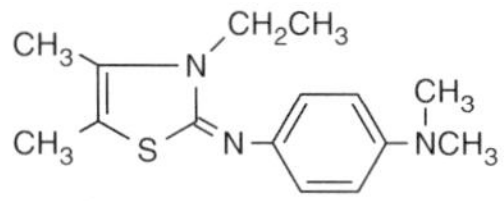

250 mg

S87,442-6
1,3-DIETHYL-2,3-DIHYDRO-2-[4-(DIMETHYLAMINO)PHENYLIMINO]-BENZIMIDAZOLE

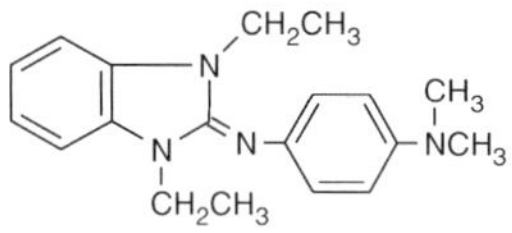

250 mg

S87,450-7
1,2-DIHYDRO-2-[4-(DIMETHYLAMINO)-PHENYLIMINO]-1-ETHYLPYRIDINE

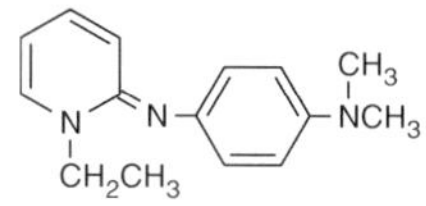

250 mg

S87,436-1
3,3-DIMETHYL-2-[4-(DIMETHYLAMINO)-PHENYLIMINO]-1-ETHYLINDOLINE HYDROCHLORIDE

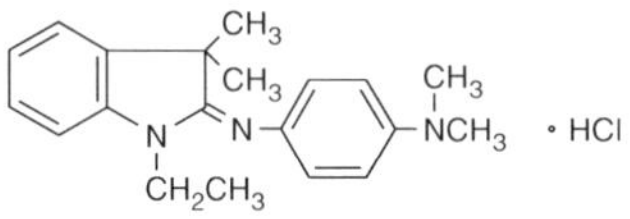

250 mg

S87,444-2
1,2-DIHYDRO-2-[4-(DIMETHYLAMINO)-PHENYLIMINO]-1-ETHYLQUINOLINE HYDROCHLORIDE

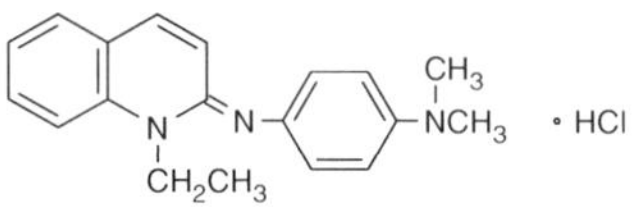

250 mg

S87,446-9
2-[4-(DIMETHYLAMINO)PHENYLIMINO]-3-ETHYLBENZOTHIAZOLINE

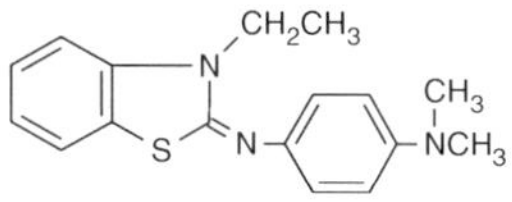

250 mg

S78,587-3
NALPHA-ACETYLNITROARGININE 4-NITROANILIDE

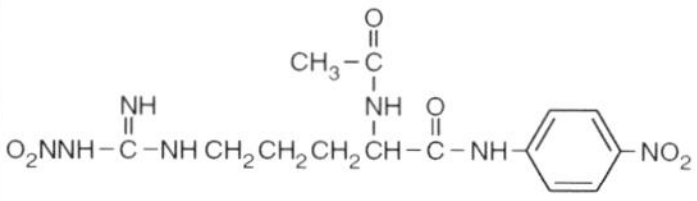

250 mg

S87,985-1
3-NITROBENZENECARBOXIMIDIC HYDRAZIDE

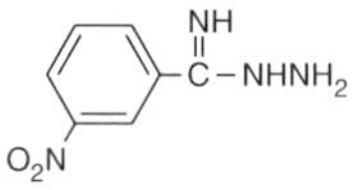

5 gm

S61,466-1
1-(4-NITROPHENYL)GUANIDINE HYDROCHLORIDE

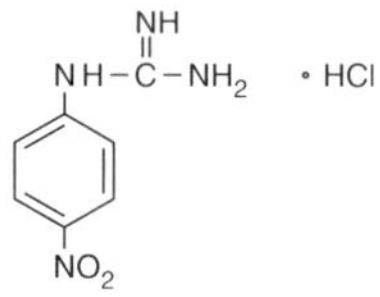

5 gm

S61,737-7
1-(4-NITROPHENYL)BIGUANIDE

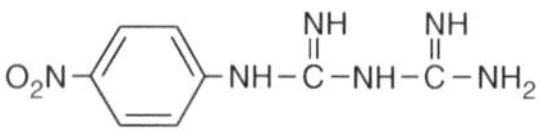

5 gm

S45,193-2
HIPPURYL-L-ARGININE ACETATE

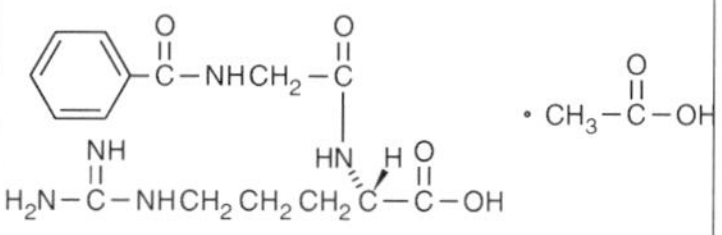

250 mg

S78,663-2
NALPHA-BENZOYL-DL-ARGININE

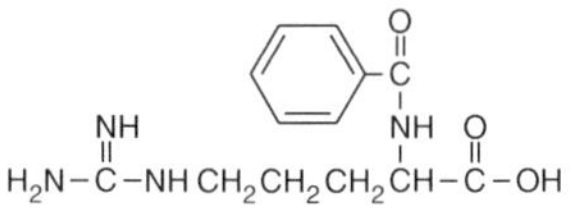

250 mg

S46,053-2
N-ALPHA-BENZOYL-L-ARGININE AMIDE HYDROCHLORIDE

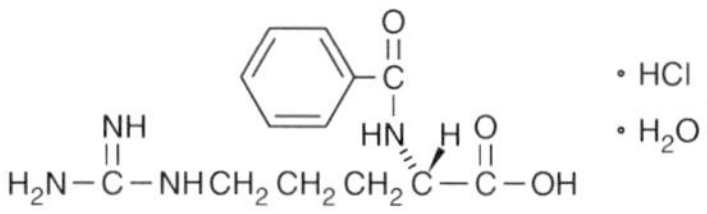

5 gm

S73,424-1
2-BENZYL-2-IMIDAZOLINE THIOSULFATE

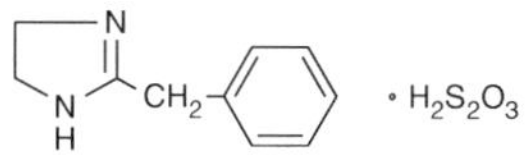

250 mg

S37,028-2
PHENETHYLGUANIDINE HEMISULFATE

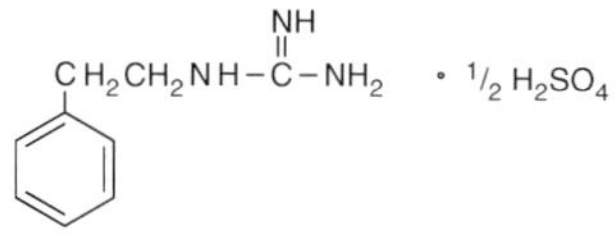

5 gm

S45,631-4
N1-PHENETHYLBIGUANIDE HYDROCHLORIDE

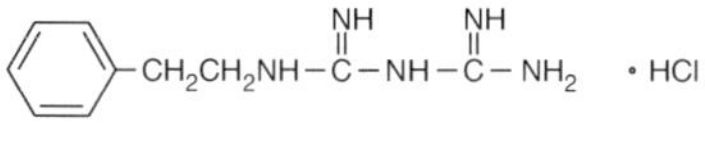

5 gm

S94,146-8
N,N'-BIS(2-ANTHRAQUINONYL)-BENZAMIDINE

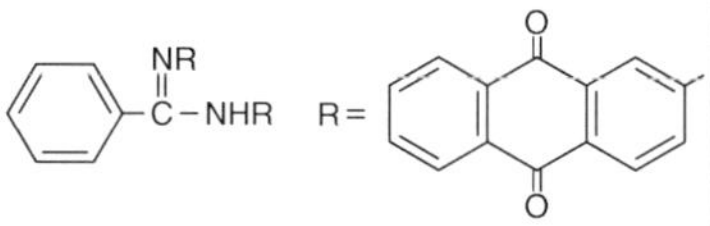

250 mg

S94,147-6
N,N'-BIS(2-ANTHRAQUINONYL)-ACETAMIDINE

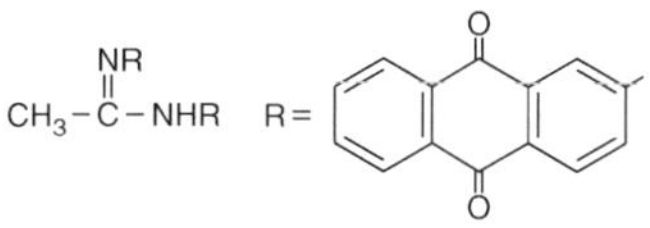

250 mg

S86,237-1
1-(P-TOLYL)GUANIDINE P-TOLUENESULFONATE

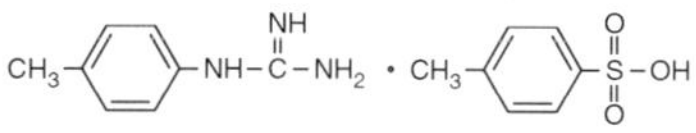

250 mg

S87,312-8
FORMIC (1,2-DIHYDRO-2-METHYL-1-ISOQUINOLYLIDENE)HYDRAZIDE

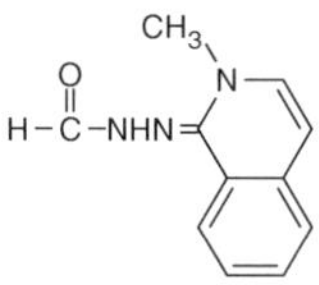

250 mg

S87,238-5
BENZOIC (3,4,5-TRIMETHYL-4-THIAZOLIN-2-YLIDENE)HYDRAZIDE

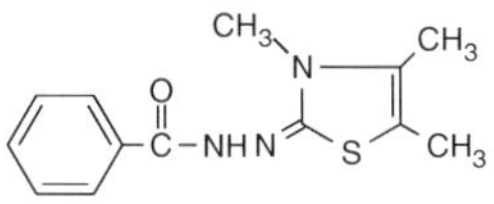

250 mg

S87,156-7
ACETIC (3-METHYL-2-BENZOTHIAZOLINYLIDENE)HYDRAZIDE

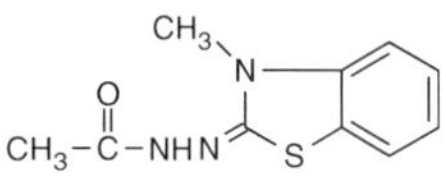

250 mg

S87,006-4
4-TERT-BUTYL-3-PHENYL-4-THIAZOLIN-2-ONE HYDRAZONE

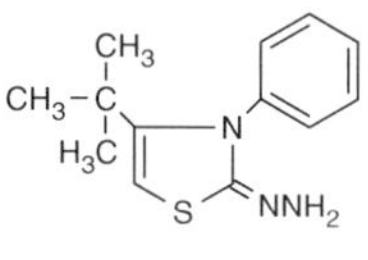

250 mg

S87,030-7
3,4-DIPHENYL-4-THIAZOLIN-2-ONE HYDRAZONE
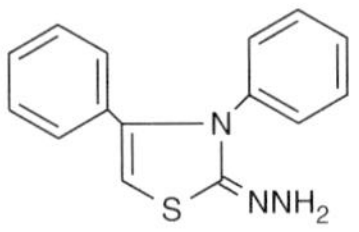
250 mg

S66,191-0
2,10-DIHYDRO-10-ETHYLIMIDAZO(2,1-B)-QUINAZOLINE-5(3H)-ONE
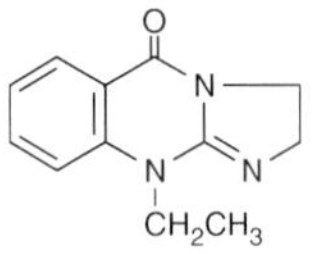
250 mg

S86,306-8
ETHYL (DIPHENYLAMIDINO)CARBAZATE
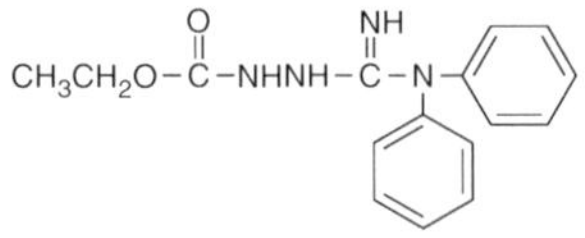
250 mg

S86,945-7
1-PHENYL-2-IMIDAZOLINE HYDRIODIDE
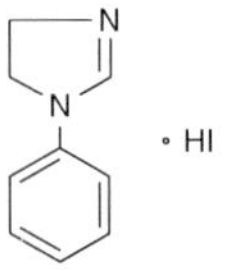
250 mg

S95,338-5
4-(DIMETHYLAMINO)-N-PHENYLBENZAMIDINE
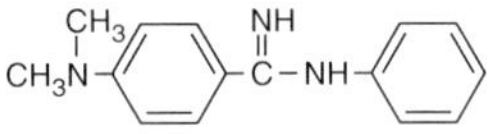
50 mg

S67,641-1
2-ETHYL-1-PHENYL-2-THIOPSEUDOUREA HYDRIODIDE
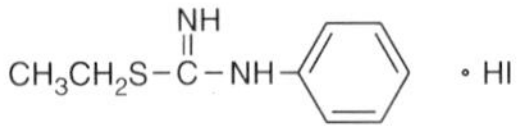
5 gm

S81,773-2
4,4'-(PHENYLIMINOMETHYLENE) BISMORPHOLINE
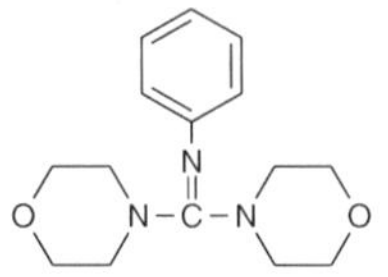
250 mg

S67,146-0
10-BENZYL-2,10-DIHYDROIMIDAZO(2,1-B)QUINAZOLIN-5(3H)-ONE METHANESULFONATE
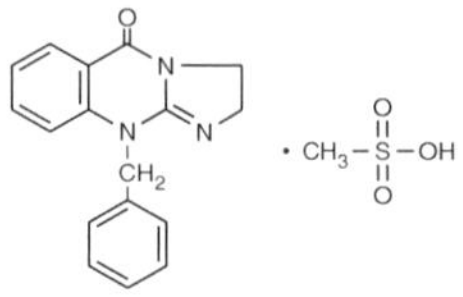
5 gm

S66,194-5
11-(4-FLUOROBENZYL)-2,3,4,11-TETRAHYDRO-6H-PYRIMIDO(2,1-B) QUINAZOLIN-6-ONE
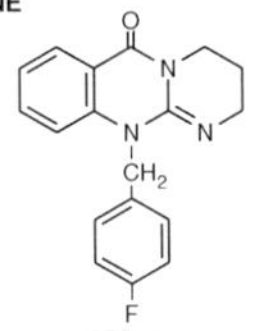
250 mg

S65,766-2
3,4-DIHYDROPYRIMIDO(2,1-A)-ISOINDOL-6(2H)-ONE
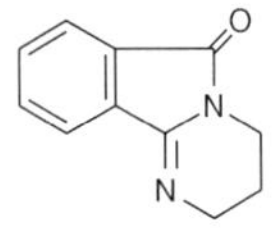
5 gm

S66,217-8
10-BENZYL-2,3-DIHYDROIMIDAZO(2,1-B)-QUINAZOLIN-5(10H)-ONE
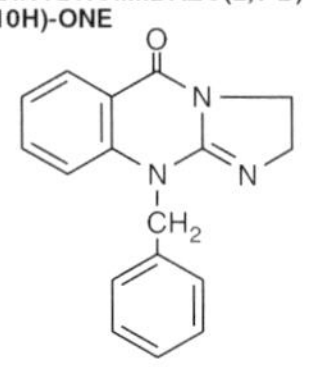
5 gm

S82,267-1
1-BENZYLGUANIDINE NITRATE
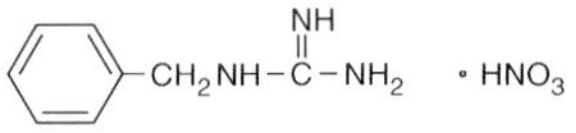
250 mg

S67,279-3
1-PHENYLBIGUANIDINE HYDROCHLORIDE
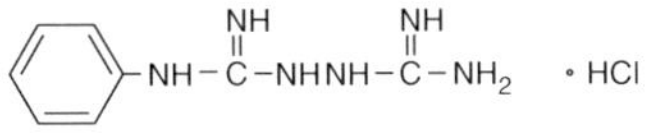
250 mg

S65,717-4
1-[N-(3-PHENYLPROPYL)AMINO]-GUANIDINE NITRATE
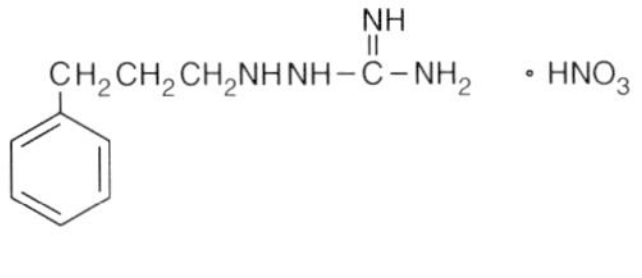
250 mg

S68,883-5
1-AMINO-3-(2,4,5-TRIMETHYLPHENYL)-GUANIDINE
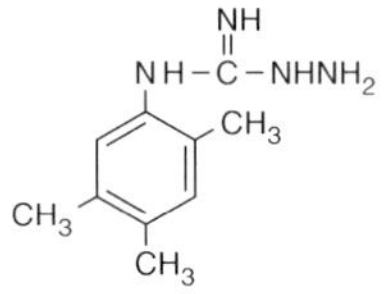
5 gm

S90,940-8
1-AMINO-1,2-DIHYDRO-2-QUINOLINIMINE
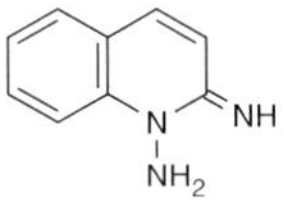
250 mg

S90,906-8
2-AMINO-1,2-DIHYDRO-1-ISOQUINOLINIMINE
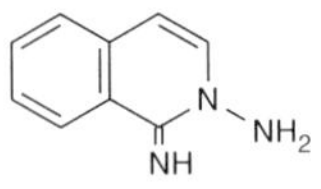
250 mg

S86,274-6
4-PHENYLSEMICARBAZIDE HYDRAZONE
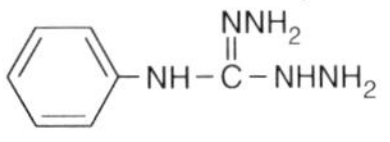
250 mg

S89,925-9
5-(2-ANTHRACENYL)-1,3-DIPHENYLFORMAZAN
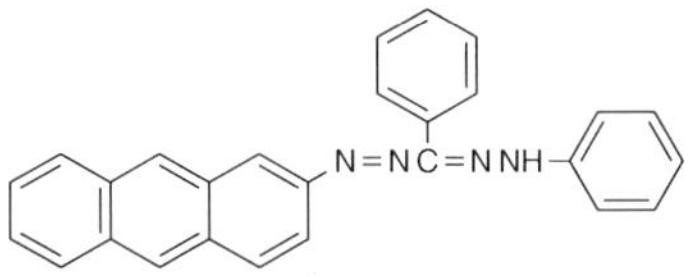
250 mg

S79,378-7
1,2,3,4,5,6-HEXAHYDRO-2-IMINO-5-PYRIMIDINYL BENZOATE HYDROCHLORIDE
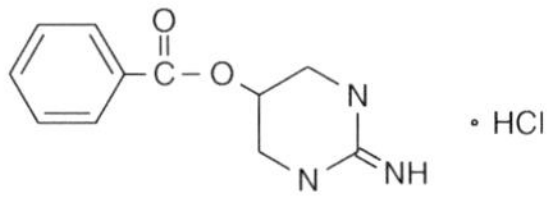
250 mg

S88,798-6
3-GUANIDINOPROPYL 4-NITROBENZOATE HYROCHLORIDE
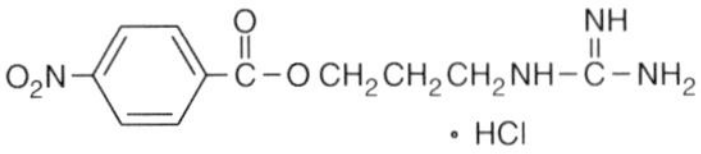
250 mg

S51,889-1
5,5-DIETHYL-1,3-DIPHENYL-2-IMINOBARBITURIC ACID
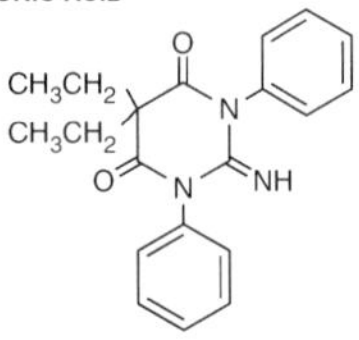
5 gm

S76,932-0
1-[4-(ETHOXYCARBONYL)PHENYL]-5-IMINO-3-PHENYLHYDANTOIN
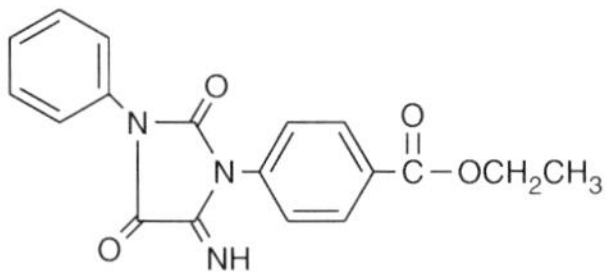
250 mg

S73,302-4
ETHYL 1-[2-(METHOXYCARBONYL)-PHENYLCARBAMOYL]-2-IMIDAZOLINE-2-ACETATE
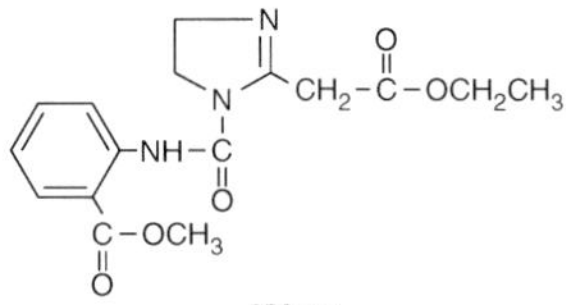
250 mg

S62,007-6
1-BENZOYLGUANIDINE
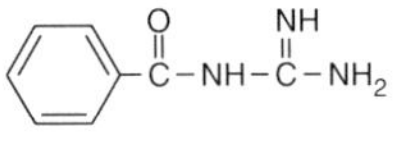
5 gm

S86,261-4
1-BENZOYLGUANIDINE HYDROCHLORIDE
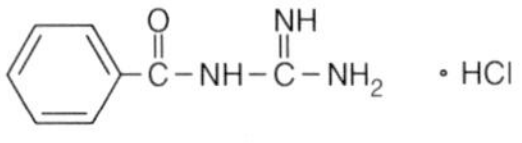
250 mg

S74,061-6
1,3-DIPHENYL-5-IMINOHYDANTOIN
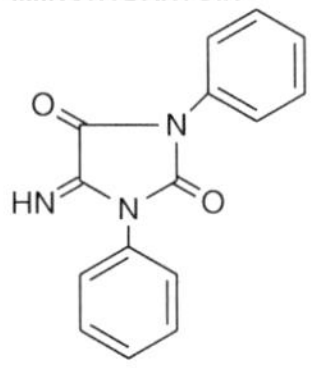
250 mg

S74,078-0
1,3-DI(P-TOLYL)-5-IMINOHYDANTOIN
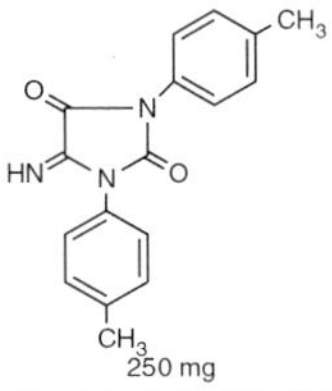
250 mg

S75,911-2
2,3-DIHYDRO-5H-IMIDAZO(2,1-A)-ISOINDOL-5-ONE
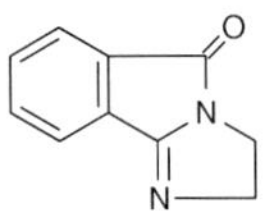
250 mg

S76,166-4
6,12-DIHYDROBENZIMIDAZO(2,1-B)-QUINAZOLIN-12-ONE
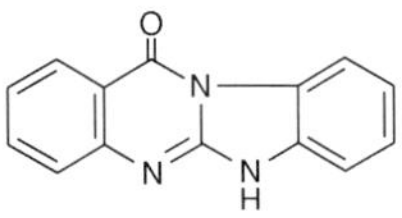
250 mg

S76,134-6
3-PHENYL-2,3,3A,10-TETRAHYDROPYRROLO(2,3-B)(1,5)-BENZODIAZEPIN-4(5H)-ONE
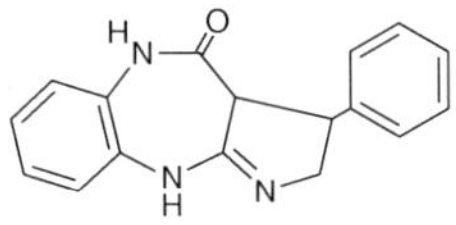
250 mg

S73,369-5
4,5-DIHYDRO-2,4,5-TRIPHENYLIMIDAZOLE
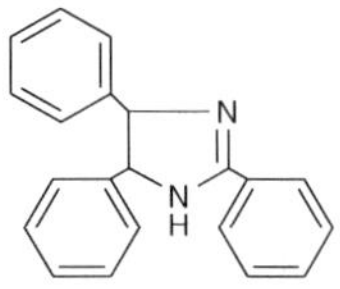
250 mg

S70,648-5
2,5-DIIMINO-3A,6A-DIPHENYL-1,3,4,6-TETRACHLOROIMIDAZO(4,5-D)IMIDAZOLE

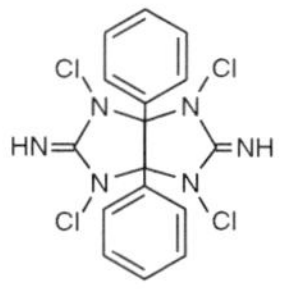

250 mg

S82,006-7
PHENYLGUANIDINE

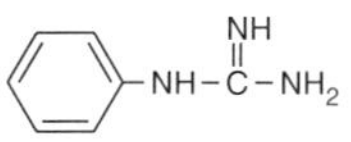

250 mg

S61,452-1
1-METHYL-1-PHENYLGUANIDINE HYDROCHLORIDE

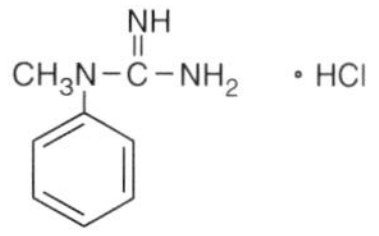

5 gm

S86,317-3
3-PHENYLPROPIONAMIDINE HYDROCHLORIDE

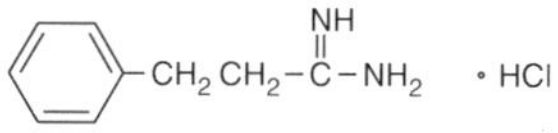

250 mg

S56,581-4
N-PHENYLBENZAMIDINE

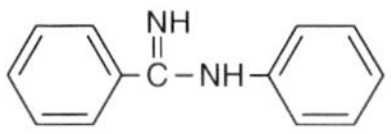

5 gm

S55,423-5
1,3-DIPHENYLGUANIDINE PHTHALATE

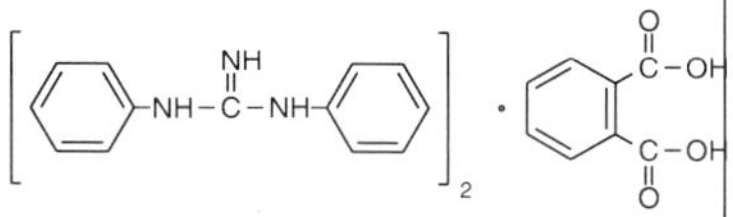

5 gm

S44,618-1
1,2,3-TRIPHENYLGUANIDINE

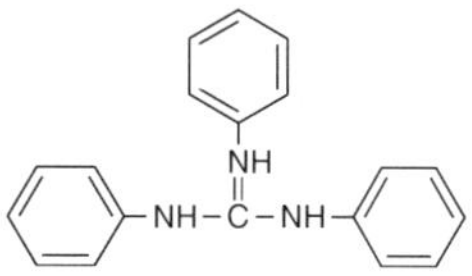

5 gm

S35,606-9
ORTHO-TOLYLBIGUANIDE HEMIHYDRATE

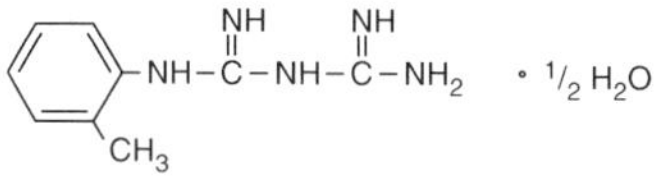

5 gm

S61,456-4
1-(4-FLUOROPHENYL)-GUANIDINE CARBONATE

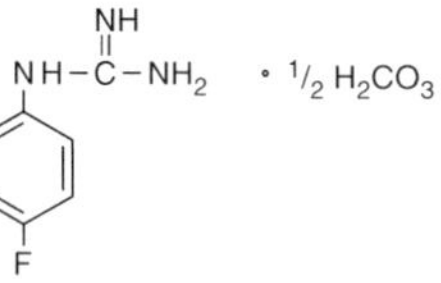

5 gm

S61,440-8
1-(4-CHLOROPHENYL)-GUANIDINE CARBONATE

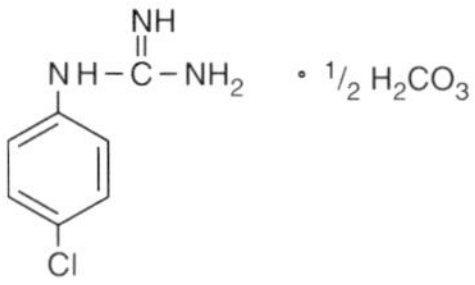

5 gm

S55,713-7
1-(P-TOLYL)GUANIDINE NITRATE

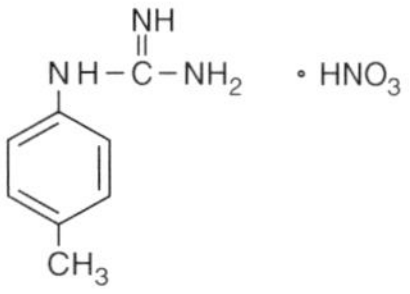

5 gm

S61,455-6
1-(P-TOLYL)GUANIDINE CARBONATE

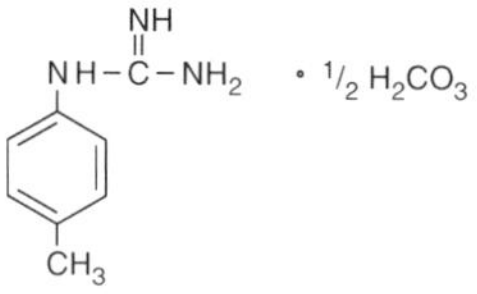

5 gm

S78,900-3
1,4-(DIAMIDINO)BENZENE DIHYDROCHLORIDE

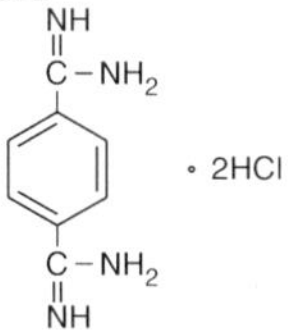

250 mg

S78,900-3
1-(4-ETHYLPHENYL)-GUANIDINE CARBONATE

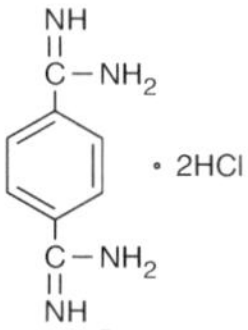

5 gm

S87,923-1
N,N'-DI(P-TOLYL)FORMAMIDINE

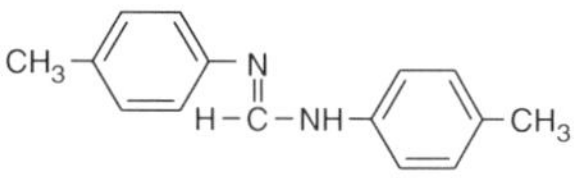

5 gm

S87,859-6
N,N'-BIS(4-BUTYLPHENYL)FORMAMIDINE

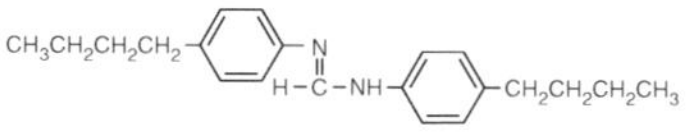

5 gm

S61,462-9
1-(3,4-DICHLOROPHENYL)-GUANIDINE CARBONATE

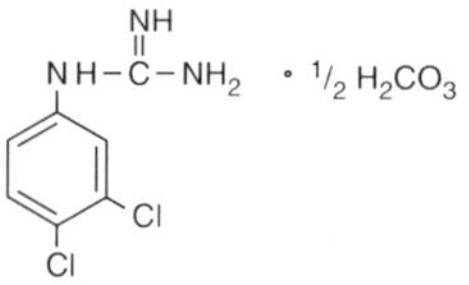

5 gm

S61,464-5
1-(2,4-XYLYL)GUANIDINE CARBONATE

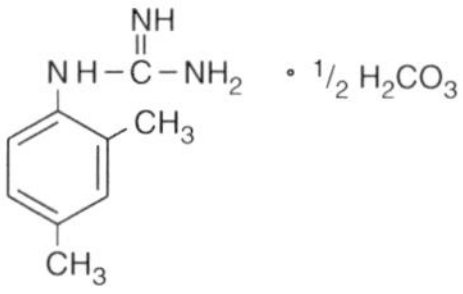

5 gm

S93,477-1
4-(3-CHLOROPHENYL)-3-THIOSEMICARBAZIDE

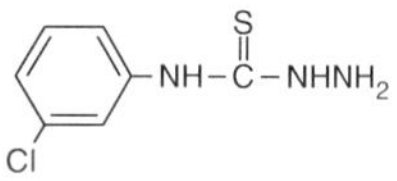

250 mg

S76,151-6
2-(2-CHLOROETHYL)-3-(4-FLUOROPHENYL)-3-HYDROXY-2,3,56-4-H-IMIDAZO(2,1-B)THIAZOLE

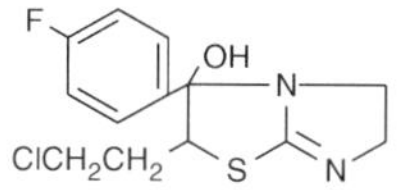

250 mg

S73,595-7
2-(2-CHLOROETHYL)-3-(4-FLUOROPHENYL)-3-HYDROXY-2,3,5,6-TETRAHYDROIMIDAZO(2,1-B)-THIAZOLE HYDROBROMIDE

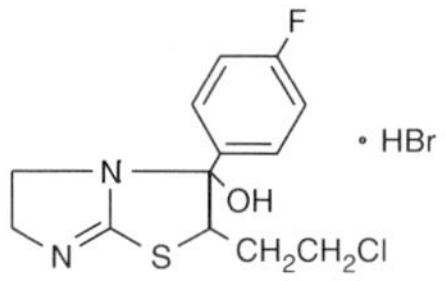

5 gm

S89,856-2
3-(2-CHLOROETHYL)-1-(4-CHLOROPHENYL)-5-THIOXOHYDANTOIN

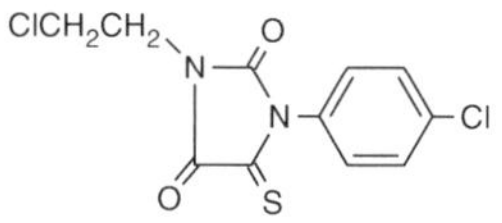

250 mg

S89,850-3
3-(2-CHLOROPHENYL)-1-(3-CHLOROPROPYL)-5-IMINO-4-THIOHYDANTOIN

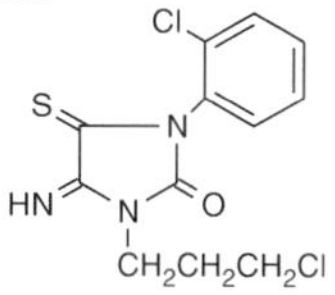

250 mg

S89,851-1
3-(4-CHLOROPHENYL)-1-(3-CHLOROPROPYL)-5-IMINO-4-THIOHYDANTOIN

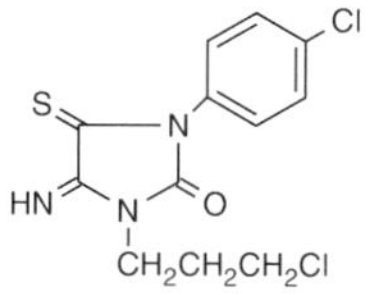

250 mg

S65,685-2
2-(2-CHLOROETHYL)-3-HYDROXY-3-PHENYL-2,3,6,7-TETRAHYDRO-5H-THIAZOLO[3,2-A]PYRIMIDINE HYDROBROMIDE

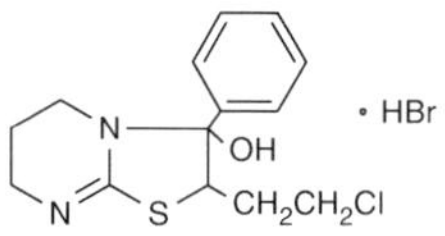

5 gm

S89,814-7
1-(3-CHLOROPROPYL)-5-IMINO-3-PHENYL-4-THIOHYDANTOIN

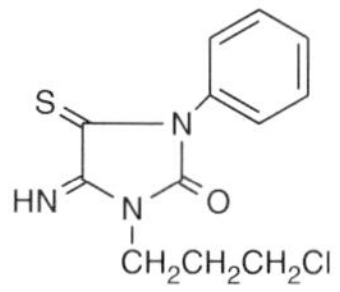

250 mg

S57,844-4
5-(DIFLUOROMETHYLSULFONYL)-3-METHYL-2(3H)-BENZOTHIAZOLETHIONE

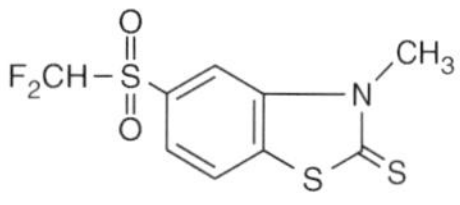

5 gm

S76,887-1
1-(HEXAMETHYLENEIMINO)-3-[3-(TRIFLUOROMETHYL)PHENYL]-2-THIOUREA

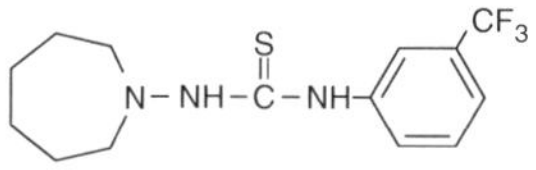

250 mg

S68,416-3
1-(5-CHLORO-2,4-DIMETHOXYPHENYL)-3-[3-(TRIFLUOROMETHYL)PHENYL]-2-THIOUREA

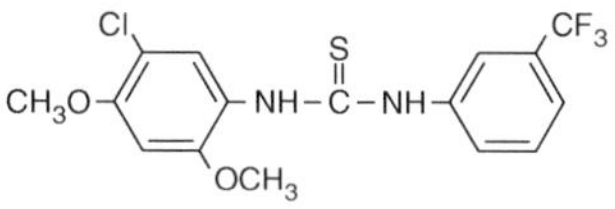

250 mg

S90,307-8
1-[2-CHLORO-5-(TRIFLUOROMETHYL) PHENYL]-3-(4-NITROPHENYL)-2-THIOUREA

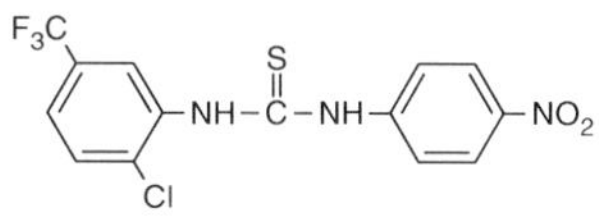

250 mg

S89,860-0
3-(4-CHLOROPHENYL)-5-IMINO-2-(TRICHLOROMETHYL)-4-OXAZOLIDINETHIONE

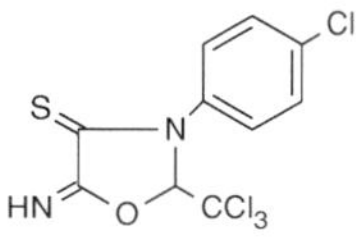

250 mg

S72,091-7
1-[4-BROMO-2-(TRIFLUOROMETHYL)-PHENYL]-6-(O-TOLYL)-5-THIOBIUREA

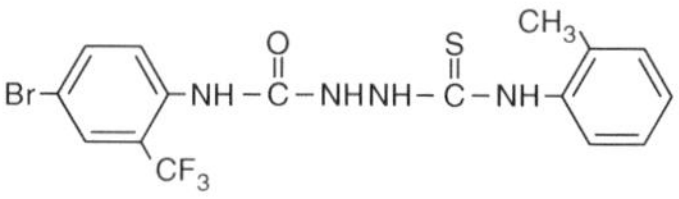

250 mg

S69,098-8
4-[4-BROMO-2-(TRIFLUOROMETHYL)-PHENYL]-1-(N-CYCLOHEXYLTHIO-CARBAMOYL)SEMICARBAZIDE

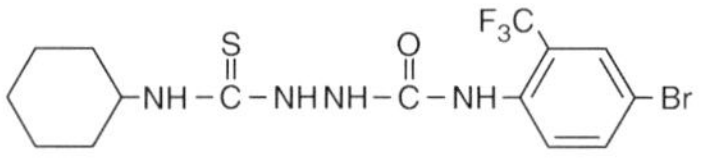

250 mg

S91,301-4
6-[2-CHLORO-5-(TRIFLUOROMETHYL)-PHENYL]-1-PHENYL-2-THIOBIUREA

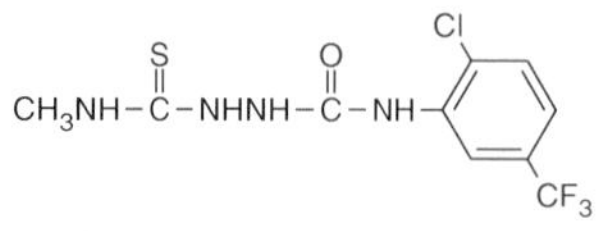

250 mg

S90,490-2
1-BENZYL-3-[4-CHLORO-2-(TRIFLUOROMETHYL)PHENYL]-2-THIOUREA

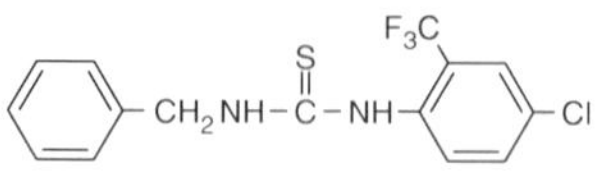

250 mg

S93,376-7
1-(2-CHLOROPHENYL)-3-[3,5-BIS (TRIFLUOROMETHYL)PHENYL]-2-THIOUREA

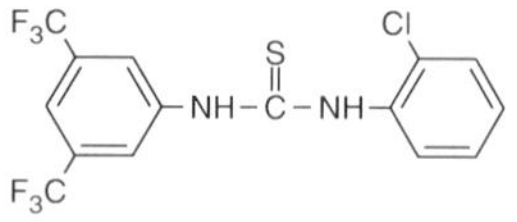

250 mg

S72,768-7
S-(2-HYDROXYPHENYL)-N-[2-(TRIFLUOROMETHYL)PHENYL]-THIOCARBAMATE

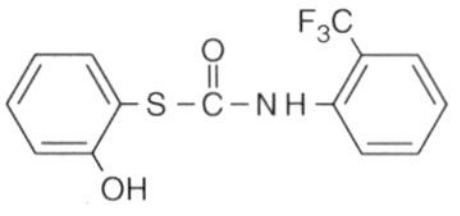

250 mg

S74,409-3
5-IMINO-3-(4-METHOXYPHENYL)-2-(TRICHLOROMETHYL)-4-OXAZOLIDINETHIONE

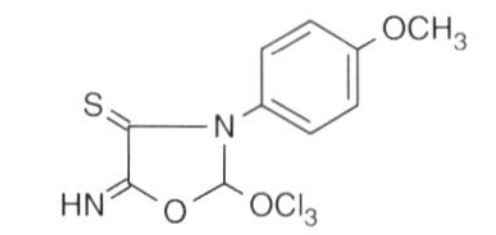

250 mg

S64,926-0
1-(4-PHENOXYPHENYL)-2-THIO-3-[3-(TRIFLUOROMETHYL)PHENYL]UREA

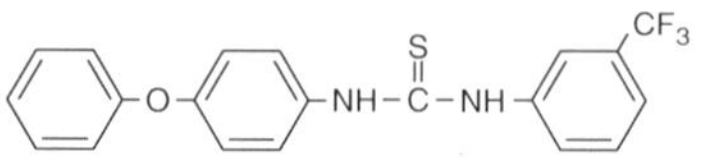

250 mg

S73,244-3
S-METHYL-1-[2,5-BIS(TRIFLUOROMETHYL)]-BENZYLIDENEISOTHIOSEMICARBAZIDE HYDRIODIDE

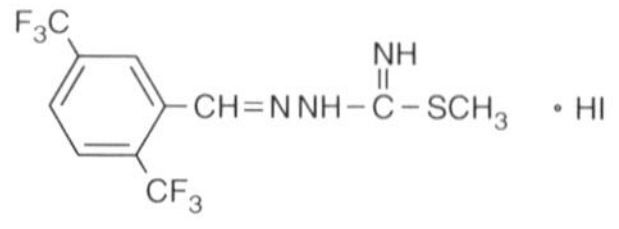

250 mg

S85,136-1
1-(4-ANILINOPHENYL)-3-[2-(TRIFLUOROMETHYL)PHENYL]-2-THIOUREA

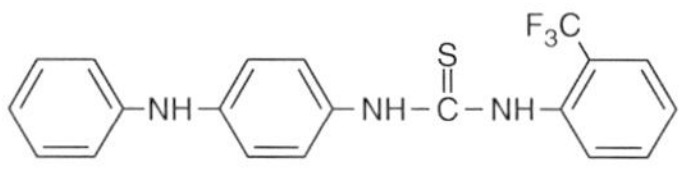

250 mg

S72,083-6
1-(4-NITROPHENYL)-3-[3-(TRIFLUOROMETHYL)PHENYL]-2-THIOUREA

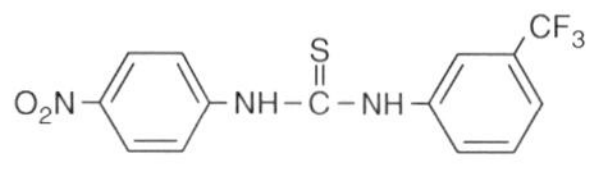

5 gm

S85,153-1
1,1-DIBENZYL-3-[2-(TRIFLUOROMETHYL)-PHENYL]-2-THIOUREA

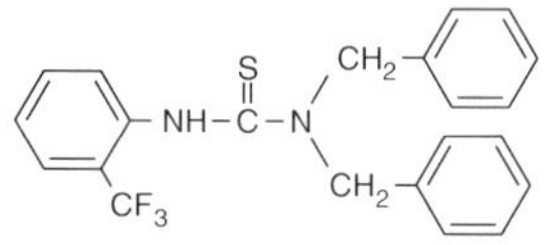

250 mg

S77,337-9
1-(1-THIOACETAMIDO)-4-[3-(TRIFLUOROMETHYL)PHENYL]OXAMIDE

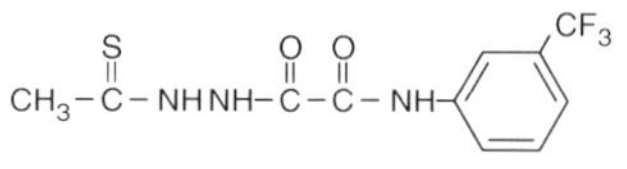

250 mg

S72,496-3
1-(4-ETHOXYCARBONYL)-3-[3-(TRIFLUOROMETHYL)PHENYL]-2-THIOUREA

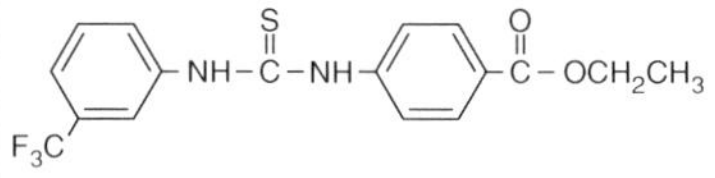

5 gm

S55,769-2
1-BENZOYL-3-(TRIFLUOROMETHYLPHENYL)-2-THIOUREA

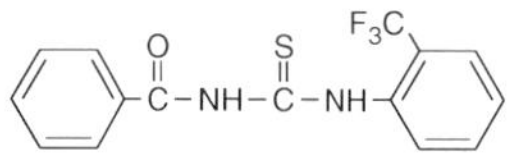

250 mg

S91,328-6
1-METHYL-6-[3-(TRIFLUOROMETHYL)-PHENYL]-2-THIOBIUREA

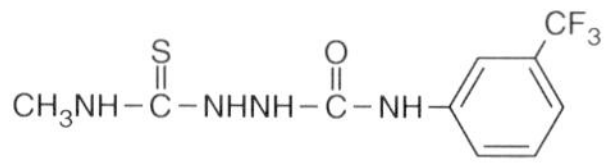

250 mg

S91,324-3
1-(N-[2-(TRIFLUOROMETHYL)PHENYL] CARBAMOYL)-3-THIOCARBOHYDRAZIDE

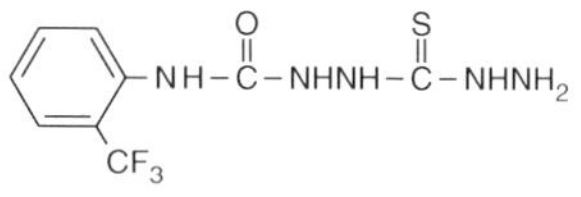

250 mg

S84,758-5
1-DODECANOYL-3-[3-(TRIFLUOROMETHYL)-PHENYL]-2-THIOUREA

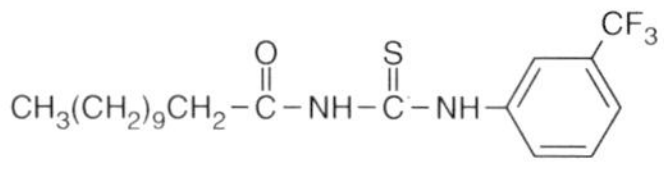

250 mg

S84,768-2
1-DODECANOYL-3-[2-(TRIFLUOROMETHYL)-PHENYL]-2-THIOUREA

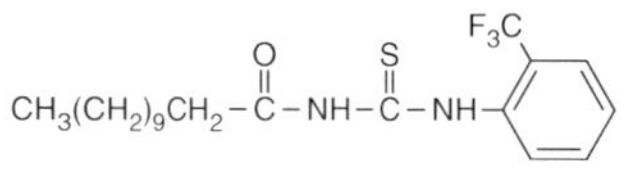

250 mg

S86,259-2
1-[3-(TRIFLUOROMETHYL)PHENYL]-2-THIOUREA

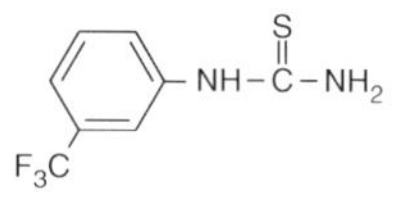

250 mg

S69,363-4
1-ISOPROPYL-3-[3-(TRIFLUOROMETHYL)-PHENYL]-2-THIOUREA

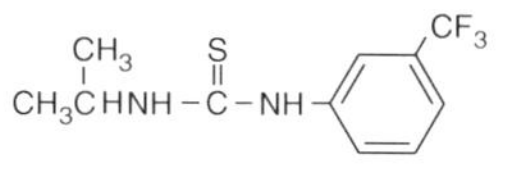

250 mg

S89,959-3
4-METHYL-2'-(TRIFLUOROMETHYL)-1-PIPERAZINETHIOCARBOXANILIDE

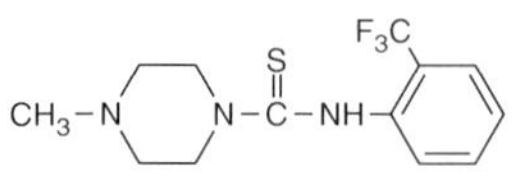

250 mg

S76,105-2
4-(4-CHLOROPHENYL)-4-HYDROXY-3-METHYL-2-(METHYLIMINO)THIAZOLIDINE HYDROBROMIDE

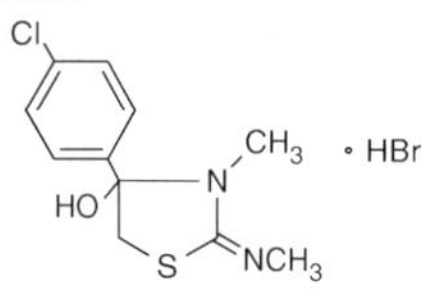

250 mg

S86,839-6
1-(2,4-DICHLOROBENZOYL)-3-(7-HYDROXY-1-NAPHTHYL)-2-THIOUREA

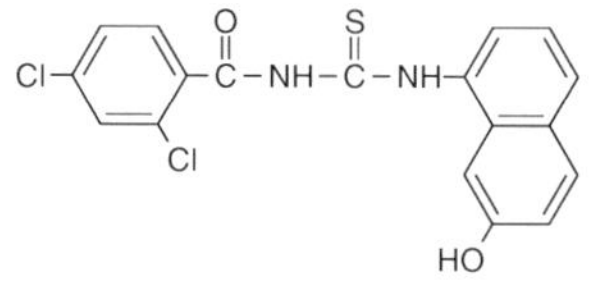

250 mg

S67,560-1
4-(4-CHLOROPHENYL)-1-(4-HYDROXYBENZOYL)THIOSEMICARBAZIDE

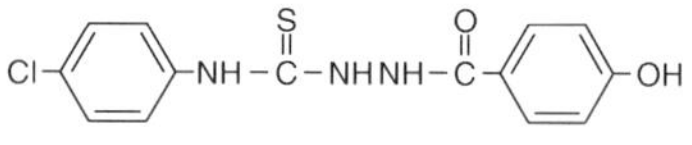

250 mg

S75,848-5
5-(4-CHLORO-3-HYDROXY-ALPHA-METHYLBENZYLIDENE)RHODANINE

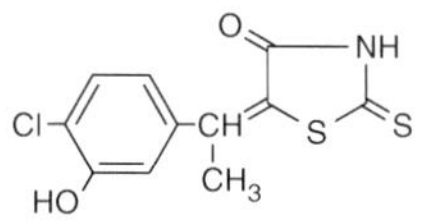

250 mg

S91,734-6
S-(2-HYDROXYPHENYL) N-(2,5-DIFLUOROPHENYL)THIOCARBAMATE

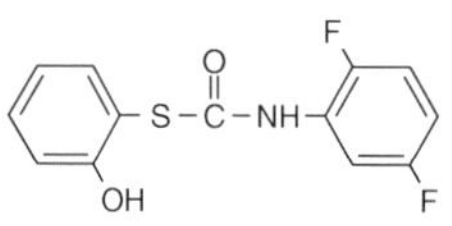

250 mg

S76,214-8
4-(4-CHLOROPHENYL)-4-HYDROXY-2-IMINOTHIAZOLIDINE HYDROBROMIDE

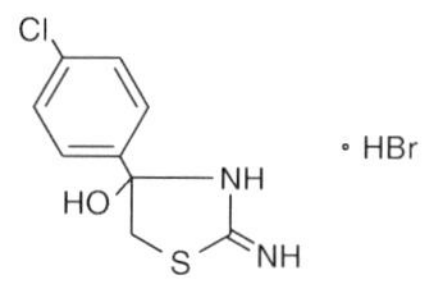

250 mg

S68,414-7
1-(5-CHLORO-2,4-DIMETHOXYPHENYL)-3-(2,5-DIMETHOXYPHENYL)-2-THIOUREA

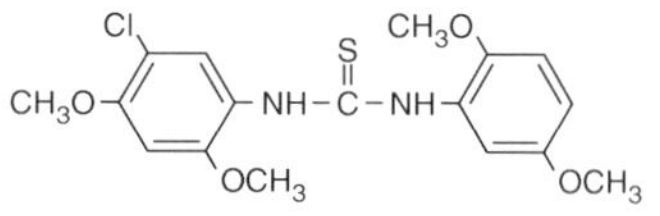

250 mg

S69,420-7
1-(2-CHLOROBENZOYL)-3-(2,4-DIMETHOXYPHENYL)-2-THIOUREA

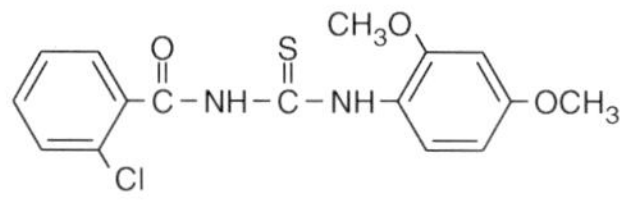

5 gm

S68,470-8
1-(4-ACETAMIDOPHENYL)-3-(5-CHLORO-2,4-DIMETHOXYPHENYL)-2-THIOUREA

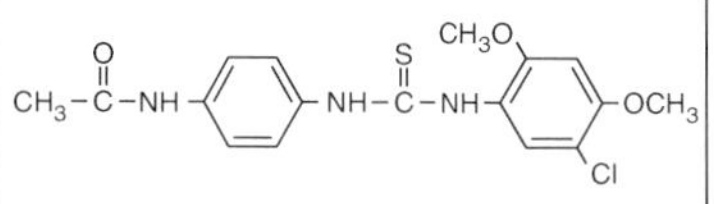

250 mg

S68,350-7
1-(5-CHLORO-2,4-DIMETHOXYPHENYL)-3-(3-FLUOROPHENYL)-2-THIOUREA

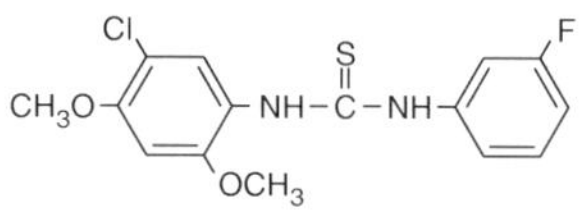

250 mg

S68,444-9
1-[N-(5-CHLORO-2,4-DIMETHOXYPHENYL)-THIOCARBAMOYL]PIPERIDINE

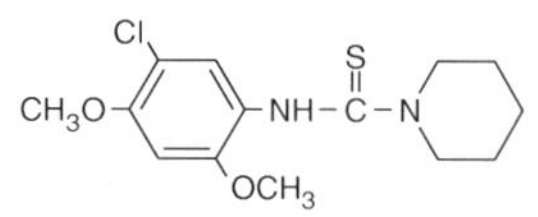

250 mg

S69,373-1
4-[N-(5-CHLORO-2,4-DIMETHOXYPHENYL)-THIOCARBAMOYL]MORPHOLINE

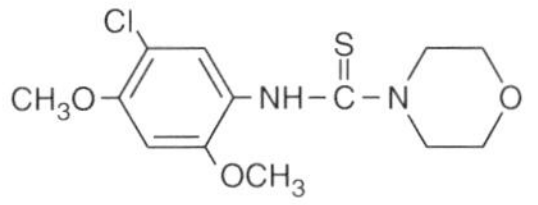

250 mg

S76,977-0
1-(5-CHLORO-2,4-DIMETHOXYPHENYL)-3-(2-FLUOROPHENYL)-3-METHYL-2-THIOUREA

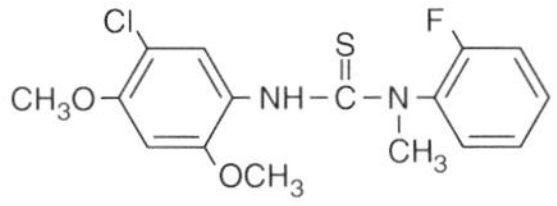

250 mg

S68,501-1
1-(5-CHLORO-2,4-DIMETHOXYPHENYL)-3-(4-CHLOROPHENYL)-2-THIOUREA

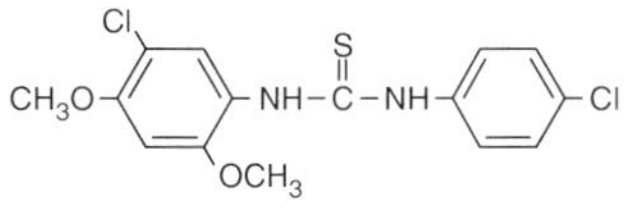

5 gm

S68,635-2
1-(5-CHLORO-2,4-DIMETHOXYPHENYL)-3-(2,4-DIMETHYLPHENYL)-2-THIOUREA

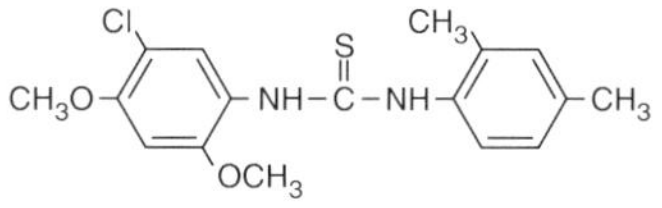

250 mg

S73,836-0
1-(5-CHLORO-2,4-DIMETHOXYPHENYL)-3-METHYL-2-THIOUREA

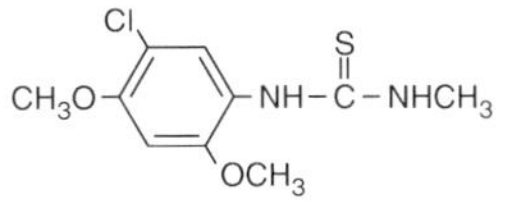

5 gm

S80,172-0
1-(2,5-DIMETHOXYPHENYL)-3-(2-FLUOROBENZOYL)-2-THIOUREA

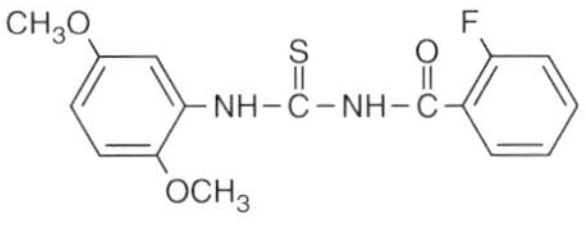

250 mg

S91,377-4
1-(4-CHLORO-2,5-DIMETHOXYPHENYL)-3-METHYL-2-THIOUREA

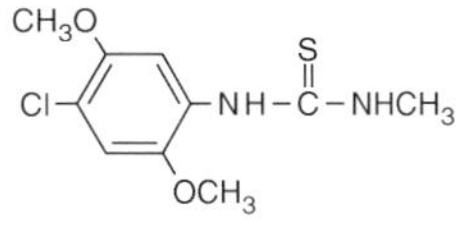

250 mg

S91,737-0
S-(4-CHLOROPHENYL) N-(2,5-DIMETHOXYPHENYL)THIOCARBAMATE

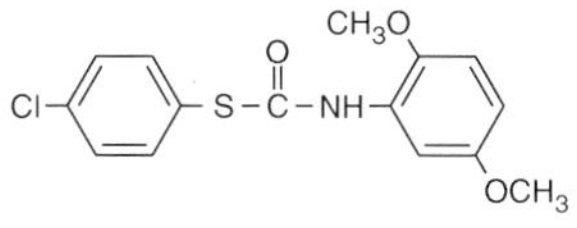

250 mg

S90,308-6
1-(4-CHLORO-2,5-DIMETHOXYPHENYL)-3-(4-NITROPHENYL)-2-THIOUREA

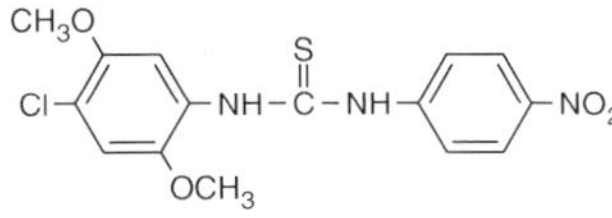

250 mg

S68,591-7
1-(5-CHLORO-2,4-DIMETHOXYPHENYL)-3-(2,4-DIFLUOROPHENYL)-2-THIOUREA

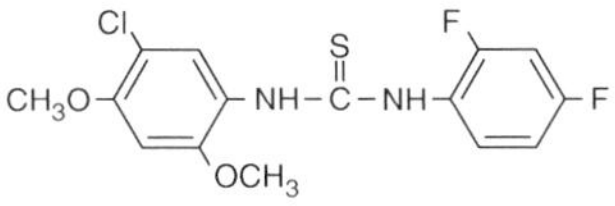

250 mg

S74,743-2
1-(2-CHLOROBENZOYL)-3-(2-METHOXY-5-METHYLPHENYL)-2-THIOUREA

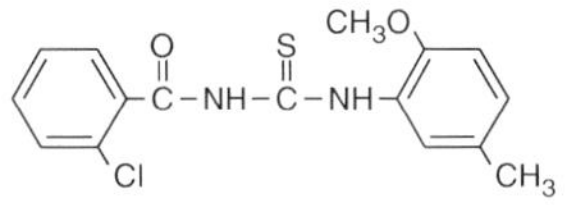

5 gm

S73,843-3
1-(5-CHLORO-2-METHOXYPHENYL)-3-(3,4-DICHLOROBENZOYL)-2-THIOUREA

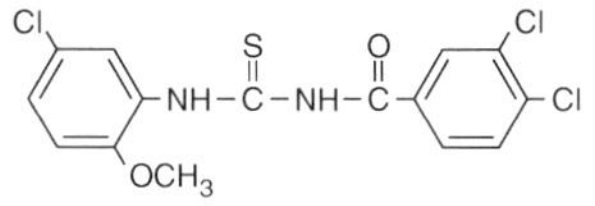

5 gm

S72,508-0
1-(5-CHLORO-2-METHOXYPHENYL)-5-THIOBIUREA

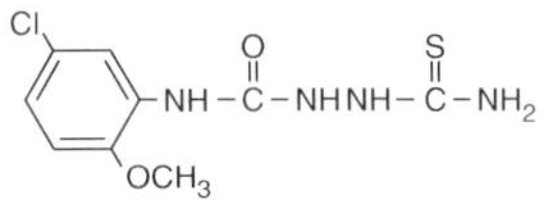

5 gm

S91,264-6
1-[N-(5-CHLORO-2-METHOXYPHENYL)-CARBAMOYL]-3-THIOCARBOHYDRAZIDE

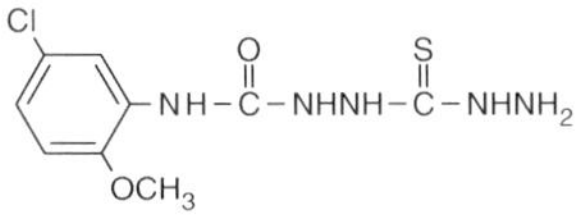

250 mg

S72,296-0
1-(3-CHLOROPHENYL)-6-(2-METHOXY-5-METHYLPHENYL)-5-THIOBIUREA

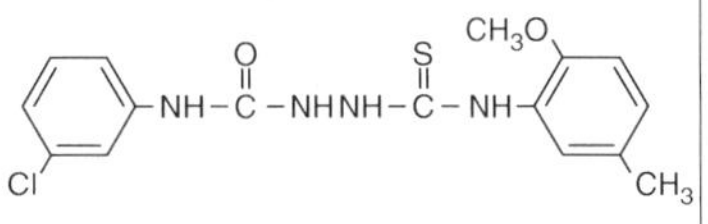

250 mg

S73,615-5
1-[N-(5-CHLORO-2-METHOXYPHENYL)-CARBAMOYL]-4-METHYL-3-THIOSEMICARBAZIDE

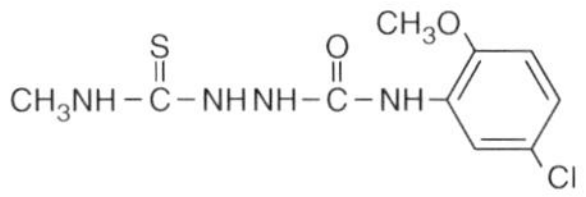

5 gm

S89,184-3
1-(5-CHLORO-2-METHOXYPHENYL)-3-PHENYL-2-THIOUREA

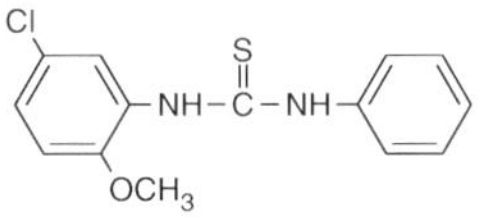

250 mg

S73,156-0
1-(2-FLUOROBENZOYL)-3-(4-ETHOXYPHENYL)-2-THIOUREA

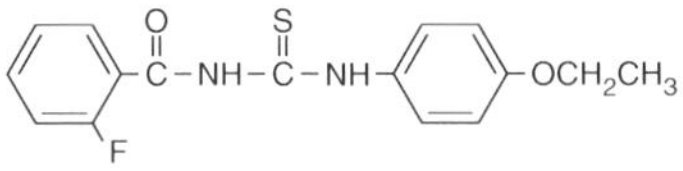

5 gm

S73,840-9
1-(2-CHLOROBENZOYL)-3-(4-ETHOXYPHENYL)-2-THIOUREA

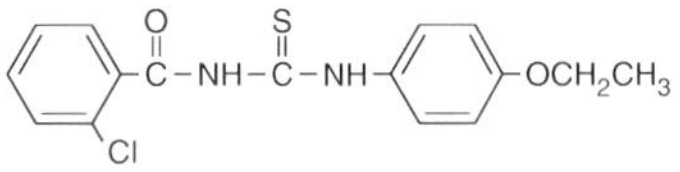

5 gm

S68,792-8
4-(3-METHOXYPROPYL)-1-(4-CHLOROBENZOYL)-3-THIOSEMICARBAZIDE

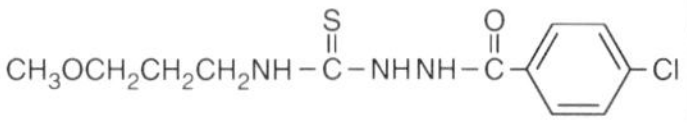

250 mg

S84,547-7
1-(2,5-DICHLOROPHENYL)-3-(2-PHENOXYPROPIONYL)-2-THIOUREA

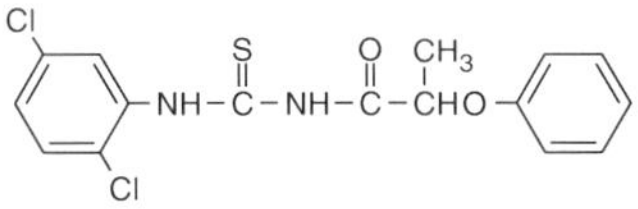

250 mg

S77,868-0
3-(4-CHLOROPHENYL)-5-(2-METHOXY-3-METHYLBENZYLIDENE)RHODANINE

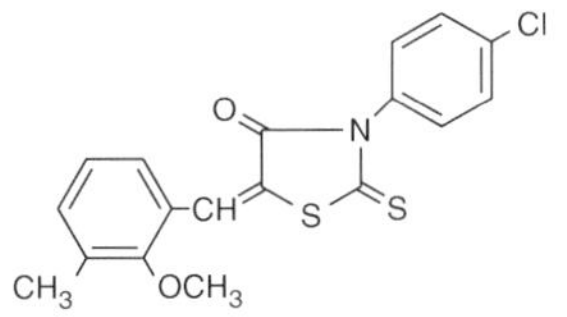

250 mg

S75,324-6
1-[2-(2,4,5-TRICHLOROPHENOXY)-ACETYL]-3-THIOSEMICARBAZIDE

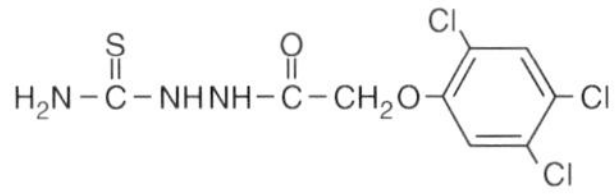

5 gm

S90,059-1
1-(2-CHLORO-4-METHOXYPHENYL)-3-CYCLOHEXYL-2-THIOUREA

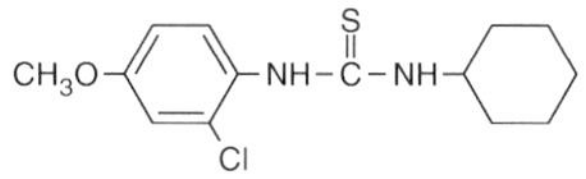

250 mg

S72,278-2
1-(2,4-DIFLUOROPHENYL)-3-(4-PHENOXYPHENYL)-2-THIOUREA

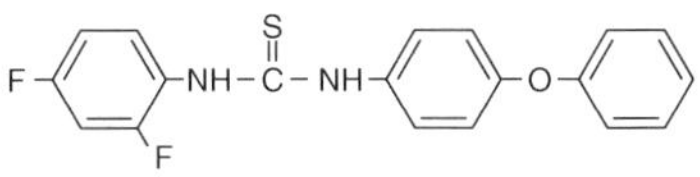

250 mg

S77,226-7
1-(3,4-DICHLOROPHENYL)-3-(4-PHENOXYPHENYL)-2-THIOUREA

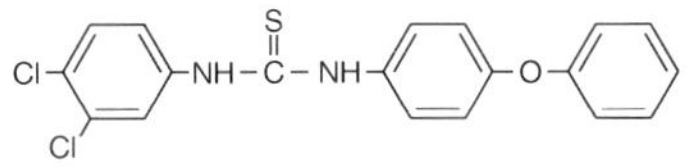

250 mg

S88,951-2
2-(2-CHLOROPHENYL)-N-ETHYL-3-OXAZOLIDINETHIOCARBOXAMIDE

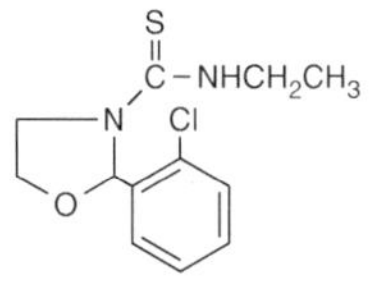

250 mg

S89,862-7
2-(2-CHLOROPHENYL)-5-IMINO-3-METHYL-4-OXAZOLIDINETHIONE

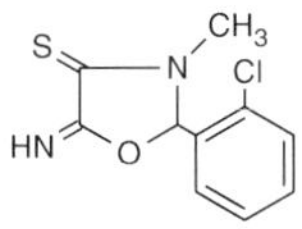

250 mg

S89,885-6
2-(4-FLUOROPHENYL)-5-IMINO-3-METHYL-4-OXAZOLIDINETHIONE

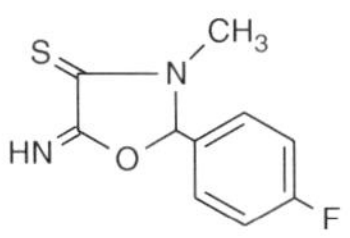

250 mg

S89,861-9
2-(4-CHLOROPHENYL)-5-IMINO-3-METHYL-4-OXAZOLIDINETHIONE

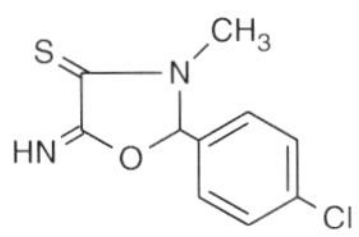

250 mg

S88,945-8
2-(4-CHLOROPHENYL)-3-OXAZOLIDINETHIOCARBOXANILIDE

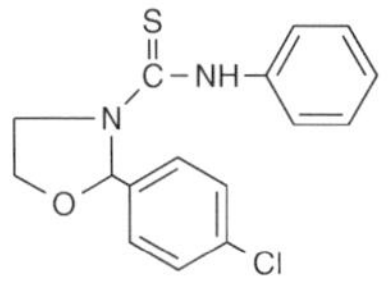

250 mg

S90,980-7
1-BENZOYL-3-(4-CHLORO-2-NITROPHENYL)-2-THIOUREA

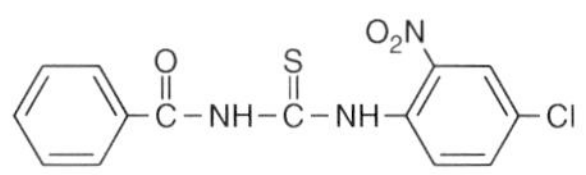

250 mg

S79,572-0
1-(4-FLUOROPHENYL)-3-(2-PYRROLIDINOETHYL)-2-THIOUREA

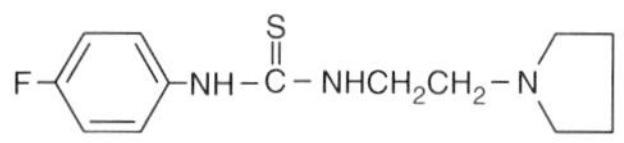

250 mg

S84,452-7
1-(3,4-DICHLOROPHENYL)-3-(3,5-DINITROBENZOYL)-2-THIOUREA

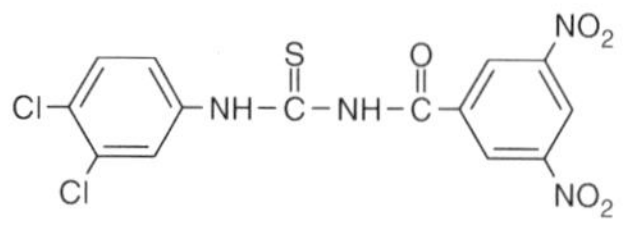

250 mg

S68,639-5
N-(4-CHLORO-3-NITROPHENYL)-N'-(3-METHYL-2-THIOUREIDO)OXAMIDE

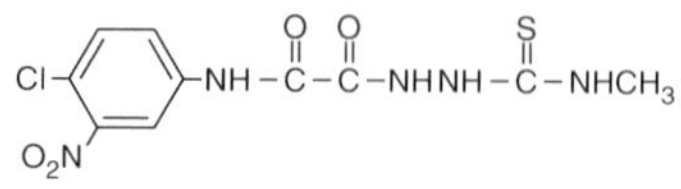

250 mg

S67,568-7
1-(2-CHLORO-5-NITROPHENYL)-3-DODECANOYL-2-THIOUREA

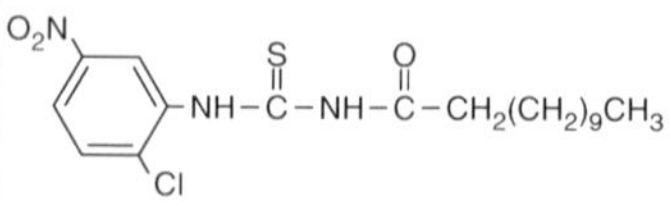

250 mg

S90,366-3
5-(4-CHLORO-3-NITROPHENYL)-2-THIOBIURET

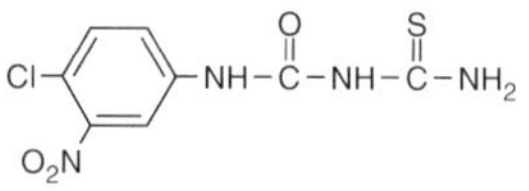

250 mg

S90,489-9
1-BENZYL-3-(4-FLUORO-3-NITROPHENYL)--2-THIOUREA

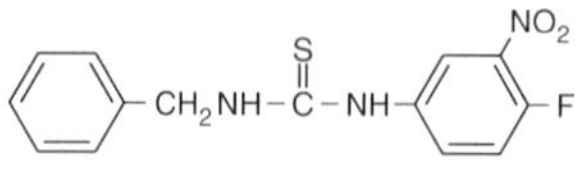

250 mg

S73,837-9
1-(2-CHLOROBENZOYL)-3-(4-NITROPHENYL)-2-THIOUREA

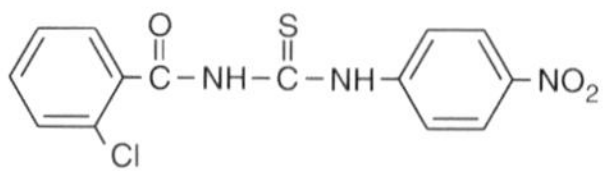

5 gm

S73,829-8
1-(3,4-DICHLOROBENZOYL)-3-(4-NITROPHENYL)-2-THIOUREA

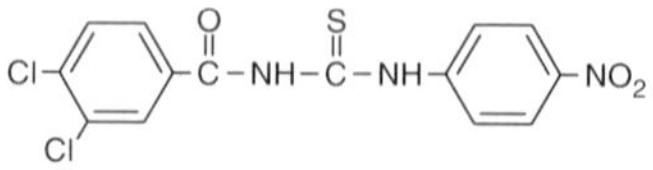

5 gm

S69,470-3
1-(2-CHLORO-4-NITROPHENYL)-3-(3,4-DICHLOROBENZOYL)-2-THIOUREA

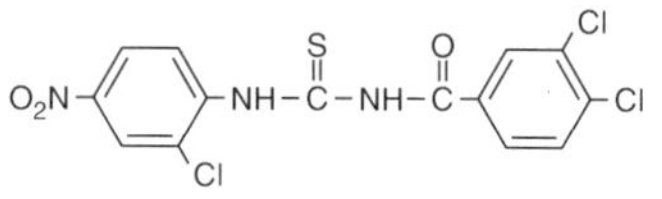

5 gm

S90,305-1
1-(3-FLUOROPHENYL)-5-(4-NITROPHENYL)-2,4-DITHIOBIURET

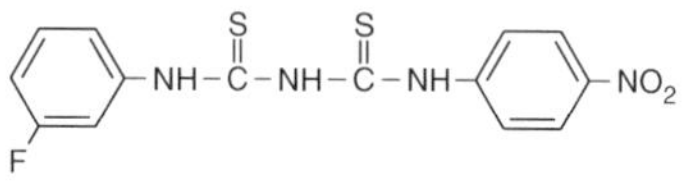

250 mg

S69,024-4
1-(2-CHLORO-4-NITROPHENYL)-3-ETHYL-2-THIOUREA

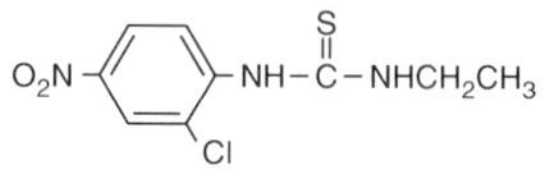

250 mg

S78,103-7
S-DODECYL N-(2-CHLORO-4-NITROPHENYL)THIOCARBAMATE

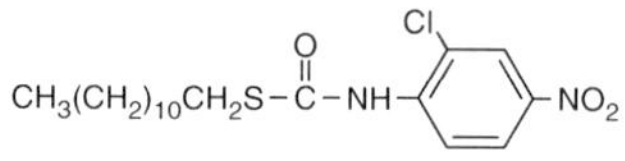

250 mg

S69,217-4
1-(4-ANILINOPHENYL)-3-(2,4-DIFLUOROPHENYL)-2-THIOUREA

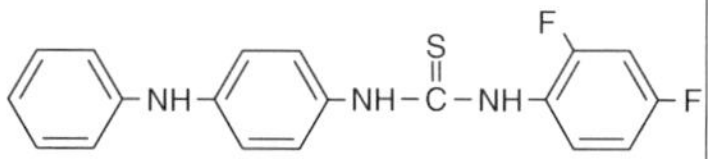

250 mg

S86,262-2
1-AMIDINO-3-(4-BROMOPHENYL)-2-THIOUREA

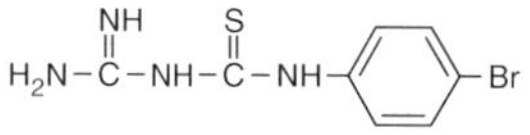

250 mg

S92,585-3
2'-CHLORO-6'-METHYLTHIOBENZANILIDE

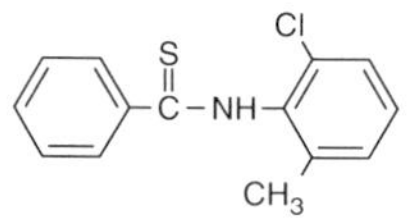

250 mg

S68,656-5
TRIETHYLAMMONIUM N-(3,4-DICHLOROPHENYL)DITHIOCARBAMATE

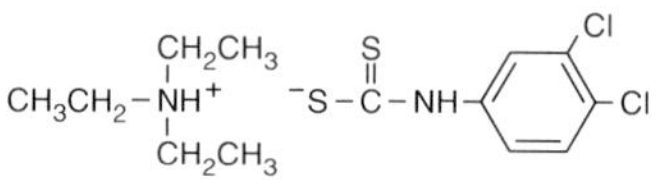

5 gm

S69,371-5
TRIETHYLAMMONIUM N-(2,4-DIFLUOROPHENYL)DITHIOCARBAMATE

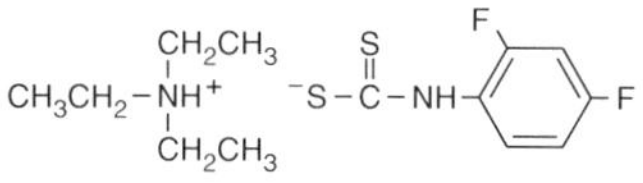

250 mg

S68,827-4
1-(N-BENZYLTHIOCARBAMOYL)-4-(3,4-DICHLOROPHENYL)SEMICARBAZIDE

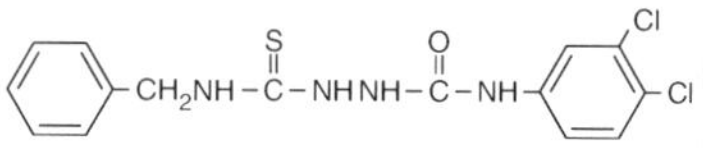

250 mg

S79,582-8
1-(2-CHLOROBENZYL)-3-CYCLOHEXYL-1-METHYL-2-THIOUREA

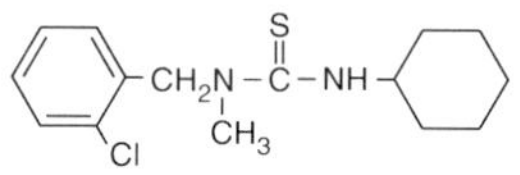

250 mg

S75,034-4
1-(4-CHLOROBENZYL)-1-METHYL-3-PHENYL-2-THIOUREA

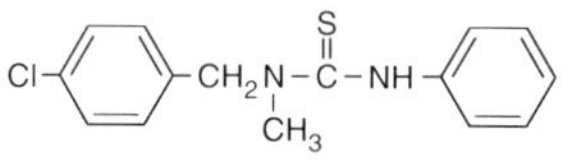

5 gm

S76,975-4
1-(3,4-DICHLOROPHENYL)-3-(DIPHENYLMETHYL)-2-THIOUREA

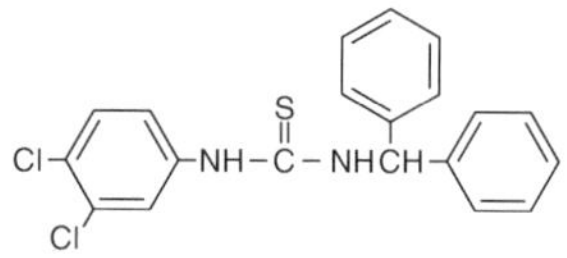

250 mg

S79,575-5
1-(3,5-DICHLOROBENZYL)-3-PHENYL-2-THIOUREA

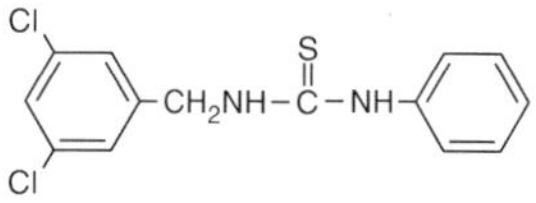

250 mg

S74,323-2
1-ALLYL-3-(4-CHLOROBENZYL)-2-THIOUREA

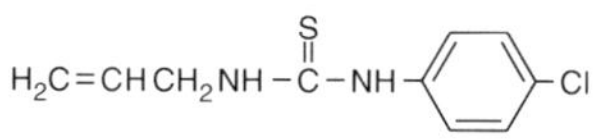

5 gm

S74,741-6
1-(3-CHLOROPHENYL)-5-THIOBIUREA

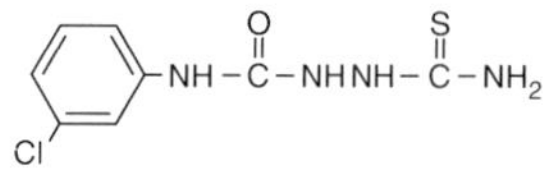

5 gm

S67,406-0
4-(3-CHLOROPHENYL)-1-(CYCLOHEXYLTHIOCARBAMOYL)-SEMICARBAZIDE

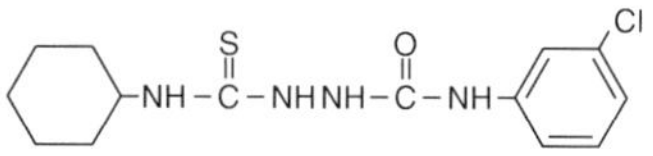

250 mg

S68,969-6
1-(2,3-DICHLOROPHENYLCARBAMOYL)-4-PHENYL-3-THIOSEMICARBAZIDE

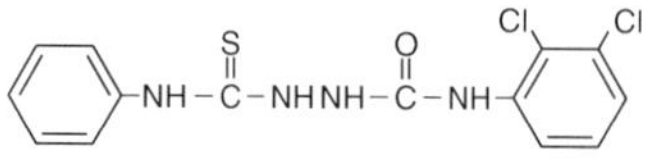

250 mg

S74,740-8
1-(2-CHLOROPHENYL)-5-THIOBIUREA

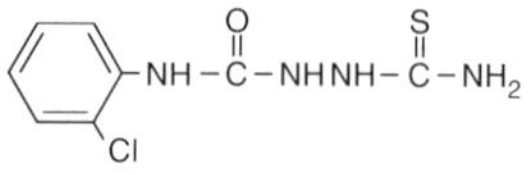

5 gm

S69,041-4
4-(2,4-DIFLUOROPHENYL)-1-(THIOCARBAMOYL)SEMICARBAZIDE

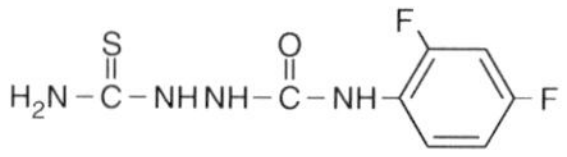

250 mg

S72,258-8
1-(3,4-DICHLOROPHENYL)-5-THIOBIUREA

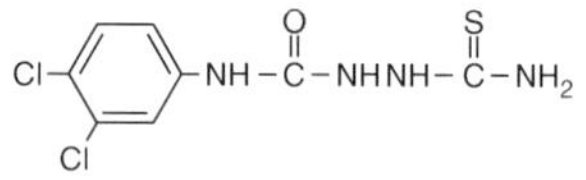

250 mg

S67,548-2
1-[N-(4-CHLORO-2-METHYLPHENYL)-CARBAMOYL]THIOSEMICARBAZIDE

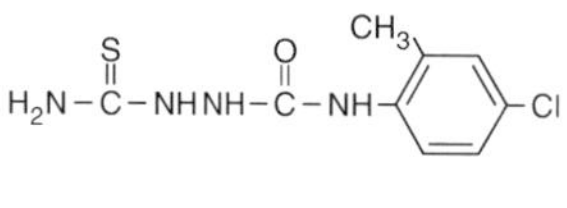

250 mg

S77,361-1
1-(2,4,5-TRICHLOROPHENYL)-5-THIOBIUREA

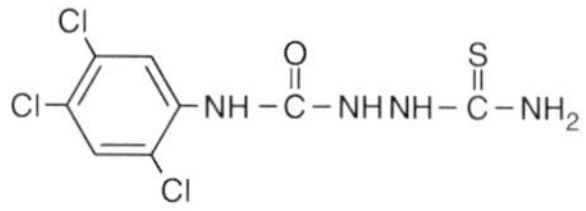

250 mg

S89,259-9
2,4-DICHLOROBENZALDEHYDE THIOSEMICARBAZONE

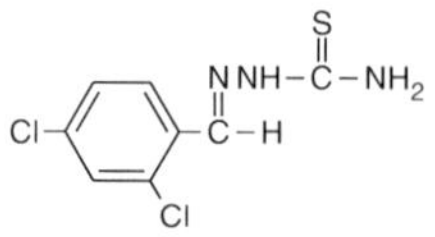

250 mg

S74,744-0
1-(2-CHLOROBENZOYL)-3-(5-CHLORO-2-METHOXYPHENYL)-2-THIOUREA

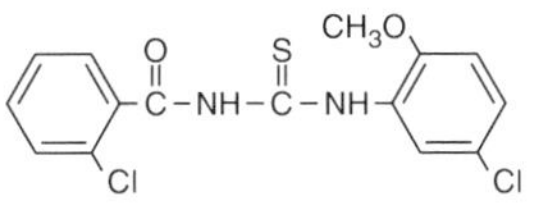

5 gm

S78,094-4
2-(2-THIOXO-1-IMIDAZOLIDINYL)ETHYL N-(2,3-DICHLOROPHENYL)CARBAMATE

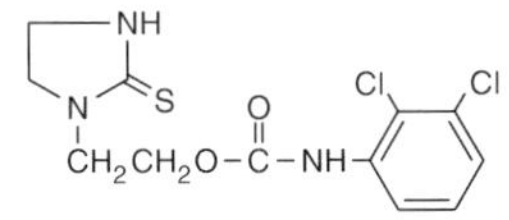

250 mg

S53,970-8
2-(4'-BROMO-3'-NITROPHENACYL)-2-THIOPSEUDOUREA HYDROBROMIDE

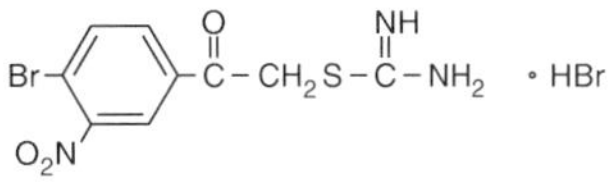

250 mg

S84,458-6
1-(3-FLUOROPHENYL)-3-(3-NITROBENZOYL)-2-THIOUREA

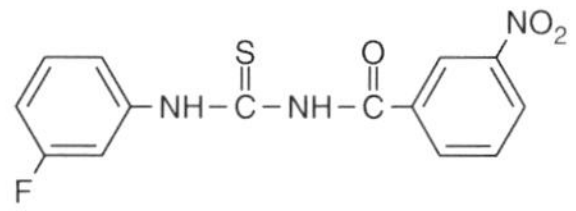

250 mg

S84,459-4
1-(4-FLUOROPHENYL)-3-(3-NITROBENZOYL)-2-THIOUREA

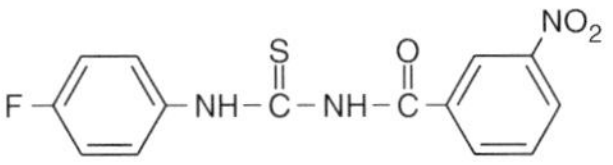

250 mg

S79,412-0
1-(2-BENZOYL-4-CHLOROPHENYL)-3-(2,4-DICHLOROBENZOYL)-2-THIOUREA

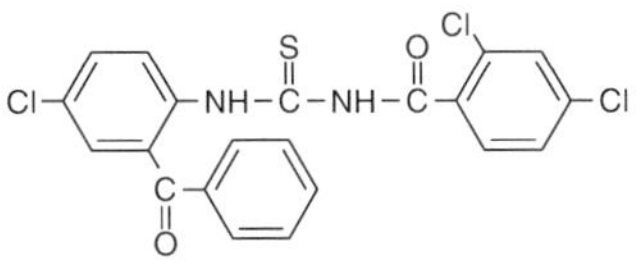

250 mg

S72,988-4
1-(3,4-DICHLOROBENZOYL)-3-(4-ETHOXYCARBONYL)-2-THIOUREA

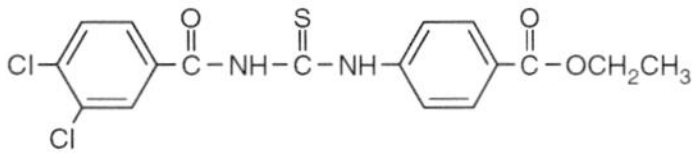

5 gm

S80,781-8
1-(2,4-DICHLOROBENZOYL)-3-(4-ETHOXYCARBONYLPHENYL)-2-THIOUREA

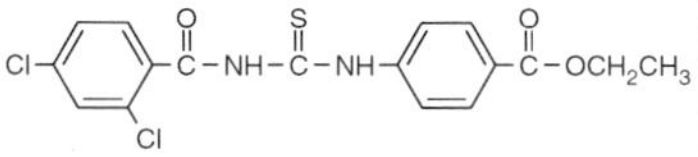

250 mg

S86,862-0
1-(2,4-DICHLOROBENZOYL)-3-(4-BUTOXYCARBONYLPHENYL)-2-THIOUREA

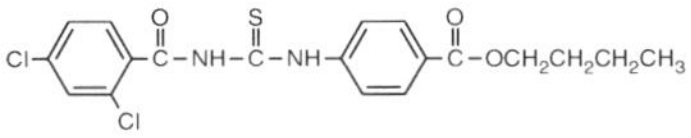

250 mg

S69,064-3
1-(2-CHLOROBENZOYL)-3-(4-ETHOXYCARBONYLPHENYL)-2-THIOUREA

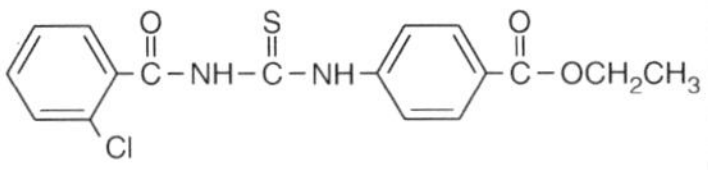

250 mg

S79,473-2
1-(3-BROMOBENZOYL)-3-(3-CHLOROPHENYL)-2-THIOUREA

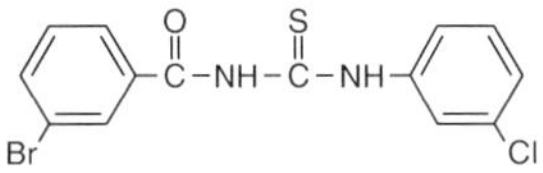

250 mg

S94,917-5
1-BENZOYL-3-(2,6-DICHLOROPHENYL)-2-THIOUREA

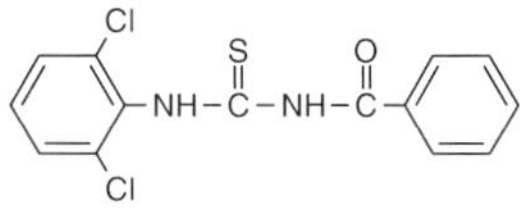

250 mg

S69,060-0
1-(2-CHLOROBENZOYL)-3-(2-FLUOROPHENYL)-2-THIOUREA

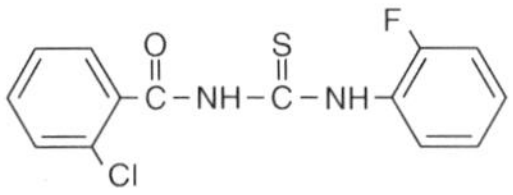

5 gm

S80,151-8
1-(3-CHLOROPHENYL)-3-(2-FLUOROBENZOYL)-2-THIOUREA

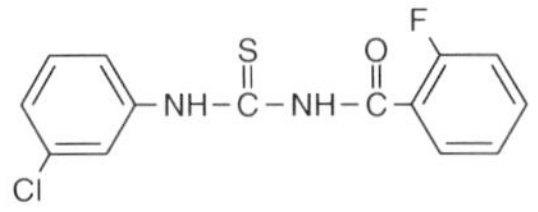

250 mg

S73,089-0
1-(2-CHLOROBENZOYL)-3-(3-CHLOROPHENYL)-2-THIOUREA

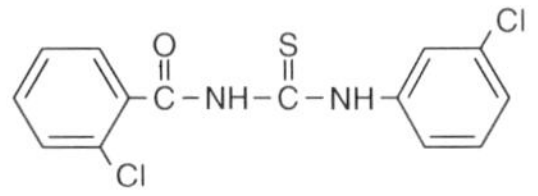

5 gm

S73,655-4
1-(2-FLUOROBENZOYL)-3-(4-FLUOROPHENYL)-2-THIOUREA

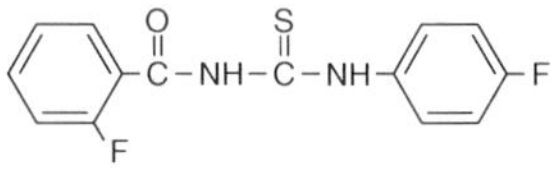

5 gm

S74,742-4
1-(2-CHLOROBENZOYL)-3-(2,4-DIFLUOROPHENYL)-2-THIOUREA

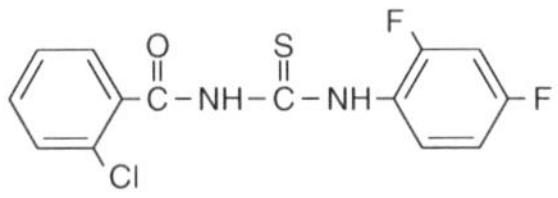

5 gm

S86,744-6
1-(2-CHLOROPHENYL)-3-(2,4-DICHLOROBENZOYL)-2-THIOUREA

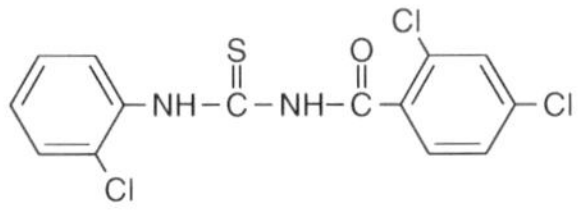

250 mg

S69,034-1
1-(3-CHLOROPHENYL)-3-(3,4-DICHLOROBENZOYL)-2-THIOUREA

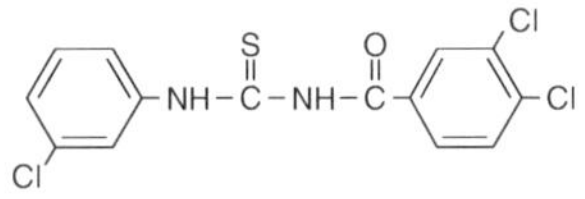

5 gm

S69,081-3
1-(2-CHLOROBENZOYL)-3-(3,4-DICHLOROPHENYL)-2-THIOUREA

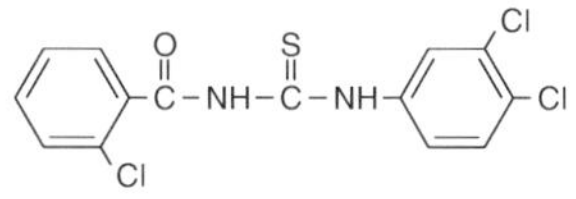

250 mg

S89,793-0
1-(3,4-DICHLOROBENZOYL)-3-(2,4-DIFLUOROPHENYL)-2-THIOUREA

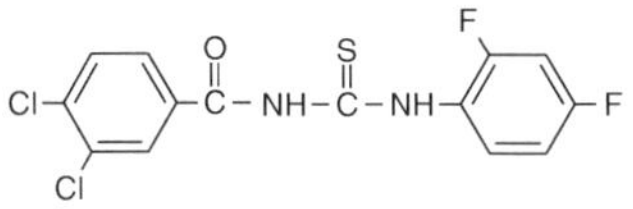

250 mg

S69,143-7
1-(3,4-DICHLOROBENZOYL)-3-(3,4-DICHLOROPHENYL)-2-THIOUREA

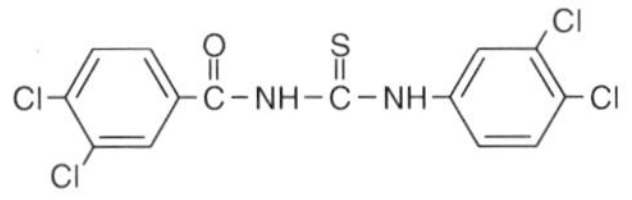

5 gm

S86,838-8
1-(2,4-DICHLOROBENZOYL)-3-(1-NAPHTHYL)-2-THIOUREA

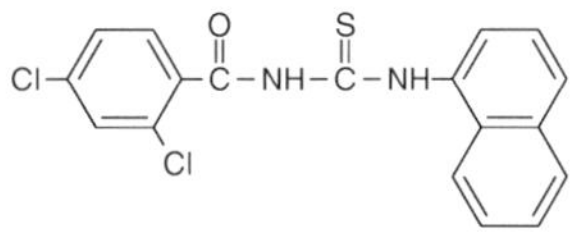

250 mg

S77,235-6
3-(4-CHLOROPHENYL)-5-(3,4-DICHLOROBENZYLIDENE)RHODANINE

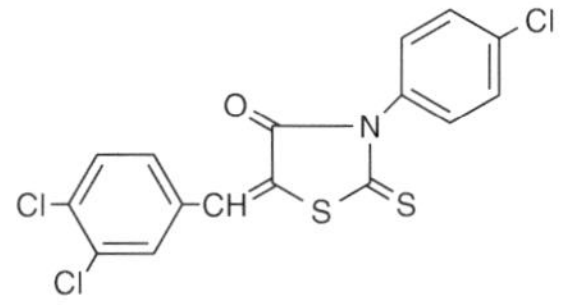

250 mg

S69,430-4
1-(3,4-DICHLOROBENZOYL)-3-(2-FLUOROPHENYL)-2-THIOUREA

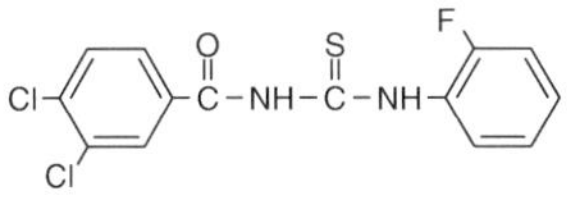

5 gm

S89,973-9
N-[3-(3-FLUOROPHENYL)-4-PHENYL-4-THIAZOLIN-2-YLIDENE]CINNAMAMIDE

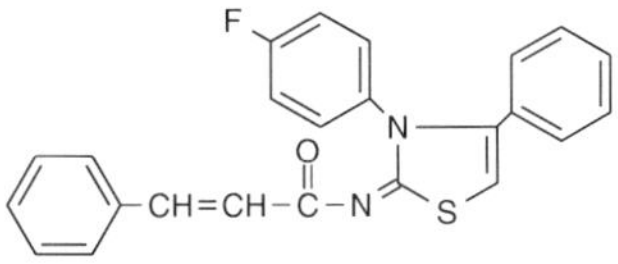

250 mg

S90,584-4
N-[3-(2-FLUOROPHENYL)-4-PHENYL-4-THIAZOLIN-2-YLIDENE]CINNAMAMIDE

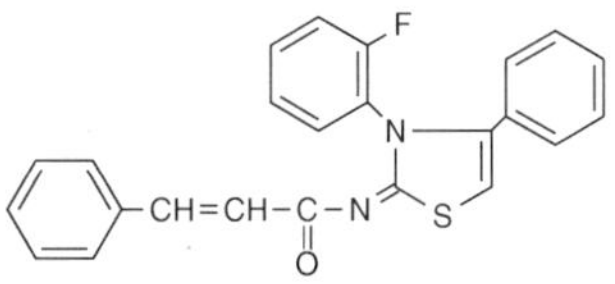

250 mg

S89,971-2
4-BROMO-N-[3-(3-CHLORO-2-METHYLPHENYL)-4-PHENYL-4-THIAZOLIN-2-YLIDENE]CINNAMAMIDE

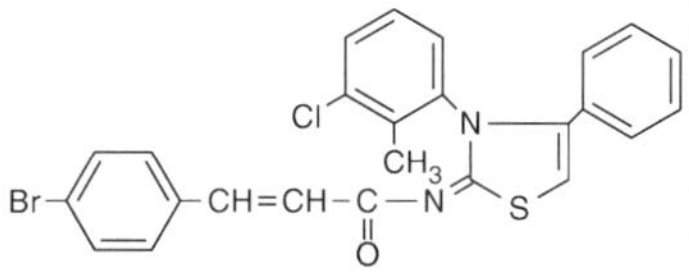

250 mg

S76,881-2
4-ALLYL-1-(2-CHLOROBENZOYL)-3-THIOSEMICARBAZIDE

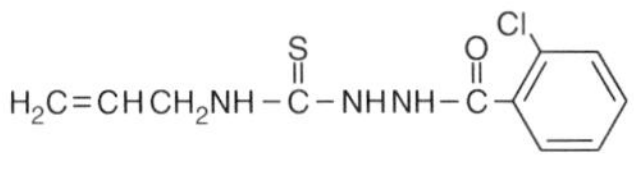

250 mg

S89,954-2
4-(3-ALLYL-2-THIOUREIDO)PHENYL N-(3,4-DICHLOROPHENYL)CARBAMATE

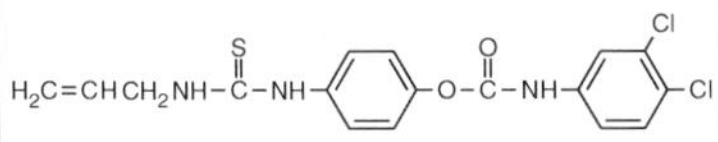

250 mg

S89,953-4
3-(3-ALLYL-2-THIOUREIDO)PHENYL N-(3,4-DICHLOROPHENYL)CARBAMATE

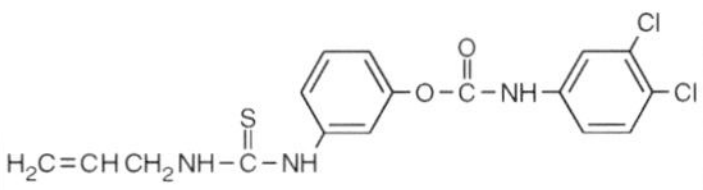

250 mg

S91,629-3
S-(PENTACHLOROPHENYL) N-(2,5-DIFLUOROPHENYL)THIOCARBAMATE

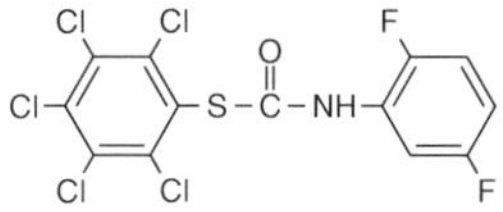

250 mg

S72,375-4
1-(3,4-DICHLOROPHENYL)-6-PHENYL-5-THIOBIUREA

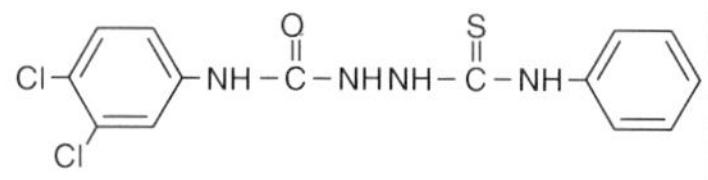

5 gm

S69,140-2
1-[N-(3-CHLOROPHENYL)CARBAMOYL)-4-(2-METHYLPHENYL]-3-THIOSEMICARBAZIDE

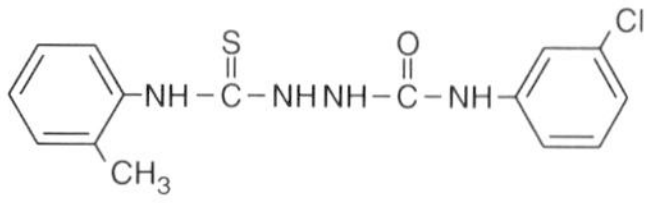

250 mg

S91,310-3
6-(3-CHLORO-4-METHYLPHENYL)-1-METHYL-2-THIOBIUREA

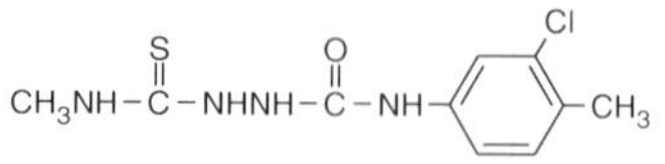

250 mg

S84,755-0
1-(3-CHLOROPHENYL)-3-DODECANOYL-2-THIOUREA

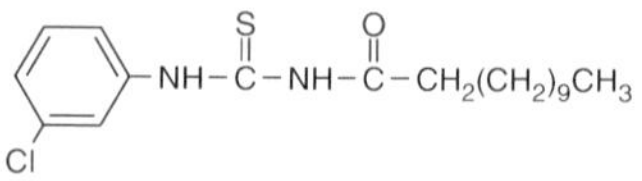

250 mg

S84,555-8
1-(4-IODOPHENYL)-3-(3-PHENYLPROPIONYL)-2-THIOUREA

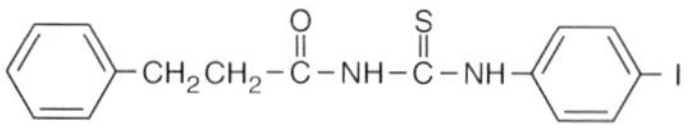

250 mg

S67,574-1
1-(2-CHLOROPHENYL)-3-DODECANOYL-2-THIOUREA

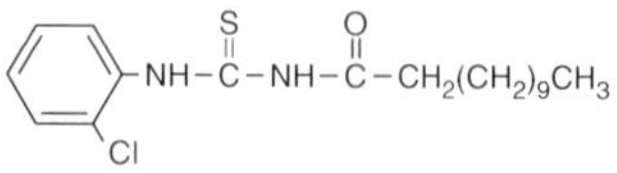

5 gm

S67,577-6
1-(2,5-DICHLOROPHENYL)-3-DODECANOYL-2-THIOUREA

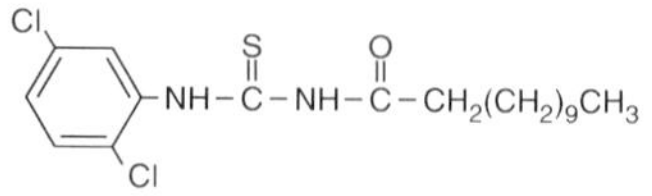

5 gm

S76,833-2
5-(4-CHLOROBENZYLIDENE)RHODANINE

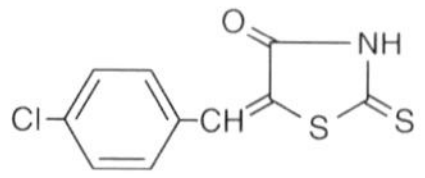

250 mg

S89,838-4
1-(3-FLUOROPHENYL)-5-OXO-3-PHENYL-2,4-DITHIOHYDANTOIN

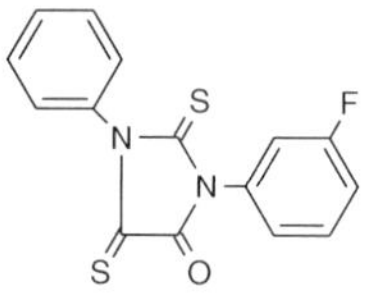

250 mg

S76,354-3
5-(4-CHLOROPHENYL)-3-METHYL-2-METHYLIMINO-4-THIAZOLIDINONE

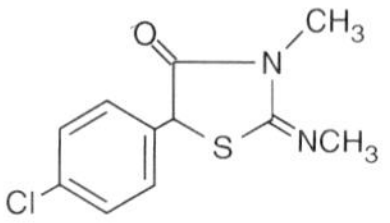

250 mg

S72,417-3
3-(3,4-DICHLOROPHENYL)RHODANINE

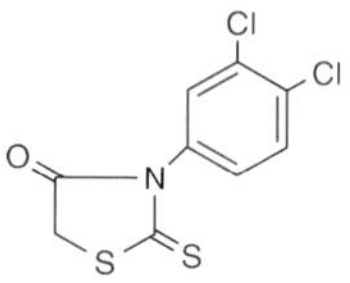

5 gm

S74,107-8
1-(3,4-DICHLOROPHENYL)-5-IMINO-3-PHENYL-4-THIOHYDANTOIN

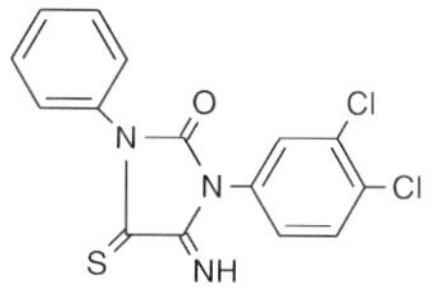

250 mg

S74,104-3
3-(4-CHLOROPHENYL)-1-(3,4-DICHLOROPHENYL)-5-IMINO-4-THIOHYDANTOIN

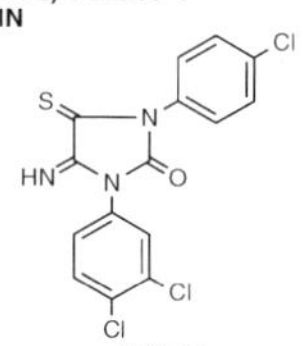

250 mg

S89,858-9
1-(3,4-DICHLOROPHENYL)-5-IMINO-3-METHYL-4-THIOHYDANTOIN

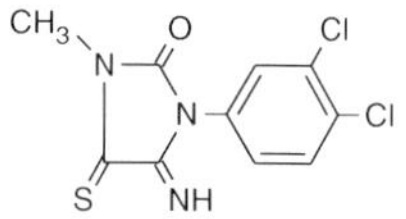

250 mg

S93,423-2
4'-CHLORO-3-MORPHOLINO-N-PHENYLTHIOCINNAMAMIDE

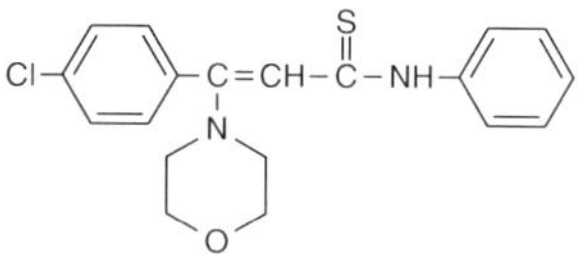

250 mg

S93,422-4
4'-CHLORO-3-MORPHOLINO-N-(4-TOLYL)-THIOCINNAMAMIDE

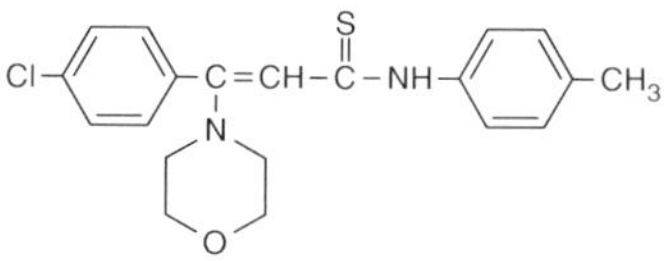

250 mg

S90,423-6
5-(3-CHLOROPHENYL)-2-THIOBIURET

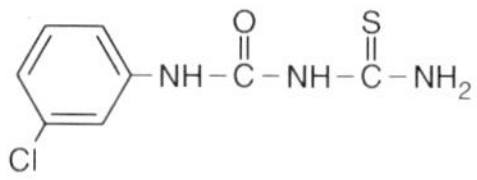

250 mg

S92,429-6
5-(4-FLUOROPHENYL)-2-THIOBIURET

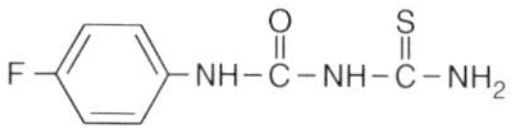

250 mg

S91,089-9
1-(2,3-DICHLOROPHENYL)-3-(2-THIAZOLIN-2-YL)UREA

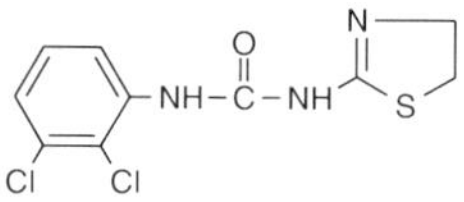

250 mg

S54,961-4
1-(3-CHLOROPHENYL)-2-THIOUREA

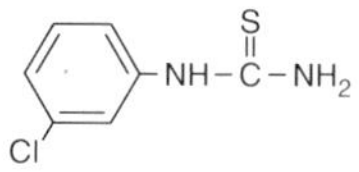

250 mg

S91,373-1
1-(3-FLUOROPHENYL)-3-METHYL-2-THIOUREA

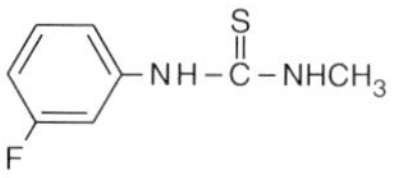

250 mg

S91,376-6
1-(3-BROMOPHENYL)-3-METHYL-2-THIOUREA

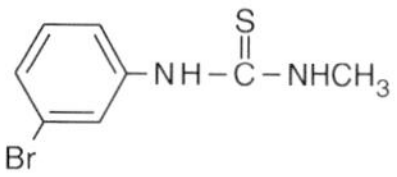

250 mg

S89,980-1
3-ALLYL-1-(3-CHLOROBENZYL)-1-METHYL-2-THIOUREA

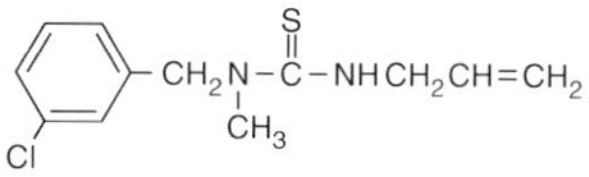

250 mg

S87,534-1
1,3-BIS(3-BROMOPHENYL)-2-THIOUREA

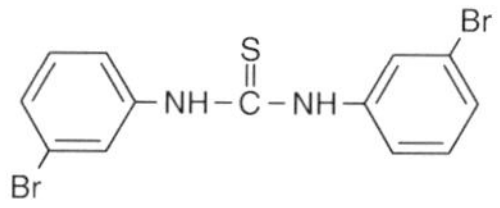

5 gm

S87,529-5
1,3-BIS(3-IODOPHENYL)-2-THIOUREA

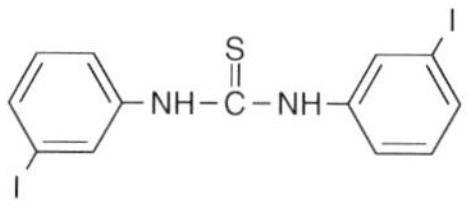

250 mg

S77,398-0
4-[N-(2,4-DIFLUOROPHENYL)-THIOCARBAMOYL]MORPHOLINE

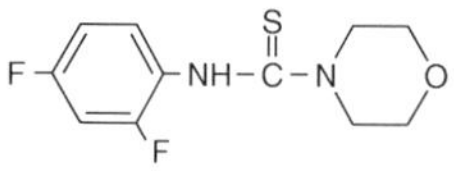

250 mg

S79,574-7
1-(3-CHLOROBENZYL)-3-PHENYL-2-THIOUREA

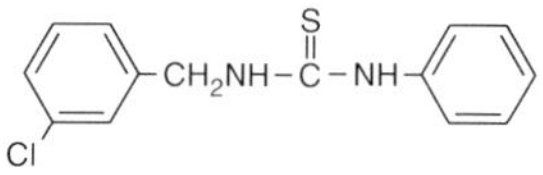

250 mg

S67,039-1
1-OCTADECYL-3-PHENYL-2-THIOUREA

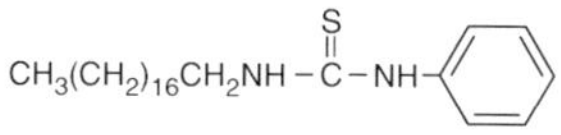

5 gm

S84,491-8
1-(2-CHLORO-6-METHYLPHENYL)-3-PHENYL-2-THIOUREA

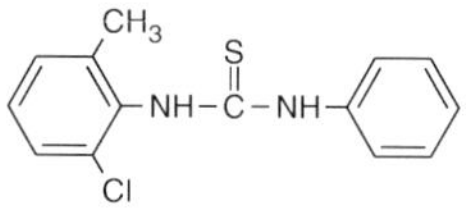

250 mg

S84,441-1
1-(3,4-DICHLOROPHENYL)-3-PHENYL-2-THIOUREA

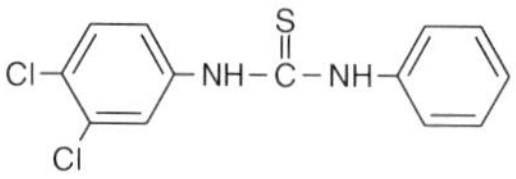

250 mg

S76,997-5
1-(3,4-DIBROMOPHENYL)-3-PHENYL-2-THIOUREA

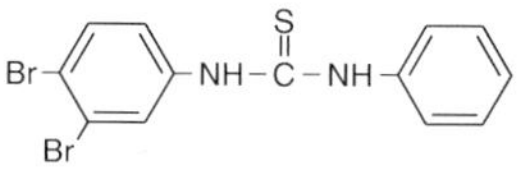

250 mg

S67,504-0
1-ETHYL-3-(2-FLUOROPHENYL)-2-THIOUREA

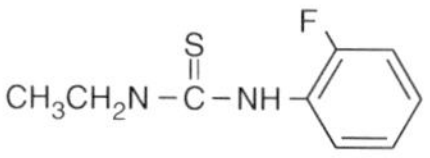

5 gm

S67,503-2
1,3-BIS(2-FLUOROPHENYL)-2-THIOUREA

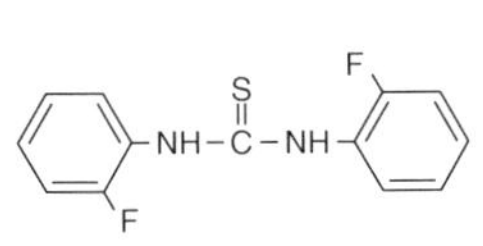

250 mg

S69,332-4
1,3-BIS(2-CHLOROPHENYL)-2-THIOUREA

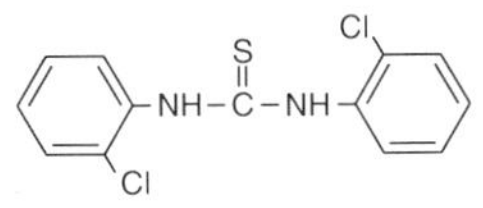

5 gm

S75,037-9
1-(2,4-DIFLUOROPHENYL)-3-(2-FLUOROPHENYL)-2-THIOUREA

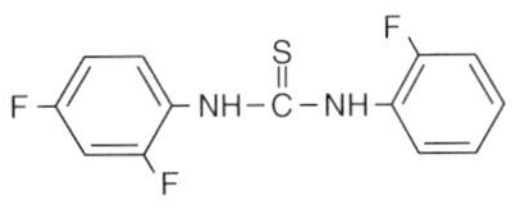

5 gm

S78,100-2
1-ALLYL-3-(4-CHLOROPHENYL)-2-THIOUREA

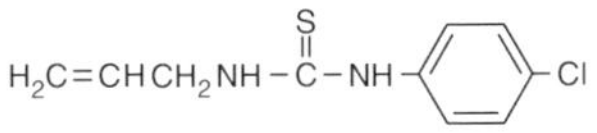

250 mg

S68,673-5
1,3-BIS(4-CHLOROPHENYL)-2-THIOUREA

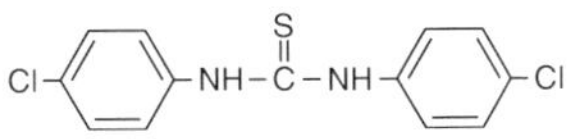

5 gm

S87,533-3
1,3-BIS(4-BROMOPHENYL)-2-THIOUREA

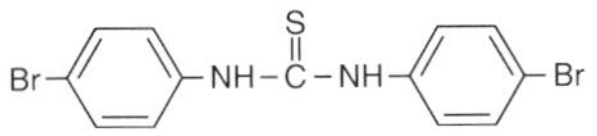

250 mg

S87,528-7
1,3-BIS(4-IODOPHENYL)-2-THIOUREA

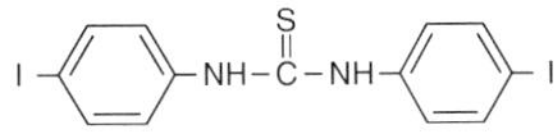

250 mg

S90,062-1
1-(4-FLUOROPHENYL)-3-(2,4,6-TRIMETHYLPHENYL)-2-THIOUREA

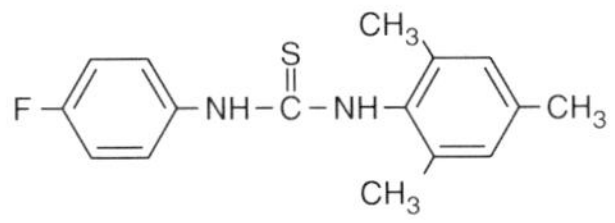

250 mg

S79,100-8
1-ALLYL-3-(3-CHLORO-2-METHYLPHENYL)-2-THIOUREA

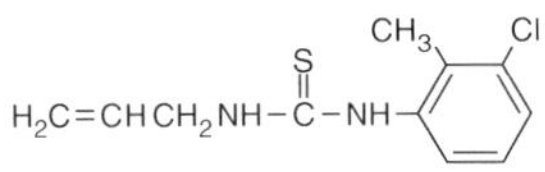

250 mg

S69,156-9
1,3-BIS(2,6-DICHLOROPHENYL)-2-THIOUREA

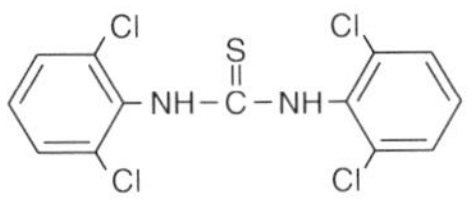

5 gm

S69,025-2
1-(2,4-DIFLUOROPHENYL)-3-ETHYL-2-THIOUREA

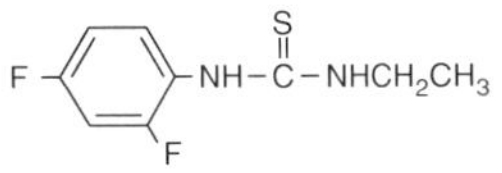

250 mg

S90,812-6
1-(3,5-DICHLOROPHENYL)-2-THIOUREA

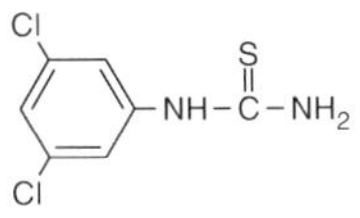

250 mg

S86,258-4
1-(2,4,6-TRIBROMOPHENYL)-2-THIOUREA

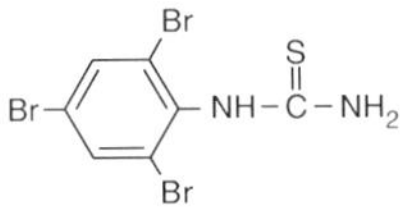

250 mg

S69,028-7
1-(2-CHLORO-4,6-DIBROMOPHENYL)-3-ETHYL-2-THIOUREA

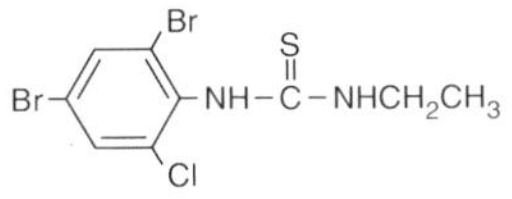

250 mg

S91,379-0
1-METHYL-3-(2,4,5-TRICHLOROPHENYL)-2-THIOUREA

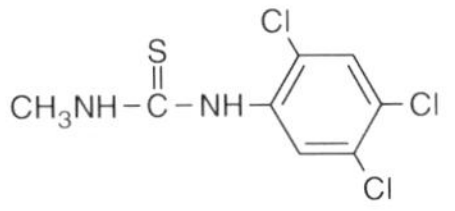

250 mg

S76,025-0
2-(4-CHLOROPHENYL)-3-IMINO-2,3,6,7-TETRAHYDRO-5H-THIAZOLO(3,2-A)-PYRIMIDINE HYDROCHLORIDE

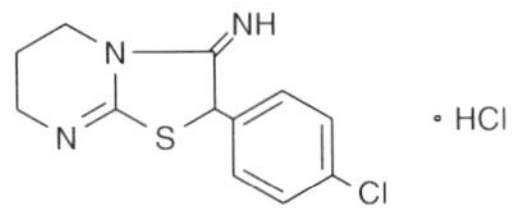

250 mg

S57,759-6
4,5-DIHYDRO-4-(HYDROXYMETHYL)-1-PHENYL-1H-TETRAZOLE-5-THIONE

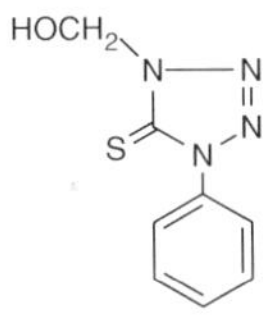

250 mg

S51,276-1
2-(2,5-DIHYDROXYPHENYL)-2-THIOPSEUDOUREA HYDROCHLORIDE

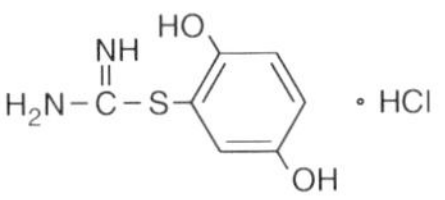

250 mg

S69,185-2
5-(3,5-DIMETHOXY-4-HYDROXYBENZYLIDENE)-3-PROPYLRHODANINE

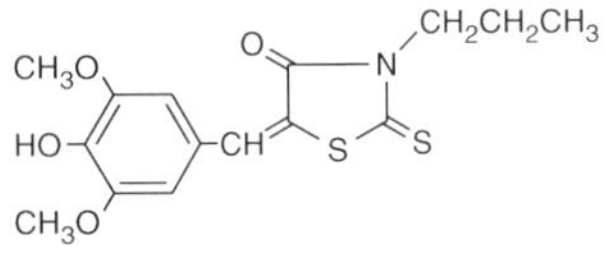

250 mg

S77,224-0
5-(4-HYDROXY-3-METHOXYBENZYLIDENE)-3-(3-METHOXYPROPYL)RHODANINE

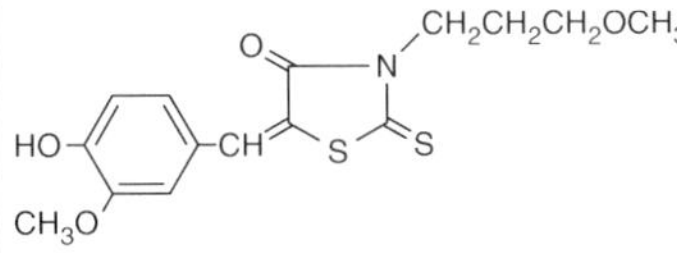

250 mg

S66,401-4
4-(4-HYDROXY-3-METHOXYPHENYLTHIOCARBONYL)-MORPHOLINE

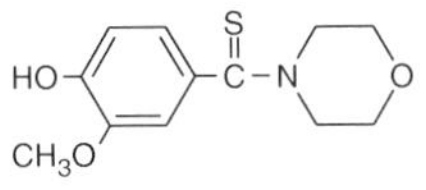

5 gm

S72,444-0
1-(2,2-DIPHENYL-2-HYDROXYACETYL)-4-(4-METHOXYPHENYL)-3-THIOSEMICARBAZIDE

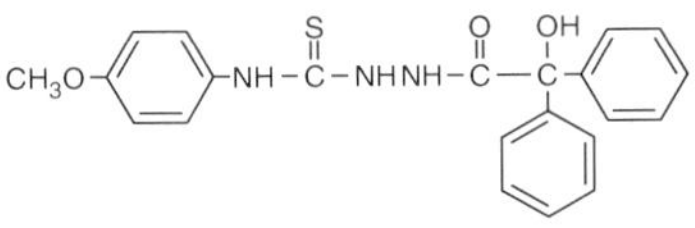

5 gm

S65,188-5
5-(ALPHA-METHYLVANILLYLIDENE)-RHODANINE

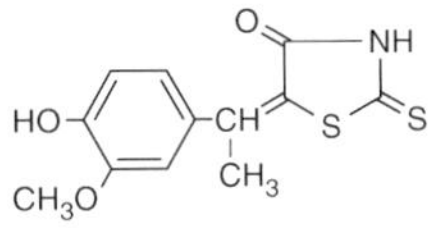

250 mg

S73,741-0
5-(4-HYDROXY-3-METHOXYBENZYLIDENE)-RHODANINE

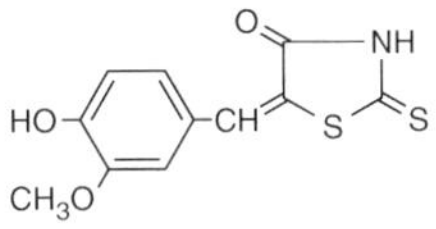

5 gm

S56,279-3
3-METHYL-2-BENZOTHIAZOLINONE, AZINE OF 4-(2-HYDROXYETHOXY)-3-METHOXYBENZALDEHYDE

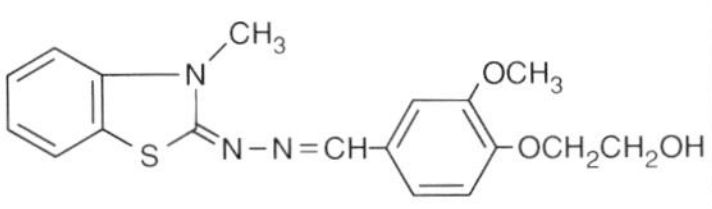

250 mg

S41,061-6
5-ANISYLIDENE-2-THIOBARBITURIC ACID

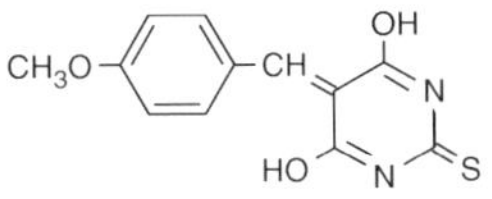

5 gm

S41,062-4
5-VANILLYLIDENE-2-THIOBARBITURIC ACID

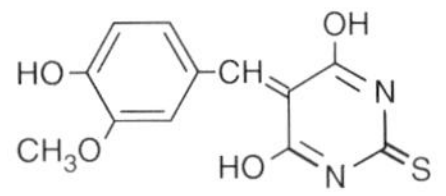

5 gm

S69,201-8
BENZILIC ACID N2-[N-(2-METHOXY-5-METHYLPHENYL)THIOCARBAMOYL]-HYDRAZIDE

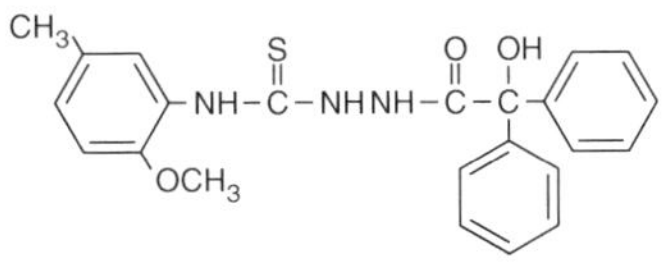

250 mg

S78,089-8
1-(2-HYDROXYBENZYLIDENE)-4-(2-METHOXY-5-METHYLPHENYL)-3-THIOSEMICARBAZIDE

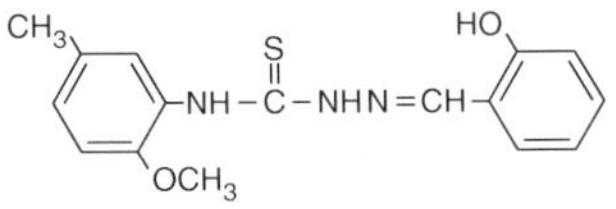

250 mg

S41,045-4
5-(O-VANILLYLIDENE)-2-THIOBARBITURIC ACID

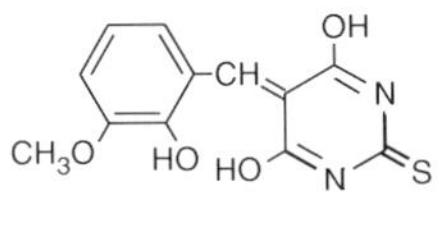

5 gm

S68,697-2
5-(2-HYDROXY-1-NAPHTHYLMETHYLENE)-3-(3-METHOXYPROPYL)RHODANINE

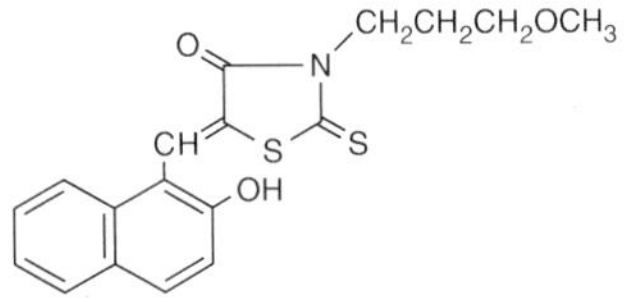

250 mg

S92,426-1
1-ALLYL-3-[3-(1-HYDROXYETHYL)-PHENYL]-2-THIOUREA

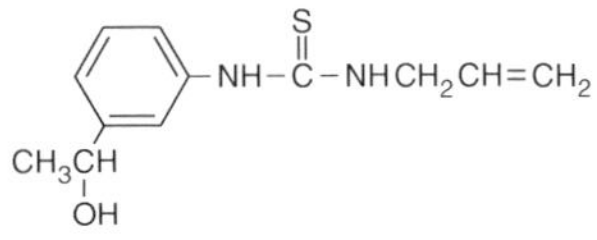

250 mg

S67,452-4
1-ALLYL-3-[2-(1-HYDROXYETHYL)-PHENYL]-2-THIOUREA
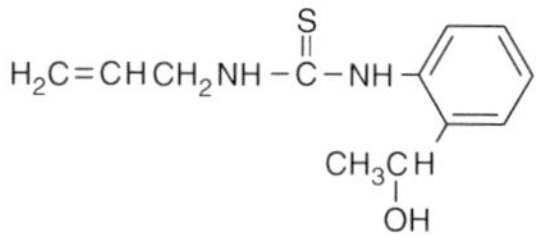
5 gm

S54,750-6
5-ETHOXYMETHYLENE-3-(4-HYDROXYPHENYL)-2-THIOXO-4-THIAZOLIDINONE
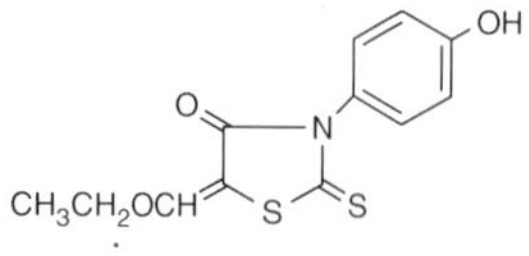
250 mg

S93,170-5
6-AMINO-1-BENZYL-4-HYDROXY-2-PYRIMIDINETHIONE
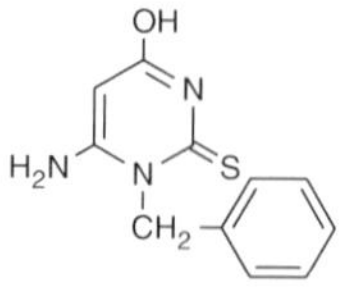
250 mg

S41,064-0
5-[4-(DIMETHYLAMINO)BENZYLIDENE]-2-THIOBARBITURIC ACID
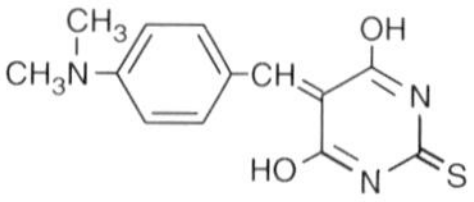
5 gm

S41,065-9
5-CINNAMYLIDENE-2-THIOBARBITURIC ACID
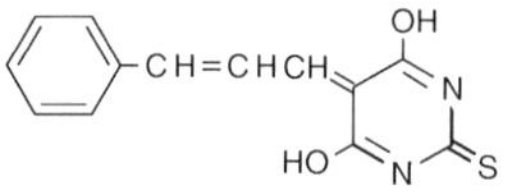
5 gm

S41,060-8
5-PIPERONYLIDENE-2-THIOBARBITURIC ACID
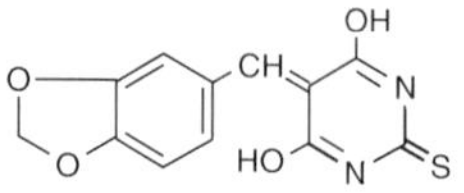
5 gm

S68,472-4
4-(4-ACETAMIDOPHENYL)-1-(4-HYDROXYBENZOYL)-3-THIOSEMICARBAZIDE
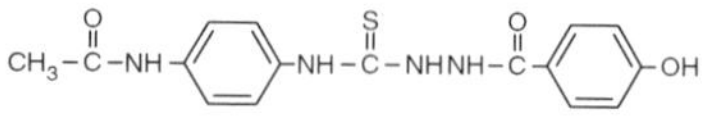
250 mg

S90,304-3
1-(3-HYDROXYPHENYL)-3-(4-NITROPHENYL)-2-THIOUREA
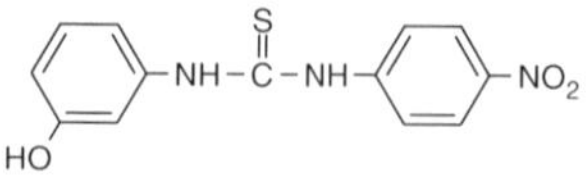
250 mg

S75,249-5
5-(4-HYDROXYBENZYL)-3-PHENYL-2-THIOHYDANTOIN
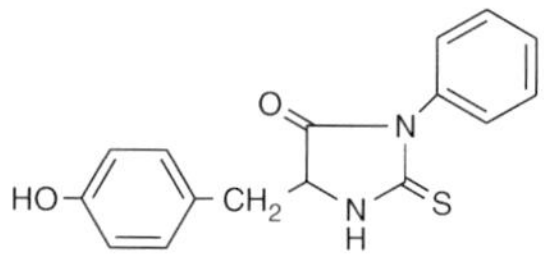
5 gm

S40,883-2
3-(4-HYDROXYANILINOMETHYL)--BENZOTHIAZOLINE-2-THIONE HYDROCHLORIDE
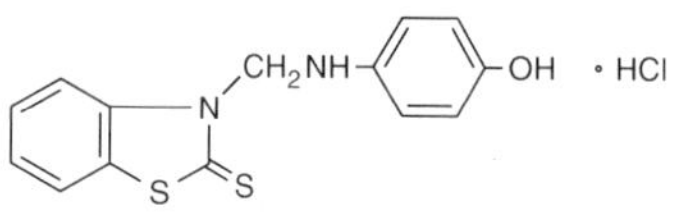
5 gm

S50,682-6
1-(P-HYDROXYANILINOMETHYL)-2-(3H)-BENZIMIDAZOLETHIONE HYDROCHLORIDE
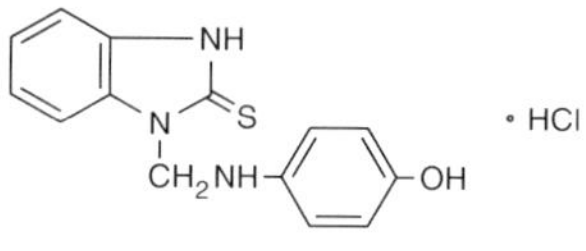
5 gm

S50,672-9
4-(2,3-DIHYDRO-2-THIOXO-1-BENZIMIDAZOLYLMETHYLAMINO)-SALICYLIC ACID
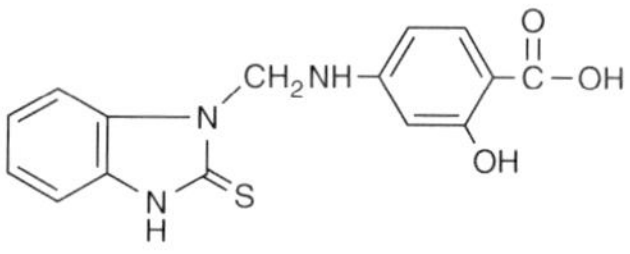
5 gm

S50,685-0
5-(2,3-DIHYDRO-2-THIOXO-1-BENZIMIDAZOLYLMETHYLAMINO)-SALICYLIC ACID
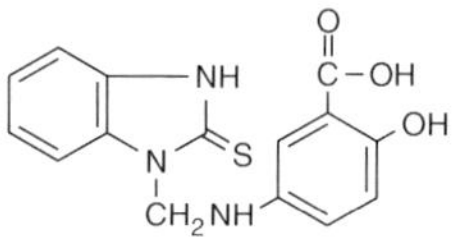
250 mg

S50,674-5
5-(2,3-DIHYDRO-2-THIOXO-3-BENZOXAZOLYLMETHYLAMINO)-SALICYLIC ACID
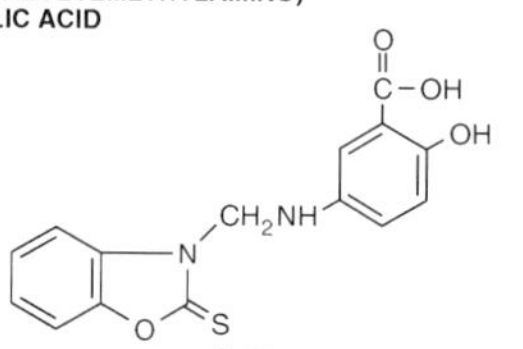
5 gm

S50,678-8
4-(2,3-DIHYDRO-2-THIOXO-3-BENZOXAZOLYLMETHYLAMINO)-SALICYLIC ACID
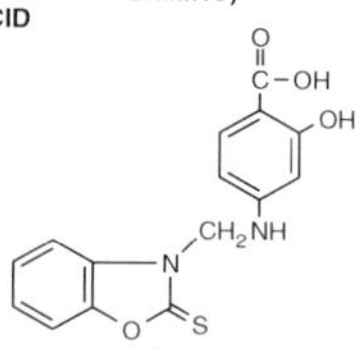
5 gm

S50,008-9
4-(4-HYDROXYTHIOBENZOYL)MORPHOLINE
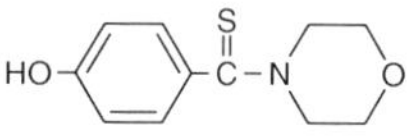
5 gm

S76,112-5
5-BENZYL-4-HYDROXY-3-METHYL-2-(METHYLIMINO)THIAZOLIDINE
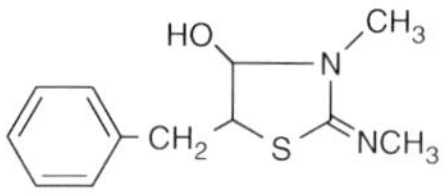
250 mg

S76,970-3
4-ETHYL-1-(4-HYDROXYBENZOYL)-3-THIOSEMICARBAZIDE
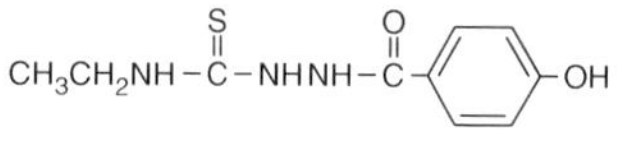
250 mg

S80,082-1
1-(3-HYDROXYBENZYLIDENE)-4-PHENYL-3-THIOSEMICARBAZIDE
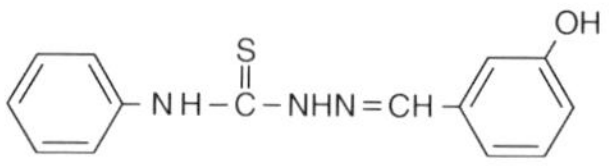
250 mg

S80,098-8
1-(2-HYDROXY-1-NAPHTHYLMETHYLENE)-4-PHENYL-3-THIOSEMICARBAZIDE
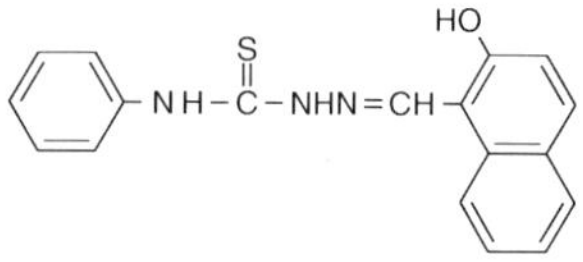
250 mg

S61,519-6
1-BENZOYL-3-(2-HYDROXYETHYL)-2-THIOUREA
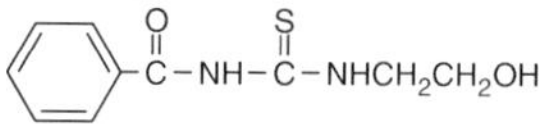
5 gm

S73,360-1
4-DIPHENYLMETHYLENE-1-(N-METHYL-THIOCARBAMOYL)-3-PIPERIDINOL
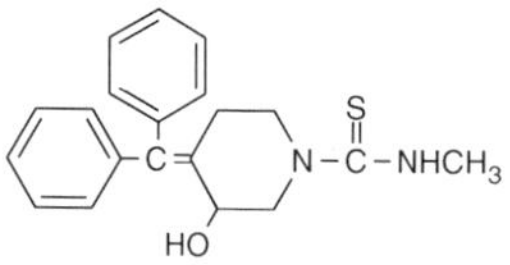
250 mg

S53,135-9
1-(2-HYDROXYETHYL)-3-PHENYL-2-THIOUREA
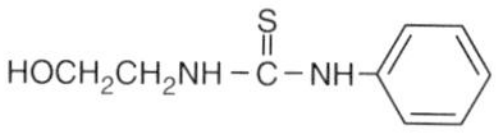
5 gm

S44,776-5
1-(2-HYDROXYPHENYL)-3-PHENYL-2-THIOUREA
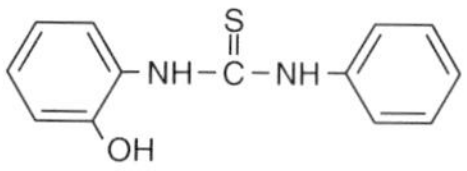
5 gm

S65,835-9
5-(3-HYDROXYBENZYLIDENE)RHODANINE
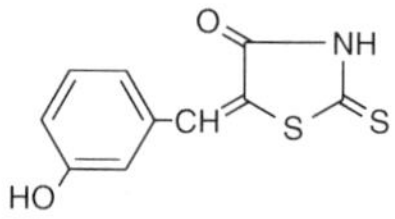
5 gm

S84,551-5
1-CINNAMOYL-3-(4-HYDROXYPHENYL)-2-THIOUREA
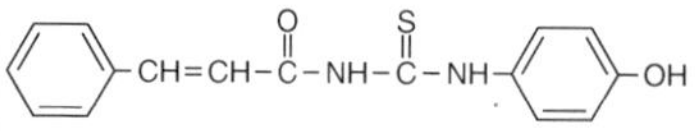
250 mg

S84,596-5
1-(3-HYDROXYPHENYL)-3-(1-NAPHTHYLCARBONYL)-2-THIOUREA
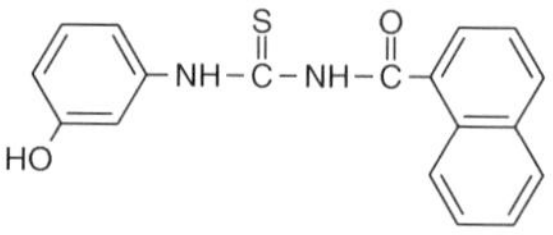
250 mg

S75,667-9
5-(3-HYDROXY-ALPHA-METHYLBENZYLIDENE)RHODANINE
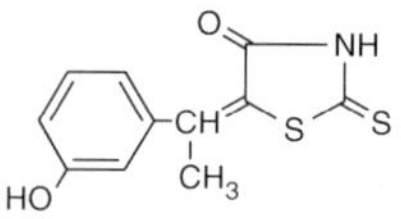
5 gm

S69,442-8
3-CYCLOPROPYL-5-(2-HYDROXYBENZYLIDENE)RHODANINE
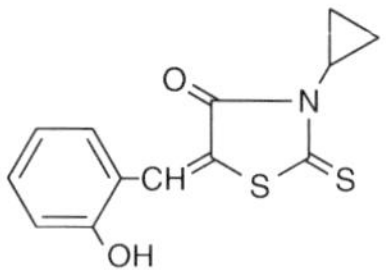
250 mg

S52,417-4
3-(3-HYDROXYPHENYL)RHODANINE
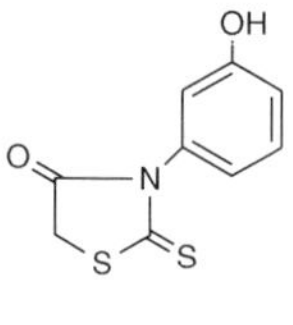
5 gm

S93,497-6
N-CIS(2-HYDROXYCYCLOPENTYL)-THIOBENZAMIDE
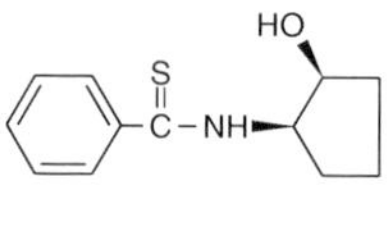
250 mg

S53,688-1
1-(3-CHLOROPHENYL)-3-(2-HYDROXYETHYL)-2-THIOUREA
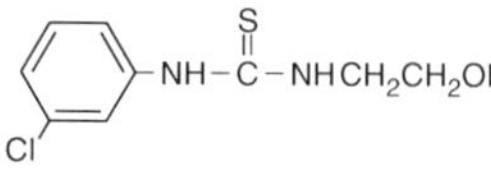
5 gm

S67,453-2
1-ALLYL-3-(3-HYDROXYPHENYL)-2-THIOUREA
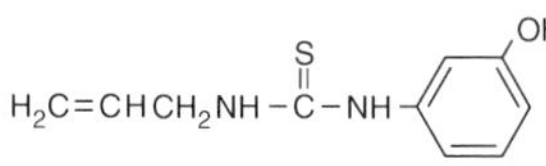
5 gm

S38,803-3
5-BENZYLIDENE-2-THIOBARBITURIC ACID
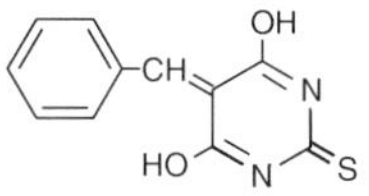
5 gm

S61,525-0
1-(3-HYDROXYPHENYL)-3-PHENYL-2-THIOUREA
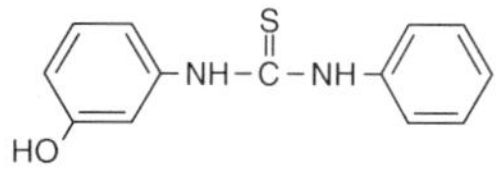
5 gm

S85,475-1
1-(5-HYDROXY-1-NAPHTHYL)-3-PHENYL-2-THIOUREA
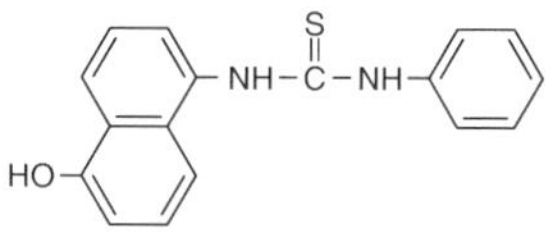
250 mg

S67,449-4
1-ALLYL-3-(4-HYDROXYPHENYL)-2-THIOUREA
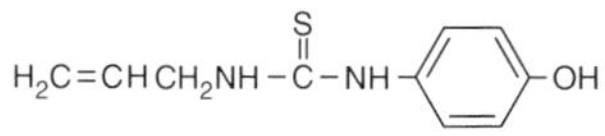
5 gm

S69,149-6
1-(4-HYDROXYPHENYL)-3-PROPYL-2-THIOUREA
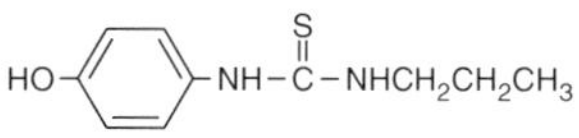
250 mg

S68,770-7
1-(2,6-DIETHYLPHENYL)-3-(4-HYDROXYPHENYL)-2-THIOUREA
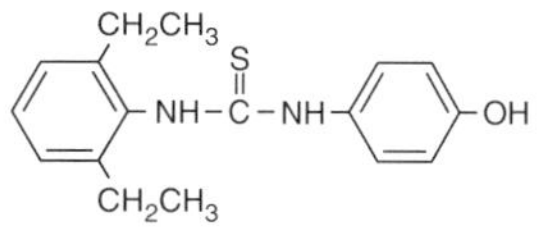
250 mg

S53,636-9
1-(2-HYDROXYETHYL)-3-(2,4-XYLYL)-2-THIOUREA
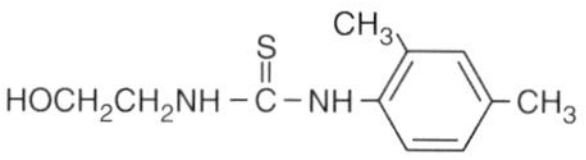
5 gm

S78,079-0
S-(2-HYDROXYPHENYL) N-(2-ETHYLPHENYL)THIOCARBAMATE
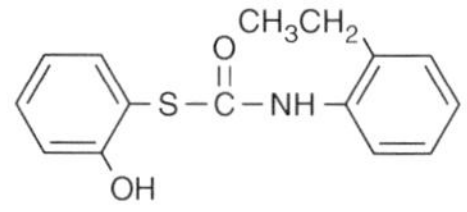
250 mg

S73,265-6
2-CHLORO-3-PHENYL-2,3,6,7-TETRAHYDRO-5H-THIAZOLO(3,2-A)-PYRIMIDIN-3-OL HYDROCHLORIDE
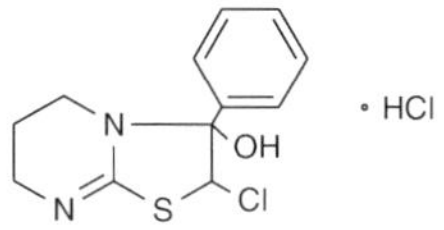
250 mg

S89,847-3
2-(3,4-DIMETHOXYPHENYL)-5-IMINO-3-METHYL-4-OXAZOLIDINETHIONE
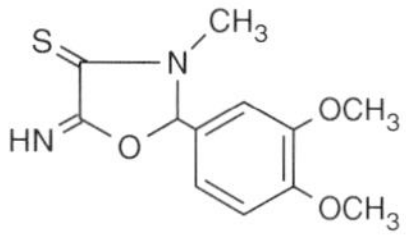
250 mg

S80,093-7
1-(3,4-DIMETHOXYBENZYLIDENE)-4-PHENYL-3-THIOSEMICARBAZIDE
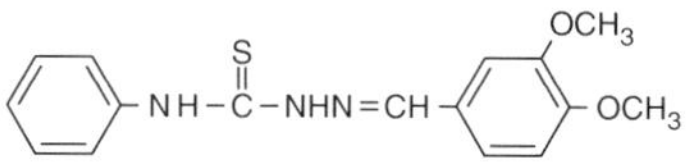
250 mg

S69,406-1
3-BUTYL-5-(2,3,4-TRIMETHOXYBENZYLIDENE)RHODANINE
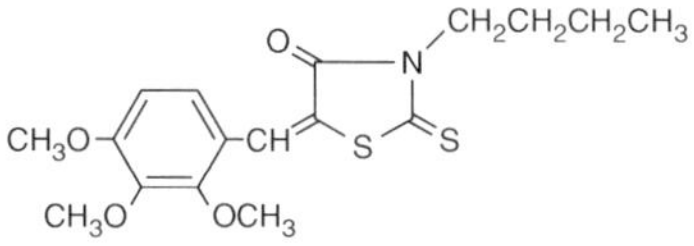
250 mg

S72,333-9
3-BUTYL-5-(3,4-DIMETHOXYBENZYLIDENE)RHODANINE
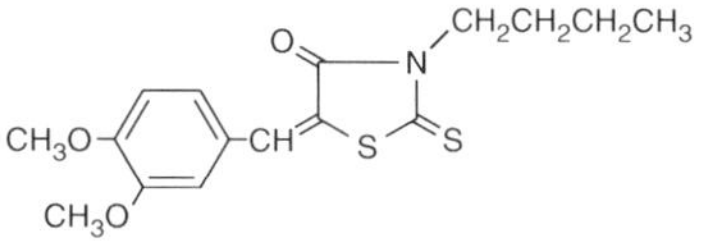
50 mg

S76,124-9
5,6-DIHYDRO-8,9-DIMETHOXY-3-(ETHYLIMINO)-3H-THIAZOLO(4,3-A)-ISOQUINOLINE HYDROCHLORIDE
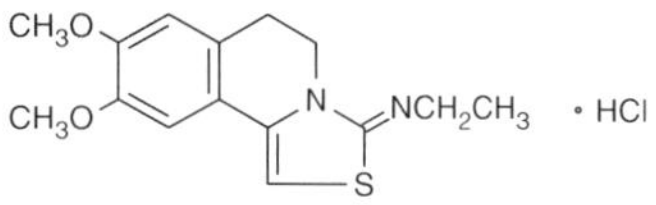
250 mg

S69,690-0
5-(2,3-DIMETHOXYBENZYLIDENE)-RHODANINE
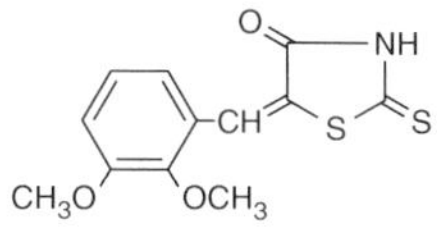
250 mg

S76,140-0
5,6-DIHYDRO-8,9-DIMETHOXY-3-(PHENYLIMINO)-3H-THIAZOLO(4,3-A)--ISOQUINOLINE HYDROCHLORIDE
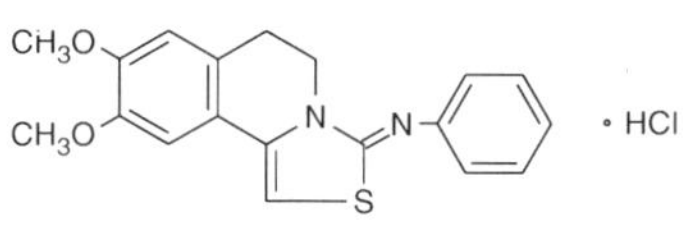
250 mg

S91,034-1
1-(2,4-DIMETHOXYPHENYL)-3-(2-THIAZOLIN-2-YL)UREA
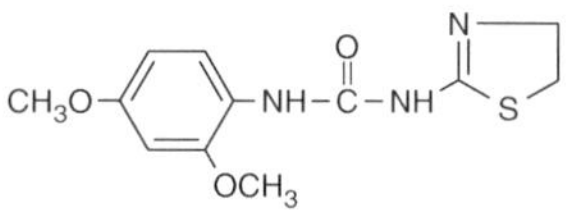
250 mg

S68,633-6
1-(3,5-DIMETHOXYPHENYL)-3-(2-ETHOXYPHENYL)-2-THIOUREA
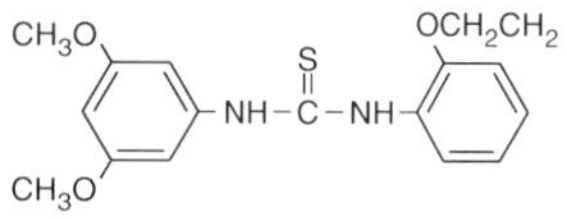
250 mg

S72,093-3
1-(2,5-DIMETHOXYPHENYL)-5-THIOBIUREA
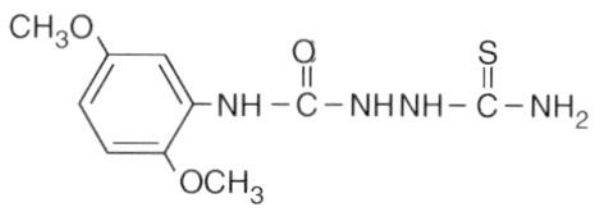
5 gm

S67,492-3
1-(2,5-DIMETHOXYPHENYL)-3-DODECANOYL-2-THIOUREA
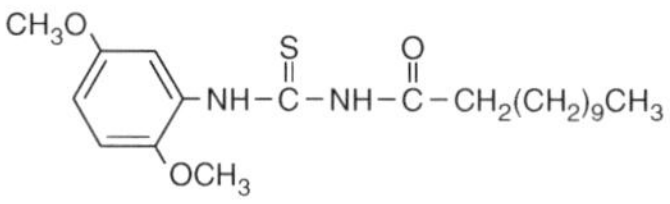
250 mg

S85,470-0
1-(2,5-DIMETHOXYPHENYL)-3-PHENYL-2-THIOUREA
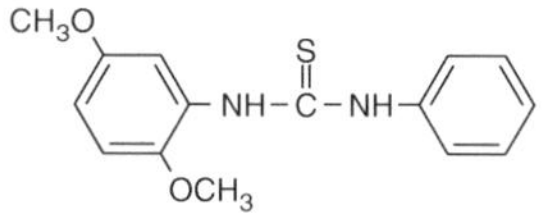
250 mg

S91,707-9
1-(2,5-DIMETHOXYPHENYL)-3-METHYL-2-THIOUREA
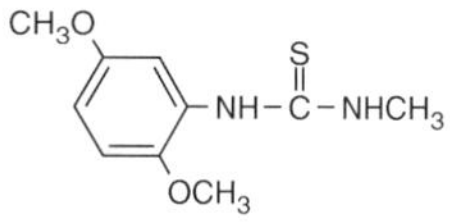
250 mg

S69,115-1
1-(2,5-DIMETHOXYPHENYL)-3-PROPYL-2-THIOUREA
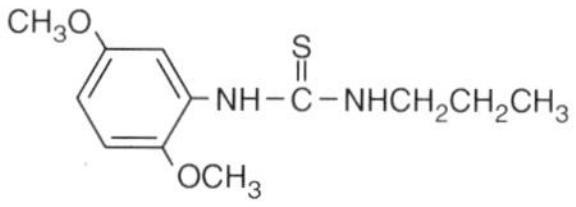
250 mg

S80,644-7
4,6-DIHYDRO-8,9-DIMETHOXY-4-(2-AMINOPHENYLIMINO)-2H-[1,3]THIAZINO (4,3-A)ISOINDOLE DIHYDROCHLORIDE
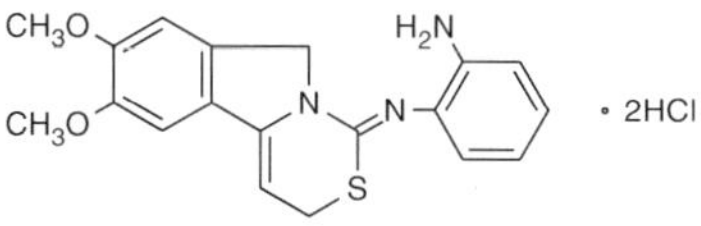
250 mg

S84,557-4
1-(2,5-DIMETHOXYPHENYL)-3-(3-NITROBENZOYL)-2-THIOUREA
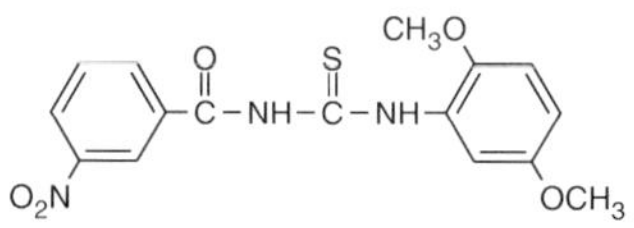
250 mg

S54,599-6
3-(3,4-DIMETHOXYPHENETHYL) -RHODANINE
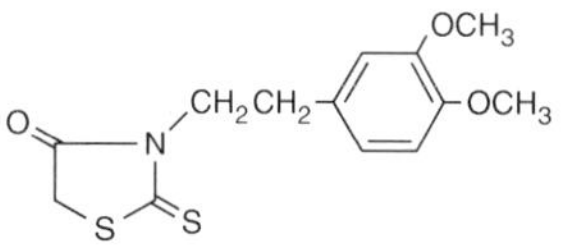
250 mg

S54,957-6
1-BENZOYL-3-(3,4-DIMETHOXYPHENETHYL)-2-THIOUREA
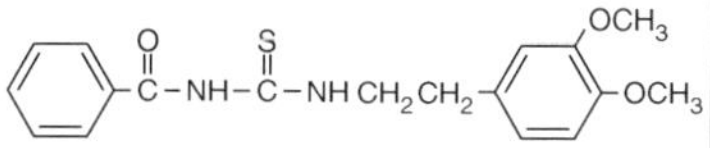
250 mg

S53,134-0
N-(3,4-DIMETHOXYPHENETHYL)-DITHIOCARBAMIC ACID, AMMONIUM SALT
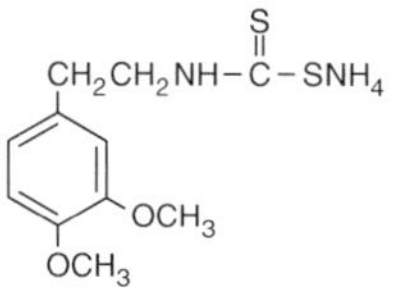
5 gm

S54,571-6
3,5-BIS(3,4-DIMETHOXYPHENETHYL)-3,4,5,6-TETRAHYDRO-2H-1,3,5-THIADIAZINE-2-THIONE
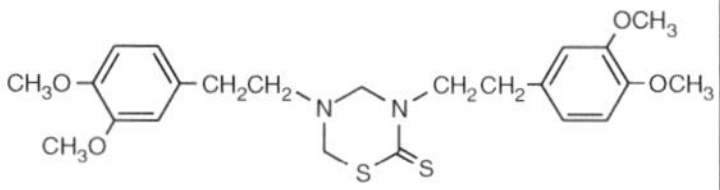
250 mg

S79,678-6
1-BENZYL-3-(2-PHENOXYPROPIONYL)-2-THIOUREA
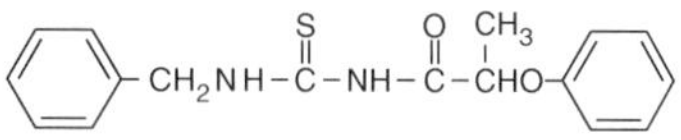
250 mg

S84,552-3
4-[3-(2-PHENOXYACETYL)-2-THIOUREIDO]BENZOIC ACID
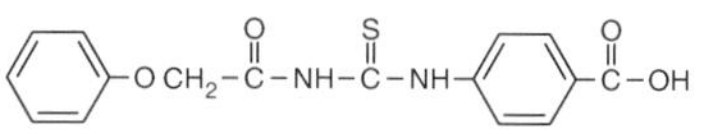
250 mg

S84,579-5
4-[3-(2-PHENOXYPROPIONYL)-2-THIOUREIDO]BENZOIC ACID
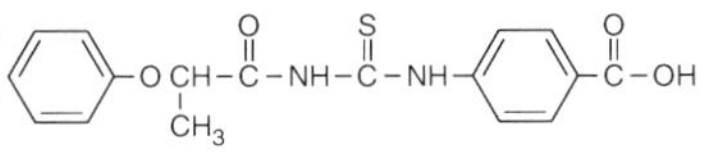
250 mg

S84,757-7
BUTYL 4-[3-(2-PHENOXYACETYL)-2-THIOUREIDO]BENZOATE
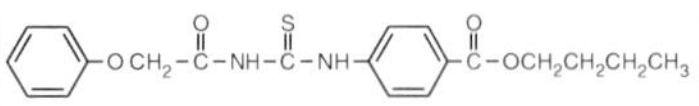
250 mg

S60,047-4
5-(3,4-DIMETHOXYBENZYLIDENE) RHODANINE
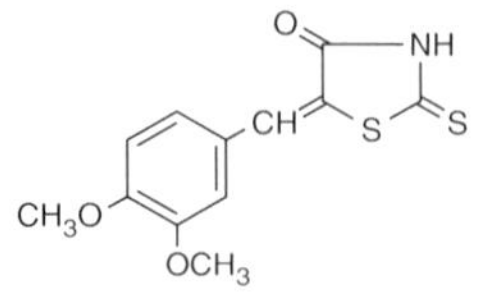
5 gm

S84,581-7
1-(4-ACETAMIDOPHENYL)-3-(2-PHENOXYACETYL)-2-THIOUREA
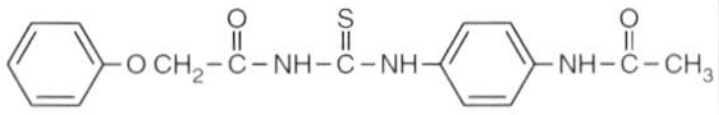
250 mg

S84,756-9
1-(1-NAPHTHYL)-3-(2-PHENOXYPROPIONYL)-2-THIOUREA
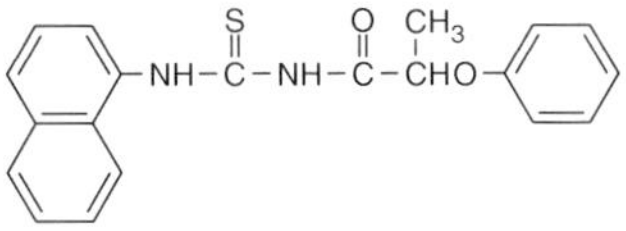
250 mg

S69,449-5
5-(3,4-DIMETHOXYBENZYLIDENE)-3-ISOPROPYLRHODANINE
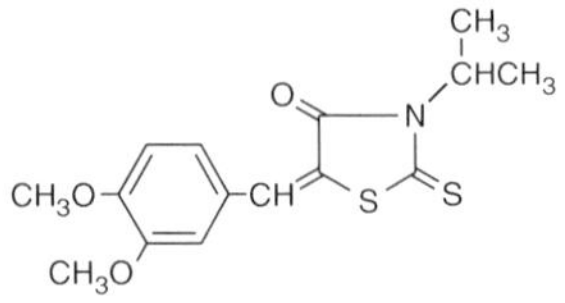
250 mg

S63,076-4
3,4-DIMETHOXY-N-HEXADECYL-N-METHYLTHIOBENZAMIDE
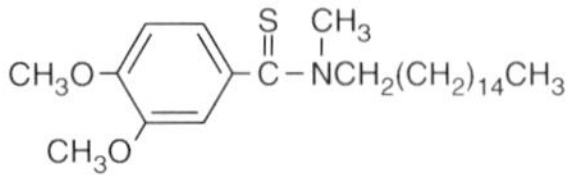
5 gm

S59,225-0
1-(3,4-DIMETHOXYPHENYL)-2-THIOUREA
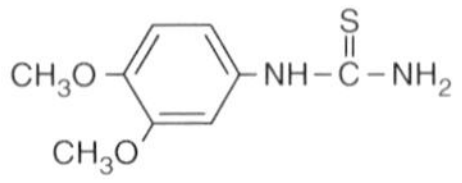
5 gm

S89,857-0
5-IMINO-2-(4-METHOXYPHENYL)-3-METHYL-4-OXAZOLIDINETHIONE
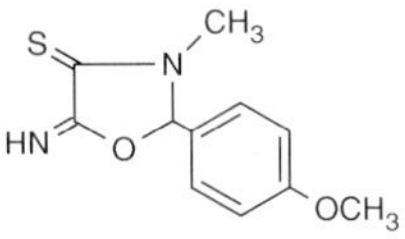
250 mg

S93,360-0
N1-(4-METHOXYPHENYL)-1-THIOOXAMIDE
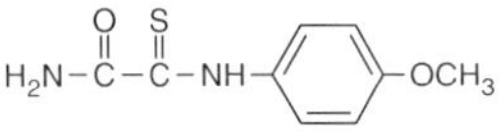
250 mg

S67,576-8
1-DODECANOYL-3-(2-METHOXY-4-NITROPHENYL)-2-THIOUREA
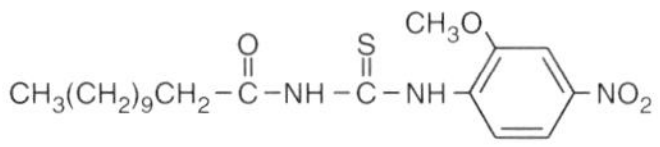
5 gm

S84,754-2
1-[4-(N-METHYLACETAMIDO)PHENYL]-3-(2-PHENOXYACETYL)-2-THIOUREA
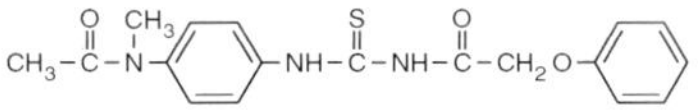
250 mg

S80,096-1
1-(4-ETHOXYPHENYL)-3-PHENETHYL-2-THIOUREA
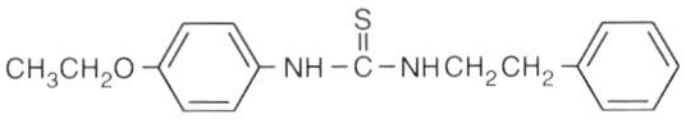
250 mg

S53,679-2
1-(1,2-DIPHENYLETHYL)-3-(4-ETHOXYPHENYL)-2-THIOUREA
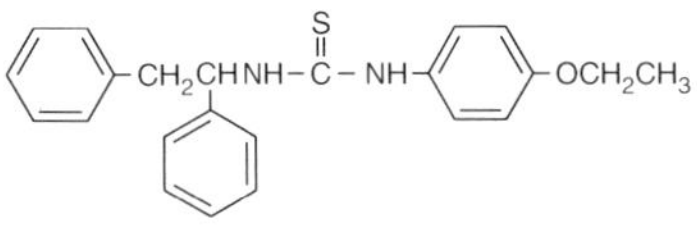
5 gm

S77,265-8
4-[4-METHOXYPHENYL)-1-(2-(METHYLAMINO)BENZOYL]-3-THIOSEMICARBAZIDE
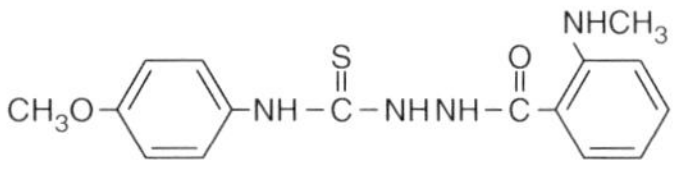
250 mg

S53,687-3
4,4'-TRIMETHYLENEBIS(1-[N-(4-ETHOXYPHENYL)THIOCARBAMOYL]-PIPERIDINE)
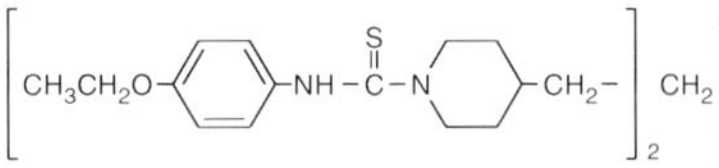
5 gm

S54,753-0
1,1-DIMETHYL-3-(2-METHOXYPHENYL)-2-THIOUREA
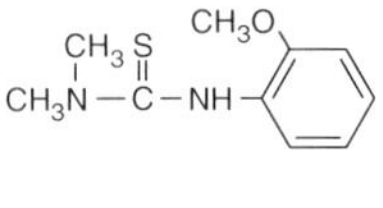
250 mg

S63,301-1
4-METHOXY-3-METHYL-2-BENZOTHIAZOLINONE HYDRAZONE HYDROCHLORIDE
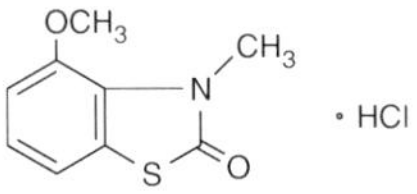
250 mg

S54,573-2
3,5-BIS(4-ETHOXYPHENYL)-3,4,5,6-TETRAHYDRO-2H-1,3,5-THIADIAZINE-2-THIONE
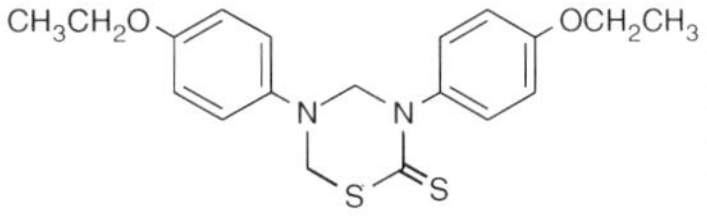
5 gm

S76,844-8
5-[4-(DIMETHYLAMINO)BENZYLIDENE]-3-(3-METHOXYPROPYL)RHODANINE
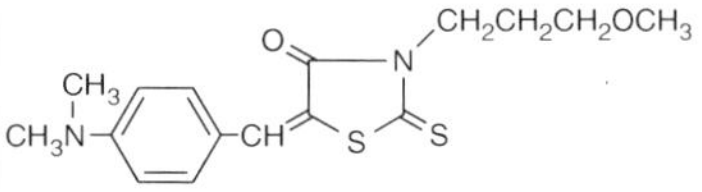
250 mg

S50,969-8
N-ETHYL-2-METHOXYTHIOBENZO-M-TOLUIDIDE
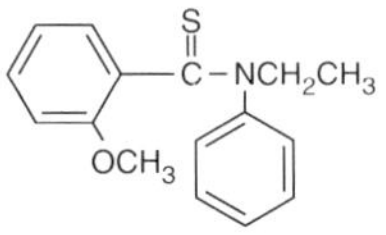
250 mg

S51,234-6
N-METHYL-P-THIOACETANISIDIDE
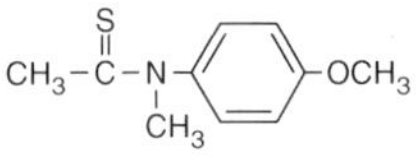
250 mg

S51,062-9
N-ETHYL-4-METHOXYTHIOBENZO-M-TOLUIDIDE
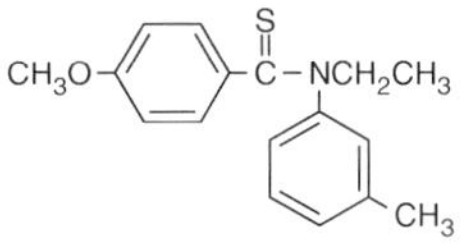
250 mg

S51,000-9
N-ISOBUTYL-4-METHOXYTHIOBENZANILIDE
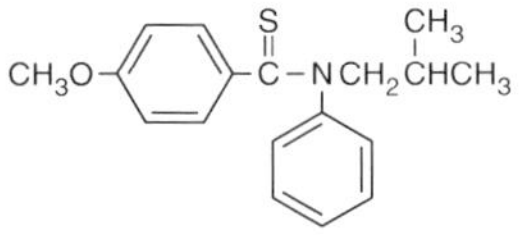
250 mg

S50,225-1
N-HEXADECYL-4-METHOXY-N-METHYLTHIOBENZAMIDE
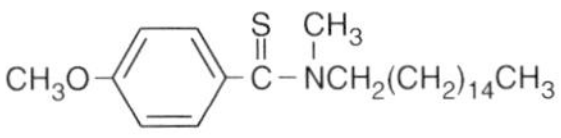
250 mg

S54,278-4
1,1-DIMETHYL-3-(4-METHOXYPHENYL)-2-THIOUREA
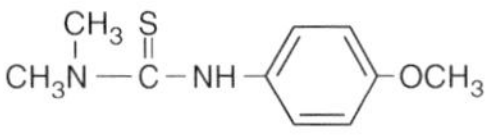
250 mg

S54,277-6
1,1-DIMETHYL-3-(4-ETHOXYPHENYL)-2-THIOUREA
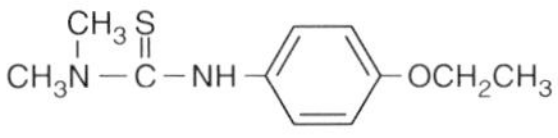
250 mg

S51,269-9
1-METHYL-1-(4-OCTYLOXYPHENYL)-2-THIOUREA
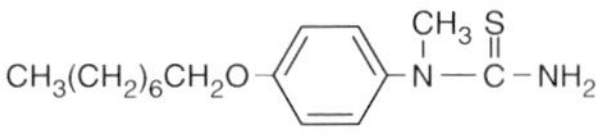
250 mg

S73,488-8
5-(4-METHOXYBENZYLIDENE)-3-METHYL-2-(METHYLIMINO)THIAZOLIDINE
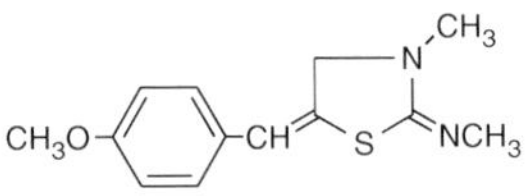
250 mg

S92,418-0
1-(3,4-DIMETHYLBENZYL)-3-(4-METHOXYPHENYL)UREA
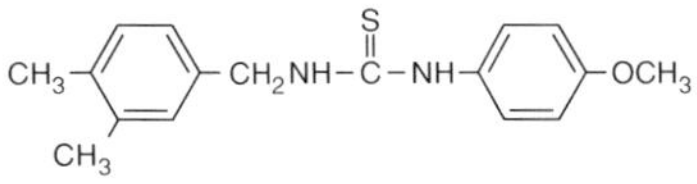
250 mg

S72,092-5
1-(2-METHOXY-5-METHYLPHENYL)-5-THIOBIUREA
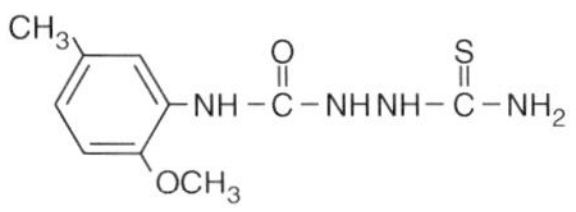
5 gm

S67,547-4
1-[N-(3-METHOXYPHENYL)CARBAMOYL]-3-THIOSEMICARBAZIDE
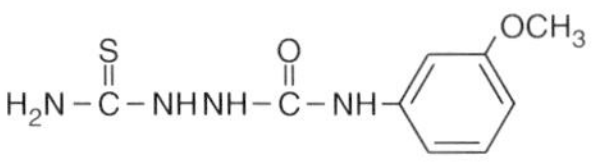
250 mg

S66,102-3
6-METHOXY-3-(3-SULFOPROPYL)-3H-BENZOTHIAZOLIN-2-ONE HYDRAZONE

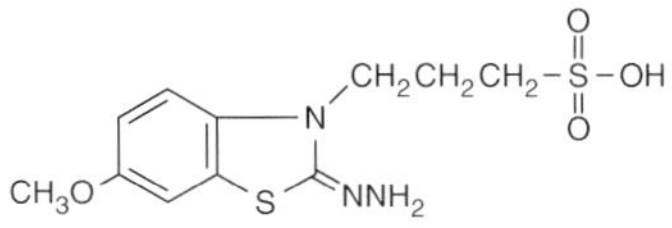

250 mg

S80,088-0
1-(4-METHOXYBENZYLIDENE)-4-PHENYL-3-THIOSEMICARBAZIDE

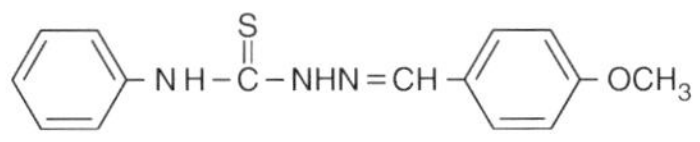

250 mg

S51,340-7
1-BENZOYL-3-(4-METHOXY-2-NITROPHENYL)-2-THIOUREA

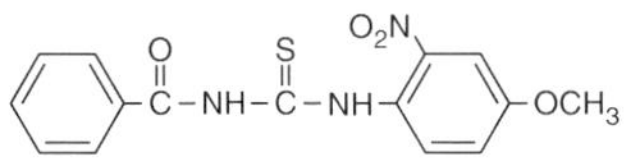

250 mg

S72,406-8
1-(2-ETHOXYPHENYL)-5-THIOBIUREA

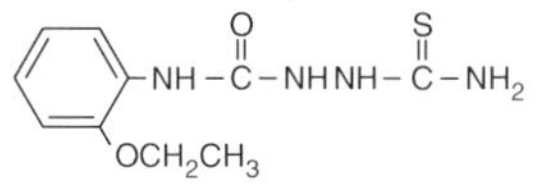

5 gm

S71,879-3
1-(DODECANOYL)-3-(2-ETHOXYPHENYL)-2-THIOUREA

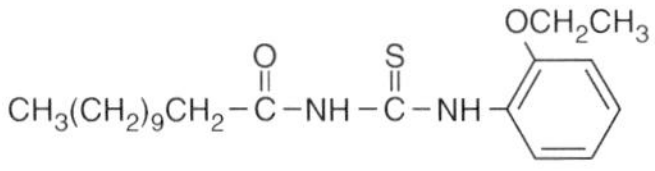

5 gm

S74,130-2
5-IMINO-1-(2-METHOXYPHENYL)-3-PHENYL-4-THIOHYDANTOIN

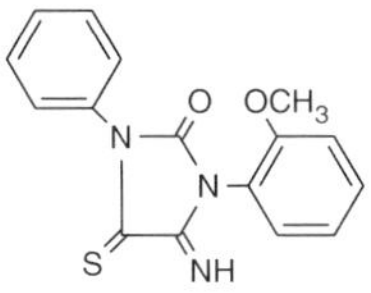

250 mg

S90,061-3
1-CYCLOHEXYL-3-(2-ETHOXYPHENYL)-2-THIOUREA

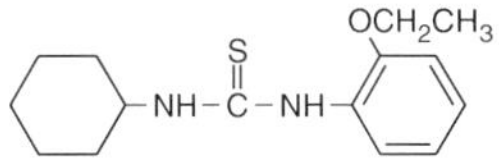

250 mg

S74,341-0
1-ALLYL-3-(2-ETHOXYPHENYL)-2-THIOUREA

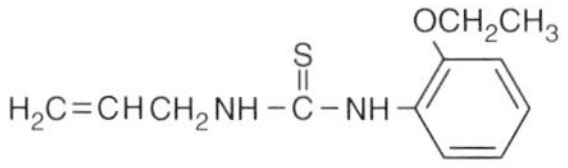

5 gm

S65,477-9
1-(2,5-DIMETHOXYPHENYL)-2-THIOUREA

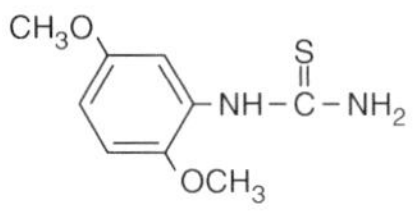

250 mg

S67,487-7
1-ETHYL-3-(2-METHOXY-5-METHYLPHENYL)-2-THIOUREA

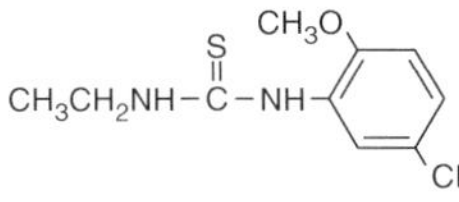

250 mg

S79,691-3
1-(2-METHOXY-5-METHYLPHENYL)-1-METHYL-3-(2-NAPHTHYL)-2-THIOUREA

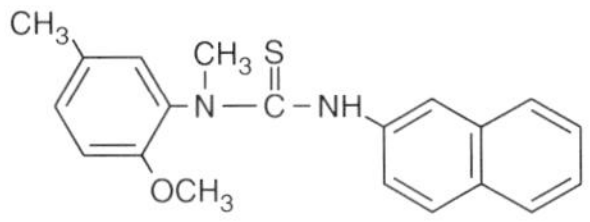

250 mg

S78,104-5
4-(2-METHOXY-5-METHYLPHENYL)-3-THIOSEMICARBAZIDE

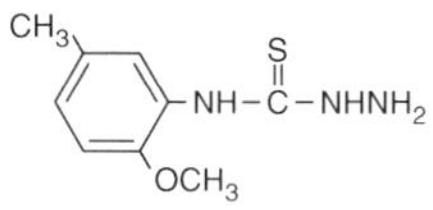

250 mg

S84,556-6
1-(4-ETHOXYPHENYL)-3-(3-NITROBENZOYL)-2-THIOUREA

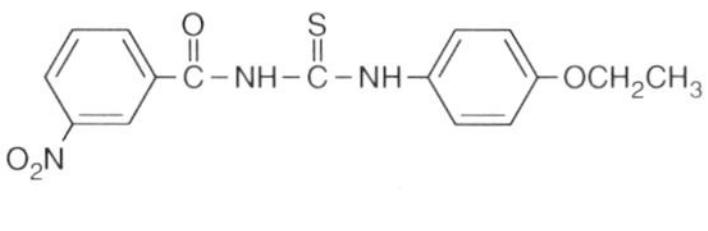

250 mg

S89,974-7
N-[3-(2-METHOXY-5-METHYLPHENYL)-4-(4-NITROPHENYL)-4-THIAZOLIN-2-YLIDENE]CINNAMAMIDE

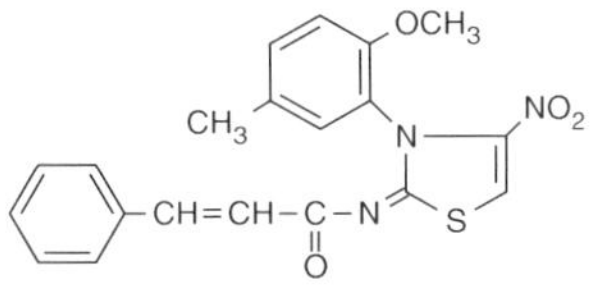

250 mg

S51,329-6
1-METHYL-3-(4-NITROBENZOYL)-1-(4-OCTYLOXYPHENYL)-2-THIOUREA

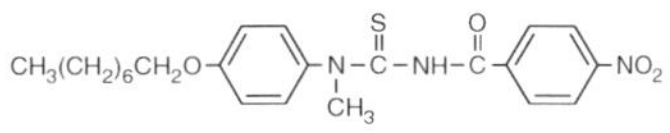

250 mg

S79,683-2
METHYL 2-[3-(2-PHENOXYPROPIONYL)-2-THIOUREIDO]BENZOATE

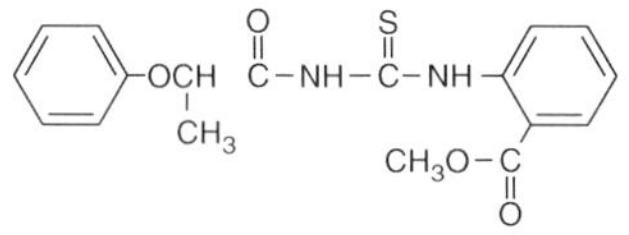

250 mg

S84,598-1
1-CINNAMOYL-3-(4-ETHOXYPHENYL)-2-THIOUREA

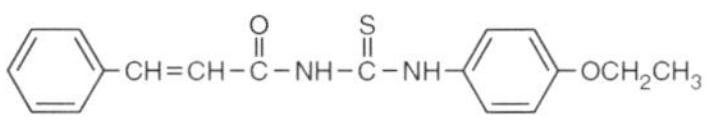

250 mg

S84,751-8
1-(4-ETHOXYPHENYL)-3-(1-NAPHTHYLCARBONYL)-2-THIOUREA

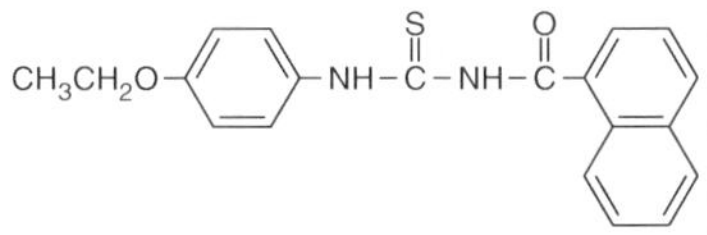

250 mg

S54,853-7
5-(ETHOXYMETHYLENE)-3-(2-METHOXYPHENYL)-RHODANINE

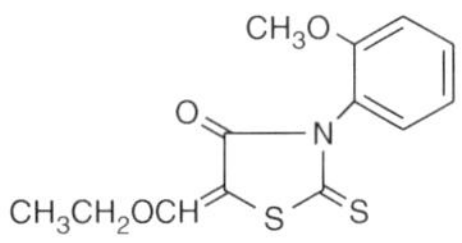

250 mg

S73,475-6
5-(4-METHOXYBENZYLIDENE)-3-METHYL-2-(METHYLIMINO)-4-THIAZOLIDINONE

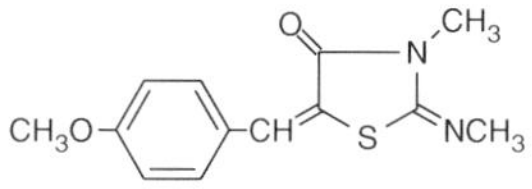

250 mg

S59,806-2
METHYL 3-(2-HEXADECYLOXY-5-METHOXYBENZOYL)DITHIOCARBAZATE

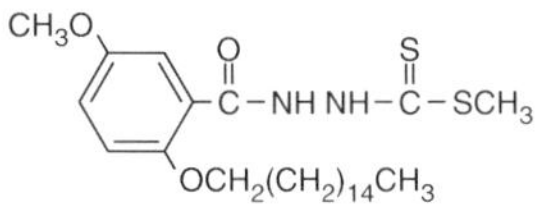

250 mg

S60,534-4
METHYL 3-[4-(HEXADECYLOXY)BENZOYL]-DITHIOCARBAZATE

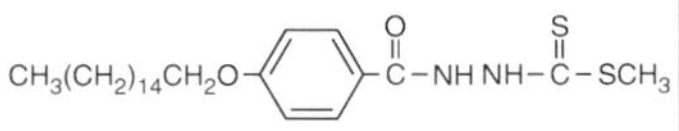

250 mg

S68,612-3
N-(4-ETHOXYPHENYL)-N'-(3-METHYL-2-THIOUREIDO)OXAMIDE

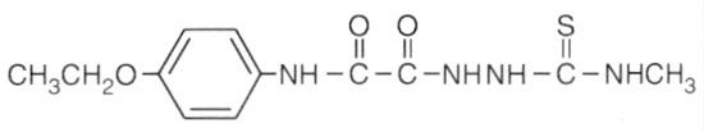

5 gm

S77,396-4
1-(4-ETHOXYPHENYL)-5-THIOBIUREA

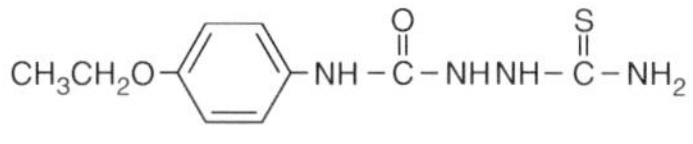

250 mg

S84,770-4
1-DODECANOYL-3-(4-METHOXYPHENYL)-2-THIOUREA

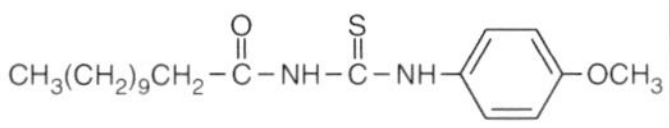

250 mg

S77,981-4
5-(3,4-DIMETHOXYBENZYLIDENE)-2-(2-METHOXYPHENYLIMINO)-3-PHENYL-4-THIAZOLIDINONE

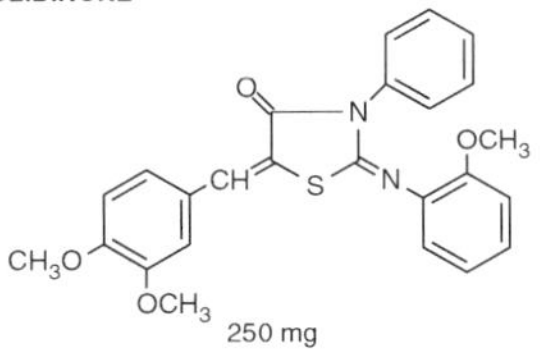

250 mg

S54,849-9
3-(2-METHOXYPHENYL)RHODANINE

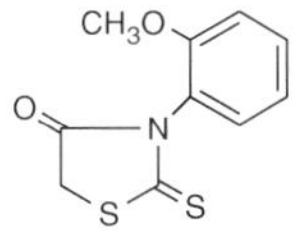

250 mg

S54,865-0
3-(3-METHOXYPHENYL)RHODANINE

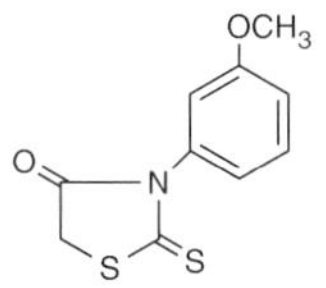

250 mg

S54,600-3
3-(4-ETHOXYPHENYL)RHODANINE

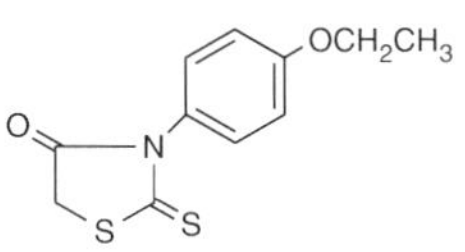

250 mg

S54,855-3
5-(ETHOXYMETHYLENE)-3-(3-METHOXYPHENYL)RHODANINE

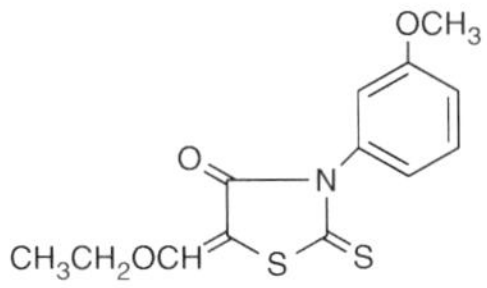

250 mg

S93,475-5
4-METHOXYTHIOBENZOIC ACID HYDRAZIDE

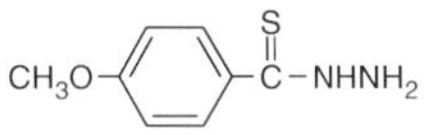

250 mg

S74,131-0
5-IMINO-1-(4-METHOXYPHENYL)-3-PHENYL-4-THIOHYDANTOIN

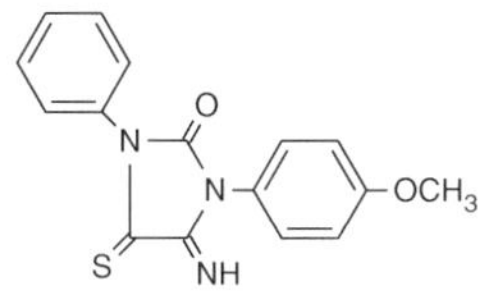

250 mg

S65,581-3
1-(4-METHOXYTHIOBENZOYL)PIPERIDINE

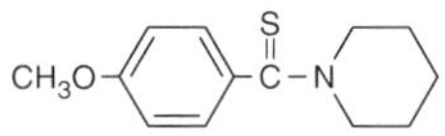

250 mg

S38,679-0
4-(4-METHOXYTHIOBENZOYL)MORPHOLINE

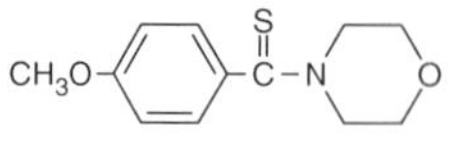

250 mg

S50,573-0
4-(2-CHLORO-5-METHOXYTHIOBENZOYL)-MORPHOLINE

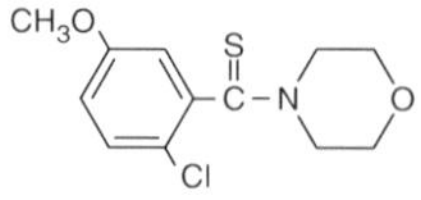

250 mg

S51,377-6
4-(4-METHOXYTHIOCINNAMOYL)-MORPHOLINE

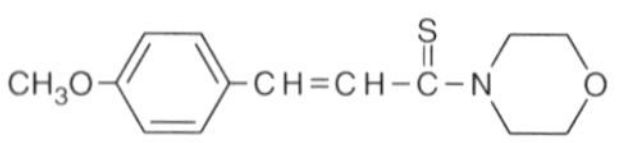

250 mg

S51,008-4
4-[3-CHLORO-4-(HEXADECYLOXY)-THIOBENZOYL]MORPHOLINE

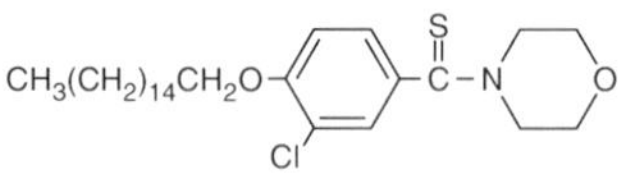

250 mg

S65,691-7
4-[2-(6-METHOXY-2-NAPHTHYL)-THIOACETYL]MORPHOLINE

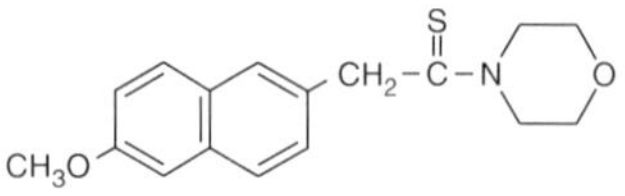

5 gm

S65,334-9
4-METHOXY-N-METHYLTHIOBENZANILIDE

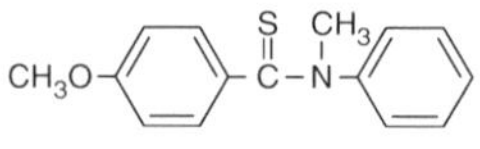

250 mg

S53,083-2
2,6-DICHLORO-N-(METHOXYMETHYL)-THIOBENZAMIDE

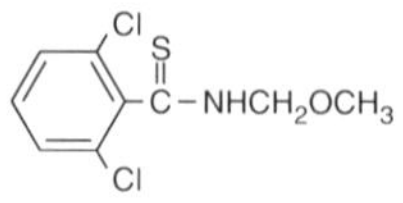

250 mg

S87,513-9
1,3-BIS(3-METHOXYPHENYL)-2-THIOUREA

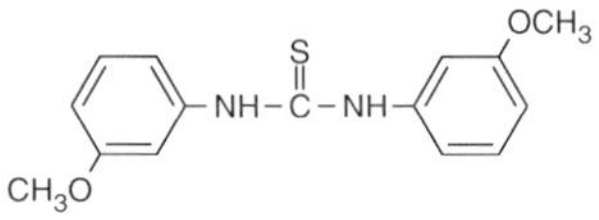

250 mg

S79,583-6
1-CYCLOHEXYL-3-(4-ETHOXYPHENYL)-2-THIOUREA

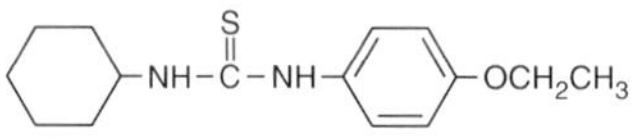

250 mg

S55,341-7
4,4'-TRIMETHYLENEBIS(1-[N-(2-METHOXYPHENYL)THIOCARBAMOYL]-PIPERIDINE)

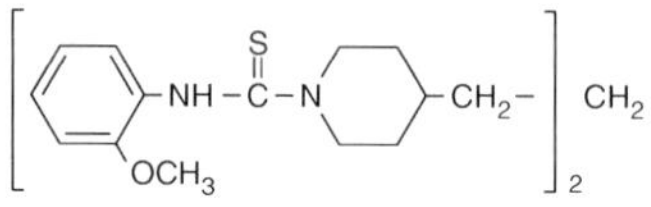

250 mg

S69,244-1
4-[N-(4-METHOXYPHENYL)-THIOCARBAMOYL]MORPHOLINE

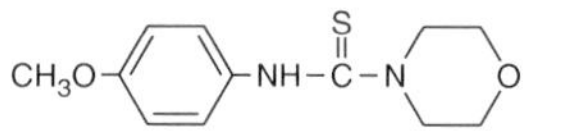

250 mg

S78,033-2
1-(3-BUTOXYPROPYL)-3-PHENYL-2-THIOUREA

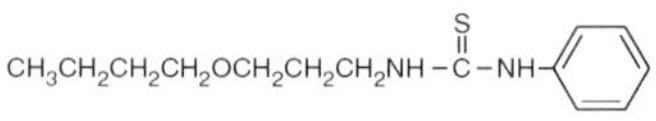

250 mg

S54,608-9
1,3-BIS(2-METHOXYPHENYL)-2-THIOUREA

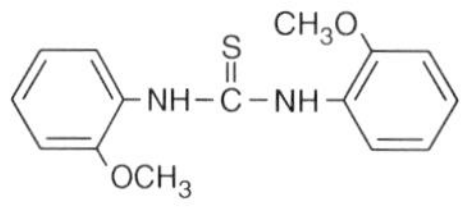

5 gm

S55,378-6
1-(4-ETHOXYPHENYL)-3-ETHYL-2-THIOUREA

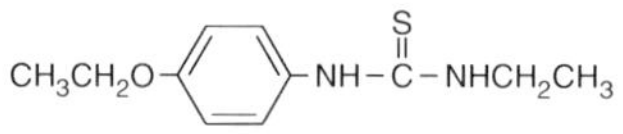

5 gm

S55,646-7
4,4'-DIMETHOXYTHIOCARBANILIDE

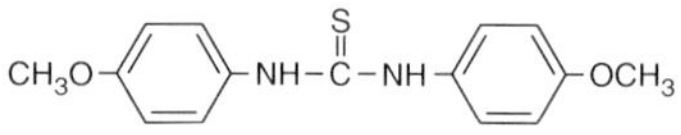

250 mg

S53,685-7
1-CYCLODODECYL-3-(4-ETHOXYPHENYL)-2-THIOUREA

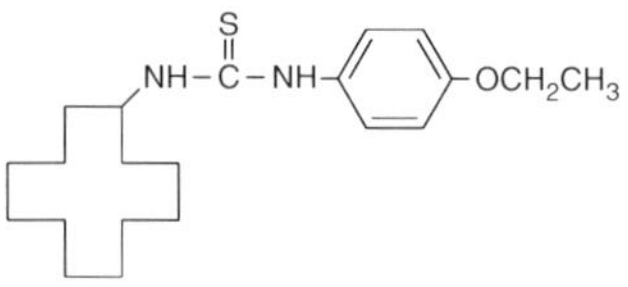

5 gm

S95,336-9
3-(2-METHOXYPHENYL)-2-METHYL-2-THIOPSEUDOUREA HYDRIODIDE

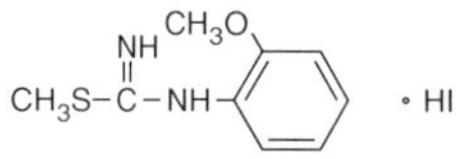

50 mg

S53,788-8
2-(4-METHOXYBENZYL)-2-THIOPSEUDOUREA HYDROCHLORIDE

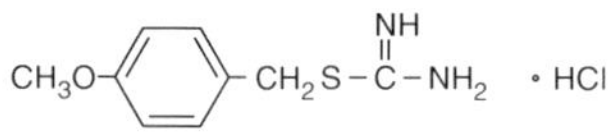

5 gm

S52,984-2
4-ETHOXYDITHIOCARBANILIC ACID, AMMONIUM SALT

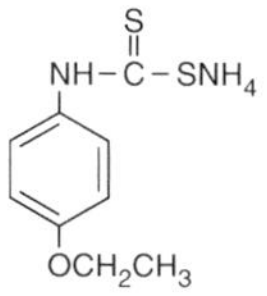

250 mg

S54,274-1
4-(4-METHOXYPHENYL)-3-THIOSEMICARBAZIDE

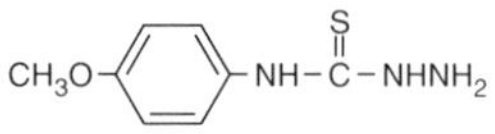

250 mg

S54,267-9
1-ANILINO-6-(4-METHOXYPHENYL)-2,5-DITHIOBIUREA

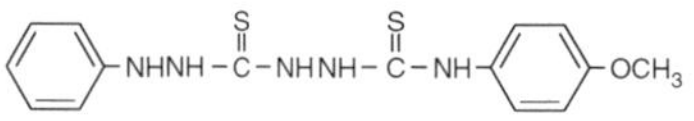

250 mg

S89,868-6
5-IMINO-3-METHYL-2-(3-NITROPHENYL)-4-OXAZOLIDINETHIONE

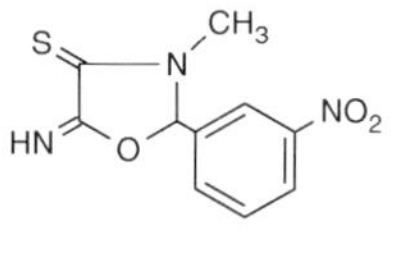

250 mg

S74,300-3
3-(3-ALLYLUREIDO)-ALPHA-METHYLBENZYL N-CYCLOHEXYLCARBAMATE

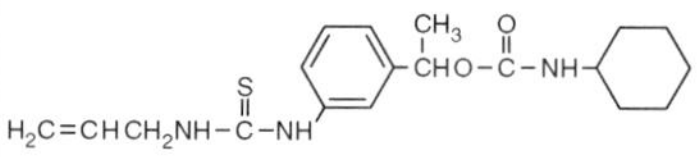

5 gm

S54,742-5
5-ETHOXYMETHYLENE-2-THIOXO-3-(M-TOLYL)-4-THIAZOLIDINONE

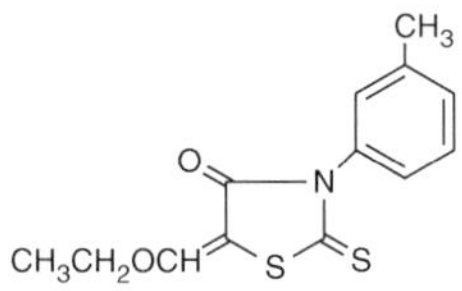

250 mg

S54,744-1
5-ETHOXYMETHYLENE-2-THIOXO-3-(O-TOLYL)-4-THIAZOLIDINONE

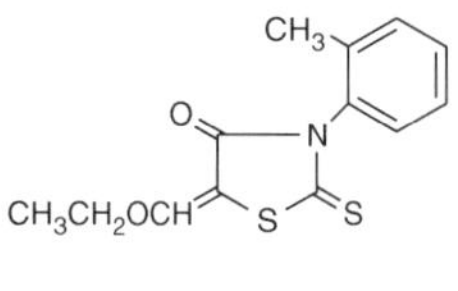

250 mg

S54,749-2
3-BENZYL-5-(ETHOXYMETHYLENE)-2-THIOXO-4-THIAZOLIDINONE

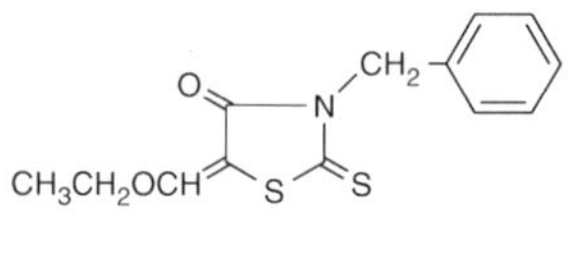

250 mg

S45,168-1
3',5'-DIBENZOYL-4-THIOTHYMIDINE

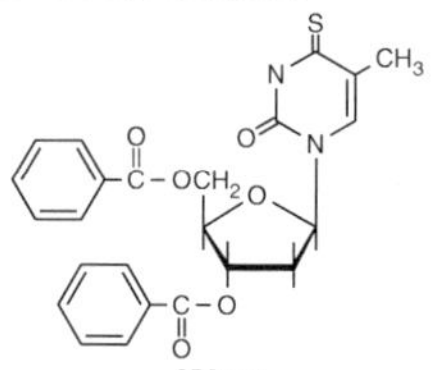

250 mg

S84,259-1
O-[2'-(DIMETHYLAMINO)CARBONYLOXY]-2-BIPHENYLYL-N,N-DIMETHYLTHIOCARBAMATE

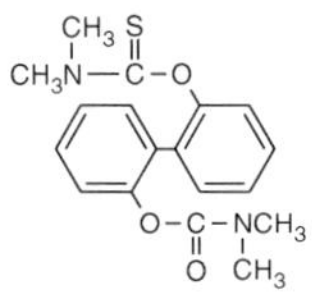

250 mg

S95,021-1
(-)-4-CYANO-5,5-DIMETHYL-6-OXOHEPTANOIC ACID BENZYL ISOTHIURONIUM SALT

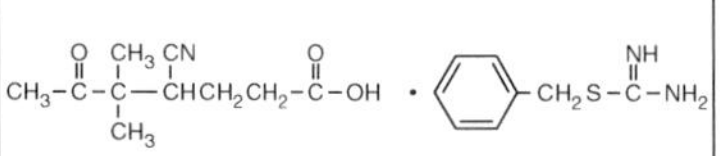

50 mg

S51,252-4
2,3-DIMETHYL-1,3-DIPHENYL-2-THIOPSEUDOUREA

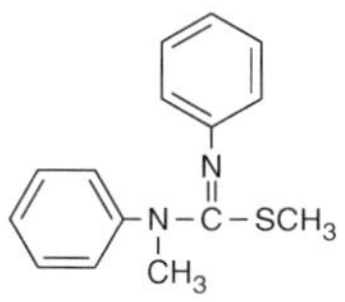

250 mg

S53,120-0
2-(2,4-DICHLOROBENZYL)-2-THIOPSEUDOUREA HYDROCHLORIDE

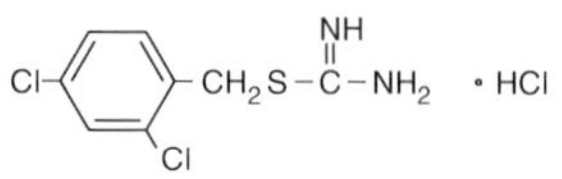

5 gm

S50,126-3
METHYL 3-(P-TOLUOYL)-DITHIOCARBAZATE

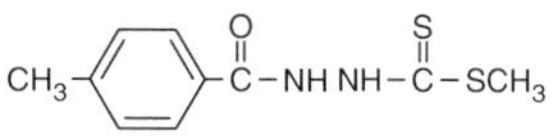

250 mg

S68,954-8
1-(N-[4-(METHYLTHIO)PHENYL]-CARBAMOYL)THIOSEMICARBAZIDE

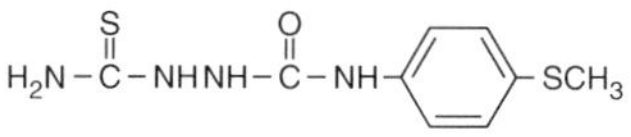

5 gm

S50,124-7
METHYL 3-BENZOYLDITHIOCARBAZATE

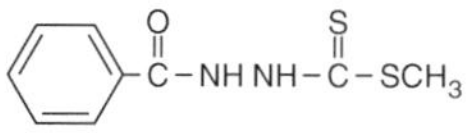

250 mg

S50,030-5
N-(1-NAPHTHYL)-2-(THIOPHENOXY)-THIOACETAMIDE

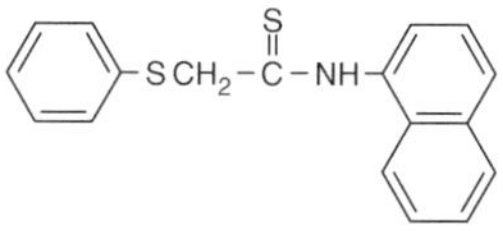

250 mg

S51,538-8
1-[2-(HEXADECYLTHIO)PHENYL]-2-THIOUREA

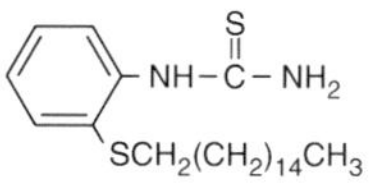

250 mg

S63,005-5
1,2-DIMETHYL-3-PHENYL-2-THIOPSEUDOUREA

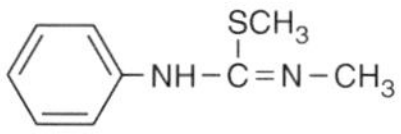

250 mg

S38,927-7
S-METHYL 5-OXO-3-PHENYL-2-PYRAZOLINE-1-THIOIMIDATE HYDRIODIDE

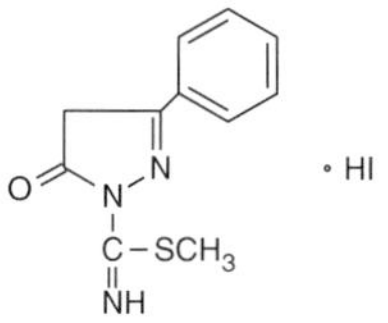

250 mg

S91,061-9
1-[4-(METHYLTHIO)PHENYL]-3-(2-THIAZOLIN-2-YL)UREA

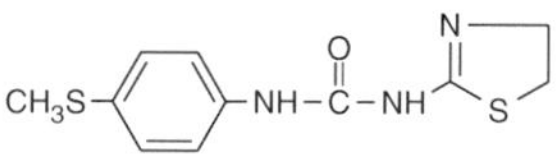

250 mg

S73,842-5
1-METHYL-3-(4-METHYLTHIOPHENYL)-2-THIOUREA

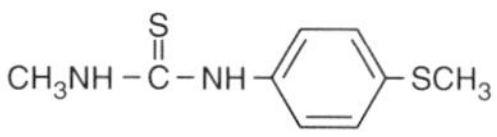

5 gm

S51,558-2
2-METHYLTHIO-4-PHENYL-5-PHENYLIMINO-1,3,4-THIADIAZOL-2-INE

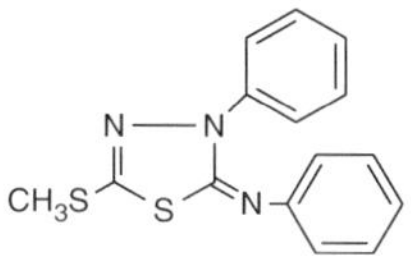

250 mg

S53,085-9
2-METHYL-3-PHENYL-2-THIOPSEUDOUREA HYDRIODIDE

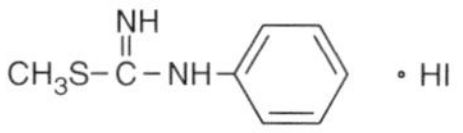

250 mg

S44,761-7
2,2'-(4,5-DIMETHYL-1,2-XYLYLENE)-BIS(2-THIOPSEUDOUREA) DIHYDROCHLORIDE

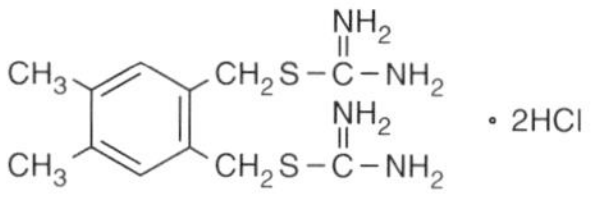

5 gm

S54,928-2
3-HEXADECYL-2-BENZOTHIAZOLINONE, AZINE WITH 2-FORMYLBENZENESULFONIC ACID, SODIUM SALT

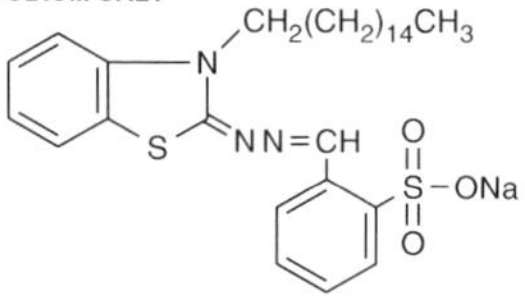

5 gm

S63,904-4
6-AMINO-3-METHYL-2-BENZOTHIAZOLINONE AZINE WITH 3-METHYL-2-BENZOTHIAZOLINONE

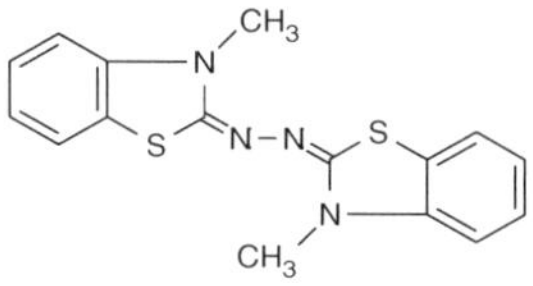

250 mg

S92,892-5
4-[(4-AMINO)THIOBENZOYL]MORPHOLINE

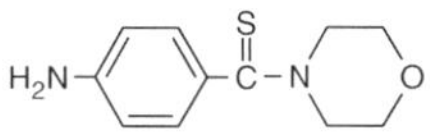

250 mg

S95,028-9
METHYL 3-AMINO-2-(N-PHENYLTHIOCARBAMOYL)-2-BUTENOATE

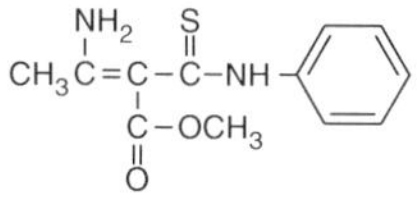

50 mg

S95,033-5
ETHYL BETA-AMINO-ALPHA-(N-PHENYLTHIOCARBAMOYL)CINNAMATE

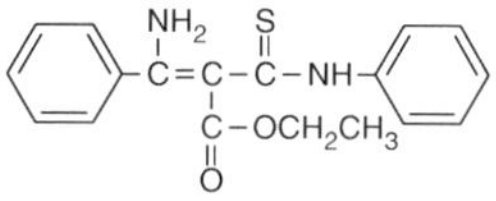

50 mg

S92,257-9
1-BENZYL-5,6-DIAMINO-2-THIOURACIL

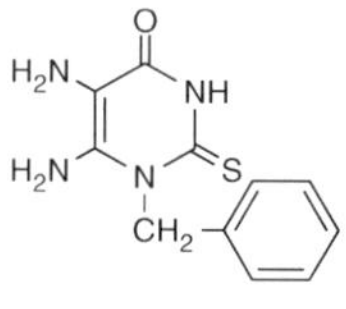

250 mg

S67,434-6
4-AMINO-5-CYANO-2,3-DIHYDRO-3-PHENYL-2-THIOXOTHIAZOLE

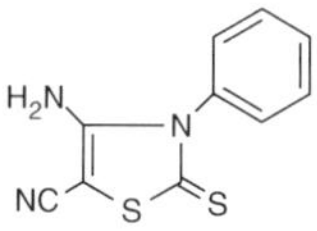

5 gm

S74,717-3
1-(2,4-DINITROPHENYL)-4-PHENYL-3-THIOSEMICARBAZIDE

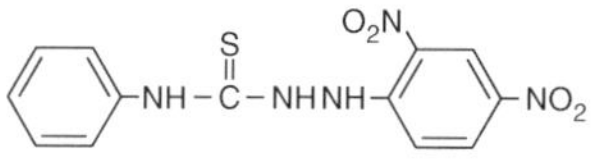

5 gm

S91,345-6
1-(1-ETHYL-3-PIPERIDINYL)-3-PHENYL-2-THIOUREA

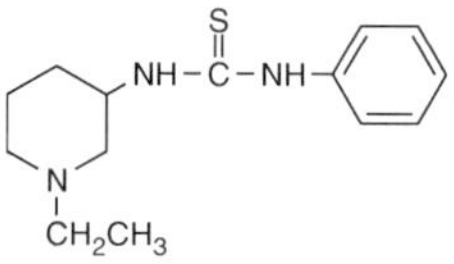

250 mg

S90,071-0
1-[2-(DIETHYLAMINO)ETHYL]-1,3-DIPHENYL-2-THIOUREA

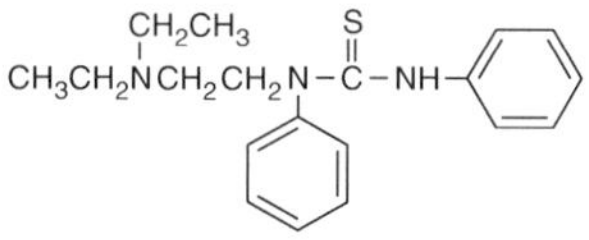

250 mg

S36,013-9
N,N'-DIBENZYLDITHIOOXAMIDE

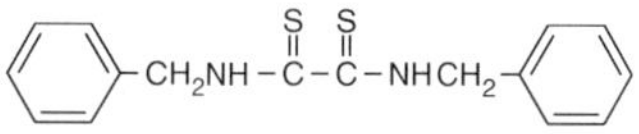

5 gm

S91,305-7
1-METHYL-6-(4-METHYL-2-NITROPHENYL)--2-THIOBIUREA

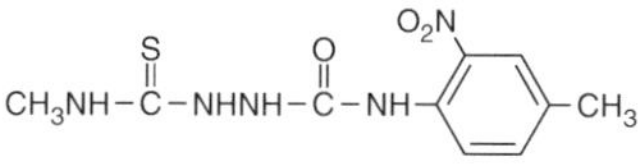

250 mg

S61,635-4
1-(2-NITROPHENYL)-3-PHENYL-2-THIOUREA

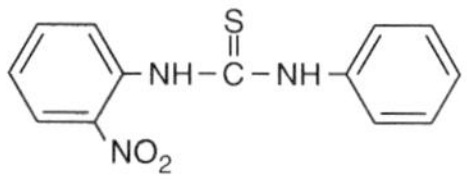

5 gm

S41,018-7
N-(THIOCARBAMOYLMETHYL)BENZAMIDE

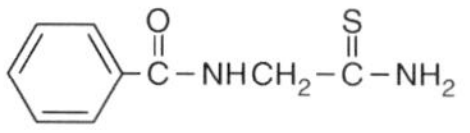

250 mg

S94,964-7
N-(2-PHTHALIMIDOACETYL)SUCCINIMIDE

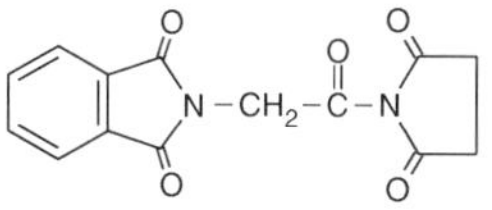

1 gm

S94,965-5
N,N-BIS(2-PHTHALIMIDOETHYL) ACETAMIDE

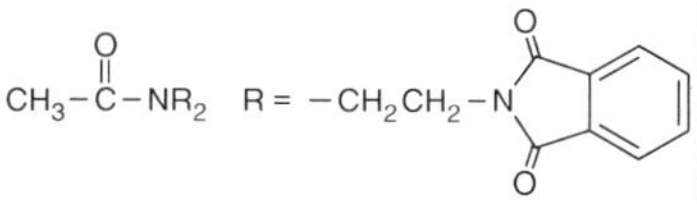

100 mg

S92,405-9
1-BENZYL-3-(2-MORPHOLINOETHYL)-2-THIOUREA

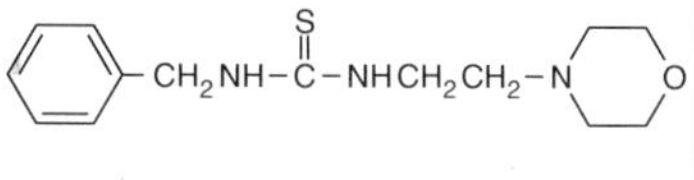

250 mg

S55,476-6
1,1'-(4,5-DIMETHYL-1,2-PHENYLENE)-BIS(3-PHENYL-2-THIOUREA)

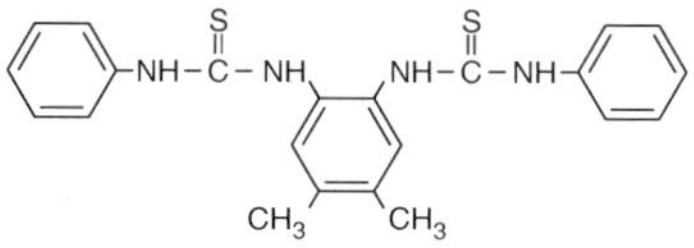

5 gm

S50,670-2
3-(3-NITROANILINOMETHYL)-2(3H)-BENZOXAZOLETHIONE

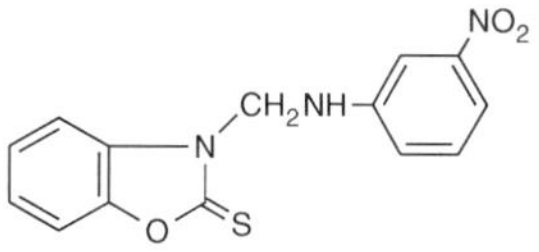

5 gm

S36,409-6
3-(3-NITROANILINOMETHYL)-2(3H)-BENZOTHIAZOLETHIONE

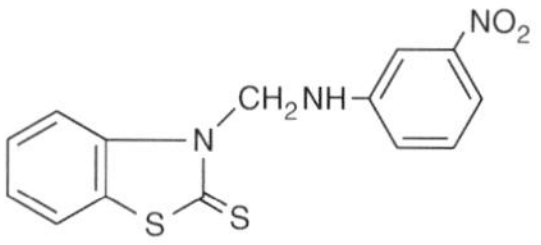

5 gm

S50,679-6
1-(3-NITROANILINOMETHYL)-2(3H)-BENZIMIDAZOLETHIONE
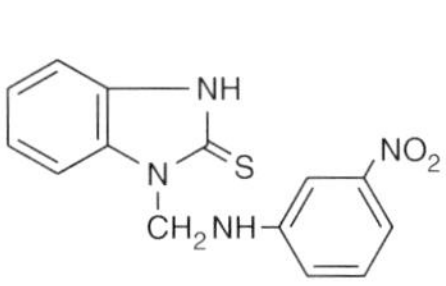
5 gm

S87,069-2
N-BENZOYL-2-MORPHOLINO-1-CYCLOPENTENE-1-THIOCARBOXAMIDE
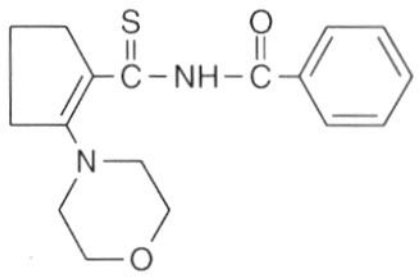
250 mg

S89,991-7
1-[3-(DIBUTYLAMINO)PROPYL]-3-PHENYL-2-THIOUREA
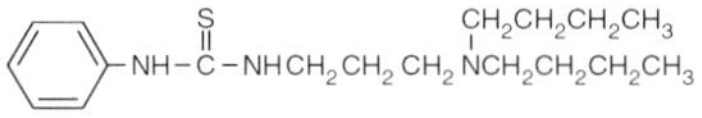
250 mg

S77,341-7
1-(2-METHYL-5-NITROPHENYL)-3-(4-OXO-2-THIAZOLIDINYLIDENE)UREA
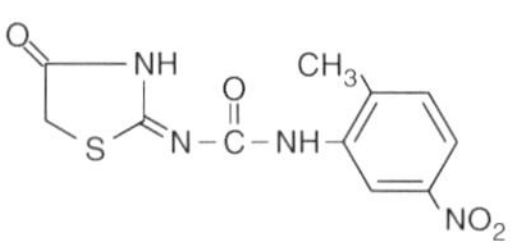
250 mg

S84,769-0
1-DODECANOYL-3-(2-METHYL-5-NITROPHENYL)-2-THIOUREA
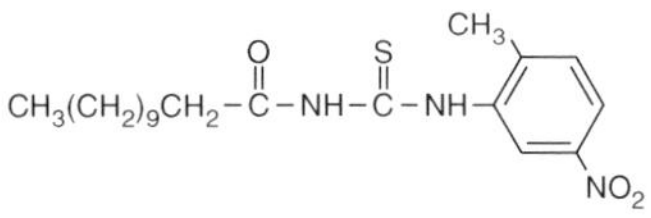
250 mg

S74,118-3
5-IMINO-1-(3-NITROPHENYL)-3-PHENYL-4-THIOHYDANTOIN
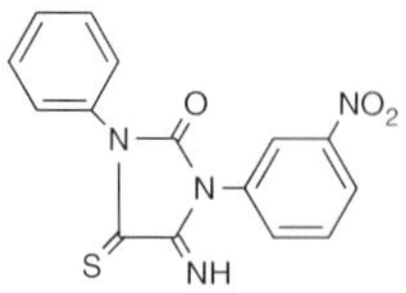
250 mg

S74,132-9
5-IMINO-1-(4-METHYL-3-NITROPHENYL)-3-PHENYL-4-THIOHYDANTOIN
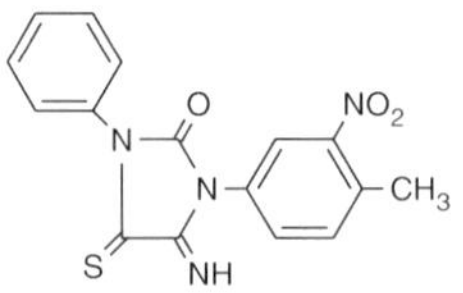
250 mg

S87,527-9
1,3-BIS(3-NITROPHENYL)-2-THIOUREA
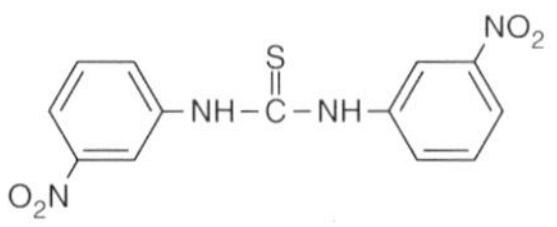
250 mg

S69,022-8
1-(ISOPROPYL)-3-(4-METHYL-3-NITROPHENYL)-2-THIOUREA
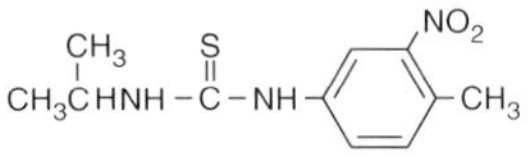
250 mg

S61,524-2
1-(3-NITROPHENYL)-3-PHENYL-2-THIOUREA
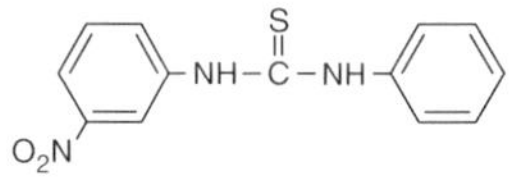
5 gm

S76,852-9
1-ANTIPYRINYL-3-(3-MORPHOLINOPROPYL)-2-THIOUREA
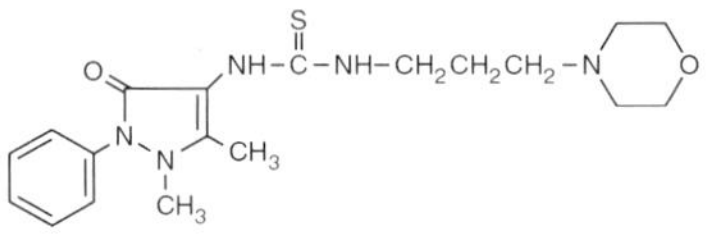
250 mg

S50,731-8
3-METHYL-6-PHENYL-4(3H)-PYRIMIDINETHIONE
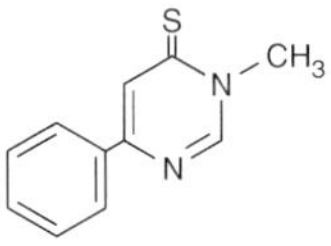
250 mg

S50,673-7
3-(4-NITROANILINOMETHYL)-2(3H)-BENZOXAZOLETHIONE
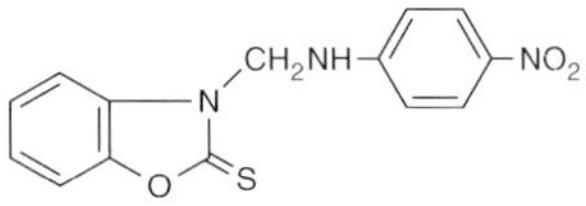
5 gm

S69,117-8
1-[4-(DIETHYLAMINO)PHENYL]-3-(2-METHOXY-5-METHYLPHENYL)-2-THIOUREA
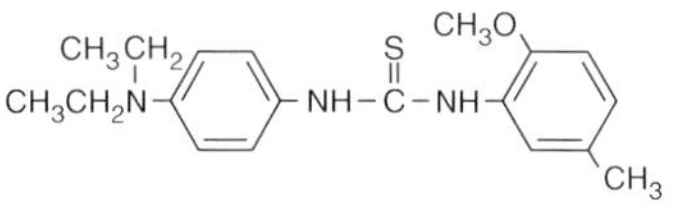
250 mg

S87,500-7
1,3-BIS[4-(DIMETHYLAMINO)PHENYL]-2-THIOUREA
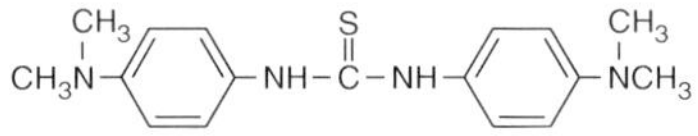
250 mg

S74,708-4
1-[4-(DIMETHYLAMINO)PHENYL]-3-(4-NITROPHENYL)-2-THIOUREA
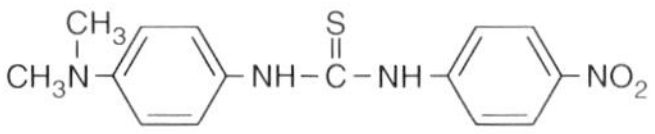
5 gm

S84,761-5
1-[4-(DIMETHYLAMINO)PHENYL]-3-DODECANOYL-2-THIOUREA
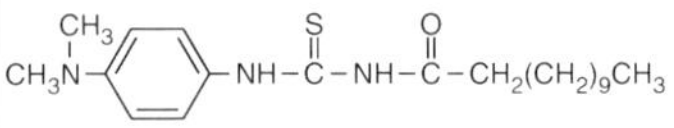
250 mg

S76,951-7
1-[4-(DIMETHYLAMINO)PHENYL]-3-(PHENYL)-2-THIOUREA
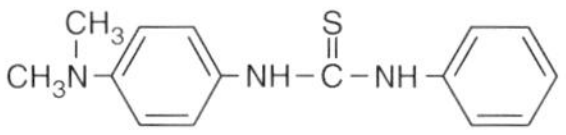
250 mg

S52,373-9
1-[4-(DIETHYLAMINO)-ORTHO-TOLYL]-2-THIO-3-(PARA-TOLYL)-UREA
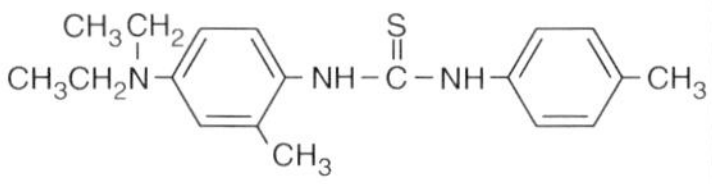
250 mg

S79,627-1
4-BENZYL-1-(4-NITROPHENYL)-3-THIOSEMICARBAZIDE
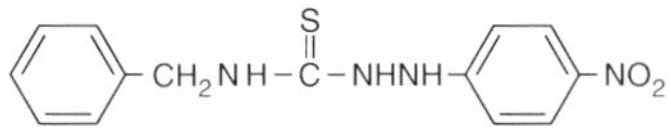
250 mg

S84,759-3
1-DODECANOYL-3-(2-METHYL-4-NITROPHENYL)-2-THIOUREA
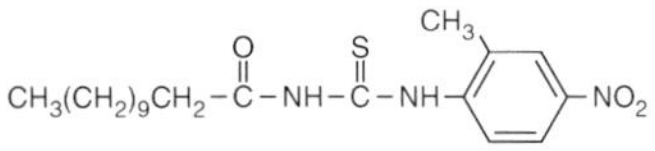
250 mg

S69,048-1
1-ETHYL-3-(4-NITROPHENYL)-2-THIOUREA
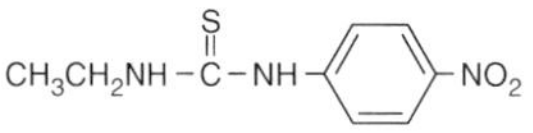
250 mg

S90,309-4
1,1-DIALLYL-3-(4-NITROPHENYL)-2-THIOUREA
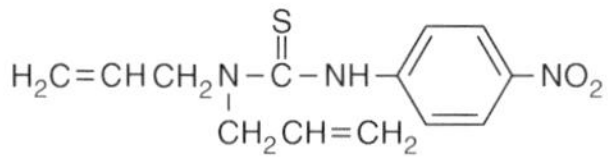
250 mg

S91,709-5
4-METHYL-1-(4-NITROPHENYL)-3-THIOSEMICARBAZIDE
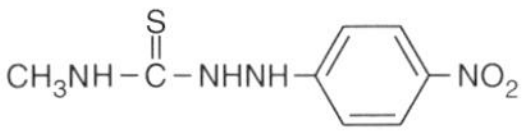
250 mg

S68,802-9
4-ISOPROPYL-1-(4-NITROPHENYL)-3-THIOSEMICARBAZIDE
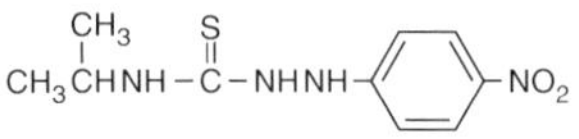
250 mg

S50,576-5
1-BENZOYL-3-(4-NITROPHENYL)-2-THIOUREA
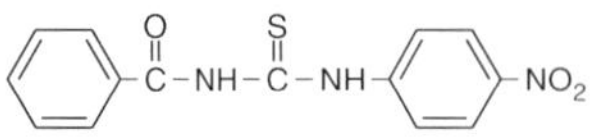
5 gm

S39,021-6
1-METHYL-3-(4-NITROPHENYL)-2-THIOUREA
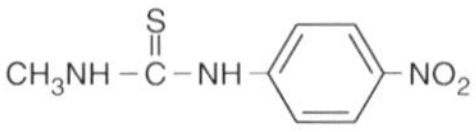
250 mg

S61,634-6
1-(4-NITROPHENYL)-3-PHENYL-2-THIOUREA
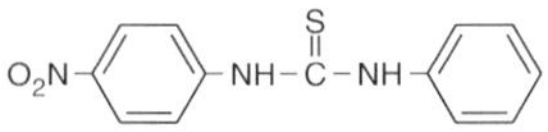
5 gm

S79,621-2
1-(4-ACETAMIDOPHENYL)-3-(2-BUTENOYL)-2-THIOUREA
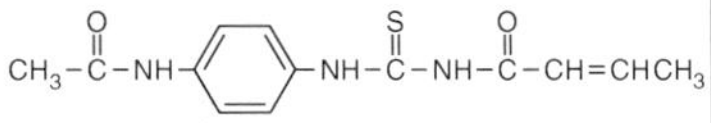
250 mg

S84,763-1
1-DODECANOYL-3-[4-(N-METHYLACETAMIDO)PHENYL]-2-THIOUREA
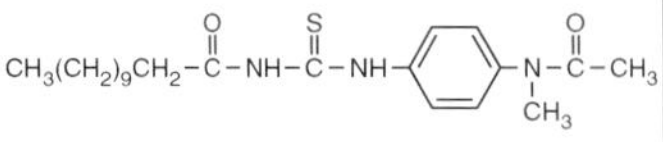
250 mg

S69,506-8
1-(4-ACETAMIDOPHENYL)-3-METHYL-2-THIOUREA
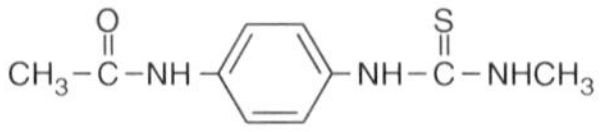
250 mg

S67,511-3
1-(4-ACETAMIDOPHENYL)-3-ETHYL-2-THIOUREA
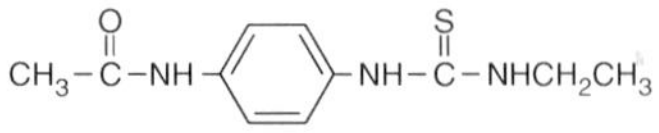
250 mg

S69,549-1
1-(4-ACETAMIDOPHENYL)-3-PROPYL-2-THIOUREA
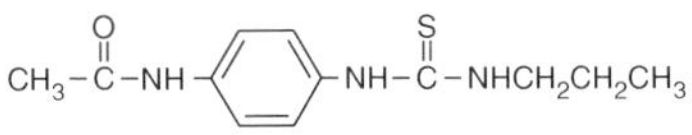
250 mg

S50,198-0
1,1'-(1,4-PHENYLENE)BIS(2-THIOUREA)
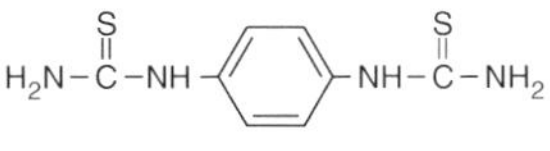
250 mg

S53,798-5
1-ALLYL-3-(1,2-DIPHENYLETHYL)-2-THIOUREA
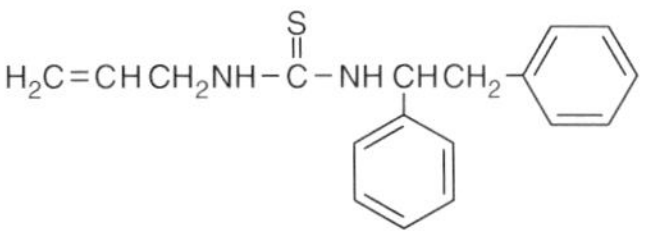
5 gm

S53,686-5
1-(1,2-DIPHENYLETHYL)-3-PHENYL-2-THIOUREA
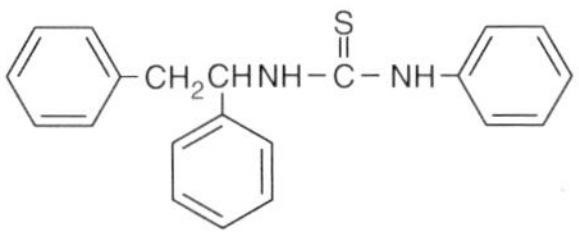
5 gm

S68,842-8
1-DIPHENYLMETHYL-3-(2-PHENETHYL)-2-THIOUREA
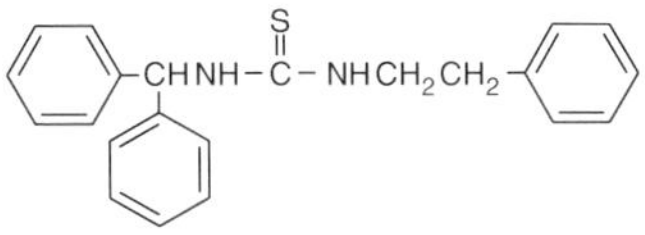
250 mg

S64,923-6
3,5-BIS(2-PHENYLETHYL)-3,4,5,6-TETRAHYDRO-2H-1,3,5-THIADIAZINE-2-THIONE
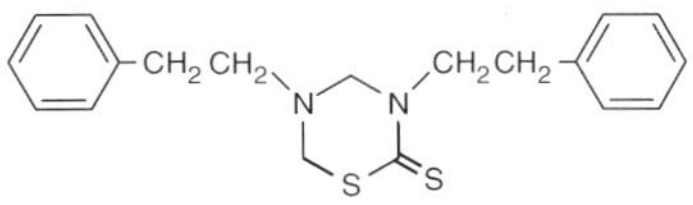
250 mg

S76,002-1
ETHYL 2-(2-IMIDAZOLIDINYLIDENE)-3(-5-OXO-2-THIOXO-4-IMIDAZOLIDINYL)-3-PHENYLPROPIONATE
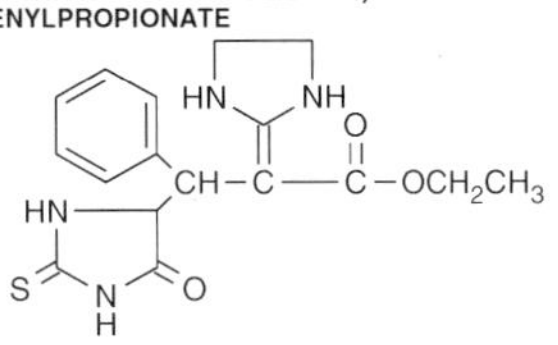
250 mg

S80,104-6
BUTYL 4-(3-PHENETHYL-2-THIOUREIDO) BENZOATE
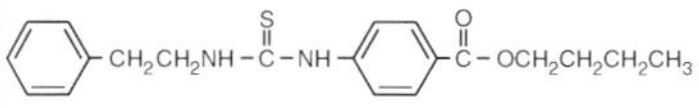
250 mg

S63,304-6
1-(PHENYLTHIOACETYL)PIPERIDINE
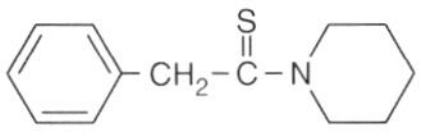
5 gm

S61,526-9
1-PHENETHYL-3-PHENYL-2-THIOUREA
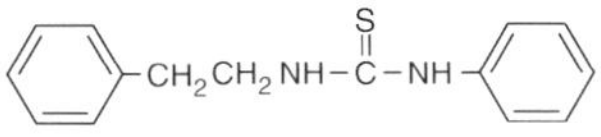
5 gm

S53,677-6
1-(1,2-DIPHENYLETHYL)-3-(ORTHO-TOLYL)-2-THIOUREA
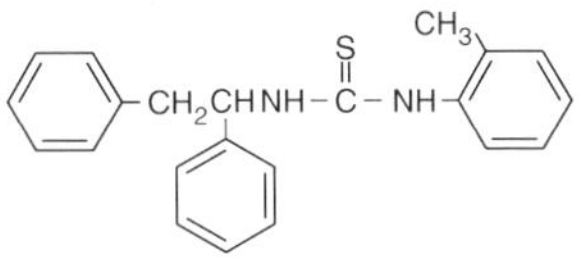
250 mg

S54,605-4
1-(1,2-DIPHENYLETHYL)-3-(2,4-XYLYL)-2-THIOUREA
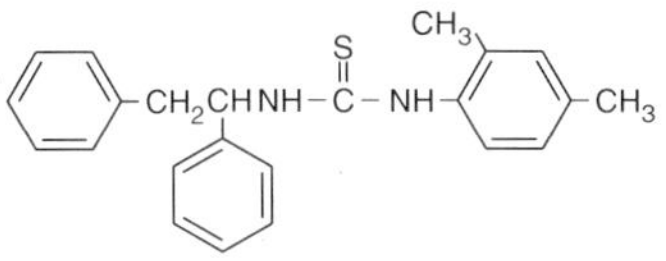
250 mg

S54,271-7
3-BENZYL-1,1-DIMETHYL-2-THIOUREA
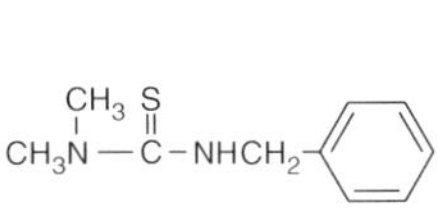
250 mg

S53,632-6
1,1-DIBENZYL-3-PHENYL-2-THIOUREA
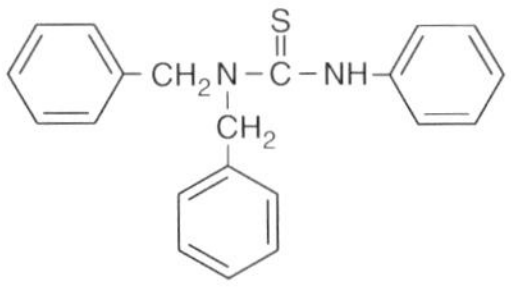
5 gm

S54,585-6
4,4'-TRIMETHYLENEBIS(1-[N-(2,4-XYLYL)THIOCARBAMOYL]PIPERIDINE)
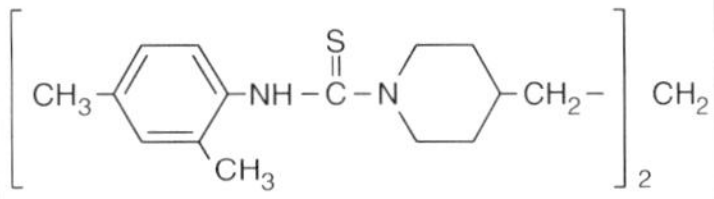
250 mg

S86,301-7
1,1,3-TRIPHENYL-2-THIOUREA
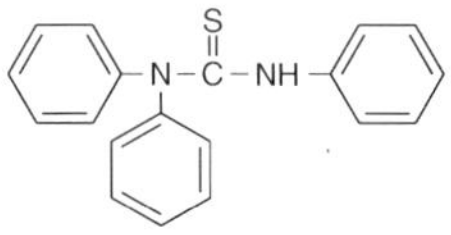
250 mg

S53,639-3
1,1-DIMETHYL-3-(PARA-TOLYL)-2-THIOUREA
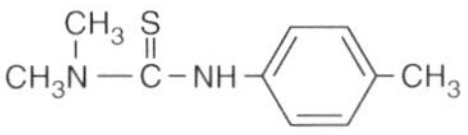
250 mg

S37,366-4
3-(4-NITROANILINOMETHYL)-2(3H)-BENZOTHIAZOLETHIONE
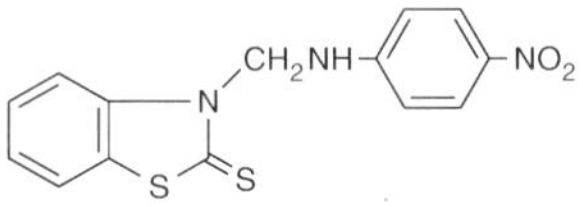
5 gm

S37,373-7
N-(2-THIOXOBENZOTHIAZOLIN-3-YLMETHYL)ANTHRANILIC ACID
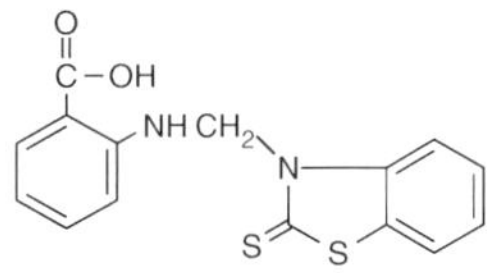
5 gm

S36,504-1
4-(2-THIOXOBENZOTHIAZOLIN-3-YLMETHYLAMINO)BENZOIC ACID
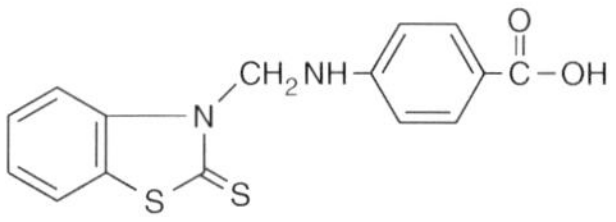
5 gm

S54,610-0
4,6-DIPHENYL-3,4,5,6-TETRAHYDRO-2H-1,3,5-THIADIAZINE-2-THIONE
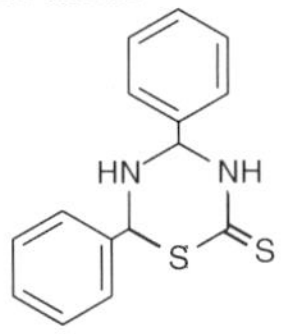
250 mg

S35,937-8
3-(2-CHLOROANILINOMETHYL)-2-BENZOTHIAZOLINETHIONE
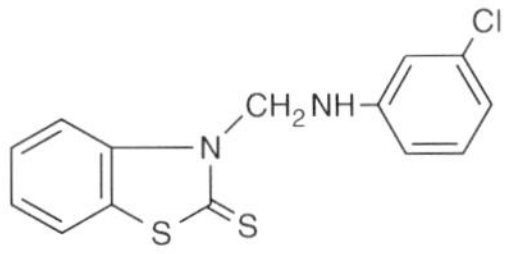
5 gm

S54,959-2
1-BENZOYL-3-(2-CHLOROPHENYL)-2-THIOUREA
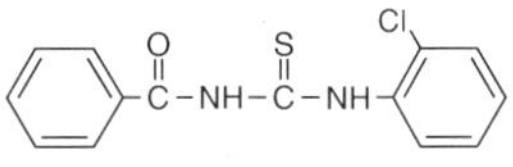
5 gm

S54,160-5
3-CYCLOHEXYL-1,2-DIHYDRO-2-THIOXO-4(3H)-QUINAZOLINONE
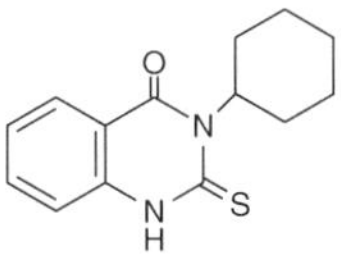
250 mg

S86,268-1
1-PHENYL-2-THIOBIURET
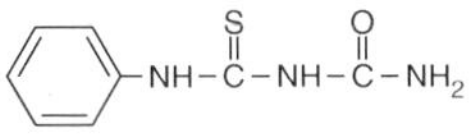
250 mg

S86,266-5
1-PHENYL-2,4-DITHIOBIURET
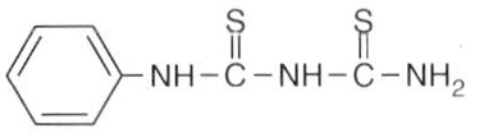
250 mg

S86,271-1
1-ACETIMIDO-3-PHENYL-2-THIOUREA
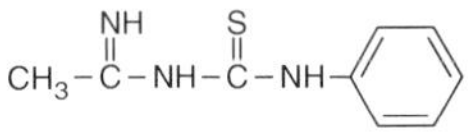
250 mg

S86,275-4
1-AMIDINO-3-(P-TOLYL)-2-THIOUREA
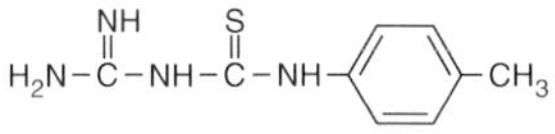
250 mg

S53,703-9
N-CYCLOHEXYLDITHIOCARBAMIC ACID, AMMONIUM SALT
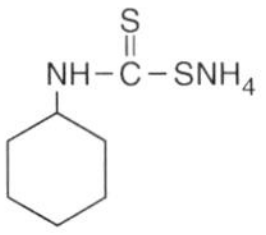
5 gm

S53,704-7
N-(2,4-XYLYL)-DITHIOCARBAMIC ACID, AMMONIUM SALT
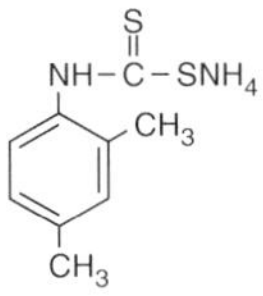
5 gm

S55,465-0
4-ETHYL-1-PHENYL-3-THIOSEMICARBAZIDE
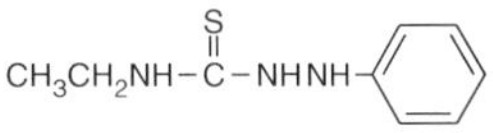
250 mg

S88,857-5
O-(M-TOLYL) N,N-DIMETHYLTHIOCARBAMATE
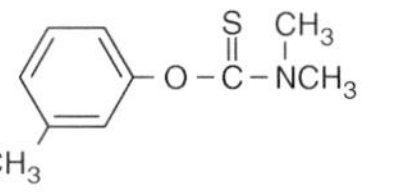
250 mg

S88,920-2
O-(M-TOLYL) N,N-DIMETHYLTHIO-CARBAMATE METHYL TRIFLUOROMETHANE-SULFONATE

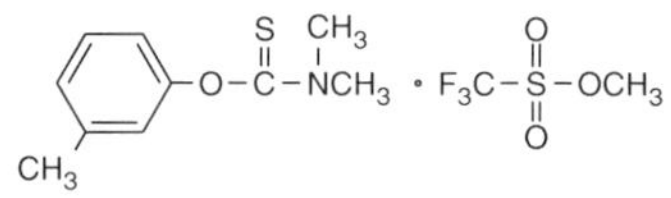

250 mg

S72,261-8
3-BUTYL-5-[4-(DIMETHYLAMINO)-BENZYLIDENE]RHODANINE

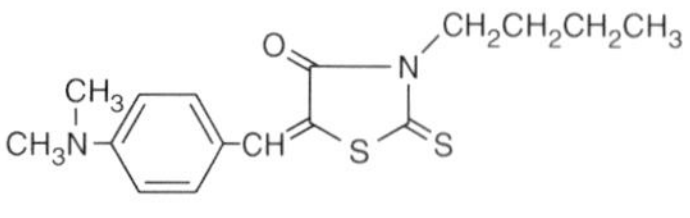

50 mg

S35,375-2
3-[N-(2-NAPHTHYL)-N-PHENYLAMINOMETHYL]BENZOTHIAZOLE-2(3H)-THIONE

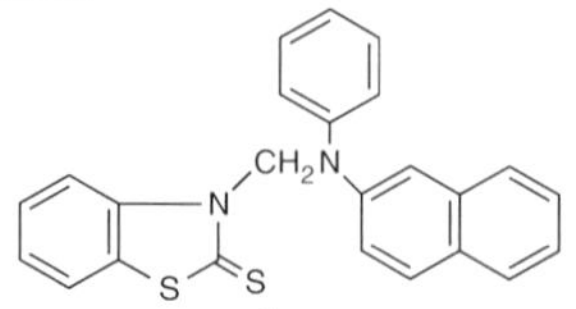

5 gm

S69,009-0
1-[4-(DIMETHYLAMINO)BENZOYL]-4-ETHYL-3-THIOSEMICARBAZIDE

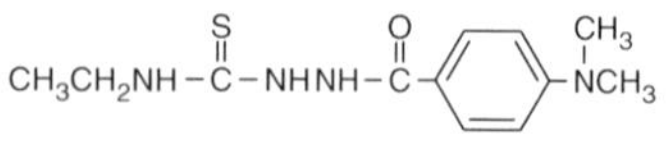

250 mg

S87,489-2
3-METHYL-2-BENZOTHIAZOLINONE [ALPHA-(4-NITROPHENYLAZO)-BENZYLIDENE]HYDRAZONE

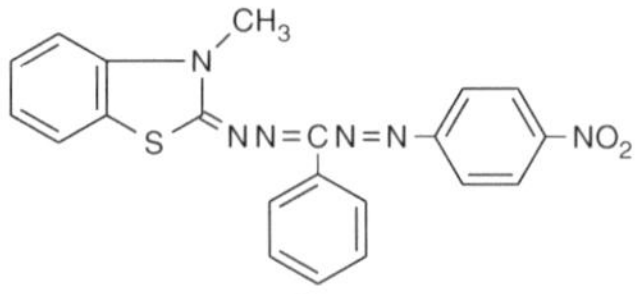

250 mg

S50,675-3
4-(2,3-DIHYDRO-2-THIOXO-1-BENZIMIDAZOLYLMETHYLAMINO)-BENZOIC ACID

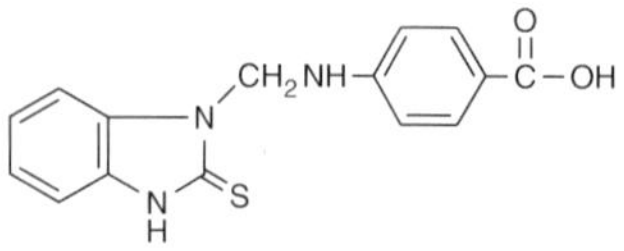

5 gm

S50,684-2
1-ANILINOMETHYL-2(3H)-BENZIMIDAZOLETHIONE

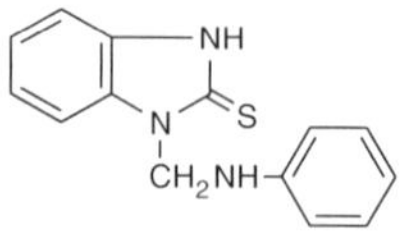

5 gm

S50,676-1
1,3-BIS(ANILINOMETHYL)-2(3H)-BENZIMIDAZOLETHIONE

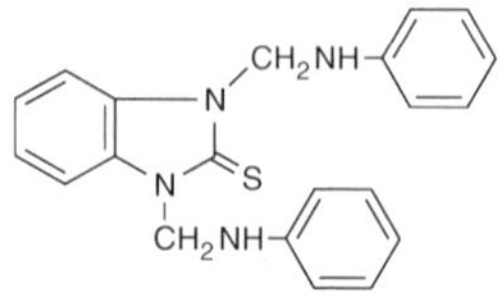

5 gm

S54,578-3
3-METHYL-5-PHENYL-3,4,5,6-TETRAHYDRO-2H-1,3,5-THIADIAZINE-2-THIONE

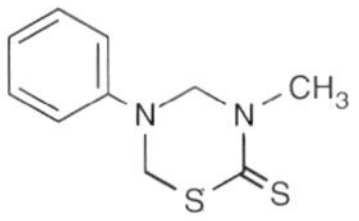

250 mg

S54,575-9
3,5-DIPHENYL-3,4,5,6-TETRAHYDRO-2H-1,3,5-THIADIAZINE-2-THIONE

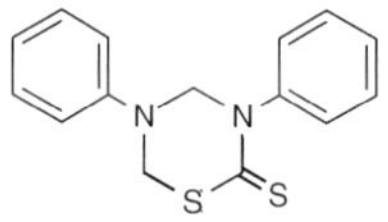

250 mg

S54,576-7
3,5-DI-P-TOLYL-3,4,5,6-TETRAHYDRO-2H-1,3,5-THIADIAZINE-2-THIONE

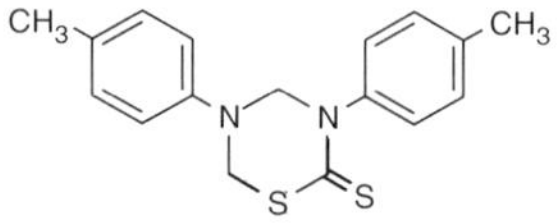

250 mg

S54,577-5
3,5-DIMETHYL-4,6-DIPHENYL-3,4,5,6-TETRAHYDRO-2H-1,3,5-THIADIAZINE-2-THIONE

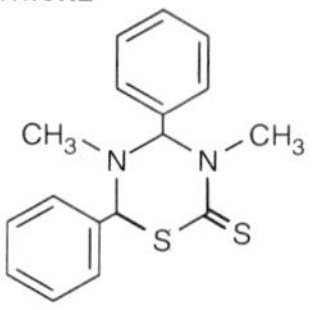

250 mg

S54,579-1
3,5-DIBENZYL-3,4,5,6-TETRAHYDRO-2H-1,3,5-THIADIAZINE-2-THIONE

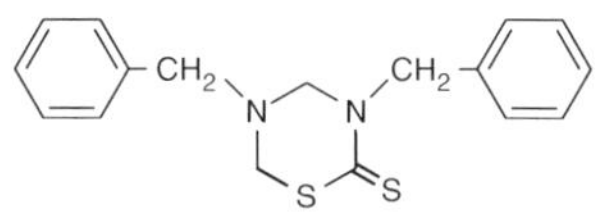

250 mg

S50,681-8
3-MORPHOLINOMETHYL-2(3H)-BENZOTHIAZOLETHIONE

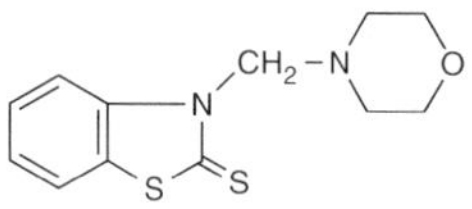

5 gm

S50,687-7
3-[N-(1-NAPHTHYL)-ANILINOMETHYL]-2-(3H)-BENZOTHIAZOLETHIONE

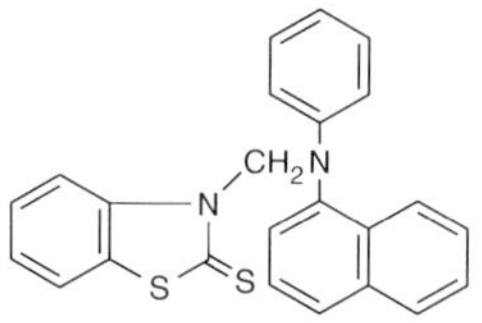

250 mg

S66,159-7
2-(N-ETHYLANILINO)ETHYL N-PHENYLTHIOCARBAMATE

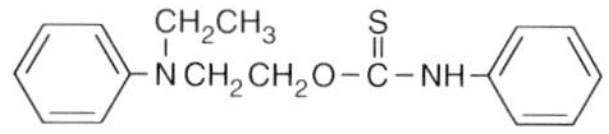

5 gm

S51,041-6
N-(1-ETHYL-3-METHYL-4-OXO-2-THIOXOIMIDAZOLIDIN-5-YLIDENEMETHYL)ACETANILIDE

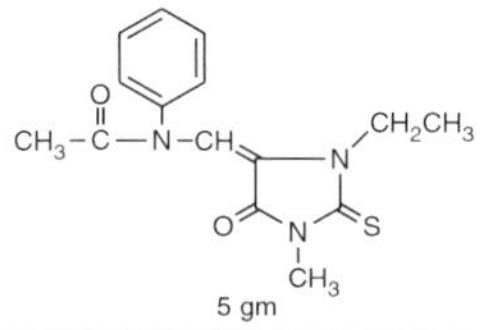

5 gm

S50,182-4
3-ALLYL-5-(5-INDANYLMETHYLENE)-RHODANINE

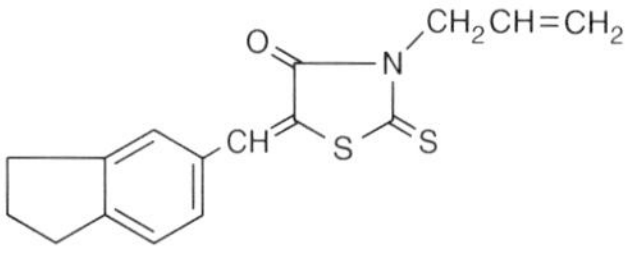

250 mg

S50,214-6
4-(MORPHOLINOTHIOCARBONYL)PHENYL PALMITATE

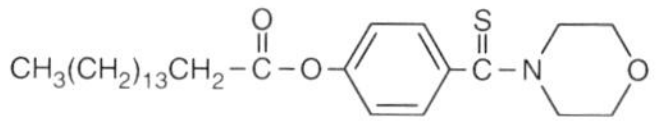

250 mg

S95,839-5
METHYL 8-BENZAMIDO-2,2-DIMETHYL-HEXAHYDRO-7-OXO-5-THIOXO-5H-THIAZOLO[3,2-C]PYRIMIDINE-3-CARBOXYLATE

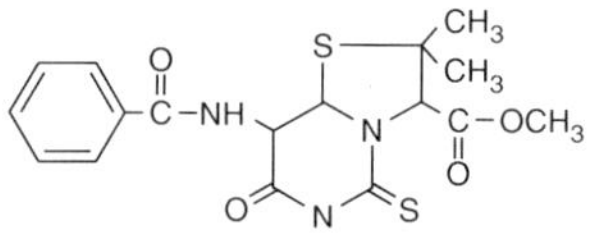

50 mg

S88,784-6
3-ETHYL-2-THIOXO-4,5-THIAZOLIDINEDIONE 5-AZINE W/3-METHYLBENZOTHIAZOL-2(3H)-ONE

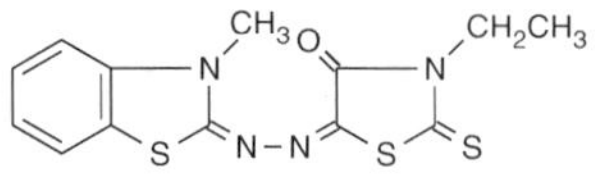

250 mg

S51,047-5
3,4-DICHLORO-N-HEXADECYL-N-METHYLTHIOBENZAMIDE

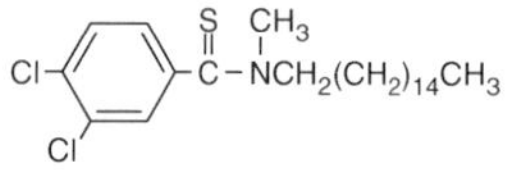

250 mg

S51,251-6
N-CYCLOHEXYLTHIOBENZANILIDE

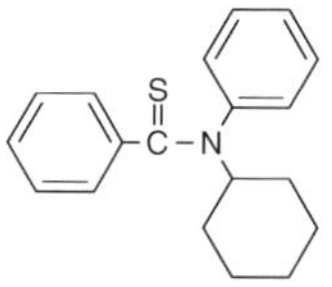

250 mg

S51,236-2
4-(THIOBENZOYL)MORPHOLINE

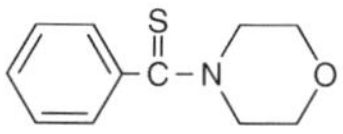

250 mg

S50,583-8
N-HEXADECYL-N-METHYLTHIOBENZAMIDE

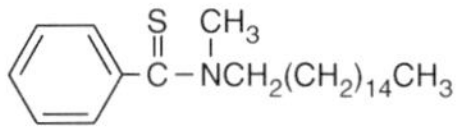

250 mg

S51,063-7
4-CHLORO-N-METHYLTHIOBENZO-ORTHO-TOLUIDIDE

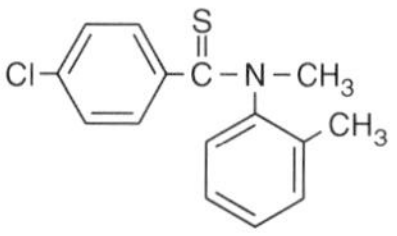

250 mg

S51,064-5
4-CHLORO-N-METHYLTHIOBENZANILIDE

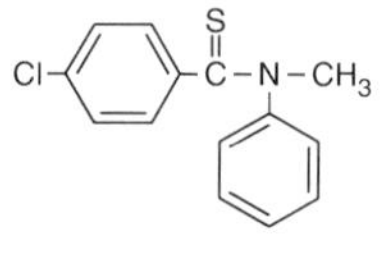

250 mg

S51,279-6
N,N,N',N'-TETRAETHYLDITHIOTEREPHTHALAMIDE

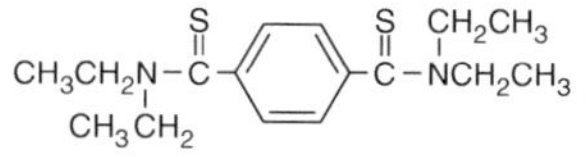

250 mg

S51,402-0
1,3,3-TRIMETHYL-2-INDOLINETHIONE

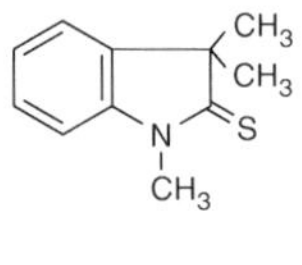

250 mg

S51,363-6
1,7-DIMETHYL-4-PENTADECYL-2(1H)-QUINOLINETHIONE

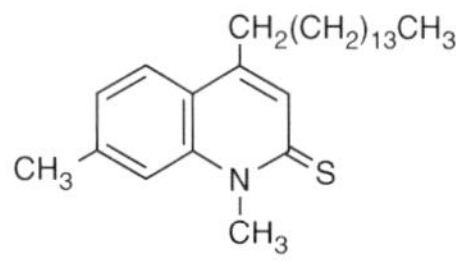

250 mg

S69,085-6
1-(N-PHENYLTHIOCARBAMOYL)-PYRROLIDINE

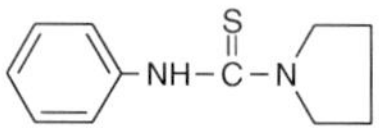

250 mg

S51,183-8
1-METHYL-1-PHENYL-2-THIOUREA

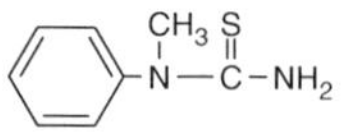

5 gm

S52,710-6
1-(4-CHLOROPHENYL)-3-(4-DIETHYLAMINO-ORTHO-TOLYL)-2-THIOUREA
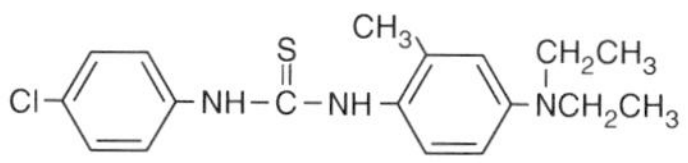
250 mg

S36,337-5
3-METHYL-2-BENZOTHIAZOLINONE
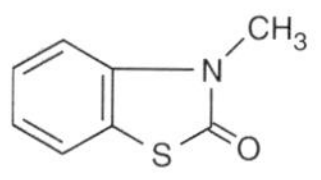
250 mg

S51,531-0
3-HEXADECYL-2-IMINOBENZOTHIAZOLINE
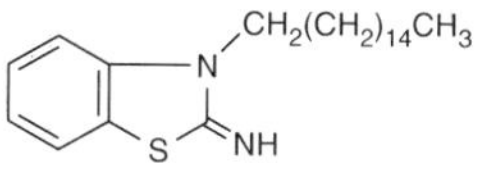
250 mg

S86,957-0
3-METHYL-4-PHENYL-4-THIAZOLIN-2-THIONE
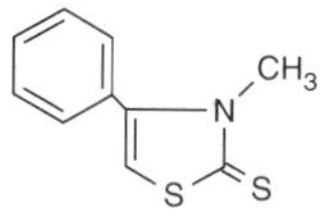
250 mg

S87,002-1
4-TERT-BUTYL-3-PHENYL-4-THIAZOLIN-2-THIONE
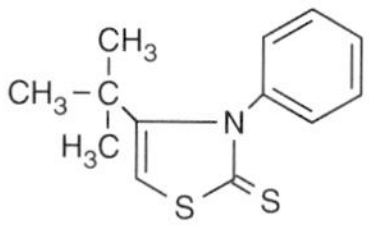
250 mg

S52,998-2
2-(4-NITROBENZYL)-2-THIOPSEUDOUREA HYDROBROMIDE
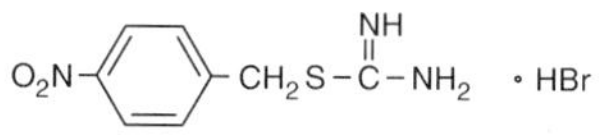
250 mg

S52,996-6
2-(2-CHLOROBENZYL)-2-THIOPSEUDOUREA HYDROCHLORIDE
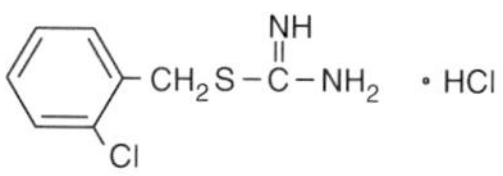
5 gm

S85,134-5
N-[3-(4-IODOPHENYL)-4-PHENYL-2-THIAZOLIDINYLIDENE]CROTONAMIDE
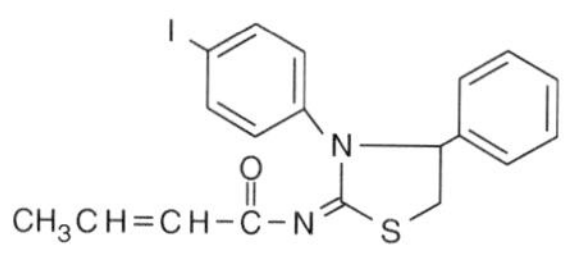
50 mg

S67,646-2
5-METHYL-3-PHENYL-2-THIOXO-4-IMIDAZOLIDINONE
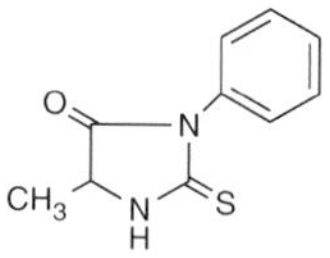
5 gm

S67,654-3
3-PHENYL-2-THIOXO-4-IMIDAZOLIDINONE
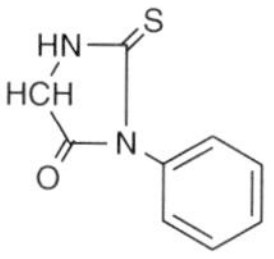
250 mg

S84,578-7
1-BENZYL-3-CINNAMOYL-2-THIOUREA
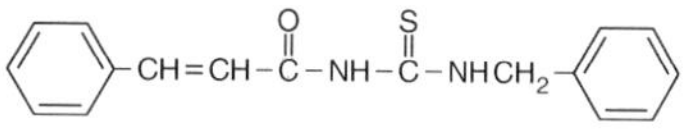
250 mg

S51,371-7
N-BENZYLTHIOBENZANILIDE
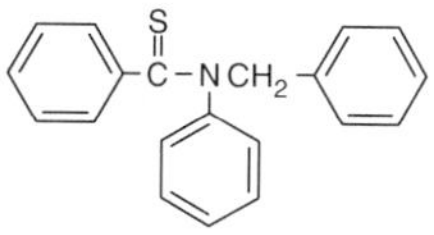
250 mg

S51,381-4
N,N-DIBENZYLTHIOBENZAMIDE
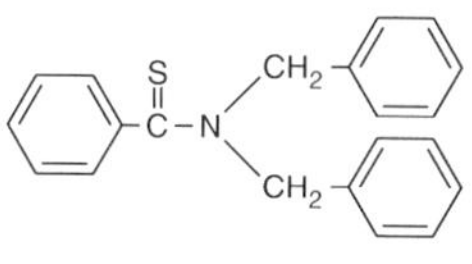
250 mg

S45,523-7
3-PHENYL-1-THIOCARBAMOYL-2-PYRAZOLIN-5-ONE
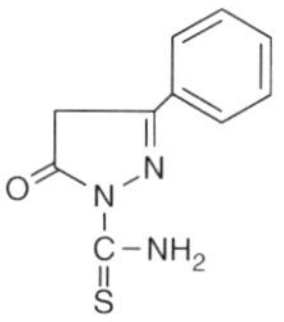
5 gm

S44,094-9
1,3-DIBENZYL-2-THIOUREA
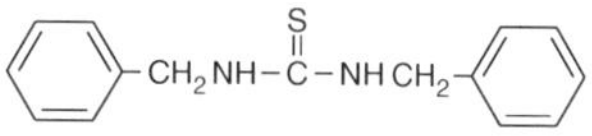
5 gm

S74,335-6
1-(4-METHYLBENZYL)-3-PHENYL-2-THIOUREA
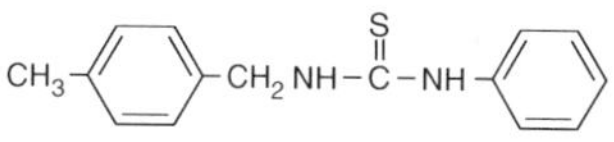
5 gm

S82,261-2
3-BENZYL-2(3H)-BENZOTHIAZOLONE
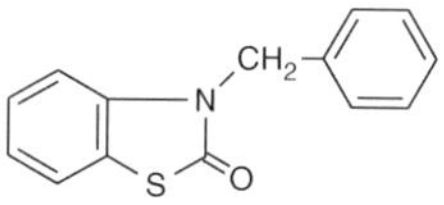
250 mg

S40,741-0
DIBENZYLDITHIOCARBAMIC ACID, ZINC SALT
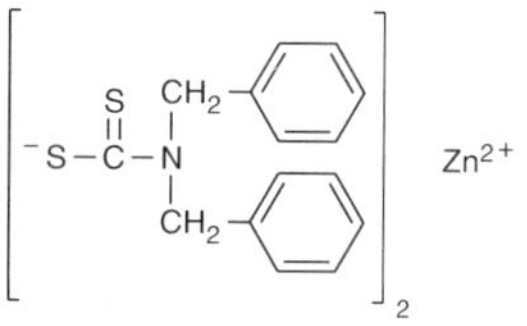
5 gm

S53,933-3
4-BENZYL-3-THIOSEMICARBAZIDE
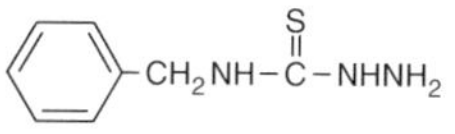
5 gm

S72,259-6
1-ETHYL-6-(O-TOLYL)-5-THIOBIUREA
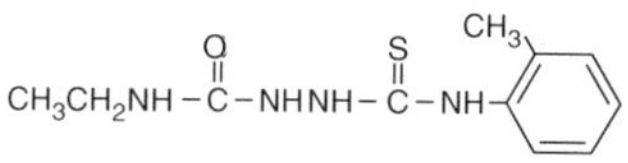
250 mg

S69,263-8
4-(2-METHYLPHENYL)-1-(N-PROPYLCARBAMOYL)-3-THIOSEMICARBAZIDE
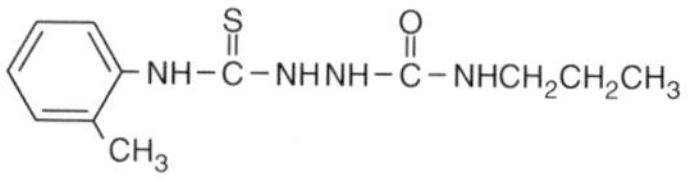
250 mg

S51,240-0
4-METHYL-1-PHENYL-3-THIOSEMICARBAZIDE
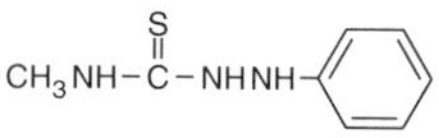
250 mg

S53,778-0
1-ANILINO-6-PHENYL-2,5-DITHIOBIUREA
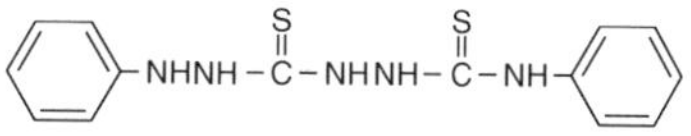
5 gm

S53,910-4
3-(PARA-ISOPROPYLBENZYLIDENEAMINO)-RHODANINE
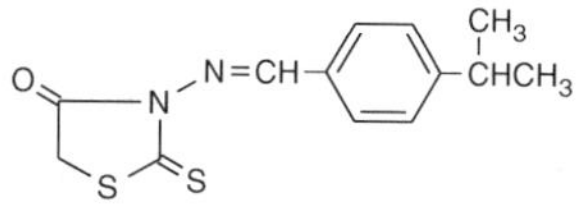
250 mg

S50,440-8
3-PHENYL-1H-1,2,4-TRIAZOLE-5(4H)-THIONE
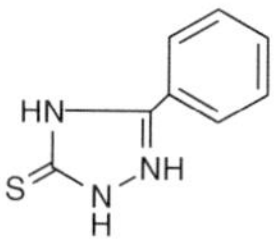
5 gm

S46,298-5
3-METHYL-2-BENZOTHIAZOLINONE HYDRAZONE
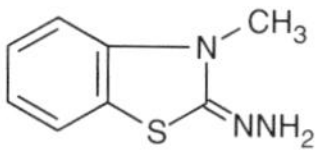
250 mg

S89,276-9
TEREPHTHALDICARBOXALDEHYDE DITHIOSEMICARBAZONE
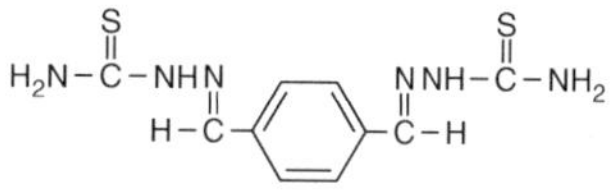
250 mg

S64,941-4
3,5-BIS(4-ISOPROPYLPHENYL)-3,4,5,6-TETRAHYDRO-2H-1,3,5-THIADIAZINE-2-THIONE
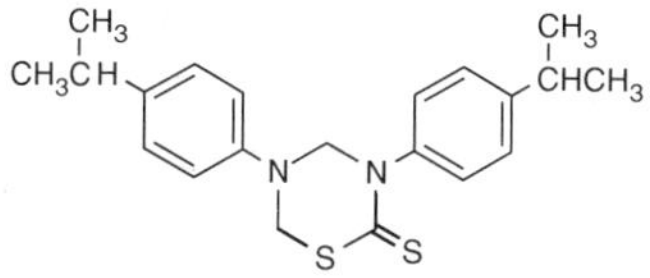
5 gm

S54,761-1
3-BENZYL-5-PROPYLTETRAHYDRO-2H-1,3,5-THIADIAZINE-2-THIONE
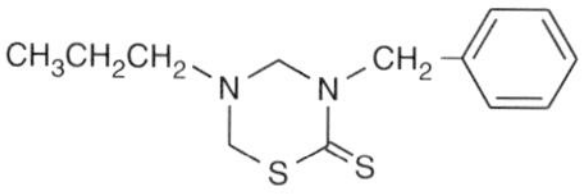
250 mg

S51,264-8
4-(BENZOYLTHIOFORMYL)MORPHOLINE
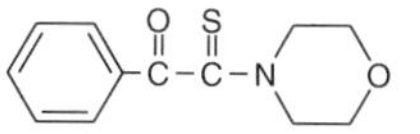
250 mg

S53,971-6
2-(3'-NITROPHENACYL)-2-THIOPSEUDOUREA HYDROBROMIDE
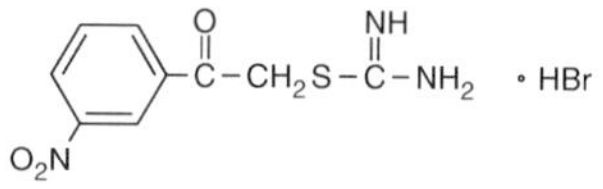
250 mg

S53,468-4
2-(4'-NITROPHENACYL)-2-THIOPSEUDOUREA HYDROBROMIDE
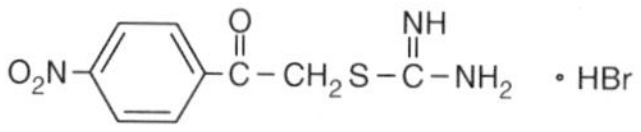
5 gm

S84,514-0
4-[3-(3-NITROBENZOYL)-2-THIOUREIDO]-BENZOIC ACID

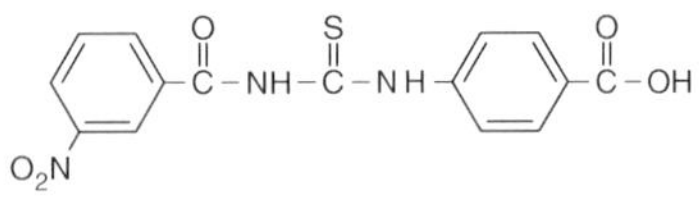

250 mg

S79,654-9
METHYL 2-[3-(3-NITROBENZOYL)-2-THIOUREIDO]BENZOATE

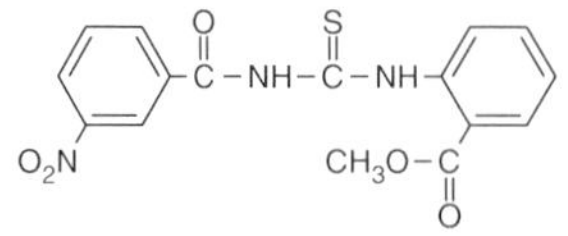

250 mg

S69,397-9
3-ISOPROPYL-5-(3-NITROBENZYLIDENE) RHODANINE

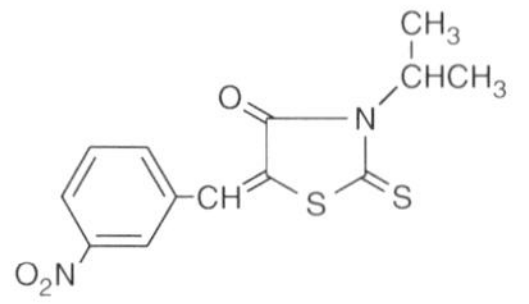

250 mg

S94,015-1
2-IMINO-3-METHYL-6-NITROBENZOTHIAZOLINE

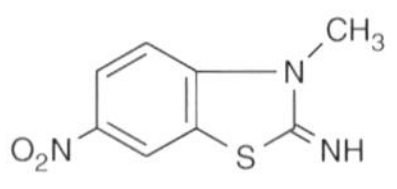

250 mg

S51,336-9
2-(4-NITROBENZYL)-2-THIOPSEUDOUREA HYDROCHLORIDE

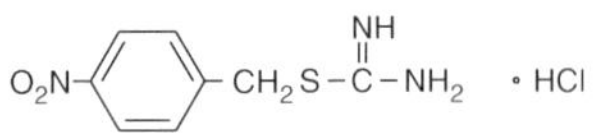

250 mg

S86,936-8
1-FORMYL-4-PHENYL-3-THIOSEMICARBAZIDE

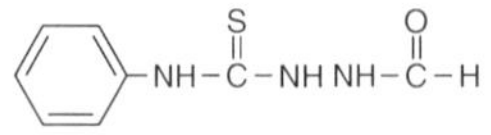

250 mg

S51,542-6
5-BENZYLIDENE-3-HEXADECYL-2-PHENYLIMINO-4-THIAZOLIDINONE

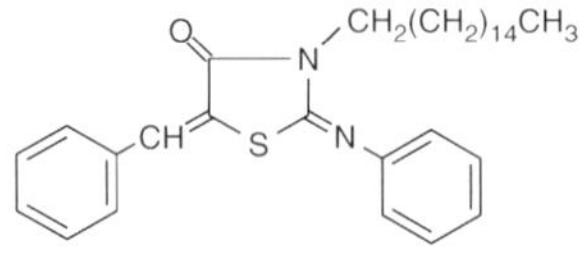

250 mg

S67,506-7
1-(3-ACETYLPHENYL)-3-ETHYL-2-THIOUREA

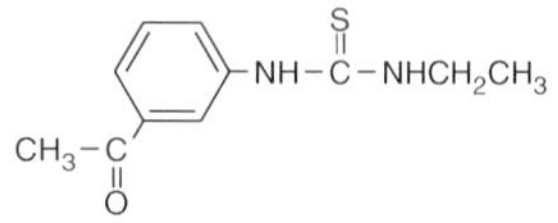

250 mg

S71,862-9
1-(3-ACETYLPHENYL)-3-ALLYL-2-THIOUREA

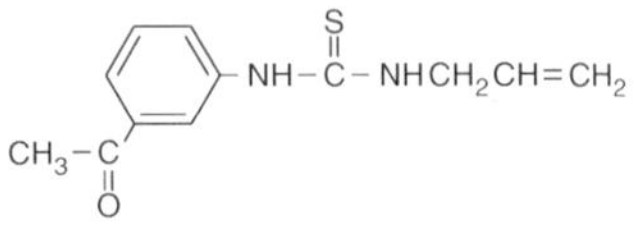

5 gm

S69,361-8
1-(3-ACETYLPHENYL)-3-PROPYL-2-THIOUREA

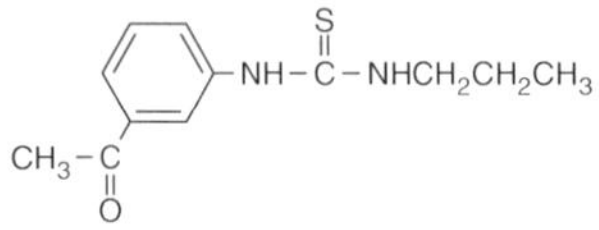

250 mg

S87,506-6
1,3-BIS(3-ACETYLPHENYL)-2-THIOUREA

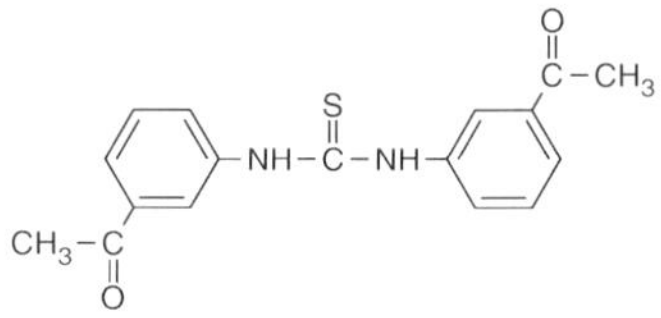

250 mg

S50,336-3
5-(5-INDANYLMETHYLENE)-2-THIOHYDANTOIN

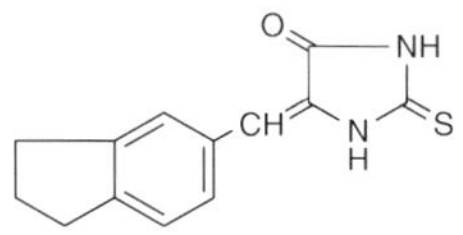

250 mg

S91,708-7
1-(2-BENZOYLPHENYL)-3-METHYL-2-THIOUREA

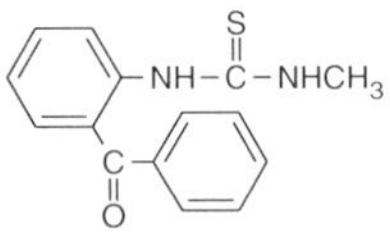

250 mg

S91,706-0
1-(4-ACETYLPHENYL)-3-METHYL-2-THIOUREA

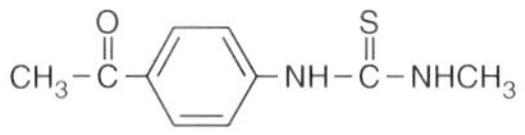

250 mg

S87,505-8
1,3-BIS(4-ACETYLPHENYL)-2-THIOUREA

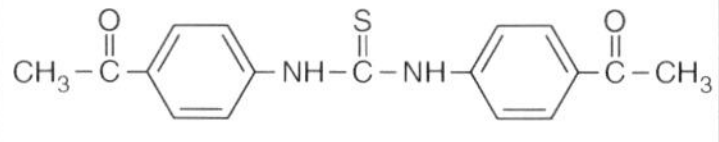

250 mg

S51,860-3
5-BENZYL-5-ETHYL-2-IMINO-4-THIAZOLIDINONE

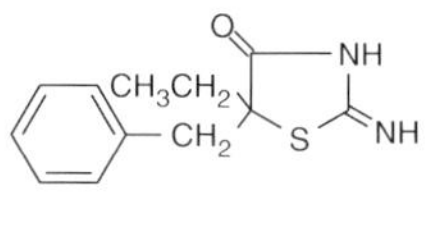

5 gm

S86,984-8
2-OXO-1-CYCLOPENTANE-THIOCARBOXANILIDE

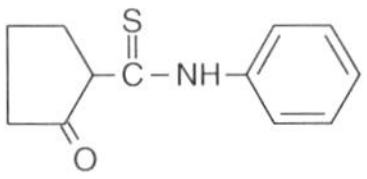

250 mg

S87,036-6
2-(CYCLOHEXYLCARBONYL) THIOACETANILIDE

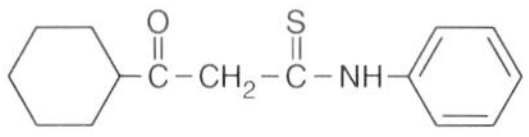

250 mg

S63,078-0
4-METHYL-1-PHENYL-2(1H)-QUINOLINETHIONE

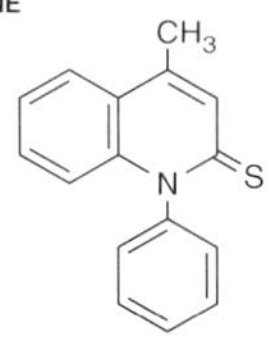

5 gm

S71,843-2
4-(3-PHENYL-2-THIOUREIDO) HIPPURIC ACID

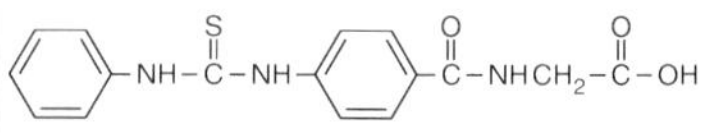

250 mg

S84,577-9
4-(3-CINNAMOYL-2-THIOUREIDO) BENZOIC ACID

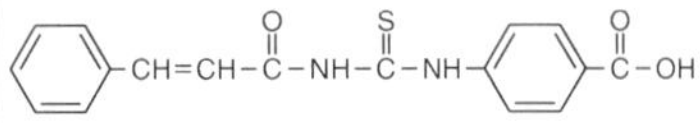

250 mg

S67,583-0
4-(3-DODECANOYL-2-THIOUREIDO) BENZOIC ACID

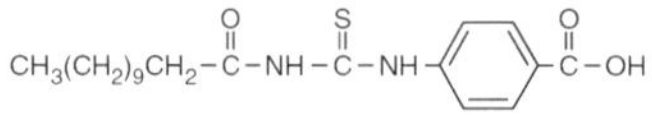

250 mg

S69,050-3
1-(3-CARBOXYPHENYL)-3-PROPYL-1-THIOUREA

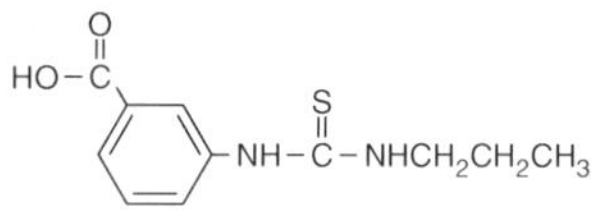

250 mg

S66,896-6
1-(2-CARBOXYPHENYL)-3-PHENYL-2-THIOUREA

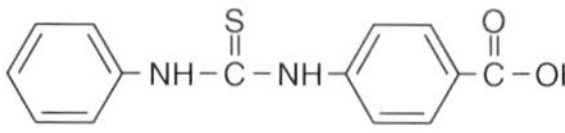

5 gm

S51,028-9
4-(3-HEXADECYL-2-THIOUREIDO)-BENZOIC ACID

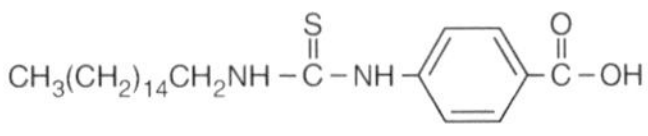

250 mg

S72,970-1
1-(4-CARBOXYPHENYL)-4-PHENYL-3-THIOSEMICARBAZIDE

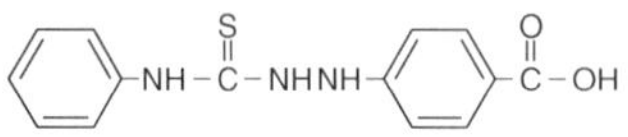

5 gm

S84,539-6
METHYL 2-[3-(3-PHENYLPROPIONYL)-2-THIOUREIDO]BENZOATE

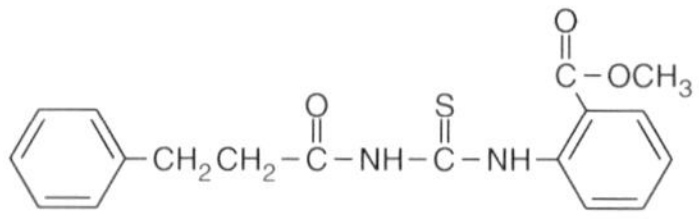

250 mg

S55,470-7
ETHYL 4-(3-BENZOYL-2-THIOUREIDO)-BENZOATE

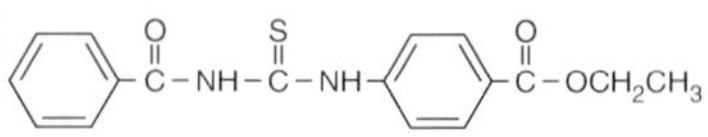

5 gm

S84,450-0
METHYL 2-[3-(2-BUTENOYL)-2-THIOUREIDO]BENZOATE

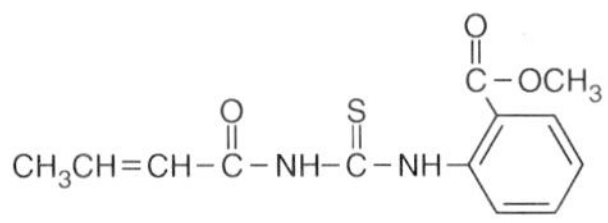

250 mg

S84,760-7
METHYL 2-(3-DODECANOYL-2-THIOUREIDO)BENZOATE

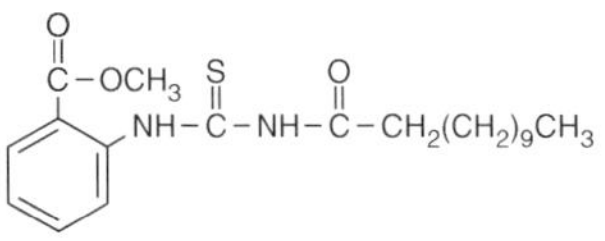

250 mg

S84,762-3
ETHYL 4-(3-DODECANOYL-2-THIOUREIDO)-BENZOATE

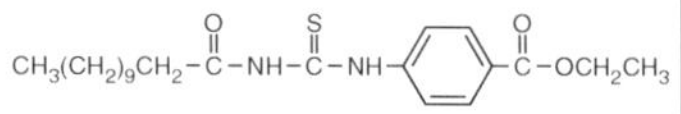

250 mg

S69,026-0
1-(4-ETHOXYCARBONYL)-3-ETHYL-2-THIOUREA

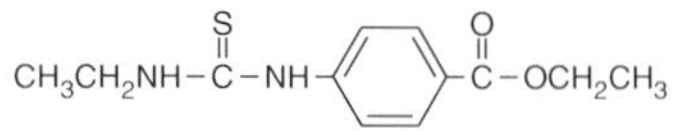

250 mg

S61,528-5
1-BENZOYL-2-THIOBIURET
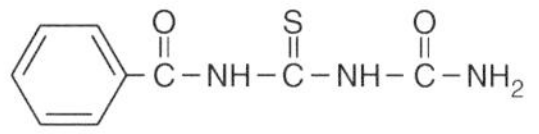
250 mg

S76,219-9
5-BENZYLIDENE-2-THIOHYDANTOIN
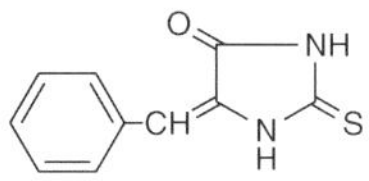
250 mg

S86,340-8
1-(BENZOYLAMIDINO)-2-THIOUREA
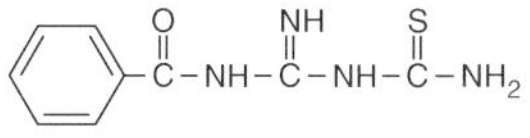
250 mg

S50,552-8
1-BENZOYL-3-PHENYL-2-THIOUREA
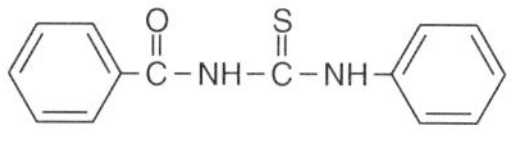
5 gm

S64,591-5
3-BENZOYL-1,1-DIALLYL-2-THIOUREA
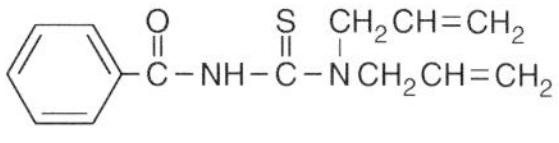
5 gm

S85,975-3
1-BENZOYL-3-METHYL-3-PHENYL-2-THIOUREA
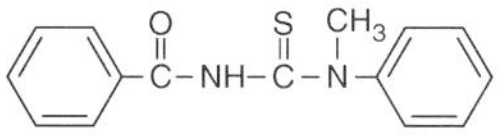
250 mg

S85,969-9
O-ETHYL N-BENZOYLTHIOCARBAMATE
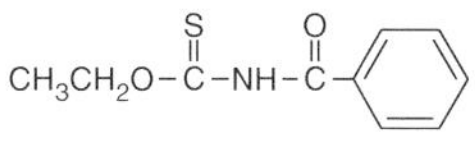
250 mg

S60,587-5
METHYL 3-(4-PENTADECYLBENZOYL)-DITHIOCARBAZATE
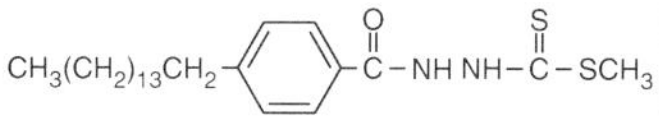
250 mg

S51,186-2
1-BENZOYL-3-THIOSEMICARBAZIDE
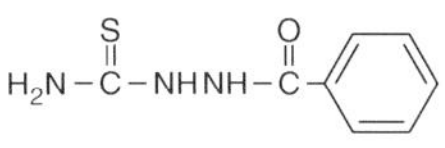
5 gm

S85,593-6
4-(3-PHENYL-2-THIOUREIDO)-BENZOIC HYDRAZIDE
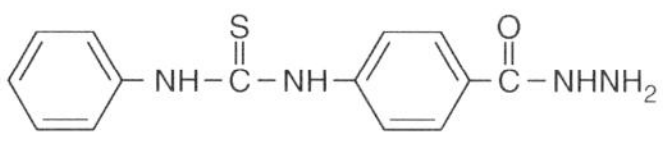
250 mg

S86,947-3
1-BENZOYL-4-METHYL-3-THIOSEMICARBAZIDE
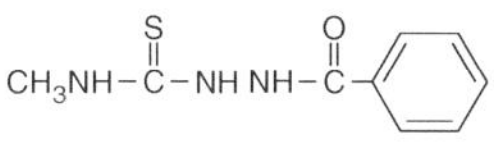
250 mg

S69,142-9
1-BENZOYL-4-PHENYL-3-THIOSEMICARBAZIDE
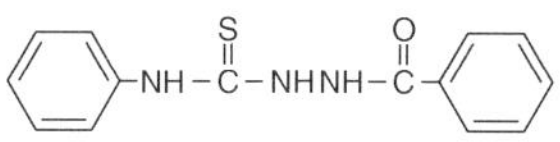
5 gm

S89,865-1
1-METHYL-5-OXO-3-PHENYL-2-THIOHYDANTOIN
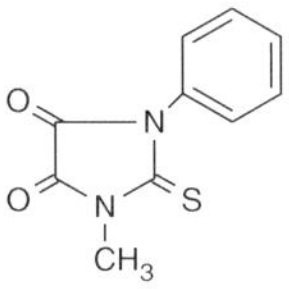
250 mg

S58,806-7
4-OXO-3-PHENYL-2-THIOXO-1-IMIDAZOLIDINEACETIC ACID
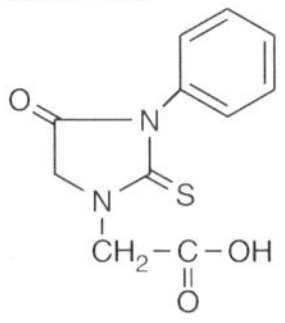
250 mg

S40,981-2
N-(N-PHENYLTHIOCARBAMOYL)GLYCINE
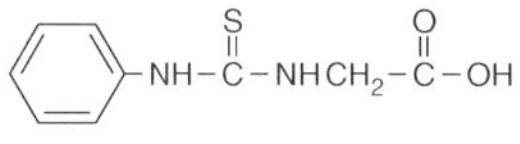
5 gm

S52,390-9
PARA-(4-OXO-2-THIOXOTHIAZOLIDINO)-BENZOIC ACID
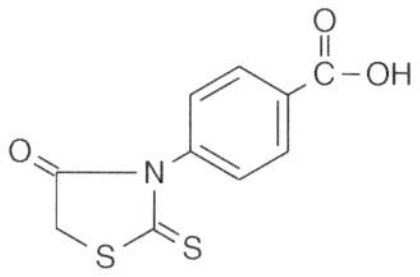
250 mg

S76,090-0
ETHYL 2-[4-(1-METHYL-6-PHENYL-1,2,3,6-TETRAHYDRO-2-THIOXO-4-PYRIMIDINYL)]PHENYL ACETATE
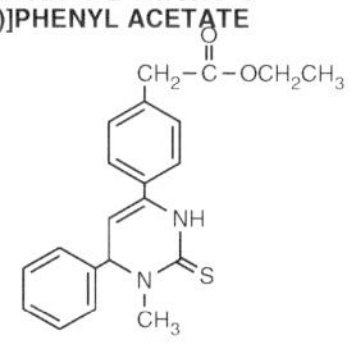
250 mg

S59,561-6
O-METHYL THIOCARBANILATE
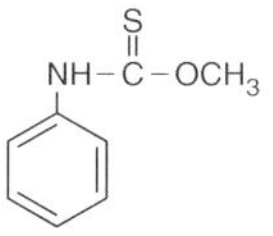
5 gm

S72,527-7
6-PHENYL-1-PROPYL-5-THIOBIUREA
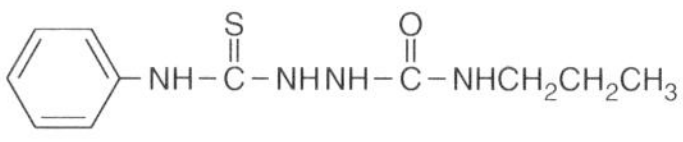
5 gm

S69,179-8
1-[N-(2,6-DIMETHYLPHENYL)-CARBAMOYL]-3-THIOSEMICARBAZIDE
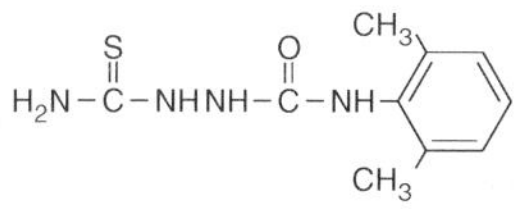
250 mg

S91,318-9
1-METHYL-6-(3,4-XYLYL)-2-THIOBIUREA
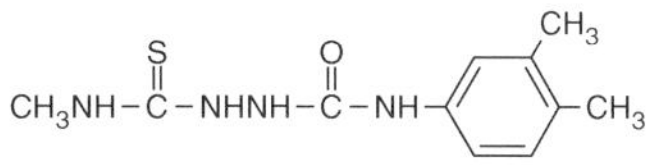
250 mg

S89,863-5
1-METHYL-3-PHENYL-5-THIOXOHYDANTOIN
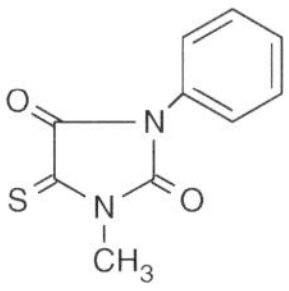
250 mg

S67,714-0
3-ETHYL-2,4-THIAZOLIDINEDIONE
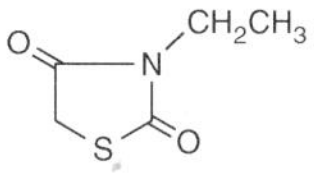
5 gm

S78,573-3
1-ACETYL-3-PHENYL-2-THIOUREA
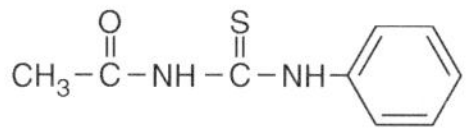
250 mg

S86,273-8
1-AMIDINO-3-BENZOYL-2-THIOUREA
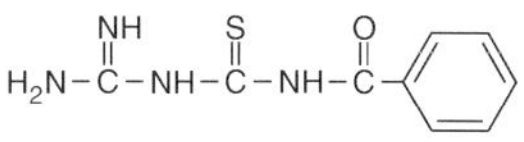
250 mg

S53,084-0
1-ISOBUTYRYL-3-PHENYL-2-THIOUREA
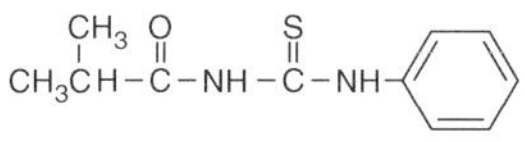
250 mg

S79,681-6
1-(3-PHENYLPROPIONYL)-3-(2,3-XYLYL)-2-THIOUREA
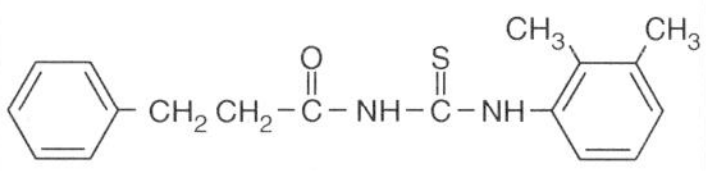
250 mg

S92,842-9
3-METHYL-8-PHENYL-2-THIOHYPOXANTHINE
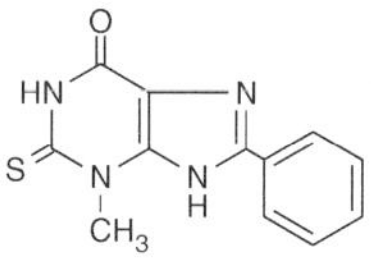
250 mg

S67,569-5
1-DODECANOYL-3-(1-NAPHTHYL)-2-THIOUREA
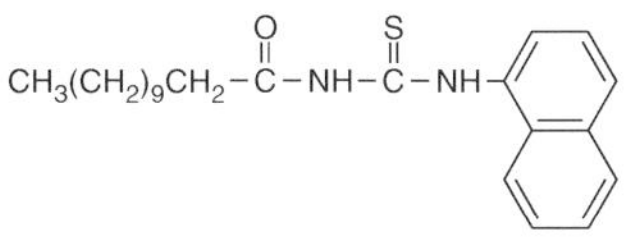
5 gm

S67,570-9
1-DODECANOYL-3-(2-NAPHTHYL)-2-THIOUREA
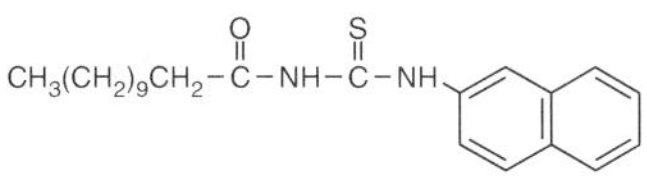
5 gm

S52,353-4
3-BENZYLRHODANINE
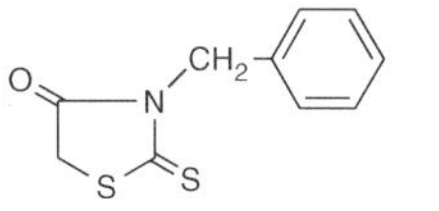
5 gm

S89,840-6
1,3-DIPHENYL-5-OXO-2,4-DITHIOHYDANTOIN
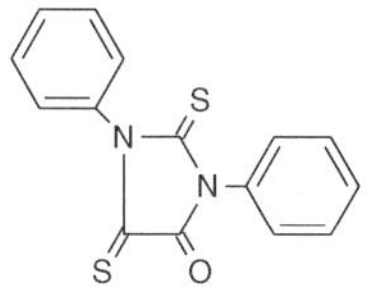
250 mg

S54,593-7
1,2-DIHYDRO-3-PROPYL-2-THIOXO-4(3H)-QUINAZOLINONE
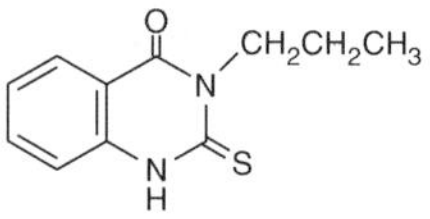
250 mg

S52,515-4
1,2-DIHYDRO-3-PHENYL-2-THIOXO-4(3H)-QUINAZOLINONE
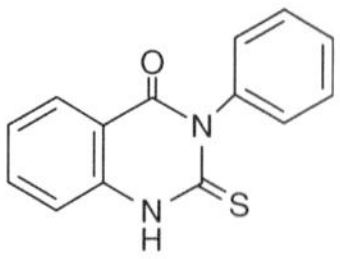
5 gm

S54,512-0
3-ALLYL-1,2-DIHYDRO-2-THIOXO-4(3H)-QUINAZOLINONE
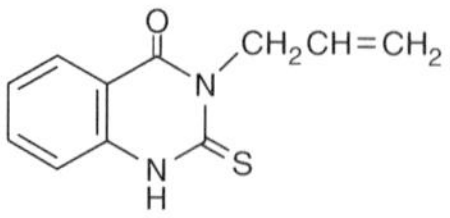
250 mg

S54,594-5
1,2-DIHYDRO-3-ISOPROPYL-2-THIOXO-4(3H)QUINAZOLINONE
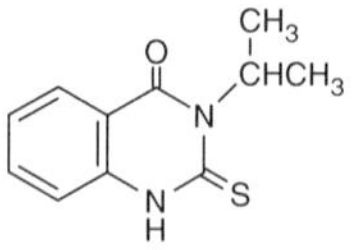
250 mg

S83,912-4
3-PHENYL-2,4-THIAZOLIDINEDIONE
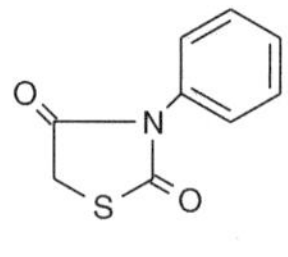
250 mg

S51,239-7
3-PHENYL-2(3H)-BENZOTHIAZOLONE
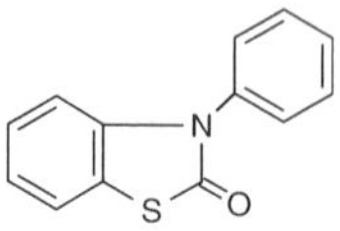
250 mg

S37,810-0
6-CHLORO-3-ETHYL-2-BENZOTHIAZOLINONE
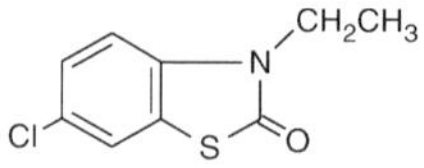
5 gm

S54,758-1
2-THIOXO-3-(O-TOLYL)-4-THIAZOLIDINONE
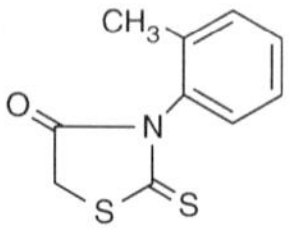
250 mg

S54,760-3
2-THIOXO-3-(M-TOLYL)-4-THIAZOLIDINONE
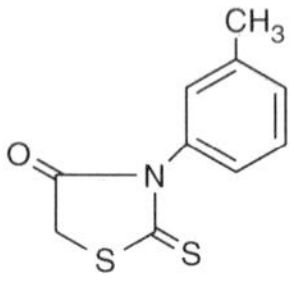
250 mg

S52,391-7
3-(4-BROMO-META-TOLYL)-RHODANINE
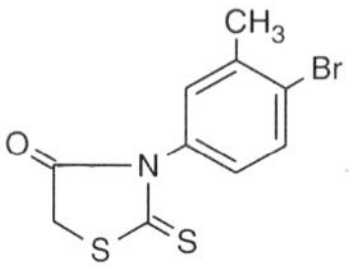
5 gm

S52,707-6
3-(5-BROMO-ORTHO-TOLYL)-RHODANINE
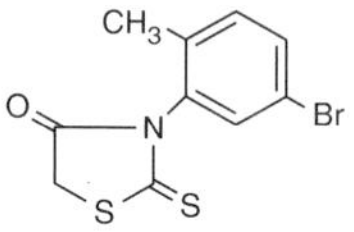
5 gm

S75,870-1
2-BENZYLIDENE-5,6-DIHYDROIMIDAZO (2,1-B)THIAZOL-3(2H)-ONE
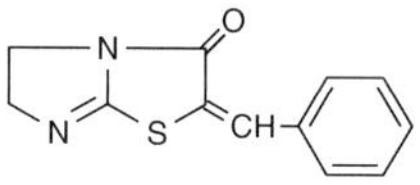
250 mg

S83,728-8
PHENYL BENZOTHIOIMIDATE
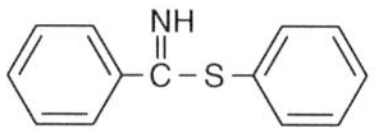
250 mg

S89,830-9
1,3-DIPHENYL-5-[N-(M-TOLYL) CARBAMOYLIMINO)DITHIOHYDANTOIN
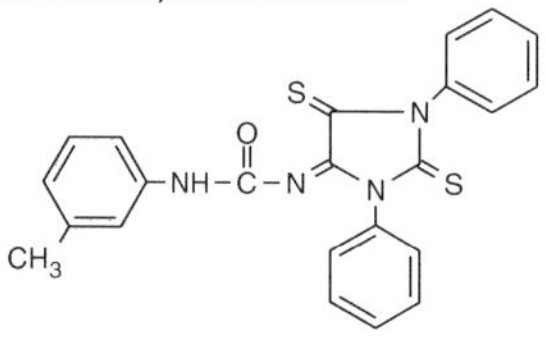
250 mg

S89,839-2
1,3-DIPHENYL-5-IMINO-4-THIOHYDANTOIN
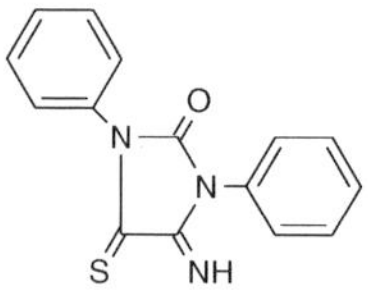
250 mg

S95,582-5
DELTA1,1'-BIISOINDOLINE-3,3'-DITHIONE
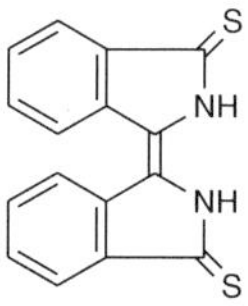
50 mg

S87,551-1
2-METHYL-1,2-BENZISOTHIAZOLE-3(2H)-THIONE
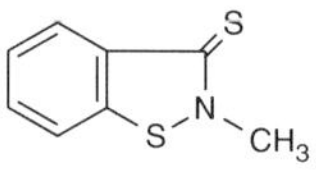
250 mg

S51,012-2
N-CYCLOHEXYL-2-PHENYLTHIOACETAMIDE
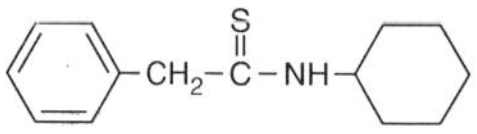
250 mg

S62,806-9
1-[(2-CHLOROPHENYL)THIOCARBAMOYL] PIPERIDINE
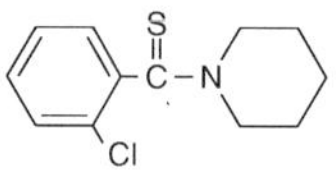
250 mg

S51,272-9
1,4-BIS(THIOBENZOYL)-PIPERAZINE
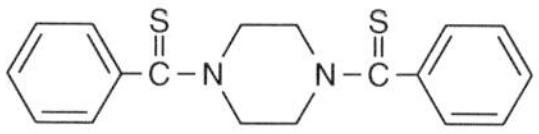
250 mg

S63,036-5
4-(THIOACETYL)MORPHOLINE
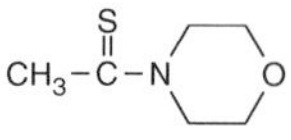
250 mg

S65,697-6
4-(4,4-DIMETHYL-4-PHENYLTHIOBUTYRYL)MORPHOLINE
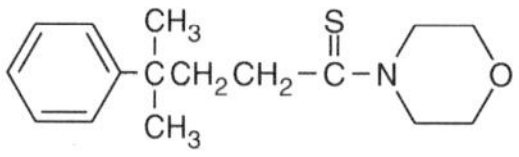
5 gm

S36,406-1
4-(2-NAPHTHYLTHIOACETYL)MORPHOLINE
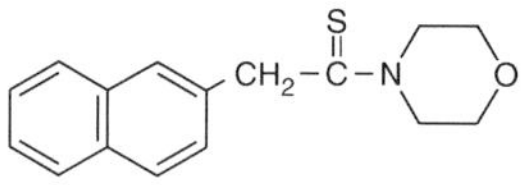
5 gm

S54,740-9
2-PHENYLTHIOACETAMIDE
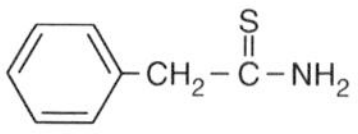
250 mg

S78,696-9
N-PHENYLTHIOBUTYRAMIDE
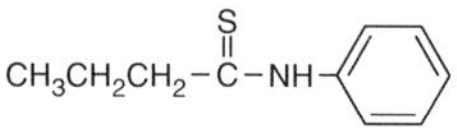
250 mg

S50,207-3
THIOBENZANILIDE
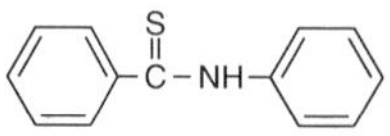
250 mg

S50,228-6
N-METHYLTHIOBENZANILIDE
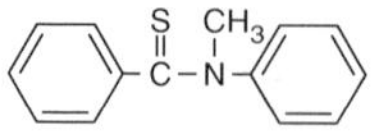
250 mg

S51,372-5
N-PHENYLTHIOACETANILIDE
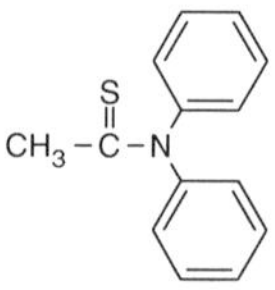
250 mg

S51,517-5
N-ALLYLTHIOBENZANILIDE
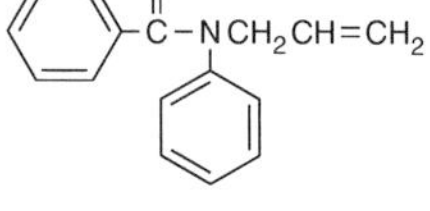
250 mg

S54,172-9
N-HEXADECYLTHIOBENZANILIDE
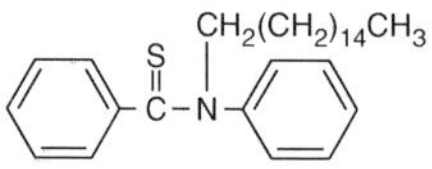
250 mg

S39,022-4
4'-CHLOROTHIOPROPIONANILIDE
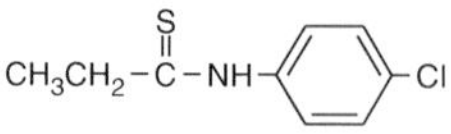
5 gm

S43,118-4
N-(1-NAPHTHYL)-THIOACETAMIDE
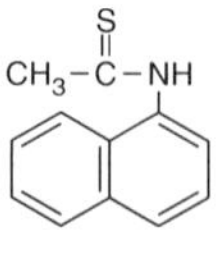
5 gm

S62,865-4
1-METHYL-2(1H)-QUINOLINETHIONE

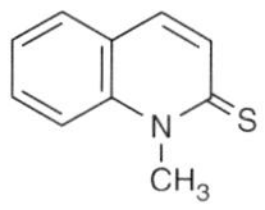

250 mg

S50,392-4
1,6-DIMETHYL-4-PENTADECYL-2(1H)-QUINOLINETHIONE

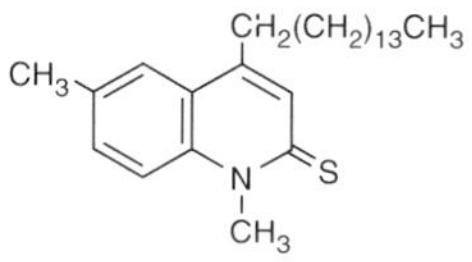

250 mg

S54,508-2
3-PHENYLDITHIOCARBAZIC ACID

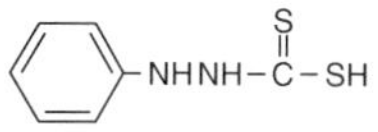

250 mg

S91,378-2
1-(1,5-DIMETHYL-3-OXO-2-PHENYL-4-PYRAZOLIN-4-YL)-3-METHYL-2-THIOUREA

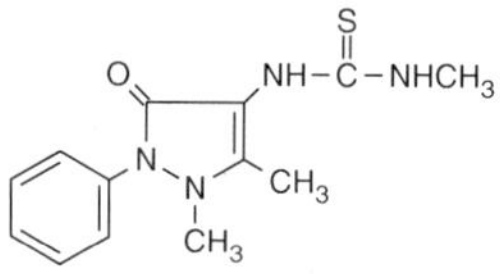

250 mg

S91,347-2
1-ACETYL-4-PHENYL-3-THIOSEMICARBAZIDE

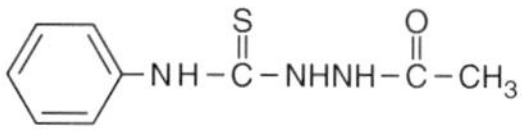

250 mg

S43,033-1
5-MERCAPTO-3-PHENYL-1,3,4-THIADIAZOLE-2-THIONE, POTASSIUM SALT

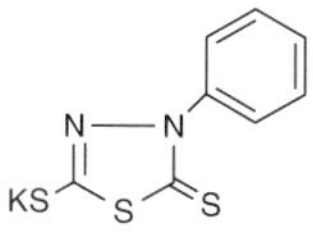

5 gm

S89,986-0
1-ALLYL-3-(1-NAPHTHYLMETHYL)-2-THIOUREA

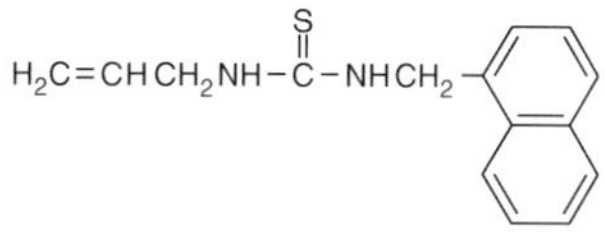

250 mg

S38,581-6
1-BENZYL-2-THIOUREA

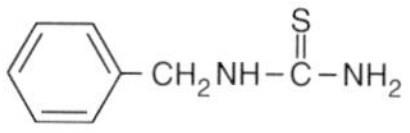

5 gm

S37,338-9
1,1-DIMETHYL-3-(META-TOLYL)-2-THIOUREA

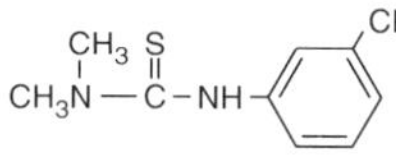

5 gm

S55,380-8
1-ETHYL-3-(M-TOLYL)-2-THIOUREA

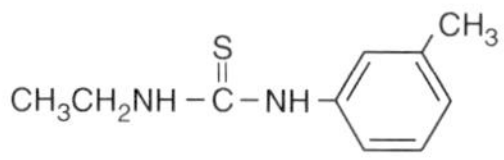

5 gm

S60,197-7
1-(3-CHLOROPHENYL)-3-PHENYL-2-THIOUREA

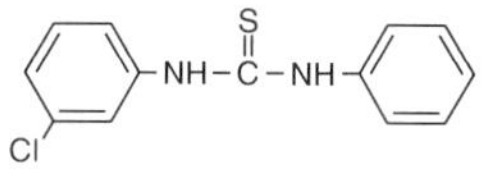

5 gm

S86,339-4
1-(CYANOAMIDINO)-3-PHENYL-2-THIOUREA

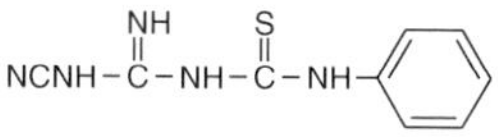

250 mg

S77,305-0
1-CYCLOPENTYL-3-PHENYL-2-THIOUREA

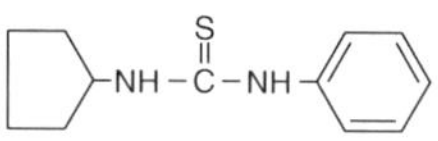

250 mg

S76,747-6
1-(N-METHYLTHIOCARBAMOYL)-2-PHENYLPYRAZOLIDINE

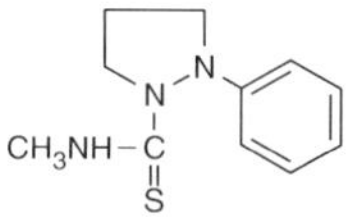

250 mg

S76,205-9
1-PROPYL-2-IMIDAZOLIDINETHIONE

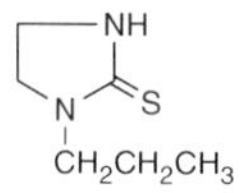

250 mg

S59,922-0
1-CYCLOHEXYL-3-PHENYL-2-THIOUREA

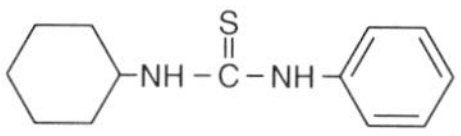

5 gm

S63,210-4
1-CYCLOHEXYL-1-METHYL-3-PHENYL-2-THIOUREA

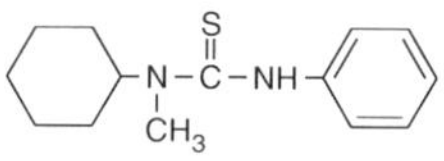

5 gm

S90,065-6
1-BENZYL-3-CYCLOHEXYL-1-ETHYL-2-THIOUREA

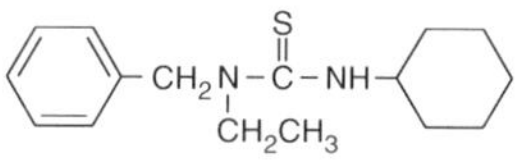

250 mg

S52,331-3
4,4'-TRIMETHYLENEBIS[1-(N-PHENYLTHIOCARBAMOYL)PIPERIDINE]

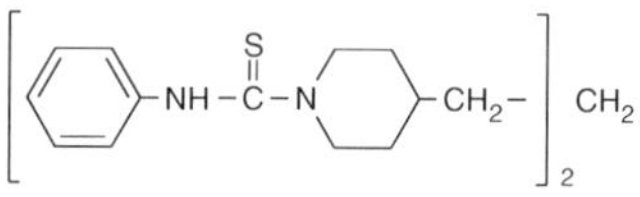

100 mg

S54,273-3
4,4'-TRIMETHYLENEBI(1-[N-(2,6-XYLYL)THIOCARBAMOYL]PIPERIDINE)

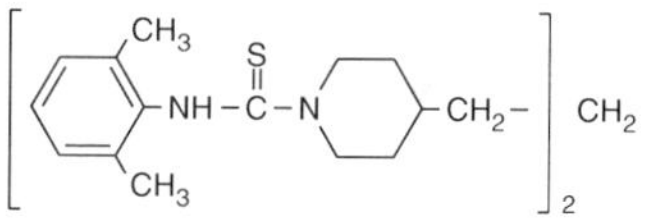

250 mg

S54,586-4
4,4'-TRIMETHYLENEBIS(1-[N-(3,4-XYLYL)THIOCARBAMOYL]PIPERIDINE)

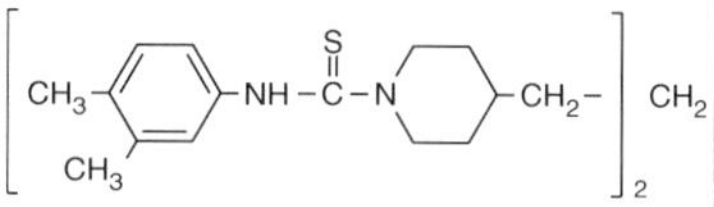

250 mg

S59,925-5
1,1-PENTAMETHYLENE-3-PHENYL-2-THIOUREA

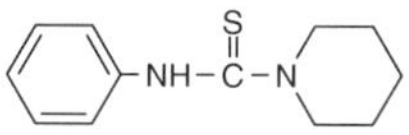

5 gm

S66,042-6
1-(2,5-DIMETHYLMORPHOLINO)-3-PHENYL-2-THIOUREA

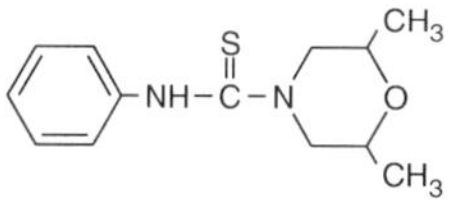

5 gm

S69,444-4
2,6-DIMETHYL-4-(3-METHYLPHENYLTHIOCARBAMOYL) MORPHOLINE

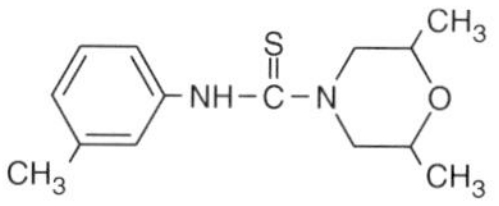

250 mg

S68,506-2
2,6-DIMETHYL-4-[N-(2-ISOPROPYLPHENYL)THIOCARBAMOYL) MORPHOLINE

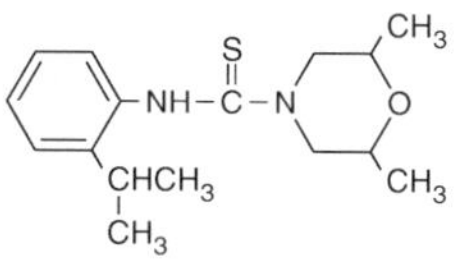

5 gm

S61,518-8
1-METHYL-3-PHENYL-2-THIOUREA

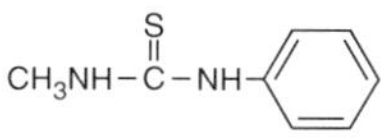

250 mg

S34,730-2
1,1-DIMETHYL-3-PHENYL-2-THIOUREA

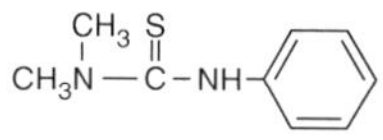

5 gm

S59,916-6
1-ETHYL-3-PHENYL-2-THIOUREA

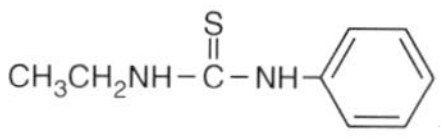

5 gm

S86,338-6
1-ACETIMIDO-3-PHENYL-2-THIOUREA P-TOLUENESULFONATE

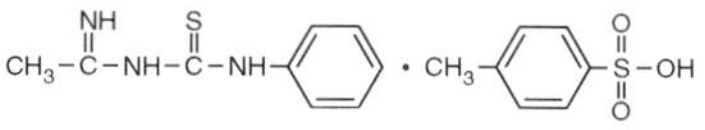

250 mg

S51,984-7
1-ALLYL-3-PHENYL-2-THIOUREA

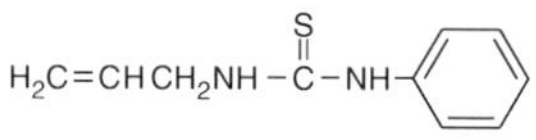

250 mg

S59,404-0
1-ISOPROPYL-3-PHENYL-2-THIOUREA

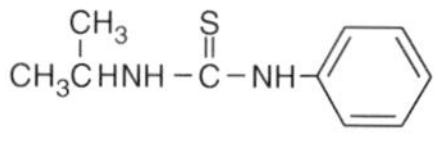

5 gm

S61,521-8
1-PHENYL-3-PROPYL-2-THIOUREA

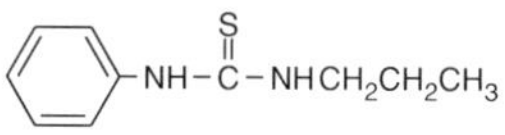

5 gm

S52,096-9
1-BUTYL-3-PHENYL-2-THIOUREA
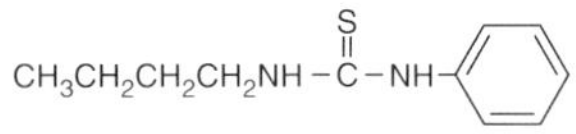
5 gm

S62,516-7
1,1-DIETHYL-3-PHENYL-2-THIOUREA
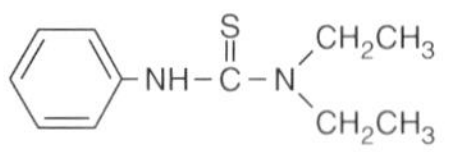
5 gm

S66,039-6
1-ISOBUTYL-3-PHENYL-2-THIOUREA
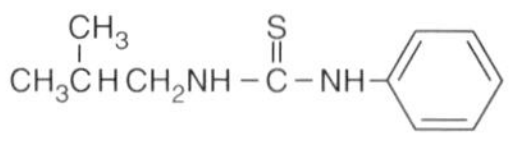
250 mg

S64,429-3
1,1-DIALLYL-3-PHENYL-2-THIOUREA
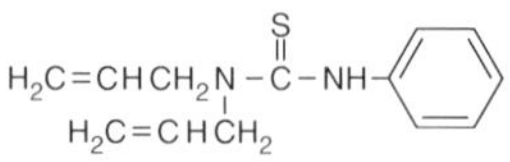
5 gm

S62,530-2
1,1-DIPROPYL-3-PHENYL-2-THIOUREA
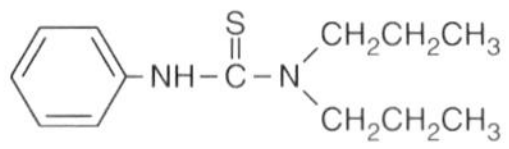
5 gm

S63,919-2
1,1-DIISOPROPYL-3-PHENYL-2-THIOUREA
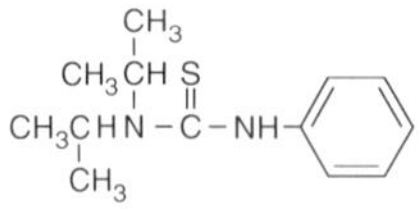
5 gm

S71,677-4
1-BENZYL-3-PHENYL-2-THIOUREA
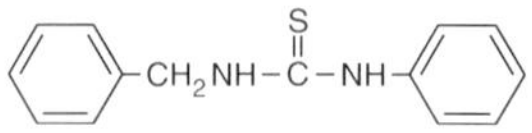
5 gm

S92,047-9
1-(ALPHA-METHYLBENZYL)-3-PHENYLTHIOUREA
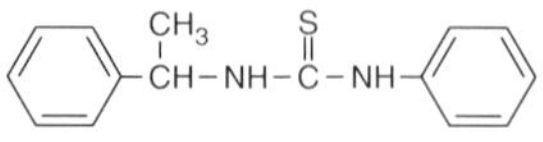
250 mg

S62,996-0
1,1-DIBUTYL-3-PHENYL-2-THIOUREA
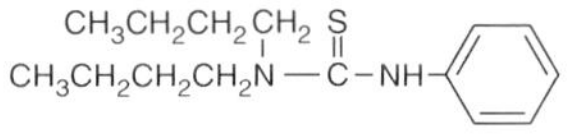
5 gm

S66,013-2
1,1-DIISOBUTYL-3-PHENYL-2-THIOUREA
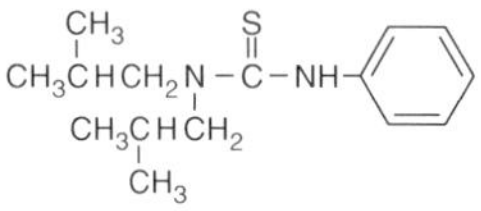
5 gm

S89,987-9
1-BENZYL-1-ETHYL-3-PHENYL-2-THIOUREA
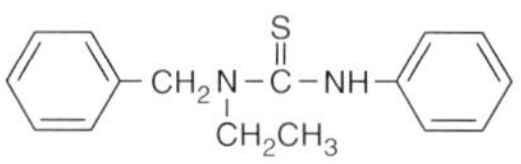
250 mg

S90,007-9
3-ALLYL-1,1-DIBENZYL-2-THIOUREA
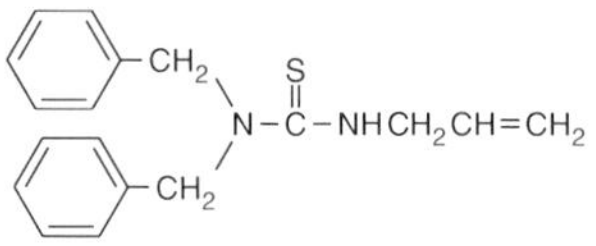
250 mg

S89,990-9
1-BENZYL-1-BUTYL-3-PHENYL-2-THIOUREA
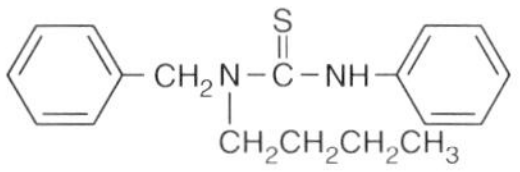
250 mg

S61,527-7
1-DODECYL-3-PHENYL-2-THIOUREA
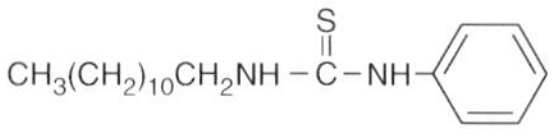
5 gm

S81,066-5
1-PHENYL-3-TETRADECYL-2-THIOUREA
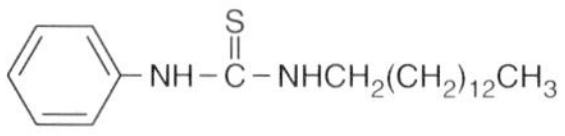
250 mg

S58,098-8
1-HEXADECYL-3-PHENYL-2-THIOUREA
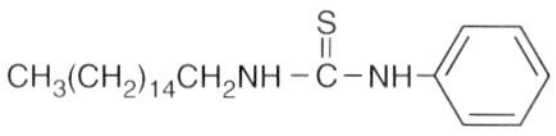
250 mg

S84,495-0
1-PHENYL-3-(P-TOLYL)-2-THIOUREA
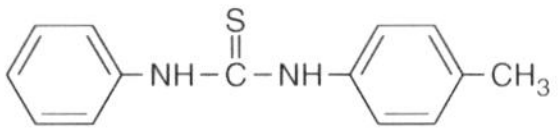
250 mg

S79,513-5
1-(4-ETHYLPHENYL)-3-PHENYL-2-THIOUREA
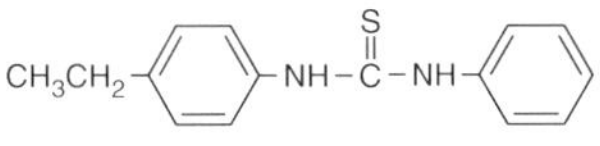
250 mg

S79,514-3
1-PHENYL-3-(2,6-XYLYL)-2-THIOUREA
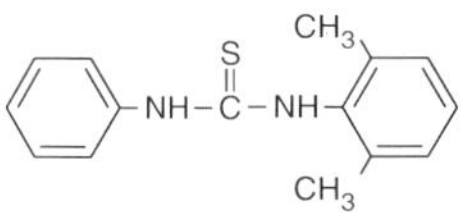
250 mg

S52,706-8
1-PHENYL-4-(2,6-XYLYL)-1H-1,2,4-TRIAZOLE-3,5(2H,4H)-DITHIONE
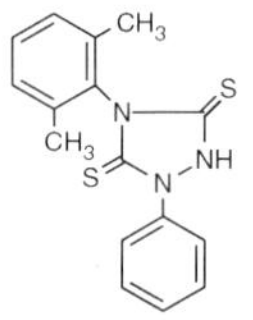
250 mg

S68,634-4
1-ALLYL-3-(DIPHENYLMETHYL)-2-THIOUREA
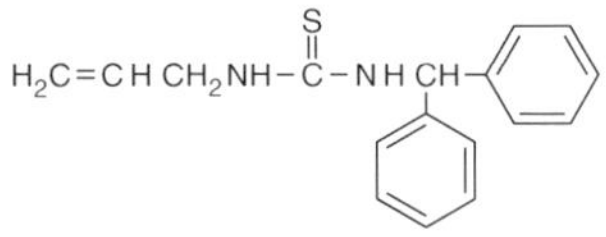
250 mg

S85,515-4
1-(2-NAPHTHYL)-3-PHENYL-2-THIOUREA
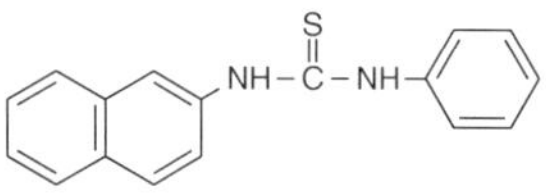
250 mg

S36,115-1
1,1-DIMETHYL-3-(ORTHO-TOLYL)-2-THIOUREA
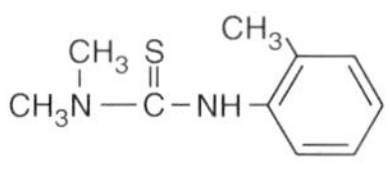
5 gm

S90,056-7
1-(2-ETHYLPHENYL)-3-METHYL-2-THIOUREA
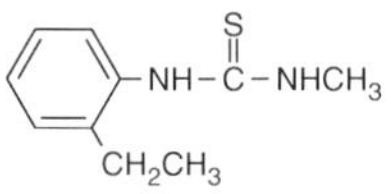
250 mg

S61,522-6
1-(2-CHLOROPHENYL)-3-PHENYL-2-THIOUREA
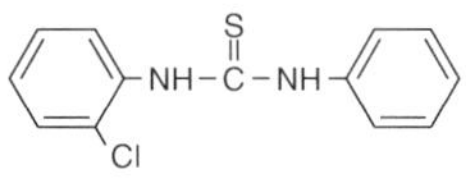
5 gm

S43,002-1
1,3-DI-O-TOLYL-2-THIOUREA
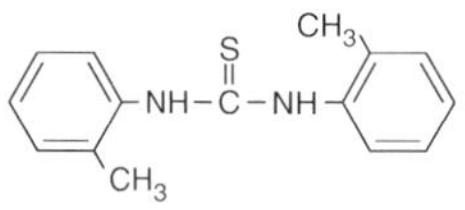
5 gm

S50,994-9
1-(4-CHLOROPHENYL)-2-THIOUREA
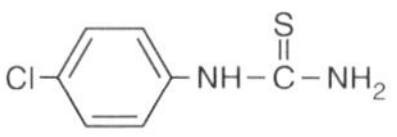
250 mg

S54,120-6
3-(4-CHLOROPHENYL)-1,1-DIMETHYL-2-THIOUREA
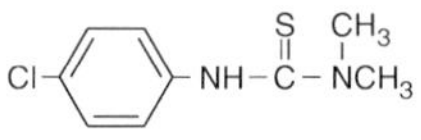
250 mg

S55,383-2
1-(4-CHLOROPHENYL)-3-ETHYL-2-THIOUREA
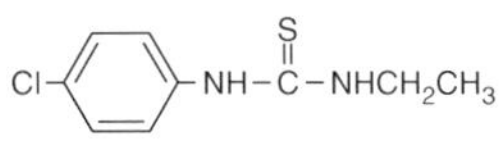
5 gm

S55,381-6
1-ETHYL-3-(P-TOLYL)-2-THIOUREA
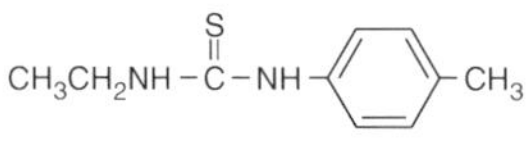
5 gm

S93,740-1
1-(4-TERT-BUTYLPHENYL)-2-THIOUREA
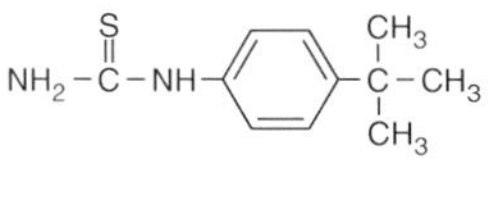
250 mg

S61,523-4
1-(4-CHLOROPHENYL)-3-PHENYL-2-THIOUREA
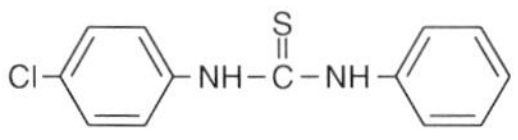
5 gm

S64,931-7
1-BUTYL-3-(4-ISOPROPYLPHENYL)-2-THIOUREA

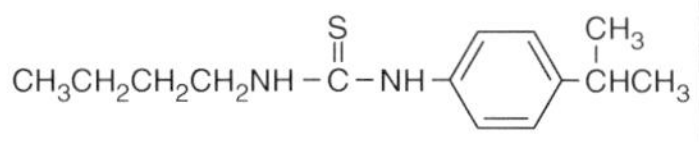

5 gm

S86,276-2
1-(2,6-XYLYL)-2-THIOUREA

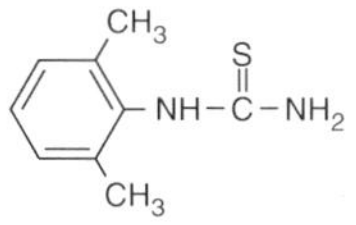

250 mg

S90,055-9
1-METHYL-3-(2,6-XYLYL)-2-THIOUREA

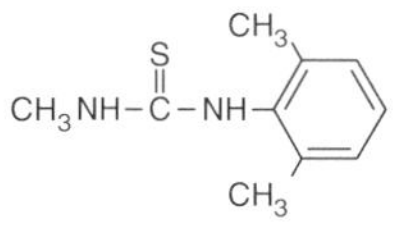

250 mg

S54,607-0
1,1-DIMETHYL-3-(2,6-XYLYL)-2-THIOUREA

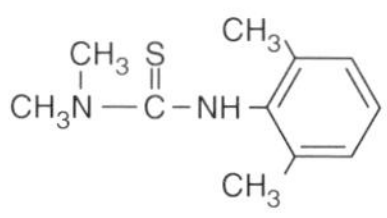

250 mg

S65,503-1
1-(2,4-DICHLOROPHENYL)-2-THIOUREA

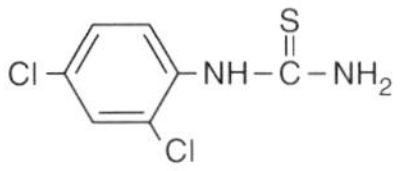

250 mg

S54,119-2
1,1-DIMETHYL-3-(3,4-XYLYL)-2-THIOUREA

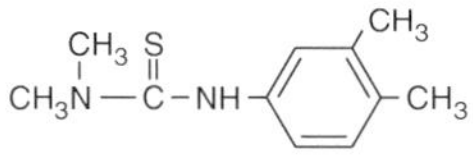

250 mg

S55,475-8
1-ETHYL-3-(3,4-XYLYL)-2-THIOUREA

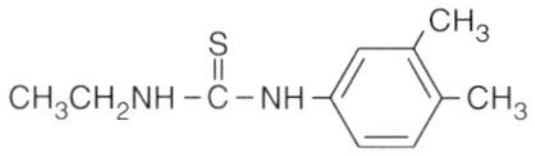

5 gm

S60,202-7
1-(2,5-DICHLOROPHENYL)-3-PHENYL-2-THIOUREA

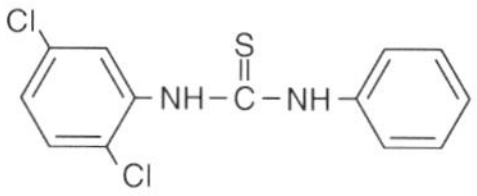

250 mg

S59,269-2
1,3-BIS(3,4-DICHLOROPHENYL)-2-THIOUREA

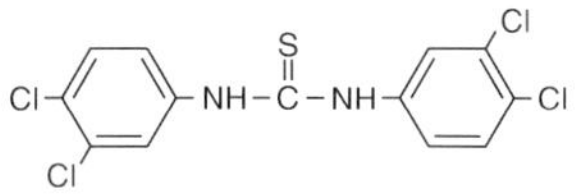

250 mg

S72,803-9
1,1'-(3,3'-DIMETHYLBIPHENYL-4,4'-DIYL)BIS[3-(2-METHYLPROPYL)-2-THIOUREA]

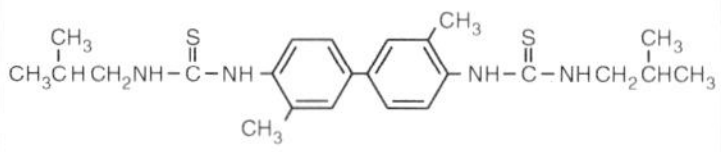

5 gm

S79,577-1
1-ALLYL-3-(1-NAPHTHYL)-2-THIOUREA

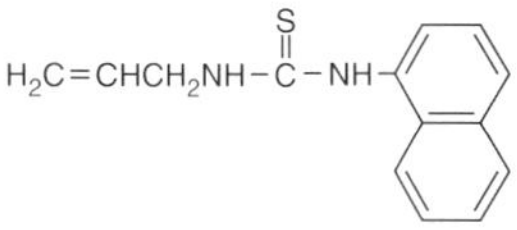

250 mg

S69,088-0
1-ISOPROPYL-3-(1-NAPHTHYL)-2-THIOUREA

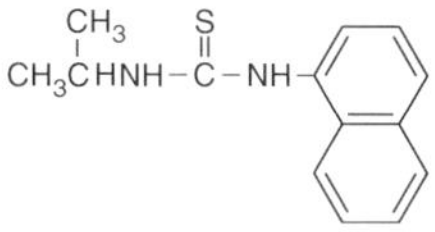

250 mg

S59,774-0
S-BUTYL N-(2,3-DICHLOROPHENYL) THIOCARBAMATE

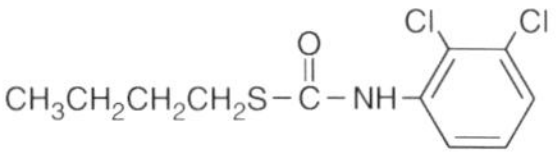

5 gm

S71,072-5
S-PROPYL N-(2-ETHYL-6-METHYLPHENYL) THIOLCARBAMATE

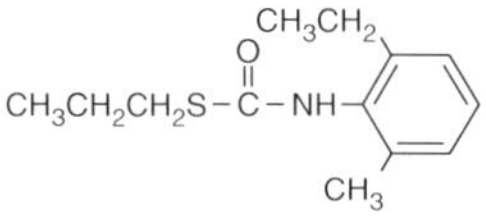

5 gm

S50,062-3
2-METHYL-2-THIOPSEUDOUREA P-TOLUENESULFONATE

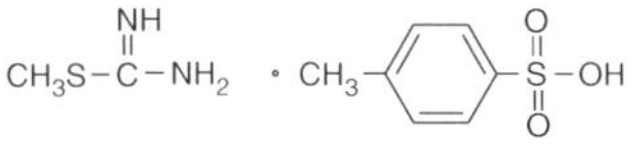

250 mg

S50,057-7
2-ETHYL-2-THIOPSEUDOUREA P-TOLUENESULFONATE

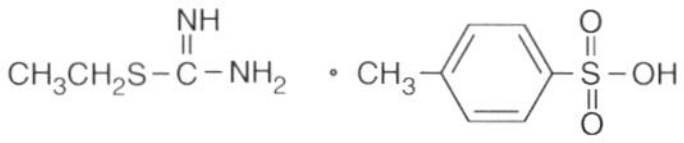

250 mg

S53,341-6
2-(3-CHLOROBENZYL)-2-THIOPSEUDOUREA HYDROCHLORIDE

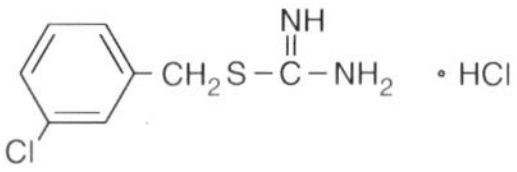

5 gm

S73,388-1
6,7-DIHYDRO-3-IMINO-1-PHENYL-3H,5H-IMIDAZO(1,2-C)THIAZOLE HYDROCHLORIDE

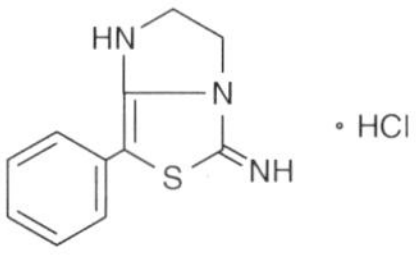

250 mg

S76,188-5
3-IMINO-2-PHENYL-2,3,6,7-TETRAHYDRO-5H-THIAZOLO(3,2-A) PYRIMIDINE HYDROCHLORIDE

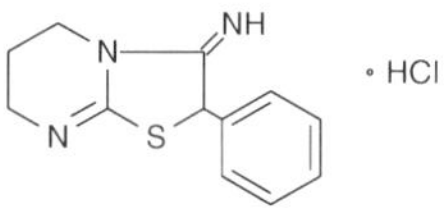

250 mg

S50,134-4
2,3-DIHYDRO-4,6-DIMETHYL-2-IMINOBENZOTHIAZOLE

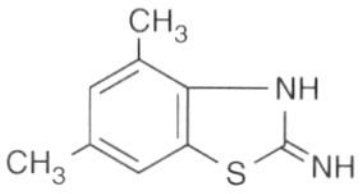

5 gm

S55,777-3
2-PENTAMETHYLBENZYL-3-PHENYL-2-THIOPSEUDOUREA HYDROCHLORIDE

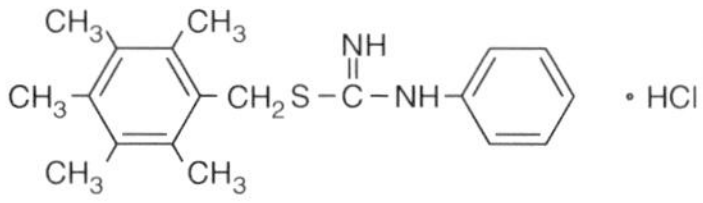

250 mg

S82,200-0
2-BENZHYDRYL-2-THIOPSEUDOUREA HYDROBROMIDE

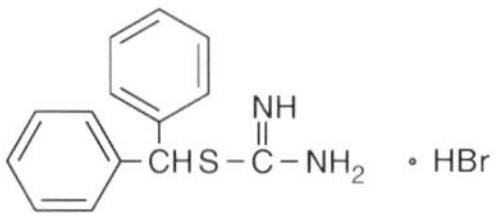

250 mg

S53,389-0
2-(4-METHYLBENZYL)-2-THIOPSEUDOUREA HYDROCHLORIDE

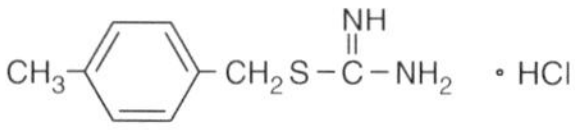

250 mg

S54,265-2
2,2'-(2,5-DIMETHYL-1,4-XYLYLENE) BIS(2-THIOPSEUDOUREA) DIHYDROCHLORIDE

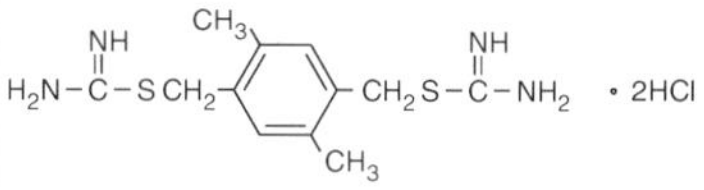

250 mg

S50,058-5
2,3-DIHYDRO-2-IMINO-3-METHYLNAPHTHO(1,2-D)THIAZOLE

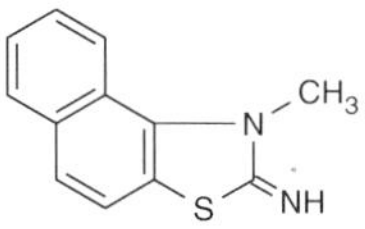

5 gm

S50,441-6
3-PHENYL-1,2,4-THIADIAZOLE-5(4H)-THIONE

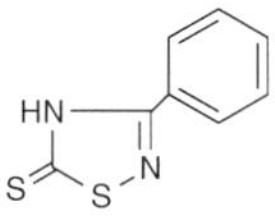

250 mg

S35,489-9
3-(PIPERIDINOMETHYL)BENZOTHIAZOLE-2(3H)-THIONE

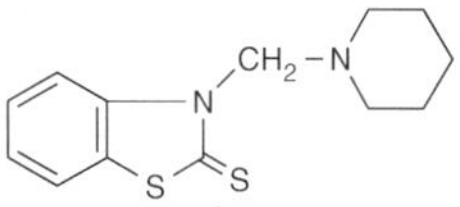

5 gm

S41,238-4
PHENYLHYDRAZINOCARBODITHIOIC ACID, AMMONIUM SALT

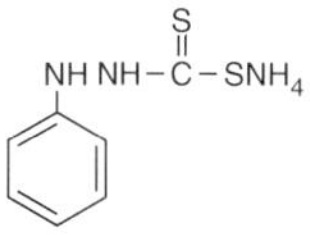

5 gm

S69,689-7
5-BENZYLIDENERHODANINE

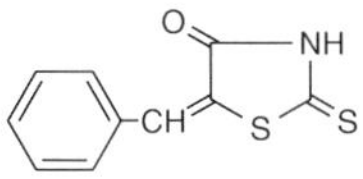

250 mg

S82,039-3
P-TOLYL N-PHENYLDITHIOCARBAMATE

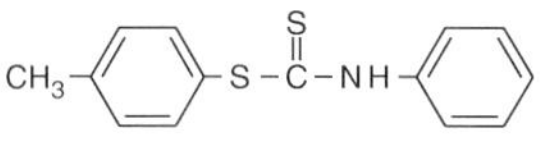

250 mg

S53,733-0
3-PHENYLBENZOTHIAZOLINE-2-THIONE

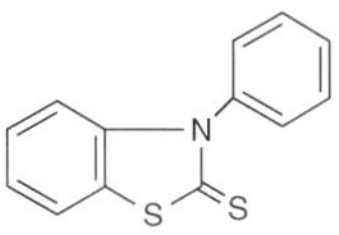

5 gm

S54,515-5
4-FLUORODITHIOCARBANILIC ACID, AMMONIUM SALT

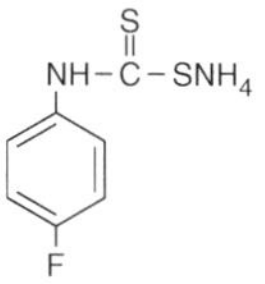

250 mg

S53,830-2
3-CHLORO-2-METHYLDITHIOCARBANILIC ACID, AMMONIUM SALT

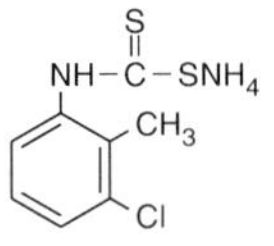

250 mg

S53,912-0
3,4-DIMETHYLDITHIOCARBANILIC ACID, AMMONIUM SALT

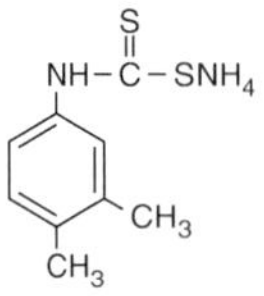

5 gm

S54,748-4
N-(1-NAPHTHYL)-DITHIOCARBAMIC ACID

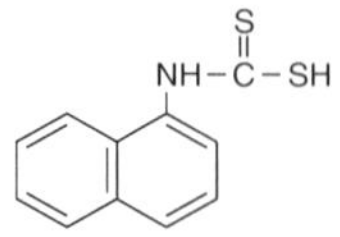

250 mg

S86,937-6
4-METHYL-1-(PHENYLSULFONYL)-3-THIOSEMICARBAZIDE

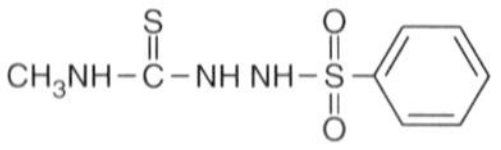

250 mg

S76,028-5
3-IMINO-2,3,5,6-TETRAHYDRO-2-(2,6-XYLYL)IMIDAZO(2,1-B)THIAZOLE HYDROCHLORIDE

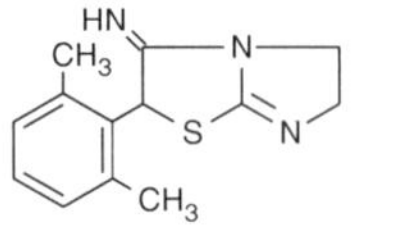

250 mg

S53,706-3
4-CYCLOHEXYL-1-PHENYL-3-THIOSEMICARBAZIDE

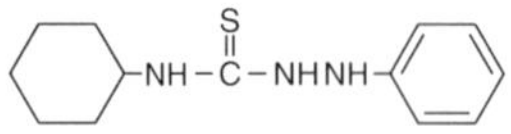

5 gm

S94,962-0
N-(CYCLOHEXYLTHIO)PHTHALIMIDE

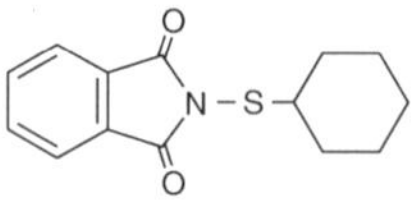

250 mg

S90,713-8
O-[4-(2,3-DIHYDRO-2,2,4-TRIMETHYL-4H-1-BENZOPYRAN-4-YL)PHENYL]-N,N-DIMETHYLTHIOCARBAMATE

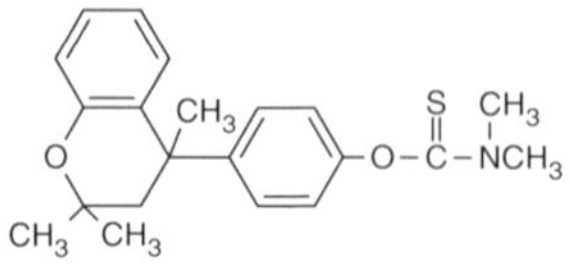

250 mg

S41,271-6
O-ETHYL THIOCARBANILATE

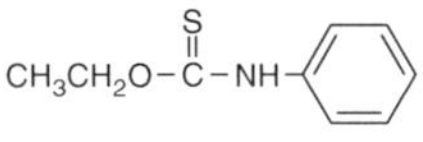

5 gm

S63,201-5
1,1-DIMETHYL-4-PHENYL-3-THIOSEMICARBAZIDE

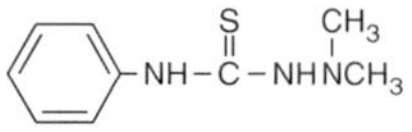

5 gm

S64,652-0
1,4-DIPHENYL-3-THIOSEMICARBAZIDE

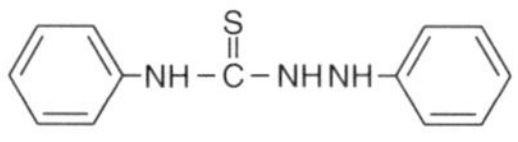

5 gm

S55,387-5
1,9-DIPHENYL-2,5,8-TRITHIOTRIUREA

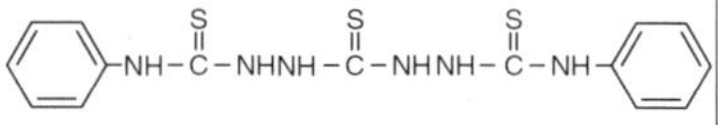

5 gm

S94,963-9
N-(BENZYLTHIO)PHTHALIMIDE

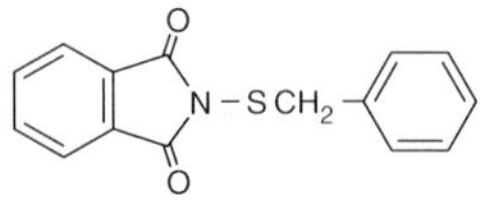

50 mg

S52,334-8
1-PHENYL-4-(ORTHO-TOLYL)-3-THIOSEMICARBAZIDE

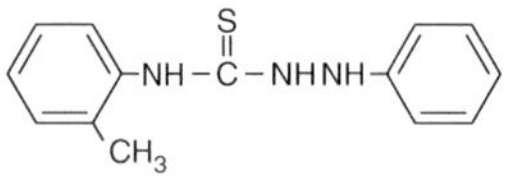

5 gm

S52,438-7
1-(2,5-DICHLOROPHENYL)-4-(ORTHO-TOLYL)-3-THIOSEMICARBAZIDE

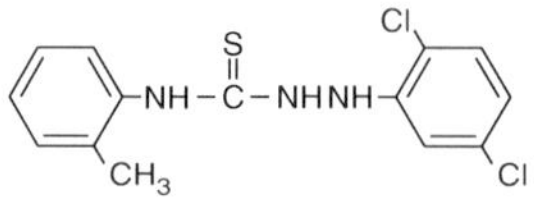

250 mg

S52,720-3
4-(4-CHLOROPHENYL)-1-PHENYL-3-THIOSEMICARBAZIDE

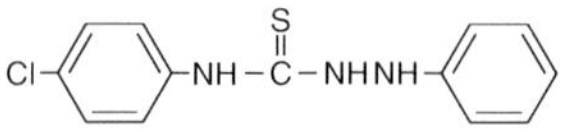

250 mg

S52,337-2
4-(3-CHLORO-ORTHO-TOLYL)-3-THIOSEMICARBAZIDE

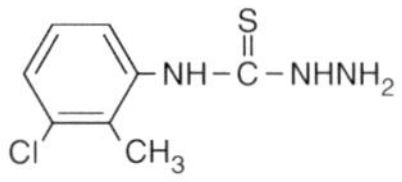

5 gm

S93,476-3
4-(3-CHLORO-4-FLUOROPHENYL)-3-THIOSEMICARBAZIDE

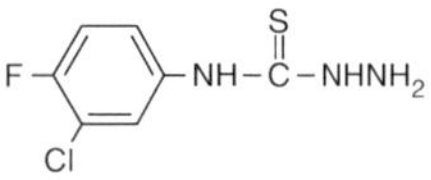

250 mg

S53,825-6
4-(4-CHLORO-ORTHO-TOLYL)-3-THIOSEMICARBAZIDE

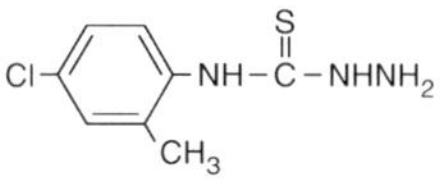

250 mg

S54,135-4
4-(3,4-XYLYL)-3-THIOSEMICARBAZIDE

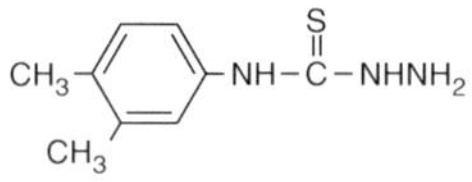

250 mg

S54,144-3
1-ANILINO-6-(3,4-XYLYL)-2,5-DITHIOBIUREA

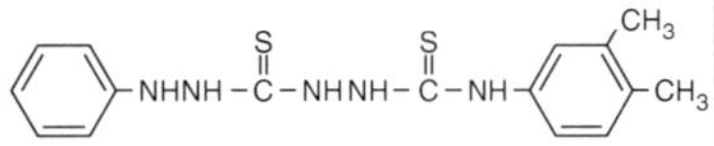

250 mg

S52,554-5
4-(4-CHLOROBUTYRYL)-4'-METHOXYDIPHENYLACETONITRILE
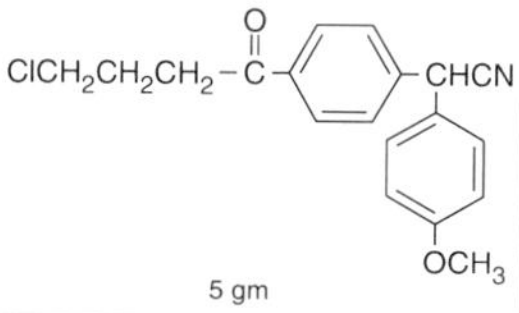
5 gm

S52,553-7
4-BROMO-2-[4-(CYCLOPROPYLFORMYL)-PHENYL]-2-(4-FLUOROPHENYL)-BUTYRONITRILE
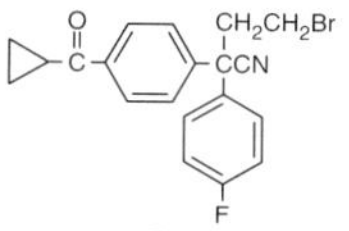
5 gm

S54,873-1
2-(2-CHLOROETHYL)-2-PHENYLPENTANENITRILE
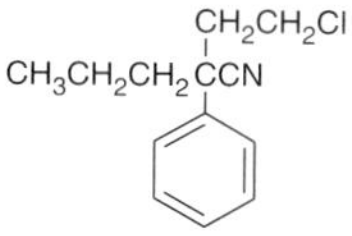
250 mg

S88,724-2
5-CHLORO-2-ISOPROPYL-2-PHENYLVALERONITRILE
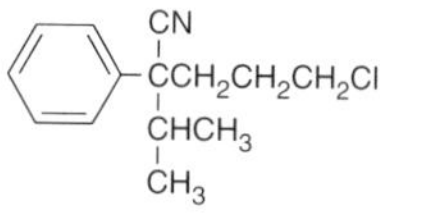
250 mg

S88,669-6
ALPHA-(3-CHLOROPROPYL)-BETA-ETHYL-ALPHA-PHENYLHYDROCINNAMONITRILE
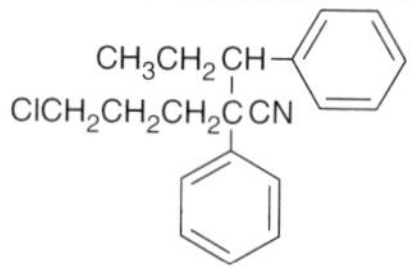
250 mg

S34,419-2
4-CHLORO-2,2-DIPHENYL-3-METHYLBUTYRONITRILE
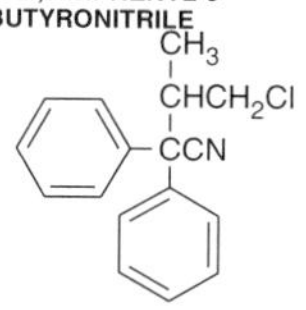
5 gm

S64,915-5
3-[2-(2-CHLOROACETAMIDO)PHENYLTHIO]-PROPIONITRILE
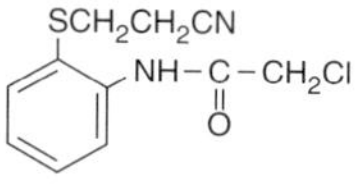
5 gm

S52,555-3
5-CHLORO-2-(ORTHO-TOLYL)-VALERONITRILE
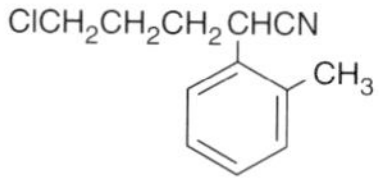
250 mg

S91,300-6
4-[2-CHLORO-5-(TRIFLUOROMETHYL)-PHENYL]-1-(2-CYANOACETYL)-SEMICARBAZIDE
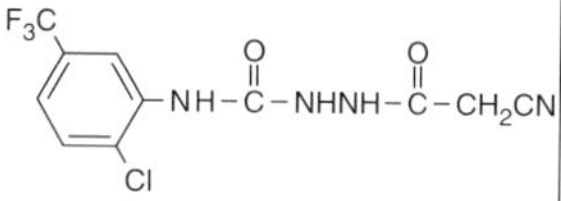
250 mg

S85,858-7
N-CYANO-2,6-DINITRO-4-(TRIFLUOROMETHYL)ANILINE
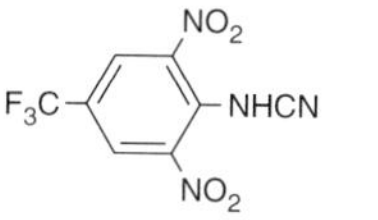
250 mg

S85,859-5
N-CYANO-2,6-DINITRO-N-METHYL-4-(TRIFLUOROMETHYL)ANILINE
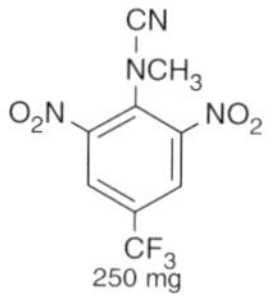
250 mg

S85,860-9
N-CYANO-2,6-DINITRO-N-PROPYL-4-(TRIFLUOROMETHYL)ANILINE
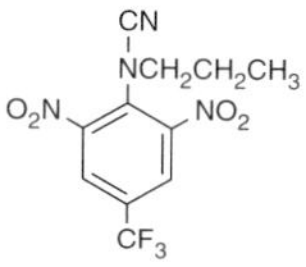
250 mg

S64,907-4
3-[3,5-BIS(TRIFLUOROMETHYL)ANILINO]-PROPIONITRILE
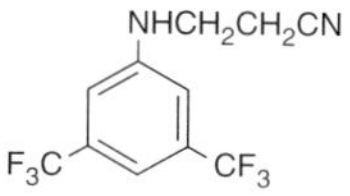
5 gm

S77,252-6
1-BENZYL-1-(2-CYANOETHYL)-3-[3-(TRIFLUOROMETHYL)PHENYL]UREA
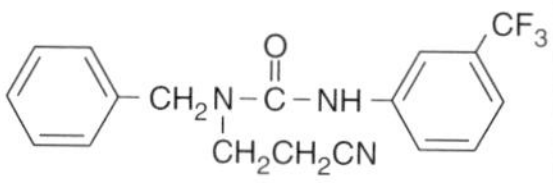
250 mg

S77,244-5
1-(2-CYANOETHYL)-1-PROPYL-3-[3-(TRIFLUOROMETHYL)PHENYL]UREA
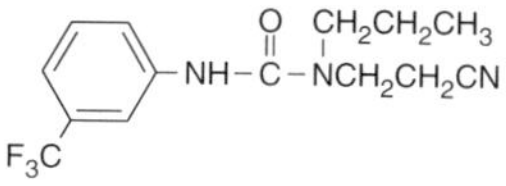
250 mg

S77,949-0
1-BUTYL-1-(2-CYANOETHYL)-3-[3-(TRIFLUOROMETHYL)PHENYL]UREA
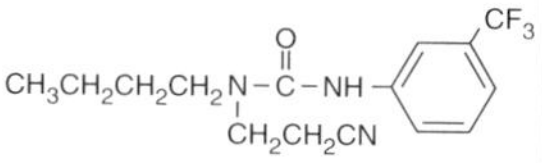
250 mg

S52,964-8
2,2,2-TRICHLOROETHYL 4-CYANO-4-PHENYL-1-PIPERIDINEPROPIONATE HYDROCHLORIDE
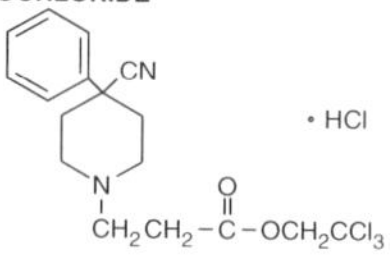
250 mg

S91,329-4
1-(2-CYANOACETYL)-4-[3-(TRIFLUOROMETHYL)PHENYL]-SEMICARBAZIDE
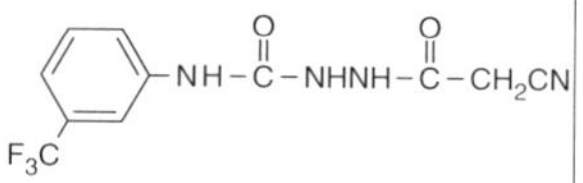
250 mg

S91,323-5
1-(2-CYANOACETYL)-4-[2-(TRIFLUOROMETHYL)PHENYL]-SEMICARBAZIDE
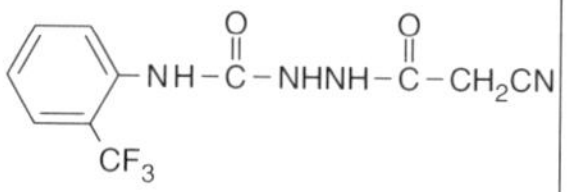
250 mg

S78,046-4
5-CYANO-6-(2-FLUOROPHENYL)-4-HYDROXY-3,4,5,6-TETRAHYDRO-2(1H)-PYRIMIDINETHIONE POTASSIUM SALT
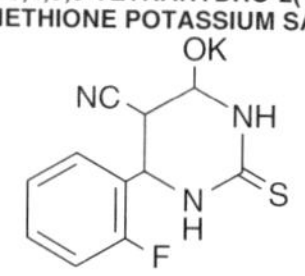
250 mg

S74,165-5
2-IODO-4,5-DIMETHOXYPHENYLACETONITRILE
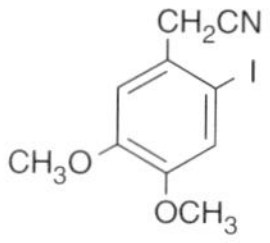
250 mg

S88,728-5
4-CHLORO-ALPHA-(4-METHOXYPHENYL)-HYDROCINNAMONITRILE
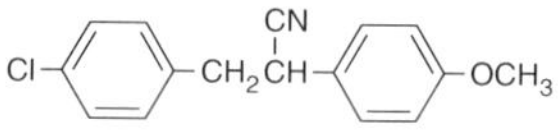
250 mg

S88,673-4
ALPHA,BETA-BIS(4-CHLOROPHENYL)-4-METHOXYHYDROCINNAMONITRILE
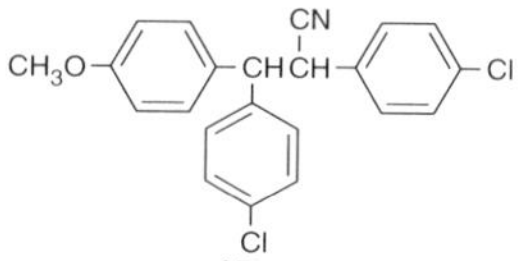
250 mg

S91,263-8
4-(5-CHLORO-2-METHOXYPHENYL)-1-(2-CYANOACETYL)SEMICARBAZIDE
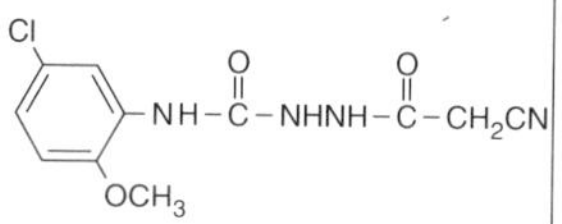
250 mg

S67,591-1
4-CHLOROPHENOXYACETONITRILE
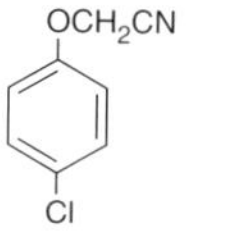
5 gm

S72,416-5
2,4-DICHLOROPHENOXYACETONITRILE
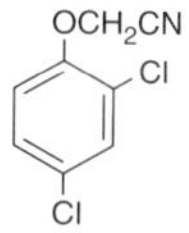
5 gm

S67,585-7
2,4,5-TRICHLOROPHENOXYACETONITRILE
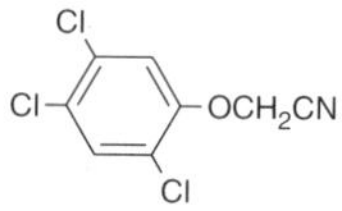
5 gm

S72,769-5
1-(4-CHLORO-2-CYANOPHENYL)-3-[4-(METHYLTHIO)PHENYL]UREA
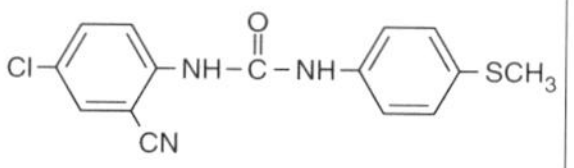
5 gm

S76,039-0
1-(4-CHLOROBENZOYL)-2-CYANO-1,2-DIHYDROQUINOLINE
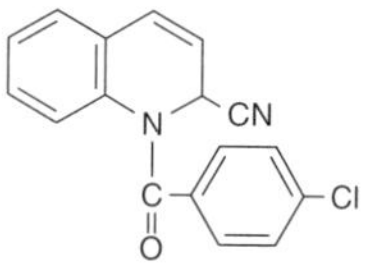
250 mg

S74,287-2
3-(4-CHLOROANILINO)PROPIONITRILE
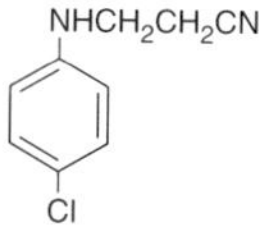
5 gm

S74,573-1
3-(4-BROMOANILINO)PROPIONITRILE
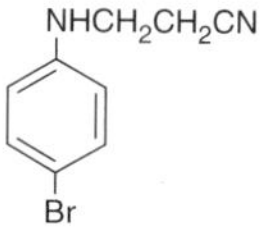
250 mg

S74,232-5
3,3'-(4-CHLOROPHENYLIMINO) DIPROPIONITRILE
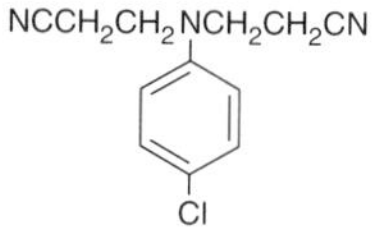
5 gm

S92,420-2
1-(4-CHLORO-3-NITROPHENYL)-3-(2-CYANOETHYL)-3-METHYLUREA
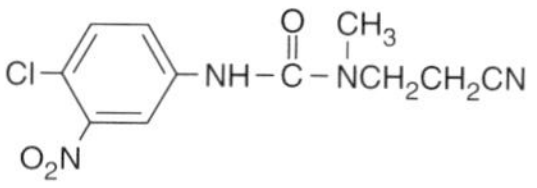
250 mg

S75,099-9
N-(2-CYANOACETAMIDO)-3,4,5,6-TETRACHLOROPHTHALAMIC ACID
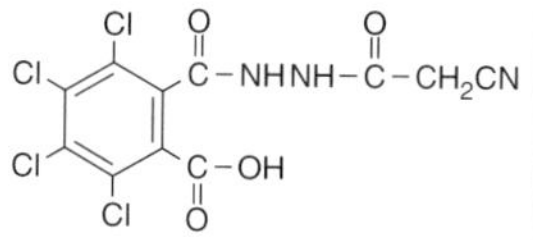
5 gm

S76,991-6
1-(2-CYANOETHYL)-3-(4-FLUOROPHENYL)-1-METHYLUREA
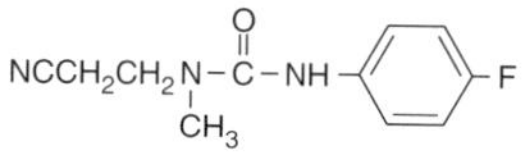
250 mg

S38,342-2
ALPHA-(4-CHLOROPHENYL)-HYDROCINNAMONITRILE
5 gm

S88,649-1
4-CHLORO-ALPHA-PHENYLHYDROCINNAMONITRILE
250 mg

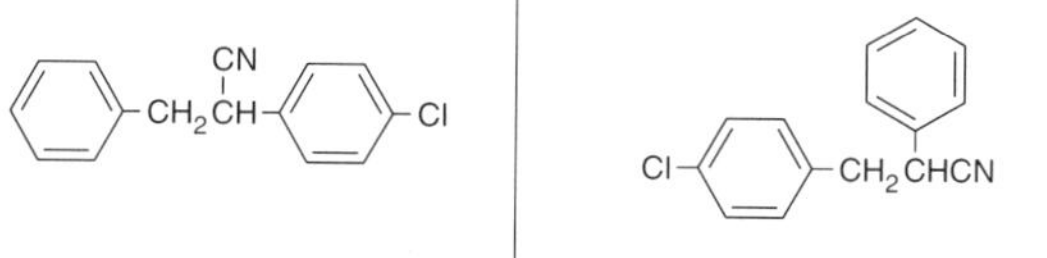

S88,651-3
4-CHLORO-BETA-METHYL-ALPHA-PHENYLHYDROCINNAMONITRILE
250 mg

S88,656-4
4-CHLORO-BETA-ETHYL-ALPHA-PHENYLHYDROCINNAMONITRILE
250 mg

S88,661-0
BETA-TERT-BUTYL-ALPHA-(4-CHLOROPHENYL)HYDROCINNAMONITRILE
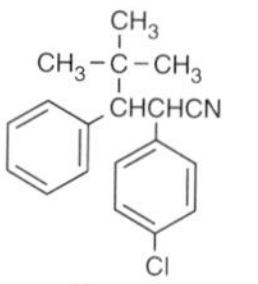
250 mg

S88,645-9
(4-CHLOROPHENYL)PHENYLACETONITRILE
250 mg

S88,644-0
(4-BROMOPHENYL)PHENYLACETONITRILE
250 mg

S45,390-0
4-CHLORO-ALPHA-METHYLPHENYLACETONITRILE
5 gm

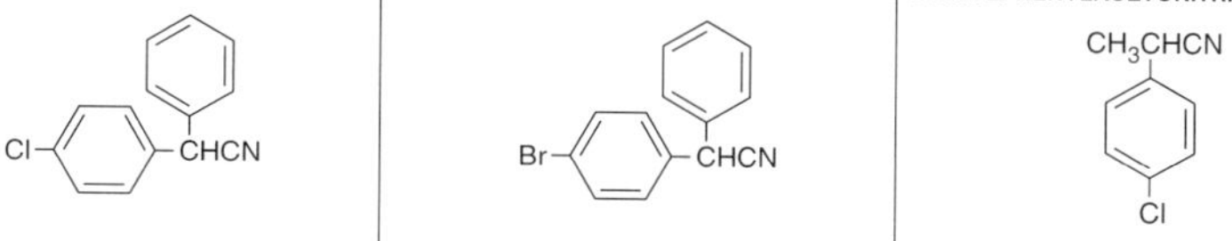

S87,943-6
4-CHLORO-ALPHA-(4-CHLOROPHENYL)-HYDROCINNAMONITRILE
5 gm

S88,664-5
2-(4-CHLOROPHENYL)-3-(1-NAPHTHYL-PROPIONITRILE)
250 mg

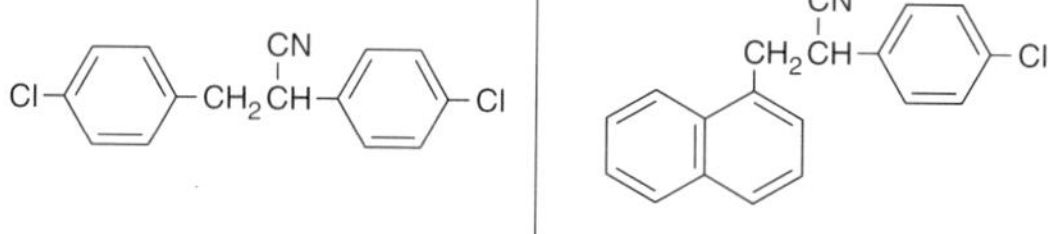

S69,321-9
1-(CYANOACETYL)-4-(2,3-DICHLOROPHENYL)SEMICARBAZIDE
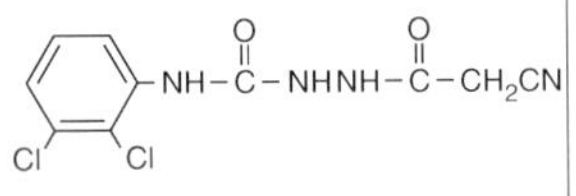
250 mg

S69,092-9
1-(CYANOACETYL)-4-(3,4-DICHLOROPHENYL)SEMICARBAZIDE
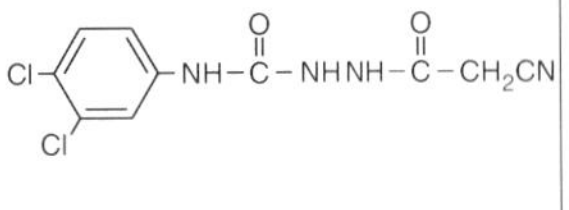
250 mg

S85,815-3
BUTYL N-(4-CHLOROPHENYL)-N-CYANOCARBAMATE
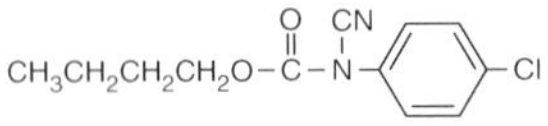
250 mg

S92,953-0
3-(4-CHLOROPHENYL)-3-OXOPROPIONITRILE
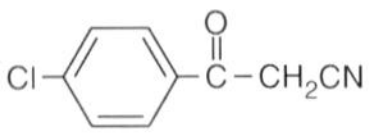
250 mg

S75,891-4
6-(4-CHLOROBENZYL)-5-OXO-6,7,8,9-TETRAHYDRO-5H-BENZOCYCLOHEPTENE-6-PROPIONITRILE
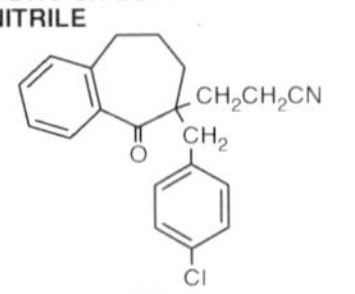
250 mg

S89,333-1
3-(2-CHLOROPHENYL)-2-CYANOPROPIONIC ACID
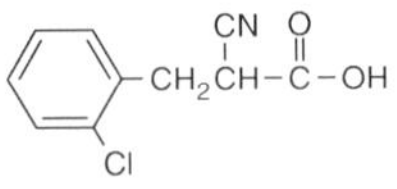
250 mg

S78,111-8
2-CYANOACETYL N-(4-FLUOROPHENYL)-CARBAMATE
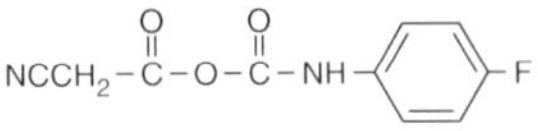
250 mg

S85,807-2
METHYL N-(3-CHLOROPHENYL)-N-CYANOCARBAMATE
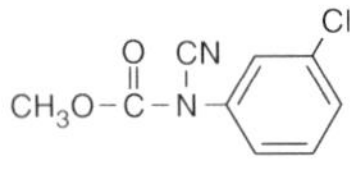
250 mg

S85,810-2
ETHYL N-(4-CHLOROPHENYL)-N-CYANOCARBAMATE
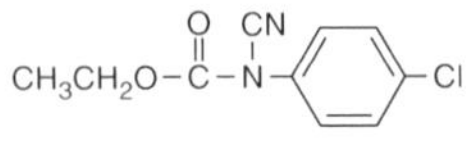
250 mg

S85,811-0
ISOPROPYL N-(4-CHLOROPHENYL)-N-CYANOCARBAMATE
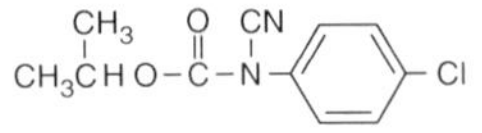
250 mg

S85,814-5
PROPYL N-(4-CHLOROPHENYL)-N-CYANOCARBAMATE
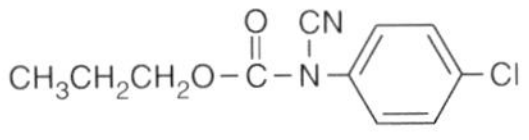
250 mg

S85,816-1
PROPYL N-CYANO-N-(3,4-DICHLOROPHENYL)CARBAMATE
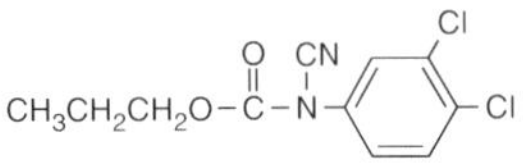
250 mg

S85,819-6
BUTYL N-CYANO-N-(3,4-DICHLOROPHENYL)CARBAMATE
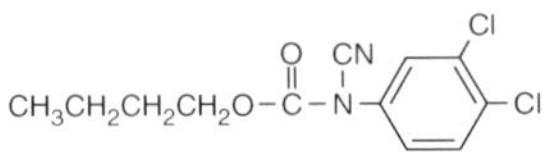
250 mg

S89,341-2
METHYL 2,2-BIS(2-CHLOROBENZYL)-2-CYANOACETATE
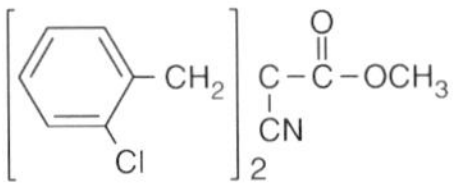
250 mg

S77,257-7
2-CYANOACETYL N-(2,3-DICHLOROPHENYL)CARBAMATE
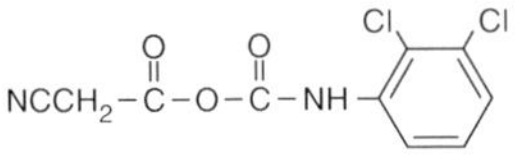
250 mg

S91,313-8
4-(3-CHLORO-4-METHYLPHENYL)-1-(2-CYANOACETYL)SEMICARBAZIDE
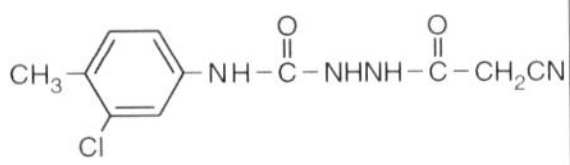
250 mg

S68,657-3
1-(3-CHLOROPHENYL)-3-[3-(2-CYANOETHYL)PHENYL]UREA
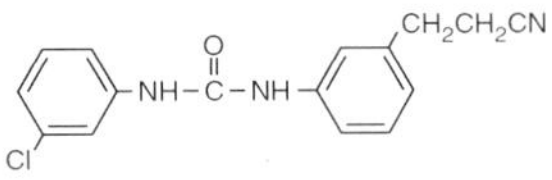
5 gm

S85,653-3
1-(4-CHLOROPHENYL)-3-CYANOUREA
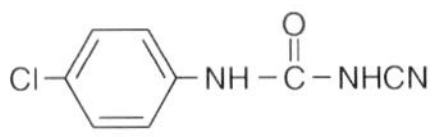
250 mg

S85,599-5
1-CYANO-1-(3,4-DICHLOROPHENYL)-3,3-DIMETHYLUREA
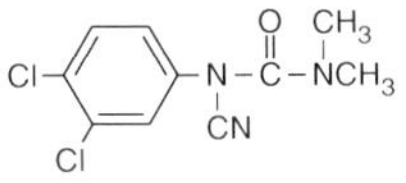
100 mg

S89,323-4
BIS(2-CHLOROBENZYL)ACETONITRILE
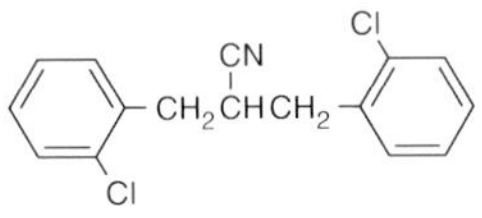
250 mg

S87,940-1
ALPHA,BETA-BIS(4-CHLOROPHENYL)-VALERONITRILE
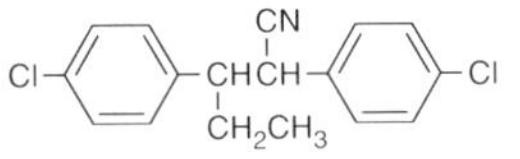
5 gm

S78,047-2
5-CYANO-4-HYDROXY-3,4,5,6-4H-6-(3,4,5-TRIMETHOXYPHENYL)-2-(1H)-PYRIMIDINETHIONE POTASSIUM SALT
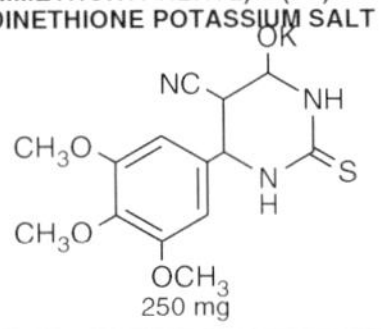
250 mg

S66,073-6
N1-ACETYL-N3-(2-CYANOETHYL)-N3-(2-HYDROXYETHYL)-1,3-PHENYLENEDIAMINE
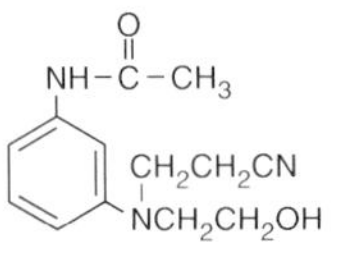
5 gm

S53,551-6
1-BENZYL-4-HYDROXYISONIPECOTONITRILE
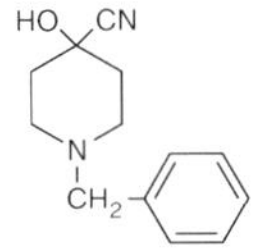
250 mg

S78,048-0
5-CYANO-6-[4-(DIMETHYLAMINO)PHENYL]-4-HYDROXY-3,4,5,6-4H-2(1H)-PYRIMIDINETHIONE POTASSIUM SALT
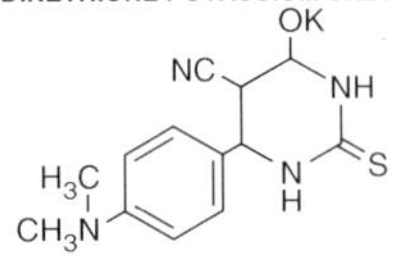
250 mg

S35,976-9
1-CYANO-1,2-DIHYDRO-2-HYDROXYQUINOLINE
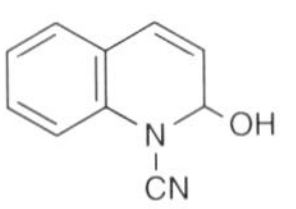
5 gm

S39,040-2
2-BENZOYL-1,2-DIHYDRO-1-ISOQUINOLINECARBONITRILE
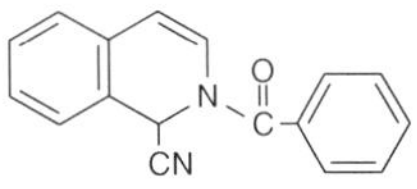
250 mg

S84,336-9
2-CYANO-3-(2-HYDROXYPHENYL)-PROPIONIC ACID
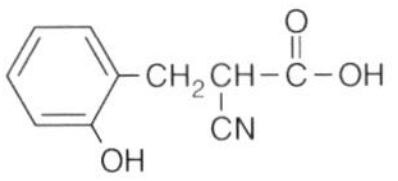
250 mg

S52,609-6
4-HYDROXY-ALPHA-METHYLHYDROCINNAMONITRILE
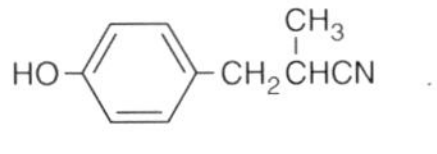
5 gm

S55,091-4
ALPHA-4-DIACETOXY-3-METHOXYPHENYLACETONITRILE
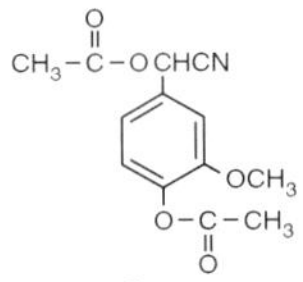
5 gm

S89,605-5
4-BENZYLOXY-1-CYANO-1-CYCLOHEXANEACETONITRILE
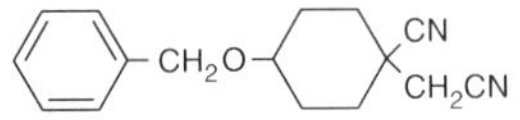
250 mg

S35,230-6
4-METHOXY-4-PHENYLBUTYRONITRILE
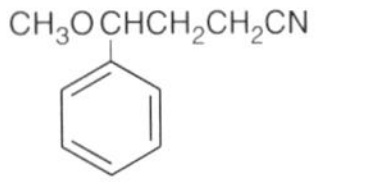
5 gm

S34,800-7
4-ETHOXY-4-PHENYLBUTYRONITRILE
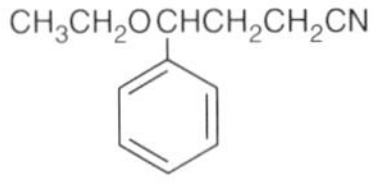
5 gm

S35,167-9
4-ISOPROPOXY-4-PHENYLBUTYRONITRILE
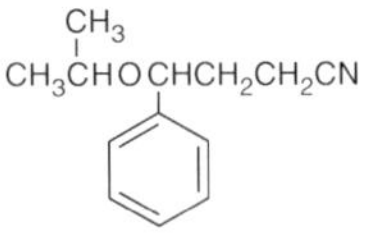
5 gm

S34,447-8
4-(4-CHLOROPHENYL)-4-METHOXYBUTYRONITRILE
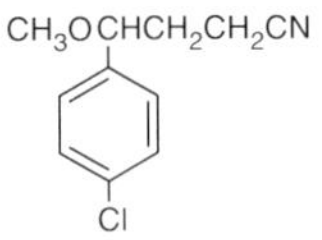
5 gm

S34,444-3
4-(4-CHLOROPHENYL)-4-ETHOXYBUTYRONITRILE
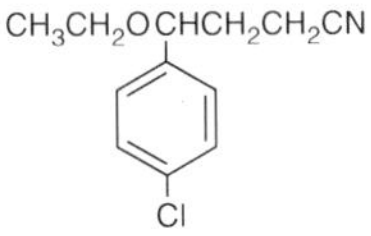
5 gm

S34,803-1
4-ETHOXY-4-(P-TOLYL)-BUTYRONITRILE
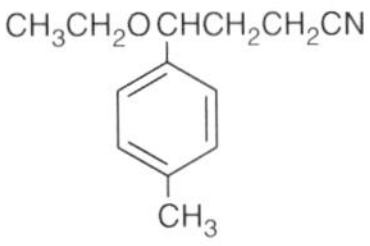
5 gm

S34,445-1
4-(4-CHLOROPHENYL)-4-ISOPROPOXYBUTYRONITRILE
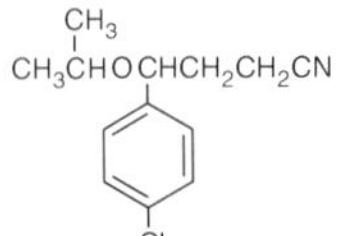
5 gm

S74,032-2
2-(2,4,5-TRIMETHOXYPHENYL)-SUCCINONITRILE
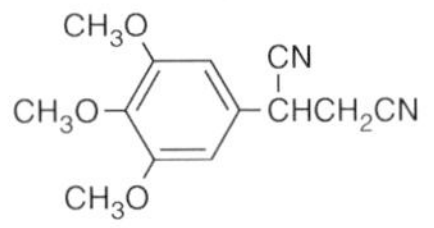
250 mg

S95,394-6
3-(3,4,5-TRIMETHOXYPHENYL)-PROPIONITRILE
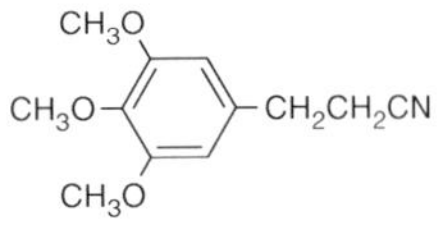
50 mg

S74,570-7
4,5-DIMETHOXY-2-NITROPHENYLACETONITRILE
250 mg

S84,954-5
(3,5-DIMETHOXYPHENYL)ACETONITRILE
250 mg

S92,954-9
3-(3,5-DIMETHOXYPHENYL)-3-OXOPROPIONITRILE
250 mg

S85,048-9
1-(2-CYANOETHYL)-1-CYCLOHEXYL-3-(2,5-DIMETHOXYPHENYL)UREA
250 mg

S93,426-7
2,5-DIMETHOXY-4-NITROPHENYLACETONITRILE
250 mg

S88,657-2
4-METHOXY-ALPHA-(4-METHOXYPHENYL)-HYDROCINNAMONITRILE
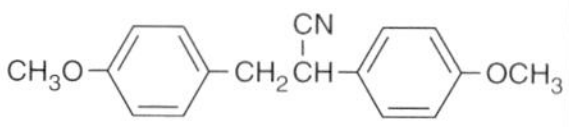
250 mg

S93,425-9
3,5-DIMETHOXYPHENYLACETONITRILE
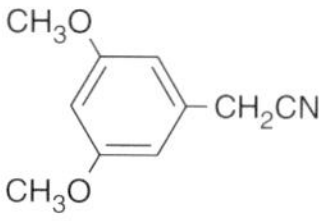
250 mg

S90,374-4
1-(2-CYANOETHYL)-3-(2,5-DIMETHOXYPHENYL)-1-METHYLUREA
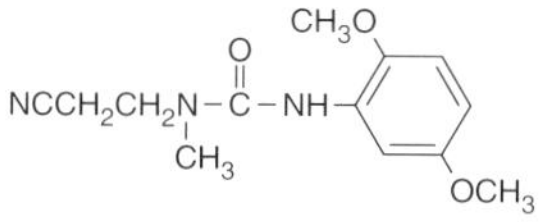
250 mg

S65,819-7
N-(ALPHA-CYANO-3,4-DICHLOROBENZYL)-3,4-DIMETHOXYPHENETHYLAMINE HYDROCHLORIDE
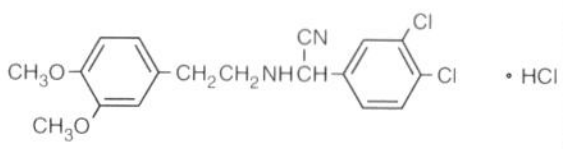
250 mg

S65,852-9
N-(ALPHA-CYANO-3,4,5-TRIMETHOXYBENZYL)-3,4-DIMETHOXYPHENETHYLAMINE HYDROCHLORIDE
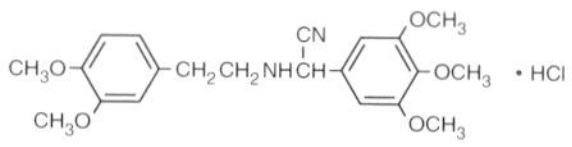
5 gm

S65,806-5
1-CYANO-N-(3,4-DIMETHOXYPHENETHYL)-CYCLOHEXYLAMINE
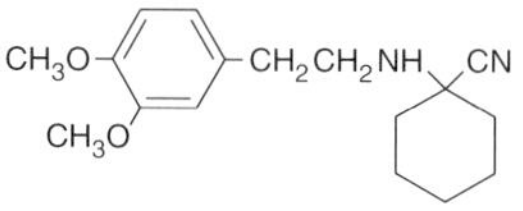
250 mg

S63,576-6
2-[N-(3,4-DIMETHOXYBENZYL)AMINO]-2-PHENYLACETONITRILE
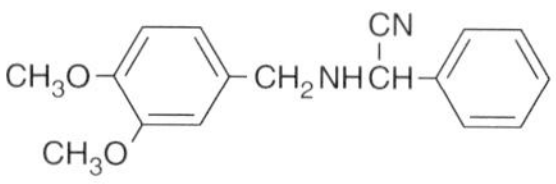
250 mg

S41,810-2
(4-AMINOPHENYL)-(PARA-ANISYL)-ACETONITRILE HYDROCHLORIDE
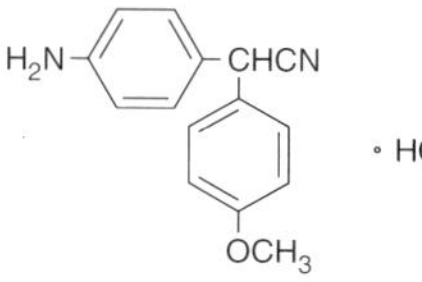
5 gm

S35,251-9
ALPHA-(4-METHOXYPHENYL)-1-PYRROLIDINEACETONITRILE
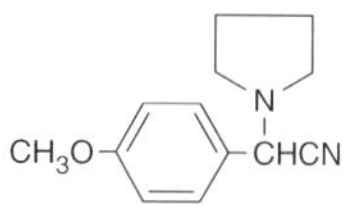
5 gm

S35,246-2
ALPHA-(4-METHOXYPHENYL)-1-PIPERIDINEACETONITRILE
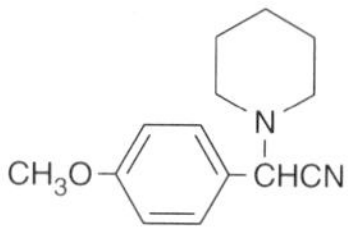
5 gm

S35,238-1
ALPHA-(4-METHOXYPHENYL)-4-MORPHOLINEACETONITRILE
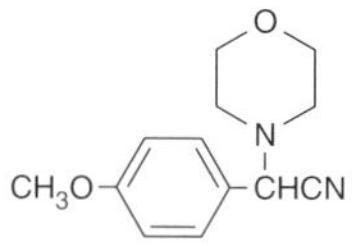
5 gm

S74,438-7
N-(2-METHOXY-5-METHYLPHENYL)-CYANOTHIOFORMAMIDE
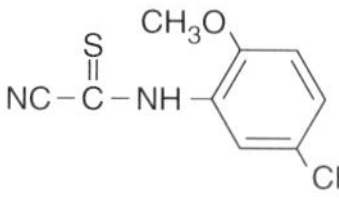
250 mg

S74,446-8
N-BENZOYL-N-(4-METHOXYPHENYL)-CYANOTHIOFORMAMIDE
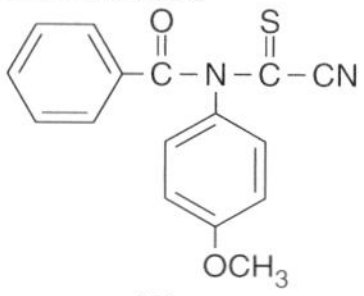
250 mg

S55,374-3
N-[2-(DODECYLOXY)PHENYL]-GLYCINONITRILE
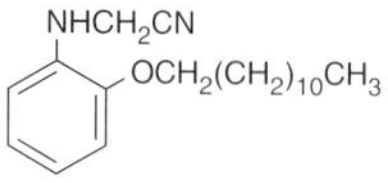
250 mg

S62,286-9
2',4'-DIMETHOXY-2-CYANOACETANILIDE
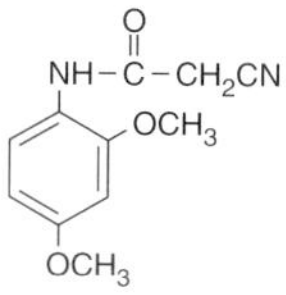
250 mg

S79,623-9
1-(2-CYANOETHYL)-3-(2-METHOXY-5-METHYLPHENYL)-1-METHYLUREA
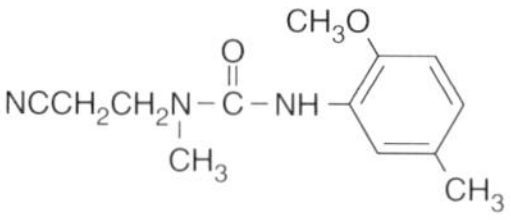
250 mg

S66,090-6
2-(P-ANISYL)-3-PHENYLSUCCINONITRILE
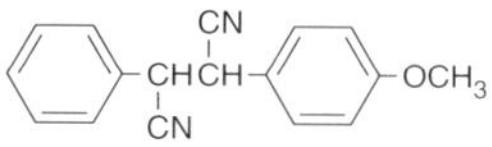
5 gm

S52,558-8
4-CYCLOPROPYLFORMYL-4'-METHOXYDIPHENYLACETONITRILE
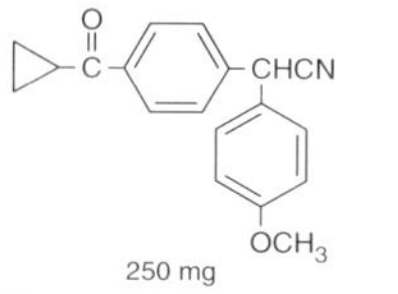
250 mg

S80,923-3
ALPHA-ACETYL-2-METHOXYPHENYLACETONITRILE
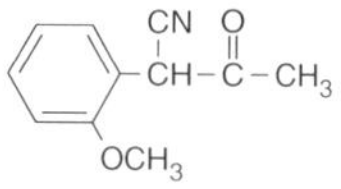
250 mg

S45,042-1
1-(4-METHOXYPHENYL)-1-CYCLOPROPANECARBONITRILE
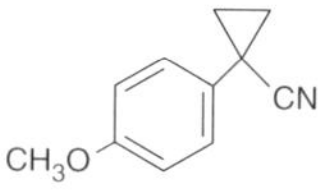
5 gm

S44,819-2
1-(4-METHOXYPHENYL)-1-CYCLOPENTANECARBONITRILE
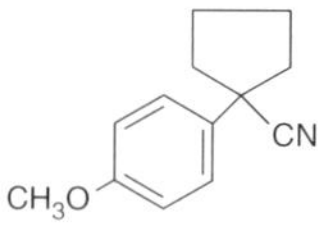
5 gm

S45,045-6
1-(4-METHOXYPHENYL)-1-CYCLOHEXANECARBONITRILE
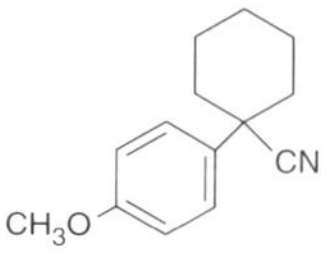
5 gm

S88,660-2
BETA-ETHYL-4-METHOXY-ALPHA-PHENYLHYDROCINNAMONITRILE
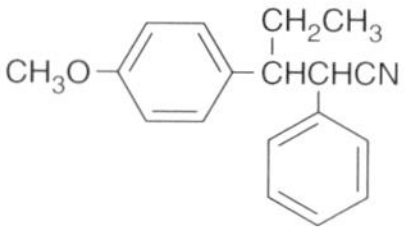
250 mg

S53,961-9
4-METHOXY-ALPHA-(4-METHOXYPHENYL)-HYDROCINNAMONITRILE
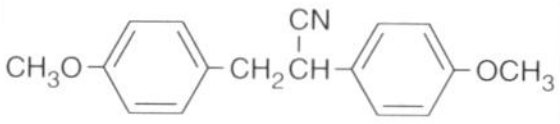
250 mg

S88,665-3
2-(4-METHOXYPHENYL)-3-(1-NAPHTHYL)-PROPIONITRILE
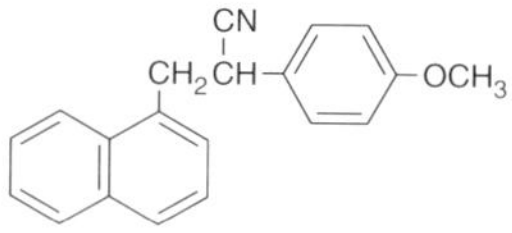
250 mg

S92,097-5
3,5-DIMETHOXYBENZYL CYANIDE
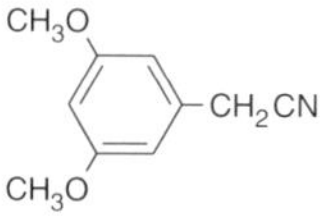
250 mg

S93,427-5
2-CHLORO-5-METHOXY-4-NITROPHENYLACETONITRILE
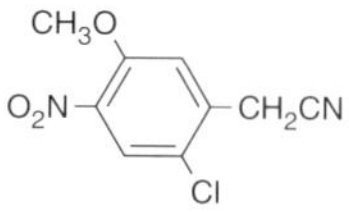
250 mg

S38,411-9
4-(META-ANISOYL)-4-PHENYLBUTYRONITRILE
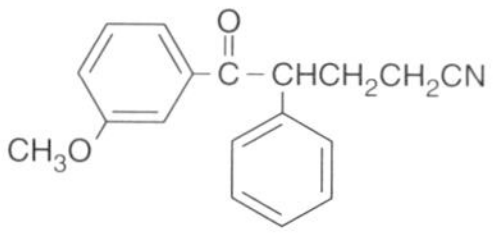
5 gm

S92,956-5
3-(4-METHOXYPHENYL)-3-OXOPROPIONITRILE
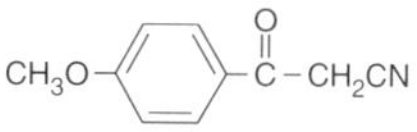
250 mg

S93,520-4
3-(4-METHOXYBENZOYL)PROPIONITRILE
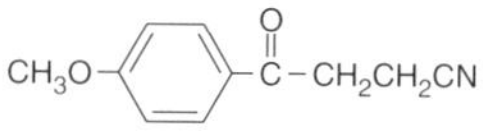
250 mg

S94,392-4
4-CYANO-4-(3-METHOXYPHENYL)-VALERIC ACID
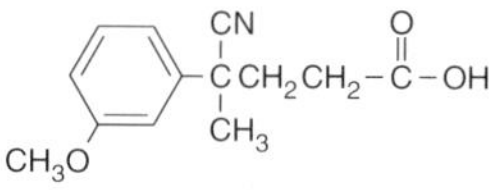
250 mg

S76,935-5
2-CYANOACETYL N-(4-METHOXYPHENYL)-CARBAMATE
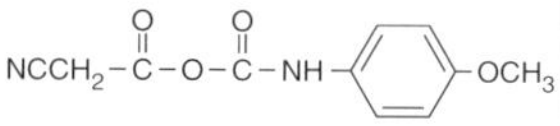
250 mg

S88,707-2
ETHYL 4-CYANO-4-(4-METHOXYPHENYL)-3-PHENYLBUTYRATE
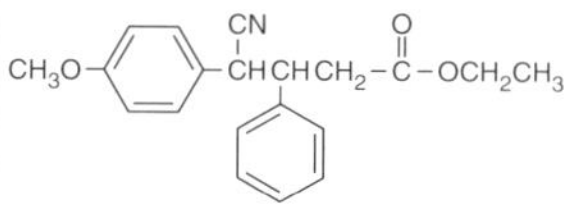
250 mg

S77,254-2
2-CYANOACETYL N-(3-METHOXYPHENYL)-CARBAMATE
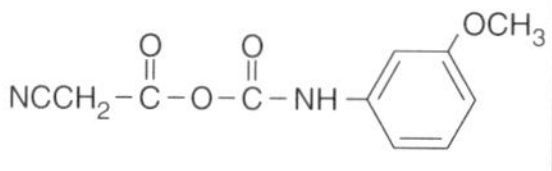
250 mg

S76,284-9
3',5'-DICYANO-3,4-DIHYDROSPIRO [NAPHTHALENE-2(1H),4'-PIPERIDINE]-2',6'-DIONE
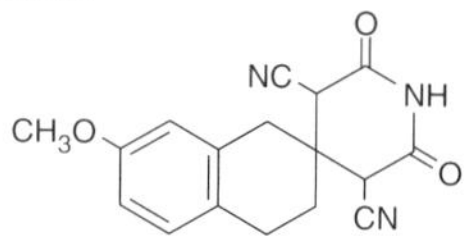
250 mg

S73,494-2
3-(CYANOMETHYL)-5-METHOXY-3-PHENYL-2(3H)-INDOLONE

S86,387-4
12-CYANO-9,10-DIHYDRO-9-METHOXY-9,10-ETHANOANTHRACENE

S38,334-1
4-BUTOXY-2-PHENYLBUTYRONITRILE

S54,367-5
1-CYANO-3-(4-METHOXYPHENYL)-GUANIDINE

S87,950-9
DL-2-(4-METHOXYPHENYL)BUTYRONITRILE

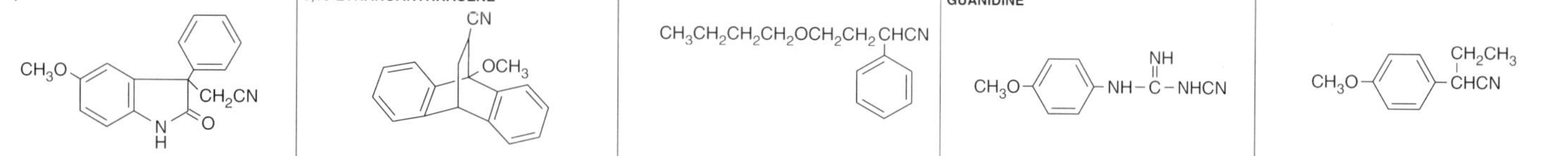

250 mg | 250 mg | 250 mg | 5 gm | 5 gm

S35,255-1
4-METHOXY-4-(P-TOLYL)-BUTYRONITRILE

S38,267-1
3,3'-(1,4-PHENYLENEDIOXY)-DIPROPIONITRILE

S35,170-9
4-ISOPROPOXY-4-(P-TOLYL)-BUTYRONITRILE

S73,394-6
CYANODIPHENYLMETHYL BENZOATE

S45,326-9
2,3,5-TRI-O-BENZOYL-BETA-D-RIBOFURANOSYL CYANIDE

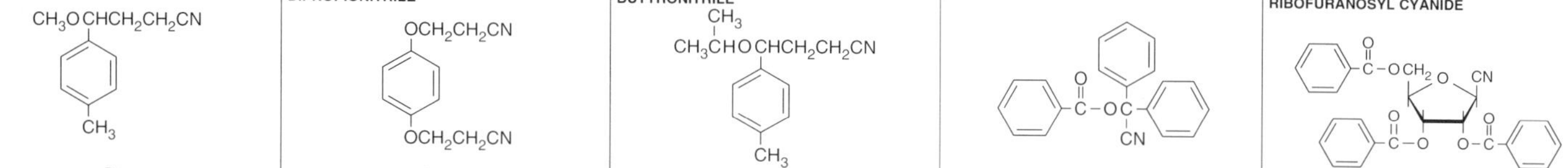

5 gm | 250 mg | 5 gm | 250 mg | 250 mg

S87,173-7
ETHYL 2-CYANO-3-OXO-5-PHENYL-4-PENTENOATE
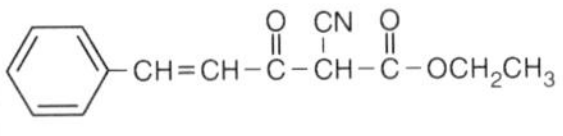
250 mg

S90,644-1
ETHYL 3-ACETOXY-2-CYANO-2-PHENYLPROPIONATE
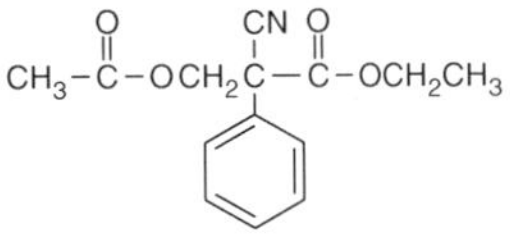
250 mg

S95,068-8
CYANOETHYL-METHYL-(N-METHYLANILINOMETHYLENE)-OXO-CYCLOPENTANECARBOXYLIC ACID
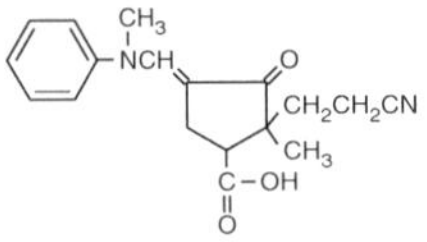
50 mg

S38,346-5
2-[2-(DIMETHYLAMINO)ETHYL]-2-[2-(METHYLTHIO)ETHYL]-2-PHENYLACETONITRILE
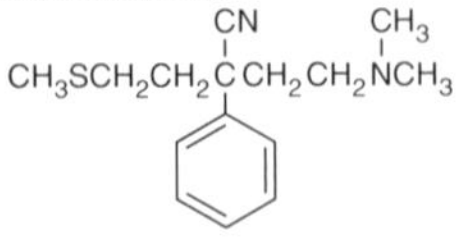
5 gm

S42,166-9
4-DIMETHYLAMINO-2-[2-(METHYLTHIO)-ETHYL]-2-PHENYLBUTYRONITRILE HYDROCHLORIDE
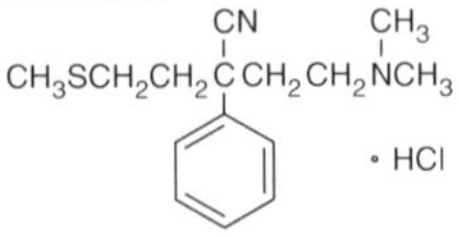
250 mg

S59,690-6
2-(PHENYLTHIO)ACETONITRILE
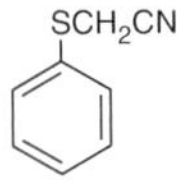
5 gm

S59,796-1
3-(PHENYLTHIO)PROPIONITRILE
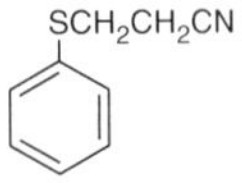
5 gm

S62,024-6
BENZYLTHIOACETONITRILE
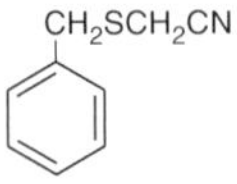
5 gm

S61,406-8
3-(BENZYLTHIO)PROPIONITRILE
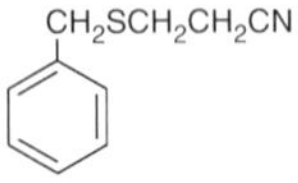
5 gm

S61,428-9
2-METHYL-3-(PHENYLTHIO)-PROPIONITRILE
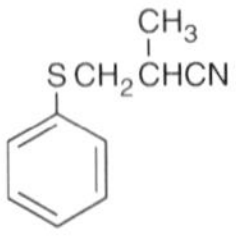
5 gm

S62,018-1
4-(PHENYLTHIO)BUTYRONITRILE
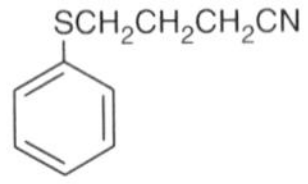
5 gm

S91,932-2
2-METHYL-2-(PHENYLTHIO)-PROPIONITRILE
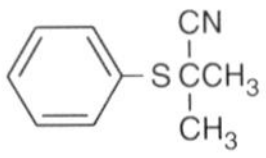
250 mg

S38,116-0
4-METHYLTHIO-2-PHENYLBUTYRONITRILE
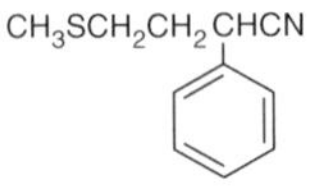
5 gm

S61,445-9
2-(4-CHLOROPHENYL)ACETONITRILE
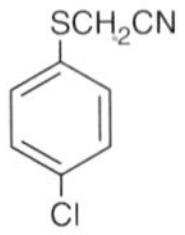
5 gm

S61,442-4
3-(4-CHLOROPHENYLTHIO)PROPIONITRILE
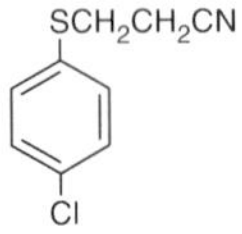
5 gm

S79,267-5
2-AMINOBENZYL CYANIDE
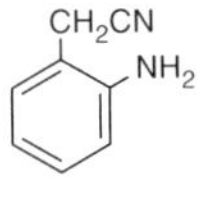
250 mg

S44,988-1
2-(4-AMINOPHENYL)-2-(PHENYL)-ACETONITRILE HYDROCHLORIDE
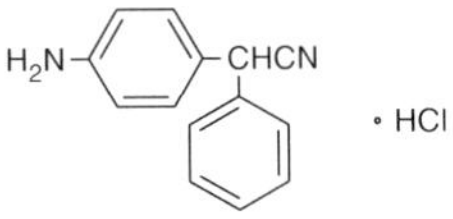
5 gm

S41,716-5
(4-AMINOPHENYL)-(META-TOLYL)-ACETONITRILE HYDROCHLORIDE
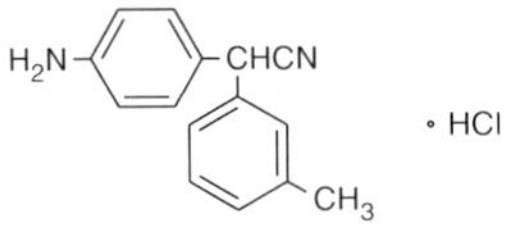
5 gm

S41,717-3
(4-AMINOPHENYL)-(PARA-TOLYL)-ACETONITRILE HYDROCHLORIDE
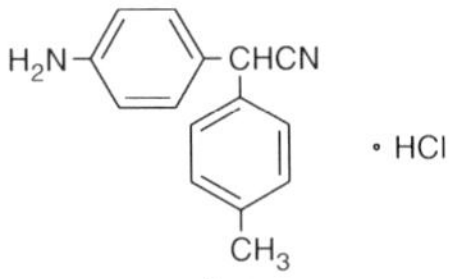
5 gm

S36,590-4
1-BENZYL-4-CYANO-4-CYCLOHEXYLAMINOPIPERIDINE
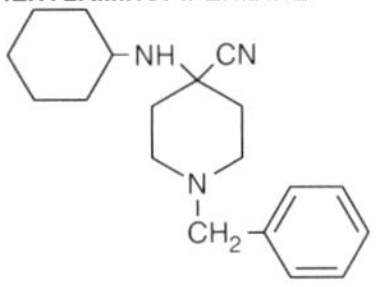
5 gm

S35,833-9
1-BENZYL-4-CYANO-4-(METHYLAMINO)-PIPERIDINE
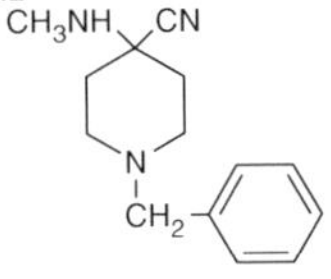
5 gm

S35,831-2
1-BENZYL-4-CYANO-4-(ETHYLAMINO)-PIPERIDINE
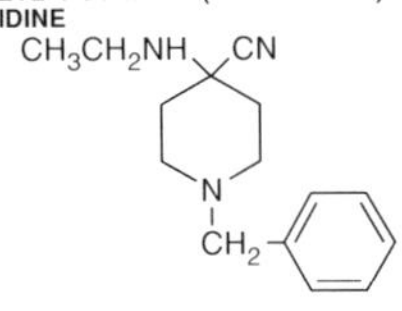
5 gm

S35,834-7
1-BENZYL-4-CYANO-4-(PROPYLAMINO)-PIPERIDINE HYDRATE
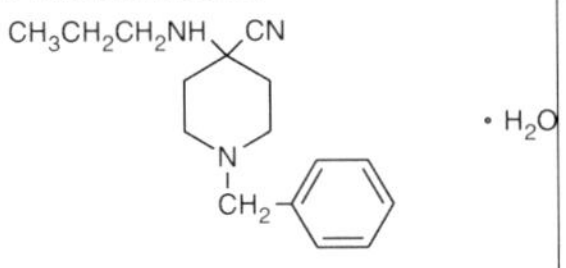
5 gm

S41,765-3
4-ANILINO-1-BENZYL-4-CYANOPIPERIDINE
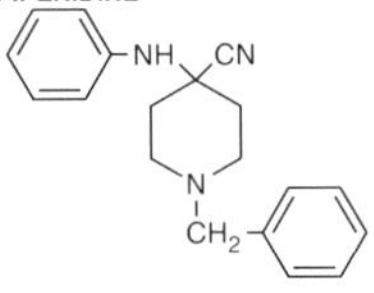
5 gm

S35,827-4
1-BENZYL-4-(4-CHLOROANILINO)-4-CYANOPIPERIDINE
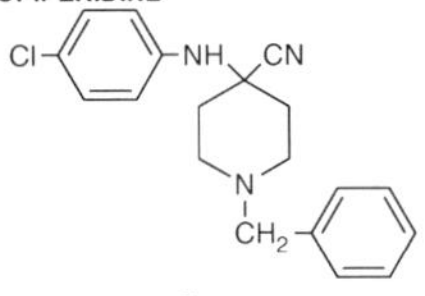
5 gm

S35,836-3
1-BENZYL-4-CYANO-4-(ORTHO-TOLUIDINO)PIPERIDINE
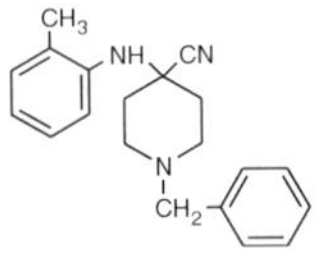
5 gm

S35,837-1
1-BENZYL-4-CYANO-4-(META-TOLUIDINO)-PIPERIDINE
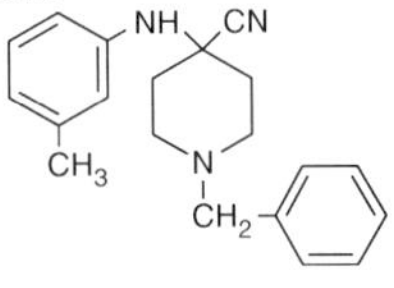
5 gm

S37,312-5
1-BENZYL-4-CYANO-4-(PARA-TOLUIDINO)-PIPERIDINE
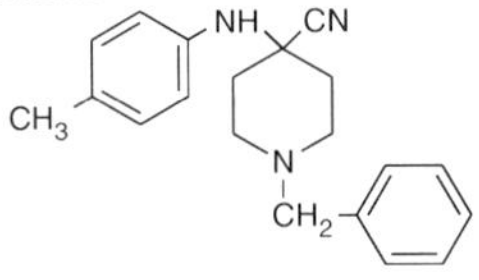
5 gm

S35,832-0
1-BENZYL-4-CYANO-4-(ISOPROPYLAMINO)-PIPERIDINE HYDRATE
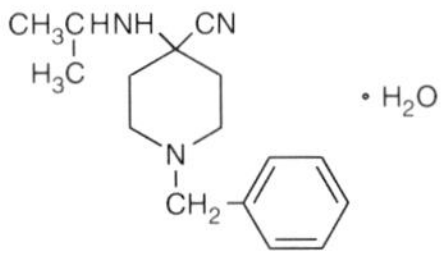
5 gm

S34,160-6
1-BENZYL-4-BUTYLAMINO-4-CYANOPIPERIDINE HYDRATE
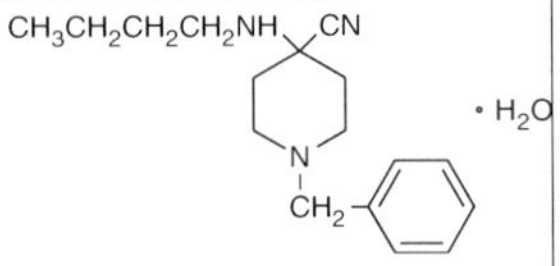
5 gm

S35,835-5
1-BENZYL-4-(1-CYANOPROPYLAMINO)-PIPERIDINE
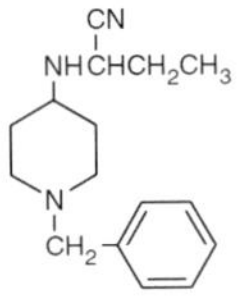
5 gm

S40,662-7
1-BENZYL-4-(3-CHLOROANILINO)-4-CYANOPIPERIDINE
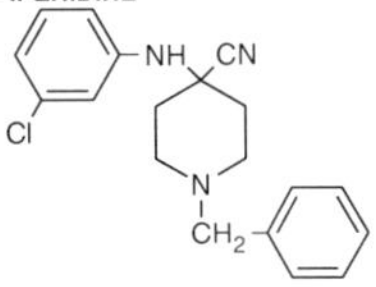
5 gm

S64,270-3
N,N'-BIS(ALPHA-CYANO-4-DIMETHYLAMINOBENZYL)ETHYLENEDIAMINE
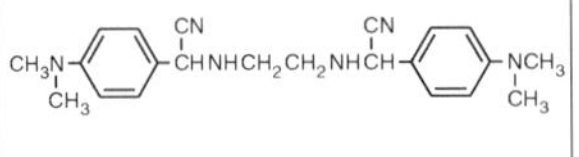
5 gm

S38,246-9
N,2-DIPHENYLGLYCINONITRILE
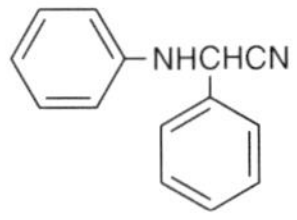
5 gm

S34,483-4
1-(A-CYANOBENZYL)-PIPERIDINE
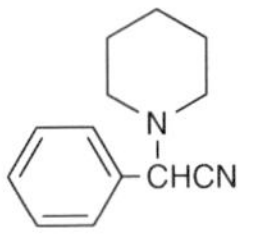
5 gm

S36,397-9
ALPHA-PHENYL-4-MORPHOLINEACETONITRILE
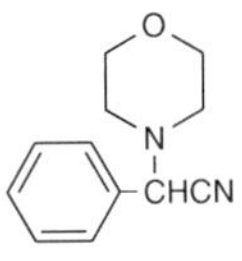
5 gm

S65,800-6
1-CYANO-N-[1-(1,2-DIPHENYL)ETHYL]-CYCLOHEXYLAMINE
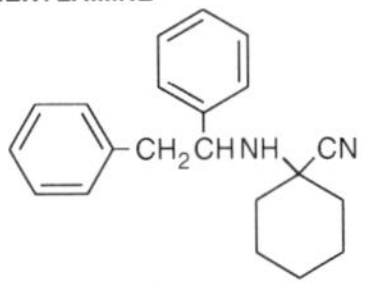
5 gm

S93,656-1
N-(5-CHLORO-2-NITROPHENYL)-CARBAMONITRILE
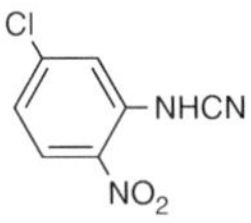
250 mg

S92,310-9
N-(2-[N-(2-CYANOETHYL)-3-METHYLANILINO]ETHYL)SUCCINIMIDE
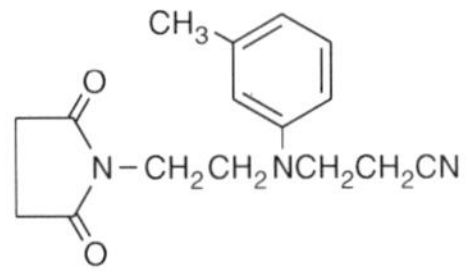
250 mg

S90,766-9
N-(2-[N-(2-CYANOETHYL)-3-METHYLANILINO]ETHYL)SUCCINIMIDE
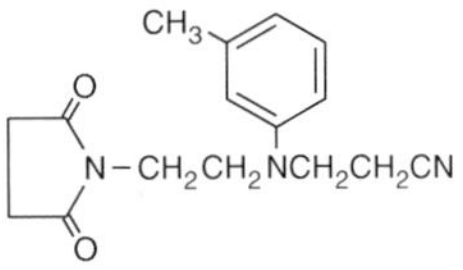
250 mg

S56,843-0
N-ETHYL-N-PHENYLGLYCINONITRILE
5 gm

S41,272-4
3-(N-ETHYLANILINO)PROPIONITRILE
5 gm

S92,254-4
1-CYANOMETHYL-1,2,3,4-TETRAHYDRO-QUINOLINE
250 mg

S91,306-5
1-(2-CYANOACETYL)-4-(4-METHYL-2-NITROPHENYL)SEMICARBAZIDE
250 mg

S61,480-7
1-BENZOYL-2-CYANO-1,2-DIHYDROQUINOLINE
5 gm

S73,233-8
BENZYL N-(CYANOMETHYL)CARBAMATE
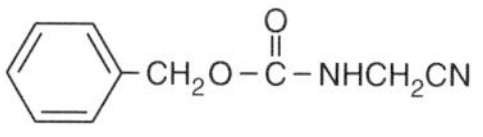
250 mg

S54,946-0
N-(CYANOMETHYL)PHTHALIMIDE
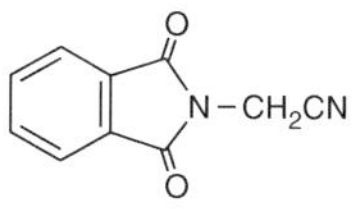
250 mg

S74,571-5
3-(P-TOLUIDINO)PROPIONITRILE
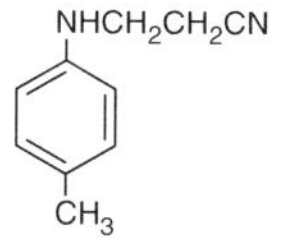
250 mg

S36,509-2
3-(O-TOLUIDINO)PROPIONITRILE
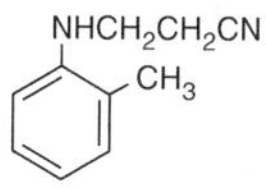
5 gm

S54,424-8
3-(3,5-XYLIDINO)PROPIONITRILE
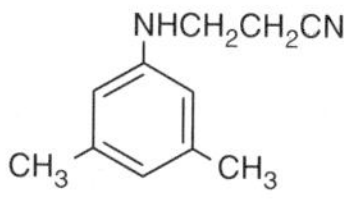
250 mg

S74,205-8
3,3'-(M-TOLYLIMINO)DIPROPIONITRILE
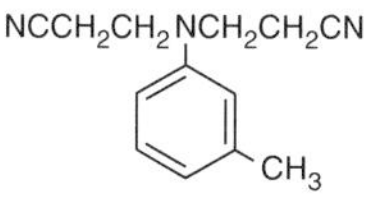
5 gm

S36,757-5
N-(2-CYANOETHYL)-N-METHYLANILINE
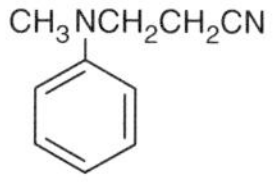
5 gm

S60,497-6
N-(2-CYANOETHYL)-N-PHENYLGLYCINE
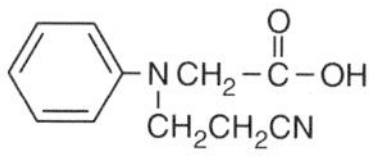
250 mg

S63,668-1
4-[N-(2-CYANOETHYL)METHYLAMINO]-BENZALDEHYDE
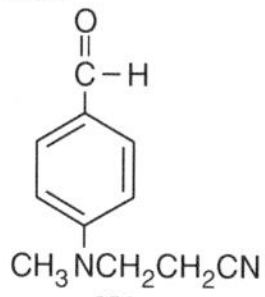
250 mg

S64,737-3
2,5-BIS[4-(N-2-CYANOETHYL-N-METHYLAMINO)BENZYLIDENE]-CYCLOPENTANONE
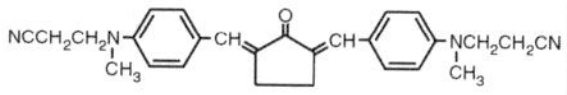
250 mg

S62,868-9
4-(4-[N,N-BIS(2-CYANOETHYL)AMINO]-BENZYLIDENE)-3-METHYL-1-PHENYL-2-PYRAZOLIN-5-ONE
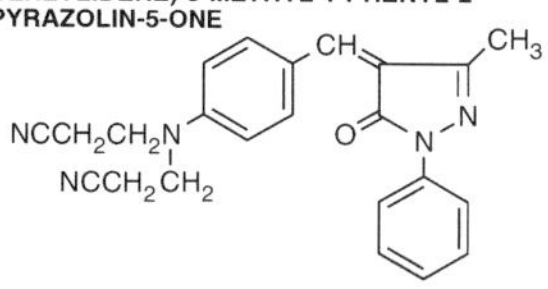
5 gm

S37,345-1
3-(N-ETHYL-META-TOLUIDINO)-PROPIONITRILE
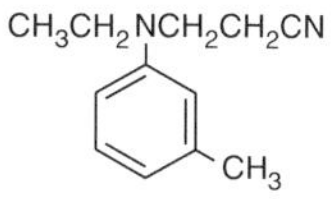
5 gm

S39,194-8
3,3'-(PHENYLIMINO)DIPROPIONITRILE
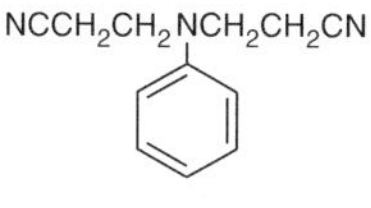
5 gm

S51,688-0
3-(N-BUTYLANILINO)PROPIONITRILE
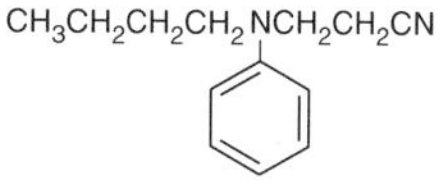
250 mg

S77,262-3
1-(2-CYANOACETYL)-4-(3-NITROPHENYL)-SEMICARBAZIDE
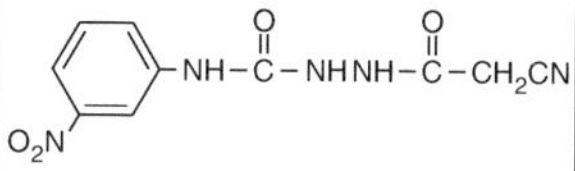
250 mg

S65,590-2
N-(2-CYANOACETYL)HYDRAZOBENZENE
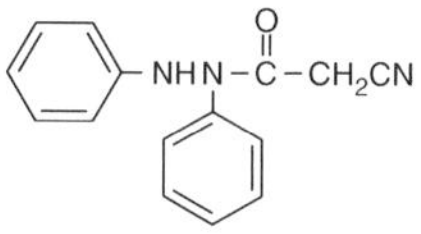
250 mg

S62,226-5
N-(2-CYANOACETYL)ANTHRANILIC ACID
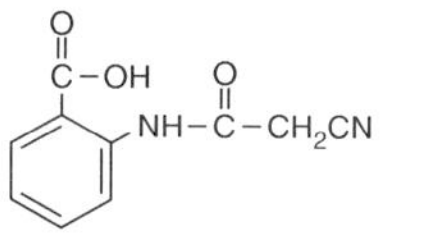
5 gm

S92,985-9
2-CYANOACETIC N2-BENZOYLHYDRAZIDE
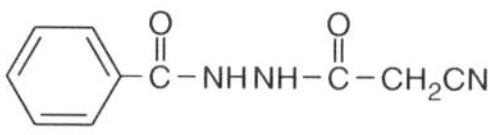
250 mg

S53,425-0
2-CYANO-2-PHENYLBUTYRAMIDE
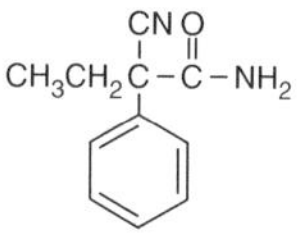
5 gm

S40,664-3
1-BENZYL-4-CYANO-4-PHENYLPIPERIDINE HYDROCHLORIDE
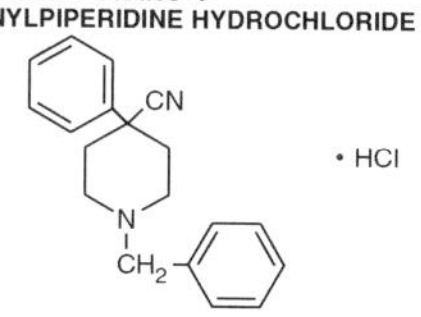
5 gm

S93,513-1
2-CHLORO-10-(2-CYANOETHYL)-10H-PHENOTHIAZINE
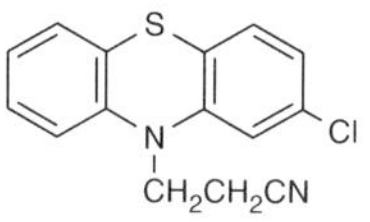
250 mg

S38,282-5
BETA-IMINO-PARA-METHYLHYDROCINNAMONITRILE
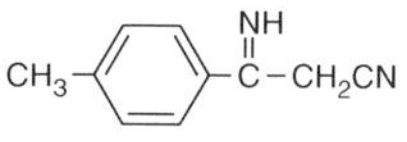
250 mg

S78,252-1
2-(2-CYANOETHYL)PHTHALIMIDE
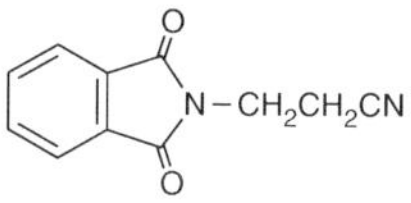
5 gm

S40,225-7
ALPHA,ALPHA-DIPHENYL-1-PIPERIDINEBUTYRONITRILE HYDROCHLORIDE
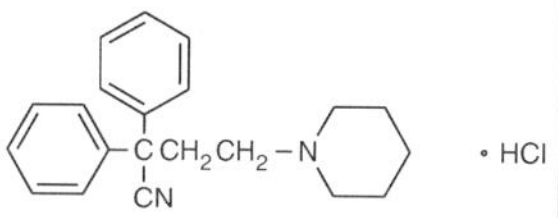
250 mg

S40,790-9
ALPHA-(2-DIMETHYLAMINOETHYL)-BENZYL CYANIDE
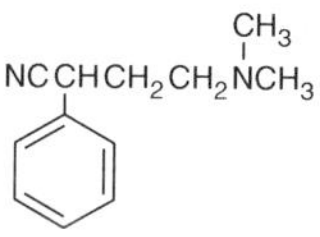
5 gm

S38,410-0
2-[2-(DIETHYLAMINO)ETHYL]-3-METHYL-2-PHENYLVALERONITRILE
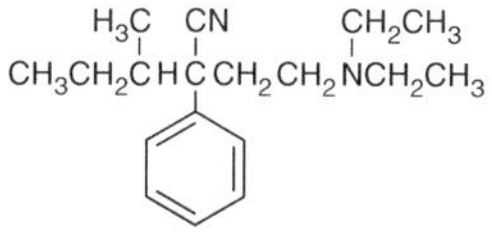
5 gm

S38,419-4
ALPHA-(2-DIMETHYLAMINOETHYL)-ALPHA-PHENYLHYDROCINNAMONITRILE
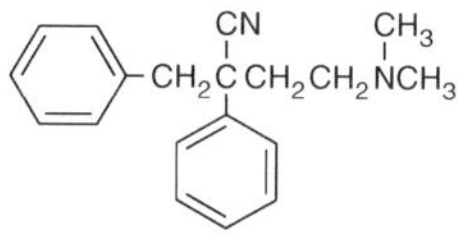
250 mg

S44,721-8
5-FLUORO-2-NITROPHENYLACETONITRILE
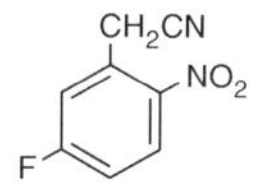
5 gm

S35,285-3
METHYL 2,3-DICYANO-3-PHENYLPROPIONATE
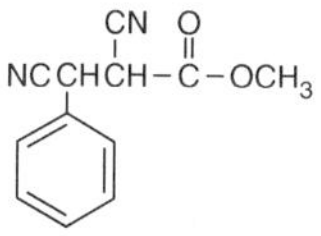
5 gm

S88,319-0
3-CYANO-N-PHENETHYLPROPIONAMIDE
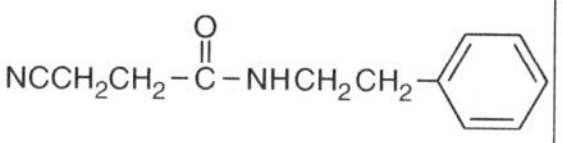
250 mg

S41,935-4
MESO-1,2-DICYANO-1,2-DIPHENYLETHANE
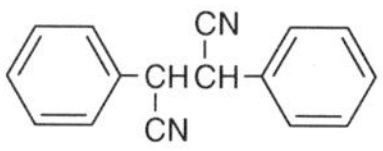
5 gm

S55,251-8
2,2,3,3-TETRAPHENYLSUCCINONITRILE
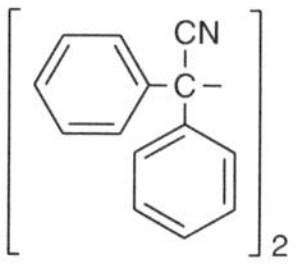
250 mg

S66,079-5
2-PHENYL-3-(P-TOLYL)SUCCINONITRILE
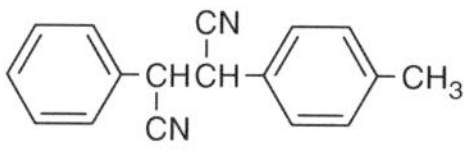
5 gm

S95,341-5
N-(2-CYANOETHYL)ANTHRANILIC ACID
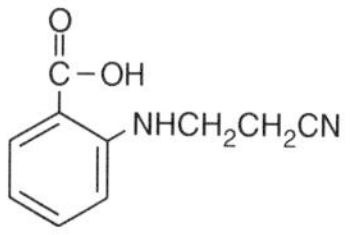
50 mg

S93,428-3
2-CYANO-4-NITROPHENYLACETONITRILE
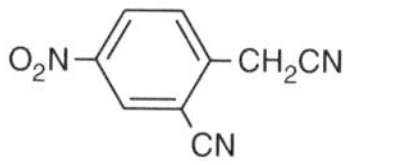
250 mg

S52,629-0
1-BENZYL-4-(4-FLUOROPHENYL)-ISONIPECOTONITRILE
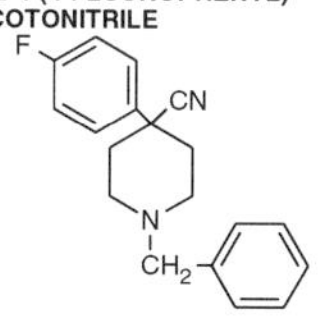
5 gm

S35,830-4
1-BENZYL-4-(ALPHA-CYANOBENZYL)-PIPERIDINE
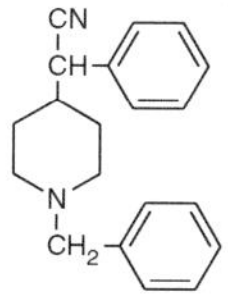
5 gm

S60,195-0
2-CHLORO-4-NITROPHENYLACETONITRILE
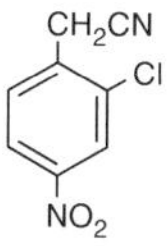
250 mg

S43,472-8
ETHYL 3-CYANO-3-PHENYLPYRUVATE
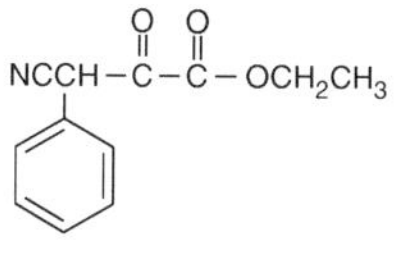
5 gm

S52,552-9
4-CYCLOPROPYLFORMYL-4'-FLUORODIPHENYLACETONITRILE
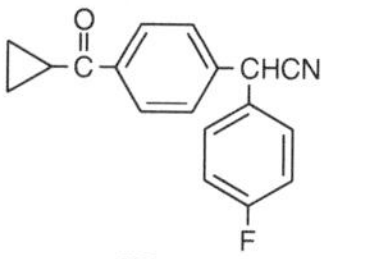
250 mg

S88,650-5
2,4-DIPHENYL-3-OXOBUTYRONITRILE
250 mg

S43,232-6
2-BENZYL-3-OXO-2-PHENYLBUTYRONITRILE
250 mg

S80,851-2
DIETHYL (ALPHA-CYANOBENZYL)-MALONATE
250 mg

S88,334-4
ETHYL ALPHA,BETA-DICYANOHYDROCINNAMATE
250 mg

S34,824-4
ETHYL 4-CYANO-4-PHENYLVALERATE
5 gm

S52,145-0
4-(TERT-PENTYL)PHENYLACETONITRILE
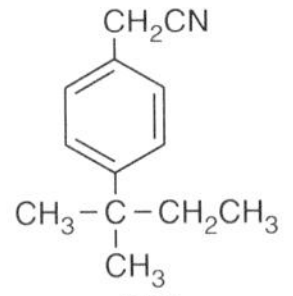
5 gm

S45,041-3
1-(4-CHLOROPHENYL)-1-CYCLOPROPANE-CARBONITRILE
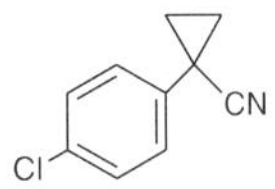
5 gm

S44,794-3
1-(P-TOLYL)-1-CYCLOPROPANECARBONITRILE
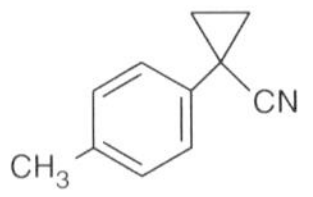
5 gm

S44,823-0
1-(4-CHLOROPHENYL)-1-CYCLOBUTANE-CARBONITRILE
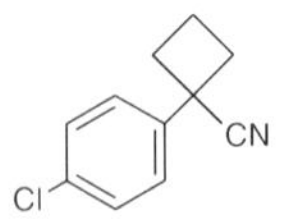
5 gm

S88,647-5
3-ETHYL-2-IMINO-3-PHENYL-1-CYCLOPENTANECARBONITRILE
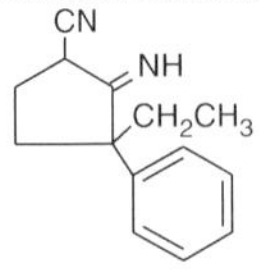
250 mg

S44,125-2
1-(4-CHLOROPHENYL)-1-CYCLOPENTANE-CARBONITRILE
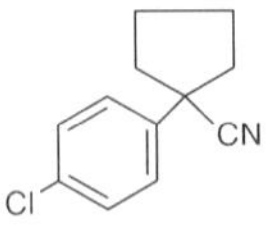
5 gm

S44,814-1
1-(P-TOLYL)-1-CYCLOPENTANECARBONITRILE
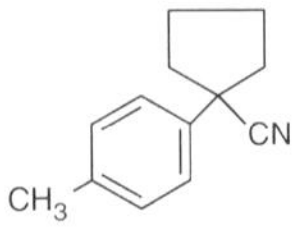
5 gm

S44,627-0
1-PHENYL-1-CYCLOHEXANECARBONITRILE
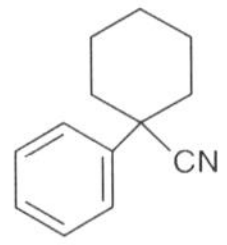
5 gm

S44,864-8
1-(4-CHLOROPHENYL)-1-CYCLOHEXANE-CARBONITRILE
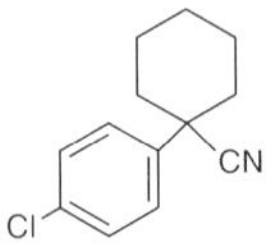
5 gm

S45,939-9
CYCLOHEXYLPHENYLACETONITRILE
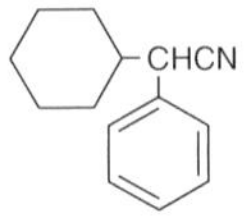
5 gm

S88,726-9
3-ETHYL-2-IMINO-3-PHENYL-1-CYCLOHEXANECARBONITRILE
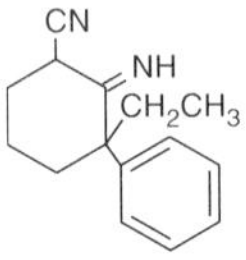
250 mg

S45,044-8
1-(P-TOLYL)-1-CYCLOHEXANECARBONITRILE
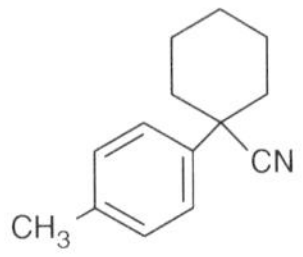
5 gm

S37,744-9
4-PHENYLTETRAHYDROPYRAN-4-CARBONITRILE
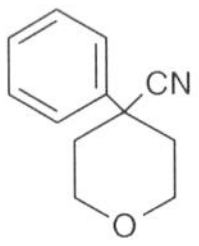
250 mg

S88,718-8
3-METHYL-2-PHENYLBUTYRONITRILE
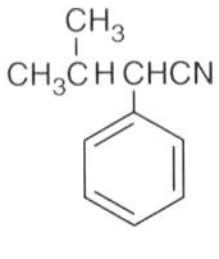
250 mg

S36,188-7
2-ETHYL-2-PHENYLBUTYRONITRILE
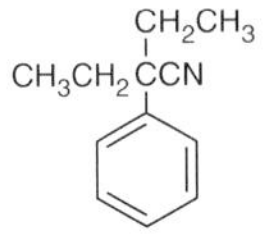
5 gm

S37,753-8
4-METHYL-2-PHENYLVALERONITRILE
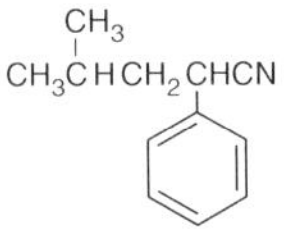
5 gm

S38,163-2
3-METHYL-2-PHENYLVALERONITRILE
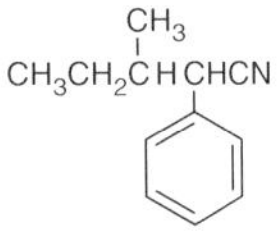
250 mg

S66,224-0
ALPHA-(TERT-BUTYL)-PHENYLACETONITRILE
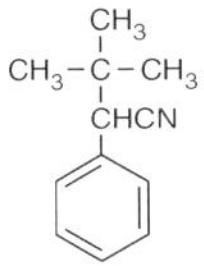
5 gm

S88,635-1
2-PHENYLHEXANENITRILE
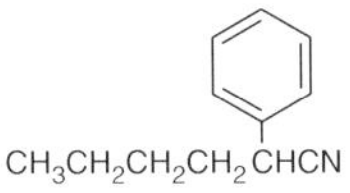
250 mg

S38,195-0
3-METHYL-2-PHENYLHEXANENITRILE
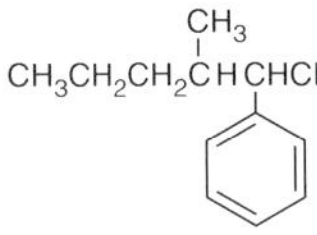
5 gm

S36,574-2
3-PHENYL-1,3,5-PENTANETRICARBONITRILE
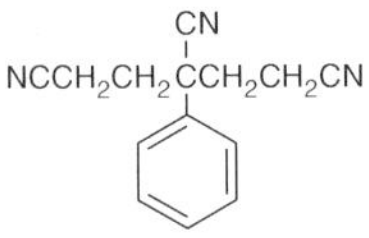
250 mg

S63,691-6
2,3-DIPHENYLPROPIONITRILE
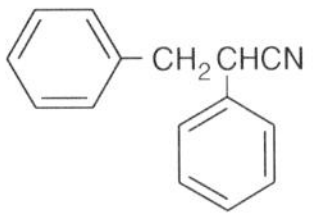
250 mg

S87,939-8
2,3-DIPHENYLVALERONITRILE
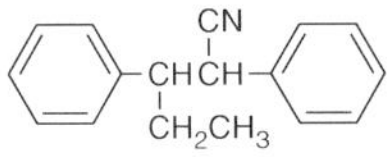
5 gm

S38,382-1
2,3-DIPHENYLGLUTARONITRILE
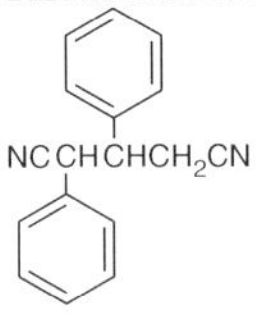
5 gm

S88,735-8
ALPHA-BENZYL-ALPHA-PHENYLHYDROCINNAMONITRILE
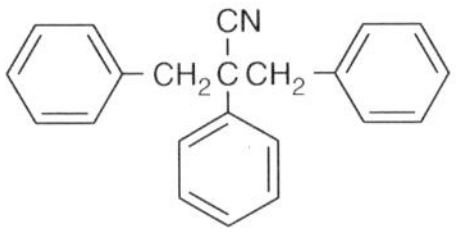
250 mg

S34,760-4
2,2-DIPHENYLBUTYRONITRILE
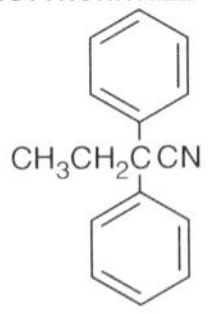
5 gm

S54,102-8
2,2-DIPHENYLVALERONITRILE
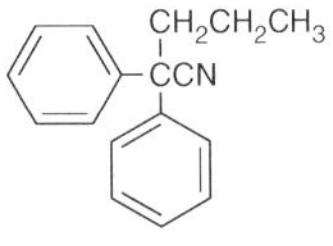
250 mg

S87,937-1
2,2-DIPHENYLISOVALERONITRILE
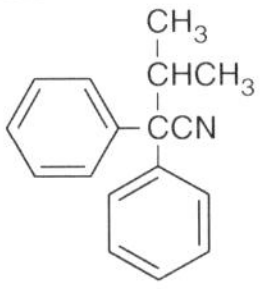
5 gm

S88,655-6
1,1-DIPHENYL-1,3-PROPANEDICARBONITRILE
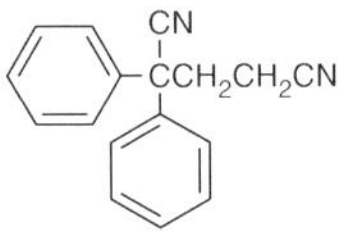
250 mg

S88,659-9
2,2-DIPHENYLHEXANENITRILE
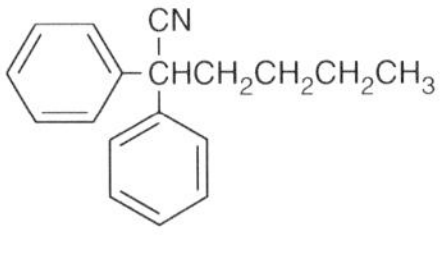
250 mg

S88,733-1
ALPHA,BETA-DIPHENYLHYDROCINNAMONITRILE

S88,736-6
ALPHA,BETA-DIPHENYL-ALPHA-METHYLHYDROCINNAMONITRILE
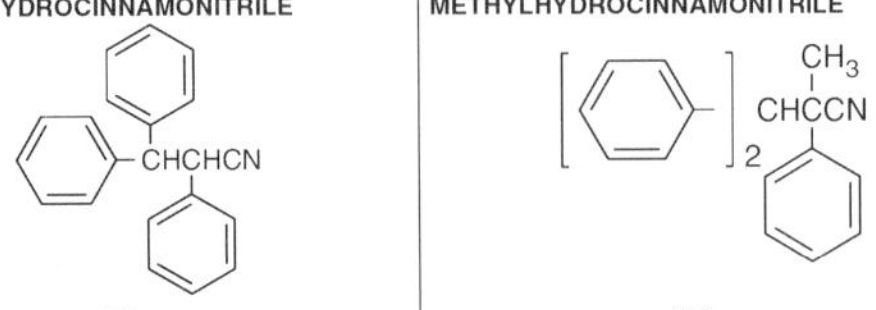
250 mg (S88,733-1)
250 mg (S88,736-6)

S88,663-7
3-(2-NAPHTHYL)-2-PHENYLPROPIONITRILE

S34,987-9
2-(4-FLUOROPHENYL)-2-PHENYLACETONITRILE
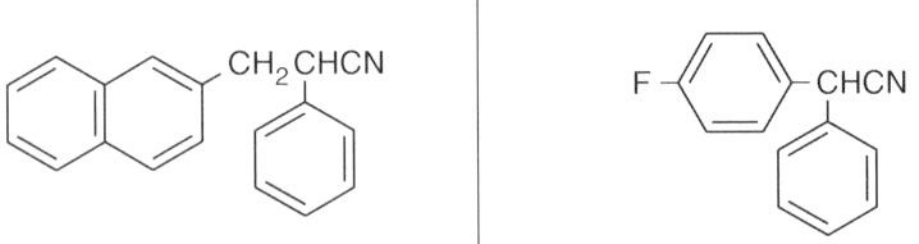
250 mg (S88,663-7)
5 gm (S34,987-9)

S64,568-0
3,4-DIMETHYLPHENYLACETONITRILE
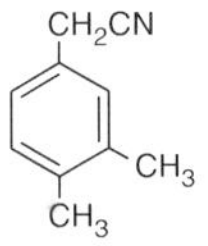
250 mg

S88,643-2
2-MESITYL-2-METHYLPROPIONITRILE

S52,636-3
N-BENZYL-N-(3-CYANOPROPYL)GLYCINE ETHYL ESTER
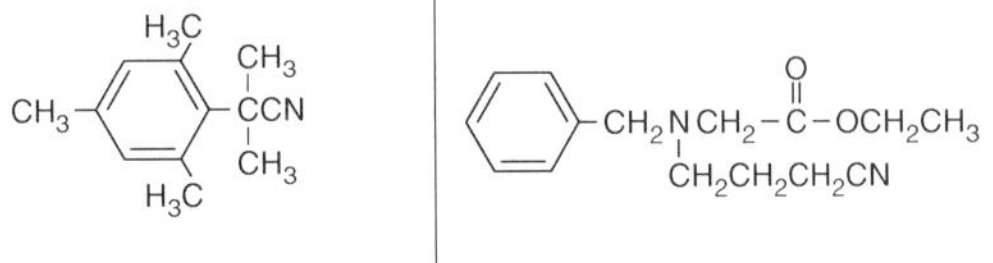
250 mg (S88,643-2)
250 mg (S52,636-3)

S83,592-7
DIBENZYLCYANAMIDE

S90,677-8
3-CYANO-1-(DIPHENYLMETHYL)AZETIDINE
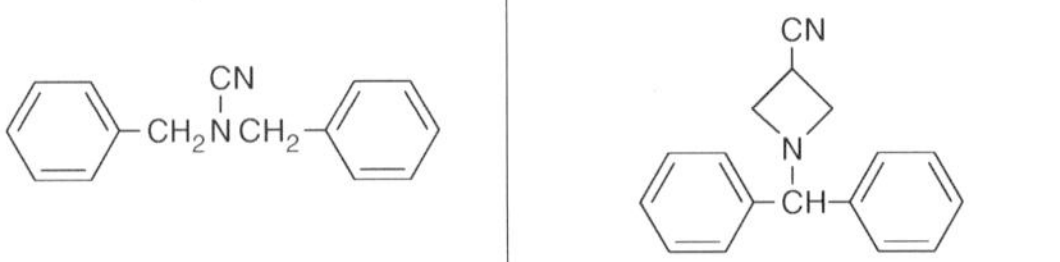
250 mg (S83,592-7)
250 mg (S90,677-8)

S92,043-6
10-PHENOTHIAZINEPROPIONITRILE
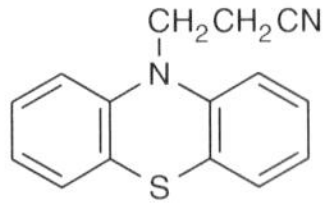
250 mg

S91,938-1
4-(1-METHYL-1-NITROPROPYL)-BENZONITRILE

S69,349-9
1-(CYANOACETYL)-4-(2-METHOXY-5-METHYLPHENYL)-3-THIOSEMICARBAZIDE
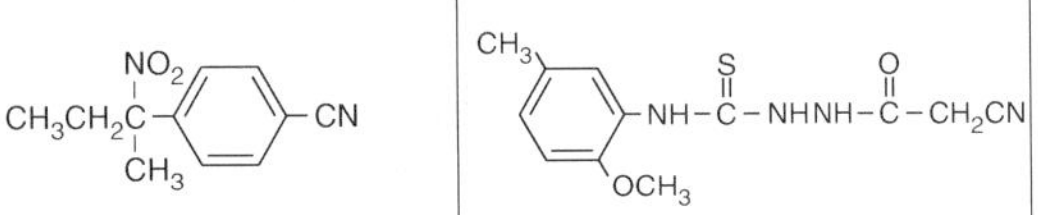
250 mg (S91,938-1)
250 mg (S69,349-9)

S69,251-4
1-(2-CYANOACETYL)-4-(4-ETHOXYCARBONYLPHENYL)-SEMICARBAZIDE

S77,209-7
1-(2-CYANOACETYL)-4-(3,4-XYLYL)-SEMICARBAZIDE
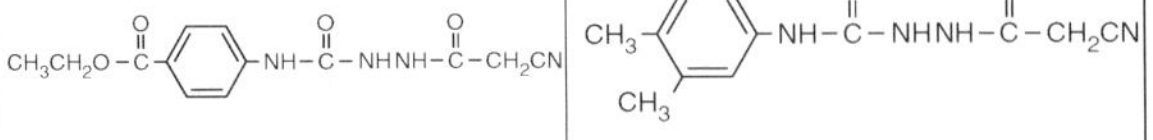
250 mg (S69,251-4)
250 mg (S77,209-7)

S86,235-5
1,2-DICYANO-1-PHENYLHYDRAZINE

250 mg

S85,823-4
BUTYL N-CYANO-N-PHENYLCARBAMATE
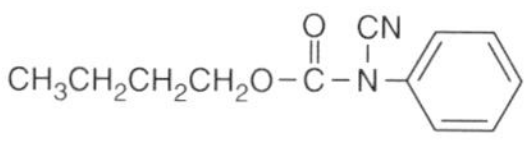
250 mg

S63,944-3
2-CYANO-2-PROPYL N-(3-CHLOROPHENYL)-CARBAMATE
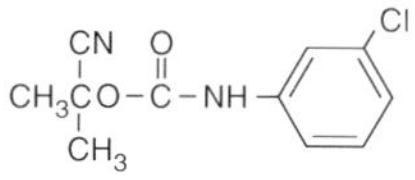
5 gm

S84,343-1
ALPHA-CHLORO-4-NITROHYDROCINNAMONITRILE
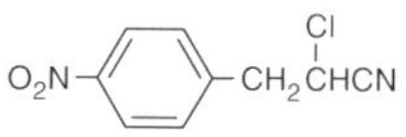
250 mg

S38,273-6
2-CYANO-1-METHYLETHYL 3-NITROBENZOATE
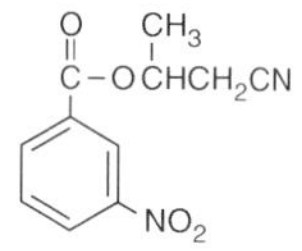
5 gm

S53,775-6
ALPHA-CYANO-3-NITROPHENETHYL BENZOATE
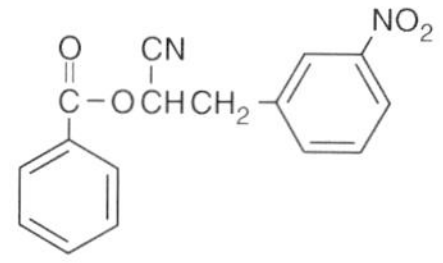
250 mg

S52,776-9
2-CHLORO-N-CYANOMETHYL-N-ETHYL-4-NITROBENZAMIDE
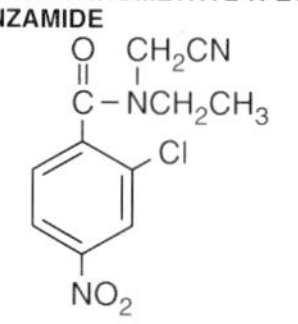
5 gm

S52,909-5
2-CHLORO-5-NITROHYDROCINNAMONITRILE
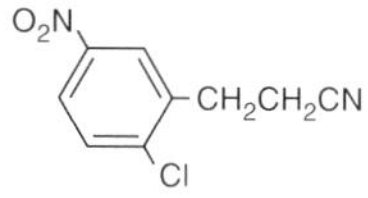
250 mg

S55,113-9
4-CYANO-2-METHYL-2-PHENYLBUTYRALDEHYDE
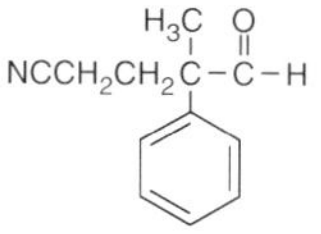
5 gm

S34,485-0
2-CYANO-1,3-INDANEDIONE
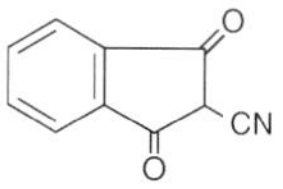
5 gm

S76,336-5
3'-(CYANOMETHYL)-2,2-DIMETHYLPROPIOPHENONE
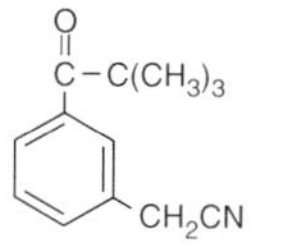
250 mg

S88,716-1
2-CYANO-1-TETRALONE
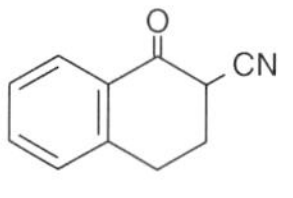
250 mg

S45,264-5
2-BENZOYLPROPIONITRILE
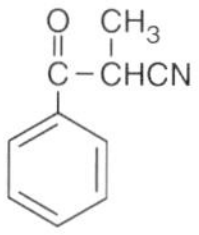
5 gm

S93,519-0
3-BENZOYLPROPIONITRILE
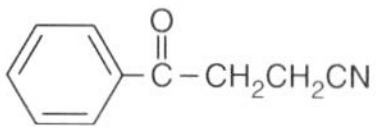
250 mg

S91,502-5
2-BENZOYLMALONONITRILE
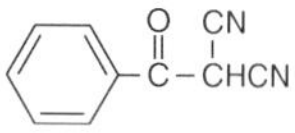
250 mg

S90,624-7
2,2-DIMETHYL-4-OXO-4-PHENYLBUTYRONITRILE
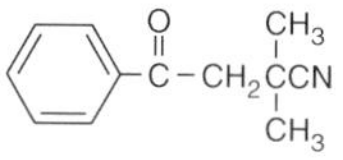
250 mg

S46,065-6
3-BENZOYL-2-PHENYLPROPIONITRILE
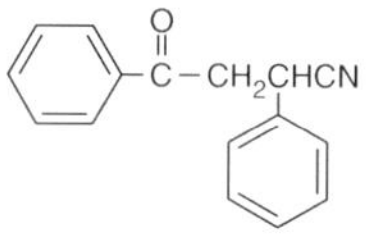
5 gm

S35,679-4
2,2,2-TRIS(2-CYANOETHYL)-ACETOPHENONE
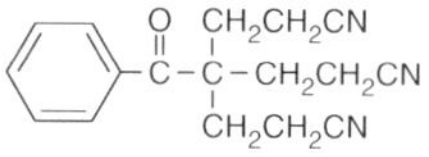
5 gm

S92,957-3
3-OXO-3-(P-TOLYL)PROPIONITRILE
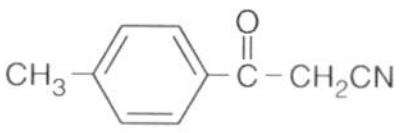
250 mg

S73,286-9
4'-(CYANOMETHYL)-2,2-DIMETHYLPROPIOPHENONE
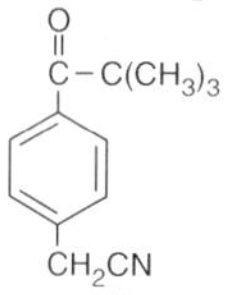
250 mg

S60,584-0
2-(2,6-DICHLOROBENZOYL)ACETONITRILE
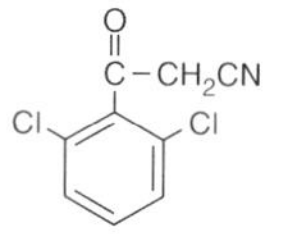
250 mg

S88,714-5
1-OXO-2-INDANCARBONITRILE
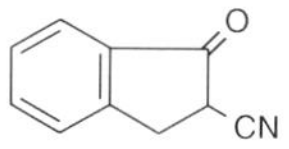
250 mg

S88,654-8
3-ISOBUTYL-2-OXO-3-PHENYL-1-CYCLOPENTANECARBONITRILE
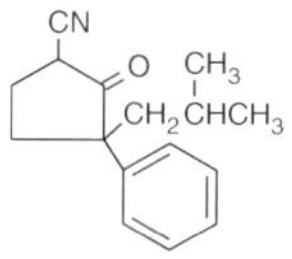
250 mg

S88,658-0
3,3-DIPHENYL-2-OXO-1-CYCLOPENTANECARBONITRILE
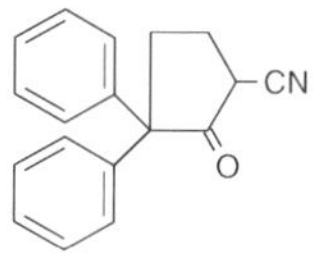
250 mg

S88,729-3
3-ISOPROPYL-2-OXO-3-PHENYL-1-CYCLOHEXANECARBONITRILE
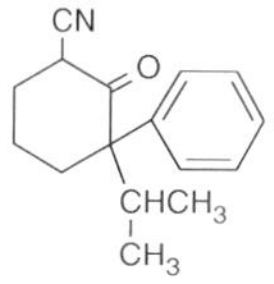
250 mg

S88,662-9
3,3-DIPHENYL-2-OXO-1-CYCLOHEXANECARBONITRILE
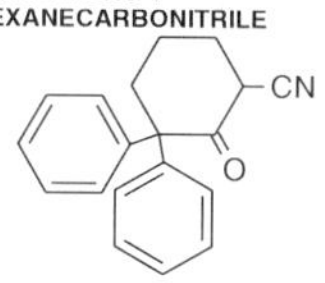
250 mg

S88,666-1
3-BENZYL-2-OXO-3-PHENYL-1-CYCLOHEXANECARBONITRILE
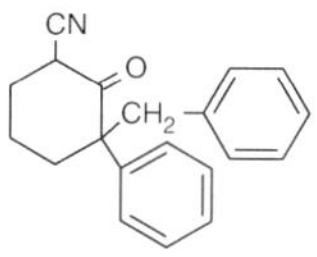
250 mg

S88,668-8
3,3-DIPHENYL-1-METHYL-2-OXO-1-CYCLOHEXANECARBONITRILE
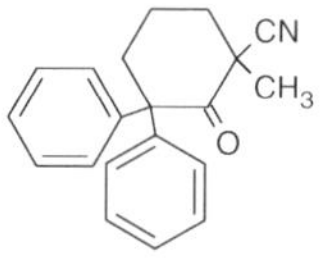
250 mg

S88,672-6
2-OXO-3-PHENETHYL-3-PHENYL-1-CYCLOHEXANECARBONITRILE
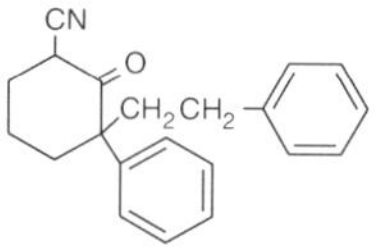
250 mg

S95,818-2
2-ACETYL-2-CYANODIBENZO(E,H)-BICYCLO[2.2.2]OCTA-5,7-DIENE
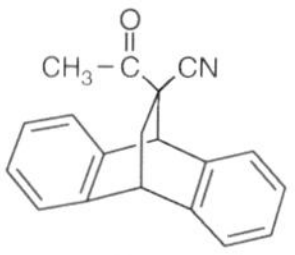
50 mg

S88,715-3
2-OXO-1-INDANCARBONITRILE
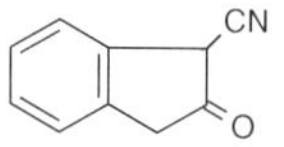
250 mg

S73,245-1
2-(CYANOMETHYL)-1-INDANONE
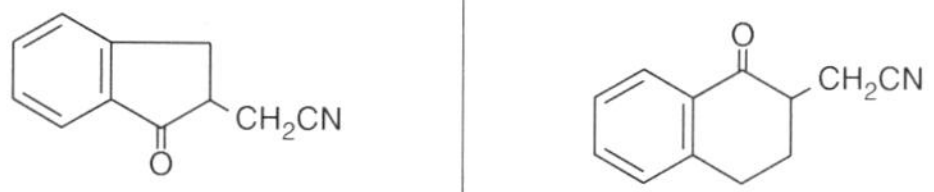
250 mg

S93,793-2
2-(1-OXO-1,2,3,4-TETRAHYDRO-2-NAPHTHYL)ACETONITRILE
250 mg

S92,592-6
3-(ALPHA-CYANOETHYL)BENZOIC ACID
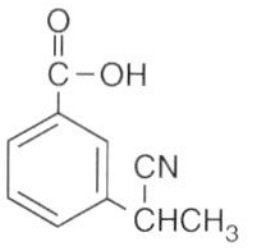
250 mg

S38,258-2
4-(2-CHLORO-2-CYANOETHYL)-BENZOIC ACID
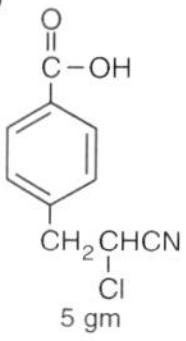
5 gm

S92,376-1
BENZOYLCYANAMIDE
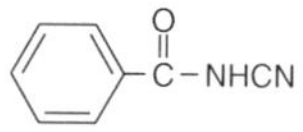
250 mg

S50,046-1
2-CHLORO-ALPHA-CYANOHYDROCINNAMIC ACID
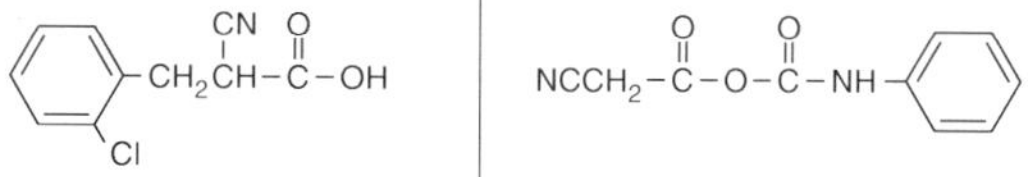
5 gm

S76,814-6
2-CYANOACETYL N-PHENYLCARBAMATE
250 mg

S64,477-3
2-CYANOETHYL N-(3-CHLOROPHENYL)-CARBAMATE
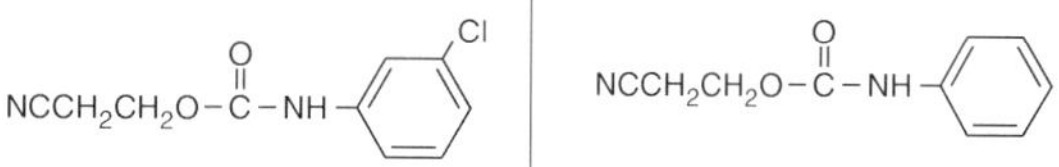
5 gm

S64,519-2
2-CYANOETHYL N-PHENYLCARBAMATE
5 gm

S85,822-6
ETHYL N-CYANO-N-PHENYLCARBAMATE
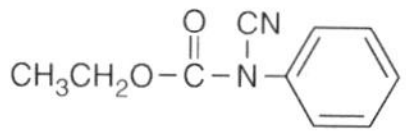
250 mg

S71,709-6
1-CYANO-1-METHYLETHYL N-PHENYLCARBAMATE
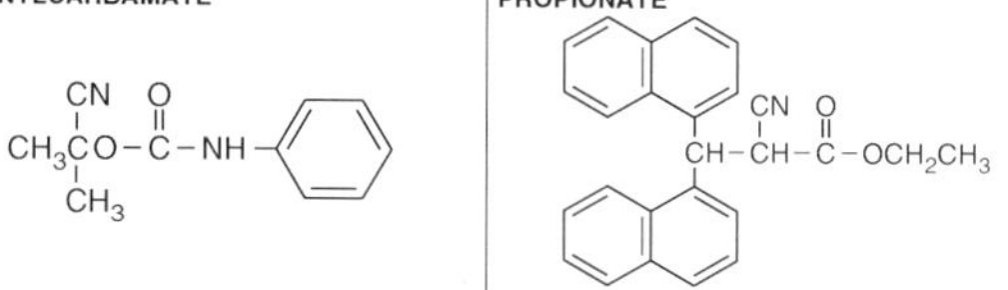
250 mg

S66,270-4
ETHYL 2-CYANO-3,3-DI(1-NAPHTHYL)-PROPIONATE
5 gm

S77,256-9
2-CYANOACETYL N-(P-TOLYL)CARBAMATE
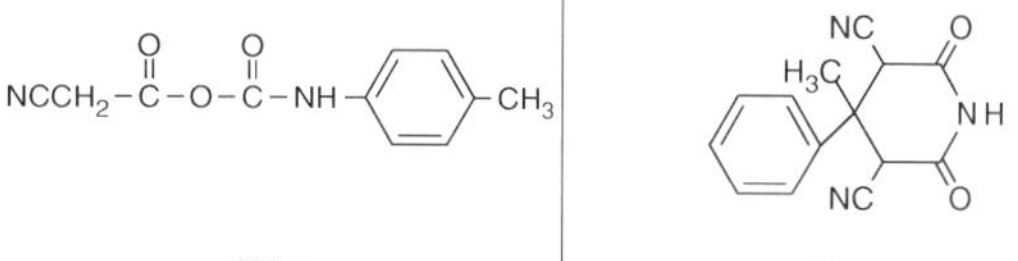
250 mg

S34,576-8
2,4-DICYANO-3-METHYL-3-PHENYLGLUTARIMIDE
5 gm

S53,416-1
2-CYANOACETANILIDE
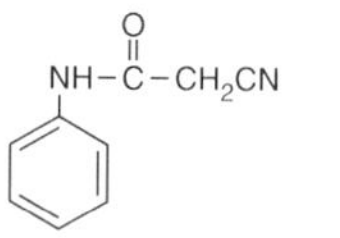
250 mg

S85,652-5
1-CYANO-3-PHENYLUREA
250 mg

S92,208-0
2,4,6-TRIMETHYLBENZYL CYANIDE
250 mg

S44,686-6
TRANS-1-CYANO-2-PHENYLCYCLOPROPANE
250 mg

S94,719-9
TRIPHENYLACRYLONITRILE EPOXIDE
250 mg

S64,936-8
ETHYL 2-CYANO-3-(4-[N-(2-CYANOETHYL)-N-METHYLAMINO]PHENYL)-PROPIONATE
250 mg

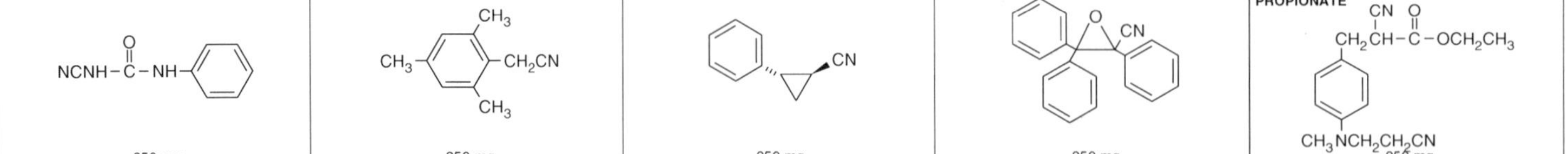

S84,318-0
6-BROMO-3,4-(METHYLENEDIOXY)-HYDROCINNAMONITRILE
250 mg

S86,386-6
11-CYANOMETHYL-9,10-DIHYDRO-9,10-ETHANOANTHRACENE
250 mg

S52,541-3
4-PHENYL-1,2,3,6-TETRAHYDRO-1-PYRIDINEVALERONITRILE HYDROCHLORIDE
250 mg

S88,632-7
ALPHA,BETA-DIBROMOHYDROCINNAMONITRILE
250 mg

S38,094-6
3-METHYL-2-PHENYLBUTYRONITRILE
250 mg

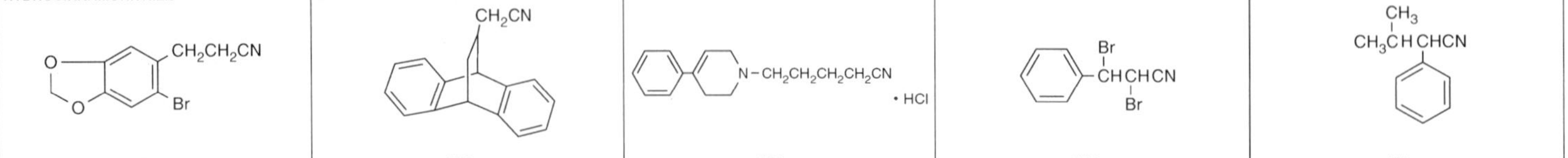

S38,100-4
2-PHENYLVALERONITRILE
250 mg

S84,227-3
2,3,4-TRIPHENYLBUTYRONITRILE
250 mg

S36,533-5
3,3,3 TRIPHENYLPROPIONITRILE
5 gm

S54,060-9
1-CYANO-3-(P-TOLYL)-GUANIDINE
5 gm

S62,063-7
2-(2-CHLORO-6-METHYLPHENYL)-ACETONITRILE
5 gm

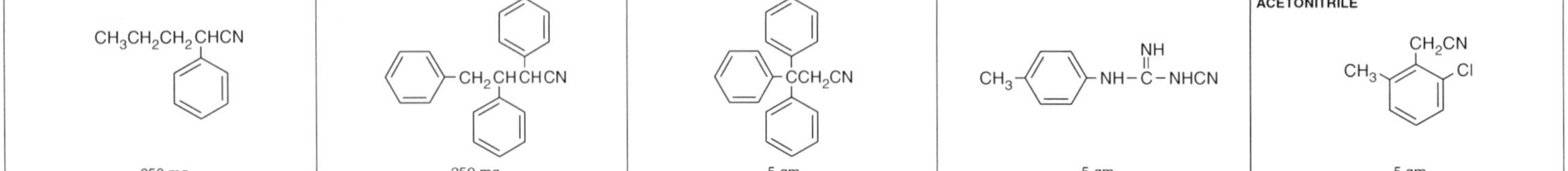

S54,621-6
1-CYANO-3-(3,4-XYLYL)-GUANIDINE
5 gm

S94,788-1
3,5,5,8,8-PENTAMETHYL-5,6,7,8-TETRAHYDRO-2-NAPHTHALENEACETONITRILE
5 gm

S34,216-5
9,9-BIS(2-CYANOETHYL)FLUORENE
5 gm

S97,388-2
AR-E48A 12-CYANOMETHYL-9-FLUOROBENZ(A)ANTHRACENE
10 mg

S92,955-7
3-OXO-3-(3,5-XYLYL)PROPIONITRILE
250 mg

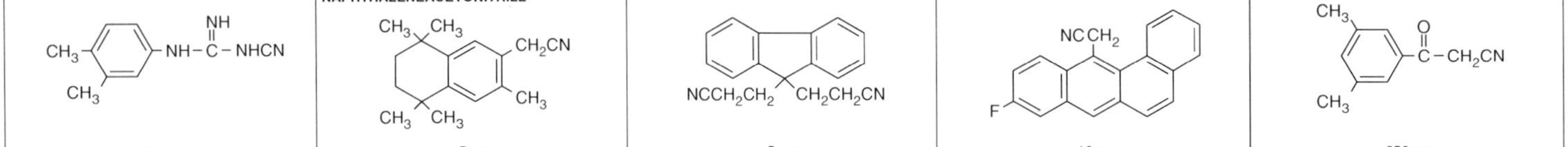

S95,806-9
BICYCLO[3.2.1]OCT-8-YL P-TOLUENESULFONATE
50 mg

S91,670-6
2'-CYANO-2,4'-DICHLOROACETANILIDE
250 mg

S74,420-4
1-(2-CHLOROETHYL)-3-(2-CYANOPHENYL)UREA
250 mg

S64,986-4
ALPHA-(3-CHLOROPROPYL)-BENZYLIDENEMALONONITRILE
250 mg

S91,381-2
2-CHLORO-2'-CYANOACETANILIDE
250 mg

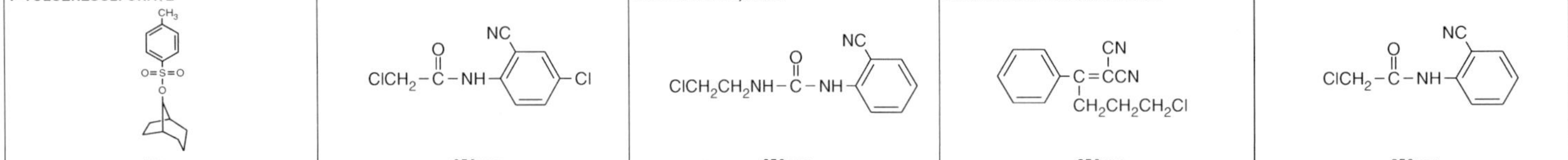

S62,774-7
2-(CHLOROMETHYL)BENZONITRILE
250 mg

S71,699-5
4-CYANOPHENYL N-(2-CHLORO-5-TRIFLUOROMETHYLPHENYL)CARBAMATE
5 gm

S90,198-9
1-[4-CHLORO-3-(TRIFLUOROMETHYL)-PHENYL]-3-(2-CYANOPHENYL)UREA
250 mg

S78,233-5
2-(PHENYLTHIO)-5-(TRIFLUOROMETHYL)-BENZONITRILE
5 gm

S74,426-3
N-(2-TRIFLUOROMETHYLPHENYL)-CYANOFORMAMIDE
250 mg

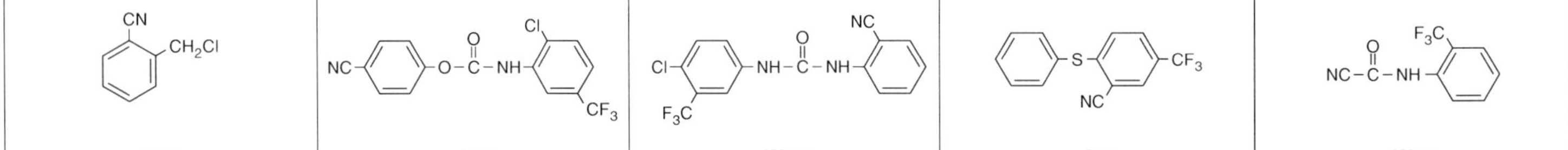

S91,446-0
4-NITRO-ALPHA-[3-(TRIFLUOROMETHYL)-PHENYL]CINNAMONITRILE
250 mg

S70,807-0
4-CYANOPHENYL N-(2-TRIFLUOROMETHYLPHENYL)-CARBAMATE
5 gm

S91,419-3
ALPHA-[3-(TRIFLUOROMETHYL)PHENYL]-CINNAMONITRILE
250 mg

S74,437-9
N-(5-CHLORO-2,4-DIMETHOXYPHENYL)-CYANOTHIOFORMAMIDE
250 mg

S88,727-7
4-CHLORO-ALPHA-(4-METHOXYPHENYL)-CINNAMONITRILE
250 mg

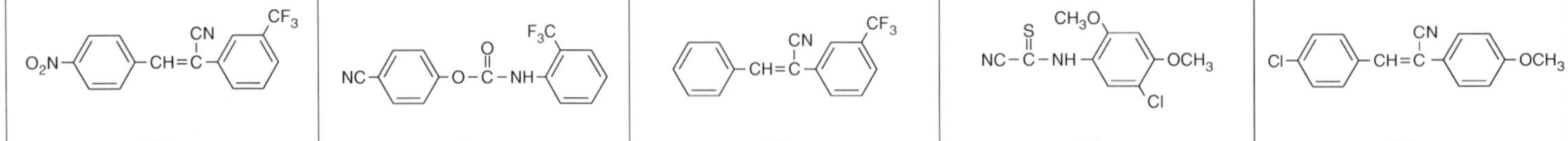

S93,135-7
ETHYL 5-BROMO-ALPHA-CYANO-2-METHOXYCINNAMATE
5 gm

S92,263-3
2-BROMO-ALPHA-CYANO-4,5-DIMETHOXYCINNAMAMIDE
5 gm

S87,942-8
ALPHA-(4-CHLOROPHENYL)-4-METHOXYCINNAMONITRILE
5 gm

S46,186-5
2-CHLORO-6-METHOXYBENZONITRILE
5 gm

S89,457-5
2,3-DICHLORO-3-(2,4,6-TRICHLOROPHENOXY)ACRYLONITRILE
250 mg

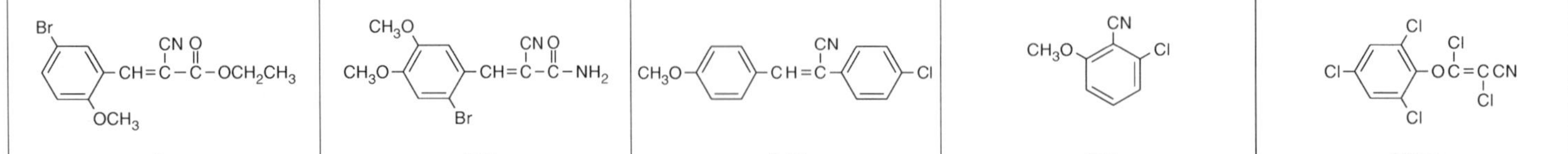

S94,701-6
6-(4-CHLOROPHENYL)-3,5-DICYANO-4-METHYL-4-PHENYL-4H-PYRAN-2-AMINE
50 mg

S91,867-9
4-CYANOBENZYL PENTACHLOROBENZOATE
250 mg

S93,483-6
3-AMINO-4-CYANO-2-(2,4-DICHLOROPHENYL)-7-NITRO-1(2H)-ISOQUINOLINONE
250 mg

S90,860-6
2-AMINO-3-CHLORO-5-NITROBENZONITRILE
250 mg

S92,336-2
4-CHLORO-2-CYANOANILINE
250 mg

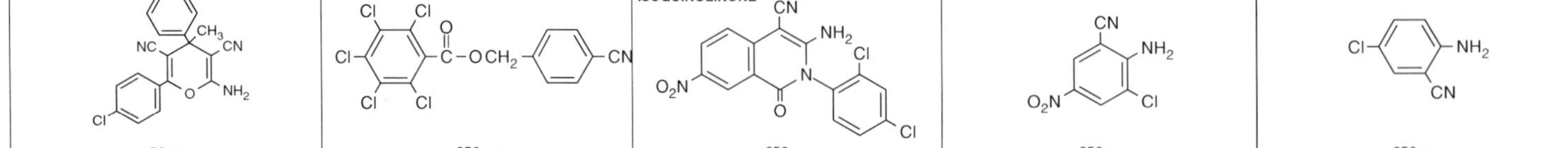

S90,815-0
2-AMINO-5-BROMOBENZONITRILE
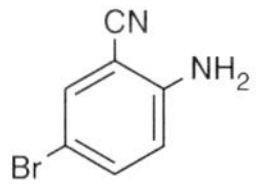
250 mg

S90,816-9
2-AMINO-3,5-DICHLOROBENZONITRILE
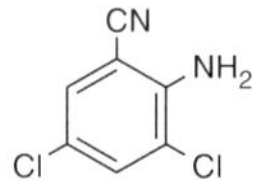
250 mg

S74,430-1
N-(2,3-DICHLOROPHENYL)-CYANOFORMAMIDE
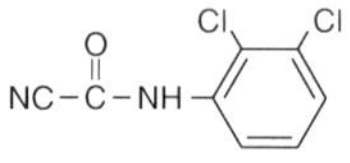
250 mg

S74,429-8
N-(2,5-DIFLUOROPHENYL)-CYANOFORMAMIDE
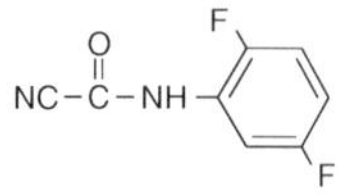
250 mg

S74,427-1
N-(3-CHLORO-4-FLUOROPHENYL)-CYANOFORMAMIDE
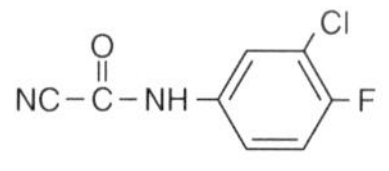
250 mg

S74,432-8
N-(2,5-DICHLOROPHENYL)-CYANOFORMAMIDE
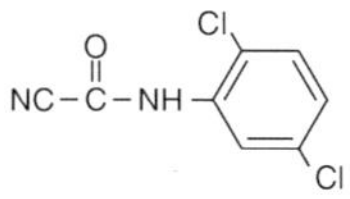
250 mg

S74,434-4
N-(2,6-DICHLOROPHENYL)-CYANOTHIOFORMAMIDE
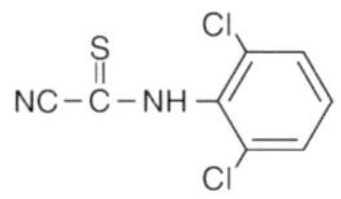
250 mg

S74,436-0
N-(2,3-DICHLOROPHENYL)-CYANOTHIOFORMAMIDE
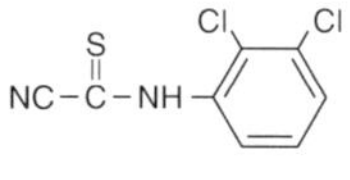
250 mg

S74,442-5
N-(3-CHLORO-2-METHYLPHENYL)-CYANOTHIOFORMAMIDE
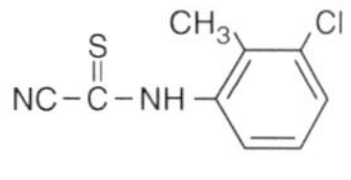
250 mg

S74,435-2
N-(2,5-DICHLOROPHENYL)-CYANOTHIOFORMAMIDE
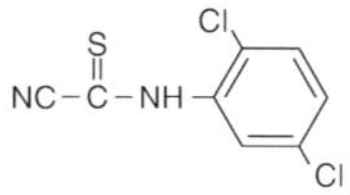
250 mg

S74,441-7
N-(5-CHLORO-2-METHYLPHENYL)-CYANOTHIOFORMAMIDE
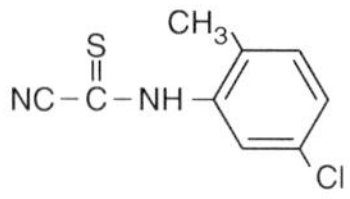
250 mg

S74,443-3
N-(2-CHLORO-6-METHYLPHENYL)-CYANOTHIOFORMAMIDE
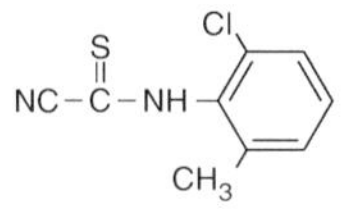
250 mg

S34,908-9
2,4-DICHLORO-3,5-DINITROBENZONITRILE
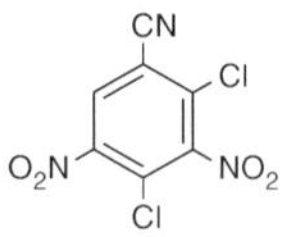
5 gm

S74,497-2
4-CHLORO-2-NITROBENZONITRILE
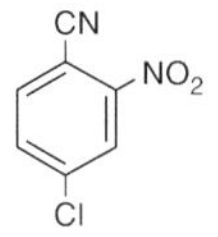
250 mg

S72,036-4
N-(2-CYANOPHENYL)-TETRACHLOROPHTHALAMIC ACID
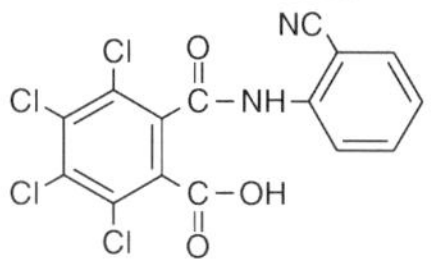
250 mg

S73,698-8
2'-CYANO-3,4,5,6-TETRABROMOPHTHALANILIC ACID
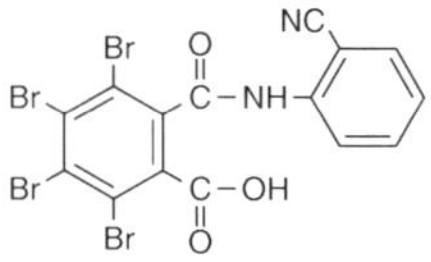
5 gm

S87,846-4
4-CHLORO-ALPHA-CYANOCINNAMAMIDE
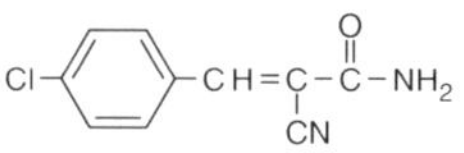
5 gm

S73,551-5
1-(2-CHLOROPHENYL)-3-(2-CYANOPHENYL)UREA
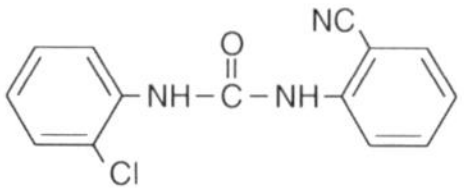
250 mg

S71,228-0
1-(2-CYANOPHENYL)-3-(2,6-DICHLOROPHENYL)UREA
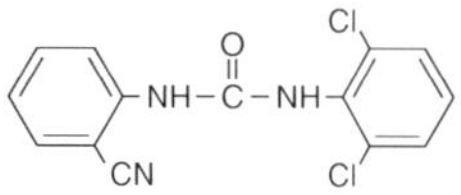
250 mg

S71,536-0
1-(2-CYANOPHENYL)-3-(2,3-DICHLOROPHENYL)UREA
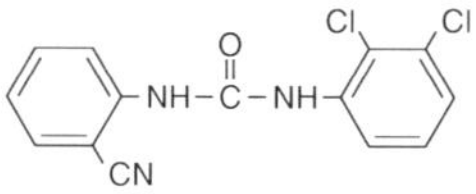
5 gm

S71,226-4
1-(2-CYANOPHENYL)-3-(2,4-DICHLOROPHENYL)UREA
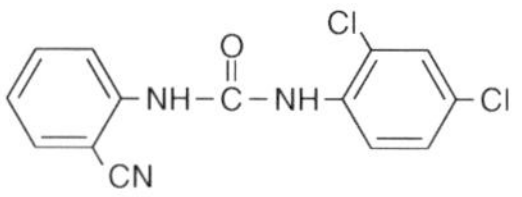
5 gm

S71,227-2
1-(2-CYANOPHENYL)-3-(2,5-DICHLOROPHENYL)UREA
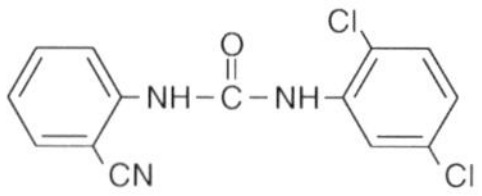
250 mg

S94,706-7
4-CHLOROBENZYLIDENEMALONONITRILE
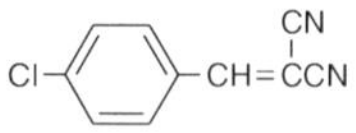
100 mg

S94,705-9
2,4-DICHLOROBENZYLIDENEMALONITRILE
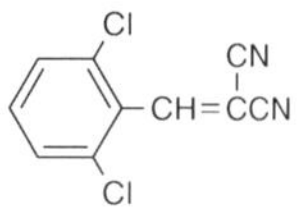
100 mg

S71,000-8
4-CYANOPHENYL N-(2-CHLORO-4-NITROPHENYL)CARBAMATE
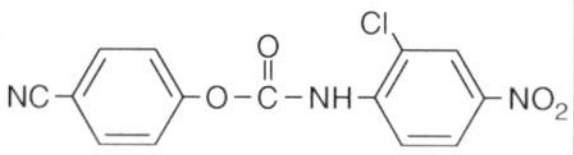
5 gm

S74,680-0
ALPHA-(4-CHLOROBENZYLIDENE)-5-NITRO-O-TOLUNITRILE
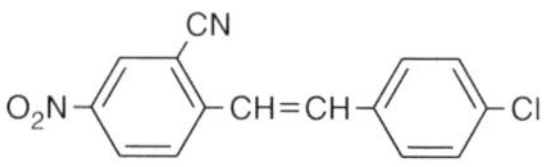
5 gm

S87,837-5
2-(2,4-DICHLOROSTYRYL)-5-NITROBENZONITRILE
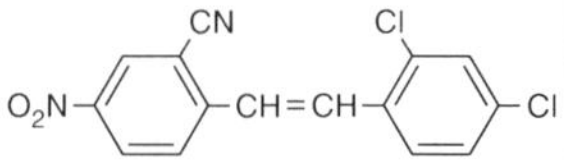
5 gm

S73,546-9
1-(2-CHLOROPHENYL)-3-(3-CYANOPHENYL)UREA
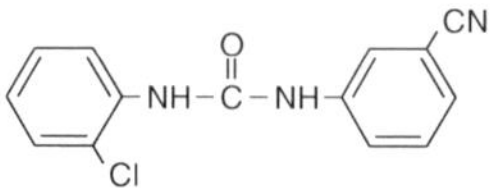
250 mg

S71,229-9
1-(3-CYANOPHENYL)-3-(2,3-DICHLOROPHENYL)UREA
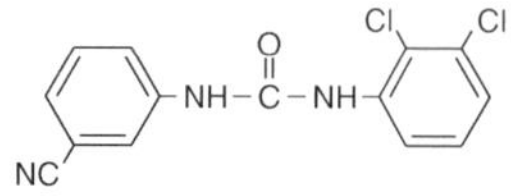
5 gm

S71,231-0
1-(3-CYANOPHENYL)-3-(2,6-DICHLOROPHENYL)UREA
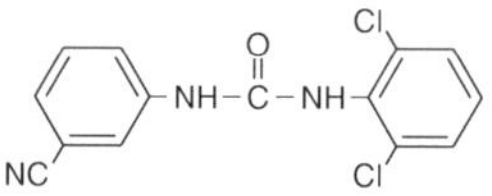
5 gm

S70,913-1
1-(2-CHLORO-6-METHYLPHENYL)-3-(3-CYANOPHENYL)UREA
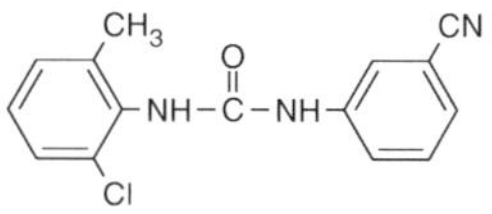
5 gm

S71,230-2
1-(3-CYANOPHENYL)-3-(2,4-DICHLOROPHENYL)UREA
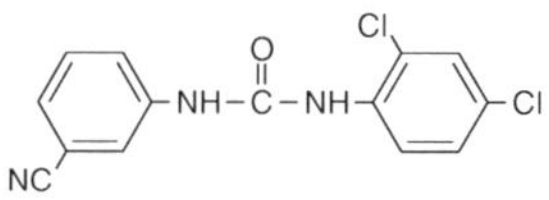
5 gm

S71,509-3
1-(3-CYANOPHENYL)-3-(2,5-DICHLOROPHENYL)UREA
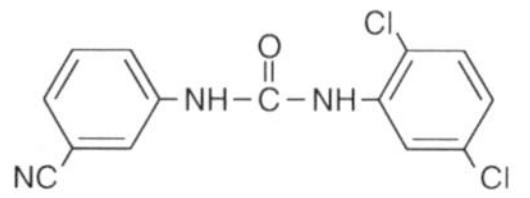
5 gm

S87,944-4
ALPHA-(4-CHLOROPHENYL)CINNAMONITRILE
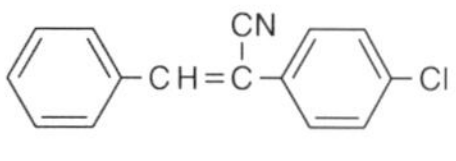
5 gm

S88,725-0
4-CHLORO-ALPHA-PHENYLCINNAMONITRILE
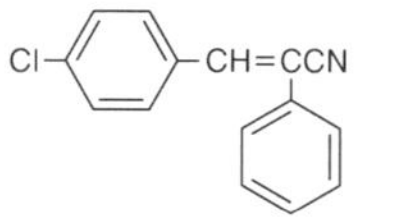
250 mg

S66,819-2
4-CHLORO-ALPHA-(4-CHLOROPHENYL)-CINNAMONITRILE

S87,936-3
ALPHA-(4-CHLOROPHENYL)-BETA-(1-NAPHTHYL)ACRYLONITRILE

S88,240-2
2-(4-BROMOANILINO)-3-CYANO-5,6-DIHYDRO-6-PHENYL-4H-THIOPYRAN-4-ONE

S88,731-5
ALPHA-(4-CHLOROPHENYL)-4-(DIMETHYLAMINO)CINNAMONITRILE

S92,228-5
4-CHLORO-ALPHA-CYANO-3-NITROCINNAMAMIDE

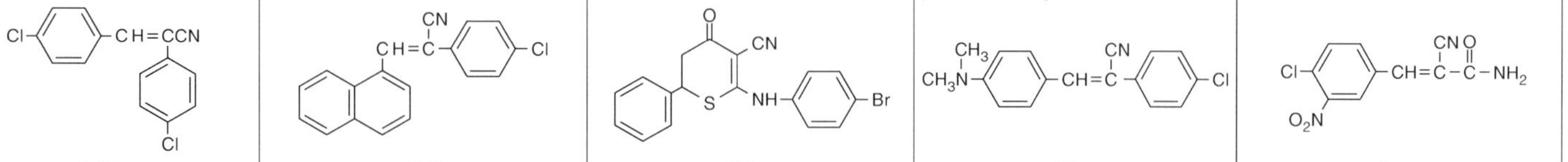

5 gm | 5 gm | 250 mg | 250 mg | 250 mg

S71,390-2
1-(2-CHLORO-4-NITROPHENYL)-3-(4-CYANOPHENYL)UREA

S93,136-5
ETHYL 4-CHLORO-ALPHA-CYANO-3-NITROCINNAMATE

S92,981-6
ALPHA-BENZOYL-4-CHLOROCINNAMONITRILE

S94,497-1
2-IODOBENZOYL CYANIDE

S93,133-0
ETHYL 2-CHLORO-ALPHA-CYANOCINNAMATE

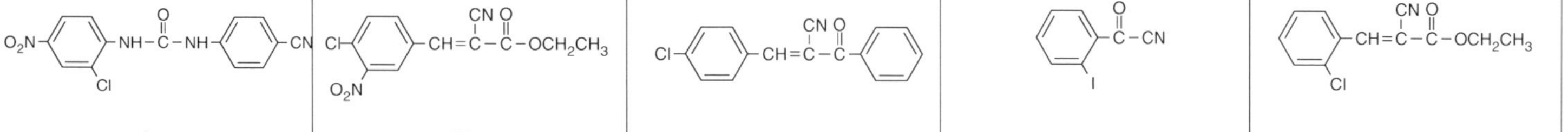

5 gm | 250 mg | 250 mg | 250 mg | 250 mg

S92,969-7
METHYL 4-CHLORO-ALPHA-CYANOCINNAMATE

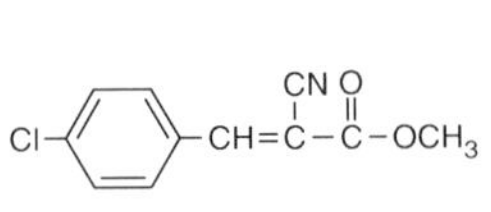

250 mg

S93,131-4
ALPHA-CYANO-2,4-DICHLOROCINNAMAMIDE

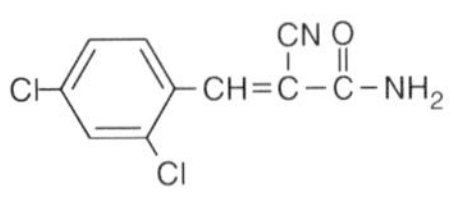

250 mg

S89,859-7
N-(4-CHLOROBENZOYL)-N-METHYL-CYANOTHIOFORMAMIDE

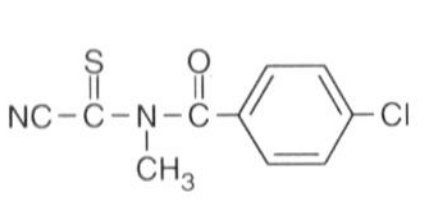

250 mg

S71,479-8
4-CYANOPHENYL N-(2-CHLOROPHENYL)-CARBAMATE

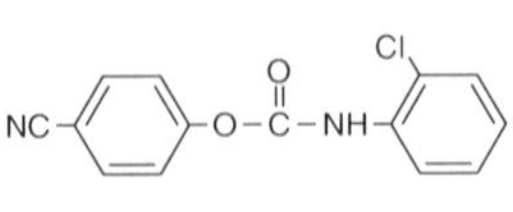

5 gm

S71,426-7
4-CYANOPHENYL N-(4-FLUOROPHENYL)-CARBAMATE

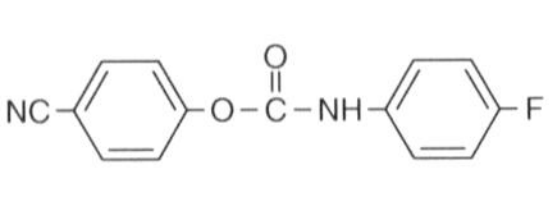

5 gm

S71,668-5
4-CYANOPHENYL N-(3-CHLOROPHENYL)-CARBAMATE

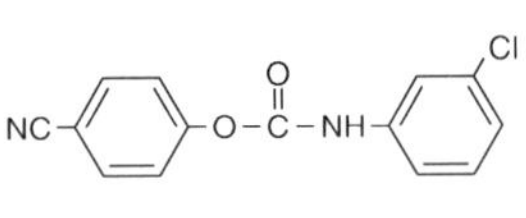

5 gm

S70,964-6
4-CYANOPHENYL N-(4-BROMOPHENYL)-CARBAMATE

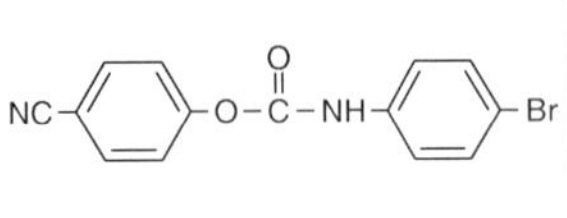

5 gm

S71,770-3
4-CYANOPHENYL N-(2,4,5-TRICHLOROPHENYL)CARBAMATE

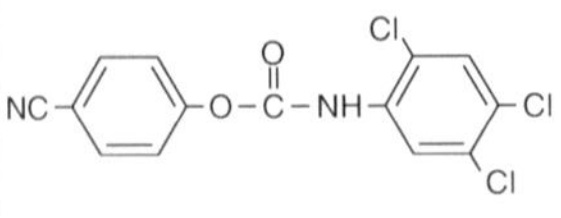

5 gm

S71,082-2
4-CYANOPHENYL N-(2-CHLORO-6-METHYLPHENYL)CARBAMATE

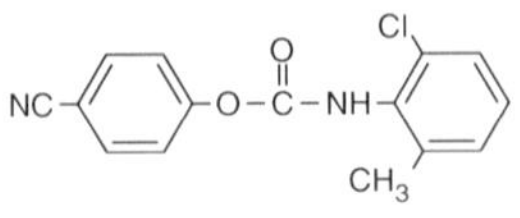

5 gm

S70,965-4
4-CYANOPHENYL N-(3,4-DICHLOROPHENYL)CARBAMATE

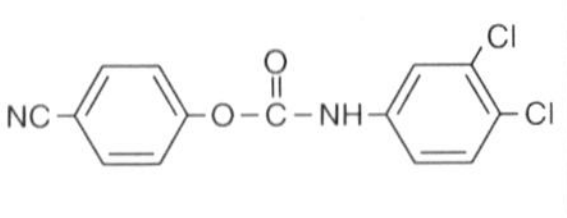

5 gm

S70,981-6
4-CYANOPHENYL N-(2,5-DICHLOROPHENYL)CARBAMATE

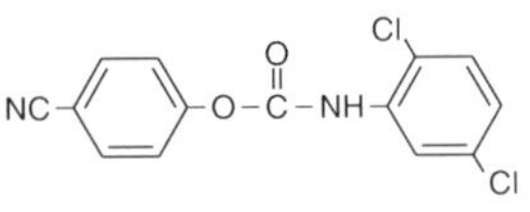

5 gm

S71,725-8
4-CYANOPHENYL N-(2,4-DICHLOROPHENYL)CARBAMATE

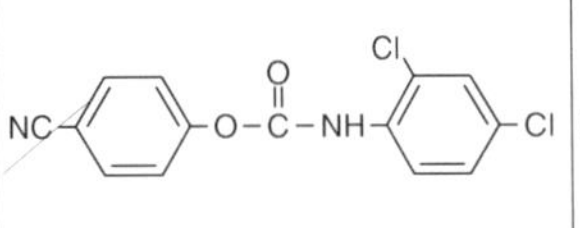

5 gm

S70,995-6
4-CYANOPHENYL N-(3-CHLORO-4-METHYLPHENYL)CARBAMATE

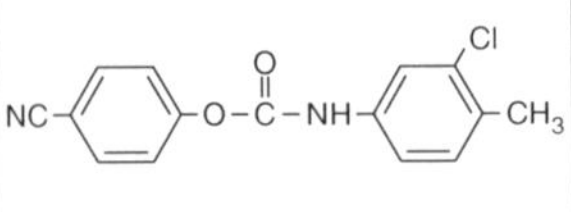

5 gm

S71,006-7
4-CYANOPHENYL N-(4-CHLORO-2-METHYLPHENYL)CARBAMATE

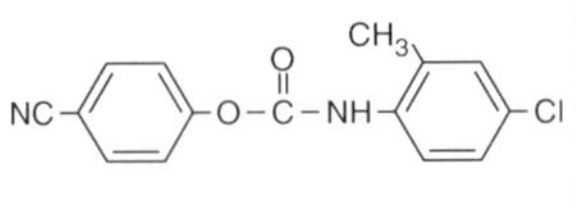

5 gm

S71,083-0
4-CYANOPHENYL N-(5-CHLORO-2-METHYLPHENYL)CARBAMATE

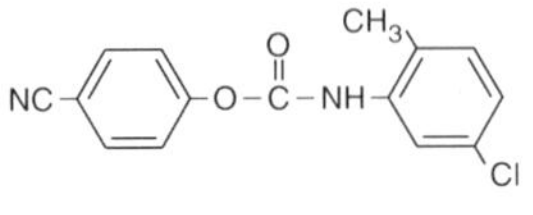

5 gm

S71,130-6
4-CYANOPHENYL N-(3,5-DICHLOROPHENYL)CARBAMATE

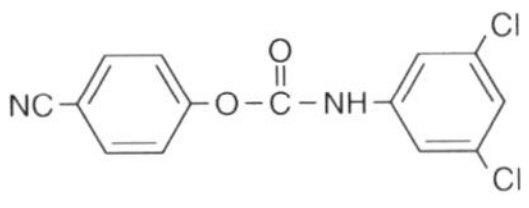

5 gm

S71,364-3
4-CYANOPHENYL N-(2,4,6-TRICHLOROPHENYL)CARBAMATE

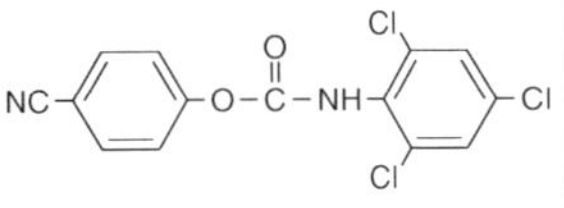

5 gm

S71,222-1
4-CYANOPHENYL N-(2,4,6-TRIBROMOPHENYL)CARBAMATE

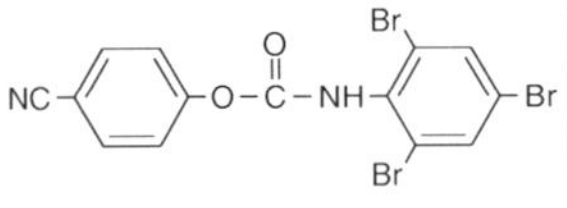

5 gm

S70,954-9
4-CYANOPHENYL N-(2,6-DICHLOROPHENYL)CARBAMATE

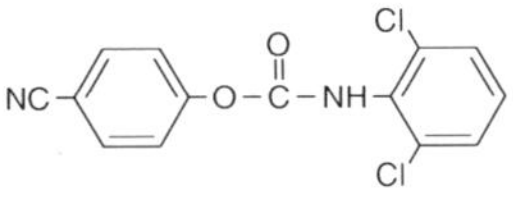

5 gm

S89,811-2
N-(3-FLUOROPHENYL)CYANOFORMAMIDE

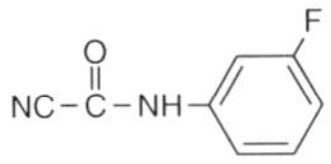

250 mg

S89,872-4
N-(2-FLUOROPHENYL)CYANOFORMAMIDE

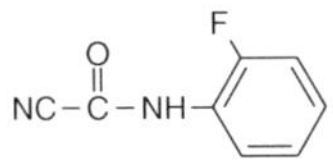

250 mg

S89,808-2
N-(2-CHLOROPHENYL)CYANOFORMAMIDE

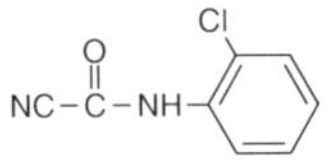

250 mg

S89,809-0
N-(4-CHLOROPHENYL)CYANOFORMAMIDE

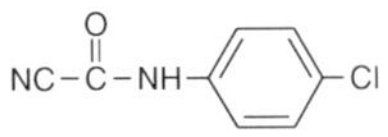

250 mg

S89,810-4
N-(4-BROMOPHENYL)CYANOFORMAMIDE

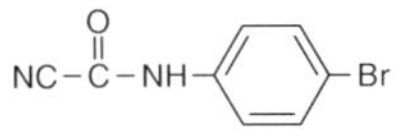

250 mg

S89,807-4
N-(3-CHLOROPHENYL)-CYANOTHIOFORMAMIDE

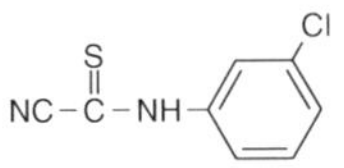

250 mg

S89,823-6
N-(4-FLUOROPHENYL)-CYANOTHIOFORMAMIDE

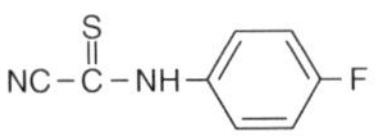

250 mg

S65,985-1
1-(4-CYANOPHENYL)-3-(2,4-DICHLOROPHENYL)UREA

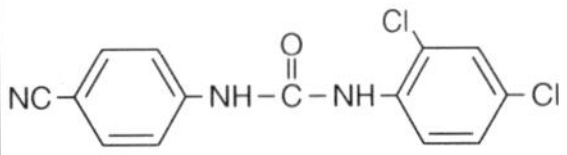

250 mg

S91,723-0
1-(4-CHLORO-2-CYANOPHENYL)-3-ISOPROPYLUREA

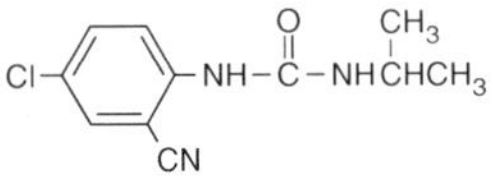

250 mg

S71,234-5
1-(3-CHLOROPHENYL)-3-(2-CYANOPHENYL)UREA

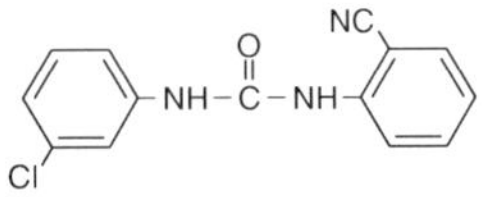

250 mg

S70,841-0
1-(4-CYANOPHENYL)-3-(2,4,5-TRICHLOROPHENYL)UREA

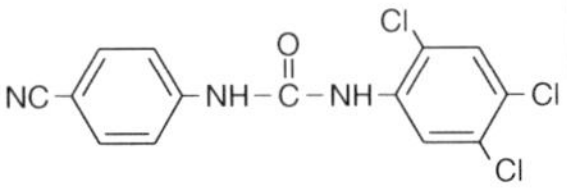

5 gm

S71,654-5
1-(4-CYANOPHENYL)-3-(2,3-DICHLOROPHENYL)UREA

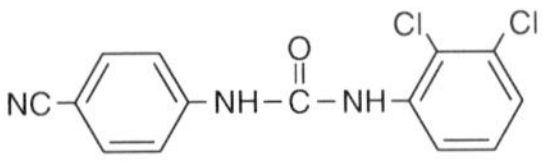

5 gm

S70,953-0
1-(2-CHLORO-6-METHYLPHENYL)-3-(4-CYANOPHENYL)UREA

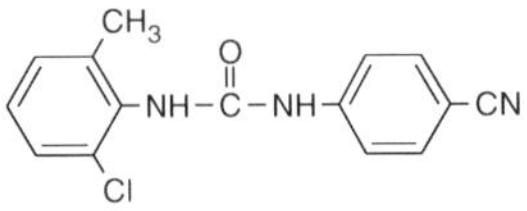

5 gm

S70,838-0
1-(4-CYANOPHENYL)-3-(2,5-DICHLOROPHENYL)UREA

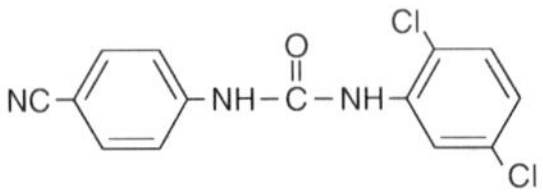

5 gm

S71,043-1
1-(4-CHLORO-2-METHYLPHENYL)-3-(4-CYANOPHENYL)UREA

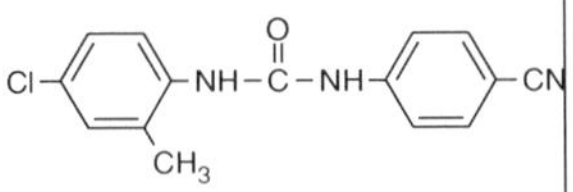

5 gm

S71,055-5
1-(5-CHLORO-2-METHYLPHENYL)-3-(4-CYANOPHENYL)UREA

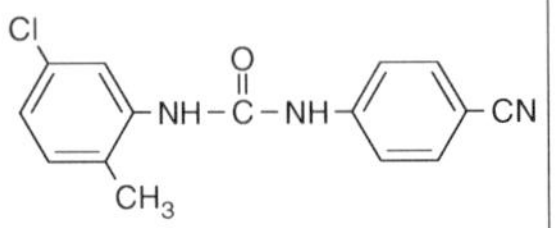

5 gm

S71,152-7
1-(3-CHLORO-4-METHYLPHENYL)-3-(4-CYANOPHENYL)UREA

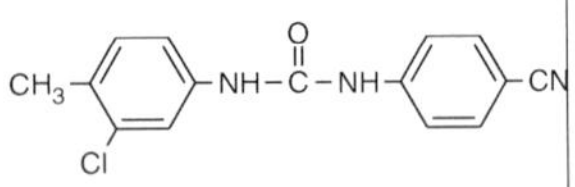

5 gm

S41,533-2
PENTAFLUOROBENZONITRILE

250 mg

S38,986-2
2,5-DICHLOROBENZONITRILE

5 gm

S93,353-8
2-BROMO-4-CHLOROBENZONITRILE

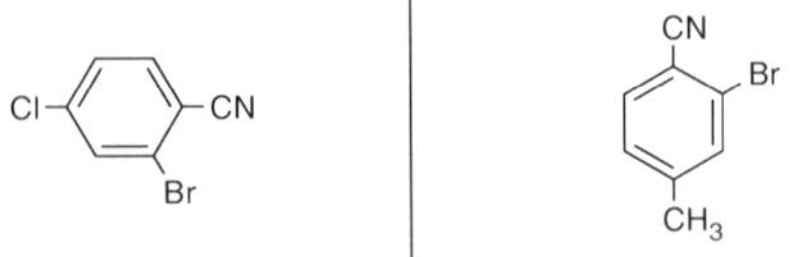

250 mg

S86,345-9
2-BROMO-4-METHYLBENZONITRILE

250 mg

S92,222-6
2-CHLORO-1,3,5-TRICYANOBENZENE

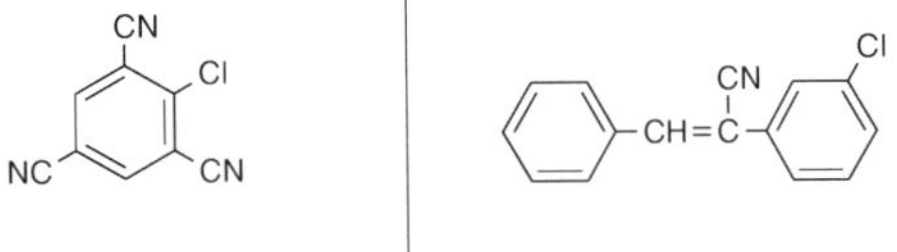

250 mg

S91,411-8
ALPHA-(3-CHLOROPHENY)-CINNAMONITRILE

250 mg

S84,950-2
2-IODOCINNAMONITRILE

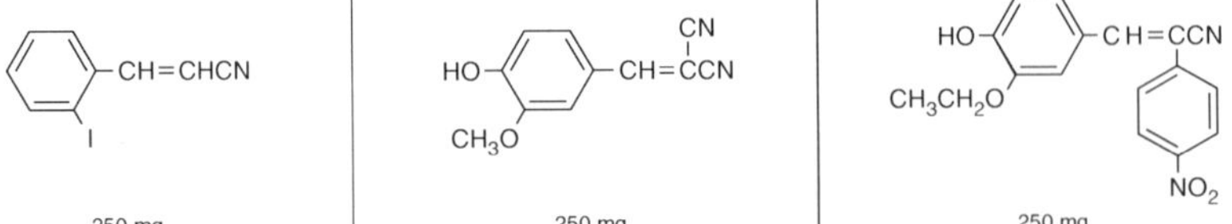

250 mg

S58,685-4
VANILLYLIDENEMALONONITRILE

250 mg

S53,987-2
3-ETHOXY-4-HYDROXY-ALPHA-(4-NITROPHENYL)CINNAMONITRILE

250 mg

S66,094-9 ALPHA-CYANO-4-HYDROXY-3-METHOXYCINNAMIC ACID — 5 gm

S66,109-0 ETHYL ALPHA-CYANO-4-HYDROXY-3-METHOXYCINNAMATE — 5 gm

S53,470-6 ALPHA,ALPHA'-PIPERONYLIDENEBIS(3,5-DICHLORO-2-HYDROXYCINNAMONITRILE) — 5 gm

S82,650-2 4-(1-HYDROXY-1-PHENYLETHYL)-BENZONITRILE — 250 mg

S82,652-9 4-[(DIPHENYL)HYDROXYMETHYL]-BENZONITRILE — 250 mg

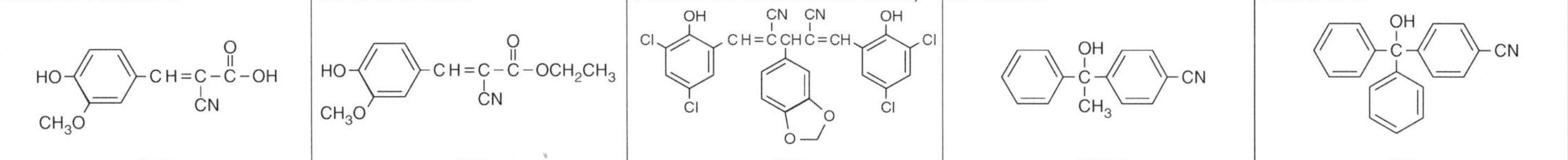

S91,856-3 4-(1-HYDROXY-1-METHYLETHYL)-BENZONITRILE — 250 mg

S51,886-7 ETHYL BETA-CYANO-2 AND 4-ETHYL-ALPHA-HYDROXYCINNAMATE — 5 gm

S51,881-6 ETHYL BETA-CYANO-ALPHA-HYDROXY-4-ISOPROPYLCINNAMATE — 5 gm

S51,883-2 METHYL BETA-CYANO-ALPHA-HYDROXYCINNAMATE — 5 gm

S51,891-3 PROPYL BETA-CYANO-ALPHA-HYDROXYCINNAMATE — 5 gm

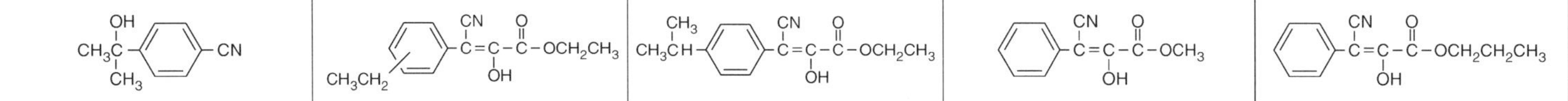

S51,880-8 BUTYL BETA-CYANO-ALPHA-HYDROXYCINNAMATE — 250 mg

S52,664-9 1-METHYLBUTYL BETA-CYANO-ALPHA-HYDROXYCINNAMATE — 250 mg

S52,678-9 HEXYL BETA-CYANO-ALPHA-HYDROXYCINNAMATE — 250 mg

S51,902-2 ETHYL BETA-CYANO-ALPHA-HYDROXY-2-METHYLCINNAMATE — 5 gm

S51,897-2 ETHYL 4-(TERT-BUTYL)-BETA-CYANO-ALPHA-HYDROXYCINNAMATE — 5 gm

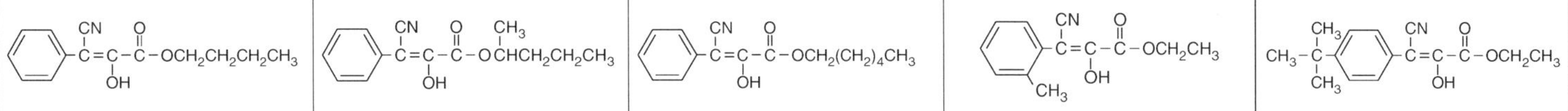

S51,892-1 ETHYL BETA-CYANO-2,3-DIMETHYL-ALPHA-HYDROXYCINNAMATE — 250 mg

S51,882-4 ETHYL BETA-CYANO-3,4-DIMETHYL-ALPHA-HYDROXYCINNAMATE — 5 gm

S51,894-8 ETHYL BETA-CYANO-2,4-DIMETHYL-ALPHA-HYDROXYCINNAMATE — 5 gm

S52,071-3 ETHYL BETA-CYANO-2,5-DIMETHYL-ALPHA-HYDROXYCINNAMATE — 5 gm

S51,759-3 ETHYL BETA-CYANO-ALPHA-HYDROXY-4-METHYLCINNAMATE — 250 mg

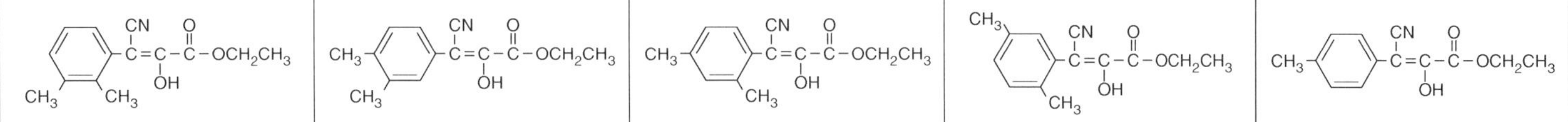

S51,871-9 ETHYL BETA-CYANO-4-ETHYL-ALPHA-HYDROXYCINNAMATE — 5 gm

S44,009-4 4-HYDROXYBENZYLIDENEMALONONITRILE — 5 gm

S64,247-9 4-CHLORO-4'-CYANO-1-HYDROXY-2-NAPHTHANILIDE — 250 mg

S92,224-2 2-(2-HYDROXYETHYLAMINO)-5-NITROBENZONITRILE — 250 mg

S51,888-3 ETHYL BETA-CYANO-ALPHA-HYDROXY-2-(1-NAPHTHYL)CINNAMATE — 5 gm

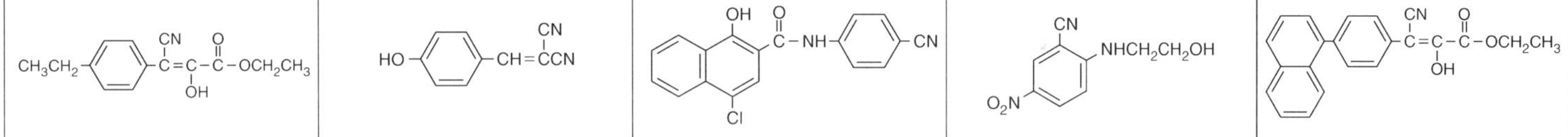

S52,369-0 ALPHA-CYANO-2,4-DICHLORO-6-HYDROXYCINNAMAMIDE — 5 gm

S52,437-9 ALPHA-CYANO-3,5-DIBROMO-2-HYDROXYCINNAMAMIDE — 5 gm

S52,506-5 ALPHA-CYANO-2-HYDROXY-1-NAPHTHALENEACRYLAMIDE — 5 gm

S53,422-6 ISOPROPYL BETA-CYANO-ALPHA-HYDROXYCINNAMATE — 250 mg

S89,623-3 4-HYDROXY-2,3,5,6-TETRAPHENYLBENZONITRILE — 250 mg

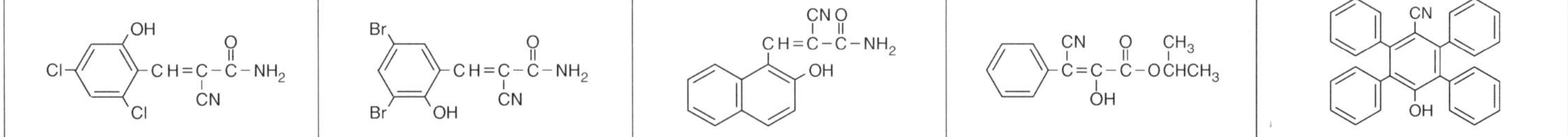

S45,233-5 3,5-DIIODO-4-HYDROXYBENZONITRILE — 5 gm

S94,702-4 3,5-DICYANO-4-(4-METHOXYPHENYL)-4-METHYL-6-PHENYL-4H-PYRAN-2-AMINE — 50 mg

S65,142-7 ETHYL 4-BENZYLOXY-ALPHA-CYANOCINNAMATE — 5 gm

S59,988-3 2,6-DIBENZYLOXYBENZONITRILE — 250 mg

S89,464-8 3,3-BIS(4-NITROPHENOXY)-2-CHLOROACRYLONITRILE — 250 mg

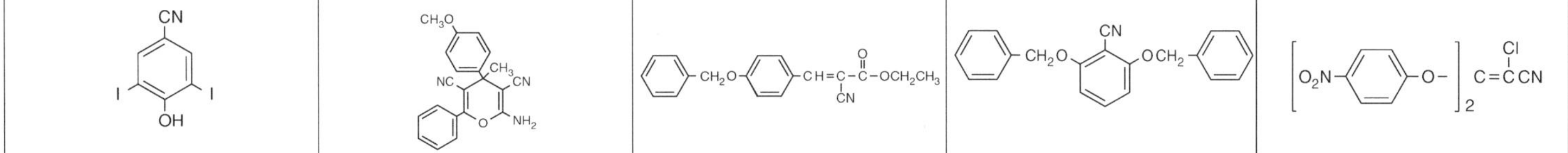

S94,707-5 3,4-DIMETHOXYBENZYLIDENE-MALONONITRILE — 100 mg

S95,387-3 2-CYANO-2-(2,3-DIMETHOXYBENZYLIDENE)ACETIC ACID — 50 mg

S71,649-9 2,5-DIMETHOXYBENZYLIDENE-MALONONITRILE — 250 mg

S71,752-5 1-(4-CYANOPHENYL)-3-(2,5-DIMETHOXYPHENYL)UREA — 5 gm

S87,716-6 3,4,5-TRIMETHOXYBENZOYL CYANIDE — 250 mg

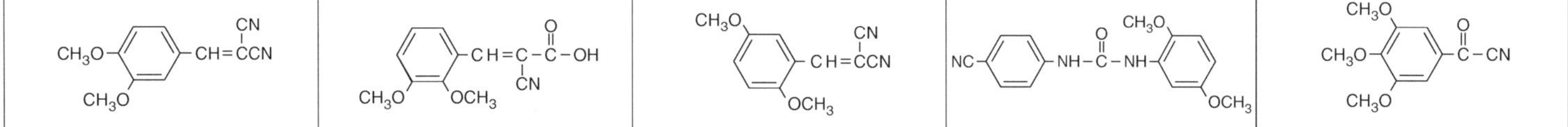

S87,855-3 ALPHA-CYANO-4,5-DIMETHOXY-2-NITROCINNAMAMIDE — 5 gm

S71,081-4 4-CYANOPHENYL N-(2,5-DIMETHOXYPHENYL)CARBAMATE — 5 gm

S69,481-9 2-CYANOPHENYL N-(2,5-DIMETHOXYPHENYL)CARBAMATE — 250 mg

S93,132-2 ALPHA-CYANO-2,3-DIMETHOXYCINNAMAMIDE — 250 mg

S95,588-4 2-BENZYL-6-CYANO-3A-(2,3-DIMETHOXY-PHENYL)-3A,4,7,7A-TETRAHYDRO-1,3-ISOINDOLINEDIONE — 50 mg

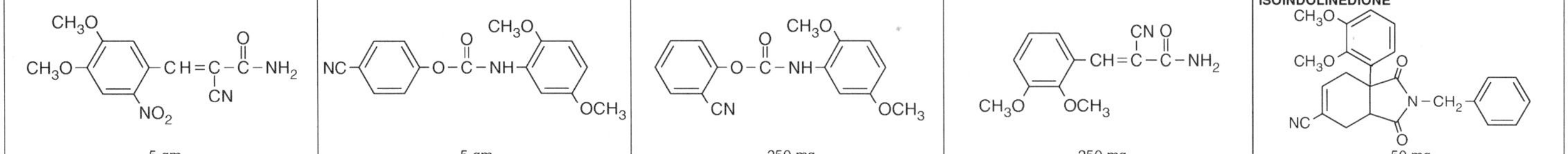

S53,180-4 (4-BENZYLOXY-3-METHOXYBENZYLIDENE)-MALONONITRILE — 5 gm

S53,258-4 ETHYL ALPHA-CYANO-2,3-DIMETHOXYCINNAMATE — 5 gm

S53,002-6 ETHYL ALPHA-CYANO-3,4-DIMETHOXYCINNAMATE — 5 gm

S53,184-7 ETHYL 4-BENZYLOXY-ALPHA-CYANO-3-METHOXYCINNAMATE — 5 gm

S52,605-3 ALPHA-(4-CHLOROPHENYL)-3,4-DIMETHOXYCINNAMONITRILE — 5 gm

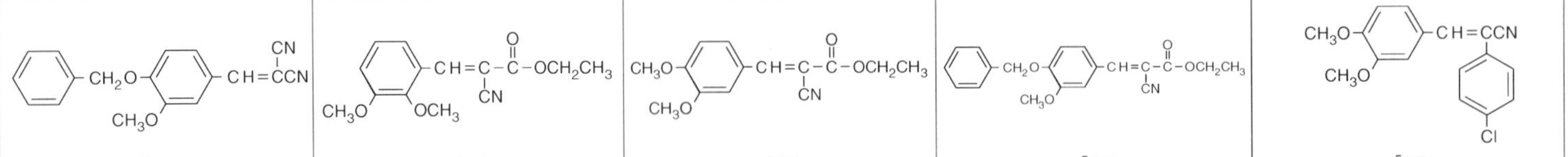

S53,236-3 3,4-DIMETHOXY-ALPHA-(PARA-TOLYL)-CINNAMONITRILE — 5 gm

S52,725-4 3,4-DIMETHOXY-ALPHA-(4-METHOXYPHENYL)CINNAMONITRILE — 250 mg

S41,151-5 3,4-DIMETHOXY-ALPHA-PHENYLCINNAMONITRILE — 5 gm

S95,370-9 ETHYL N-(2-CYANO-4-METHYL-6-NITROPHENYL)CARBAMATE — 50 mg

S55,907-5 6-NITRO-O-ANISONITRILE — 250 mg

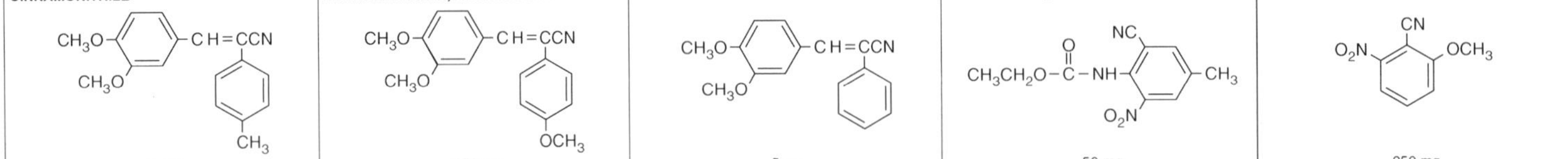

S87,852-9 ALPHA-CYANO-2-METHOXYCINNAMAMIDE — 5 gm

S53,780-2 3-BROMO-ALPHA-CYANO-4-METHOXYCINNAMAMIDE — 5 gm

S87,853-7 ALPHA-CYANO-4-METHOXYCINNAMAMIDE — 5 gm

S52,424-7 ALPHA-CYANO-3,4-DIMETHOXYCINNAMAMIDE — 5 gm

S43,435-3 4-METHOXYBENZYLIDENEMALONONITRILE — 5 gm

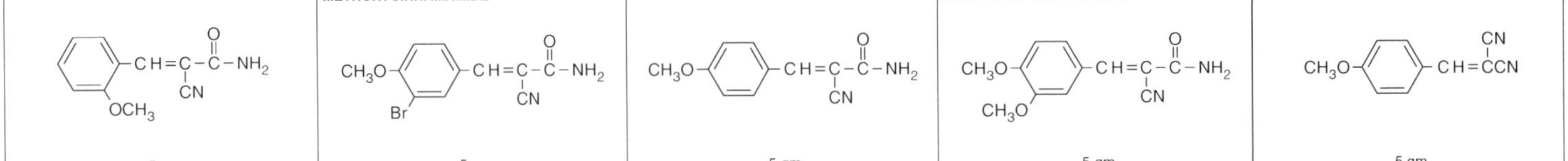

S54,945-2 (4-ETHOXYBENZYLIDENE)MALONONITRILE — 250 mg

S55,473-1 (3,5-DIBROMO-2-ETHOXYBENZYLIDENE)-MALONONITRILE — 250 mg

S74,679-7 ALPHA-(4-METHOXYBENZYLIDENE)-5-NITRO-O-TOLUNITRILE — 5 gm

S88,732-3 4-(DIMETHYLAMINO)-ALPHA-(4-METHOXYPHENYL)CINNAMONITRILE — 250 mg

S52,716-5 ALPHA-(4-METHOXYPHENYL)-3-NITROCINNAMONITRILE — 250 mg

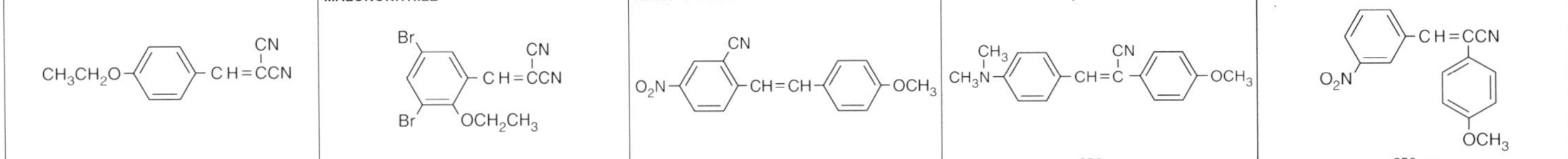

S66,055-8 2-METHOXY-ALPHA-PHENYLCINNAMONITRILE — 250 mg

S93,155-1 2-CYANO-2-(4-METHOXYPHENYLHYDRAZONO)ACETAMIDE — 250 mg

S66,086-8 ETHYL ALPHA-CYANO-4-METHOXY-3-NITROCINNAMATE — 5 gm

S71,259-0 4-CYANOPHENYL N-(2-METHOXYPHENYL)-CARBAMATE — 5 gm

S70,961-1 4-CYANOPHENYL N-(2-METHOXY-5-METHYLPHENYL)CARBAMATE — 5 gm

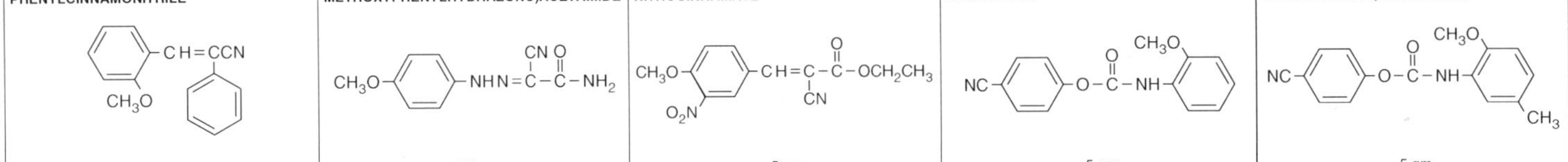

S89,806-6 N-(2-METHOXYPHENYL)CYANOFORMAMIDE — 250 mg

S70,796-1 1-(4-CYANOPHENYL)-3-(2-METHOXYPHENYL)UREA — 5 gm

S52,599-5 4-ETHOXY-ALPHA-(4-NITROPHENYL)-CINNAMONITRILE — 5 gm

S92,980-8 ALPHA-BENZOYL-4-METHOXYCINNAMONITRILE — 250 mg

S43,249-0 ETHYL 3-BENZYLOXY-ALPHA-CYANOCINNAMATE — 5 gm

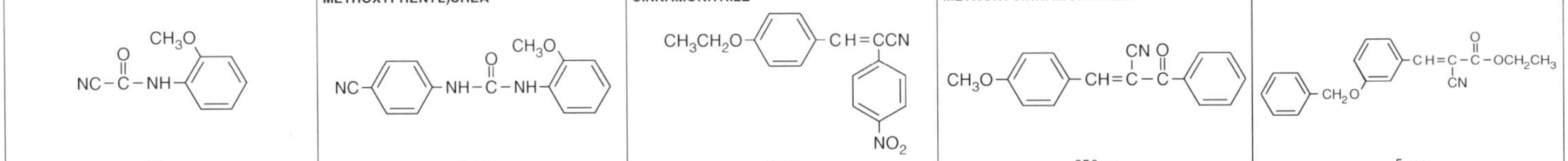

S53,189-8 ETHYL 3-BROMO-ALPHA-CYANO-4-METHOXYCINNAMATE — 5 gm

S53,260-6 ETHYL ALPHA-CYANO-2,5-DIMETHOXYCINNAMATE — 5 gm

S52,489-1 ALPHA-CYANO-2,4-DIMETHOXYCINNAMAMIDE — 5 gm

S53,828-0 ALPHA-CYANO-2,5-DIMETHOXYCINNAMAMIDE — 5 gm

S70,951-4 4-CYANOPHENYL N-(4-METHOXYPHENYL)-CARBAMATE — 5 gm

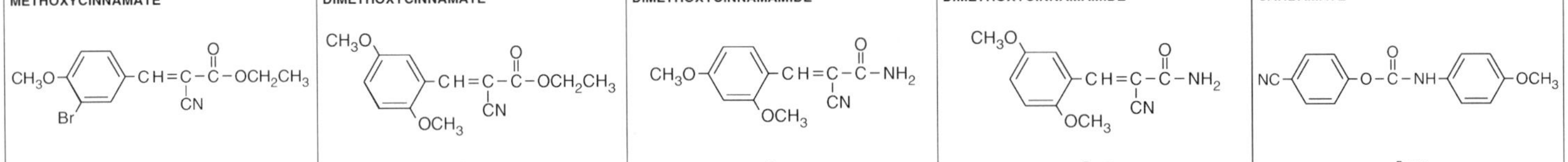

S71,714-2 4-CYANOPHENYL N-(3-METHOXYPHENYL)-CARBAMATE — 5 gm

S93,134-9 ETHYL ALPHA-CYANO-2-METHOXYCINNAMATE — 250 mg

S53,172-3 ETHYL ALPHA-CYANO-4-METHOXYCINNAMATE — 5 gm

S53,007-7 ETHYL ALPHA-CYANO-2,4-DIMETHOXYCINNAMATE — 250 mg

S53,006-9 4'-(BETA-CYANO-4-METHOXYSTYRYL)-ACETANILIDE — 250 mg

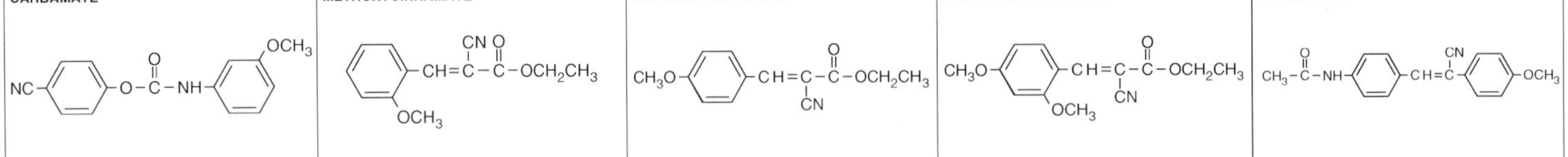

S70,795-3 1-(4-CYANOPHENYL)-3-(4-METHOXYPHENYL)UREA — 250 mg

S71,508-5 1-(4-CYANOPHENYL)-3-(3-METHOXYPHENYL)UREA — 250 mg

S44,790-0 2,5-DIMETHYL-PARA-ANISONITRILE — 5 gm

S76,340-3 5-CYANO-2-ETHOXY-6-METHYL-4-PHENYL-1,2,3,4-TETRAHYDROPYRIDINE — 250 mg

S91,600-5 BETA-METHOXY-ALPHA-PHENYLCINNAMONITRILE — 250 mg

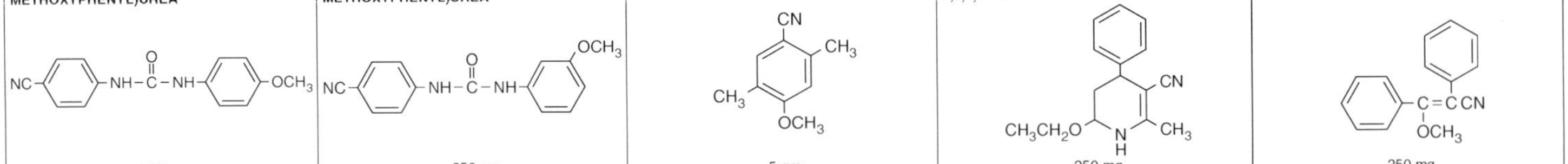

S53,073-5
ALPHA-(4-METHOXYPHENYL) CINNAMONITRILE
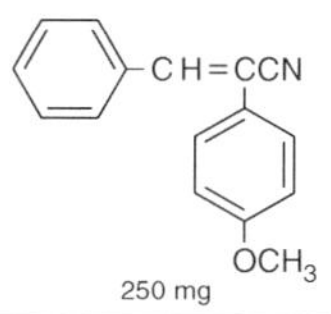
250 mg

S53,659-8
ALPHA-(3-CHLOROPHENYL)-4-METHOXYCINNAMONITRILE
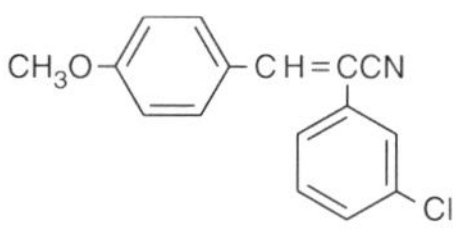
5 gm

S34,233-5
2,3-BIS(4-METHOXYPHENYL) ACRYLONTIRILE
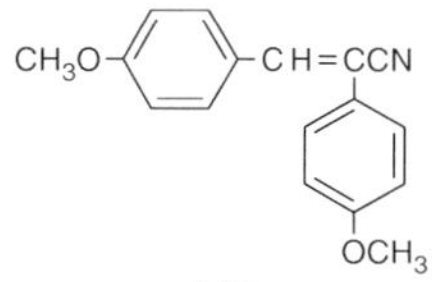
5 gm

S53,005-0
2,6-DIMETHOXY-ALPHA-(4-METHOXYPHENYL)CINNAMONITRILE
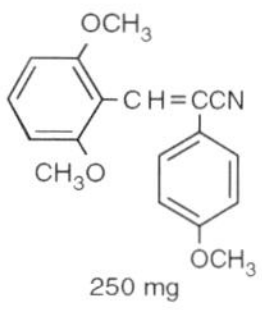
250 mg

S52,976-1
2,4-DICHLORO-ALPHA-(4-METHOXYPHENYL)CINNAMONITRILE
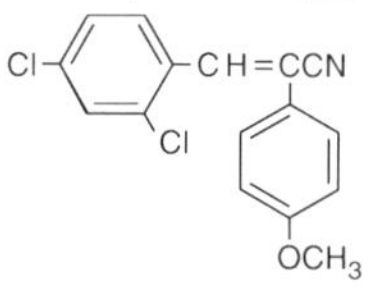
250 mg

S53,133-2
ALPHA-(5-BROMO-2-METHOXYBENZYLIDENE)-1-NAPHTHALENEACETONITRILE
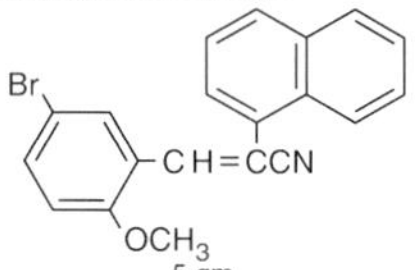
5 gm

S94,703-2
3,5-DICYANO-4-METHYL-4-(4-NITROPHENYL)-6-PHENYL-4H-PYRAN-2-AMINE
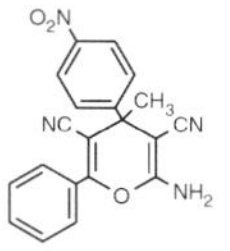
50 mg

S94,700-8
3,5-DICYANO-4,6-DIPHENYL-4-METHYL-4H-PYRAN-2-AMINE
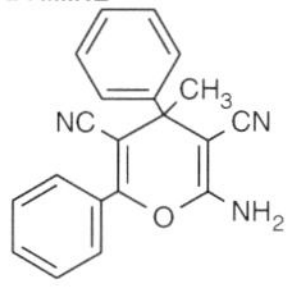
100 mg

S91,782-6
4-CYANOBENZYL 2,4,6-TRIMETHYLBENZOATE
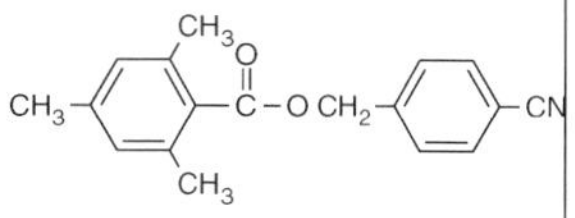
250 mg

S50,981-7
1,2-DIHYDRO-4,6-DIMETHYL-2-OXO-1-(4,5-TRIMETHOXYBENZYLIDENEAMINO)-NICOTINONITRILE
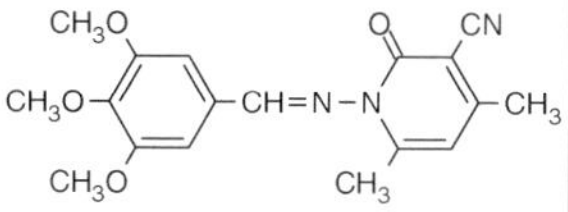
250 mg

S87,889-8
2-BROMO-ALPHA-CYANO-4,5-(METHYLENEDIOXY)CINNAMAMIDE
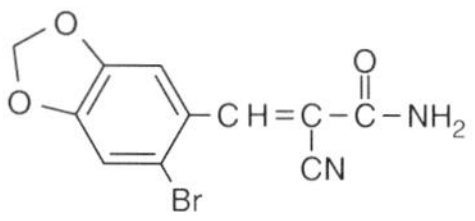
5 gm

S94,610-9
4-CYANOBENZIL
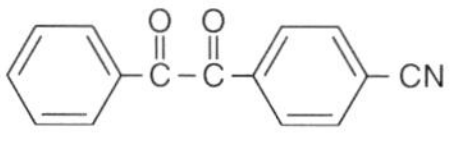
250 mg

S94,709-1
ETHYL 6-AMINO-1-BENZAMIDO-5-CYANO-4-(4-NITROPHENYL)-2-OXOTETRAHYDRO-3-PYRIDINECARBOXYLATE
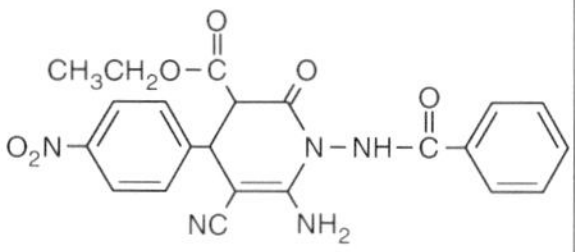
50 mg

S67,173-8
2,4-DIACETOXYBENZONITRILE
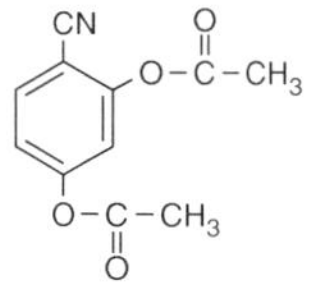
5 gm

S88,888-5
ALPHA-CYANO-BETA-METHYL-2-(METHYLTHIO)CINNAMIC ACID
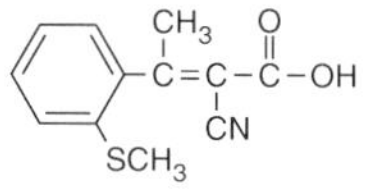
250 mg

S88,893-1
ALPHA-CYANO-BETA,5-DIMETHYL-2-(METHYLTHIO)CINNAMIC ACID
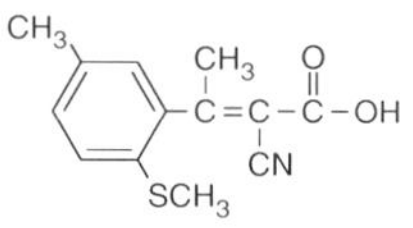
250 mg

S74,847-1
3-CYANO-1-METHYL-4-(METHYLTHIO)-6-PHENYL-2(1H)-PYRIDINONE
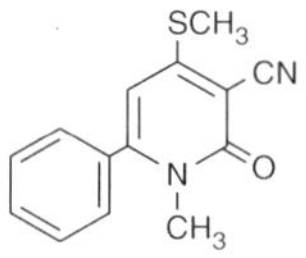
250 mg

S93,351-1
2-(PHENYLTHIO)BENZONITRILE
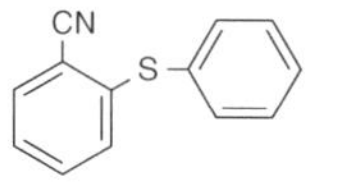
250 mg

S91,937-3
4-[1-METHYL-1-(METHYLTHIO)ETHYL]-BENZONITRILE
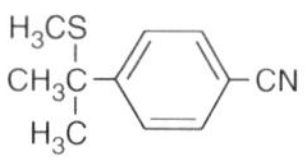
250 mg

S89,388-9
3,3-BIS(2-NAPHTHYLTHIO)-2-CHLOROACRYLONITRILE
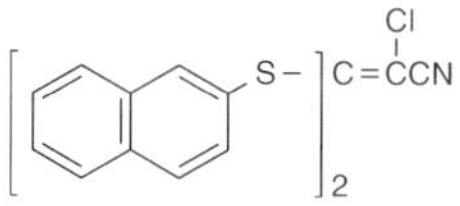
250 mg

S89,389-7
2,3,3-TRIS(2-NAPHTHYLTHIO)-ACRYLONITRILE
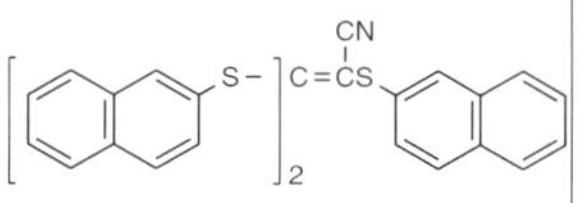
250 mg

S87,523-6
2,2'-DITHIOBIS(BENZONITRILE)-
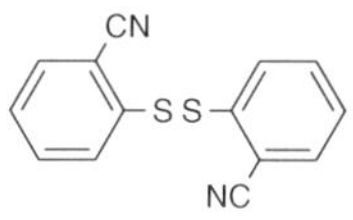
250 mg

S90,861-4
2-AMINO-3,5-DINITROBENZONITRILE
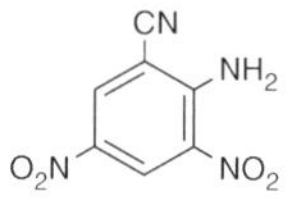
250 mg

S53,709-8
4-AMINO-1-PHENYL-3-CYCLOHEXENE-1,3-DICARBONITRILE
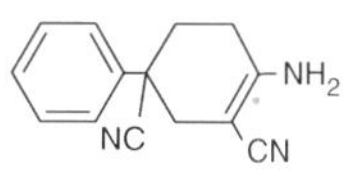
250 mg

S93,482-8
3-AMINO-4-CYANO-2-METHYL-7-NITRO-1(2H)-ISOQUINOLINONE
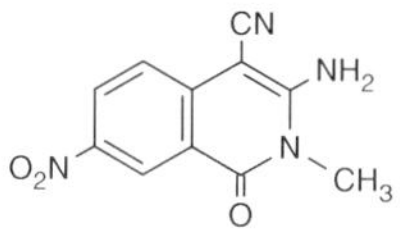
250 mg

S93,628-6
3-AMINO-2-BENZYL-4-CYANO-7-NITRO-1(2H)-ISOQUINOLINONE
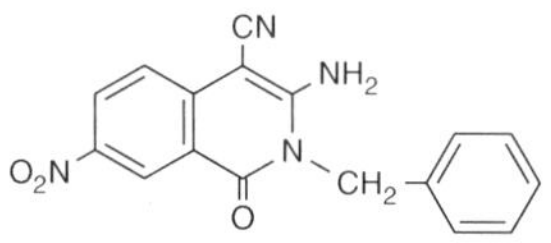
250 mg

S94,010-0
3-AMINO-4-CYANO-2-ISOPROPYL-7-NITRO-1(2H)-ISOQUINOLINONE
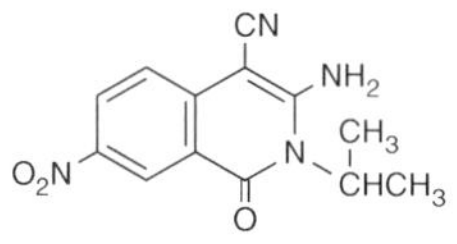
250 mg

S94,009-7
3-AMINO-4-CYANO-2-HEXYL-7-NITRO-1(2H)-ISOQUINOLINONE
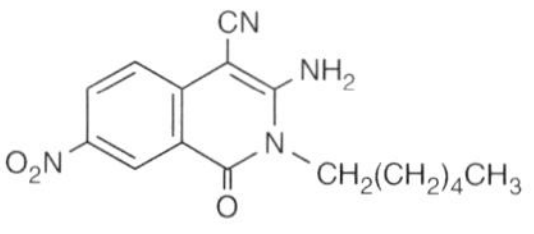
250 mg

S93,630-8
3-AMINO-4-CYANO-7-NITRO-2-PHENYL-1(2H)-ISOQUINOLINONE
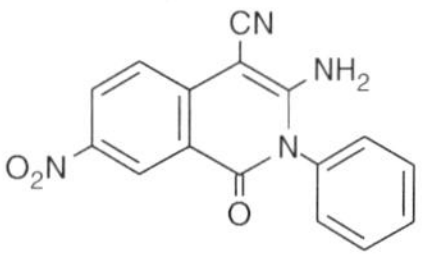
250 mg

S64,736-5
5'-AMINO-2'-CYANOACETANILIDE
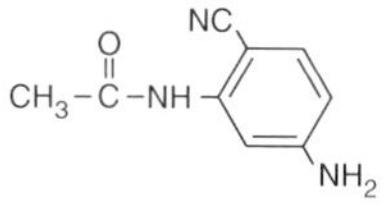
5 gm

S67,769-8
3-AMINO-4-CYANO-5,6,7,8-TETRAHYDRO-1H-2-BENZOTHIOPYRAN-1-THIONE
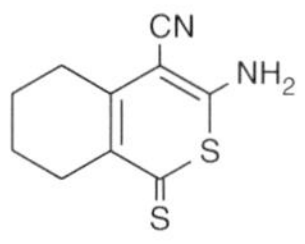
5 gm

S67,915-1
3-AMINOBENZONITRILE HYDROCHLORIDE
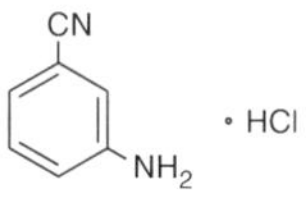
250 mg

S93,492-5
5-AMINO-7-CYANO-10-NITRO-12H-ISOQUINO(2,3-A)QUINAZOLIN-12-ONE
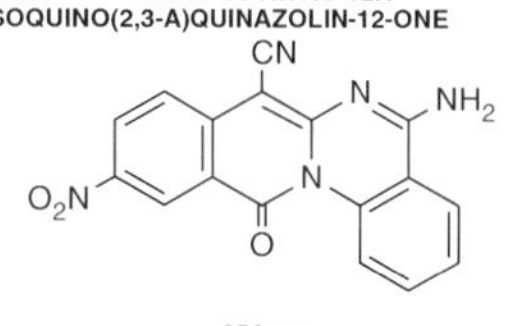
250 mg

S74,425-5
N-(4-ACETYLPHENYL)CYANOFORMAMIDE
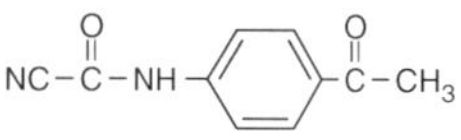
250 mg

S74,433-6
N-(4-ETHOXYCARBONYLPHENYL)-CYANOFORMAMIDE
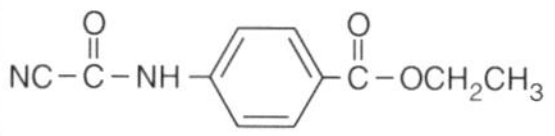
250 mg

S71,588-3
2'-CYANO-2,2-DIMETHYLPROPIONANILIDE

S74,424-7
N-(3,5-XYLYL)CYANOFORMAMIDE

S74,423-9
N-(2,3-XYLYL)CYANOFORMAMIDE

S74,439-6
N-(2,6-XYLYL)CYANOTHIOFORMAMIDE

S74,440-9
N-(2,3-XYLYL)CYANOTHIOFORMAMIDE

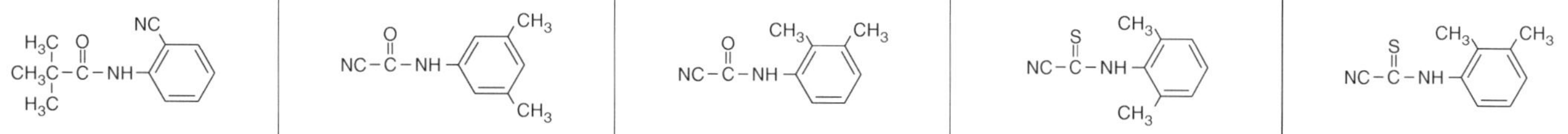

250 mg | 250 mg | 250 mg | 250 mg | 250 mg

S66,021-3
1,1'-(4-METHYL-1,3-PHENYLENE)BIS[3-(3-CYANOPHENYL)UREA]

S80,847-4
5-(ALPHA-CYANOBENZYLIDENE)-3-METHYL-4-PHENYL-2-PYRROLIDINONE

S55,873-7
(ANILINOMETHYLENE)-MALONONITRILE

S92,974-3
ALPHA-[4-(DIMETHYLAMINO)-BENZYLIDENE]MALONONITRILE

S93,771-1
1-(4-DIMETHYLAMINOPHENYL)-1,2,2-TRICYANOETHYLENE

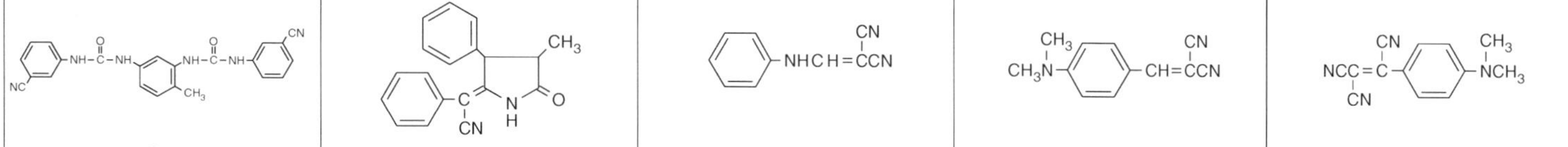

5 gm | 250 mg | 250 mg | 250 mg | 250 mg

S71,721-5
4-CYANOPHENYL N-(3-NITROPHENYL) CARBAMATE
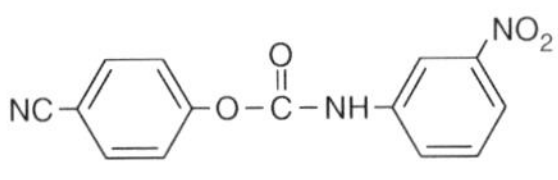
5 gm

S89,812-0
N-(3-NITROPHENYL)CYANOFORMAMIDE
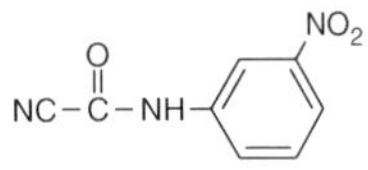
250 mg

S77,111-2
2-NITRO-P-TOLUNITRILE
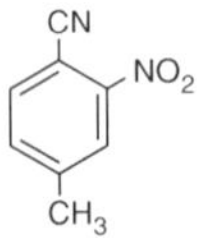
250 mg

S87,899-5
ALPHA-CYANO-4,5-(METHYLENEDIOXY)-2-NITROCINNAMAMIDE
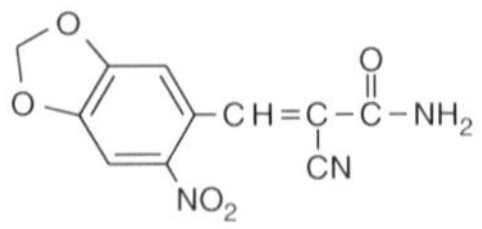
5 gm

S53,169-3
(3-NITROBENZYLIDENE)MALONONITRILE
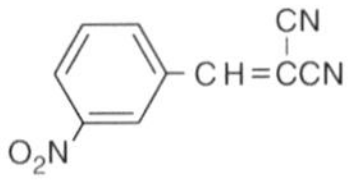
5 gm

S53,186-3
(4-NITROBENZYLIDENE)MALONONITRILE
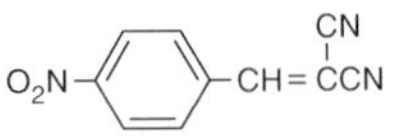
250 mg

S54,722-0
2-CYANO-5-PHENYL-2,4-PENTADIENOAMIDE
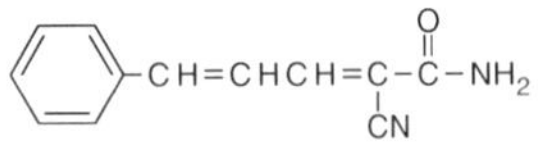
5 gm

S87,851-0
ALPHA-CYANO-4-METHYLCINNAMAMIDE
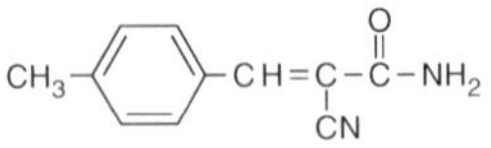
5 gm

S64,768-3
ALPHA-CYANO-3,4,5-TRIMETHOXYCINNAMIDE
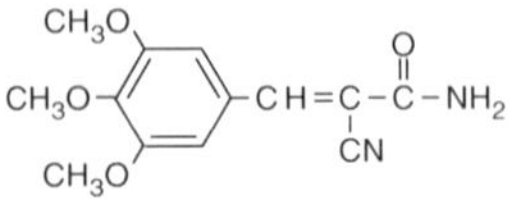
5 gm

S50,186-7
ALPHA-CYANO-5-INDANACRYLAMIDE
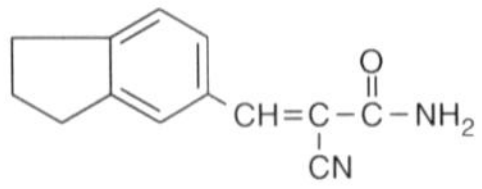
250 mg

S87,858-8
2-CYANO-3-(1-NAPHTHYL)ACRYLAMIDE
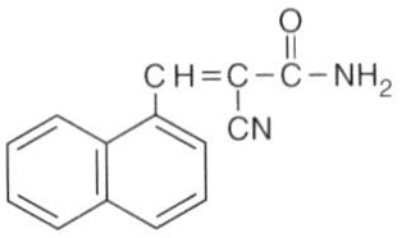
5 gm

S74,387-9
2'-CYANO-2,4-HEXADIENANILIDE
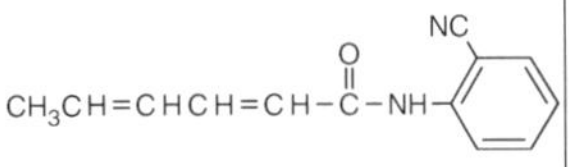
5 gm

S78,041-3
2'-CYANOGLUTARANILIC ACID
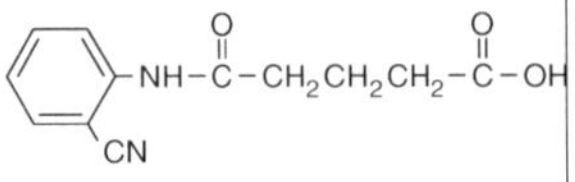
250 mg

S53,181-2
ALPHA-DICYANOMETHYLENE-PARA-ACETOTOLUIDIDE
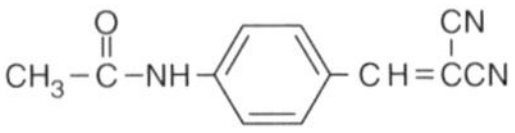
5 gm

S51,039-4
N-(2,2-DICYANOVINYL)-ACETANILIDE
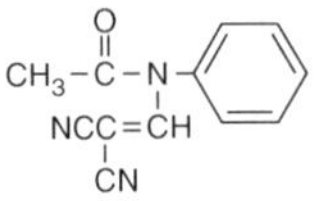
250 mg

S74,948-6
1-(2-CYANOPHENYL)-3-PHENYLUREA
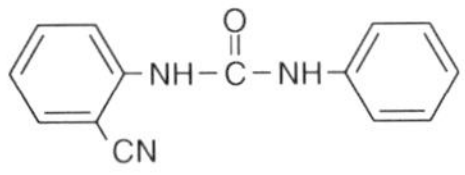
250 mg

S65,965-7
1,1'-(4-METHYL-1,3-PHENYLENE)BIS[3-(4-CYANOPHENYL)UREA]
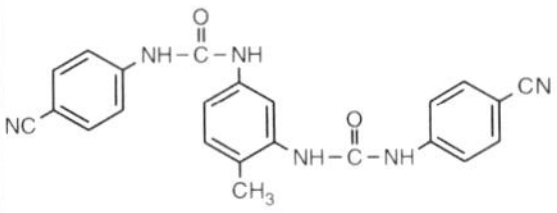
5 gm

S71,093-8
1,1'-(METHYLENEDI-4,1-PHENYLENE)-BIS[3-(2-CYANOPHENYL)UREA]
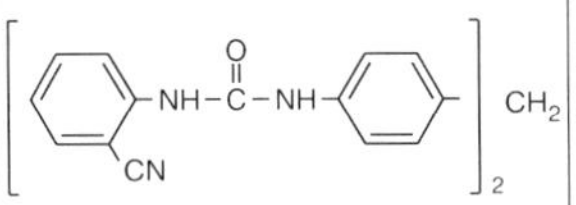
5 gm

S71,758-4
1,1'-(4-METHYL-1,3-PHENYLENE)BIS[3-(2-CYANOPHENYL)UREA]
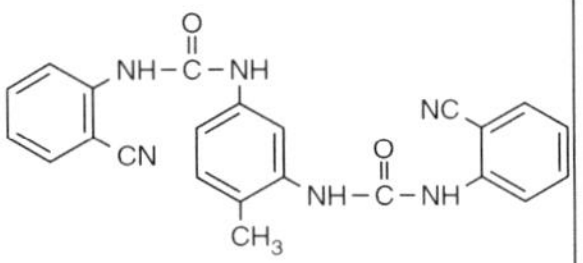
5 gm

S65,200-8
ALPHA-CYCLOPROPYLBENZYLIDENE-MALONONITRILE
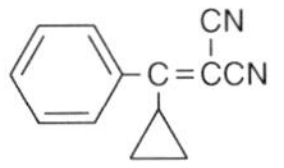
250 mg

S43,436-1
PIPERONYLIDENEMALONONITRILE
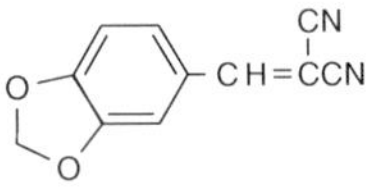
5 gm

S92,975-1
(CYCLOHEXANE-1,4-DIYLIDENE)-DIMALONONITRILE
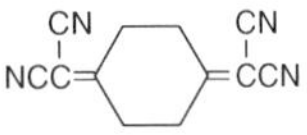
250 mg

S94,708-3
2,5-DIBENZYL(CYCLOHEXANE-1,4-DIYLIDENE)DIMALONONITRILE
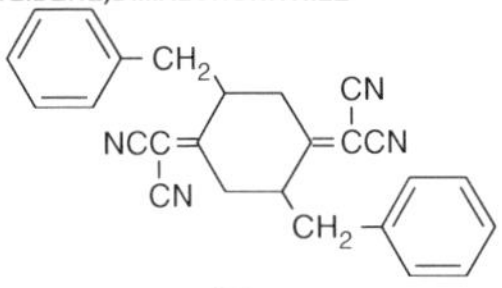
250 mg

S64,763-2
3-CYANO-2-IMINO-2H-1-BENZOPYRAN
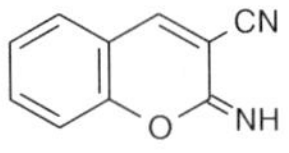
250 mg

S94,322-3
ALPHA-METHYLBENZYLIDENE-MALONONITRILE
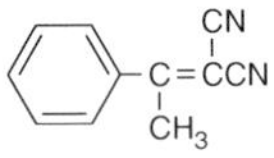
250 mg

S64,988-0
ALPHA-ETHYLBENZYLIDENEMALONONITRILE
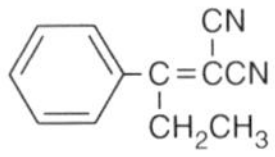
250 mg

S65,199-0
ALPHA-PROPYLBENZYLIDENE-MALONONITRILE
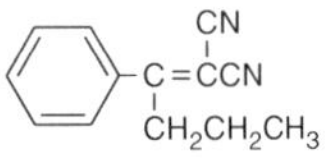
5 gm

S64,987-2
ALPHA-ISOBUTYLBENZYLIDENE-MALONONITRILE
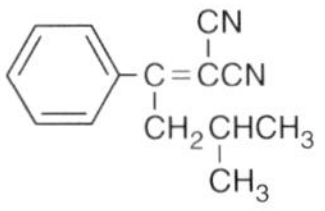
250 mg

S65,156-7
ALPHA-BUTYLBENZYLIDENE-MALONONITRILE
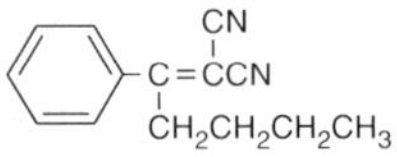
5 gm

S41,880-3
DIPHENYLMETHYLENEMALONONITRILE
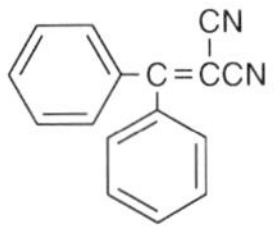
5 gm

S94,704-0
2-METHYLBENZYLIDENEMALONONITRILE

100 mg

S43,438-8
4-METHYLBENZYLIDENEMALONONITRILE
5 gm

S53,171-5
(2,4-DICHLOROBENZYLIDENE)-MALONONITRILE
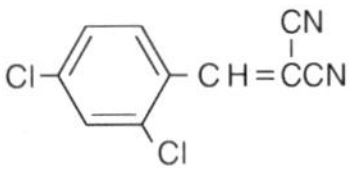
5 gm

S50,180-8
(5-INDANYLMETHYLENE)-MALONONITRILE
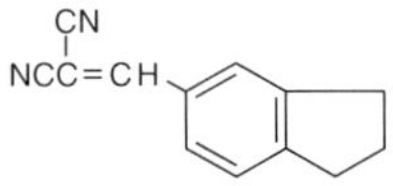
250 mg

S65,073-0
1-(DICYANOMETHYLENE)TETRALIN
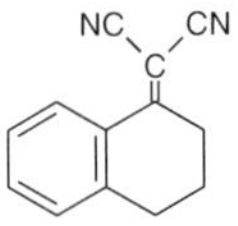
250 mg

S84,298-2
ALPHA,BETA-DIPHENYLFUMARONITRILE
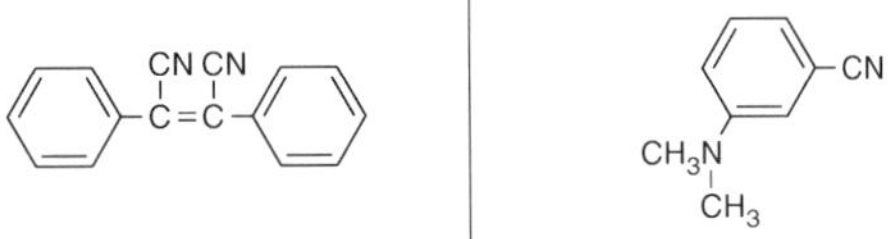
250 mg

S86,944-9
3-(DIMETHYLAMINO)BENZONITRILE
250 mg

S52,507-3
ALPHA-CYANO-4-(DIMETHYLAMINO)-CINNAMAMIDE
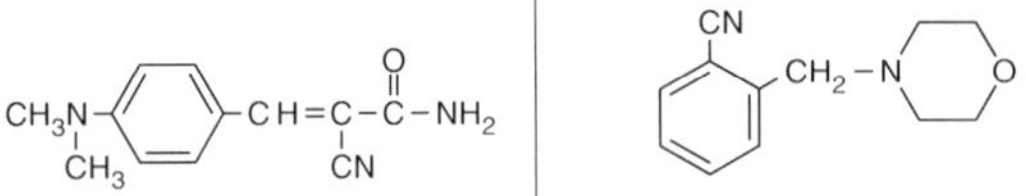
5 gm

S74,515-4
2-(MORPHOLINOMETHYL)BENZONITRILE
250 mg

S87,898-7
2-[2-(2-BROMO-4,5-METHYLENEDIOXYPHENYL)VINYL]-5-NITROBENZONITRILE
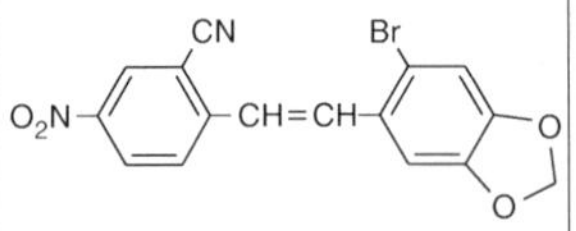
250 mg

S87,862-6
5-NITRO-2-STYRYLBENZONITRILE
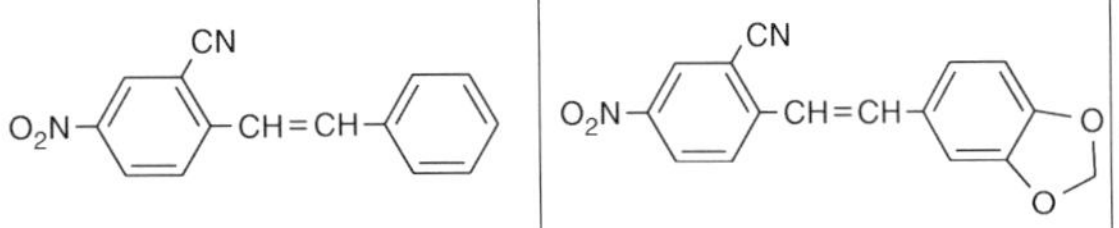
5 gm

S74,683-5
5-NITRO-ALPHA-PIPERONYLIDENE-O-TOLUNITRILE
5 gm

S74,681-9
ALPHA-(4-METHYLBENZYLIDENE)-5-NITRO-O-TOLUNITRILE
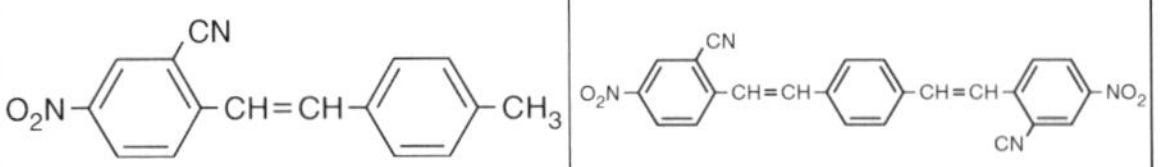
5 gm

S74,685-1
ALPHA,ALPHA'-TEREPHTHALYLIDENEBIS (5-NITRO-O-TOLUNITRILE)
5 gm

S74,682-7
5-NITRO-ALPHA-(3-NITROBENZYLIDENE)-O-TOLUNITRILE
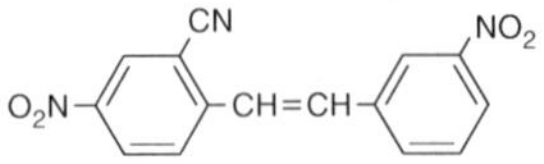
5 gm

S52,336-4 ALPHA-CYANO-3-NITROCINNAMAMIDE — 5 gm

S93,491-7 6-CYANO-5-METHYL-9-NITROBENZIMIDAZO(1,2-B)ISOQUINOLIN-11(5H)-ONE — 250 mg

S57,195-4 1-(3-CYANOPHENYL)-3-METHYL-2-PYRAZOLIN-5-ONE — 250 mg

S65,488-4 1,1'-(METHYLENEDI-4,1-PHENYLENE)-BIS[3-(3-CYANOPHENYL)UREA] — 5 gm

S71,104-7 1-(3-CYANOPHENYL)-3-(4-CYANOPHENYL)UREA — 250 mg

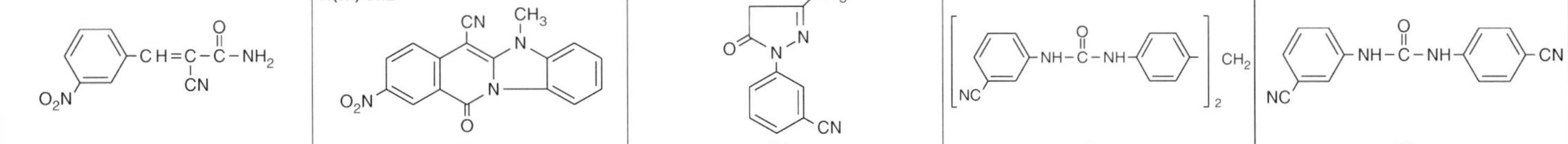

S70,790-2 1-(3-CYANOPHENYL)-3-(2,6-DIMETHYLPHENYL)UREA — 5 gm

S87,519-8 1,3-BIS(3-CYANOPHENYL)-2-THIOUREA — 250 mg

S64,758-6 ALPHA-[4-(DIMETHYLAMINO)-BENZYLIDENE]-4-NITROPHENYLACETONITRILE — 5 gm

S35,829-0 1-BENZYL-4-(ALPHA-CYANOBENZYLIDENE)-PIPERIDINE HYDROCHLORIDE — 5 gm

S94,672-9 4-NITROBENZOYL CYANIDE — 250 mg

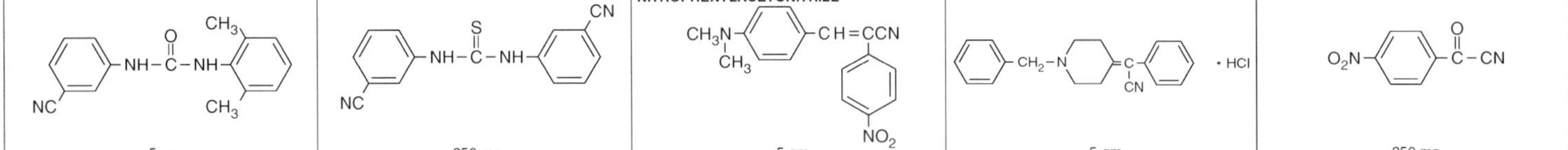

S34,139-8 BETA-(4-ANTIPYRYL)-ALPHA-CYANO-4-NITROSTYRENE — 5 gm

S91,762-1 4-(1,1-DIMETHYL-2-NITROETHYL)-BENZONITRILE — 250 mg

S91,860-1 4-(2-METHYL-2-NITROPROPYL)-BENZONITRILE — 250 mg

S91,756-7 4-(2-NITRO-1,1,2-TRIMETHYLPROPYL)-BENZONITRILE — 250 mg

S91,753-2 4-(1,2-DIMETHYL-1-ETHYL-2-NITROPROPYL)BENZONITRILE — 250 mg

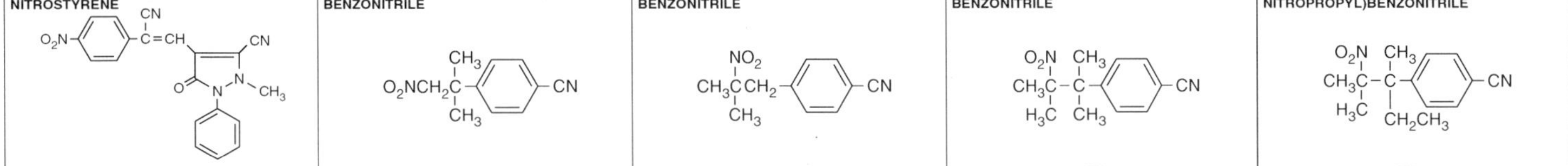

S91,754-0 4-(2-NITRO-1,1,2-TRIMETHYLBUTYL)-BENZONITRILE — 250 mg

S66,063-9 2-NITRO-ALPHA-PHENYLCINNAMONITRILE — 250 mg

S65,822-7 2-CHLORO-ALPHA-(4-NITROPHENYL)-CINNAMONITRILE — 5 gm

S94,501-3 2,2'-BIS(BENZOYLCYANIDE) — 250 mg

S55,760-9 4'-(BETA-CYANOSTYRYL)ACETANILIDE — 250 mg

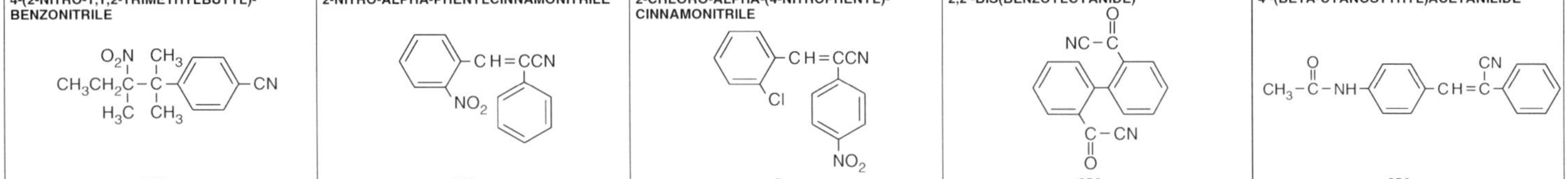

S44,816-8 2,3,5,6-TETRAMETHYL-1,4-BENZENEDIACETONITRILE — 5 gm

S53,234-7 3,4-METHYLENEDIOXY-ALPHA-(PARA-TOLYL)CINNAMONITRILE — 250 mg

S43,774-3 TRIPHENYLACRYLONITRILE — 5 gm

S44,806-0 4-FLUORO-ALPHA-METHYLPHENYLACETONITRILE — 5 gm

S65,797-2 3-CHLORO-ALPHA-(1-NAPHTHYLMETHYLENE)-PHENYLACETONITRILE — 5 gm

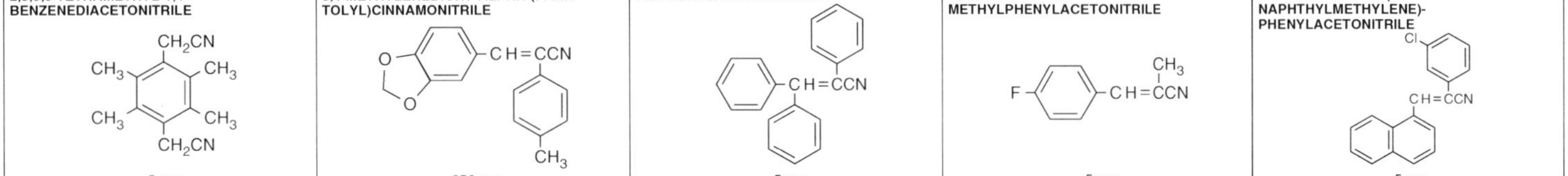

S92,967-0 ALPHA-CYANO-4-(DIMETHYLAMINO)-CINNAMAMIDE — 250 mg

S34,591-1 4-(DIETHYLAMINO)BENZONITRILE — 250 mg

S87,987-8 3-(DIMETHYLAMINOMETHYL)BENZONITRILE — 5 gm

S68,311-6 4-(DIMETHYLAMINOMETHYL)BENZONITRILE — 5 gm

S93,764-9 4-CYANO-7-DIETHYLAMINO-3-(METHOXYCARBONYLACETYL)COUMARIN — 250 mg

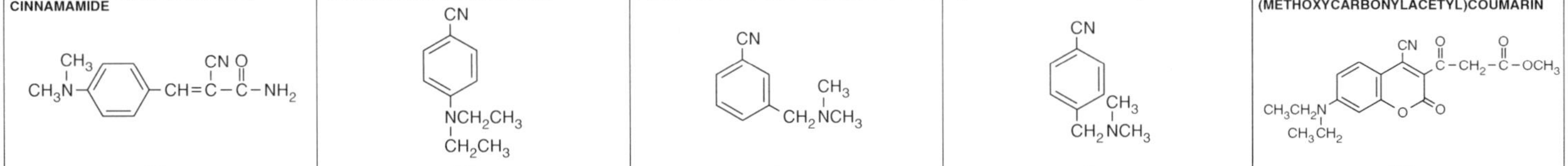

S52,993-1 ETHYL ALPHA-CYANO-4-(DIMETHYLAMINO) CINNAMATE — 5 gm

S52,989-3 ETHYL ALPHA-CYANO-4-(DIETHYLAMINO)-CINNAMATE — 5 gm

S52,990-7 ETHYL 2-CHLORO-ALPHA-CYANO-4-(DIMETHYLAMINO)CINNAMATE — 5 gm

S53,217-7 4-DIMETHYLAMINO-ALPHA-(PARA-TOLYL)-CINNAMONITRILE — 5 gm

S87,465-5 4-CYANO-N-METHYL-N-NITROSOANILINE — 250 mg

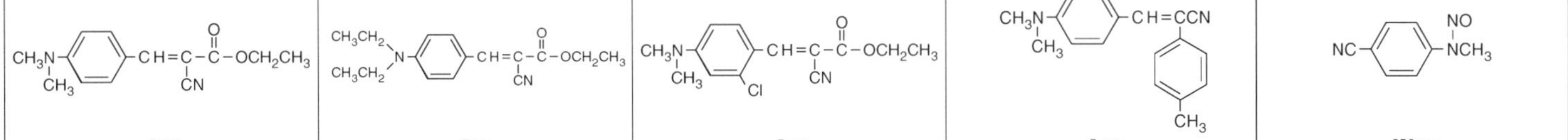

S73,538-8 4-CYANOANILINO N-(3,5-XYLYL)-CARBAMATE — 250 mg

S91,855-5 4-(1-METHYL-1-NITROETHYL)-BENZONITRILE — 250 mg

S88,621-1 4-CYANOBENZALDEHYDE METHYLPHENYLHYDRAZONE — 250 mg

S93,152-7 2-CYANO-2-(PHENYLHYDRAZONO)-ACETAMIDE — 250 mg

S93,153-5 2-CYANO-2-(P-TOLYLHYDRAZONO)-ACETAMIDE — 250 mg

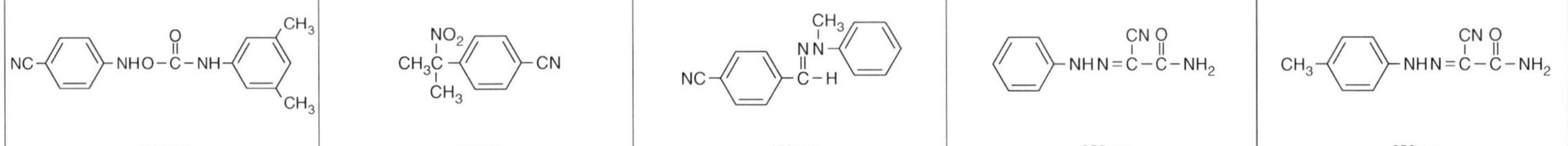

S84,931-6 4-CYANOBENZALDEHYDE PHENYLHYDRAZONE — 250 mg

S92,961-1 ALPHA-BENZOYL-4-NITROCINNAMONITRILE — 250 mg

S92,971-9 METHYL ALPHA-CYANO-4-NITROCINNAMATE — 250 mg

S92,754-6 2-(ALPHA-CYANO-4-NITROBENZYLIDENE)-3-METHYLBENZOTHIAZOLINE — 250 mg

S91,428-2 ALPHA-(4-NITROPHENYL)CINNAMONITRILE — 250 mg

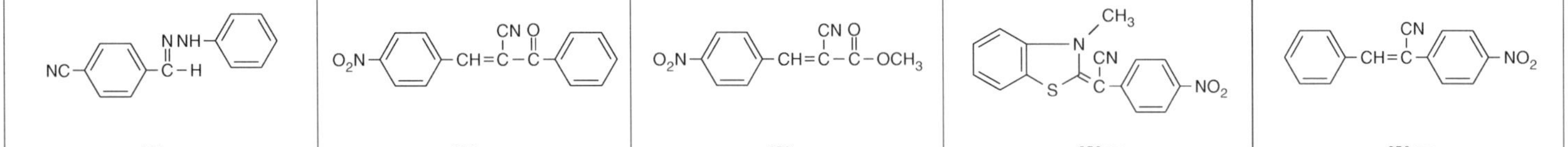

S84,342-3 4-NITROCINNAMONITRILE — 250 mg

S91,413-4 4-METHYL-ALPHA-(4-NITROPHENYL)-CINNAMONITRILE — 250 mg

S74,687-8 ALPHA-(1-NAPHTHYLMETHYLENE)-5-NITRO-O-TOLUNITRILE — 5 gm

S53,968-6 ETHYL ALPHA-CYANO-3-NITROCINNAMATE — 5 gm

S37,736-8 ETHYL ALPHA-CYANO-4-NITROCINNAMATE — 5 gm

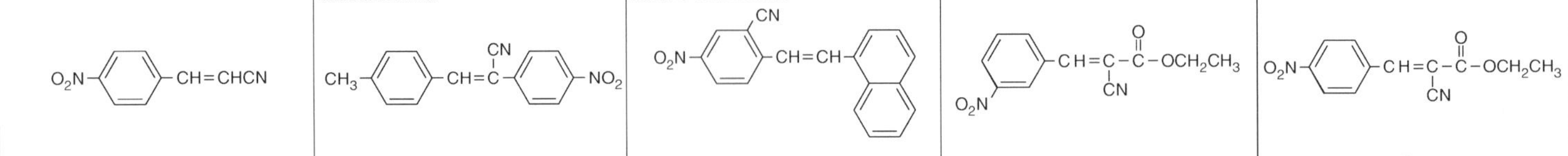

S52,513-8 ALPHA-CYANO-4-NITROCINNAMAMIDE — 5 gm

S52,726-2 2-CHLORO-ALPHA-CYANO-5-NITROCINNAMAMIDE — 250 mg

S36,106-2 2,5-DIMETHYL-4-NITROBENZONITRILE — 250 mg

S53,232-0 3-NITRO-ALPHA-(PARA-TOLYL)-CINNAMONITRILE — 250 mg

S67,943-7 ETHYL 2-(4-CYANOBENZOYL)ACETATE — 5 gm

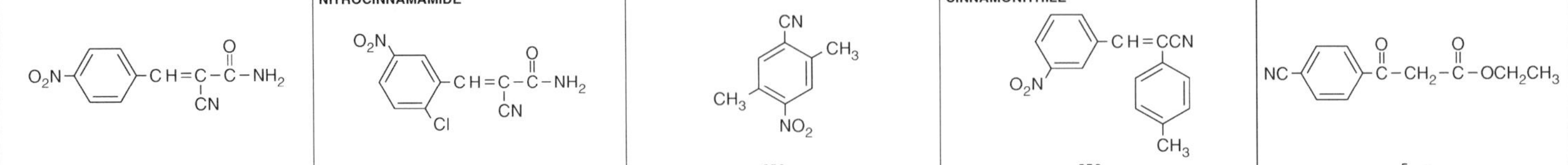

S87,679-8 4-CYANOBENZOPHENONE — 250 mg

S92,979-4 ALPHA-BENZOYL-4-METHYLCINNAMONITRILE — 250 mg

S92,963-8 4-METHYL-ALPHA-(4-METHYLBENZOYL)-CINNAMONITRILE — 250 mg

S87,026-9 2-(DICYANOMETHYLENE)-1-ACENAPHTHENONE — 250 mg

S92,566-7 3-CYANOBENZANTHRONE — 250 mg

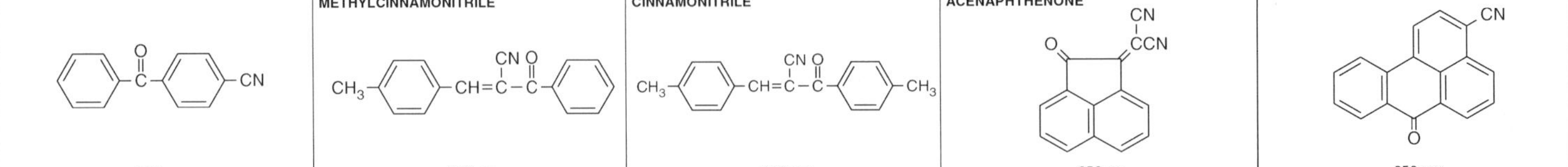

S87,049-8 10-(DICYANOMETHYLENE)-9(10H)-PHENANTHRENONE — 250 mg

S93,748-7 6-CYANOBENZOTHIAZOLO(3,2-A)-QUINOLIN-5-ONE — 250 mg

S93,746-0 6-CYANO-7-METHYLBENZIMIDAZO(1,2-A)-QUINOLINE-5(7H)-THIONE — 250 mg

S80,944-6 6-BROMO-ALPHA-CYANO-1,3-BENZODIOXOLE-5-ACRYLIC ACID — 250 mg

S50,037-2 2-CYANO-5-PHENYL-2,4-PENTADIENOIC ACID — 250 mg

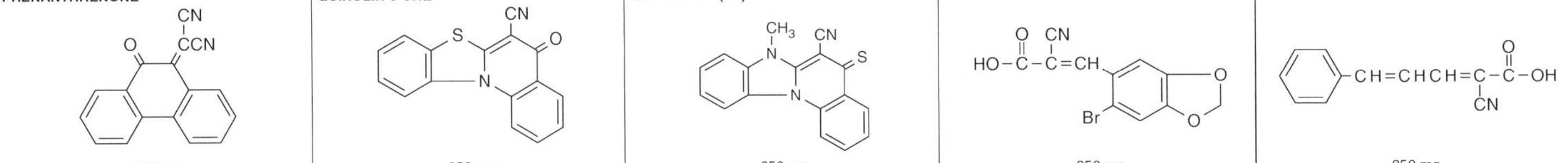

S66,139-2 DIETHYL 3,3'-(1,4-PHENYLENE)BIS-(2-CYANOACRYLATE) — 5 gm

S53,116-2 ETHYL 4-ACETAMIDO-ALPHA-CYANOCINNAMATE — 5 gm

S74,864-1 4-CYANOPHENYL BENZOATE — 250 mg

S44,539-8 METHYL 2-CYANOBENZOATE — 5 gm

S66,095-7 ETHYL ALPHA-CYANO-3,4-METHYLENEDIOXYCINNAMATE — 250 mg

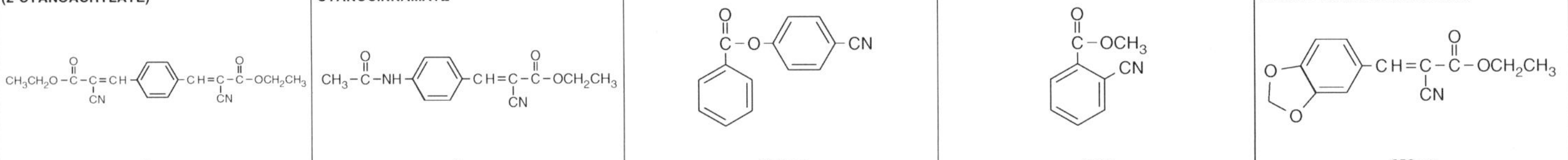

S92,973-5 METHYL ALPHA-CYANO-3,4-(METHYLENEDIOXY)CINNAMATE — 250 mg

S84,316-4 ETHYL 6-BROMO-ALPHA-CYANO-(3,4-METHYLENEDIOXY)CINNAMATE — 250 mg

S35,276-4 METHYL ALPHA-CYANOCINNAMATE — 5 gm

S38,158-6 ETHYL ALPHA-CYANOCINNAMATE — 5 gm

S66,113-9 ETHYL 2-CINNAMYLIDENE-2-CYANOACETATE — 250 mg

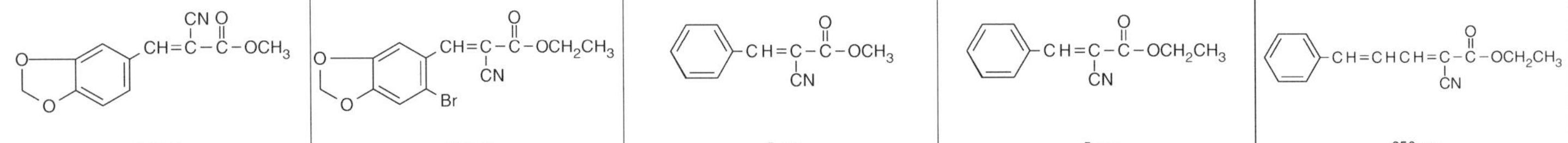

S66,141-4 ETHYL ALPHA-CYANO-BETA-ETHYLCINNAMATE — 250 mg

S70,151-3 ETHYL ALPHA-CYANO-BETA-PHENYLCINNAMATE — 5 gm

S92,970-0 METHYL ALPHA-CYANO-4-METHYLCINNAMATE — 250 mg

S38,007-5 ETHYL 4-CHLORO-ALPHA-CYANOCINNAMATE — 250 mg

S66,100-7 ETHYL ALPHA-CYANO-4-METHYLCINNAMATE — 250 mg

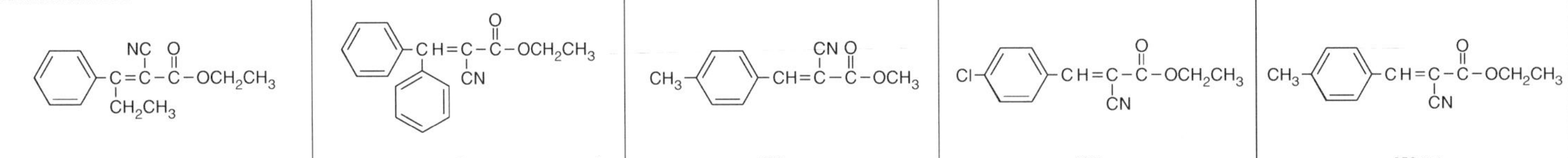

S53,008-5 ETHYL ALPHA-CYANO-4-PHENYLCINNAMATE — 5 gm

S53,187-1 ETHYL ALPHA-CYANO-3,4-DICHLOROCINNAMATE — 5 gm

S51,461-6 ETHYL ALPHA-CYANO-5-INDANACRYLATE — 250 mg

S53,183-9 ETHYL ALPHA-CYANO-1-NAPHTHALENEACRYLATE — 5 gm

S92,694-9 METHYL 10-CYANO-9-ANTHRACENECARBOXYLATE — 250 mg

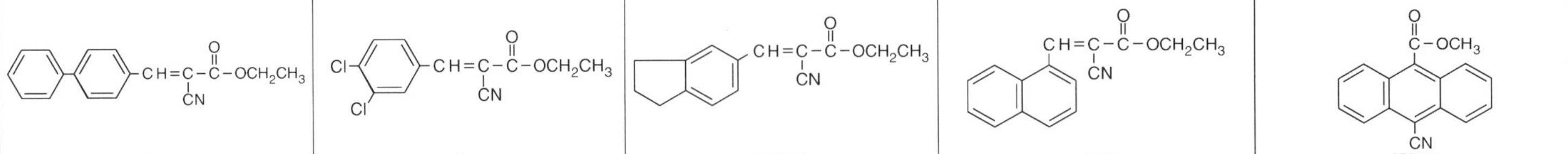

S76,078-1 TERT-BUTYL 2-CYANO-2-(1,2-DIHYDRO-2-QUINOLYLIDENE)ACETATE — 250 mg

S52,374-7 ALPHA-CYANO-3,4-(METHYLENEDIOXY)-CINNAMAMIDE — 5 gm

S92,966-2 ALPHA-CYANOCINNAMAMIDE — 250 mg

S52,367-4 2-CHLORO-ALPHA-CYANOCINNAMAMIDE — 5 gm

S89,832-5 N-BENZOYL-N-(P-TOLYL)-CYANOTHIOFORMAMIDE — 250 mg

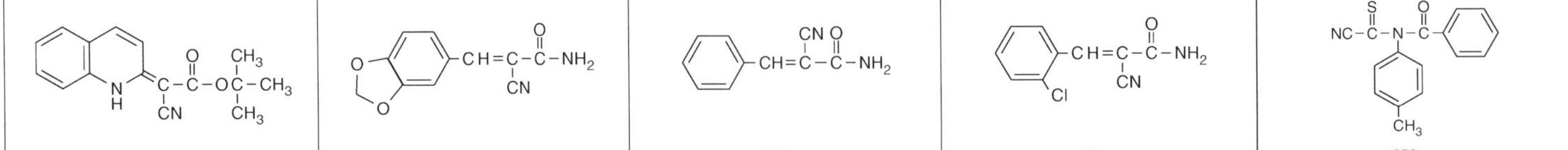

S53,215-0 4'-(BETA-CYANO-4-METHYLSTYRYL)-ACETANILIDE — 5 gm

S74,421-2 ETHYL 2-(2-CYANOPHENYLUREIDO)-ACETATE — 250 mg

S64,459-5 4-CYANOPHENYL N-PHENYLCARBAMATE — 5 gm

S64,460-9 4-CYANOPHENYL N-(4-CHLOROPHENYL)-CARBAMATE — 5 gm

S65,000-5 4-CYANOPHENYL N-(2,3-XYLYL)-CARBAMATE — 5 gm

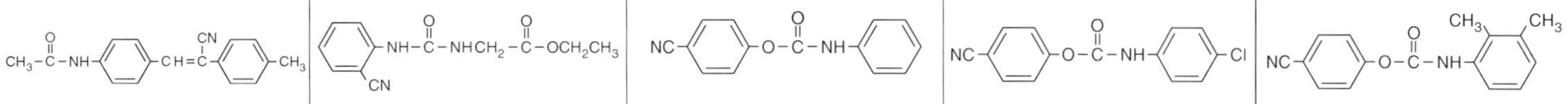

S66,648-3 4-CYANOPHENYL N-(2,6-XYLYL)-CARBAMATE — 5 gm

S66,755-2 4-CYANOPHENYL N-(2-ETHYL-6-METHYLPHENYL)CARBAMATE — 5 gm

S74,597-9 4-CYANOPHENYL N-(2,6-DIISOPROPYLPHENYL)CARBAMATE — 250 mg

S66,793-5 4-CYANOPHENYL N-(3,5-XYLYL)--CARBAMATE — 5 gm

S80,600-5 METHYL N-(2-CYANO-5-METHYLPHENYL)-N-ISOPROPYLCARBAMATE — 250 mg

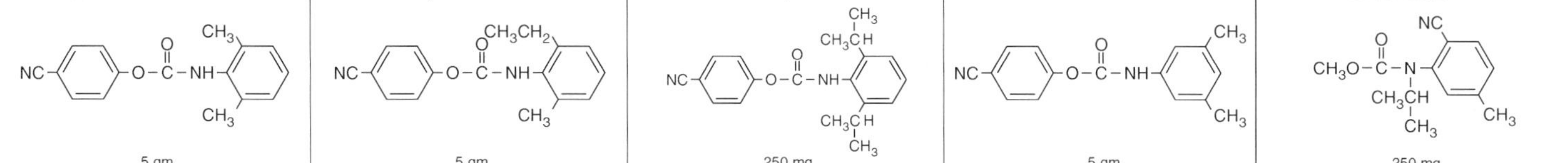

S71,546-8 4-CYANOPHENYL N-(O-TOLYL)CARBAMATE — 5 gm

S71,476-3 4-CYANOPHENYL N-(P-TOLYL)CARBAMATE — 5 gm

S71,477-1 4-CYANOPHENYL N-(M-TOLYL)CARBAMATE — 5 gm

S71,359-7 4-CYANOPHENYL N-(2,3-DICHLOROPHENYL)CARBAMATE — 5 gm

S71,762-2 4-CYANOPHENYL N-(2,6-DIETHYLPHENYL) CARBAMATE — 5 gm

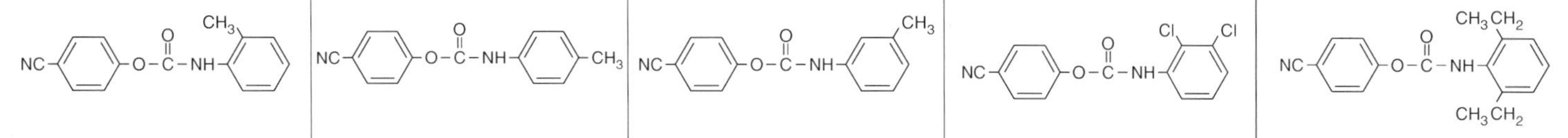

S71,138-1 4-CYANOPHENYL N-(2,4-DIMETHYLPHENYL)CARBAMATE — 5 gm

S71,001-6 4-CYANOPHENYL N-(1-NAPHTHYL)-CARBAMATE — 5 gm

S34,489-3 3-CYANOPHENYL ACETATE — 5 gm

S65,746-8 ETHYL 2-(4-CYANOPHENYL)PROPIONATE — 5 gm

S94,920-5 2-CYANOBENZAMIDE — 250 mg

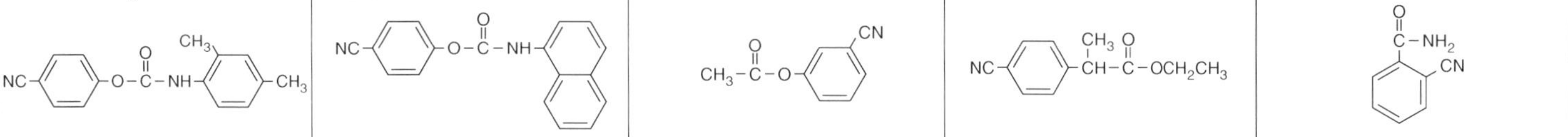

S90,539-9 N-(2-CYANOPHENYL)-1,4,5,6,7,7-HEXACHLORO-5-NORBORNENE-2,3-DICARBOXIMIDE — 250 mg

S89,819-8 N-PHENYLCYANOFORMAMIDE — 250 mg

S89,800-7 N-(O-TOLYL)CYANOFORMAMIDE — 250 mg

S52,355-0 4'-(4-BROMO-BETA-CYANOSTYRYL)-ACETANILIDE — 5 gm

S89,803-1 N-(3,4-XYLYL)CYANOFORMAMIDE — 250 mg

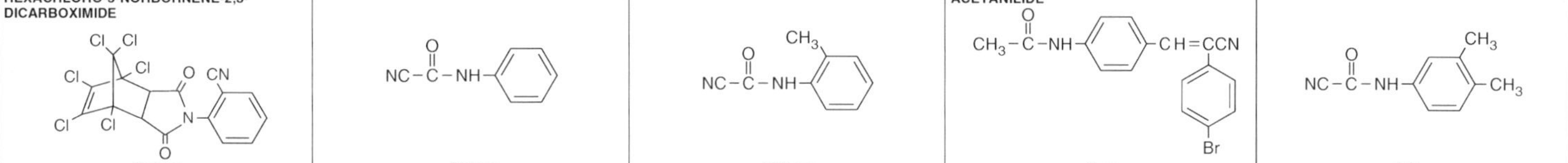

S53,645-8 N,N'-[4-(ALPHA-CYANO-2,4-DICHLOROSTYRYL)-1,2-PHENYLENE]-BISACETAMIDE — 5 gm

S39,133-6 N-(4-CYANO-1-NAPHTHYL)-ACETAMIDE — 5 gm

S89,791-4 N-(P-TOLYL)CYANOTHIOFORMAMIDE — 250 mg

S89,869-4 N-(2,4-XYLYL)CYANOTHIOFORMAMIDE — 250 mg

S71,428-3 1-(4-CYANOPHENYL)-3-PHENYLUREA — 250 mg

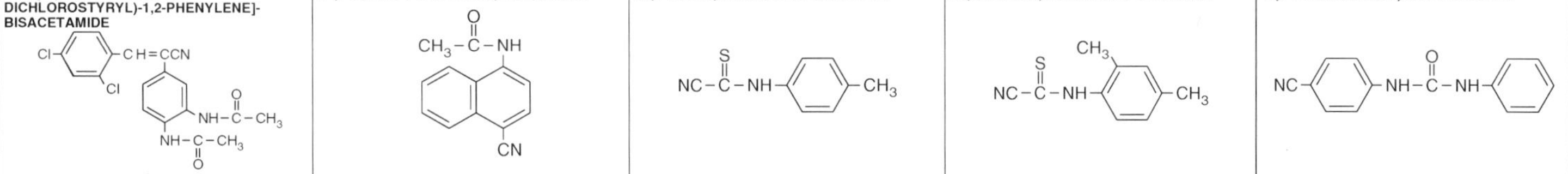

S73,567-1 1-(4-CYANOPHENYL)-3-(P-TOLYL)UREA — 250 mg

S87,517-1 1,3-BIS(4-CYANOPHENYL)-2-THIOUREA — 250 mg

S69,776-1 2-(ALPHA-METHYLBENZYL)BENZONITRILE — 250 mg

S44,530-4 2,3-DIMETHYLBENZONITRILE — 5 gm

S65,436-1 2,4-DICHLOROBENZONITRILE — 250 mg

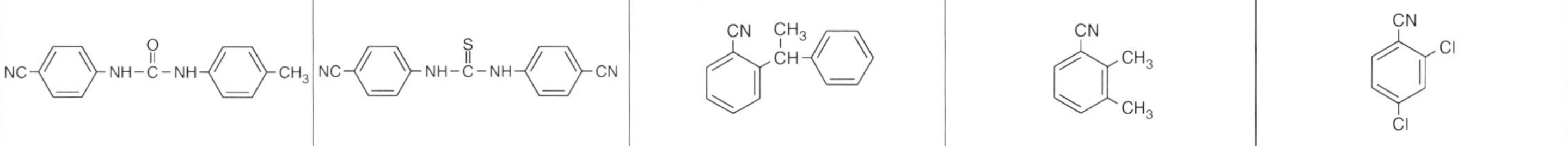

S44,141-4 2,4-DIMETHYLBENZONITRILE — 5 gm

S45,400-1 3,4-DIMETHYLBENZONITRILE — 5 gm

S44,592-4 2,4,6-TRIMETHYLBENZONITRILE — 250 mg

S87,935-5 BETA-(1-NAPHTHYL)-ALPHA-PHENYLACRYLONITRILE — 5 gm

S42,810-8 2-CHLOROCINNAMONITRILE (MIXTURE OF CIS AND TRANS) — 5 gm

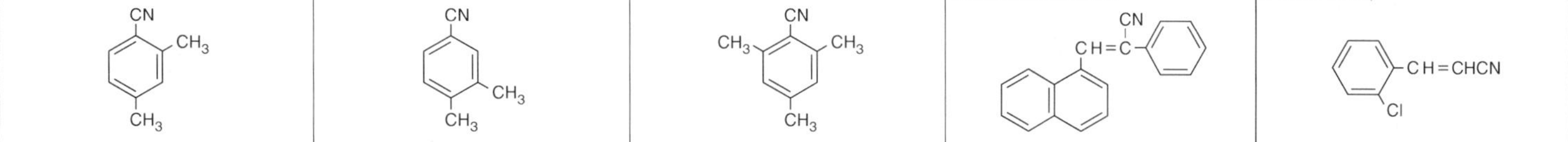

S53,669-5 2-CHLORO-ALPHA-(3-CHLOROPHENYL)-CINNAMONITRILE — 5 gm

S43,413-2 4-METHYLCINNAMONITRILE (MIXTURE OF CIS AND TRANS) — 5 gm

S43,415-9 4-ISOPROPYLCINNAMONITRILE (MIXTURE OF CIS AND TRANS) — 5 gm

S53,235-5 4-CHLORO-ALPHA-(PARA-TOLYL)-CINNAMONITRILE — 5 gm

S53,956-2 ALPHA-(3-CHLOROPHENYL)-2,6-DICHLOROCINNAMONITRILE — 250 mg

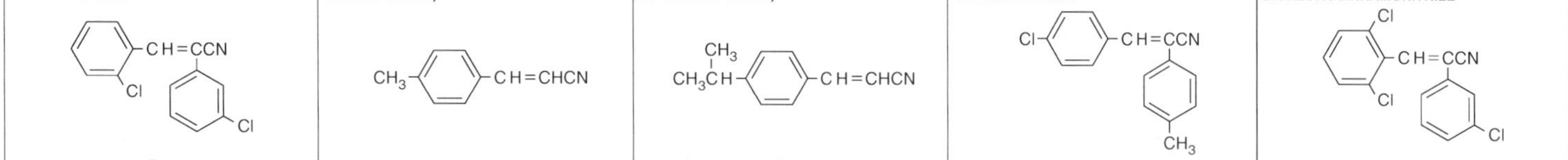

S94,794-6
6-CYANO-1,1,4,4,7-PENTAMETHYL-1,2,3,4-TETRAHYDRONAPHTHALENE
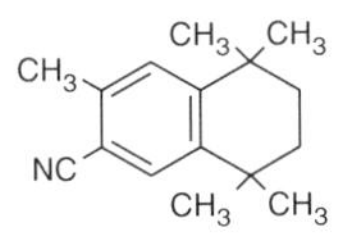
5 gm

S94,795-4
6-CYANO-7-ETHYL-1,2,3,4-TETRAHYDRO-1,1,4,4-TETRAMETHYLNAPHTHALENE
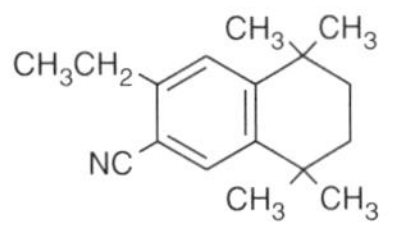
250 mg

S85,954-0
2,7-DICYANONAPHTHALENE
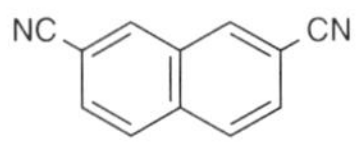
250 mg

S97,389-0
AR-E50 5-CYANO-7-METHYLBENZ(A)-ANTHRACENE
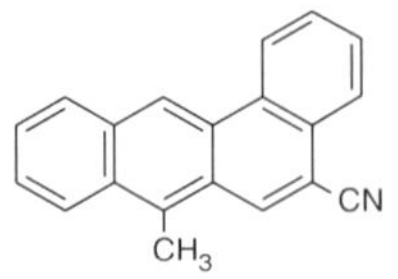
10 mg

S97,418-8
AR-E80 5-CYANO-7,12-DIMETHYLBENZ(A)ANTHRACENE
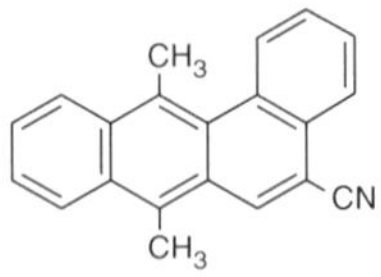
10 mg

S67,104-5
1,3,5-TRICYANOBENZENE
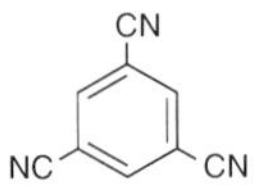
250 mg

S44,025-6
4-HYDROXYPHENYL THIOCYANATE MONOHYDRATE
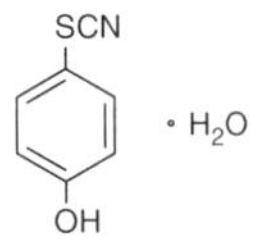
5 gm

S51,158-7
4-HYDROXY-M-TOLYL THIOCYANATE
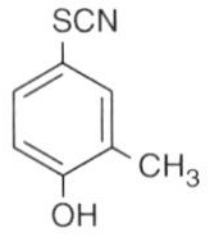
5 gm

S36,860-1
2,3-DIMETHYL-4-THIOCYANOPHENOL
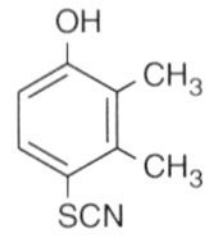
250 mg

S75,663-6
3,4-DIMETHOXYPHENYL THIOCYANATE
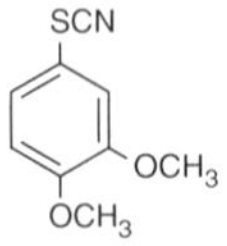
5 gm

S58,714-1
4-AMINO-3-METHOXYPHENYL THIOCYANATE HYDROCHLORIDE
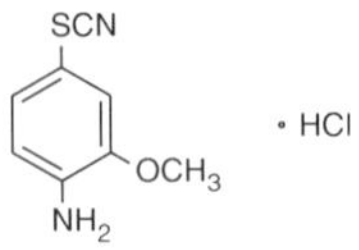
250 mg

S63,342-9
4-ACETAMIDO-3-METHOXYPHENYL THIOCYANATE
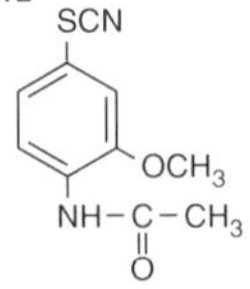
5 gm

S38,757-6
4-(HEXADECYLOXY)BENZOYLMETHYL THIOCYANATE
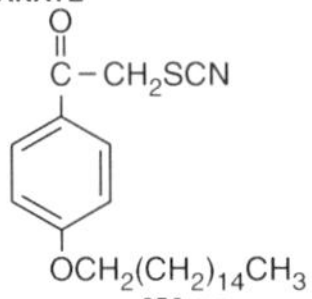
250 mg

S92,909-3
1-AMINO-2,4-NAPHTHALENEDITHIOCYANATE
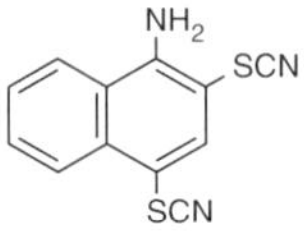
250 mg

S83,726-1
4-(DIMETHYLAMINO)PHENYL THIOCYANATE
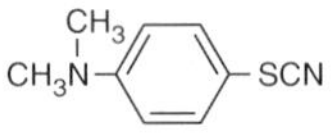
250 mg

S35,599-2
4-THIOCYANATOPHENYLHYDRAZINE HYDROCHLORIDE
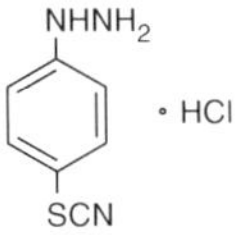
5 gm

S43,205-9
4-NITROBENZYL THIOCYANATE
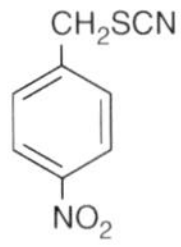
5 gm

S90,606-9
3-NITRO-1-AZULENYL THIOCYANATE
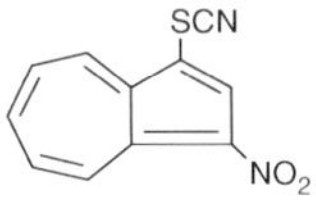
250 mg

S43,179-6
4-CHLOROBENZOYLMETHYL THIOCYANATE
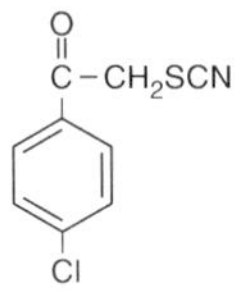
5 gm

S42,803-5
2-NAPHTHOYLMETHYL THIOCYANATE
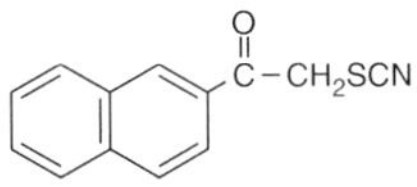
5 gm

S41,148-5
3-METHYLBENZYL THIOCYANATE
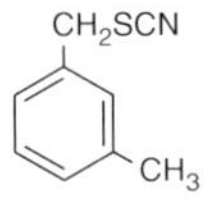
5 gm

S81,655-8
TRIPHENYLMETHYL THIOCYANATE
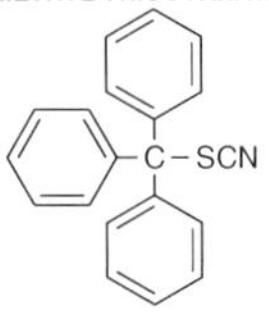
250 mg

S42,775-6
P-XYLYLENE DITHIOCYANATE
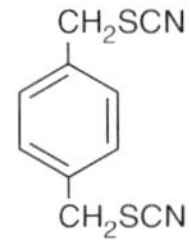
5 gm

S42,893-0
2,6-DICHLOROBENZYL THIOCYANATE
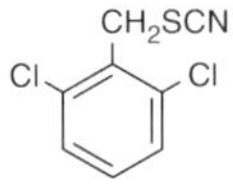
5 gm

S77,827-3
MESITYL THIOCYANATE
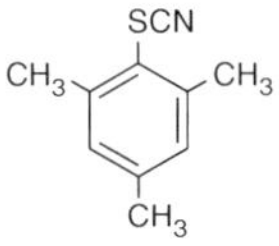
250 mg

S65,686-0
3,5-DI-TERT-BUTYL-4-HYDROXYPHENYL ISOTHIOCYANATE
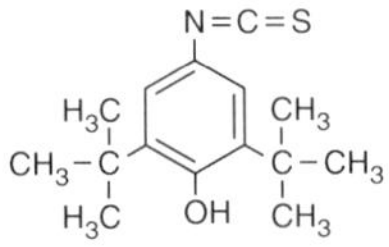
5 gm

S64,547-8
ALPHA-METHYLBENZYL ISOTHIOCYANATE
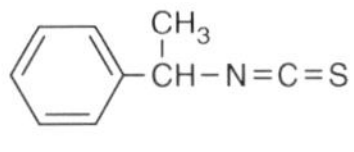
5 gm

S39,144-1
2-BIPHENYLYL ISOTHIOCYANATE
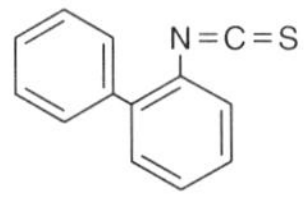
250 mg

S93,564-6
4-BIPHENYLYL ISOTHIOCYANATE
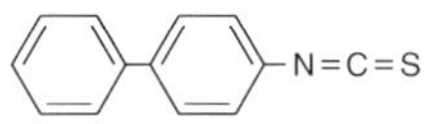
250 mg

S64,884-1
2-ISOPROPYLPHENYL ISOTHIOCYANATE
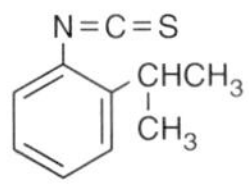
5 gm

S64,886-8
4-ISOPROPYLPHENYL ISOTHIOCYANATE

250 mg

S39,367-3
4-(TERT-BUTYL)-PHENYL ISOTHIOCYANATE
5 gm

S39,364-9
3-CHLORO-PARA-TOLYL ISOTHIOCYANATE
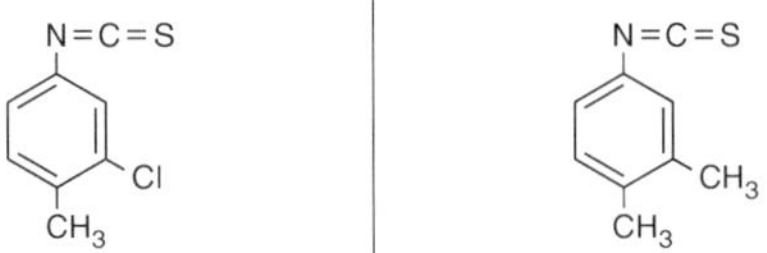
250 mg

S39,363-0
3,4-XYLYL ISOTHIOCYANATE
250 mg

S64,545-1
2,4-XYLYL ISOTHIOCYANATE
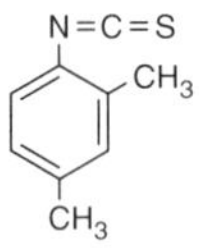
250 mg

S64,883-3
MESITYL ISOTHIOCYANATE
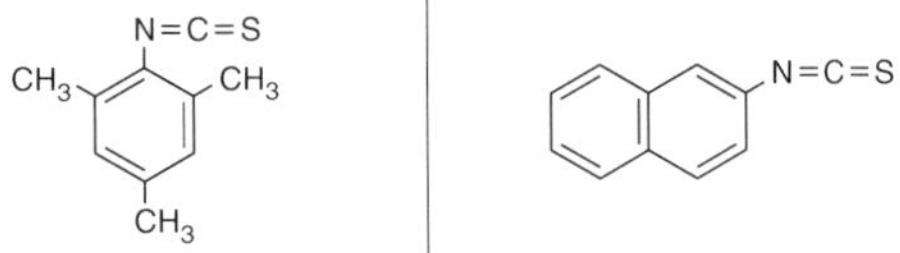
250 mg

S43,615-1
2-NAPHTHYL ISOTHIOCYANATE
5 gm

S67,701-9
3-ETHYLPHENYL ISOCYANATE
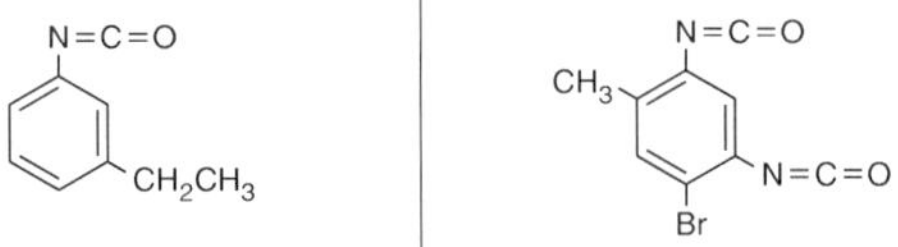
5 gm

S67,746-9
4-BROMO-6-METHYL-1,3-PHENYLENE DIISOCYANATE
5 gm

S53,846-9
3-CHLORO-4-FLUOROPHENYL ISOCYANATE
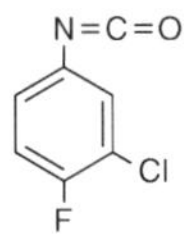
250 mg

S67,745-0
2,6-DIBROMO-4-ETHYLPHENYL ISOCYANATE
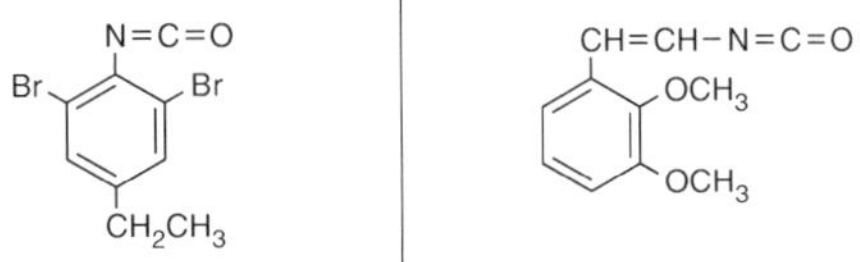
5 gm

S95,367-9
2,3-DIMETHOXYSTYRYL ISOCYANATE
50 mg

S67,725-6
4-PHENOXYPHENYL ISOCYANATE
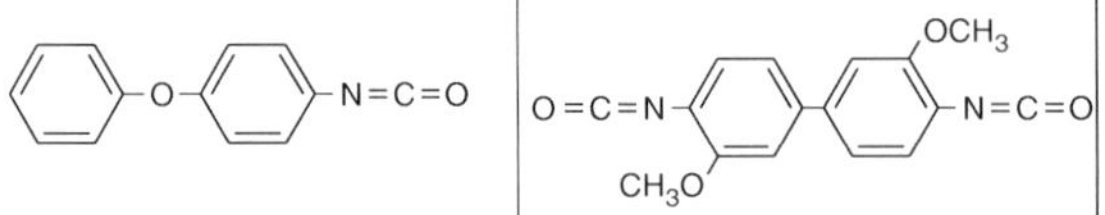
250 mg

S36,077-5
3,3'-DIMETHOXY-4,4'-BIPHENYLENE DIISOCYANATE
5 gm

S55,022-1
5-CHLORO-2,4-DIMETHOXYPHENYL ISOCYANATE
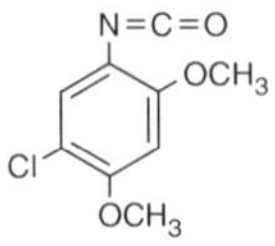
250 mg

S66,865-6
4-ETHOXYPHENYL ISOCYANATE
5 gm

S53,844-2
5-CHLORO-2-METHOXYPHENYL ISOCYANATE
5 gm

S55,024-8
4-CHLORO-2-METHOXYPHENYL ISOCYANATE
250 mg

S55,025-6
6-METHOXY-M-TOLYL ISOCYANATE
250 mg

S53,839-6
4-CHLORO-2-NITROPHENYL ISOCYANATE
250 mg

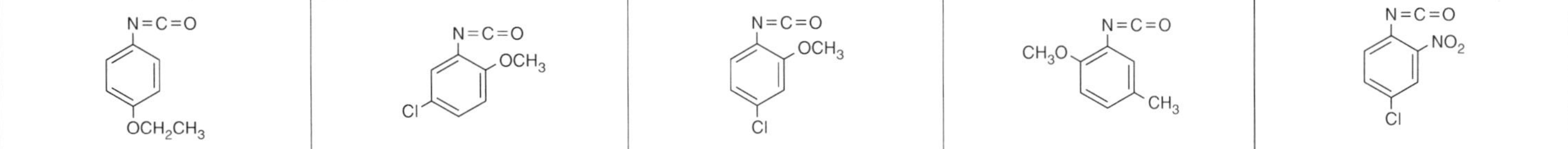

S51,696-1
ALPHA,ALPHA-DIMETHYL-ALPHA,4-PHENETHYL DIISOCYANATE
250 mg

S67,719-1
2-ISOPROPYL-6-METHYLPHENYL ISOCYANATE
5 gm

S56,616-0
TRIPHENYLMETHYL ISOCYANATE
5 gm

S53,834-5
2-NITRO-PARA-TOLYL ISOCYANATE
5 gm

S67,700-0
4-ETHYLPHENYL ISOCYANATE
5 gm

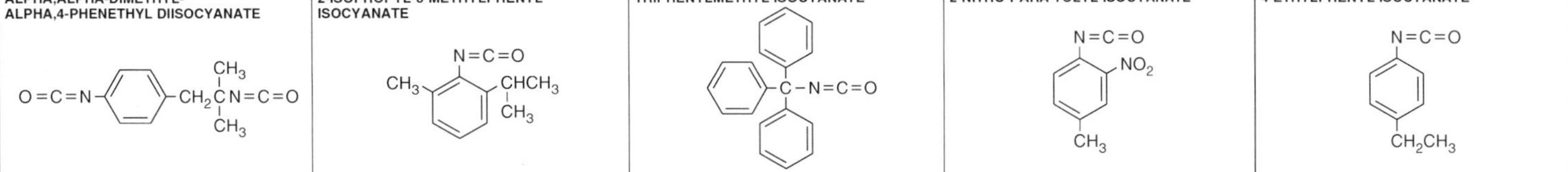

S67,702-7
2,3-DIMETHYLPHENYL ISOCYANATE
5 gm

S53,850-7
3-CHLORO-PARA-TOLYL ISOCYANATE
250 mg

S55,017-5
5-CHLORO-O-TOLYL ISOCYANATE
250 mg

S55,018-3
4-CHLORO-O-TOLYL ISOCYANATE
250 mg

S55,011-6
2,4-XYLYL ISOCYANATE
250 mg

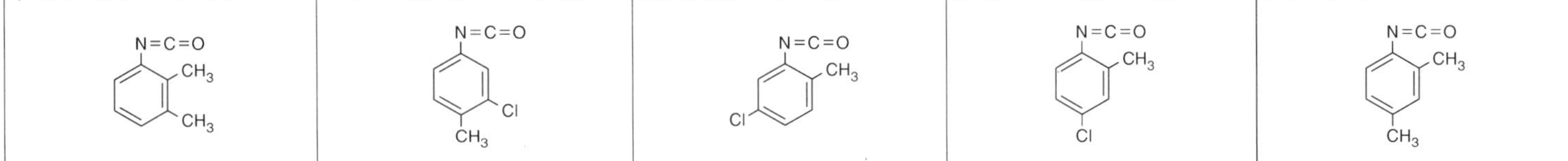

S34,692-6
3,3'-DIMETHYLDIPHENYLMETHANE-4,4'-DIISOCYANATE
5 gm

S36,090-2
3,3'-DIMETHYL-4,4'-BIPHENYLENE DIISOCYANATE
5 gm

S38,333-3
MESITYL ISOCYANATE
250 mg

S87,595-3
7-ISOPROPYL-1,2,3,4,4A,9,10,10A-OCTAHYDRO-1-PHENANTHRYLMETHYL ISOCYANATE
250 mg

S87,538-4
BIS(4-CHLOROPHENYL)CARBODIIMIDE
250 mg

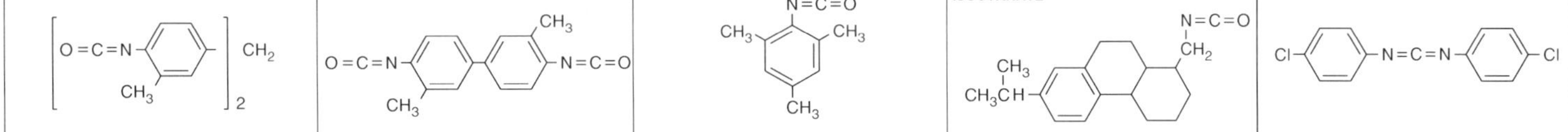

S87,539-2
BIS(4-BROMOPHENYL)CARBODIIMIDE
250 mg

S87,537-6
BIS(4-IODOPHENYL)CARBODIIMIDE
250 mg

S87,516-3
BIS(3-METHOXYPHENYL)CARBODIIMIDE
250 mg

S87,501-5
BIS[4-(DIMETHYLAMINO)PHENYL]-CARBODIIMIDE
250 mg

S45,819-8
N,N'-BIS(4-NITROPHENYL)-CARBODIIMIDE
5 gm

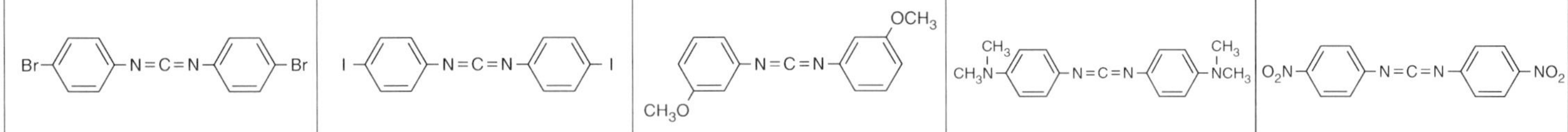

S34,244-0
BIS(TRIPHENYLMETHYL)CARBODIIMIDE
5 gm

S45,000-6
1-TERT-BUTYL-3-(TRIPHENYLMETHYL)-CARBODIIMIDE
5 gm

S42,940-6
N,N'-BIS(2,6-DIETHYLPHENYL)-CARBODIIMIDE
5 gm

S89,824-4
N-(AZIDOCARBONYL)-2-FLUOROANILINE
250 mg

S89,825-2
N-(AZIDOCARBONYL)-4-FLUOROANILINE
250 mg

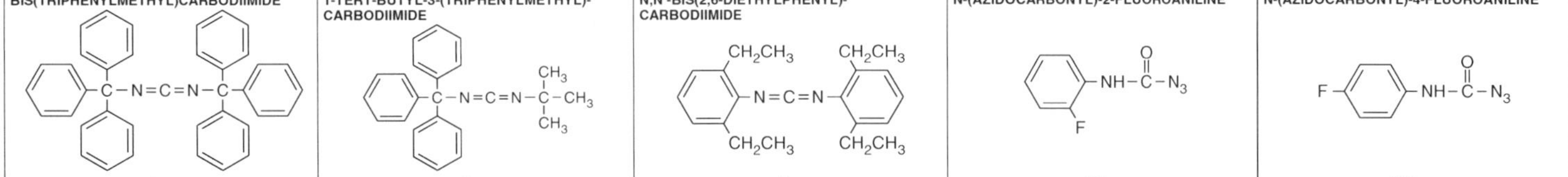

S88,579-7
1,2-BIS(2-HYDROXY-3-METHOXYBENZYLIDENE)HYDRAZINE
250 mg

S53,678-4
SALICYLALDEHYDE AZINE
250 mg

S87,349-7
2,6-DIPHENYL-1-METHYL-4(1H)-PYRIDONE AZINE
250 mg

S34,854-6
4-DIAZO-N,N-DIMETHYLANILINE, ZINC CHLORIDE DOUBLE SALT
5 gm

S36,009-0
4-DIAZO-N,N-DIETHYL-META-TOLUIDINE FLUOROBORATE
5 gm

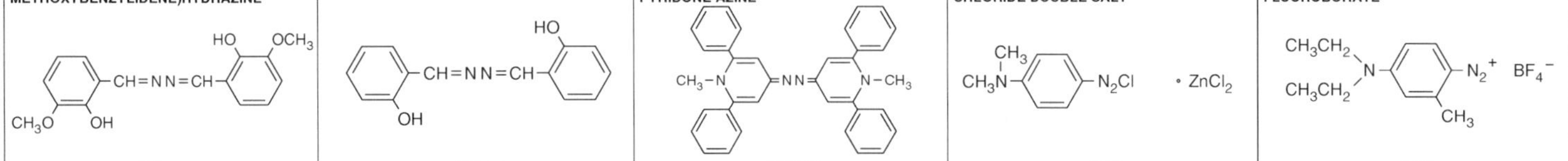

S34,856-2
4-DIAZO-N-ETHYL-O-TOLUIDINE FLUOROBORATE
5 gm

S95,317-2
1-(AZIDOCARBONYL)-2-BENZOYLHYDRAZINE
50 mg

S74,003-9
5,6-DIHYDRAZONOCYCLOOCTENE
250 mg

S88,792-7
2,6-DIPHENYL-4H-THIOPYRAN-4-ONE AZINE
250 mg

S88,761-7
1H-PHENALEN-1-ONE AZINE
250 mg

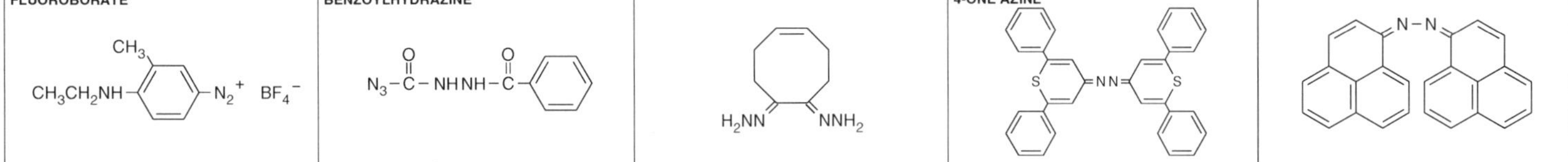

S87,334-9
1,4-BIS(AZIDOMETHYL)-1,4-BIS(P-TOLYL)-2-TETRAZENE
250 mg

S95,286-9
2-AZIDO-5-(4-NITROPHENYL)-1,3,4-OXADIAZOLE
50 mg

S91,920-9
1-(1-AZIDO-1-METHYLETHYL)-4-NITROBENZENE
250 mg

S95,088-2
2-(AZIDOCARBONYL)-3,5-DIMETHYLPYRROLE
50 mg

S89,826-0
N-(AZIDOCARBONYL)ANILINE
250 mg

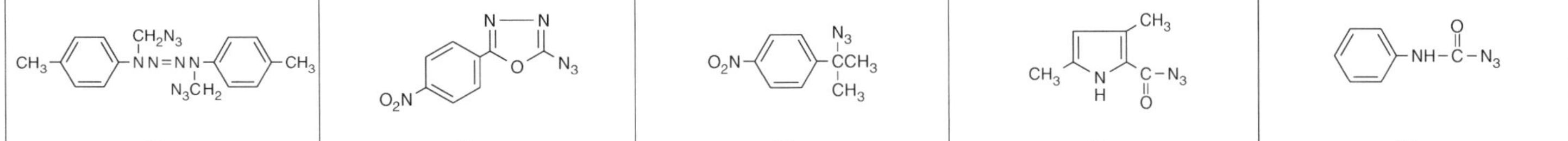

S89,822-8
N-(AZIDOCARBONYL)-4-METHYLANILINE

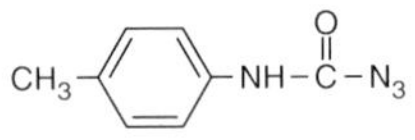

250 mg

S86,217-7
2-METHYLBENZALDEHYDE AZINE

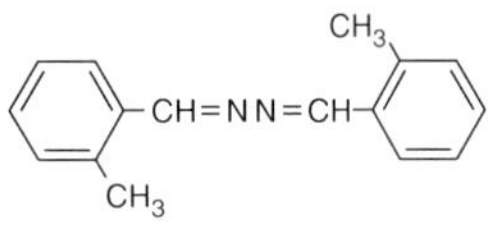

250 mg

S88,776-5
BENZOYL CHLORIDE AZINE

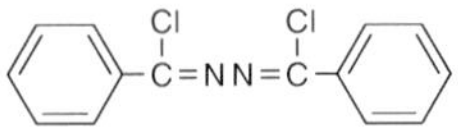

250 mg

S91,829-6
9-ANTHRACENYLMETHYL AZIDE

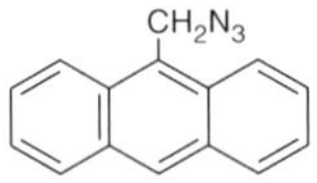

250 mg

S77,124-4
2-(4-CHLOROPHENYL)-1-METHYLETHYL 2,4-DINITROBENZENESULFENATE

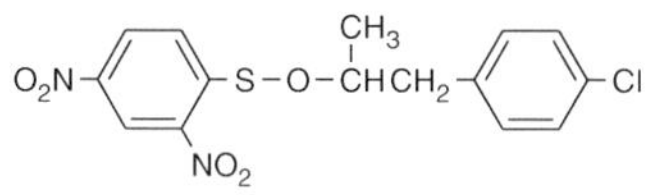

250 mg

S91,771-0
4-BROMOPHENYL METHYL SULFOXIDE

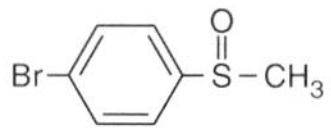

250 mg

S86,245-2
P-TOLYL DISULFOXIDE

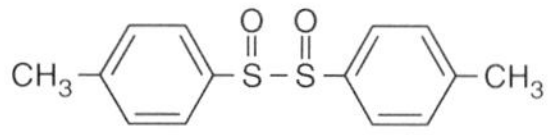

250 mg

S94,661-3
10-(3-DIMETHYLAMINOPROPYL)-10H-PHENOTHIAZINE-5-OXIDE HYDROCHLORIDE

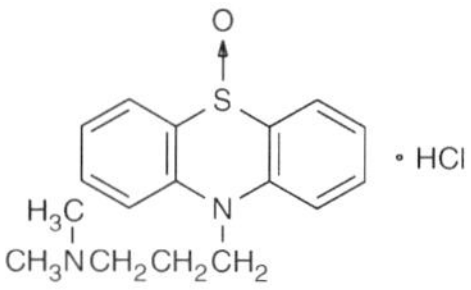

250 mg

S77,123-6
1-METHYL-2-PHENYLPROPYL 2,4-DINITROBENZENESULFENATE

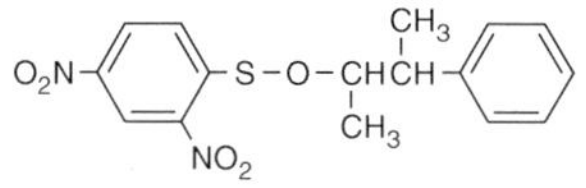

250 mg

S74,829-3
TRANS-4-TERT-BUTYLCYCLOHEXYL 2,4-DINITROBENZENESULFENATE

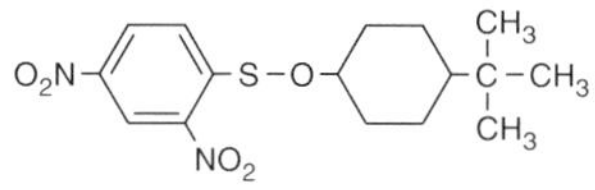

250 mg

S83,607-9
10-ETHYL-3-NITROPHENOTHIAZINE-5-OXIDE

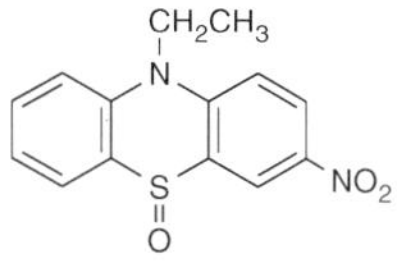

250 mg

S71,024-5
4-NITRO-N-SULFINYLANILINE

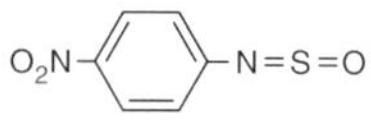

250 mg

S82,095-4
10-ACETYL-3,7-DINITROPHENOTHIAZINE-5-OXIDE

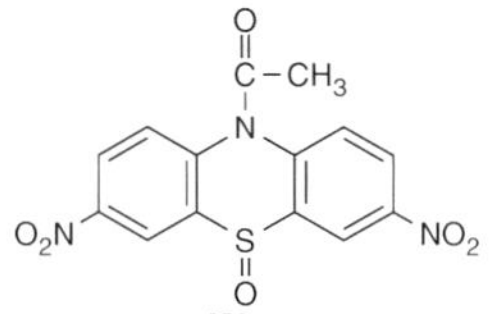

250 mg

S81,194-7
PHENOTHIAZINE-5-OXIDE

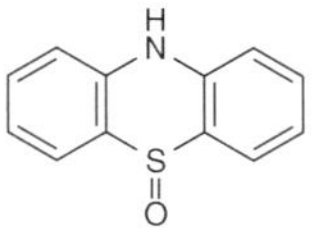

250 mg

S93,619-7
2-CHLOROPHENOTHIAZINE-5-OXIDE

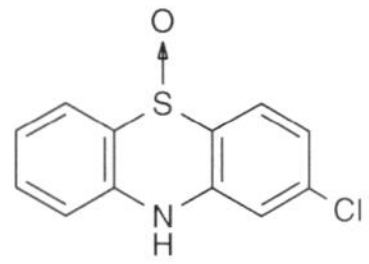

250 mg

S82,413-5
10-PHENYLPHENOTHIAZINE-5-OXIDE

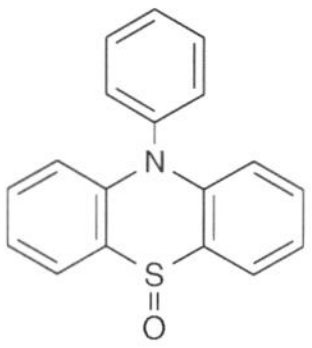

250 mg

S82,412-7
10-DECYLPHENOTHIAZINE-5-OXIDE

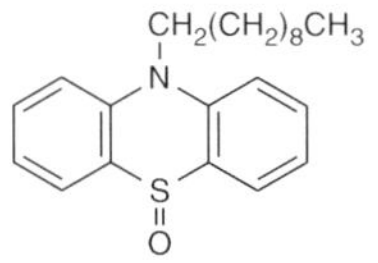

250 mg

S83,611-7
10-OCTADECYLPHENOTHIAZINE-5-OXIDE

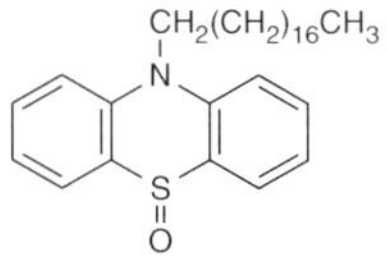

250 mg

S72,183-2
BENZYL 2-NITROPHENYL SULFOXIDE

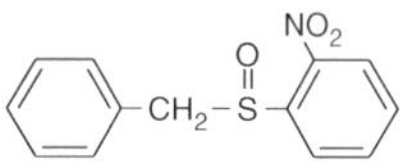

250 mg

S67,974-7
2-(BENZOYLMETHYLSULFINYL)-ACETANILIDE

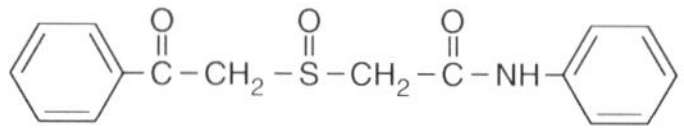

250 mg

S36,119-4
3,7-DINITROPHENOTHIAZINE-5-OXIDE

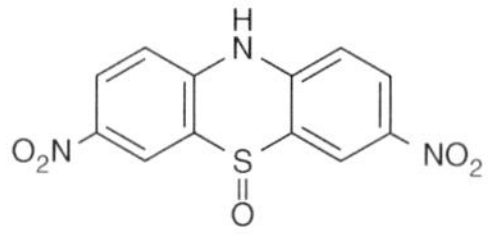

5 gm

S70,712-0
5'-METHYL-2'-(METHYLSULFINYL)-ACETOPHENONE

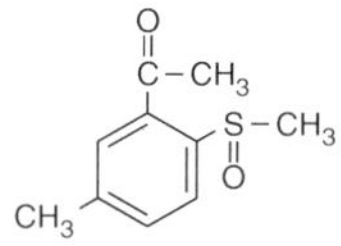

5 gm

S82,331-7
PHENOXATHIIN-10-OXIDE

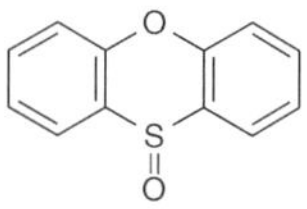

250 mg

S62,004-1
BENZYL METHYL SULFOXIDE

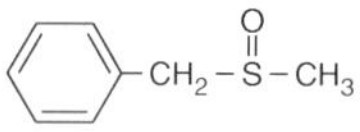

5 gm

S62,021-1
ALLYL PHENYL SULFOXIDE

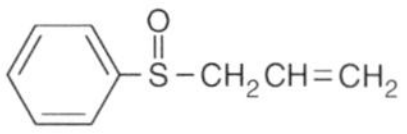

5 gm

S79,985-8
PHENYL PROPYL SULFOXIDE

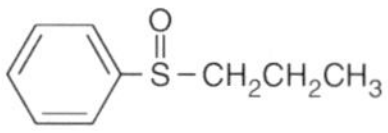

250 mg

S86,323-8
2-PHENYL-1,3-DITHIANE 1-OXIDE

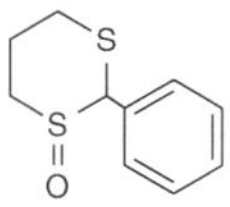

250 mg

S62,003-3
BENZYL PHENYL SULFOXIDE

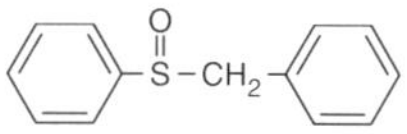

250 mg

S62,010-6
BIS(PHENYLSULFINYL)METHANE

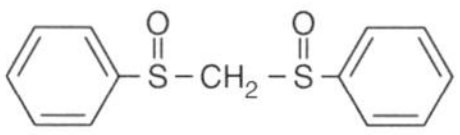

250 mg

S77,098-1
BENZYL 2-TERT-BUTYLPHENYL SULFOXIDE

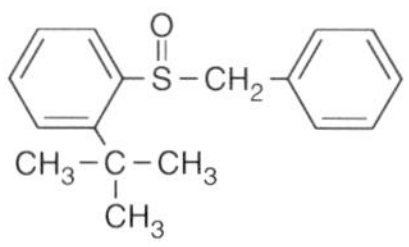

250 mg

S86,085-9
ETHYL P-TOLYL SULFOXIDE

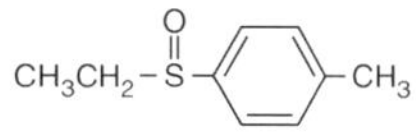

250 mg

S79,908-4
METHYL 1-NAPHTHYLMETHYL SULFOXIDE

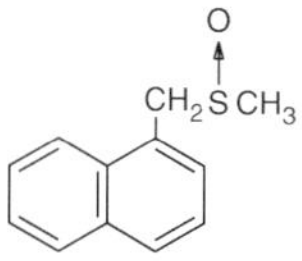

250 mg

S77,837-0
ETHYL 2-NAPHTHYL SULFOXIDE

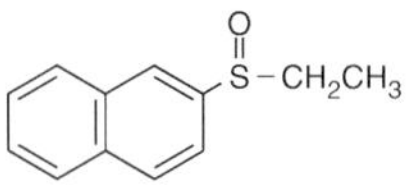

250 mg

S83,479-3
THIANTHRENE-5,10-DIOXIDE

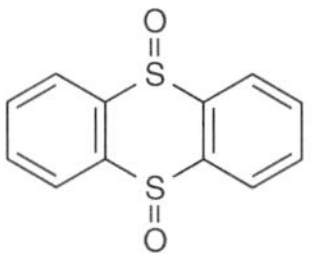

250 mg

S86,963-5
1-ISOPROPYLIDENE-4-PHENYL-3-THIOSEMICARBAZIDE

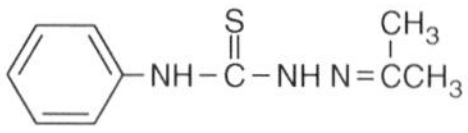

250 mg

S74,533-2
CHLOROMETHYL 4-CHLOROPHENYL SULFONE

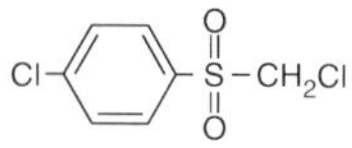

250 mg

S82,198-5
3-(3-CHLOROPROPYLSULFONYL)PROPYL 4-NITROPHENYL ETHER
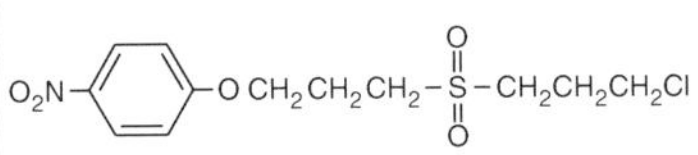
250 mg

S94,590-0
4-(PHENYLSULFONYL)BUTYL 11-BROMOUNDECANOATE
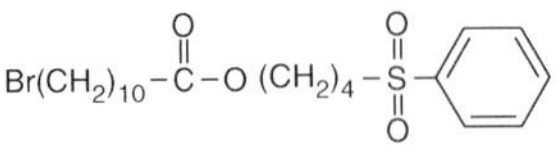
250 mg

S71,811-4
4',4''-SULFONYLBIS(2-CHLOROACETANILIDE)
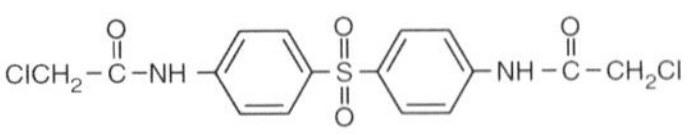
250 mg

S67,994-1
2-CHLORO-4'-HEXADECYLSULFONYLACETANILIDE
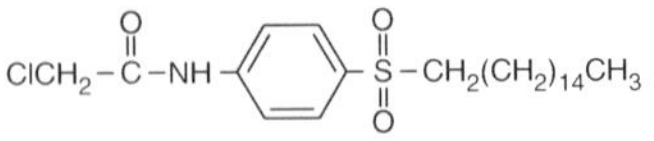
250 mg

S57,698-0
2-CHLOROETHYL PHENYL SULFONE
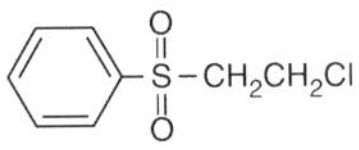
250 mg

S61,833-0
2-BROMOETHYL PHENYL SULFONE
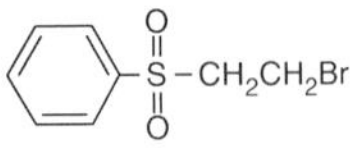
250 mg

S61,441-6
BENZYL 2-CHLOROETHYL SULFONE
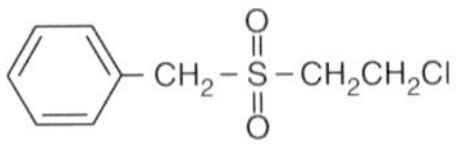
5 gm

S79,946-7
3-BROMO-1-PROPENYL PHENYL SULFONE
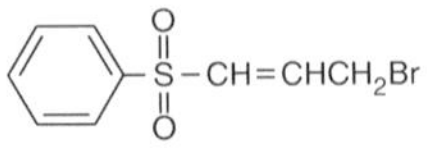
250 mg

S79,911-4
BROMOMETHYL 2-CHLORO-2-PHENYLETHYL SULFONE
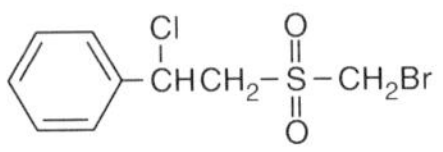
250 mg

S79,995-5
BENZYL 3-CHLOROPROPYL SULFONE
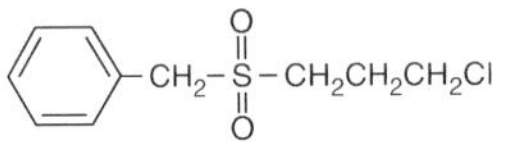
250 mg

S75,957-0
4-BROMO-2-BUTENYL PHENYL SULFONE
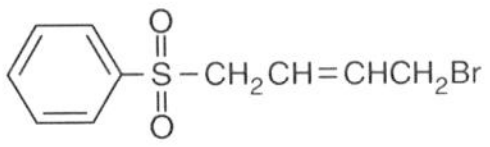
250 mg

S79,915-7
BENZYL 4-CHLOROBUTYL SULFONE
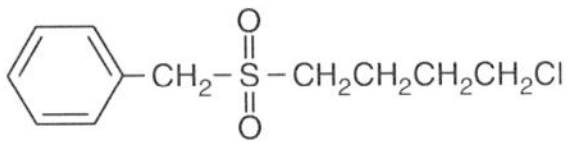
250 mg

S73,220-6
2-(BROMOMETHYL)BENZYL METHYL SULFONE
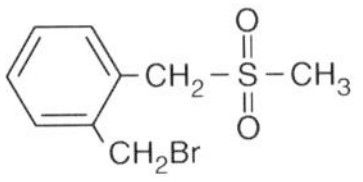
250 mg

S79,947-5
BROMOMETHYL P-TOLYL SULFONE
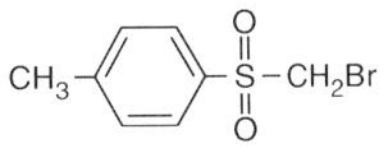
250 mg

S62,006-8
P-CHLOROBENZYL 2-CHLOROETHYL SULFONE
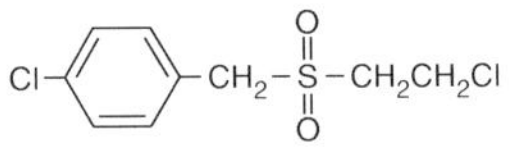
250 mg

S75,955-4
3-CHLORO-1-PROPENYL P-TOLYL SULFONE
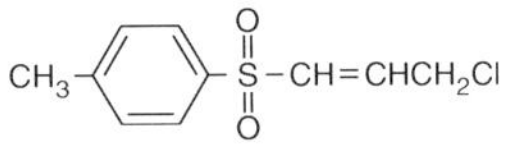
250 mg

S77,564-9
3-CHLOROPROPYL P-TOLYL SULFONE
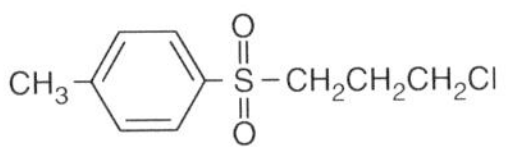
250 mg

S61,778-4
3-AMINO-4-CHLOROPHENYL DIFLUOROMETHYL SULFONE
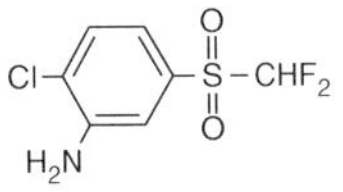
5 gm

S54,183-4
5-(DIFLUOROMETHYLSULFONYL)-ORTHO-ANISIDINE
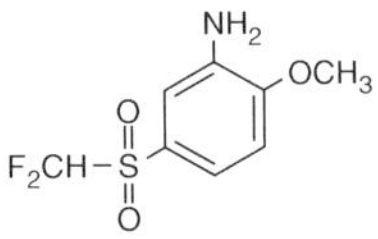
250 mg

S63,228-7
3-AMINO-1-[2-AMINO-4-(DIFLUOROMETHYLSULFONYL)PHENYL]-2-PYRAZOLIN-5-ONE
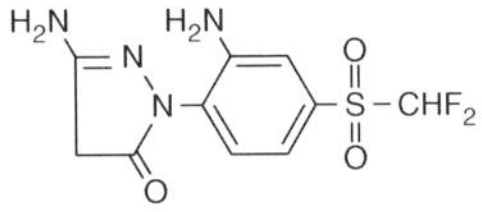
250 mg

S50,518-8
4-DIFLUOROMETHYLSULFONYL-2-NITROPHENYLHYDRAZINE
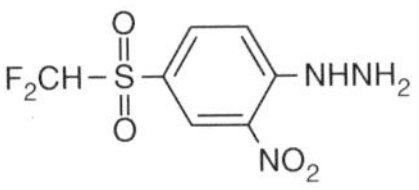
5 gm

S62,223-0
4-(2-CHLORO-1,1,2-TRIFLUOROETHYLSULFONYL)-2-NITROPHENYLHYDRAZINE
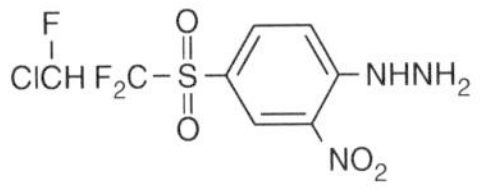
5 gm

S64,202-9
3-CHLORO-4-HYDRAZINOPHENYL DIFLUOROMETHYL SULFONE
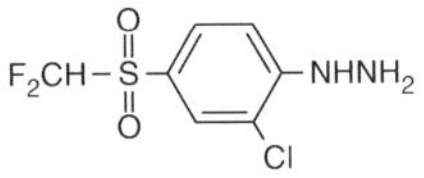
250 mg

S38,604-9
4-CHLORO-3-NITROPHENYL 2-CHLORO-1,1,2-TRIFLUOROETHYL SULFONE
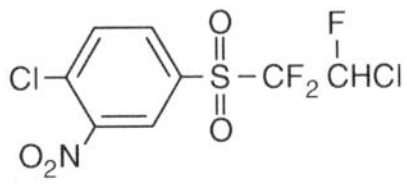
250 mg

S68,861-4
4-(2,2-DICHLOROACETAMIDO)PHENYL SULFONE
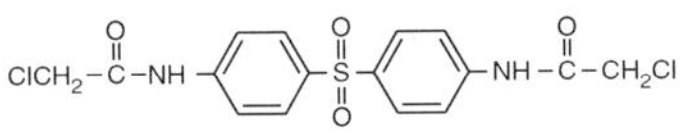
250 mg

S38,887-4
4-CHLOROPHENYL DIFLUOROMETHYL SULFONE
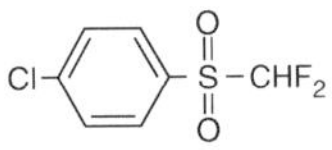
5 gm

S62,216-8
2,5-DICHLOROPHENYL DIFLUOROMETHYL SULFONE
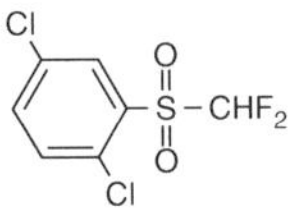
5 gm

S51,475-6
6-ETHYLSULFONYL-ALPHA,ALPHA,ALPHA-TRIFLUORO-M-TOLUIDINE
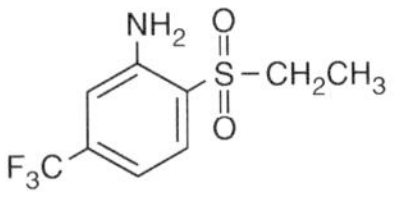
250 mg

S64,204-5
4-HEXADECYLSULFONYL-2-TRIFLUOROMETHYLPHENYLHYDRAZINE
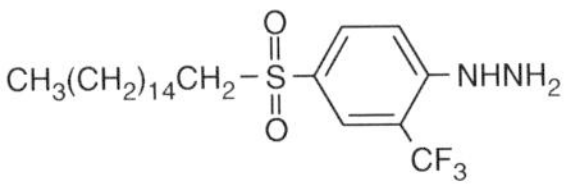
250 mg

S50,975-2
4-HEXADECYLSULFONYL-N-(TRIFLUOROACETYL)-METANILYL FLUORIDE
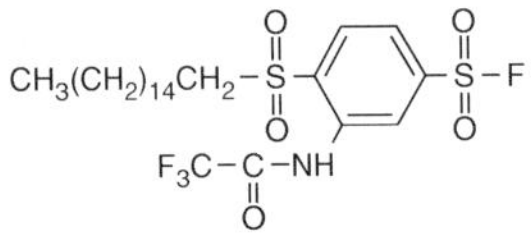
250 mg

S77,227-5
1,1'-(SULFONYLDI-1,4-PHENYLENE)BIS[3-(2-TRIFLUOROMETHYLPHENYL)UREA]
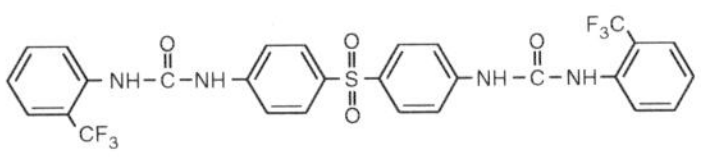
250 mg

S93,347-3
PHENYL TRIFLUOROMETHYL SULFONE
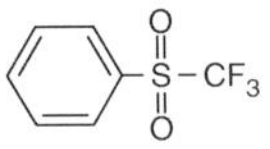
250 mg

S93,346-5
4-CHLOROPHENYL TRIFLUOROMETHYL SULFONE

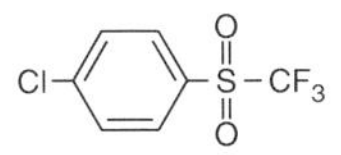

250 mg

S79,891-6
4,6-DICHLORO-ALPHA-[2,4-DIMETHYL-6-(3-METHYLBENZYL)PHENYLSULFONYL]-O-CRESOL

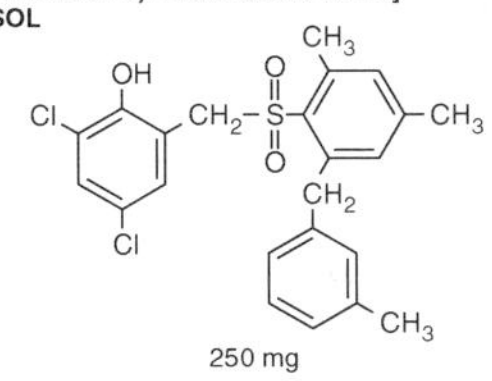

250 mg

S88,950-4
3-[2-(2-CHLOROPHENOXY)ACETOXY]-2,3-DIHYDROTHIOPENE-1,1-DIOXIDE

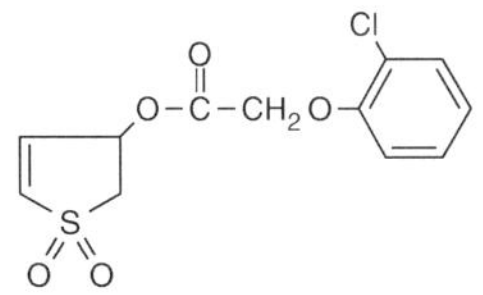

250 mg

S88,949-0
3-[2-(2,4-DICHLOROPHENOXY)ACETOXY]-2,3-DIHYDROTHIOPHENE-1,1-DIOXIDE

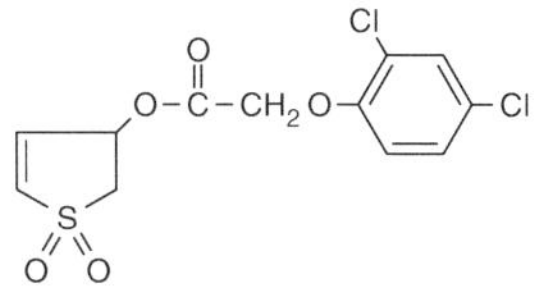

250 mg

S83,581-1
4'-[3-(4-CHLOROPHENYL)-1-(4-METHOXYPHENYL)-3-OXOPROPYLSULFONYL]-ACETANILIDE

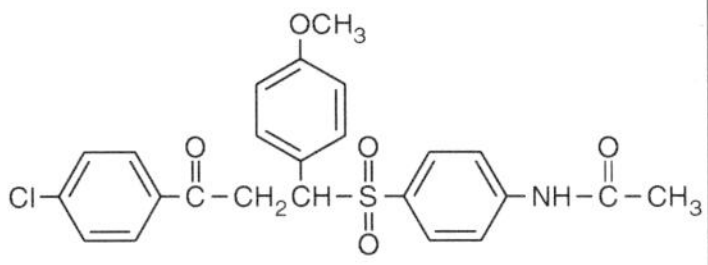

250 mg

S84,032-7
3-(2-CHLOROPHENYL)-4'-METHOXY-3-(P-TOSYL)PROPIOPHENONE

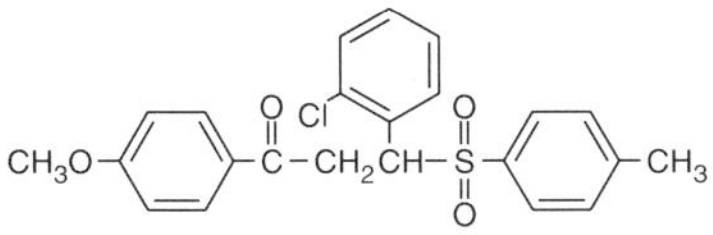

250 mg

S81,241-2
4'-CHLORO-3-(4-METHOXYPHENYL)-3-(P-TOSYL)PROPIOPHENONE

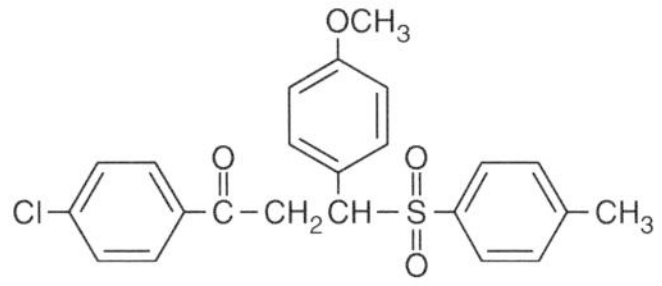

250 mg

S94,614-1
4-(4-[(4-CHLOROPHENYL)SULFONYL]-PHENOXY)BENZIL

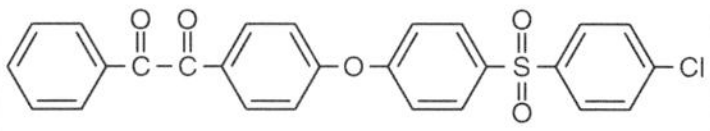

250 mg

S77,804-4
METHYL 2-FLUORO-4-METHOXY-6-(PHENYLSULFONYL)BENZOATE

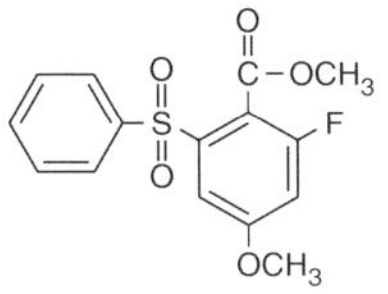

250 mg

S81,240-4
3-[4-(ACETAMIDO)PHENYLSULFONYL]-3-(2-CHLOROPHENYL)-4'-METHOXYPROPIOPHENONE

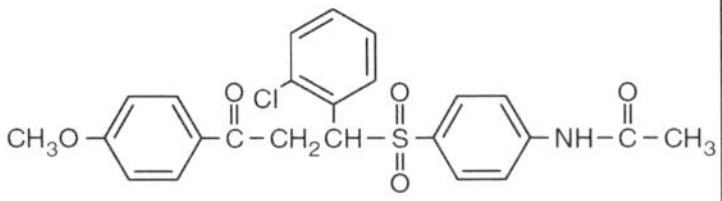

250 mg

S94,566-8
4-CHLORO-4'-ETHOXYDIPHENYL SULFONE

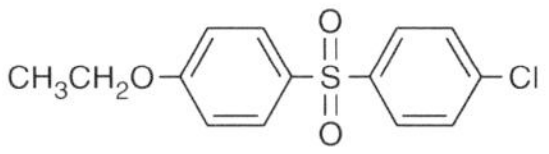

250 mg

S66,170-8
2-AMINO-4-CHLORO-4'-METHYLDIPHENYL SULFONE

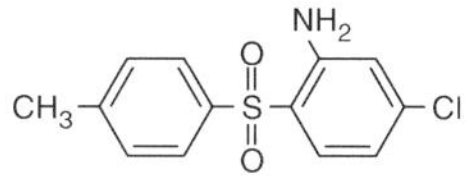

5 gm

S45,524-5
5-CHLORO-2-(HEXADECYLSULFONYL)-ANILINE

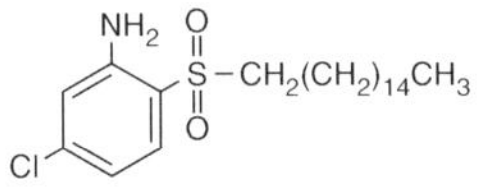

5 gm

S65,479-5
4-AMINO-4'-CHLORODIPHENYL SULFONE

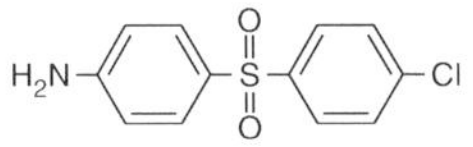

250 mg

S83,626-5
4-(3-CHLOROPHENYLSULFONYL)ANILINE

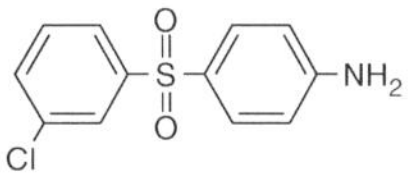

250 mg

S38,571-9
4-(4-BROMOPHENYLSULFONYL)-ANILINE

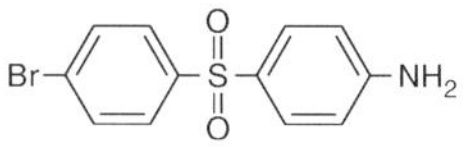

5 gm

S44,828-1
2-CHLORO-4-(METHYLSULFONYL)-ANILINE

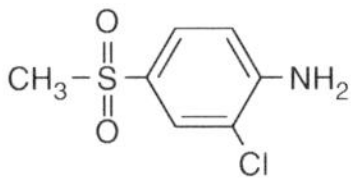

5 gm

S58,264-6
2-CHLORO-5-(METHYLSULFONYL)ANILINE

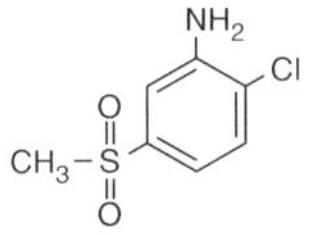

250 mg

S43,238-5
N-[4-(4-CHLOROPHENYLSULFONYL)-PHENYL]GLYCINE

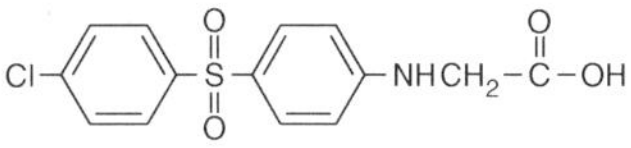

5 gm

S93,857-2
3-(3-BROMOANILINO)-2,3-DIHYDROTHIOPHENE-1,1-DIOXIDE

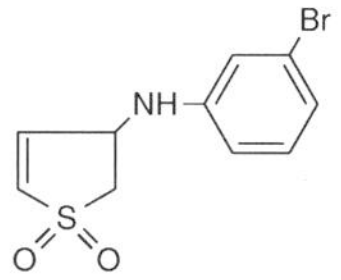

250 mg

S93,860-2
3-(4-BROMOANILINO)-2,3-DIHYDROTHIOPHENE-1,1-DIOXIDE

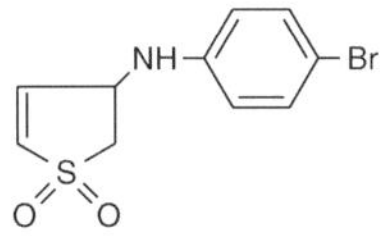

250 mg

S90,079-6
4-CHLOROBENZALDEHYDE N-PHENYL-N-(3-SULFOLANYL)HYDRAZONE

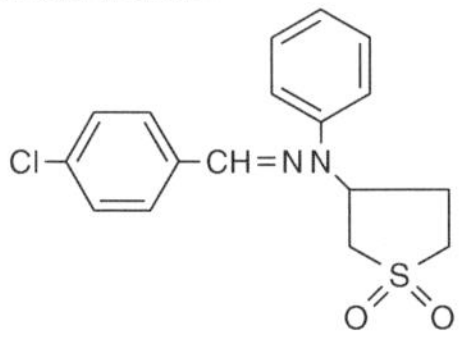

250 mg

S82,798-3
2-CHLOROPHENYL 4-NITROPHENYL SULFONE

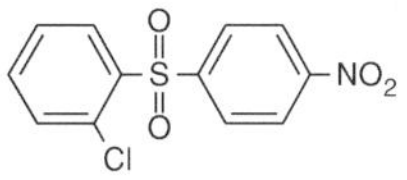

250 mg

S39,769-5
4-CHLORO-3-NITROPHENYL METHYL SULFONE

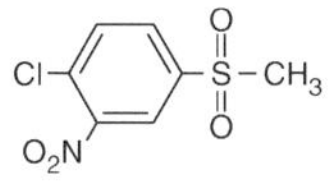

5 gm

S92,894-1
4,4'-CHLORO-3,3'-NITROPHENYL SULFONE

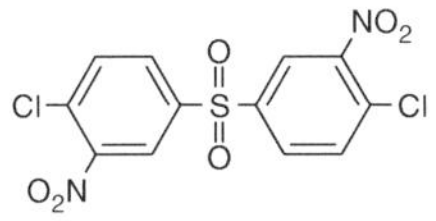

250 mg

S69,637-4
4-(HEXADECYLSULFONYL)-3-NITROCHLOROBENZENE

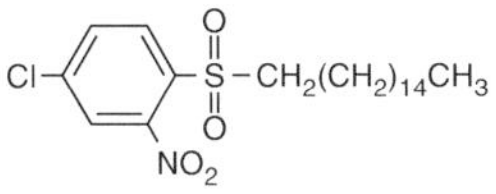

250 mg

S77,805-2
2-CHLORO-6-(PHENYLSULFONYL)-BENZOIC ACID

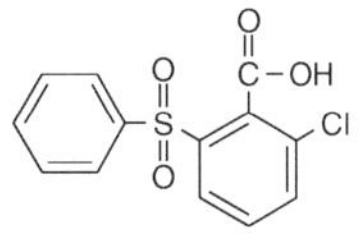

250 mg

S45,522-9
4-(4-CHLOROPHENYLSULFONYL)-BENZOIC ACID

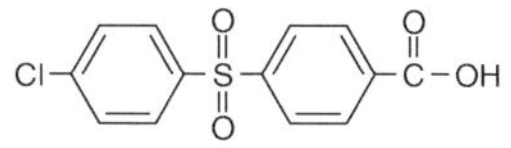

5 gm

S76,718-2
2-(4-CHLOROPHENYL)-4-PHENYLSULFONYL-1,3(2H,4H)-ISOQUINOLINEDIONE

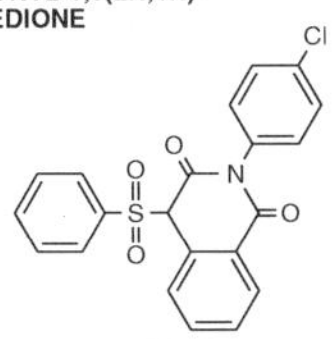

250 mg

S34,015-4
4-ACETAMIDO-4'-CHLORODIPHENYL SULFONE

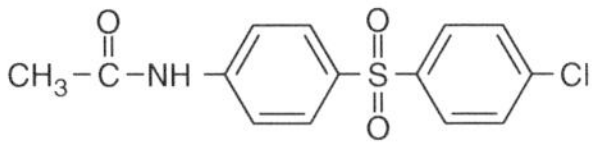

5 gm

S76,961-4
1,1'-(SULFONYLDI-1,4-PHENYLENE)BIS[3-(4-BROMOPHENYL)UREA]

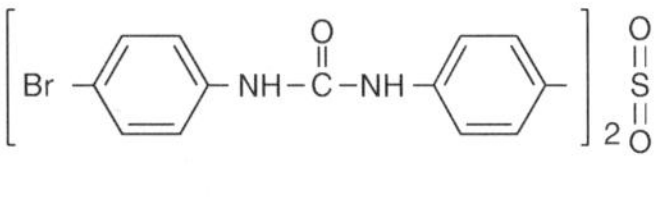

250 mg

S72,805-5
1,1'-(SULFONYLDI-1,4-PHENYLENE)BIS[3-(4-CHLORO-2-METHYLPHENYL]UREA)

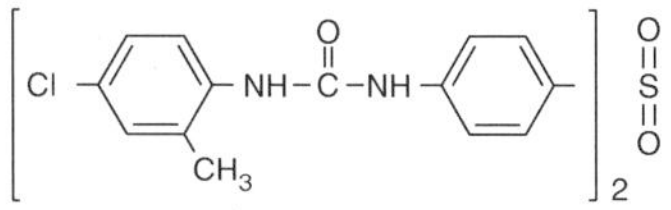

100 mg

S76,962-2
1,1'-(SULFONYLDI-1,4-PHENYLENE)BIS[3-(3-CHLORO-4-METHYLPHENYL]UREA)

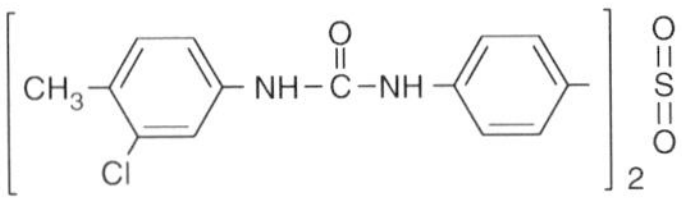

250 mg

S94,277-4
TRANS-3-(4-BROMOPHENYLSULFONYL)-ACRYLONITRILE

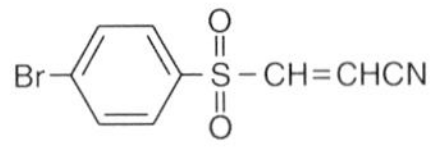

250 mg

S79,942-4
3-BROMOPHENYL ISOPROPYL SULFONE

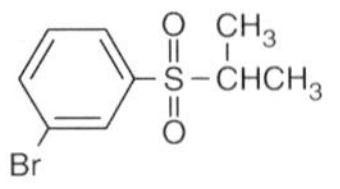

250 mg

S80,591-2
4-BROMOPHENYL 2-PHENYLCYCLOPROPYL SULFONE

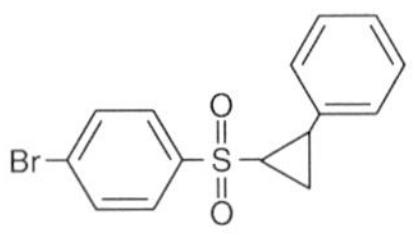

250 mg

S80,904-7
3-(4-CHLOROPHENYL)-3,4-DIHYDRO-2-PHENYL-2H-1,4-BENZOTHIAZINE-1,1-DIOXIDE

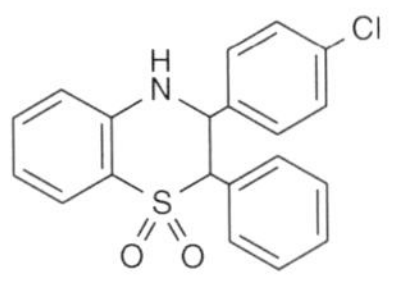

250 mg

S77,835-4
4-BROMOPHENYL PHENYL SULFONE

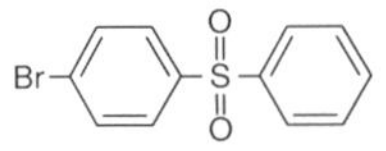

250 mg

S86,330-0
BENZYL 4-CHLOROPHENYL SULFONE

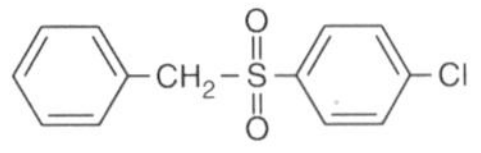

250 mg

S94,276-6
4-CHLOROPHENYL TRANS-STYRYL SULFONE

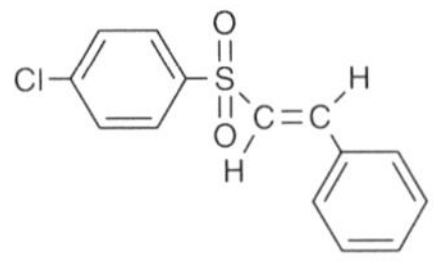

250 mg

S88,041-8
3-BROMO-4-METHYLPHENYL PHENYL SULFONE

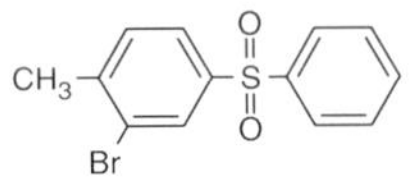

250 mg

S67,167-3
4-BROMOPHENYL SULFONE

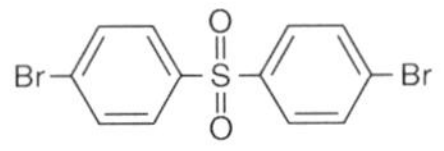

5 gm

S38,280-9
4-BROMOPHENYL PARA-TOLYL SULFONE

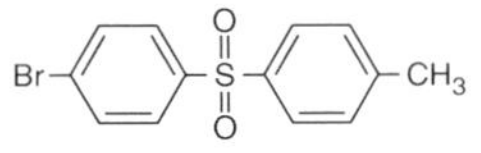

250 mg

S77,849-4
4-FLUOROPHENYL MESITYL SULFONE

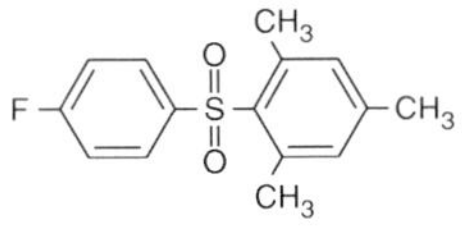

250 mg

S80,723-0
4-CHLOROPHENYL MESITYL SULFONE

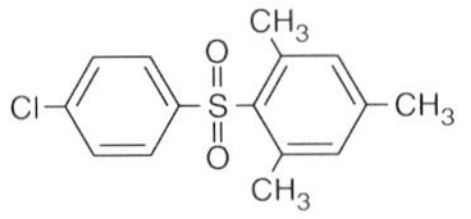

250 mg

S39,305-3
3,4-DICHLOROPHENYL HEXADECYL SULFONE

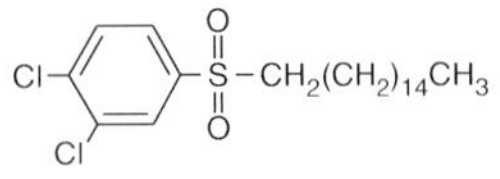

250 mg

S39,318-5
3,4-DICHLOROPHENYL METHYLSULFONYLMETHYL SULFONE

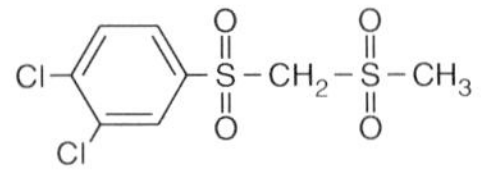

250 mg

S80,909-8
2,5-BIS(DIPHENYLHYDROXYMETHYL)-TETRAHYDROTHIOPHENE-1,1-DIOXIDE

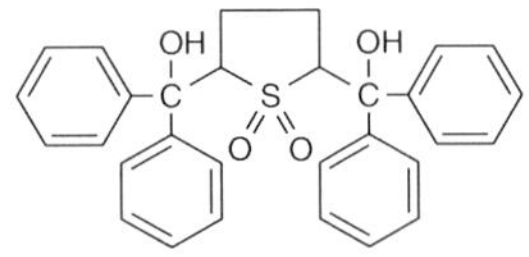

250 mg

S91,953-5
2-(2-NITROPHENYLSULFONYL)-HYDROQUINONE

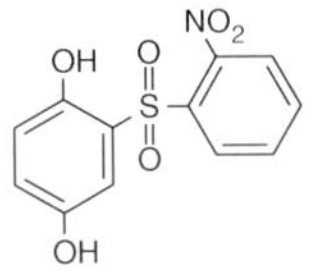

250 mg

S92,116-5
2-(3-NITROPHENYLSULFONYL)-HYDROQUINONE

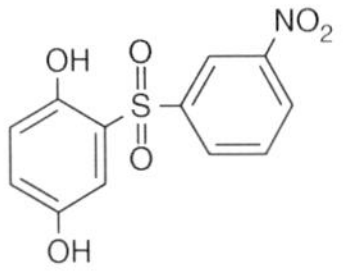

250 mg

S92,030-4
2-(PHENYLSULFONYL)HYDROQUINONE

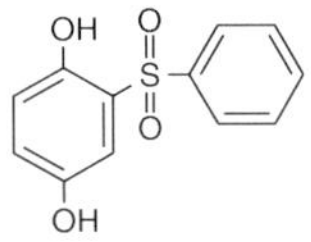

250 mg

S44,107-4
3-AMINO-4-HYDROXYPHENYL SULFONE

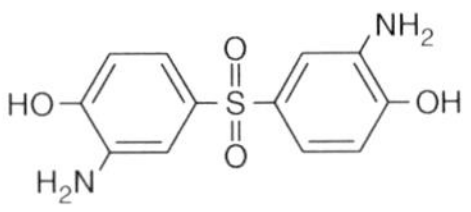

5 gm

S50,431-9
2,2'-SULFONYLBIS[4-(1,1,3,3-TETRAMETHYLBUTYL)PHENOL]

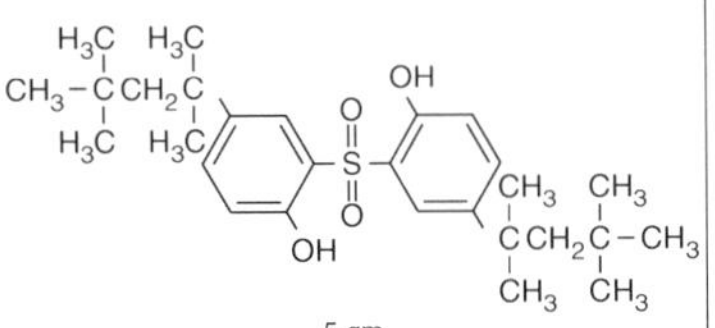

5 gm

S50,692-3
4,4'-SULFONYLBIS[6-(TERT-BUTYL)-M-CRESOL]

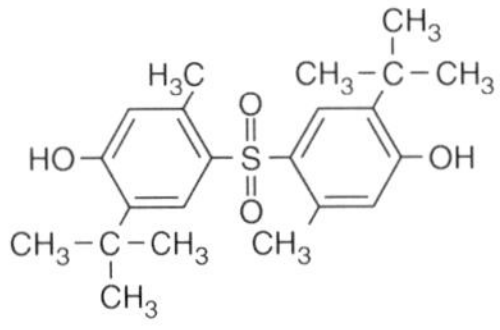

5 gm

S90,083-4
3-ETHOXY-4-HYDROXYBENZALDEHYDE N-PHENYL-N-(3-SULFOLANYL)HYDRAZONE

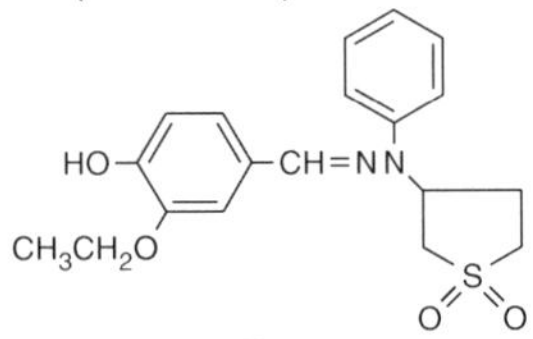

250 mg

S77,551-7
ALPHA,ALPHA-DIPHENYL-2-SULFOLANEMETHANOL

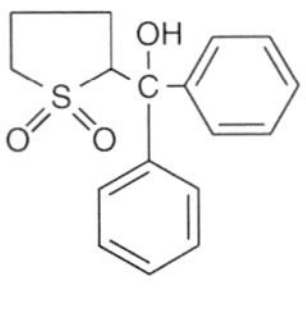

250 mg

S82,509-3
2-(METHYLSULFONYL)-1-PHENYLETHANOL

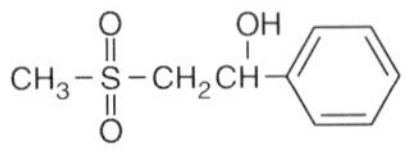

250 mg

S38,916-1
1-HYDROXY-4'-(METHYLSULFONYL)-2-NAPHTHANILIDE

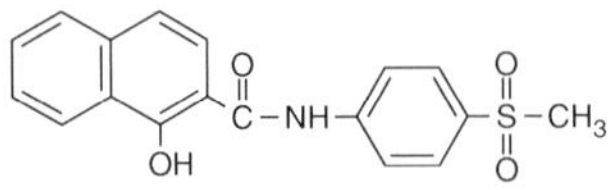

250 mg

S76,730-1
3-(1-HYDROXYETHYLIDENE)-4-(PHENYLSULFONYLMETHYL)-1-ISOPROPYL-2-PYRROLIDINONE

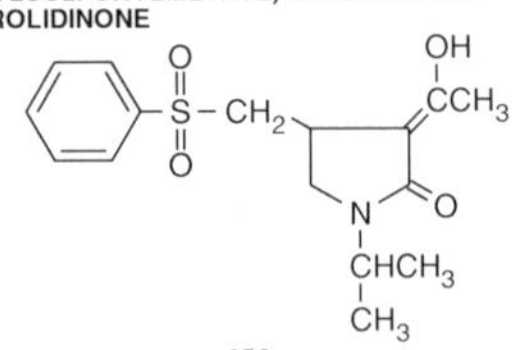

250 mg

S38,252-3
2-(4-SULFANILYLANILINO)ETHANOL

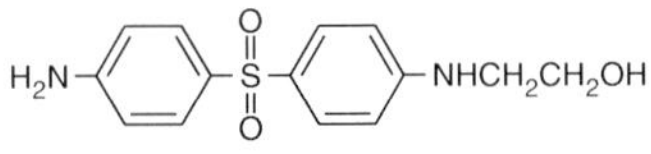

250 mg

S93,232-9
CIS-3-BENZYLAMINO-4-HYDROXY-2,3,4,5-TETRAHYDROTHIOPHENE-1,1-DIOXIDE

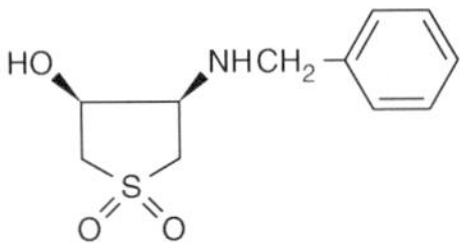

250 mg

S93,239-6
TRANS-3-(BENZYLAMINO)-4-HYDROXYTETRAHYDROTHIOPHENE-1,1-DIOXIDE

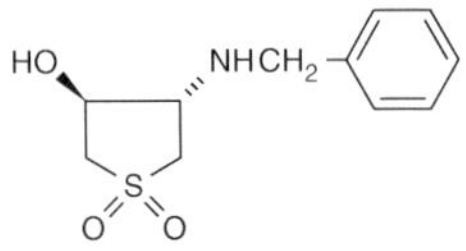

250 mg

S58,161-5
3-CARBOXY-4-HYDROXYBENZENESULFONIC (METHYL-PENTADECYL-PYRIMIDINYLIDENE)HYDRAZIDE

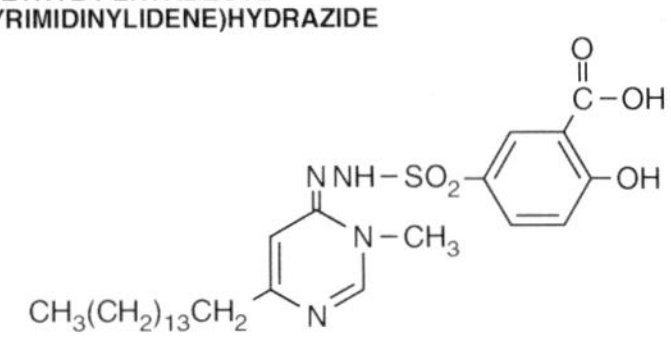

250 mg

S90,082-6
SALICYLALDEHYDE N-PHENYL-N-(3-SULFOLANYL)HYDRAZONE

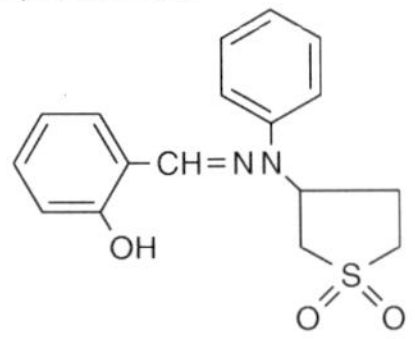

250 mg

S90,084-2
3,5-DI-TERT-BUTYL-4-HYDROXYBENZALDEHYDE N-PHENYL-N-(3-SULFOLANYL)HYDRAZONE
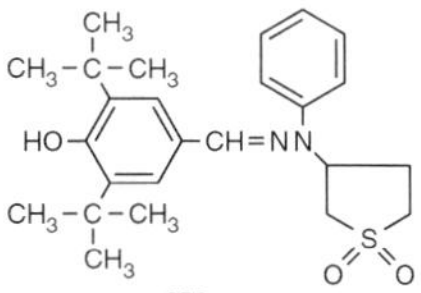
250 mg

S77,228-3
2-HYDROXY-1-NAPHTHALDEHYDE N-PHENYL-N-(3-SULFOLANYL)HYDRAZONE
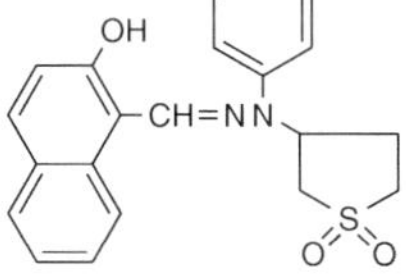
250 mg

S39,067-4
4-METHYLSULFONYL-2-NITROPHENOL
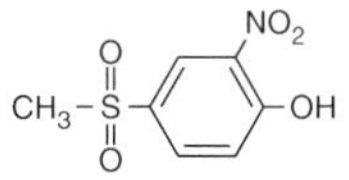
250 mg

S64,239-8
PHENYL 4-(4-CHLOROPHENYLSULFONYL)-1-HYDROXY-2-NAPHTHOATE
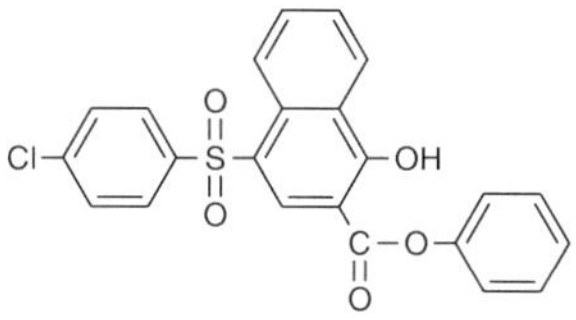
250 mg

S64,217-7
PHENYL 4-(PHENYLSULFONYL)-1-HYDROXY-2-NAPHTHOATE
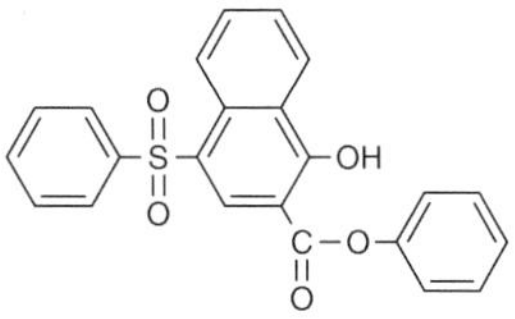
250 mg

S93,876-9
TRANS-1-(4-HYDROXY-3-SULFOLANYL)-3-PHENYL-2-THIOUREA
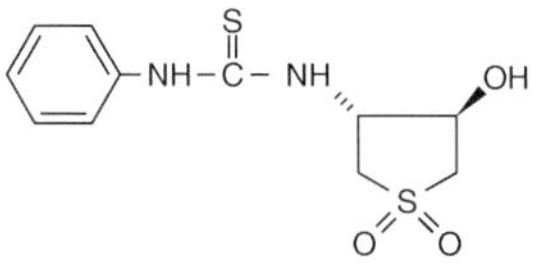
250 mg

S39,054-2
(PARA-TOLYLSULFONYL)-METHANOL
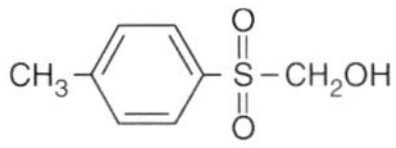
250 mg

S86,318-1
2-(P-TOSYL)-1-ETHANOL
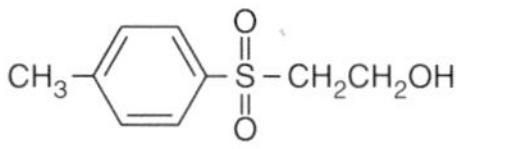
250 mg

S73,328-8
3,4-DIHYDRO-6,7-DIMETHOXY-1-(PHENYLSULFONYLMETHYL)ISOQUINOLINE
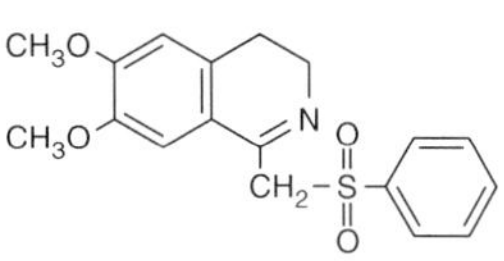
250 mg

S94,617-6
1,1'-(SULFONYLBIS)4,1-PHENYLENEOXY-4,1-(PHENYLENE)BIS(2-PHENYLETHANEDIONE)
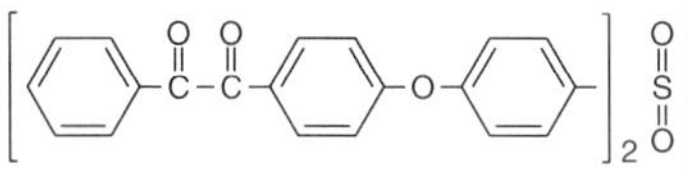
250 mg

S64,215-0
3-AMINO-4-(HEXADECYLOXY)PHENYL PHENYL SULFONE
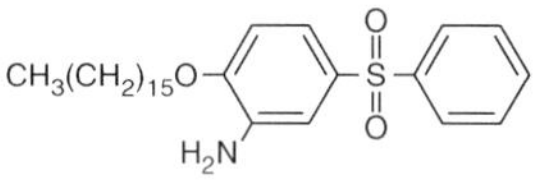
250 mg

S35,775-8
3-AMINO-4-METHOXYPHENYL ETHYL SULFONE
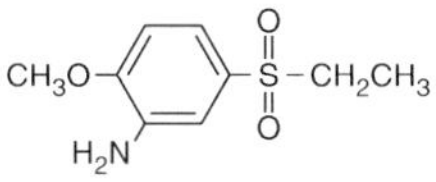
5 gm

S58,253-0
2-(HEXADECYLOXY)-5-(METHYLSULFONYL) ANILINE
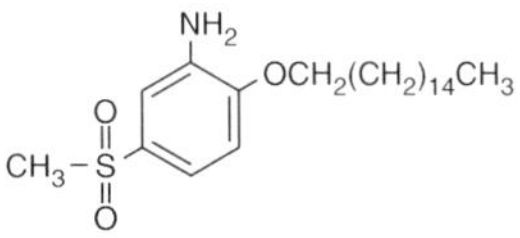
5 gm

S61,870-5
2-(4-AMINOPHENOXY)ETHYL PHENYL SULFONE
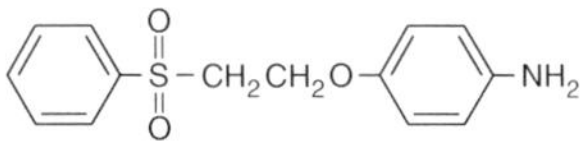
250 mg

S93,858-0
2,3-DIHYDRO-3-(3-METHOXYANILINO)-THIOPHENE-1,1-DIOXIDE
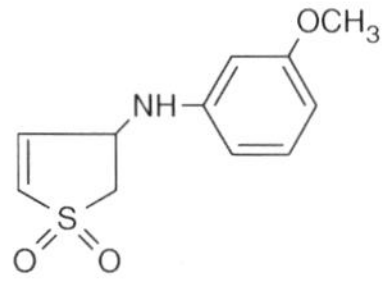
250 mg

S93,856-4
2,3-DIHYDRO-3-(2-METHOXYANILINO)-THIOPHENE-1,1-DIOXIDE
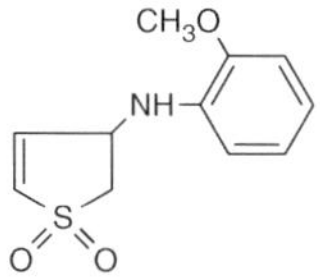
250 mg

S83,584-6
3-[4-(DIMETHYLAMINO)PHENYL]-4'-METHOXY-3-(PHENYLSULFONYL) PROPIOPHENONE
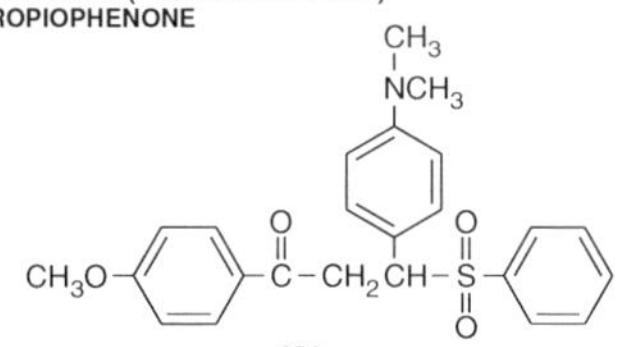
250 mg

S90,556-9
4-METHOXYBENZALDEHYDE N2-PHENYL-N2-(3-SULFOLANYL)HYDRAZONE
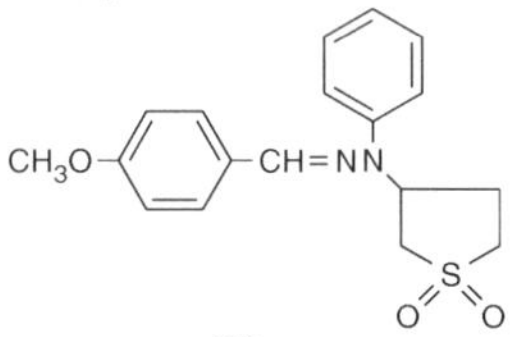
250 mg

S58,693-5
4-(HEXADECYLOXY)-3-NITROPHENYL METHYL SULFONE
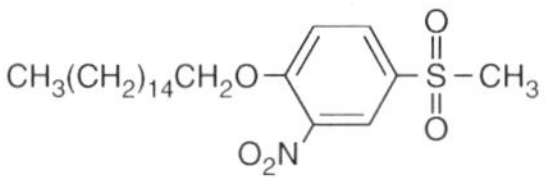
250 mg

S68,264-0
2,3-DIHYDRO-S,S-DIOXIDE-5-ETHOXY-1,4-BENZOTHIAZEPINE
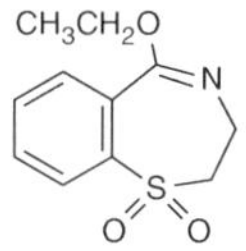
250 mg

S80,722-2
4'-METHOXY-2-(PHENYLSULFONYL)-ACETOPHENONE
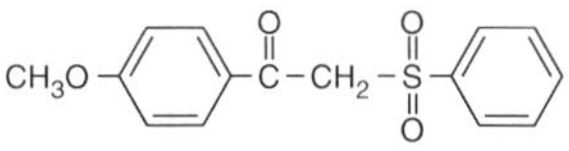
250 mg

S94,616-8
4-(4-[(4-HYDROXYPHENYL)SULFONYL]-PHENOXY)BENZIL
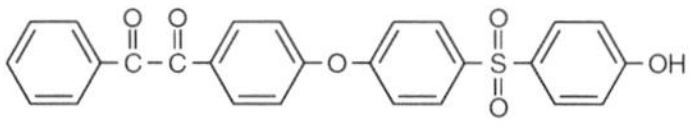
250 mg

S60,097-0
3-(3-METHYLPHENOXY)SULFOLANE
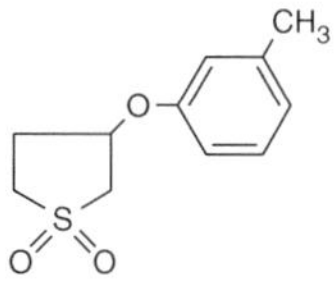
5 gm

S94,092-5
3-(PHENOXY)TETRAHYDROTHIOPHENE-1,1-DIOXIDE
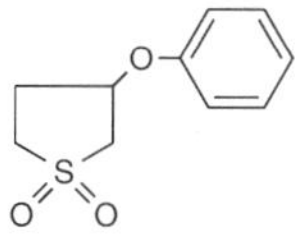
250 mg

S83,333-9
3-METHOXYPHENYL PHENYL SULFONE
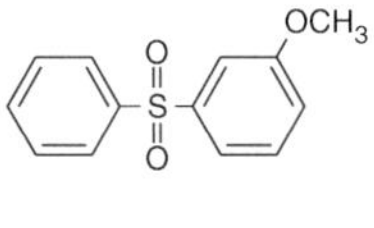
250 mg

S83,334-7
4-METHOXYPHENYL PHENYL SULFONE
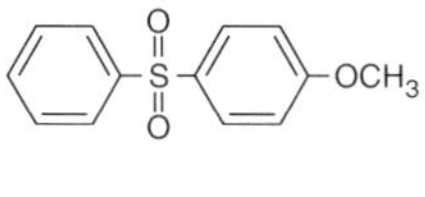
250 mg

S45,473-7
4-METHOXYPHENYL METHYL SULFONE
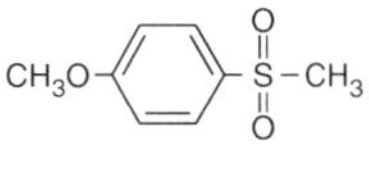
5 gm

S43,354-3
4-FLUOROPHENYL 4-METHOXYPHENYL SULFONE
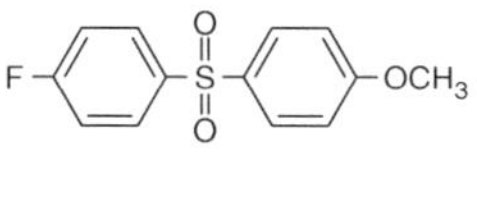
5 gm

S45,376-5
4,4'-SULFONYLDIBENZYL ALCOHOL
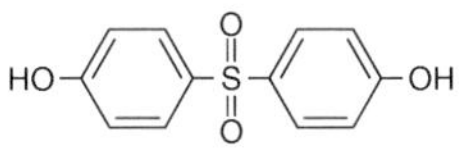
5 gm

S83,112-3
1,9-DICARBETHOXYPHENOXATHIIN-10,10-DIOXIDE
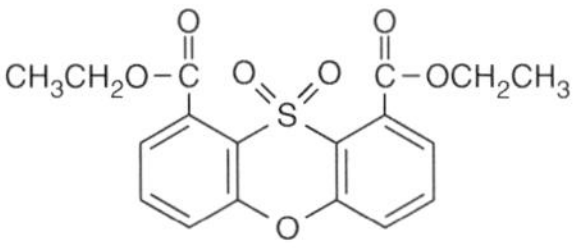
250 mg

S40,050-5
2-(4-AMINOPHENYLSULFONYL)-N-[2-(METHYLTHIO)ETHYL]ACETAMIDE
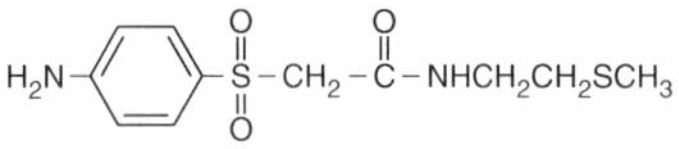
250 mg

S40,071-8
N-[2-(METHYLTHIO)ETHYL)-2-(PARA-TOLYLSULFONYL]ACETAMIDE
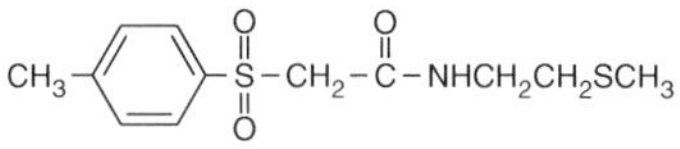
250 mg

S65,586-4
4'-METHYLSULFONYL-2-(4-NITROPHENYLTHIO)ACETOPHENONE
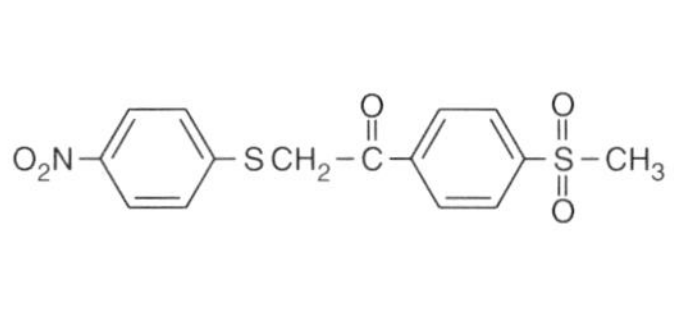

5 gm

S95,510-8
2-[2-(PHENYLSULFONYL)ETHYLTHIO]-TETRAHYDROTHIOPYRAN-4-ONE
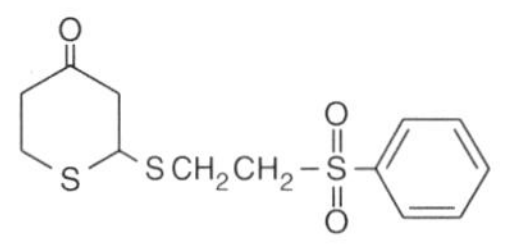

50 mg

S95,511-6
2-[1,4-DIOXA-8-THIASPIRO(4.5)DEC-7-YLTHIO]ETHYL PHENYL SULFONE
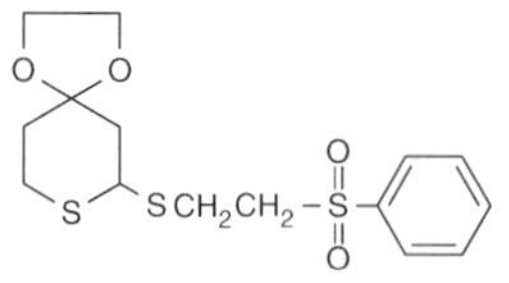

50 mg

S80,758-3
CIS-P-TOLYL 2-(P-TOSYL)-CYCLOHEXYL SULFIDE
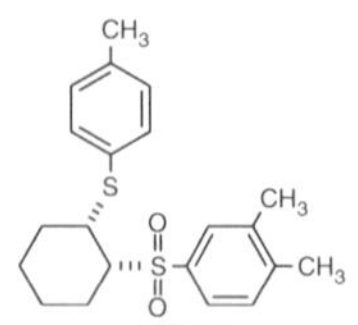

250 mg

S80,756-7
P-TOLYL 2-(3,4-XYLYLSULFONYL)-CYCLOHEXYL SULFIDE
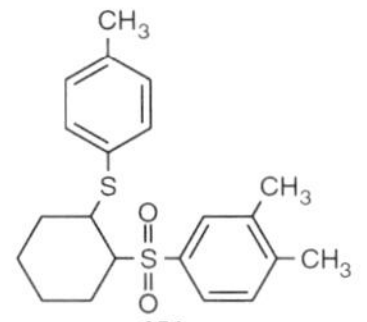

250 mg

S91,797-4
METHYL 4-(PHENYLSULFONYL)PHENYL SULFIDE
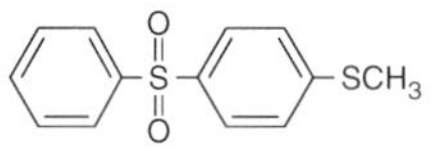

250 mg

S91,784-2
ALPHA,ALPHA-DIMETHYL-4-(PHENYLSULFONYL)BENZYL METHYL SULFIDE
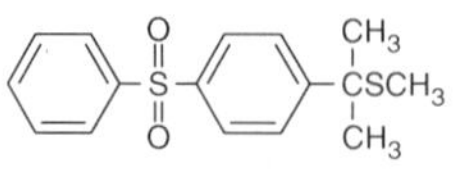

250 mg

S59,273-0
4-METHYLSULFONYL-2-NITROPHENYL DISULFIDE
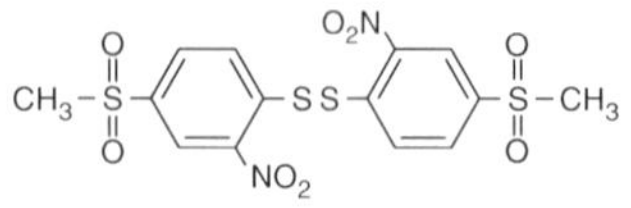

250 mg

S86,246-0
P-TOLYL DISULFONE
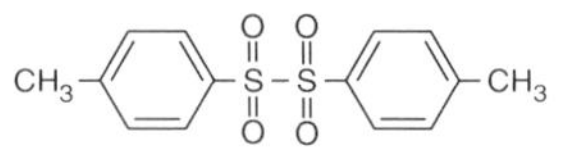

250 mg

S64,225-8
3-AMINO-1-(2-AMINO-4-METHYLSULFONYLPHENYL)-2-PYRAZOLIN-5-ONE
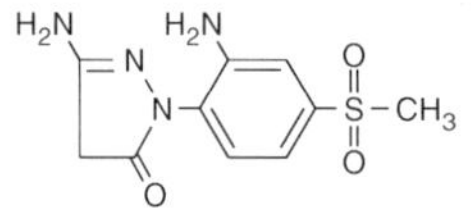

5 gm

S64,236-3
3-AMINO-1-(2-AMINO-4-METHYLSULFONYLPHENYL)-2-PYRAZOLIN-5-ONE HYDROCHLORIDE
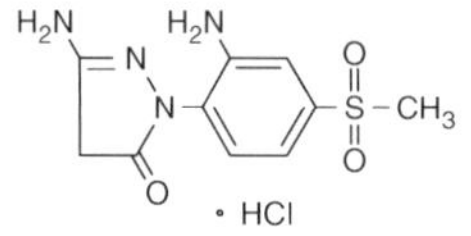

5 gm

S59,257-9
N1-HEXADECYL-N1-METHYL-4-METHYLSULFONYL-1,2-PHENYLENEDIAMINE
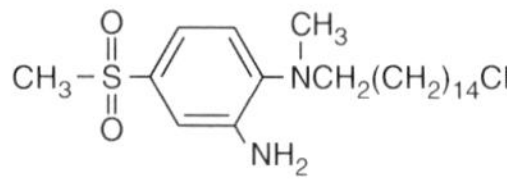

5 gm

S82,897-1
4-(M-TOSYL)-1,3-PHENYLENEDIAMINE
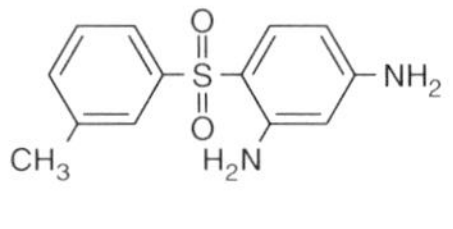

250 mg

S63,910-9
3-AMINO-1-(2-CHLORO-4-METHYLSULFONYLPHENYL)-2-PYRAZOLIN-5-ONE
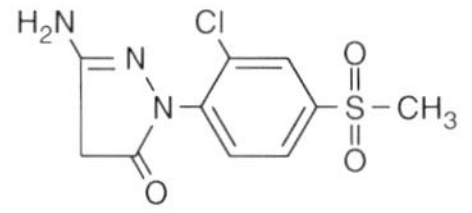

250 mg

S60,517-4
3-AMINO-1-(4-HEXADECYL-SO2)PH-2-PYRAZOLIN-5-ONE, MIXTURE OF BRANCHED ISOMERS
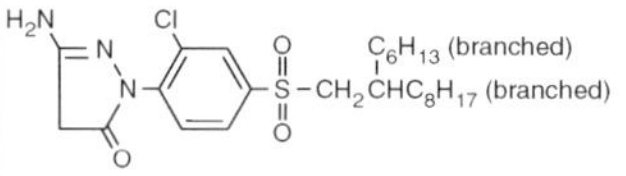

5 gm

S65,977-0
4-[4-(P-TOLYLSULFONYL)BENZYLAMINO]-ANILINE
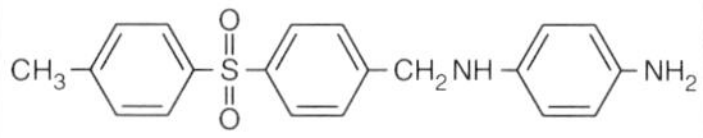

5 gm

S83,661-3
2-(METHYLSULFONYL)-4-NITROANILINE
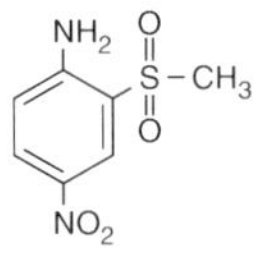

250 mg

S40,159-5
N-[4-(4-AMINOPHENYLSULFONYL)-PHENYL]GLYCINE, MORPHOLINE SALT
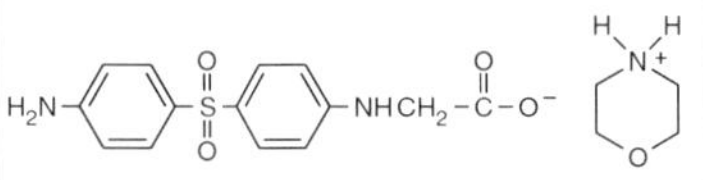

250 mg

S54,064-1
N-[4-(4-AMINOPHENYLSULFONYL)-PHENYL]GLYCINE HYDROCHLORIDE
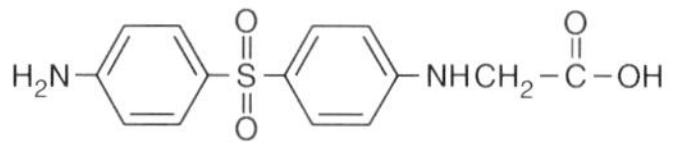

5 gm

S37,742-2
4-(2-DIETHYLAMINOETHYLSULFONYL)-ANILINE
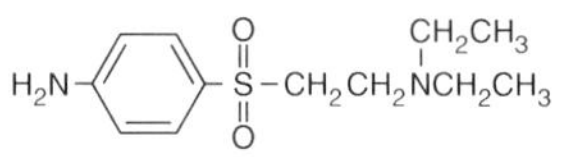

250 mg

S38,307-4
METHYL 3-AMINO-4-(HEXADECYLSULFONYL)BENZOATE
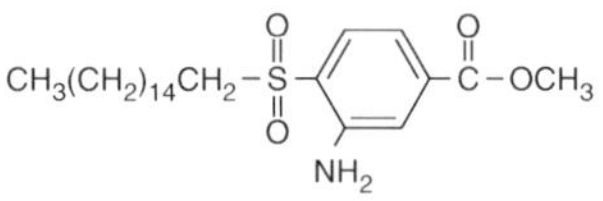

250 mg

S40,096-3
2-(4-AMINOPHENYLSULFONYL)-N-BUTYLPROPIONAMIDE
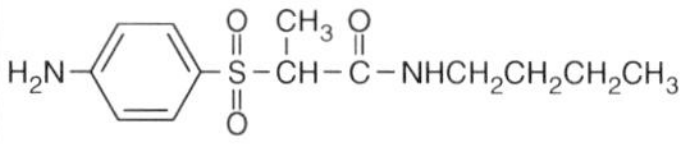

250 mg

S62,775-5
2-AMINOPHENYL HEXADECYL SULFONE
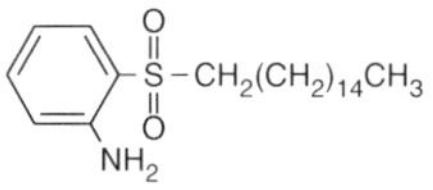

5 gm

S90,721-9
2-(P-TOSYL)ANILINE
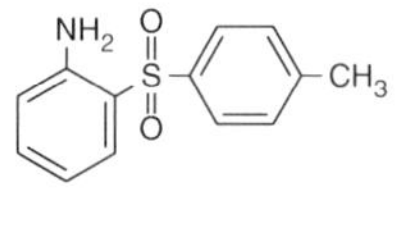

250 mg

S92,670-1
N,N-DIMETHYL-2-(P-TOSYL)-1,4-PHENYLENEDIAMINE
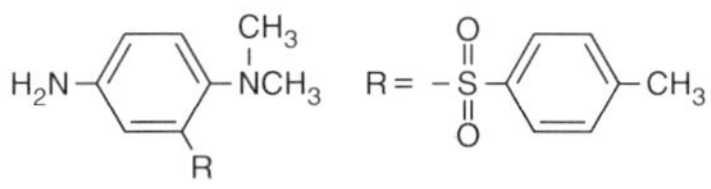

250 mg

S68,888-6
4-AMINO-4'-NITRODIPHENYL SULFONE
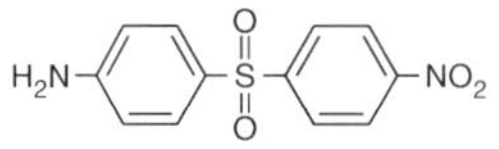

5 gm

S85,861-7
2,6-DINITRO-4-(METHYLSULFONYL)-ANILINE
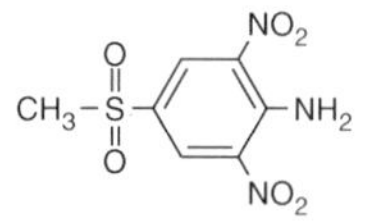

250 mg

S64,491-9
3-AMINODIPHENYL SULFONE
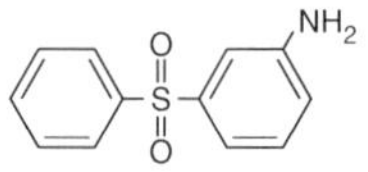

250 mg

S82,908-0
4-(4-ISOPROPYLPHENYLSULFONYL)-ANILINE
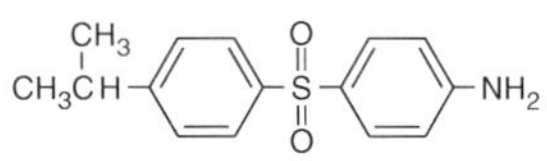

250 mg

S81,247-1
4-(DODECYLSULFONYL)ANILINE
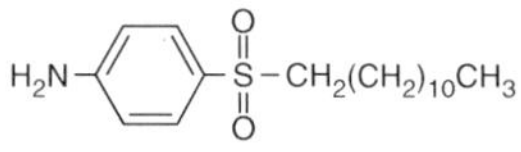

250 mg

S83,662-1
2,4-BIS(METHYLSULFONYL)ANILINE
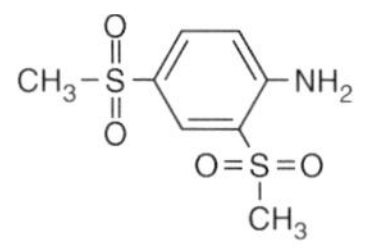

250 mg

S40,076-9
4-[2-(4-AMINOPHENYLSULFONYL)-ACETYL]MORPHOLINE
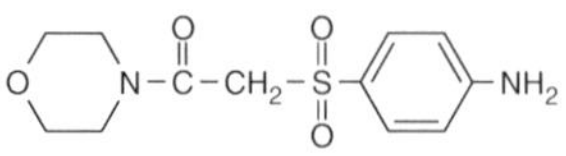

250 mg

S88,889-3
3,7-DIAMINODIBENZOTHIOPHENE-5,5-DIOXIDE

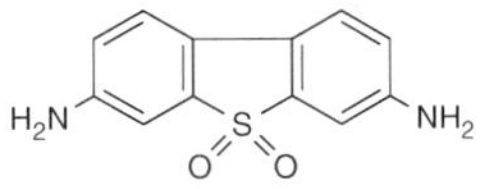

250 mg

S79,012-5
D-METHIONINE SULFONE

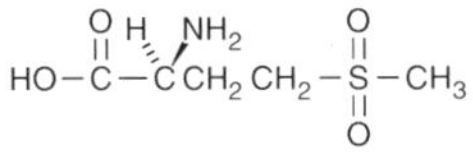

250 mg

S40,018-1
2-(4-AMINOPHENYLSULFONYL)-ACETIC ACID

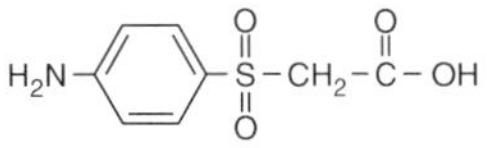

250 mg

S83,997-3
2-AMINOPHENOXATHIIN-10,10-DIOXIDE

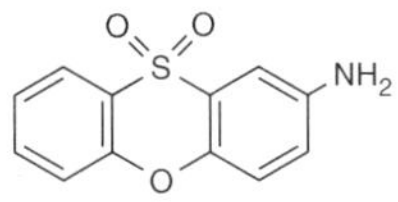

250 mg

S94,151-4
4-AMINOPHENYL METHYL SULFONE

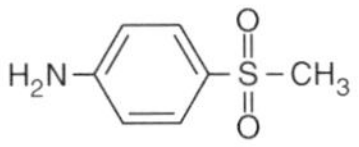

250 mg

S85,906-0
N-CYANO-2,6-DINITRO-4-(METHYLSULFONYL)ANILINE

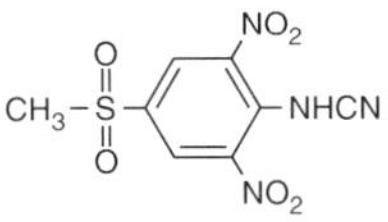

250 mg

S85,862-5
2,6-DINITRO-N-METHYL-4-(METHYLSULFONYL)ANILINE

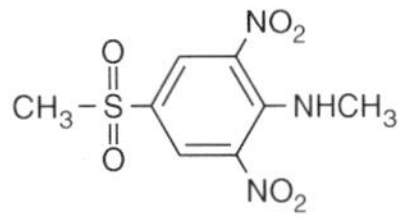

250 mg

S85,863-3
2,6-DINITRO-N-ETHYL-4-(METHYLSULFONYL)ANILINE

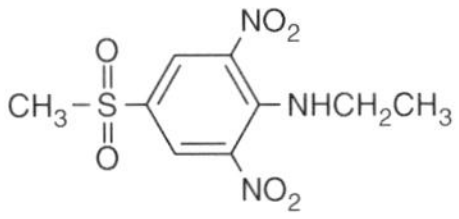

250 mg

S85,864-1
2,6-DINITRO-4-(METHYLSULFONYL)-N-PROPYLANILINE

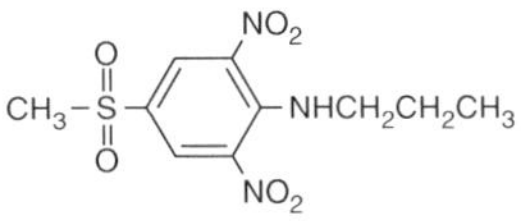

250 mg

S85,866-8
N-BUTYL-2,6-DINITRO-4-(METHYLSULFONYL)ANILINE

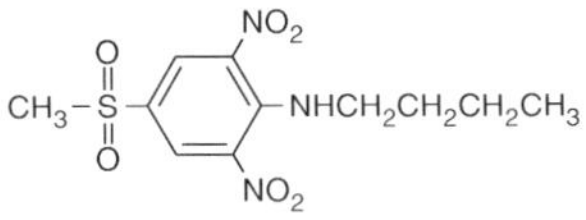

250 mg

S93,877-7
TRANS-3,4-BIS(BENZYLAMINO) TETRAHYDROTHIOPHENE-1,1-DIOXIDE

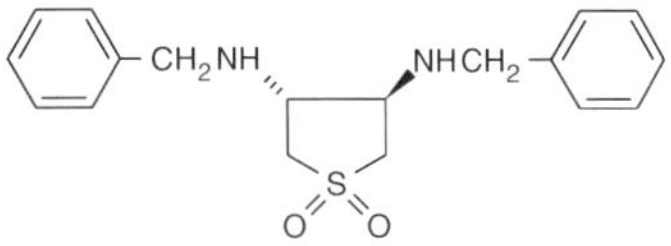

250 mg

S85,907-9
N-BUTYL-N-CYANO-2,6-DINITRO-4-(METHYLSULFONYL)ANILINE

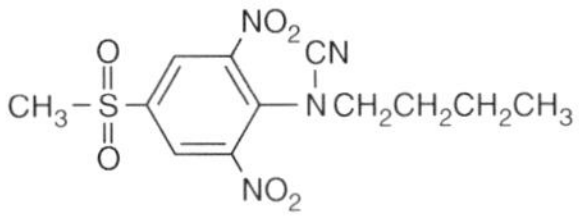

250 mg

S59,237-4
N-HEXADECYL-N-METHYL-4-METHYLSULFONYL-2-NITROANILINE

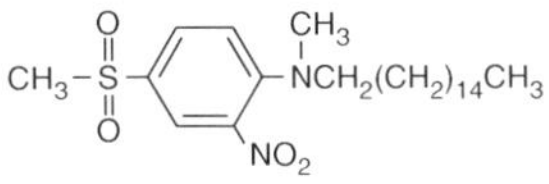

5 gm

S53,106-5
1-[4-(METHYLSULFONYL)BENZYL]-HEXAMETHYLENETETRAMINIUM BROMIDE

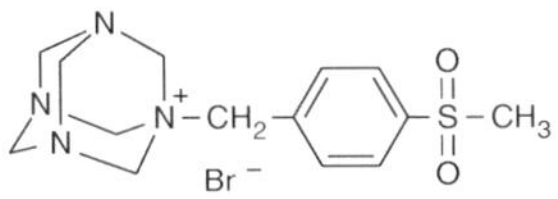

250 mg

S58,242-5
4-(METHYLSULFONYL)-2-NITROPHENYLHYDRAZINE

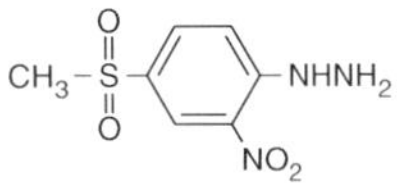

5 gm

S62,222-2
2-CARBETHOXYACETAMIDE 4-METHYLSULFONYL-2-NITROPHENYLHYDRAZONE

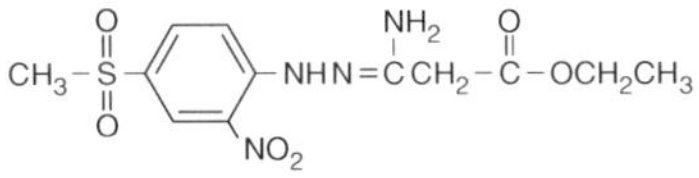

250 mg

S85,867-6
METHYL N-[2,6-DINITRO-4-(METHYLSULFONYL)PHENYL]CARBAMATE

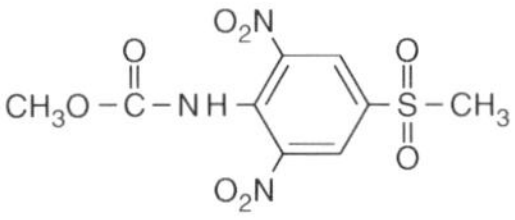

250 mg

S85,869-2
PROPYL N-[2,6-DINITRO-4-(METHYLSULFONYL)PHENYL]CARBAMATE

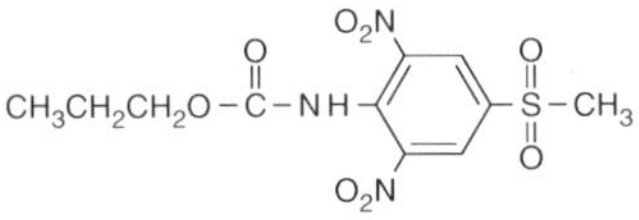

250 mg

S85,870-6
ISOPROPYL N-[2,6-DINITRO-4-(METHYLSULFONYL)PHENYL]CARBAMATE

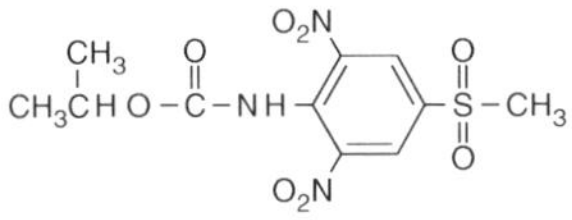

250 mg

S85,897-8
ETHYL N-[2,6-DINITRO-4-(METHYLSULFONYL)PHENYL]-N-METHYLCARBAMATE

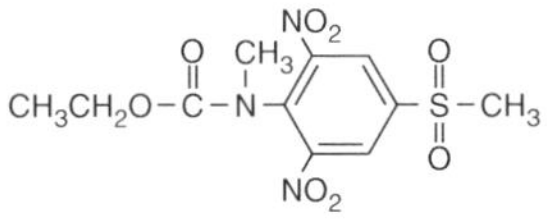

250 mg

S85,894-3
ISOPROPYL N-[2,6-DINITRO-4-(METHYLSULFONYL)PHENYL]-N-METHYLCARBAMATE

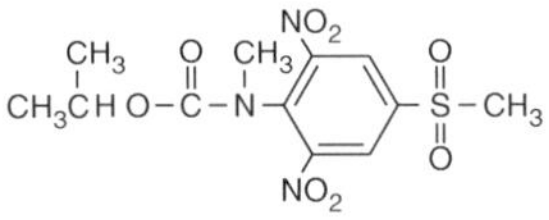

250 mg

S85,898-6
ETHYL N-[2,6-DINITRO-4-(METHYLSULFONYL)PHENYL]-N-ETHYLCARBAMATE

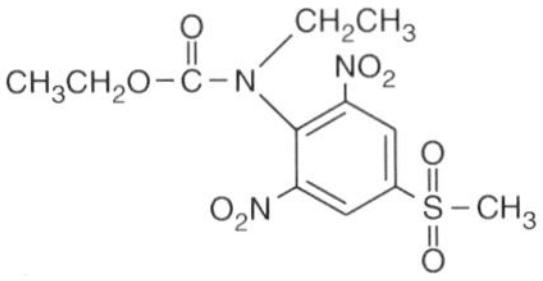

250 mg

S85,900-1
METHYL N-[2,6-DINITRO-4-(METHYLSULFONYL)PHENYL]-N-PROPYLCARBAMATE

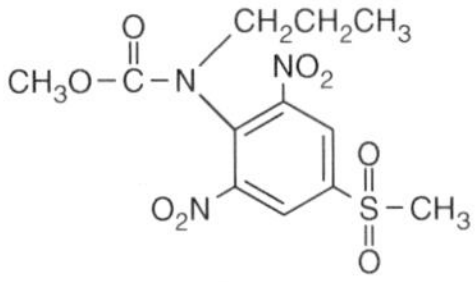

250 mg

S85,892-7
PROPYL N-[2,6-DINITRO-4-(METHYLSULFONYL)PHENYL]-N-ETHYLCARBAMATE

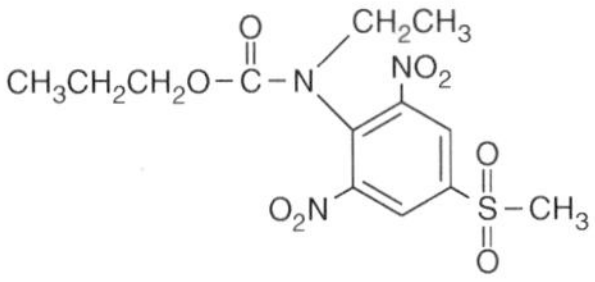

250 mg

S85,893-5
PROPYL N-[2,6-DINITRO-4-(METHYLSULFONYL)PHENYL]-N-PROPYLCARBAMATE

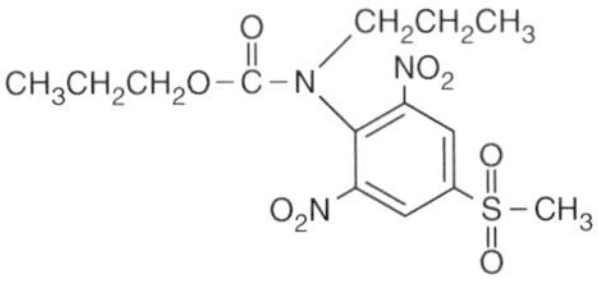

250 mg

S85,895-1
ISOPROPYL N-[2,6-DINITRO-4-(METHYLSULFONYL)PHENYL]-N-PROPYLCARBAMATE

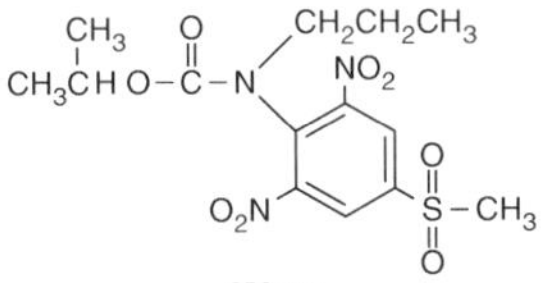

250 mg

S85,899-4
ETHYL N-BUTYL-N-[2,6-DINITRO-4-(METHYLSULFONYL)PHENYL]CARBAMATE

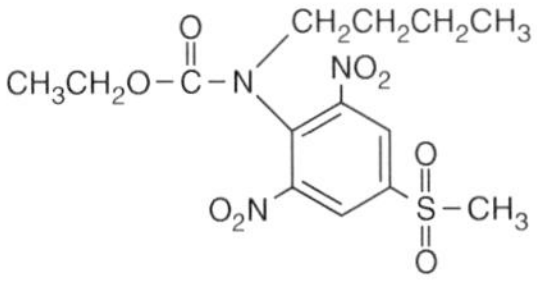

250 mg

S85,871-4
2',6'-DINITRO-4'-(METHYLSULFONYL)-ACETANILIDE

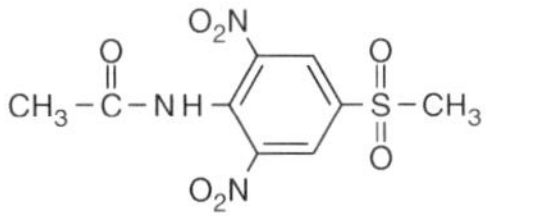

250 mg

S85,872-2
2',6'-DINITRO-4'-(METHYLSULFONYL)-BUTYRANILIDE

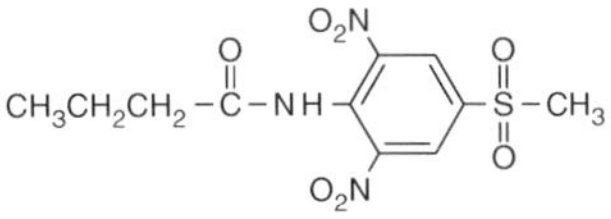

250 mg

S85,889-7
2',6'-DINITRO-N-METHYL-4'-(METHYLSULFONYL)ACETANILIDE

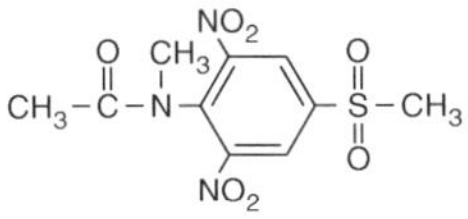

250 mg

S85,885-4
2',6'-DINITRO-N-METHYL-4'-(METHYLSULFONYL)PROPIONANILIDE

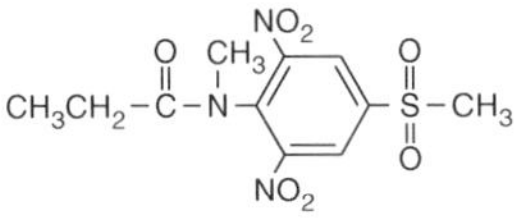

250 mg

S85,880-3
2',6'-DINITRO-N-METHYL-4'-(METHYLSULFONYL)BUTYRANILIDE

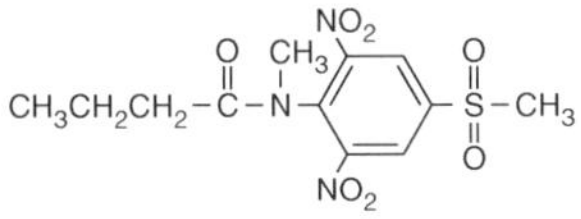

250 mg

S85,886-2
2',6'-DINITRO-N-ETHYL-4'-(METHYLSULFONYL)PROPIONANILIDE
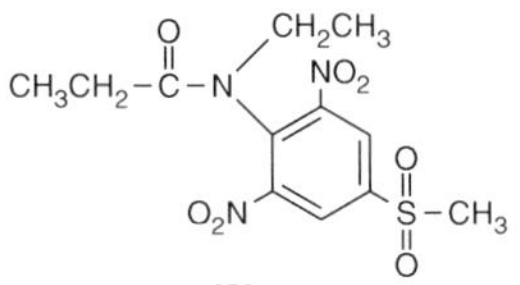
250 mg

S85,890-0
2',6'-DINITRO-4'-(METHYLSULFONYL)-N-PROPYLACETANILIDE
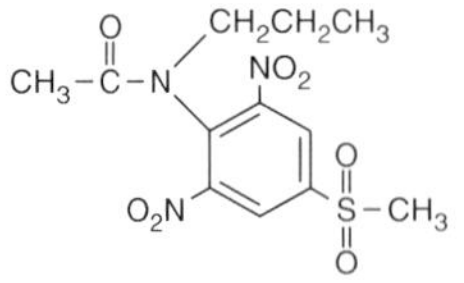
250 mg

S85,881-1
2',6'-DINITRO-N-ETHYL-4'-(METHYLSULFONYL)BUTYRANILIDE
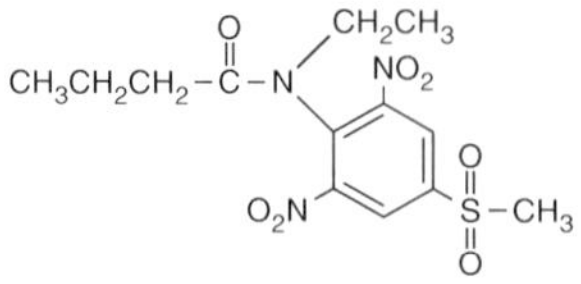
250 mg

S85,887-0
2',6'-DINITRO-4'-(METHYLSULFONYL)-N-PROPYLPROPIONANILIDE
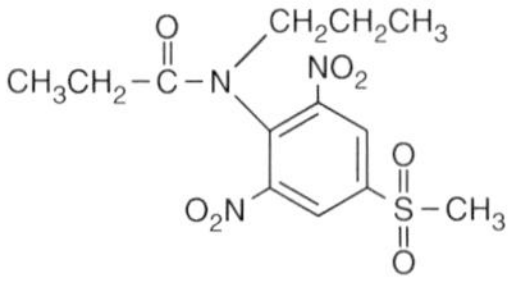
250 mg

S85,891-9
N-BUTYL-2',6'-DINITRO-4'-(METHYLSULFONYL)ACETANILIDE
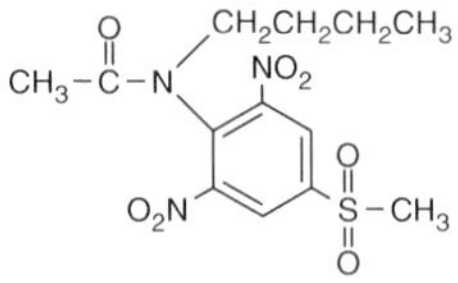
250 mg

S85,883-8
2',6'-DINITRO-4'-(METHYLSULFONYL)-N-PROPYLBUTYRANILIDE
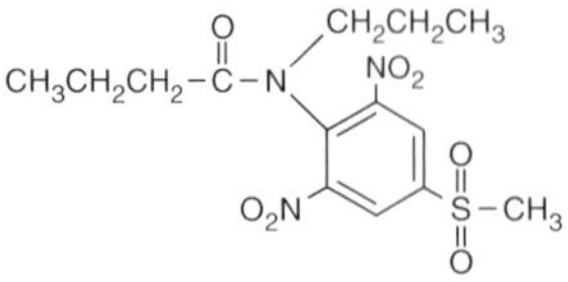
250 mg

S85,888-9
N-BUTYL-2',6'-DINITRO-4'-(METHYLSULFONYL)PROPIONANILIDE
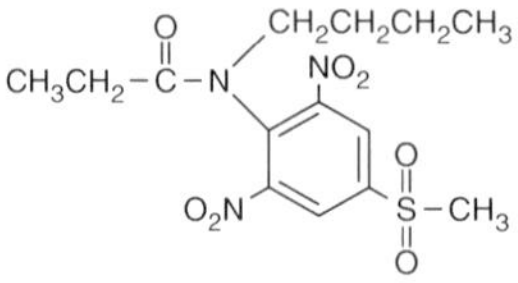
250 mg

S85,884-6
N-BUTYL-2',6'-DINITRO-4'-(METHYLSULFONYL)BUTYRANILIDE
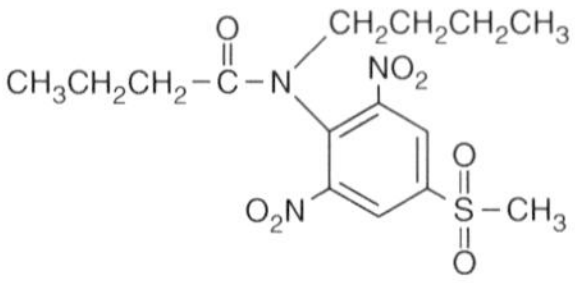
250 mg

S77,803-6
N,N-DIMETHYL-4-(PHENYLSULFONYL)-3-PIPERIDINOANILINE
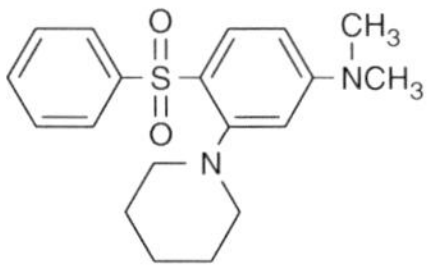
250 mg

S82,893-9
2,4-DINITROPHENYL M-TOLYL SULFONE
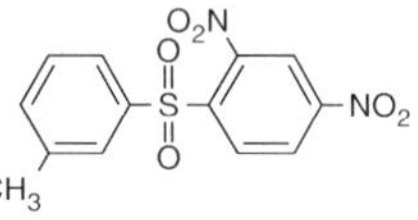
250 mg

S91,785-0
4-(1,1-DIMETHYL-2-NITROETHYL)PHENYL PHENYL SULFONE
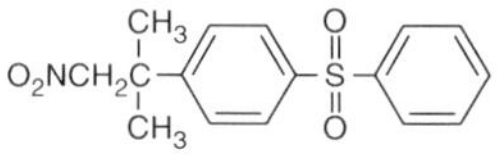
250 mg

S55,366-2
4-CHLORO-N-[4-(HEXADECYLSULFONYL)-PHENYL]-BENZYLAMINE
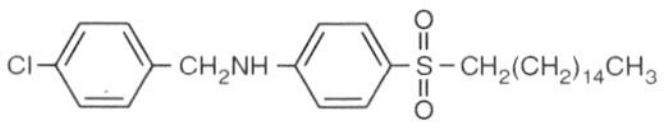
250 mg

S93,706-1
N-(4-SULFAMOYLPHENYL)-BETA-ALANINE
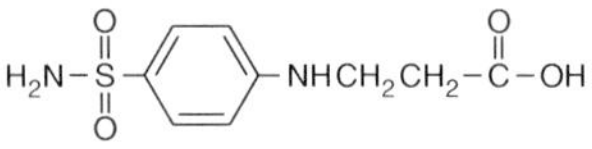
250 mg

S93,861-0
3-ANILINOTETRAHYDROTHIOPHENE-1,1-DIOXIDE
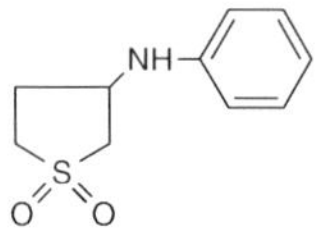
250 mg

S93,859-9
2,3-DIHYDRO-3-(4-METHYLANILINO)THIOPHENE-1,1-DIOXIDE
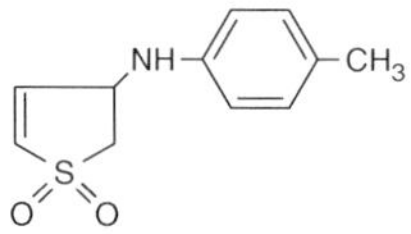
250 mg

S75,453-6
N-METHYL-2-(PHENYLSULFONYLMETHYL)BENZYLAMINE HYDROBROMIDE
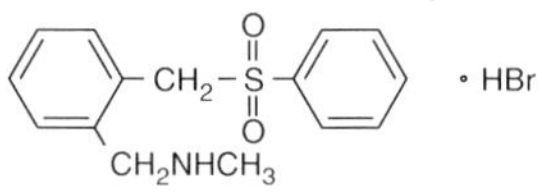
250 mg

S86,331-9
2,2-DIMETHYL-3-(DIMETHYLAMINO)-4-(4-NITROPHENYL)THIETANE-1,1-DIOXIDE
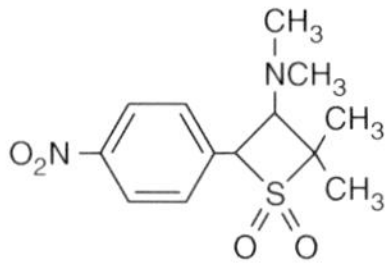
250 mg

S73,444-6
DIHYDRO-1'-METHYL-4-PHENYLSULFONYLSPIRO(3H-2-BENZOPYRAN-3,4'-PIPERIDIN)-1-ONE
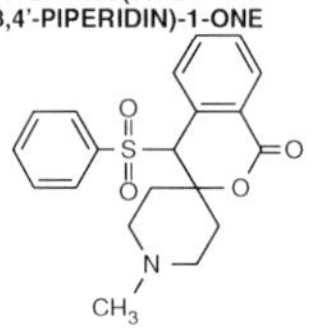
250 mg

S89,345-5
4-MORPHOLINOPHENYL SULFONE
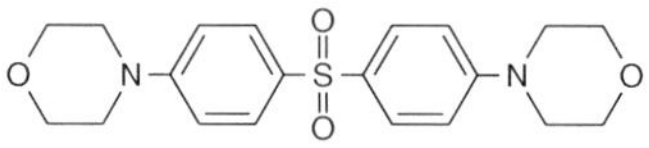
250 mg

S90,786-3
4-PHENYLTHIOMORPHOLINE-1,1-DIOXIDE
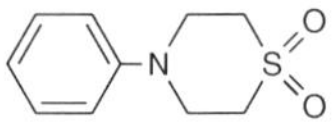
250 mg

S82,415-1
10-PHENYLPHENOTHIAZINE-5,5-DIOXIDE
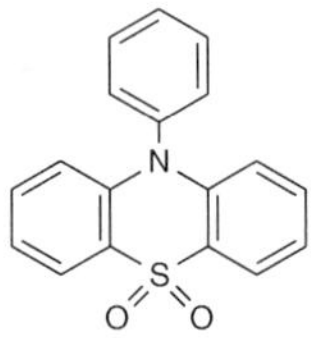
250 mg

S81,535-7
10-ETHYLPHENOTHIAZINE-5,5-DIOXIDE
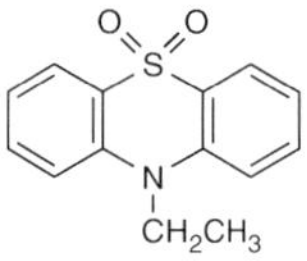
250 mg

S82,925-0
3,7-DICHLORO-10-ETHYLPHENOTHIAZINE-5,5-DIOXIDE
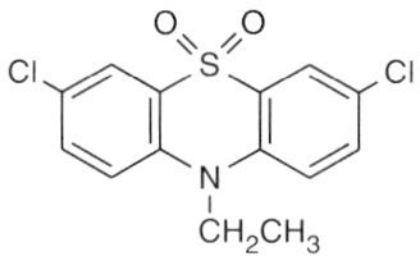
250 mg

S82,414-3
10-DECYLPHENOTHIAZINE-5,5-DIOXIDE
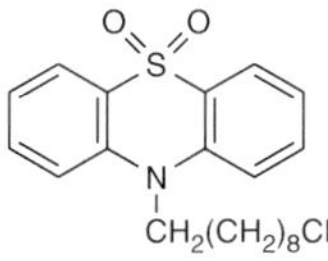
250 mg

S83,612-5
10-OCTADECYLPHENOTHIAZINE-5,5-DIOXIDE
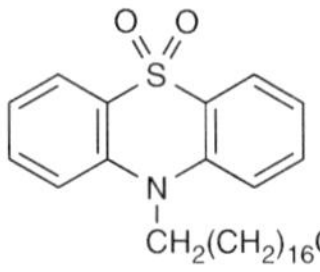
250 mg

S73,465-9
N,N-DIMETHYL-2-(PHENYLSULFONYLMETHYL)BENZYLAMINE
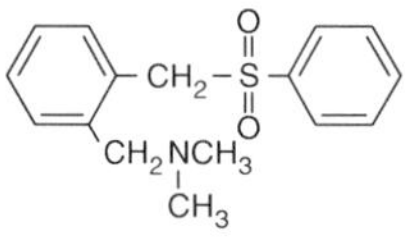
250 mg

S73,515-9
1-METHYL-4-(METHYLSULFONYLMETHYL)-PYRIDINIUM CHLORIDE
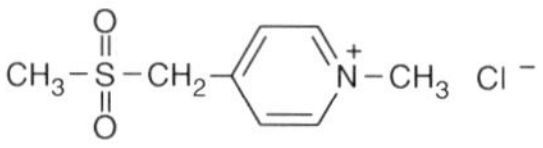
250 mg

S80,716-8
ALPHA-NITROBENZYL PHENYL SULFONE
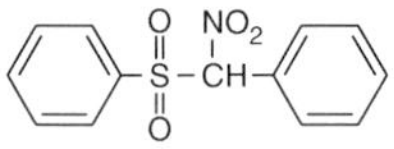
250 mg

S38,300-7
4-(HEXADECYLSULFONYL)-PHENYLHYDRAZINE
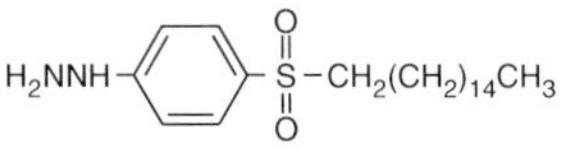
5 gm

S39,308-8
2-CHLORO-4-(METHYLSULFONYL)-PHENYLHYDRAZINE
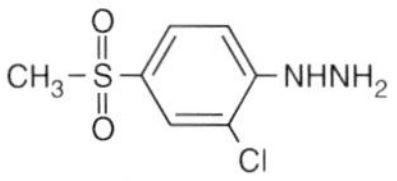
250 mg

S64,246-0
3-CHLORO-4-HYDRAZINOPHENYL METHYLSULFONYLMETHYL SULFONE
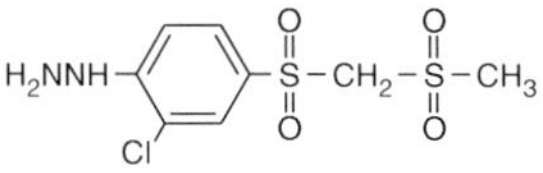
5 gm

S65,575-9
2-CHLORO-4-HEXADECYLSULFONYLPHENYLHYDRAZINE
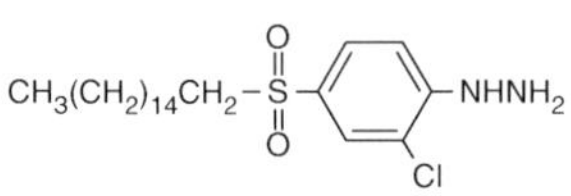
250 mg

S65,632-1
N-(4-HEXADECYLSULFONYLPHENYL)-PROPIONALDEHYDE HYDRAZONE

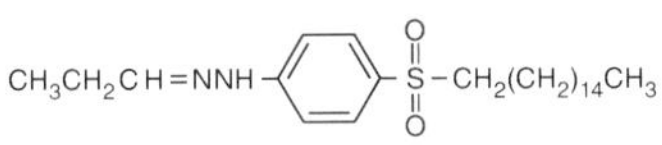

250 mg

S59,664-7
4-METHOXY-3-NITROPHENYL METHYL SULFONE

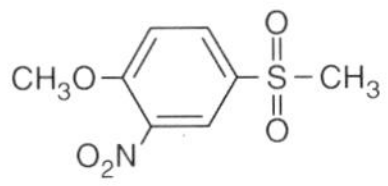

250 mg

S46,156-3
4-METHYLSULFONYL-3-NITROBENZOIC ACID

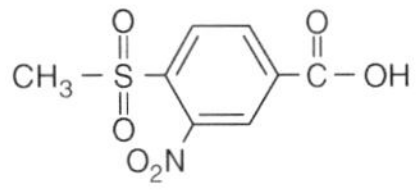

5 gm

S91,934-9
2-NITROPHENYL PHENYL SULFONE

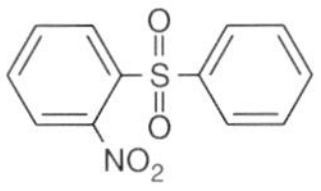

250 mg

S83,635-4
2-NITROPHENYL O-TOLYL SULFONE

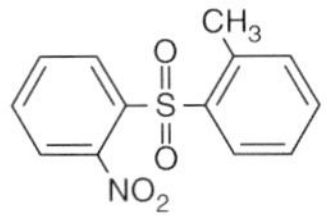

250 mg

S83,578-1
HEXADECYL 2-NITROPHENYL SULFONE

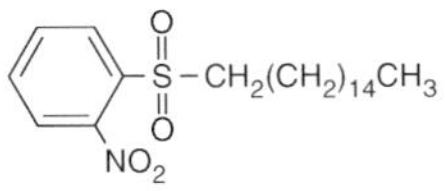

250 mg

S60,598-0
4-HEXADECYLSULFONYL-3-NITROBENZOIC ACID

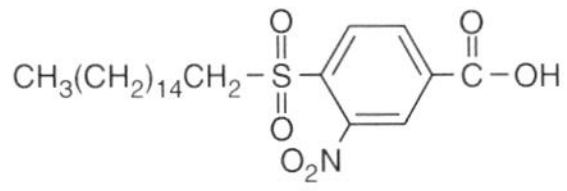

250 mg

S38,926-9
METHYL 4-HEXADECYLSULFONYL-3-NITROBENZOATE

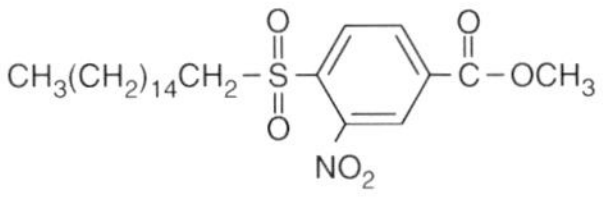

250 mg

S61,481-5
2-(4-ACETAMIDOPHENYLSULFONYL)-ACETAMIDE

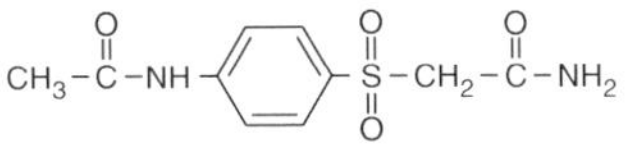

5 gm

S40,010-6
2-(PARA-TOLYLSULFONYL)-ACETAMIDE

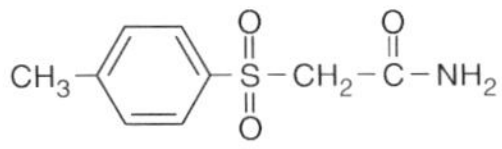

250 mg

S40,082-3
N-ALLYL-2-(PARA-TOLYLSULFONYL)-ACETAMIDE

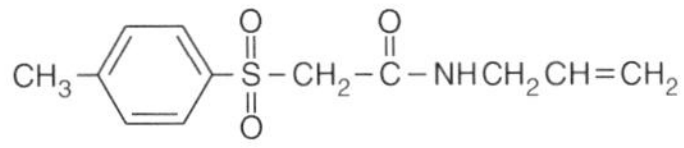

250 mg

S40,072-6
N-PROPYL-2-(PARA-TOLYLSULFONYL)-ACETAMIDE

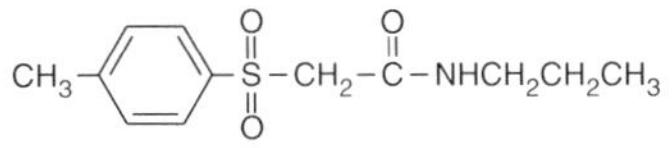

250 mg

S40,073-4
N-ISOPROPYL-2-(PARA-TOLYLSULFONYL)-ACETAMIDE

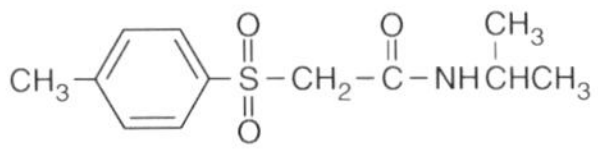

250 mg

S40,376-8
N-BUTYL-2-(PARA-TOLYLSULFONYL)-ACETAMIDE

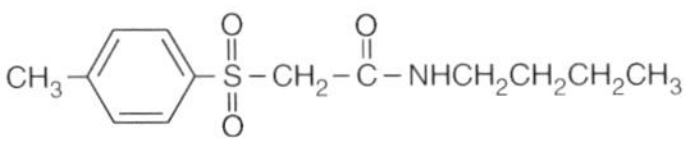

250 mg

S40,106-4
N-BUTYL-2-(PARA-TOLYLSULFONYL)-PROPIONAMIDE

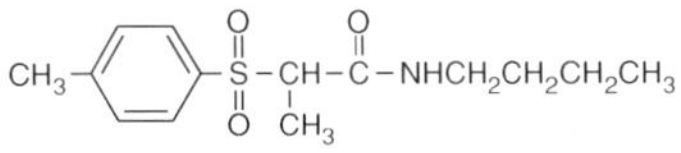

250 mg

S68,346-9
2,3-DIHYDRO-S,S-DIOXIDE-1,4-BENZOTHIAZEPIN-5(4H)-ONE

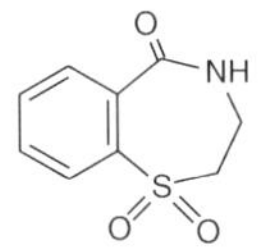

5 gm

S40,101-3
N,N-DIETHYL-2-(PARA-TOLYLSULFONYL)-ACETAMIDE

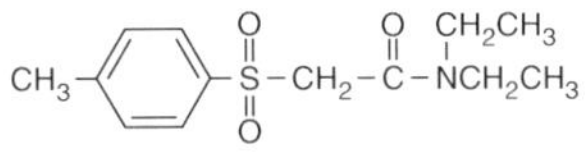

250 mg

S53,110-3
2-(PARA-TOLYLSULFONYL)-ACETIC ACID HYDRAZIDE

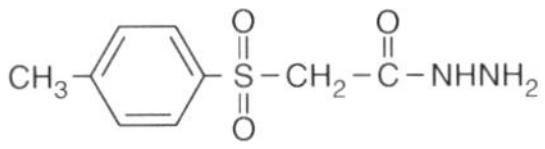

250 mg

S90,066-4
1,4-DIPHENYL-1-(3-SULFOLANYL)-3-THIOSEMICARBAZIDE

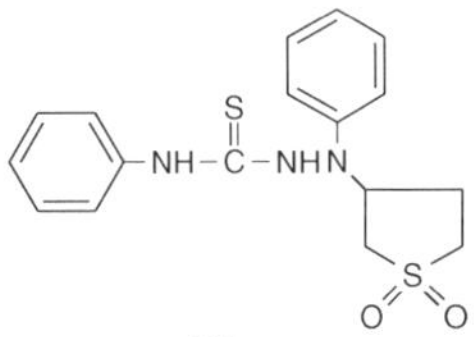

250 mg

S75,902-3
5-METHYL-7-PHENYL-8-(PHENYLSULFONYL)-1,2,3,7-TETRAHYDROIMIDAZO(1,2-A)PYRIDINE

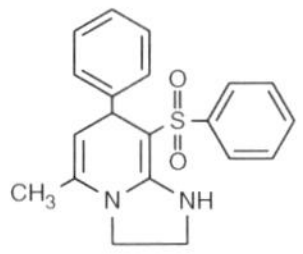

250 mg

S92,025-8
2-(4-NITROPHENYLSULFONYL)-1,4-BENZOQUINONE

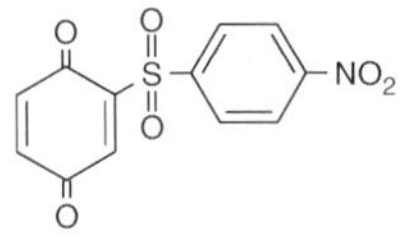

250 mg

S79,918-1
METHYL 3-NITROPHENYL SULFONE

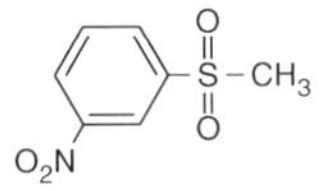

250 mg

S79,943-2
TERT-BUTYL 3-NITROPHENYL SULFONE

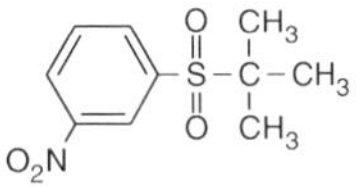

250 mg

S79,939-4
3-NITROPHENYL SULFONE

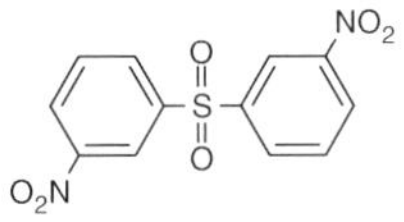

250 mg

S79,996-3
4-NITROPHENYL PHENYL SULFONE

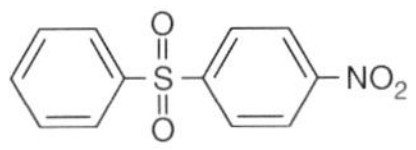

250 mg

S91,793-1
ALPHA,ALPHA-DIMETHYL-4-NITROBENZYL PHENYL SULFONE

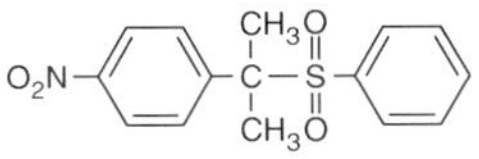

250 mg

S51,331-8
ETHYL 4-NITROPHENYL SULFONE

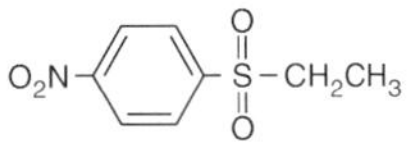

250 mg

S82,894-7
4-NITROPHENYL M-TOLYL SULFONE

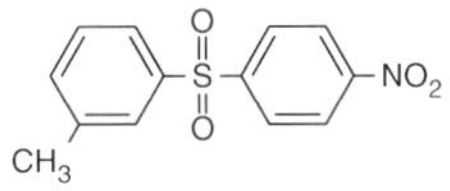

250 mg

S67,271-8
4-NITROBENZYL P-TOLYL SULFONE

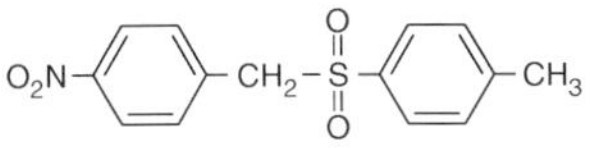

250 mg

S82,907-2
4-ISOPROPYLPHENYL 4-NITROPHENYL SULFONE

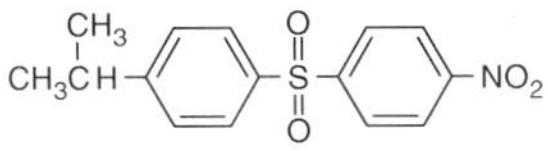

250 mg

S83,575-7
DODECYL 4-NITROPHENYL SULFONE

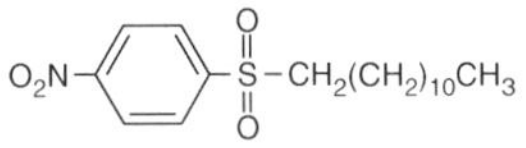

250 mg

S61,813-6
2-HEXADECYLSULFONYL-5-NITROBENZOIC ACID

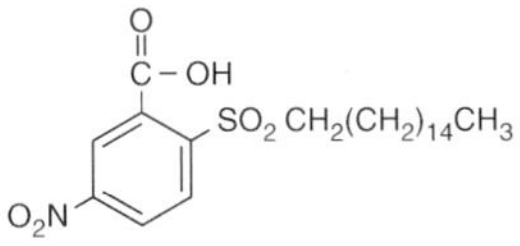

250 mg

S36,621-8
4-ACETAMIDO-4'-NITRODIPHENYL SULFONE

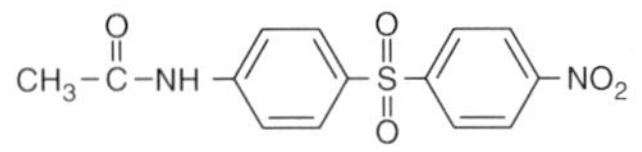

5 gm

S91,799-0
1-NITROCYCLOHEXYL PHENYL SULFONE

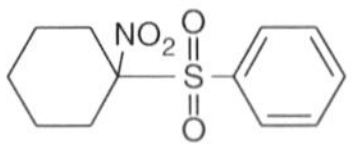

250 mg

S91,796-6
1-NITROCYCLOHEXYL P-TOLYL SULFONE

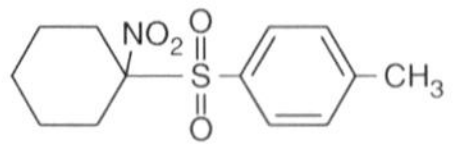

250 mg

S83,110-7
2-NITROPHENOXATHIIN-10,10-DIOXIDE

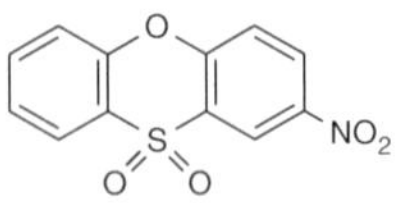

250 mg

S79,954-8
2-NITRO-2-PROPYL PHENYL SULFONE

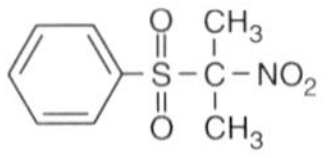

250 mg

S91,929-2
1-NITROPROPYL PHENYL SULFONE

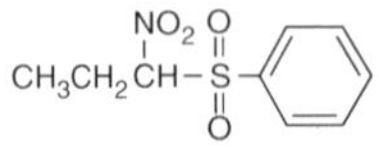

250 mg

S91,781-8
1-NITROCYCLODODECYL PHENYL SULFONE

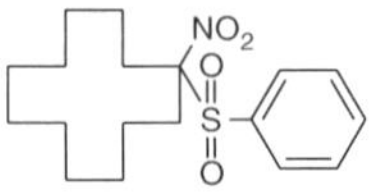

250 mg

S91,789-3
4-(1-METHYL-1-NITROETHYL)PHENYL PHENYL SULFONE

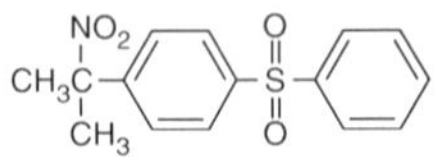

250 mg

S91,940-3
1-METHYL-1-NITROETHYL P-TOLYL SULFONE

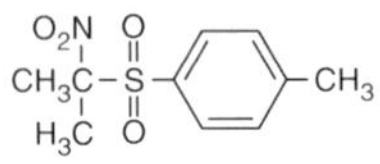

250 mg

S91,936-5
1-METHYL-1-NITROPROPYL P-TOLYL SULFONE

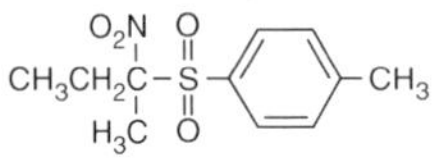

250 mg

S62,219-2
HEXADECYL 4-NITROPHENYL SULFONE

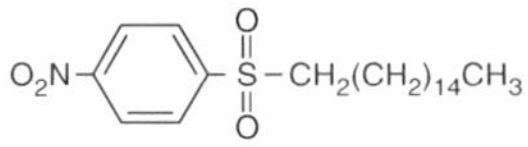

5 gm

S91,775-3
1-NITROHEXADECYL PHENYL SULFONE

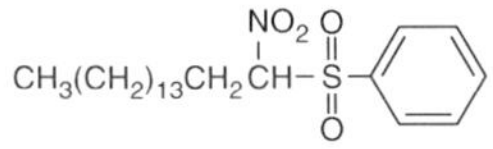

250 mg

S59,672-8
4-FLUORO-3-NITROPHENYL METHYL SULFONE

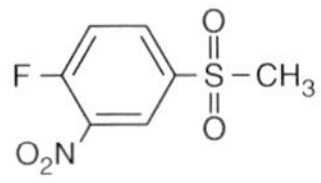

250 mg

S44,775-7
4-FLUORO-3-NITROPHENYL 3-NITROPHENYL SULFONE

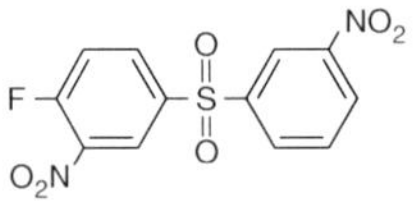

5 gm

S77,852-4
3-PHENYL-3-(P-TOSYL)PROPIONALDEHYDE

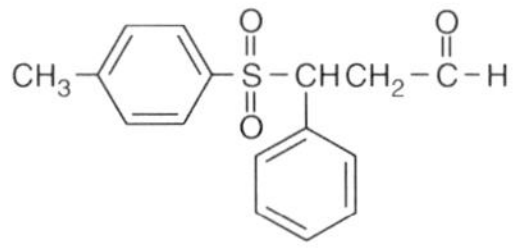

250 mg

S60,551-4
ETHYL 2-(METHYLSULFONYL)-BENZOYLACETATE

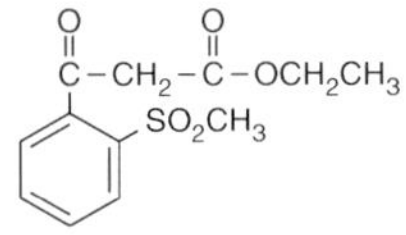

250 mg

S53,067-0
2-BENZOYL-2-(PHENYLSULFONYL)-ACETANILIDE

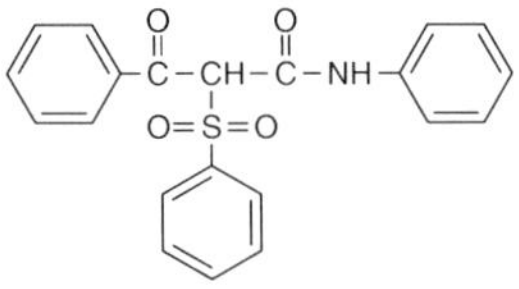

250 mg

S84,029-7
3-(4-ACETAMIDOPHENYLSULFONYL)-3-[(3,4-METHYLENEDIOXY)PHENYL] PROPIOPHENONE

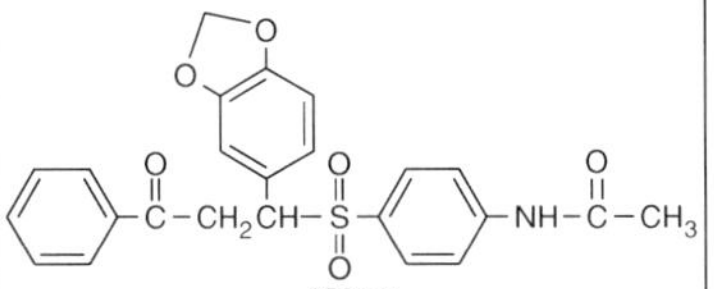

250 mg

S83,580-3
4'-(1,3-DIPHENYL-3-OXOPROPYLSULFONYL)ACETANILIDE

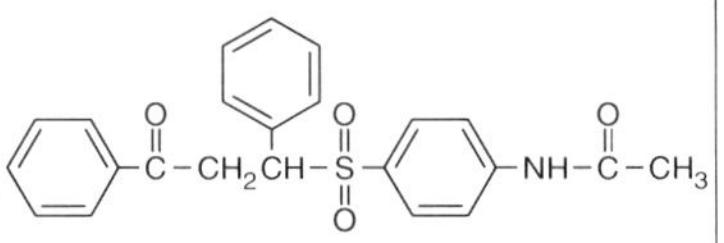

250 mg

S84,033-5
3-[(3,4-METHYLENEDIOXY)PHENYL]-3-(P-TOSYL)PROPIOPHENONE

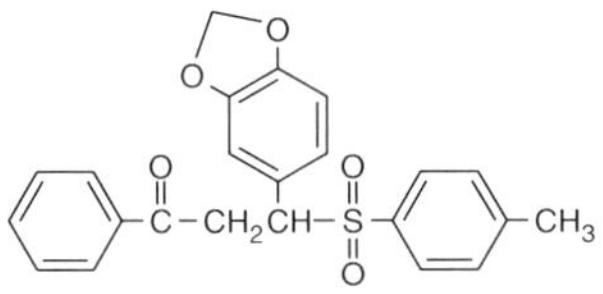

250 mg

S86,379-3
THIOXANTHEN-9-ONE-10,10-DIOXIDE

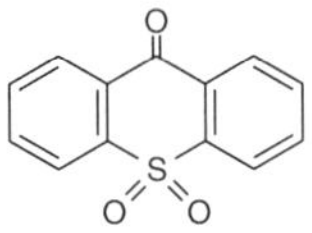

250 mg

S59,686-8
2-(METHYLSULFONYL)ACETOPHENONE

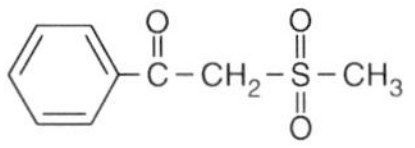

5 gm

S80,739-7
2-(BENZYLSULFONYL)ACETOPHENONE

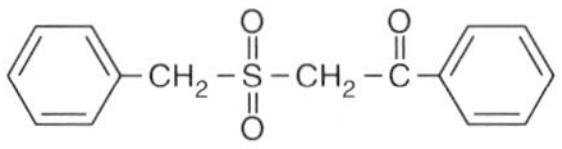

250 mg

S83,770-9
3-(METHYLSULFONYL)-3-PHENYLPROPIOPHENONE

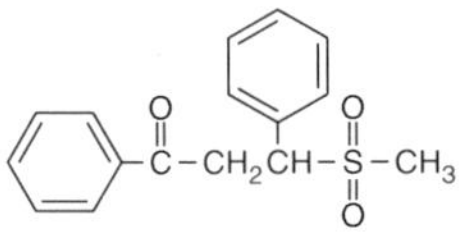

250 mg

S91,786-9
2-METHYL-2-(PHENYLSULFONYL)-PROPIOPHENONE

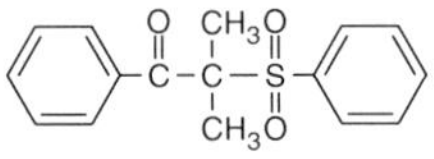

250 mg

S92,984-0
ALPHA-(PHENYLSULFONYL)CHALCONE

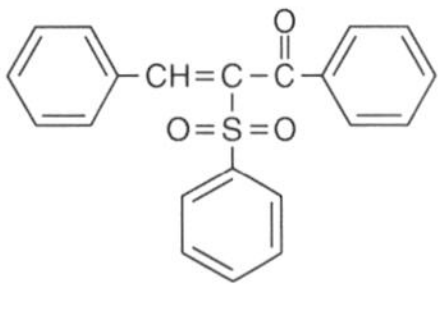

250 mg

S81,244-7
3-PHENYL-3-(PHENYLSULFONYL)-PROPIOPHENONE

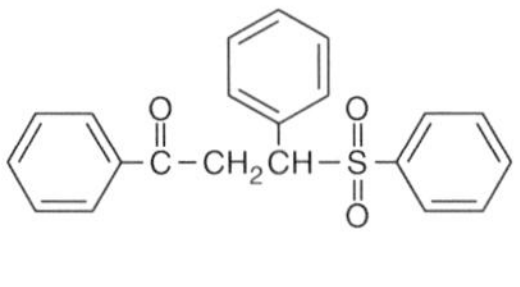

250 mg

S73,336-9
2-(PHENYLSULFONYL)BENZOPHENONE

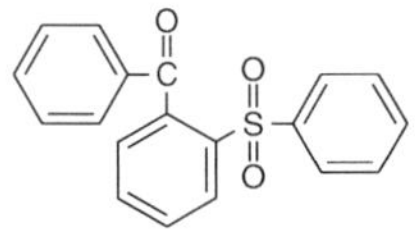

250 mg

S81,242-0
3-PHENYL-3-(P-TOSYL)PROPIOPHENONE

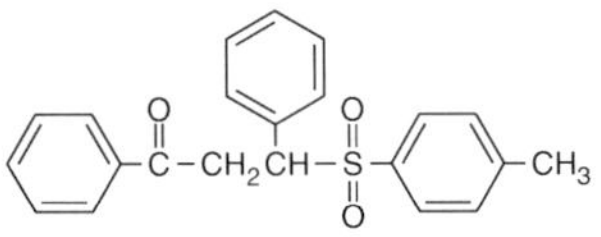

250 mg

S59,657-4
2-(P-TOLYLSULFONYL)ACETOPHENONE

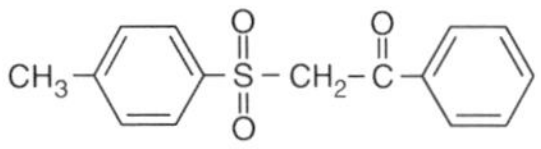

5 gm

S61,474-2
4'-METHYL-2-(P-TOLYLSULFONYL)-ACETOPHENONE

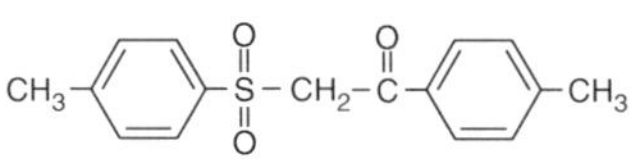

5 gm

S50,507-2
4'-(HEXADECYLSULFONYL)ACETOPHENONE

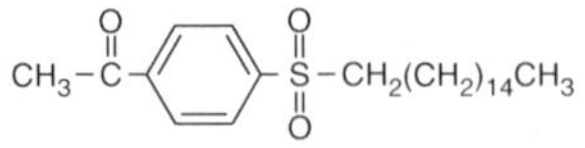

5 gm

S92,045-2
3,6-DIMETHYL-2-(4-TOLYLSULFONYL)-1,4-BENZOQUINONE
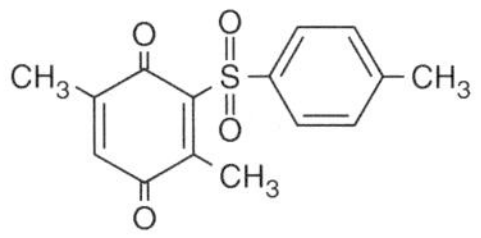
250 mg

S80,621-8
3-(P-TOSYL)ALLYL PYRUVATE
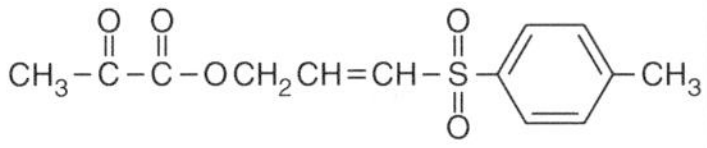
250 mg

S61,912-4
METHYL 4-(PHENYLSULFONYL)-ACETOACETATE
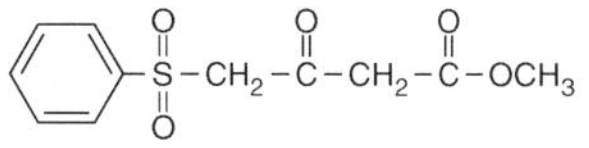
250 mg

S61,482-3
(4-ACETAMIDOPHENYLSULFONYL)ACETONE
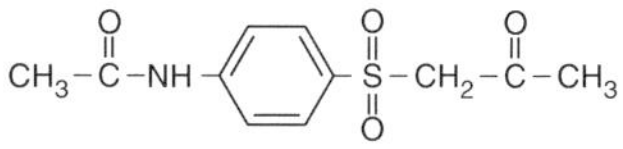
250 mg

S93,105-5
2,3-DIHYDROBENZO(B)THIOPHEN-3-ONE 1,1-DIOXIDE
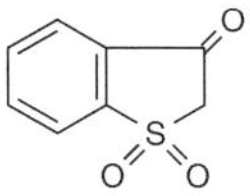
250 mg

S77,831-1
METHYL 3-(PHENYLSULFONYL)PROPYL KETONE
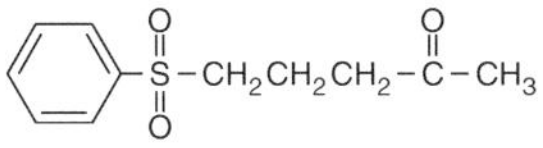
250 mg

S59,670-1
1-(PHENYLSULFONYL)PINACOLONE
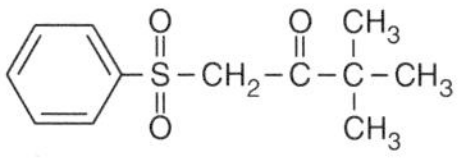
250 mg

S94,327-4
2-OXO-1-CYCLODODECYL PHENYL SULFONE
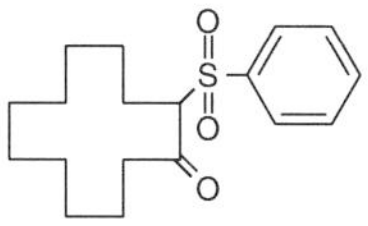
250 mg

S61,622-2
4-CHLOROPHENYLSULFONYLACETONE
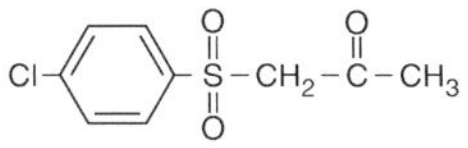
5 gm

S45,474-5
1-(P-TOLYLSULFONYL)-2-PROPANONE
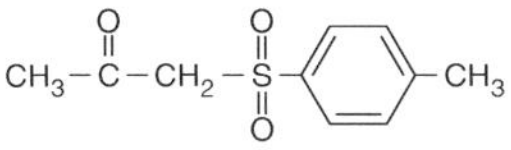
5 gm

S59,684-1
1-(P-TOLYLSULFONYL)PINACOLONE
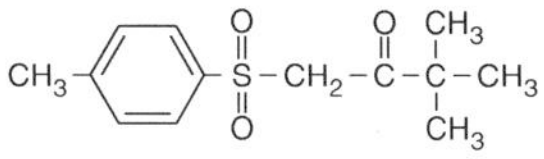
250 mg

S80,767-2
4-METHYL-4-(P-TOSYL)-2-PENTANONE
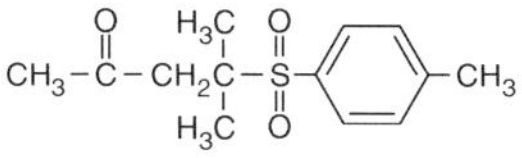
250 mg

S69,842-3
4',4''-SULFONYLDIMALEANILIC ACID
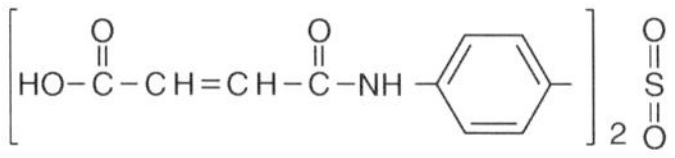
5 gm

S83,118-2
1,9-DICARBOXYPHENOXATHIIN-10,10-DIOXIDE
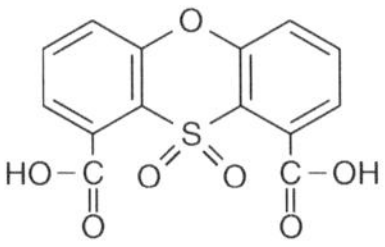
250 mg

S39,298-7
2-(HEXADECYLSULFONYL)BENZOIC ACID
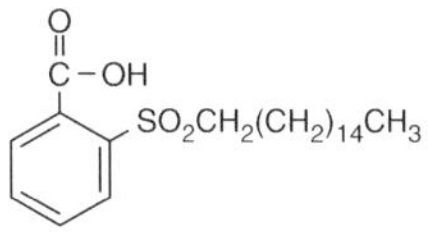
250 mg

S79,937-8
4-(TERT-BUTYLSULFONYL)BENZOIC ACID HYDRATE
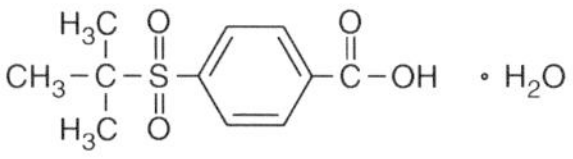
250 mg

S38,704-5
4-(HEXADECYLSULFONYL)BENZOIC ACID
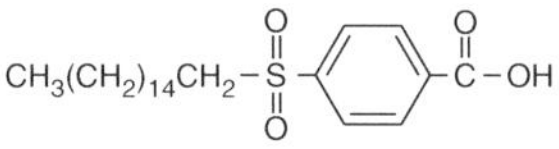
250 mg

S93,852-1
3-BENZOYLOXY-2,3-DIHYDROTHIOPHENE-1,1-DIOXIDE
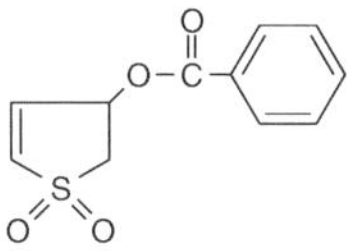
250 mg

S88,948-2
3-(BENZOYLOXY)TETRAHYDROTHIOPHENE-1,1-DIOXIDE
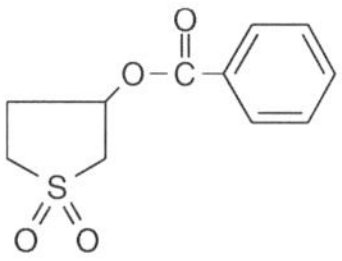
250 mg

S90,901-7
METHYL 4-(METHYLSULFONYL)BENZOATE
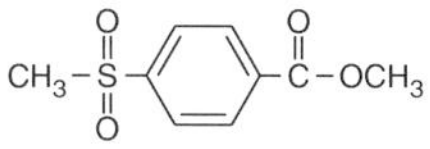
250 mg

S76,386-1
2-PHENYL-4-(PHENYLSULFONYL)-ISOQUINOLINE-1,3(2H,4H)-DIONE
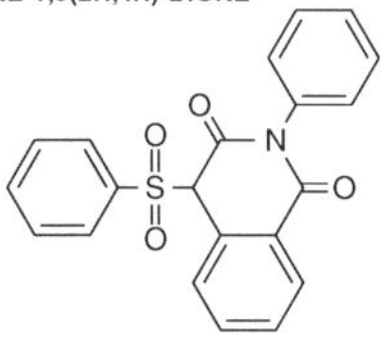
250 mg

S40,379-2
2-(PHENYLSULFONYL)ACETIC ACID
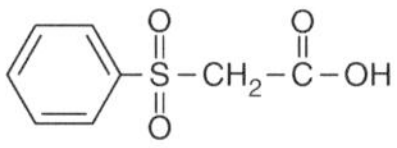
250 mg

S38,014-8
2-(PARA-TOLYLSULFONYL)ACETIC ACID
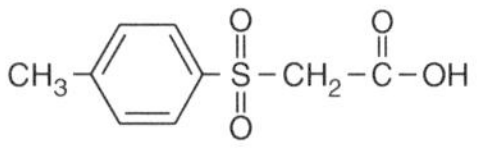
250 mg

S90,842-8
[4-(METHYLSULFONYL)PHENYL]-METHYLENE DIACETATE
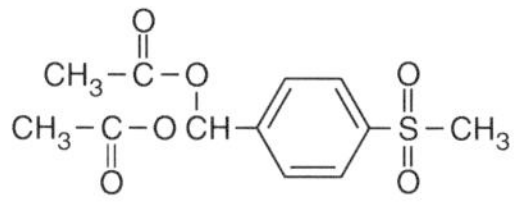
250 mg

S61,458-0
ETHYL 2-(4-ACETAMIDOPHENYLSULFONYL) ACETATE
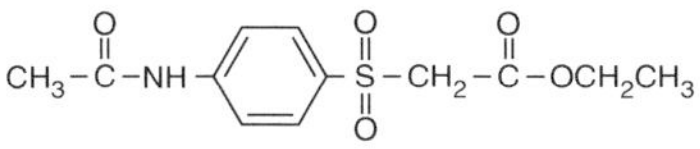
5 gm

S76,369-1
3-PHENYL-4-(PHENYLSULFONYLMETHYL)-2-OXAZOLIDINONE
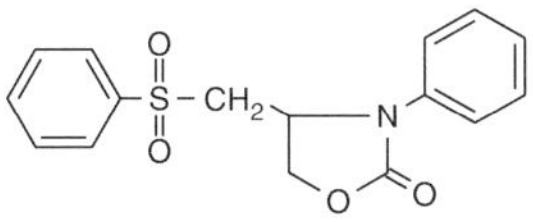
250 mg

S80,385-5
ETHYL N-(PHENYLSULFONYLMETHYL)-CARBAMATE
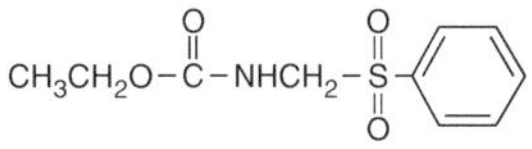
250 mg

S59,687-6
ETHYL 2-(PHENYLSULFONYL)ACETATE
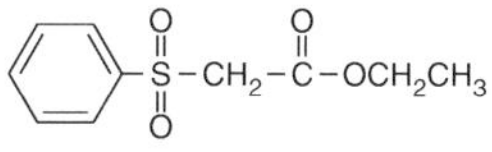
5 gm

S61,454-8
METHYL 3-(PHENYLSULFONYL)PROPIONATE
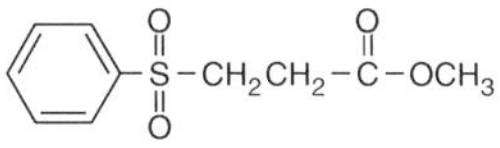
5 gm

S77,836-2
ETHYL 3-(BENZYLSULFONYL)PROPIONATE
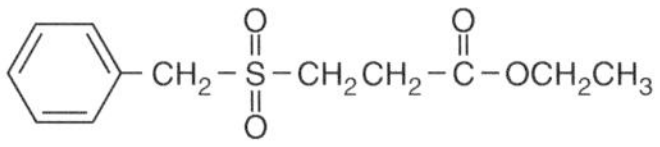
250 mg

S94,273-1
METHYL 11-(PHENYLSULFONYL)-UNDECANOATE
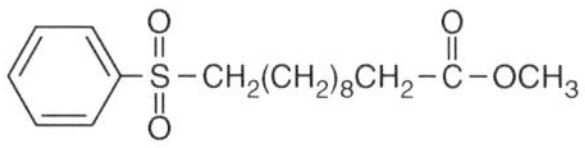
5 gm

S94,589-7
11-(PHENYLSULFONYL)UNDECYL ACETATE
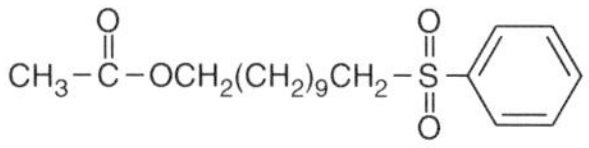
250 mg

S61,413-0
ETHYL 2-(4-CHLOROPHENYLSULFONYL)-ACETATE
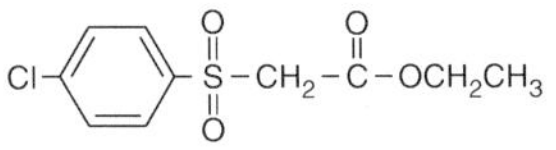
250 mg

S53,113-8
ETHYL 2-(PARA-TOLYLSULFONYL)-ACETATE
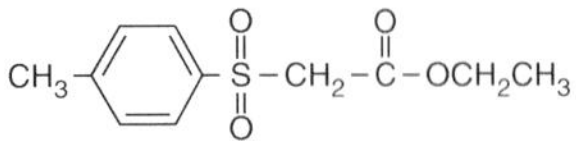
250 mg

S38,221-3
BUTYL 2-(PARA-TOLYLSULFONYL)-ACETATE
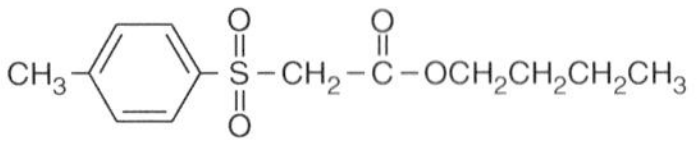
250 mg

S80,714-1
2-PHENYL-2-(PHENYLSULFONYL)ACETYL CHLORIDE
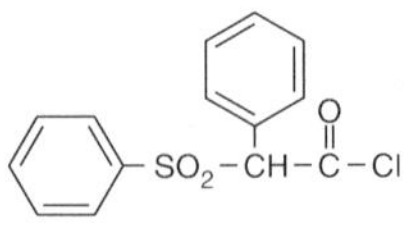
250 mg

S53,111-1
2-(PARA-TOLYLSULFONYL)-ACETYL CHLORIDE
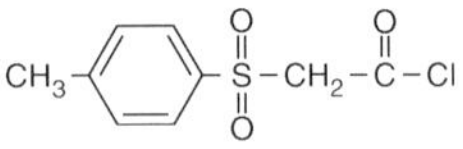
250 mg

S63,795-5
3-BENZYLSULFONYLPROPIONAMIDE
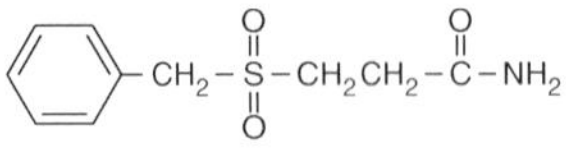
5 gm

S61,446-7
2-(4-CHLOROPHENYLSULFONYL)ACETAMIDE
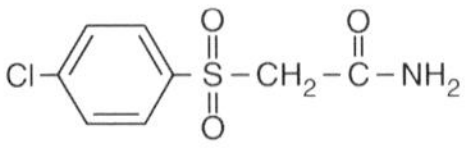
250 mg

S59,594-2
4-ACETAMIDOPHENYLSULFONYLACETONITRILE
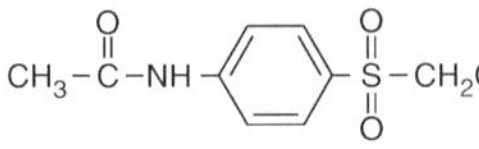
5 gm

S37,362-1
4'-(METHYLSULFONYL)ACETANILIDE
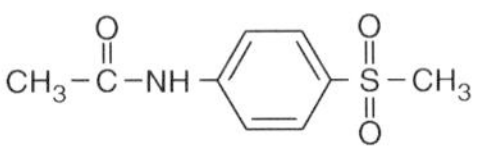
5 gm

S69,743-5
N-(P-TOLYLSULFONYLMETHYL)FORMAMIDE
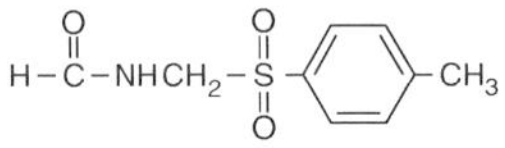
5 gm

S53,173-1
4',4''-SULFONYLDIFORMANILIDE
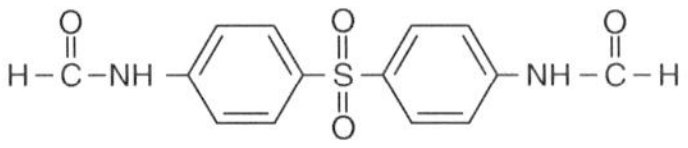
5 gm

S38,735-5
4'-(HEXADECYLSULFONYL)ACETANILIDE
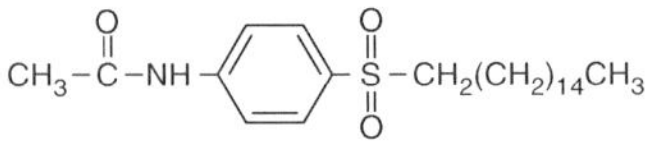
250 mg

S72,804-7
1,1'-(SULFONYLDI-4,1-PHENYLENE)BIS-(3-PHENYLUREA)
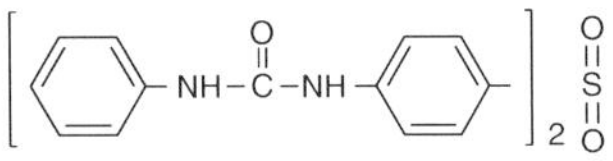
100 mg

S77,203-8
1,1'-(SULFONYLDI-1,4-PHENYLENE)BIS-[3-(2,4-XYLYL)UREA]
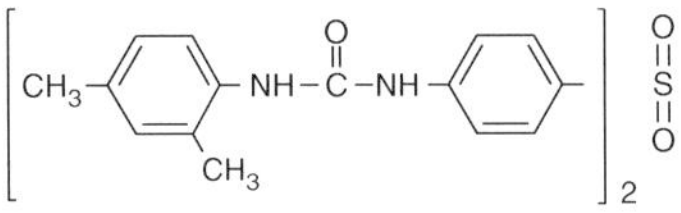
250 mg

S51,262-1
1-[4-(HEXADECYLSULFONYL)PHENYL]-2-THIOUREA
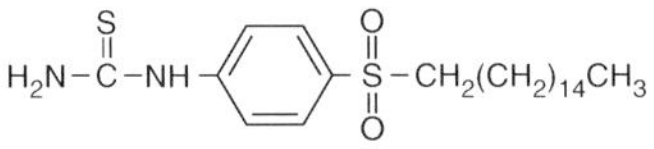
250 mg

S76,671-2
2-ETHYLIMINO-3-ISOPROPYL-5-(PHENYLSULFONYLMETHYL)THIAZOLIDINE HYDROCHLORIDE
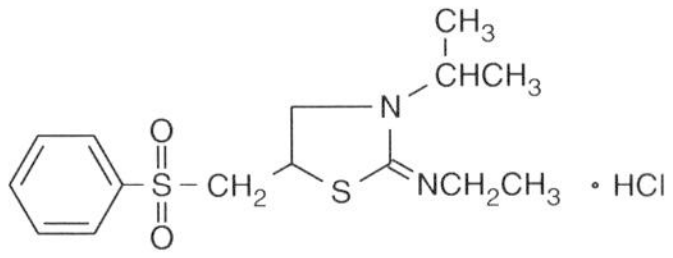
250 mg

S91,798-2
4-(PHENYLSULFONYL)BENZONITRILE
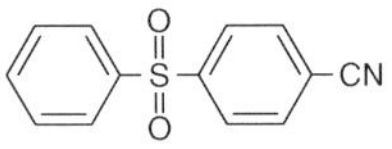
250 mg

S91,872-5
2,4-DICYANO-1-[1-METHYL-1-(PHENYLSULFONYL)ETHYL]BENZENE
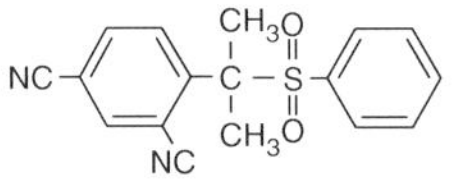
250 mg

S78,691-8
4,4'-SULFONYLBIS(BENZONITRILE)
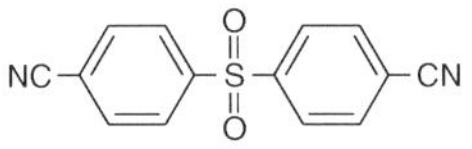
250 mg

S80,736-2
PHENYL 3-(PHENYLSULFONYL)PHENYL SULFOXIDE
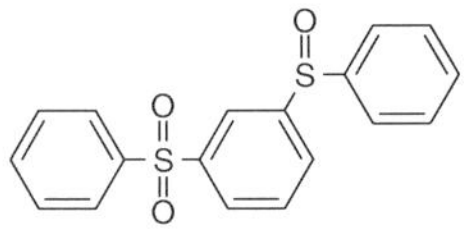
250 mg

S85,219-8
3-BROMO-5-(PHENYLSULFONYL)TRICYCLO[2.2.1.0(2,6)]HEPTANE
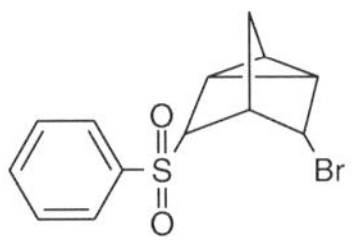
250 mg

S77,563-0
TRANS-PHENYL 2-PHENYLCYCLOPROPYL SULFONE
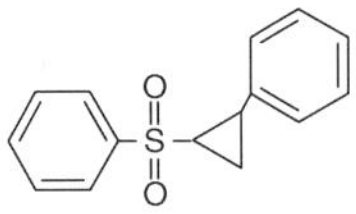
250 mg

S80,586-6
2-NAPHTHYL 2-PHENYLCYCLOPROPYL SULFONE
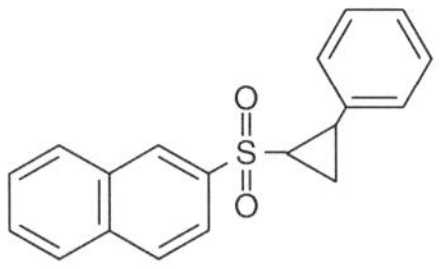
250 mg

S79,935-1
1-METHYLCYCLOPROPYL P-TOLYL SULFONE
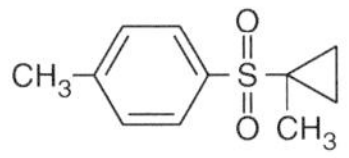
250 mg

S80,901-2
2-(1-AZIRIDINYL)VINYL P-TOLYL SULFONE
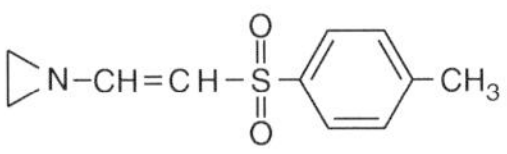
250 mg

S86,528-1
2-(P-TOSYL)-5-NORBORNENE
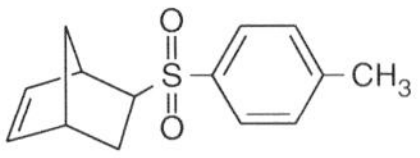
250 mg

S86,328-9
TRANS-2-CHLOROCYCLOPENTYL P-TOLYL SULFONE
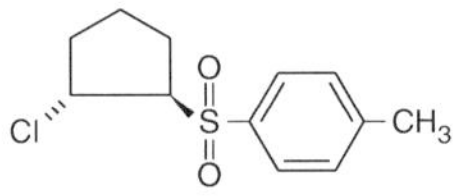
250 mg

S86,442-0
3-CHLORO-2-NORBORNYL P-TOLYL SULFONE
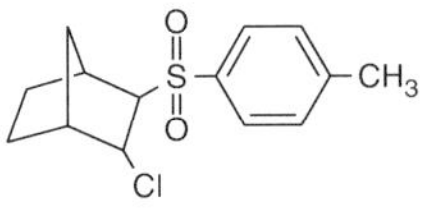
250 mg

S88,862-1
3,7-DIFLUORODIBENZOTHIOPHENE-5,5-DIOXIDE
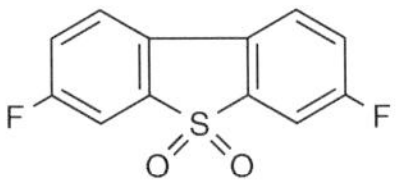
250 mg

S93,250-7
2,5-DIHYDRO-3-PHENYLTHIOPHENE-1,1-DIOXIDE
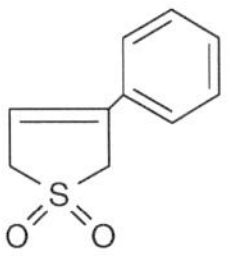
250 mg

S73,459-4
PHENYL (3-PYRROLIDINO-1-PROPENYL)-SULFONE HYDROCHLORIDE
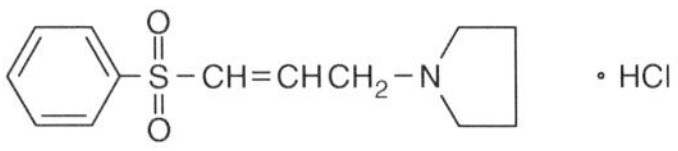
250 mg

S79,941-6
CYCLOHEXYL PHENYL SULFONE
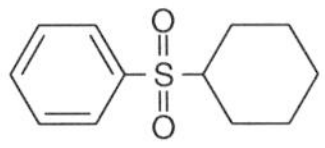
250 mg

S77,901-6
PHENYL 3,4,5-TRIMETHYLCYCLOHEXYLMETHYL SULFONE

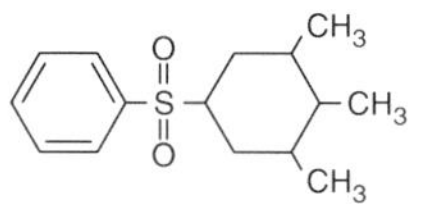

250 mg

S80,706-0
BENZYL 4-(CYCLOHEXYLMETHYL)PHENYL SULFONE

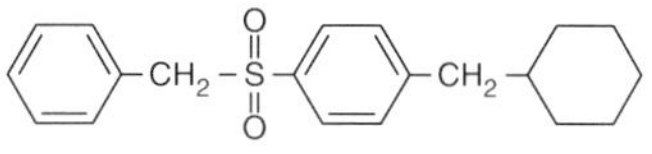

250 mg

S77,857-5
BENZYL 4-(1-METHYLCYCLOHEXYLMETHYL)PHENYL SULFONE

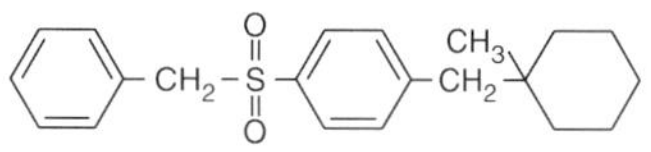

250 mg

S80,751-6
CYCLOHEXYL P-TOLYL SULFONE

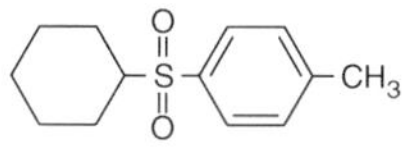

250 mg

S77,909-1
CYCLOHEXYL 4-METHYLBENZYL SULFONE

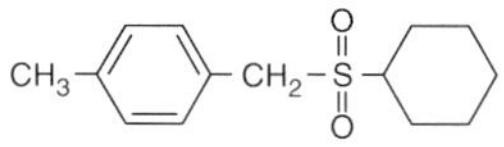

250 mg

S80,727-3
2-CYCLOHEXYLVINYL P-TOLYL SULFONE

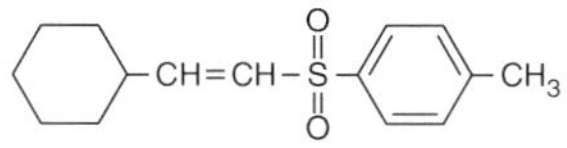

250 mg

S80,766-4
1-CYCLOHEXENYL P-TOLYL SULFONE

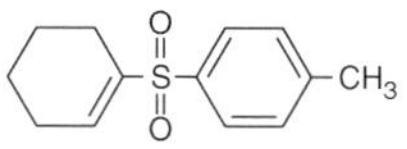

250 mg

S80,757-5
CIS-4,5-DI-P-TOSYL-1-CYCLOHEXENE

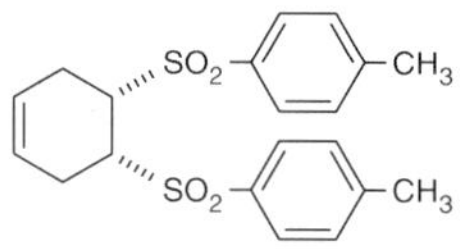

250 mg

S79,887-8
5-CHLORO-3,4-DIHYDRO-6-PHENYL-2H-THIOPYRAN-1,1-DIOXIDE

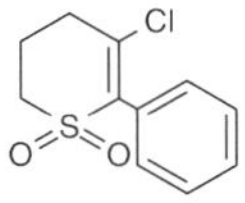

250 mg

S79,898-3
1,5,7-TRIMETHYLTHIOXANTHENE-10,10-DIOXIDE

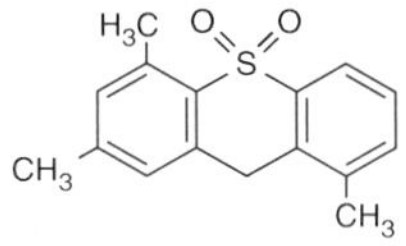

250 mg

S79,899-1
2,4,7-TRIMETHYLTHIOXANTHENE-10,10-DIOXIDE

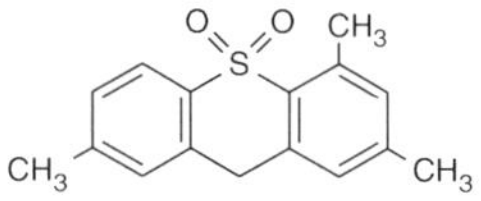

250 mg

S90,894-0
2H-1-BENZOTHIOPYRAN-1,1-DIOXIDE

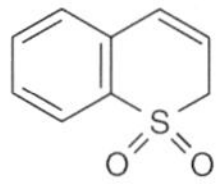

250 mg

S83,113-1
PHENOXATHIIN-10,10-DIOXIDE

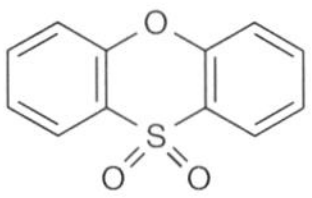

250 mg

S83,109-3
2-IODOPHENOXATHIIN-10,10-DIOXIDE

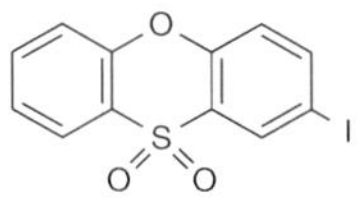

250 mg

S41,550-2
METHYL PHENYL SULFONE

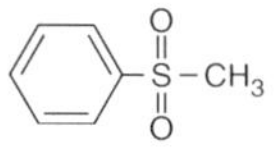

250 mg

S57,696-4
ETHYL PHENYL SULFONE

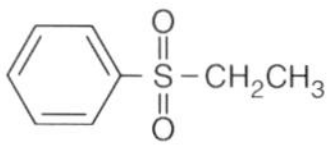

5 gm

S62,008-4
BENZYL METHYL SULFONE

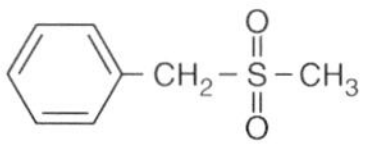

5 gm

S77,824-9
METHYL STYRYL SULFONE

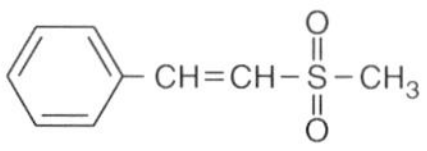

250 mg

S57,691-3
ISOPROPYL PHENYL SULFONE

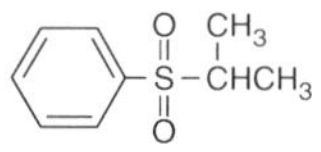

5 gm

S77,896-6
2-IODO-1-PROPENYL PHENYL SULFONE

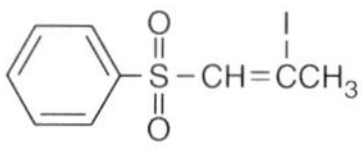

250 mg

S79,836-3
TERT-BUTYL PHENYL SULFONE

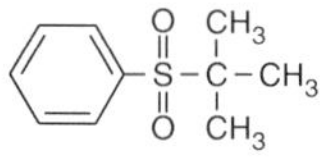

250 mg

S79,926-2
BUTYL PHENYL SULFONE

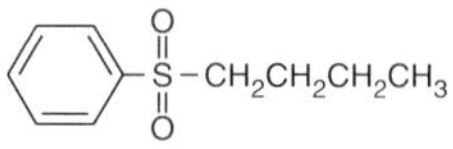

250 mg

S77,900-8
ETHYL ALPHA-IODOSTYRYL SULFONE

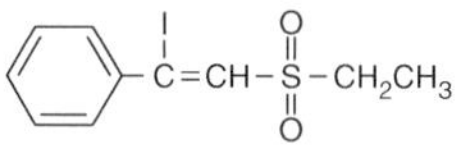

250 mg

S86,324-6
1-PENTEN-1-YL PHENYL SULFONE

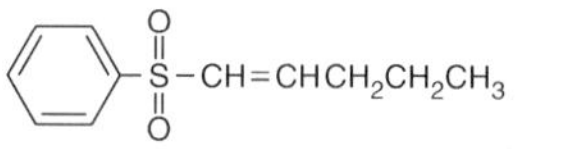

250 mg

S43,123-0
BENZYL TERT.-BUTYL SULFONE

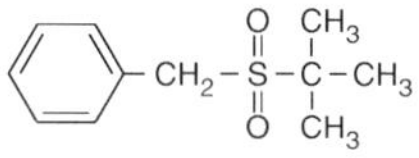

5 gm

S79,945-9
BENZYL BUTYL SULFONE

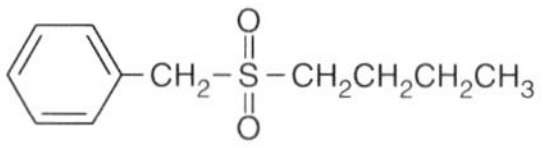

250 mg

S77,838-9
2-CHLORO-1-HEXEN-1-YL PHENYL SULFONE

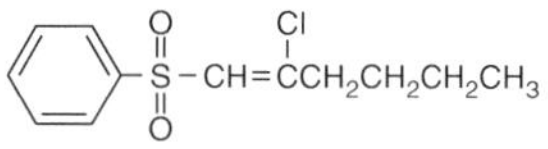

250 mg

S77,842-7
2-IODO-1-HEXEN-1-YL PHENYL SULFONE

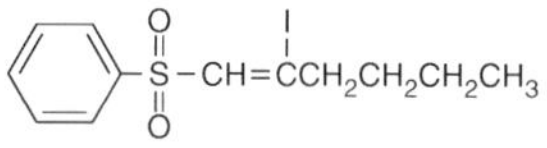

250 mg

S61,443-2
BENZYL PHENYL SULFONE

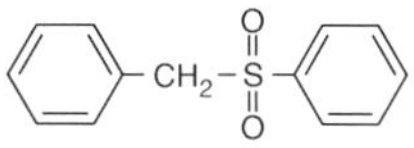

5 gm

S94,275-8
PHENYL TRANS-STYRYL SULFONE

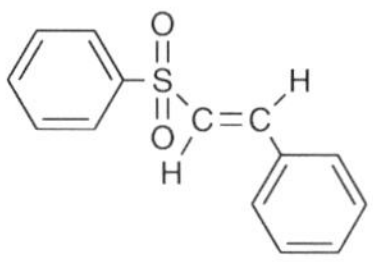

250 mg

S88,360-3
PHENETHYL PHENYL SULFONE

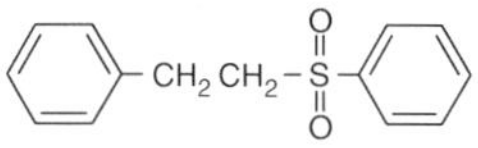

250 mg

S63,369-0
1,2-BIS(PHENYLSULFONYL)ETHANE

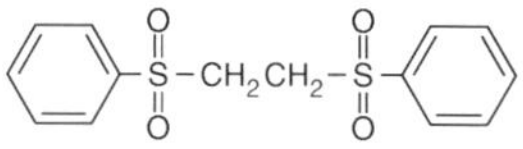

5 gm

S77,561-4
1,2-BIS(PHENYLSULFONYL)-1,2-DICHLOROETHYLENE

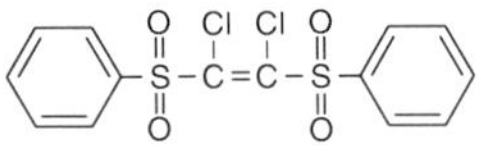

250 mg

S77,576-2
PHENYL 3-PHENYLPROPYL SULFONE

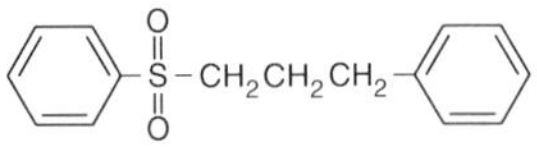

250 mg

S80,728-1
BENZYL OCTYL SULFONE

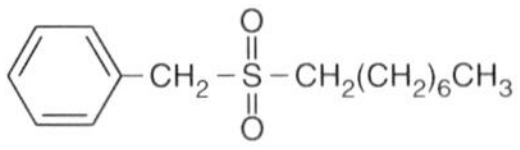

250 mg

S88,699-8
2,7-DIHYDRO-3,6-DIPHENYL-1,4,5-THIADIAZEPINE-1,1-DIOXIDE

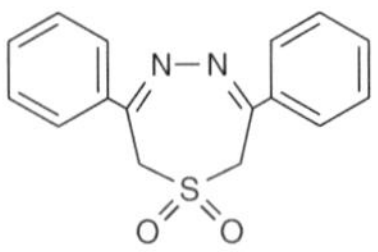

250 mg

S79,900-9
1,2-BIS(BENZYLSULFONYL)ETHENE

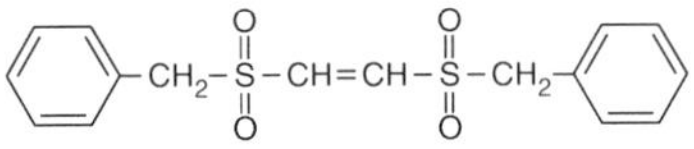

250 mg

S80,710-9
PHENYL ALPHA-PHENYLSTYRYL SULFONE

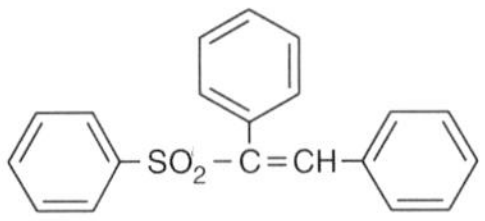

250 mg

S77,888-5
DIBENZYLMETHYL PHENYL SULFONE

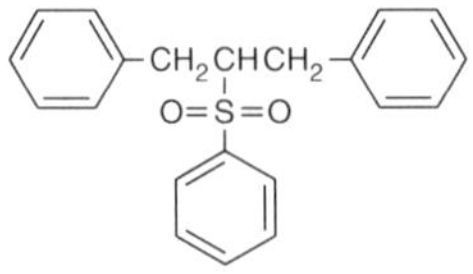

250 mg

S77,889-3
1,1-DIBENZYLETHYL PHENYL SULFONE

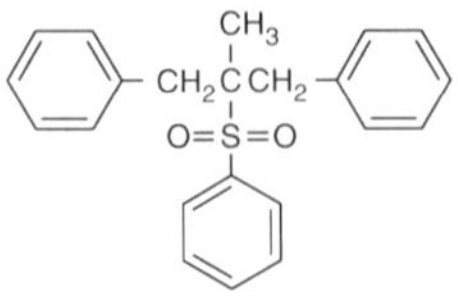

250 mg

S77,898-2
3,5-DIPHENYL-2-METHYL-4-(PHENYLSULFONYL)ISOXAZOLIDINE

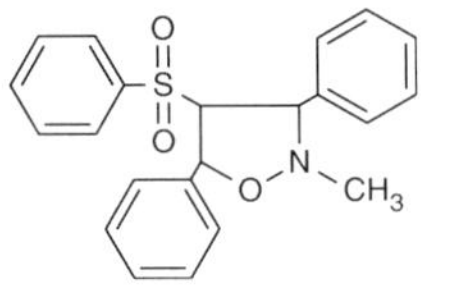

250 mg

S84,895-6
PHENYL O-TOLYL SULFONE

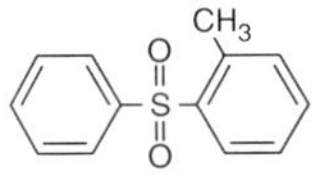

250 mg

S77,554-1
1-METHYL-1-PHENYLETHYL O-TOLYL SULFONE

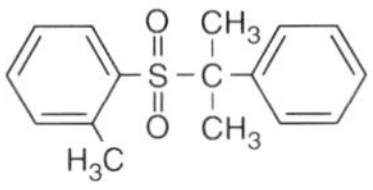

250 mg

S80,596-3
2-BENZYLPHENYL TERT-BUTYL SULFONE

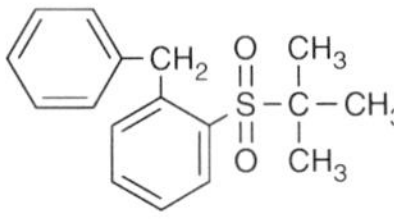

250 mg

S44,733-1
4-FLUOROPHENYL PHENYL SULFONE

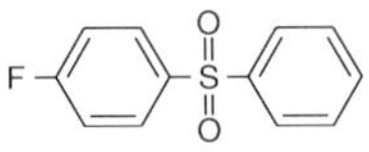

5 gm

S77,906-7
BENZYL P-TOLYL SULFONE

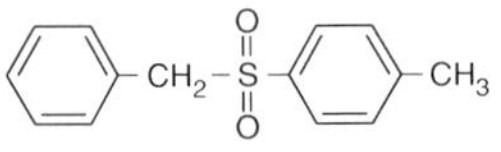

250 mg

S80,749-4
4-ETHYLPHENYL PHENYL SULFONE

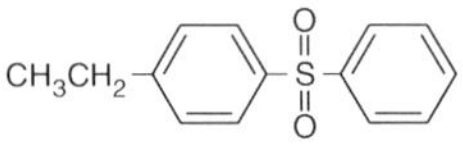

250 mg

S80,712-5
STYRYL P-TOLYL SULFONE

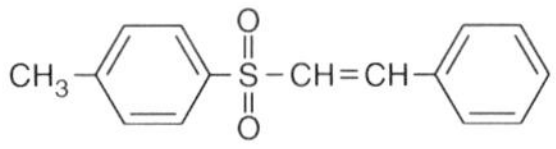

250 mg

S81,075-4
PHENETHYL P-TOLYL SULFONE

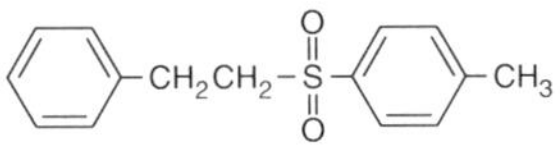

250 mg

S91,788-5
4-ISOPROPYLPHENYL PHENYL SULFONE

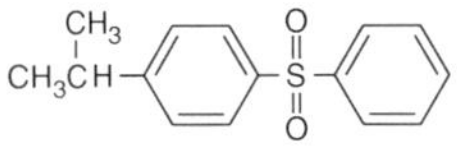

250 mg

S91,791-5
4-(1-CHLORO-1-METHYLETHYL)PHENYL PHENYL SULFONE

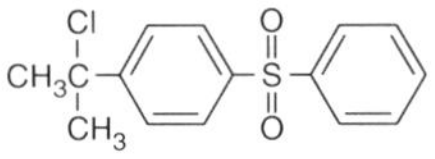

250 mg

S77,897-4
ALPHA-IODOSTYRYL P-TOLYL SULFONE

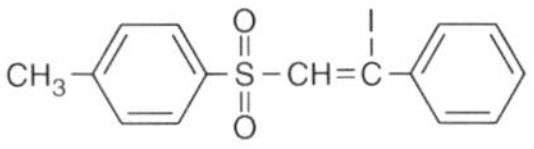

250 mg

S80,915-2
2-IODO-2-PHENYLVINYL P-TOLYL SULFONE

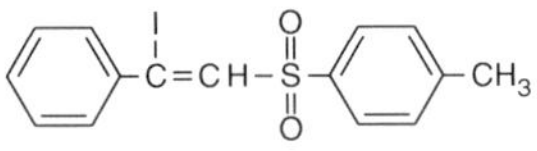

250 mg

S80,916-0
1-METHYL-2-PHENYLVINYL P-TOLYL SULFONE

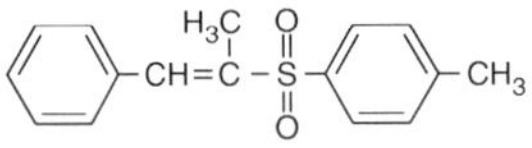

250 mg

S80,593-9
BENZYL 4-(NEOPENTYL)PHENYL SULFONE

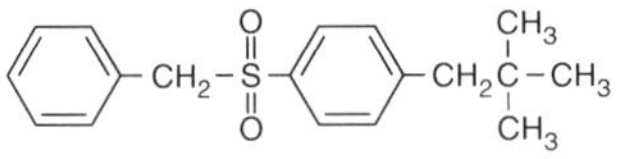

250 mg

S86,334-3
BENZYL 4-TERT-PENTYLPHENYL SULFONE

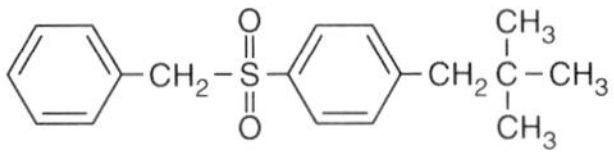

250 mg

S91,783-4
1,4-BIS(PHENYLSULFONYL)BENZENE

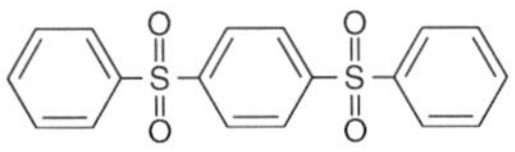

250 mg

S77,571-1
2,2-DIPHENYLETHYL P-TOLYL SULFONE

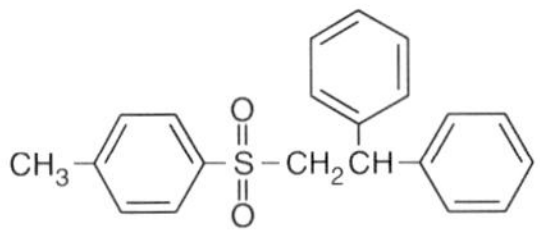

250 mg

S86,372-6
2-CHLORO-1,2-DIPHENYLETHYL P-TOLYL SULFONE

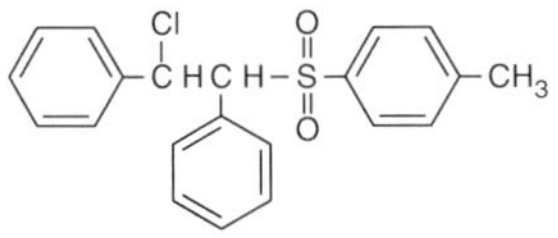

250 mg

S80,705-2
P-TOLYL TRIPHENYLMETHYL SULFONE

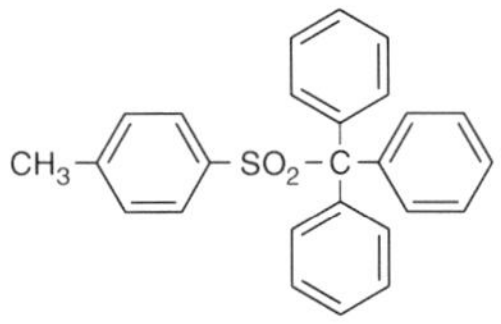

250 mg

S80,760-5
2,4-DIETHYLPHENYL PHENYL SULFONE

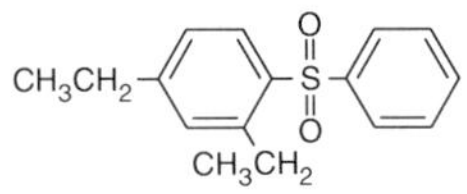

250 mg

S79,892-4
4-BENZYLSULFONYL-3-(PHENYLSULFONYLMETHYL)TOLUENE

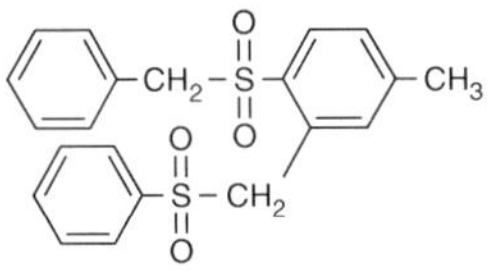

250 mg

S77,904-0
1,2-DIPHENYLETHENYL MESITYL SULFONE

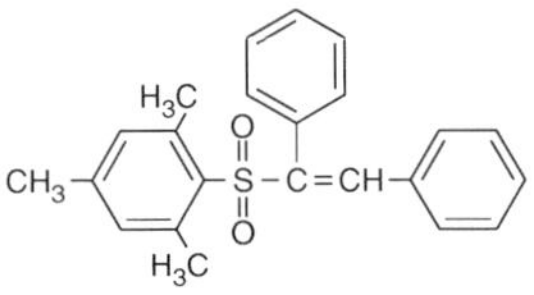

250 mg

S79,901-7
MESITYL STYRYL SULFONE

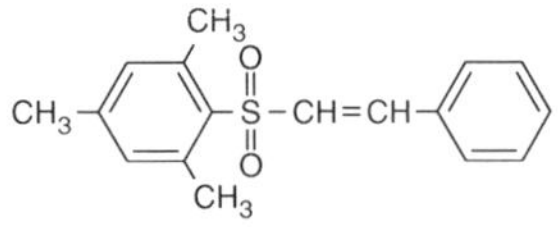

250 mg

S81,780-5
DIPHENYLMETHYL PHENYL SULFONE
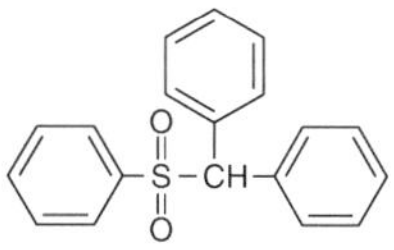
250 mg

S83,772-5
4,5-DIHYDRO-5,5-DIPHENYL-2H-3,1-BENZOXATHIEPIN-1,1-DIOXIDE
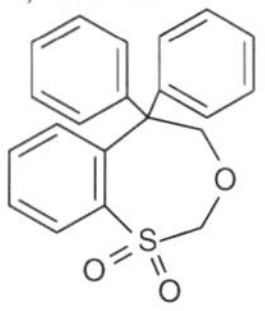
250 mg

S80,588-2
1-NAPHTHYL PHENYL SULFONE
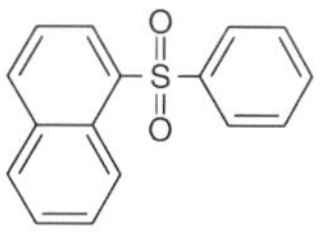
250 mg

S77,570-3
2-NAPHTHYL PHENYL SULFONE
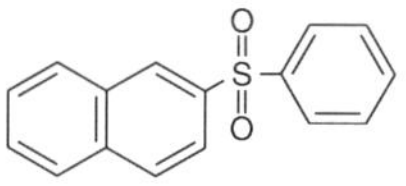
250 mg

S80,371-5
BUTYL O-TOLYL SULFONE
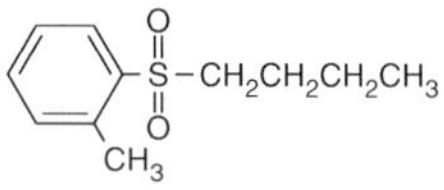
250 mg

S80,717-6
TERT-BUTYL 2-ISOPROPYLPHENYL SULFONE
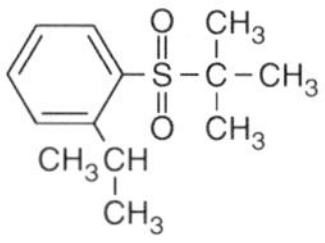
250 mg

S80,587-4
MESITYL O-TOLYL SULFONE
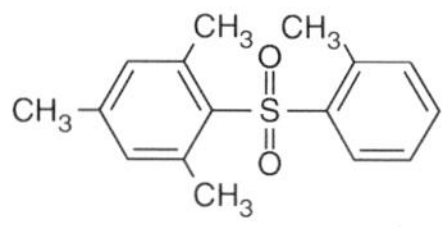
250 mg

S38,598-0
4-CHLOROPHENYL METHYL SULFONE
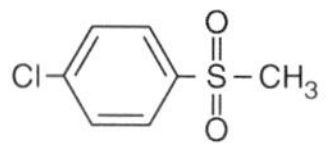
250 mg

S59,683-3
4-BROMOPHENYL METHYL SULFONE
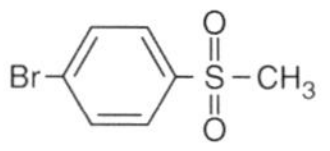
5 gm

S81,258-7
METHYL P-TOLYL SULFONE
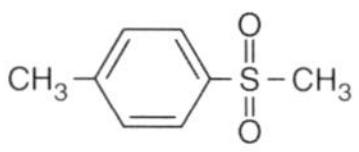
250 mg

S61,444-0
2-CHLOROETHYL 4-CHLOROPHENYL SULFONE
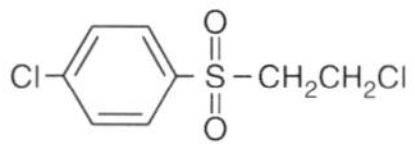
5 gm

S58,834-2
(4-CHLOROPHENYLSULFONYL)-(METHYLSULFONYL)METHANE
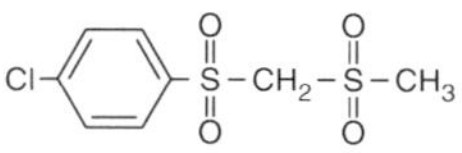
250 mg

S77,823-0
P-TOLYL VINYL SULFONE
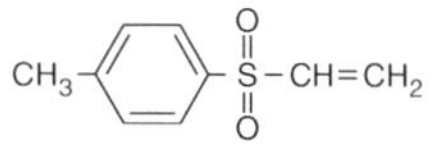
250 mg

S82,269-8
ETHYL P-TOLYL SULFONE
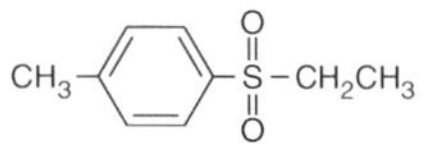
250 mg

S86,321-1
ALLYL P-TOLYL SULFONE
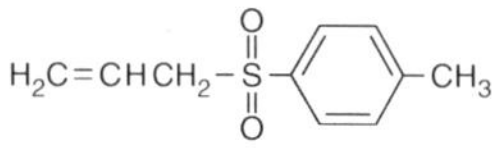
250 mg

S79,925-4
ISOPROPYL P-TOLYL SULFONE
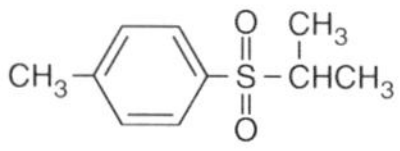
250 mg

S77,910-5
2-IODO-1-PROPENYL P-TOLYL SULFONE
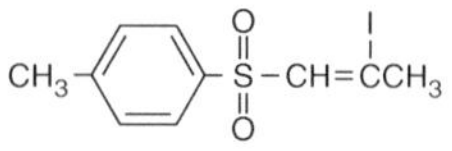
250 mg

S77,112-0
2-METHYLALLYL P-TOLYL SULFONE
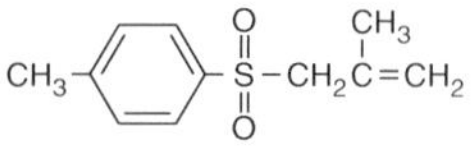
250 mg

S77,830-3
TERT-BUTYL P-TOLYL SULFONE
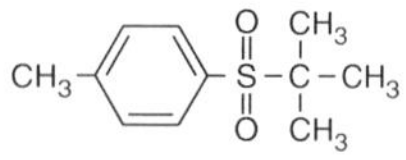
250 mg

S82,012-1
BUTYL P-TOLYL SULFONE
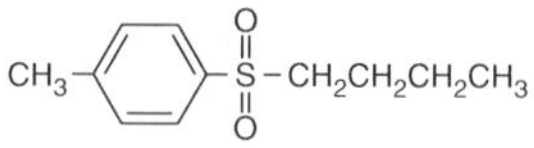
250 mg

S77,902-4
2,2-DIMETHYLPROPYL P-TOLYL SULFONE
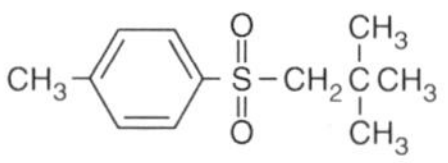
250 mg

S80,902-0
PENTYL P-TOLYL SULFONE
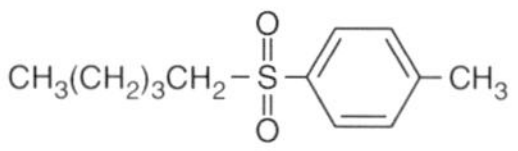
250 mg

S43,352-7
4-CHLOROPHENYL 4-FLUOROPHENYL SULFONE
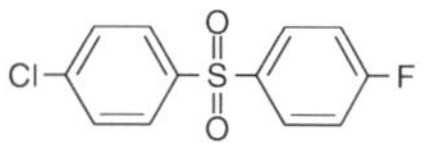
5 gm

S42,834-5
2,4,4',5-TETRACHLORODIPHENYL SULFONE
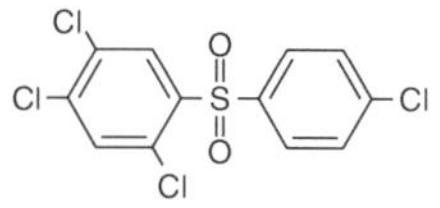
5 gm

S45,050-2
PHENYL P-TOLYL SULFONE
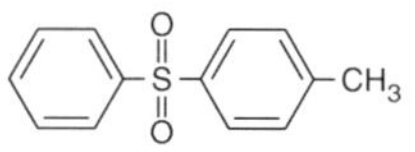
5 gm

S45,002-2
4-FLUOROPHENYL P-TOLYL SULFONE
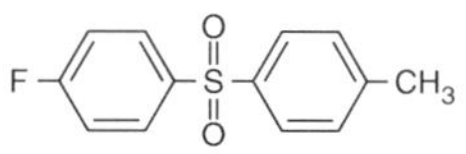
5 gm

S43,361-6
4-CHLOROPHENYL P-TOLYL SULFONE
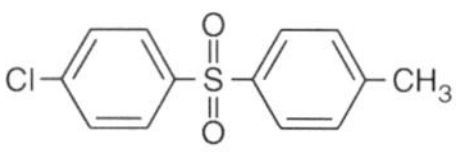
5 gm

S80,713-3
4-METHYLBENZYL P-TOLYL SULFONE
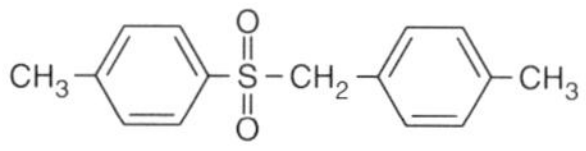
250 mg

S77,577-0
BIS(P-TOSYL)METHANE
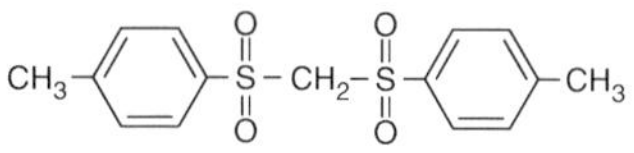
250 mg

S77,566-5
1,2-DICHLORO-1,2-DI-P-TOSYLETHYLENE
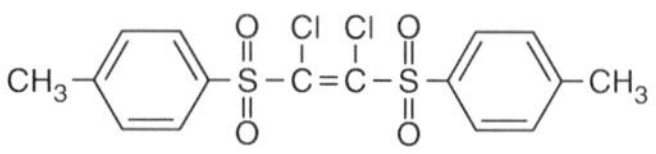
250 mg

S80,910-1
2,3-BIS(P-TOSYL)-1-PROPENE
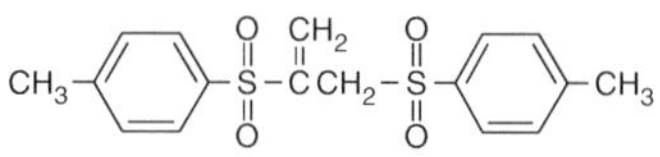
250 mg

S38,888-2
4-CHLOROPHENYL HEXADECYL SULFONE
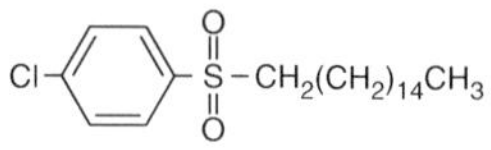
5 gm

S50,513-7
HEXADECYL P-TOLYL SULFONE
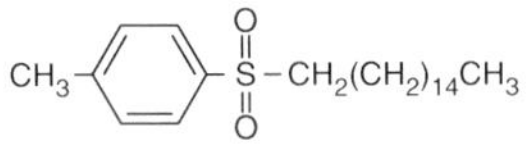
5 gm

S64,248-7
4-CHLOROPHENYL 4-(TERT-DODECYL)BENZYL SULFONE
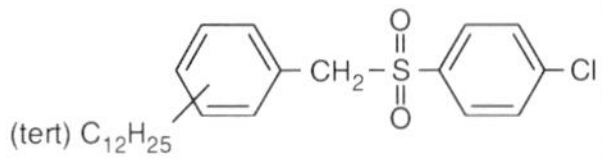
250 mg

S79,897-5
MESITYL P-TOLYL SULFONE
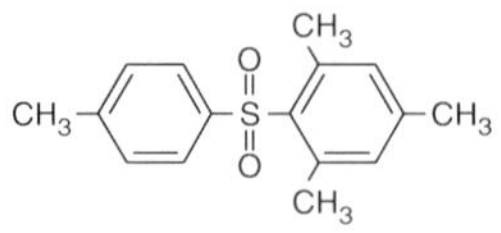
250 mg

S76,963-0
1,1'-(SULFONYLDI-4,1-PHENYLENE)BIS-[3-(1-NAPHTHYL)UREA]
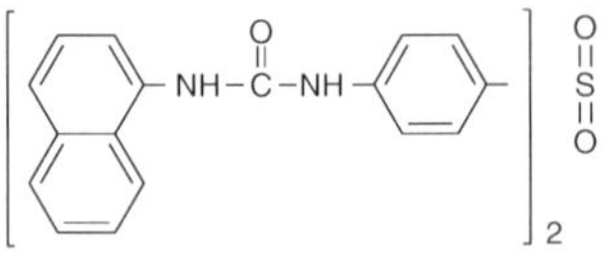
250 mg

S62,857-3
3,4-DICHLOROPHENYL METHYL SULFONE
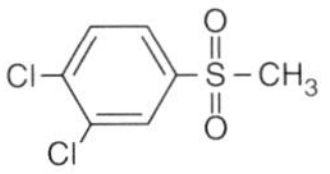
5 gm

S45,020-0
2,4-BIS(METHYLSULFONYL)-1-CHLOROBENZENE
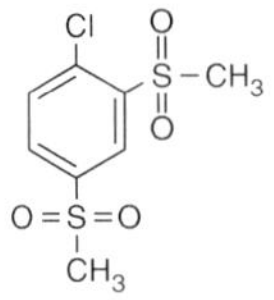
5 gm

S64,213-4
2,5-DICHLOROPHENYL METHYLSULFONYLMETHYL SULFONE
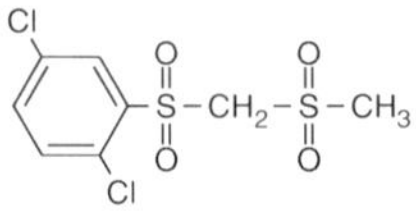
5 gm

S77,560-6
TERT-BUTYL 2,4-XYLYL SULFONE
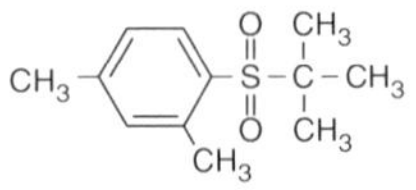
250 mg

S62,858-1
3,3',4,4'-TETRACHLORODIPHENYL SULFONE
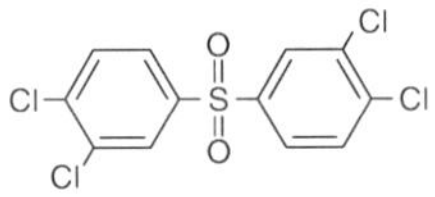
250 mg

S79,959-9
ISOPROPYL MESITYL SULFONE
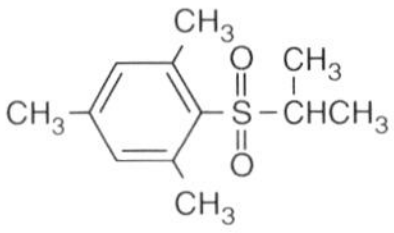
250 mg

S77,558-4
TERT-BUTYL 2,4,6-TRIMETHYLPHENYL SULFONE
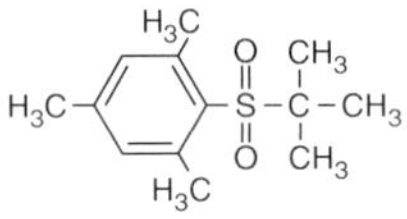
250 mg

S77,851-6
MESITYL M-TOLYL SULFONE
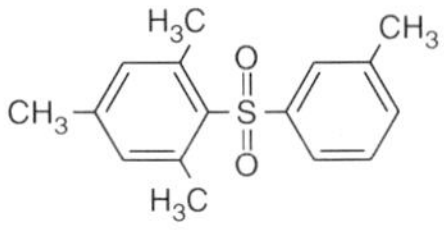
250 mg

S72,197-2
DIMESITYL SULFONE
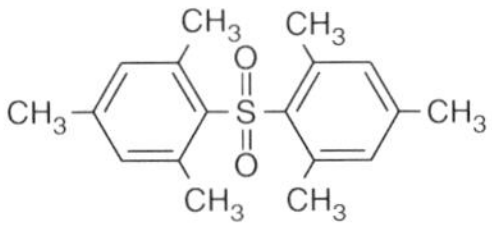
5 gm

S79,895-9
MESITYL 2-NAPHTHYL SULFONE
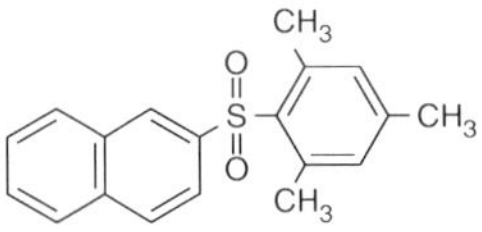
250 mg

S86,335-1
1-NAPHTHYL SULFONE
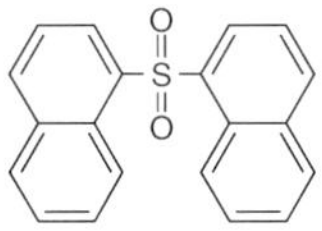
250 mg

S79,940-8
METHYL 2-NAPHTHYL SULFONE
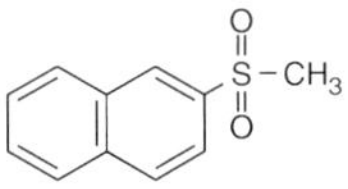
250 mg

S79,907-6
METHYL 1-NAPHTHYLMETHYL SULFONE
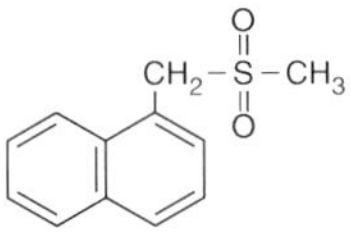
250 mg

S34,147-9
1-PHENYLSULFONYL-2,3-DIMETHYLINDOLINE (MIXTURE OF ISOMERS)
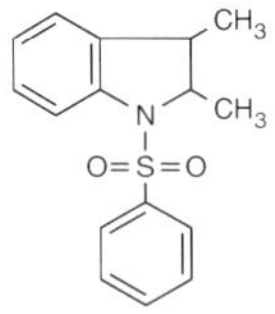
5 gm

S81,259-5
3-CHLORO-4-METHOXYBENZENESULFINIC ACID, SODIUM SALT
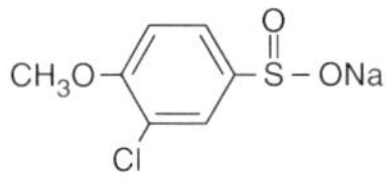
250 mg

S81,340-0
4-CHLORO-3-NITROBENZENESULFINIC ACID, BARIUM SALT
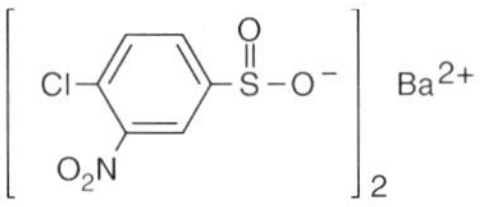
250 mg

S83,268-5
2-CHLORO-5-NITROBENZENESULFINIC ACID, SODIUM SALT
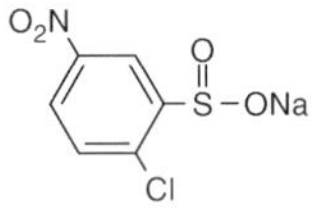
250 mg

S63,899-4
4-(4-AMINO-3-METHYLPHENOXY)-1-BUTANESULFINIC ACID
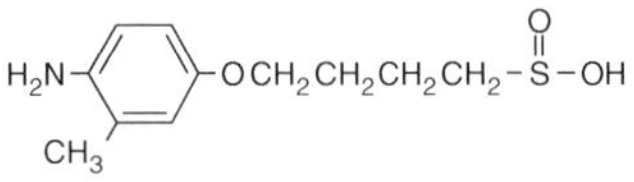
250 mg

S45,039-1
3-ACETAMIDO-4-METHOXYBENZENESULFINIC ACID, HYDRATE
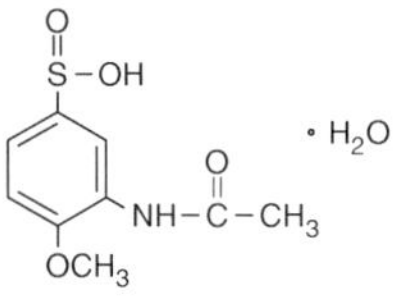
5 gm

S81,260-9
3-ACETAMIDO-4-METHOXYBENZENESULFINIC ACID, SODIUM SALT
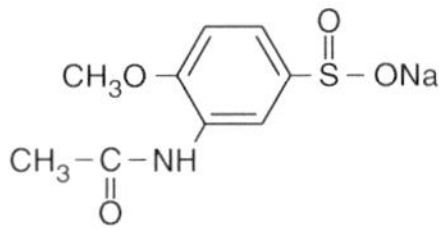
250 mg

S84,112-9
5-NITRO-2-(PHENYLTHIO)-BENZENESULFINIC ACID
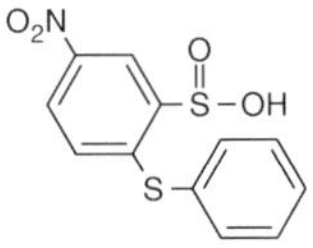
250 mg

S65,527-9
2-CHLORO-5-NITROBENZENESULFINIC ACID
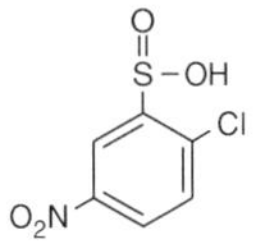
5 gm

S43,613-5
4-ACETAMIDOBENZENESULFINIC ACID
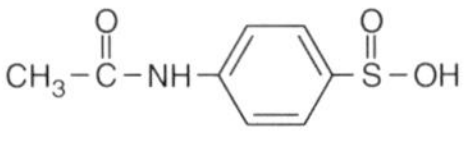
5 gm

S83,572-2
4-ACETAMIDOBENZENESULFINIC ACID, SODIUM SALT
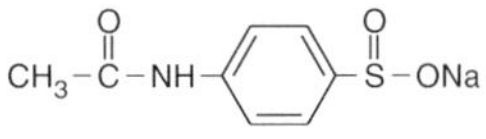
250 mg

S86,522-2
9,10-DIHYDRO-9,10-ETHANOANTHRACEN-11-YL-METHYL-1,1-DIMETHYL-ETHANESULFINATE
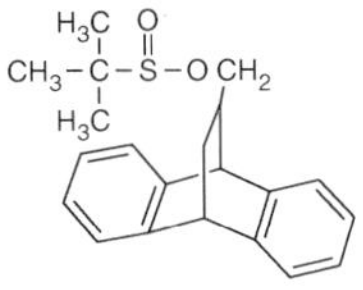
250 mg

S81,261-7
BENZENESULFINIC ACID NITRATE
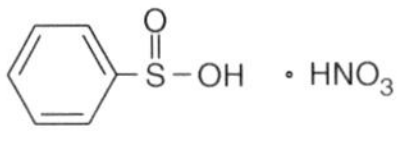
250 mg

S86,998-8
P-TOLYL BENZENESULFINATE
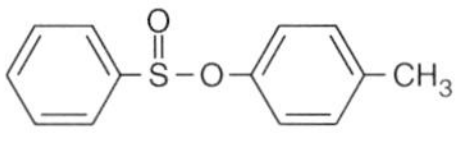
250 mg

S61,439-4
4-CHLOROBENZENESULFINIC ACID SODIUM SALT
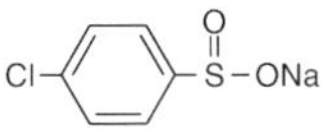
5 gm

S83,549-8
P-TOLUENESULFINIC ACID

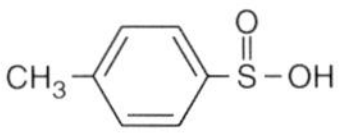

250 mg

S79,932-7
4-(2,2-DIMETHYLPROPYL)-BENZENESULFINIC ACID

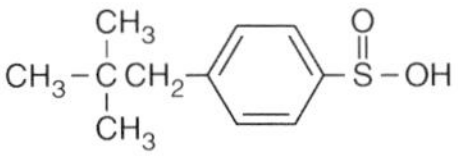

250 mg

S79,933-5
4-(2,2-DIMETHYLPROPYL)-BENZENESULFINIC ACID, SODIUM SALT

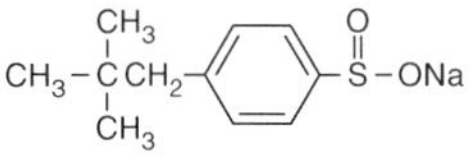

250 mg

S79,889-4
2,4,6-TRIMETHYLBENZENESULFINIC ACID

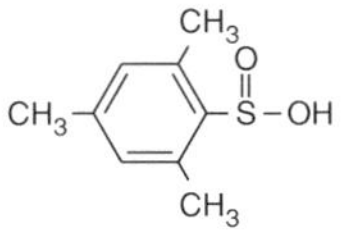

250 mg

S79,896-7
2,4-DIMETHYL-6-(3-METHYLBENZYL)-BENZENESULFINIC ACID, SODIUM SALT

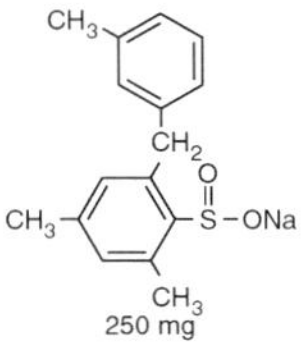

250 mg

S65,208-3
2-NAPHTHYLSULFINIC ACID

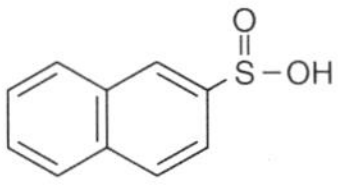

250 mg

S50,847-0
6-(2,4,5-TRICHLOROPHENOXY)-METANILIC ACID

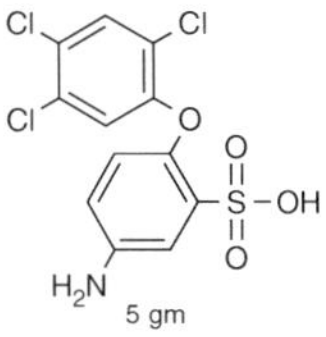

5 gm

S42,947-3
2-AMINO-5-CHLORO-P-TOLUENESULFONIC ACID

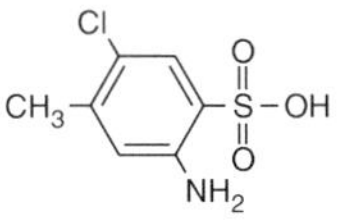

5 gm

S43,203-2
6-AMINO-4-CHLORO-META-TOLUENESULFONIC ACID

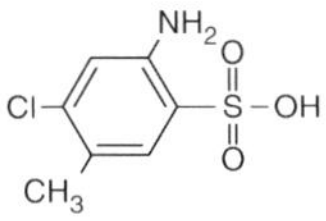

5 gm

S35,753-7
2-AMINO-5-CHLORO-4-ETHYLBENZENESULFONIC ACID

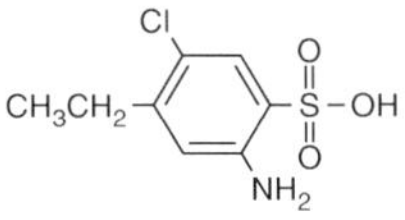

5 gm

S41,934-6
5-AMINO-2-CHLOROBENZENESULFONIC ACID

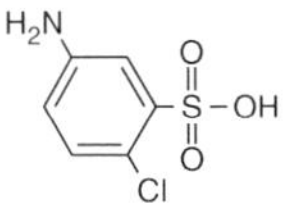

5 gm

S43,877-4
4-AMINO-3-CHLOROBENZENESULFONIC ACID

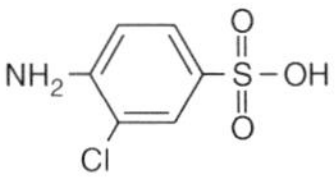

5 gm

S60,824-6
3,5-DIBROMOSULFANILIC ACID

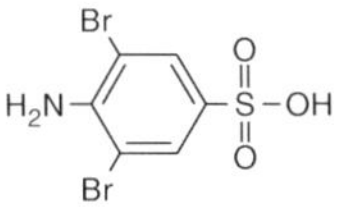

5 gm

S35,957-2
4-AMINO-5-CHLORO-M-TOLUENESULFONIC ACID

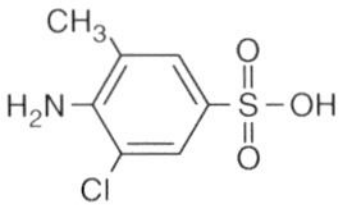

5 gm

S38,744-4
2,5-DICHLOROSULFANILIC ACID

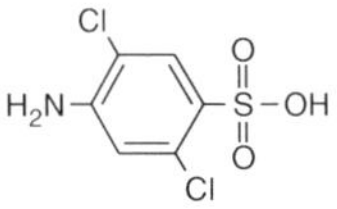

5 gm

S42,381-5
2,4-DIHYDROXY-5-(ISOBUTYLTHIO)-BENZENESULFONIC ACID

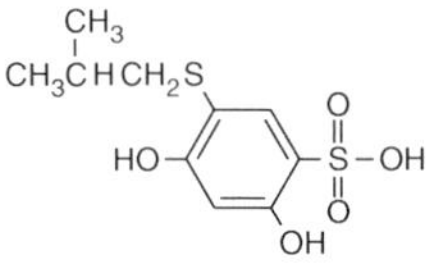

5 gm

S45,466-4
5-BENZOYL-4-HYDROXY-2-METHOXY-BENZENESULFONIC ACID

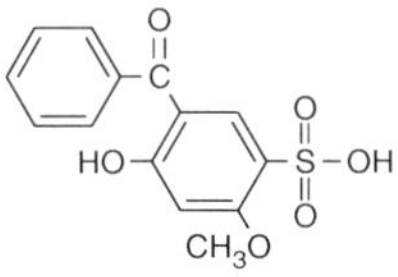

5 gm

S78,805-8
L-TYROSINE BENZYL ESTER PARA-TOLUENESULFONATE

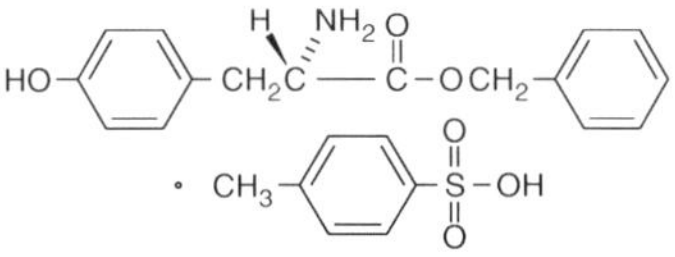

250 mg

S35,772-3
5-AMINO-2-HYDROXY-3-SULFOBENZOIC ACID

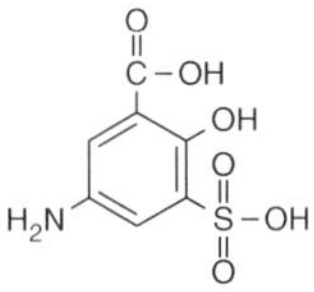

5 gm

S55,429-4
2-AMINO-6-HYDROXYBENZENESULFONIC ACID

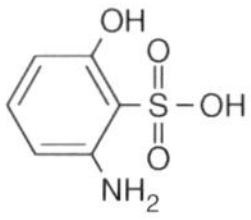

5 gm

S35,782-0
5-AMINO-2-HYDROXYBENZENESULFONIC ACID

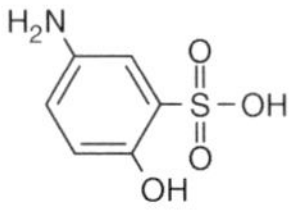

5 gm

S44,088-4
4-AMINO-3-HYDROXY-1-NAPHTHALENE-SULFONIC ACID HEMIHYDRATE

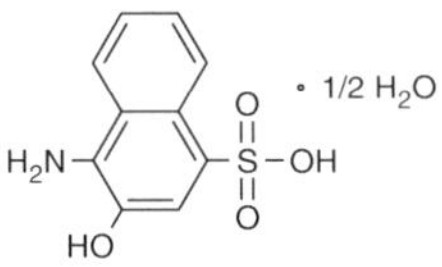

5 gm

S34,133-9
6-ANILINO-4-HYDROXY-2-NAPHTHALENE-SULFONIC ACID

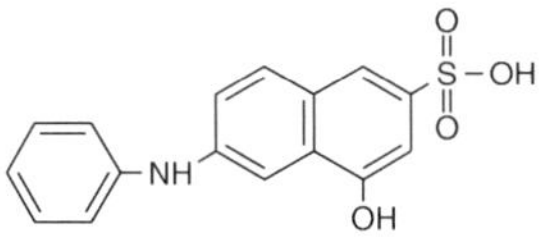

5 gm

S35,129-6
1-HYDROXY-5-SULFO-2-NAPHTHALENEDIAZONIUM CHLORIDE

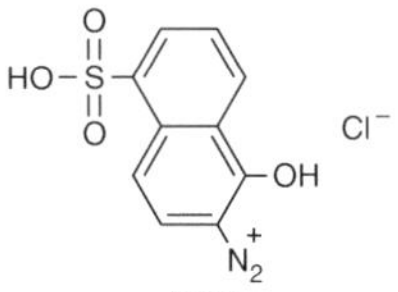

5 gm

S42,796-9
BROMOPHENOL RED

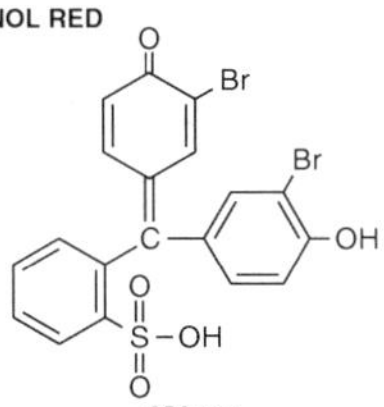

250 mg

S43,983-5
QUINIZARIN-2-SULFONIC ACID

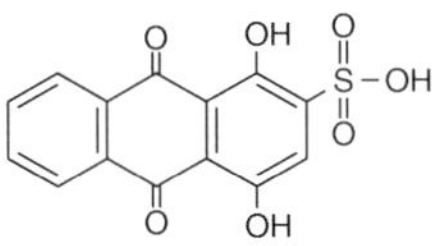

250 mg

S36,036-8
3,5-DICHLORO-4-HYDROXYBENZENESULFONIC ACID

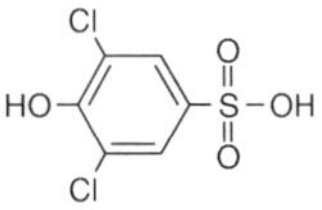

5 gm

S56,346-3
7-HYDROXY-1,3-NAPHTHALENEDISULFONIC ACID

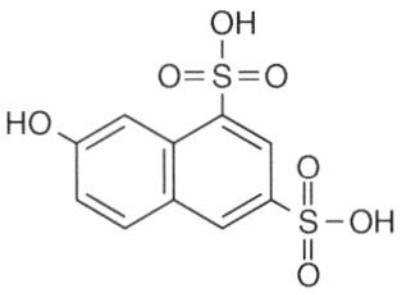

5 gm

S56,369-2
4,5-DIHYDROXY-2,7-NAPHTHALENEDISULFONIC ACID

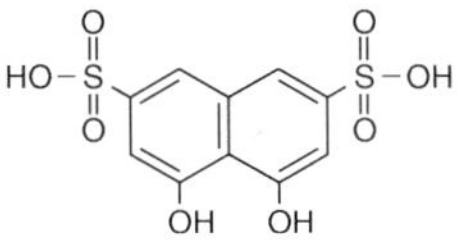

5 gm

S44,789-7
4-METHOXYMETANILIC ACID

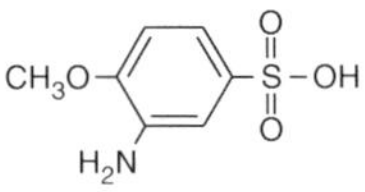

5 gm

S65,856-1
2-AMINO-5-(4-CYCLOHEXYLPHENOXY)-BENZENESULFONIC ACID

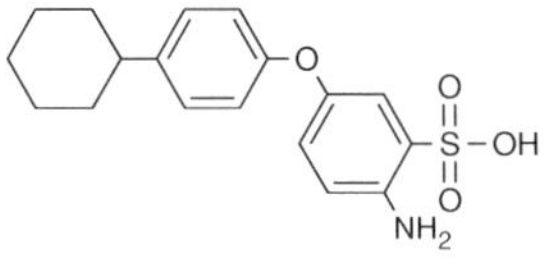

5 gm

S81,098-3
3-NITRO-4-PHENOXYBENZENESULFONIC ACID

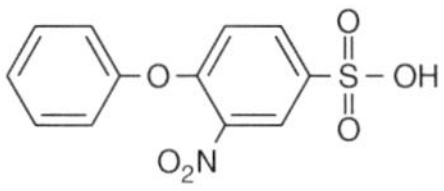

250 mg

S58,199-2
4-(HEXADECYLOXY)-3-NITROBENZENESULFONIC ACID

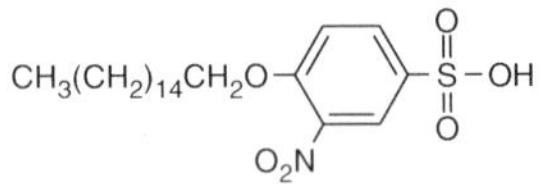

250 mg

S62,794-1
3-NITRO-4-OCTADECYLOXYBENZENESULFONIC ACID

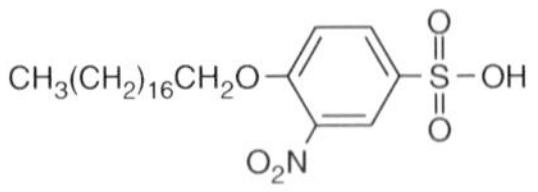

250 mg

S50,111-5
5-NITRO-2-[4-(1,1,3,3-TETRAMETHYLBUTYL)PHENOXY]-BENZENESULFONIC ACID

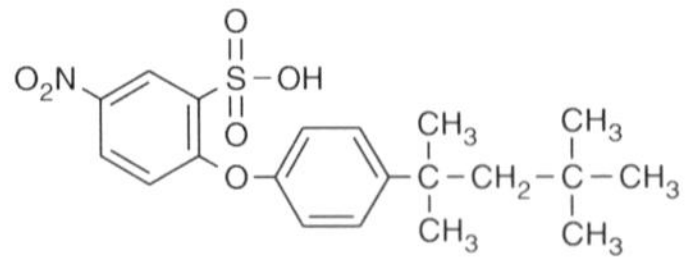

250 mg

S50,505-6
2-HEXADECYLOXY-5-SULFOBENZOIC ACID

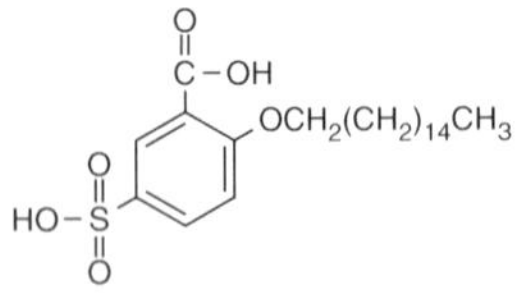

250 mg

S61,764-4
3-(2-HEXADECYLOXY-5-SULFOPHENYL)-PROPIONIC ACID

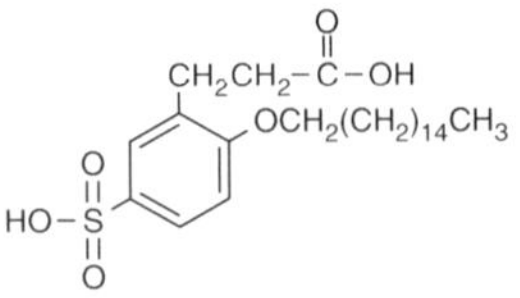

250 mg

S60,525-5
4-(4-HEXADECYLOXY-3-SULFOPHENYL)-BUTYRIC ACID

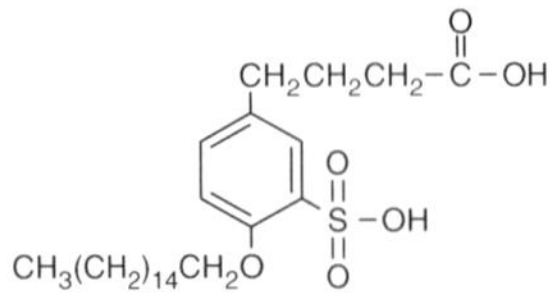

250 mg

S42,374-2
2-(2,4,6-TRIIODOPHENOXY)-ETHANESULFONIC ACID

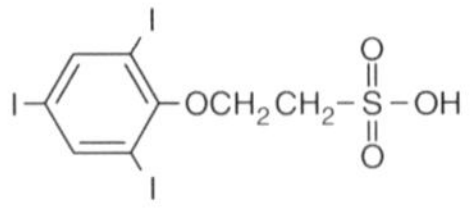

5 gm

S61,798-9
2,6-DIMETHYL-4-HEXADECYLOXYBENZENESULFONIC ACID

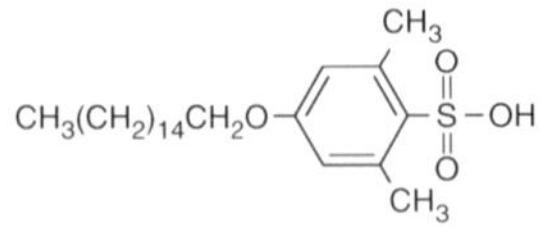

5 gm

S38,302-3
4-HEXADECYLTHIO-3-NITROBENZENESULFONIC ACID

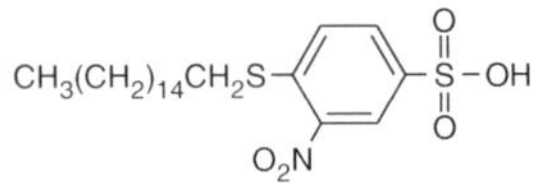

250 mg

S59,293-5
2-HEXADECYLTHIO-5-SULFOBENZOIC ACID

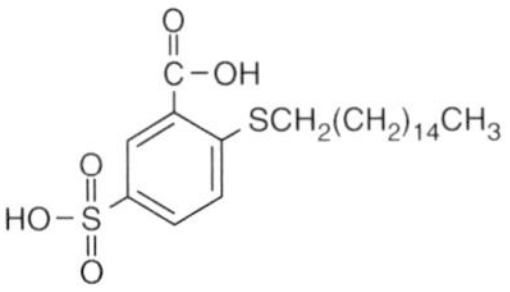

5 gm

S59,272-2
4-(HEXADECYLTHIO)-BENZENESULFONIC ACID

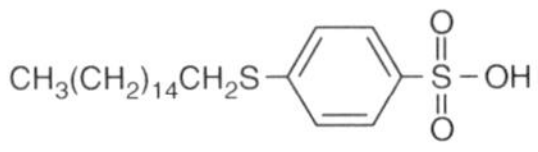

5 gm

S65,858-8
4-(3-AMINO-4-METHYLANILINO)-3-NITROBENZENESULFONIC ACID

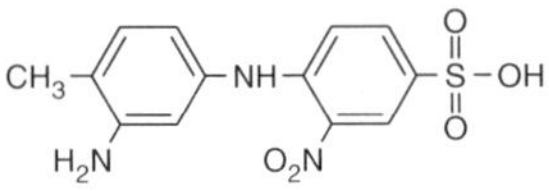

5 gm

S57,209-8
4-(4-AMINOANILINO)-3-NITRO-BENZENESULFONIC ACID

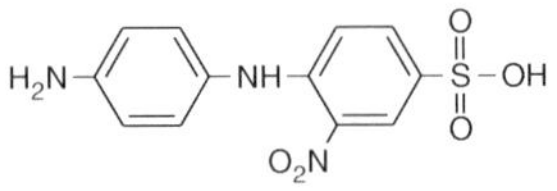

5 gm

S55,459-6
3-AMINO-4-(DODECYLAMINO)-BENZENESULFONIC ACID

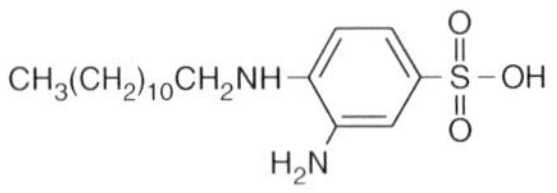

250 mg

S92,005-3
5-AMINO-4-METHOXY-2-NITROBENZENESULFONIC ACID

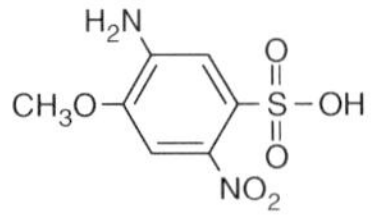

250 mg

S57,203-9
5-AMINO-2-(4-AMINOANILINO)-BENZENESULFONIC ACID

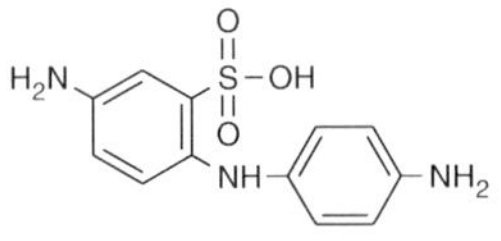

5 gm

S34,111-8
5-AMINO-2-(ANILINO)BENZENESULFONIC ACID

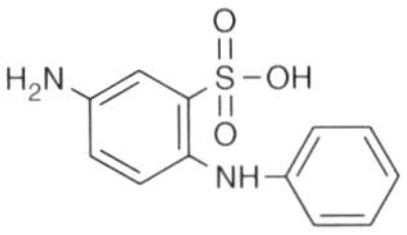

5 gm

S65,763-8
3-(4-AMINO-N1-BUTYLANILINO)PROPANE-SULFONIC ACID

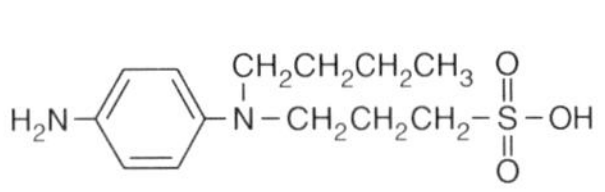

5 gm

S35,786-3
5-AMINO-2-(4-METHYLANILINO)BENZENE-SULFONIC ACID

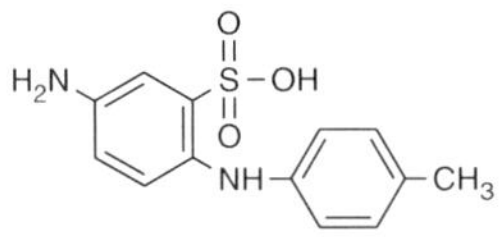

5 gm

S37,328-1
4,4'-DIAMINOSTILBENE-2,2'-DISULFONIC ACID

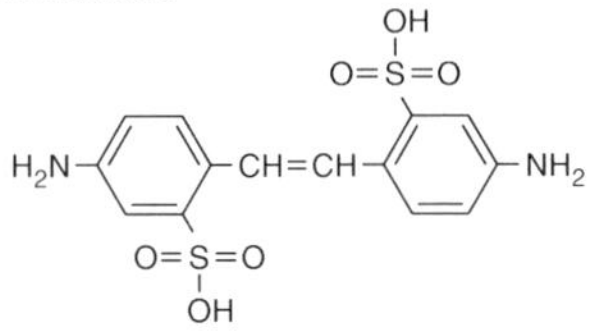

5 gm

S68,042-7
2,2'-BENZIDINEDISULFONIC ACID

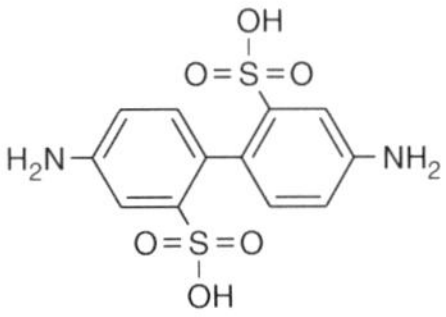

250 mg

S42,832-9
4-AMINO-M-TOLUENESULFONIC ACID

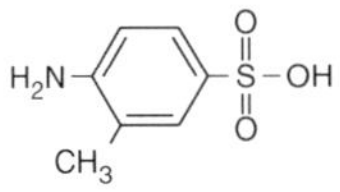

5 gm

S51,850-6
4-METHYLMETANILIC ACID

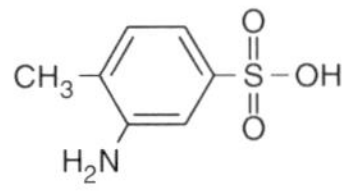

5 gm

S55,439-1
5,6-DIMETHYLMETANILIC ACID

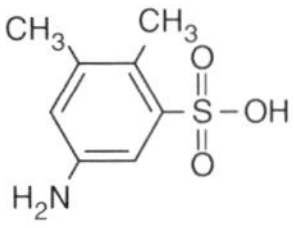

5 gm

S35,790-1
4,6-DIMETHYLMETANILIC ACID

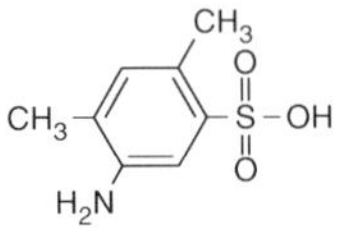

5 gm

S35,746-4
1-AMINOANTHRAQUINONE-2-SULFONIC ACID

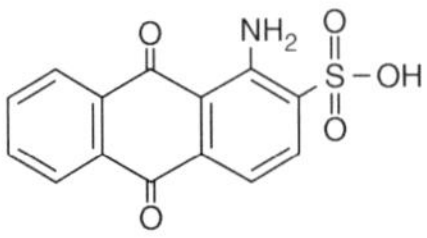

5 gm

S44,258-5
1-AMINO-2-NAPHTHALENESULFONIC ACID

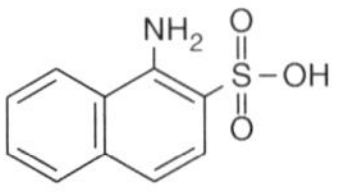

5 gm

S45,881-3
2-AMINO-1,5-NAPHTHALENEDISULFONIC ACID

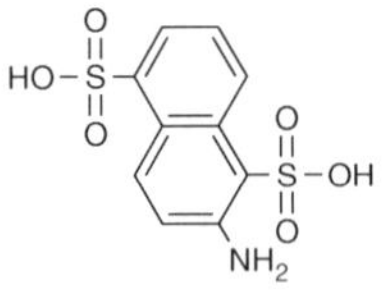

5 gm

S61,739-3
DL-LEUCINE BENZYL ESTER, P-TOLUENESULFONATE

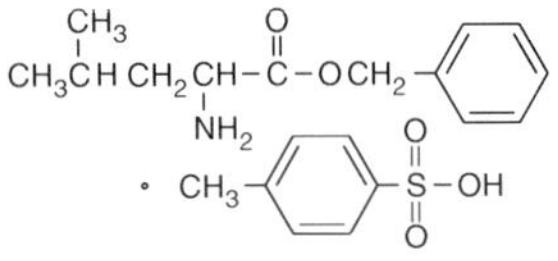

5 gm

S93,104-7
MONO-2-(4-AMINOPHENYLSULFONYL)-ETHYL SULFATE

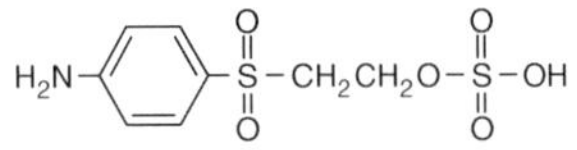

250 mg

S69,299-9
3-AMINO-1-NAPHTHALENESULFONIC ACID HYDRATE

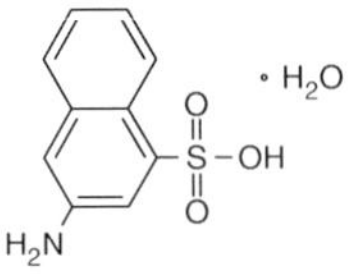

5 gm

S51,843-3
6-AMINO-1,3-NAPHTHALENEDISULFONIC ACID

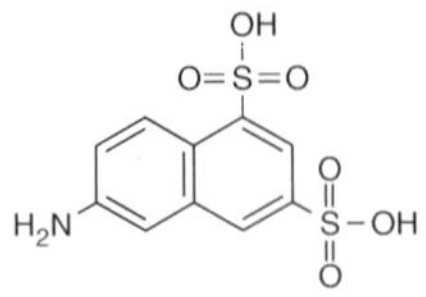

5 gm

S56,402-8
4-AMINO-2,7-NAPHTHALENEDISULFONIC ACID

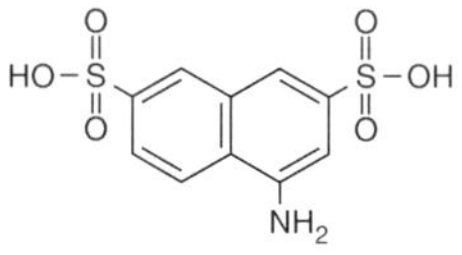

5 gm

S65,846-4
4-AMINO-1,7-NAPHTHALENEDISULFONIC ACID

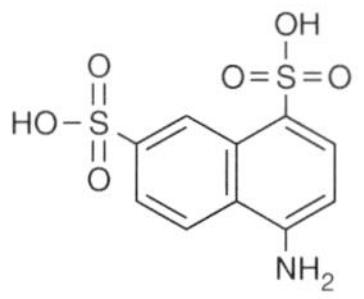

5 gm

S57,908-4
5-AMINO-1-NAPHTHALENESULFONIC ACID

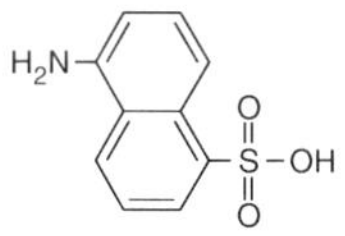

5 gm

S59,091-6
6-AMINO-2-NAPHTHALENESULFONIC ACID

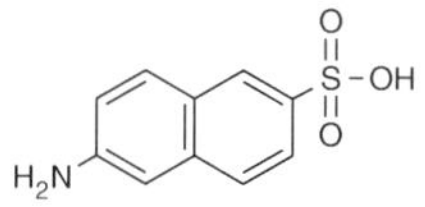

250 mg

S70,678-7
7-AMINO-2-NAPHTHALENESULFONIC ACID

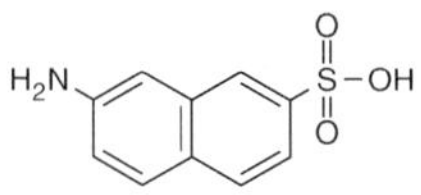

250 mg

S51,644-9
N,N-DIMETHYL-3-NITROSULFANILIC ACID

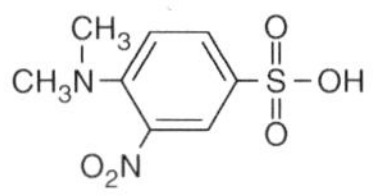

5 gm

S57,436-8
4-CHLORO-2,6-DINITROBENZENESULFONIC ACID

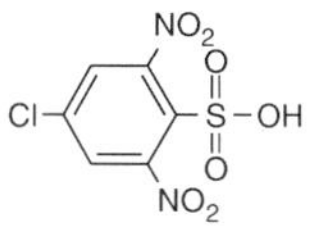

250 mg

S50,883-7
4-CHLORO-3,5-DINITROBENZENESULFONIC ACID

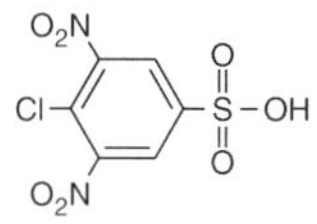

5 gm

S34,758-2
DIPHENYLAMINE-4-SULFONIC ACID

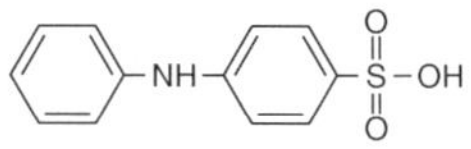

5 gm

S43,297-0
N,4-DIMETHYLMETANILIC ACID

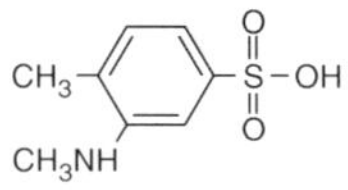

5 gm

S56,235-1
N-PHENYLTAURINE

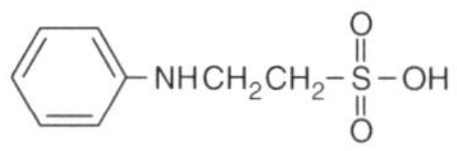

250 mg

S53,414-5
N-(SULFOMETHYL)ANTHRANILIC ACID

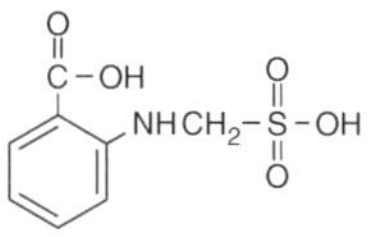

250 mg

S55,425-1
N-ETHYL-5-SULFOANTHRANILIC ACID

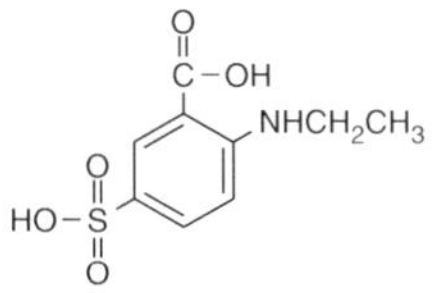

5 gm

S56,268-8
3-ANILINO-1-PROPANESULFONIC ACID

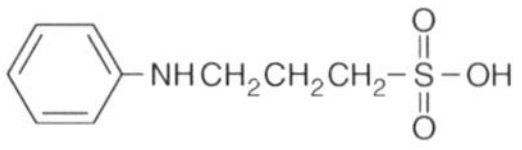

250 mg

S57,687-5
N-ETHYL-4-METHYLMETANILIC ACID

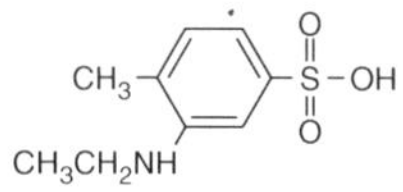

5 gm

S45,769-8
N,N-DIISOPROPYLETHYLAMINE P-TOLUENESULFONATE

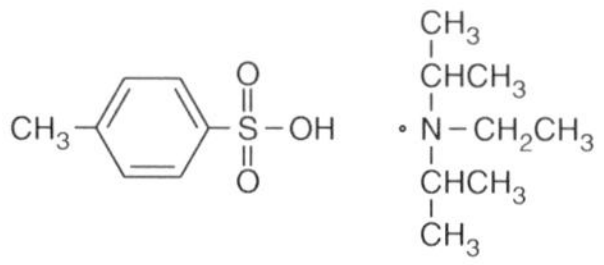

250 mg

S63,332-1
4-(4-AMINO-N1-HEXADECYLANILINO)-BUTANESULFONIC ACID

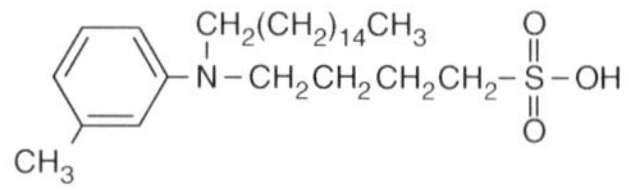

5 gm

S53,418-8
3-(N-METHYLANILINO)-1-PROPANESULFONIC ACID

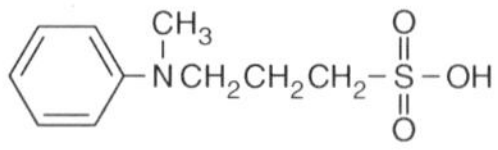

250 mg

S62,233-8
3-(N-OCTYLANILINO)-1-PROPANESULFONIC ACID

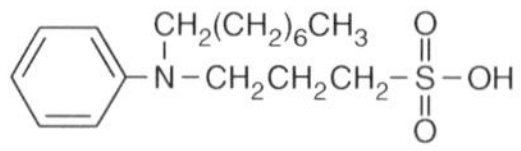

250 mg

S57,200-4
4,4'-DIAMINO-5,5'-DIMETHYL-2,2'-BIPHENYLDISULFONIC ACID

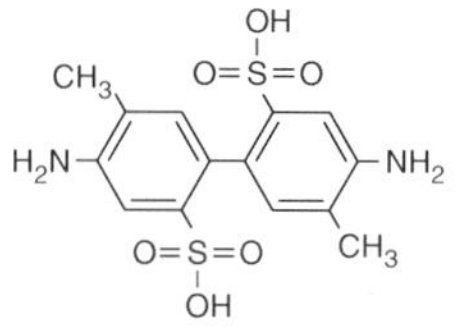

5 gm

S57,103-2
4-HYDRAZINOBENZENESULFONIC ACID

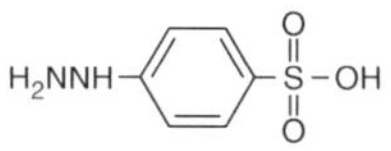

5 gm

S35,965-3
3-CHLORO-4-HYDRAZINOBENZENE-SULFONIC ACID

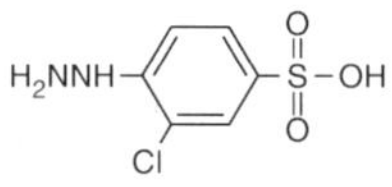

5 gm

S63,774-2
4,4'-DIHYDRAZINO-2,2'-BIPHENYLDISULFONIC ACID

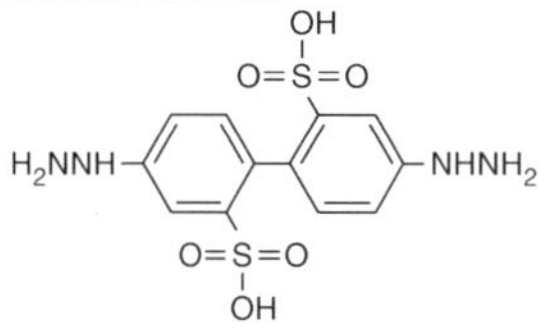

250 mg

S58,857-1
3-NITRO-2-NAPHTHALENESULFONIC ACID-BIS(HEXACHLOROCYCLOPENTADIENE) ADDUCT

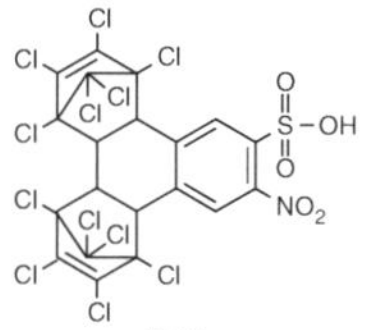

5 gm

S71,459-3
5-NITROANTHRAQUINONE-1-SULFONIC ACID

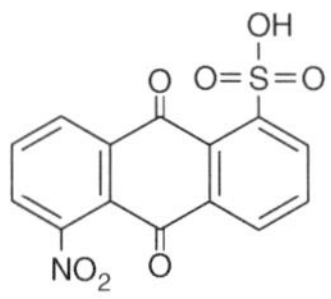

250 mg

S59,844-5
3-CHLORO-4-NITROBENZENESULFONIC ACID

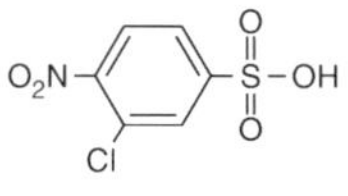

250 mg

S42,726-8
5-NITRO-O-TOLUENESULFONIC ACID

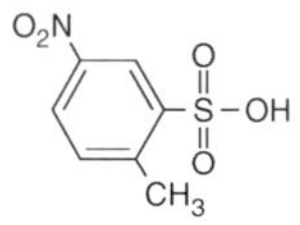

5 gm

S51,332-6
4-CHLORO-5-NITRO-O-TOLUENESULFONIC ACID

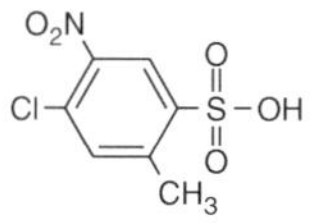

250 mg

S62,869-7
4-[4-(4-CHLOROBENZYLIDENE)-3-METHYL-5-OXO-2-PYRAZOLIN-2-YL]-BENZENESULFONIC ACID PYRIDINE SALT

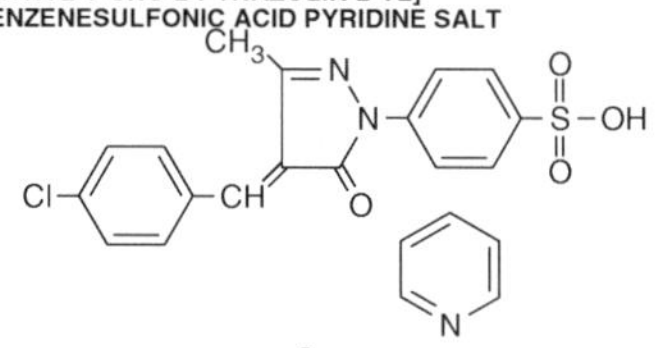

5 gm

S66,419-7
SULFOPHTHALIC ACID, MIXTURE OF ISOMERS

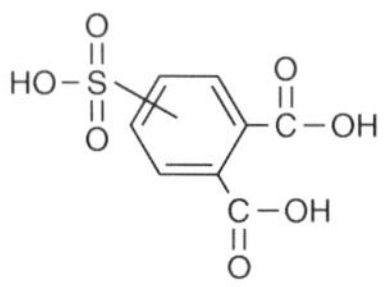

5 gm

S37,189-0
4-SULFOPHTHALIC ACID, SODIUM SALT

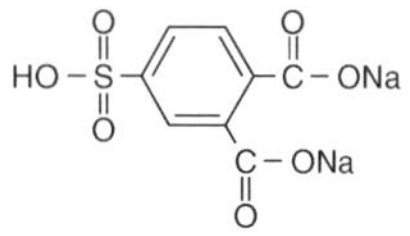

5 gm

S58,256-5
2-(HEXADECYLSULFONYL)-5-SULFOBENZOIC ACID

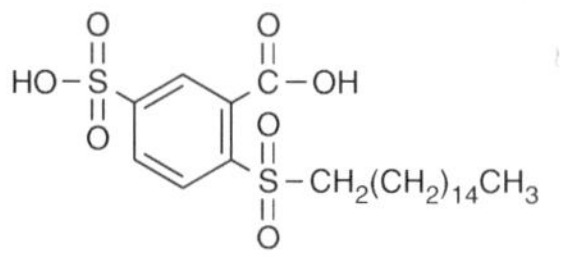

5 gm

S38,690-1
3-SULFOBENZOIC ACID
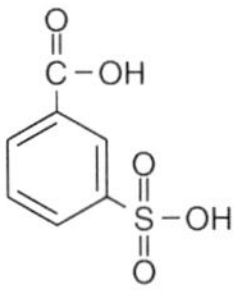
250 mg

S59,834-8
2-BROMO-5-SULFOBENZOIC ACID
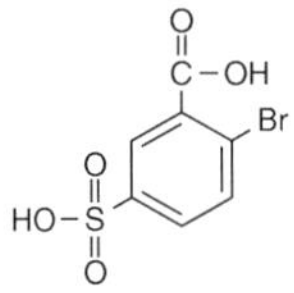
250 mg

S35,552-6
5-OXO-1-(4-SULFOPHENYL)-2-PYRAZOLINE-3-CARBOXYLIC ACID
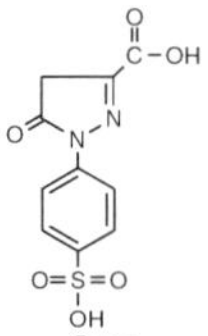
5 gm

S34,570-9
1-(2,5-DICHLORO-4-SULFOPHENYL)-5-PYRAZOLONE-3-CARBOXYLIC ACID
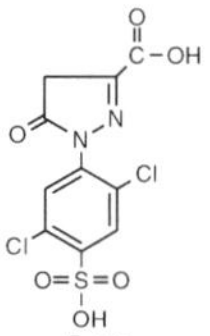
5 gm

S63,675-4
2-(4-SULFOPHENYL)ACETIC ACID, BARIUM SALT
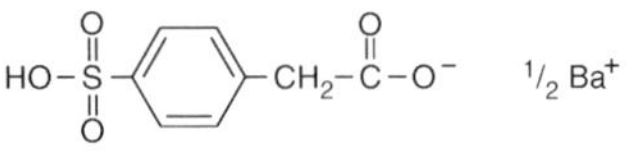
250 mg

S65,513-9
1-ACETOXYNAPHTHALENE-6-SULFONIC ACID
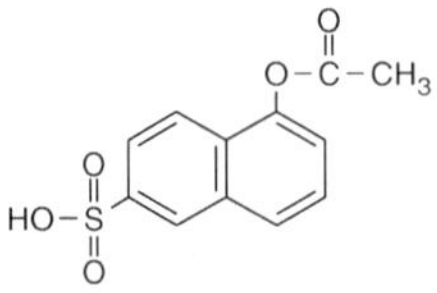
5 gm

S50,515-3
5-SULFO-1-NAPHTHYL ACETATE
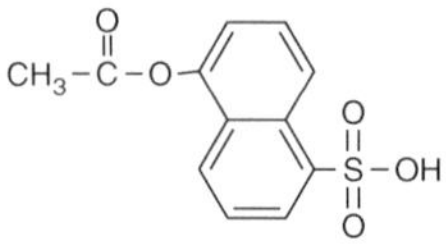
250 mg

S38,732-0
ETHYL 4-SULFO-1-NAPHTHYL CARBONATE
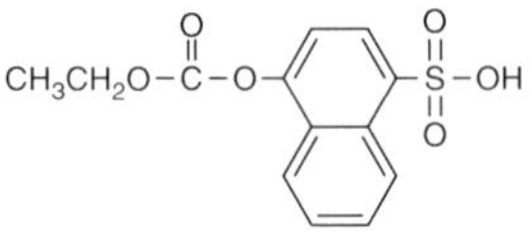
250 mg

S68,858-4
4-(3-ETHYLPHENYL)-1-(4-SULFOPHENYL) SEMICARBAZIDE
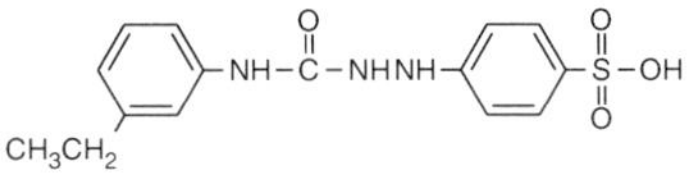
250 mg

S36,609-9
4-(3-METHYL-5-OXO-2-PYRAZOLIN-1-YL) -BENZENESULFONIC ACID
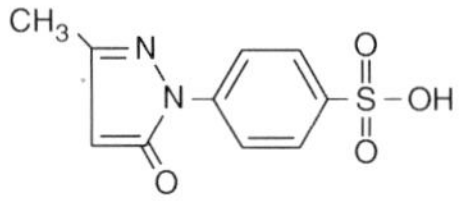
5 gm

S60,715-0
4-CHLORO-3-(3-METHYL-5-OXO-2-PYRAZOLIN-1-YL)-BENZENESULFONIC ACID
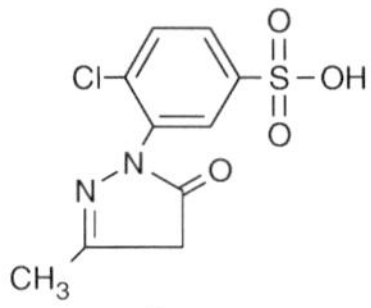
5 gm

S58,856-3
2-NAPHTHALENESULFONIC ACID-BIS-(HEXACHLOROCYCLOPENTADIENE) ADDUCT
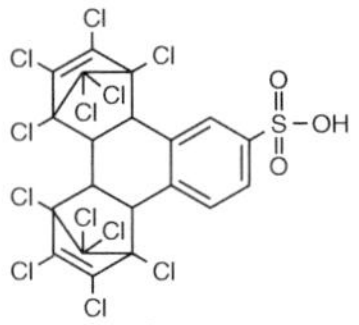
5 gm

S58,850-4
3-METHYL-2-NAPHTHALENESULFONIC ACID-BIS(HEXACHLOROCYCLOPENTADIENE) ADDUCT
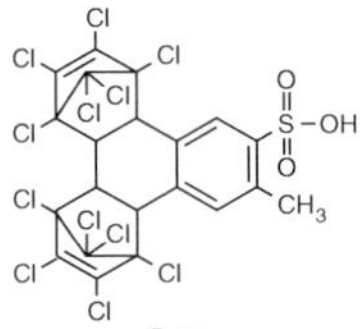
5 gm

S81,101-7
2-DIBENZOFURANSULFONIC ACID
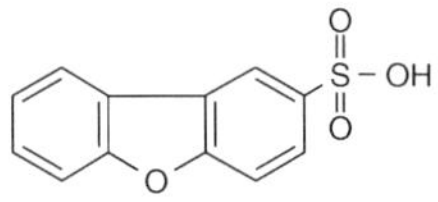
250 mg

S60,256-6
N-(AMIDINOAMIDINO)SULFANILIC ACID
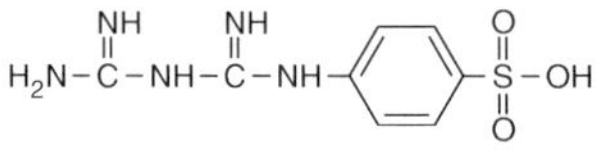
5 gm

S44,651-3
4-BIPHENYLSULFONIC ACID
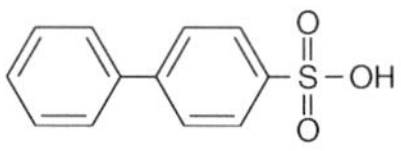
5 gm

S41,515-4
3,4-DICHLOROBENZENESULFONIC ACID
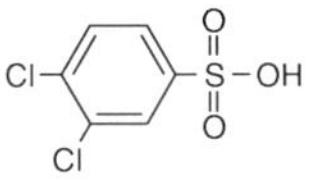
5 gm

S56,360-9
2,5-XYLENESULFONIC ACID DIHYDRATE
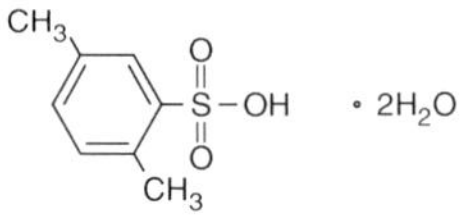
5 gm

S37,048-7
2,4,5-TRIMETHYLBENZENESULFONIC ACID
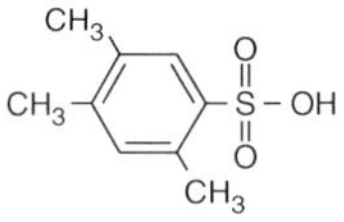
250 mg

S94,644-3
INDOLINE-2-SULFONIC ACID
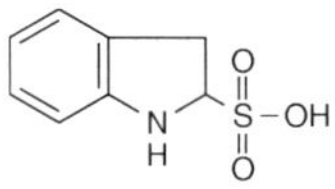
250 mg

S50,870-5
2-(2-AMINO-A,A,A-TRIFLUORO-P-TOLYLOXY)-3,5-DICHLOROBENZENE SULFONIC ACID, SODIUM SALT
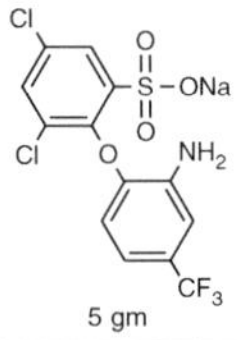
5 gm

S39,867-5
5-CHLORO-2-[4-CHLORO-2-(4-CHLOROTRIFLUORO-M-TOLYLUREIDO]-PHENOXY)BENZENESULFONIC ACID NASALT
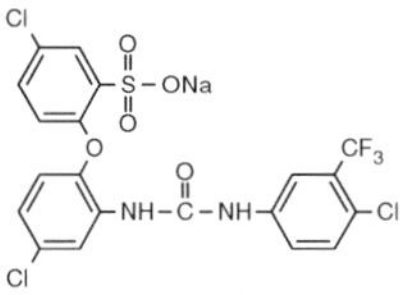
250 mg

S39,896-9
2-(4-CHLORO-2-SULFOPHENYLTHIO)-4,5-' DICHLORO-3-(TRIFLUOROMETHYL)-CARBANILIDE, SODIUM SALT
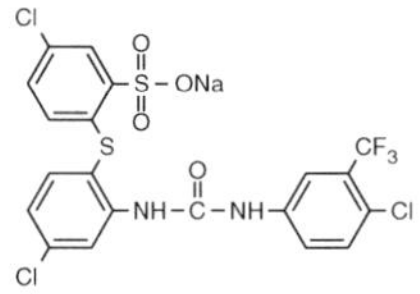
250 mg

S69,462-2
ANILINIUM TRIFLUOROMETHANESULFONATE
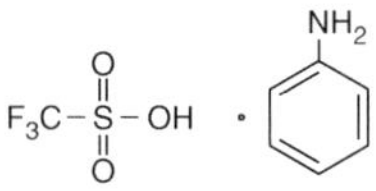
5 gm

S50,824-1
2-AMINO-6-CHLORO-3-(2,4-DICHLOROPHENOXY)BENZENESULFONIC ACID, SODIUM SALT
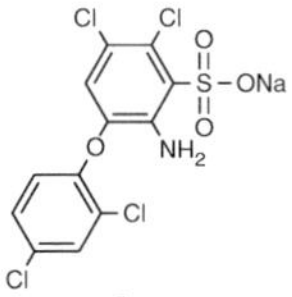
5 gm

S50,803-9
4-(2,4-DICHLOROPHENOXY)METANILIC ACID, SODIUM SALT
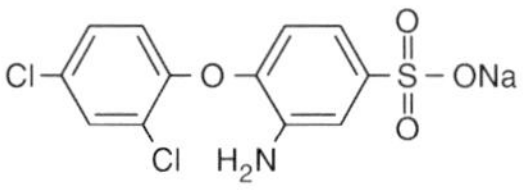
5 gm

S50,806-3
4-(2-AMINO-4-CHLOROPHENOXY)-3-CHLOROBENZENESULFONIC ACID, SODIUM SALT
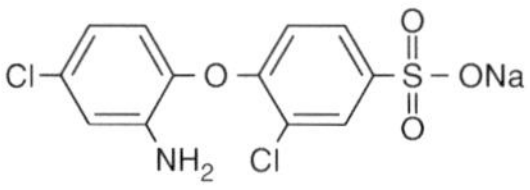
5 gm

S50,860-8
2-(2-AMINO-4-CHLOROPHENOXY)-3,5-DICHLOROBENZENESULFONIC ACID, SODIUM SALT
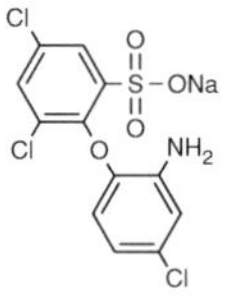
5 gm

S50,862-4
2-(2-AMINO-4-CHLOROPHENOXY)-4,5-DICHLOROBENZENESULFONIC ACID, SODIUM SALT
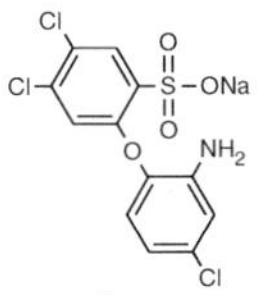
5 gm

S50,866-7
2-AMINO-5-(4-CHLOROPHENOXY)-BENZENESULFONIC ACID, SODIUM SALT
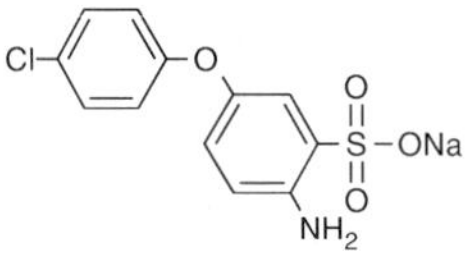
5 gm

S50,793-8
2-AMINO-4-CHLORO-5-(2,4-DICHLOROPHENOXY)BENZENESULFONIC ACID, SODIUM SALT
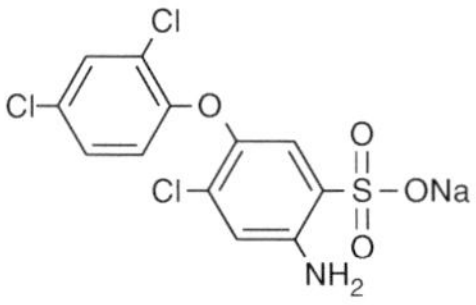
5 gm

S50,805-5
6-(2-CHLOROPHENOXY)METANILIC ACID, SODIUM SALT
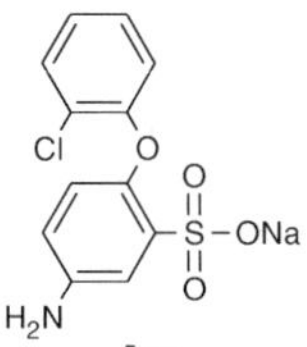
5 gm

S50,820-9
6-(4-CHLOROPHENOXY)METANILIC ACID, SODIUM SALT

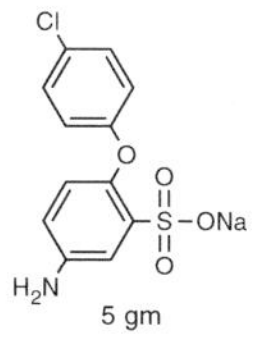

5 gm

S50,794-6
2-(4-AMINOPHENOXY)-3,5-DICHLOROBENZENESULFONIC ACID, SODIUM SALT

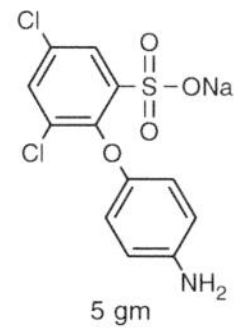

5 gm

S50,816-0
2-(4-AMINO-2-CHLOROPHENOXY)-5-CHLOROBENZENESULFONIC ACID, SODIUM SALT

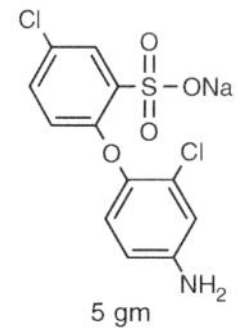

5 gm

S50,859-4
6-(4-CHLORO-M-TOLYLOXY)METANILIC ACID, SODIUM SALT

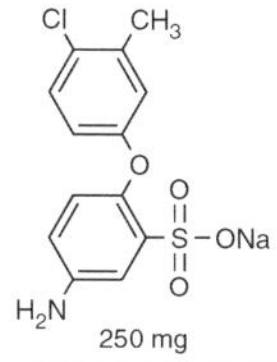

250 mg

S50,826-8
2-(4-AMINO-2-CHLOROPHENOXY)-4,5-DICHLOROBENZENESULFONIC ACID, SODIUM SALT

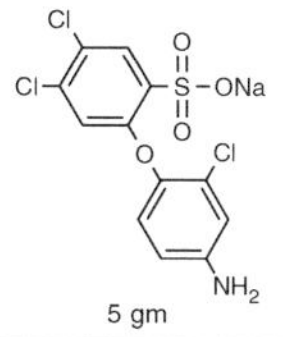

5 gm

S39,875-6
A,A-BIS(3,5-DICHLORO-2-ETHOXYPHENYL)-ORTHO-TOLUENESULFONIC ACID, SODIUM SALT

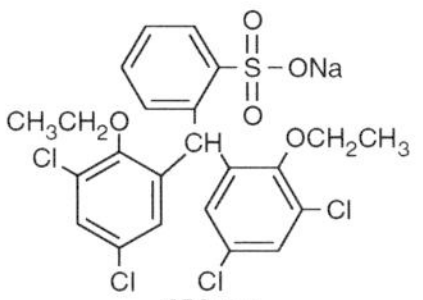

250 mg

S39,877-2
A,A-BIS(3,5-DICHLORO-2-ETHOXYPHENYL)-5-CHLORO-O-TOLUENESULFONIC ACID, SODIUM SALT

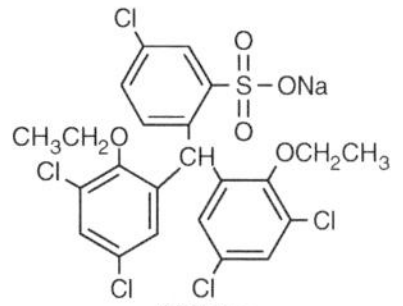

250 mg

S92,893-3
4-CHLORO-3-NITROBENZENESULFONIC ACID, SODIUM SALT

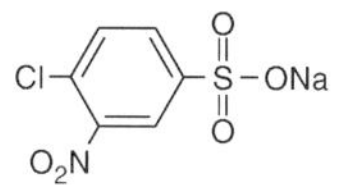

250 mg

S74,930-3
(3,5-DIFLUOROANILINO) METHANESULFONIC ACID, SODIUM SALT

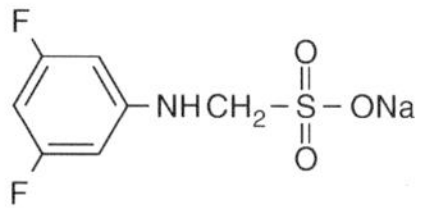

250 mg

S84,108-0
2-CHLORO-3-NITROBENZENESULFONIC ACID, POTASSIUM SALT

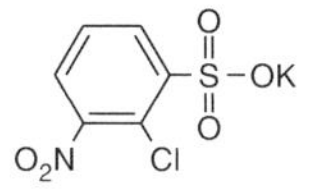

250 mg

S82,888-2
4-BROMO-3-NITROBENZENESULFONIC ACID, THALLIUM SALT

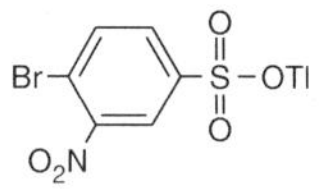

250 mg

S71,452-6
4-BROMOBENZENESULFONIC ACID, SODIUM SALT

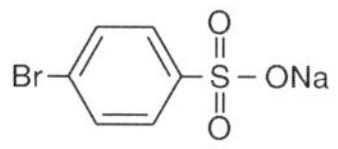

5 gm

S76,489-2
2,4,5-TRICHLOROBENZENESULFONIC ACID, POTASSIUM SALT

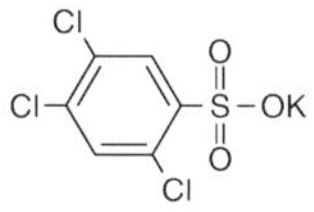

5 gm

S43,036-6
HYDROQUINONESULFONIC ACID, SODIUM SALT

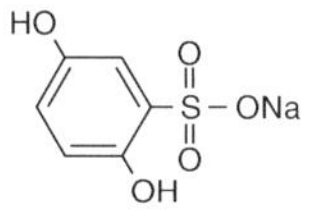

5 gm

S92,113-0
5,6-DIHYDROXY-8-(2-NITROPHENYLTHIO)-2-NAPHTHALENESULFONIC ACID, SODIUM SALT

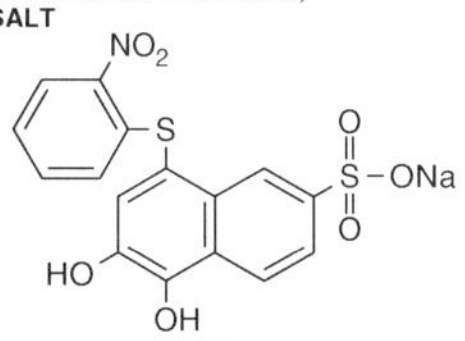

250 mg

S51,037-8
ALPHA,2-DIHYDROXY-ALPHA-TOLUENESULFONIC ACID, SODIUM SALT

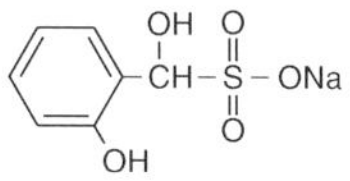

250 mg

S45,591-1
6,7-DIHYDROXYNAPHTHALENE-2-SULFONIC ACID, SODIUM SALT

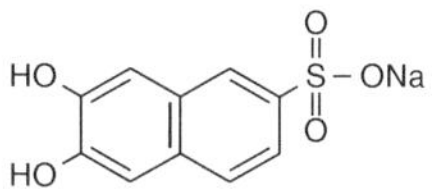

5 gm

S35,093-1
5-BENZOYL-4-HYDROXY-2-METHOXY-BENZENESULFONIC ACID, SODIUM SALT

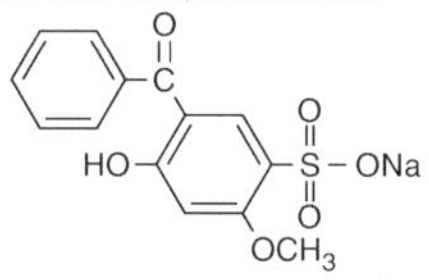

5 gm

S68,051-6
2-METHOXYBENZALDEHYDE SODIUM BISULFITE ADDITION COMPOUND

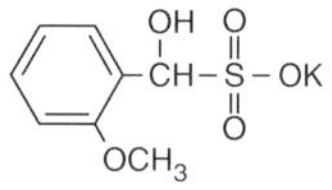

250 mg

S35,801-0
P-ANISALDEHYDE SODIUM BISULFITE ADDITION COMPOUND

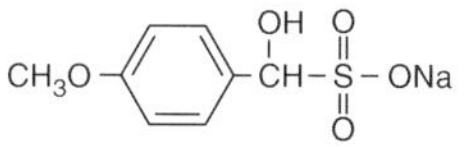

5 gm

S56,748-5
[ALPHA-(4-AMINO-M-TOLYL)-ALPHA-(ANILINOPHENYL)-ALPHA-HYDROXY-P-TOLYL]-SULFANILIC ACID, SODIUM SALT

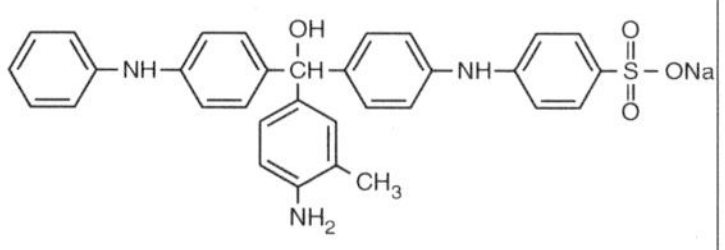

5 gm

S67,898-8
3-AMINO-5-HYDROXY-2,7-NAPHTHALENE-DISULFONIC ACID, MONOSODIUM SALT
80-90%

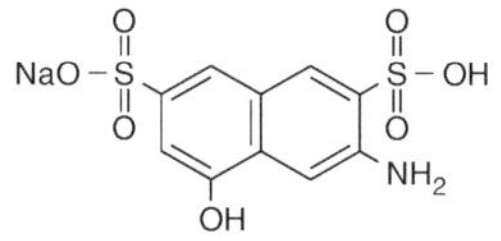

5 gm

S41,776-9
5-(2,4-DINITROANILINO)-2-HYDROXY-BENZENESULFONIC ACID, SODIUM SALT

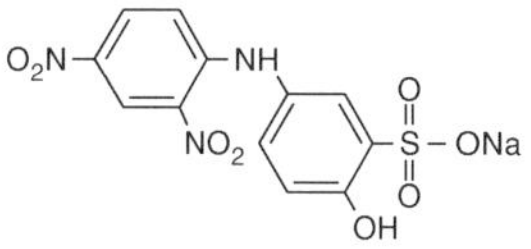

5 gm

S36,422-3
4-HYDROXY-3-NITROSO-2,7-NAPHTHALENE NAPHTHALENEDISULFONIC, DISODIUM SALT

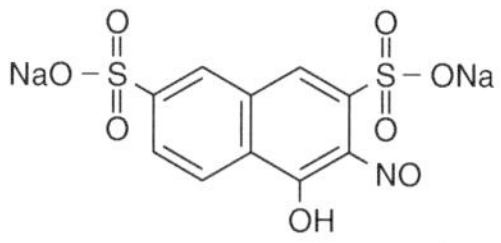

5 gm

S93,984-6
4-HYDROXY-3-NITROSO-1-NAPHTHALENESULFONIC ACID, SODIUM SALT

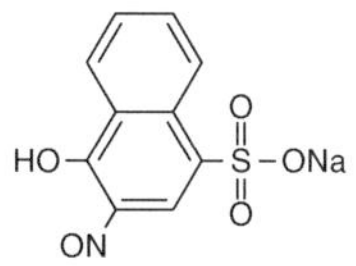

250 mg

S63,685-1
4-HYDROXY-3-NITROBENZENESULFONIC ACID, DISODIUM SALT

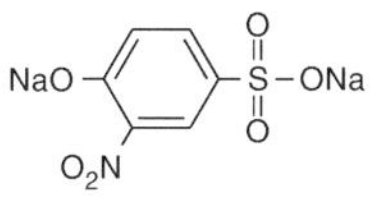

250 mg

S38,896-3
4-HYDROXY-5-NITRO-1,3-BENZENEDISULFONIC ACID, DIPOTASSIUM SALT

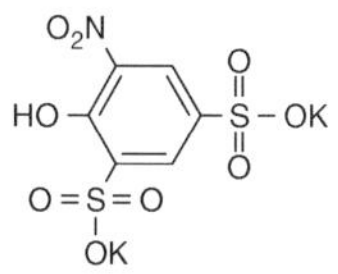

5 gm

S36,831-8
4,5-DIHYDROXY-3-NITROSO-2,7-NAPHTHALENEDISULFONIC ACID, COPPER COMPLEX

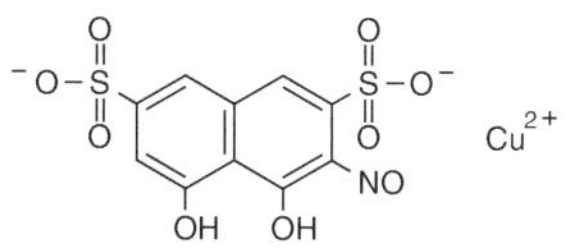

5 gm

S36,416-9
2-HYDROXY-5-NITROBENZENESULFONIC ACID, DISODIUM SALT

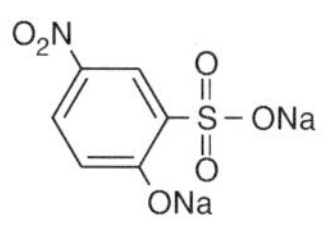

5 gm

S34,021-9
7-ACETAMIDO-4-HYDROXYNAPHTHALENE-2-SULFONIC ACID, SODIUM SALT

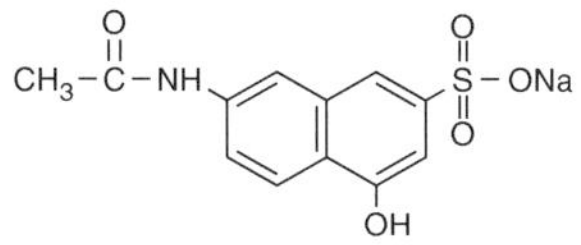

5 gm

S35,705-7
4-ACETAMIDO-5-HYDROXYNAPHTHALENE-2,7-DISULFONIC ACID DISODIUM SALT

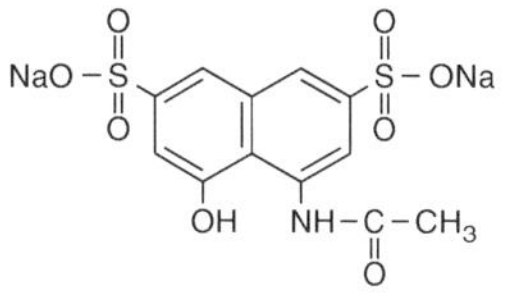

5 gm

S51,218-4
ALPHA-HYDROXY-ALPHA-TOLUENESULFONIC ACID, SODIUM SALT

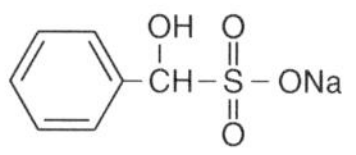

250 mg

S70,652-3
BENZYLACETONE SODIUM BISULFITE ADDUCT

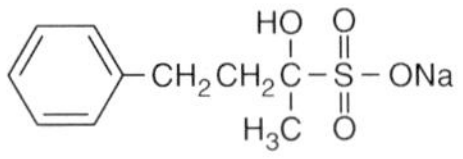

5 gm

S77,593-2
2-HYDROXY-ALPHA-TOLUENESULFONIC ACID, SODIUM SALT

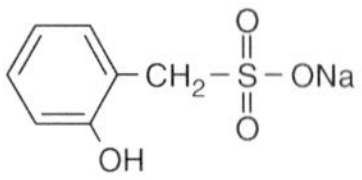

250 mg

S39,872-1
ALPHA,ALPHA-BIS(3,5-DICHLORO-2-HYDROXYPHENYL)-O-TOLUENESULFONIC ACID, SODIUM SALT

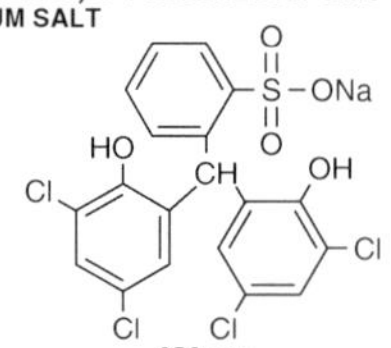

250 mg

S57,321-3
4-HYDROXYBENZENESULFONIC ACID, POTASSIUM SALT

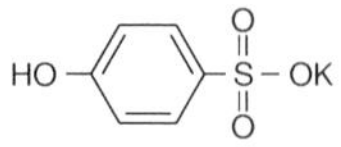

5 gm

S51,033-5
ALPHA,ALPHA'-DIHYDROXY-P-XYLENE-ALPHA,ALPHA'-DISULFONIC ACID, DISODIUM SALT

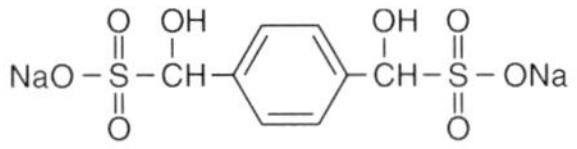

250 mg

S57,993-9
7-HYDROXY-1-NAPHTHALENESULFONIC ACID, SODIUM SALT

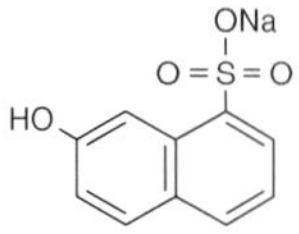

5 gm

S52,203-1
ALPHA-HYDROXY-1-NAPHTHALENEMETHANESULFONIC ACID,SODIUM SALT

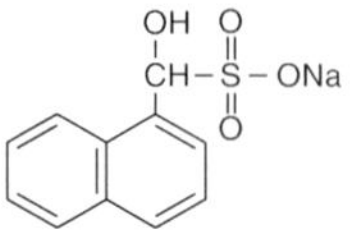

5 gm

S64,875-2
5-HYDROXY-1-NAPHTHALENESULFONIC ACID, SODIUM SALT

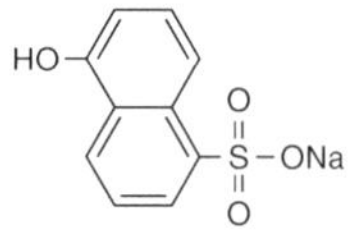

5 gm

S64,877-9
8-HYDROXY-1-NAPHTHALENESULFONIC ACID, SODIUM SALT

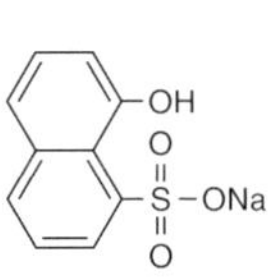

250 mg

S57,592-5
7-HYDROXY-2-NAPHTHALENESULFONIC ACID, SODIUM SALT

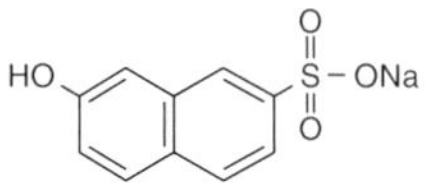

5 gm

S59,298-6
4-[2-(4-ETHOXY-3-METHOXYBENZOYL)-ACETAMIDO]BENZENESULFONIC ACID, BARIUM SALT

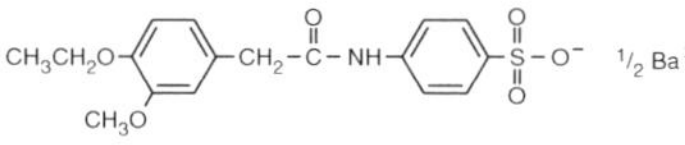

250 mg

S62,853-0
5-AMINO-2-PHENOXYBENZENESULFONIC ACID SODIUM SALT

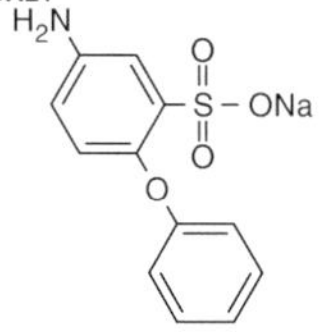

250 mg

S34,112-6
5-AMINO-6-ETHOXY-2-NAPHTHALENE-SULFONIC ACID, SODIUM SALT

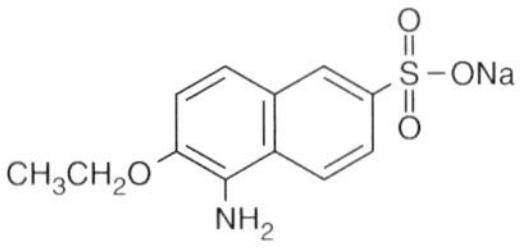

5 gm

S62,732-1
2-METHOXYANILINOMETHANESULFONIC ACID, SODIUM SALT DIHYDRATE

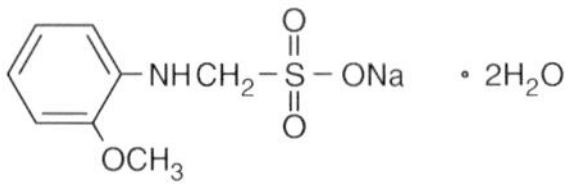

250 mg

S52,691-6
4-ACETAMIDO-1-NAPHTHALENE-SULFONIC ACID, SODIUM SALT

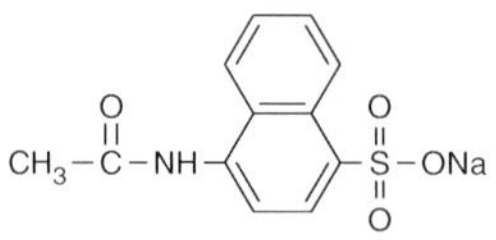

250 mg

S67,183-5
4-[4-(2-HEXADECYLOXYPHENYL)-5-OXO-2-PYRAZOLIN-1-YL]BENZENESULFONIC ACID SODIUM

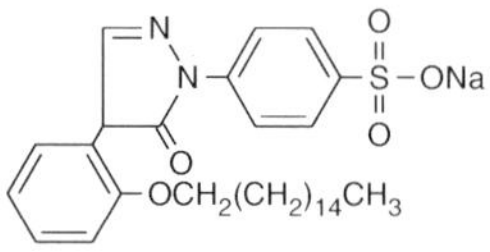

5 gm

S39,863-2
5-CHLORO-2-[4-CHLORO-2-(3,4-DICHLOROPHENYLUREIDO)PHENOXY]-BENZENESULFON.ACID NA

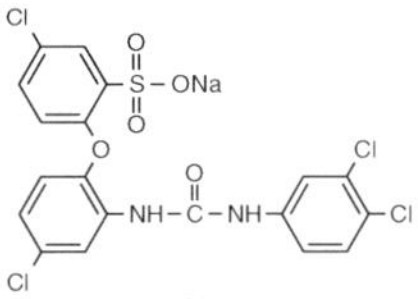

250 mg

S65,541-4
4-(3-METHYL-4-NITROPHENOXY)-1-BUTANESULFONIC ACID, SODIUM SALT

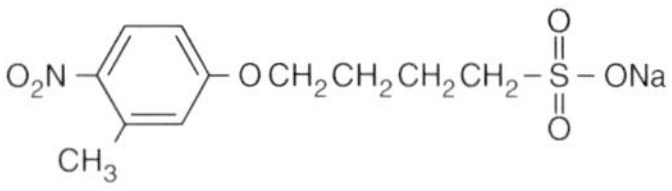

250 mg

S67,182-7
4-(3-METHYL-4-NITROPHENOXY)-1-BUTANESULFONIC ACID, POTASSIUM SALT

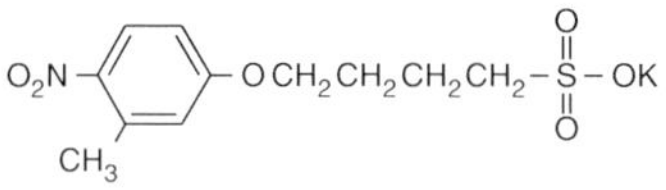

5 gm

S50,141-7
5-NITRO-2-PHENOXYBENZENESULFONIC ACID, SODIUM SALT

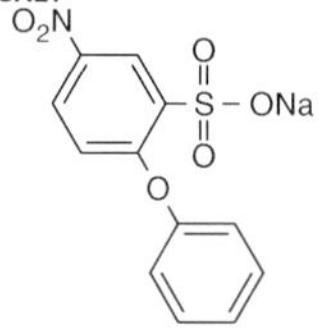

5 gm

S59,295-1
4-[2-(P-ANISOYL)ACETAMIDO]-BENZENESULFONIC ACID, BARIUM SALT

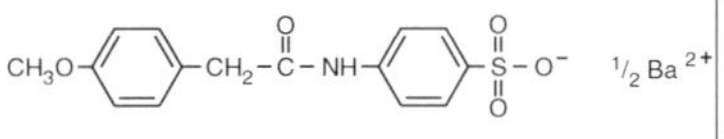

250 mg

S59,789-9
2-(2-FLUOROBENZOYL)-4'-(2-SULFOBENZAMIDO)ACET-O-ANISIDIDE, BARIUM SALT

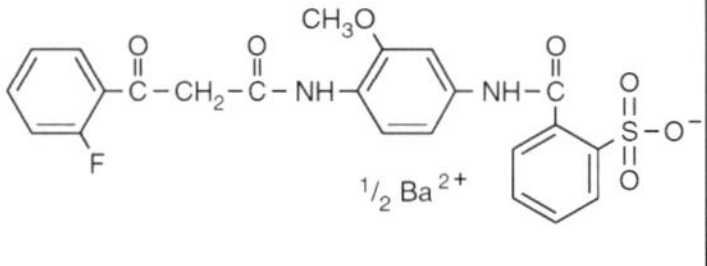

250 mg

S59,255-2
5-ACETYL-2-(HEXADECYLOXY) BENZENESULFONIC ACID, POTASSIUM SALT

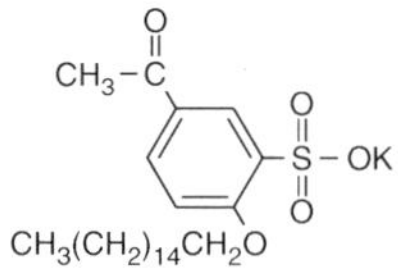

5 gm

S51,580-9
4-(4-SULFOPHENOXY)BENZOIC ACID, MAGNESIUM SALT

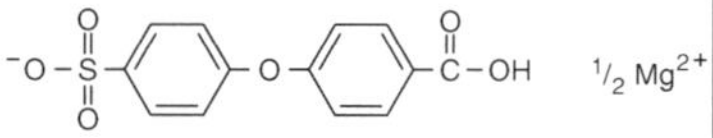

250 mg

S39,876-4
A,A-BIS(3,5-DICHLORO-2-METHOXYPHENYL)-O-TOLUENESULFONIC ACID, NA SALT

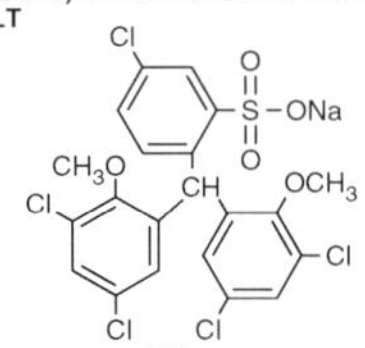

250 mg

S58,843-1
2-[4-(ISO-NONYL)PHENOXY]-1-ETHANESULFONIC ACID, SODIUM SALT, MIXTURE OF ISOMERS

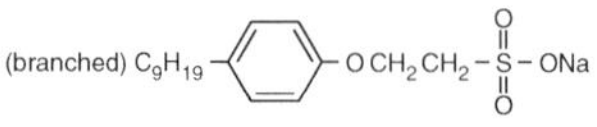

250 mg

S43,385-3
4-METHOXY-1-NAPHTHALENESULFONIC ACID, SODIUM SALT

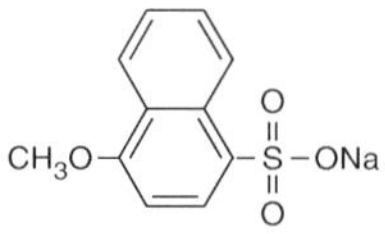

5 gm

S52,785-8
DL-PHENYLALANINE BENZYL ESTER PARA-TOLUENESULFONATE

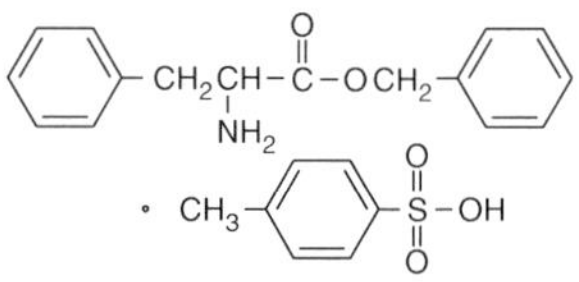

5 gm

S76,558-9
L-PHENYLALANINE BENZYL ESTER P-TOLUENESULFONATE

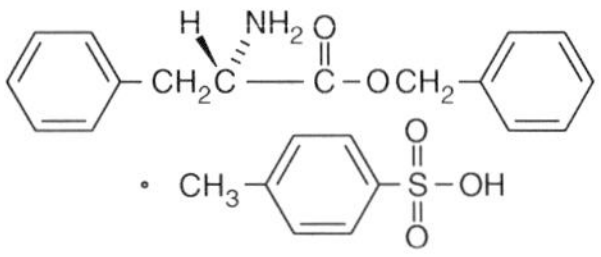

5 gm

S93,314-7
D-ASPARTIC ACID DIBENZYL ESTER P-TOLUENESULFONATE

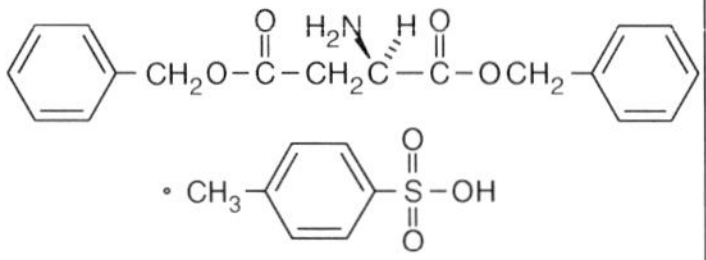

250 mg

S78,925-9
L-2-AMINOBUTYRIC ACID BENZYL ESTER PARA-TOLUENESULFONATE

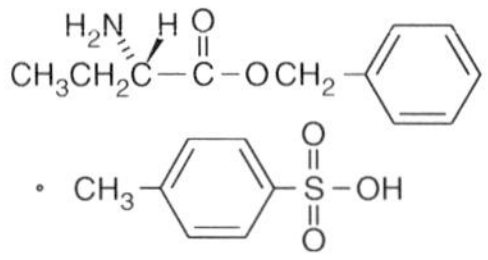

250 mg

S78,335-8
L-ISOLEUCINE BENZYL ESTER P-TOLUENESULFONATE

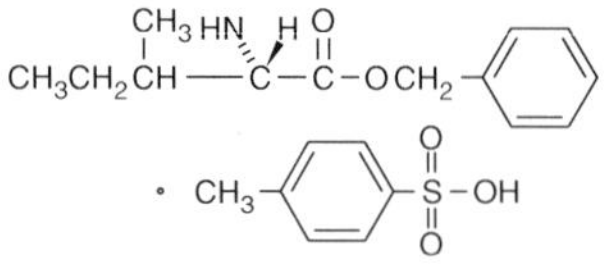

5 gm

S70,415-6
S-BENZYLOXYCARBONYLTHIOSULFONIC ACID SODIUM SALT

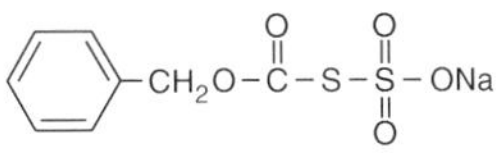

5 gm

S38,765-7
4-(HEXADECYLTHIO)METANILIC ACID, POTASSIUM SALT

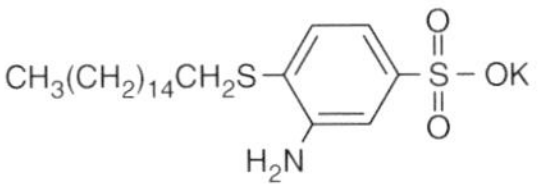

250 mg

S59,787-2
2-HEXADECYLTHIO-5-SULFOBENZOIC ACID, PYRIDINE SALT

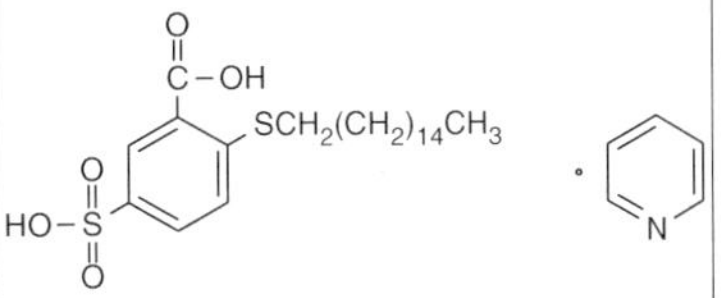

250 mg

S60,560-3
3-(2-NITROPHENYLTHIO)-1-PROPANESULFONIC ACID, SODIUM SALT

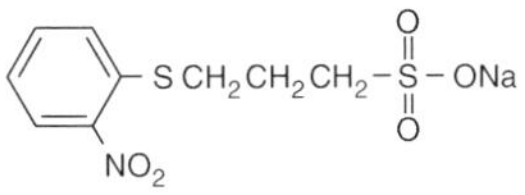

250 mg

S92,114-9
4,4'-DITHIOBIS(BENZENESULFONIC ACID, POTASSIUM SALT)

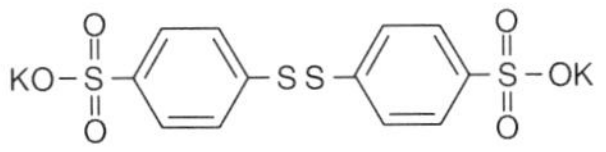

250 mg

S65,674-7
ETHYLENEDIAMINE P-TOLUENESULFONATE

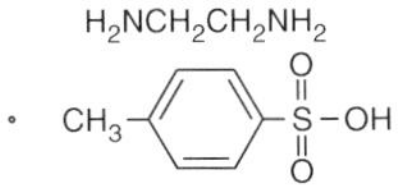

5 gm

S66,012-4
1-AMINO-4-(P-TOLUENESULFONAMIDO)-ANTHRAQUINONE-2-SULFONIC ACID SODIUM SALT

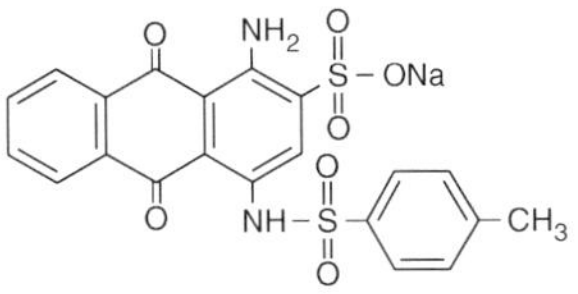

5 gm

S90,929-7
5,6-DIAMINOPHENANTHRIDINIUM P-TOLUENESULFONATE

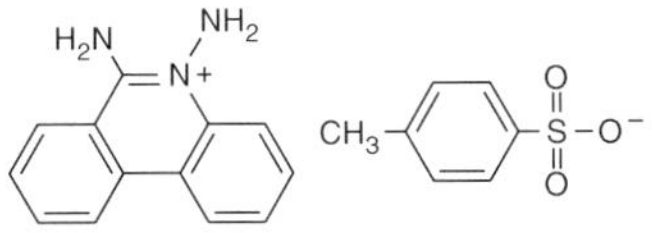

250 mg

S43,687-9
METANILIC ACID, SODIUM SALT

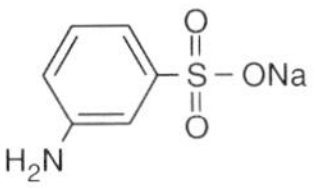

5 gm

S82,613-8
SULFANILIC ACID, THALLIUM(I) SALT

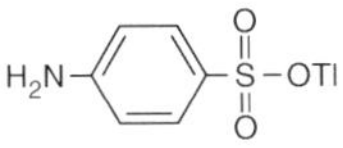

250 mg

S53,410-2
4-AMINO-ALPHA-TOLUENESULFONIC ACID, SODIUM SALT

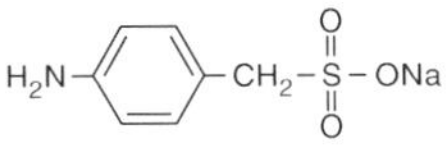

250 mg

S82,838-6
3-AMINO-4-METHYLBENZENESULFONIC ACID, CESIUM SALT

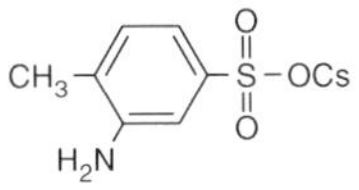

250 mg

S35,747-2
1-AMINO-4-BROMOANTHRAQUINONE-2-SULFONIC ACID, SODIUM SALT

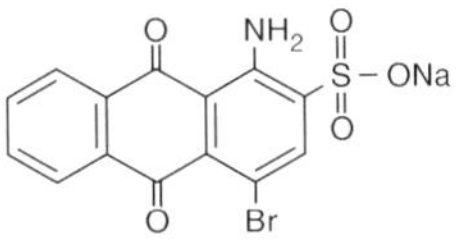

5 gm

S41,702-5
8-AMINO-2-NAPHTHALENESULFONIC ACID, SODIUM SALT, TECH., 65%

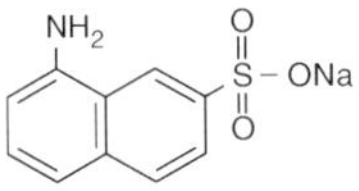

5 gm

S45,991-7
5-AMINO-1-NAPHTHALENESULFONIC ACID, SODIUM SALT

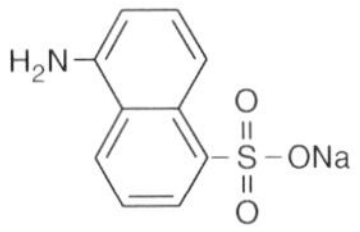

5 gm

S41,910-9
4-AMINO-1,5-NAPHTHALENEDISULFONIC ACID, MONOSODIUM SALT

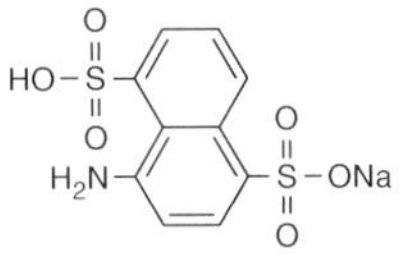

5 gm

S71,457-7
8-AMINO-1,6-NAPHTHALENEDISULFONIC ACID, MONOSODIUM SALT

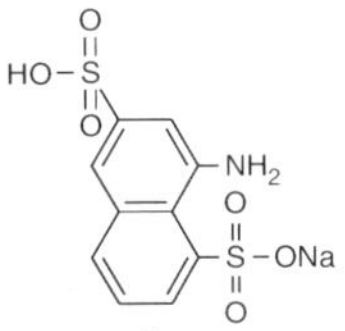

5 gm

S75,486-2
9-AMINO-9-FLUORENESULFONIC ACID, SODIUM SALT

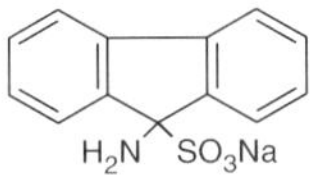

250 mg

S89,668-3
2-AMINO-1,4,5,6-TETRAHYDROPYRIMIDINE P-TOLUENESULFONATE

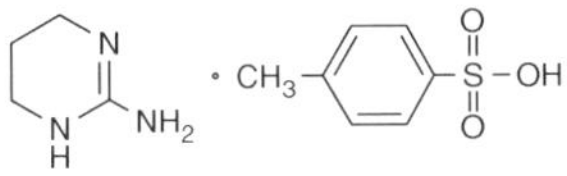

250 mg

S38,715-0
N-ETHYL-3-NITROSULFANILIC ACID, POTASSIUM SALT

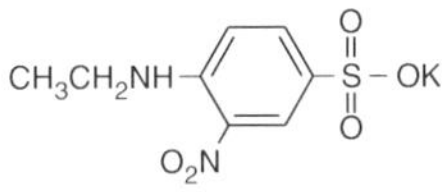

250 mg

S51,677-5
N-DODECYL-3-NITROSULFANILIC ACID, SODIUM SALT

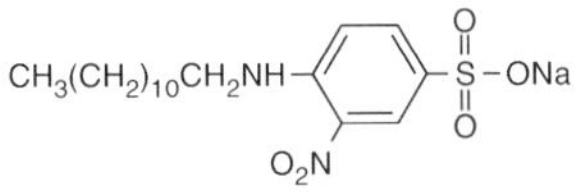

250 mg

S85,006-3
4-ACETAMIDO-3-NITROBENZENESULFONIC ACID, SODIUM SALT

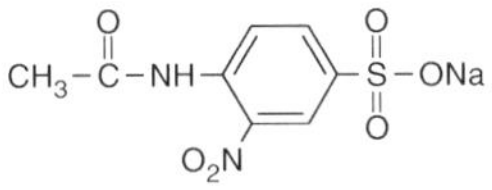

250 mg

S59,249-8
4-(1-BENZYL-5-OXO-2-PYRAZOLIN-3-YLCARBAMOYL)BENZENESULFONIC ACID, BARIUM SALT

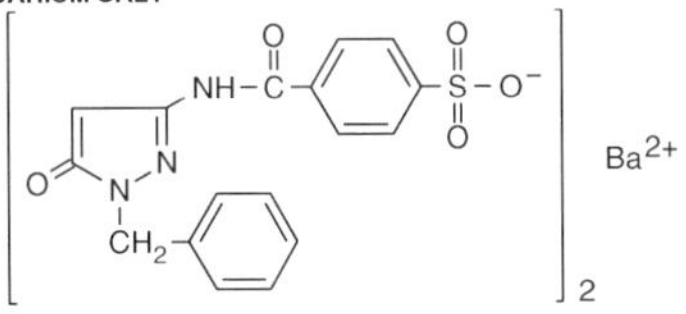

5 gm

S36,737-0
4-CHLORO-3,5-DINITROBENZENESULFONIC ACID, POTASSIUM SALT

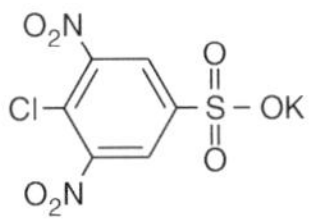

5 gm

S63,330-5
1-PHENYL-3-(3-SULFOBENZAMIDO)-2-PYRAZOLIN-5-ONE, BARIUM SALT

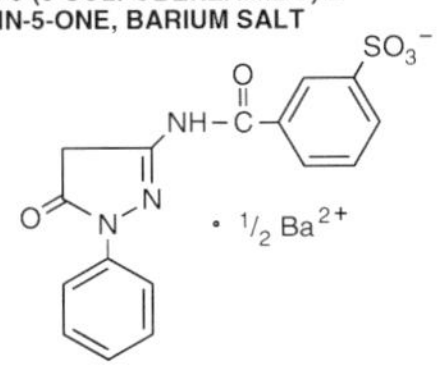

250 mg

S62,302-4
3-(N-DODECYL-3-METHYL-4-NITROANILINO)-1-PROPANESULFONIC ACID, AMMONIUM SALT

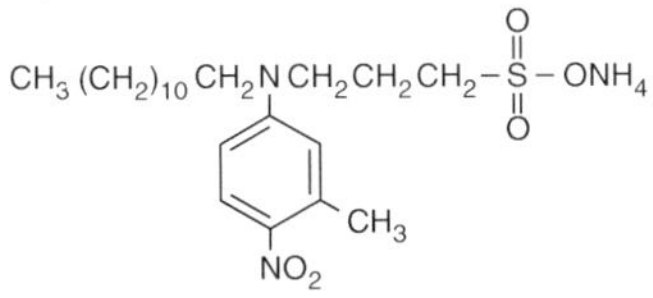

250 mg

S71,455-0
2,5-DINITROBENZENESULFONIC ACID, SODIUM SALT

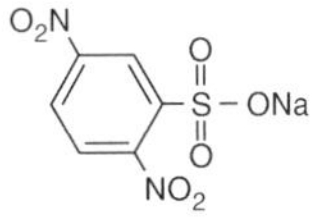

5 gm

S97,203-7
Q255 4-(4-SULFOANILINO)-1,2-NAPHTHOQUINONE, SODIUM SALT

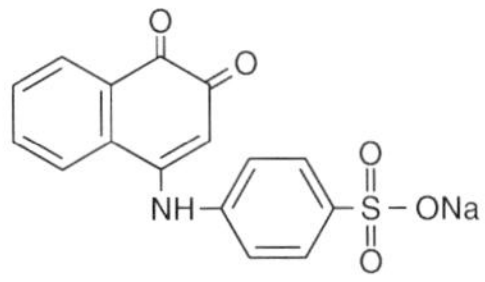

100 mg

S50,559-5
ANILINOMETHANESULFONIC ACID, SODIUM SALT

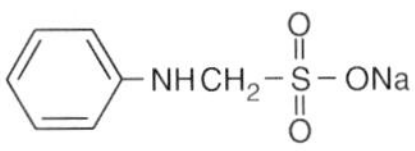

5 gm

S36,125-9
DIPHENYLAMINE-4-SULFONIC ACID, MAGNESIUM SALT

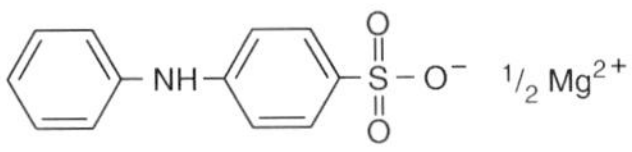

5 gm

S62,730-5
N-(O-TOLYL)AMINOMETHANESULFONIC ACID, SODIUM SALT MONOHYDRATE

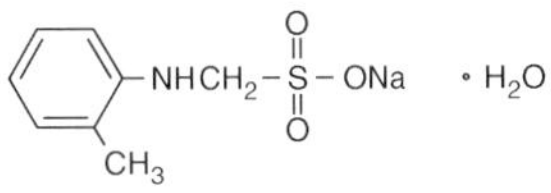

250 mg

S90,890-8
2-(N-ETHYLANILINO)-1-ETHANESULFONIC ACID, POTASSIUM SALT

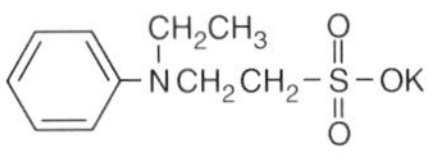

250 mg

S55,069-8
4-[BIS(CYANOETHYL)AMINOBENZYLIDENE]-5-OXO-1-(P-SULFOPHENYL-2-PYRAZOLINE)-3-CARBOXYLIC ACID, SODIUM SALT

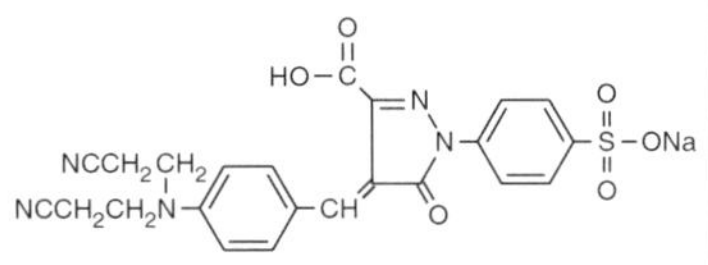

250 mg

S40,087-4
N-ANTIPYRINYL-N-METHYLAMINOMETHANESULFONIC ACID, SODIUM SALT MONOHYDRATE

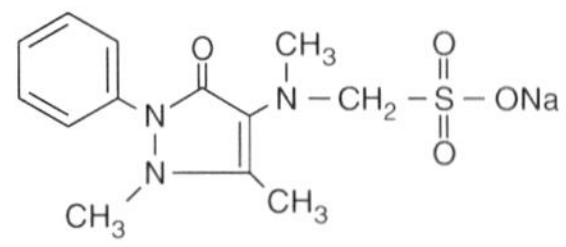

250 mg

S60,526-3
3-(N-DODECYL-M-TOLUIDINO)-1-PROPANESULFONIC ACID, SODIUM SALT

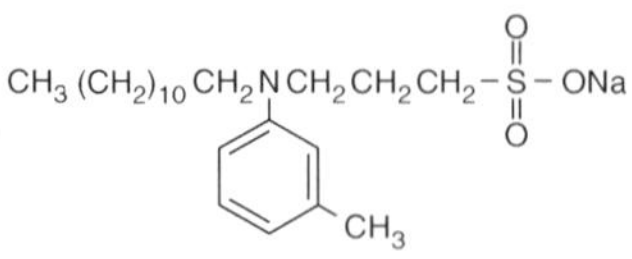

250 mg

S65,533-3
4-(N-DODECYL-3-METHYLANILINO)-1-BUTANESULFONIC ACID, SODIUM SALT

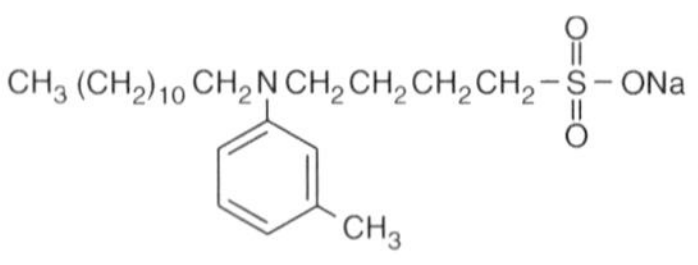

5 gm

S51,231-1
N-(4-CHLOROBENZYLIDENEAMINO)-SULFANILIC ACID, PYRIDINIUM SALT

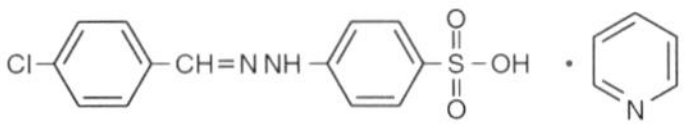

5 gm

S42,064-6
[2-(2,4,6-TRIIODOPHENOXY)ETHYL]-TRIMETHYLAMMONIUM METHANESULFONATE

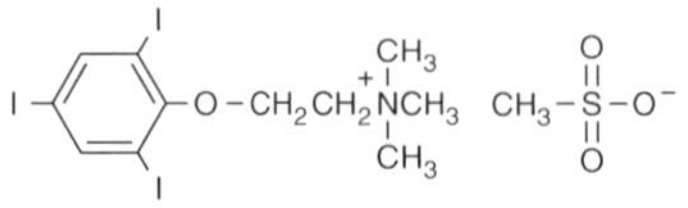

5 gm

S70,538-1
METHYLTRIBUTYLAMMONIUM P-TOLUENESULFONATE

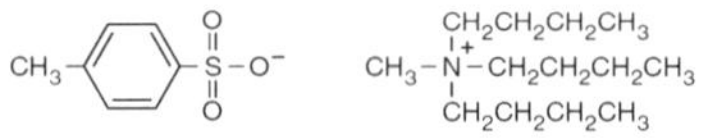

5 gm

S63,867-6
4-(4-HYDRAZINOPHENYL)-BENZENESULFONIC ACID, SODIUM SALT

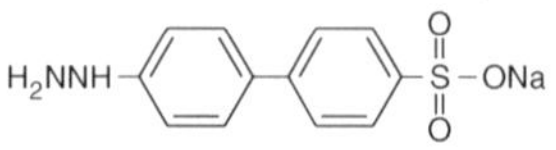

250 mg

S53,426-9
ACETIC ACID (2-SULFOBENZYLIDENE)-HYDRAZIDE, SODIUM SALT

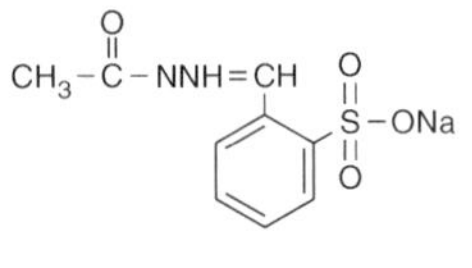

5 gm

S63,868-4
5-OXO-1-(4-SULFOPHENYL)-2-PYRAZOLINE-3-CARBOXYLIC ACID, LEAD SALT

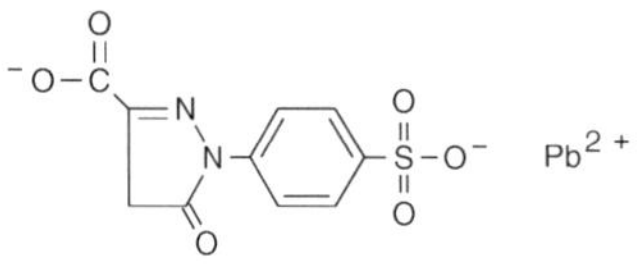

250 mg

S71,454-2
2-ACETAMIDOBENZENESULFONIC ACID, BARIUM SALT

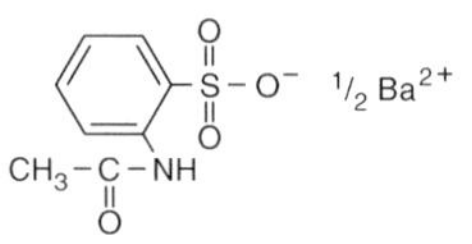

5 gm

S55,852-4
1,3-DIOXO-2-ISOINDOLINEETHANESULFONIC ACID, POTASSIUM SALT

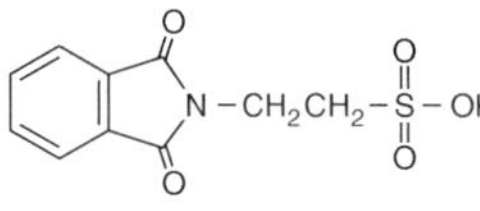

250 mg

S71,453-4
4-NITROBENZENESULFONIC ACID, AMMONIUM SALT

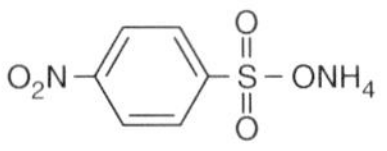

5 gm

S86,430-7
4-NITROBENZENESULFONIC ACID, SODIUM SALT

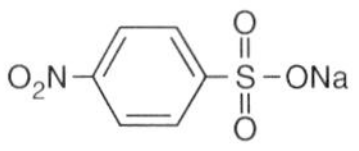

250 mg

S39,044-5
4'-NITRO-4-BIPHENYLSULFONIC ACID, SODIUM SALT

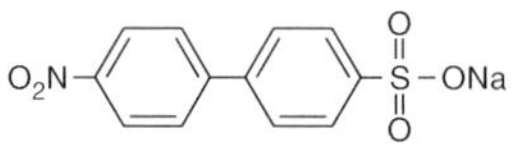

250 mg

S62,271-0
2-(2-NITRO-4-SULFOPHENYL)ACETIC ACID SODIUM SALT

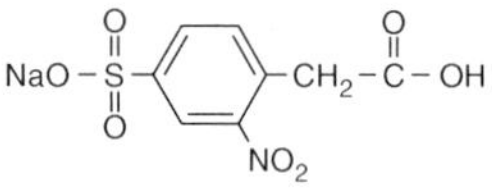

250 mg

S62,278-8
2-BROMO-3-NITRO-ALPHA-TOLUENESULFONIC ACID SODIUM SALT

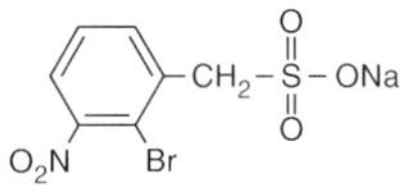

250 mg

S54,098-6
2-CHLORO-5-NITROBENZENESULFONIC ACID, SODIUM SALT

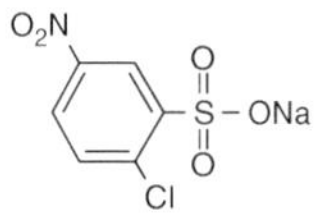

5 gm

S34,752-3
4,4'-DINITROSTILBENE-2,2'-DISULFONIC ACID, DISODIUM SALT

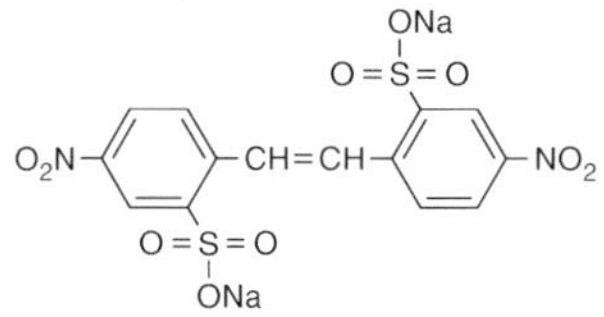

5 gm

S51,930-8
4-CHLORO-5-NITRO-META-TOLUENESULFONIC ACID, SODIUM SALT

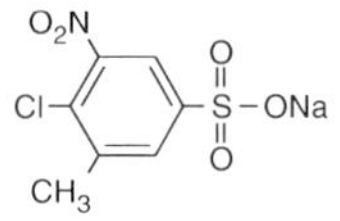

250 mg

S55,478-2
5-NITRO-2,3-XYLENESULFONIC ACID, SODIUM SALT

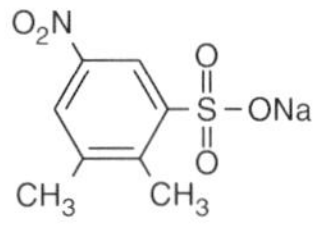

250 mg

S55,479-0
5-NITRO-2,4-XYLENESULFONIC ACID, SODIUM SALT

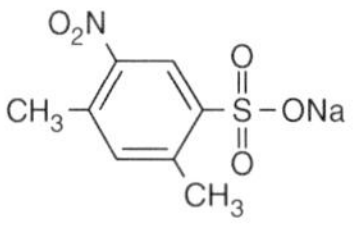

5 gm

S58,885-7
6-NITRO-1-NAPHTHALENESULFONIC ACID, SODIUM SALT

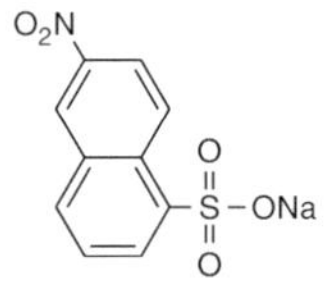

5 gm

S81,254-4
1,3-DIPHENYL-3-OXO-1-PROPANESULFONIC ACID, SODIUM SALT

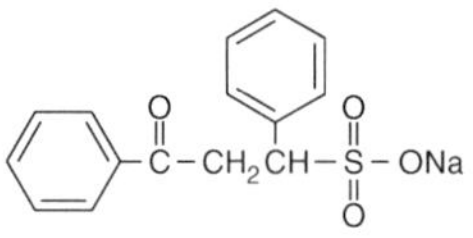

250 mg

S82,901-3
ANTHRAQUINONE-2-SULFONIC ACID, THALLIUM SALT

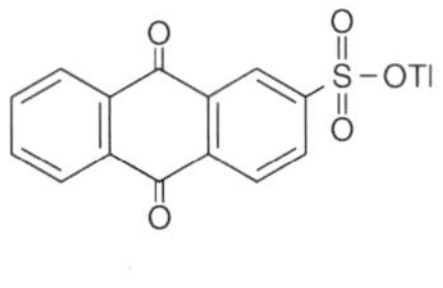

250 mg

S59,528-4
N-(4-SULFOPHENYL)ACETOACETAMIDE, POTASSIUM SALT

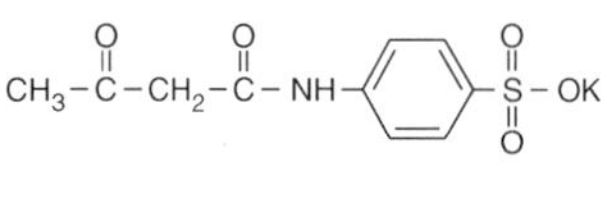

5 gm

S79,912-2
3-OXO-1-PHENYL-1-BUTANESULFONIC ACID SODIUM SALT

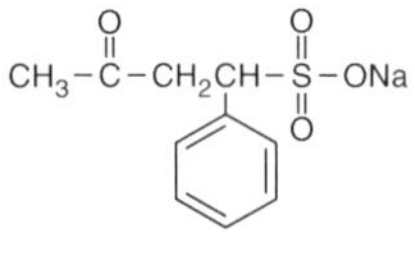

250 mg

S46,166-0
4-SULFOPHTHALIC ACID, TRISODIUM SALT DIHYDRATE

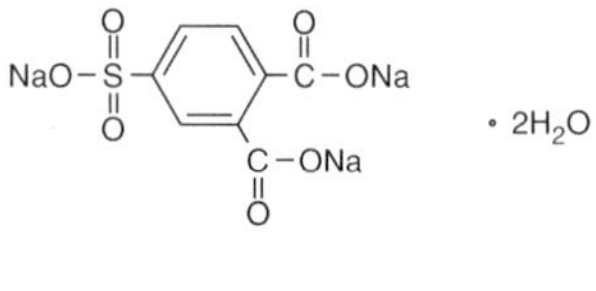

5 gm

S59,241-2
2-HEXADECYLSULFONYL-5-SULFOBENZOIC ACID, LEAD SALT

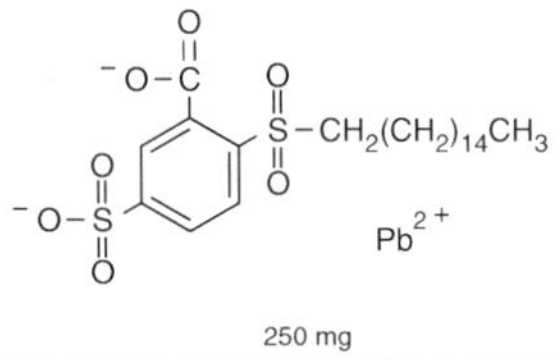

250 mg

S81,967-0
3-SULFOBENZOIC ACID, DISODIUM SALT

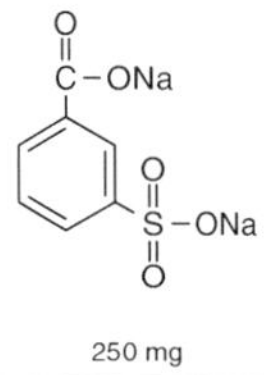

250 mg

S58,852-0
3-SULFO-2-NAPHTHOIC ACID-BIS (HEXACHLOROCYCLOPENTADIENE) ADDUCT, DISODIUM SALT

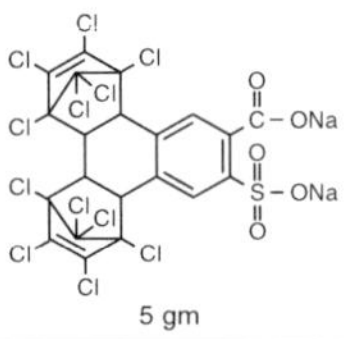

5 gm

S83,735-0
2-CARBOXYBENZENESULFONIC ACID, AMMONIUM SALT

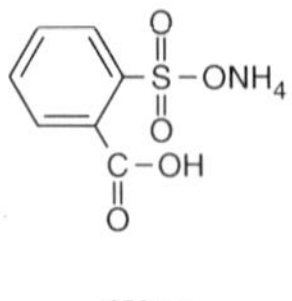

250 mg

S65,296-2
4-CARBOXYBENZENESULFONIC ACID, SODIUM SALT

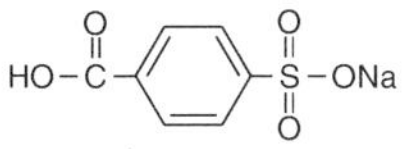

250 mg

S39,349-5
ETHYL 2-SULFOBENZOATE, SODIUM SALT

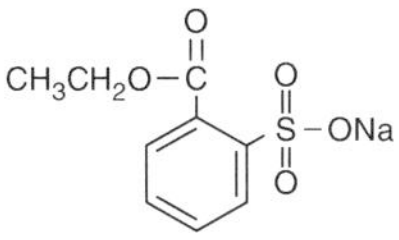

250 mg

S58,845-8
ISO-OCTADECYL 2-SULFOBENZOATE, SODIUM SALT, MIXTURE OF BRANCHED ISOMERS

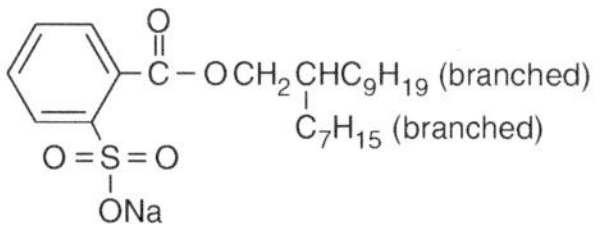

250 mg

S62,305-9
4-(4,4,4-TRIMETHYLACETOACETAMIDO) BENZENESULFONIC ACID, BARIUM SALT

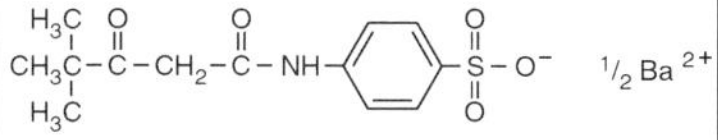

250 mg

S56,887-2
ALPHA-SULFOPHENYLACETIC ACID, PYRIDINE SALT

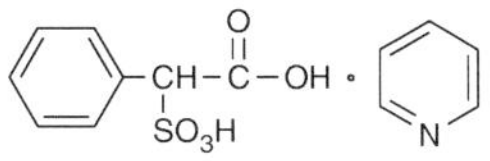

5 gm

S80,731-1
ALPHA-SULFOPHENYLACETIC ACID, SODIUM SALT

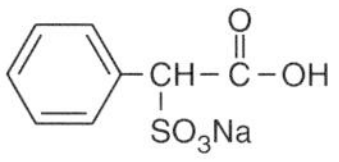

250 mg

S86,326-2
ALPHA-(ETHOXYCARBONYL)-ALPHA-TOLUENESULFONIC ACID, BARIUM SALT

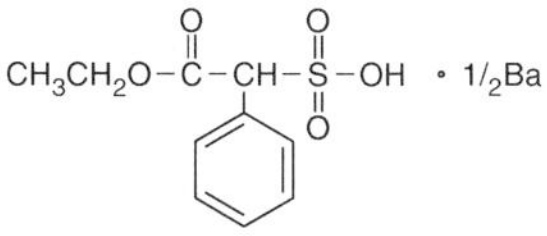

250 mg

S82,794-0
3-ACETAMIDOBENZENESULFONIC ACID, SODIUM SALT DIHYDRATE

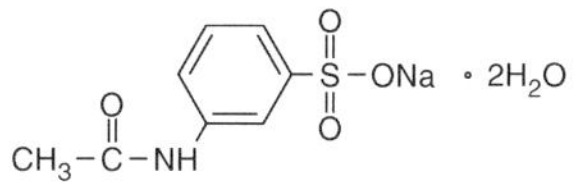

250 mg

S60,570-0
3-ACETAMIDO-4-CHLOROBENZENESULFONIC ACID, POTASSIUM SALT

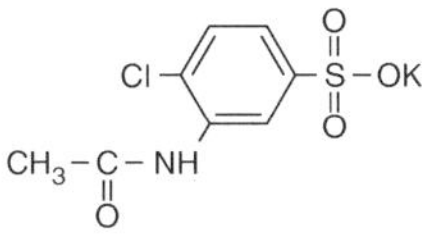

250 mg

S59,264-1
4-(5-OXO-3-PENTADECYL-2-PYRAZOLIN-1-YL)BENZENESULFONIC ACID, SODIUM SALT

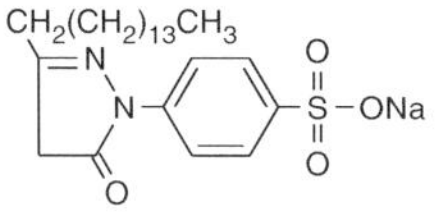

5 gm

S86,229-0
1-AMIDINOUREA BENZENESULFONATE

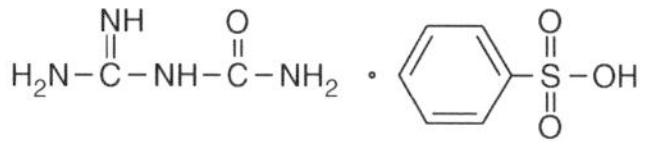

250 mg

S86,251-7
AMIDINOUREA P-TOLUENESULFONATE

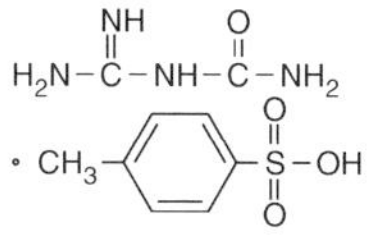

250 mg

S63,284-8
3-(3-BUTYLUREIDO)BENZENESULFONIC ACID SODIUM SALT

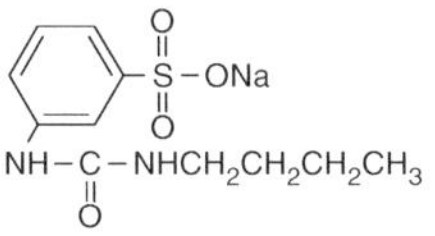

250 mg

S86,256-8
1-AMIDINO-3-METHYL-2-THIOUREA P-TOLUENESULFONATE

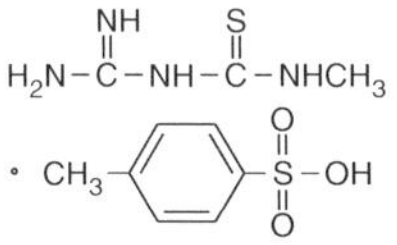

250 mg

S86,230-4
1-AMIDINO-3-PROPYL-2-THIOUREA P-TOLUENESULFONATE

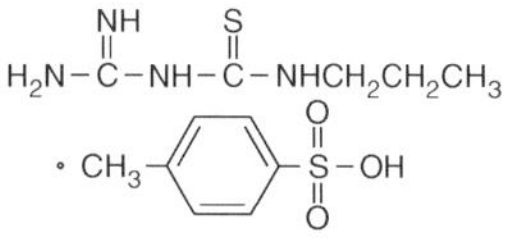

250 mg

S56,890-2
N-PHENYLCARBAMOYLSULFONIC ACID, POTASSIUM SALT

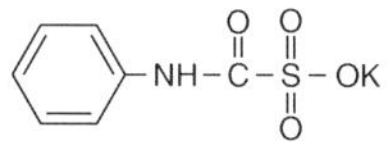

250 mg

S82,266-3
2-BUTYL-2-THIOPSEUDOUREA P-TOLUENESULFONATE

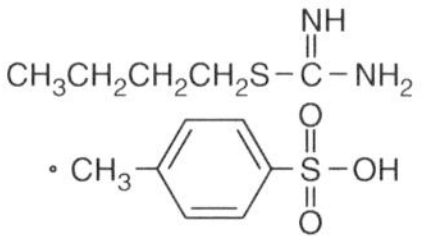

250 mg

S59,224-2
2-CYANO-ALPHA-TOLUENESULFONIC ACID, SODIUM SALT

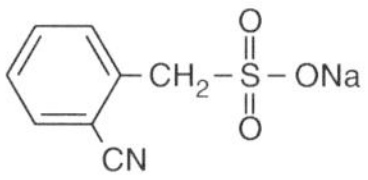

5 gm

S50,780-6
4-(HEXADECYLSULFONYL)-BENZENESULFONIC ACID, SODIUM SALT

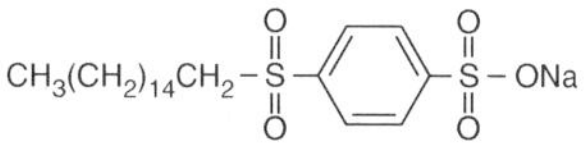

250 mg

S80,754-0
2'-(N-METHYL-N-PHENYLSULFAMOYL)-3,3',5,5'TETRAMETHYL-2-BIBENZYLSULFONIC ACID SODIUM SALT

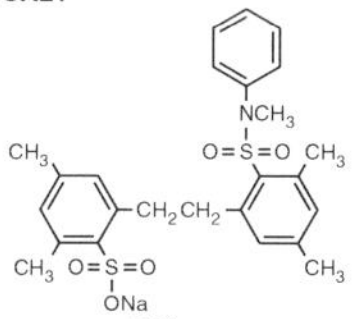

250 mg

S58,851-2
3-METHYL-2-NAPHTHALENESULFONIC ACID BIS(HEXACHLOROCYCLOPENTADIENE) ADDUCT, CALCIUM SALT

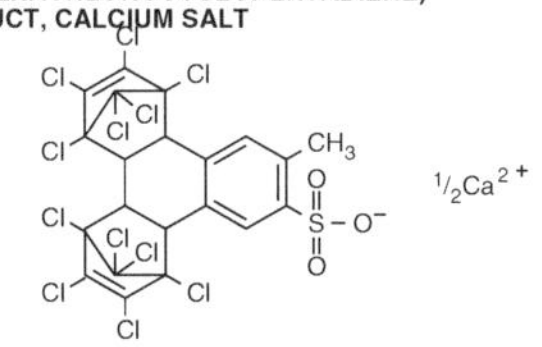

5 gm

S82,833-5
BENZENESULFONIC ACID, THALLIUM SALT

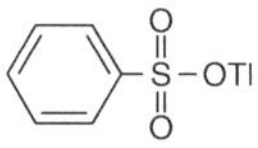

250 mg

S79,982-3
ALPHA-TOLUENESULFONIC ACID, TRIETHYLAMINE SALT

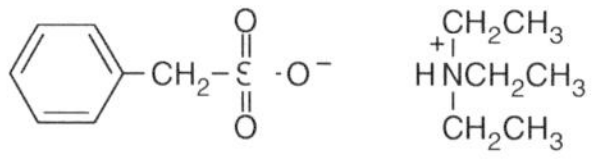

250 mg

S80,896-2
ALPHA-TOLUENESULFONIC ACID, ETHYLAMINE SALT

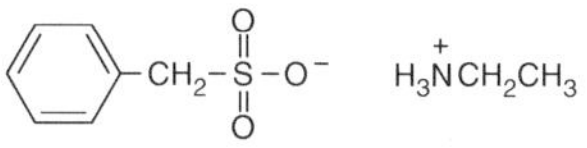

250 mg

S57,021-4
BETA-STYRENESULFONIC ACID, AMMONIUM SALT

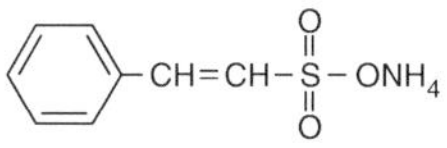

250 mg

S79,955-6
4-PHENYL-1-BUTANESULFONIC ACID SODIUM SALT

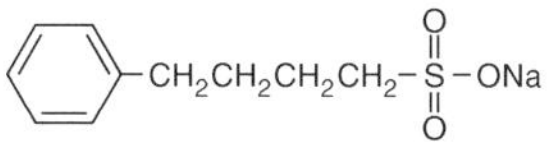

250 mg

S79,958-0
6-PHENYL-1-HEXANESULFONIC ACID, SODIUM SALT

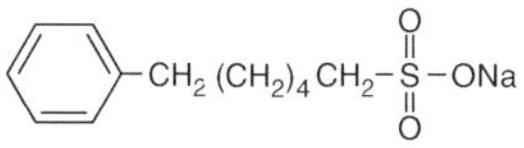

250 mg

S59,793-7
1,2-BENZENEDISULFONIC ACID, DISODIUM SALT

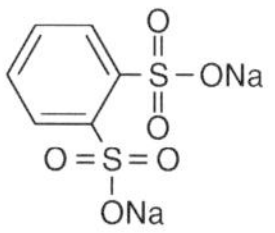

5 gm

S86,250-9
GUANIDINE P-TOLUENESULFONATE

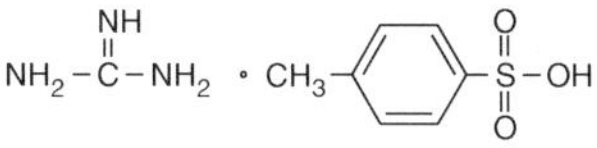

250 mg

S41,551-0
4-CHLOROBENZENESULFONIC ACID, SODIUM SALT

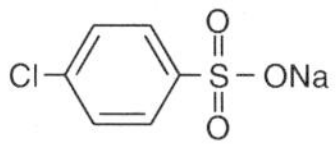

250 mg

S82,615-4
P-TOLUENESULFONIC ACID, THALLIUM (I)SALT

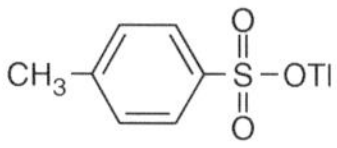

250 mg

S92,289-7
LITHIUM P-TOLUENESULFONATE

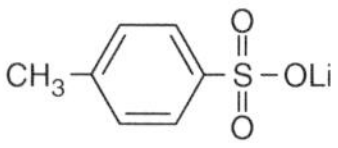

250 mg

S74,270-8
4-METHYL-1,2-BENZENEDISULFONIC ACID, DISODIUM SALT

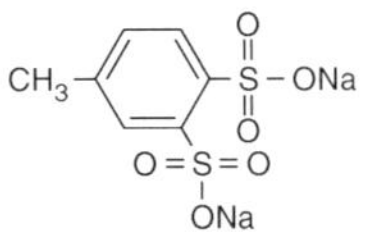

5 gm

S60,541-7
3,5-DICHLORO-P-TOLUENESULFONIC ACID, SODIUM SALT

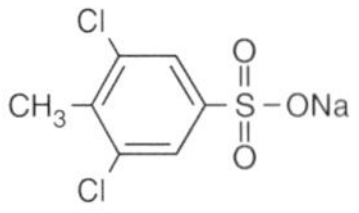

250 mg

S52,758-0
2,4,5-TRICHLOROBENZENESULFONIC ACID, SODIUM SALT

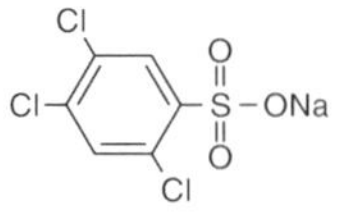

5 gm

S41,279-1
5-INDANSULFONIC ACID, AMMONIUM SALT

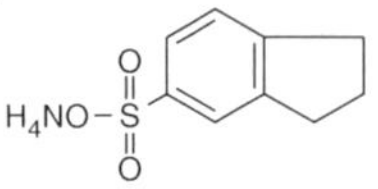

5 gm

S59,256-0
2-BROMO-1-NAPHTHALENESULFONIC ACID

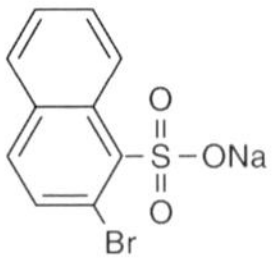

250 mg

S58,883-0
1,6-NAPHTHALENEDISULFONIC ACID, DISODIUM SALT DIHYDRATE

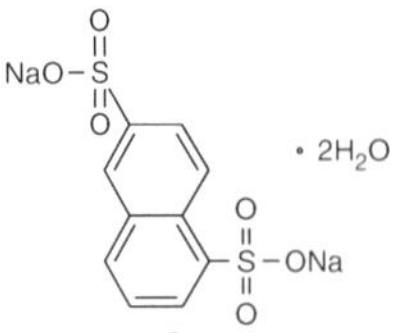

5 gm

S36,400-2
1,3,6-NAPHTHALENETRISULFONIC ACID, SODIUM SALT

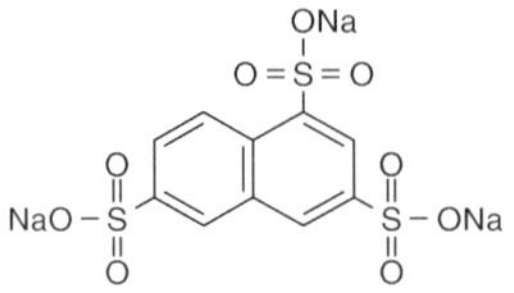

5 gm

S77,584-3
2-ANTHRACENESULFONIC ACID, SODIUM SALT

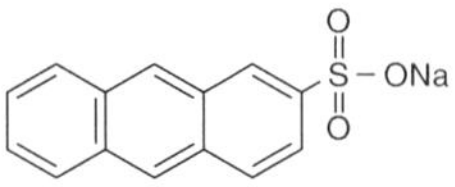

250 mg

S77,585-1
1-ANTHRACENESULFONIC ACID, SODIUM SALT

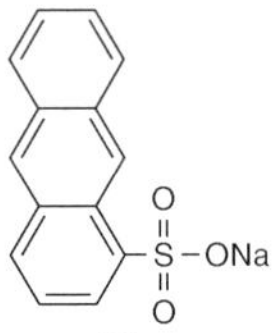

250 mg

S75,702-0
3-PHENANTHRENESULFONIC ACID, POTASSIUM SALT

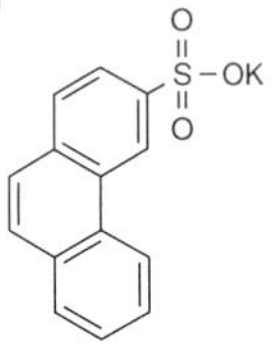

5 gm

S72,251-0
5,6,7,8-TETRAHYDRO-2-PHENANTHRENESULFONIC ACID SODIUM SALT

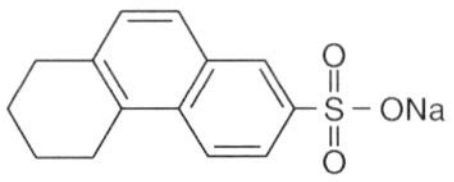

5 gm

S40,620-1
ACENAPHTHENE-5-SULFONIC ACID, POTASSIUM SALT

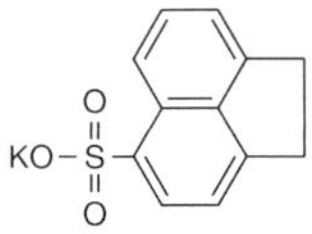

5 gm

S83,272-3
2-SULFOTHIANTHRENE-5,5,10,10-TETRAOXIDE, SODIUM SALT

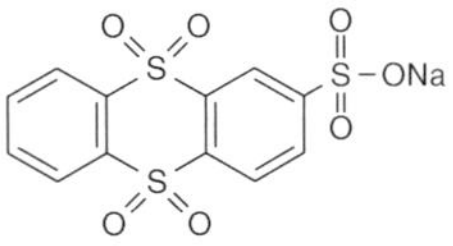

250 mg

S40,911-1
MENADIOL DISULFATE DISODIUM SALT TRIHYDRATE

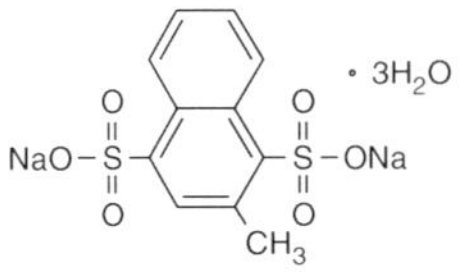

5 gm

S51,395-4
2-CHLOROETHYL 2,4-DINITROBENZENESULFONATE

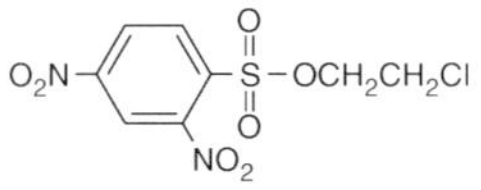

250 mg

S50,130-1
2-CHLOROETHYL N-ACETYLSULFANILATE

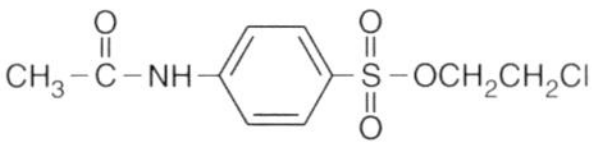

250 mg

S83,278-2
2-BROMOETHYL BENZENESULFONATE

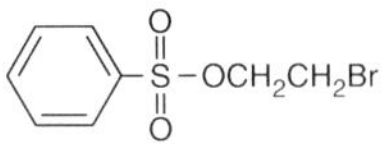

250 mg

S83,279-0
2-BROMOETHYL P-TOLUENESULFONATE

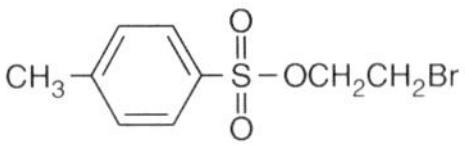

250 mg

S82,015-6
3-CHLOROPROPYL P-TOLUENESULFONATE

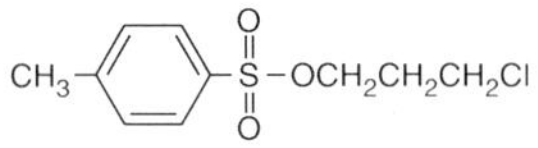

250 mg

S51,533-7
2,2,3,3-TETRAFLUOROPROPYL BENZENESULFONATE

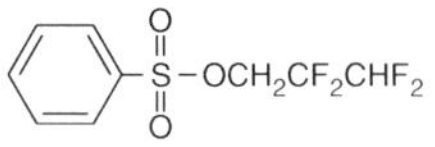

250 mg

S39,157-3
1-METHYL-2,2,3,3-TETRAFLUOROPROPYL BENZENESULFONATE

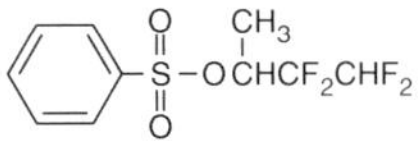

250 mg

S60,505-0
2,2,3,3,4,4,5,5,6,6,7,7,8,8,9,9-HEXADECYLFLUORONONYL BENZENESULFONATE

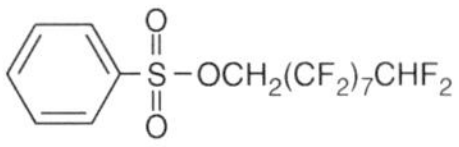

5 gm

S38,685-5
2,2,3,3,4,4,5,5,6,6,7,7,8,8,9,9-HEXADECAFLUORONONYL PARA-TOLUENESULFONATE

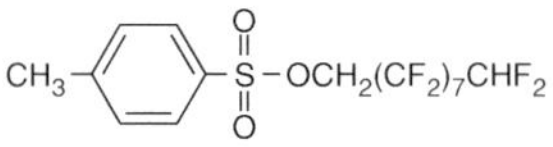

250 mg

S90,808-8
2,2,2-TRIFLUOROETHYL 4-AMINO-3,5-DIBROMOBENZENESULFONATE

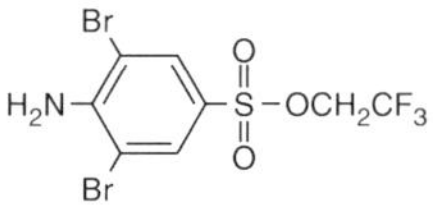

250 mg

S90,867-3
1,1,1,3,3,3-HEXAFLUOROISOPROPYL 4-AMINO-3,5-DIBROMOBENZENESULFONATE

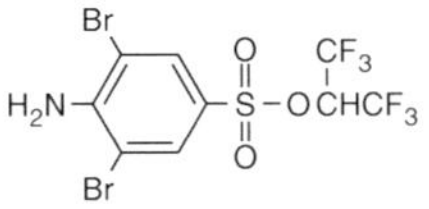

250 mg

S60,506-9
ALPHA-(TRIFLUOROMETHYL)BENZYL BENZENESULFONATE

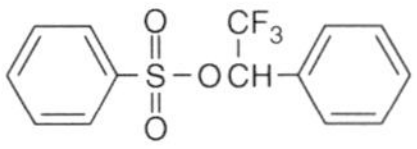

250 mg

S39,166-2
2,2,2-TRICHLOROETHYL BENZENESULFONATE

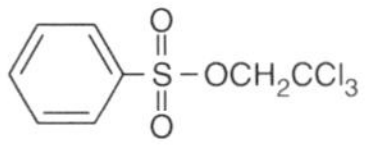

250 mg

S80,899-7
2,2,2-TRIFLUOROETHYL ALPHA-TOLUENESULFONATE

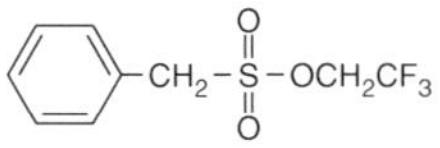

250 mg

S60,508-5
1,1,1,3,3,3-HEXAFLUOROISOPROPYL BENZENESULFONATE

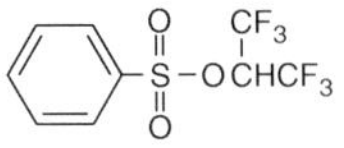

250 mg

S67,501-6
4'-BROMOPHENACYL P-TOLUENESULFONATE

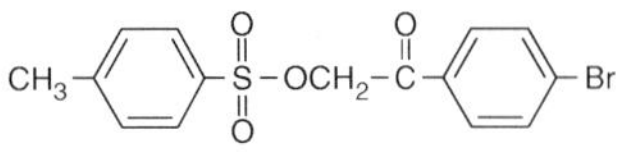

5 gm

S86,519-2
BICYCLO[2.2.2]OCTANE-2,3-DIOL BIS(4-BROMOBENZENESULFONATE)

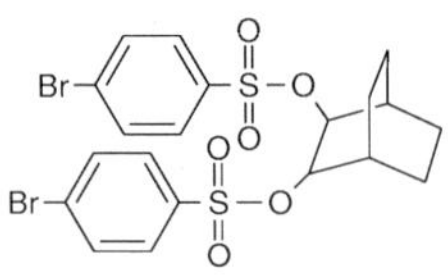

250 mg

S87,573-2
9,10-DIHYDRO-9,10-PROPANOANTHRACEN-11-YL 4-BROMOBENZENESULFONATE

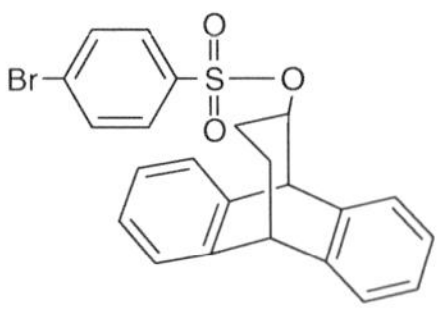

250 mg

S91,487-8
1,2,3,4-TETRAHYDRO-1-NAPHTHYL 4-BROMOBENZENESULFONATE

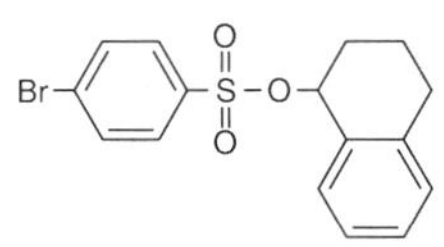

250 mg

S39,401-7
2,4,6-TRICHLOROPHENYL 3,4-DICHLOROBENZENESULFONATE

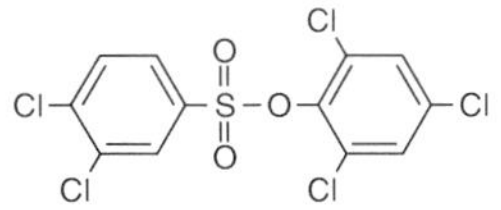

250 mg

S67,197-5
2,4,6-TRICHLOROPHENYL METHANESULFONATE

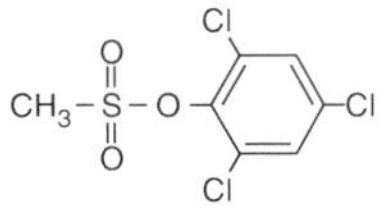

5 gm

S93,932-3
2-O-(P-TOSYL)-D-ARABINOSE

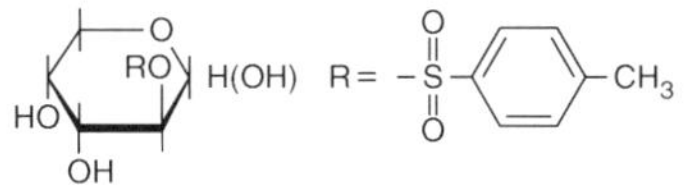

250 mg

S53,295-9
METHYL 4,6-O-BENZYLIDENEGLUCOSIDE, 2-PARA-TOLUENESULFONATE

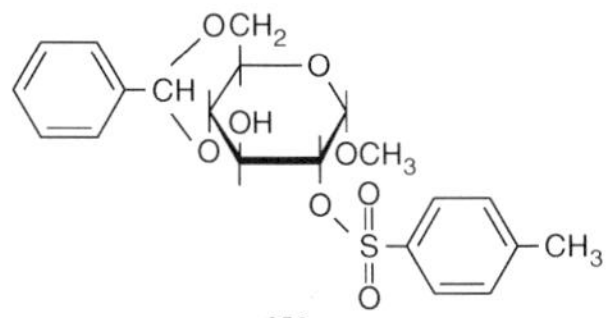

250 mg

S52,533-2
3-CYCLOHEXYL-3-HYDROXY-3-PHENYLPROPYL PARA-TOLUENESULFONATE

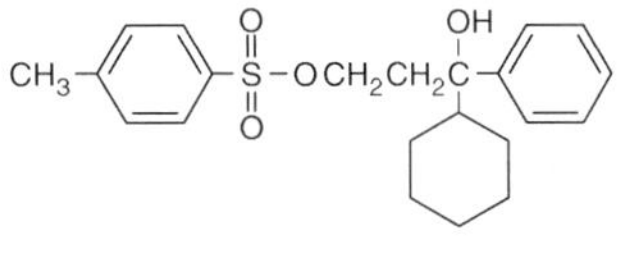

250 mg

S80,670-6
6-HYDROXYHEXYL P-TOLUENESULFONATE

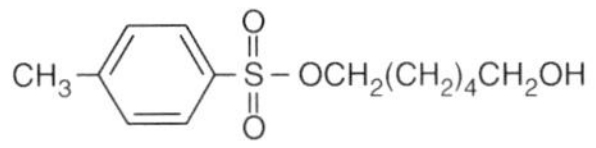

250 mg

S50,039-9
2-PHENOXYETHYL METANILATE

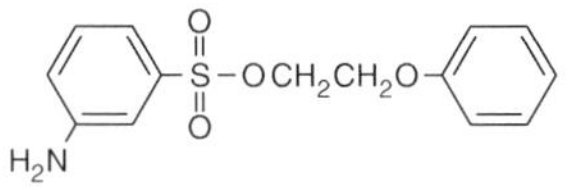

250 mg

S62,792-5
2-PHENOXYETHYL 5-ACETAMIDO-2-METHOXY-3-NITROBENZENESULFONATE

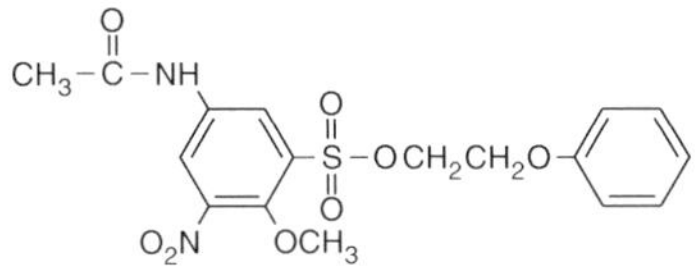

250 mg

S63,019-5
2-PHENOXYETHYL 4-(4-[2-(4-ISOOCTYLPHENOXY)ETHOXY]-3-NITROBENZAMIDO)BENZENESULFONATE

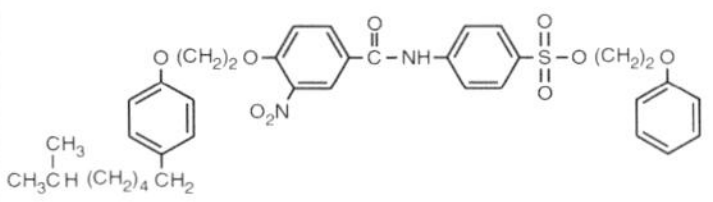

5 gm

S62,910-3
2-(4-TERT-OCTYLPHENOXY)ETHYL 3-NITROBENZENESULFONATE

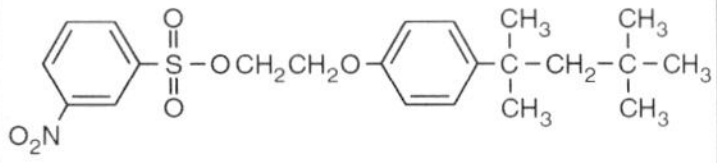

5 gm

S59,870-4
2-(4-NITROPHENOXY)ETHYL P-TOLUENESULFONATE

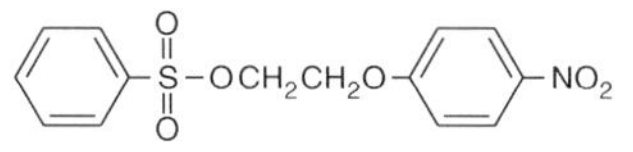

250 mg

S61,973-6
2-CARBOXY-6-METHOXYPHENYL BENZENESULFONATE

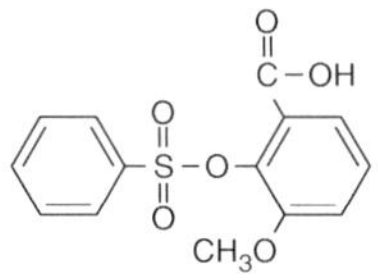

250 mg

S87,723-9
2-PHENOXYETHYL METHANESULFONATE

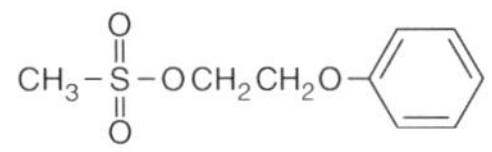

250 mg

S39,149-2
2-METHOXYETHYL PARA-TOLUENESULFONATE

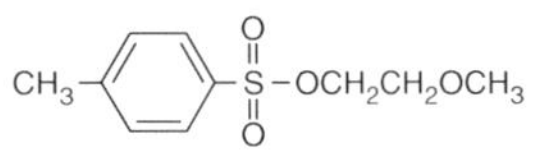

250 mg

S60,511-5
2-ETHOXYETHYL P-TOLUENESULFONATE

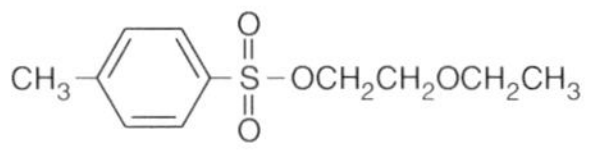

250 mg

S51,616-3
2-BUTOXYETHYL PARA-TOLUENESULFONATE

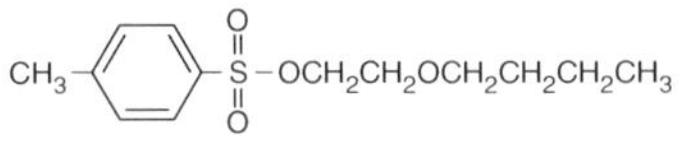

250 mg

S60,513-1
2-[2-(2-ETHOXY)ETHOXY]ETHYL P-TOLUENESULFONATE

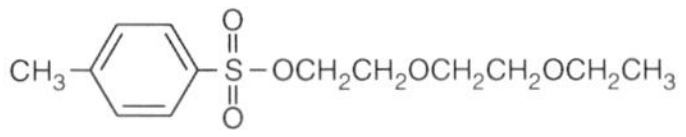

250 mg

S62,413-6
2-PHENOXYETHYL P-TOLUENESULFONATE

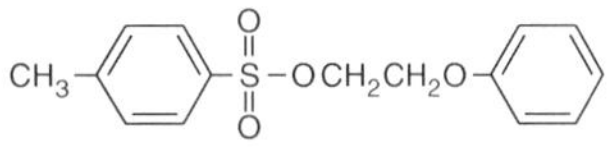

5 gm

S62,968-5
2-(4-TERT-AMYLPHENOXY)ETHYL P-TOLUENESULFONATE

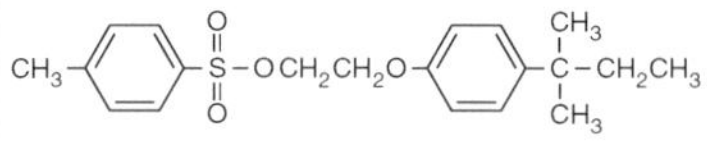

5 gm

S62,929-4
2-(4-TERT-OCTYLPHENOXY)ETHYL P-TOLUENESULFONATE

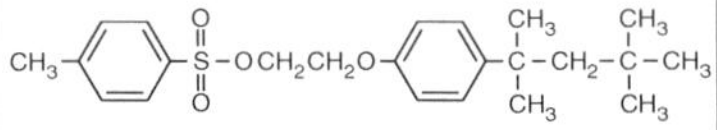

250 mg

S59,820-8
2-(2-ETHYLBUTOXY)ETHYL P-TOLUENESULFONATE

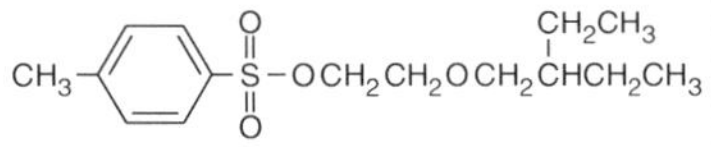

250 mg

S90,689-1
N-(3-METHOXYBENZYLIDENE)-6-O-(P-TOSYL)-1,3,4-TRI-O-ACETYL-BETA-D-GALACTOSAMINE

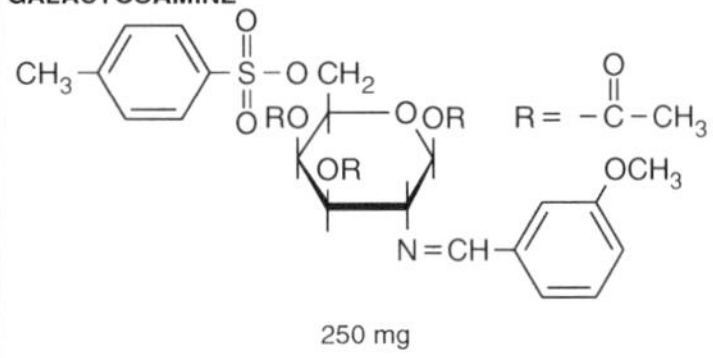

250 mg

S61,976-0
2-[2-(2-CARBOXYETHYL)-5-METHOXY-6-TOSYLOXYPHENYL]SUCCINIC ACID

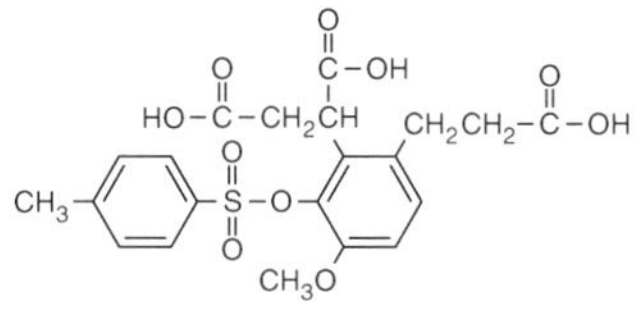

250 mg

S59,876-3
2,4-DI(TERT-PENTYL)PHENYL 3-ACETAMIDO-4-METHOXYBENZENESULFONATE

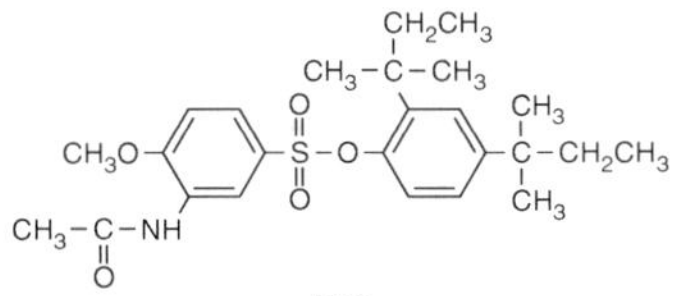

5 gm

S59,261-7
6-ETHOXY-3H-2,1-BENZOTHIOL-3-ONE-1,1-DIOXIDE

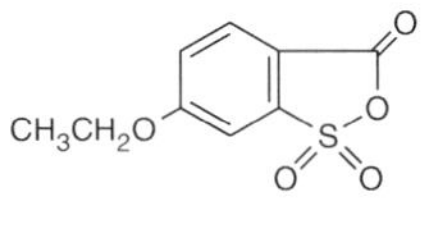

5 gm

S62,924-3
P-TOLYL 2-PHENOXYETHANESULFONATE

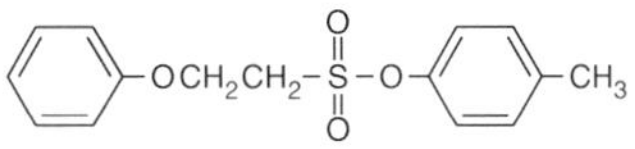

5 gm

S59,810-0
2-METHOXYBUTYL P-TOLUENESULFONATE

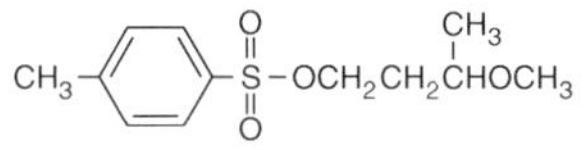

5 gm

S70,722-8
2-(BENZYLOXYETHYL)-P-TOLUENESULFONATE

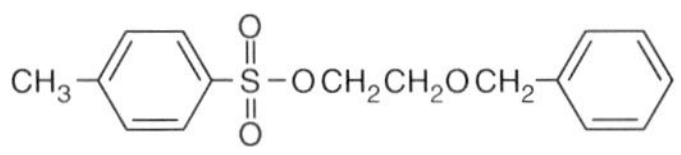

5 gm

S87,651-8
BENZYL N-[1-PHENYL-2-(P-TOSYLOXY)-ETHYL]CARBAMATE

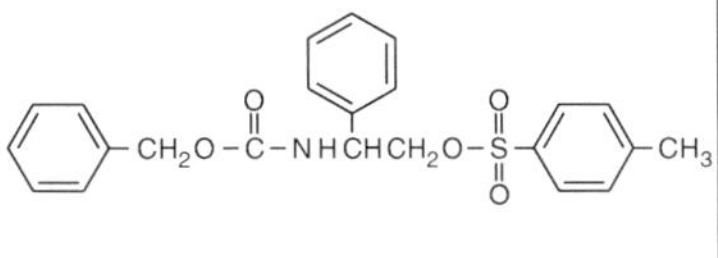

250 mg

S66,844-3
1-CARBOBENZYLOXY-4-PIPERIDINYL P-TOLUENESULFONATE

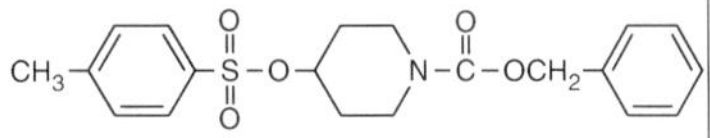

5 gm

S79,346-9
ETHYL ALPHA-(P-TOLYLSULFONYLOXY)-PHENYLACETATE

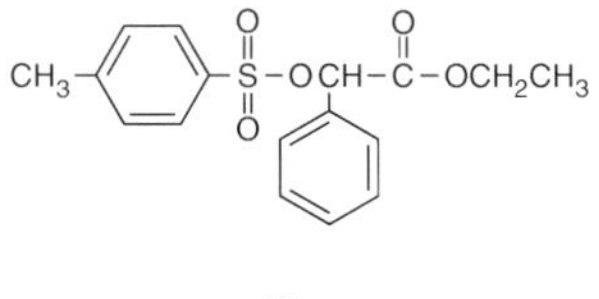

250 mg

S93,910-2
1,3-O-BENZYLIDENE-2,4,5-TRIS-O-(METHYLSULFONYL)-D-ARABITOL
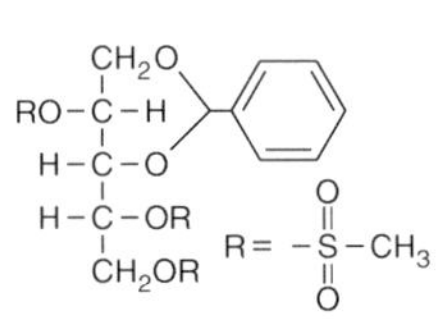
250 mg

S93,913-7
2,4:3,5-DI-O-METHYLENE-1,6-BIS-O-(P-TOSYL)-L-IDITOL
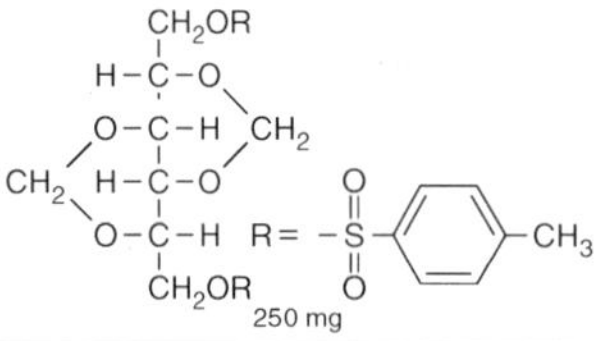
250 mg

S93,933-1
2-O-(METHYLSULFONYL)-1,3,4-TRI-O-BENZOYL-D-ARABINOSE
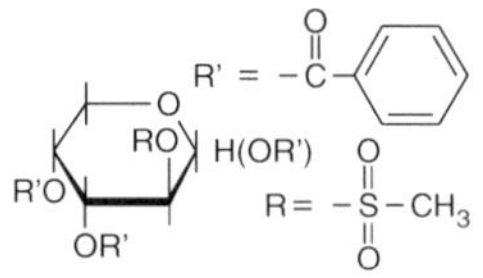
250 mg

S86,513-3
2-(P-TOSYLOXY)CYCLOHEXYL NITRATE
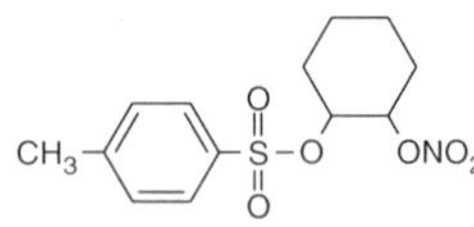
250 mg

S65,742-5
2,2-DIMETHYL-1,3-DIOXOLAN-4-YLMETHYL P-TOLUENESULFONATE
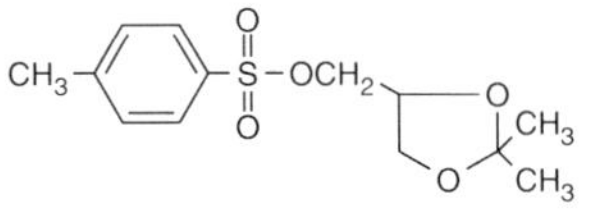
5 gm

S85,275-9
TRANS-1-CYCLOHEXENE-1,2-DIOL DI-P-TOSYLATE
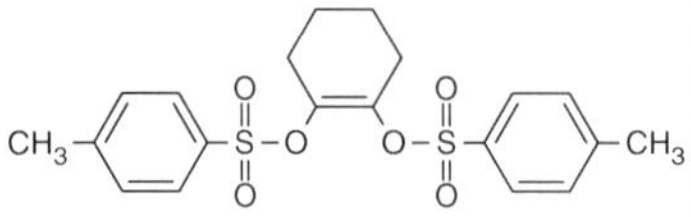
250 mg

S85,321-6
CIS-1-CYCLOHEXENE-1,2-DIOL DI-P-TOSYLATE
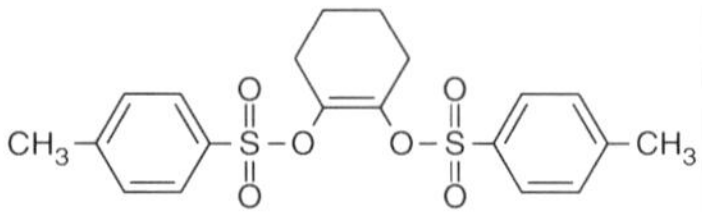
250 mg

S69,734-6
1,2-PROPANEDIOL DITOSYLATE
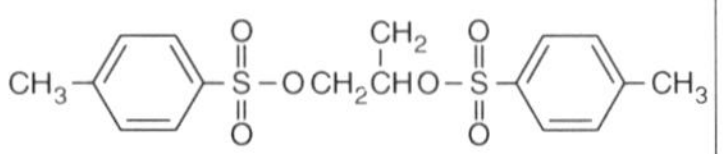
250 mg

S67,951-8
6-DIAZO-5,6-DIHYDRO-5-OXO-1-NAPHTHALENESULFONIC 4-BENZOYL-1,3-PHENYLENE ESTER
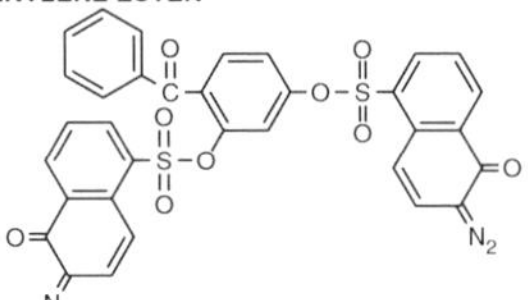
250 mg

S86,452-8
2-OXO-1,3,3-TRIMETHYLCYCLOHEXYLMETHYL P-TOLUENESULFONATE
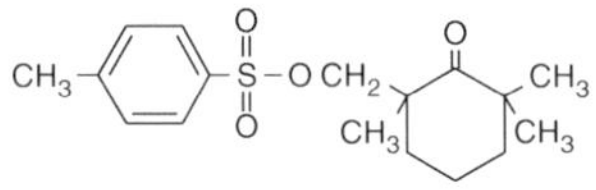
250 mg

S91,493-2
DIETHYL [ALPHA-(P-TOSYLOXY)-BENZYLIDENE]MALONATE
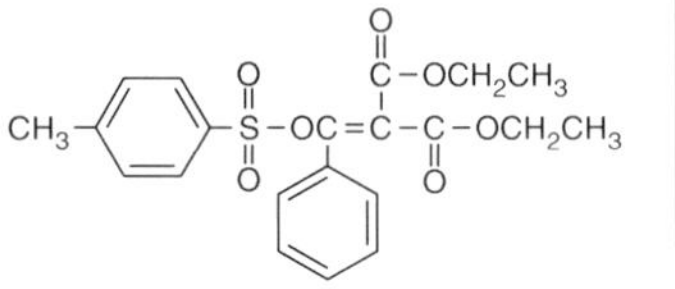
250 mg

S66,271-2
2,6-BIS[2,2(DIHYDROXYMETHYL)PROPYL DIMETHANESULFONATE]-1,3,5-TRIMETHYLBENZENE
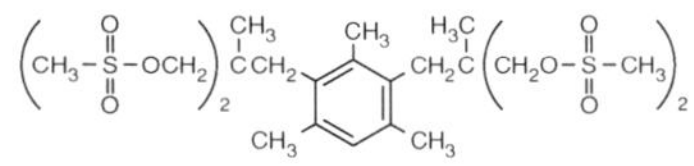
250 mg

S87,277-6
2,2-DIMETHYL-1,3-PROPANEDIOL BIS(BENZENESULFONATE)
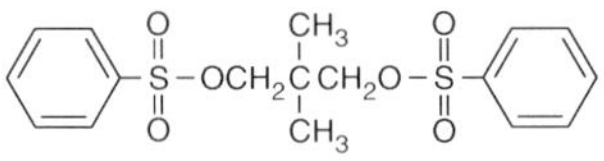
250 mg

S93,900-5
PENTAERYTHRITOL TETRAKIS(P-TOLUENESULFONATE)
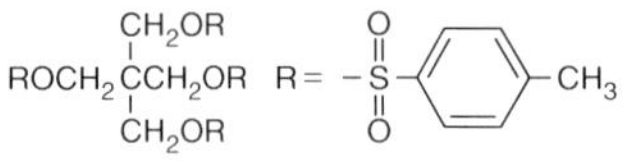
250 mg

S90,692-1
N,1,3,4-TETRAACETYL-6-O-(P-TOSYL)-BETA-D-GALACTOSAMINE
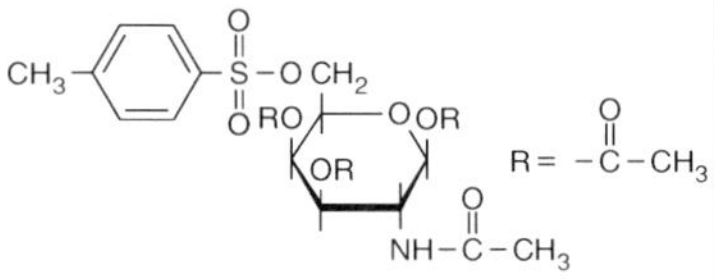
250 mg

S84,699-6
BENZENESULFONOTHIOIC ACID ANHYDROSULFIDE
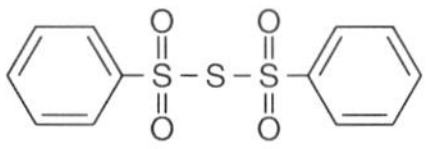
250 mg

S83,785-7
BIS(P-TOLYLSULFONYL) TRISULFIDE
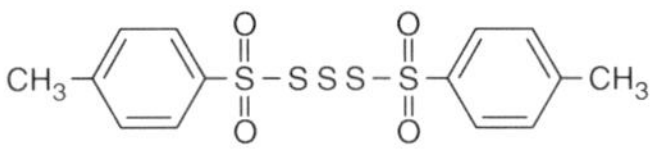
250 mg

S86,337-8
BIS(P-TOSYL) TRISULFIDE
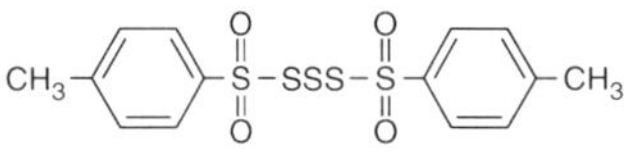
250 mg

S62,913-8
2-CYANOETHYL 4-AMINO-3-NITROBENZENESULFONATE
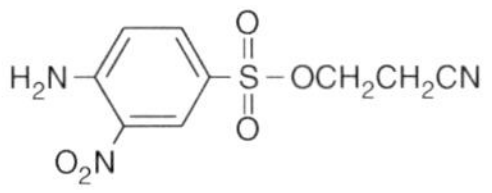
250 mg

S50,201-4
4-AMINO-3-NITROPHENYL 1-BUTANESULFONATE
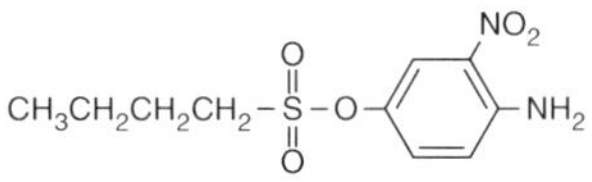
250 mg

S50,022-4
PHENYL SULFANILATE
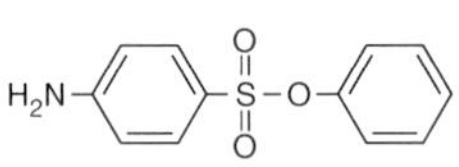
5 gm

S93,345-7
PHENYL 2,4-DINITROBENZENESULFONATE
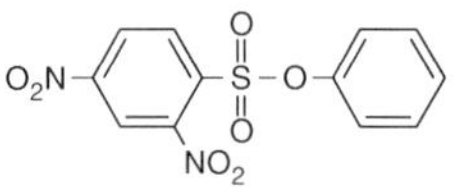
250 mg

S90,876-2
3-(4-NITROANILINO)PROPYL P-TOLUENESULFONATE
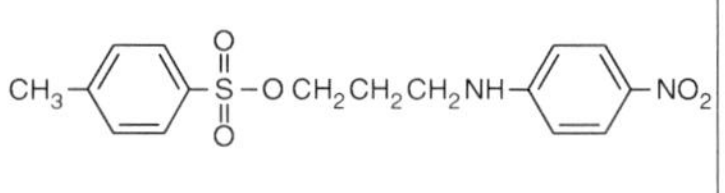
250 mg

S90,739-1
2-(BIS[4-(DIMETHYLAMINO)PHENYL]-METHYL)PHENYL METHANESULFONATE
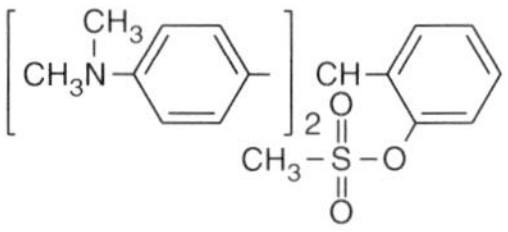
250 mg

S90,670-0
1-(DIPHENYLMETHYL)-3-AZETIDINYL METHANESULFONATE
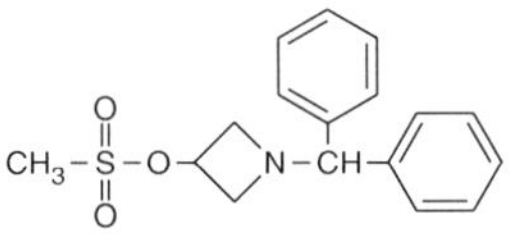
250 mg

S44,708-0
3-QUINUCLIDYL P-TOLUENESULFONATE HYDROCHLORIDE
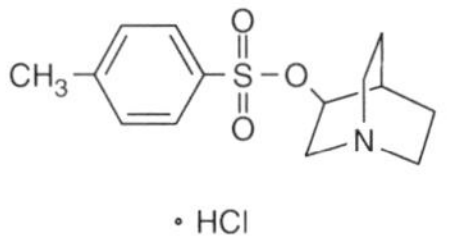
250 mg

S63,886-2
CYCLOHEXYLTRIMETHYLAMMONIUM P-TOLUENESULFONATE
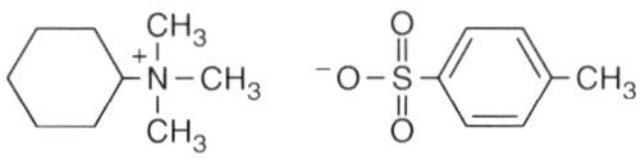
250 mg

S63,876-5
PHENYLTRIMETHYLAMMONIUM P-TOLUENESULFONATE
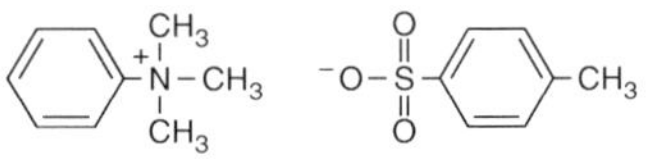
250 mg

S58,692-7
2-NITROPHENYL METHANESULFONATE
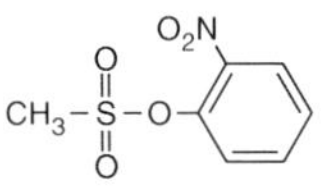
5 gm

S91,760-5
2-METHYL-2-NITROPROPYL P-TOLUENESULFONATE
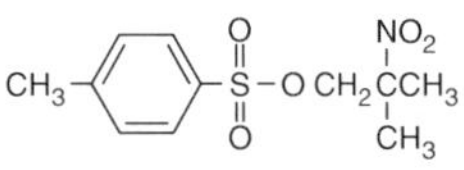
250 mg

S61,859-4
6-NITRO-M-TOLYL METHANESULFONATE
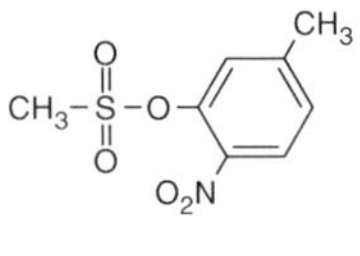
5 gm

S44,423-5
CYANOMETHYL P-TOLUENESULFONATE
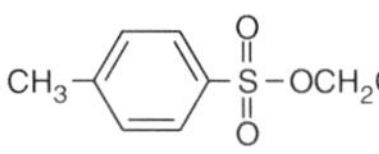
5 gm

S73,271-0
1-CYANO-2,2-DIMETHYLPROPYL BENZENESULFONATE
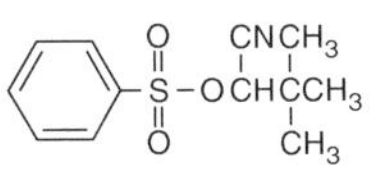
250 mg

S36,454-1
2-PHTHALIMIDOETHYL P-TOLUENESULFONATE
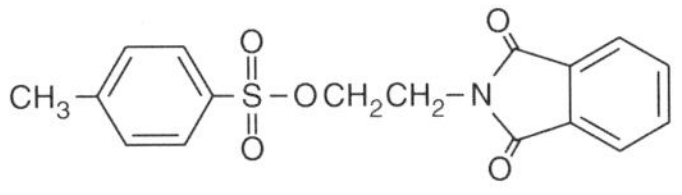
5 gm

S67,196-7
2-PHTHALIMIDOETHYL BENZENESULFONATE
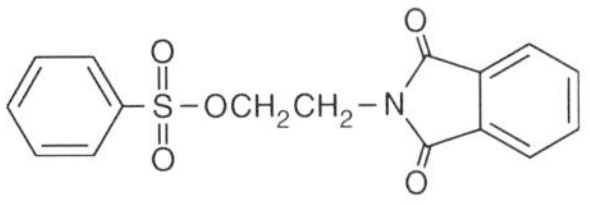
5 gm

S46,295-0
TERT-BUTYL N-(MESITYLENESULFONYLOXY)CARBAMATE
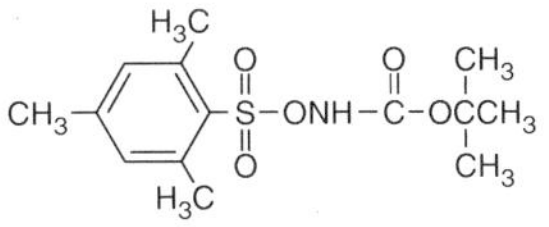
5 gm

S95,421-7
2-(2-NITROPHENYL)ETHYL P-TOLUENESULFONATE
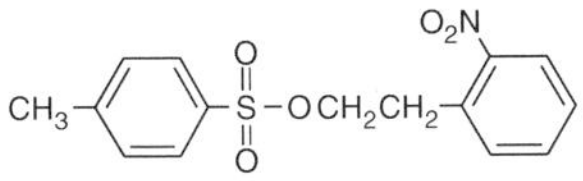
50 mg

S80,765-6
4-NITROPHENYL P-TOLUENESULFONATE
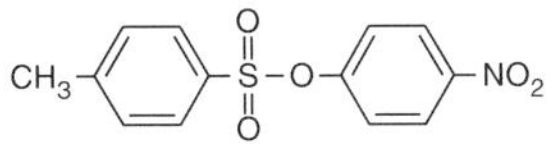
250 mg

S56,264-5
3-(METHYLSULFONYLOXY)-4-NITROBENZOIC ACID
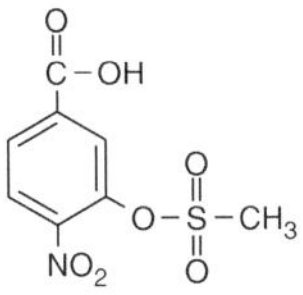
250 mg

S50,013-5
2-CYANOETHYL 3-NITROBENZENESULFONATE
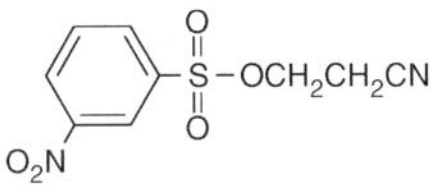
250 mg

S35,454-6
PHENYL 4-CHLORO-3-NITROBENZENESULFONATE
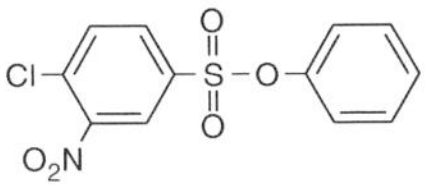
5 gm

S38,290-6
ETHYL 3-NITROBENZENESULFONATE
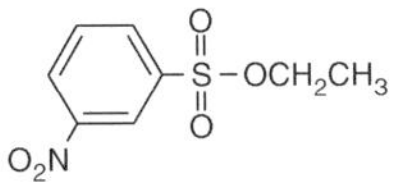
250 mg

S77,844-3
3-BENZOYL-3H-1,2-BENZOXATHIOLE 2,2-DIOXIDE
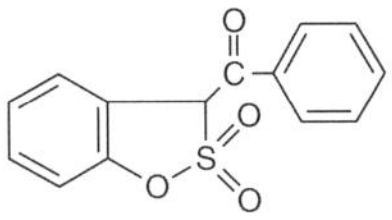
250 mg

S80,721-4
3-BENZOYL-3,3A-DIHYDROCYCLOHEPT(D)-1,2-OXATHIOLE-2,2-DIOXIDE
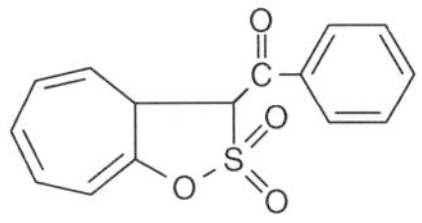
250 mg

S67,200-9
PHENYL 2-(3-CYCLOHEXYLHEXAHYDRO-2,4,6-TRIOXOPYRIMIDIN-1-YL) BENZENESULFONATE
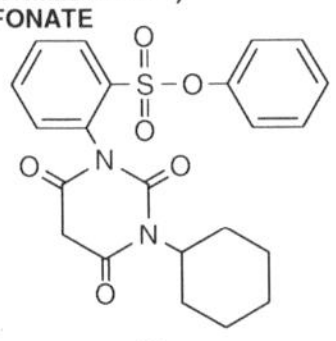
50 mg

S66,158-9
1-METHYL-2-OXOCYCLOPENTYLMETHYL P-TOLUENESULFONATE
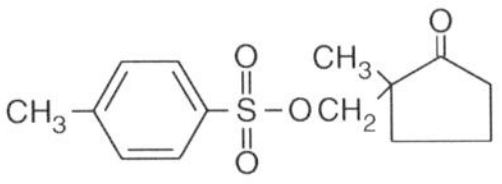
250 mg

S58,853-9
3-SULFO-2-NAPHTHOIC ACID, CYCLIC ANHYDRIDE-BIS (HEXACHLOROCYCLOPENTADIENE) ADDUCT
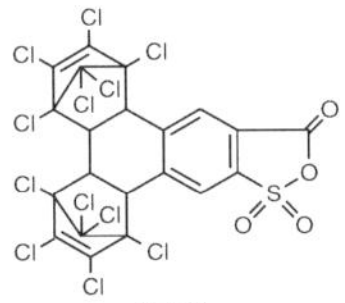
5 gm

S95,333-4
S-(P-TOSYL)MERCAPTOACETIC ACID
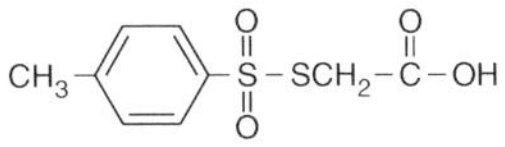
50 mg

S86,460-9
BENZYL N-[4-(P-TOSYLOXY)CYCLOHEXYL]-CARBAMATE

250 mg

S95,796-8
TRICYCLO[5.2.1.0(2,6)]DEC-10-YL P-TOLUENESULFONATE
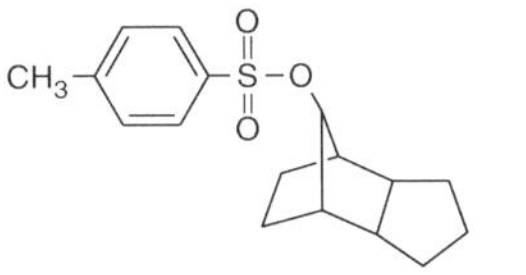
50 mg

S95,794-1
TRICYCLO(5.2.1.02,6)DECA-4,8-DIEN-10-YL P-TOLUENESULFONATE
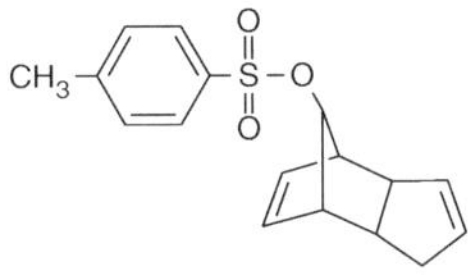
50 mg

S93,866-1
PHENYL 3-SULFOLANESULFONATE
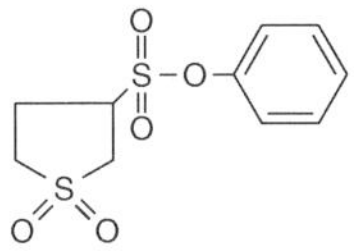
250 mg

S86,707-1
7,7-DIMETHYL-2-METHYLENEBICYCLO (3.1.1)HEPTAN-3-YLMETHYL P-TOSYLATE
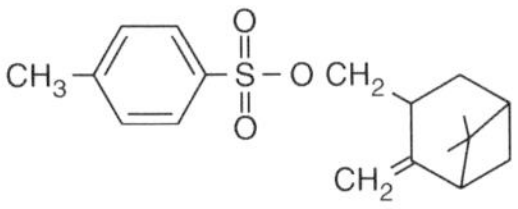
250 mg

S43,810-3
ALPHA-HYDROXY-O-TOLUENESULFONIC ACID GAMMA-SULTONE
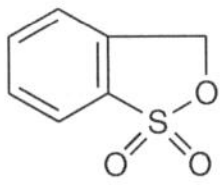
5 gm

S66,822-2
CYCLOHEXYL P-TOLUENESULFONATE
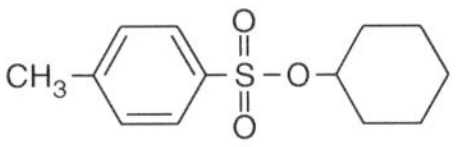
5 gm

S85,320-8
CIS-2-BROMOCYCLOHEXANOL P-TOSYLATE
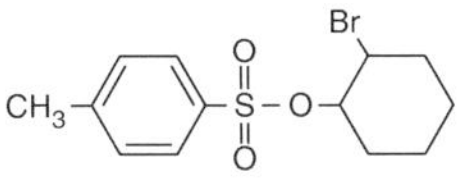
250 mg

S85,271-6
9,10-DIHYDRO-9,10-ETHANOANTHRACEN-11-OL P-TOSYLATE
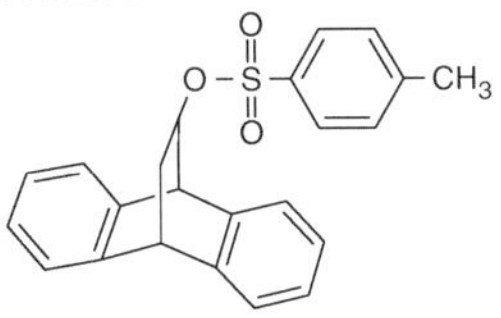
250 mg

S85,272-4
9,10-DIHYDRO-9,10-ETHANOANTHRACENE-11-METHANOL P-TOSYLATE
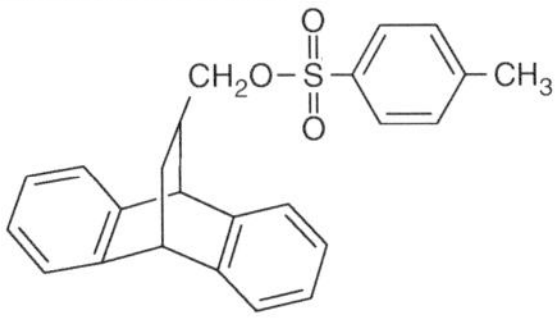
250 mg

S85,273-2
9,10-DIHYDRO-ALPHA-METHYL-9,10-ETHANOANTHRACENE-11-METHANOL P-TOSYLATE
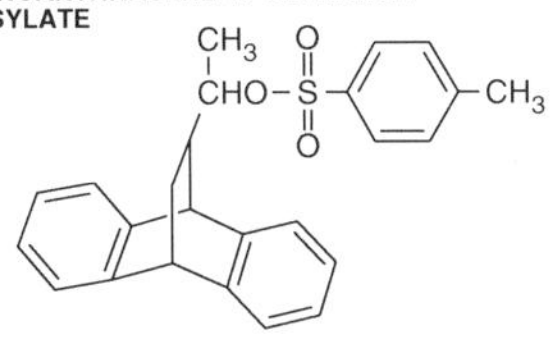
250 mg

S85,274-0
9,10-DIHYDRO-9,10-ETHANOANTHRACENE-11-ETHANOL P-TOSYLATE
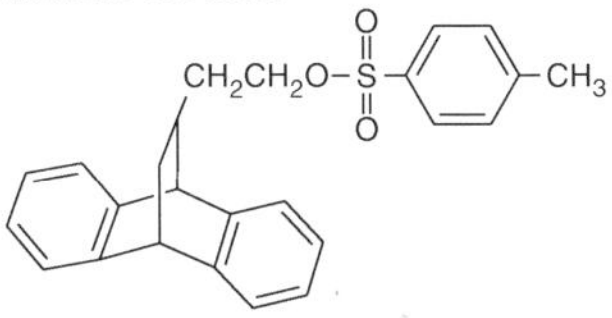
250 mg

S86,461-7
9,10-DIHYDRO-11-METHYL-9,10-ETHANOANTHRACENE-11-METHANOL P-TOLUENESULFONATE
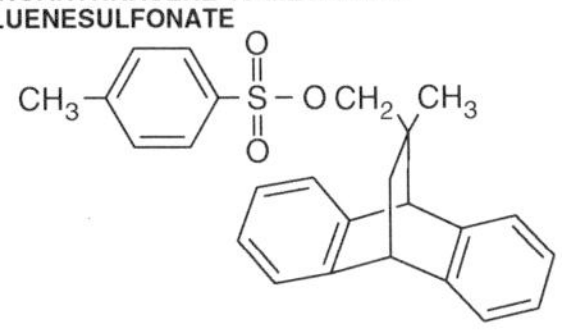
250 mg

S86,545-1
9,10-DIHYDRO-12-METHYL-9,10-ETHANOANTHRACENE-11-METHANOL P-TOSYLATE
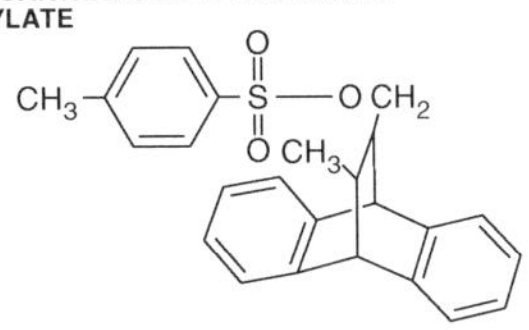
250 mg

S86,418-8
9,10-DIHYDRO-9,10-ETHANOANTHRACEN-11-YL METHANESULFONATE
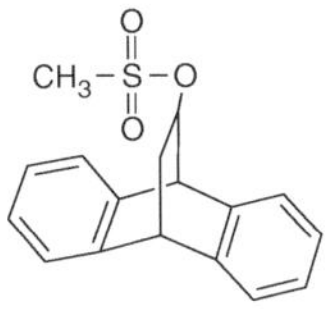
250 mg

S86,389-0
9,10-DIHYDRO-9,10-ETHANOANTHRACENE-11-METHANOL METHANESULFONATE
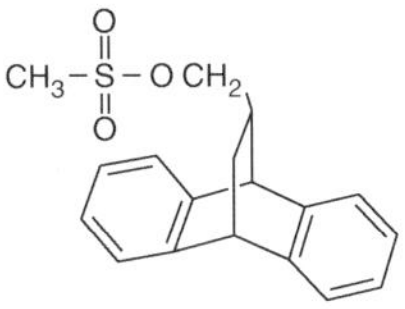
250 mg

S85,270-8
9,10-DIHYDRO-9,10-ETHANOANTHRACENE-11-METHANOL 1-BUTANESULFONATE
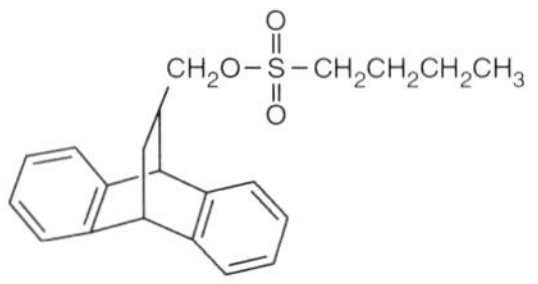
250 mg

S52,542-1
1-BENZYL-4-PIPERIDYL METHANESULFONATE HYDROCHLORIDE
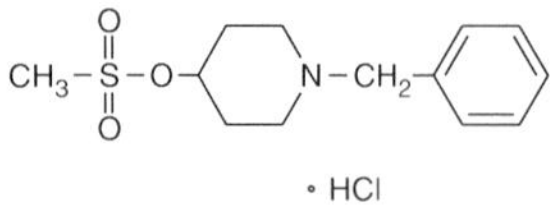
250 mg

S54,479-5
METHYL BENZENESULFONATE
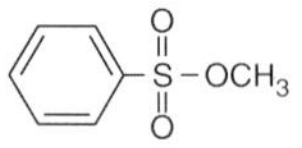
5 gm

S85,197-3
PHENYL METHANESULFONATE
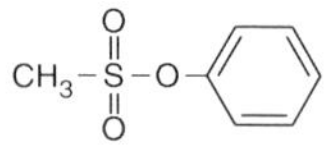
250 mg

S54,478-7
ETHYL BENZENESULFONATE
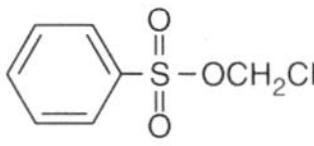
5 gm

S70,614-0
BUTYL BENZENESULFONATE
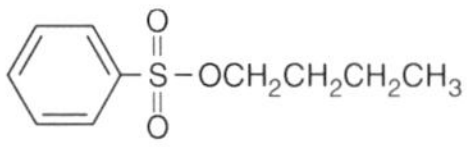
5 gm

S86,660-1
4-PHENYLBUTYL METHANESULFONATE
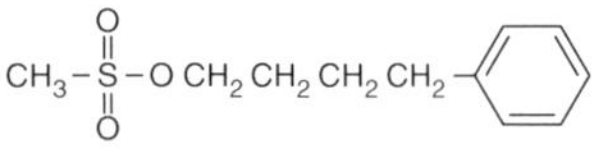
250 mg

S55,446-4
DIPHENYL 1,2-ETHANEDISULFONATE
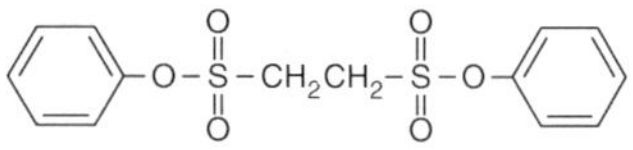
5 gm

S77,848-6
2,2-DIMETHYLPROPYL 2-METHYL-1-PHENYL-1-PROPANE SULFONATE
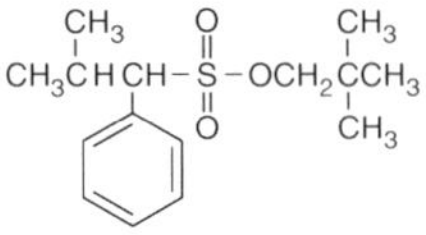
250 mg

S51,866-2
2,2-DIETHYL-1,3-PROPANEDIOL BIS(BENZENESULFONATE)
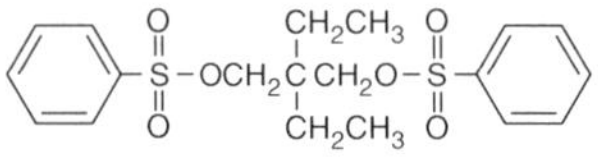
250 mg

S51,904-9
2-BUTYL-2-ETHYL-1,3-PROPANEDIOL BIS(TOLUENESULFONATE)
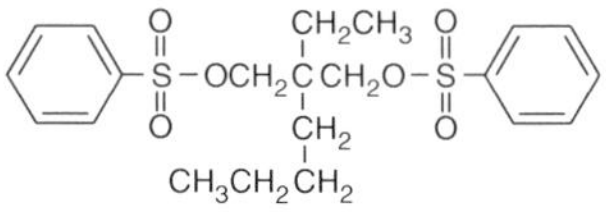
250 mg

S58,303-0
ISO-HEXADECYL BENZENESULFONATE (MIXTURE OF ISOMERS)
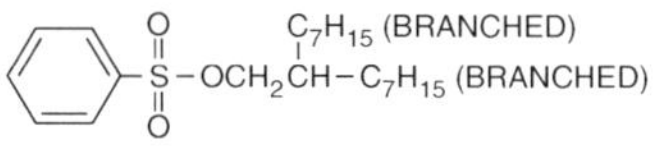
5 gm

S62,308-3
HEXADECYL BENZENESULFONATE
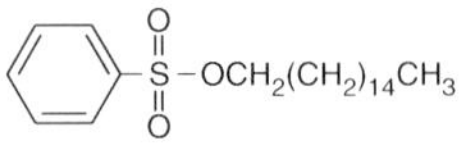
250 mg

S35,424-4
PENTAERYTHRITOL TETRAKIS (BENZENESULFONATE)
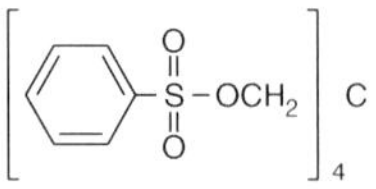
5 gm

S59,804-6
2-PHENYLETHYL P-TOLUENESULFONATE
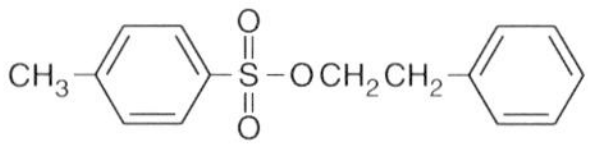
250 mg

S86,700-4
3-PHENYLPROPYL P-TOLUENESULFONATE
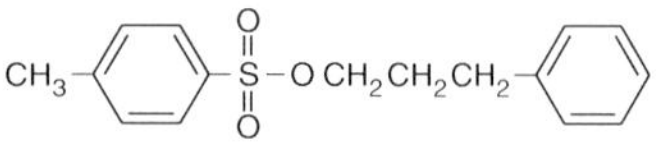
250 mg

S86,701-2
4-PHENYLBUTYL P-TOLUENESULFONATE
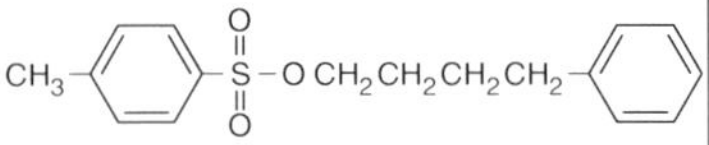
250 mg

S86,162-6
2,3-XYLYL BENZENESULFONATE
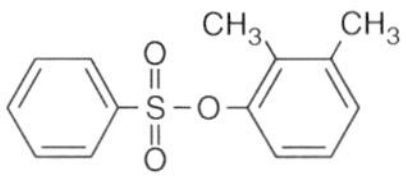
250 mg

S80,740-0
PHENYL MESITYLENESULFONATE
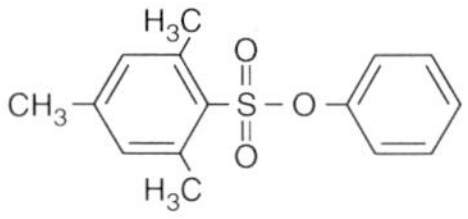
250 mg

S66,246-1
2-BENZHYDRYL-1,3-PROPANEDIOL BIS(METHANESULFONATE)
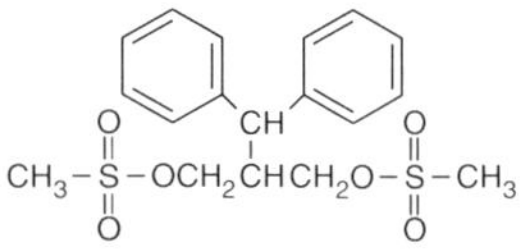
250 mg

S44,540-1
ACETYL P-TOLUENESULFONATE
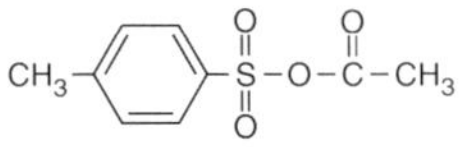
5 gm

S53,427-7
BUTYL PARA-TOLUENESULFONATE
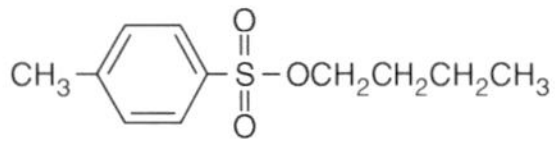
250 mg

S69,782-6
ISOBUTYL TOSYLATE
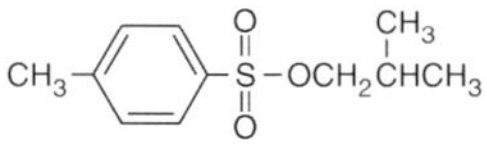
250 mg

S86,438-2
4-TERT-BUTYLPHENYL METHANESULFONATE
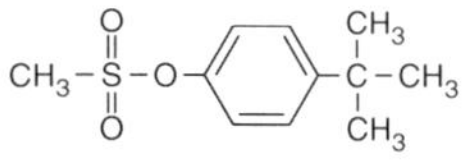
250 mg

S77,907-5
2,2-DIMETHYLPROPYL P-TOLUENESULFONATE
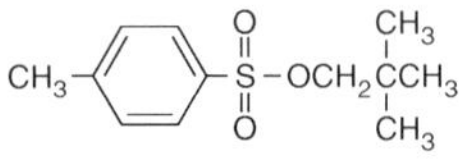
250 mg

S83,284-7
PENTYL P-TOLUENESULFONATE
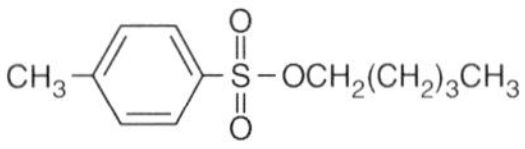
250 mg

S87,674-7
NEOPENTYL P-TOLUENESULFONATE
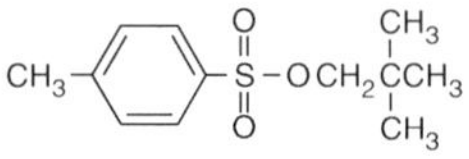
250 mg

S50,933-7
PHENYL PARA-TOLUENESULFONATE
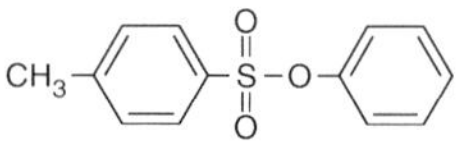
250 mg

S77,878-8
2,2-DIMETHYLPROPYL 2-(P-TOLYL) ETHANESULFONATE
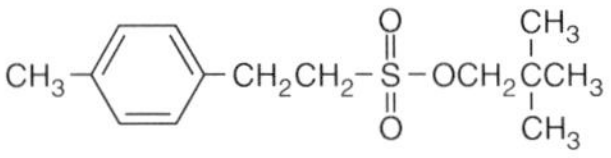
250 mg

S77,854-0
4-TERT-BUTYLPHENYL P-TOLUENESULFONATE
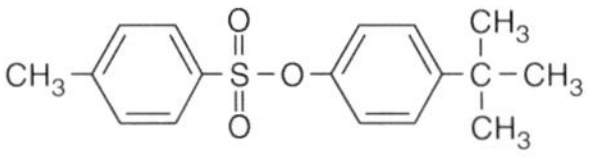
250 mg

S50,508-0
4-(1,1,3,3-TETRAMETHYLBUTYL)PHENYL P-TOLUENESULFONATE
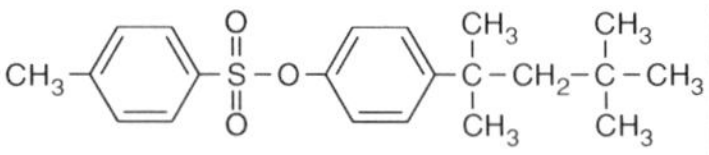
250 mg

S60,716-9
TETRADECYL P-TOLUENESULFONATE
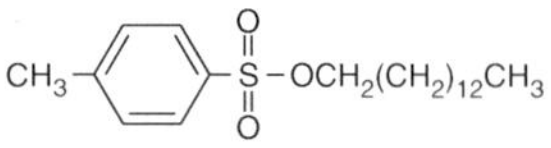
5 gm

S38,446-1
2,2-DIETHYL-1,3-PROPANEDIOL DI-P-TOLUENESULFONATE
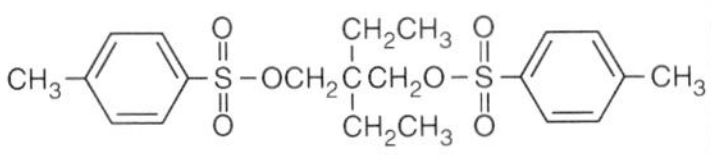
5 gm

S50,142-5
HEXADECYL P-TOLUENESULFONATE
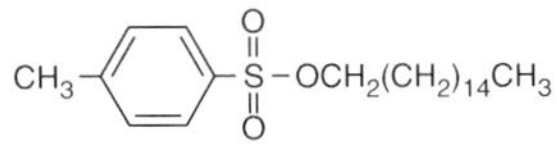
5 gm

S51,684-8
2-HEXYLDECYL PARA-TOLUENESULFONATE
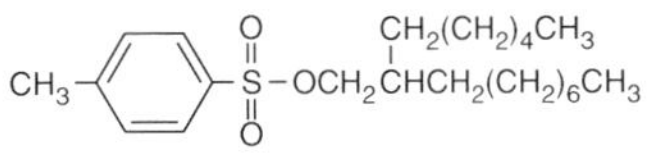
250 mg

S64,254-1
2-(OCTYL)DECYL P-TOLUENESULFONATE, BRANCHED ISOMERS
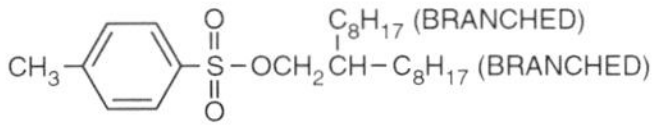

250 mg

S51,959-6
2,4-DICHLOROPHENYL PARA-TOLUENESULFONATE
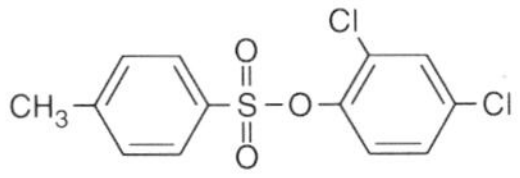
5 gm

S80,725-7
P-TOLYL 2,4-XYLENESULFONATE
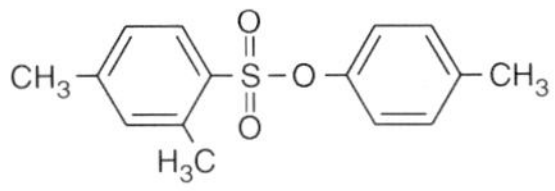
250 mg

S79,890-8
P-TOLYL 2-[2-(5-METHYL-2-SULFINOPHENYL)ETHYL]-P-TOLUENESULFONATE, SODIUM SALT
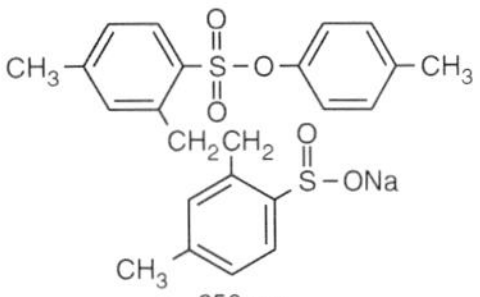
250 mg

S77,850-8
P-TOLYL MESITYLSULFONATE
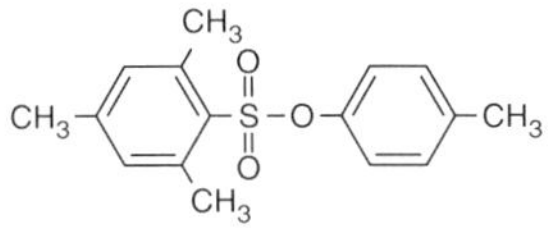
250 mg

S80,755-9
P-TOLYL 2,4,6-TRIISOPROPYLBENZENESULFONATE
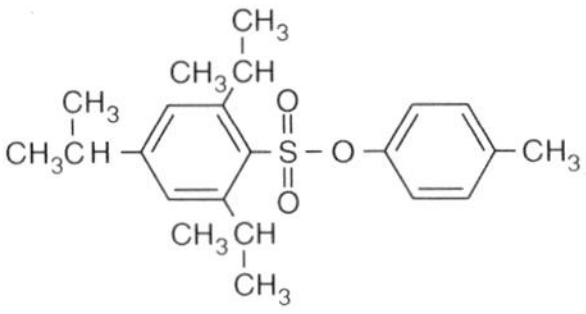
250 mg

S80,709-5
4-BIPHENYLYL MESITYLENESULFONATE
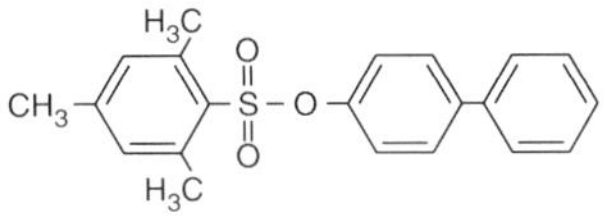
250 mg

S79,894-0
1-NAPHTHYL MESITYLENESULFONATE
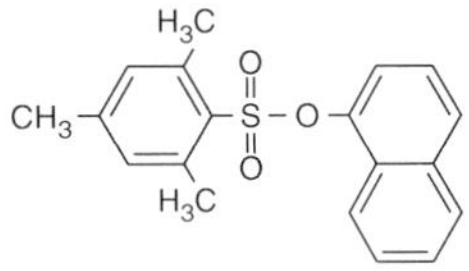
250 mg

S59,807-0
ETHYLENE GLYCOL BIS(BENZENESULFONATE)
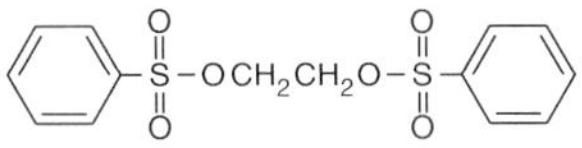
5 gm

S56,297-1
2-MESITYLENESULFONIC ANHYDRIDE
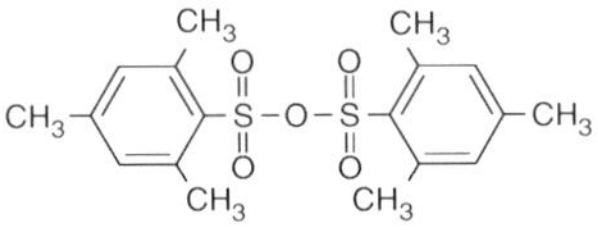
5 gm

S41,157-4
N-(METHYLSULFONYLOXY)PHTHALIMIDE
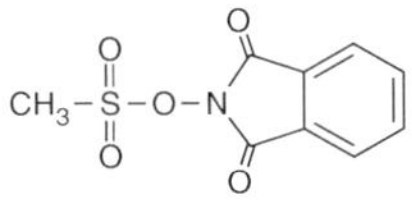
5 gm

S77,578-9
S-PHENYL P-TOLUENETHIOSULFONATE
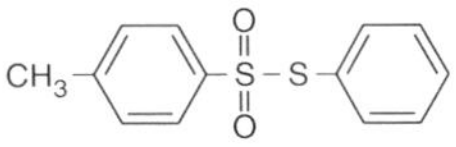
250 mg

S77,575-4
S-BENZYL P-TOLUENETHIOSULFONATE
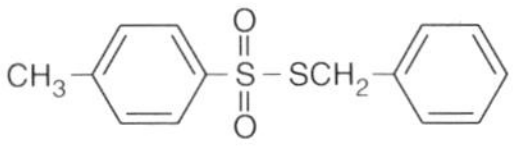
250 mg

S76,509-0
S-METHYL P-TOLUENETHIOSULFONATE
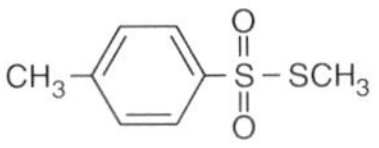
5 gm

S80,764-8
S-(P-TOLYL) P-TOLUENETHIOSULFONATE
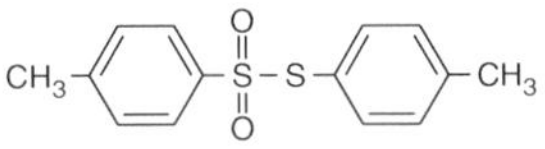
250 mg

S80,906-3
4,5-DIHYDRO-4-PHENYL-1,3,2,5-BENZODIOXATHIAZEPINE-2,2-DIOXIDE
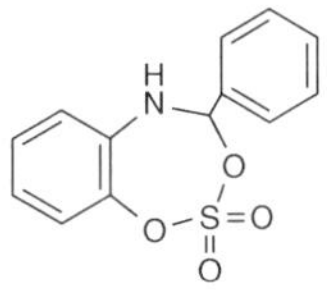
250 mg

S46,146-6
2,3,4,-TRICHLOROBENZENESULFONYL CHLORIDE
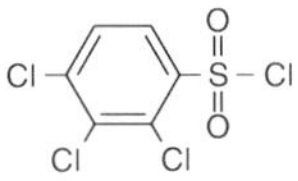
5 gm

S60,579-4
3-(BROMOACETYL)BENZENESULFONYL FLUORIDE
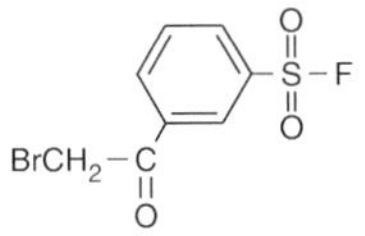
250 mg

S60,207-8
4-AMINO-6-CHLORO-1,3-BENZENEDISULFONYL CHLORIDE
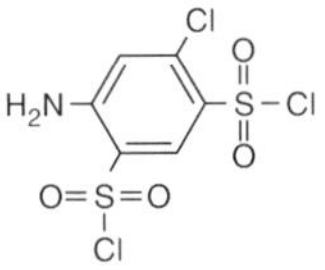
250 mg

S38,702-9
3-BROMOSULFANILYL FLUORIDE
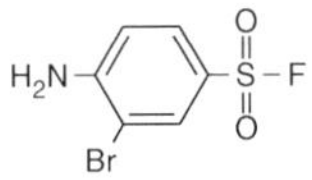
250 mg

S76,207-5
4-CHLORO-3-(CHLOROSULFONYL)-5-NITROBENZOIC ACID
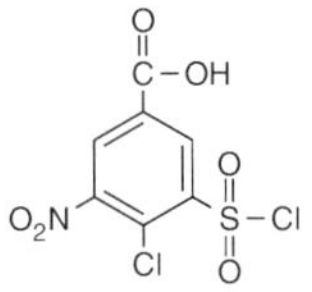
250 mg

S65,626-7
1-(4-FLUOROSULFONYLPHENYL)-3-[2-(2-HYDROXYETHOXY)ETHYL]UREA
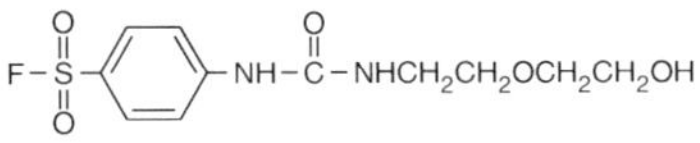
250 mg

S59,300-1
1-HYDROXY-5'-FLUOROSULFONYL-2-NAPHTH-O-ANISIDIDE
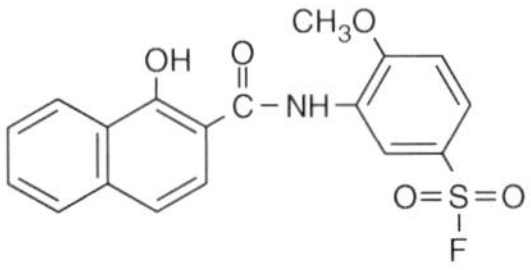
250 mg

S63,322-4
4'-AMINO-3-(FLUOROSULFONYL)-2'-HYDROXYBENZANILIDE
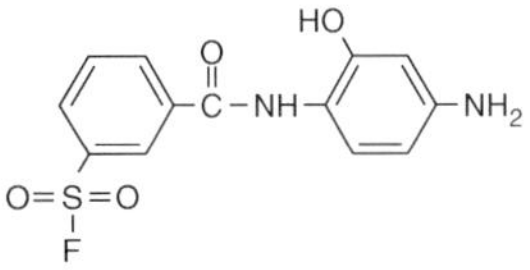
250 mg

S62,283-4
4'-ACETAMIDO-4-FLUOROSULFONYL-2'-HYDROXYBENZANILIDE
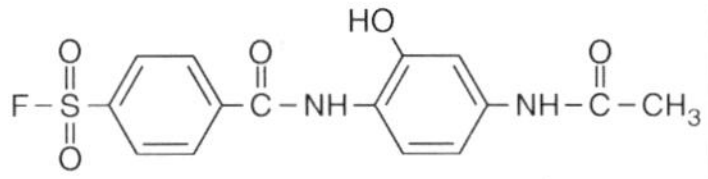
250 mg

S62,937-5
4'-FLUOROSULFONYL-1-HYDROXY-2'-NITRO-2-NAPHTHANILIDE
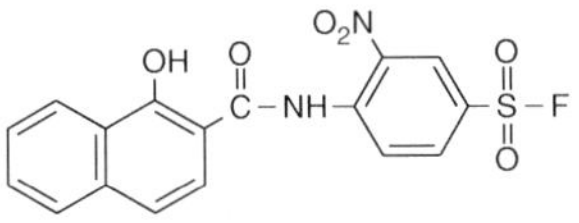
5 gm

S65,630-5
4-(3-[3-(FLUOROSULFONYL)PHENYL]-UREIDO]SALICYLIC ACID
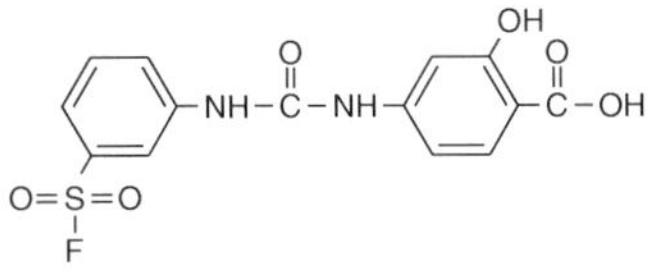
250 mg

S93,319-8
4-FLUOROSULFONYL-1-HYDROXY-2-NAPHTHOIC ACID
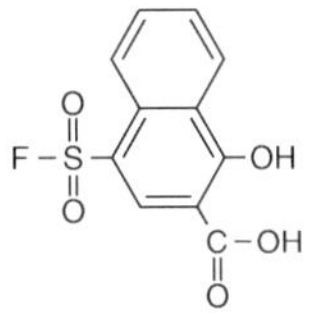
250 mg

S38,320-1
PHENYL 4-FLUOROSULFONYL-1-HYDROXY-2-NAPHTHOATE
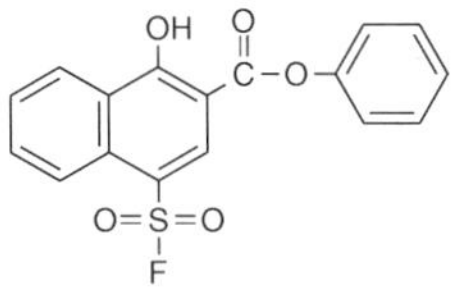
5 gm

S38,322-8
PHENYL 4,7-BIS(FLUOROSULFONYL)-1-HYDROXY-2-NAPHTHOATE
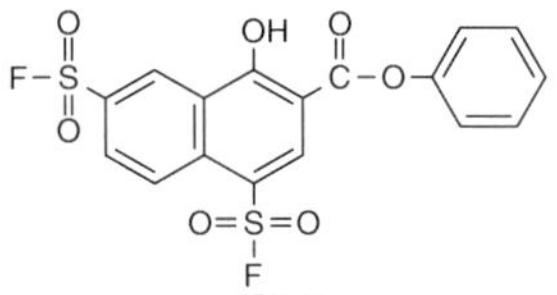
250 mg

S58,026-0
N-(1-HYDROXY-2-NAPHTHOYL)-SULFANILYL FLUORIDE
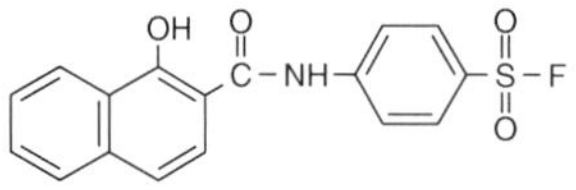
250 mg

S51,547-7
3-(4-CHLOROPHENYLCARBAMOYL)-4-HYDROXY-1-NAPHTHALENESULFONYL FLUORIDE
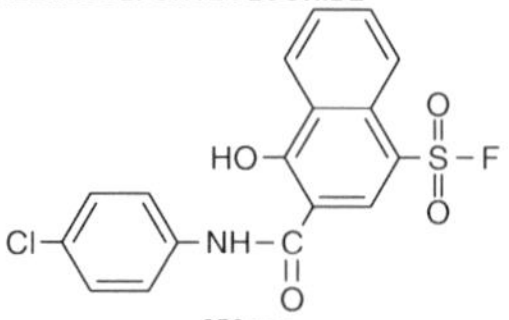
250 mg

S58,784-2
2-(4-ETHOXY-3-METHOXYBENZOYL)-4'-(FLUOROSULFONYL)ACETANILIDE
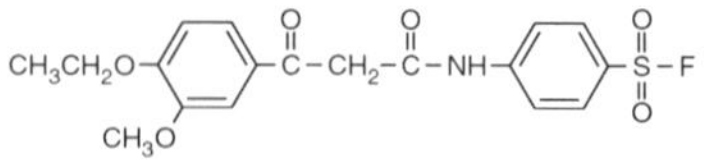
250 mg

S55,998-9
2-[4-(FLUOROSULFONYL)-PHENOXY]ACETIC ACID
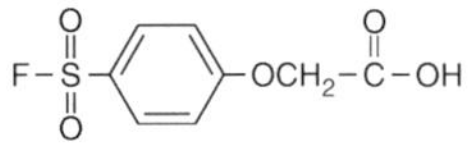
5 gm

S62,777-1
3-AMINO-4-ETHOXYBENZENESULFONYL FLUORIDE
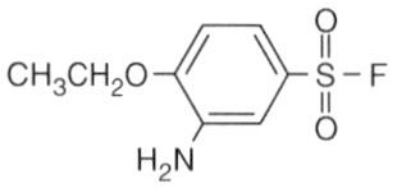
250 mg

S58,197-6
4-(HEXADECYLOXY)METANILYL FLUORIDE
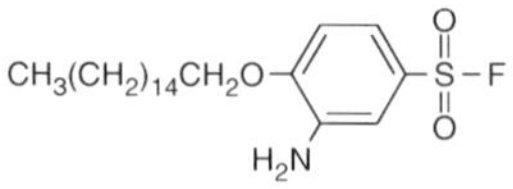
250 mg

S54,433-7
3-[2-(HEXADECYLOXY)BENZYLAMINO]-BENZENESULFONYL FLUORIDE
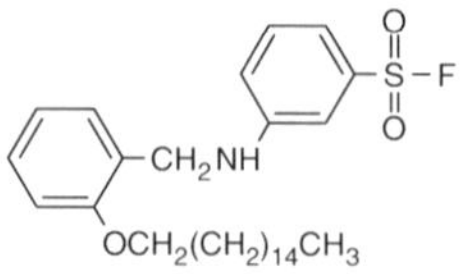
5 gm

S58,271-9
5'-(FLUOROSULFONYL)-2-(3-NITROBENZOYL)-O-ACETANISIDIDE
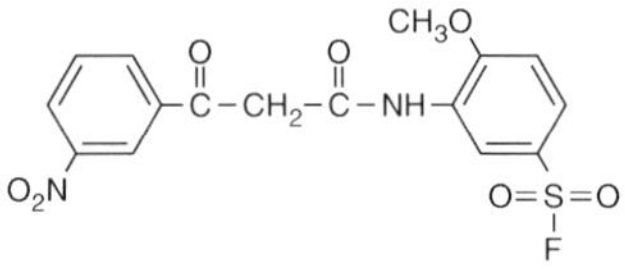
250 mg

S58,289-1
4-(HEXADECYLOXY)-3-NITROBENZENESULFONYL FLUORIDE
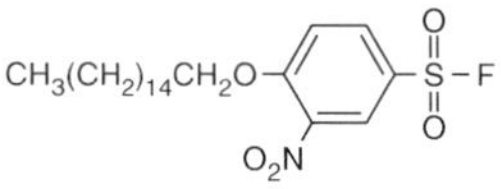
250 mg

S51,391-1
3-NITRO-4-(OCTADECYLOXY)-BENZENESULFONYL FLUORIDE
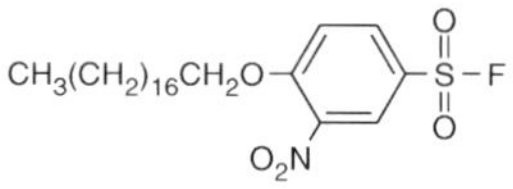
250 mg

S60,516-6
2,5-DIMETHOXY-4-NITROBENZENESULFONYL CHLORIDE
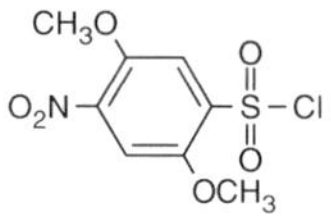
250 mg

S38,719-3
N-ACETYL-4-ETHOXYMETANILYL FLUORIDE
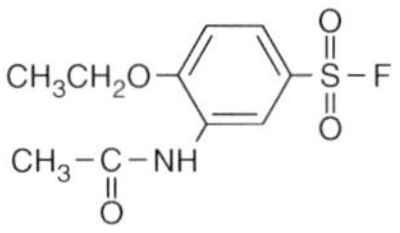
250 mg

S59,258-7
METHYL 2-[3-FLUOROSULFONYL-4-(HEXADECYLOXY)BENZOYL]ACETATE
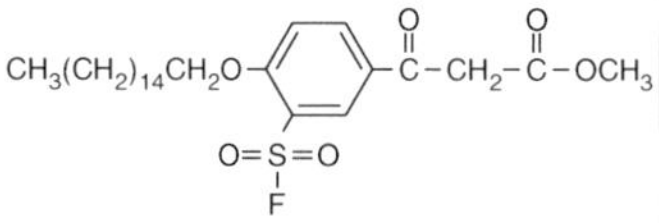
250 mg

S64,190-1
METHYL 2-[5-FLUOROSULFONYL-2-(HEXADECYLOXY)BENZOYL]ACETATE
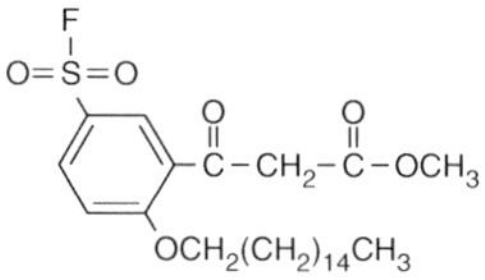
250 mg

S50,510-2
METHYL 3-CHLOROSULFONYL-4-(HEXADECYLOXY)BENZOATE
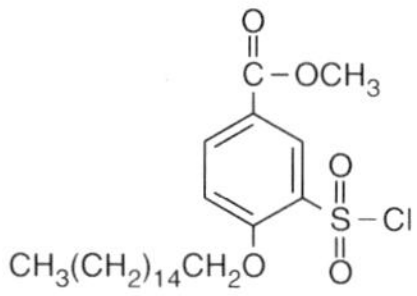
250 mg

S44,747-1
5-FLUOROSULFONYL-O-ANISIC ACID
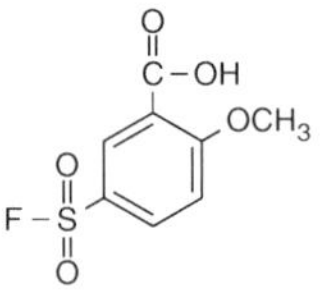
5 gm

S56,882-1
3-(FLUOROSULFONYL)-P-ANISIC ACID
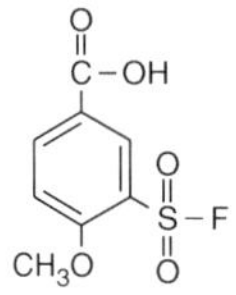
250 mg

S56,837-6
3-(CHLOROSULFONYL)-P-ANISIC ACID
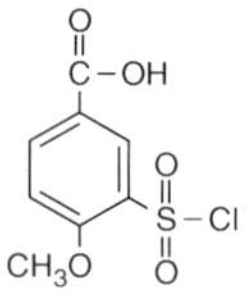
250 mg

S58,155-0
5-(FLUOROSULFONYL)-2-(HEXADECYLOXY)-BENZOIC ACID
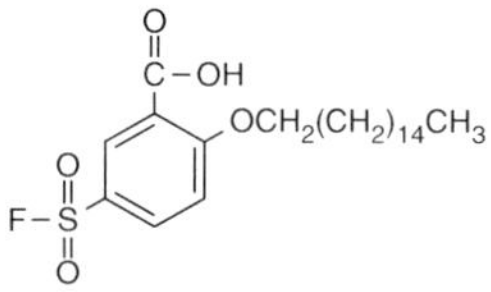
250 mg

S58,186-0
5-(CHLOROSULFONYL)-2-(HEXADECYLOXY)-BENZOIC ACID
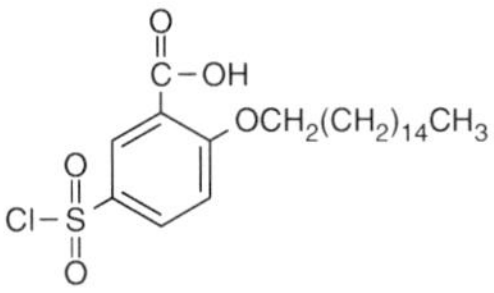
250 mg

S59,849-6
3-CHLOROSULFONYL-4-HEXADECYLOXYBENZOIC ACID
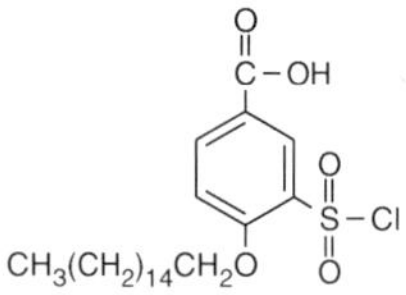
250 mg

S58,826-1
5-FLUOROSULFONYL-2-(HEXADECYLOXY)-BENZOYL CHLORIDE
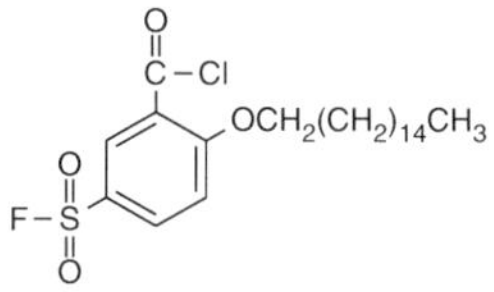
5 gm

S50,586-2
N-(P-ANISOYL)SULFANILYL FLUORIDE
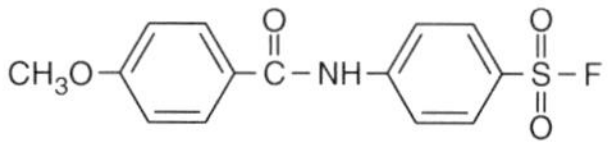
250 mg

S55,447-2
N-(4-ISOPROPOXYBENZOYLACETYL)-SULFANILYL FLUORIDE
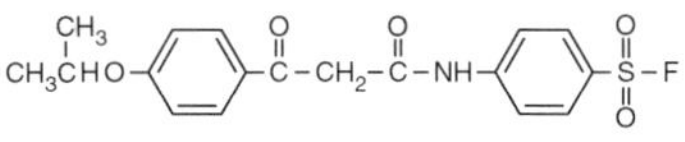
250 mg

S38,295-7
N-ACETYL-4-METHOXYMETANILYL FLUORIDE
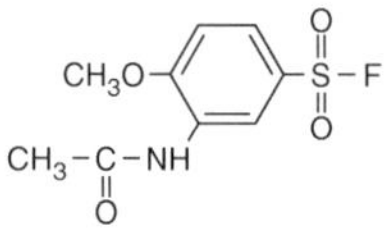
250 mg

S54,049-8
4,4'-OXYBIS-(BENZENESULFONYL CHLORIDE)
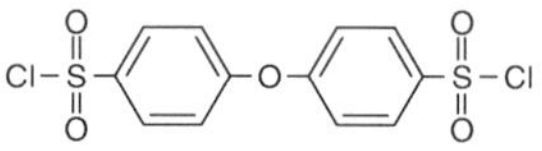
5 gm

S58,270-0
4-(HEXADECYLOXY)-BENZENESULFONYL CHLORIDE
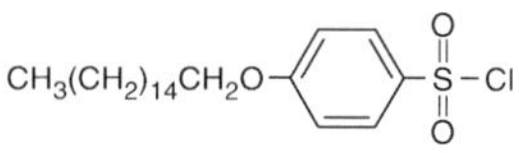
5 gm

S51,582-5
4-(OCTADECYLOXY)-BENZENESULFONYL FLUORIDE
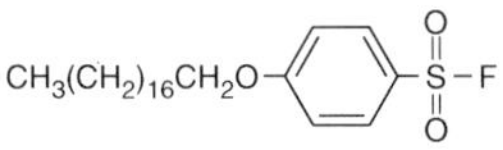
250 mg

S38,669-3
2,5-DIMETHOXYBENZENESULFONYL CHLORIDE
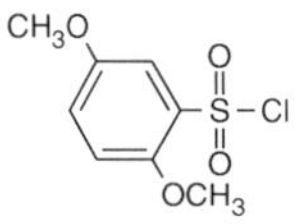
250 mg

S58,756-7
4-(HEXADECYLTHIO)METANILYL FLUORIDE
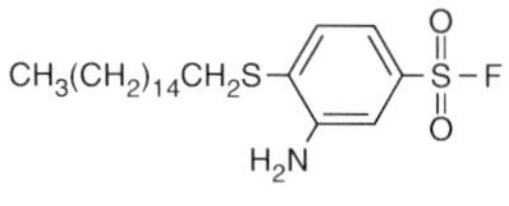
5 gm

S58,822-9
5-FLUOROSULFONYL-2-(HEXADECYLTHIO)-BENZOIC ACID
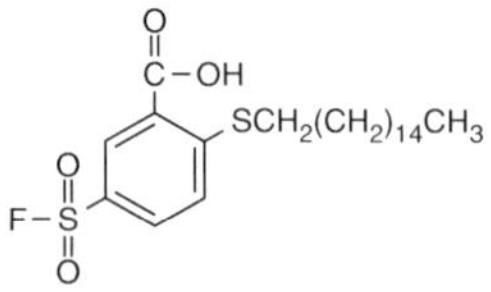
5 gm

S62,282-6
4'-AMINO-4-(FLUOROSULFONYL)-BENZANILIDE
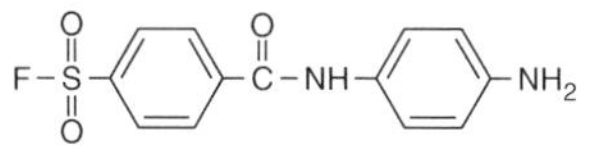
250 mg

S38,326-0
4-(METHYLSULFONYL)-METANILYL FLUORIDE
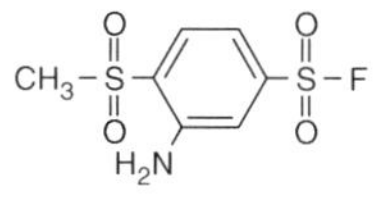
250 mg

S51,562-0
4'-AMINO-4-BIPHENYLSULFONYL FLUORIDE
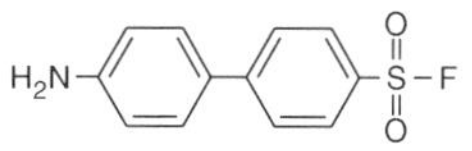
250 mg

S67,192-4
3-AMINO-4-METHYLBENZENESULFONYL FLUORIDE
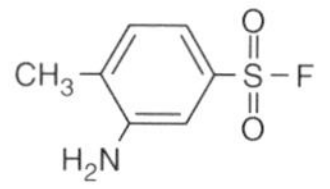
5 gm

S38,687-1
4-(HEXADECYLSULFONYL)-METANILYL FLUORIDE
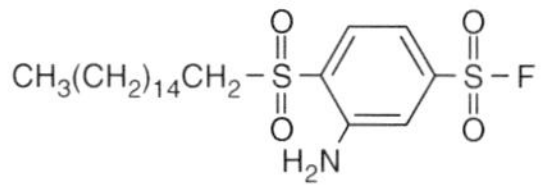
250 mg

S63,232-5
3-NITRO-4-(OCTADECYLAMINO)-BENZENESULFONYL CHLORIDE
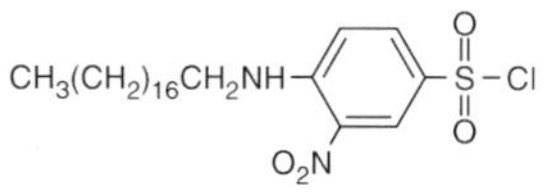
250 mg

S63,119-1
4-DIMETHYLAMINO-3-NITROBENZENESULFONYL CHLORIDE
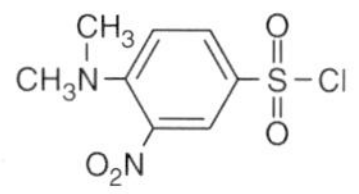
250 mg

S65,631-3
1,2-BIS(2-FLUOROSULFONYLPHENYLUREIDO)ETHANE
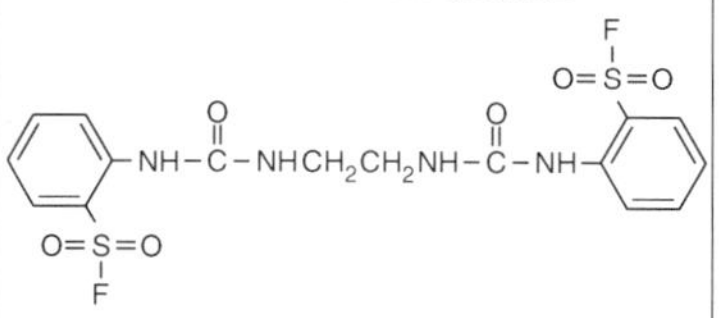
5 gm

S58,811-3
N-(1-BENZYL-5-OXO-2-PYRAZOLIN-3-YL)-4-(FLUOROSULFONYL)BENZAMIDE
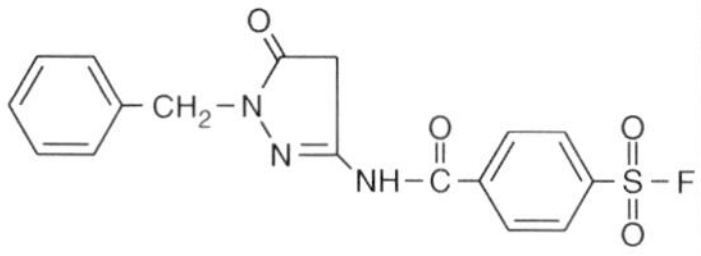
5 gm

S65,596-1
3-[3-(3-NITROPHENYL)UREIDO]-BENZENESULFONYL FLUORIDE
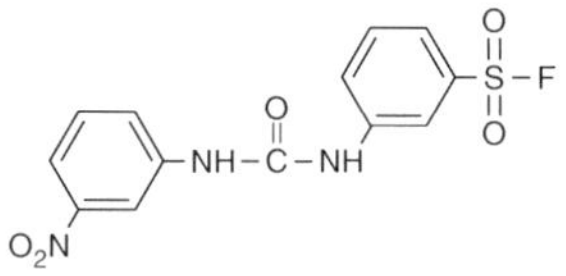
250 mg

S63,248-1
4-[N-(4-NITROPHENYL)CARBAMOYL]-BENZENESULFONYL FLUORIDE
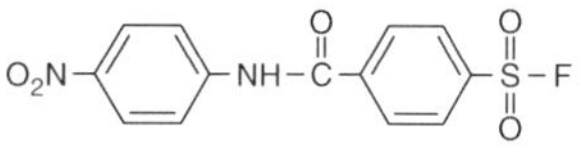
250 mg

S63,303-8
4'-(3-FLUOROSULFONYLBENZAMIDO)-SUCCINANILIC ACID
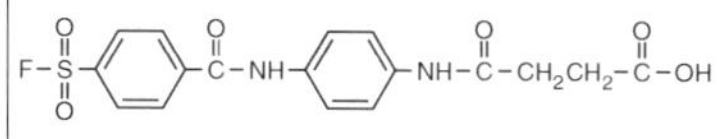
5 gm

S58,074-0
N-DODECYLSULFANILYL FLUORIDE
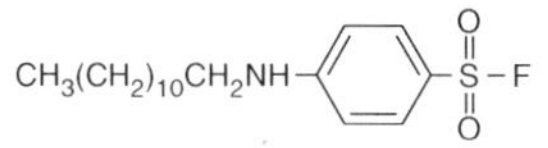
250 mg

S63,282-1
BENZOIC [2-(4-FLUOROSULFONYLPHENYL)-4-METHYL-1,3,4-THIADIAZOL-2-IN-5-YLIDENE]-HYDRAZIDE
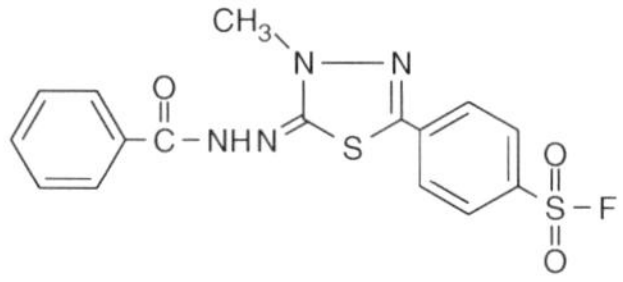
250 mg

S65,635-6
3-(3-BENZYLUREIDO)BENZENE-SULFONYL FLUORIDE
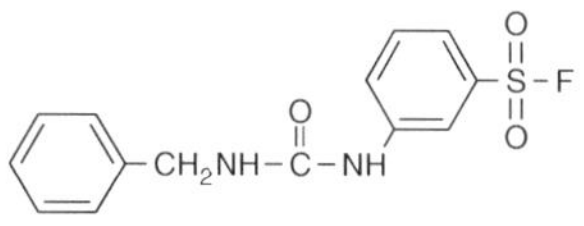
250 mg

S62,272-9
2-(4-FLUOROSULFONYLPHENYL)-5-HYDRAZONO-4-METHYL-1,3,4-THIADIAZOL-2-INE DIHYDROCHLORIDE
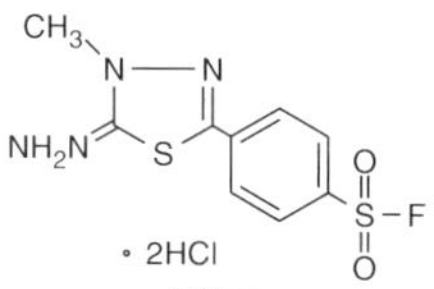
250 mg

S63,275-9
5-FLUOROSULFONYL-3-METHYL-2-BENZOTHIAZOLINONE HYDRAZONE HYDROCHLORIDE
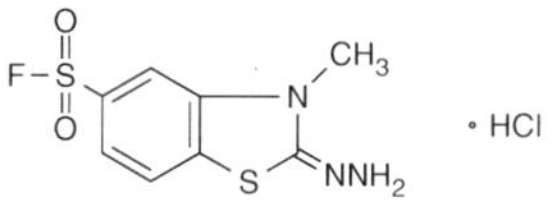
250 mg

S38,694-4
4-BENZENESULFONYL-3-NITROBENZENESULFONYL FLUORIDE
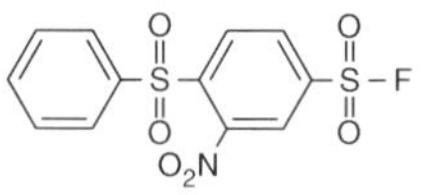
250 mg

S38,766-5
4-HEXADECYLSULFONYL-3-NITROBENZENESULFONYL FLUORIDE
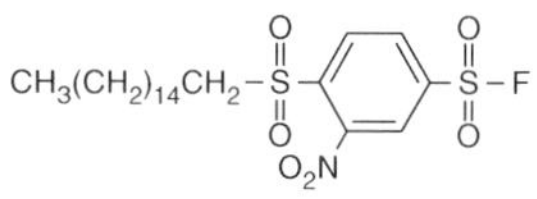
250 mg

S62,881-6
3-ACETAMIDO-4-HEXADECYLSULFONYLBENZENESULFONYL FLUORIDE
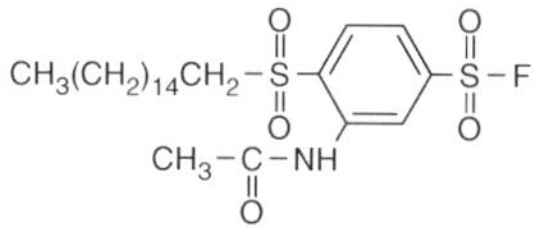
5 gm

S65,576-7
2-[3-(1-NAPHTHYLUREIDO)]BENZENE-SULFONYL FLUORIDE
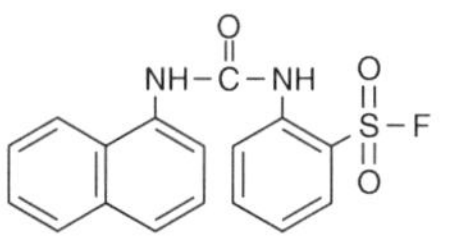
250 mg

S74,007-1
6-DIAZO-5,6-DIHYDRO-5-OXO-1-NAPHTHALENESULFONYL CHLORIDE
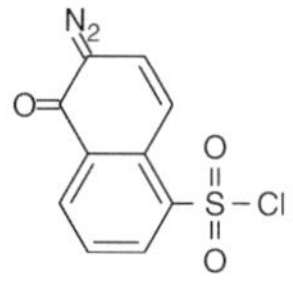
250 mg

S71,161-6
3,5-DINITRO-P-TOLUENESULFONYL CHLORIDE
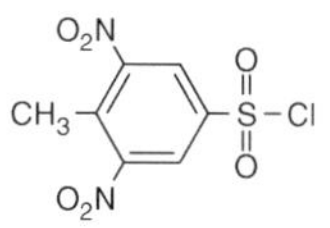
250 mg

S38,312-0
3-FLUOROSULFONYL-5-NITROBENZOIC ACID
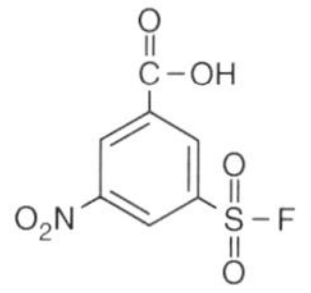
250 mg

S36,411-8
3-NITROBENZENESULFONYL FLUORIDE
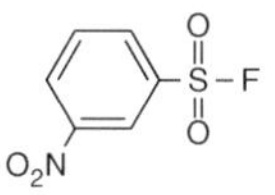
5 gm

S35,960-2
4-CHLORO-3-NITROBENZENESULFONYL FLUORIDE
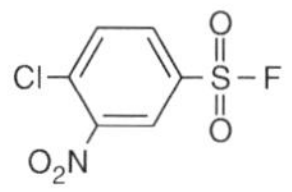
5 gm

S67,190-8
4-METHYL-3-NITROBENZENESULFONYL FLUORIDE
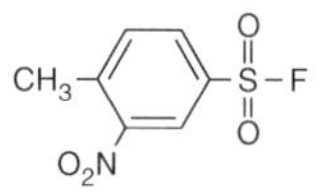
5 gm

S38,591-3
4-(TERT-BUTYL)-3-NITROBENZENESULFONYL FLUORIDE
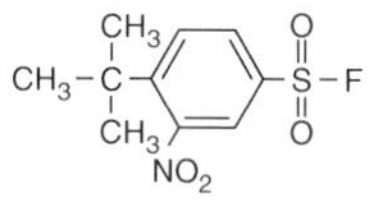
5 gm

S59,838-0
2,4-DIMETHYL-5-NITROBENZENESULFONYL FLUORIDE
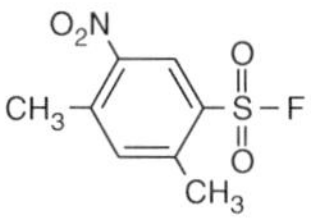
250 mg

S60,554-9
5-NITRO-1-NAPHTHALENESULFONYL FLUORIDE
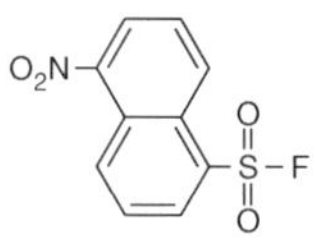
250 mg

S39,195-6
ETHYL 3-(FLUOROSULFONYL)-BENZOYLACETATE
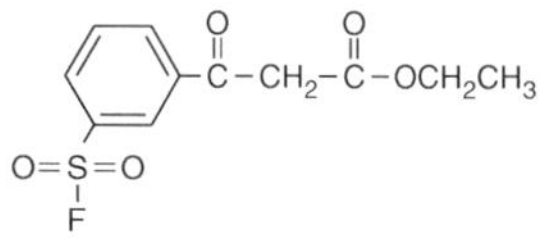
5 gm

S60,542-5
3-(2-BENZOYLACETAMIDO)-BENZENESULFONYL FLUORIDE
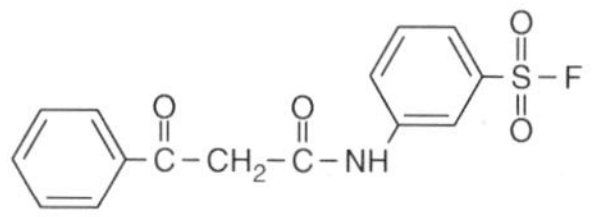
5 gm

S78,700-0
5-(CHLOROSULFONYL)ISOPHTHALIC ACID

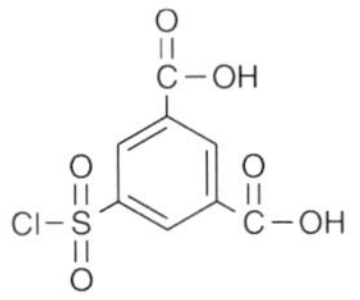

250 mg

S38,314-7
4-(FLUOROSULFONYL)PHTHALIC ACID

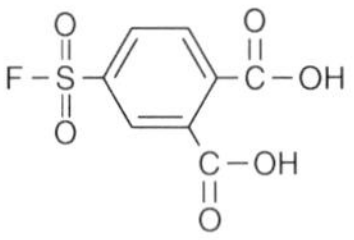

250 mg

S34,903-8
5-(FLUOROSULFONYL)ISOPHTHALIC ACID

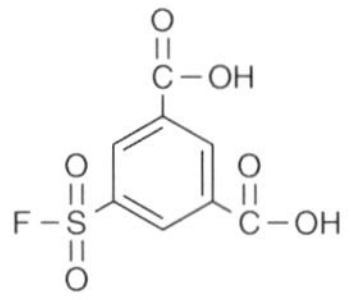

5 gm

S38,588-3
3-(FLUOROSULFONYL)BENZOIC ACID, COPPER(II) SALT

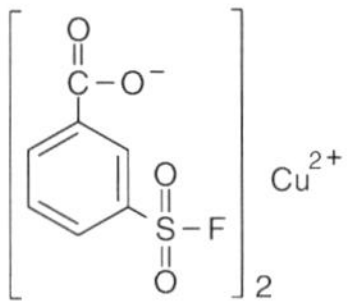

250 mg

S65,595-3
2-[3-(3-FLUOROSULFONYLPHENYL)-UREIDO]BENZOIC ACID

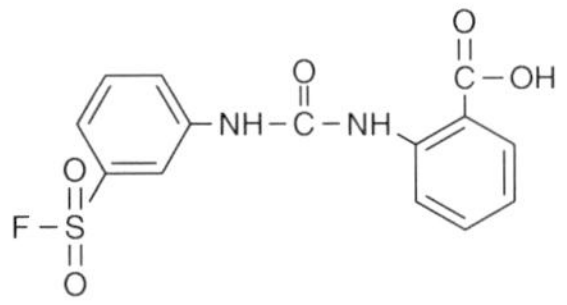

5 gm

S58,268-9
5-(CHLOROSULFONYL)-2-(HEXADECYLSULFONYL)BENZOIC ACID

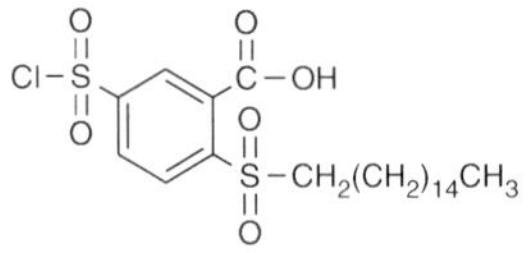

250 mg

S36,752-4
3-(CHLOROSULFONYL)BENZOIC ACID

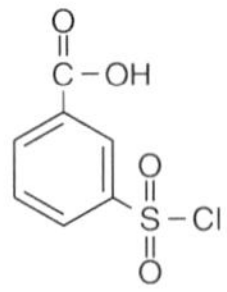

250 mg

S62,863-8
3,5-BIS(FLUOROSULFONYL)BENZOIC ACID

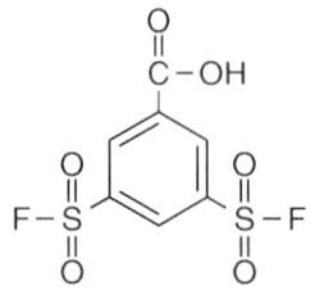

250 mg

S65,628-3
ETHYL 4-[3-(FLUOROSULFONYL)-PHENYLUREIDO]BENZOATE

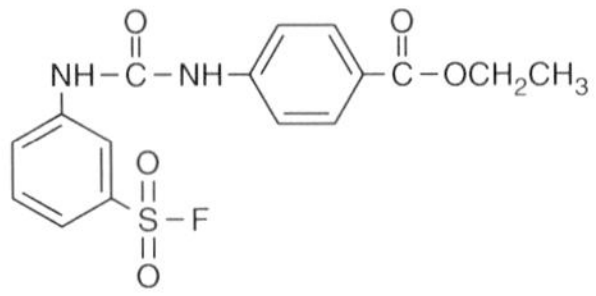

250 mg

S60,509-3
METHYL 2-(FLUOROSULFONYL)BENZOATE

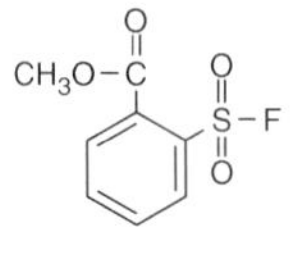

250 mg

S58,825-3
5-CHLOROSULFONYL-2-(HEXADECYLSULFONYL)BENZOYL CHLORIDE

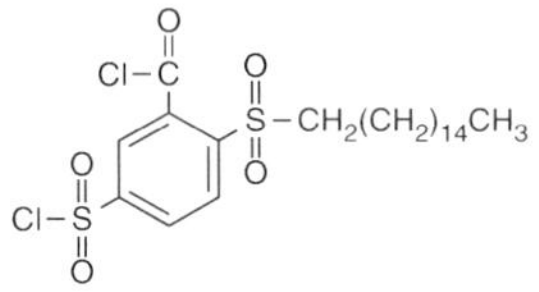

5 gm

S59,836-4
1-CHLORO-4-FLUOROSULFONYL-2-NAPHTHOYL CHLORIDE

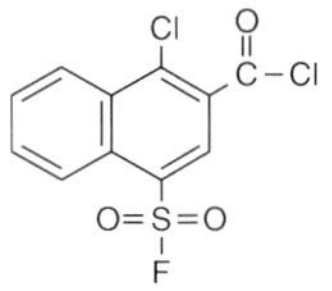

5 gm

S65,388-8
4-FLUOROSULFONYL-N-HEXADECYLBENZAMIDE

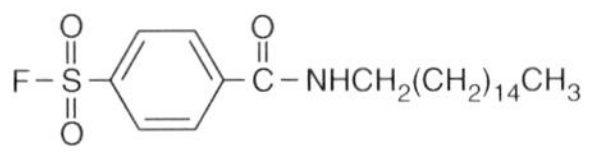

250 mg

S39,297-9
BIS[4-(FLUOROSULFONYL)PHENYL]-SUCCINATE

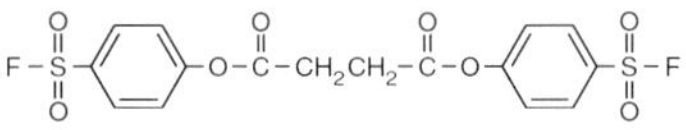

5 gm

S51,586-8
5-FLUOROSULFONYL-2-(HEXADECYLSULFONYL)BENZOIC ACID

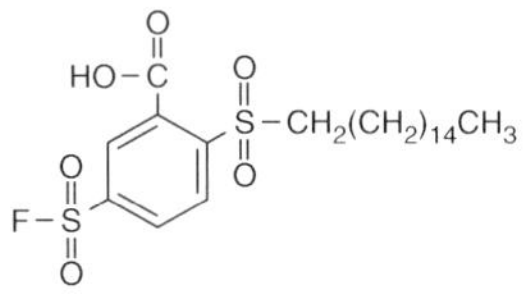

250 mg

S43,141-9
N-ACETYLSULFANILYL FLUORIDE

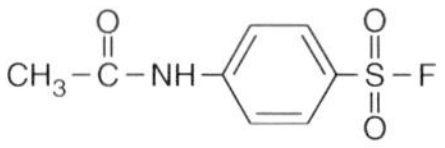

5 gm

S53,749-7
3-(PIPERIDINOCARBONYL)-BENZENESULFONYL FLUORIDE

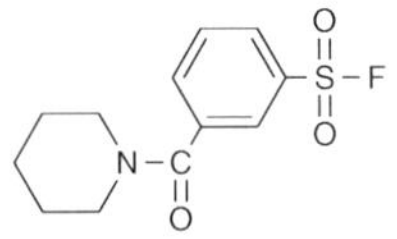

250 mg

S65,594-5
3-(3-ETHYLUREIDO)-BENZENESULFONYL FLUORIDE

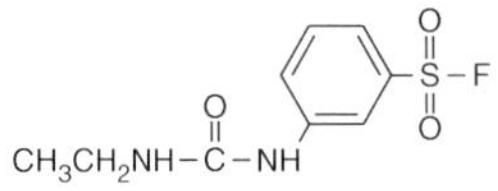

5 gm

S65,637-2
1-(3-FLUOROSULFONYLPHENYL)-3-PHENYLUREA

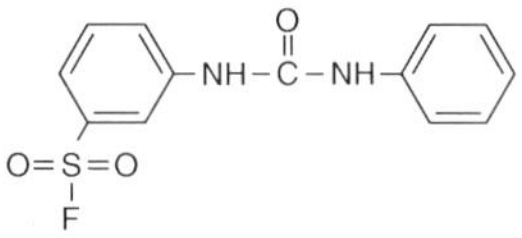

250 mg

S65,629-1
1-(3-FLUOROSULFONYLPHENYL)-3-HEXYLUREA

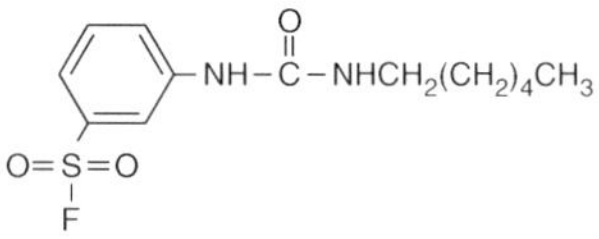

250 mg

S65,599-6
1-(2-FLUOROSULFONYLPHENYL)-3-(3-FLUOROSULFONYLPHENYL)UREA

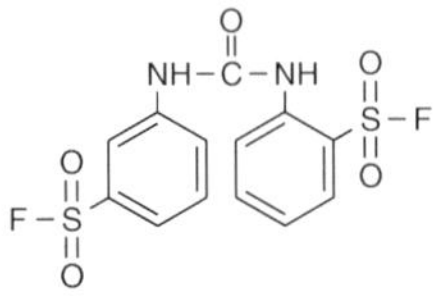

250 mg

S39,068-2
5-METHYLSULFONYL-ORTHO-TOLUENESULFONYL CHLORIDE

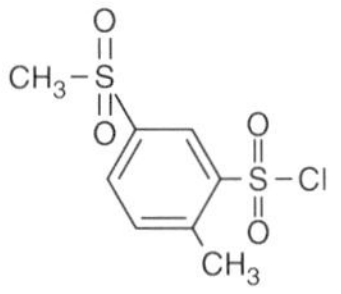

250 mg

S62,687-2
PENTAMETHYLBENZENESULFONYL CHLORIDE

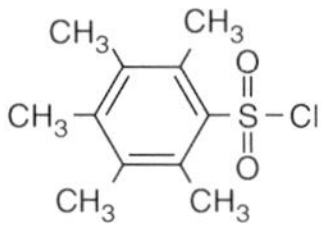

5 gm

S46,134-2
O-FLUOROSULFONYLBENZENESULFONYL CHLORIDE

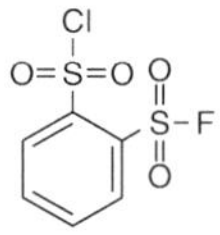

5 gm

S68,574-7
O-TOLUENESULFONYL CHLORIDE

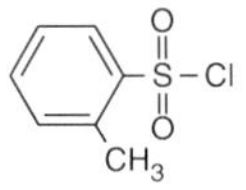

5 gm

S59,275-7
4-(1-BROMOETHYL)-BENZENESULFONYL FLUORIDE

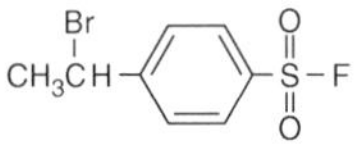

5 gm

S41,771-8
2-CHLORO-5-FLUOROSULFONYLBENZENESULFONYL CHLORIDE

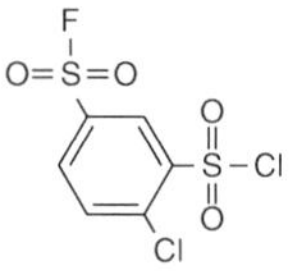

5 gm

S59,729-5
2,5-DIMETHYLBENZENESULFONYL CHLORIDE

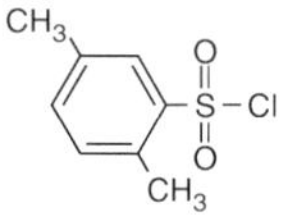

5 gm

S38,601-4
2-MESITYLENESULFONYL FLUORIDE

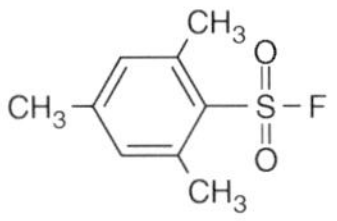

250 mg

S38,289-2
2,6-NAPHTHALENEDISULFONYL CHLORIDE

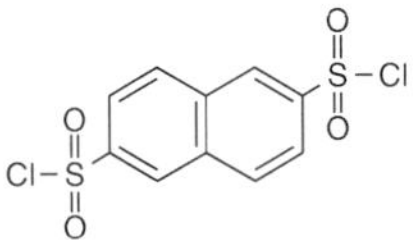

250 mg

S78,125-8
L-1-(P-TOSYLAMINO)ISOBUTYL CHLOROMETHYL KETONE

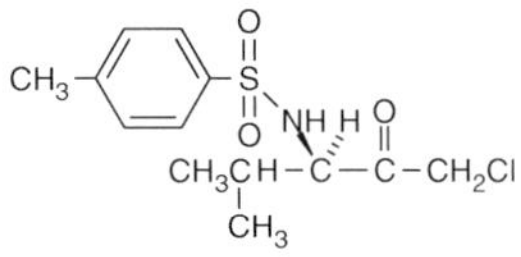

250 mg

S87,157-5
2-(3-BROMOPROPYL)-1,2-BENZISOTHIAZOL-3(2H)-ONE-1,1-DIOXIDE

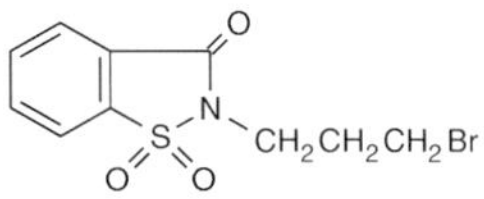

250 mg

S73,623-6
N4-CHLOROACETYLSULFANILAMIDE
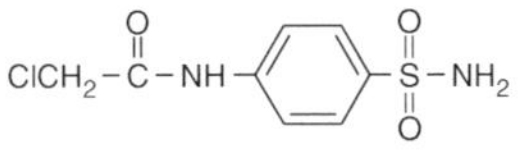
5 gm

S39,812-8
1-BUTYL-3-(ALPHA-CHLORO-PARA-TOLYLSULFONYL)UREA
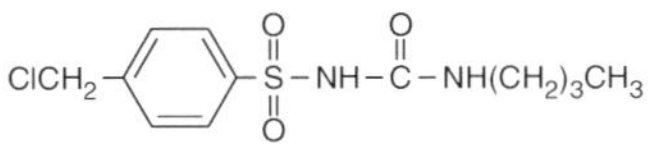
5 gm

S35,863-0
N,N-BIS(2-CHLOROETHYL)-P-TOLUENESULFONAMIDE
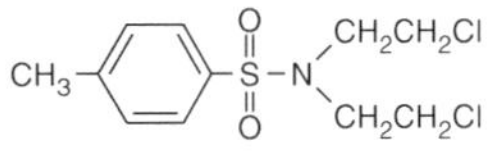
5 gm

S57,868-1
3-(TERT-DODECYLCARBAMOYL)BENZENESULFONIC ACID (5-DIFLUOROMETHYLSULFONYL-3-METHYL-2-BENZOTHIAZOLINYLIDENE)HYDRAZIDE
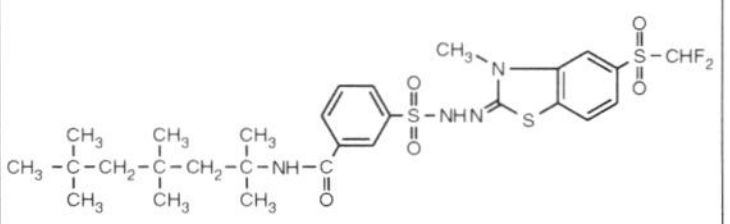
5 gm

S58,234-4
4-(TERT-DODECYLCARBAMOYL)BENZENESULFONIC ACID (5-DIFLUOROMETHYLSULFONYL-3-METHYL-2-BENZOTHIAZOLINYLIDENE)HYDRAZIDE
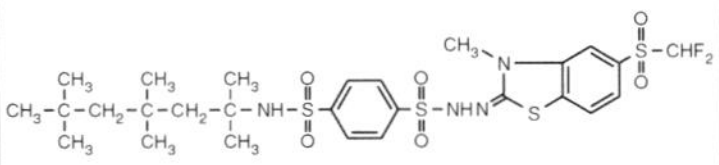
250 mg

S58,175-5
4-(HEXADECYLSULFAMOYL)BENZENESULFONIC ACID (5-DIFLUOROMETHYLSULFONYL-3-METHYL-2-BENZOTHIAZOLINYLIDENE)HYDRAZIDE
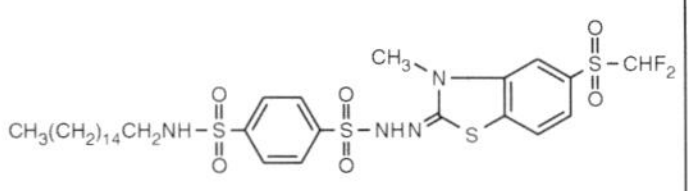
250 mg

S57,751-0
3-(HEXADECYLSULFAMOYL)BENZENESULFONIC [5-(DIFLUOROMETHYLSULFONYL)-3-METHYL-2-BENZOTHIAZOLINYLIDENE]HYDRAZIDE
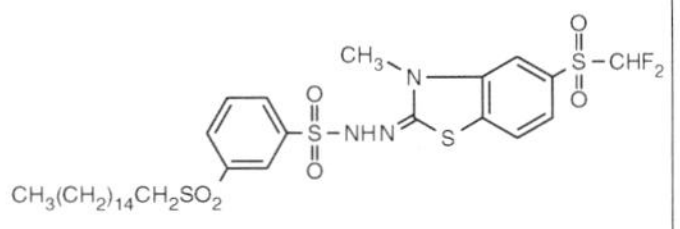
250 mg

S56,869-4
ALKYLCARBAMOYLBENZENESULFONIC ACID [5-(DIFLUOROMETHYLSULFONYL)-3-METHYL-2-BENZOTHIAZOLINYLIDENE]HYDRAZIDE
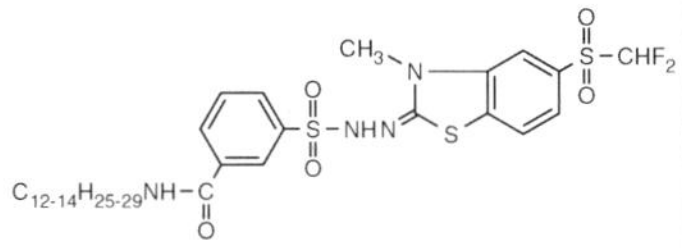
250 mg

S90,531-3
2'-CHLORO-2-NITRO-5'-(TRIFLUOROMETHYL)-BENZENESULFONANILIDE
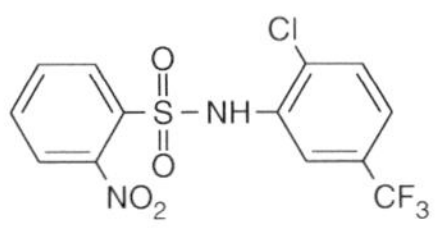
250 mg

S68,401-5
N-ACETYL-4-[4-CHLORO-2-(TRIFLUOROMETHYL)PHENYLUREIDO]-BENZENESULFONAMIDE
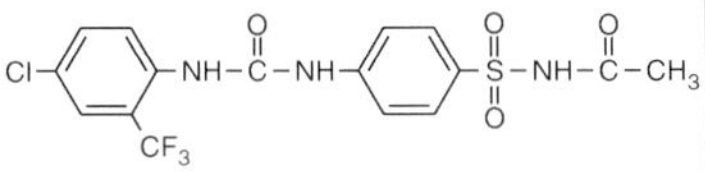
250 mg

S90,575-5
1-[2-CHLORO-5-(TRIFLUOROMETHYL)-PHENYL]-3-(4-SULFAMOYLPHENYL)UREA
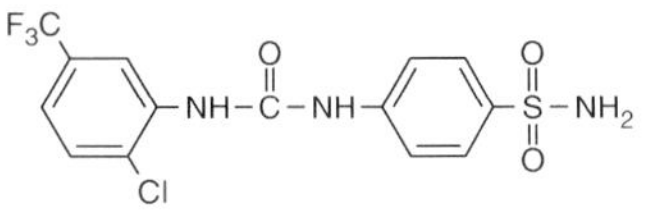
250 mg

S67,590-3
4-CHLORO-N-[2-CHLORO-5-(TRIFLUOROMETHYL)PHENYL]-BENZENESULFONAMIDE
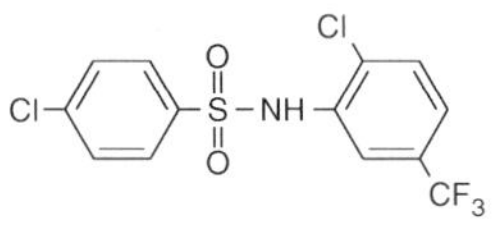
5 gm

S90,526-7
3,4,4'-TRICHLORO-2'-(TRIFLUOROMETHYL)-BENZENESULFONANILIDE
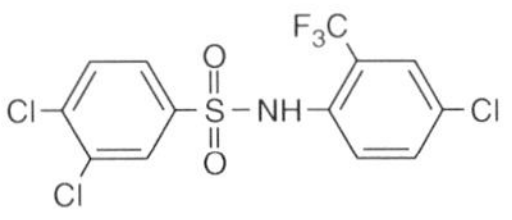
250 mg

S79,571-2
2,4,5-TRICHLORO-3'-(TRIFLUOROMETHYL)-BENZENESULFONANILIDE
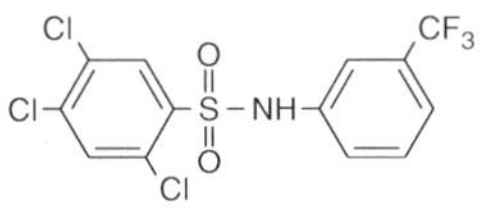
250 mg

S86,234-7
4-(TRIFLUOROMETHYLTHIO)-BENZENESULFONAMIDE
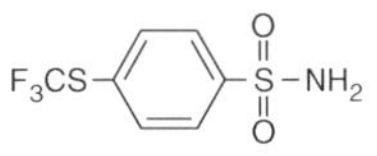
250 mg

S91,688-9
2'-NITRO-4'-(TRIFLUOROMETHYL)-METHANESULFONANILIDE
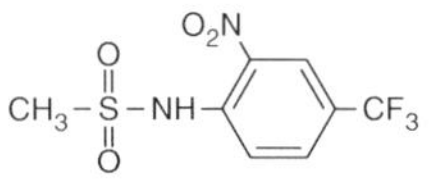
250 mg

S85,790-4
2',6'-DINITRO-4'-(TRIFLUOROMETHYL)-METHANESULFONANILIDE
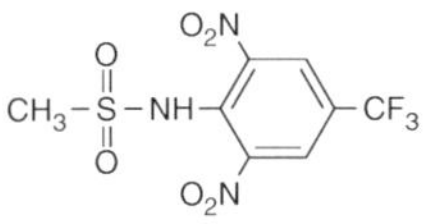
250 mg

S85,791-2
2',6'-DINITRO-N-METHYL-4'-(TRIFLUOROMETHYL)-METHANESULFONANILIDE
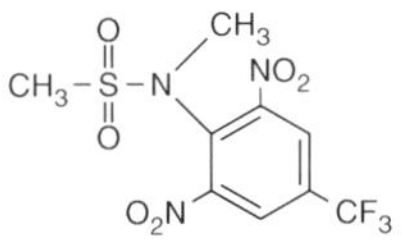
250 mg

S85,792-0
2',6'-DINITRO-N-ETHYL-4'-(TRIFLUOROMETHYL)-METHANESULFONANILIDE
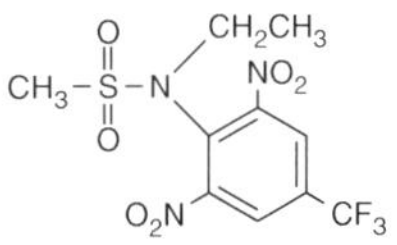
250 mg

S85,793-9
2',6'-DINITRO-N-PROPYL-4'-(TRIFLUOROMETHYL)-METHANESULFONANILIDE
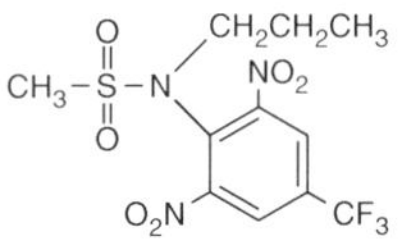
250 mg

S85,878-1
2',6'-DINITRO-N-HEXYL-4'-(TRIFLUOROMETHYL)-METHANESULFONANILIDE
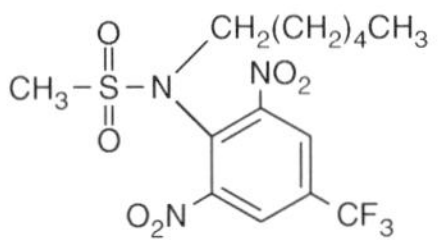
250 mg

S79,567-4
3-NITRO-3'-(TRIFLUOROMETHYL)-BENZENESULFONANILIDE
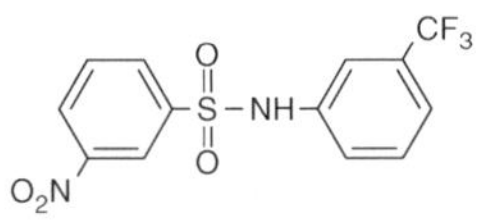
250 mg

S72,172-7
N3-TRICHLOROACETYLMETANILAMIDE
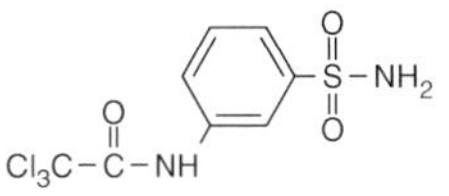
250 mg

S90,122-9
N4-ACETYL-2'-TRIFLUOROMETHYL-SULFANILANILIDE
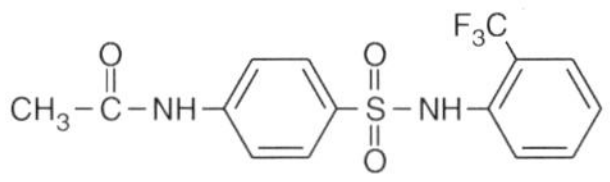
250 mg

S80,170-4
1-(4-SULFAMOYLPHENYL)-3-[2-(TRIFLUOROMETHYL)PHENYL]UREA
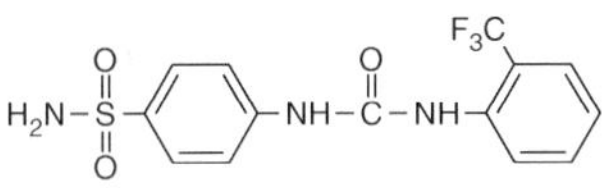
250 mg

S89,123-1
1-(P-TOSYL)-3-[3-(TRIFLUOROMETHYL)-PHENYL]UREA
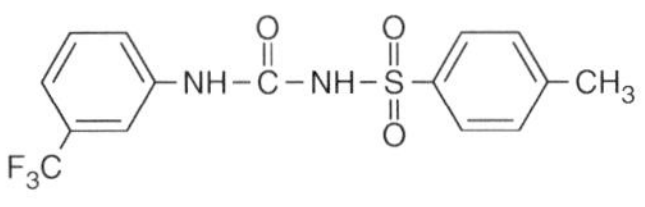
250 mg

S86,233-9
4-(TRIFLUOROMETHYLSULFONYL)-BENZENESULFONAMIDE
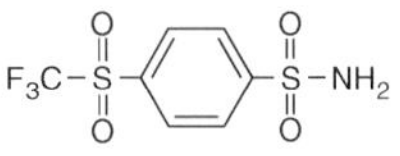
250 mg

S86,232-0
ALPHA,ALPHA,ALPHA-TRIFLUORO-P-TOLUENESULFONAMIDE
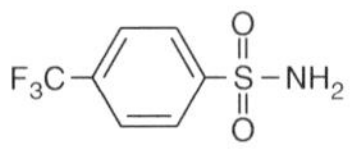
250 mg

S94,974-4
2-(TRICHLOROMETHYLTHIO)-1,2-BENZISOTHIAZOL-3(2H)-ONE-1,1-DIOXIDE
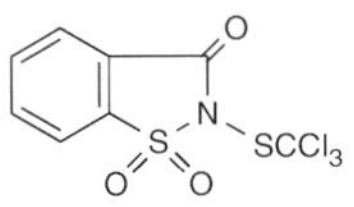
250 mg

S94,723-7
BENZENESULFONIC ACID 2-[2-BROMO-5-(TRIFLUOROMETHYL)PHENYL]HYDRAZIDE
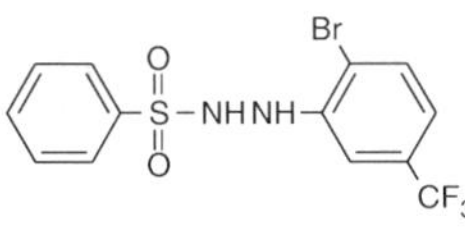
250 mg

S90,185-7
3'-(TRIFLUOROMETHYL)-2,4,6-TRIMETHYLBENZENESULFONANILIDE
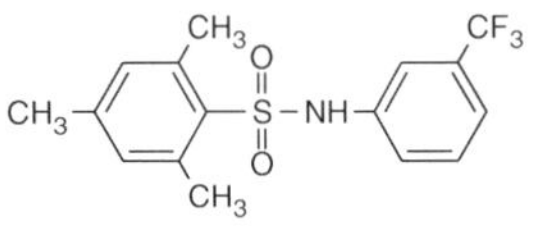
250 mg

S92,460-1
3'-TRIFLUOROMETHYL-2-NAPHTHALENESULFONANILIDE
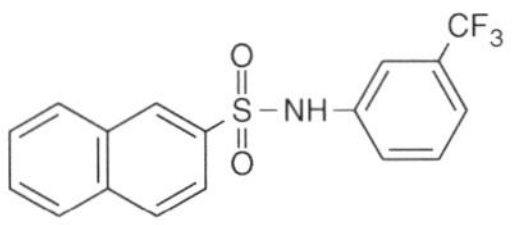
250 mg

S78,019-7
N-[3-(4-CHLOROPHENOXY)-2-HYDROXYPROPYL]SACCHARIN
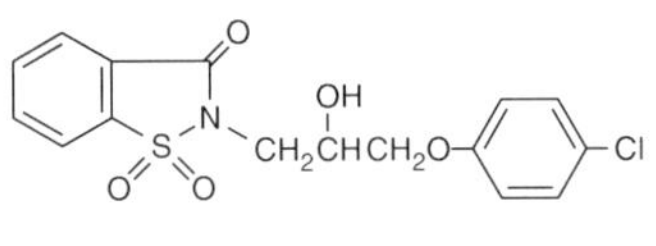
250 mg

S90,589-5
5'-CHLORO-2'-HYDROXYMETHANESULFONANILIDE
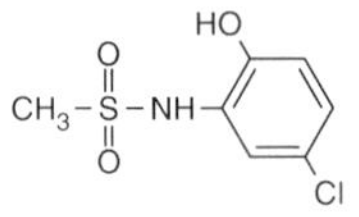
250 mg

S67,614-4
4-CHLORO-N-(3-HYDROXYPHENYL)-BENZENESULFONAMIDE
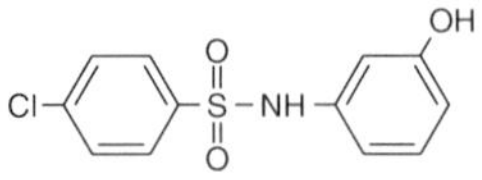
5 gm

S72,123-9
1-(5-CHLORO-2,4-DIMETHOXYPHENYL)-3-(4-SULFAMOYLPHENYL)UREA
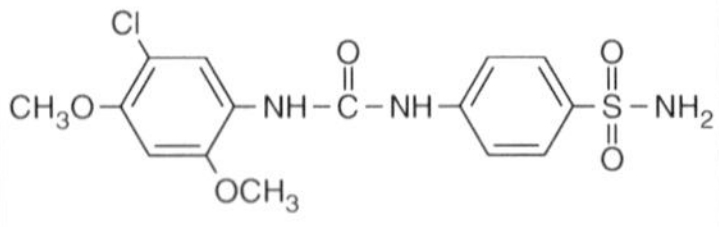
250 mg

S68,823-1
4-BROMO-N-(5-CHLORO-2,4-DIMETHOXYPHENYL)BENZENESULFONAMIDE
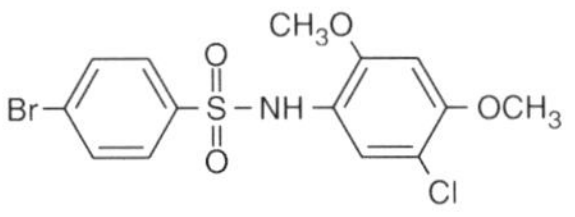
250 mg

S90,527-5
2',5'-DIMETHOXY-3,4,4'-TRICHLOROBENZENESULFONANILIDE
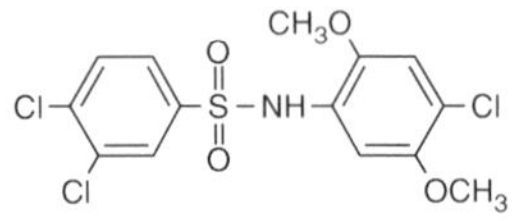
250 mg

S90,596-8
4'-CHLORO-2',5'-DIMETHOXYMETHANESULFONANILIDE
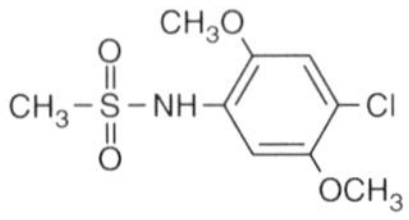
250 mg

S92,469-5
5'-CHLORO-2',4'-DIMETHOXYMETHANESULFONANILIDE
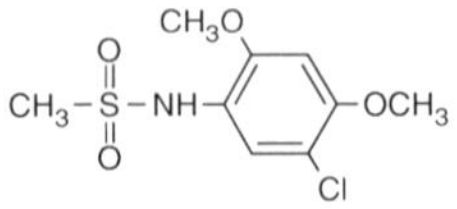
250 mg

S90,135-0
4-CHLORO-2'-METHOXY-5'-NITROBENZENESULFONANILIDE
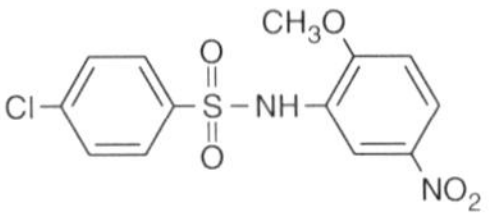
250 mg

S76,914-2
4-CHLORO-3,5'-DINITRO-2'-METHOXYBENZENESULFONANILIDE
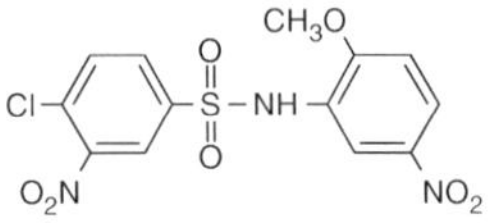
250 mg

S90,110-5
2'-METHOXY-5'-NITRO-2,4,5-TRICHLOROBENZENESULFONANILIDE
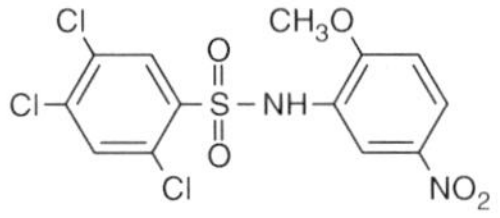
250 mg

S71,833-5
N-(3-CHLORO-2-METHOXYPHENYL)-N'-DIMETHYLSULFAMIDE
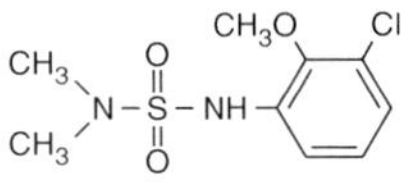
25 mg

S79,568-2
4,5'-DICHLORO-2'-METHOXY-2-NITROBENZENESULFONANILIDE
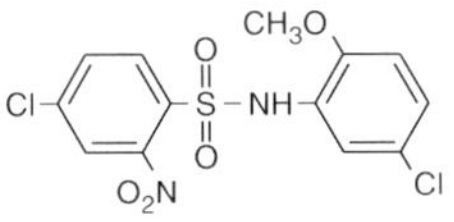
250 mg

S74,367-4
5'-CHLORO-2'-METHOXY-3-NITROBENZENESULFONANILIDE
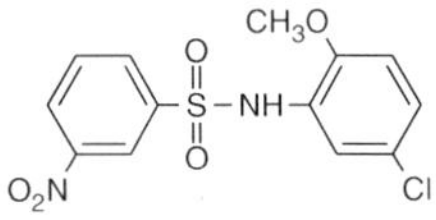
250 mg

S90,112-1
4,5'-DICHLORO-2'-METHOXY-3-NITROBENZENESULFONANILIDE
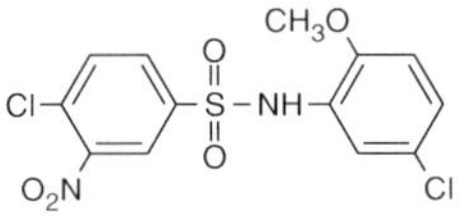
250 mg

S72,674-5
4-(5-CHLORO-2-METHOXYPHENYL)-1-(P-TOSYL)SEMICARBAZIDE
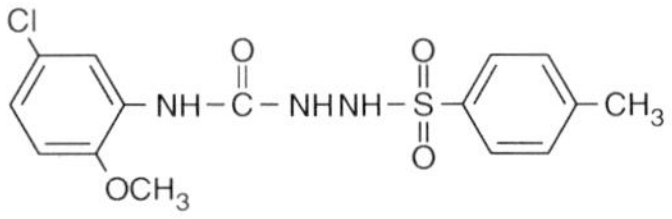
5 gm

S67,457-5
N4-ACETYL-5'-CHLORO-2'-METHOXY-SULFANILANILIDE
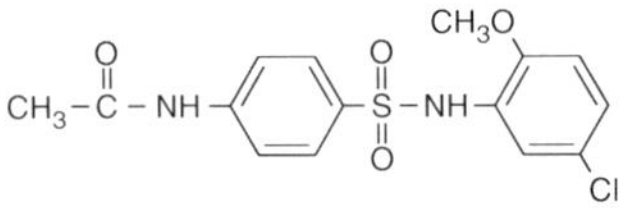
250 mg

S91,684-6
4'-CHLORO-2'-(PHENOXY)-METHANESULFONANILIDE
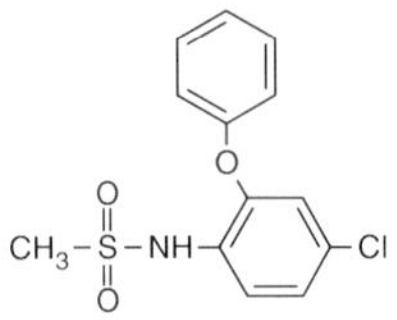
250 mg

S75,057-3
4-CHLORO-2'-METHOXYBENZENESULFONANILIDE
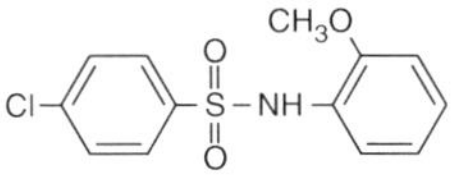
5 gm

S90,136-9
4,5'-DICHLORO-2'-METHOXYBENZENESULFONANILIDE
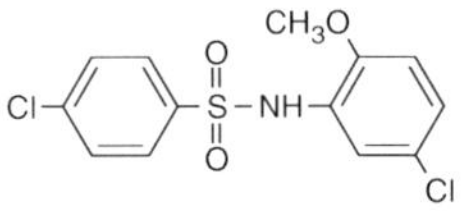
250 mg

S71,831-9
3,4-DICHLORO-2'-METHOXY-5'-METHYLBENZENESULFONANILIDE
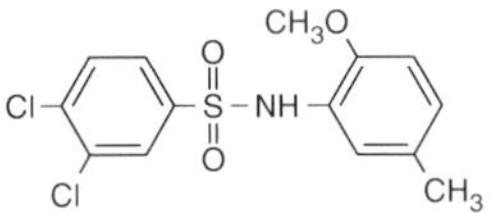
5 gm

S90,116-4
2'-METHOXY-2,4,5,5'-TETRACHLOROBENZENESULFONANILIDE
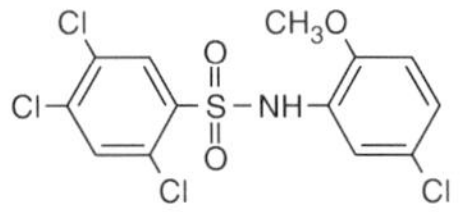
250 mg

S79,563-1
5'-CHLORO-2'-METHOXY-2-NAPHTHALENESULFONANILIDE
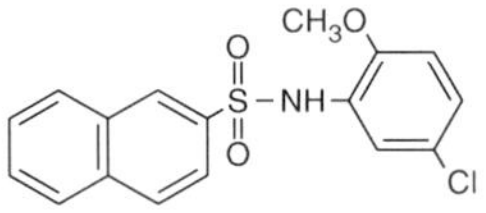
250 mg

S68,708-1
4-BROMO-4'-PHENOXYBENZENE-SULFONANILIDE
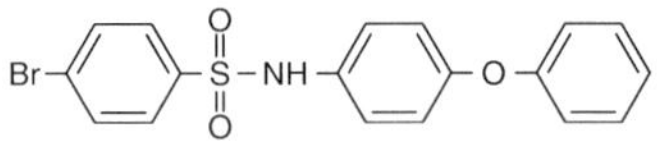
250 mg

S68,478-3
4-BROMO-4'-CHLORO-2'-PHENOXY-BENZENESULFONANILIDE
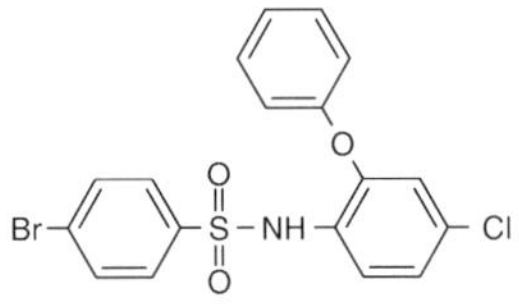
250 mg

S90,137-7
4-CHLORO-4'-ETHOXYBENZENESULFONANILIDE
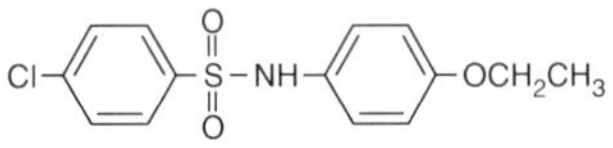
250 mg

S67,617-9
2',4-DICHLORO-4'-METHOXYBENZENE-SULFONANILIDE
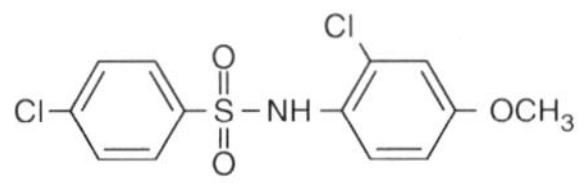
5 gm

S79,569-0
2',5'-DICHLORO-4-METHOXYBENZENESULFONANILIDE
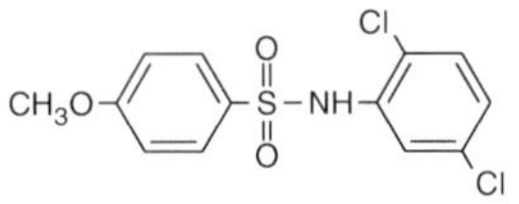
250 mg

S90,097-4
5'-CHLORO-4-METHOXY-2'-METHYLBENZENESULFONANILIDE
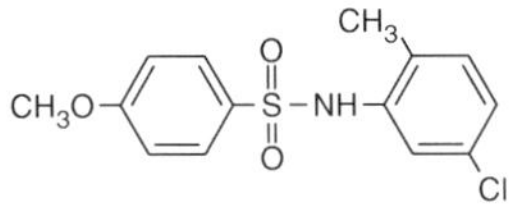
250 mg

S75,060-3
3,4-DICHLORO-4'-ETHOXYBENZENESULFONANILIDE
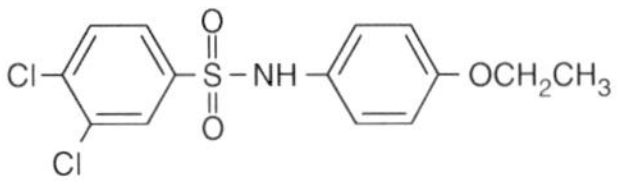
5 gm

S77,574-6
4-(4-CHLOROPHENYLTHIOMETHYL-SULFONYL)MORPHOLINE
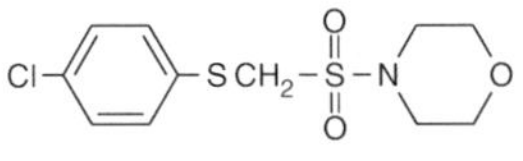
250 mg

S64,760-8
2'-AMINO-4-CHLORO-4'-METHYLBENZENESULFONANILIDE
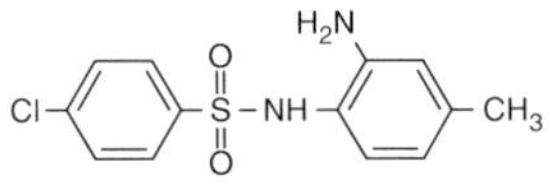
5 gm

S51,614-7
4-CHLORO-N-HEXADECYL-N-PHENETHYLMETANILAMIDE
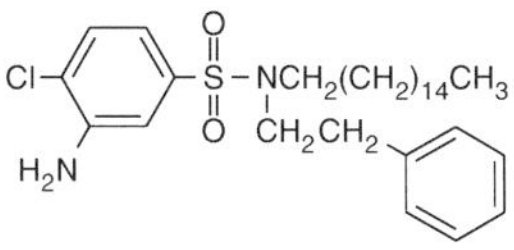
250 mg

S35,148-2
4'-IODOSULFANILANILIDE
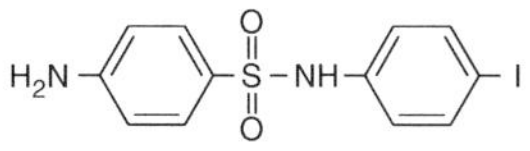
5 gm

S66,081-7
5-AMINO-2-CHLORO-2',4'-DIMETHYLBENZENESULFONANILIDE
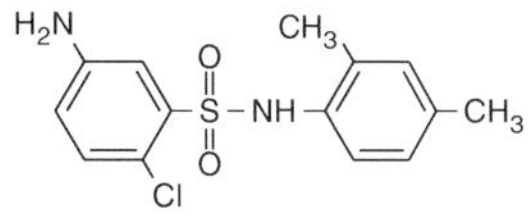
5 gm

S56,638-1
2,6-DIBROMOSULFANILAMIDE
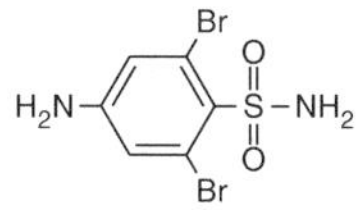
5 gm

S65,952-5
3,5-DIBROMOSULFANILAMIDE
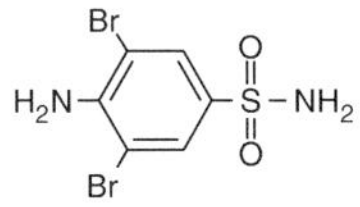
250 mg

S65,861-8
3,5-DICHLORO-N,N-DIMETHYLSULFANILAMIDE
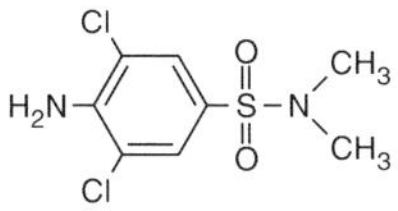
5 gm

S57,199-7
2,5-DICHLORO-N,N-DIMETHYLSULFANILAMIDE
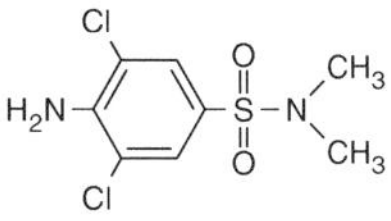
5 gm

S50,543-9
METANILIC ACID 2-(5-CHLORO-ORTHO-TOLYL)-HYDRAZIDE
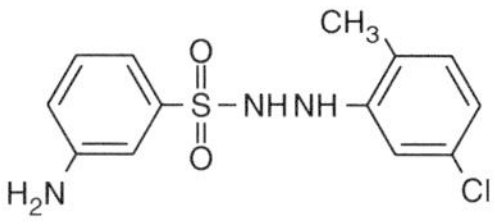
250 mg

S86,734-9
1-(4-CHLORO-2-NITROPHENYL)-3-(4-SULFAMOYLPHENYL)UREA
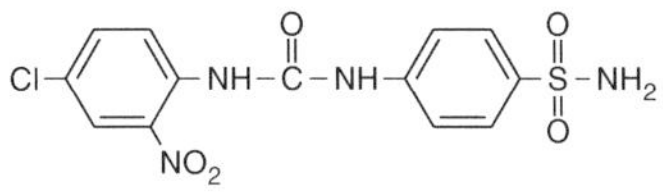
250 mg

S39,415-7
N,N'-ETHYLENEBIS(4-CHLOROBENZENESULFONAMIDE)
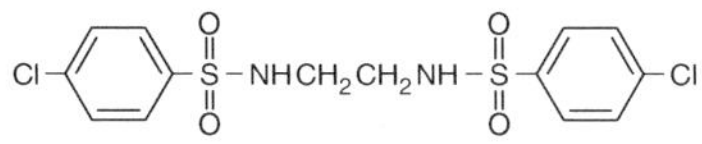
250 mg

S68,658-1
N-(4-CHLORO-3-NITROPHENYL)-2-NITROBENZENESULFONAMIDE
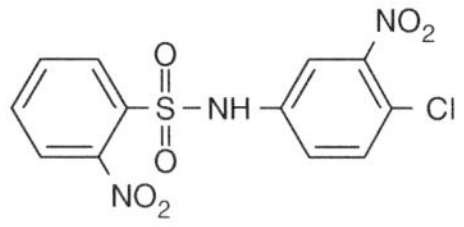
250 mg

S84,448-9
1-(2-FLUORO-5-NITROPHENYL)-3-(4-SULFAMOYLPHENYL)UREA
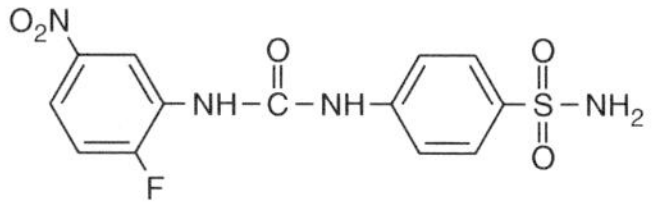
250 mg

S90,529-1
3,4-DICHLORO-3'-NITROBENZENESULFONANILIDE
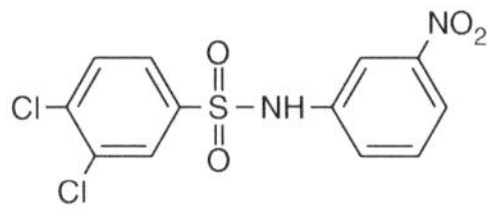
250 mg

S74,371-2
4-CHLORO-3,5'-DINITRO-2'-METHYLBENZENESULFONANILIDE
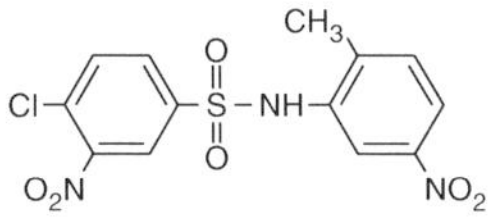
250 mg

S71,551-4
1,1'-(4,6-DIBROMO-1,3-PHENYLENE)-BIS[3-(4-SULFAMOYLPHENYL)UREA]
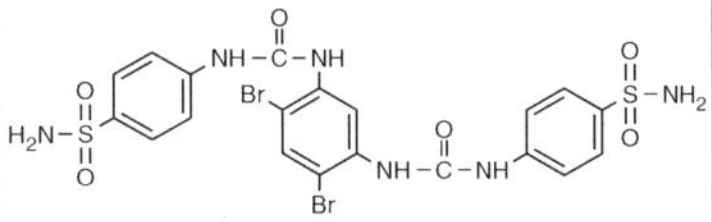
250 mg

S76,235-0
4'-CHLORO-2'-(3,4-DIHYDRO-4,4-DIMETHYL-1-ISOQUINOLINYL)-N,4-DIMETHYL-BENZENESULFONANILIDE HYDROCHLORIDE
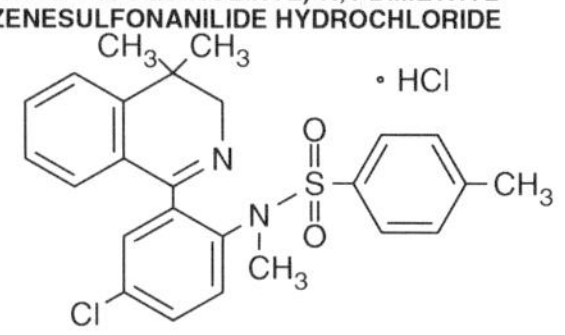
250 mg

S89,054-5
2'-CHLORO-4'-NITROBENZENESULFONANILIDE
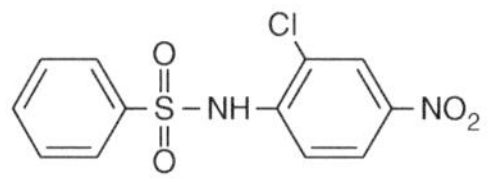
250 mg

S68,432-5
4-CHLORO-2'-METHYL-4'-NITRO-BENZENESULFONANILIDE
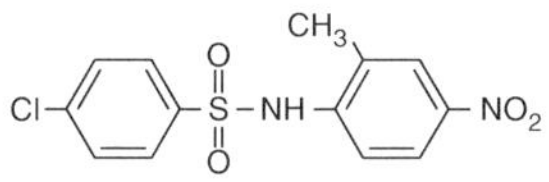
250 mg

S72,160-3
4'-NITRO-2',3,4-TRICHLOROSULFONANILIDE
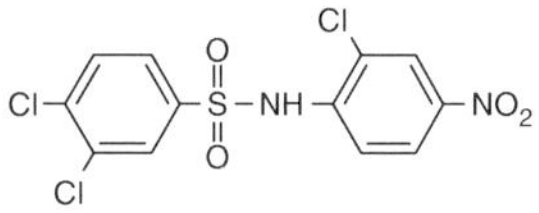
5 gm

S92,402-4
N'-(3-CHLORO-4-METHYLPHENYL)-N,N-DIMETHYLSULFAMIDE
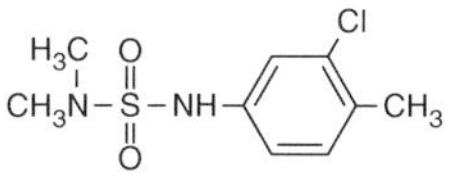
250 mg

S74,366-6
ETHYL 4-(4-CHLORO-3-NITROBENZENESULFONAMIDO)BENZOATE
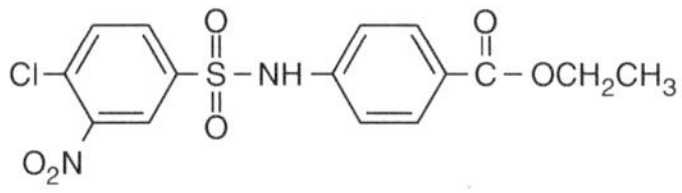
5 gm

S90,108-3
2'-FLUORO-3-NITROBENZENESULFONANILIDE
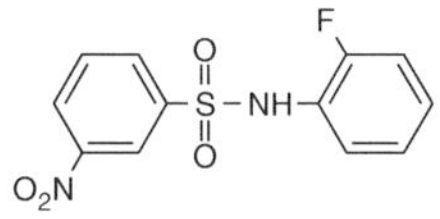
250 mg

S79,178-4
2'-CHLORO-3-NITROBENZENESULFONANILIDE
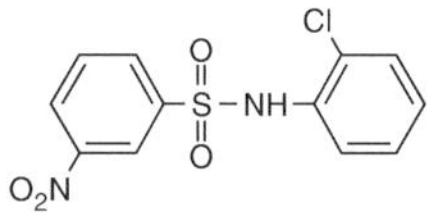
250 mg

S90,107-5
4-CHLORO-2'-FLUORO-3-NITROBENZENESULFONANILIDE
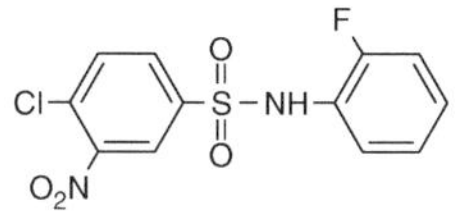
250 mg

S79,180-6
3'-CHLORO-2'-METHYL-3-NITROBENZENESULFONANILIDE
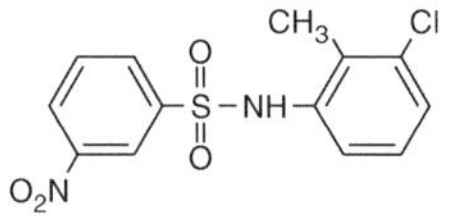
250 mg

S63,079-9
4-CHLORO-N,N-DIMETHYL-3-NITROBENZENESULFONAMIDE
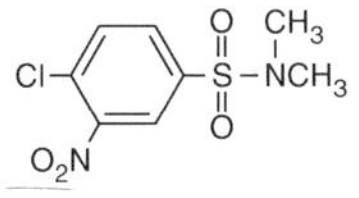
250 mg

S77,366-2
5'-CHLORO-2'-METHYL-3-NITROBENZENESULFONANILIDE
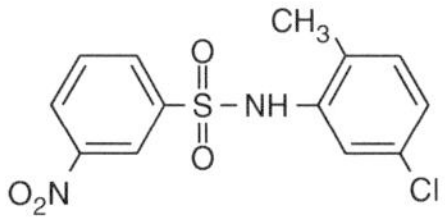
250 mg

S79,181-4
4'-CHLORO-2'-METHYL-3-NITROBENZENESULFONANILIDE
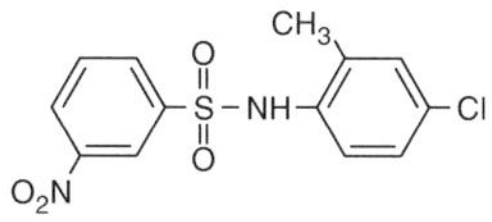
250 mg

S90,111-3
4,5'-DICHLORO-2'-METHYL-3-NITROBENZENESULFONANILIDE
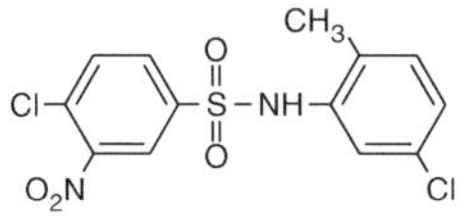
250 mg

S71,830-0
2'-BENZOYL-4'-CHLOROMETHANE-SULFONANILIDE
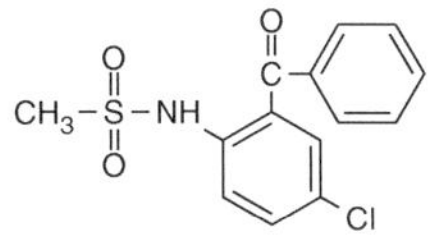
250 mg

S67,643-8
3'-ACETYL-4-CHLOROBENZENE-SULFONANILIDE
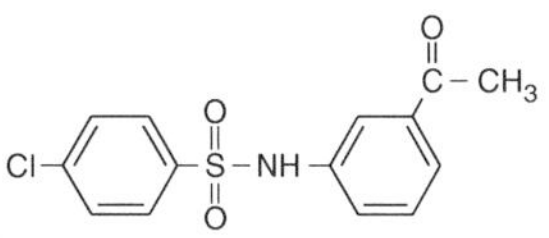
250 mg

S74,710-6
METHYL 2-(4-CHLOROPHENYLSULFONAMIDO]-BENZOATE
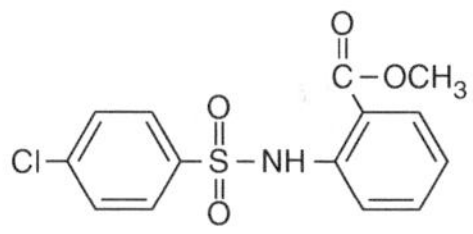
5 gm

S75,010-7
ETHYL 4-[(3,4-DICHLOROPHENYL)-SULFONAMIDO]BENZOATE
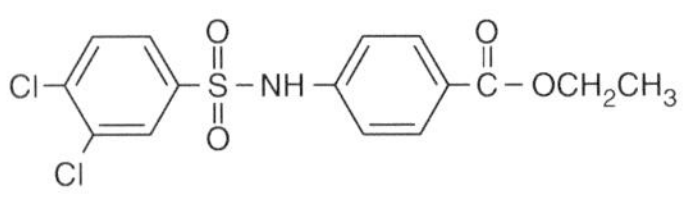
5 gm

S90,471-6
3,4-DICHLOROBENZOIC N2(PHENYLSULFONYL)HYDRAZIDE
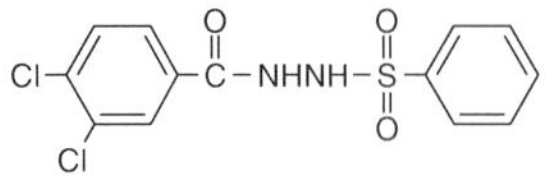
250 mg

S68,471-6
N2-(BUTYLSULFONYL)-4-CHLOROBENZOIC HYDRAZIDE
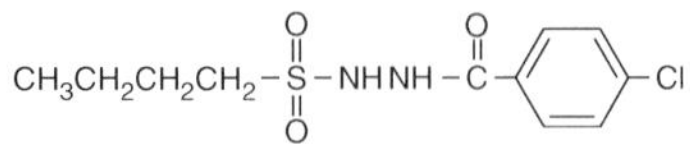
250 mg

S72,247-2
N-(4-CHLOROPHENYL)-N-TOSYL-BETA-ALANINE
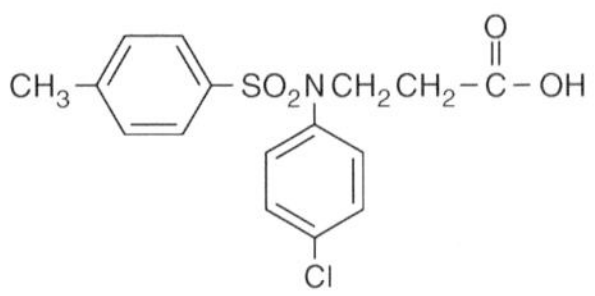
5 gm

S75,575-3
N-(3-CHLOROPHENYL)-N-TOSYL-BETA-ALANINE
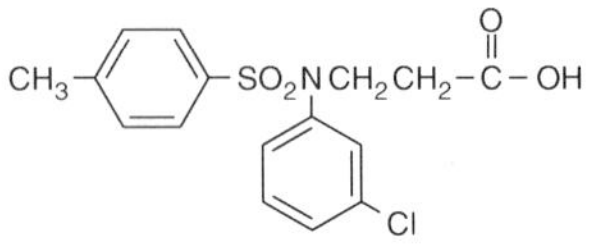
250 mg

S85,805-6
METHYL N-(3-CHLOROPHENYL)-N-(METHYLSULFONYL)CARBAMATE
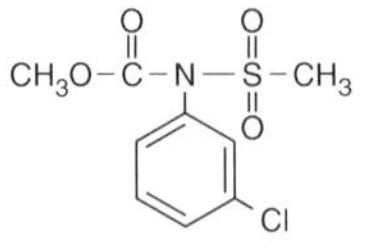
250 mg

S85,801-3
PROPYL N-(3-CHLOROPHENYL)-N-(METHYLSULFONYL)CARBAMATE
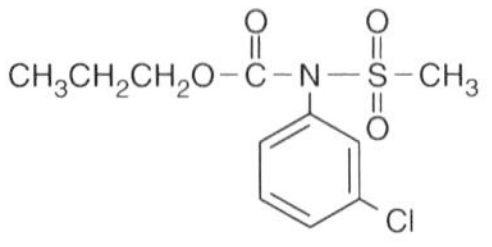
250 mg

S85,802-1
ISOPROPYL N-(3-CHLOROPHENYL)-N-(METHYLSULFONYL)CARBAMATE
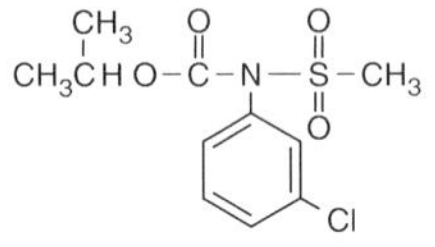
250 mg

S85,804-8
BUTYL N-(3-CHLOROPHENYL)-N-(METHYLSULFONYL)CARBAMATE
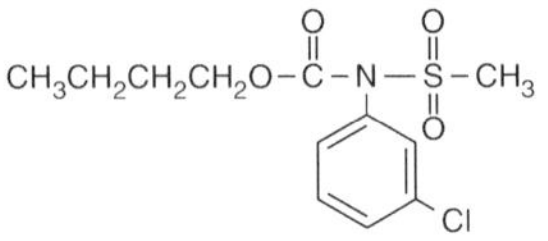
250 mg

S85,762-9
METHYL N-(4-CHLOROPHENYL)-N-(METHYLSULFONYL)CARBAMATE
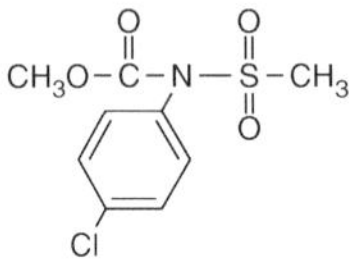
250 mg

S85,763-7
ETHYL N-(4-CHLOROPHENYL)-N-(METHYLSULFONYL)CARBAMATE
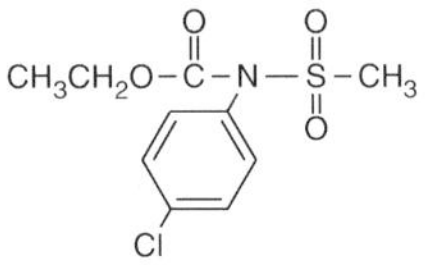
250 mg

S85,771-8
PROPYL N-(4-CHLOROPHENYL)-N-(METHYLSULFONYL)CARBAMATE
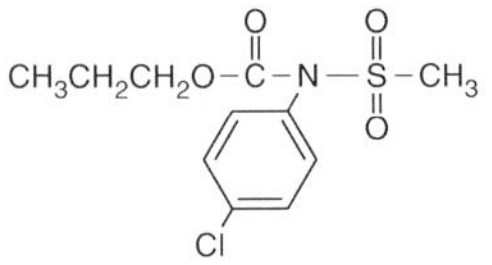
250 mg

S85,772-6
ISOPROPYL N-(4-CHLOROPHENYL)-N-(METHYLSULFONYL)CARBAMATE
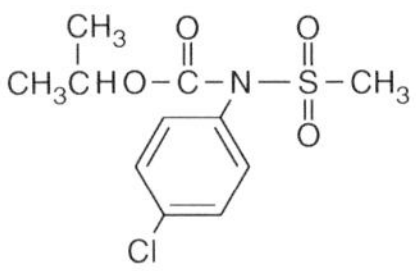
250 mg

S85,776-9
BUTYL N-(4-CHLOROPHENYL)-N-(METHYLSULFONYL)CARBAMATE
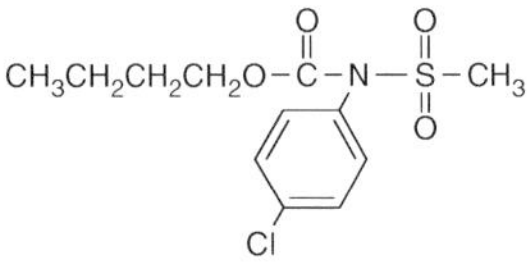
250 mg

S79,599-2
2,4-DICHLOROPHENYL N-(P-TOSYL)-CARBAMATE
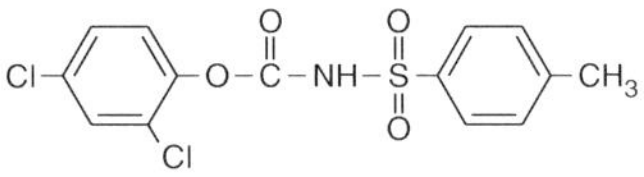
250 mg

S85,766-1
METHYL N-(3,4-DICHLOROPHENYL)-N-(METHYLSULFONYL)CARBAMATE
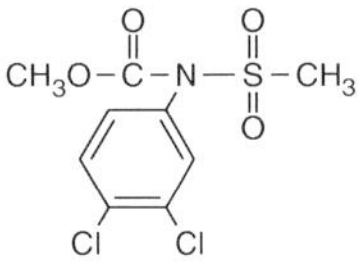
250 mg

S85,768-8
ETHYL N-(3,4-DICHLOROPHENYL)-N-(METHYLSULFONYL)CARBAMATE
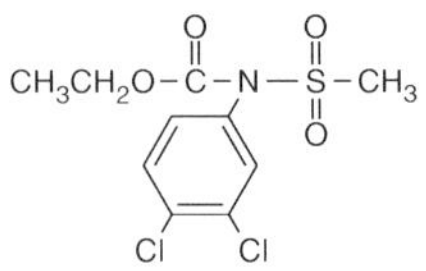
250 mg

S85,769-6
PROPYL N-(3,4-DICHLOROPHENYL)-N-(METHYLSULFONYL)CARBAMATE
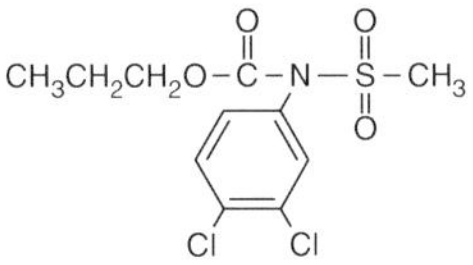
250 mg

S85,773-4
ISOPROPYL N-(3,4-DICHLOROPHENYL)-N-(METHYLSULFONYL)CARBAMATE
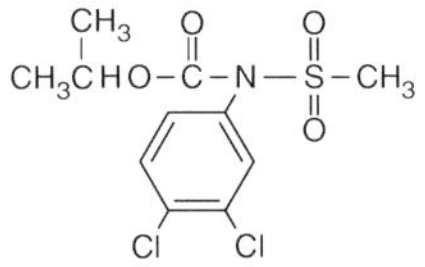
250 mg

S85,774-2
BUTYL N-(3,4-DICHLOROPHENYL)-N-(METHYLSULFONYL)CARBAMATE
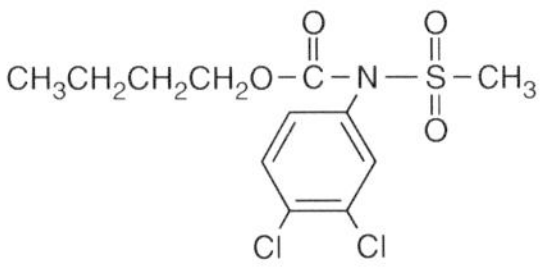
250 mg

S90,599-2
4-(4-CHLOROPHENYLSULFONYL)-1-(P-TOSYL)SEMICARBAZIDE
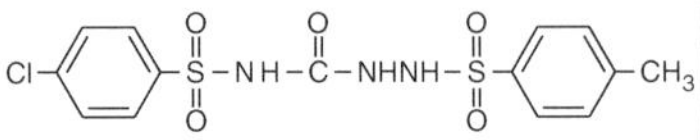
250 mg

S72,991-4
4-(2-FLUOROPHENYL)-1-(4-TOLYLSULFONYL)SEMICARBAZIDE
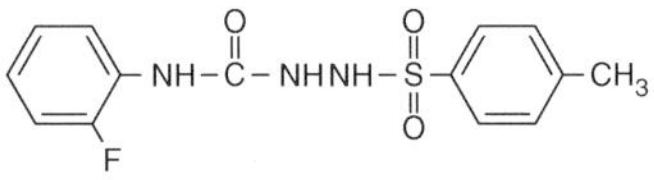
5 gm

S86,290-8
4-ACETAMIDO-4'-BROMO-N-CYANOBENZENESULFONANILIDE
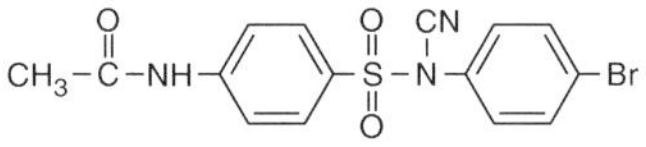
250 mg

S90,121-0
N4-ACETYL-2'-FLUOROSULFANILANILIDE
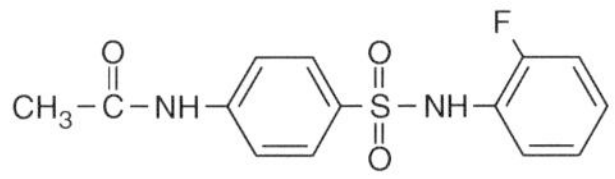
250 mg

S75,063-8
N4-ACETYL-4'-FLUOROSULFANILANILIDE
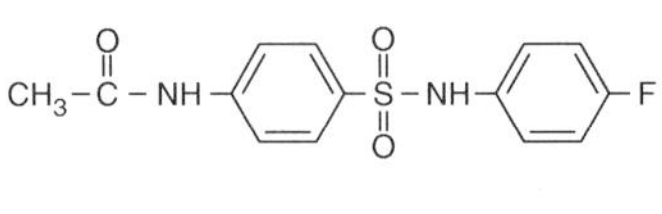
5 gm

S71,604-9
N4-ACETYL-3'-CHLOROSULFANILANILIDE
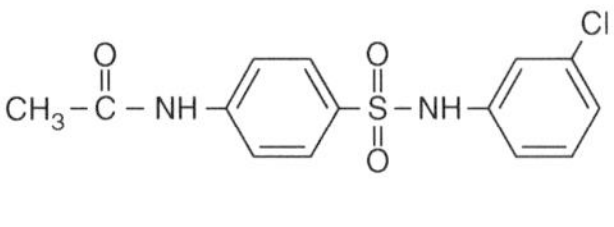
250 mg

S90,125-3
N4-ACETYL-2',5'-DICHLORO-SULFANILANILIDE
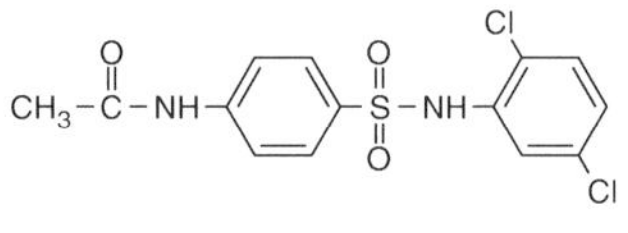
250 mg

S69,056-2
N4-ACETYL-2',6'-DIBROMO-4'-FLUORO-SULFANILANILIDE
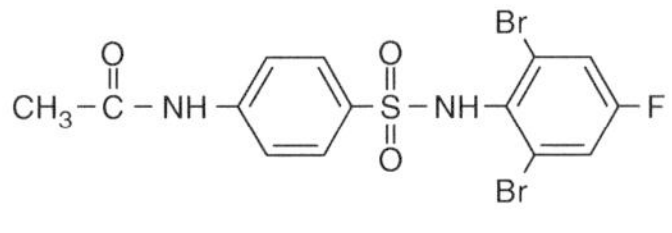
250 mg

S77,392-1
N4-ACETYL-4'-CHLORO-2,6'-DIBROMO-SULFANILANILIDE
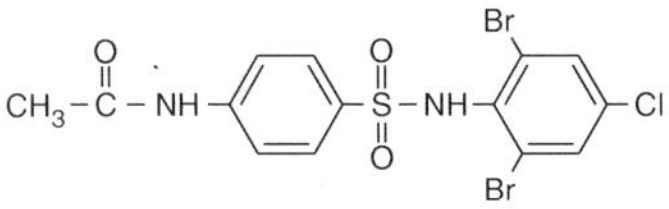
250 mg

S85,627-4
1-(4-CHLOROPHENYL)-3-(METHYLSULFONYL)UREA
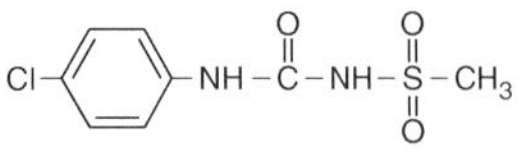
250 mg

S73,638-4
1-(4-FLUOROPHENYL)-3-(4-SULFAMOYLPHENYL)UREA
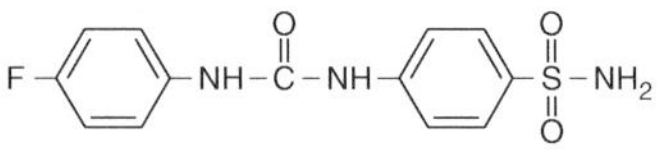
250 mg

S84,449-7
1-(3-FLUOROPHENYL)-3-(4-SULFAMOYLPHENYL)UREA
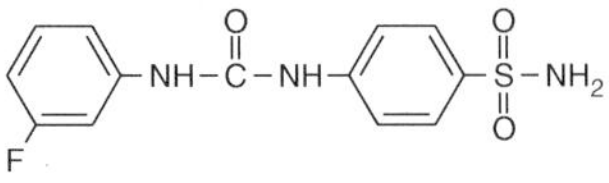
250 mg

S86,736-5
1-(3-CHLOROPHENYL)-3-(4-SULFAMOYLPHENYL)UREA

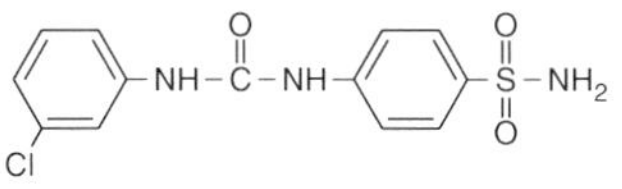

250 mg

S84,489-6
1-(4-BROMOPHENYL)-3-(P-TOSYL)UREA

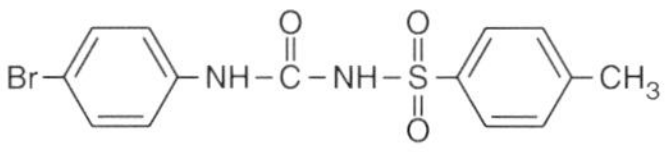

250 mg

S79,617-4
1-(2,3-DICHLOROPHENYL)-3-(4-SULFAMOYLPHENYL)UREA

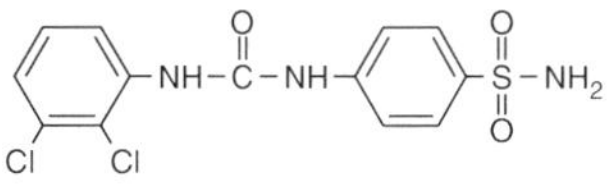

250 mg

S80,772-9
1-(3-CHLORO-2-METHYLPHENYL)-3-(4-SULFAMOYLPHENYL)UREA

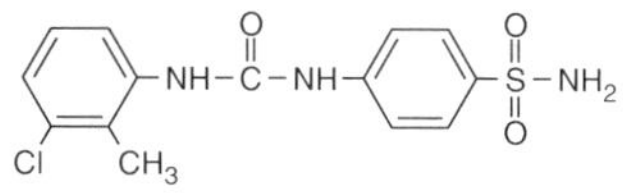

250 mg

S77,866-4
N-ACETYL-4-(2,3-DICHLOROPHENYLUREIDO)-BENZENESULFONAMIDE

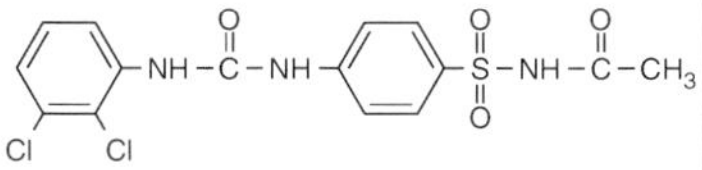

250 mg

S92,414-8
1-(3,4-DICHLOROPHENYL)-3-(P-TOSYL)UREA

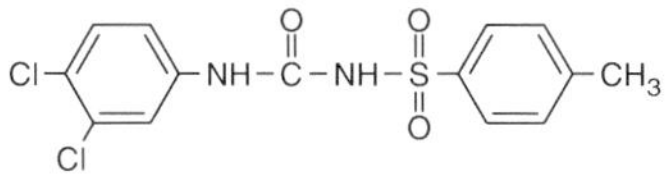

250 mg

S85,628-2
1-(3,4-DICHLOROPHENYL)-3-(METHYLSULFONYL)UREA

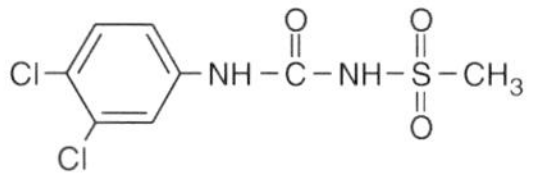

250 mg

S85,619-3
1-(3,4-DICHLOROPHENYL)-3,3-DIMETHYL-1-(METHYLSULFONYL)UREA

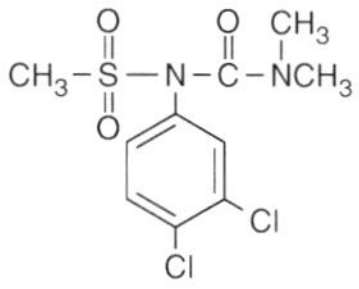

250 mg

S85,620-7
1-(3,4-DICHLOROPHENYL)-3,3-DIETHYL-1-(METHYLSULFONYL)UREA

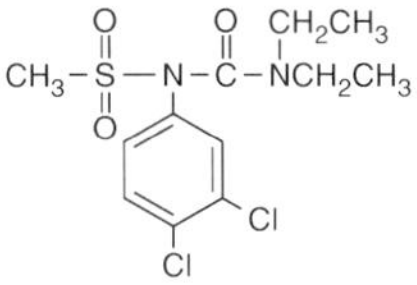

250 mg

S86,277-0
3'-BROMO-N-CYANOBENZENESULFONANILIDE

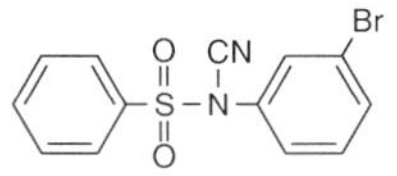

250 mg

S86,278-9
4'-CHLORO-N-CYANOBENZENESULFONANILIDE

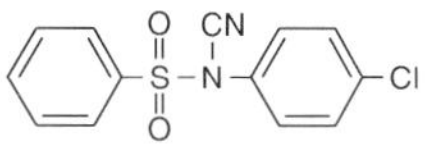

250 mg

S76,856-1
4-CHLORO-4'-CYANOBENZENESULFONANILIDE

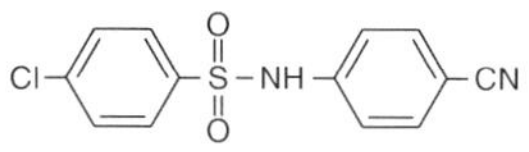

250 mg

S86,279-7
4'-CHLORO-N-CYANO-P-TOLUENESULFONANILIDE

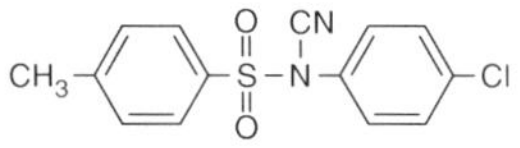

250 mg

S86,280-0
4'-BROMO-N-CYANO-P-TOLUENESULFONANILIDE

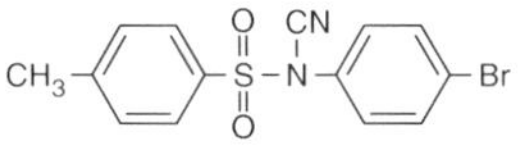

250 mg

S86,281-9
3'-BROMO-N-CYANO-P-TOLUENESULFONANILIDE

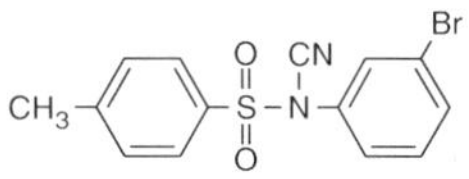

250 mg

S79,179-2
4'-FLUOROBENZENESULFONANILIDE

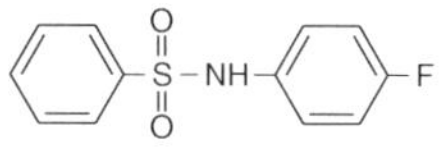

250 mg

S90,609-3
4'-BROMO-N-METHYLBENZENESULFONANILIDE

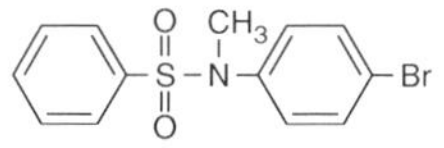

250 mg

S79,183-0
3'-CHLORO-2'-METHYLBENZENESULFONANILIDE

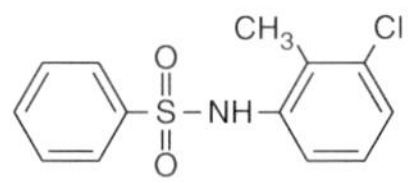

250 mg

S34,437-0
N-(5-CHLORO-2-METHYLPHENYL)-BENZENESULFONAMIDE

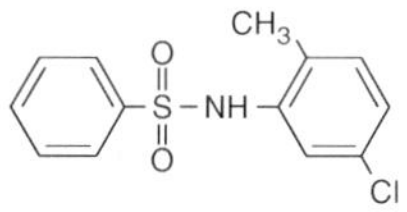

5 gm

S79,184-9
4'-CHLORO-2'-METHYLBENZENESULFONANILIDE

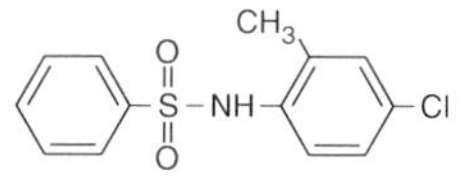

250 mg

S50,026-7
N-(2-CHLOROPHENYL)-N-METHYLMETHANESULFONAMIDE

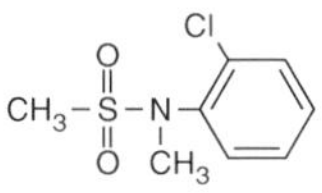

250 mg

S89,036-7
4-CHLORO-2'-ETHYLBENZENESULFONANILIDE

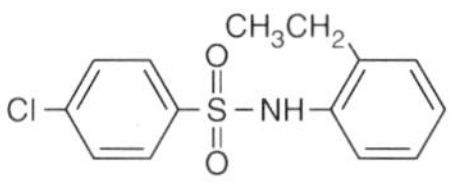

250 mg

S90,106-7
2'-FLUORO-2,4,5-TRICHLOROBENZENESULFONANILIDE

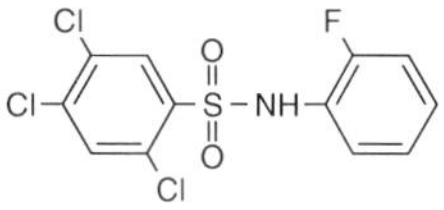

250 mg

S82,022-9
4-IODOBENZENESULFONAMIDE

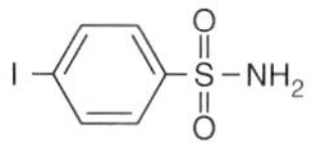

250 mg

S71,829-7
N-(4-BROMOPHENYL)METHANESULFONAMIDE

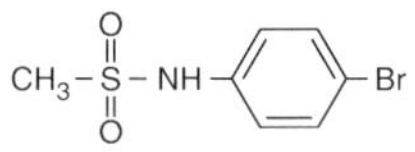

250 mg

S90,133-4
1-(4-CHLOROPHENYLSULFONYL)-4-METHYLPIPERIDINE

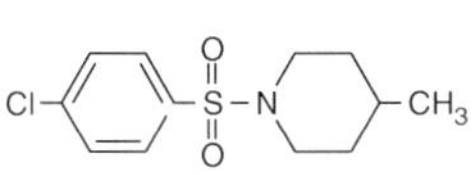

250 mg

S90,131-8
3',4-DICHLOROBENZENESULFONANILIDE

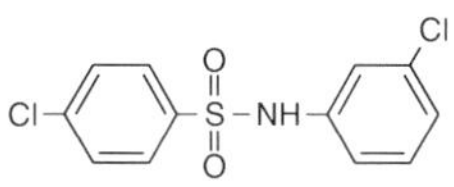

250 mg

S34,910-0
4,4'-DICHLORODIPHENYLDISULFIMIDE

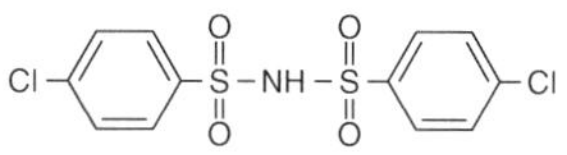

5 gm

S77,217-8
4-CHLORO-N-METHYLBENZENESULFONO-M-TOLUIDIDE

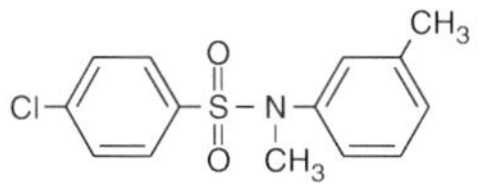

250 mg

S79,185-7
4-CHLORO-4'-ETHYLBENZENESULFONANILIDE

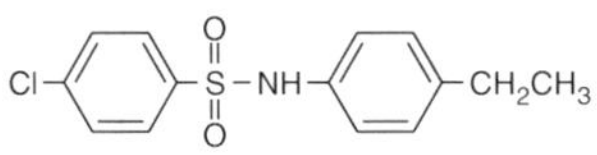

250 mg

S76,834-0
2',3',4-TRICHLOROBENZENESULFONANILIDE

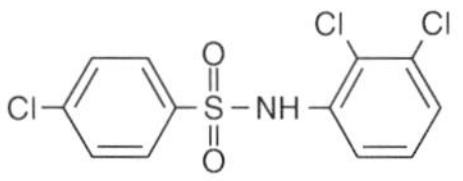

250 mg

S89,031-6
3',4-DICHLORO-2'-METHYLBENZENESULFONANILIDE

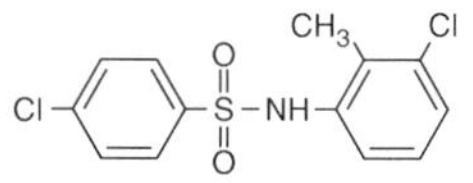

250 mg

S68,752-9
4-CHLORO-N-(3,4-DICHLOROPHENYL)-BENZENESULFONAMIDE
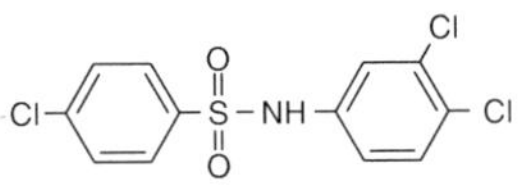
250 mg

S90,129-6
2',4,5'-TRICHLOROBENZENESULFONANILIDE
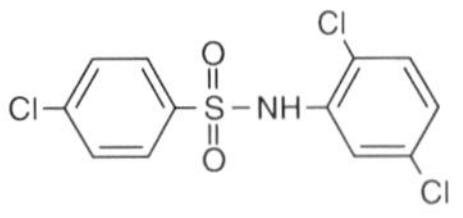
250 mg

S90,528-3
3,4-DICHLORO-4'-IODOBENZENESULFONANILIDE
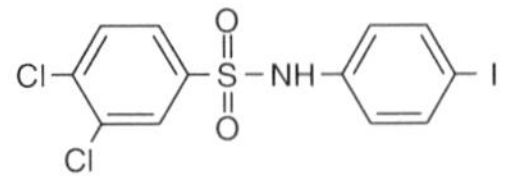
250 mg

S79,182-2
4,5'-DICHLORO-2'-METHYLBENZENESULFONANILIDE
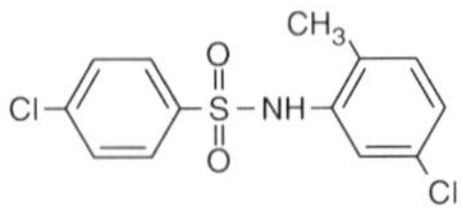
250 mg

S79,188-1
4-CHLORO-2',4',6'-TRIMETHYLBENZENESULFONANILIDE
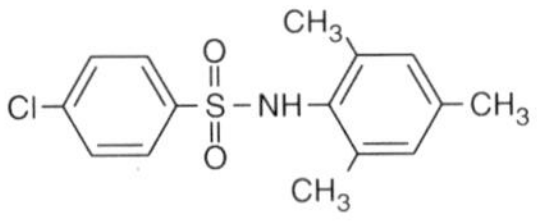
250 mg

S90,143-1
4-CHLORO-2',4',5'-TRIMETHYLBENZENESULFONANILIDE
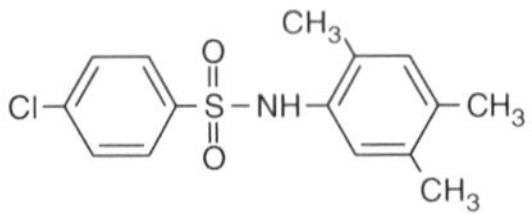
250 mg

S75,058-1
4-CHLORO-N-(1-NAPHTHYL)-BENZENESULFONAMIDE
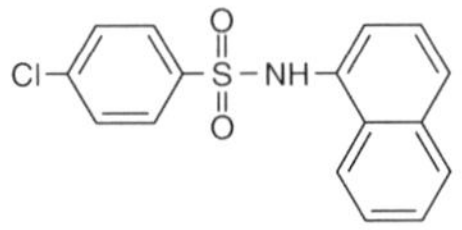
5 gm

S91,686-2
2',5'-DIFLUOROMETHANESULFONANILIDE
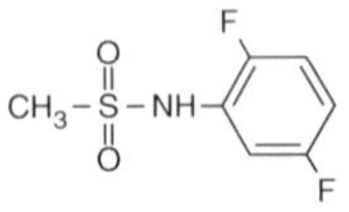
250 mg

S83,763-6
2-IODO-P-TOLUENESULFONAMIDE
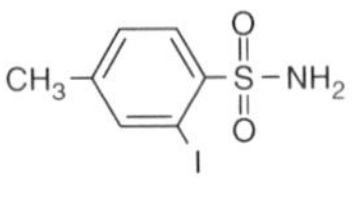
250 mg

S90,591-7
5'-CHLORO-2'-METHYLMETHANESULFONANILIDE
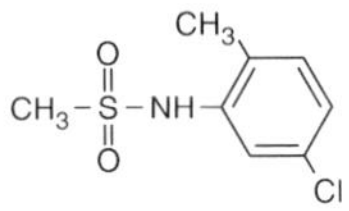
250 mg

S74,292-9
3,3',4,4'-TETRACHLOROBENZENESULFONANILIDE
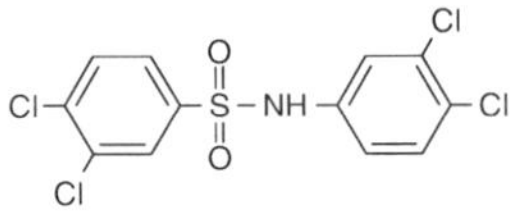
5 gm

S74,293-7
3,3',4,5'-TETRACHLOROBENZENESULFONANILIDE
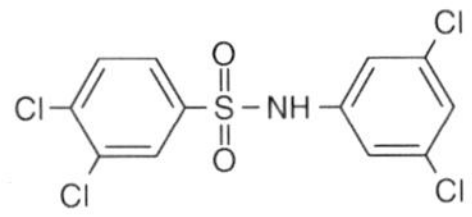
250 mg

S91,394-4
2',3,4,4',5'-PENTACHLOROBENZENESULFONANILIDE
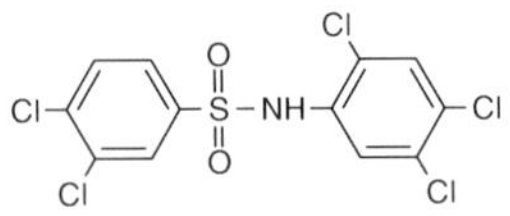
250 mg

S90,109-1
2'-METHYL-2,4,5,5'-TETRACHLOROBENZENESULFONANILIDE
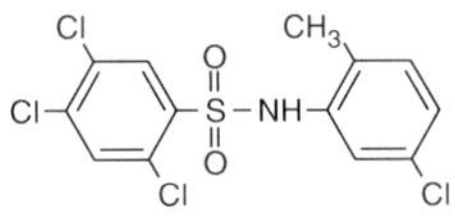
250 mg

S90,114-8
2',5'-DIMETHYL-2,4,5-TRICHLOROBENZENESULFONANILIDE
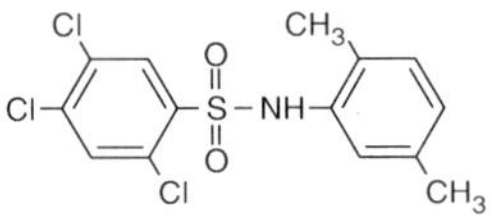
250 mg

S74,295-3
3,4-DICHLORO-N-(1-NAPHTHYL)-BENZENESULFONAMIDE
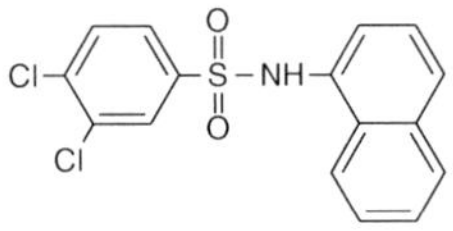
5 gm

S77,375-1
5'-CHLORO-2'-METHYL-2-NAPHTHALENESULFONANILIDE
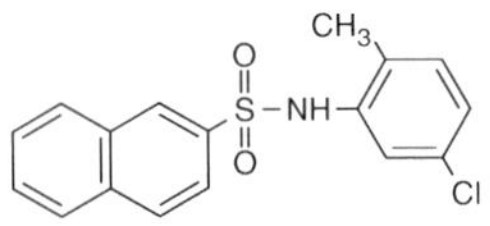
250 mg

S71,828-9
N-(3,5-DICHLOROPHENYL)-METHANESULFONAMIDE
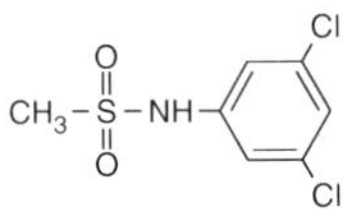
5 gm

S71,827-0
N-(4-BROMO-2,6-DIMETHYLPHENYL)-METHANESULFONAMIDE
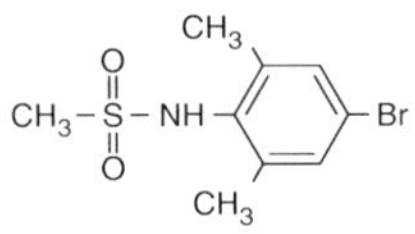
250 mg

S77,357-3
2',6'-DIBROMO-4'-ETHYLMETHANESULFONANILIDE
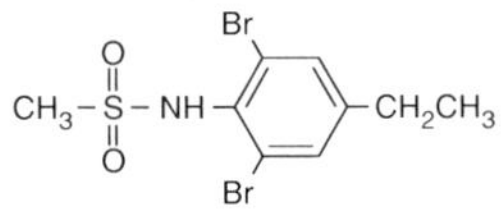
250 mg

S86,995-3
4-(4-CHLOROPHENYL)-1-(PHENYLSULFONYL)-3-THIOSEMICARBAZIDE
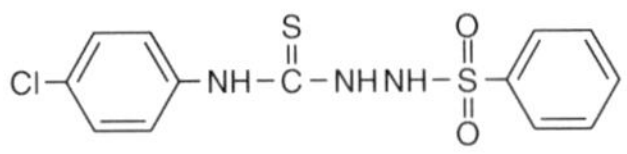
250 mg

S84,138-2
4-CHLOROBENZENESULFONIC HYDRAZIDE
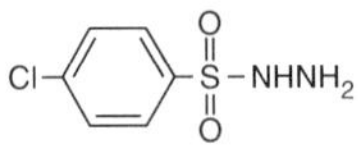
250 mg

S95,310-5
4-BROMOBENZENESULFONIC HYDRAZIDE
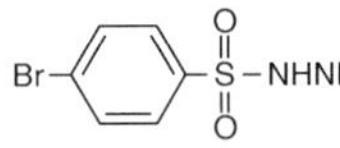
50 mg

S94,499-8
4-IODOBENZENESULFONIC ACID HYDRAZIDE
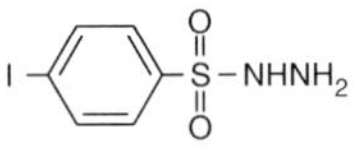
250 mg

S87,999-1
2,5-DICHLOROBENZENESULFONIC HYDRAZIDE
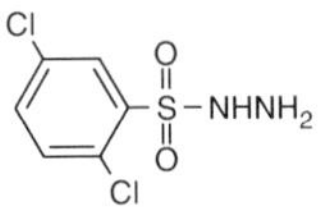
5 gm

S71,028-8
4-CHLOROBENZENESULFONYL AZIDE
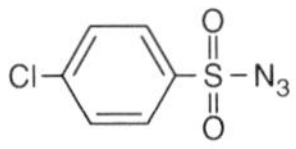
5 gm

S62,977-4
3-AMINO-N,N-BIS(2-HYDROXYETHYL)-BENZENESULFONAMIDE
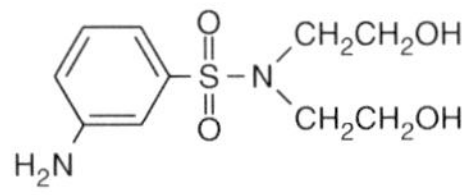
5 gm

S62,899-9
N1,N1-BIS(2-HYDROXYETHYL)BENZENE-SULFANILAMIDE
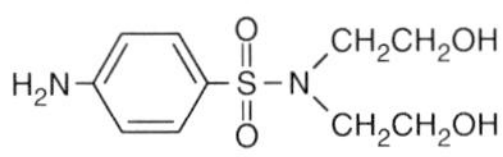
5 gm

S69,012-0
N4-(2,3-DIHYDROXYPROPYL)-SULFANILAMIDE
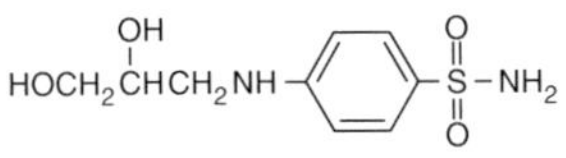
5 gm

S58,314-6
4'-AMINO-4-(HEXADECYLOXY)-2'-HYDROXYBENZENESULFONANILIDE
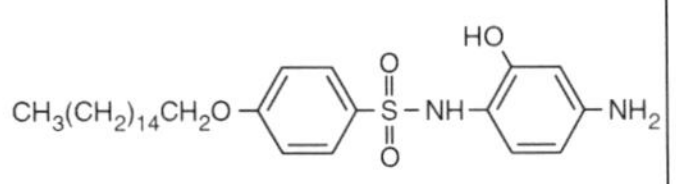
5 gm

S60,596-4
4'-AMINO-2'-HYDROXY-6-METHOXY-3,4-XYLENESULFONANILIDE
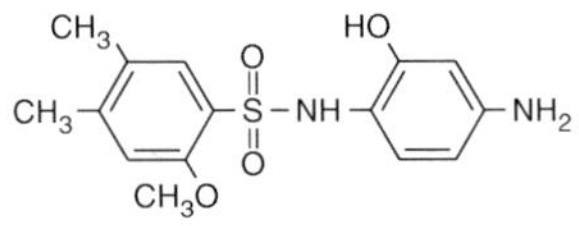
250 mg

S54,693-3
SULFANILAMIDE-N4-L-ARABINOSIDE
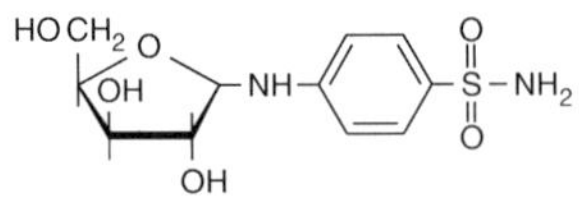
5 gm

S73,333-4
2-[ALPHA-(1,1-DIMETHYLETHYL)-ALPHA-HYDROXYBENZYL]-N-METHYLBENZENESULFONAMIDE

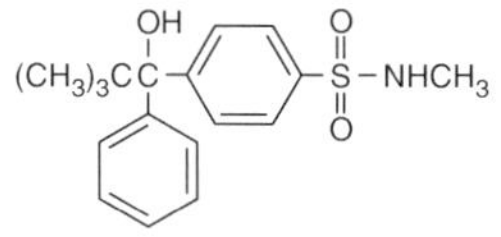

250 mg

S80,753-2
2-(DIPHENYLHYDROXYMETHYL)-N-METHYL-N-PHENYL-P-TOLUENESULFONAMIDE

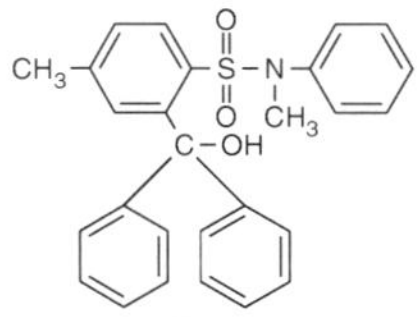

250 mg

S73,507-8
2-(2,2-DIMETHYL-1-HYDROXY)PROPYL-N-METHYLBENZENESULFONAMIDE

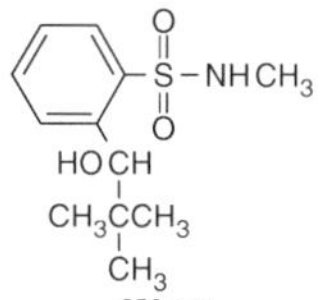

250 mg

S57,963-7
SULFANILAMIDE-N4-D-GLUCOSIDE

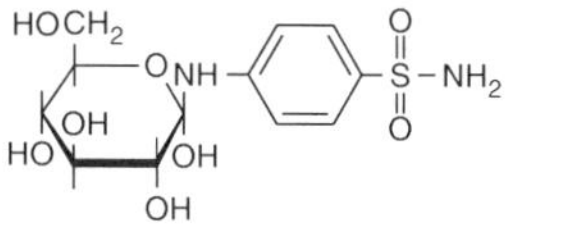

5 gm

S63,768-8
5-HYDRAZINOSULFONYLSALICYLIC ACID

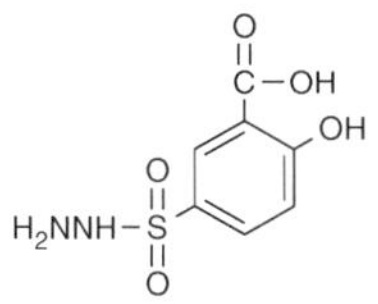

5 gm

S58,730-3
4'-AMINO-2'-HYDROXY-N4-PROPIONYL-SULFANILANILIDE

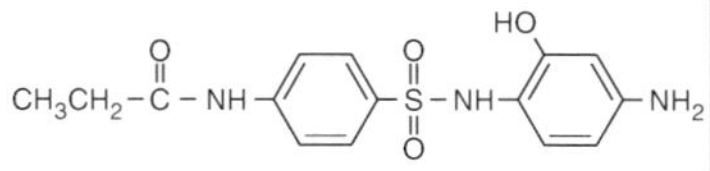

250 mg

S58,793-1
4'-AMINO-N4-BUTYRYL-2'-HYDROXYSULFANILANILIDE

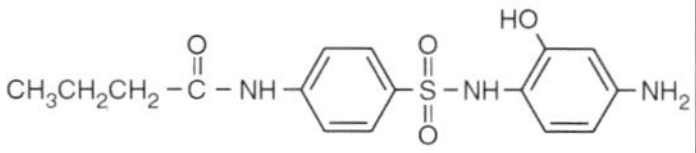

5 gm

S58,739-7
4'-AMINO-2'-HYDROXY-N4-PIVALOYLSULFANILANILIDE

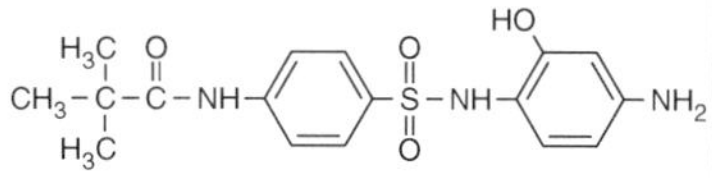

250 mg

S58,772-9
4'-AMINO-2'-HYDROXY-N4-VALERYLSULFANILANILIDE

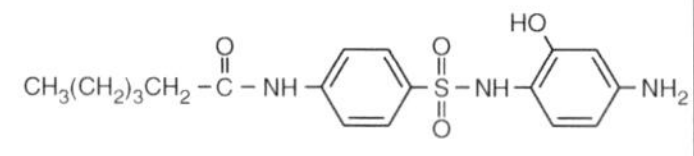

250 mg

S58,795-8
4'-AMINO-2'-HYDROXY-N4-ISOVALERYLSULFANILANILIDE

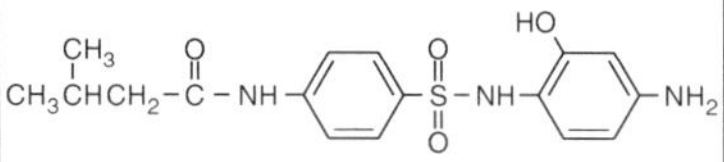

250 mg

S58,797-4
4'-AMINO-2'-HYDROXY-N4-ISONONANOYLSULFANILANILIDE

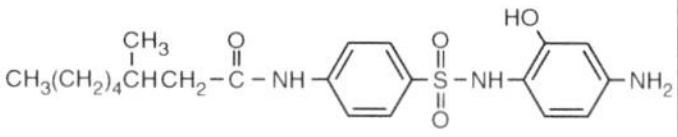

250 mg

S58,800-8
4'-AMINO-2'-HYDROXY-N4-NONANOYLSULFANILANILIDE

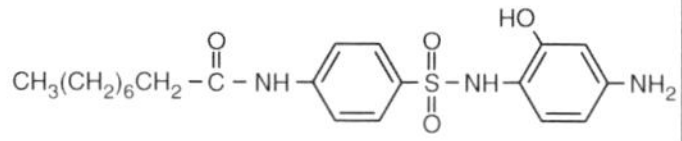

250 mg

S63,906-0
4'-AMINO-N4-DECANOYL-2'-HYDROXYSULFANILANILIDE

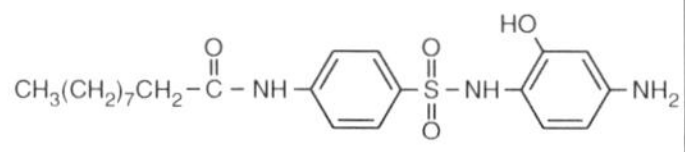

250 mg

S64,339-4
4'-AMINO-2'-HYDROXYBENZENESULFONANILIDE

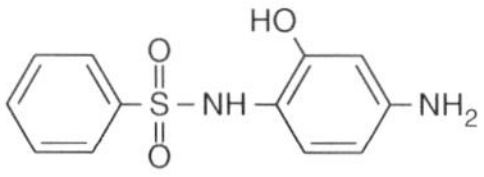

5 gm

S58,796-6
4'-AMINO-5'-CHLORO-2'-HYDROXYBENZENESULFONANILIDE

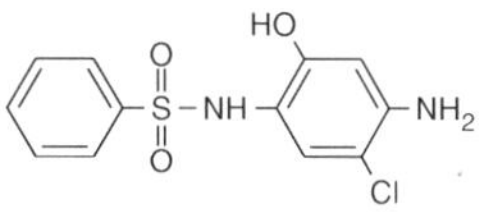

250 mg

S58,755-9
4'-AMINO-4-(TERT-BUTYL)-2'-HYDROXYBENZENESULFONANILIDE

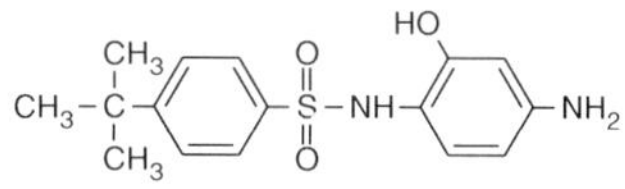

250 mg

S60,603-0
4'-AMINO-N-BENZYL-2'-HYDROXY-P-TOLUENESULFANILIDE

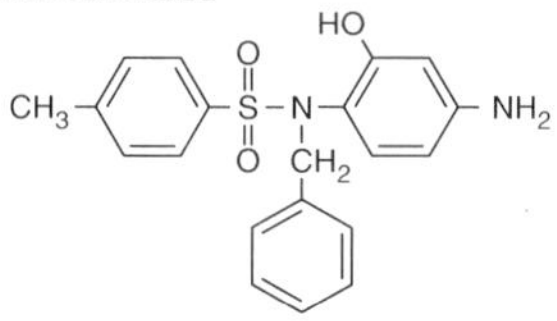

250 mg

S60,573-5
4'-AMINO-2'-HYDROXY-2-NAPHTHALENESULFONANILIDE

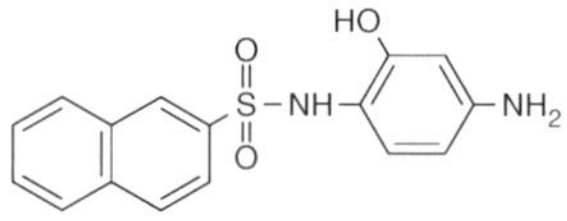

250 mg

S57,205-5
N-(3-AMINO-4-HYDROXYPHENYLSULFONYL)-ANTHRANILIC ACID

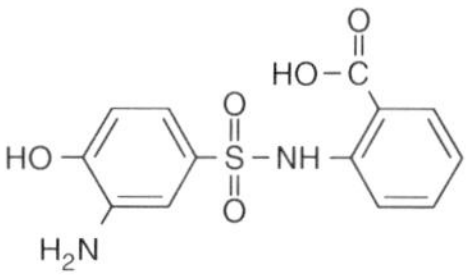

5 gm

S61,822-5
4'-AMINO-N4-ISOBUTYRYL-2'-HYDROXY-SULFANILANILIDE

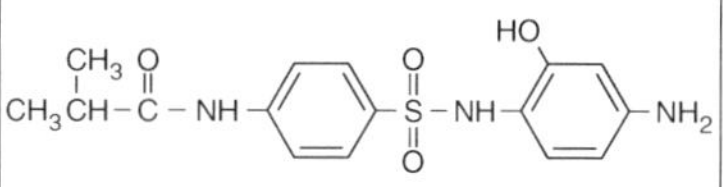

5 gm

S57,207-1
3-AMINO-4-HYDROXYBENZENESULFONANILIDE

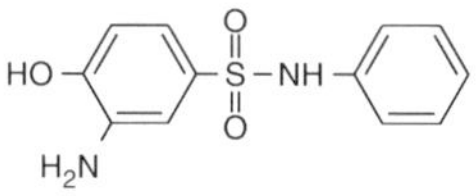

5 gm

S67,175-4
N1-(2-HYDROXYETHYL)SULFANILAMIDE

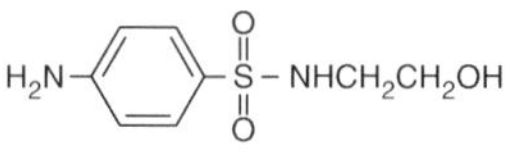

5 gm

S44,802-8
4-HYDROXYMETANILAMIDE

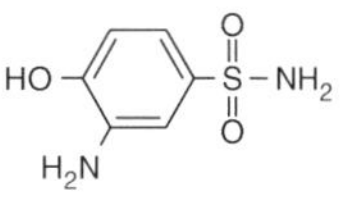

5 gm

S58,773-7
4'-AMINO-N4-(2-ETHYLHEXANOYL)-2'-HYDROXYSULFANILANILIDE

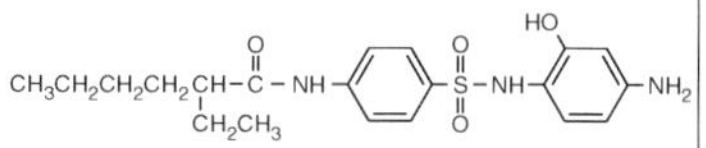

250 mg

S50,011-9
N1-(2-HYDROXYETHYL)-3-NITROSULFANILAMIDE

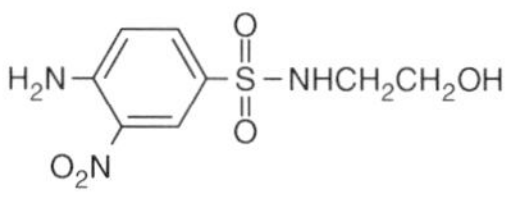

250 mg

S90,588-7
2'-HYDROXY-5'-NITROMETHANESULFONANILIDE

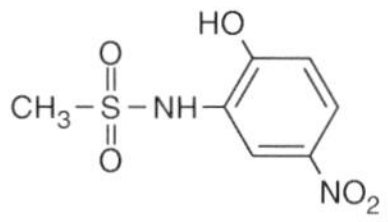

250 mg

S63,041-1
2'-HYDROXY-4'-NITRO-P-TOLUENESULFONANILIDE

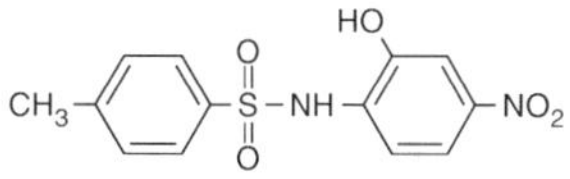

5 gm

S58,799-0
4-(SEC-BUTYL)-2'-HYDROXY-4'-NITROBENZENESULFONANILIDE

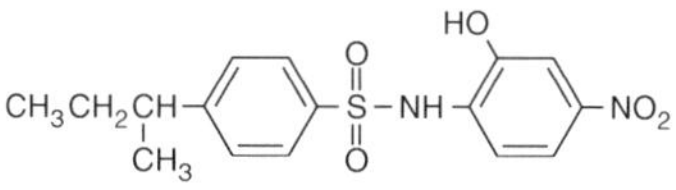

250 mg

S62,409-8
2'-HYDROXY-4'-NITROBENZENESULFONANILIDE

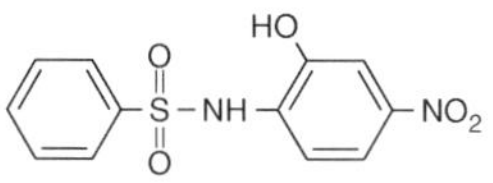

250 mg

S58,318-9
4-(SEC-BUTYL)-5'-CHLORO-2'-HYDROXY-4'-NITROBENZENESULFONANILIDE

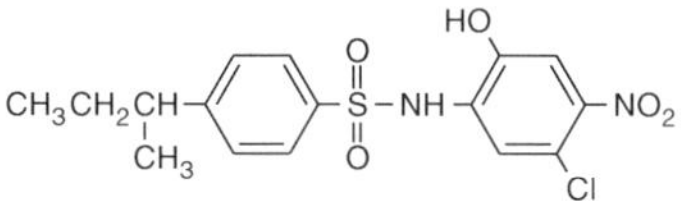

5 gm

S63,878-1
4-FLUOROSULFONYL-3'-HYDROXY-4'-METHANESULFONAMIDOBENZANILIDE

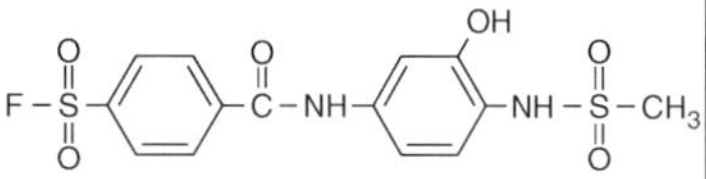

250 mg

S63,318-6
4-HYDRAZINOSULFONYL-3'-HYDROXY-4'-METHYLSULFONAMIDOBENZANILIDE

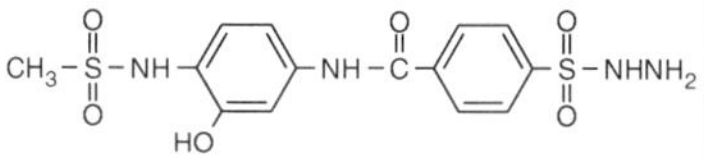

250 mg

S63,250-3
4'-ACETAMIDO-4-FLUOROSULFONYL-2'-HYDROXYBENZENESULFONANILIDE

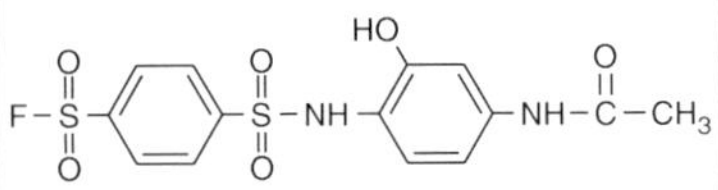

5 gm

S60,547-6
3'-HYDROXY-4'-(PHENYLSULFONYLAMINO)-METHACRYLANILIDE

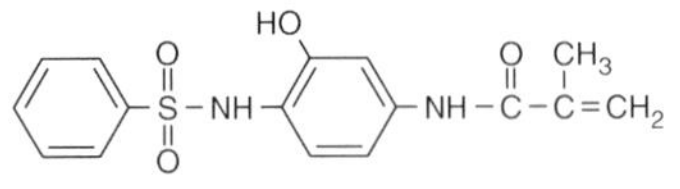

250 mg

S63,254-6
2'-HYDROXY-4'-TRIMETHYLACETAMIDOBENZENESULFONANILIDE

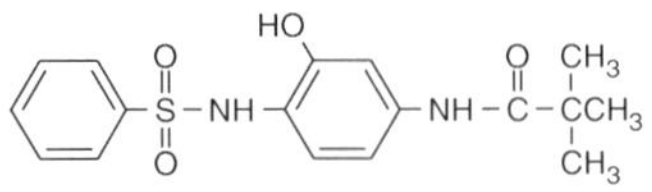

5 gm

S87,109-5
BENZENESULFONIC N'-(3,5-DIETHYL-4-HYDROXYPHENYL)-N2-(3-METHYL-2-BENZOTHIAZOLINYLIDENE)HYDRAZIDE

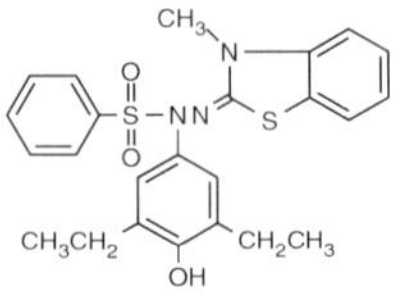

250 mg

S56,963-1
1-HEXADECANESULFONIC ACID (M-HYDROXY-ALPHA-MORPHOLINOBENZYLIDENE)HYDRAZIDE

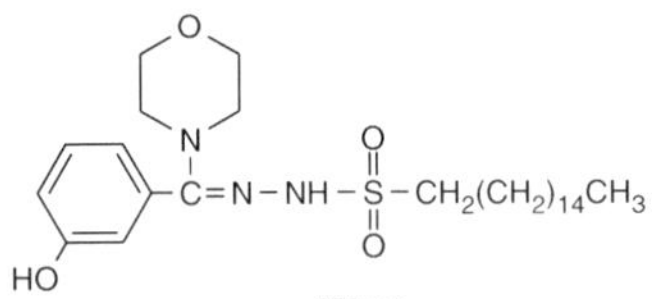

250 mg

S59,242-0
(HYDROXY-NAPHTHAMIDO)-BENZENESULFONIC (ALPHA-CYCLOHEXYLAMINO-2-CHLORO-BENZYLIDEN)HYDRAZIDE

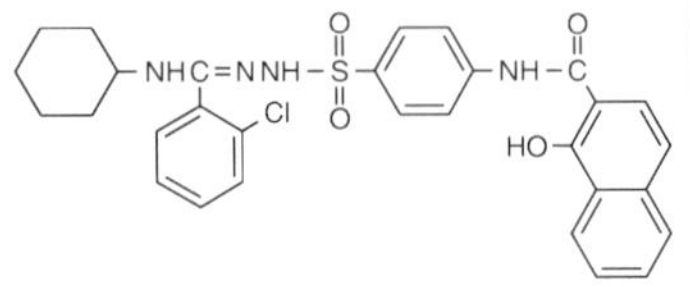

250 mg

S86,561-3
2-(DIMETHYLHYDROXYMETHYL)-CYCLOHEXANONE (P-TOSYL)HYDRAZONE

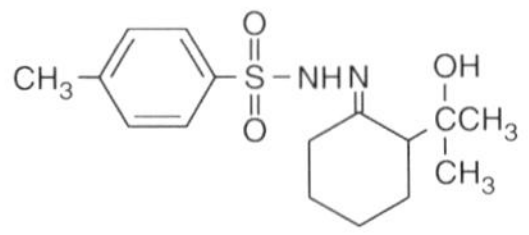

250 mg

S86,392-0
3-HYDROXYISOVALEROPHENONE (P-TOSYL)-HYDRAZONE

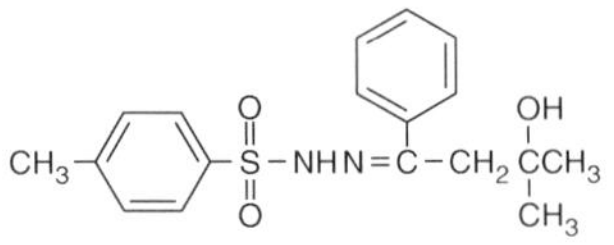

250 mg

S34,225-4
N,N-BIS(2-HYDROXYETHYL)-3-CARBOXYBENZENESULFONAMIDE

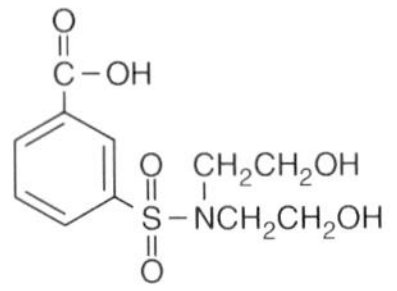

5 gm

S36,243-3
N-(2-HYDROXYETHYL)-N-(2-HYDROXYPROPYL)-P-TOLUENESULFONAMIDE

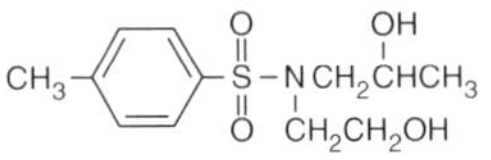

5 gm

S56,983-6
2'-HYDROXY-4'-(P-NITROBENZENESULFONAMIDO)-HEXADECANANILIDE

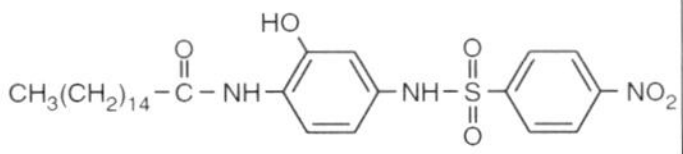

250 mg

S50,191-3
4'-BUTYLSULFAMOYL-1-HYDROXY-2'-NITRO-2-NAPHTHANILIDE

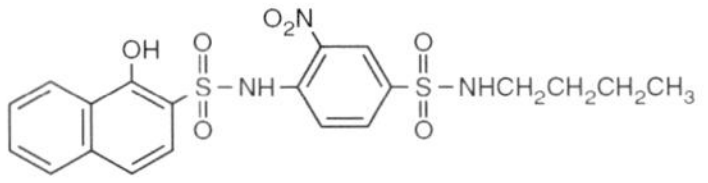

5 gm

S42,085-9
4-(BENZENESULFONAMIDO)-SALICYLIC ACID

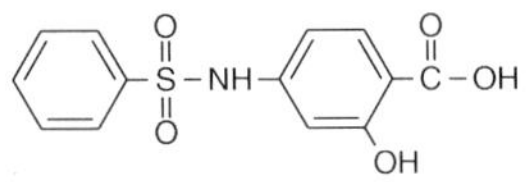

250 mg

S63,273-2
3'-(HYDRAZINOSULFONYL)-1-HYDROXY-2-NAPHTHANILIDE

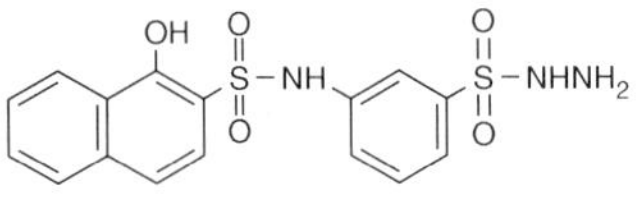

5 gm

S64,233-9
2'-CHLORO-5'-[N-(2-CYANOETHYL)-SULFAMOYL]-1-HYDROXY-2-NAPHTHANILIDE

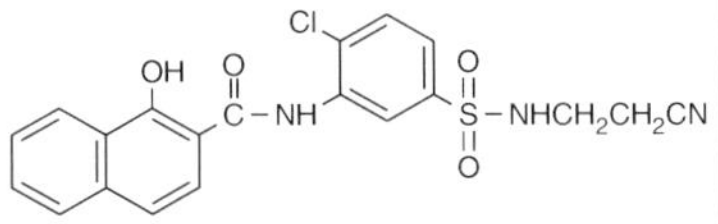

250 mg

S59,515-2
1-(HYDROXYMETHYL)-3-(4-PHTHALIMIDOPHENYLSULFONYL)UREA

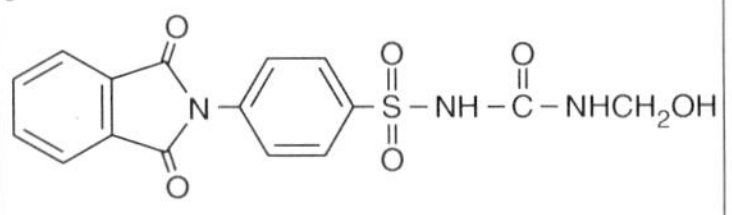

5 gm

S35,124-5
4-HYDROXY-4-PHENYL-1-P-TOLYLSULFONYLPIPERIDINE

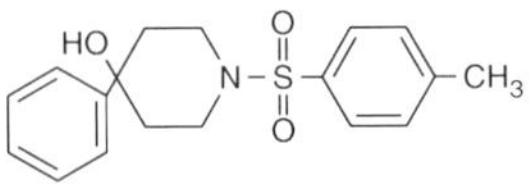

5 gm

S45,777-9
3'-HYDROXYBENZENESULFONANILIDE

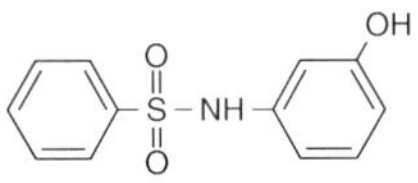

5 gm

S38,815-7
4'-HYDROXYBENZENESULFONANILIDE

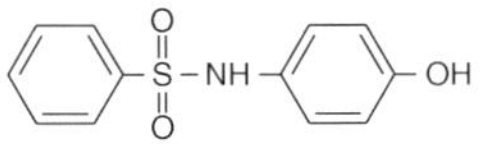

250 mg

S40,884-0
N-(2-HYDROXYETHYL)-N-(3-HYDROXYPROPYL)-PARA-TOLUENESULFONAMIDE

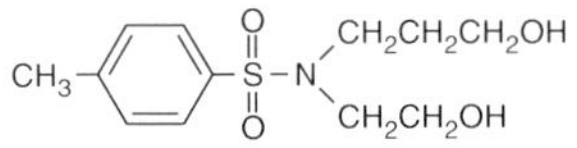

5 gm

S59,854-2
N,N-DIMETHYL-8-HYDROXYNAPHTHALENESULFONAMIDE

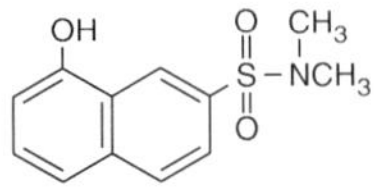

250 mg

S85,541-3
4-METHOXYBENZYL N-(P-TOSYL)-CARBAMATE

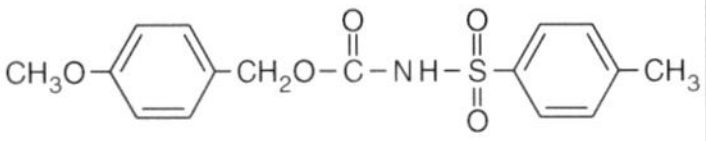

250 mg

S62,972-3
4'-[2-(4-TERT-OCTYLPHENOXY)ETHOXY]-SULFANILANILIDE

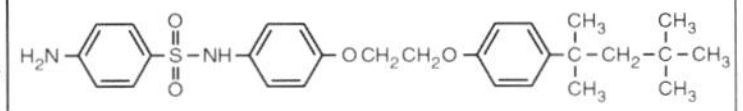

5 gm

S85,122-1
1-(2,4-DIMETHOXYPHENYL)-3-(4-SULFAMOYLPHENYL)UREA

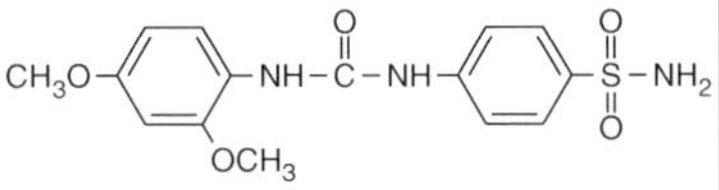

250 mg

S55,437-5
2,5-DIMETHOXYSULFANILANILIDE

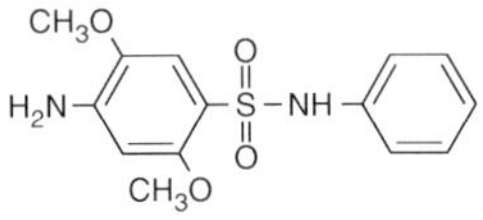

5 gm

S40,789-5
2,5-DIMETHOXYSULFANILAMIDE

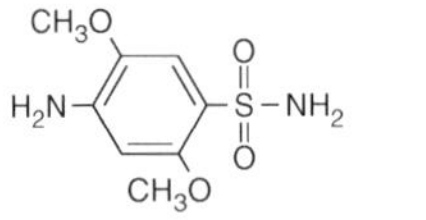

5 gm

S79,566-6
2',4-DIMETHOXY-5'-NITROBENZENESULFONANILIDE

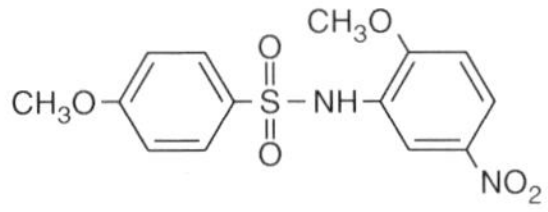

250 mg

S73,521-3
N-(3,4-DIMETHOXYBENZYL)-10-CAMPHORSULFONAMIDE

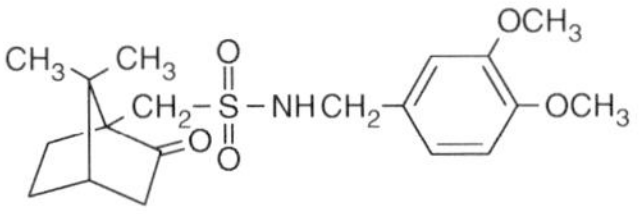

250 mg

S51,025-4
3-NITROBENZENESULFONIC ACID [ALPHA-(N-METHYLHEXADECYLAMINO)VERATRYLIDENE]-HYDRAZIDE

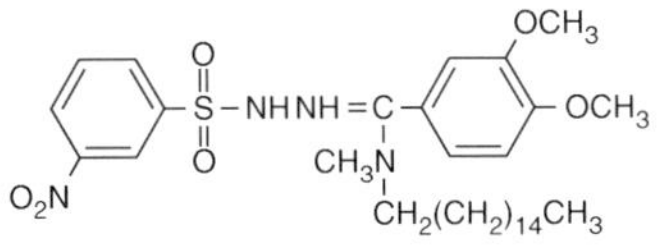

250 mg

S56,871-6
4-SULFOBENZOIC ACID 4-[ALPHA-(N-HEXADECYL-N-METHYLAMINO)-VERATRYLIDENE]HYDRAZIDE

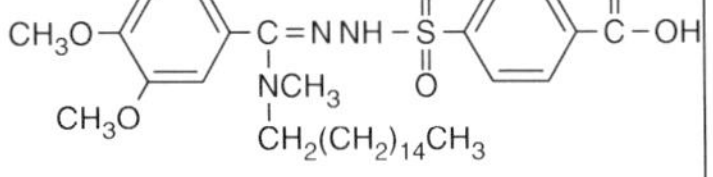

250 mg

S58,752-4
3,5-DICARBOXYBENZENESULFONIC-4-(HEXADECYLOXY-3-METHOXY-ALPHA-PIPERIDINOBENZYLIDENE)HYDRAZIDE

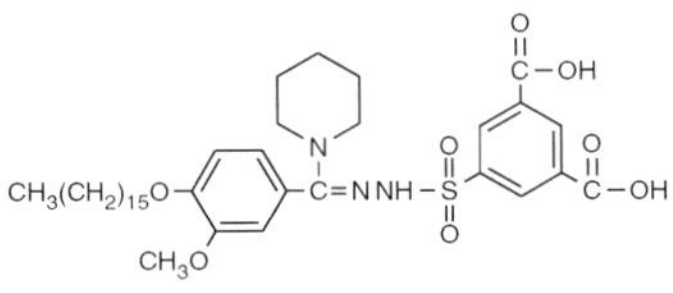

250 mg

S58,774-5
4-(CARBOXYMETHOXY)BENZENESULFONIC (1-METHYL-4-PENTADECYL-2-QUINOLYLIDENE)-HYDRAZIDE

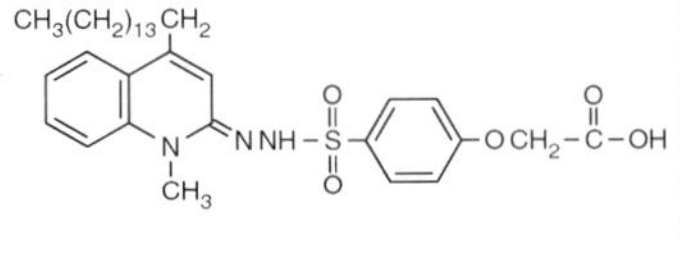

5 gm

S57,766-9
1-HEXADECANESULFONIC ACID (ALPHA-PIPERIDINOVERATRYLIDENE)HYDRAZIDE

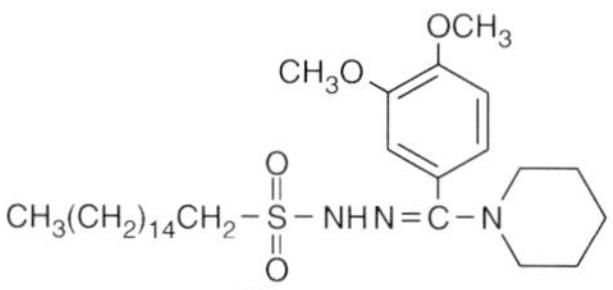

250 mg

S51,061-0
4-[2-(METHYLSULFONAMIDO)ETHOXY]-META-ANISIC ACID

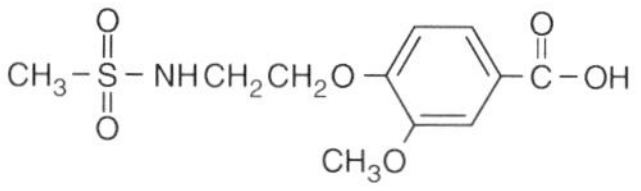

250 mg

S50,067-4
4-NITRO-4'-(2-[4-(1,1,3,3-TETRAMETHYLBUTYL)PHENOXY]ETHOXY)-BENZENESULFONANILIDE

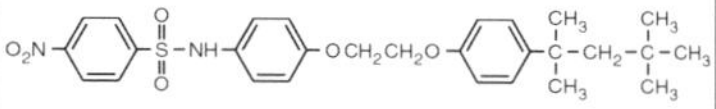

5 gm

S38,286-8
2-[4-(HYDRAZINOSULFONYL)-PHENOXY]ACETIC ACID

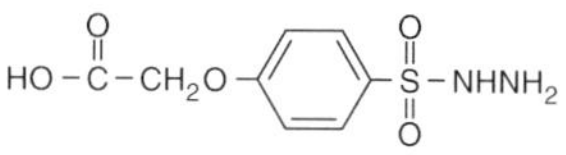

5 gm

S90,558-5
4-ALLYL-2-METHOXYPHENYL N-(P-TOSYL)-CARBAMATE

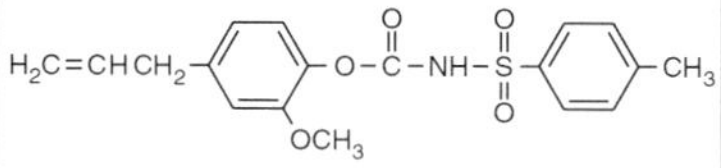

250 mg

S50,939-6
2'-AMINO-5'-METHOXY-P-TOLUENE-SULFONANILIDE

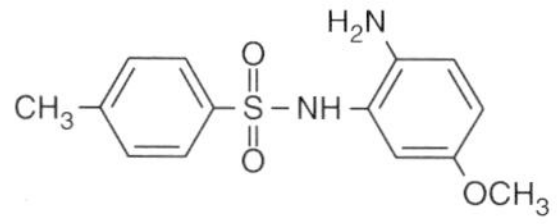

5 gm

S35,773-1
N-(4-AMINO-3-METHOXYANTHRAQUINON-1-YL)-P-TOLUENESULFONAMIDE

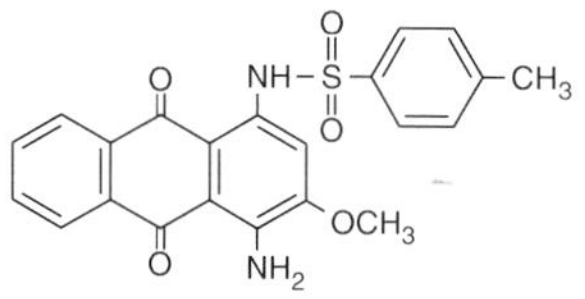

5 gm

S39,300-2
N1-HEXADECYL-4-METHOXY-N2-(ALPHA-METHYLBENZYL)METANILAMIDE

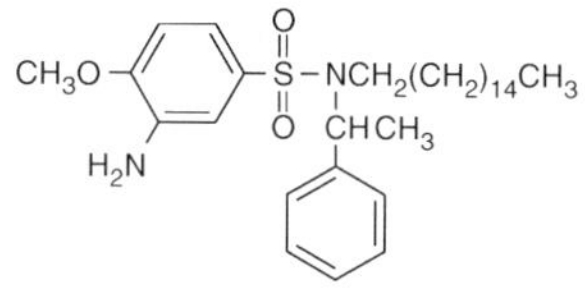

250 mg

S59,288-9
N1,N1-DIBENZYL-4-(HEXADECYLOXY)-METANILAMIDE

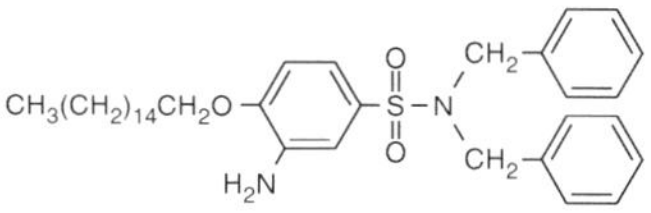

5 gm

S50,194-8
DIMETHYL 5-[4-(HEXADECYLOXY)-METANILAMIDO]ISOPHTHALATE

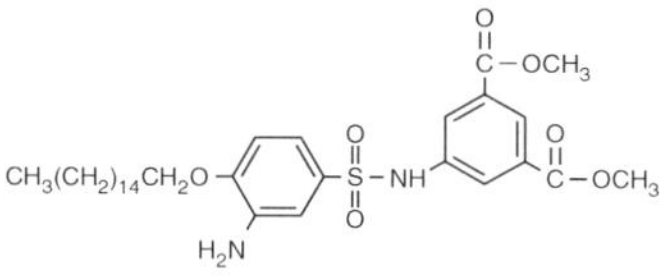

250 mg

S38,600-6
3-[3-AMINO-4-(HEXADECYLOXY)-BENZENESULFONAMIDO]-BENZENESULFONYL FLUORIDE

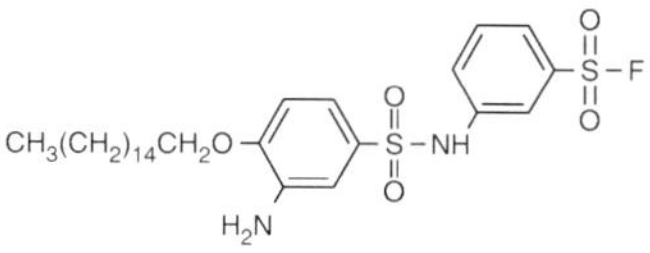

250 mg

S38,606-5
4-METHOXYMETANILANILIDE

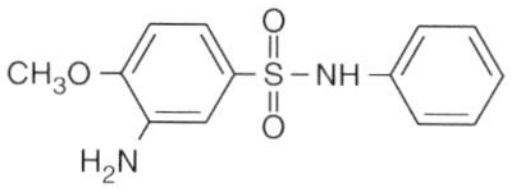

250 mg

S35,765-0
3-AMINO-N-ETHYL-4-PHENOXYBENZENESULFONANILIDE

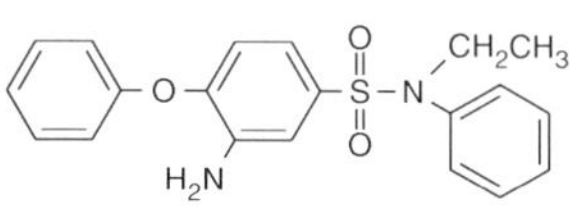

5 gm

S55,895-8
2'-(DODECYLOXY)METANILANILIDE

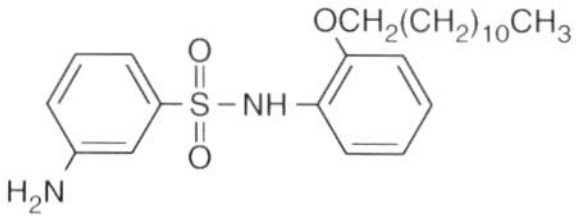

250 mg

S34,094-4
3-AMINO-N-BUTYL-4-METHOXYBENZENESULFONAMIDE

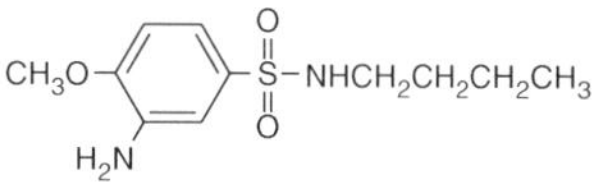

5 gm

S38,611-1
N1-HEXADECYL-4-METHOXYMETANILAMIDE

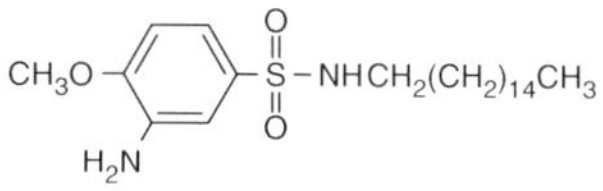

250 mg

S60,504-2
N1,N1-DIMETHYL-4-HEXADECYLOXYMETANILAMIDE

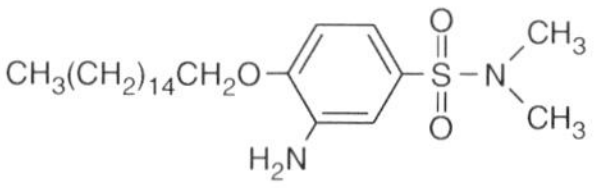

250 mg

S59,236-6
N1-(TERT-BUTYL)-4-(HEXADECYLOXY)-METANILAMIDE

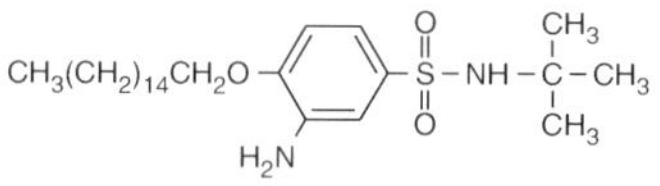

5 gm

S65,517-1
4'-HEXADECYLOXYSULFANILANILIDE

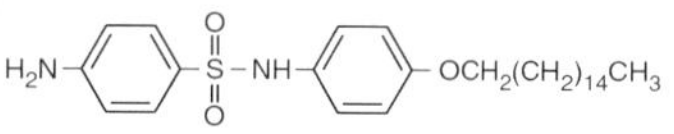

250 mg

S59,235-8
N1-CYCLOHEXYL-4-(HEXADECYLOXY)-METANILAMIDE

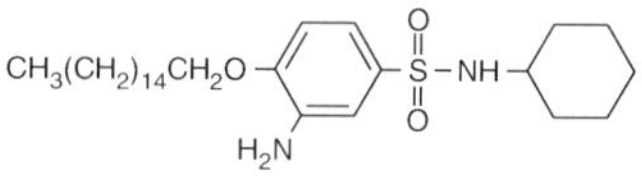

5 gm

S58,828-8
2-HEXADECYLOXY-5-(MORPHOLINOSULFONYL)ANILINE

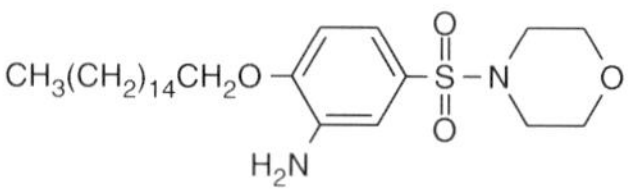

250 mg

S68,541-0
O-TERT-BUTYL-L-SERINE TERT-BUTYL ESTER DIBENZENESULFIMIDE SALT

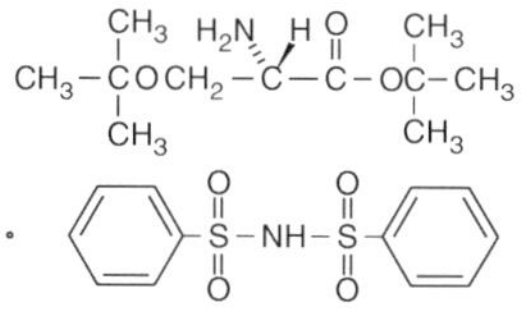

5 gm

S67,882-1
O-TERT-BUTYL-L-THREONINE TERT-BUTYL ESTER DIBENZENESULFONAMIDE SALT

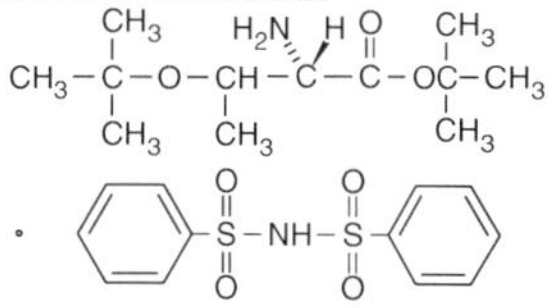

250 mg

S51,356-3
N-[2-(2,5-DIMETHOXY-N-HEXADECYLANILINO]ETHYL)-METHANESULFONAMIDE

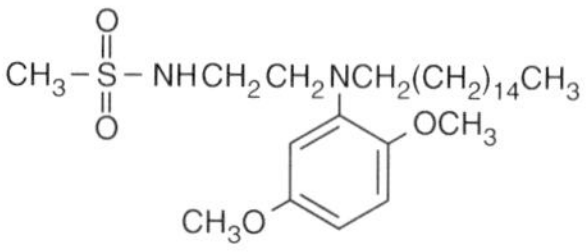

250 mg

S91,687-0
4'-METHOXY-2'-NITROMETHANESULFONANILIDE

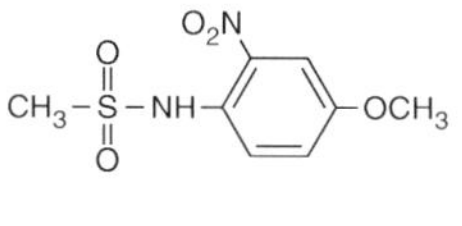

250 mg

S57,863-0
1-HEXADECANESULFONIC [2-(4-METHOXY-PHENYL)-1-METHYL-6-SULFO-4(1H)-QUINOLYLIDENE]HYDRAZIDE

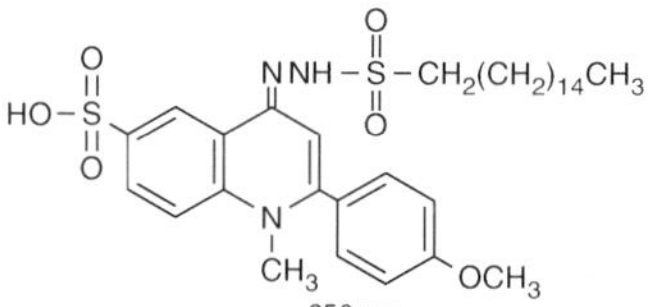

250 mg

S75,061-1
N4-ACETYL-2'-METHOXY-5'-NITRO-SULFANILANILIDE

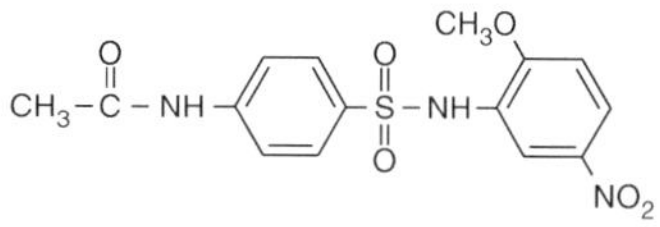

5 gm

S69,916-0
N-(4-METHOXY-3-NITROPHENYL)-METHANESULFONAMIDE

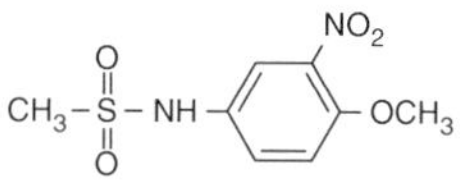

250 mg

S74,370-4
2'-METHOXY-5'-NITROMETHANESULFONANILIDE

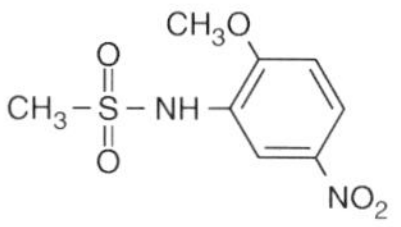

5 gm

S51,316-4
5'-CHLORO-2'-METHOXY-4'-NITRO-P-TOLUENESULFONANILIDE

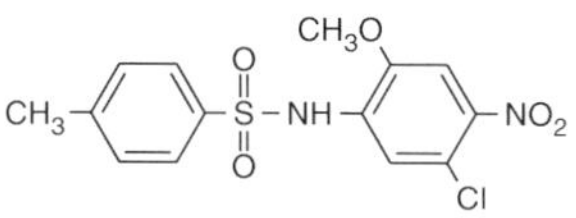

250 mg

S53,061-1
2',4'-DIMETHOXY-PARA-TOLUENESULFONANILIDE

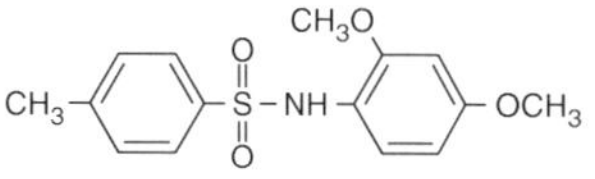

250 mg

S87,057-9
4-METHOXYBENZENESULFONIC (3-METHYL-2-BENZOTHIAZOLINYLIDENE)HYDRAZIDE

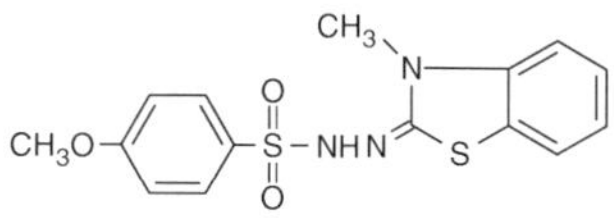

250 mg

S57,858-4
5-CARBOXY-2-(HEXADECYLOXY)-BENZENESULFONIC ACID (3-METHYL-2-BENZOTHIAZOLINYLIDENE)HYDRAZIDE

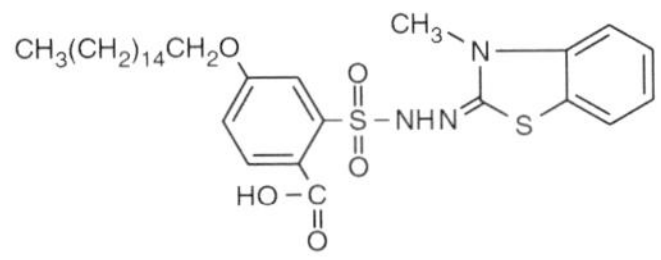

5 gm

S56,965-8
4-(HEXADECYLOXY)-3-SULFOBENZOIC ACID 3-[(4,6-DICHLORO-3-METHYL-2-BENZOTHIAZOLINYLIDENE)HYDRAZIDE]

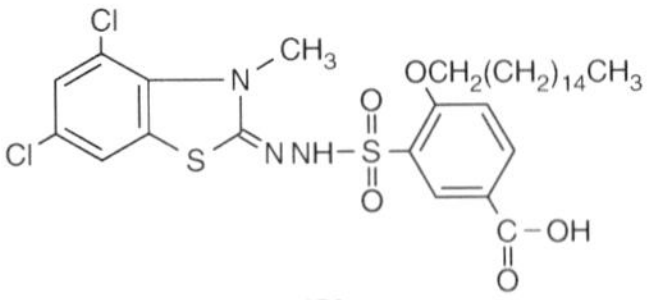

250 mg

S57,771-5
3-CARBOXYBENZENESULFONIC (4-HEXADECYLOXY-ALPHA-PIPERIDINOBENZYLIDENE)HYDRAZIDE

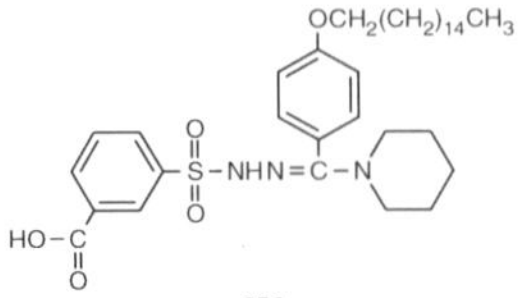

250 mg

S58,342-1
5-CARBOXY-2-(HEXADECYLOXY)-BENZENESULFONIC (3-METHYL-5-SULFO-2-BENZOTHIAZOLINYLIDENE)HYDRAZID

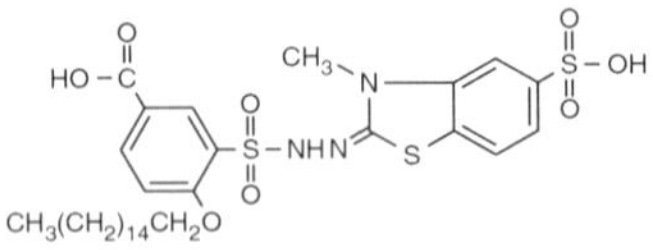

5 gm

S57,792-8
4-(HEXADECYLOXY)BENZENESULFONIC ACID (3-METHYL-2-BENZOTHIAZOLINYLIDENE)HYDRAZIDE

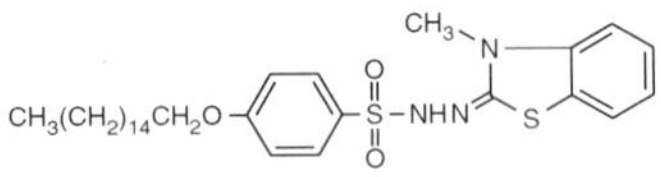

250 mg

S50,205-7
N,N-DIBENZYL-4-HEXADECYLOXY-3-NITROBENZENESULFONAMIDE

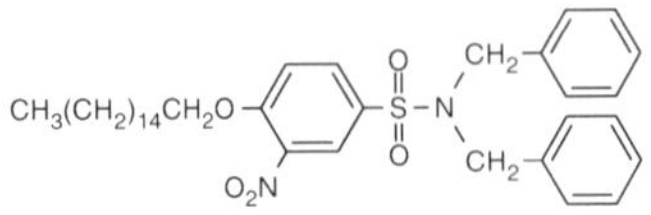

250 mg

S86,304-1
4-ACETAMIDO-N-CYANO-4'-ETHOXYBENZENESULFONANILIDE

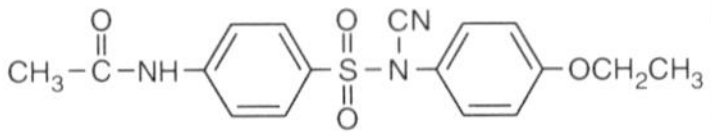

250 mg

S72,772-5
1-(2-METHOXY-4-NITROPHENYL)-3-(4-SULFAMOYLPHENYL)UREA

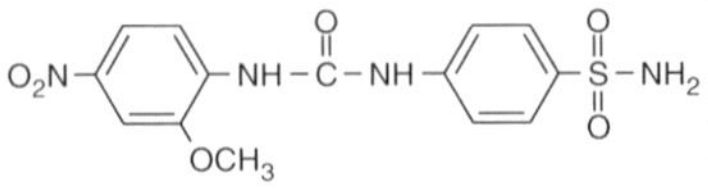

5 gm

S73,105-6
2'-METHOXY-4'-NITROMETHANESULFONANILIDE

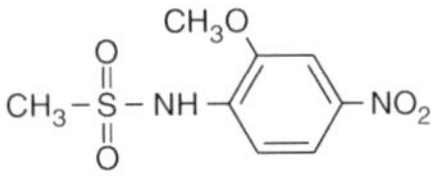

5 gm

S74,365-8
2'-METHOXY-5'-NITRO-2-NAPHTHALENESULFONANILIDE

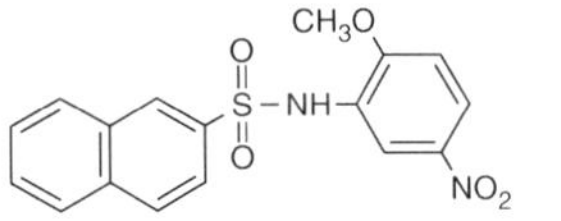

250 mg

S51,038-6
4-(4-HEXADECYLOXY-3-NITROPHENYLSULFONYL)MORPHOLINE

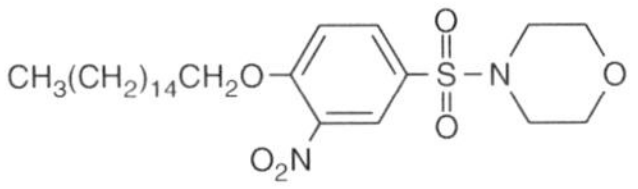

250 mg

S55,868-0
2'-DODECYLOXY-3-NITROBENZENESULFONANILIDE

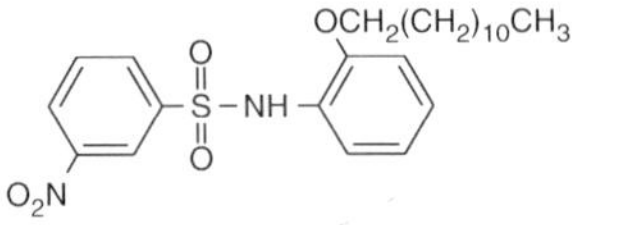

250 mg

S51,518-3
4-HEXADECYLOXY-3-NITROBENZENESULFONAMIDE

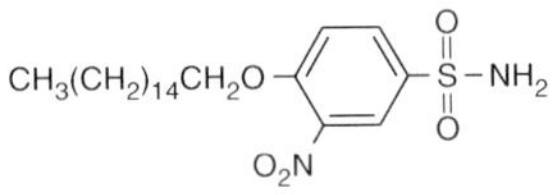

250 mg

S50,020-8
N-(TERT-BUTYL)-4-HEXADECYLOXY-3-NITROBENZENESULFONAMIDE

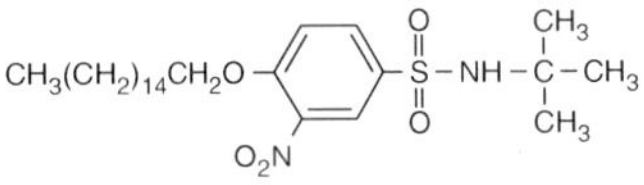

5 gm

S51,051-3
3-NITROBENZENESULFONIC ACID 2-(2-METHOXYPHENYL)HYDRAZIDE

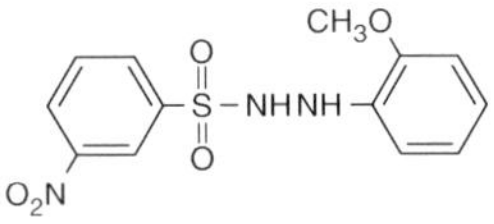

250 mg

S63,007-1
METHYL 3-(4-CARBOXYPHENYLSULFONAMIDO)-4-HEXADECYLOXYBENZOATE

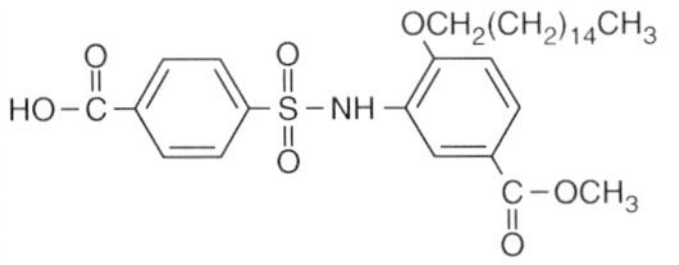

250 mg

S51,020-3
4-[2-(METHYLSULFONAMIDO)ETHOXY] BENZOIC ACID

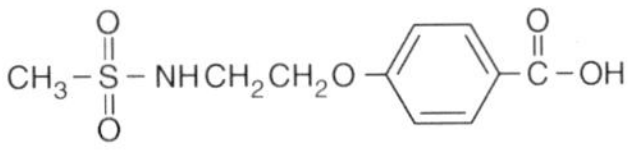

250 mg

S67,907-0
TRANS-O-TERT-BUTYL-HYDROXY-L-PROLINE TERT-BUTYL ESTER DIBENZENZESULFONAMIDE SALT

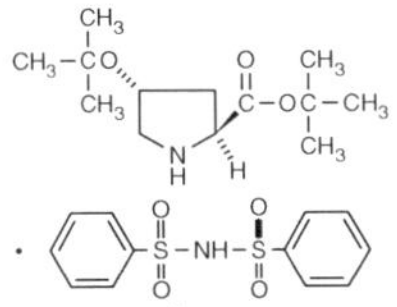

5 gm

S75,064-6
N4-ACETYL-2'-METHOXYSULFANILANILIDE

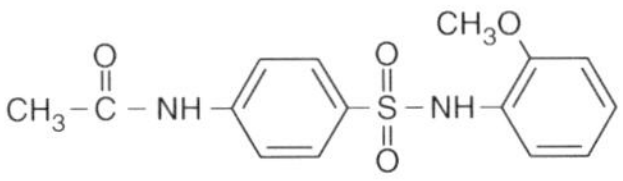

5 gm

S39,299-5
N3-ACETYL-N1,N1-BIS(2-ETHYLHEXYL)-4-METHOXYMETANILAMIDE

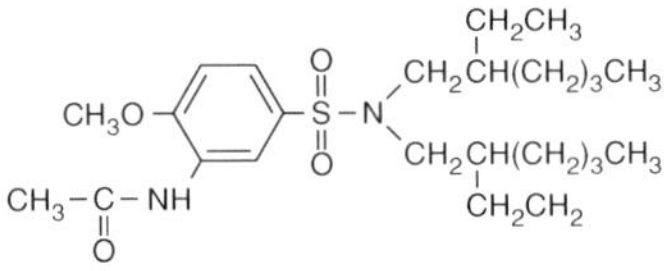

250 mg

S60,578-6
N3-ACETYL-N1-HEXADECYL-4-METHOXY-N1-METHYLMETANILAMIDE

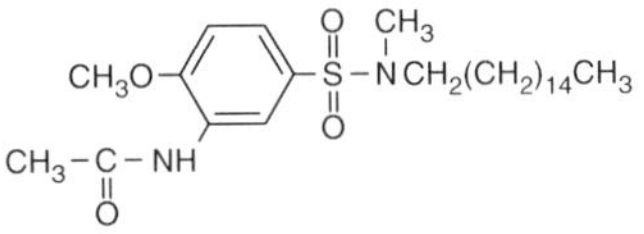

250 mg

S58,278-6
5'-(N-BENZYL-N-HEXADECYLSULFAMOYL)-O-ACETANISIDIDE

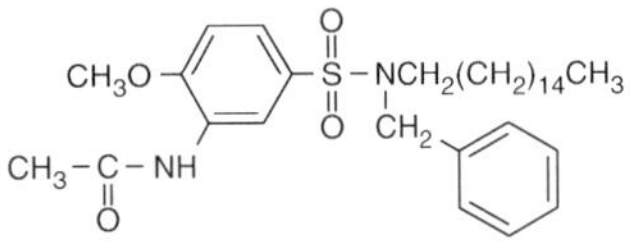

5 gm

S86,763-2
1-(2-METHOXYPHENYL)-3-(4-SULFAMOYLPHENYL)UREA

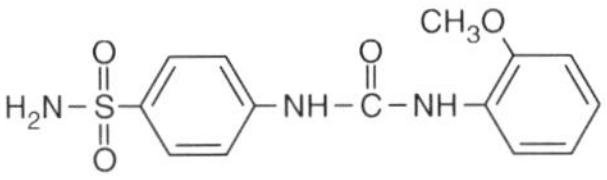

250 mg

S74,368-2
2'-ETHOXYMETHANESULFONANILIDE

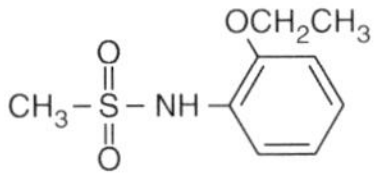

5 gm

S90,594-1
2'-METHOXY-5'-METHYLMETHANESULFONANILIDE

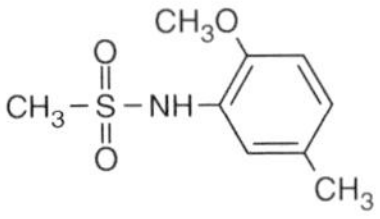

250 mg

S35,218-7
2'-METHOXY-5'-METHYLBENZENESULFONANILIDE

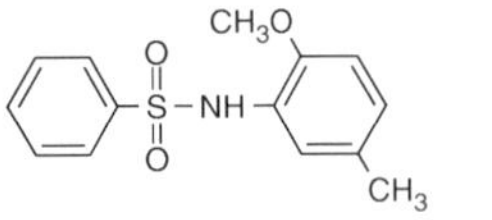

5 gm

S51,354-7
3-NITROBENZENESULFONIC ACID 2-(3-METHOXYPHENYL)HYDRAZIDE

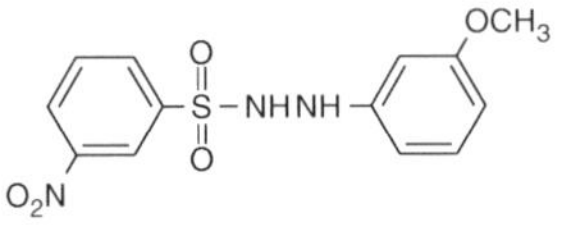

250 mg

S89,910-0
2'-(4-METHOXYBENZOYL)-P-TOLUENESULFONANILIDE

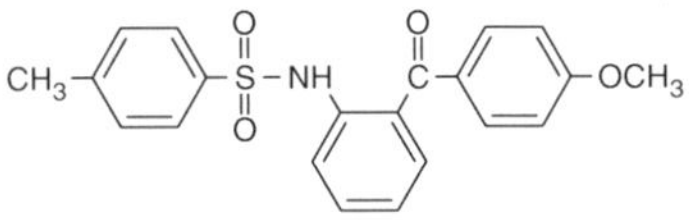

250 mg

S58,753-2
5-CARBOXY-2-(HEXADECYLOXY)-BENZENESULFONIC [5-(4-METHOXYPHENYL)-3-METHYL-1,3,4-THIADIAZOL-2(3H)-YLIDENE]HYDRAZIDE

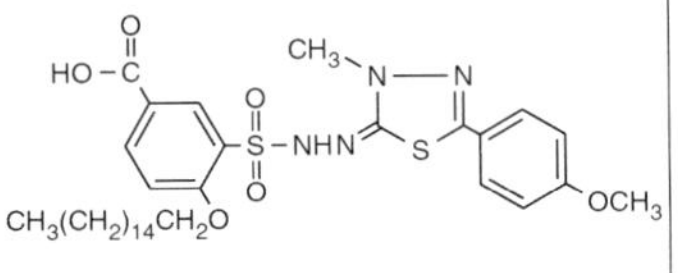

5 gm

S90,103-2
ETHYL 4-(4-METHOXYPHENYLSULFONAMIDO)-BENZOATE

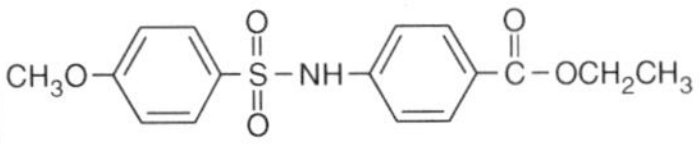

250 mg

S62,264-8
5-(DIMETHYLSULFAMOYL)-2-METHOXYBENZOYL CHLORIDE

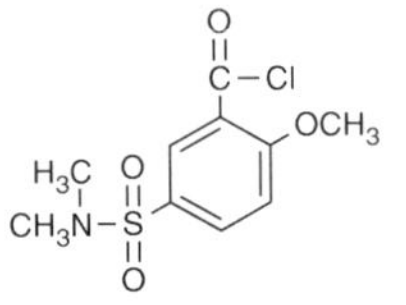

5 gm

S35,657-3
1-(PHENYLSULFONYL)-2-(3,4,5-TRIMETHOXYBENZOYL)HYDRAZINE

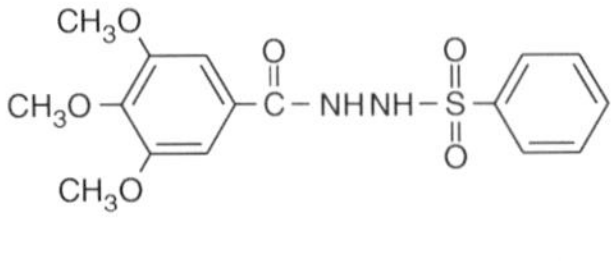

5 gm

S68,851-7
N-(4-ETHOXYPHENYL)-N'-(4-TOLYLSULFONAMIDO)OXAMIDE

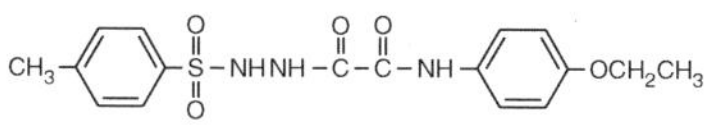

250 mg

S51,521-3
4-HEXADECYLOXY-3-(HYDRAZINOSULFONYL)BENZOIC ACID

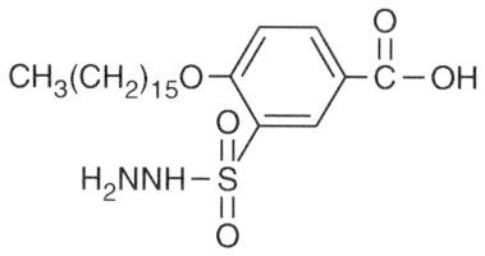

250 mg

S54,142-7
2-[N-(2-PHENOXYETHYL)-ANILINOSULFONYL]BENZOYL CHLORIDE

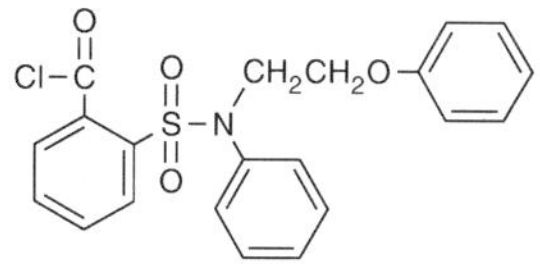

250 mg

S57,850-9
N4-(3-SULFO-4-HEXADECYLOXYBENZOYL)-N1-NAPHTHYLSULFANILAMIDE SODIUM SALT

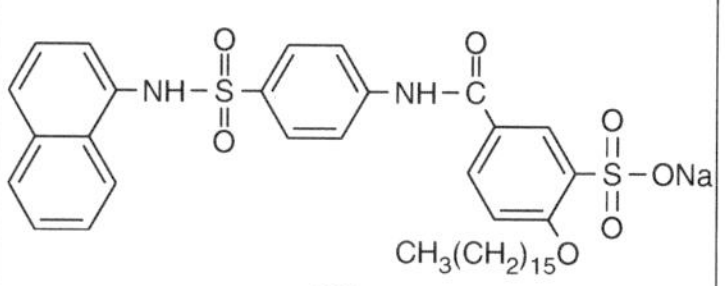

250 mg

S58,113-5
N1-NAPHTHYL-N4-(3-SULFO-2-HEXADECYLOXYBENZOYL)SULFANILAMIDE POTASSIUM SALT

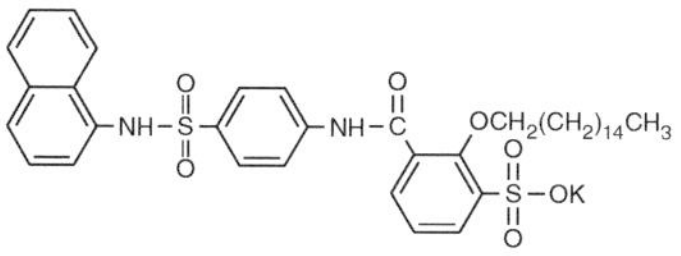

5 gm

S50,587-0
4-HEXADECYLOXY-3-[4-(1-NAPHTHYLSULFAMOYL)PHENYLCARBAMOYL]-BENZENESULFONYL FLUORIDE

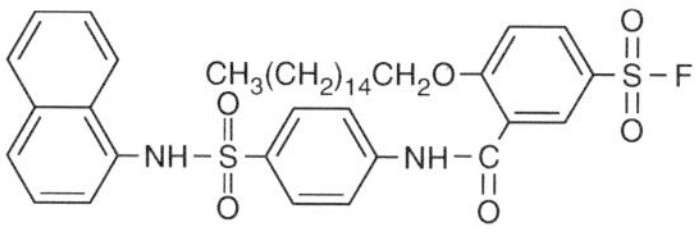

5 gm

S51,011-4
N4-(3-FLUOROSULFONYL-4-HEXADECYLOXYBENZOYL)-N1-NAPHTHYLSULFANILAMIDE

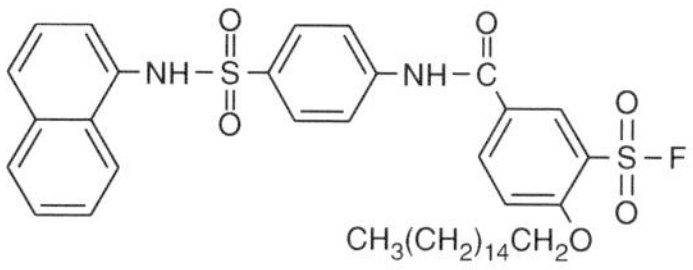

5 gm

S71,656-1
N4-ACETYL-4'-METHOXYSULFANILANILIDE

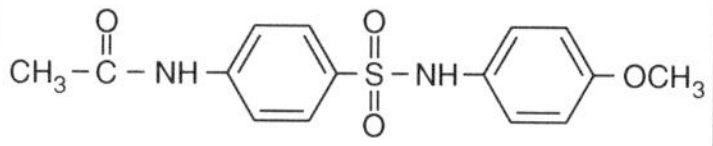

5 gm

S63,017-9
N4-ACETYL-4'-HEXADECYLOXY-SULFANILANILIDE

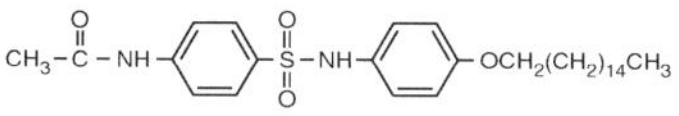

5 gm

S55,377-8
N-[2-(2-DODECYLOXYPHENYLSULFAMOYL)-ETHYL]PHTHALIMIDE

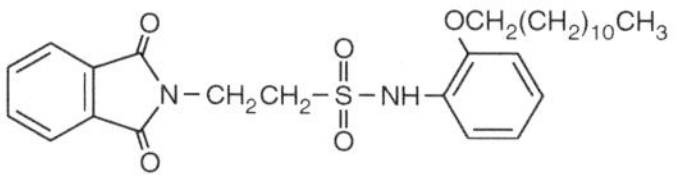

250 mg

S43,530-9
1-(3,5-DIMETHOXYBENZOYL)-2-(PHENYLSULFONYL)HYDRAZINE

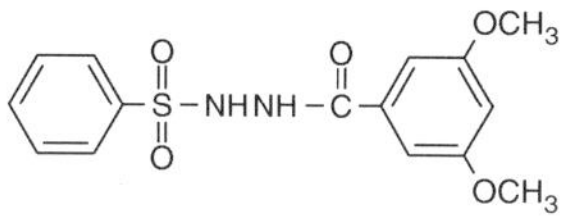

5 gm

S40,104-8
1-(3-ISOPROPOXYPROPYL)-3-(PARA-TOLYLSULFONYL)UREA

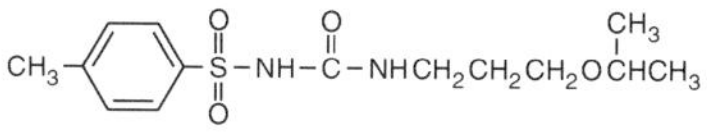

250 mg

S39,822-5
1-(3-METHOXYPROPYL)-3-(PARA-TOLYLSULFONYL)UREA

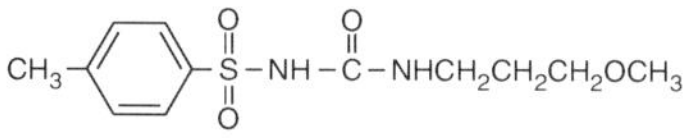

5 gm

S86,312-2
N-CYANO-4'-ETHOXYBENZENESULFONANILIDE

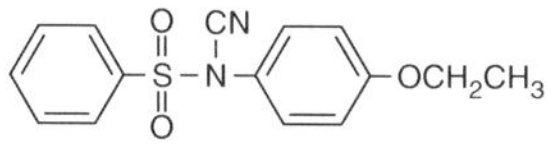

250 mg

S36,308-1
4-METHOXYBENZENESULFONAMIDE

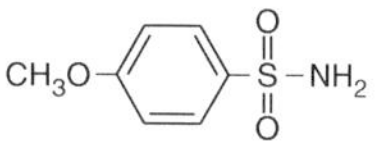

5 gm

S35,219-5
N-(3-METHOXY-4-METHYLPHENYL)-BENZENESULFONAMIDE

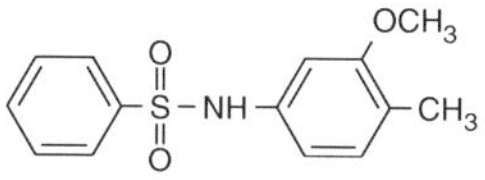

5 gm

S73,106-4
4,4'-OXYBIS(METHANESULFONANILIDE)

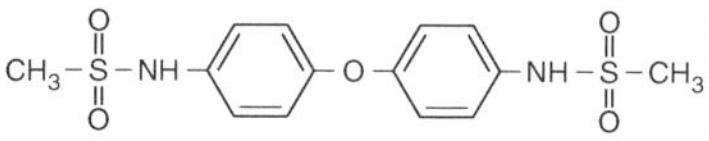

250 mg

S61,535-8
4,4'-OXYBIS(N-ALLYLBENZENESULFONAMIDE)

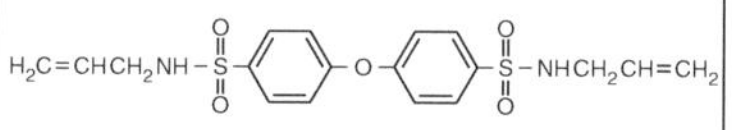

5 gm

S87,033-1
1-(BENZYLSULFONYL)-4-(4-METHOXYPHENYL)-3-THIOSEMICARBAZIDE

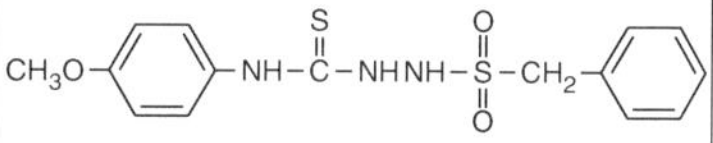

250 mg

S61,779-2
5-AMINO-2-(P-TOLUENESULFONAMIDO)PHENYL P-TOLUENESULFONATE

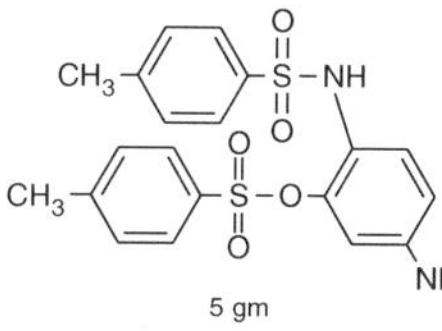

5 gm

S78,259-9
N-DELTA-CARBOBENZYLOXY-N-ALPHA-(P-TOSYL)-L-ORNITHINE

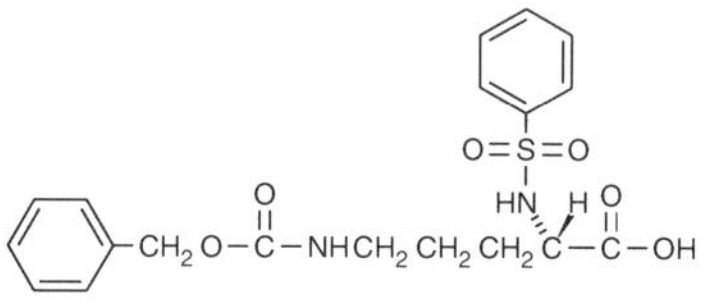

5 gm

S76,587-2
CARBOBENZYLOXY-N-EPSILON-TOSYL-L-LYSINE

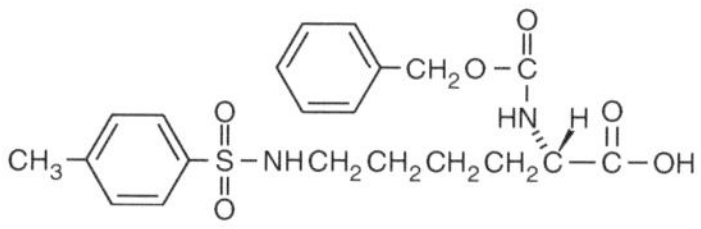

5 gm

S95,699-6
N-BENZYL-9,10-DIHYDRO-9,10-DIOXO-2-ANTHRACENESULFONAMIDE

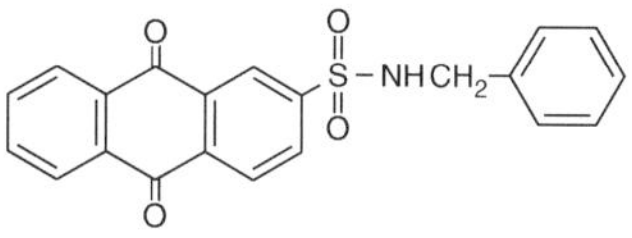

50 mg

S39,861-6
1-(2-MERCAPTOETHYL)-3-(PARA-TOLYLSULFONYL)UREA

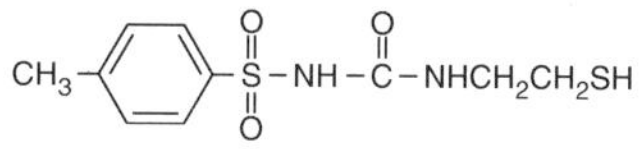

250 mg

S51,516-7
1-HEXADECANESULFONIC ACID (ALPHA-ANILINO-ALPHA-MERCAPTOMETHYLENE)-HYDRAZIDE

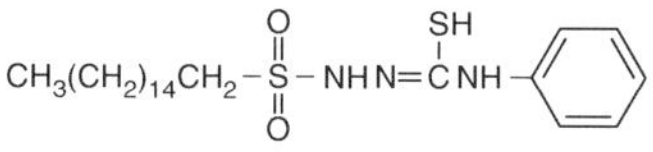

250 mg

S38,673-1
N1,N1-DIMETHYL-4-(HEXADECYLTHIO)-METANILAMIDE

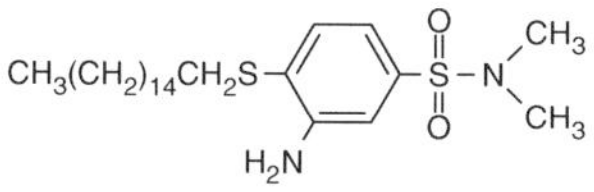

250 mg

S72,271-5
1-[2-(2,4-DINITROPHENYLTHIO)PHENYL]-3-(P-TOSYL)UREA

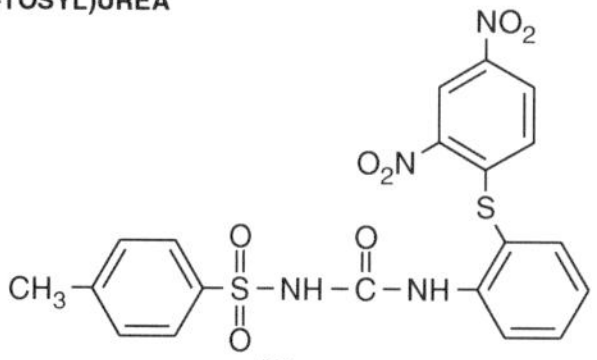

50 mg

S59,823-2
4-(HEXADECYLTHIO)-3-NITROBENZENESULFONAMIDE

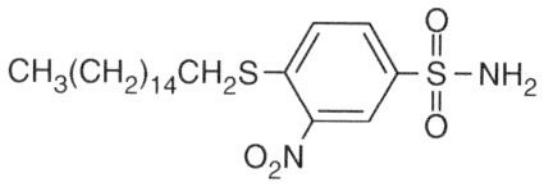

5 gm

S34,203-3
S-BENZYL-N-(4-TOLUENESULFONYL)-L-CYSTEINE

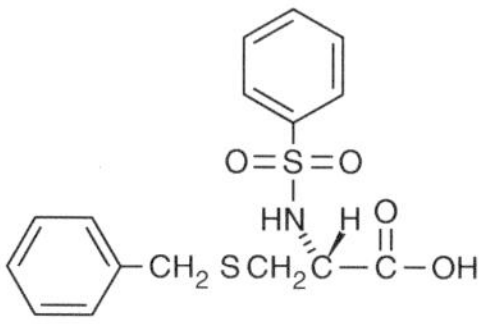

5 gm

S40,063-7
1-[2-(ETHYLTHIO)ETHYL]-3-(PARA-TOLYLSULFONYL)UREA

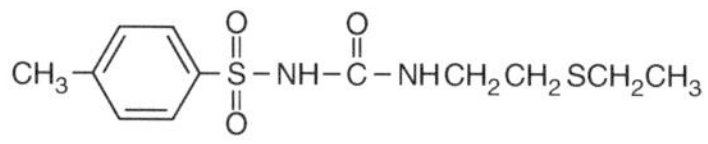

250 mg

S63,914-1
5-(DIMETHYLSULFAMOYL)-2-(HEXADECYLTHIO)BENZOYL CHLORIDE

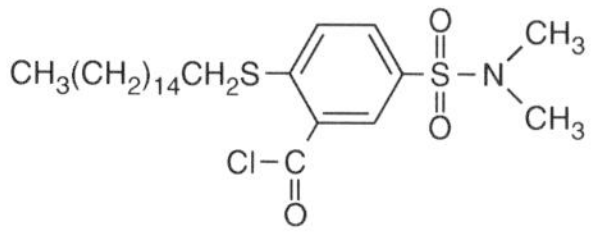

5 gm

S40,083-1
4-METHYLTHIO-2-[(3-P-TOLYLSULFONYL)UREIDO]BUTYRIC ACID

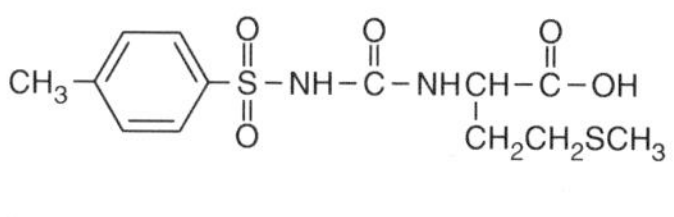

250 mg

S39,835-7
1-[2-(METHYLTHIO)ETHYL]-3-(PARA-TOLYLSULFONYL)UREA

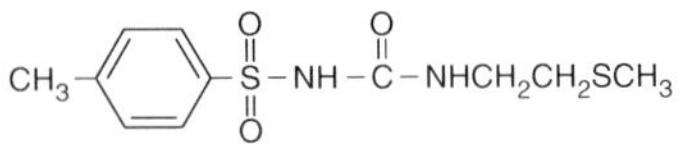

5 gm

S79,468-6
1-[4-(METHYTHIO)PHENYL]-3-(4-SULFAMOYLPHENYL)UREA

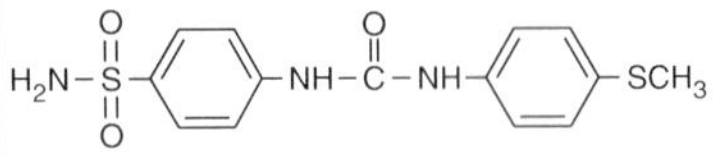

250 mg

S51,364-4
1-HEXADECANESULFONIC ACID [(METHYLTHIO)(P-TOLUIDINO)-METHYLENE]HYDRAZIDE

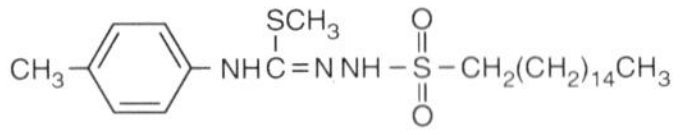

250 mg

S90,593-3
4'-(METHYLTHIO)METHANESULFONANILIDE

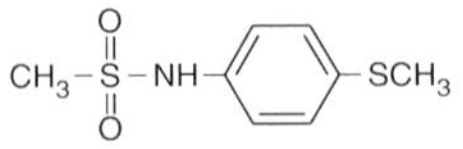

250 mg

S39,837-3
N,N'-BIS(PARA-TOLYLSULFONYLCARBAMOYL)CYSTAMINE

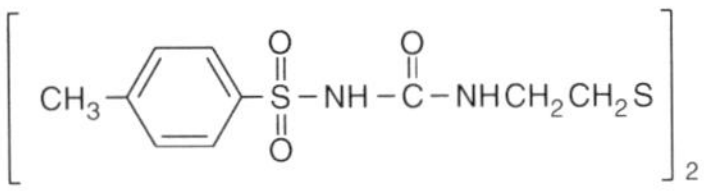

5 gm

S42,314-9
1,1'-DITHIODIETHYLBIS[3-(PARA-TOLYLSULFONYL)UREA], SODIUM DERIVATIVE

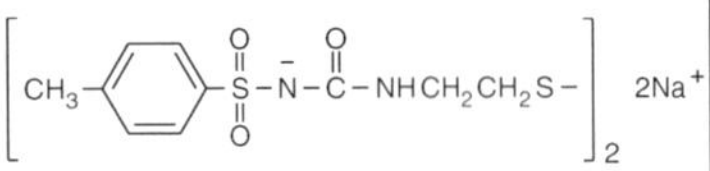

250 mg

S40,257-5
1,1'-[DITHIOBIS(TRIMETHYLENE)]BIS-[3-(PARA-TOLYLSULFONYL)UREA]

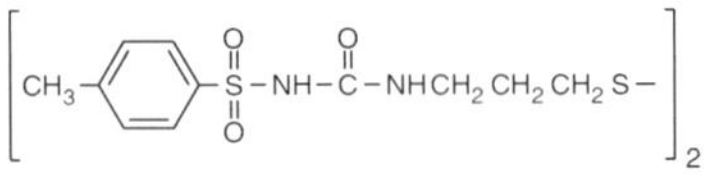

250 mg

S91,689-7
4-(METHYLSULFONAMIDO)PHENYL DISULFIDE

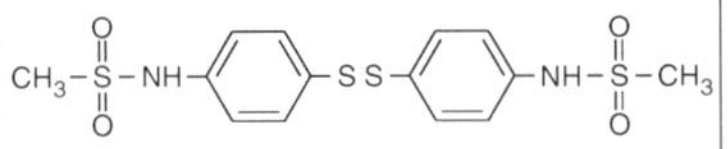

250 mg

S61,845-4
2,6-DIMETHYL-N4-HEXADECYL-N4-[2-(METHYLSULFONYLAMINO)ETHYL]-1,4-PHENYLENEDIAMINE

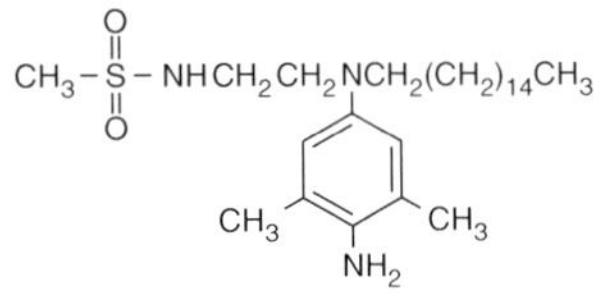

250 mg

S50,512-9
3-AMINO-N2,N4-DIMETHYLSULFANILAMIDE

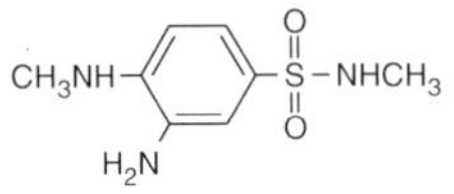

250 mg

S64,199-5
3-AMINO-N4-OCTADECYL-N1,N1,N4-TRIMETHYLSULFANILAMIDE

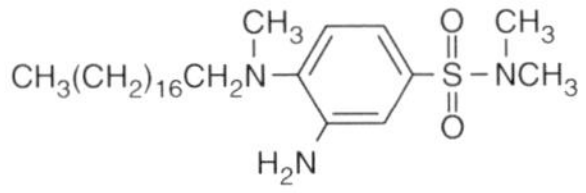

250 mg

S64,226-6
3-AMINO-N4,N4-DIETHYL-N1-OCTADECYLSULFANILAMIDE

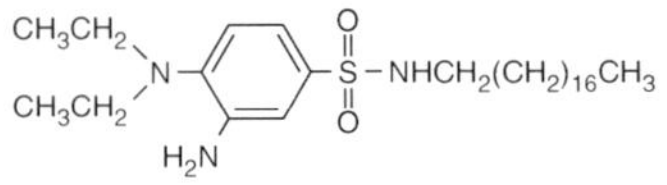

250 mg

S51,001-7
N1-HEXADECYL-4-MORPHOLINOMETANILAMIDE

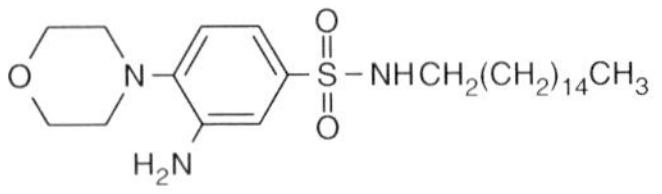

250 mg

S65,415-9
4-AMINO-3-NITROBENZENESULFONYL HYDRAZIDE

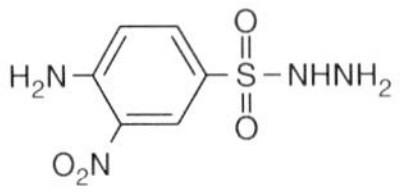

250 mg

S67,960-7
1,2-DIMETHANILYLETHANE

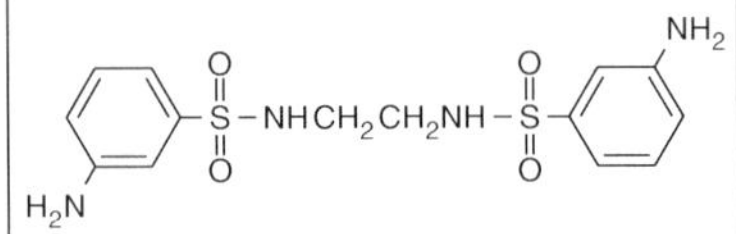

5 gm

S50,732-6
2'-AMINO-P-TOLUENESULFONANILIDE

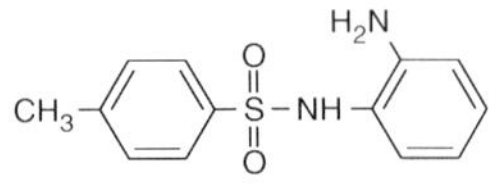

5 gm

S63,126-4
3-AMINO-N-HEXADECYL-4-(4-MORPHOLINO)-N-METHYLBENZENESULFONAMIDE

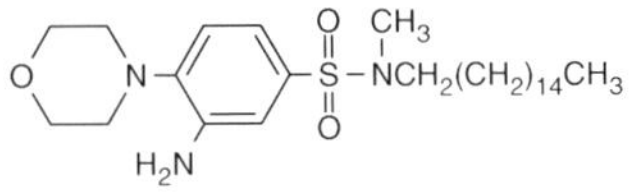

5 gm

S66,223-2
N-(2-AMINOETHYL)-P-TOLUENESULFONAMIDE

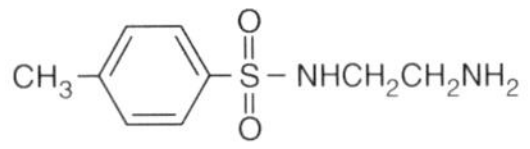

5 gm

S64,765-9
3'-AMINO-4'-METHYLMETHANESULFONANILIDE

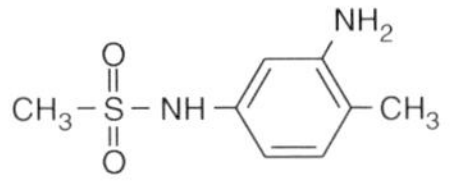

5 gm

S61,787-3
2-[N-(3-AMINOPHENYL)-N-METHYLSULFAMOYL]ETHYL SULFATE, POTASSIUM SALT

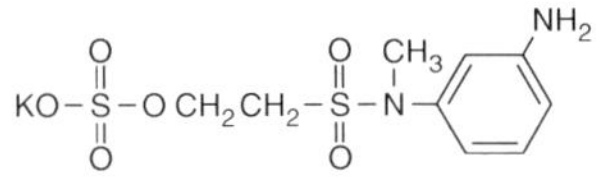

5 gm

S56,973-9
4-MORPHOLINO-3-(MORPHOLINOSULFONYL)-ANILINE

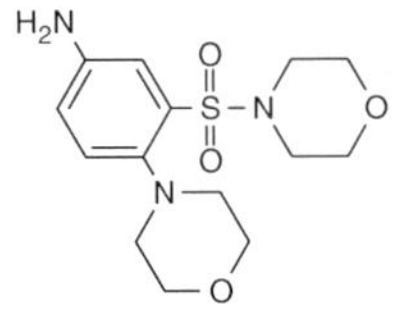

250 mg

S50,585-4
1-HEXADECANESULFONIC ACID 2-(3-AMINO-ALPHA-PIPERIDINO-BENZYLIDENE)HYDRAZIDE HYDROCHLORIDE

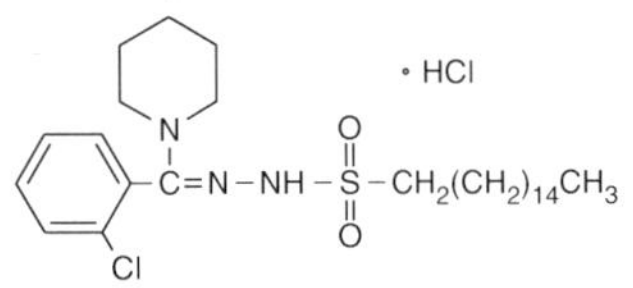

5 gm

S51,373-3
HEXADECANESULFONIC ACID (5-AMINO-2-CHLORO-ALPHA-PIPERIDINOBENZYLIDENE)-HYDRAZIDE

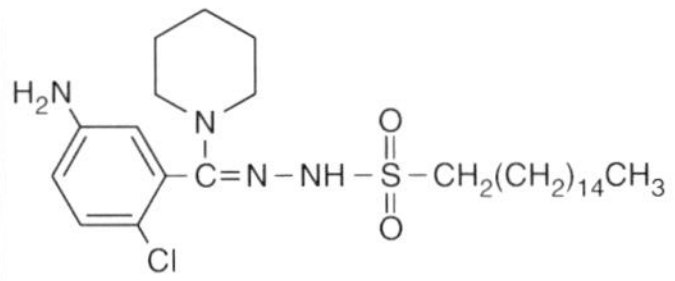

250 mg

S50,547-1
1-HEXADECANESULFONIC ACID 2-(3-AMINO-ALPHA-MORPHOLINOBENZYLIDENE)HYDRAZIDE

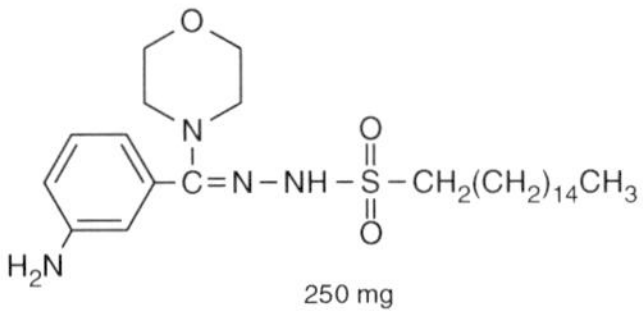

250 mg

S35,618-2
N6-(P-TOSYL)-L-LYSINE

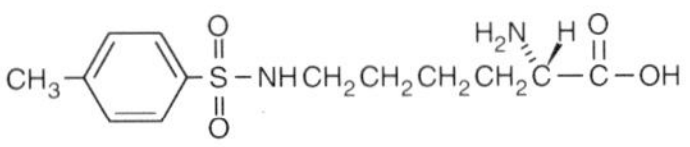

5 gm

S76,590-2
N-DELTA-TOSYL-L-ORNITHINE TERT-BUTYL ESTER

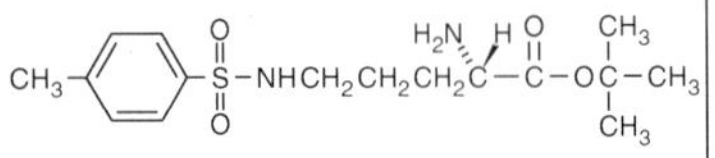

5 gm

S76,513-9
L-PHENYLALANINE TERT-BUTYL ESTER DIBENZENESULFIMIDE SALT

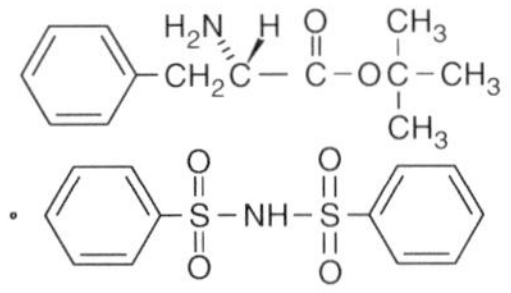

5 gm

S51,373-3
HEXADECANESULFONIC ACID (4-AMINO-2-CHLORO-ALPHA-PIPERIDINOBENZYLIDENE)-HYDRAZIDE

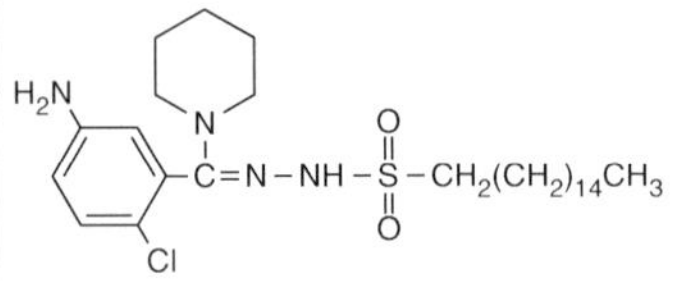

250 mg

S61,857-8
N1-BENZYL-N1-HEXADECYLMETANILAMIDE

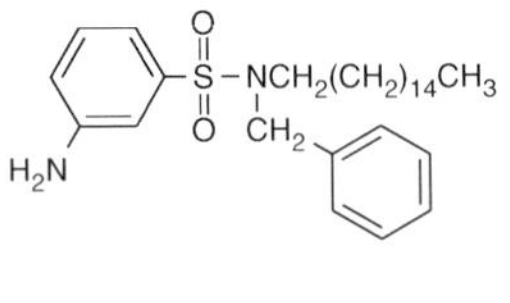

250 mg

S59,785-6
N1-HEXADECYL-N1-METHYL-4-METHYLSULFONYLMETANILAMIDE

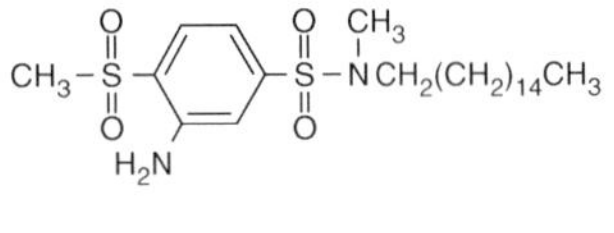

250 mg

S76,707-7
2-AMINO-N-METHYL-N-(2-METHYL-1,3-DIOXOLAN-2-YLMETHYL)-BENZENESULFONAMIDE

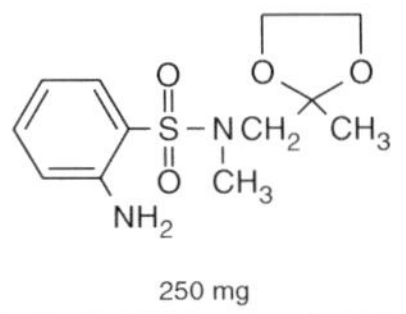

250 mg

S35,762-6
2-AMINO-N-ETHYLBENZENESULFONANILIDE

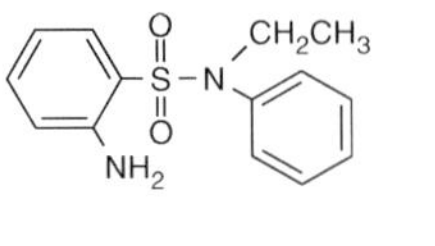

5 gm

S63,670-3
4-AMINO-3-CARBOXY-N-(2-CARBOXYPHENYL)BENZENESULFONAMIDE
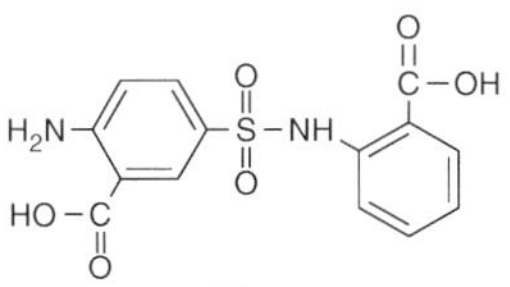
250 mg

S38,701-0
DIMETHYL 5-SULFANILAMIDOISOPHTHALATE
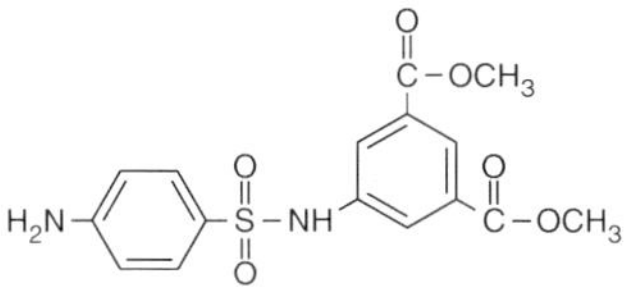
250 mg

S40,314-8
1-(PARA-AMINOPHENYLSULFONYL)-3-ISOPENTYLUREA
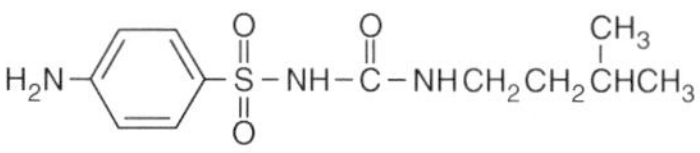
250 mg

S44,110-4
1-SULFANILYLUREA
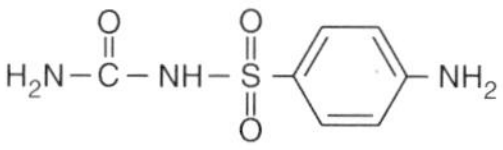
5 gm

S50,146-8
N-SULFANILYLSULFANILIC ACID
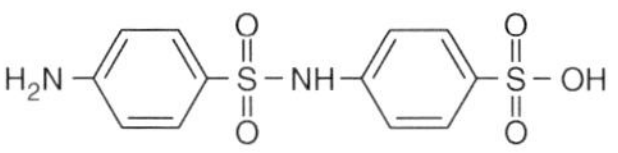
250 mg

S63,889-7
N1,N1-BIS(2-ETHYLHEXYL)METANILAMIDE
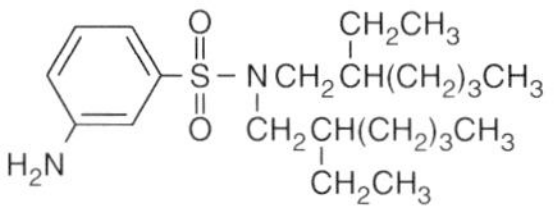
5 gm

S59,842-9
N1-HEXADECYL-N1-METHYLMETANILAMIDE
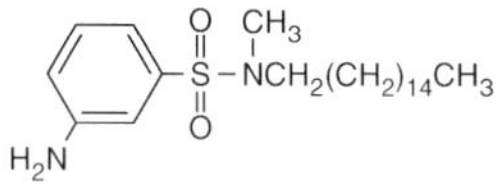
250 mg

S62,240-0
N1,N1-DIDODECYLMETANILAMIDE
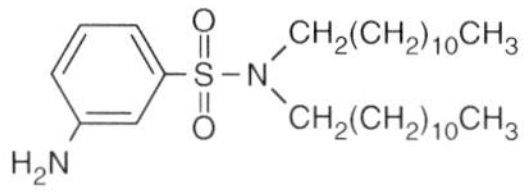
250 mg

S65,253-9
N1-METHYLMETANILANILIDE
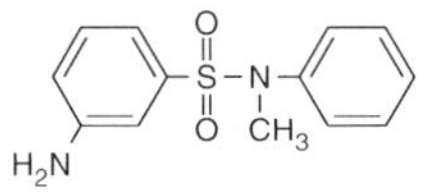
5 gm

S58,721-4
N1-HEXADECYLMETANILANILIDE
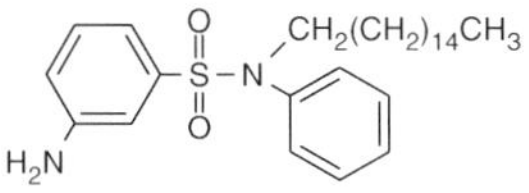
250 mg

S58,707-9
N1-HEXADECYLSULFANILANILIDE
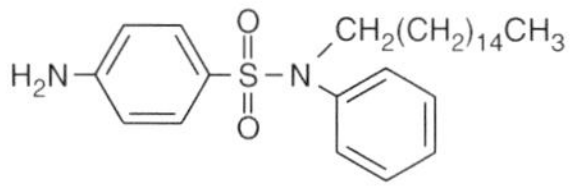
250 mg

S35,785-5
5-AMINO-ORTHO-TOLUENESULFONANILIDE
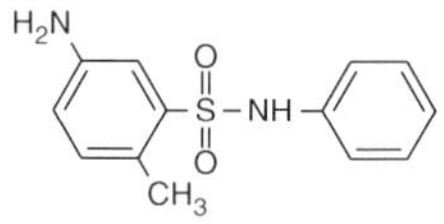
5 gm

S35,766-9
5-AMINO-N-ETHYL-ORTHO-TOLUENESULFONANILIDE
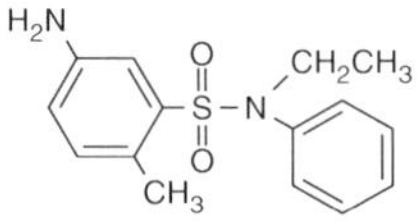
5 gm

S82,817-3
N1-OCTADECYLSULFANILAMIDE HYDROCHLORIDE
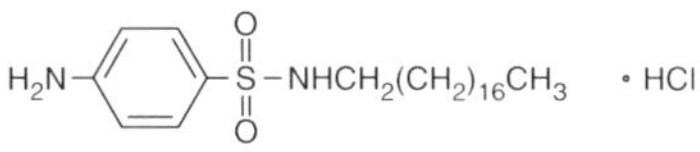
250 mg

S45,314-5
3-AMINO-1-(P-TOLUENESULFONYL)-3-PYRAZOLINE
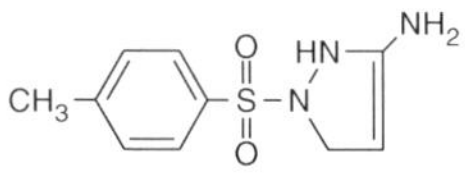
5 gm

S38,297-3
2-AMINO-N-HEXADECYLBENZENESULFONAMIDE
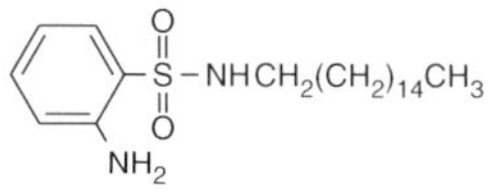
5 gm

S75,674-1
6-METHYL-N-EPSILON-(P-TOSYL)LYSINE HYDROCHLORIDE
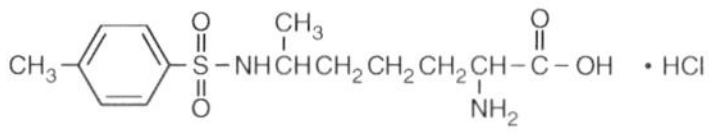
5 gm

S76,538-4
L-ALANINE TERT-BUTYL ESTER DIBENZENESULFIMIDE
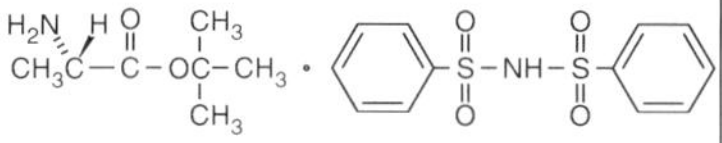
5 gm

S76,514-7
L-LEUCINE TERT-BUTYL ESTER DIBENZENESULFIMIDE SALT
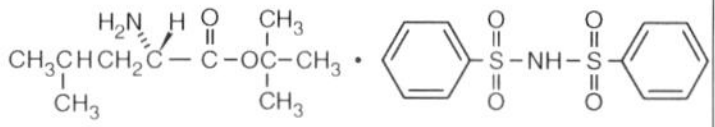
5 gm

S78,449-4
N-ALPHA-(P-TOSYL)LYSINAMIDE ACETATE MONOHYDRATE
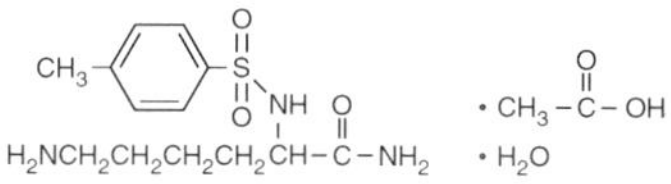
250 mg

S62,300-8
METHANESULFONIC ACID [2-(4-AMINOPHENYL)-4-3-SULFOPROPYL-1,3,4-THIADIAZOL-2-IN-5-YLIDENE]HYDRAZIDE
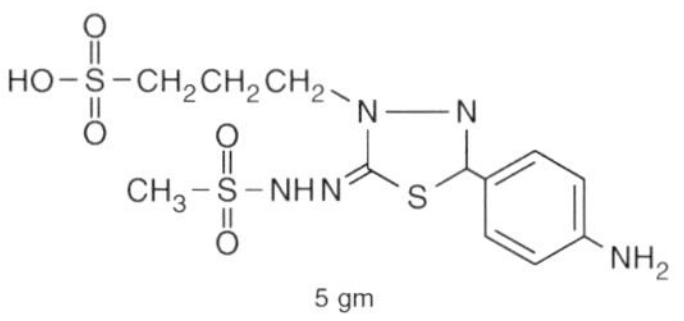
5 gm

S65,601-1
N1-CYCLOHEXYLMETANILAMIDE
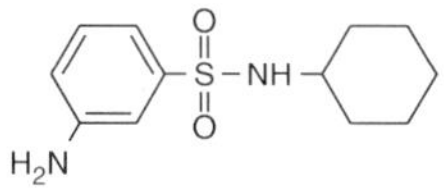
5 gm

S71,347-3
N1,N1-DIMETHYLSULFANILAMIDE
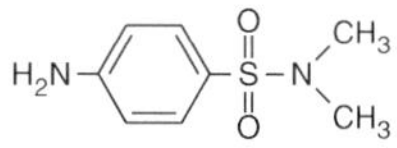
5 gm

S38,700-2
N1-HEXADECYLSULFANILAMIDE
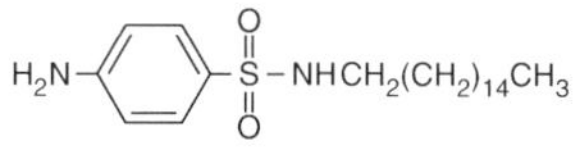
5 gm

S34,105-3
3-AMINO-N,N-DIMETHYL-4-METHYL-BENZENESULFONAMIDE
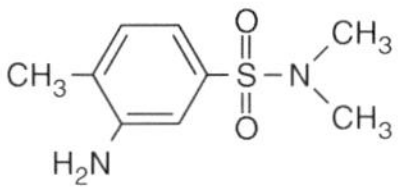
5 gm

S34,569-5
3,5-DICHLOROSULFANILAMIDE
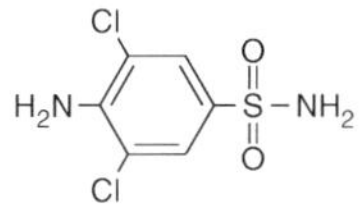
5 gm

S34,109-6
N-(2,4-DINITRO-1-NAPHTHYL)-P-TOLUENESULFONAMIDE
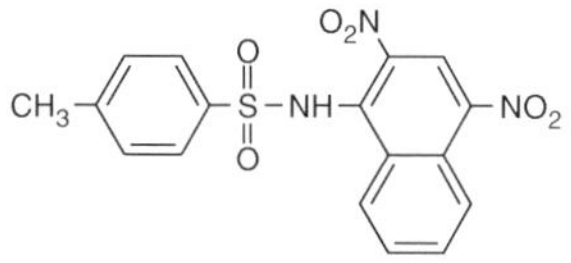
5 gm

S67,188-6
N-[2-(N-ETHYL-3-METHYL-4-NITRO-ANILINO)ETHYL]METHANESULFONAMIDE
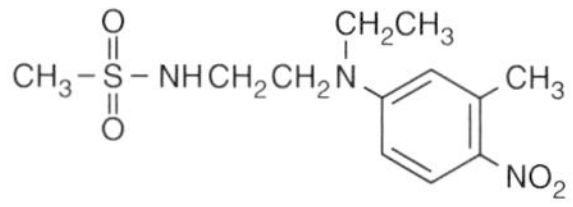
5 gm

S46,174-1
DISPERSE YELLOW 42 (C.I. 10338)
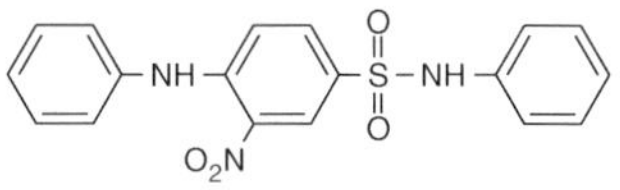
5 gm

S64,493-5
4-(METHYLAMINO)-N-METHYL-3-NITROBENZENESULFONAMIDE
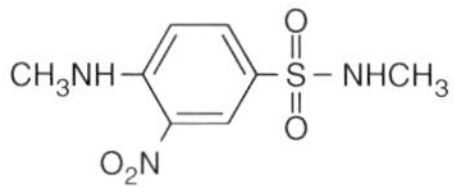
5 gm

S63,081-0
N-ETHYL-4-ETHYLAMINO-3-NITROBENZENESULFONAMIDE
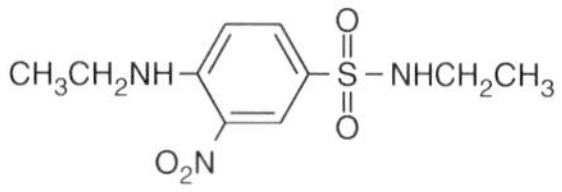
250 mg

S51,685-6
3-NITRO-N1,N1,N4,N4-TETRAKIS(2-ETHYLHEXYL)SULFANILAMIDE
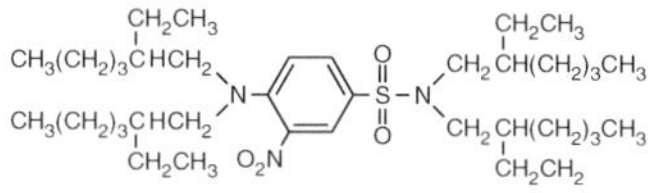
250 mg

S50,782-2
N-[2-(N-HEXADECYLANILINO)ETHYL]-METHANESULFONAMIDE
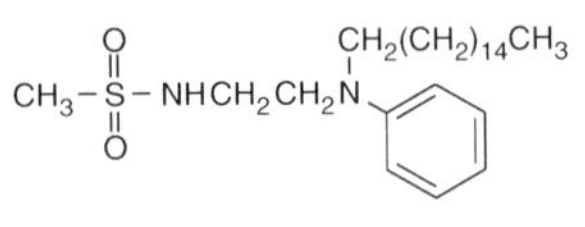
250 mg

S72,084-4
1-(4-METHYL-2-NITROPHENYL)-3-(4-SULFAMOYLPHENYL)UREA
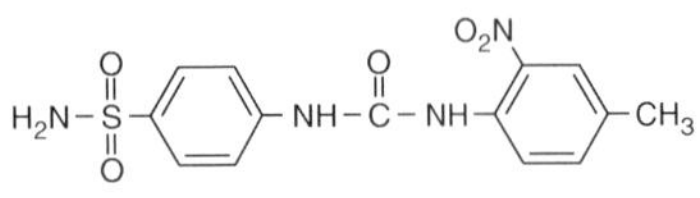
250 mg

S85,874-9
2',6'-DINITRO-N-ETHYL-4'-(METHYLSULFONYL)-METHANESULFONANILIDE
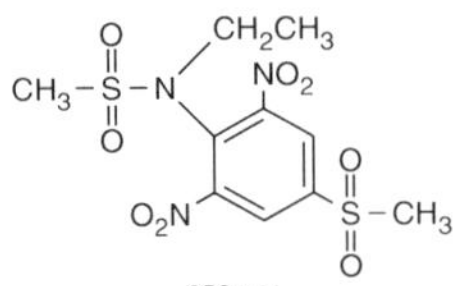
250 mg

S85,875-7
N-BUTYL-2',6'-DINITRO-4'-(METHYLSULFONYL)-METHANESULFONANILIDE
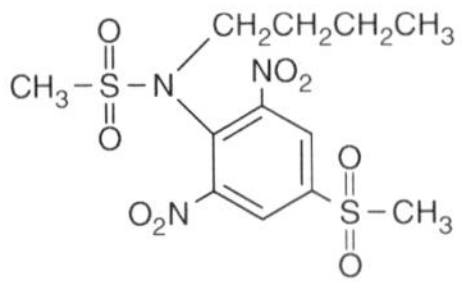
250 mg

S85,876-5
2',6'-DINITRO-N-PENTYL-4'-(METHYLSULFONYL)-METHANESULFONANILIDE
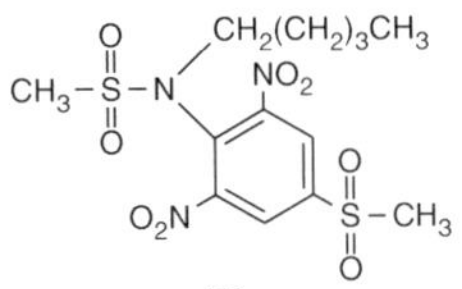
250 mg

S74,496-4
N-METHYL-2'-NITRO-P-TOLUENESULFONANILIDE
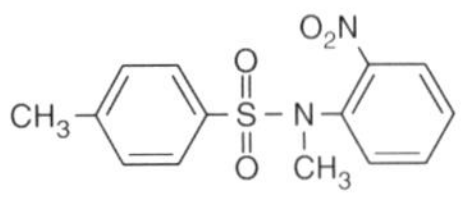
250 mg

S90,590-9
4'-METHYL-2'-NITROMETHANESULFONANILIDE
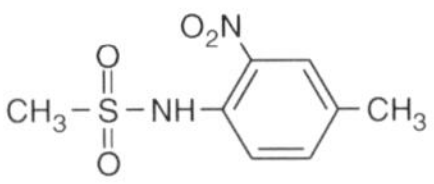
250 mg

S85,873-0
2',6'-DINITRO-N-METHYL-4'-(METHYLSULFONYL)-METHANESULFONANILIDE
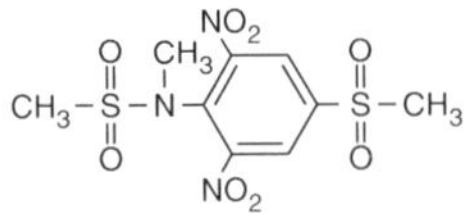
250 mg

S63,046-2
4-DODECANAMIDO-3-NITROBENZENESULFONIC 2-PHENYLHYDRAZIDE
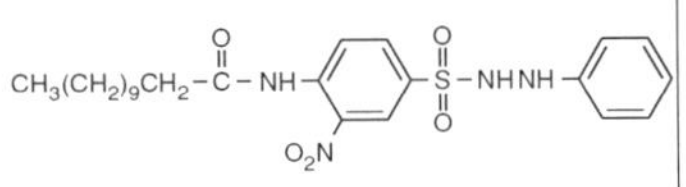
5 gm

S63,144-2
N,N-DIDODECYL-4-MORPHOLINO-3-NITROBENZENESULFONAMIDE
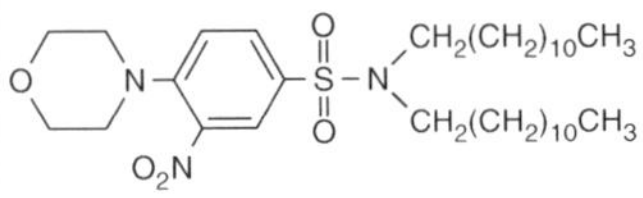
5 gm

S64,252-5
N,N-BIS(2-ETHYLHEXYL)-4-MORPHOLINO-3-NITROBENZENESULFONAMIDE
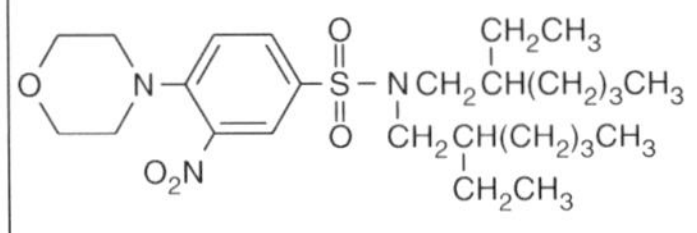
250 mg

S62,866-2
4-METHYL-2'-NITROBENZENESULFONANILIDE
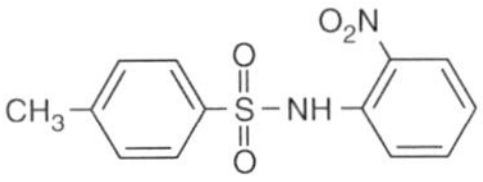
250 mg

S38,636-7
1-NITRO-2-PARA-TOLUENESULFONAMIDONAPHTHALENE
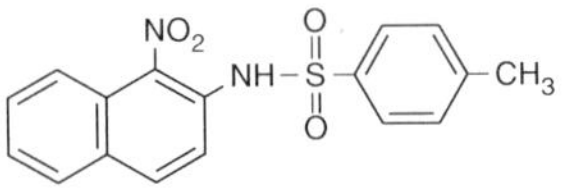
250 mg

S78,246-7
1-CYANO-1,2-DIHYDRO-2-(PHENYLSULFONYL)ISOQUINOLINE
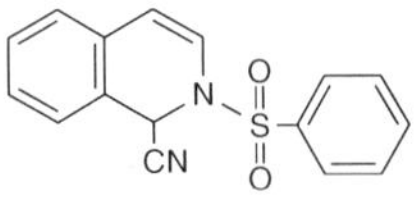
5 gm

S52,634-7
1-METHYL-4-(PARA-TOLYLSULFONYL)-PIPERAZINE
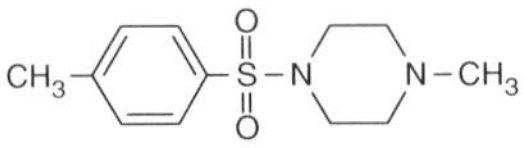
250 mg

S80,581-5
N,N'-ETHYLENEBIS-(BENZENESULFONAMIDE)
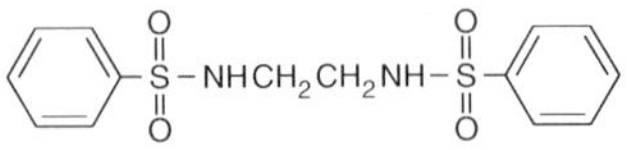
250 mg

S73,935-9
N,N'-(4-METHYL-1,2-PHENYLENE)BIS-(BENZENESULFONAMIDE)
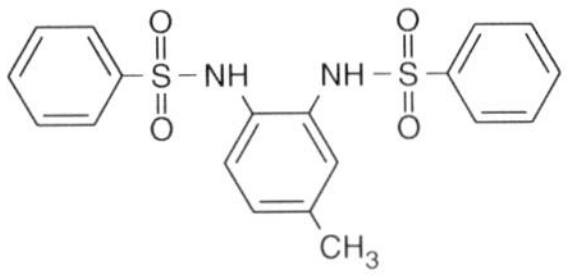
5 gm

S94,290-1
N,N'-(1,2-PHENYLENE)BIS(P-TOLUENESULFONAMIDE)
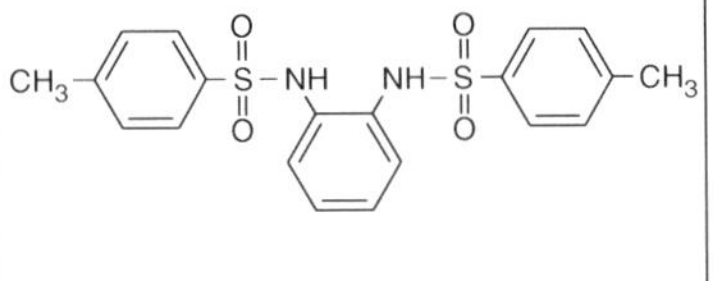
250 mg

S56,281-5
N,N'-ETHYLENEBIS(N-METHYL-P-TOLUENESULFONAMIDE)
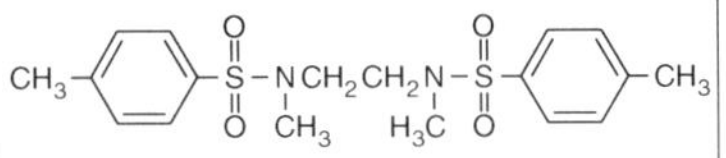
250 mg

S57,856-8
1-HEXADECANESULFONIC [2-(3,5-DI-TERT-BUTYLPHENYL)-1-METHYL-6-SULFO-4(1H)-QUINOLYLIDENE]HYDRAZIDE
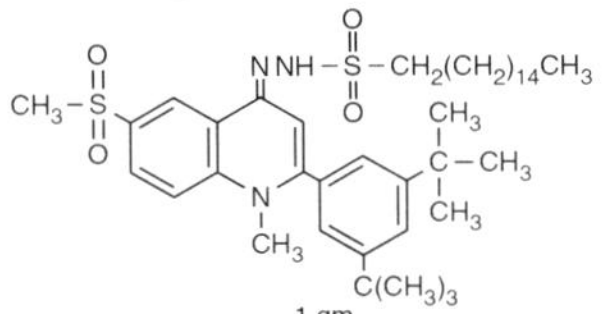
1 gm

S65,255-5
4-(DIMETHYLAMINO)-N,N-DIDODECYL-3-NITROBENZENESULFONAMIDE
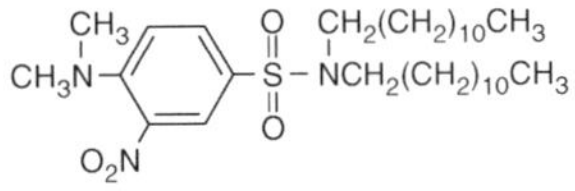
250 mg

S76,503-1
(3-DANSYLAMIDO)-PROPYLTRIMETHYLAMMONIUM IODIDE
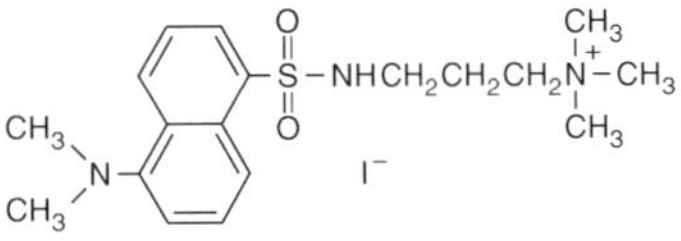
5 gm

S87,175-3
2-(3-ANILINOPROPYL)-1,2-BENZISOTHIAZOL-3(2H)-ONE-1,1-DIOXIDE
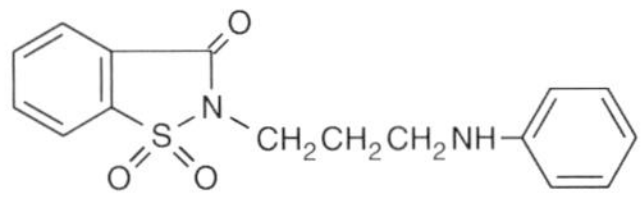
250 mg

S65,973-8
N-METHYL-4'-CHLORO-2'-(3,4-DIHYDRO-1-ISOQUINOLINYL)-P-TOLUENESULFONANILIDE
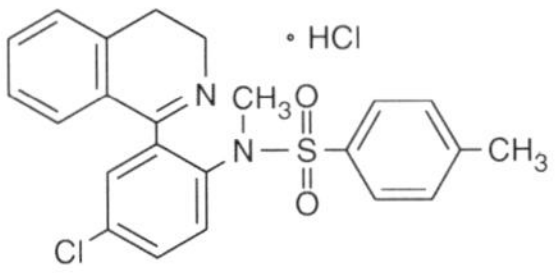
5 gm

S69,902-0
2-(METHYLSULFONAMIDO)-4-NITROBENZOIC ACID
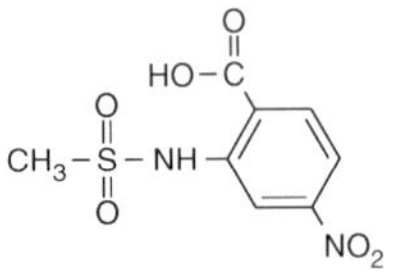
250 mg

S78,114-2
3,5-DINITROBENZOYL N-(M-TOLYLSULFONYL)CARBAMATE
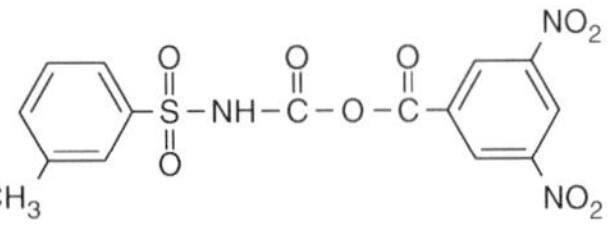
250 mg

S79,629-8
1-(2-METHYL-5-NITROPHENYL)-3-(4-SULFAMOYLPHENYL)UREA
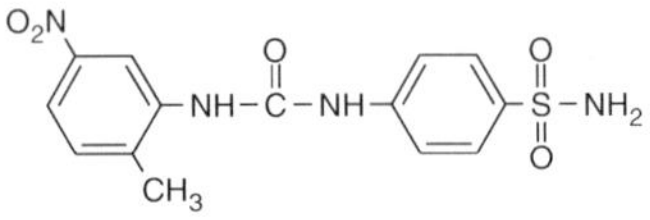
250 mg

S79,570-4
3,5'-DINITRO-2'-METHYLBENZENESULFONANILIDE
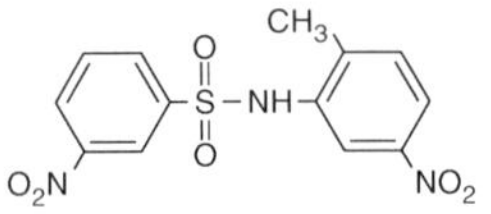
250 mg

S79,565-8
5'-NITRO-2,2',4,6-TETRAMETHYLBENZENESULFONANILIDE
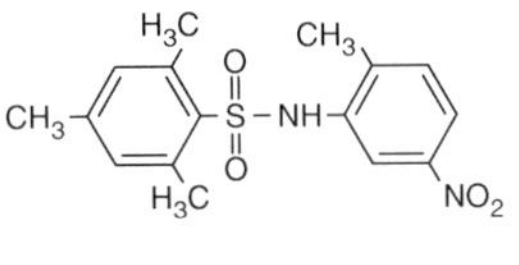
250 mg

S87,019-6
1-(BENZYLSULFONYL)-4-(3-NITROPHENYL)-3-THIOSEMICARBAZIDE
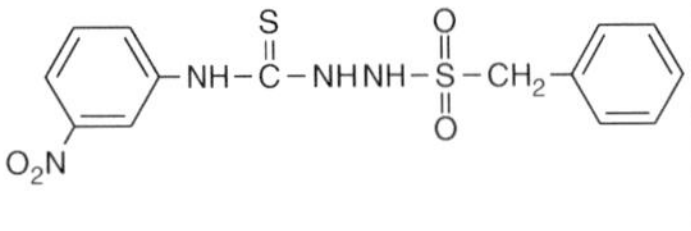
250 mg

S62,877-8
4-CHLORO-N-(2-CYANOETHYL)-3-NITROBENZENESULFONAMIDE
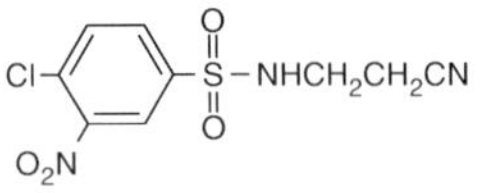
5 gm

S64,747-0
4'-METHYL-3'-NITROMETHANESULFONANILIDE
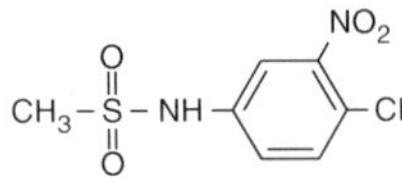
5 gm

S53,710-1
1-ANTHRANILOYL-2-(PARA-TOLYLSULFONYL)HYDRAZINE

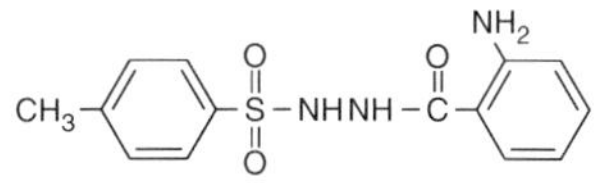

5 gm

S61,540-4
N,N-BIS(2-CYANOETHYL)-P-TOLUENE-SULFONANILIDE

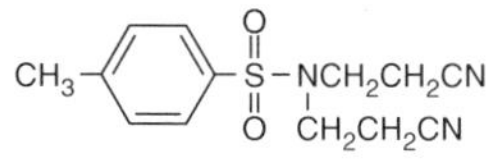

5 gm

S76,202-4
3,4-DIHYDRO-1-[2-(DI-P-TOSYLAMINO)-PHENYL]ISOQUINOLINE

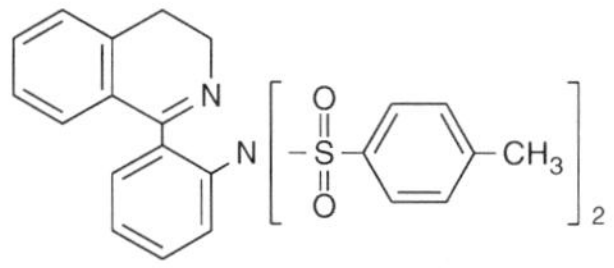

250 mg

S93,096-2
4'-ANILINO-P-TOLUENESULFONANILIDE

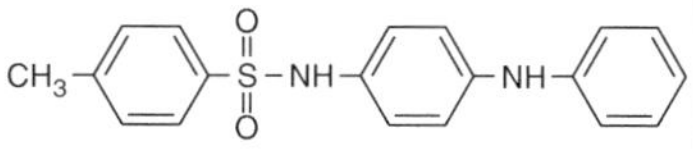

250 mg

S90,994-7
4'-(DIETHYLAMINO)-METHANESULFONANILIDE HYDROCHLORIDE

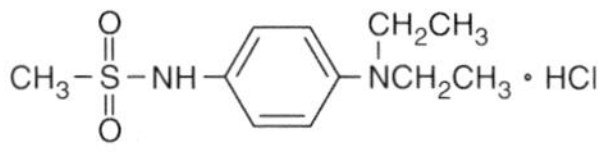

250 mg

S87,035-8
4-[4-(DIMETHYLAMINO)PHENYL]-1-(PHENYLSULFONYL)-3-THIOSEMICARBAZIDE

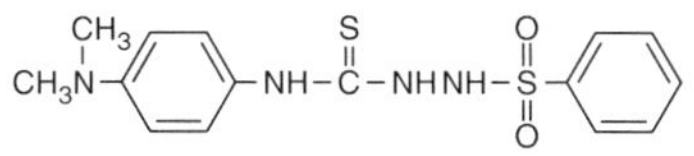

250 mg

S56,979-8
1-MORPHOLINO-2-(MORPHOLINOSULFONYL)-4-NITROBENZENE

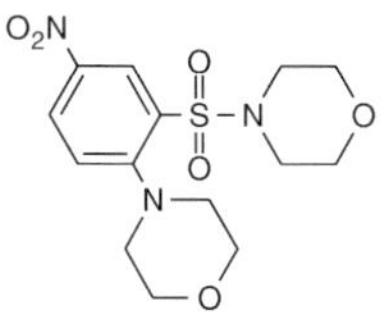

250 mg

S56,892-9
4'-(DIETHYLAMINO)-METHANESULFONANILIDE

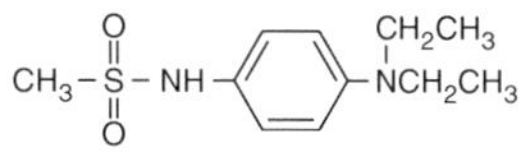

250 mg

S63,040-3
5-NITRO-2-(P-TOLUENESULFONAMIDO)PHENYL P-TOLUENESULFONATE

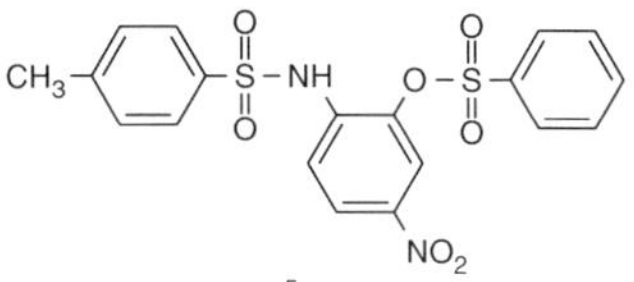

5 gm

S70,633-7
4'-NITROBENZENESULFONANILIDE

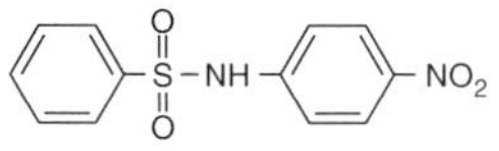

5 gm

S80,734-6
N-(4-NITROPHENYL)DI-ALPHA-TOLUENESULFONAMIDE

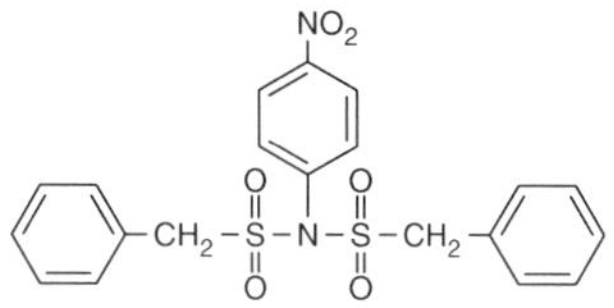

250 mg

S71,603-0
4'-NITROMETHANESULFONANILIDE

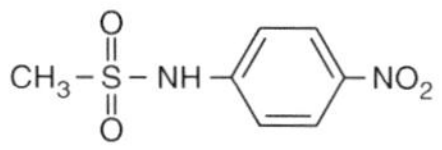

250 mg

S72,171-9
2'-METHYL-4'-NITROMETHANESULFONANILIDE

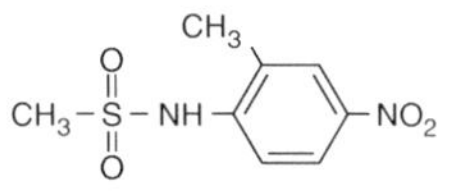

250 mg

S34,051-0
N4-ACETYL-4'-NITROSULFANILANIDE

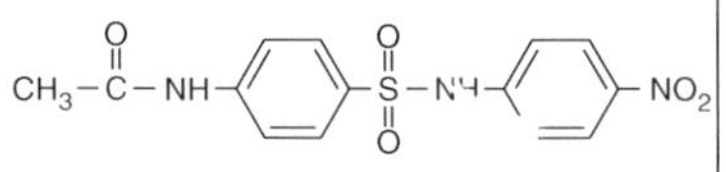

5 gm

S50,021-6
N,N'-BIS(4-NITROPHENYL)SULFAMIDE

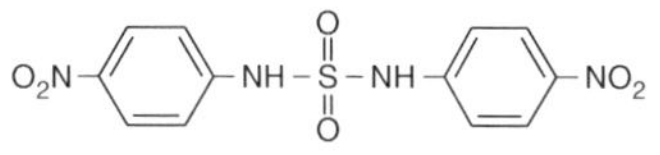

250 mg

S34,436-2
N-(5-CHLORO-2-METHYL-4-NITROPHENYL)-BENZENESULFONAMIDE

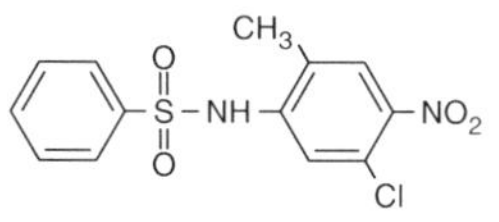

5 gm

S35,312-4
N-(2-METHYL-4-NITROPHENYL)-BENZENESULFONAMIDE

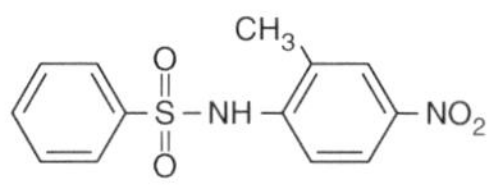

5 gm

S51,124-2
N-ACETYL-4'-(P-TOLYLSULFONAMIDO)-BENZANILIDE

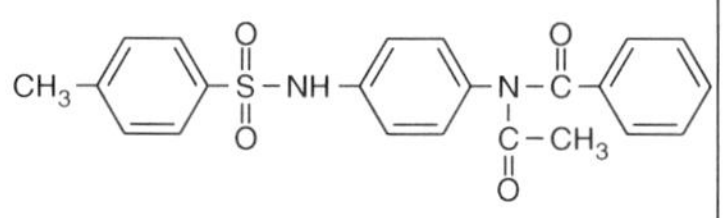

5 gm

S76,321-7
N-(3-CYANOPROPYL)SACCHARIN

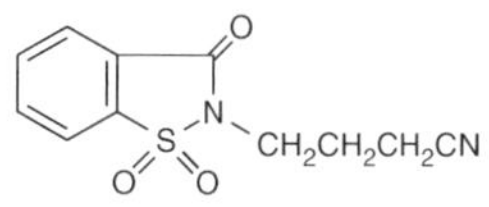

250 mg

S51,359-8
3'-(2-METHYL-5-OXO-1-PHENYL-3-PYRAZOLIN-3-YL)-1-HEXADECANESULFONANILIDE

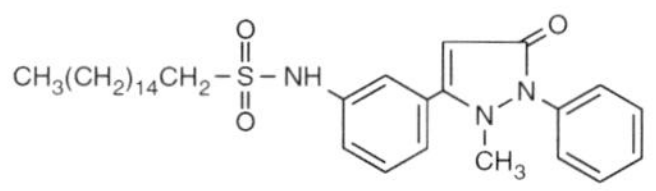

250 mg

S36,515-7
N-ALPHA-P-TOSYL-L-ARGININE HYDROCHLORIDE

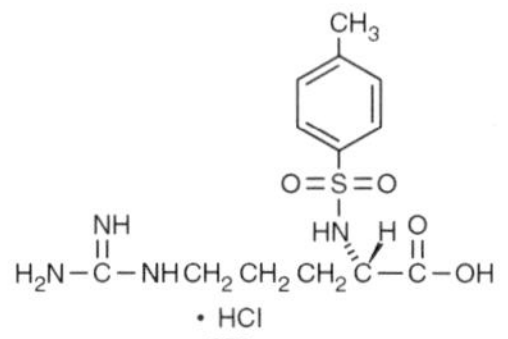

250 mg

S36,516-5
N-ALPHA-P-TOSYL-L-ARGININE TRIHYDRATE

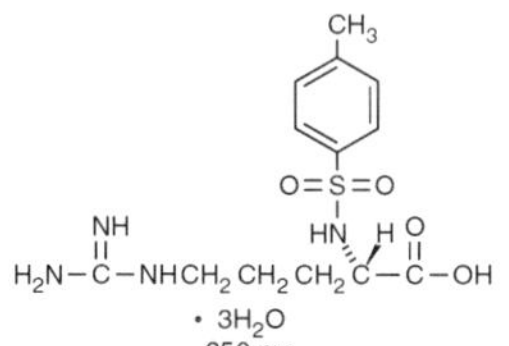

250 mg

S78,741-8
N-ALPHA-(P-TOSYL)-D-ARGININE

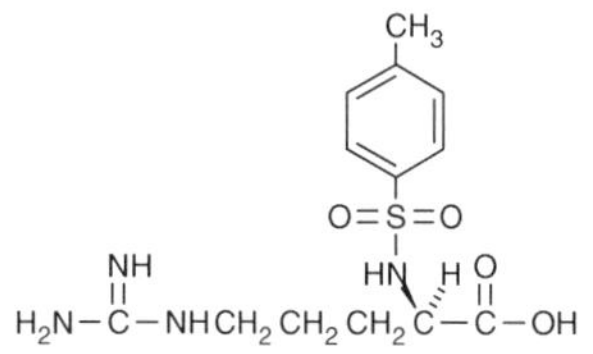

250 mg

S52,530-8
N,N-TETRAMETHYLENE-4-(META-TOLYL)-1-(PARA-TOLYLSULFONYL)-ISONIPECOTAMIDE

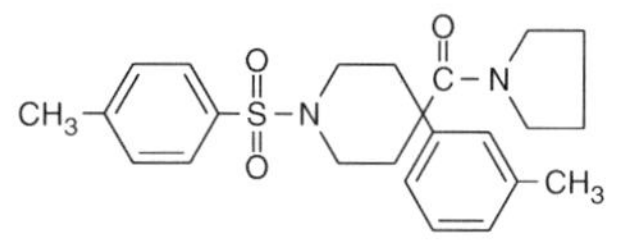

250 mg

S70,496-2
N-ALPHA-TOSYL-L-ARGININAMIDE HYDROCHLORIDE

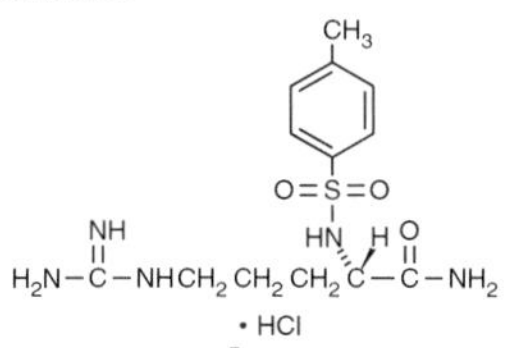

5 gm

S73,939-1
N,N'-(2-METHYL-1,4-PHENYLENE)BIS-(BENZENESULFONAMIDE)

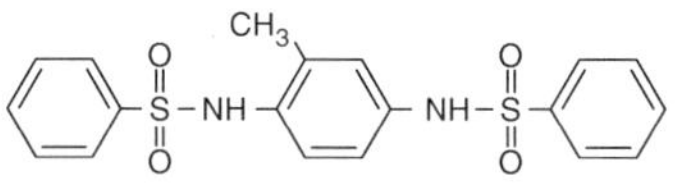

5 gm

S51,353-9
N,N'-(1,4-PHENYLENE)BIS(P-TOLUENESULFONAMIDE)

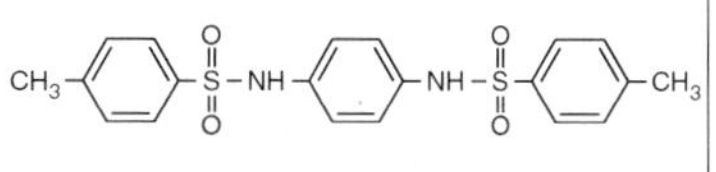

250 mg

S50,995-7
N,N'-(1,4-PHENYLENE)BIS(N-METHYL-P-TOLUENESULFONAMIDE)

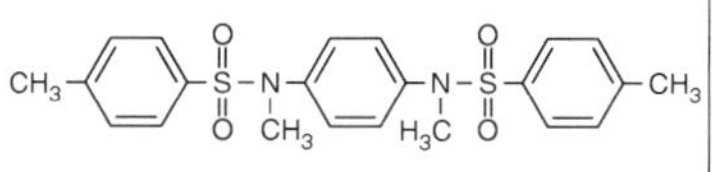

250 mg

S86,397-1
DEOXYBENZOIN (P-TOSYL)HYDRAZONE

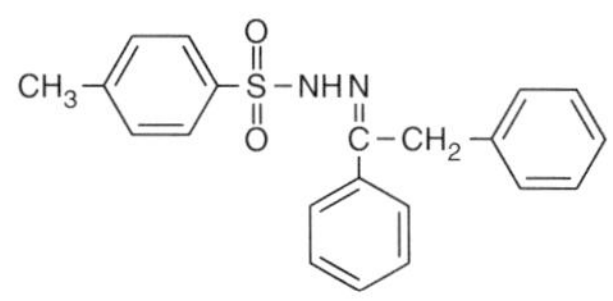

250 mg

S65,833-2
2-(N-METHYL-P-TOLUENESULFONAMIDO)-N-PHENETHYLBENZAMIDE

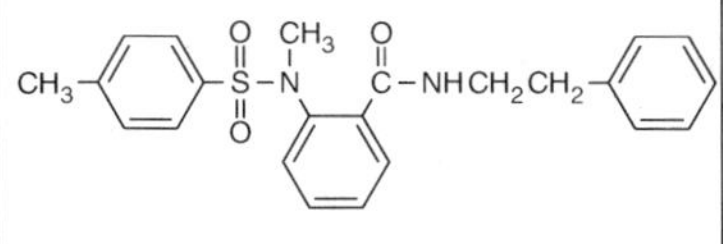

5 gm

S68,891-6
4-SULFAMOYLANILINOMETHANESULFONIC ACID, SODIUM SALT

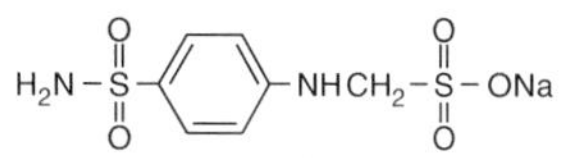

5 gm

S86,302-5
1-(4-BIPHENYLYL)-3-(2-NITROPHENYLSULFONYL)UREA

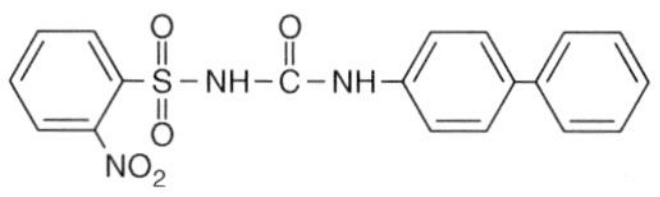

250 mg

S36,194-1
2-(N-ETHYL-P-TOLUENESULFONAMIDO)-9-FLUORENONE
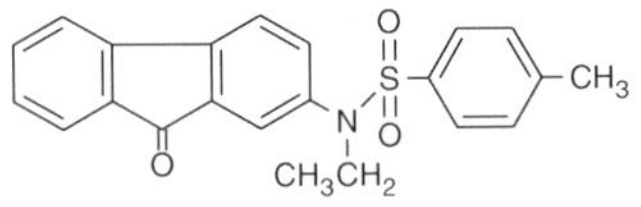
250 mg

S86,283-5
1-PHENYL-3-(P-TOSYL)UREA
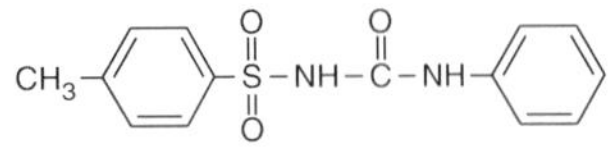
250 mg

S53,747-0
4-(N-HEXADECYL-N-METHYLAMINOSULFONYL)-BENZENESULFONIC ACID HYDRAZIDE
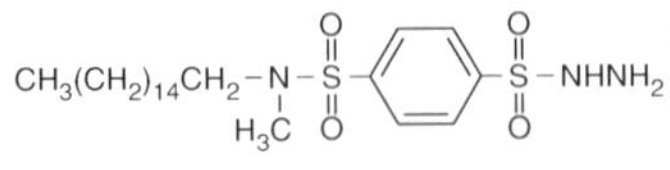
250 mg

S53,121-9
1,3,5-TRIS(PARA-TOLYLSULFONYL)-HEXAHYDRO-1,3,5-TRIAZINE
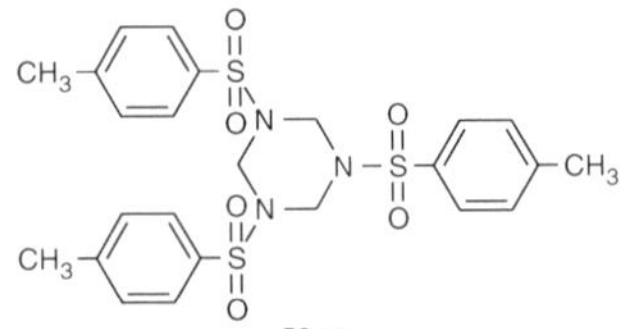
50 mg

S87,219-9
1-PHENYL-1-(3-[2-(PYRROLIDINO CARBONYL)PHENYLSULFONAMIDO]PROPYL)-PYRROLIDINIUM PERCHLORATE
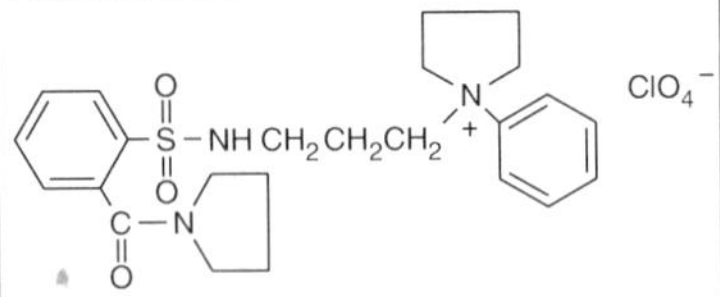
250 mg

S86,309-2
N-ACETYL-P-TOLUENESULFONANILIDE
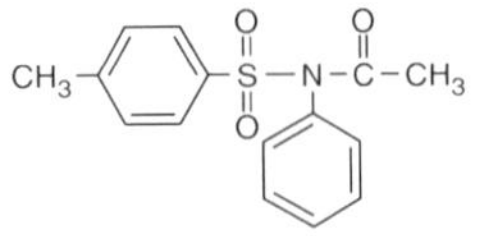
250 mg

S50,585-4
1-HEXADECANESULFONIC ACID (2-CHLORO-ALPHA-PIPERIDINOBENZYLIDENE)HYDRAZIDE HCL
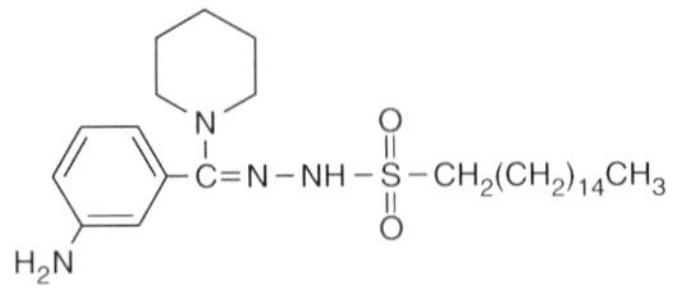
250 mg

S87,264-4
2-NITROBENZENESULFONIC (3-METHYL-2-BENZOTHIAZOLINYLIDENE)HYDRAZIDE
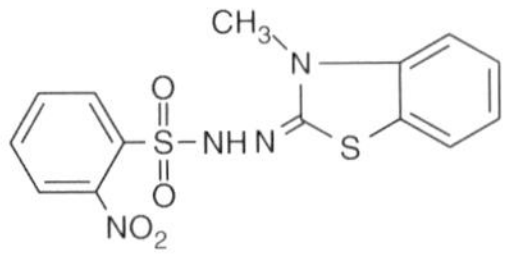
250 mg

S87,016-1
4-NITROBENZENESULFONIC (3-METHYL-2-BENZOTHIAZOLINYLIDENE)HYDRAZIDE
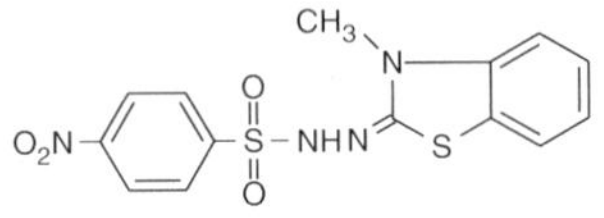
250 mg

S87,263-6
2,4-DINITROBENZENESULFONIC (3-METHYL-2-BENZOTHIAZOLINYLIDENE)-HYDRAZIDE
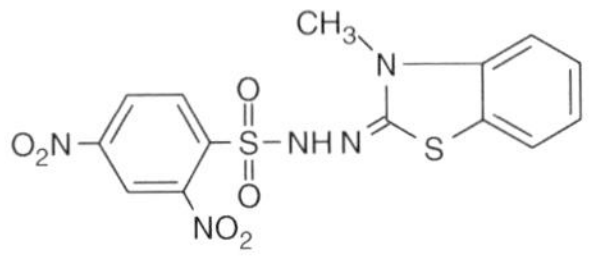
250 mg

S87,260-1
3-NITROBENZENESULFONIC (3-METHYL-2-BENZOTHIAZOLINYLIDENE)HYDRAZIDE
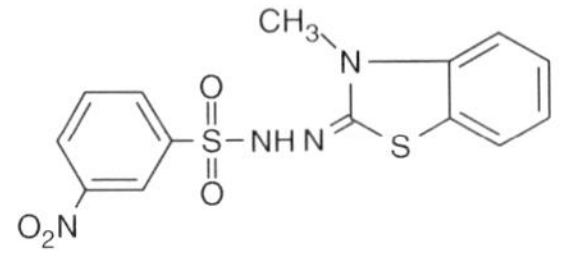
250 mg

S55,354-9
N4-ACETYLSULFANILIC ACID (3-METHYL-2-BENZOTHIAZOLINYLIDENE)HYDRAZIDE
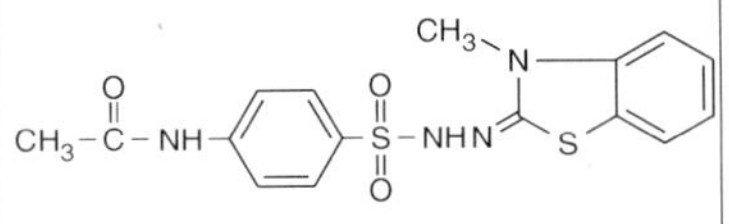
250 mg

S87,222-9
2-(3-METHYL-2-BENZOTHIAZOLINYLIDENE)-1-HYDRAZINESULFONIC ACID
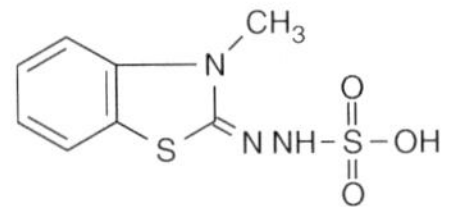
250 mg

S87,259-8
BENZENESULFONIC (3-METHYL-4,5,6,7-TETRAHYDRO-2-BENZOTHIAZOLINYLIDENE)-HYDRAZIDE
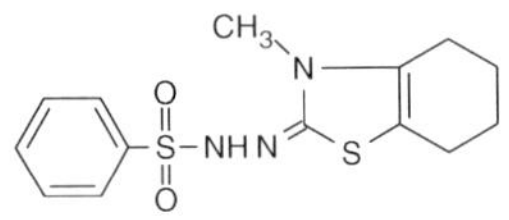
250 mg

S87,058-7
BENZENESULFONIC (3-METHYL-4-PHENYL-4-THIAZOLIN-2-YLIDENE)HYDRAZIDE
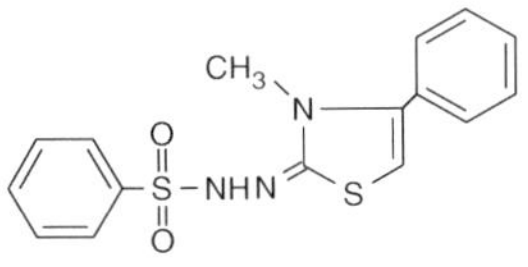
250 mg

S87,018-8
BENZENESULFONIC (3-METHYL-2-BENZOTHIAZOLINYLIDENE)HYDRAZIDE
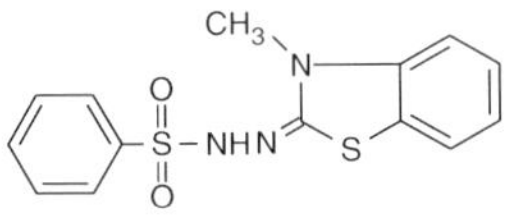
250 mg

S87,056-0
P-TOLUENESULFONIC (3-METHYL-2-BENZOTHIAZOLINYLIDENE)HYDRAZIDE
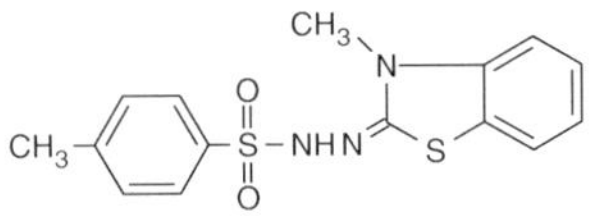
250 mg

S87,233-4
METHANESULFONIC (3-METHYL-2-BENZOTHIAZOLINYLIDENE)HYDRAZIDE
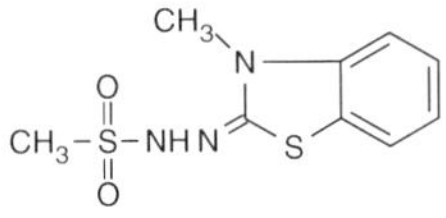
250 mg

S51,222-2
P-TOLUENESULFONIC ACID (1,2-DIHYDRO-1,4-DIMETHYL-2-QUINOLYLIDENE)HYDRAZIDE
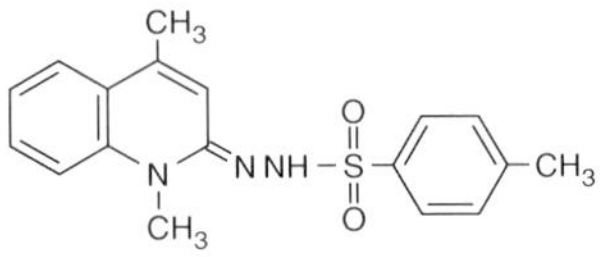
250 mg

S87,269-5
BENZENESULFONIC (1-METHYL-2-INDOLINYLIDENE)HYDRAZIDE
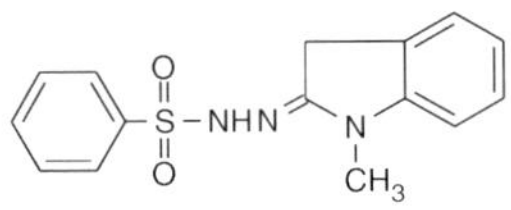
250 mg

S58,138-0
3-ACETAMIDO-4-HEXADECYLSULFONYLBENZ-ENESULFONIC ACID (3-METHYL-5-PIPERIDINO-SULFONYL-2-BENZOTHIAZOLIDENE)HYDRAZIDE
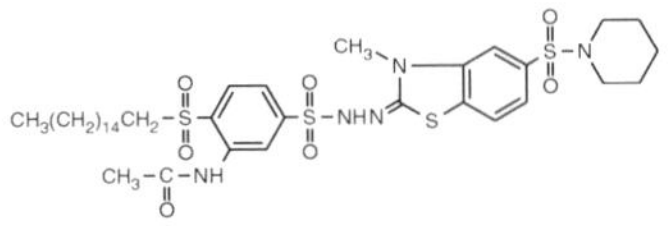
5 gm

S56,951-8
2-(HEXADECYLSULFONYL)-5-SULFOBENZOIC ACID 5-(3-METHYL-2-BENZOTHIAZOLINYLIDENE)HYDRAZIDE
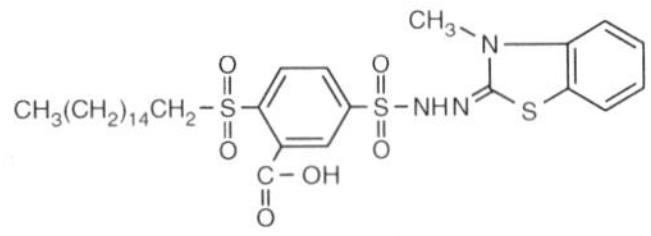
250 mg

S58,237-9
3-ACETAMID-4-(HEXADECYLSULFONYL)BENZENE SULFONIC ACID (3-METHYL-5-SULFO-2-BENZOTHIAZOLINYLIDENE)HYDRAZIDE
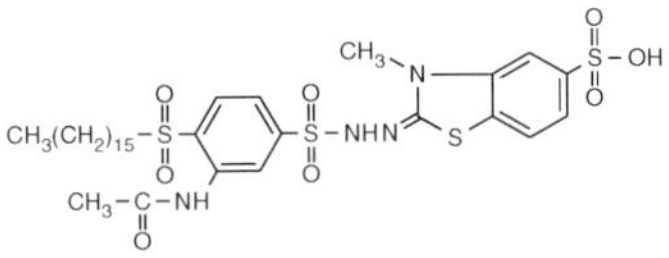
250 mg

S50,635-4
1-(2-CHLOROTHIOBENZOYL)-4-(P-TOLYLSULFONYL)PIPERAZINE
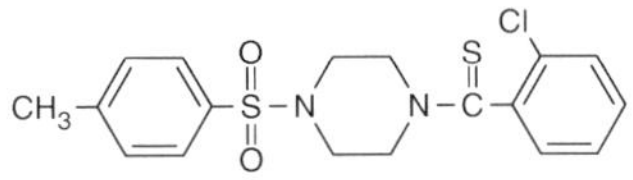
250 mg

S87,637-2
N-(P-TOSYL)-1-PIPERIDINECARBOXAMIDE
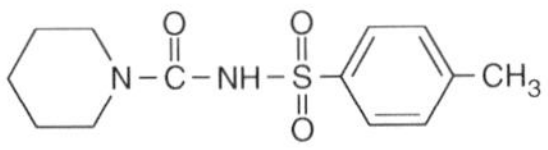
250 mg

S57,893-2
4-(HEXADECYLSULFAMOYL)BENZENE-SULFONIC (3-METHYL-5-SULFO-2-BENZO-THIAZOLINYLIDENE)HYDRAZIDE
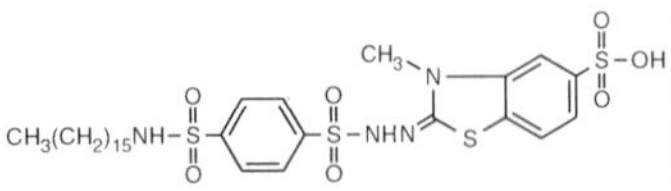
250 mg

S57,896-7
(N-HEXADECYL-N-METHYLSULFAMOYL)BENZENE-SULFONIC ACID(3-METHYL-5-SULFO-2-BENZOTHIAZOLINYLIDEN)HYDRAZIDE
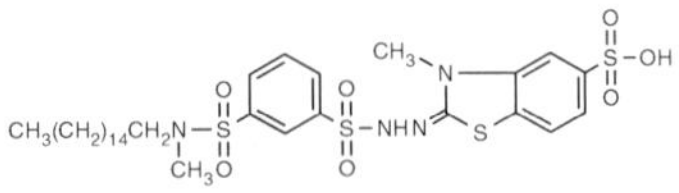
250 mg

S58,349-9
1-HEXADECANESULFONIC (METHYL-SULFO-BENZOTHIAZOLINYLIDENE)HYDRAZIDE, PYRIDINE SALT
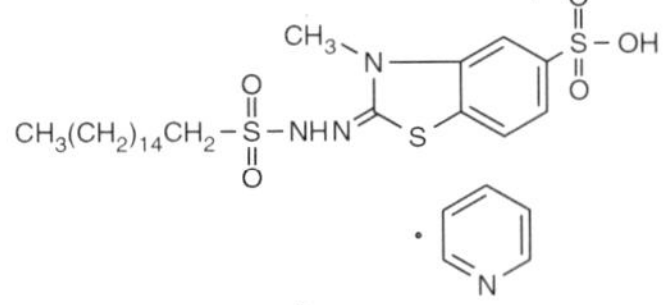
5 gm

S58,359-6
1-HEXADECANESULFONIC ACID [3-(4-SULFOBUTYL)-2-BENZOTHIAZOLINYLIDENE]HYDRAZIDE
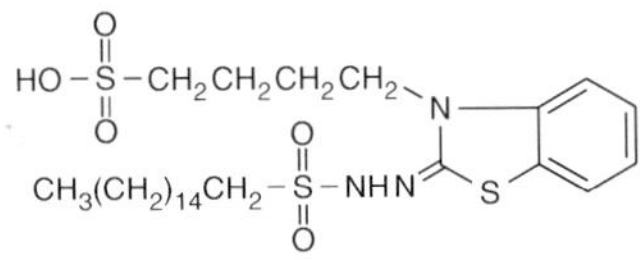
250 mg

S57,875-4
4-(N-HEXADECYL-N-METHYLSULFAMOYL)-BENZENESULFONIC ACID (3-METHYL-2-BENZOTHIAZOLINYLIDENE)HYDRAZIDE
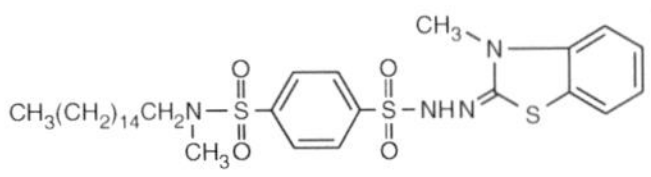
250 mg

S57,848-7
4-(HEXADECYLSULFAMOYL)BENZENE-SULFONIC ACID (3-METHYL-2-BENZOTHIAZOLINYLIDENE)HYDRAZIDE
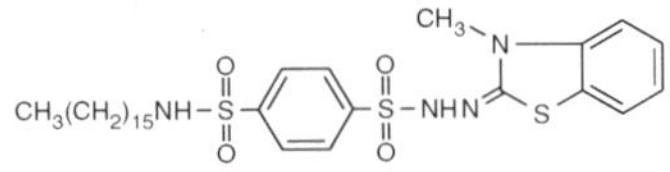
250 mg

S87,024-2
BENZENESULFONIC (4-TERT-BUTYL-3-METHYL-4-THIAZOLIN-2-YLIDENE)-HYDRAZIDE HYDROBROMIDE
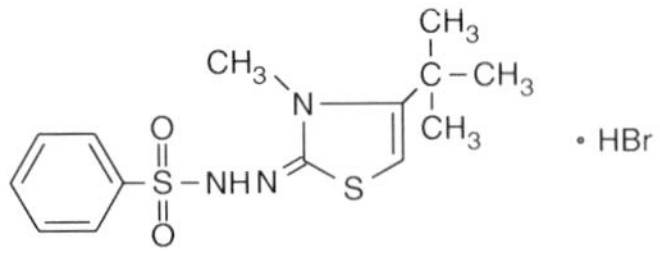
250 mg

S87,025-0
BENZENESULFONIC (4-TERT-BUTYL-3-METHYL-4-THIAZOLIN-2-YLIDENE)-HYDRAZIDE
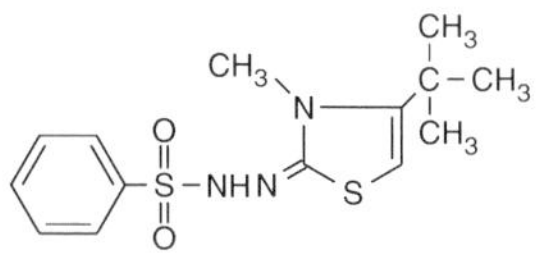
250 mg

S57,880-0
1-HEXADECANESULFONIC ACID (3-METHYL-2-BENZOTHIAZOLINYLIDENE)-HYDRAZIDE
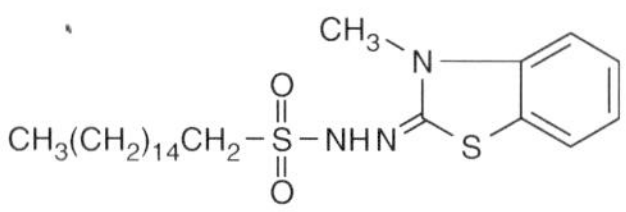
5 gm

S56,970-4
1-HEXADECANESULFONIC ACID (2-CHLORO-ALPHA-PIPERIDINOBENZYLIDENE)HYDRAZIDE
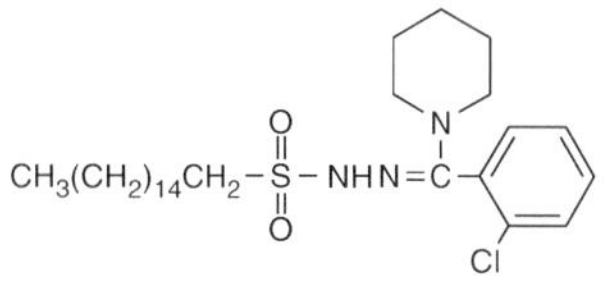
250 mg

S57,770-7
1-HEXADECANESULFONIC ACID (4-CHLORO-ALPHA-PIPERIDINOBENZYLIDENE)HYDRAZIDE
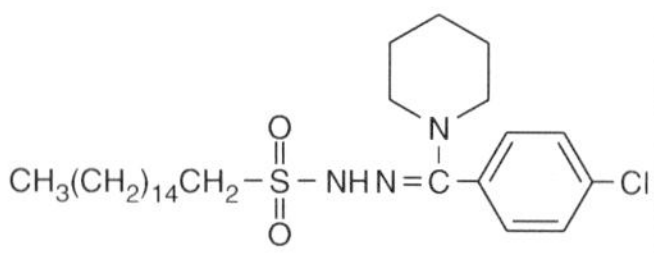
250 mg

S58,174-7
1-HEXADECANESULFONIC ACID-[3,6-DIMETHYL-4(3H)-PYRIMIDINYLIDENE]HYDRAZIDE
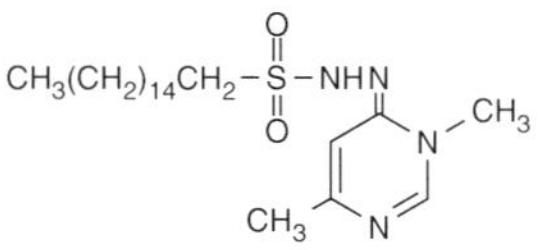
250 mg

S57,862-2
1-DODECANESULFONIC ACID (2-CHLORO-ALPHA-MORPHOLINOBENZYLIDENE)-HYDRAZIDE
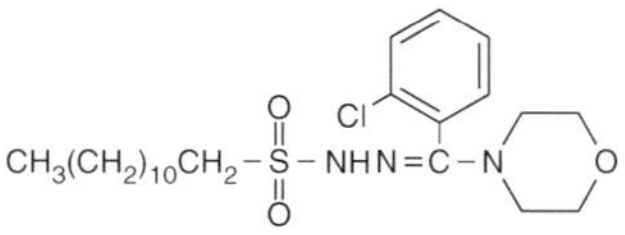
5 gm

S58,819-9
1-HEXADECANESULFONIC [4-(DIETHYLAMINO)-ALPHA-MORPHOLINOBENZYLIDENE]HYDRAZIDE
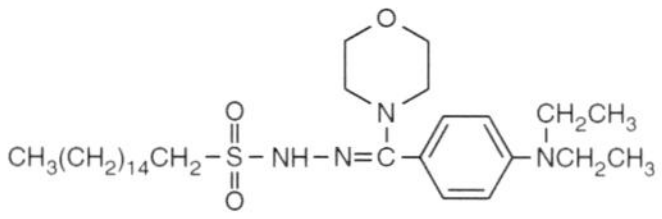
5 gm

S87,089-7
P-TOLUENESULFONIC N2-(1,2-DIHYDRO-1-ETHYLPYRIDYLIDENE)-N1-PHENYLHYDRAZIDE
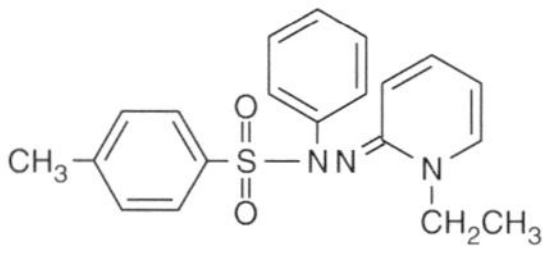
250 mg

S58,741-9
1-HEXADECANESULFONIC ACID [4-METHYL-1-PHENYL-2(1H)-QUINOLYLIDENE]HYDRAZIDE
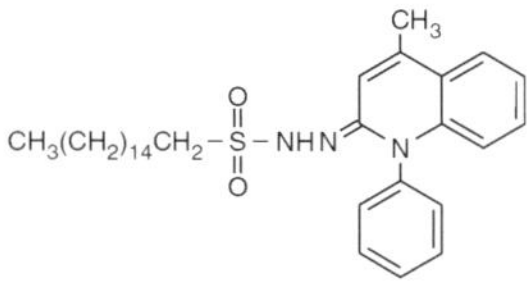
5 gm

S58,351-0
1-HEXADECANESULFONIC ACID [1-METHYL-2(1H)-QUINOLYLIDENE]-HYDRAZIDE
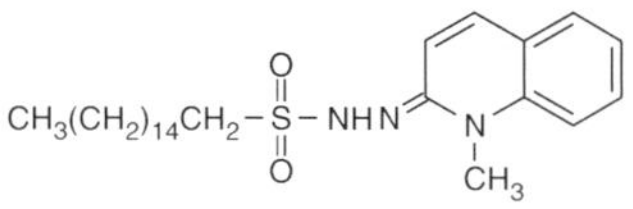
250 mg

S67,895-3
N-METHYL-N-(P-TOLUENESULFONYL)-PYRROLIDINIUM HEXACHLOROANTIMONATE
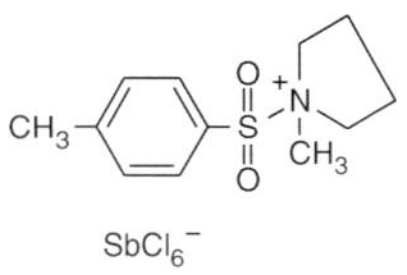
5 gm

S63,897-8
1-[4-(ACETYLSULFAMOYL)BUTYL]-3,3-DIMETHYL-3H-INDOLIUM BROMIDE
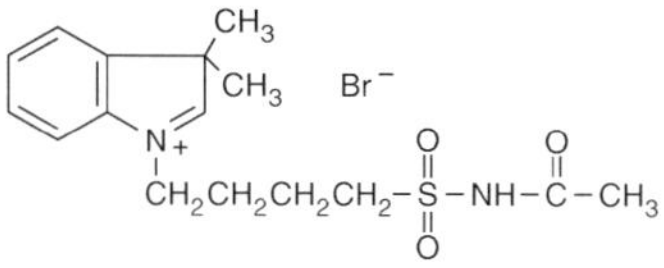
250 mg

S77,597-5
3,4-DIHYDRO-2-(METHYLSULFONAMIDO)-ISOQUINOLINIUM BROMIDE

250 mg

S67,786-8
4'-(ACETYLSULFAMOYL)PHTHALANILIC ACID, SODIUM SALT
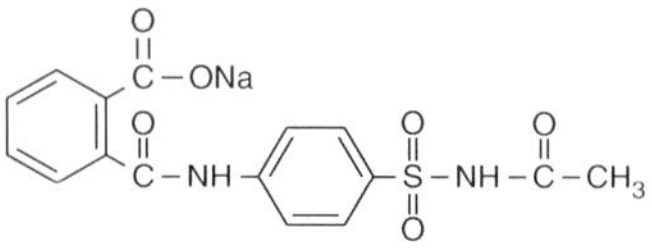
5 gm

S62,225-7
1-HEXADECANESULFONIC ACID (2-CHLORO-ALPHA-MORPHOLINOBENZYLIDENE)HYDRAZIDE
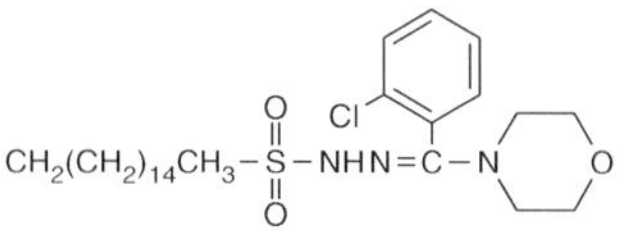
5 gm

S56,231-9
2-(N-BENZYL-P-TOLUENESULFONAMIDO)-5-NITROPHENYL P-TOLUENESULFONATE
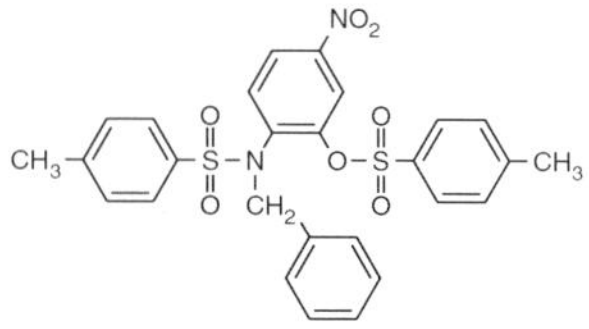
250 mg

S60,577-8
N4-ACETYL-N1-BENZYL-N1-HEXADECYL-SULFANILAMIDE
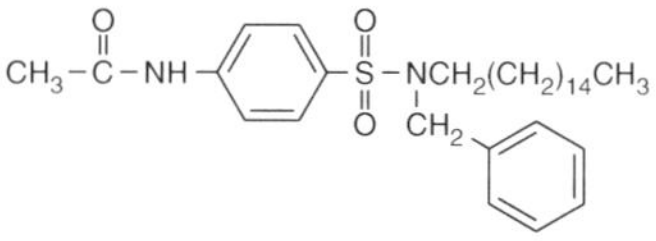
5 gm

S58,372-3
1-HEXADECANESULFONIC-5-(2-FLUORO PHENYL)-3-(3-SULFOPROPYL)THIADIAZOL-2-YLIDENE HYDRAZIDE
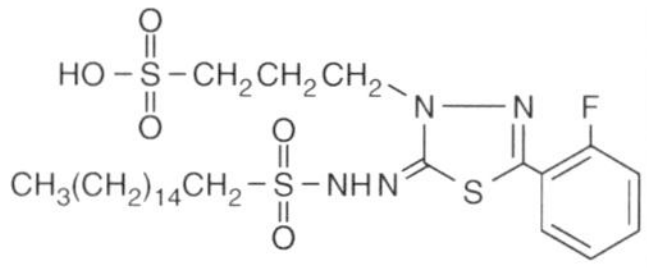
250 mg

S65,672-0
4-CHLORO-N-METHYL-N-SULFAMOYLBENZYLAMINE
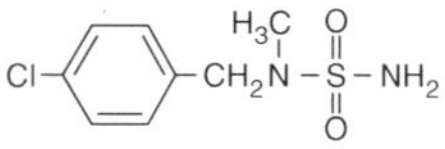
5 gm

S81,687-6
N-(BENZYL)BENZENESULFONAMIDE
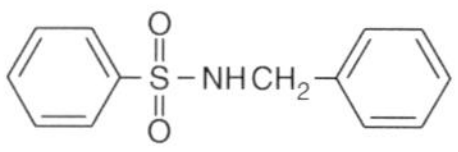
250 mg

S94,976-0
N-BENZYL-P-TOLUENESULFONAMIDE
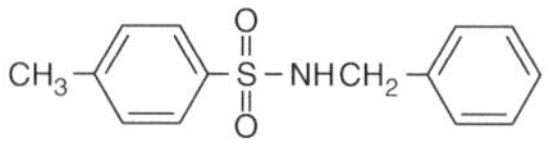
1 gm

S59,832-1
4-HYDRAZINO-N1-HEXADECYL-N1-METHYLBENZENESULFONAMIDE HYDROCHLORIDE
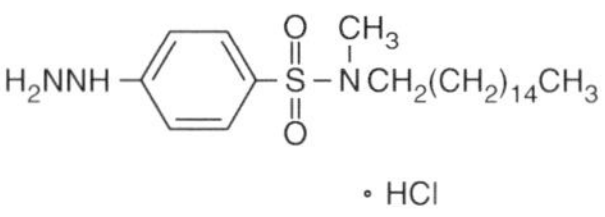
5 gm

S54,170-2
N-HEXADECYL-4-HYDRAZINOBENZENESULFONAMIDE
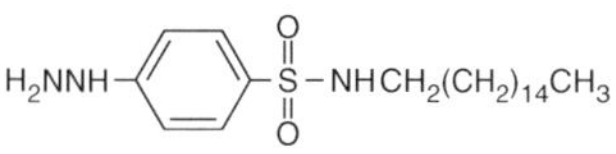
250 mg

S59,856-9
3-CHLORO-N1-HEXADECYL-4-HYDRAZINOBENZENESULFONAMIDE
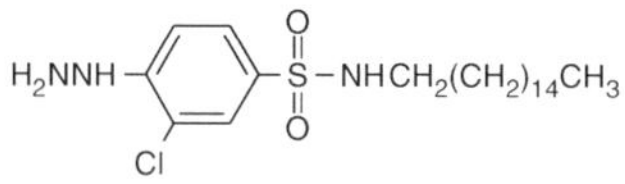
250 mg

S59,860-7
3-CARBOXYBENZENESULFONIC ACID (3-HEXADECYL-2-BENZOTHIAZOLINYLIDENE)HYDRAZIDE
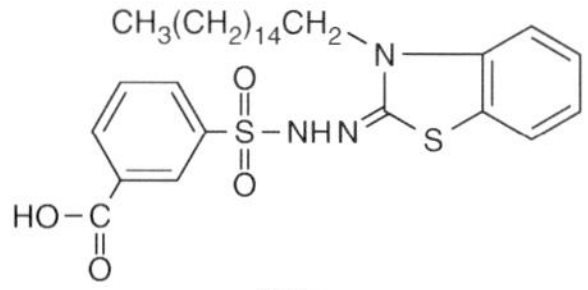
250 mg

S87,550-3
BENZENESULFONIC (4-THIAZOLIN-2-YLIDENE)HYDRAZIDE
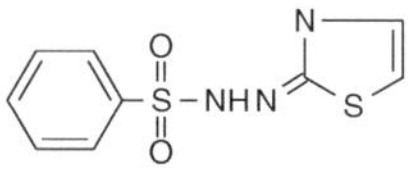
250 mg

S87,054-4
BENZENESULFONIC ISOPROPYLIDENEHYDRAZIDE
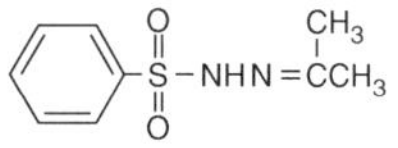
250 mg

S86,996-1
BENZENESULFONIC BENZYLIDENEHYDRAZIDE
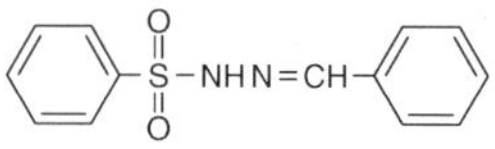
250 mg

S87,174-5
BENZENESULFONIC N1-ALLYL-N2-BENZYLIDENEHYDRAZIDE
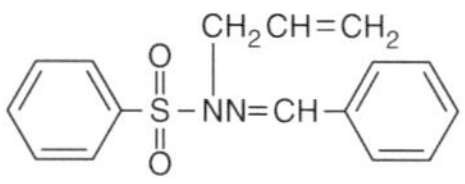
250 mg

S86,521-4
2-PHENYLPROPIOPHENONE-(PHENYLSULFONYL)HYDRAZONE
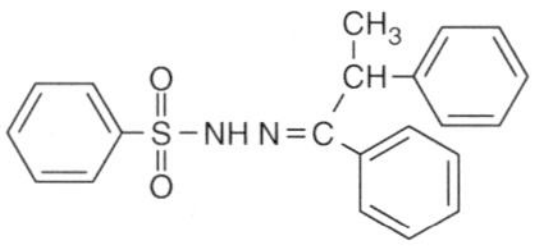
250 mg

S92,464-4
PHENYLACETALDEHYDE P-TOSYLHYDRAZONE
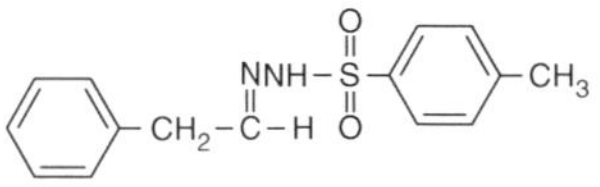
250 mg

S86,365-3
PROPIOPHENONE (P-TOSYL)HYDRAZONE
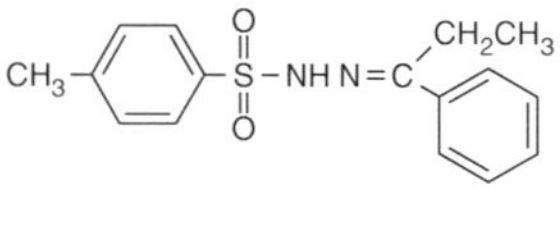
250 mg

S86,571-0
2-METHYLPROPIOPHENONE (P-TOSYL)-HYDRAZONE

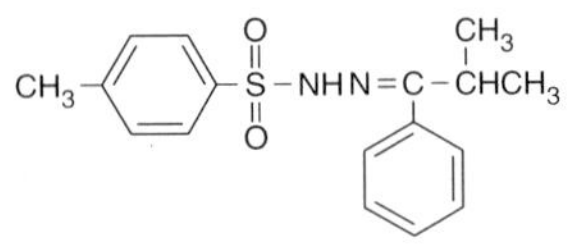

250 mg

S86,391-2
VALEROPHENONE (P-TOSYL)HYDRAZONE

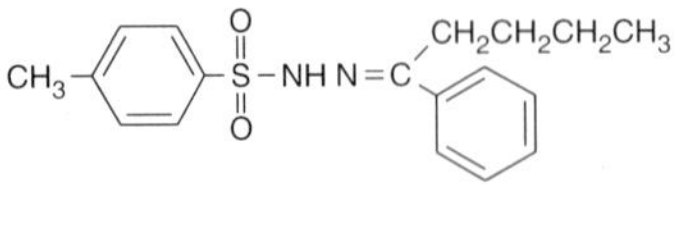

250 mg

S87,241-5
MESITYLENESULFONIC ISOPROPYLIDENEHYDRAZIDE

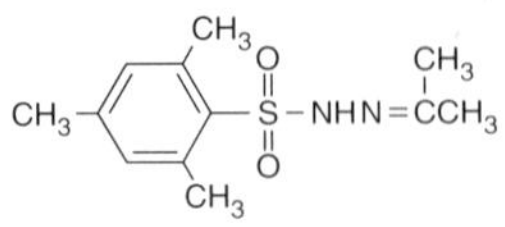

250 mg

S95,018-1
N-METHYL-N-(PHENYLSULFONYL)GLYCINE

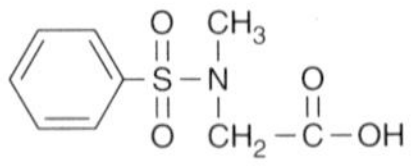

50 mg

S51,877-8
2,2'-(PHENYLSULFONYLIMINO)DIETHYL-BIS(BENZENESULFONATE)

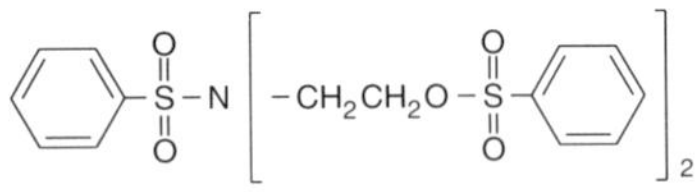

5 gm

S86,315-7
2-(METHYLSULFONYLAMINO)PHENYL METHANESULFONATE

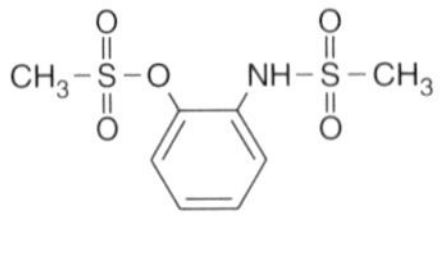

250 mg

S59,837-2
4-HEXADECYLSULFONYL-3-NITROBENZENESULFONAMIDE

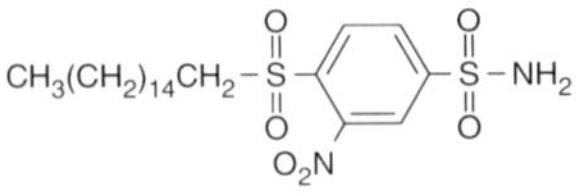

250 mg

S67,176-2
N-HEXADECYL-2-NITROBENZENESULFONAMIDE

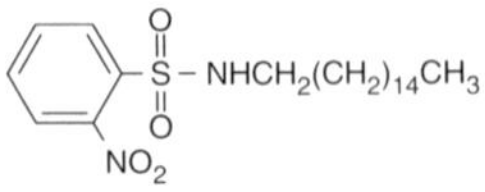

5 gm

S40,109-9
1-BUTYL-3-[2-(PARA-TOLYLSULFONYL)-ACETYL]UREA

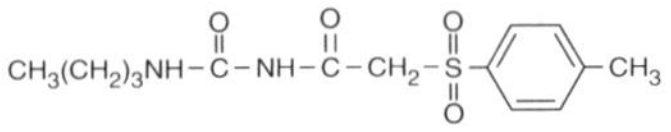

250 mg

S56,263-7
N-DODECYL-N-METHYL-2-PHTHALIMIDOETHANESULFONAMIDE

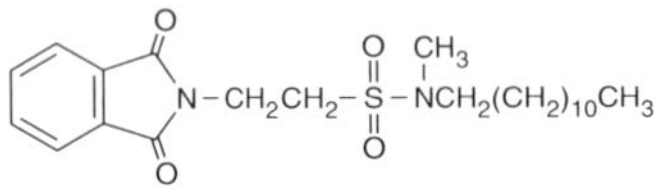

250 mg

S40,060-2
1-[2-(ETHYLSULFINYL)ETHYL]-3-(PARA-TOLYLSULFONYL)UREA

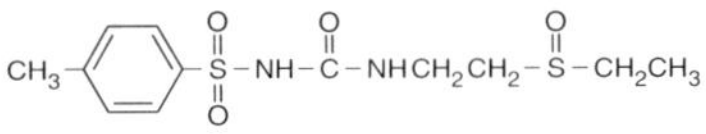

250 mg

S65,604-6
1-CYCLOHEXYL-3-(2-SULFAMOYLPHENYL)UREA

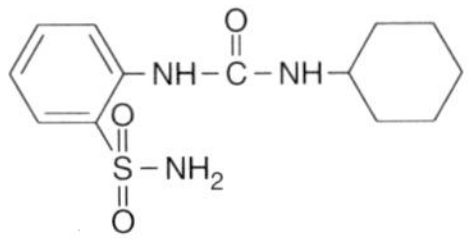

250 mg

S73,671-6
1-(3-NITROBENZOYL)-3-(4-SULFAMOYLPHENYL)-2-THIOUREA

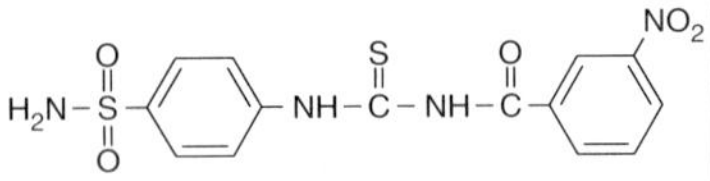

250 mg

S67,456-7
N4-ACETYL-3'-NITROSULFANILANILIDE

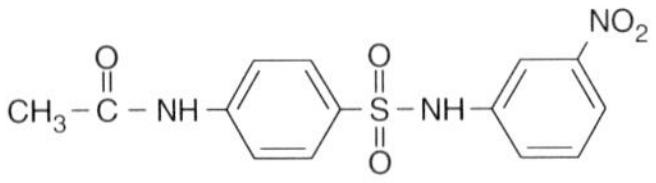

250 mg

S86,282-7
N-CYANO-4'-METHYL-3-NITROBENZENESULFONANILIDE

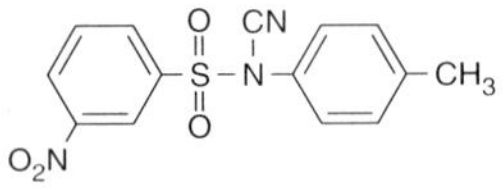

250 mg

S55,375-1
2-(DI-P-TOSYLAMINO)-5-NITROPHENYL P-TOLUENESULFONATE

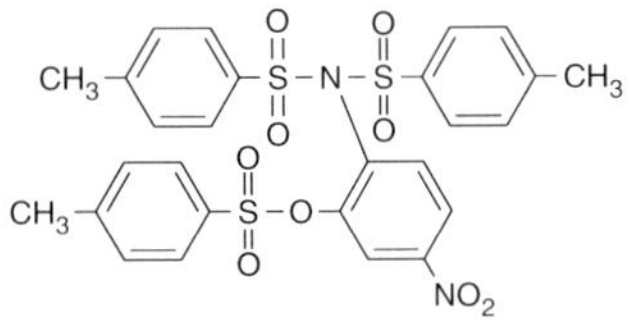

250 mg

S37,131-9
2-CHLORO-5-NITROBENZENESULFONAMIDE

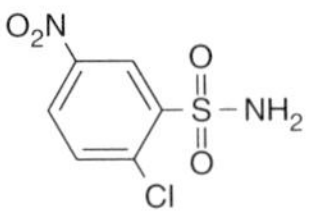

5 gm

S35,959-7
4-CHLORO-3-NITROBENZENESULFONANILIDE

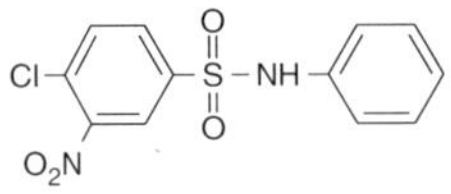

5 gm

S58,187-9
N-HEXADECYL-3-NITROBENZENESULFONAMIDE

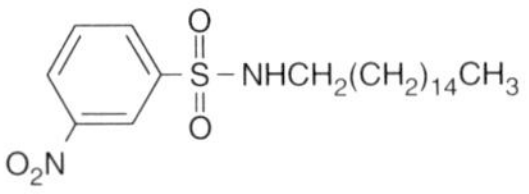

250 mg

S58,717-6
N-HEXADECYL-3-NITROBENZENESULFONANILIDE

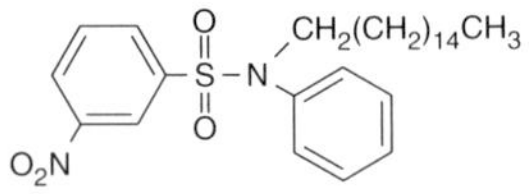

250 mg

S58,762-1
N,N-DIDODECYL-3-NITROBENZENESULFONAMIDE

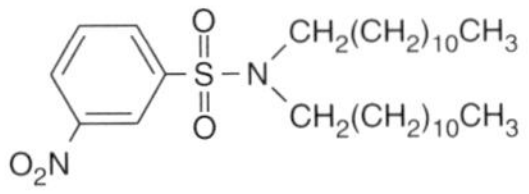

250 mg

S37,969-7
1-(4-NITROPHENYLSULFONYL)GUANIDINE

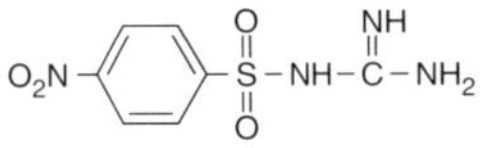

250 mg

S59,196-3
N-HEXADECYL-N-METHYL-4-NITROBENZENESULFONAMIDE

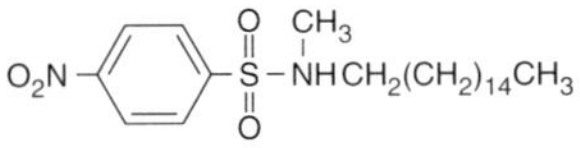

250 mg

S58,712-5
N-HEXADECYL-4-NITROBENZENESULFONANILIDE

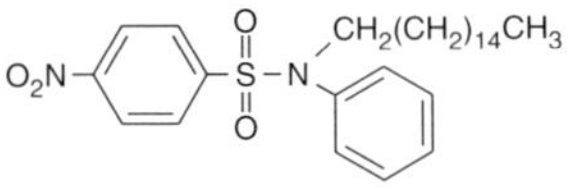

250 mg

S74,958-3
N-(7-NITRO-2-FLUORENYL)-P-TOLUENESULFONAMIDE

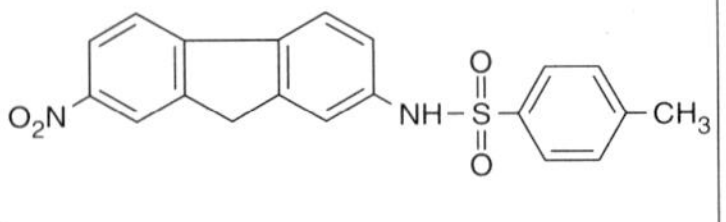

250 mg

S38,714-2
4-CHLORO-3-NITROBENZENESULFONAMIDE

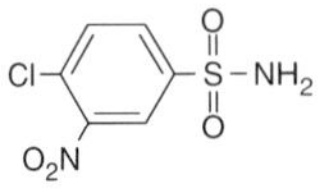

250 mg

S58,167-4
4-CHLORO-N-HEXADECYL-N-METHYL-3-NITROBENZENESULFONAMIDE

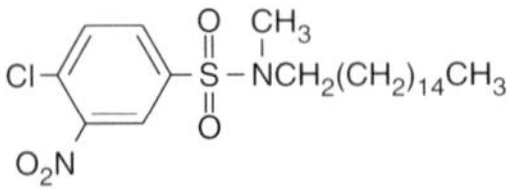

250 mg

S51,004-1
4-CHLORO-N,N-DIDODECYL-3-NITROBENZENESULFONAMIDE

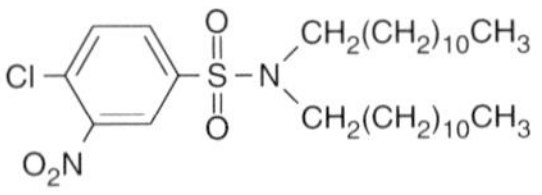

250 mg

S51,054-8
3-NITROBENZENESULFONIC ACID 2-(2-BROMO-P-TOLYL)HYDRAZIDE

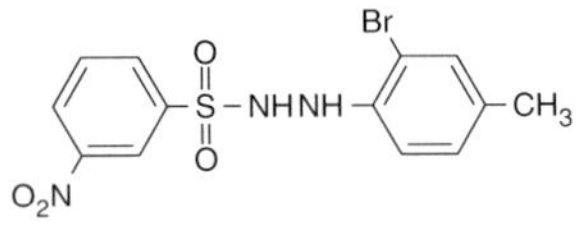

250 mg

S51,343-1
3-NITROBENZENESULFONIC ACID 2-(4-BROMO-O-TOLYL)HYDRAZIDE

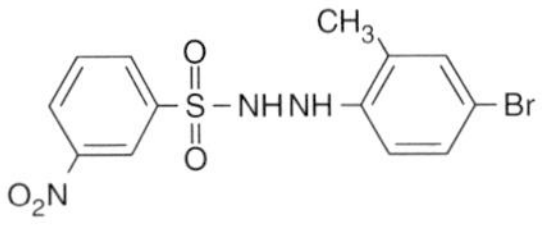

250 mg

S84,024-6
3-PHENYL-3-(PIPERIDINOSULFONYL)-PROPIOPHENONE

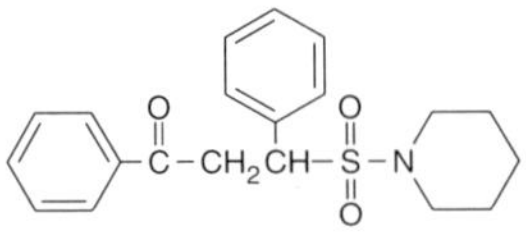

250 mg

S84,020-3
N,N-DIETHYL-1,3-DIPHENYL-3-OXOPROPANESULFONAMIDE

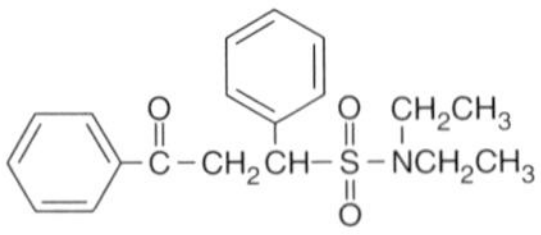

250 mg

S84,021-1
N-BUTYL-1,3-DIPHENYL-3-OXOPROPANESULFONAMIDE

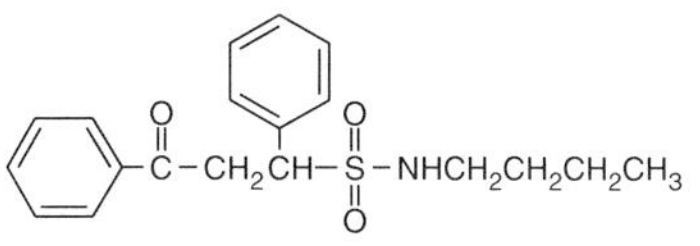

250 mg

S79,860-6
2'-BENZOYL-P-TOLUENESULFONANILIDE

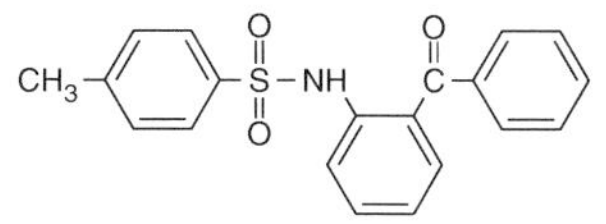

250 mg

S91,685-4
4'-ACETYLMETHANESULFONANILIDE

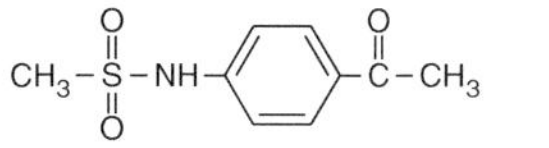

250 mg

S36,508-4
2-(PARA-TOLUENESULFONAMIDO)-9-FLUORENONE

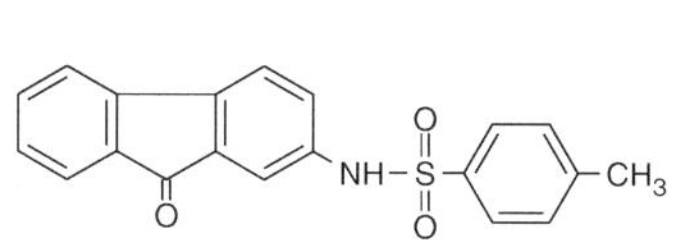

250 mg

S36,393-6
2-(N-METHYL-P-TOLUENESULFONAMIDO)-9-FLUORENONE

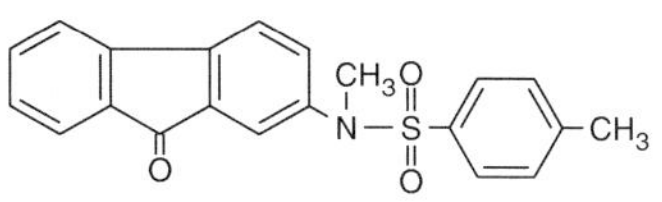

5 gm

S77,452-9
N-(9-OXO-2-FLUORENYL)DI-P-TOLUENESULFONIMIDE

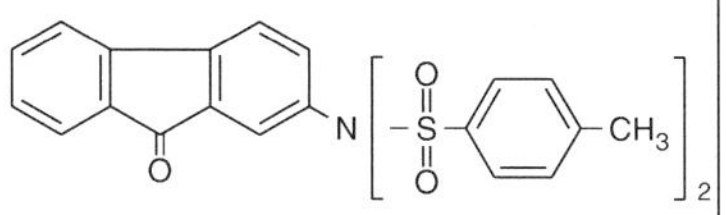

250 mg

S87,023-4
2-OXO-N-(P-TOSYL)-1-CYCLOHEXANECARBOXAMIDE

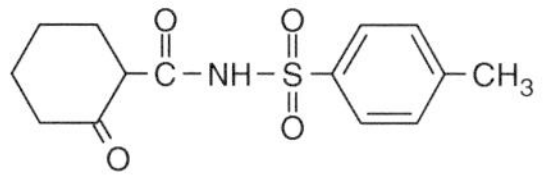

250 mg

S35,548-8
4'-SULFAMOYLSUCCINANILIC ACID

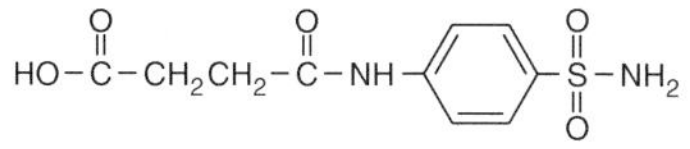

5 gm

S64,731-4
5-METHYL-1-(P-TOLUENESULFONYL)-2-THIOXOBENZIMIDAZOLINE

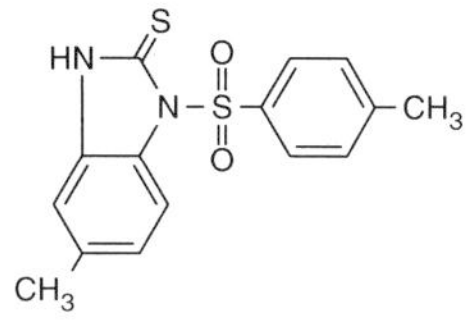

250 mg

S39,351-7
1-(P-TOLYLSULFONYL)-2(3H)-BENZIMIDAZOLETHIONE

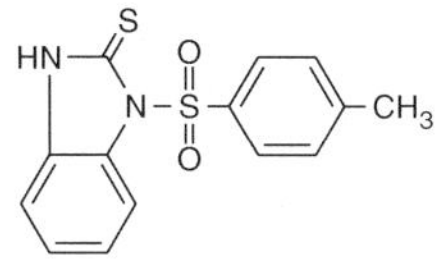

250 mg

S35,197-0
4'-(ACETYLSULFAMOYL)MALEANILIC ACID

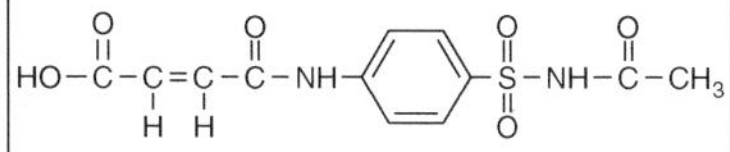

5 gm

S34,568-7
4-(N,N-DICHLOROSULFAMYL)BENZOIC ACID, SODIUM SALT

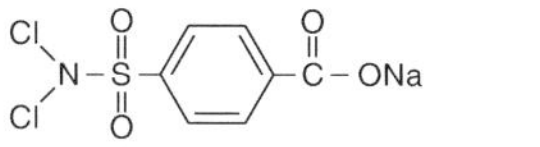

5 gm

S37,039-8
4'-(ACETYLSULFAMOYL)-PHTHALANILIC ACID

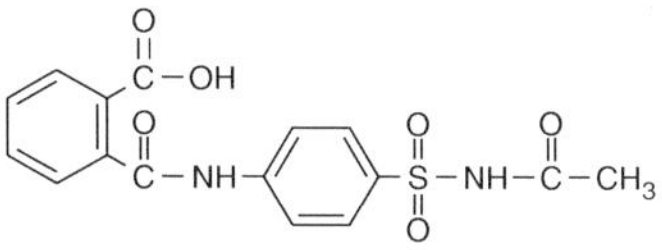

5 gm

S35,481-3
4'-SULFAMOYLPHTHALANILIC ACID

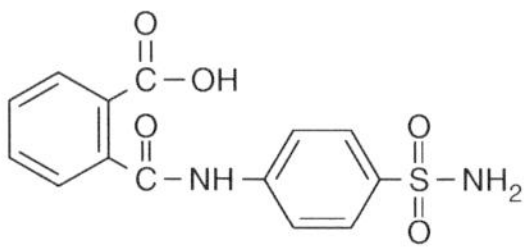

5 gm

S35,198-9
4'-SULFAMOYLMALEANILIC ACID

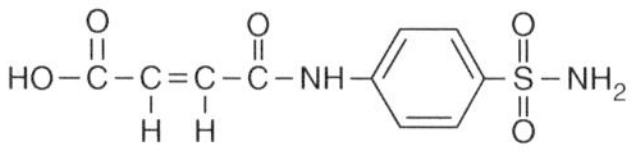

5 gm

S56,868-6
2-HEXADECYLSULFONYL-5-SULFOBENZOIC ACID 5-[(3-METHYL-5-SULFO-2-BENZOTHIAZOLINYLIDENE)HYDRAZIDE]

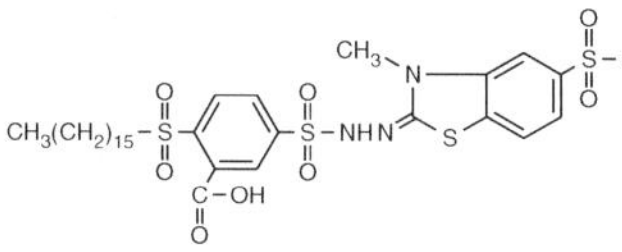

5 gm

S34,201-7
4-(BENZYLSULFONAMIDO)BENZOIC ACID

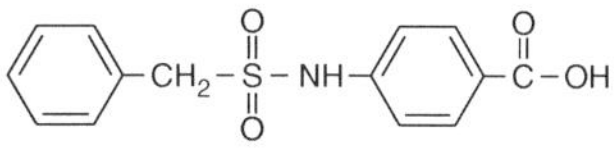

5 gm

S79,878-9
N-(P-TOSYL)ANTHRANILIC ACID

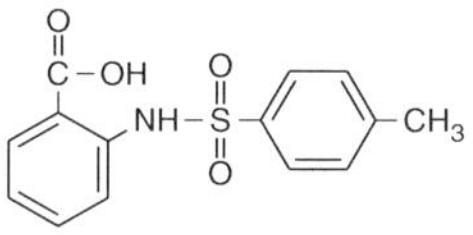

250 mg

S67,033-2
N-ETHYL-N-(P-TOLUENESULFONYL)-ANTHRANILIC ACID

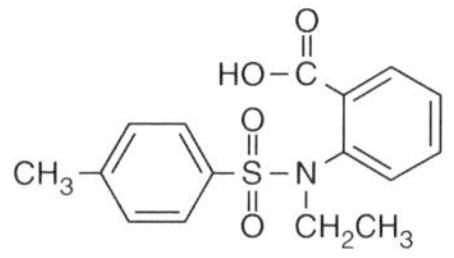

5 gm

S54,638-0
4-(METHYLSULFONAMIDO)BENZOIC ACID

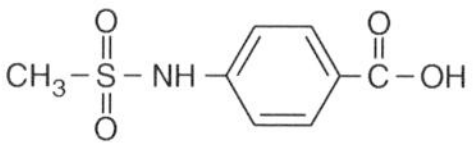

5 gm

S34,364-1
N,N-DIETHYL-4-SULFAMOYLBENZOIC ACID

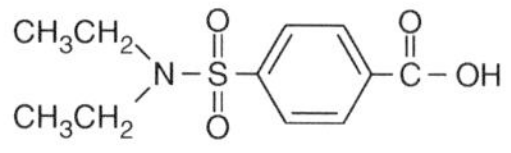

5 gm

S57,204-7
2-CHLORO-5-(METHYLSULFAMOYL)-BENZOIC ACID

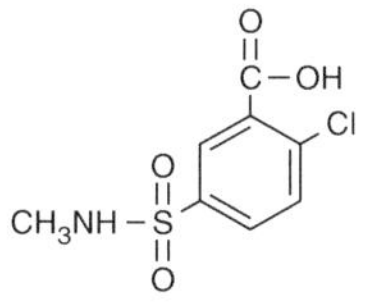

5 gm

S41,693-2
4-(HYDRAZINOSULFONYL)BENZOIC ACID

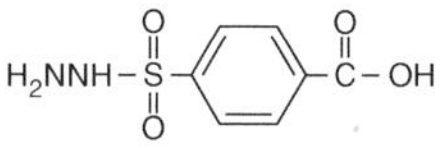

5 gm

S85,540-5
4-(METHOXYCARBONYL)PHENYL N-(P-TOSYL)CARBAMATE

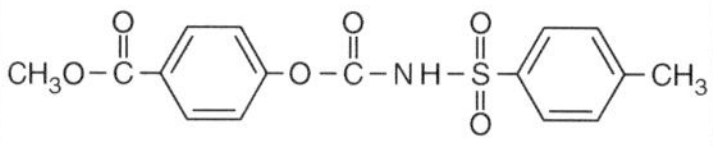

250 mg

S67,675-6
ETHYL 4-[(4-ACETAMIDOPHENYL)-SULFONAMIDO]BENZOATE

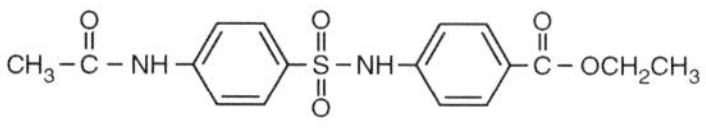

250 mg

S71,835-1
ETHYL 4-(METHYLSULFONAMIDO)BENZOATE

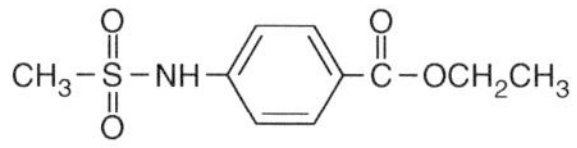

5 gm

S69,891-1
N3-(4-ETHYLBENZOYL)METANILAMIDE

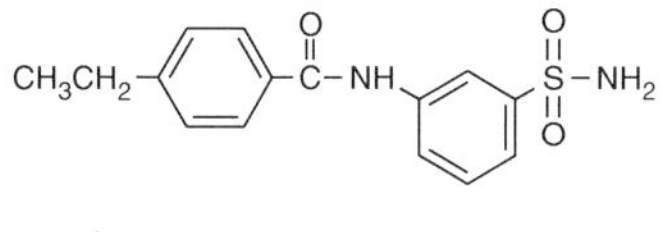

250 mg

S51,512-4
N-(PARA-TOLYLSULFONYL)BENZANILIDE

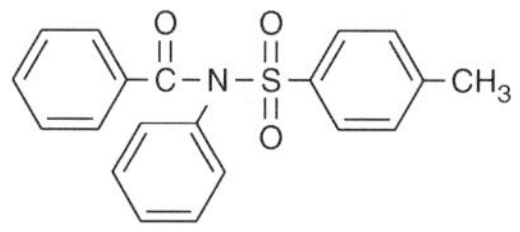

250 mg

S82,617-0
2-(DIMETHYLTHALLYL)SACCHARIN

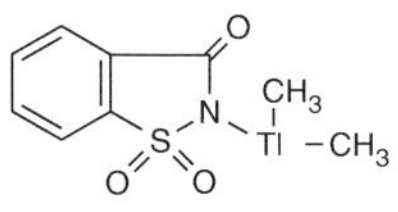

250 mg

S82,618-9
2-(DIETHYLTHALLYL)SACCHARIN

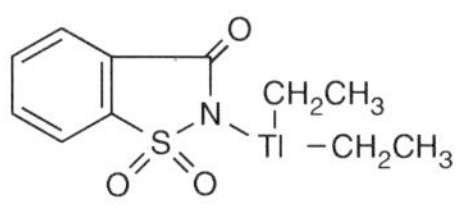

250 mg

S51,868-9
1-PHENYLSULFONYL-4,4-PIPERIDINEDICARBOXYLIC ACID

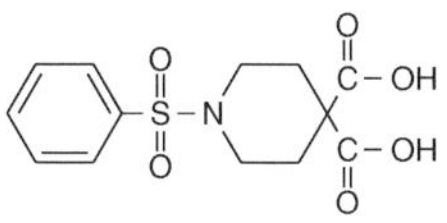

250 mg

S51,907-3
DIETHYL 1-PHENYLSULFONYL-4,4-PIPERIDINEDICARBOXYLATE

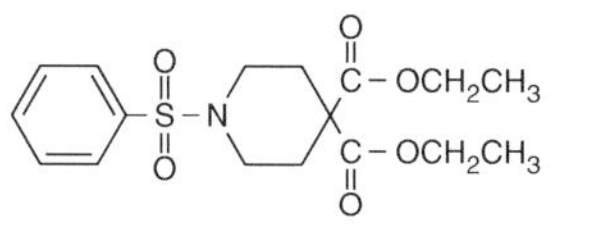

5 gm

S91,279-4
HEXAHYDRO-4'-SULFAMOYLPHTHALANILIC ACID
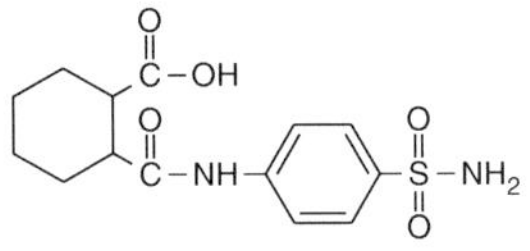
250 mg

S91,278-6
4'-SULFAMOYLGLUTARANILIC ACID
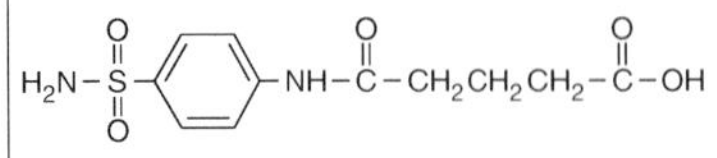
250 mg

S59,941-7
N-(P-TOSYL)-L-GLUTAMIC ACID
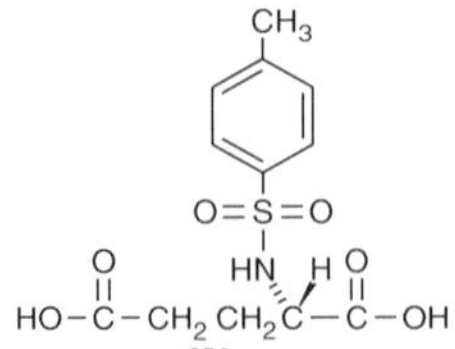
250 mg

S59,942-5
N-(P-TOSYL)-D-GLUTAMIC ACID
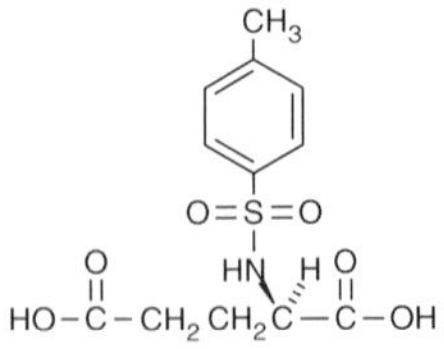
250 mg

S78,739-6
N-(P-TOSYL)-L-ISOLEUCINE
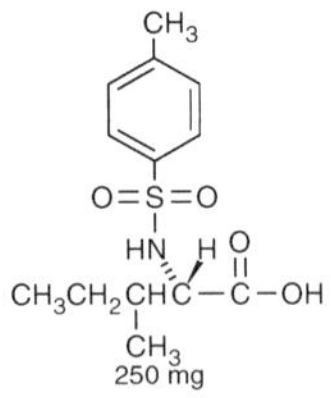
250 mg

S35,546-1
N1-ACETYL-N4-SUCCINOYLSULFANILAMIDE
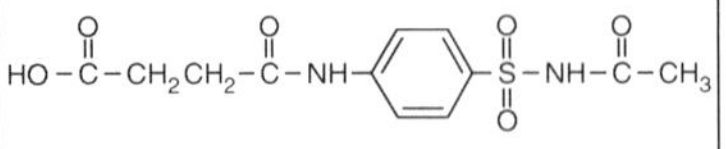
5 gm

S34,376-5
4-CARBOXY-4-PHENYL-1-P-TOSYLPIPERIDINE
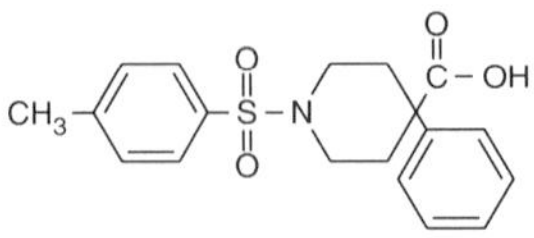
5 gm

S53,753-5
4-[N-METHYL(METHYLSULFONAMIDO)]-BENZOIC ACID
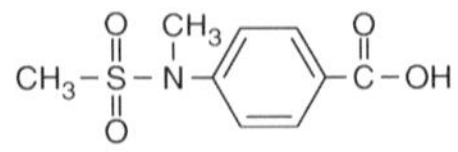
250 mg

S59,946-8
N-(P-TOSYL)-L-GLUTAMINE ISOPROPYL ESTER
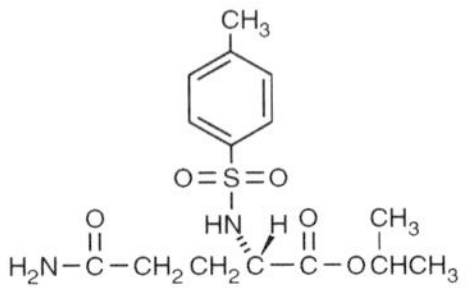
250 mg

S61,475-0
N-(P-TOSYL)-L-LEUCINE
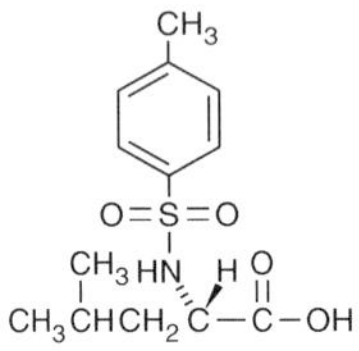
5 gm

S85,757-2
METHYL N-(METHYLSULFONYL)-N-PHENYLCARBAMATE
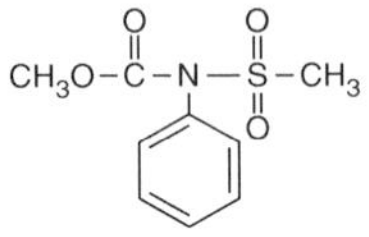
250 mg

S85,758-0
ETHYL N-(METHYLSULFONYL)-N-PHENYLCARBAMATE
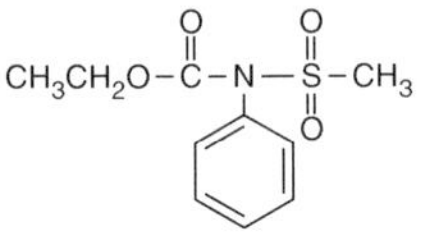
250 mg

S85,759-9
PROPYL N-(METHYLSULFONYL)-N-PHENYLCARBAMATE
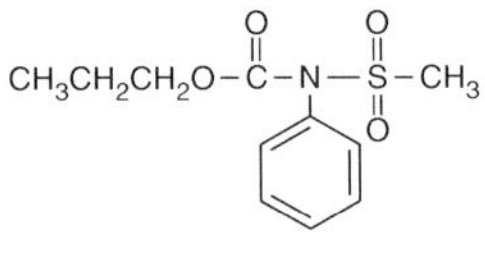
250 mg

S85,760-2
ISOPROPYL N-(METHYLSULFONYL)-N-PHENYLCARBAMATE
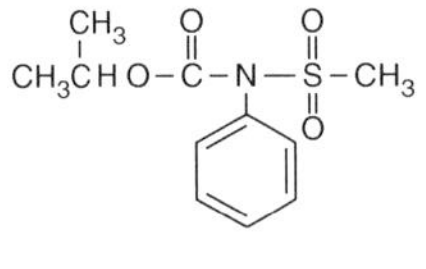
250 mg

S85,761-0
BUTYL N-(METHYLSULFONYL)-N-PHENYLCARBAMATE
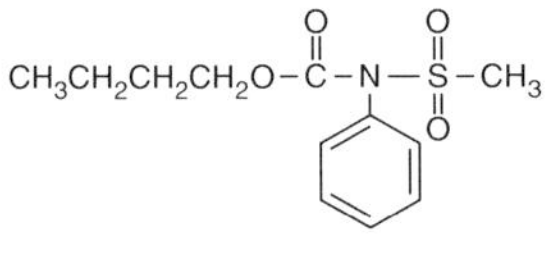
250 mg

S90,536-4
PROPYL N-(4-SULFAMOYLPHENYL)-CARBAMATE
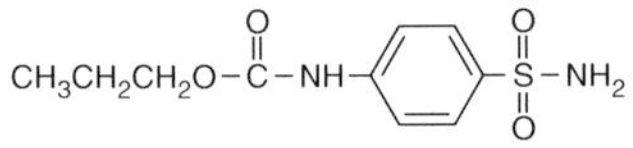
250 mg

S52,754-8
ETHYL 4-(METHYLSULFONAMIDO)BENZOATE
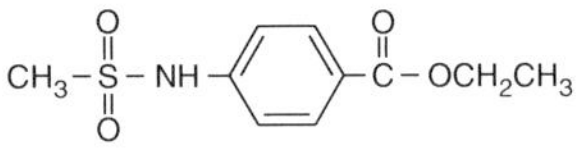
5 gm

S87,670-4
5-OXO-1-(P-TOSYL)-2-PYRROLIDINECARBOXYLIC ACID CHLORIDE
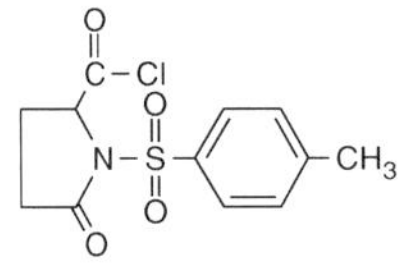
250 mg

S94,511-0
4-(SEMICARBAZIDOSULFONYL)-ACETANILIDE
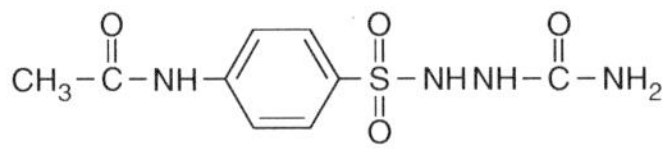
250 mg

S67,268-8
1,2-BENZISOTHIAZOL-3(2H)-ONE-1,1-DIOXIDE DIETHYLAMINE SALT
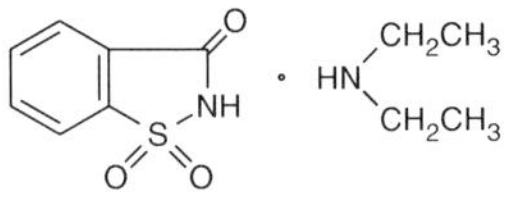
250 mg

S67,269-6
1,2-BENZISOTHIAZOL-3(2H)-ONE,1,1-DIOXIDE TRIBENZYLAMINE SALT
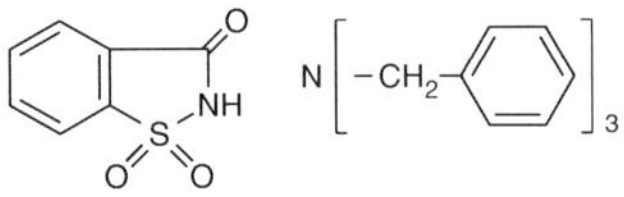
250 mg

S87,000-5
4-PHENYL-1-(PHENYLSULFONYL)-3-THIOSEMICARBAZIDE
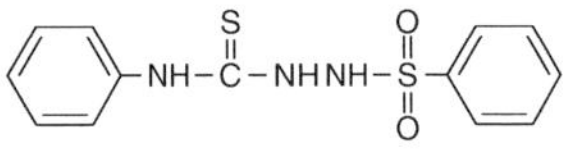
250 mg

S91,026-0
4-ISOPROPYL-1-(P-TOSYL)-SEMICARBAZIDE
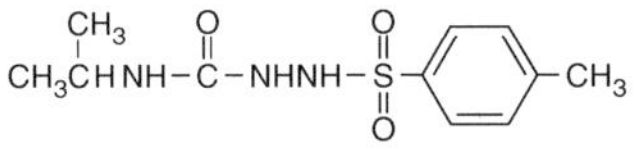
250 mg

S90,541-0
1,4,5,6,7,7-HEXACHLORO-N-(4-SULFAMOYLPHENYL)-5-NORBORNENE-2,3-DICARBOXIMIDE
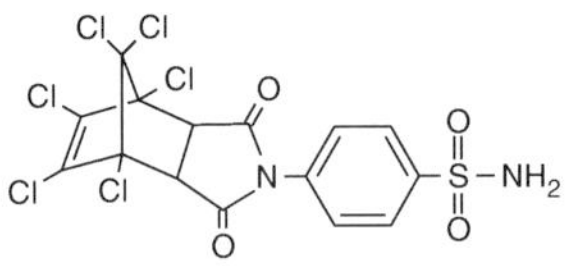
250 mg

S40,105-6
1-(4-ACETAMIDOPHENYLSULFONYL)-3-ISOPENTYLUREA
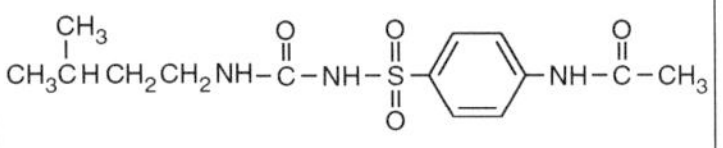
250 mg

S34,058-8
1-(N-ACETYLSULFANILYL)UREA
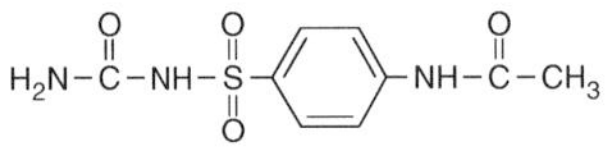
5 gm

S65,556-2
1-(4-[N-(ACETYL)SULFAMOYL]PHENYL)-3-ETHYLUREA
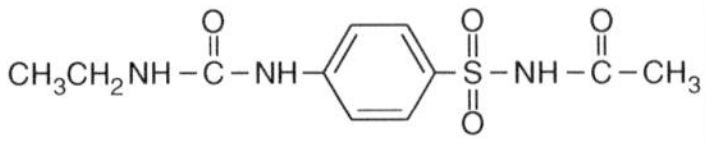
5 gm

S40,068-8
1-(4-ACETAMIDOPHENYLSULFONYL)-3-PROPYLUREA
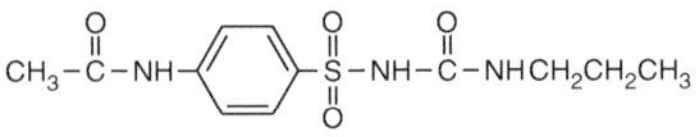
250 mg

S40,069-6
1-(4-ACETAMIDOPHENYLSULFONYL)-3-ISOPROPYLUREA
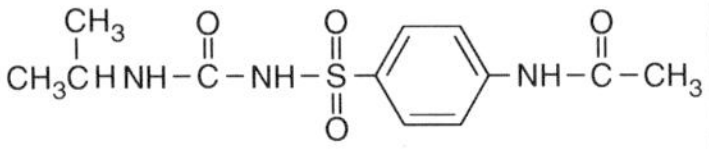
250 mg

S86,296-7
4-ACETAMIDO-N-CYANOBENZENESULFONANILIDE
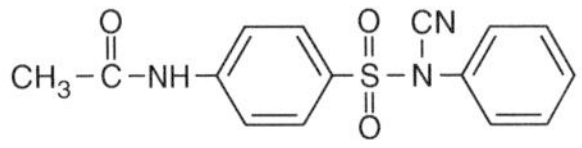
250 mg

S53,736-5
N3-ACETYL-4-(HEXADECYLSULFONYL)-METANILIC ACID HYDRAZIDE
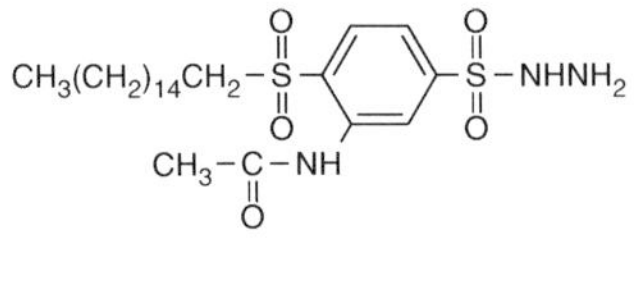
5 gm

S71,813-0
N3-BUTYRYLMETANILAMIDE
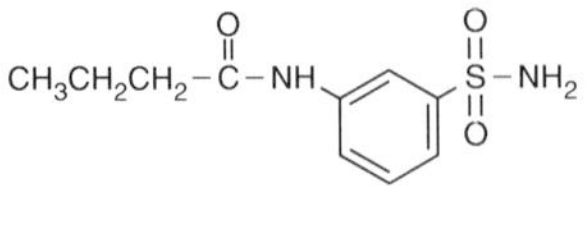
250 mg

S61,876-4
N4-ACETYL-N1-CYCLOHEXYL-N1-HEXADECYLSULFANILAMIDE

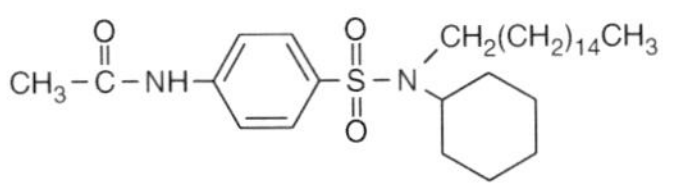

250 mg

S71,605-7
N4-ACETYLSULFANILANILIDE

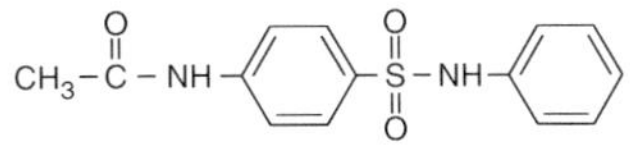

250 mg

S90,123-7
N4-ACETYL-N1-BENZYL-N1-METHYLSULFANILAMIDE

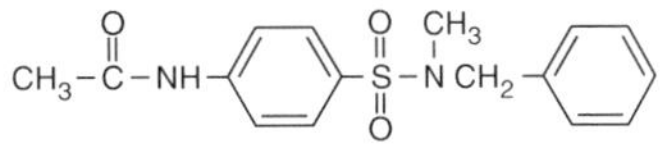

250 mg

S90,278-0
N4-PIVALOYLSULFANILAMIDE

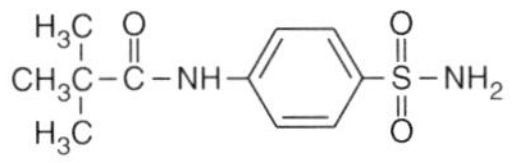

250 mg

S38,251-5
N4-ACETYL-4'-SULFAMOYL-SULFANILANILIDE

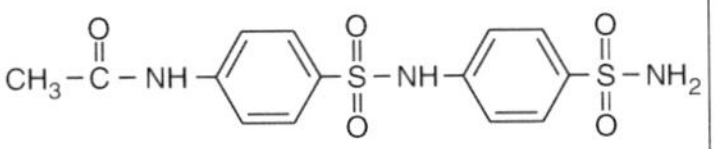

5 gm

S67,677-2
N4-ACETYL-4'-ETHYLSULFANILANILIDE

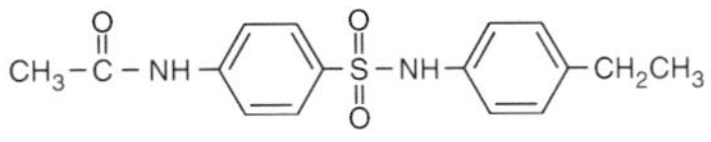

5 gm

S91,399-5
N4-DECANOYLSULFANILANILIDE

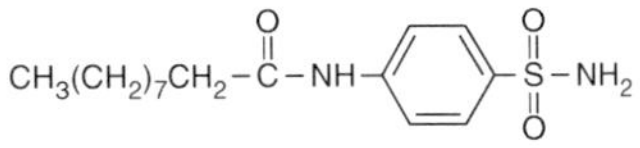

250 mg

S74,294-5
N4-ACETYL-2',4',5'-TRIMETHYL-SULFANILANILIDE

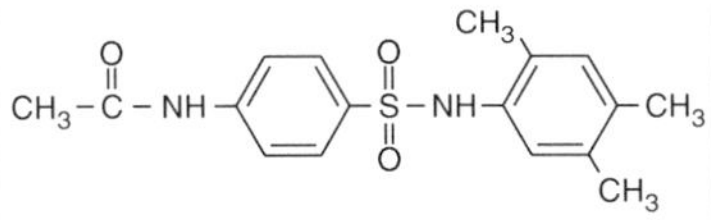

5 gm

S38,824-6
N4-ACETYL-N1-HEXADECYLSULFANILAMIDE

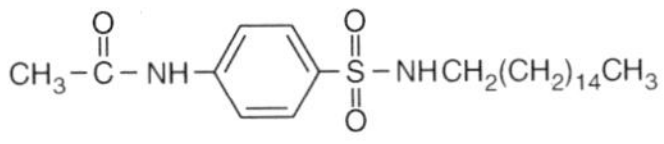

250 mg

S67,458-3
N4-ACETYL-2',3'-DIMETHYL-SULFANILANILIDE

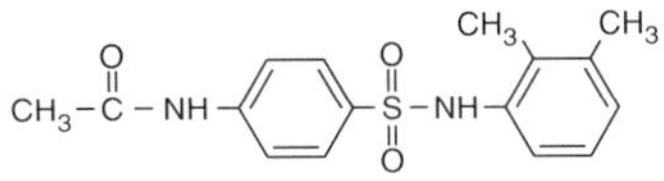

250 mg

S75,065-4
N4-ACETYL-2',6'-DIMETHYL-SULFANILANILIDE

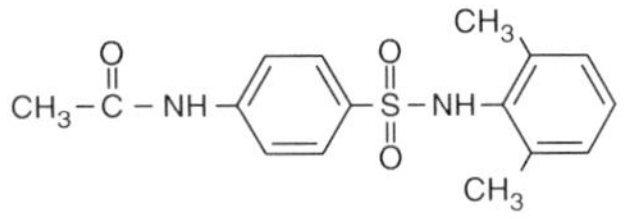

5 gm

S67,596-2
N4-ACETYL-2',5'-DIMETHYL-SULFANILANILIDE

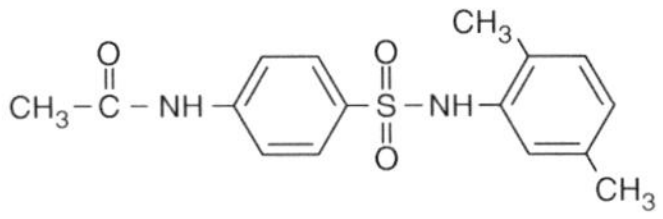

5 gm

S90,124-5
N4-ACETYL-N1-(1-NAPHTHYL)-SULFANILAMIDE

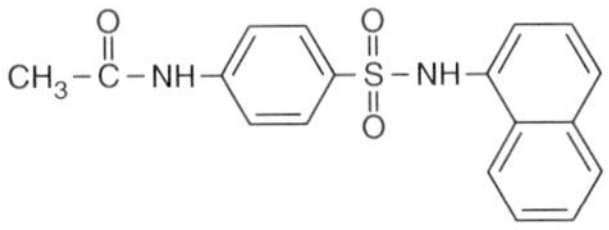

250 mg

S69,488-6
N3-(1-NAPHTHYLACETYL)METANILAMIDE

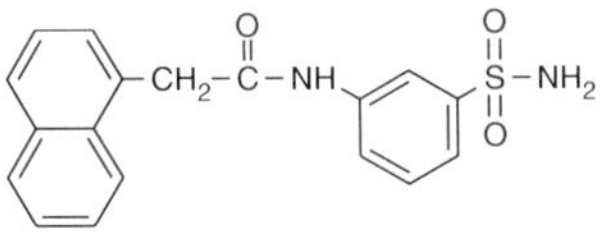

250 mg

S40,059-9
1-[3-(METHYLSULFONYL)PROPYL]-3-(PARA-TOLYLSULFONYL)UREA

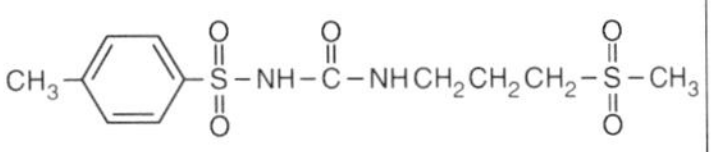

250 mg

S40,312-1
1-ISOBUTYL-3-(PARA-TOLYLSULFONYL)UREA

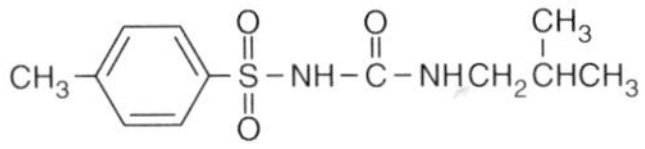

250 mg

S72,032-1
1-METHYL-3-(3-SULFAMOYLPHENYL)UREA

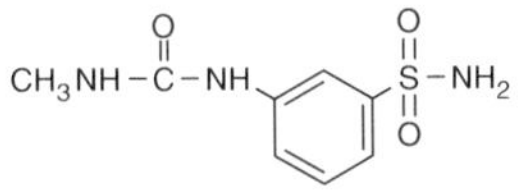

250 mg

S69,856-3
1-ETHYL-3-(3-SULFAMOYLPHENYL)UREA

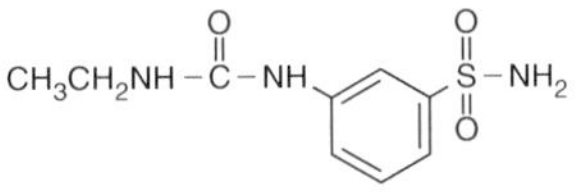

250 mg

S77,287-9
1-CYCLOPENTYL-3-(P-TOSYL)UREA

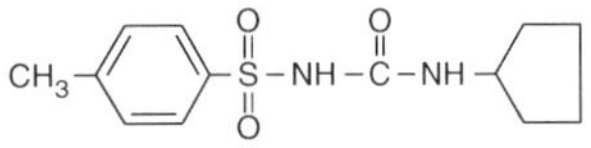

250 mg

S65,545-7
1-CYCLOHEXYL-3-(4-SULFAMOYLPHENYL)UREA

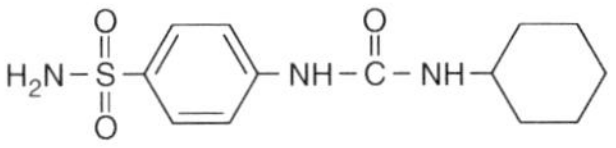

5 gm

S75,634-2
1-(PHENYLSULFONYL)UREA

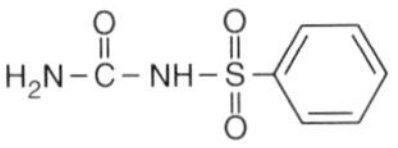

5 gm

S85,626-6
1-(METHYLSULFONYL)-3-PHENYLUREA

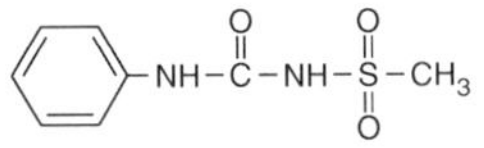

250 mg

S40,051-3
1-BUTYL-3-(PHENYLSULFONYL)UREA

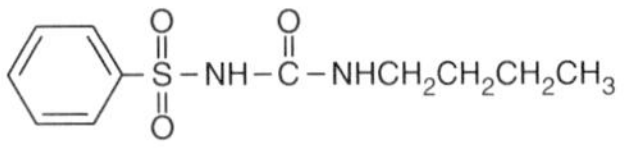

250 mg

S85,625-8
1,1-DIETHYL-3-(METHYLSULFONYL)-3-PHENYLUREA

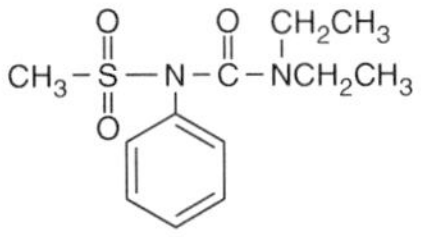

250 mg

S40,065-3
1-BENZYLSULFONYL-3-BUTYLUREA

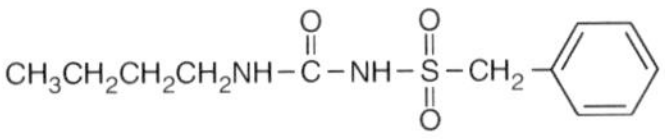

250 mg

S40,208-7
1,1'-HEXAMETHYLENEBIS(3-PHENYLSULFONYLUREA)

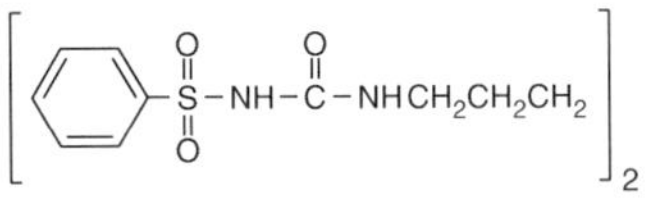

250 mg

S73,659-7
1-(4-SULFAMOYLPHENYL)-3-(O-TOLYL)UREA

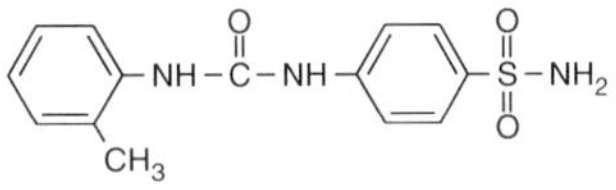

5 gm

S71,784-3
1-(2-ETHYLPHENYL)-3-(4-SULFAMOYLPHENYL)UREA

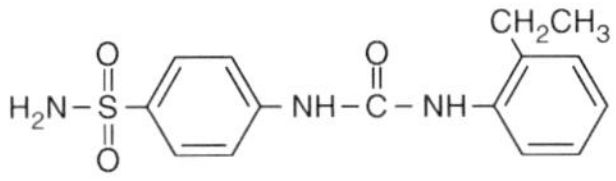

250 mg

S72,088-7
1-METHYL-3-(4-SULFAMOYLPHENYL)UREA

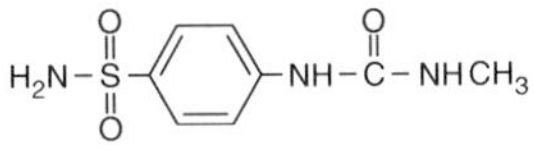

5 gm

S65,563-5
1-ETHYL-3-(4-SULFAMOYLPHENYL)UREA

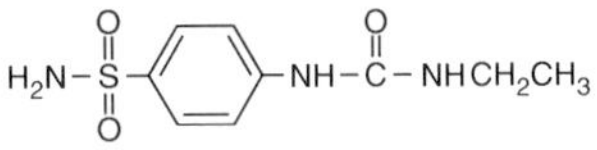

5 gm

S38,062-8
1-ETHYL-3-(PARA-TOLYLSULFONYL)UREA

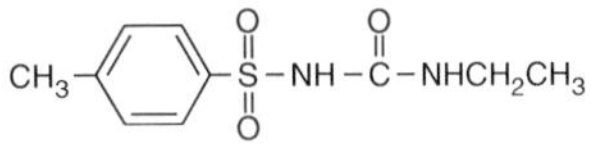

250 mg

S71,045-8
1,1-DIMETHYL-3-(P-TOLYLSULFONYL)UREA

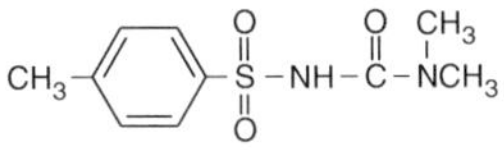

250 mg

S90,559-3
1-PROPYL-3-(4-SULFAMOYLPHENYL)UREA
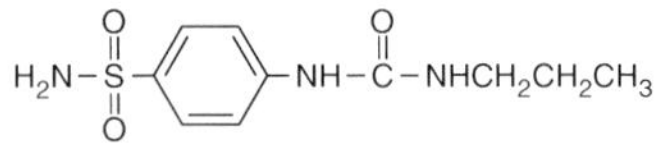
250 mg

S71,394-5
1,1-DIMETHYL-3-[4-(N,N-DIMETHYLSULFAMOYL)PHENYL]UREA
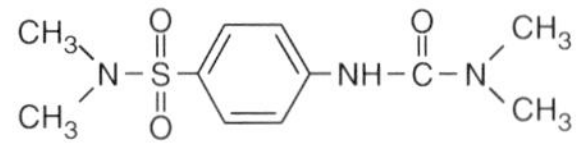
5 gm

S39,834-9
1-BUTYL-3-(PARA-TOLYLSULFONYL)UREA
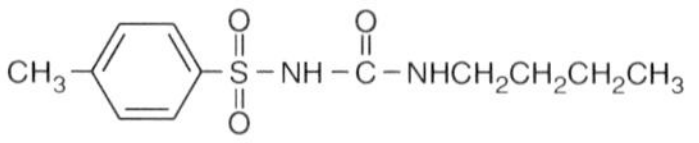
5 gm

S40,064-5
1-(1-METHYLPROPYL)-3-(PARA-TOLYLSULFONYL)UREA
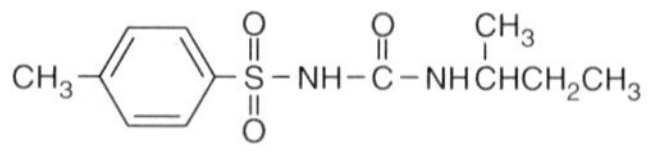
250 mg

S60,767-3
1,1-DIETHYL-3-(P-TOLYLSULFONYL)UREA
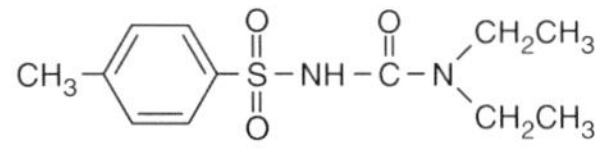
5 gm

S40,074-2
1-(4-CHLOROPHENYLSULFONYL)-3-ISOPENTYLUREA
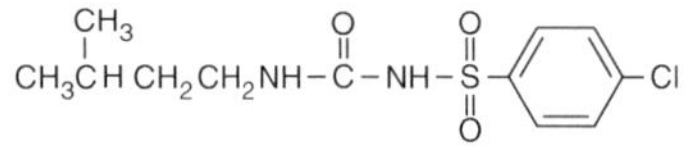
250 mg

S65,567-8
1-PHENYL-3-(4-SULFAMOYLPHENYL)UREA
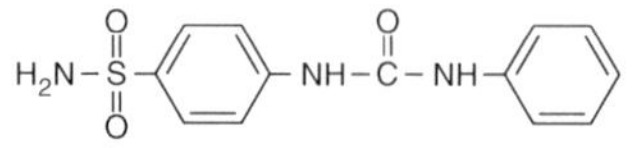
250 mg

S65,558-9
1-[(4-ACETYLSULFAMOYL)PHENYL]-3-BUTYLUREA
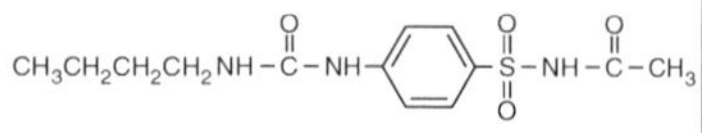
5 gm

S60,766-5
1,1-DIPROPYL-3-(P-TOLYLSULFONYL)UREA
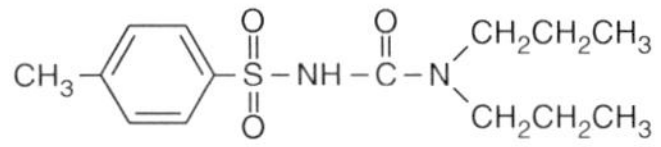
5 gm

S86,762-4
1-(4-SULFAMOYLPHENYL)-3-(M-TOLYL)UREA
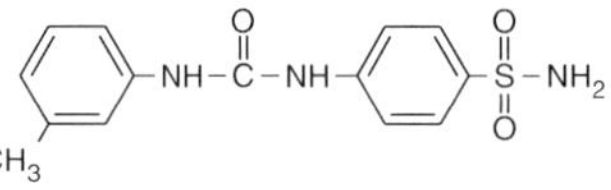
250 mg

S71,534-4
1-[6-(P-TOLUENESULFONAMIDO)HEXYL]-3-(P-TOLUENESULFONYL)UREA
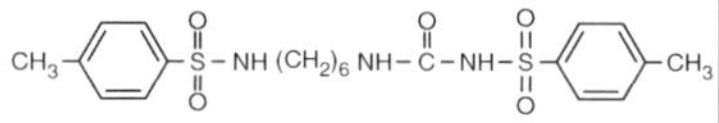
5 gm

S65,465-5
1,1'-HEXAMETHYLENEBIS[3-(P-TOLUENESULFONAMIDO)UREA]
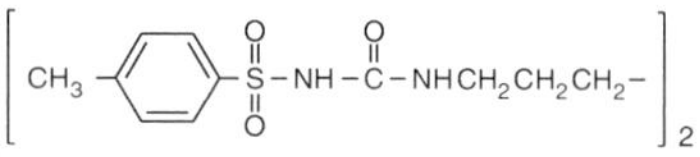
5 gm

S77,948-2
1-(4-SULFAMOYLPHENYL)-3-(2,6-XYLYL)UREA
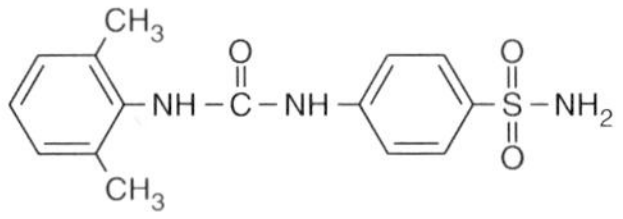
250 mg

S75,273-8
1-(2,6-DIETHYLPHENYL)-3-(P-TOSYL)UREA
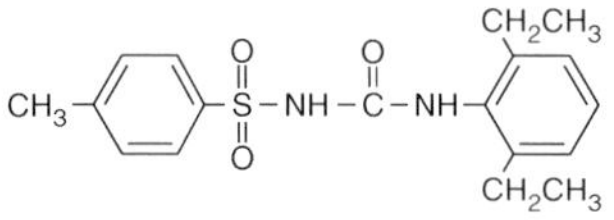
250 mg

S73,703-8
1-(4-SULFAMOYLPHENYL)-3-(2,5-XYLYL)UREA
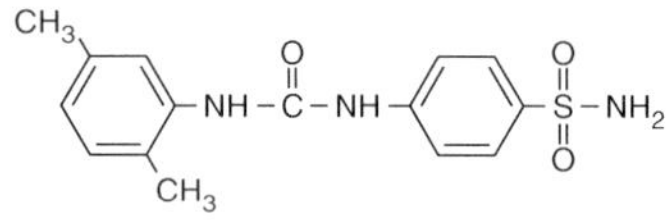
5 gm

S67,661-6
4,4'-BIS[3-(P-TOLUENESULFONYL)-UREIDO]-3,3'-DIMETHYLBIPHENYL
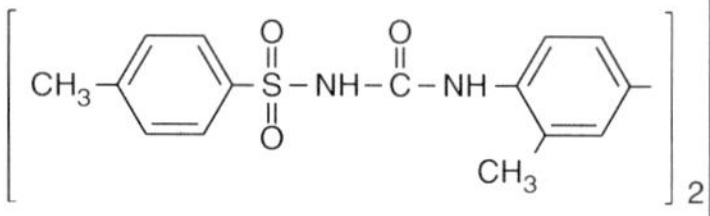
250 mg

S72,946-9
1-(1-NAPHTHYL)-3-(4-SULFAMOYLPHENYL)UREA
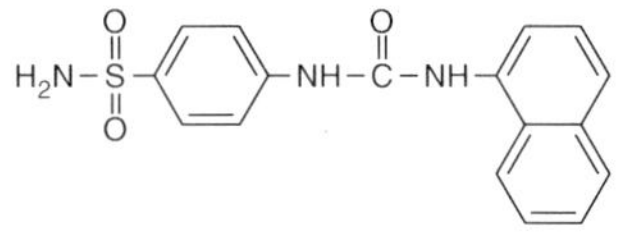
5 gm

S38,141-1
1-(2-NAPHTHYLSULFONYL)UREA
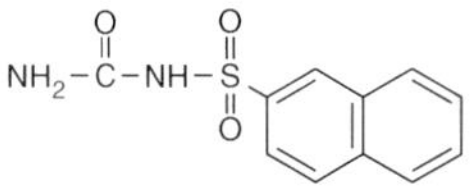
250 mg

S89,892-9
4-(P-TOSYL)-1H-TETRAZOL-5(4H)-ONE
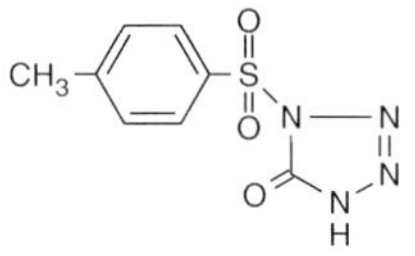
250 mg

S67,507-5
1-ETHYL-3-(3-SULFAMOYLPHENYL)-2-THIOUREA
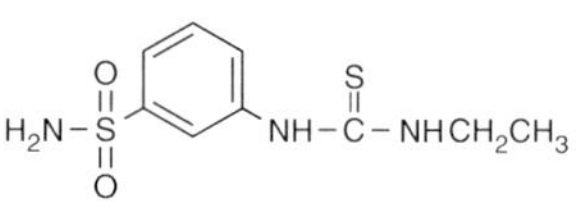
250 mg

S69,120-8
1-ETHYL-3-(4-SULFAMOYLPHENYL)-2-THIOUREA
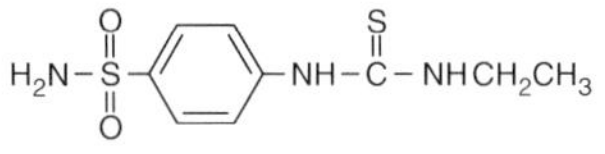
250 mg

S55,892-3
1-ETHYL-3-(P-TOLYLSULFONYL)-2-THIOUREA
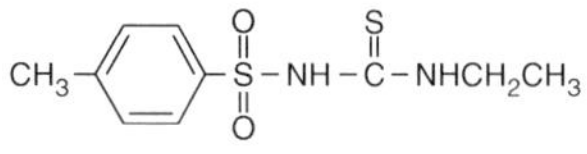
250 mg

S65,755-7
1-PHENYL-3-(P-TOLYLSULFONYL)-2-THIOUREA
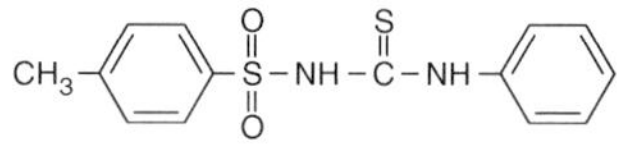
5 gm

S86,288-6
N-CYANO-2'-METHYLBENZENESULFONANILIDE
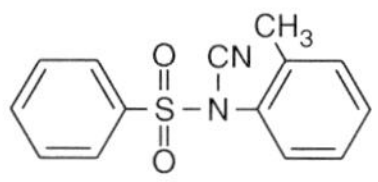
250 mg

S86,300-9
N-CYANO-2'-PHENYL-P-TOLUENESULFONANILIDE
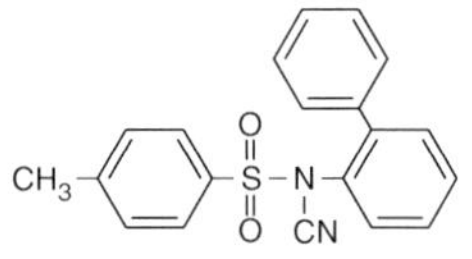
250 mg

S86,285-1
N-CYANO-P-TOLUENESULFONANILIDE
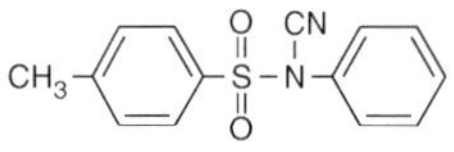
250 mg

S86,287-8
N-CYANO-4'-METHYLBENZENESULFONANILIDE
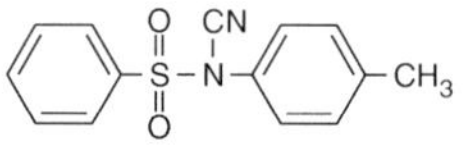
250 mg

S86,303-3
N-CYANO-4'-PHENYLBENZENESULFONANILIDE
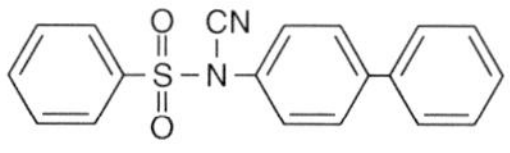
250 mg

S86,299-1
N-CYANO-4'-PHENYL-P-TOLUENESULFONANILIDE
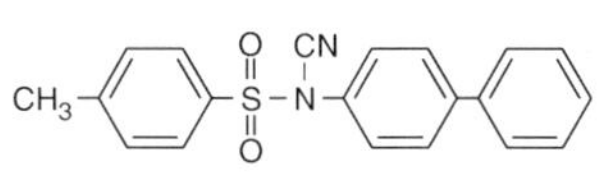
250 mg

S86,294-0
N-CYANO-2',6'-DIMETHYLBENZENESULFONANILIDE
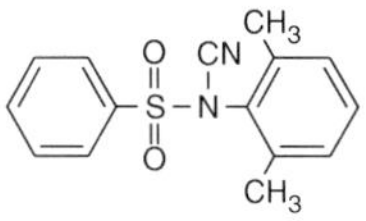
250 mg

S86,293-2
N-CYANO-3'-METHYL-O-TOLUENESULFONANILIDE
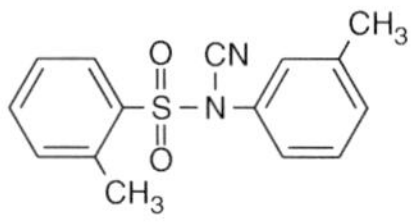
250 mg

S86,248-7
N-CYANO-2'-METHYL-P-TOLUENESULFONANILIDE
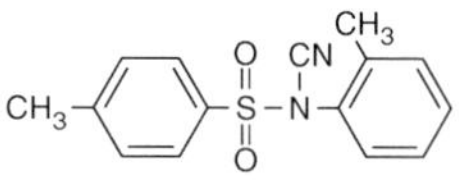
250 mg

S86,292-4
N-CYANO-3'-METHYL-P-TOLUENESULFONANILIDE

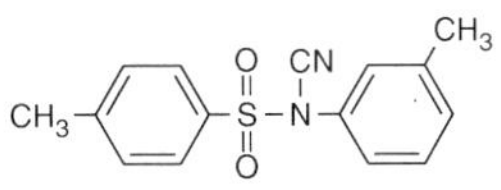

250 mg

S86,297-5
N-CYANO-4'-METHYL-P-TOLUENESULFONANILIDE

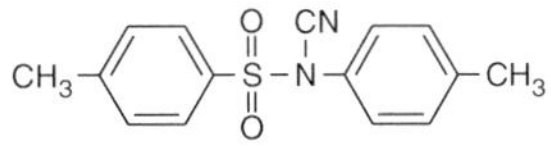

250 mg

S86,307-6
N-CYANO-2',6'-DIMETHYL-P-TOLUENESULFONANILIDE

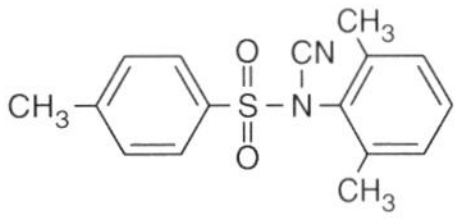

250 mg

S44,387-5
1-(4-ISOTHIOCYANATOPHENYLSULFONYL)-PIPERIDINE

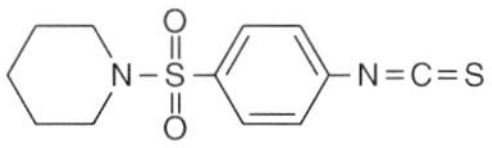

5 gm

S62,082-3
3,3'-SULFONYLDIBENZENESULFONIC ACID DIHYDRAZIDE

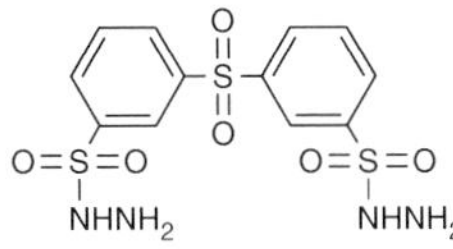

5 gm

S91,510-6
1-METHYL-3-(P-TOSYLAMINO)PROPYL P-TOLUENESULFONATE

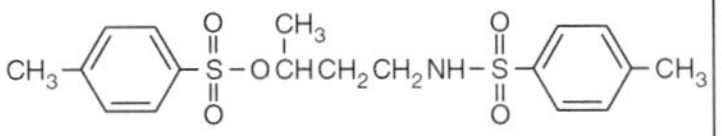

250 mg

S50,221-9
3-(HEXADECYLSULFAMOYL)-BENZENESULFONYL FLUORIDE

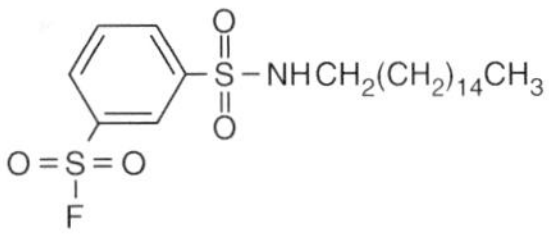

250 mg

S50,123-9
3-(N-HEXADECYL-N-METHYLAMINOSULFONYL)-BENZENESULFONYL FLUORIDE

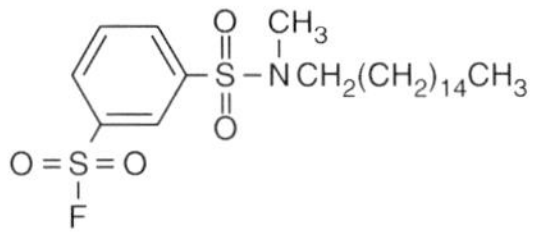

250 mg

S63,255-4
N-(1-ANTHRACENYL)-4-FLUOROSULFONYLBENZENESULFONAMIDE

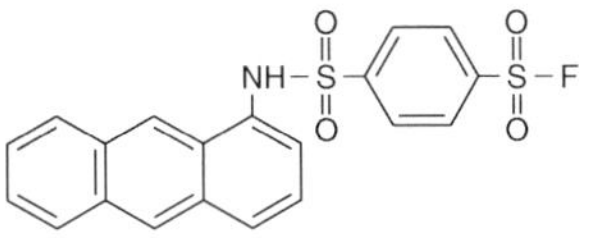

250 mg

S34,495-8
N-CYCLOHEXYLBENZENESULFONAMIDE

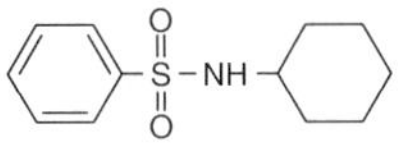

5 gm

S69,664-1
N-(2-METHYL-1-NAPHTHYL)-BENZENESULFONAMIDE

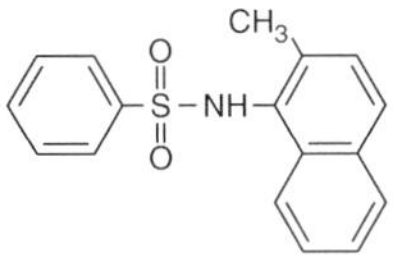

5 gm

S36,511-4
N-(ORTHO-TOLYL)BENZENESULFONAMIDE

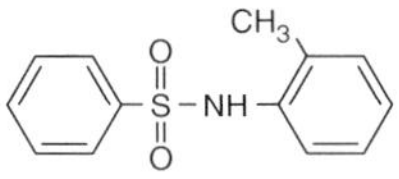

5 gm

S44,212-7
4-FLUOROBENZENESULFONAMIDE

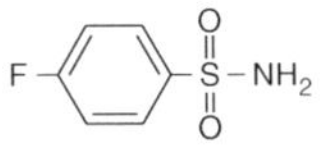

5 gm

S54,038-2
1,3-BENZENEDISULFONAMIDE

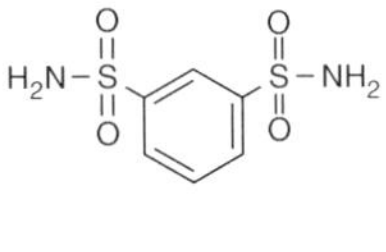

5 gm

S88,911-3
2,3,5,6-TETRAMETHYLBENZENESULFONAMIDE

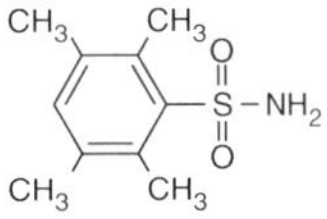

250 mg

S53,728-4
4-(N-HEXADECYLAMINOSULFONYL)-BENZENESULFONIC ACID HYDRAZIDE

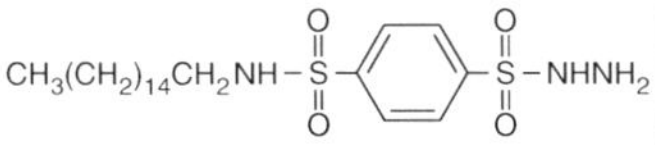

5 gm

S53,724-1
4-SULFAMOYLBENZENESULFONIC ACID HYDRAZIDE

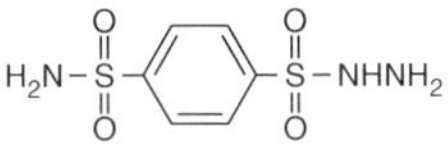

5 gm

S52,546-4
2-(PARA-TOLYLSULFONYL)ISOINDOLINE

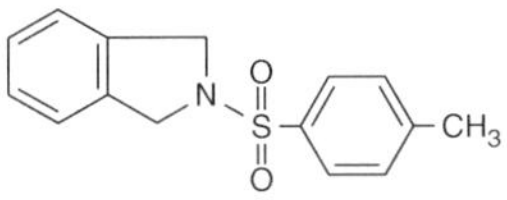

250 mg

S70,724-4
N-CYCLOHEXYL-P-TOLUENESULFONAMIDE

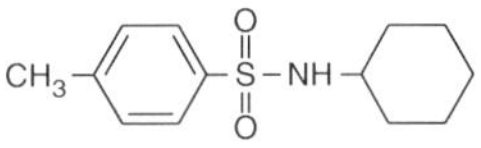

5 gm

S84,110-2
1-(BENZYLSULFONYL)PIPERIDINE

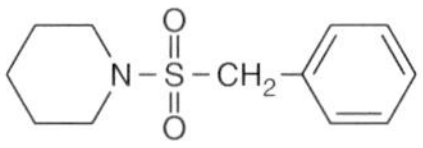

250 mg

S94,473-4
N-P-TOSYLPIPERIDINE

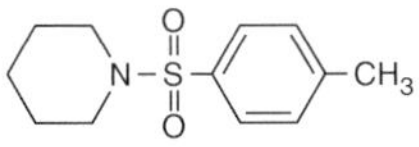

250 mg

S52,537-5
4-PHENYL-1,2,3,6-TETRAHYDRO-1-(PARA-TOLYLSULFONYL)PYRIDINE

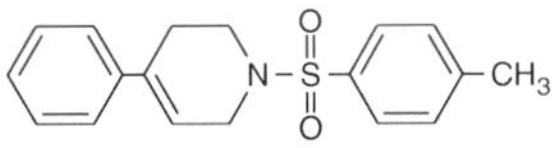

250 mg

S64,590-7
1,3,5-TRIS(PHENYLSULFONYL)-HEXAHYDRO-1,3,5-TRIAZINE

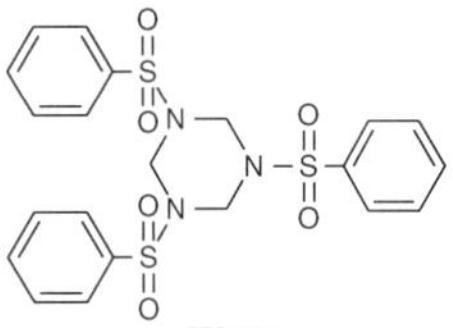

250 mg

S77,908-3
4-(BENZYLSULFONYL)MORPHOLINE

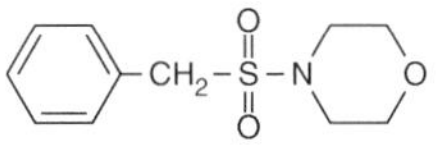

250 mg

S77,553-3
4-(3-PHENYLPROPYLSULFONYL)-MORPHOLINE

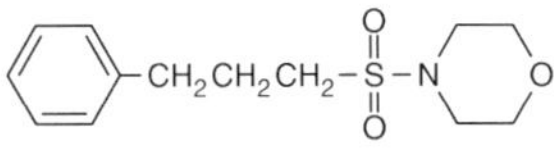

250 mg

S51,954-5
N,N-DICHLOROBENZENESULFONAMIDE

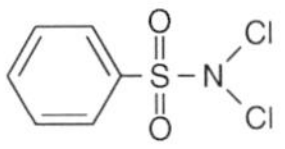

5 gm

S79,981-5
METHANESULFONANILIDE

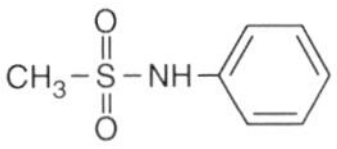

250 mg

S35,911-4
N-BUTYLBENZENESULFONAMIDE

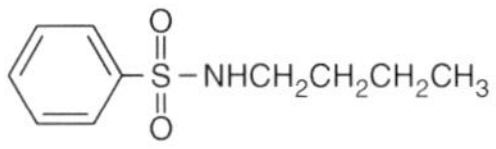

5 gm

S81,545-4
BENZENESULFONANILIDE

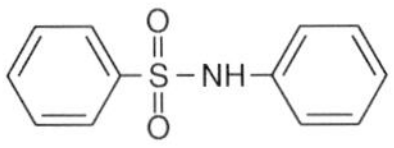

250 mg

S44,087-6
DIBENZENESULFONAMIDE

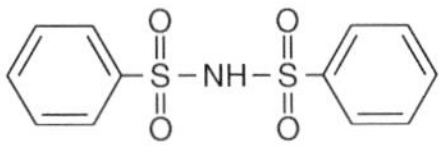

5 gm

S79,902-5
N-METHYL-ALPHA-TOLUENESULFONANILIDE

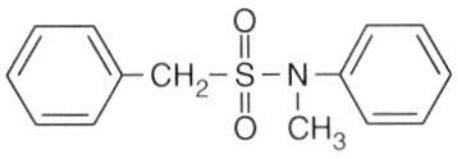

250 mg

S83,088-7
N-TETRADECYLBENZENESULFONAMIDE

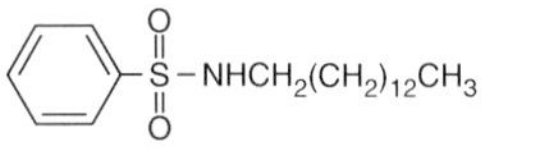

250 mg

S79,190-3
N-ETHYL-2'-METHYLBENZENESULFONANILIDE

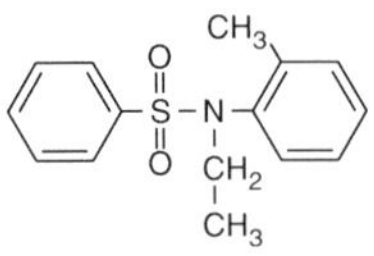

250 mg

S80,595-5
N-METHYL-P-TOLUENESULFONANILIDE

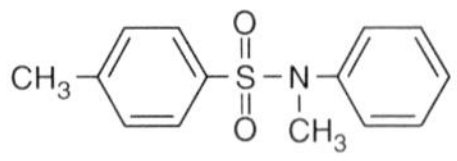

250 mg

S82,534-4
N-PHENYL-P-TOLUENESULFONANILIDE

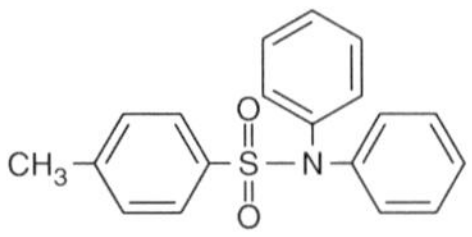

250 mg

S77,545-2
4-(2-[4-(N-METHYL-N-PHENYLSULFAMOYL)PHENYL]ETHYL)-BENZENESULFINIC ACID, SODIUM SALT

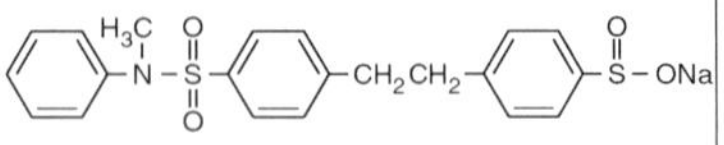

250 mg

S79,186-5
2',3'-DIMETHYLBENZENESULFONANILIDE

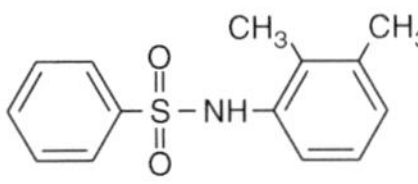

250 mg

S79,187-3
2',6'-DIMETHYLBENZENESULFONANILIDE

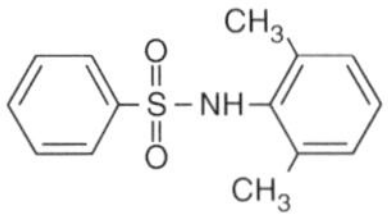

250 mg

S89,065-0
2',4'-DIMETHYLBENZENESULFONANILIDE

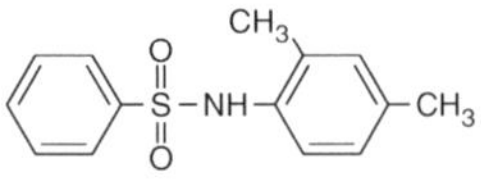

250 mg

S80,762-1
N-METHYL-2,4-XYLENESULFONANILIDE

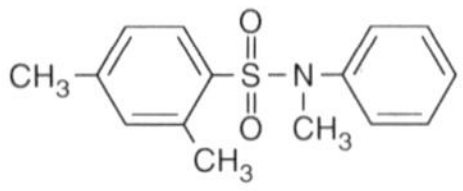

250 mg

S79,893-2
N,N-DIPHENYL-2,4-XYLENESULFONAMIDE

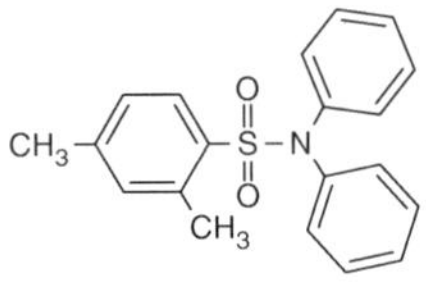

250 mg

S80,594-7
2,4,6,N-TETRAMETHYLBENZENESULFONANILIDE

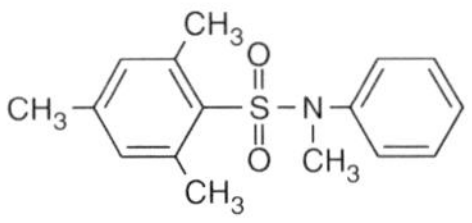

250 mg

S67,817-1
P-TOLUENESULFONO-O-TOLUIDIDE

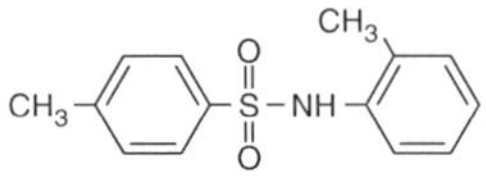

250 mg

S86,270-3
1-(P-TOSYL)GUANIDINE

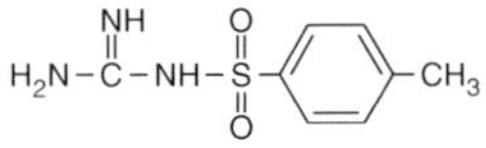

250 mg

S42,013-1
(PARA-TOLUENESULFONAMIDOMETHYL)-PHOSPHONIC ACID

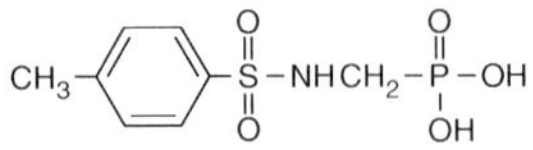

250 mg

S91,476-2
N-(P-TOSYL)ACETAMIDE

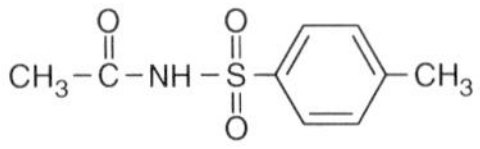

250 mg

S71,021-0
S,S-DIMETHYL-N-(P-TOLYLSULFONYL)-SULFILIMINE

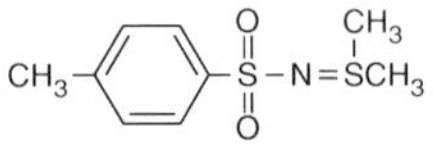

250 mg

S66,148-1
1-(P-TOSYL)AZETIDINE

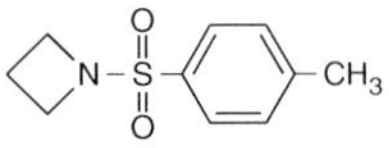

250 mg

S41,057-8
N-BUTYL-PARA-TOLUENESULFONAMIDE

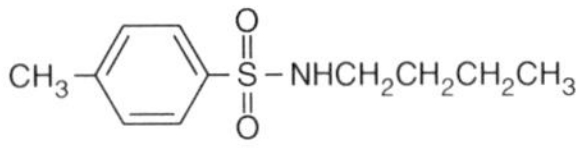

5 gm

S79,931-9
N,N-DIETHYL-P-TOLUENESULFONAMIDE

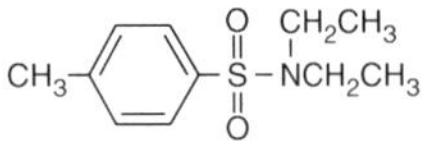

250 mg

S45,038-3
4'-BROMOBENZENESULFONANILIDE

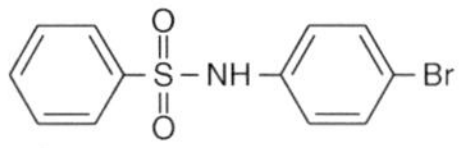

5 gm

S56,370-6
4,4'-DICHLORODIBENZENESULFONAMIDE, SODIUM SALT

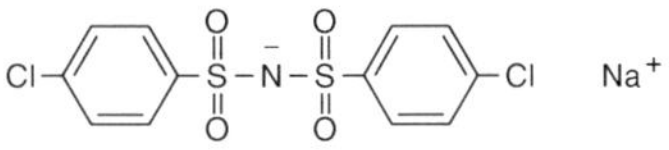

5 gm

S56,000-6
P-TOLUENESULFONANILIDE

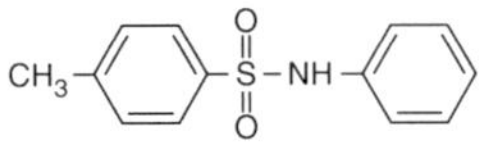

5 gm

S40,315-6
N-(PARA-TOLYLSULFONYL)HEXANAMIDE

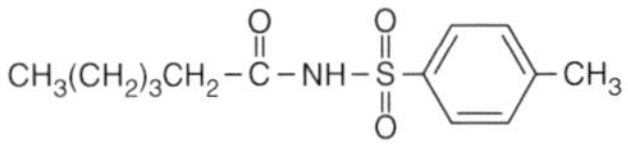

250 mg

S73,409-8
S,S-DIISOPROPYL-N-(P-TOLYLSULFONYL)-SULFOXIMINE

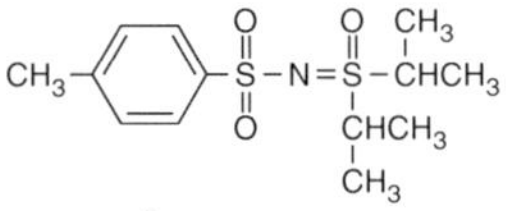

250 mg

S51,588-4
4'-CHLORO-N-METHYL-PARA-TOLUENESULFONANILIDE

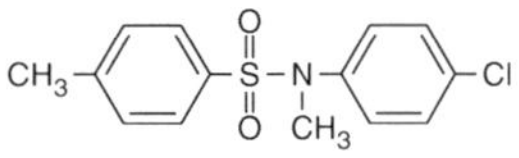

250 mg

S67,896-1
DI-P-TOLUENESULFONAMIDE

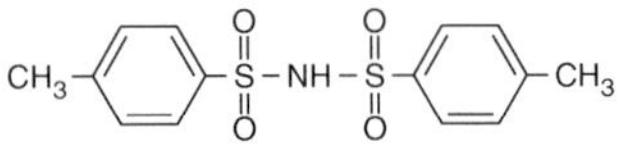

5 gm

S38,615-4
4-FLUORO-N-HEXADECYLBENZENESULFONAMIDE

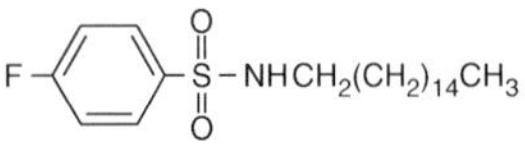

250 mg

S38,613-8
4-CHLORO-N-HEXADECYLBENZENESULFONAMIDE

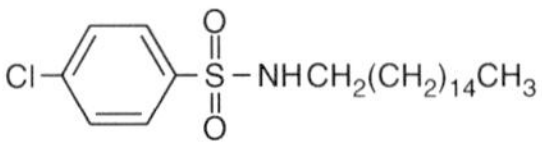

5 gm

S51,076-9
4-BROMO-N-HEXADECYLBENZENESULFONAMIDE

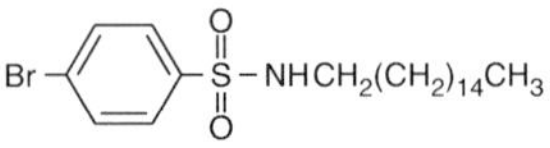

5 gm

S51,561-2
N-HEXADECYL-PARA-TOLUENESULFONAMIDE

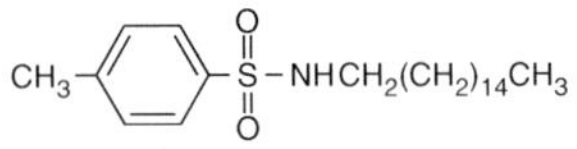

250 mg

S59,848-8
4-BROMO-N-HEXADECYL-N-METHYLBENZENESULFONAMIDE

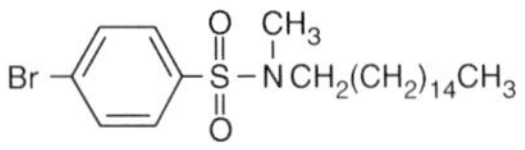

5 gm

S82,818-1
N-OCTADECYL-P-TOLUENESULFONAMIDE

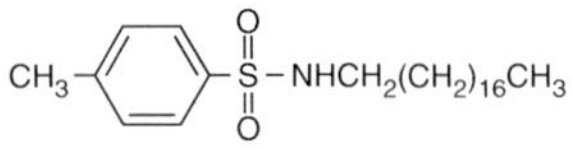

250 mg

S82,141-1
N-(P-TOSYL)OCTADECANAMIDE

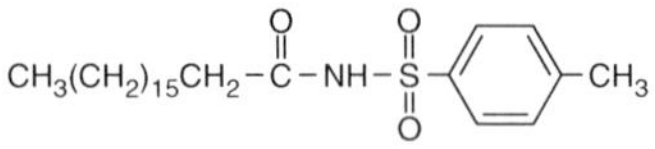

250 mg

S38,610-3
4-BROMO-N,N-DIDODECYLBENZENESULFONAMIDE
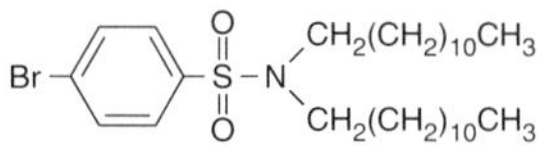
5 gm

S74,274-0
TETRAKIS(P-TOSYLAMINOMETHYL)METHANE
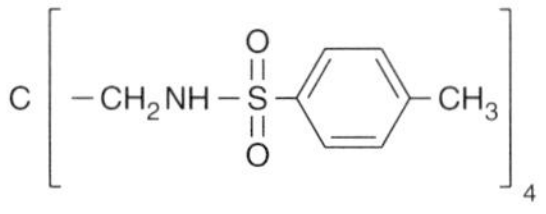
5 gm

S65,207-5
2',6'-DIETHYL-P-TOLUENESULFONANILIDE
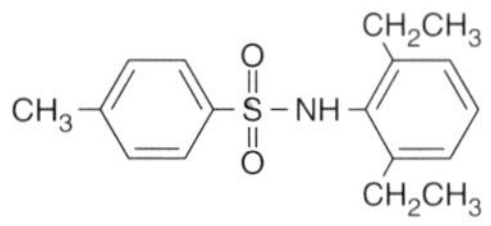
5 gm

S90,118-0
4'-ETHYL-2,4,6-TRIMETHYLBENZENESULFONANILIDE
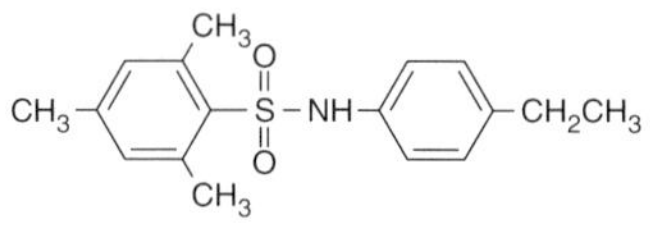
250 mg

S77,110-4
N-(2-FLUORENYL)-N-METHYL-P-TOLUENESULFONAMIDE
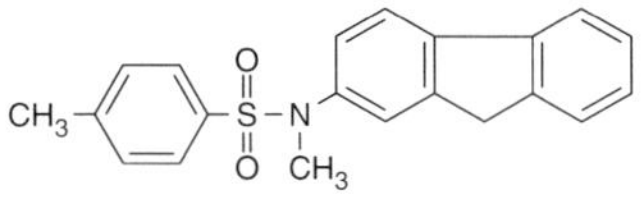
250 mg

S75,416-1
N-ETHYL-N-(2-FLUORENYL)-P-TOLUENESULFONAMIDE
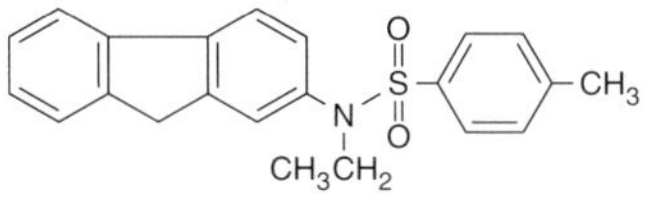
250 mg

S37,326-5
3-CHLORO-P-TOLUENESULFONAMIDE
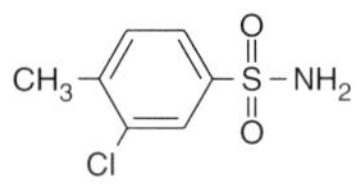
5 gm

S50,550-1
2,5-XYLENESULFONAMIDE
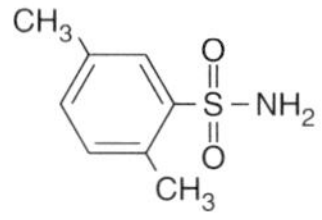
250 mg

S58,248-4
3,4-DICHLORO-N,N-DIMETHYLBENZENESULFONAMIDE
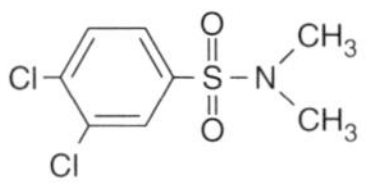
5 gm

S51,400-4
2,4-DIMETHYL-N-METHANESULFONANILIDE
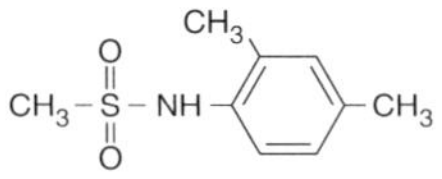
250 mg

S65,613-5
3,4-DICHLORO-N-HEXADECYL-N-METHYLBENZENESULFONAMIDE
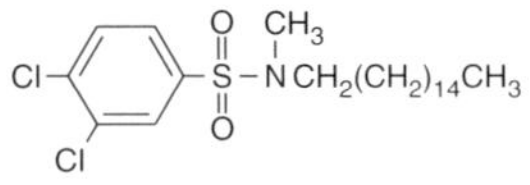
250 mg

S56,265-3
3,4-DICHLORO-N-METHYL-N-OCTADECYLBENZENESULFONAMIDE
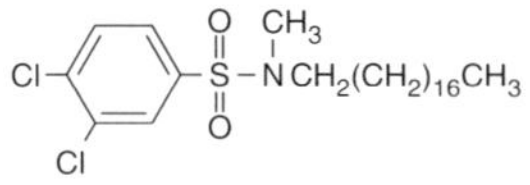
250 mg

S80,708-7
N,N-DIETHYL-2,4,6-TRIMETHYLBENZENESULFONAMIDE
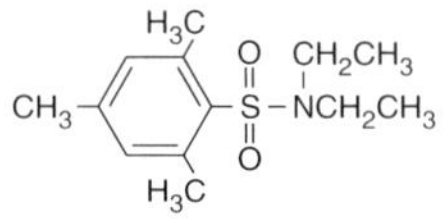
250 mg

S90,592-5
2',4',5'-TRIMETHYLMETHANESULFONANILIDE
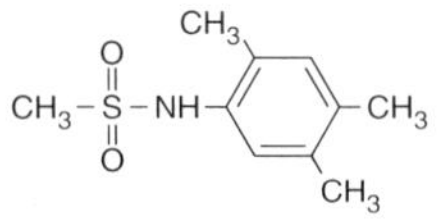
250 mg

S63,027-6
N-HEXADECYL-1,2-BENZENEDISULFONIMIDE
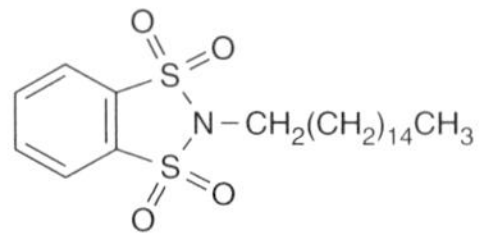
250 mg

S51,406-3
METHANESULFONIC ACID HYDRAZIDE
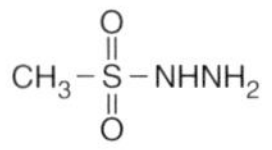
5 gm

S87,525-2
4-PHENYL-1-(PHENYLSULFONYL)-3-THIOSEMICARBAZIDE
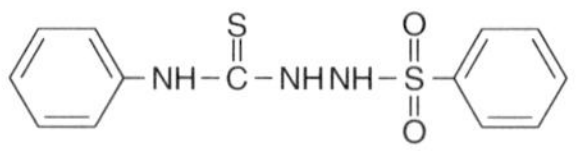
250 mg

S87,021-8
1-(PHENYLSULFONYL)-4-(P-TOLYL)-3-THIOSEMICARBAZIDE
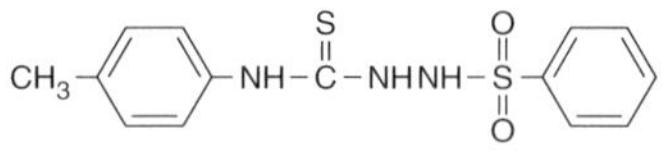
250 mg

S80,737-0
3-PHENYL-1,5,6,10B-TETRAHYDRO-3H-(1,2,3)THIADIAZOLO(4,3-A)-ISOQUINOLINE-2,2-DIOXIDE
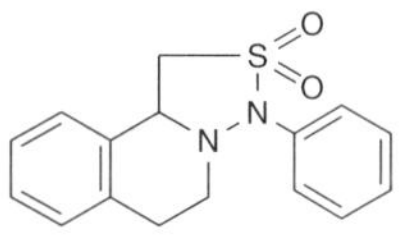
250 mg

S64,280-0
1,2-BIS(P-TOLYLSULFONYL)HYDRAZINE
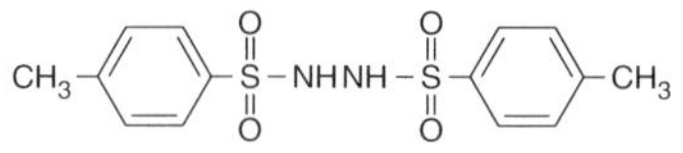
250 mg

S74,197-3
TRIS(3-CHLOROPHENYL)PHOSPHINE

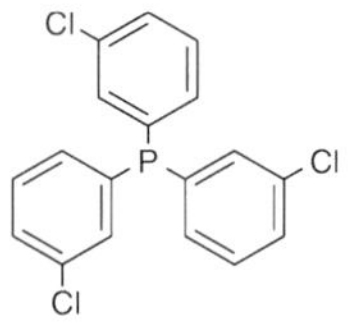

250 mg

S84,131-5
ETHYLENEBIS[(2-METHOXYPHENYL)-PHENYLPHOSPHINE] (DIPAMP)

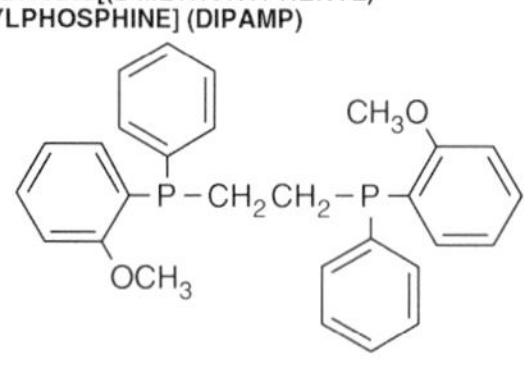

250 mg

S54,316-0
TRIS(2-METHOXYPHENYL)PHOSPHINE

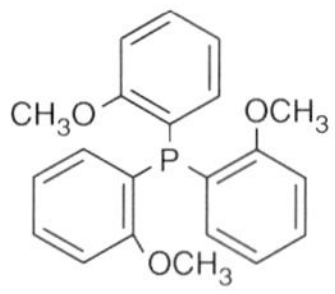

250 mg

S74,192-2
3-[PHENYL(3,5-XYLYL)PHOSPHINO]-PROPIONIC ACID

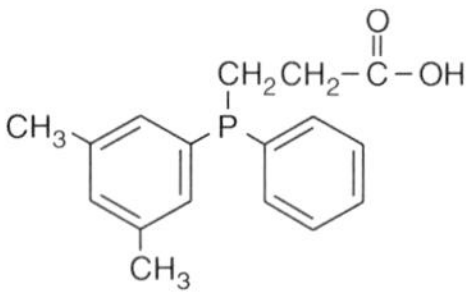

250 mg

S54,361-6
(DIPROPYL)PHENYLPHOSPHINE

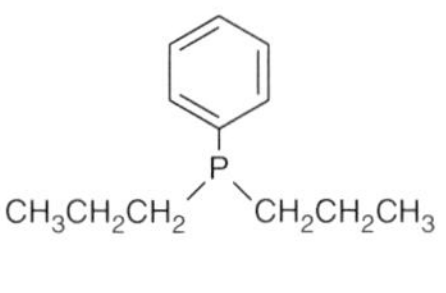

5 gm

S74,288-0
DI-M-TOLYLPHENYLPHOSPHINE

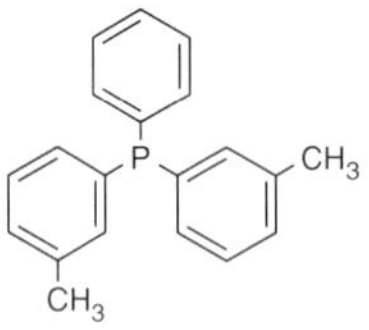

5 gm

S83,591-9
1,2-DIHYDROPHOSPHINOLINE

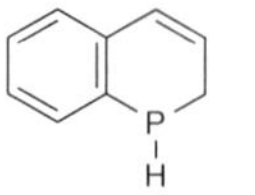

250 mg

S55,003-5
ETHYLTRIS(3-CHLOROPHENYL)-PHOSPHONIUM IODIDE

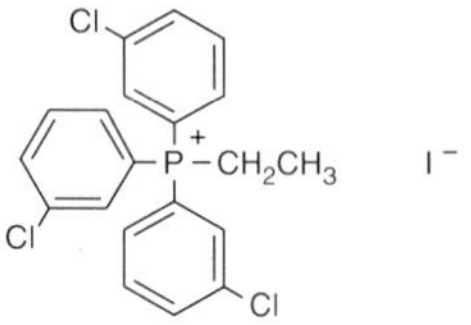

250 mg

S95,039-4
TETRAMETHYLENEBIS-(TRIPHENYLPHOSPHONIUM BROMIDE)

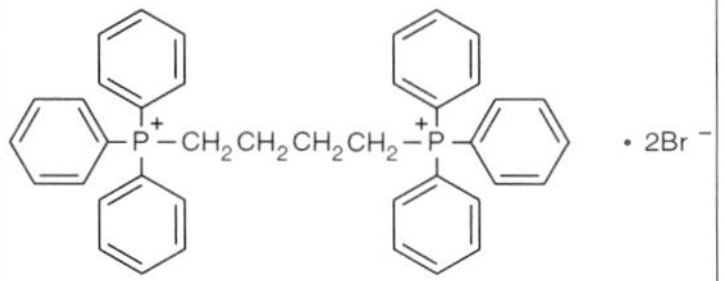

50 mg

S36,706-0
3-(BROMOMETHYL)-BENZYLTRIPHENYLPHOSPHONIUM BROMIDE

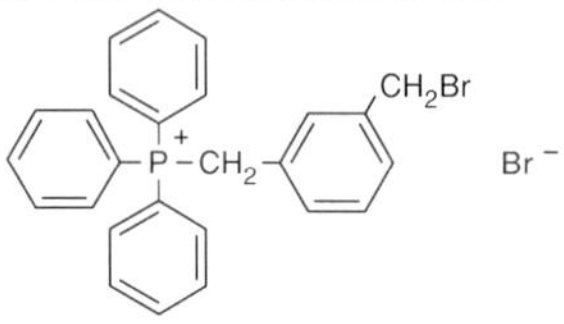

5 gm

S43,511-2
2-(BROMOMETHYL)-BENZYLTRIPHENYLPHOSPHONIUM BROMIDE

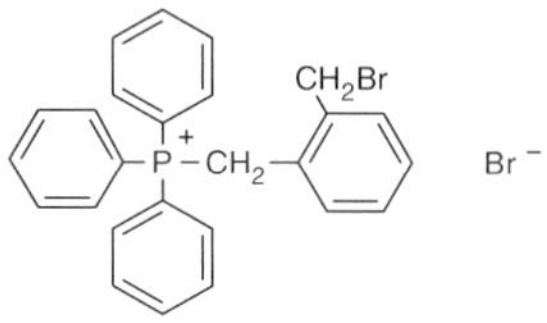

5 gm

S35,894-0
4-(BROMOMETHYL)-BENZYLTRIPHENYLPHOSPHONIUM BROMIDE

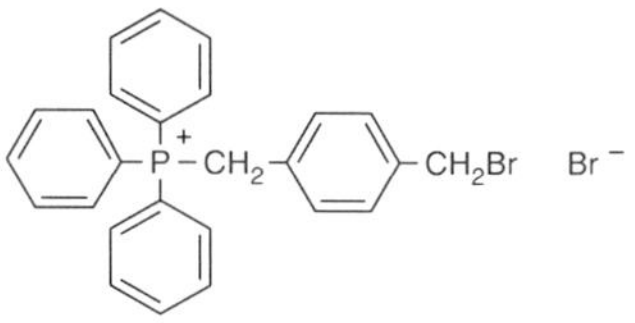

5 gm

S78,676-4
METHYLTRIPHENOXYPHOSPHONIUM TRIFLUOROMETHANESULFONATE

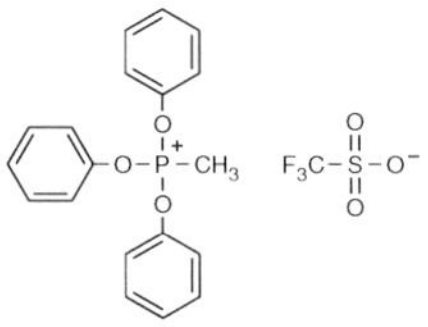

250 mg

S44,447-2
(4-CHLOROPHENOXYMETHYL)-TRIPHENYLPHOSPHONIUM CHLORIDE

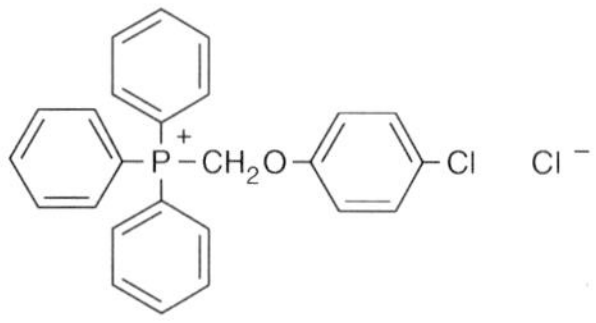

250 mg

S77,415-4
(4-FLUOROBENZYL)-TRIPHENYLPHOSPHONIUM CHLORIDE

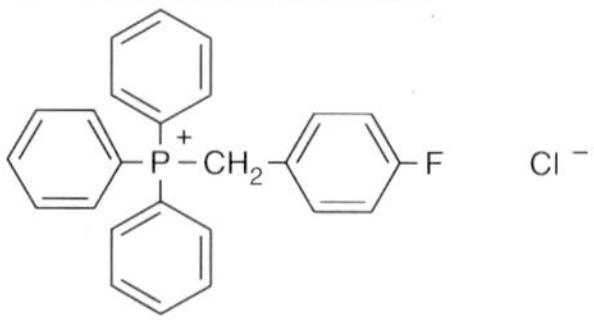

250 mg

S77,414-6
(4-FLUOROBENZYL)-TRIPHENYLPHOSPHONIUM BROMIDE

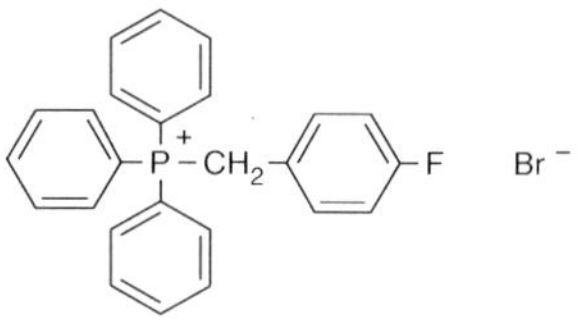

250 mg

S61,671-0
(HYDROXYMETHYL)-TRIPHENYLPHOSPHONIUM CHLORIDE

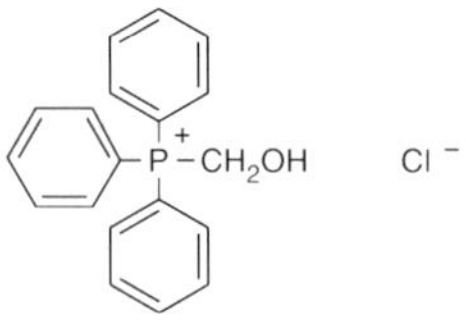

5 gm

S86,016-6
(3-HYDROXYPROPYL)-TRIPHENYLPHOSPHONIUM BROMIDE

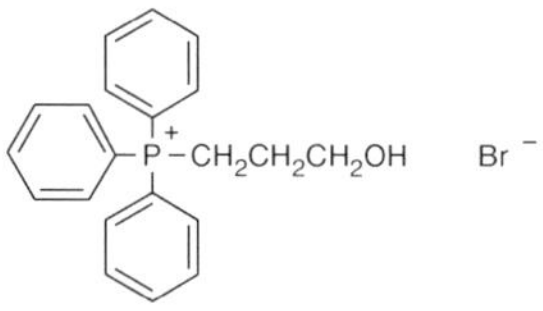

250 mg

S54,327-6
(3-HYDROXYPROPYL)-TRIPHENYLPHOSPHONIUM IODIDE

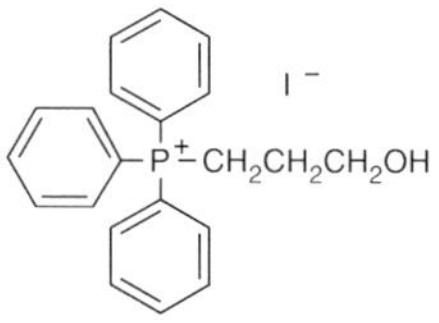

250 mg

S54,581-3
(3,5-DI-TERT-BUTYL-4-HYDROXYBENZYL)-TRIPHENYLPHOSPHONIUM BROMIDE

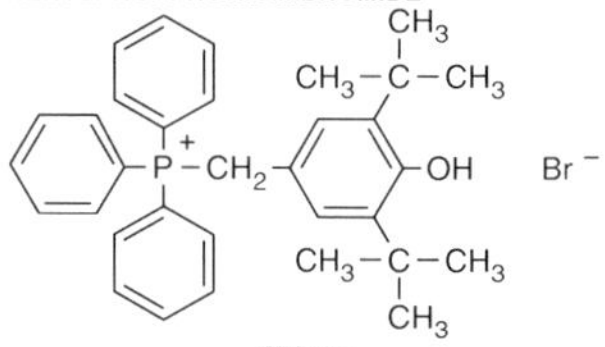

250 mg

S89,615-2
[ETHOXY(ETHOXYCARBONYL)METHYL]-TRIPHENYLPHOSPHONIUM CHLORIDE

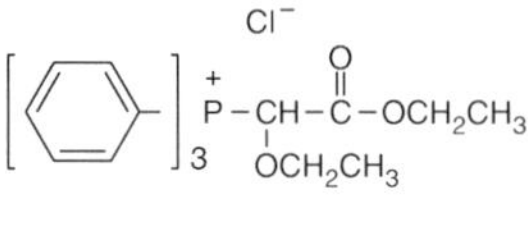

250 mg

S76,515-5
(2-ETHOXY-1-PROPENYL)-TRIPHENYLPHOSPHONIUM IODIDE

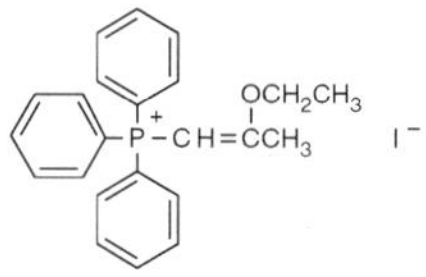

5 gm

S67,676-4
(2,5-DIMETHYL-4-METHOXYBENZYL)-TRIPHENYLPHOSPHONIUM BROMIDE

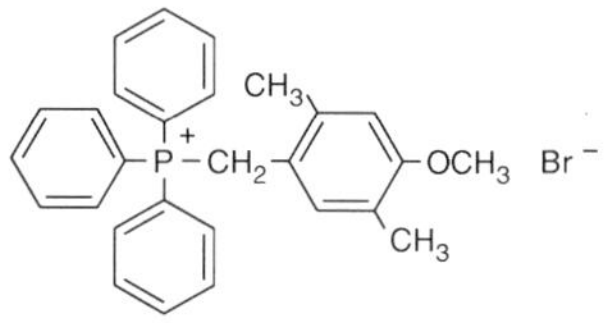

250 mg

S88,546-0
[3-(3-METHOXYPHENYL)PROPYL]-TRIPHENYLPHOSPHONIUM BROMIDE

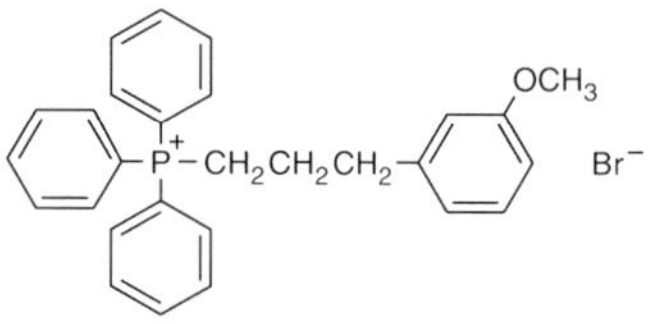

250 mg

S53,683-0
(4-METHOXYBENZYL)-TRIPHENYLPHOSPHONIUM CHLORIDE

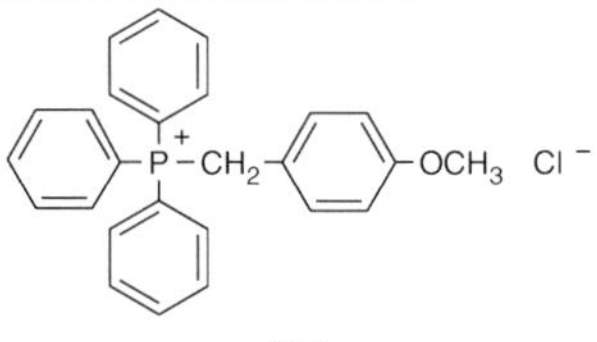

5 gm

S54,974-6
(4-ETHOXYBENZYL)-TRIPHENYLPHOSPHONIUM CHLORIDE

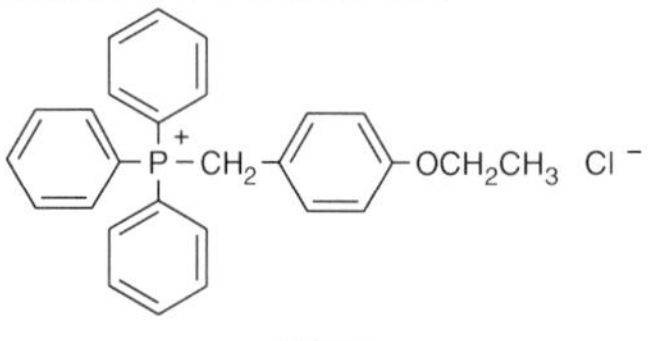

250 mg

S62,667-8
(METHYLTHIOMETHYL)-TRIPHENYLPHOSPHONIUM CHLORIDE

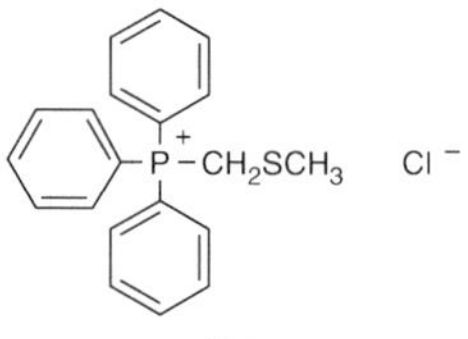

5 gm

S61,670-2
(PHENYLTHIOMETHYL)-TRIPHENYLPHOSPHONIUM CHLORIDE

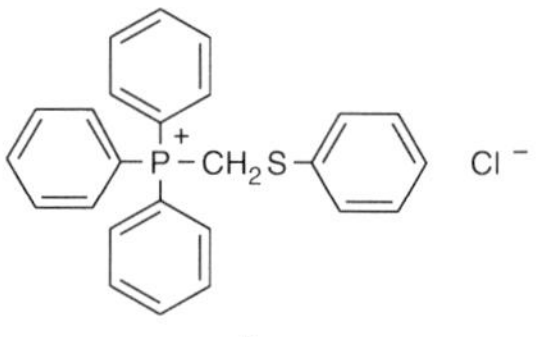

5 gm

S55,463-4
2-HYDROXY-1-PHENYL-3-(TRIPHENYLPHOSPHONIUM)-2-PYRROLIN-5-ONE, INNER SALT

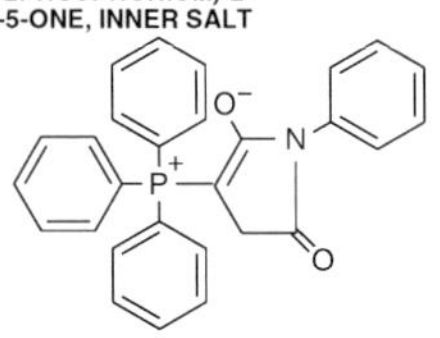

250 mg

S63,374-7
(4'-NITROPHENACYL)-TRIPHENYLPHOSPHONIUM BROMIDE

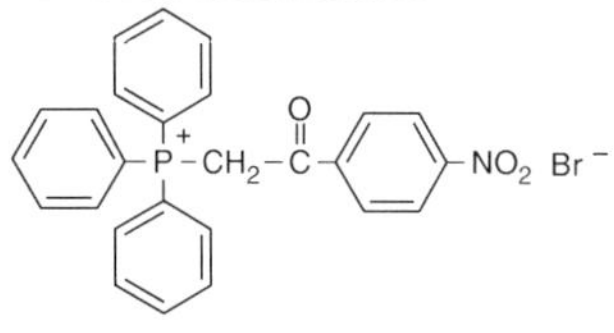

5 gm

S54,346-2
DIMETHYL(4-NITROBENZYL)-PHENYLPHOSPHONIUM BROMIDE

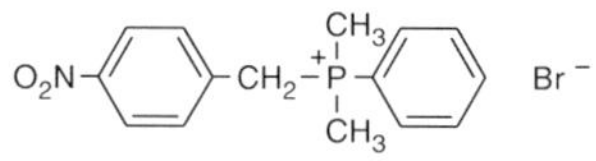

250 mg

S52,425-5
(4-NITROBENZYL)-TRIBUTYLPHOSPHONIUM BROMIDE

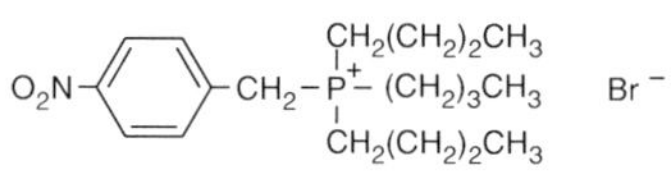

5 gm

S54,324-1
(DIPHENYL)METHYL(4-NITROBENZYL)-PHOSPHONIUM BROMIDE

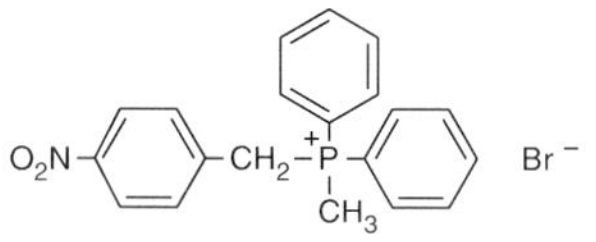

250 mg

S95,037-8
(4-OXO-2-BUTEN-1-YL)-TRIPHENYLPHOSPHONIUM BROMIDE

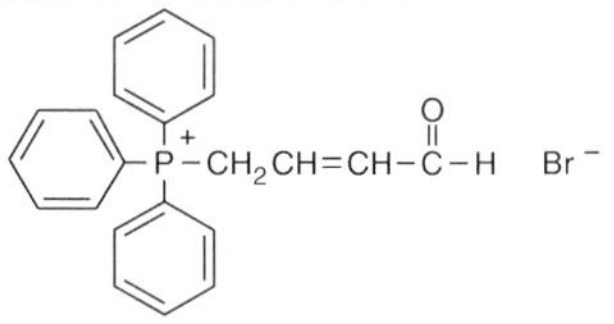

50 mg

S94,757-1
(ACETONYL)-TRIPHENYLPHOSPHONIUM BROMIDE

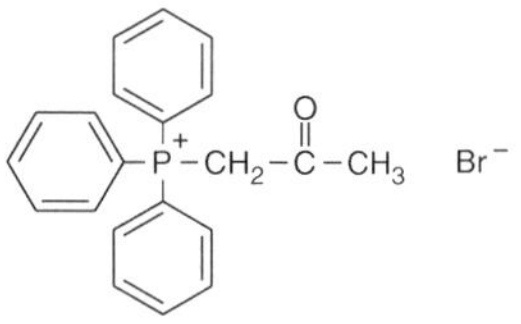

5 gm

S95,038-6
[3-(METHOXYCARBONYL)-2-PROPEN-1-YL]-TRIPHENYLPHOSPHONIUM BROMIDE

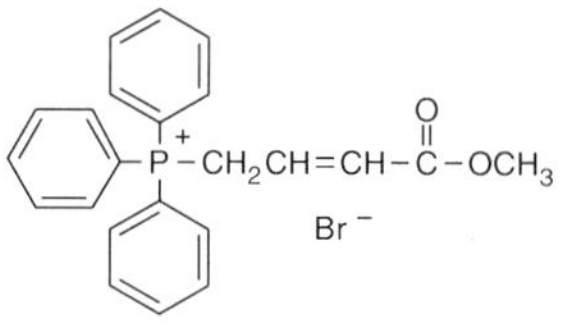

50 mg

S55,467-7
[1-(CARBOXYMETHYL)-2,2-DIHYDROXYVINYL]-TRIPHENYLPHOSPHONIUM HYDROXIDE

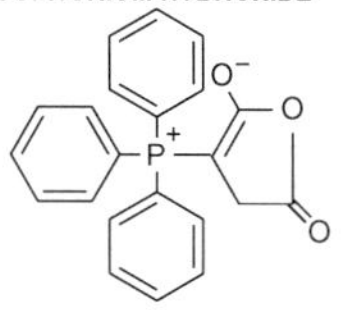

5 gm

S84,083-1
[2-(ETHOXYCARBONYL)ETHYL]-TRIPHENYLPHOSPHONIUM CHLORIDE

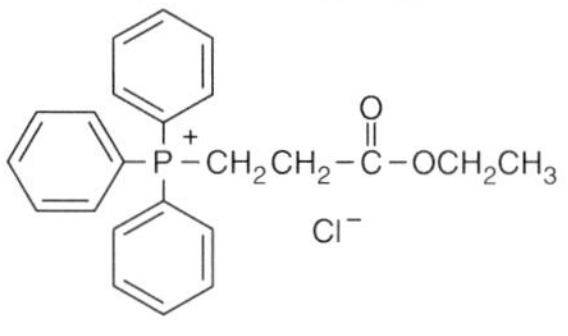

250 mg

S70,253-6
[1-(ETHOXYCARBONYL)ETHYL]-TRIPHENYLPHOSPHONIUM BROMIDE

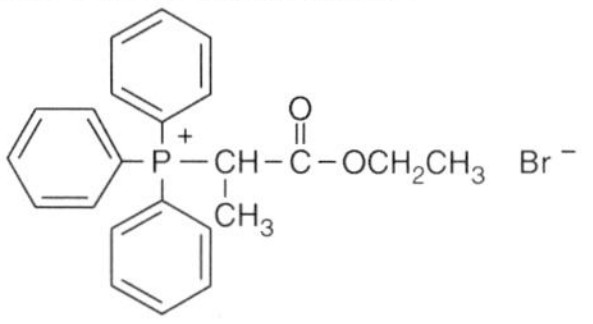

250 mg

S93,882-3
[2-(ETHOXYCARBONYL)ALLYL]-TRIPHENYLPHOSPHONIUM BROMIDE

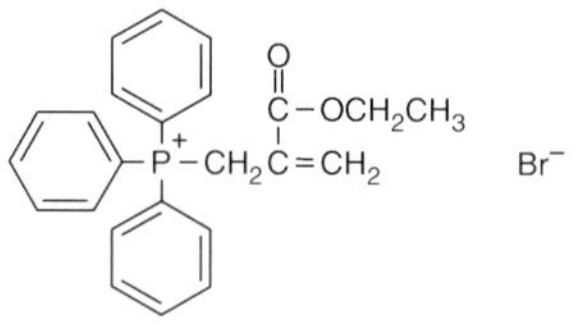

250 mg

S53,641-5
[1-(ETHOXYCARBONYL)PROPYL]-TRIPHENYLPHOSPHONIUM BROMIDE

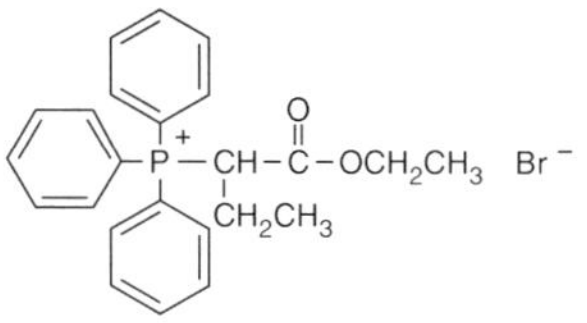

5 gm

S63,789-0
CARBAMOYLMETHYLTRIPHENYLPHOSPHONIUM CHLORIDE

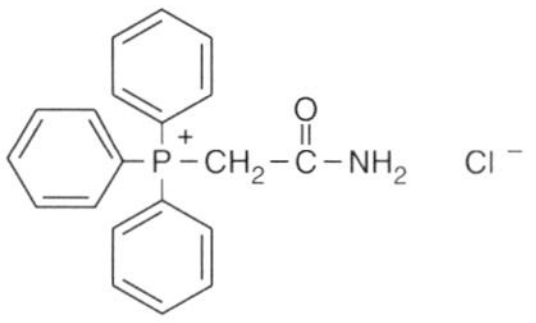

250 mg

S89,536-9
[1-(DIMETHYLCARBAMOYL)ETHYL]-TRIPHENYLPHOSPHONIUM BROMIDE

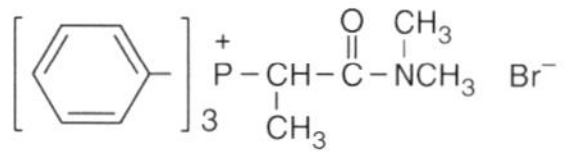

250 mg

S63,371-2
(CYANOMETHYL)-TRIPHENYLPHOSPHONIUM CHLORIDE

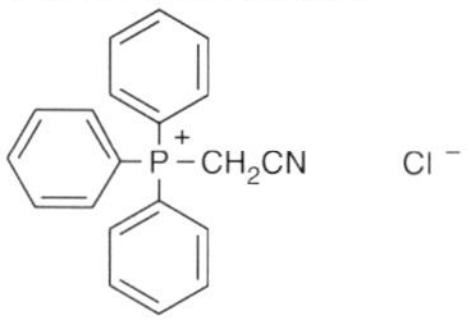

5 gm

S91,507-6
[4-(4,4,5,5-TETRAMETHYL-1,3-DIOXOLAN-2-YL)BUTYL]-TRIPHENYLPHOSPHONIUM BROMIDE

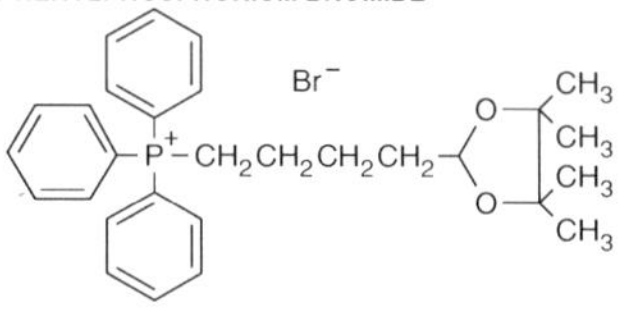

250 mg

S86,139-1
(1,3-BENZODIOXOL-5-YLMETHYL)-TRIPHENYLPHOSPHONIUM CHLORIDE

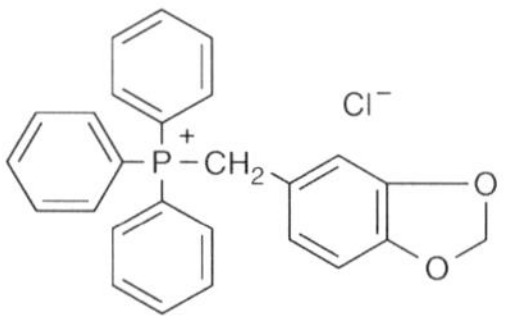

250 mg

S92,765-1
BIS(1-ETHYL-3-METHYL-2-BENZIMIDAZOLINYL)PHOSPHONIUM TETRAFLUOROBORATE

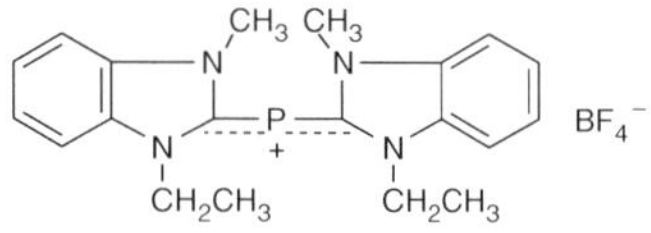

250 mg

S61,649-4
CYCLOHEXYLTRIPHENYLPHOSPHONIUM BROMIDE

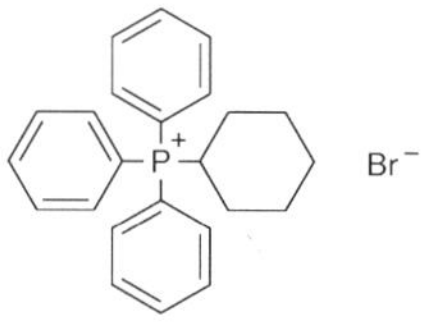

5 gm

S89,649-7
(2-PIPERIDINYLMETHYL)-TRIPHENYLPHOSPHONIUM IODIDE

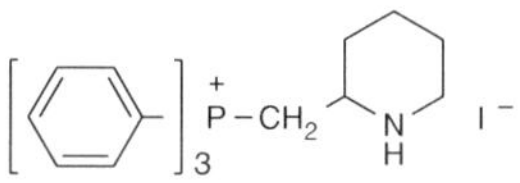

250 mg

S54,336-5
PHENYLTRIETHYLPHOSPHONIUM IODIDE

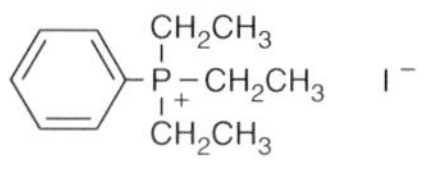

250 mg

S74,554-5
BENZYLTRIETHYLPHOSPHONIUM CHLORIDE

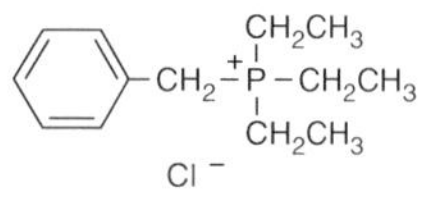

250 mg

S74,553-7
BENZYLTRIETHYLPHOSPHONIUM BROMIDE

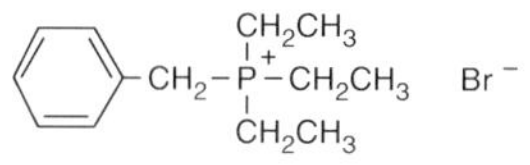

250 mg

S54,613-5
AMINOTRIPHENYLPHOSPHONIUM CHLORIDE

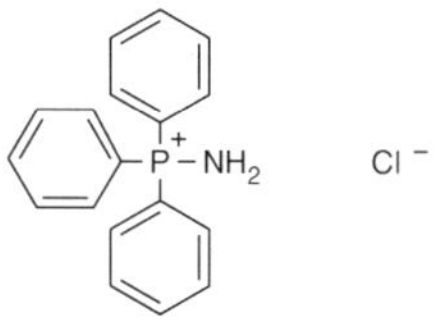

250 mg

S54,345-4
AMINOTRIPHENYLPHOSPHONIUM BROMIDE

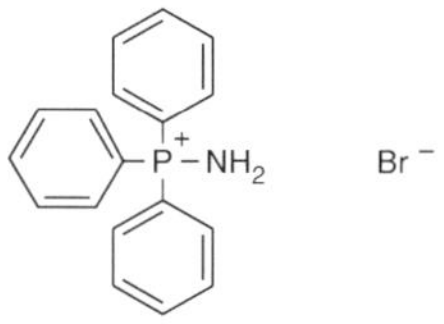

250 mg

S69,461-4
ETHYLTRIPHENYLPHOSPHONIUM FLUOROBORATE

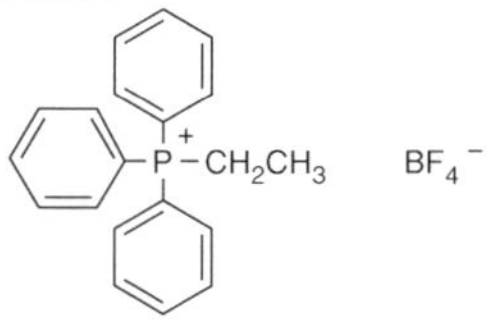

250 mg

S65,412-4
PROPYLAMINOTRIPHENYLPHOSPHONIUM BROMIDE

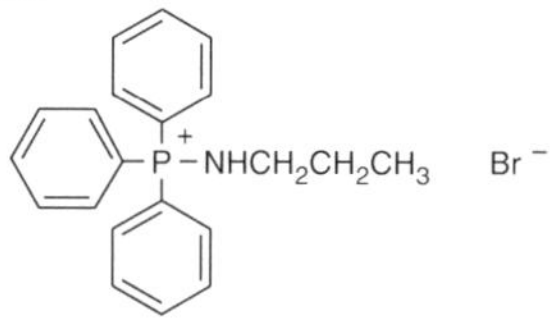

250 mg

S88,098-1
(3-BROMOALLYL)-TRIPHENYLPHOSPHONIUM BROMIDE

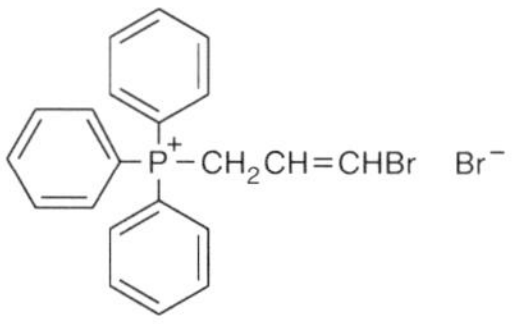

250 mg

S55,002-7
(2-METHYLALLYL)-TRIPHENYLPHOSPHONIUM CHLORIDE

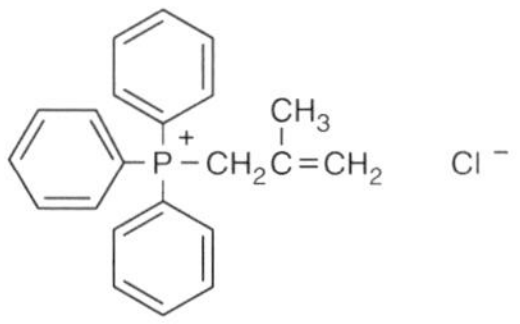

5 gm

S88,001-9
(3-BUTEN-1-YL)-TRIPHENYLPHOSPHONIUM BROMIDE

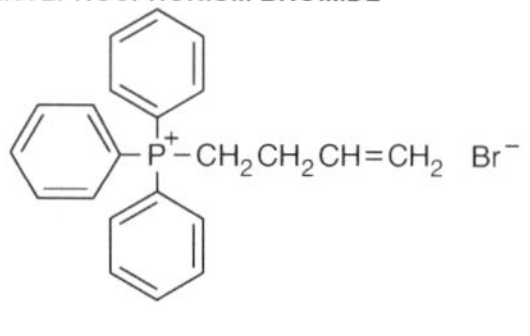

250 mg

S65,327-6
BUTYLAMINOTRIPHENYLPHOSPHONIUM BROMIDE

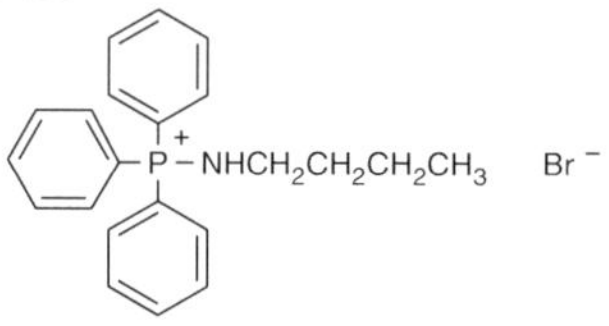

250 mg

S88,551-7
(3-METHYL-2-BUTEN-1-YL)-TRIPHENYLPHOSPHONIUM CHLORIDE

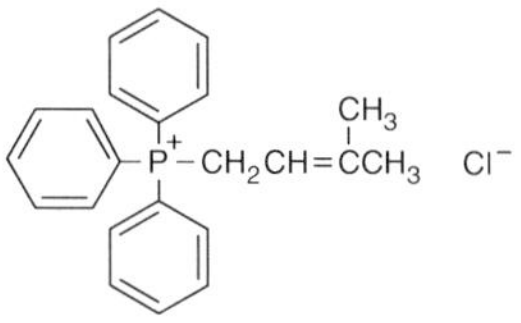

250 mg

S88,232-1
(4-PENTEN-1-YL)-TRIPHENYLPHOSPHONIUM BROMIDE

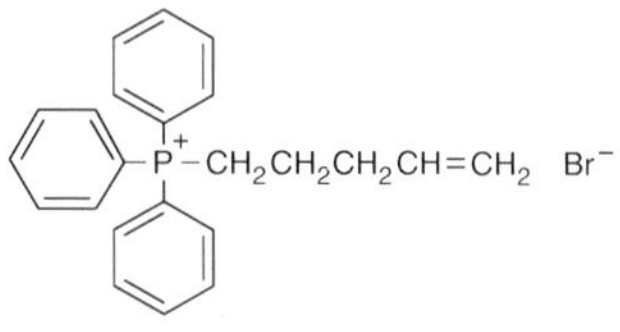

250 mg

S94,726-1
(ISOAMYL)-TRIPHENYLPHOSPHONIUM IODIDE

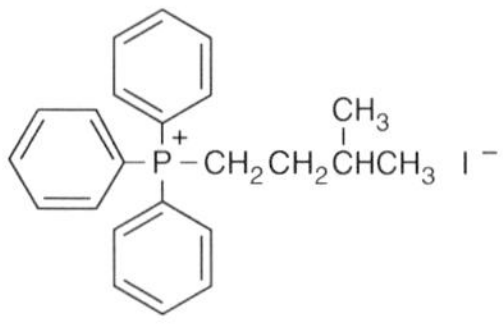

250 mg

S54,335-7
(3-CHLOROBENZYL)-TRIPHENYLPHOSPHONIUM CHLORIDE

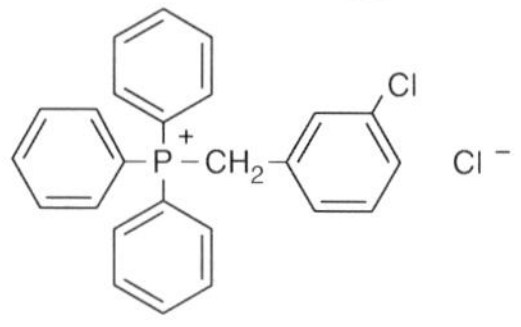

250 mg

S60,814-9
BENZYLTRIPHENYLPHOSPHONIUM BROMIDE
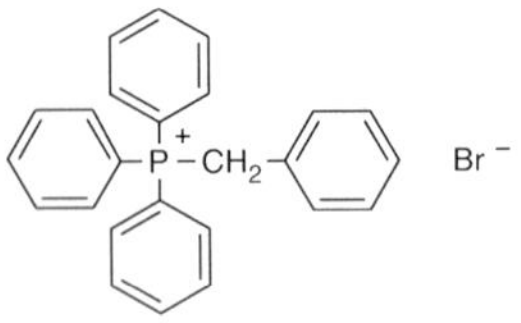
250 mg

S62,677-5
(3-METHYLBENZYL)-TRIPHENYLPHOSPHONIUM CHLORIDE
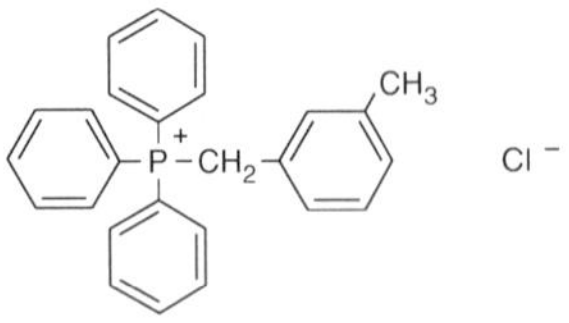
5 gm

S87,112-5
[2-(2,5-CYCLOHEXADIEN-1-YLIDENE)-ETHYL]TRIPHENYLPHOSPHONIUM TETRAFLUOROBORATE
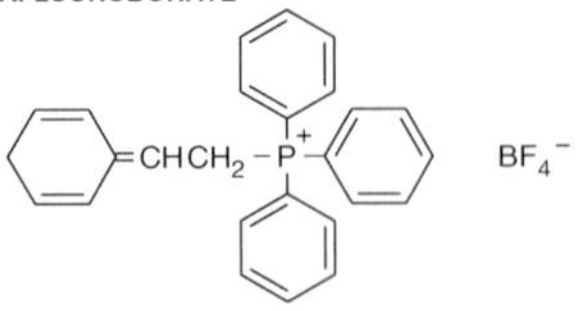
250 mg

S87,111-7
[2-(2,5-CYCLOHEXADIEN-1-YLIDENE)-VINYL]TRIPHENYLPHOSPHONIUM PERCHLORATE
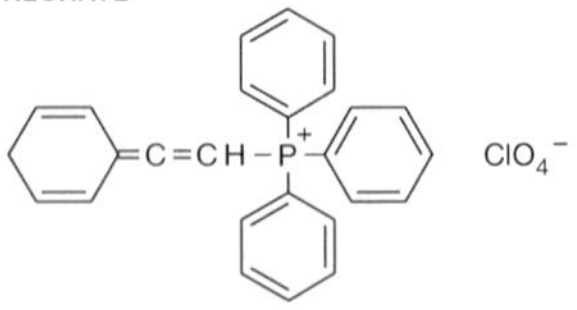
250 mg

S87,113-3
[2-(2,5-CYCLOHEXADIEN-1-YLIDENE)-ETHYL]TRIPHENYLPHOSPHONIUM PERCHLORATE
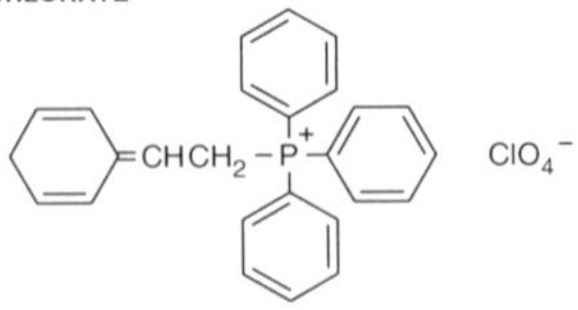
250 mg

S88,516-9
(3-PHENYLPROPYL)-TRIPHENYLPHOSPHONIUM BROMIDE
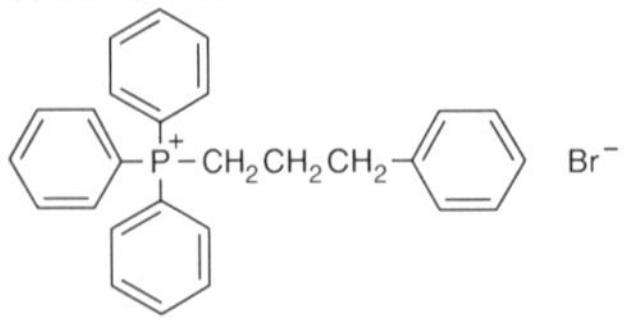
250 mg

S87,116-8
[2-(1-METHYL-2,5-CYCLOHEXADIEN-1-YL)ETHYL]TRIPHENYLPHOSPHONIUM PERCHLORATE
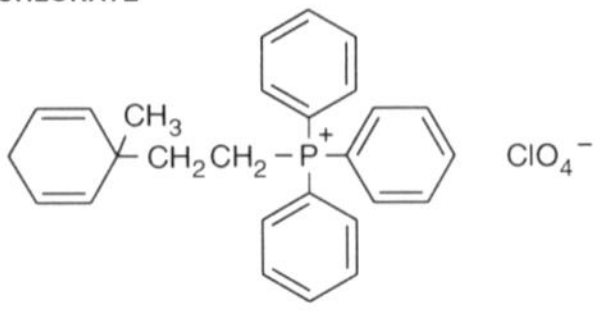
250 mg

S87,119-2
[4-(2,5-CYCLOHEXADIEN-1-YLIDENE)-2-BUTEN-1-YL]TRIPHENYLPHOSPHONIUM TETRAFLUOROBORATE
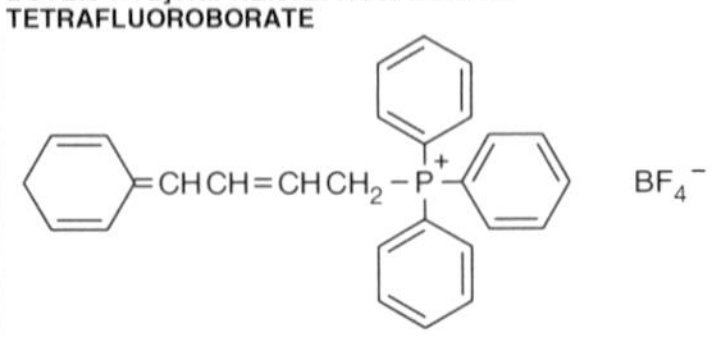
250 mg

S87,120-6
[4-(2,5-CYCLOHEXADIEN-1-YLIDENE)-2-BUTEN-1-YL]-TRIPHENYLPHOSPHONIUM IODIDE
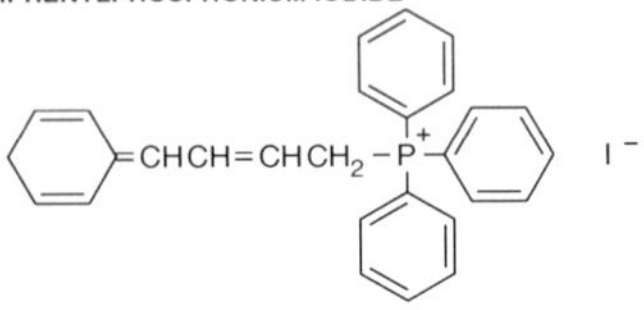
250 mg

S87,121-4
(PENTAMETHYLBENZYL)-TRIPHENYLPHOSPHONIUM CHLORIDE
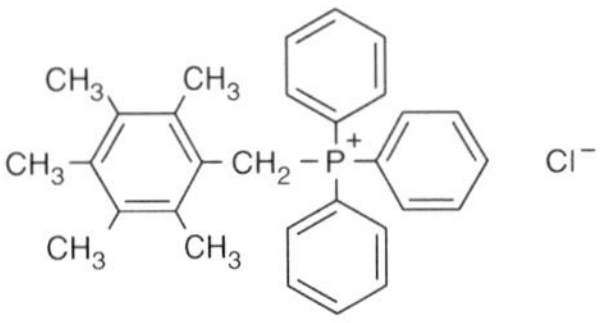
250 mg

S53,682-2
ETHYLENEBIS-(TRIPHENYLPHOSPHONIUM BROMIDE)
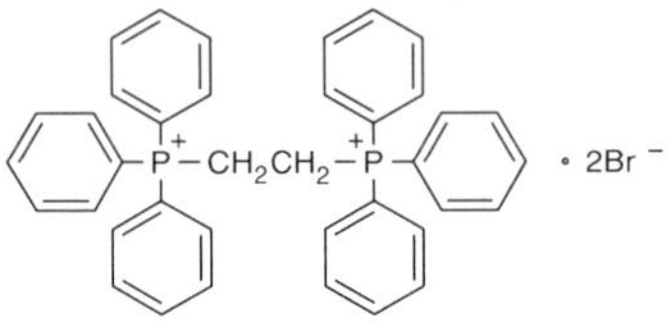
5 gm

S61,882-9
TRANS-2-BUTYLENE-1,4-BIS-(TRIPHENYLPHOSPHONIUM CHLORIDE)
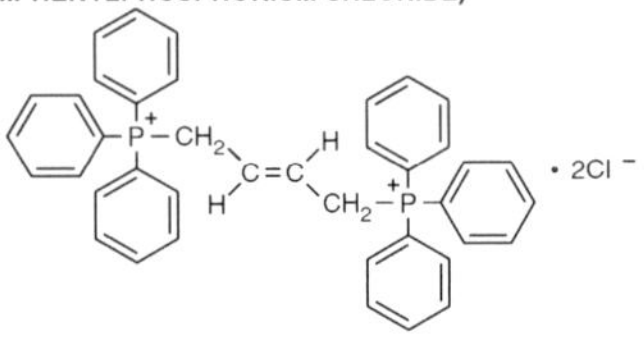
5 gm

S62,676-7
(2-METHYLBENZYL)-TRIPHENYLPHOSPHONIUM CHLORIDE
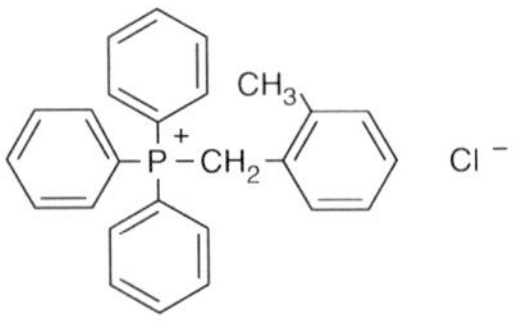
5 gm

S62,669-4
(4-METHYLBENZYL)-TRIPHENYLPHOSPHONIUM CHLORIDE
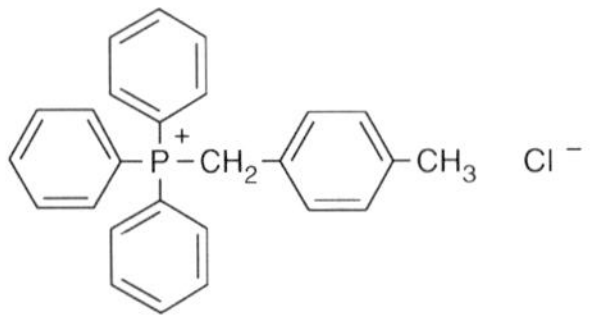
5 gm

S36,345-6
(4-METHYLBENZYL)-TRIPHENYLPHOSPHONIUM BROMIDE
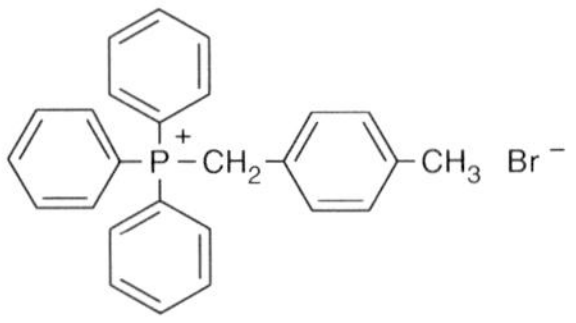
5 gm

S57,222-5
P-XYLYLENEBIS-(TRIPHENYLPHOSPHONIUM CHLORIDE)
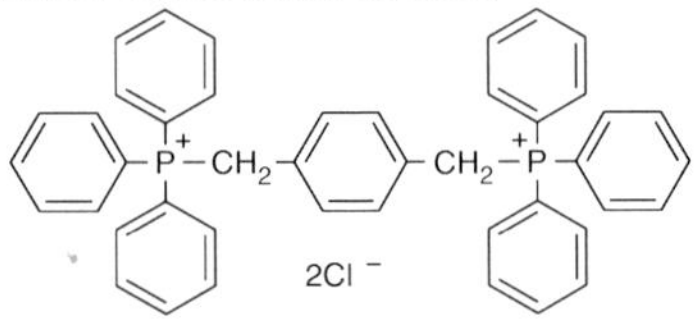
250 mg

S62,009-2
2,4-DICHLOROBENZYL-TRIPHENYLPHOSPHONIUM CHLORIDE
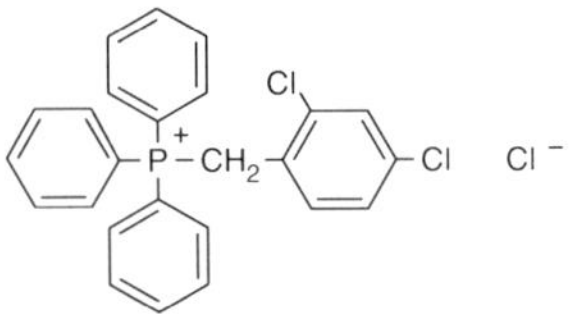
5 gm

S53,820-5
(2,5-DIMETHYL-1,4-XYLYLENE)BIS-(TRIPHENYLPHOSPHONIUM CHLORIDE)
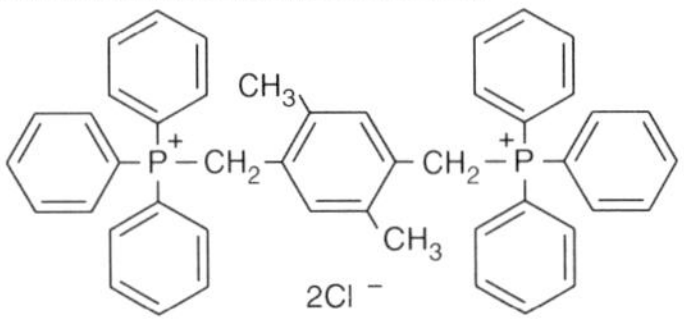
5 gm

S54,350-0
(4,5-DIMETHYL-1,2-XYLYLENE)BIS-(TRIPHENYLPHOSPHONIUM CHLORIDE)
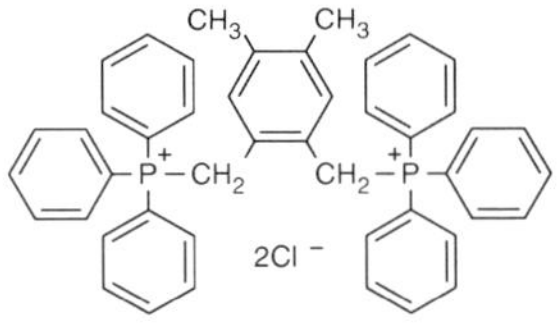
50 mg

S61,880-2
1-NAPHTHYLMETHYL-TRIPHENYLPHOSPHONIUM CHLORIDE
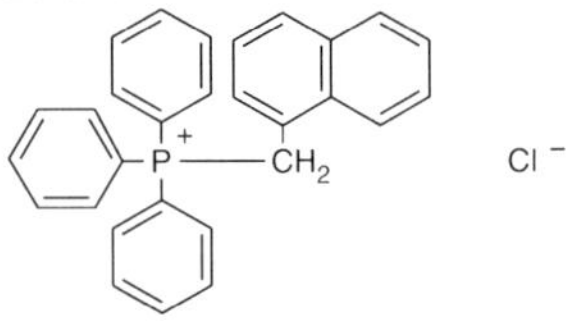
5 gm

S61,883-7
2-NAPHTHYLMETHYL-TRIPHENYLPHOSPHONIUM CHLORIDE
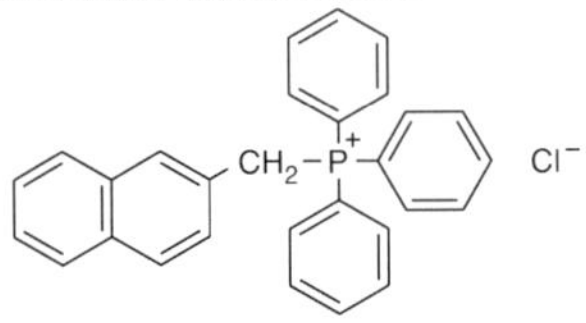
5 gm

S94,773-3
(2-NAPHTHYLMETHYL)-TRIPHENYLPHOSPHONIUM BROMIDE
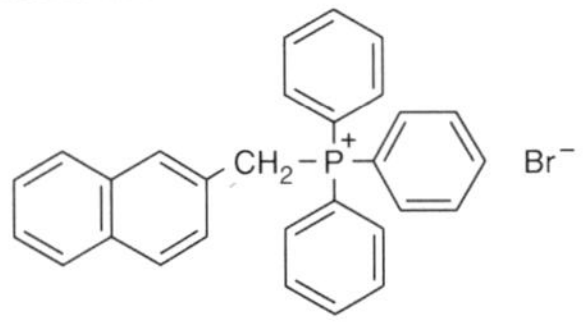
5 gm

S75,503-6
9-FLUORENYLTRIPHENYLPHOSPHONIUM BROMIDE
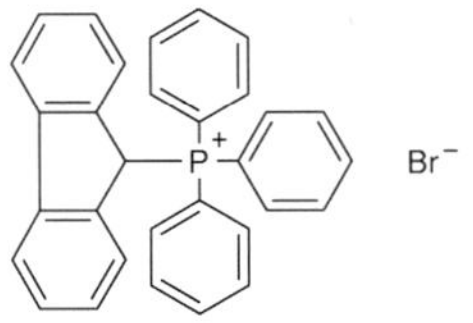
250 mg

S54,343-8
METHYLTRIS(4-FLUOROPHENYL)-PHOSPHONIUM IODIDE
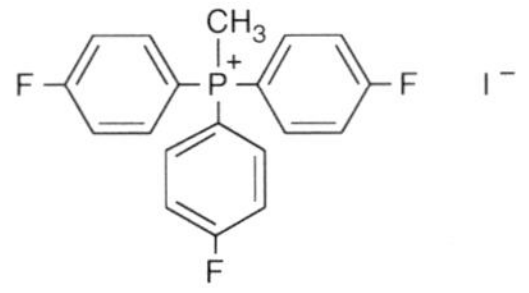
250 mg

S54,321-7
METHYLTRIS(4-CHLOROPHENYL)-PHOSPHONIUM IODIDE
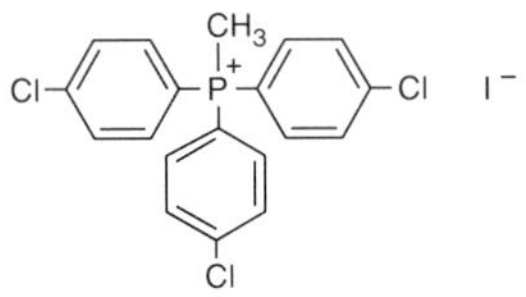
250 mg

S54,994-0
(4-METHYLBENZYL)-TRIBUTYLPHOSPHONIUM CHLORIDE
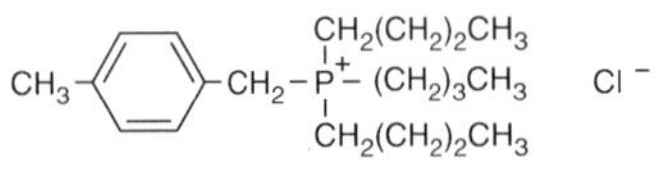
5 gm

S65,640-2
1,1'-BIPHENYL-4,4'-DIYLBIS(METHYL)-BIS(TRIPHENYLPHOSPHONIUM CHLORIDE)
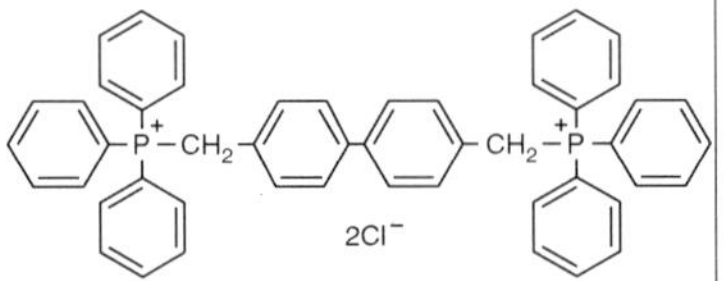
5 gm

S84,258-3
1,1-DIETHYL-1H-PHOSPHINDOLIUM PICRATE
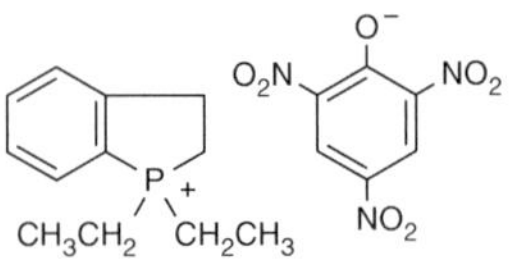
250 mg

S75,418-8
[2-(DIMETHYLAMINO)-9-FLUORENYLIDENE]-TRIPHENYLPHOSPHORANE
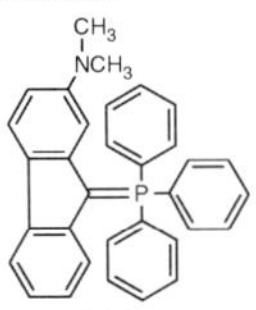
250 mg

S75,152-9
(2-NITRO-9-FLUORENYLIDENE)-TRIPHENYLPHOSPHORANE
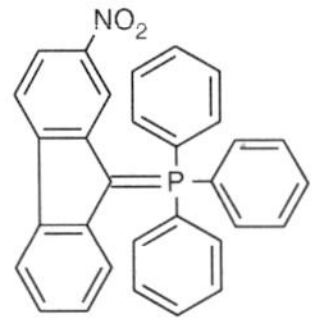
5 gm

S61,650-8
(PHENACYLIDENE)-TRIPHENYLPHOSPHORANE
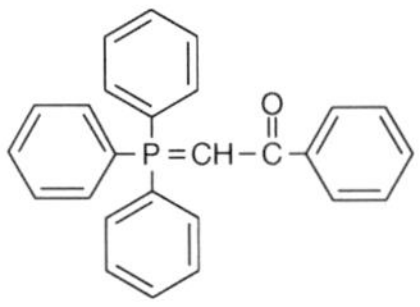
5 gm

S43,757-3
2-(TRIPHENYLPHOSPHORANYLIDENE)-GAMMA-BUTYROLACTONE
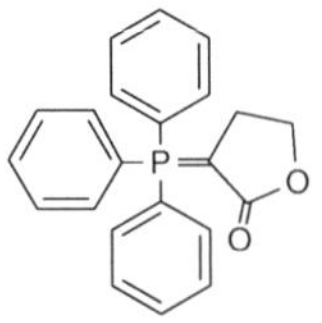
250 mg

S74,907-9
(2,4,7-TRICHLORO-9-FLUORENYLIDENE)-TRIPHENYLPHOSPHORANE
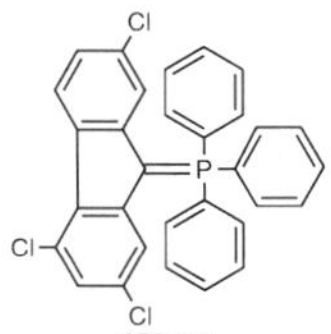
250 mg

S41,352-6
PHENYL ETHYLENE PHOSPHITE
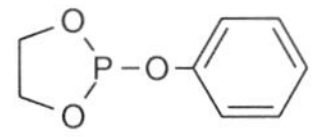
5 gm

S40,583-3
2,2-DIMETHYLTRIMETHYLENE PHENYL PHOSPHITE
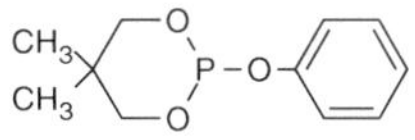
5 gm

S42,004-2
4-DIMETHYLAMINO-META-TOLYLPHOSPHINIC ACID, MONOSODIUM SALT
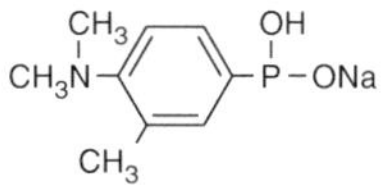
250 mg

S61,719-9
DIISOPROPYL PHENYLPHOSPHONITE
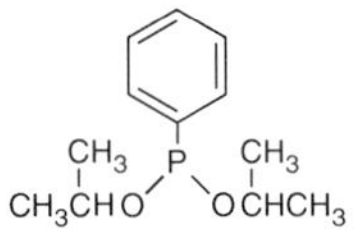
250 mg

S43,493-0
BUTYL DIPHENYLPHOSPHINITE
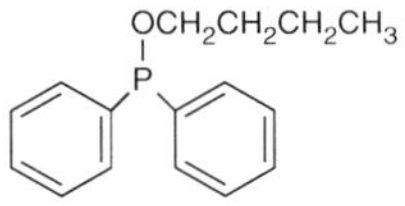
5 gm

S62,641-4
1,2-PHENYLENE PHOSPHOROBROMIDITE
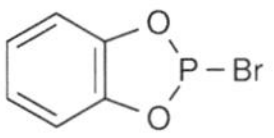
5 gm

S91,575-0
(M-TOLYL)PHOSPHONIC ACID
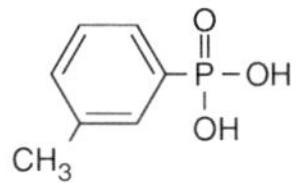
250 mg

S91,453-3
(DICHLOROMETHYL)PHOSPHONIC DICHLORIDE
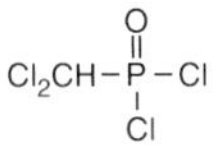
250 mg

S77,346-8
O,O'-BIS(4-TERT-BUTYLPHENYL)-[3-(TRIFLUOROMETHYL)PHENYLTHIOUREIDO]-THIOPHOSPHORAMIDATE
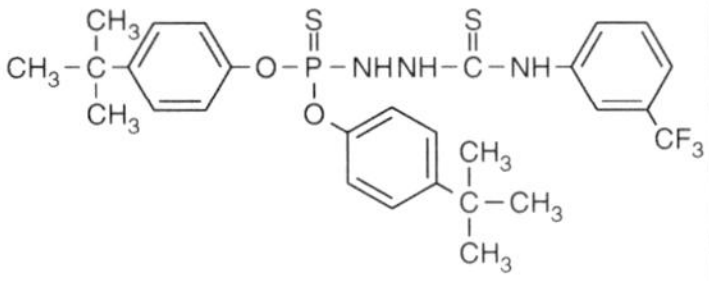
250 mg

S91,564-5
(3-FLUOROPHENYL)PHOSPHONIC ACID
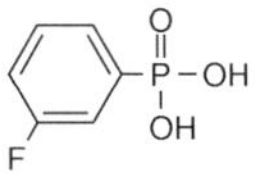
250 mg

S91,597-1
(3-CHLOROPHENYL)PHOSPHONIC ACID
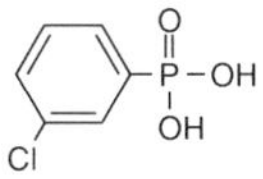
250 mg

S53,038-7
PENTACHLOROPHENYL TETRAMETHYLPHOSPHORODIAMIDATE
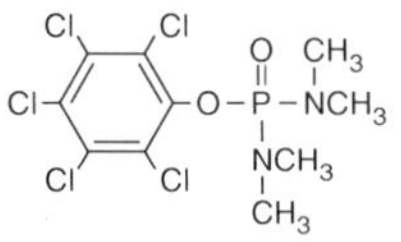
250 mg

S67,344-7
BIS(4-T-BUTYL-PHENYL) N-(5-CHLORO-2,4-DIMETHOXY-PHENYL)-2-THIOUREIDOTHIOPHOSPHORAMIDATE
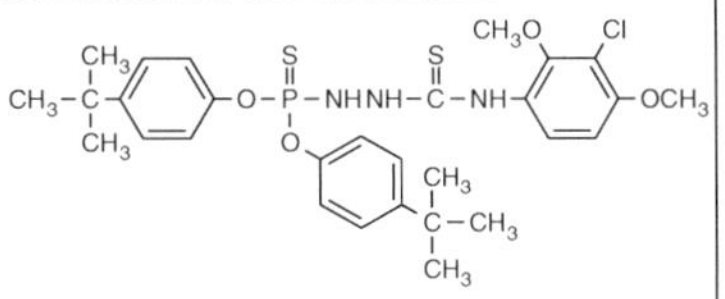
5 gm

S77,995-4
O,O'-BIS(4-TERT-BUTYLPHENYL)-(2-CHLORO-6-METHOXYPHENYLUREIDO)-THIOPHOSPHORAMIDATE
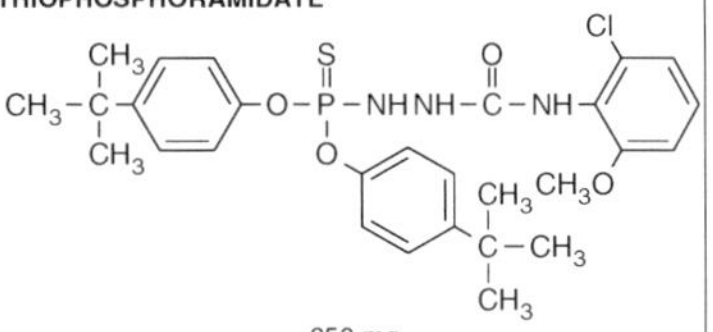
250 mg

S78,645-4
4'-CHLORO-3-HYDROXY-2'-METHYL-2-NAPHTHANILIDE MONOPHOSPHATE ESTER
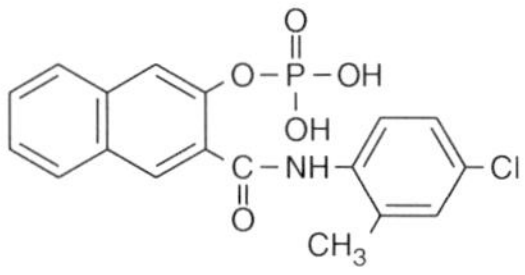
250 mg

S67,343-9
BIS(4-TERT-BUTYLPHENYL) N-(2-CHLORO-4-METHYLTHIO-6-METHYLPHENYLUREYLENE)-THIOPHOSPHORAMIDATE
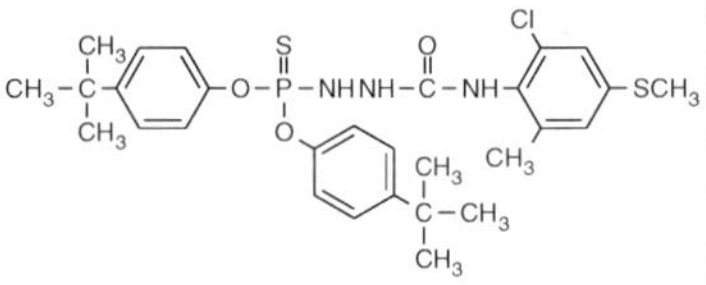
5 gm

S69,203-4
O,O'-BIS(4-TERT-BUTYLPHENYL)-(3,4-DICHLOROPHENYLUREIDO)-THIOPHOSPHORAMIDATE
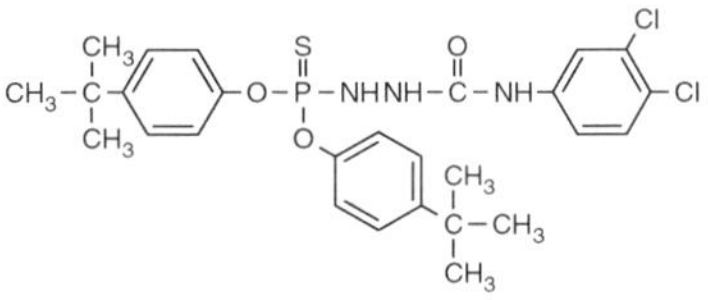
250 mg

S92,874-7
N-[N-(4-CHLOROPHENYL)CARBAMOYL]-N',N',N'',N''-TETRAETHYLPHOSPHORAMIDE
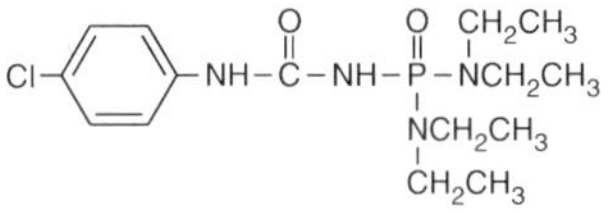
250 mg

S83,120-4
DIPHENYL(PENTACHLOROPHENYL)-PHOSPHINE OXIDE
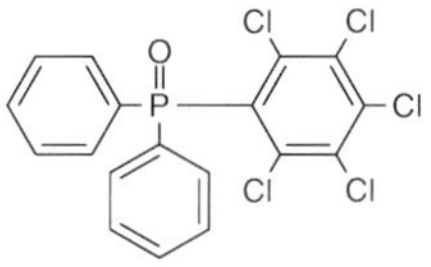
250 mg

S72,706-7
(4-CHLOROPHENYL)-PHENYLPHOSPHINIC ACID
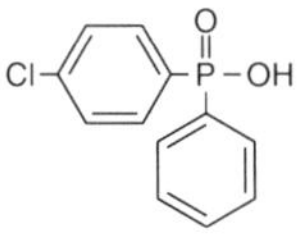
5 gm

S91,582-3
(2-FLUOROPHENYL)PHOSPHONIC ACID
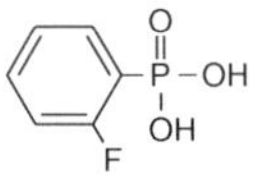
250 mg

S91,576-9
(2-CHLOROPHENYL)PHOSPHONIC ACID
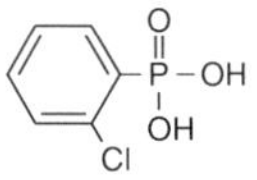
250 mg

S91,566-1
(2-BROMOPHENYL)PHOSPHONIC ACID
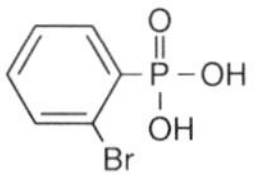
250 mg

S91,579-3
(2-IODOPHENYL)PHOSPHONIC ACID
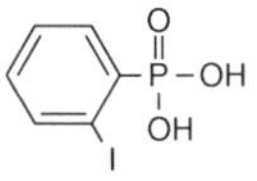
250 mg

S65,947-9
TRIS(2-CHLOROPHENYL) PHOSPHATE
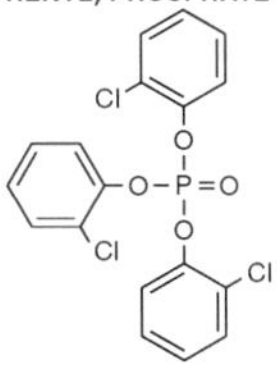
5 gm

S85,919-2
(4-CHLOROPHENYL)PHOSPHONIC ACID
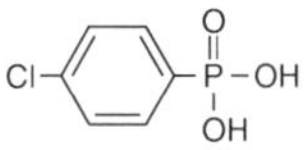
250 mg

S91,587-4
(4-BROMOPHENYL)PHOSPHONIC ACID
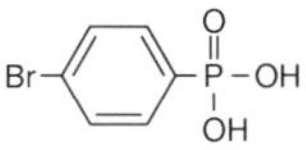
250 mg

S45,685-3
DIETHYL 4-FLUOROPHENYLAMIDOPHOSPHATE
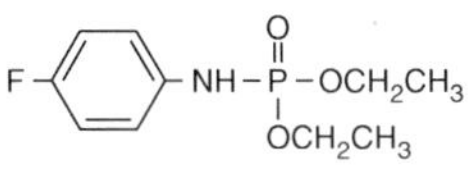
5 gm

S55,770-6
(DIPHENYLPHOSPHINYL)METHANOL
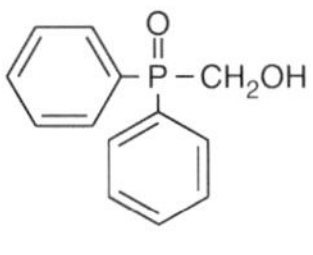
250 mg

S86,029-8
TRANS-1,2-CYCLOHEXANEDIOL MONO(DIBENZYL PHOSPHATE)
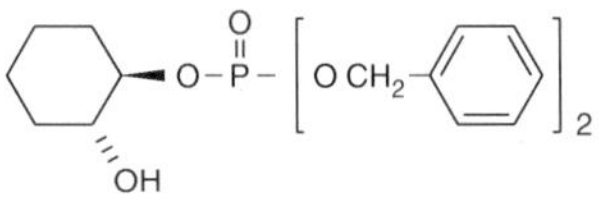
250 mg

S35,992-0
ALPHA-HYDROXYBENZYLPHOSPHINIC ACID, SODIUM SALT
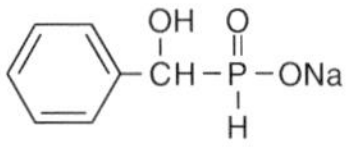
5 gm

S60,260-4
ALPHA-HYDROXY-ALPHA-PHENYLMETHANEPHOSPHONIC ACID DISODIUM SALT
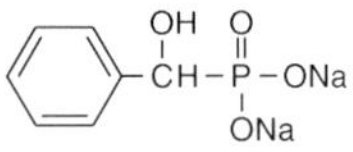
5 gm

S41,265-1
DIETHYL 4-[BIS(2-HYDROXYETHYL)-AMINO]BENZYLPHOSPHONATE
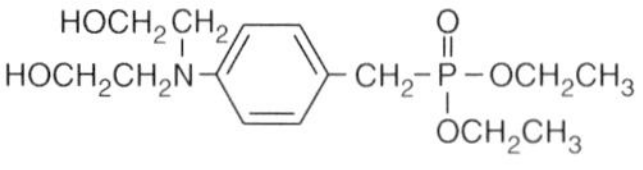
5 gm

S67,345-5
BIS(4-TERT-BUTYLPHENYL) N-(2-HYDROXYBENZYLIDENEAMINO)-THIOPHOSPHORAMIDATE
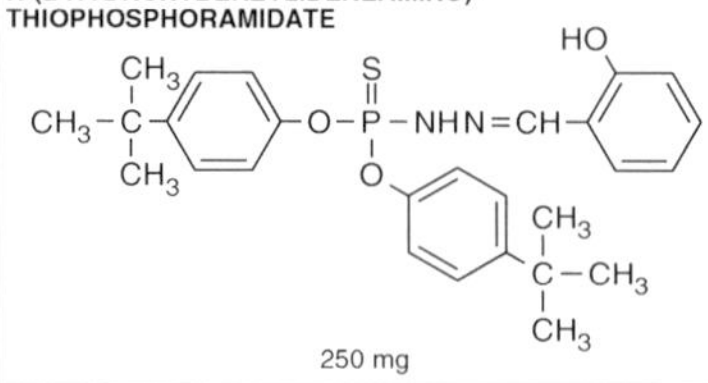
250 mg

S57,626-3
ALPHA-HYDROXYBENZYLPHOSPHINIC ACID
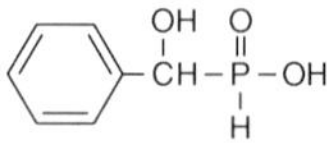
5 gm

S55,778-1
ALPHA-(DIPHENYLPHOSPHINYL)ETHANOL
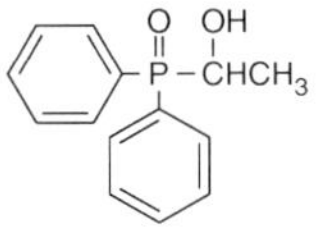
5 gm

S91,585-8
(3-METHOXYPHENYL)PHOSPHONIC ACID
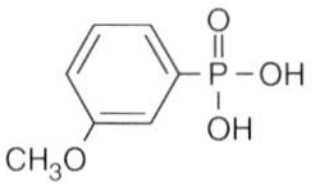
250 mg

S67,399-4
BIS(4-TERT-BUTYLPHENYL) N-(2,3-DIMETHOXYBENZYLIDENEAMINO)-THIOPHOSPHORAMIDATE
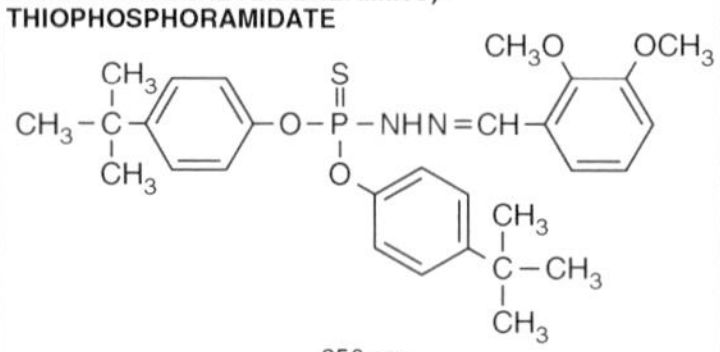
250 mg

S69,390-1
O,O'-BIS(4-TERT-BUTYLPHENYL)-(2-METHOXY-4-NITROPHENYLUREIDO)-THIOPHOSPHORAMIDATE
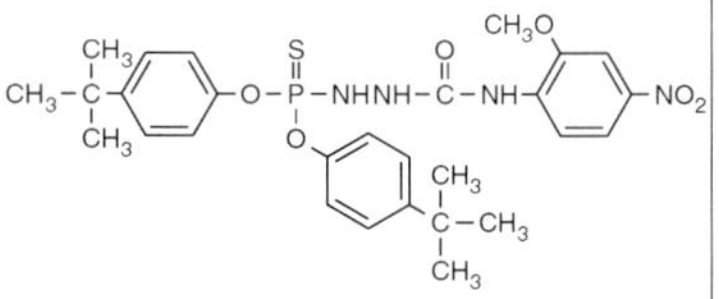
250 mg

S69,292-1
O,O'-BIS(4-TERT-BUTYLPHENYL)-(2-METHOXY-5-METHYLPHENYLUREIDO)-THIOPHOSPHORAMIDATE
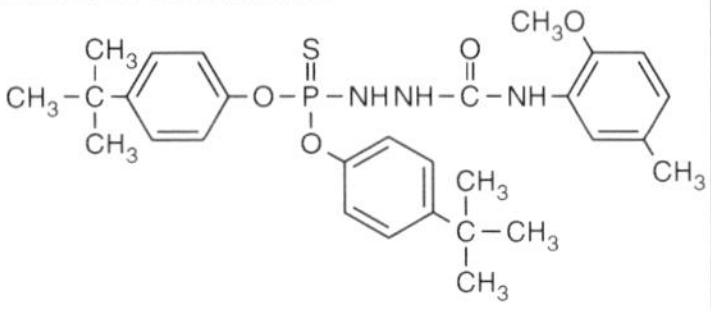
250 mg

S74,198-1
3-[(4-METHOXYPHENYL)-PHENYLPHOSPHINYL]PROPIONIC ACID
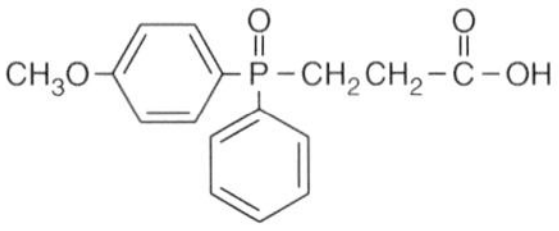
250 mg

S72,707-5
(3-METHOXYPHENYL)-PHENYLPHOSPHINIC ACID
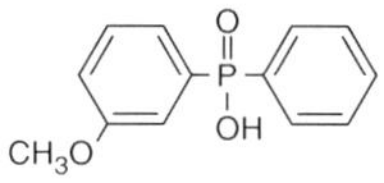
5 gm

S74,238-4
(2-METHOXYPHENYL)-PHENYLPHOSPHINIC ACID
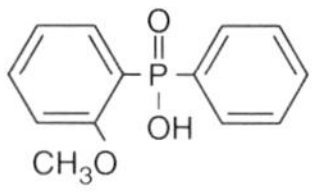
5 gm

S72,705-9
(4-METHOXYPHENYL)-PHENYLPHOSPHINIC ACID
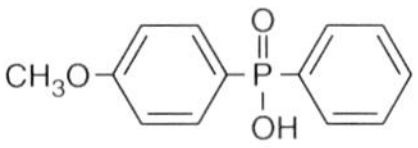
5 gm

S72,729-6
(4-ETHOXYPHENYL)-PHENYLPHOSPHINIC ACID
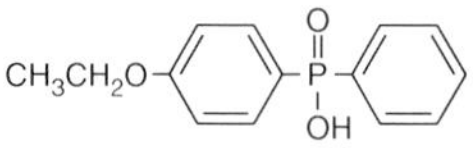
5 gm

S91,568-8
(2-METHOXYPHENYL)PHOSPHONIC ACID
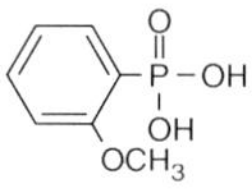
250 mg

S91,586-6
(2-ETHOXYPHENYL)PHOSPHONIC ACID
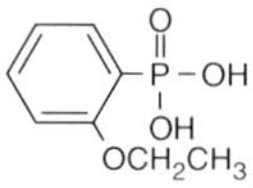
250 mg

S54,582-1
4-METHOXYPHENYLPHOSPHONIC ACID
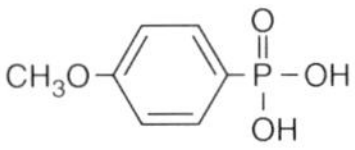
250 mg

S55,830-3
4-ETHOXYPHENYLPHOSPHONIC ACID
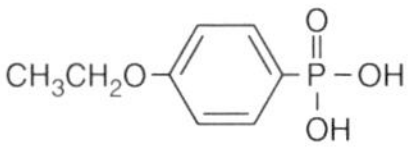
5 gm

S67,395-1
BIS(4-TERT-BUTYLPHENYL) N-[3-(3-METHOXYPROPYL)-2-THIOUREIDO]-THIOPHOSPHORAMIDATE
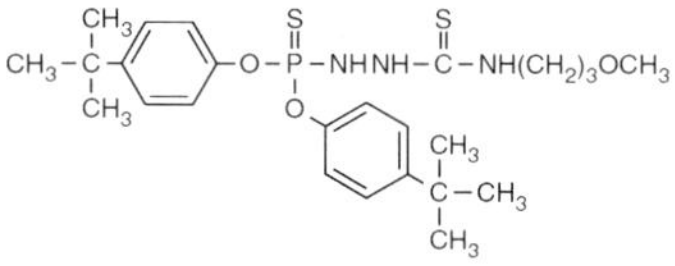
250 mg

S95,029-7
3-(BENZYLOXY)PROPYL ETHYL PHOSPHORAMIDATE
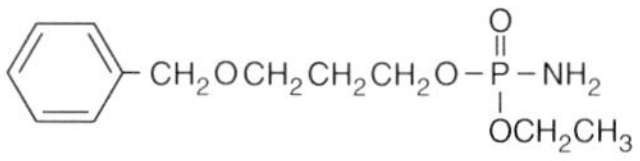
50 mg

S85,937-0
2-AMINOETHYL BENZYL PHOSPHATE
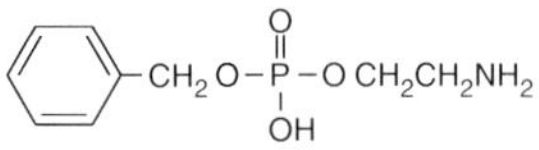
250 mg

S39,901-9
DIBENZYL (ALPHA-ANILINOBENZYL)-PHOSPHONATE
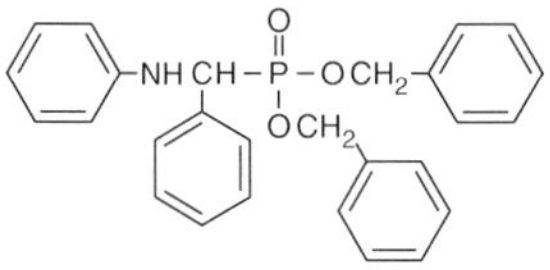
5 gm

S85,973-7
DIBENZYL PHOSPHORAMIDATE
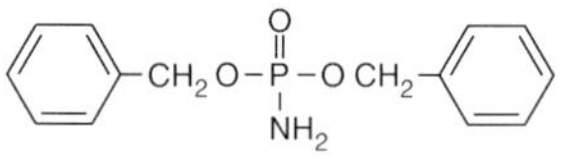
250 mg

S86,025-5
(+/-)-1,2-CYCLOHEXANEDIOL BIS(DIPHENYL PHOSPHATE)
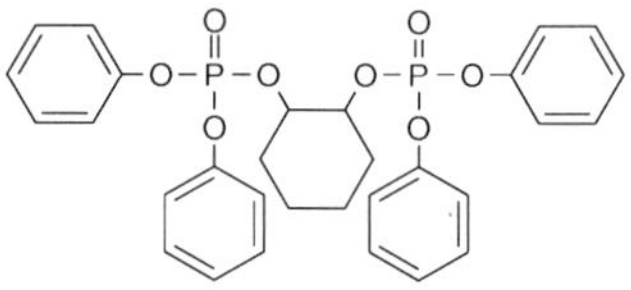
250 mg

S86,130-8
5-ETHYL-5-METHYL-2-PHENYL-1,3,2-DIOXAPHOSPHORINANE OXIDE
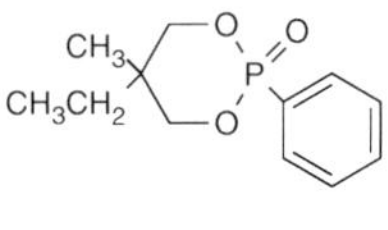
250 mg

S86,026-3
TRANS-1,4-CYCLOHEXANEDIOL BIS(DIPHENYL PHOSPHATE)
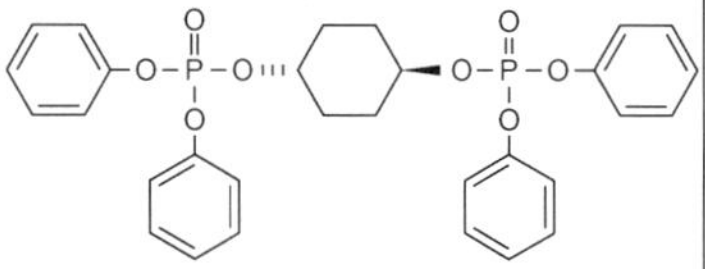
250 mg

S67,423-0
BIS(4-TERT-BUTYLPHENYL) N-[4-(METHYLTHIO)PHENYLUREIDO]-THIOPHOSPHORAMIDATE
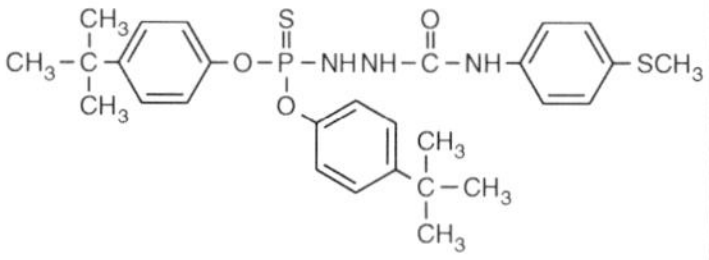
250 mg

S54,989-4
TRIS(3-AMINOPHENYL)PHOSPHINE OXIDE
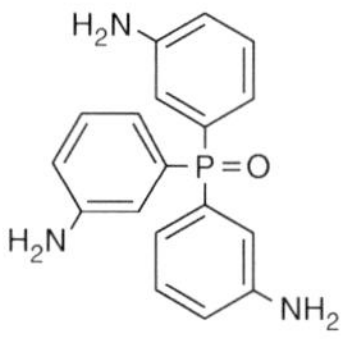
250 mg

S50,728-8
DIPHENYL (ALPHA-ANILINO-2-PYRIDYLMETHYL)PHOSPHONATE
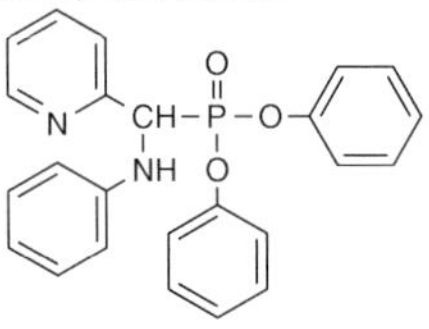
5 gm

S51,056-4
1-[2-(N-ETHYLANILINO)ETHYLAMINO]-ISOPROPYLPHOSPHINIC ACID
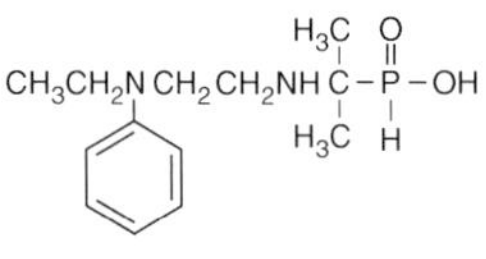
250 mg

S73,304-0
DIMORPHOLINOPHENETHYLAMINO PHOSPHINE OXIDE
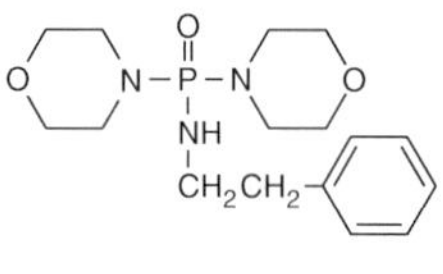
250 mg

S50,092-5
DIPHENYL(ALPHA-ANILINOBENZYL)-PHOSPHONATE
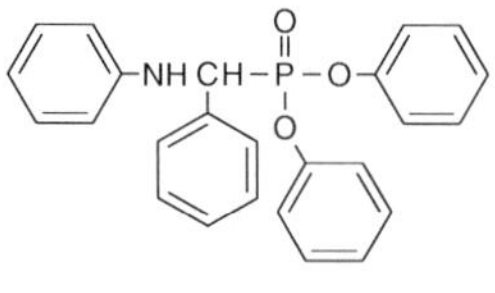
5 gm

S53,984-8
TRIS[4-(DIMETHYLAMINO)PHENYL]-PHOSPHINE OXIDE

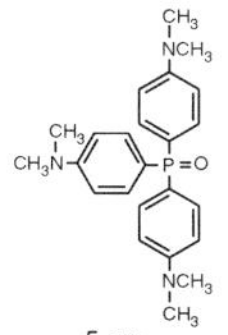

5 gm

S86,173-1
DIISOPROPYL (N,N-DIBENZYLCARBAMOYLMETHYL)-PHOSPHONATE

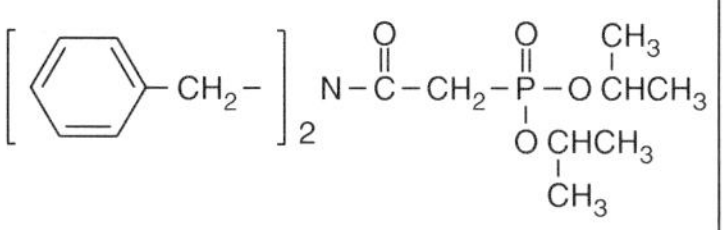

250 mg

S94,978-7
DIPHENYL N-BENZYLPHOSPHORAMIDATE

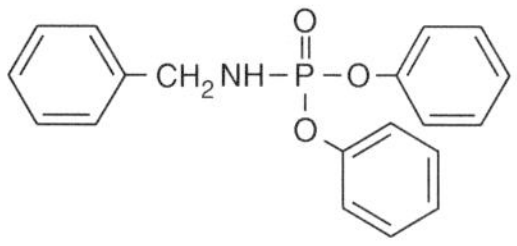

250 mg

S43,198-2
DIETHYL DIPHENYLMETHYLAMIDOPHOSPHATE

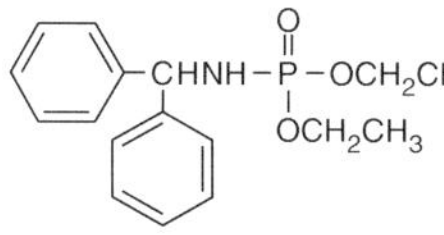

5 gm

S69,207-7
O,O'-BIS(4-TERT-BUTYLPHENYL)-(PIPERONYLIDENEAMINO)-THIOPHOSPHORAMIDATE

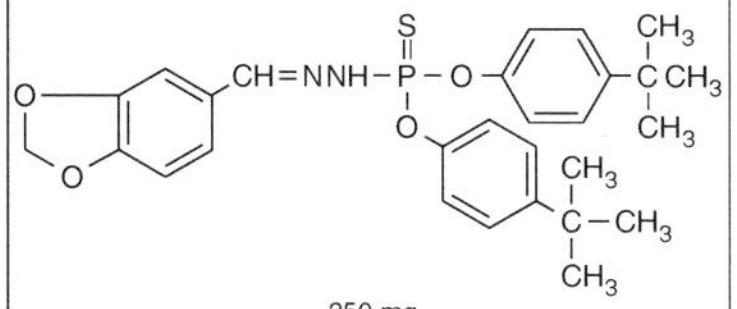

250 mg

S91,573-4
(3-NITROPHENYL)PHOSPHONIC ACID

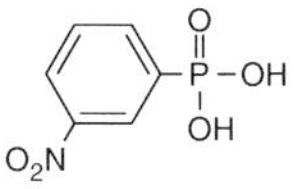

250 mg

S91,572-6
(2-NITROPHENYL)PHOSPHONIC ACID

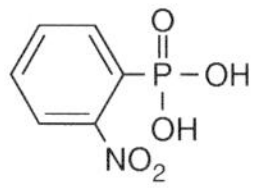

250 mg

S91,574-2
(4-NITROPHENYL)PHOSPHONIC ACID

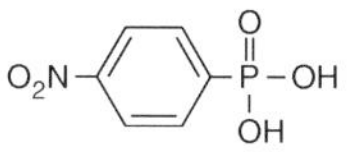

250 mg

S91,580-7
(4-NITROBENZYL)PHOSPHONIC ACID

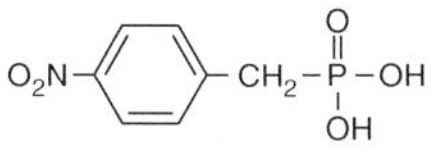

250 mg

S67,451-6
TRIS(5-METHYL-2-NITROPHENYL)-PHOSPHATE

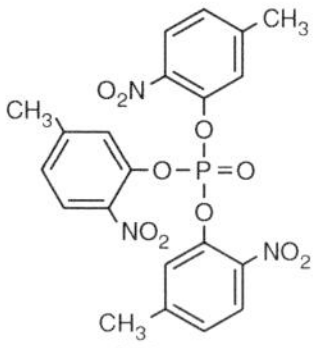

250 mg

S89,797-3
TRIS(3-METHYL-4-NITROPHENYL)-PHOSPHATE

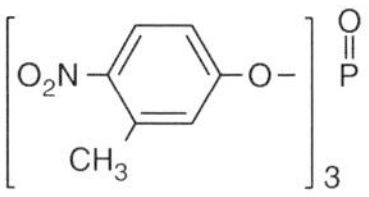

250 mg

S55,598-3
BIS(3-NITROPHENYL)-METHYLPHOSPHINE OXIDE

5 gm

S55,595-9
BIS(3-NITROPHENYL)-BUTYLPHOSPHINE OXIDE

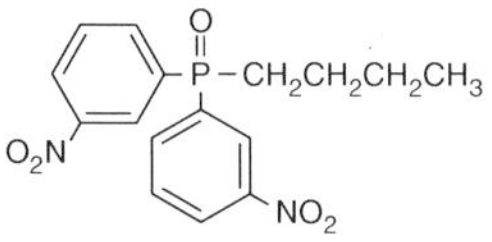

5 gm

S54,589-9
TRIS(3-NITROPHENYL)PHOSPHINE OXIDE

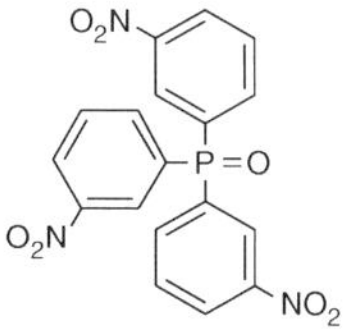

250 mg

S54,353-5
N-(3-NITROPHENYL)-TRIPHENYLPHOSPHINE IMIDE

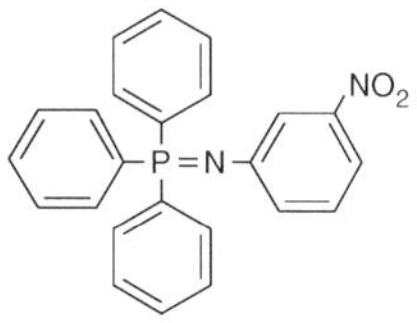

250 mg

S54,354-3
N-(2-NITROPHENYL)-TRIPHENYLPHOSPHINE IMIDE

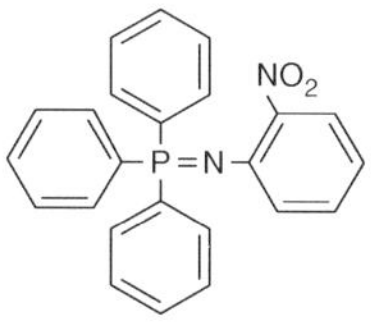

250 mg

S54,352-7
N-(4-NITROPHENYL)-TRIPHENYLPHOSPHINE IMIDE

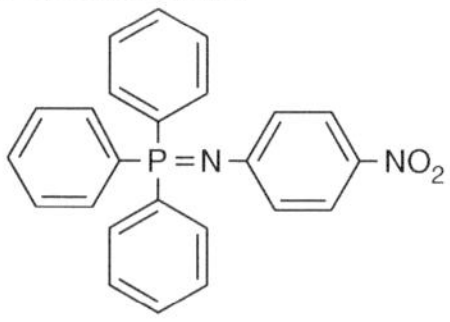

250 mg

S91,569-6
(3-CARBOXYPHENYL)PHOSPHONIC ACID

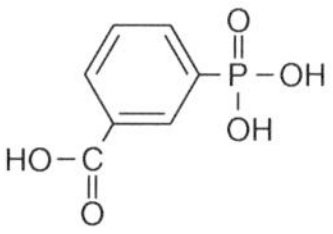

250 mg

S42,797-7
PHOSPHOENOLPYRUVIC ACID, SILVER BARIUM SALT, DIHYDRATE

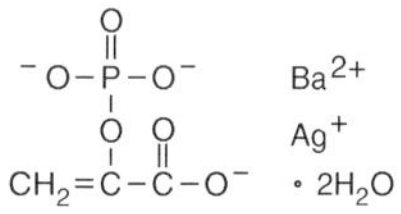

5 gm

S91,590-4
4-PHOSPHONOBENZOIC ACID

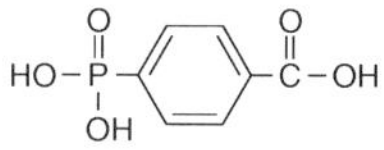

250 mg

S93,147-0
DIMETHYL N-BENZOYLPHOSPHORAMIDATE

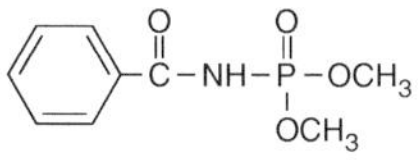

250 mg

S74,988-5
DIETHYL (2-ACETAMIDO-9-FLUORENYL)-PHOSPHONATE

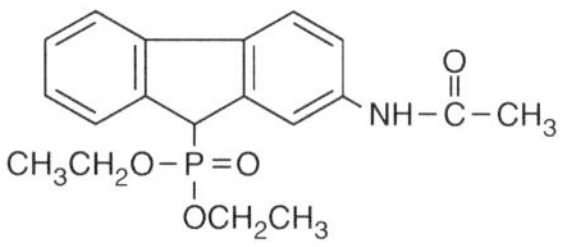

250 mg

S86,179-0
ETHYL (N,N-DIETHYLCARBAMOYLMETHYL)-PHENYLPHOSPHINATE

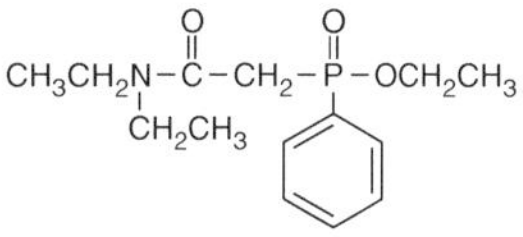

250 mg

S85,948-6
DIPHENYL N-[N-(PHENYL)-THIOCARBAMOYL]PHOSPHORAMIDATE

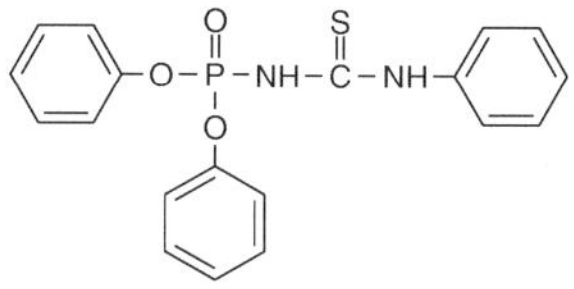

250 mg

S93,854-8
2,3-DIHYDRO-3-(DIPHENYLPHOSPHINYLOXY)THIOPHENE-1,1-DIOXIDE

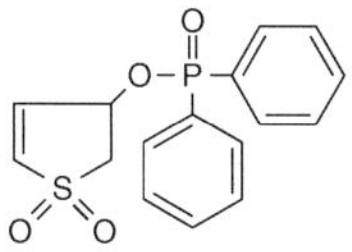

250 mg

S85,950-8
CYCLOHEXYL DIPHENYL PHOSPHATE

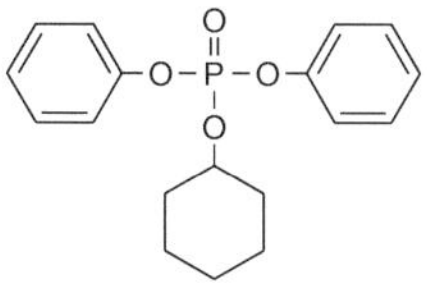

250 mg

S86,704-7
1-CYCLOHEXEN-1-YL DIPHENYL PHOSPHATE

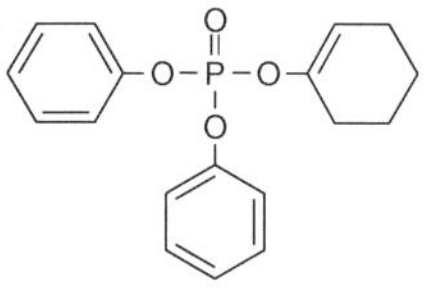

250 mg

S36,449-5
PHENYLPHOSPHINIC ACID, SODIUM SALT

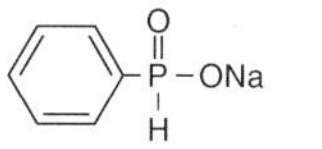

5 gm

S82,835-1
PHENYLPHOSPHONIC ACID, DITHALLIUM SALT

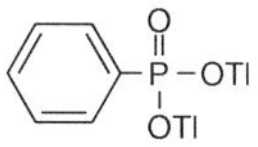

250 mg

S50,163-8
PHENYL PHOSPHORODICHLORIDOTHIONATE

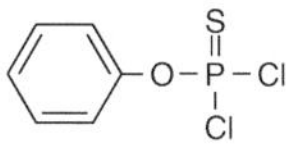

5 gm

S43,501-5
BENZYL PHOSPHITE, AMMONIUM SALT

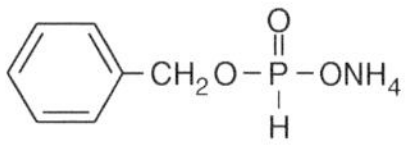

5 gm

S44,313-1
DIMETHYL BENZYLPHOSPHONATE

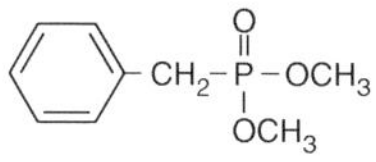

5 gm

S54,333-0
DIMETHYLPHENYLPHOSPHINE SULFIDE METHIODIDE
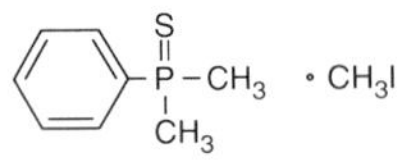
250 mg

S41,227-9
DIETHYL PHENYLAMIDOPHOSPHATE
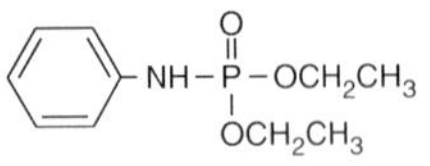
5 gm

S86,128-6
DIISOPROPYL PHENYLPHOSPHONATE
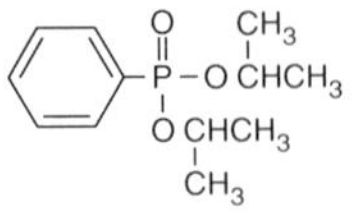
250 mg

S44,407-3
DIPHENYLPHOSPHINYL AZIDE
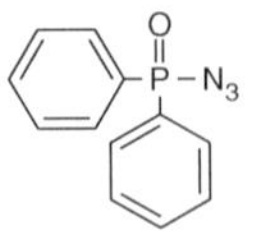
250 mg

S50,571-4
DIPHENYL AMINOPHOSPHORAMIDATE
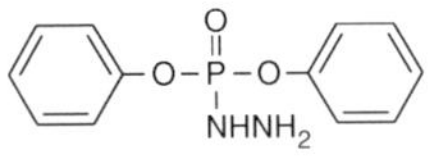
250 mg

S59,514-4
DI-SEC-BUTYL PHENYLPHOSPHONATE
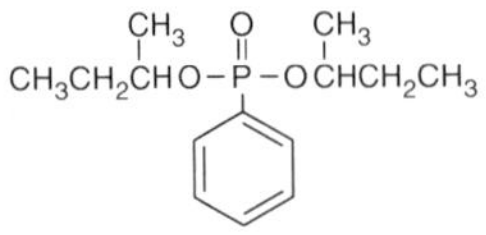
5 gm

S54,329-2
1,2-DIMETHYL-1,2-DIPHENYLDIPHOSPHINE DISULFIDE
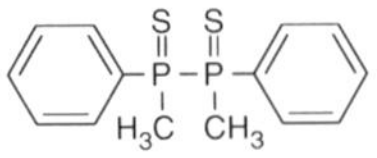
250 mg

S86,194-4
BIS(1-METHYLBUTYL)-PHENYLPHOSPHONATE
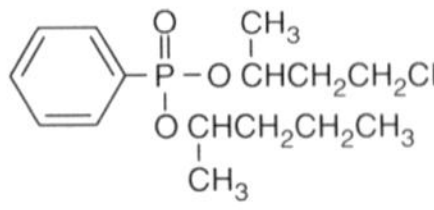
250 mg

S79,811-8
DIPHENYL N-PHENYLPHOSPHORAMIDATE
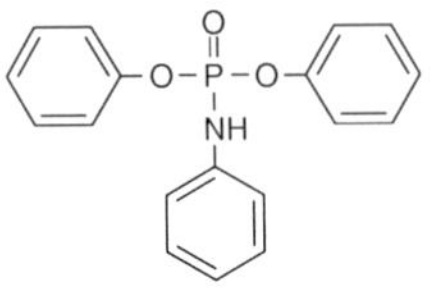
250 mg

S85,978-8
DIPHENYL N-(N-METHYLANILINO)-PHOSPHORAMIDATE
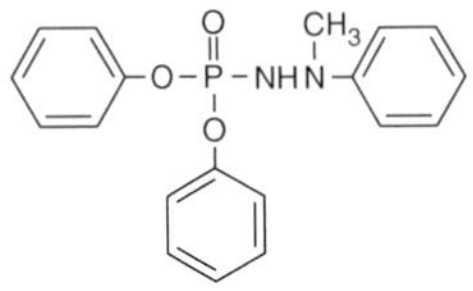
250 mg

S94,759-8
DIPHENYL ALPHA-CHLOROBENZYLPHOSPHONATE
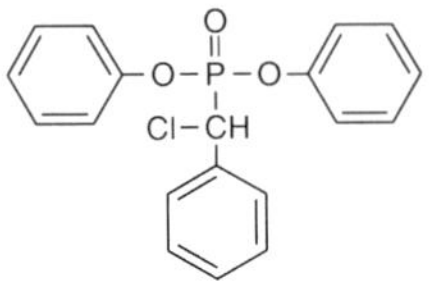
250 mg

S53,640-7
TRIBENZYLPHOSPHINE OXIDE
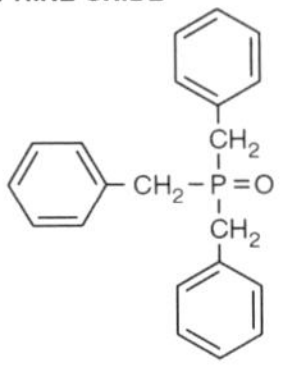
5 gm

S54,357-8
N-(3-TOLYL)TRIPHENYLPHOSPHINE IMIDE
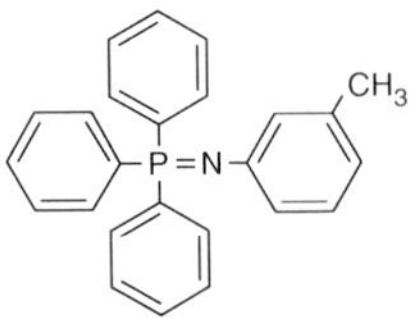
250 mg

S81,035-5
TRIS(3-PHENYLPROPYL)PHOSPHINE OXIDE
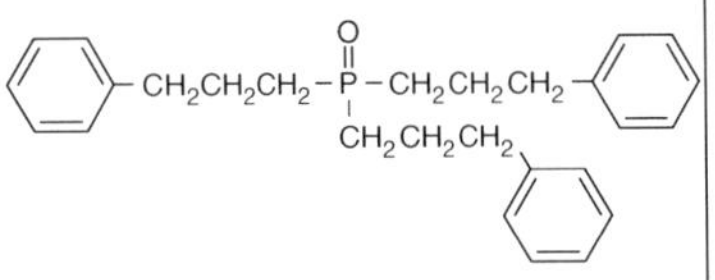
250 mg

S55,101-5
DIPHENYL (TRIPHENYLPHOSPHORANYLIDENEMETHYL)-PHOSPHONATE
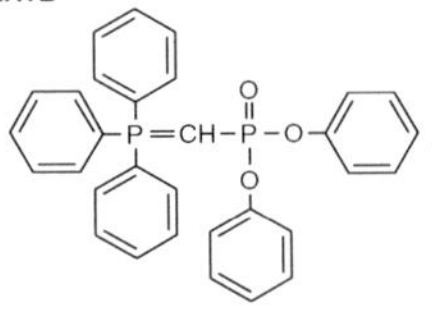
5 gm

S54,358-6
N-(2-TOLYL)TRIPHENYLPHOSPHINE IMIDE
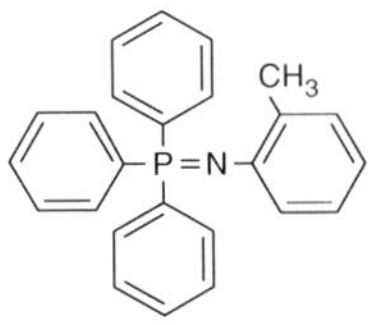
250 mg

S54,359-4
N-(4-CHLOROPHENYL)-TRIPHENYLPHOSPHINE IMIDE
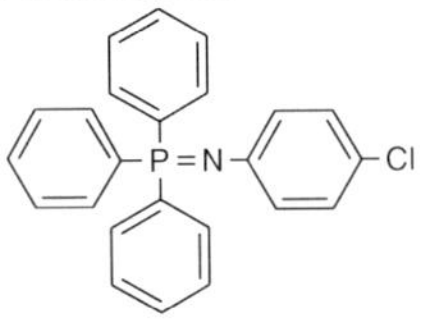
250 mg

S54,355-1
N-(P-TOLYL)TRIPHENYLPHOSPHINE IMIDE
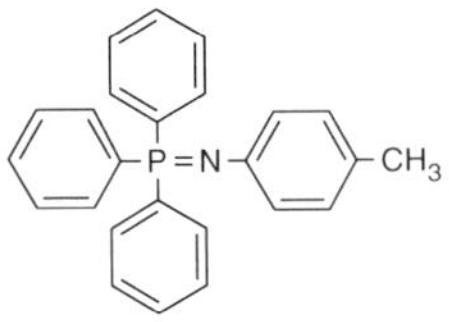
250 mg

S74,600-2
PHENYL(3,5-XYLYL)PHOSPHINIC ACID
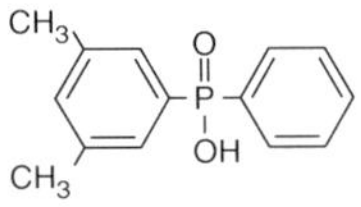
5 gm

S74,174-4
ETHYL PHENYL(3,5-XYLYL)PHOSPHINATE
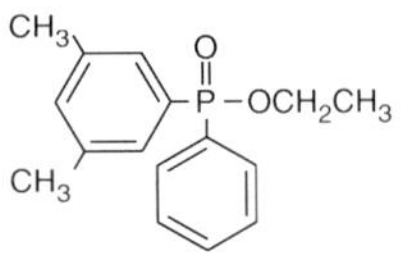
250 mg

S91,581-5
(O-TOLYL)PHOSPHONIC ACID
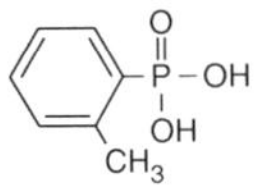
250 mg

S91,565-3
(2-ETHYLPHENYL)PHOSPHONIC ACID
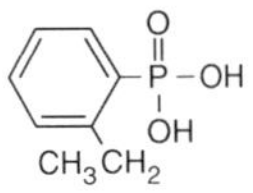
250 mg

S91,577-7
(2-ISOPROPYLPHENYL)PHOSPHONIC ACID
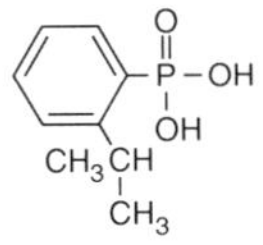
250 mg

S40,832-8
DI-ORTHO-TOLYL CHLOROPHOSPHATE
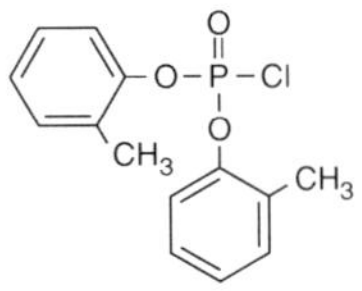
5 gm

S34,215-7
2,2'-DICHLOROTRIPHENYL PHOSPHATE
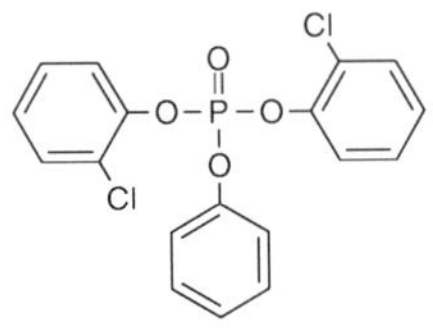
5 gm

S50,161-1
P-TOLYL PHOSPHORODICHLORIDOTHIONATE
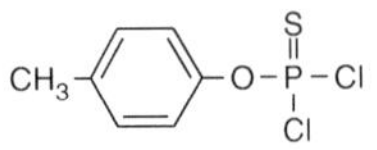
5 gm

S91,595-5
(4-ETHYLPHENYL)PHOSPHONIC ACID
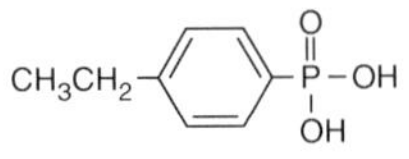
250 mg

S56,624-1
4-(TERT-PENTYL)PHENYL PHOSPHATE
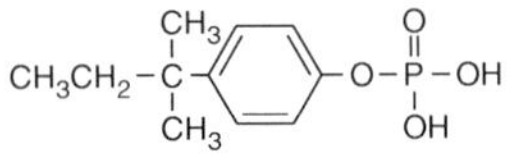
5 gm

S43,977-0
DI-P-TOLYL CHLOROPHOSPHATE
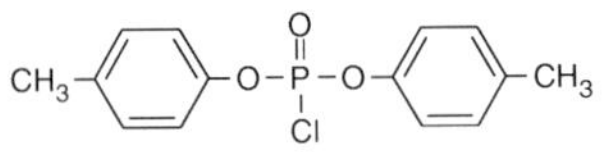
5 gm

S54,966-5
TRI-P-TOLYLPHOSPHINE SULFIDE
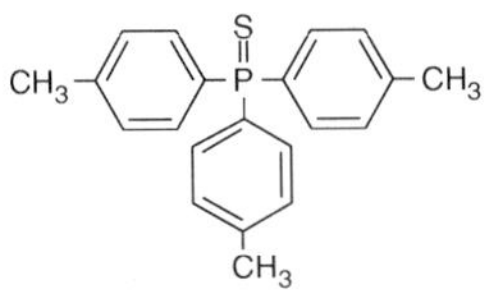
250 mg

S35,633-6
TRIS(4-TERT-BUTYLPHENYL) PHOSPHATE
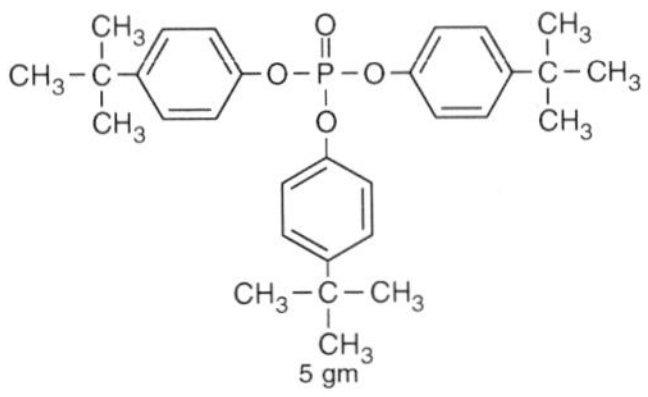
5 gm

S36,689-7
4-ETHYL-3',3'',5',5''-TETRAMETHYLTRIPHENYL PHOSPHATE
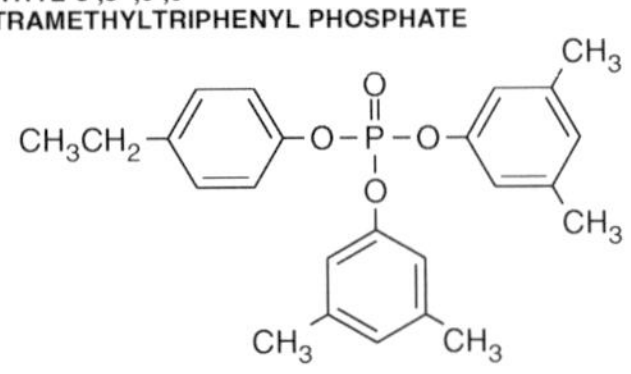
5 gm

S91,578-5
(2,3-XYLYL)PHOSPHONIC ACID
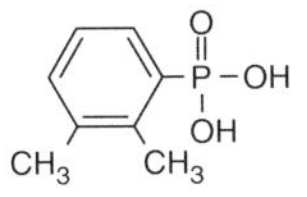
250 mg

S91,583-1
(2,6-XYLYL)PHOSPHONIC ACID
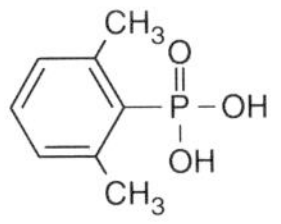
250 mg

S55,679-3
TRIS(2,6-DICHLOROPHENYL)PHOSPHATE
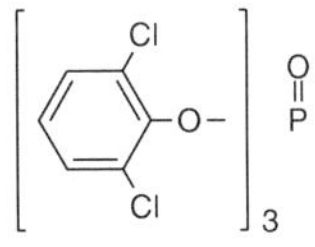
5 gm

S36,534-3
TRIS(2,3-DIMETHYLPHENYL)PHOSPHATE
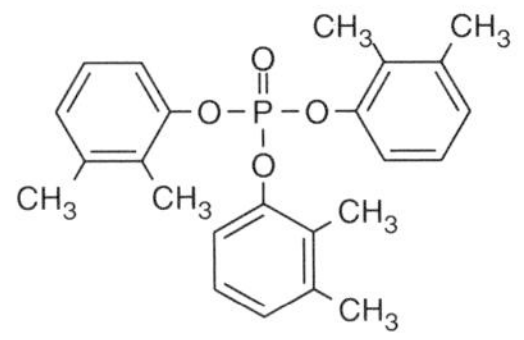
5 gm

S37,376-1
TRIS-(2,6-DIMETHYLPHENYL)PHOSPHATE
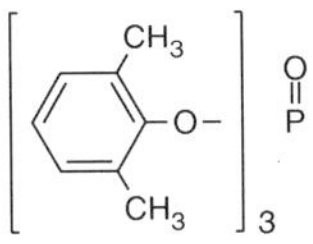
5 gm

S54,584-8
3,4-XYLYLPHOSPHONIC ACID
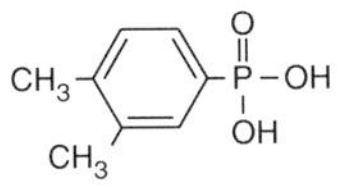
250 mg

S91,571-8
(2,4-XYLYL)PHOSPHONIC ACID
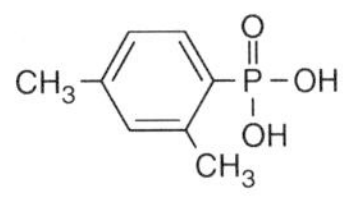
250 mg

S91,594-7
(2,5-XYLYL)PHOSPHONIC ACID
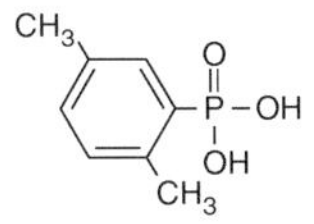
250 mg

S36,535-1
TRIS(2,5-DIMETHYLPHENYL)PHOSPHATE
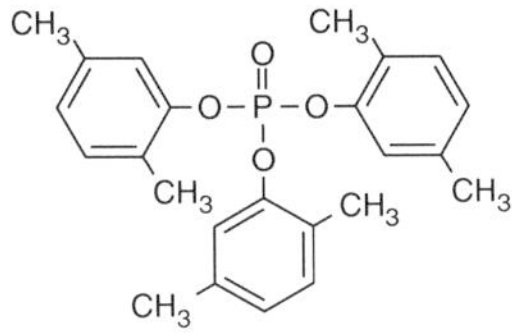
5 gm

S36,537-8
TRIS(3,4-DIMETHYLPHENYL)PHOSPHATE
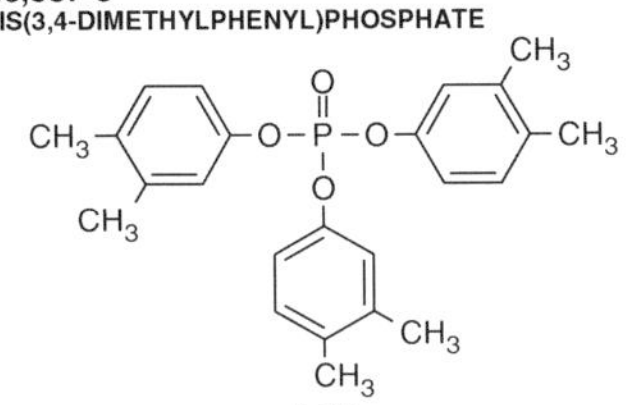
5 gm

S40,575-2
TRIS(2,4-DIMETHYLPHENYL)PHOSPHATE
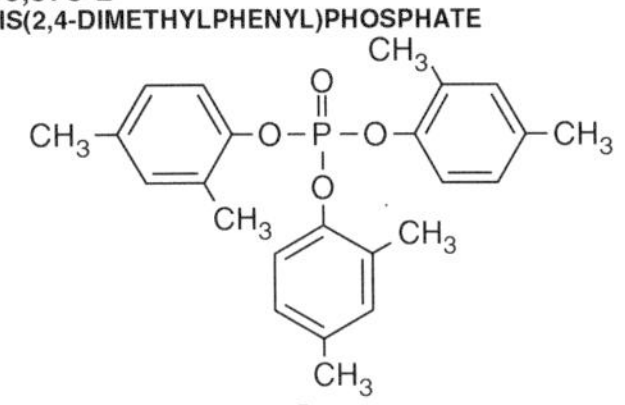
5 gm

S91,596-3
(3,5-XYLYL)PHOSPHONIC ACID
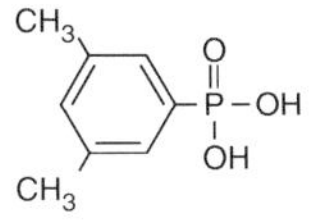
250 mg

S35,870-3
BIS(3,5-DIMETHYLPHENYL)-
CHLOROPHOSPHATE
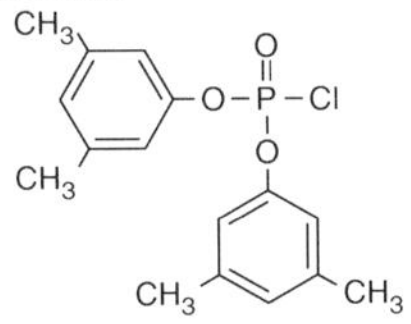
5 gm

S91,604-8
(2,4,6-TRIMETHYLPHENYL)-
PHOSPHONIC ACID
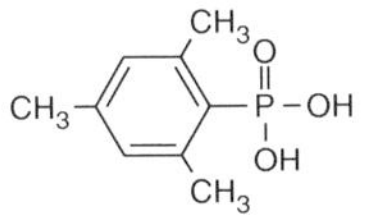
250 mg

S88,585-1
MESITYL DICHLOROPHOSPHATE
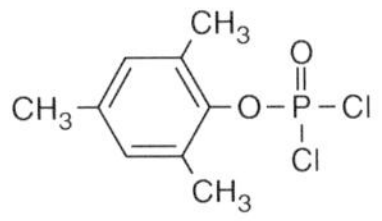
250 mg

S91,601-3
(1-NAPHTHYL)PHOSPHONIC ACID
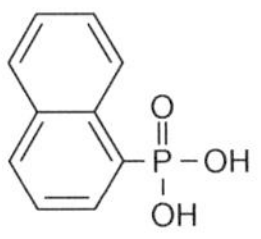
250 mg

S75,547-8
DIETHYL N-(7-FLUORO-2-FLUORENYL)-
PHOSPHORAMIDATE
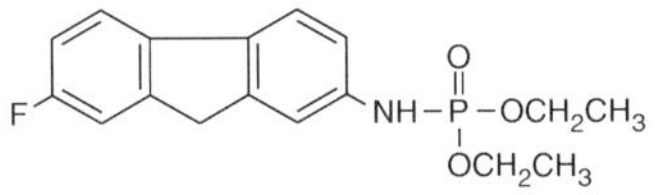
250 mg

S44,676-9
9-FLUORENYLIDENE
TRIPHENYLPHOSPHORANE
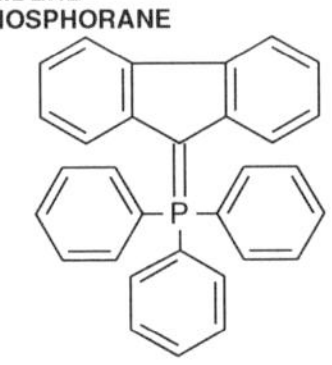
250 mg

S76,005-6
1,3-DIMETHYL-1,2,3,4-TETRAHYDRO-
1,3,2-BENZODIAZAPHOSPHORINE-
2,4-DIONE
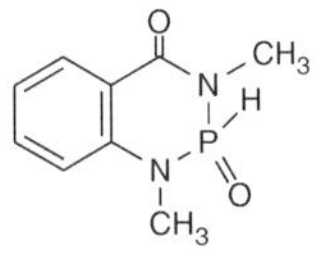
250 mg

S75,974-0
2-BENZYL 4-ETHYL 5-CHLOROMETHYL-3-METHYL-2,4-PYRROLEDICARBOXYLATE

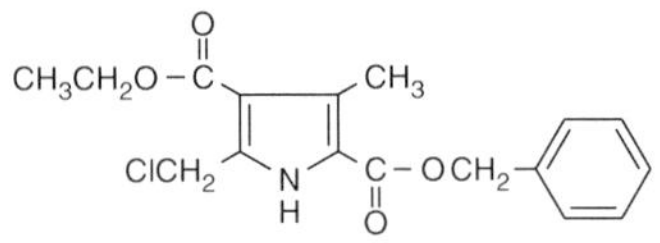

250 mg

S75,680-6
DIETHYL 2-(BROMOMETHYL)-4-METHYL-3,5-PYRROLEDICARBOXYLATE

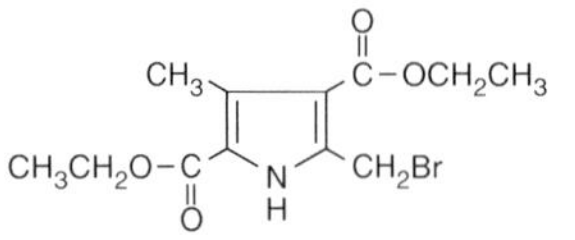

5 gm

S84,407-1
N-(2-FLUORO-5-NITROPHENYL)-2-PYRROLECARBOXAMIDE

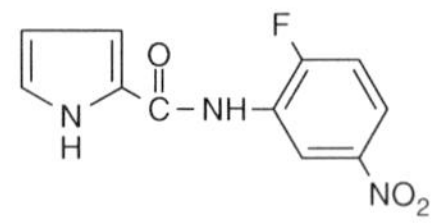

250 mg

S83,537-4
1-(4-BROMOPHENYL)-2-METHYL-5-PHENYLPYRROLE

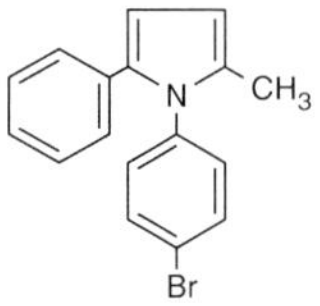

250 mg

S76,413-2
1-(4-CHLOROPHENYL)-2,5-DIMETHYLPYRROLE

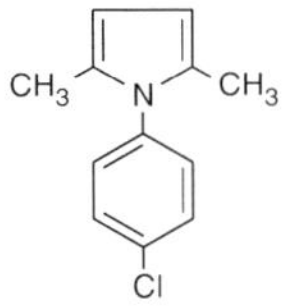

5 gm

S83,011-9
2,5-BIS(PENTAFLUOROPHENYL)PYRROLE

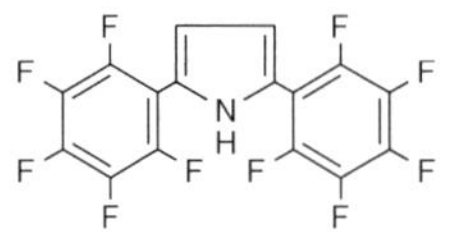

250 mg

S76,057-9
4-CHLOROPHENYL 2-PYRROLYL KETONE

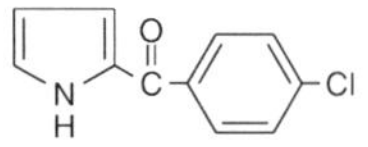

250 mg

S80,058-9
1-[N-(4-FLUOROPHENYL)CARBAMOYL]-PYRROLE

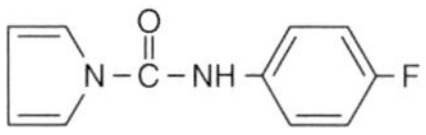

250 mg

S89,538-5
ALPHA-(3,4-DIMETHYL-5-ETHOXY-2H-PYRROLYLIDENE)-BETA-BENZYL ALCOHOL

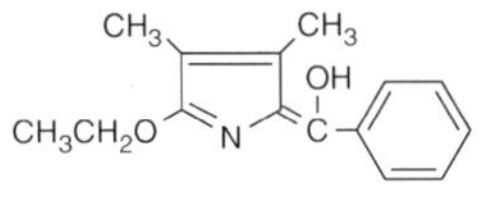

250 mg

S80,804-0
BENZYL 3,5-DIMETHYL-4-(2-HYDROXYETHYL)-2-PYRROLECARBOXYLATE

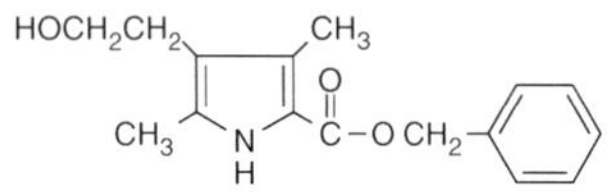

250 mg

S92,815-1
ETHYL-4-HYDROXY-2-METHYL-3-PYRROLECARBOXYLATE

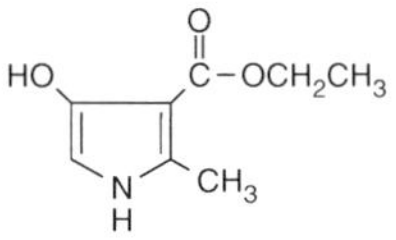

250 mg

S62,149-8
5-METHOXY-1-PYRROLIDINO-1-PENTENE

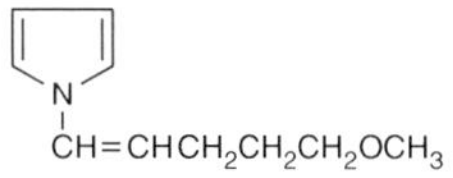

250 mg

S74,596-0
TERT-BUTYL 3,5-DIMETHYL-4-METHOXY-2-PYRROLECARBOXYLATE

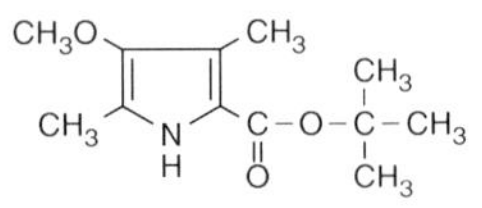

250 mg

S89,598-9
5-ETHOXY-3-METHYL-2-PYRROLECARBOXANILIDE

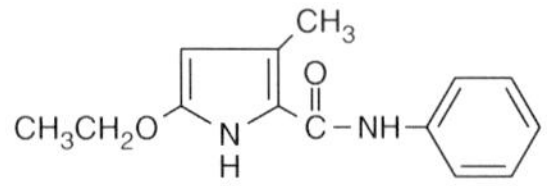

250 mg

S80,803-2
DIBENZYL 5,5'-METHYLENEBIS(3,4-DIMETHYL-2-PYRROLECARBOXYLATE)

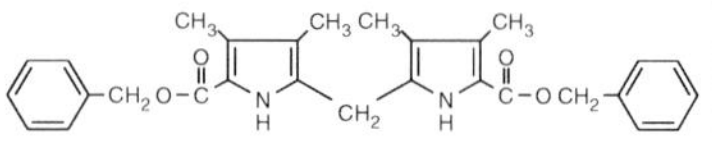

250 mg

S94,118-2
DIBENZYL 5,5'-METHYLENEBIS[3-METHYL-4-(2-CYANOETHYL)-2-PYRROLECARBOXYLATE]

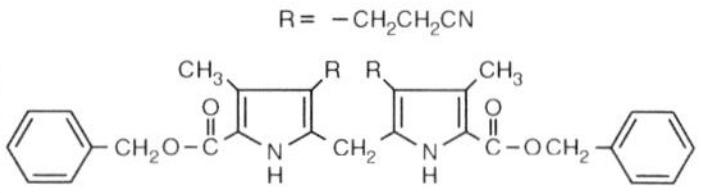

250 mg

S94,102-6
BENZYL 4-ETHYL-5-FORMYL-3-METHYL-2-PYRROLECARBOXYLATE

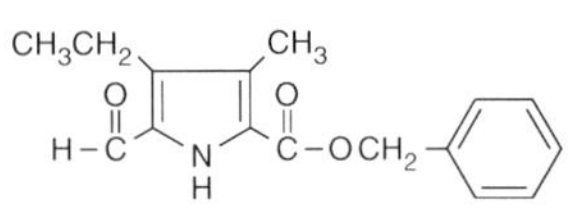

50 mg

S94,104-2
BENZYL 4-ACETYL-3,5-DIMETHYL-2-PYRROLECARBOXYLATE

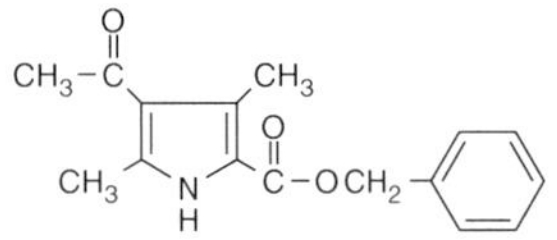

250 mg

S95,190-0
BENZYL 3,5-DIMETHYL-4-ETHYL-2-PYRROLECARBOXYLATE

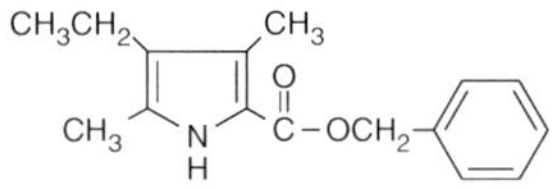

50 mg

S95,194-3
BENZYL 5-ACETOXYMETHYL-3,4-DIMETHYL-2-PYRROLECARBOXYLATE

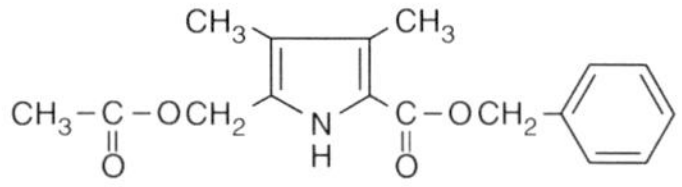

50 mg

S95,205-2
BENZYL 5-ACETOXYMETHYL-4-ETHYL-3-METHYL-2-PYRROLECARBOXYLATE

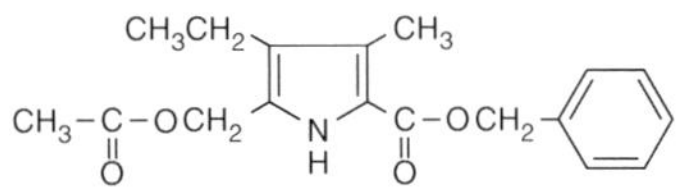

50 mg

S95,224-9
DIBENZYL 2,2'-METHYLENEBIS(5-PYRROLECARBOXYLATE)

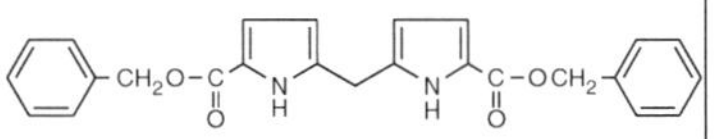

50 mg

S64,810-8
BENZYL 3,4,5-TRIMETHYLPYRROLE-2-CARBOXYLATE

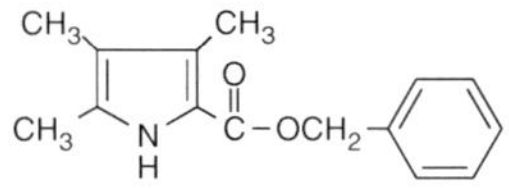

250 mg

S94,342-8
ETHYL 5-(BENZYLOXYCARBONYL)-2,4-DIMETHYL-3-PYRROLEACETATE

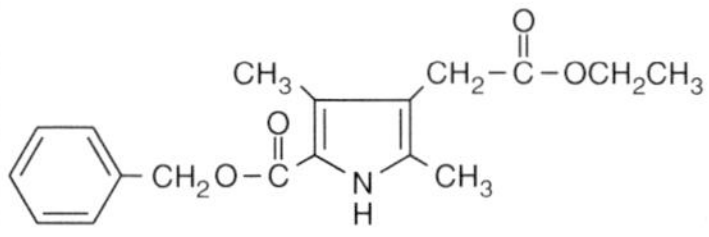

250 mg

S95,230-3
DIBENZYL 2,2'-METHYLENEBIS(3-ETHYL-4-METHYL-5-PYRROLECARBOXYLATE)

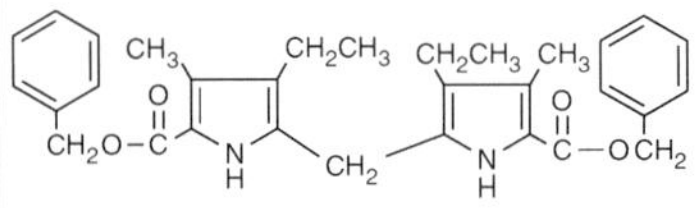

50 mg

S95,197-8
ETHYL 2-[2,2-BIS(ETHOXYCARBONYL)-VINYL]-5-CARBOXY-4-METHYL-3-PYRROLECARBOXYLATE

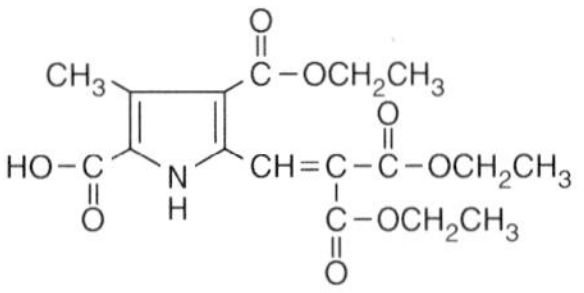

50 mg

S95,159-5
DIETHYL 5-FORMYL-3-METHYL-2,4-PYRROLEDICARBOXYLATE

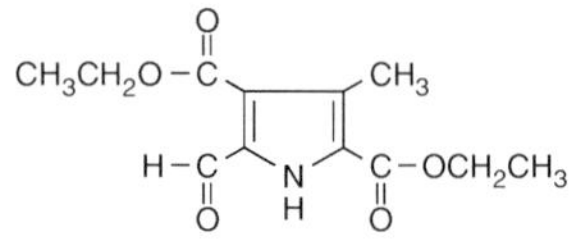

50 mg

S95,177-3
DIETHYL 5-(1,3-DIOXOLAN-2-YL)-3-METHYL-2,4-PYRROLEDICARBOXYLATE

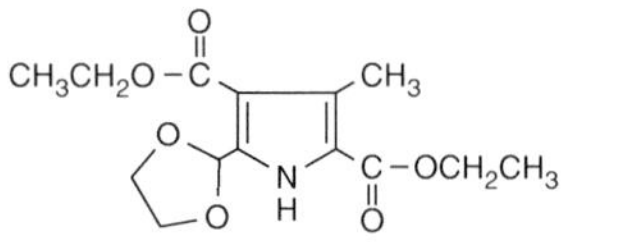

50 mg

S95,112-9
ETHYL 2-CARBOXY-4-METHYL-3-PYRROLECARBOXYLATE

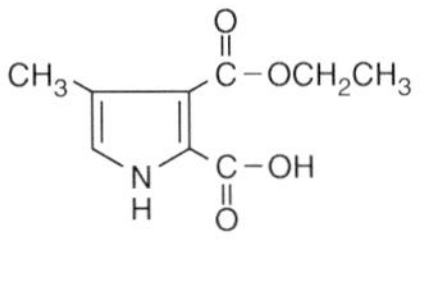

50 mg

S95,130-7
ETHYL 2-CARBOXY-4,5-DIMETHYL-3-PYRROLECARBOXYLATE

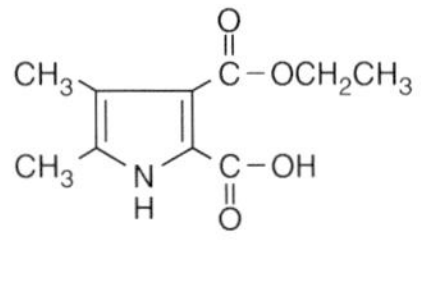

50 mg

S95,128-5
3-ETHYL 2-METHYL 4-METHYL-2,3-PYRROLEDICARBOXYLATE

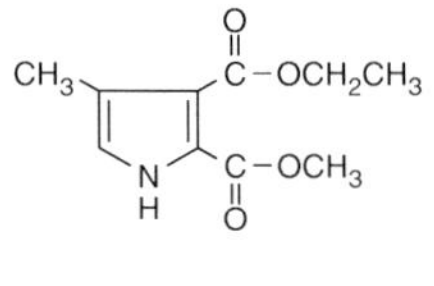

50 mg

S82,912-9
2-(2,5-DIMETHYL-1-PYRROLYL)PHENYL O-TOLYL SULFIDE

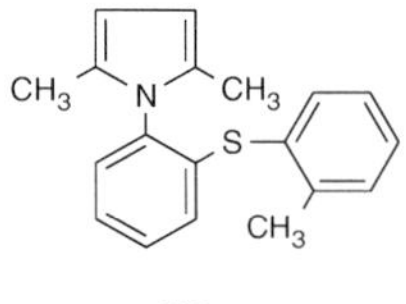

250 mg

S82,911-0
4-(2,5-DIMETHYL-1-PYRROLYL)PHENYL PHENYLSULFIDE
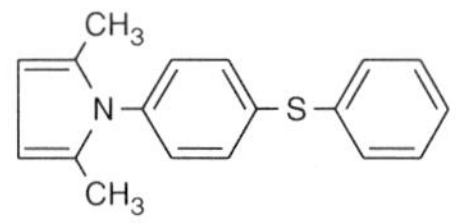
250 mg

S83,640-0
4-(2,5-DIMETHYL-1-PYRROLYL)PHENYL M-TOLYL SULFIDE
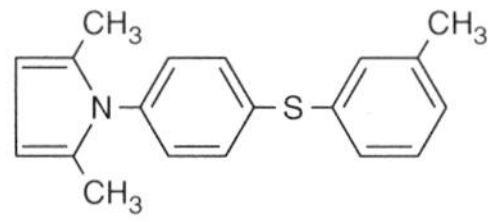
250 mg

S92,576-4
2-CYANO-3,4-DIPHENYL-1-(P-TOLYL)-PYRROLE
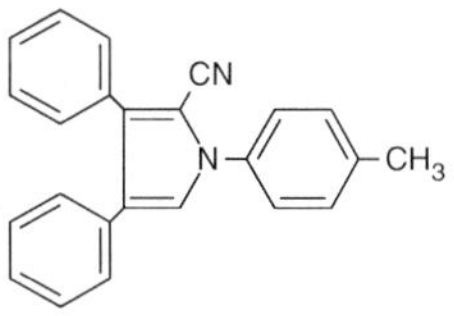
250 mg

S75,968-6
5,5'-DIETHYL-3,3',4,4'-TETRAMETHYL-2,2'-BIPYRROLE
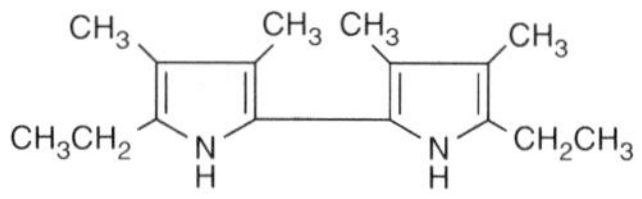
250 mg

S73,998-7
2,5-DIMETHYL-1-(3-NITROPHENYL)-PYRROLE
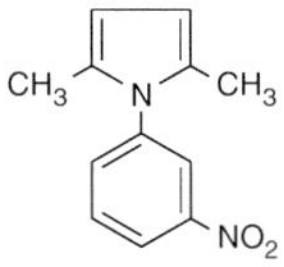
5 gm

S94,344-4
DIMETHYL 2,2'-METHYLENEBIS(5-CARBOXY-4-METHYL-3-PYRROLEPROPIONATE)
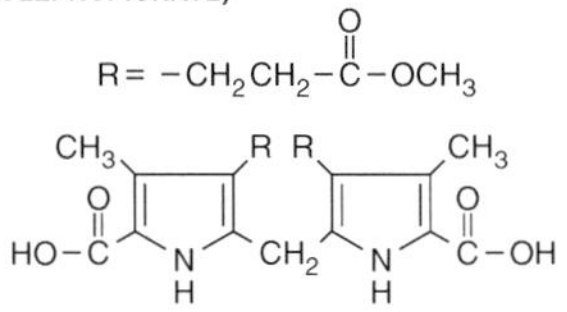
250 mg

S75,962-7
DIETHYL 5-(3-ETHOXYCARBONYL-4-METHYL-PYRROL-2-YLMETHYL)3-METHYL-2,4-PYRROLEDICARBOXYLATE
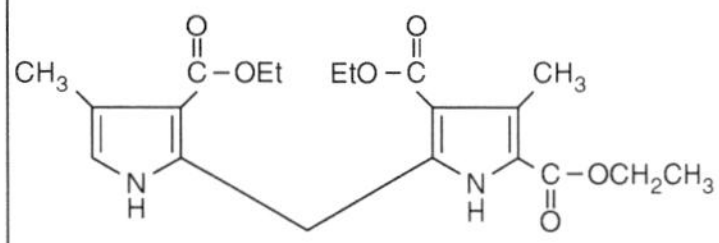
250 mg

S94,345-2
DIMETHYL 2,2'-METHYLENEBIS[5-(BENZYLOXYCARBONYL)-4-METHYL-3-PYRROLEACETATE]
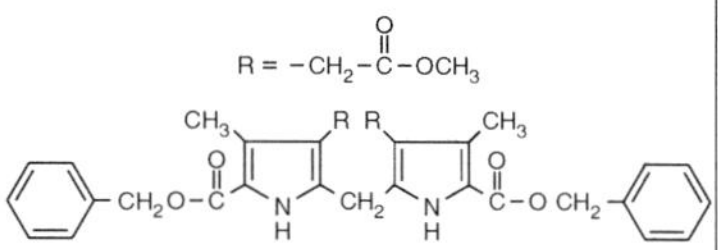
250 mg

S95,185-4
5-BROMO-2-(5-BROMO-3-ETHYL-4-METHYL-PYRROL-2-YLMETHYLENE)-3-ETHYL-4-METHYL-2H-PYRROLE
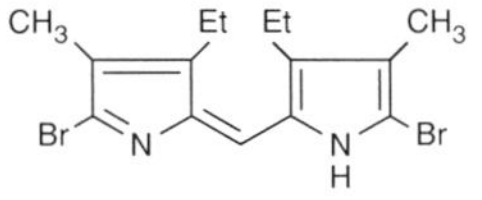
50 mg

S75,965-1
3,5-DIMETHYL-2-(3,5-DIMETHYL-4-ETHYLPYRROL-2-YLMETHYLENE)-4-ETHYL-2H-PYRROLE HYDROBROMIDE
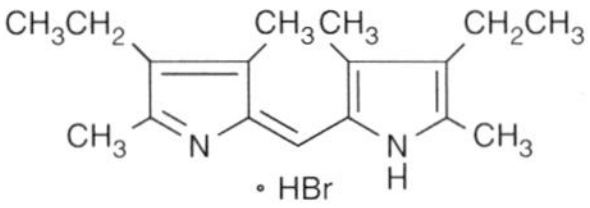
250 mg

S82,079-2
1,1'-(1,3-PHENYLENE)BIS(2,5-DIMETHYLPYRROLE)
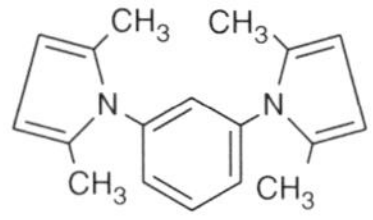
250 mg

S67,872-4
5,10,15,20-TETRAPHENYL-21H,23H-PORPHINE HYDROCHLORIDE
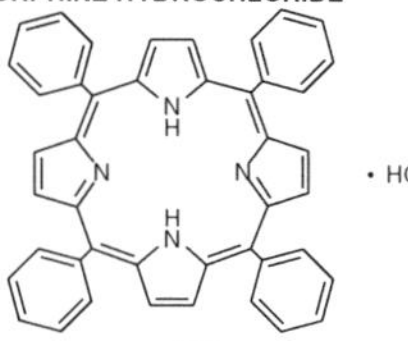
250 mg

S94,105-0
2-(2,2-DICYANOVINYL)-5-METHYLPYRROLE
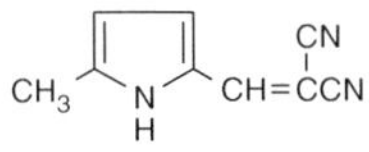
250 mg

S94,106-9
2-(2,2-DICYANOVINYL)-3,4,5-TRIMETHYLPYRROLE
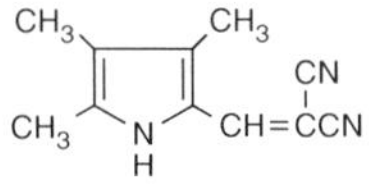
250 mg

S75,978-3
ETHYL 2-(2,2-DICYANOVINYL)-5-IODO-4-METHYL-3-PYRROLECARBOXYLATE
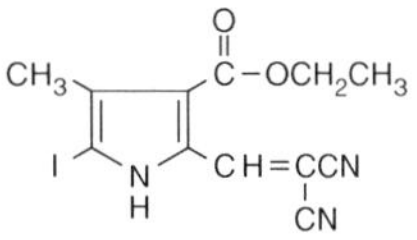
250 mg

S95,212-5
METHYL 5[4(2-ACETAMIDOETHYL)-3-METHYL-2H-PYRROL-2-YLIDENE-METHYL]-4-METHYL-3-PYRROLEPROPIONATE HYDROBROMIDE
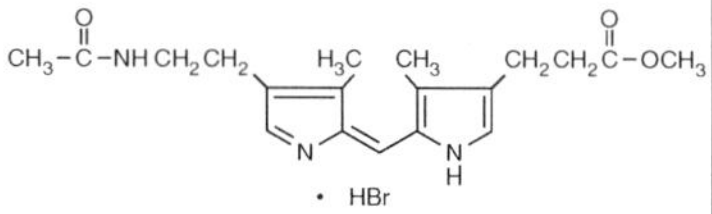
50 mg

S95,200-1
2-(4,5-DIMETHYL-3-ETHYLPYRROL-2-YLMETHYLENE)-3,5-DIMETHYL-4-ETHYL-2H-PYRROLE HYDROBROMIDE
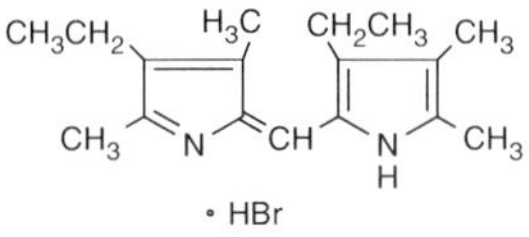
50 mg

S95,223-0
2,5-DI(2-NAPHTHYL)PYRROLE
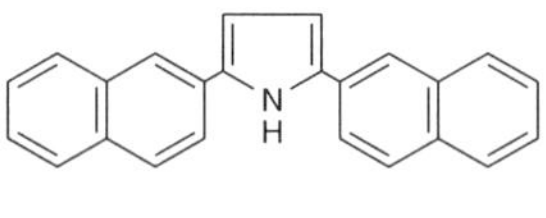
50 mg

S94,101-8
ETHYL 5-FORMYL-2-PYRROLECARBOXYLATE
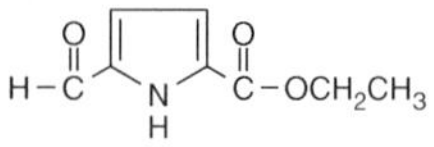
250 mg

S94,103-4
3,5-DIIODO-4-METHYL-2-PYRROLECARBOXALDEHYDE
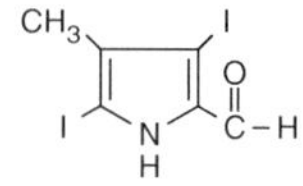
250 mg

S94,108-5
METHYL 5-METHOXYCARBONYL-4-METHOXYCARBONYLMETHYL-2-METHYL-3-PYRROLEACRYLATE
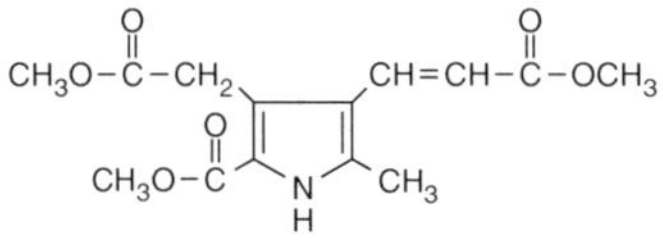
50 mg

S95,142-0
ETHYL 5-FORMYL-3-METHYL-4-(2-NITROETHYL)-2-PYRROLECARBOXYLATE
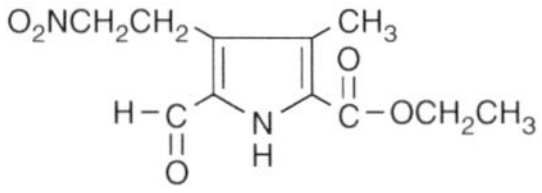
50 mg

S95,123-4
ETHYL 5-BROMO-3-METHYL-4-(2-NITROVINYL)-2-PYRROLECARBOXYLATE
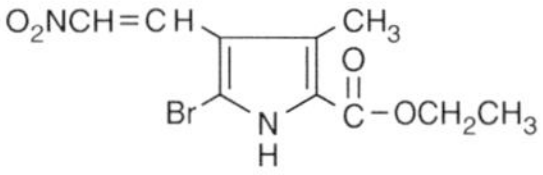
50 mg

S95,105-6
3-METHYL-4-(2-NITROVINYL)-2,5-PYRROLEDICARBOXYLIC ACID
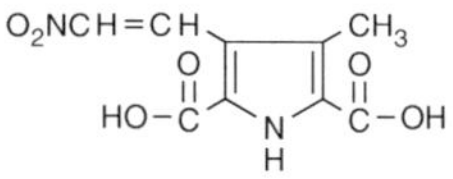
50 mg

S95,107-2
4-METHYL-3-(2-NITROETHYL)-2,5-PYRROLEDICARBOXYLIC ACID
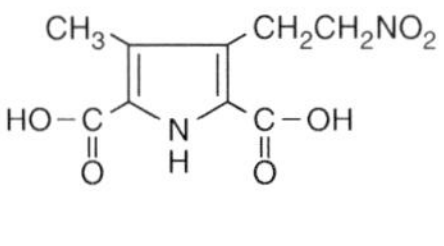
50 mg

S95,141-2
ETHYL 3,5-DIMETHYL-4-(2-NITROVINYL)-2-PYRROLECARBOXYLATE
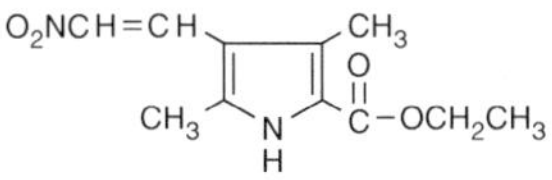
50 mg

S95,124-2
ETHYL 5-BROMO-3-METHYL-4-(2-NITROETHYL)-2-PYRROLECARBOXYLATE
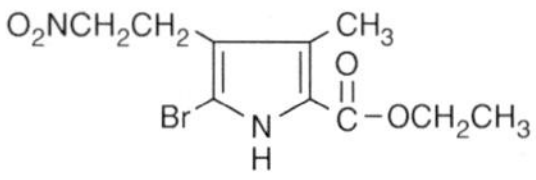
50 mg

S95,110-2
ETHYL 5-FORMYL-4-METHYL-3-PYRROLECARBOXYLATE
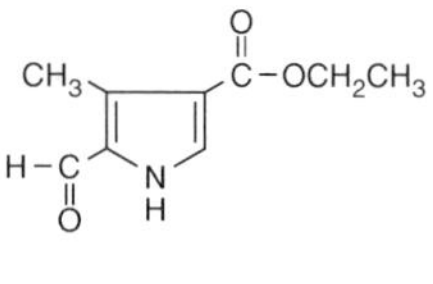
50 mg

S95,125-0
METHYL 5-FORMYL-4-METHYL-3-PYRROLEPROPIONATE
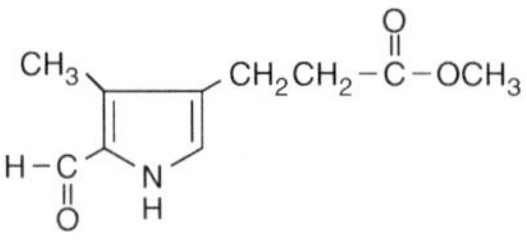
50 mg

S95,127-7
ETHYL 3,5-DIMETHYL-4-FORMYL-2-PYRROLECARBOXYLATE
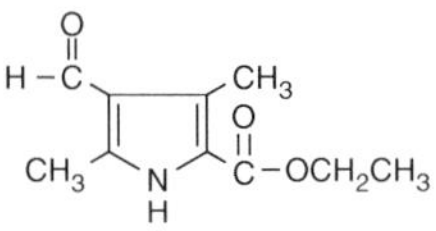
50 mg

S95,181-1
BENZYL 3,4-DIMETHYL-5-FORMYL-2-PYRROLECARBOXYLATE
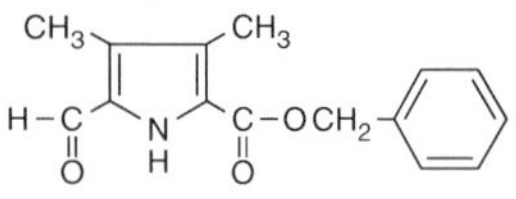
50 mg

S95,093-9
3,5-DIMETHYL-4-FORMYL-2-PYRROLECARBOXYLIC ACID
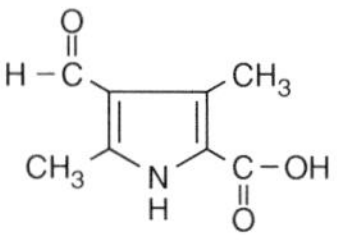
50 mg

S95,108-0
5-FORMYL-4-METHYL-3-PYRROLEPROPIONIC ACID

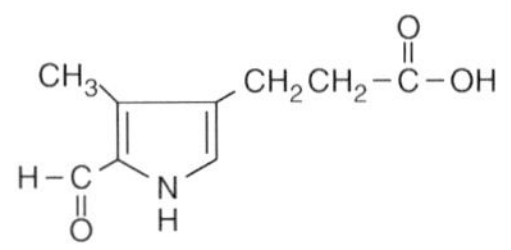

50 mg

S95,109-9
2-FORMYL-4-METHYL-3-PYRROLEPROPIONIC ACID

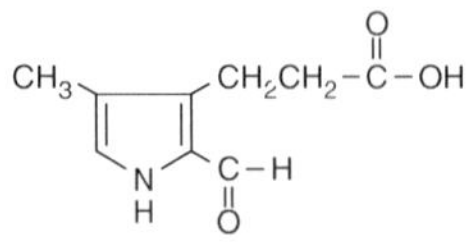

50 mg

S95,082-3
3-METHYL-2-PYRROLECARBOXALDEHYDE

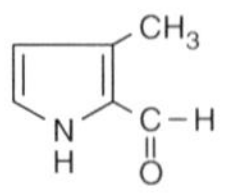

50 mg

S95,092-0
3,4-DIMETHYL-2,5-PYRROLEDICARBOXALDEHYDE

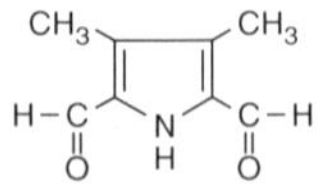

50 mg

S95,116-1
4,5-DIMETHYL-3-ETHYL-2-PYRROLECARBOXALDEHYDE

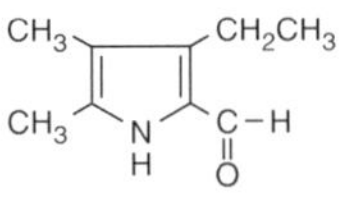

50 mg

S95,225-7
METHYL 5-[-(2,2-DICYANOVINYL)-3-ETHYL-4-METHYL-PYRROL-2-YLMETHYL]-3-ETHOXYCARBONYL-4-METHYL-8-OXO-2-PYRROLEBUTYRATE

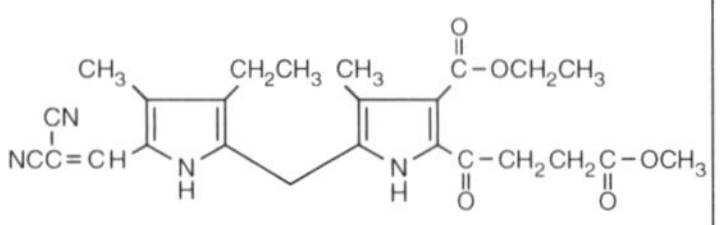

50 mg

S95,202-8
BIS(3,5-DIMETHYL-4-ETHYLPYRROL-2-YL) KETONE

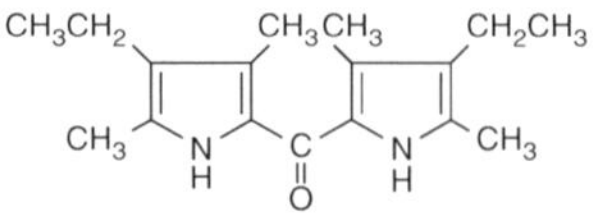

50 mg

S40,628-7
3-ACETYL-2,4-DIMETHYLPYRROLE-5-CARBOXYLIC ACID

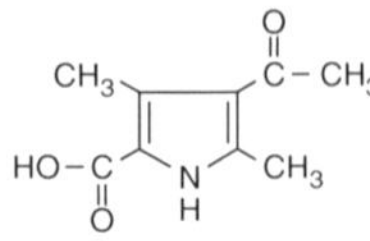

250 mg

S95,172-2
ETHYL 5-(CHLOROFORMYL)-3-(ETHOXYCARBONYL)-4-METHYL-2-PYRROLEACRYLATE

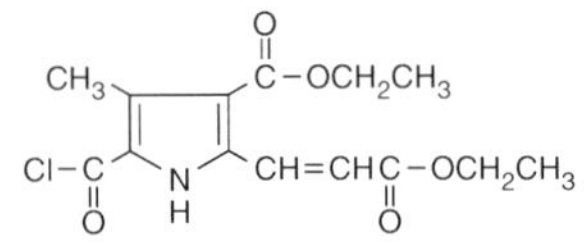

50 mg

S95,174-9
3,3',4,4'-TETRAMETHYL-(2,2'-BIPYRROLE)-5,5'-DICARBOXYLIC ACID

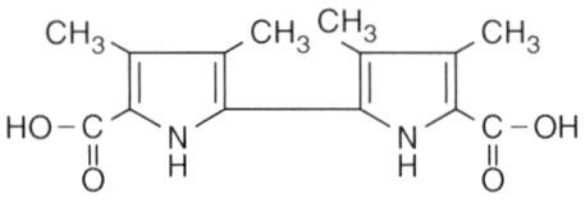

50 mg

S95,186-2
2,2'-METHYLENEBIS(3,4-DIMETHYL-5-PYRROLECARBOXYLIC ACID)

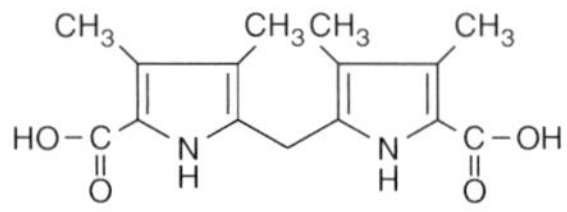

50 mg

S95,198-6
5,5'-METHYLENEBIS(4-ETHYL-3-METHYL-2-PYRROLECARBOXYLIC ACID)

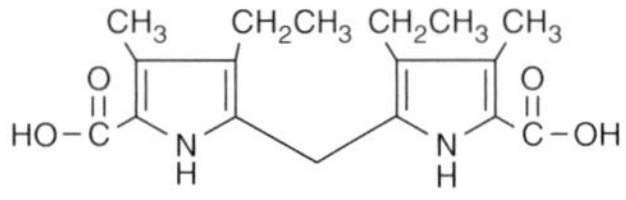

50 mg

S95,161-7
2,4-DIMETHYL-5-ETHOXYCARBONYL-3-PYRROLEPROPIONIC ACID

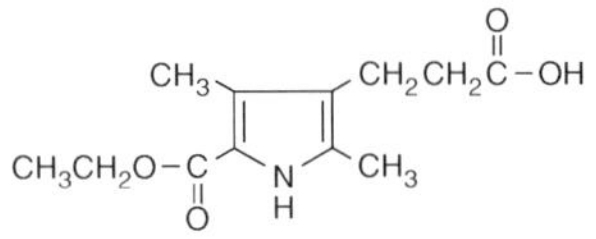

50 mg

S95,162-5
ETHYL 5-(ACETOXYMETHYL)-3,4-DIMETHYL-2-PYRROLECARBOXYLATE

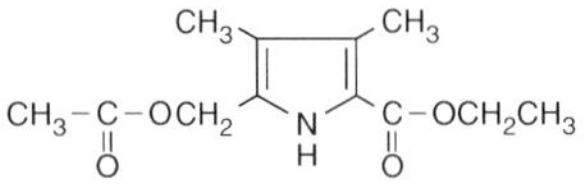

50 mg

S95,168-4
DIETHYL 3-ETHYL-5-METHYL-2,4-PYRROLEDICARBOXYLATE

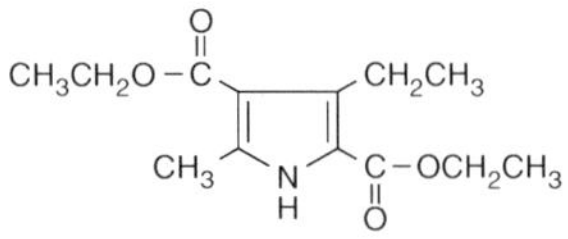

250 mg

S95,176-5
ETHYL 2,4-DIMETHYL-5-(ETHOXYCARBONYL)-3-PYRROLEACRYLATE

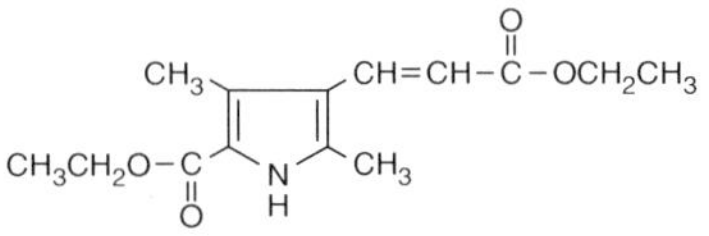

50 mg

S95,208-7
TERT-BUTYL 3,5-BIS(ETHOXYCARBONYL)-4-METHYL-2-PYRROLEACRYLATE

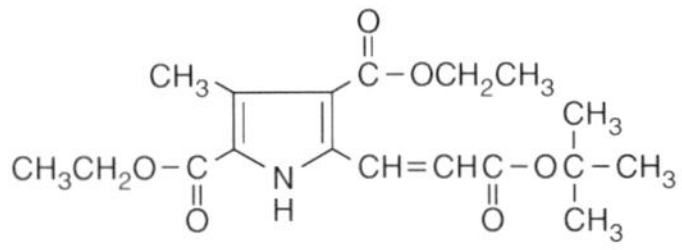

50 mg

S95,213-3
ETHYL 5-[4-(ETHOXYCARBONYL)-3-METHYL-PYRROL-2-YLMETHYL]-4-ETHYL-3-METHYL-2-PYRROLECARBOXYLATE

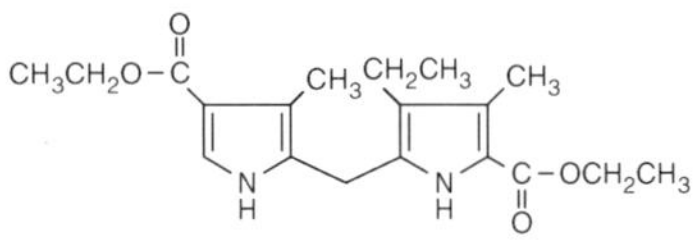

50 mg

S95,214-1
DIETHYL 2,2'-METHYLENEBIS(3,4-DIMETHYL-5-PYRROLECARBOXYLATE)

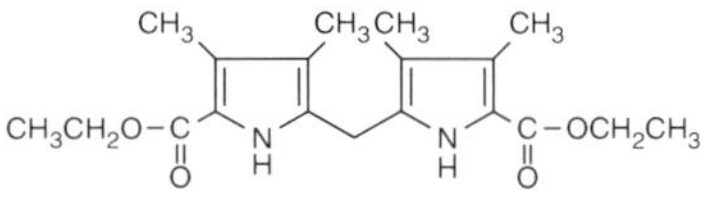

50 mg

S95,222-2
DI-TERT-BUTYL 2,2'-METHYLENEBIS-(3,4-DIMETHYL-5-PYRROLECARBOXYLATE)

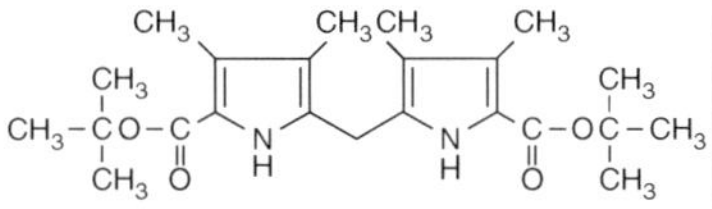

50 mg

S95,217-6
ETHYL 5-(DICYANOVINYL-ETHYL-METHYLPYRROLYLMETHYL)-2,4-DIMETHYL-PYRROLECARBOXYLATE

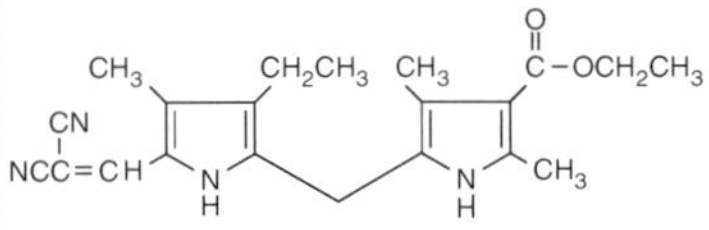

50 mg

S95,218-4
ETHYL 2-(DICYANOVINYL-ETHYL-METHYLPYRROLYLMETHYL)-4,5-DIMETHYL-PYRROLECARBOXYLATE

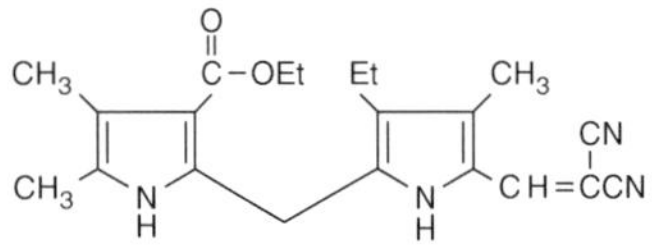

50 mg

S95,166-8
ETHYL 4-(2,2-DICYANOVINYL)-3,5-DIMETHYL-2-PYRROLECARBOXYLATE

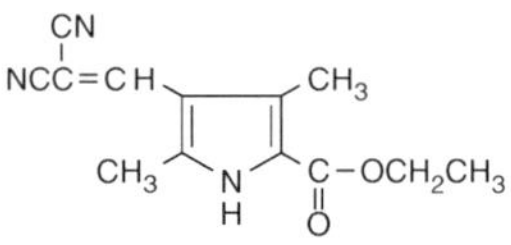

50 mg

S94,107-7
ETHYL 2-METHYL-3-PYRROLECARBOXYLATE

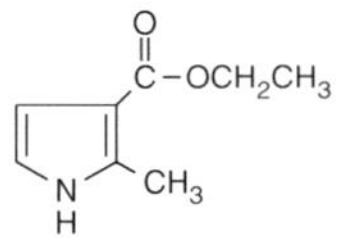

250 mg

S95,119-6
ETHYL 3,4-DIMETHYL-2-PYRROLECARBOXYLATE

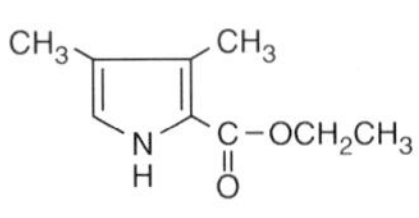

50 mg

S95,126-9
ETHYL 2,4-DIMETHYL-5-FORMYL-3-PYRROLECARBOXYLATE

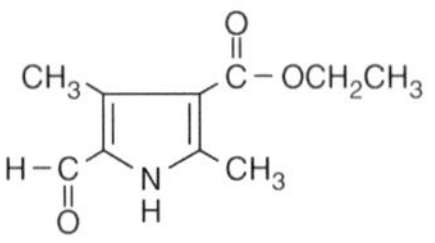

50 mg

S95,175-7
ETHYL 5-CARBOXY-3-(ETHOXYCARBONYL)-4-METHYL-2-PYRROLEACRYLATE

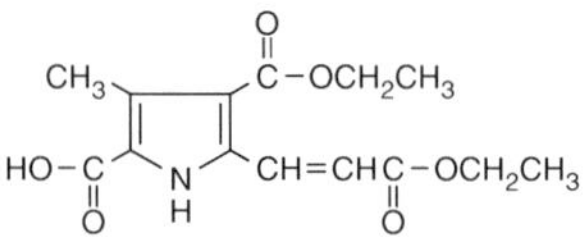

50 mg

S95,094-7
3,4-DIMETHYL-2,5-PYRROLEDICARBOXYLIC ACID

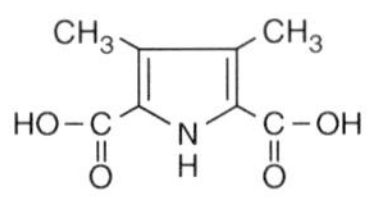

50 mg

S95,146-3
ETHYL 5-CARBOXY-4-ETHYL-3-METHYL-2-PYRROLECARBOXYLATE

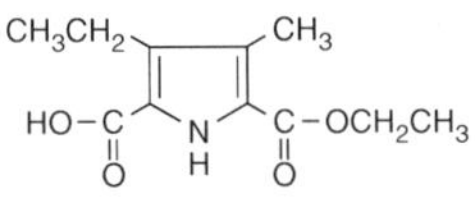

50 mg

S95,106-4
ETHYL 5-BROMO-4-CARBOXY-3-METHYL-2-PYRROLECARBOXYLATE

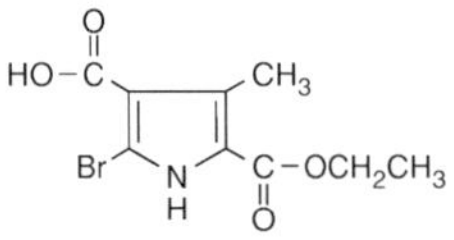

50 mg

S95,131-5
ETHYL 5-CARBOXY-2,4-DIMETHYL-3-PYRROLECARBOXYLATE

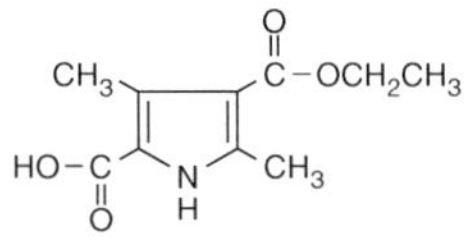

50 mg

S95,083-1
4-METHYL-3-PYRROLECARBOXYLIC ACID

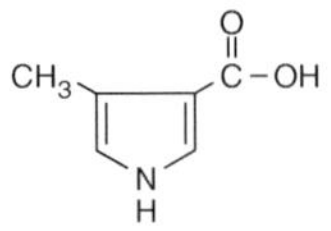

50 mg

S95,099-8
4-ETHYL-3-METHYL-2-PYRROLECARBOXYLIC ACID

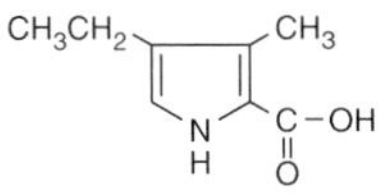

50 mg

S45,079-0
5-PHENYL-2-PYRROLEPROPIONIC ACID

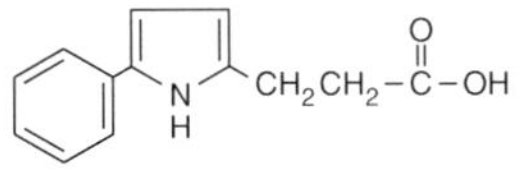

5 gm

S95,145-5
DIETHYL 3-METHYL-2,4-PYRROLEDICARBOXYLATE

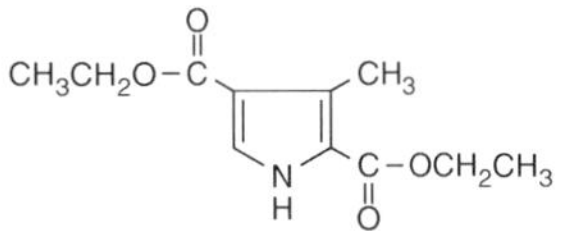

50 mg

S95,139-0
DIETHYL 5-BROMO-3-METHYL-2,4-PYRROLEDICARBOXYLATE

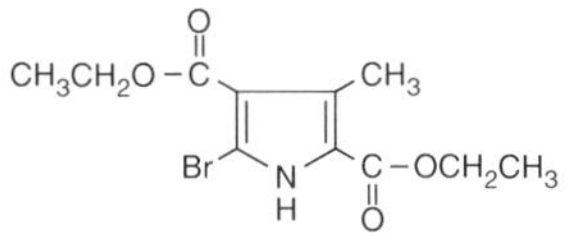

50 mg

S95,132-3
ETHYL 3-ETHYL-5-METHYL-2-PYRROLECARBOXYLATE

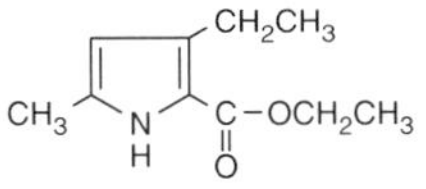

50 mg

S95,152-8
ETHYL 3,5-DIMETHYL-4-ETHYL-2-PYRROLECARBOXYLATE

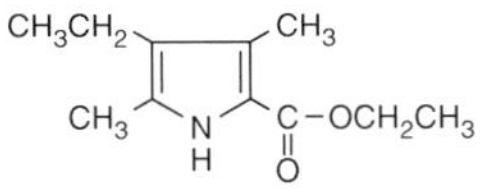

50 mg

S95,153-6
ETHYL 4,5-DIMETHYL-3-ETHYL-2-PYRROLECARBOXYLATE

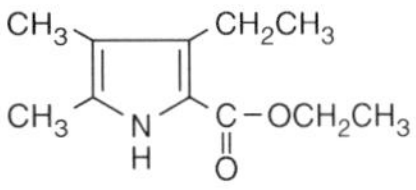

50 mg

S95,140-4
ETHYL 2-(2-PYRROLYL)-1-PYRROLINE-5-CARBOXYLATE

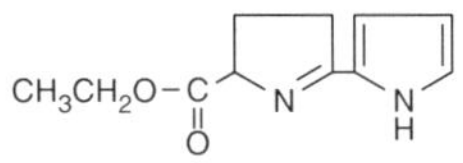

50 mg

S95,147-1
TERT-BUTYL 3,4-DIMETHYL-5-IODO-2-PYRROLECARBOXYLATE

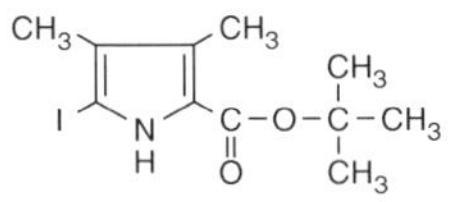

50 mg

S95,129-3
ETHYL 5-CARBOXY-3,4-DIMETHYL-2-PYRROLECARBOXYLATE

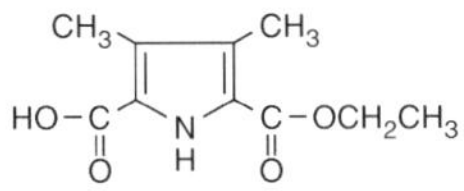

50 mg

S95,169-2
METHYL 2,4-DIMETHYL-5-(ETHOXYCARBONYL)-3-PYRROLEPROPIONATE

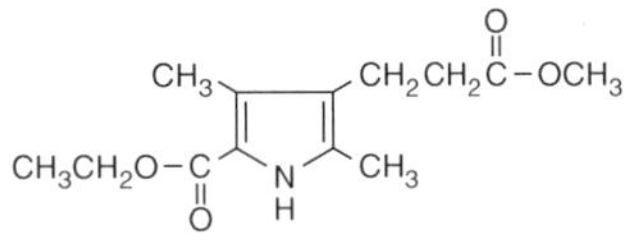

50 mg

S95,207-9
DIETHYL 3,3',4,4'-TETRAMETHYL-(2,2'-BIPYRROLE)-5,5'-DICARBOXYLATE

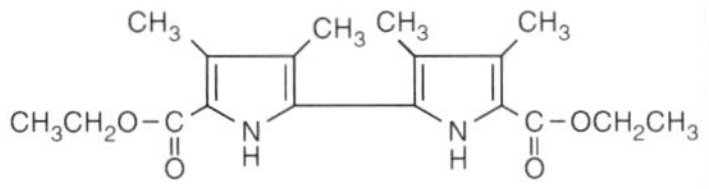

50 mg

S95,104-8
METHYL ALPHA-CYANO-2-PYRROLEACRYLATE

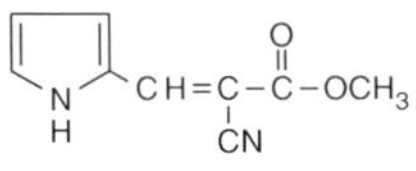

50 mg

S95,138-2
METHYL ALPHA-CYANO-3,4-DIMETHYL-2-PYRROLEACRYLATE

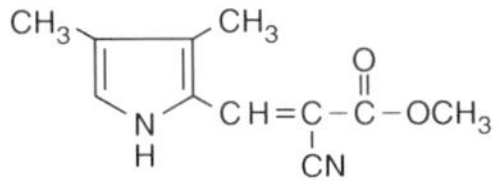

50 mg

S95,097-1
ETHYL 4-METHYL-3-PYRROLECARBOXYLATE

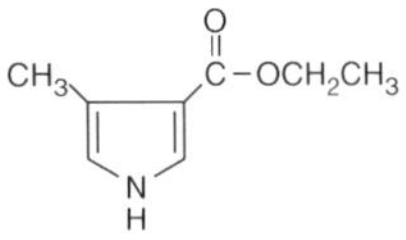

50 mg

S95,113-7
2,3-DIMETHYL-4,5,6,7-TETRAHYDRO-1H-PYRROLO(3,2-C)PYRIDIN-4-ONE

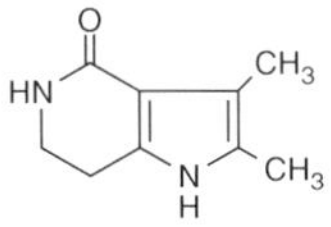

50 mg

S95,101-3
3,4,5-TRIMETHYL-2-PYRROLECARBOXYLIC HYDRAZIDE

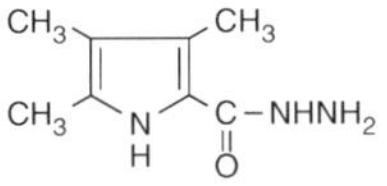

50 mg

S95,089-0
3,4-DIMETHYL-2-PYRROLECARBOXALDEHYDE

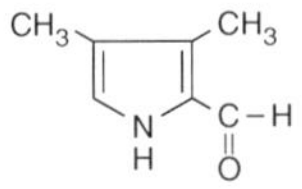

50 mg

S95,096-3
2-(2-PYRROLYL)-1-PYRROLINE

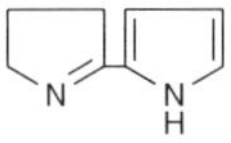

50 mg

S83,841-1
2,5-DIMETHYL-1-[4-(PHENYLSULFONYL)-PHENYL]PYRROLE

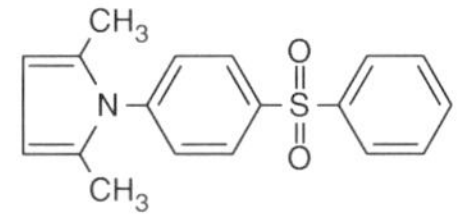

250 mg

S83,638-9
4-(2,5-DIMETHYL-1-PYRROLYL)PHENYL M-TOLYL SULFONE

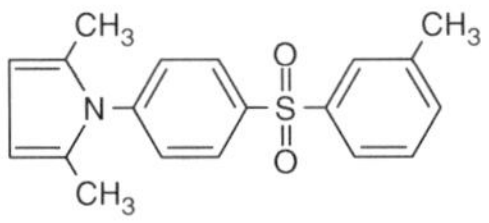

250 mg

S83,639-7
4-(2,5-DIMETHYL-1-PYRROLYL)PHENYL P-TOLYL SULFONE

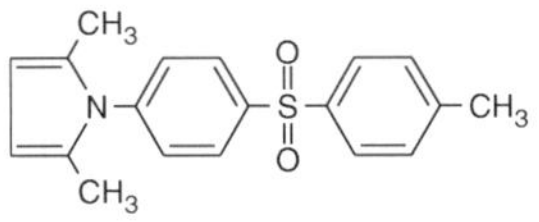

250 mg

S56,864-3
4-(2,5-DIMETHYL-3-FORMYL-1-PYRROLYL)BENZOIC ACID

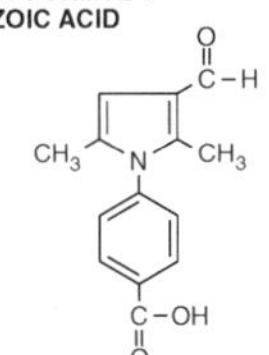

250 mg

S95,144-7
ETHYL 5-FORMYL-1,2,4-TRIMETHYL-3-PYRROLECARBOXYLATE

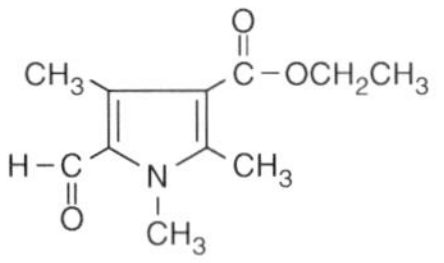

50 mg

S60,546-8
2,5-DIMETHYL-1-(4-ETHOXYCARBONYLPHENYL)-3-FORMYLPYRROLE

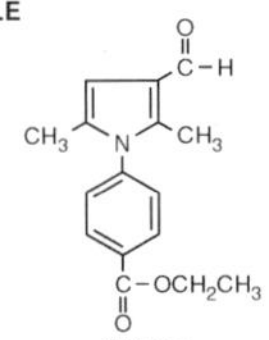

250 mg

S87,873-1
N-(4-ACETYLPHENYL)-2,5-DIMETHYLPYRROLE

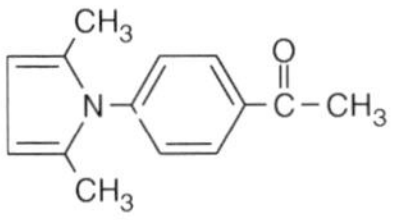

5 gm

S60,550-6
2,5-DIMETHYL-1-(4-ETHOXYCARBONYLPHENYL)PYRROLE

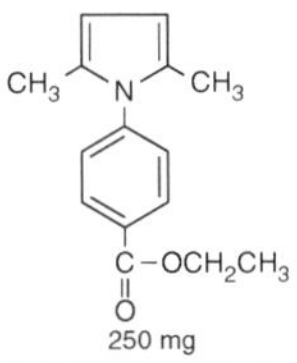

250 mg

S90,672-7
MONO-METHYL 3-(1-PYRROLYL)GLUTARATE

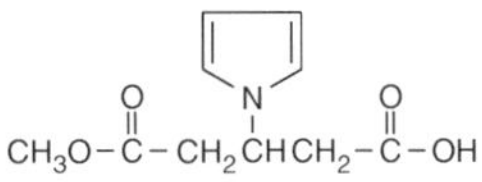

250 mg

S90,664-6
DIMETHYL 3-(1-PYRROLYL)GLUTARATE

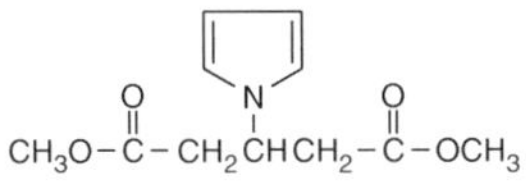

250 mg

S38,205-1
6'-CHLORO-1-PYRROLEACETO-ORTHO-TOLUIDIDE

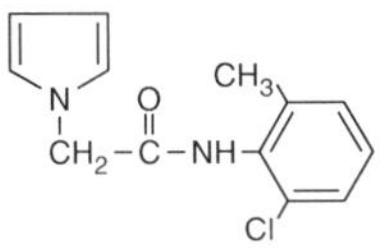

250 mg

S64,912-0
1-(2,3,5,6-TETRACHLOROPHENYL)-PYRROLE

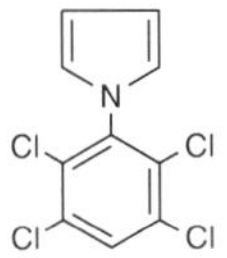

250 mg

S34,729-9
2,5-DIMETHYL-1-PHENYLPYRROLE

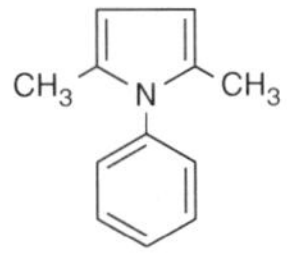

5 gm

S55,588-6
1-DECYL-2,5-DIMETHYLPYRROLE

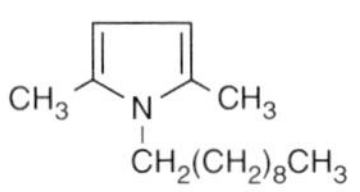

250 mg

S81,245-5
2,5-DIMETHYL-1-DODECYLPYRROLE

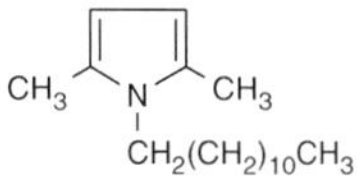

250 mg

S84,028-9
2,5-DIMETHYL-1-OCTADECYLPYRROLE

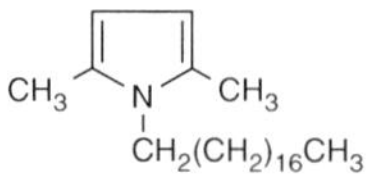

250 mg

S95,199-4
METHYL 2-[4-(2-AMINOETHYL)-3-METHYLPYRROL-2-YLMETHYLENE]-3-METHYL-2H-PYRROLE-4-PROPIONATE DIHYDROBROMIDE DIHYDRATE

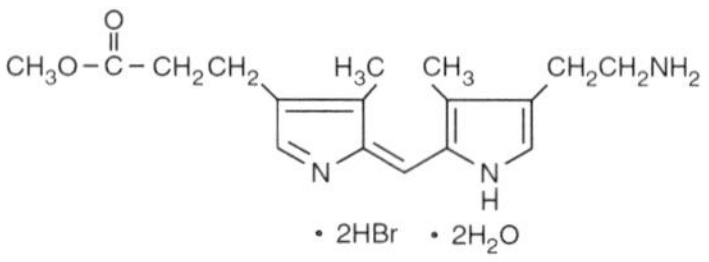

50 mg

S84,058-0
DIETHYL 1-AMINO-2,5-DIMETHYL-3,4-PYRROLEDICARBOXYLATE

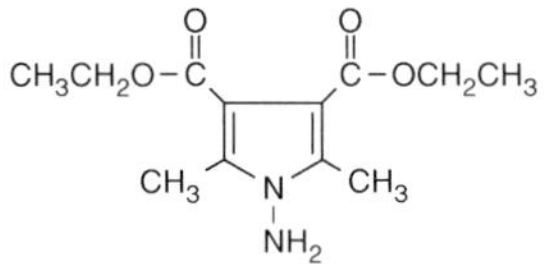

250 mg

S76,770-0
ETHYL 4-ACETYL-3-ETHYL-5-METHYL-2-PYRROLECARBOXYLATE

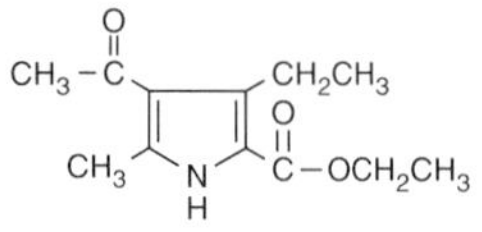

250 mg

S76,276-8
ETHYL 3,4,5-TRIMETHYL-2-PYRROLECARBOXYLATE

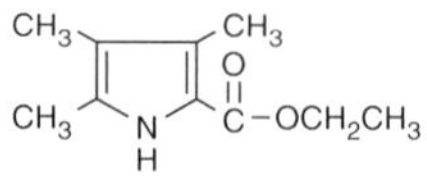

250 mg

S76,768-9
TERT-BUTYL 3,5-DIMETHYL-4-ETHYL-2-PYRROLECARBOXYLATE

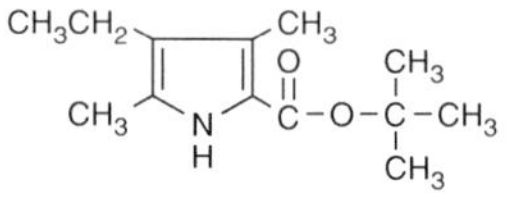

250 mg

S76,769-7
ETHYL 4-BENZYL-3,5-DIMETHYL-2-PYRROLECARBOXYLATE

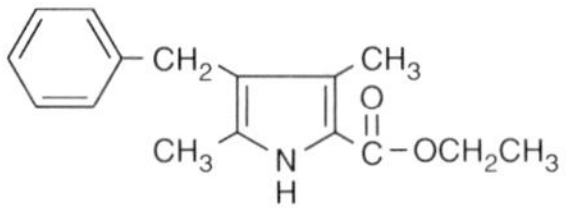

250 mg

S95,086-6
2,5-DIIODO-4-METHYL-3-(2-NITROVINYL)PYRROLE

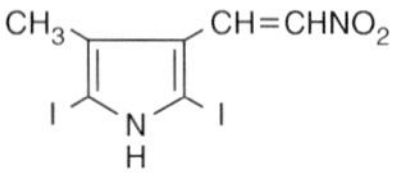

50 mg

S89,491-5
ETHYL 2-(ETHOXYCARBONYL)-4-(ETHOXYCARBONYLMETHYL)-5-FORMYL-3-PYRROLEPROPIONATE

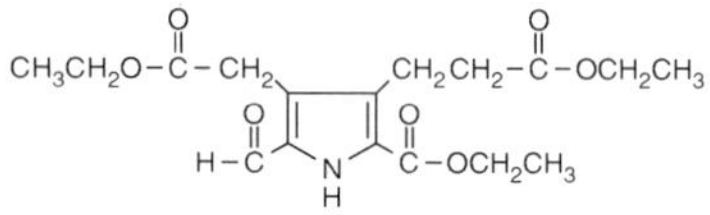

250 mg

S75,958-9
DIETHYL 3-ETHYL-5-FORMYL-2,4-PYRROLEDICARBOXYLATE

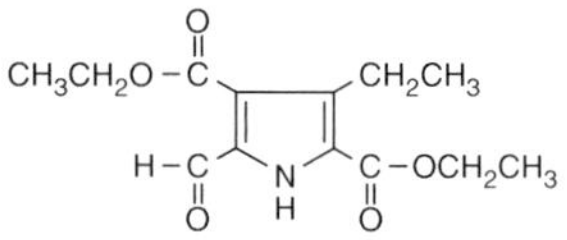

250 mg

S86,142-1
METHYL 4-FORMYL-1-(PHENYLSULFONYL)-2-PYRROLECARBOXYLATE

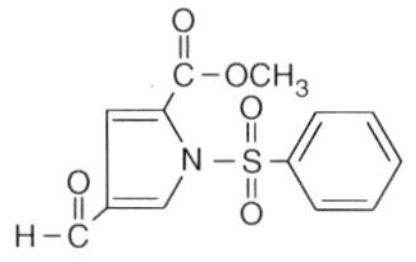

250 mg

S75,959-7
4-ETHYL-5-FORMYL-3-METHYL-2-PYRROLECARBOXYLIC ACID

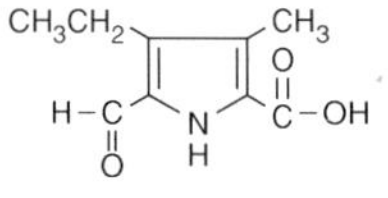

250 mg

S74,038-1
3,4-DIBROMO-5-METHYL-2-PYRROLECARBOXALDEHYDE

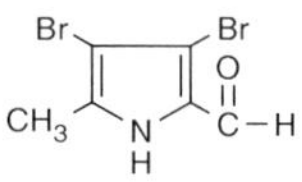

250 mg

S95,156-0
METHYL ALPHA-CYANO-3,4-DIMETHYL-5-FORMYL-2-PYRROLEACRYLATE

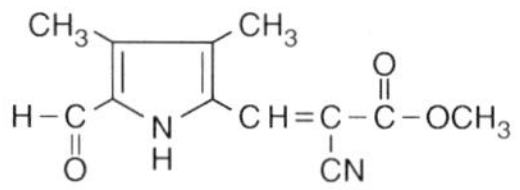

50 mg

S34,041-3
ETHYL 3-ACETYL-2,4-DIMETHYL-5-PYRROLECARBOXYLATE

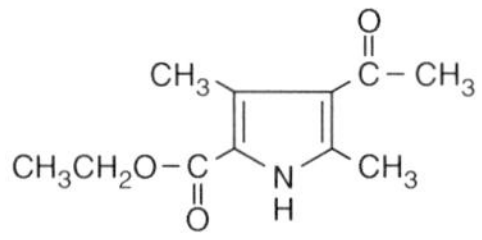

5 gm

S83,655-9
2-BENZOYLPYRROLE

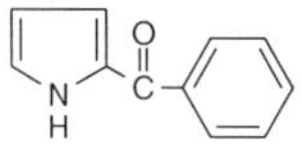

250 mg

S65,320-9
4'-METHYL-3-PYRROLINOPROPIOPHENONE

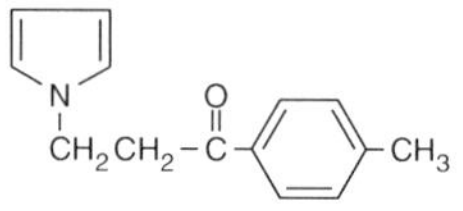

5 gm

S89,586-5
4-(CARBOXYMETHYL)-2-(ETHOXYCARBONYL)-5-METHYL-3-PYRROLEPROPIONIC ACID

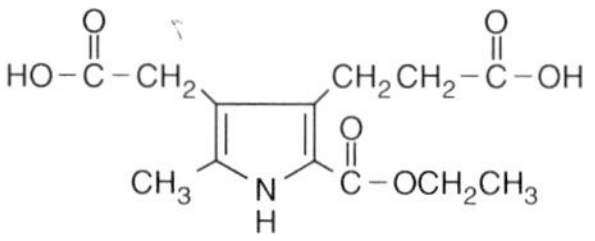

250 mg

S89,585-7
ETHYL 2,4-DIMETHYL-5-(ETHOXYCARBONYL)-3-PYRROLEACETATE

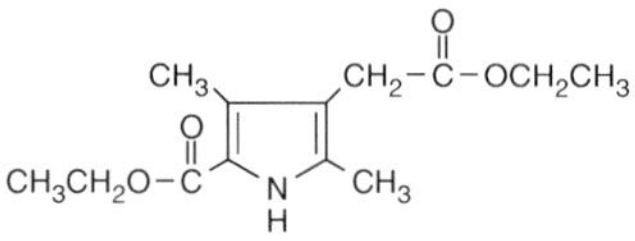

250 mg

S89,492-3
ETHYL 2-(ETHOXYCARBONYL)-4-(ETHOXYCARBONYLMETHYL)-5-METHYL-3-PYRROLEPROPIONATE

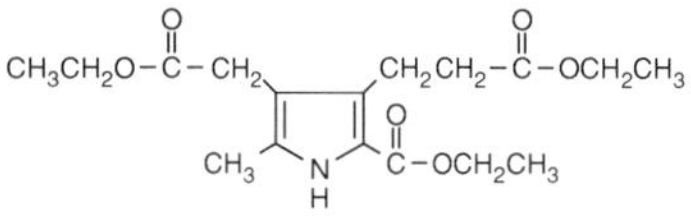

250 mg

S89,496-6
ETHYL 4-(TERT-BUTOXYCARBONYL)-2-(ETHOXYCARBONYL)-5-METHYL-3-PYRROLEPROPIONATE

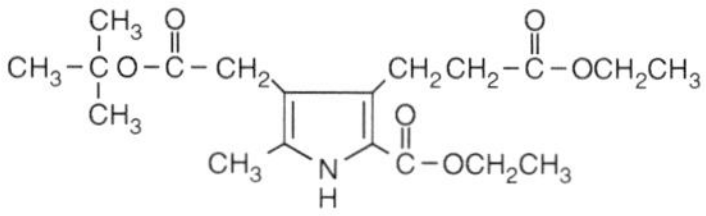

250 mg

S73,924-3
METHYL 4-(TERT-BUTOXYCARBONYL)-2-(METHOXYCARBONYL)-5-METHYL-3-PYRROLEACETATE

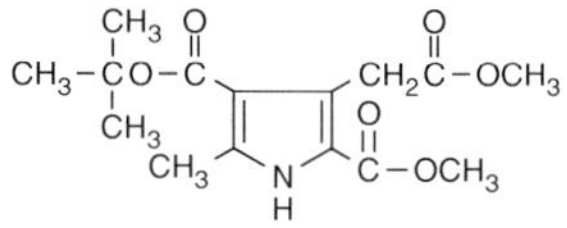

5 gm

S76,713-1
2,2'-METHYLENEBIS(DIETHYL 4-METHYL-3,5-PYRROLEDICARBOXYLATE)

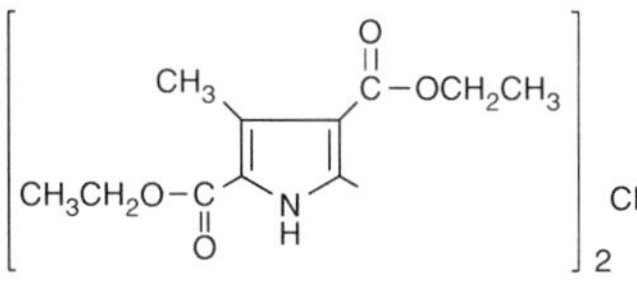

250 mg

S80,806-7
2,4-DIMETHYL-5-(TERT-BUTOXYCARBONYL)-3-PYRROLEPROPIONIC ACID

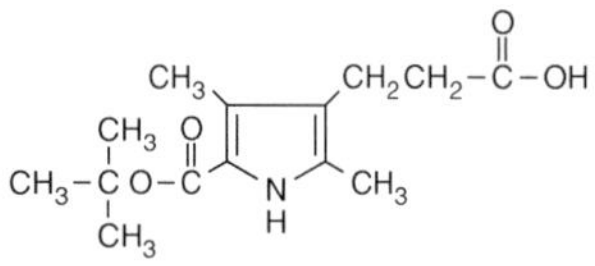

250 mg

S94,330-4
METHYL 5-CARBOXY-2-(DIMETHYLCARBAMOYL)-4-METHYL-3-PYRROLEPROPIONATE

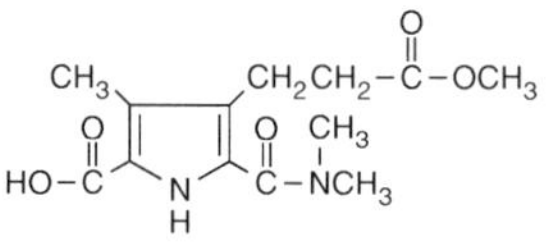

250 mg

S73,928-6
5-(BENZYLOXYCARBONYL)-2,4-DIMETHYL-3-PYRROLEPROPIONIC ACID

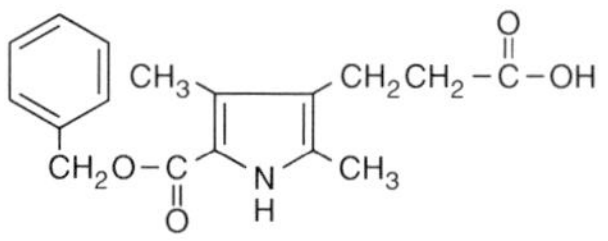

5 gm

S80,805-9
METHYL 2,4-DIMETHYL-5-(TERT-BUTOXYCARBONYL)-3-PYRROLEPROPIONATE

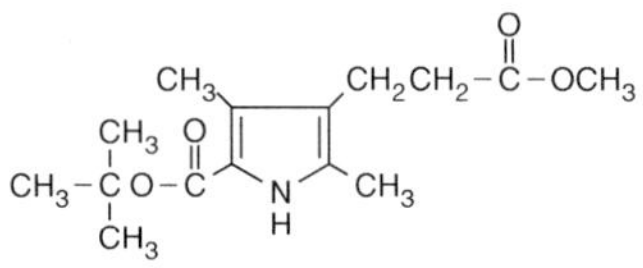

250 mg

S94,339-8
ETHYL 4-(TERT-BUTOXYCARBONYL)-2-(ETHOXYCARBONYL)-5-METHYL-3-PYRROLEACETATE

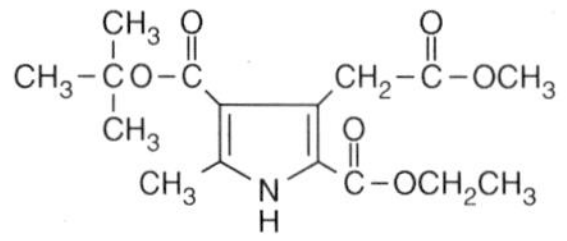

5 gm

S73,927-8
BENZYL 3,5-DIMETHYL-4-(METHOXYCARBONYL)-2-PYRROLECARBOXYLATE

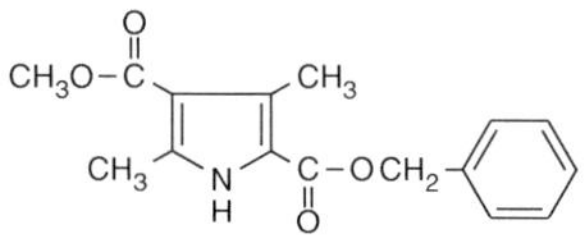

250 mg

S73,925-1
ETHYL 5-(BENZYLOXYCARBONYL)-2,4-DIMETHYL-3-PYRROLEPROPIONATE

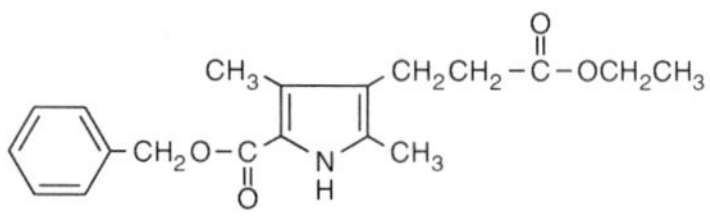

5 gm

S94,336-3
2-BENZYL 4-TERT-BUTYL 3,5-DIMETHYL-2,4-PYRROLEDICARBOXYLATE

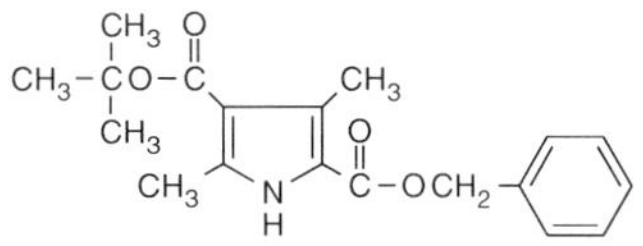

250 mg

S94,335-5
4-BENZYL 2-ETHYL 3-ETHYL-2-METHYL-2,4-PYRROLEDICARBOXYLATE

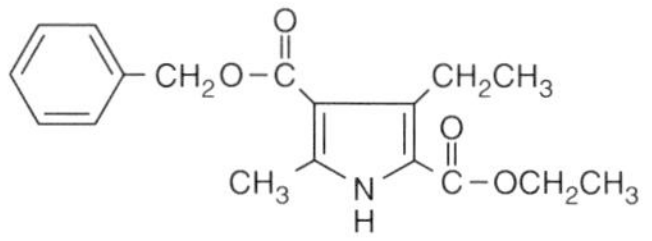

5 gm

S94,343-6
ETHYL 2,4-DIMETHYL-5-(DIMETHYLCARBAMOYL)-3-PYRROLECARBOXYLATE

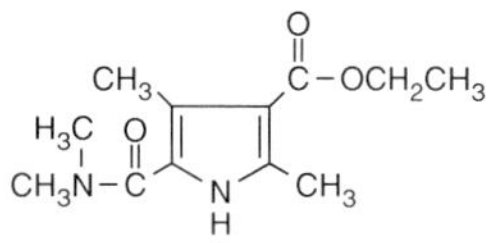

5 gm

S94,333-9
ETHYL 2,4-DIMETHYL-5-ETHYL-3-PYRROLECARBOXYLATE

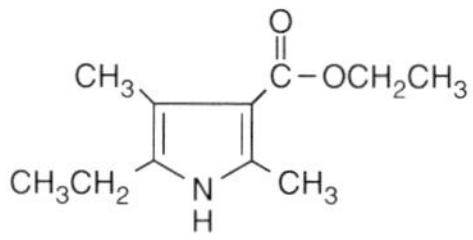

250 mg

S34,836-8
ETHYL 4,5-DIPHENYL-2-METHYL-3-PYRROLECARBOXYLATE

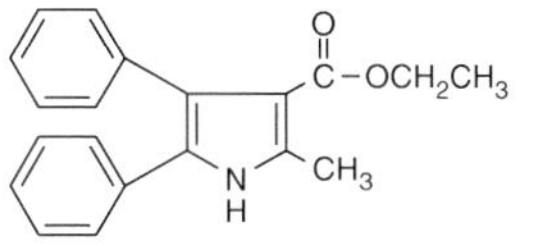

5 gm

S94,331-2
METHYL 5-(BENZYLOXYCARBONYL)-2-CARBOXY-4-METHYL-3-PYRROLEPROPIONATE

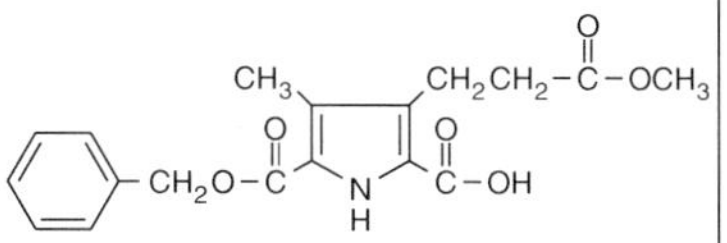

250 mg

S75,960-0
ETHYL 5-CARBOXY-2-IODO-4-METHYL-3-PYRROLECARBOXYLATE

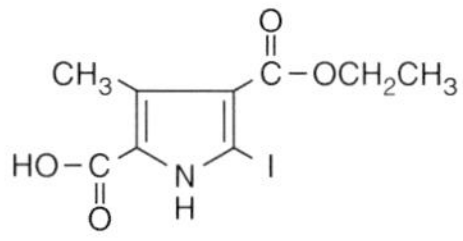

250 mg

S82,231-0
1-ACETYLPYRROLE

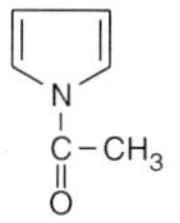

250 mg

S94,251-0
4-ETHYL-N,N,3,5-TETRAMETHYL-2-PYRROLECARBOXAMIDE

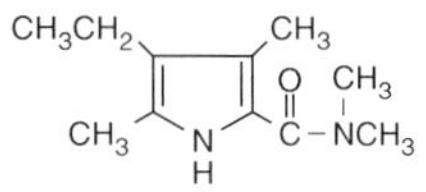

5 gm

S60,566-2
2,5-DIMETHYL-1-HEXYLPYRROLE

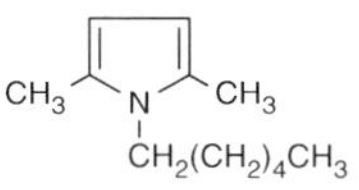

250 mg

S43,391-8
METHYL 5-CHLOROMETHYL-2-FURANCARBOXYLATE

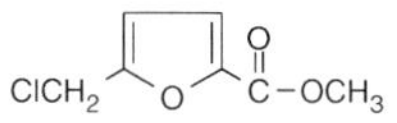

250 mg

S83,645-1
FURFURYL 2-CHLOROACETATE

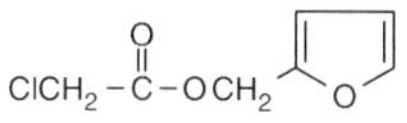

250 mg

S83,644-3
FURFURYL 2,2-DICHLOROACETATE

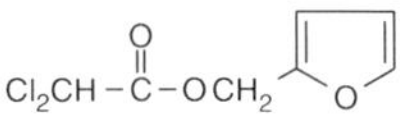

250 mg

S91,698-6
2'-CHLORO-5'-(TRIFLUOROMETHYL)-2-FURANILIDE

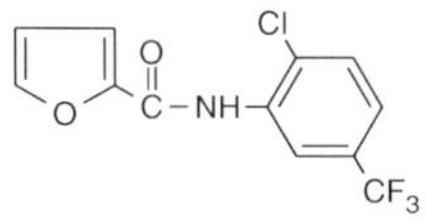

250 mg

S90,197-0
1-[4-CHLORO-3-(TRIFLUOROMETHYL)-PHENYL]-3-FURFURYLUREA

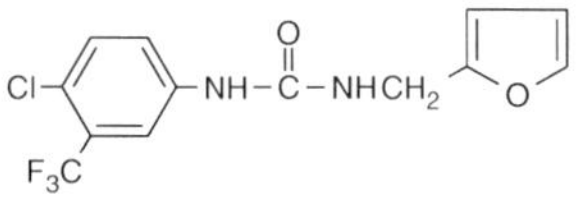

250 mg

S90,260-8
1-[4-CHLORO-2-(TRIFLUOROMETHYL)-PHENYL]-3-FURFURYLUREA

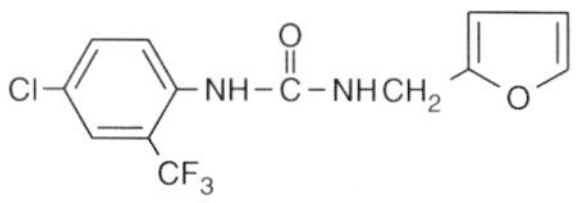

250 mg

S72,345-2
2-(FURFURYLUREIDO)ACETIC N2-[N-(2-CHLORO-5-TRIFLUOROMETHYLPHENYL)-CARBAMOYL]HYDRAZIDE

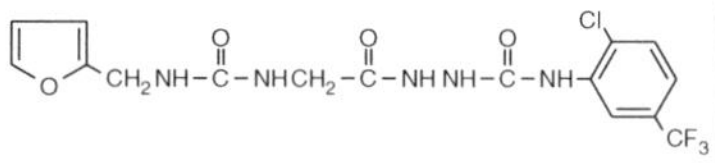

5 gm

S75,320-3
N-(1-HYDROXY-2,2,2-TRICHLOROETHYL)-2-FURAMIDE

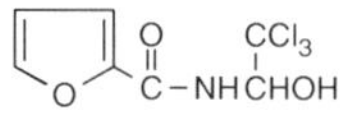

5 gm

S64,335-1
5-[2-NITRO-4-(TRIFLUOROMETHYL)-PHENYL]-2-FURALDEHYDE

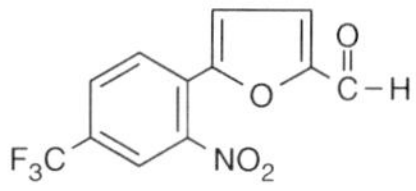

5 gm

S69,236-0
2-(FURFURYLUREIDO)ACETIC N2-(N-[2-(TRIFLUOROMETHYL)PHENYL]CARBAMOYL)-HYDRAZIDE

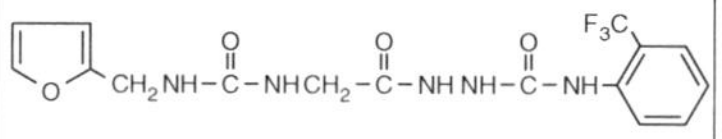

250 mg

S61,169-7
1-FURFURYL-3-(3-TRIFLUOROMETHYLPHENYL)UREA

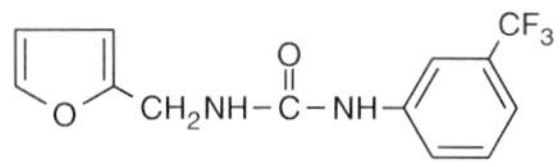

5 gm

S91,327-8
1-FURFURYL-1-METHYL-3-[2-(TRIFLUOROMETHYL)PHENYL]UREA

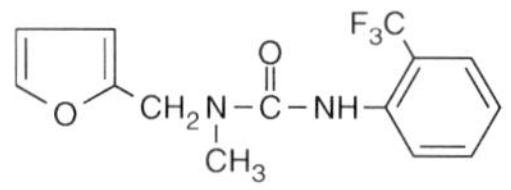

250 mg

S66,614-9
1-[2-CHLORO-5-(TRIFLUOROMETHYL)PHENYL]-3-FURFURYLUREA

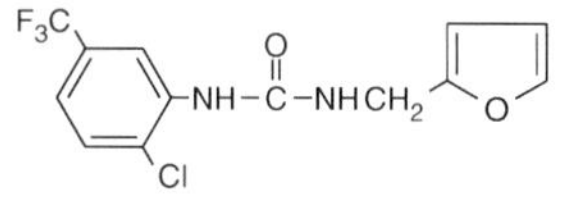

5 gm

S55,193-7
1-(2-FURYL)-4,4,4-TRIFLUORO-1,3-BUTANEDIONE

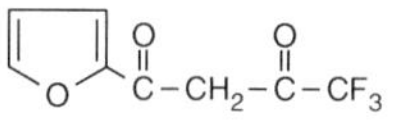

250 mg

S71,774-6
FURFURYL N-(3-TRIFLUOROMETHYLPHENYL)-CARBAMATE

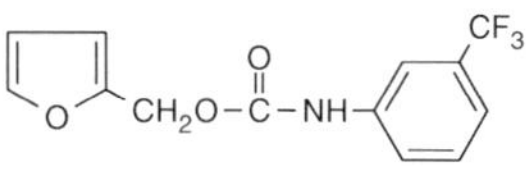

5 gm

S76,427-2
N-FURFURYL-2,2,2-TRICHLOROACETAMIDE

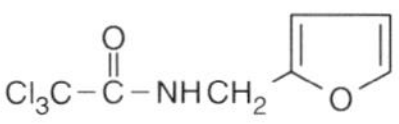

5 gm

S77,966-0
2'-(TRIFLUOROMETHYL)-2-FURANILIDE

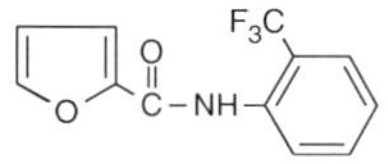

250 mg

S70,863-1
1-FURFURYL-3-(2-TRIFLUOROMETHYLPHENYL)UREA

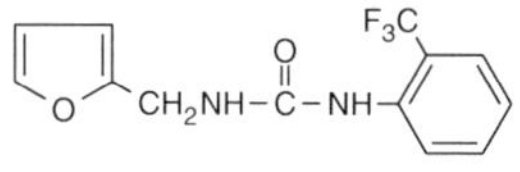

5 gm

S85,404-2
1-(5-CHLORO-2,4-DIMETHOXYPHENYL)-3-FURFURYLUREA

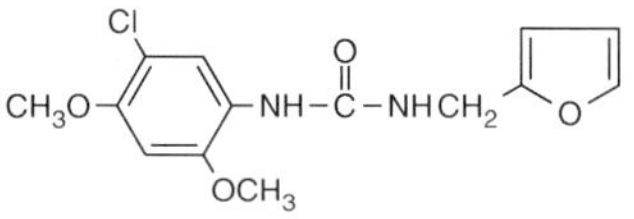

250 mg

S70,058-4
4'-CHLORO-2',5'-DIMETHOXY-2-FURANILIDE

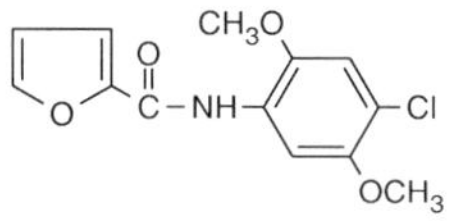

250 mg

S79,475-9
FURFURYL N-(5-CHLORO-2-METHOXYPHENYL)CARBAMATE

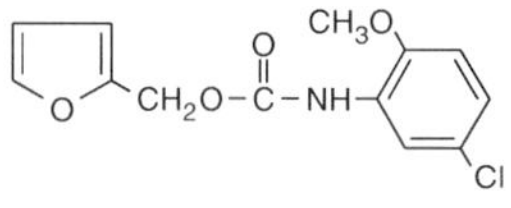

250 mg

S85,100-0
FURFURYL N-(4-CHLORO-2-METHOXYPHENYL)CARBAMATE

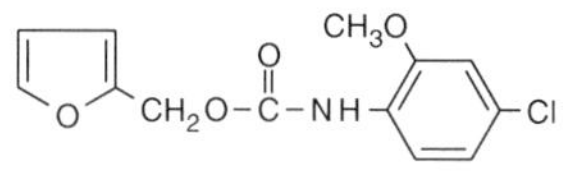

250 mg

S91,695-1
4'-CHLORO-2'-PHENOXY-2-FURANILIDE

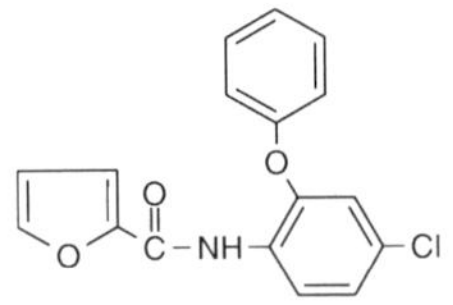

250 mg

S73,846-8
4'-CHLORO-2'-METHOXY-2-FURANILIDE

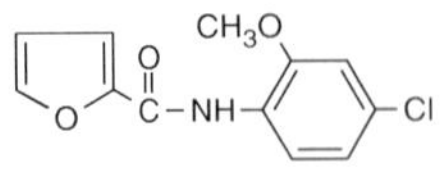

5 gm

S90,153-9
5'-CHLORO-2'-METHOXY-2-FURANILIDE

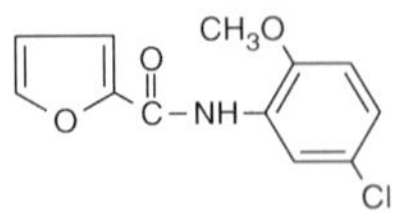

250 mg

S91,282-4
3-(5-CHLORO-2-METHOXYPHENYL)-1-FURFURYL-1-METHYLUREA

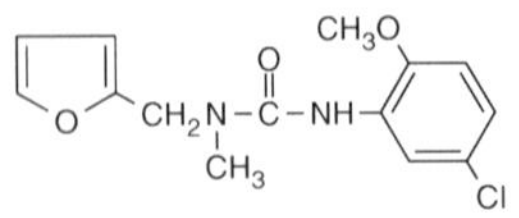

250 mg

S84,413-6
FURFURYL N-(2-CHLORO-4-NITROPHENYL)-CARBAMATE

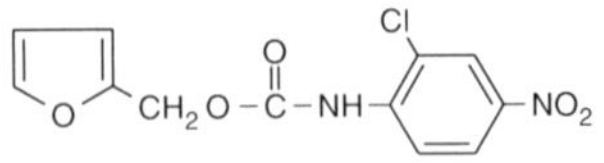

250 mg

S89,215-7
3'-CHLORO-4'-METHYL-2-FURANILIDE

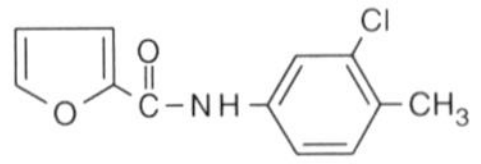

250 mg

S93,124-1
3-(4-CHLORO-3-NITROPHENYL)-1-(2-FURYL)-2-PROPEN-1-ONE

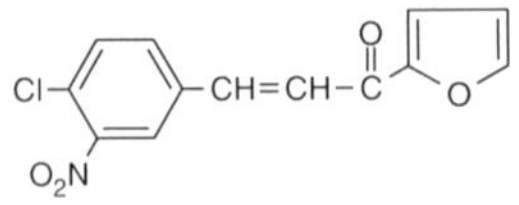

250 mg

S84,414-4
1-(4-FLUOROPHENYL)-3-FUROYL-2-THIOUREA

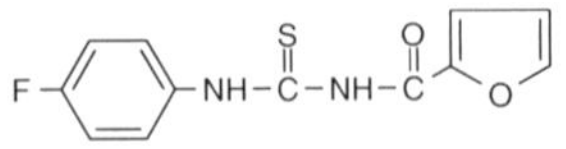

250 mg

S73,832-8
2',4'-DICHLORO-2-FURANILIDE

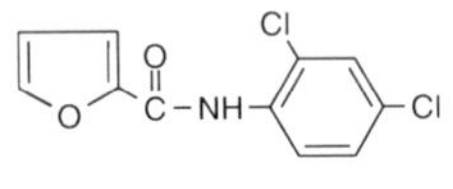

5 gm

S79,199-7
3',4'-DICHLORO-2-FURANILIDE

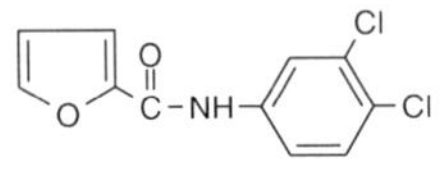

250 mg

S77,221-6
3',5'-DICHLORO-2-FURANILIDE

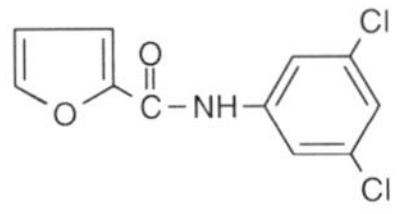

250 mg

S79,618-2
1-(3-CHLORO-4-METHYLPHENYL)-3-(2-FUROYL)-2-THIOUREA

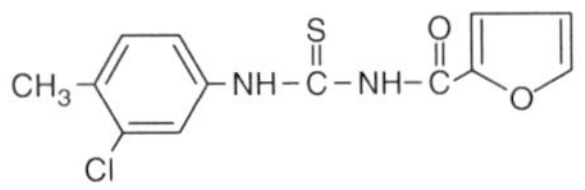

250 mg

S69,816-4
1-(2-FUROYL)-4-(4-FLUORO-2-NITROPHENYL)SEMICARBAZIDE

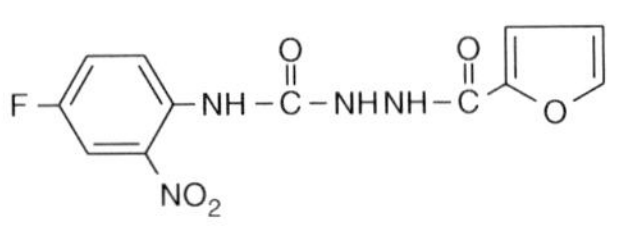

250 mg

S73,849-2
4'-CHLORO-3'-NITRO-2-FURANILIDE

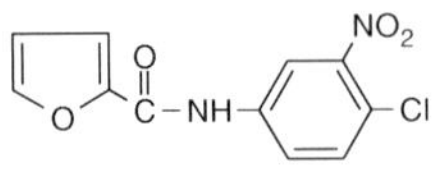

5 gm

S72,802-0
FURFURYL N-(2-CHLORO-5-NITROPHENYL)-CARBAMATE

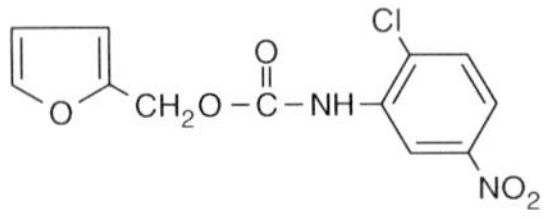

5 gm

S88,719-6
ALPHA-(4-CHLOROPHENYL)-BETA-(2-FURYL)ACRYLONITRILE

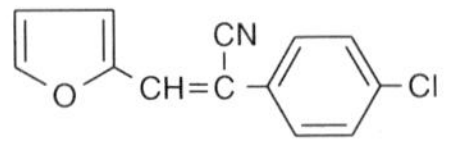

250 mg

S73,450-0
2,6-DICHLORO-N-FURFURYL-N-METHYLBENZYLAMINE

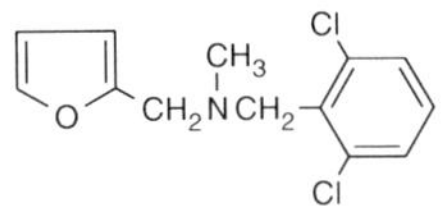

250 mg

S67,422-2
4-(3-CHLOROPHENYL)-1-(3-FURFURYLUREIDOACETYL)SEMICARBAZIDE

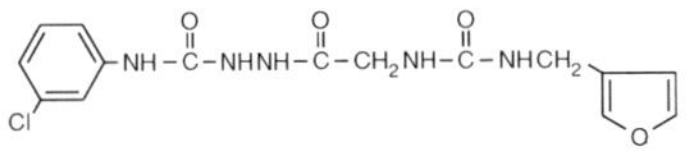

250 mg

S91,607-2
N-FURFURYL-3,4,5,6-TETRABROMOPHTHALAMIC ACID

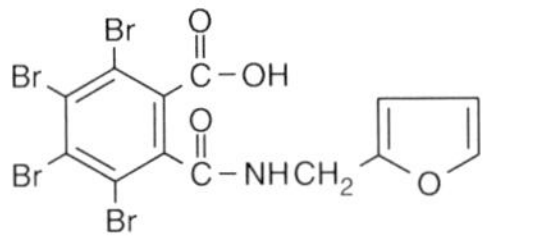

250 mg

S90,424-4
4-(4-FLUOROPHENYL)-1-[3-(FURFURYL)-UREIDOACETYL]SEMICARBAZIDE

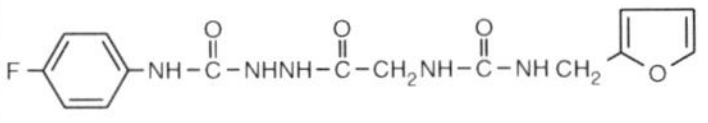

250 mg

S85,135-3
1-(2,3-DICHLOROPHENYL)-3-FURFURYLUREA

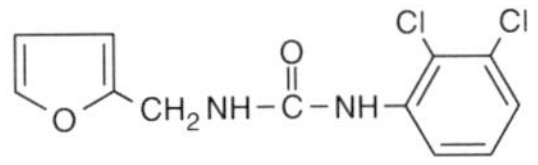

250 mg

S69,493-2
2-CHLORO-4-NITROPHENYL 2-FUROATE

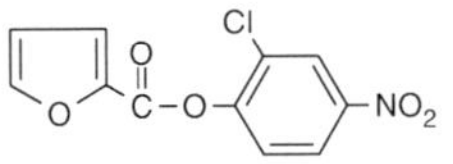

250 mg

S93,123-3
3-(2-CHLORO-5-NITROPHENYL)-1-(2-FURYL)-2-PROPEN-1-ONE

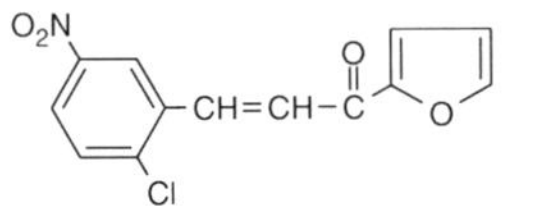

250 mg

S77,260-7
N-FURFURYLTETRACHLOROPHTHALAMIC ACID

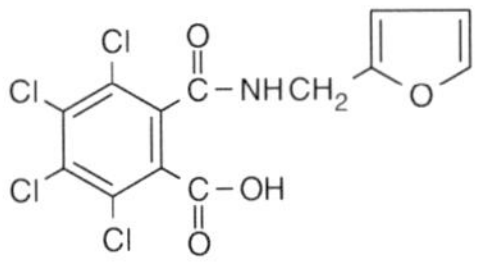

250 mg

S69,985-3
4'-BROMO-2-FURANILIDE

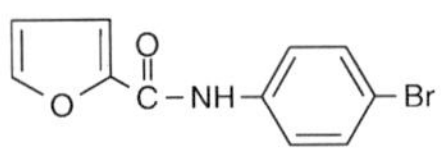

250 mg

S72,038-0
2',6'-DICHLORO-2-FURANILIDE

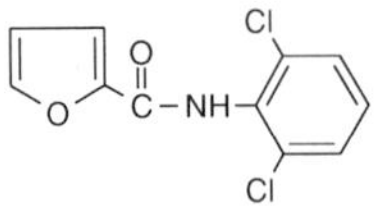

250 mg

S73,882-4
2',5'-DICHLORO-2-FURANILIDE

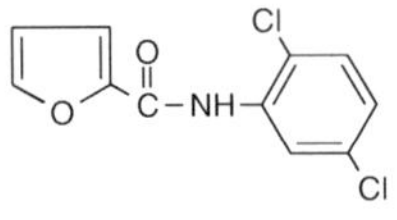

5 gm

S73,788-7
2'-CHLORO-4',6'-DIBROMO-2-FURANILIDE

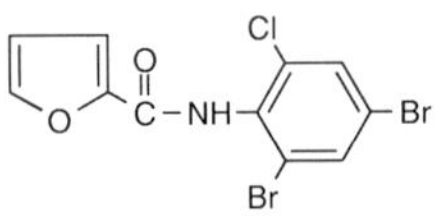

5 gm

S73,789-5
4'-CHLORO-2',6'-DIBROMO-2-FURANILIDE

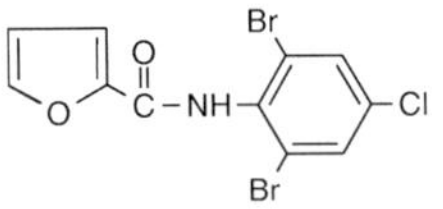

5 gm

S84,415-2
FURFURYL N-(3-FLUOROPHENYL)-CARBAMATE

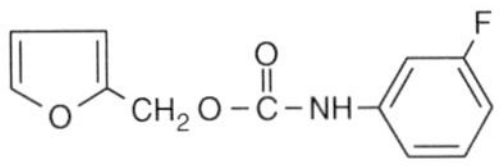

250 mg

S71,010-5
FURFURYL N-(3,4-DICHLOROPHENYL)-CARBAMATE

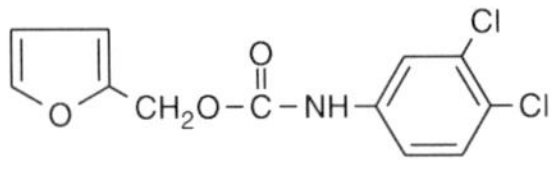

5 gm

S70,942-5
FURFURYL N-(2,4,5-TRICHLOROPHENYL)-CARBAMATE

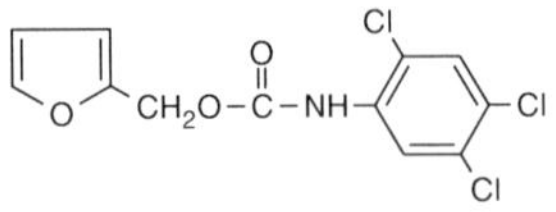

5 gm

S72,515-3
1-(3-CHLORO-4-METHYLPHENYL)-3-FURFURYL-3-METHYLUREA

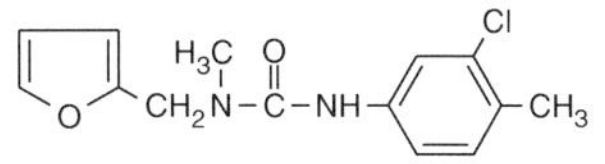

5 gm

S76,956-8
4-CHLORO-N-FURFURYLBENZENESULFONAMIDE

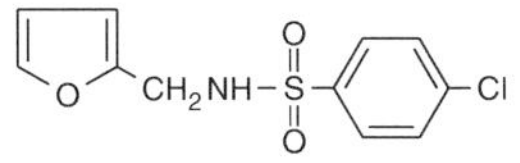

250 mg

S84,012-2
2,5-BIS(PENTAFLUOROPHENYL)FURAN

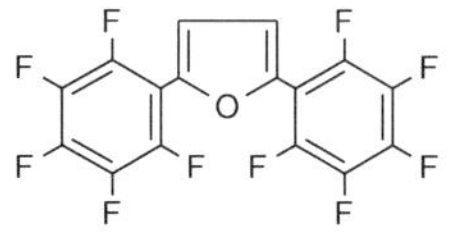

250 mg

S81,169-6
DIMETHYL 3,4-DIHYDROXY-2,5-FURANDICARBOXYLATE, DISODIUM SALT

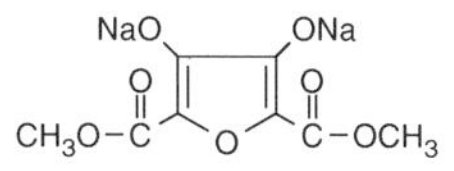

250 mg

S81,171-8
DIETHYL 3,4-DIHYDROXY-2,5-FURANDICARBOXYLATE, DISODIUM SALT

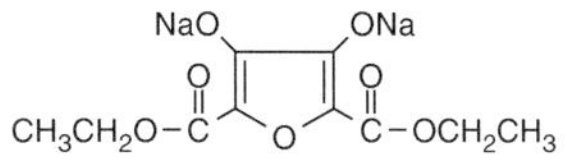

250 mg

S81,172-6
DIETHYL 3,4-DIHYDROXY-2,5-FURANDICARBOXYLATE

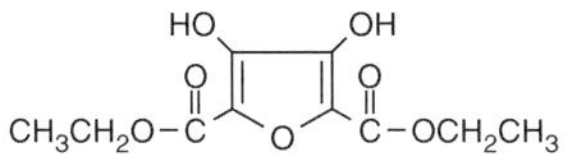

250 mg

S77,965-2
1-(2-FUROYL)-2-(3,4,5-TRIHYDROXYBENZOYL)HYDRAZINE

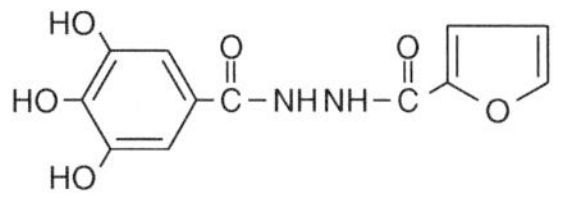

250 mg

S77,951-2
4-(2,3-DIHYDROXYBENZYLIDENE)-2-FURFURYL-5-THIOXO-3-ISOTHIAZOLIDINONE

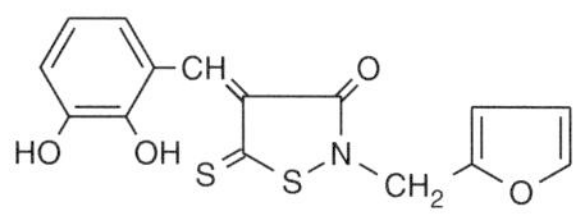

250 mg

S81,459-8
DIMETHYL 3-HYDROXY-4-METHOXY-2,5-FURANDICARBOXYLATE

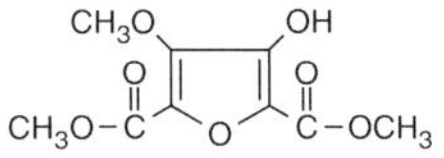

250 mg

S81,053-3
4,5-DIBROMO-2-(HYDROXYMETHYL)-3-FUROIC ACID

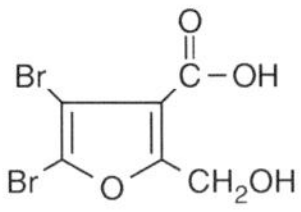

250 mg

S60,591-3
4'-AMINO-5'-CHLORO-2'-HYDROXY-2-FURANILIDE

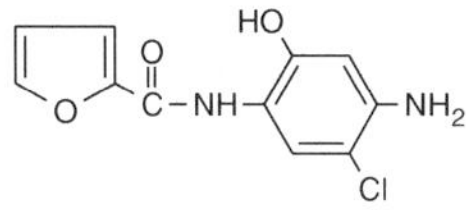

5 gm

S58,313-8
2'-HYDROXY-4'-PIVALAMIDO-2-FURANILIDE

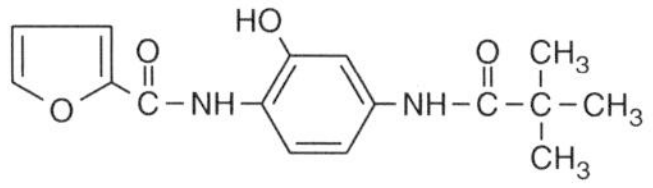

250 mg

S73,865-4
4'-HYDROXY-2-FURANILIDE

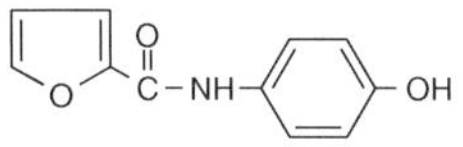

5 gm

S73,851-4
N-(5-HYDROXY-1-NAPHTHYL)-2-FURAMIDE

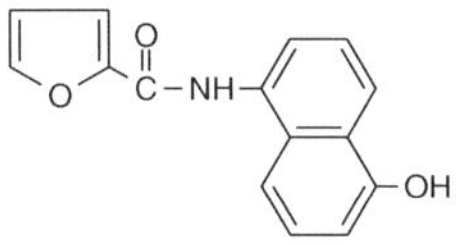

50 mg

S81,139-4
ALPHA,ALPHA-DIMETHYLFURFURYL ALCOHOL

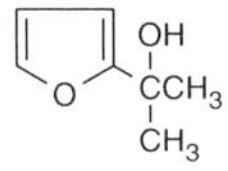

250 mg

S95,240-0
ETHYL 2-HYDROXY-3-FURANCARBOXYLATE

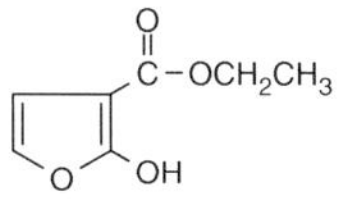

250 mg

S69,169-0
4-FURFURYL-1-(2-HYDROXYBENZOYL)-3-THIOSEMICARBAZIDE

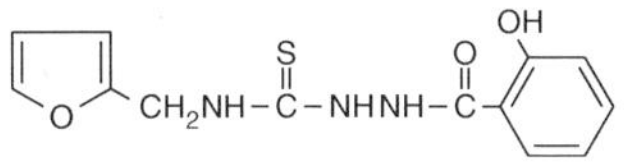

250 mg

S86,102-2
2-METHYL-3-FURANMETHANOL

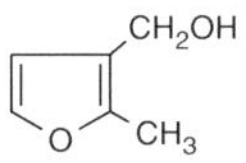

250 mg

S41,063-2
5-FURFURYLIDENE-2-THIOBARBITURIC ACID

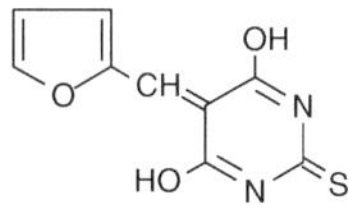

5 gm

S59,252-8
5'-CHLORO-2'-HYDROXY-4'-NITRO-2-FURANILIDE

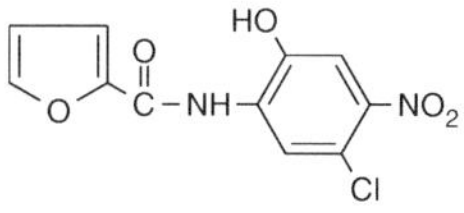

250 mg

S60,538-7
4'-(2-FUROYLAMINO)-3'-HYDROXYMALONANILIC ACID

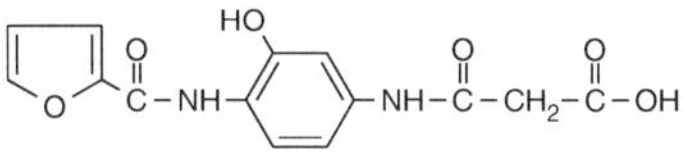

250 mg

S63,343-7
4'-[3-(HYDRAZINOSULFONYL)BENZAMIDO]-2'-HYDROXY-2-FURANILIDE

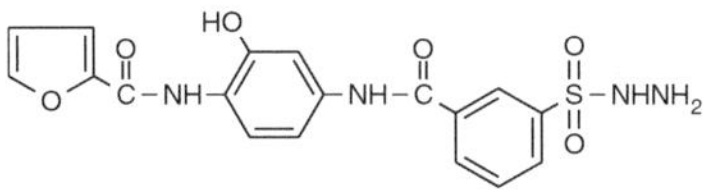

250 mg

S63,769-6
2'-HYDROXY-4'-METHACRYLAMIDO-2-FURANILIDE

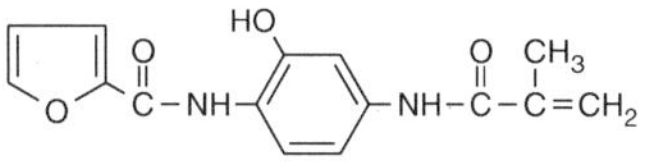

5 gm

S46,281-0
3,4-FURANDIMETHANOL

250 mg

S81,141-6
ALPHA,ALPHA-DIPHENYLFURFURYL ETHER

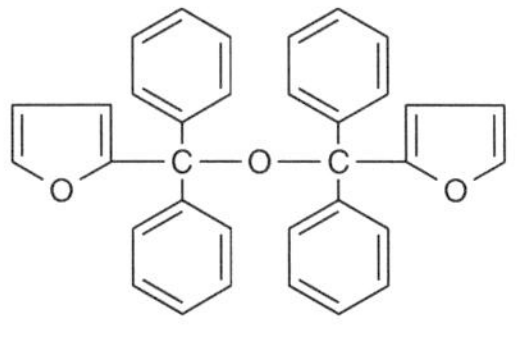

250 mg

S63,979-6
3-(4-BENZYLOXYPHENYL)-1-(2-FURYL)-2-PROPEN-1-ONE

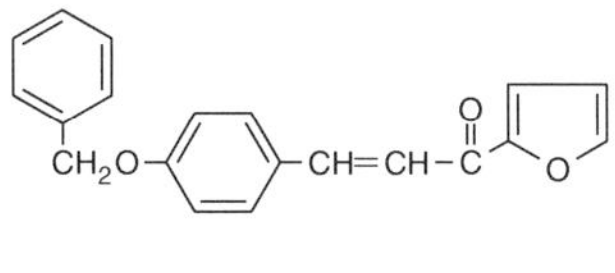

5 gm

S82,019-9
3,4-DIMETHOXY-2,5-FURANDICARBOXYLIC ACID

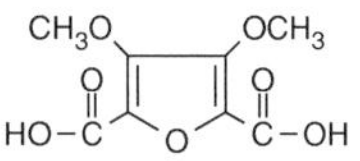

250 mg

S74,857-9
4',5'-DIMETHOXY-2-FURFURYLIDENE-2'-NITROACETOPHENONE

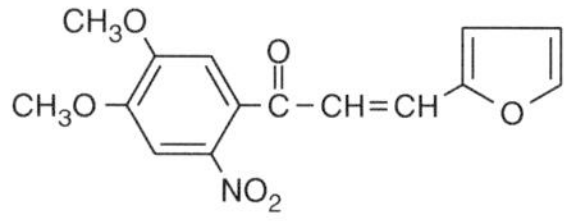

250 mg

S95,250-8
METHYL 5-(3,4-DIMETHOXYPHENYL)-2-METHYL-3-FURANCARBOXYLATE

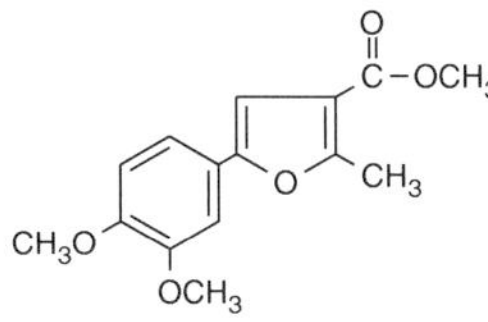

50 mg

S70,050-9
4-(2,5-DIMETHOXYPHENYL)-1-(2-FUROYL)SEMICARBAZIDE

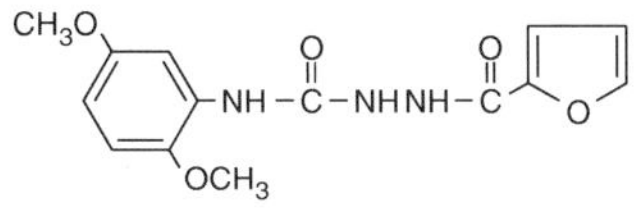

250 mg

S73,834-4
2',5'-DIMETHOXY-2-FURANILIDE

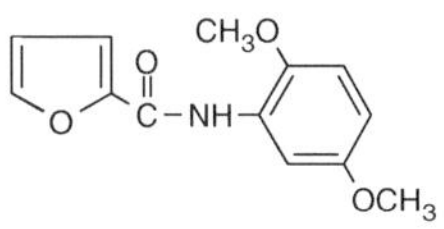

5 gm

S90,364-7
4-(2,5-DIMETHOXYPHENYL)-1-(3-FURFURYLUREIDOACETYL)SEMICARBAZIDE

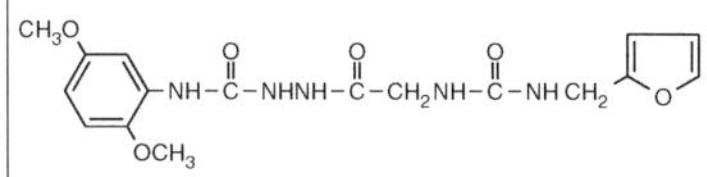

250 mg

S69,188-7
4,4'-BIS(FURFURYLUREIDO)-3,3'-DIMETHOXYBIPHENYL

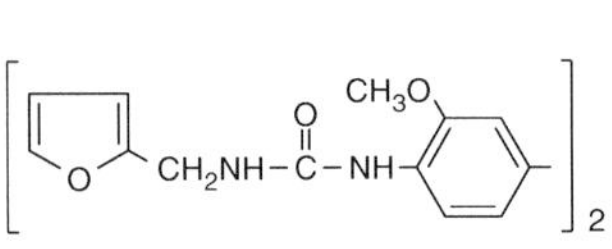

250 mg

S89,216-5
2'-METHOXY-2-FURANILIDE

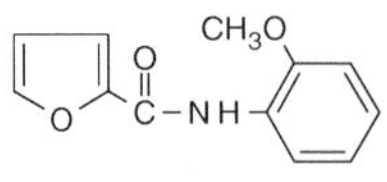

250 mg

S89,211-4
4'-METHOXY-2-FURANILIDE

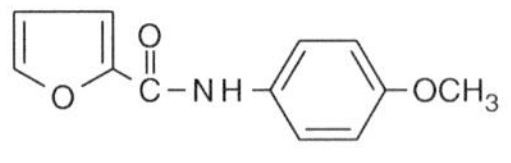

250 mg

S81,419-9
2-(4-METHOXYBENZOYL)-5-NITROFURAN

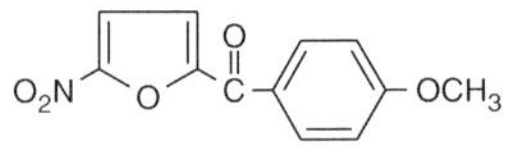

250 mg

S83,945-0
4-ETHOXYPHENYL 2-FURYL KETONE

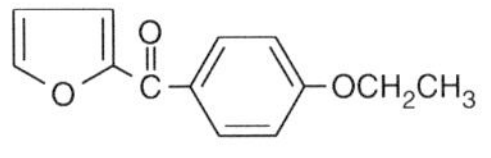

250 mg

S84,494-2
1-(4-ETHOXYPHENYL)-3-FUROYL-2-THIOUREA

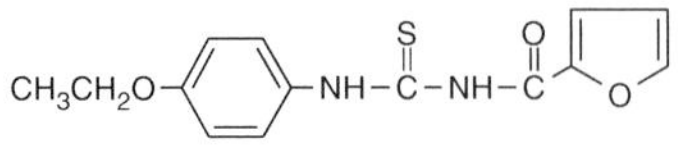

250 mg

S73,838-7
3'-METHOXY-2-FURANILIDE

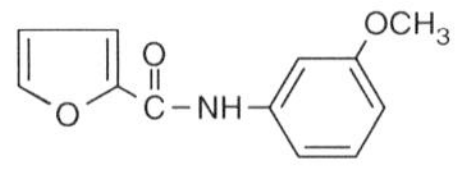

5 gm

S70,068-1
2'-METHOXY-4'-NITRO-2-FURANILIDE

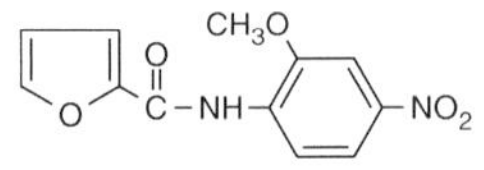

250 mg

S78,030-8
4'-ACETAMIDO-3'-METHOXY-2-FURANILIDE

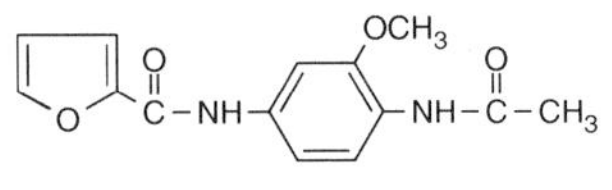

250 mg

S88,723-4
BETA-(2-FURYL)-ALPHA-(4-METHOXYPHENYL)ACRYLONITRILE

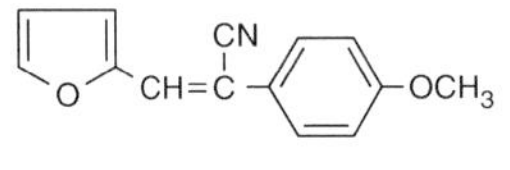

250 mg

S77,933-4
1-(FURFURYLUREIDOACETYL)-4-(2-METHOXY-5-NITROPHENYL)SEMICARBAZIDE

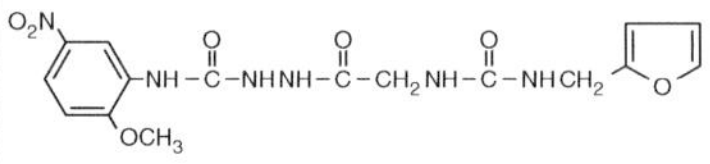

250 mg

S72,365-7
2-(FURFURYLUREIDO)ACETIC ACID N2-[N-(4-ETHOXYPHENYL)CARBAMOYL]-HYDRAZIDE

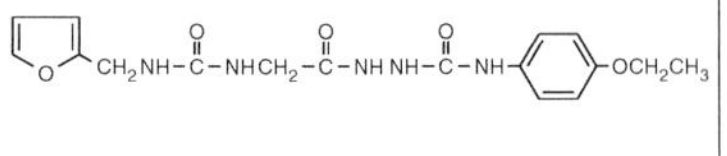

5 gm

S91,396-0
2'-METHOXY-5'-METHYL-2-FURANILIDE

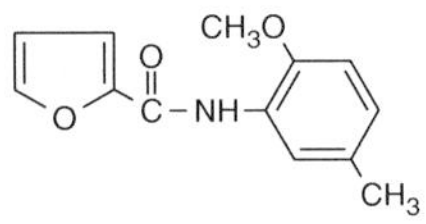

250 mg

S63,643-6
FURFURYL N-(2-METHOXYPHENYL)-CARBAMATE

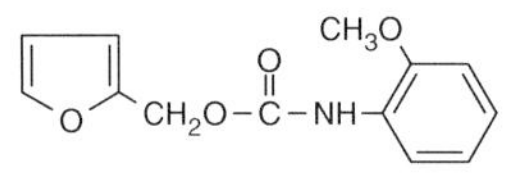

5 gm

S79,576-3
FURFURYL N-(2-METHOXY-5-METHYLPHENYL)CARBAMATE

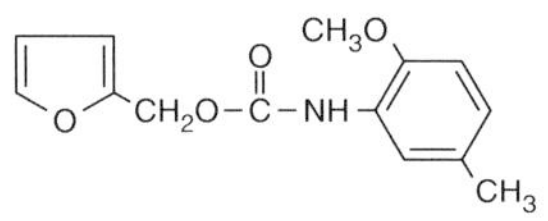

250 mg

S60,940-4
1-(FURFURYL)-3-(3-METHOXYPHENYL)UREA

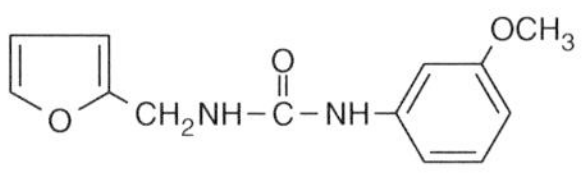

5 gm

S61,057-7
1-(O-ANISYL)-3-FURFURYLUREA

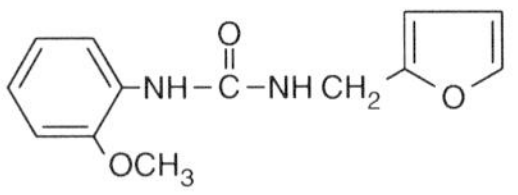

5 gm

S61,290-1
1-(P-ANISYL)-3-FURFURYLUREA

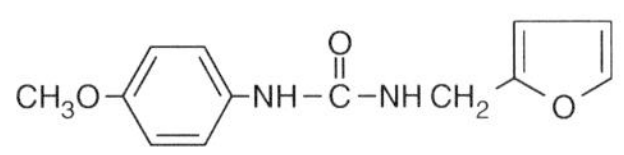

5 gm

S61,093-3
1-(2,5-DIMETHOXYPHENYL)-3-FURFURYLUREA

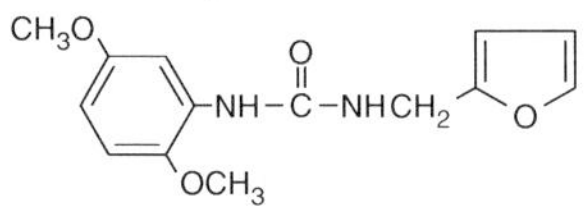

5 gm

S65,026-9
1-(3-CHLORO-2-METHOXYPHENYL)-3-(FURFURYL)UREA

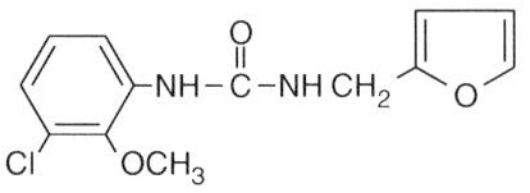

5 gm

S55,827-3
1-(2-FURYL)-3-(4-METHOXYPHENYL)-2-PROPEN-1-ONE

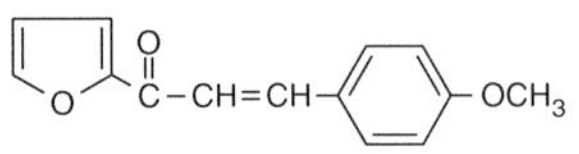

5 gm

S74,372-0
4'-ETHOXY-2-FURANILIDE

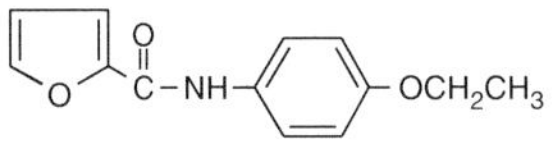

5 gm

S73,798-4
N,N'-(OXYDI-4,1-PHENYLENE)BIS-(2-FURAMIDE)

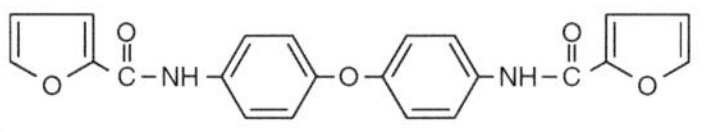

50 mg

S70,023-1
1-(4-BUTOXYPHENYL)-3-FURFURYLUREA

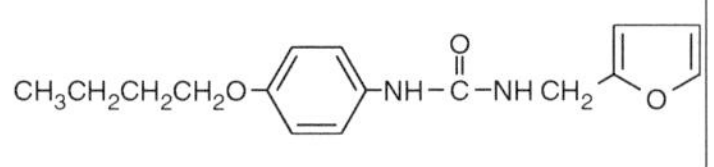

250 mg

S69,492-4
3-(4-BUTOXYPHENYL)-1-FURFURYL-1-METHYLUREA

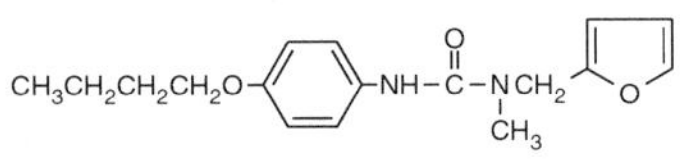

250 mg

S69,089-9
1-FURFURYL-3-(4-METHOXY-2-METHYLPHENYL)UREA

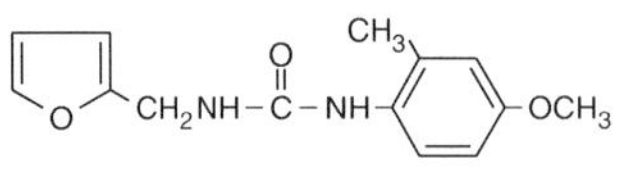

250 mg

S84,236-2
3,4,5-TRIPHENYL-2(3H)-FURANONE

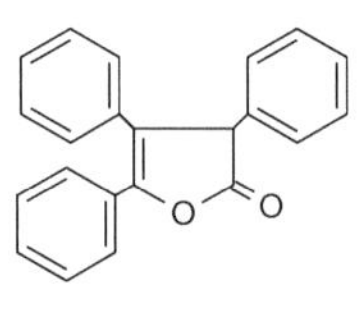

250 mg

S95,596-5
4,7-DIHYDRO-1,3-DIPHENYLISOBENZOFURAN

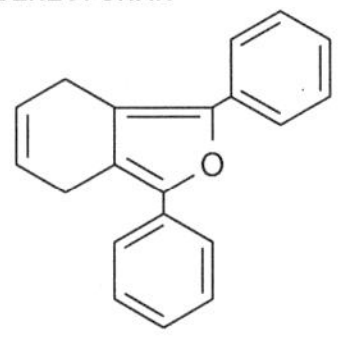

50 mg

S84,679-1
TETRAPHENYLFURAN

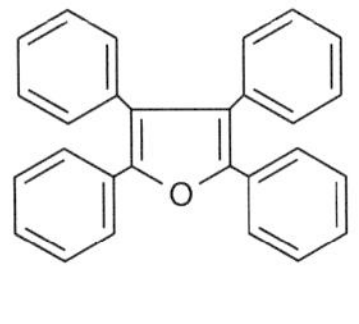

250 mg

S45,019-7
2-ACETYL-3,4-BIS(ACETOXYMETHYL)-FURAN

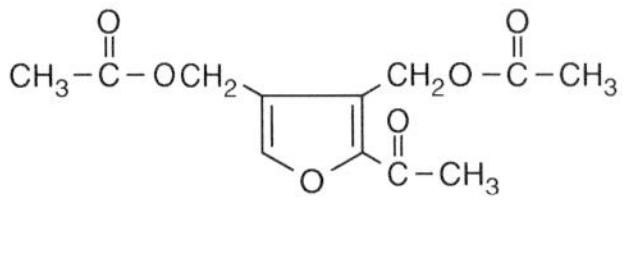

5 gm

S82,020-2
DIMETHYL 3,4-DIACETOXY-2,5-FURANDICARBOXYLATE

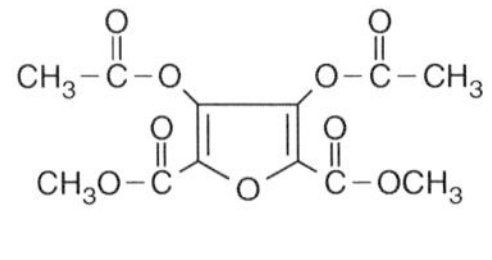

250 mg

S81,342-7
DIETHYL 3,4-BIS(BENZOYLOXY)-2,5-FURANDICARBOXYLATE

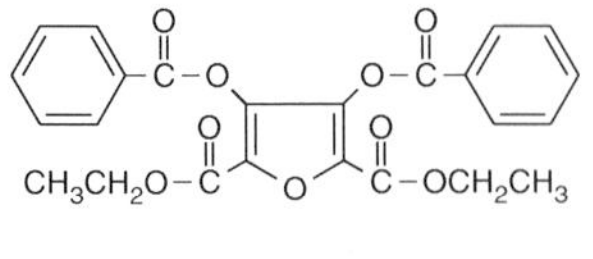

250 mg

S73,833-6
4'-METHYL-2'-NITRO-2-FURANILIDE
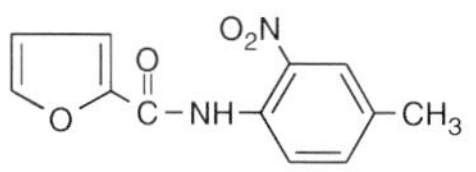
5 gm

S85,191-4
FURFURYL N-(4-METHYL-2-NITROPHENYL)-CARBAMATE
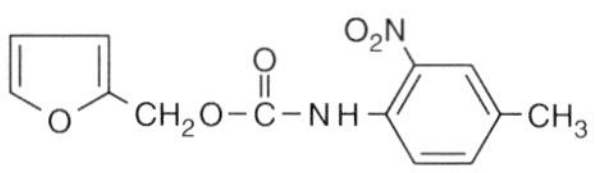
250 mg

S59,270-6
2-FURAMIDO-4-NITROPHENYL 2-FUROATE
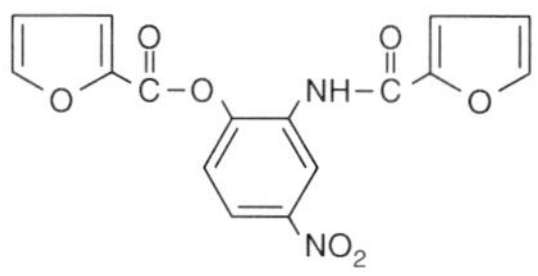
5 gm

S78,093-6
2-FURAMIDO-4-NITROBENZOIC ACID
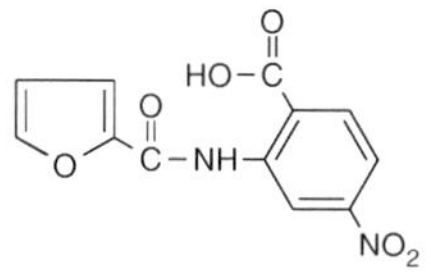
250 mg

S73,881-6
4'-METHYL-3'-NITRO-2-FURANILIDE
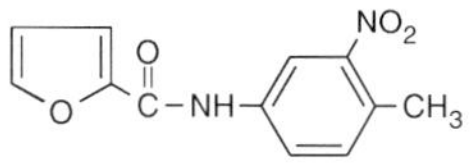
5 gm

S84,816-6
FURFURYL N-(2-METHYL-5-NITROPHENYL)-CARBAMATE
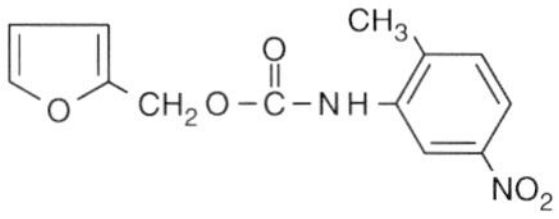
250 mg

S63,639-8
FURFURYL N-(3-NITROPHENYL)CARBAMATE
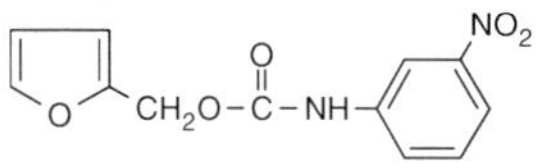
5 gm

S66,728-5
N,N'-BIS(2-FURYLMETHOXYCARBONYL)-4-METHYL-1,3-PHENYLENEDIAMINE
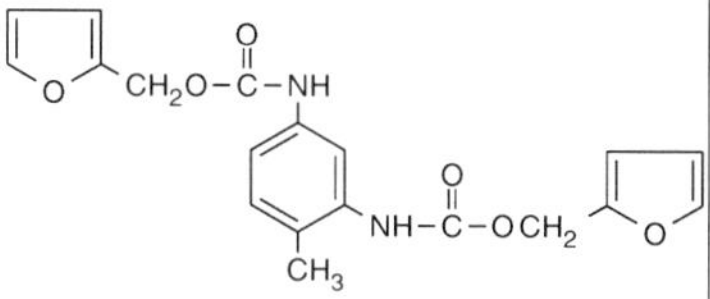
5 gm

S73,864-6
4'-ACETAMIDO-2-FURANILIDE
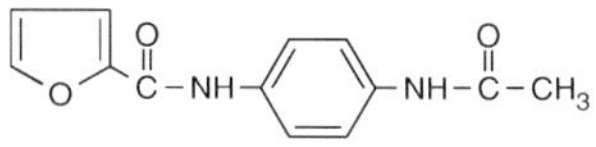
5 gm

S51,765-8
4-(METHYLAMINO)PHENYL 2-FUROATE
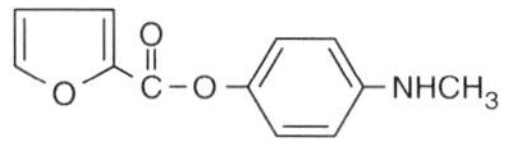
5 gm

S94,628-1
5-(DIMETHYLAMINO)-2-FURALDEHYDE
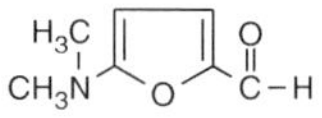
250 mg

S65,403-5
4,4'-METHYLENEBIS(FURFURYL-N-PHENYLCARBAMATE)
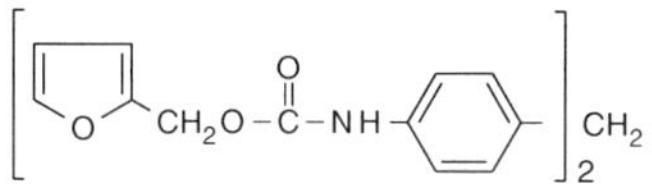
250 mg

S89,213-0
2',4'-DIMETHYL-2-FURANILIDE
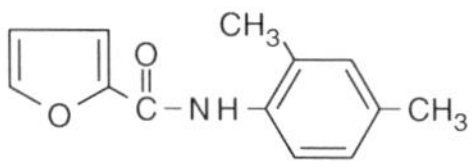
250 mg

S89,209-2
N-(1-NAPHTHYL)-2-FURAMIDE
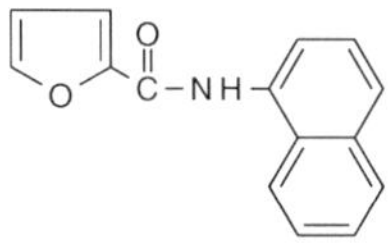
250 mg

S81,422-9
2,5-BIS(3-NITROBENZOYL)-3-NITROFURAN
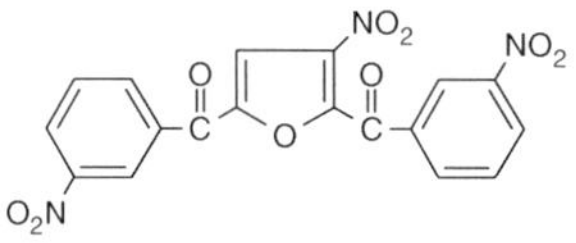
250 mg

S83,976-0
5-(4-NITROPHENYL)-2-FUROIC ACID
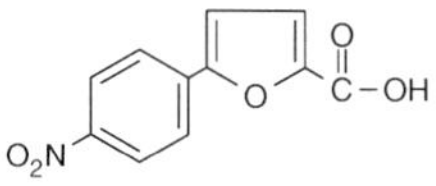
250 mg

S81,417-2
2-BENZOYL-5-NITROFURAN
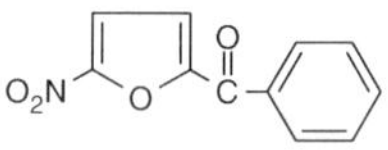
250 mg

S81,416-4
METHYL 5-NITRO-2-FUROATE
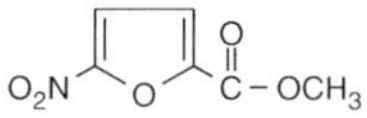
250 mg

S81,463-6
ETHYL 5-NITRO-2-FUROATE
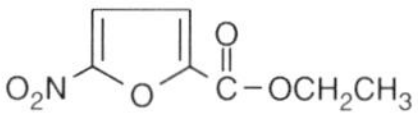
250 mg

S83,939-6
5-NITRO-2-FUROYL CHLORIDE
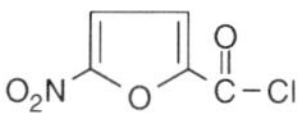
250 mg

S82,208-6
5-NITRO-2-FURANILIDE
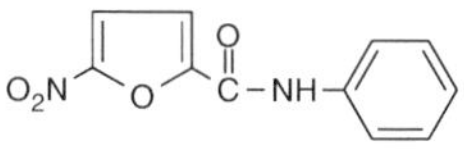
50 mg

S43,136-2
ETHYL 5-FORMYL-2-METHYL-3-FURANCARBOXYLATE
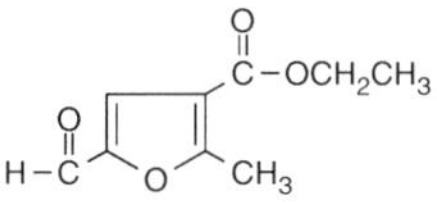
250 mg

S94,627-3
5-BROMO-2-FURALDEHYDE
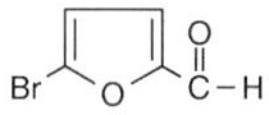
250 mg

S94,626-5
5-IODO-2-FURALDEHYDE
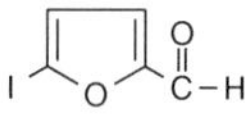
250 mg

S81,688-4
GAMMA-OXO-2-FURANBUTYRIC ACID
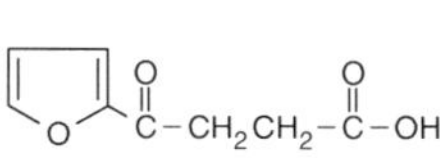
250 mg

S64,279-7
4-METHYL-4-(2-FUROYL)PIMELONITRILE
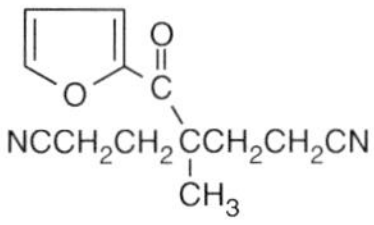
5 gm

S59,368-0
4-(2-CYANOETHYL)-4-(5-METHYL-2-FUROYL)PIMELONITRILE
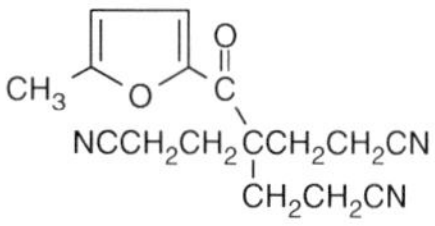
250 mg

S59,323-0
2-BUTYRYLFURAN
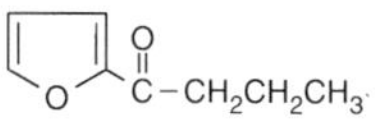
5 gm

S81,140-8
2,5-DIACETYLFURAN
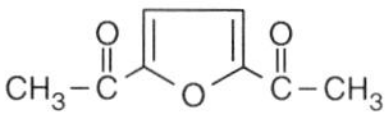
250 mg

S63,859-5
3-(4-CHLOROPHENYL)-1-(2-FURYL)-2-PROPEN-1-ONE
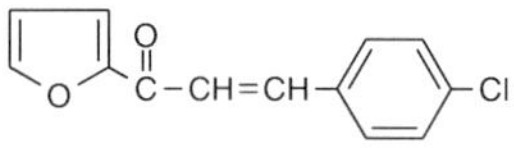
5 gm

S63,860-9
3-(2-CHLOROPHENYL)-1-(2-FURYL)-2-PROPEN-1-ONE
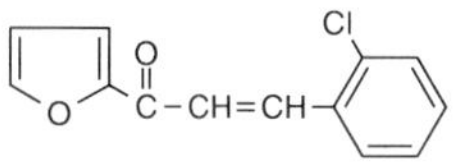
5 gm

S83,683-4
1-(2-FURYL)-1-OCTADECANONE
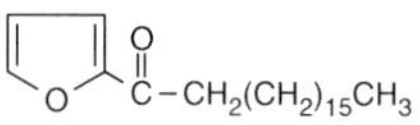
250 mg

S83,650-8
2-FURYL 2-PYRROLYL KETONE

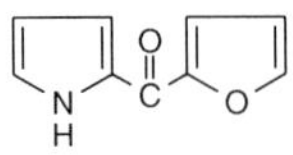

250 mg

S83,850-0
2,3,4,5-FURANTETRACARBOXYLIC ACID, MONOSODIUM SALT

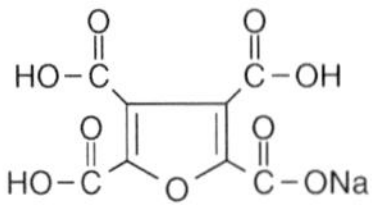

250 mg

S84,045-9
5-BROMO-2,4-FURANDICARBOXYLIC ACID

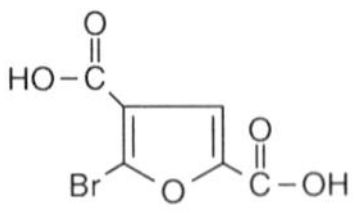

250 mg

S83,849-7
2,3,4,5-FURANTETRACARBOXYLIC ACID

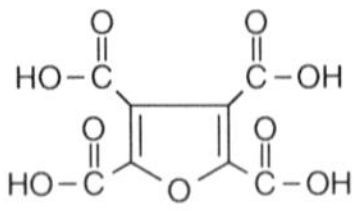

250 mg

S92,722-8
2-FUROIC ACID, ANILINE SALT

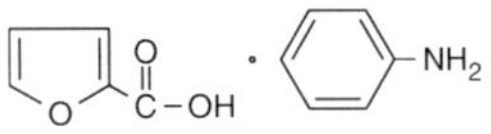

250 mg

S81,374-5
3,5-DIBROMO-2-FUROIC ACID

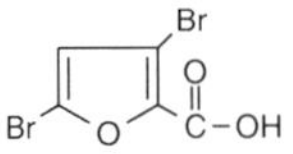

250 mg

S81,436-9
3,5-DIMETHYL-2,4-FURANDICARBOXYLIC ACID

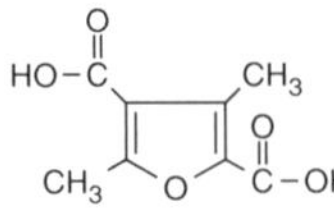

250 mg

S82,595-6
2-FUROIC ACID, DODECYLAMINE SALT

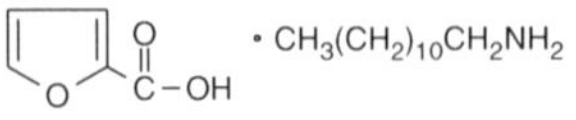

250 mg

S82,600-6
5-CHLORO-2-FUROIC ACID

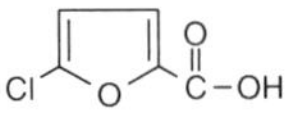

250 mg

S81,050-9
3,4-DICHLORO-2-FUROIC ACID

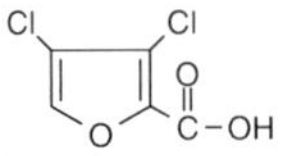

250 mg

S81,334-6
4-BROMO-2-FUROIC ACID

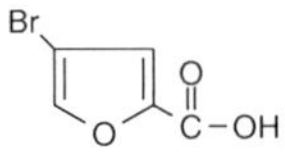

250 mg

S81,819-4
3-BROMO-2-FUROIC ACID

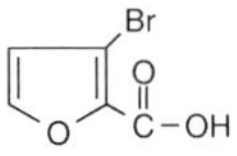

250 mg

S81,132-7
5-BROMO-3,4-DICHLORO-2-FUROIC ACID

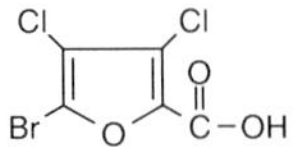

250 mg

S81,051-7
4,5-DIBROMO-2-FUROIC ACID

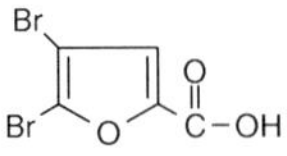

250 mg

S81,481-4
3,4,5-TRIBROMO-2-FUROIC ACID

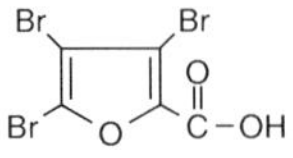

250 mg

S81,134-3
5-METHYL-2-FUROIC ACID

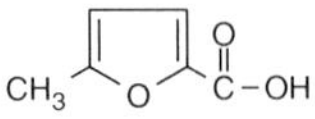

250 mg

S84,003-3
5-TERT-BUTYL-2-FUROIC ACID

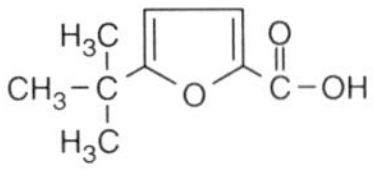

250 mg

S79,459-7
FURFURYL N-[4-(ETHOXYCARBONYL)-PHENYL]CARBAMATE

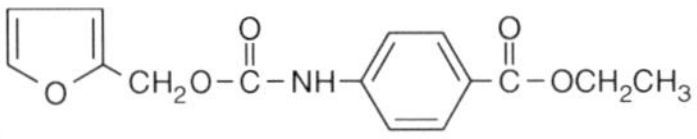

250 mg

S81,453-9
DIMETHYL 3,4-DIIODO-2,5-FURANDICARBOXYLATE

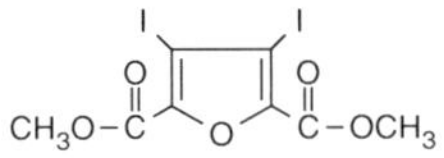

250 mg

S81,054-1
METHYL 5-BROMO-2-FUROATE

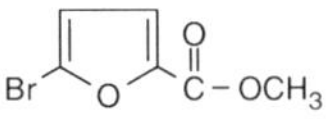

250 mg

S81,450-4
METHYL 4-BROMO-2-FUROATE

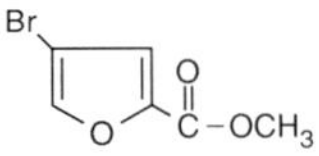

250 mg

S81,137-8
ETHYL 3,4-DICHLORO-2-FUROATE

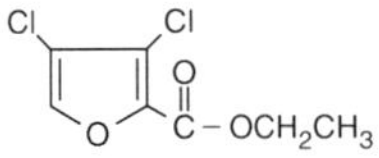

250 mg

S81,138-6
ETHYL 5-BROMO-2-FUROATE

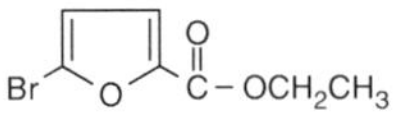

250 mg

S81,135-1
ETHYL 3,4,5-TRICHLORO-2-FUROATE

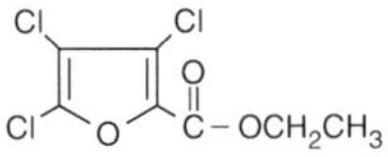

250 mg

S81,452-0
ETHYL 4,5-DIBROMO-2-FUROATE

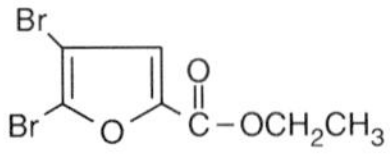

250 mg

S81,462-8
ETHYL 3,5-DIBROMO-2-FUROATE

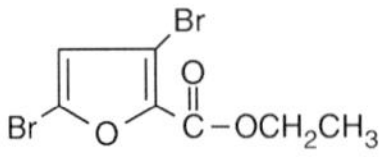

250 mg

S83,916-7
ETHYL 3,4-DIBROMO-2-FUROATE

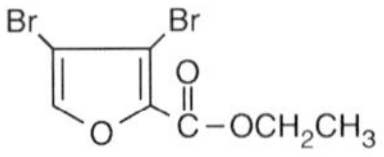

250 mg

S34,076-6
ALLYL FUROATE

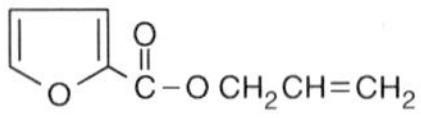

5 gm

S82,704-5
N-DODECYL-2-FURAMIDE

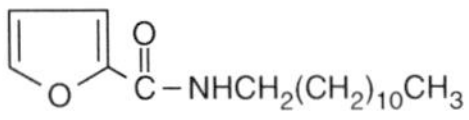

250 mg

S82,705-3
N-OCTADECYL-2-FURAMIDE

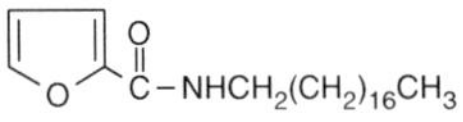

250 mg

S83,974-4
N,N-DIETHYL-5-PENTYL-2-FURAMIDE

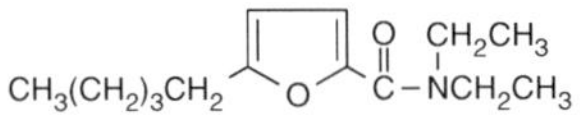

250 mg

S63,841-2
FURFURYL N-PHENYLCARBAMATE

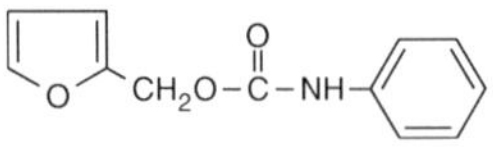

5 gm

S63,825-0
FURFURYL N-(4-CHLOROPHENYL)-CARBAMATE

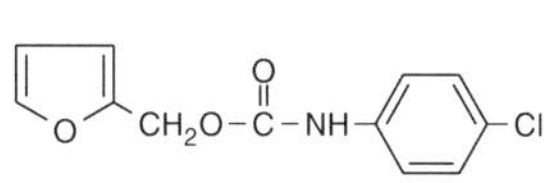

5 gm

S63,836-6
FURFURYL N-(2-CHLOROPHENYL)-CARBAMATE

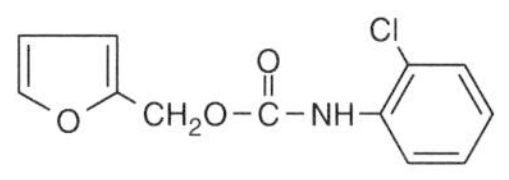

5 gm

S63,951-6
FURFURYL N-(3-CHLOROPHENYL)-CARBAMATE

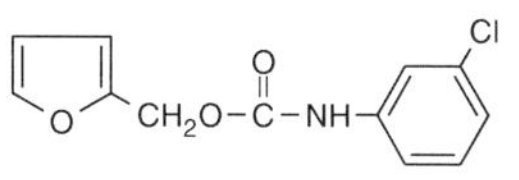

5 gm

S63,651-7
FURFURYL N-(2,3-DICHLOROPHENYL)-CARBAMATE

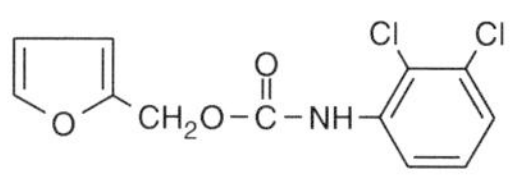

5 gm

S64,480-3
FURFURYL N-(M-TOLYL)CARBAMATE

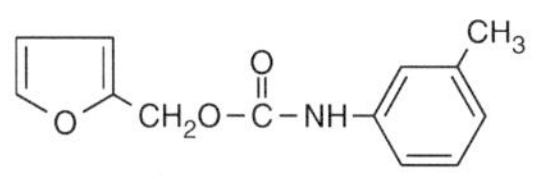

5 gm

S66,759-5
FURFURYL N-(O-TOLYL)CARBAMATE

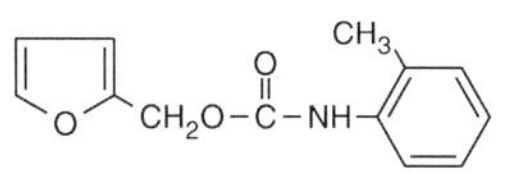

5 gm

S81,478-4
FURFURYL N,N-DIPHENYLCARBAMATE

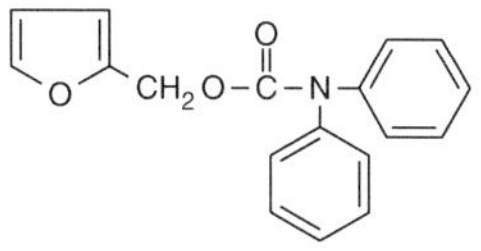

250 mg

S66,790-0
FURFURYL N-(2-ETHYLPHENYL)CARBAMATE

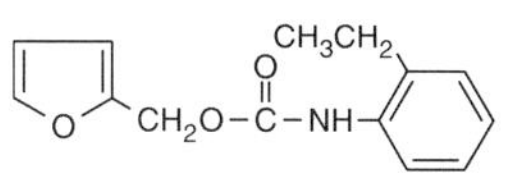

5 gm

S66,721-8
FURFURYL N-(4-BROMOPHENYL)CARBAMATE

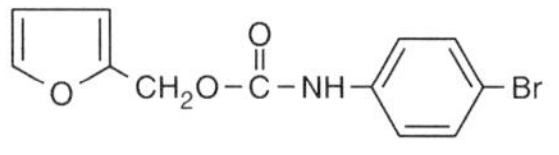

5 gm

S65,236-9
FURFURYL N-(P-TOLYL)CARBAMATE

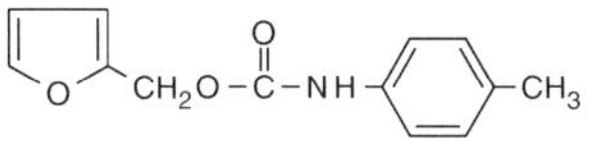

5 gm

S66,515-0
FURFURYL N-(2,3-XYLYL)CARBAMATE

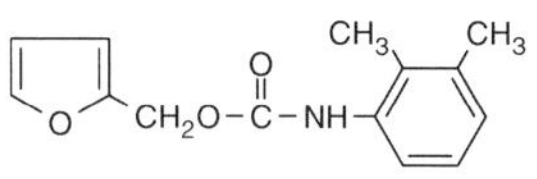

5 gm

S66,654-8
FURFURYL N-(2,6-XYLYL)CARBAMATE

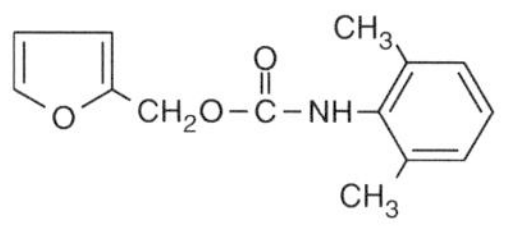

5 gm

S65,158-3
FURFURYL N-(2,4-DICHLOROPHENYL)-CARBAMATE

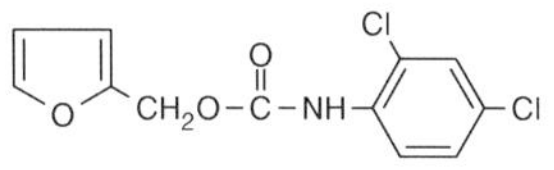

5 gm

S66,574-6
FURFURYL N-(4-CHLORO-2-METHYLPHENYL)CARBAMATE

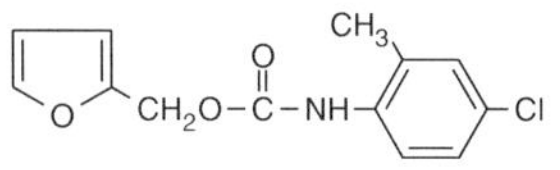

5 gm

S66,723-4
FURFURYL N-(5-CHLORO-2-METHYLPHENYL)CARBAMATE

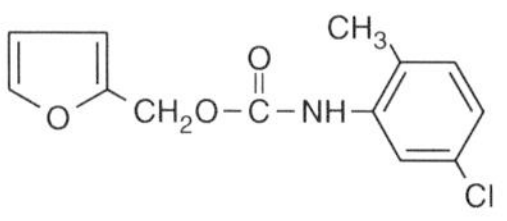

5 gm

S66,554-1
FURFURYL N-(2,4-XYLYL)CARBAMATE

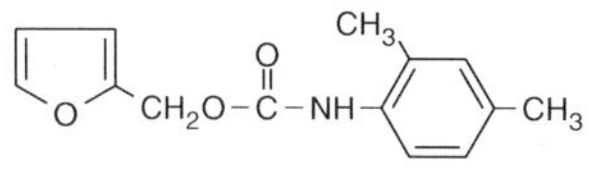

5 gm

S66,650-5
FURFURYL N-(3,4-XYLYL)CARBAMATE

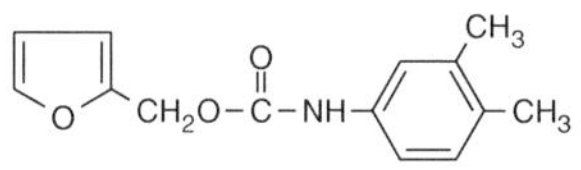

5 gm

S66,802-8
FURFURYL N-(3,5-XYLYL)CARBAMATE

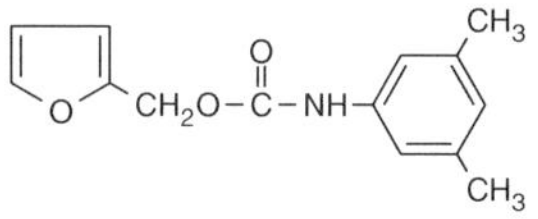

5 gm

S86,795-0
FURFURYL N-(1-NAPHTHYL)CARBAMATE

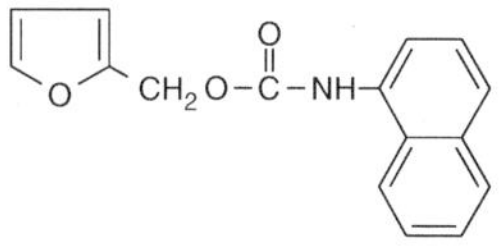

250 mg

S83,654-0
FURFURYL 2,2-DIMETHYLPROPIONATE

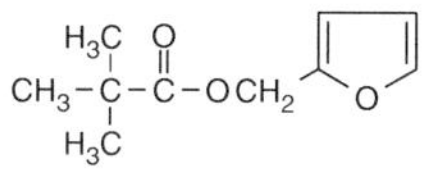

250 mg

S83,684-2
FURFURYL OCTADECANOATE

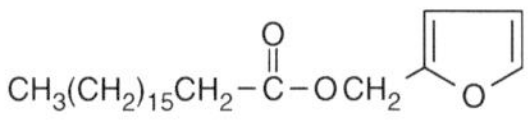

250 mg

S75,683-0
3-CARBOXY-2-FURANACETIC ACID

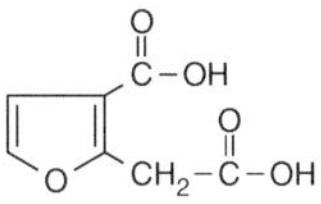

5 gm

S94,650-8
ETHYL 5-AMINO-4-CYANO-2-METHYL-3-FURANCARBOXYLATE

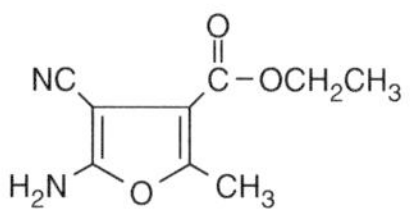

250 mg

S86,141-3
2-METHYL-3-FURYLMETHYL 4-NITROBENZOATE

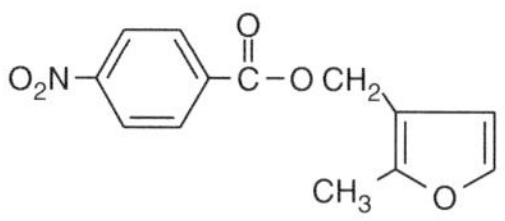

250 mg

S84,000-9
2,4-DIMETHYL-5-NITRO-3-FUROIC ACID

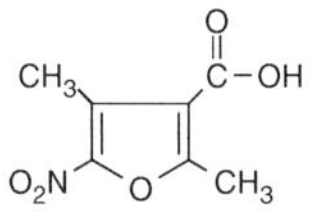

250 mg

S95,244-3
METHYL 5-FURFURYLIDENE-4-OXO-3-TETRAHYDROTHIOPHENECARBOXYLATE

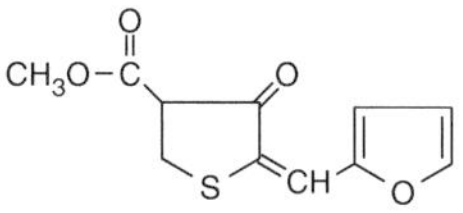

50 mg

S75,772-1
3,4-FURANDICARBOXYLIC ACID

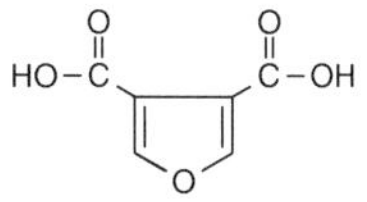

5 gm

S83,848-9
METHYL 4-CARBOXY-3-FUROATE

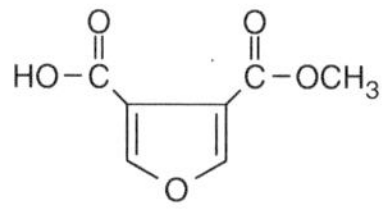

250 mg

S86,144-8
2-METHYL-3-FURFURYL BENZOATE

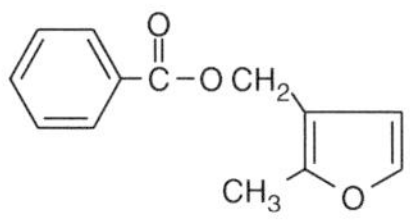

250 mg

S82,539-5
2,5-DIMETHYL-3-FURAMIDE

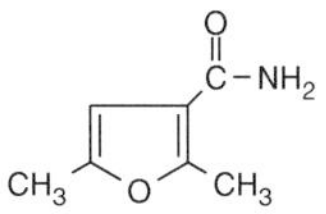

250 mg

S86,095-6
2-(2-FURYL)ETHYL ACETATE

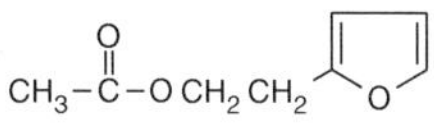

250 mg

S82,602-2
2-FURANACRYLIC ACID, SODIUM SALT

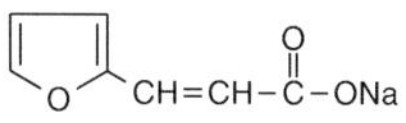

250 mg

S45,203-3
2-(FURFURYLAMINO)ETHANETHIOL
850

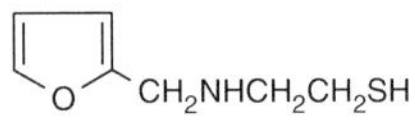

5 gm

S72,362-2
2-(FURFURYLUREIDO)ACETIC ACID N2-(N-[4-(METHYLTHIO)PHENYL]-CARBAMOYL)HYDRAZIDE

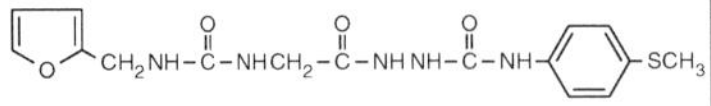

5 gm

S85,345-3
1-FURFURYL-3-[3-(METHYLTHIO)-PHENYL]UREA

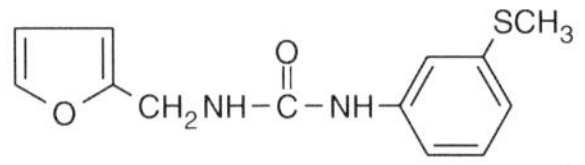

250 mg

S83,660-5
3-(2-FURYL)-3-(P-TOLYLTHIO)-PROPIOPHENONE

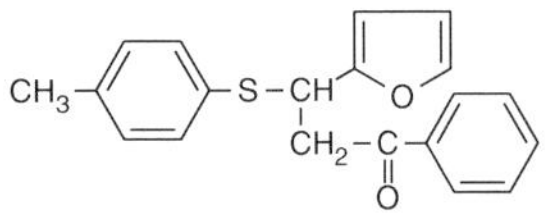

250 mg

S80,489-4
6-AMINO-3-ETHYL-1-FURFURYLURACIL

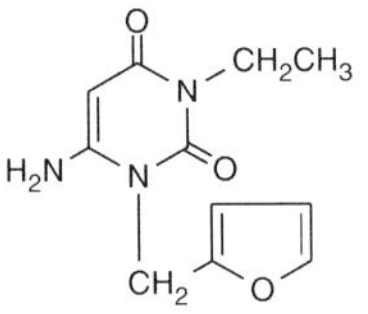

250 mg

S69,833-4
2-(2-FURYL)GLYCINE

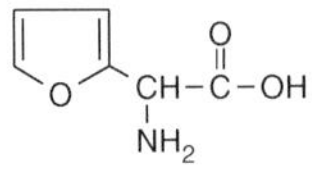

250 mg

S73,013-0
N-FURFURYLPICRYLAMINE

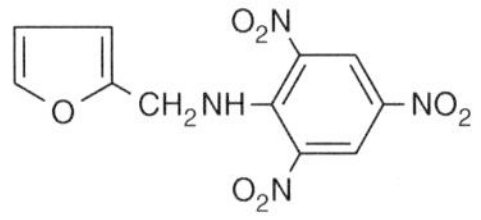

5 gm

S45,037-5
ALPHA-METHYLAMINO-2-FURANACETONITRILE HYDROCHLORIDE

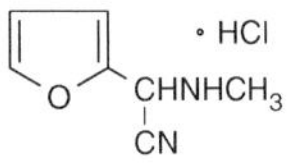

5 gm

S67,050-2
1-FURFURYL-3-(2-NITROPHENYL)UREA

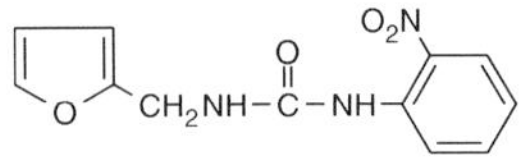

250 mg

S76,800-6
1-FURFURYL-1-METHYL-3-(2-METHYL-6-NITROPHENYL)UREA

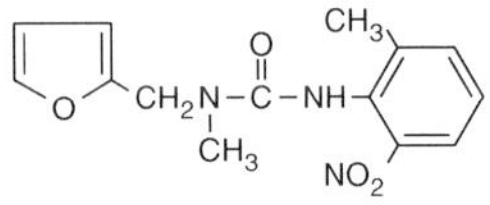

250 mg

S69,854-7
N-(4,5-DIMETHYL-2-NITROPHENYL)-2-FURAMIDE

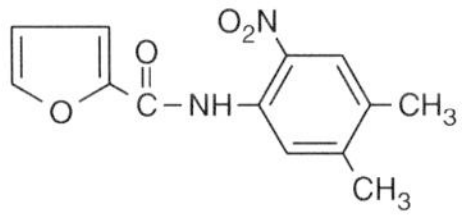

250 mg

S76,702-6
1-FURFURYL-1-METHYL-3-[2-(N-METHYLFURFURYLAMINOMETHYL)-PHENYL]UREA

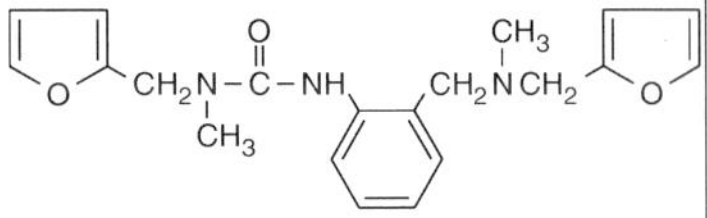

250 mg

S90,151-2
3'-NITRO-2-FURANILIDE

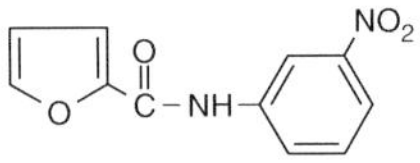

250 mg

S91,350-2
1-FURFURYL-1-METHYL-3-(3-NITROPHENYL)UREA

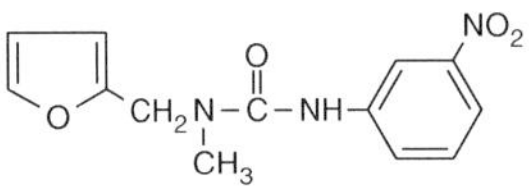

250 mg

S71,099-7
1-FURFURYL-3-(4-METHYL-3-NITROPHENYL)UREA

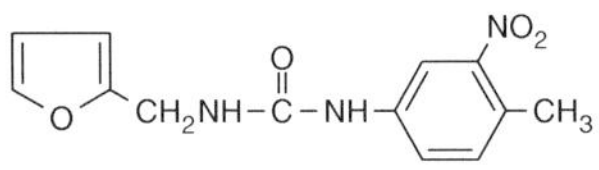

5 gm

S63,826-9
1-FURFURYL-3-(3-NITROPHENYL)UREA

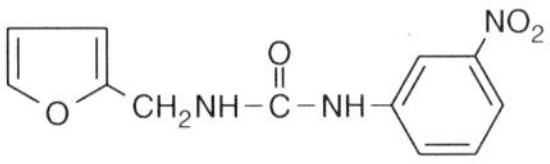

5 gm

S64,443-9
1,1'-(4-METHYL-1,3-PHENYLENE)BIS(3-FURFURYLUREA)

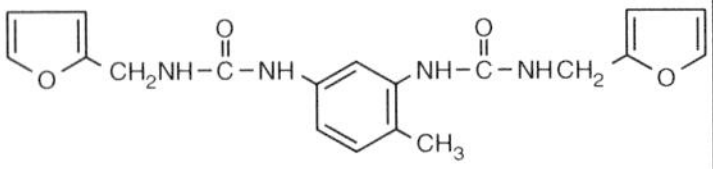

5 gm

S74,489-1
ALPHA-FURFURYLIDENE-2,4-DINITROTOLUENE

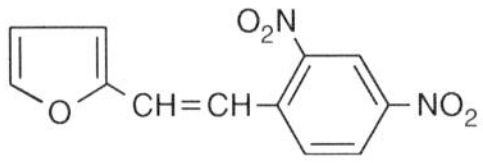

250 mg

S91,699-4
2'-CYANO-4'-NITRO-2-FURANILIDE

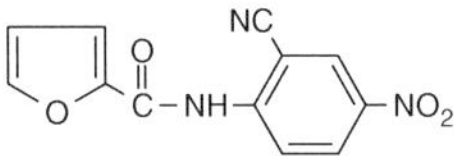

250 mg

S66,895-8
1-FURFURYL-3-(4-NITROPHENYL)UREA

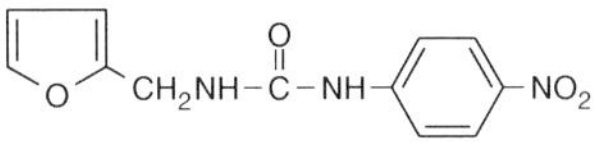

5 gm

S78,098-7
1-FURFURYL-3-(2-METHYL-4-NITROPHENYL)UREA

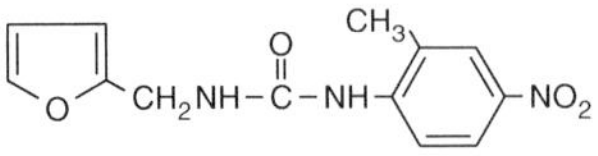

250 mg

S69,992-6
2'-METHYL-4'-NITRO-2-FURANILIDE

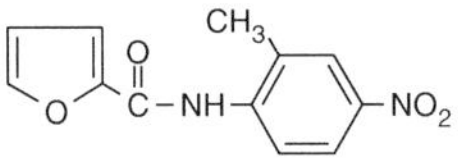

250 mg

S71,106-3
1,1'-(1,5-NAPHTHALENEDIYL)BIS(3-FURFURYLUREA)

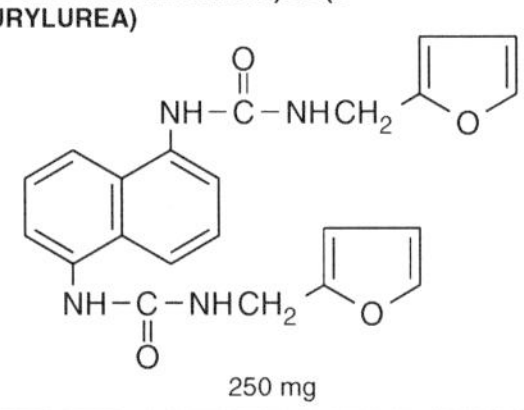

250 mg

S81,420-2
ALPHA-PHENYL-2-FURANACRYLONITRILE

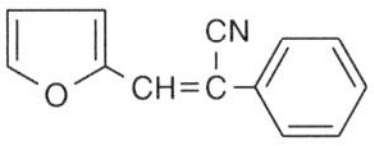

250 mg

S88,642-4
3-(2-FURYL)-2-PHENYLPROPIONITRILE

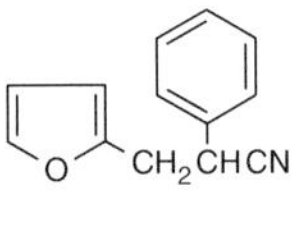

250 mg

S73,982-0
2-CHLORO-3-(FURFURYLAMINO)-1,4-NAPHTHOQUINONE

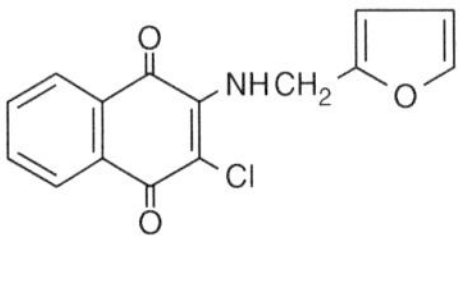

5 gm

S41,626-6
N-METHYLFURFURYLAMINE

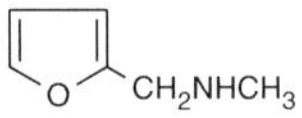

5 gm

S40,665-1
N-(ALPHA-BENZYLFURFURYL)ETHYLAMINE

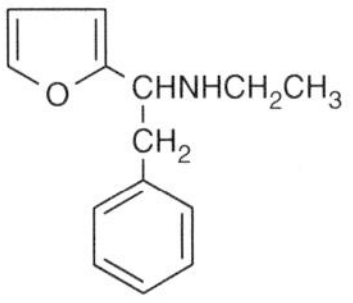

5 gm

S83,602-8
N,N,N',N'-TETRAMETHYL-4,4'-FURFURYLIDENEDIANILINE

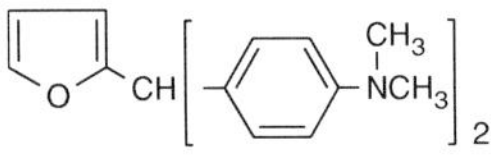

250 mg

S50,899-3
N,N,5-TRIMETHYLFURFURYLAMINE

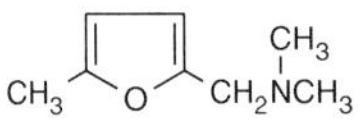

5 gm

S55,822-2
3-[4-(DIMETHYLAMINO)PHENYL]-1-(2-FURYL)-2-PROPEN-1-ONE

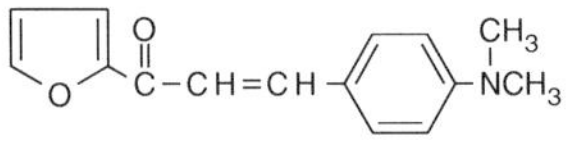

5 gm

S54,859-6
N-FURFURYLMALEIMIDE

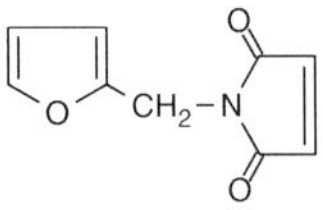

250 mg

S72,366-5
2-(FURFURYLUREIDO)ACETIC ACID N2-[N-(2,4-XYLYL)CARBAMOYL]-HYDRAZIDE

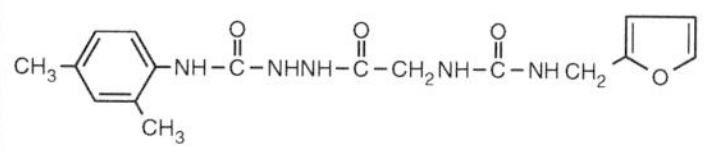

5 gm

S69,240-9
2-(FURFURYLUREIDO)ACETIC ACID N2-[N-(1-NAPHTHYL)CARBAMOYL]-HYDRAZIDE

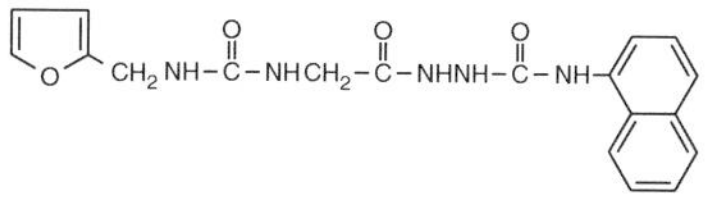

250 mg

S90,228-4
2-(3-FURFURYLUREIDO)ACETIC (2-BUTEN-1-YLIDENE)HYDRAZIDE

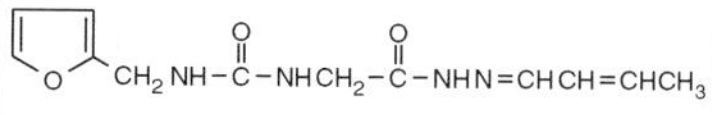

250 mg

S59,259-5
ACETIC ACID FURFURYLIDENEHYDRAZIDE

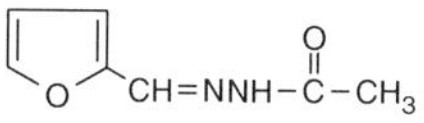

250 mg

S91,362-6
N-FURFURYLPHTHALAMIC ACID

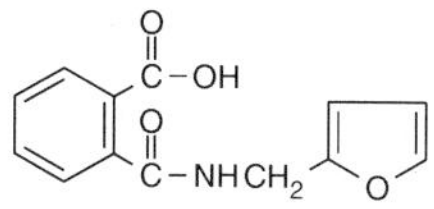

250 mg

S91,700-1
DIMETHYL 5-(FURAMIDO)ISOPHTHALATE

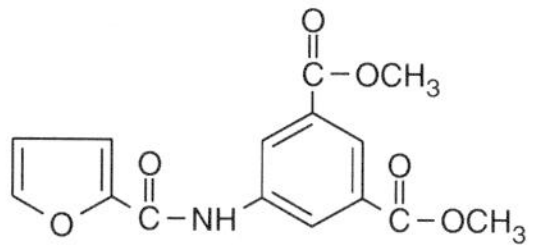

250 mg

S72,602-8
ETHYL 3-(FURFURYLUREIDO)BENZOATE

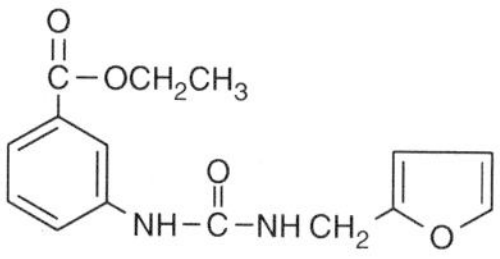

50 mg

S38,958-7
ALPHA-CYANOFURFURYL BENZOATE

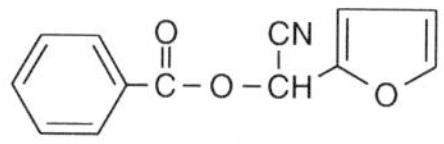

250 mg

S61,494-7
5-BROMO-2-FUROAMIDE

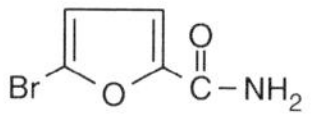

5 gm

S91,702-8
4',4''-SULFONYLBIS(2-FURANILIDE)

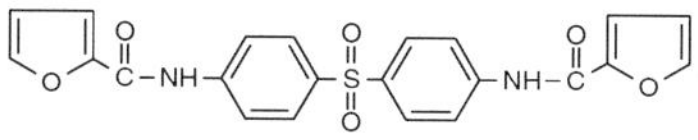

250 mg

S91,697-8
N3-FUROYLMETANILAMIDE

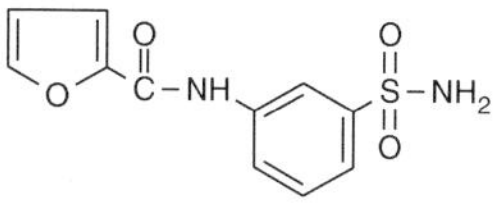

250 mg

S90,997-1
N-FURFURYL-3-METHYLBENZAMIDE

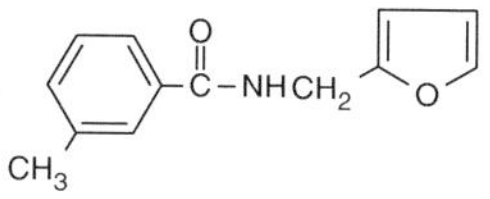

250 mg

S90,154-7
4'-ETHYL-2-FURANILIDE

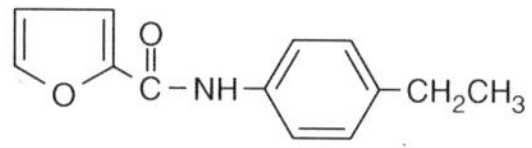

250 mg

S91,012-0
N-FURFURYL-4-METHYLBENZAMIDE

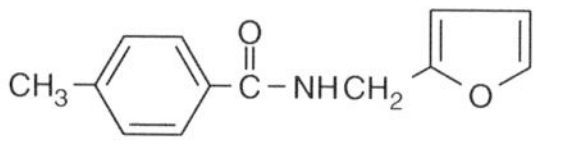

250 mg

S90,159-8
2',4',5'-TRIMETHYL-2-FURANILIDE

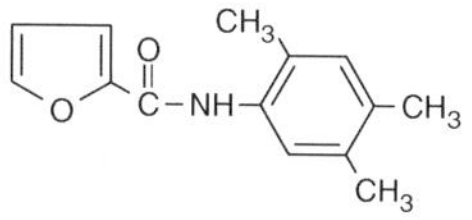

250 mg

S90,532-1
N-FURFURYL-1-NAPHTHAMIDE

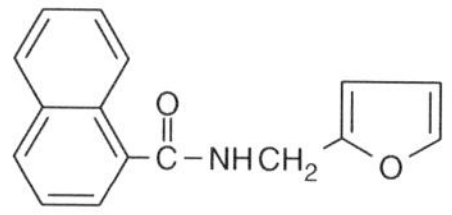

250 mg

S79,573-9
N-FURFURYL-1,2,3,6-TETRAHYDROPHTHALAMIC ACID

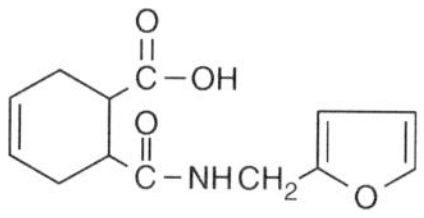

250 mg

S38,960-9
ALPHA-CYANOFURFURYL ACETATE

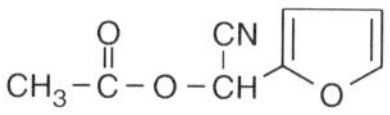

5 gm

S38,935-8
N-FURFURYLFORMAMIDE

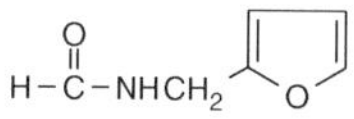

250 mg

S90,493-7
N-FURFURYL-1-CYCLOHEXANECARBOXAMIDE

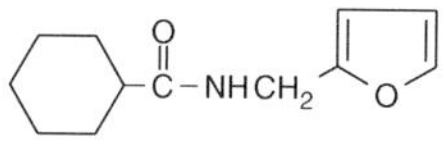

250 mg

S67,413-3
2-(FURFURYLUREIDO)ACETIC ACID HYDRAZIDE

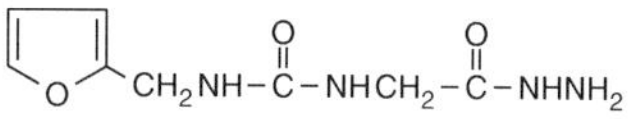

250 mg

S39,199-9
1,3-DIFURFURYLUREA

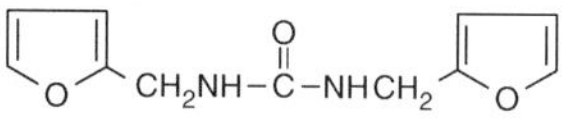

5 gm

S63,396-8
1-(2-CHLOROPHENYL)-3-FURFURYLUREA

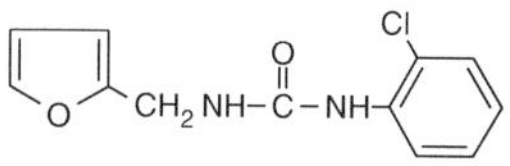

5 gm

S61,173-5
1-(2,5-DICHLOROPHENYL)-3-FURFURYLUREA

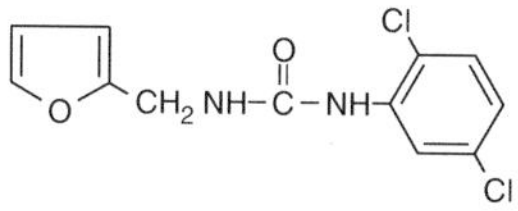

5 gm

S61,178-6
1-(2,6-DICHLOROPHENYL)-3-FURFURYLUREA

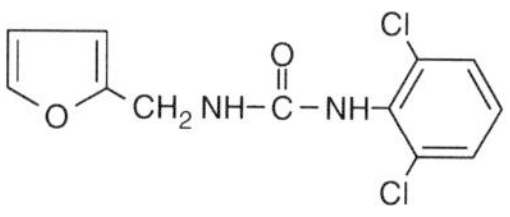

5 gm

S61,603-6
1-(2,3-DICHLOROPHENYL)-3-FURFURYLUREA

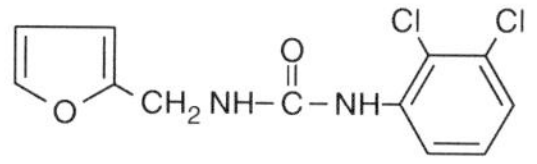

5 gm

S63,397-6
1-FURFURYL-3-(2,4,5-TRICHLOROPHENYL)UREA

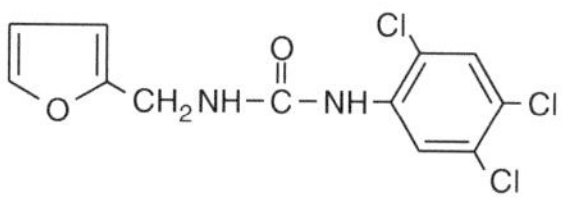

5 gm

S61,359-2
1-FURFURYL-3-(P-TOLYL)UREA

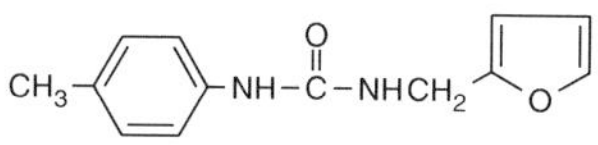

5 gm

S61,000-3
1-(5-CHLORO-2-METHYLPHENYL)-3-FURFURYLUREA

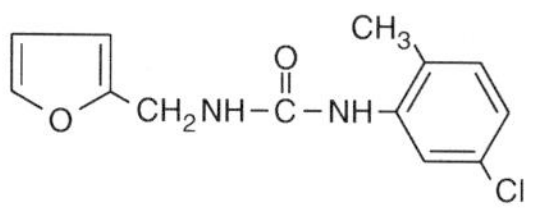

5 gm

S63,634-7
1-(3-CHLORO-4-METHYLPHENYL)-3-FURFURYLUREA

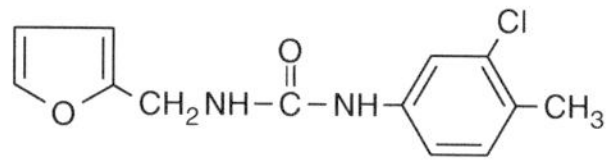

5 gm

S61,041-0
1-FURFURYL-3-(2,4-XYLYL)UREA

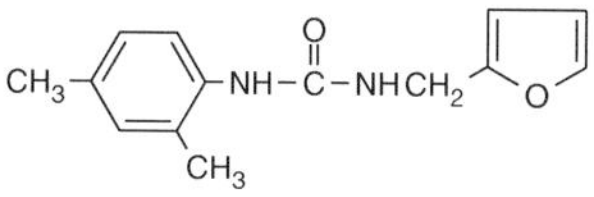

5 gm

S64,998-8
1-FURFURYL-3-(2,5-XYLYL)UREA

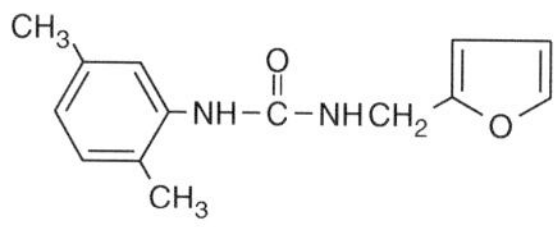

5 gm

S63,632-0
1,1'-HEXAMETHYLENEBIS(3-FURFURYLUREA)

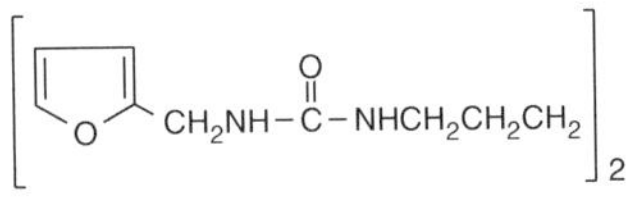

5 gm

S91,340-5
3-CYCLOHEXYL-1-FURFURYL-1-METHYLUREA

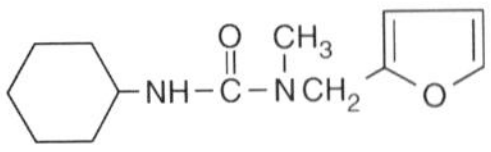

250 mg

S65,775-1
1-(2-ETHYLPHENYL)-3-FURFURYLUREA

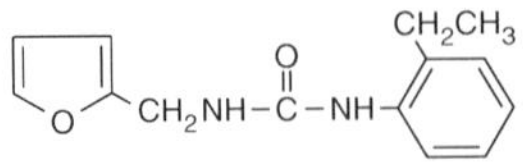

5 gm

S64,036-0
1-(4-BROMOPHENYL)-3-FURFURYLUREA

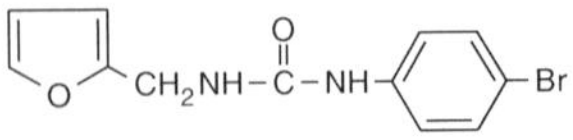

5 gm

S65,225-3
1-(2-CHLORO-6-METHYLPHENYL)-3-FURFURYLUREA

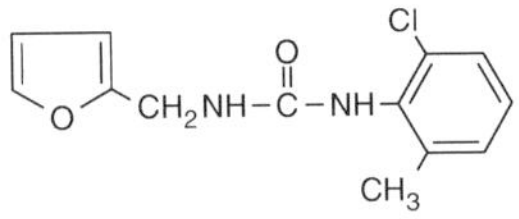

5 gm

S64,392-0
1-FURFURYL-3-(2,3-XYLYL)UREA

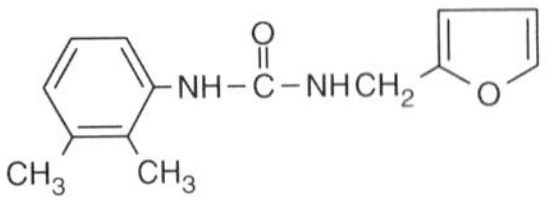

5 gm

S65,828-6
1-FURFURYL-3-(2,6-XYLYL)UREA

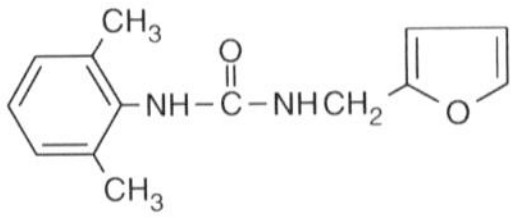

5 gm

S66,091-4
1-(2-ETHYL-6-METHYLPHENYL)-3-FURFURYLUREA

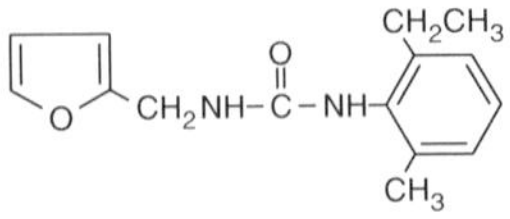

5 gm

S66,599-1
1-(2,6-DIETHYLPHENYL)-3-FURFURYLUREA

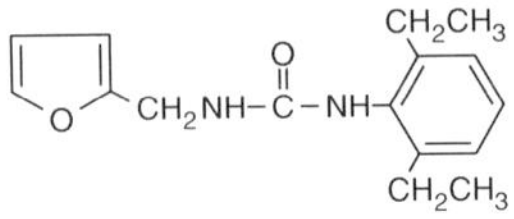

5 gm

S66,602-5
1-(2,6-DIISOPROPYLPHENYL)-3-FURFURYLUREA

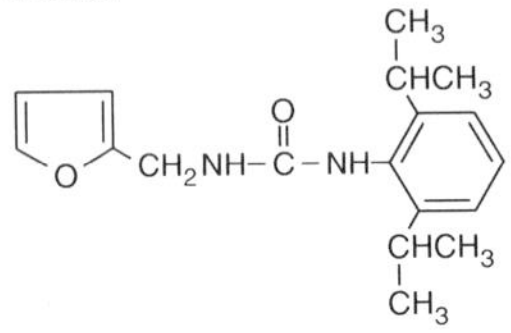

5 gm

S64,022-0
1-(2,4-DICHLOROPHENYL)-3-FURFURYLUREA

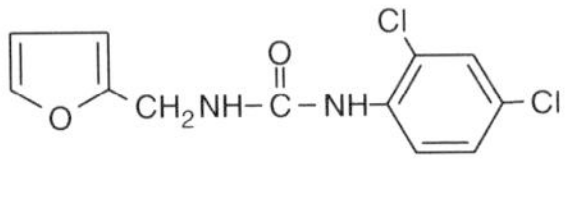

5 gm

S66,363-8
1-FURFURYL-3-(3,4-XYLYL)UREA

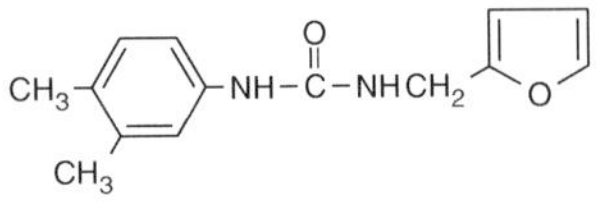

5 gm

S76,967-3
1-FURFURYL-1-METHYL-3-(3,4-XYLYL)UREA

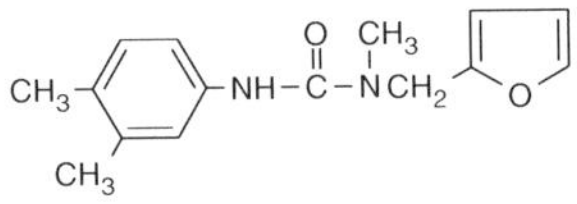

250 mg

S66,153-8
1-FURFURYL-3-(3,5-XYLYL)UREA

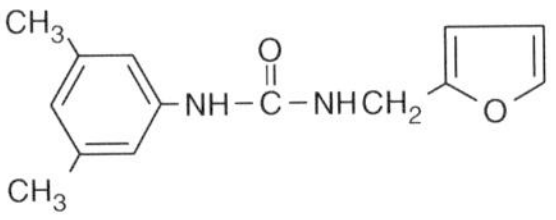

5 gm

S63,905-2
1-FURFURYL-3-(1-NAPHTHYL)UREA

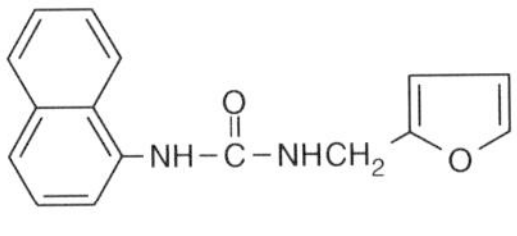

5 gm

S61,728-8
1,3-DIFURFURYL-2-THIOUREA

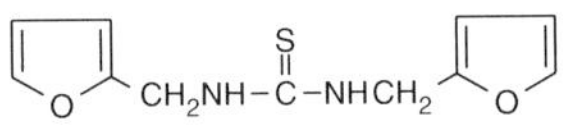

5 gm

S63,955-9
1-FURFURYL-3-PHENYL-2-THIOUREA

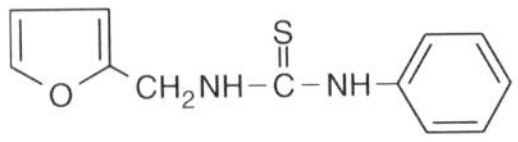

5 gm

S90,530-5
1-FURFURYL-3-(1-NAPHTHYL)-2-THIOUREA

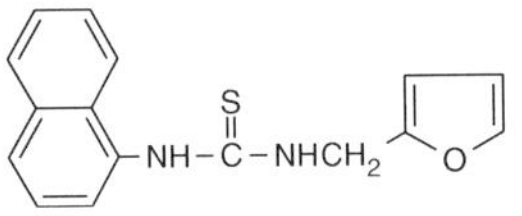

250 mg

S67,696-9
3-FURFURYLRHODANINE

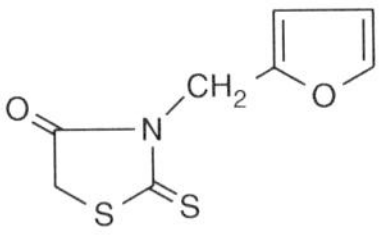

5 gm

S61,559-5
N-(FURFURYLIDENE)-P-TOLUIDINE

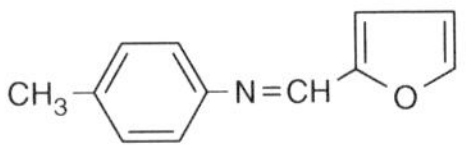

250 mg

S36,525-4
2,4,5-TRI-(2-FURYL)-2-IMIDAZOLINE

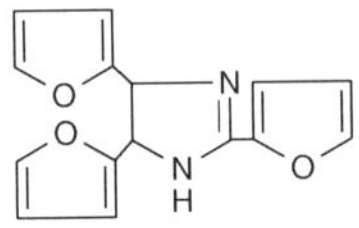

5 gm

S79,619-0
1-BENZYL-3-FURFURYL-2-THIOUREA

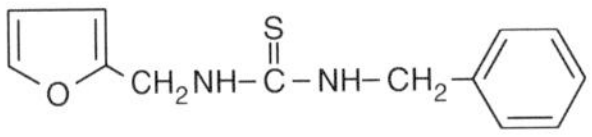

250 mg

S63,619-3
1-(2-FURYL)-3-(3-NITROPHENYL)-2-PROPEN-1-ONE

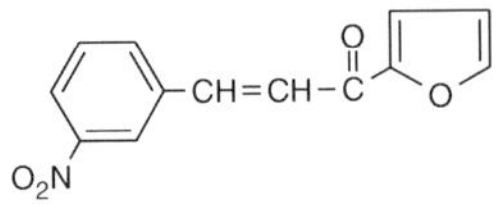

5 gm

S51,762-3
1,5-BIS(5-NITRO-2-FURYL)-1,4-PENTADIEN-3-ONE

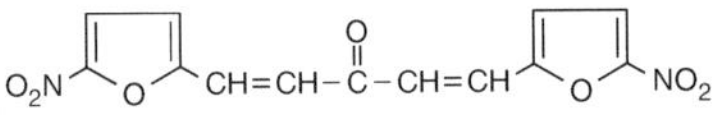

5 gm

S81,230-7
2-BENZOYL-5-NITROFURAN

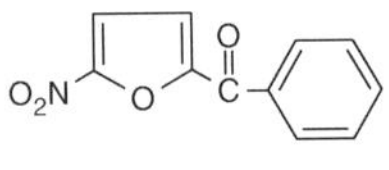

250 mg

S81,434-2
2-METHYL-5-NITRO-3-FUROIC ACID

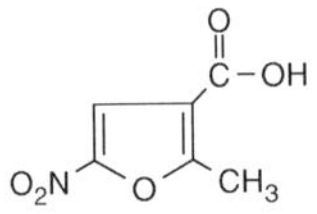

250 mg

S82,207-8
2,5-DINITROFURAN

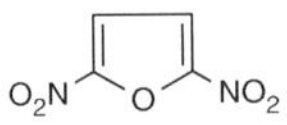

250 mg

S37,551-9
2-(2-NITROVINYL)FURAN

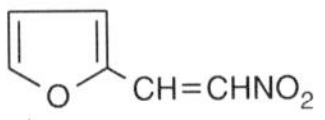

5 gm

S69,859-8
2'-BENZOYL-2-FURANILIDE

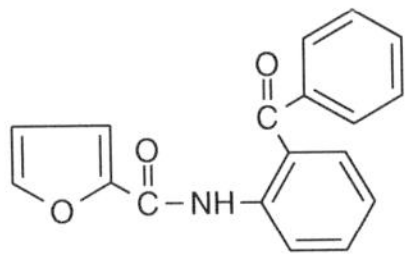

250 mg

S83,659-1
3-(2-FURYL)-3-(PHENYLSULFONYL)-PROPIOPHENONE

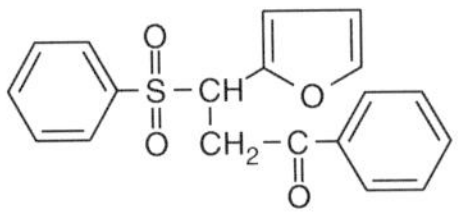

250 mg

S56,658-6
2-FURFURYLIDENEACETOPHENONE

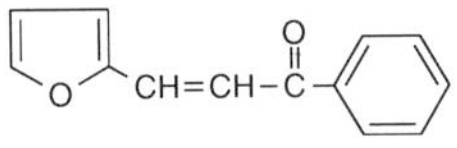

5 gm

S63,620-7
3-(2,4-DICHLOROPHENYL)-1-(2-FURYL)-2-PROPEN-1-ONE

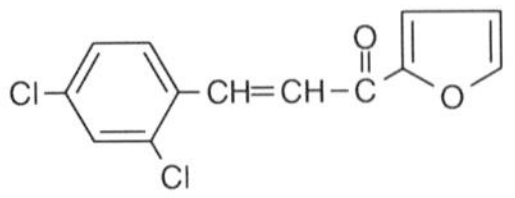

5 gm

S67,671-3
1,5-BIS(5-METHYL-2-FURYL)-1,4-PENTADIEN-3-ONE

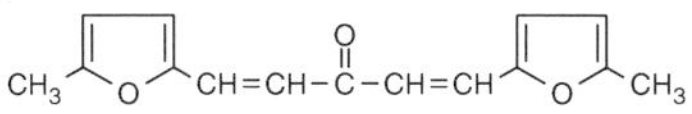

5 gm

S59,776-7
3,3'-(1,4-PHENYLENE)BIS[1-(2-FURYL)-2-PROPEN-1-ONE]

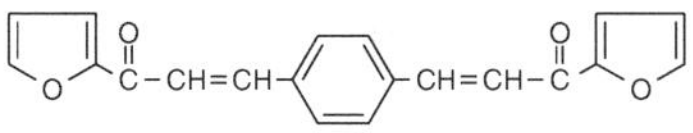

250 mg

S81,148-3
2-BENZOYLFURAN

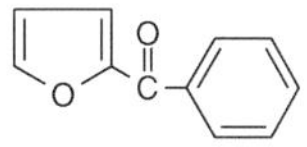

250 mg

S83,941-8
3-METHYL-2,3,4,5-TETRAHYDRO-2,4-FURANDIONE

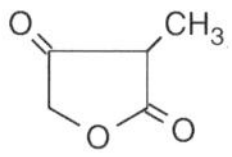

250 mg

S55,209-7
1-(2-FURYL)-1-PROPANONE

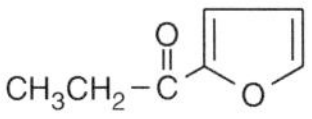

250 mg

S54,161-3
1,5-DIPHENYL-3-(2-FURYL)-1,5-PENTANEDIONE

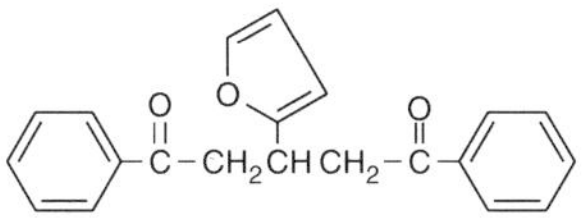

250 mg

S68,011-7
2-CYCLOHEXYLMETHYLENE-6-FURFURYLIDENE-1-CYCLOHEXANONE

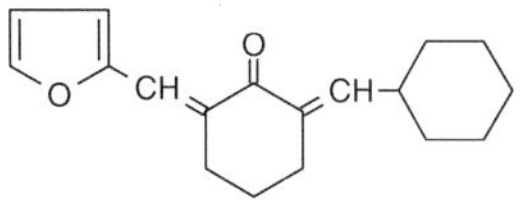

250 mg

S36,941-1
3,4,6,7,8,-HEXAHYDRO-1(2H)-DIBENZOFURAN-1-ONE

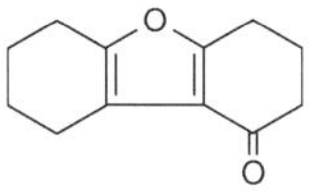

250 mg

S37,334-6
DIETHYL FURFURYLIDENEMALONATE

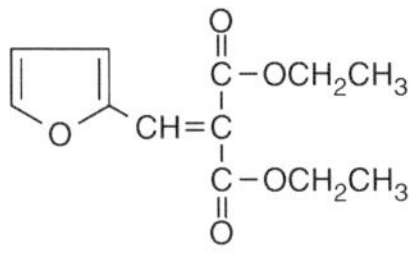

5 gm

S81,480-6
2,4-FURANDICARBOXYLIC ACID

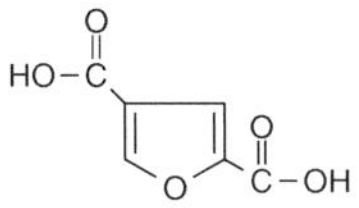

250 mg

S81,375-3
2,5-DIMETHYL-3,4-FURANDICARBOXYLIC ACID

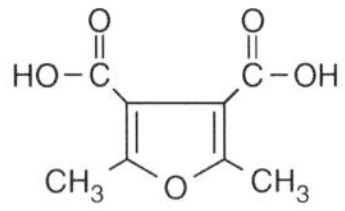

250 mg

S81,149-1
2,5-DIMETHYL-4-(ETHOXYCARBONYL)-3-FUROIC ACID

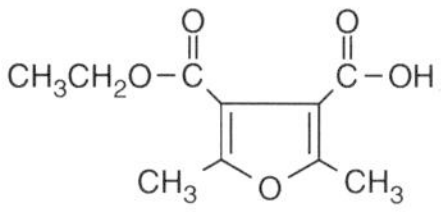

250 mg

S81,059-2
DIMETHYL 2,5-FURANDICARBOXYLATE

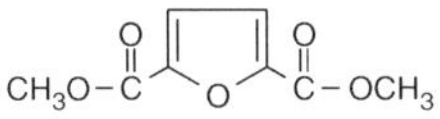

250 mg

S81,147-5
DIETHYL 2,5-DIMETHYL-3,4-FURANDICARBOXYLATE

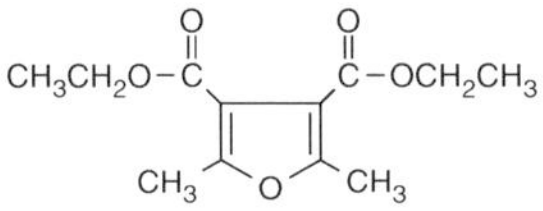

250 mg

S81,479-2
2-FURANACRYLIC ACID, POTASSIUM SALT

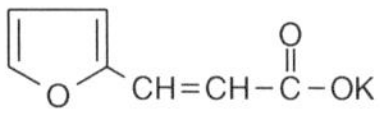

250 mg

S54,852-9
N-FURFURYLMALEAMIC ACID

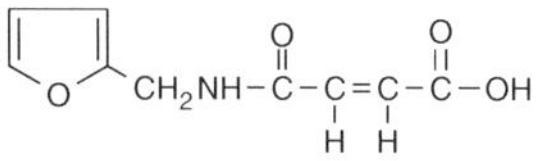

250 mg

S81,056-8
2-METHYL-3-FUROIC ACID

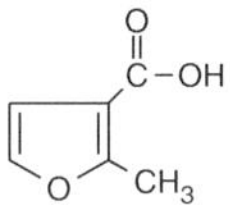

250 mg

S82,603-0
2-FURANACRYLIC ACID, MERCURY (II) SALT

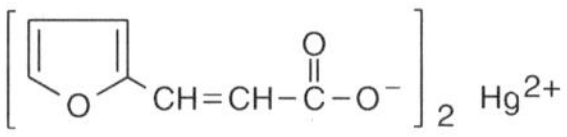

250 mg

S80,835-0
ALPHA-METHYL-2-FURANACRYLIC ACID

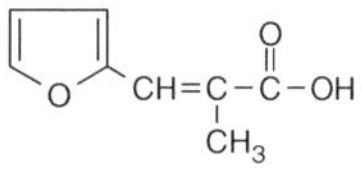

250 mg

S54,494-9
TRIETHYL 2,3,4-FURANTRICARBOXYLATE

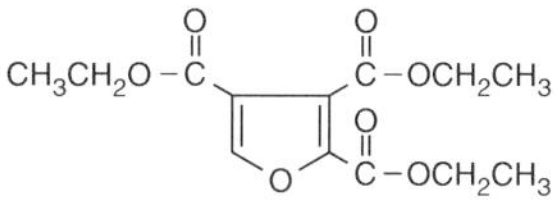

250 mg

S38,940-4
DIBUTYL 3,4-FURANDICARBOXYLATE

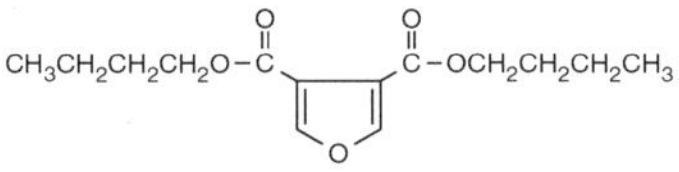

5 gm

S66,067-1
ETHYL ALPHA-CYANO-2-FURANACRYLATE

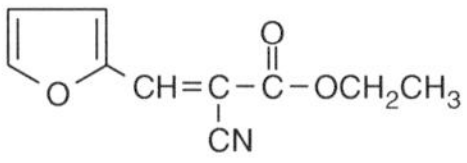

250 mg

S83,944-2
ETHYL 5-BROMO-ALPHA-CYANO-2-FURANACRYLATE

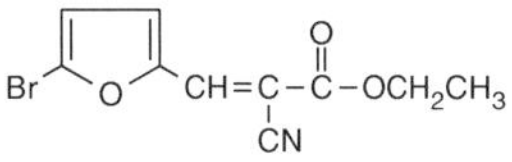

250 mg

S82,004-0
METHYL 5-CHLORO-2-FUROATE

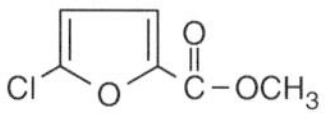

250 mg

S81,333-8
METHYL 4,5-DIBROMO-2-FUROATE

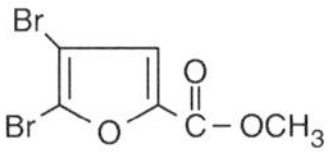

250 mg

S81,084-3
ETHYL 4,5-BIS(CHLOROMERCURI)-2-METHYL-3-FUROATE

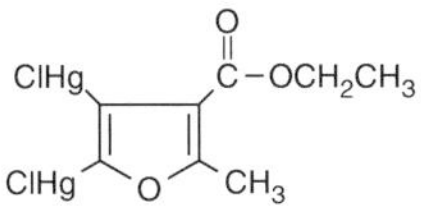

250 mg

S76,840-5
N-FURFURYL-2-FURAMIDE

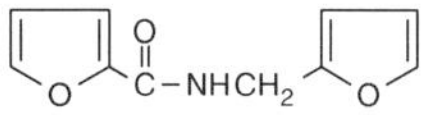

250 mg

S70,941-7
2',6'-DIMETHYL-2-FURANILIDE

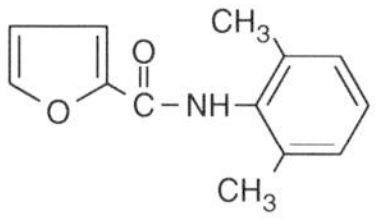

250 mg

S67,339-0
3,4-DIHYDRO-3-FURFURYL-4-OXO-1,2,3-BENZOTRIAZINE

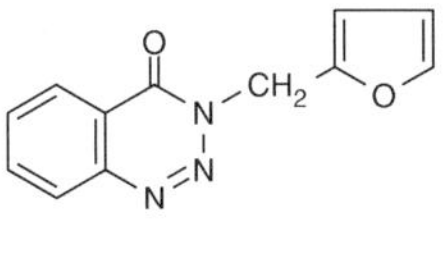

250 mg

S35,008-7
FURFURYLIDENEACETONE

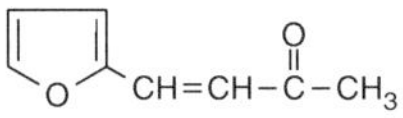

5 gm

S55,151-1
1,3-DI-(2-FURYL)-2-PROPEN-1-ONE

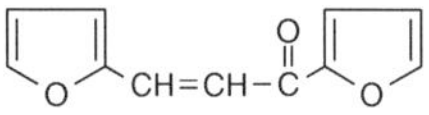

5 gm

S42,691-1
4'-FLUORO-3-(2-FURYL)ACRYLOPHENONE

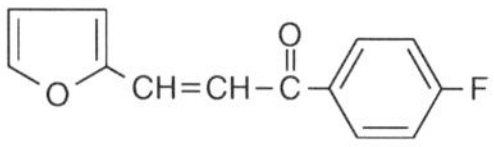

5 gm

S53,118-9
4-FURFURYLIDENE-2-PHENYL-2-OXAZOLIN-5-ONE
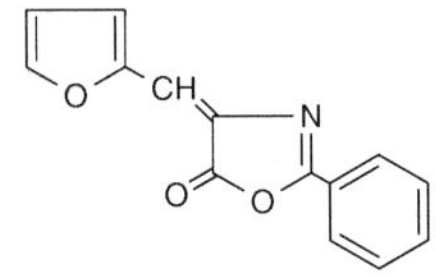
250 mg

S93,708-8
3-FURYL-1-(2-NAPHTHYL)-2-PROPEN-1-ONE
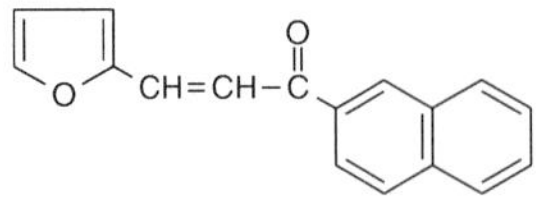
250 mg

S76,401-9
N,N'-DIFURFURYLOXAMIDE
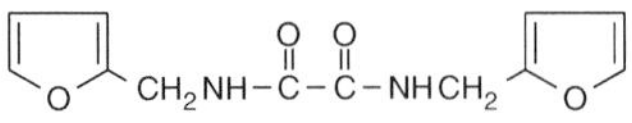
5 gm

S73,064-5
3-[N-(3-FURFURYL)CARBAMOYL]-5-NORBORNENE-2-CARBOXYLIC ACID
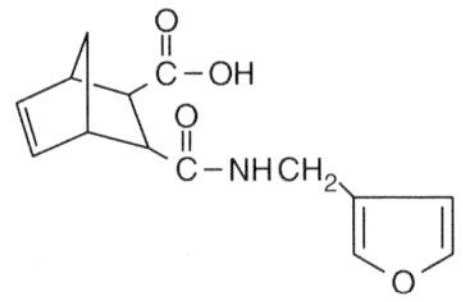
5 gm

S73,014-9
N-FURFURYLGLUTARAMIC ACID
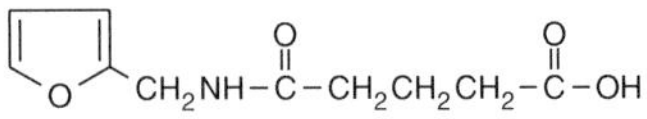
5 gm

S71,916-1
CIS-N-FURFURYLHEXAHYDRO-PHTHALAMIC ACID
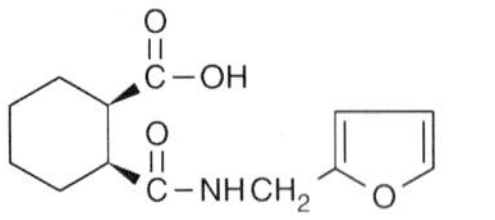
5 gm

S74,393-3
2-(N-FURFURYLCARBAMOYL)-CYCLOHEXANECARBOXYLIC ACID
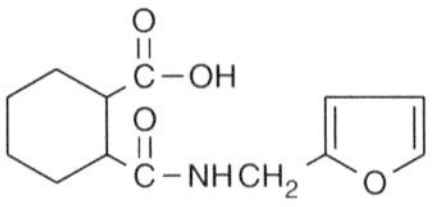
5 gm

S93,181-0
METHYL 3-(METHOXYCARBONYL)-2-FURANACETATE
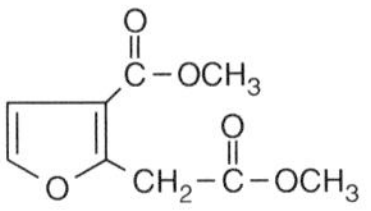
250 mg

S54,306-3
2-(2-FURYL)-3,1-BENZOXAZIN-4-(4H)-ONE
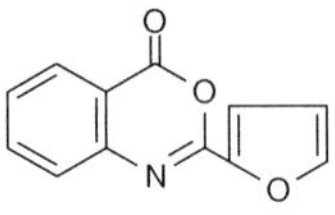
5 gm

S83,904-3
2-FUROIC ANHYDRIDE
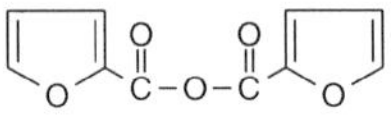
250 mg

S92,723-6
5-(2-FURYL)HYDANTOIN
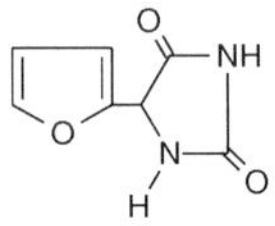
250 mg

S75,682-2
5-FURFURYLIDENERHODANINE
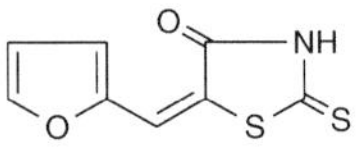
5 gm

S54,601-1
3-(2-FURYL)RHODANINE
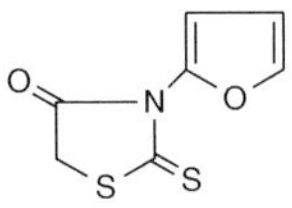
250 mg

S40,872-7
1-FURFURYL-3-(P-TOLYLSULFONYL)UREA
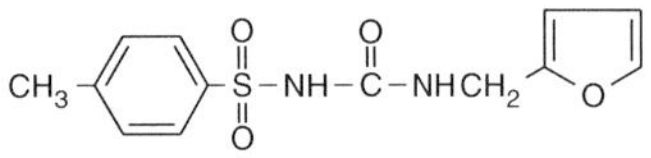
5 gm

S60,737-1
1-BUTYL-3-FURFURYLUREA
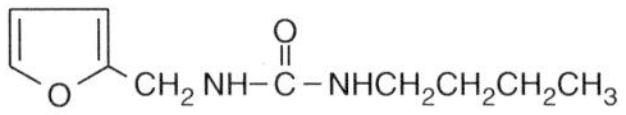
5 gm

S79,560-7
1-TERT-BUTYL-3-FURFURYLUREA
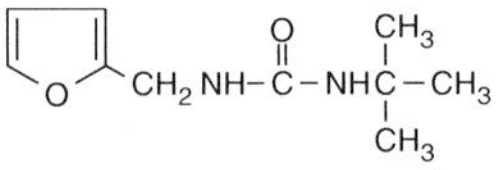
250 mg

S60,696-0
1-(4-CHLOROPHENYL)-3-FURFURYLUREA
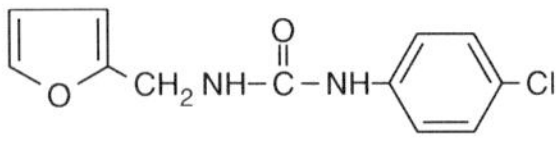
5 gm

S61,007-0
1-(3-CHLOROPHENYL)-3-FURFURYLUREA
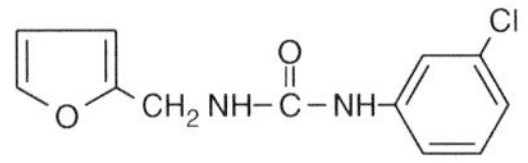
5 gm

S60,675-8
1-(3,4-DICHLOROPHENYL)-3-FURFURYLUREA
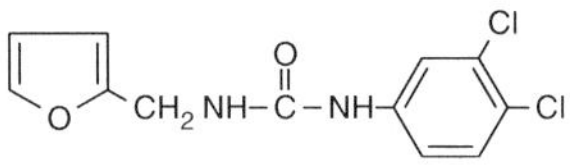
5 gm

S60,636-7
1-FURFURYL-3-(M-TOLYL)UREA
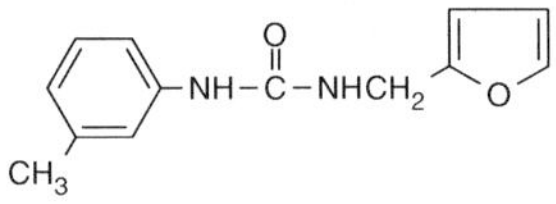
5 gm

S60,648-0
1-FURFURYL-3-(O-TOLYL)UREA
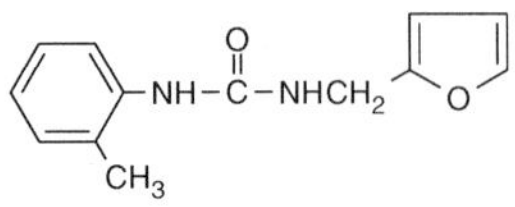
5 gm

S61,543-9
1-(4-CHLORO-2-METHYLPHENYL)-3-FURFURYLUREA
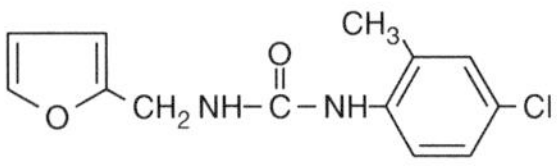
5 gm

S60,732-0
1-CYCLOHEXYL-3-FURFURYLUREA
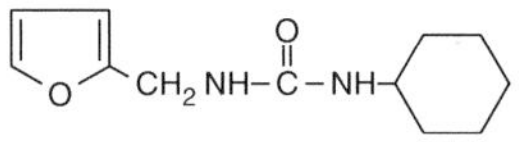
5 gm

S60,686-3
1-FURFURYL-3-PHENYLUREA
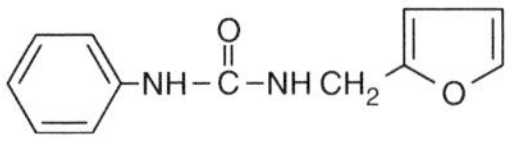
5 gm

S63,473-5
4,4'-METHYLENEBIS(1-FURFURYL-3-PHENYLUREA)
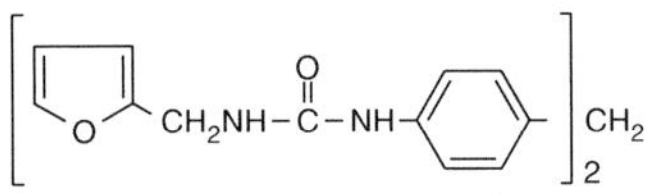
5 gm

S88,730-7
ALPHA-(2-FURYL)-BETA-(1-NAPHTHYL)-ACRYLONITRILE
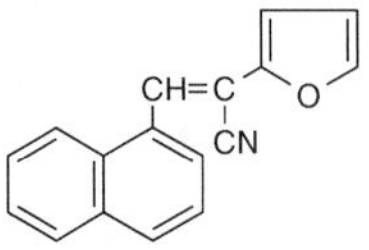
250 mg

S77,886-9
1-(2-FURYLSULFONYL)-2-(METHYLSULFONYL)ETHANE
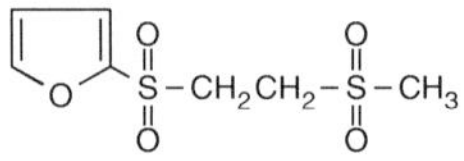
250 mg

S59,334-6
2-BUTYL-5-METHYLFURAN
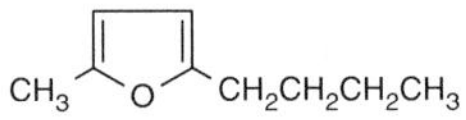
250 mg

S44,987-3
DIETHYL FURFURYLAMIDOPHOSPHATE
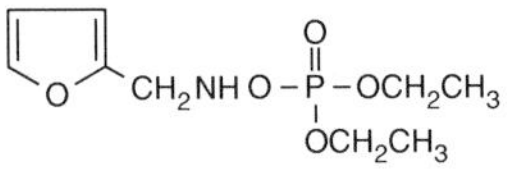
5 gm

S66,237-2
1-(4-METHYL-2-FURYL)-2,6,6-TRIMETHYL-1-CYCLOHEXENE
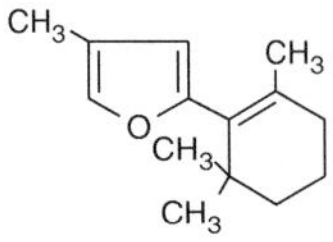
5 gm

S81,440-7
3,4-DIPHENYLFURAN
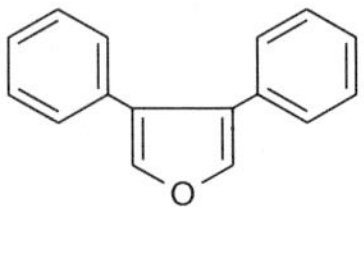
250 mg

S81,405-9
3-BROMO-2,4,5-TRIPHENYLFURAN
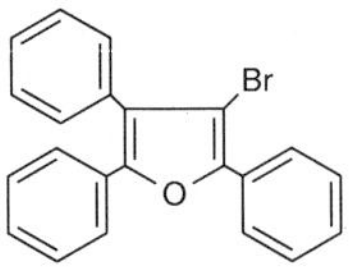
250 mg

S92,586-1
9-(2-FURFURYLIDENE)FLUORENE

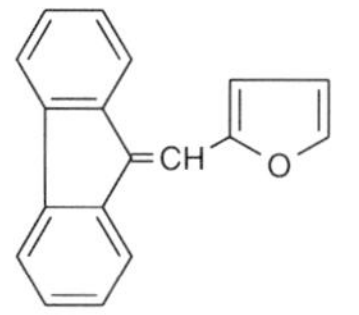

250 mg

S72,029-1
ETHYL 2-(2-CHLOROACETAMIDO)-4,5,6,7-TETRAHYDROBENZO(B)-THIOPHENE-3-CARBOXYLATE

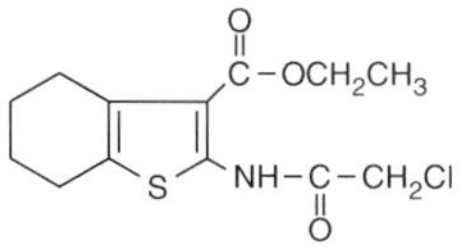

250 mg

S91,680-3
N-[3-CYANO-4,5,6,7-TETRAHYDROBENZO-(B)THIOPHEN-2-YL]-2,2-DICHLOROACETAMIDE

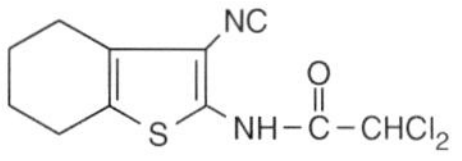

250 mg

S91,645-5
ETHYL 2-[-3-[2-CHLORO-5-(TRIFLUOROMETHYL)-PHENYL]UREIDO]-4,5,6,7-TETRAHYDRO-BENZO[B]THIOPHENE-3-CARBOXYLATE

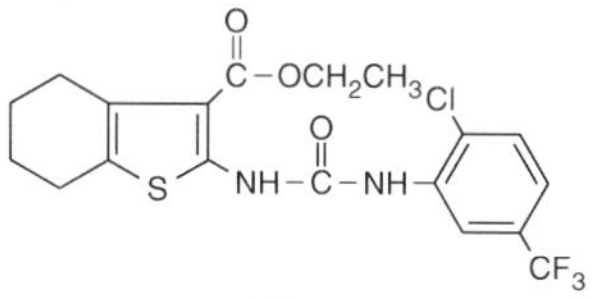

250 mg

S91,640-4
1-(3-CYANO-4,5,6,7-TETRAHYDROBENZO[B]THIOPHENE-2-YL)-3-[3-(TRIFLUOROMETHYL)PHENYL]UREA

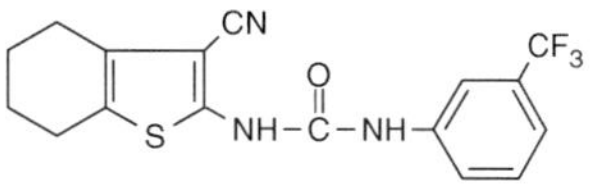

250 mg

S93,779-7
2-AMINO-5-(4-BROMOPHENYL)THIOPHENE HYDROCHLORIDE

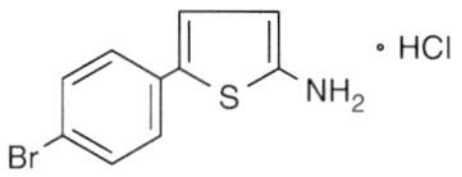

250 mg

S77,999-7
1-(2-CHLORO-4-NITROPHENYL)-3-[3-CYANO-4,5,6,7-TETRAHYDROBENZO(B)-THIOPHEN-2-YL]UREA

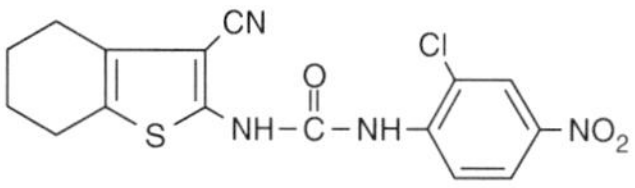

250 mg

S88,640-8
2-(4-CHLOROPHENYL)-3-(2-THIENYL)-PROPIONITRILE

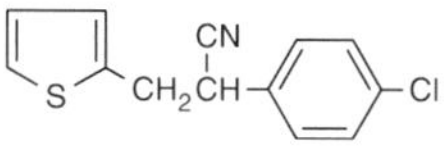

250 mg

S82,976-5
2-(PENTAFLUOROBENZOYL)-3,4,5-TRICHLOROTHIOPHENE

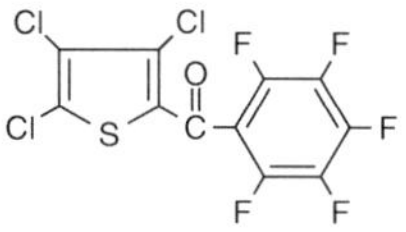

250 mg

S70,338-9
4-CHLORO-2-(2-THIENYLETHYL)-BENZOIC ACID

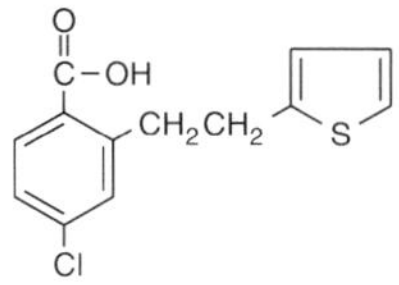

5 gm

S91,633-1
ETHYL 2-[3-(3-CHLOROPHENYL)UREIDO]-4,5,6,7-TETRAHYDROBENZO(B)-THIOPHENE-3-CARBOXYLATE

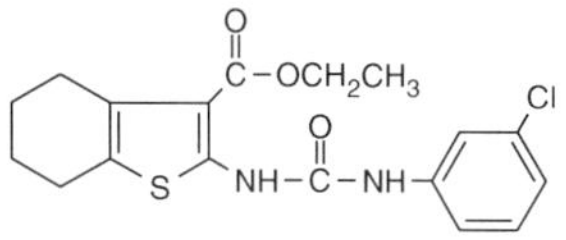

250 mg

S92,468-7
ETHYL 2-(4-FLUOROPHENYLUREIDO)-4,5,6,7-TETRAHYDRO-3-BENZO(B)-THIOPHENECARBOXYLATE

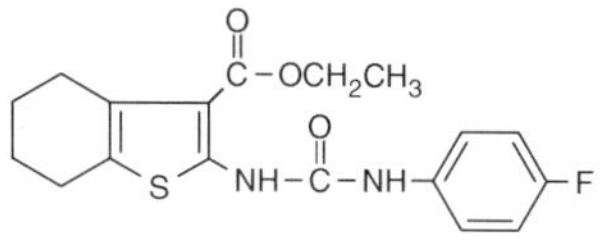

250 mg

S91,636-6
1-[3-CYANO-4,5,6,7-TETRAHYDROBENZO-(B)THIOPHEN-2-YL]-3-(4-FLUOROPHENYL)UREA

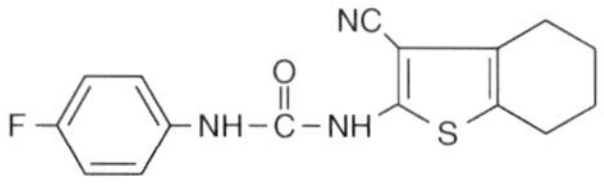

250 mg

S91,632-3
1-(3-CHLOROPHENYL)-3-[3-CYANO-4,5,6,7-TETRAHYDROBENZO(B)-THIOPHEN-2-YL]UREA

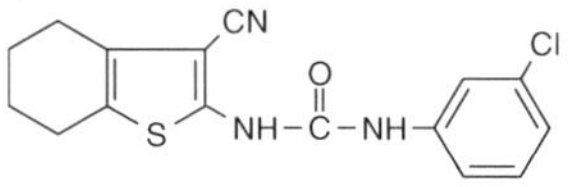

250 mg

S88,721-8
ALPHA-(4-CHLOROPHENYL)-BETA-(2-THIENYL)ACRYLONITRILE

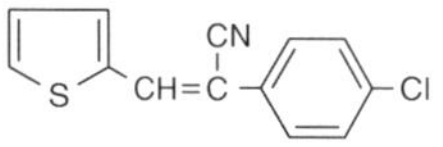

250 mg

S83,177-8
3,4-DICHLORO-2-(PENTACHLOROPHENYL)-THIOPHENE

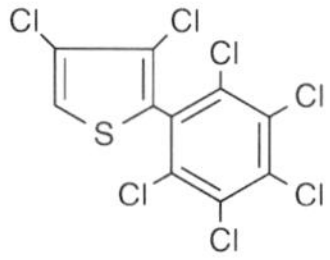

250 mg

S83,178-6
2-(2,3,5,6-TETRACHLOROPHENYL)-3,4,5-TRICHLOROTHIOPHENE

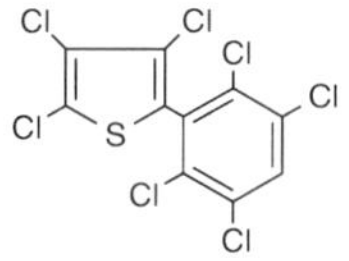

250 mg

S83,179-4
2-(PENTACHLOROPHENYL)-3,4,5-TRICHLOROTHIOPHENE

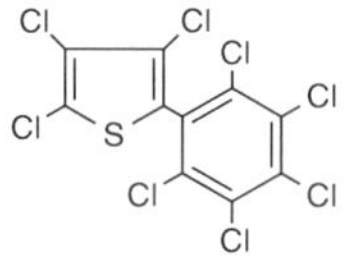

250 mg

S86,121-9
DIETHYL 3,4-DIHYDROXY-2,5-THIOPHENEDICARBOXYLATE, DISODIUM SALT

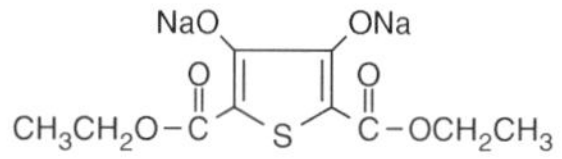

250 mg

S44,596-7
3-(2-THIENYL)-L-SERINE MONOHYDRATE

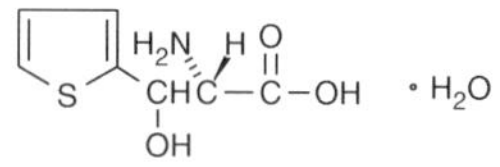

250 mg

S58,104-6
1-PHENYL-3-(2-THIENYL)-1,3-PROPANEDIONE, COPPER(I) DERIVATIVE

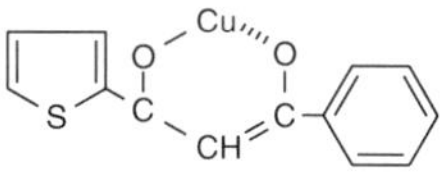

250 mg

S55,217-8
THENOIN

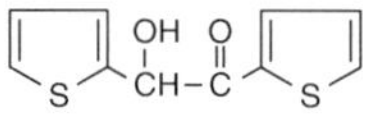

250 mg

S56,742-6
ALPHA-CYCLOPROPYL-2-THIOPHENEMETHANOL

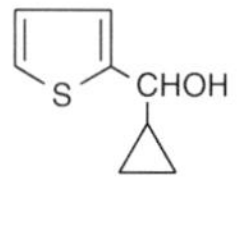

5 gm

S52,478-6
ALPHA-CYCLOPROPYL-ALPHA-(4-FLUOROPHENYL)-2-THIOPHENEMETHANOL

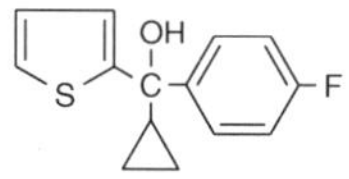

250 mg

S36,501-7
1-(2-THIENYL)-2-PROPANOL

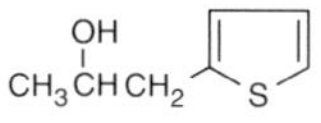

5 gm

S94,740-7
5-THENYL-2-THIOPHENEMETHANOL

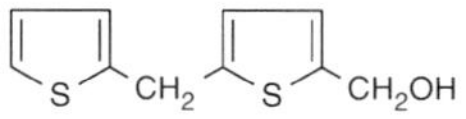

100 mg

S59,819-4
2',5'-DIMETHOXY-2-(2-THENOYL)-ACETANILIDE

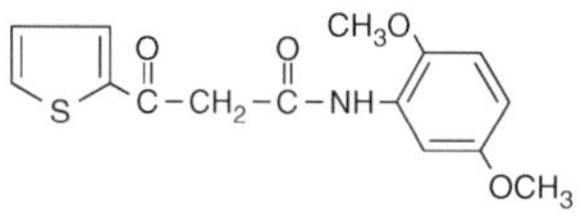

250 mg

S91,642-0
ETHYL 2-[3-(2-METHOXY-5-METHYLPHENYL)UREIDO]TETRAHYDROBENZO-(B)-THIOPHENE-3-CARBOXYLATE

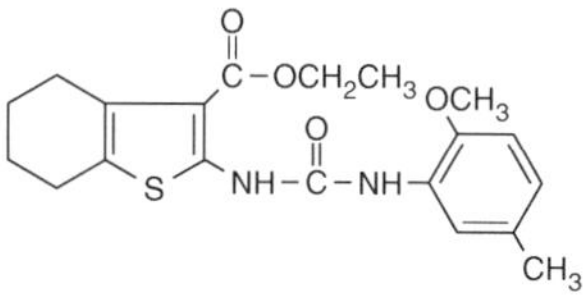

250 mg

S58,002-3
3-(4-METHOXYPHENYL)-1-(2-THIENYL)-2-PROPEN-1-ONE

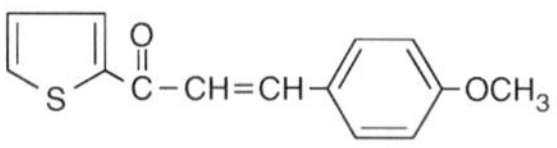

5 gm

S83,649-4
2-FURYL 2-THIENYL KETONE

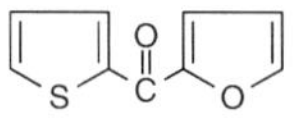

250 mg

S94,577-3
DIETHYL 2,2'-(2-THENYLIDENE)BIS-ACETOACETATE

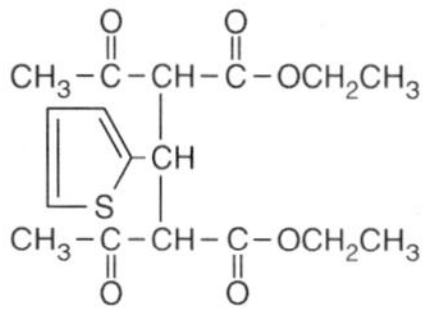

250 mg

S88,822-2
3-BENZOYL-2-THIOPHENECARBOXYLIC ACID

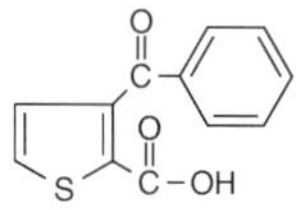

250 mg

S41,016-0
1-(2-THIENYL)-1-BUTANONE

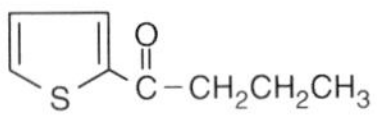

5 gm

S79,916-5
2,5-DIMETHYL-3-(PHENYLTHIO)-THIOPHENE

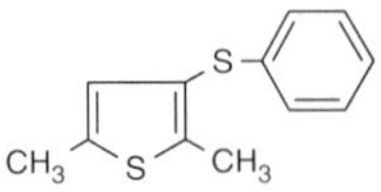

250 mg

S79,904-1
2-METHYL-5-(O-TOLYLTHIO)THIOPHENE

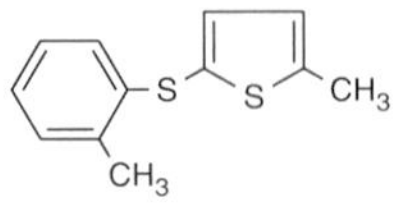

250 mg

S93,526-3
2-AMINO-3,5-DINITROTHIOPHENE

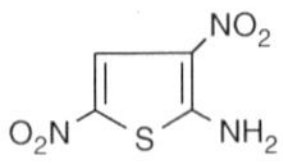

250 mg

S90,820-7
5-ACETYL-2-AMINO-3-NITROTHIOPHENE

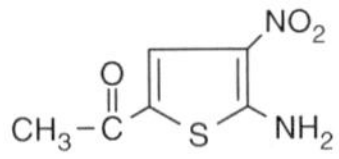

250 mg

S81,103-3
4-(2-THIENYL)ANILINE HYDROCHLORIDE

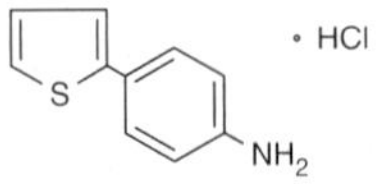

250 mg

S90,793-6
METHYL 2-AMINO-5-(CYCLOPROPYLCARBONYL)-3-THIOPHENECARBOXYLATE

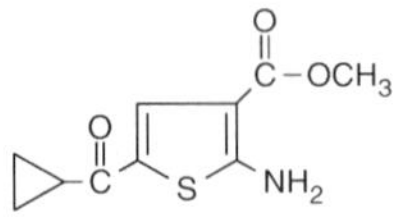

250 mg

S90,866-5
METHYL 5-ACETYL-2-AMINO-3-THIOPHENECARBOXYLATE

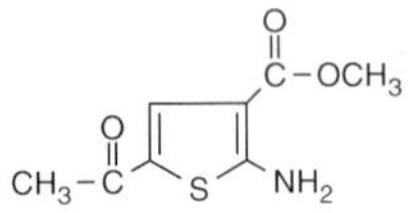

250 mg

S90,785-5
ETHYL 5-ACETYL-2-AMINO-4-METHYL-3-THIOPHENECARBOXYLATE

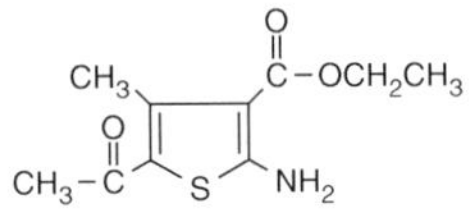

250 mg

S90,830-4
METHYL 2-AMINO-5-BENZOYL-3-THIOPHENECARBOXYLATE

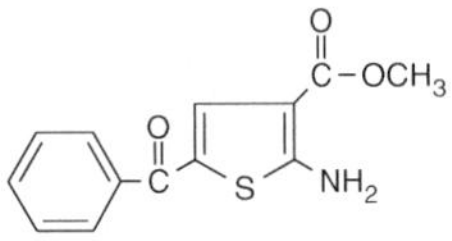

250 mg

S94,857-8
2-(AMINOACETYL)THIOPHENE HYDROCHLORIDE

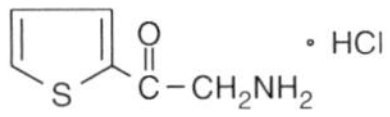

250 mg

S90,868-1
DIMETHYL 5-AMINO-3-METHYL-2,4-THIOPHENEDICARBOXYLATE

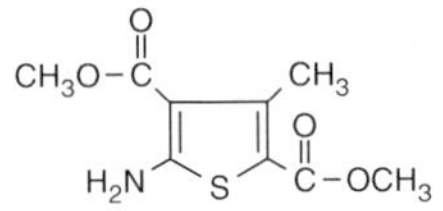

250 mg

S90,851-7
METHYL 2-AMINO-4-METHYL-5-(N-PHENYLCARBAMOYL)-3-THIOPHENECARBOXYLATE

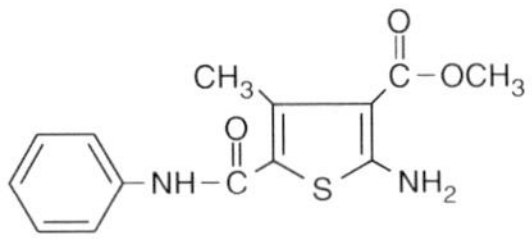

250 mg

S93,535-2
METHYL 2-AMINO-3-THIOPHENECARBOXYLATE

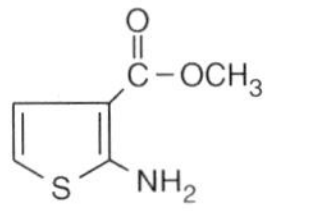

250 mg

S90,828-2
5-AMINO-4-CYANO-3-METHYL-2-THIOPHENECARBOXANILIDE

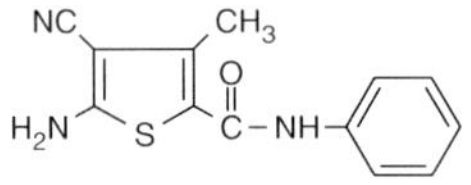

250 mg

S52,545-6
4-(2-THIENYL)-1-(PARA-TOLYLSULFONYL)ISONIPECOTONITRILE

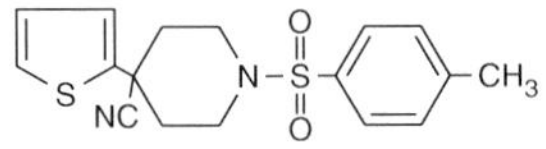

250 mg

S58,393-6
2-(2,4-DINITROSTYRYL)THIOPHENE

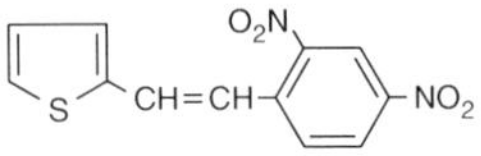

5 gm

S87,849-9
2-CYANO-3-(2-THIENYL)ACRYLAMIDE

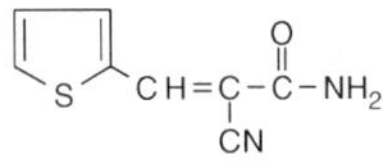

5 gm

S76,259-8
4,5,6,7-TETRAHYDROBENZO(B)-THIOPHENE-4-YLIDENEMALONONITRILE

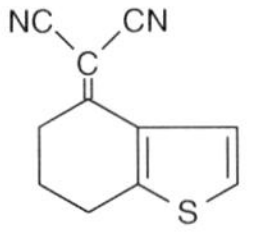

250 mg

S74,684-3
5-NITRO-2-[2-(2-THIENYL)VINYL]-BENZONITRILE

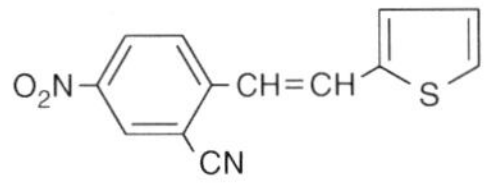

5 gm

S65,821-9
4-NITRO-ALPHA-2-THENYLIDENE-PHENYLACETONITRILE

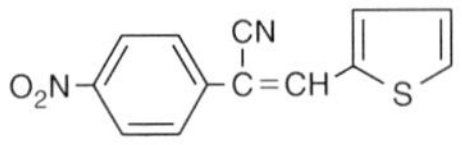

5 gm

S88,722-6
ALPHA-PHENYL-BETA-(2-THIENYL)-ACRYLONITRILE

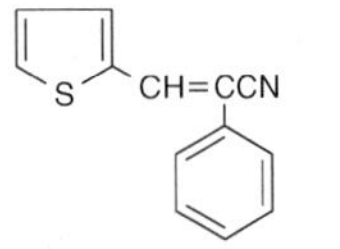

250 mg

S56,475-3
3-[4-(DIMETHYLAMINO)PHENYL]-1-(2-THIENYL)-2-PROPEN-1-ONE

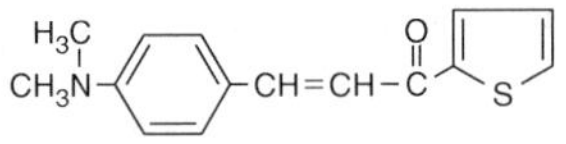

5 gm

S91,115-1
4-[4-(ETHOXYCARBONYL)PHENYL]-1-(2-THENOYL)SEMICARBAZIDE

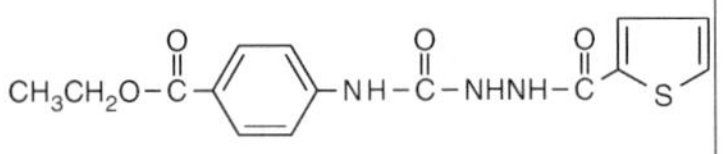

250 mg

S93,671-5
N-TERT-BUTYL-2-THIOPHENECARBOXAMIDE

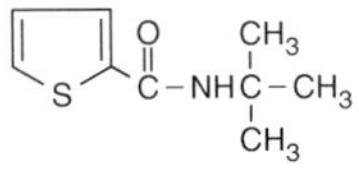

250 mg

S88,063-9
2-THIOPHENECARBOXANILIDE

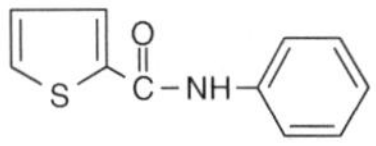

250 mg

S51,248-6
4-(2-THIENYLTHIOFORMYL)MORPHOLINE

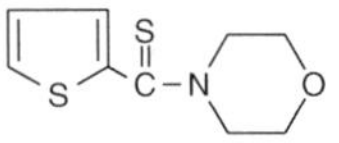

250 mg

S86,216-9
3,5-DIMETHYL-2-(DIMORPHOLINOMETHYL)-THIOPHENE

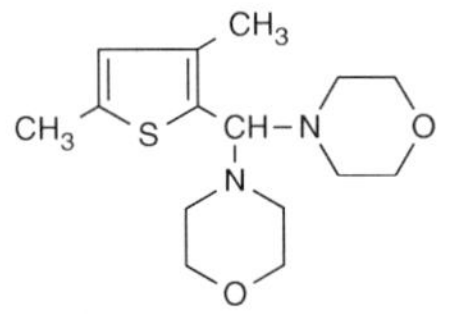

250 mg

S61,955-8
4'-NITRO-3-(2-THIENYL)ACRYLOPHENONE

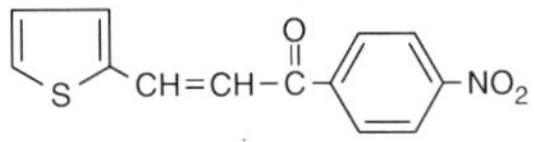

250 mg

S93,529-8
TERT-BUTYL 5-NITRO-2-THIOPHENECARBOXYLATE

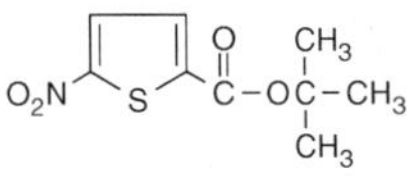

250 mg

S93,536-0
5-NITRO-2-THIOPHENECARBOXYLIC ACID

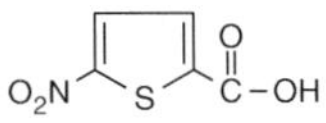

250 mg

S55,188-0
2-BROMO-5-NITROTHIOPHENE

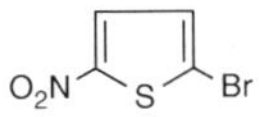

250 mg

S95,253-2
3-METHYL-2-NITROTHIOPHENE

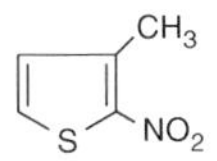

50 mg

S93,527-1
3-NITRO-2-PHENYLTHIOPHENE

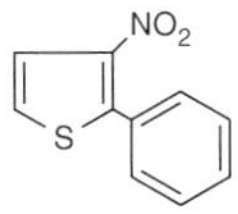

250 mg

S92,004-5
4-METHYL-2-THIOPHENECARBOXALDEHYDE

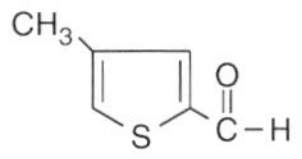

250 mg

S56,488-5
2-THIOPHENEACROLEIN

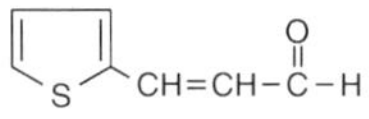

5 gm

S88,905-9
2-ACETYL-3-THIOPHENECARBOXYLIC ACID

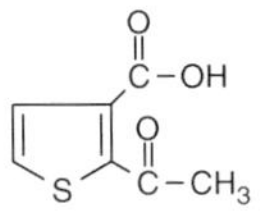

250 mg

S90,806-1
4-OXO-4-(2-THIENYL)-2-BUTENOIC ACID

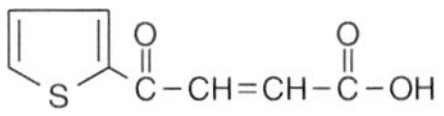

250 mg

S84,712-7
4-BENZOYL-2,5-DICHLORO-3-THIOPHENECARBOXYLIC ACID

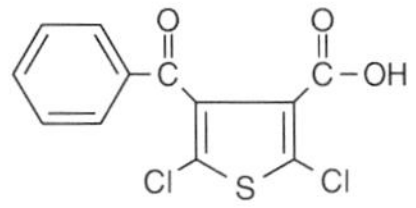

250 mg

S90,845-2
2-(2-THIENYLCARBONYL)BENZOIC ACID

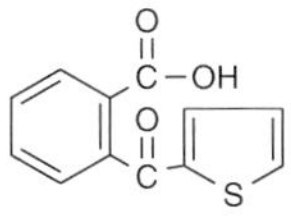

250 mg

S83,163-8
2,5-DIACETYL-3,4-DICHLOROTHIOPHENE

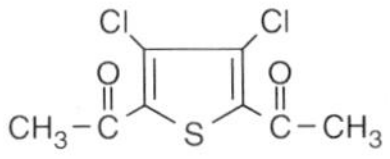

250 mg

S82,770-3
1,5-BIS(3,4,5-TRICHLORO-2-THIENYL)-2,2,3,3,4,4-HEXAFLUORO-1,5-PENTANEDIONE

250 mg

S92,814-3
2-ACETAMIDO-5-ACETYL-3-THIOPHENECARBOXYLIC ACID, SODIUM SALT

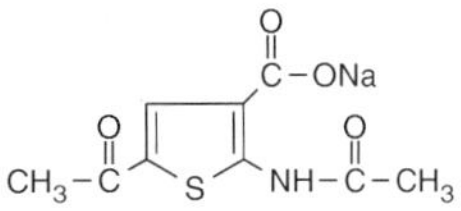

250 mg

S44,219-4
3-(2-THENOYL)PROPIONIC ACID

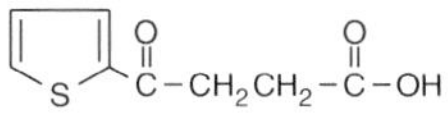

5 gm

S56,481-8
4-[2-(2-THENOYL)VINYL]ACETANILIDE

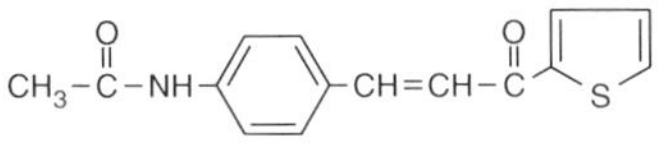

5 gm

S93,525-5
4-OXO-4-(2-THIENYL)BUTYRONITRILE

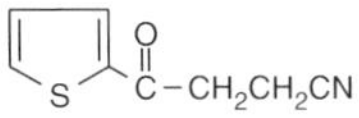

250 mg

S84,382-2
1,3-DICHLORO-5,6-DIHYDRO-4H-CYCLOPENTA(C)THIOPHEN-4-ONE

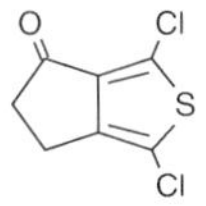

250 mg

S82,809-2
2-ACETYL-3,4,5-TRICHLOROTHIOPHENE

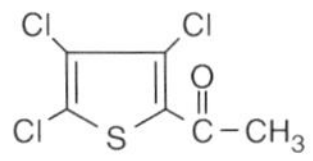

250 mg

S59,331-1
2-BUTYRYL-3-METHYLTHIOPHENE

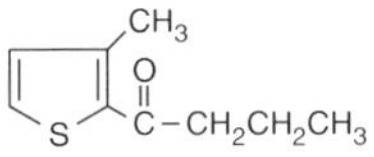

5 gm

S58,402-9
4'-METHYL-3-(2-THIENYL)-ACRYLOPHENONE

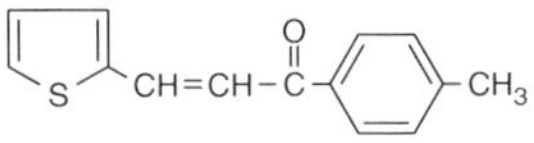

5 gm

S63,777-7
5-PHENYL-1-(2-THIENYL)-2,4-PENTADIEN-1-ONE

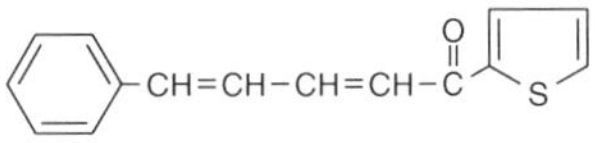

250 mg

S81,270-6
2-OCTADECANOYLTHIOPHENE

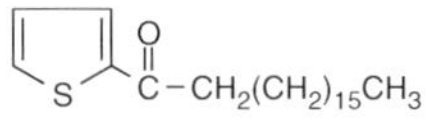

250 mg

S56,485-0
3-(3,4-METHYLENEDIOXYPHENYL)-1-(2-THIENYL)-2-PROPEN-1-ONE

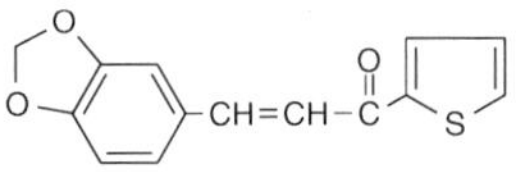

5 gm

S87,878-2
2,6-BIS(2-THIENYLIDENE)-CYCLOHEXANONE

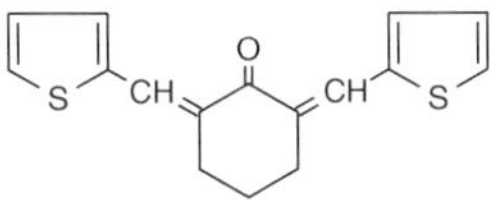

5 gm

S93,532-8
2-CHLOROACETYL-5-PHENYLTHIOPHENE

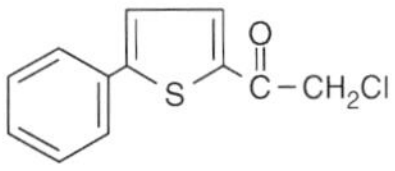

250 mg

S87,880-4
1-(2-NAPHTHYL)-3-(2-THIENYL)-2-PROPEN-1-ONE

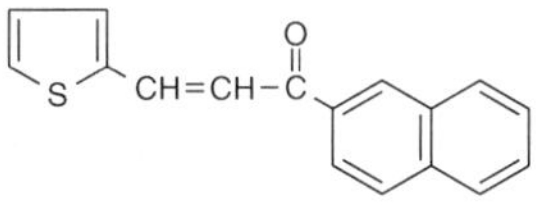

5 gm

S93,780-0
2-ACETYL-5-PHENYLTHIOPHENE

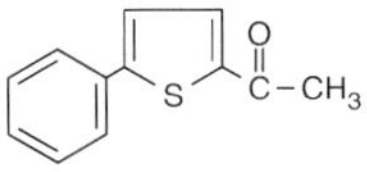

250 mg

S55,182-1
5-CHLORO-2-THIENYL METHYL KETONE

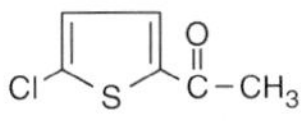

250 mg

S55,189-9
2,5-DICHLORO-3-THIENYL METHYL KETONE

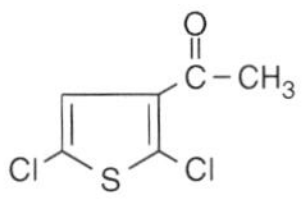

250 mg

S55,203-8
2-ACETYL-5-METHYLTHIOPHENE

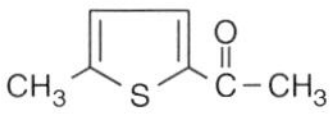

5 gm

S53,204-5
3-MORPHOLINO-1-(2-THIENYL)-1-PROPANONE HYDROCHLORIDE

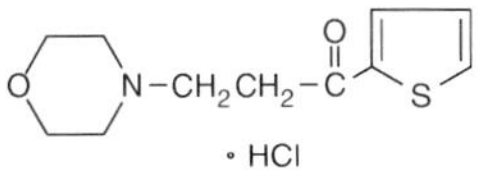

5 gm

S70,139-4
1-PHENYL-5-(2-THIENYL)-1,4-PENTADIEN-3-ONE

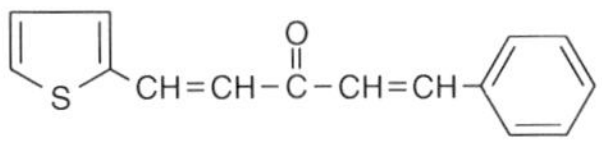

5 gm

S92,720-1
2,4-THIOPHENEDICARBOXYLIC ACID

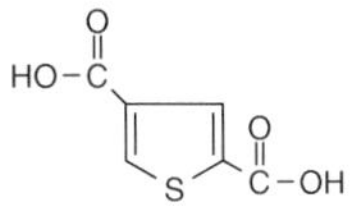

250 mg

S81,228-5
3,4,5-TRICHLORO-2-THIOPHENECARBOXYLIC ACID, SODIUM SALT

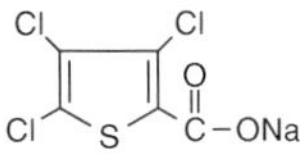

250 mg

S77,923-7
2-(2-THIENYL)HIPPURIC ACID

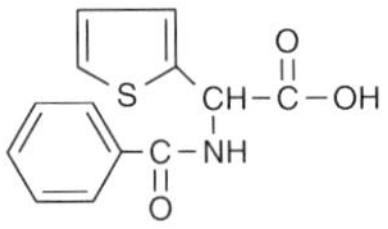

250 mg

S90,811-8
2-ACETAMIDO-3-THIOPHENECARBOXYLIC ACID

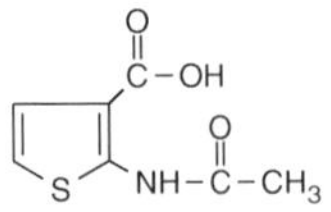

250 mg

S81,229-3
3,4-DICHLORO-2-THIOPHENECARBOXYLIC ACID

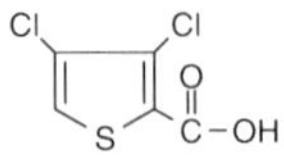

250 mg

S84,365-2
2,5-DICHLORO-3-THIOPHENECARBOXYLIC ACID

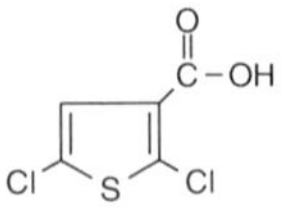

250 mg

S81,227-7
3,4,5-TRICHLORO-2-THIOPHENECARBOXYLIC ACID

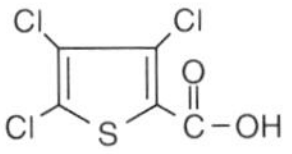

250 mg

S65,174-5
4-METHYL-2-THIOPHENECARBOXYLIC ACID

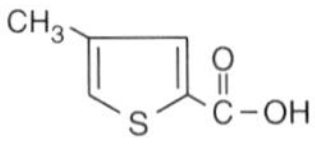

250 mg

S80,871-7
ALPHA-METHYL-2-THIOPHENEACRYLIC ACID

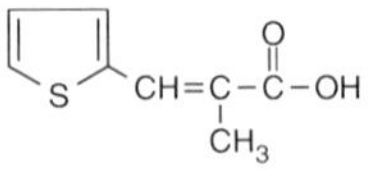

250 mg

S84,390-3
2,5-DICHLORO-ALPHA-METHYL-3-THIOPHENEACRYLIC ACID

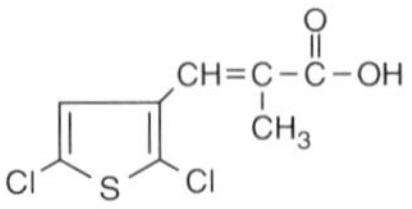

250 mg

S88,817-6
4H-DITHIENO(2,3-B:2',3'-D)-THIOPYRAN-2-CARBOXYLIC ACID

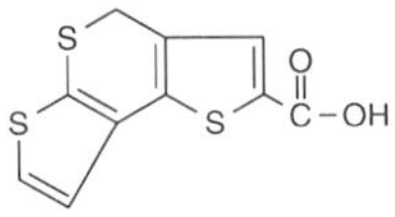

250 mg

S54,544-9
DIETHYL 2-ACETAMIDO-4-METHYL-3,5-THIOPHENEDICARBOXYLATE

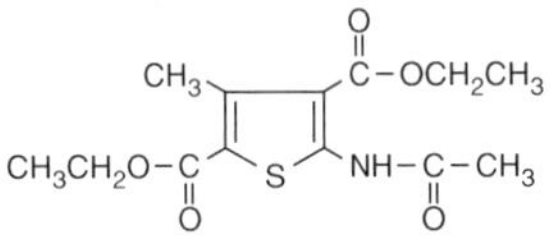

5 gm

S54,458-2
ETHYL 2-ACETAMIDO-4,5,6,7-TETRAHYDROBENZO(B)THIOPHENE-3-CARBOXYLATE

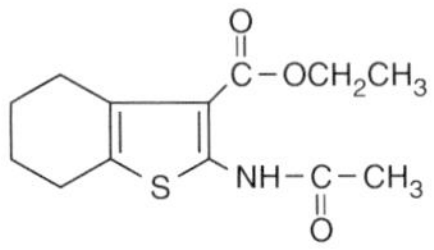

5 gm

S93,137-3
ETHYL 2-CYANO-3-(2-THIENYL)ACRYLATE

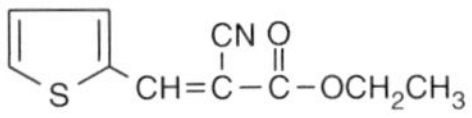

250 mg

S66,476-6
4,5-DIHYDRO-2-(ETHOXYCARBONYL)-NAPHTHO(1,2-B)THIOPHENE

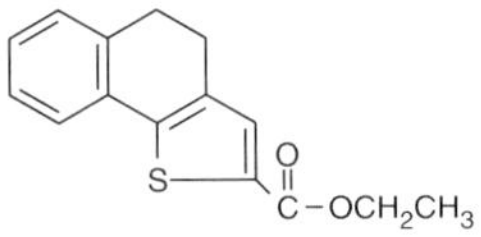

5 gm

S82,381-3
3,4-DICHLORO-2,5-THIOPHENEDICARBONYL CHLORIDE

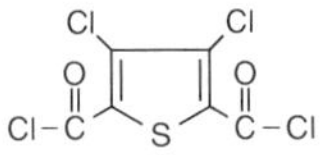

250 mg

S55,155-4
1-(2-FURYL)-3-(2-THIENYL)-2-PROPEN-1-ONE

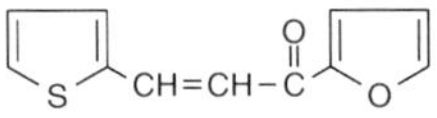

5 gm

S55,218-6
3-(2-FURYL)-1-(2-THIENYL)-2-PROPEN-1-ONE

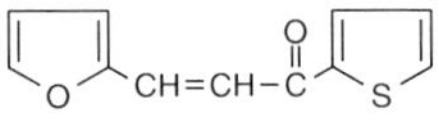

250 mg

S94,244-8
2,2-DIMETHYL-5-(2-THENYLIDENE)-1,3-DIOXANE-4,6-DIONE

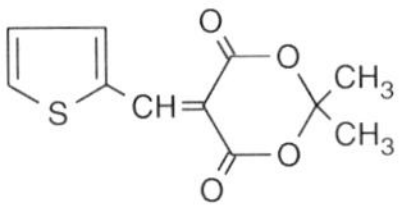

5 gm

S94,247-2
1,3-DIMETHYL-5-(2-THENYLIDENE)-BARBITURIC ACID

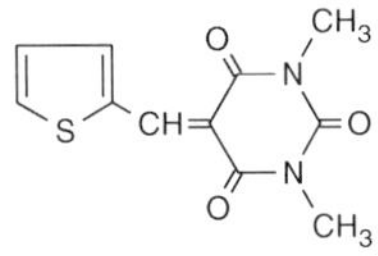

5 gm

S93,534-4
2-FORMAMIDO-3-THIOPHENECARBOXYLIC ACID

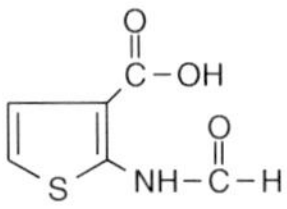

250 mg

S93,778-9
2,2'-BITHIOPHENE-5-CARBOXYLIC ACID

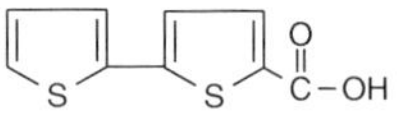

250 mg

S61,637-0
5-ISOPROPYL-2-THIOPHENEBUTYRIC ACID

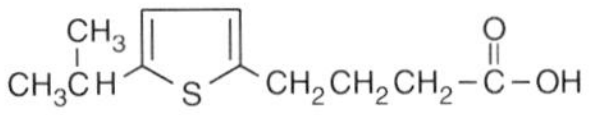

250 mg

S76,029-3
5,6-DIHYDRO-3-(2-THIENYL)IMIDAZO-(2,1-B)THIAZOLE-2-ACETIC ACID HYDROBROMIDE

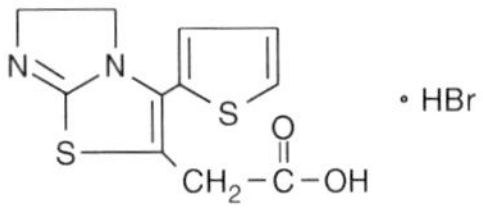

250 mg

S60,502-6
2',5'-DICHLORO-3-THIOPHENEACETAMIDE

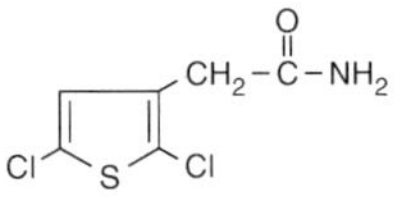

5 gm

S52,630-4
6,7-DIHYDRO-2',6'-DIOXOSPIRO-[THIANAPHTHENE-4(5H)-4'-PIPERIDINE]-3', 5'-DICYANIDE

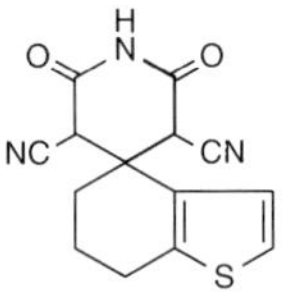

250 mg

S93,517-4
2-PHTHALIMIDOMETHYLTHIOPHENE

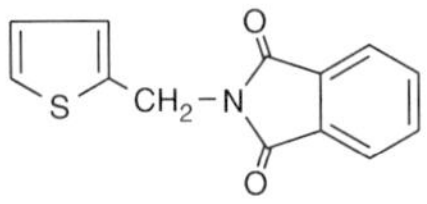

250 mg

S76,256-3
N-(2-THIENYL)ACETAMIDE

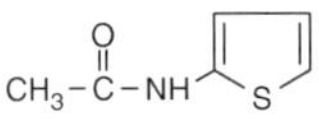

250 mg

S94,388-6
N-(3-THIENYL)ACETAMIDE

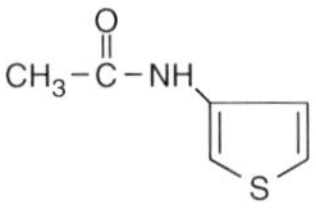

250 mg

S67,904-6
2-THIOPHENEACRYLONITRILE

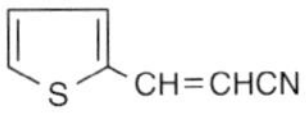

5 gm

S94,509-9
(2-THENYLIDENE)MALONONITRILE

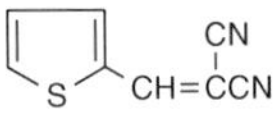

5 gm

S79,988-2
2-(PHENYLSULFONYL)THIOPHENE

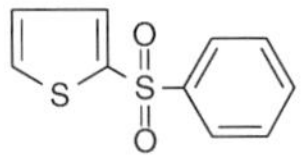

250 mg

S79,990-4
2-METHYL-5-(PHENYLSULFONYL)-THIOPHENE

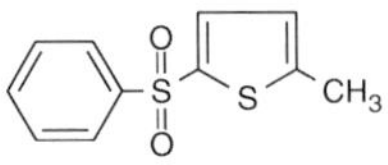

250 mg

S79,906-8
2,5-DIMETHYL-3-(PHENYLSULFONYL)-THIOPHENE

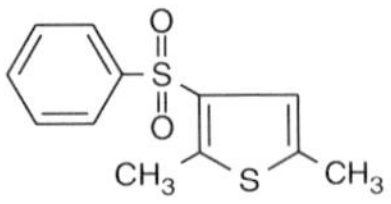

250 mg

S80,726-5
2,5-DIMETHYL-3-(MESITYLSULFONYL)-THIOPHENE
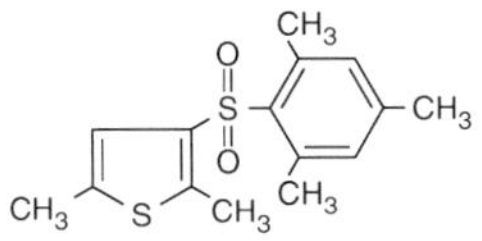
250 mg

S80,741-9
3,5-DIMETHYL-2-(MESITYLSULFONYL)-THIOPHENE
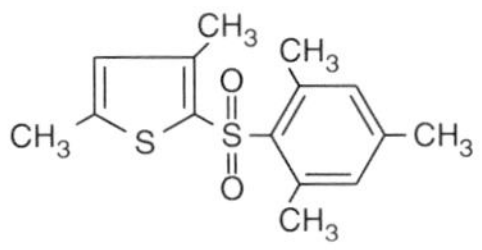
250 mg

S43,563-5
2-THIOPHENESULFONIC ACID, SODIUM SALT MONOHYDRATE
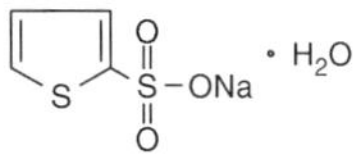
5 gm

S79,884-3
2,5-DIMETHYL-3-THIOPHENESULFONIC ACID, SODIUM SALT
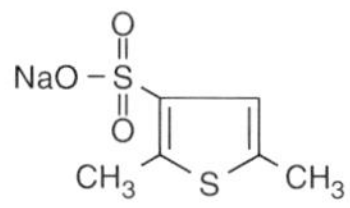
250 mg

S82,492-5
2-IODO-3,4,5-TRICHLOROTHIOPHENE
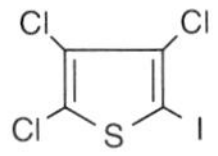
250 mg

S55,205-4
2,3,5-TRIBROMOTHIOPHENE
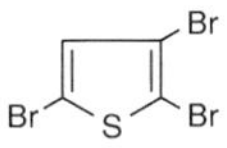
250 mg

S84,363-6
2,5-DIBROMO-3,4-DIIODOTHIOPHENE
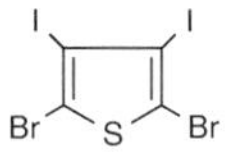
250 mg

S35,582-8
TETRAIODOTHIOPHENE
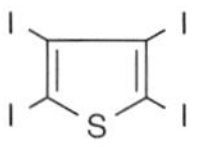
250 mg

S65,193-1
2,5-DICHLORO-3-METHYLTHIOPHENE
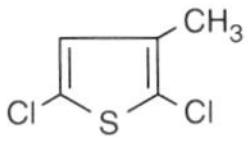
5 gm

S55,200-3
2-IODO-5-METHYLTHIOPHENE
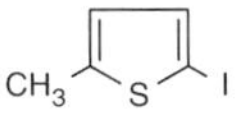
250 mg

S65,294-6
3-METHYL-2,4,5-TRIIODOTHIOPHENE
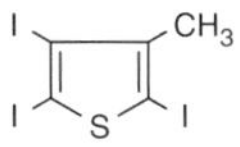
250 mg

S82,811-4
2-ALLYL-3,4,5-TRICHLOROTHIOPHENE
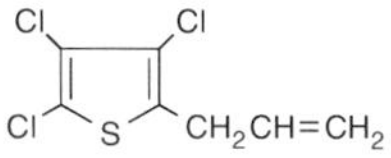
250 mg

S64,970-8
2-TERT-BUTYLTHIOPHENE
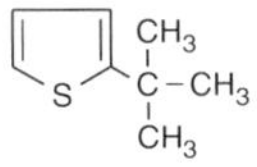
5 gm

S82,494-1
3,3',4,4',5,5'-HEXACHLORO-2,2'-BITHIOPHENE
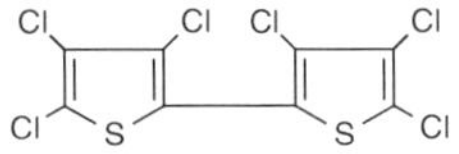
250 mg

S93,661-8
5',5'-DIIODO-2,2'-BITHIOPHENE
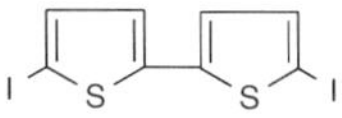
250 mg

S79,888-6
2-BUTYL-5-METHYLTHIOPHENE
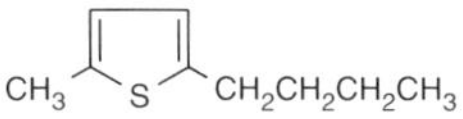
250 mg

S93,777-0
5'-IODO-5-METHYL-2,2'-BITHIOPHENE
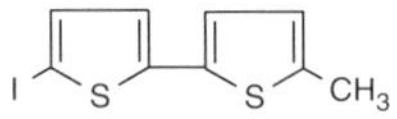
250 mg

S65,302-0
2-HEXYLTHIOPHENE
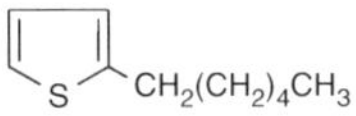
5 gm

S39,158-1
2-HEXADECYLTHIOPHENE
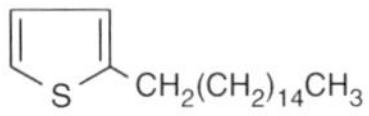
250 mg

S94,119-0
2-(3,5-DIMETHYL-4-ETHYL-2H-PYRROL-2-YLIDENEMETHYL)THIOPHENE TETRA-FLUOROBORATE
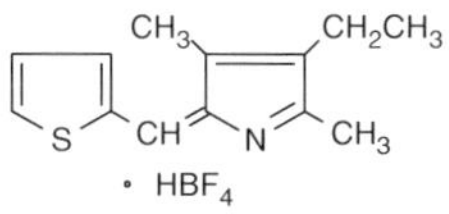
50 mg

S73,252-4
2-(5-BROMO-2-THIENYL)-2,5,5-TRIMETHYL-1,3-DIOXANE
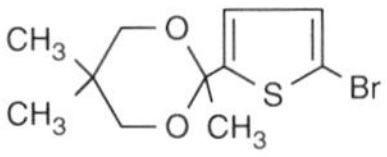
250 mg

S93,530-1
2-BROMO-5-PHENYLTHIOPHENE
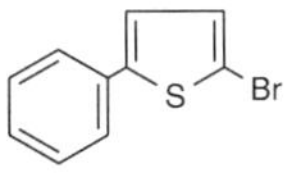
250 mg

S82,493-3
2-PHENYL-3,4,5-TRICHLOROTHIOPHENE
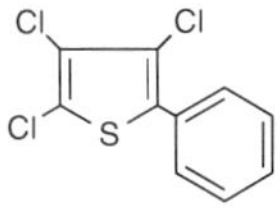
250 mg

S92,719-8
TETRAPHENYLTHIOPHENE
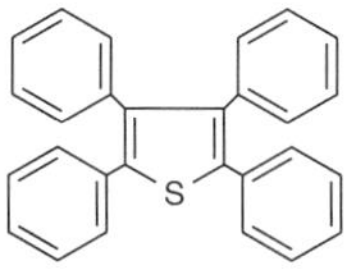
250 mg

S92,588-8
9-(2-THENYLIDENE)FLUORENE
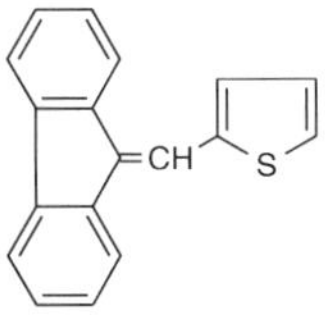
250 mg

S34,528-8
DIACENAPHTHO(1,2-B 1',2'-D)-THIOPHENE
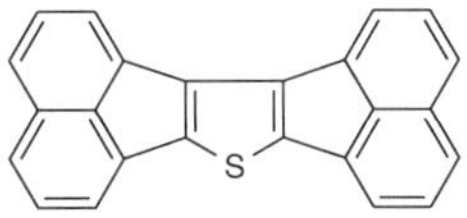
5 gm

S89,434-6
4,5-DICHLORO-3-(DICHLOROMETHYL)-1-PHENYLPYRAZOLE
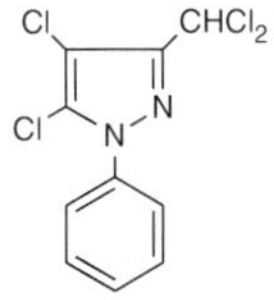
250 mg

S76,075-7
METHYL 3-HYDROXY-1-METHYL-5[3-(TRIFLUOROMETHYL)BENZYL]-4-PYRAZOLEACETATE
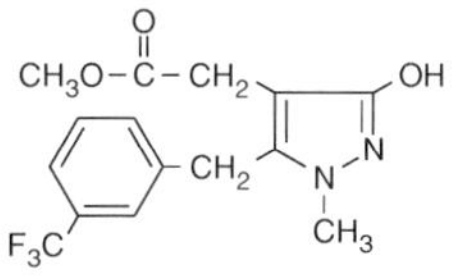
250 mg

S89,376-5
1-PHENYL-3-(TRICHLOROMETHYL)-PYRAZOLE
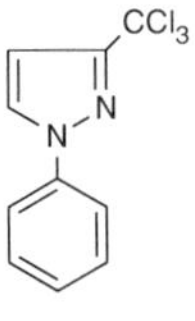
250 mg

S70,720-1
1-(4-FLUOROPHENYL)-5-HYDROXY-3-METHYLPYRAZOLE
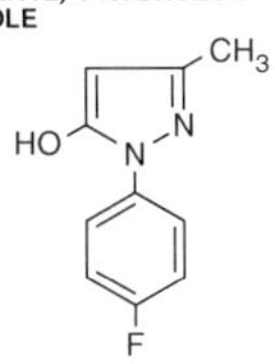
5 gm

S91,127-5
1-[N-(5-CHLORO-2,4-DIMETHOXYPHENYL)-CARBAMOYL]PYRAZOLE
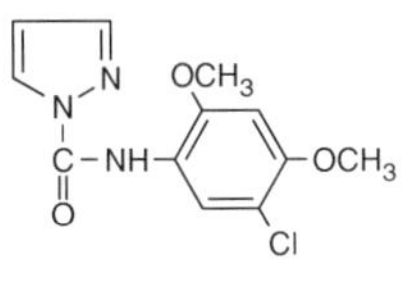
250 mg

S91,693-5
1-(N-[4-CHLORO-2-(PHENOXY)PHENYL]-CARBAMOYL)-3,5-DIMETHYLPYRAZOLE
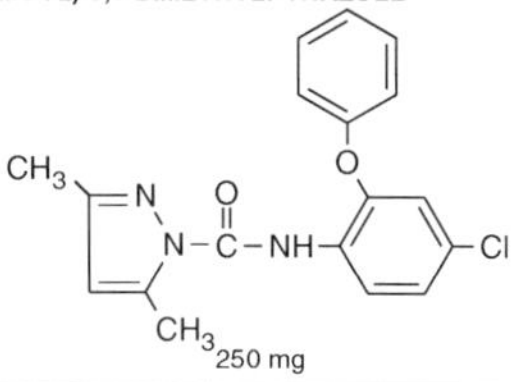
250 mg

S91,165-8
1-[N-(2-FLUORO-5-NITROPHENYL)-CARBAMOYL]PYRAZOLE

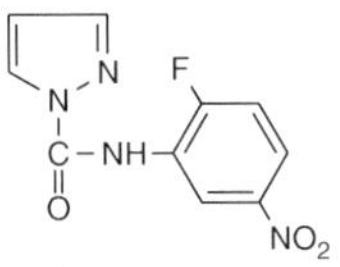

250 mg

S91,096-1
1-(2,3-DICHLOROPHENYL)-3-(1-PHENYL-5-PYRAZOLYL)UREA

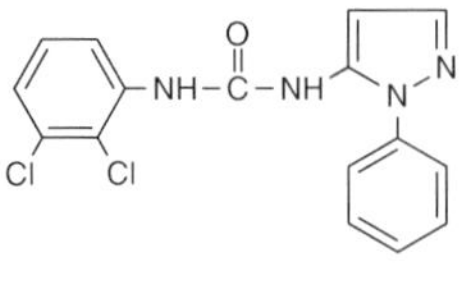

250 mg

S91,234-4
1-[N-(4-FLUOROPHENYL)CARBAMOYL]-PYRAZOLE

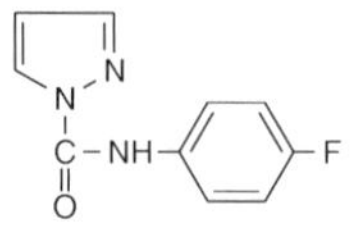

250 mg

S91,088-0
1-[N-(2,3-DICHLOROPHENYL)CARBAMOYL]-3,5-DIMETHYLPYRAZOLE

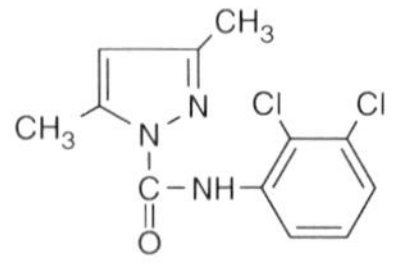

250 mg

S69,742-7
4-(4-CHLOROPHENYL)DIHYDRO-DIMETHYL-1H-PYRAZOLO(3,4-E)(1,4)-THIAZEPIN-7(6H)-ONE

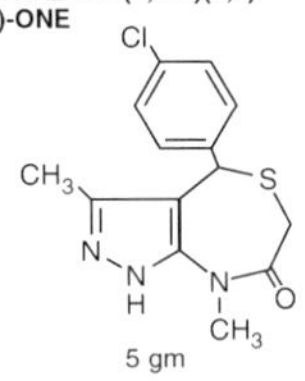

5 gm

S91,057-0
1-(3-FLUOROPHENYL)-3-(1-PHENYL-5-PYRAZOLYL)UREA

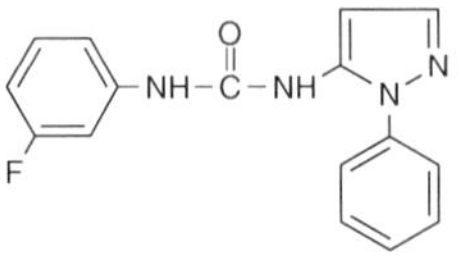

250 mg

S77,640-8
5-AMINO-4-CYANO-1-(2-HYDROXYETHYL)-PYRAZOLE

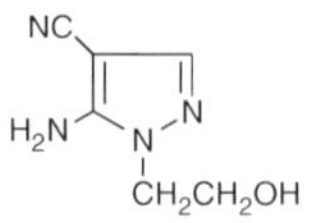

250 mg

S78,506-7
5-AMINO-1-(2-HYDROXYETHYL)-4-PYRAZOLECARBOXAMIDE

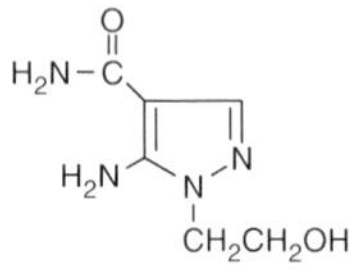

250 mg

S94,052-6
5-METHYL-4-NITRO-1H-PYRAZOLE-3-METHANOL

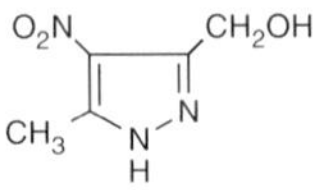

250 mg

S60,057-1
5-(DIMETHYLAMINOMETHYLENEAMINO)-1,3-DIPHENYL-4-PYRAZOLEMETHANOL

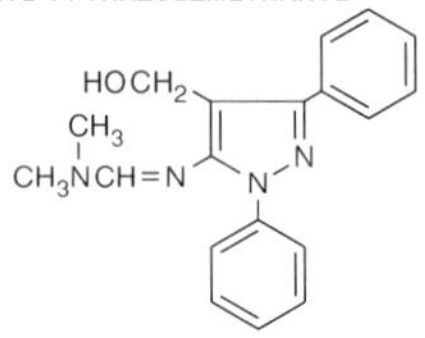

5 gm

S35,076-1
4-(2-HYDROXYETHYL)-3-METHYL-2-PYRAZOLIN-5-ONE

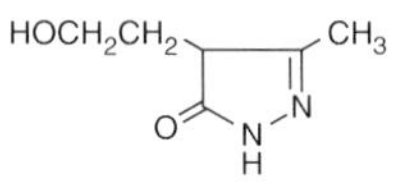

5 gm

S72,042-9
3,5-DIMETHYL-4-HYDROXYPYRAZOLE

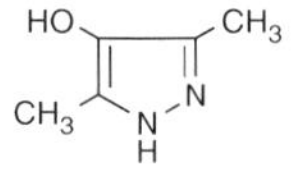

250 mg

S94,532-3
3-METHYL-1H-PYRAZOLE-5-METHANOL

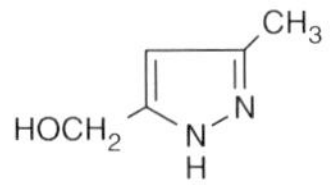

250 mg

S35,890-8
4-BROMO-3,5-DIMETHYL-1-PYRAZOLEMETHANOL

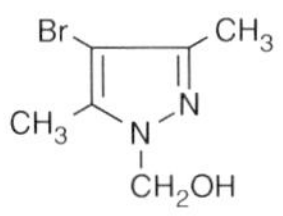

250 mg

S94,538-2
3,5-DIMETHYL-1H-PYRAZOLE-4-METHANOL

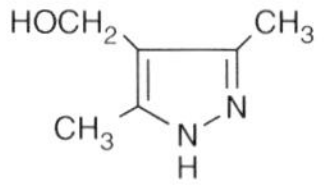

100 mg

S94,038-0
4-HYDROXYMETHYL-1-PHENYL-1H-PYRAZOLE

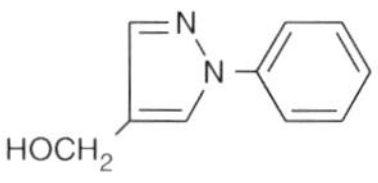

250 mg

S76,358-6
2-(3-METHYL-2-PYRAZOLIN-5-YL)PHENOL

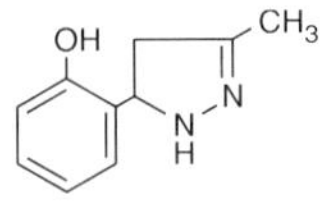

250 mg

S91,033-3
1-[N-(2,4-DIMETHOXYPHENYL)-CARBAMOYL]PYRAZOLE

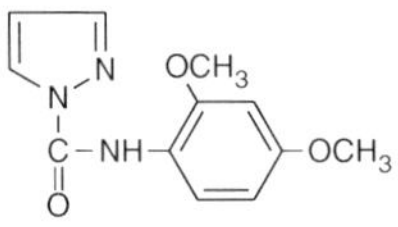

250 mg

S91,044-9
1-[N-(2,5-DIMETHOXYPHENYL)-CARBAMOYL]-3,5-DIMETHYLPYRAZOLE

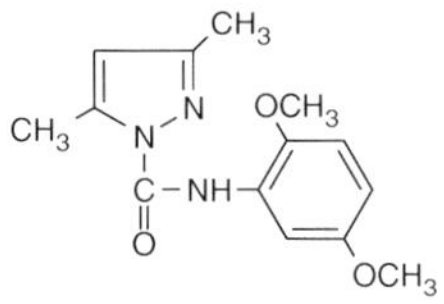

250 mg

S93,466-6
ETHYL 5-(4-METHOXYANILINO)-3-METHYL-4-PYRAZOLECARBOXYLATE

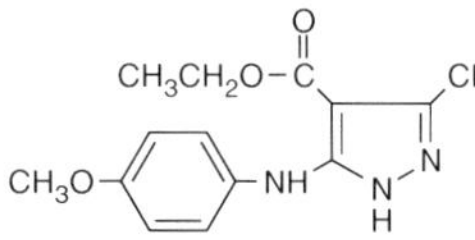

250 mg

S94,531-5
5-(4-METHOXYPHENYL)-1H-PYRAZOLE

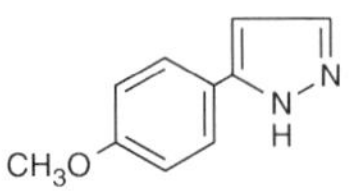

250 mg

S93,470-4
3,9-DIMETHYL-6-METHOXY-1H-PYRAZOLO-(3,4-B)QUINOL-4(9H)-ONE

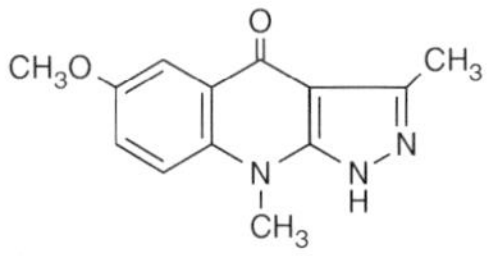

250 mg

S93,468-2
5-(4-METHOXYPHENYL)-3-METHYL-4-PYRAZOLECARBOXYLIC ACID

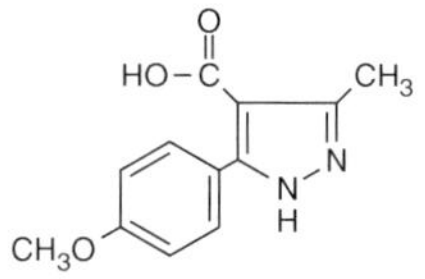

250 mg

S89,421-4
METHYL 4-(1-CHLOROVINYL)-3-PYRAZOLECARBOXYLATE

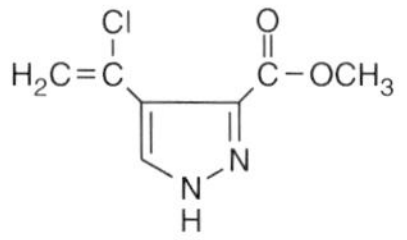

250 mg

S91,067-8
1-[4-(METHYLTHIO)PHENYL]-3-(1-PHENYL-5-PYRAZOLYL)UREA

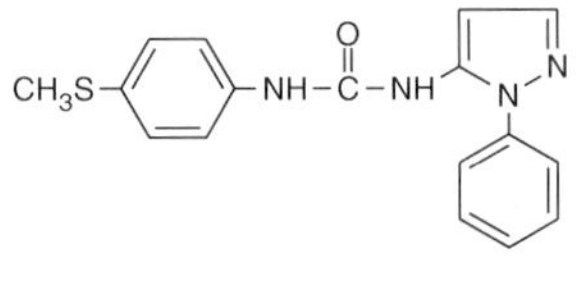

250 mg

S80,305-7
5-AMINO-4-NITROSO-4-PYRAZOLIN-3-ONE

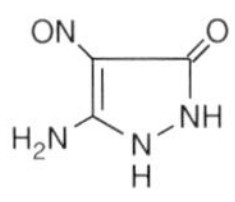

250 mg

S45,313-7
5-AMINO-4-CYANO-1-PHENYL-3-PYRAZOLEACETONITRILE

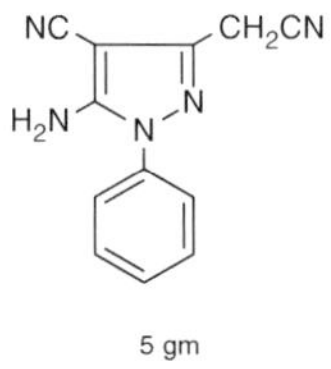

5 gm

S34,097-9
5-AMINO-4-CYANO-3-CYANOMETHYLPYRAZOLE

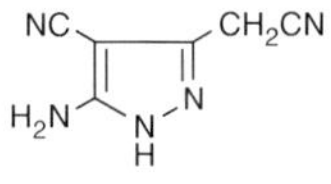

5 gm

S43,421-3
3-AMINO-2,4,5,6-TETRAHYDROCYCLOPENTAPYRAZOLE-2-PROPIONITRILE

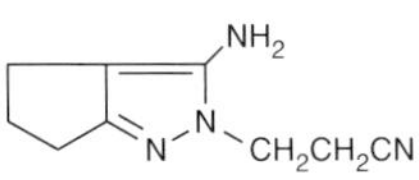

5 gm

S34,098-7
5-AMINO-4-CYANO-1-PHENYLPYRAZOLE

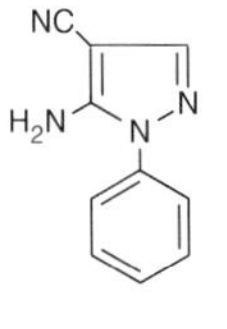

5 gm

S43,807-3
5-AMINO-1-PHENYLPYRAZOLE

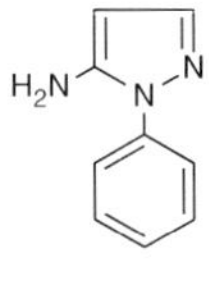

5 gm

S78,339-0
5-AMINO-4-PYRAZOLECARBONITRILE

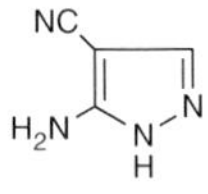

5 gm

S59,302-8
5-AMINO-3-METHYL-1-(P-TOLYL)-PYRAZOLE

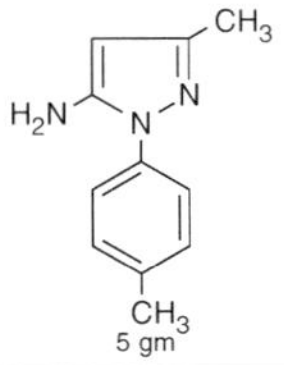

5 gm

S66,403-0
5-(2-AMINO-5-CHLOROPHENYL)-3-METHYLPYRAZOLE

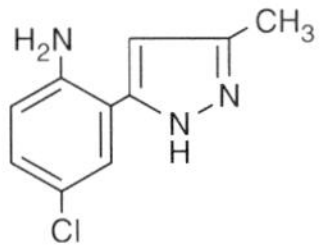

5 gm

S50,094-1
2-(4-AMINOPHENYL)-3-METHYL-3-PYRAZOLIN-5-ONE

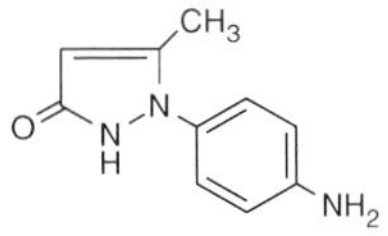

5 gm

S34,808-2
ETHYL 5-AMINO-1-PHENYLPYRAZOLE-4-CARBOXYLATE

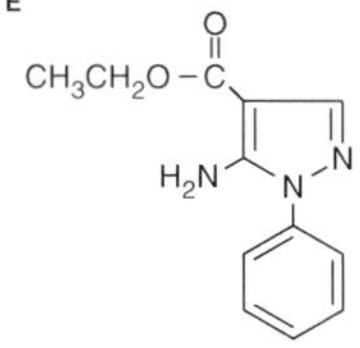

5 gm

S35,751-0
5-AMINO-4-CARBAMYL-3-PYRAZOLEACETIC ACID

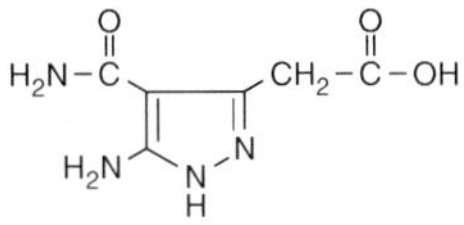

250 mg

S78,502-4
4-AMINO-5-PYRAZOLECARBOXAMIDE HEMISULFATE

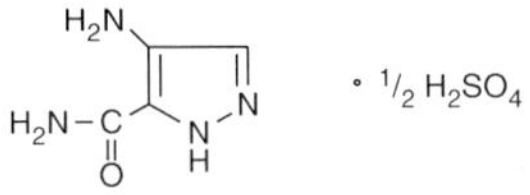

250 mg

S78,304-8
5-AMINO-1-METHYL-4-PYRAZOLECARBOXAMIDE SULFATE

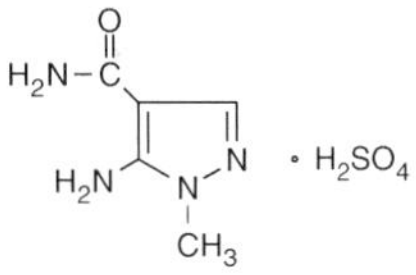

5 gm

S78,407-9
4-AMINO-3-METHYL-5-PYRAZOLECARBOXAMIDE SULFATE

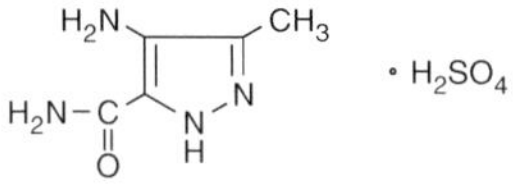

250 mg

S56,103-7
3-AMINO-5-PYRAZOLONE

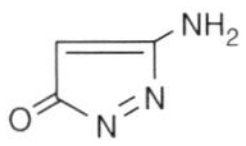

5 gm

S94,034-8
5-AMINO-1,3-DIMETHYL-1H-PYRAZOLE

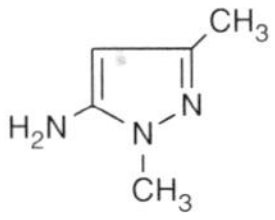

250 mg

S70,715-5
5-AMINO-3-METHYL-1-PHENYLPYRAZOLE

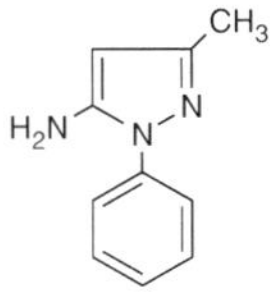

5 gm

S43,290-3
3-METHYL-4-PHENYLHYDRAZINO-2-PYRAZOLIN-5-ONE

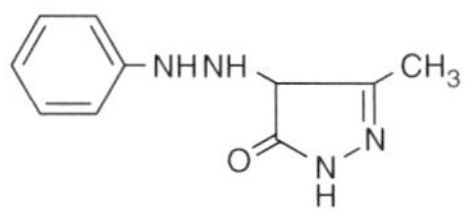

5 gm

S94,539-0
3,5-DI(TERT-BUTYL)-4-NITROSO-1H-PYRAZOLE

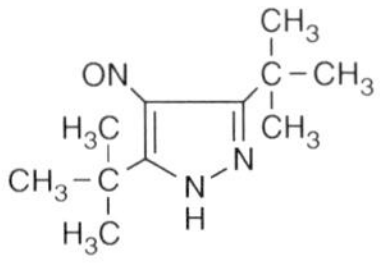

100 mg

S94,051-8
3,5-DIMETHYL-4-NITROSO-1H-PYRAZOLE

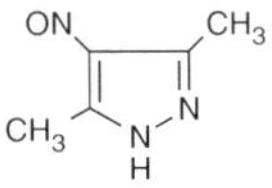

250 mg

S94,044-5
3,4-DINITRO-1H-PYRAZOLE

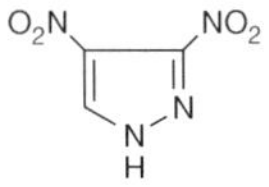

250 mg

S94,035-6
4-NITRO-1,3,5-TRIMETHYL-1H-PYRAZOLE

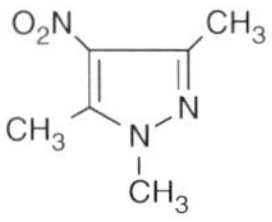

250 mg

S94,115-8
4-NITRO-5-(1H-PYRAZOL-1-YL)-1H-PYRAZOLE

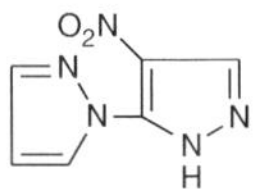

250 mg

S94,553-6
3,5-DIMETHYL-4'-NITRO-1,5'-BI-1H-PYRAZOLE

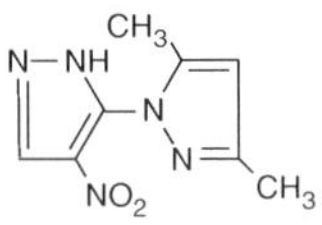

250 mg

S94,554-4
4'-NITRO-3,3',5-TRIMETHYL-1,5'-BI-1H-PYRAZOLE

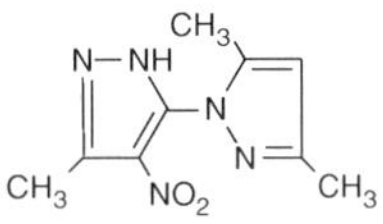

250 mg

S94,552-8
1,1'-DIMETHYL-3,3'-BI-1H-PYRAZOLE

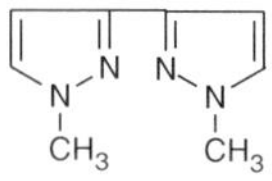

100 mg

S41,044-6
3,5-DIMETHYL-1-(2,4-DINITROPHENYL)-PYRAZOLE

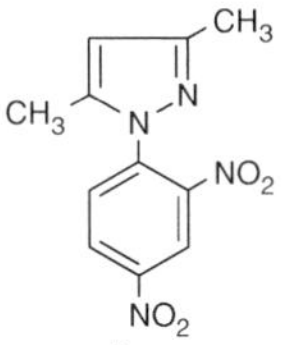

5 gm

S94,875-6
3,5-DIANILINO-1H-PYRAZOLE

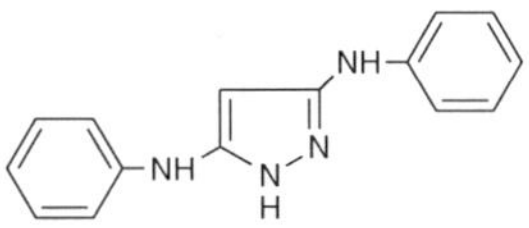

250 mg

S64,564-8
N-(1,3-DIMETHYL-5-PYRAZOLYL)-ANTHRANILIC ACID

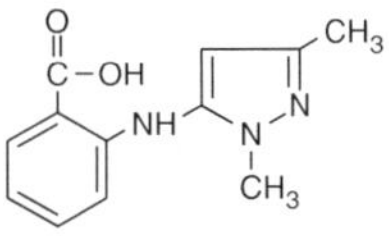

5 gm

S53,300-9
4-(1-PYRAZOLYL)-2-BUTANONE

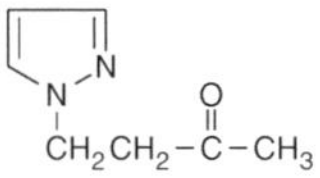

5 gm

S56,617-9
5-METHYL-3-PHENYL-4-PYRAZOLECARBOXAMIDE

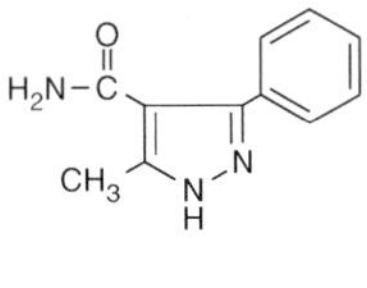

5 gm

S63,251-1
3-(5-IMINO-3-METHYL-2-PYRAZOLIN-1-YL)BENZENESULFONIC ACID

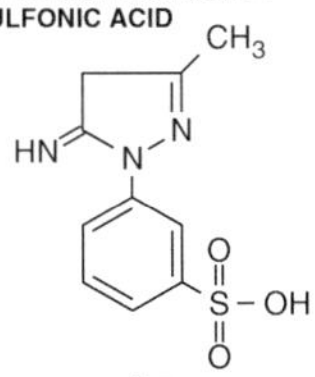

5 gm

S50,213-8
ETHYL 9-[4-(3-METHYL-5-OXO-3-PYRAZOLIN-2-YL)-PHENYLCARBAMOYL]-NONANOATE

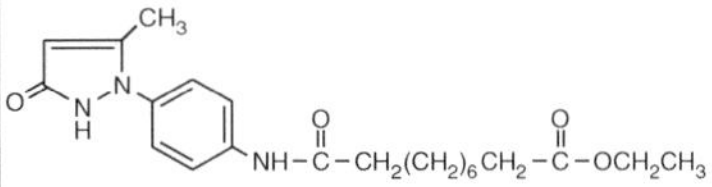

250 mg

S50,233-2
3-METHYL-2-(4-NITROPHENYL)-3-PYRAZOLIN-5-ONE

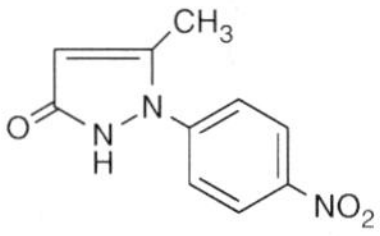

5 gm

S93,950-1
4-ETHYL-1-(4-NITROPHENYL)PYRAZOLE

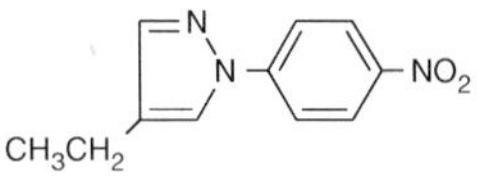

250 mg

S50,688-5
1-ANILINOMETHYL-3,5-DIMETHYLPYRAZOLE

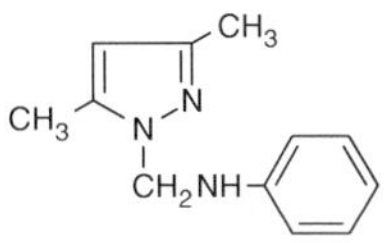

5 gm

S93,469-0
1,3-DIMETHYL-1H-PYRAZOLO(3,4-B)-QUINOL-4(9H)-ONE

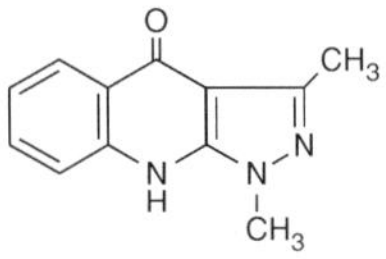

250 mg

S93,463-1
5-ANILINO-3-METHYL-4-PYRAZOLECARBOXYLIC ACID

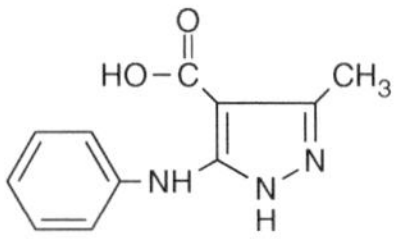

250 mg

S93,467-4
DIETHYL 5-(3-CHLOROANILINO)-3,4-PYRAZOLEDICARBOXYLATE
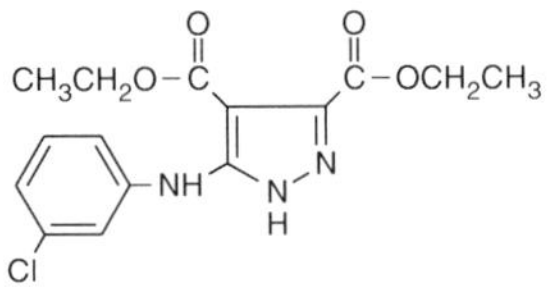
250 mg

S93,465-8
ETHYL 5-ANILINO-3-METHYL-4-PYRAZOLECARBOXYLATE
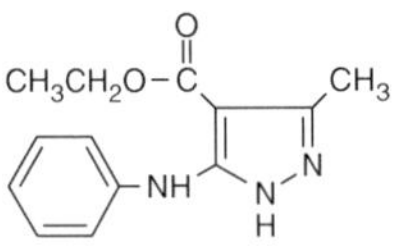
250 mg

S92,869-0
1,5-DIMETHYL-3-[2-(DIMETHYLAMINO)-VINYL]-4-NITRO-1H-PYRAZOLE
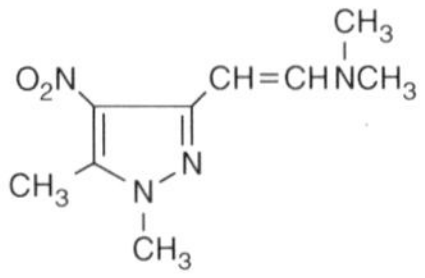
250 mg

S89,374-9
4,5-DICHLORO-1-PHENYL-4-PYRAZOLIN-3-ONE
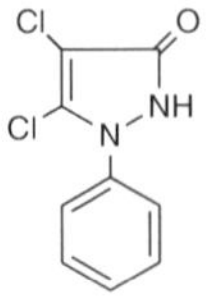
250 mg

S87,004-8
3-ANILINO-4,5,6,7-TETRAHYDRO-1H-INDAZOLE
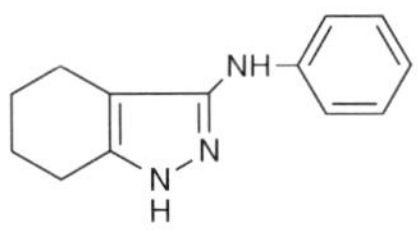
250 mg

S87,029-3
3-ANILINO-5-PHENYLPYRAZOLE
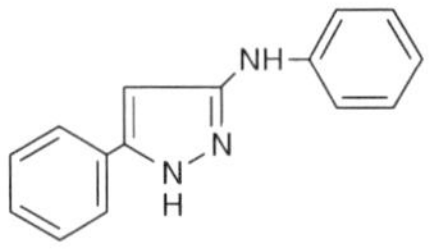
250 mg

S73,499-3
METHYL 2-(5-METHYL-1-PYRAZOLYLMETHYL)BENZOATE
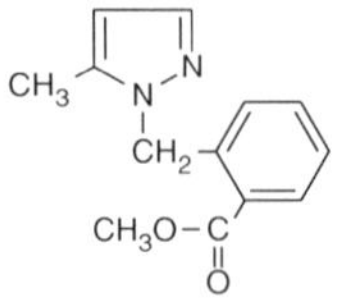
250 mg

S94,466-1
3-PHENYL-1H-PYRAZOL-5(4H)-ONE
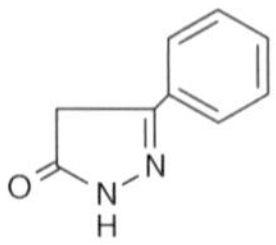
250 mg

S94,461-0
4-BENZYL-1,3-DIPHENYL-1H-PYRAZOL-5-(4H)-ONE
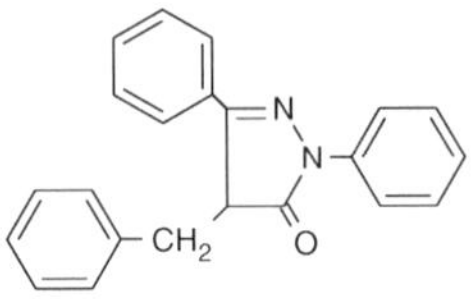
250 mg

S35,729-4
4-ACETYL-5-METHYL-1-PHENYLPYRAZOLE
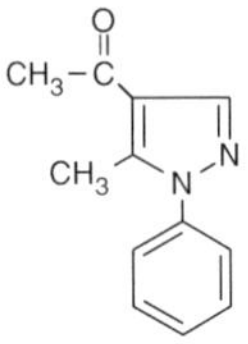
5 gm

S34,938-0
ETHYL 5-METHYL-1-PHENYLPYRAZOLE-4-CARBOXYLATE
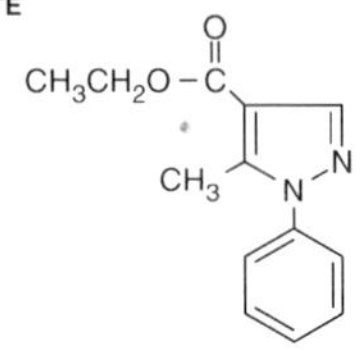
5 gm

S54,087-0
3-METHYL-2-PYRAZOLIN-5-ONE
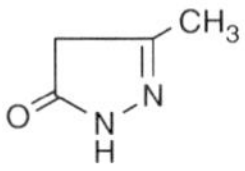
5 gm

S36,986-1
3-METHYL-5-PHENYLPYRAZOLE
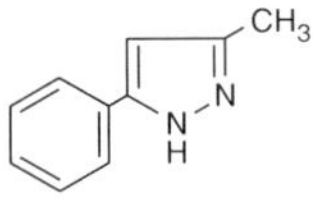
250 mg

S92,729-5
1,3,5-TRIPHENYLPYRAZOLE
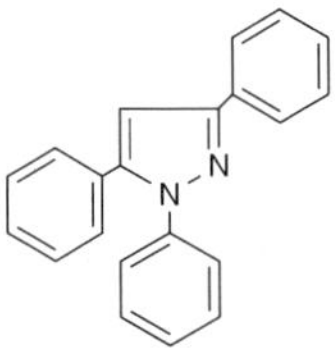
250 mg

S94,544-7
ETHYL 1H-PYRAZOLE-1-ACETATE
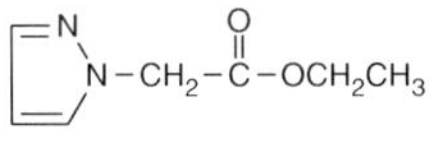
250 mg

S89,373-0
METHYL 4-(TRICHLOROVINYL)-3-PYRAZOLECARBOXYLATE
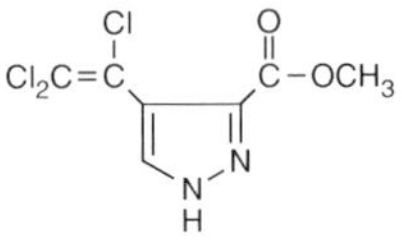
250 mg

S94,464-5
ETHYL 1-ETHYL-5-METHYL-1H-PYRAZOLE-3-CARBOXYLATE
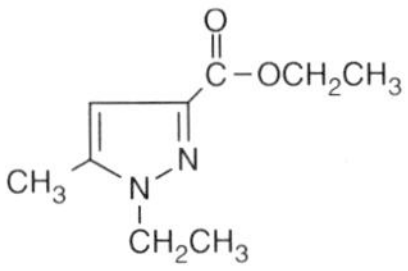
250 mg

S78,949-6
4-NITRO-5-PYRAZOLECARBOXYLIC ACID
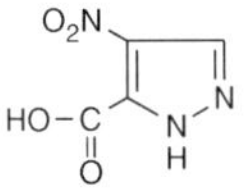
250 mg

S93,939-0
3-NITRO-1H-PYRAZOLE
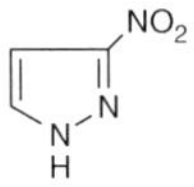
250 mg

S93,940-4
4-NITRO-1H-PYRAZOLE
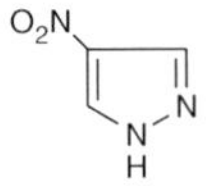
250 mg

S94,028-3
1-NITRO-1H-PYRAZOLE
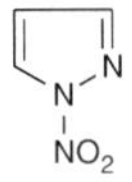
250 mg

S93,941-2
3-METHYL-4-NITRO-1H-PYRAZOLE
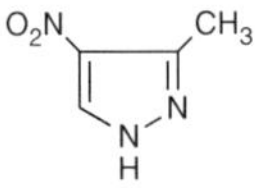
250 mg

S93,945-5
3,5-DIMETHYL-4-NITRO-1H-PYRAZOLE
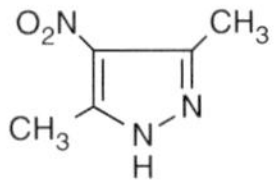
250 mg

S94,037-2
4-ETHYL-1-METHYL-3-NITRO-1H-PYRAZOLE
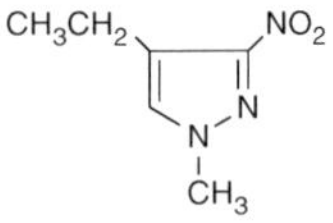
250 mg

S93,943-9
5-TERT-BUTYL-3-NITRO-1H-PYRAZOLE
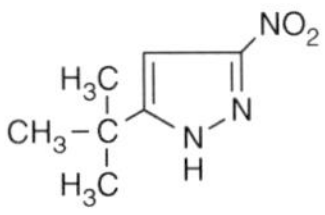
250 mg

S94,234-0
1-NITRO-3-TERT-BUTYL-1H-PYRAZOLE
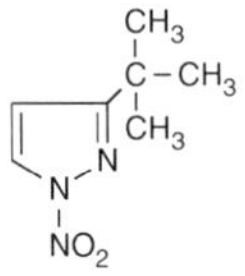
250 mg

S93,942-0
5-TERT-BUTYL-3-METHYL-4-NITRO-1H-PYRAZOLE
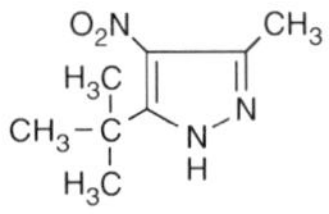
250 mg

S94,057-7
3-NITRO-5-PHENYL-1H-PYRAZOLE
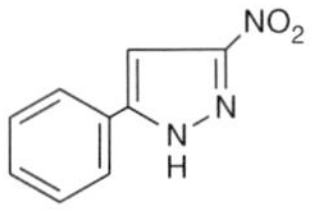
250 mg

S94,236-7
1-NITRO-3-PHENYL-1H-PYRAZOLE
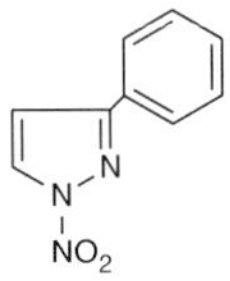
250 mg

S94,462-9
1-NITRO-4-PHENYL-1H-PYRAZOLE
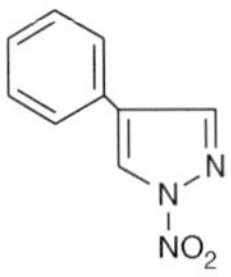
100 mg

S94,537-4
1,3,5-TRIMETHYL-1H-PYRAZOLE-4-CARBOXALDEHYDE
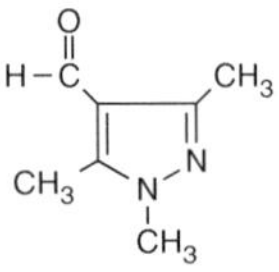
100 mg

S93,944-7
3-METHYL-1H-PYRAZOLE-5-CARBOXALDEHYDE
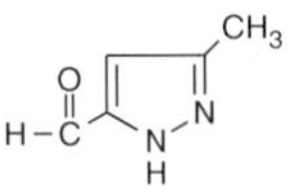
250 mg

S94,029-1
1,5-DIMETHYL-1H-PYRAZOLE-4-CARBOXYALDEHYDE

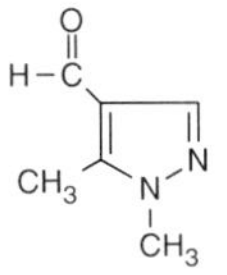

250 mg

S94,235-9
1,3-DIMETHYL-1H-PYRAZOLE-4-CARBOXALDEHYDE

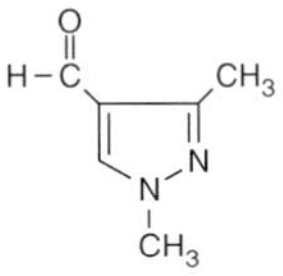

250 mg

S94,540-4
4-ACETYL-1-METHYL-1H-PYRAZOLE

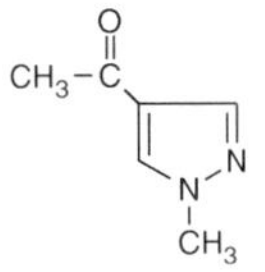

250 mg

S75,459-5
2-(3,5-DIMETHYL-1-PYRAZOLYL)-ACETOPHENONE

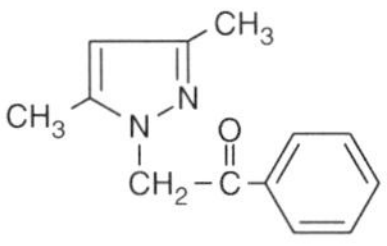

250 mg

S94,039-9
4-ACETYL-1-PHENYL-1H-PYRAZOLE

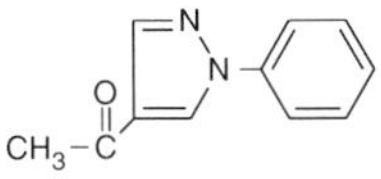

250 mg

S78,332-3
3-METHYL-5-PYRAZOLECARBOXYLIC ACID

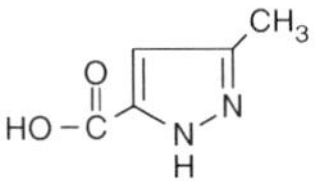

5 gm

S93,947-1
1-METHYL-4-PYRAZOLECARBOXYLIC ACID

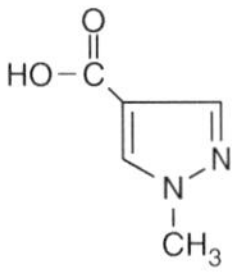

250 mg

S93,462-3
ETHYL 4-CHLORO-5-METHYL-3-PYRAZOLECARBOXYLATE

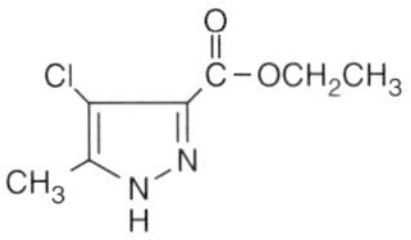

250 mg

S35,183-0
4-KETO-6-METHYL-1-PHENYLPYRAZOLO-(3,4-D)-5,7-OXAZINE

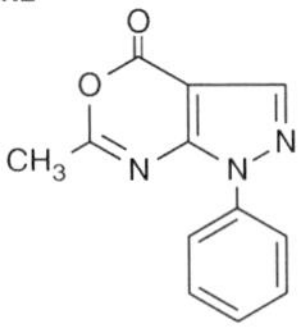

250 mg

S58,091-0
3,5-PYRAZOLEDICARBOXYLIC ACID, MONOPOTASSIUM SALT

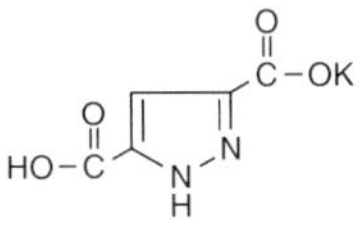

5 gm

S94,529-3
1H-PYRAZOLE-3-CARBOXYLIC ACID

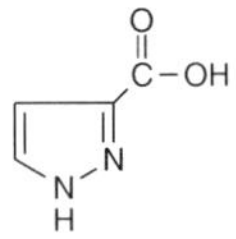

250 mg

S94,056-9
4-METHYL-1H-PYRAZOLE-5-CARBOXYLIC ACID

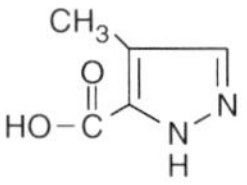

250 mg

S94,459-9
4-BROMO-1H-PYRAZOLE-1-ACETIC ACID

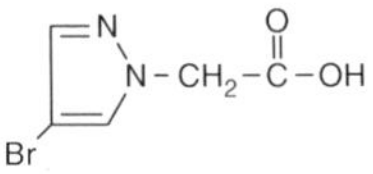

250 mg

S94,032-1
1,3-DIMETHYL-1H-PYRAZOLE-5-CARBOXYLIC ACID

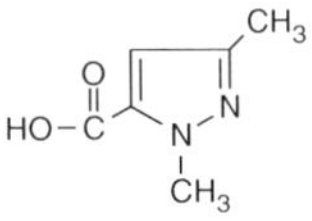

250 mg

S94,053-4
3,5-DIMETHYL-1H-PYRAZOLE-4-CARBOXYLIC ACID

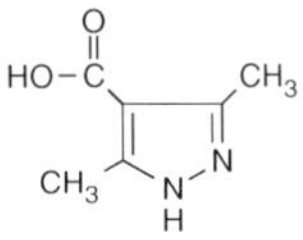

250 mg

S94,031-3
1,3,5-TRIMETHYL-1H-PYRAZOLE-4-CARBOXYLIC ACID

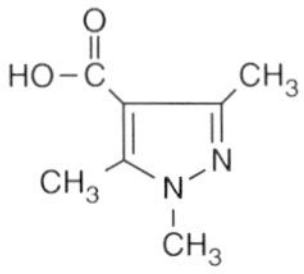

250 mg

S82,214-0
5-METHYL-1-PHENYL-3-PYRAZOLECARBOXYLIC ACID MONOHYDRATE

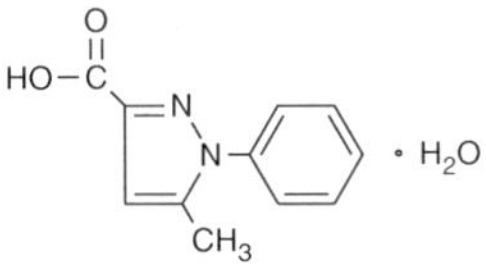

250 mg

S94,040-2
4-(DIACETOXYMETHYL)-1-PHENYL-1H-PYRAZOLE

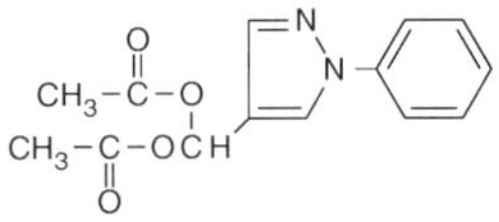

250 mg

S94,043-7
ETHYL 1H-PYRAZOLE-3-CARBOXYLATE

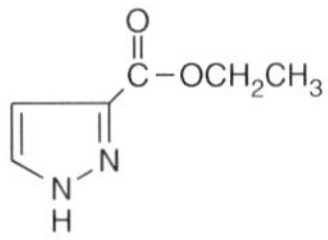

250 mg

S94,055-0
METHYL 4-METHYL-1H-PYRAZOLE-3-CARBOXYLATE

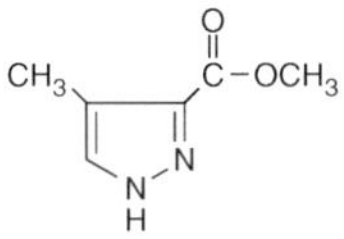

250 mg

S94,036-4
ETHYL 1,5-DIMETHYL-1H-PYRAZOLE-3-CARBOXYLATE

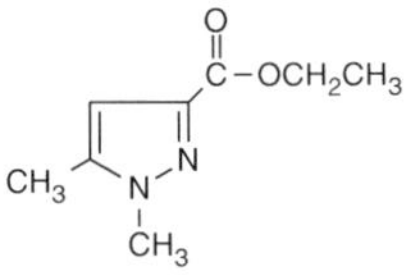

250 mg

S94,054-2
ETHYL 3,5-DIMETHYL-1H-PYRAZOLE-4-CARBOXYLATE

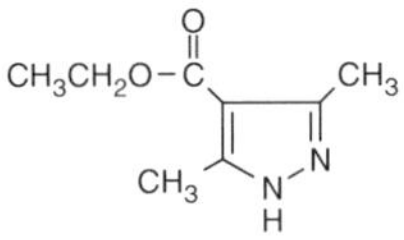

250 mg

S94,030-5
ETHYL 1,3,5-TRIMETHYL-1H-PYRAZOLE-4-CARBOXYLATE

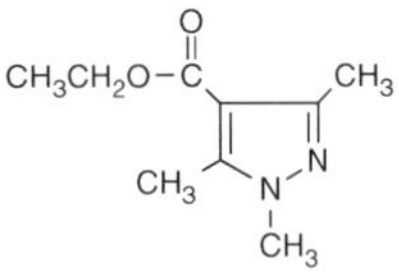

250 mg

S94,463-7
ETHYL 1-ETHYL-3-METHYL-1H-PYRAZOLE-5-CARBOXYLATE

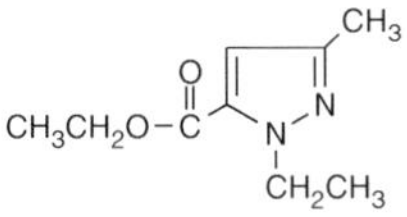

250 mg

S87,041-2
N-(5-PHENYL-3-PYRAZOLYL)BENZAMIDE

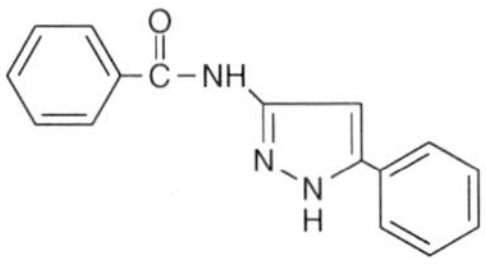

250 mg

S35,346-9
3-METHYL-1-PYRAZOLECARBOXAMIDE

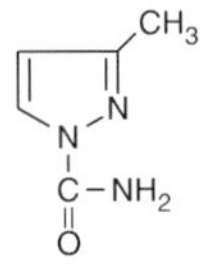

5 gm

S36,113-5
3,4-DIMETHYL-5-PYRAZOLONE

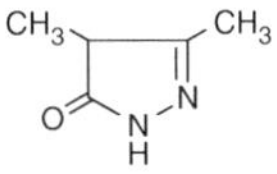

5 gm

S93,949-8
1,3-DIMETHYL-4-PHENYL-2-PYRAZOLIN-5-ONE

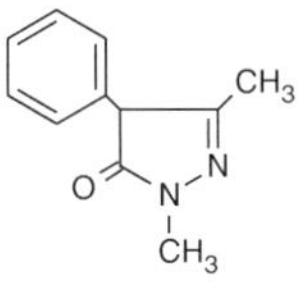

250 mg

S94,530-7
4-CHLORO-1H-PYRAZOLE

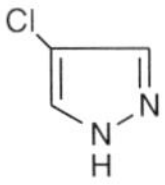

250 mg

S94,045-3
3,4-DIBROMO-1H-PYRAZOLE

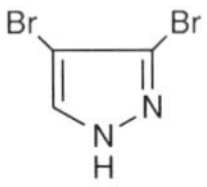

250 mg

S94,046-1
3,5-DIBROMO-1H-PYRAZOLE

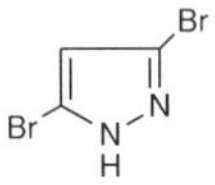

250 mg

S94,042-9
3,4,5-TRIBROMO-1H-PYRAZOLE

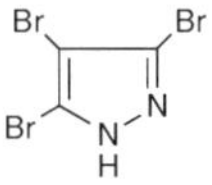

250 mg

S93,946-3
3,4-DIBROMO-5-METHYL-1H-PYRAZOLE
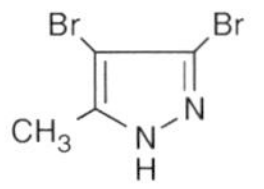
250 mg

S56,680-2
3-METHYL-1-PYRAZOLETHIOCARBOXAMIDE
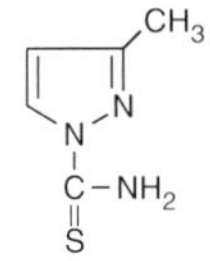
5 gm

S76,778-6
1-ALLYLPYRAZOLE
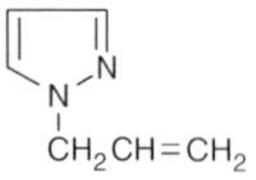
250 mg

S72,206-5
1-ALLYL-3,4,5-TRIBROMOPYRAZOLE
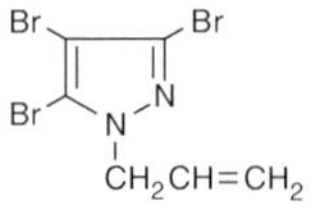
5 gm

S94,049-6
5-TERT-BUTYL-1H-PYRAZOLE
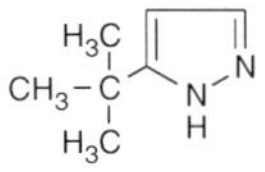
250 mg

S76,791-3
1-ALLYL-3,5-DIMETHYLPYRAZOLE
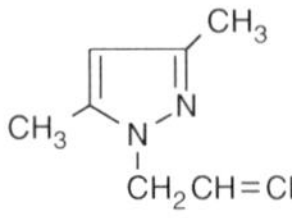
250 mg

S45,214-9
3,5-DIMETHYL-1-PHENYLPYRAZOLE
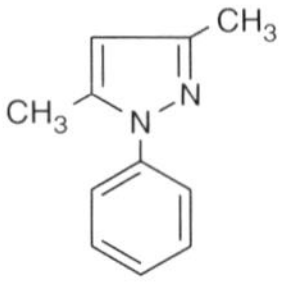
5 gm

S94,048-8
3,5-DI-TERT-BUTYL-1H-PYRAZOLE
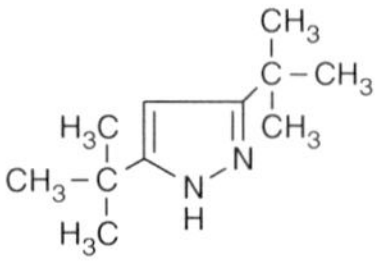
250 mg

S94,528-5
3-PHENYL-1H-PYRAZOLE
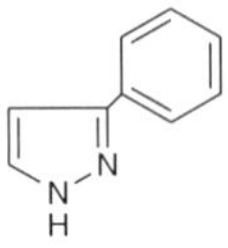
250 mg

S94,542-0
1-PHENYL-3,4,5-TRIBROMO-1H-PYRAZOLE
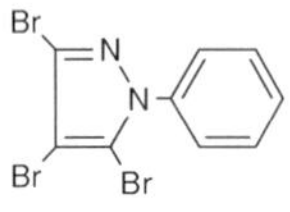
250 mg

S94,041-0
1-BENZYL-3,4,5-TRIBROMO-1H-PYRAZOLE
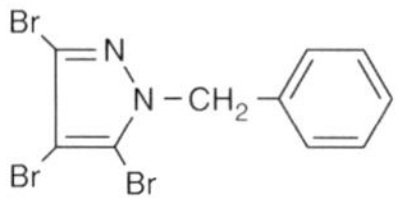
250 mg

S94,536-6
3,5-DIMETHYL-4-PHENYL-1H-PYRAZOLE
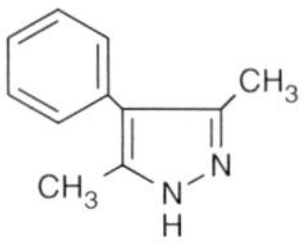
250 mg

S77,091-4
4-BROMO-1-CINNAMYLPYRAZOLE
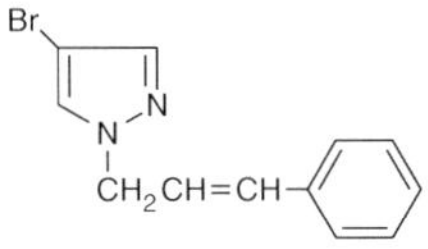
250 mg

S94,465-3
3,5-DIPHENYL-1H-PYRAZOLE
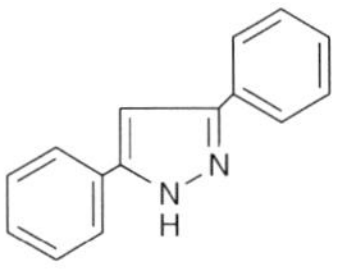
250 mg

S79,227-6
3,3-DIPHENYL-4,5,6,7,8,9-HEXAHYDRO-3H-CYCLOOCTAPYRAZOLE
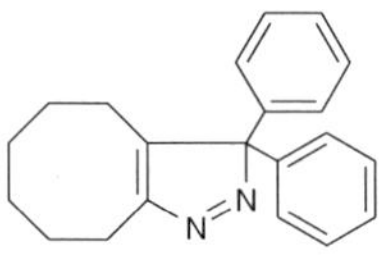
250 mg

S79,228-4
2,3-DIPHENYL-4,5,6,7,8,9-HEXAHYDRO-2H-CYCLOOCTAPYRAZOLE
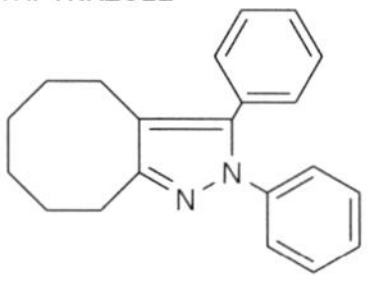
250 mg

S94,237-5
1-METHYL-4-(P-TOLYL)-1H-PYRAZOLE
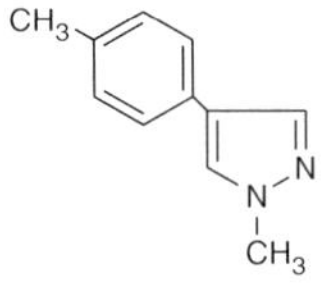
250 mg

S76,104-4
2,4-DIHYDROINDENO(1,2-C)PYRAZOLE
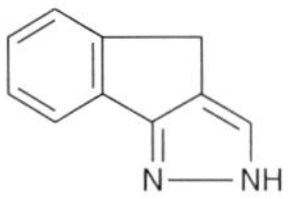
250 mg

S76,081-1
4,5-DIHYDRO-3-(M-TOLYL)-2H-BENZ(G)-INDAZOLE
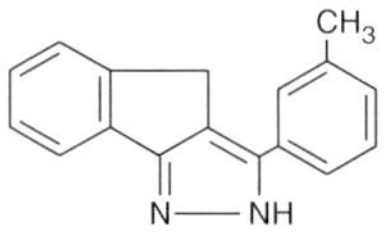
250 mg

S93,988-9
5-CHLOROMETHYL-1-(4-FLUOROPHENYL)-4-ISOPROPYL-2-PHENYLIMIDAZOLE HYDROCHLORIDE
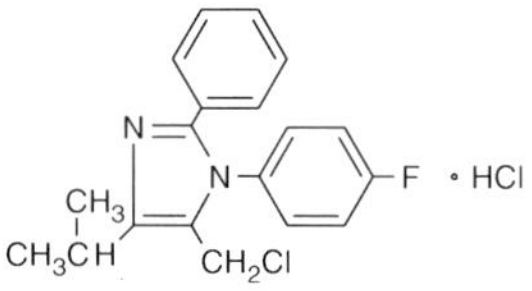
250 mg

S94,027-5
5-CHLOROMETHYL-1-(4-FLUOROPHENYL)-4-ISOPROPYL-2-PHENYLIMIDAZOLE
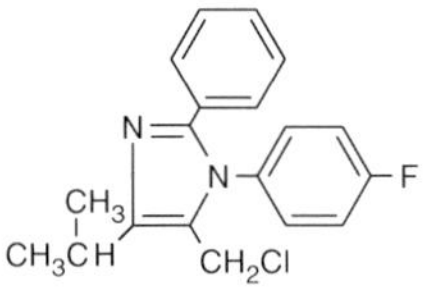
250 mg

S88,504-5
1-(2,3-DICHLOROPROPYL)IMIDAZOLE
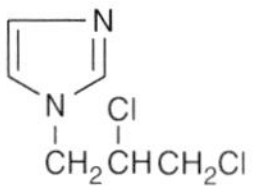
250 mg

S92,653-1
5-CHLOROMETHYL-4-METHYLIMIDAZOLE HYDROCHLORIDE
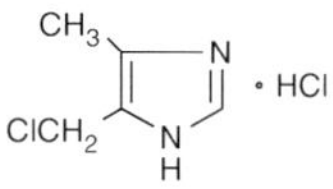
250 mg

S88,341-7
1-[3-(TRIFLUOROMETHYL)PHENYL]-DIPHENYLMETHYLIMIDAZOLE HYDROCHLORIDE
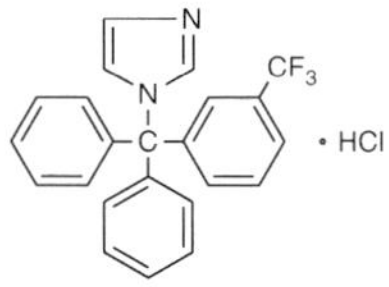
250 mg

S69,459-2
IMIDAZOLE TRIFLUOROMETHANESULFONATE
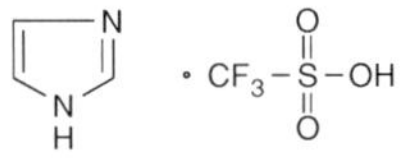
250 mg

S76,015-3
ALPHA-(4-CHLOROPHENYL)-1-METHYL-ALPHA-[2-(2-TETRAHYDROPYRANYLOXY)PHENYL]-2-IMIDAZOLEMETHANOL
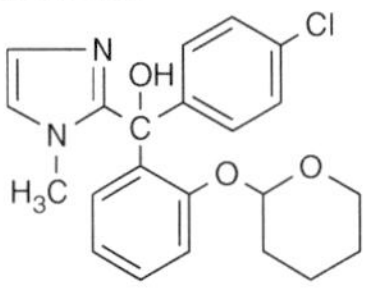
250 mg

S73,314-8
BETA-(4-CHLOROPHENYL)-BETA-HYDROXY-2-(2-IMIDAZOLYL)HYDROCINNAMONITRILE
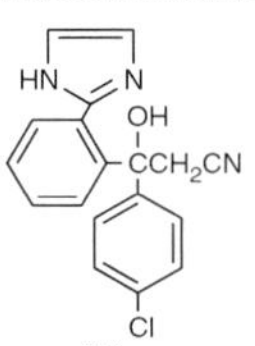
250 mg

S76,053-6
5-(4-CHLOROPHENYL)-5-HYDROXY-5H-IMIDAZO(2,1-A)ISOINDOLE P-TOLUENESULFONATE
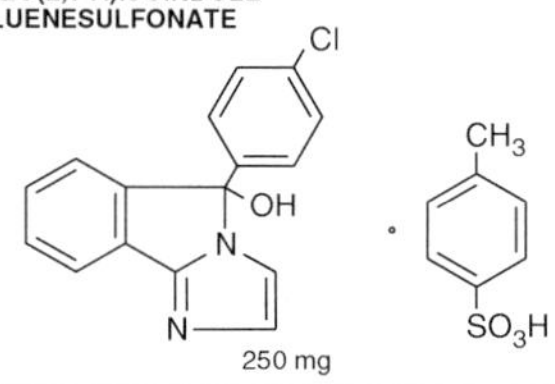
250 mg

S76,696-8
4-CHLORO-2'-(2-IMIDAZOLYL)-ALPHA-(1-METHYL-2-IMIDAZOLIN-2-YL)-BENZHYDROL
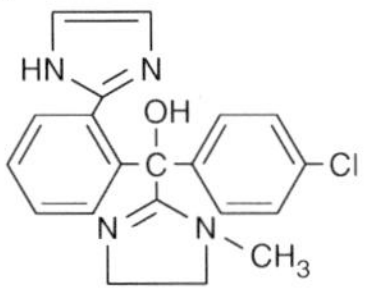
250 mg

S73,520-5
6-[BETA-(4-CHLOROPHENYL)-2-BETA-2-HYDROXY-2-(2-IMIDAZOLYL)PHENETHYL]-PYRIDIN-2(1H)-ONE DIHYDROCHLORIDE
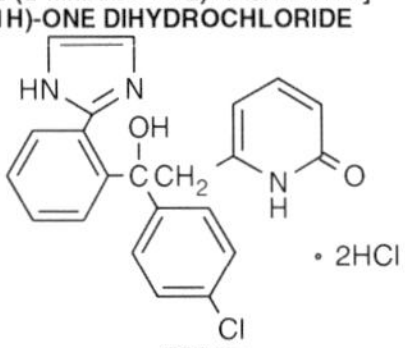
250 mg

S94,021-6
1-(4-FLUOROPHENYL)-5-(HYDROXYMETHYL)-4-ISOPROPYL-2-PHENYLIMIDAZOLE
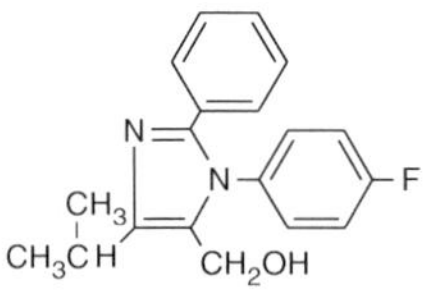
250 mg

S76,987-8
1-[N-(5-CHLORO-2,4-DIMETHOXYPHENYL)-CARBAMOYL]IMIDAZOLE
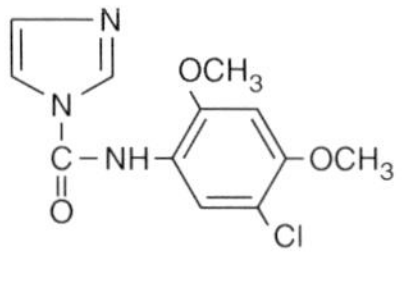
250 mg

S91,539-4
3-CHLORO-N-(ALPHA-METHOXY-1-METHYL-2-IMIDAZOLYLMETHYLENE)ANILINE
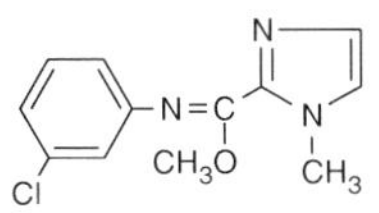
250 mg

S91,525-4
3-CHLORO-N-(ALPHA-METHOXY-2-IMIDAZOLYLMETHYLENE)ANILINE TETRAFLUOROBORATE
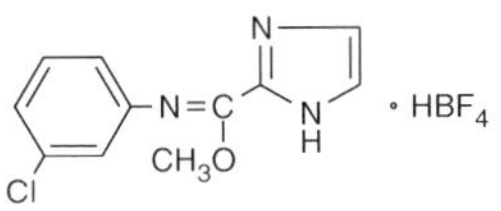
250 mg

S76,822-7
1-[N-(5-CHLORO-2-METHOXYPHENYL)-CARBAMOYL]IMIDAZOLE
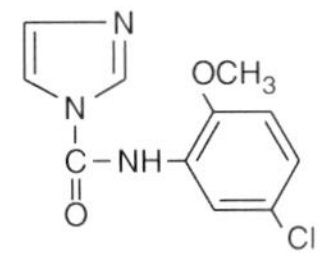
250 mg

S77,210-0
1-[N-(4-CHLORO-2-METHOXYPHENYL)-CARBAMOYL]IMIDAZOLE
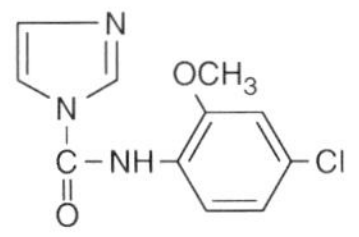
250 mg

S91,694-3
1-[N-(4-CHLORO-2-ETHOXYPHENYL)-CARBAMOYL]IMIDAZOLE
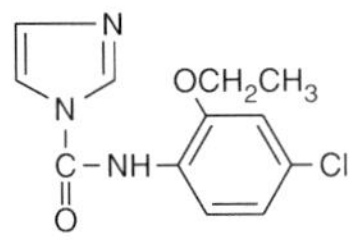
250 mg

S91,531-9
3'-CHLORO-1-METHYL-5-NITRO-2-IMIDAZOLECARBOXANILIDE
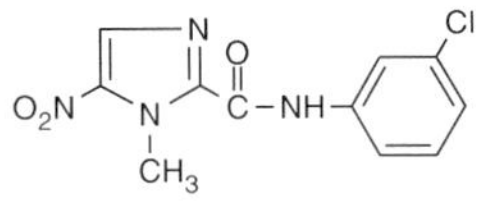
250 mg

S76,726-3
1-(4-CHLORO-2-NITROBENZYL)IMIDAZOLE
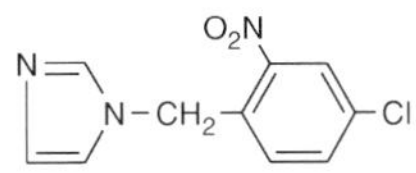
250 mg

S91,196-8
1-[N-(4-FLUORO-3-NITROPHENYL)-CARBAMOYL]IMIDAZOLE
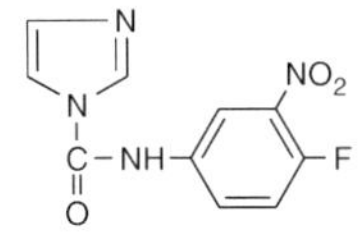
250 mg

S94,023-2
3-[1-(4-FLUOROPHENYL)-4-ISOPROPYL-2-PHENYLIMIDAZOL-5-YL]ACRYLONITRILE
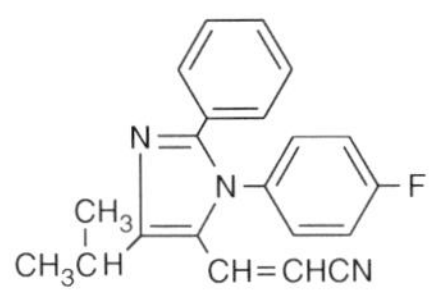
5 gm

S76,640-2
5-(4-CHLOROBENZYL)-5-(4-CHLOROPHENYL)-5H-IMIDAZO(2,1-A)-ISOINDOLE
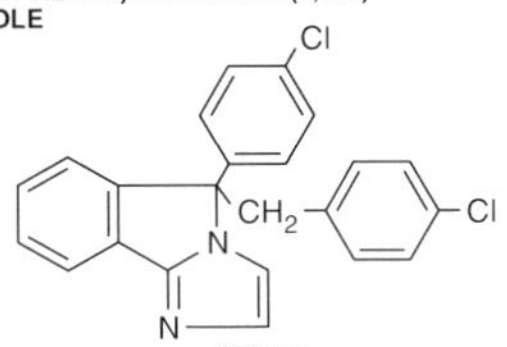
250 mg

S91,533-5
3'-CHLORO-1-METHYL-2-IMIDAZOLECARBOXANILIDE
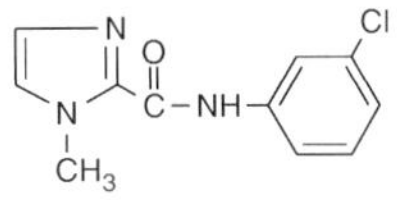
250 mg

S80,601-3
5-(DIACETYLPHOSPHINYLMETHYL)-1-(4-FLUOROPHENYL)-4-ISOPROPYL-2-PHENYLIMIDAZOLE
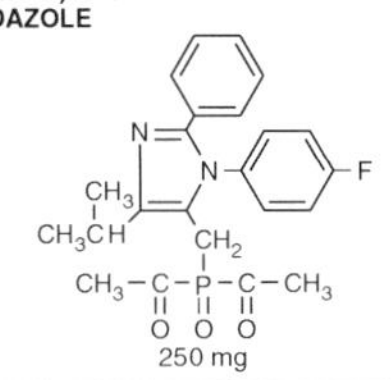
250 mg

S93,987-0
1-(4-FLUOROPHENYL)-4-ISOPROPYL-2-PHENYLIMIDAZOLE-5-CARBOXALDEHYDE
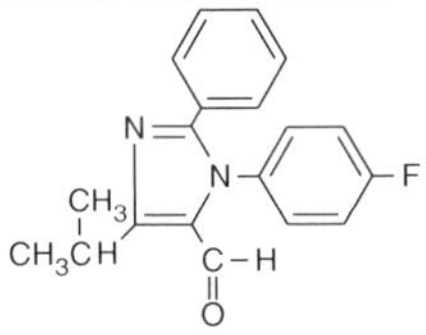
5 gm

S76,659-3
4'-CHLORO-2-(2-METHYLIMIDAZOL-1-YLMETHYL)BENZANILIDE
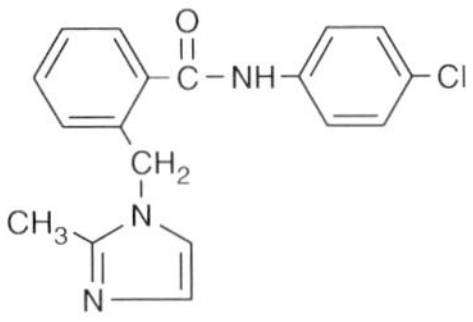
250 mg

S93,989-7
[1-(4-FLUOROPHENYL)-4-ISOPROPYL-2-PHENYLIMIDAZOL-5-YL]METHYL ACETATE
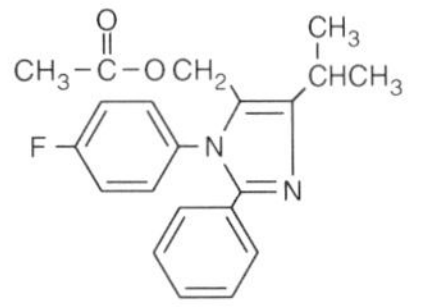
250 mg

S93,986-2
ETHYL 3-[1-(4-FLUOROPHENYL)-4-ISOPROPYL-2-PHENYLIMIDAZOL-5-YL]-ACRYLATE
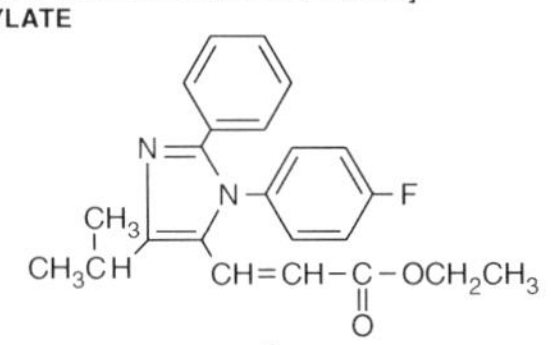
5 gm

S73,243-5
4-CHLOROPHENYL 1-METHYL-2-IMIDAZOLYL KETONE
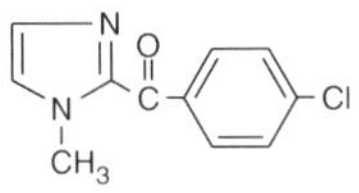
250 mg

S72,085-2
2-(IMIDAZOL-1-YLCARBONYL)-3,4,5,6-TETRACHLOROBENZOIC ACID
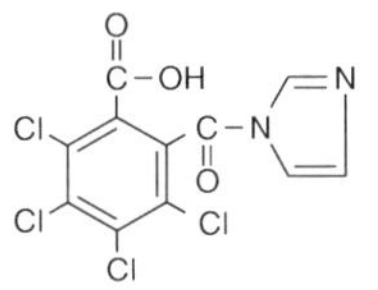
5 gm

S78,044-8
2-(1-IMIDAZOLYLCARBONYL)-3,4,5,6-TETRABROMOBENZOIC ACID
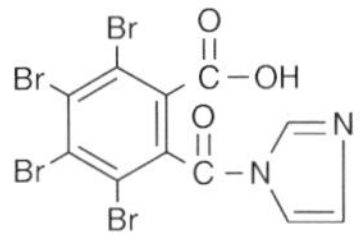
250 mg

S89,817-1
2'-FLUORO-2-IMIDAZOLECARBOXANILIDE
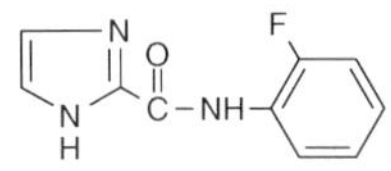
250 mg

S72,322-3
2-(2-CHLOROPHENYL)-4,5-DIPHENYLIMIDAZOLE
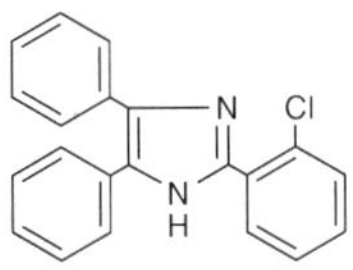
50 mg

S94,022-4
1-(4-FLUOROPHENYL)-4-ISOPROPYL-5-METHYL-2-PHENYLIMIDAZOLE
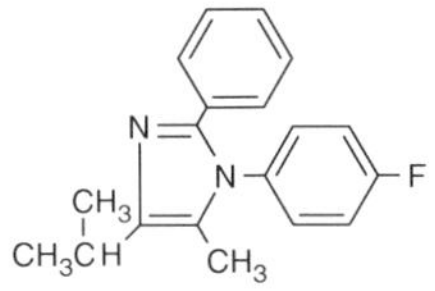
5 gm

S69,230-1
2-(4-CHLOROPHENYL)-4,5-DIPHENYLIMIDAZOLE
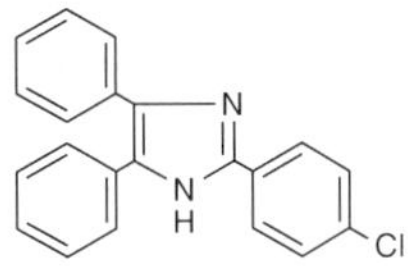
250 mg

S77,338-7
3-(4,5-DIPHENYL-2-IMIDAZOLYL)-CATECHOL
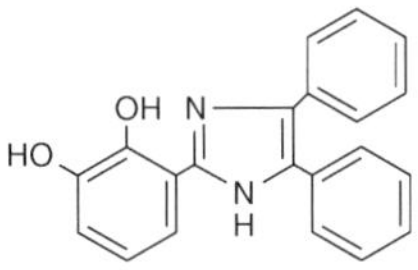
250 mg

S93,992-7
ETHYL (4-FLUOROPHENYL-4-ISOPROPYL-2-PHENYLIMIDAZOL-5-YL)-3,5-DIHYDROXY-6-HEPTENOATE
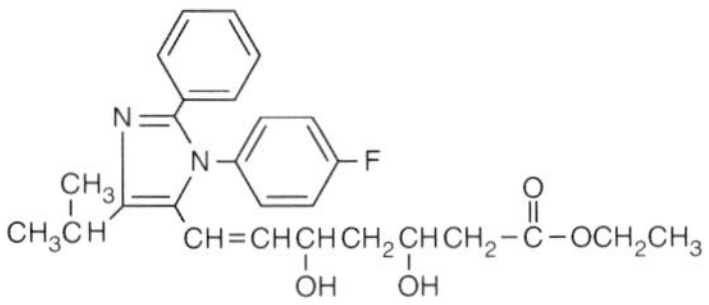
5 gm

S68,586-0
2,2-DIMETHYL-4'-[4-(ALPHA-HYDROXY-1-METHYL-2-IMIDAZOLYLMETHYL)PHENOXY]-PROPIOPHENONE ETHYLENE KETAL
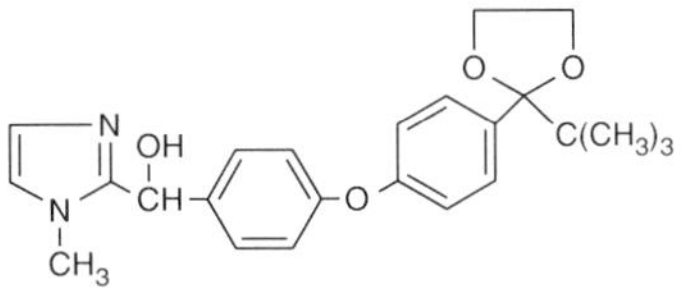
250 mg

S91,521-1
2-(1-METHYL-5-NITRO-2-IMIDAZOLYL)-1-PHENYL-1-ETHANOL
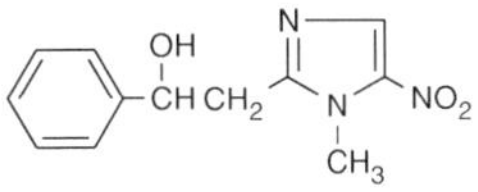
250 mg

S75,889-2
6-[ALPHA-(4-CHLOROPHENYL)-ALPHA-HYDROXY-2-(2-IMIDAZOLYL)BENZYL]-PYRIDIN-2(1H)-ONE
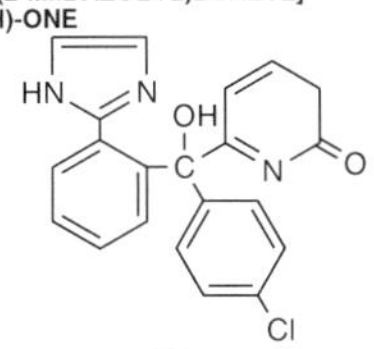
250 mg

S91,541-6
2-(5-NITRO-1-IMIDAZOLYL)-1-ETHANOL
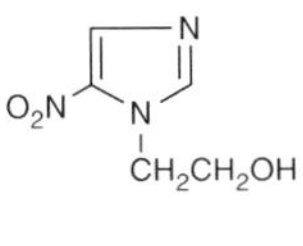
250 mg

S78,485-0
CARBOBENZYLOXY-L-HISTIDYL-L-TYROSINAMIDE
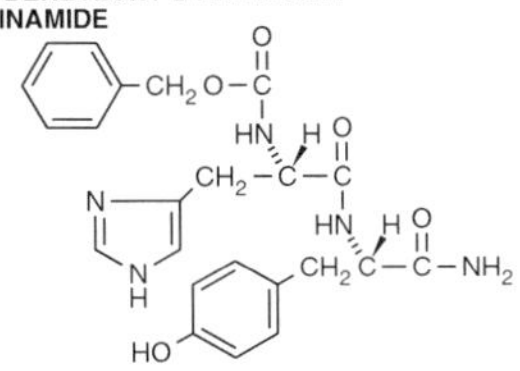
250 mg

S91,515-7
1-METHYL-5-NITRO-2-IMIDAZOLEMETHANOL
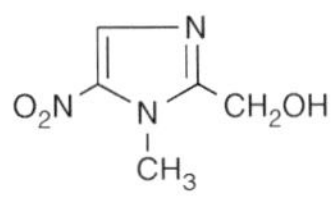
250 mg

S93,331-7
N-BENZOYL-L-HISTIDINOL
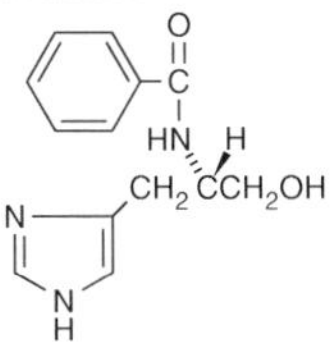
250 mg

S35,128-8
7-(2-HYDROXYPROPYL)THEOPHYLLINE
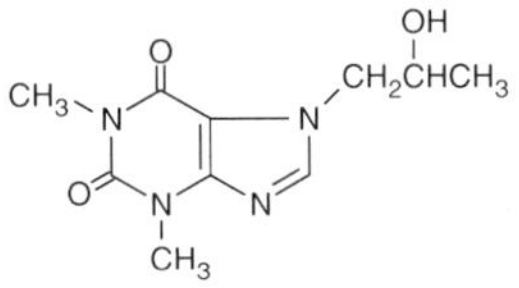
5 gm

S87,900-2
N-(2-HYDROXY-3-METHOXY-5-NITROBENZYLIDENE)HISTIDINE
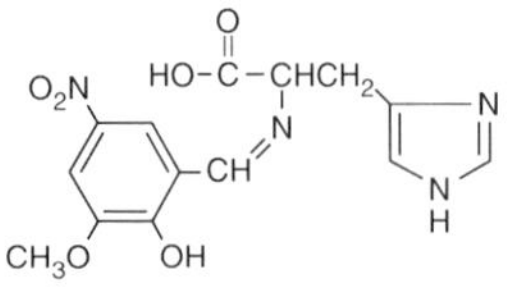
5 gm

S78,142-8
CARBOBENZYLOXY-L-HISTIDYL-L-TYROSINE MONOHYDRATE
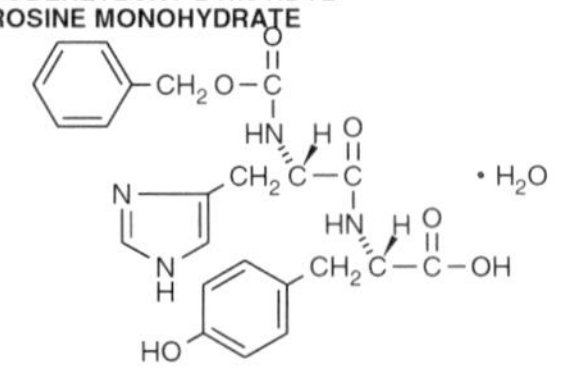
250 mg

S78,769-8
CARBOBENZYLOXY-L-HISTIDYL-L-TYROSINE METHYL ESTER
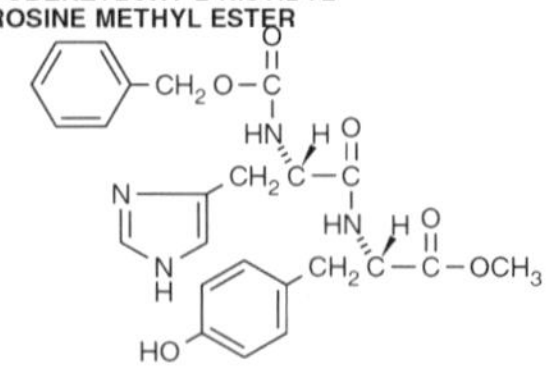
250 mg

S72,043-7
4-(HYDROXYMETHYL)IMIDAZOLE PICRATE
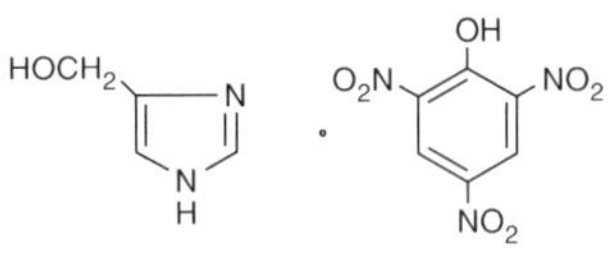
5 gm

S76,675-5
1-[2-(BENZYLOXY)ETHYL]IMIDAZOLE
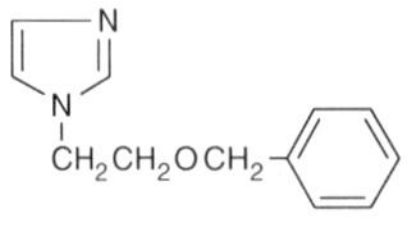
250 mg

S75,122-7
2-(2,5-DIMETHOXYPHENYL)-4,5-DIPHENYLIMIDAZOLE
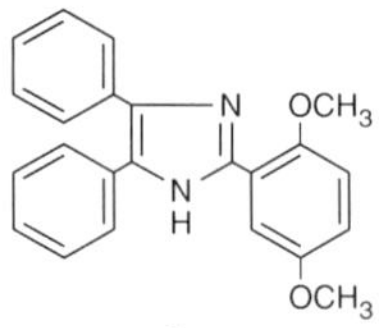
5 gm

S84,989-8
1-(2-METHOXYBENZOYL)IMIDAZOLE
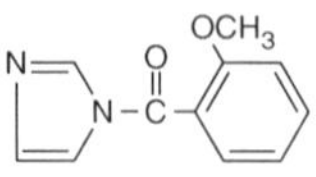
250 mg

S36,994-2
5-METHYL-5-(3,4,5-TRIMETHOXYPHENYL)-HYDANTOIN
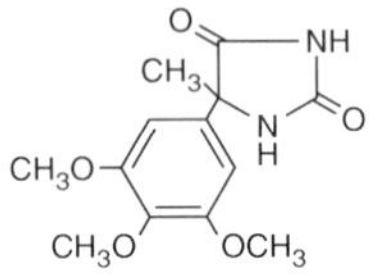
5 gm

S72,272-3
4,5-DIPHENYL-2-(4-METHOXYPHENYL)-IMIDAZOLE
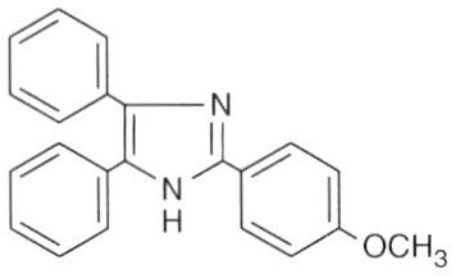
250 mg

S79,023-0
N',N-ALPHA-BIS(CARBOBENZYLOXY)-L-HISTIDINE METHYL ESTER
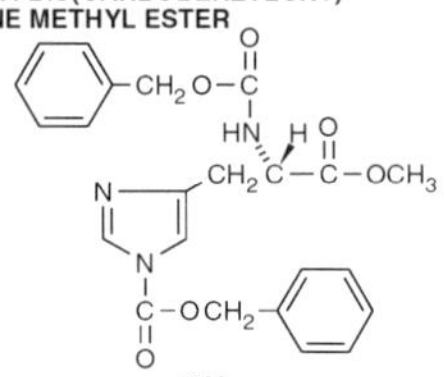
250 mg

S78,584-9
CARBOBENZYLOXYHISTIDINE HYDRAZIDE
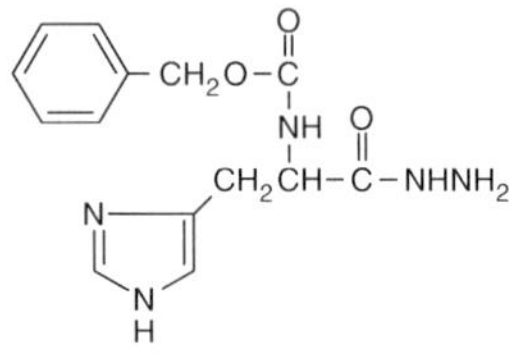
250 mg

S78,622-5
CARBOBENZYLOXYHISTIDYLPHENYLALANINE METHYL ESTER
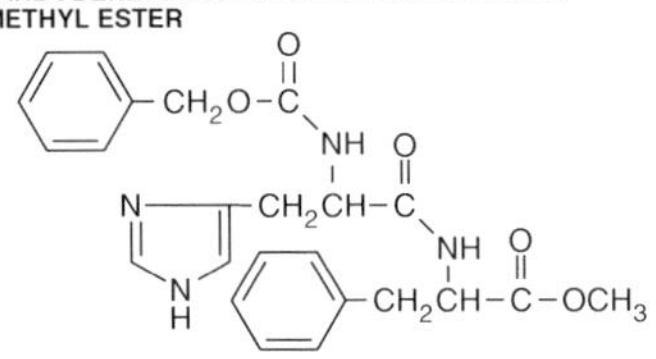
250 mg

S78,460-5
CARBOBENZYLOXY-L-HISTIDYLGLYCINE
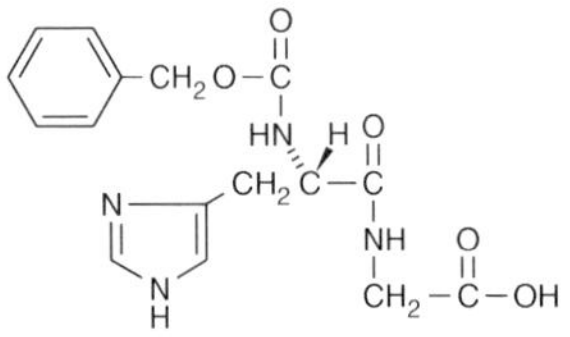
250 mg

S77,795-1
CARBOBENZYLOXY-L-HISTIDYL-L-LEUCINAMIDE
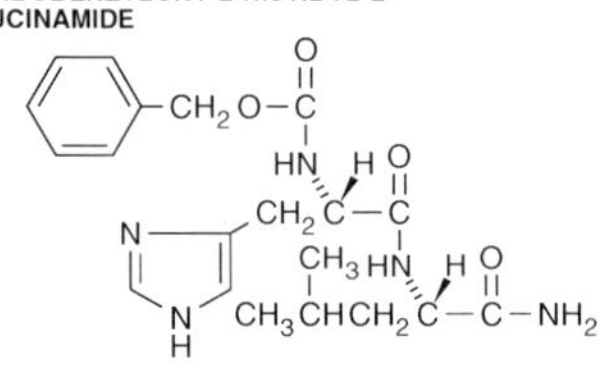
250 mg

S88,499-5
2-(1-IMIDAZOLYL)ACETALDEHYDE DIETHYL ACETAL
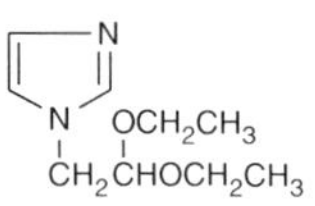
250 mg

S76,024-2
1-BENZYL-2-MERCAPTOIMIDAZOLE
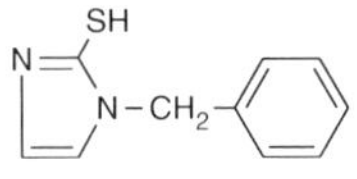
250 mg

S46,232-2
1-(2-AMINOETHYL)-2-METHYL-5-NITROIMIDAZOLE DIHYDROCHLORIDE MONOHYDRATE
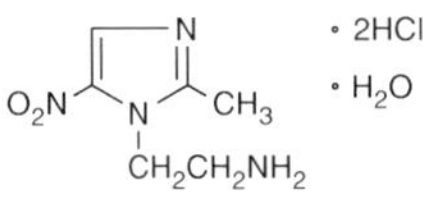
5 gm

S88,381-6
1-(2,3-DIAMINOPROPYL)IMIDAZOLE TRIHYDROCHLORIDE
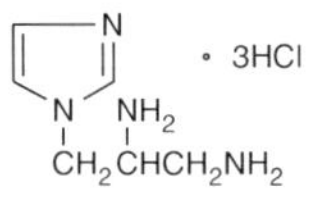
250 mg

S45,187-8
GLYCYL-L-HISTIDYLGLYCINE
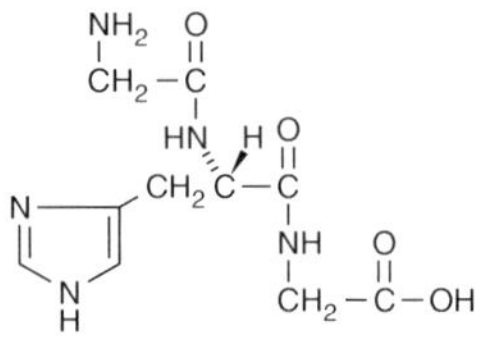
250 mg

S58,094-5
DL-HISTIDINE DIHYDROCHLORIDE
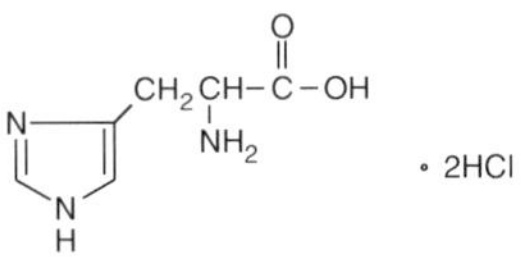
5 gm

S58,364-2
DL-HISTIDINE HYDROCHLORIDE HYDRATE
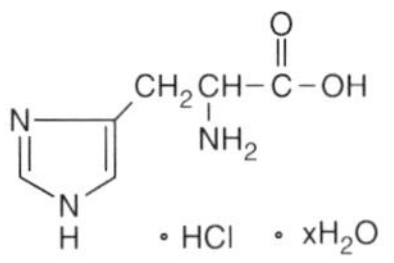
5 gm

S78,934-8
L-HISTIDYL-L-ISOLEUCINE HEMIHYDRATE
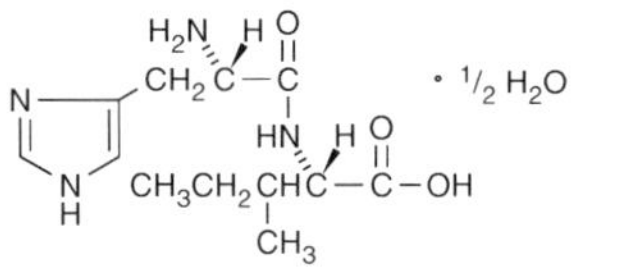
250 mg

S45,717-5
L-(+)-HOMOCARNOSINE SULFATE
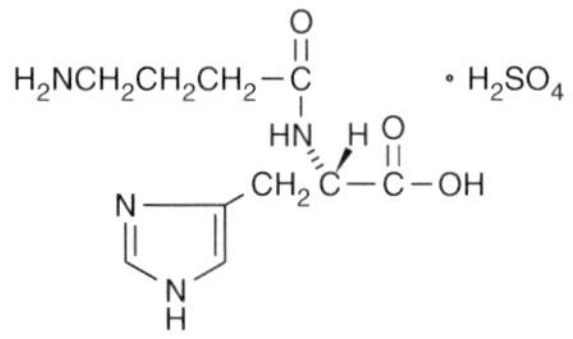
250 mg

S78,719-1
L-HISTIDINAMIDE DIHYDROCHLORIDE
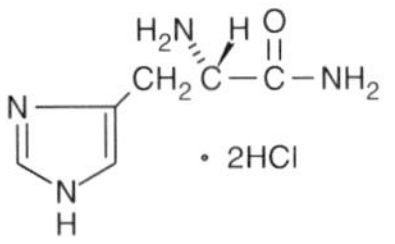
250 mg

S91,471-1
2-AMINOIMIDAZOLE HYDROGEN SULFATE
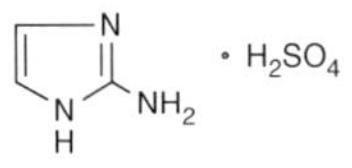
250 mg

S94,634-6
2-AMINO-1-PHENYL-2-IMIDAZOLINE HYDROBROMIDE
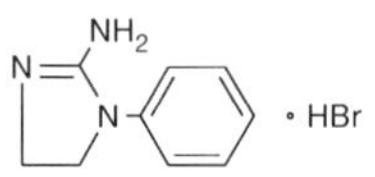
250 mg

S41,015-2
THEOPHYLLINE-7-ACETIC ACID
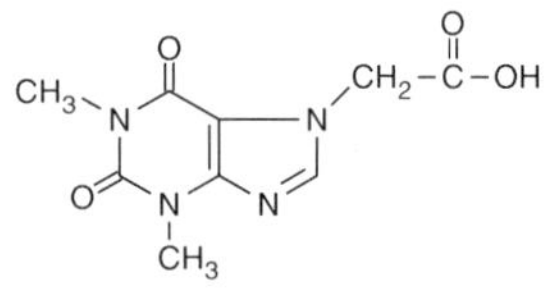
5 gm

S40,985-5
N-PHTHALOYL-DL-HISTIDINE
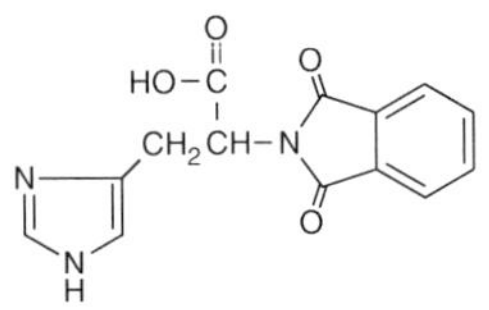
5 gm

S41,912-5
ETHYL 7-THEOPHYLLINEACETATE
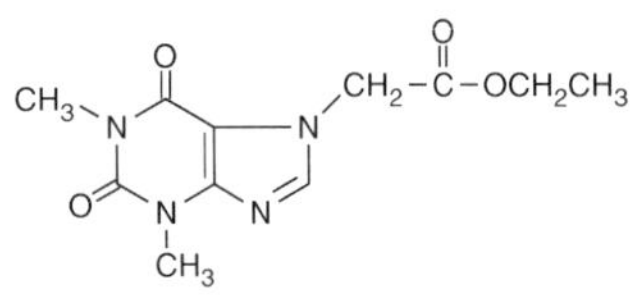
5 gm

S84,701-1
5-(BENZYLAMINO)-1-METHYL-4-NITROIMIDAZOLE
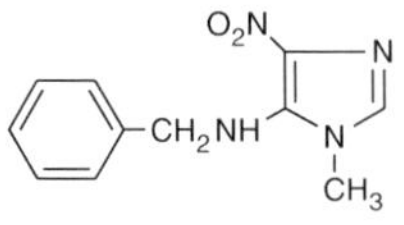
250 mg

S94,116-6
2,2'-BI-1H-IMIDAZOLE
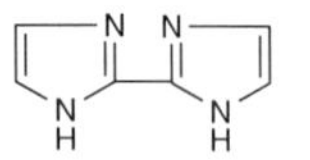
250 mg

S41,816-1
4,5-DIBROMO-2-PHENYLIMIDAZOLE

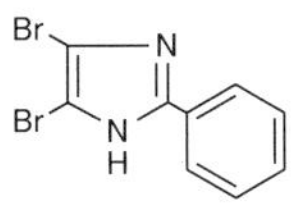

5 gm

S42,911-2
THEOBROMINE

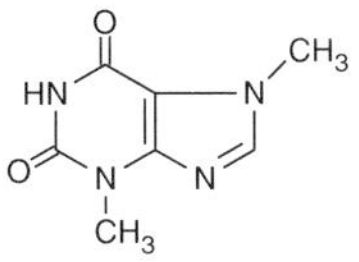

5 gm

S88,508-8
2-(4,5-DIMETHYL-1-IMIDAZOLYL)-ACETAMIDE

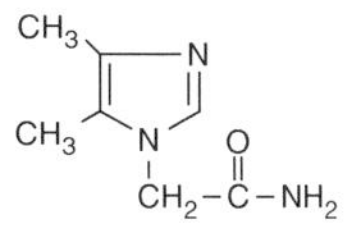

250 mg

S91,518-1
2-(1-METHYL-5-NITRO-2-IMIDAZOLYL)-ACETOPHENONE

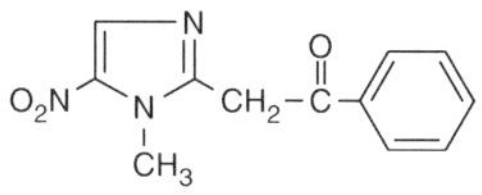

250 mg

S91,529-7
1-METHYL-4-NITROIMIDAZOLE

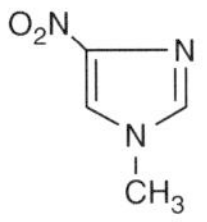

250 mg

S91,483-5
1-METHYL-5-NITRO-2-STYRYLIMIDAZOLE

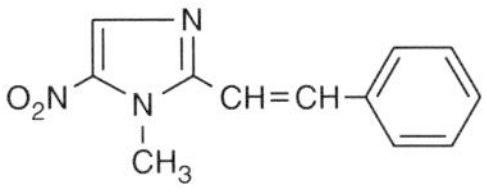

250 mg

S80,227-1
5-NITROIMIDAZOLE

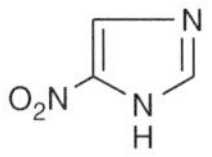

250 mg

S41,671-1
METHYL 1-PHENYLIMIDAZOLE-5-CARBOXYLATE

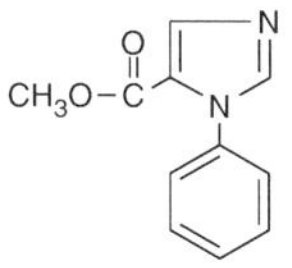

250 mg

S88,487-1
N-[3-(1-IMIDAZOLYL)-2-OXOPROPYL]-PHTHALIMIDE

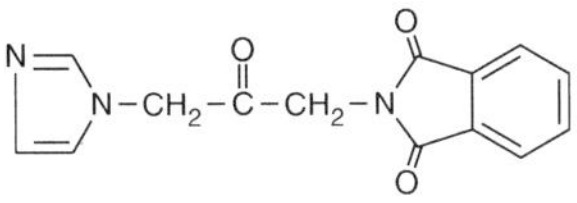

250 mg

S78,733-7
N-TERT-BUTOXYCARBONYLHISTIDINE

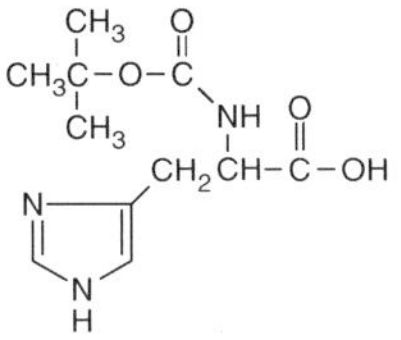

250 mg

S78,917-8
L-HISTIDYL-L-2-AMINOBUTYRIC ACID

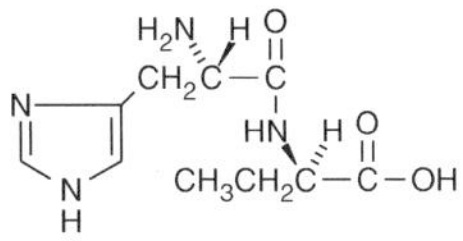

250 mg

S51,768-2
1,1'-(1,4-PHENYLENE)BIS[2-(5-INDANYL)-4,5-DIPHENYLIMIDAZOLE]

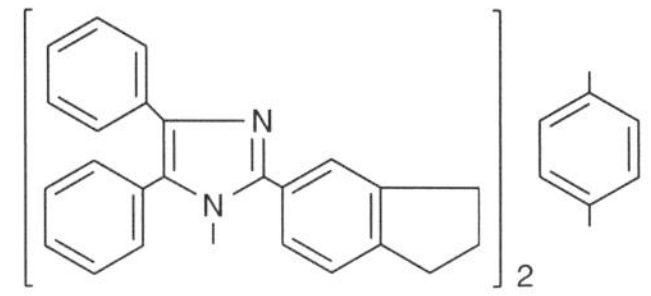

250 mg

S72,404-1
5-[4-(DIMETHYLAMINO)BENZYLIDENE]-HYDANTOIN

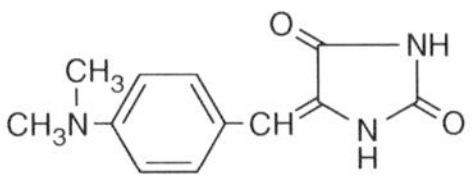

50 mg

S66,397-2
4,5-DIPHENYL-4-IMIDAZOLIN-2-ONE

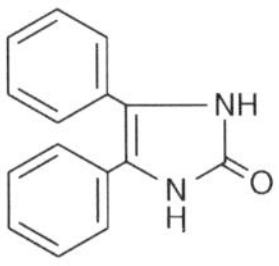

5 gm

S78,781-7
CARBOBENZYLOXYHISTIDYLPHENYLALANINE BENZYL ESTER

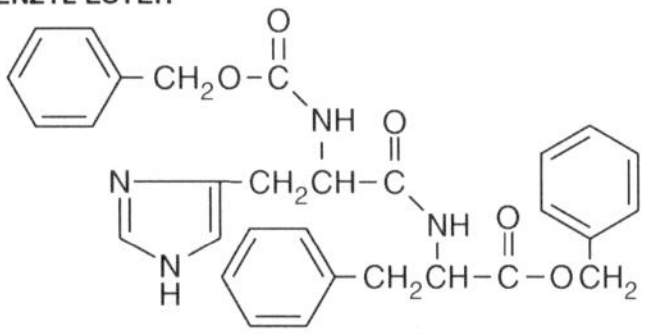

250 mg

S50,653-2
4,5-DIPHENYL-2-(5-INDANYL)-IMIDAZOLE

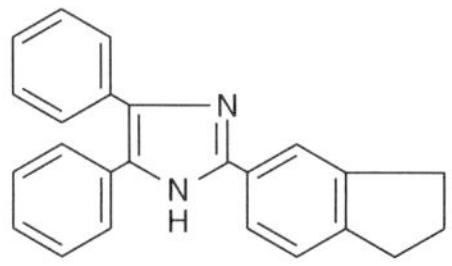

250 mg

S76,493-0
N,1-DIBENZYL-L-HISTIDINE

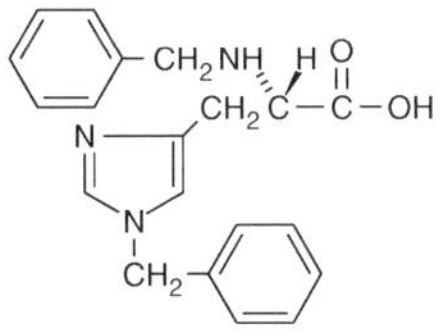

250 mg

S92,819-4
ETHYL 4,5-DIPHENYL-2-IMIDAZOLECARBOXYLATE

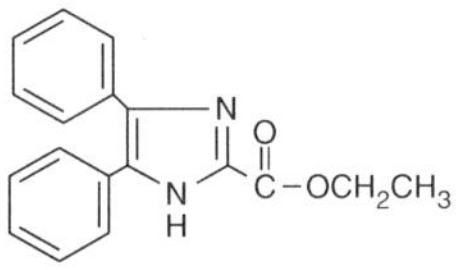

250 mg

S45,015-4
4,5-DIPHENYL-2-METHYLIMIDAZOLE

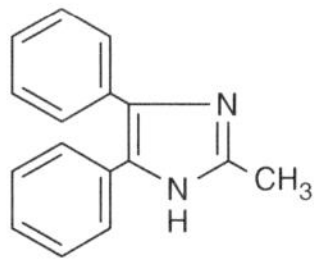

250 mg

S73,586-8
3-(ALPHA-CARBOXYBENZYL)-1,2-DIMETHYLIMIDAZOLIUM BROMIDE

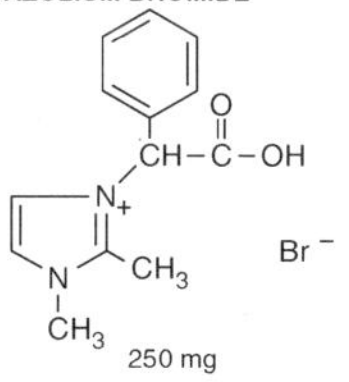

250 mg

S88,501-0
1,3-DIMETHYLIMIDAZOLIUM IODIDE

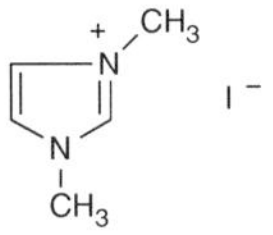

250 mg

S88,848-6
1,2-DIMETHYL-5-NITROIMIDAZOLE

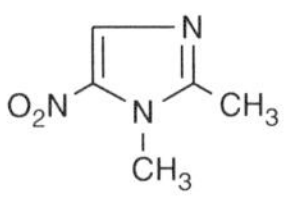

5 gm

S39,251-0
7-(PIPERIDINOMETHYL)THEOPHYLLINE

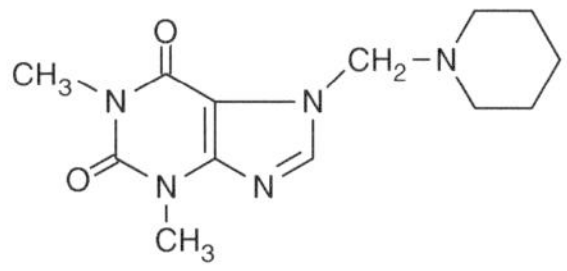

5 gm

S88,382-4
1,1'-METHYLENEBIS(IMIDAZOLE)

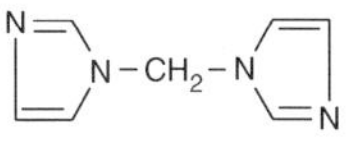

250 mg

S88,383-2
1,1'-METHYLENEBIS(IMIDAZOLE)-DIHYDROCHLORIDE

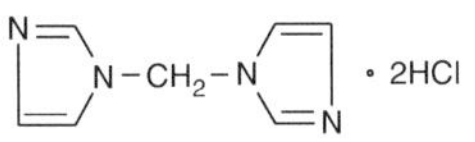

250 mg

S68,305-1
1-(3,7-DIMETHYL-2,6-OCTADIEN-1-YL)-IMIDAZOLE

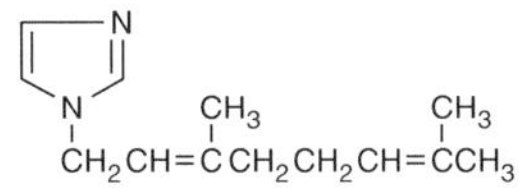

250 mg

S91,516-5
1-METHYL-2-STYRYLIMIDAZOLE

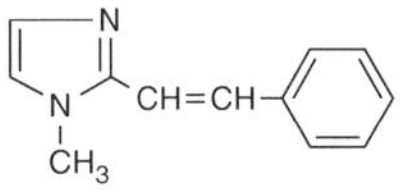

250 mg

S88,502-9
3-(2-CHLORO-2-CYANOETHYL)-IMIDAZOLIUM CHLORIDE

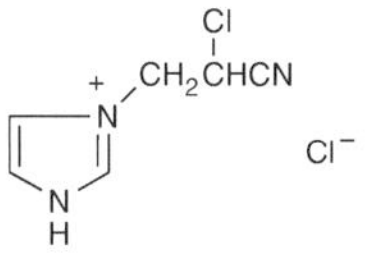

250 mg

S93,990-0
DIMETHYL [1-(4-FLUOROPHENYL)-4-ISOPROPYL-2-PHENYLIMIDAZOL-5-YL]-METHYLPHOSPHONATE

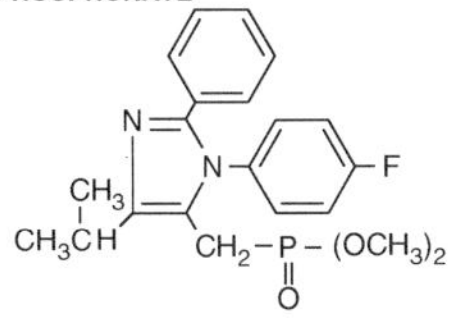

250 mg

S39,252-9
7-(MORPHOLINOMETHYL)THEOPHYLLINE

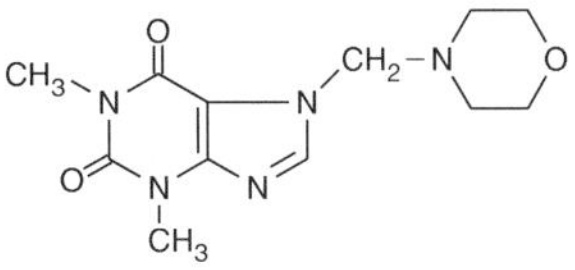

5 gm

S39,247-2
7,7'-(1,4-PIPERAZINEDIYLDIMETHYL)-DITHEOPHYLLINE

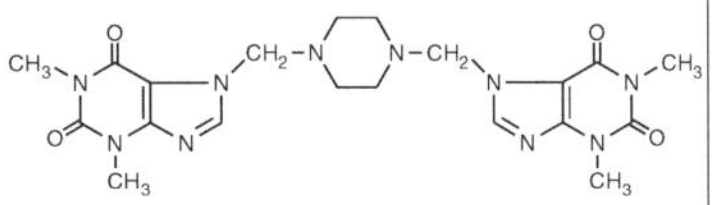

5 gm

S42,638-5
7-(PYRROLIDINOMETHYL)THEOPHYLLINE

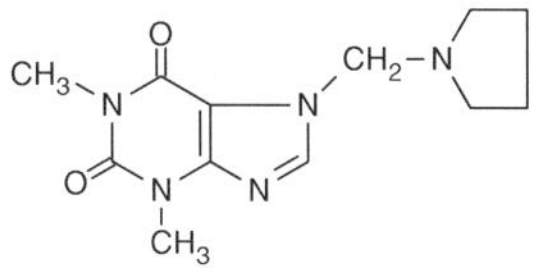

5 gm

S95,259-1
1-METHYL-5-[ALPHA-(PHENYLSULFONYL)-2-NITROHYDROCINNAMOYL]IMIDAZOLE

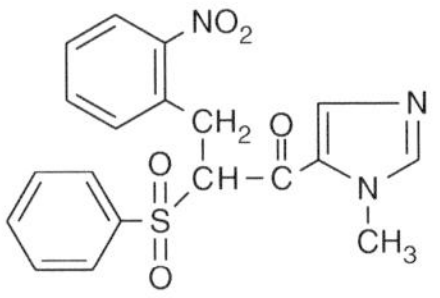

50 mg

S78,516-4
4-NITRO-1-(P-TOSYL)IMIDAZOLE

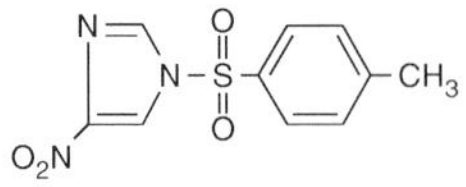

250 mg

S77,683-1
1-(MESITYLENESULFONYL)-4-NITROIMIDAZOLE

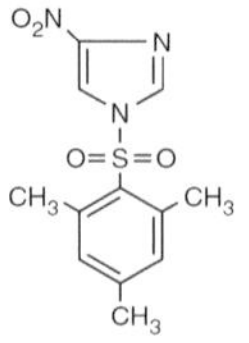

250 mg

S74,950-8
5-BROMO-4-NITROIMIDAZOLE

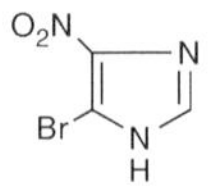

250 mg

S80,308-1
5-METHYL-4-NITROIMIDAZOLE

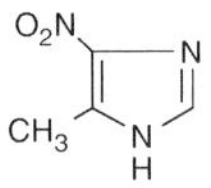

250 mg

S75,845-0
4-(2-METHYLIMIDAZOL-1-YL)-BENZOPHENONE

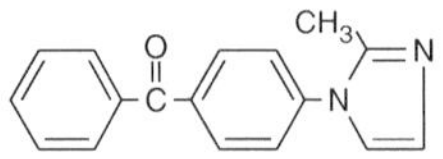

250 mg

S75,108-1
2-(1-IMIDAZOLYLCARBONYL)-BENZOIC ACID

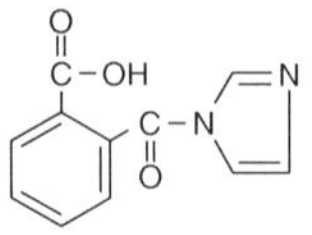

5 gm

S91,119-4
ETHYL 4-(1-IMIDAZOLYLCARBONYLAMINO)-BENZOATE

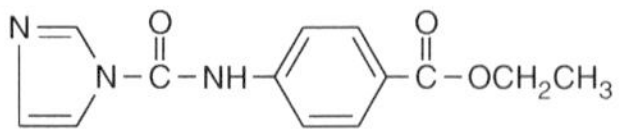

250 mg

S93,176-4
3-METHYL-8-PHENYL-2-THIOHYPOXANTHINE

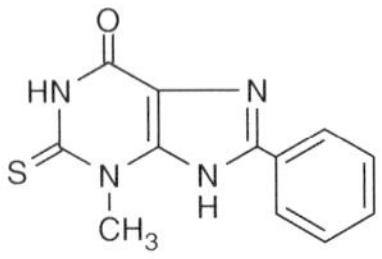

250 mg

S61,905-1
N-BENZOYLIMIDAZOLE

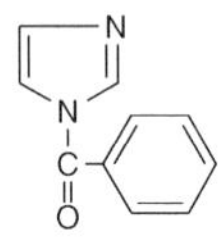

250 mg

S78,535-0
CARBOBENZYLOXYGLYCYL-L-HISTIDINE HYDRATE

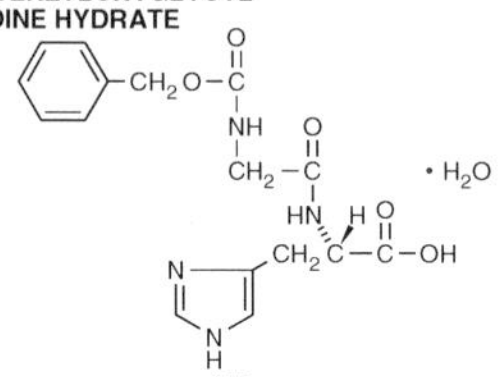

250 mg

S78,472-9
CARBOBENZYLOXY-L-HISTIDYLGLYCYLGLYCINE

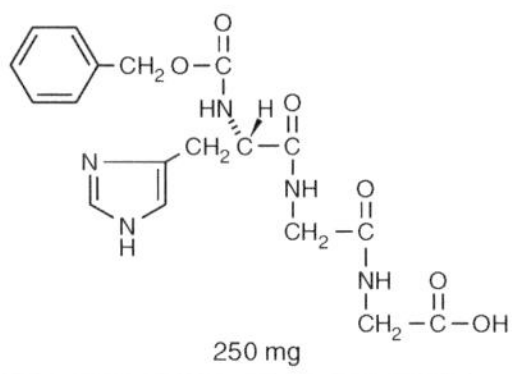

250 mg

S78,980-1
N-ALPHA-ACETYL-DL-HISTIDINE HYDRATE

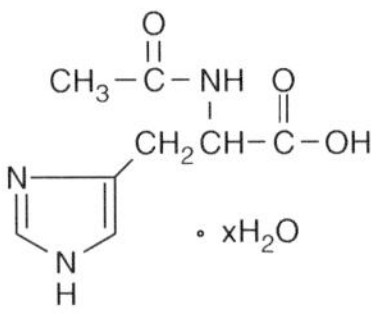

250 mg

S78,258-0
L-ASPARTIC ACID HYDANTOIN

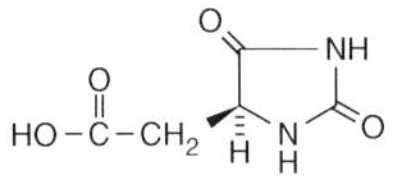

5 gm

S78,539-3
CARBOBENZYLOXY-L-HISTIDYL-L-ALANINE METHYL ESTER

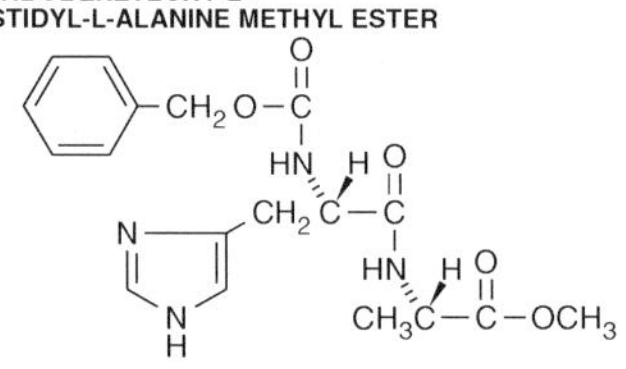

250 mg

S78,773-6
CARBOBENZYLOXY-L-HISTIDYL-L-VALINE BENZYL ESTER

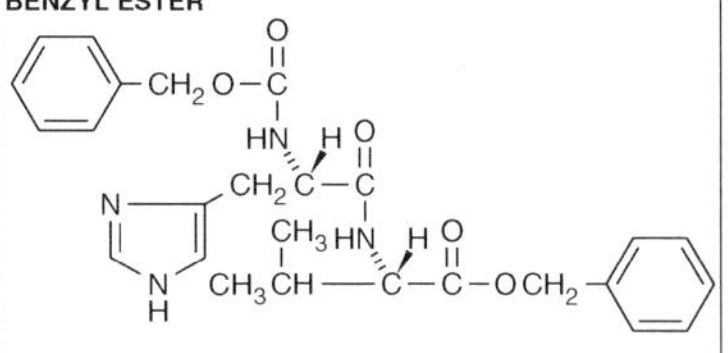

250 mg

S42,890-6
2-CARBETHOXYTHIO-1-METHYLIMIDAZOLE

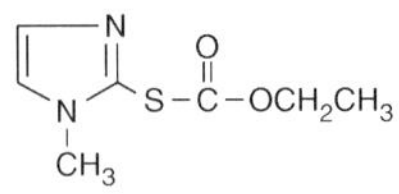

5 gm

S35,904-1
8-BROMOTHEOPHYLLINE

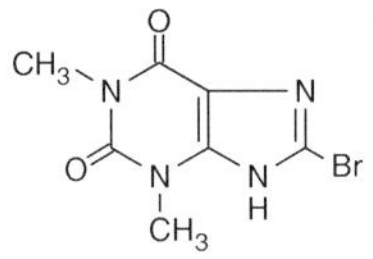

5 gm

S92,793-7
3-METHYL-2-THIOXANTHINE

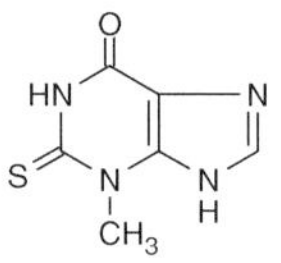

250 mg

S91,465-7
2-IMIDAZOLECARBOXANILIDE

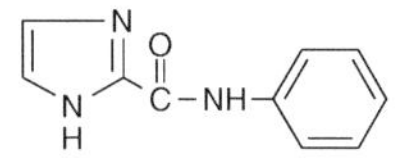

250 mg

S93,461-5
2'-ETHYL-6'-METHYL-1-IMIDAZOLECARBOXANILIDE

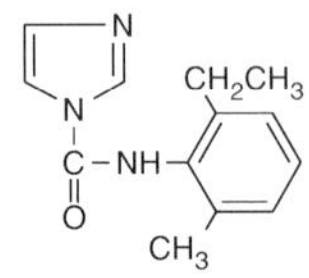

250 mg

S93,177-2
3,8-DIMETHYL-2-THIOXANTHINE

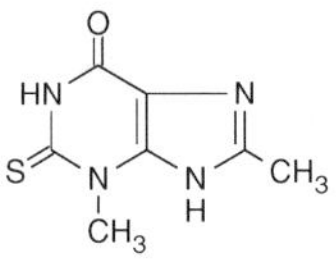

250 mg

S91,216-6
1-[N-(2,6-DIETHYLPHENYL)CARBAMOYL]-IMIDAZOLE

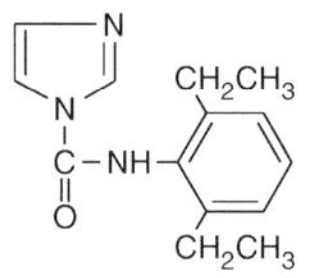

250 mg

S61,767-1
IMIDAZOLE PERFLUOROPROPIONIC ACID SALT

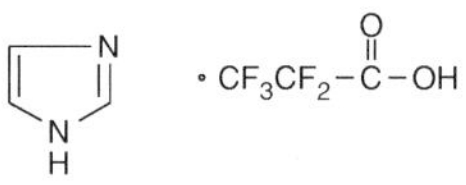

5 gm

S61,697-4
IMIDAZOLE PERFLUOROBUTYRIC ACID SALT

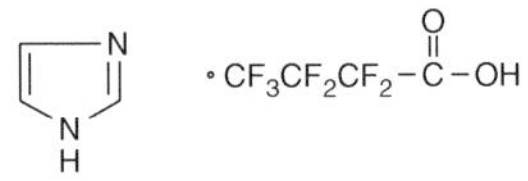

5 gm

S80,304-9
5-BROMOIMIDAZOLE

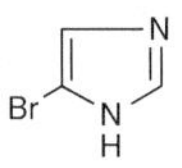

250 mg

S80,450-9
4-BROMOIMIDAZOLE

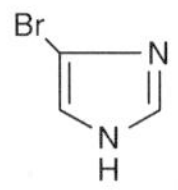

250 mg

S88,384-0
2-IODOIMIDAZOLE

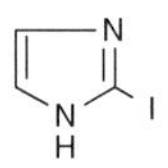

250 mg

S78,381-1
2-BROMO-4,5-DICHLOROIMIDAZOLE

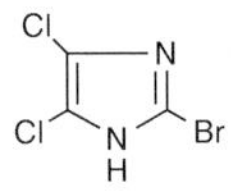

5 gm

S80,226-3
4,5-DIBROMOIMIDAZOLE

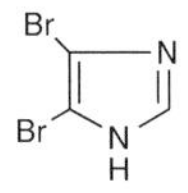

250 mg

S61,970-1
IMIDAZOLE TRIFLUOROACETIC ACID SALT

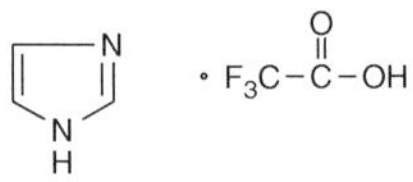

5 gm

S34,761-2
4,5-DIPHENYL-2-ETHYLIMIDAZOLE

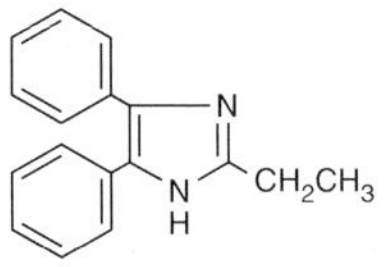

5 gm

S34,772-8
4,5-DIPHENYL-2-PROPYLIMIDAZOLE

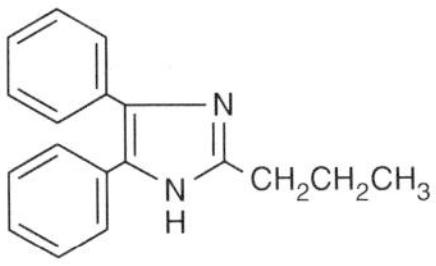

250 mg

S76,212-1
2-PHENYLIMIDAZOLE HYDROCHLORIDE

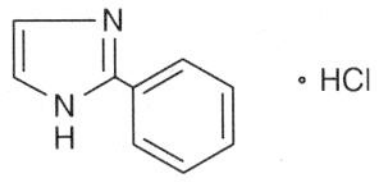

250 mg

S91,535-1
2-STYRYLIMIDAZOLE

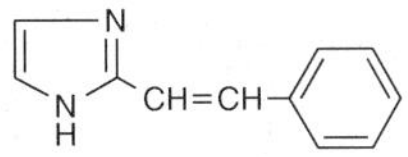

250 mg

S92,756-2
4,5-DIPHENYL-1-TRITYLIMIDAZOLE

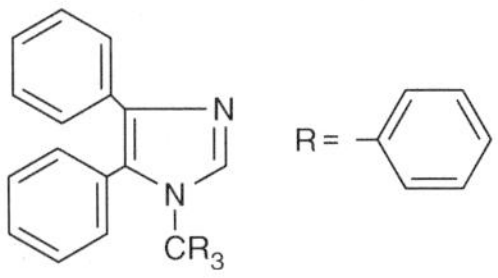

250 mg

S90,648-4
1,4,5,6,7,8-HEXAHYDRO-2-PHENYLCYCLOHEPTIMIDAZOLE

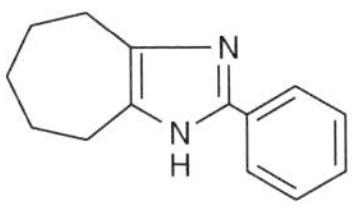

250 mg

S94,637-0
3-CHLORO-1H-1,2,4-TRIAZOLE

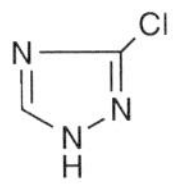

250 mg

S46,313-2
4-METHYL-5-(TRIFLUOROMETHYL)-4H-1,2,4-TRIAZOLINE-3(2H)-THIONE

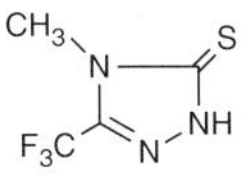

250 mg

S72,884-5
3-[4-CHLORO-2-(TRIFLUOROMETHYL)-ANILINO]-5-MERCAPTO-4H-1,2,4-TRIAZOLE

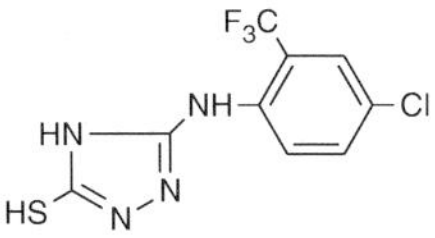

250 mg

S91,186-0
1-(5-PHENYL-1H-1,2,4-TRIAZOL-3-YL)-3-[2-(TRIFLUOROMETHYL)PHENYL]UREA

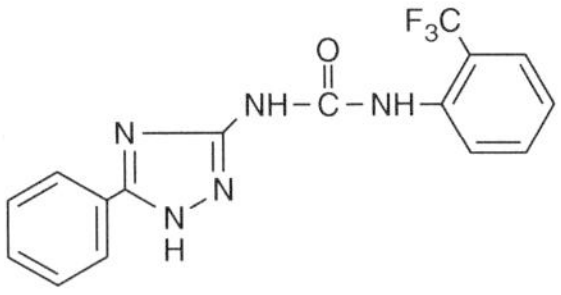

250 mg

S78,248-3
5,6-DIHYDRO-3-(TRIFLUOROMETHYL)-1,2,4-TRIAZOLO(3,4-A)ISOQUINOLINE

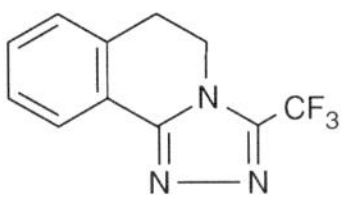

5 gm

S87,466-3
1-(4-CHLOROPHENYL)-1,2,4-TRIAZOL-3-OL

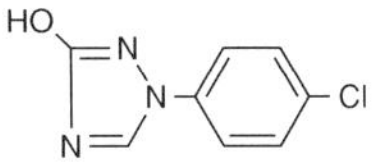

250 mg

S68,598-4
5-(2-AMINO-5-CHLOROPHENYL)-3-PHENYL-1H-1,2,4-TRIAZOLE

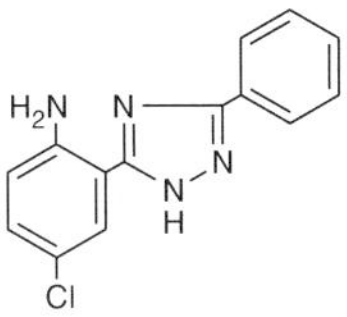

250 mg

S91,056-2
1-(3-FLUOROPHENYL)-3-(5-PHENYL-1,2,4-TRIAZOL-3-YL)UREA

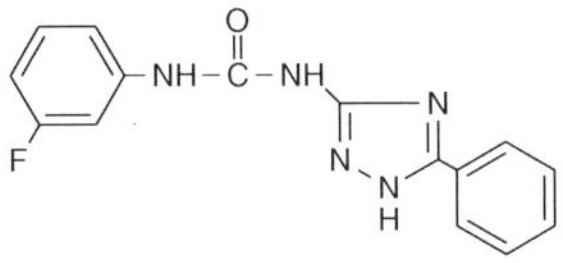

250 mg

S91,054-6
1-(2,5-DIFLUOROPHENYL)-3-(5-PHENYL-1,2,4-TRIAZOL-3-YL)UREA

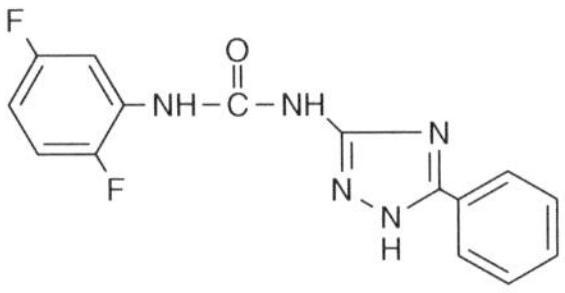

250 mg

S87,463-9
1-(P-TOLYL)-1,2,4-TRIAZOL-3-OL

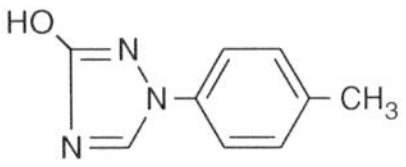

250 mg

S54,712-3
4-AMINO-4H-1,2,4-TRIAZOLE-3,5-DIOL

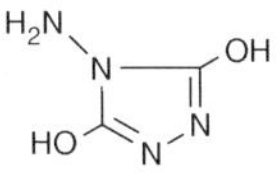

5 gm

S91,941-1
METHYL 5-HYDROXY-1-METHYL-1H-1,2,3-TRIAZOLE-4-CARBOXYLATE

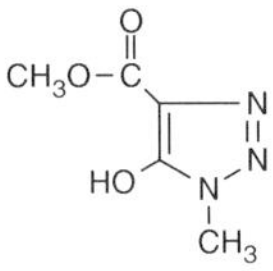

250 mg

S86,934-1
1-PHENYL-1,2,4-TRIAZOL-3-OL

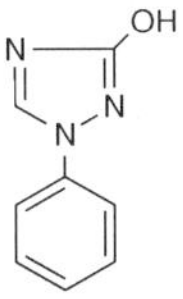

250 mg

S68,284-5
3-BENZYL-4-HYDROXY-4H-1,2,4-TRIAZOLE

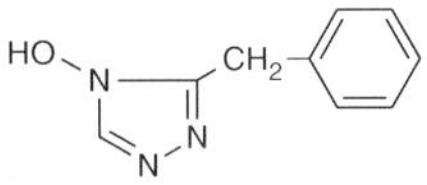

5 gm

S72,885-3
3-(2-ETHOXYANILINO)-5-MERCAPTO-4H-1,2,4-TRIAZOLE

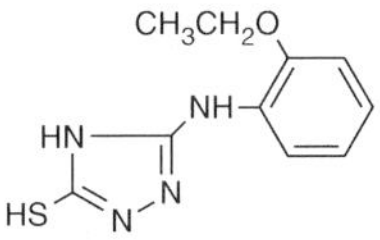

250 mg

S91,068-6
1-[4-(METHYLTHIO)PHENYL]-3-(5-PHENYL-1,2,4-TRIAZOL-3-YL)UREA

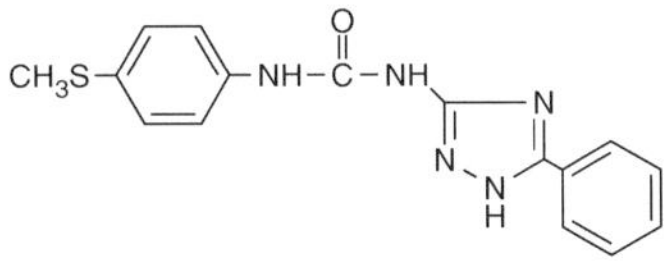

250 mg

S43,724-7
3-AMINO-5-PHENYL-1H-1,2,4-TRIAZOLE

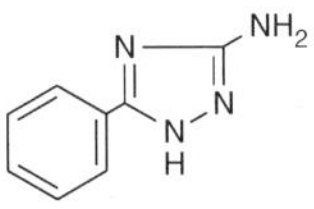

5 gm

S66,212-7
2-(2-AMINOPHENYL)-1H-1,3,4-TRIAZOLE

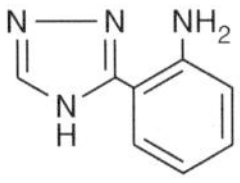

5 gm

S92,104-1
5-AMINO-1,4-DIPHENYL-1,2,3-TRIAZOLE

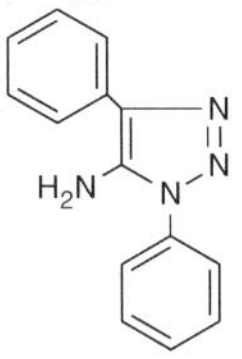

250 mg

S94,638-9
3-AMINO-1-BENZYL-1H-1,2,4-TRIAZOLE

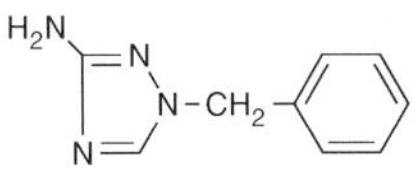

250 mg

S35,789-8
4-AMINO-1,2,4-TRIAZOLE HYDROCHLORIDE

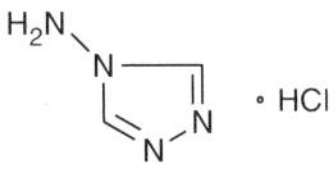

5 gm

S35,478-3
3-PHENYL-5-UREIDO-1,2,4-TRIAZOLE

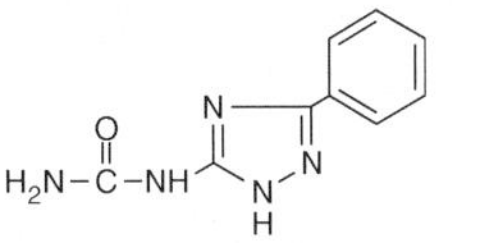

5 gm

S92,349-4
5-ANILINO-4-PHENYL-1H-1,2,3-TRIAZOLE

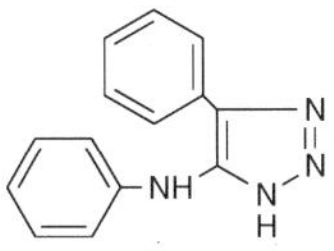

250 mg

S93,456-9
3,5-DIBROMO-4H-1,2,4-TRIAZOLE

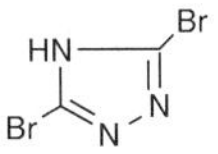

5 gm

S59,627-2
3-METHYLAMINO-5-PHENYL-1,2,4-TRIAZOLE

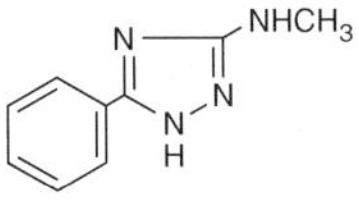

250 mg

S92,231-5
5-ANILINO-1-PHENYL-1H-1,2,4-TRIAZOLE

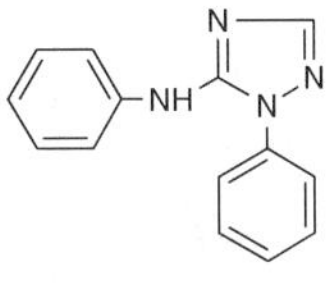

250 mg

S94,555-2
3-PHENYL-1H-1,2,4-TRIAZOLE

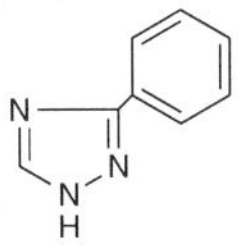

250 mg

S94,876-4
4-(BENZYLAMINO)-4H-1,2,4-TRIAZOLE

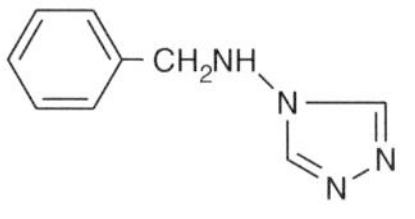

250 mg

S93,432-1
1,1-BIS(1H-1,2,4-TRIAZOL-4-YLAMINO)-PROPANE

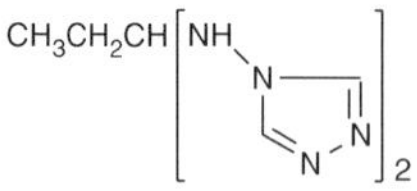

250 mg

S93,436-4
4-(DIPROPYLAMINO)-4H-1,2,4-TRIAZOLE

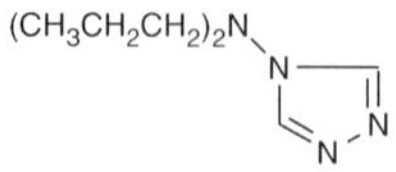

250 mg

S93,434-8
1,1-BIS(1H-1,2,4-TRIAZOL-4-YLAMINO)-HEPTANE

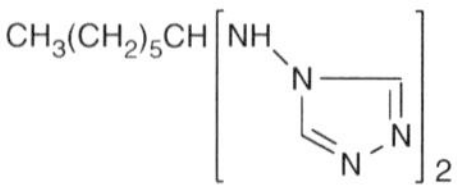

250 mg

S93,435-6
4-[BIS(CARBOXYETHYL)AMINO]-4H-1,2,4-TRIAZOLE

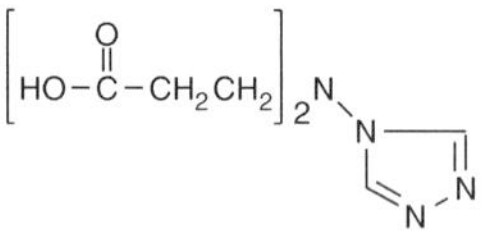

250 mg

S58,334-0
4-PHENYL-1,2,4-TRIAZOLIDINE-3,5-DIONE

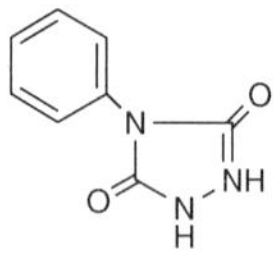

250 mg

S58,673-0
4,6-DIMETHYL-1H-V-TRIAZOLO(4,5-D)-PYRIMIDINE-5,7(4H,6H)-DIONE

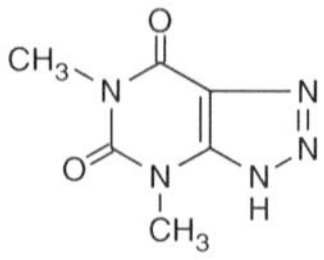

250 mg

S56,577-6
3-ACETAMIDO-1H-1,2,4-TRIAZOLE

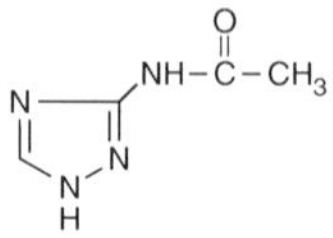

5 gm

S89,798-1
4'-METHYL-1,2,4-TRIAZOLE-5-CARBOXANILIDE

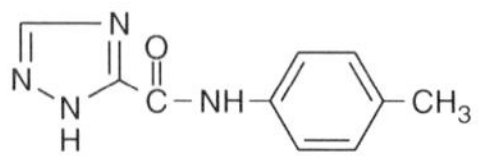

250 mg

S91,946-2
5-CHLORO-1-PHENYL-1H-1,2,3-TRIAZOLE

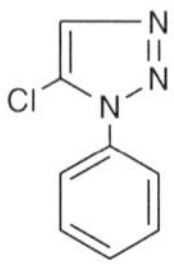

250 mg

S93,457-7
3,4-DIPHENYL-4H-1,2,4-TRIAZOLE

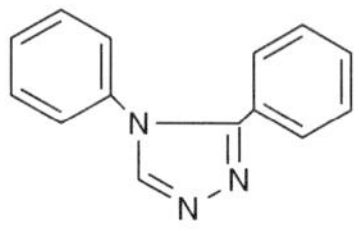

250 mg

S51,698-8
5-TRIFLUOROMETHYL-2H-1,2,3,4-TETRAZOLE

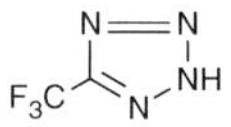

250 mg

S95,307-5
5-PHENOXY-1H-TETRAZOLE

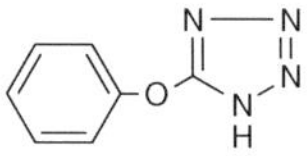

50 mg

S93,642-1
5-(METHYLTHIO)-1H-TETRAZOLE

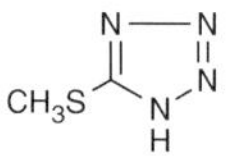

250 mg

S89,472-9
1-METHYL-5-(METHYLTHIO)-1H-TETRAZOLE

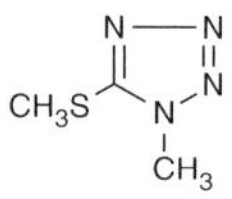

250 mg

S94,878-0
2-ALLYL-5-AMINO-2H-TETRAZOLE

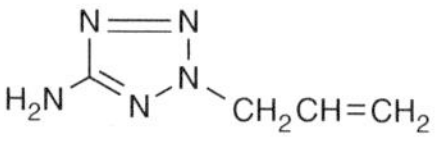

250 mg

S45,024-3
6,7-DIHYDRO-5H-PYRROLOTETRAZOLE

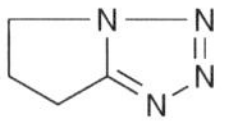

5 gm

S95,306-7
1-AMINO-4,5-DIHYDRO-1H-TETRAZOL-5-ONE

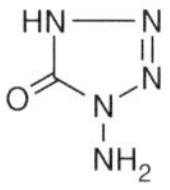

50 mg

S54,168-0
5-BROMO-1-PHENYL-1H-TETRAZOLE

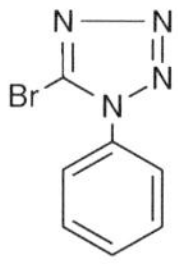

250 mg

S53,138-3
2,3,5-TRIPHENYL-2H-TETRAZOLIUM BROMIDE

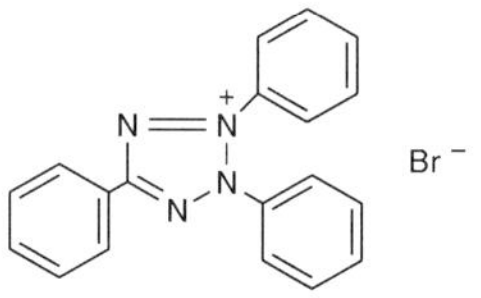

250 mg

S76,400-0
2-(FORMYLAMINO)-1H-TETRAZOLE

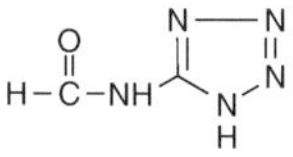

250 mg

S95,309-1
4-(4-CYCLOHEXYLCYCLOHEXYL)-4,5-DIHYDRO-1H-TETRAZOL-5-ONE

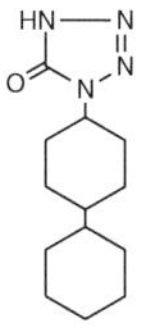

50 mg

S51,384-9
2,3-DIHYDRO-3-METHYL-2-(1-PHENYL-1H-TETRAZOL-5-YLTHIOIMINO)-BENZOTHIAZOLE

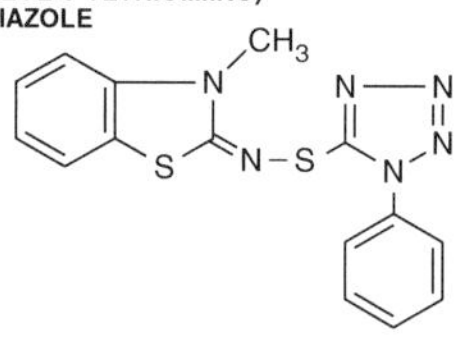

250 mg

S39,701-6
5-METHYL-1-PHENYL-1H-TETRAZOLE

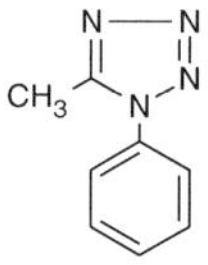

250 mg

S44,534-7
3-METHYLISOXAZOLE

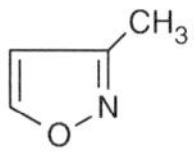

250 mg

S95,260-5
5-TERT-BUTYLISOXAZOLE

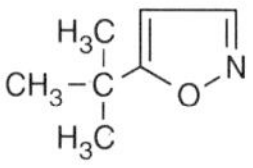

50 mg

S88,531-2
5-(BROMOMETHYL)-3-PHENYLISOXAZOLE

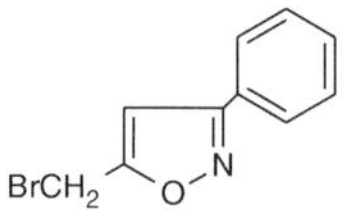

250 mg

S76,036-6
2-(5-PHENYL-3-ISOXAZOLYL)BENZYL BROMIDE

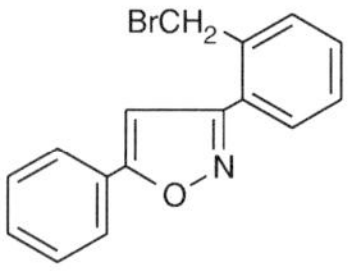

250 mg

S85,403-4
1-(5-CHLORO-2,4-DIMETHOXYPHENYL)-3-(5-METHYL-3-ISOXAZOLYL)UREA

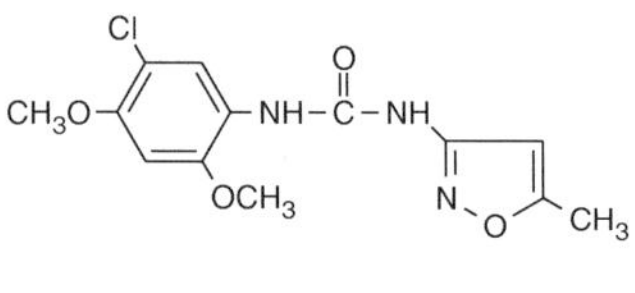

250 mg

S76,690-9
1-[2-(4-FLUOROPHENYL)-3-(3-PHENYL-5-ISOXAZOLYL)PROPIONYL]-4-METHYLPIPERAZINE HYDROCHLORIDE

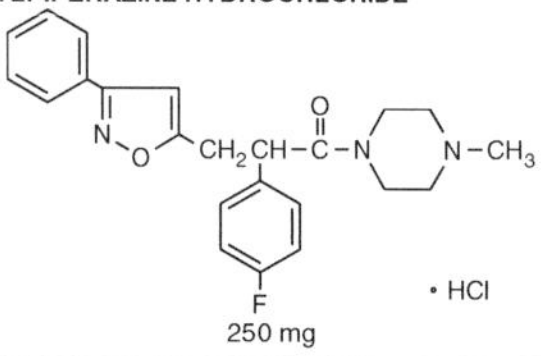

250 mg

S74,964-8
N-(3,4-DIMETHYL-5-ISOXAZOLYL)-TETRACHLOROPHTHALAMIC ACID

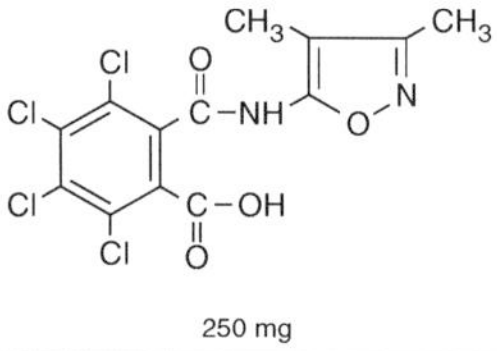

250 mg

S91,368-5
N-(5-METHYL-3-ISOXAZOLYL)-3,4,5,6-TETRACHLOROPHTHALAMIC ACID

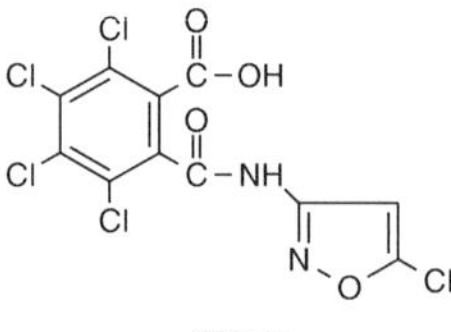

250 mg

S76,990-8
3,4-DICHLORO-N-(5-METHYL-3-ISOXAZOLYL)BENZAMIDE
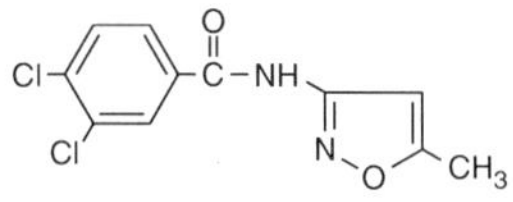
250 mg

S71,794-0
1-(4-FLUOROPHENYL)-3-(5-METHYLISOXAZOL-3-YL)UREA
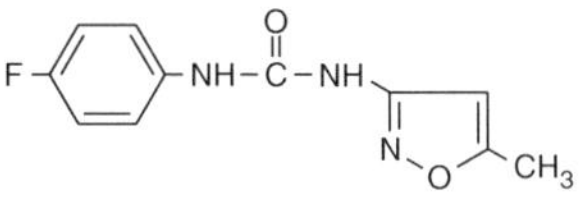
250 mg

S72,798-9
1-(2,4-DIFLUOROPHENYL)-3-(5-METHYL-3-ISOXAZOLYL)UREA
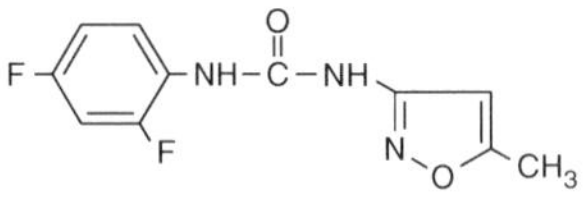
5 gm

S73,383-0
4-(DIPHENYLHYDROXYMETHYL)-N-METHYL-3-PHENYL-5-ISOXAZOLECARBOXAMIDE
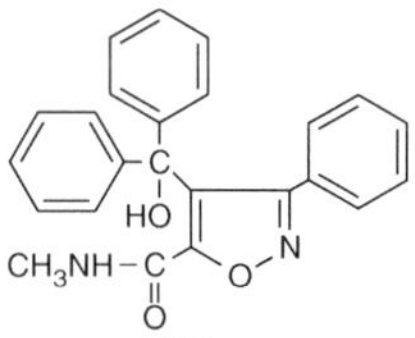
250 mg

S76,232-6
5-[3-(DIMETHYLAMINO)PROPYL]-N-(2-METHOXYBENZYL)-3-PHENYL-4-ISOXAZOLECARBOXAMIDE
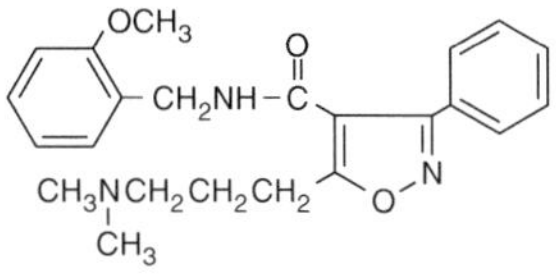
250 mg

S65,837-5
2-(TERT-BUTYL)-5-PHENYLISOXAZOLIUM TETRAFLUOROBORATE
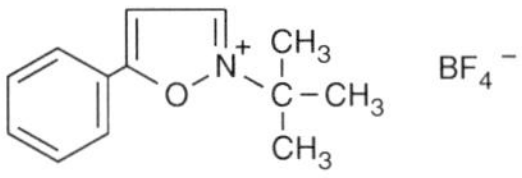
5 gm

S95,262-1
2-ETHYL-5-PHENYLISOXAZOLIUM TETRAFLUOROBORATE
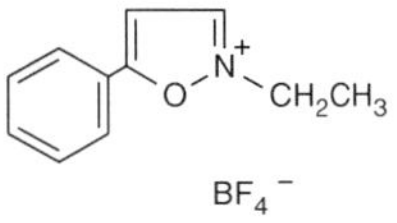
50 mg

S95,261-3
5-(P-TOLYL)ISOXAZOLE
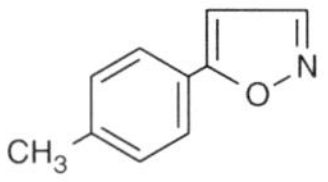
50 mg

S87,976-2
5-METHYL-3-PHENYL-4-ISOXAZOLECARBONYL CHLORIDE
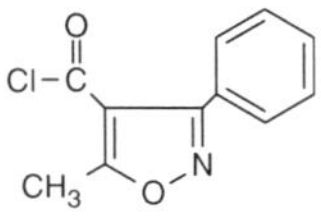
5 gm

S73,240-0
3-TERT-BUTYL-5-ISOXAZOLEPROPIONIC ACID
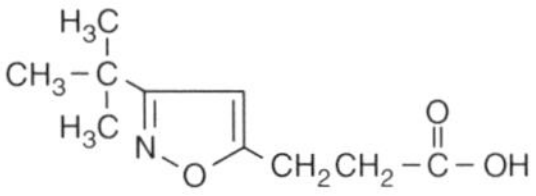
250 mg

S91,075-9
1-(3,4-DIMETHYL-5-ISOXAZOLYL)-3-[4-(METHYLTHIO)PHENYL]UREA
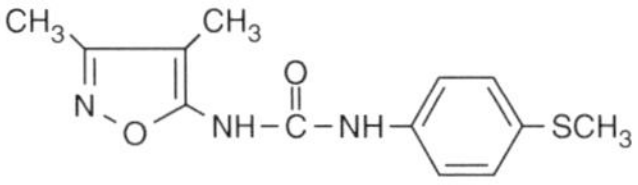
250 mg

S70,334-6
5-(ALPHA-AMINOSTYRYL)-N-METHYL-3-PHENYLISOXAZOLE-4-CARBOXAMIDE
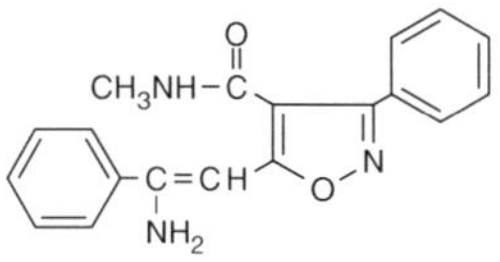
5 gm

S89,789-2
1-(5-METHYL-3-ISOXAZOLYL)-3-(2-METHYL-6-NITROPHENYL)UREA
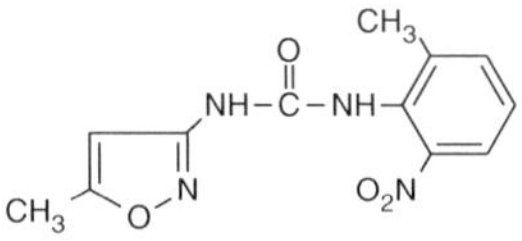
250 mg

S95,264-8
2,5-DI-TERT-BUTYLISOXAZOLIUM-PERCHLORATE
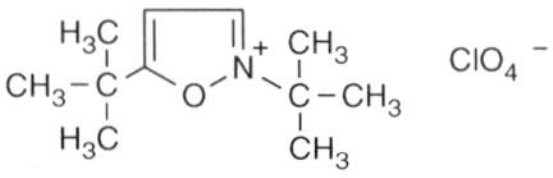
50 mg

S44,996-2
2-ETHYL-5-PHENYLISOXAZOLIUM-4'-SULFONATE MONOHYDRATE
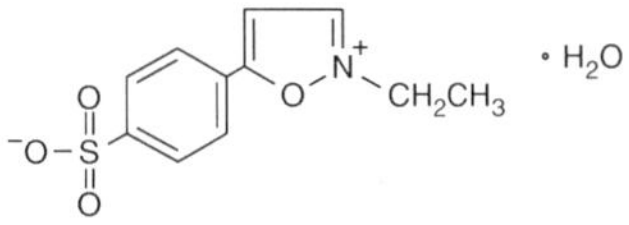
5 gm

S67,224-6
DIETHYL [(5-METHYL-3-PHENYL-4-ISOXAZOLYL)FORMYL]MALONATE
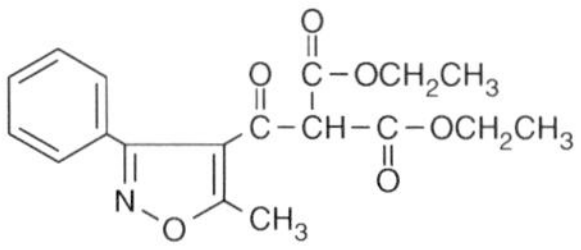
250 mg

S95,265-6
3-PHENYL-4,5,6,7-TETRAHYDRO-1,2-BENZISOXAZOLE
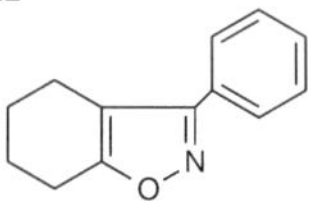
50 mg

S73,627-9
N-(5-METHYL-3-ISOXAZOLYL)-PHTHALAMIC ACID
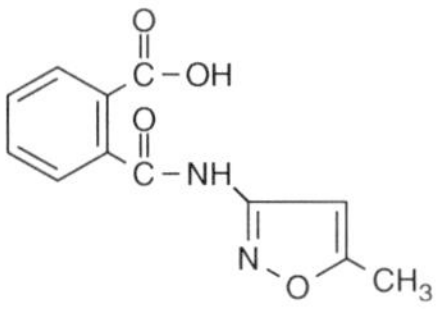
5 gm

S73,639-2
N-(3,4-DIMETHYL-5-ISOXAZOLYL)-PHTHALAMIC ACID
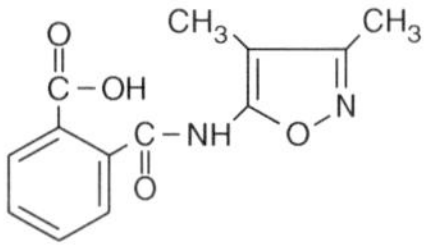
5 gm

S76,303-9
ETHYL 3-(1-BROMOETHYL)-5-METHYL-4-ISOXAZOLECARBOXYLATE
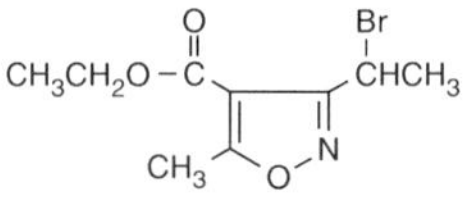
250 mg

S90,194-6
N,N'-BIS(5-METHYL-3-ISOXAZOLYL)-OXAMIDE
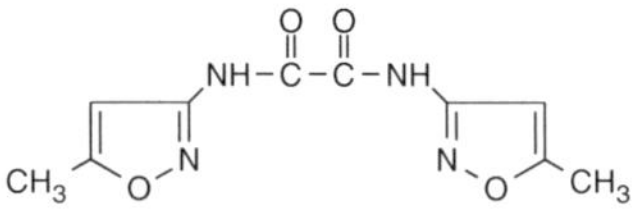
250 mg

S52,766-1
3-(2-CHLOROPHENYL)-5-METHYL-4-ISOXAZOLECARBOXYLIC ACID
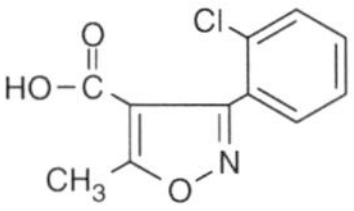
5 gm

S73,275-3
3-TERT-BUTYL-4,5,6,7-TETRAHYDRO-2,1-BENZISOXAZOLE-7-CARBOXYLIC ACID
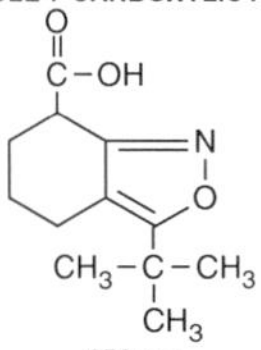
250 mg

S37,325-7
3-(2-CHLOROPHENYL)-5-METHYL-4-ISOXAZOLECARBOXYLIC ACID CHLORIDE
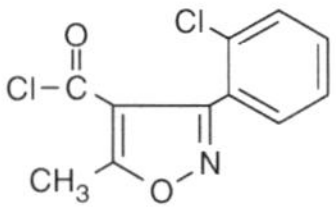
5 gm

S86,900-7
1,1'-HEXAMETHYLENEBIS[3-(5-METHYL-3-ISOXAZOLYL)UREA]
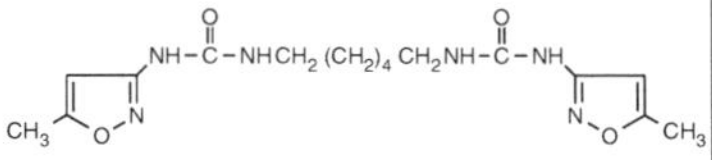
250 mg

S73,428-4
1,3-BIS(3,4-DIPHENYL-5-ISOXAZOLYLETHYL)UREA
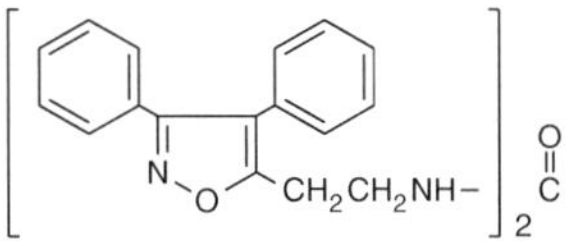
250 mg

S91,218-2
1-(2,6-DIETHYLPHENYL)-3-(5-METHYL-3-ISOXAZOLYL)UREA
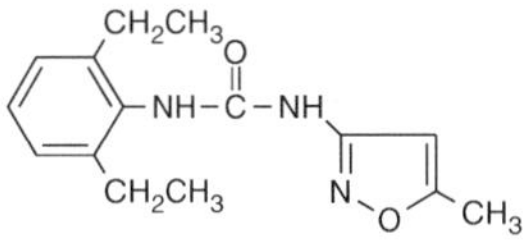
250 mg

S86,778-0
1-(5-METHYL-3-ISOXAZOLYL)-3-(1-NAPHTHYL)UREA
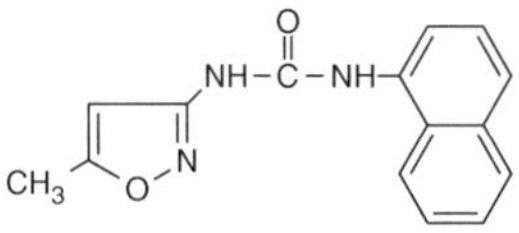
250 mg

S73,216-8
4-(CYANOMETHYL)-3-ETHYL-5-METHYLISOXAZOLE
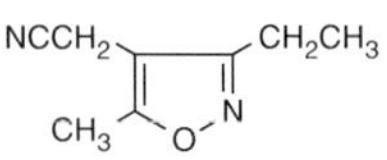
250 mg

S67,404-4
5-PHENYLISOXAZOLE
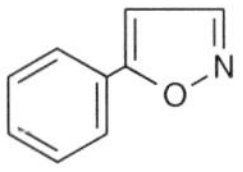
5 gm

S43,338-1
3-METHYL-5-PHENYLISOXAZOLE
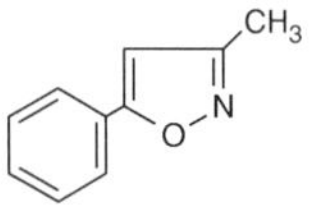
5 gm

S73,438-1
3-METHYL-5-(3-PHENYLPROPYL)-ISOXAZOLE
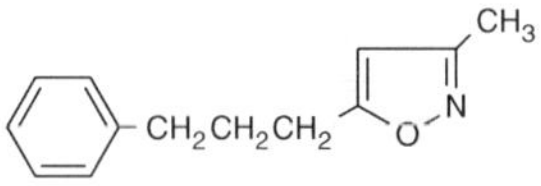
250 mg

S76,225-3
N-(3-ETHYL-5-METHYL-4-ISOXAZOLYLMETHYL)PHTHALIMIDE

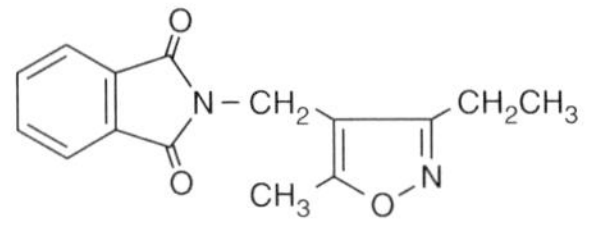

250 mg

S92,725-2
2-(3-CHLOROPHENYL)-4-METHYL-5-PHENYLOXAZOLE

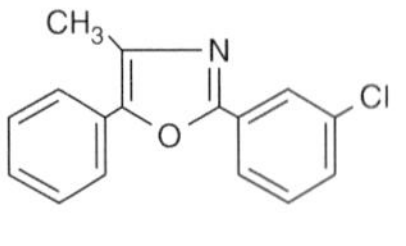

250 mg

S73,435-7
4-(4,5-DIHYDRO-4,4-DIMETHYL-2-OXAZOLYL)-5-(2-HYDROXY-2-PHENYLETHYL)-3-PHENYLISOXAZOLE

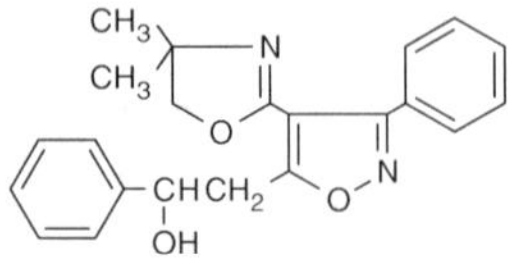

250 mg

S88,330-1
4-(1-METHOXY-1-METHYLETHYL)-2-PHENYL-2-OXAZOLIN-5-ONE

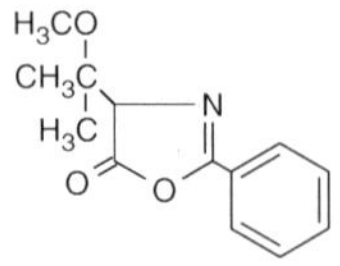

250 mg

S34,989-5
2-(3-FLUOROPHENYL)-5-PHENYLOXAZOLE

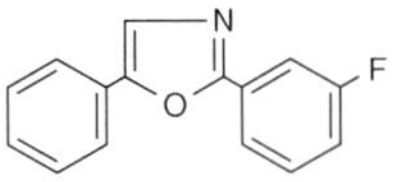

250 mg

S35,243-8
2-(3-METHOXYPHENYL)-5-PHENYLOXAZOLE

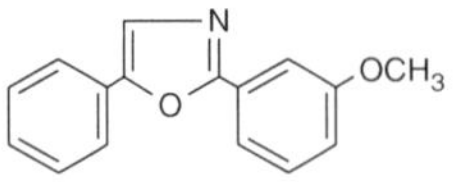

250 mg

S92,824-0
2-(4-NITROPHENYL)-4,5-DIPHENYLOXAZOLE

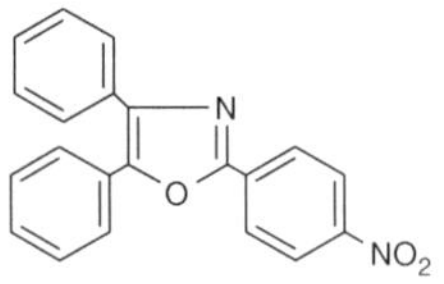

250 mg

S92,724-4
2,4-DIPHENYL-5-METHYLOXAZOLE

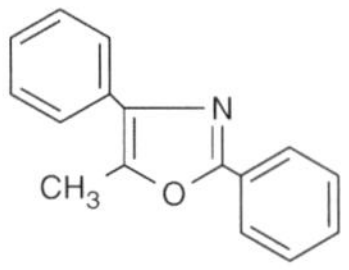

250 mg

S87,234-2
2-(2-OXAZOLYL)BENZOIC ACID, POTASSIUM SALT

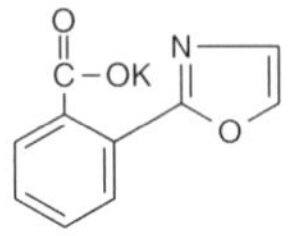

250 mg

S93,255-8
3-(4-METHYL-5-PHENYL-2-OXAZOLYL)-ACRYLIC ACID

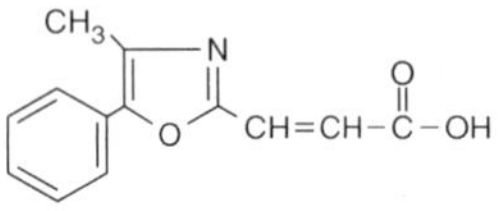

250 mg

S59,245-5
2-AMINO-5-PHENYLOXAZOLE

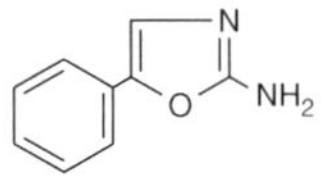

250 mg

S50,190-5
2-AMINO-5-ETHYL-4-PHENYLOXAZOLE

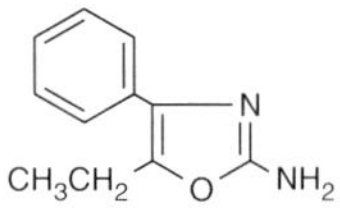

250 mg

S35,313-2
2-METHYL-5-(4-NITROPHENYL)OXAZOLE

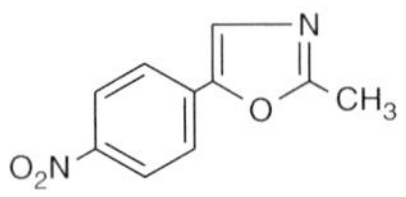

1 gm

S92,822-4
2-METHYL-4,5-BIS(4-NITROPHENYL)-OXAZOLE

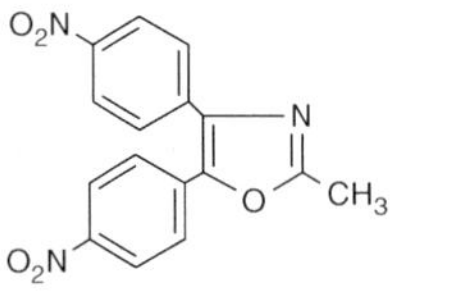

250 mg

S92,821-6
3-(4-METHYL-5-PHENYL-2-OXAZOLYL)-ACRYLIC ACID, POTASSIUM SALT

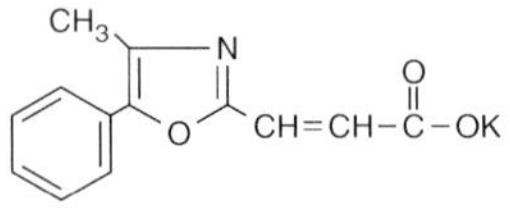

250 mg

S92,825-9
4-(5-METHYL-4-PHENYL-2-OXAZOLYL)-BENZOIC ACID

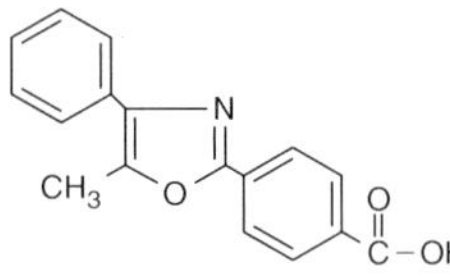

250 mg

S43,375-6
N4-ACETYL-N1-(4,5-DIMETHYLOXAZOL-2-YL)SULFANILAMIDE

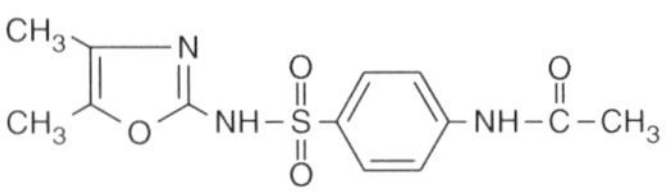

5 gm

S34,455-9
2-(2-CHLOROPHENYL)-5-PHENYLOXAZOLE

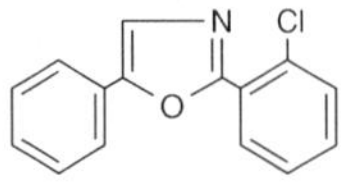

250 mg

S34,566-0
2-(3,4-DICHLOROPHENYL)-5-PHENYLOXAZOLE

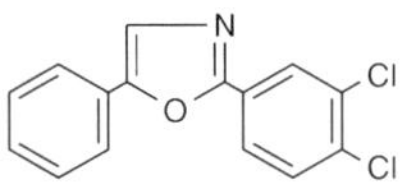

250 mg

S35,901-7
2-(2-BROMOPHENYL)-5-PHENYLOXAZOLE

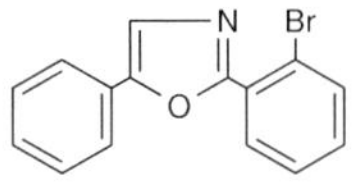

250 mg

S35,332-9
2-(2-METHYLPHENYL)-5-PHENYLOXAZOLE

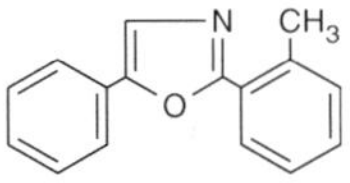

250 mg

S35,333-7
2-(3-METHYLPHENYL)-5-PHENYLOXAZOLE

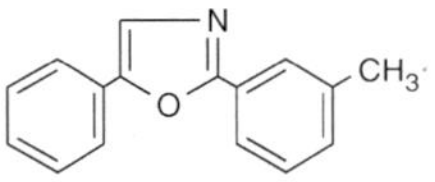

250 mg

S42,747-0
5-PHENYL-2-(P-TOLYL)OXAZOLE

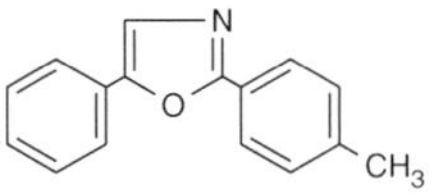

250 mg

S36,765-6
2-CYCLOHEXYL-5-PHENYLOXAZOLE

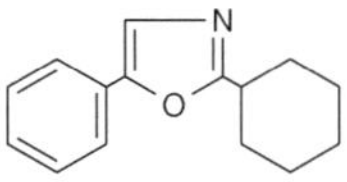

250 mg

S35,377-9
2-(1-NAPHTHYL)-5-PHENYLOXAZOLE

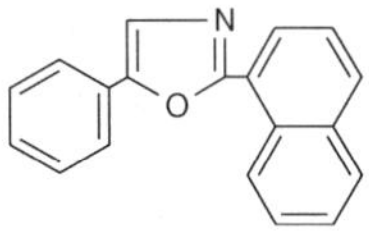

250 mg

S89,854-6
5-PHENYL-1,2,4-OXADIAZOL-3-YL-N-(2-CHLOROETHYL)CARBAMATE

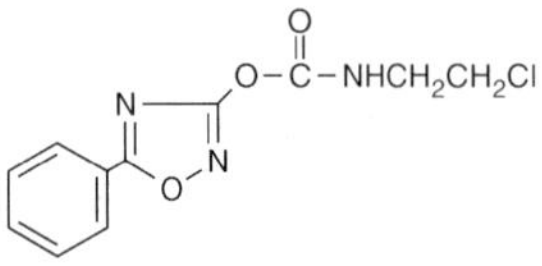

250 mg

S89,842-2
5-PHENYL-1,2,4-OXADIAZOL-3-YL-N-(2-CHLOROPHENYL)CARBAMATE

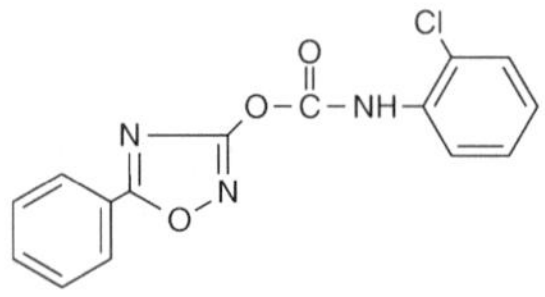

250 mg

S89,841-4
5-PHENYL-1,2,4-OXADIAZOL-3-YL-N-(4-FLUOROPHENYL)CARBAMATE

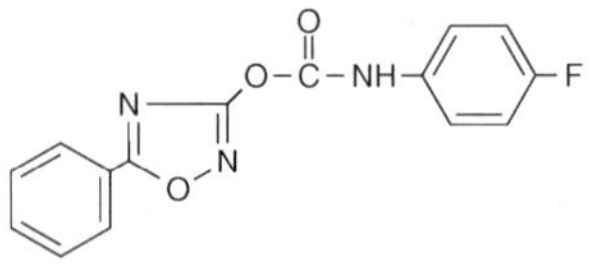

250 mg

S89,864-3
5-PHENYL-1,2,4-OXADIAZOL-3-YL-N-(3,4-DICHLOROPHENYL)CARBAMATE

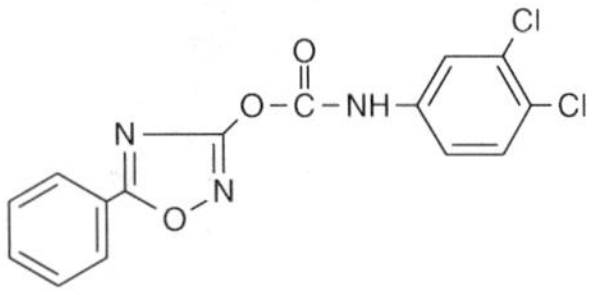

250 mg

S77,869-9
S-(5-HEPTADECYL-1,3,4-OXADIAZOL-2-YL) N-(3,4-DICHLOROPHENYL)-THIOCARBAMATE

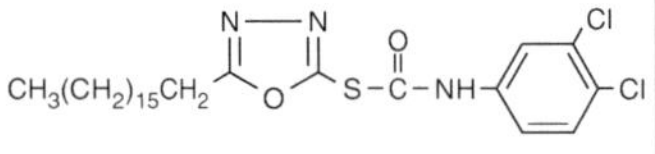

250 mg

S52,617-7
3-(PARA-CHLOROPHENYL)-5-METHYL-1,2,4-OXADIAZOLE

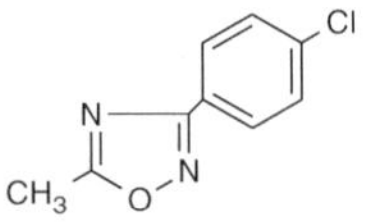

250 mg

S41,107-8
2-(1,3,4-OXADIAZOL-2-YL)PHENOL

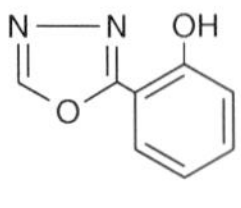

5 gm

S34,983-6
2-(4-FLUOROPHENYL)-5-(2-NAPHTHYL)-1,3,4-OXADIAZOLE
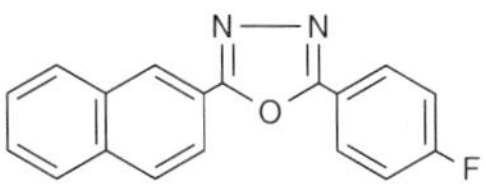
250 mg

S95,303-2
TERT-BUTYL N-[N-(5-AZIDO-ALPHA-BENZYL-1,3,4-OXADIAZOL-2-YLMETHYL)-CARBAMOYLMETHYL]CARBAMATE
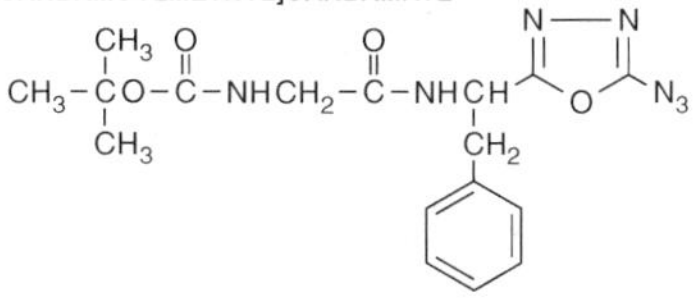
50 mg

S95,302-4
TERT-BUTYL N-(5-AZIDO-ALPHA-BENZYL-1,3,4-OXADIAZOL-2-YLMETHYL)-CARBAMATE
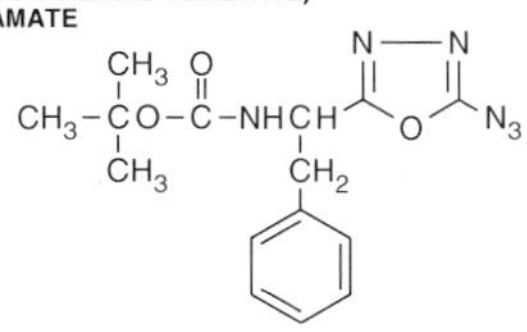
50 mg

S34,984-4
2-(2-FLUOROPHENYL)-5-PHENYL-1,3,4-OXADIAZOLE
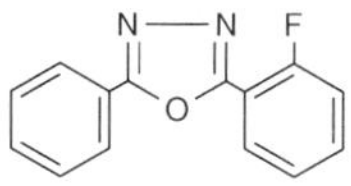
250 mg

S36,323-5
2-(4-METHOXYPHENYL)-5-(1-NAPHTHYL)-1,3,4-OXADIAZOLE
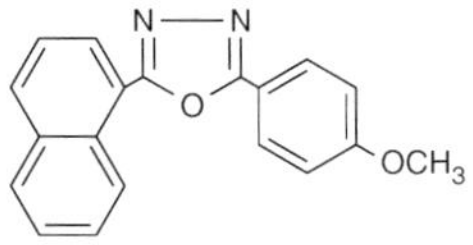
250 mg

S34,450-8
2-(3-CHLOROPHENYL)-5-(1-NAPHTHYL)-1,3,4-OXADIAZOLE
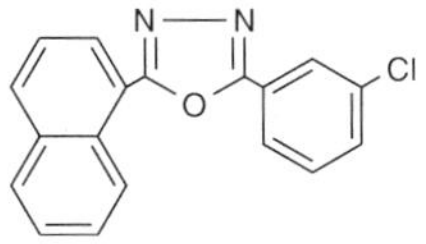
250 mg

S36,322-7
2-(3-METHOXYPHENYL)-5-(2-NAPHTHYL)-1,3,4-OXADIAZOLE
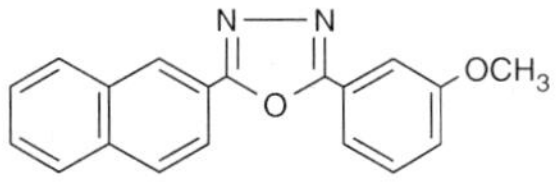
250 mg

S34,452-4
2-(4-CHLOROPHENYL)-5-(2-NAPHTHYL)-1,3,4-OXADIAZOLE
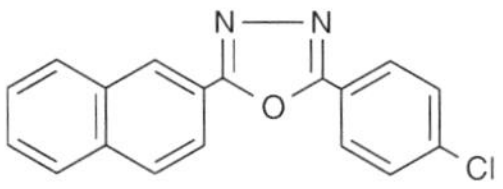
250 mg

S94,635-4
2-AMINO-5-PHENYL-1,3,4-OXADIAZOLE
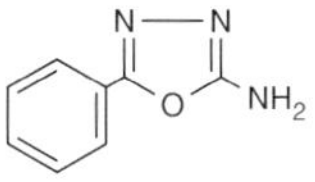
250 mg

S88,842-7
BENZYL N-[3-(3-PHENYL-1,2,4-OXADIAZOL-5-YL)PROPYL]CARBAMATE
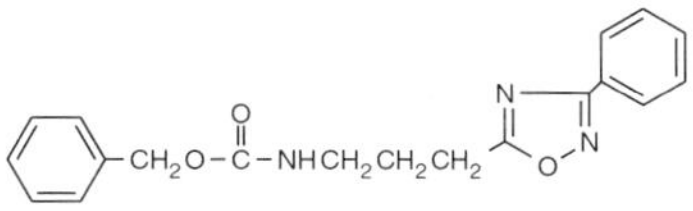
250 mg

S89,867-8
5-PHENYL-1,2,4-OXADIAZOL-3-YL-N-METHYLCARBAMATE
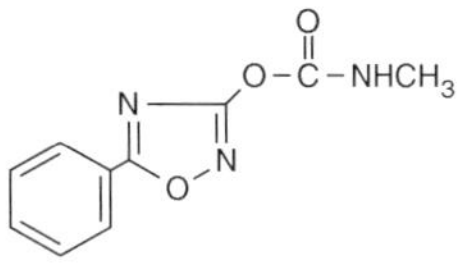
250 mg

S89,836-8
5-PHENYL-1,2,4-OXADIAZOL-3-YL-N-(P-TOLYL)CARBAMATE
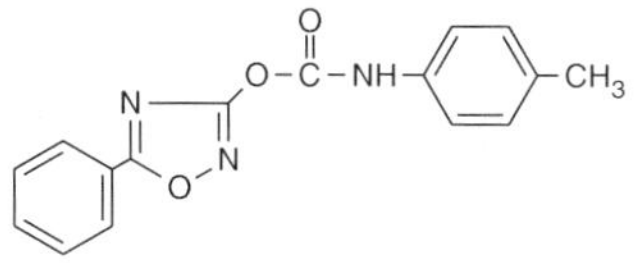
250 mg

S95,301-6
2-(METHYLTHIO)-5-UNDECYL-1,3,4-OXADIAZOLE
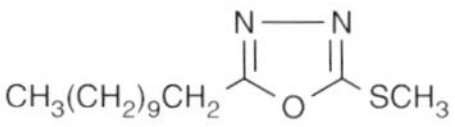
50 mg

S35,477-5
3-PHENYL-5-UREIDO-1,2,4-OXADIAZOLE
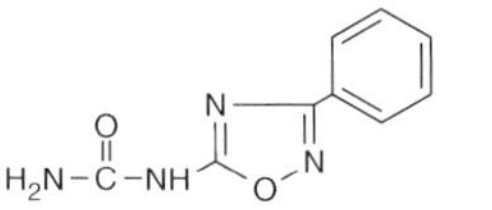
5 gm

S94,636-2
2-AMINO-5-BENZYL-1,3,4-OXADIAZOLE
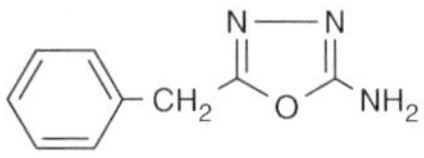
250 mg

S95,290-7
5-BENZYL-2,3-DIHYDRO-1,3,4-OXADIAZOL-2-ONE
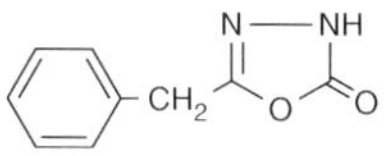
50 mg

S95,289-3
2-(METHYLSULFONYL)-5-(4-NITROPHENYL)-1,3,4-OXADIAZOLE
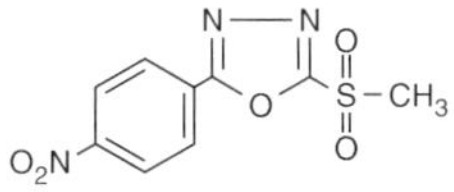
50 mg

S95,288-5
2-(METHYLTHIO)-5-(4-NITROPHENYL)-1,3,4-OXADIAZOLE
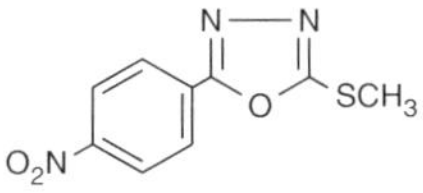
50 mg

S38,274-4
5-METHYL-3-(PARA-TOLYL)-1,2,4-OXADIAZOLE
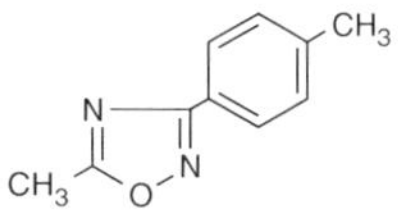
250 mg

S50,439-4
4'-NITRO-2-[3-(P-TOLYL)-1,2,4-OXADIAZOL-5-YL]ACETOPHENONE
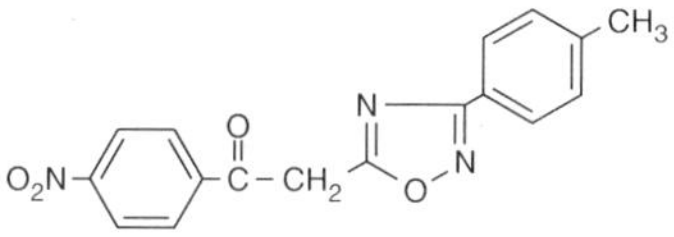
250 mg

S40,961-8
2-(5-NITRO-2-FURYL)-1,3,4-OXADIAZOL-2-IN-5-ONE
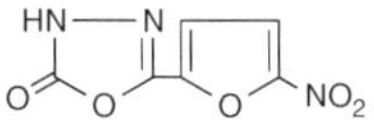
5 gm

S39,236-7
2-PHENYL-1,3,4-OXADIAZOL-2-IN-5-ONE
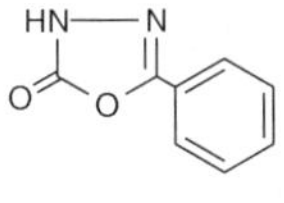
250 mg

S95,299-0
2-(METHYLSULFONYL)-5-(1-NAPHTHYL)-1,3,4-OXADIAZOLE
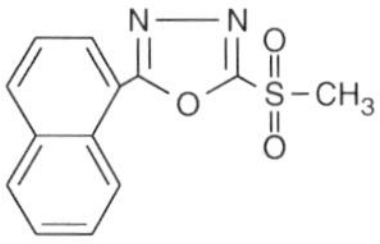
50 mg

S95,293-1
N-(5-AZIDO-1,3,4-OXADIAZOL-2-YLMETHYL)-P-TOLUENESULFONAMIDE
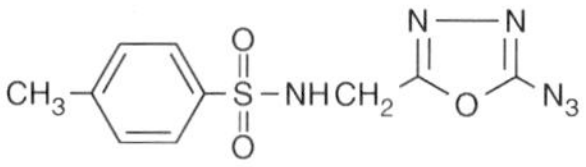
50 mg

S95,300-8
2-AZIDO-5-UNDECYL-1,3,4-OXADIAZOLE
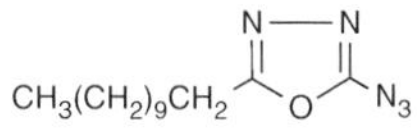
50 mg

S34,559-8
2,5-BIS-(2-CHLOROPHENYL)-1,3,4-OXADIAZOLE
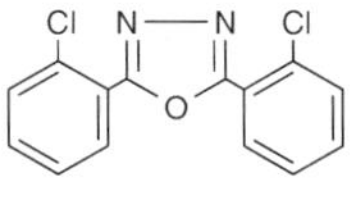
250 mg

S34,564-4
2-(3,4-DICHLOROPHENYL)-5-PHENYL-1,3,4-OXADIAZOLE
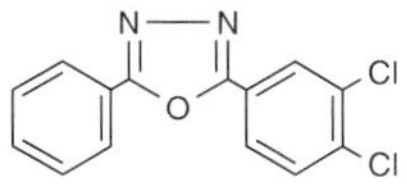
250 mg

S36,784-2
2,5-BIS(4-BROMOPHENYL)-1,3,4-OXADIAZOLE
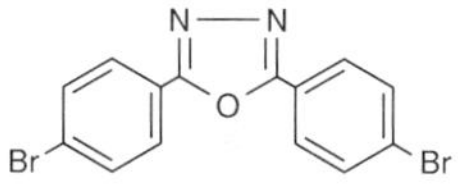
250 mg

S35,331-0
2-(3-METHYLPHENYL)-5-PHENYL-1,3,4-OXADIAZOLE
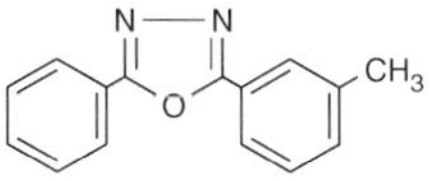
250 mg

S34,454-0
2-(4-CHLOROPHENYL)-5-PHENYL-1,3,4-OXADIAZOLE
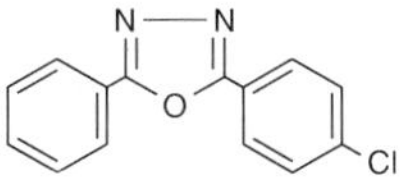
250 mg

S42,749-7
2-(2-FLUOROPHENYL)-5-(1-NAPHTHYL)-1,3,4-OXADIAZOLE
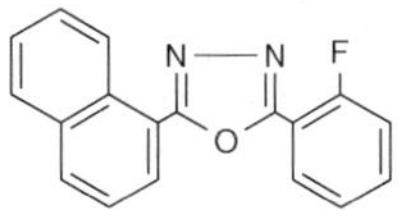
250 mg

S35,325-6
2-(2-METHYLPHENYL)-5-(1-NAPHTHYL)-1,3,4-OXADIAZOLE
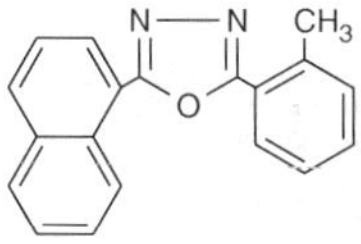
250 mg

S35,328-0
2-(4-METHYLPHENYL)-5-(1-NAPHTHYL)-1,3,4-OXADIAZOLE
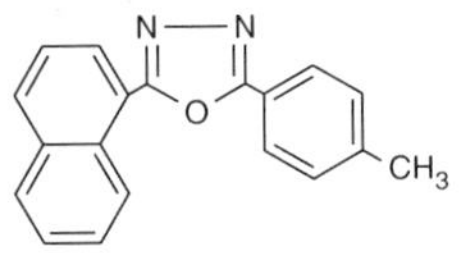
250 mg

S34,557-1
2-(3,4-DICHLOROPHENYL)-5-(2-NAPHTHYL)-1,3,4-OXADIAZOLE
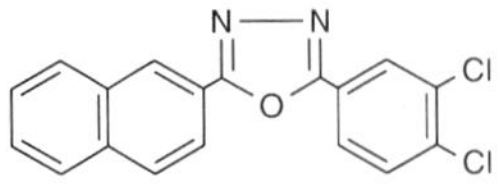
250 mg

S34,555-5
2-(2,4-DICHLOROPHENYL)-5-(2-NAPHTHYL)-1,3,4-OXADIAZOLE
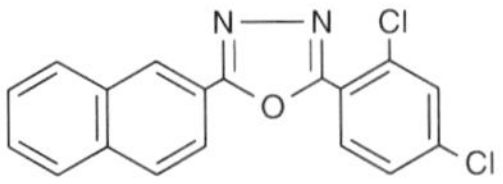
250 mg

S95,298-2
2-AZIDO-5-(1-NAPHTHYL)-1,3,4-OXADIAZOLE
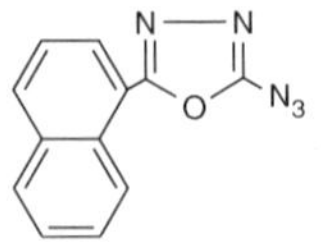
50 mg

S35,326-4
2-(3-METHYLPHENYL)-5-(1-NAPHTHYL)-1,3,4-OXADIAZOLE
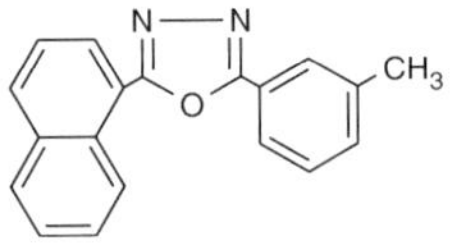
250 mg

S35,327-2
2-(3-METHYLPHENYL)-5-(2-NAPHTHYL)-1,3,4-OXADIAZOLE
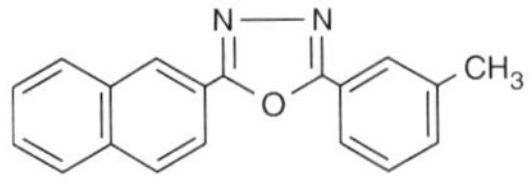
250 mg

S40,610-4
2-(2-BROMOPHENYL)-5-PHENYL-1,3,4-OXADIAZOLE
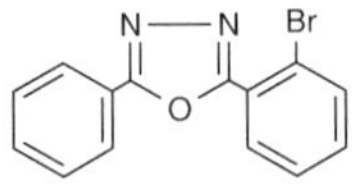
250 mg

S76,350-0
6-[4-(BENZYLOXY)PHENYL]-5-METHYL-3-PHENYLISOXAZOLO(4,5-C)PYRIDIN-4(5H)-ONE
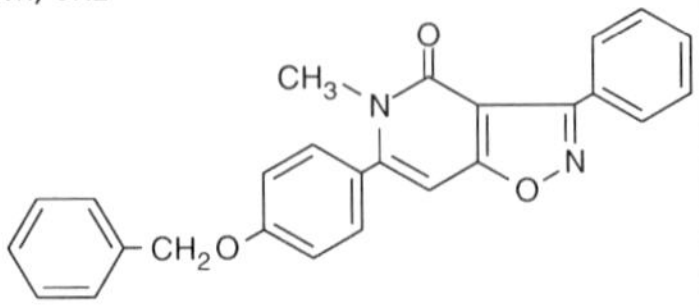
250 mg

S67,100-2
3-ETHYL-5-METHYL-6-PHENYL-4(5H)-ISOXAZOLO(4,5-C)PYRIDINONE
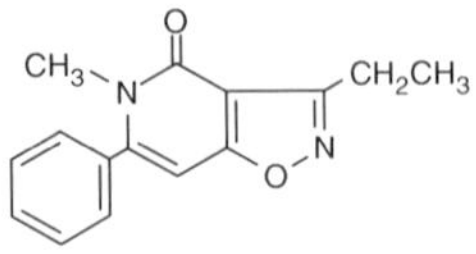
5 gm

S67,603-9
4-(DIMETHYLAMINOMETHYL)PHENYL-5-METHYL-3-PHENYLISOXAZOLO(4,5-C)-PYRIDIN-4(5H)-ONE
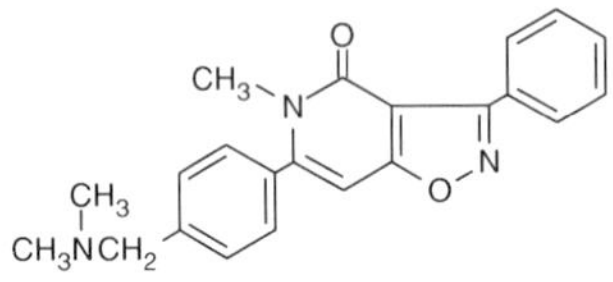
5 gm

S75,764-0
3,6-DIPHENYL-5-METHYLISOXAZOLO(4,5-C)PYRIDIN-4(5H)-ONE
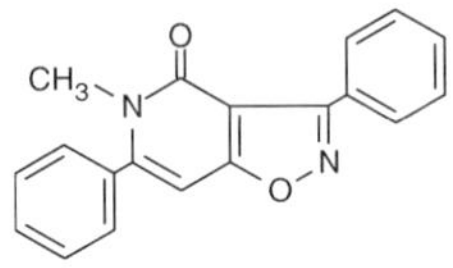
5 gm

S75,473-0
3-CYCLOHEXYL-5-HYDROXY-1,2,3,4-OXATRIAZOL-3-IUM, INNER SALT
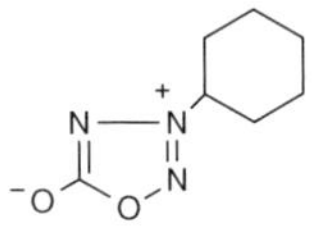
250 mg

S76,238-5
3-CHLORO-4,5-DIHYDRO(1)-BENZOTHIEPINO(5,4-C)ISOXAZOLE
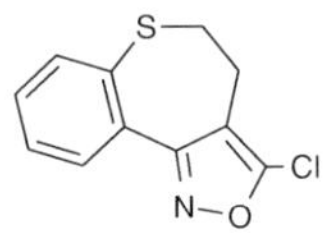
250 mg

S95,273-7
3-[2-(3,4,5-TRIMETHOXYPHENYL)ETHYL]-4-ISOTHIAZOLECARBOXYLIC ACID
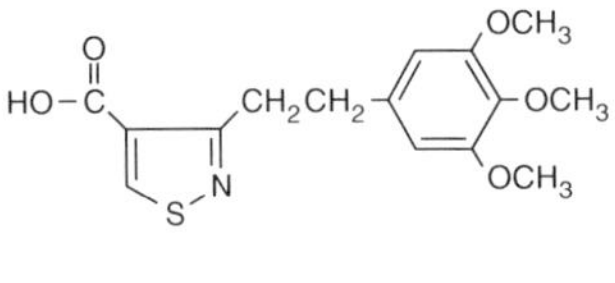
50 mg

S95,274-5
METHYL 3-[2-(3,4,5-TRIMETHOXYPHENYL)ETHYL]-4-ISOTHIAZOLECARBOXYLATE
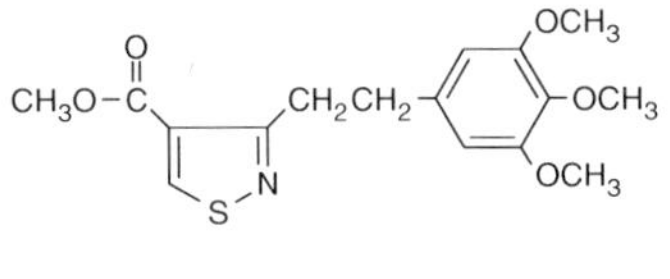
50 mg

S90,856-8
5-AMINO-4-BROMO-3-METHYLISOTHIAZOLE
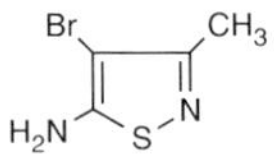
250 mg

S76,381-0
4-CYANOISOTHIAZOLE
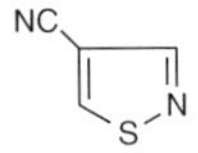
250 mg

S95,272-9
5-ACETAMIDO-2,3-DIMETHYLISOTHIAZOLIUM IODIDE
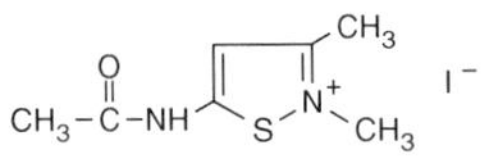
50 mg

S95,271-0
3-METHYL-4-ISOTHIAZOLECARBOXYLIC ACID
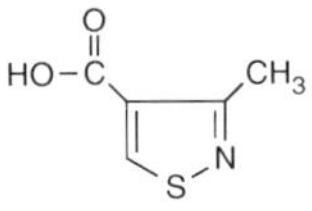
50 mg

S90,859-2
N-(3-METHYL-5-ISOTHIAZOLYL)-ACETAMIDE
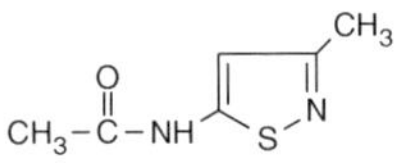
250 mg

S55,499-5
2,4-DICHLORO-N-(3-ISOTHIAZOLYL)-BENZAMIDE
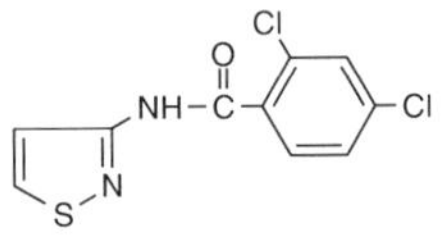
250 mg

S91,677-3
N-[4-(CHLOROMETHYL)-2-THIAZOLYL]-3,4-DICHLOROBENZAMIDE
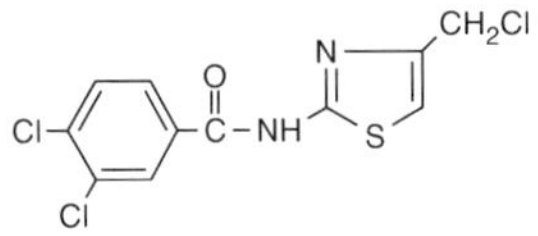
250 mg

S75,469-2
2-AMINO-5-(2-CHLOROETHYL)-4-PHENYLTHIAZOLE HYDROBROMIDE
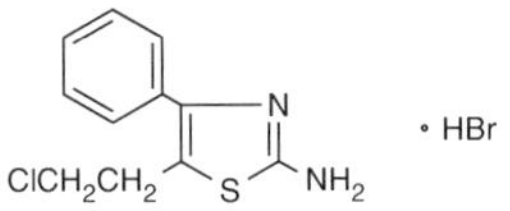
250 mg

S88,588-6
3-BENZYL-5-(2-CHLOROETHYL)-4-METHYLTHIAZOLIUM CHLORIDE
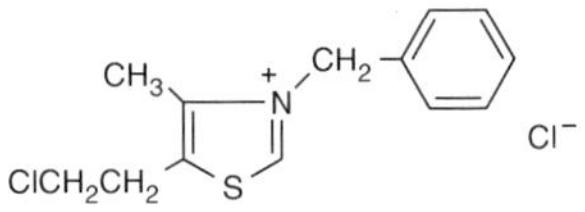
250 mg

S57,826-6
2-ACETAMIDO-4-(CHLOROMETHYL)-THIAZOLE
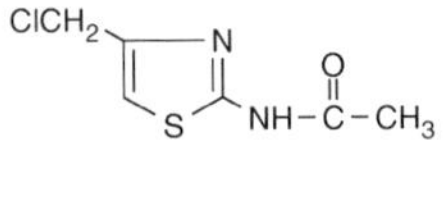
5 gm

S69,235-2
2-[3-(2-THIAZOLYL)UREIDO]ACETIC ACID N2-(N-[2-CHLORO-5-(TRIFLUOROMETHYL)-PHENYL]CARBOMOYL)HYDRAZIDE
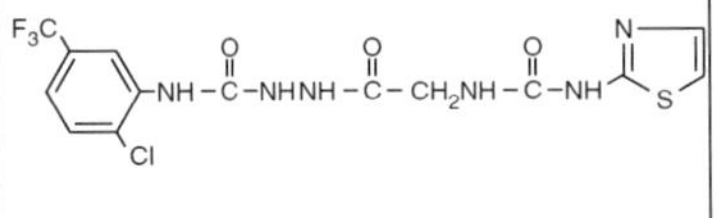
250 mg

S90,261-6
1-[4-CHLORO-2-(TRIFLUOROMETHYL)-PHENYL]-3-(2-THIAZOLYL)UREA
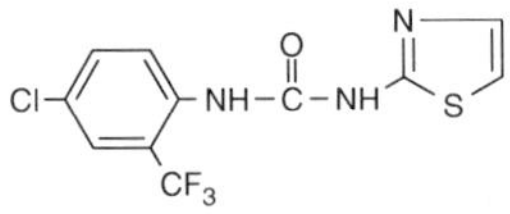
250 mg

S88,343-3
2-PHENYL-2-[4-(TRIFLUOROMETHYL)-2-THIAZOLYLAMINO]-1-ETHANOL
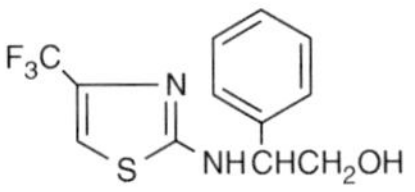
250 mg

S72,876-4
2-[3-(2-THIAZOLYL)UREIDO]ACETIC N2-[N-(2-TRIFLUOROMETHYLPHENYL)-CARBAMOYL]HYDRAZIDE
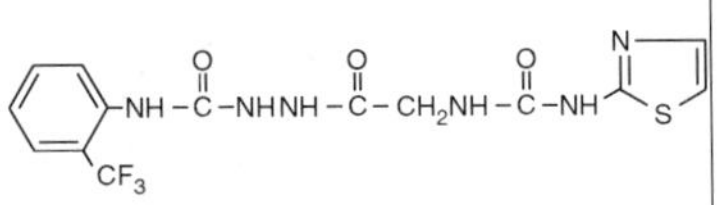
250 mg

S88,344-1
2-ACETAMIDO-4-(TRIFLUOROMETHYL)-THIAZOLE
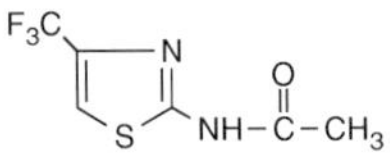
250 mg

S80,110-0
1-(2-THIAZOLYL)-3-[3-(TRIFLUOROMETHYL)PHENYL]UREA
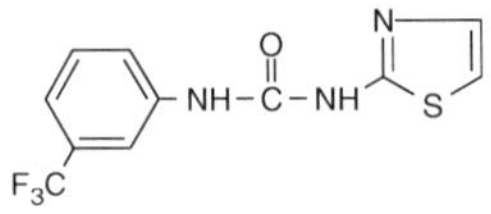
250 mg

S80,166-6
1-(2-THIAZOLYL)-3-[2-(TRIFLUOROMETHYL)PHENYL]UREA
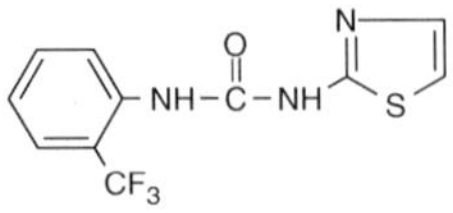
250 mg

S89,790-6
2-(2,4-DICHLOROPHENOXY)-N-(2-THIAZOLYL)ACETAMIDE

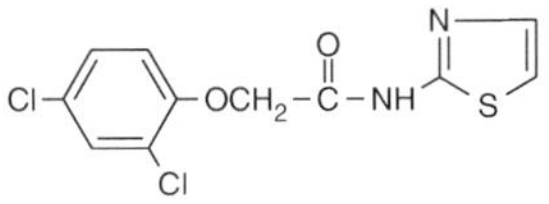

250 mg

S91,179-8
1-(5-CHLORO-2-METHOXYPHENYL)-3-(4-PHENYL-2-THIAZOLYL)UREA

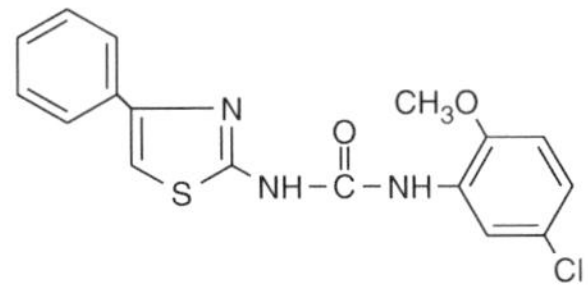

250 mg

S79,697-2
1-(5-CHLORO-2-METHOXYPHENYL)-3-(2-THIAZOLYL)UREA

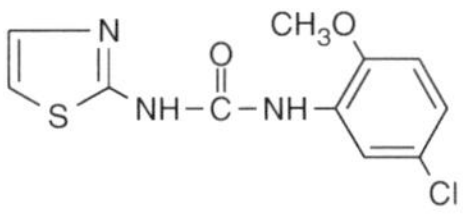

250 mg

S79,607-7
1-(2-CHLORO-4-METHOXYPHENYL)-3-(2-THIAZOLYL)UREA

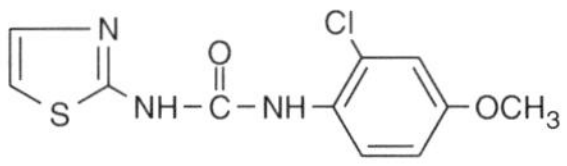

250 mg

S93,479-8
2-(2-CHLOROBENZOYL)-4-METHYL-2-THIAZOLYLACETONITRILE

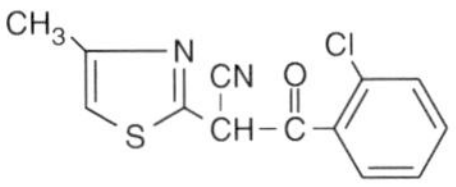

250 mg

S77,232-1
1-(4-FLUORO-3-NITROPHENYL)-3-(4-PHENYL-2-THIAZOLYL)UREA

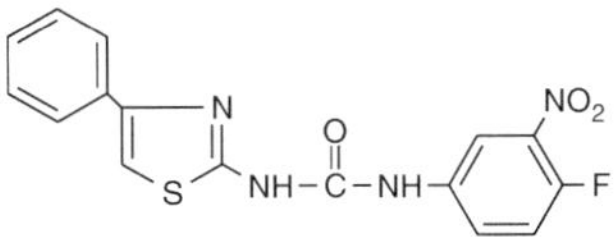

250 mg

S79,509-7
1-(2-CHLORO-5-NITROPHENYL)-3-(2-THIAZOLYL)UREA

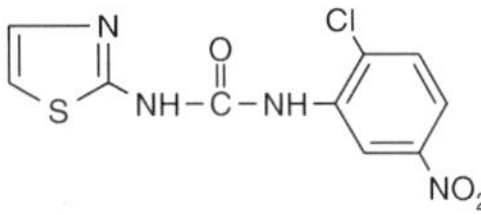

250 mg

S90,988-2
1-(2-CHLORO-4-NITROPHENYL)-3-(4-METHYL-2-THIAZOLYL)UREA

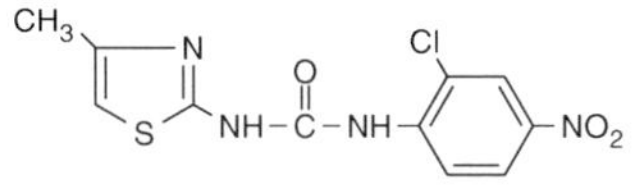

250 mg

S91,608-0
3,4,5,6-TETRABROMO-N-(2-THIAZOLYL) PHTHALAMIC ACID

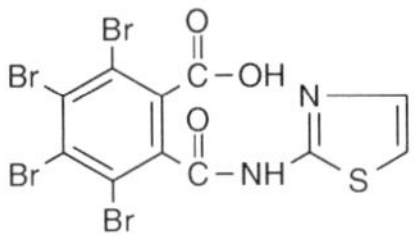

250 mg

S74,694-0
1-(3,4-DICHLOROBENZOYL)-3-(2-THIAZOLYL)-2-THIOUREA

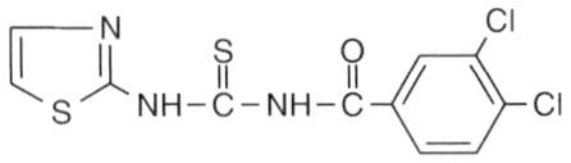

5 gm

S90,467-8
3,4-DICHLORO-N-(2-THIAZOLYL)-BENZAMIDE

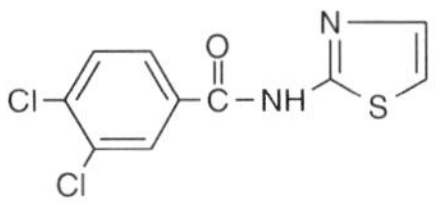

250 mg

S77,225-9
1-(2-FLUOROPHENYL)-3-(4-PHENYL-2-THIAZOLYL)UREA

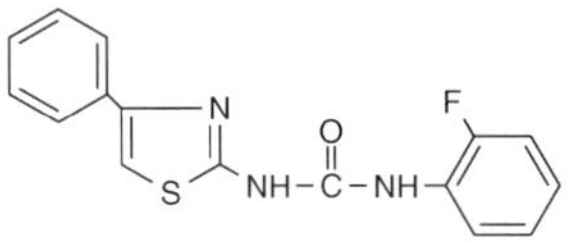

250 mg

S84,402-0
1-(3-FLUOROPHENYL)-3-(2-THIAZOLYL)UREA

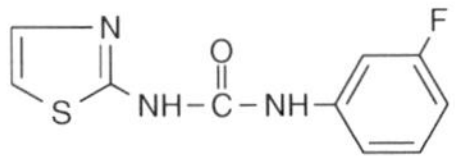

250 mg

S91,235-2
1-(4-FLUOROPHENYL)-3-(4-PHENYL-2-THIAZOLYL)UREA

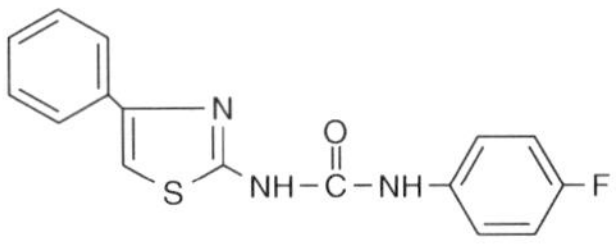

250 mg

S91,106-2
1-(2,3-DICHLOROPHENYL)-3-(4-PHENYL-2-THIAZOLYL)UREA

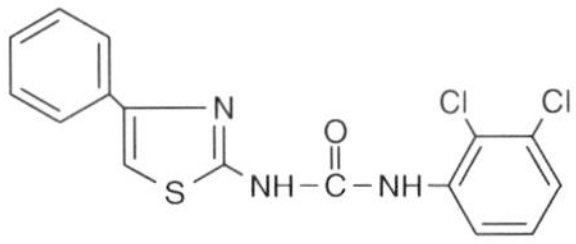

250 mg

S85,151-5
1-(2,3-DICHLOROPHENYL)-3-(2-THIAZOLYL)UREA

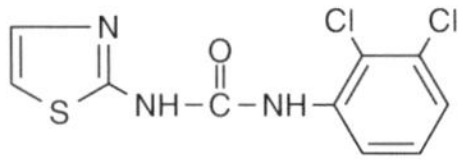

250 mg

S79,606-9
1-(3-CHLORO-4-FLUOROPHENYL)-3-(2-THIAZOLYL)UREA

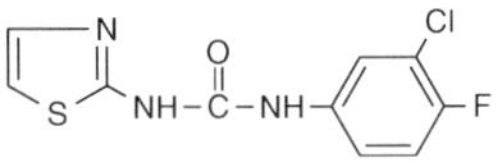

250 mg

S75,654-7
2-AMINO-4-(3,4-DIHYDROXYPHENYL)-THIAZOLE HYDROCHLORIDE

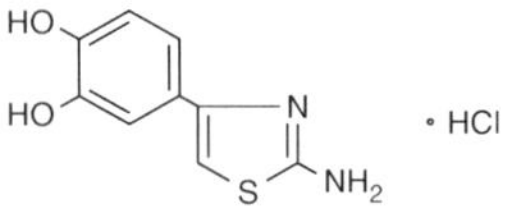

5 gm

S69,237-9
2-[3-(2-THIAZOLYL)UREIDO]ACETIC ACID N2-(3,4-DIHYDROXYBENZYLIDENE)-HYDRAZIDE

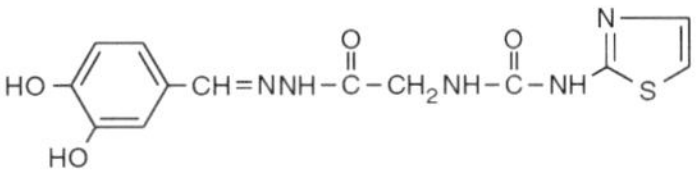

250 mg

S73,971-5
6-METHOXY-ALPHA-(2-THIAZOLYLAMINO)-O-CRESOL

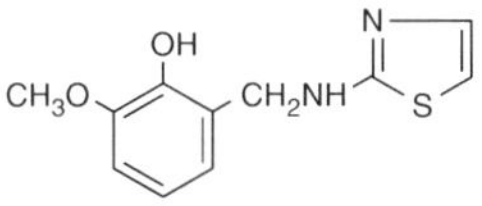

5 gm

S94,866-7
N-(5-BROMO-2-THIAZOLYL)-N'-(2-HYDROXYETHYL)OXAMIDE

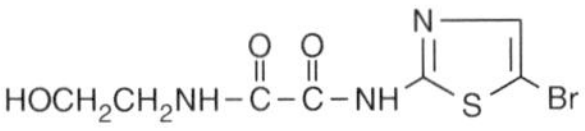

250 mg

S64,879-5
ALPHA-(2-THIAZOLYLAMINO)-O-CRESOL

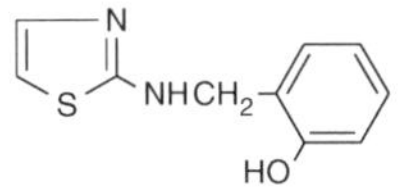

5 gm

S34,641-1
2,4-DIHYDROXYTHIAZOLE-5-ACETIC ACID

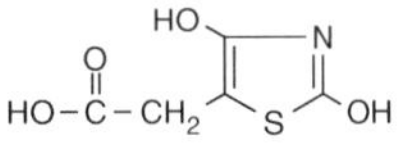

5 gm

S67,368-4
4,4'-BIS(2-THIAZOLYLUREIDO)-3,3'-DIMETHOXYBIPHENYL

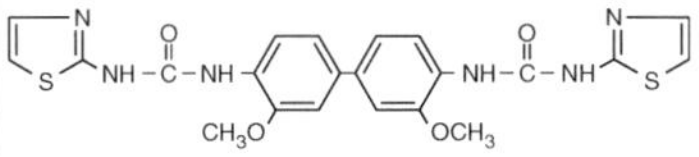

250 mg

S58,279-4
2-AMINO-4-(4-METHOXYPHENYL)-5-TETRADECYLTHIAZOLE

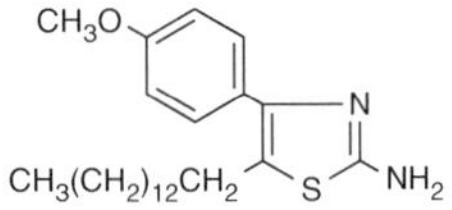

250 mg

S94,870-5
N-(2-THIAZOLYL)OXAMIC N2-(4-METHOXYPHENYLSULFONYL)HYDRAZIDE

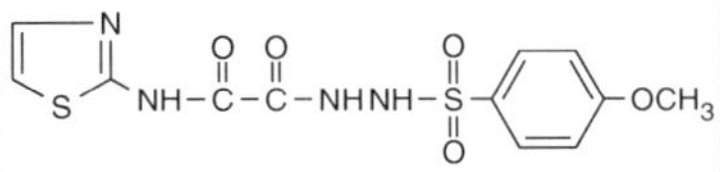

250 mg

S79,508-9
1-(2-METHOXY-5-NITROPHENYL)-3-(2-THIAZOLYL)UREA

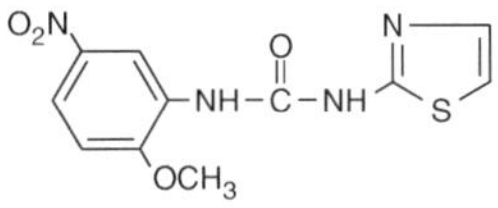

250 mg

S64,237-1
2-(2-METHOXYBENZOYL)-N-[3-(3-NITROPHENYL)-2-THIAZOLYL]ACETAMIDE

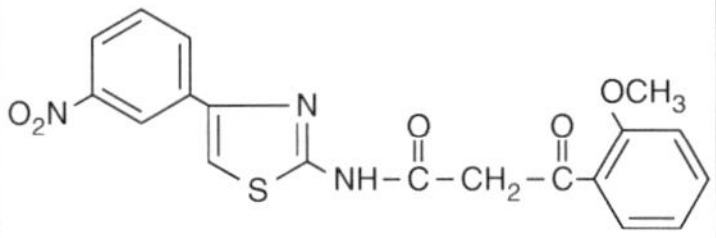

5 gm

S69,205-0
2-[3-(2-THIAZOLYL)UREIDO]ACETIC ACID N2-(2-METHOXYBENZYLIDENE)-HYDRAZIDE

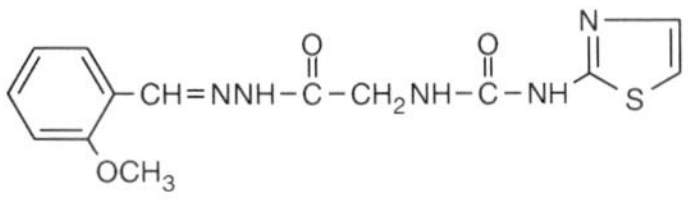

250 mg

S90,549-6
1-(2-METHOXY-5-METHYLPHENYL)-3-(4-METHYL-2-THIAZOLYL)UREA

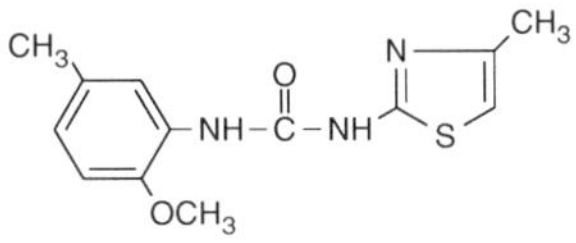

250 mg

S38,756-8
2-BROMO-4-[4-(HEXADECYLOXY)PHENYL]-THIAZOLE

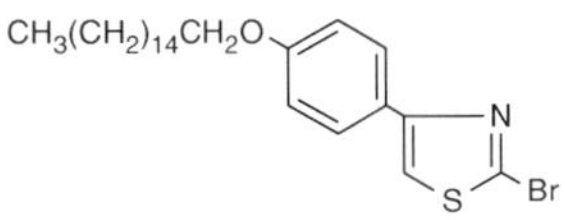

250 mg

S76,035-8
4,5-DIPHENYL-2-MERCAPTOTHIAZOLE

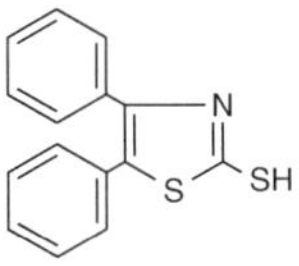

250 mg

S92,002-9
2-AMINO-3-(2-CHLOROACETYL)-4-THIAZOLINE

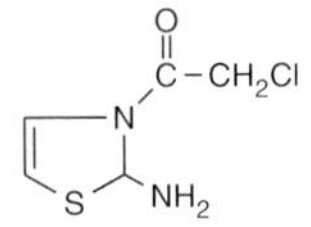

250 mg

S35,711-1
2-ACETAMIDO-5-(4-NITROPHENYLTHIO)-THIAZOLE

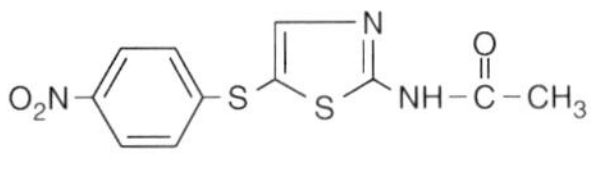

5 gm

S91,074-0
1-[4-(METHYLTHIO)PHENYL]-3-(4-PHENYL-2-THIAZOLYL)UREA

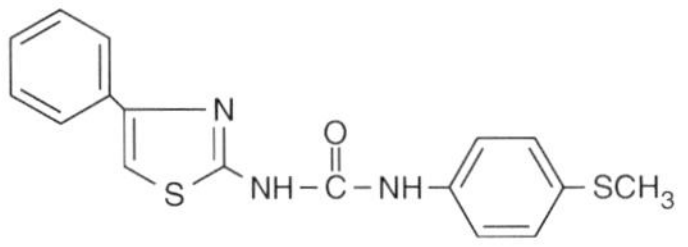

250 mg

S62,951-0
2-AMINO-4-(3-NITROPHENYL)THIAZOLE

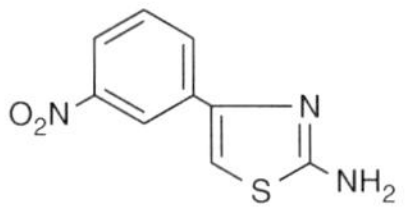

5 gm

S90,792-8
4-AMINO-2-(N-METHYLANILINO)THIAZOLE

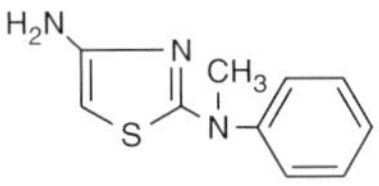

250 mg

S64,544-3
2-AMINO-4-PHENYL-5-THIAZOLYLACETIC ACID

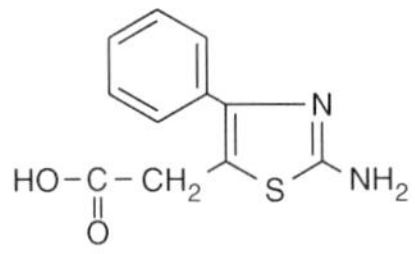

250 mg

S64,494-3
2-(3-AMINOBENZOYL)-N-(2-THIAZOLYL)-ACETAMIDE

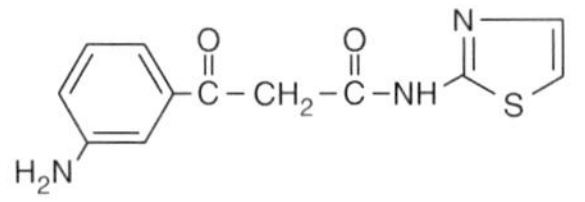

5 gm

S90,854-1
2-AMINO-5-THIAZOLESULFONYL FLUORIDE

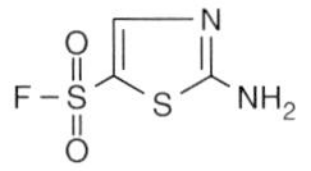

250 mg

S38,678-2
3-(2-AMINO-4-THIAZOLYL)-BENZENESULFONYL FLUORIDE

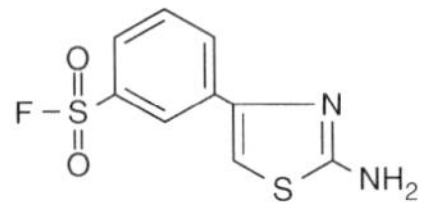

250 mg

S45,475-3
2-AMINO-5,6-DIHYDRO-4H-CYCLOPENTATHIAZOLE HYDROCHLORIDE

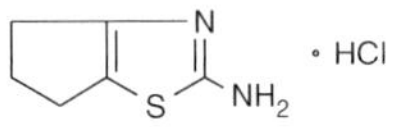

5 gm

S87,149-4
2-AMINO-5,6-DIHYDRO-4H-CYCLOPENTATHIAZOLE

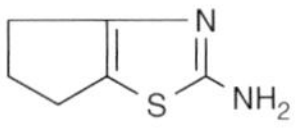

250 mg

S94,631-1
2-AMINO-5-CHLOROTHIAZOLE HYDROCHLORIDE

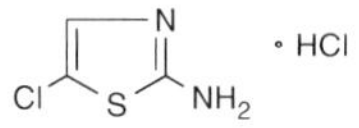

250 mg

S36,653-6
2-AMINO-4-METHYLTHIAZOLE HYDROCHLORIDE

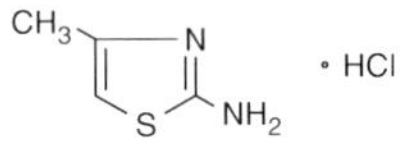

5 gm

S45,752-3
2-AMINO-4,5-DIMETHYLTHIAZOLE HYDROBROMIDE

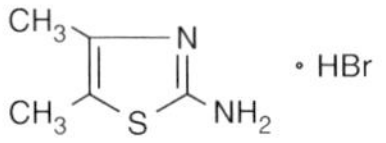

5 gm

S84,136-6
2-AMINO-4-TERT-BUTYLTHIAZOLE

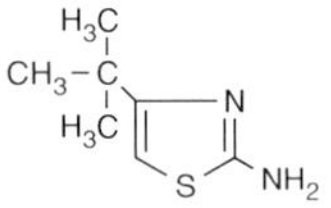

250 mg

S56,371-4
2-AMINO-4-PHENETHYLTHIAZOLE HYDROCHLORIDE

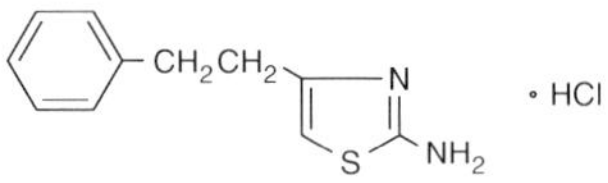

5 gm

S81,913-1
2-AMINO-4-PHENYLTHIAZOLE

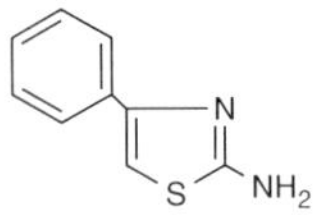

250 mg

S64,208-8
2-AMINO-4,5,6,7-TETRAHYDROBENZOTHIAZOLE

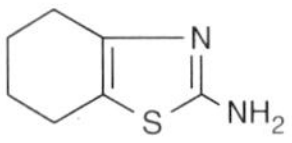

5 gm

S62,446-2
2-(PICRYLAMINO)THIAZOLE

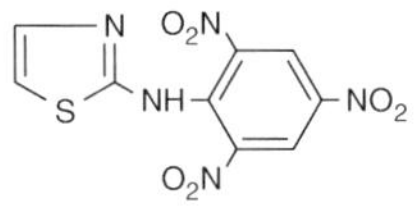

5 gm

S94,867-5
N-(5-CHLORO-2-THIAZOLYL)OXAMIC HYDRAZIDE

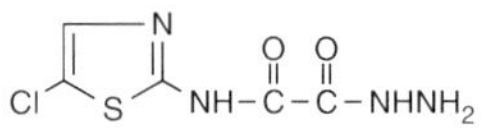

250 mg

S94,868-3
N-(2-THIAZOLYL)OXAMIC N2-ACETYLHYDRAZIDE

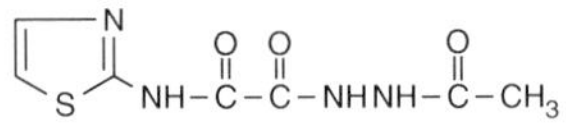

250 mg

S58,716-8
2-HEPTADECYL-4-(HYDRAZINOMETHYL)-THIAZOLE HYDROCHLORIDE

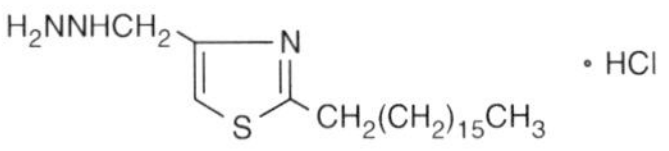

5 gm

S85,338-0
1-(4-METHYL-2-NITROPHENYL)-3-(2-THIAZOLYL)UREA

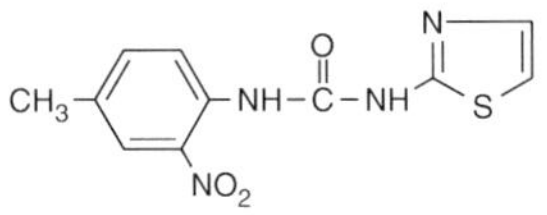

250 mg

S36,423-1
N-(5-NITROTHIAZOL-2-YL)BENZAMIDE

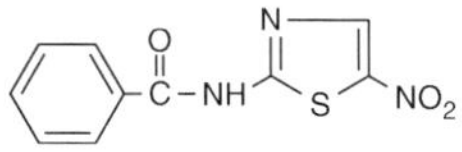

5 gm

S94,871-3
N-(2-THIAZOLYL)OXAMIC-N2-4-TOLUENESULFONYLHYDRAZIDE

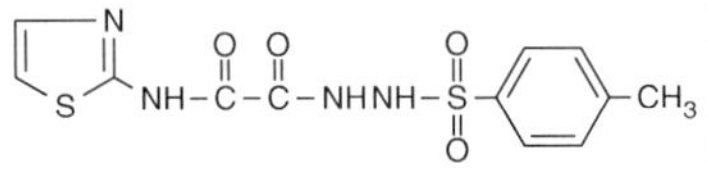

250 mg

S84,409-8
1-(2-METHYL-5-NITROPHENYL)-3-(2-THIAZOLYL)UREA

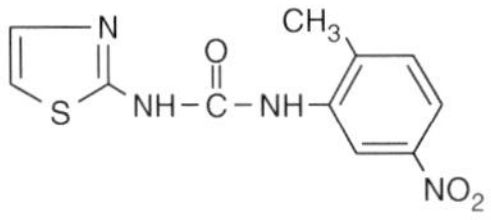

250 mg

S90,763-4
2-(N-ETHYLANILINO)-4-(4-NITROPHENYL)THIAZOLE

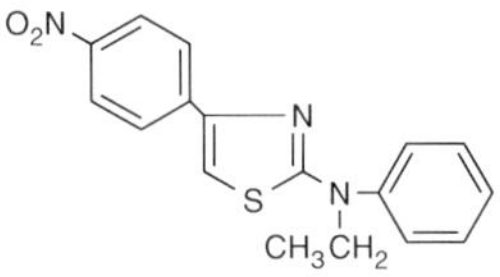

250 mg

S90,898-3
4-METHYL-2-(N-METHYLANILINO)-THIAZOLE

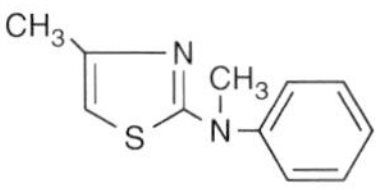

250 mg

S90,753-7
2-(DIMETHYLAMINO)-4-PHENYLTHIAZOLE

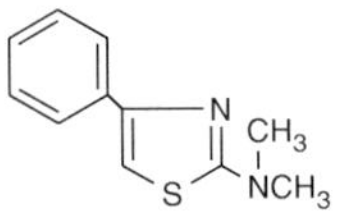

250 mg

S39,015-1
2,4-DIMETHYL-3-ETHYLTHIAZOLIUM IODIDE

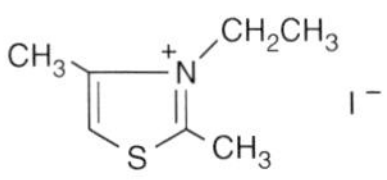

250 mg

S63,874-9
2-HYDRAZINO-4-PHENYLTHIAZOLE

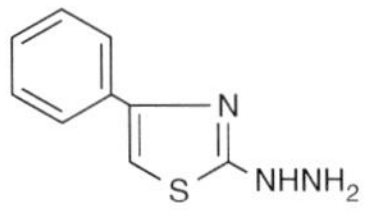

250 mg

S72,875-6
2-[3-(2-THIAZOLYL)UREIDO]ACETIC N2-[N-(2,4-XYLYL)CARBAMOYL]-HYDRAZIDE

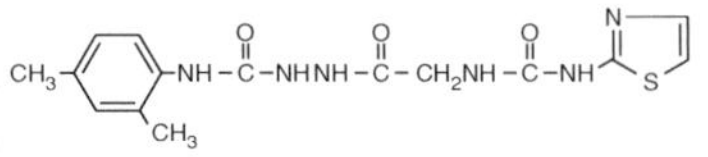

100 mg

S76,251-2
ACETONE (4-METHYL-2-THIAZOLYL)-HYDRAZONE HYDROCHLORIDE
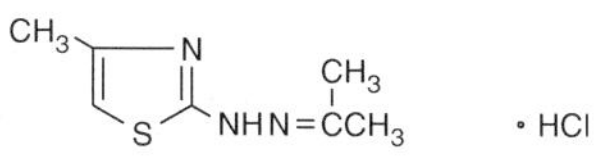
250 mg

S77,967-9
ETHYL 2-[3-(5-NITRO-2-THIAZOLYL)-UREIDO]ACETATE
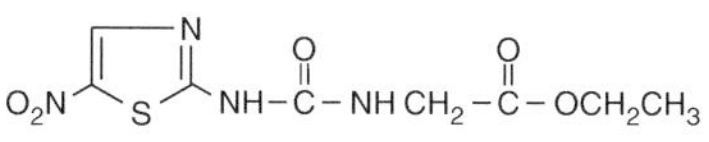
250 mg

S88,879-6
ETHYL 2,5-DIMETHYL-4-THIAZOLECARBOXYLATE
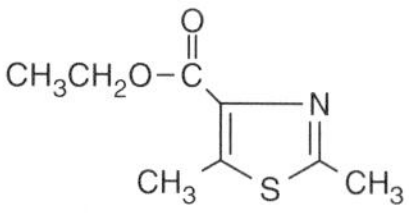
250 mg

S37,300-1
2-ACETAMIDO-5-(4-NITROPHENYLSULFONYL)THIAZOLE
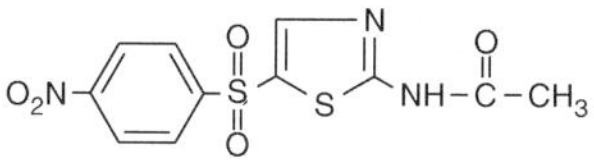
5 gm

S89,928-3
4-NITRO-N-(2-THIAZOLYL)BENZAMIDE
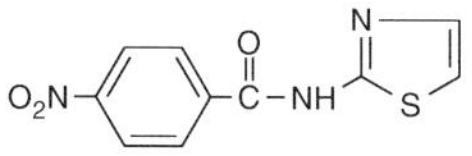
250 mg

S94,869-1
N-(2-THIAZOLYL)OXAMIC N2-(4-NITROBENZOYL)HYDRAZIDE
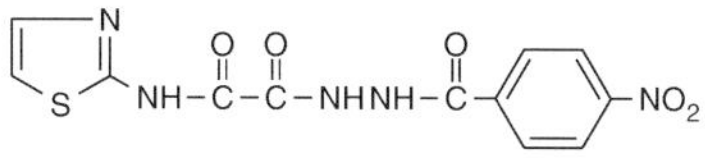
250 mg

S91,359-6
N-(5-NITRO-2-THIAZOLYL)-PHTHALAMIC ACID
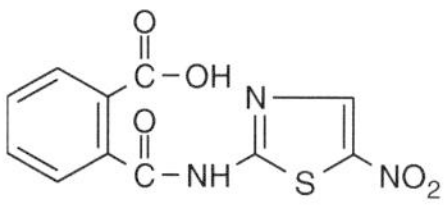
250 mg

S54,084-6
2-ACETAMIDO-5-NITROTHIAZOLE
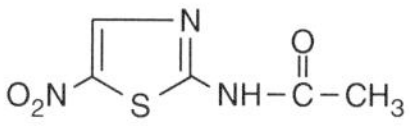
5 gm

S79,593-3
1-(3-ACETYLPHENYL)-3-(2-THIAZOLYL)UREA
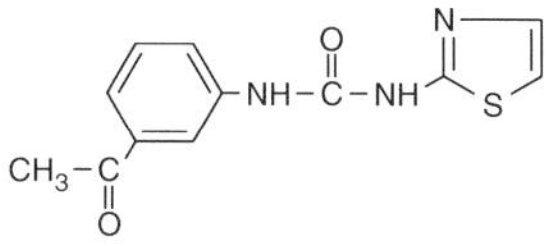
250 mg

S43,138-9
N-(2-THIAZOLYL)ACETOACETAMIDE
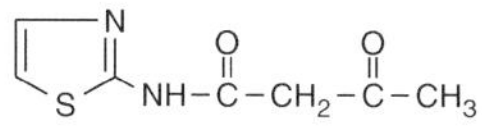
5 gm

S43,865-0
2-ACETOACETAMIDO-4-METHYLTHIAZOLE
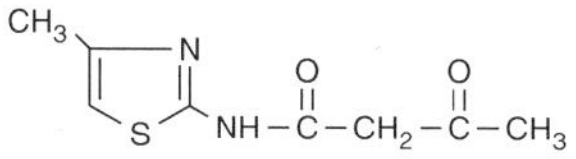
5 gm

S94,865-9
2-AMINO-5-ACETYL-4-METHYLTHIAZOLE HYDROCHLORIDE
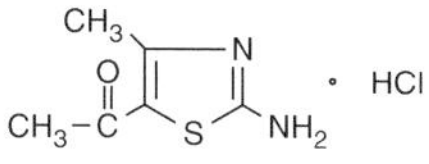
250 mg

S71,918-8
N-(2-THIAZOLYL)PHTHALAMIC ACID
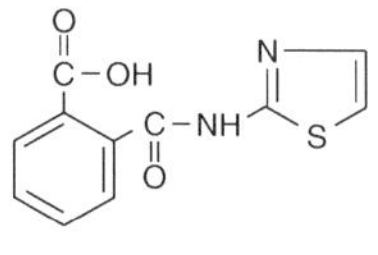
5 gm

S74,792-0
N,N'-BIS(2-THIAZOLYL)ISOPHTHALAMIDE
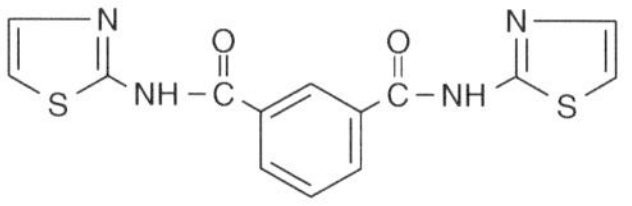
5 gm

S77,925-3
ETHYL 2-(2-BENZAMIDO-4-THIAZOLYL)-4-THIAZOLECARBOXYLATE
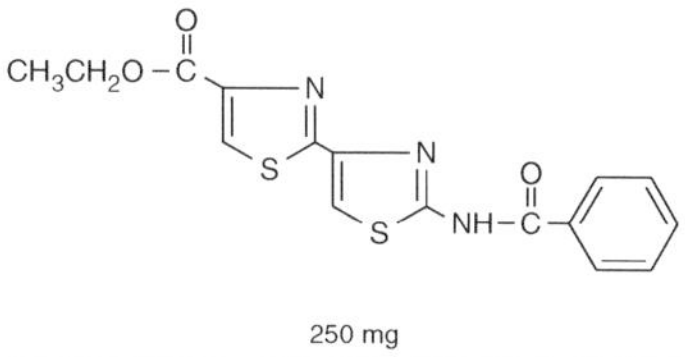
250 mg

S72,927-2
ETHYL 4-(2-THIAZOLYLUREIDO)BENZOATE
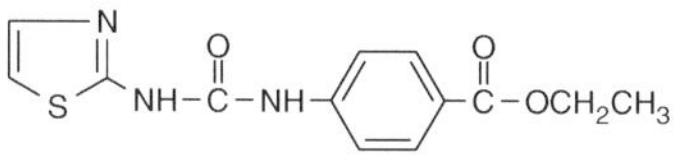
5 gm

S84,446-2
1-CINNAMOYL-3-(2-THIAZOLYL)-2-THIOUREA
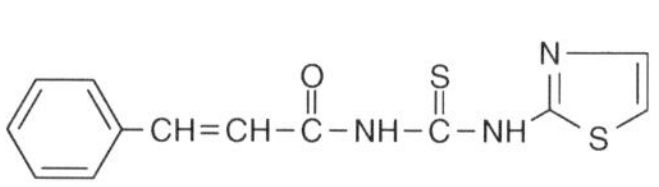
250 mg

S90,999-8
3-METHYL-N-(2-THIAZOLYL)BENZAMIDE
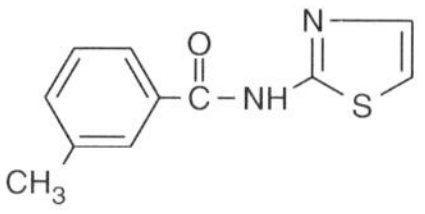
250 mg

S55,521-5
4-CHLORO-N-(2-THIAZOLYL)-BENZAMIDE
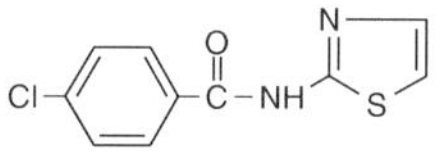
250 mg

S72,072-0
N-(2-THIAZOLYL)GLUTARAMIC ACID
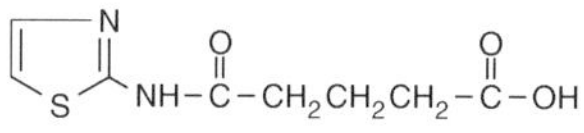
5 gm

S74,654-1
N-(2-THIAZOLYL)-1,2,3,6-TETRAHYDROPHTHALAMIC ACID
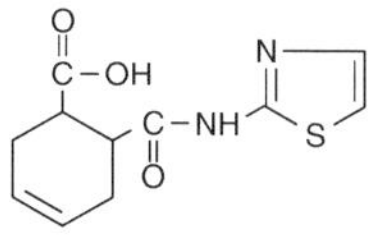
5 gm

S72,179-4
CIS-HEXAHYDRO-N-(2-THIAZOLYL)-PHTHALAMIC ACID
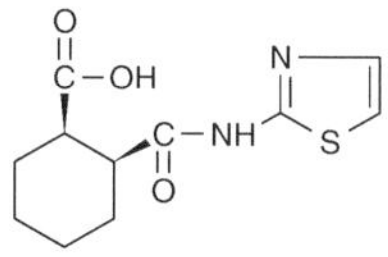
5 gm

S36,484-3
N1-(2-THIAZOYL)-N4-SUCCINOYL-SULFANILAMIDE MONOHYDRATE
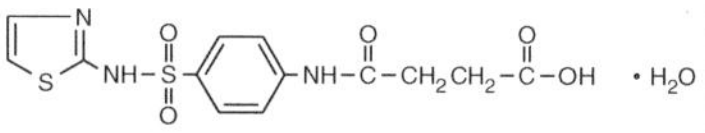
5 gm

S92,209-9
N-(2-THIAZOLYL)SUCCINAMIC ACID
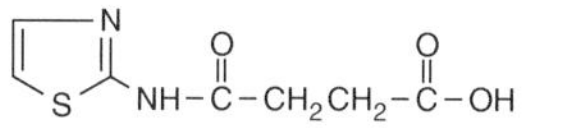
5 gm

S38,317-1
ETHYL 2-(2-AMINO-4-THIAZOLYL)-OCTADECANOATE
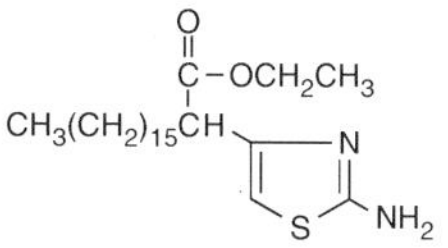
250 mg

S41,396-8
2-ACETAMIDOTHIAZOLE
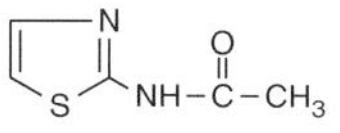
5 gm

S38,828-9
2-ACETAMIDO-4-PHENYLTHIAZOLE
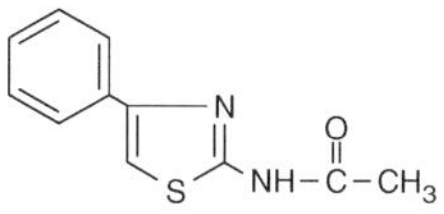
250 mg

S76,243-1
ACETIC N2-(4-METHYL-2-THIAZOLYL)-HYDRAZIDE HYDROCHLORIDE
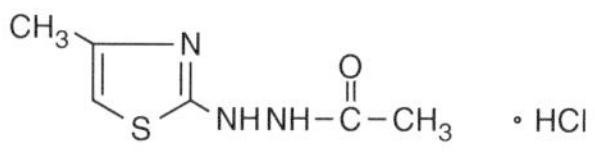
250 mg

S89,929-1
1-CYCLOHEXYL-3-(2-THIAZOLYL)UREA
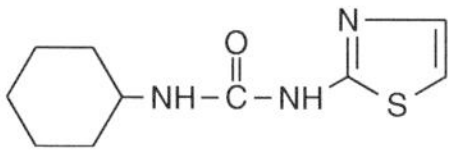
250 mg

S91,219-0
1-(2,6-DIETHYLPHENYL)-3-(2-THIAZOLYL)UREA
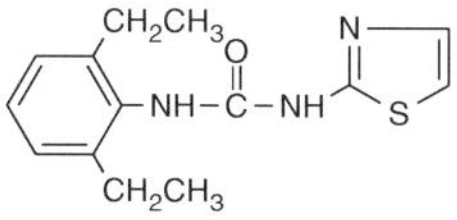
250 mg

S90,190-3
1-(2-THIAZOLYL)-3-(3,4-XYLYL)UREA
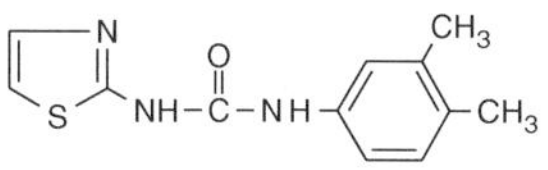
250 mg

S76,913-4
1-(2-THIAZOLYL)-3-(P-TOLYLSULFONYL)UREA
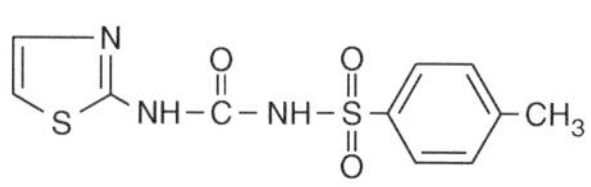
250 mg

S69,394-4
4,4'-METHYLENEBIS[1-PHENYL-3-(2-THIAZOLYL)UREA]

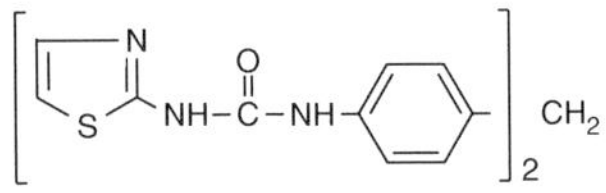

5 gm

S59,516-0
2-MERCAPTOTHIAZOLE

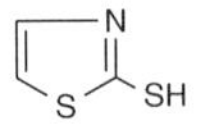

5 gm

S44,523-1
2,4-DIMETHYLTHIAZOLE

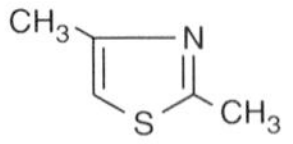

5 gm

S62,304-0
4-PHENYL-5-TETRADECYLTHIAZOLE

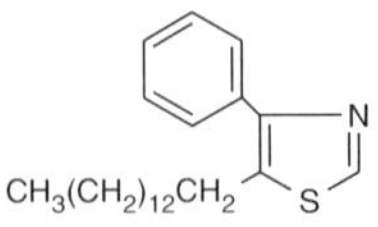

250 mg

S81,912-3
4-PHENYLTHIAZOLE

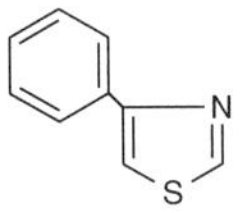

250 mg

S92,726-0
2,4,5-TRIPHENYLTHIAZOLE

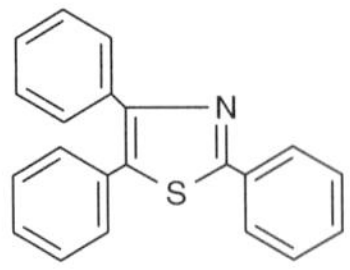

250 mg

S90,422-8
1-(3-CHLOROPHENYL)-3-(5-METHYL-1,3,4-THIADIAZOL-2-YL)UREA

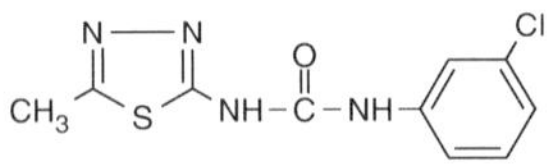

250 mg

S90,355-8
1-(2-FLUOROPHENYL)-3-(5-METHYL-1,3,4-THIADIAZOL-2-YL)UREA

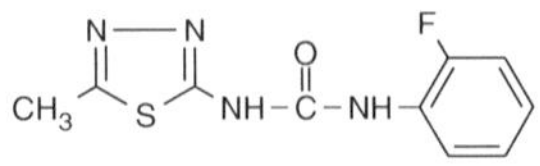

250 mg

S74,325-9
1-(3-CHLORO-4-FLUOROPHENYL)-3-(5-METHYL-1,3,4-THIADIAZOL-2-YL)-UREA

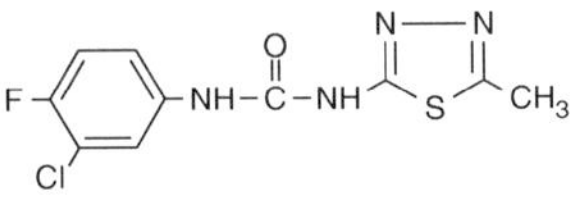

250 mg

S94,658-3
5-AMINO-1,3,4-THIADIAZOL-2-OL

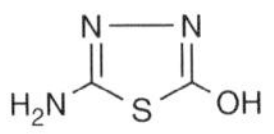

250 mg

S86,260-6
5-ANILINO-3-HYDROXY-1,2,4-THIADIAZOLE

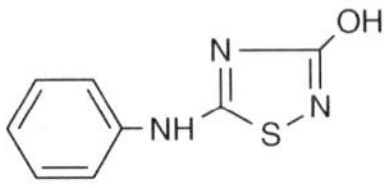

250 mg

S86,240-1
5-ANILINO-3-METHOXY-1,2,4-THIADIAZOLE

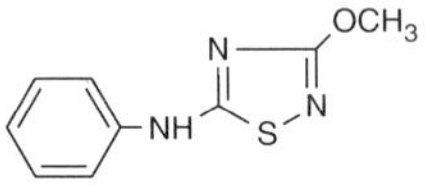

250 mg

S86,243-6
5-ANILINO-3-ETHOXY-1,2,4-THIADIAZOLE

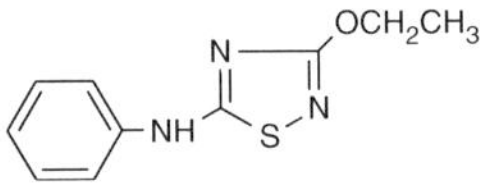

250 mg

S90,441-4
1-(2-METHOXY-5-METHYLPHENYL)-3-(1,3,4-THIADIAZOL-2-YL)UREA

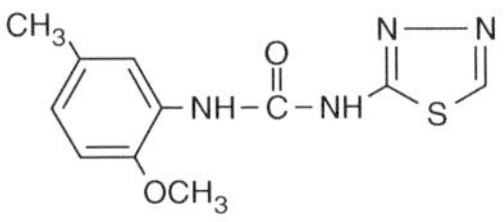

250 mg

S92,415-6
1-(2,5-DIMETHOXYPHENYL)-3-(1,3,4-THIADIAZOL-2-YL)UREA

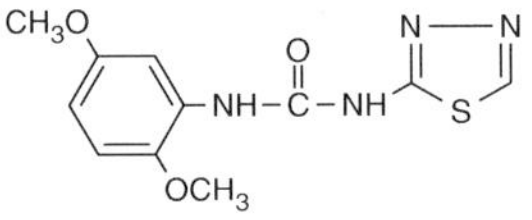

250 mg

S65,579-1
2-MERCAPTO-5-METHYLAMINO-1,3,4-THIADIAZOLE

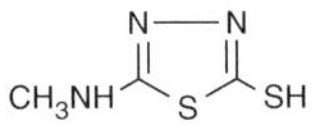

250 mg

S60,610-3
2-METHYLTHIO-5-(4-PENTADECYLPHENYL)-1,3,4-THIADIAZOLE

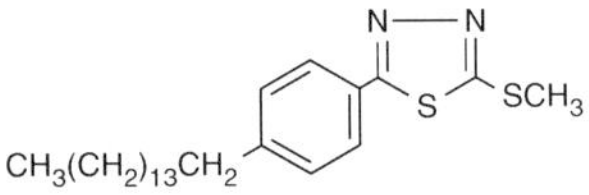

250 mg

S38,616-2
2-AMINO-5-HEXADECYLTHIO-1,3,4-THIADIAZOLE

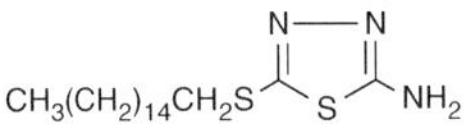

5 gm

S90,086-9
N-[5-(2,4-DINITROPHENYLTHIO)-1,3,4-THIADIAZOL-2-YL]ACETAMIDE

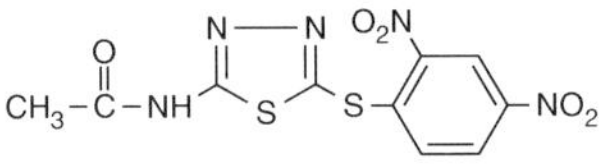

250 mg

S59,223-4
3-BENZYLTHIO-5-CHLORO-1,2,4-THIADIAZOLE

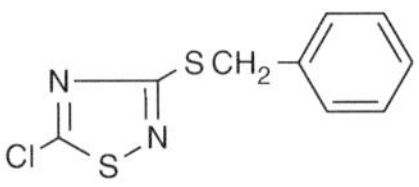

250 mg

S60,601-4
2-(4-HEXADECYLSULFONYLPHENYL)-5-METHYLTHIO-1,3,4-THIADIAZOLE

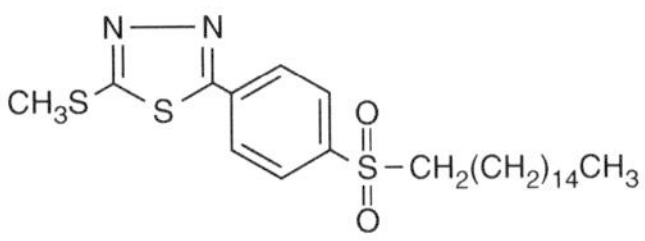

250 mg

S57,798-7
2-(METHYLTHIO)-5-(P-TOLYL)-1,3,4-THIADIAZOLE

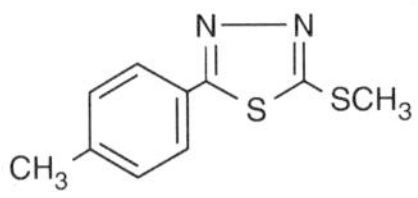

5 gm

S58,815-6
3-AMINO-1-(5-PHENYL-1,3,4-THIADIAZOL-2-YLMETHYL)-2-PYRAZOLIN-5-ONE

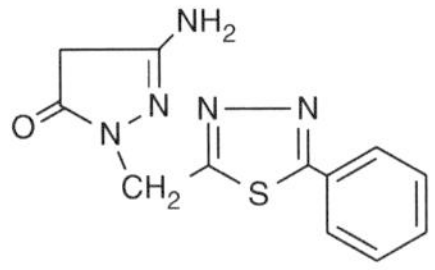

250 mg

S86,253-3
3-AMINO-5-(METHYLAMINO)-1,2,4-THIADIAZOLE

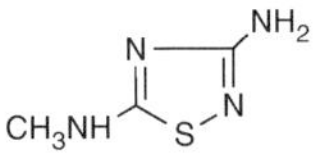

250 mg

S86,254-1
3-AMINO-5-(METHYLAMINO)-1,2,4-THIADIAZOLE P-TOLUENESULFONATE

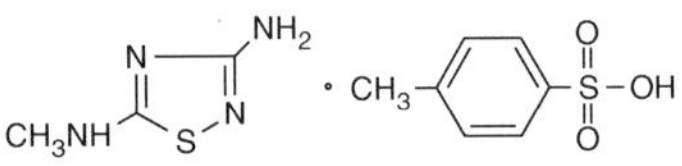

250 mg

S86,263-0
5-AMINO-3-ANILINO-1,2,4-THIADIAZOLE HYDROCHLORIDE

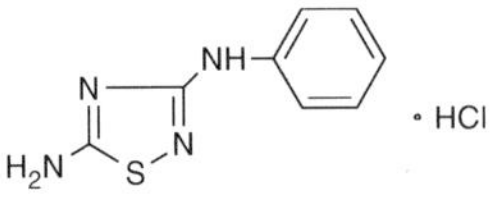

250 mg

S86,264-9
5-AMINO-3-ANILINO-1,2,4-THIADIAZOLE

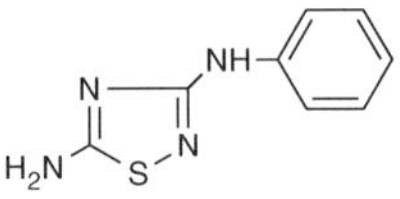

250 mg

S86,265-7
3-AMINO-5-ANILINO-1,2,4-THIADIAZOLE

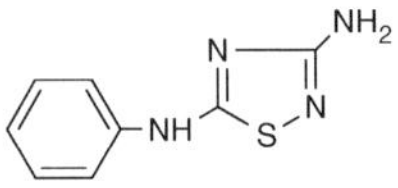

250 mg

S86,311-4
5-AMINO-3-(BENZYLPHENYLAMINO)-1,2,4-THIADIAZOLE

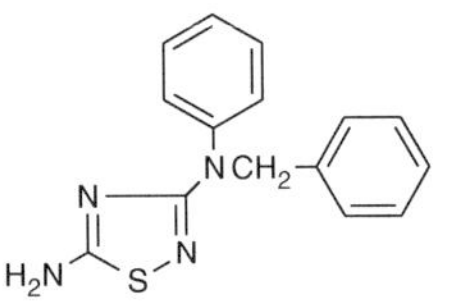

250 mg

S86,249-5
3,5-DIAMINO-1,2,4-THIADIAZOLE P-TOLUENESULFONATE

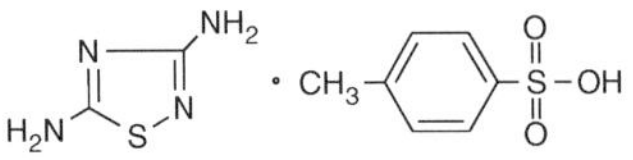

250 mg

S94,678-8
3,5-DIAMINO-1,2,4-THIADIAZOLE

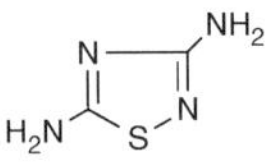

250 mg

S86,236-3
3-AMINO-5-PHENYL-1,2,4-THIADIAZOLE

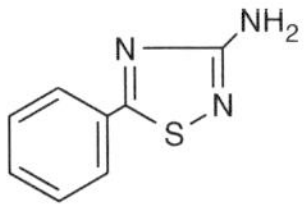

250 mg

S90,819-3
2-AMINO-5-(BUTYLTHIO)-1,3,4-THIADIAZOLE

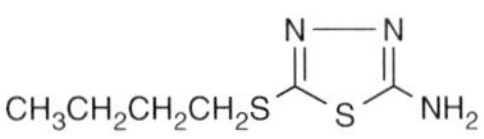

250 mg

S43,578-3
2-AMINO-5-METHYL-1,3,4-THIADIAZOLE HYDROCHLORIDE

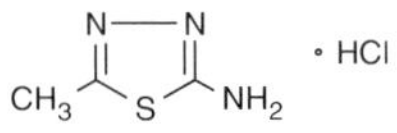

5 gm

S53,832-9
5-(3-CHLORO-O-TOLUIDINO)-1,3,4-THIADIAZOLE-2(3H)-THIONE

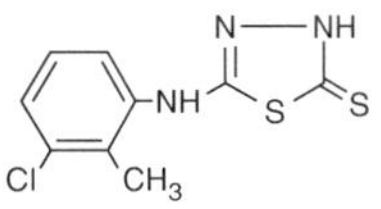

5 gm

S86,291-6
1-PHENYL-3-(5-PHENYL-1,2,4-THIADIAZOL-3-YL)-2-THIOUREA

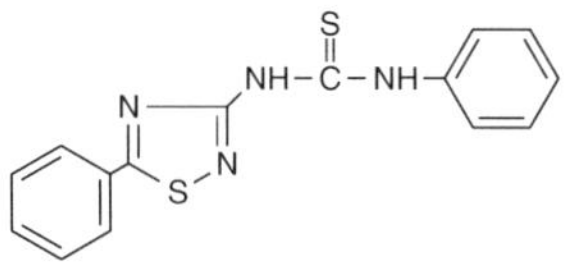

250 mg

S52,508-1
5-(PARA-TOLUIDINO)-1,3,4-THIADIAZOLE-2(3H)-THIONE

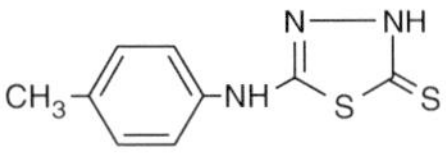

5 gm

S54,592-9
5-CYCLOHEXYLAMINO-1,3,4-THIADIAZOLE-2(3H)-THIONE

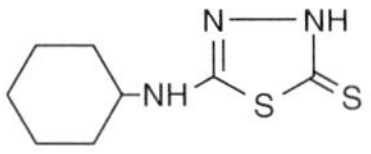

250 mg

S86,295-9
1-(5-ANILINO-1,2,4-THIADIAZOL-3-YL)-3-PHENYL-2-THIOUREA

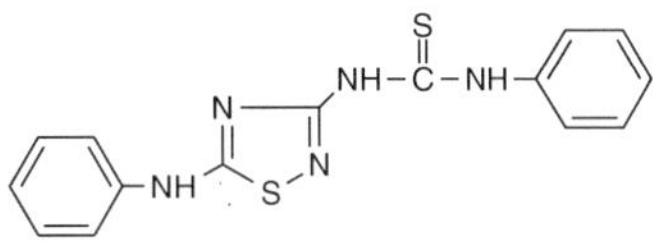

250 mg

S54,511-2
5-(ORTHO-TOLUIDINO)-1,3,4-THIADIAZOLE-2(3H)-THIONE

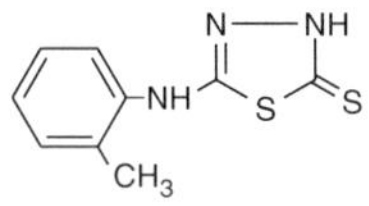

250 mg

S86,244-4
5-ANILINO-3-PHENYL-1,2,4-THIADIAZOLE

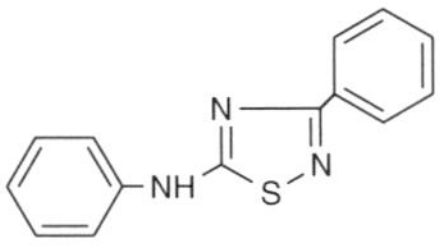

250 mg

S95,305-9
2-(METHYLSULFONYL)-5-PHENYL-1,3,4-THIADIAZOLE

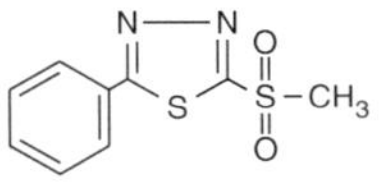

50 mg

S59,835-6
5-NITROSOAMINO-3-PHENYL-1,2,4-THIADIAZOLE

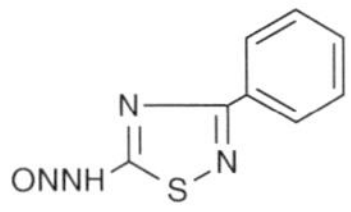

5 gm

S86,310-6
1-BENZOYL-3-(5-PHENYL-1,2,4-THIADIAZOL-3-YL)-2-THIOUREA

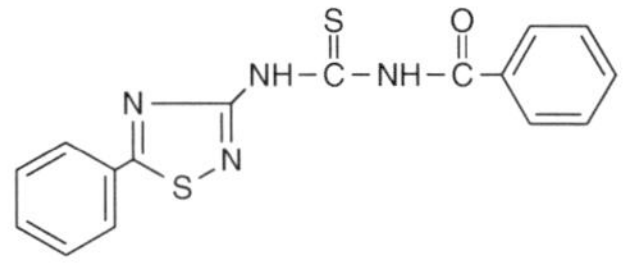

250 mg

S90,537-2
1,4,5,6,7,7-HEXACHLORO-N-(2-THIAZOLYL)-5-NORBORNENE-2,3-DICARBOXIMIDE

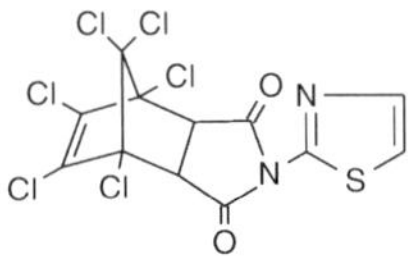

250 mg

S50,406-8
2,5-DIACETAMIDO-1,3,4-THIADIAZOLE

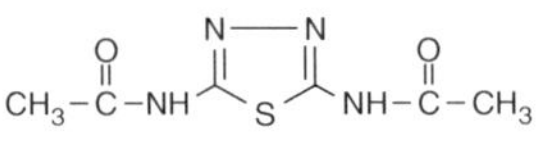

5 gm

S86,231-2
3,5-DIACETAMIDO-1,2,4-THIADIAZOLE

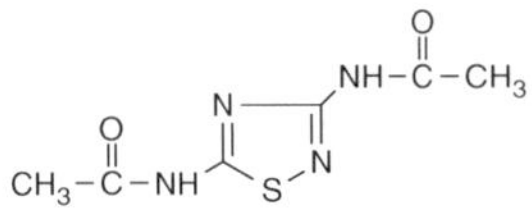

250 mg

S90,492-9
N-(1,3,4-THIADIAZOL-2-YL)-1-CYCLOHEXANECARBOXAMIDE

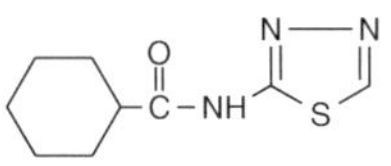

250 mg

S86,289-4
1-(5-PHENYL-1,2,4-THIADIAZOL-3-YL)-2-THIOUREA

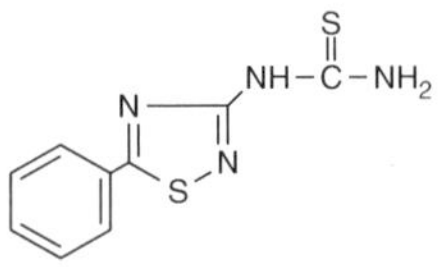

250 mg

S90,360-4
1-METHYL-3-(5-METHYL-1,3,4-THIADIAZOL-2-YL)UREA

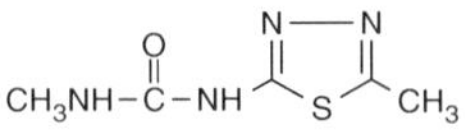

250 mg

S37,555-1
2-BUTYL-5-(PARA-TOLUENESULFONAMIDO)-1,3,4-THIADIAZOLE

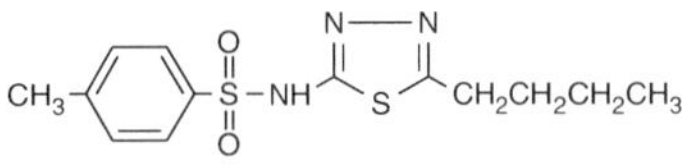

5 gm

S37,557-8
2-ISOBUTYL-5-(PARA-TOLUENESULFONAMIDO)-1,3,4-THIADIAZOLE

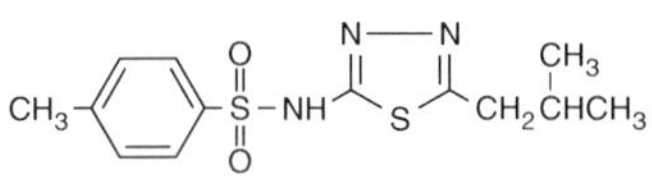

250 mg

S94,003-8
6-ETHYL-2-METHYLTHIAZOLO(5,4-D)-THIAZOLE-5(6H)-THIONE

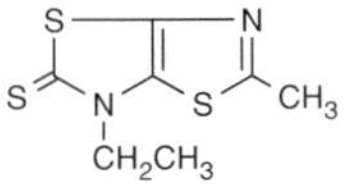

250 mg

S72,167-0
5,6-DIHYDRO-3-PHENYLIMIDAZO(2,1-B)-THIAZOLE

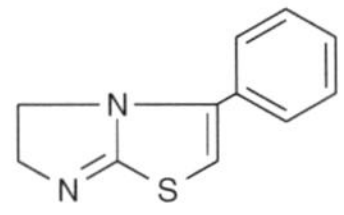

5 gm

S87,220-2
2,2'-BIS(THIAZOLE) DIHYDROCHLORIDE

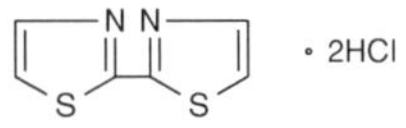

250 mg

S89,802-3
6,7-DIHYDRO-3-PHENYL-5H-THIAZOLO-(3,2-A)PYRIMIDINE HYDROBROMIDE

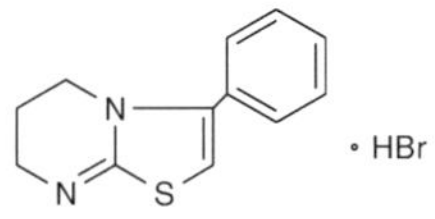

5 gm

S46,106-7
2,5-DIMETHYL-1,3,4-THIADIAZOLE

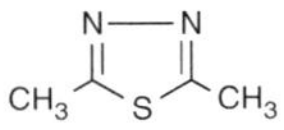

5 gm

S54,152-4
4-PHENYL-1,2,3-THIADIAZOLE

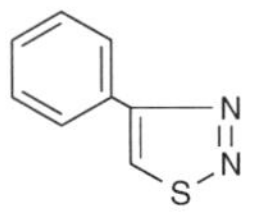

250 mg

S89,904-6
5,6-DIHYDRO-3-(3,4-DIHYDROXYPHENYL)-IMIDAZO(2,1-B)THIAZOLE HYDROCHLORIDE

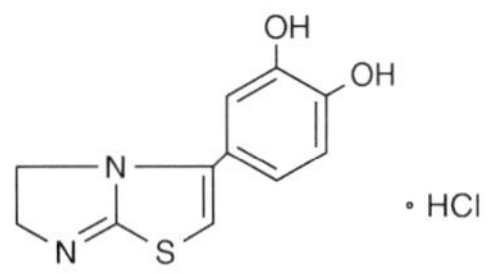

250 mg

S94,877-2
5-AMINO-1,2,3,4-THIATRIAZOLE

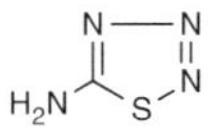

250 mg

S92,343-5
4,4'-BIS(4-FLUOROPHENYL)-TETRATHIAFULVALENE

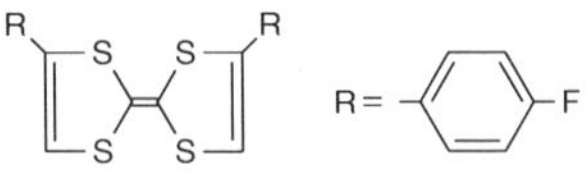

250 mg

S92,352-4
4,4'-BIS(4-CHLOROPHENYL)-TETRATHIAFULVALENE

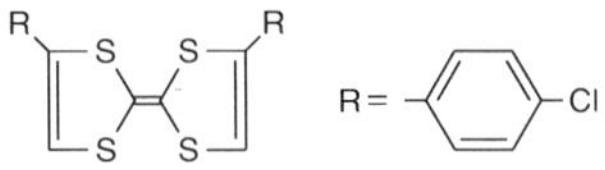

250 mg

S92,601-9
4,4'-BIS(4-BROMOPHENYL)-TETRATHIAFULVALENE

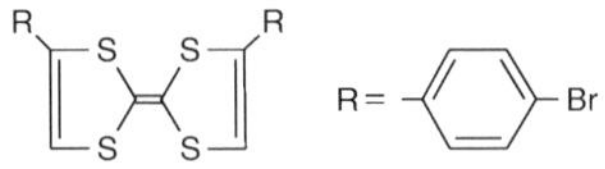

250 mg

S92,353-2
4,4'-BIS(4-IODOMETHYL)-TETRATHIAFULVALENE

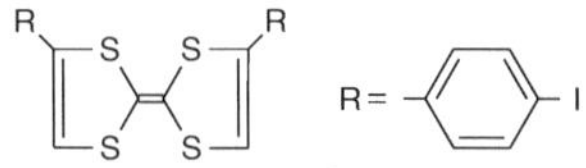

250 mg

S92,359-1
4,4'-BIS(4-CHLOROPHENYL)-5,5'-DIMETHYLTETRATHIAFULVALENE

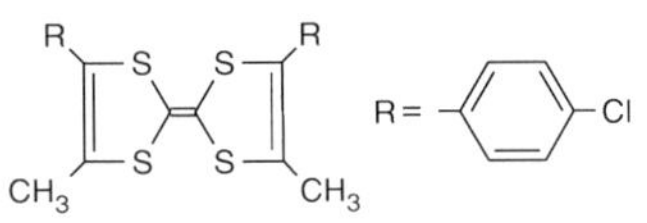

250 mg

S92,603-5
4,4'-BIS(2,5-DIBUTOXY-4-ETHYLPHENYL)TETRATHIAFULVALENE

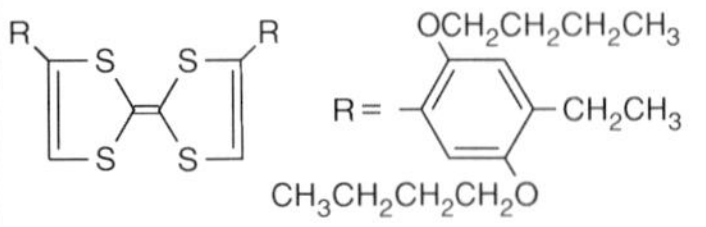

250 mg

S92,589-6
4,4'-BIS(4-METHOXYPHENYL)-TETRATHIAFULVALENE

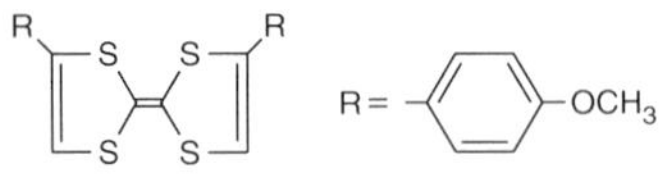

250 mg

S92,572-1
4,4'-BIS(4-METHOXYPHENYL)-5,5'-DIMETHYLTETRATHIAFULVALENE

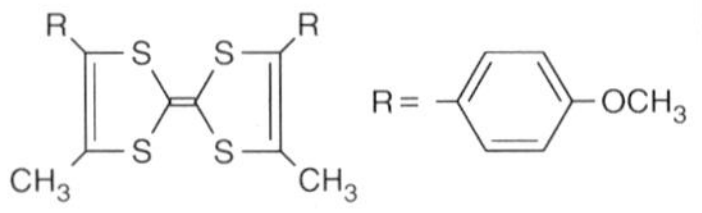

250 mg

S92,345-1
4,4'-BIS(4-ACETAMIDOPHENYL)-TETRATHIAFULVALENE

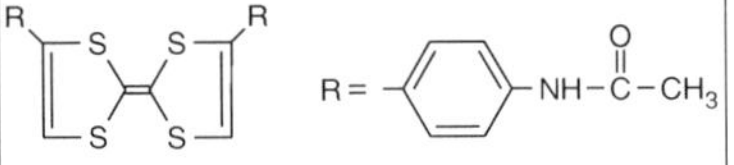

250 mg

S92,253-6
4',4,5',5-TETRACYANOTETRATHIAFULVALENE

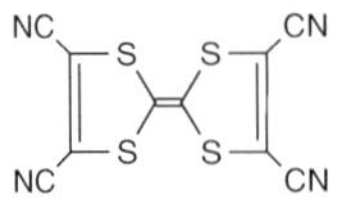

250 mg

S92,604-3
4,4',5,5'-BIS(PENTAMETHYLENE)-TETRATHIAFULVALENE

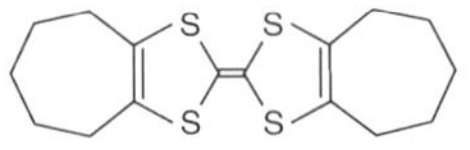

250 mg

S92,578-0
4,4',5,5'-TETRAPHENYLTETRATHIAFULVALENE

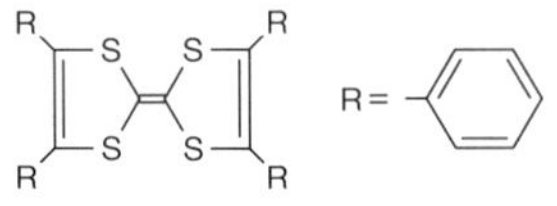

250 mg

S92,562-4
4,4'-BIS(BIPHENYLYL)-TETRATHIAFULVALENE

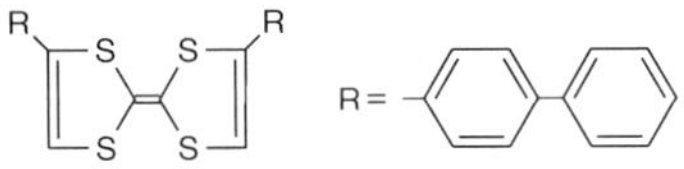

250 mg

S92,243-9
4,4'-(4-BIPHENYLYL)-5,5'-DIMETHYL-TETRATHIAFULVALENE

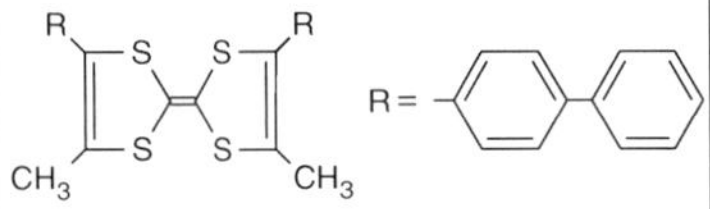

250 mg

S92,571-3
4,4'-BIS(P-TOLYL)TETRATHIAFULVALENE

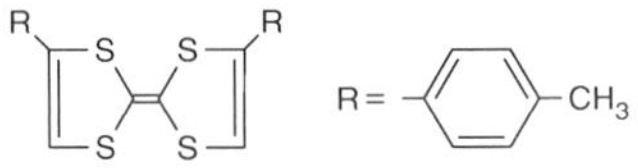

250 mg

S92,361-3
4,4'-BIS[4-(METHYLTHIO)PHENYL]-TETRATHIAFULVALENE

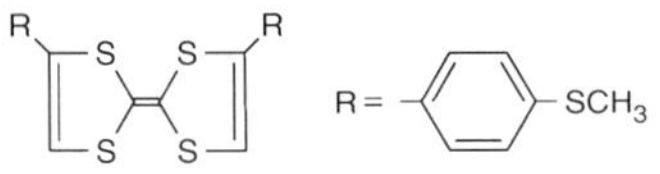

250 mg

S92,574-8
4,4'-BIS(P-TOLYL)-5,5'-DIMETHYLTETRATHIAFULVALENE

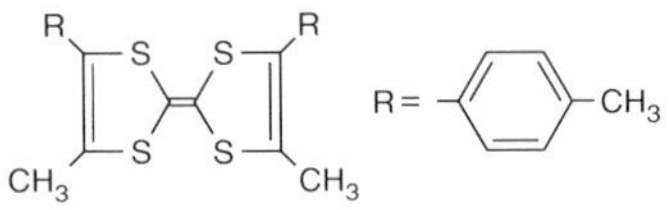

250 mg

S93,429-1
4,4'-BIS(2-NAPHTHYL)-5,5'-DIMETHYLTETRATHIAFULVALENE

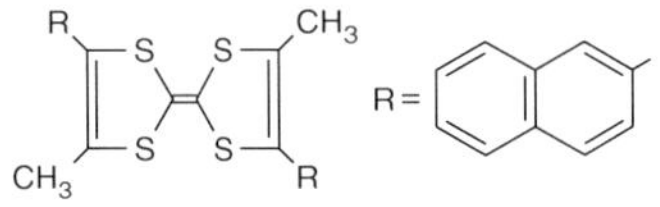

250 mg

S77,935-0
1-PHENYL-1-(5-MERCAPTO-1,3,4-THIADIAZOL-2-YL)-3-[3-(TRIFLUORO-METHYL)PHENYL]UREA

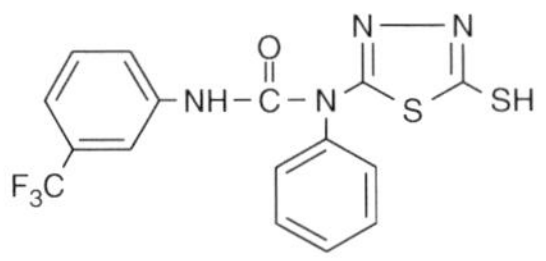

250 mg

S79,605-0
1-(5-BROMO-1,3,4-THIADIAZOL-2-YL)-3-[3-(TRIFLUOROMETHYL)PHENYL]UREA

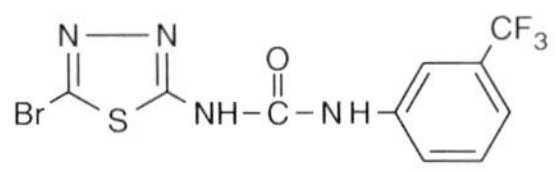

250 mg

S87,954-1
2-(1,3-DIMETHYLUREIDO)-5-(3-FLUOROBENZYLTHIO)-1,3,4-THIADIAZOLE

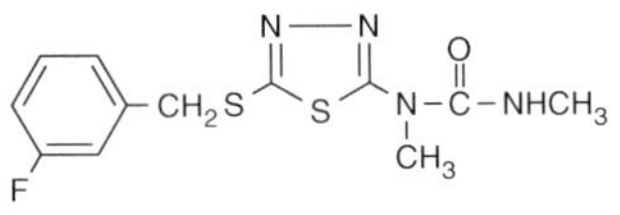

5 gm

S76,637-2
2-AMINO-5-(4-CHLOROBENZYL)-1,3,4-THIADIAZOLE

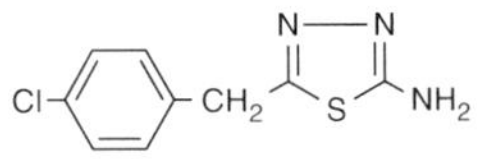

250 mg

S90,458-9
4-CHLORO-N-(5-METHYL-1,3,4-THIADIAZOL-2-YL)BENZAMIDE

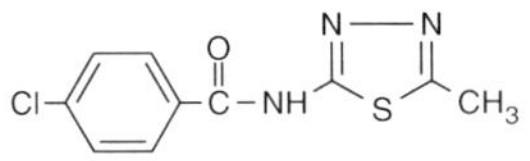

250 mg

S90,464-3
3,4-DICHLORO-N-(5-METHYL-1,3,4-THIADIAZOL-2-YL)BENZAMIDE

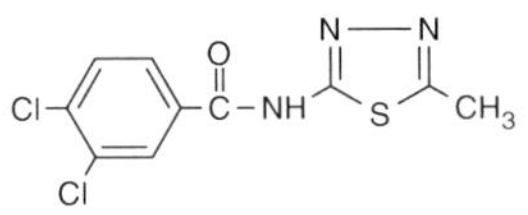

250 mg

S45,080-4
N-ALPHA-CHLOROACETYL-DL-TRYPTOPHAN

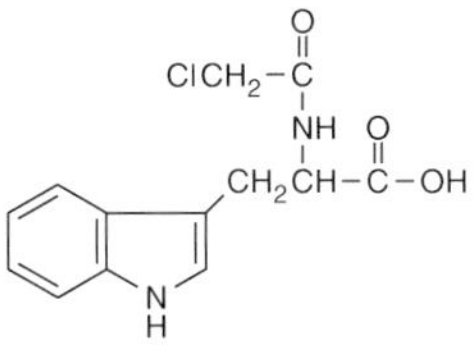

5 gm

S79,016-8
N-ALPHA-CHLOROACETYL-L-TRYPTOPHAN

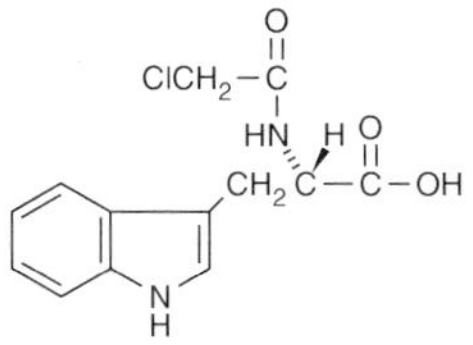

250 mg

S34,423-0
2-[4-N(2-CHLOROETHYL)METHYLAMINO]-STYRYL-1,3,3-TRIMETHYL-3H-PSEUDOINDOLIUM CHLORIDE

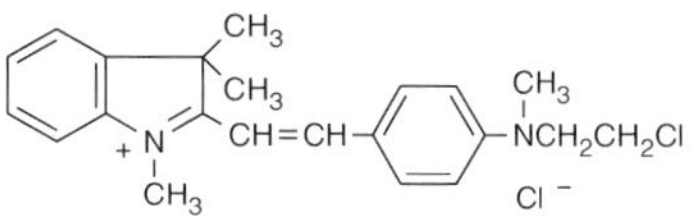

5 gm

S88,293-3
METHYL 3,5-DIHYDROXY-7-[3-(4-FLUOROPHENYL)-1-ISOPROPYL-2-INDOLYL]-6-HEPTENOATE

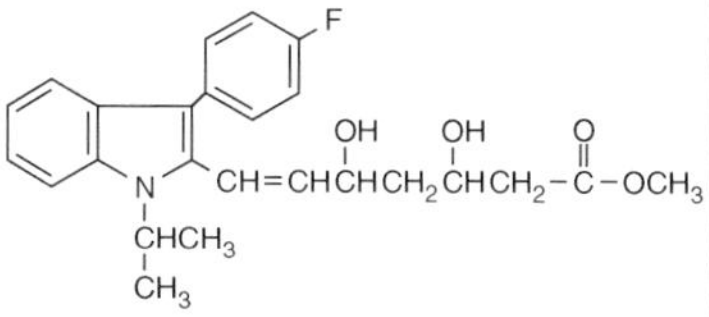

250 mg

S75,892-2
3-(4-CHLOROPHENYLTHIO)-1-METHYLINDOLE

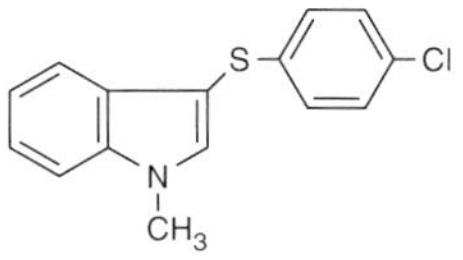

250 mg

S91,197-6
1-[N-(4-FLUORO-3-NITROPHENYL)-CARBAMOYL]INDOLE

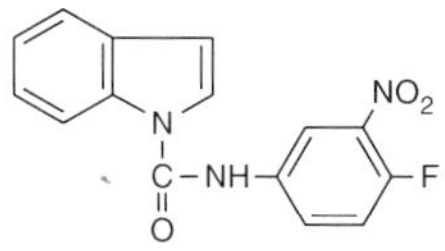

250 mg

S74,565-0
5-CHLORO-3-(4-CHLOROPHENYL)-1-ETHYLINDOLE

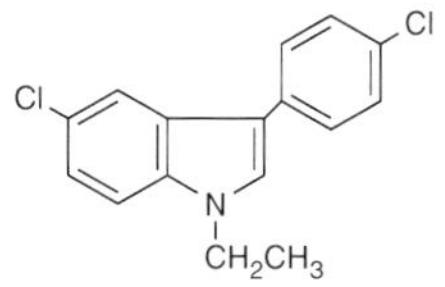

250 mg

S80,664-1
3-(4-CHLOROBENZOYL)-2-METHYLINDOLE

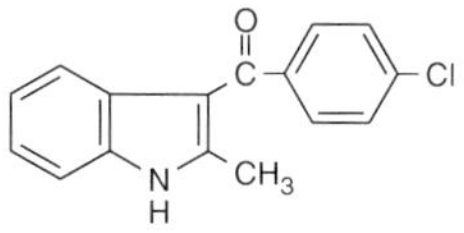

250 mg

S77,936-9
2-(1-INDOLYLCARBONYL)-3,4,5,6-TETRACHLOROBENZOIC ACID

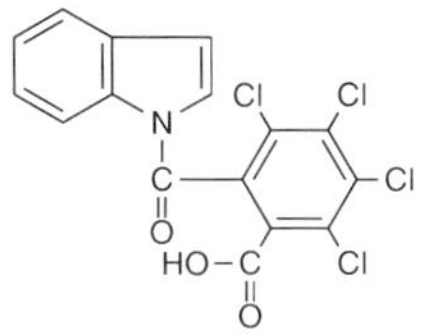

250 mg

S72,704-0
2-(4-CHLOROPHENYL)-1-ETHYLINDOLE

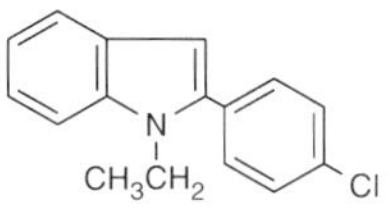

5 gm

S72,726-1
3-(4-CHLOROPHENYL)-1-ETHYL-5-METHYLINDOLE

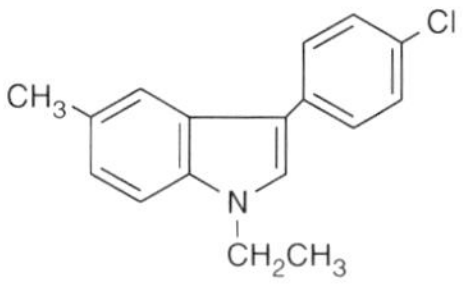

5 gm

S72,725-3
1-TERT-BUTYL-3-(4-CHLOROPHENYL)-INDOLE

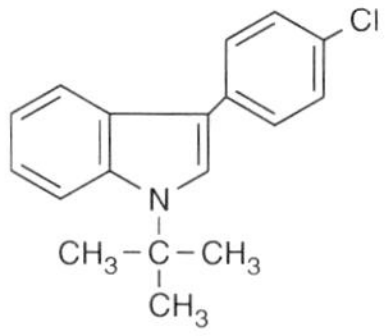

5 gm

S72,745-8
1-TERT-BUTYL-2-(4-CHLOROPHENYL)-INDOLE

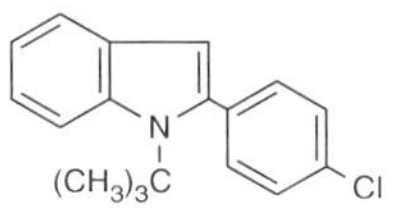

5 gm

S89,602-0
METHYL 3-(3-HYDROXY-3-METHYLBUTOXY)-2-INDOLECARBOXYLATE

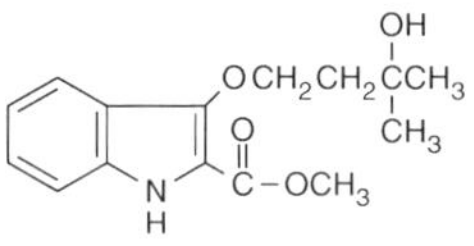

250 mg

S76,026-9
ETHYL 2,3-DIHYDRO-ALPHA-HYDROXY-2-OXO-3-INDOLEACETATE

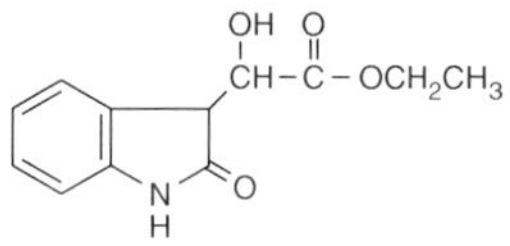

250 mg

S95,552-3
METHYL 1-ALLYL-3-HYDROXY-2-INDOLECARBOXYLATE

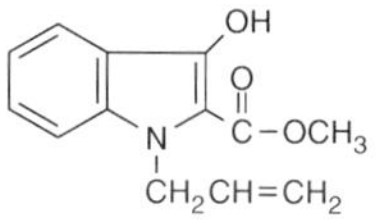

50 mg

S95,553-1
METHYL 3-HYDROXY-2-(METHOXYCARBONYL)-1-INDOLEACETATE

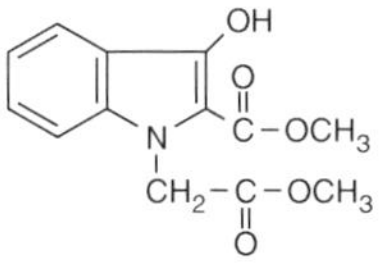

50 mg

S87,975-4
N,N-DIMETHYL-2-(3-ETHYL-5-METHYL-4-ISOXAZOLYL)-BETA-HYDROXYTRYPTAMINE

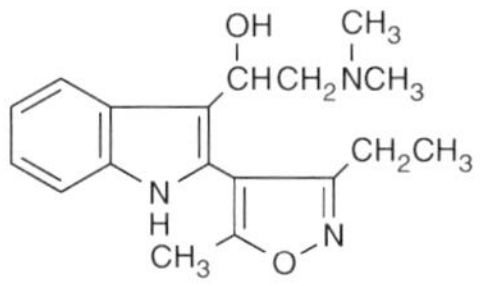

5 gm

S60,331-7
2,4-DI(3-INDOLYL)-1-BUTANOL

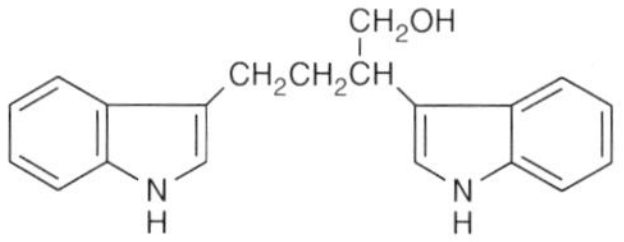

5 gm

S87,322-5
2-(4-HYDROXYSTYRYL)-1,3,3-TRIMETHYL-3H-INDOLIUM IODIDE

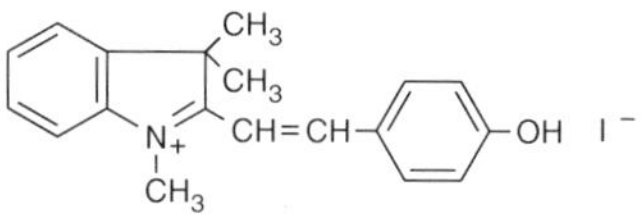

250 mg

S75,924-4
4,5-DIHYDRO-2-(2-HYDROXYBUTYL)-6-(3-INDOLYL)-3(2H)-PYRIDAZINONE

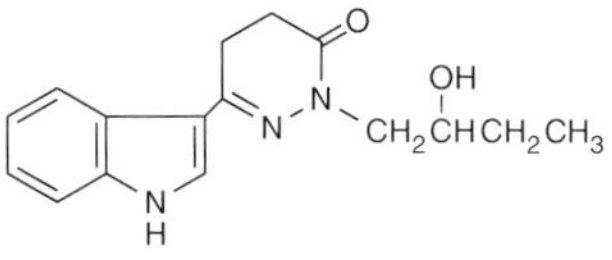

250 mg

S89,529-6
ETHYL 5-HYDROXY-3-(2-OXOCYCLOHEXYLMETHYL)-2-INDOLECARBOXYLATE

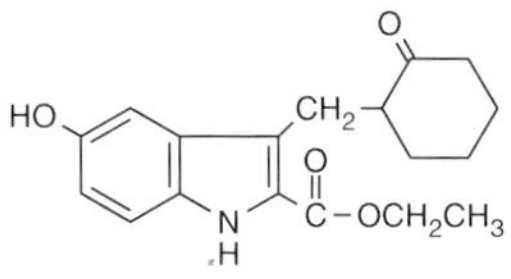

250 mg

S78,626-8
CARBOBENZYLOXYTRYPTOPHYLSERINE METHYL ESTER

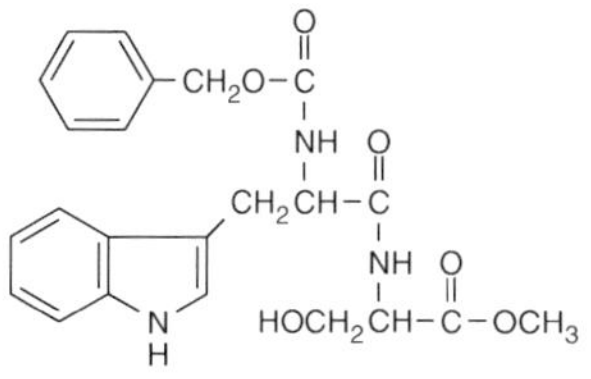

250 mg

S78,780-9
CARBOBENZYLOXYTRYPTOPHYLTYROSINE METHYL ESTER

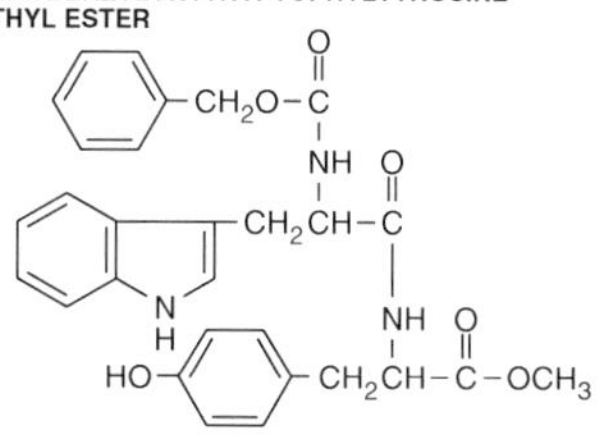

250 mg

S78,784-1
CARBOBENZYLOXYTRYPTOPHYLTYROSINE BENZYL ESTER

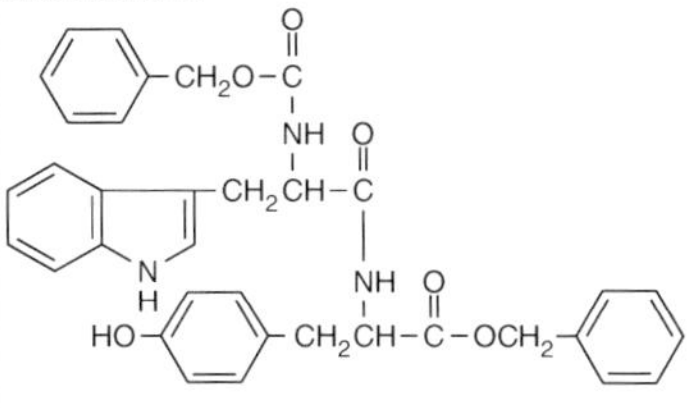

250 mg

S78,499-0
CARBOBENZYLOXY-L-TYROSYL-L-TRYPTOPHANAMIDE

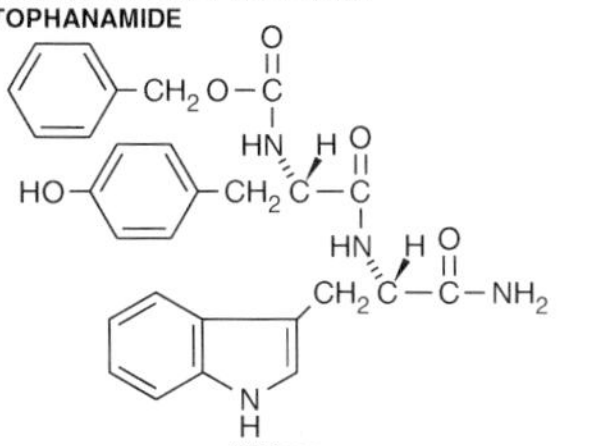

250 mg

S41,710-6
5-BENZYLOXY-N,N-DIMETHYLTRYPTAMINE OXALATE

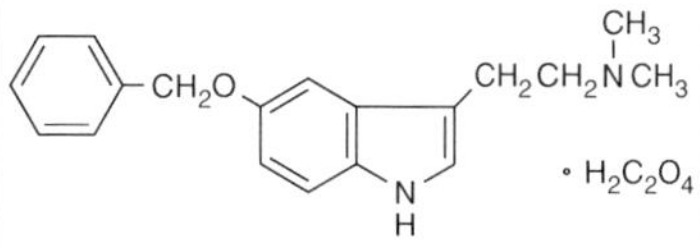

250 mg

S40,671-6
5-BENZYLOXY-N,N-DIMETHYL-3-INDOLEGLYOXYLAMIDE

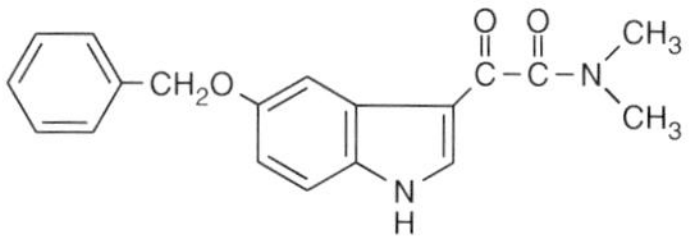

5 gm

S35,851-7
5-BENZYLOXYINDOLE-3-ACETAMIDE

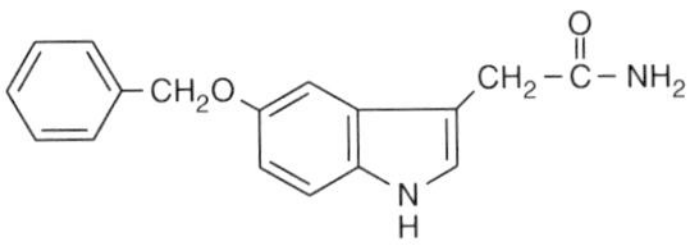

250 mg

S35,852-5
5-BENZYLOXYINDOLE-3-ACETONITRILE

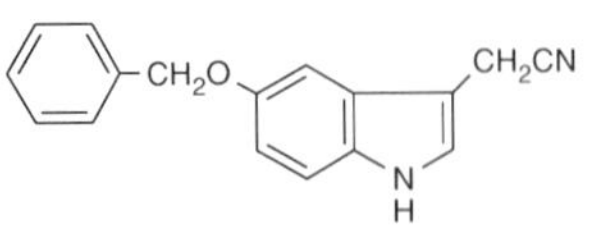

250 mg

S36,342-1
N-METHYL-5-BENZYLOXYINDOLE

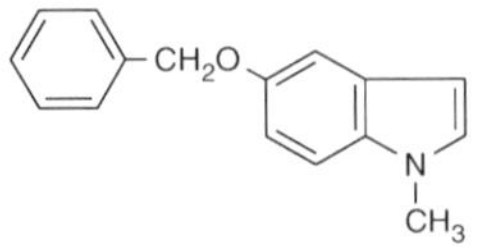

250 mg

S36,151-8
ETHYL 5-BENZYLOXYINDOLE-2-CARBOXYLATE

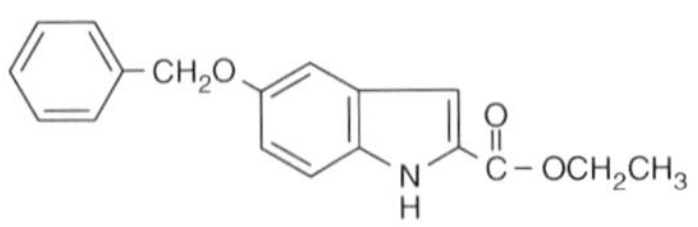

250 mg

S45,452-4
5-BENZYLOXYINDOLE-3-ACETIC ACID

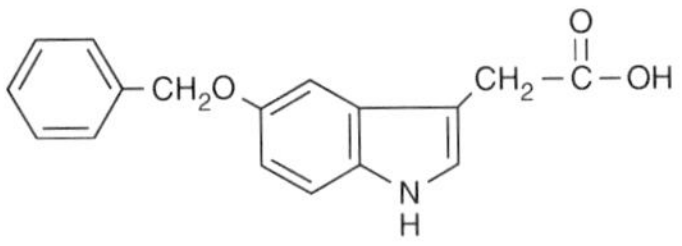

250 mg

S75,679-2
3-(2-AMINOETHYL)-2-(3,4-DIMETHOXYPHENYL)INDOLE

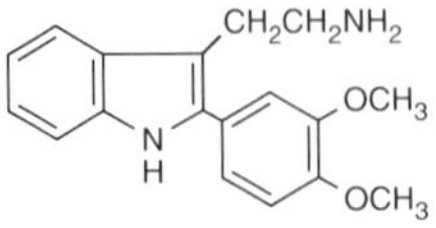

5 gm

S95,562-0
3-(CYANOMETHYL)-2-(3,4-DIMETHOXYPHENYL)-3-ACETONITRILE

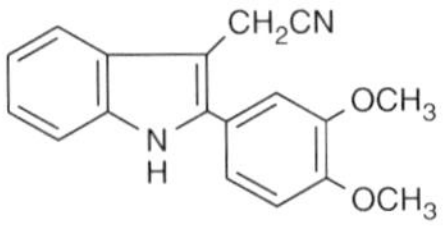

50 mg

S91,738-9
1-[N-(2,5-DIMETHOXYPHENYL)-CARBAMOYL]INDOLE

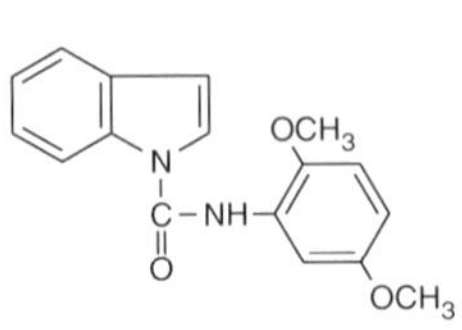

250 mg

S89,610-1
ETHYL 3-(2,2-DIETHOXYETHOXY)-2-INDOLECARBOXYLATE

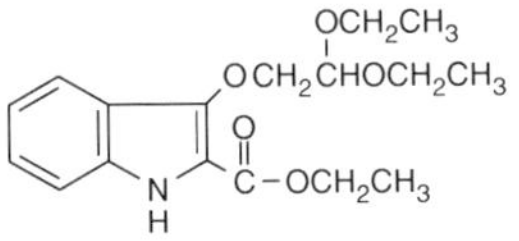

250 mg

S89,579-2
ETHYL 3-(FORMYLMETHOXY)-2-INDOLECARBOXYLATE

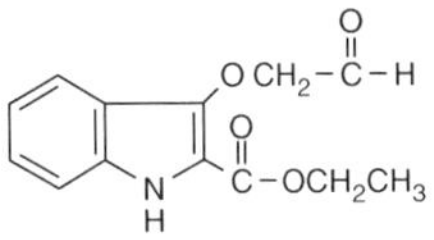

250 mg

S93,999-4
4-GLYCIDYLOXY-2-INDOLECARBOXAMIDE, COMPOUND WITH 1,4-DIOXANE

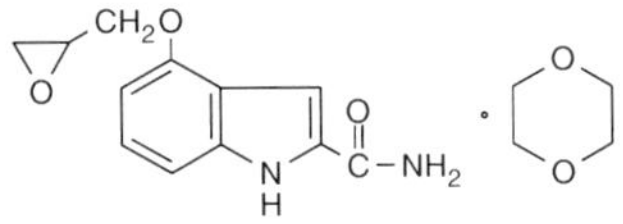

5 gm

S89,488-5
DIMETHYL 2-(2-ETHOXY-3-INDOLYL)-MALEATE

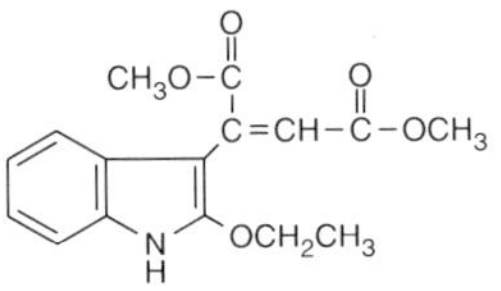

250 mg

S95,560-4
2-(4-METHOXYPHENYL)TRYPTAMINE

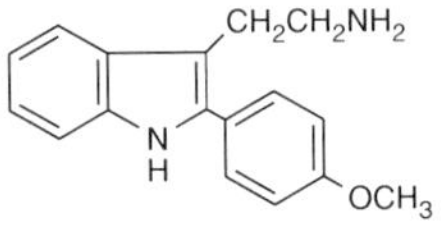

50 mg

S95,561-2
2-(4-METHOXYPHENYL)TRYPTAMINE HYDROCHLORIDE

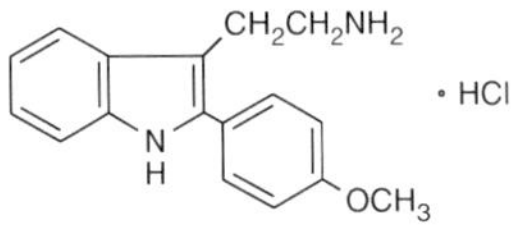

50 mg

S42,488-9
HARMALINE METHOSULFATE

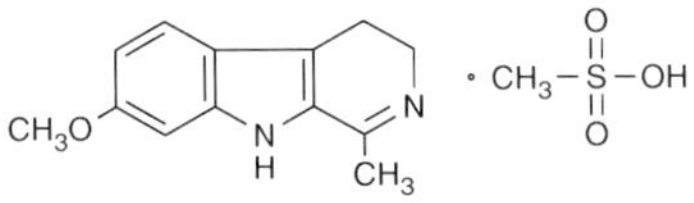

250 mg

S40,997-9
2-PROPYL-1,2,3,4-TETRAHYDROHARMINE HYDROCHLORIDE

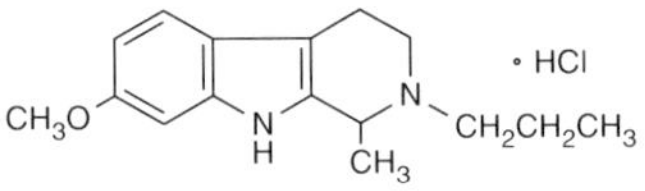

250 mg

S41,772-6
6-METHOXY-1-TETRAHYDRONORHARMANONE

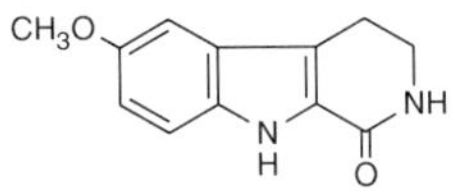

250 mg

S40,672-4
5-BENZYLOXY-N-ETHYL-N-METHYLTRYPTAMINE OXALATE

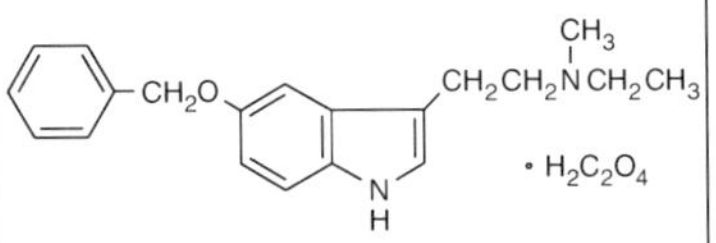

5 gm

S37,180-7
5-METHOXY-3-INDOLEACETAMIDE

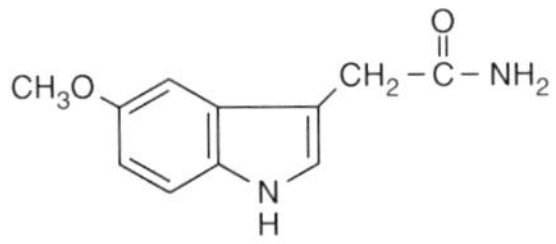

250 mg

S56,755-8
5-BENZYLOXY-3-INDOLECARBOXALDEHYDE

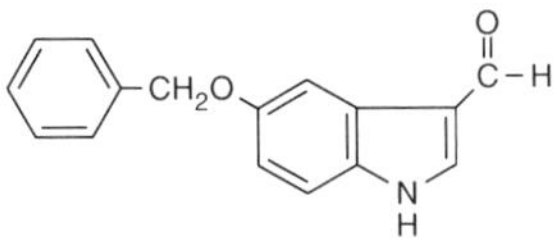

5 gm

S34,192-4
5-BENZYLOXY-N,N-DIBENZYLTRYPTAMINE HYDROCHLORIDE

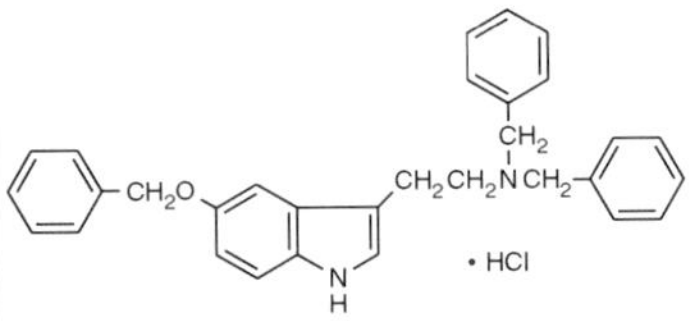

250 mg

S87,246-6
1-ETHOXY-2-ETHYLISOINDOLIUM TETRAFLUOROBORATE

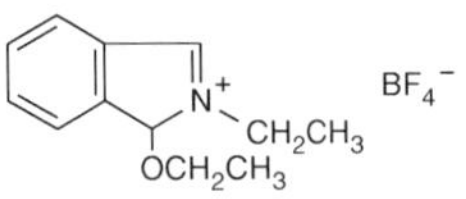

250 mg

S95,558-2
3-(CYANOMETHYL)-2-(4-METHOXYPHENYL)-INDOLE

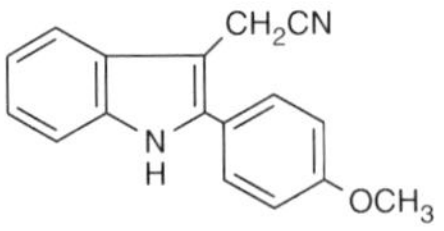

50 mg

S89,572-5
2-(3-INDOLYLOXY)ACETALDEHYDE SEMICARBAZONE

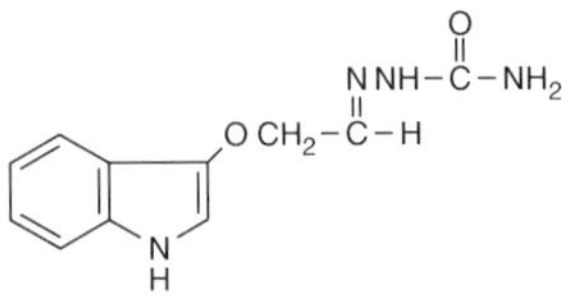

250 mg

S95,546-9
6-METHOXY-3-INDOLEGLYOXYLAMIDE

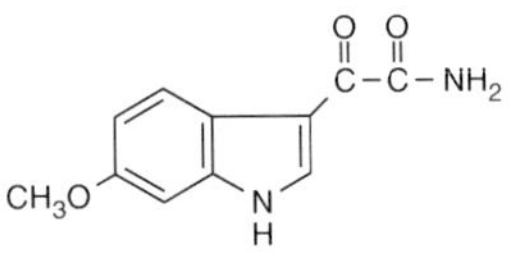

50 mg

S56,780-9
ETHYL 5-METHOXY-2-INDOLECARBOXYLATE

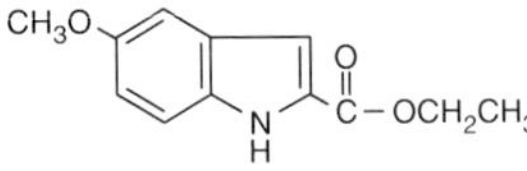

250 mg

S42,410-2
N-ACETYL-2,3-DIHYDROHARMALINE

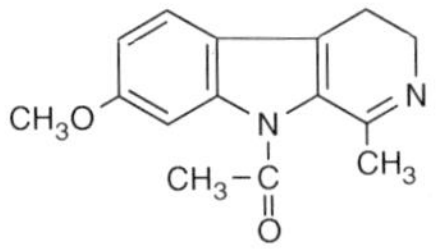

50 mg

S95,573-6
N-(2-[2-(4-METHOXYPHENYL)INDOL-3-YL]ETHYL)-P-TOLUENESULFONAMIDE

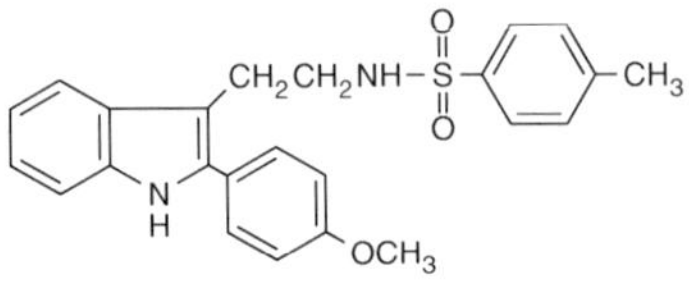

50 mg

S54,710-7
5-METHOXY-2-PHENYLINDOLE
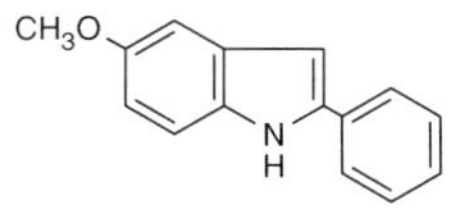
5 gm

S95,548-5
2,3-DIMETHYL-6-METHOXYINDOLE
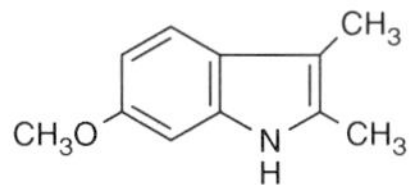
5 gm

S95,549-3
2,3-DIMETHYL-7-METHOXYINDOLE
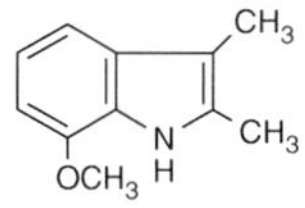
50 mg

S88,321-2
4-METHOXY-1,2,3,4-TETRAHYDROCARBAZOLE
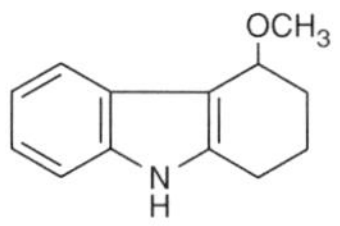
250 mg

S78,851-1
CARBOBENZYLOXYMETHIONYLTRYPTOPHAN METHYL ESTER
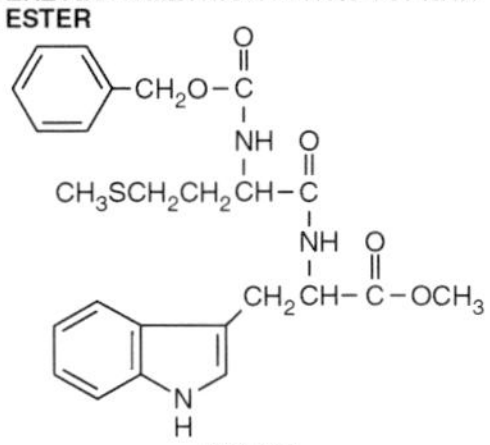
250 mg

S57,987-4
CARBOBENZYLOXY-DL-TRYPTOPHAN
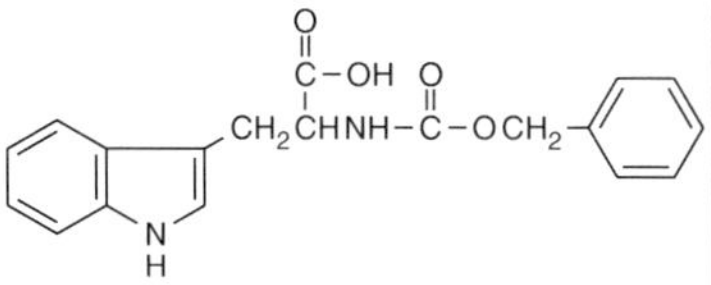
5 gm

S79,084-2
CARBOBENZYLOXYNORVALYLTRYPTOPHAN METHYL ESTER
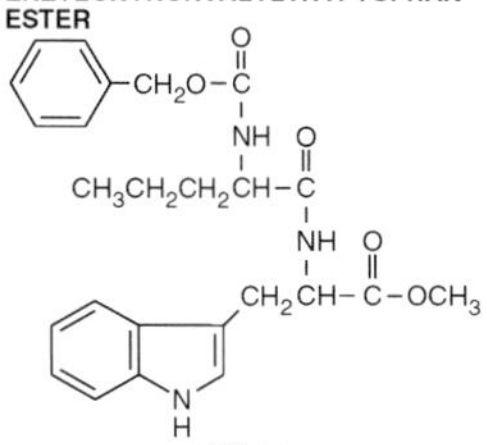
250 mg

S79,065-6
CARBOBENZYLOXYNORVALYL-TRYPTOPHANAMIDE
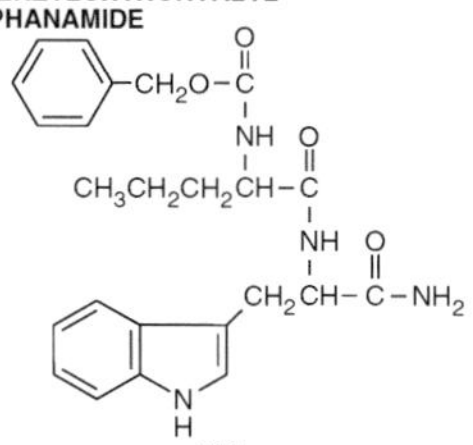
250 mg

S78,147-9
CARBOBENZYLOXY-L-TRYPTOPHANAMIDE
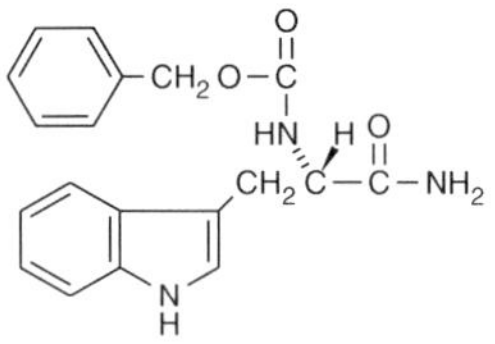
250 mg

S89,535-0
BENZYL 1-(3-INDOLYLMETHYL)-2-OXO-1-CYCLOHEXANECARBOXYLATE
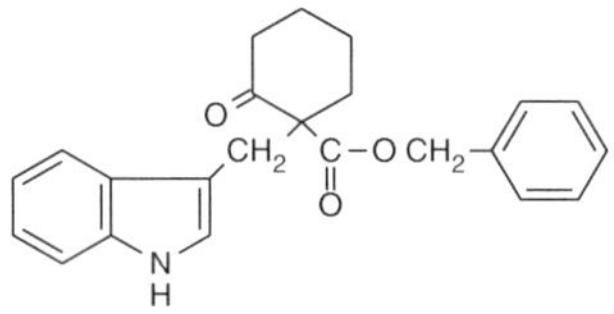
250 mg

S78,975-5
CARBOBENZYLOXYTRYPTOPHYLALANINE METHYL ESTER
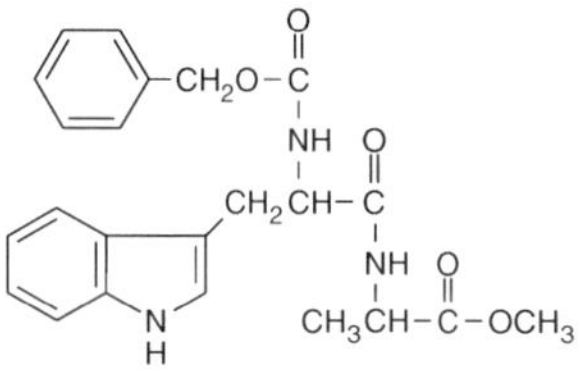
250 mg

S78,636-5
CARBOBENZYLOXYTRYPTOPHYLVALINE METHYL ESTER
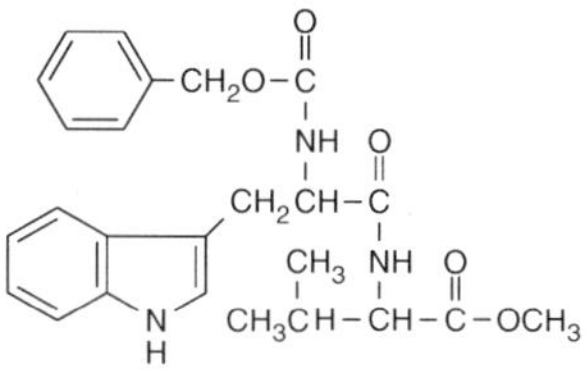
250 mg

S78,637-3
CARBOBENZYLOXY-L-TRYPTOPHYLGLYCINE METHYL ESTER
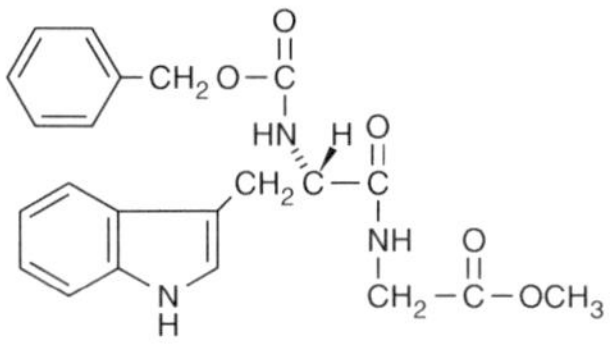
250 mg

S79,073-7
CARBOBENZYLOXYGLYCYL-L-TRYPTOPHAN METHYL ESTER
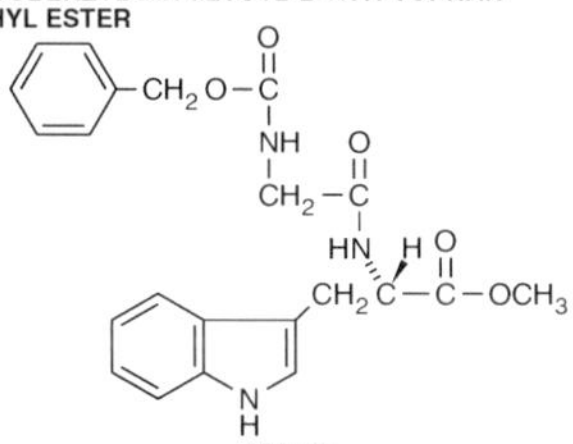
250 mg

S78,627-6
CARBOBENZYLOXYSARCOSYL-L-TRYPTOPHAN METHYL ESTER
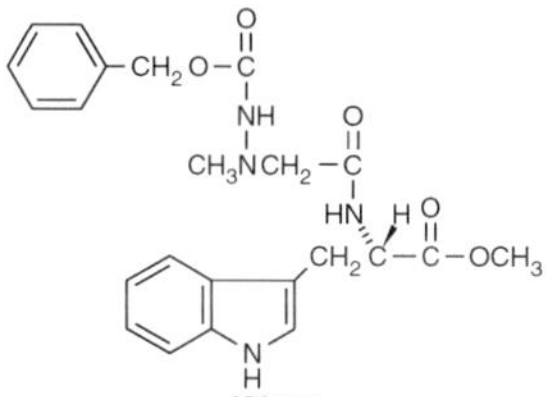
250 mg

S78,767-1
CARBOBENZYLOXYGLYCYLGLYCYL-L-TRYPTOPHAN METHYL ESTER
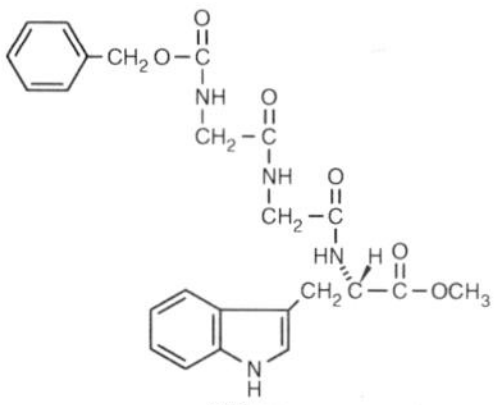
250 mg

S78,614-4
CARBOBENZYLOXYTRYPTOPHYLLEUCINE METHYL ESTER
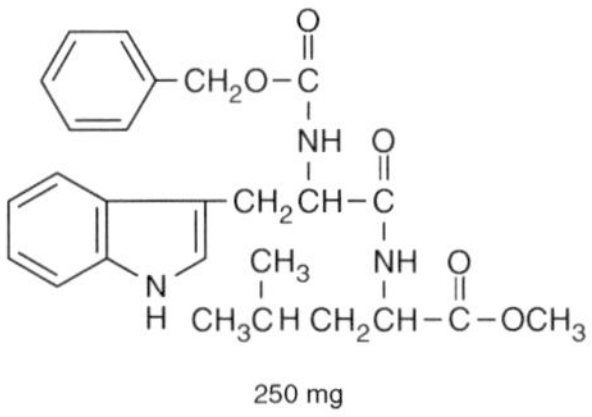
250 mg

S79,075-3
CARBOBENZYLOXYTRYPTOPHYLGLYCYL-GLYCINE BENZYL ESTER
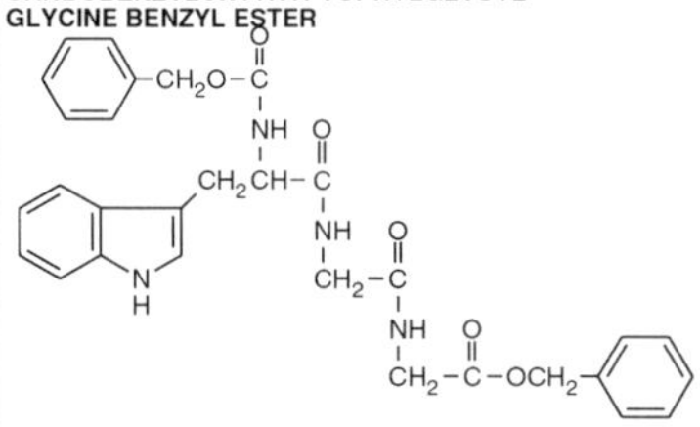
250 mg

S78,948-8
CARBOBENZYLOXY-L-NORLEUCYL-L-TRYPTOPHAN METHYL ESTER
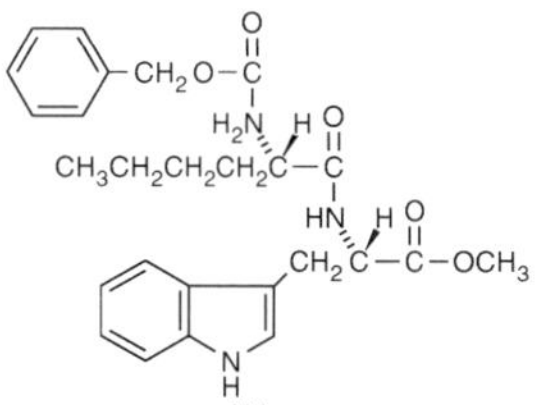
250 mg

S77,797-8
CARBOBENZYLOXYGLYCYL-L-TRYPTOPHANAMIDE
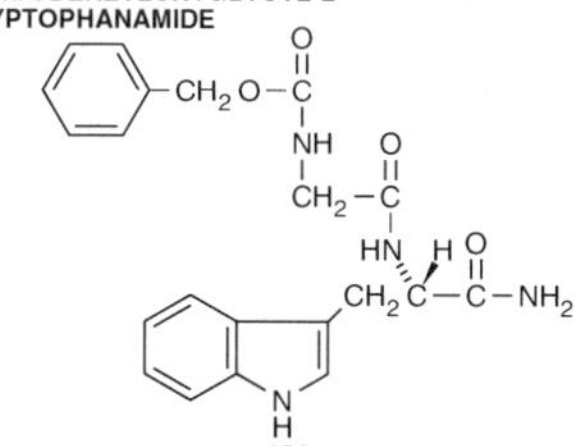
250 mg

S78,262-9
CARBOBENZYLOXYSARCOSYL-L-TRYPTOPHANAMIDE
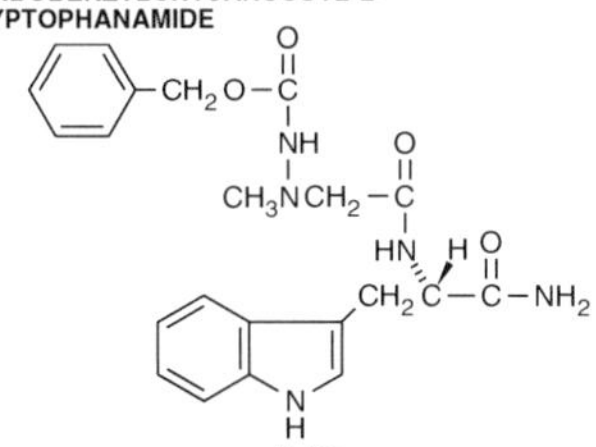
5 gm

S78,490-7
CARBOBENZYLOXY-L-TRYPTOPHYL-L-LEUCINAMIDE
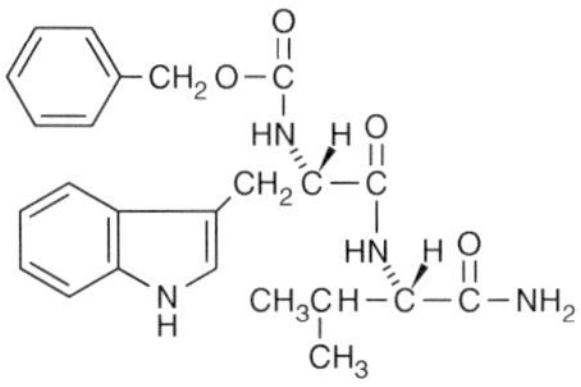
250 mg

S95,566-3
DIETHYL 2-ACETAMIDO-2-(1-METHYLINDOL-3-YLMETHYL)MALONATE
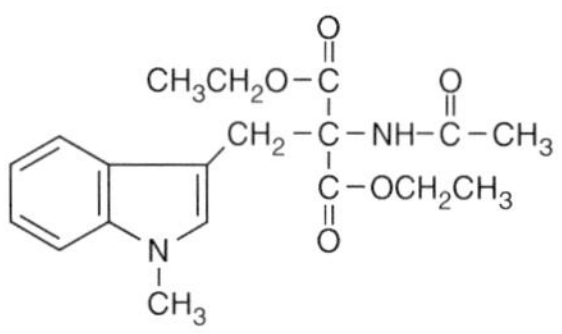
50 mg

S89,594-6
2-(5-BROMO-1-METHYL-3-INDOLYLMETHYL)SUCCINIC ACID
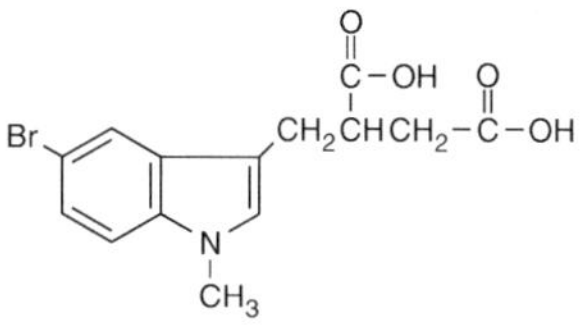
250 mg

S77,663-7
CARBOBENZYLOXY-L-METHIONYL-L-TRYPTOPHAN
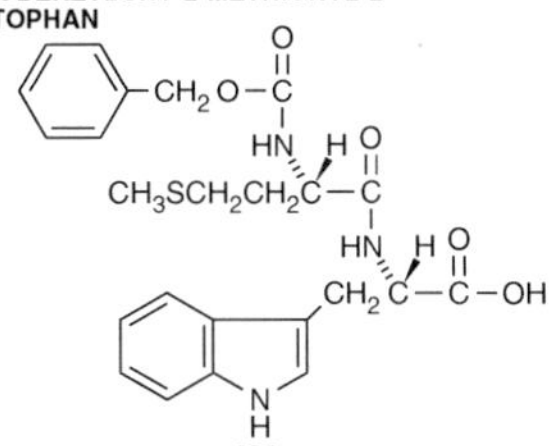
250 mg

S77,662-9
CARBOBENZYLOXY-L-METHIONYL-L-TRYPTOPHANAMIDE
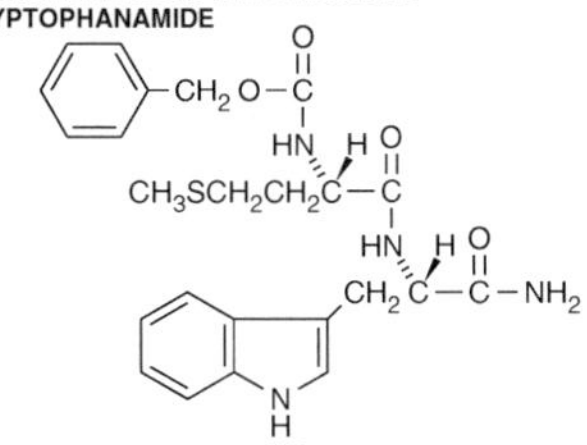
250 mg

S45,123-1
BETA-ALANYL-L-TRYPTOPHAN
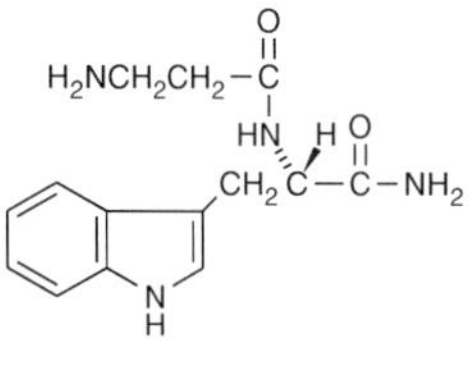
250 mg

S76,366-7
ETHYL 1-(3-AMINOPROPYL)-3-(ETHOXYCARBONYLMETHYL)-2-INDOLEPROPIONATE HYDROCHLORIDE
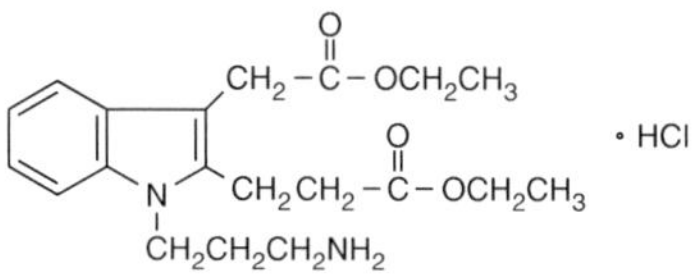
250 mg

S78,650-0
N-ALPHA-L-ISOLEUCYL-L-TRYPTOPHANAMIDE HYDROCHLORIDE

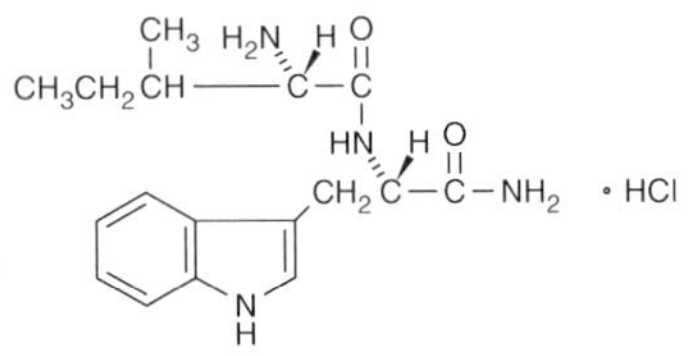

250 mg

S44,918-0
GLYCYL-L-TRYPTOPHAN

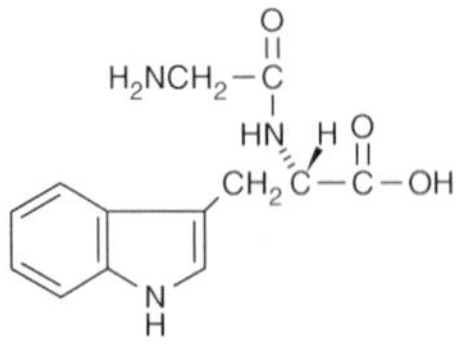

250 mg

S74,927-3
DL-6-METHYLTRYPTOPHAN

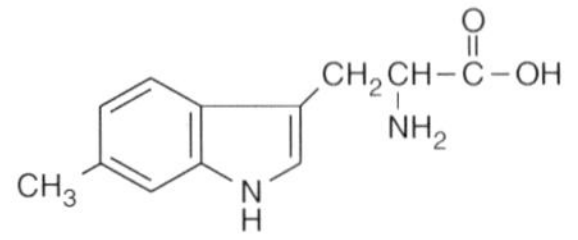

250 mg

S76,578-3
DL-TRYPTOPHAN METHYL ESTER

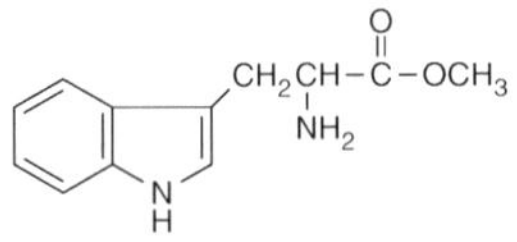

5 gm

S45,152-5
5-FLUORO-ALPHA-METHYLTRYPTAMINE HYDROCHLORIDE

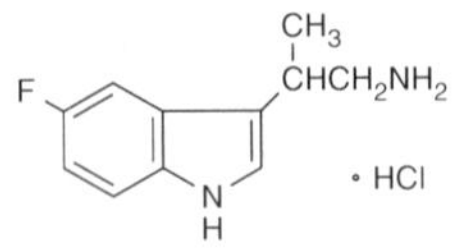

250 mg

S45,153-3
6-FLUORO-ALPHA-METHYLTRYPTAMINE

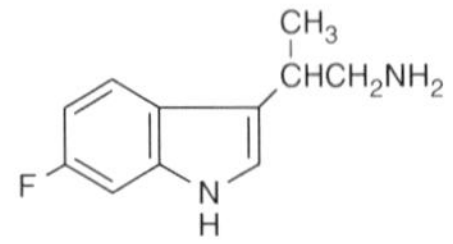

250 mg

S42,859-0
5-AMINO-2,3-DIMETHYLINDOLE

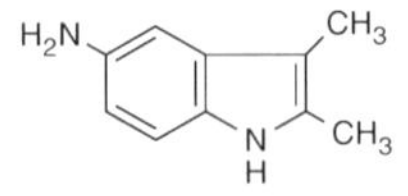

250 mg

S78,165-7
L-NORLEUCYL-L-TRYPTOPHAN

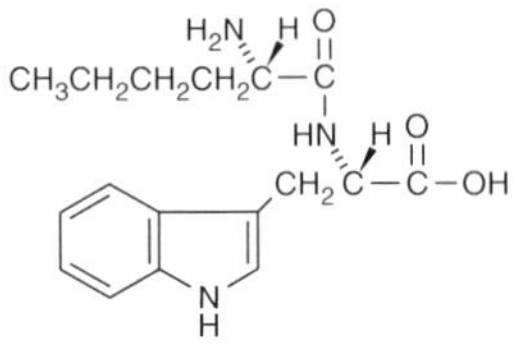

250 mg

S35,028-1
GLYCYL-DL-TRYPTOPHAN

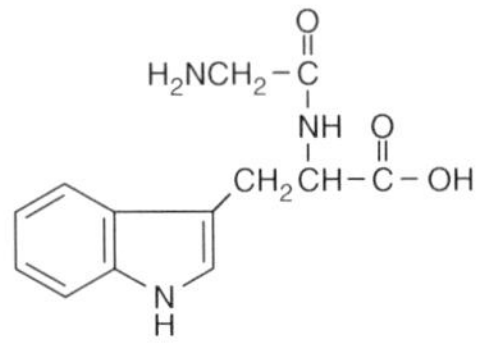

250 mg

S78,149-5
L-TRYPTOPHYL-L-LEUCINAMIDE ACETATE

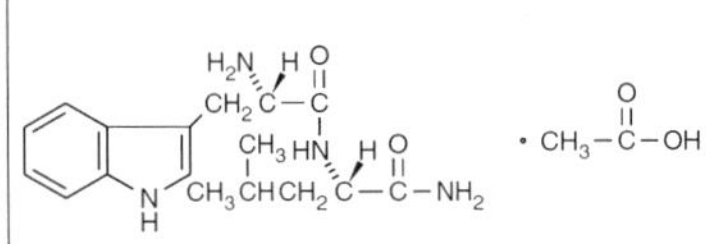

250 mg

S77,737-4
N-(ALPHA)-L-LEUCYL-L-TRYPTOPHANAMIDE ACETATE

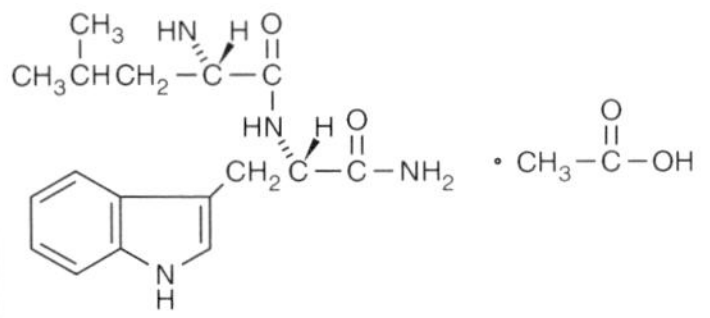

250 mg

S41,475-1
3-INDOLEPROPIONIC ACID HYDRAZIDE

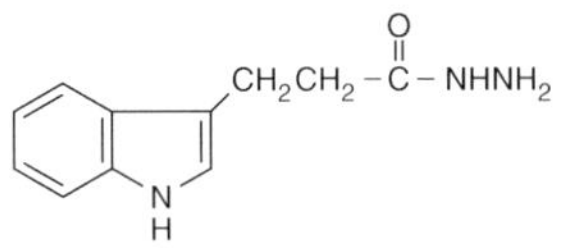

5 gm

S40,903-0
LEPTOCLADINE HYDROCHLORIDE

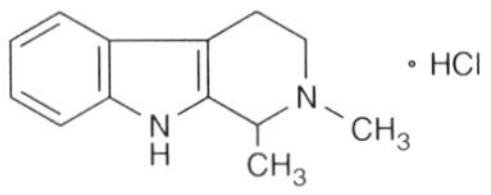

1 gm

S43,326-8
8-CHLORO-1-TETRAHYDRONORHARMANONE

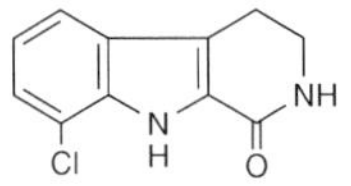

250 mg

S44,753-6
1-TETRAHYDRONORHARMANONE

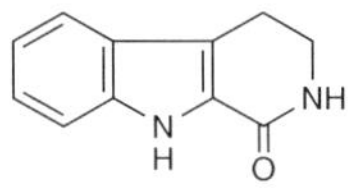

250 mg

S43,286-5
8-METHYL-1-TETRAHYDRONORHARMANONE

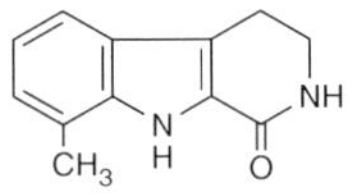

250 mg

S40,996-0
2-PROPYL-1,2,3,4-TETRAHYDROHARMANE HYDROCHLORIDE

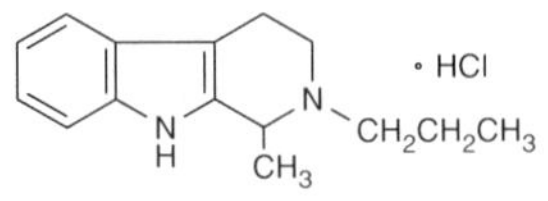

250 mg

S78,465-6
N-(2,4-DINITROPHENYL)-L-TRYPTOPHAN HYDRATE

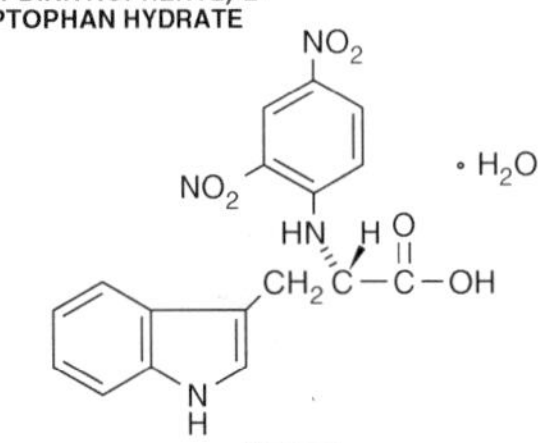

250 mg

S41,679-7
1,2,3,4-TETRAHYDROHARMANE-3-CARBOXYLIC ACID DIHYDRATE

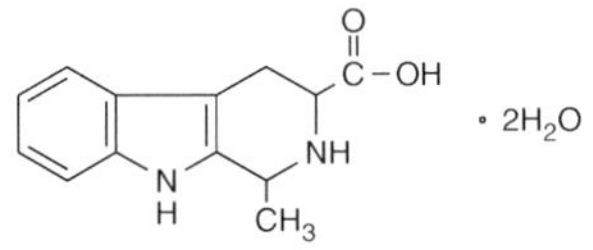

250 mg

S74,566-9
1-ETHYL-3-NITROSO-2-PHENYLINDOLE

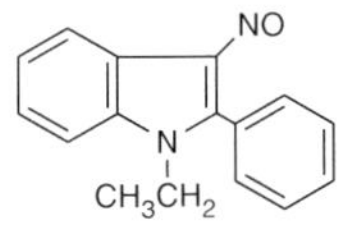

250 mg

S42,560-5
3-(N-METHYLBENZYLAMINOMETHYL)-INDOLE

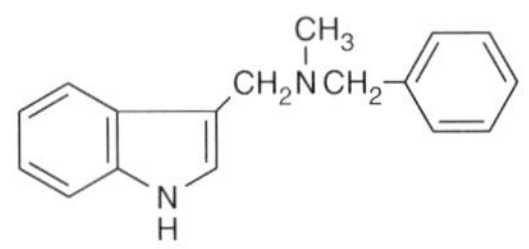

5 gm

S42,572-9
3-(3-METHYLPIPERIDINOMETHYL)INDOLE

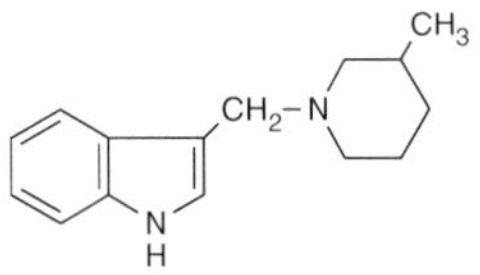

5 gm

S42,573-7
3-(4-METHYLPIPERIDINOMETHYL)INDOLE

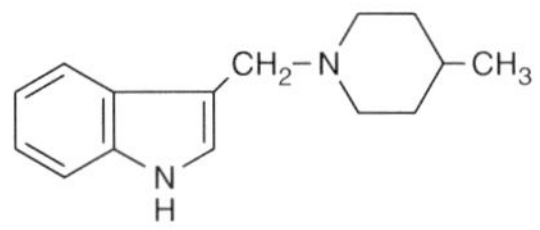

5 gm

S42,584-2
2-METHYL-3-(MORPHOLINOMETHYL)-INDOLE

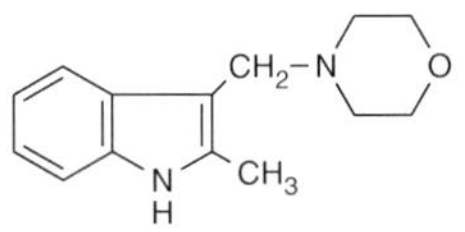

5 gm

S57,044-3
2-METHYLGRAMINE

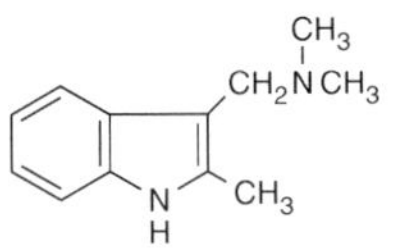

5 gm

S38,203-5
3-(DIETHYLAMINOMETHYL)INDOLE HYDROCHLORIDE

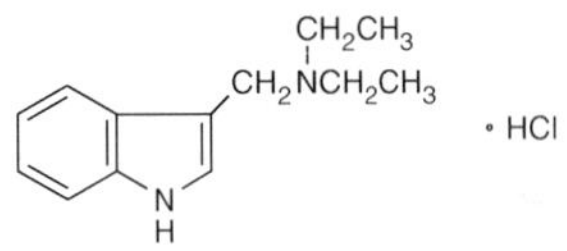

250 mg

S42,467-6
3-(DIALLYLAMINOMETHYL)INDOLE

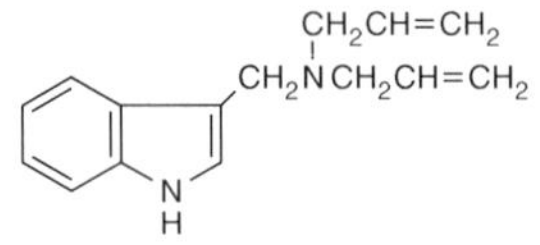

5 gm

S34,493-1
3-(CYCLOHEXYLAMINOMETHYL)INDOLE

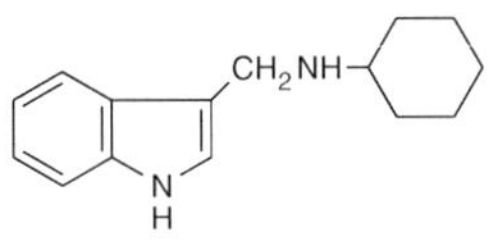

5 gm

S88,298-4
3-[2-(DIMETHYLAMINO)ACETYL]-2-(3-ETHYL-5-METHYL-4-ISOXAZOLYL)INDOLE

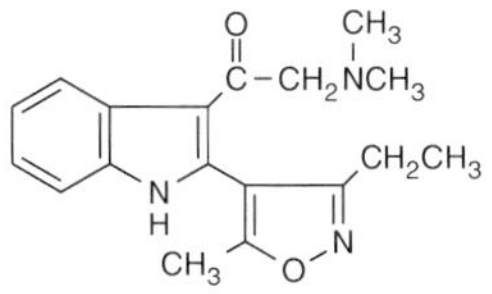

250 mg

S91,543-2
3-(DIMETHYLAMINOMETHYL)-6-NITROINDOLE

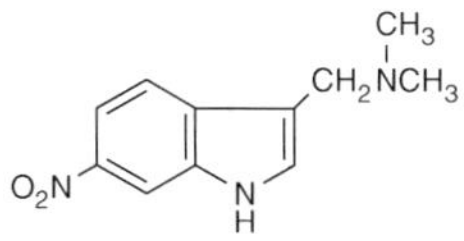

250 mg

S36,279-4
1-INDOLEPROPIONAMIDE

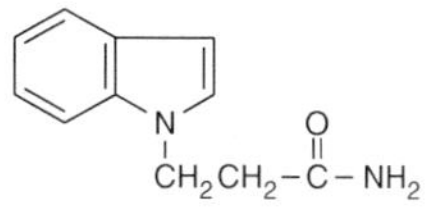

5 gm

S36,926-8
1-INDOLEPROPIONITRILE

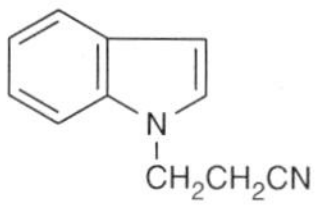

5 gm

S36,005-8
3-1,5-DIAZABICYCLO[3.2.1]OCT-8-YL-INDOLE

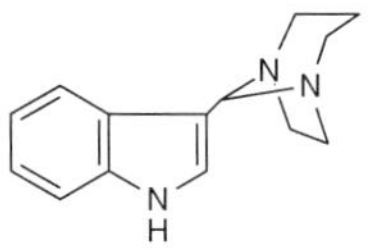

250 mg

S89,549-0
4-BROMO-3-(PIPERIDINOMETHYL)INDOLE

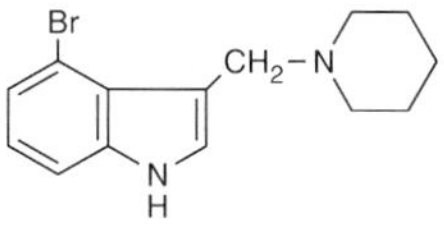

250 mg

S45,151-7
5-FLUOROGRAMINE

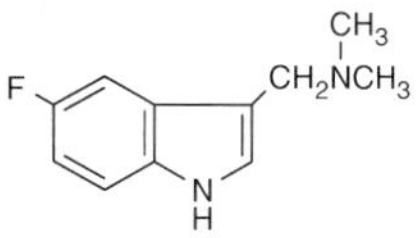

250 mg

S67,204-1
5-BROMOGRAMINE

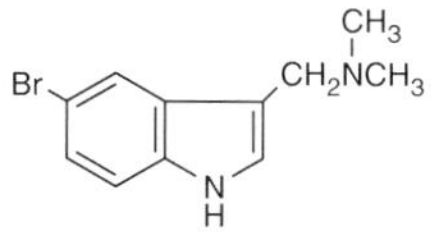

250 mg

S35,294-2
5-METHYLGRAMINE

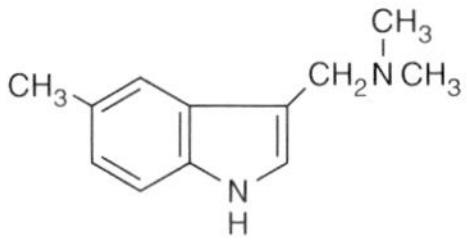

250 mg

S36,846-6
1-(3-DIMETHYLAMINOPROPYL)INDOLE

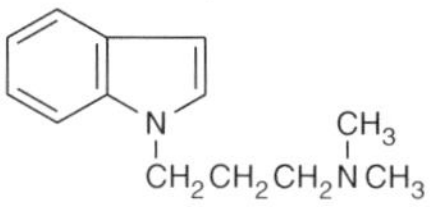

250 mg

S89,514-8
(3-INDOLYLMETHYL)-TRIMETHYLAMMONIUM IODIDE

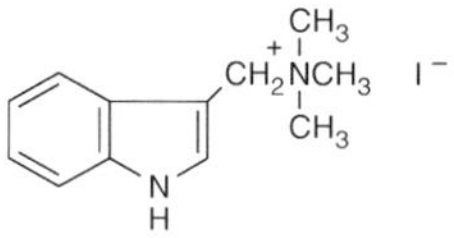

250 mg

S74,845-5
2,3-DIMETHYL-4-NITROINDOLE

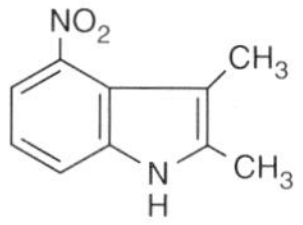

250 mg

S76,195-8
N,N-DIMETHYL-2-(5-METHYL-3-PHENYL-4-ISOXAZOLYL)-ALPHA-OXO-3-INDOLEACETAMIDE

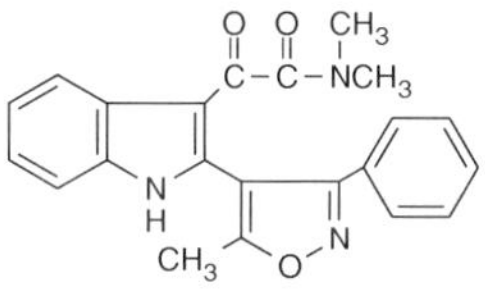

250 mg

S87,982-7
3-ACETYL-2-(3-ETHYL-5-METHYL-4-ISOXAZOLYL)INDOLE

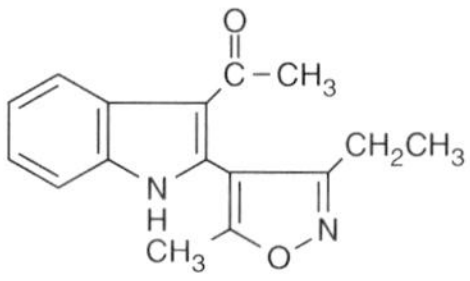

5 gm

S34,486-9
ALPHA-CYANO-3-INDOLEACRYLAMIDE

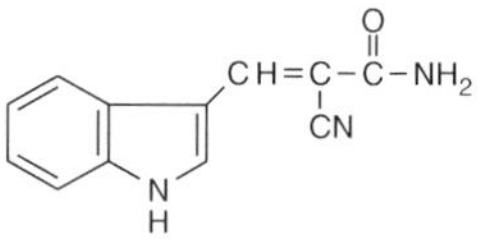

5 gm

S34,487-7
ALPHA-CYANO-3-INDOLEACRYLONITRILE

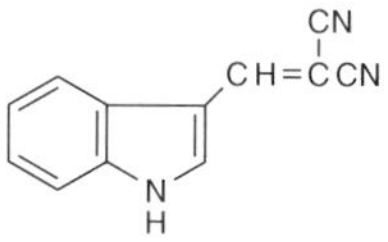

5 gm

S87,984-3
2-(3-ETHYL-5-METHYL-4-ISOXAZOLYL)-INDOLE

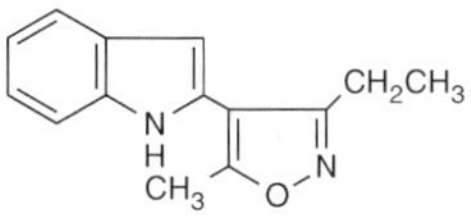

5 gm

S74,016-0
N-BETA-BENZYLTRYPTAMINE HYDROCHLORIDE

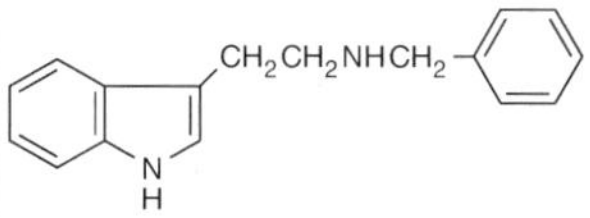

250 mg

S42,903-1
3-(2-NITROVINYL)INDOLE

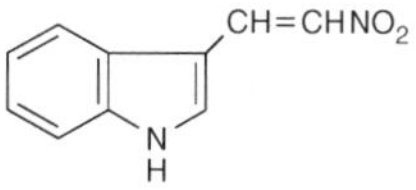

250 mg

S36,980-2
3-(2-NITRO-1-PROPENYL)INDOLE

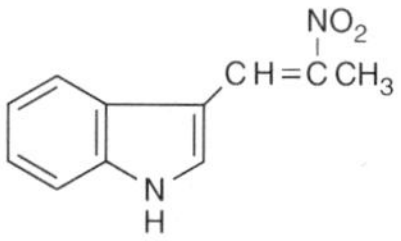

5 gm

S74,222-8
N-OMEGA-FORMYLTRYPTAMINE

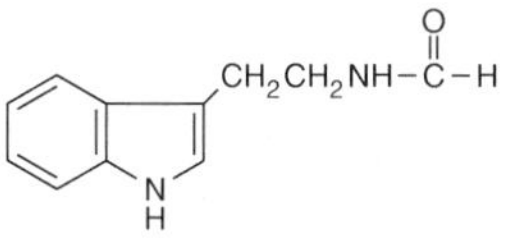

250 mg

S75,992-9
N-OMEGA-METHYLTRYPTAMINE HYDROCHLORIDE

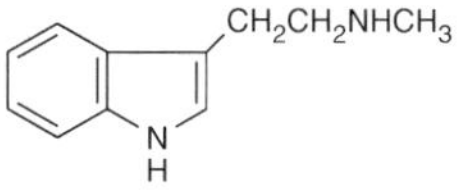

250 mg

S75,841-8
5-(1-METHYLHEPTYLAMINO)INDOLINE DIHYDROCHLORIDE

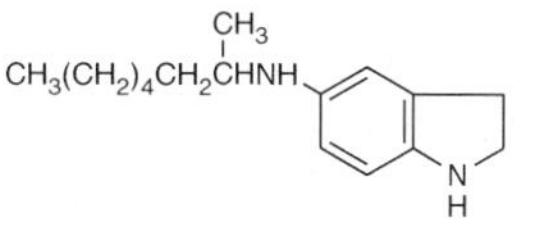

250 mg

S44,705-6
N/ALPHA-ACETYL-N-METHYLTRYPTOPHAN METHYL ESTER

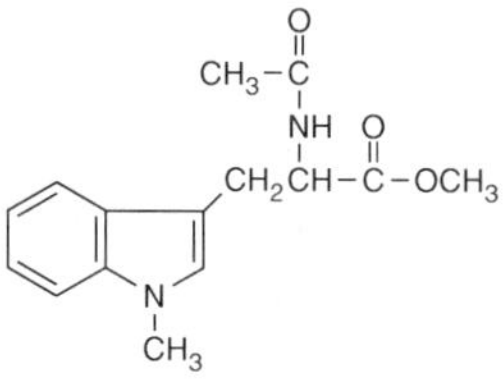

100 mg

S89,506-7
1-METHYL-5-NITRO-2-INDOLINONE

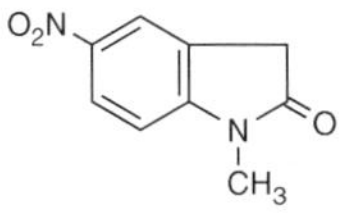

250 mg

S76,157-5
2-METHYL-5-NITROINDOLE

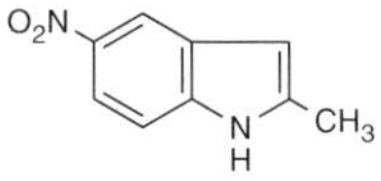

250 mg

S74,453-0
3-ETHYL-2-METHYL-5-NITROINDOLE

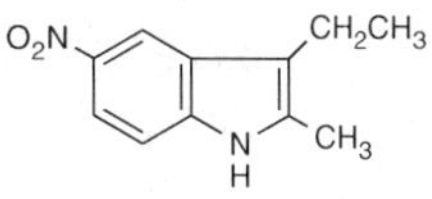

250 mg

S89,497-4
DIETHYL 2-[4-(CYANOMETHYL)-3-INDOLYLMETHYL]-2-FORMAMIDOMALONATE

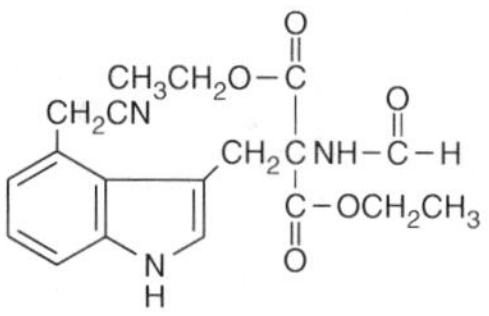

250 mg

S74,534-0
N-OMEGA-(ACETOACETYL)TRYPTAMINE

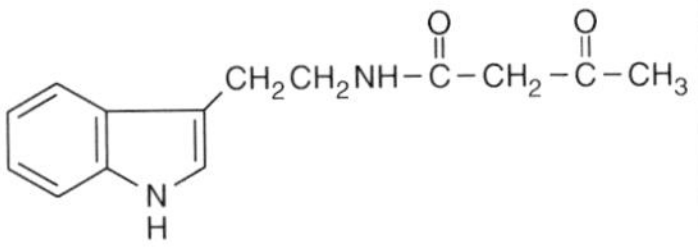

250 mg

S78,657-8
N-ALPHA-BENZOYL-DL-TRYPTOPHAN

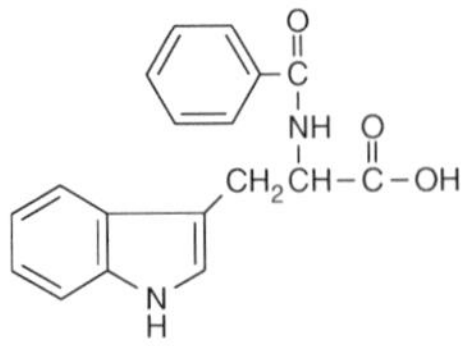

250 mg

S74,548-0
DIETHYL (N-[2-(3-INDOLYL)ETHYL] CARBAMOYLMETHYL)MALONATE

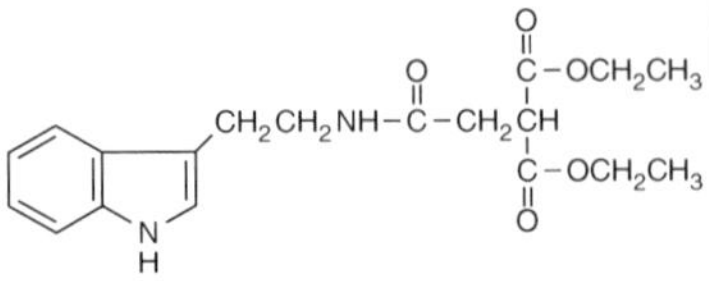

250 mg

S78,823-6
CARBOBENZYLOXYGLYCYL-D-TRYPTOPHAN

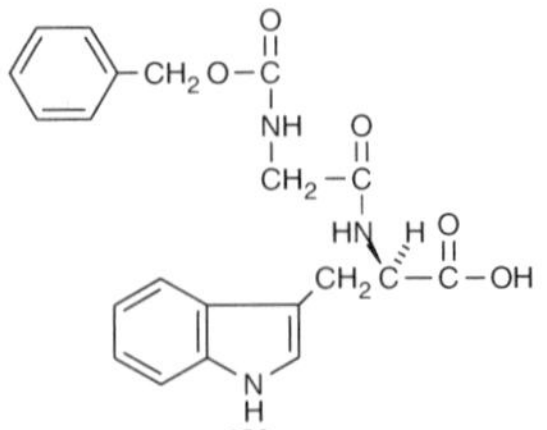

250 mg

S78,279-3
N-(TERT-BUTOXYCARBONYL)-L-TRYPTOPHAN

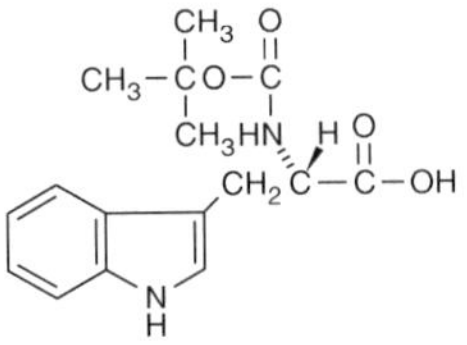

5 gm

S78,662-4
N-ALPHA-ACETYL-5-METHYL-DL-TRYPTOPHAN

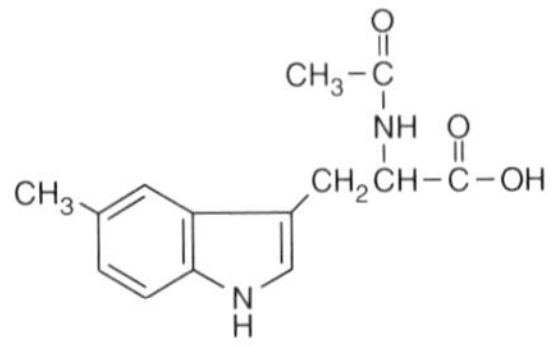

250 mg

S45,130-4
N-ACETYL-L-TRYPTOPHAN METHYL ESTER

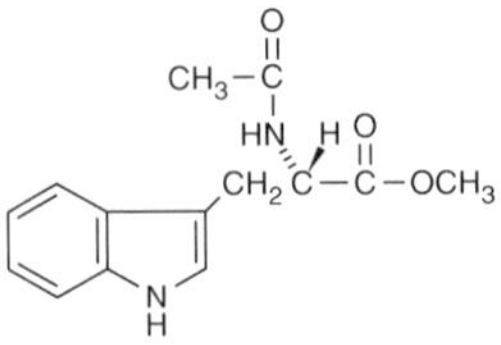

250 mg

S78,312-9
N-ALPHA-ACETYL-DL-TRYPTOPHAN METHYL ESTER

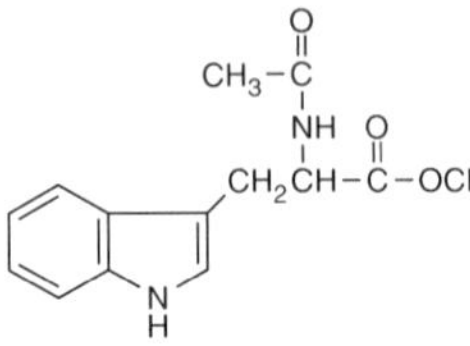

5 gm

S78,193-2
DL-TRYPTOPHAN METHYL ESTER HYDROCHLORIDE

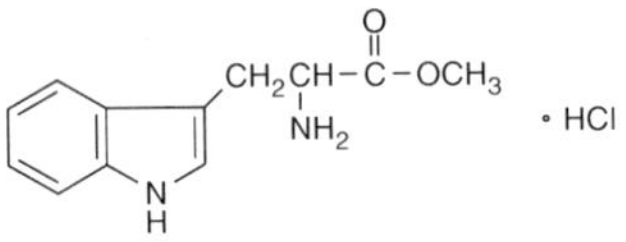

250 mg

S78,127-4
CARBOBENZYLOXY-L-ISOLEUCYL-L-TRYPTOPHANAMIDE

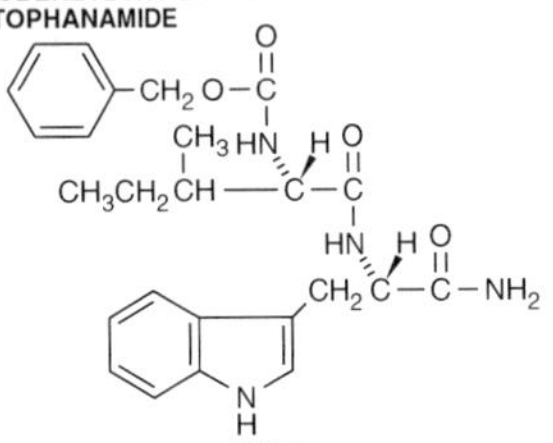

250 mg

S77,600-9
N-ACETYL-D-TRYPTOPHANAMIDE

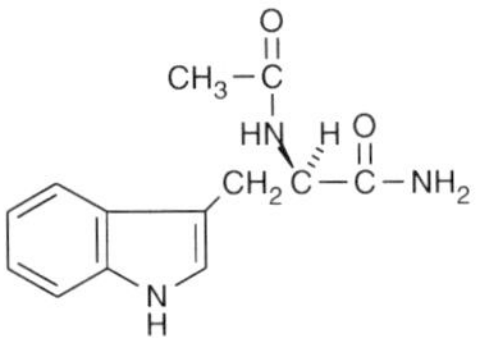

250 mg

S36,608-0
5-METHYLINDOLE-3-ACETAMIDE

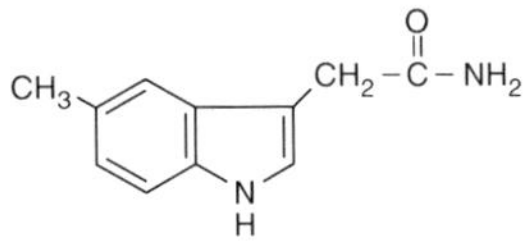

250 mg

S74,177-9
2-(2-PHENYL-3-INDOLYL)ACRYLONITRILE

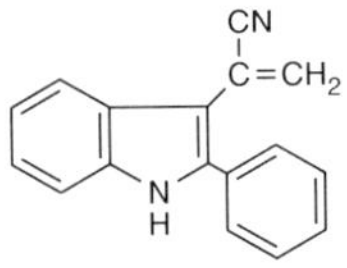

250 mg

S74,017-9
N-[2-(3-INDOLYL)ETHYL]-P-TOLUENESULFONAMIDE

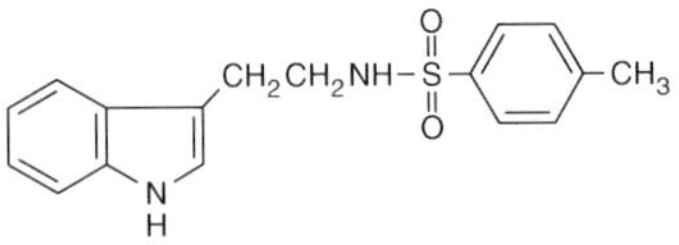

250 mg

S95,557-4
2-BENZOYL-3-METHYLINDOLE

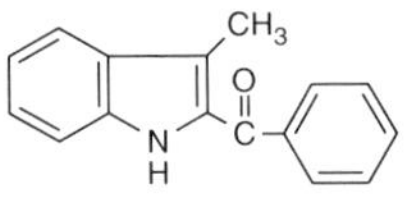

50 mg

S78,779-5
CARBOBENZYLOXYTRYPTOPHYLPHENYLALANINE METHYL ESTER

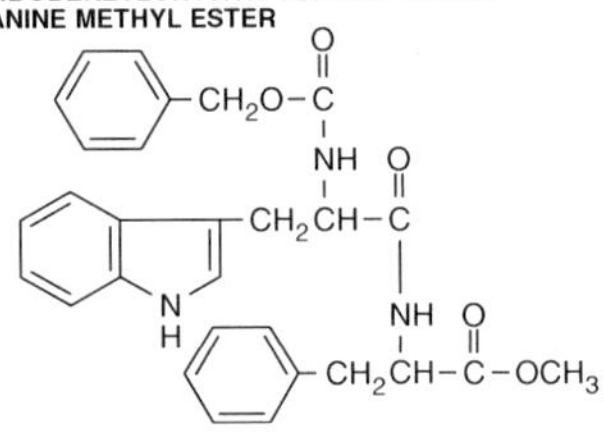

250 mg

S79,078-8
CARBOBENZYLOXYPHENYLALANYLTRYPTOPHAN METHYL ESTER

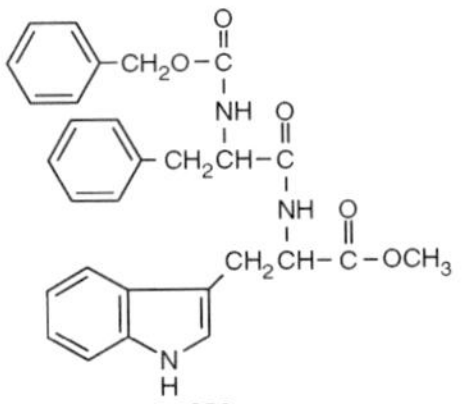

250 mg

S77,707-2
CARBOBENZYLOXY-L-PHENYLALANYL-L-TRYPTOPHANAMIDE

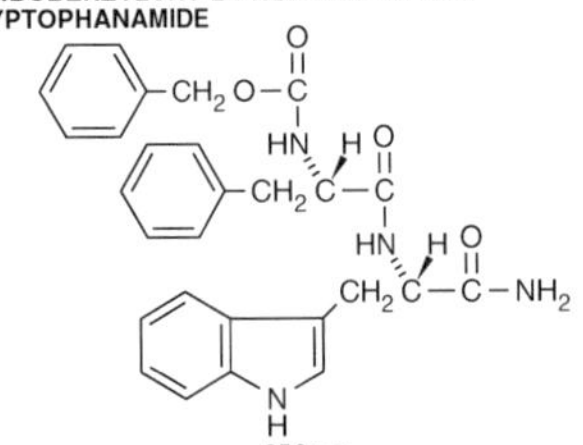

250 mg

S78,181-9
CARBOBENZYLOXY-L-TRYPTOPHYL-L-PHENYLALANINAMIDE

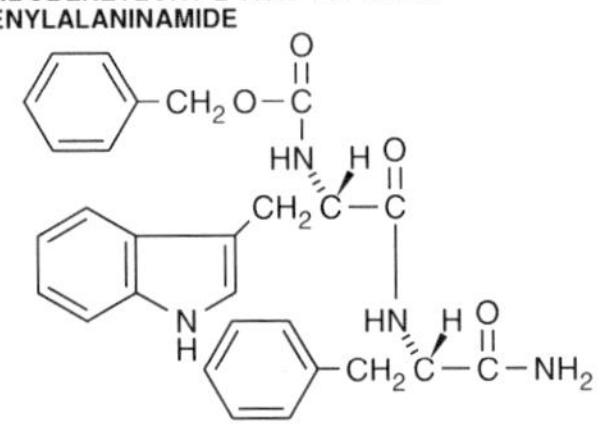

250 mg

S37,336-2
3-[3-(DIMETHYLAMINO)PROPIONYL] INDOLE

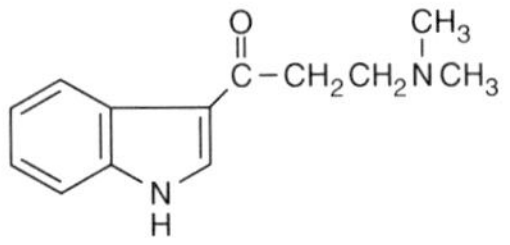

5 gm

S93,995-1
2-(3-ETHYL-5-METHYL-4-ISOXAZOLYL)-N,N-DIMETHYL-3-INDOLEGLYOXYLAMIDE

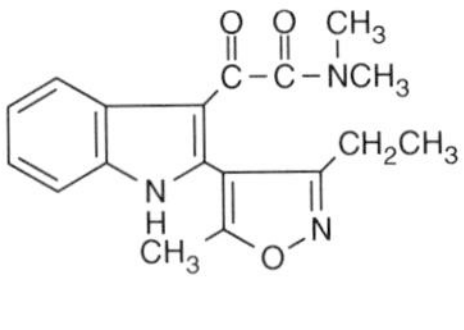

5 gm

S95,691-0
3,4-DIHYDRO-2-METHYL-9H-PYRIDO(3,4-B)INDOLIUM IODIDE

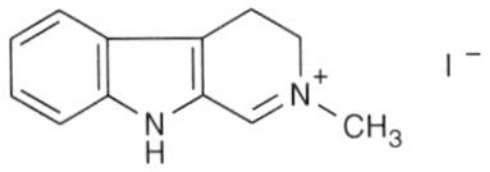

50 mg

S40,811-5
2,3-DIMETHYL-5-NITROINDOLE

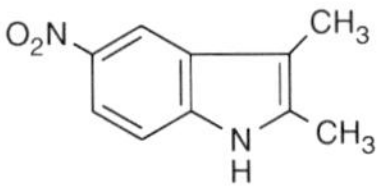

5 gm

S40,812-3
2,3-DIMETHYL-7-NITROINDOLE

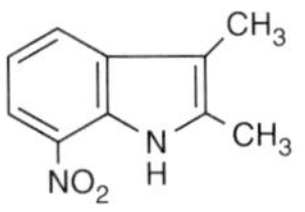

250 mg

S44,849-4
5-METHYLINDOLE-2-CARBOXYLIC ACID

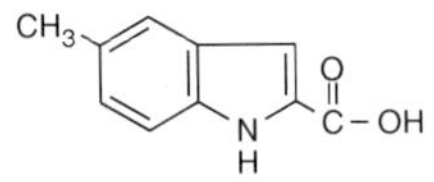

250 mg

S94,508-0
4-(3-INDOLYLMETHYLAMINO)-BENZOIC ACID

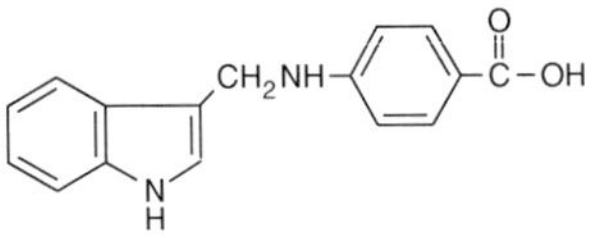

250 mg

S93,270-1
2,3-INDOLEDICARBOXYLIC ACID

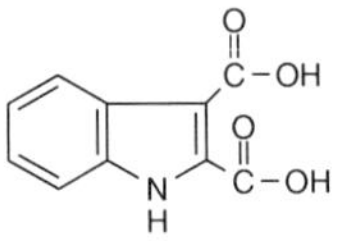

250 mg

S42,679-2
5-BROMOINDOLE-2-CARBOXYLIC ACID

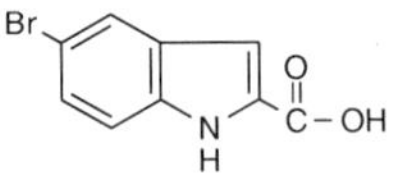

250 mg

S94,250-2
2-METHYLINDOLE-3-ACETIC ACID
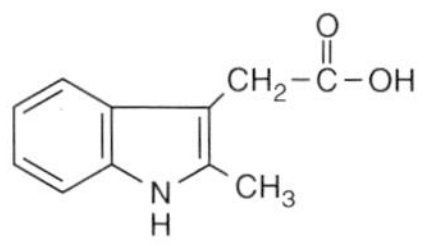
250 mg

S93,994-3
N-OLEOYL-DL-TRYPTOPHAN ETHYL ESTER
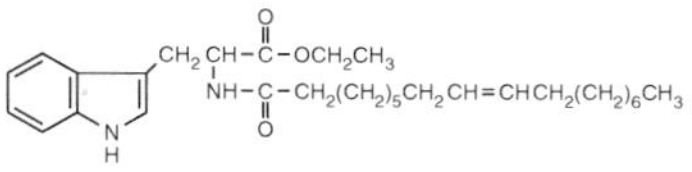
5 gm

S45,302-1
METHYL 3-INDOLYLACETATE
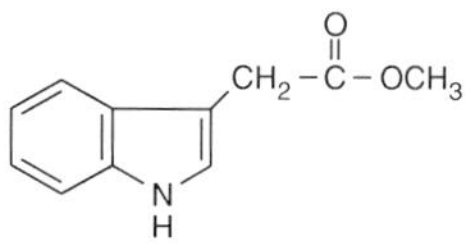
5 gm

S36,976-4
METHYL 3-INDOLEPROPIONATE
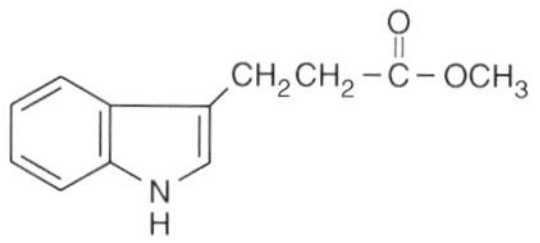
5 gm

S54,467-1
METHYL 3-INDOLEBUTYRATE
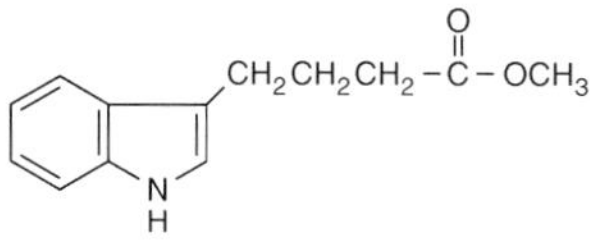
5 gm

S54,466-3
ETHYL 3-INDOLEBUTYRATE
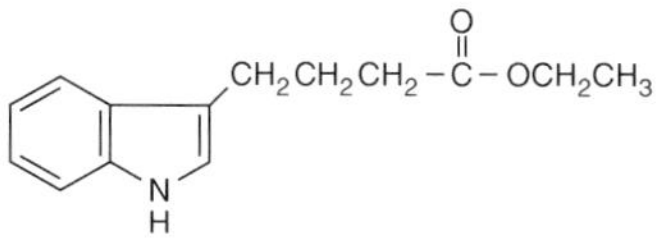
250 mg

S40,688-0
5-BROMOINDOLE-3-ACETAMIDE
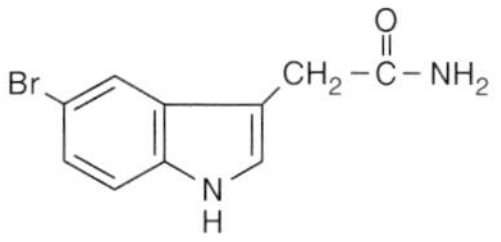
250 mg

S54,288-1
2-(PARA-TOLYL)-INDOLE
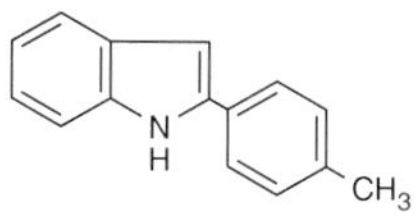
250 mg

S42,575-3
3-[4-(METHYLAMINO)BENZYL]INDOLE
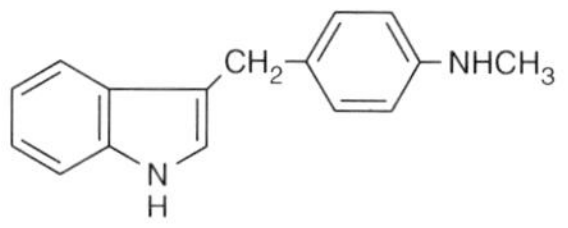
250 mg

S41,612-6
2,3-DIPHENYLINDOLE
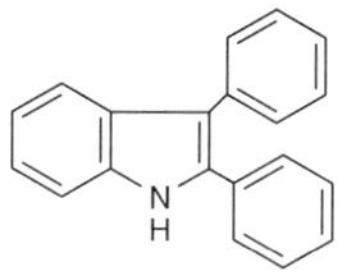
5 gm

S85,012-8
6-BROMO-1,2,3,4-TETRAHYDROCARBAZOLE
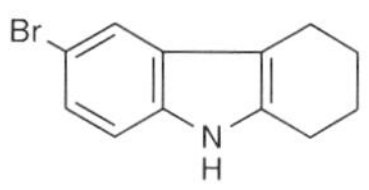
250 mg

S41,076-4
6-METHYL-1,2,3,4-TETRAHYDROCARBAZOLE
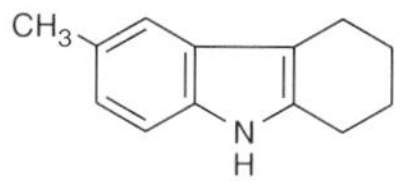
5 gm

S43,712-3
1-BENZYLINDOLE-3-CARBOXALDEHYDE
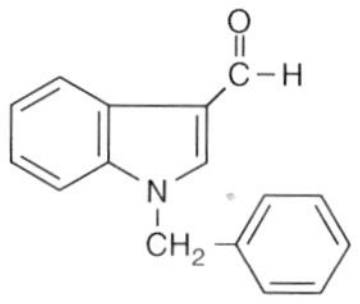
250 mg

S95,556-6
1-BENZOYL-3-METHYLINDOLE
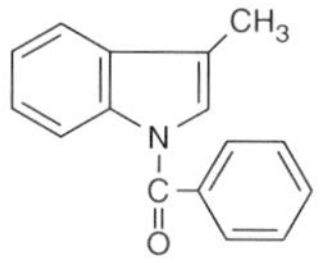
50 mg

S95,565-5
[2-(4-METHOXYPHENYL)INDOL-3-YLMETHYL]TRIMETHYLAMMONIUM IODIDE
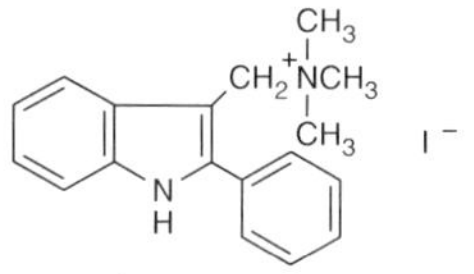
50 mg

S36,363-4
METHYL 1-METHYLINDOLE-2-CARBOXYLATE
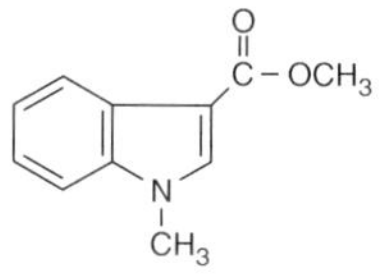
250 mg

S54,039-0
1-INDOLEPROPIONIC ACID
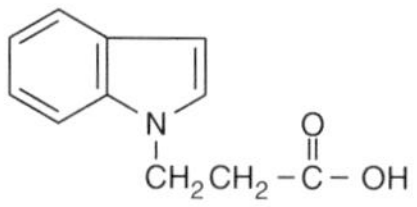
5 gm

S54,042-0
METHYL 1-INDOLEPROPIONATE
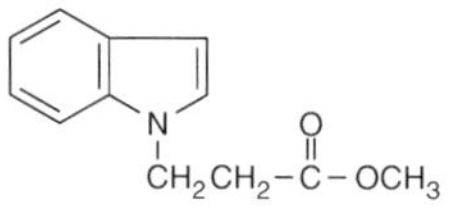
5 gm

S54,041-2
METHYL 1-INDOLEBUTYRATE
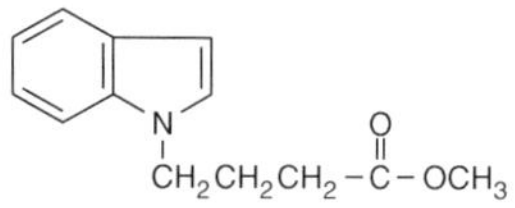
5 gm

S54,043-9
ISOPROPYL 1-INDOLEBUTYRATE
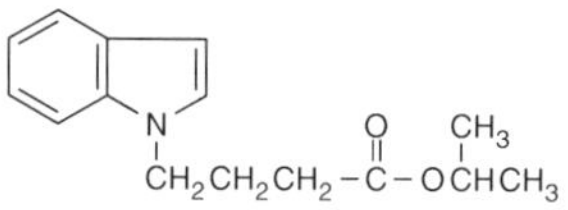
5 gm

S36,626-9
N-ACETYLINDOLE
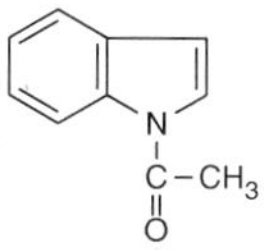
250 mg

S40,979-0
2-PHENYLINDOLIZINE
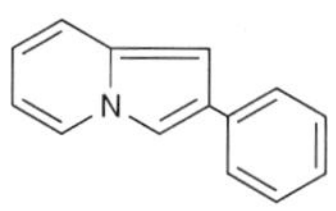
250 mg

S74,279-1
1-ETHYL-2-PHENYLINDOLE
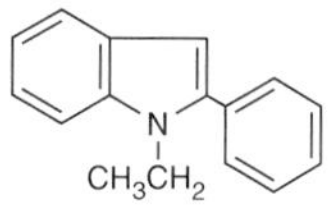
5 gm

S80,643-9
1-ISOPROPYL-3-(3,5-XYLYL)INDOLE
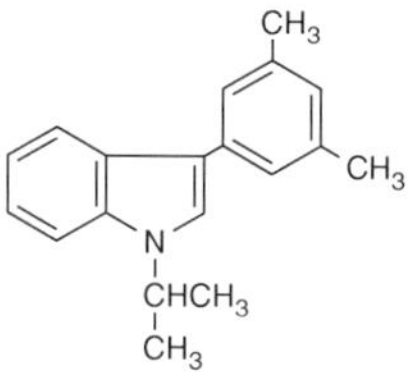
250 mg

S88,816-8
5-BROMO-1-METHYLINDOLE
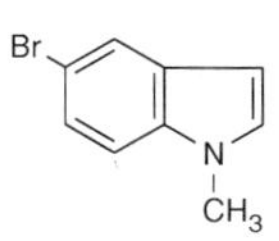
250 mg

S36,100-3
N,N-DIMETHYLHOMOTRYPTAMINE
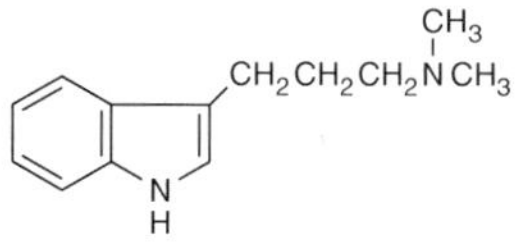
250 mg

S40,809-3
5,N,N-TRIMETHYL TRYPTAMINE
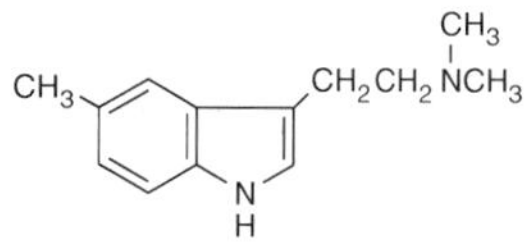
250 mg

S51,754-2
(3-[2-(METHOXYCARBONYL)ETHYL]-2-INDOLYL)TRIMETHYLAMMONIUM IODIDE
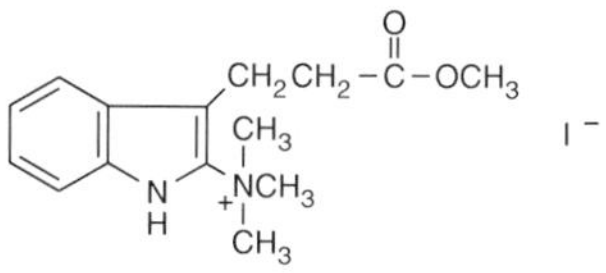
5 gm

S87,161-3
1,2,3,3-TETRAMETHYL-3H-INDOLIUM PERCHLORATE

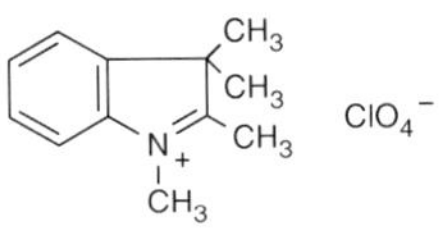

250 mg

S88,743-9
2,2'-(1,3-BUTADIENE-1,4-DIYL)BIS-1,3,3-TRIMETHYL-3H-INDOLIUM IDOIDE

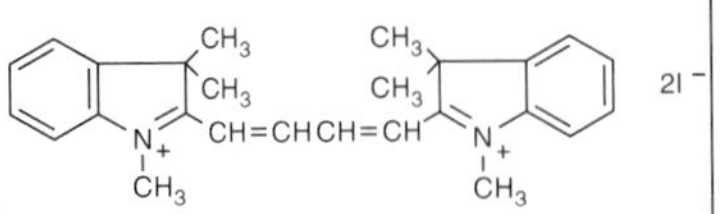

250 mg

S44,547-9
1-BENZYLINDOLE-3-CARBOXYLIC ACID

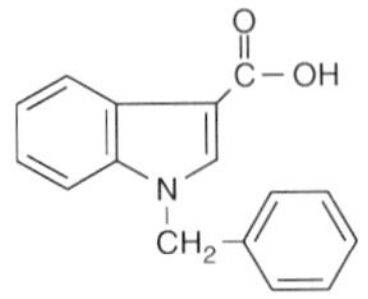

250 mg

S72,939-6
4-BUTYL-1-(3-INDOLYLACETYL)-SEMICARBAZIDE

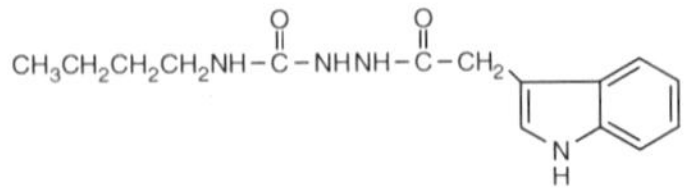

5 gm

S89,576-8
1-ETHYLPYRANO(3,4-B)INDOL-3(9H)-ONE

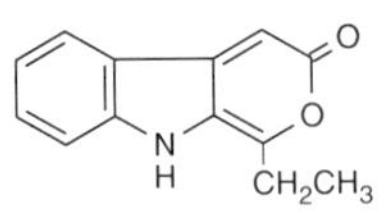

250 mg

S89,591-1
1-PROPYLPYRANO(3,4-B)INDOL-3(9H)-ONE

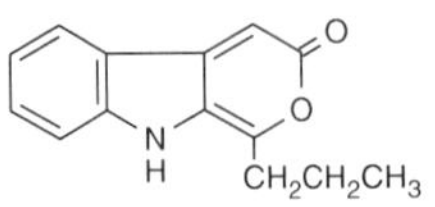

250 mg

S78,278-5
5-BROMO-6-CHLORO-3-INDOLYL ACETATE

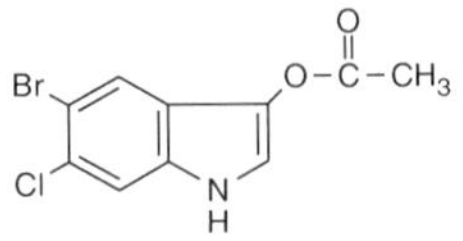

5 gm

S76,519-8
3-INDOLYL BUTYRATE

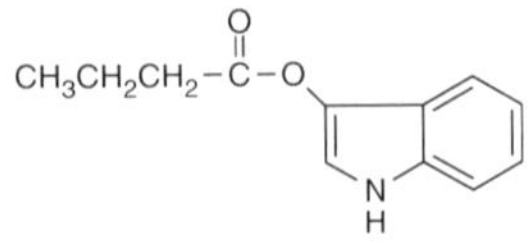

5 gm

S76,527-9
N-ALPHA-T-BOC-L-TRYPTOPHAN 4-NITROPHENYL ESTER

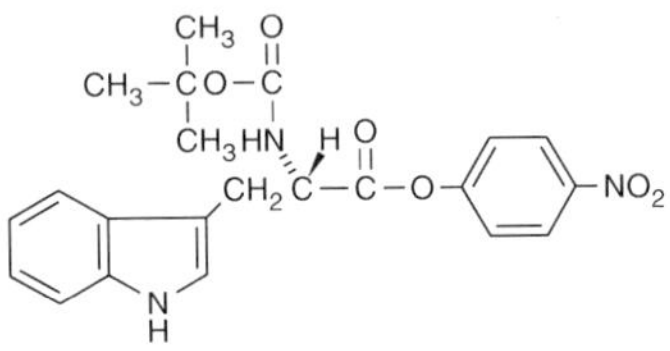

5 gm

S79,085-0
N-ALPHA-CARBOBENZYLOXY-L-TRYPTOPHAN 2-NITROPHENYL ESTER

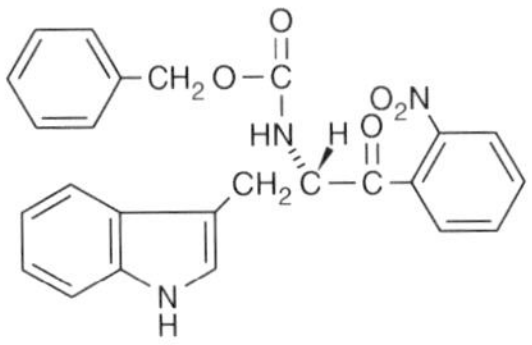

250 mg

S67,887-2
N-ALPHA-CARBOBENZYLOXY-DL-TRYPTOPHAN 4-NITROPHENYL ESTER

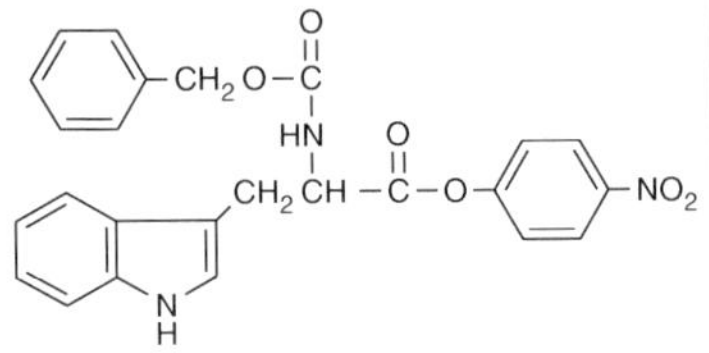

5 gm

S36,206-9
5-FLUOROINDOLE-3-CARBOXALDEHYDE

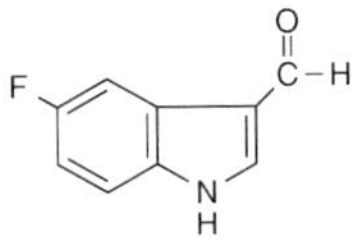

250 mg

S43,816-2
5-METHYLINDOLE-3-CARBOXALDEHYDE

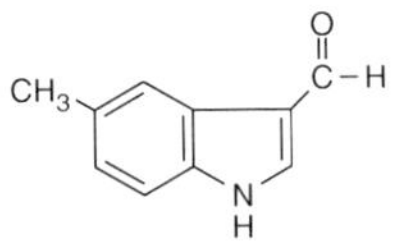

250 mg

S34,934-8
ETHYL 2-METHYL-3-INDOLYLGLYOXYLATE

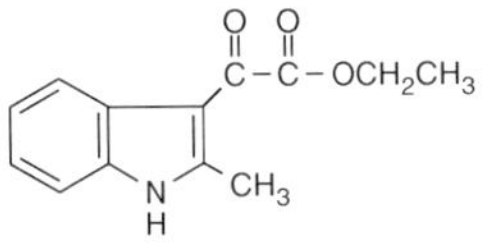

5 gm

S44,903-2
INDOLE-3-GLYOXYLAMIDE

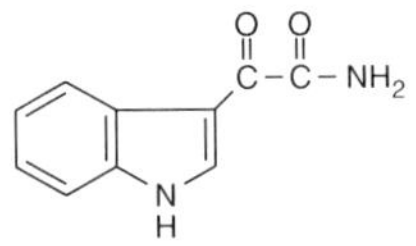

250 mg

S89,532-6
ETHYL 6-BROMO-3-[1-(ETHOXYCARBONYL)-2-OXOCYCLOHEXYLMETHYL]-2-INDOLECARBOXYLATE

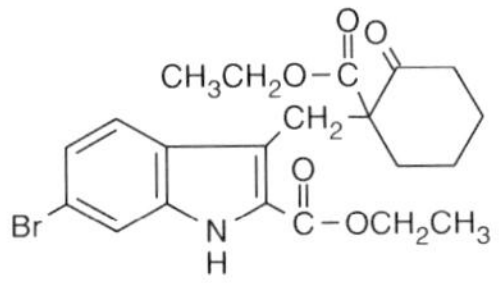

250 mg

S89,614-4
ETHYL 4-BROMO-3-[1-(ETHOXYCARBONYL)-2-OXOCYCLOHEXYLMETHYL]-2-INDOLECARBOXYLATE

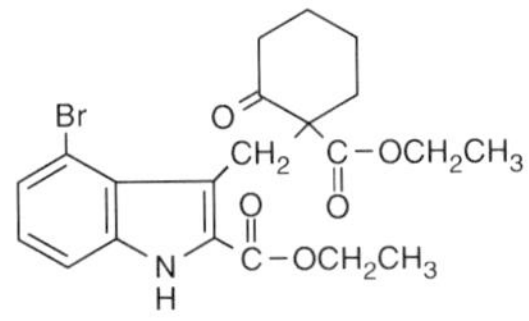

250 mg

S88,896-6
2-(3-INDOLYLCARBONYL)BENZOIC ACID

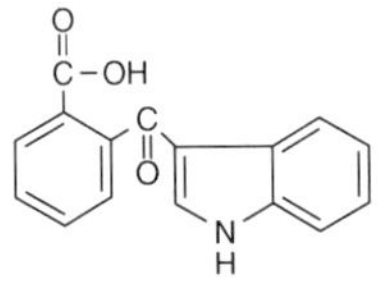

250 mg

S89,581-4
2-PROPIONYL-3-INDOLEACETIC ACID

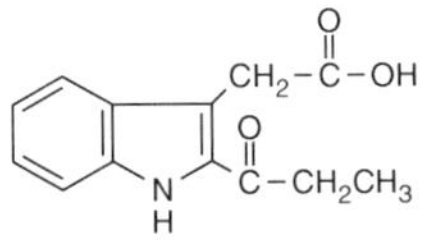

250 mg

S76,091-9
N-[2-(3-INDOLYL)ETHYL]-6-OXO-6-PHENYLHEXANAMIDE

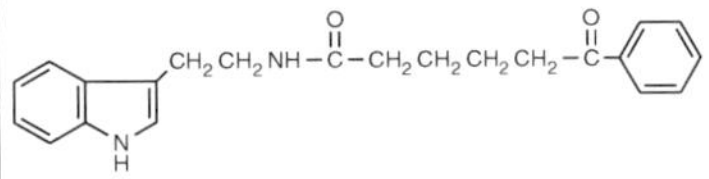

250 mg

S89,599-7
N-METHYL-2-PROPIONYL-3-INDOLEACETAMIDE

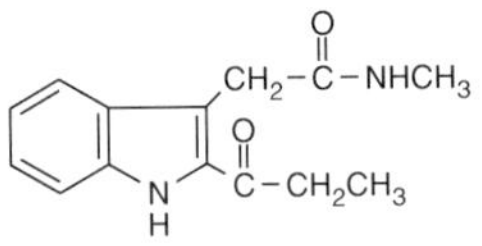

250 mg

S95,574-4
ALPHA-(1,4-DIACETYL-1,4-DIHYDROPYRIDIN-3-YL)-3,3'-METHYLENEDIINDOLE

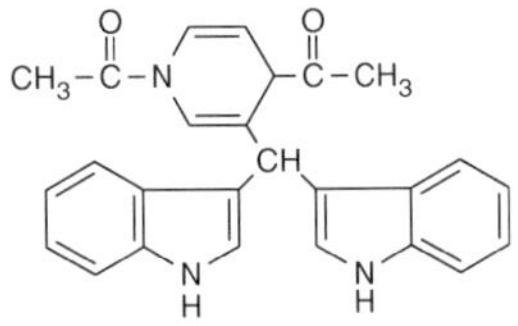

50 mg

S87,162-1
3-METHYL-1-PROPIONYLINDOLIZINE

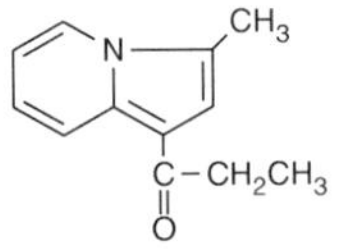

250 mg

S36,675-7
3-BENZOYLINDOLE

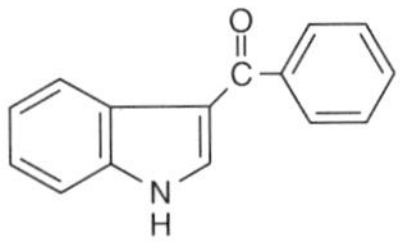

250 mg

S37,045-2
3-PROPIONYLINDOLE

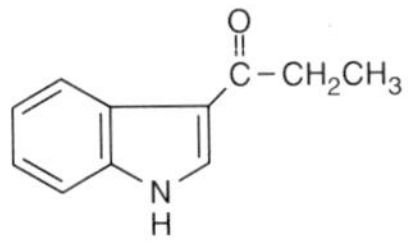

250 mg

S35,345-0
2-METHYL-3-PROPIONYLINDOLE

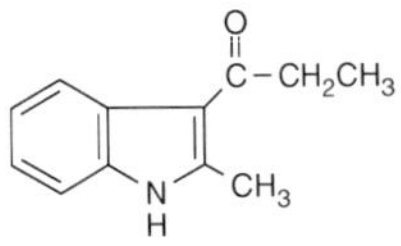

1 gm

S92,839-9
3-ACETYL-2-ETHYLINDOLE

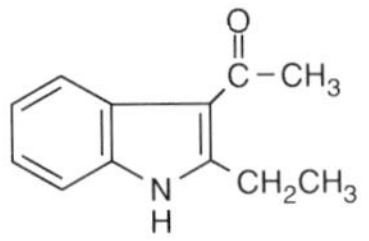

250 mg

S95,488-8
3-[1,1-BIS(3-INDOLYL)ETHYL]-1,4-DIACETYL-1,4-DIHYDROPYRIDINE

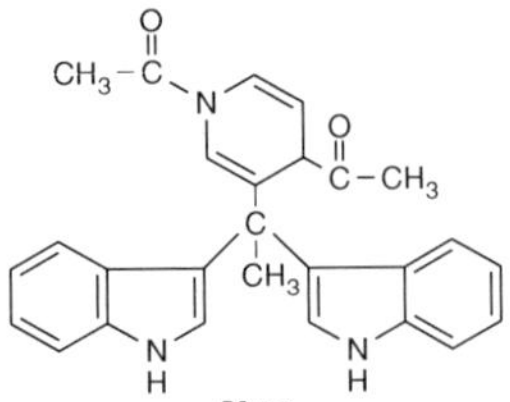

50 mg

S40,700-3
3-BUTYRYLINDOLE

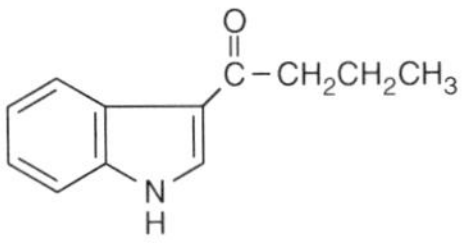

5 gm

S94,252-9
2-INDOLYL KETONE

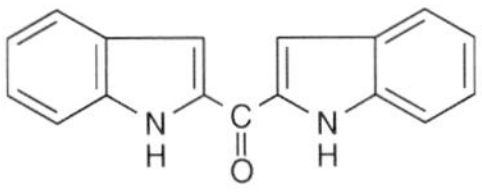

250 mg

S34,922-4
ETHYL 3-INDOLYLGLYOXYLATE

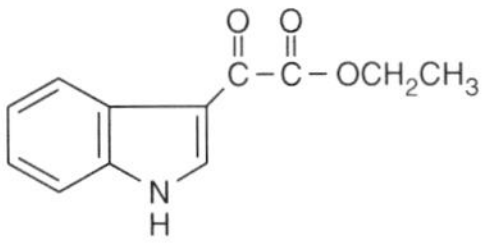

5 gm

S89,547-4
S-ETHYL 3-INDOLETHIOGLYOXYLATE

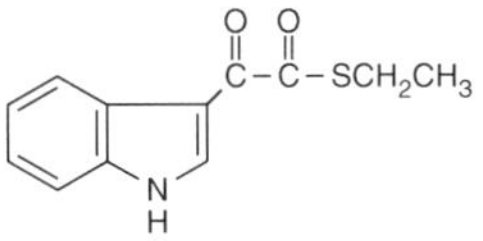

250 mg

S89,607-1
N-METHYL-N-PHENYL-3-INDOLEGLYOXYLAMIDE

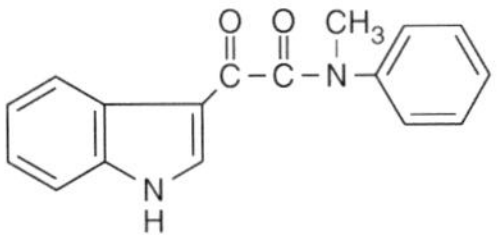

250 mg

S34,817-1
ETHYL ALPHA-(ETHOXYCARBONYL)-3-INDOLEACRYLATE

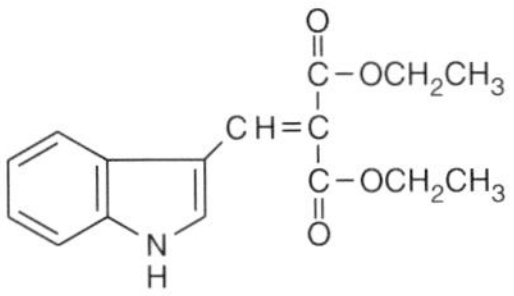

5 gm

S34,814-7
ETHYL 5-BROMO-ALPHA-(ETHOXYCARBONYL)-3-INDOLEACRYLATE

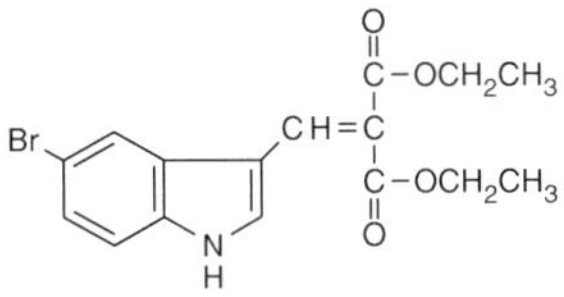

5 gm

S89,580-6
2-CARBOXY-3-INDOLEBUTYRIC ACID

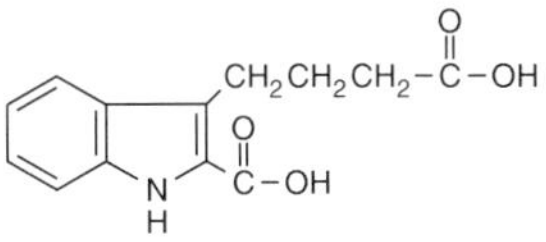

250 mg

S89,481-8
5-BROMO-2-CARBOXY-3-INDOLEBUTYRIC ACID

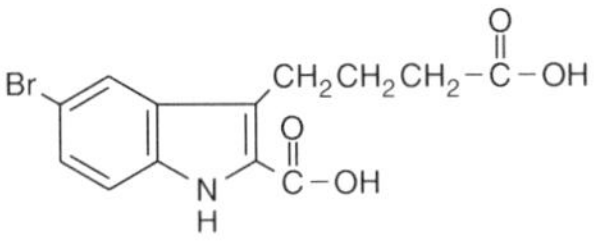

250 mg

S44,773-0
5-ETHYL-2-INDOLECARBOXYLIC ACID

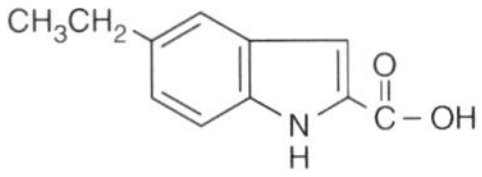

5 gm

S76,058-7
2,3-DIMETHYL-7-INDOLECARBOXYLIC ACID

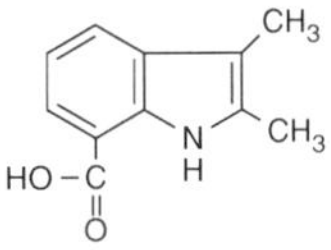

250 mg

S88,819-2
METHYL 3-METHYL-2-INDOLECARBOXYLATE

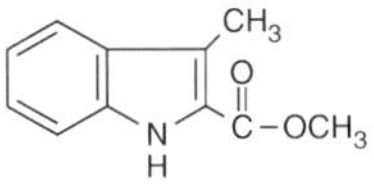

250 mg

S35,277-2
METHYL ALPHA-CYANO-3-INDOLEACRYLATE

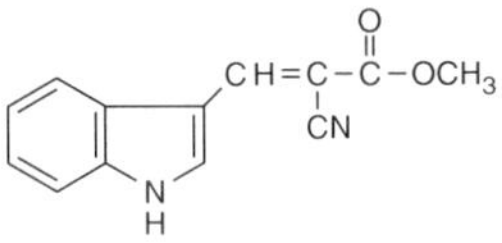

5 gm

S34,815-5
ETHYL 5-BROMO-ALPHA-CYANO-3-INDOLEACRYLATE

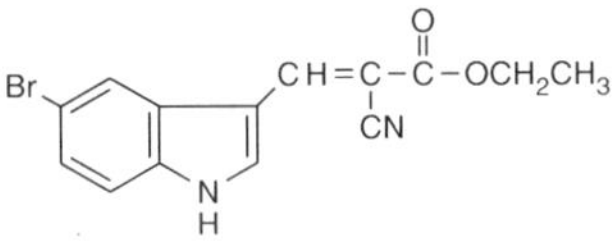

5 gm

S78,871-6
N-ALPHA-FURYLACRYLOYL-L-TRYPTOPHANAMIDE

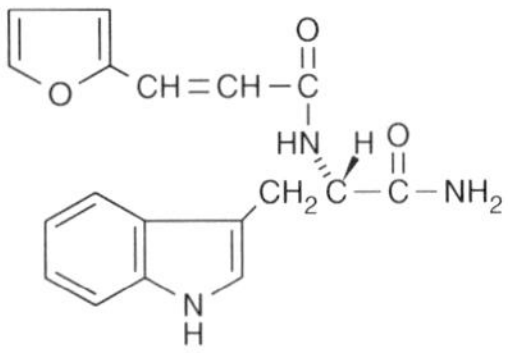

250 mg

S95,570-1
6-BENZAMIDO-N-[2-(3-INDOLYL)ETHYL] HEXANAMIDE

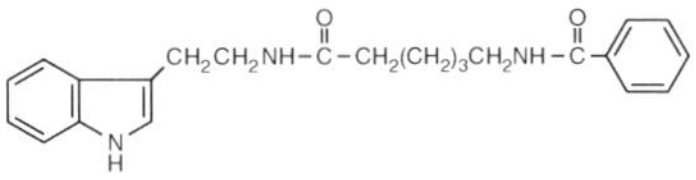

50 mg

S55,768-4
1-BENZOYLINDOLE

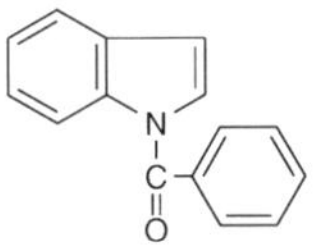

250 mg

S78,028-6
3-BUTYL-5-(3-INDOLYLMETHYLENE)-RHODANINE

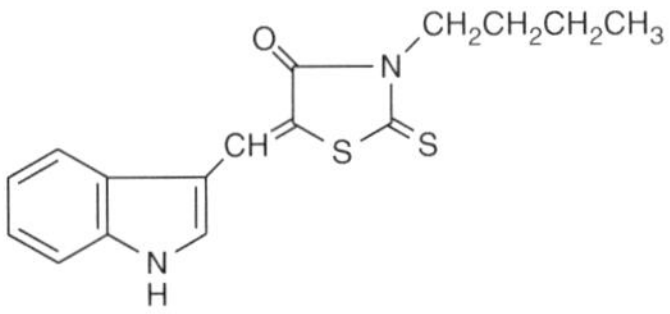

250 mg

S94,246-4
1,3-DIMETHYL-5-(3-INDOLYLMETHYLENE)-BARBITURIC ACID

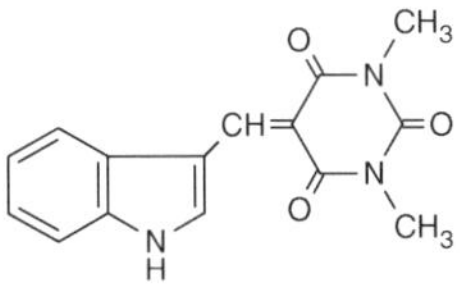

250 mg

S89,609-8
3-(4-INDOLYL)-5-METHYL-4-CYCLOHEXENE-1,2-DICARBOXYLIC ACID

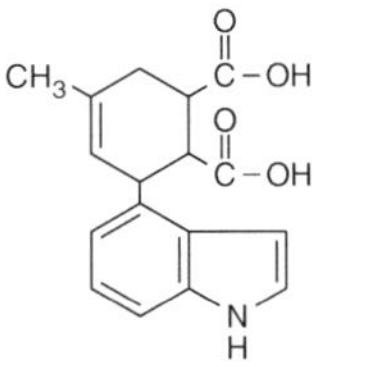

250 mg

S89,486-9
5-BROMO-2-(ETHOXYCARBONYL)-3-INDOLEBUTYRIC ACID

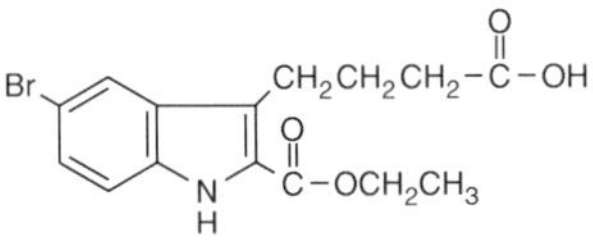

250 mg

S78,479-6
CARBOBENZYLOXY-L-TRYPTOPHYLGLYCINE

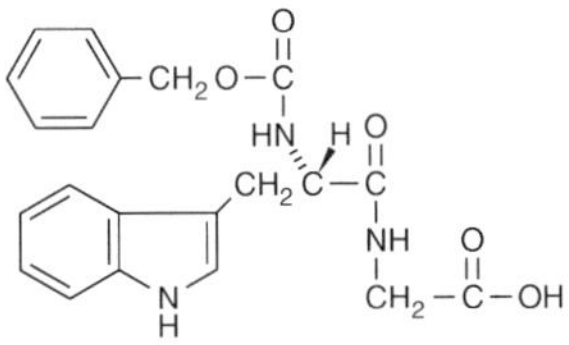

250 mg

S77,631-9
CARBOBENZYLOXY-SARCOSYL-L-TRYPTOPHAN

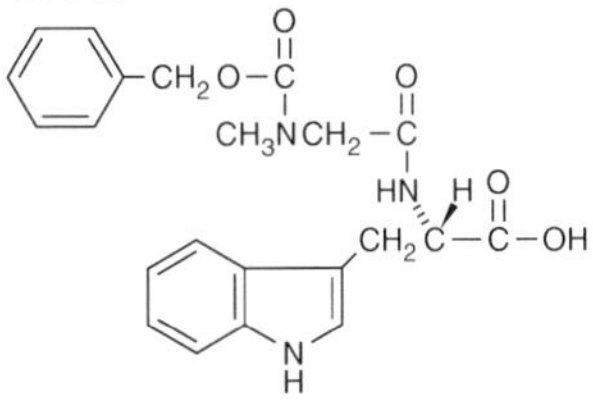

250 mg

S77,632-7
CARBOBENZYLOXY-L-ALANYL-L-TRYPTOPHAN

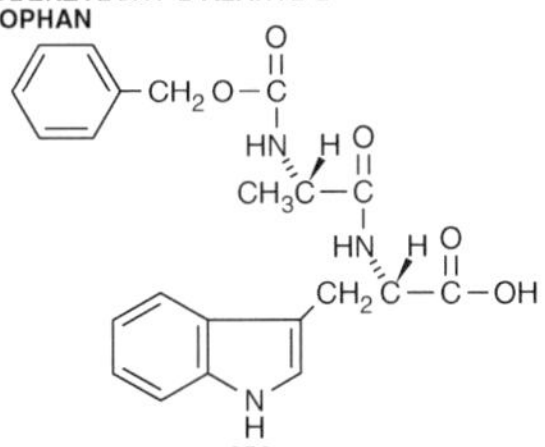

250 mg

S78,967-4
CARBOBENZYLOXY-L-TRYPTOPHYLGLYCYLGLYCINE

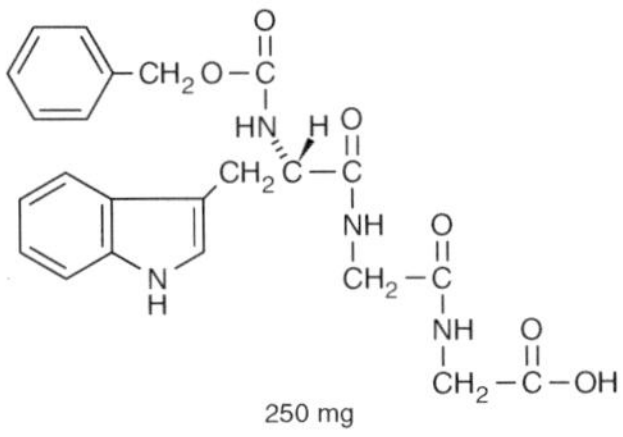

250 mg

S77,769-2
CARBOBENZYLOXY-L-NORVALYL-L-TRYPTOPHAN

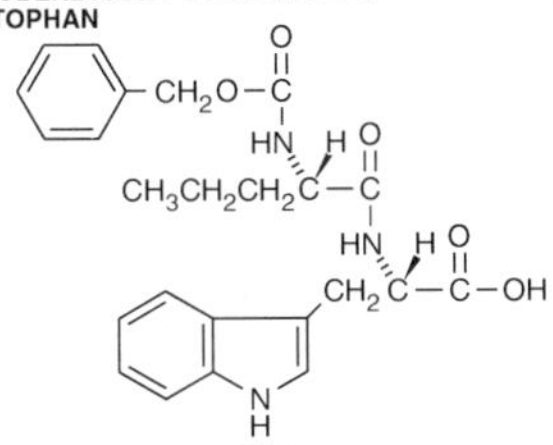

250 mg

S78,139-8
CARBOBENZYLOXY-L-ISOLEUCYL-L-TRYPTOPHAN

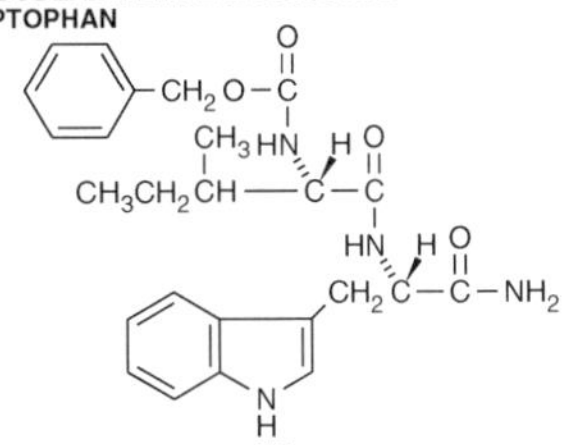

250 mg

S78,606-3
CARBOBENZYLOXY-L-TRYPTOPHYL-L-PHENYLALANINE

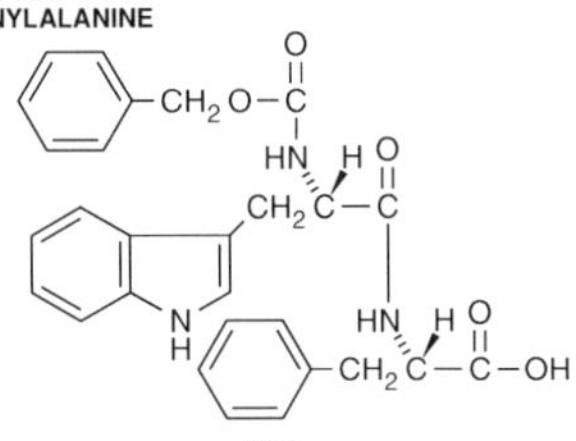

250 mg

S77,745-5
N-ACETYL-D-TRYPTOPHAN

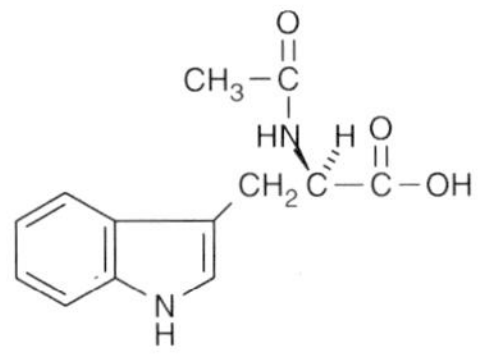

250 mg

S77,719-6
N-ALPHA-ACETYL-6-METHYL-DL-TRYPTOPHAN

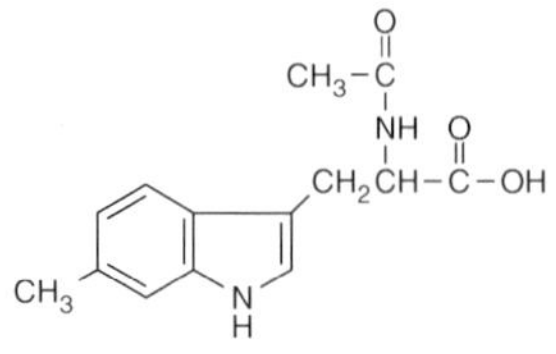

250 mg

S77,643-2
CARBOBENZYLOXY-L-TRYPTOPHYL-L-TRYPTOPHAN

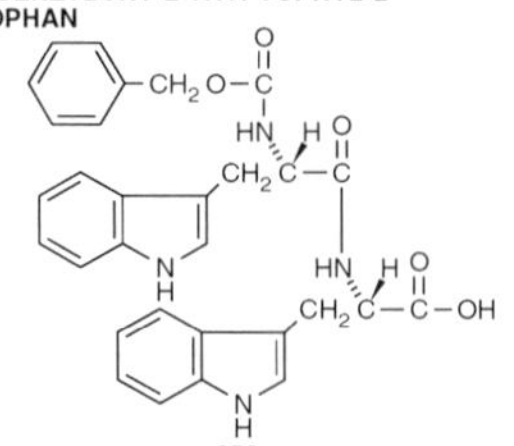

250 mg

S54,711-5
1-INDOLEACETIC ACID

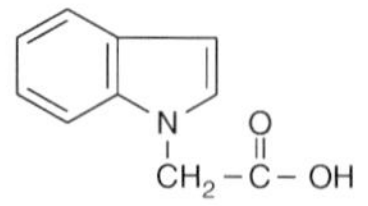

5 gm

S43,990-8
5-CHLORO-2-METHYL-3-INDOLEACETIC ACID

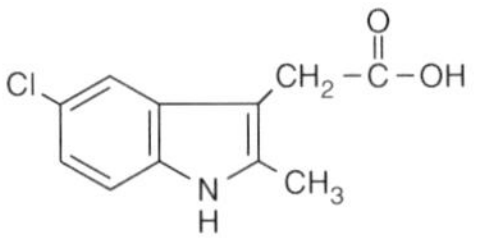

250 mg

S89,482-6
5-BROMO-3-INDOLEBUTYRIC ACID

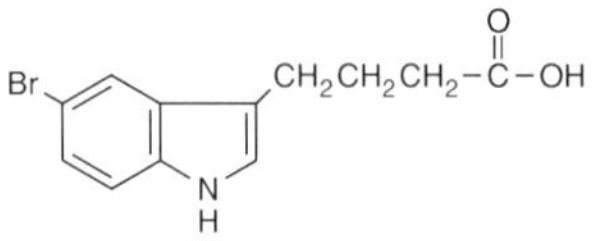

250 mg

S78,621-7
CARBOBENZYLOXY-L-TRYPTOPHYLGLYCYLGLYCINE METHYL ESTER

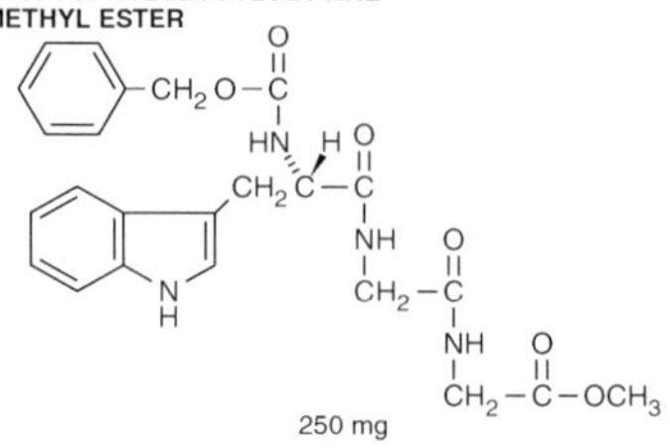

250 mg

S78,854-6
CARBOBENZYLOXY-L-ISOLEUCYL-L-TRYPTOPHAN METHYL ESTER

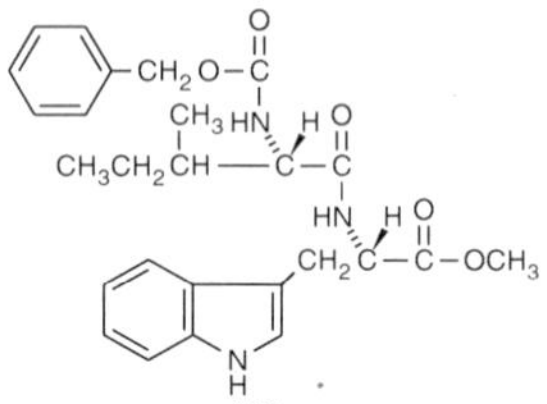

250 mg

S78,604-7
CARBOBENZYLOXYTRYPTOPHYLALANINE BENZYL ESTER

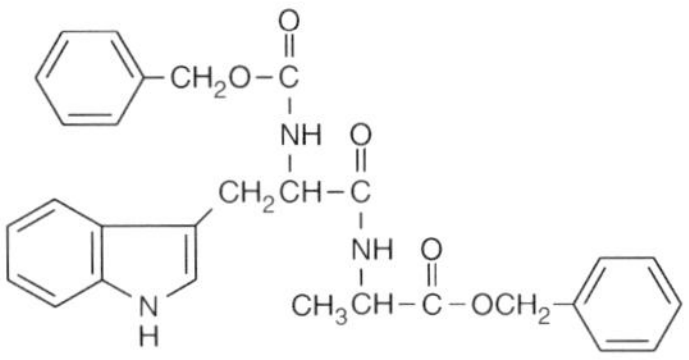

250 mg

S78,783-3
CARBOBENZYLOXYTRYPTOPHYLVALINE BENZYL ESTER

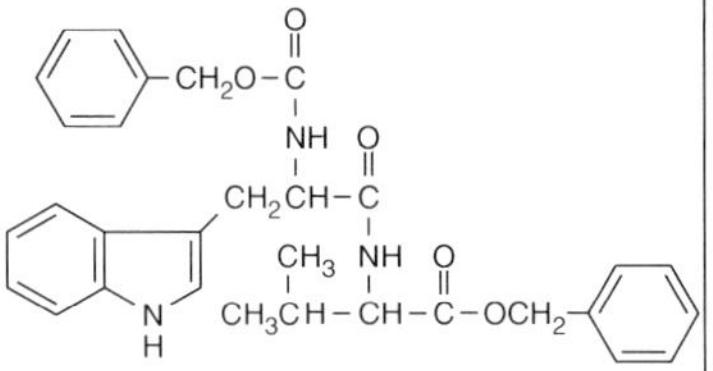

250 mg

S78,782-5
CARBOBENZYLOXYTRYPTOPHYLTRYPTOPHAN METHYL ESTER

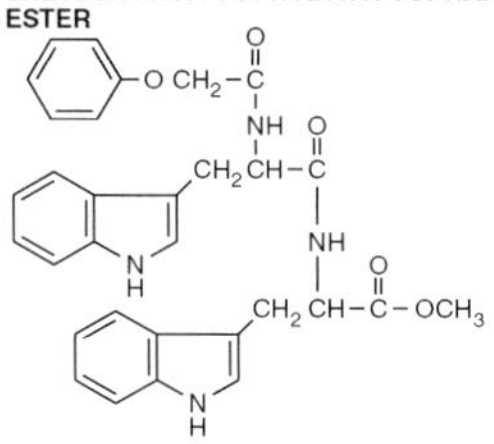

250 mg

S78,601-2
CARBOBENZYLOXY-L-TRYPTOPHYL-L-PHENYLALANINE BENZYL ESTER

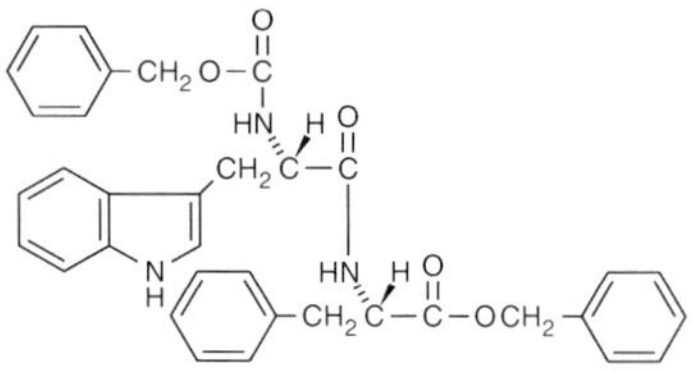

250 mg

S78,602-0
CARBOBENZYLOXY-L-TRYPTOPHYL-L-LEUCINE BENZYL ESTER

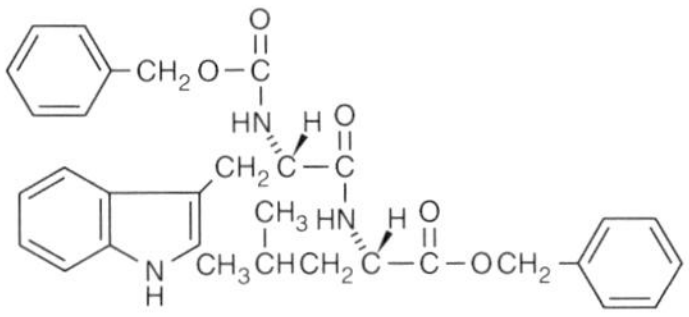

250 mg

S78,455-9
N-ACETYL-D-TRYPTOPHAN METHYL ESTER

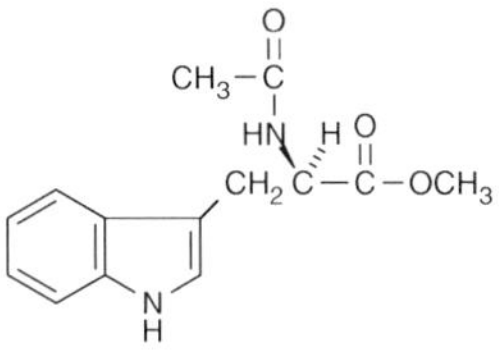

250 mg

S36,179-8
ETHYL 5-METHYLINDOLE-2-CARBOXYLATE

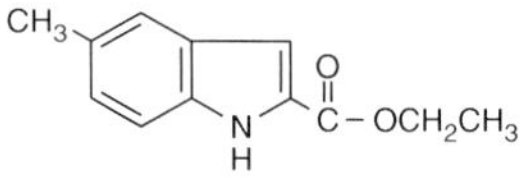

250 mg

S77,721-8
N-ALPHA-(TERT-BUTOXYCARBONYL)-L-TRYPTOPHAN-N-HYDROXYSUCCINIMIDE ESTER

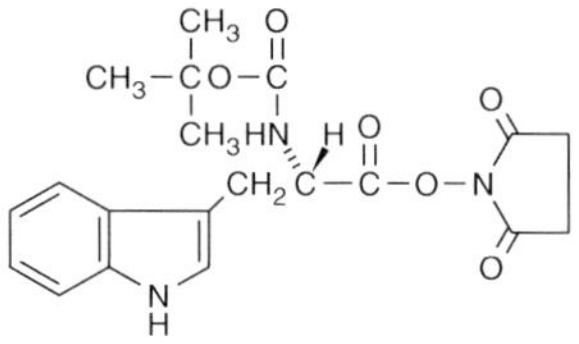

250 mg

S79,003-6
CARBOBENZYLOXYTRYPTOPHAN BETA-NAPHTHYLAMIDE

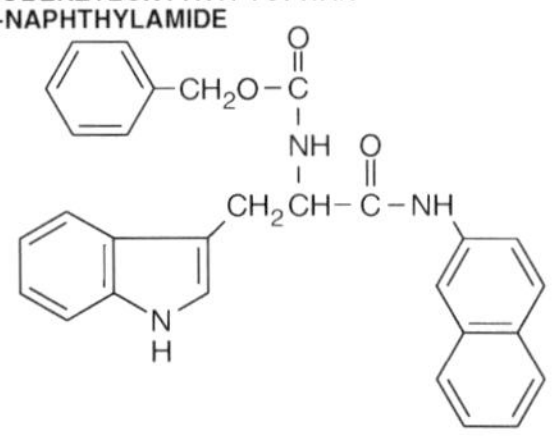

250 mg

S74,029-2
1-ACETYL-3-INDOLYL ACETATE

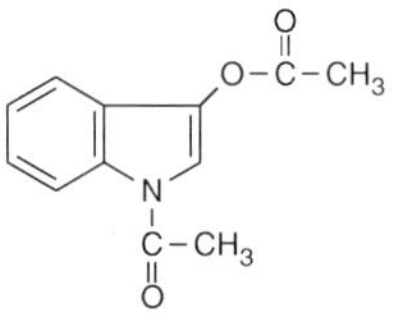

250 mg

S88,587-8
METHYL 3-METHYL-1-INDOLECARBOXYLATE

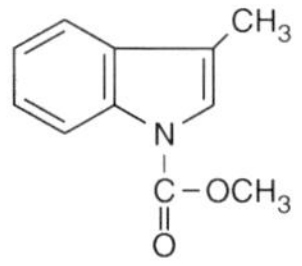

250 mg

S89,485-0
ETHYL 5-BROMO-3-INDOLEBUTYRATE

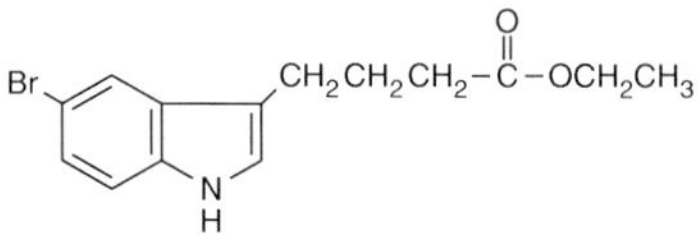

250 mg

S78,189-4
HEXYL 3-INDOLEACETATE

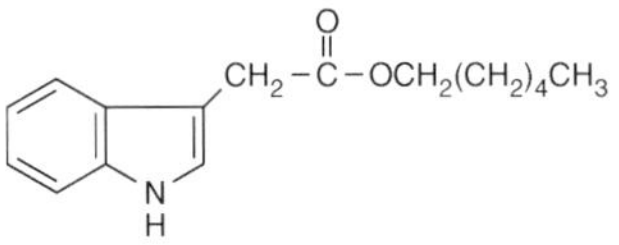

250 mg

S76,257-1
5-(3-INDOLYLMETHYLENE)RHODANINE

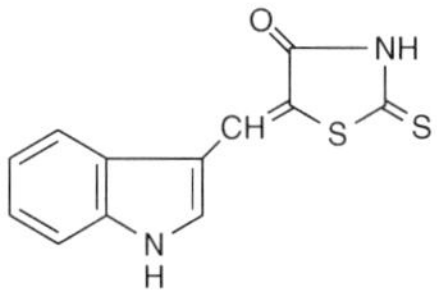

250 mg

S76,829-4
5-(3-INDOLYLMETHYLENE)-3-ISOPROPYLRHODANINE

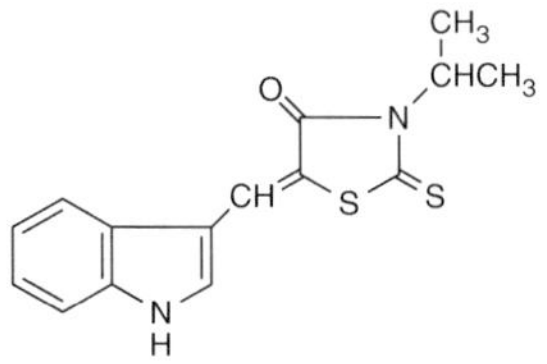

250 mg

S95,726-7
1,2,3,3A,4,5-HEXAHYDROPYRIDO(3,2,1-JK)CARBAZOL-6(6H)-ONE

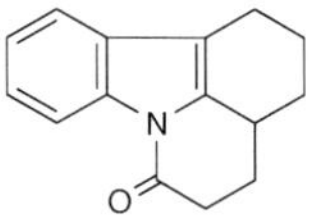

50 mg

S88,176-7
1-ACETYL-3-BROMOINDOLE

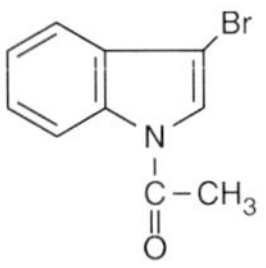

250 mg

S36,428-2
1-PROPIONYLINDOLE

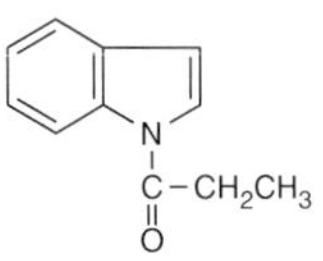

5 gm

S74,688-6
1-ACETYL-2,3-DIMETHYLINDOLE

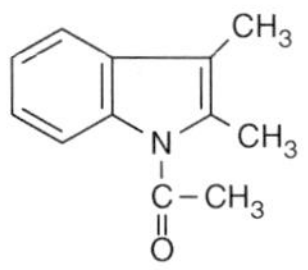

5 gm

S91,136-4
1-[N-(P-TOLYL)CARBAMOYL]INDOLE

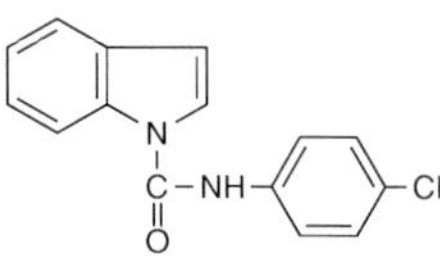

250 mg

S37,250-1
3-INDOLEPROPIONITRILE

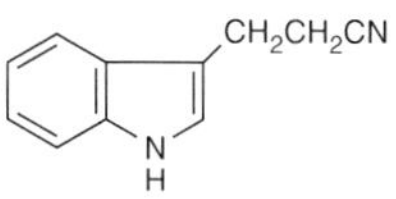

250 mg

S89,571-7
3-METHYL-4-INDOLEACETONITRILE

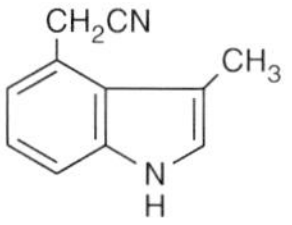

250 mg

S76,169-9
2-(1-METHYLCYCLOHEXYL)INDOLE

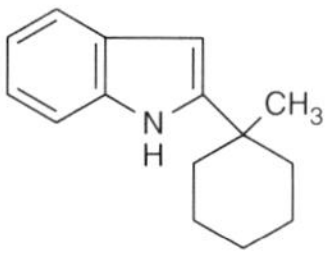

250 mg

S74,562-6
3-PHENYLINDOLE

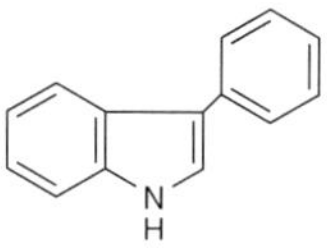

250 mg

S37,361-3
3-METHYL-2-PHENYLINDOLE

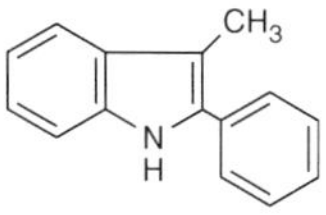

5 gm

S43,256-3
5-METHYL-2-PHENYLINDOLE

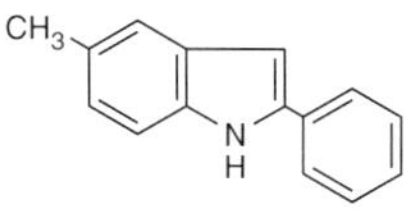

5 gm

S79,864-9
5,6-DIHYDROINDOLO(2,1-A)-ISOQUINOLINE

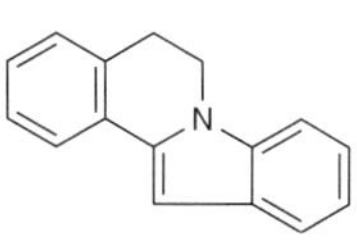

250 mg

S41,793-9
2,7-DIMETHYLINDOLE

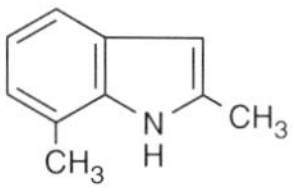

250 mg

S36,381-2
2-METHYL-3-PROPYLINDOLE

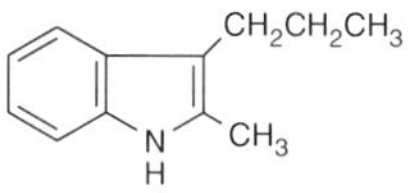

5 gm

S74,957-5
1-(1H-INDOL-3-YL)-1-(3H-INDOL-3-YL)-ETHENE

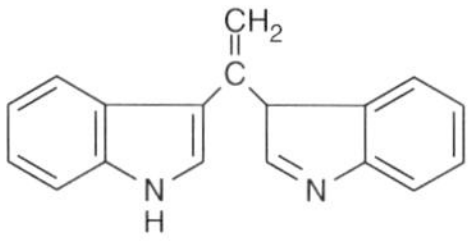

250 mg

S45,231-9
2,3-CYCLOOCTENOINDOLE

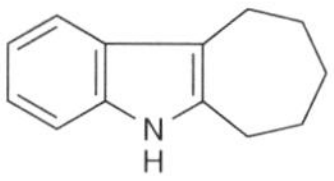

250 mg

S83,490-4
9-(CHLOROACETYL)-3-NITROCARBAZOLE

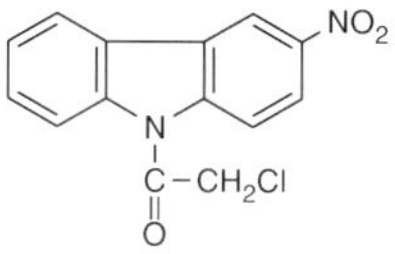

250 mg

S83,489-0
9-(BROMOACETYL)-3-NITROCARBAZOLE

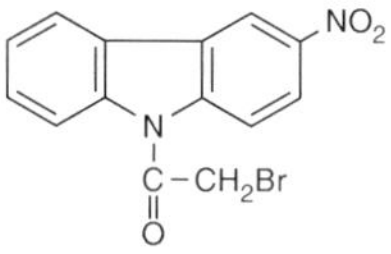

250 mg

S83,491-2
9-(CHLOROACETYL)CARBAZOLE

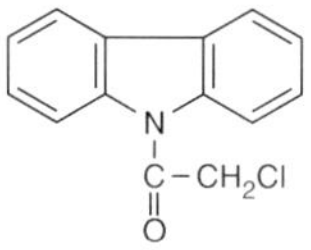

250 mg

S89,633-0
9-(2-IODOACETYL)CARBAZOLE

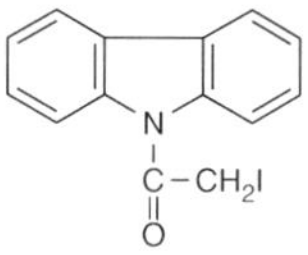

250 mg

S83,846-2
4'-CHLORO-2-HYDROXY-3-CARBAZOLECARBOXANILIDE

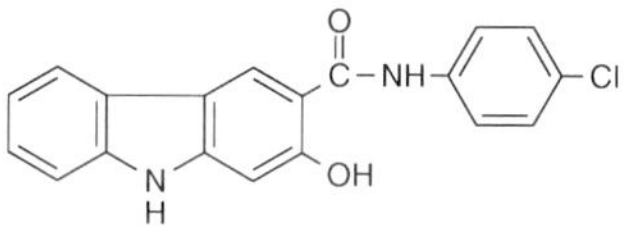

250 mg

S78,000-6
9-[N-(4-CHLOROPHENYL)CARBAMOYL] CARBAZOLE

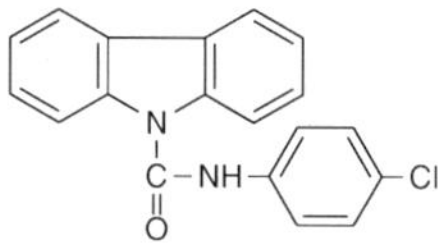

250 mg

S41,850-1
9-CARBAZOLEMETHANOL

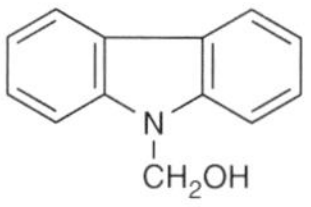

5 gm

S82,849-1
9-(2-NITROPHENYL)CARBAZOLE

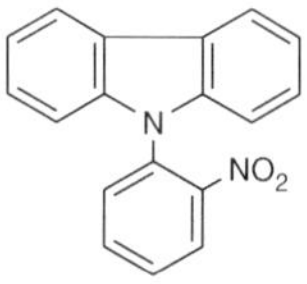

250 mg

S83,559-5
2,7-DINITRO-9-PHENYLCARBAZOLE

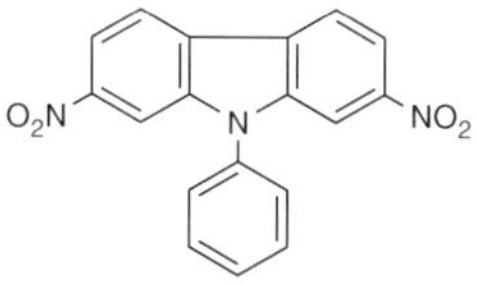

250 mg

S38,288-4
9-CARBAZOLEPROPIONITRILE

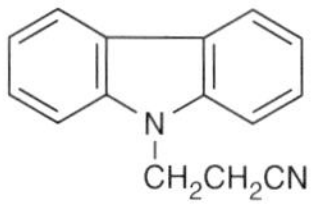

250 mg

S82,804-1
N,N'-(1,4-PHENYLENE)DICARBAZOLE

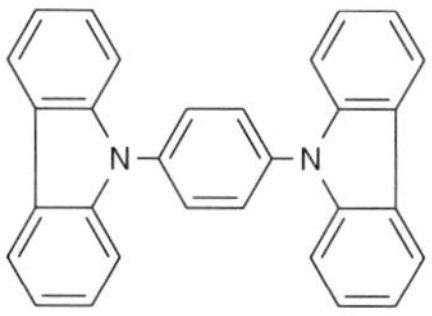

250 mg

S37,546-2
9-ETHYL-3-NITROCARBAZOLE

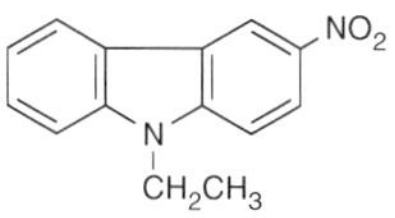

5 gm

S57,761-8
9-(P-NITROPHENYL)CARBAZOLE

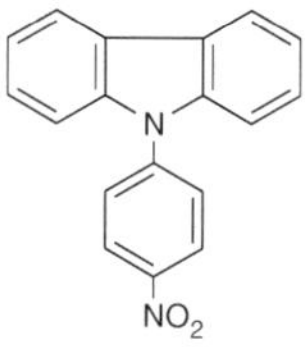

250 mg

S82,802-5
3-BENZAMIDO-9-ETHYLCARBAZOLE

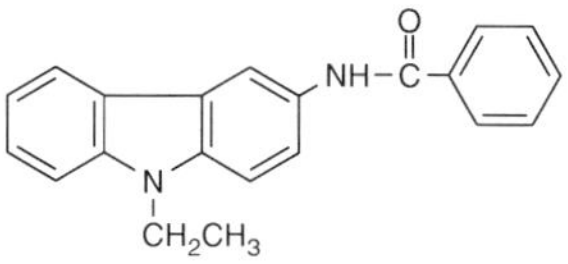

250 mg

S83,845-4
3-ACETAMIDO-9-ETHYLCARBAZOLE

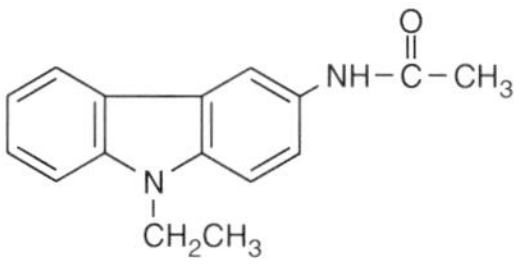

250 mg

S67,421-4
9-ETHYL-3-(4-METHYLBENZAMIDO)-CARBAZOLE

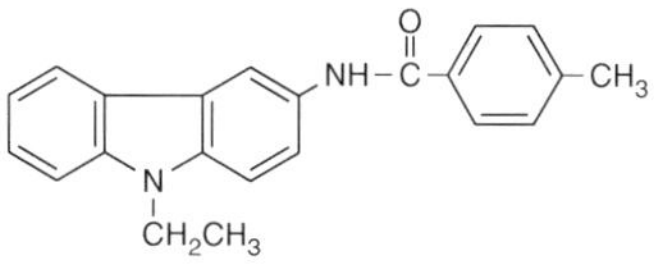

250 mg

S41,942-7
9-NITROSOCARBAZOLE

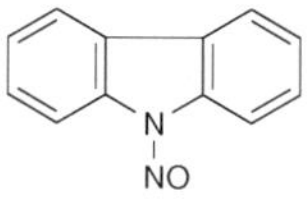

5 gm

S42,582-6
9-(MORPHOLINOMETHYL)CARBAZOLE

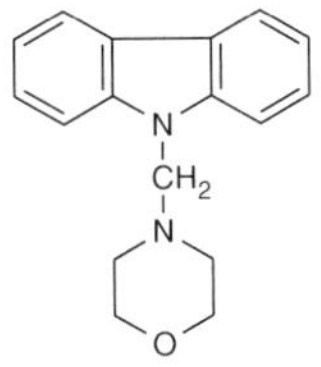

250 mg

S42,476-5
9-(DIMETHYLAMINOMETHYL)CARBAZOLE

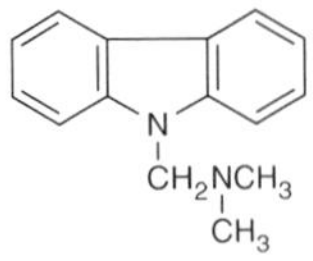

250 mg

S42,462-5
9-(DIETHYLAMINOMETHYL)CARBAZOLE

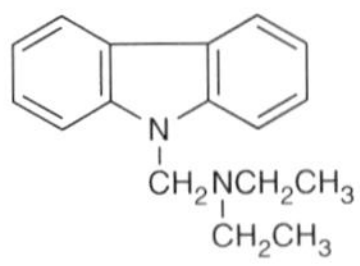

250 mg

S63,002-0
3,6-DIACETYL-9-ETHYLCARBAZOLE

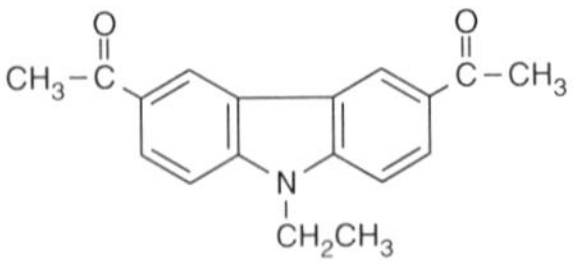

250 mg

S81,862-3
3-CHLORO-9-ETHYLCARBAZOLE

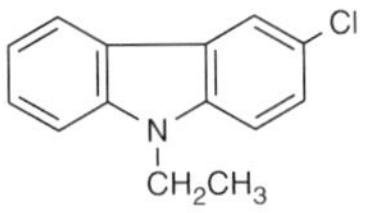

250 mg

S81,178-5
3-BROMO-9-ETHYLCARBAZOLE

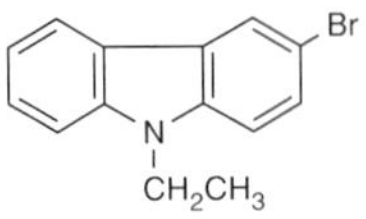

250 mg

S81,179-3
3-BROMO-6-CHLORO-9-ETHYLCARBAZOLE

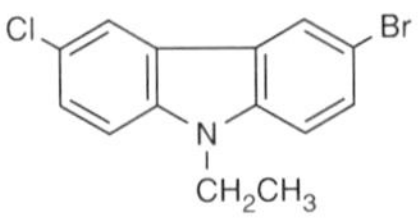

250 mg

S81,736-8
9-ETHYL-3-IODOCARBAZOLE

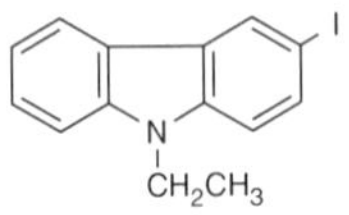

250 mg

S83,593-5
9-ETHYL-3-CARBAZOLEPHOSPHONOUS ACID

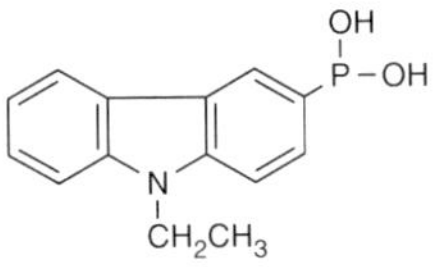

250 mg

S82,032-6
2,8-DIBROMO-9-ETHYLCARBAZOLE

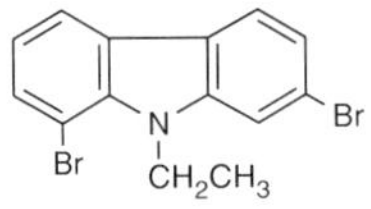

250 mg

S82,031-8
2,8-DIIODO-9-ETHYLCARBAZOLE

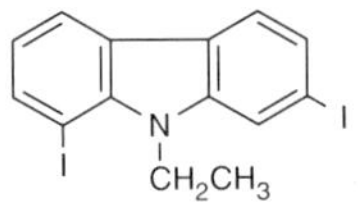

250 mg

S82,033-4
3,6-DIIODO-9-ETHYLCARBAZOLE

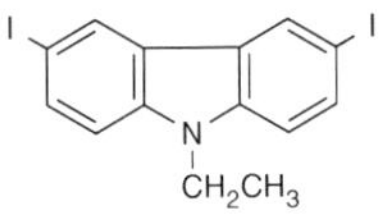

250 mg

S82,800-9
9-ALLYLCARBAZOLE

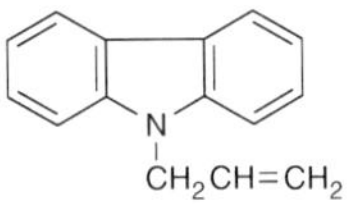

250 mg

S79,245-4
11-METHYL-11H-BENZO(A)CARBAZOLE

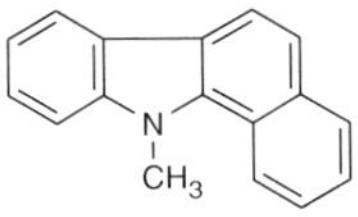

250 mg

S65,589-9
9-(ETHOXYCARBONYLMETHYL)CARBAZOLE

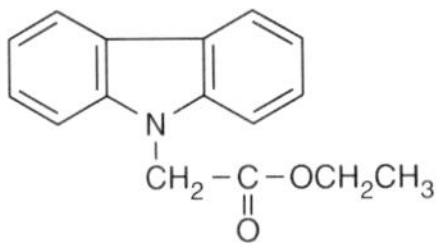

250 mg

S72,829-2
9-ETHYL-3-(3-NITROPHENYLCARBOXAMIDO)CARBAZOLE

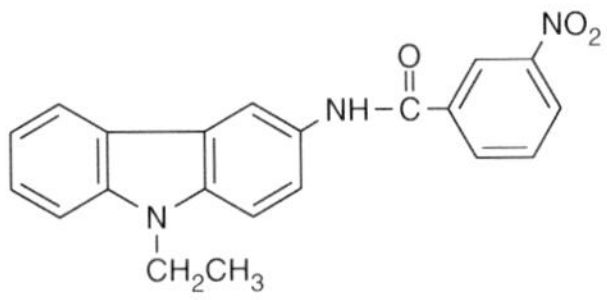

250 mg

S75,827-2
3-NITROCARBAZOLE

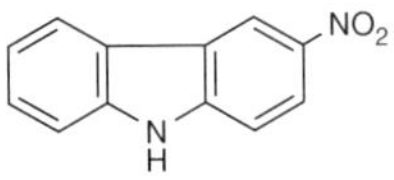

250 mg

S64,173-1
2,9-DIACETYLCARBAZOLE

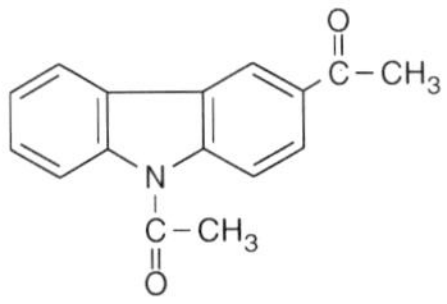

5 gm

S63,998-2
2-ACETYLCARBAZOLE

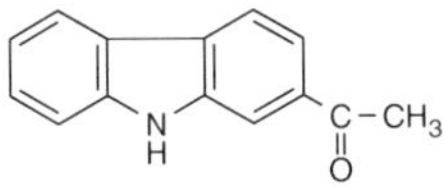

250 mg

S64,057-3
3-ACETYLCARBAZOLE

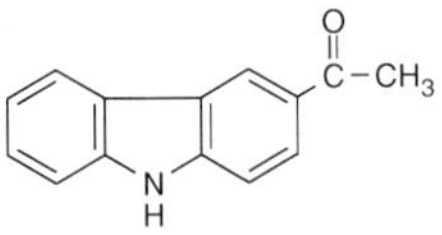

250 mg

S38,889-0
9-CARBAZOLEPROPIONIC ACID

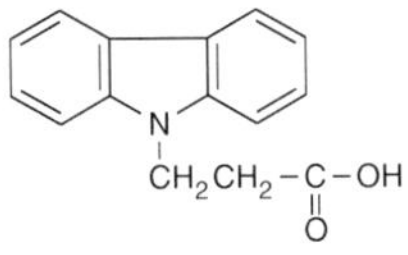

5 gm

S52,943-5
9-ACETYLCARBAZOLE

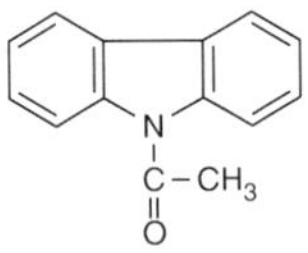

250 mg

S81,353-2
9-OCTADECANOYLCARBAZOLE

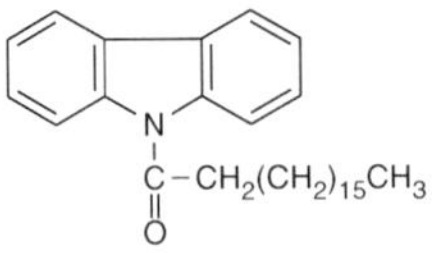

250 mg

S82,099-7
9-OCTANOYLCARBAZOLE

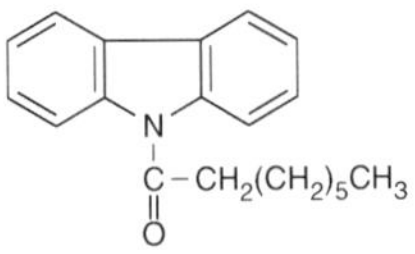

250 mg

S63,995-8
1,3,6,8-TETRA-TERT-BUTYLCARBAZOLE

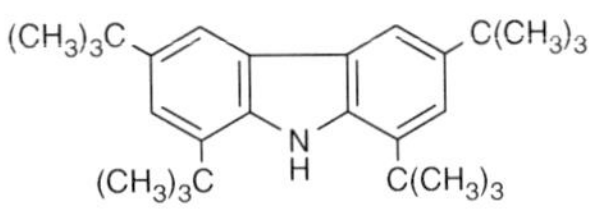

250 mg

S82,806-8
N,N'-(4,4'-BIPHENYLENE)DICARBAZOLE

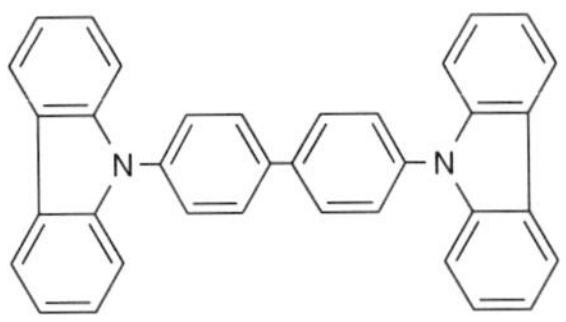

250 mg

S94,396-7
13H-DIBENZO(A,I)CARBAZOLE

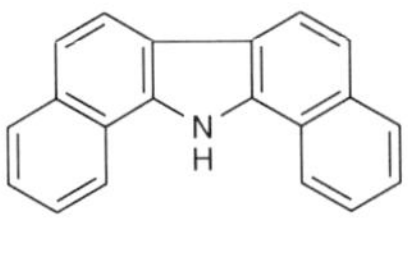

100 mg

S81,163-7
2-(CHLOROACETYL)DIBENZOFURAN

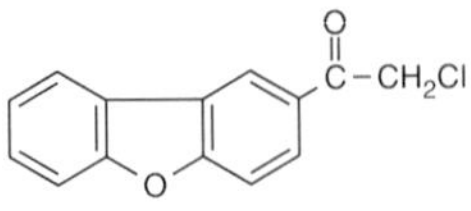

250 mg

S81,164-5
2-(BROMOACETYL)DIBENZOFURAN

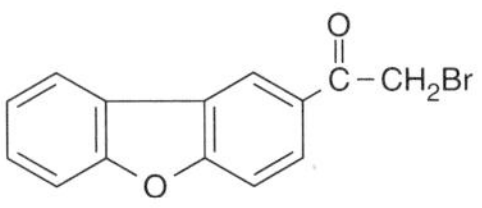

250 mg

S81,206-4
2-(2-BROMOETHYL)DIBENZOFURAN

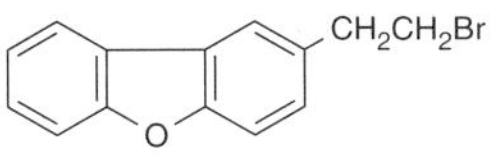

250 mg

S81,494-6
1,9-DIBROMO-3-NITRO-2,8-DIBENZOFURANDIOL

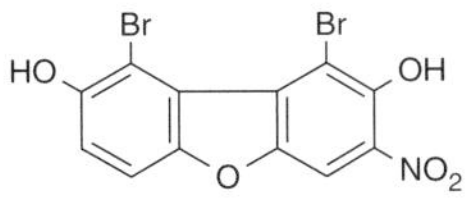

250 mg

S83,753-9
3-HYDROXY-4-METHOXYDIBENZOFURAN

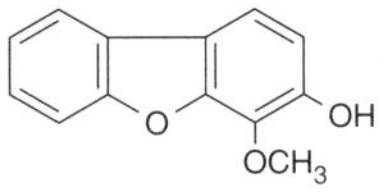

250 mg

S83,751-2
1,3-DIBROMO-4-HYDROXY-6-METHOXYDIBENZOFURAN

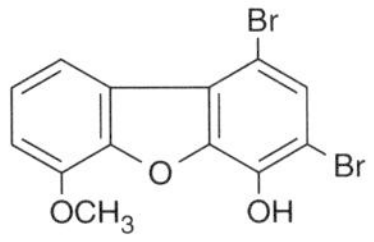

250 mg

S83,679-6
3-ALLYL-4-HYDROXYDIBENZOFURAN

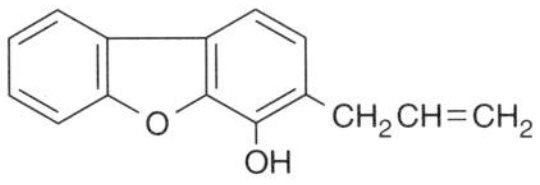

250 mg

S83,874-8
TRIS(4-DIBENZOFURYL)METHANOL

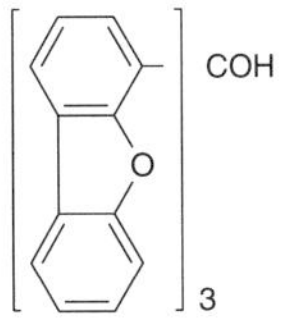

250 mg

S81,205-6
4-(2-HYDROXYETHYL)DIBENZOFURAN

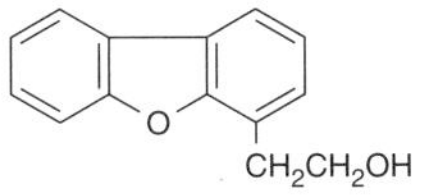

250 mg

S35,072-9
2-HYDROXY-3-DIBENZOFURANCARBOXYLIC ACID

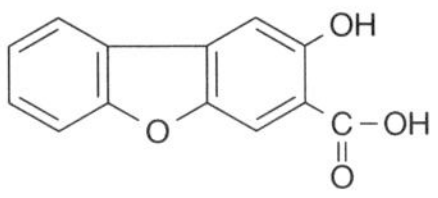

5 gm

S83,820-9
2-(2-HYDROXYETHYL)DIBENZOFURAN

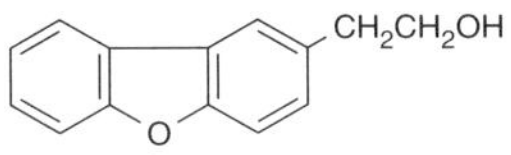

250 mg

S83,762-8
1,1'-OXALYLBIS(4-METHOXYDIBENZOFURAN)

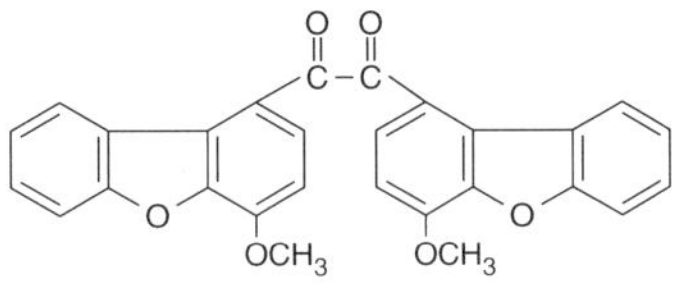

250 mg

S83,854-3
4,6-DIMETHOXYDIBENZOFURAN

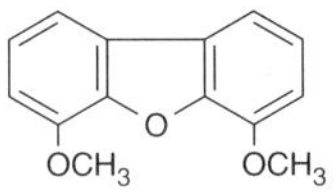

250 mg

S83,755-5
1-BROMO-3,4-DIMETHOXYDIBENZOFURAN

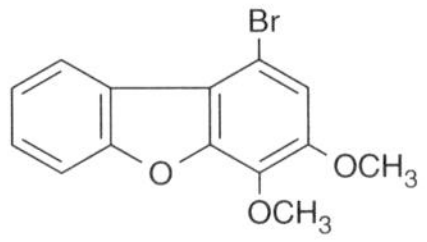

250 mg

S83,759-8
4,4'-BI(6-METHOXYDIBENZOFURAN)

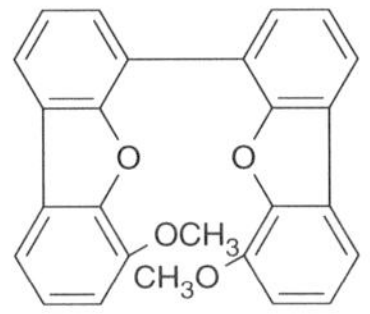

250 mg

S83,760-1
1,1'-METHYLENEBIS(4-METHOXYDIBENZOFURAN)

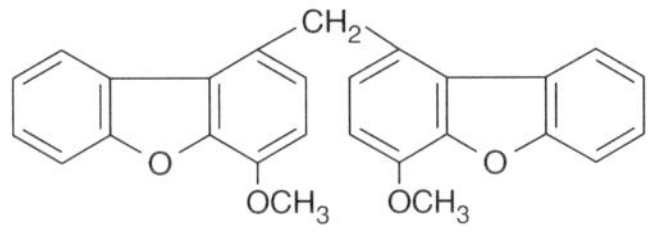

250 mg

S81,493-8
1,9-DIBROMO-2,8-DIMETHOXY-3-NITRODIBENZOFURAN

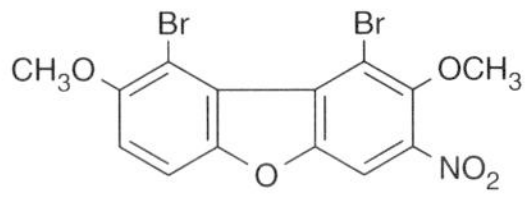

250 mg

S81,487-3
2,2'-(DIBENZOFURAN-2,8-DIYLDIOXY)-DIACETIC ACID

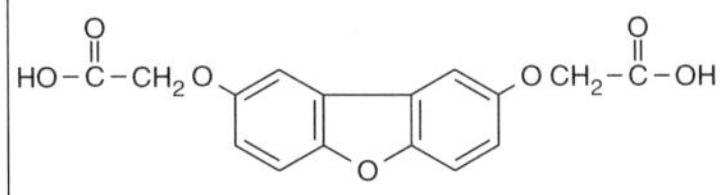

250 mg

S83,853-5
2,8-DIMETHOXYDIBENZOFURAN

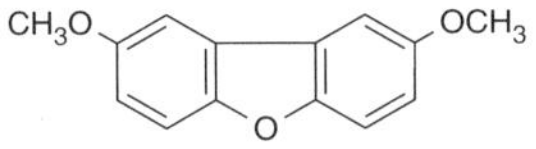

250 mg

S81,491-1
1,9-DIBROMO-2,8-DIMETHOXYDIBENZOFURAN

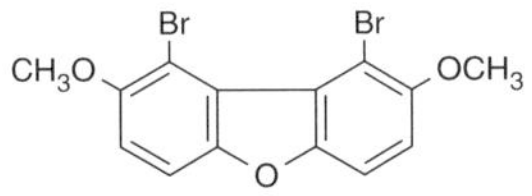

250 mg

S81,486-5
2,8-DIALLYLOXYDIBENZOFURAN

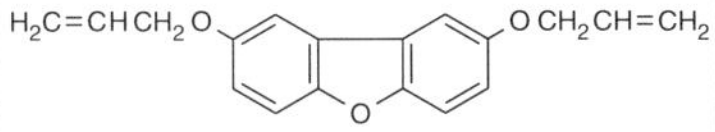

250 mg

S81,490-3
2,8-DIMETHOXY-1,3,7,9-TETRANITRODIBENZOFURAN

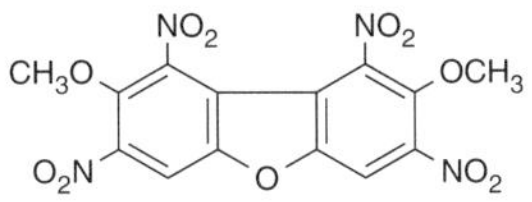

250 mg

S83,752-0
3-AMINO-1-BROMO-4-METHOXYDIBENZOFURAN

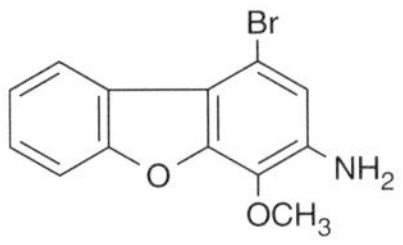

250 mg

S81,097-5
4-METHOXYDIBENZOFURAN PICRATE

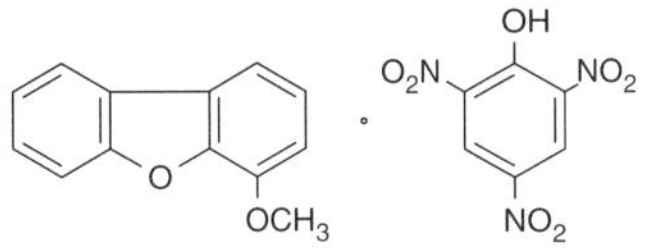

250 mg

S81,391-5
2-METHOXY-1-DIBENZOFURANCARBOXYLIC ACID

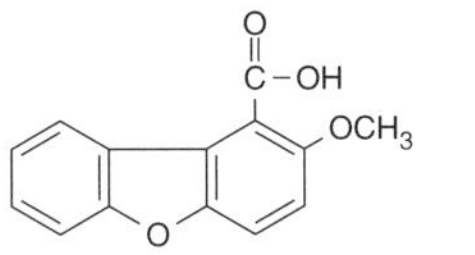

250 mg

S83,675-3
3-METHOXYDIBENZOFURAN

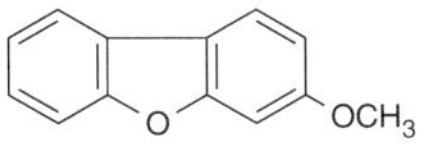

250 mg

S89,329-3
1-(3-DIBENZOFURYL)-3-[4-(2-METHOXYPHENYL)PIPERAZINO]-1-PROPANONE

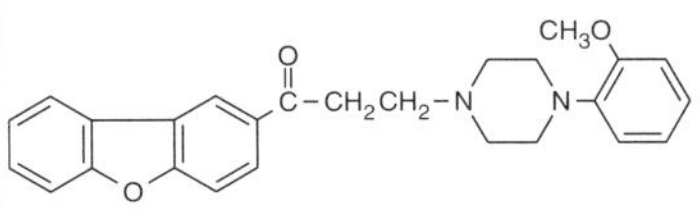

250 mg

S83,852-7
1,3-DIBROMO-4-METHOXYDIBENZOFURAN

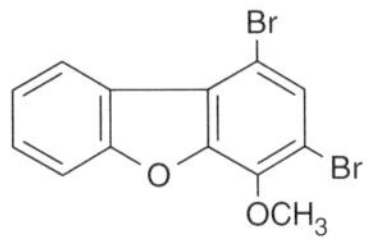

250 mg

S83,546-3
5-BROMO-2-BENZOFURANCARBOXYLIC ACID

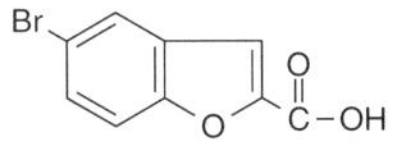

250 mg

S84,052-1
5,7-DIBROMO-2-BENZOFURANCARBOXYLIC ACID
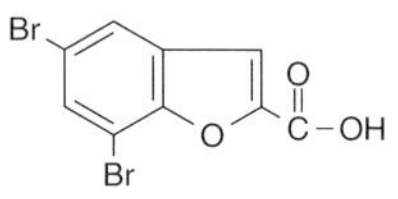
250 mg

S81,527-6
2-BENZOFURANCARBOXAMIDE
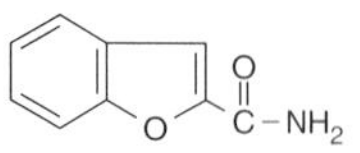
250 mg

S81,125-4
4-DIBENZOFURYL ACETATE
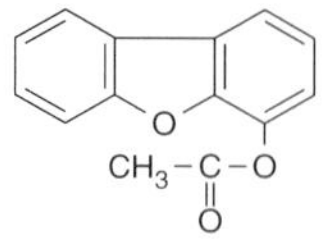
250 mg

S81,387-7
7-NITRO-4-DIBENZOFURANCARBOXAMIDE
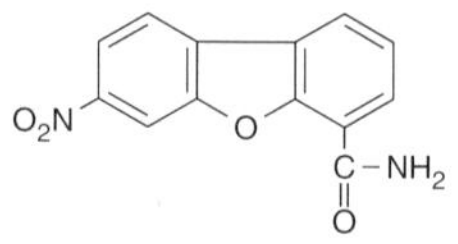
250 mg

S81,015-0
8-CHLORO-4-DIBENZOFURANCARBOXYLIC ACID
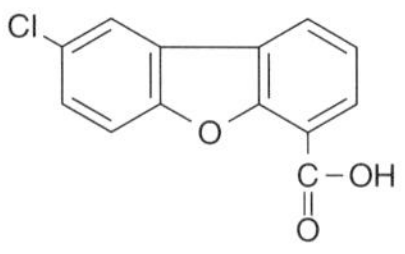
250 mg

S81,022-3
METHYL 4-DIBENZOFURANCARBOXYLATE
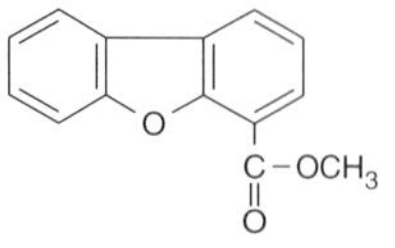
250 mg

S81,021-5
METHYL 2-CHLORO-4-DIBENZOFURANCARBOXYLATE
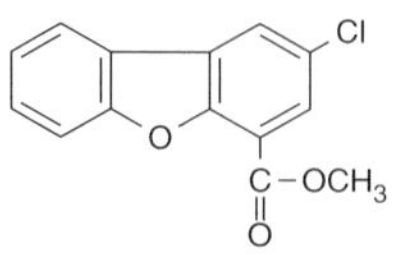
250 mg

S81,020-7
METHYL 8-BROMO-4-DIBENZOFURANCARBOXYLATE
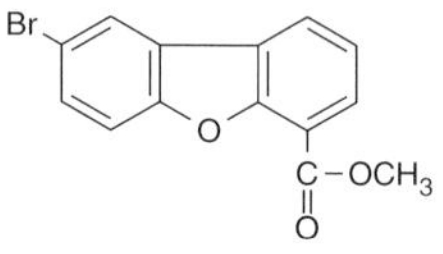
250 mg

S83,872-1
N,N-DIMETHYL-4-DIBENZOFURANCARBOXAMIDE
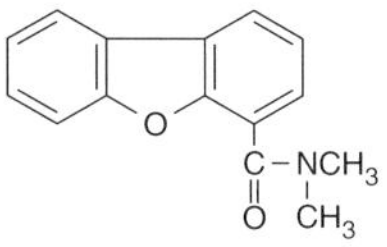
250 mg

S83,927-2
METHYL 1-DIBENZOFURANCARBOXYLATE
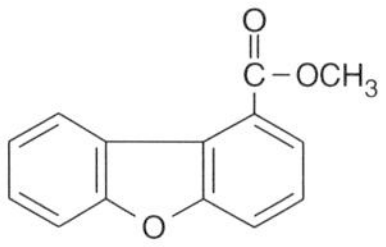
250 mg

S81,207-2
4-DIBENZOFURANACETIC ACID
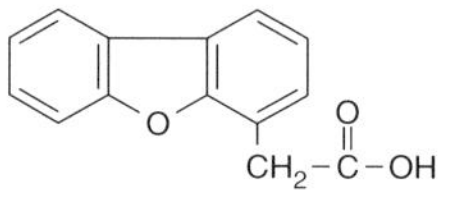
250 mg

S81,966-2
3-AMINO-2-NITRODIBENZOFURAN
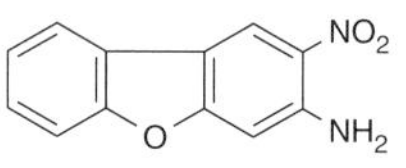
250 mg

S81,235-8
7-AMINO-2-NITRODIBENZOFURAN
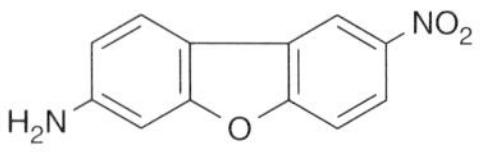
250 mg

S81,012-6
2,7-DIAMINODIBENZOFURAN
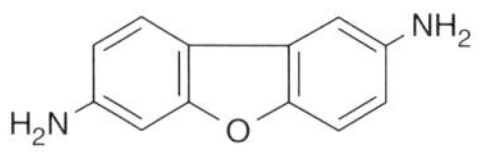
250 mg

S81,094-0
3-AMINODIBENZOFURAN
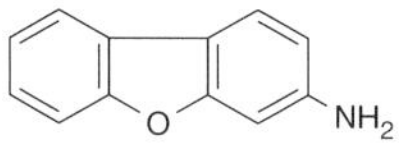
250 mg

S81,544-6
3-AMINODIBENZOFURAN HYDROCHLORIDE
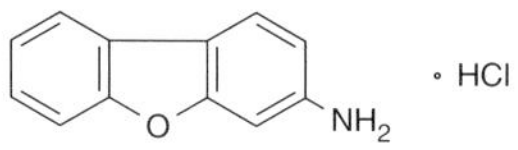
250 mg

S82,255-8
2-AMINODIBENZOFURAN HYDROCHLORIDE
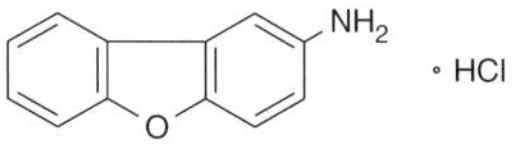
250 mg

S83,746-6
7-AMINO-1,4-DIHYDRODIBENZOFURAN HYDROCHLORIDE
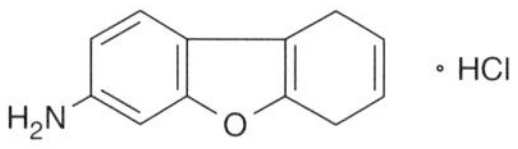
250 mg

S81,095-9
3-AMINO-2-CHLORODIBENZOFURAN
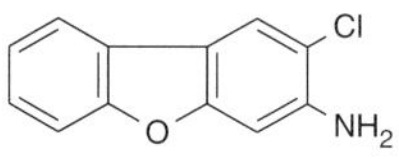
250 mg

S83,271-5
3-AMINO-2-BROMODIBENZOFURAN HYDROCHLORIDE
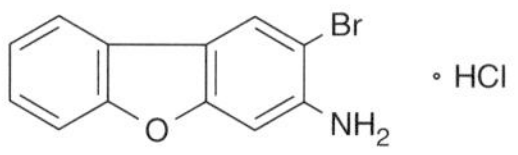
250 mg

S83,764-4
3-AMINO-2-BROMODIBENZOFURAN
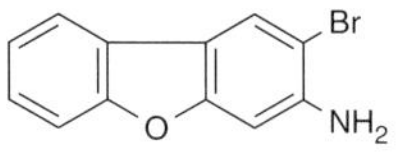
250 mg

S83,767-9
4-(2-AMINOETHYL)DIBENZOFURAN HYDROCHLORIDE
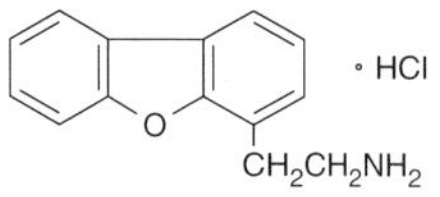
250 mg

S84,074-2
3-ACETAMIDO-8-ACETYL-2-NITRODIBENZOFURAN
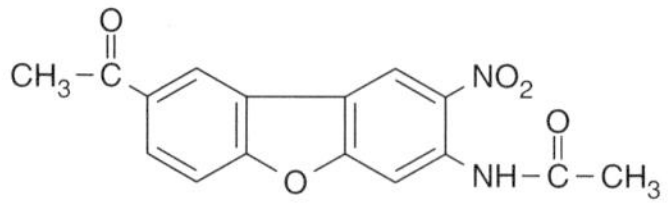
250 mg

S83,913-2
3-(DIACETYLAMINO)-2-NITRODIBENZOFURAN
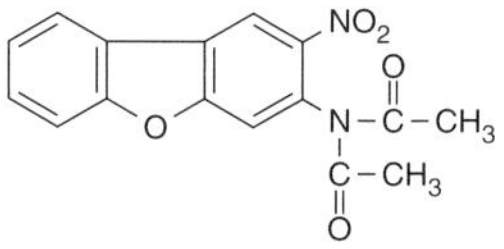
250 mg

S82,257-4
3-ACETAMIDO-2-NITRODIBENZOFURAN
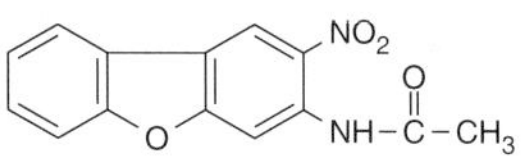
250 mg

S81,204-8
3-(ETHYLAMINO)DIBENZOFURAN HYDROCHLORIDE
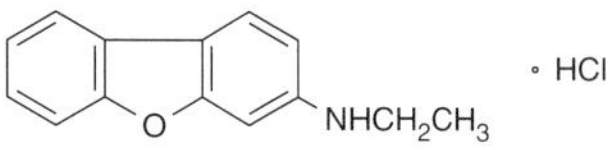
250 mg

S83,768-7
3-(PROPYLAMINO)DIBENZOFURAN
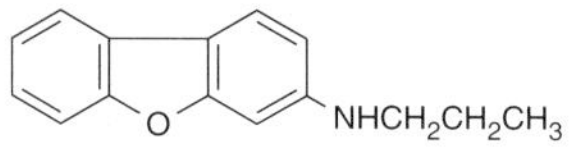
250 mg

S83,851-9
2-(2-PIPERIDINOACETYL)DIBENZOFURAN
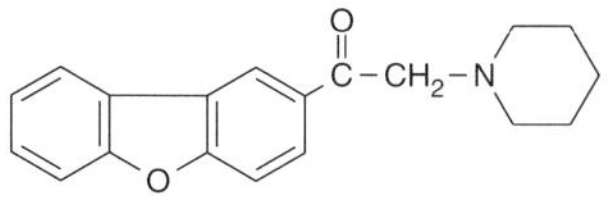
250 mg

S81,202-1
2-[3-(DIMETHYLAMINO)PROPIONYL]-DIBENZOFURAN HYDROCHLORIDE

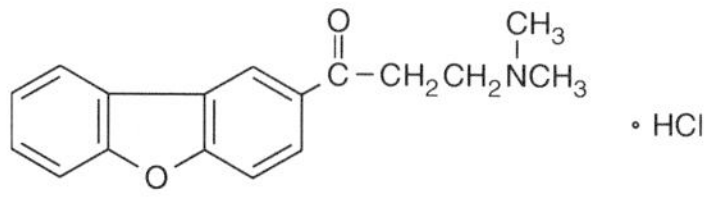

250 mg

S83,748-2
3-(DIETHYLAMINO)DIBENZOFURAN

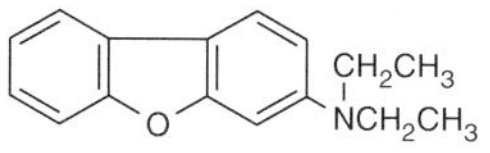

250 mg

S81,200-5
2-[2-(DIETHYLAMINO)ETHYL]-DIBENZOFURAN HYDROCHLORIDE

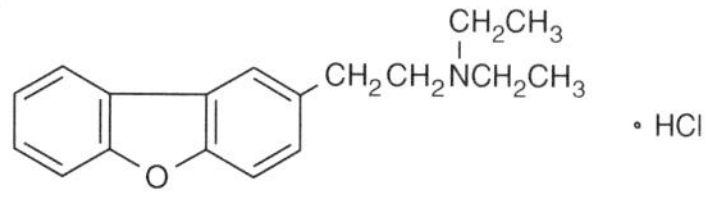

250 mg

S83,673-7
2-[2-(DIETHYLAMINO)ETHYL]-DIBENZOFURAN

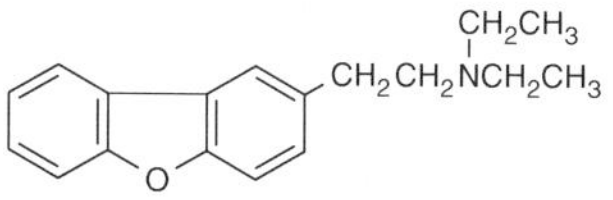

250 mg

S81,201-3
4-[2-(DIETHYLAMINO)ETHYL]-DIBENZOFURAN HYDROCHLORIDE

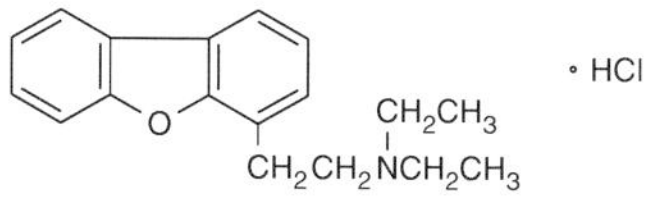

250 mg

S81,208-0
2-ACETYL-7-NITRODIBENZOFURAN

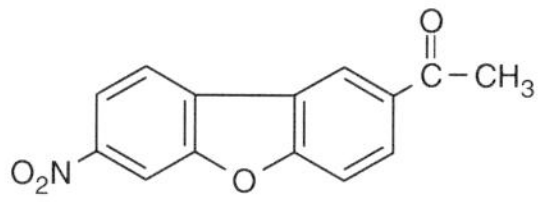

250 mg

S88,259-3
8-NITRO-2-DIBENZOFURANSULFONIC ACID

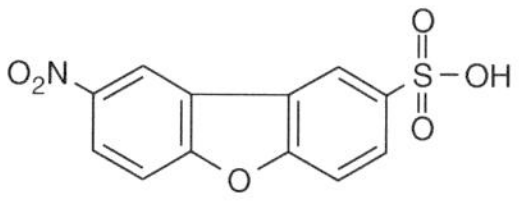

250 mg

S81,092-4
3-NITRODIBENZOFURAN

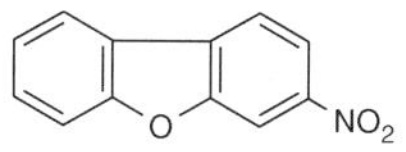

250 mg

S83,765-2
7-NITRO-1,2,3,4-TETRAHYDRODIBENZOFURAN

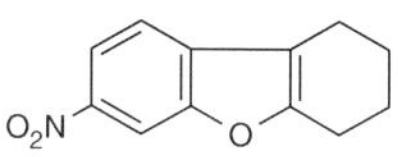

250 mg

S81,005-3
2,7-DINITRODIBENZOFURAN

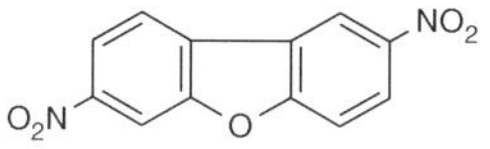

250 mg

S81,591-8
2-BROMO-7-NITRODIBENZOFURAN

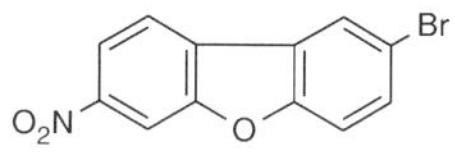

250 mg

S83,813-6
3-BROMO-7-NITRODIBENZOFURAN

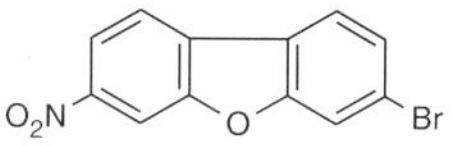

250 mg

S62,055-6
DIBENZOFURANCARBOXALDEHYDE

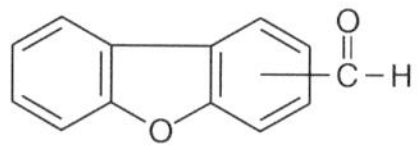

250 mg

S84,978-2
2-(2-BENZOFURANYLCARBONYL)-BENZOIC ACID

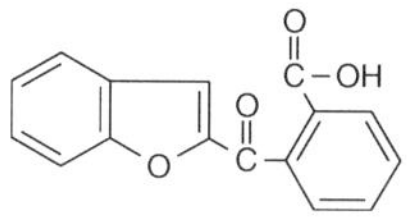

250 mg

S84,122-6
2,8-DIBENZOYLDIBENZOFURAN

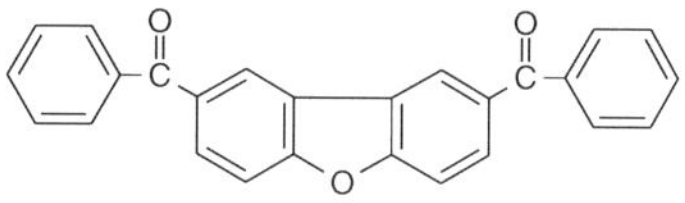

250 mg

S83,743-1
2-BENZOYLDIBENZOFURAN

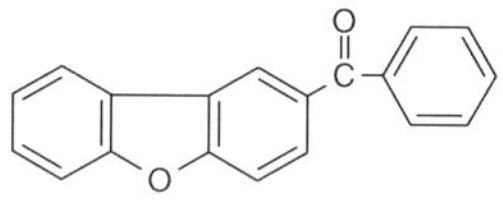

250 mg

S81,162-9
2-ACETYLDIBENZOFURAN

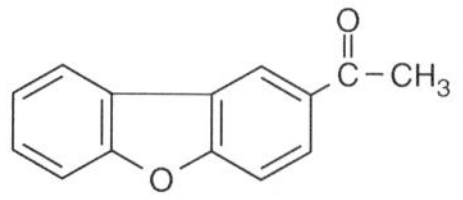

250 mg

S97,200-2
DINAPHTHO(1,2-B:2',3'-D)FURAN-7,12-DIONE

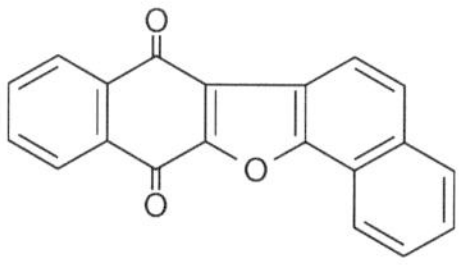

100 mg

S57,076-1
DIBENZOFURAN-2-CARBOXYLIC ACID

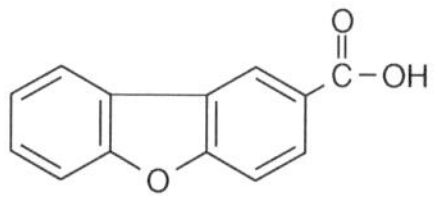

250 mg

S83,740-7
METHYL 2-DIBENZOFURANCARBOXYLATE

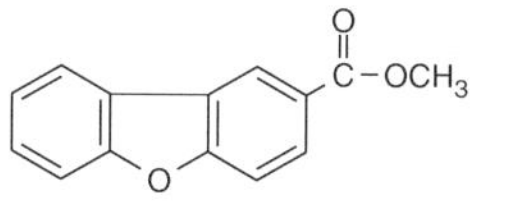

250 mg

S81,401-6
ETHYL 2-DIBENZOFURANCARBOXYLATE

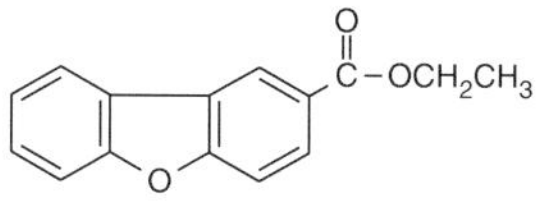

250 mg

S81,644-2
8-BROMO-2-DIBENZOFURANCARBOXAMIDE

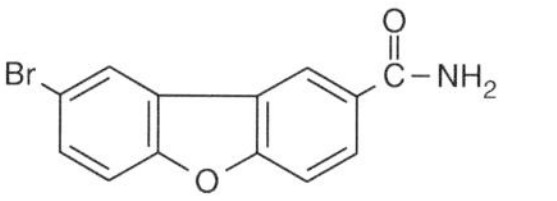

250 mg

S83,343-6
2-DIBENZOFURANCARBOXYLIC HYDRAZIDE

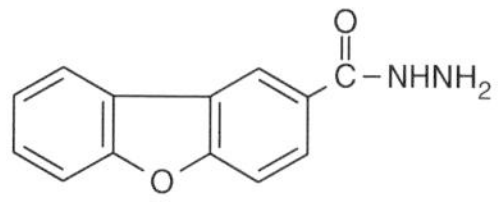

250 mg

S75,703-9
2-ACETAMIDODIBENZOFURAN

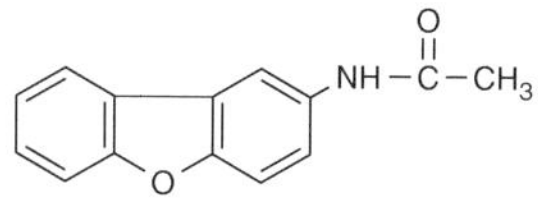

5 gm

S81,111-4
3-ACETAMIDODIBENZOFURAN

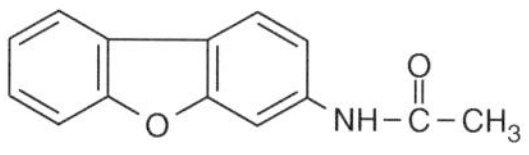

250 mg

S82,256-6
3-ACETAMIDO-2-BROMODIBENZOFURAN

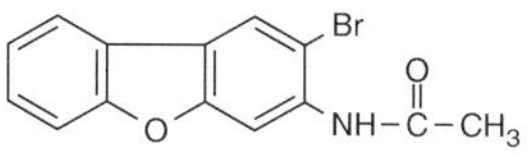

250 mg

S82,445-3
7-ACETAMIDO-2-BROMODIBENZOFURAN

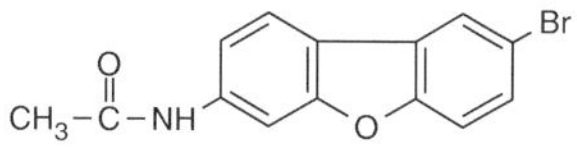

250 mg

S82,276-0
ETHYL 2-DIBENZOFURANCARBODITHIOATE

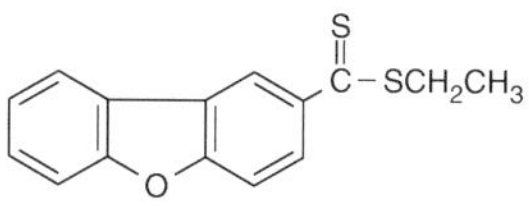

250 mg

S81,165-3
2-CYANODIBENZOFURAN

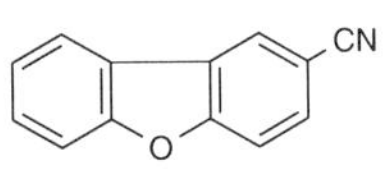

250 mg

S83,766-0
2-(CYANOMETHYL)DIBENZOFURAN

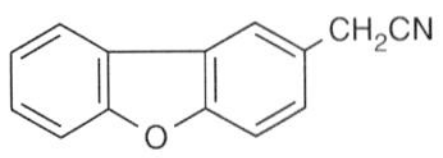

250 mg

S84,047-5
2-BENZOFURANSULFONIC ACID

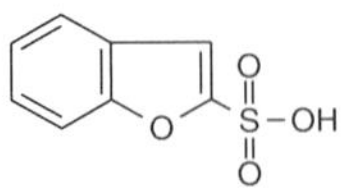

250 mg

S45,449-4
4-DIBENZOFURANSULFONIC ACID, SODIUM SALT MONOHYDRATE

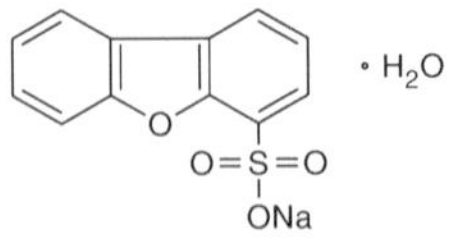

250 mg

S81,209-9
2-DIBENZOFURANSULFONIC ACID, SODIUM SALT

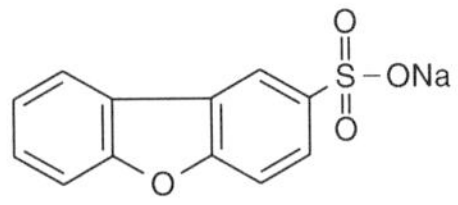

250 mg

S81,100-9
2,8-DIBENZOFURANDISULFINIC ACID, DISODIUM SALT

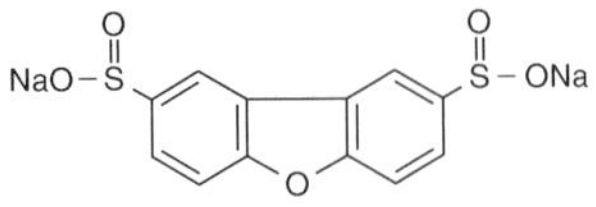

250 mg

S83,547-1
8-BROMO-2-DIBENZOFURANSULFONIC ACID, SODIUM SALT

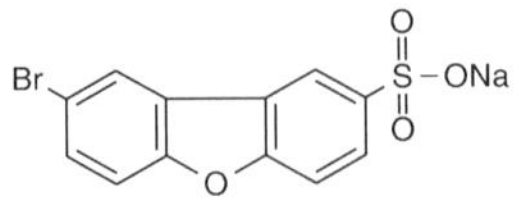

250 mg

S81,173-4
4,6-DIBENZOFURANDISULFONIC ACID

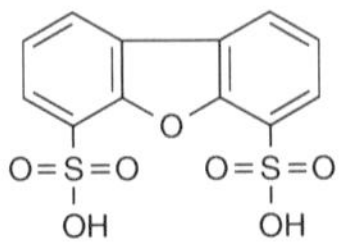

250 mg

S83,983-3
2,8-DIBENZOFURANDISULFONIC ACID, DISODIUM SALT

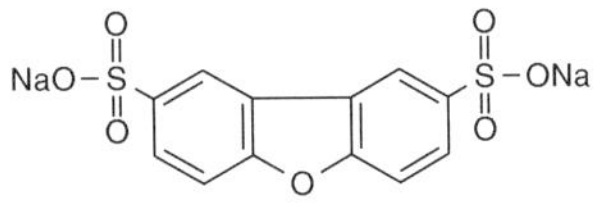

250 mg

S83,981-7
8-(CHLOROMERCURI)-2-DIBENZOFURANSULFONIC ACID, SODIUM SALT

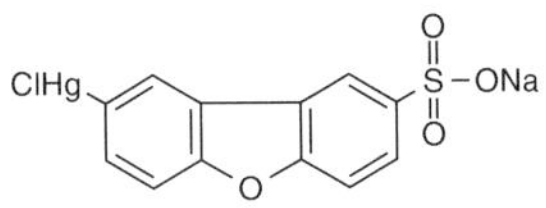

250 mg

S81,233-1
2-DIBENZOFURANSULFONIC ACID, BARIUM SALT

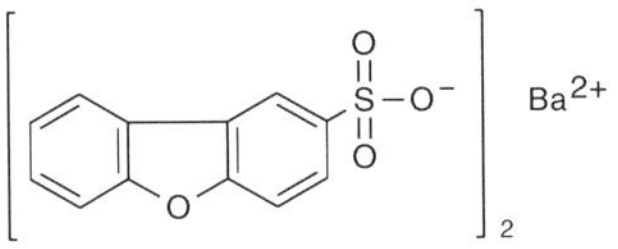

250 mg

S83,903-5
2-DIBENZOFURANSULFONYL CHLORIDE

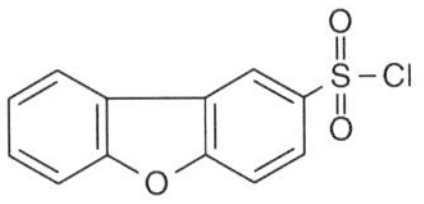

250 mg

S81,155-6
2,8-DIBENZOFURANDISULFONYL CHLORIDE

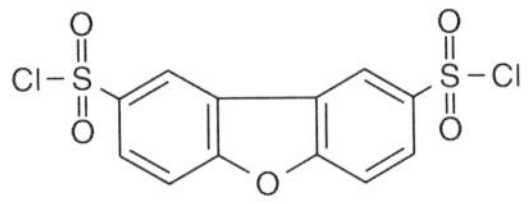

250 mg

S81,099-1
1-DIBENZOFURANSULFONAMIDE

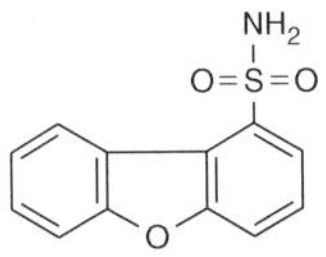

250 mg

S45,666-7
2,3-BENZODIPHENYLENE OXIDE

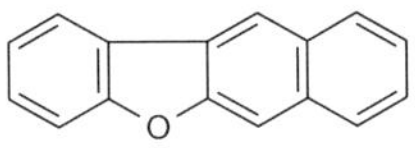

5 gm

S84,069-6
6-(4-DIBENZOFURYL)-5,6-DIHYDROPHENANTHRIDINE

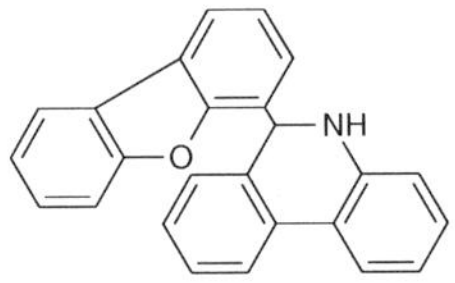

250 mg

S84,068-8
2-(4-DIBENZOFURYL)-1,2-DIHYDROQUINOLINE

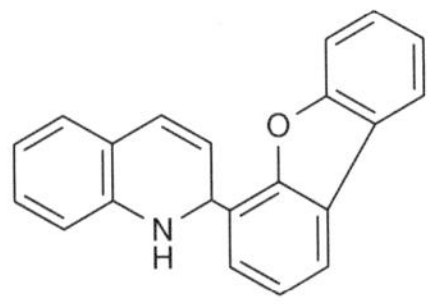

250 mg

S83,270-7
3-FLUORODIBENZOFURAN

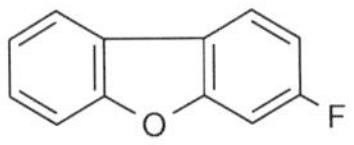

250 mg

S81,006-1
2-CHLORODIBENZOFURAN

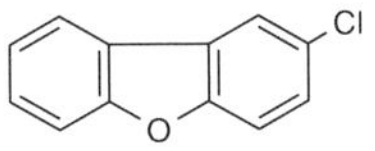

250 mg

S81,102-5
2-DIBENZOFURANSULFINIC ACID, SODIUM SALT

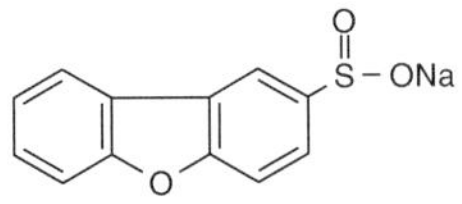

250 mg

S81,003-7
2,8-DICHLORODIBENZOFURAN

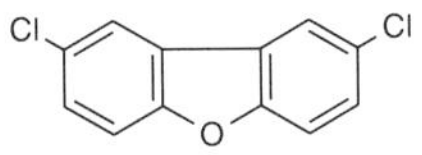

250 mg

S81,093-2
2-BROMODIBENZOFURAN

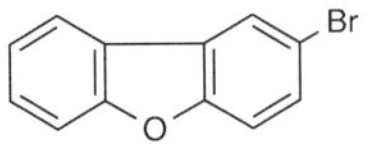

250 mg

S83,925-6
1-BROMODIBENZOFURAN

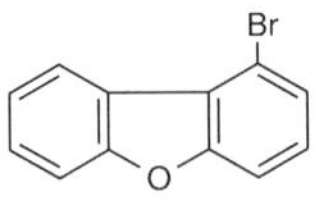

250 mg

S81,236-6
3-IODODIBENZOFURAN

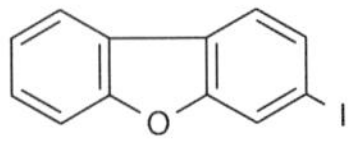

250 mg

S81,589-6
2-IODODIBENZOFURAN

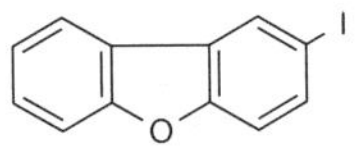

250 mg

S84,128-5
4-IODODIBENZOFURAN

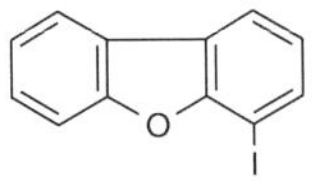

250 mg

S83,165-4
2,8-DIBROMODIBENZOFURAN

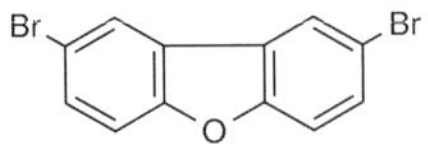

250 mg

S84,127-7
2,8-DIIODODIBENZOFURAN

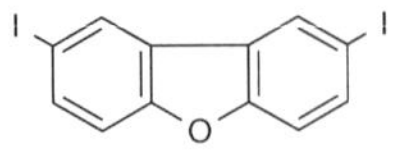

250 mg

S83,273-1
4-METHYLDIBENZOFURAN

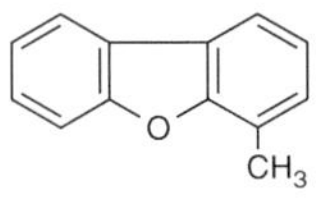

250 mg

S83,674-5
2-METHYLDIBENZOFURAN

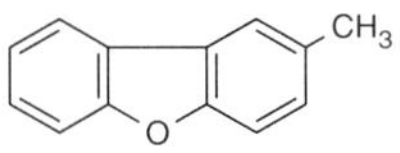

250 mg

S83,676-1
2-ETHYLDIBENZOFURAN

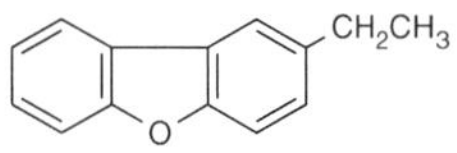

250 mg

S76,265-2
3-(BROMOMETHYL)BENZO(B)THIOPHENE-5-YL BENZOATE

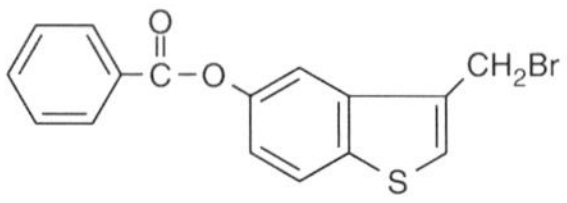

250 mg

S36,876-8
6-ETHOXY-3-HYDROXYBENZO(B)THIOPHENE

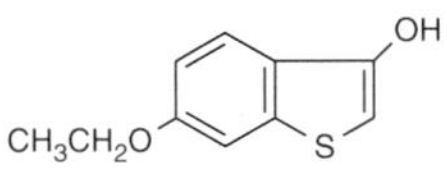

5 gm

S76,055-2
2-ACETYL-3-HYDROXYBENZO(B)THIOPHENE

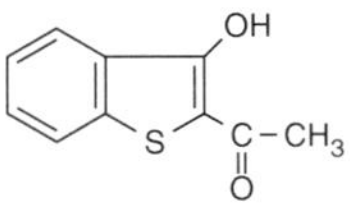

250 mg

S36,745-1
6-CHLORO-4-METHYLBENZO(B)-THIOPHENE-3-OL

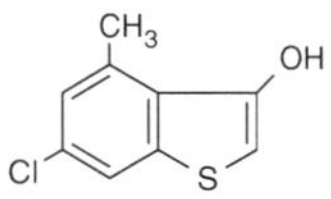

250 mg

S76,267-9
4,5-DIMETHOXYBENZO(B)THIOPHENE-2-CARBOXYLIC ACID

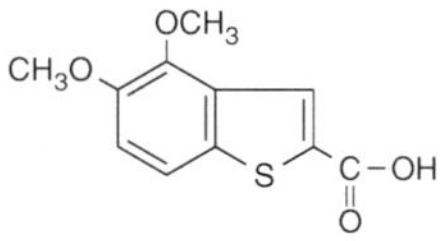

250 mg

S75,569-9
5,6-DIMETHOXYBENZO(B)THIOPHENE-3-ACETAMIDE

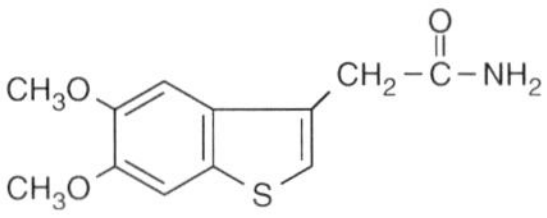

250 mg

S75,586-9
6-METHOXY-5-(METHOXYMETHOXY)BENZO-(B)THIOPHENE-3-ACETAMIDE

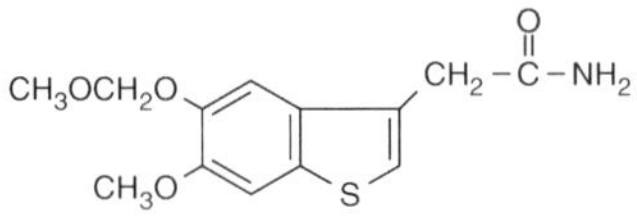

250 mg

S75,987-2
3-(2-AMINOETHYL)-6-METHOXYBENZO(B)-THIOPHENE HYDROCHLORIDE

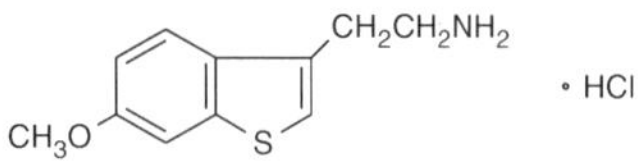

250 mg

S76,248-2
3-(2-AMINOETHYL)-4-METHOXYBENZO(B)-THIOPHENE HYDROCHLORIDE

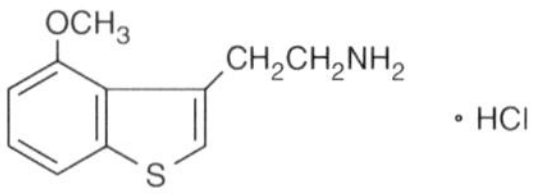

250 mg

S76,249-0
3-(2-AMINOPROPYL)-4-METHOXYBENZO(B)-THIOPHENE HYDROCHLORIDE

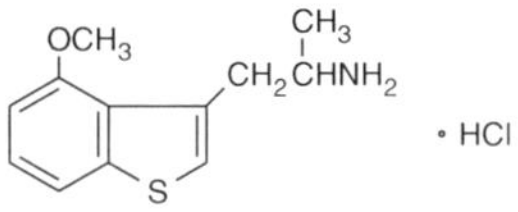

250 mg

S76,247-4
3-[2-(DIMETHYLAMINO)ETHYL]-4-METHOXYBENZO(B)THIOPHENE HYDROCHLORIDE

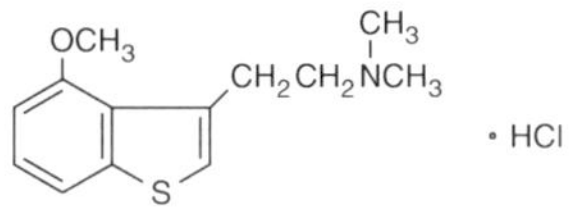

250 mg

S75,800-0
6-METHOXYBENZO(B)THIOPHENE-3-ACETAMIDE

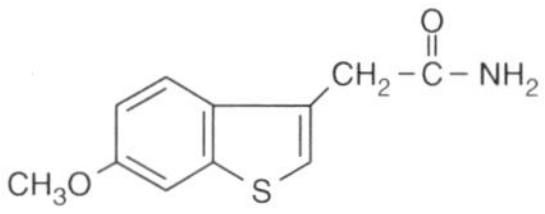

250 mg

S83,536-6
4-METHOXYDIBENZOTHIOPHENE-5,5-DIOXIDE

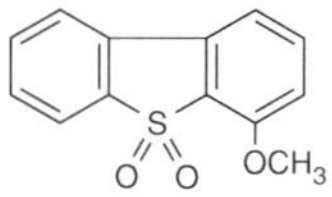

250 mg

S81,634-5
4-METHOXYDIBENZOTHIOPHENE

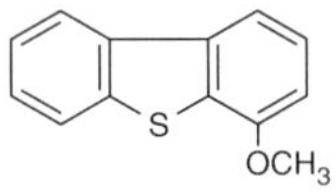

250 mg

S53,489-7
7-BROMOTHIENO(3,2-C)PYRID-4(5H)-ONE

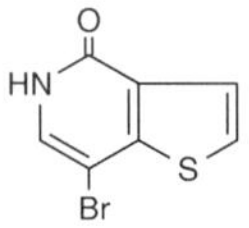

5 gm

S82,344-9
4-AMINODIBENZOTHIOPHENE

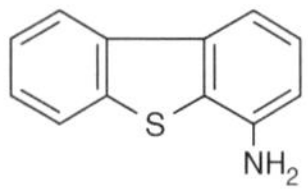

250 mg

S81,599-3
3,7-DIAMINODIBENZOTHIOPHENE-SULFONE DIHYDROCHLORIDE

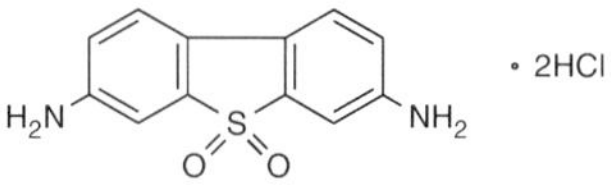

250 mg

S77,062-0
2-AMINODIBENZOTHIOPHENE SULFONE

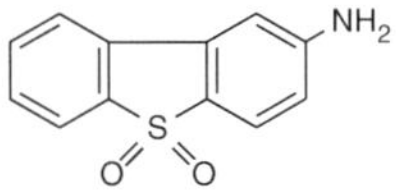

250 mg

S75,986-4
3-[2-(DIMETHYLAMINO)ETHYL]BENZO(B)-THIOPHENE HYDROCHLORIDE

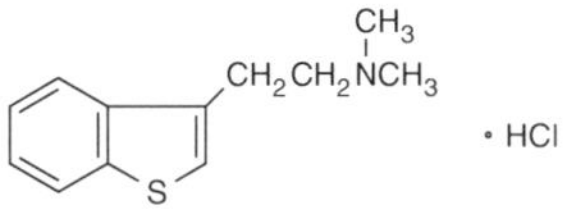

250 mg

S76,266-0
N-(2-[BENZO(B)THIOPHENE-3-YLCARBONYL]ETHYL)-N-METHYLMORPHOLINIUM IODIDE

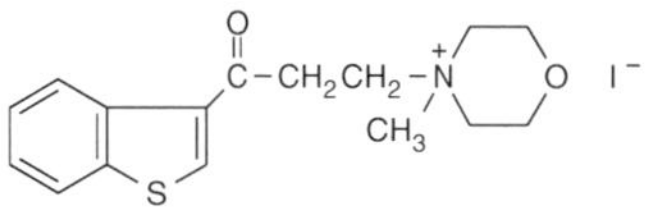

250 mg

S75,583-4
DIETHYL 2-([BENZO(B)THIOPHENE-3-YLMETHYL]-2-FORMAMIDO)MALONATE

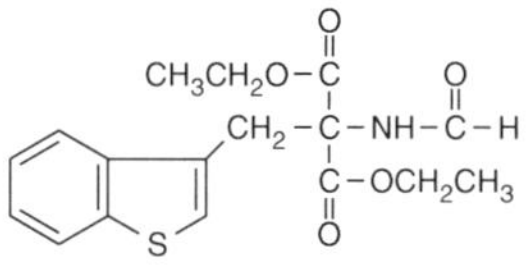

250 mg

S75,599-0
BENZO(B)THIOPHENE-4-YL 4-NITROBENZOATE

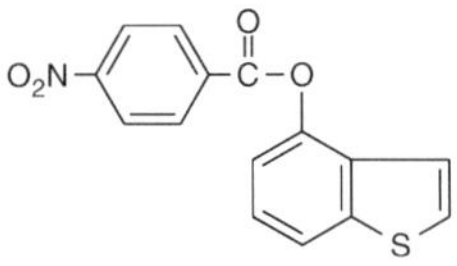

250 mg

S55,219-4
4-NITROTHIANAPHTHENE-2-CARBOXYLIC ACID

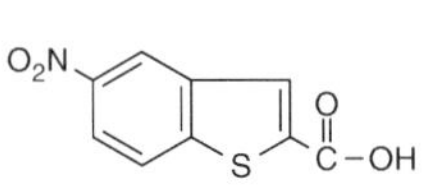

250 mg

S81,096-7
2-NITRODIBENZOTHIOPHENE

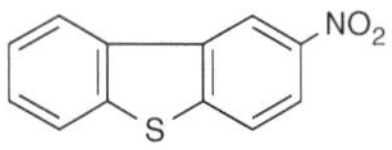

250 mg

S82,444-5
2,8-DINITRODIBENZOTHIOPHENE

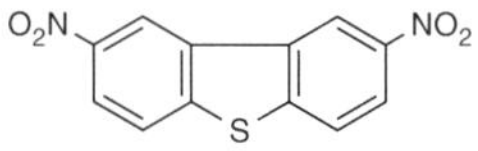

250 mg

S84,116-1
2-(DIBENZOTHIOPHENE-2-YLCARBONYL)-BENZOIC ACID

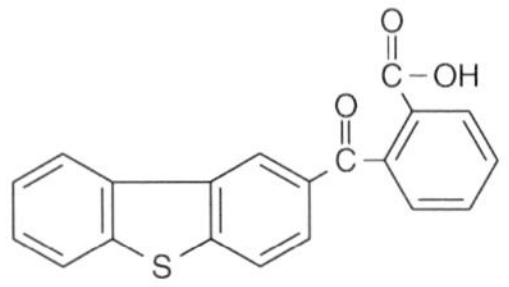

250 mg

S88,858-3
2-CHLORO-2H-THIOPYRANO(2,3-B)-1-BENZOTHIOPHEN-4(3H)-ONE

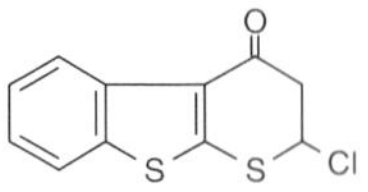

250 mg

S75,590-7
BENZO(B)THIOPHENE-3-YL 2-PIPERIDINOETHYL KETONE

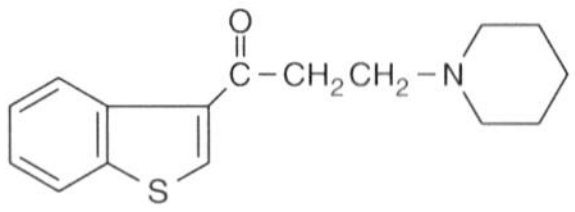

250 mg

S76,263-6
3-(3-MORPHOLINOPROPIONYL)BENZO(B)-THIOPHENE HYDROCHLORIDE

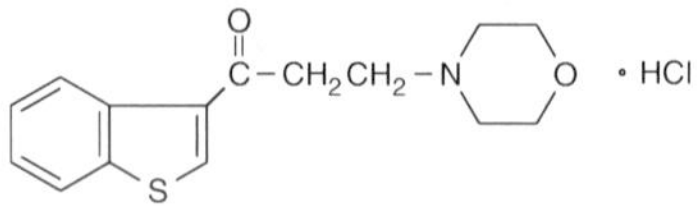

250 mg

S81,800-3
4-BENZOYLDIBENZOTHIOPHENE

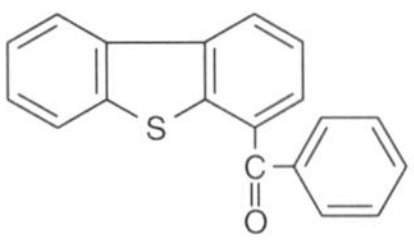

250 mg

S75,991-0
2-ACETYLBENZO(B)THIOPHENE

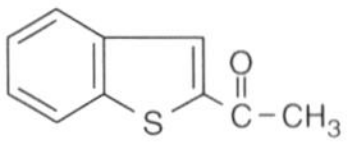

250 mg

S75,595-8
BENZO(B)THIOPHENE-3-YL ETHYL KETONE

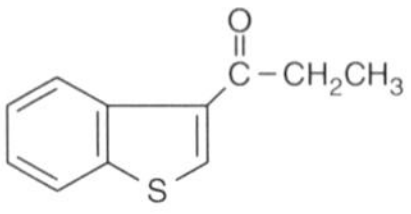

250 mg

S81,633-7
2-ACETYLDIBENZOTHIOPHENE

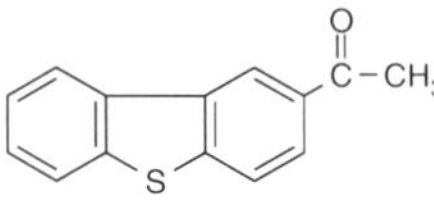

250 mg

S81,536-5
4-CARBOXYDIBENZOTHIOPHENE-5,5-DIOXIDE

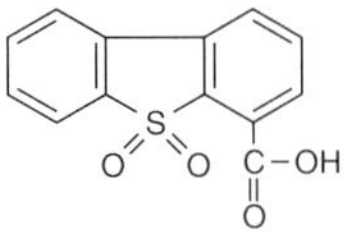

250 mg

S76,268-7
BENZO(B)THIOPHENE-3-ACRYLIC ACID

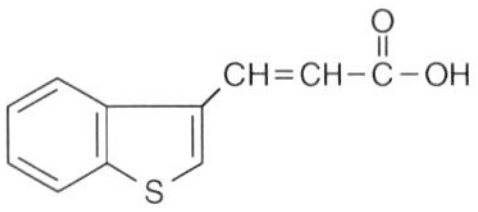

250 mg

S81,805-4
4-DIBENZOTHIOPHENECARBOXYLIC ACID

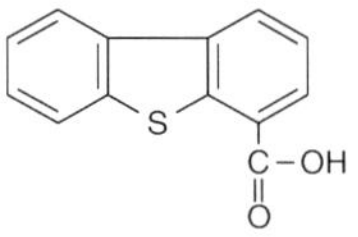

250 mg

S75,572-9
BENZO(B)THIOPHENE-4-YL BENZOATE

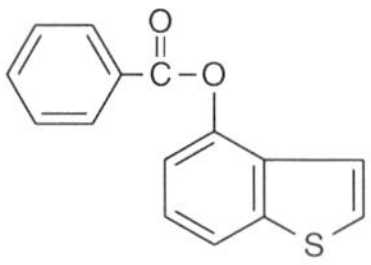

250 mg

S75,585-0
3-BROMOBENZO(B)THIOPHENE-4-YL BENZOATE

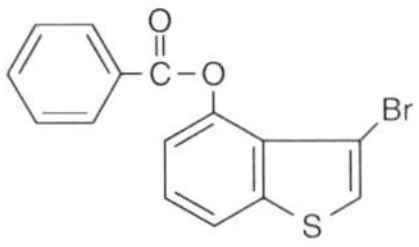

250 mg

S81,810-0
METHYL 4-DIBENZOTHIOPHENECARBOXYLATE

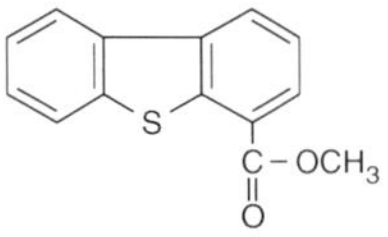

250 mg

S83,557-9
4-DIBENZOTHIOPHENECARBOXAMIDE

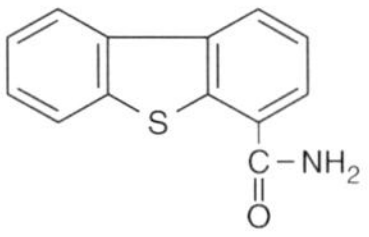

250 mg

S82,342-2
N-(3-DIBENZOTHIENYL)BENZAMIDE

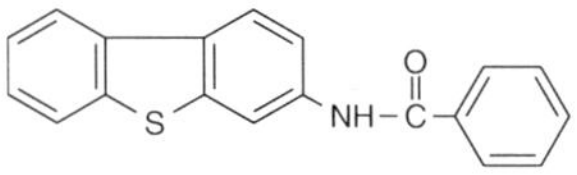

250 mg

S81,632-9
2-DIBENZOTHIOPHENEPROPIONIC ACID

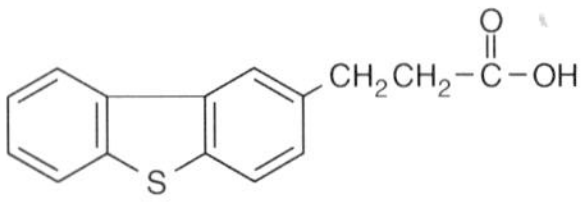

250 mg

S81,631-0
2-DIBENZOTHIOPHENEBUTYRIC ACID

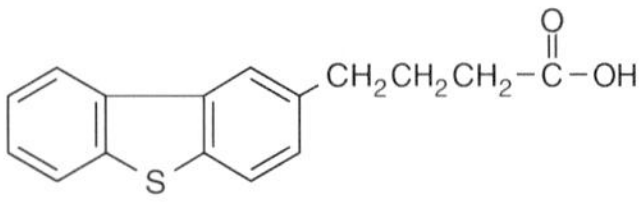

250 mg

S76,264-4
DIETHYL 2-ACETAMIDO-2-[BENZO(B)-THIOPHENE-3-YLMETHYL]MALONATE

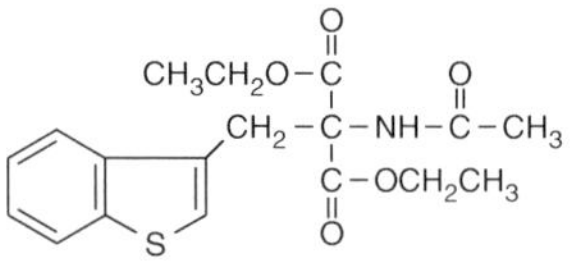

250 mg

S75,652-0
ETHYL 2,2-DIMETHYLTHIENO(2,3-F)-1,3-BENZODIOXOLE-7-ACETATE

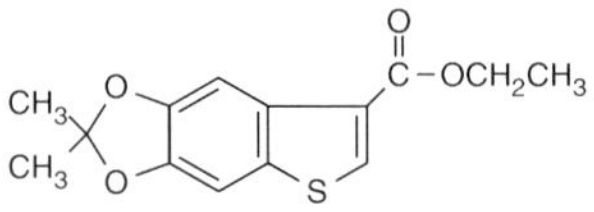

5 gm

S81,804-6
2-(DIACETYLAMINO)DIBENZOTHIOPHENE

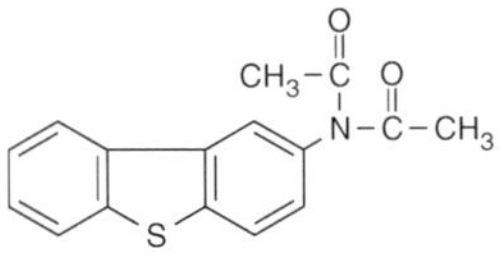

250 mg

S92,453-9
N-[3-CYANO-2-BENZO(B)THIOPHEN-2-YL] PROPIONAMIDE

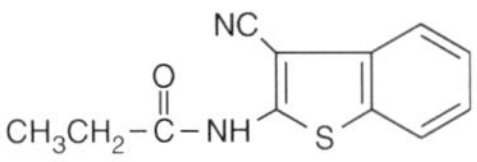

250 mg

S82,334-1
2-ACETAMIDODIBENZOTHIOPHENE

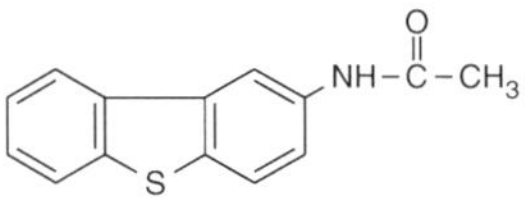

250 mg

S75,660-1
BENZO(B)THIOPHENE-3-ACETONITRILE

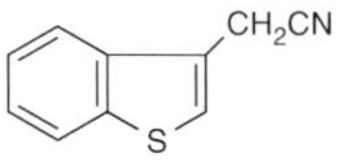

5 gm

S81,803-8
DIBENZOTHIOPHENE-5-OXIDE

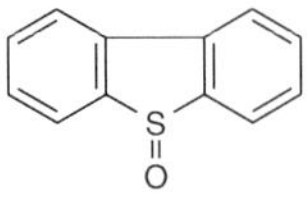

250 mg

S85,241-4
THIONAPHTHENE-1,1-DIOXIDE

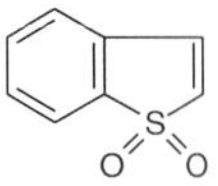

250 mg

S82,375-9
2,8-DIBROMODIBENZOTHIOPHENE-5,5-DIOXIDE

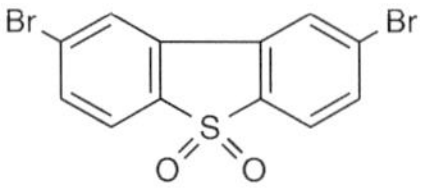

250 mg

S75,582-6
BENZO(B)THIOPHENE-4-YL METHANESULFONATE

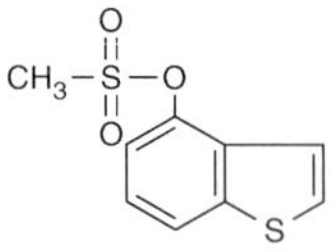

250 mg

S65,295-4
8,9,10,11-TETRAHYDROBENZO(B)-NAPHTHO(1,2-D)-THIOPHENE PICRATE

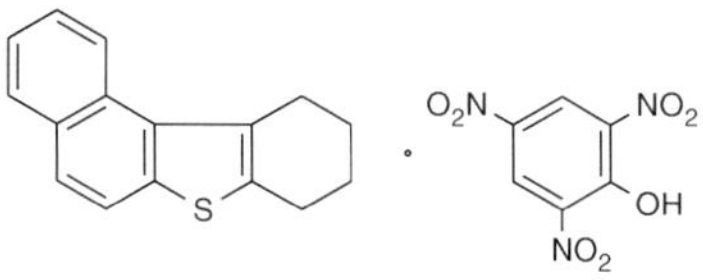

250 mg

S36,405-3
2-(2-NAPHTHYL)THIANAPHTHENE

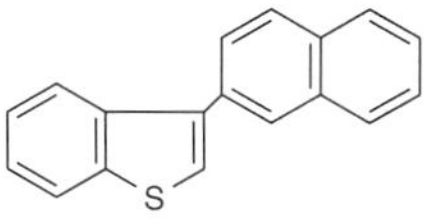

5 gm

S65,187-7
3-BROMODIBENZOTHIOPHENE

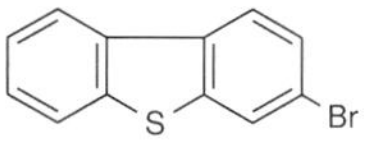

5 gm

S83,535-8
2-BROMODIBENZOTHIOPHENE

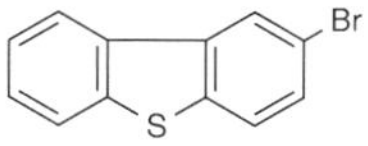

250 mg

S81,806-2
4-IODODIBENZOTHIOPHENE

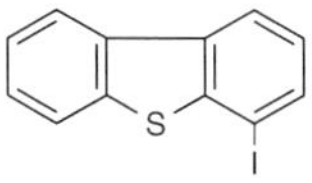

250 mg

S81,807-0
2,8-DIBROMODIBENZOTHIOPHENE

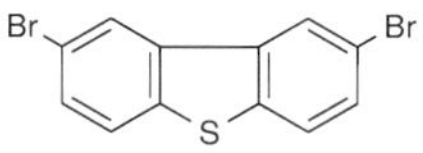

250 mg

S82,337-6
4,4'-BI(DIBENZOTHIOPHENE)

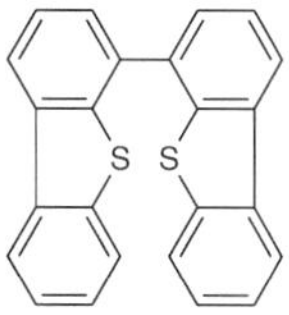

250 mg

S82,338-4
2,2'-METHYLENEBIS-(DIBENZOTHIOPHENE)

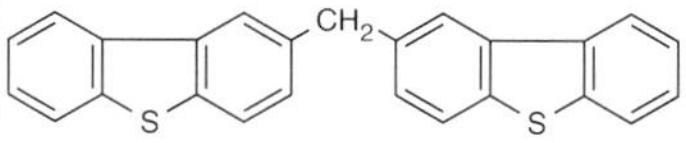

250 mg

S55,195-3
3-BROMO-1-BENZOTHIOPHENE

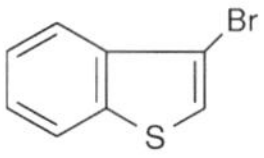

250 mg

S91,209-3
1-(N-[4-CHLORO-3-(TRIFLUOROMETHYL)-PHENYL]CARBAMOYL)-6-NITRO-1H-INDAZOLE

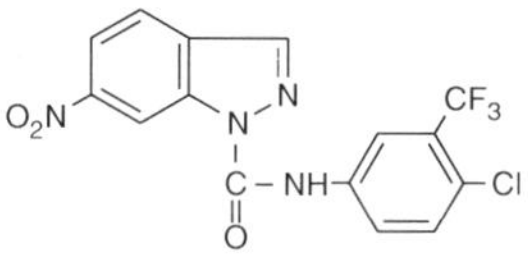

250 mg

S91,187-9
1-(1H-INDAZOL-5-YL)-3-[2-(TRIFLUOROMETHYL)PHENYL]UREA

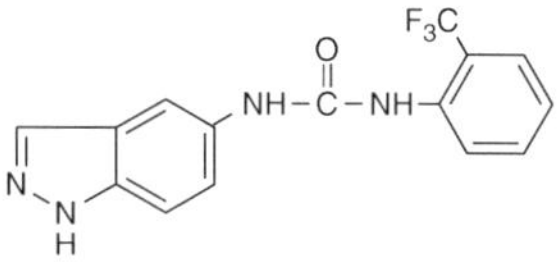

250 mg

S77,206-2
1-(5-CHLORO-2,4-DIMETHOXYPHENYL)-3-(1H-INDAZOL-5-YL)UREA

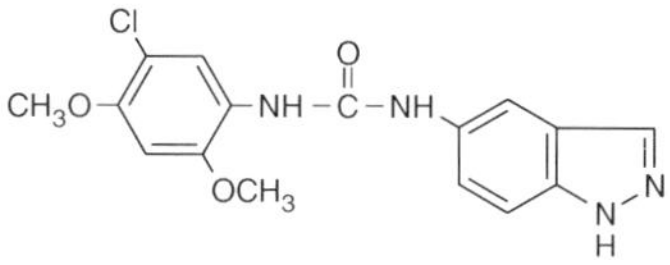

250 mg

S75,004-2
1-[N-(5-CHLORO-2,4-DIMETHOXYPHENYL)-CARBAMOYL]-6-NITRO-1H-INDAZOLE

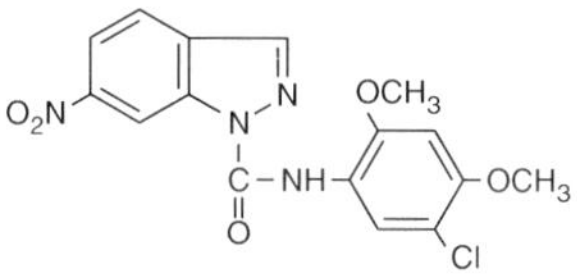

5 gm

S72,904-3
5-CHLORO-1-[N-(5-CHLORO-2-METHOXYPHENYL)CARBAMOYL]-1H-INDAZOLE

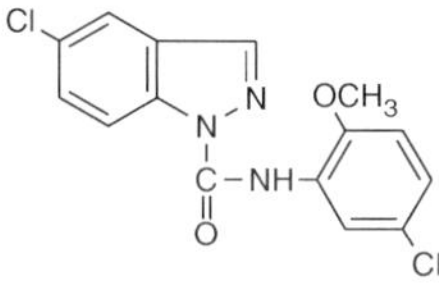

5 gm

S91,199-2
1-(4-FLUORO-3-NITROPHENYL)-3-(1H-INDAZOL-5-YL)UREA

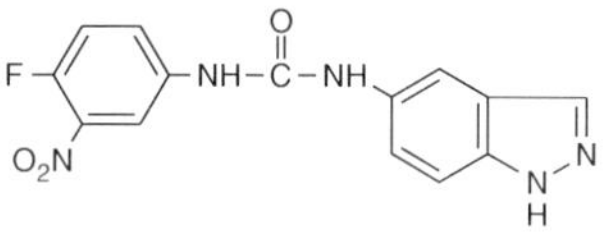

250 mg

S76,878-2
5-CHLORO-1-[N-(4-FLUORO-3-NITROPHENYL)CARBAMOYL]-1H-INDAZOLE

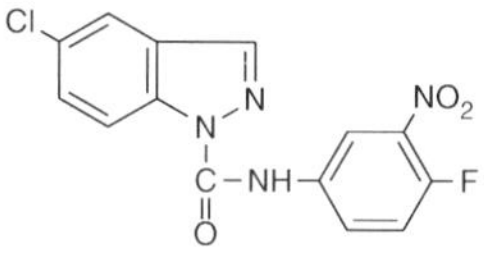

250 mg

S91,169-0
5-CHLORO-1-[N-(2-FLUORO-5-NITROPHENYL)CARBAMOYL]-1H-INDAZOLE

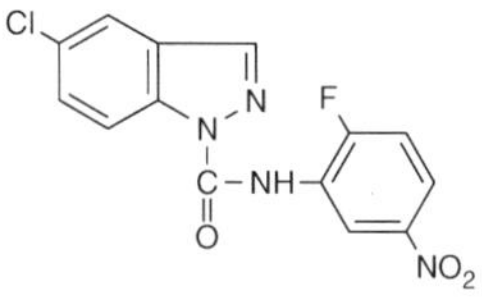

250 mg

S76,805-7
1-[N-(4-FLUORO-3-NITROPHENYL)-CARBAMOYL]-6-NITRO-1H-INDAZOLE

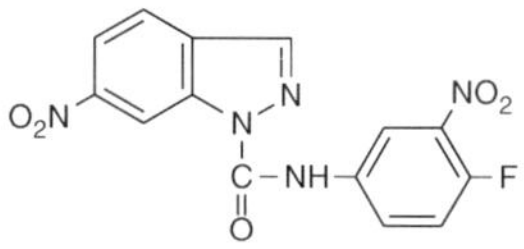

250 mg

S76,883-9
1-(4-FLUORO-3-NITROPHENYL)-3-(1H-INDAZOL-6-YL)UREA

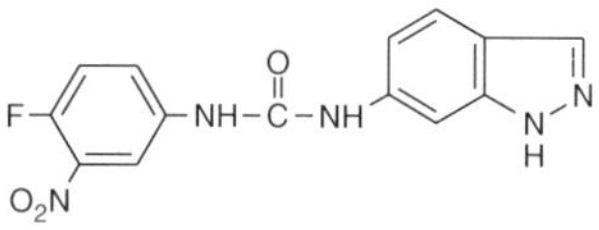

250 mg

S91,230-1
3-CHLORO-1-[N-(2-FLUOROPHENYL)-CARBAMOYL]-6-NITRO-1H-INDAZOLE

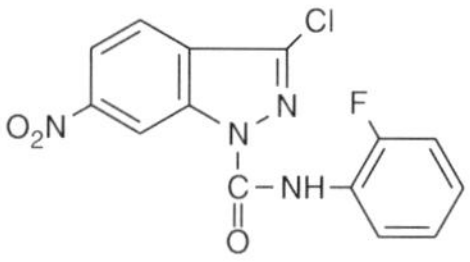

250 mg

S91,250-6
1-[N-(3-CHLORO-4-METHYLPHENYL)-CARBAMOYL]-6-NITRO-1H-INDAZOLE

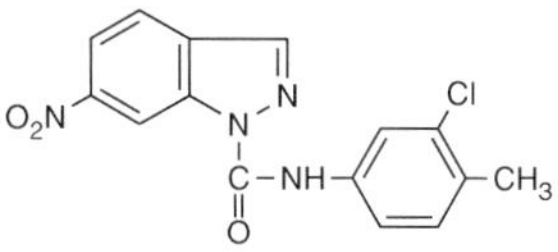

250 mg

S91,251-4
3-CHLORO-1-[N-(3-CHLORO-4-METHYLPHENYL)CARBAMOYL]-6-NITRO-1H-INDAZOLE

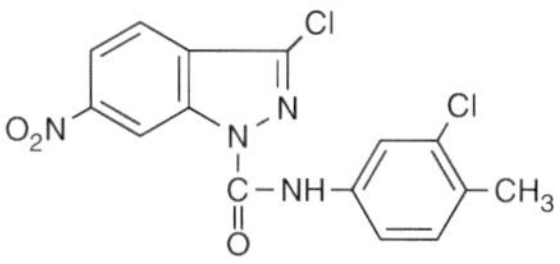

250 mg

S78,023-5
N-(1H-INDAZOL-6-YL)-3,4,5,6-TETRACHLOROPHTHALAMIC ACID

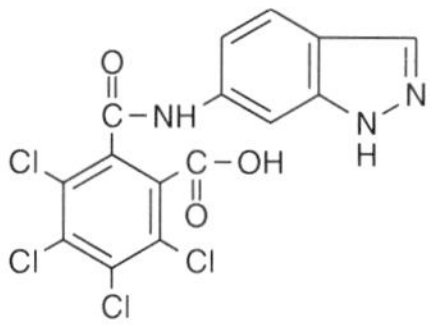

250 mg

S91,105-4
1-(2,3-DICHLOROPHENYL)-3-(1H-INDAZOL-5-YL)UREA

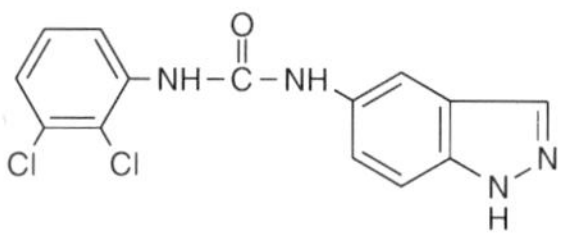

250 mg

S73,700-3
1-[N-(5-CHLORO-2-METHYLPHENYL)-CARBAMOYL]-6-NITRO-1H-INDAZOLE

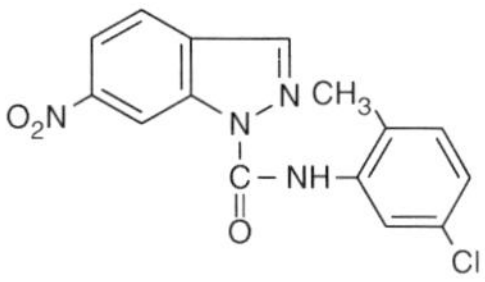

5 gm

S76,989-4
5-CHLORO-1-[N-(5-CHLORO-2-METHYLPHENYL)CARBAMOYL]-1H-INDAZOLE

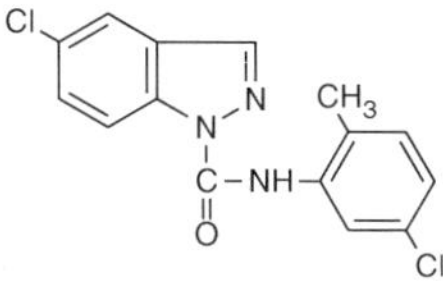

250 mg

S61,607-9
3,6-DIHYDROXY-4-METHYL-1-PYRAZOLO-(3,4-B)PYRIDINE

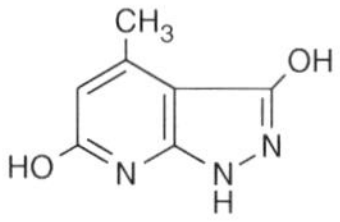

5 gm

S91,073-2
1-(1H-INDAZOL-5-YL)-3-[4-(METHYLTHIO)PHENYL]UREA

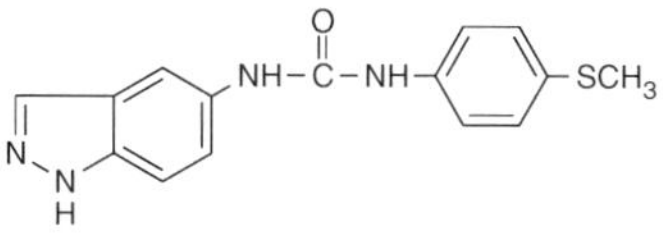

250 mg

S91,072-4
1-(N-[4-(METHYLTHIO)PHENYL]-CARBAMOYL)-1H-INDAZOLE

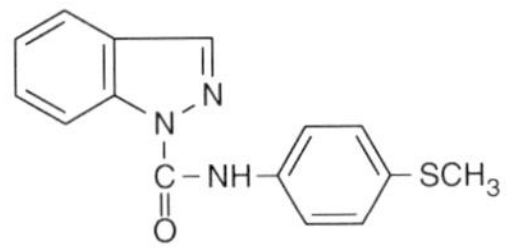

250 mg

S43,581-3
7-AMINOINDAZOLE

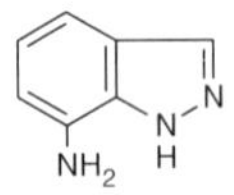

250 mg

S45,632-2
7-NITROINDAZOLE

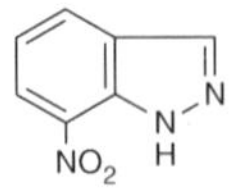

5 gm

S94,110-7
2,7-DINITRO-2H-INDAZOLE

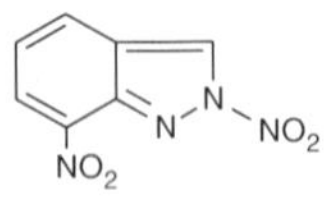

250 mg

S94,546-3
3-BROMO-7-NITRO-1H-INDAZOLE

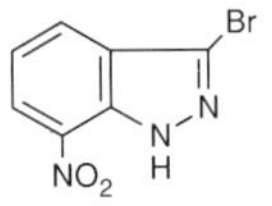

50 mg

S91,140-2
1-[N-(4-METHYL-3-NITROPHENYL)-CARBAMOYL]-6-NITRO-1H-INDAZOLE

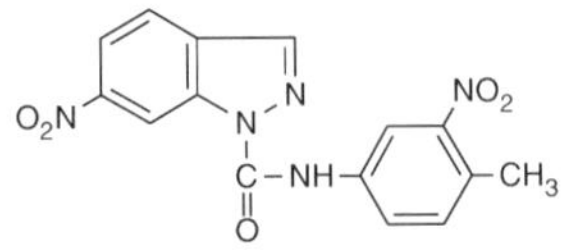

250 mg

S91,223-9
ETHYL 2-(6-NITRO-1H-INDAZOL-1-YLCARBONYLAMINO)ACETATE

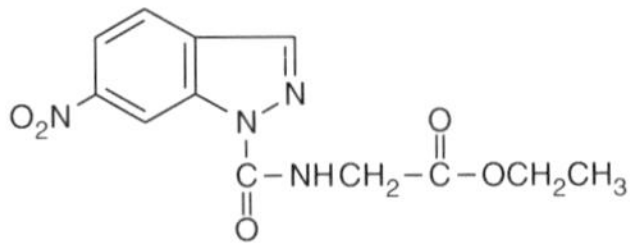

250 mg

S91,242-5
6-NITRO-1-[N-(3,4-XYLYL)CARBAMOYL]-1H-INDAZOLE

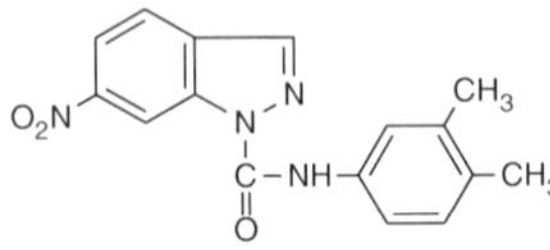

250 mg

S91,244-1
3-CHLORO-6-NITRO-1-[N-(3,4-XYLYL)-CARBAMOYL]-1H-INDAZOLE

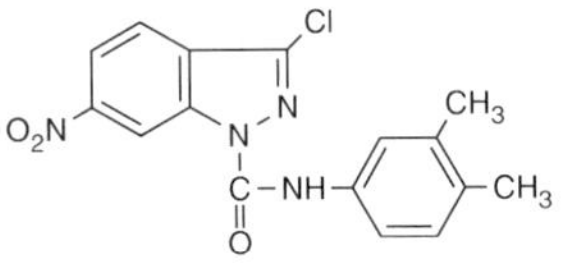

250 mg

S94,550-1
5,7-DINITRO-1H-INDAZOL-3(2H)-ONE

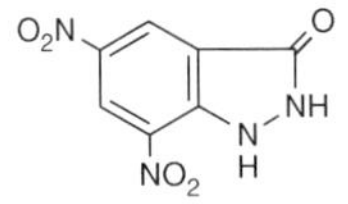

250 mg

S40,962-6
6-NITRO-3-INDAZOLINONE

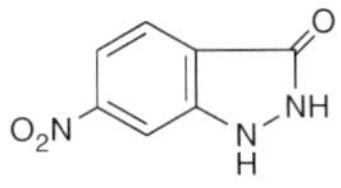

5 gm

S92,768-6
2',6'-DIETHYL-6-NITRO-1H-INDAZOLE-1-CARBOXANILIDE

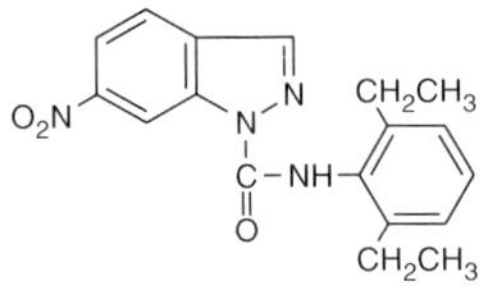

250 mg

S94,111-5
1,6-DINITRO-1H-INDAZOLE-3-CARBONITRILE

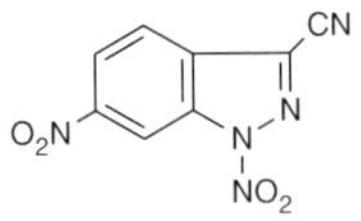

250 mg

S94,549-8
6-NITRO-1-PHENYL-1H-INDAZOLE

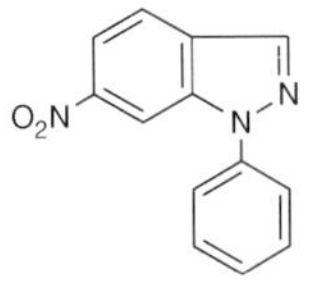

50 mg

S94,547-1
3-BROMO-1,5-DINITRO-1H-INDAZOLE

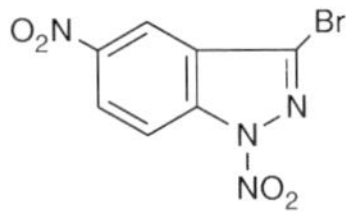

250 mg

S77,223-2
ETHYL 2-[3-(1H-INDAZOL-5-YL)UREIDO] ACETATE

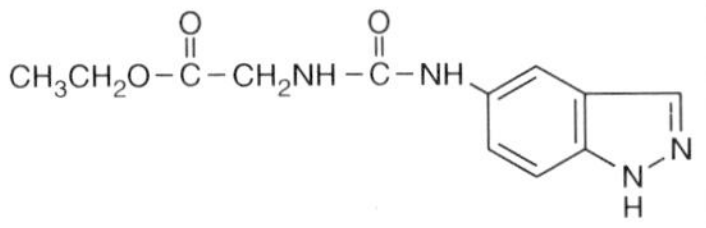

250 mg

S93,953-6
2-PHENYLINDAZOLE

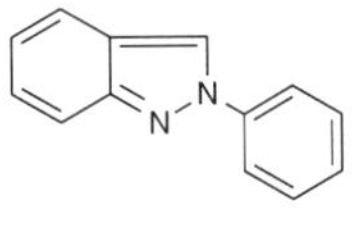

250 mg

S94,271-5
3,5-DIBROMOINDAZOLE

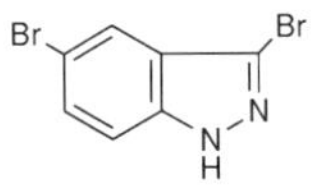

250 mg

S93,952-8
6-NITRO-1-PHENYL-1H-INDAZOLE-3-CARBOXYLIC ACID

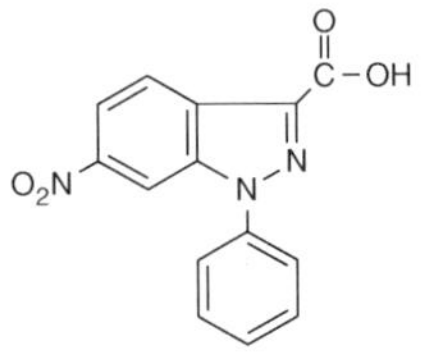

250 mg

S40,999-5
1,9-PYRAZOLOANTHRONE

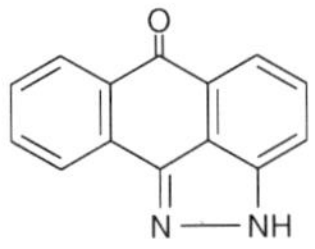

5 gm

S91,148-8
ETHYL 4-(1H-INDAZOL-1-YLCARBONYLAMINO)BENZOATE

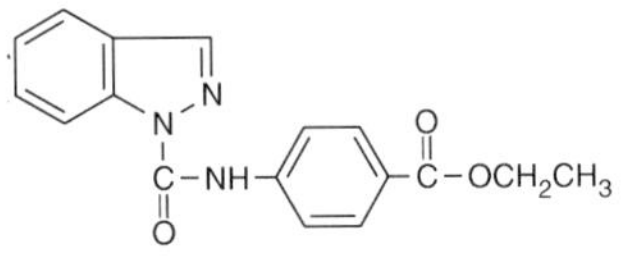

250 mg

S91,146-1
ETHYL 4-(5-CHLORO-1H-INDAZOL-1-YLCARBONYLAMINO)BENZOATE

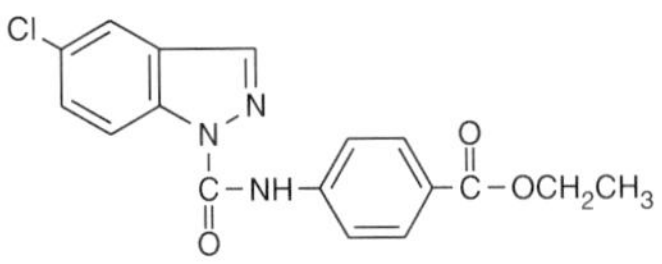

250 mg

S92,559-4
ETHYL 4-[3-(5-1H-INDAZOLYL)UREIDO]-BENZOATE

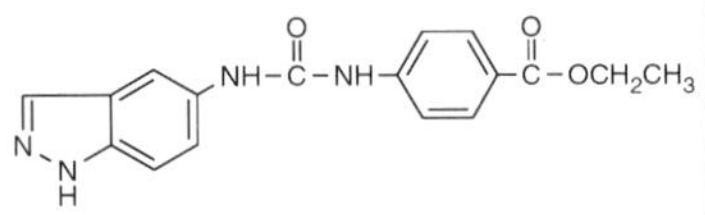

250 mg

S72,429-7
ETHYL 2-(5-CHLORO-1-INDAZOLYLCARBOXAMIDO)ACETATE

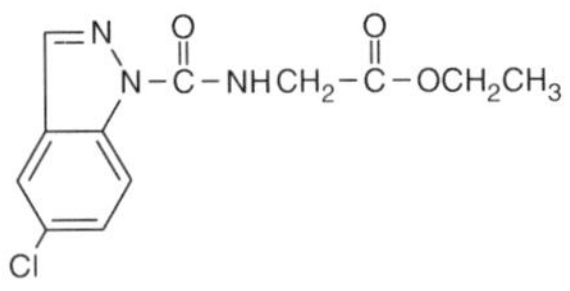

5 gm

S94,109-3
1-PHENYL-1H-INDAZOLE

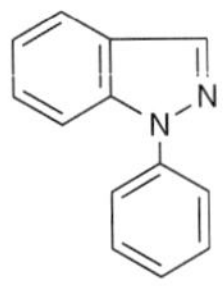

250 mg

S94,113-1
3-BROMO-2-PHENYL-2H-INDAZOLE

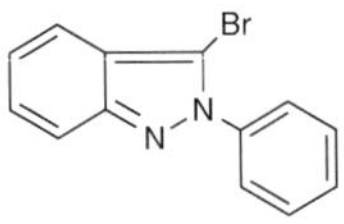

250 mg

S94,112-3
3-BROMO-1H-INDAZOLE

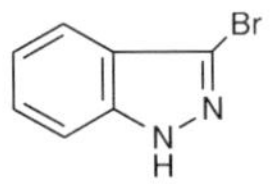

250 mg

S87,152-4
3-ETHYL-1H-INDAZOLE

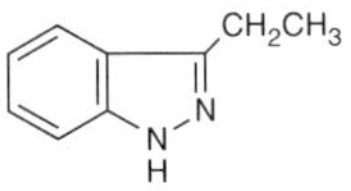

250 mg

S58,750-8
2-(CHLOROMETHYL)-5-NITROBENZIMIDAZOLE

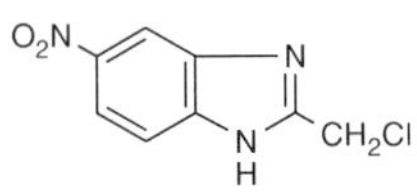

250 mg

S93,485-2
2-(CHLOROMETHYL)BENZIMIDAZOLE HYDROCHLORIDE

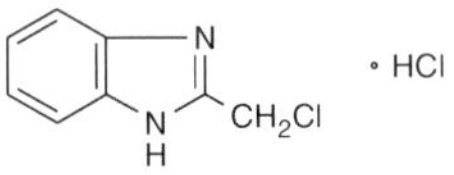

250 mg

S91,204-2
1-(N-[4-CHLORO-2-(TRIFLUOROMETHYL)-PHENYL]CARBAMOYL)BENZIMIDAZOLE

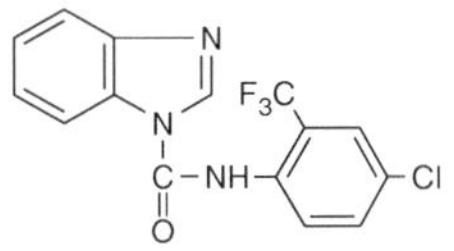

250 mg

S64,938-4
2,3-DIHYDRO-2-THIOXO-6-TRIFLUOROMETHYLBENZIMIDAZOLE

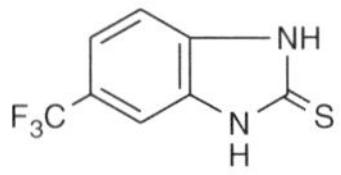

250 mg

S45,346-3
2-(TRICHLOROMETHYL)BENZIMIDAZOLE

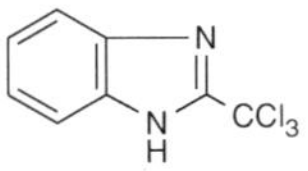

5 gm

S39,523-4
1,3-DIETHYL-2-METHYL-5-(TRIFLUOROMETHYL)-BENZIMIDAZOLIUM IODIDE

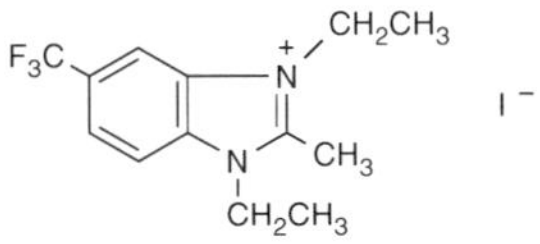

250 mg

S91,183-6
1-(N-[2-(TRIFLUOROMETHYL)PHENYL]-CARBAMOYL)BENZIMIDAZOLE

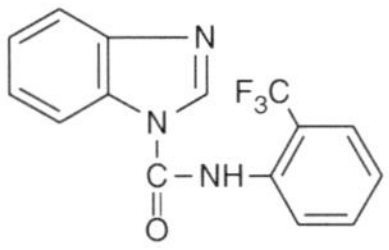

250 mg

S73,086-6
1-(2-BENZIMIDAZOLYL)-3-[2-(TRIFLUOROMETHYL)PHENYL]UREA

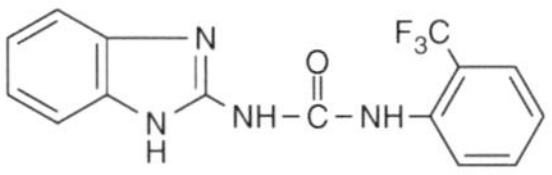

5 gm

S46,012-5
2-(TRIFLUOROMETHYL)BENZIMIDAZOLE

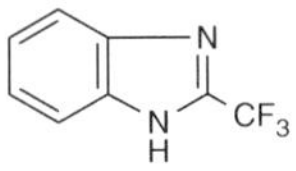

250 mg

S76,030-7
3-(4-CHLOROPHENYL)-2,3-DIHYDRO-3-HYDROXYTHIAZOLO(3,2-A)-BENZIMIDAZOLE HYDROBROMIDE

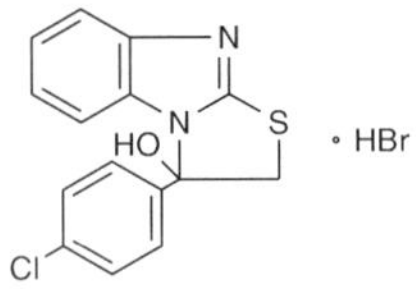

250 mg

S73,725-9
1-[(5-CHLORO-2,4-DIMETHOXYPHENYL)-CARBAMOYL]-6-NITROBENZIMIDAZOLE

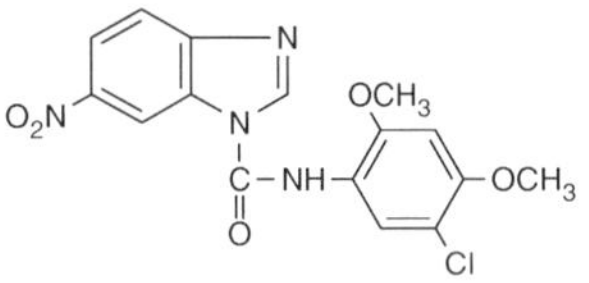

5 gm

S94,075-5
2-(4-CHLOROPHENOXYMETHYL)-1-METHYLBENZIMIDAZOLE

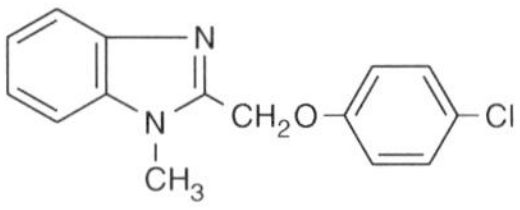

250 mg

S94,074-7
2-(4-BROMOPHENOXYMETHYL)-1-METHYLBENZIMIDAZOLE

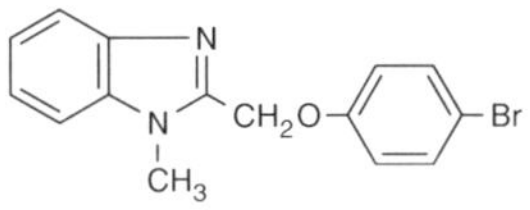

250 mg

S91,176-3
1-[N-(5-CHLORO-2-METHOXYPHENYL)-CARBAMOYL]BENZIMIDAZOLE

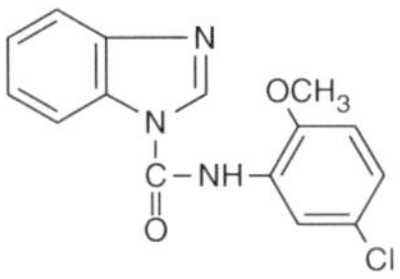

250 mg

S77,275-5
1-[N-(4-FLUORO-3-NITROPHENYL)-CARBAMOYL]BENZIMIDAZOLE

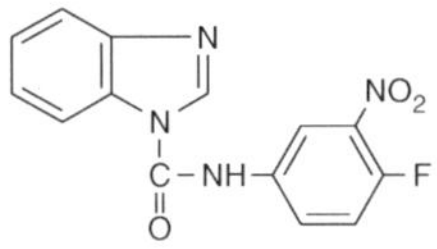

250 mg

S91,198-4
1-[N-(4-FLUORO-3-NITROPHENYL)-CARBAMOYL]-6-NITROBENZIMIDAZOLE

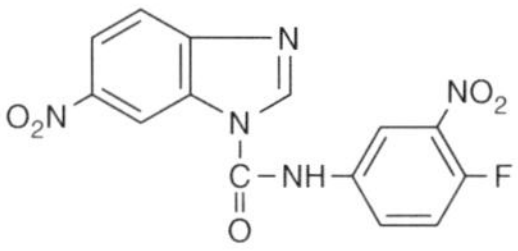

250 mg

S91,098-8
1-[N-(2,3-DICHLOROPHENYL)CARBAMOYL]-6-NITROBENZIMIDAZOLE

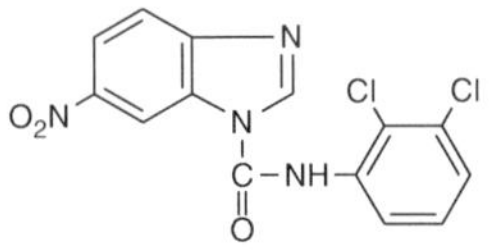

250 mg

S93,665-0
2-(2-FLUOROBENZOYL)-2-(1-METHYL-2-BENZIMIDAZOLYL)ACETONITRILE

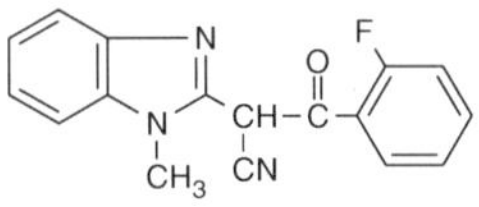

250 mg

S93,753-3
2-(2-IODOBENZOYL)-2-(1-METHYL-2-BENZIMIDAZOLYL)ACETONITRILE

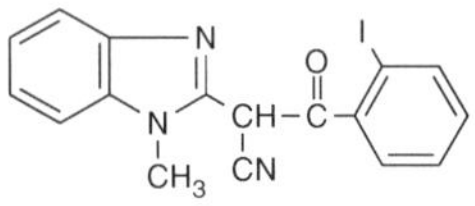

250 mg

S91,091-0
1-[N-(2,3-DICHLOROPHENYL)CARBAMOYL]-BENZIMIDAZOLE

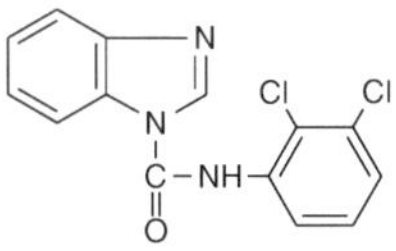

250 mg

S91,051-1
1-[N-(2,5-DIFLUOROPHENYL)CARBAMOYL]-BENZIMIDAZOLE

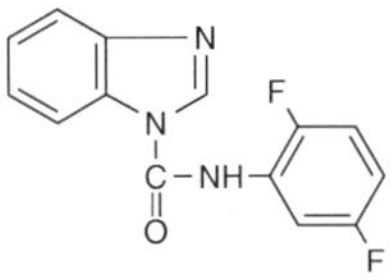

250 mg

S77,955-5
1-(2-BENZIMIDAZOLYL)-3-(2,5-DIFLUOROPHENYL)UREA

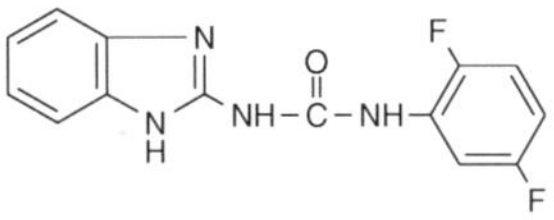

250 mg

S76,996-7
1-[N-(5-CHLORO-2-METHYLPHENYL)-CARBAMOYL]BENZIMIDAZOLE

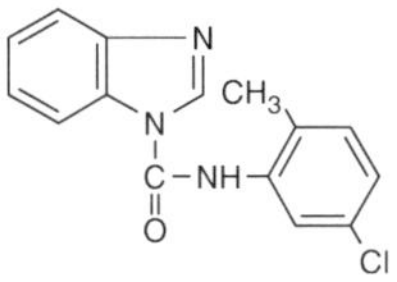

250 mg

S76,926-6
1-[N-(3-FLUOROPHENYL)CARBAMOYL]-BENZIMIDAZOLE

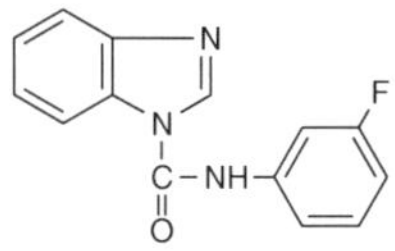

250 mg

S69,146-1
3-(4-BROMOPHENYL)THIAZOLO(3,2-A)-BENZIMIDAZOLE

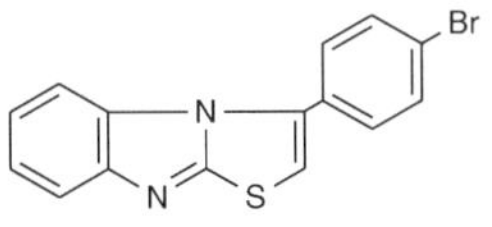

250 mg

S94,262-6
2-(4-CHLOROPHENYL)-1-METHYLBENZIMIDAZOLE

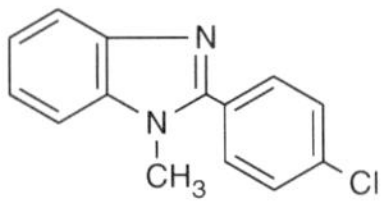

250 mg

S94,263-4
2-(4-BROMOPHENYL)-1-METHYLBENZIMIDAZOLE

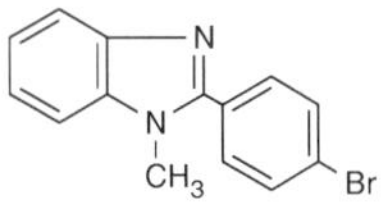

250 mg

S92,764-3
2-(3-BROMOPHENYL)BENZIMIDAZOLE

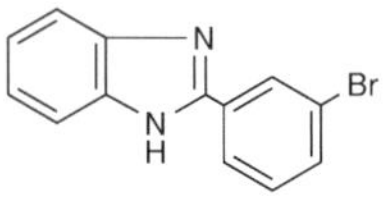

250 mg

S36,239-5
2-(ALPHA-HYDROXYBENZYL)-5,6-DIMETHYLBENZIMIDAZOLE

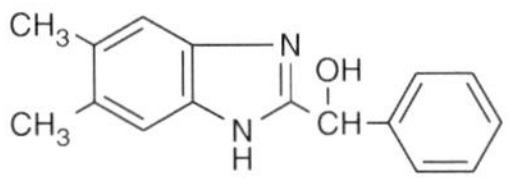

250 mg

S80,275-1
5,6-DIAMINO-2-HYDROXYBENZIMIDAZOLE DIHYDROCHLORIDE

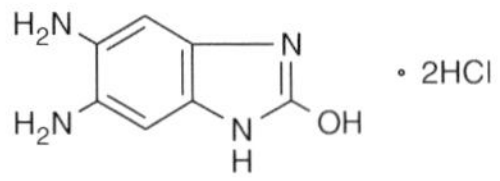

250 mg

S80,235-2
5,6-DINITRO-2-HYDROXYBENZIMIDAZOLE

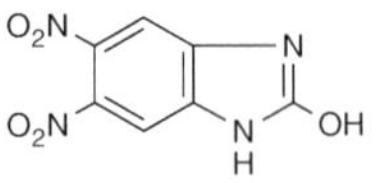

250 mg

S59,867-4
5,6-DICHLORO-1-(2-HYDROXYETHYL)-2-METHYLBENZIMIDAZOLE

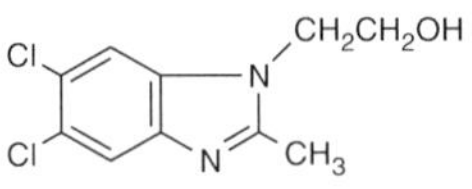

250 mg

S80,205-0
2-HYDROXY-5-NITROBENZIMIDAZOLE

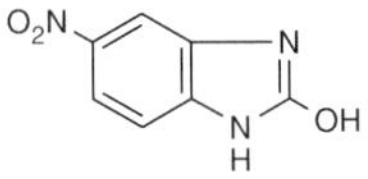

250 mg

S65,815-4
3,5-DICHLORO-ALPHA-(5,6-DIMETHYL-2-BENZIMIDAZOLYL)-2-HYDROXYCINNAMONITRILE

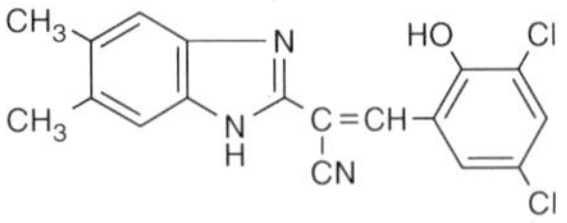

250 mg

S35,868-1
1,2-BIS(5,6-DIMETHYL-2-BENZIMIDAZOLYL)ETHANOL HYDROCHLORIDE

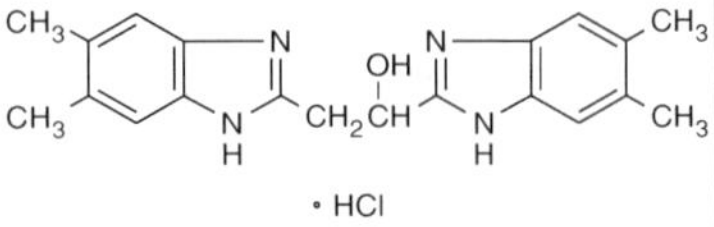

250 mg

S92,763-5
ALPHA,6-DIMETHYLBENZIMIDAZOLEMETHANOL

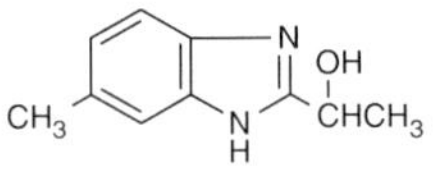

250 mg

S94,080-1
2-(2-BENZIMIDAZOLYLMETHYL)PHENOL

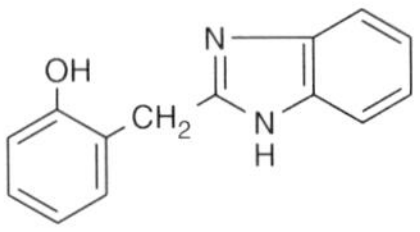

250 mg

S93,150-0
2-(2-BENZIMIDAZOLYLAMINO)-1-ETHANOL

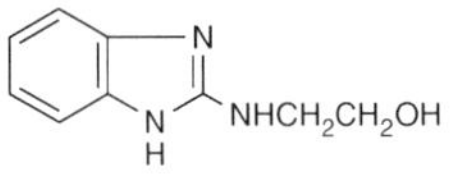

250 mg

S44,150-3
2-(3-HYDROXYPROPYL)BENZIMIDAZOLE

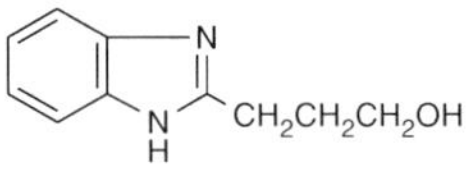

5 gm

S35,585-2
ALPHA,ALPHA,5,6-TETRAMETHYL-2-BENZIMIDAZOLEMETHANOL

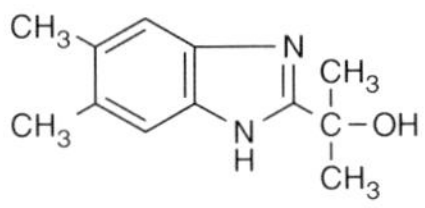

5 gm

S34,675-6
5,6-DIMETHYL-1-BENZIMIDAZOLEMETHANOL

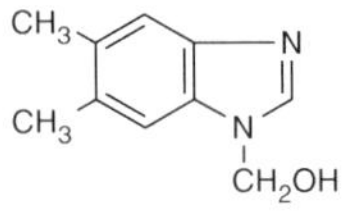

5 gm

S37,368-0
5,6-DIMETHYL-2-BENZIMIDAZOLEETHANOL

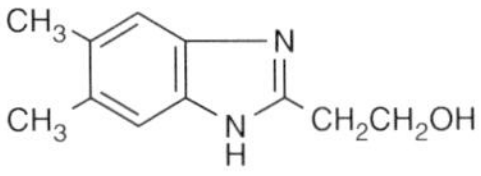

250 mg

S92,762-7
ALPHA-PHENYL-2-BENZIMIDAZOLEMETHANOL

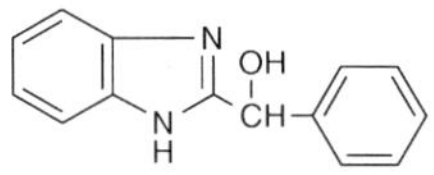

250 mg

S45,328-5
2-BENZIMIDAZOLEMETHANOL

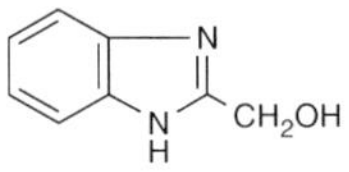

250 mg

S36,104-6
5,6-DIMETHYL-2-BENZIMIDAZOLEMETHANOL

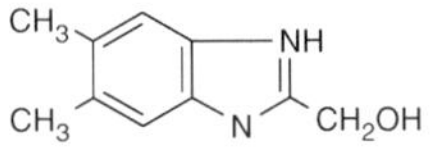

5 gm

S35,663-8
ALPHA,5,6-TRIMETHYL-2-BENZIMIDAZOLEMETHANOL

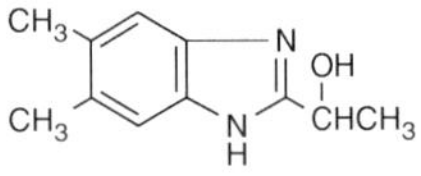

5 gm

S80,204-2
5-CHLORO-2-HYDROXYBENZIMIDAZOLE

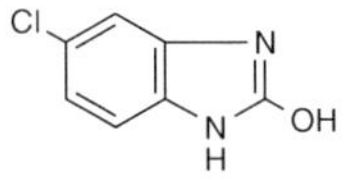

250 mg

S80,236-0
6-CHLORO-2-HYDROXYBENZIMIDAZOLE

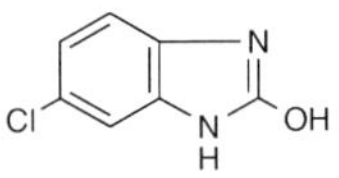

250 mg

S34,701-9
5,6-DIMETHYL-2-HYDROXYBENZIMIDAZOLE

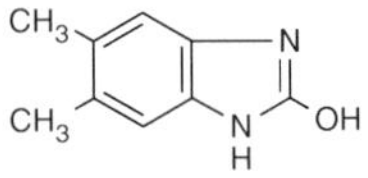

5 gm

S93,961-7
5-HYDROXYMETHYL-2-METHYL-1-PHENYLBENZIMIDAZOLE

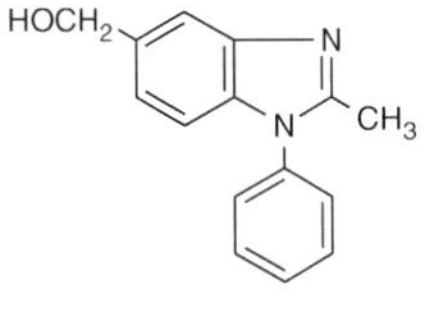

250 mg

S36,924-1
2-(2-BENZIMIDAZOLYL)PHENOL

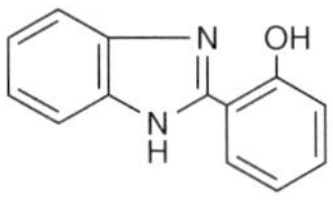

250 mg

S34,678-0
2-(5,6-DIMETHYL-2-BENZIMIDAZOLYL)-PHENOL

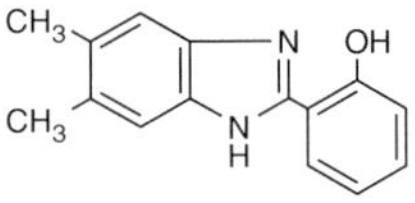

5 gm

S35,808-8
2-BENZIMIDAZOLEETHANOL

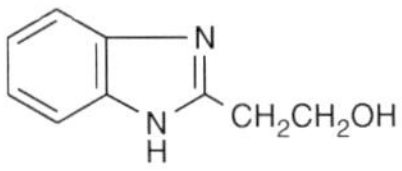

5 gm

S94,897-7
1-ETHYL-5-HYDROXY-2-METHYLBENZIMIDAZOLE

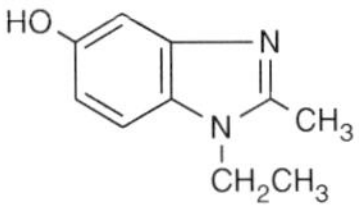

250 mg

S34,676-4
5,6-DIMETHYL-2-BENZIMIDAZOLEPROPANOL

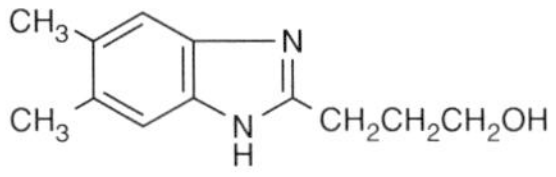

250 mg

S76,013-7
2-(3-BENZYLOXYPHENYL)BENZIMIDAZOLE

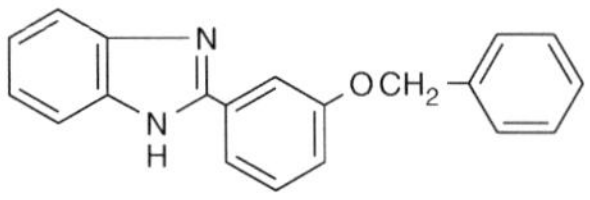

250 mg

S91,050-3
1-[N-(2,4-DIMETHOXYPHENYL)-CARBAMOYL]BENZIMIDAZOLE
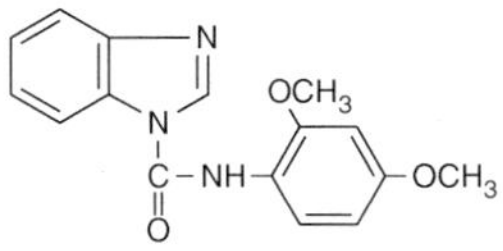
250 mg

S80,353-7
4,7-DIMETHOXYBENZIMIDAZOLE
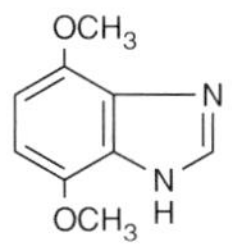
250 mg

S93,770-3
2-(2-CHLORO-1-BENZIMIDAZOLYL)-4'-METHOXYACETANILIDE
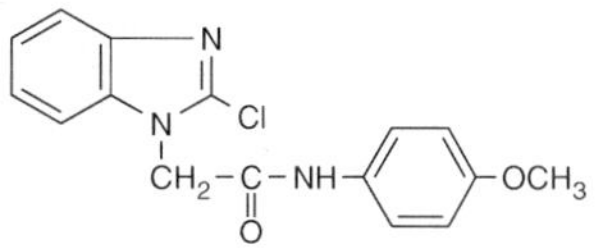
250 mg

S94,076-3
2-(PHENOXYMETHYL)BENZIMIDAZOLE
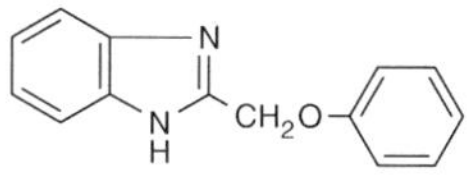
250 mg

S94,073-9
1-METHYL-2-(P-TOLYLOXYMETHYL)-BENZIMIDAZOLE
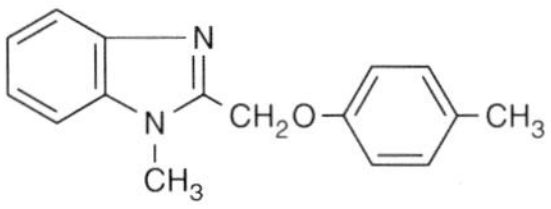
250 mg

S94,261-8
2-(2-METHOXYPHENYL)-1-METHYLBENZIMIDAZOLE
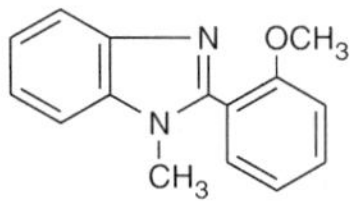
250 mg

S94,264-2
2-(4-METHOXYPHENYL)-1-METHYLBENZIMIDAZOLE
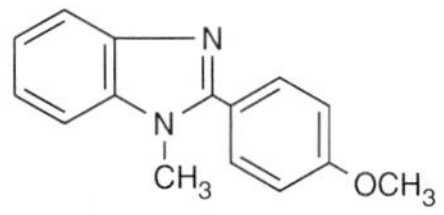
250 mg

S50,982-5
2-(3,4,5-TRIMETHOXYPHENYL)DIBENZO-(E,G)BENZIMIDAZOLE
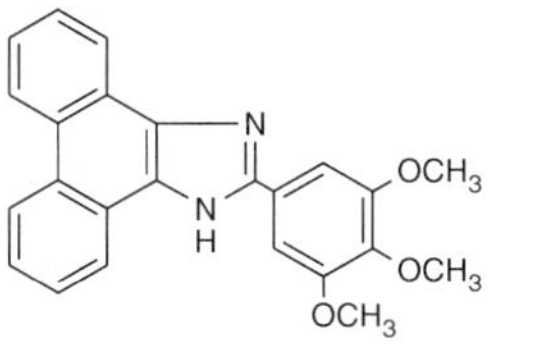
250 mg

S91,058-9
1-(N-ETHYLCARBAMOYL)-2-MERCAPTO-5-NITROBENZIMIDAZOLE
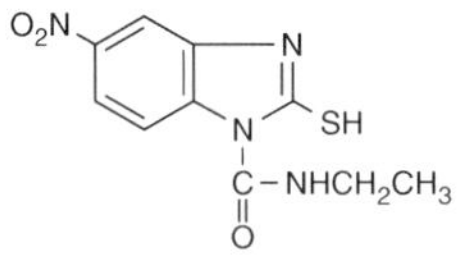
250 mg

S54,635-6
N,1-DIMETHYL-2-MERCAPTO-5-BENZIMIDAZOLESULFONAMIDE
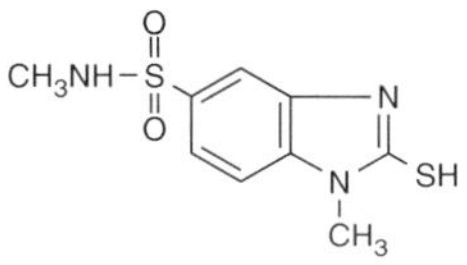
5 gm

S58,769-9
2-(METHYLTHIOMETHYL)-5-NITROBENZIMIDAZOLE
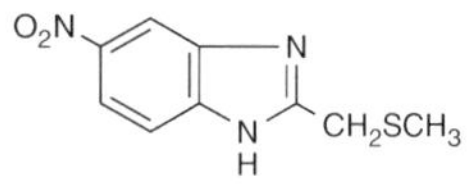
250 mg

S91,069-4
1-(N-[4-(METHYLTHIO)PHENYL]-CARBAMOYL)-6-NITROBENZIMIDAZOLE
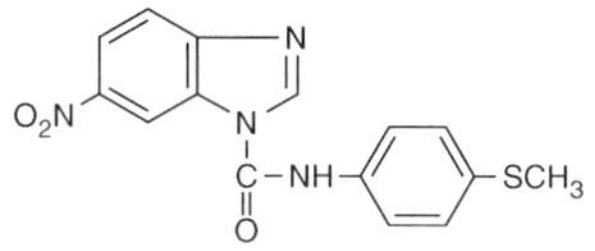
250 mg

S94,017-8
2-(BENZYLTHIO)BENZIMIDAZOLE
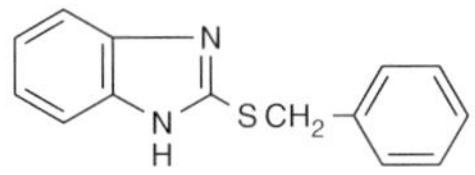
250 mg

S93,275-2
2-(PHENYLTHIOMETHYL)BENZIMIDAZOLE
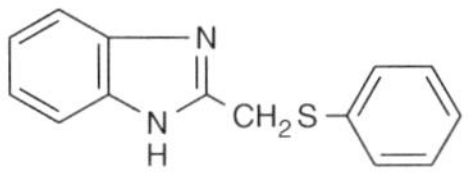
250 mg

S93,486-0
2-(ISOPROPYLTHIOMETHYL)-BENZIMIDAZOLE
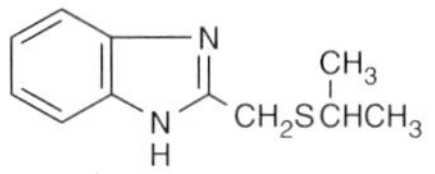
250 mg

S39,460-2
1,N-DIETHYL-2-METHYLTHIO-5-BENZIMIDAZOLESULFONAMIDE
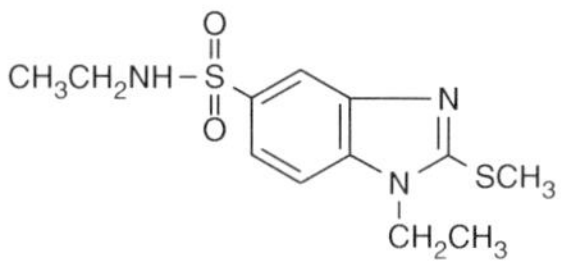
250 mg

S50,724-5
1-(BUTYLTHIOMETHYL)BENZIMIDAZOLE
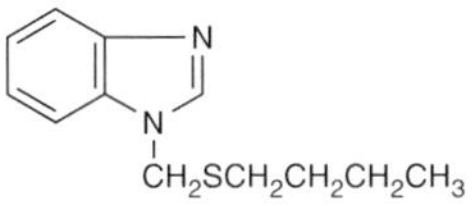
5 gm

S54,639-9
N,1-DIMETHYL-2-METHYLTHIO-5-BENZIMIDAZOLESULFONAMIDE
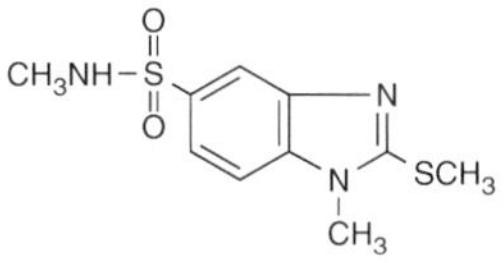
250 mg

S39,464-5
(2-BENZIMIDAZOLYLTHIO)ACETIC ACID
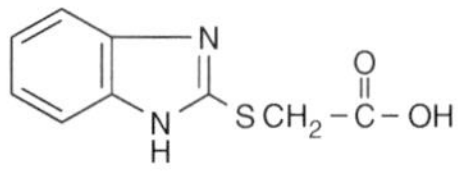
5 gm

S34,679-9
(5,6-DIMETHYL-2-BENZIMIDAZOLYLTHIO)-ACETIC ACID
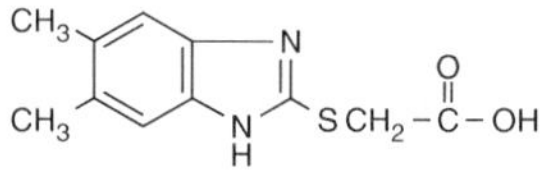
5 gm

S46,131-8
(2-BENZIMIDAZOLYLMETHYLTHIO)-ACETIC ACID
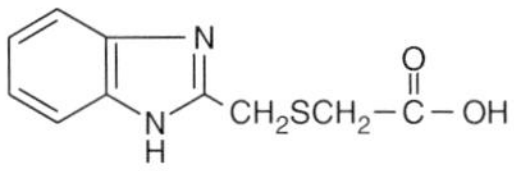
5 gm

S91,062-7
1-(N-[4-(METHYLTHIO)PHENYL]-CARBAMOYL)BENZIMIDAZOLE
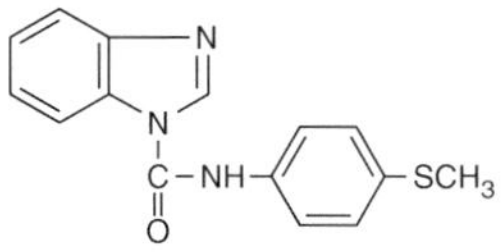
250 mg

S34,711-6
5,6-DIMETHYL-2-(METHYLTHIO)-BENZIMIDAZOLE
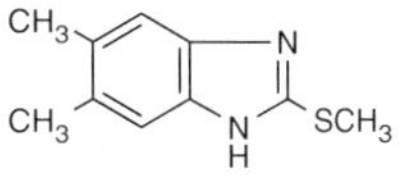
5 gm

S65,605-4
2-METHYLTHIOBENZIMIDAZOLESULFONIC ACID
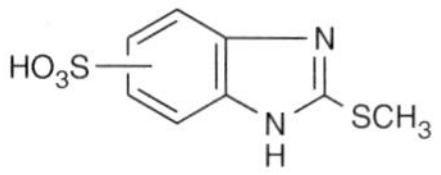
250 mg

S92,759-7
2-(ETHYLTHIOMETHYL)BENZIMIDAZOLE
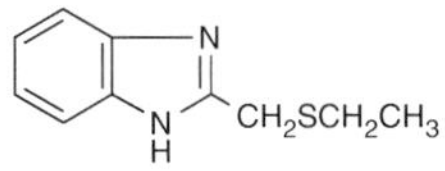
250 mg

S92,760-0
2-(ALLYLTHIOMETHYL)BENZIMIDAZOLE
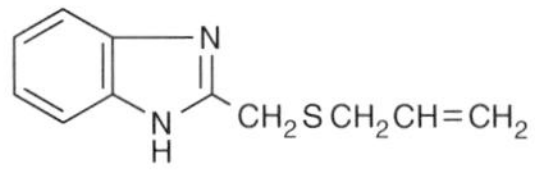
250 mg

S94,268-5
2-(OCTYLTHIO)BENZIMIDAZOLE
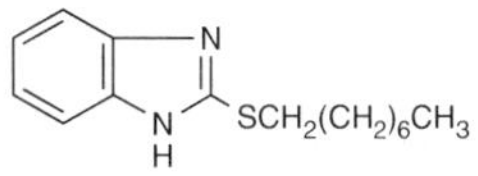
250 mg

S40,639-2
5-AMINO-1-ETHYL-2-METHYLBENZIMIDAZOLE DIHYDROCHLORIDE
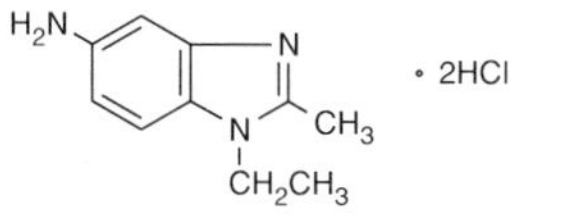
5 gm

S92,596-9
1-ACETAMIDO-2-AMINOBENZIMIDAZOLE HYDROBROMIDE

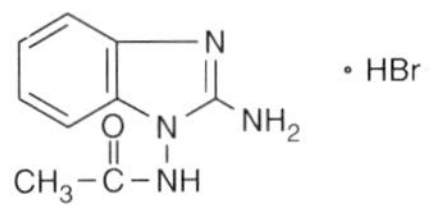

250 mg

S93,962-5
5-AMINO-2-METHYL-1-PHENYLBENZIMIDAZOLE

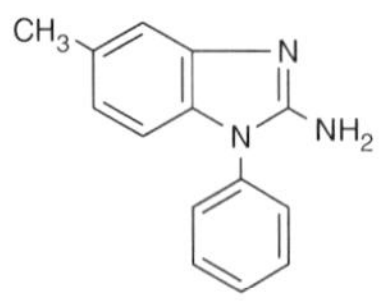

250 mg

S34,118-5
2-AMINOMETHYL-5,6-DIMETHYLBENZIMIDAZOLE HYDROCHLORIDE

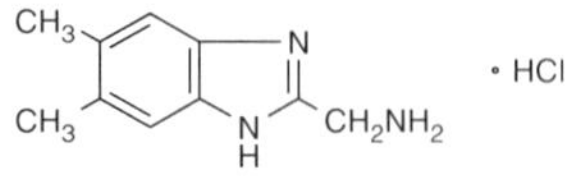

5 gm

S51,700-3
2-GUANIDINO-5-NITROBENZIMIDAZOLE

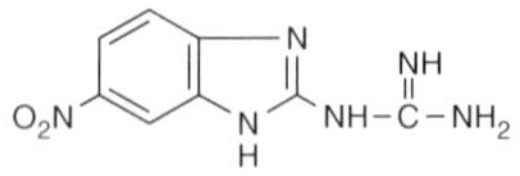

5 gm

S56,996-8
2-(3-AMINO-3-METHYLPENTYL)-1-OCTADECYL-5-BENZIMIDAZOLESULFONIC ACID, SODIUM SALT

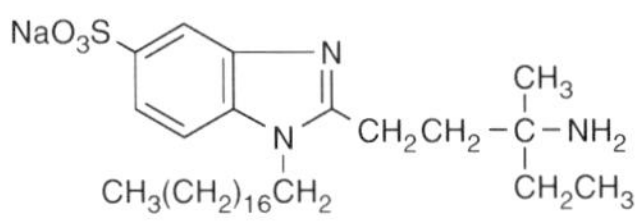

250 mg

S61,784-9
2-(M-AMINOPHENYL)-1-DODECYLBENZIMIDAZOLE-5-SULFONIC ACID, SODIUM SALT

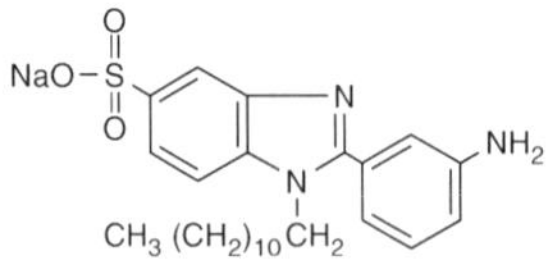

250 mg

S94,077-1
2-AMINO-1-METHYLBENZIMIDAZOLE

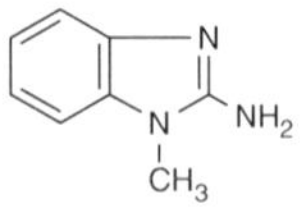

250 mg

S90,987-4
2-AMINO-1-ETHYLBENZIMIDAZOLE

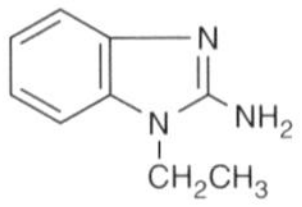

250 mg

S94,072-0
2-AMINO-1-ISOPROPYLBENZIMIDAZOLE

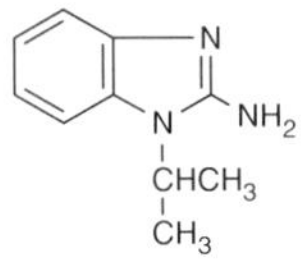

250 mg

S92,757-0
1-ACETYL-2-AMINOBENZIMIDAZOLE

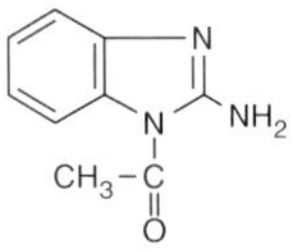

250 mg

S51,703-8
5-CHLORO-2-GUANIDINOBENZIMIDAZOLE

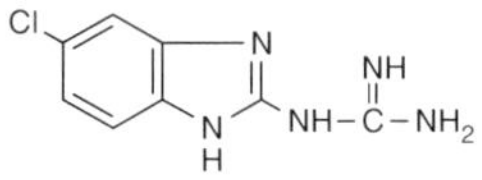

5 gm

S92,018-5
2,5-DIMETHYL-7-AMINOBENZIMIDAZOLE

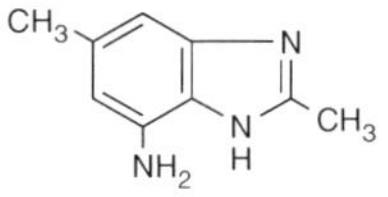

250 mg

S94,895-0
2-AMINO-1-BUTYLBENZIMIDAZOLE

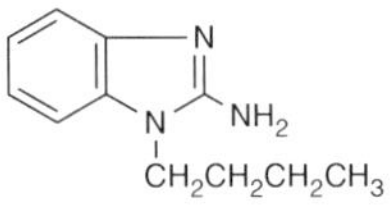

250 mg

S40,900-6
4-(2-KETO-1-BENZIMIDAZOLINYL)-1,2,3,6-TETRAHYDROPYRIDINE

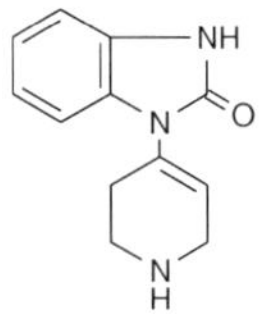

250 mg

S80,233-6
5,6-DINITROBENZIMIDAZOLE

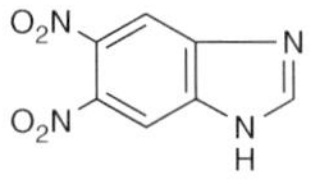

250 mg

S80,234-4
5,7-DINITROBENZIMIDAZOLE

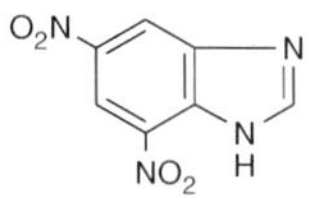

250 mg

S93,276-0
2-(PIPERIDINOMETHYL)BENZIMIDAZOLE

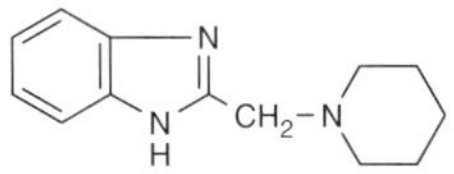

250 mg

S41,684-3
5,6-DIMETHYL-2-TRIFLUOROMETHYLBENZIMIDAZOLE

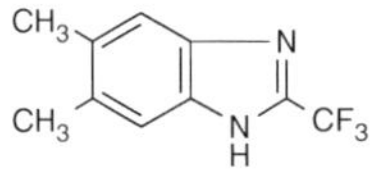

5 gm

S92,401-6
1-(CYCLOPROPYLCARBONYL)-5-NITROBENZIMIDAZOLE

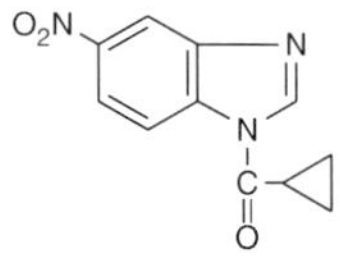

250 mg

S93,768-1
2-(2-CHLORO-1-BENZIMIDAZOLYL)-ACETAMIDE

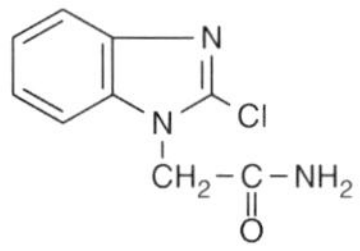

250 mg

S52,566-9
1-(4-CHLOROBENZYL)-2-METHYLBENZIMIDAZOLE HYDROCHLORIDE

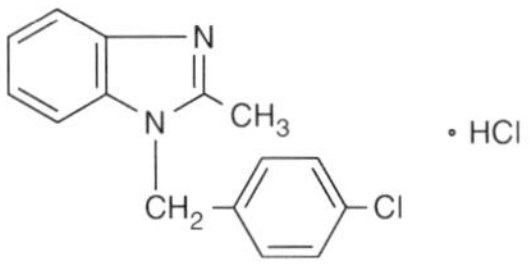

250 mg

S36,497-5
5,5',6,6'-TETRAMETHYL-2,2'-BIBENZIMIDAZOLE

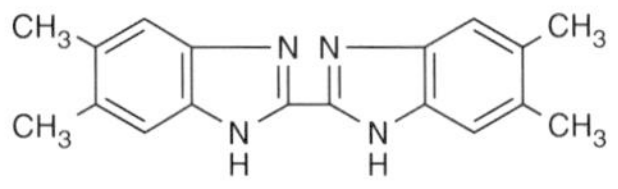

5 gm

S39,456-4
5-CHLORO-1-ETHYL-2-METHYL-6-NITROBENZIMIDAZOLE

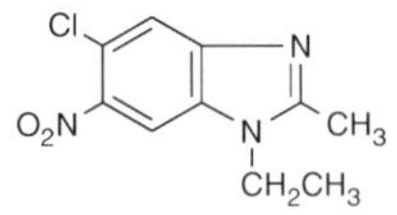

5 gm

S39,510-2
5-CHLORO-1,3-DIETHYL-2-METHYL-6-NITROBENZIMIDAZOLIUM IODIDE

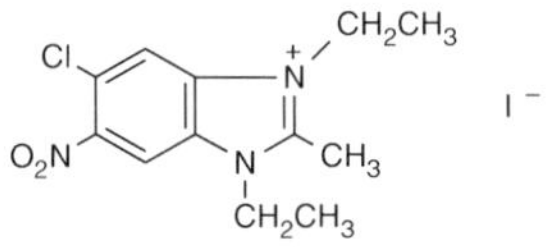

250 mg

S74,346-1
1-(2,2-DIMETHYLPROPIONYL)-5-NITROBENZIMIDAZOLE

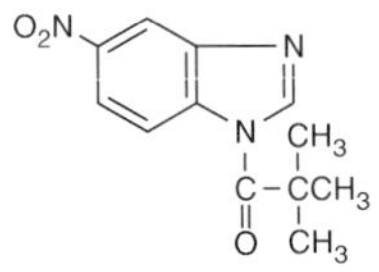

5 gm

S76,379-9
6-NITROBENZIMIDAZOLE

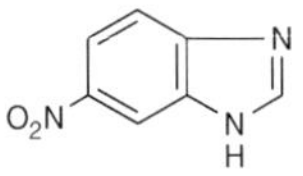

250 mg

S76,380-2
6-NITROBENZIMIDAZOLE NITRATE

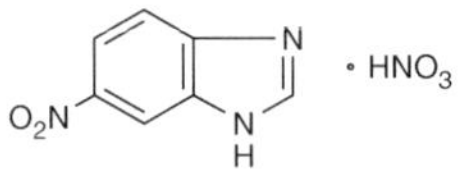

250 mg

S57,304-3
5-NITROBENZIMIDAZOLE SULFATE

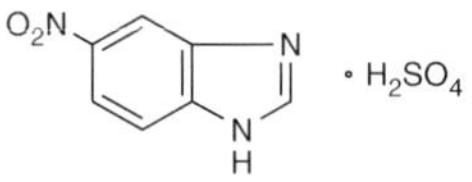

5 gm

S93,757-6
2-(2-BENZIMIDAZOYL)-2-VALERYLACETONITRILE

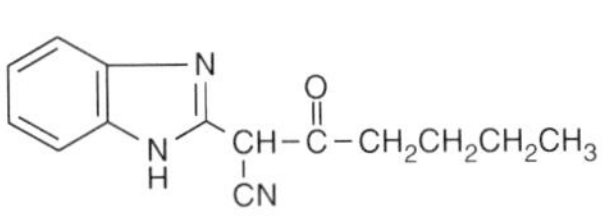

250 mg

S93,754-1
2-(2-CHLORO-5-NITROBENZOYL)-2-(1-METHYL-2-BENZIMIDAZOLYL)-ACETONITRILE

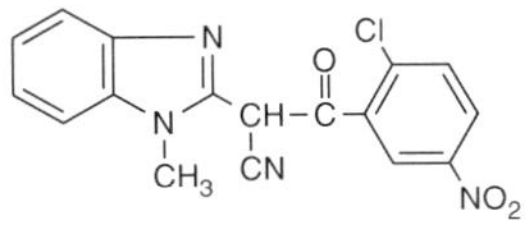

250 mg

S91,226-3
ETHYL 2-(5-NITRO-1-BENZIMIDAZOLYLCARBONYLAMINO)ACETATE

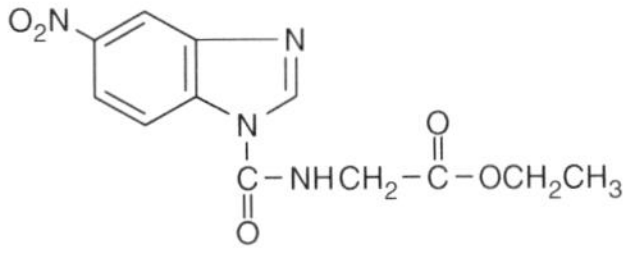

250 mg

S44,704-8
6-NITROBENZIMIDAZOLE, SODIUM SALT

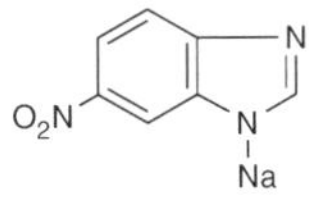

5 gm

S80,241-7
5-METHYL-6-NITROBENZIMIDAZOLE

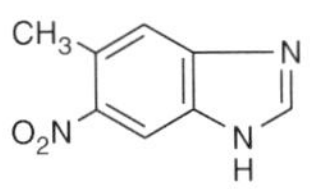

250 mg

S93,756-8
2-(1-METHYL-2-BENZIMIDAZOLYL)-2-VALERYLACETONITRILE

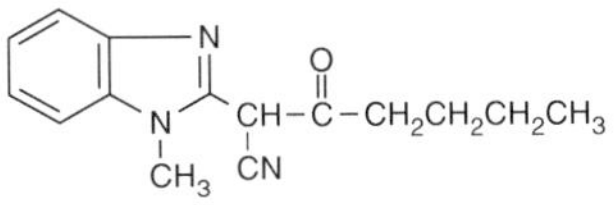

250 mg

S93,488-7
4-CYANO-3,5-DIMETHYLPYRIDO(1,2-A)-BENZIMIDAZOL-1(5H)-ONE

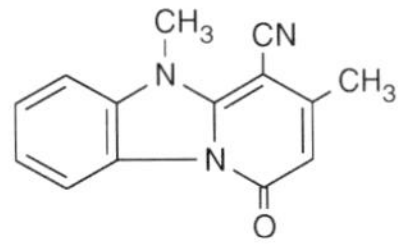

250 mg

S93,490-9
6-CYANO-7,8,9,10-TETRAHYDROBENZIMIDAZOLO(1,2-B)-ISOQUINOLIN-11(5H)-ONE

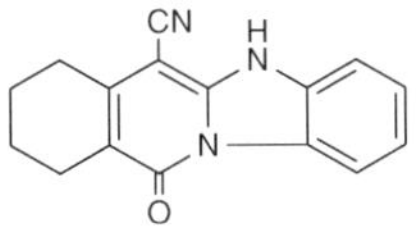

250 mg

S93,487-9
4-CYANO-3,5-DIMETHYL-2-ETHYLPYRIDO-(1,2-A)BENZIMIDAZOL-1(5H)-ONE

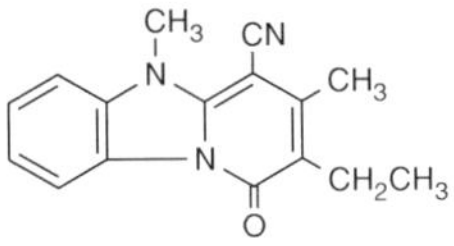

250 mg

S67,189-4
N-ACETYL-4-(5-CYANO-1-ETHYL-2-METHYL-3-BENZIMIDAZOLIUM)-BUTANESULFONAMIDE BROMIDE

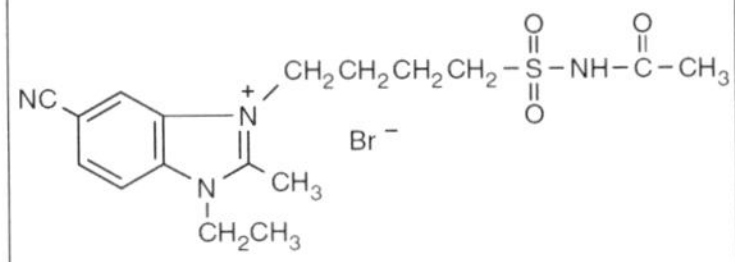

250 mg

S50,689-3
1-(4-NITROANILINOMETHYL)-2(3H)-BENZIMIDAZOLETHIONE

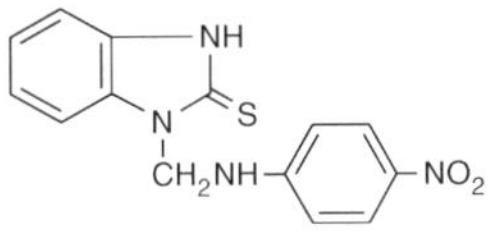

250 mg

S38,563-8
5-CYANO-1-ETHYL-2-METHYLBENZIMIDAZOLE

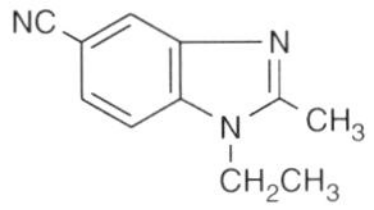

5 gm

S80,573-4
1-[4-(DIMETHYLAMINO)PHENYL]-6-ETHYLBENZIMIDAZOLE

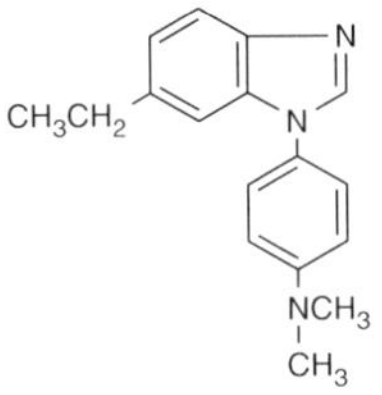

250 mg

S89,843-0
5-CHLORO-2-(CYANOAMINO)-1-(4-NITROPHENYL)BENZIMIDAZOLE

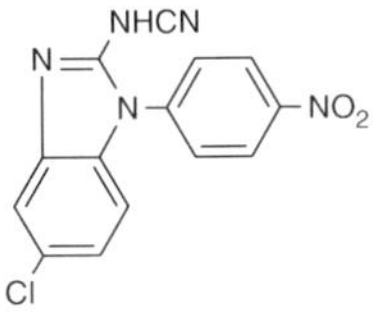

250 mg

S51,155-2
TETRAKIS(5,6-DIMETHYL-2-BENZIMIDAZOLYL)ETHYLENE

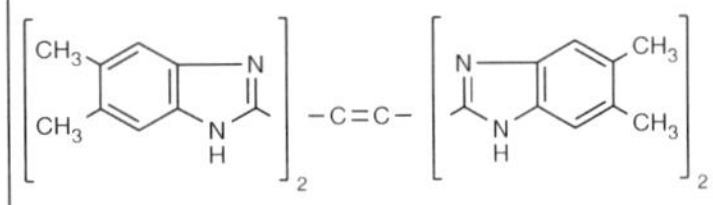

5 gm

S35,825-8
2-BENZYLBENZIMIDAZOLE

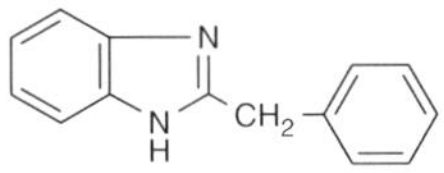

5 gm

S34,180-0
2-BENZYL-5,6-DIMETHYLBENZIMIDAZOLE

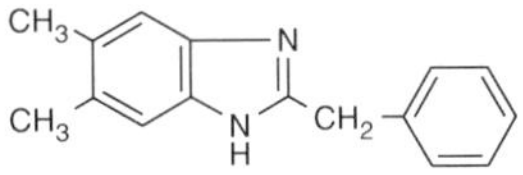

5 gm

S44,964-4
2-BENZIMIDAZOLECARBAMONITRILE

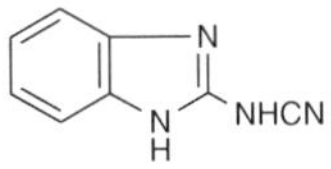

5 gm

S94,892-6
2-BENZOYLBENZIMIDAZOLE

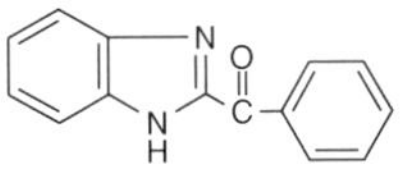

250 mg

S94,893-4
2-(ISOPROPYLTHIO)BENZIMIDAZOLE

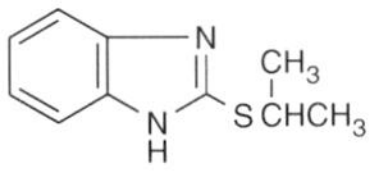

250 mg

S43,210-5
4-(2-BENZIMIDAZOLYL)-BENZENESULFONIC ACID

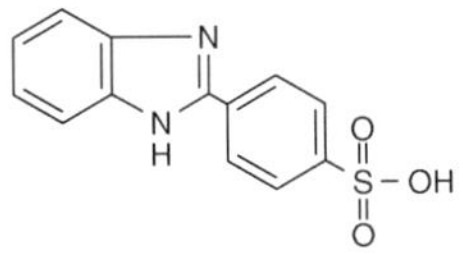

5 gm

S93,148-9
2-BENZIMIDAZOLESULFONIC ACID

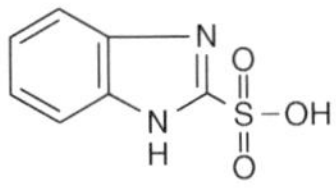

250 mg

S94,894-2
2-ANILINOBENZIMIDAZOLE

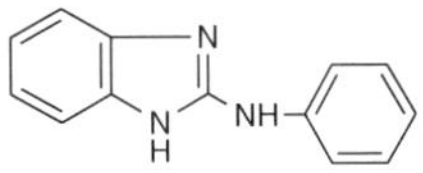

250 mg

S93,151-9
2-CHLOROBENZIMIDAZOLE

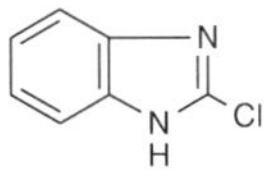

250 mg

S44,710-2
2-METHYLBENZIMIDAZOLE HYDROCHLORIDE

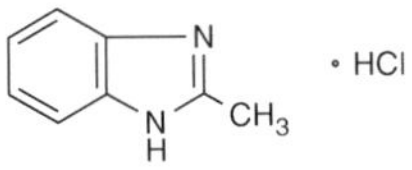

5 gm

S41,832-3
2-GUANIDINO-5-METHYLBENZIMIDAZOLE

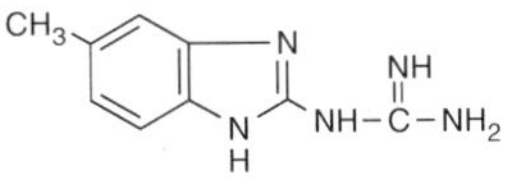

5 gm

S55,674-2
2-PROPYLBENZIMIDAZOLE

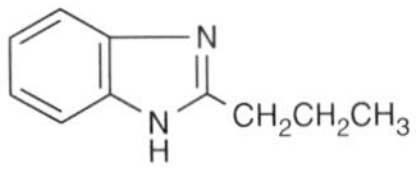

5 gm

S42,721-7
2-ETHYLBENZIMIDAZOLE

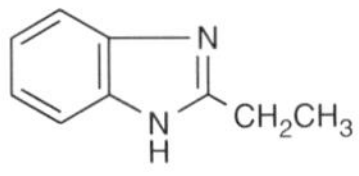

5 gm

S39,524-2
5-ACETYL-1,3-DIETHYL-2-METHYLBENZIMIDAZOLIUM IODIDE

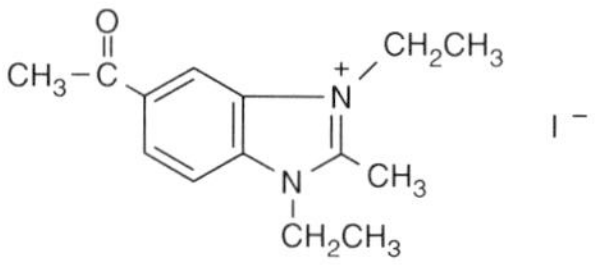

250 mg

S53,064-6
1-(2-ACETOXYETHYL)-5,6-DICHLORO-3-ETHYL-2-METHYLBENZIMIDAZOLIUM BROMIDE

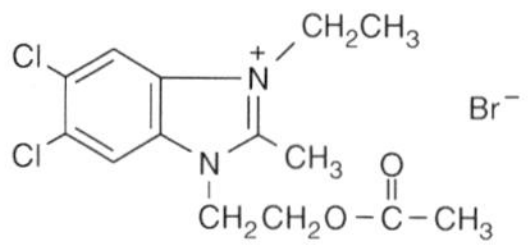

250 mg

S39,501-3
1-ETHYL-2-METHYL-3-(3-SULFOOXYPROPYL)BENZIMIDAZOLIUM HYDROXIDE, INNER SALT

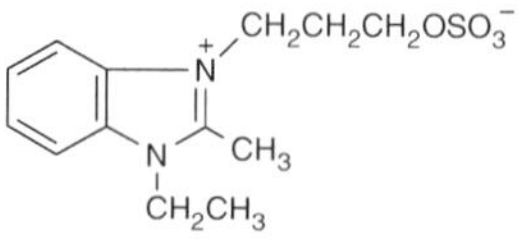

250 mg

S36,112-7
5,6-DIMETHYL-1-(PIPERIDINOMETHYL)-BENZIMIDAZOLE

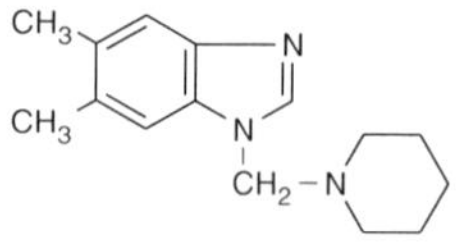

250 mg

S92,886-0
3-NITRO-IMIDAZO(1,2-A)PYRIDINE

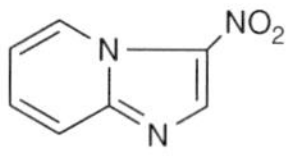

250 mg

S39,458-0
1-ETHYL-2-METHYLBENZIMIDAZOL-5-YL METHYL KETONE

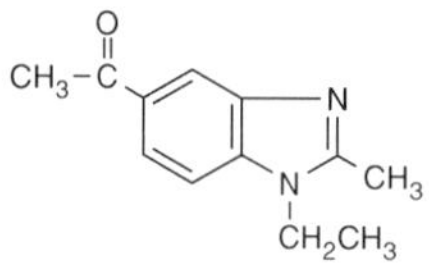

250 mg

S92,022-3
2-ACETYL-1-METHYLBENZIMIDAZOLE

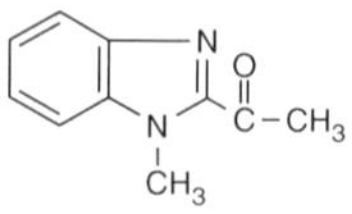

250 mg

S56,879-1
1-ETHYL-2-METHYL-5-(METHYLSULFONYL)BENZIMIDAZOLE

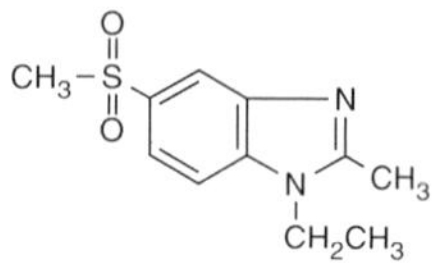

250 mg

S65,791-3
1-ETHYL-2-METHYL-3-(3-SULFOPROPYL)-BENZIMIDAZOLE

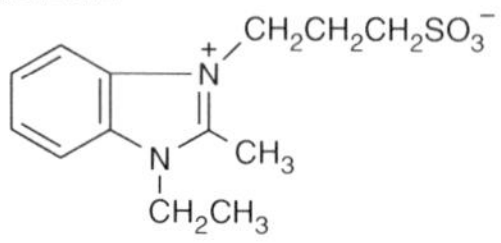

5 gm

S56,991-7
1-ETHYL-2-UNDECYL-5-BENZIMIDAZOLESULFONIC ACID

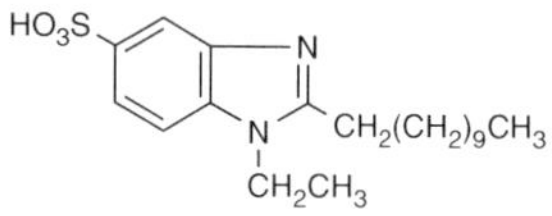

250 mg

S55,882-6
2-METHYL-1-OCTADECYL-5-BENZIMIDAZOLESULFONIC ACID

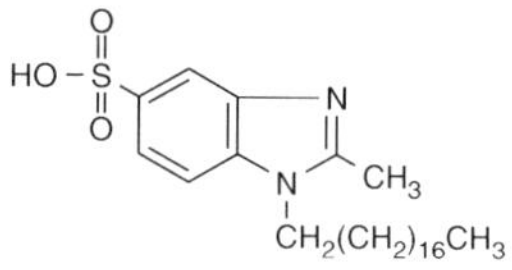

250 mg

S44,835-4
N,1-DIETHYL-2-METHYL-5-BENZIMIDAZOLE-SULFONAMIDE

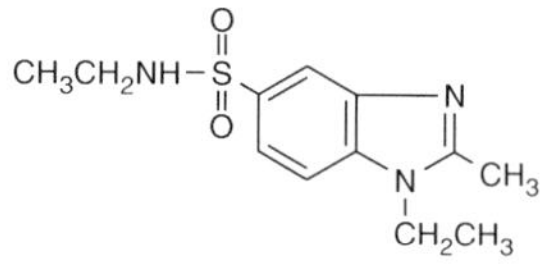

5 gm

S39,489-0
5,6-DICHLORO-2-METHYL-1-PHENYLBENZIMIDAZOLE

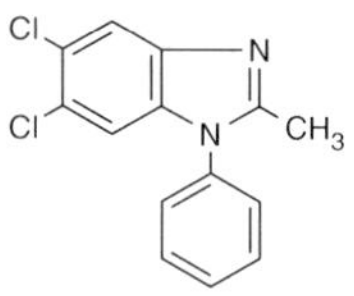

250 mg

S39,511-0
5,6-DICHLORO-1-METHYL-2-BENZIMIDAZOLETHIOL

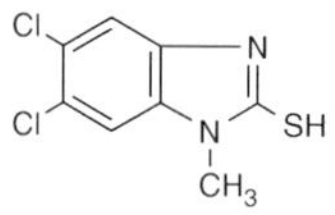

250 mg

S94,079-8
1,2-DIMETHYLBENZIMIDAZOLE

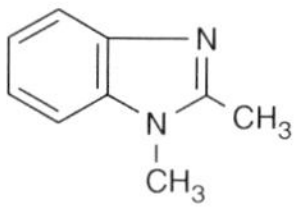

250 mg

S39,472-6
5,6-DICHLORO-1,2-DIMETHYLBENZIMIDAZOLE

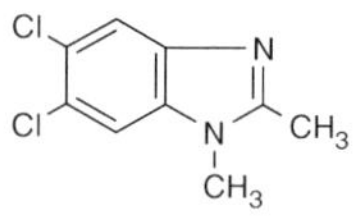

250 mg

S45,291-2
5-CHLORO-1-ETHYL-2-METHYLBENZIMIDAZOLE

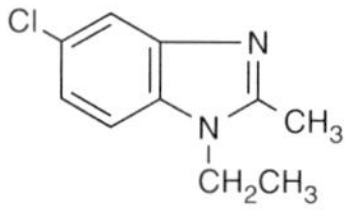

250 mg

S39,480-7
5,6-DICHLORO-1-ETHYL-2-METHYLBENZIMIDAZOLE

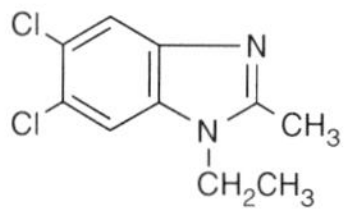

250 mg

S93,692-8
2-PHENYL-1,3-DIMETHYLBENZIMIDAZOLIUM IODIDE

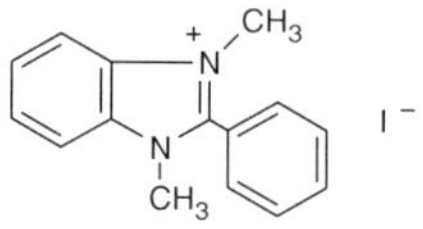

250 mg

S93,693-6
2-HEXYL-1,3-DIMETHYLBENZIMIDAZOLIUM IODIDE

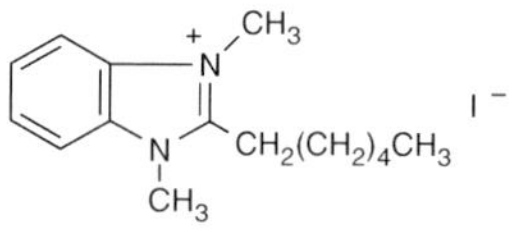

250 mg

S54,236-9
2-(5-CHLORO-2-BENZIMIDAZOLYLIMINO)-4,5-IMIDAZOLIDINEDIONE

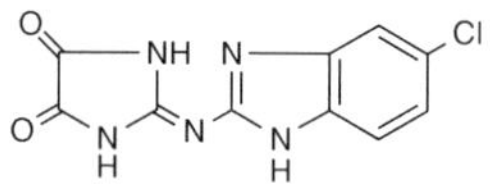

250 mg

S54,636-4
1-ETHYL-2-METHYL-5-BENZIMIDAZOLECARBOXAMIDE

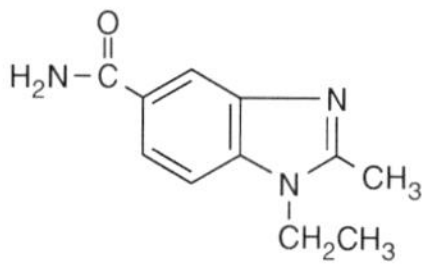

5 gm

S54,237-7
2-(5-NITRO-2-BENZIMIDAZOLYLIMINO)-4,5-IMIDAZOLIDINEDIONE

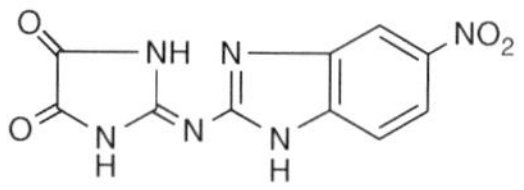

250 mg

S94,899-3
6-NITRO-1,2,3-TRIMETHYLBENZIMIDAZOLIUM METHYL SULFATE

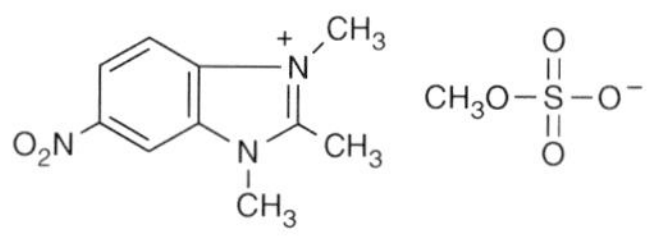

250 mg

S92,014-2
2-ACETYLBENZIMIDAZOLE

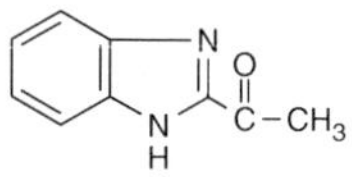

250 mg

S62,898-0
2-(2-OXOPROPYL)BENZIMIDAZOLE

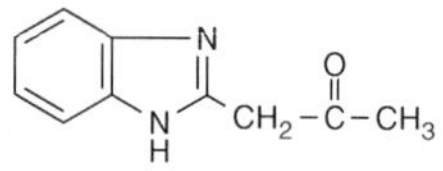

250 mg

S91,149-6
ETHYL 4-(1-BENZIMIDAZOLYLCARBONYLAMINO)-BENZOATE

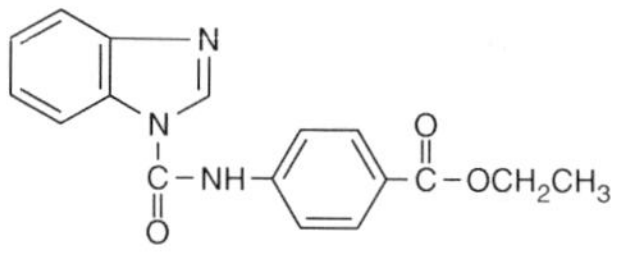

250 mg

S51,085-8
2-(2-BENZIMIDAZOLYLIMINO)-4,5-IMIDAZOLIDINEDIONE

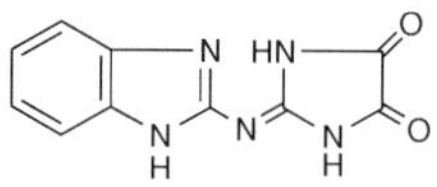

250 mg

S50,777-6
2-(ACETOACETAMIDO)BENZIMIDAZOLE

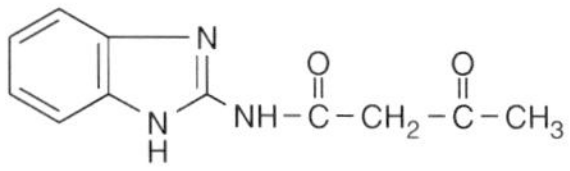

5 gm

S34,672-1
5,6-DIMETHYLBENZIMIDAZOLE-1-ACETIC ACID HYDROCHLORIDE

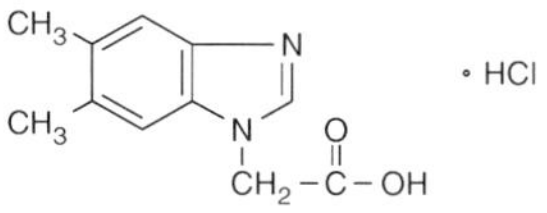

5 gm

S92,809-7
1-METHYL-2-BENZIMIDAZOLEACETIC ACID

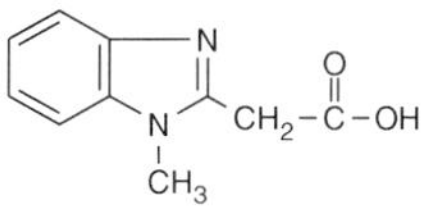

250 mg

S34,833-3
ETHYL 5,6-DIMETHYLBENZIMIDAZOLE-1-ACETATE

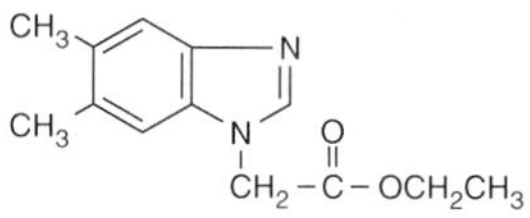

5 gm

S79,845-2
2,3-DIHYDRO-2-OXO-5-BENZIMIDAZOLYL ACETATE

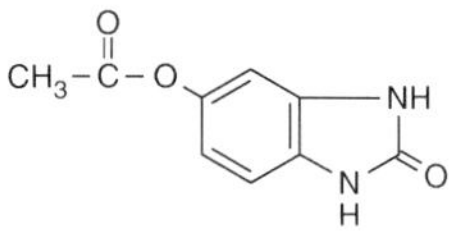

250 mg

S94,267-7
2-METHYL-1-PHENYL-5-BENZIMIDAZOLYL PROPIONATE

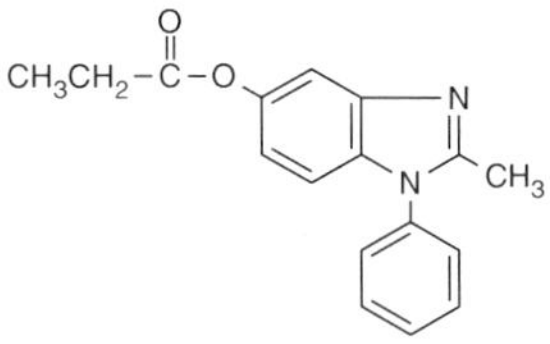

250 mg

S93,767-3
METHYL 2-(2-CHLORO-1-BENZIMIDAZOLYL)ACETATE

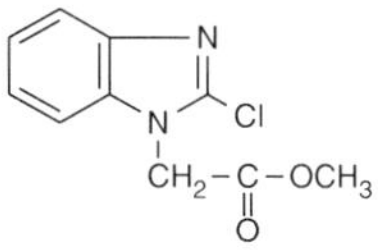

250 mg

S34,671-3
5,6-DIMETHYLBENZIMIDAZOLE-1-ACETHYDRAZIDE

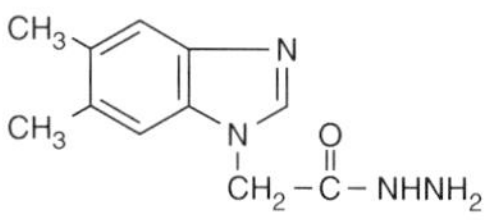

5 gm

S34,674-8
5,6-DIMETHYLBENZIMIDAZOLE-1-CARBOXANILIDE

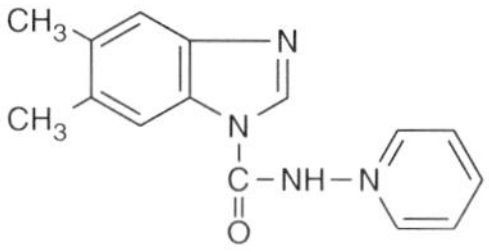

5 gm

S92,991-3
(1-METHYL-2-BENZIMIDAZOLYL)-ACETONITRILE

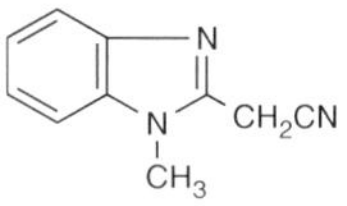

250 mg

S39,107-7
(5,6-DIMETHYL-2-BENZIMIDAZOLYL)-CARBAMONITRILE

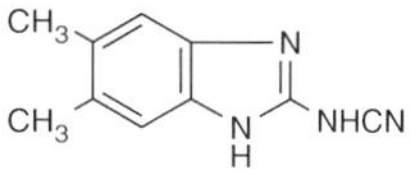

250 mg

S39,103-4
2-AMINO-1,4-DIHYDRO-4,4,7,8-TETRAMETHYL-S-TRIAZINO(1,2-A)-BENZIMIDAZOLE

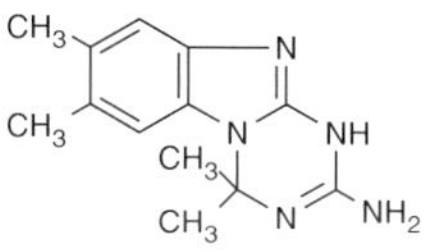

250 mg

S73,300-8
1-ALLYL-2-PHENYL-1H-BENZIMIDAZOLE

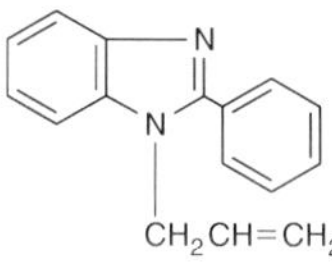

250 mg

S94,265-0
1-METHYL-2-(P-TOLYL)BENZIMIDAZOLE

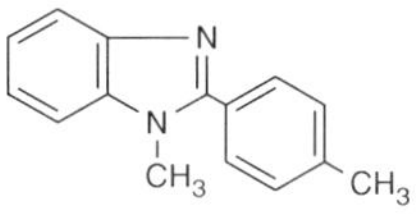

250 mg

S80,274-3
2,5-DICHLOROBENZIMIDAZOLE

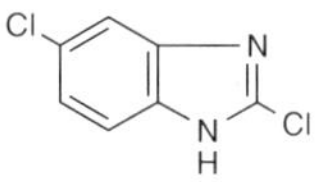

250 mg

S88,380-8
2-BROMOBENZIMIDAZOLE

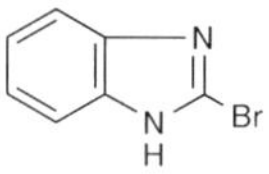

250 mg

S94,269-3
5-BROMOBENZIMIDAZOLE

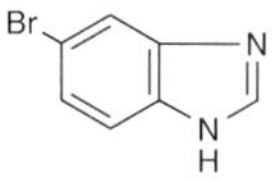

250 mg

S43,881-2
5-CHLORO-2-METHYLBENZIMIDAZOLE

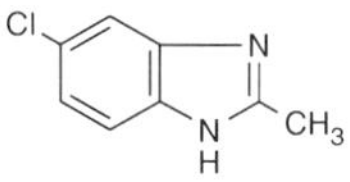

5 gm

S91,544-0
2-(1-PROPENYL)BENZIMIDAZOLE

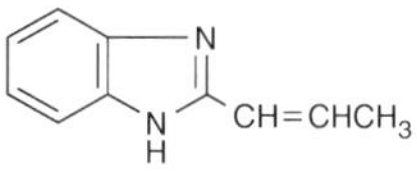

250 mg

S51,720-8
2-ETHYL-4-METHYLBENZIMIDAZOLE

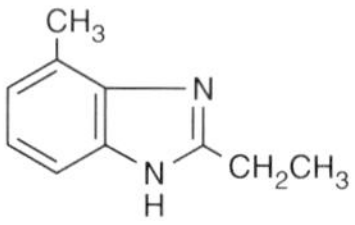

5 gm

S94,266-9
2-ETHYL-1-METHYLBENZIMIDAZOLE

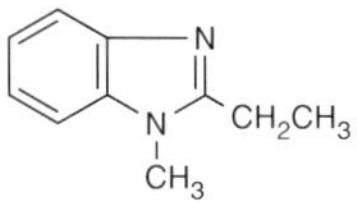

250 mg

S34,699-3
5,6-DIMETHYL-2-ETHYLBENZIMIDAZOLE

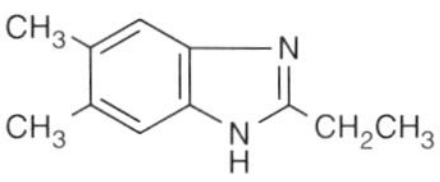

5 gm

S52,024-1
2-(2-BENZIMIDAZOLYLAMINO)-4,7-DIMETHYL-5H-1,3-DIAZEPINE

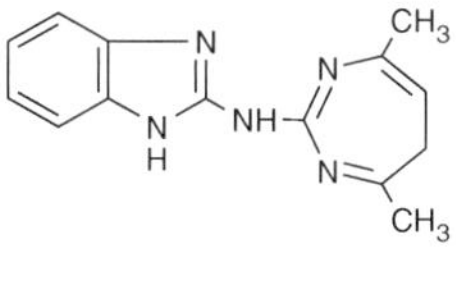

250 mg

S37,415-6
DI-2-BENZIMIDAZOLYLMETHANE

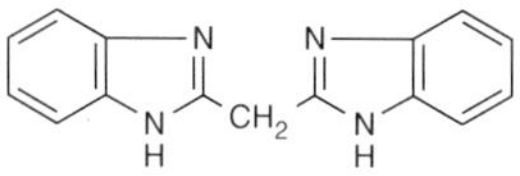

5 gm

S56,984-4
2-NONYLBENZIMIDAZOLE

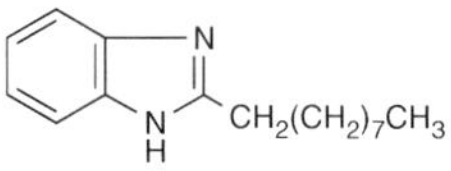

250 mg

S55,763-3
2,2'-ETHYLENEBIBENZIMIDAZOLE

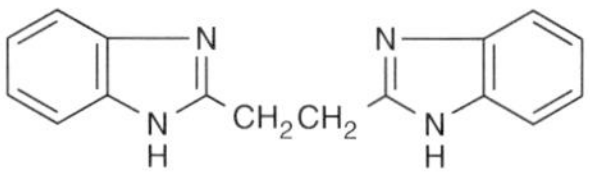

250 mg

S37,406-7
2,2'-METHYLENEBIS(5,6-DIMETHYLBENZIMIDAZOLE)

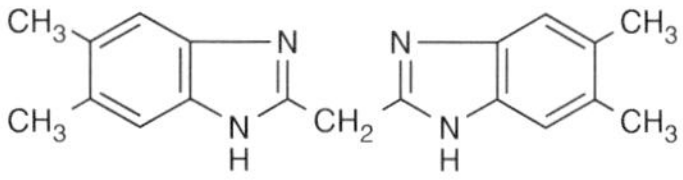

5 gm

S56,898-8
2-PENTADECYLBENZIMIDAZOLE

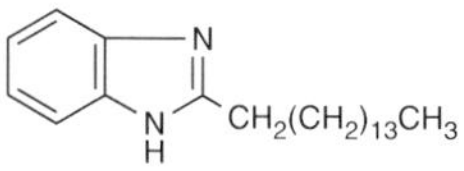

250 mg

S51,208-7
2-(4,5-DIPHENYL-2-IMIDAZOLYLIDENEAMINO)BENZIMIDAZOLE

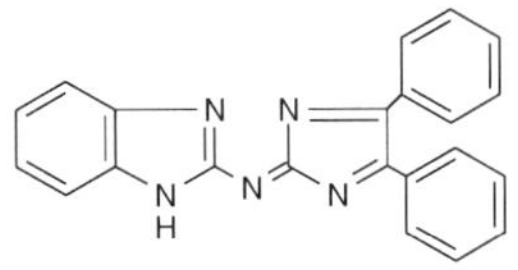

250 mg

S41,766-1
12H-BENZ(5,6)ISOINDOLO(2,1-A)-BENZIMIDAZOLE

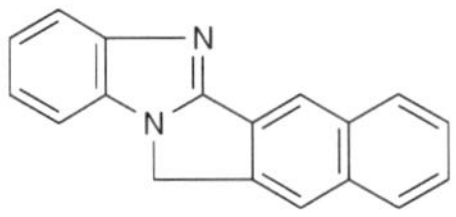

250 mg

S92,758-9
2-STYRYLBENZIMIDAZOLE

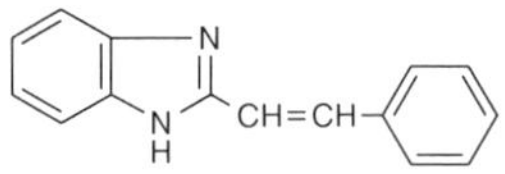

250 mg

S76,838-3
1-(2,2-DICHLOROACETYL)-1H-BENZOTRIAZOLE

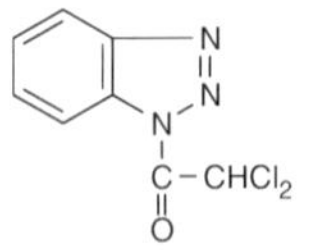

250 mg

S91,325-1
1-(N-[2-(TRIFLUOROMETHYL)PHENYL]-CARBAMOYL)-1H-BENZOTRIAZOLE

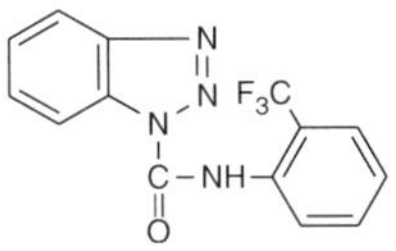

250 mg

S91,330-8
1-(N-[3-(TRIFLUOROMETHYL)PHENYL]-CARBAMOYL)-1H-BENZOTRIAZOLE

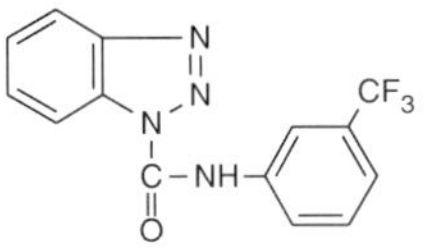

250 mg

S91,331-6
5-CHLORO-1-(N-[3-(TRIFLUOROMETHYL)-PHENYL]CARBAMOYL)-1H-BENZOTRIAZOLE

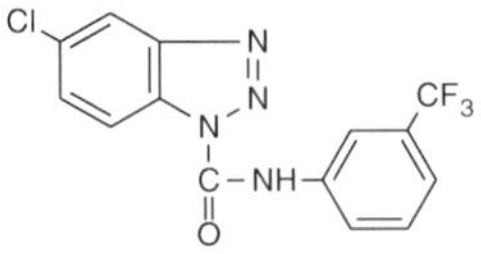

250 mg

S78,249-1
3-(TRIFLUOROMETHYL)-1,2,4-TRIAZOLO(3,4-A)ISOQUINOLINE

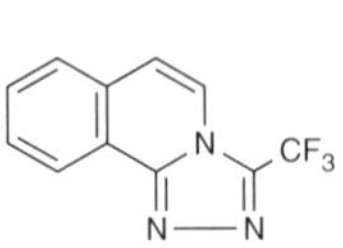

5 gm

S91,266-2
5-CHLORO-1-[N-(5-CHLORO-2-METHOXYPHENYL)CARBAMOYL]-1H-BENZOTRIAZOLE

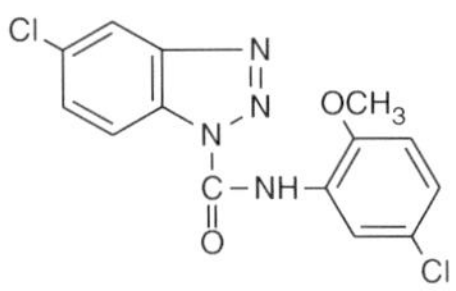

250 mg

S91,267-0
1-[N-(5-CHLORO-2-METHOXYPHENYL)-CARBAMOYL]-5,6-DIMETHYL-1H-BENZOTRIAZOLE

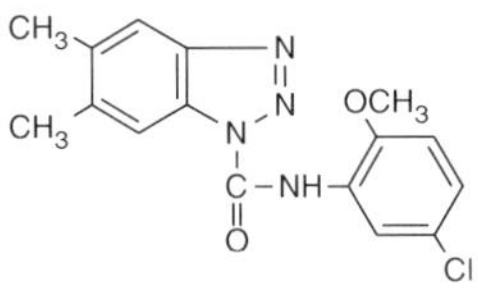

250 mg

S73,651-1
2-(1-BENZOTRIAZOLYLCARBONYL)-3,4,5,6-TETRACHLOROBENZOIC ACID

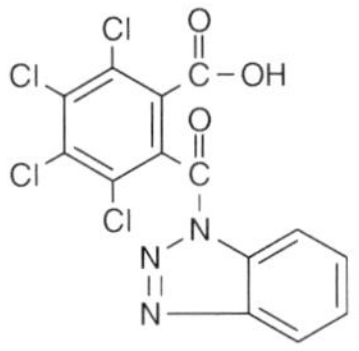

5 gm

S91,295-6
5-CHLORO-1-[N-(2-FLUOROPHENYL)-CARBAMOYL]-1H-BENZOTRIAZOLE

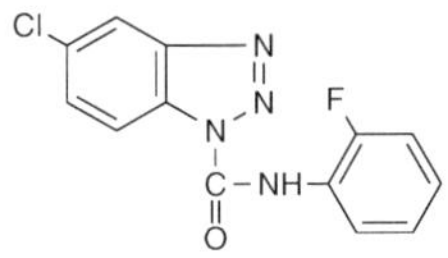

250 mg

S91,314-6
1-[N-(3-CHLORO-4-METHYLPHENYL)-CARBAMOYL]-1H-BENZOTRIAZOLE

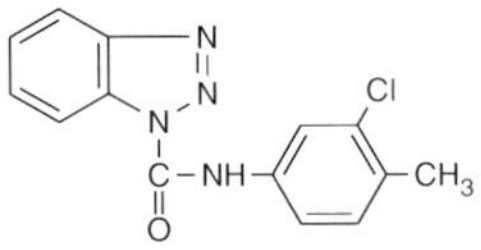

250 mg

S91,315-4
5-CHLORO-1-[N-(3-CHLORO-4-METHYLPHENYL)CARBAMOYL]-1H-BENZOTRIAZOLE

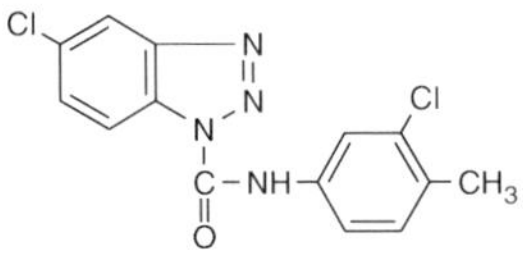

250 mg

S58,319-7
2-AMINO-6-(2H-BENZOTRIAZOL-2-YL)-P-CRESOL

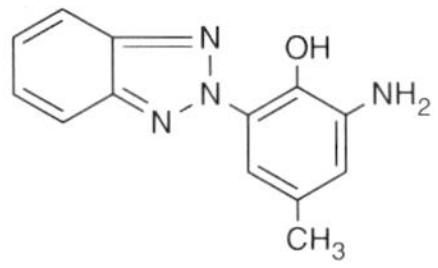

250 mg

S36,254-9
2-(2-HYDROXY-5-METHYLPHENYL)-BENZOTRIAZOLE

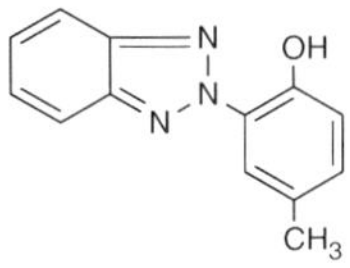

5 gm

S58,306-5
2-(2H-BENZOTRIAZOL-2-YL)-4-(TERT-BUTYL)-6-NITROPHENOL

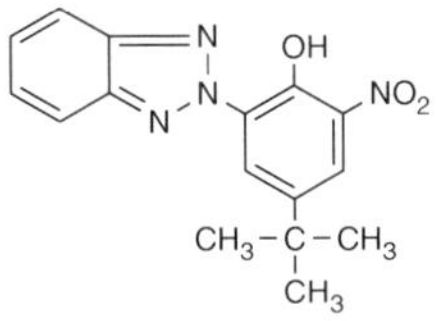

250 mg

S58,817-2
2-(2H-BENZOTRIAZOL-2-YL)-HYDROQUINONE

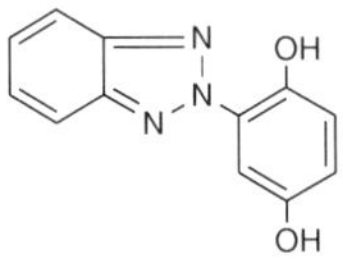

25 mg

S79,866-5
8,9-DIMETHOXY-1,2,4-TRIAZOLO(3,4-A)-ISOQUINOLINE

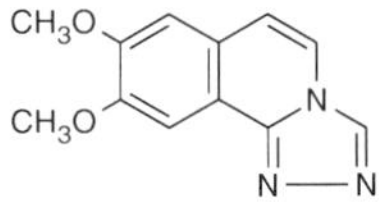

250 mg

S79,880-0
8,9-DIMETHOXY-3-METHYL-1,2,4-TRIAZOLO(3,4-A)ISOQUINOLINE

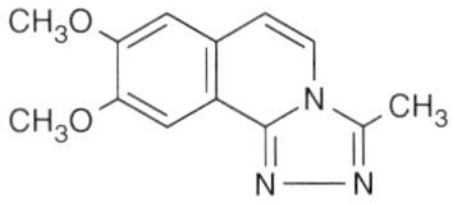

250 mg

S90,976-9
2,3-BIS(4-METHOXYPHENYL)-3H-NAPHTHO(1,2-D)TRIAZOLIUM TETRAFLUOROBORATE

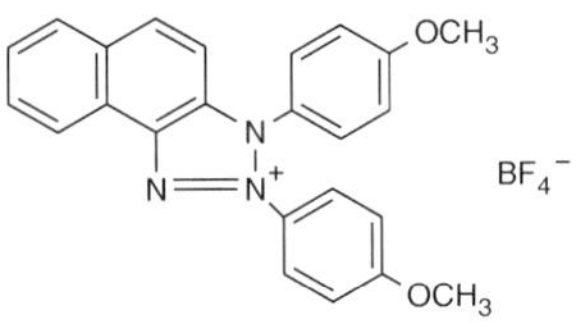

250 mg

S55,581-9
2-(O-AMINOPHENYL)-5,6-DIMETHOXY-2H-BENZOTRIAZOLE

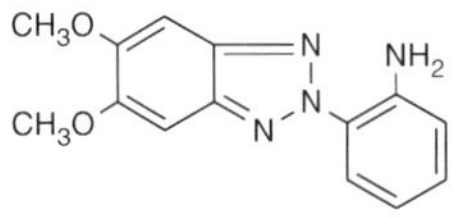

250 mg

S90,975-0
2-(4-METHOXYPHENYL)-3-(P-TOLYL)-3H-NAPHTHO(1,2-D)TRIAZOLIUM BROMIDE

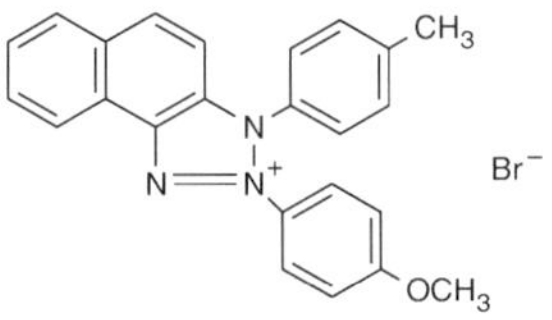

250 mg

S84,374-1
5,6-DIAMINO-1H-BENZOTRIAZOLE DIHYDROCHLORIDE

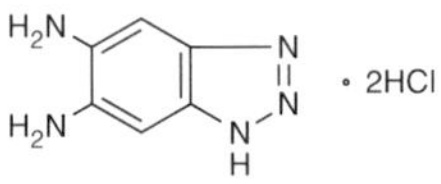

250 mg

S84,373-3
6,7-DIAMINO-1H-BENZOTRIAZOLE DIHYDROCHLORIDE

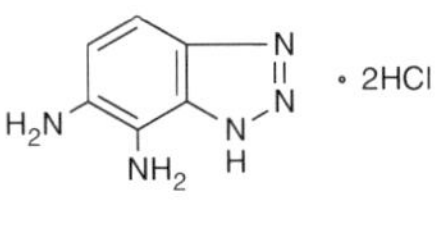

250 mg

S54,449-3
2-(2-AMINOPHENYL)-2H-NAPHTHO(1,2-D)-TRIAZOLE

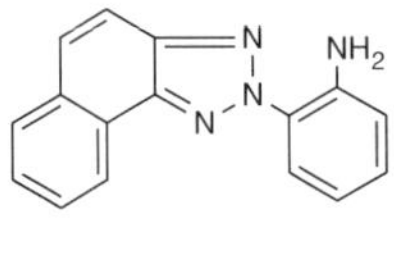

5 gm

S91,307-3
5-CHLORO-1-[N-(4-METHYL-2-NITROPHENYL)CARBAMOYL]-1H-BENZOTRIAZOLE

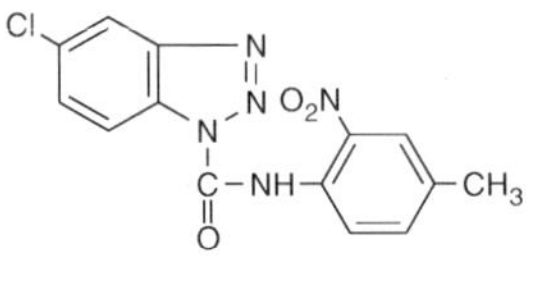

250 mg

S91,348-0
1-[N-(3-NITROPHENYL)CARBAMOYL]-1H-BENZOTRIAZOLE

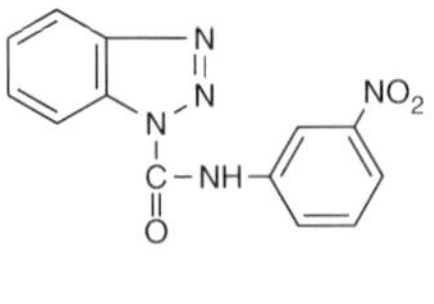

250 mg

S91,349-9
5-CHLORO-1-[N-(3-NITROPHENYL)-CARBAMOYL]-1H-BENZOTRIAZOLE

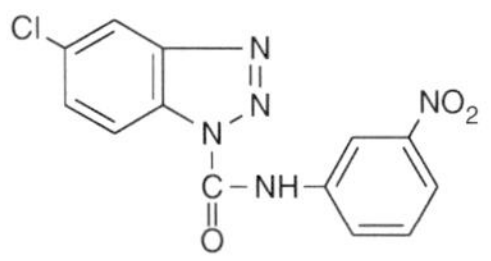

250 mg

S80,390-1
1-(3-PHTHALAMIDOPROPYL)-1H-BENZOTRIAZOLE

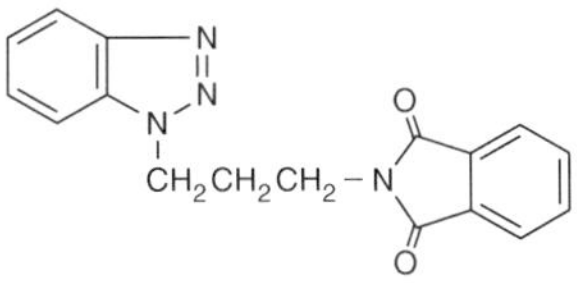

250 mg

S80,599-8
1,1'-TRIMETHYLENEBIS(BENZOTRIAZOLE)

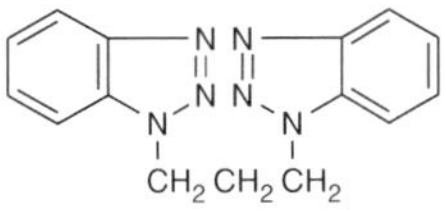

250 mg

S94,507-2
4-(1H-BENZOTRIAZOL-1-YLMETHYLAMINO)BENZOIC ACID

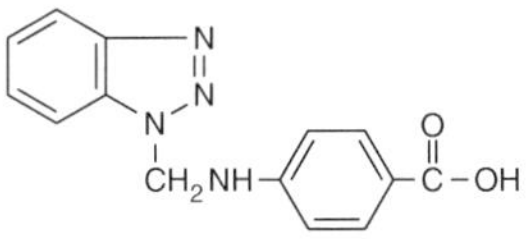

250 mg

S94,506-4
1-(ANILINOMETHYL)-1H-BENZOTRIAZOLE

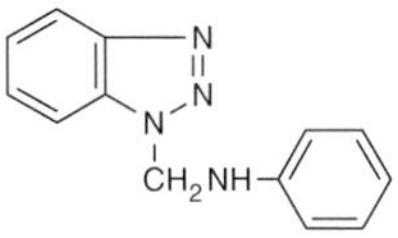

250 mg

S93,430-5
1-BENZYL-8-MERCAPTO-1H-1,2,4-TRIAZOLO(4,3-B)PYRIDAZIN-4-IUM HYDROXIDE, INNER SALT

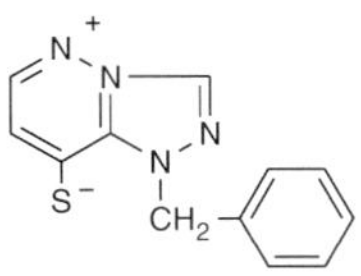

250 mg

S61,892-6
DIETHYL 1-(BENZOTRIAZOLYLIMINO)-MALONATE

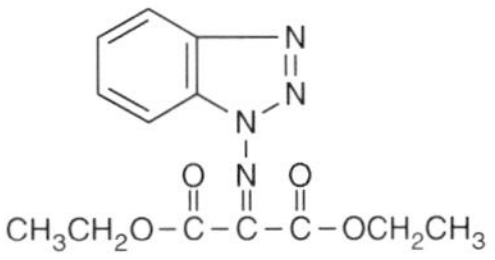

250 mg

S93,954-4
4-NITROBENZOTRIAZOLE

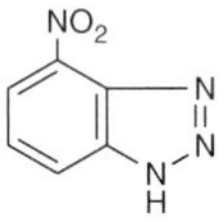

250 mg

S74,345-3
1-[N-(2,3-XYLYL)CARBAMOYL]-1H-BENZOTRIAZOLE

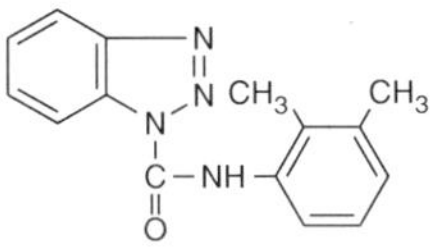

50 mg

S91,320-0
5-CHLORO-1-[N-(3,4-XYLYL)-CARBAMOYL]-1H-BENZOTRIAZOLE

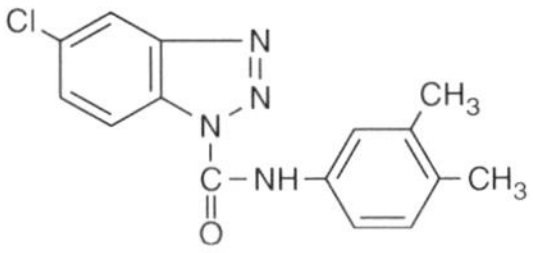

250 mg

S76,821-9
1-[N-(3,4-XYLYL)CARBAMOYL]-1H-BENZOTRIAZOLE

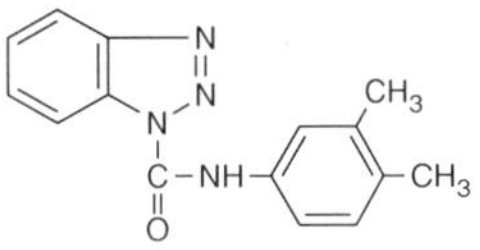

250 mg

S92,858-5
3-PHENYL-1H-1,2,4-TRIAZOLO(4,3-B)-ISOQUINOLINE PERCHLORATE

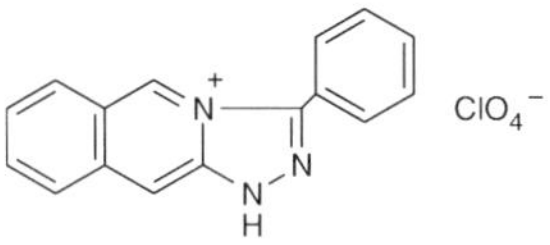

250 mg

S92,856-9
TETRAZOLO(1,5-A)QUINOLINE

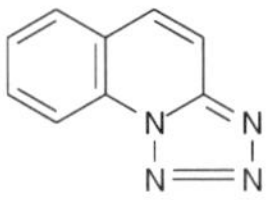

250 mg

S76,555-4
2-ETHYL-7-HYDROXY-1,2-BENZISOXAZOLIUM TETRAFLUOROBORATE

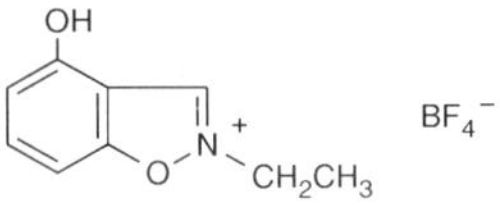

5 gm

S83,858-6
2-[2-(DIETHYLAMINO)ETHOXY]-BENZOXAZOLE HYDROCHLORIDE

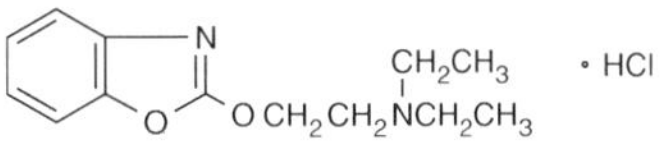

250 mg

S54,435-3
5-CHLORO-2,6-DIMETHYLBENZOXAZOLE

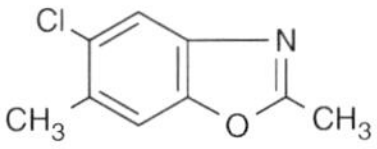

250 mg

S65,730-1
3-PHENYL-2,1-BENZISOXAZOLE

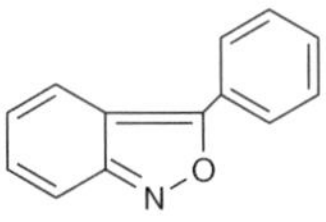

5 gm

S68,282-9
5-CHLORO-6-METHYL-3-PHENYL-2,1-BENZISOXAZOLE

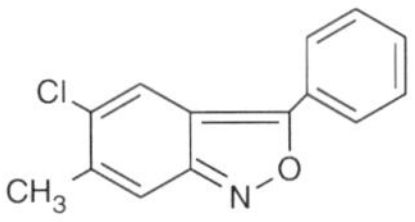

5 gm

S41,783-1
5-PHENYL-2-BENZOXAZOLETHIOL

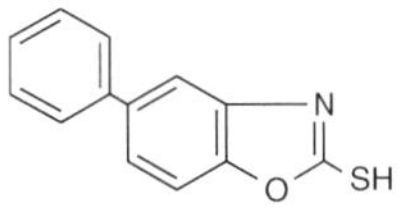

5 gm

S53,209-6
6-METHYL-2-BENZOXAZOLETHIOL

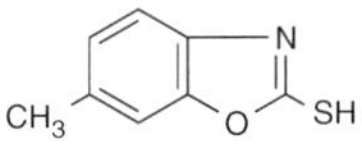

5 gm

S87,298-9
2-(METHYLTHIO)BENZOXAZOLE

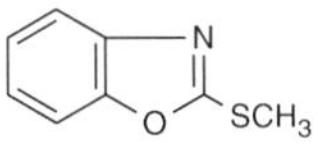

250 mg

S94,885-3
2-(BENZYLTHIO)BENZOXAZOLE

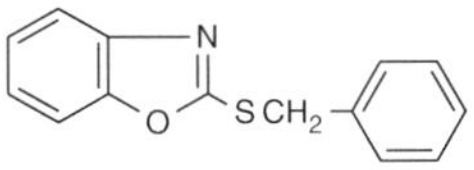

250 mg

S81,839-9
6-AMINO-2-PHENYLBENZOXAZOLE

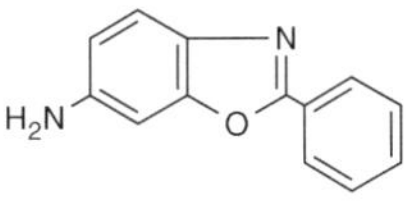

250 mg

S34,028-6
2-(2-N-ACETANILINOVINYL)-BENZOXAZOLE ETHIODIDE

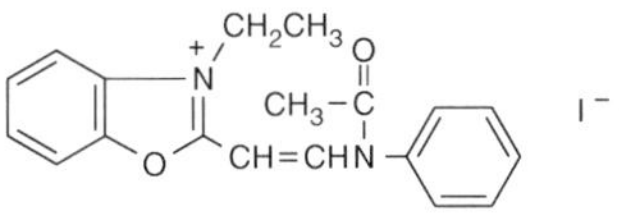

250 mg

S84,939-1
2-(ACI-NITROMETHYL)NAPHTH(1,2-D)-OXAZOLE, LITHIUM SALT

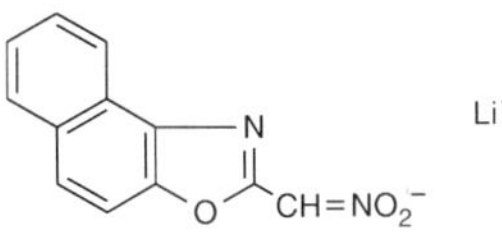

250 mg

S39,485-8
5-ACETAMIDO-2-METHYLBENZOXAZOLE

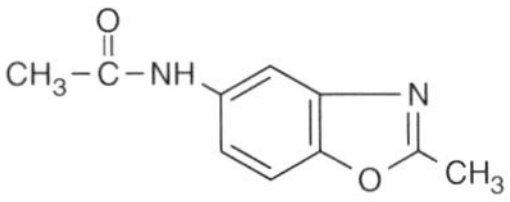

250 mg

S93,640-5
2-BENZOXAZOLEACETONITRILE

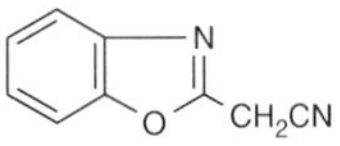

250 mg

S94,001-1
2'-[NAPHTH(2,3-D)OXAZOL-2-YL]-P-TOLUENESULFONANILIDE

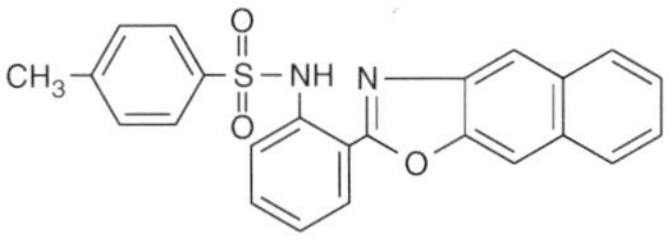

250 mg

S60,530-1
2-METHYL-6-NITROBENZOXAZOLE

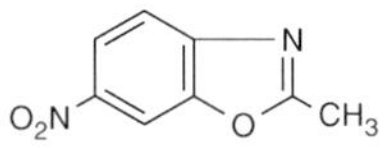

250 mg

S81,540-3
6-NITRO-2(3H)-BENZOXAZOLONE

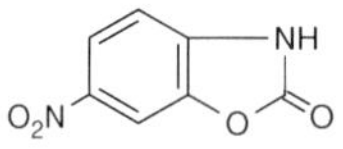

250 mg

S50,239-1
N-(2-METHYL-6-BENZOXAZOLYL)-P-TOLUENESULFONAMIDE

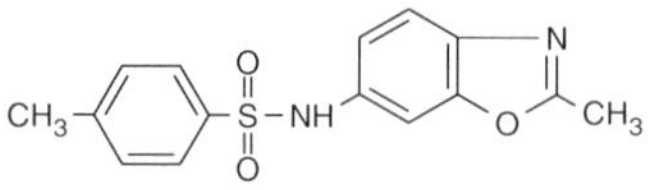

250 mg

S46,177-6
3-(2-CARBOXYETHYL)-2,5-DIMETHYLBENZOXAZOLIUM BROMIDE

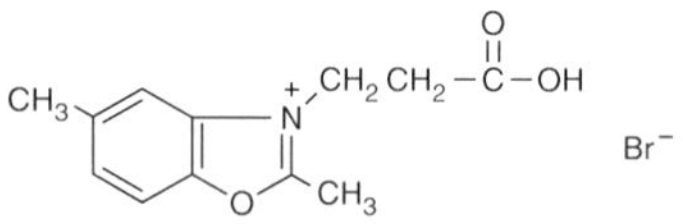

5 gm

S39,462-9
3-ACETAMIDOSULFONYLBUTYL-2-METHYL-5-PHENYLBENZOXAZOLIUM BROMIDE

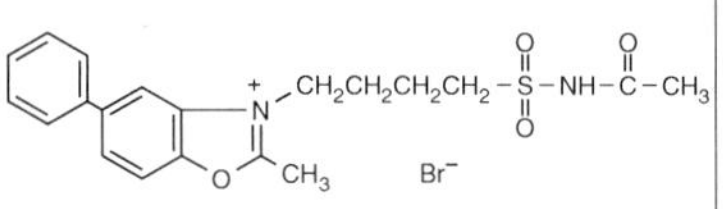

250 mg

S38,990-0
2-(2-ACETANILINOVINYL)-5,6-DIMETHYL-3-ETHYLBENZOXAZOLIUM IODIDE

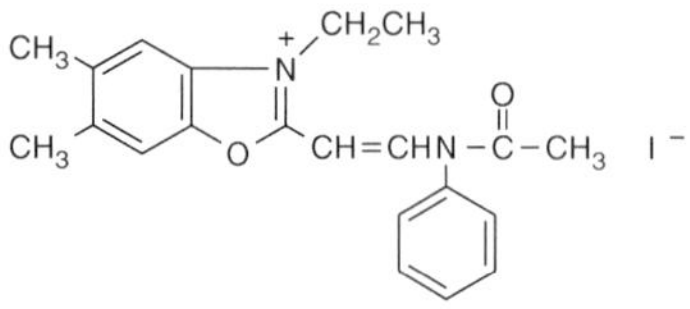

250 mg

S44,811-7
3-ETHYL-2-METHYL-5-PHENYLBENZOXAZOLIUM ETHYL SULFATE

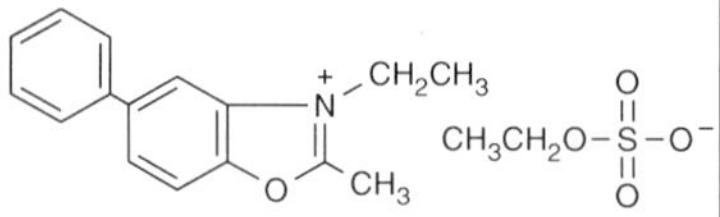

5 gm

S80,808-3
2,3-DIMETYL-5-PHENYLBENZOXAZOLIUM METHYL SULFATE

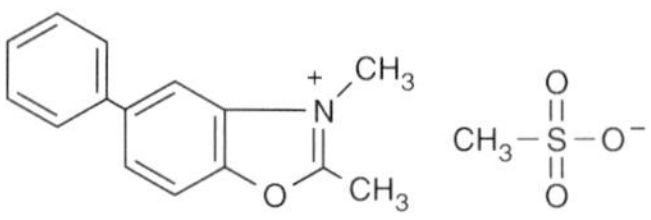

250 mg

S39,516-1
2-METHYL-5-PHENYL-3-PROPYLBENZOXAZOLIUM IODIDE

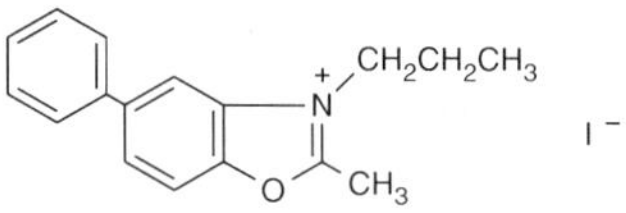

250 mg

S38,810-6
2-HYDRAZINOBENZOXAZOLE

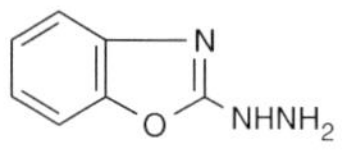

250 mg

S83,914-0
DIETHYL 2-(2-BENZOXAZOLYL)-2-ETHYLMALONATE

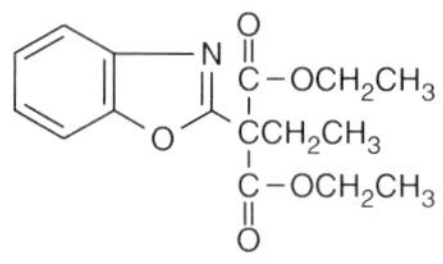

250 mg

S90,754-5
5-CHLORO-2-(CYANOMETHYL)BENZOXAZOLE

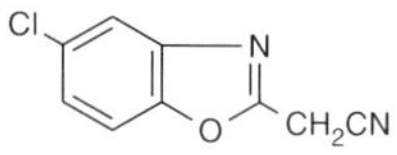

250 mg

S43,942-8
2,5,6-TRIMETHYLBENZOXAZOLE

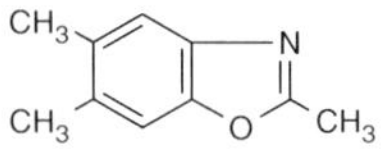

5 gm

S39,945-0
5-CYCLOHEXYL-2-METHYLBENZOXAZOLE

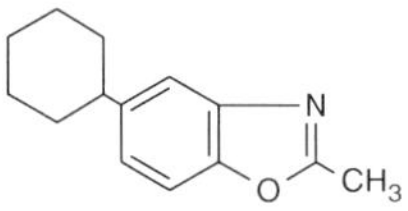

250 mg

S45,649-7
2-METHYLNAPHTH(2,1-D)OXAZOLE

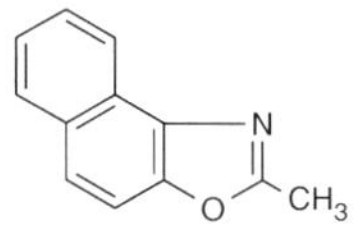

5 gm

S46,024-9
2-METHYLNAPHTH(1,2-D)OXAZOLE

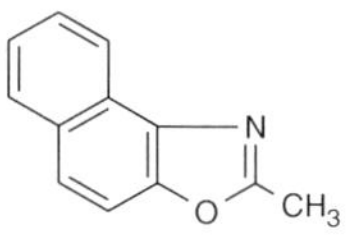

5 gm

S35,962-9
5-CHLORO-3-PHENYLANTHRANIL

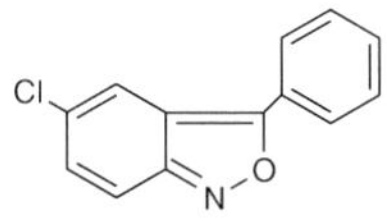

5 gm

S77,167-8
3-PHENYL-1,2-BENZISOXAZOLE

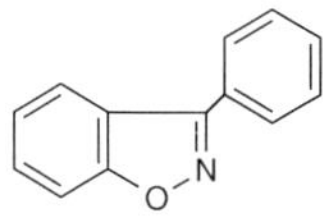

250 mg

S37,776-7
2-(2-CHLOROACETAMIDO)BENZOTHIAZOLE

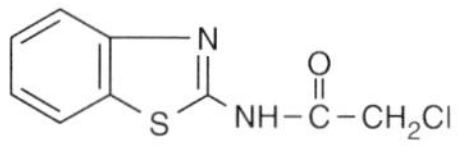

5 gm

S58,300-6
5-(DIFLUOROMETHYLSULFONYL)-2-(METHYLTHIO)BENZOTHIAZOLE

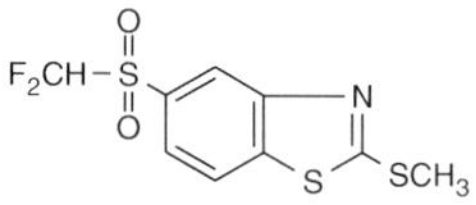

5 gm

S88,494-4
3-(DIBROMOMETHYL)-1,2-BENZISOTHIAZOLE

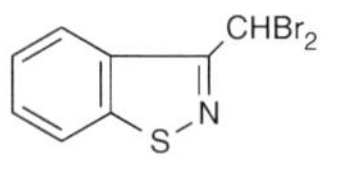

250 mg

S91,207-7
1-(2-BENZOTHIAZOLYL)-3-[4-CHLORO-2-(TRIFLUOROMETHYL)PHENYL]UREA

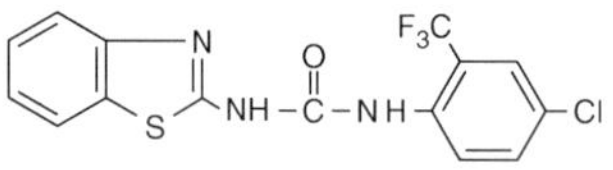

250 mg

S73,699-6
1-(4-CHLORO-2-BENZOTHIAZOLYL)-3-[2-CHLORO-5-(TRIFLUOROMETHYL)-PHENYL]UREA

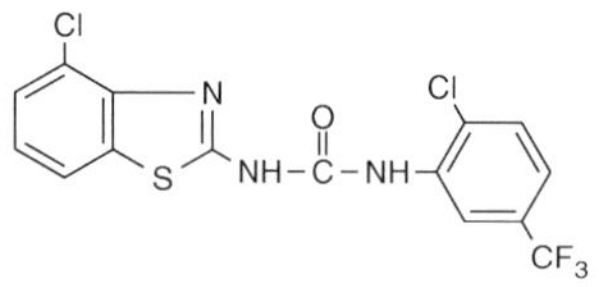

5 gm

S91,205-0
1-(4-CHLORO-2-BENZOTHIAZOLYL)-3-[4-CHLORO-2-(TRIFLUOROMETHYL)-PHENYL]UREA

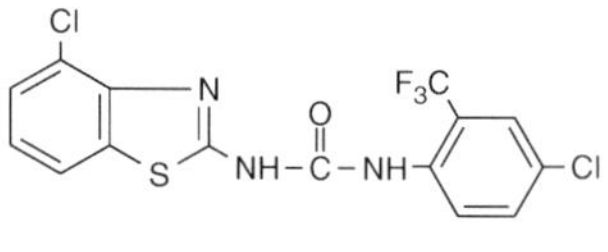

250 mg

S72,587-0
1-(4-CHLORO-2-BENZOTHIAZOLYL)-3-[4-CHLORO-3-(TRIFLUOROMETHYL)-PHENYL]UREA

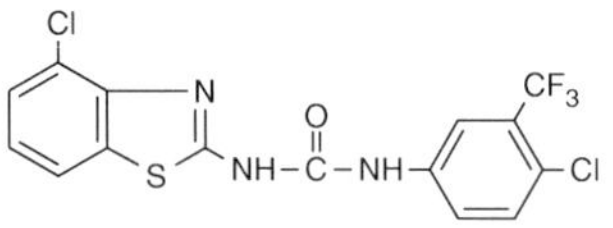

5 gm

S90,809-6
2-AMINO-4-(TRIFLUOROMETHYL)-BENZOTHIAZOLE

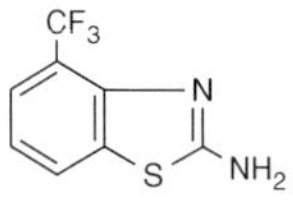

250 mg

S90,864-9
2-AMINO-6-BROMO-4-(TRIFLUOROMETHYL)BENZOTHIAZOLE

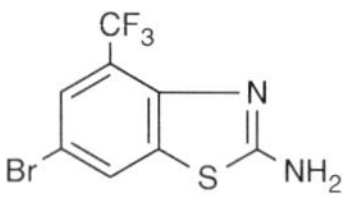

250 mg

S71,589-1
5,6-DIMETHYL-2-(2,2,2-TRIFLUOROACETAMIDO)BENZOTHIAZOLE

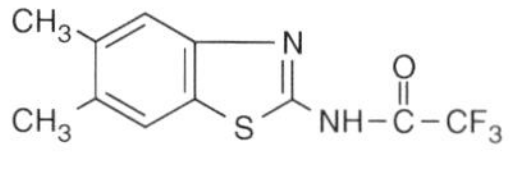

250 mg

S91,185-2
1-(2-BENZOTHIAZOLYL)-3-[2-(TRIFLUOROMETHYL)PHENYL]UREA

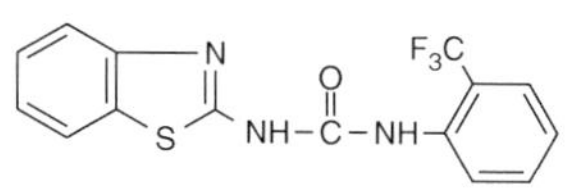

250 mg

S91,152-6
1-(5,6-DIMETHYL-2-BENZOTHIAZOLYL)-3-[2-(TRIFLUOROMETHYL)PHENYL]UREA

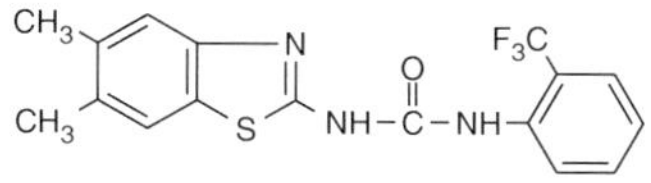

250 mg

S64,039-5
1-(2-BENZOTHIAZOLYL)-3-(3-TRIFLUOROMETHYLPHENYL)UREA

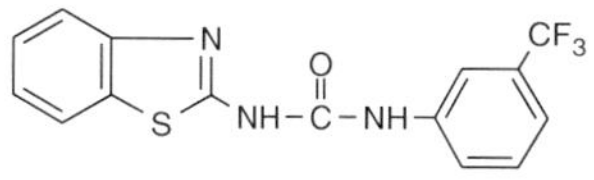

5 gm

S77,971-7
1-(5,6-DIMETHYL-2-BENZOTHIAZOLYL)-3-[3-(TRIFLUOROMETHYL)PHENYL]UREA

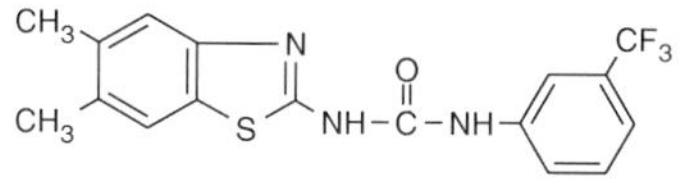

250 mg

S45,013-8
2-(TRIFLUOROMETHYL)BENZOTHIAZOLE

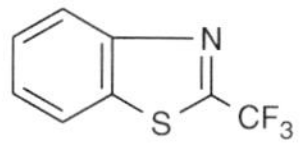

5 gm

S83,290-1
ALPHA-(4-CHLOROPHENYL)-ALPHA-METHYL-2-BENZOTHIAZOLEMETHANOL

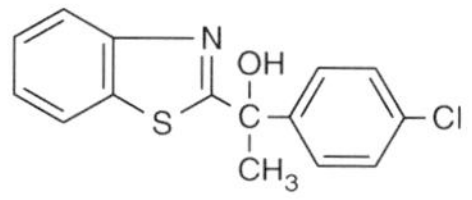

250 mg

S72,975-2
1-(2-BENZOTHIAZOLYL)-3-(5-CHLORO-2,4-DIMETHOXYPHENYL)UREA

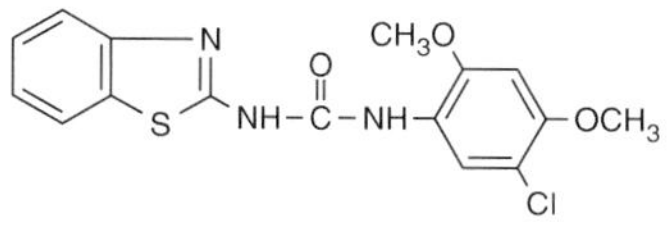

5 gm

S73,674-0
1-(2-BENZOTHIAZOLYL)-3-(5-CHLORO-2-METHOXYPHENYL)UREA

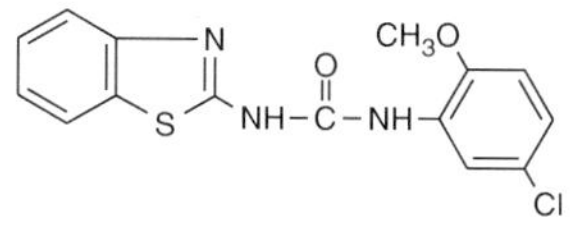

5 gm

S72,133-6
1-(4-CHLORO-2-BENZOTHIAZOLYL)-3-(5-CHLORO-2-METHOXYPHENYL)UREA

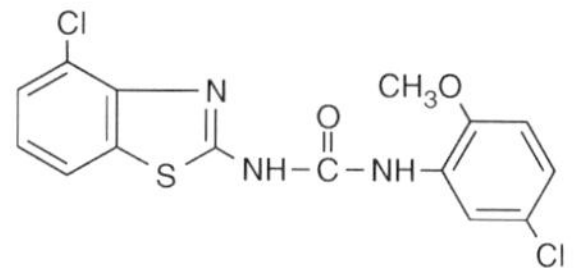

250 mg

S71,567-0
1-(5-CHLORO-2-METHOXYPHENYL)-3-(5,6-DIMETHYL-2-BENZOTHIAZOLYL)UREA

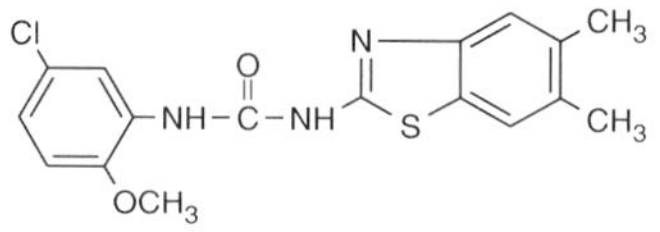

50 mg

S91,094-5
1-(2,3-DICHLOROPHENYL)-3-(6-METHOXY-2-BENZOTHIAZOLYL)UREA

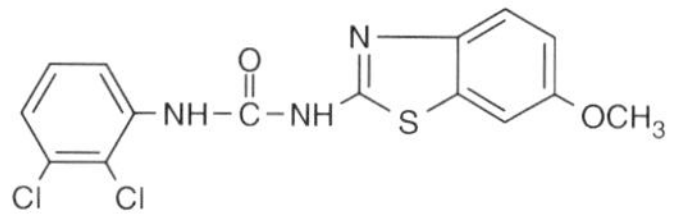

250 mg

S72,993-0
1-(3-FLUOROPHENYL)-3-(6-METHOXY-2-BENZOTHIAZOLYL)UREA

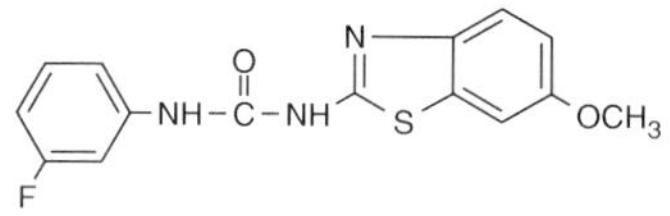

5 gm

S71,521-2
1-(4-CHLOROPHENYL)-3-(6-METHOXY-2-BENZOTHIAZOLYL)UREA

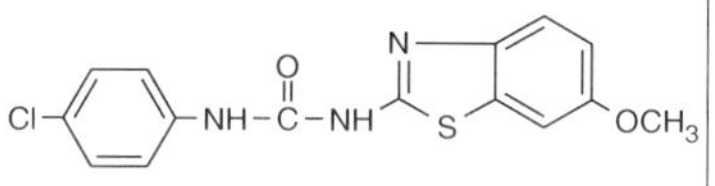

5 gm

S77,259-3
2-[(2-CHLORO-4-NITROPHENYL)THIO]-BENZOTHIAZOLE

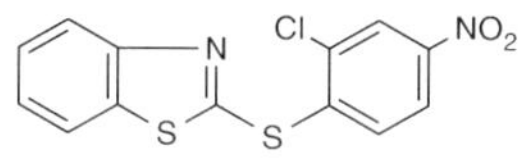

250 mg

S91,192-5
1-(4-CHLORO-2-BENZOTHIAZOLYL)-3-(4-FLUORO-3-NITROPHENYL)UREA

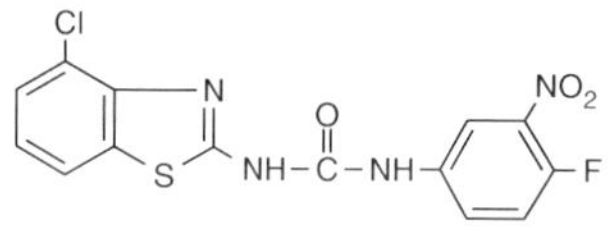

250 mg

S72,134-4
1-(2-BENZOTHIAZOLYL)-3-(2-FLUORO-5-NITROPHENYL)UREA

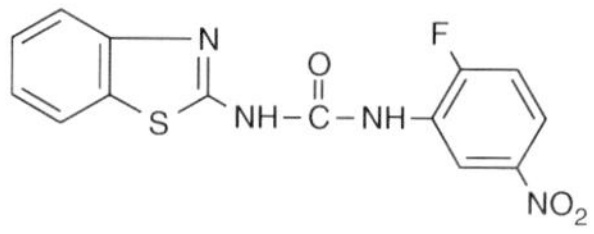

250 mg

S87,009-9
2-(2,5-DICHLOROPHENYLSULFONYLAZO)-3-METHYLBENZOTHIAZOLIUM TETRAFLUOROBORATE

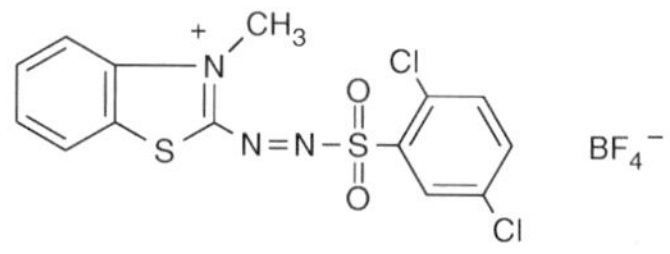

250 mg

S67,149-5
3-(4-CHLOROPHENYL)THIAZOLO(3,2-A)-PYRIDINIUM BISULFATE

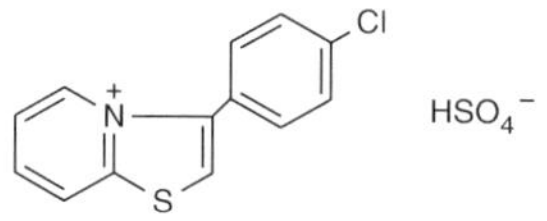

5 gm

S89,788-4
1-(2-BENZOTHIAZOLYL)-3-(4-CHLORO-3-NITROPHENYL)UREA

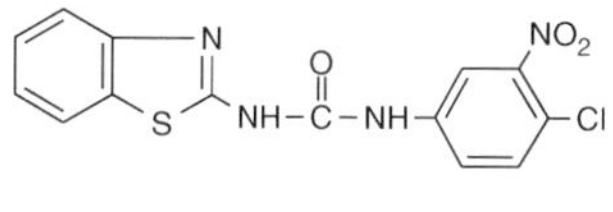

250 mg

S74,416-6
3',4'-DICHLORO-2-BENZOTHIAZOLECARBOXANILIDE

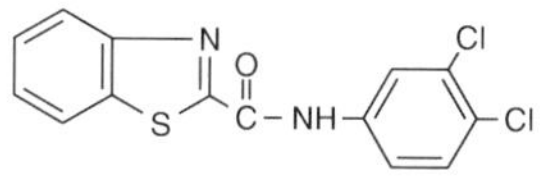

250 mg

S77,234-8
1-(2-BENZOTHIAZOLYL)-3-(2-FLUOROPHENYL)UREA

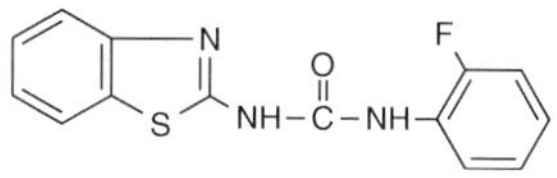

250 mg

S72,994-9
1-(4-CHLORO-2-BENZOTHIAZOLYL)-3-(2-FLUOROPHENYL)UREA

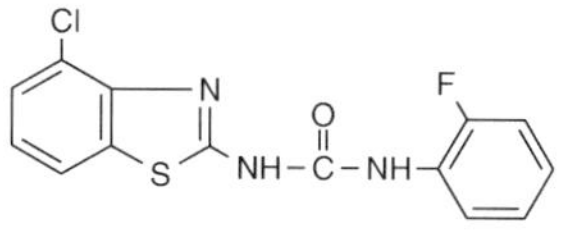

5 gm

S72,979-5
1-(4-CHLORO-2-BENZOTHIAZOLYL)-3-(4-FLUOROPHENYL)UREA

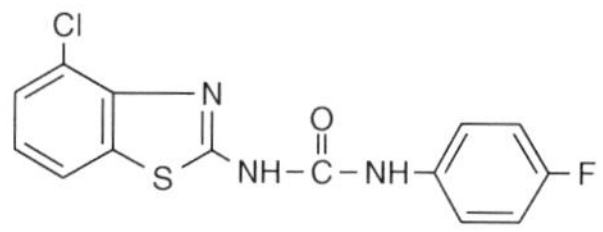

5 gm

S91,095-3
1-(2-BENZOTHIAZOLYL)-3-(2,3-DICHLOROPHENYL)UREA

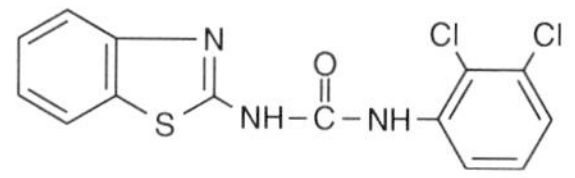

250 mg

S91,093-7
1-(4-CHLORO-2-BENZOTHIAZOLYL)-3-(2,3-DICHLOROPHENYL)UREA

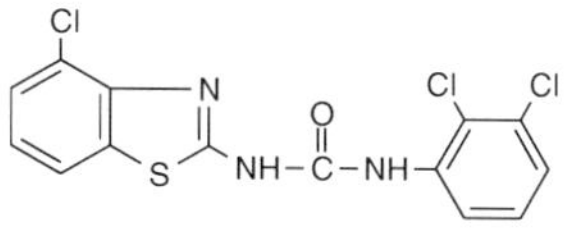

250 mg

S77,998-9
1-(7-CHLORO-2-BENZOTHIAZOLYL)-3-(2,5-DIFLUOROPHENYL)UREA

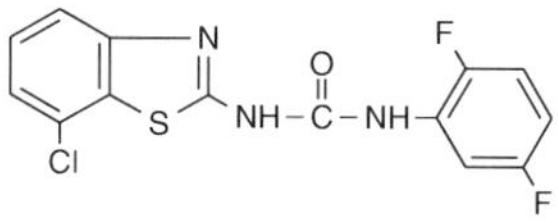

250 mg

S91,053-8
1-(4-CHLORO-2-BENZOTHIAZOLYL)-3-(2,5-DIFLUOROPHENYL)UREA

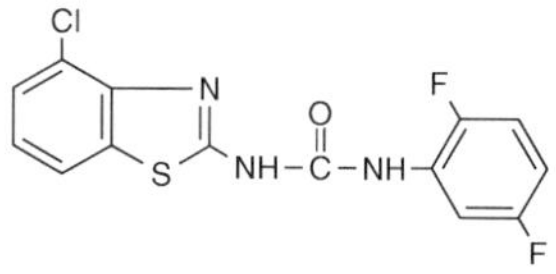

250 mg

S91,249-2
1-(4-CHLORO-2-BENZOTHIAZOLYL)-3-(3-CHLORO-4-METHYLPHENYL)UREA

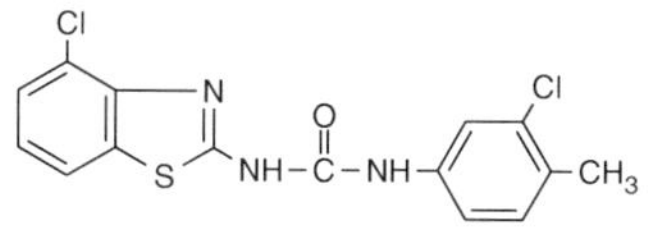

250 mg

S77,926-1
1-(2,5-DIFLUOROPHENYL)-3-(5,6-DIMETHYL-2-BENZOTHIAZOLYL)UREA

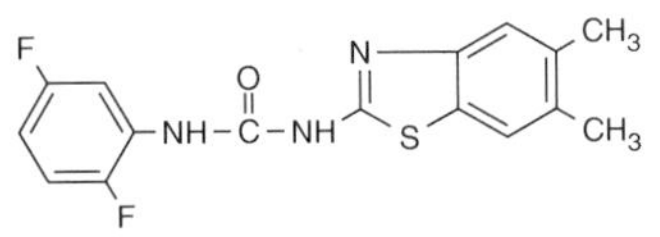

250 mg

S91,055-4
1-(5,6-DIMETHYL-2-BENZOTHIAZOLYL)-3-(3-FLUOROPHENYL)UREA

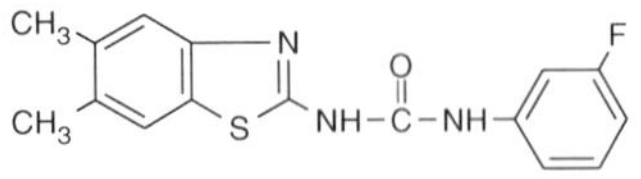

250 mg

S71,515-8
1-(2-BENZOTHIAZOLYL)-3-(4-CHLOROPHENYL)UREA

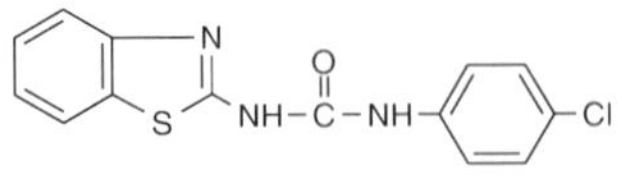

5 gm

S70,731-7
1-(2-BENZOTHIAZOLYL)-3-(2,6-DICHLOROPHENYL)UREA

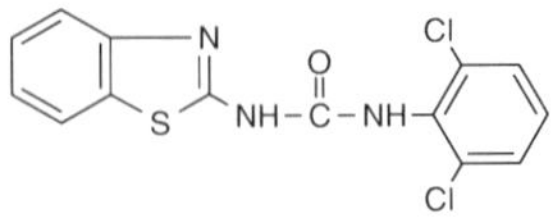

250 mg

S72,135-2
1-(2-BENZOTHIAZOLYL)-3-(2,5-DICHLOROPHENYL)UREA

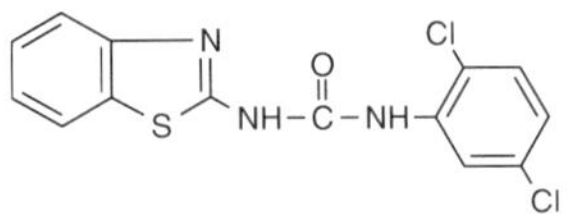

5 gm

S72,136-0
1-(4-CHLORO-2-BENZOTHIAZOLYL)-3-(3-FLUOROPHENYL)UREA

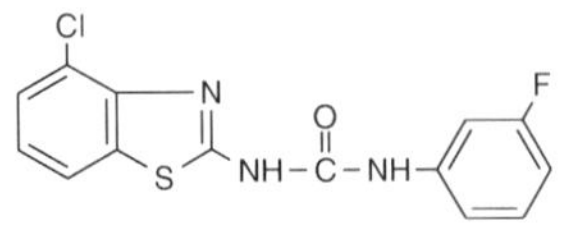

250 mg

S72,137-9
1-(2-BENZOTHIAZOLYL)-3-(2,4,5-TRICHLOROPHENYL)UREA

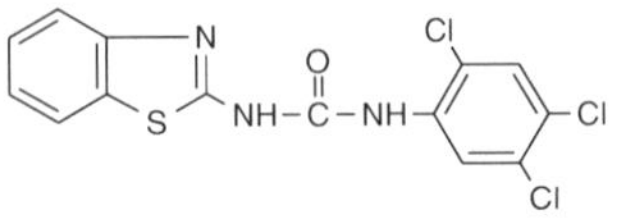

250 mg

S67,460-5
3-(4-BROMOPHENYL)IMIDAZO(2,1-B)-BENZOTHIAZOLE

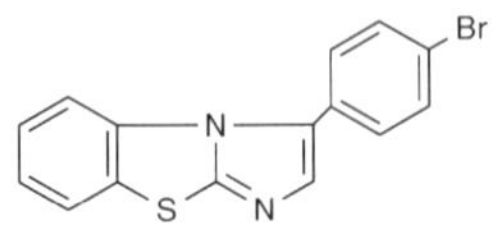

250 mg

S71,895-5
2-(4-CHLOROPHENYL)BENZOTHIAZOLE

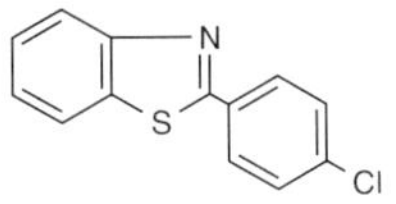

250 mg

S83,926-4
2-(2,4-DIHYDROXYPHENYL)-BENZOTHIAZOLE

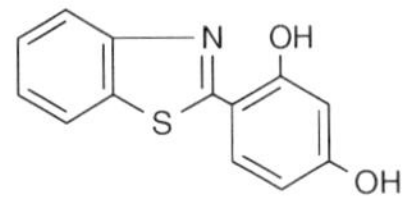

250 mg

S89,782-5
2,6-DIMETHOXY-4-(2-BENZOTHIAZOLYL)-PHENOL

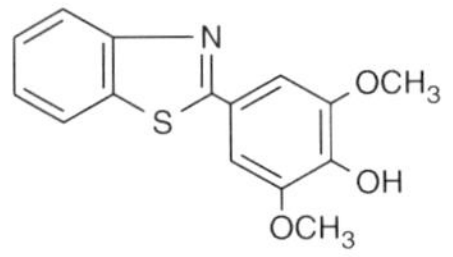

250 mg

S71,894-7
2-(4-HYDROXY-3-METHOXYPHENYL)-BENZOTHIAZOLE

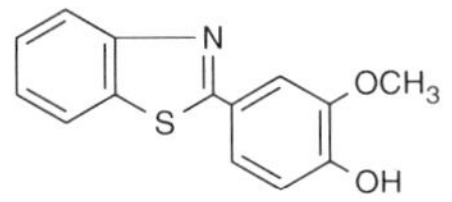

250 mg

S83,294-4
ALPHA-[4-(DIMETHYLAMINO)PHENYL]-2-BENZOTHIAZOLEMETHANOL

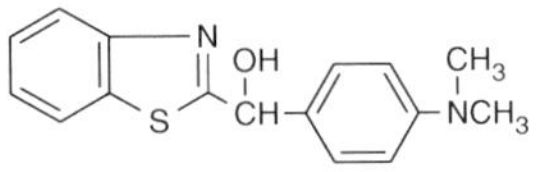

250 mg

S90,920-3
ALPHA-PHENYL-2-BENZOTHIAZOLEMETHANOL

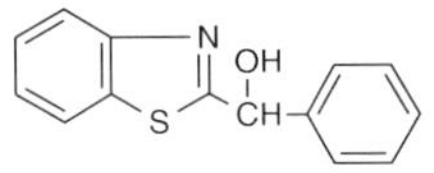

250 mg

S83,295-2
ALPHA,ALPHA-DIPHENYL-2-BENZOTHIAZOLEMETHANOL

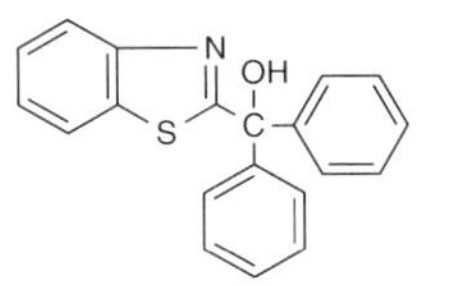

250 mg

S93,955-2
2-AMINO-4-HYDROXYBENZOTHIAZOLE

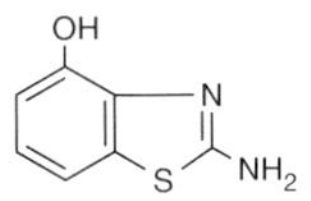

250 mg

S57,846-0
3-(2-BENZOTHIAZOLYLAMINO)-5-NITROPHENOL

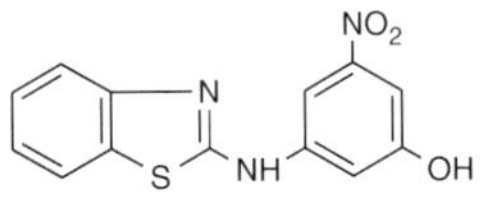

5 gm

S64,881-7
ALPHA-(2-BENZOTHIAZOLYLAMINO)-O-CRESOL

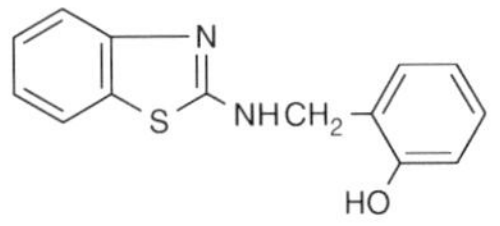

250 mg

S82,793-2
SACCHARIN, SILVER SALT

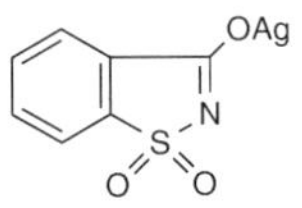

250 mg

S45,467-2
2-(2-HYDROXYPHENYL)BENZOTHIAZOLE

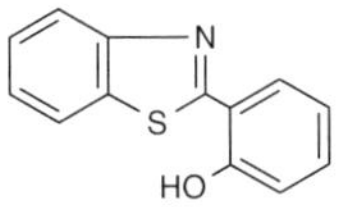

5 gm

S82,260-4
2-(BENZYLOXY)BENZOTHIAZOLE

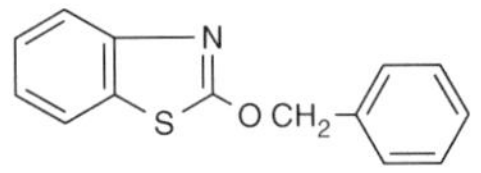

250 mg

S67,549-0
2-(2,3,4-TRIMETHOXYPHENYL)-BENZOTHIAZOLE

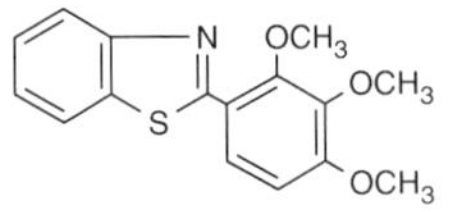

250 mg

S68,876-2
2-(4-ETHOXY-3-METHOXYPHENYL)-BENZOTHIAZOLE

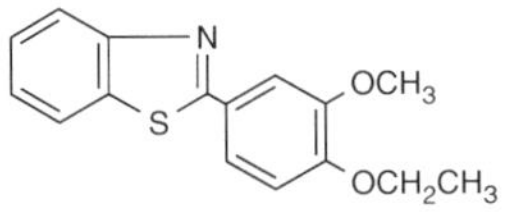

250 mg

S91,037-6
1-(2-BENZOTHIAZOLYL)-3-(2,4-DIMETHOXYPHENYL)UREA

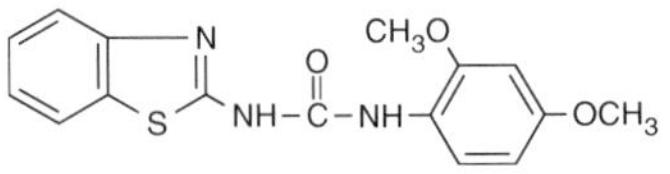

250 mg

S91,038-4
1-(2,4-DIMETHOXYPHENYL)-3-[4-(6-METHYL-2-BENZOTHIAZOLYL)PHENYL]UREA

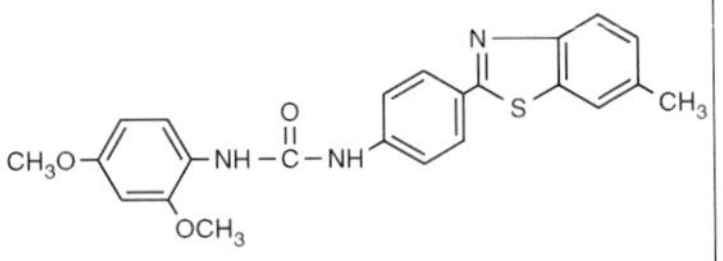

250 mg

S91,047-3
1-(4-CHLORO-2-BENZOTHIAZOLYL)-3-(2,5-DIMETHOXYPHENYL)UREA

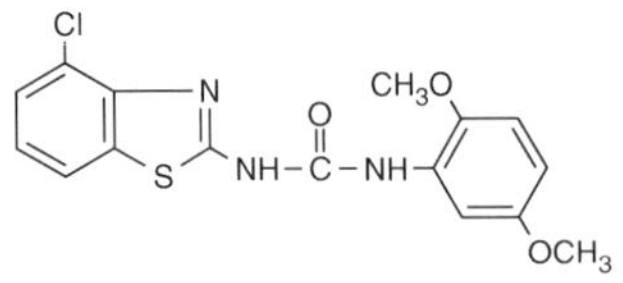

250 mg

S91,048-1
1-(2,5-DIMETHOXYPHENYL)-3-(6-METHOXY-2-BENZOTHIAZOLYL)UREA

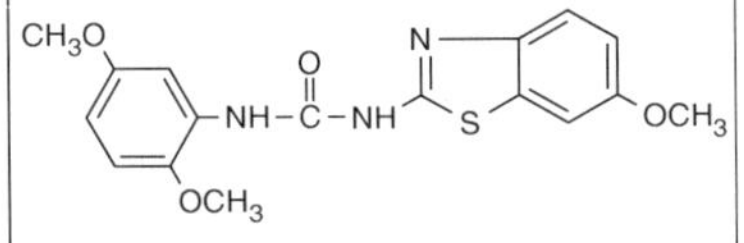

250 mg

S72,933-7
1-(2-BENZOTHIAZOLYL)-3-(2,5-DIMETHOXYPHENYL)UREA

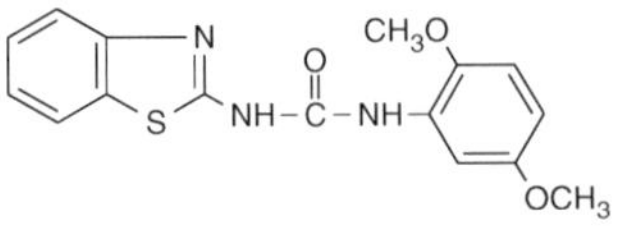

5 gm

S65,532-5
6-HEXADECYLOXY-2-MERCAPTOBENZOTHIAZOLE

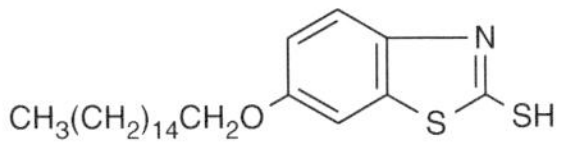

250 mg

S91,064-3
1-(6-METHOXY-2-BENZOTHIAZOLYL)-3-[4-(METHYLTHIO)PHENYL]UREA

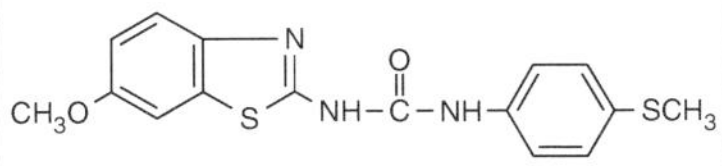

250 mg

S94,013-5
5-METHOXY-2-METHYL-6-(METHYLTHIO)-BENZOTHIAZOLE

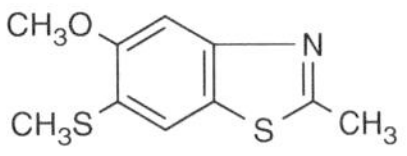

250 mg

S92,569-1
5,6-DIBUTOXY-2-(METHYLTHIO)-BENZOTHIAZOLE

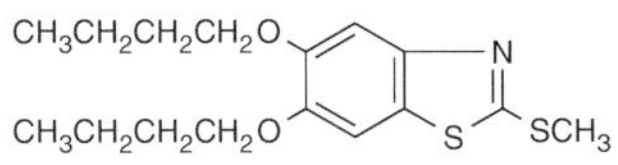

250 mg

S65,993-2
1,1'-(4-METHYL-1,3-PHENYLENE)BIS[3-(6-METHOXY-2-BENZOTHIAZOLYL)UREA]

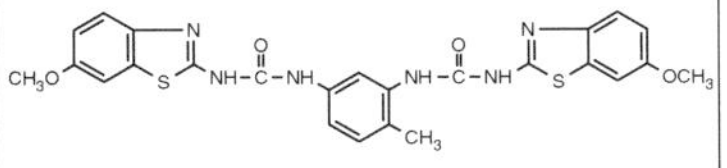

5 gm

S87,190-7
6-METHOXY-3-METHYL-2-[(3-METHYL-2-BENZOTHIAZOLINYLIDENE)AMINOAZO]-BENZOTHIAZOLIUM TETRAFLUOROBORATE

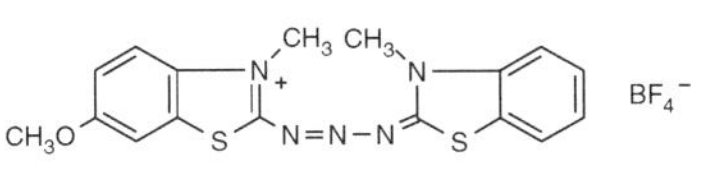

250 mg

S87,187-7
2-[(6-METHOXY-3-METHYL-2-BENZOTHIAZOLINYLIDENE)AMINOAZO]-3-METHYLBENZOTHIAZOLIUM PERCHLORATE

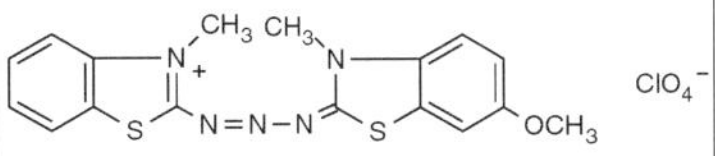

250 mg

S87,216-4
3-ISOBUTYL-2-[(6-METHOXY-3-METHYL-2-BENZOTHIAZOLINYLIDENE)AMINOAZO]-BENZOTHIAZOLIUM PERCHLORATE

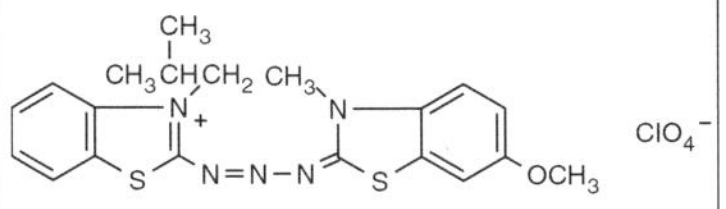

250 mg

S35,943-2
5-CHLORO-2-[4-(2-DIETHYLAMINOETHOXY)PHENYL]-BENZOTHIAZOLE HYDROCHLORIDE

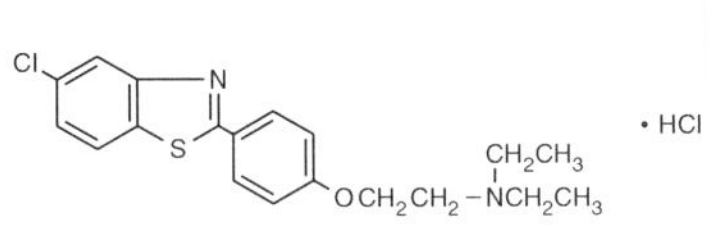

5 gm

S87,374-8
2-(3-ETHOXY-1H-PHENALEN-1-YLIDENEMETHYL)-3-ETHYL-BENZOTHIAZOLIUM TETRAFLUOROBORATE

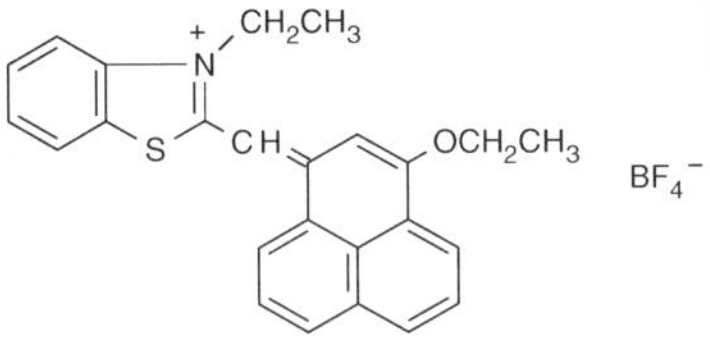

250 mg

S92,213-7
2-CHLORO-5-METHOXYBENZOTHIAZOLE

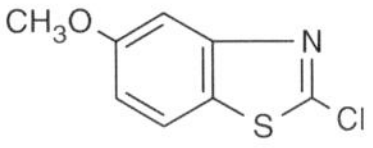

250 mg

S35,715-4
2-ACETOACETAMIDO-6-ETHOXYBENZOTHIAZOLE

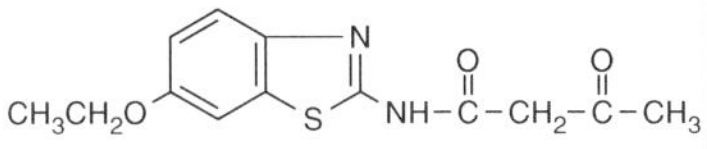

5 gm

S91,121-6
ETHYL 4-[3-(6-METHOXY-2-BENZOTHIAZOLYL)UREIDO]BENZOATE

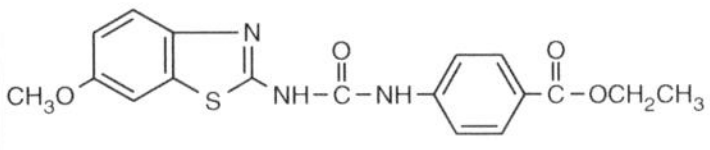

250 mg

S65,994-0
4,4'-METHYLENEBIS[3-(6-METHOXY-2-BENZOTHIAZOLYL)-1-PHENYLUREA]

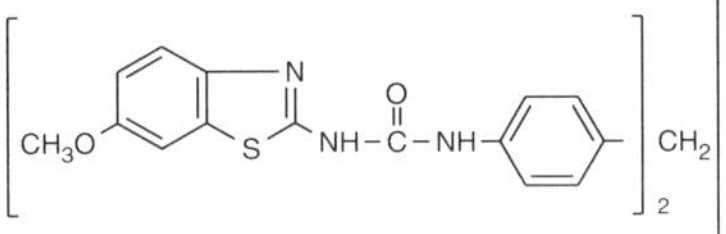

5 gm

S64,436-6
2-[N-(3,4-DICHLOROPHENYL)-CARBAMOYL]AMINO-6-METHOXYBENZOTHIAZOLE

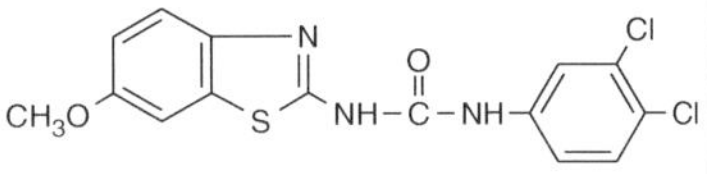

5 gm

S71,516-6
1-(6-METHOXY-2-BENZOTHIAZOLYL)-3-PHENYLUREA

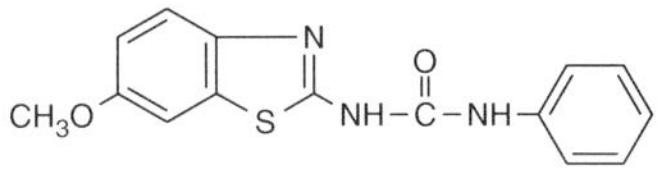

250 mg

S82,253-1
2-ETHOXYBENZOTHIAZOLE

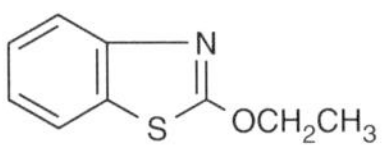

250 mg

S39,488-2
6-ETHOXY-2-METHYLBENZOTHIAZOLE

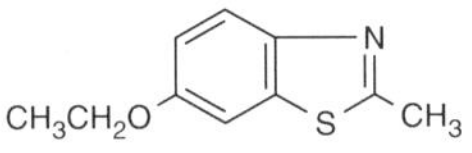

250 mg

S94,889-6
2,5-DIMETHYL-6-METHOXYBENZOTHIAZOLE

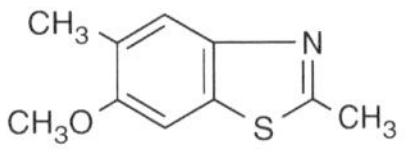

250 mg

S69,635-8
2-MERCAPTOBENZOTHIAZOLE-5-SULFONIC ACID, SODIUM SALT

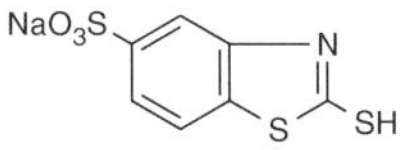

250 mg

S50,554-4
6-AMINO-2-(METHYLTHIO)-BENZOTHIAZOLE

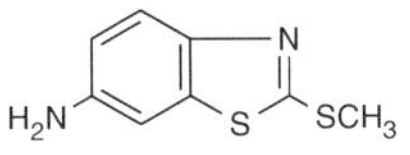

250 mg

S50,546-3
3-METHYL-2-METHYLTHIO-6-NITRO-5-SULFOBENZOTHIAZOLIUM METHYL SULFATE

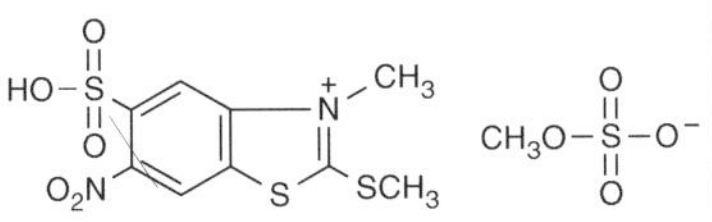

250 mg

S58,832-6
2-(CARBOXYMETHYLTHIO)-3-METHYLBENZOTHIAZOLIUM CHLORIDE

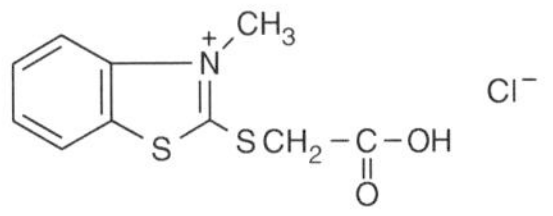

250 mg

S62,841-7
2-(CARBOXYMETHYLTHIO)-3-METHYLBENZOTHIAZOLINIUM BROMIDE

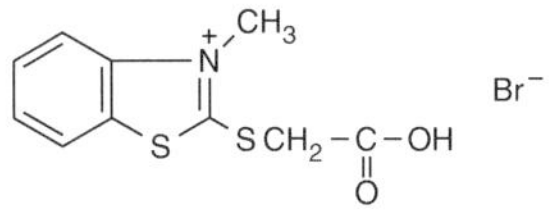

250 mg

S63,772-6
3-HEXADECYL-2-METHYLTHIOBENZOTHIAZOLIUM P-TOLUENESULFONATE

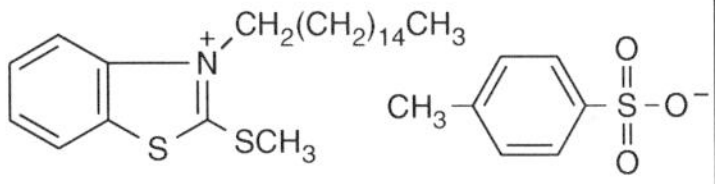

250 mg

S50,115-8
N-BENZYL-2-METHYLTHIO-5-BENZOTHIAZOLESULFONAMIDE

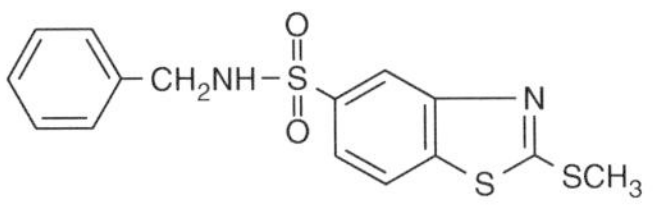

250 mg

S50,636-2
N-BENZYL-2-METHYLTHIO-6-BENZOTHIAZOLESULFONAMIDE

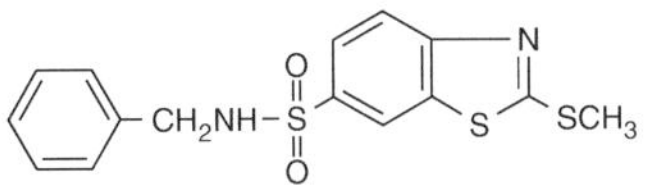

250 mg

S34,743-4
4-(2-BENZOTHIAZOLYLTHIO)-BUTYRIC ACID

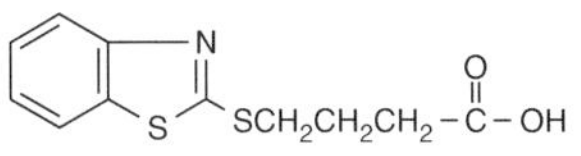

5 gm

S91,065-1
1-(2-BENZOTHIAZOLYL)-3-[4-(METHYLTHIO)PHENYL]UREA

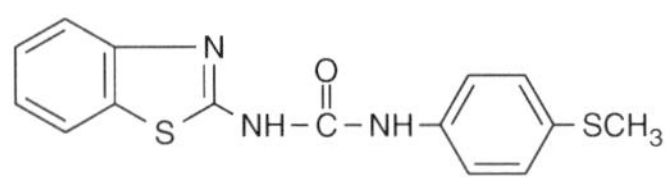

250 mg

S90,128-8
3-(2-BENZOTHIAZOLYLTHIO)-PROPIONITRILE

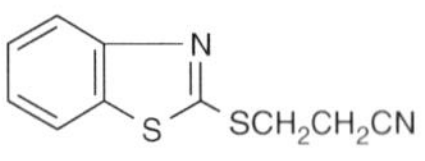

250 mg

S50,015-1
2-METHYLTHIO-5-BENZOTHIAZOLESULFONIC ACID

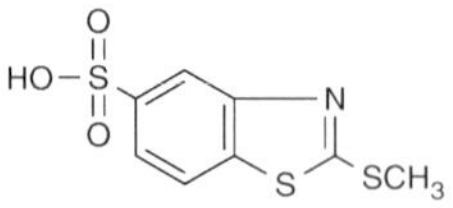

5 gm

S50,109-3
2-METHYLTHIO-N,N-PENTAMETHYLENE-5-BENZOTHIAZOLESULFONAMIDE

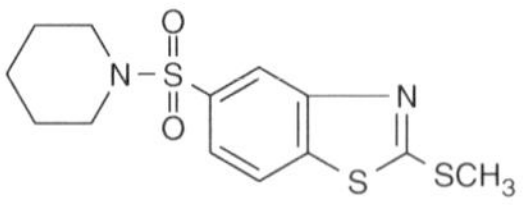

250 mg

S50,637-0
N-HEXADECYL-2-METHYLTHIO-6-BENZOTHIAZOLESULFONAMIDE

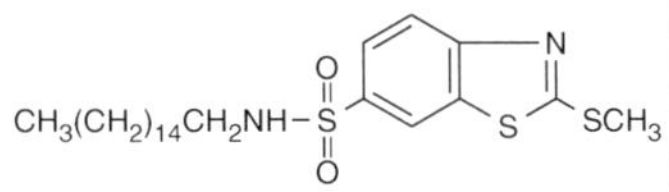

250 mg

S83,552-8
2-(PHENYLTHIO)BENZOTHIAZOLE

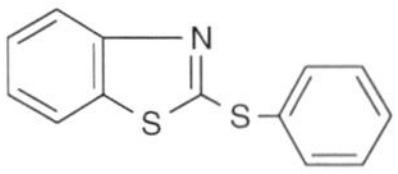

250 mg

S83,816-0
2-(P-TOLYLTHIO)BENZOTHIAZOLE

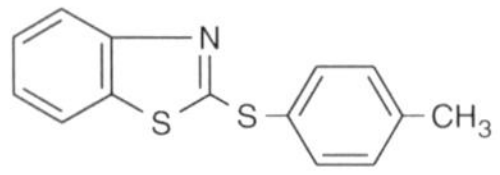

250 mg

S82,446-1
2-BENZOTHIAZOLYL SULFIDE

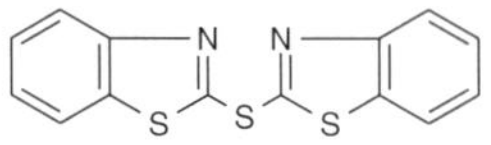

250 mg

S84,034-3
2-(OCTADECYLTHIO)BENZOTHIAZOLE

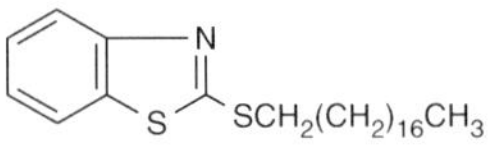

250 mg

S39,702-4
3-METHYL-2-(METHYLTHIO)-BENZOTHIAZOLIUM IODIDE

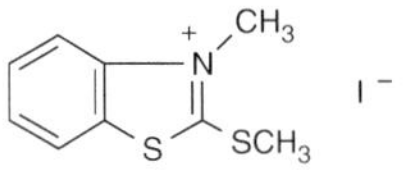

250 mg

S39,504-8
6-AMINO-2-METHYL-7-NITROBENZOTHIAZOLE

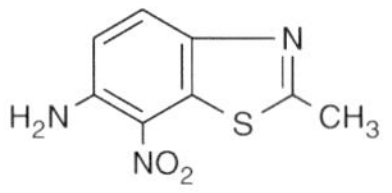

250 mg

S94,891-8
4,5-DIAMINO-2-METHYLBENZOTHIAZOLE

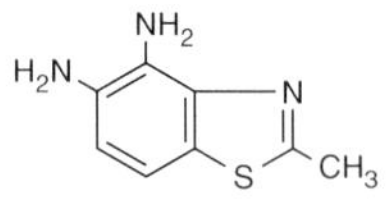

250 mg

S90,814-2
2-AMINO-5-NITROBENZOTHIAZOLE

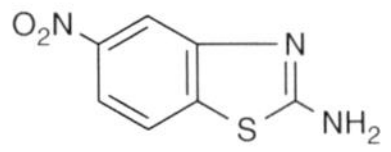

250 mg

S88,533-9
5-AMINO-3-METHYL-1,2-BENZISOTHIAZOLE

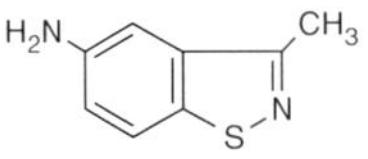

250 mg

S84,129-3
6-AMINO-2-PHENYLBENZOTHIAZOLE

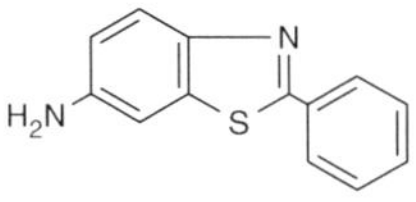

250 mg

S94,645-1
2-AMINOBENZOTHIAZOLE-5-CARBOXYLIC ACID

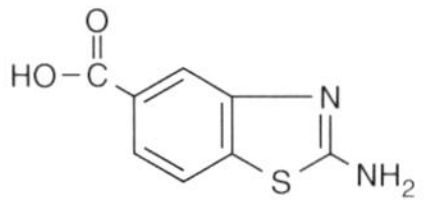

250 mg

S90,865-7
2-AMINO-6-BENZOTHIAZOLYL-THIOCYANATE

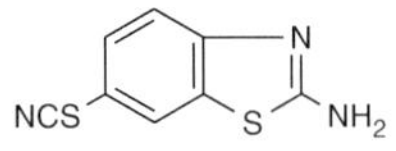

250 mg

S42,948-1
2-AMINO-4-CHLOROBENZOTHIAZOLE HYDROBROMIDE

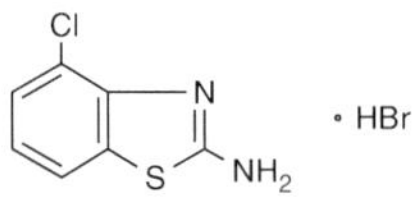

5 gm

S94,014-3
2-AMINO-6-CHLOROBENZOTHIAZOLE HYDROCHLORIDE

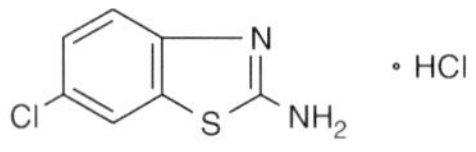

250 mg

S89,473-7
2-AMINO-6-BROMOBENZOTHIAZOLE

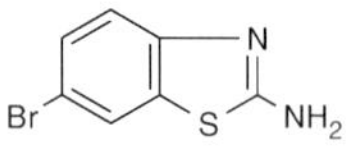

250 mg

S40,529-9
2-(P-AMINOPHENYL)-6-METHYLBENZOTHIAZOLE-7-SULFONIC ACID

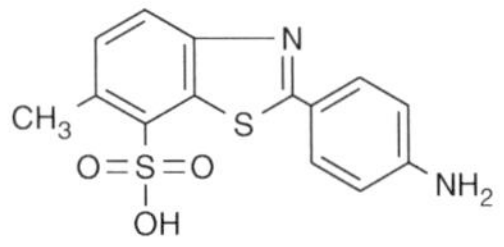

5 gm

S94,886-1
6-AMINO-2-METHYLBENZOTHIAZOLE

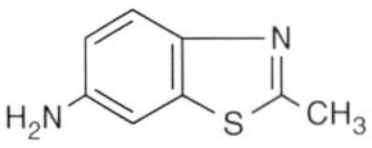

250 mg

S62,610-4
2-(2,4,6-TRINITROPHENYLAMINO)-BENZOTHIAZOLE

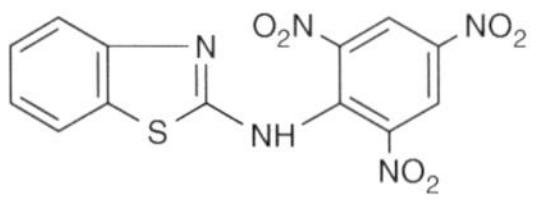

5 gm

S57,976-9
2-BENZOTHIAZOLECARBOXAMIDINE

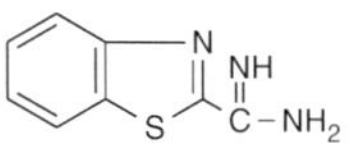

5 gm

S93,956-0
5-CHLORO-2-METHYL-4-NITROBENZOTHIAZOLE

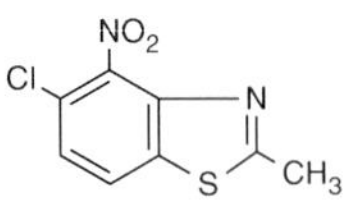

250 mg

S81,193-9
2,2'-BIBENZOTHIAZOLE

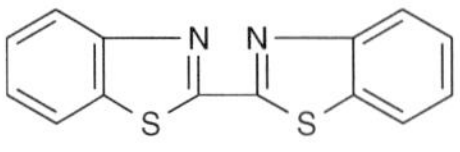

250 mg

S62,785-2
NITROBENZOTHIAZOLE (MIXTURE OF ISOMERS)

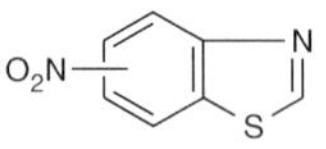

250 mg

S94,011-9
4'-AMINO-2'-(2-BENZOTHIAZOLYL)-P-TOLUENESULFONANILIDE

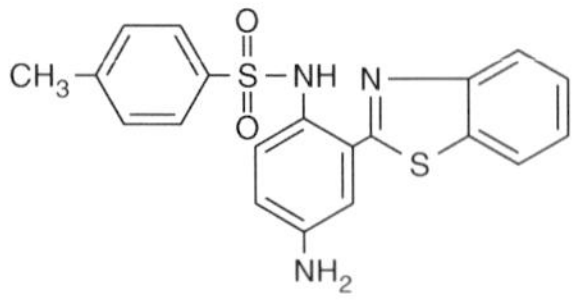

250 mg

S87,202-4
2-[(1,4-DIHYDRO-1-METHYL-4-QUINOLYLIDENE)AMINOAZO]-3-METHYLBENZOTHIAZOLIUM PERCHLORATE

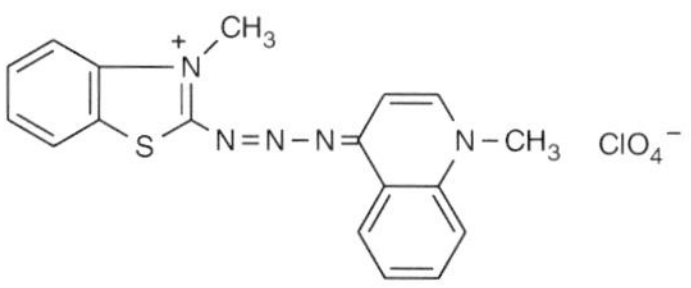

250 mg

S53,721-7
6(N-HEXADECYL-N-METHYLAMINO-SULFONYL)-3-METHYL-2-(METHYLTHIO)BENZO-THIAZOLIUM-2,4-DINITROBZENESULONATE

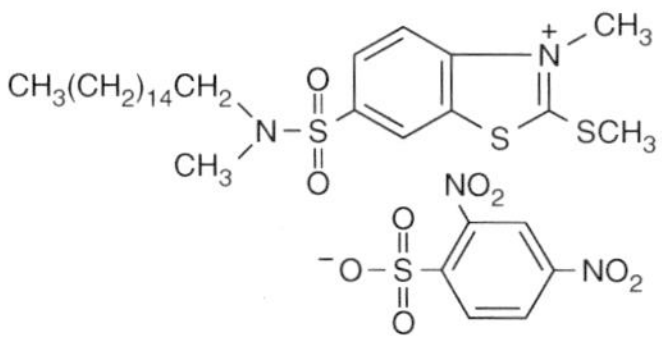

5 gm

S94,000-3
2'-(2-BENZOTHIAZOLYL)-4'-NITRO-P-TOLUENESULFONANILIDE

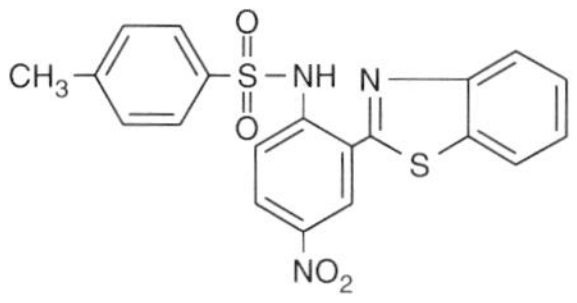

250 mg

S66,024-8
1,1'-(4-METHYL-1,3-PHENYLENE)BIS[3-(2-BENZOTHIAZOLYL)UREA]

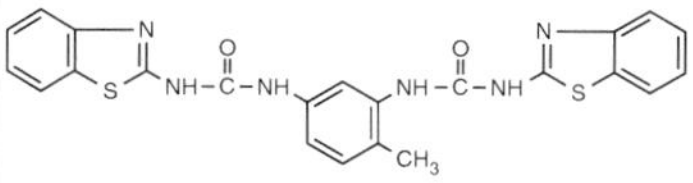

5 gm

S94,012-7
2'-(2-BENZOTHIAZOLYL)-P-TOLUENESULFONANILIDE

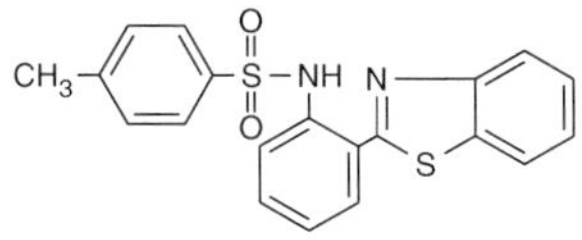

250 mg

S39,692-3
2,3-DIMETHYL-6-NITROBENZOTHIAZOLIUM P-TOLUENESULFONATE

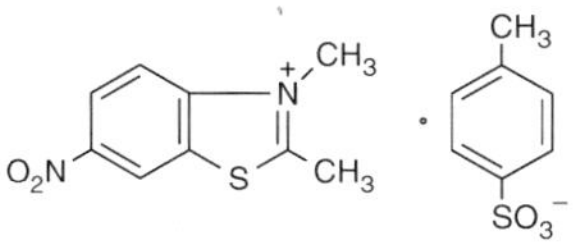

250 mg

S50,219-7
2-HYDRAZINO-6-NITROBENZOTHIAZOLE

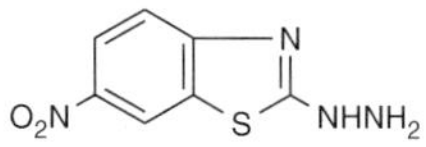

250 mg

S91,542-4
N-(6-NITRO-2-BENZOTHIAZOLYL)-ACETAMIDE

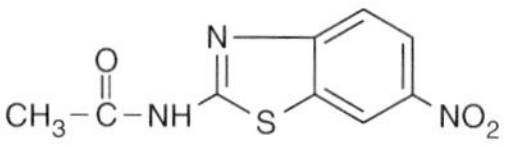

250 mg

S87,223-7
6-NITRO-2-BENZOTHIAZOLINONE METHYLENEHYDRAZONE

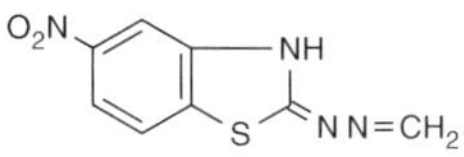

50 mg

S87,184-2
2-[(1,3-DIMETHYL-2-BENZIMIDAZOLINYLIDENE)AMINOAZO]-3-METHYLBENZOTHIAZOLIUM PERCHLORATE

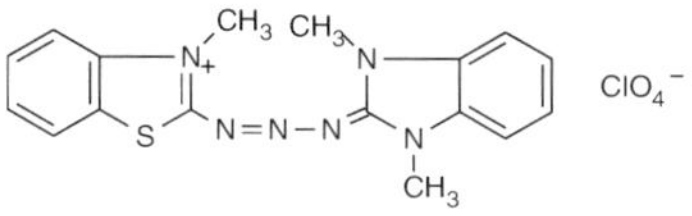

250 mg

S87,214-8
2-[(1,3-DIMETHYL-2-BENZIMIDAZOLINYLIDENE)AMINOAZO]-3-ISOBUTYLBENZOTHIAZOLIUM PERCHLORATE

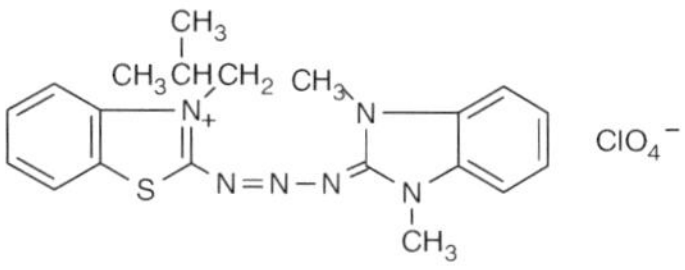

250 mg

S87,201-6
2-[(1,2-DIHYDRO-1-METHYL-2-QUINOLYLIDENE)AMINOAZO]-3-METHYLBENZOTHIAZOLIUM PERCHLORATE

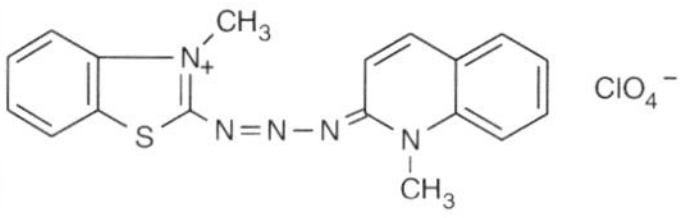

250 mg

S87,218-0
2-[(1,4-DIHYDRO-1-METHYL-4-QUINOLYLIDENE)AMINOAZO]-3-ISOBUTYLBENZOTHIAZOLIUM PERCHLORATE

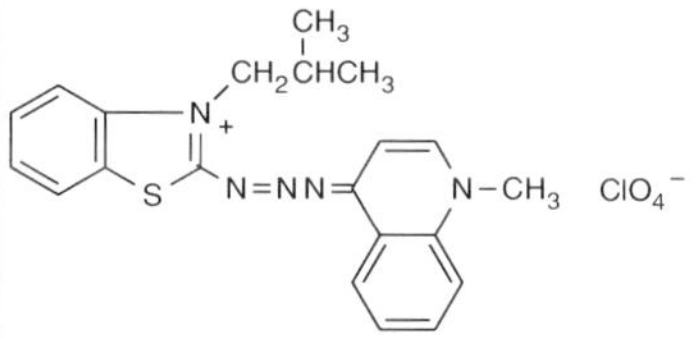

250 mg

S87,178-8
3-METHYL-2-[(3-METHYL-2-BENZOTHIAZOLINYLIDENE)AMINOAZO]-BENZOTHIAZOLIUM TETRAFLUOROBORATE

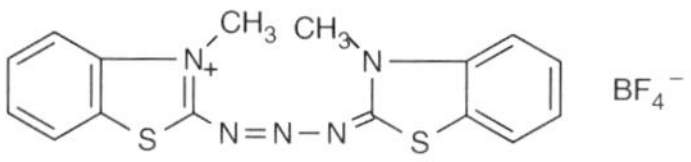

250 mg

S87,176-1
3-METHYL-2-[(3-METHYL-2-BENZOTHIAZOLINYLIDENE)AMINOAZO]-BENZOTHIAZOLIUM PERCHLORATE

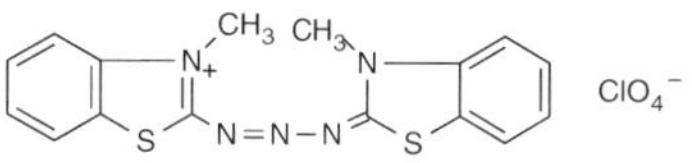

250 mg

S87,211-3
3-ISOBUTYL-2-[(3-METHYL-2-BENZOTHIAZOLINYLIDENE)AMINOAZO]-BENZOTHIAZOLIUM PERCHLORATE

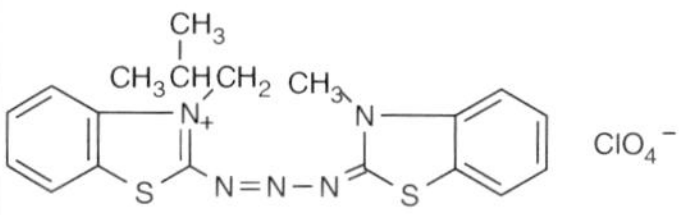

250 mg

S36,087-2
2-[4-(DIMETHYLAMINO)STYRYL]NAPHTHO(1,2-D)THIAZOLE METHIODIDE

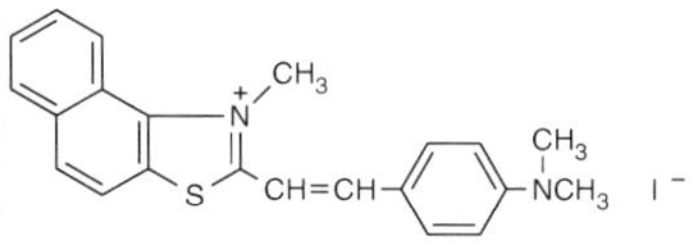

250 mg

S43,911-8
DITHIAZANINE IODIDE

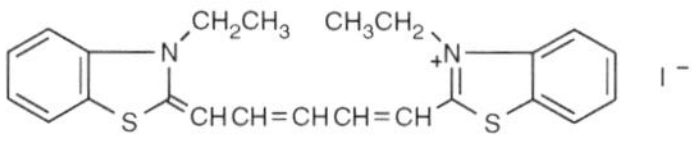

5 gm

S87,296-2
2,3-DIHYDRO-3-METHYL-2-BENZOSELENAZOLONE HYDRAZONE

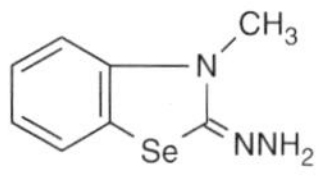

250 mg

S87,305-5
2,3-DIHYDRO-3-METHYL-2-BENZOSELENAZOLONE HYDRAZONE P-TOLUENESULFONATE

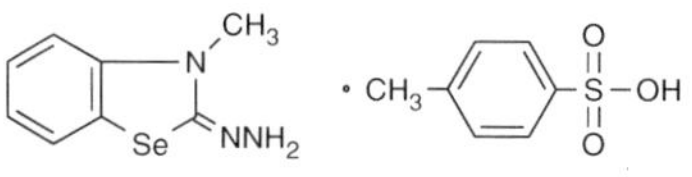

250 mg

S87,503-1
2,3-DIHYDRO-3-ETHYL-2-[2-(PHENYLIMINO)ETHYLIDENE]-BENZOSELENAZOLE

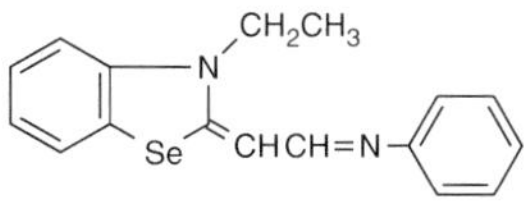

250 mg

S39,469-6
3-BENZYL-2-METHYLBENZOTHIAZOLIUM BROMIDE

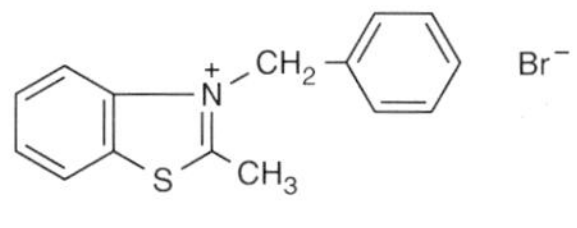

250 mg

S87,515-5
3-(2-BENZOYLHYDRAZINO)-2-METHYL-1,2-BENZISOTHIAZOLE PERCHLORATE

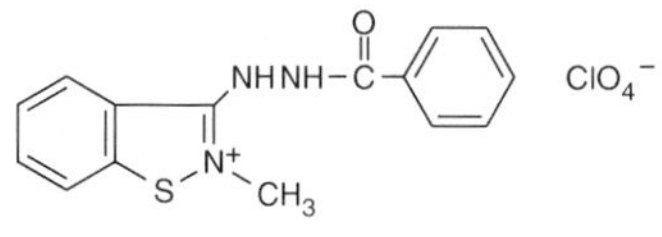

250 mg

S39,005-4
3-(2-CARBOXYETHYL)-5-CHLORO-2-METHYLBENZOTHIAZOLIUM IODIDE

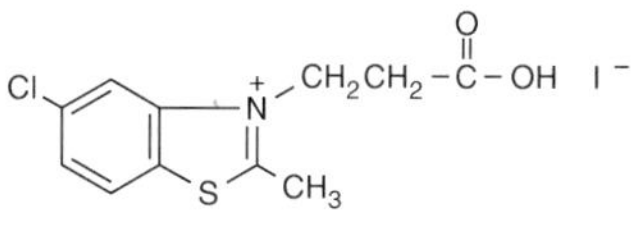

5 gm

S51,382-2
2-AMINO-6-ETHOXYCARBONYLMETHYL-3-METHYLBENZOTHIAZOLIUM IODIDE

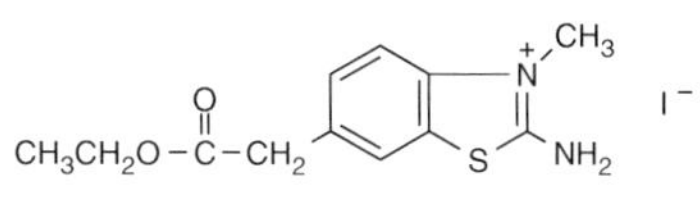

250 mg

S38,980-3
2-(2-ACETANILINOVINYL)-3-ETHYLBENZOTHIAZOLIUM IODIDE

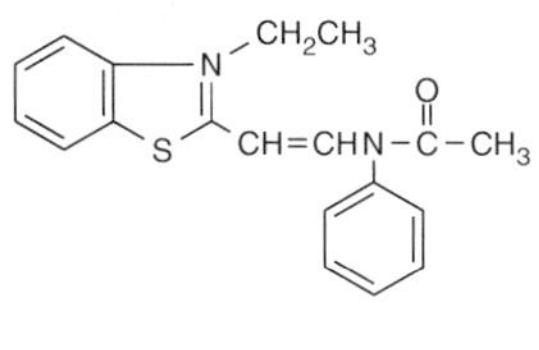

250 mg

S87,282-2
3-ETHYL-2-[2-(4,6,8-TRIMETHYL-1-AZULENYL)VINYL]-BENZOTHIAZOLIUM IODIDE

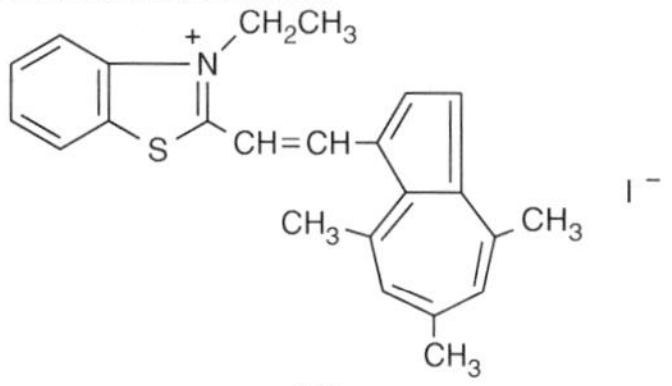

250 mg

S43,011-0
3-METHYL-2-(METHYLTHIO)-BENZOTHIAZOLIUM P-TOLUENESULFONATE

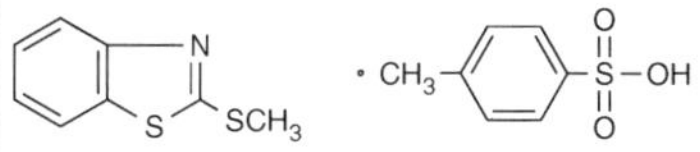

5 gm

S42,985-6
2-(METHYLTHIO)BENZOTHIAZOLE ETHYL P-TOLUENESULFONATE

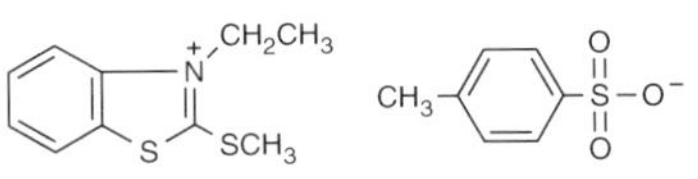

5 gm

S39,494-7
3-(3-SULFOOXYPROPYL)-2,5,6-TRIMETHYLBENZOTHIAZOLIUM HYDROXIDE, INNER SALT

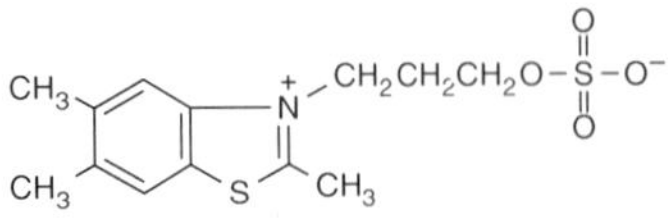

250 mg

S53,734-9
2-METHYLTHIO-3-PHENYLBENZOTHIAZOLIUM P-TOLUENESULFONATE

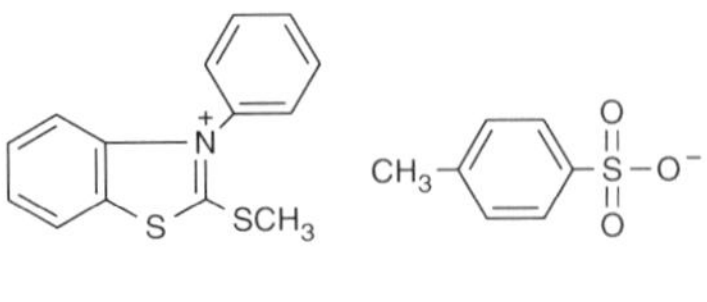

5 gm

S57,085-0
2,3-DIMETHYLBENZOTHIAZOLIUM METHYL SULFATE

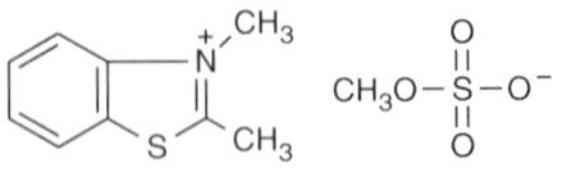

5 gm

S44,872-9
1,2-DIMETHYLNAPHTHO(1,2-D)-THIAZOLIUM METHYL SULFATE

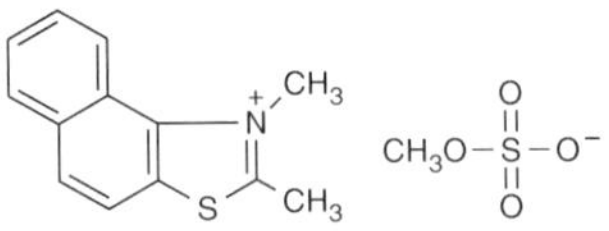

250 mg

S63,309-7
3-METHYLBENZOTHIAZOLIUM P-TOLUENESULFONATE

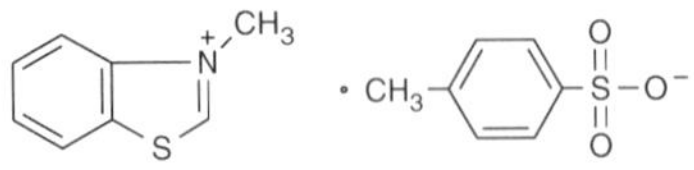

250 mg

S43,976-2
2,3-DIMETHYLBENZOTHIAZOLIUM IODIDE

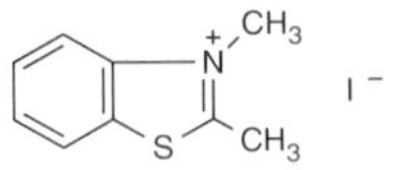

5 gm

S76,497-3
2-CHLORO-3-METHYLBENZOTHIAZOLIUM TRIFLUOROMETHANESULFONATE

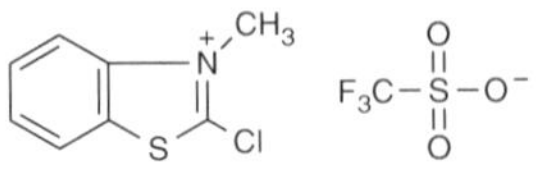

5 gm

S62,293-1
2,5-DIMETHYL-3-ETHYLBENZOTHIAZOLIUM ETHYL SULFATE

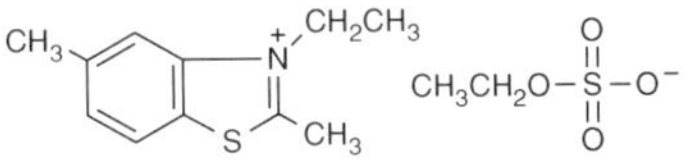

5 gm

S38,978-1
2,3-DIMETHYLNAPHTHO(1,2-D)-THIAZOLIUM IODIDE

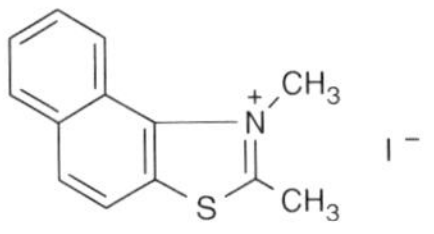

250 mg

S87,268-7
2-BENZOTHIAZOLINONE (ALPHA-ACETYLBENZYLIDENE)HYDRAZONE

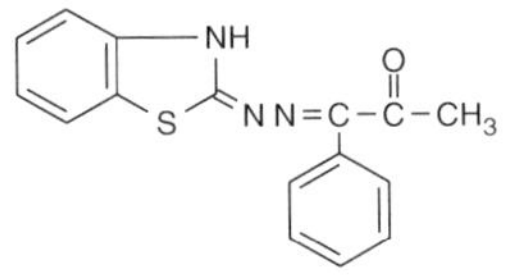

250 mg

S58,790-7
2-(2-THIENYLMETHYLENEHYDRAZINO)-BENZOTHIAZOLE

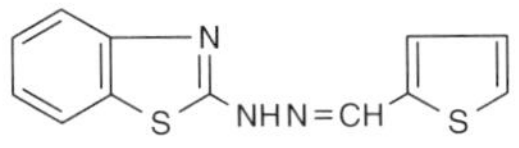

250 mg

S94,888-8
2-METHYL-6-NITROBENZOTHIAZOLE

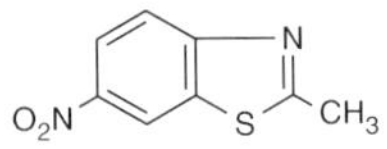

250 mg

S38,647-2
5-NITROBENZOTHIAZOLE

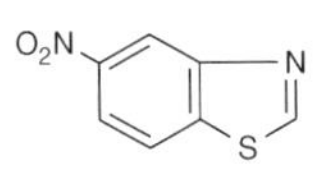

250 mg

S83,287-1
2-BENZOYLBENZOTHIAZOLE

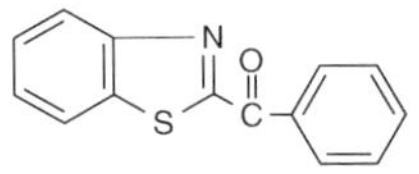

250 mg

S90,799-5
1-(2-BENZOTHIAZOLYL)ACETONE

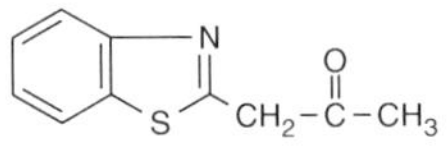

250 mg

S91,120-8
ETHYL 4-[3-(4-CHLORO-2-BENZOTHIAZOLYL)UREIDO]BENZOATE

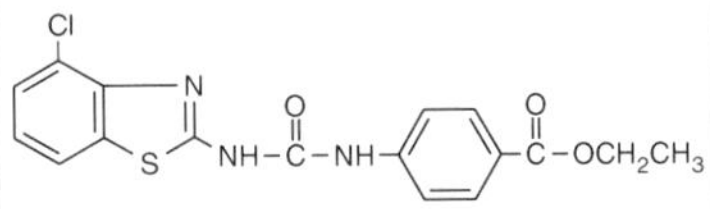

250 mg

S91,122-4
ETHYL 4-[3-(5,6-DIMETHYL-2-BENZOTHIAZOLYL)UREIDO]BENZOATE

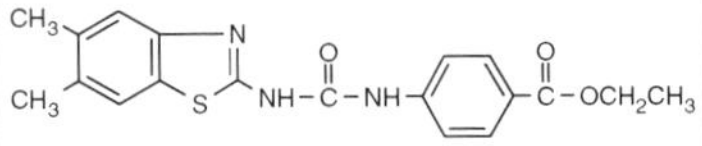

250 mg

S83,915-9
DIETHYL 2-(2-BENZOTHIAZOLYL)-2-ETHYLMALONATE

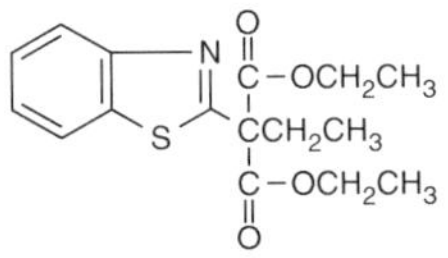

250 mg

S35,811-8
N-(2-BENZOTHIAZOLYL)-ACETOACETAMIDE

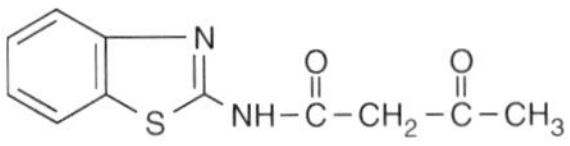

5 gm

S36,674-9
3-(2-BENZOTHIAZOLYLTHIO)-PROPIONIC ACID

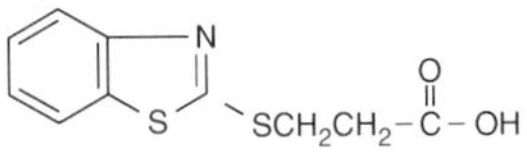

250 mg

S94,647-8
2-BENZOTHIAZOLEBUTYRIC ACID

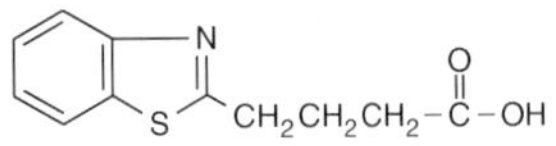

250 mg

S91,227-1
ETHYL 2-(3-[4-(6-METHYL-2-BENZOTHIAZOLYL)PHENYL]UREIDO)-ACETATE

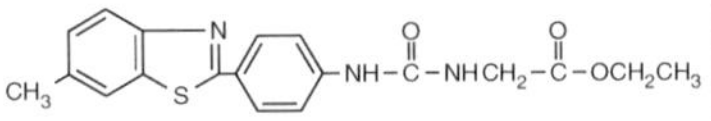

250 mg

S91,523-8
N-(2-BENZOTHIAZOLYL)ACETAMIDE

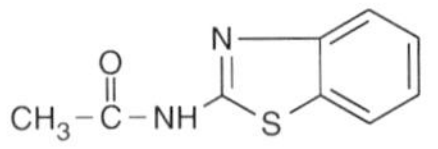

250 mg

S74,689-4
4'-(6-METHYL-2-BENZOTHIAZOLYL)-ACETANILIDE

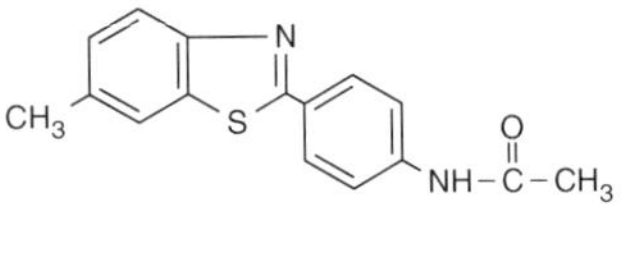

5 gm

S65,318-7
2-FORMAMIDOBENZOTHIAZOLE

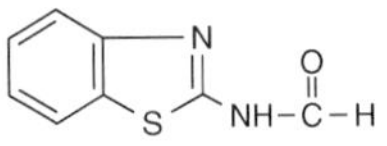

5 gm

S64,461-7
1-(2-BENZOTHIAZOLYL)-3-PHENYLUREA

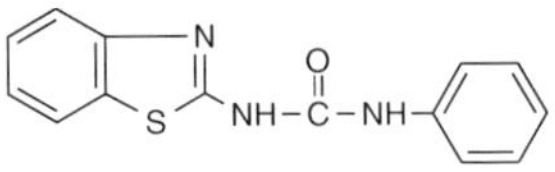

5 gm

S65,780-8
4,4'-METHYLENEBIS(N-[N-(2-BENZOTHIAZOLYL)CARBAMOYL]ANILINE)

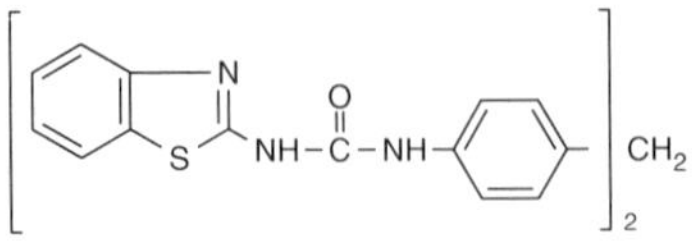

5 gm

S64,458-7
1-(2-BENZOTHIAZOLYL)-3-(3,4-DICHLOROPHENYL)UREA

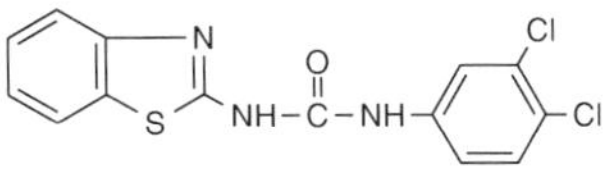

5 gm

S91,059-7
1-ETHYL-3-[NAPHTHO(1,2-D)-THIAZOL-2-YL]UREA

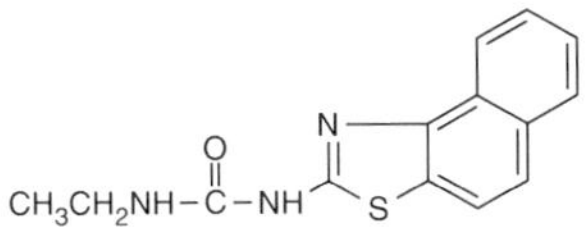

250 mg

S40,659-7
(2-BENZOTHIAZOLYL)GUANIDINE

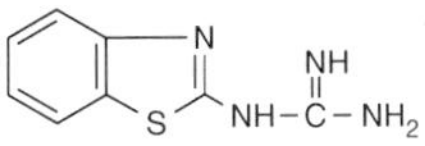

5 gm

S87,014-5
2-(P-TOSYLAZO)BENZOTHIAZOLE

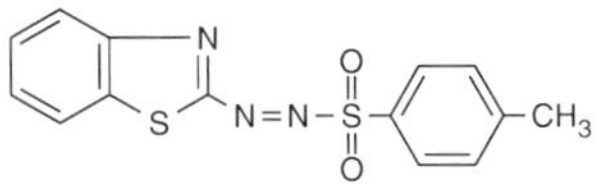

250 mg

S84,073-4
2-PIPERIDINOBENZOTHIAZOLE

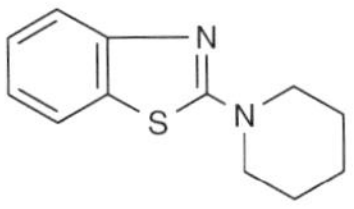

250 mg

S76,236-9
4,4'-TRIMETHYLENEBIS[1-(2-BENZOTHIAZOLYL)PIPERIDINE]

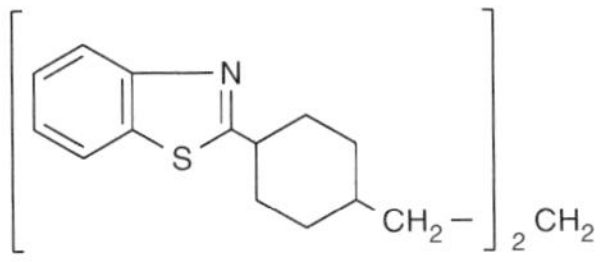

250 mg

S92,990-5
2-(2-BENZOTHIAZOLYL)ACETONITRILE

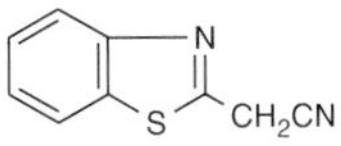

250 mg

S62,849-2
BENZOTHIAZOLE-2,5-DISULFONIC ACID DISODIUM SALT

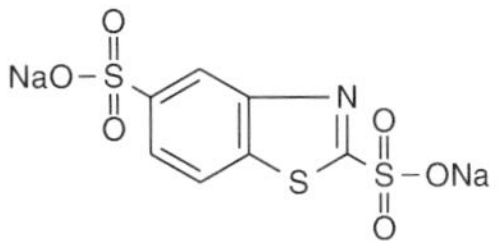

250 mg

S36,597-1
N-CYCLOHEXYL-2-BENZOTHIAZOLESULFENAMIDE

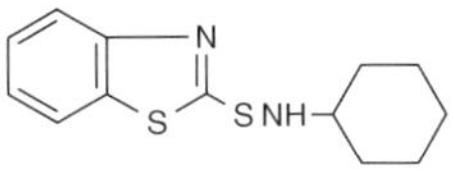

5 gm

S36,399-5
2-(MORPHOLINOTHIO)BENZOTHIAZOLE

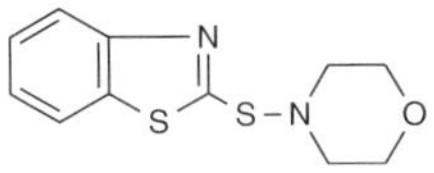

5 gm

S92,577-2
4-METHYL-2-PHENYLBENZOTHIAZOLE

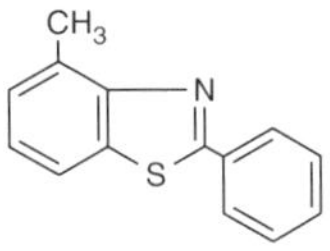

250 mg

S94,384-3
6-METHYL-2-PHENYLBENZOTHIAZOLE

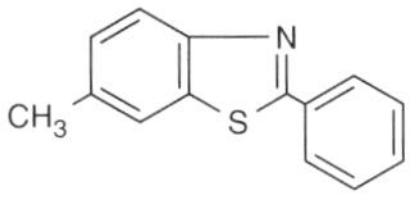

250 mg

S39,519-6
2-METHYL-6-PHENYLBENZOTHIAZOLE

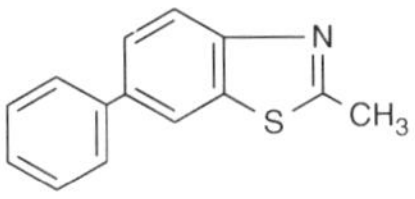

250 mg

S92,755-4
2-(O-TOLYL)BENZOTHIAZOLE

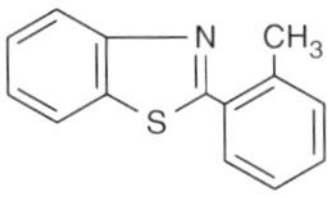

250 mg

S68,537-2
2-(P-TOLYL)BENZOTHIAZOLE

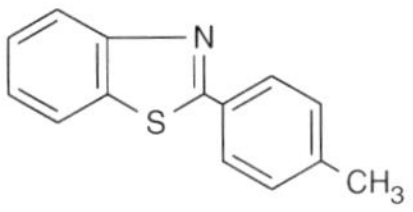

250 mg

S93,744-4
2-(3-INDENYL)BENZOTHIAZOLE

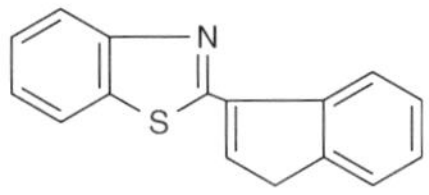

250 mg

S81,328-1
BENZOTHIAZOLE PICRATE

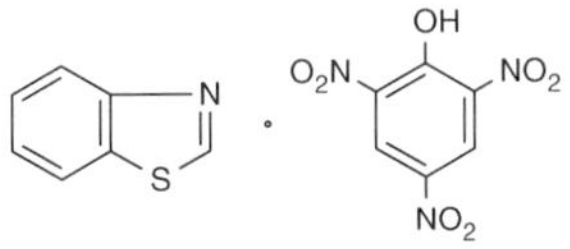

250 mg

S37,356-7
2-METHOXYBENZOTHIAZOLE

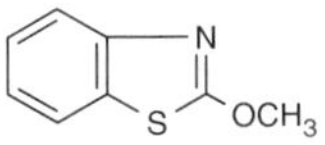

5 gm

S38,296-5
5-CHLORO-2-METHYLBENZOTHIAZOLE

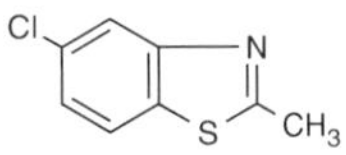

250 mg

S39,686-9
4,6-DICHLORO-2-METHYLBENZOTHIAZOLE

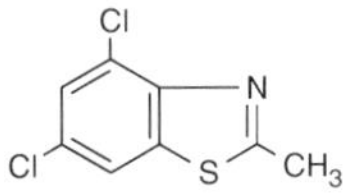

250 mg

S94,256-1
4,5-DICHLORO-2-METHYLBENZOTHIAZOLE

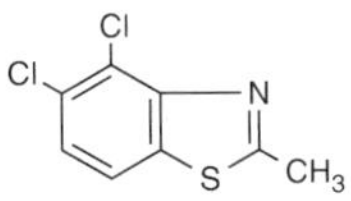

250 mg

S43,898-7
2,6-DIMETHYLBENZOTHIAZOLE

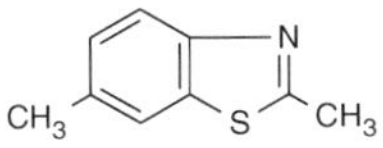

250 mg

S50,723-7
ETHYL (2-BENZOTHIAZOLYLTHIO)-FORMATE

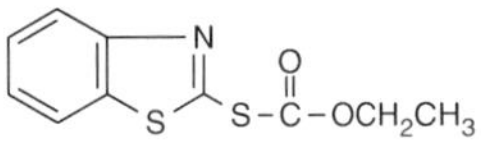

250 mg

S43,032-3
N-TERT-BUTYL-2-BENZOTHIAZOLESULFENAMIDE

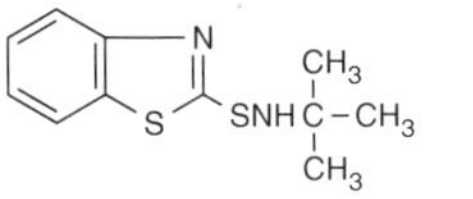

5 gm

S73,557-4
2-(4-CHLOROPHENYL)OXAZOLO(4,5-B)-PYRIDINE

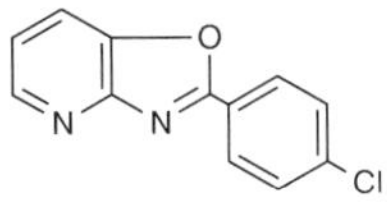

250 mg

S73,558-2
2-(3,4-DICHLOROPHENYL)OXAZOLO(4,5-B)-PYRIDINE

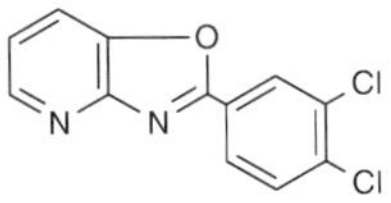

250 mg

S84,372-5
5-AMINO-4-NITROBENZOFURAZAN

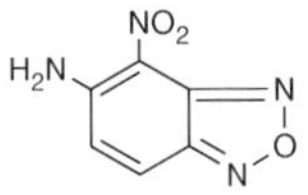

250 mg

S40,851-4
ETHYL 7-HYDROXY-S-TRIAZOLO(1,5-A)-PYRIMIDINE-6-CARBOXYLATE

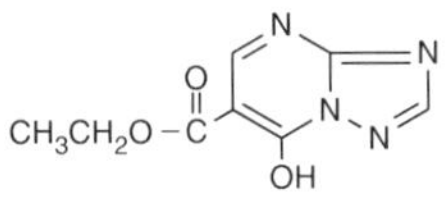

5 gm

S51,579-5
2-BENZYLTHIO-5-METHYL-S-TRIAZOLO (1,5-A)PYRIMIDIN-7-OL

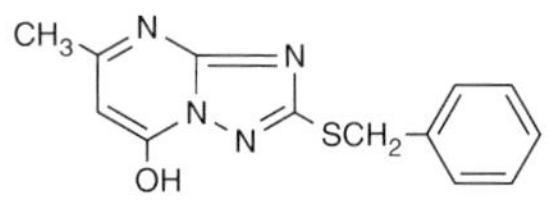

250 mg

S51,112-9
3-AMINOPYRAZOLO(4,5-C)-PYRIDINE-4,6-DIOL

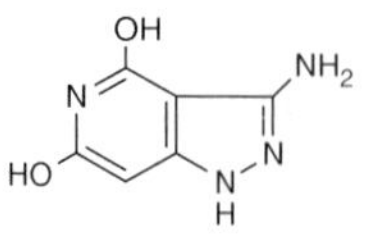

5 gm

S60,064-4
5,7-DIHYDROXY-S-TRIAZOLO-(1,5-A)-PYRIMIDINE

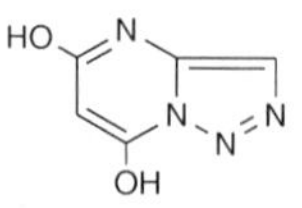

5 gm

S59,311-7
3,6-DIMETHYL-1-PHENYL-1H-PYRAZOLO-(3,4-B)-PYRIDIN-4-OL

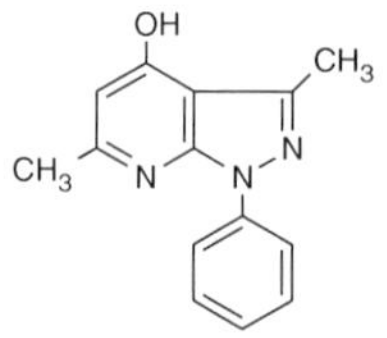

250 mg

S40,983-9
5-PHENYL-S-TRIAZOLO(1,5-A)-PYRIMIDIN-7-OL

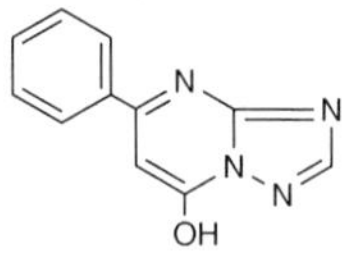

5 gm

S34,634-9
3,4-DIHYDROXY-6-METHYLPYRAZOLO(3,4-B)-PYRIDINE

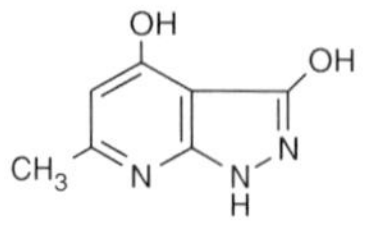

5 gm

S34,707-8
4,6-DIMETHYL-3-HYDROXYPYRAZOLO-(3,4-B)-PYRIDINE

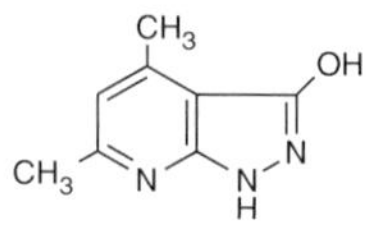

5 gm

S59,306-0
2,3-DIHYDRO-3-IMINO-5-METHYL-2-PHENYL-S-TRIAZOLO(4,3-A)-PYRIMIDIN-6-OL

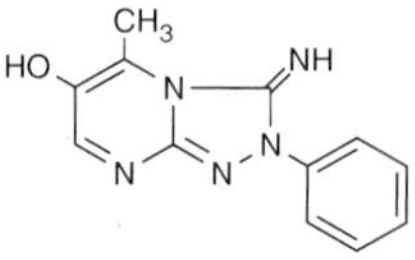

250 mg

S51,027-0
4-METHYLPYRIMIDO(1,2-A)-BENZIMIDAZOL-2-OL

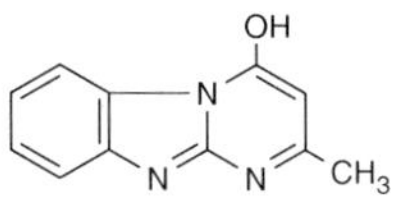

5 gm

S56,681-0
5-METHYL-S-TRIAZOLO(4,3-A)-S-TRIAZIN-7-OL

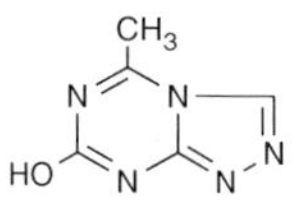

5 gm

S36,258-1
5-HYDROXY-7-METHYL-1,2,3,8-TETRAAZAINDOLIZINE

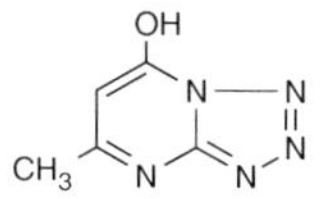

5 gm

S42,566-4
HARMALINE METHOCHLORIDE

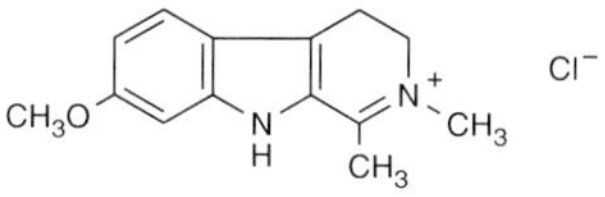

250 mg

S42,495-1
HARMINE METHOSULFATE

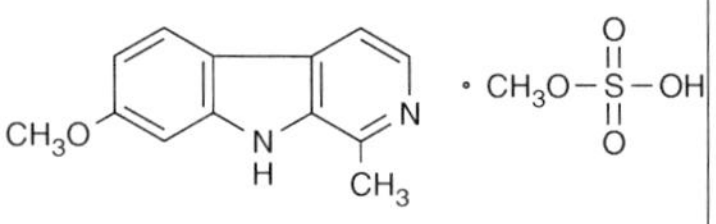

250 mg

S42,497-8
HARMINE METHIODIDE

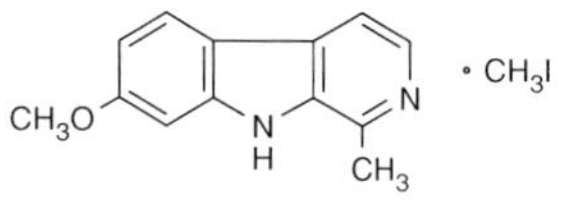

250 mg

S42,592-3
6-NITROHARMALINE NITRATE

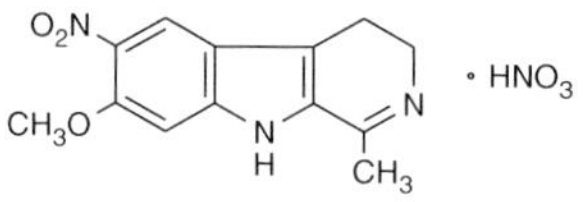

250 mg

S42,565-6
N-METHYLHARMINE

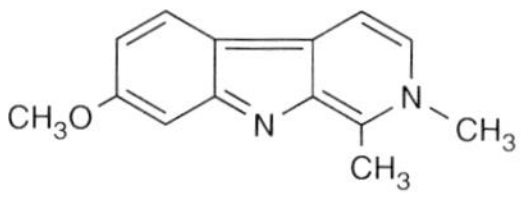

250 mg

S42,579-6
N-METHYLTETRAHYDROHARMINE

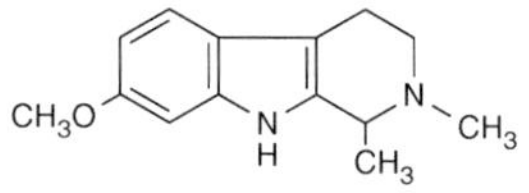

250 mg

S42,595-8
2-NITROSO-1,2-DIHYDROHARMALINE

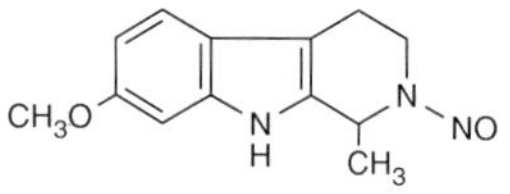

250 mg

S42,411-0
N-ACETYLHARMALINE

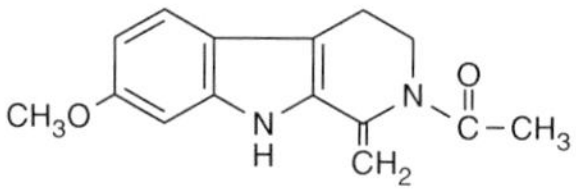

250 mg

S41,757-2
BROMOHARMINE HYDROBROMIDE

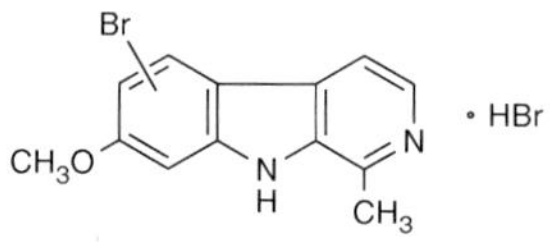

250 mg

S42,591-5
ALPHA-(4-NITROBENZYLIDENE)HARMINE

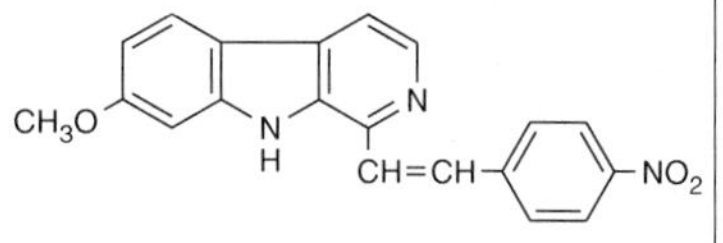

250 mg

S42,491-9
HARMALOL HYDROBROMIDE

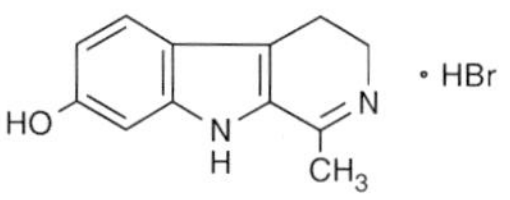

250 mg

S42,625-3
2-PROPYLHARMINE IODIDE

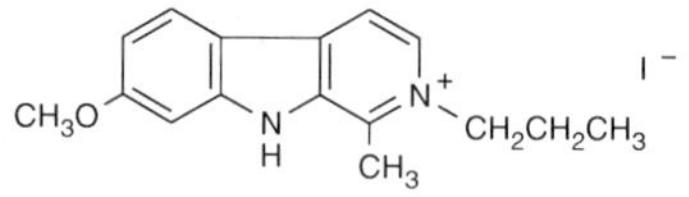

250 mg

S42,574-5
9-PROPYLHARMINE METHIODIDE

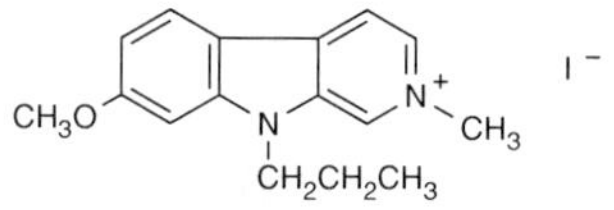

250 mg

S42,492-7
HARMAN METHOSULFATE

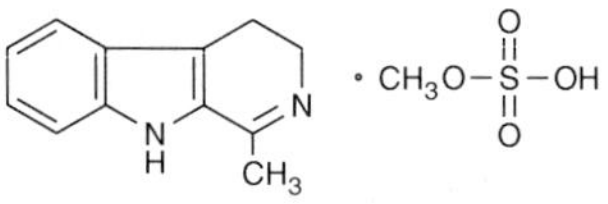

250 mg

S42,433-1
BROMONORHARMAN (BR POSITION UNSPECIFIED)

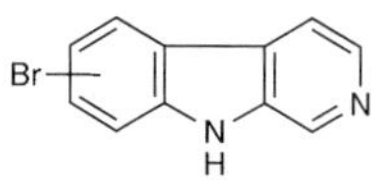

250 mg

S42,493-5
HARMAN-N-SULFONIC ACID

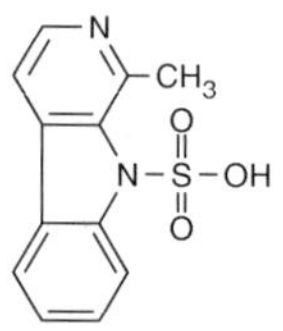

250 mg

S42,734-9
4-AZA-2-MERCAPTOBENZIMIDAZOLE

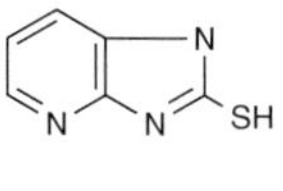

5 gm

S54,936-3
4-(METHYLTHIO)-1H-PYRAZOLO(3,4-D)-PYRIMIDINE

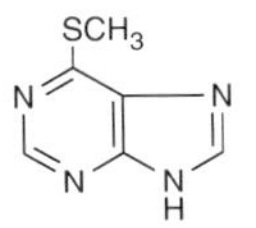

250 mg

S92,033-9
ETHYL 2-[7-METHYL(1,2,4)TRIAZOLO-(3,4-C)(1,2,4)BENZOTRIAZIN-1-YLTHIO]ACETATE

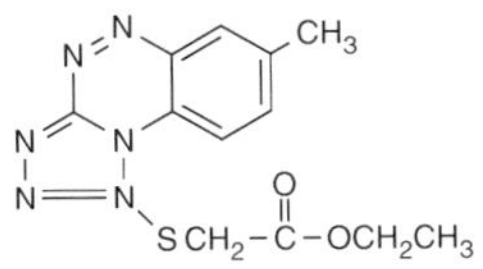

250 mg

S78,390-0
7-DEAZA-6-(METHYLTHIO)PURINE

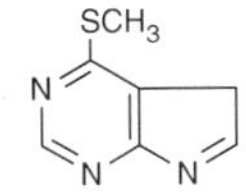

5 gm

S41,258-9
7-AMINO-S-TRIAZOLO(1,5-A)-PYRIMIDIN-5(4H)-ONE

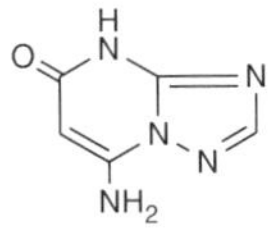

5 gm

S42,606-7
1-OCTYL-1,2,3,4-TETRAHYDRONORHARMAN-3-CARBOXYLIC ACID

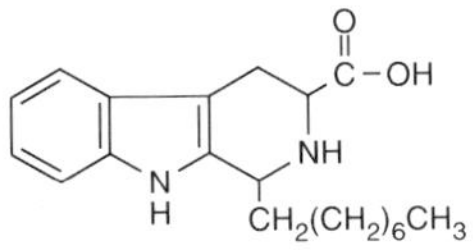

250 mg

S42,626-1
1-PROPYL-1,2,3,4-TETRAHYDRONORHARMAN-3-CARBOXYLIC ACID

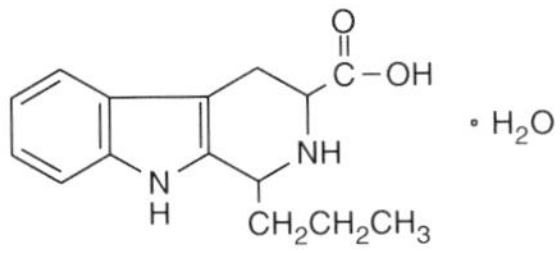

250 mg

S42,530-3
1-ISOBUTYL-1,2,3,4-TETRAHYDRONORHARMAN-3-CARBOXYLIC ACID

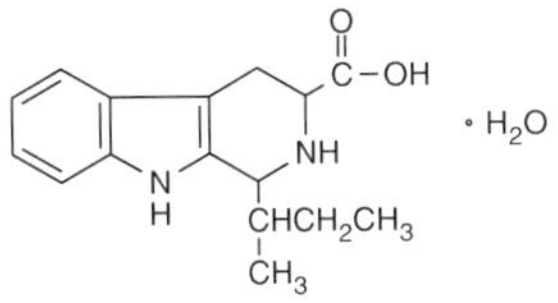

250 mg

S42,466-8
1-(DIETHYLMETHYL)-1,2,3,4-TETRAHYDRONORHARMAN-3-CARBOXLIC ACID

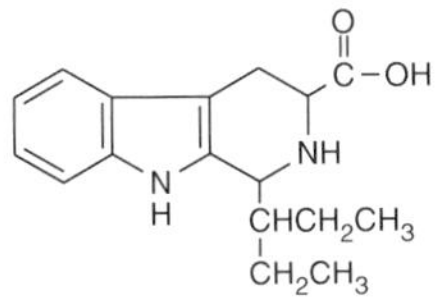

250 mg

S42,593-1
6-NITROHARMAN NITRATE

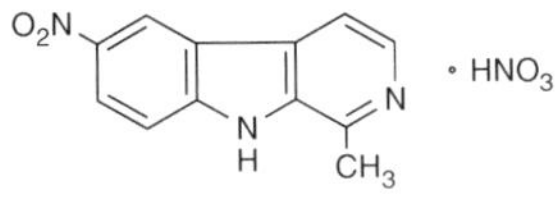

250 mg

S42,621-0
1-PHENYL-1,2,3,4-TETRAHYDRONORHARMAN-3-CARBOXYLIC ACID

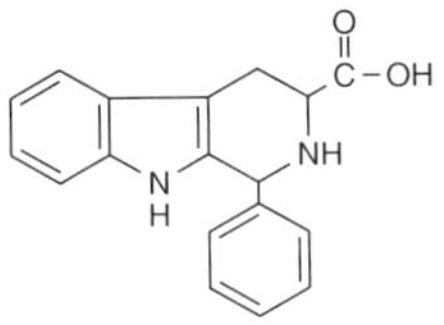

250 mg

S78,212-2
8-BROMO-5-METHYL-5H-1,2,4-TRIAZINO(5,6-B)INDOLE-3(2H)-THIONE

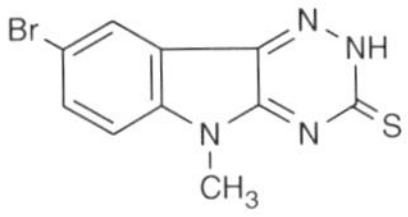

5 gm

S80,448-7
1,3-DIETHYL-5-NITROSO-6-PHENYL-7H-PYRROLO-(2,3-D)PYRIMIDINE-2,4-(1H,3H)-DIONE

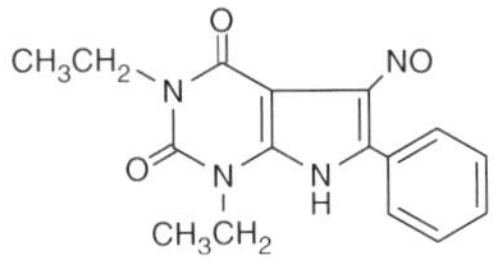

250 mg

S40,937-5
METHYL 2-PHENYLIMIDAZO(1,2-A)-PYRIDINE-7-CARBOXYLATE

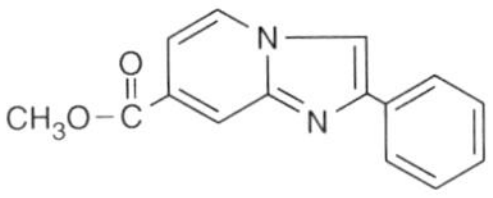

100 mg

S41,801-3
5-METHYL-2,1,3-BENZOSELENADIAZOLE

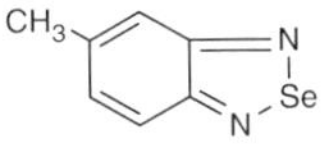

5 gm

S42,558-3
1-(4-METHOXYPHENYL)-1,2,3,4-TETRAHYDRONORHARMAN-3-CARBOXYLIC ACID

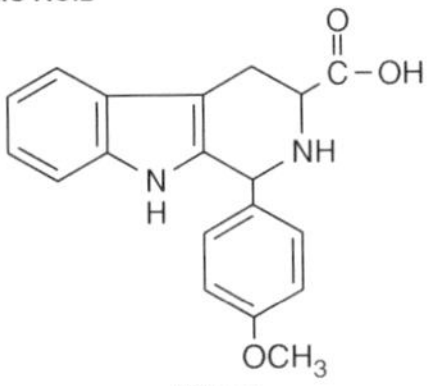

250 mg

S42,597-4
NORHARMAN METHIODIDE

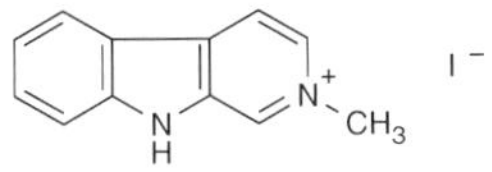

250 mg

S42,598-2
6-NORHARMANSULFONIC ACID

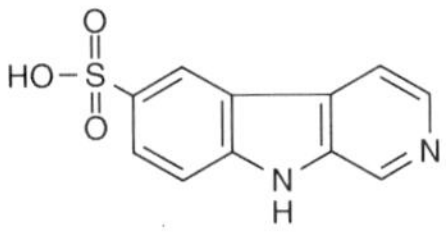

250 mg

S78,698-5
5-METHYL-7-(1H-1,2,4-TRIAZOL-5-YLAMINO)-S-TRIAZOLO(1,5-A)-PYRIMIDINE

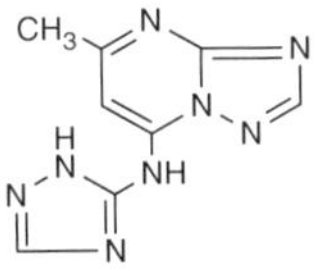

250 mg

S40,978-2
2-PHENYLIMIDAZO(1,2-A)PYRIDINE-6-CARBOXYLIC ACID

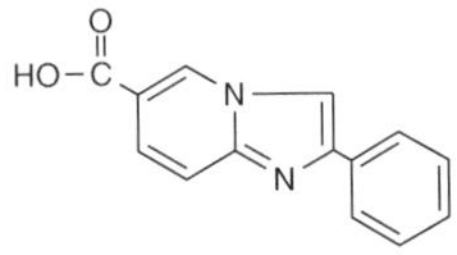

250 mg

S69,716-8
3-METHYLTHIAZOLO(3,2-A)PYRIDINIUM PERCHLORATE

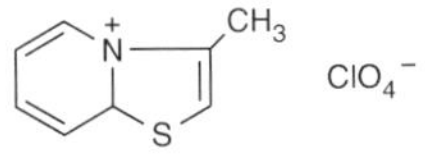

250 mg

S94,515-3
5,6-DICHLORO-1,2,3-BENZOTHIADIAZOLE

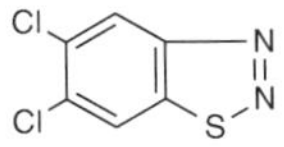

250 mg

S64,753-5
2-ETHOXYCARBONYL-3,4,5,6,7,8-HEXAHYDRO(1)BENZOTHIENO(2,3-D)-PYRIMIDIN-4-ONE

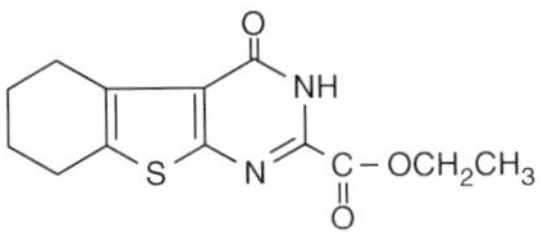

5 gm

S64,754-3
8-BROMO-2,3-DIHYDRO-5H-AS-TRIAZINO-(5,6-B)INDOLE-3-THIONE

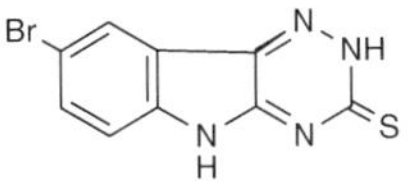

5 gm

S80,253-0
6-CHLORO-1,3-DIETHYL-7H-PYRROLO-(2,3-D)PYRIMIDINE-2,4(1H,3H)-DIONE

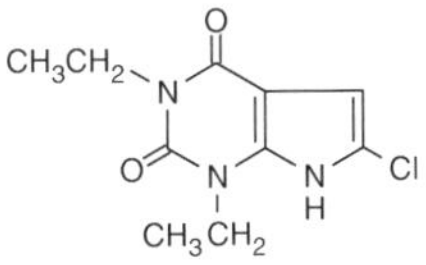

250 mg

S58,782-6
4-METHYLBENZOFURO(3,2-G)QUINOL-2-(1H)-ONE

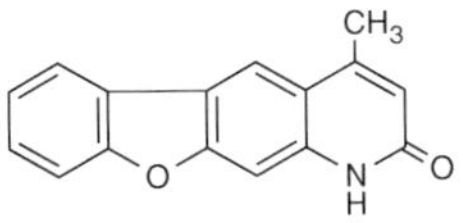

250 mg

S76,041-2
3,6-DIPHENYLISOXAZOLO(4,5-C)-PYRIDIN-4(5H)-ONE

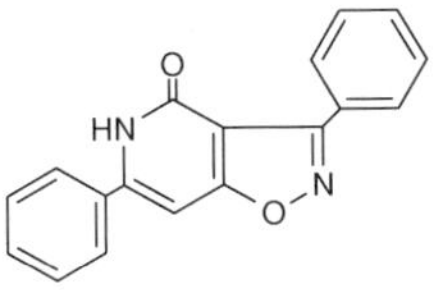

250 mg

S92,036-3
TETRAZOLO(5,1-C)(1,2,4)-BENZOTRIAZINE

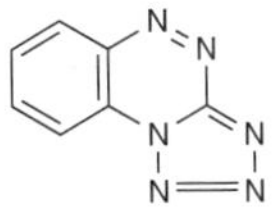

250 mg

S73,225-7
5-CHLORO-S-TRIAZOLO(3,4-A)-ISOQUINOLINE

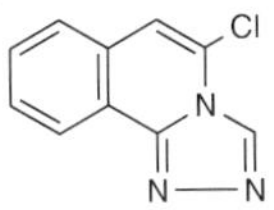

250 mg

S40,945-6
5-METHYL-S-TRIAZOLO(1,5-A)-PYRIMIDINE

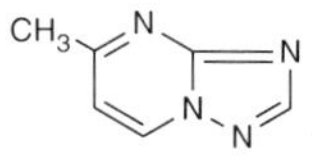

5 gm

S80,432-0
7-(CHLOROMETHYL)-1,3-DIMETHYLXANTHINE

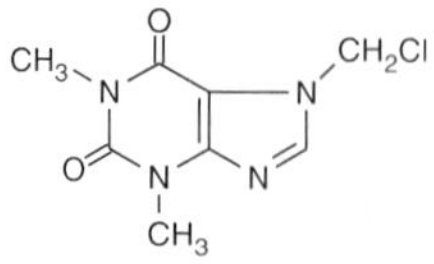

250 mg

S80,431-2
8-CHLORO-7-(CHLOROMETHYL)-1,3-DIMETHYLXANTHINE

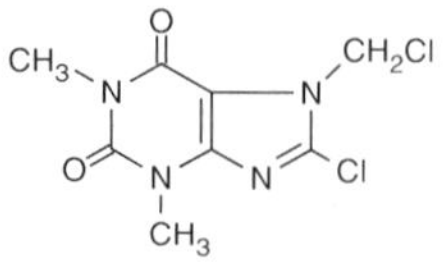

250 mg

S77,775-7
6-(TRICHLOROMETHYL)PURINE

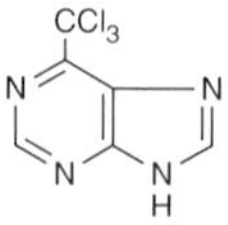

250 mg

S45,565-2
ADENOSINE-5'-MONOPHOSPHORIC ACID, DISODIUM SALT

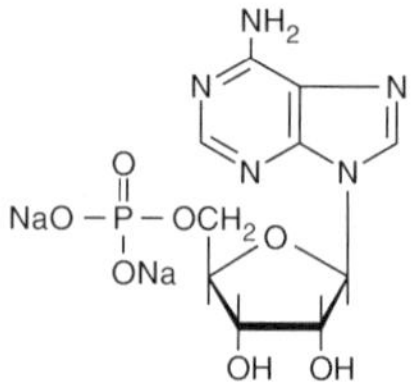

250 mg

S46,014-1
ADENOSINE 2'(3')-MONOPHOSPHORIC ACID

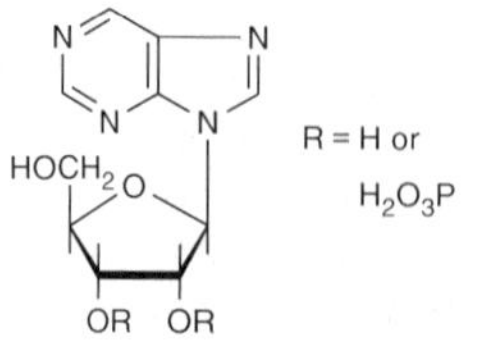

250 mg

S45,501-6
PURINE RIBOSIDE

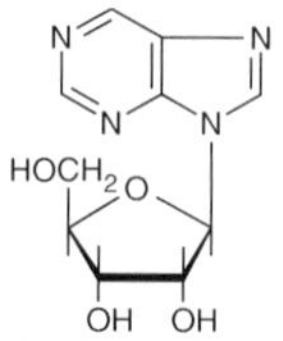

250 mg

S46,016-8
6-DIMETHYLAMINOPURINE-9-RIBOSIDE

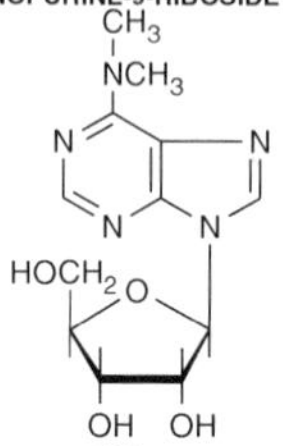

250 mg

S46,151-2
ADENOSINE 2'-MONOPHOSPHORIC ACID

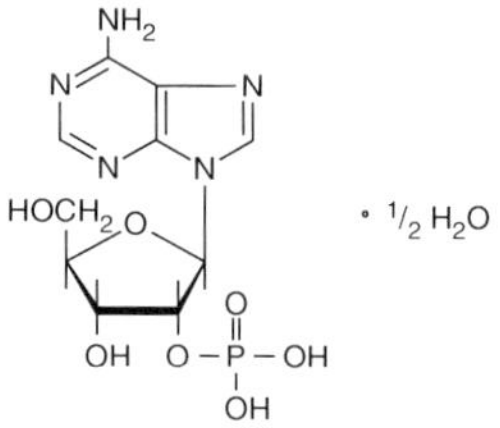

5 gm

S78,295-5
2,6-DIHYDROXY-8-(METHYLTHIO)PURINE

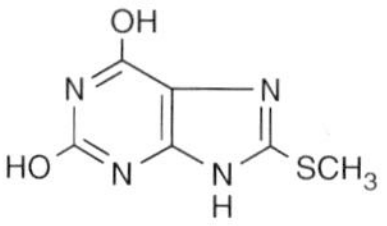

5 gm

S45,614-4
2,8-DIHYDROXYADENINE

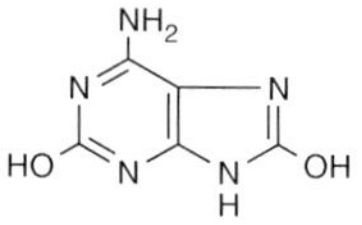

250 mg

S80,322-7
8-HYDROXY-3-(2-HYDROXYETHYL)-XANTHINE

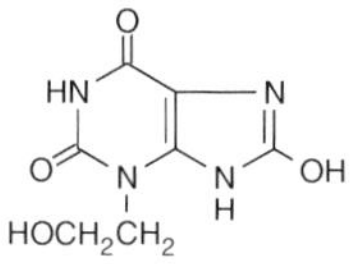

250 mg

S45,171-1
2',3'-ISOPROPYLIDENE-6-MERCAPTOPURINE RIBOSIDE

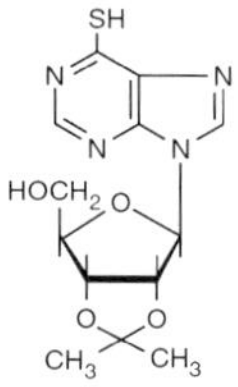

250 mg

S45,011-1
8-BROMOINOSINE

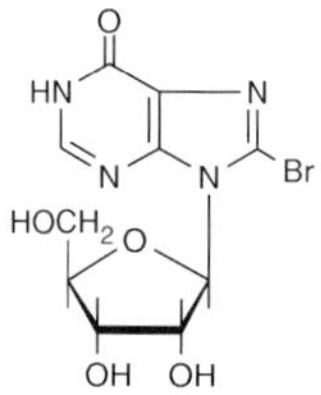

250 mg

S43,825-1
8-BROMOXANTHOSINE

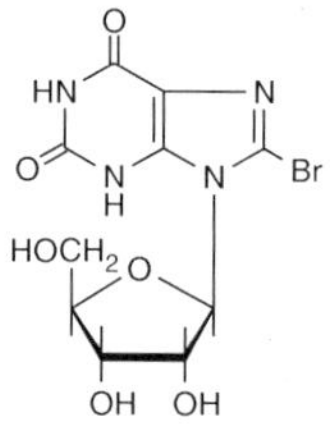

250 mg

S45,165-7
2'-DEOXYINOSINE 5'-MONOPHOSPHATE, DISODIUM SALT DIHYDRATE

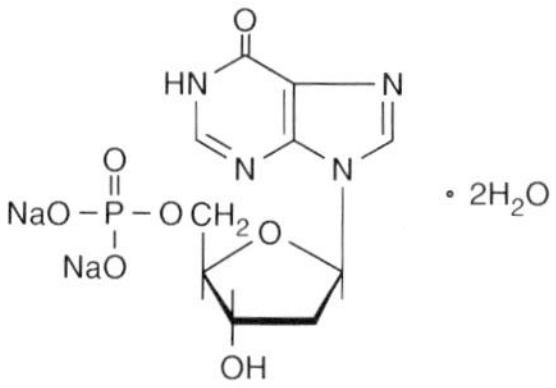

250 mg

S46,033-8
2',3',5'-TRIACETYLINOSINE

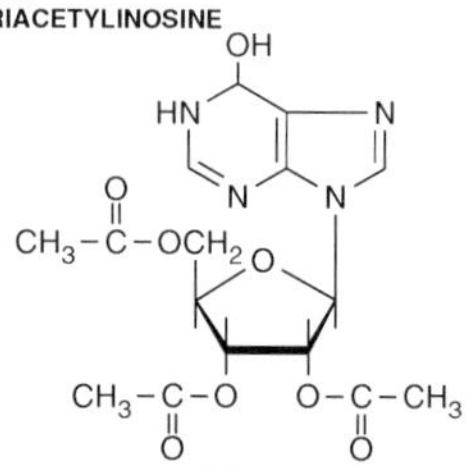

250 mg

S78,328-5
2',3',5'-TRIBENZOYLTHIOGUANOSINE

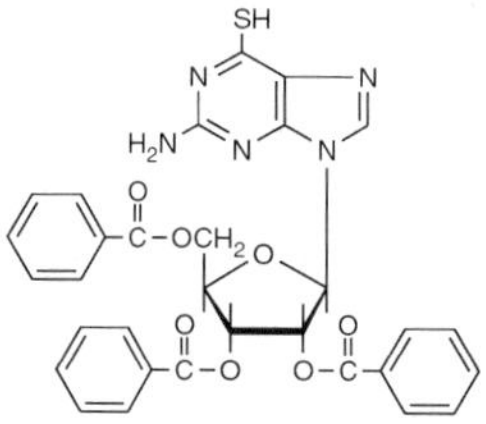

5 gm

S78,272-6
8-MERCAPTOGUANOSINE

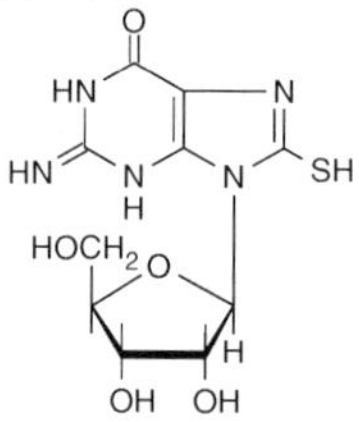

5 gm

S45,500-8
2',3'-DIACETYLADENOSINE

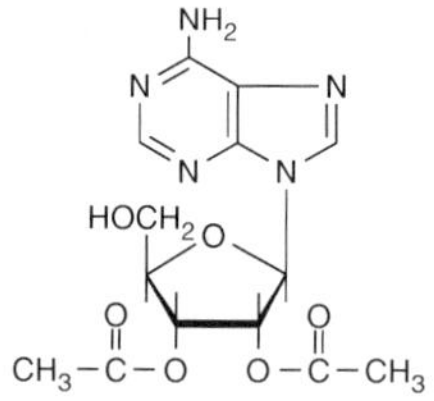

250 mg

S86,042-5
2',3'-O-BENZYLIDENEADENOSINE

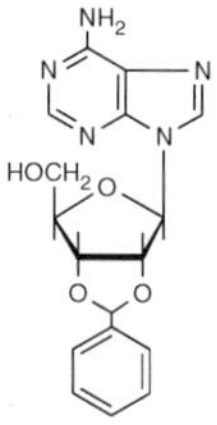

250 mg

S45,071-5
2',3'-ISOPROPYLIDENEINOSINE

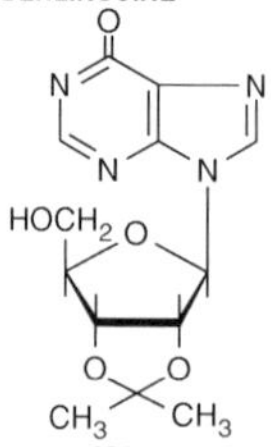

250 mg

S35,091-5
4-HYDROXY-6-MERCAPTOPYRAZOLO-(3,4-D)-PYRIMIDINE

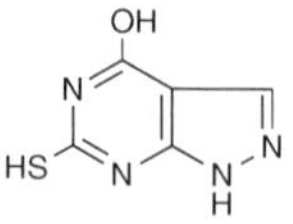

250 mg

S44,852-4
6-HYDROXY-2-MERCAPTOPURINE

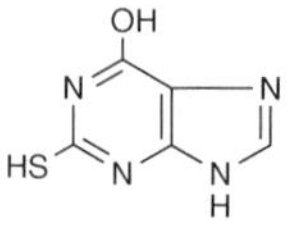

5 gm

S78,504-0
6,8-DIMERCAPTO-2-HYDROXYPURINE

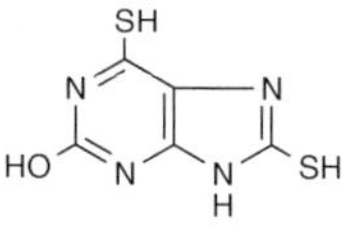

250 mg

S78,714-0
2,8-DIMERCAPTO-6-HYDROXYPURINE

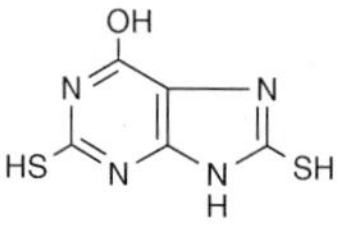

250 mg

S45,812-0
6-HYDROXY-8-MERCAPTOPURINE MONOHYDRATE

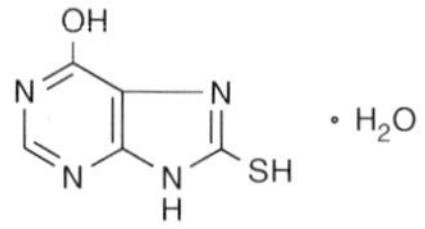

5 gm

S40,777-1
6,8-DIHYDROXY-2-METHYLMERCAPTOPURINE

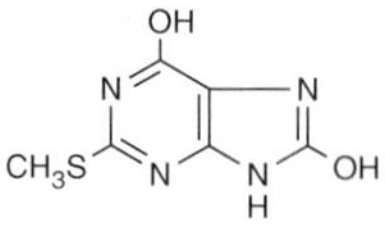

5 gm

S42,321-1
2-AMINO-6-PURINOL

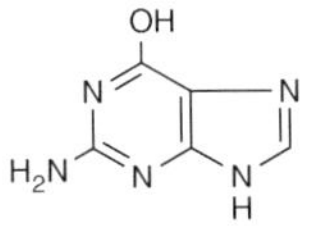

5 gm

S78,889-9
6-AMINO-2-HYDROXYPURINE SULFATE

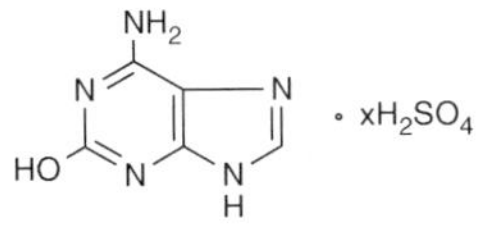

250 mg

S40,650-3
3-AMINO-2-PHENYL-2H-PYRAZOLO-(4,3-C)-PYRIDINE-4,6-DIOL

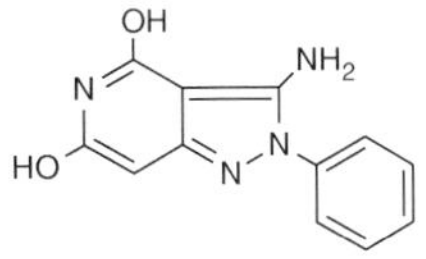

5 gm

S78,179-7
2-AMINO-9-BUTYL-6-HYDROXYPURINE

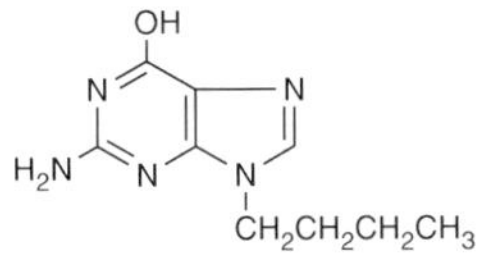

250 mg

S78,791-4
6-AMINO-2-HYDROXY-7-METHYLPURINE

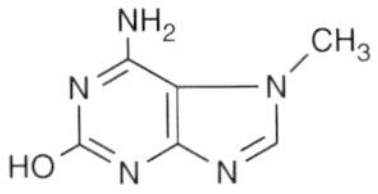

250 mg

S36,690-0
6-[BIS(2-HYDROXYETHYL)AMINO]PURINE

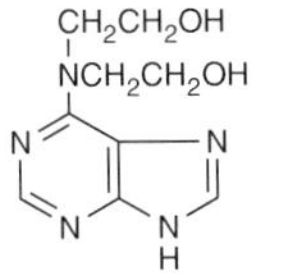

250 mg

S42,180-4
7-(2,3-DIHYDROXYPROPYL)-THEOPHYLLINE NICOTINATE

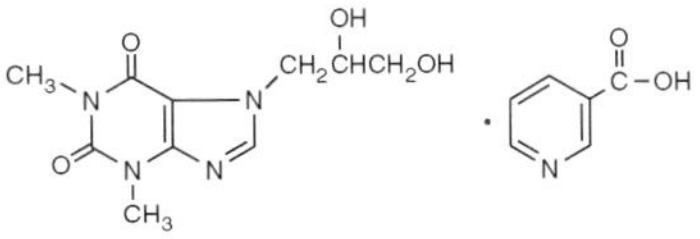

250 mg

S92,210-2
6-(2-HYDROXYETHYLAMINO)PURINE

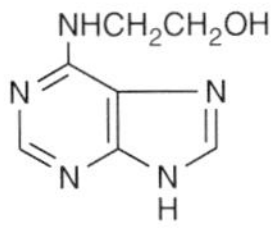

250 mg

S92,843-7
2-MERCAPTO-6,8-PURINEDIOL

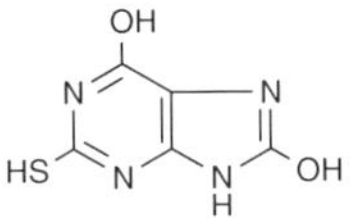

250 mg

S78,409-5
2-HYDROXY-6-METHYLPURINE

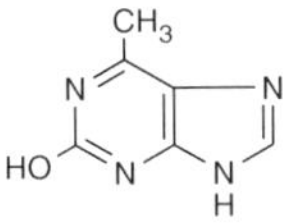

250 mg

S78,512-1
2-(DIMETHYLAMINO)-6-HYDROXY-9-METHYLPURINE

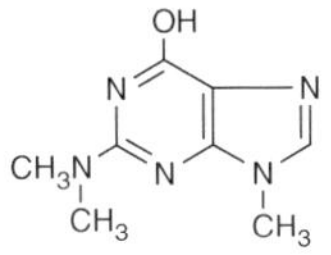

250 mg

S35,074-5
3-HYDROXY-5,6-DIMETHYL-1H-PYRAZOLO-(3,4-B)-PYRAZINE

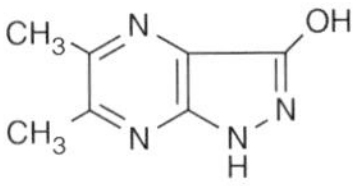

5 gm

S80,358-8
1-ETHYL-3-(2-HYDROXYETHYL)URIC ACID

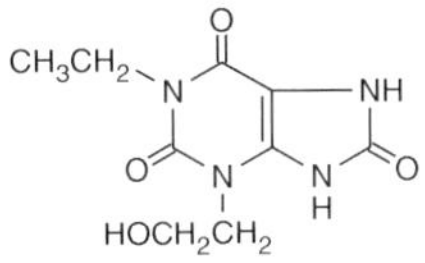

250 mg

S78,408-7
6-HYDROXY-2-METHYLPURINE

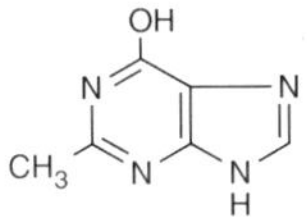

250 mg

S80,292-1
1-ETHYL-8-HYDROXY-3-METHYLXANTHINE

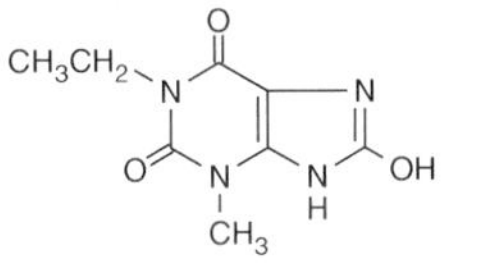

250 mg

S80,338-3
3-ETHYL-8-HYDROXY-1-METHYLXANTHINE

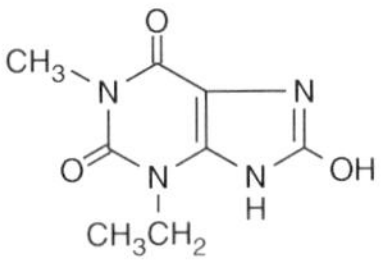

250 mg

S80,462-2
3-ALLYL-8-HYDROXY-1-METHYLXANTHINE

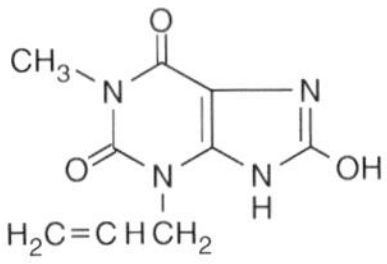

50 mg

S80,463-0
1,3-DIETHYL-8-HYDROXYXANTHINE

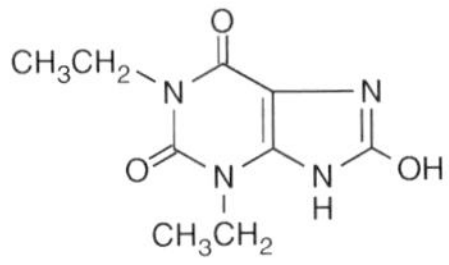

250 mg

S80,366-9
3-ALLYL-1-ETHYL-8-HYDROXYXANTHINE

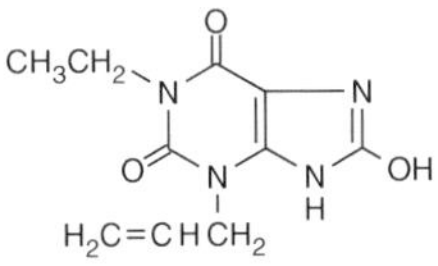

250 mg

S80,221-2
1-BUTYL-4,5-DIHYDRO-3-ETHYL-8-HYDROXYXANTHINE

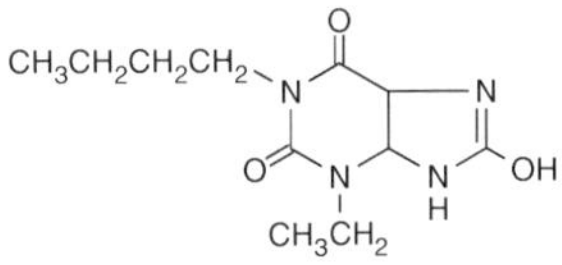

250 mg

S35,122-9
4-HYDROXY-1-PHENYLPYRAZOLO(3,4-D)-PYRIMIDINE

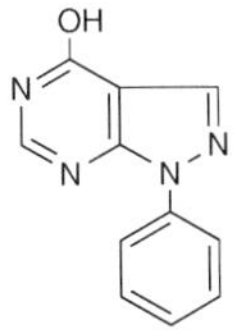

5 gm

S45,159-2
2',3'-ISOPROPYLIDENE-5'-TRITYLINOSINE

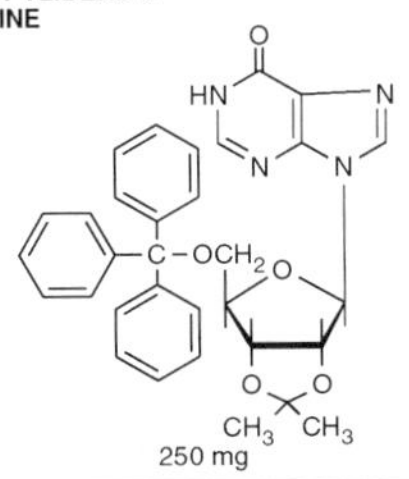

250 mg

S86,044-1
N6-(4-METHOXYBENZOYL)ADENOSINE

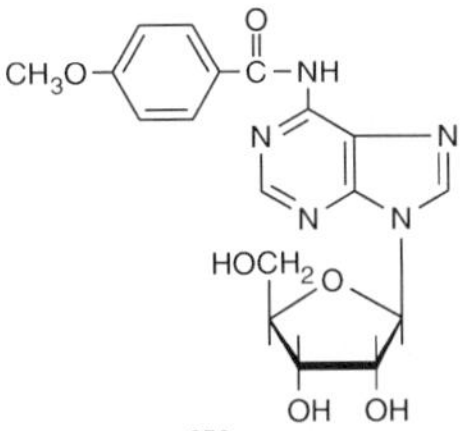

250 mg

S35,174-1
2',3'-ISOPROPYLIDENE-5'-TRITYLADENOSINE

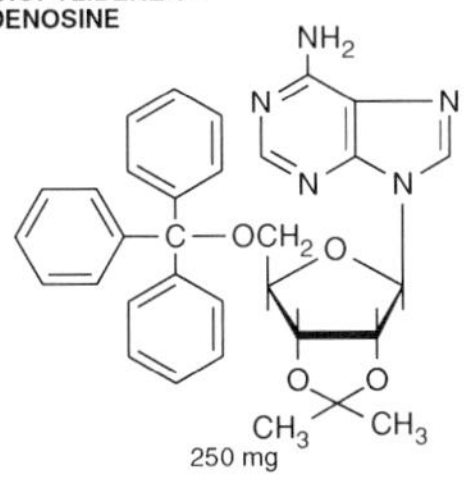

250 mg

S43,720-4
6-CHLORO-9-METHOXYMETHYLPURINE

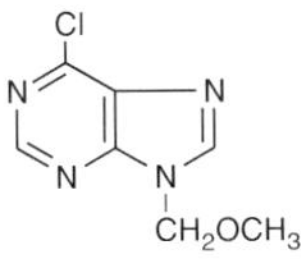

250 mg

S80,296-4
1-ETHYL-3-(2-METHOXYETHYL)XANTHINE

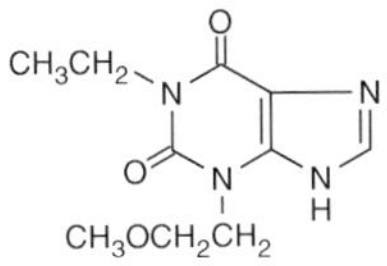

250 mg

S92,272-2
6-METHOXY-3-METHYL-8-PHENYLPURINE

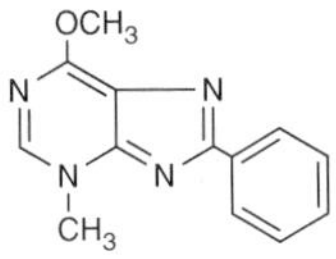

250 mg

S45,082-0
6-PROPOXYPURINE

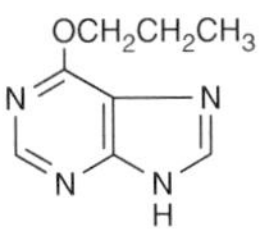

250 mg

S42,795-0
8-BROMOADENOSINE 2',3',5'-TRIACETATE
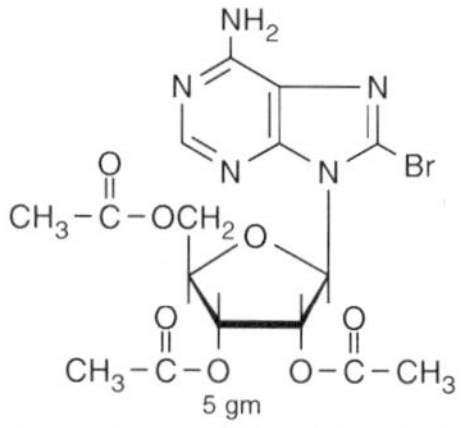
5 gm

S46,257-8
2',3',5'-TRIBENZOYLINOSINE
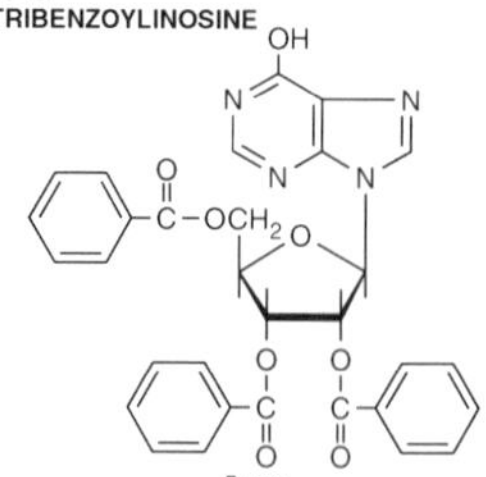
5 gm

S94,399-1
GUANOSINE 2'(3')-MONOPHOSPHATE, DISODIUM SALT MONOHYDRATE (MIXTURE OF ISOMERS)
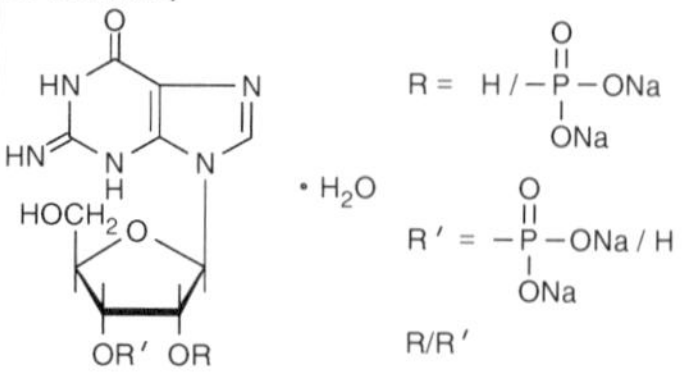

250 mg

S78,370-6
BETA-NICOTINAMIDE ADENINE DINUCLEOTIDE, OXIDIZED, FREE ACID (NAD,DPN)
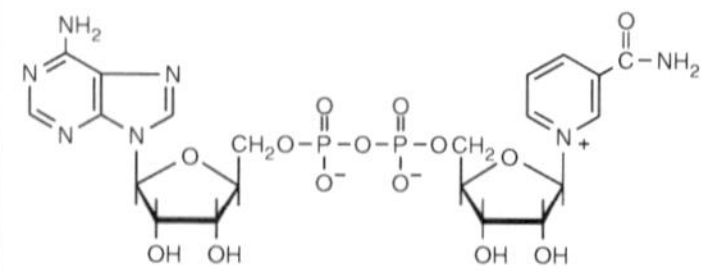
5 gm

S44,974-1
8-BROMOGUANOSINE 2',3',5'-TRIACETATE
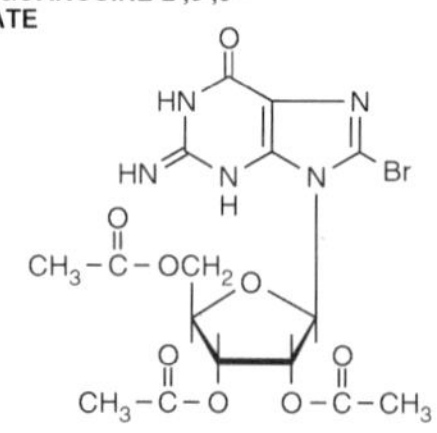
5 gm

S35,142-3
INOSINE 2',3',5'-TRIS(N-CYCLOHEXYLCARBAMATE)
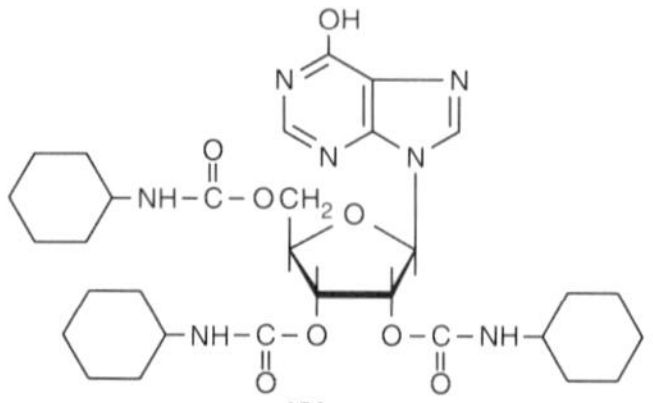
250 mg

S86,049-2
5'-O-ACETYL-2',3'-O-ISOPROPYLIDENEGUANOSINE
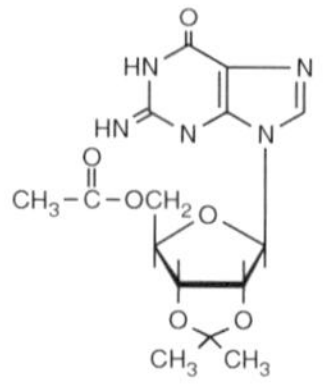
250 mg

S78,394-3
2,8-DIAMINO-6-MERCAPTOPURINE
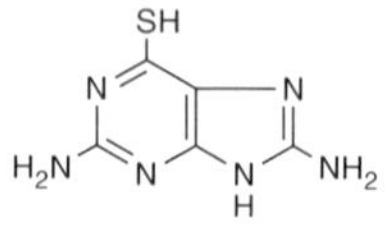
250 mg

S78,344-7
6-AMINO-2-MERCAPTOPURINE SULFATE
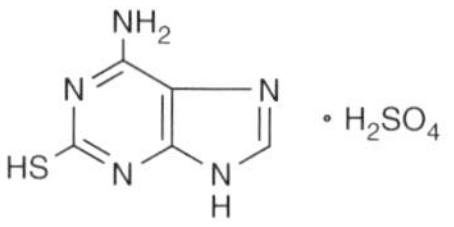
5 gm

S78,410-9
2,6-DIMERCAPTO-7-METHYLPURINE
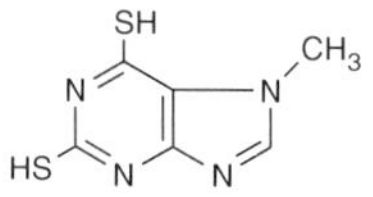
250 mg

S78,508-3
6-MERCAPTO-9-METHYLPURINE
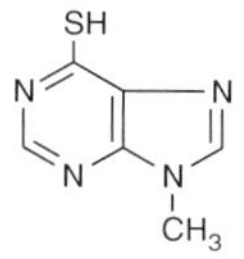
250 mg

S77,723-4
7-METHYL-2,6,8-TRIMERCAPTOPURINE
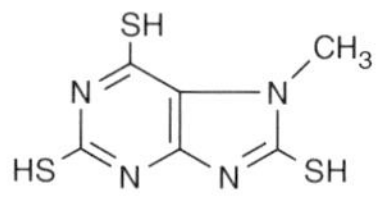
250 mg

S78,574-1
2-AMINO-9-BUTYL-6-MERCAPTOPURINE
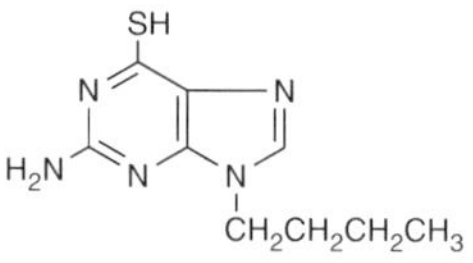
250 mg

S78,274-2
6,8-DIMERCAPTOPURINE
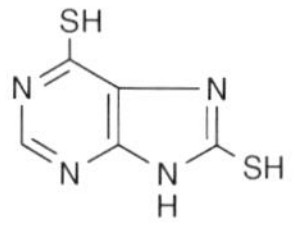
5 gm

S80,321-9
1,3-DIMETHYL-8-MERCAPTOXANTHINE
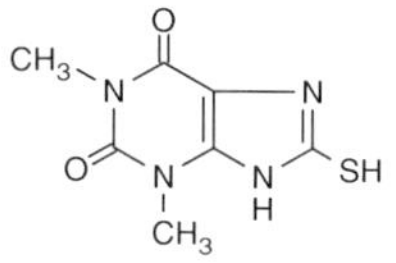
250 mg

S80,395-2
3-ISOBUTYL-8-MERCAPTO-1-METHYLXANTHINE
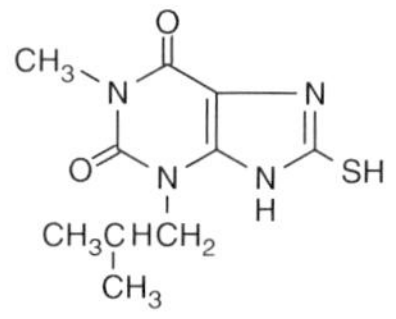
250 mg

S77,773-0
2-(METHYLTHIO)-6-THIOPURINE
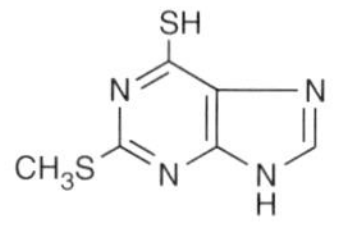
250 mg

S45,907-0
2,6-DIMERCAPTOPURINE
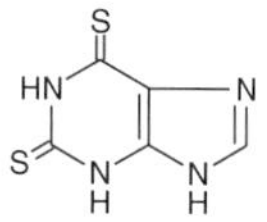
250 mg

S42,736-5
2,6,8-TRIMERCAPTOPURINE
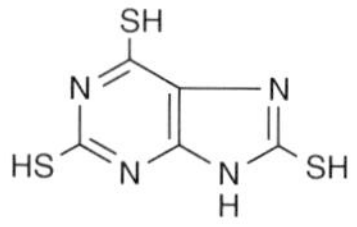
250 mg

S34,089-8
2-AMINO-6-BENZYLTHIOPURINE
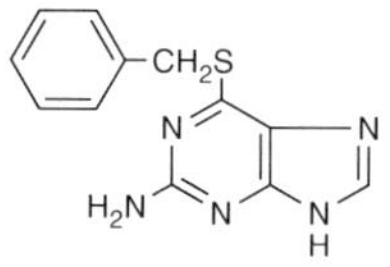
250 mg

S45,182-7
6-(4-CARBOXYBUTYLTHIO)PURINE
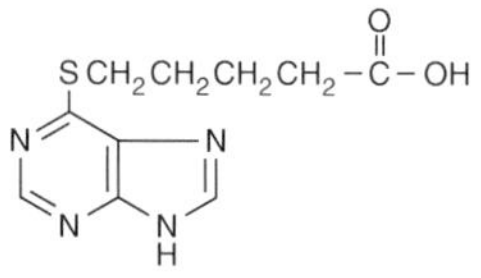
5 gm

S35,058-3
6-(HEXYLTHIO)PURINE
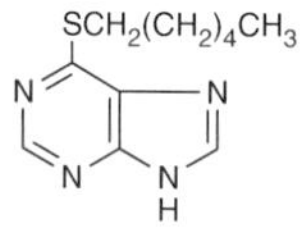
250 mg

S65,953-3
6-(CARBOXYMETHYLTHIO)PURINE HYDRAZIDE
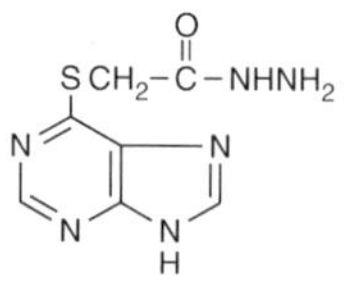
250 mg

S34,928-3
6-ETHYLTHIOPURINE
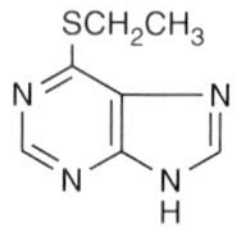
250 mg

S35,503-8
6-PROPYLMERCAPTOPURINE
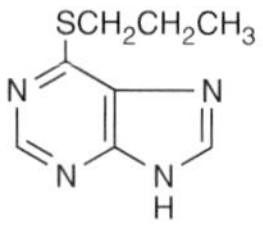
250 mg

S35,430-9
6-(PENTYLTHIO)PURINE
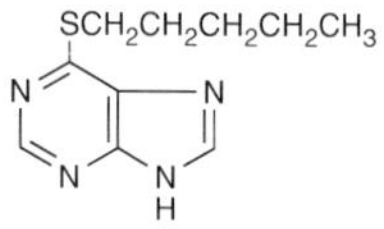
250 mg

S35,413-9
6-(OCTYLTHIO)PURINE
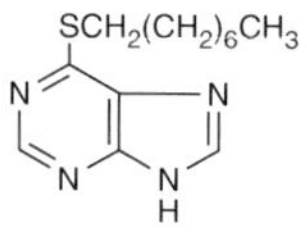
250 mg

S88,245-3
ETHYL 5-METHYL-2-(METHYLTHIO)-(1,2,4)TRIAZOLO(1,5-A)PYRIMIDINE-7-CARBOXYLATE
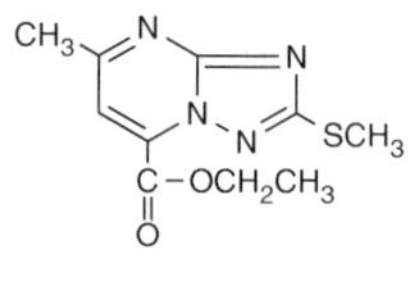
250 mg

S37,317-6
6-(BUTYLTHIO)PURINE

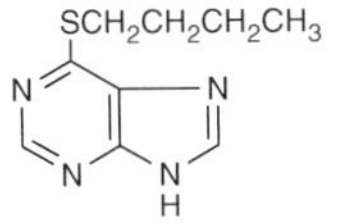

5 gm

S88,251-8
5,7-DIMETHYL-2-(METHYLTHIO)-1,2,4-TRIAZOLO(1,5-A)PYRIMIDINE

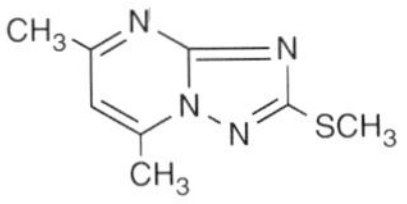

250 mg

S92,791-0
3-BENZYL-2,6,8-TRIS(METHYLTHIO)-PURINE

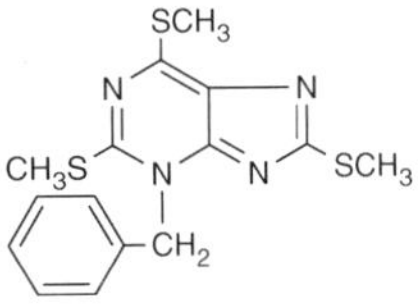

250 mg

S78,895-3
2,8-DI(METHYLTHIO)PURINE

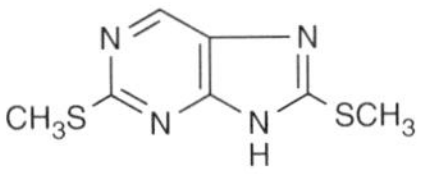

250 mg

S92,792-9
2,6,8-TRIS(METHYLTHIO)PURINE

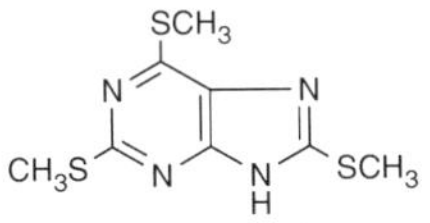

250 mg

S35,041-9
6-HEPTYLMERCAPTOPURINE

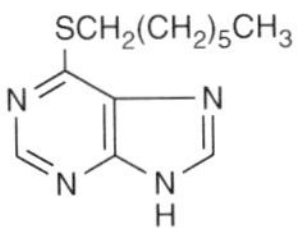

5 gm

S75,669-5
6-(HEPTYLTHIO)PURINE

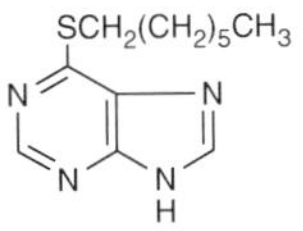

5 gm

S78,135-5
2,6-DIAMINO-7-ETHYL-7H-PURINE

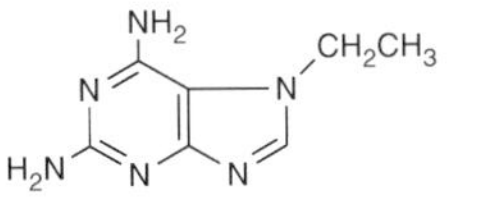

250 mg

S93,178-0
2,6-DIAMINO-8-PHENYLPURINE

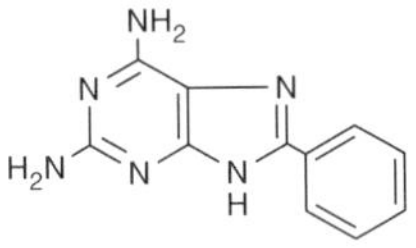

250 mg

S35,784-7
4-AMINO-1-PHENYLPYRAZOLO(3,4-D)-PYRIMIDINE

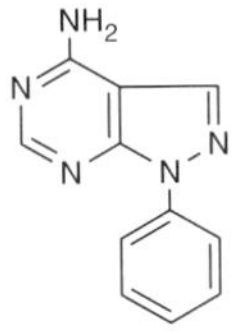

250 mg

S78,970-4
9-BENZYLGUANINE

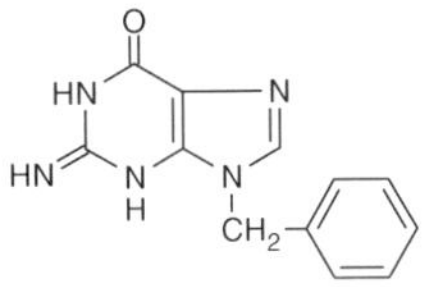

250 mg

S78,368-4
2-AMINO-6-(DIMETHYLAMINO)PURINE

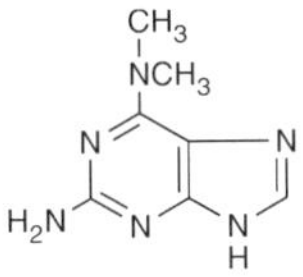

250 mg

S78,893-7
6-AMINO-2-(DIMETHYLAMINO)PURINE

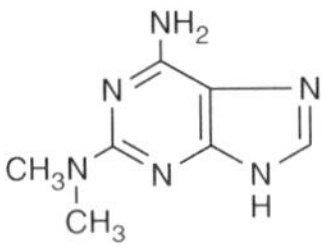

250 mg

S80,324-3
8-AMINOTHEOPHYLLINE

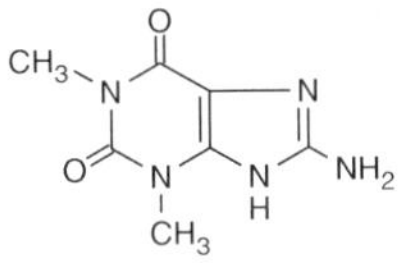

250 mg

S80,403-7
8-AMINOTHEOPHYLLINE HYDROCHLORIDE

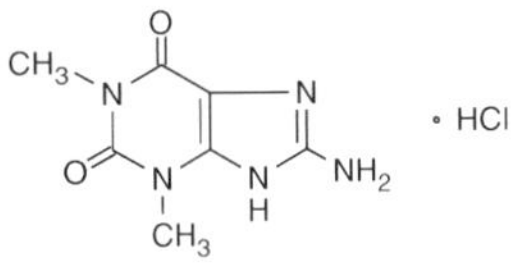

250 mg

S45,101-0
ADENINE PHOSPHATE

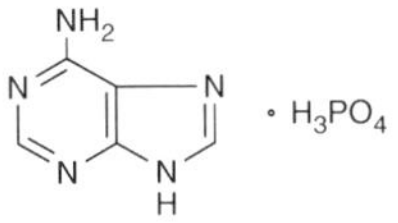

5 gm

S45,162-2
TRIACANTHINE

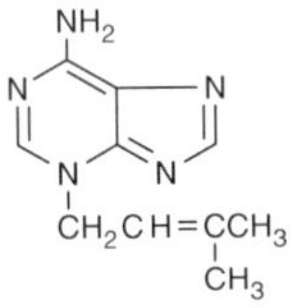

250 mg

S46,017-6
N6-(DELTA-2-ISOPENTENYL)ADENINE

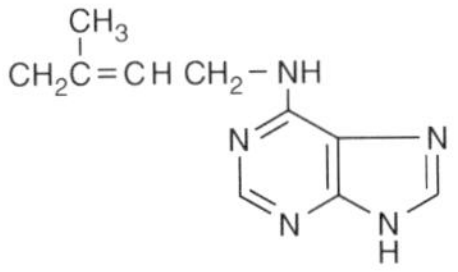

250 mg

S74,967-2
9-METHYL-2-(METHYLAMINO)-6(1H)-PURINONE

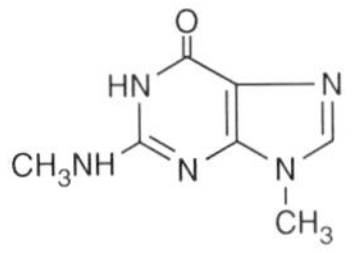

250 mg

S92,794-5
2-(METHYLTHIO)HYPOXANTHINE

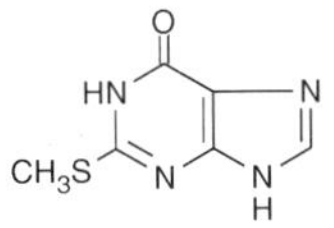

250 mg

S44,879-6
2-HYDROXYPURINE MONOHYDRATE

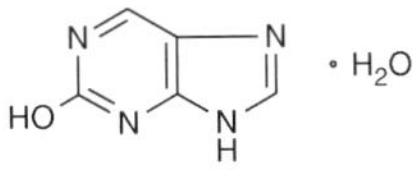

250 mg

S86,476-5
1,7-DIETHYL-3-METHYLXANTHINE

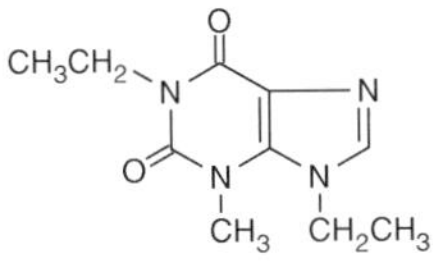

250 mg

S86,474-9
1,3-DIMETHYL-7-ETHYLXANTHINE

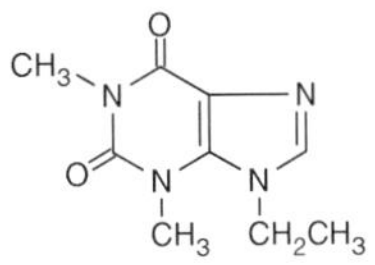

250 mg

S85,203-1
3,7-DIMETHYL-1-ETHYLXANTHINE

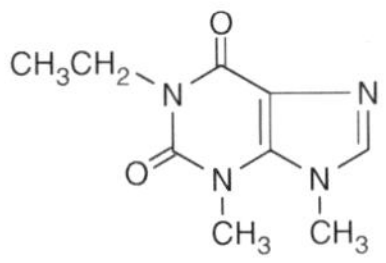

250 mg

S78,198-3
6-(DIMETHYLAMINO)-9-ETHYLPURINE

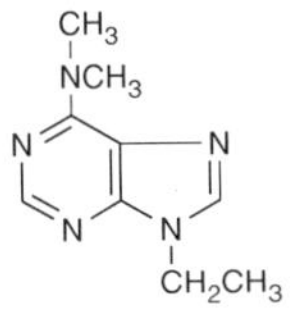

250 mg

S80,341-3
8-HYDRAZINOCAFFEINE

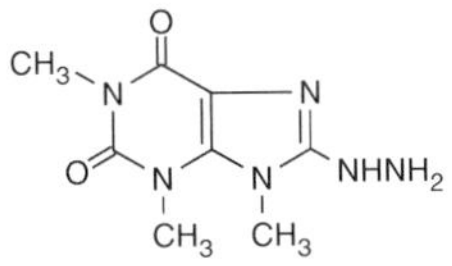

250 mg

S34,503-2
6-CYCLOHEXYLUREIDOPURINE

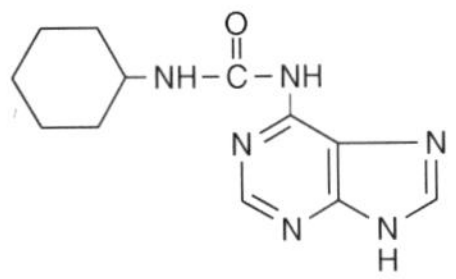

250 mg

S80,383-9
1-ETHYL-3-ISOBUTYLXANTHINE

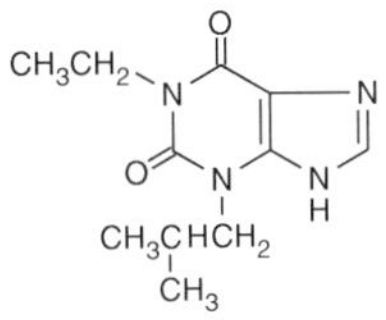

250 mg

S44,973-3
8-BROMOGUANINE HYDROBROMIDE

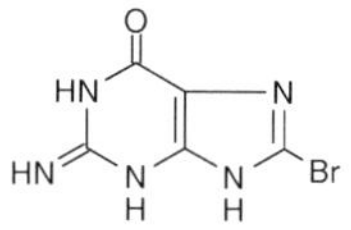

250 mg

S78,894-5
8-NITROTHEOPHYLLINE

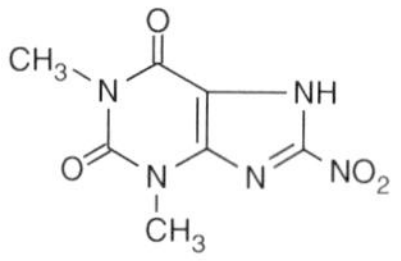

250 mg

S80,332-4
8-CHLOROCAFFEINE

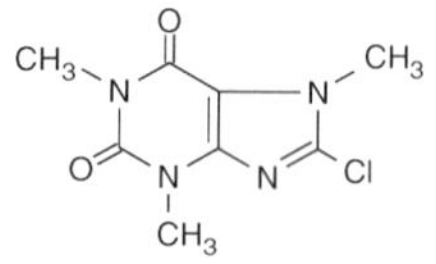

250 mg

S80,415-0
8-BROMOCAFFEINE

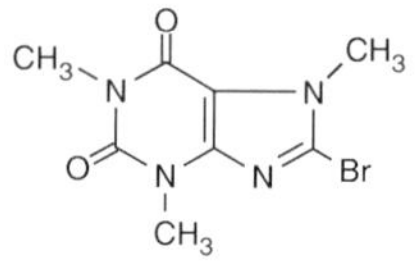

250 mg

S80,269-7
3-ETHYL-1-PROPYLXANTHINE

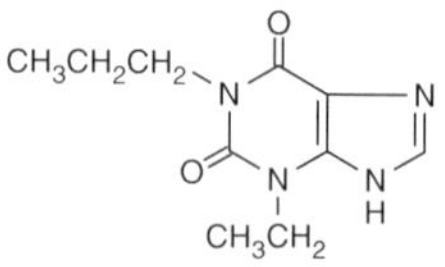

250 mg

S80,369-3
1,3-DIETHYL-7-METHYLXANTHINE

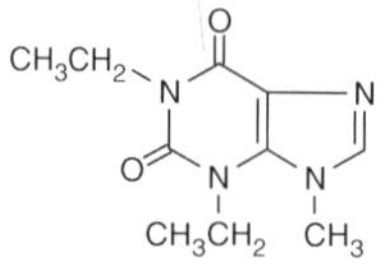

250 mg

S80,456-8
3-ALLYL-8-CHLORO-1-ETHYLXANTHINE

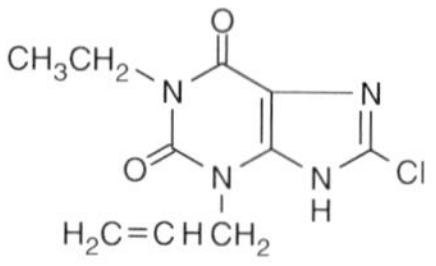

250 mg

S78,153-3
9-CYCLOPENTYLHYPOXANTHINE

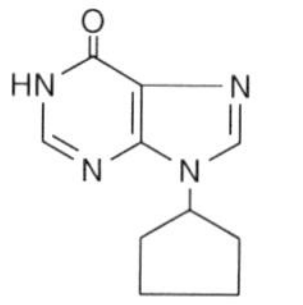

250 mg

S77,633-5
3-METHYL-8-AZAGUANINE

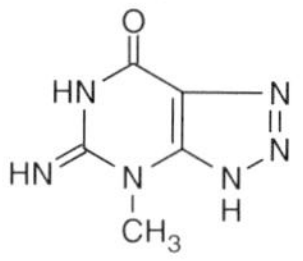

250 mg

S45,297-1
GUANINE HYDROCHLORIDE

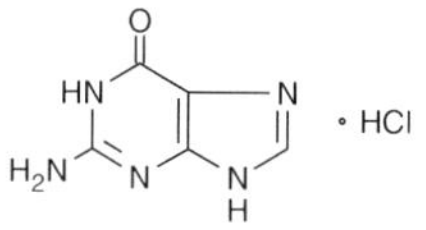

5 gm

S89,912-7
8-BROMO-1,3-DIETHYLXANTHINE

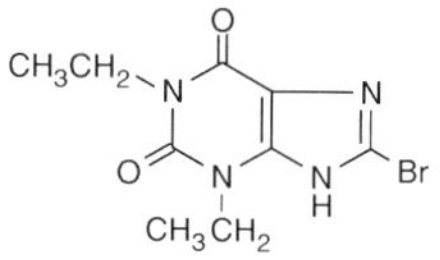

250 mg

S89,911-9
1,3-DIETHYL-8-IODOXANTHINE

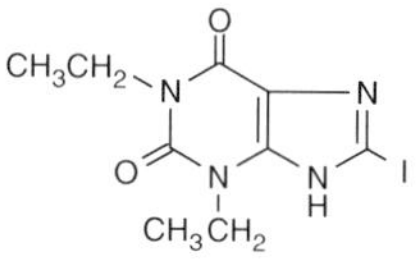

250 mg

S42,394-7
CAFFEINE NICOTINATE

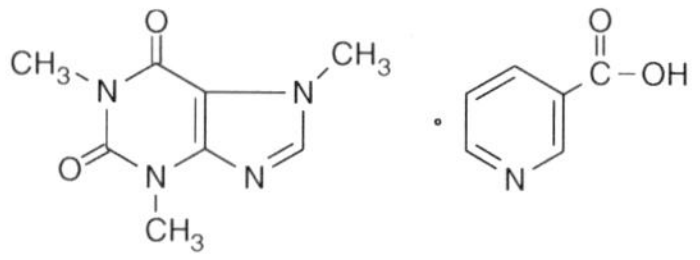

5 gm

S89,914-3
8-IODOTHEOPHYLLINE

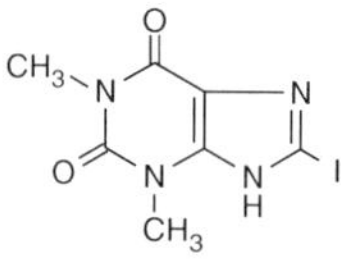

250 mg

S87,031-5
8-(2,4-DICHLOROPHENYLAZO)-1,3-DIETHYLXANTHINE

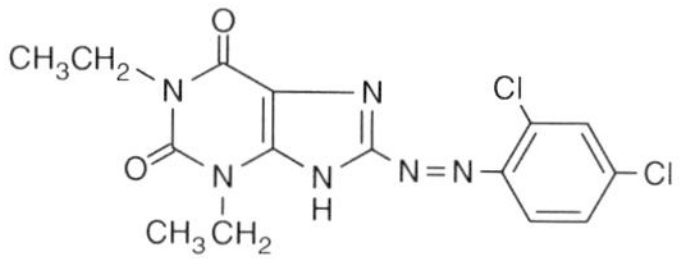

250 mg

S78,565-2
8-BROMOGUANINE

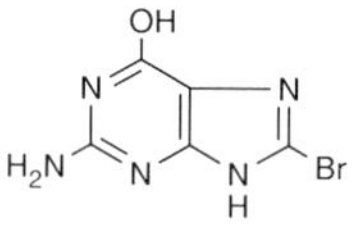

250 mg

S44,628-9
2-ACETAMIDO-6-HYDROXYPURINE HEMIHYDRATE

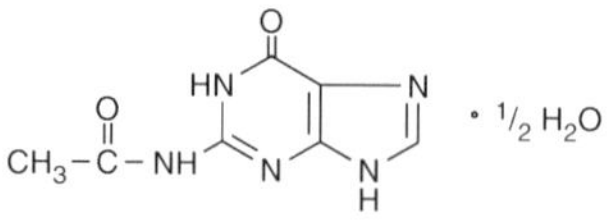

250 mg

S78,122-3
2-AMINO-1,7-DIHYDRO-1-METHYL-6H-PURINE-6-THIONE

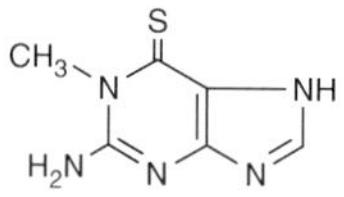

250 mg

S77,750-1
6,8-DICHLORO-2-(METHYLTHIO)PURINE

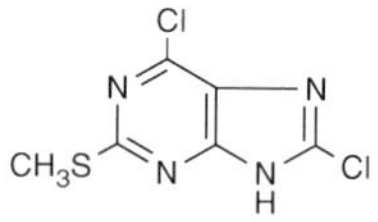

250 mg

S40,451-9
6-PIPERIDINOPURINE

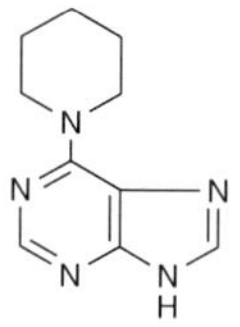

100 mg

S44,891-5
GUANINE SULFATE

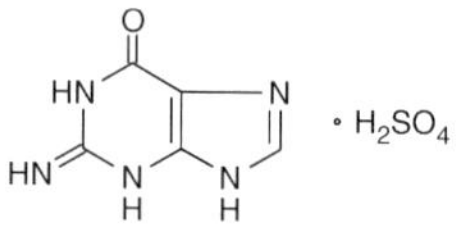

5 gm

S46,255-1
2,6,8-TRICHLOROPURINE, AMMONIUM SALT

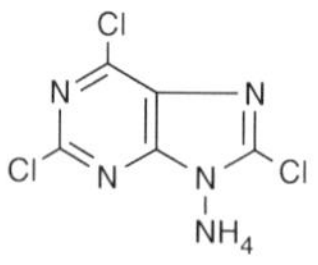

5 gm

S39,908-6
2-(2-CHLOROACETAMIDO)PYRIDINE

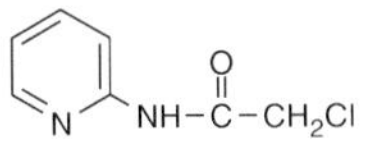

250 mg

S83,167-0
4-(CHLOROACETYL)-2,3,5,6-TETRACHLOROPYRIDINE

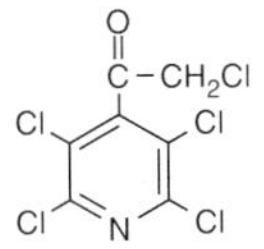

250 mg

S82,194-2
2-(3-CHLOROPROPYL)PYRIDINE

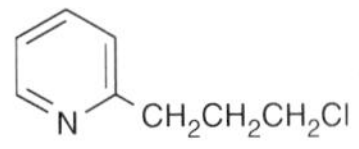

250 mg

S91,210-7
1-[4-CHLORO-3-(TRIFLUOROMETHYL)-PHENYL]-3-(3-HYDROXY-2-PYRIDYL)UREA

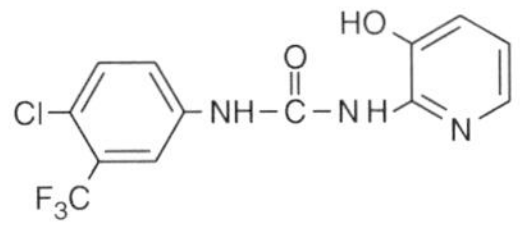

250 mg

S91,253-0
1-[2-CHLORO-5-(TRIFLUOROMETHYL)-PHENYL]-3-(3-PYRIDYL)UREA

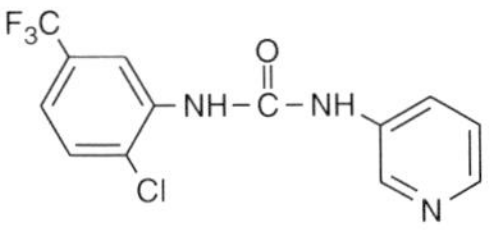

250 mg

S68,348-5
1-[4-CHLORO-2-(TRIFLUOROMETHYL)-PHENYL]-3-(3-PYRIDYLMETHYL)UREA

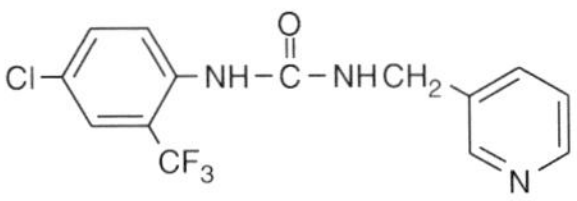

250 mg

S72,157-3
1-[4-CHLORO-3-(TRIFLUOROMETHYL)-PHENYL]-3-(5-NITRO-2-PYRIDYL)UREA

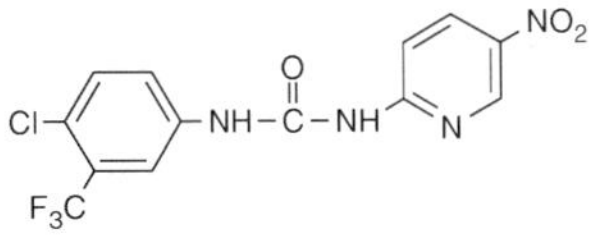

250 mg

S91,208-5
1-[4-CHLORO-2-(TRIFLUOROMETHYL)-PHENYL]-3-(5-NITRO-2-PYRIDYL)UREA

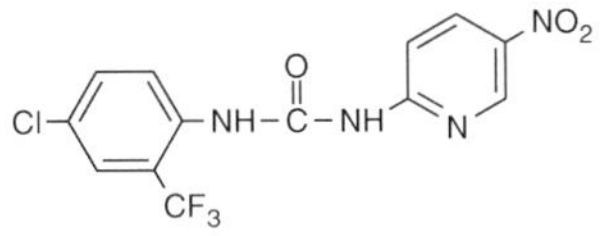

250 mg

S91,254-9
1-[2-CHLORO-5-(TRIFLUOROMETHYL)-PHENYL]-3-(5-NITRO-2-PYRIDYL)UREA

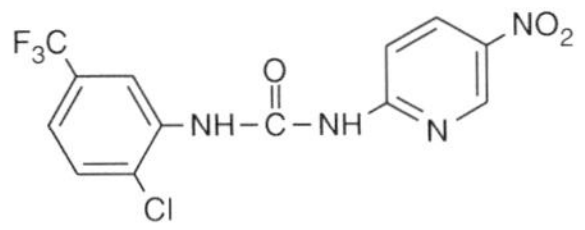

250 mg

S67,443-5
2-PYRIDYL N-[4-BROMO-2-(TRIFLUOROMETHYL)PHENYL]CARBAMATE

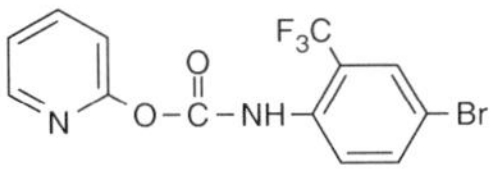

250 mg

S68,837-1
2-(2-PYRIDYL)ETHYL N-[4-CHLORO-2-(TRIFLUOROMETHYL)PHENYL]CARBAMATE

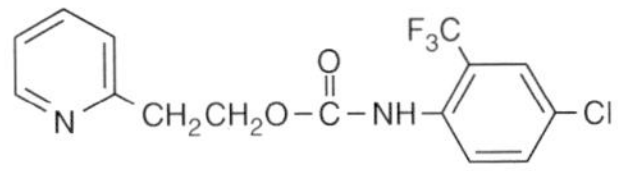

250 mg

S76,905-3
4-PYRIDYL N-[4-CHLORO-2-(TRIFLUOROMETHYL)PHENYL]CARBAMATE

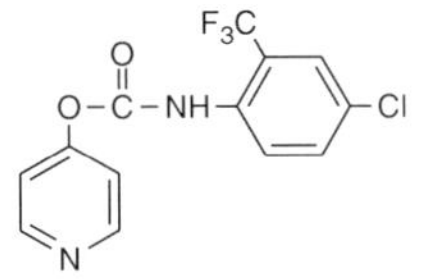

250 mg

S72,358-4
2-(5-METHYL-2-PYRIDYLUREIDO)ACETIC N2-[N-(4-CHLORO-3-TRIFLUOROMETHYL-PHENYL)CARBAMOYL]HYDRAZIDE

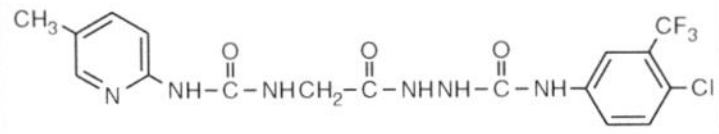

5 gm

S91,206-9
1-(5-CHLORO-2-PYRIDYL)-3-[4-CHLORO-2-(TRIFLUOROMETHYL)PHENYL]UREA

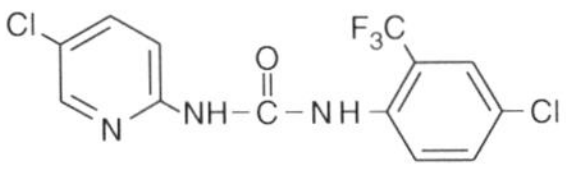

250 mg

S65,257-1
1-(2-PYRIDYL)-3,3,3-TRICHLORO-2-PROPANOL

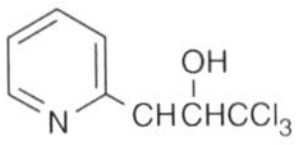

250 mg

S95,471-3
1-(6-METHYL-2-PYRIDYL)-3,3,3-TRICHLORO-2-PROPANOL

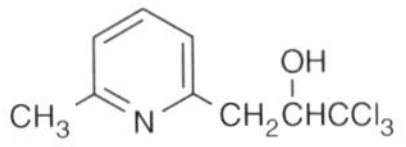

50 mg

S94,519-6
4-AMINO-3,5-DICHLORO-2-(TRICHLOROMETHYL)PYRIDINE

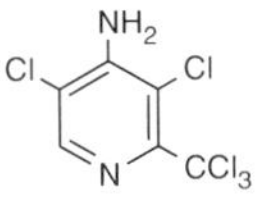

250 mg

S93,368-6
2,2,2-TRIFLUOROACETIC N2-(3-NITRO-2-PYRIDYL)HYDRAZIDE

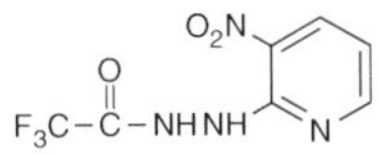

250 mg

S69,843-1
N-(3-PYRIDYL)-2,2,2-TRICHLOROACETAMIDE

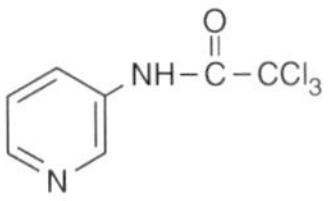

5 gm

S65,105-2
2-[2-(2-CHLORO-4-TRIFLUOROMETHYLANILINO)ETHYL]-PYRIDINE

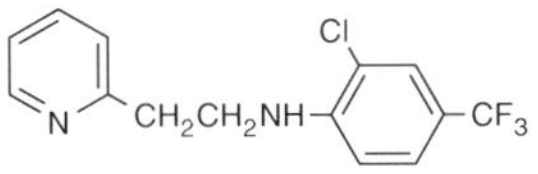

5 gm

S65,130-3
2-(2-[3,5-BIS(TRIFLUOROMETHYL)-ANILINO]ETHYL)PYRIDINE

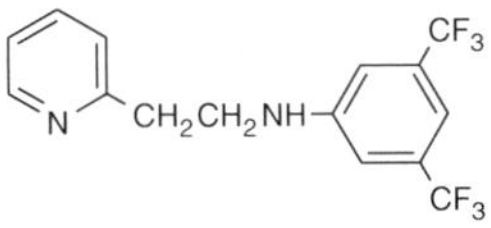

250 mg

S65,095-1
2-(2-[3-(TRIFLUOROMETHYL)ANILINO]-ETHYL)PYRIDINE

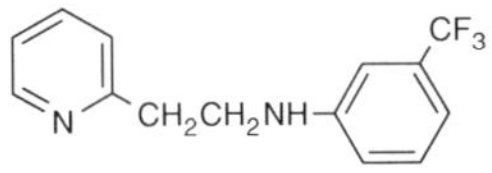

5 gm

S61,611-7
1-(3-PYRIDYLMETHYL)-3-[3-(TRIFLUOROMETHYL)PHENYL]UREA

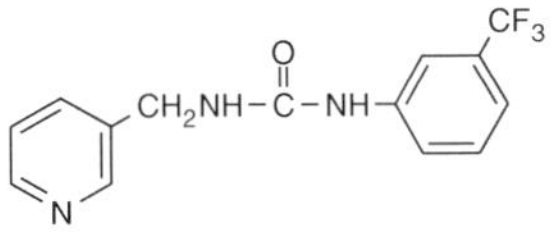

5 gm

S90,515-1
1-(4-PYRIDYL)-3-[2-(TRIFLUOROMETHYL)PHENYL]UREA

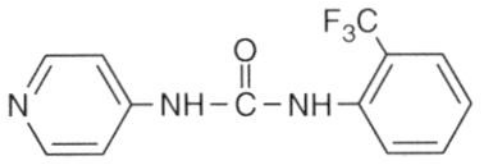

250 mg

S66,625-4
1-(2-CHLORO-5-TRIFLUOROMETHYLPHENYL)-3-(3-PYRIDYLMETHYL)UREA

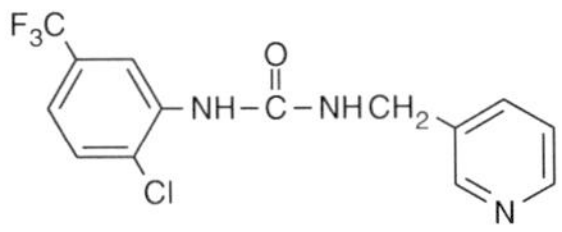

5 gm

S90,443-0
1-(4-PYRIDYLCARBONYL)-4-[2-(TRIFLUOROMETHYL)PHENYL]-SEMICARBAZIDE

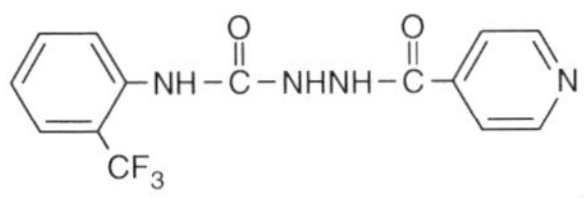

250 mg

S65,016-1
1-(4-PYRIDYLMETHYL)-3-[3-(TRIFLUOROMETHYL)PHENYL]UREA

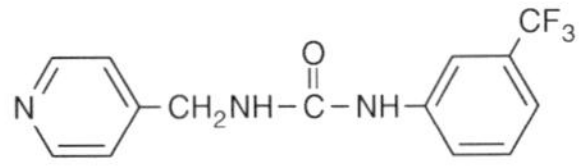

5 gm

S89,848-1
2-[3-(TRIFLUOROMETHYL)ANILINO]-PYRIDINE

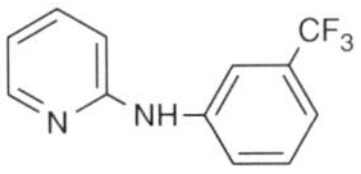

250 mg

S77,354-9
1-(5-METHYL-2-PYRIDYLUREIDOACETYL)-4-[3-(TRIFLUOROMETHYL)PHENYL]-SEMICARBAZIDE

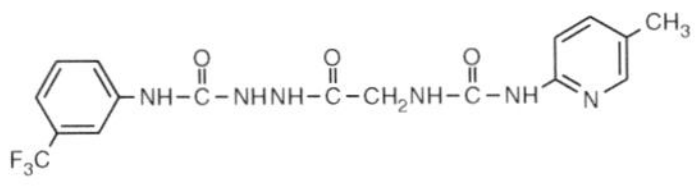

250 mg

S84,772-0
4-(3-[3-(TRIFLUOROMETHYL)PHENYL]-UREIDO)BENZOIC (4-PYRIDYLMETHYLENE)-HYDRAZIDE

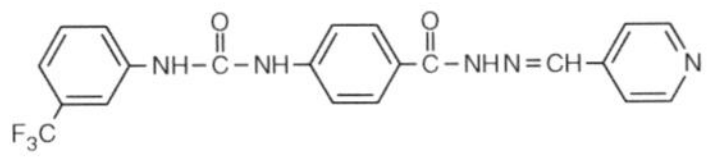

250 mg

S69,670-6
N-NITROSO-N-(2-PYRIDYL)-3-(TRIFLUOROMETHYL)ANILINE

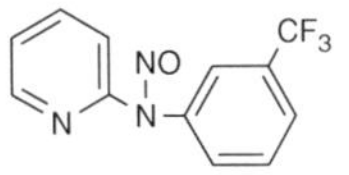

5 gm

S71,706-1
3-PYRIDYL N-(3-TRIFLUOROMETHYLPHENYL)-CARBAMATE

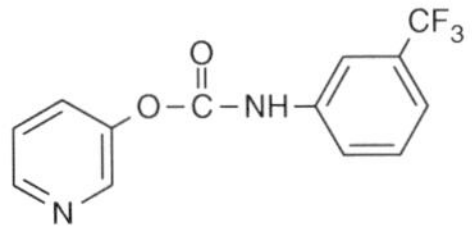

5 gm

S69,630-7
N-(3-METHYL-2-PYRIDYL)-2,2,2-TRICHLOROACETAMIDE

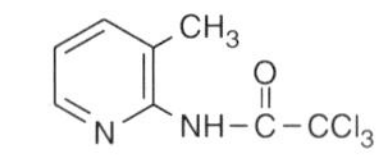

250 mg

S69,631-5
N-(4-METHYL-2-PYRIDYL)-2,2,2-TRICHLOROACETAMIDE

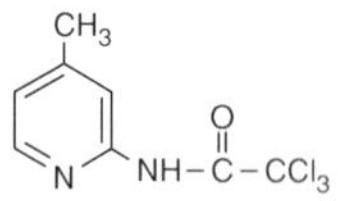

250 mg

S69,629-3
N-(2-PYRIDYL)-2,2,2-TRICHLOROACETAMIDE

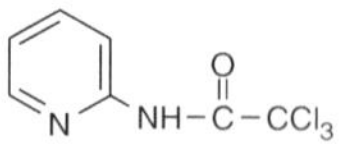

250 mg

S71,977-3
N-(3-PYRIDYLMETHYL)-2,2,2-TRICHLOROACETAMIDE

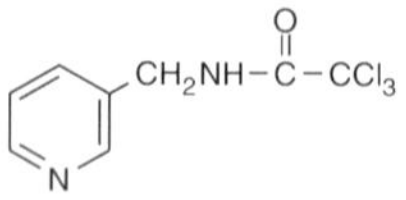

5 gm

S69,551-3
1-(2-CHLORO-3-PYRIDYL)-3-[2-(TRIFLUOROMETHYL)PHENYL]UREA

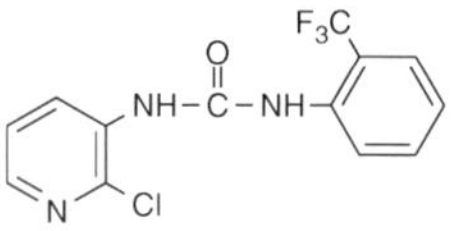

250 mg

S91,154-2
1-(5-CHLORO-2-PYRIDYL)-3-[2-(TRIFLUOROMETHYL)PHENYL]UREA

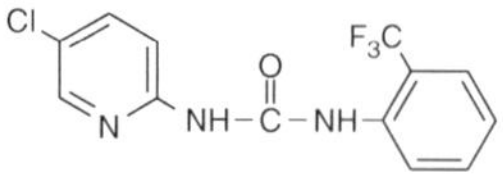

250 mg

S77,271-2
1-(5-CHLORO-2-PYRIDYL)-3-[3-(TRIFLUOROMETHYL)PHENYL]UREA

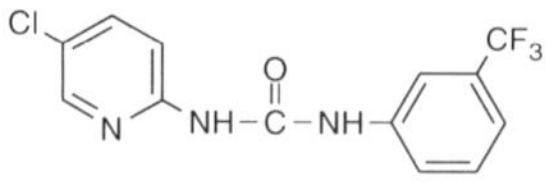

250 mg

S80,113-5
1-(4,6-DIMETHYL-2-PYRIDYL)-3-[3-(TRIFLUOROMETHYL)PHENYL]UREA

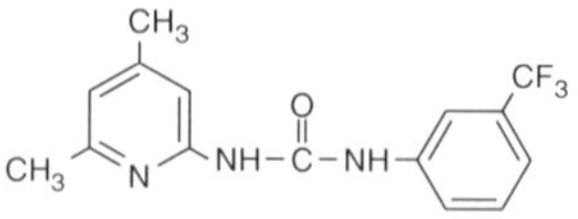

250 mg

S76,206-7
2-CHLORO-3-CYANO-5-(TRICHLOROMETHYL)PYRIDINE

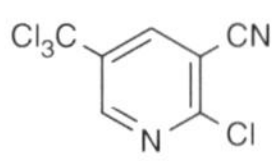

250 mg

S94,527-7
2-CHLORO-5-(TRICHLOROMETHYL)-PYRIDINE

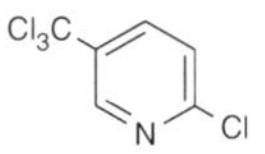

250 mg

S94,518-8
3,4,5-TRICHLORO-2-(TRICHLOROMETHYL)-PYRIDINE

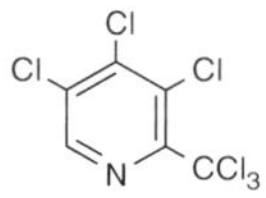

250 mg

S94,526-9
2-CHLORO-3-(TRICHLOROMETHYL)-PYRIDINE

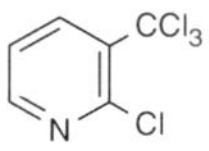

250 mg

S94,516-1
2-(TRICHLOROMETHYL)PYRIDINE HYDROCHLORIDE

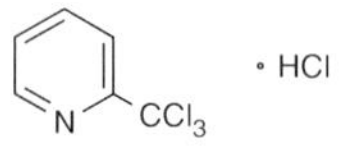

250 mg

S91,153-4
1-(5-CHLORO-2,4-DIMETHOXYPHENYL)-3-(3-HYDROXY-2-PYRIDYL)UREA

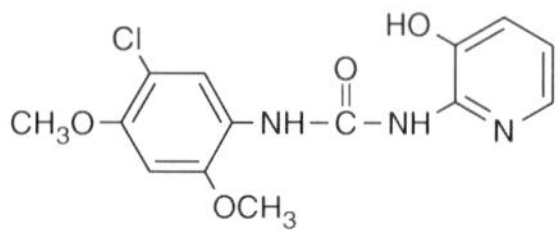

250 mg

S85,533-2
4-PYRIDINECARBOXYLIC (3-BROMO-4-HYDROXY-5-METHOXYBENZYLIDENE)-HYDRAZIDE

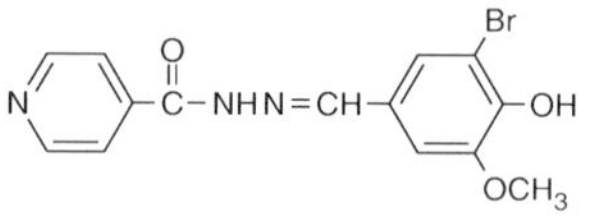

250 mg

S76,222-9
ALPHA-(4-CHLOROPHENYL)-ALPHA-METHYL-2-PYRIDINEMETHANOL

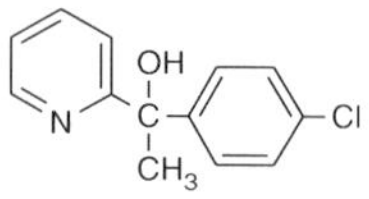

250 mg

S91,101-1
1-(2,3-DICHLOROPHENYL)-3-(3-HYDROXY-2-PYRIDYL)UREA

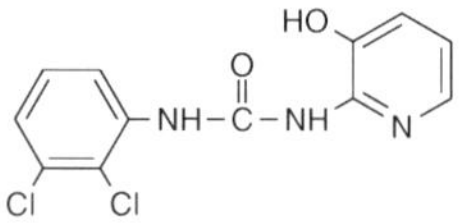

250 mg

S77,284-4
1-(4-FLUORO-3-NITROPHENYL)-3-(3-HYDROXY-2-PYRIDYL)UREA

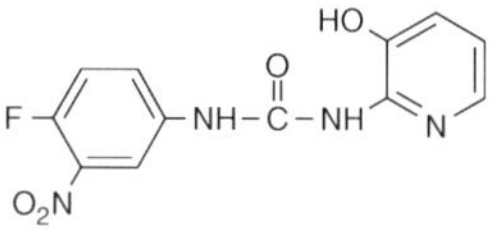

250 mg

S92,101-7
2,6-DIPHENYL-4-(4-FLUOROPHENYL)-PYRYLIUM TETRAFLUOROBORATE

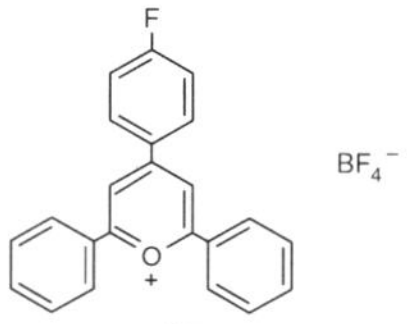

250 mg

S92,103-3
2,4,6-TRIS(4-FLUOROPHENYL)PYRYLIUM TETRAFLUOROBORATE

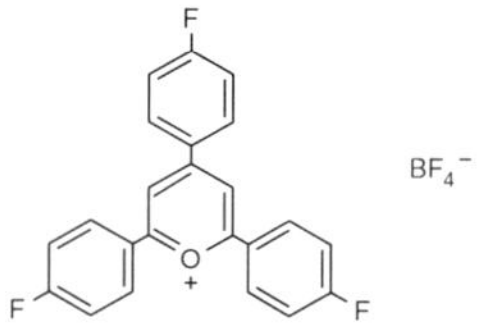

250 mg

S70,091-6
7-(4-CHLOROPHENYL)-5,6,8,9-TETRAHYDRODIBENZO(C,H)XANTHYLIUM TETRAFLUOROBORATE

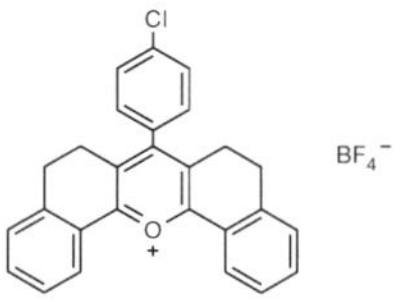

5 gm

S70,144-0
7-(2,4-DICHLOROPHENYL)-5,6,8,9-TETRAHYDRODIBENZO(C,H)XANTHYLIUM PERCHLORATE

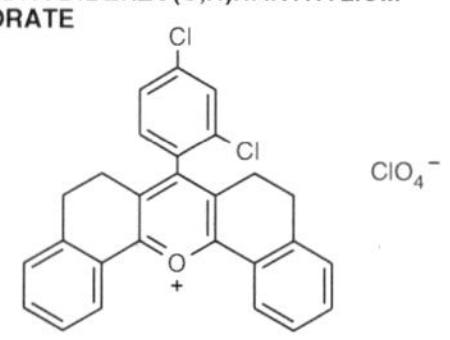

5 gm

S71,418-6
1-(4-FLUOROBENZYL)-3-(4-FLUOROBENZYLOXY)PYRIDINIUM CHLORIDE

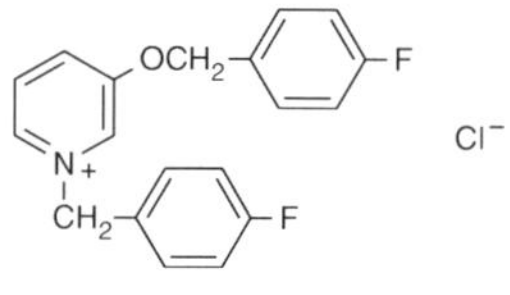

250 mg

S76,982-7
1-(5-CHLORO-2,4-DIMETHOXYPHENYL)-3-(3-PYRIDYL)UREA

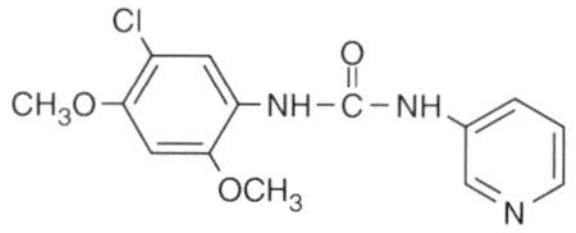

250 mg

S76,867-7
4-PYRIDYL N-[5-CHLORO-2,4-(DIMETHOXY)PHENYL]CARBAMATE

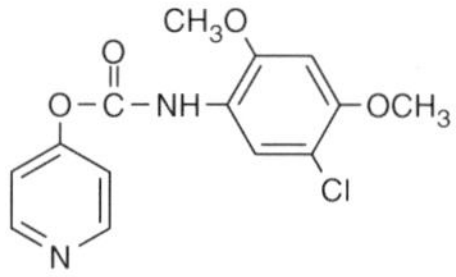

250 mg

S69,344-8
4-PYRIDYLMETHYL N-(5-CHLORO-2,4-DIMETHOXYPHENYL)CARBAMATE

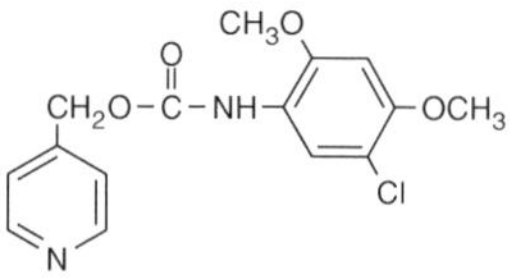

250 mg

S72,992-2
4-(5-CHLORO-2,4-DIMETHOXYPHENYL)-1-(3-PYRIDYLCARBONYL)SEMICARBAZIDE

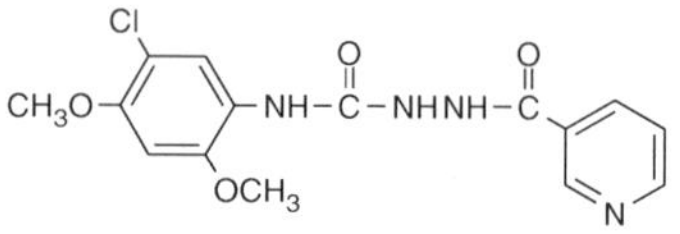

5 gm

S77,268-2
2-PYRIDYL N-(5-CHLORO-2,4-DIMETHOXYPHENYL)CARBAMATE

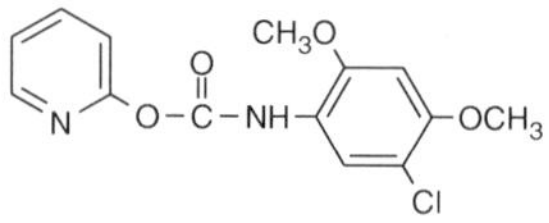

250 mg

S91,151-8
1-(5-CHLORO-2,4-DIMETHOXYPHENYL)-3-(5-CHLORO-2-PYRIDYL)UREA

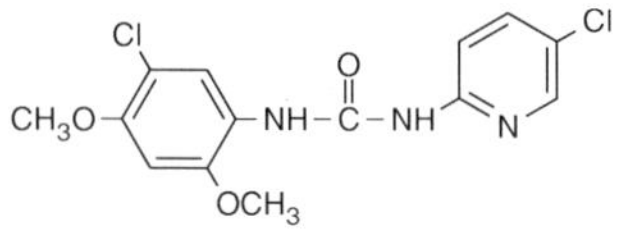

250 mg

S91,175-5
1-(5-CHLORO-2-METHOXYPHENYL)-3-(2-PYRIDYLMETHYL)UREA

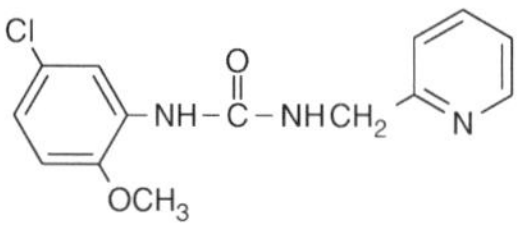

250 mg

S76,988-6
1-(5-CHLORO-2-METHOXYPHENYL)-3-(3-PYRIDYL)UREA

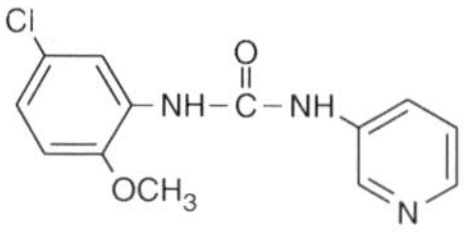

250 mg

S69,305-7
4-(3-CHLORO-6-METHOXYPHENYL)-1-(4-PYRIDYLCARBONYL)SEMICARBAZIDE

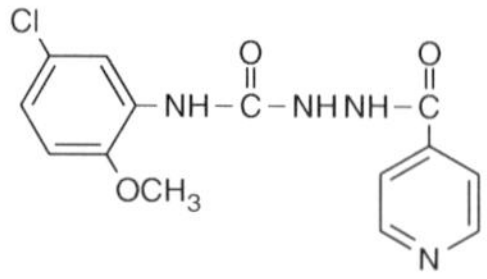

250 mg

S72,524-2
2-PYRIDYL N-(5-CHLORO-2-METHOXYPHENYL)CARBAMATE

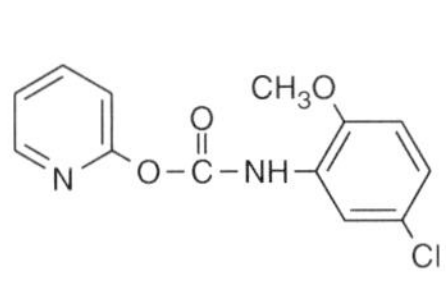

5 gm

S72,538-2
2-PYRIDYL N-(4-CHLORO-2-METHOXYPHENYL)CARBAMATE

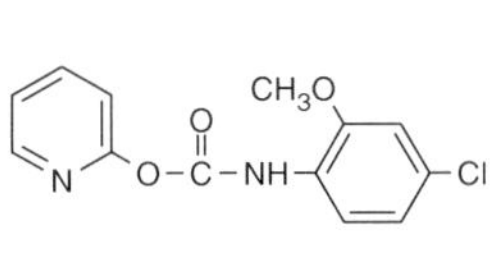

5 gm

S69,125-9
2-PYRIDYLMETHYL N-(5-CHLORO-2-METHOXYPHENYL)CARBAMATE

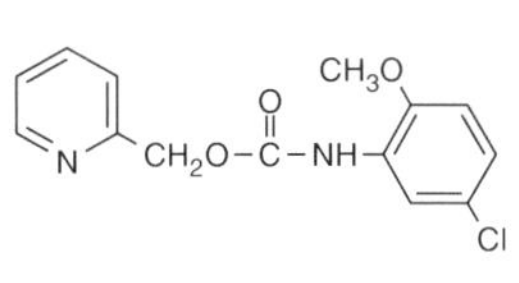

250 mg

S69,335-9
3-PYRIDYLMETHYL N-(5-CHLORO-2-METHOXYPHENYL)CARBAMATE

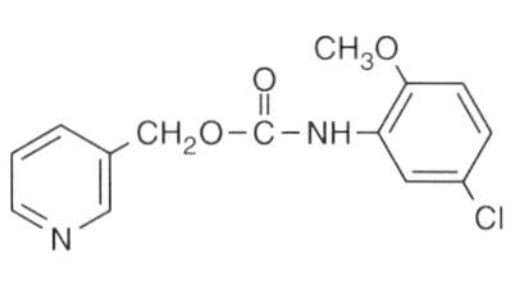

250 mg

S73,826-3
4-PYRIDYLMETHYL N-(5-CHLORO-2-METHOXYPHENYL)CARBAMATE

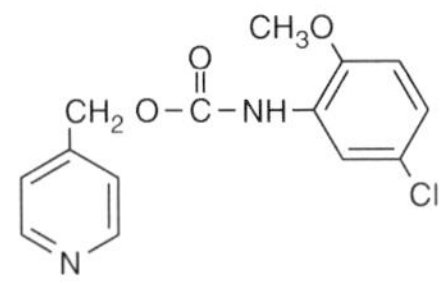

5 gm

S77,204-6
4-(5-CHLORO-2-METHOXYPHENYL)-1-(2-PYRIDYL)SEMICARBAZIDE

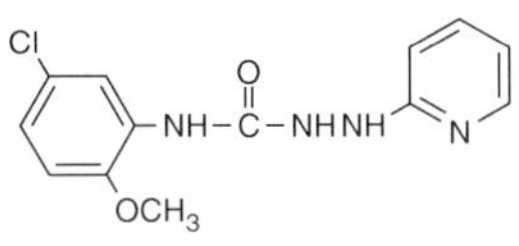

250 mg

S91,134-8
1-(4-CHLORO-2-METHOXYPHENYL)-3-(3,5-DICHLORO-2-PYRIDYL)UREA

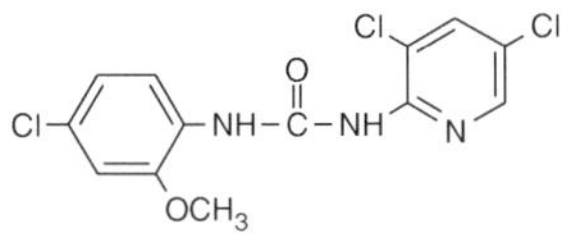

250 mg

S91,177-1
1-(5-CHLORO-2-METHOXYPHENYL)-3-(5-CHLORO-2-PYRIDYL)UREA

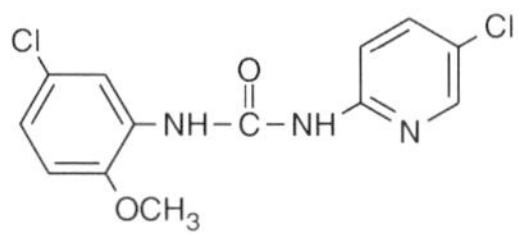

250 mg

S91,132-1
1-(4-CHLORO-2-METHOXYPHENYL)-3-(5-CHLORO-2-PYRIDYL)UREA

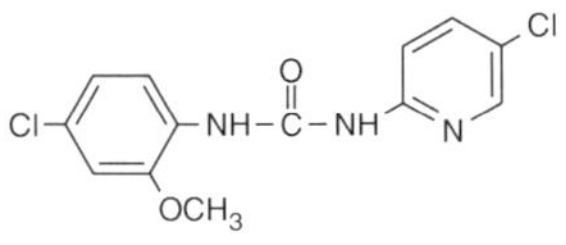

250 mg

S73,657-0
1-(5-CHLORO-2-METHOXYPHENYL)-3-(4-METHYL-2-PYRIDYL)UREA

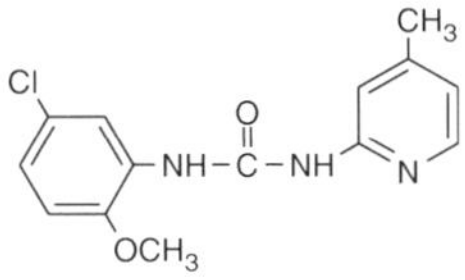

5 gm

S91,135-6
1-(4-CHLORO-2-METHOXYPHENYL)-3-(2-PYRIDYL)UREA

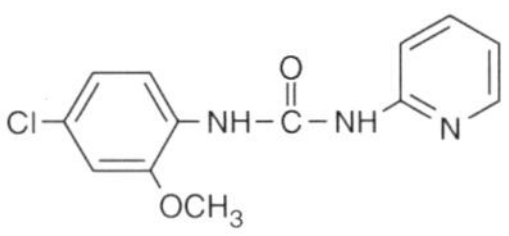

250 mg

S72,521-8
1-(4-CHLORO-2-METHOXYPHENYL)-3-(3-PYRIDYL)UREA

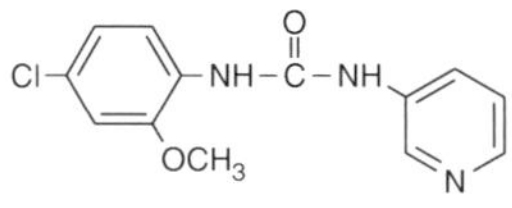

5 gm

S76,855-3
1-(4-CHLORO-2-METHOXYPHENYL)-3-(3-PYRIDYLMETHYL)UREA

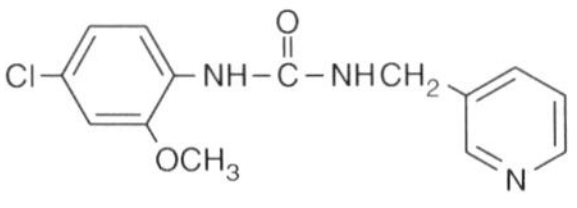

250 mg

S91,129-1
5-NITRO-2-PYRIDYL N-(4-CHLORO-2-METHOXYPHENYL)CARBAMATE

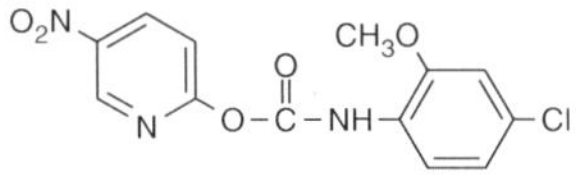

250 mg

S91,155-0
5-NITRO-2-PYRIDYL N-(5-CHLORO-2,4-DIMETHOXYPHENYL)CARBAMATE

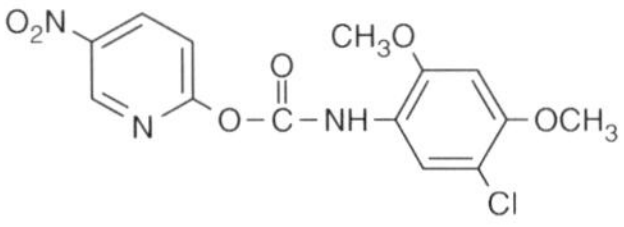

250 mg

S91,131-3
1-(4-CHLORO-2-METHOXYPHENYL)-3-(5-NITRO-2-PYRIDYL)UREA

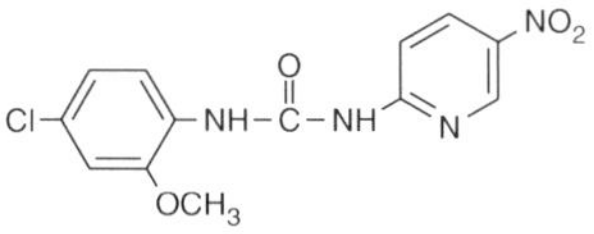

250 mg

S91,158-5
4-(5-CHLORO-2,4-DIMETHOXYPHENYL)-1-(2-PYRIDYLCARBONYL)SEMICARBAZIDE

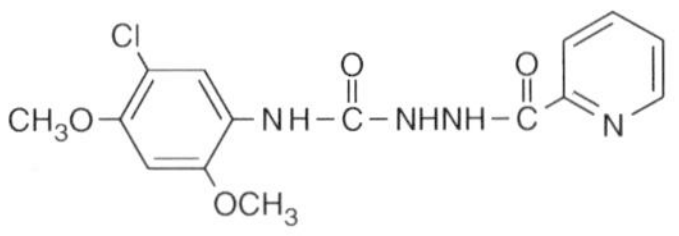

250 mg

S93,382-1
4'-CHLORO-2-(2-PYRIDYLTHIO)-ACETOPHENONE

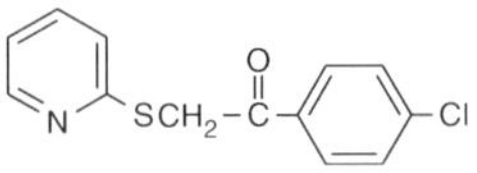

250 mg

S68,270-5
2-AMINO-5-CHLORO-N-(2-PYRIDYLMETHYL)BENZAMIDE

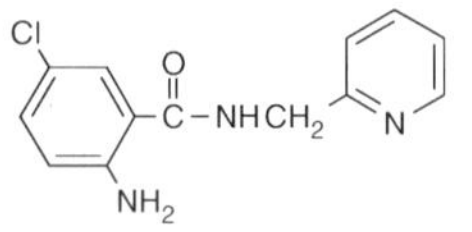

250 mg

S92,733-3
5-BROMO-2,3-DIAMINO-6-METHYLPYRIDINE

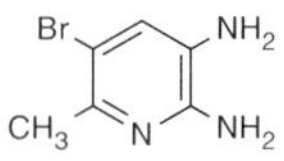

250 mg

S91,166-6
1-(2-FLUORO-5-NITROPHENYL)-3-(3-PYRIDYL)UREA

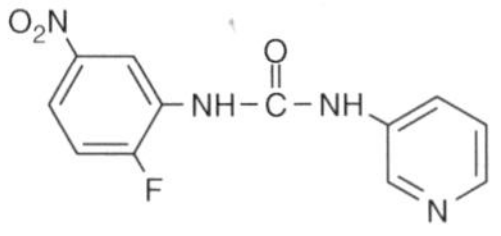

250 mg

S91,191-7
1-(4-FLUORO-3-NITROPHENYL)-3-(3-PYRIDYL)UREA

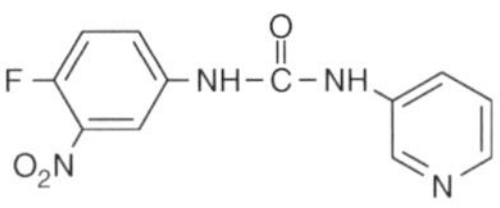

250 mg

S80,121-6
1-(2-FLUOROPHENYL)-3-(5-NITRO-2-PYRIDYL)UREA

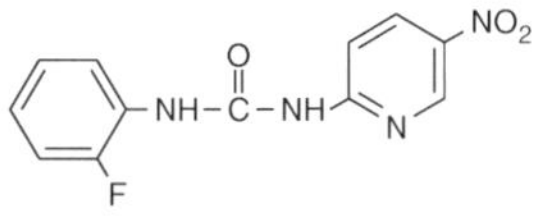

250 mg

S70,221-8
2-PYRIDYL N-(4-FLUORO-2-NITROPHENYL)CARBAMATE

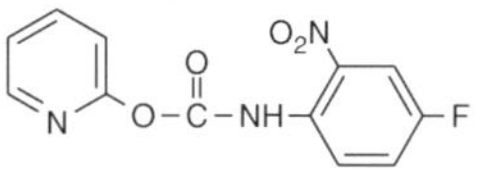

250 mg

S86,721-7
3-PYRIDYL N-(4-CHLORO-2-NITROPHENYL)CARBAMATE

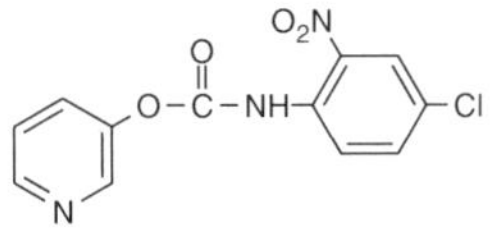

250 mg

S73,630-9
1-(4-CHLORO-2-NITROPHENYL)-3-(6-METHYL-2-PYRIDYL)UREA

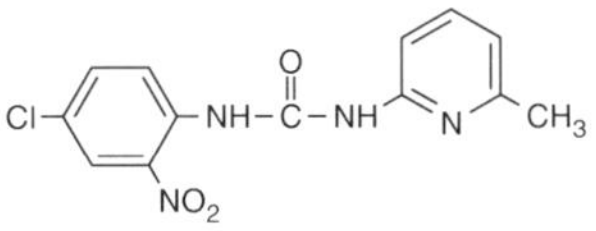

5 gm

S91,163-1
5-NITRO-2-PYRIDYL N-(5-CHLORO-2-METHYLPHENYL)CARBAMATE

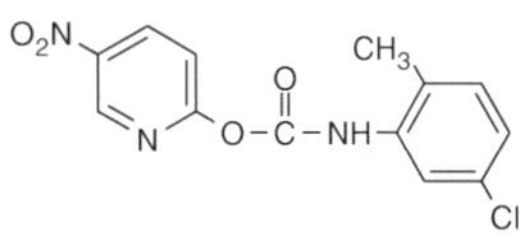

250 mg

S91,104-6
1-(2,3-DICHLOROPHENYL)-3-(5-NITRO-2-PYRIDYL)UREA

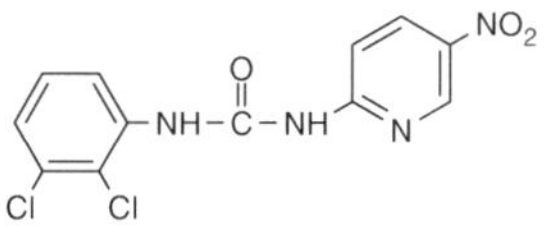

250 mg

S91,161-5
1-(5-CHLORO-2-METHYLPHENYL)-3-(5-NITRO-2-PYRIDYL)UREA

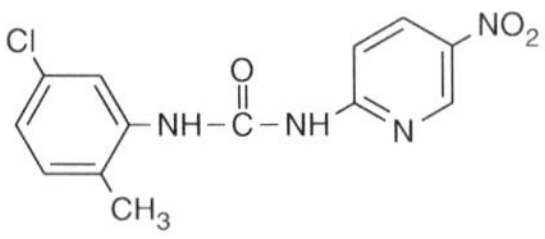

250 mg

S91,092-9
1-(2,3-DICHLOROPHENYL)-3-(3-PYRIDYL)UREA

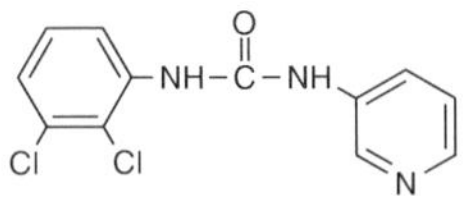

250 mg

S72,408-4
1-(3,4-DICHLOROPHENYL)-3-(3-PYRIDYL)UREA

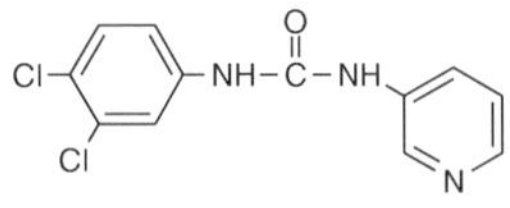

5 gm

S91,248-4
1-(3-CHLORO-4-METHYLPHENYL)-3-(3-PYRIDYL)UREA

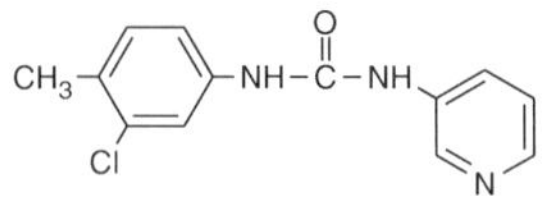

250 mg

S67,098-7
2-DIMETHYLAMINO-1-(4-FLUOROBENZYL)-4(1H)-PYRIDO(2,3-D)PYRIMIDINONE
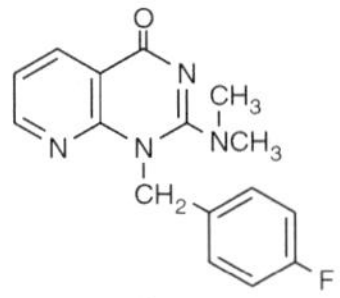
5 gm

S86,722-5
2-PYRIDYL N-(4-FLUORO-3-NITROPHENYL)CARBAMATE
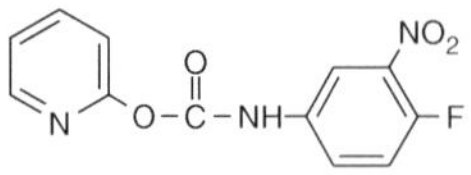
250 mg

S79,477-5
3-PYRIDYL N-(4-FLUORO-3-NITROPHENYL)CARBAMATE
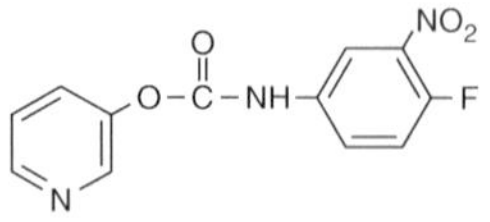
250 mg

S91,201-8
4-PYRIDYL N-(4-FLUORO-3-NITROPHENYL)CARBAMATE
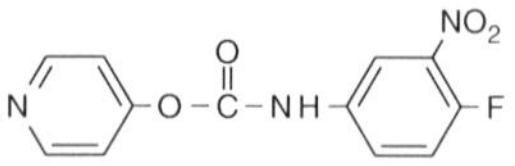
250 mg

S86,735-7
1-(4-FLUORO-3-NITROPHENYL)-3-(4-METHYL-2-PYRIDYL)UREA
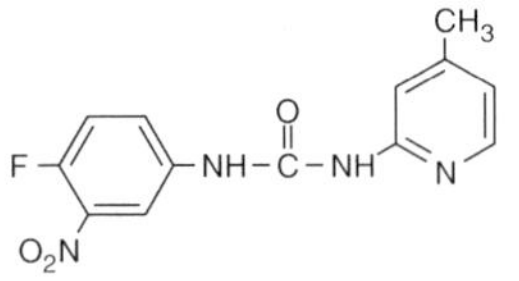
250 mg

S84,447-0
1-(2-FLUORO-5-NITROPHENYL)-3-(3-PYRIDYLMETHYL)UREA
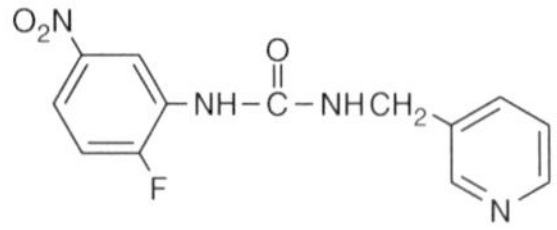
250 mg

S93,761-4
2-(2-CHLOROBENZOYL)-2-PYRIDYLACETONITRILE
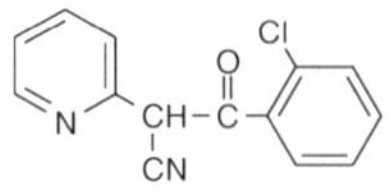
250 mg

S91,367-7
N-[2-(2-PYRIDYL)ETHYL]-3,4,5,6-TETRACHLOROPHTHALAMIC ACID
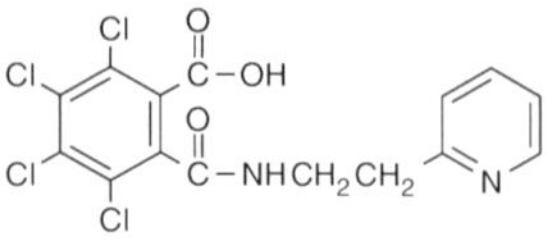
250 mg

S90,598-4
4-(4-CHLOROPHENYLSULFONYL)-1-NICOTINOYLSEMICARBAZIDE
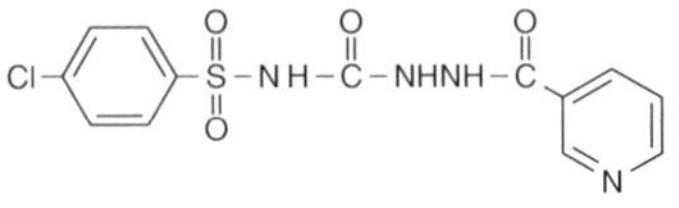
250 mg

S79,540-2
1-(2-CHLORO-6-METHYLPHENYL)-3-[2-(2-PYRIDYL)ETHYL]UREA
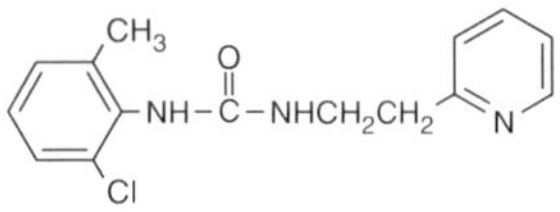
250 mg

S71,175-6
1-(2,6-DICHLOROPHENYL)-3-(4-PYRIDYL)UREA
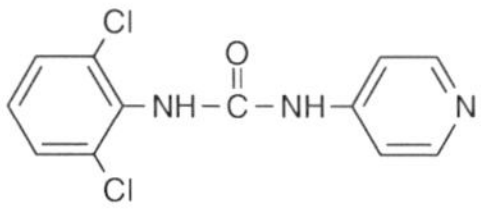
250 mg

S71,116-0
1-(3-BROMOPHENYL)-3-(3-PYRIDYLMETHYL)UREA
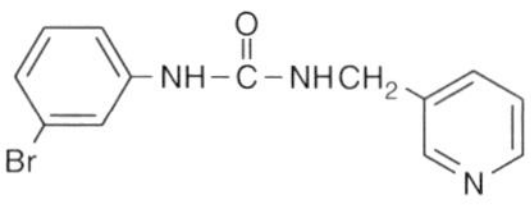
250 mg

S91,292-1
4-PYRIDINECARBOXYLIC (2,4-DICHLOROBENZYLIDENE)HYDRAZIDE
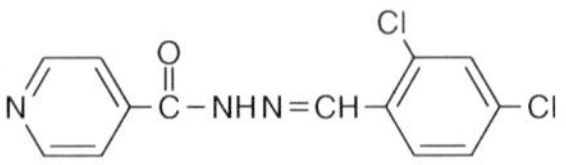
250 mg

S68,698-0
1-(2-CHLORO-4-NITROPHENYL)-3-(5-CHLORO-2-PYRIDYL)UREA
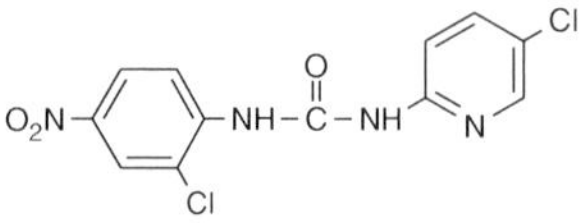
250 mg

S70,035-5
S-(2,3,5,6-TETRACHLORO-4-PYRIDYL) N-(2-CHLORO-4-NITROPHENYL)-THIOLCARBAMATE
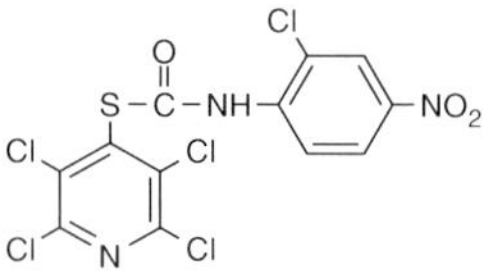
250 mg

S68,407-4
1-(2-BROMO-4,5-DIMETHYLPHENYL)-3-(4-PYRIDYLMETHYL)UREA
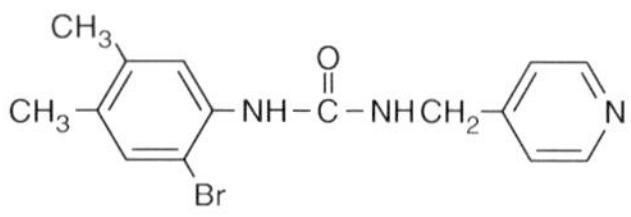
250 mg

S67,062-6
1-(4-FLUOROBENZYL)-2-(METHYLAMINO)-PYRIDO(2,3-D)PYRIMIDIN-4(1H)-ONE
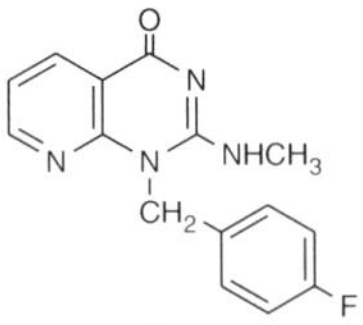
5 gm

S74,418-2
2-FLUORO-N-(2-PYRIDYL)ANILINE
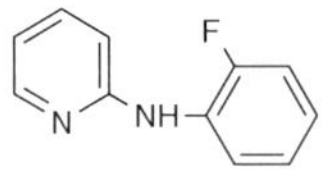
250 mg

S94,586-2
3-(4-BROMO-2-CHLOROANILINOMETHYL)-PYRIDINE
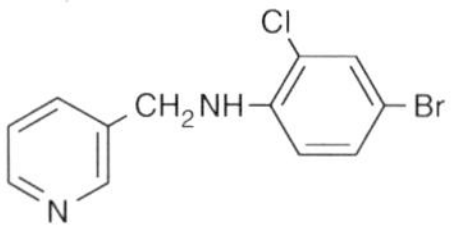
250 mg

S79,686-7
2-[3-(6-METHYL-2-PYRIDYL)UREIDO]ACETIC [2-CHLORO-4-(DIMETHYLAMINO)]-BENZYLIDENE HYDRAZIDE
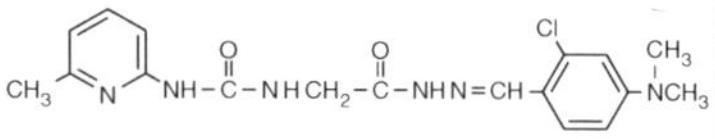
250 mg

S75,457-9
1-(4-FLUOROBENZYL)-4-(METHYLSULFONYLMETHYL)-PYRIDINIUM CHLORIDE
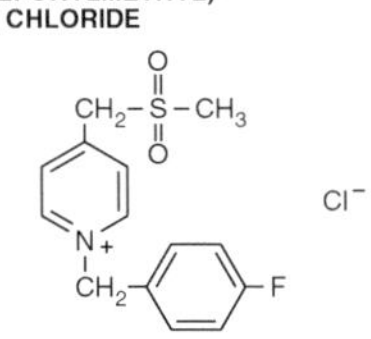
250 mg

S73,513-2
1-(4-CHLOROBENZOYLMETHYL)-4-(METHYLSULFONYLMETHYL)-PYRIDINIUM BROMIDE
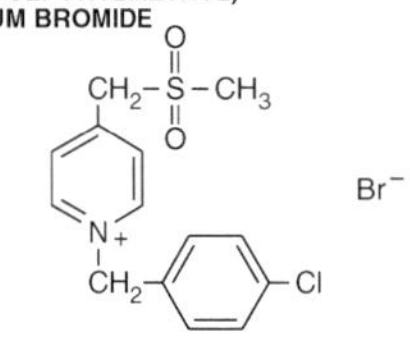
250 mg

S68,390-6
2,4-BIS(4-CHLOROPHENYL)-6-CHLOROPYRIDINE
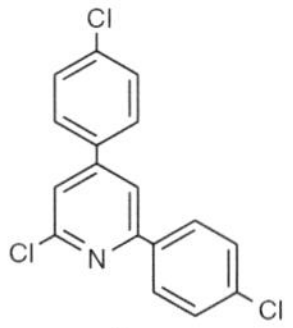
5 gm

S69,496-7
4-(2,4-DIFLUOROPHENYL)-1-(4-PYRIDYLCARBONYL)SEMICARBAZIDE
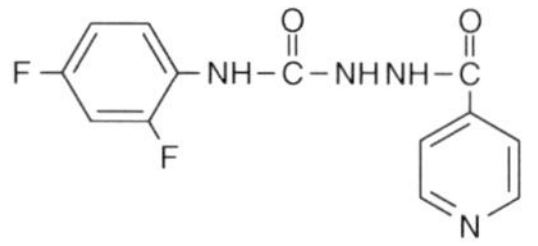
250 mg

S77,277-1
4-(2-FLUOROPHENYL)-1-(4-PYRIDYLCARBONYL)-3-THIOSEMICARBAZIDE
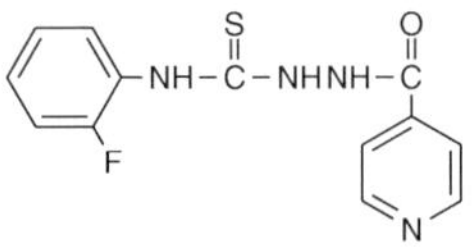
250 mg

S69,187-9
4-(4-CHLOROPHENYL)-1-(5-METHYL-2-PYRIDYLUREIDOACETYL)SEMICARBAZIDE
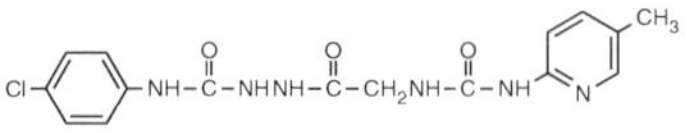
250 mg

S69,241-7
2-(5-METHYL-2-PYRIDYLUREIDO)ACETIC N2-[N-(3-CHLOROPHENYL)CARBAMOYL]-HYDRAZIDE
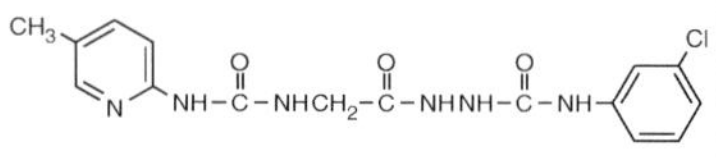
250 mg

S72,360-6
2-(5-METHYL-2-PYRIDYLUREIDO)ACETIC ACID N2-[N-(3-BROMOPHENYL)-CARBAMOYL]HYDRAZIDE
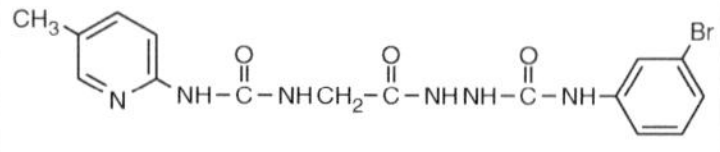
5 gm

S77,932-6
4-(3-CHLOROPHENYL)-1-(6-METHYL-2-PYRIDYLUREIDOACETYL)SEMICARBAZIDE
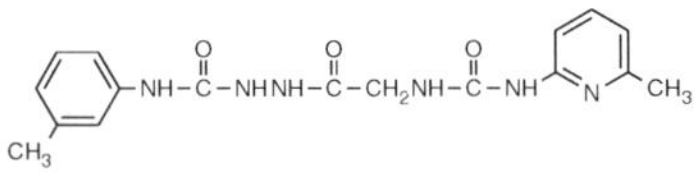
250 mg

S90,310-8
2-[3-(2-CHLORO-4-NITROPHENYL)-UREIDO]ACETIC (2-PYRIDYLMETHYLENE)-HYDRAZIDE
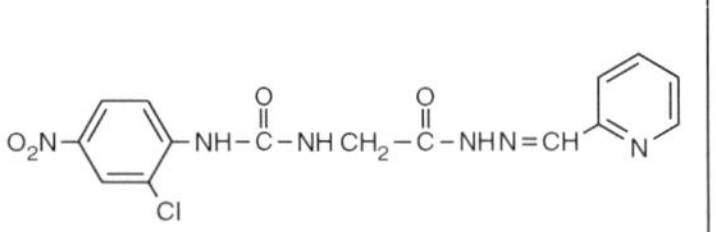
250 mg

S90,315-9
2-[3-(2-CHLORO-4-NITROPHENYL)-UREIDO]ACETIC (4-PYRIDYLMETHYLENE)-HYDRAZIDE
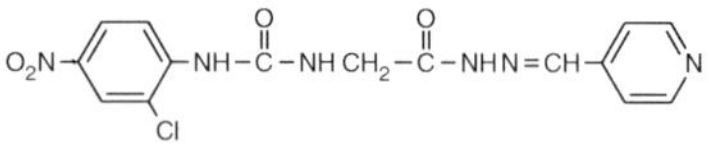
250 mg

S84,766-6
4-[3-(2-FLUOROPHENYL)UREIDO]-BENZOIC (2-PYRIDYLMETHYLENE)-HYDRAZIDE
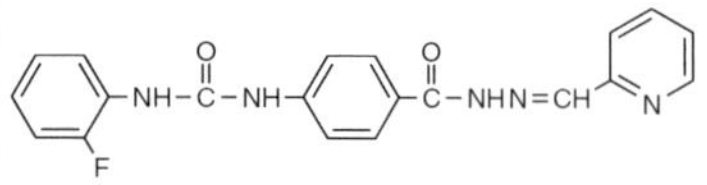
250 mg

S84,767-4
4-[3-(2-FLUOROPHENYL)UREIDO]-BENZOIC (3-PYRIDYLMETHYLENE)-HYDRAZIDE

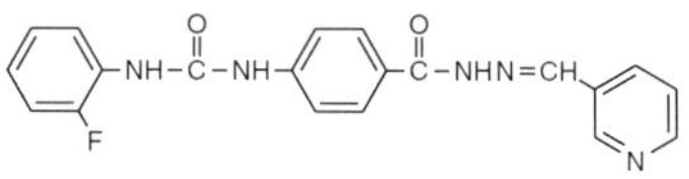

250 mg

S72,451-3
2-PYRIDYLMETHYL N-(2,5-DICHLOROPHENYL)CARBAMATE

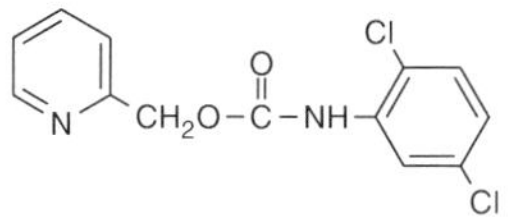

5 gm

S79,596-8
3-PYRIDYL N-(2-BROMO-4-METHYLPHENYL)CARBAMATE

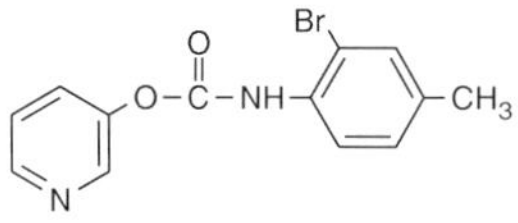

250 mg

S73,827-1
4-PYRIDYLMETHYLENEAMINO N-(5-CHLORO-2-METHOXYPHENYL)-CARBAMATE

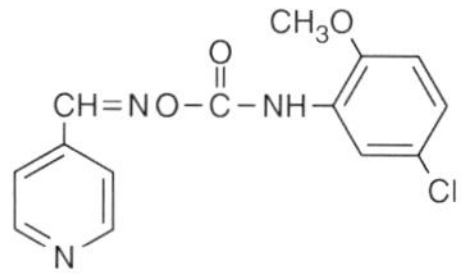

5 gm

S69,582-3
4-(2-CHLORO-4-NITROPHENYL)-1-(4-PYRIDYLCARBONYL)SEMICARBAZIDE

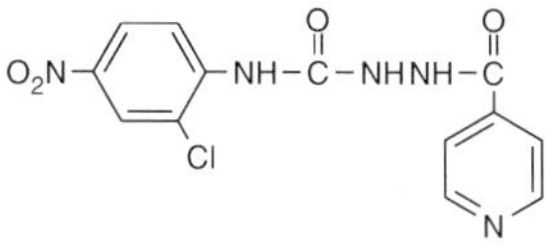

250 mg

S91,171-2
5-NITRO-2-PYRIDYL N-(2-FLUORO-5-NITROPHENYL)CARBAMATE

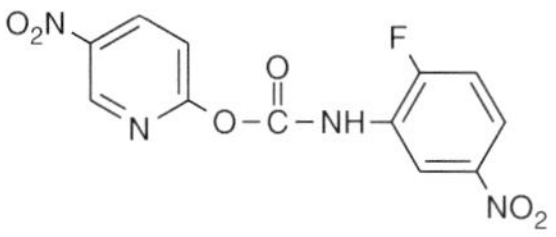

250 mg

S91,202-6
5-NITRO-2-PYRIDYL N-(4-FLUORO-3-NITROPHENYL)CARBAMATE

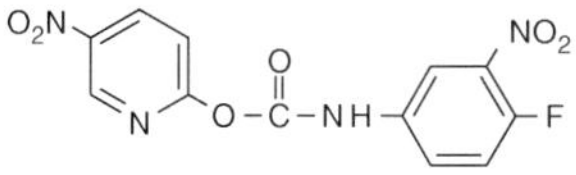

250 mg

S68,832-0
5-CHLORO-2-PYRIDYL N-(2-CHLORO-4-NITROPHENYL)CARBAMATE

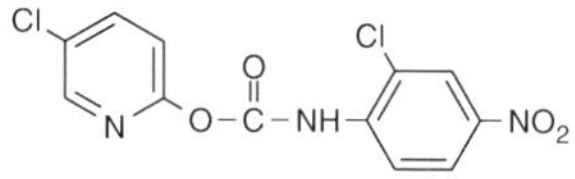

250 mg

S91,170-4
2-PYRIDYL N-(2-FLUORO-5-NITROPHENYL)CARBAMATE

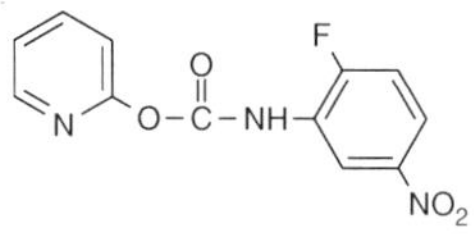

250 mg

S68,687-5
3-PYRIDYLMETHYL N-(2-CHLORO-4-NITROPHENYL)CARBAMATE

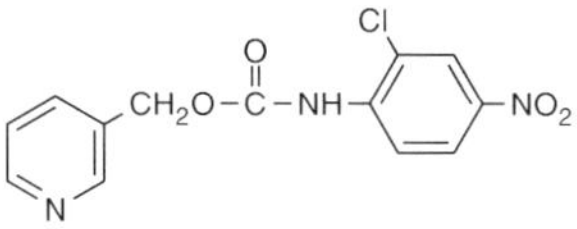

5 gm

S91,167-4
1-(5-CHLORO-2-PYRIDYL)-3-(2-FLUORO-5-NITROPHENYL)UREA

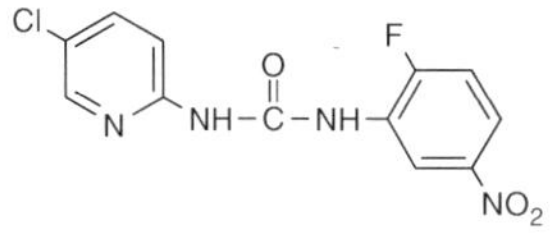

250 mg

S91,193-3
1-(5-CHLORO-2-PYRIDYL)-3-(4-FLUORO-3-NITROPHENYL)UREA

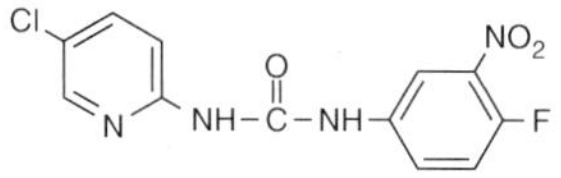

250 mg

S91,194-1
1-(4-FLUORO-3-NITROPHENYL)-3-(5-NITRO-2-PYRIDYL)UREA

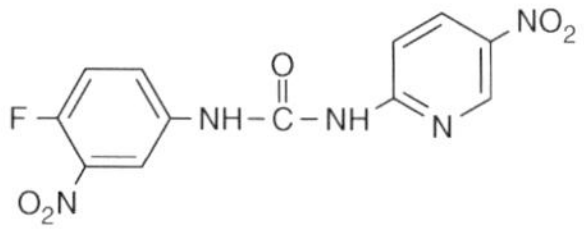

250 mg

S91,168-2
1-(5-BROMO-2-PYRIDYL)-3-(2-FLUORO-5-NITROPHENYL)UREA

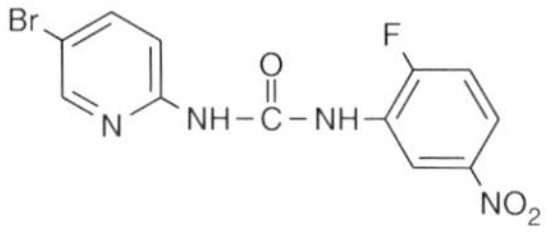

250 mg

S91,369-3
N-(5-NITRO-2-PYRIDYL)-3,4,5,6-TETRACHLOROPHTHALAMIC ACID

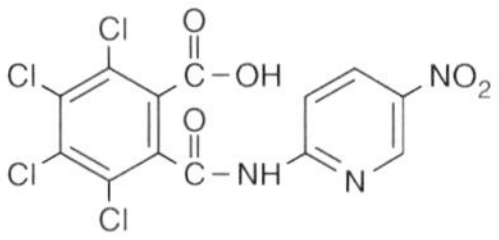

250 mg

S91,255-7
4-BROMO-N-(5-NITRO-2-PYRIDYL)-BENZAMIDE

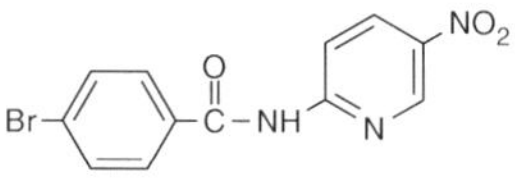

250 mg

S46,240-3
4-CHLOROPHENYL 3-PYRIDYL KETONE

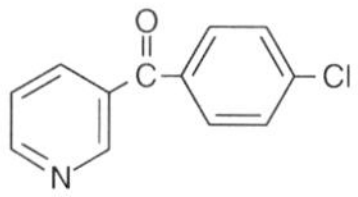

5 gm

S92,452-0
4-(4-CHLOROPHENYLSULFONYL)-1-(4-PYRIDYLCARBONYL)SEMICARBAZIDE

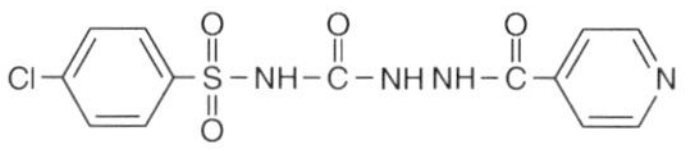

250 mg

S77,438-3
2',4',6'-TRIBROMONICOTINANILIDE

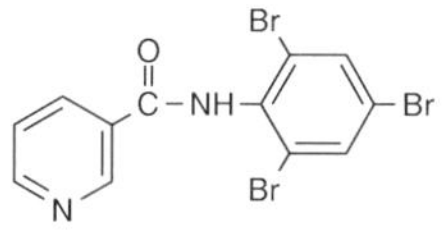

250 mg

S76,815-4
3-(4-PYRIDYL)PROPYL N-(2-CHLOROPHENYL)CARBAMATE

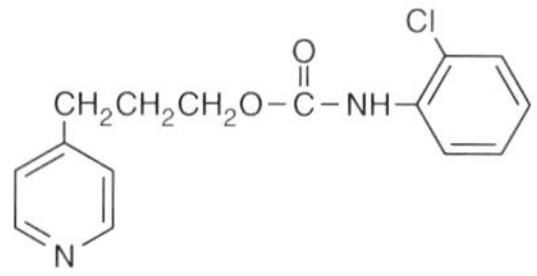

250 mg

S73,821-2
3-PYRIDYLMETHYL N-(4-FLUOROPHENYL)-CARBAMATE

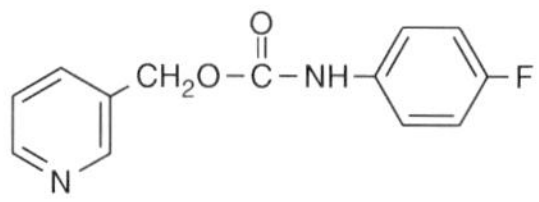

5 gm

S71,773-8
2-PYRIDYL N-(2,3-DICHLOROPHENYL)-CARBAMATE

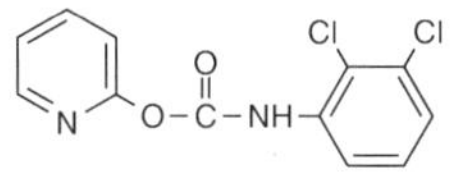

5 gm

S68,776-6
4-PYRIDYLMETHYL N-(2,3-DICHLOROPHENYL)CARBAMATE

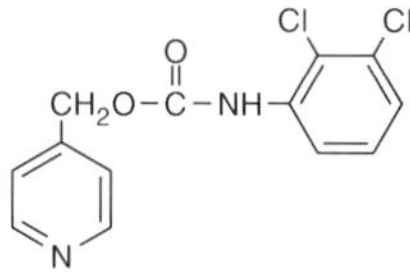

250 mg

S72,314-2
3-(4-PYRIDYL)PROPYL N-(2,3-DICHLOROPHENYL)CARBAMATE

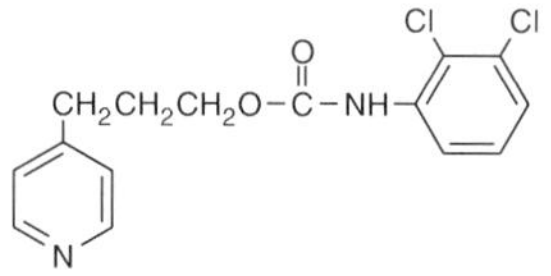

250 mg

S73,753-4
2-PYRIDYL N-(2,5-DIFLUOROPHENYL)-CARBAMATE

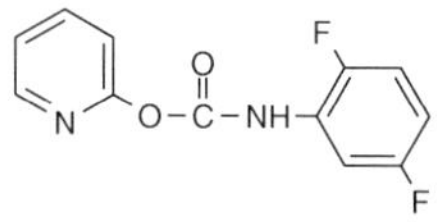

5 gm

S72,604-4
3-PYRIDYLMETHYL N-(3,4-DICHLOROPHENYL)CARBAMATE

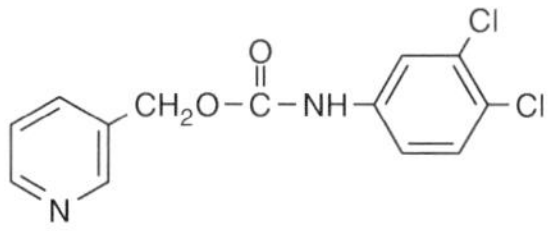

5 gm

S73,904-9
3-(4-PYRIDYL)PROPYL N-(3,4-DICHLOROPHENYL)CARBAMATE

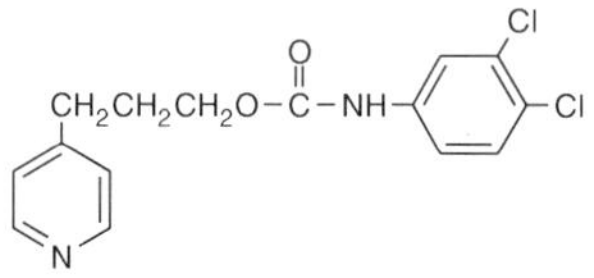

5 gm

S68,616-6
5-CHLORO-2-PYRIDYL N-(2,5-DICHLOROPHENYL)CARBAMATE

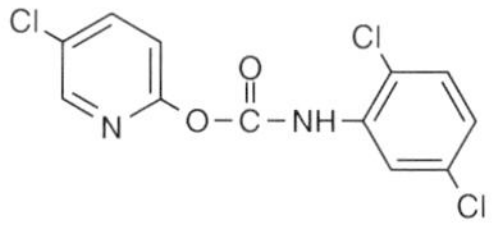

250 mg

S71,472-0
3-PYRIDYL N-(3-CHLOROPHENYL)-CARBAMATE

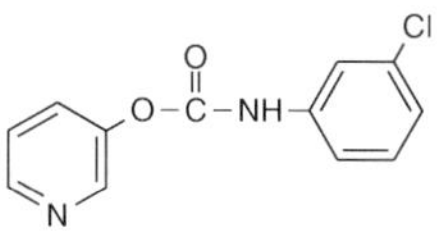

250 mg

S68,701-4
3-PYRIDYLMETHYL N-(3-FLUOROPHENYL)-CARBAMATE

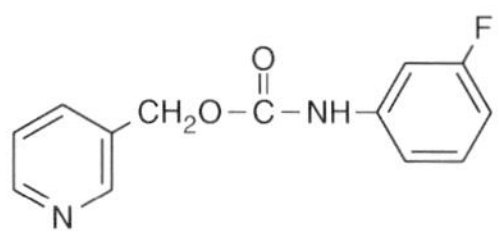

250 mg

S73,847-6
3-PYRIDYLMETHYL N-(3-CHLOROPHENYL)-CARBAMATE

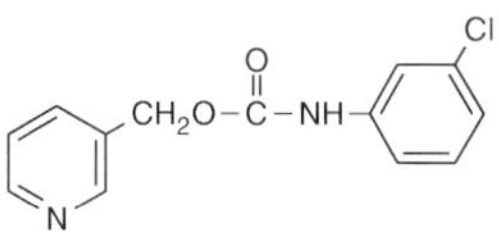

5 gm

S69,328-6
4-PYRIDYLMETHYL N-(3-CHLOROPHENYL)-CARBAMATE

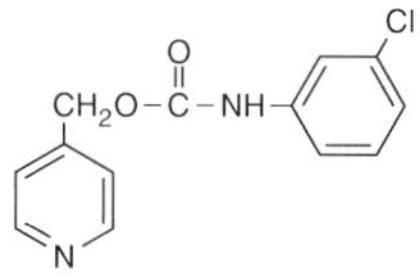

250 mg

S73,787-9
4-PYRIDYLPROPYL N-(3-CHLOROPHENYL)-CARBAMATE

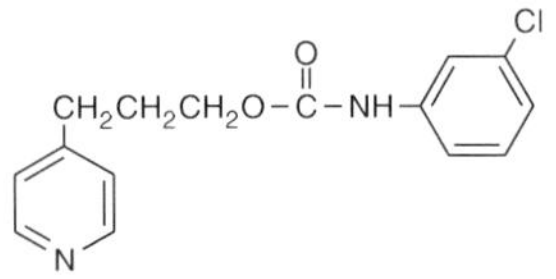

1 gm

S75,939-2
10-(4-FLUOROBENZYL)-1,2,3,5,10,10A-HEXAHYDROIMIDAZO(1,2-A)PYRIDO(2,3-D)-PYRIMIDIN-5-ONE

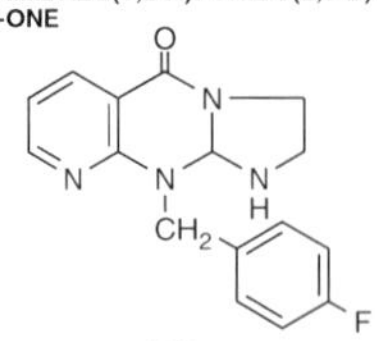

250 mg

S93,369-4
4-CHLOROBENZOIC N2-(3-NITRO-2-PYRIDYL)HYDRAZIDE

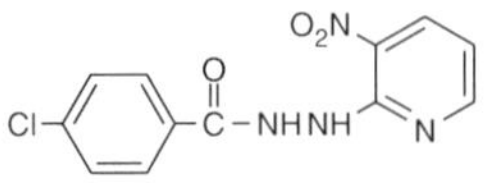

250 mg

S69,341-3
1-(2-CHLOROPHENYL)-3-(5-CHLORO-2-PYRIDYL)UREA

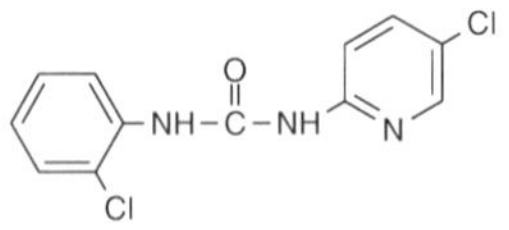

250 mg

S71,882-3
1-(2-FLUOROPHENYL)-3-(4-METHYL-2-PYRIDYL)UREA

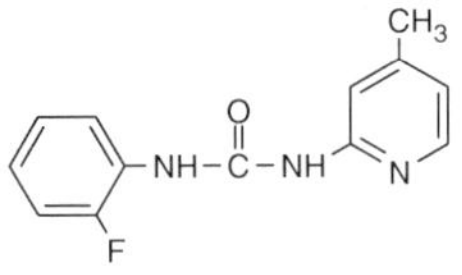

5 gm

S89,231-9
1-(2-CHLOROPHENYL)-3-(4-METHYL-2-PYRIDYL)UREA

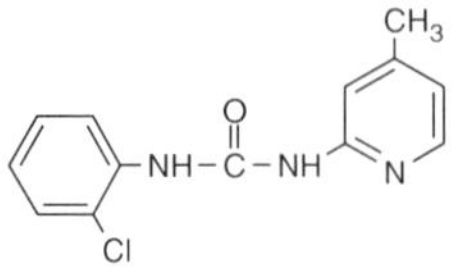

250 mg

S72,153-0
1-(2-CHLOROPHENYL)-3-(6-METHYL-2-PYRIDYL)UREA

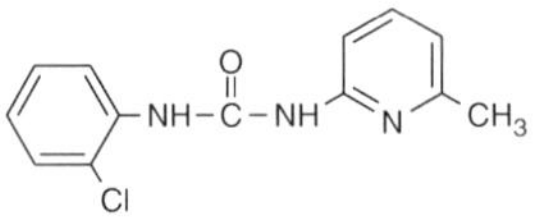

5 gm

S91,090-2
1-(2,3-DICHLOROPHENYL)-3-(3,5-DICHLORO-2-PYRIDYL)UREA

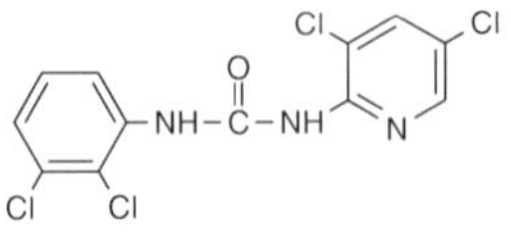

250 mg

S69,250-6
1-(2-CHLORO-3-PYRIDYL)-3-(2,3-DICHLOROPHENYL)UREA

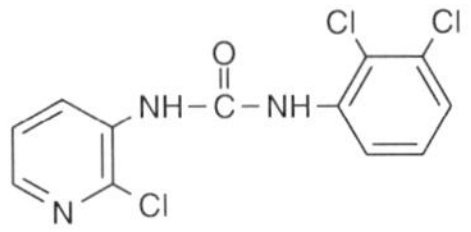

250 mg

S91,100-3
1-(5-CHLORO-2-PYRIDYL)-3-(2,3-DICHLOROPHENYL)UREA

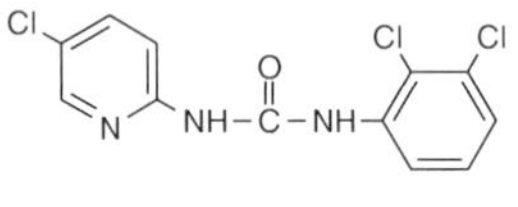

250 mg

S91,099-6
1-(5-BROMO-2-PYRIDYL)-3-(2,3-DICHLOROPHENYL)UREA

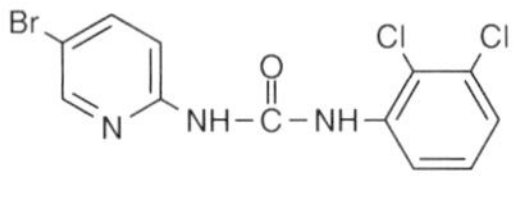

250 mg

S71,389-9
1-(2,6-DICHLOROPHENYL)-3-(6-METHYL-3-PYRIDYL)UREA

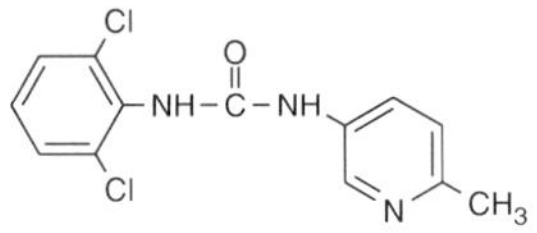

250 mg

S91,087-2
1-(2,3-DICHLOROPHENYL)-3-(2-PYRIDYL)UREA

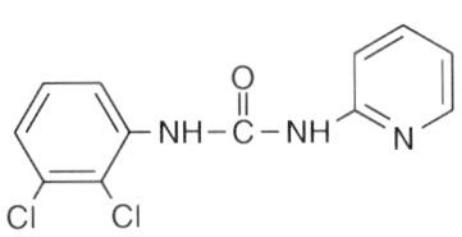

250 mg

S79,616-6
1-(2,3-DICHLOROPHENYL)-3-(6-METHYL-2-PYRIDYL)UREA

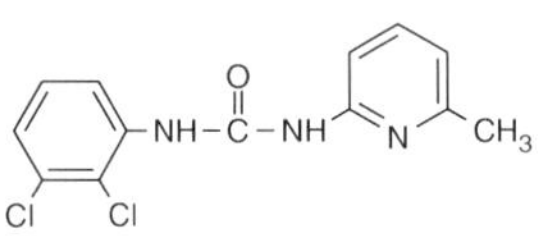

250 mg

S71,548-4
1-(2-CHLORO-6-METHYLPHENYL)-3-(3-PYRIDYL)UREA

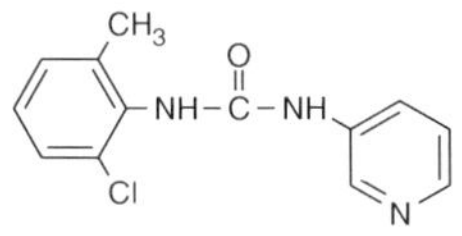

5 gm

S71,174-8
1-(2,6-DICHLOROPHENYL)-3-(3-PYRIDYL)UREA

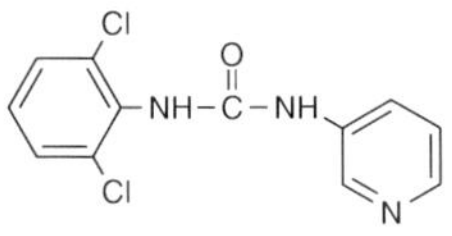

5 gm

S76,885-5
1-(2-CHLORO-3-PYRIDYL)-3-(2,4-DIFLUOROPHENYL)UREA

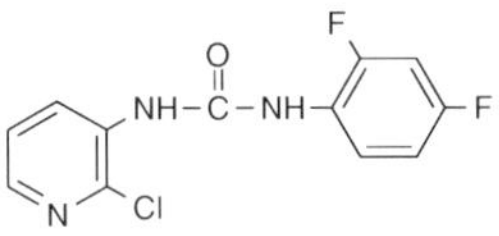

250 mg

S77,272-0
1-(5-CHLORO-2-PYRIDYL)-3-(5-CHLORO-2-METHYLPHENYL)UREA

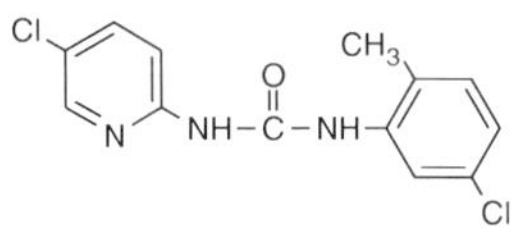

250 mg

S91,162-3
1-(5-BROMO-2-PYRIDYL)-3-(5-CHLORO-2-METHYLPHENYL)UREA

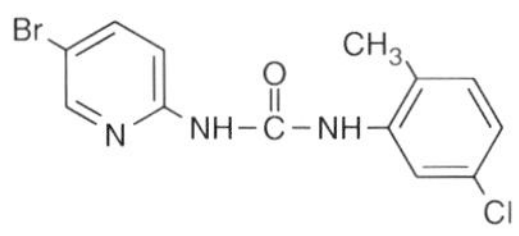

250 mg

S73,091-2
1-(4,6-DIMETHYL-2-PYRIDYL)-3-(2,5-DIFLUOROPHENYL)UREA

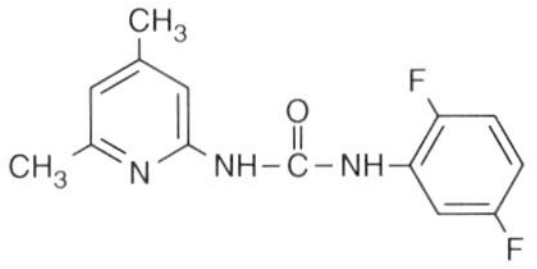

5 gm

S72,630-3
1-(3,4-DICHLOROPHENYL)-3-(2-PYRIDYL)UREA

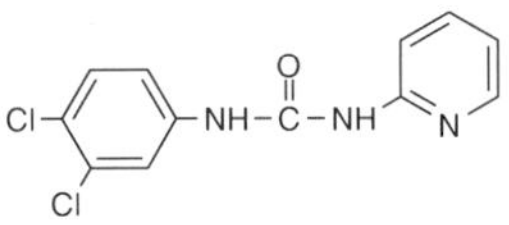

5 gm

S92,423-7
1-(3-CHLORO-4-METHYLPHENYL)-3-[2-(2-PYRIDYL)ETHYL]UREA

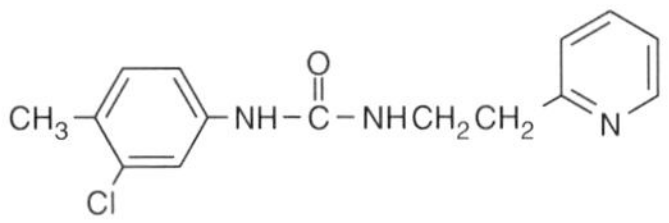

250 mg

S72,972-8
1-(2,5-DIFLUOROPHENYL)-3-(6-METHYL-2-PYRIDYL)UREA

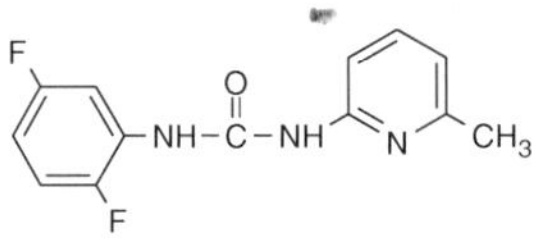

5 gm

S90,320-5
1-(2,4-DICHLOROPHENYL)-3-(6-METHYL-2-PYRIDYL)UREA

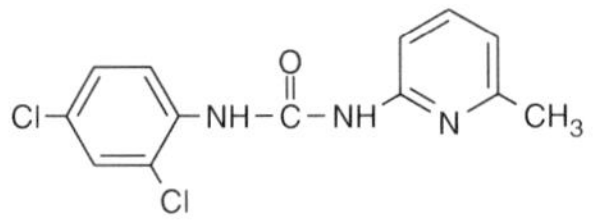

250 mg

S71,883-1
1-(2,5-DIFLUOROPHENYL)-3-(3-PYRIDYL)UREA

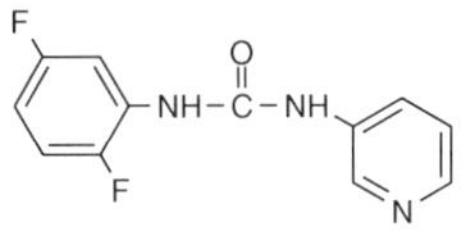

250 mg

S71,489-5
1-(3,4-DICHLOROPHENYL)-3-(4-PYRIDYLMETHYL)UREA

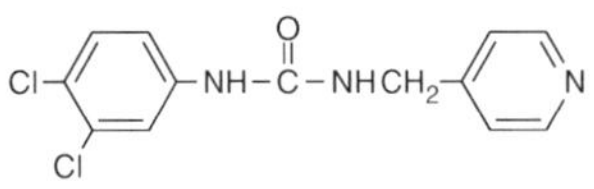

5 gm

S69,389-8
1-(5-CHLORO-2-PYRIDYL)-3-(2,4,5-TRICHLOROPHENYL)UREA

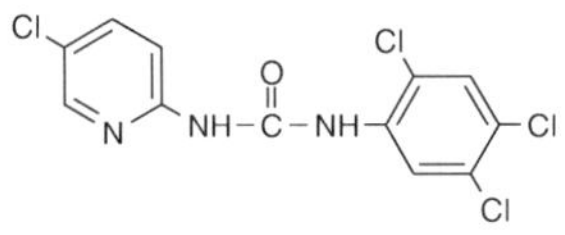

250 mg

S72,031-3
1-(3-BROMOPHENYL)-3-(4,6-DIMETHYL-2-PYRIDYL)UREA

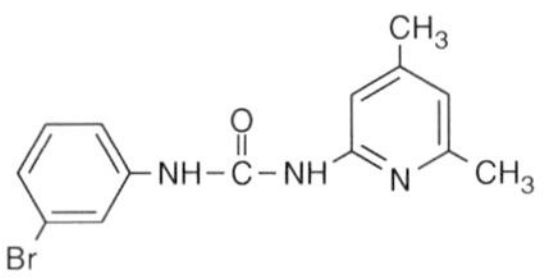

5 gm

S85,051-9
1-(3-CHLOROPHENYL)-3-(4-METHYL-2-PYRIDYL)UREA

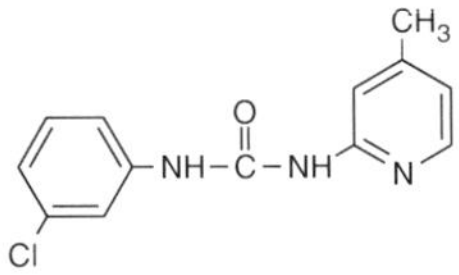

250 mg

S76,062-5
1-(3-FLUOROPHENYL)-3-(3-PYRIDYL)UREA

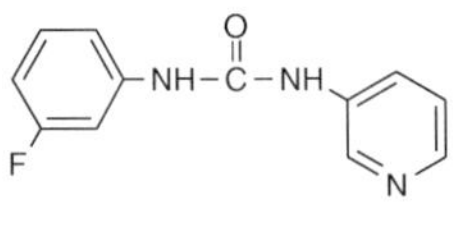

250 mg

S71,661-8
1-(3-PYRIDYLMETHYL)-3-(2,4,5-TRICHLOROPHENYL)UREA

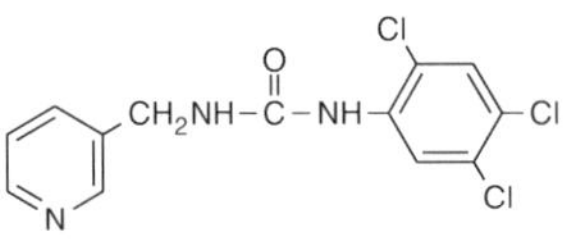

5 gm

S84,422-5
S-(2,3,5,6-TETRACHLORO-4-PYRIDYL) N-(3-CHLORO-2-METHYLPHENYL)-THIOCARBAMATE

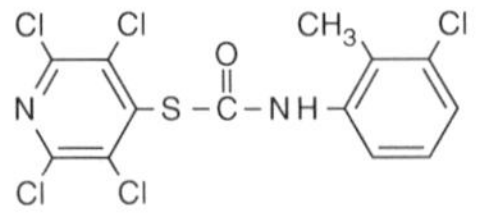

250 mg

S69,541-6
7-(4-CHLOROPHENYL)-5,6,8,9-TETRAHYDRODIBENZO(C,H)XANTHYLIUM PERCHLORATE

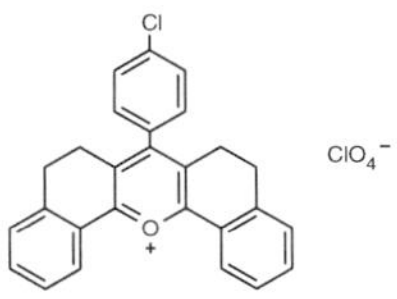

5 gm

S92,831-3
2-(2-CHLOROSTYRYL)PYRIDINE

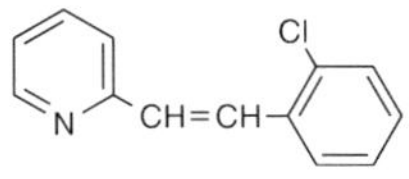

250 mg

S92,056-8
2,6-BIS(4-FLUOROPHENYL)-4-METHYLPHOSPHORIN

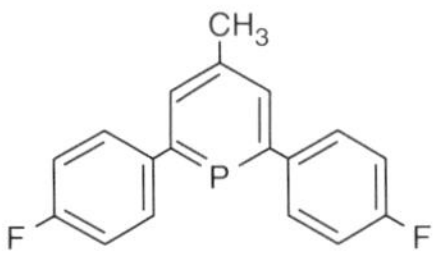

250 mg

S92,066-5
2,4,6-TRIS(4-FLUOROPHENYL)-PHOSPHORIN

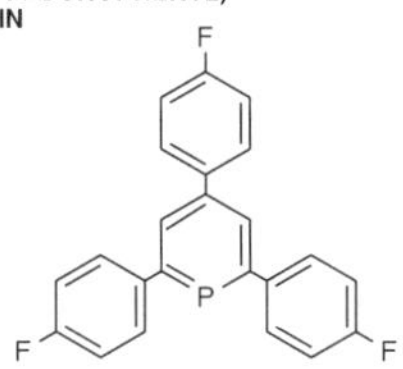

250 mg

S92,065-7
2,6-BIS(4-FLUOROPHENYL)-4-(4-CHLOROPHENYL)PHOSPHORIN

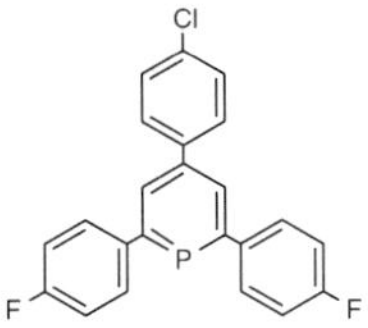

250 mg

S89,855-4
2-(4-FLUOROANILINO)PYRIDINE

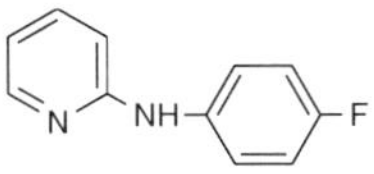

250 mg

S87,140-0
4-CHLORO-N-(2-PYRIDYLAZO)ANILINE

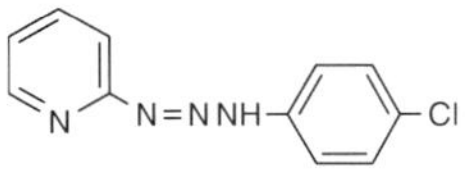

250 mg

S90,973-4
3-(4-CHLOROPHENYL)-1-(5-METHYL-2-PYRIDYL)TRIAZENE

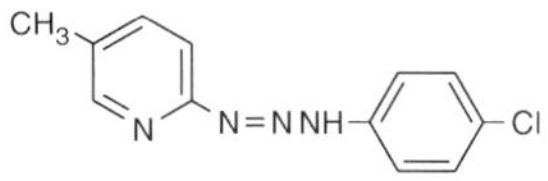

250 mg

S87,704-2
4-(4-CHLOROSTYRYL)PYRIDINE

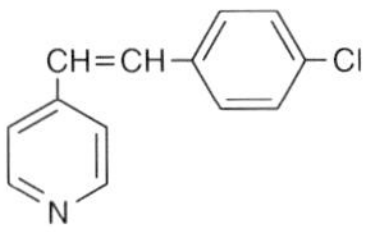

250 mg

S89,853-8
2-(3,4-DICHLOROANILINO)PYRIDINE

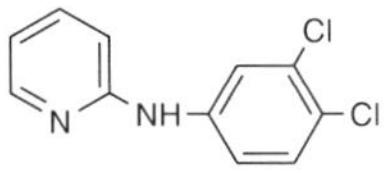

250 mg

S89,815-5
2-(3,5-DICHLOROANILINO)PYRIDINE

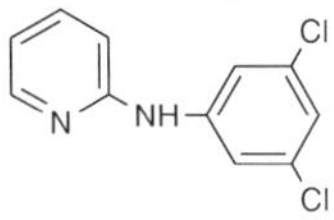

250 mg

S83,166-2
2-(PENTACHLOROPHENYL)PYRIDINE

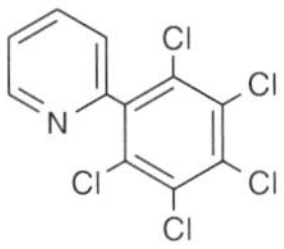

250 mg

S45,850-3
MESO-1,2-BIS(4-PYRIDYL)-1,2-ETHANEDIOL TRIHYDRATE

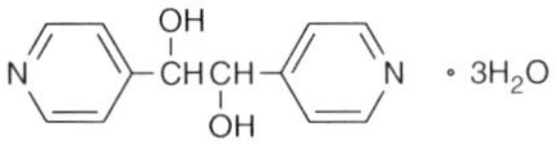

5 gm

S87,621-6
3,4-PYRIDINEDIMETHANOL

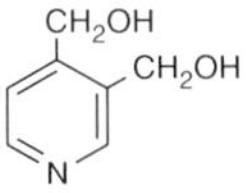

250 mg

S80,624-2
2,2'-(4-PYRIDYLMETHYLENE)BIS(5,5-DIMETHYL-3-HYDROXY-2-CYCLOHEXEN-1-ONE)

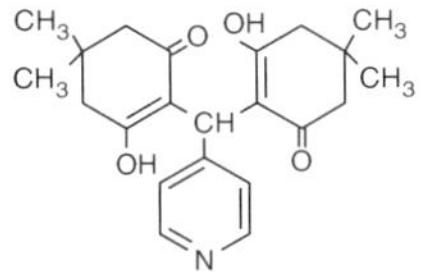

250 mg

S45,867-8
2,6-DIHYDROXYPYRIDINE HEMISULFATE

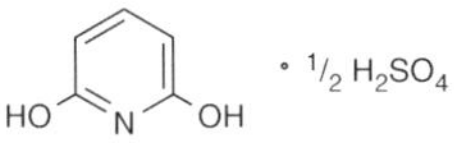

5 gm

S74,842-0
ALPHA-(2,3-DIMETHOXY-1-NAPHTHYL)-ALPHA-(3-PYRIDYL)BENZYL ALCOHOL

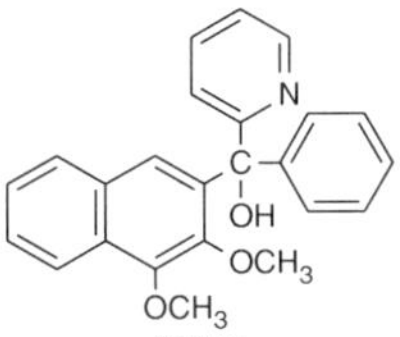

250 mg

S69,655-2
6-METHOXY-ALPHA-(2-PYRIDYLAMINO)-O-CRESOL

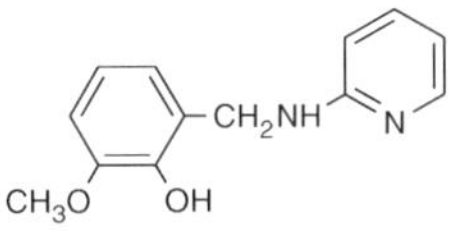

5 gm

S39,606-0
1-METHYL-2-(VANILLYLIDENEMETHYL)-PYRIDINIUM IODIDE

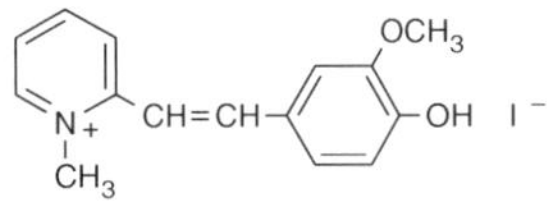

5 gm

S69,809-1
2-HYDROXY-3-METHOXY-N-(4-METHYL-2-PYRIDYL)BENZYLAMINE

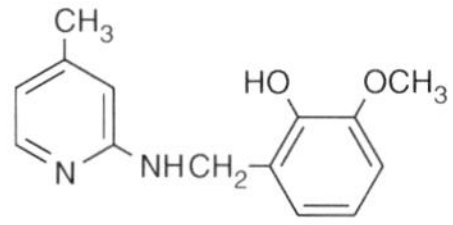

5 gm

S68,191-1
3-ETHOXY-ALPHA-PHENYL-ALPHA-(2-PYRIDYL)-2-NAPHTHALENEMETHANOL

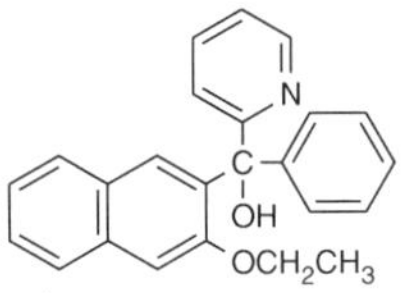

250 mg

S35,249-7
4-METHOXYPHENYL 2-PYRIDYL CARBINOL

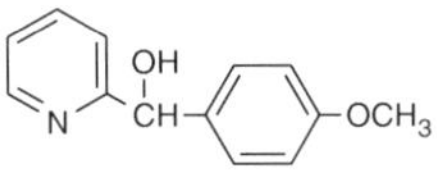

5 gm

S73,353-9
4-HYDROXY-3-(ALPHA-IMINOBENZYL)-6-(4-METHOXYPHENYL)-1-METHYLPYRIDIN-2(1H)-ONE

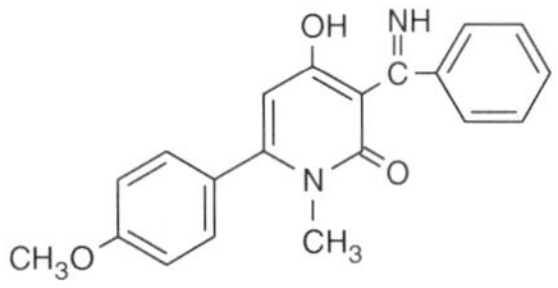

250 mg

S73,460-8
2,3-DIETHOXY-ALPHA-PHENYL-ALPHA-(2-PYRIDYL)-1-NAPHTHALENEMETHANOL

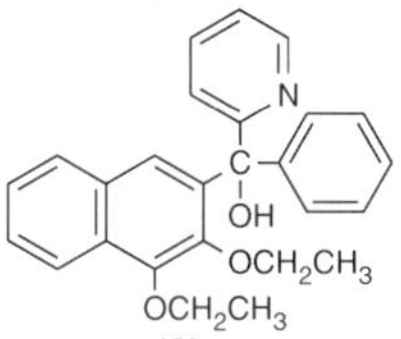

250 mg

S76,231-8
3-METHOXY-ALPHA-PHENYL-ALPHA-(4-PYRIDYL)-2-NAPHTHALENEMETHANOL

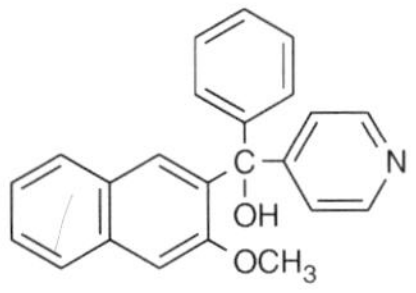

250 mg

S66,839-7
1,2-DIPHENYL-2-(2-PYRIDYL)ETHANOL

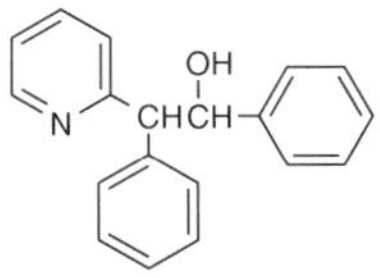

250 mg

S73,370-9
N,5-DIMETHYL-2-[2-HYDROXY-2-PHENYL-2-(2-PYRIDYL)ETHYL]-BENZENESULFONAMIDE

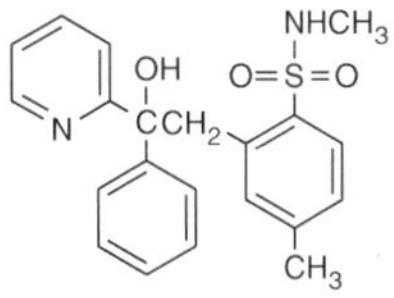

250 mg

S73,334-2
ALPHA-(PIPERIDINOMETHYL)-ALPHA-(2-PYRIDYL)BENZYL ALCOHOL

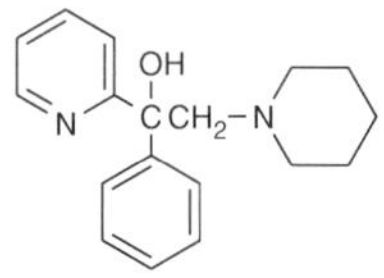

250 mg

S68,181-4
6-(2,2-DIPHENYL-2-HYDROXYETHYL)-2-(1H)PYRIDONE

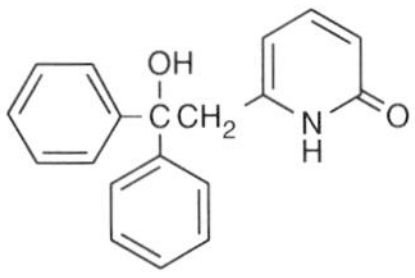

250 mg

S76,143-5
1-HYDROXY-1-PHENYL-2-(4-PYRIDYLMETHYL)-1H-ISOINDOL-3-(2H)-ONE

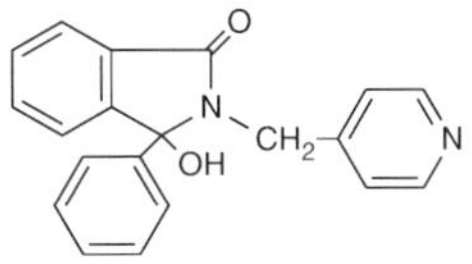

250 mg

S74,838-2
4-(MORPHOLINOMETHYL)-ALPHA-(2-PYRIDYL)BENZHYDROL

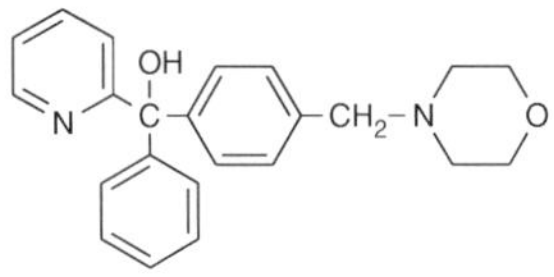

250 mg

S76,084-6
ALPHA-(4-BIPHENYLYL)-ALPHA-METHYL-3-PYRIDINEMETHANOL

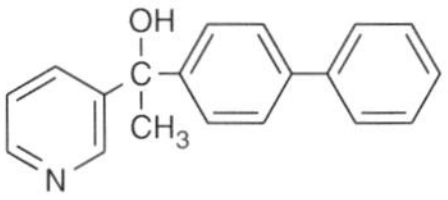

250 mg

S82,390-2
2,5-DICHLORO-ALPHA-PHENYL-4-PYRIDINEMETHANOL

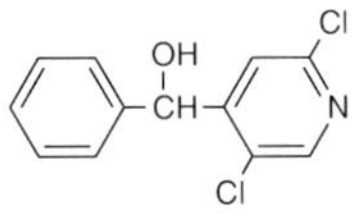

250 mg

S82,981-1
ALPHA-(2,3,5,6-TETRACHLORO-4-PYRIDYL)BENZYL ALCOHOL

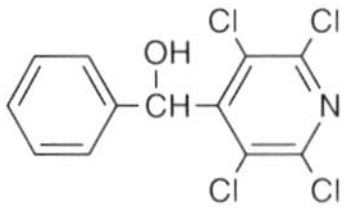

250 mg

S82,391-0
2,5-DICHLORO-ALPHA,ALPHA-DIPHENYL-4-PYRIDINEMETHANOL

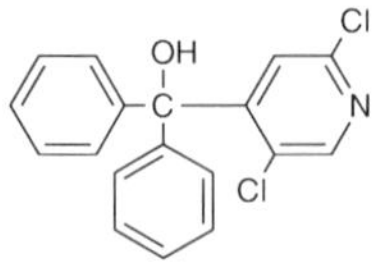

250 mg

S59,376-1
ALPHA,ALPHA-DIPHENYL-2-PYRIDINEETHANOL

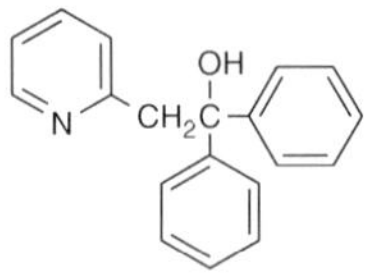

5 gm

S59,379-6
ALPHA,ALPHA-DIPHENYL-2-PYRIDINEETHANOL HYDROCHLORIDE

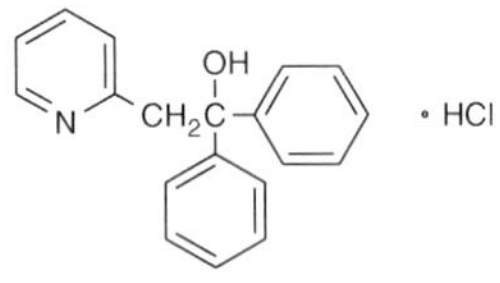

5 gm

S36,695-1
1,2-DI-2-PYRIDYL-1,2-ETHANEDIOL

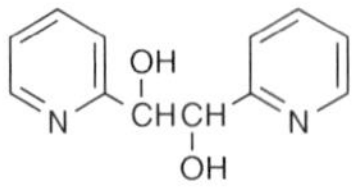

250 mg

S56,052-9
1,2-BIS(6-METHYL-2-PYRIDYL)-1,2-ETHANEDIOL

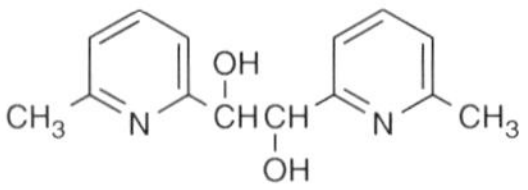

5 gm

S51,726-7
3-HYDROXY-3-(6-METHYL-2-PYRIDYLMETHYL)-2-INDOLINONE

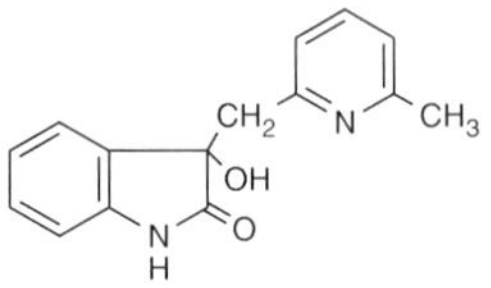

5 gm

S36,588-2
6,6'-DIMETHYL-2,2'-PYRIDOIN

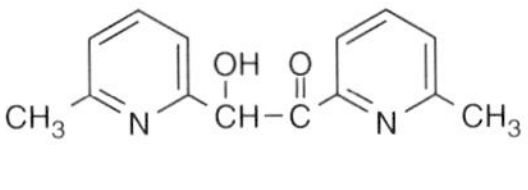

5 gm

S37,147-5
1-(2-PYRIDYL)-1,2-ETHANEDIOL

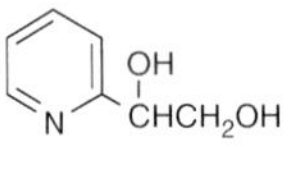

5 gm

S82,620-0
[2,2'-DIPYRIDYLTHALLYL(III)]-LACTATE

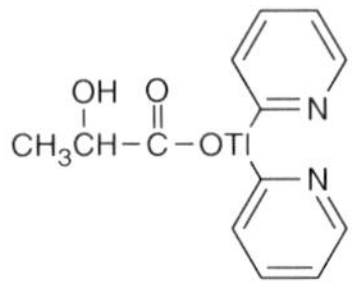

250 mg

S42,517-6
ALPHA-(3-HYDROXY-2-OXO-3-INDOLINYL)-2-METHYLPYRIDINE

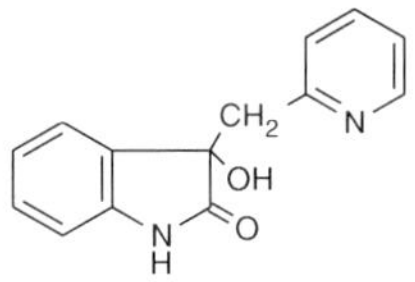

250 mg

S69,931-4
5H-(1)-BENZOPYRANO(2,3-B)-PYRIDIN-5-OL

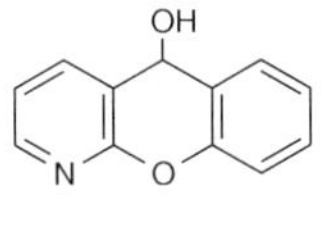

250 mg

S91,071-6
1-(3-HYDROXY-2-PYRIDYL)-3-[4-(METHYLTHIO)PHENYL]UREA

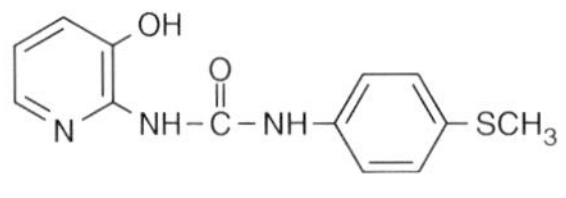

250 mg

S83,400-9
2-(2-PYRIDYLTHIO)ETHANOL

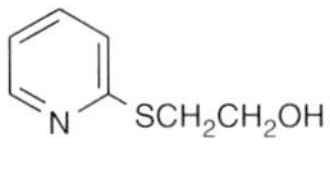

250 mg

S44,655-6
PYRIDOXAMINE 5-PHOSPHATE MONOHYDROCHLORIDE

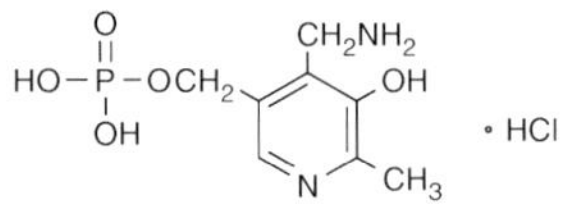

250 mg

S56,479-6
ALPHA-(3-PYRIDYLAMINO)-O-CRESOL

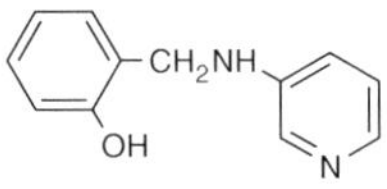

5 gm

S35,527-5
3-(2-PYRIDYLMETHYLAMINO)-1-PROPANOL

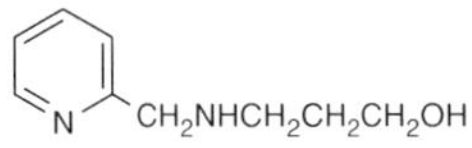

1 gm

S43,277-6
2-[N-BENZYL-N-(2-HYDROXYETHYL)-AMINOMETHYL]-3-HYDROXYPYRIDINE

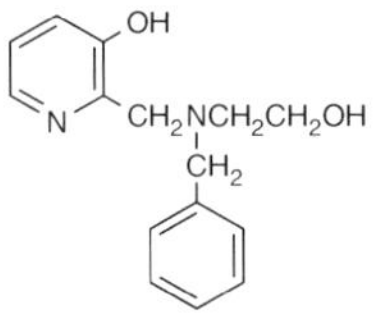

5 gm

S44,714-5
2-[N-METHYL-N-(5-NITRO-2-PYRIDYL)-AMINO]ETHANOL

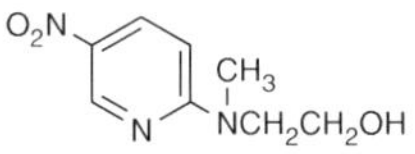

5 gm

S36,274-3
3-HYDROXY-2-(PYRROLIDINOMETHYL)-PYRIDINE

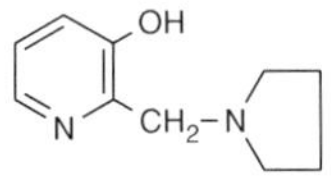

5 gm

S35,125-3
3-HYDROXY-2-(PIPERIDINOMETHYL)-PYRIDINE

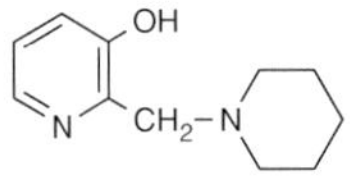

5 gm

S39,244-8
1-(2-PYRIDYLMETHYL)-3-PIPERIDINOL

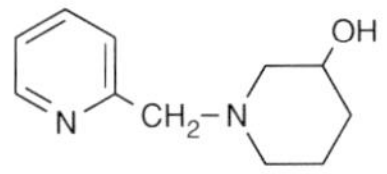

250 mg

S35,105-9
3-HYDROXY-2-(MORPHOLINOMETHYL)-PYRIDINE

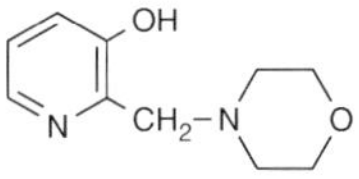

5 gm

S36,687-0
2,6-BIS(DIMETHYLAMINOMETHYL)-3-PYRIDINOL

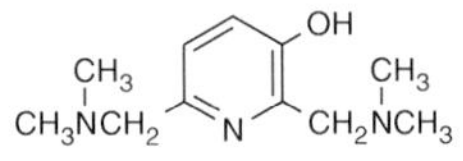

5 gm

S34,595-4
2-(DIETHYLAMINOMETHYL)-3-HYDROXYPYRIDINE

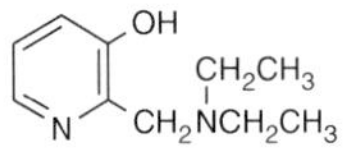

5 gm

S50,708-3
2-[BIS(2-PYRIDYLMETHYL)AMINO]-ETHANOL

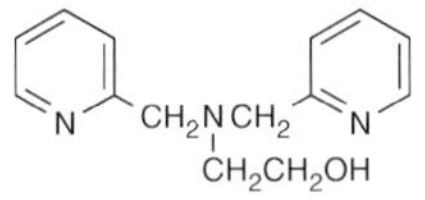

5 gm

S41,104-3
N-(5-NITRO-2-PYRIDYL)-IMINODIETHANOL

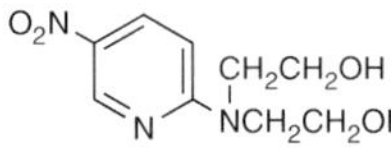

5 gm

S45,227-0
5-BROMO-3-NITRO-2-PYRIDINOL

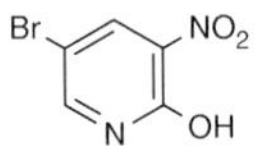

250 mg

S80,550-5
6-HYDROXY-2-METHYL-3-NITROPYRIDINE

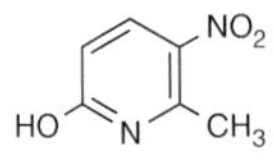

250 mg

S94,880-2
4-HYDROXY-3-NITROPYRIDINE

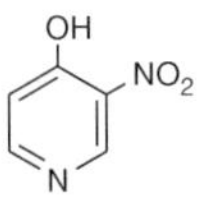

250 mg

S52,796-3
2-(4-PHENYL-1,2,3,6-TETRAHYDRO-1-PYRIDINYLMETHYL)-3-PYRIDINOL HYDROCHLORIDE

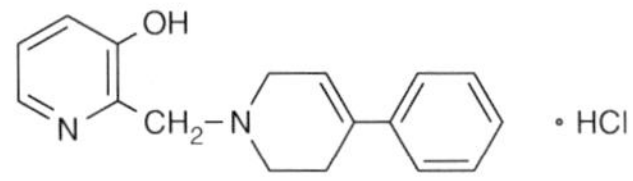

5 gm

S41,058-6
5-PICOLINYLIDENE-2-THIOBARBITURIC ACID

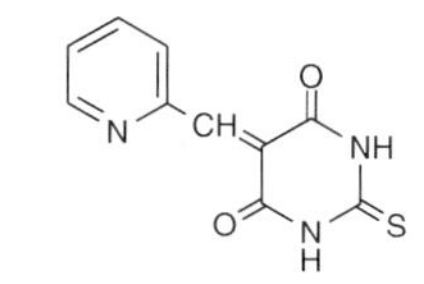

5 gm

S35,528-3
3-(3-PYRIDYLMETHYLAMINO)-1-PROPANOL

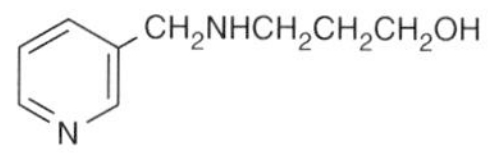

1 gm

S91,443-6
3-CARBAMOYL-1-(2-HYDROXYETHYL)-PYRIDINIUM CHLORIDE

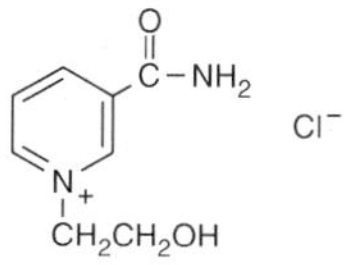

250 mg

S43,250-4
5-[N-(HYDROXYMETHYL)CARBAMOYL]-NICOTINIC ACID, SODIUM SALT

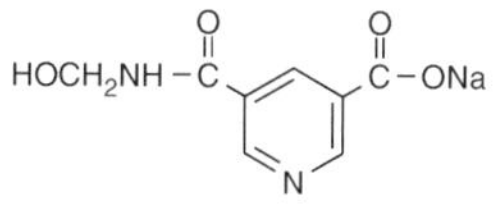

5 gm

S40,723-2
5-CYANO-2,6-DIHYDROXY-3,4-DIMETHYLPYRIDINE

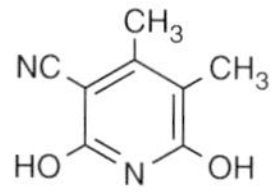

5 gm

S44,706-4
2-(5-NITRO-2-PYRIDYLAMINO)ETHANOL

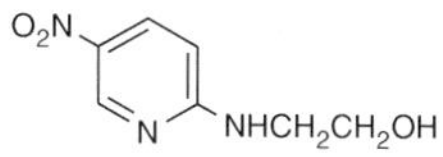

5 gm

S35,529-1
3-(4-PYRIDYLMETHYLAMINO)-1-PROPANOL

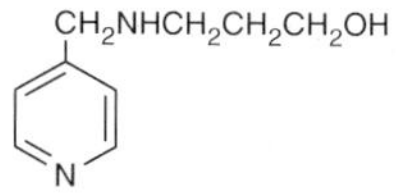

1 gm

S64,187-1
ALPHA-(6-METHYL-2-PYRIDYLAMINO)-O-CRESOL

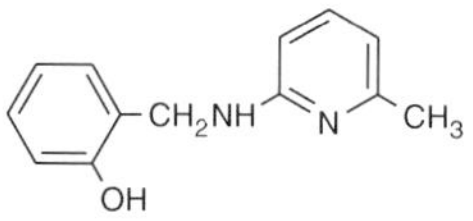

5 gm

S61,953-1
ALPHA-(3,5-DICHLORO-2-PYRIDYLAMINO)-O-CRESOL

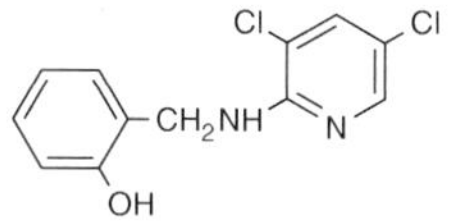

250 mg

S55,952-0
ALPHA-(5-CHLORO-2-PYRIDYLAMINO)-O-CRESOL

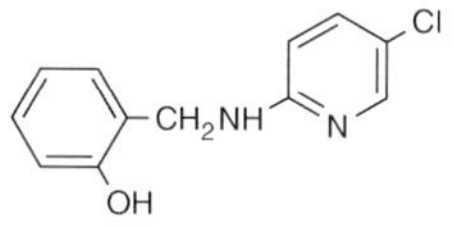

250 mg

S55,929-6
ALPHA-(3-METHYL-2-PYRIDYLAMINO)-O-CRESOL

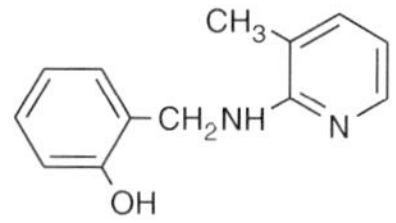

5 gm

S63,327-5
1-(2-HYDROXY-5-NITROBENZYL)-PYRIDINIUM CHLORIDE

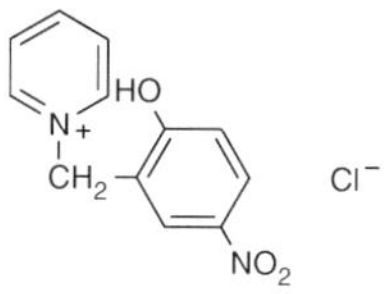

250 mg

S90,688-3
1-(2-HYDROXY-5-METHYLBENZOYLMETHYL)-PYRIDINIUM IODIDE

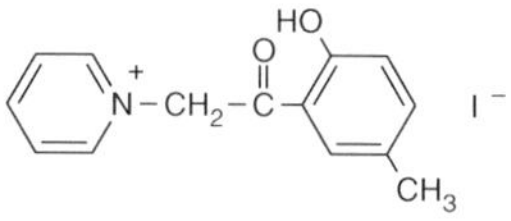

250 mg

S53,015-8
2-CARBAMOYL-3-HYDROXY-1-METHYLPYRIDINIUM CHLORIDE

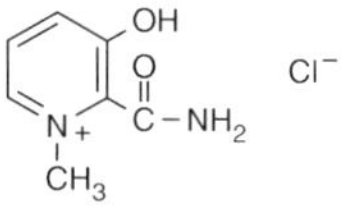

5 gm

S44,739-0
2-(2-HYDROXYSTYRYL)-1-METHYLPYRIDINIUM IODIDE

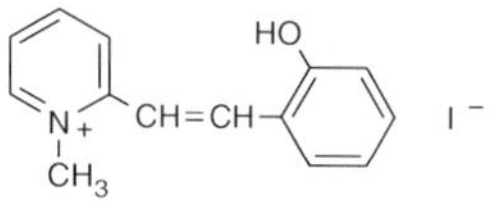

5 gm

S50,564-1
1-ETHYL-4-(2-HYDROXYSTYRYL)-PYRIDINIUM IODIDE

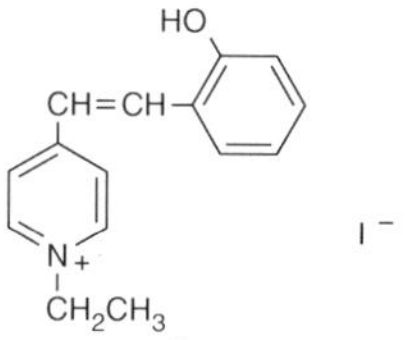

5 gm

S58,992-6
2-(3-HYDROXYSTYRYL)-1-METHYLPYRIDINIUM IODIDE

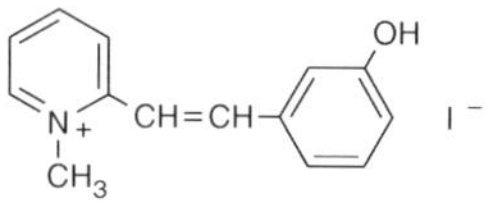

5 gm

S35,104-0
3-HYDROXY-1-METHYLPYRIDINIUM BROMIDE

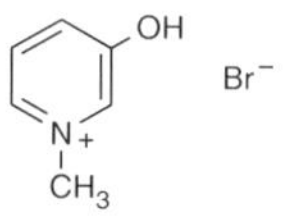

5 gm

S92,523-3
4-HYDROXYBENZOIC (3-PYRIDYLMETHYLENE)HYDRAZIDE

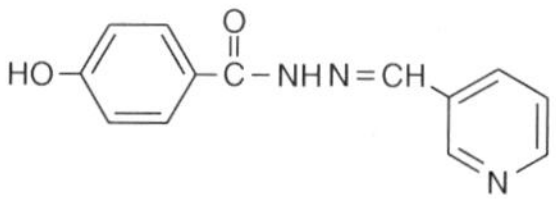

250 mg

S73,158-7
ISONICOTINIC (2-HYDROXY-1-NAPHTHYLMETHYLENE)HYDRAZIDE

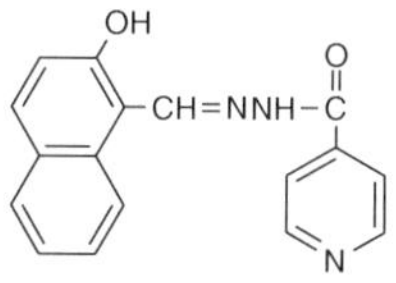

100 mg

S42,730-6
6-METHYL-2-NITRO-3-PYRIDINOL

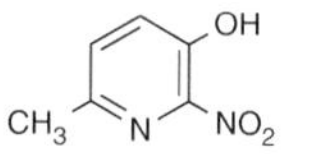

250 mg

S50,397-5
2'-HYDROXY-4'-NITRONICOTINANILIDE

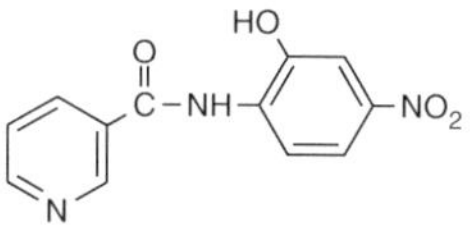

250 mg

S91,497-5
3-(3-PYRIDYL)-DL-SERINE

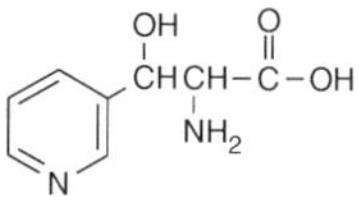

250 mg

S91,524-6
3-(4-PYRIDYL)-DL-SERINE

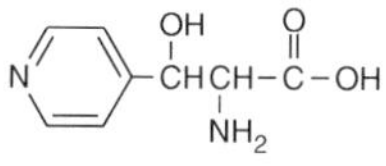

250 mg

S91,243-3
1-(3-HYDROXY-2-PYRIDYL)-3-(3,4-XYLYL)UREA

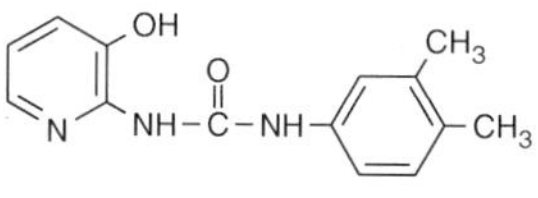

250 mg

S57,045-1
ALPHA,ALPHA-DIPHENYL-2-PYRIDINEMETHANOL

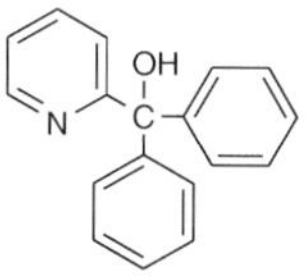

5 gm

S43,427-2
PYRIDOXINE SULFATE

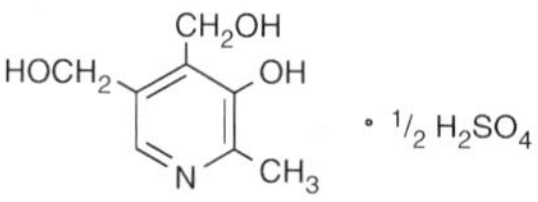

5 gm

S94,748-2
2-(P-TOLYL)PYRIDINE PICRATE

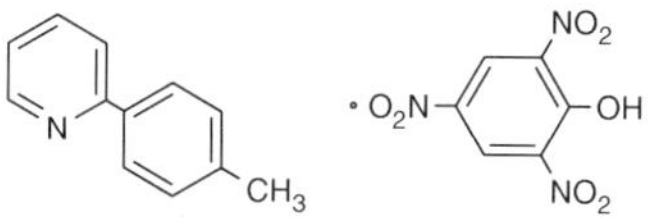

250 mg

S91,126-7
ETHYL 4-[3-(3-HYDROXY-2-PYRIDYL)-UREIDO]BENZOATE

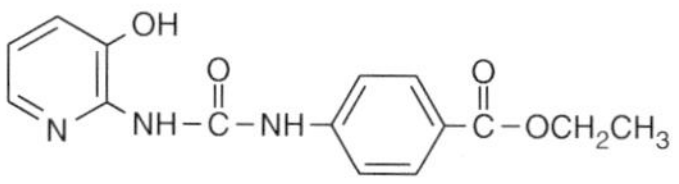

250 mg

S44,546-0
N-(HYDROXYMETHYL)ISONICOTINAMIDE

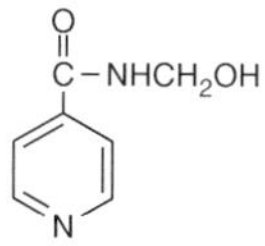

250 mg

S38,801-7
5-(4-PYRIDYLMETHYLENE)-2-THIOBARBITURIC ACID

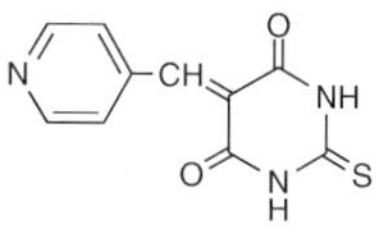

5 gm

S38,802-5
5-(3-PYRIDYLMETHYLENE)-2-THIOBARBITURIC ACID

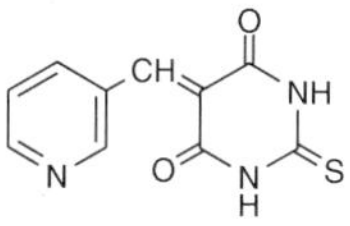

5 gm

S37,372-9
4-PYRIDYLHYDROXYMETHANESULFONIC ACID
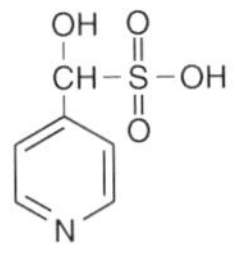
5 gm

S87,439-6
2,3-DIPHENYL-1-(4-PYRIDYL)-1-CYCLOPROPANOL
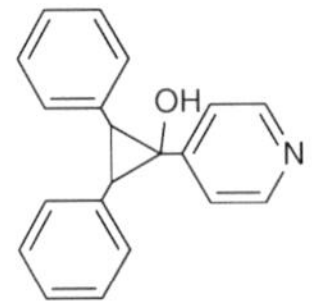
250 mg

S76,677-1
4-(4-PYRIDYL)-4-TETRAHYDROPYRANETHANOL
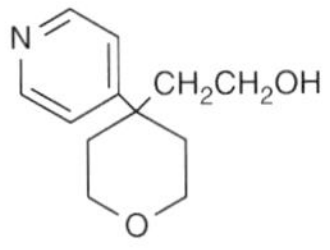
250 mg

S76,019-6
ALPHA-(4-BIPHENYLYL)-ALPHA-(MORPHOLINOMETHYL)-2-PYRIDINEMETHANOL
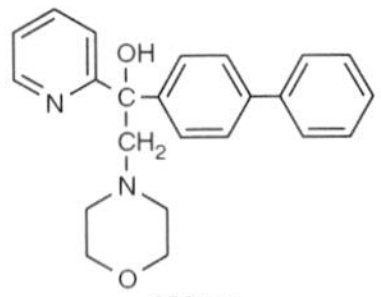
250 mg

S66,478-2
2,6-DIPHENYL-4-(4-HYDROXYPHENYL)-PYRIDINE
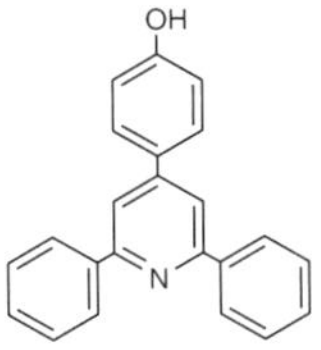
5 gm

S76,066-8
ALPHA-(2-PYRIDYLMETHYLENE)-O-CRESOL
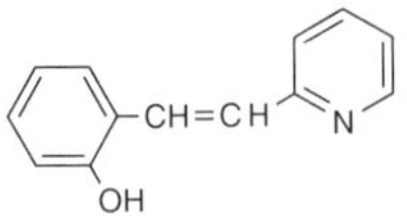
250 mg

S81,814-3
4-(2-PYRIDYL)PHENOL HYDROBROMIDE
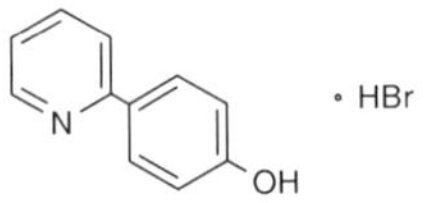
250 mg

S37,747-3
ALPHA-(4-CHLOROPHENYL)-2-PYRIDINEMETHANOL
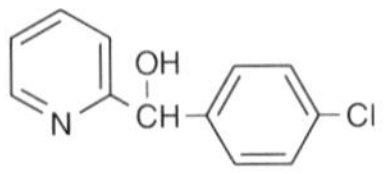
250 mg

S51,763-1
4,4'-(2-PYRIDYLMETHYLENE)DIPHENOL
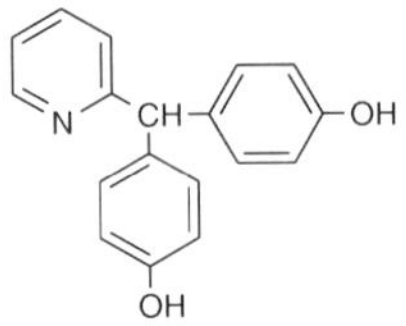
5 gm

S37,698-1
3,5-DIBROMO-2-PYRIDINOL
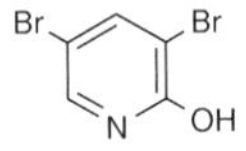
250 mg

S40,891-3
3-HYDROXY-2,6-PYRIDINEDIMETHANOL HYDROCHLORIDE
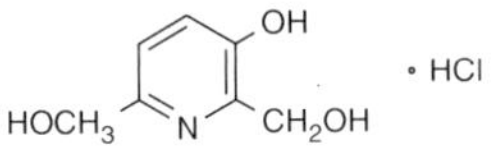
5 gm

S43,252-0
ALPHA,2,4,6-TETRAMETHYL-3-PYRIDINEMETHANOL
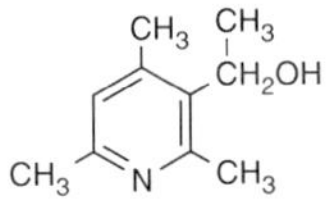
5 gm

S46,132-6
2-IODO-3-PYRIDINOL
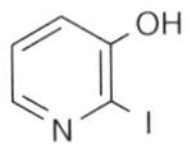
5 gm

S95,464-0
2-METHYL-3-PYRIDINEMETHANOL
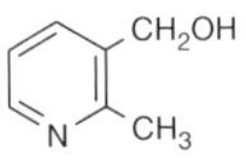
50 mg

S81,708-2
1,1-BIS(2-PYRIDYLMETHYL)ETHANOL
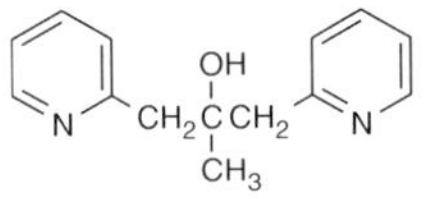
250 mg

S40,942-1
6-METHYL-2-PYRIDINEMETHANOL
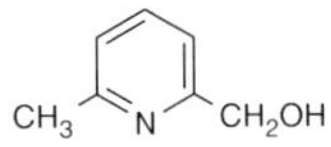
5 gm

S58,117-8
2,2'-BIS(6-METHYL-2-PYRIDYL)-TERT-BUTANOL
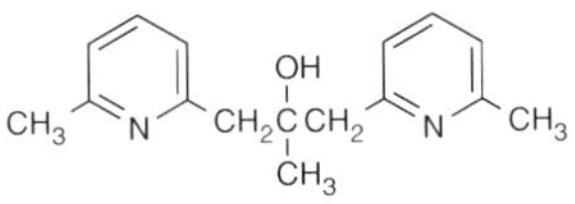
250 mg

S66,827-3
2-(4-PYRIDYL)-2-PROPANOL
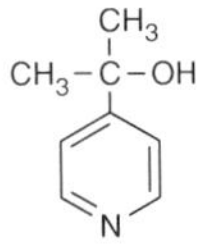
250 mg

S92,073-8
2,4-BIS(4-METHOXYPHENYL)-6-PHENYLPYRYLIUM TETRAFLUOROBORATE
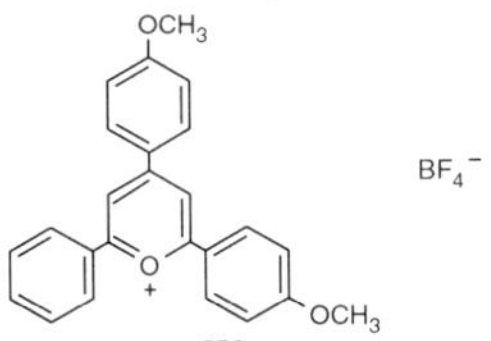
250 mg

S92,367-2
2,4-BIS(4-METHOXYPHENYL)-6-METHYLPYRYLIUM PERCHLORATE
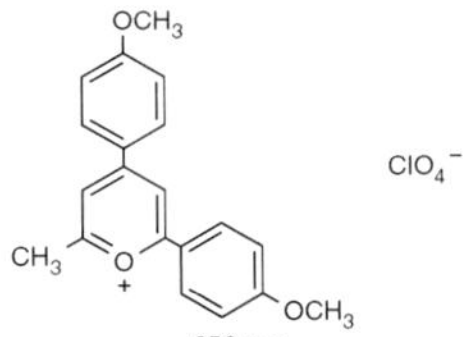
250 mg

S92,188-2
5,6-DIHYDRO-2,4-(4-METHOXYPHENYL)-NAPHTHO(1,2-B)PYRYLIUM TETRAFLUOROBORATE
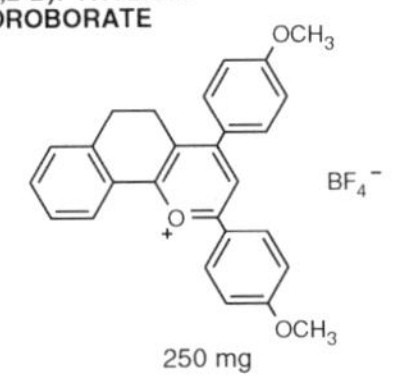
250 mg

S92,190-4
5,6-DIHYDRO-2-(4-METHOXYPHENYL)-4-PHENYLNAPHTHO(1,2-B)PYRYLIUM TETRAFLUOROBORATE
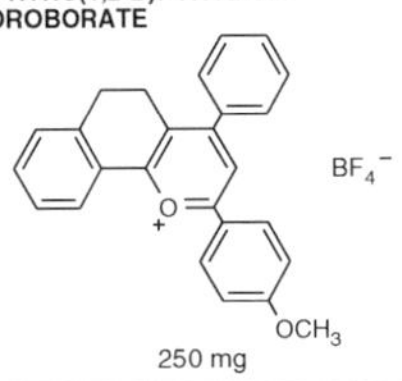
250 mg

S93,188-8
2,4-DI-TERT-BUTYL-6-(2-METHOXYPHENYL)PYRYLIUM TETRAFLUOROBORATE
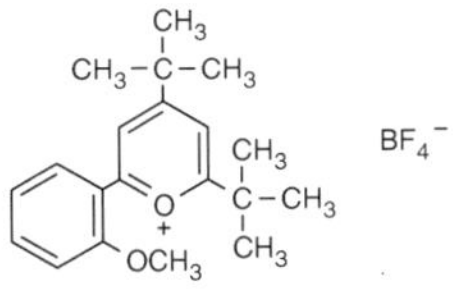
250 mg

S92,187-4
2,4,6-TRIS(4-METHOXYPHENYL)-PYRYLIUM TETRAFLUOROBORATE
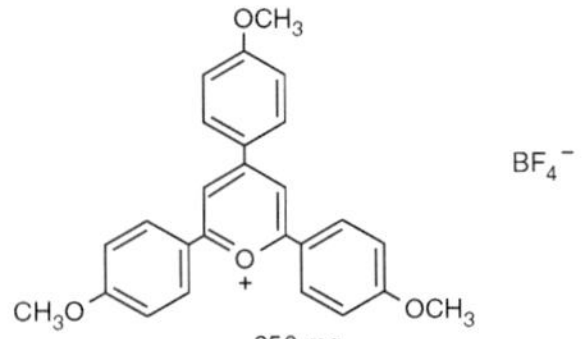
250 mg

S92,683-3
2,4-DIPHENYL-6-(4-METHOXYPHENYL)-PYRYLIUM TETRAFLUOROBORATE
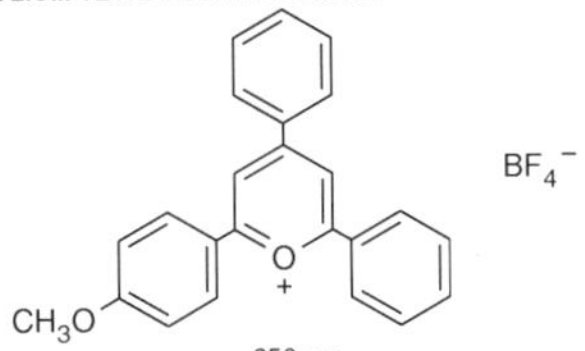
250 mg

S92,684-1
2,4-DIPHENYL-6-(4-METHOXYPHENYL)-PYRYLIUM PERCHLORATE
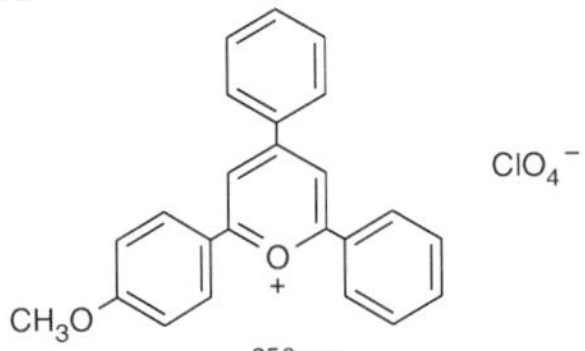
250 mg

S92,070-3
2,6-DIPHENYL-4-(4-METHOXYPHENYL)-PYRYLIUM TETRACHLOROFERRATE
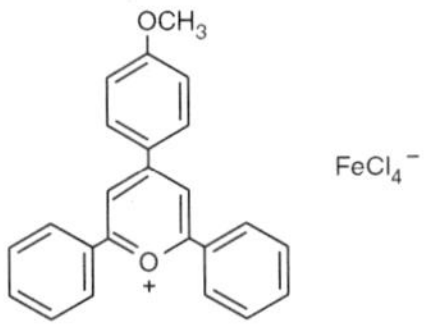
250 mg

S92,100-9
2,6-DIPHENYL-6-(4-METHOXYPHENYL)-PYRYLIUM TETRACHLOROFERRATE
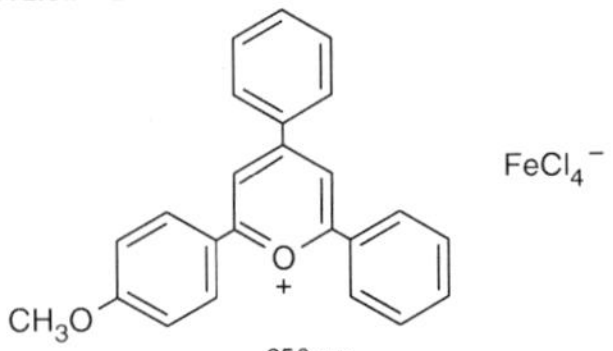
250 mg

S70,145-9
7-(2-METHOXYPHENYL)-5,6,8,9-TETRAHYDRODIBENZO(C,H)XANTHYLIUM PERCHLORATE
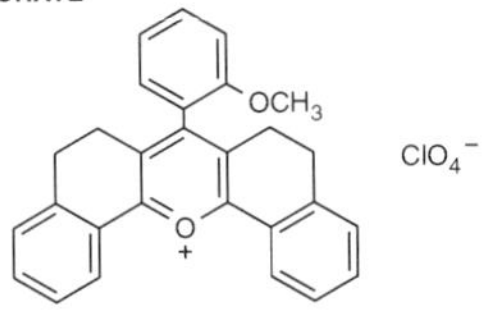
5 gm

S70,143-2
7-(4-METHOXYPHENYL)-5,6,8,9-TETRAHYDRODIBENZO(C,H)XANTHYLIUM TETRAFLUOROBORATE
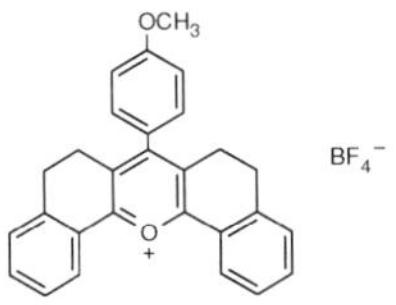
250 mg

S69,273-5
7-(4-METHOXYPHENYL)-5,6,8,9-TETRAHYDRODIBENZO(C,H)XANTHYLIUM PERCHLORATE
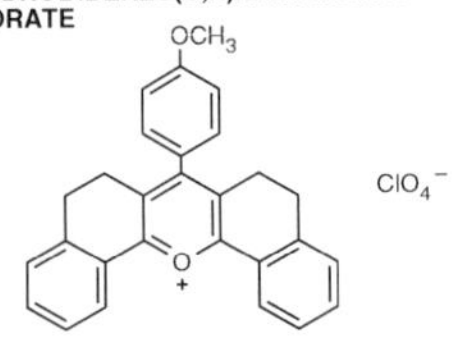
5 gm

S92,129-7
2,4-DIPHENYL-6-METHYLPYRYLIUM PERCHLORATE
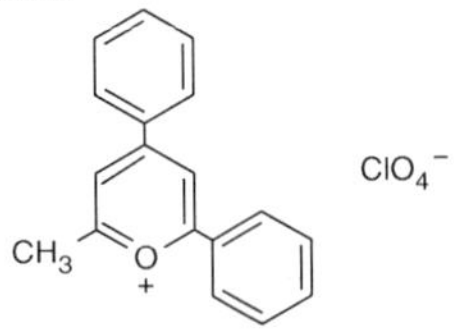
250 mg

S92,057-6
2,4-DIPHENYL-6-METHYLPYRYLIUM IODIDE
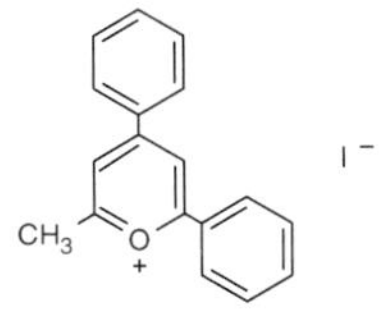

250 mg

S92,135-1
2,6-DIPHENYL-4-ISOPROPYLPYRYLIUM TETRAFLUOROBORATE
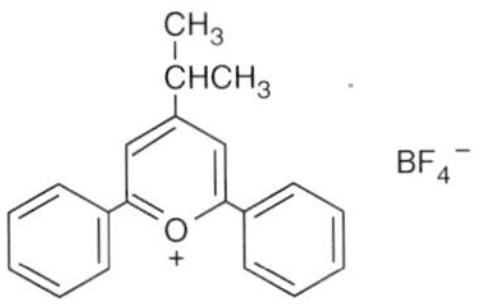

250 mg

S92,369-9
2-TERT-BUTYL-4,6-DIPHENYLPYRYLIUM TETRACHLOROFERRATE
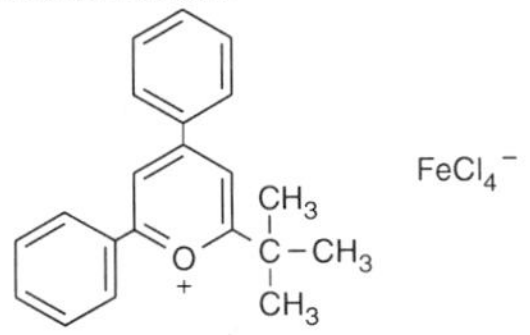

250 mg

S92,390-7
2,4,6-TRIPHENYLPYRYLIUM TETRACHLOROFERRATE
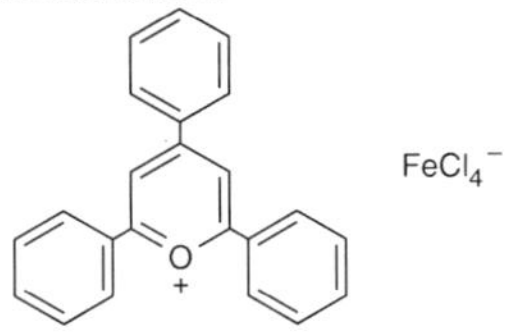

250 mg

S92,375-3
4-BENZYL-2,6-DIPHENYLPYRYLIUM TETRAFLUOROBORATE
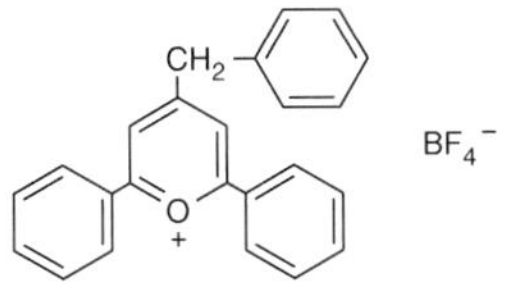

250 mg

S92,398-2
2,3,4,6-TETRAPHENYLPYRYLIUM TETRAFLUOROBORATE
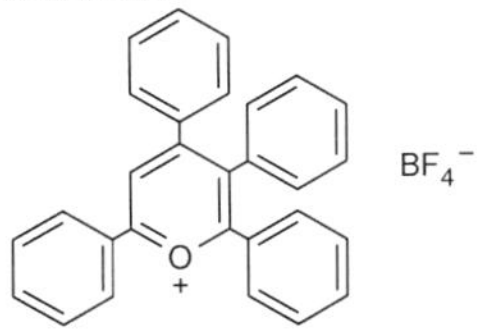

250 mg

S87,395-0
4,4'-BIS(2,6-DIPHENYLPYRYLIUM TETRAFLUOROBORATE)
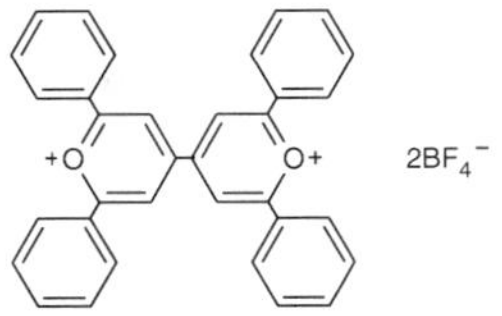

250 mg

S93,186-1
4-(P-TOLYL)-2,3,6-TRIPHENYLPYRYLIUM TETRACHLOROFERRATE
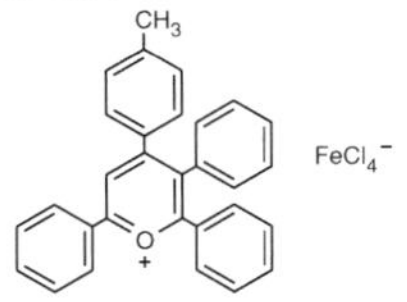

250 mg

S87,392-6
1,4-PHENYLENE-4,4'-BIS(2,6-DIPHENYL-4-PYRYLIUM TETRAFLUOROBORATE)
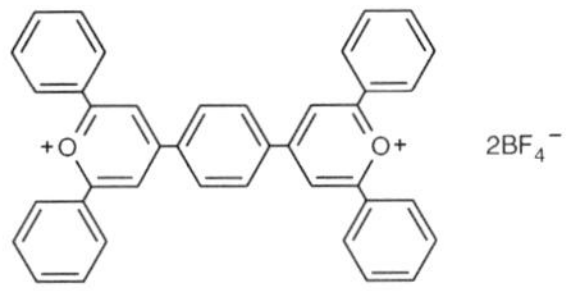

250 mg

S92,075-4
5,6-DIHYDRO-4-(4-METHOXYPHENYL)-2-PHENYLNAPHTHO(1,2-B)PYRYLIUM TETRAFLUOROBORATE
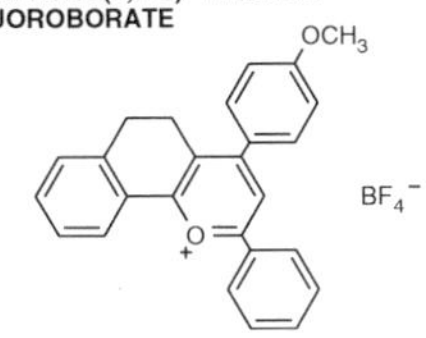

250 mg

S94,005-4
2-(4-CYANOPHENYL)-5-(4-PENTYLPHENYL)PYRYLIUM PERCHLORATE
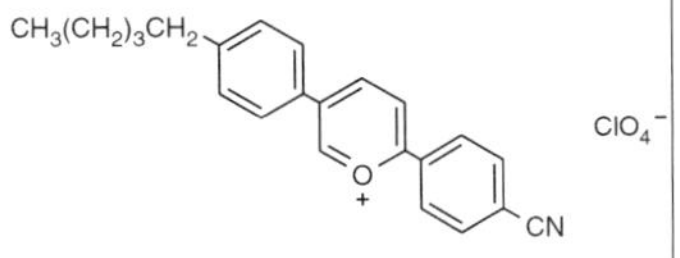

250 mg

S92,689-2
2,6-DICYCLOPROPYLPYRYLIUM PERCHLORATE
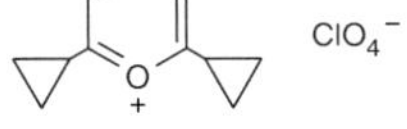

250 mg

S92,013-4
2,4,6-TRIMETHYLPYRYLIUM TETRAFLUOROBORATE
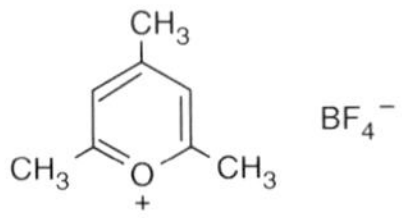

250 mg

S92,019-3
2,3,4,6-TETRAMETHYLPYRYLIUM PERCHLORATE
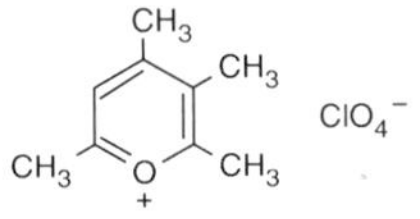

250 mg

S92,035-5
2,6-DI-TERT-BUTYLPYRYLIUM PERCHLORATE
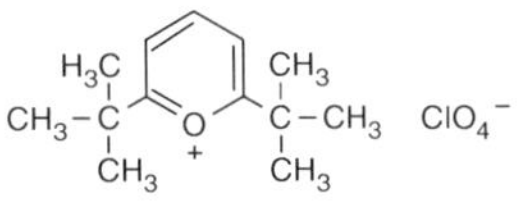

250 mg

S92,335-4
2,6-DI-TERT-BUTYL-4-METHYLPYRYLIUM TETRAFLUOROBORATE
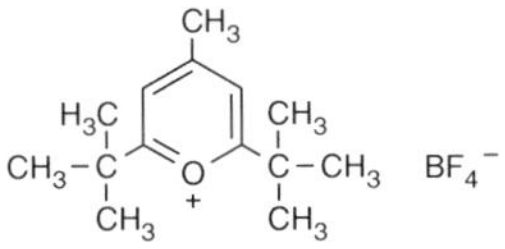

5 gm

S92,356-7
2,6-DI-TERT-BUTYL-4-METHYLPYRYLIUM PERCHLORATE
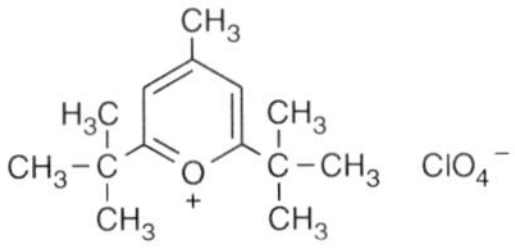

250 mg

S92,049-5
2,6-DI-TERT-BUTYL-4-ISOPROPYLPYRYLIUM TETRACHLOROFERRATE
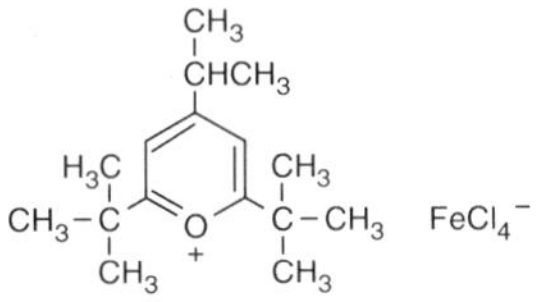

250 mg

S92,096-7
2,6-DIPHENYLPYRYLIUM PERCHLORATE
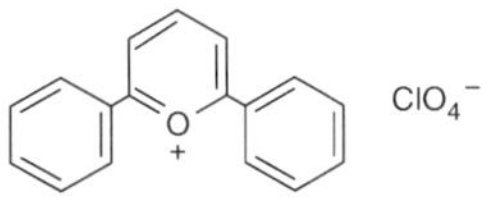

5 gm

S92,090-8
2,4,6-TRI-TERT-BUTYLPYRYLIUM TETRAFLUOROBORATE
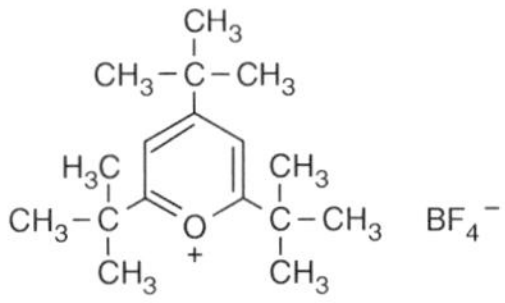

250 mg

S92,341-9
2,4,6-TRI-TERT-BUTYLPYRYLIUM IODIDE
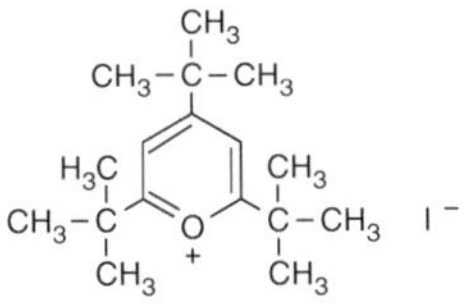

250 mg

S92,218-8
2,4,6-TRI-TERT-BUTYL-3-METHYLPYRYLIUM TETRAFLUOROBORATE
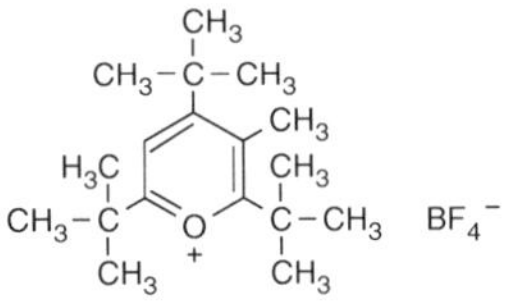

250 mg

S92,687-6
2,4-DIPHENYL-6-ISOPROPYLPYRYLIUM PERCHLORATE
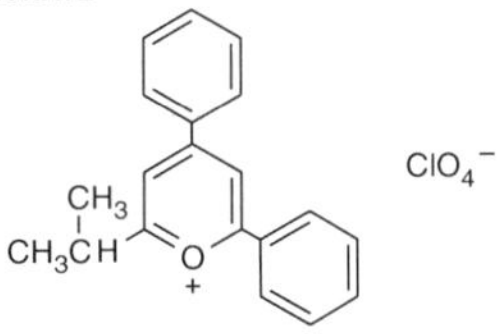

250 mg

S92,688-4
2,6-DIMETHYL-4-PHENYLPYRYLIUM PERCHLORATE
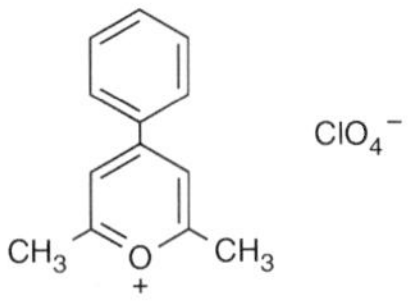

250 mg

S92,261-7
2,6-DIPHENYLPYRYLIUM PERCHLORATE
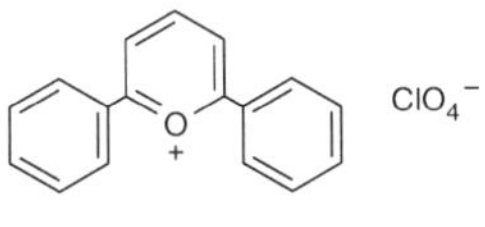

5 gm

S92,052-5
2,6-DIPHENYLPYRYLIUM TETRAFLUOROBORATE
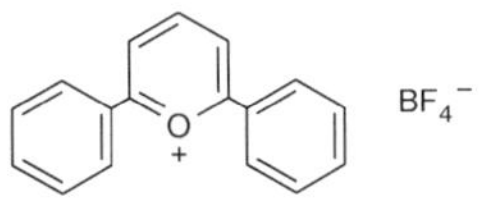

250 mg

S92,259-5
2,4-DIPHENYL-6-METHYLPYRYLIUM TETRAFLUOROBORATE
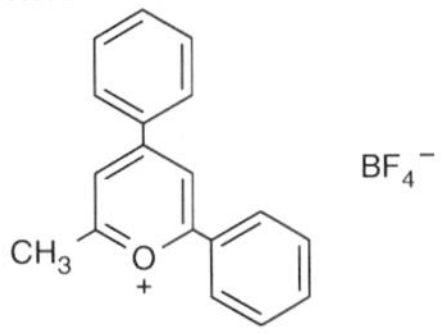

250 mg

S91,980-2
2,6-DIPHENYL-4-METHYLPYRYLIUM TETRAFLUOROBORATE
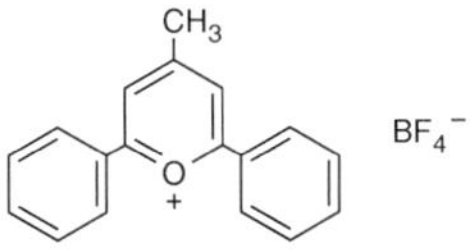

250 mg

S92,088-6
4-ETHYL-2,6-DIPHENYLPYRYLIUM TETRAFLUOROBORATE
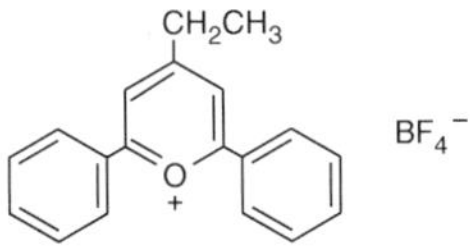

250 mg

S92,385-0
2,6-DI-TERT-BUTYL-4-PHENYLPYRYLIUM TETRAFLUOROBORATE
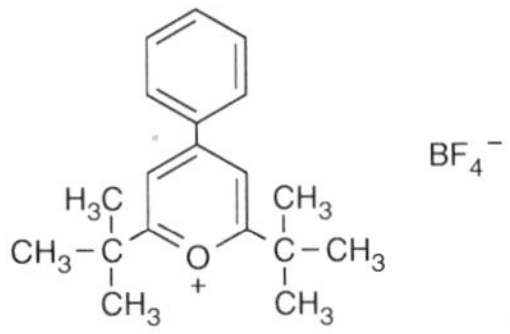

250 mg

S92,102-5
2,4-DIPHENYL-6-ETHYLPYRYLIUM PERCHLORATE
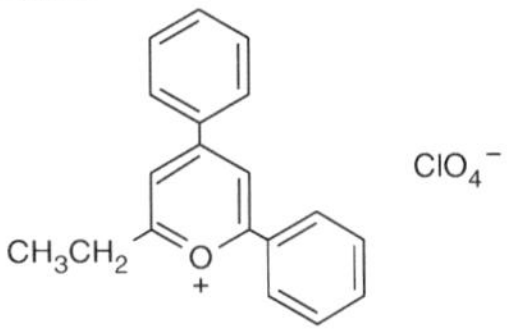

250 mg

S92,060-6
4-BENZYL-2,6-DI-TERT-BUTYLPYRYLIUM TETRAFLUOROBORATE
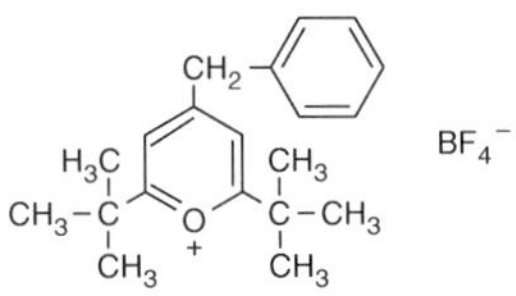

250 mg

S93,185-3
2,6-DIPHENYL-4-TERT-BUTYLPYRYLIUM PERCHLORATE
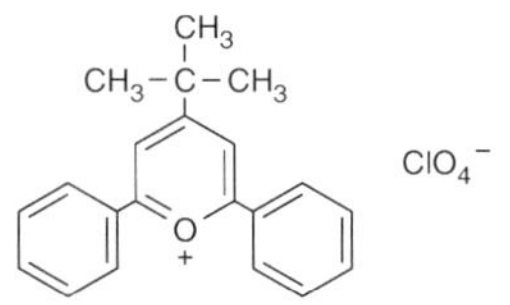
250 mg

S92,185-8
2,4-DIPHENYL-6-STYRYLPYRYLIUM PERCHLORATE
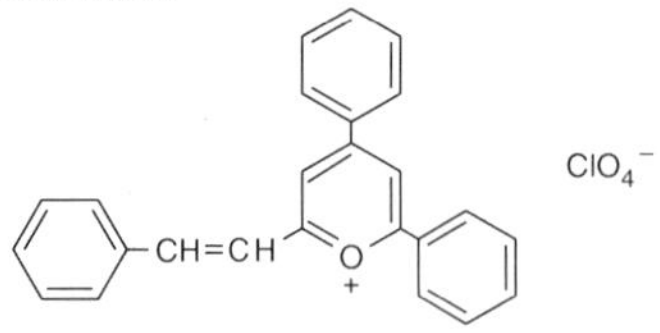
250 mg

S92,686-8
2,3,5,6-TETRAPHENYLPYRYLIUM TETRAFLUOROBORATE
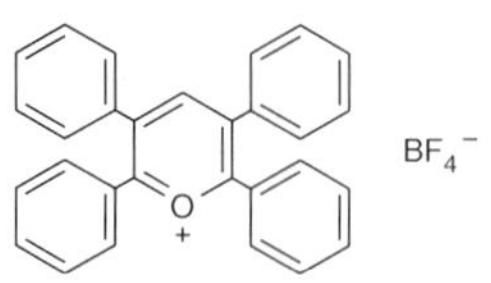
250 mg

S92,068-1
2,6-DIPHENYL-4-(P-TOLYL)PYRYLIUM TETRAFLUOROBORATE
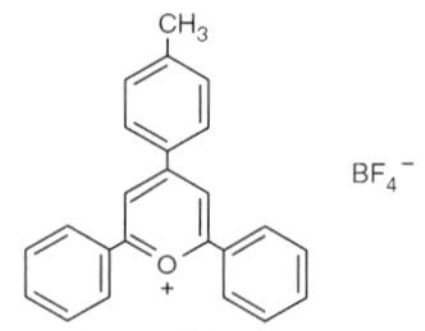
250 mg

S92,191-2
2,4-DIPHENYL-6-(P-TOLYL)PYRYLIUM TETRACHLOROFERRATE
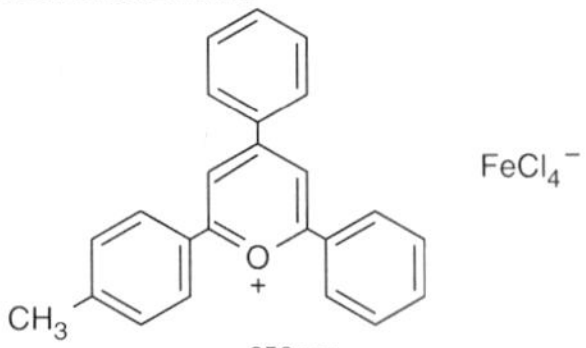
250 mg

S92,074-6
2,6-BIS(P-TOLYL)-4-PHENYLPYRYLIUM TETRACHLOROFERRATE
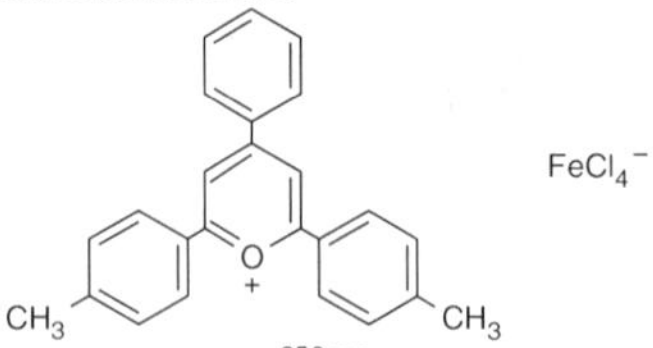
250 mg

S92,196-3
2,6-BIS(P-TOLYL)-4-PHENYLPYRYLIUM TETRAFLUOROBORATE
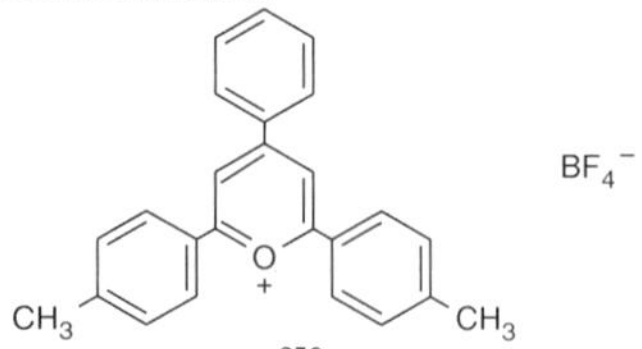
250 mg

S92,077-0
2,4-DIPHENYL-6-(2,4,6-TRIMETHYLPHENYL)PYRYLIUM TETRACHLOROFERRATE
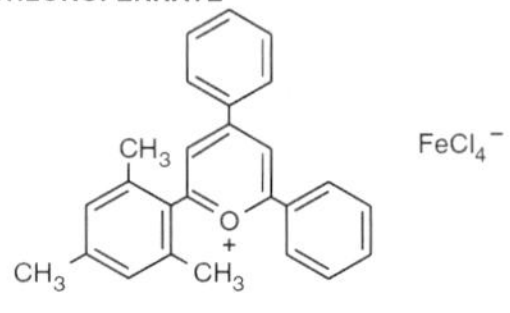
250 mg

S93,184-5
2,4-DIPHENYL-5,6,7,8-TETRAHYDROBENZOPYRYLIUM TETRAFLUOROBORATE
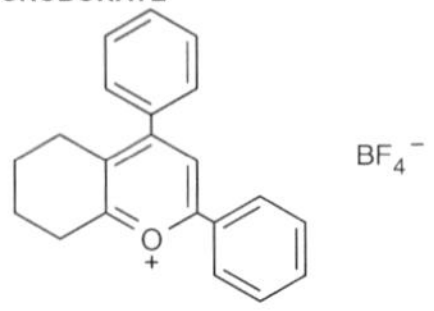
250 mg

S68,921-1
7-PHENYL-5,6,8,9-TETRAHYDRODIBENZO-(C,H)XANTHYLIUM PERCHLORATE
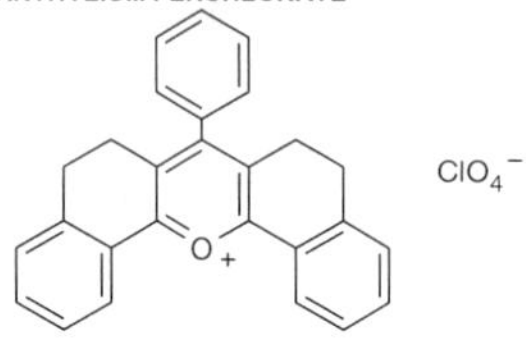
5 gm

S69.659-5
7-(4-METHYLPHENYL)-5,6,8,9-TETRAHYDRODIBENZO(C,H)XANTHYLIUM TETRAFLUOROBORATE
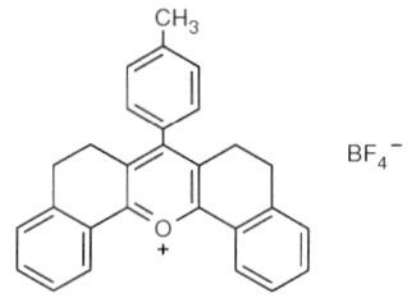
5 gm

S92,319-2
8-BENZYLIDENE-2,4-DIPHENYL-5,6,7,8-TETRAHYDRO-BENZOPYRYLIUM TETRAFLUOROBORATE
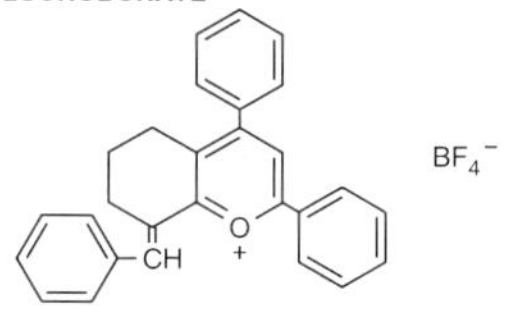
250 mg

S81,645-0
N,N-DIETHYL-BETA-(2-PYRIDYLOXY)-PHENETHYLAMINE
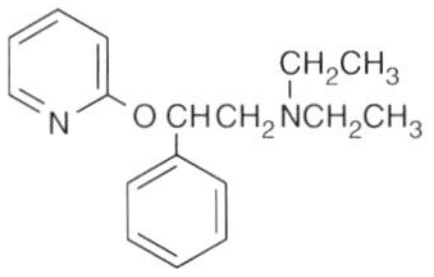
250 mg

S71,419-4
1-(4-METHYLBENZYL)-3-(4-METHYLBENZYLOXY)PYRIDINIUM CHLORIDE
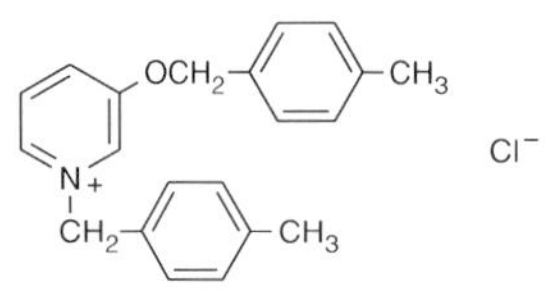
5 gm

S62,468-3
2-(4-BENZYLOXYSTYRYL)-1-METHYLPYRIDINIUM IODIDE
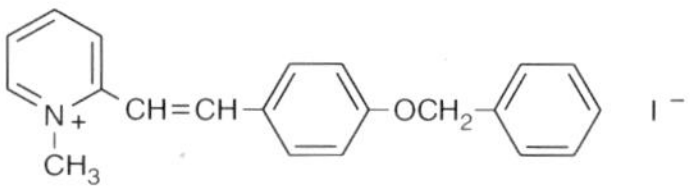
250 mg

S64,180-4
2-(4-BENZYLOXYSTYRYL)-1-ETHYLPYRIDINIUM IODIDE
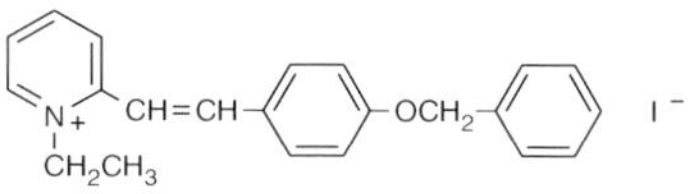
5 gm

S79,853-3
3,4-DIMETHOXYPHENYL 2-PYRIDINECARBOXYLATE
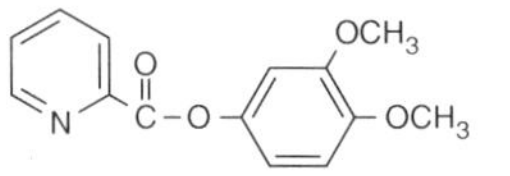
250 mg

S90,353-1
1-[2-(3,4-DIMETHOXYBENZYLIDENE)-HYDRAZINOCARBONYLMETHYL]-PYRIDINIUM CHLORIDE
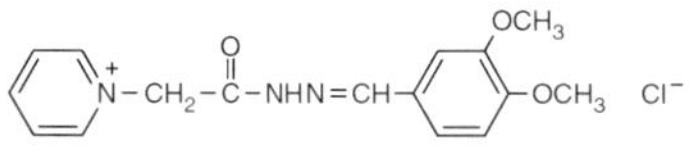
250 mg

S84,081-5
2,6-BIS(2,3-DIMETHOXYSTYRYL)-PYRIDINE
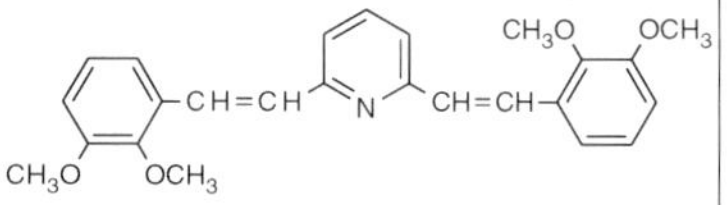
250 mg

S79,339-6
4,6-DIETHOXY-2-PYRIDINECARBOXYLIC ACID
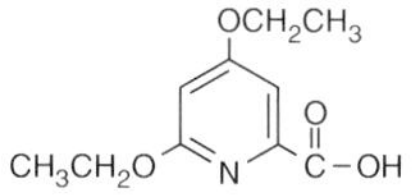
250 mg

S85,117-5
2-(3-PYRIDYL)ETHYL N-(2,4-DIMETHOXYPHENYL)CARBAMATE
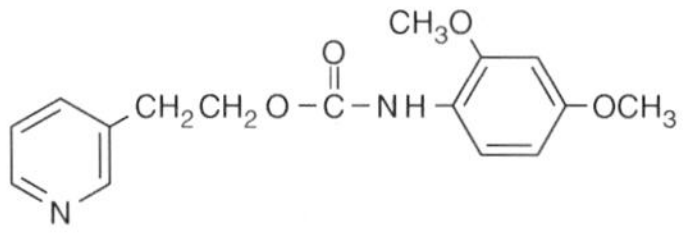
250 mg

S91,036-8
1-(3,5-DICHLORO-2-PYRIDYL)-3-(2,4-DIMETHOXYPHENYL)UREA
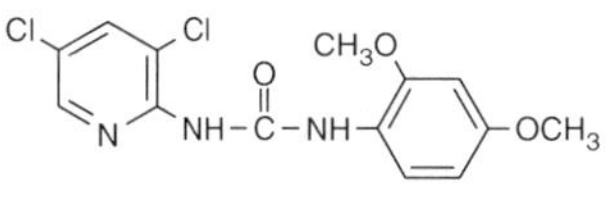
250 mg

S75,044-1
1-(2,4-DIMETHOXYPHENYL)-3-(5-METHYL-2-PYRIDYL)UREA
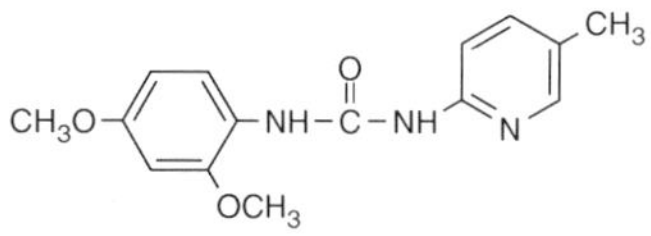
5 gm

S72,980-9
1-(2,4-DIMETHOXYPHENYL)-3-(3-PYRIDYLMETHYL)UREA
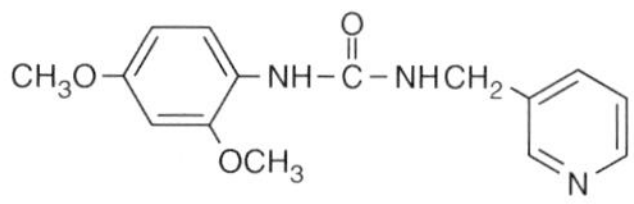
5 gm

S91,045-7
1-(2,5-DIMETHOXYPHENYL)-3-(2-PYRIDYLMETHYL)UREA
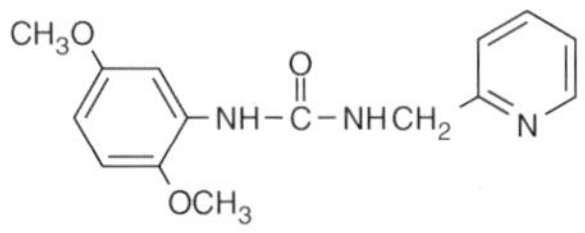
250 mg

S71,874-2
3-PYRIDYLMETHYL N-(2,5-DIMETHOXYPHENYL)CARBAMATE
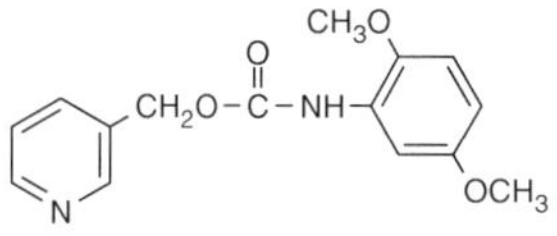
250 mg

S45,815-5
3,5-DIAMINO-2,6-DIMETHOXYPYRIDINE DIHYDROCHLORIDE
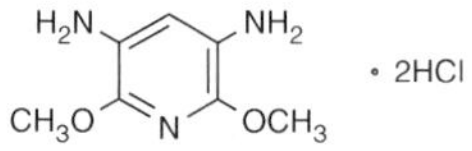
5 gm

S90,291-8
2-[3-(6-METHYL-2-PYRIDYL)-UREIDO]ACETIC (3,4-DIMETHOXYBENZYLIDENE)HYDRAZIDE
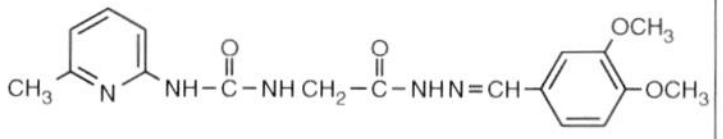
250 mg

S91,046-5
1-(2,5-DIMETHOXYPHENYL)-3-(3-METHYL-2-PYRIDYL)UREA
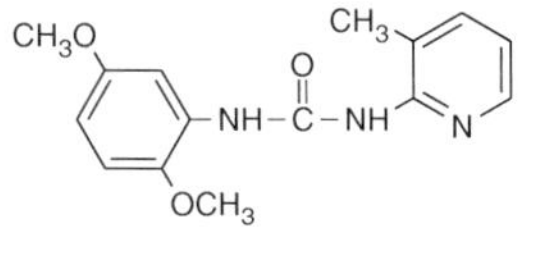
250 mg

S68,826-6
5-CHLORO-2-PYRIDYL N-(2,5-DIMETHOXYPHENYL)CARBAMATE
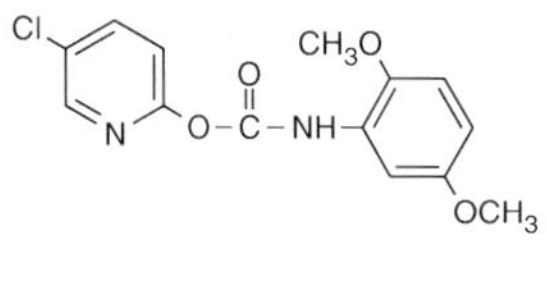
250 mg

S90,514-3
1-(2-CHLORO-3-PYRIDYL)-3-(2,5-DIMETHOXYPHENYL)UREA
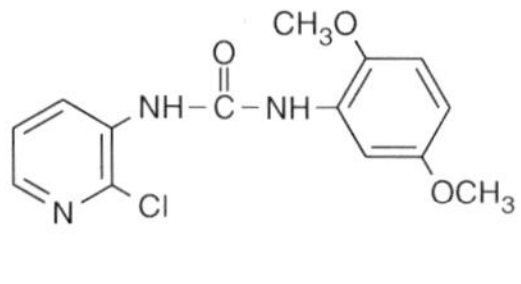
250 mg

S95,482-9
METHYL 6-CHLORO-5-CYANO-4-(DIMETHOXYPHENYL)-2-PYRIDINECARBOXYLATE
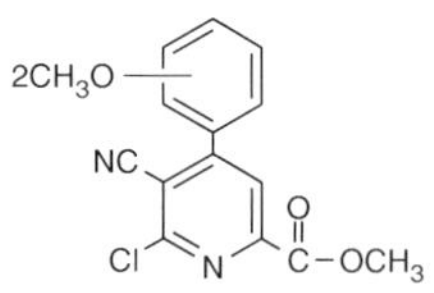
50 mg

S72,361-4
2-(5-METHYL-2-PYRIDYLUREIDO)ACETIC N2-[N-(2,5-DIMETHOXYPHENYL)CARBAMOYL]-HYDRAZIDE
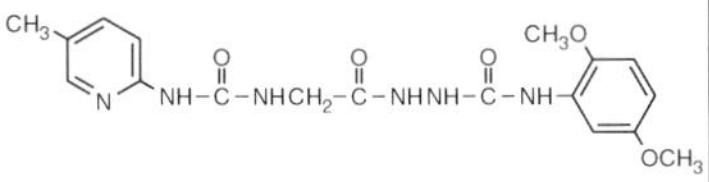
5 gm

S68,360-4
1-(2,5-DIMETHOXYPHENYL)-3-(3-PYRIDYL)UREA
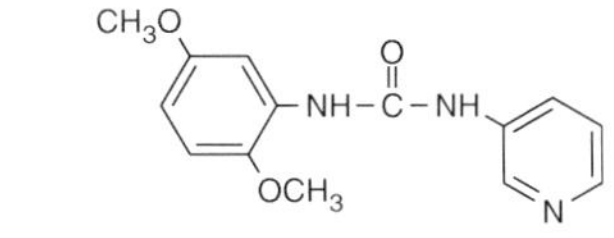
250 mg

S92,076-2
2,4,6-TRIS(4-METHOXYPHENYL)-THIOPYRYLIUM TETRAFLUOROBORATE
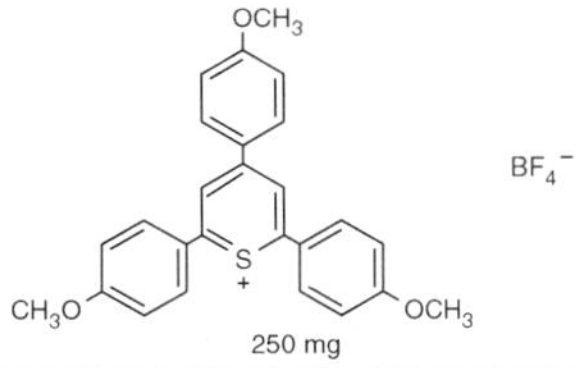
250 mg

S92,379-6
2,6-BIS(4-METHOXYPHENYL)-4-PHENYLPHOSPHORIN
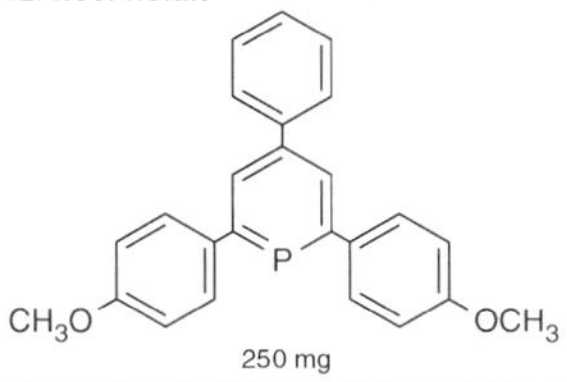
250 mg

S92,079-7
2,4,6-(4-METHOXYPHENYL)PHOSPHORIN
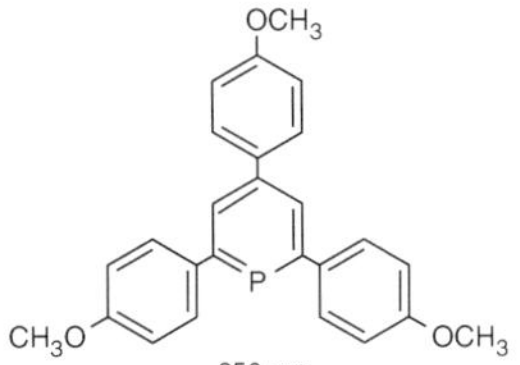
250 mg

S41,485-9
(2-PYRIDYLMETHYLAMINO)-ACETALDEHYDE DIETHYL ACETAL
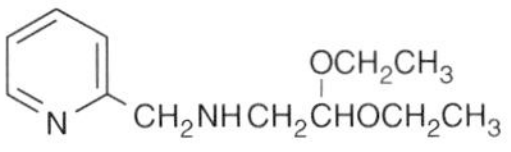
5 gm

S59,084-3
2-(2,3-DIMETHOXYSTYRYL)-1-METHYLPYRIDINIUM IODIDE
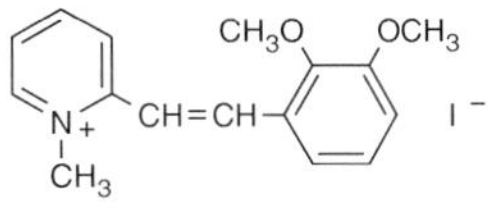
250 mg

S50,569-2
2-(3,4-DIMETHOXYSTYRYL)-1-METHYLPYRIDINIUM IODIDE
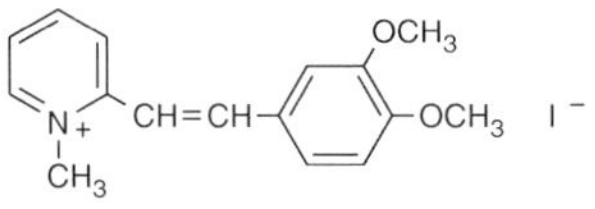
5 gm

S63,966-4
2-(3,4-DIMETHOXYSTYRYL)-1-ETHYLPYRIDINIUM IODIDE
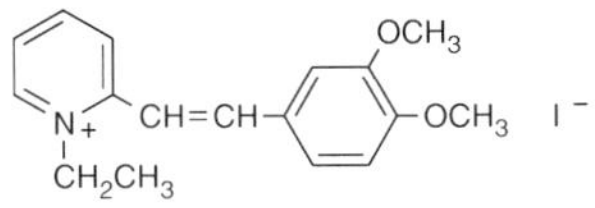
250 mg

S50,771-7
1-ALLYL-2-(3,4-DIMETHOXYSTYRYL)-PYRIDINIUM IODIDE
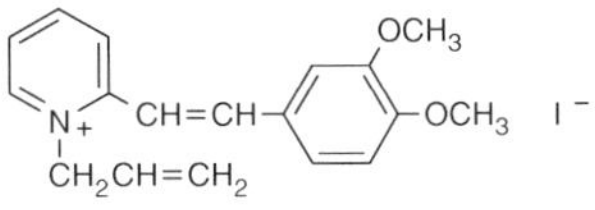
5 gm

S50,456-4
4-(3,4-DIMETHOXYSTYRYL)-1-METHYLPYRIDINIUM IODIDE
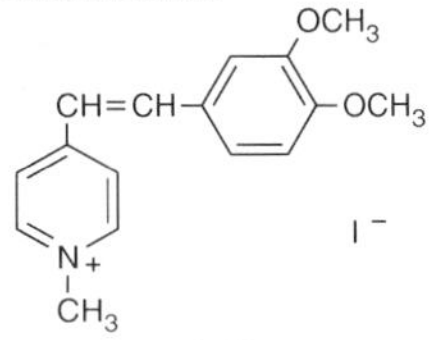
5 gm

S50,570-6
4-(3,4-DIMETHOXYSTYRYL)-1-ETHYLPYRIDINIUM IODIDE
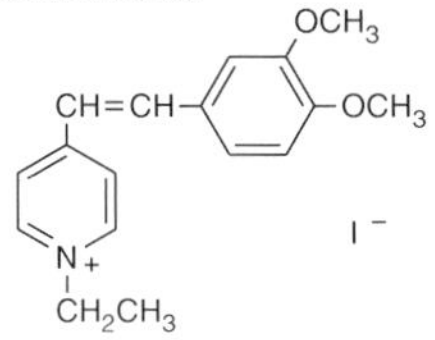
250 mg

S44,780-3
ETHYL 6-METHYL-3-PYRIDYLOXYACETATE
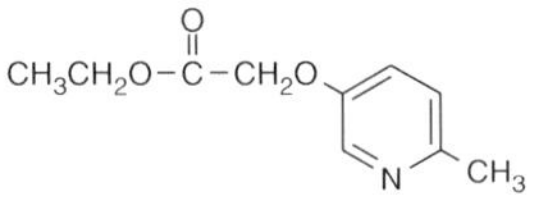
5 gm

S61,726-1
2-METHOXYPHENYL NICOTINATE
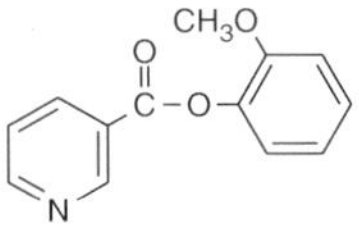
5 gm

S83,528-5
ETHYL 2-(3-PYRIDYLOXY)ACETATE
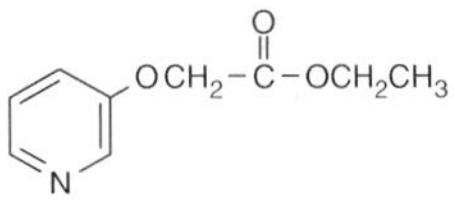
250 mg

S73,322-9
8-METHOXY-2-(4-PYRIDYLMETHYLENE)-1-BENZOSUBERONE
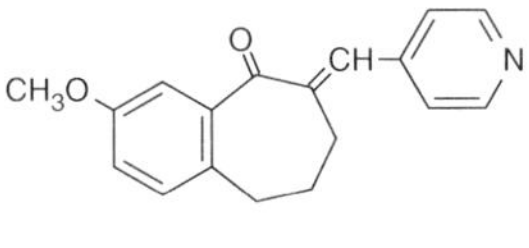
250 mg

S92,354-0
2-(4-METHOXYPHENYL)-6-PHENYLTHIOPYRYLIUM TETRAFLUOROBORATE
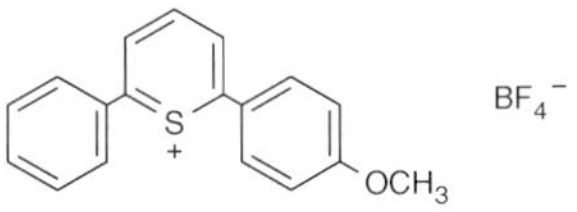
250 mg

S35,750-2
5-AMINO-2-BUTOXYPYRIDINE HYDROCHLORIDE
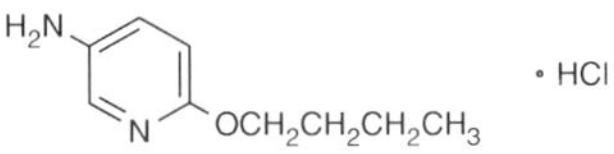
5 gm

S81,080-0
2-(3-AMINO-4-METHOXYPHENYL)PYRIDINE
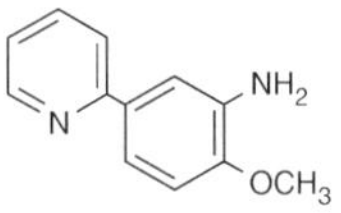
250 mg

S78,684-5
2-(2-AMINOETHOXY)PYRIDINE
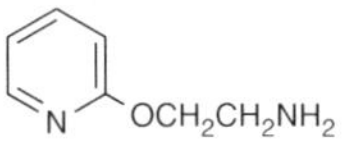
250 mg

S44,731-5
1,2-DIHYDRO-4-(METHOXYMETHYL)-6-METHYL-5-NITRO-2-OXONICOTINONITRILE
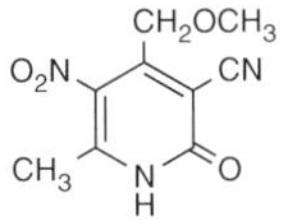
5 gm

S44,994-6
2-CHLORO-4-(METHOXYMETHYL)-6-METHYL-5-NITRONICOTINONITRILE
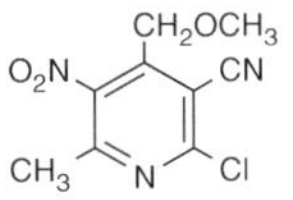
250 mg

S43,176-1
2-(3-METHOXYPROPYLAMINO)-5-NITROPYRIDINE
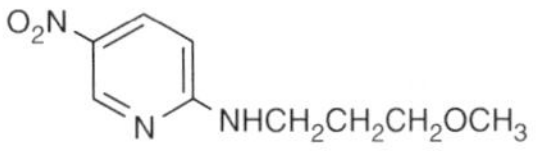
5 gm

S34,140-1
4-CHLORO-N-(6-METHOXY-3-PYRIDYL)-ANTHRANILIC ACID
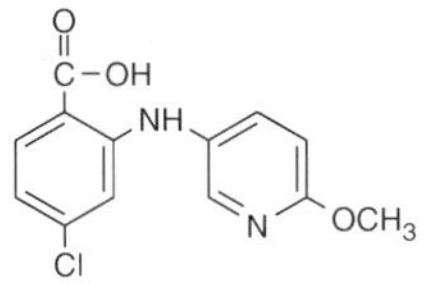
5 gm

S67,127-4
PYRILAMINE MALEATE
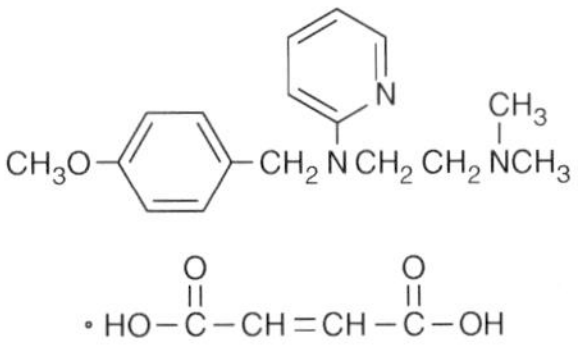
5 gm

S34,710-8
N,N-DIMETHYL-N'-(4-METHOXYBENZYL)-N'-(2-PYRIDYL)ETHYLENEDIAMINE
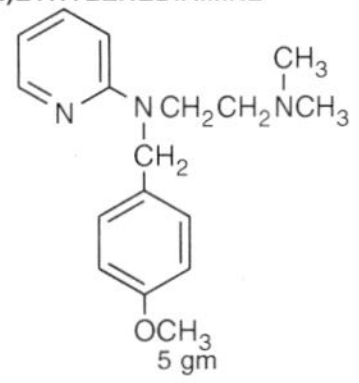
5 gm

S38,355-4
N-(6-BUTOXY-3-PYRIDYL)-2-(DIETHYLAMINO)ACETAMIDE
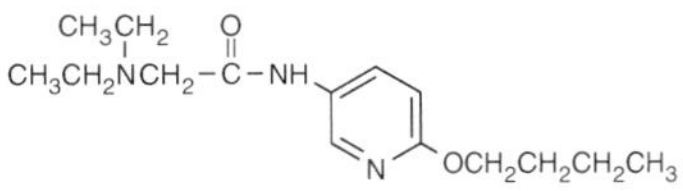
250 mg

S88,498-7
2-ETHOXY-1-ETHYL-5-NITROPYRIDINIUM TETRAFLUOROBORATE
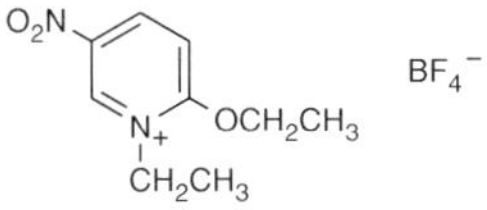
250 mg

S90,295-0
2-[3-(2-ETHOXYPHENYL)UREIDO]ACETIC (2-PYRIDYLMETHYLENE)HYDRAZIDE
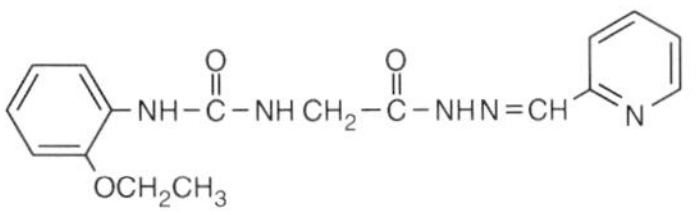
250 mg

S43,550-3
2-METHOXY-3-NITROPYRIDINE
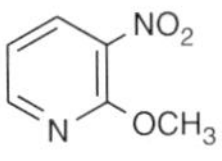
5 gm

S37,813-5
2-BUTOXY-5-NITROPYRIDINE
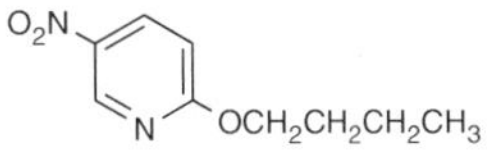
5 gm

S69,033-3
1-(2-METHOXY-5-METHYLPHENYL)-3-(3-PYRIDYL)-2-THIOUREA
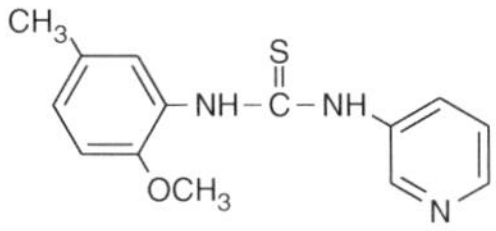
250 mg

S63,237-6
3-CYANO-1-(2-METHOXYETHYL)-PYRIDINIUM P-TOLUENESULFONATE

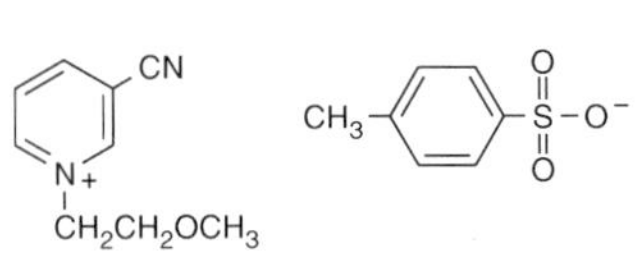

250 mg

S91,426-6
3-CARBAMOYL-1-(METHOXYMETHYL)-PYRIDINIUM CHLORIDE

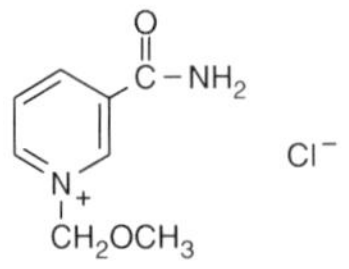

250 mg

S90,414-7
NICOTINIC (4-METHOXYBENZYLIDENE)-HYDRAZIDE

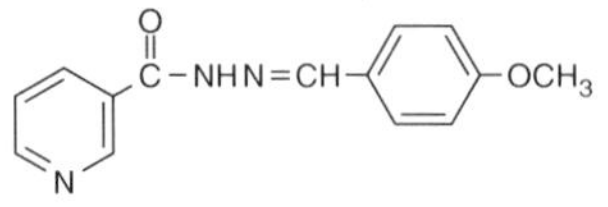

250 mg

S43,394-2
3-(2,4-DINITROPHENOXY)PYRIDINE

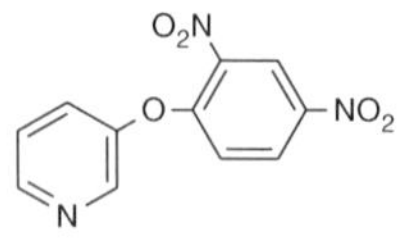

250 mg

S60,939-0
1-(2-METHOXYPHENYL)-3-(3-PYRIDYLMETHYL)UREA

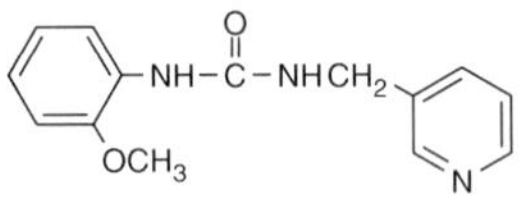

5 gm

S65,087-0
1-(3-CHLORO-2-METHOXYPHENYL)-3-(3-PYRIDYLMETHYL)UREA

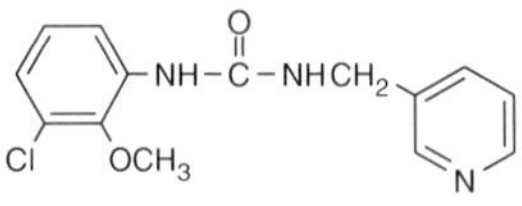

5 gm

S66,584-3
1-(2-METHOXY-5-METHYLPHENYL)-3-(3-PYRIDYLMETHYL)UREA

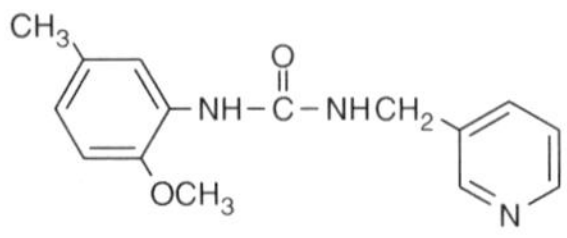

5 gm

S61,015-1
1-(2,5-DIMETHOXYPHENYL)-3-(3-PYRIDYLMETHYL)UREA

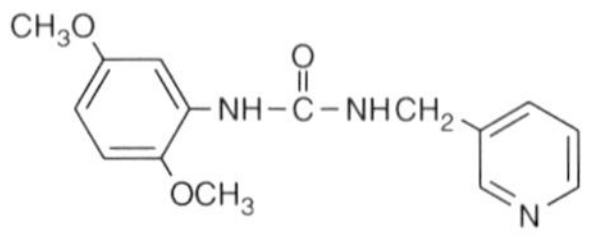

5 gm

S40,193-5
N-(DIPHENYLMETHOXYMETHYL)-NICOTINAMIDE

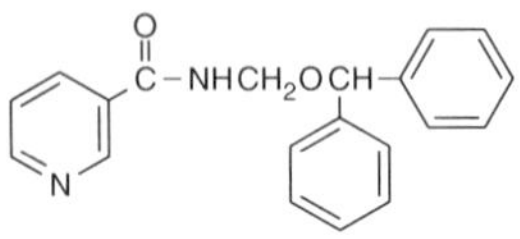

250 mg

S60,955-2
1-(3-METHOXYPHENYL)-3-(3-PYRIDYLMETHYL)UREA

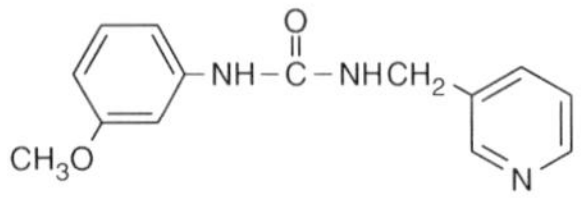

5 gm

S61,335-5
1-(P-ANISYL)-3-(3-PYRIDYLMETHYL)UREA

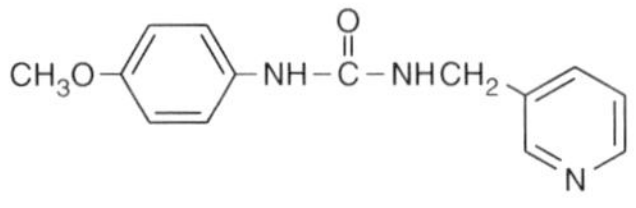

5 gm

S85,518-9
1-(4-METHOXYPHENYL)-3-(3-PYRIDYLMETHYL)UREA

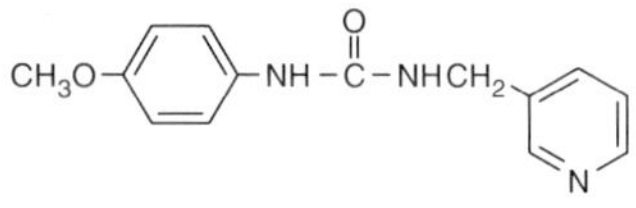

250 mg

S92,017-7
3-AMINO-6-CHLORO-5-CYANO-4-(METHOXYMETHYL)-2-METHYLPYRIDINE

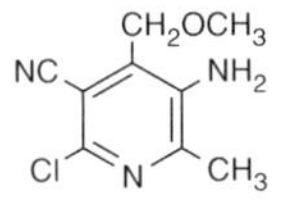

250 mg

S80,008-2
1-(2-METHOXY-4-NITROPHENYL)-3-(4-PYRIDYLMETHYL)UREA

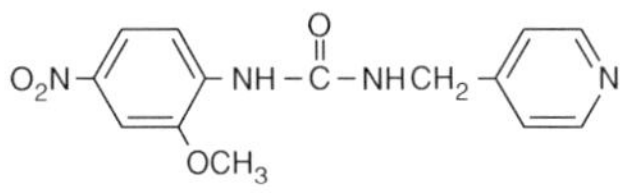

250 mg

S79,706-5
1-(2-METHOXY-4-NITROPHENYL)-3-(4-METHYL-2-PYRIDYL)UREA

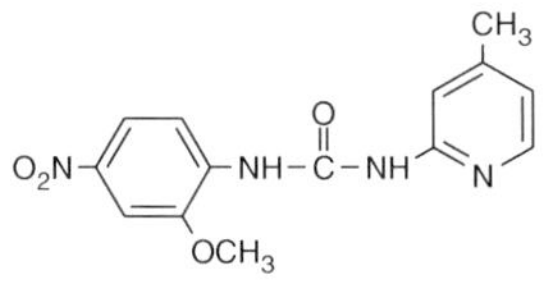

250 mg

S88,899-0
2-(2,6-DIETHYL-ALPHA-[2-(DIMETHYLAMINO)ETHOXY]BENZYL)-PYRIDINE HYDROCHLORIDE

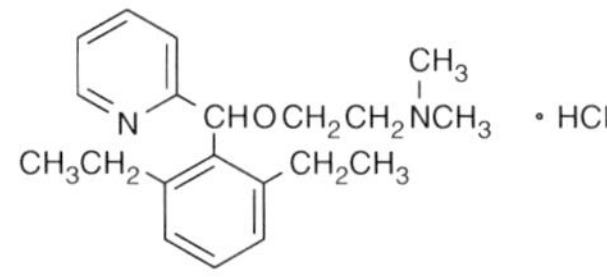

250 mg

S34,583-0
2,6-BIS(2-DIMETHYLAMINOETHOXY)-PYRIDINE

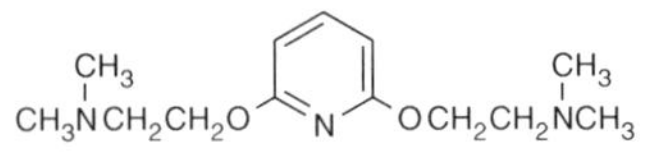

5 gm

S63,101-9
1-ALLYL-2-(4-METHOXYSTYRYL)-PYRIDINIUM IODIDE

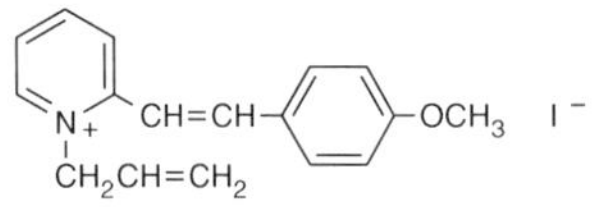

250 mg

S51,555-8
1-(4'-PHENOXYPHENACYL)-PYRIDINIUM IODIDE

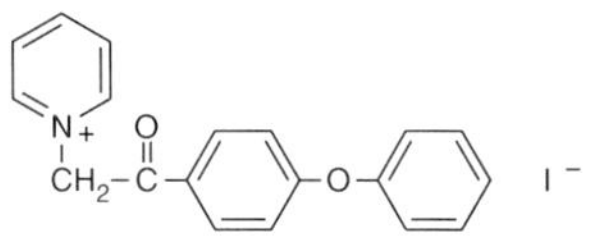

250 mg

S54,645-3
1-(2',5'-DIMETHOXYPHENACYL)-PYRIDINIUM IODIDE

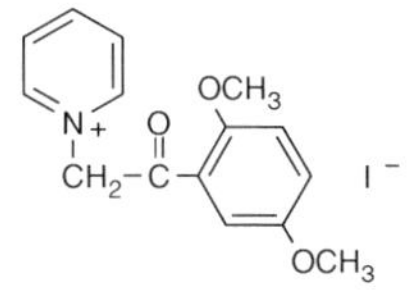

250 mg

S73,362-8
1-CINNAMYL-3-(N-[3-(1-METHYLETHOXY)-PROPYL]SULFAMOYL)PYRIDINIUM BROMIDE

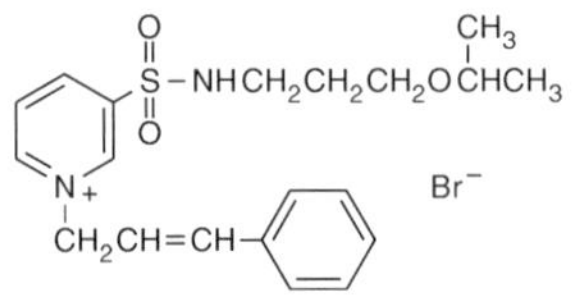

250 mg

S87,514-7
1-(METHOXYMETHYL)-4-STYRYLPYRIDINIUM BROMIDE

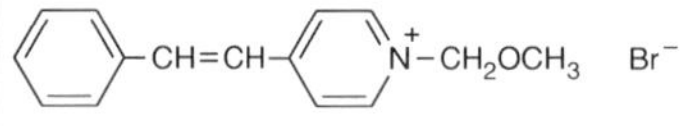

250 mg

S62,711-9
2-(2-ETHOXYSTYRYL)-1-METHYLPYRIDINIUM IODIDE

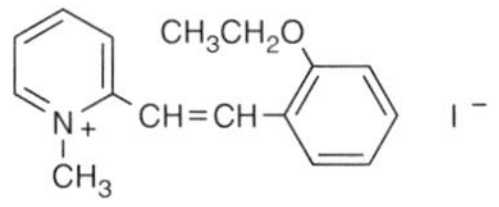

250 mg

S63,615-0
2-(2-ETHOXYSTYRYL)-1-ETHYLPYRIDINIUM IODIDE

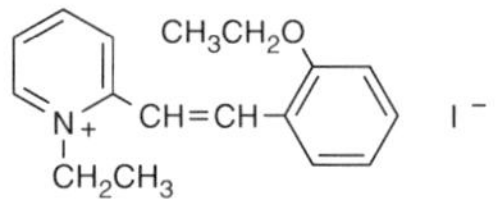

250 mg

S62,324-5
1-ETHYL-4-(2-METHOXYSTYRYL)-PYRIDINIUM IODIDE

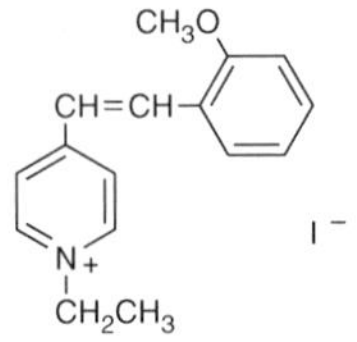

5 gm

S63,984-2
1-ETHYL-2-(4-METHOXYSTYRYL)-PYRIDINIUM IODIDE

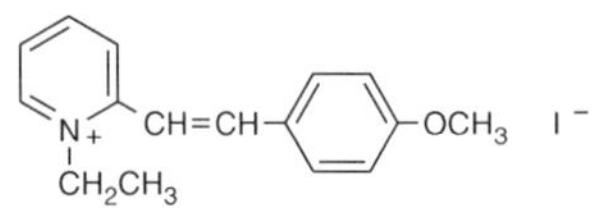

5 gm

S62,315-6
2-(4-ETHOXYSTYRYL)-1-METHYLPYRIDINIUM IODIDE

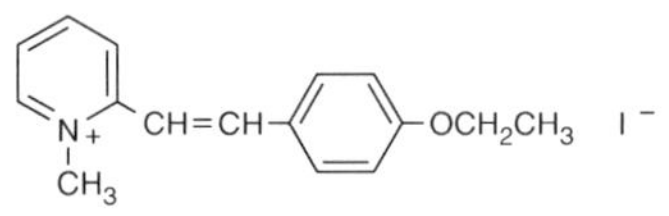

5 gm

S50,446-7
2-(4-ACETAMIDOSTYRYL)-1-METHYLPYRIDINIUM IODIDE

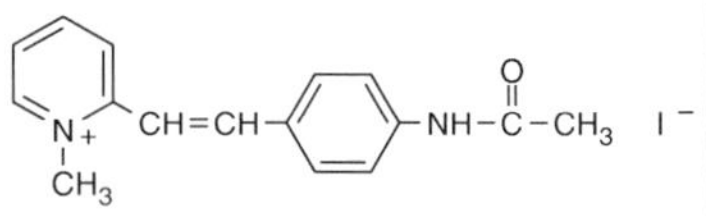

5 gm

S50,099-2
4-(4-METHOXYSTYRYL)-1-METHYLPYRIDINIUM IODIDE

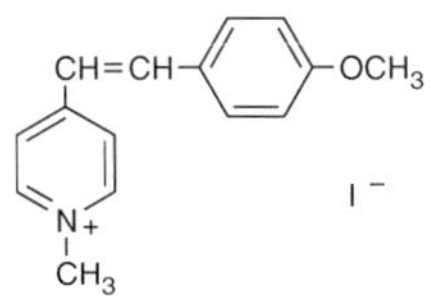

5 gm

S62,312-1
2-(2,5-DIMETHOXYSTYRYL)-1-METHYLPYRIDINIUM IODIDE

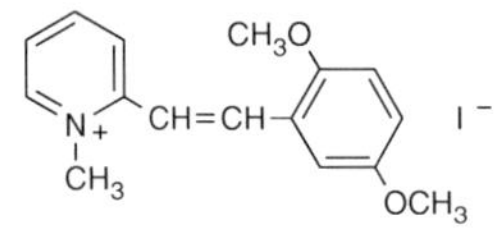

5 gm

S67,782-5
1-ETHYL-2-ETHOXYPYRIDINIUM TETRAPHENYLBORATE

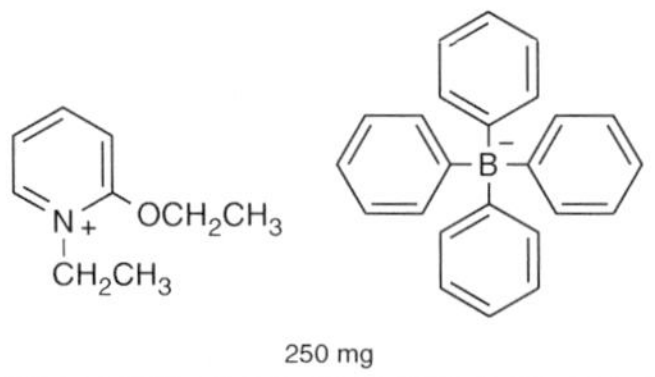

250 mg

S58,044-9
4-METHOXY-1,3-DIMETHYLPYRIDINIUM IODIDE

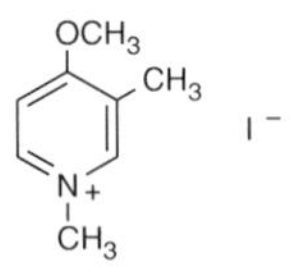

5 gm

S52,321-6
4-METHOXY-1-METHYLPYRIDINIUM IODIDE
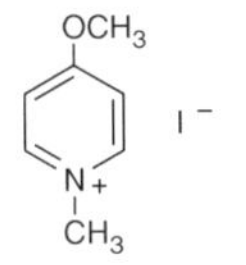
250 mg

S69,227-1
1-(4-PYRIDYLCARBONYL)-2-(CARBOXYMETHOXYACETYL)HYDRAZINE
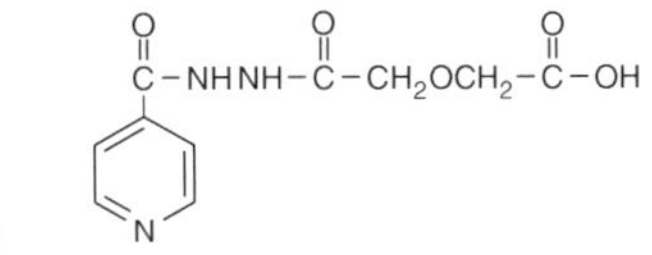
250 mg

S53,911-2
1-(4-ETHOXYPHENYL)-3-(2-PYRIDYL)-2-THIOUREA
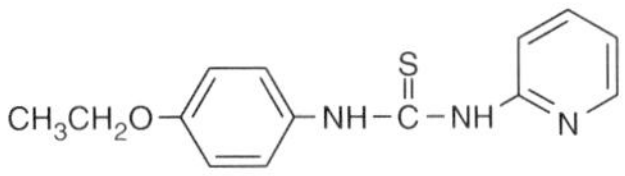
5 gm

S90,978-5
3-(2-METHOXY-5-NITROPHENYL)-1-(2-PYRIDYL)TRIAZENE
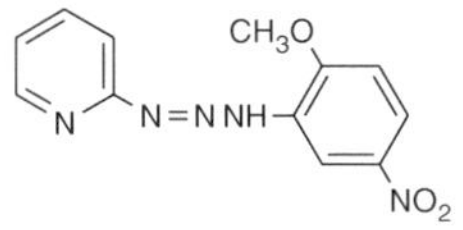
250 mg

S90,437-6
4-(2-METHOXY-5-METHYLPHENYL)-1-NICOTINOYLSEMICARBAZIDE
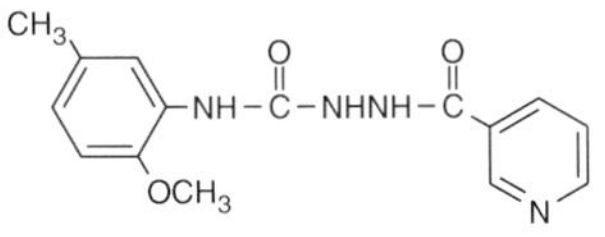
250 mg

S90,438-4
4-(2-METHOXY-5-METHYLPHENYL)-1-(4-PYRIDYLCARBONYL)SEMICARBAZIDE
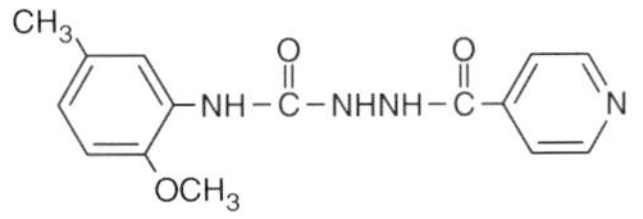
250 mg

S60,970-6
2-PYRIDYL N-(2,5-DIMETHOXYPHENYL)-CARBAMATE
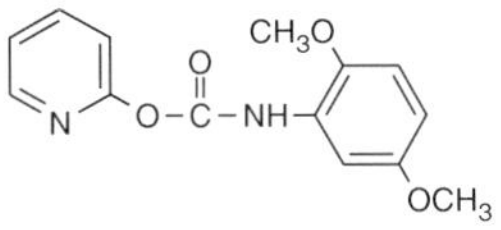
5 gm

S71,863-7
5-CHLORO-2-PYRIDYL N-(2-METHOXYPHENYL)CARBAMATE
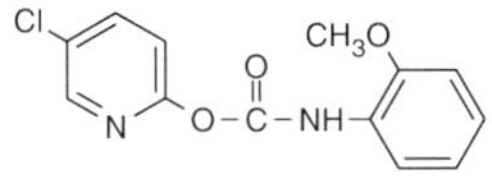
5 gm

S59,753-8
2-PYRIDYL N-(2-METHOXYPHENYL)-CARBAMATE
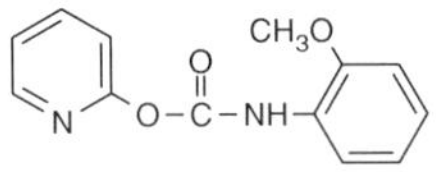
5 gm

S59,767-8
3-PYRIDYL N-(2-METHOXYPHENYL)-CARBAMATE
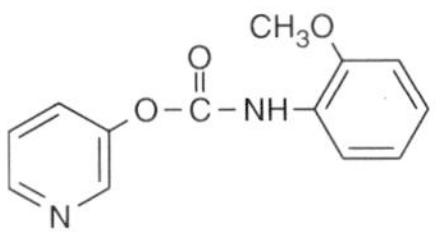
5 gm

S68,730-8
4-PYRIDYLMETHYL N-(2-METHOXYPHENYL)-CARBAMATE
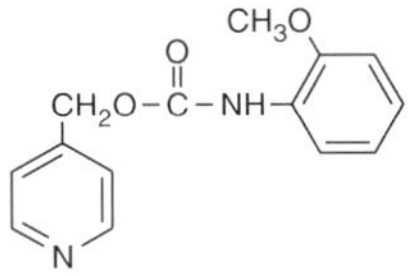
250 mg

S60,993-5
3-PYRIDYL N-(2,5-DIMETHOXYPHENYL)-CARBAMATE
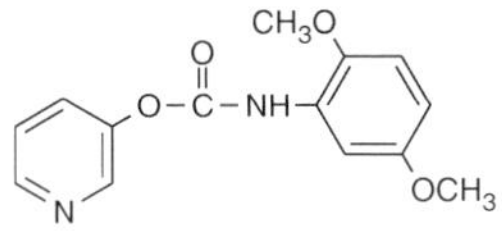
5 gm

S61,091-7
3-PYRIDYL N-(M-ANISYL)CARBAMATE
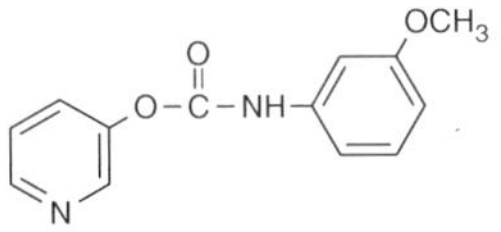
5 gm

S79,139-3
1-(4,6-DIMETHYL-2-PYRIDYL)-3-(2-METHOXYPHENYL)UREA
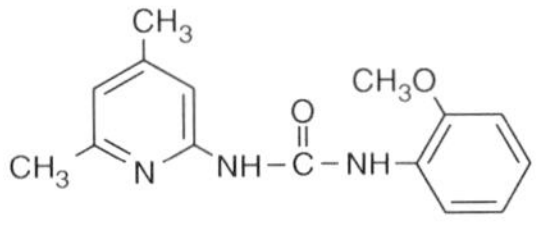
250 mg

S72,624-9
1-(5-CHLORO-2-PYRIDYL)-3-(2-METHOXY-5-METHYLPHENYL)-2-THIOUREA
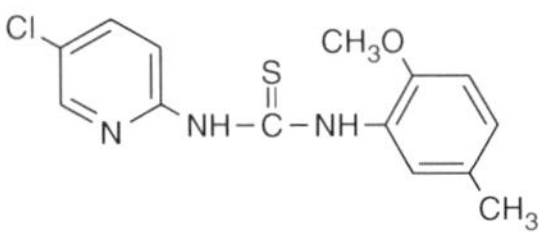
5 gm

S36,314-6
2-(METHOXYMETHYL)PYRIDINE
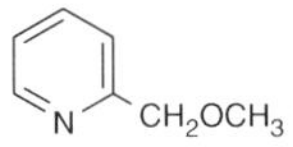
5 gm

S35,216-0
2-METHOXYMETHYL-6-METHYLPYRIDINE
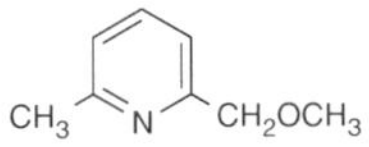
5 gm

S42,383-1
N-(PROPOXYMETHYL)NICOTINAMIDE
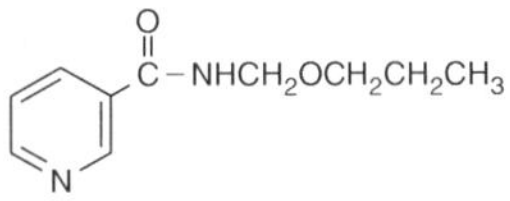
250 mg

S69,172-0
1-(4-METHOXY-2-METHYLPHENYL)-3-(2-PYRIDYL)UREA
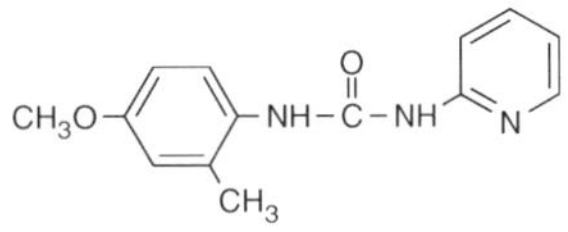
250 mg

S69,032-5
1-(4-METHOXY-2-METHYLPHENYL)-3-(3-PYRIDYLMETHYL)UREA
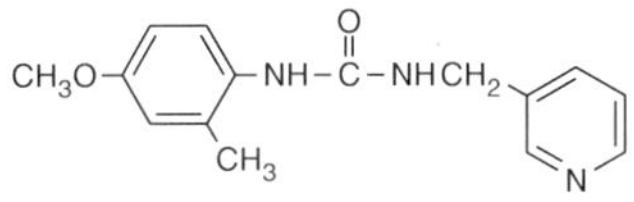
250 mg

S54,270-9
1-(4-METHOXYPHENYL)-3-(2-PYRIDYL)-2-THIOUREA
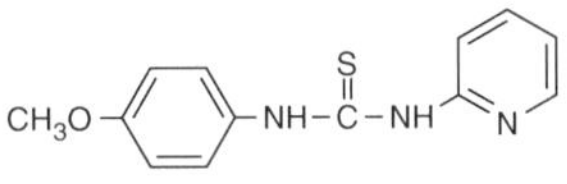
250 mg

S73,758-5
1-(4-METHOXYPHENYL)-3-(2-PYRIDYL)-2-THIOUREA
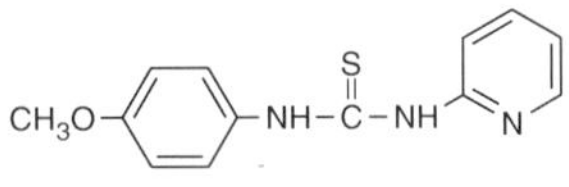
5 gm

S93,959-5
2-CHLORO-6-PHENOXYPYRIDINE
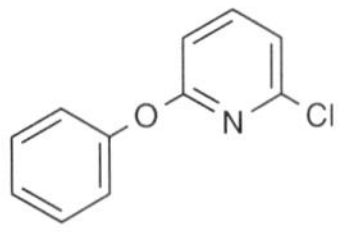
250 mg

S81,083-5
2-(4-METHOXYPHENYL)PYRIDINE
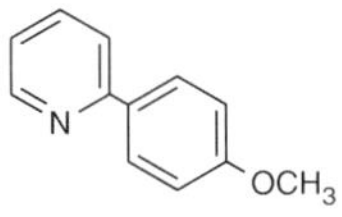
250 mg

S81,081-9
4-(4-METHOXYPHENYL)PYRIDINE
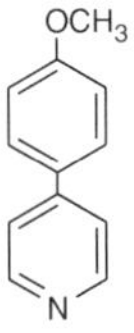
250 mg

S70,126-2
7-(4-METHOXYPHENYL)-5,6,8,9-TETRAHYDRODIBENZO(C,H)ACRIDINE
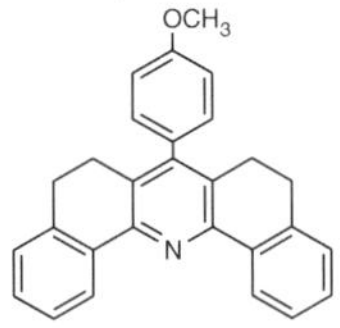
5 gm

S44,226-7
2-ETHOXYPYRIDINE
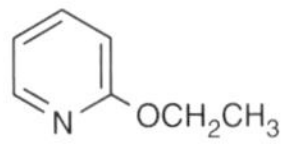
5 gm

S34,081-2
2-(ALLYLOXY)PYRIDINE
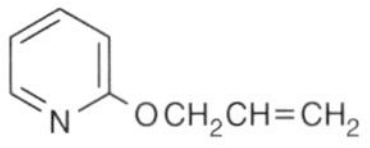
5 gm

S94,223-5
2-ALLYLOXYPYRIDINE
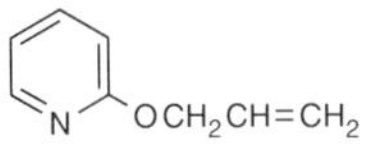
5 gm

S43,384-5
2-(2-METHOXYETHYL)PYRIDINE
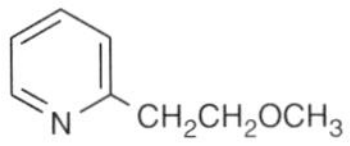
5 gm

S36,140-2
2-(2-ETHOXYETHYL)PYRIDINE
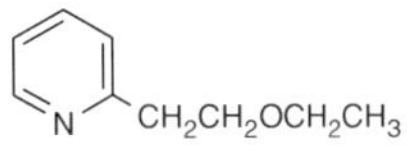
5 gm

S35,414-7
2-(OCTYLOXY)PYRIDINE
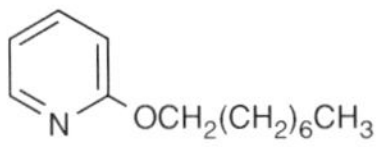
5 gm

S36,873-3
2-DODECYLOXYPYRIDINE
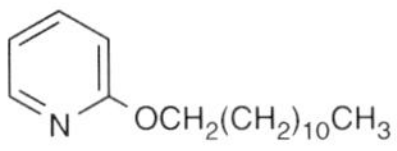
5 gm

S35,047-8
2-(HEXADECYLOXY)PYRIDINE
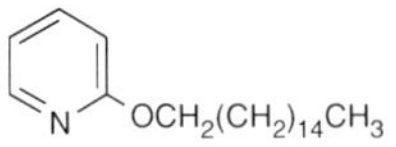
5 gm

S35,408-2
2-(OCTADECYLOXY)PYRIDINE
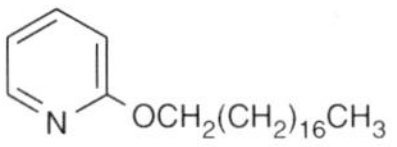
5 gm

S61,250-2
2-BROMO-6-METHOXYPYRIDINE
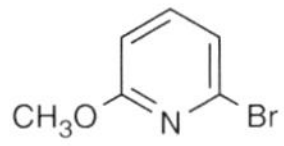
5 gm

S44,633-5
2-CHLORO-6-ETHOXYPYRIDINE
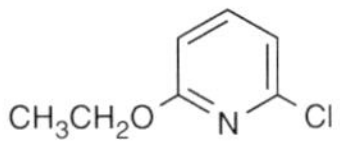
5 gm

S36,954-3
2-(2-METHOXYETHYL)-6-METHYLPYRIDINE
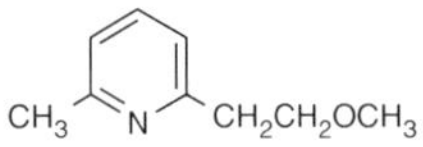
5 gm

S36,139-9
2-(2-ETHOXYETHYL)-6-METHYLPYRIDINE
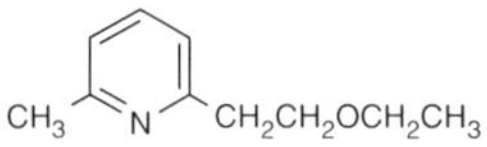
5 gm

S41,750-5
3-PENTYLOXYPYRIDINE
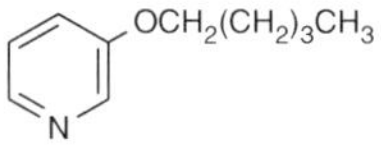
250 mg

S37,261-7
4-(2-METHOXYETHYL)PYRIDINE
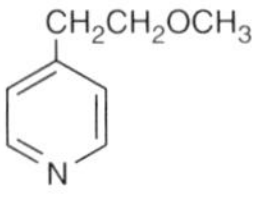
250 mg

S36,874-1
4-(2-ETHOXYETHYL)PYRIDINE
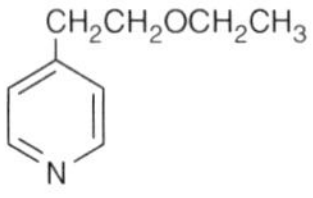
5 gm

S65,694-1
1-BENZYLOXY-4-(TERT-BUTYL)-PYRIDINIUM BROMIDE
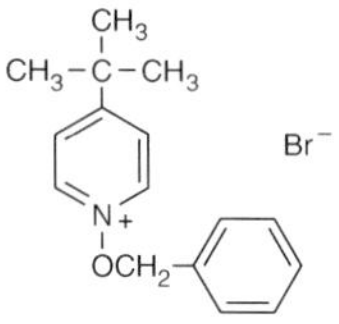
5 gm

S76,004-8
ALPHA-(ALPHA-MORPHOLINOBENZYL)-ALPHA-PHENYL-2-PYRIDINEMETHANOL
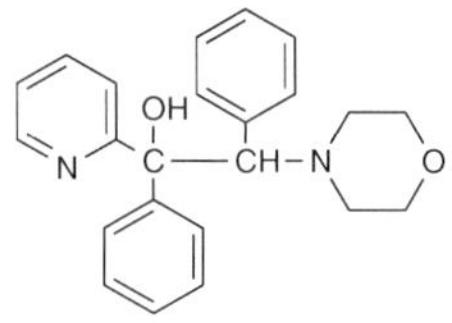
250 mg

S89,381-1
3,5-DICHLORO-2-PHENYL-4H-PYRAN-4-ONE
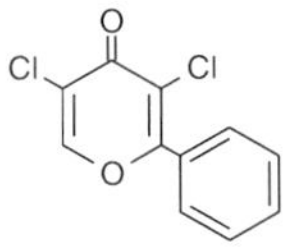
250 mg

S92,051-7
2,6-DIPHENYL-4H-PYRAN-4-ONE
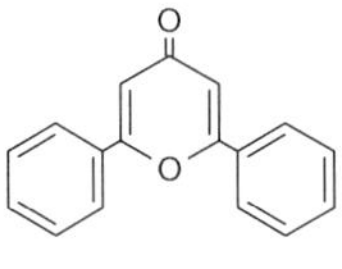
250 mg

S39,836-5
4-METHYLBENZYL NICOTINATE
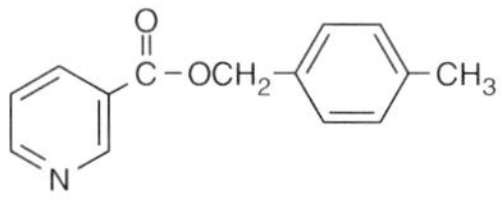
5 gm

S92,083-5
4-(1-BENZOTHIOPHEN-3-YLMETHYL)-2,4,6-TRIPHENYL-4H-PYRAN
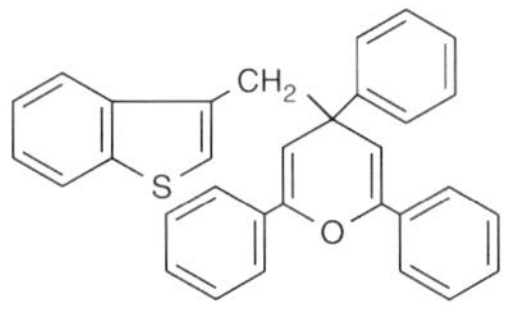
250 mg

S36,473-8
(3-PYRIDYLMETHYLAMINO)-ACETALDEHYDE DIETHYL ACETAL
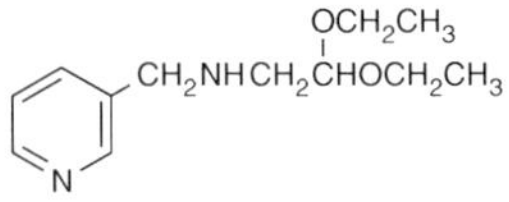
5 gm

S36,474-6
(4-PYRIDYLMETHYLAMINO)-ACETALDEHYDE DIETHYL ACETAL
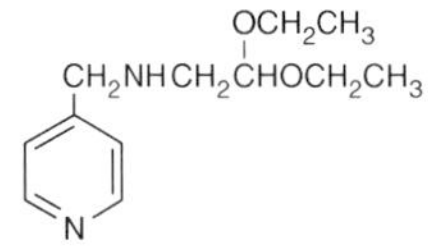
5 gm

S53,087-5
1,2,3-PROPANETRIYL TRINICOTINATE
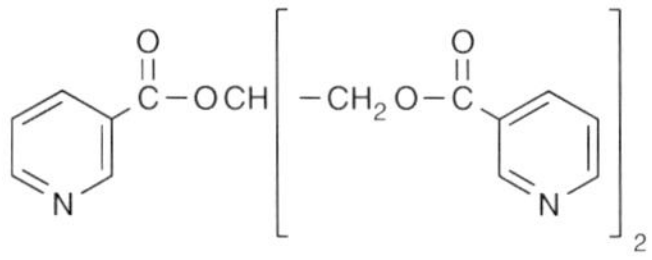
250 mg

S63,871-4
ETHYLENE GLYCOL BIS(3-PYRIDINIUMPROPIONATE) DICHLORIDE
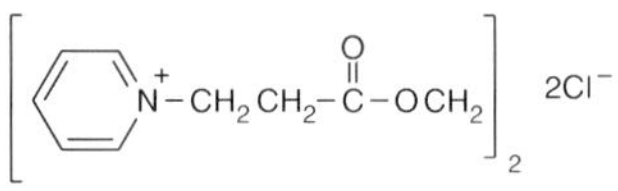
250 mg

S74,014-4
ALSTONINE PERCHLORATE
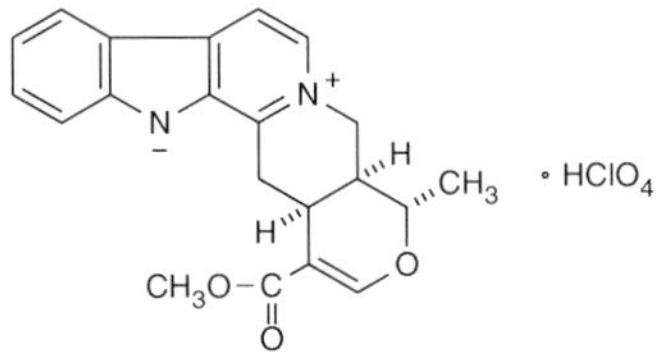
250 mg

S35,718-9
2-(ACETOXYMETHYL)-3-ACETOXYPYRIDINE
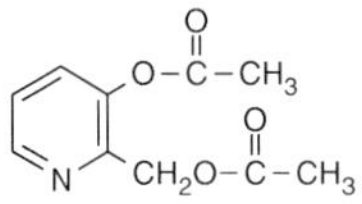
5 gm

S94,640-0
4,6-DIMETHYL-2-OXO-2H-PYRAN-5-CARBOXYLIC ACID
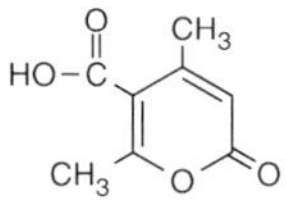
250 mg

S87,951-7
2-METHYL-3,4,6-PYRIDINETRICARBOXYLIC ACID
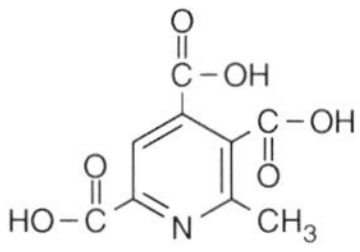
5 gm

S72,560-9
N-(6-METHYL-2-PYRIDYL)-3,4,5,6-TETRACHLOROPHTHALAMIC ACID
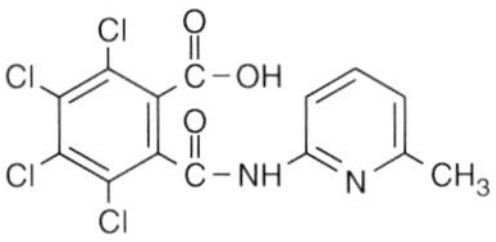
5 gm

S87,848-0
N-(3-METHYL-2-PYRIDYL)-SUCCINAMIC ACID
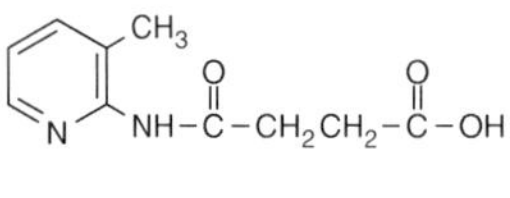
5 gm

S87,801-4
N-(6-METHYL-2-PYRIDYL)-SUCCINAMIC ACID
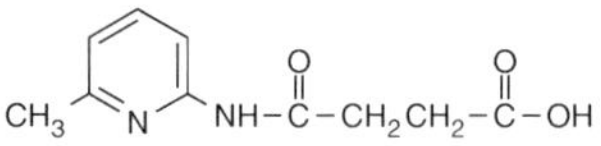
5 gm

S87,724-7
DIMETHYL 3,4-PYRIDINEDICARBOXYLATE
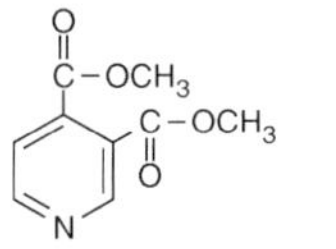
250 mg

S93,175-6
6-MERCAPTO-2-(4-PYRIDYL)PURINE
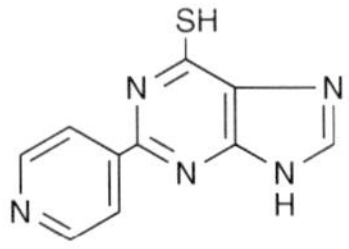
250 mg

S69,691-9
2-MERCAPTO-5-NITROPYRIDINE
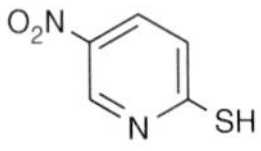
250 mg

S78,200-9
2-(2-MERCAPTO-4,5,6-TRIMETHYL-3-PYRIDYL)BENZOIC ACID

5 gm

S92,299-4
2,4,6-TRIPHENYLTHIOPYRYLIUM IODIDE
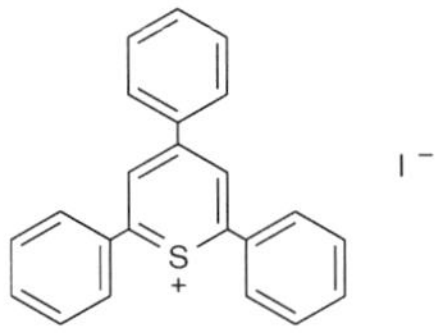
250 mg

S92,300-1
2,3,4,5,6-PENTAPHENYLTHIOPYRYLIUM TETRAFLUOROBORATE

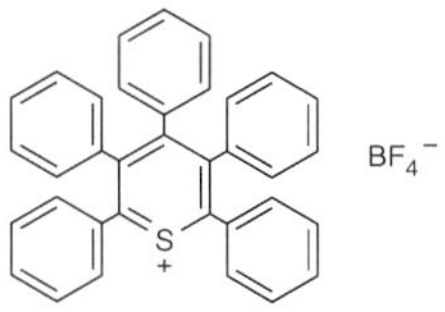

250 mg

S91,076-7
5-NITRO-2-PYRIDYL N-[4-(METHYLTHIO)-PHENYL]CARBAMATE

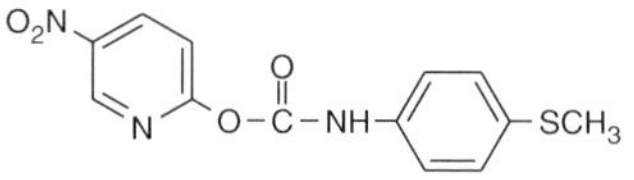

250 mg

S65,995-9
4-ETHYLTHIO-N-[2-(2-PYRIDYL)ETHYL]-1,8-NAPHTHALIMIDE

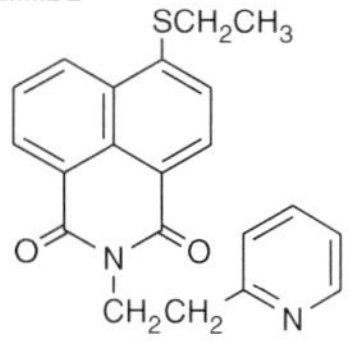

250 mg

S51,583-3
1-[4'-(METHYLTHIO)PHENACYL]-PYRIDINIUM IODIDE

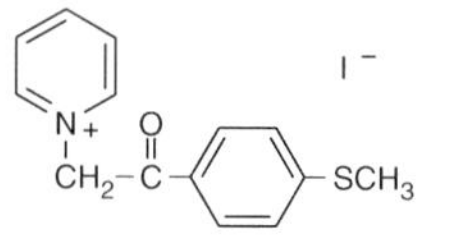

250 mg

S91,077-5
4-PYRIDYL N-[4-(METHYLTHIO)PHENYL]-CARBAMATE

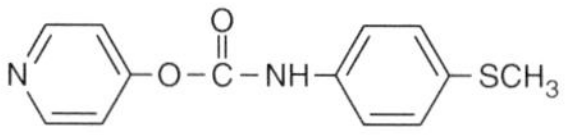

250 mg

S71,859-9
4-PYRIDYLMETHYL N-[4-(METHYLTHIO)-PHENYL]CARBAMATE

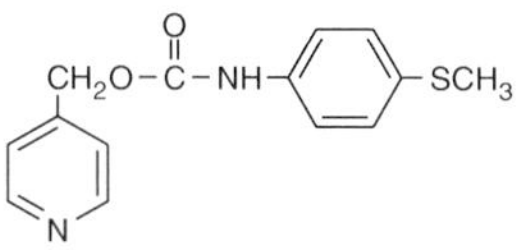

250 mg

S71,844-0
3-(4-PYRIDYL)PROPYL N-[4-(METHYLTHIO)PHENYL]CARBAMATE

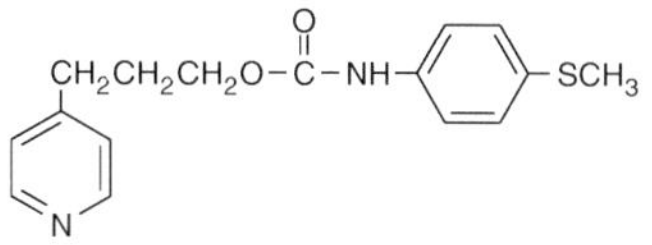

250 mg

S91,063-5
1-(3,5-DICHLORO-2-PYRIDYL)-3-[4-(METHYLTHIO)PHENYL]UREA

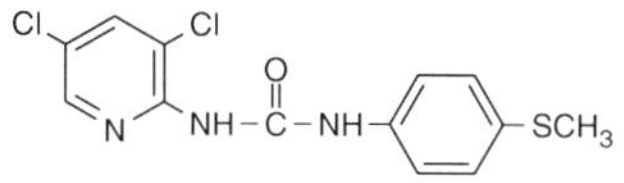

250 mg

S91,070-8
1-(5-BROMO-2-PYRIDYL)-3-[4-(METHYLTHIO)PHENYL]UREA

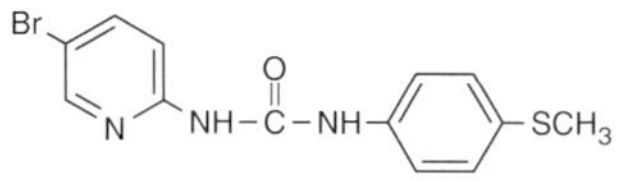

250 mg

S75,042-5
1-(3-METHYL-2-PYRIDYL)-3-[4-(METHYLTHIO)PHENYL]UREA

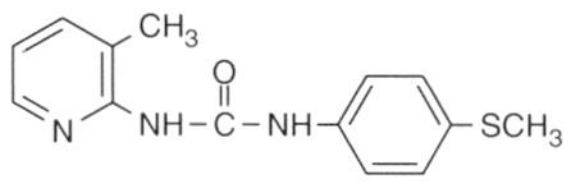

5 gm

S79,729-4
1-(4,6-DIMETHYL-2-PYRIDYL)-3-[4-(METHYLTHIO)PHENYL]UREA

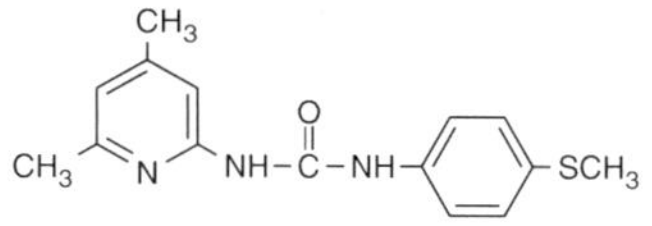

250 mg

S79,469-4
1-(4-METHYL-2-PYRIDYL)-3-[4-(METHYLTHIO)PHENYL]UREA

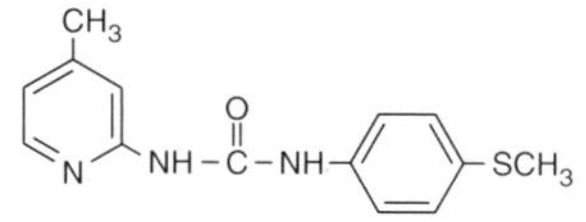

250 mg

S75,030-1
1-[4-(METHYLTHIO)PHENYL]-3-(2-PYRIDYL)UREA

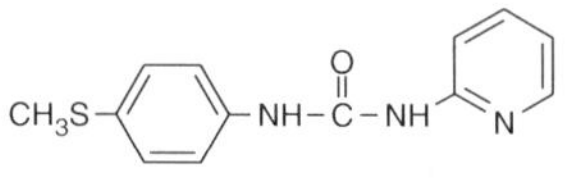

5 gm

S90,572-0
1-(6-METHYL-2-PYRIDYL)-3-[4-(METHYLTHIO)PHENYL]UREA

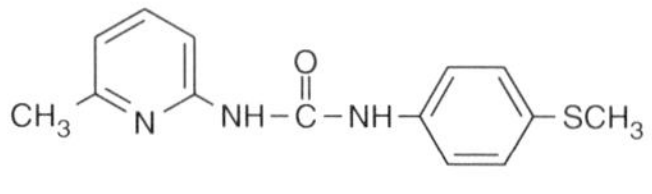

250 mg

S77,236-4
1-(5-CHLORO-2-PYRIDYL)-3-[4-(METHYLTHIO)PHENYL]UREA

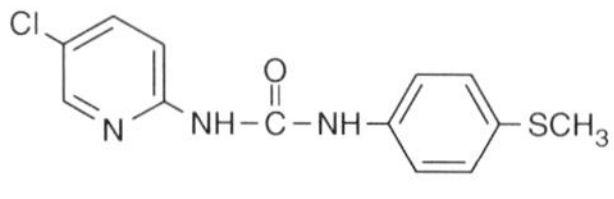

250 mg

S73,824-7
1-(4-METHYL-2-PYRIDYL)-3-[3-(METHYLTHIO)PHENYL]UREA

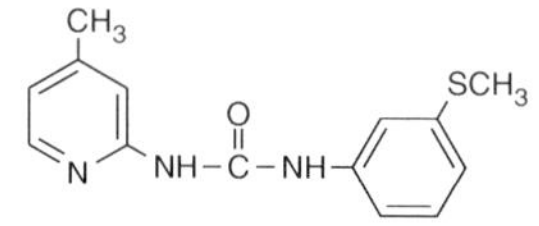

5 gm

S72,555-2
1-(4,6-DIMETHYL-2-PYRIDYL)-3-[3-(METHYLTHIO)PHENYL]UREA

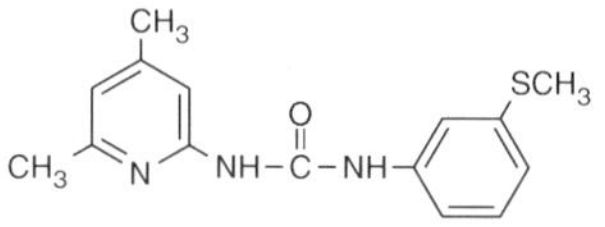

5 gm

S82,196-9
2-[2-(DIPHENYLMETHYLTHIO)ETHYL]-PYRIDINE PICRATE

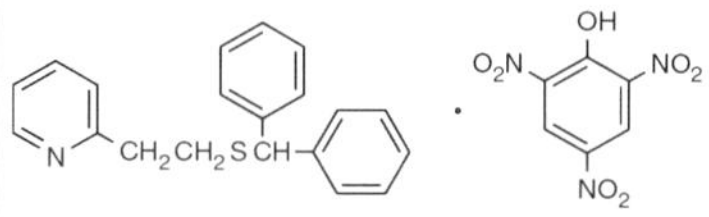

250 mg

S88,783-8
4-(DIMETHYLSULFONIO)-1-METHYLPYRIDINIUM DIBROMIDE

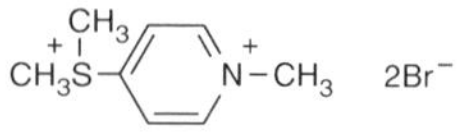

250 mg

S65,584-8
5-AMINO-2-(2-PYRIDYL)BENZIMIDAZOLE

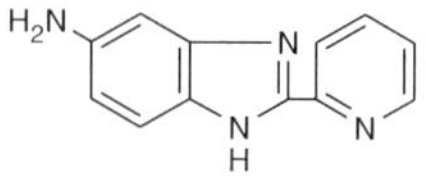

250 mg

S94,225-1
2-AMINO-3,5-DINITROPYRIDINE

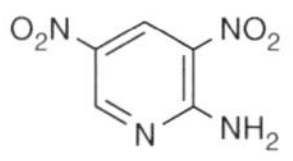

250 mg

S80,314-6
2-AMINO-3,5-DINITRO-6-METHYLPYRIDINE

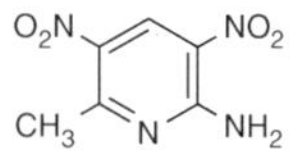

250 mg

S91,429-0
4-AMINO-3,5-DINITROPYRIDINE

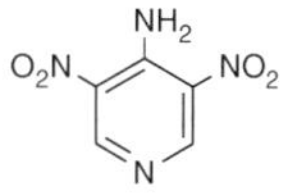

250 mg

S93,258-2
4-AMINO-3-NITROPYRIDINE

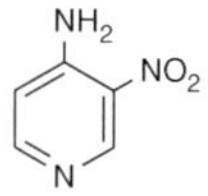

250 mg

S38,127-6
N1-(5-NITRO-2-PYRIDYL)-SULFANILAMIDE

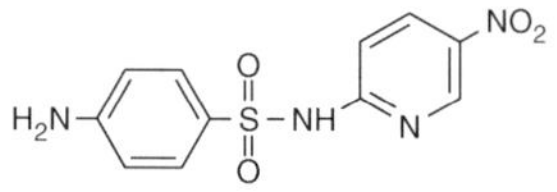

250 mg

S92,732-5
2-AMINO-5-METHYL-3-NITROPYRIDINE

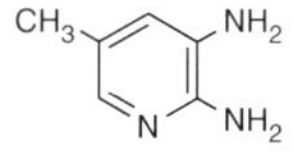

250 mg

S52,061-6
2,6-PYRIDINEDITHIOCARBOXAMIDE

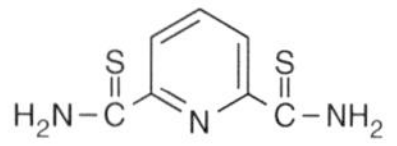

5 gm

S80,551-3
2-AMINO-5-CHLORO-3-NITROPYRIDINE

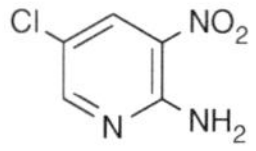

250 mg

S90,949-1
2-AMINO-3-CHLORO-5-NITROPYRIDINE

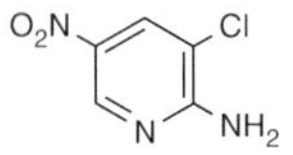

250 mg

S80,229-8
2-AMINO-5-BROMO-6-METHYL-3-NITROPYRIDINE

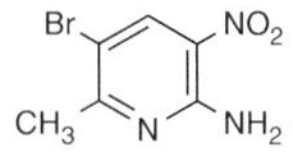

250 mg

S35,341-8
6-METHYLPICOLINIC THIONAMIDE

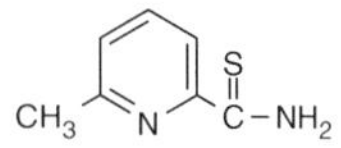

5 gm

S93,374-0
ACETIC N2-(3-AMINO-4-PYRIDYL)-HYDRAZIDE HYDROCHLORIDE

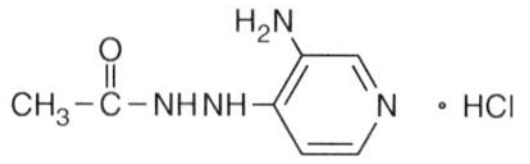

250 mg

S92,731-7
3,4,5-TRIAMINOPYRIDINE TRIHYDROCHLORIDE
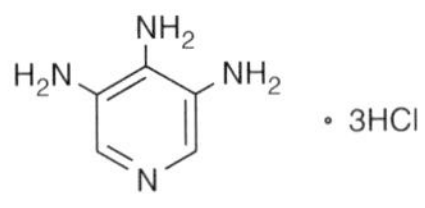
250 mg

S89,730-2
4-AMINO-1-BENZYLPYRIDINIUM BROMIDE
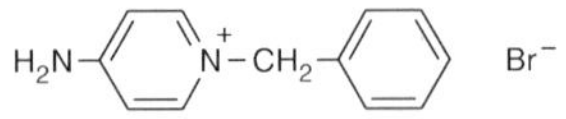
250 mg

S37,926-3
2-AMINO-3-AMINOMETHYL-6-METHYLPYRIDINE DIHYDROCHLORIDE
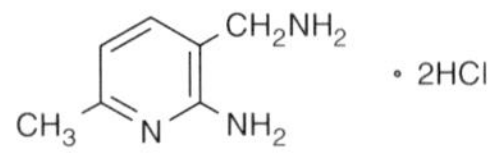
250 mg

S46,123-7
6-AMINO-2,4-DIMETHYL-3-PYRIDINECARBONITRILE
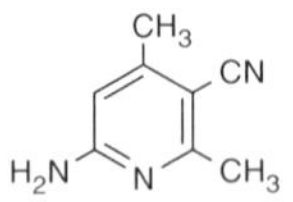
5 gm

S94,521-8
4-AMINO-3,5-DICHLORO-2-METHYLPYRIDINE
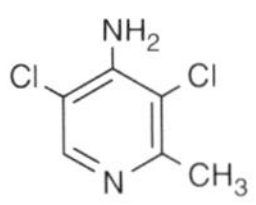
250 mg

S95,465-9
4-AMINO-2,3-LUTIDINE
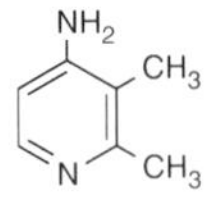
50 mg

S76,378-0
3-(AMINOMETHYL)PYRIDINE DIHYDROCHLORIDE
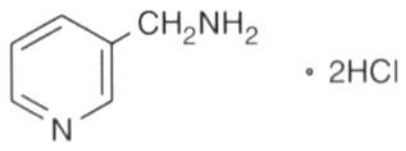
250 mg

S80,318-9
6-AMINO-2-METHYL-3-NITROPYRIDINE
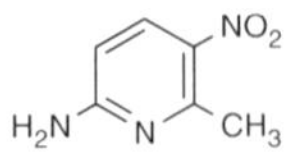
250 mg

S38,178-0
DIETHYL (6-AMINO-2-PYRIDYLAMINOMETHYLENE)MALONATE
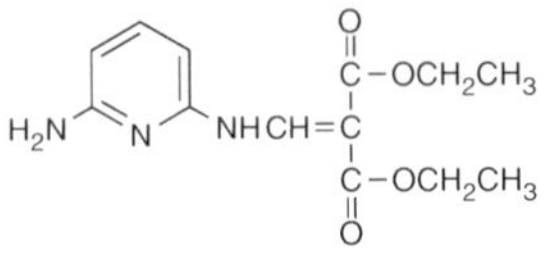
5 gm

S93,459-3
5-AMINO-3-(2-PYRIDYL)-4H-1,2,4-TRIAZOLE
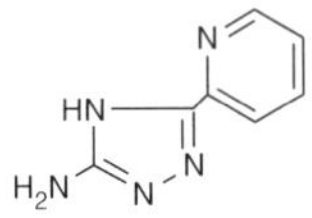
250 mg

S90,983-1
1,2-DIAMINOPYRIDINIUM P-TOLUENESULFONATE
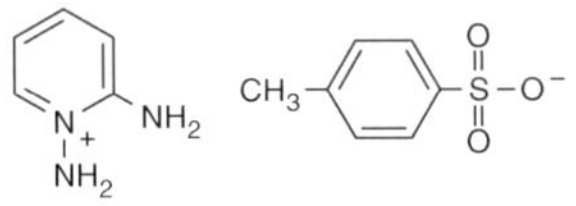
250 mg

S74,473-5
2-[2-(2-PYRIDYL)VINYLENE]ANILINE
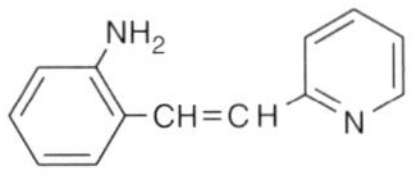
250 mg

S93,256-6
2-AMINO-4-BENZYLPYRIDINE
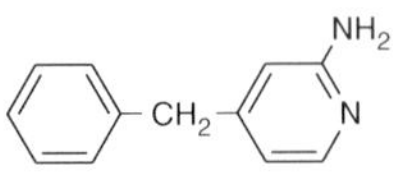
250 mg

S80,316-2
2-AMINO-6-BROMO-3-METHYL-5-NITROPYRIDINE
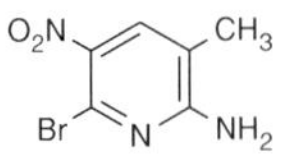
250 mg

S90,607-7
METHYL 2-AMINONICOTINATE
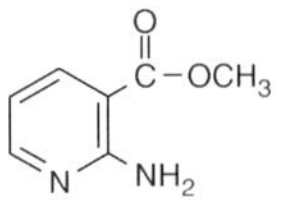
250 mg

S92,204-8
3,4-DIAMINO-5-METHYL-2(5H)-PYRIDONE
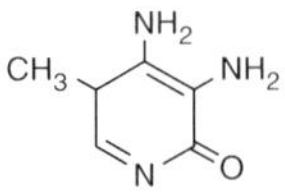
250 mg

S75,804-3
3-AMINOPICOLINIC ACID
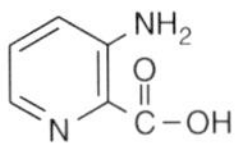
250 mg

S94,228-6
N-(5-AMINO-2-PYRIDYL)ACETAMIDE
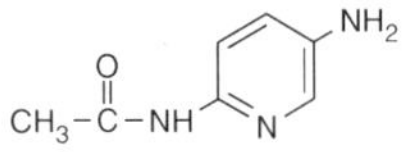
250 mg

S76,223-7
2-AMINO-5-(3-PYRIDYLMETHYL)-1,3,4-THIADIAZOLE
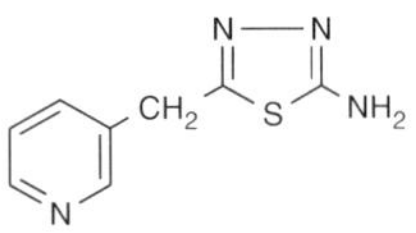
250 mg

S94,659-1
6-AMINO-3-PYRIDINESULFONIC ACID
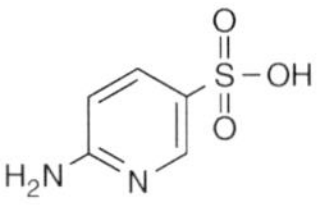
250 mg

S80,317-0
6-AMINO-3-BROMO-2-METHYLPYRIDINE
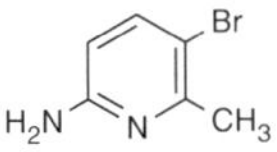
250 mg

S94,018-6
2-AMINO-5-IODOPYRIDINE
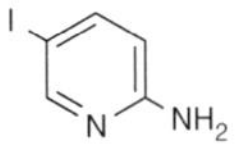
250 mg

S81,719-8
2-AMINOPYRIDINE PICRATE
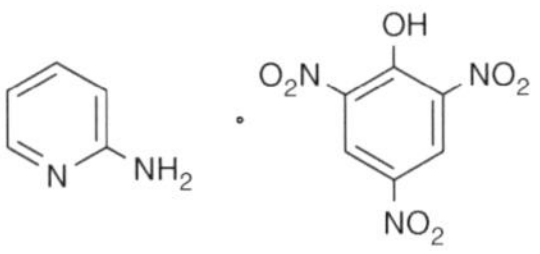
250 mg

S82,232-9
2-AMINOPYRIDINE HYDROCHLORIDE
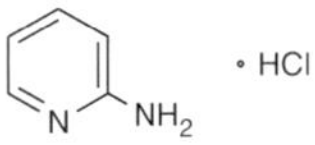
250 mg

S55,996-2
N-(5-CHLORO-2-PYRIDYL)PICRYLAMINE
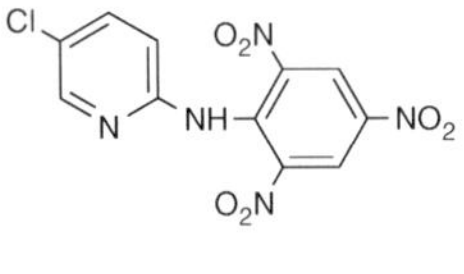
250 mg

S55,999-7
N-(4-METHYL-2-PYRIDYL)PICRYLAMINE
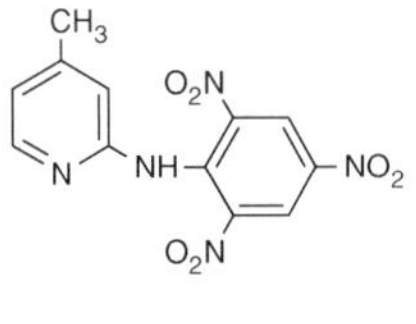
5 gm

S55,993-8
N-(6-METHYL-2-PYRIDYL)PICRYLAMINE
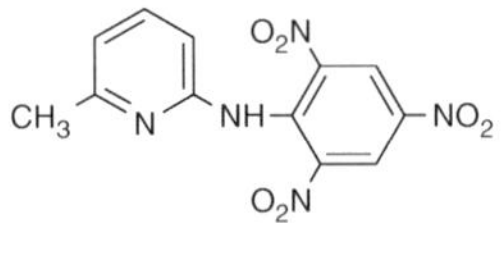
250 mg

S62,204-4
2-(2,4-DINITROPHENYL)AMINOPYRIDINE
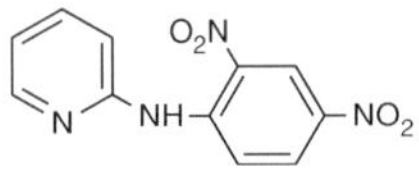
250 mg

S40,091-2
3-CARBAMOYL-1-(PHENYLCARBAMOYLMETHYL)-PYRIDINIUM CHLORIDE
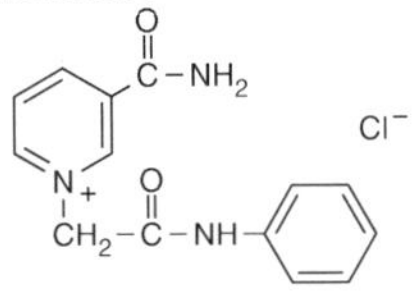
250 mg

S61,477-7
3-NITROSO-2,4,6-TRIACETAMIDOPYRIDINE
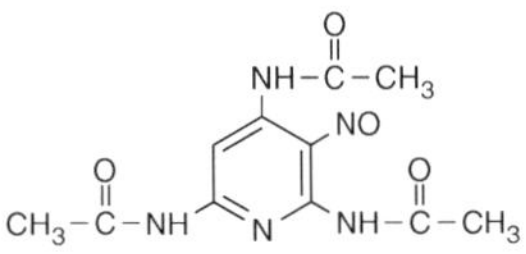
5 gm

S61,763-6
6-NITRO-2-(2-PYRIDYL)BENZIMIDAZOLE
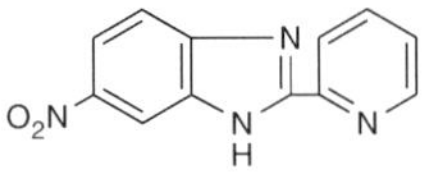
250 mg

S42,325-4
N-[2-(DIETHYLAMINO)ETHYL]-N-(1,2-DIPHENYLETHYL)NICOTINAMIDE HYDROCHLORIDE
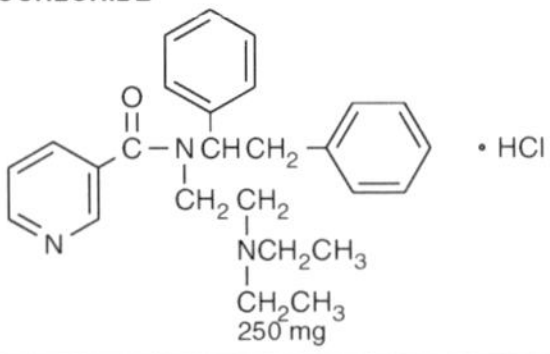
250 mg

S35,525-9
1-[2-(3-PYRIDYLMETHYLAMINO)ETHYL]-PIPERAZINE
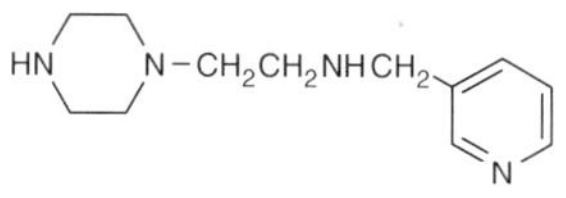
1 gm

S91,409-6
4-(2,4-DINITROANILINO)PYRIDINE
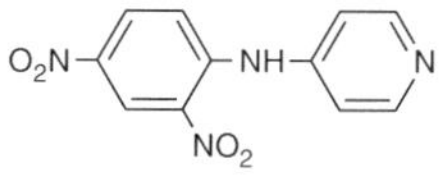
250 mg

S55,988-1
N-(3-METHYL-2-PYRIDYL)PICRYLAMINE
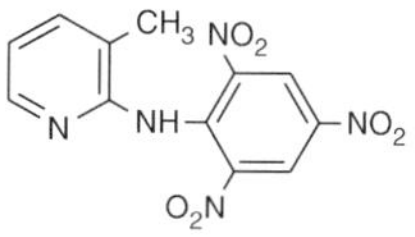
5 gm

S55,985-7
N-(3-PYRIDYL)PICRYLAMINE
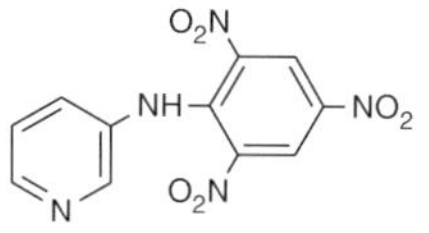
5 gm

S55,982-2
N-(4-PYRIDYL)PICRYLAMINE
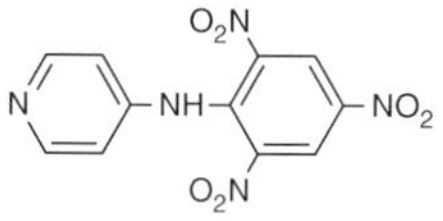
250 mg

S40,043-2
4-CARBAMOYL-1-(2,4-DINITROPHENYL)-PYRIDINIUM CHLORIDE
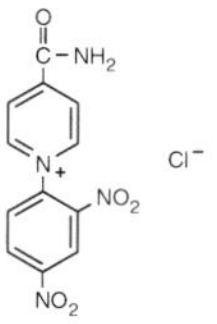
250 mg

S91,138-0
1-(4-METHYL-3-NITROPHENYL)-3-(2-PYRIDYLMETHYL)UREA
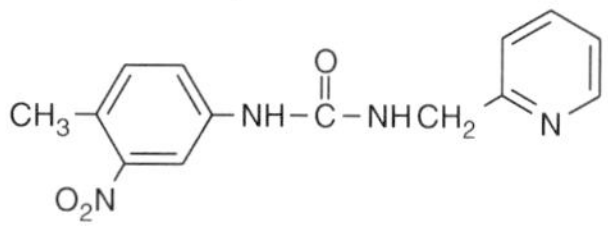
250 mg

S91,142-9
1-(4-METHYL-3-NITROPHENYL)-3-(3-PYRIDYL)UREA
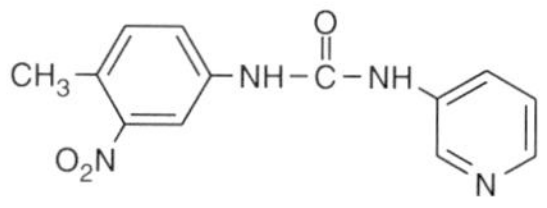
250 mg

S66,034-5
1,1'-(4-METHYL-1,3-PHENYLENE)BIS[3-(3-PYRIDYL)UREA]
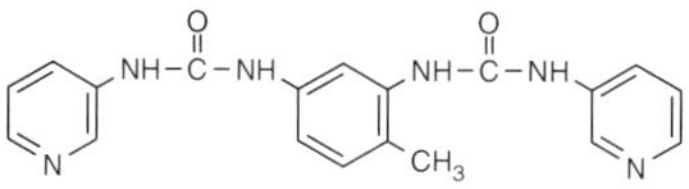
5 gm

S41,038-1
2-CYCLOHEXYLAMINO-5-NITROPYRIDINE
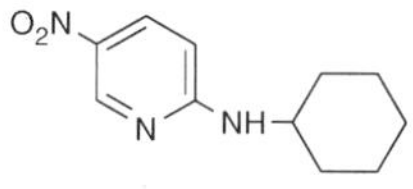
5 gm

S50,649-4
ALPHA-(2-NITROBENZYLIDENE)-N-PHENYL-2-PYRIDINEMETHYLAMINE
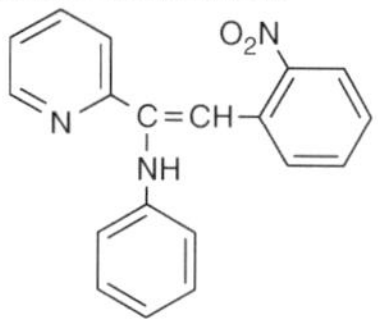
5 gm

S44,746-3
2-METHYLAMINO-5-NITROPYRIDINE
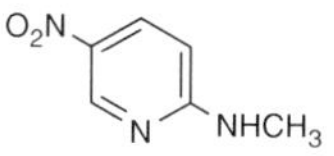
5 gm

S43,178-8
2-ETHYLAMINO-5-NITROPYRIDINE
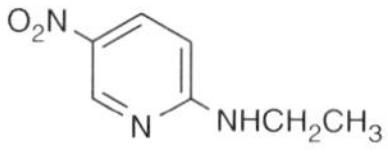
5 gm

S76,116-8
N-(5-NITRO-2-PYRIDYL)OCTYLAMINE
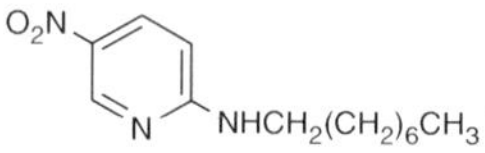
250 mg

S34,240-8
1,2-BIS(4-PYRIDYL)-N,N'-DIMETHYLETHYLENEDIAMINE
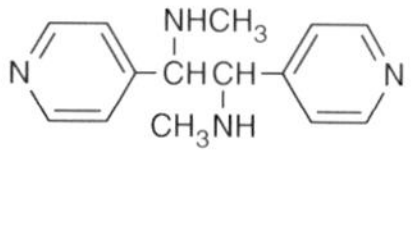
5 gm

S45,064-2
4-[CYANO(HYDROXIMINO)METHYL]-1-METHYLPYRIDINIUM IODIDE
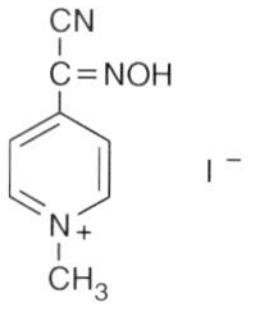
5 gm

S80,423-1
3,7-DINITROBENZO(B)(1,8)-NAPHTHYRIDIN-5(10H)-ONE
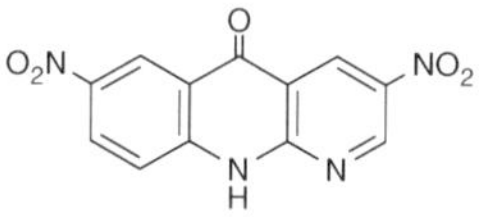
250 mg

S79,105-9
1-(5-NITRO-2-PYRIDYL)-3-PHENYLUREA
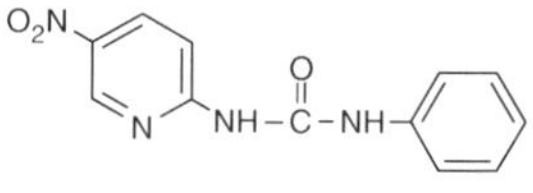
250 mg

S41,211-2
5-NITRO-2-PIPERIDINOPYRIDINE
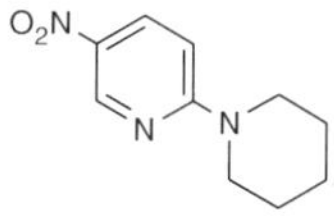
5 gm

S64,745-4
4-(3-CHLOROPHENYL)-1-PICOLINOYLTHIOSEMICARBAZIDE
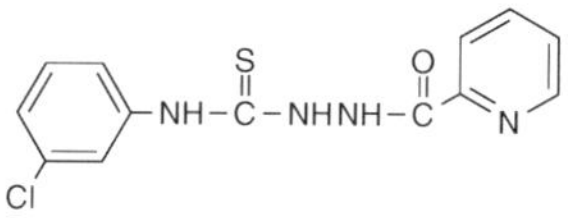
5 gm

S66,700-5
1-(2-PYRIDYLMETHYL)-3-(2,4,5-TRICHLOROPHENYL)UREA
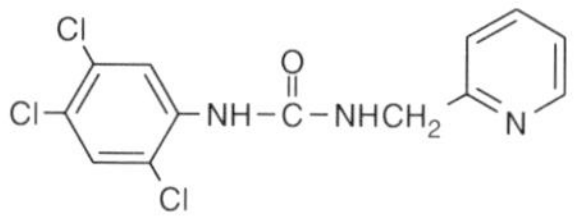
5 gm

S64,601-6
1-(3-PYRIDYL)-2-THIOUREA
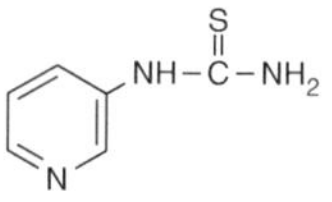
5 gm

S38,420-8
ALPHA-PHENYL-N-(3-PYRIDYL)-PHENETHYLAMINE
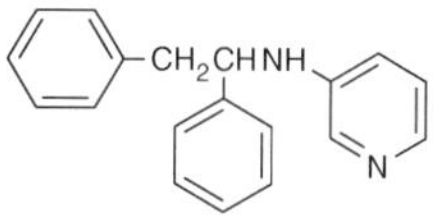
250 mg

S35,524-0
1-[2-(2-PYRIDYLMETHYLAMINO)ETHYL]-PIPERAZINE
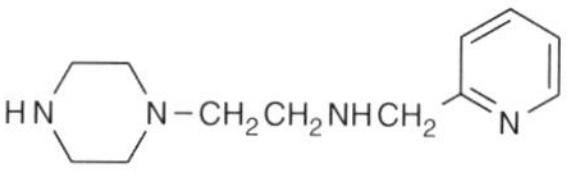
1 gm

S43,098-6
2-BENZYLAMINO-5-NITROPYRIDINE
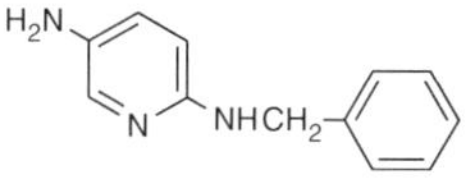
5 gm

S53,323-8
2-(ALPHA-ETHYLBENZYLAMINOMETHYL)-PYRIDINE
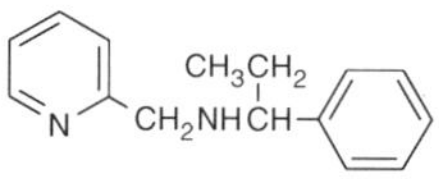
5 gm

S53,320-3
2-(ALPHA-ETHYLBENZYLAMINOMETHYL)-6-METHYLPYRIDINE
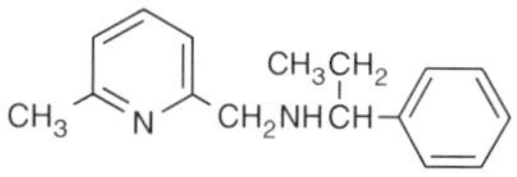
5 gm

S44,982-2
2-ANILINO-5-NITROPYRIDINE
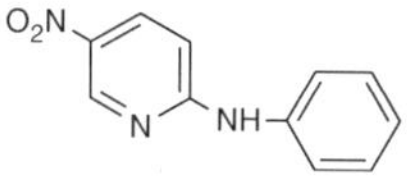
5 gm

S93,524-7
3-(3-PYRIDYLAMINO)PROPIOPHENONE
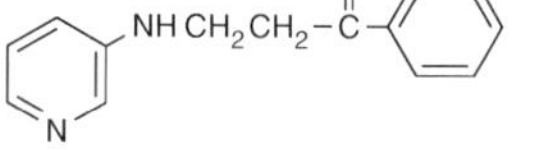
250 mg

S54,856-1
3-(4-METHYLBENZYLAMINO)PYRIDINE
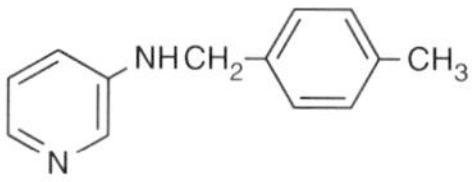
250 mg

S36,377-4
6-METHYL-2-PICOLYLMETHYLAMINE
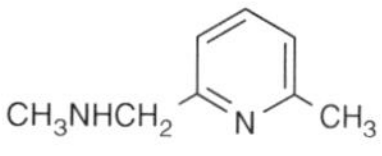
5 gm

S89,312-9
N,N-DIMETHYL-N'-FURFURYL-N'-(2-PYRIDYL)ETHYLENEDIAMINE FUMARATE
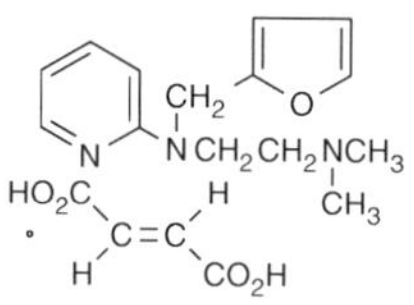
250 mg

S76,706-9
2-(2-CHLORO-5-NITRO-4-PYRIDYL)-3-(DIMETHYLAMINO)ACROLEIN

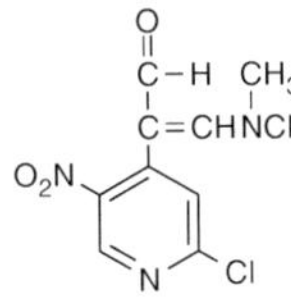

250 mg

S53,326-2
ETHYL 4-(2-PYRIDYLMETHYL)-1-PIPERAZINECARBOXYLATE

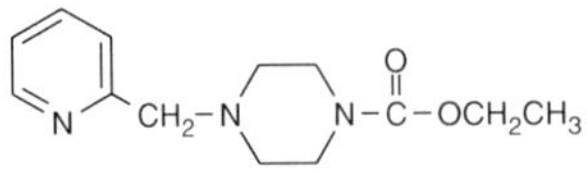

5 gm

S35,491-0
2-(PIPERIDINOMETHYL)PYRIDINE

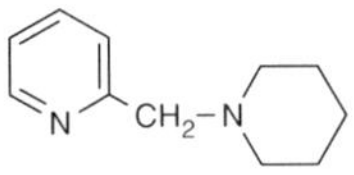

5 gm

S43,349-7
2-(MORPHOLINOMETHYL)PYRIDINE

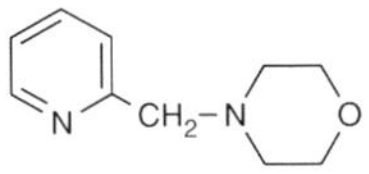

5 gm

S36,082-1
2-(DIMETHYLAMINOMETHYL)PYRIDINE

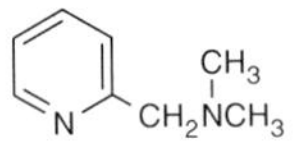

5 gm

S34,596-2
2-(DIETHYLAMINOMETHYL)PYRIDINE

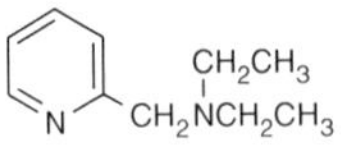

5 gm

S91,440-1
3-ACETYL-1-(CARBAMOYLMETHYL)-PYRIDINIUM CHLORIDE

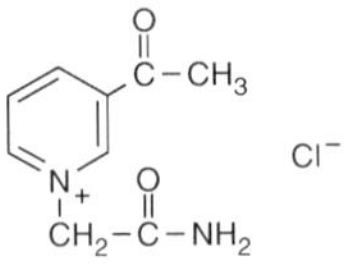

250 mg

S91,416-9
1-(CARBAMOYLMETHYL)-3-FLUOROPYRIDINIUM CHLORIDE

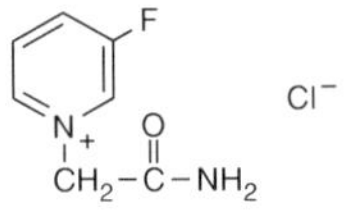

250 mg

S45,311-0
1-METHYL-4-CYANOFORMYLPYRIDINIUM OXIMATE

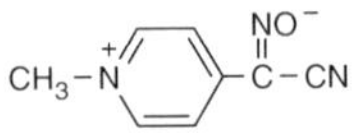

250 mg

S58,770-2
1,1'-ETHYLENEDIPYRIDINIUM DI-P-TOLUENESULFONATE

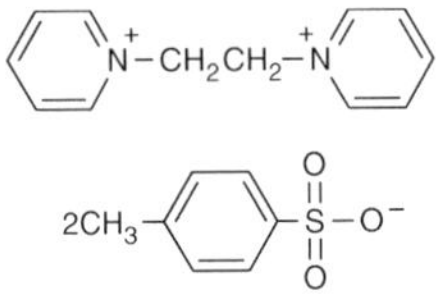

250 mg

S86,968-6
1-(2-PYRIDYLMETHYL)-PYRIDINIUM BROMIDE

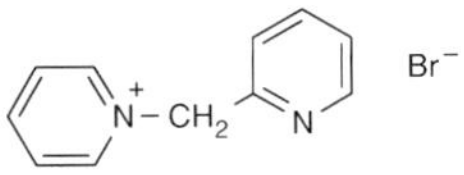

250 mg

S63,544-8
1,1'-ETHYLENEBIS-(PYRIDINIUM BROMIDE)

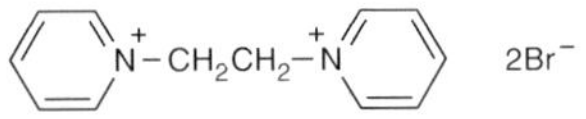

250 mg

S50,528-5
1,1'-(IMINODIETHYLENE)BIS-(PYRIDINIUM CHLORIDE)

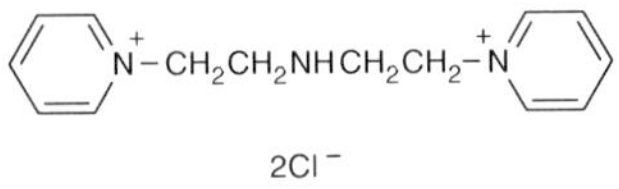

250 mg

S38,924-2
2-HYDRAZINO-3-NITROPYRIDINE

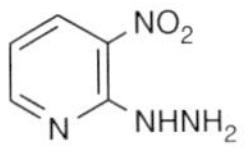

250 mg

S90,219-5
2-(3-FURFURYLUREIDO)ACETIC (3-PYRIDYLMETHYLENE)HYDRAZIDE

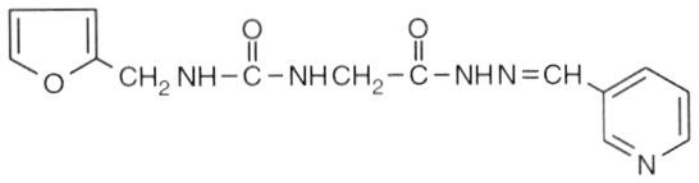

250 mg

S91,474-6
2-PYRIDINECARBOXALDEHYDE SEMICARBAZONE

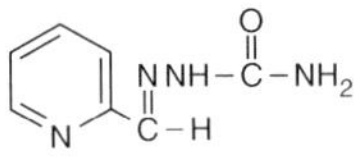

250 mg

S87,620-8
2-PYRIDINECARBOXALDEHYDE THIOSEMICARBAZONE

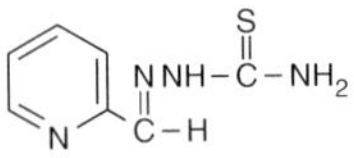

250 mg

S37,213-7
PICOLINAMIDE

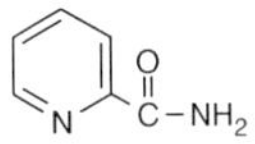

5 gm

S85,326-7
1-(4-METHYL-2-NITROPHENYL)-3-(6-METHYL-2-PYRIDYL)UREA

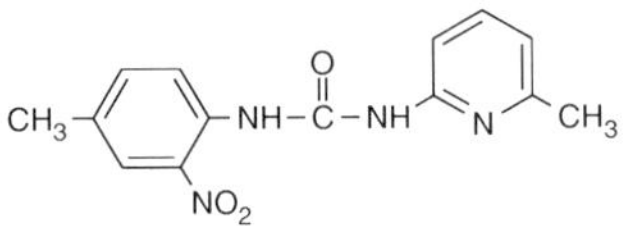

250 mg

S90,945-9
3-NITRO-4(1H)-PYRIDONE

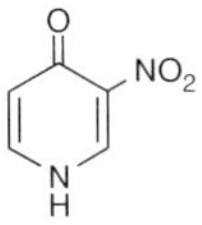

250 mg

S51,078-5
2-BENZOYLACETAMIDO-5-NITROPYRIDINE

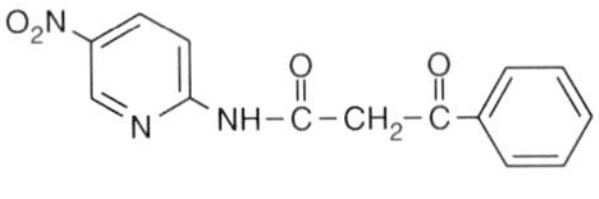

250 mg

S91,356-1
N-(5-NITRO-3-PYRIDYL)-PHTHALAMIC ACID

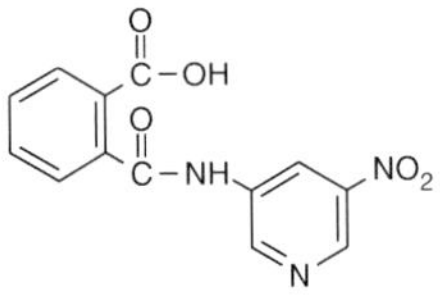

250 mg

S87,131-1
1-BENZOYL-3-(3-NITRO-4-PYRIDYL)UREA

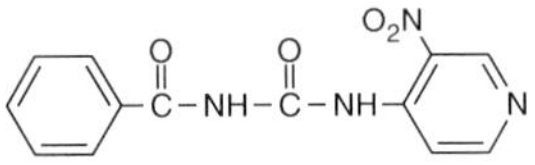

250 mg

S80,854-7
N-(4-METHYL-3-NITRO-2-PYRIDYL)-BENZAMIDE

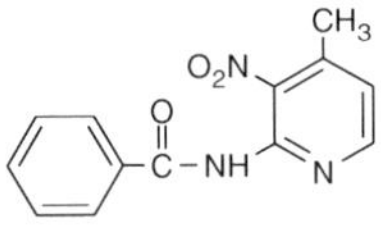

250 mg

S80,333-2
2-ACETAMIDO-3-METHYL-5-NITROPYRIDINE

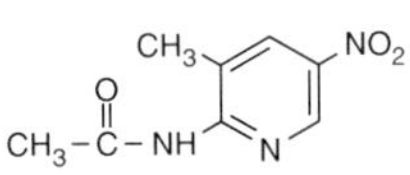

250 mg

S40,963-4
3-NITRO-2,6-LUTIDINE

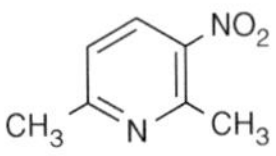

5 gm

S67,226-2
3-METHYL-2-(NITROMETHYL)-PYRIDINE PICRATE

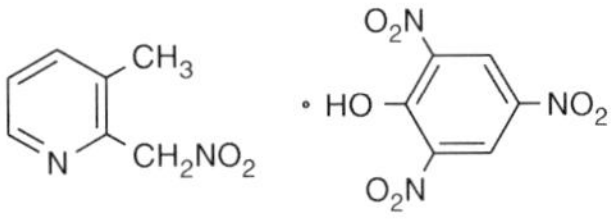

250 mg

S89,686-1
5,5-DIMETHYL-3-(3-PYRIDYLAMINO)-2-CYCLOHEXEN-1-ONE

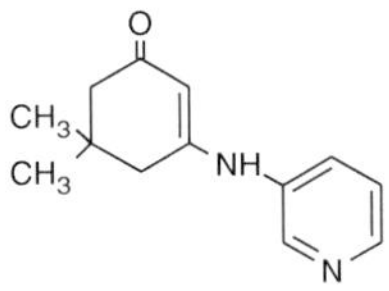

250 mg

S39,082-8
N-(6-METHYL-2-PYRIDYL)-ACETOACETAMIDE

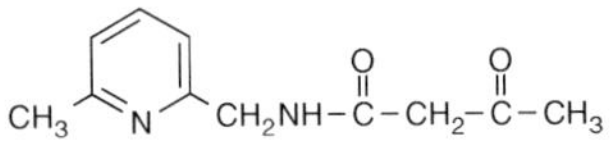

5 gm

S72,073-9
N-(3-PYRIDYL)PHTHALAMIC ACID

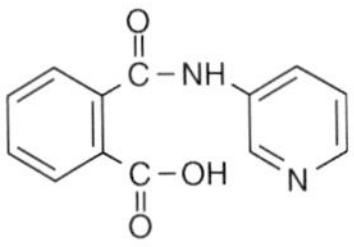

5 gm

S91,358-8
N-METHYL-N-(2-PYRIDYLMETHYL)-PHTHALAMIC ACID

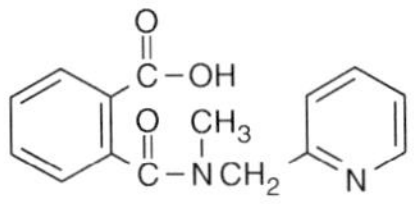

250 mg

S55,490-1
N-(3-PYRIDYL)MALEAMIC ACID

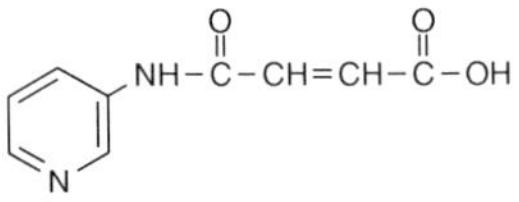

5 gm

S91,118-6
ETHYL 4-[3-(3-PYRIDYL)UREIDO]-BENZOATE
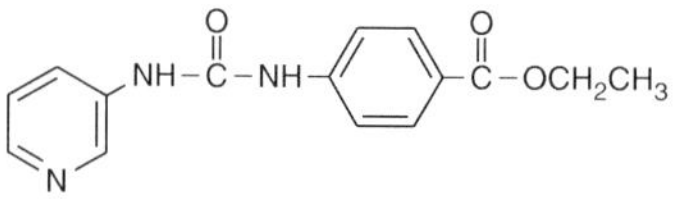
250 mg

S72,069-0
N-(3-PYRIDYL)GLUTARAMIC ACID
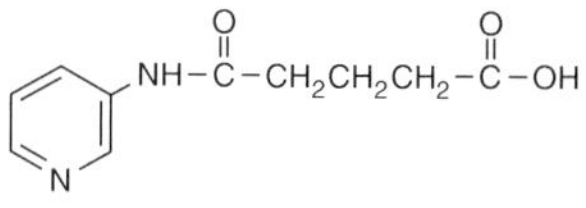
5 gm

S94,227-8
3,4-BIS(FORMAMIDO)PYRIDINE
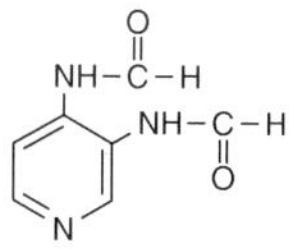
250 mg

S64,426-9
1-PHENYL-3-(3-PYRIDYL)UREA
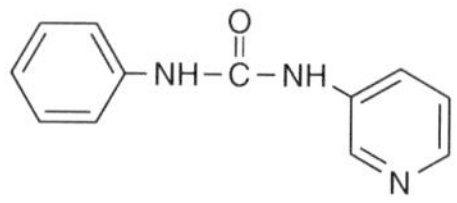
5 gm

S64,404-8
1-(2-PYRIDYLMETHYL)-3-(2,3-XYLYL)UREA
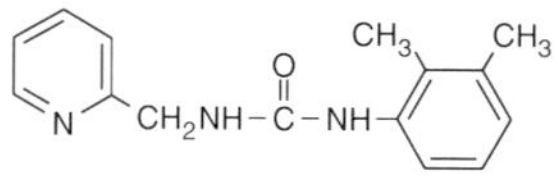
250 mg

S65,496-5
1-(2-PYRIDYLMETHYL)-3-(2,6-XYLYL)UREA
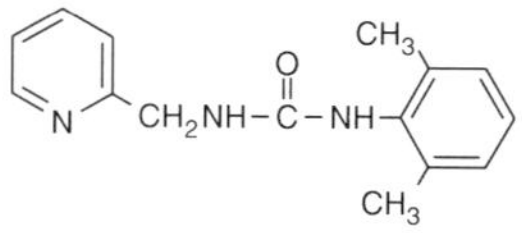
5 gm

S91,213-1
1-(2,6-DIETHYLPHENYL)-3-(2-PYRIDYLMETHYL)UREA
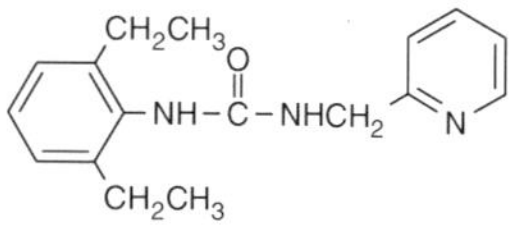
250 mg

S91,241-7
1-(2-PYRIDYLMETHYL)-3-(3,4-XYLYL)UREA
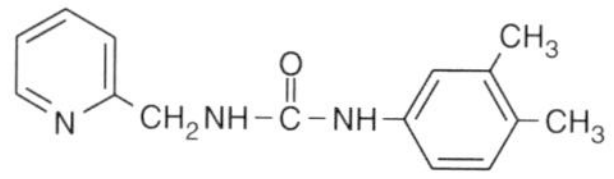
250 mg

S68,987-4
1-(4-CHLORO-2-METHYLPHENYL)-3-(3-PYRIDYL)UREA
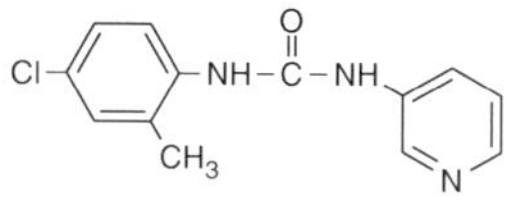
250 mg

S91,353-7
1-PHENYL-3-(3-PYRIDYL)-2-THIOUREA
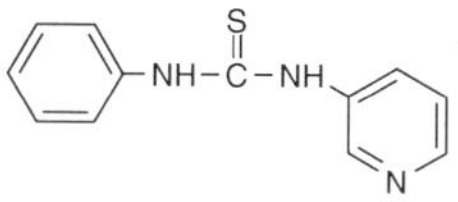
250 mg

S91,375-8
1-METHYL-3-(2-PYRIDYLMETHYL)-2-THIOUREA
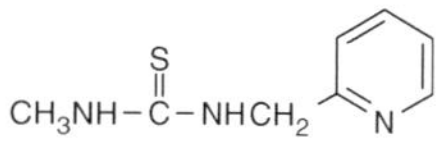
250 mg

S52,538-3
4-BENZYL-ALPHA-(3-PYRIDYL)-1-PIPERAZINEACETONITRILE
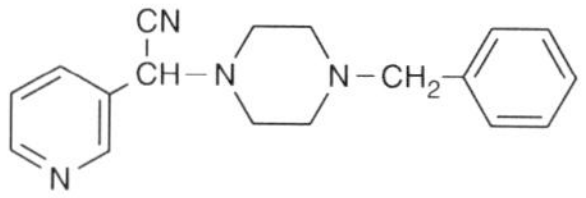
250 mg

S94,522-6
3,4,5-TRICHLORO-2-PYRIDINECARBONITRILE
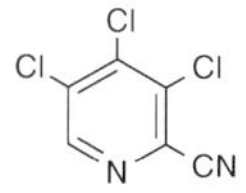
250 mg

S59,731-7
3,4,5,6-TETRACHLOROPICOLINONITRILE
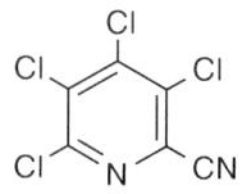
5 gm

S70,887-9
1-(2,6-DIMETHYLPHENYL)-3-(3-PYRIDYL)UREA
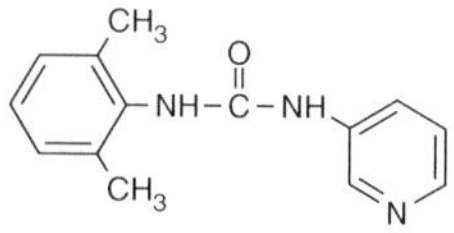
5 gm

S84,928-6
2-(DIBROMONITROMETHYL)-3-METHYLPYRIDINE
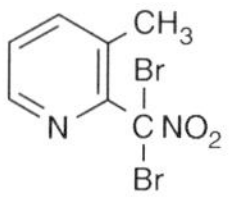
250 mg

S73,508-6
2-(2-PYRIDYL)INDOLE
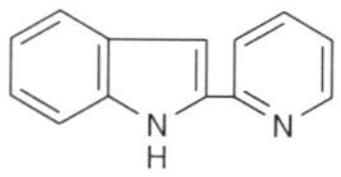
250 mg

S84,927-8
2-(DIBROMONITROMETHYL)-6-METHYLPYRIDINE
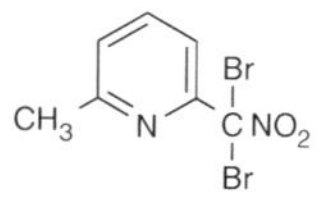
250 mg

S79,336-1
4-CYANO-1-METHYL-4-(2-PYRIDYL)-PIPERIDINE
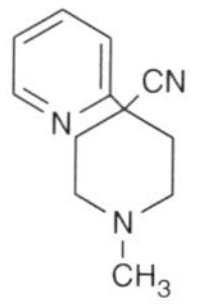
250 mg

S66,892-3
1-(4-NITROPHENYL)-3-(3-PYRIDYLMETHYL)UREA
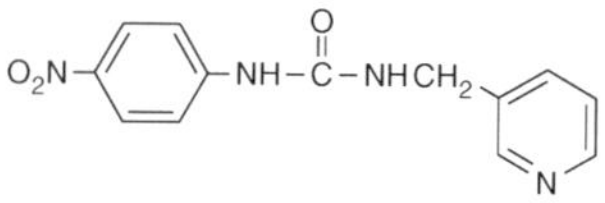
5 gm

S60,734-7
1-(2-CHLORO-4-NITROPHENYL)-3-(3-PYRIDYLMETHYL)UREA
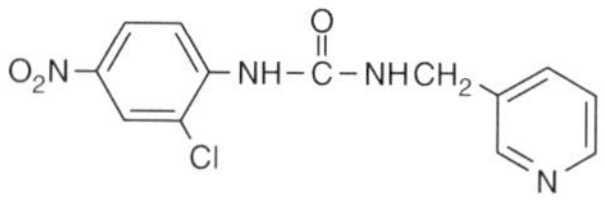
5 gm

S53,349-1
N-(2-PYRIDYL)PICRYLAMINE
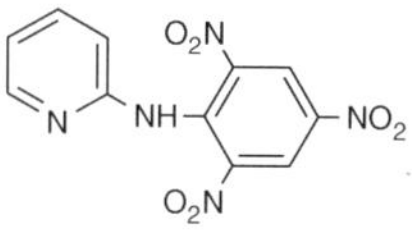
5 gm

S41,215-5
1,2-DIHYDRO-1-[2-(2-PYRIDYL)ETHYL]-3,6-PYRIDAZINEDIONE
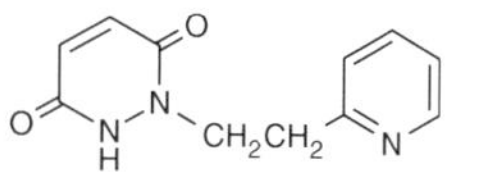
5 gm

S66,028-0
1,1'-(4-METHYL-1,3-PHENYLENE)BIS[3-(4-PYRIDYLMETHYL)UREA]
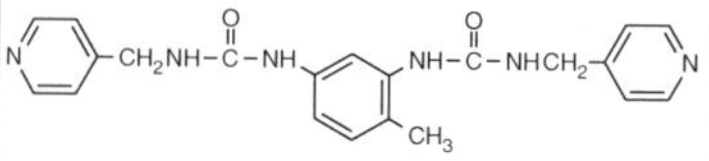
5 gm

S41,153-1
2-PICOLINYLIDENEMETHYL-3-O-TOLYL-4(3H)-QUINAZOLINONE
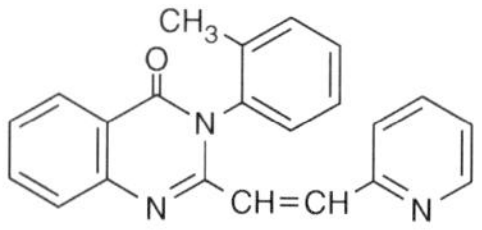
5 gm

S65,108-7
2-[2-(PHENETHYLAMINO)ETHYL]PYRIDINE
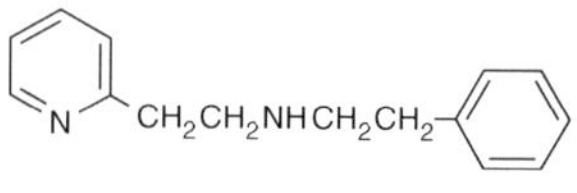
5 gm

S38,188-8
ALPHA-(4-CHLOROPHENYL)-2-PYRIDINEACETONITRILE
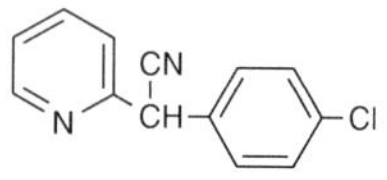
5 gm

S55,096-5
2-[2-(PROPYLAMINO)ETHYL]PYRIDINE
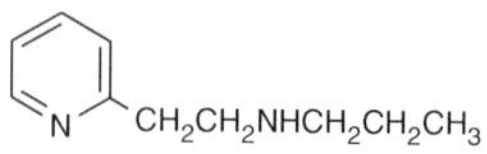
5 gm

S65,093-5
2-(2-BENZYLAMINOETHYL)PYRIDINE
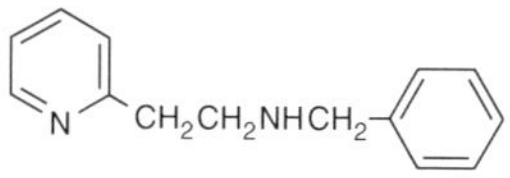
5 gm

S65,754-9
3-(4-PYRIDYLAMINO)PHTHALIDE
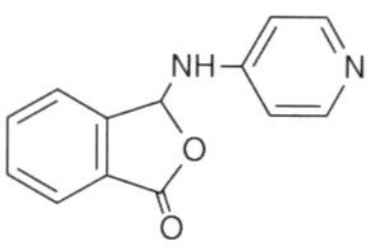
5 gm

S53,809-4
PICOLINIC ACID (3-INDOLYLMETHYLENE)-HYDRAZIDE
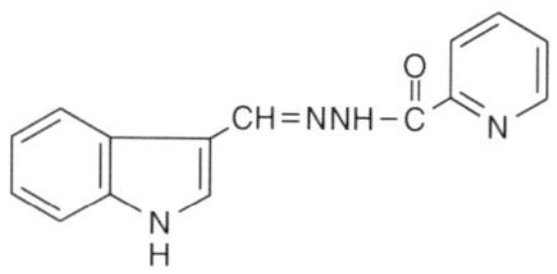
5 gm

S59,356-7
2-(2-ANILINOETHYL)PYRIDINE
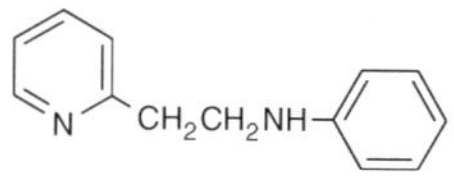
5 gm

S50,287-1
DIPHENYL (ALPHA-ANILINO-3-PYRIDYLMETHYL)PHOSPHONATE

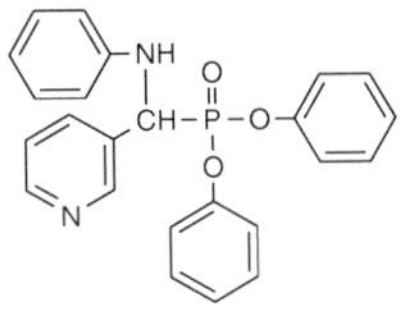

5 gm

S65,090-0
2-[2-(2-CHLOROANILINO)ETHYL]-PYRIDINE

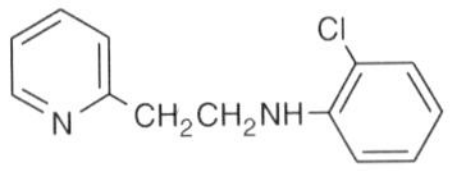

5 gm

S65,097-8
2-[2-(2-METHYLANILINO)ETHYL]-PYRIDINE

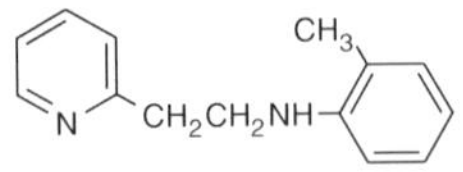

5 gm

S65,094-3
2-[2-(4-FLUOROANILINO)ETHYL]-PYRIDINE

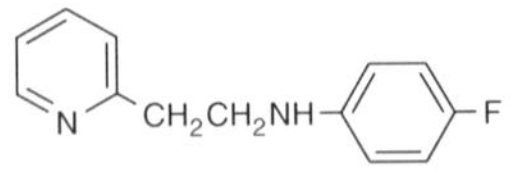

5 gm

S65,084-6
2-[2-(4-CHLOROANILINO)ETHYL]-PYRIDINE

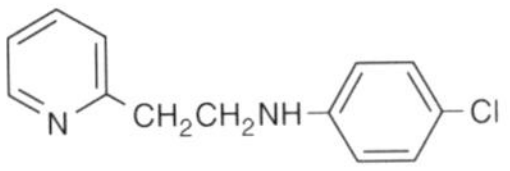

5 gm

S65,131-1
2-[2-(4-METHYLANILINO)ETHYL]-PYRIDINE

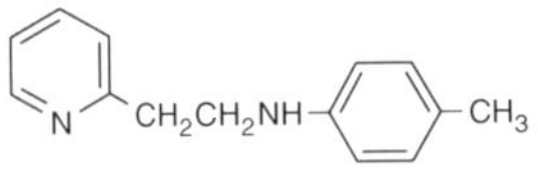

5 gm

S65,092-7
2-[2-(4-ISOPROPYLANILINO)ETHYL]-PYRIDINE

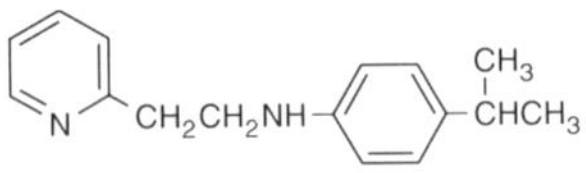

5 gm

S65,128-1
2-[2-(2,4-DIMETHYLANILINO)ETHYL]-PYRIDINE

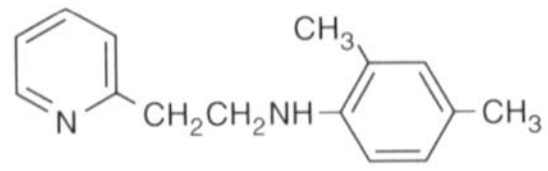

5 gm

S36,145-3
2-(2-ETHYLAMINOETHYL)PYRIDINE

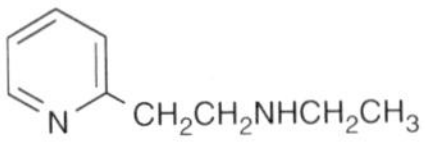

5 gm

S55,093-0
2-[2-(ISOPROPYLAMINO)ETHYL]-PYRIDINE

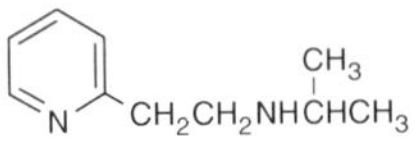

250 mg

S65,083-8
2-[2-(3-CHLOROANILINO)ETHYL]-PYRIDINE

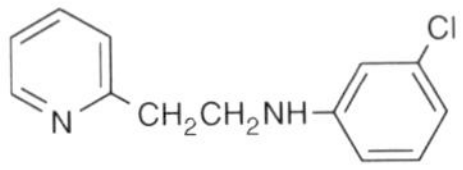

5 gm

S65,091-9
2-[2-(3-METHYLANILINO)ETHYL]-PYRIDINE

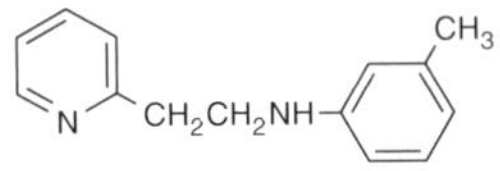

5 gm

S75,429-3
N-[2-(2-PYRIDYL)ETHYL]-2-FLUORENAMINE

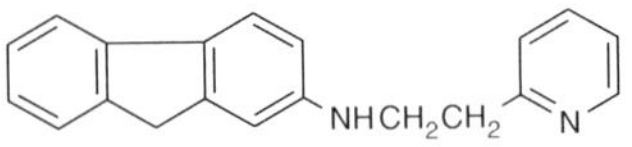

250 mg

S65,104-4
N-METHYL-N-[2-(2-PYRIDYL)ETHYL]-2-TETRAHYDROFURYLAMINE

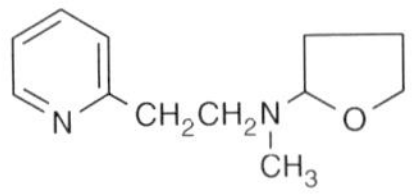

250 mg

S35,536-4
2-(2-PYRROLIDINOETHYL)PYRIDINE

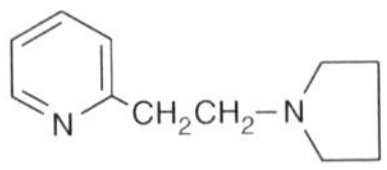

5 gm

S89,430-3
4-MORPHOLINO-2,3,5,6-TETRACHLOROPYRIDINE

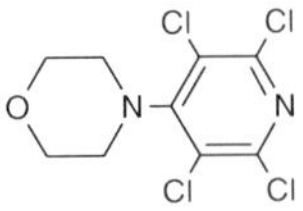

250 mg

S35,367-1
2-(2-MORPHOLINOETHYL)PYRIDINE

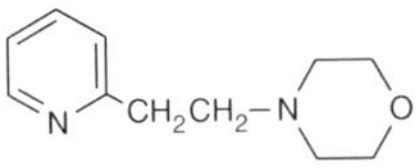

5 gm

S66,173-2
2-[2-(N-ETHYL-N-PHENYLAMINO)ETHYL]-PYRIDINE HYDROCHLORIDE

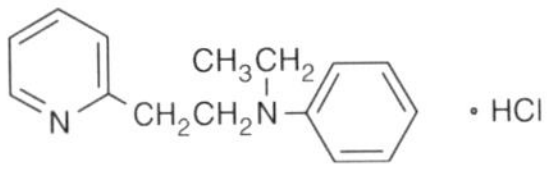

5 gm

S58,160-7
2-(2-PYRIDYL)-N,N,N',N'-TETRAPHENYL-1,3-PROPANEDIAMINE

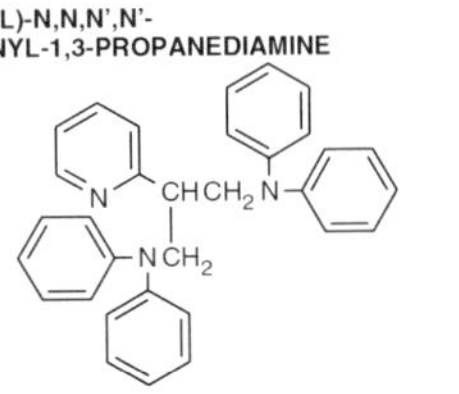

5 gm

S37,259-5
2-(2-DIETHYLAMINOETHYL)PYRIDINE

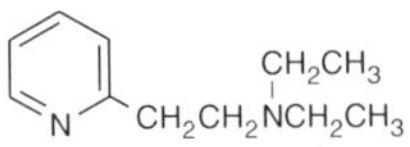

5 gm

S59,365-6
2-[2-(DIBUTYLAMINO)ETHYL]PYRIDINE

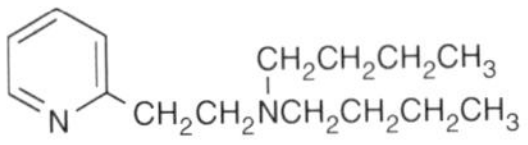

5 gm

S60,544-1
1-(2-CHLORO-6-NITROBENZYL)-PYRIDINIUM BROMIDE

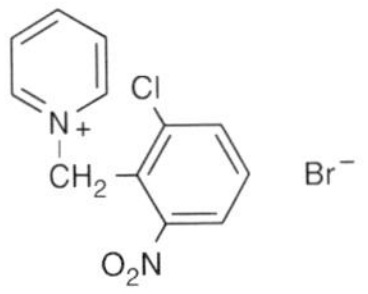

250 mg

S59,188-2
1-(2,4-DINITROBENZYL)-PYRIDINIUM CHLORIDE

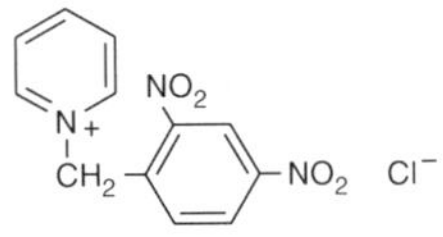

250 mg

S40,288-5
3-BENZYLCARBAMOYL-1-(DECYLCARBAMOYLMETHYL)-PYRIDINIUM CHLORIDE

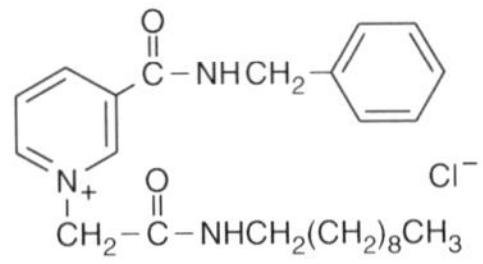

250 mg

S40,286-9
1-(3,4-DICHLOROBENZYL)-3-(DODECYLCARBAMOYL)-PYRIDINIUM CHLORIDE

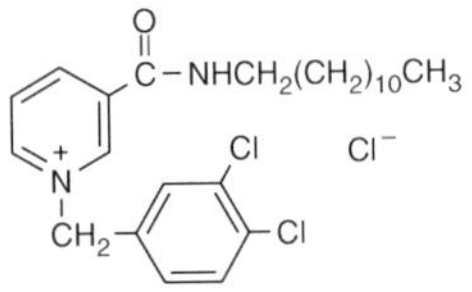

250 mg

S40,125-0
1-(3,4-DICHLOROBENZYL)-3-(DIMETHYLCARBAMOYL)-PYRIDINIUM CHLORIDE

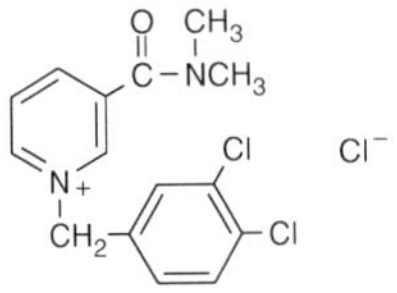

250 mg

S67,539-3
4-METHYL-4-[2-PHENYLSULFONYL-2-(2-PYRIDYL)ETHYL]MORPHOLINIUM IODIDE

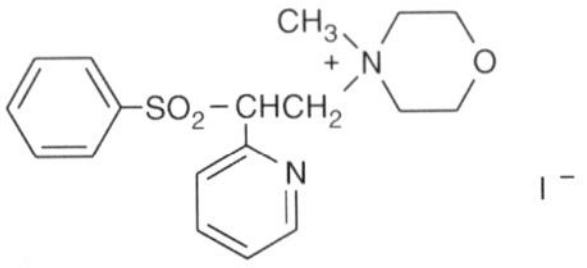

250 mg

S91,437-1
3-CARBAMOYL-1-(4-NITROBENZYL)-PYRIDINIUM CHLORIDE

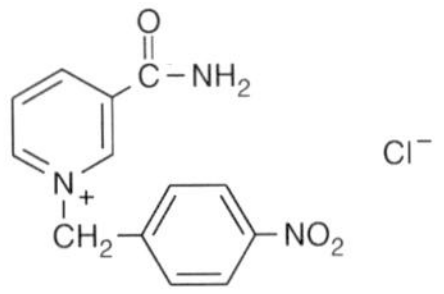

250 mg

S91,530-0
3-CARBAMOYL-1-(4-NITROBENZYL)-PYRIDINIUM BROMIDE

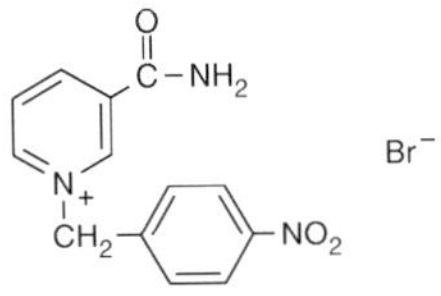

250 mg

S91,444-4
1-ACETONYL-3-CARBAMOYLPYRIDINIUM CHLORIDE

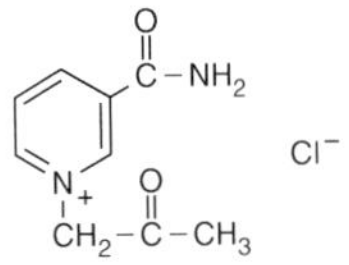

250 mg

S91,425-8
3-CARBAMOYL-1-(ISOPROPOXYCARBONYLMETHYL)-PYRIDINIUM CHLORIDE

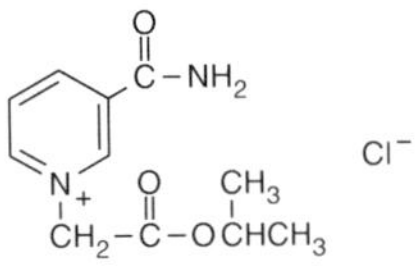

250 mg

S91,557-2
3-CARBAMOYL-1-(METHOXYCARBONYLMETHYL)-PYRIDINIUM CHLORIDE

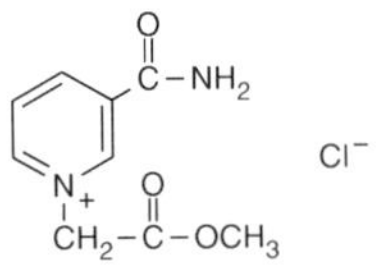

250 mg

S40,108-0
3-CARBAMOYL-1-(HEXYLOXYCARBONYLMETHYL)-PYRIDINIUM CHLORIDE
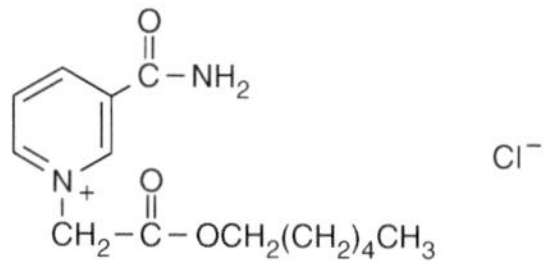
250 mg

S40,402-0
3-CARBAMOYL-1-(DODECYLOXYCARBONYLMETHYL)-PYRIDINIUM CHLORIDE
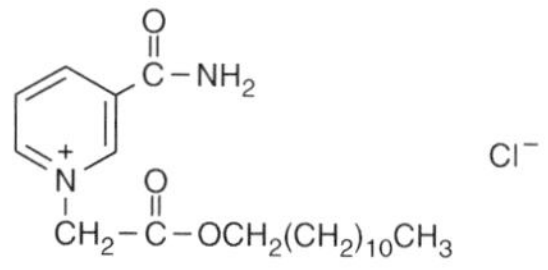
250 mg

S40,267-2
3-CARBAMOYL-1-(HEXADECYLOXYCARBONYLMETHYL)-PYRIDINIUM CHLORIDE
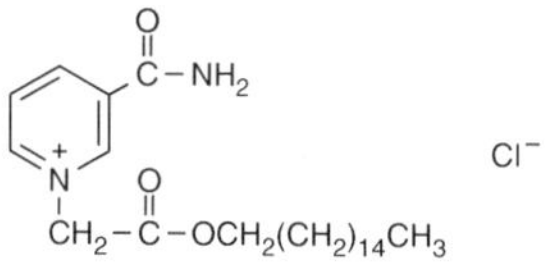
250 mg

S40,270-2
1-DECYLOXYCARBONYLMETHYL-3-(PHENYLCARBAMOYL)-PYRIDINIUM CHLORIDE
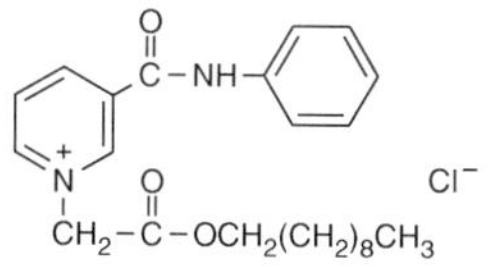
250 mg

S40,285-0
3-(N-PHENYLCARBAMOYL)-1-(UNDECYLOXYCARBONYLMETHYL)-PYRIDINIUM CHLORIDE
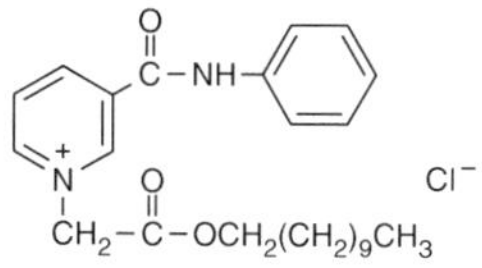
250 mg

S40,297-4
1-DODECYLOXYCARBONYLMETHYL-3-(PHENYLCARBAMOYL)-PYRIDINIUM CHLORIDE
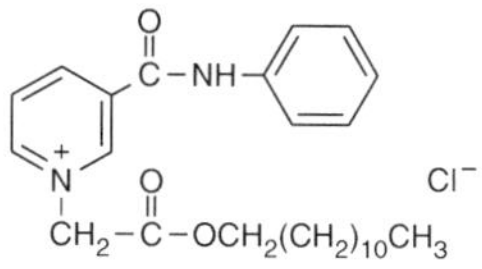
250 mg

S40,271-0
3-(4-CHLOROPHENYLCARBAMOYL)-1-(DECYLOXYCARBONYLMETHYL)-PYRIDINIUM CHLORIDE
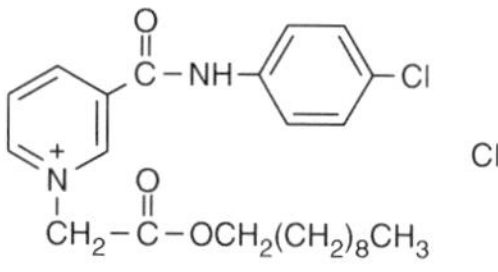
250 mg

S40,090-4
1-CARBAMOYLMETHYL-3-(PHENYLCARBAMOYL)-PYRIDINIUM CHLORIDE
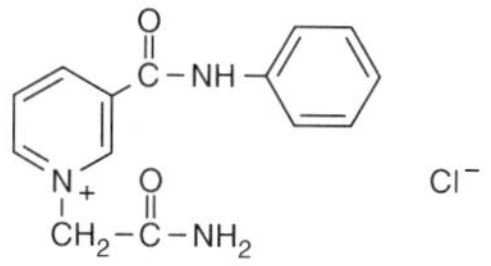
250 mg

S40,093-9
1-CARBAMOYLMETHYL-3-(4-CHLOROPHENYLCARBAMOYL)-PYRIDINIUM CHLORIDE
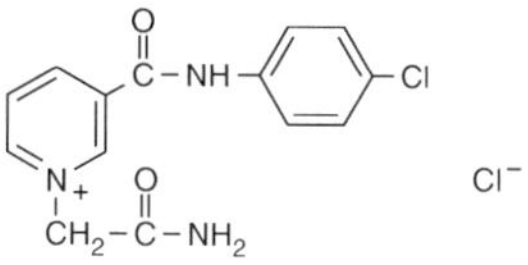
250 mg

S40,406-3
3-CARBAMOYL-1-(DECYLCARBAMOYLMETHYL)-PYRIDINIUM CHLORIDE
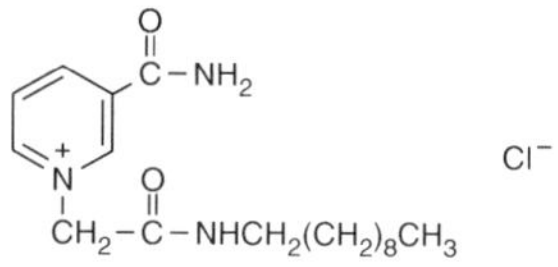
250 mg

S34,327-7
1-(2-CARBAMOYLETHYL)-3-(DIETHYLCARBAMOYL)-PYRIDINIUM CHLORIDE
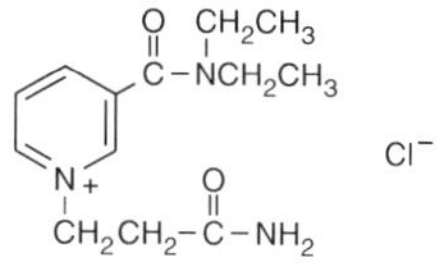
5 gm

S40,016-5
3-CARBAMOYL-1-(CARBAMOYLMETHYL)-PYRIDINIUM CHLORIDE
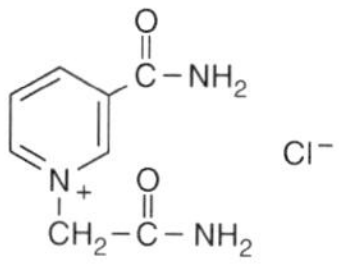
250 mg

S91,427-4
3-CARBAMOYL-1-(CYANOMETHYL)-PYRIDINIUM CHLORIDE
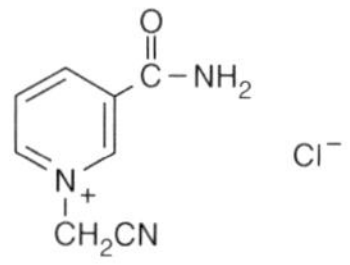
250 mg

S91,517-3
1-BENZYL-3-CARBAMOYLPYRIDINIUM BROMIDE
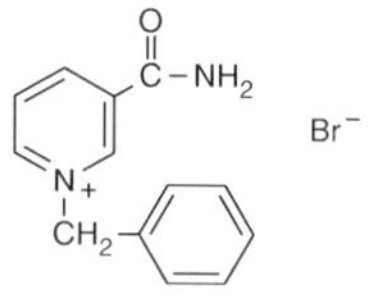
250 mg

S90,623-9
3-CARBAMOYL-1-PROPYLPYRIDINIUM BROMIDE
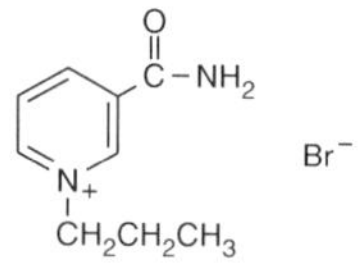
250 mg

S62,682-1
3-CARBAMOYL-1-METHYLPYRIDINIUM IODIDE
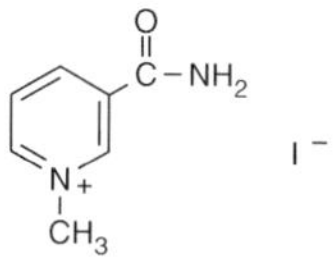
5 gm

S40,056-4
3-CARBAMOYL-1-HEXYLPYRIDINIUM BROMIDE
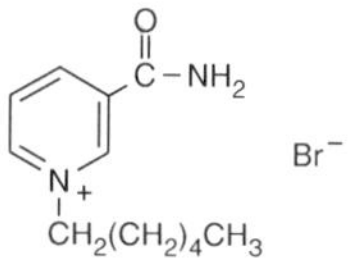
250 mg

S40,133-1
3-CARBAMOYL-1-NONYLPYRIDINIUM BROMIDE
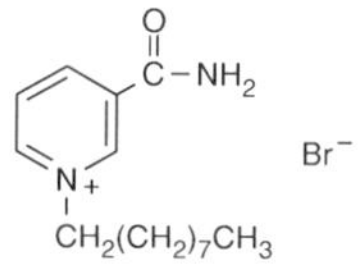
250 mg

S40,383-0
3-CARBAMOYL-1-DECYLPYRIDINIUM BROMIDE
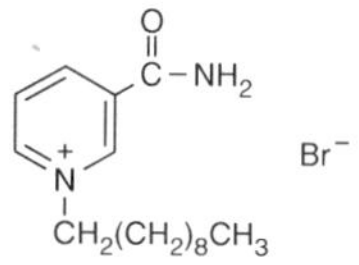
250 mg

S40,147-1
3-CARBAMOYL-1-UNDECYLPYRIDINIUM BROMIDE
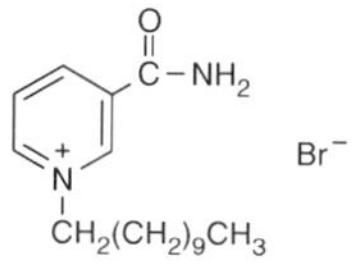
250 mg

S40,170-6
3-CARBAMOYL-1-DODECYLPYRIDINIUM BROMIDE
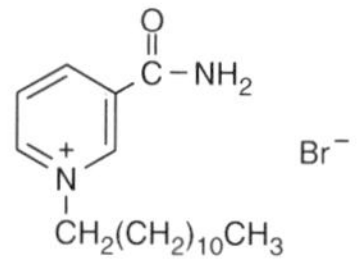
250 mg

S40,140-4
3-DIMETHYLCARBAMOYL-1-(PHENYLCARBAMOYLMETHYL)-PYRIDINIUM CHLORIDE
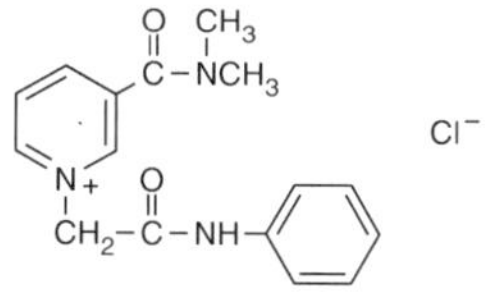
250 mg

S40,129-3
1-(4-CHLOROPHENYLCARBAMOYLMETHYL)-3-(DIMETHYLCARBAMOYL)-PYRIDINIUM CHLORIDE
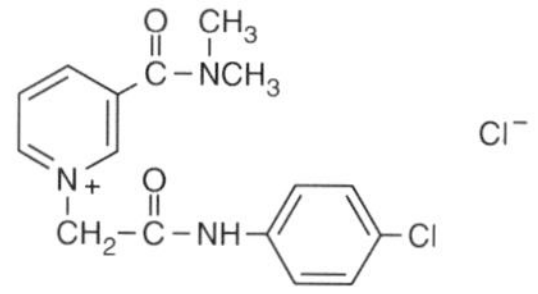
250 mg

S40,163-3
1-(4-CHLOROPHENYLCARBAMOYLMETHYL)-3-(DIETHYLCARBAMOYL)-PYRIDINIUM CHLORIDE
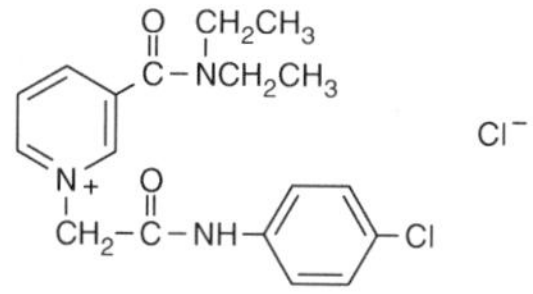
250 mg

S40,161-7
3-DIETHYLCARBAMOYL-1-(PHENYLCARBAMOYLMETHYL)-PYRIDINIUM CHLORIDE
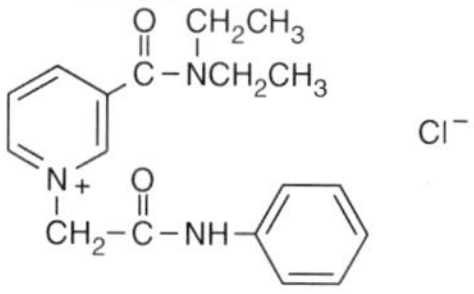
250 mg

S40,218-4
1-DECYLCARBAMOYLMETHYL-3-(DIMETHYLCARBAMOYL)-PYRIDINIUM CHLORIDE
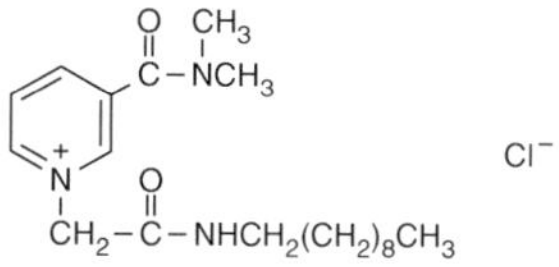
250 mg

S40,253-2
3-DIMETHYLCARBAMOYL-1-(DODECYLCARBAMOYLMETHYL)-PYRIDINIUM CHLORIDE
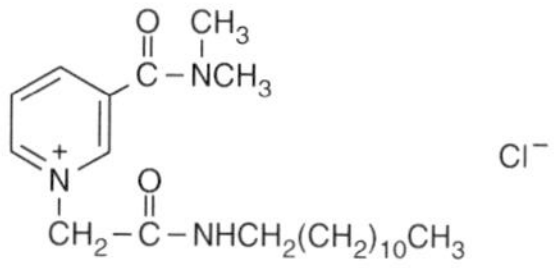
250 mg

S40,254-0
1-DECYLCARBAMOYLMETHYL-3-(DIETHYLCARBAMOYL)-PYRIDINIUM CHLORIDE
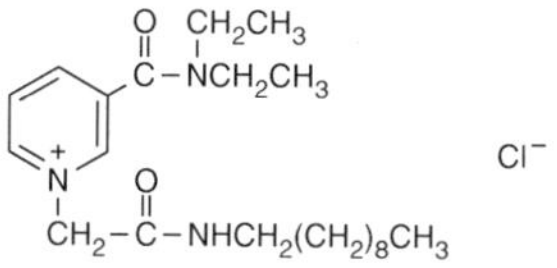
250 mg

S40,282-6
3-DIETHYLCARBAMOYL-1-(DODECYLCARBAMOYLMETHYL)-PYRIDINIUM CHLORIDE
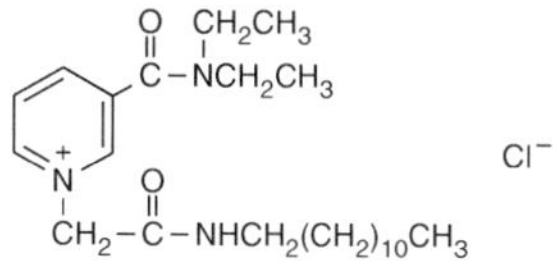
250 mg

S40,142-0
1-ALLYLCARBAMOYLMETHYL-3-(PHENYLCARBAMOYL)-PYRIDINIUM CHLORIDE
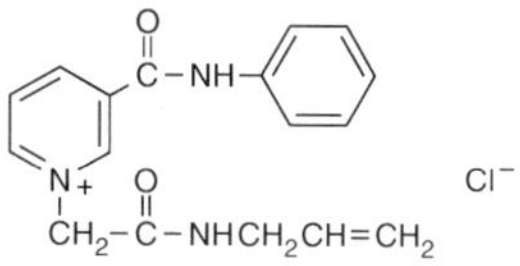
250 mg

S40,194-3
3-(N-PHENYLCARBAMOYL)-1-(N-PHENYLCARBAMOYLMETHYL)-PYRIDINIUM CHLORIDE
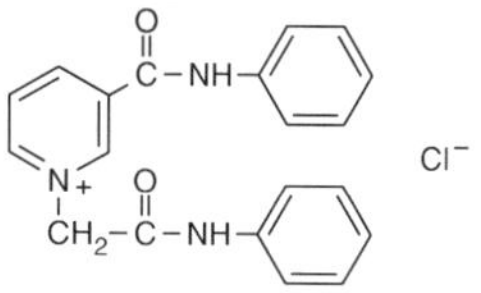
250 mg

S40,268-0
1-DECYLCARBAMOYLMETHYL-3-(PHENYLCARBAMOYL)-PYRIDINIUM CHLORIDE
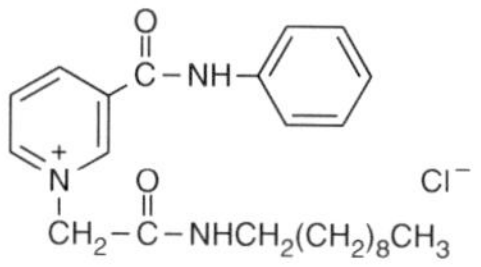
250 mg

S40,299-0
1-DODECYLCARBAMOYLMETHYL-3-(PHENYLCARBAMOYL)-PYRIDINIUM CHLORIDE
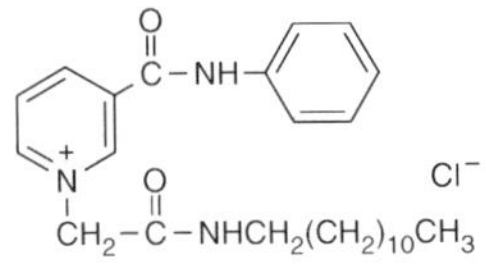
250 mg

S40,126-9
1-(4-CHLOROPHENYLCARBAMOYLMETHYL)-3-(METHYLCARBAMOYL)-PYRIDINIUM CHLORIDE
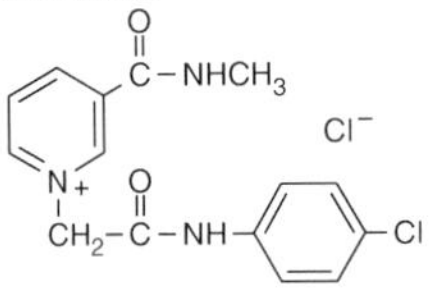
250 mg

S40,269-9
3-(4-CHLOROPHENYLCARBAMOYL)-1-(DECYLCARBAMOYLMETHYL)-PYRIDINIUM CHLORIDE
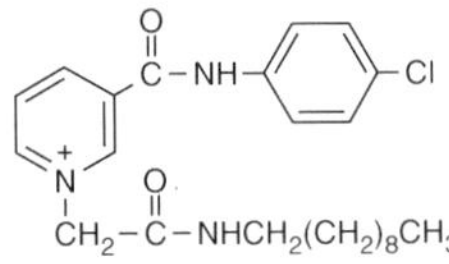
250 mg

S40,298-2
3-(4-CHLOROPHENYLCARBAMOYL)-1-(DODECYLCARBAMOYLMETHYL)-PYRIDINIUM CHLORIDE
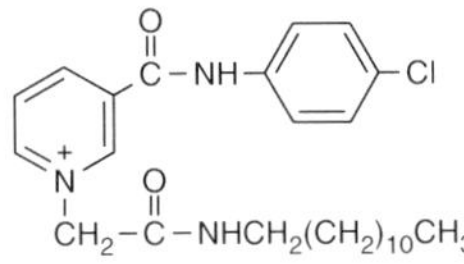
250 mg

S40,240-0
3-ALLYLCARBAMOYL-1-(DECYLCARBAMOYLMETHYL)-PYRIDINIUM CHLORIDE
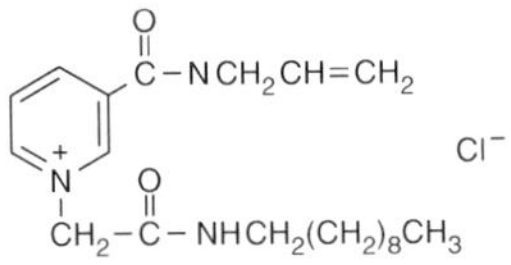
250 mg

S40,235-4
1-DODECYLCARBAMOYLMETHYL-3-(METHYLCARBAMOYL)-PYRIDINIUM CHLORIDE
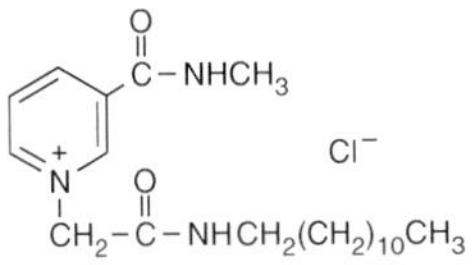
250 mg

S40,203-6
3-DODECYLCARBAMOYL-1-METHYLPYRIDINIUM CHLORIDE
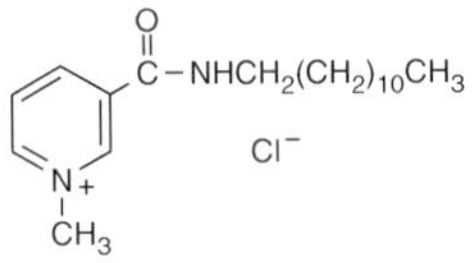
250 mg

S40,205-2
3-DODECYLCARBAMOYL-1-METHYLPYRIDINIUM BROMIDE
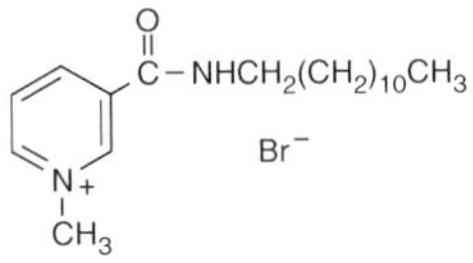
250 mg

S40,200-1
3-DODECYLCARBAMOYL-1-METHYLPYRIDINIUM IODIDE
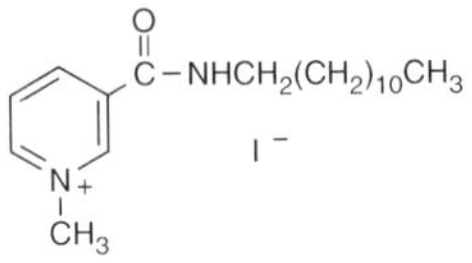
250 mg

S40,219-2
3-DODECYLCARBAMOYL-1-METHYLPYRIDINIUM METHANESULFONATE
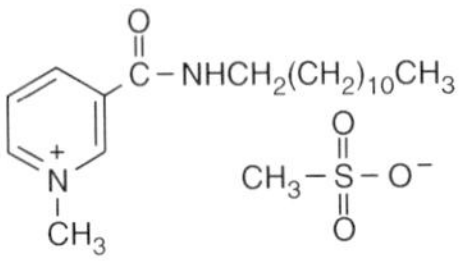
250 mg

S40,156-0
1-(3,4-DICHLOROBENZYL)-3-(DIETHYLCARBAMOYL)-PYRIDINIUM CHLORIDE
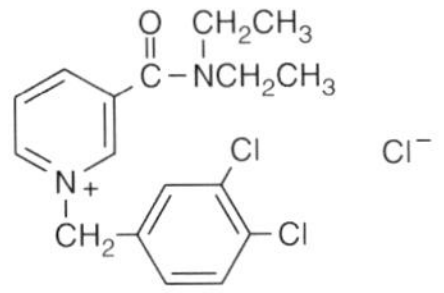
250 mg

S34,326-9
1-(2-CARBAMOYLETHYL)-3-CARBOXYPYRIDINIUM CHLORIDE
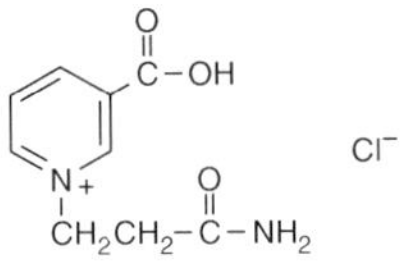
5 gm

S36,722-2
1-(2-CARBAMOYLETHYL)-3-CYANOPYRIDINIUM CHLORIDE
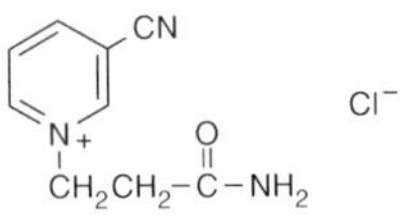
250 mg

S34,331-5
1-(2-CARBAMOYLETHYL)-PYRIDINIUM CHLORIDE
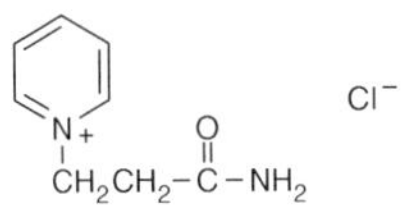
5 gm

S34,328-5
1-(2-CARBAMOYLETHYL)-2-METHYLPYRIDINIUM CHLORIDE
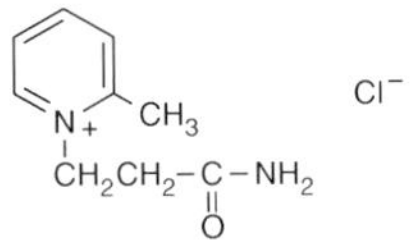
5 gm

S34,329-3
1-(2-CARBAMOYLETHYL)-3-METHYLPYRIDINIUM CHLORIDE
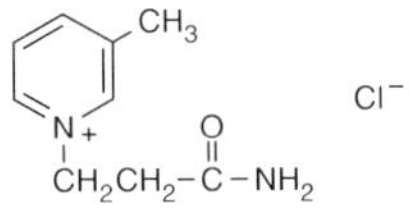
5 gm

S34,330-7
1-(2-CARBAMOYLETHYL)-4-METHYLPYRIDINIUM CHLORIDE
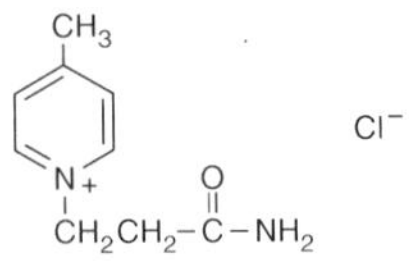
5 gm

S87,266-0
1-(3-PHTHALIMIDOPROPYL)PYRIDINIUM PERCHLORATE
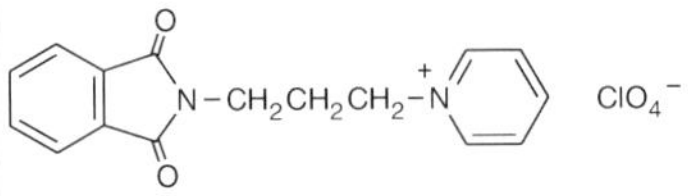
250 mg

S40,183-8
1-(4-CHLOROBENZYLCARBAMOYLMETHYL)-3-(DIETHYLCARBAMOYL)-PYRIDINIUM CHLORIDE
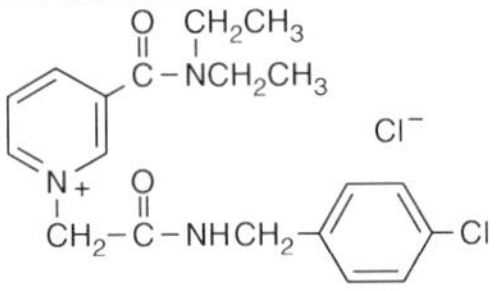
250 mg

S51,575-2
1,1'-[N,N'-HEXAMETHYLENEBIS(2-CARBAMOYLETHYL)]BIS-(PYRIDINIUM CHLORIDE)
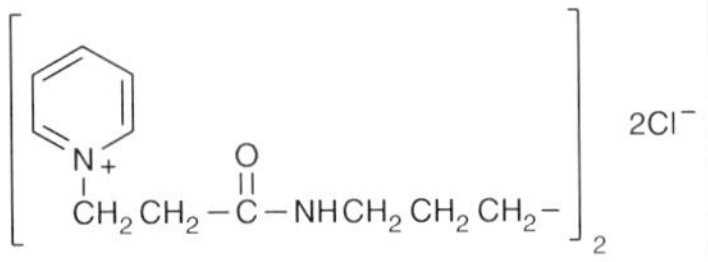
250 mg

S62,382-2
1,1'-TRIMETHYLENEBIS(2,4-DIMETHYLPYRIDINIUM BROMIDE)
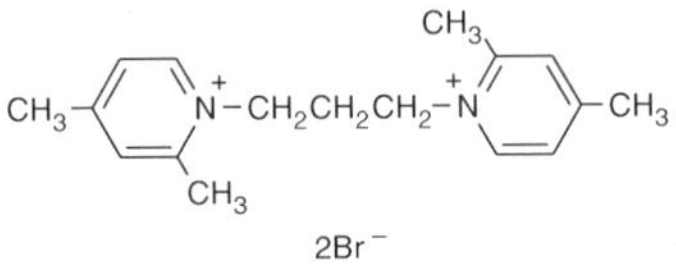
5 gm

S62,380-6
1,1'-TRIMETHYLENEBIS(3,5-DIMETHYLPYRIDINIUM BROMIDE)
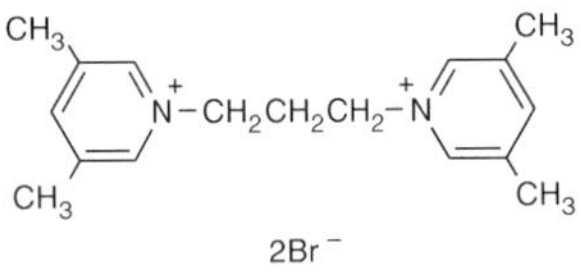
250 mg

S62,390-3
1,1'-TRIMETHYLENEBIS(3,4-DIMETHYLPYRIDINIUM BROMIDE)
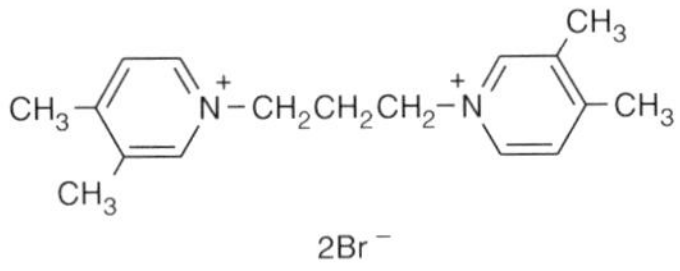
250 mg

S62,397-0
1,1'-TRIMETHYLENEBIS(3-METHYLPYRIDINIUM BROMIDE)
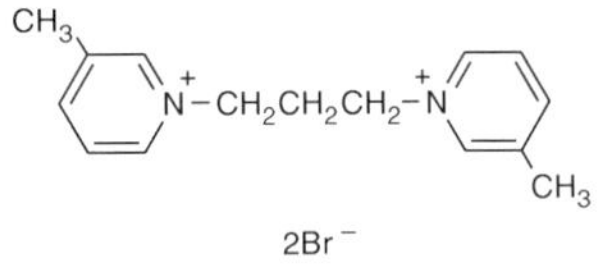
250 mg

S62,391-1
1,1'-TRIMETHYLENEBIS(4-METHYLPYRIDINIUM BROMIDE)
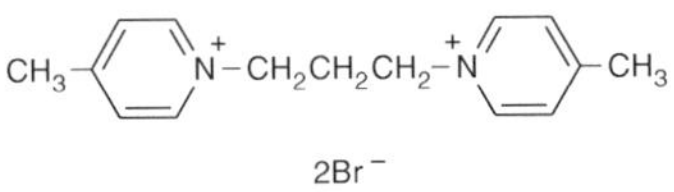
250 mg

S62,385-7
1,1'-TRIMETHYLENEBIS-(PYRIDINIUM BROMIDE)
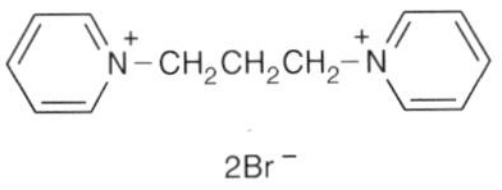
250 mg

S65,490-6
N,N'-(4-METHYL-1,3-PHENYLENE)BIS-(2-PYRIDYL CARBAMATE)
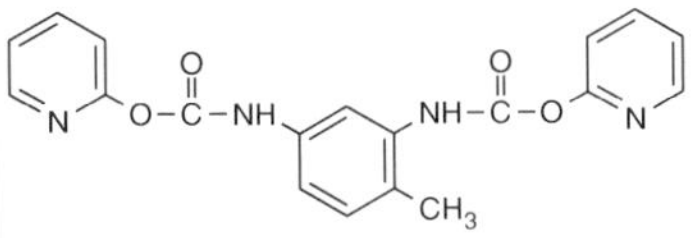
5 gm

S42,304-1
N-BENZYL-N-(2-CYCLOHEXYL-1-PHENYLETHYL)NICOTINAMIDE
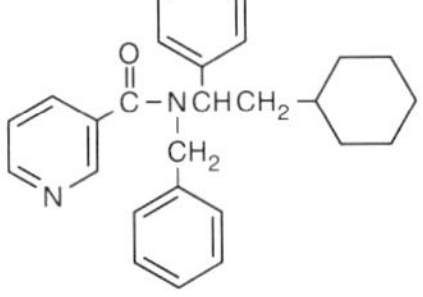
250 mg

S91,526-2
3-PYRIDINECARBOXALDEHYDE SEMICARBAZONE
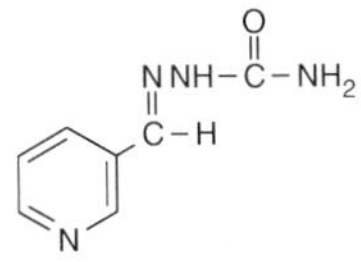
250 mg

S79,646-8
2-[3-(3-NITROPHENYL)UREIDO]ACETIC (3-PYRIDYLMETHYLENE)HYDRAZIDE
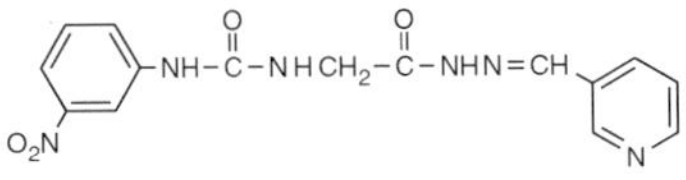
250 mg

S93,788-6
3-(BUTYRYL)PYRIDINE SEMICARBAZONE
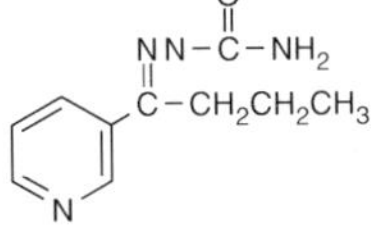
250 mg

S36,475-4
3-(3-PYRIDYLMETHYLAMINO)-1,2,4-TRIAZOLE
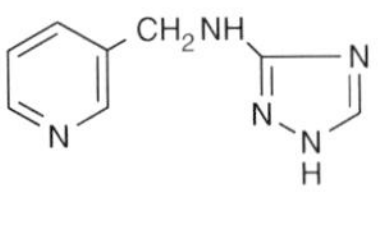
5 gm

S55,368-9
2-(3-PYRIDYL)BENZIMIDAZOLE

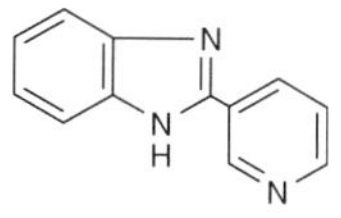

250 mg

S65,483-3
N,N'-(4-METHYL-1,3-PHENYLENE)BIS-(3-PYRIDYL CARBAMATE)

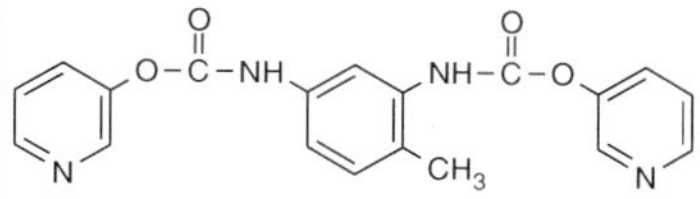

5 gm

S60,277-9
3-PYRIDYL N-(3-NITROPHENYL)-CARBAMATE

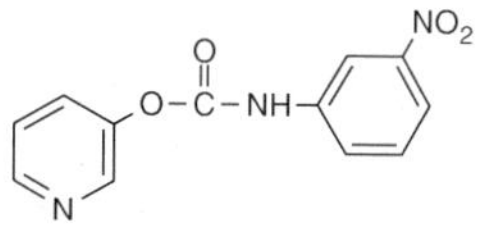

5 gm

S87,141-9
4-NITRAMINOPYRIDINE

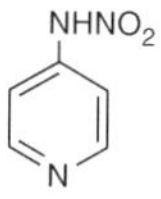

250 mg

S69,377-4
4-PYRIDYLMETHYL N-(2-METHYL-5-NITROPHENYL)CARBAMATE

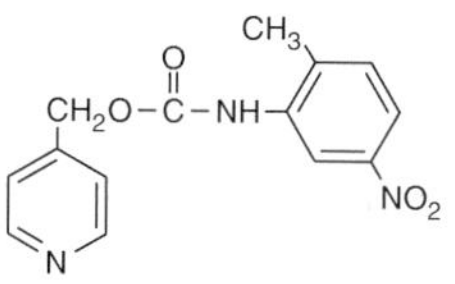

250 mg

S91,141-0
1-(3,5-DICHLORO-2-PYRIDYL)-3-(4-METHYL-3-NITROPHENYL)UREA

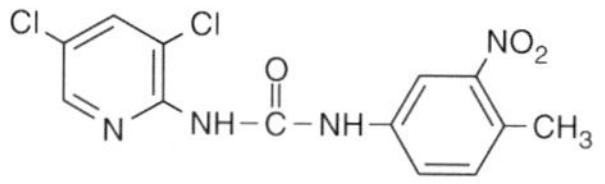

250 mg

S91,139-9
1-(5-CHLORO-2-PYRIDYL)-3-(4-METHYL-3-NITROPHENYL)UREA

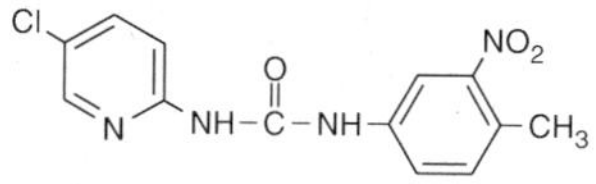

250 mg

S84,793-3
1-(2-METHYL-5-NITROPHENYL)-3-(3-PYRIDYLMETHYL)UREA

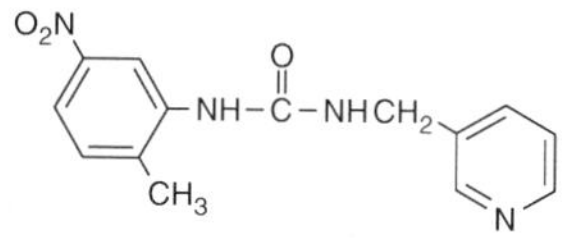

250 mg

S84,300-8
2-(2-NITROPHENYL)PYRIDINE PICRATE

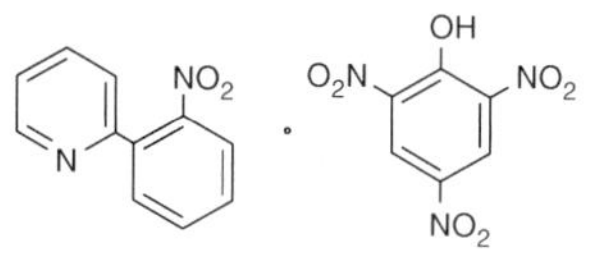

250 mg

S76,516-3
2-(2,4-DINITROBENZYL)PYRIDINE

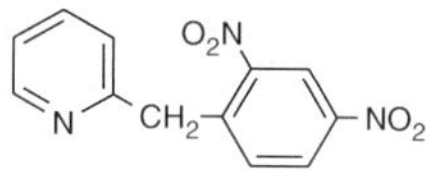

5 gm

S60,163-2
2-PYRIDYL N-(3-NITROPHENYL)-CARBAMATE

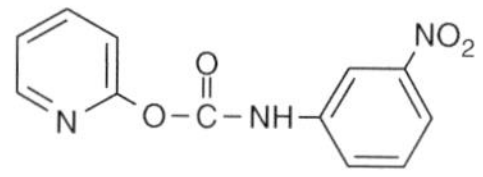

5 gm

S63,407-7
1-(3-NITROPHENYL)-3-(3-PYRIDYLMETHYL)UREA

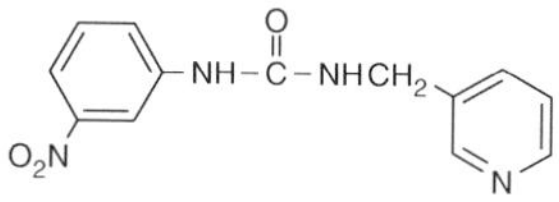

5 gm

S39,077-1
N-(3-PYRIDYLMETHYL)ACETOACETAMIDE

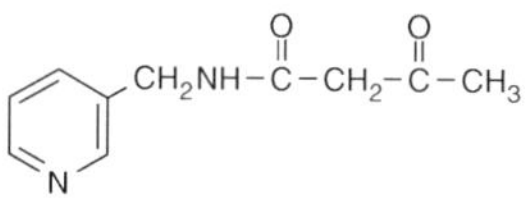

5 gm

S72,074-7
N-(3-PYRIDYLMETHYL)PHTHALAMIC ACID

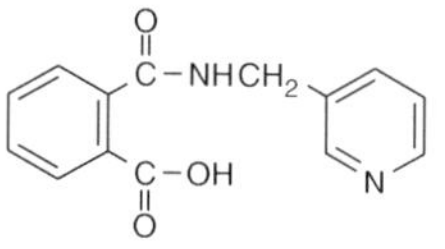

5 gm

S37,678-7
ETHYL 2-CHLORO-3-CYANO-6-METHYLISONICOTINATE

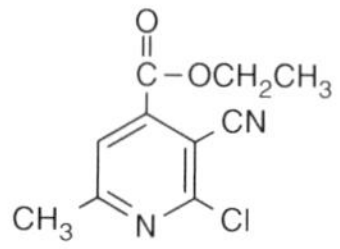

250 mg

S91,714-1
1-PROPYL-3-NICOTINOYLUREA

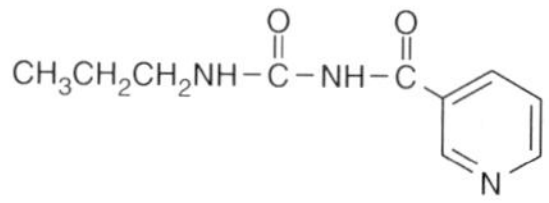

250 mg

S37,724-4
NICOTINANILIDE

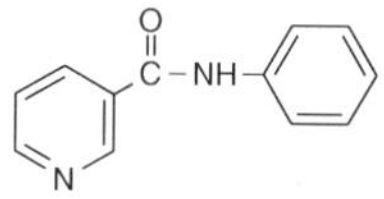

5 gm

S40,386-5
NICOTINAMIDOMETHYLPHOSPHONIC ACID, DISODIUM SALT MONOHYDRATE

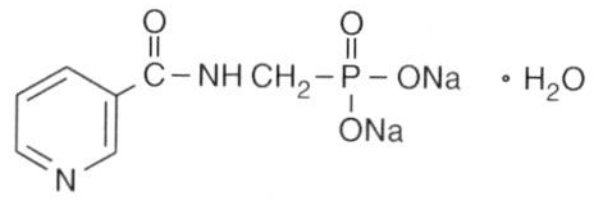

250 mg

S93,659-6
N-(4-PYRIDYL)BENZAMIDE

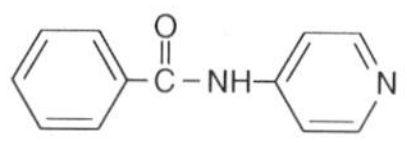

250 mg

S61,725-3
(NICOTINAMIDOMETHYL)PHOSPHONIC ACID, DISODIUM SALT

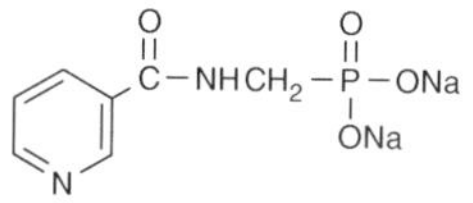

5 gm

S43,542-2
N-ETHYLNICOTINAMIDE

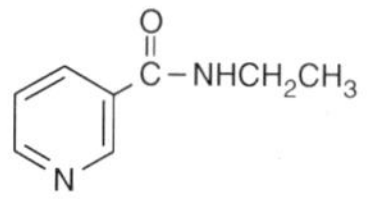

5 gm

S39,888-8
1-NICOTINOYLPYRROLIDINE

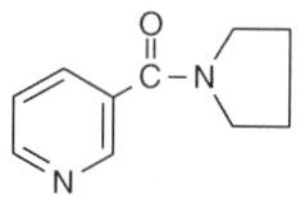

250 mg

S39,893-4
4-NICOTINOYLMORPHOLINE

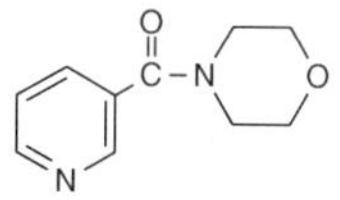

5 gm

S72,008-9
CIS-HEXAHYDRO-N-(3-PYRIDYLMETHYL)-PHTHALAMIC ACID

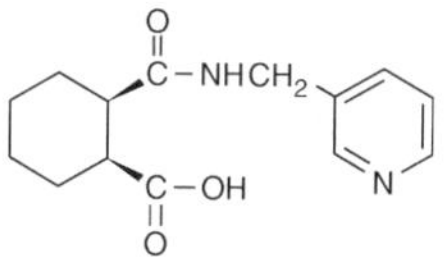

5 gm

S75,023-9
N-[2-(2-PYRIDYL)ETHYL]-GLUTARAMIC ACID

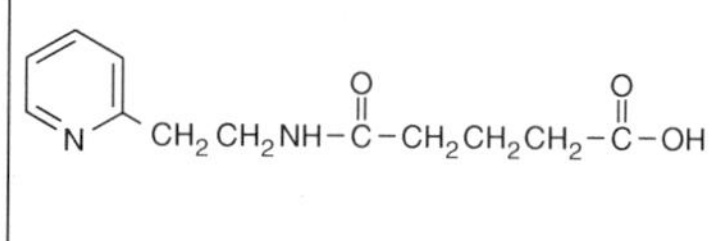

5 gm

S75,024-7
N-[2-(2-PYRIDYL)ETHYL]-3,4,5,6-TETRABROMOPHTHALAMIC ACID

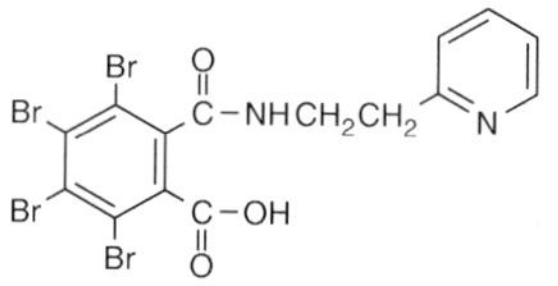

5 gm

S91,222-0
ETHYL 2-(3-[2-(2-PYRIDYL)ETHYL]-UREIDO)ACETATE

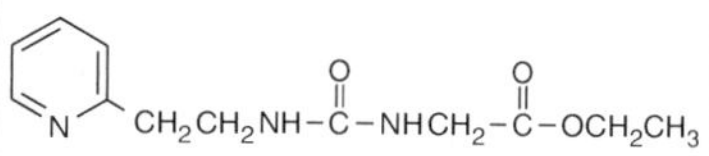

250 mg

S92,826-7
ETHYL 4-PYRIDYLCARBAMATE

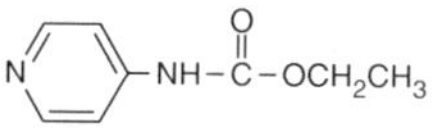

250 mg

S87,732-8
5-FLUORO-3-PYRIDINECARBOXAMIDE

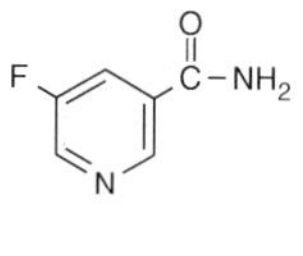

250 mg

S37,757-0
3',4'-DICHLORONICOTINANILIDE

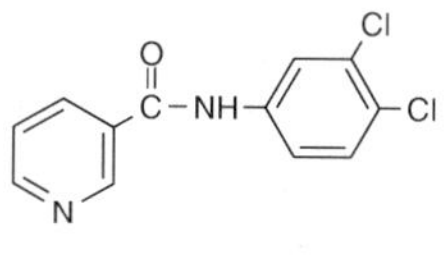

250 mg

S42,393-9
(NICOTINAMIDOMETHYL)PHOSPHONIC ACID, IRON(II) SALT

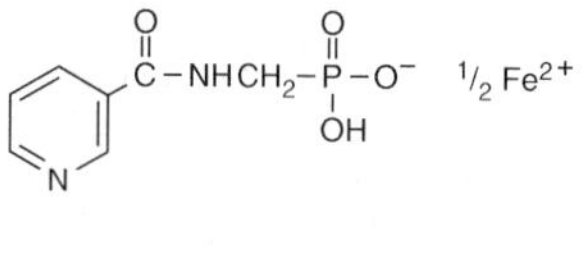

5 gm

S38,207-8
N,N'-DIALLYL-3,5-PYRIDINEDICARBOXAMIDE

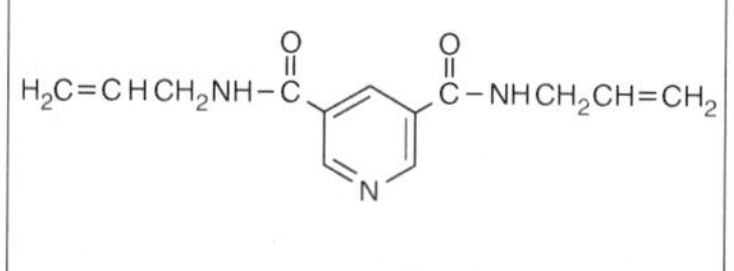

250 mg

S37,761-9
N,N-DIPROPYLNICOTINAMIDE

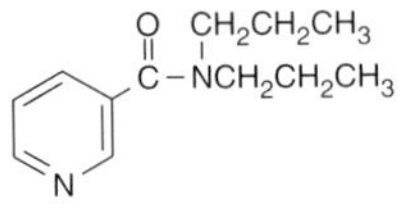

5 gm

S38,167-5
N,N-DIISOPROPYLNICOTINAMIDE

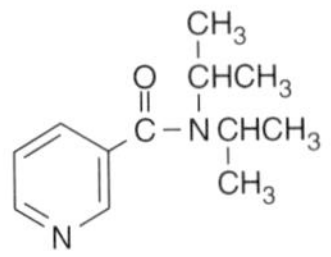

5 gm

S41,216-3
2,3-DIHYDRO-2-[2-(2-PYRIDYL)ETHYL]-1,4-PHTHALAZINEDIONE

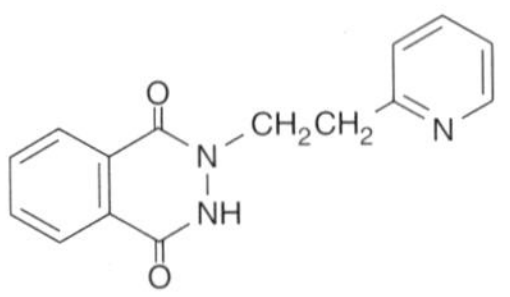

5 gm

S60,795-9
1-CYCLOHEXYL-3-(3-PYRIDYLMETHYL)UREA

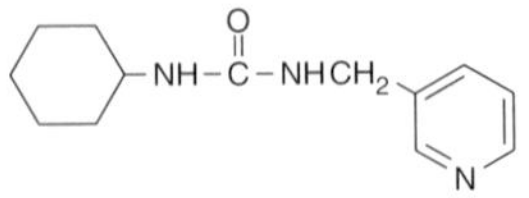

5 gm

S64,433-1
1-PHENYL-3-(4-PYRIDYL)UREA

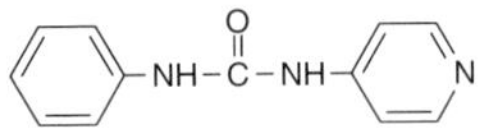

5 gm

S63,375-5
1-(2-CHLOROPHENYL)-3-(3-PYRIDYLMETHYL)UREA

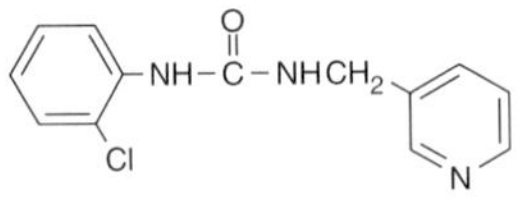

5 gm

S61,097-6
1-(3-PYRIDYLMETHYL)-3-(O-TOLYL)UREA

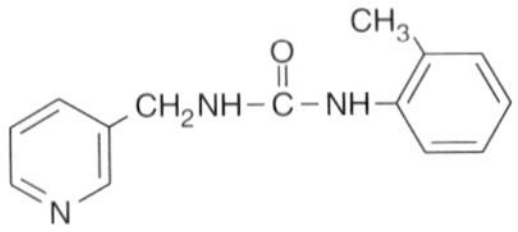

5 gm

S66,357-3
1-(2-ETHYLPHENYL)-3-(3-PYRIDYLMETHYL)UREA

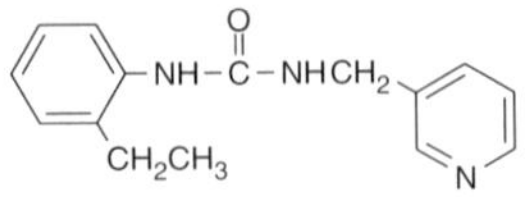

5 gm

S60,695-2
1-(4-CHLOROPHENYL)-3-(3-PYRIDYLMETHYL)UREA

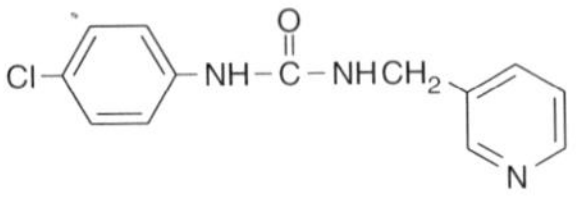

5 gm

S64,029-8
1-(4-BROMOPHENYL)-3-(3-PYRIDYLMETHYL)UREA

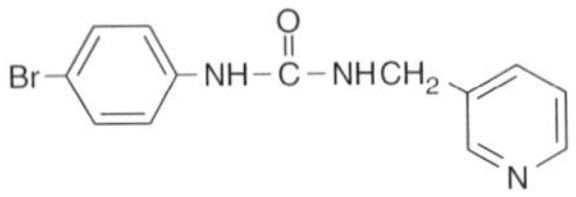

5 gm

S61,271-5
1-(3-PYRIDYLMETHYL)-3-(P-TOLYL)UREA

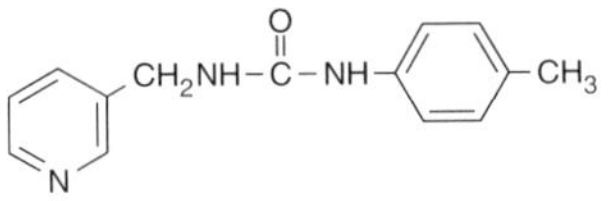

5 gm

S65,851-0
4,4'-BIS[3-(3-PYRIDYL)UREIDO]-DIPHENYLMETHANE

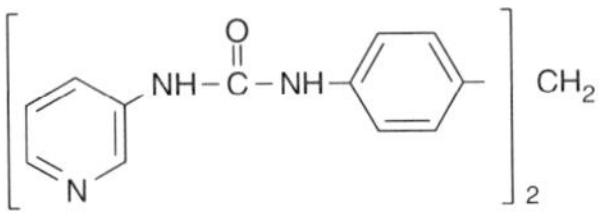

5 gm

S63,495-6
4,4'-METHYLENEBIS[1-PHENYL-3-(3-PYRIDYLMETHYL)UREA]

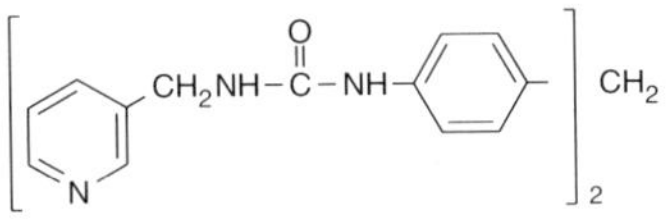

5 gm

S91,215-8
1-(2,6-DIETHYLPHENYL)-3-[2-(2-PYRIDYL)ETHYL]UREA

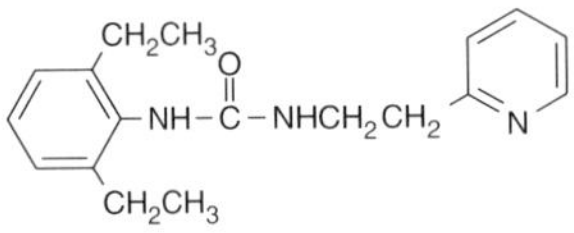

250 mg

S61,595-1
1-(2,3-DICHLOROPHENYL)-3-(3-PYRIDYLMETHYL)UREA

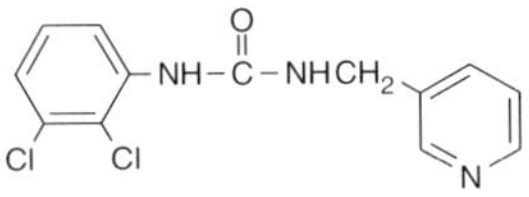

5 gm

S63,628-2
1-(2,6-DICHLOROPHENYL)-3-(3-PYRIDYLMETHYL)UREA

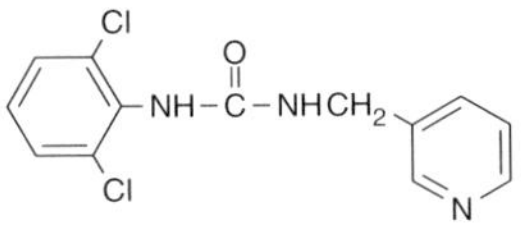

5 gm

S65,234-2
1-(2-CHLORO-6-METHYLPHENYL)-3-(3-PYRIDYLMETHYL)UREA

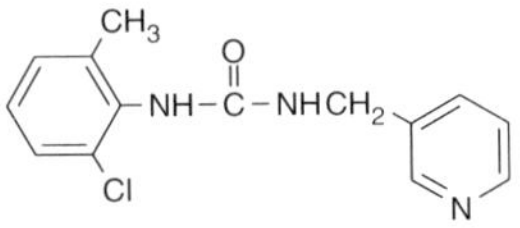

5 gm

S71,756-8
1-(3-PYRIDYLMETHYL)-3-(2,6-XYLYL)UREA

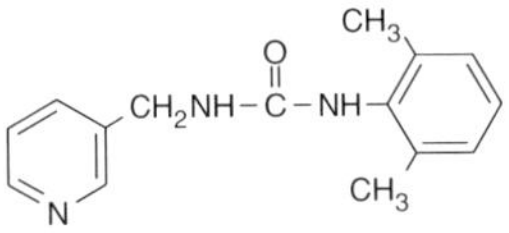

250 mg

S66,383-2
1-(2-ETHYL-6-METHYLPHENYL)-3-(3-PYRIDYLMETHYL)UREA

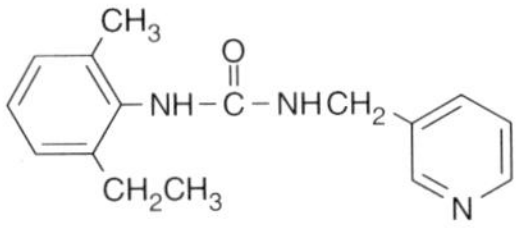

5 gm

S66,626-2
1-(2,6-DIETHYLPHENYL)-3-(3-PYRIDYLMETHYL)UREA

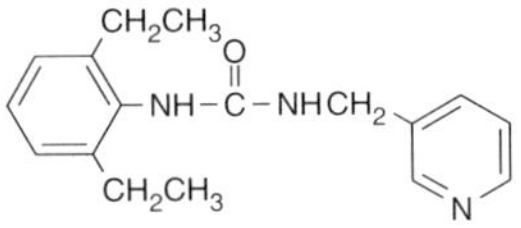

5 gm

S66,618-1
1-(2,6-DIISOPROPYLPHENYL)-3-(3-PYRIDYLMETHYL)UREA

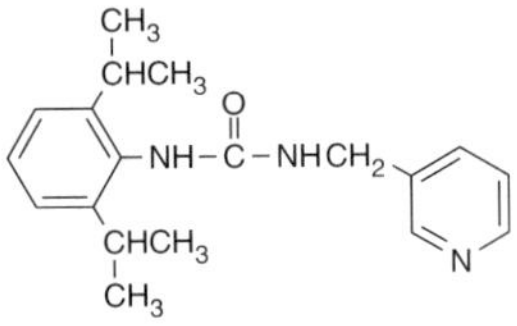

5 gm

S91,240-9
1-[2-(2-PYRIDYL)ETHYL]-3-(3,4-XYLYL)UREA

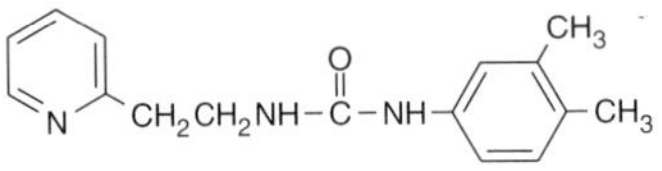

250 mg

S60,763-0
1-(3,4-DICHLOROPHENYL)-3-(3-PYRIDYLMETHYL)UREA

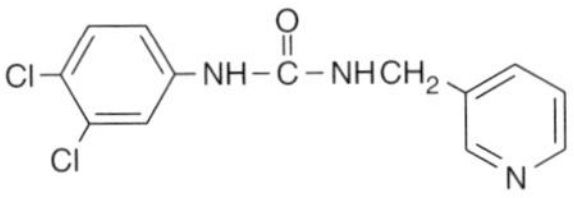

5 gm

S61,198-0
1-(2,5-DICHLOROPHENYL)-3-(3-PYRIDYLMETHYL)UREA

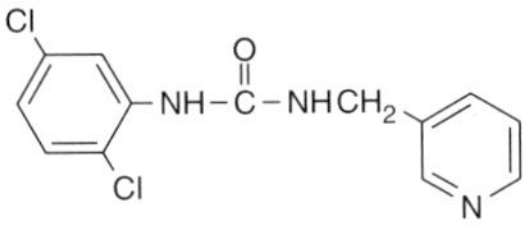

5 gm

S61,357-6
1-(2,4-DICHLOROPHENYL)-3-(3-PYRIDYLMETHYL)UREA

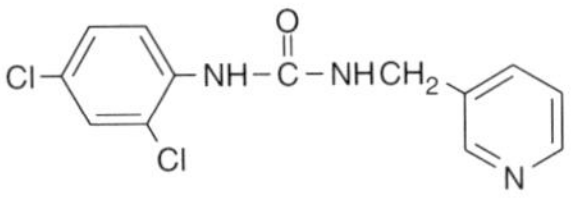

5 gm

S60,793-2
1-(4-CHLORO-2-METHYLPHENYL)-3-(3-PYRIDYLMETHYL)UREA

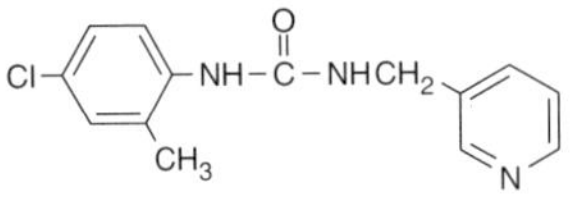

5 gm

S60,942-0
1-(5-CHLORO-2-METHYLPHENYL)-3-(3-PYRIDYLMETHYL)UREA

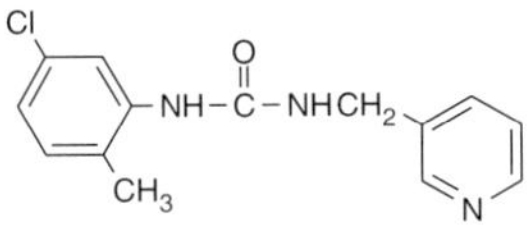

5 gm

S63,849-8
1-(3-CHLORO-4-METHYLPHENYL)-3-(3-PYRIDYLMETHYL)UREA

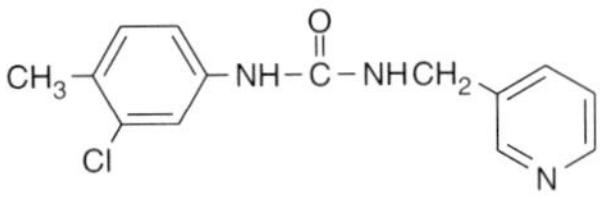

5 gm

S61,118-2
1-(3-PYRIDYLMETHYL)-3-(2,4-XYLYL)UREA

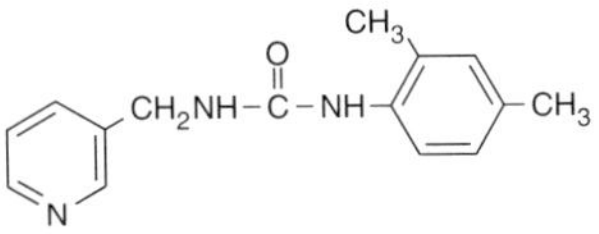

5 gm

S64,996-1
1-(3-PYRIDYLMETHYL)-3-(2,5-XYLYL)UREA

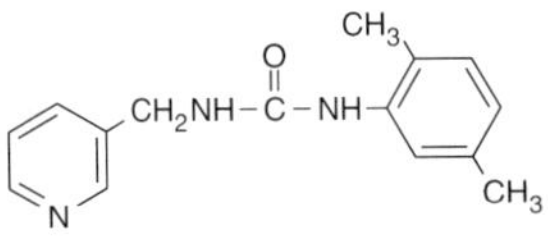

5 gm

S66,101-5
1-(3-PYRIDYLMETHYL)-3-(3,4-XYLYL)UREA

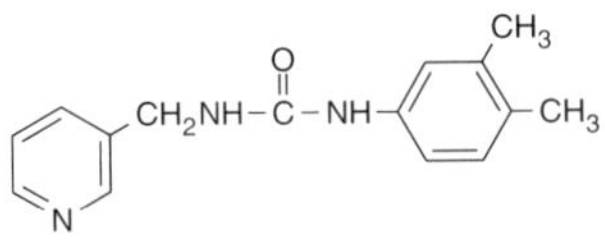

5 gm

S64,454-4
1,1'-(4-METHYL-1,3-PHENYLENE)BIS[3-(3-PYRIDYLMETHYL)UREA]

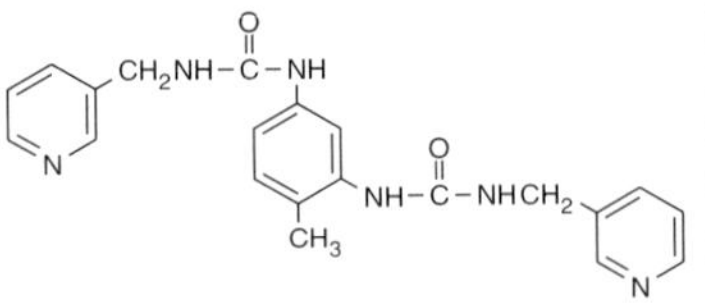

5 gm

S66,386-7
1-(3-PYRIDYLMETHYL)-3-(3,5-XYLYL)UREA

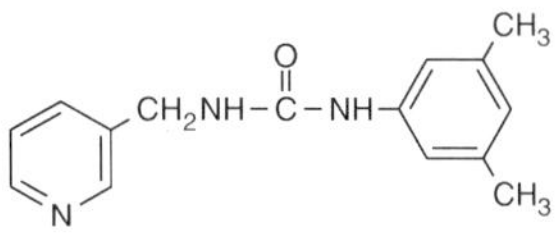

5 gm

S60,799-1
1-BUTYL-3-(3-PYRIDYLMETHYL)UREA

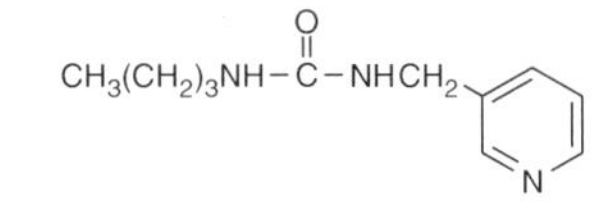

5 gm

S85,508-1
1-TERT-BUTYL-3-(3-PYRIDYLMETHYL)UREA

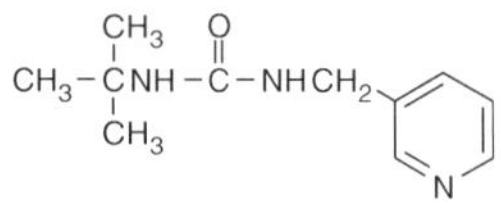

250 mg

S66,785-4
1,3-BIS(3-PYRIDYLMETHYL)UREA

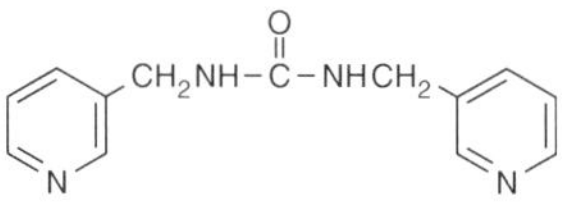

5 gm

S60,949-8
1-(3-CHLOROPHENYL)-3-(3-PYRIDYLMETHYL)UREA

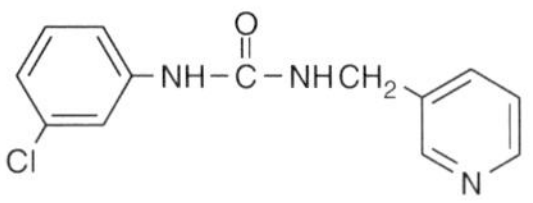

5 gm

S60,662-6
1-(3-PYRIDYLMETHYL)-3-(M-TOLYL)UREA

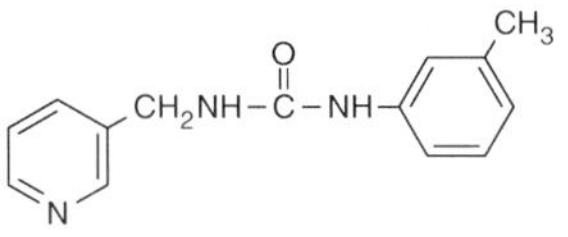

5 gm

S63,845-5
1,1'-HEXAMETHYLENEBIS[3-(3-PYRIDYLMETHYL)UREA]

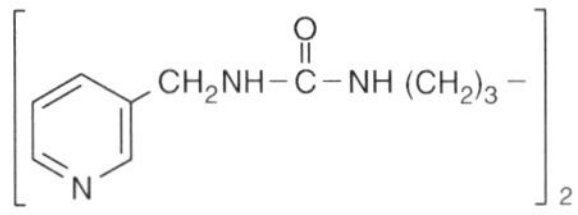

5 gm

S63,956-7
1-(1-NAPHTHYL)-3-(3-PYRIDYLMETHYL)UREA

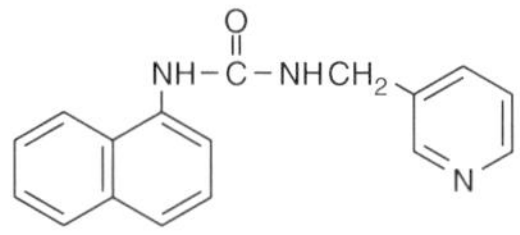

5 gm

S90,517-8
1-ETHYL-3-(4-PYRIDYL)UREA

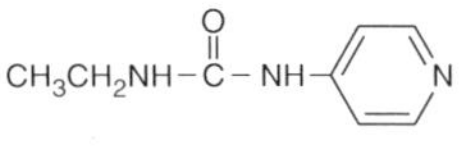

250 mg

S73,660-0
1-PHENYL-3-[2-(2-PYRIDYL)ETHYL]-2-THIOUREA

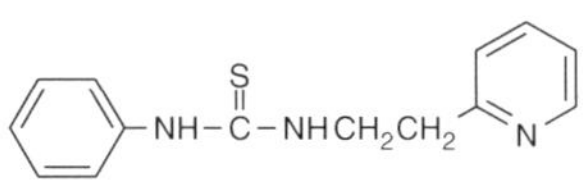

5 gm

S62,510-8
1-PHENYL-3-(3-PYRIDYLMETHYL)-2-THIOUREA

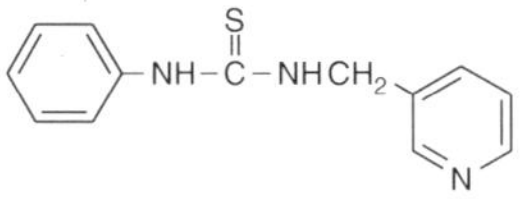

5 gm

S55,379-4
N-[2-(2-PYRIDYL)ETHYL]-DITHIOCARBAMIC ACID, AMMONIUM SALT

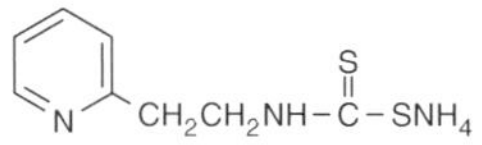

250 mg

S44,552-5
ALPHA-PHENYL-2-PYRIDINEACETONITRILE

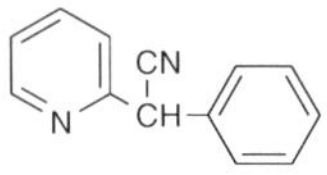

250 mg

S76,102-8
3-(4,5-DIHYDRO-2-IMIDAZOLYL)-PYRIDINE

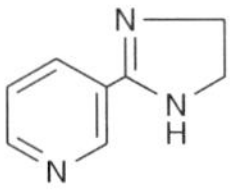

250 mg

S42,254-1
N-(1,2-DIPHENYLETHYL)-ISONICOTINAMIDE

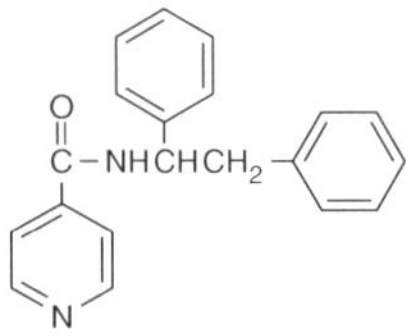

250 mg

S53,324-6
4-(ALPHA-ETHYLBENZYLAMINOMETHYL)-PYRIDINE

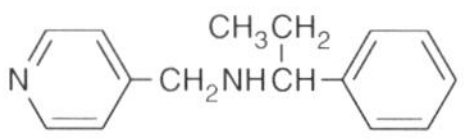

5 gm

S36,857-1
5,6-DIMETHYL-2-(4-PYRIDYL)-BENZIMIDAZOLE

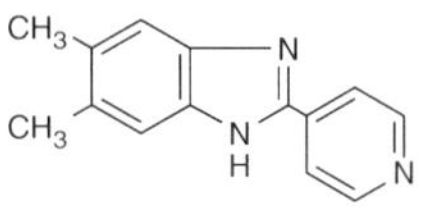

250 mg

S60,527-1
1-(3-NITROBENZYL)-PYRIDINIUM CHLORIDE

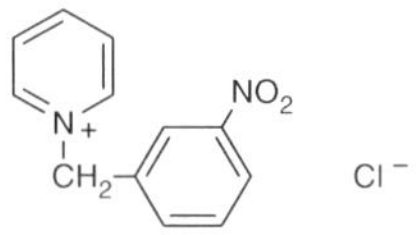

5 gm

S40,290-7
1-BENZYL-4-(DODECYLCARBAMOYL)-PYRIDINIUM CHLORIDE

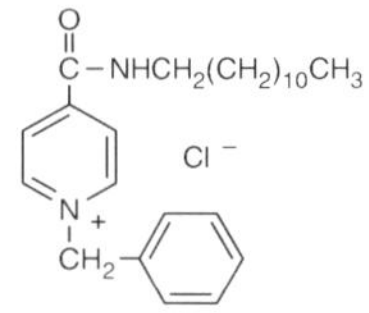

250 mg

S40,287-7
1-(3,4-DICHLOROBENZYL)-4-(DODECYLCARBAMOYL)-PYRIDINIUM CHLORIDE

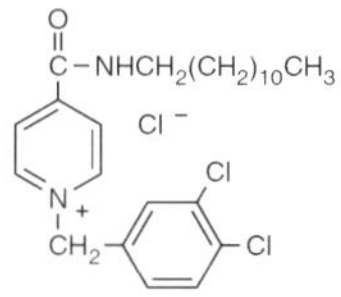

250 mg

S61,785-7
1-(3-NITROBENZYL)PYRIDINIUM P-TOLUENESULFONATE

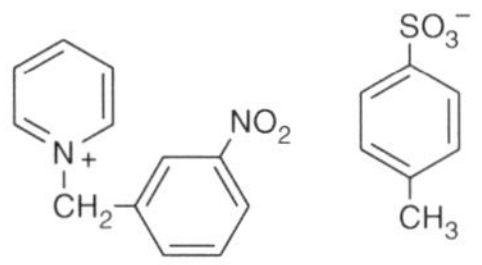

250 mg

S40,107-2
4-CARBAMOYL-1-(HEXYLOXYCARBONYLMETHYL)-PYRIDINIUM CHLORIDE

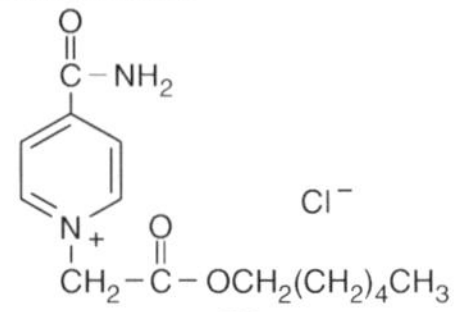

250 mg

S40,171-4
4-CARBAMOYL-1-(DECYLOXYCARBONYLMETHYL)-PYRIDINIUM CHLORIDE

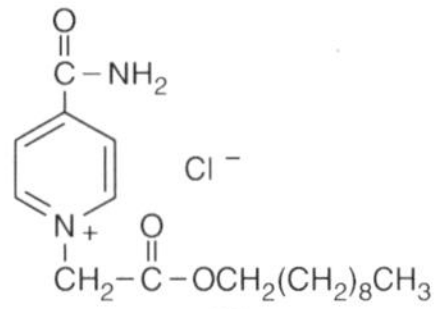

250 mg

S40,401-2
4-CARBAMOYL-1-(DODECYLOXYCARBONYLMETHYL)-PYRIDINIUM CHLORIDE

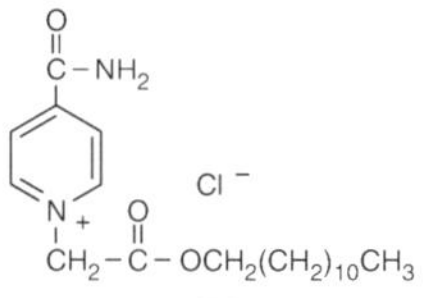

250 mg

S40,266-4
4-CARBAMOYL-1-(HEXADECYLOXYCARBONYLMETHYL)-PYRIDINIUM CHLORIDE

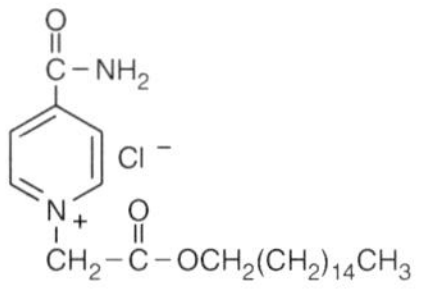

250 mg

S40,217-6
4-CARBAMOYL-1-(DODECYLCARBAMOYLMETHYL)-PYRIDINIUM CHLORIDE

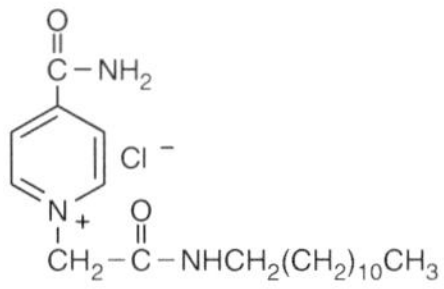

250 mg

S40,146-3
4-CARBAMOYL-1-UNDECYLPYRIDINIUM BROMIDE

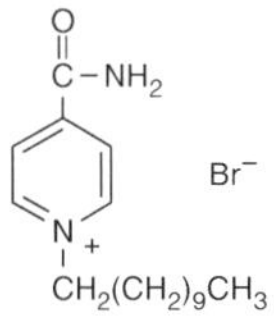

250 mg

S40,033-5
1-BUTYL-4-CARBAMOYLPYRIDINIUM BROMIDE

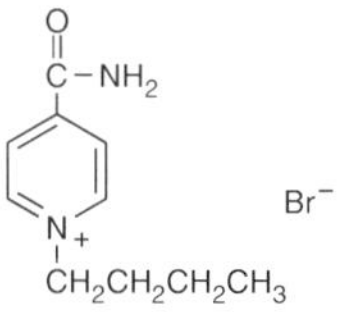

250 mg

S40,046-7
4-CARBAMOYL-1-ISOPENTYLPYRIDINIUM BROMIDE

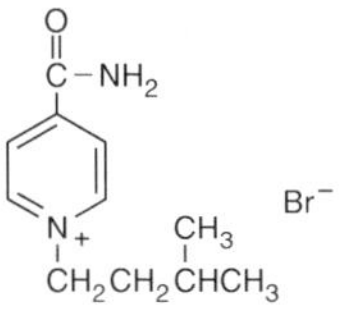

250 mg

S40,048-3
4-CARBAMOYL-1-PENTYLPYRIDINIUM BROMIDE

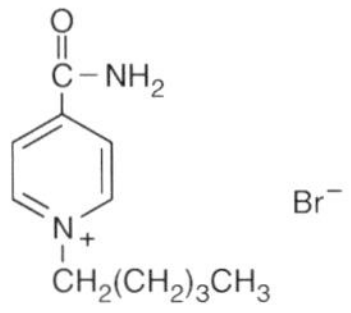

250 mg

S40,057-2
4-CARBAMOYL-1-HEXYLPYRIDINIUM BROMIDE

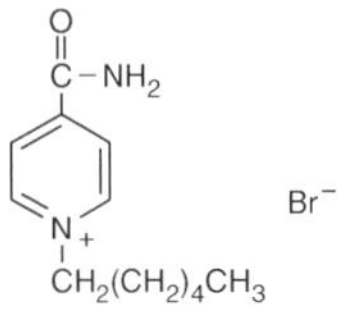

250 mg

S40,132-3
4-CARBAMOYL-1-NONYLPYRIDINIUM BROMIDE

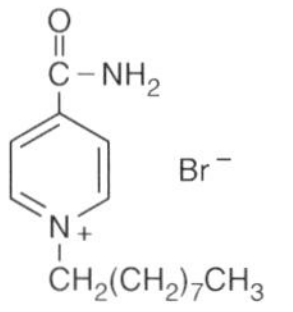

250 mg

S40,144-7
4-CARBAMOYL-1-DECYLPYRIDINIUM BROMIDE

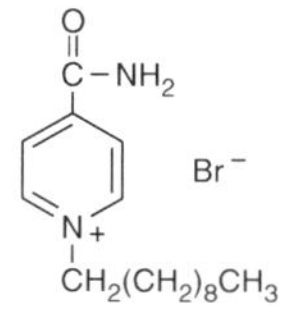

250 mg

S40,404-7
4-CARBAMOYL-1-HEXADECYLPYRIDINIUM BROMIDE

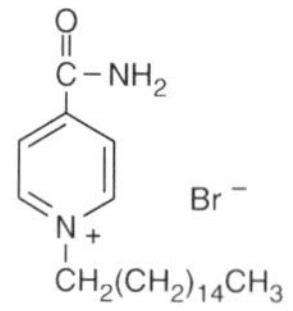

250 mg

S40,283-4
4-CARBAMOYL-1-OCTADECYLPYRIDINIUM BROMIDE

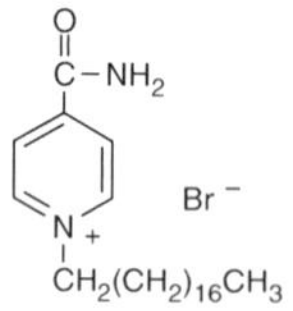

250 mg

S40,202-8
4-DODECYLCARBAMOYL-1-METHYLPYRIDINIUM CHLORIDE

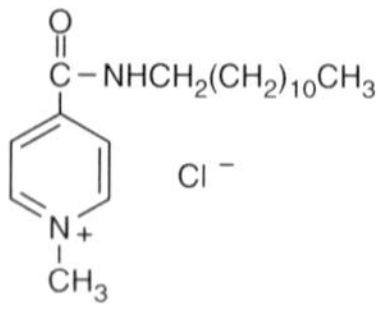

250 mg

S40,204-4
4-DODECYLCARBAMOYL-1-METHYLPYRIDINIUM BROMIDE

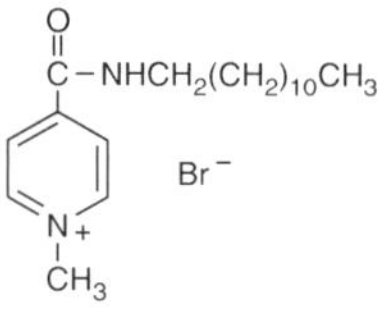

250 mg

S40,199-4
4-DODECYLCARBAMOYL-1-METHYLPYRIDINIUM IODIDE

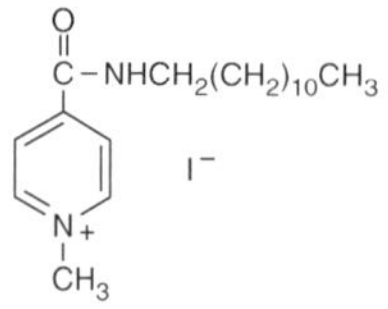

250 mg

S40,220-6
4-DODECYLCARBAMOYL-1-METHYLPYRIDINIUM METHANESULFONATE

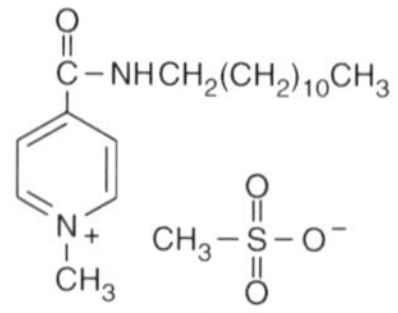

250 mg

S40,155-2
1-(3,4-DICHLOROBENZYL)-4-(DIETHYLCARBAMOYL)-PYRIDINIUM CHLORIDE

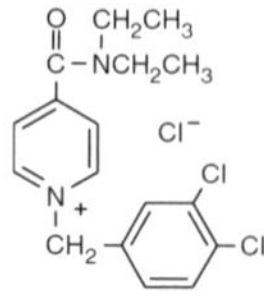

250 mg

S64,908-2
4-CYANO-1-METHYLPYRIDINIUM IODIDE

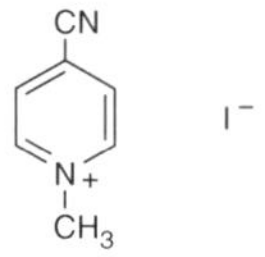

250 mg

S87,362-4
2,2'-VINYLENEBIS(1-METHYLPYRIDINIUM TETRAFLUOROBORATE)

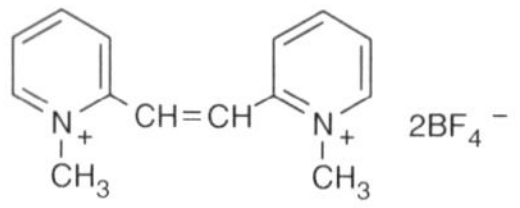

250 mg

S62,347-4
1,1'-TETRAMETHYLENEBIS(2,5-DIMETHYLPYRIDINIUM PERCHLORATE)

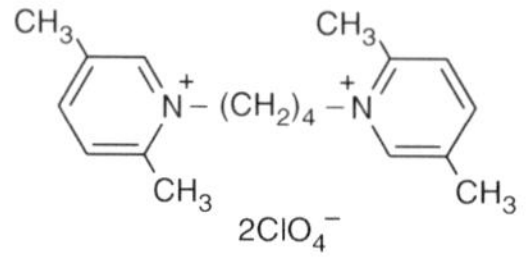

5 gm

S50,331-2
1-METHYL-2-[2-(2-PYRIDYL)VINYL]-PYRIDINIUM IODIDE

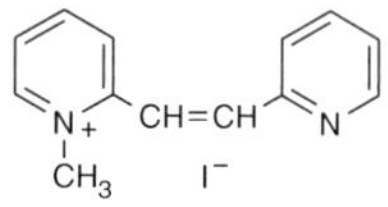

250 mg

S62,348-2
1,1'-TETRAMETHYLENEBIS(3,5-DIMETHYLPYRIDINIUM BROMIDE)

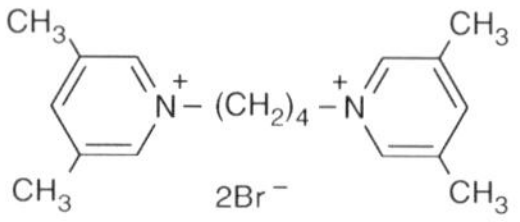

250 mg

S62,353-9
1,1'-TETRAMETHYLENEBIS(3,4-DIMETHYLPYRIDINIUM BROMIDE)

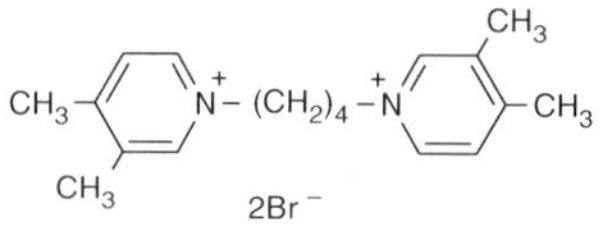

250 mg

S62,349-0
1,1'-TETRAMETHYLENEBIS(3,4-DIMETHYLPYRIDINIUM PERCHLORATE)

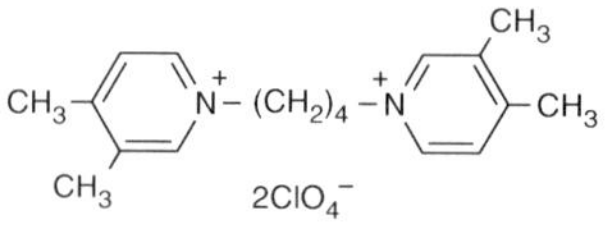

250 mg

S43,562-7
1,1'-TETRAMETHYLENEBIS-(PYRIDINIUM CHLORIDE)

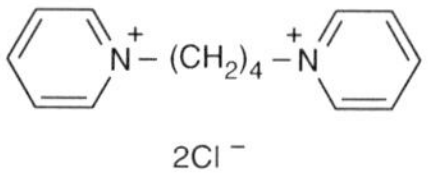

5 gm

S64,917-1
1-(ALPHA-METHYLBENZYL)-3-(4-PYRIDYLMETHYL)-2-THIOUREA

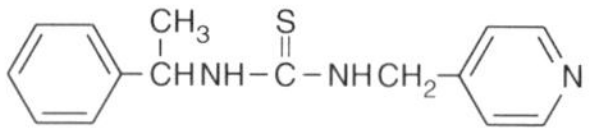

5 gm

S42,048-4
ISONICOTINIC ACID (1-METHYL-2-OXOPROPYLIDENE)HYDRAZIDE

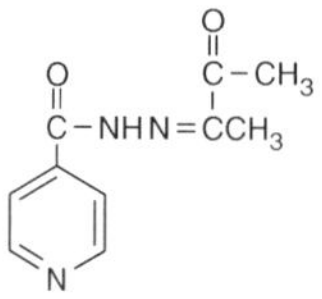

5 gm

S91,470-3
4-PYRIDINECARBOXALDEHYDE SEMICARBAZONE

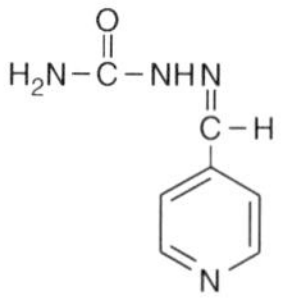

250 mg

S42,001-8
ISONICOTINIC ACID ISOPROPYLIDENEHYDRAZIDE

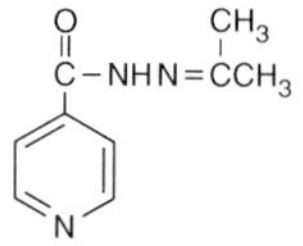

5 gm

S36,286-7
ISONICOTINIC ACID 2,2'-METHYLENEDIHYDRAZIDE

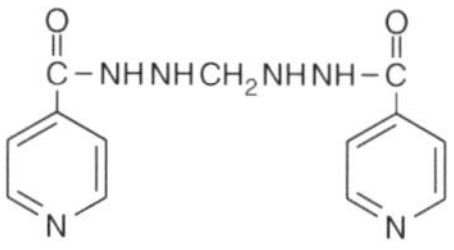

5 gm

S36,856-3
6,6'-DIMETHYL-2,2'-PYRIDIL

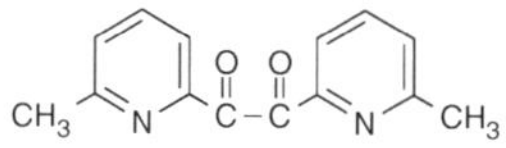

250 mg

S72,152-2
1-(4-NITROPHENYL)-3-(6-METHYL-2-PYRIDYL)-2-THIOUREA

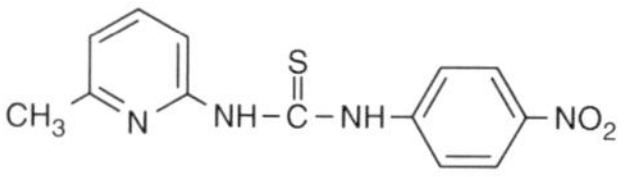

5 gm

S81,107-6
3-(2-NITROPHENYL)PYRIDINE PICRATE

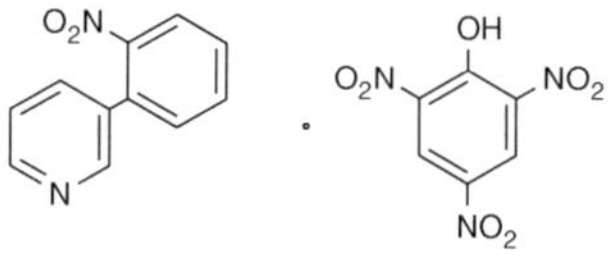

250 mg

S83,962-0
3-(2-NITROPHENYL)PYRIDINE

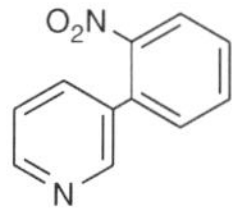

250 mg

S81,104-1
2-(3-NITROPHENYL)PYRIDINE

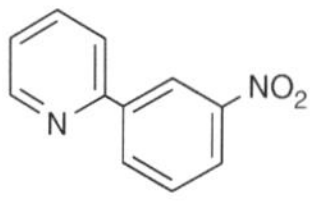

250 mg

S81,106-8
2-(3-NITROPHENYL)PYRIDINE PICRATE

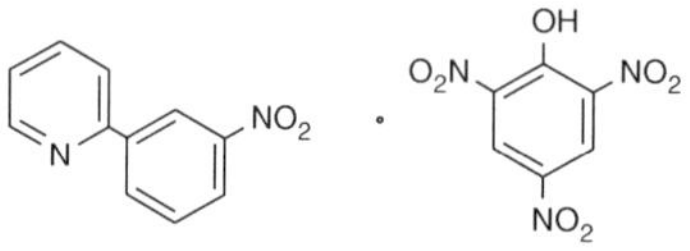

250 mg

S80,209-3
6-ACETAMIDO-2-METHYL-3-NITROPYRIDINE

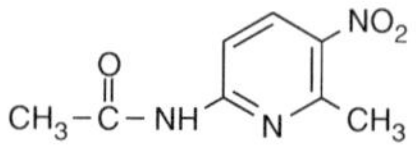

250 mg

S91,887-3
4-(1-METHYL-1-NITROETHYL)PYRIDINE

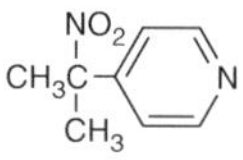

250 mg

S75,954-6
4-PHENYL-2-(4-PYRIDYL)-1,4,7,8-TETRAHYDROQUINOLIN-5(6H)-ONE

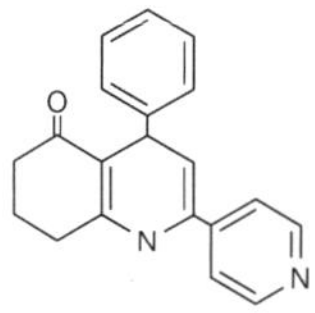

250 mg

S91,024-4
4-ISOPROPYL-1-(4-PYRIDYLCARBONYL)-SEMICARBAZIDE

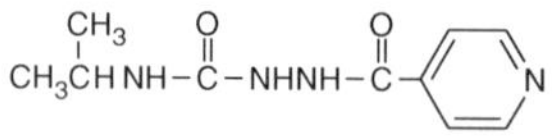

250 mg

S37,726-0
N,N-DIISOPROPYLISONICOTINAMIDE

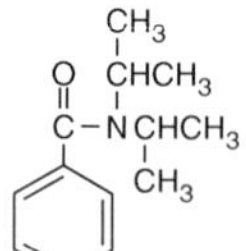

250 mg

S39,094-1
N,N-DIPROPYLISONICOTINAMIDE
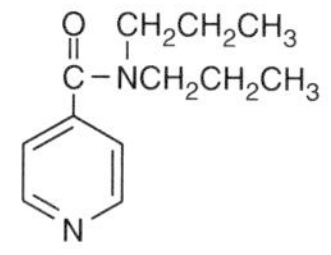
5 gm

S66,724-2
2-PYRIDYL N-(4-CHLOROPHENYL)-CARBAMATE
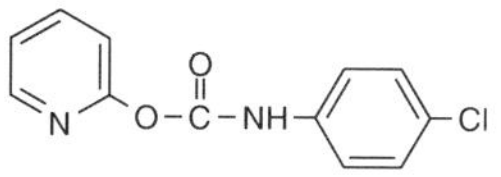
5 gm

S66,702-1
2-PYRIDYL N-(2,4-DICHLOROPHENYL)-CARBAMATE
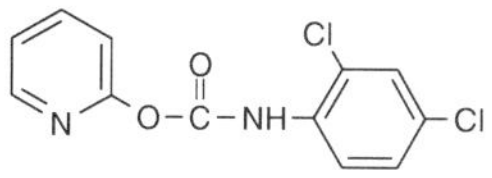
5 gm

S66,693-9
2-PYRIDYL N-(2,4,5-TRICHLOROPHENYL)-CARBAMATE
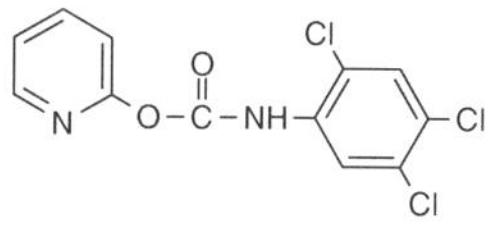
5 gm

S42,113-8
4-CARBAMOYL-1-HEPTYLPYRIDINIUM BROMIDE
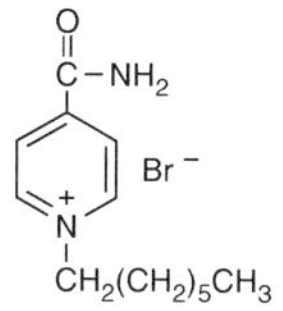
5 gm

S42,118-9
4-CARBAMOYL-1-OCTYLPYRIDINIUM BROMIDE
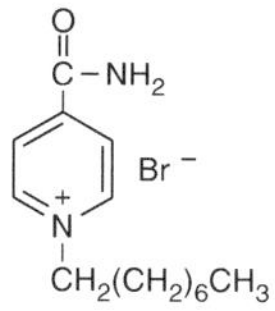
5 gm

S42,096-4
ISONICOTINIC ACID 2-(SULFOMETHYL)-HYDRAZIDE, CALCIUM SALT DIHYDRATE
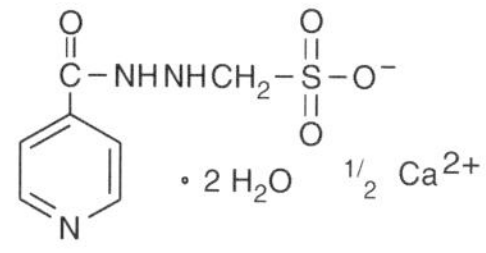
5 gm

S62,819-0
1-PHENYL-3-(4-PYRIDYLMETHYL)UREA
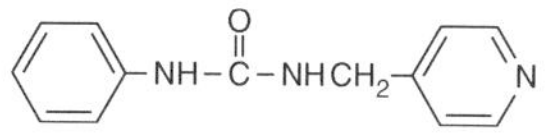
5 gm

S62,822-0
1-(4-CHLOROPHENYL)-3-(4-PYRIDYLMETHYL)UREA
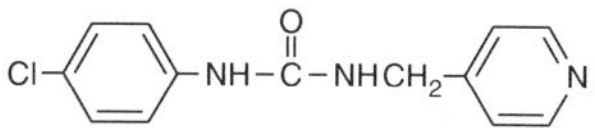
5 gm

S66,027-2
4,4'-METHYLENEBIS[1-PHENYL-3-(4-PYRIDYLMETHYL)UREA]
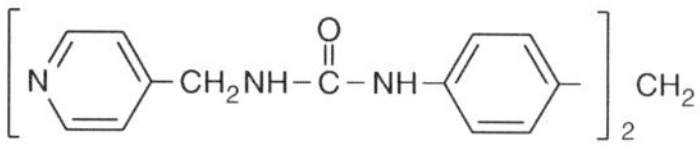
5 gm

S71,395-3
1,1'-(1,5-NAPHTHALENEDIYL)BIS[3-(3-PYRIDYLMETHYL)UREA]
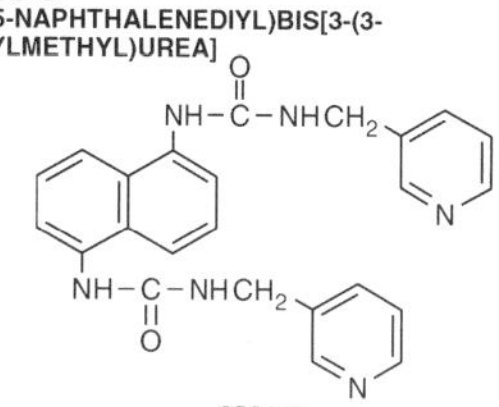
250 mg

S65,794-8
1-(4-PYRIDYLMETHYL)-3-(2,6-XYLYL)-2-THIOUREA
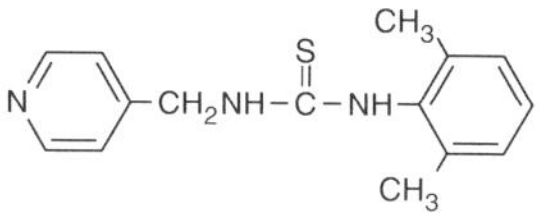
5 gm

S87,295-4
4-CYANO-3-(CYANOMETHYL)PYRIDINE
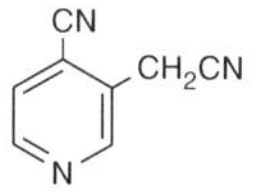
250 mg

S65,725-5
3-(4-PYRIDYL)-2H-BENZ(G)INDAZOLE
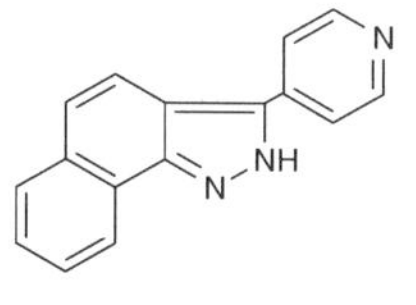
5 gm

S67,145-2
3-(4-PYRIDYL)-2H-BENZ(G)INDAZOLE METHANESULFONATE
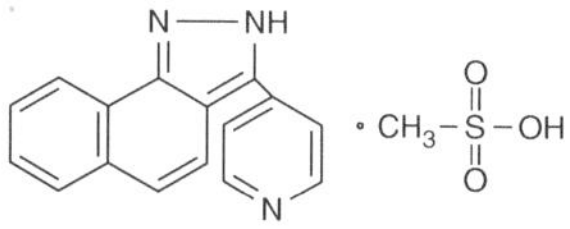
250 mg

S36,456-8
2-(3-PICOLYL)-2-IMIDAZOLINE HYDROCHLORIDE
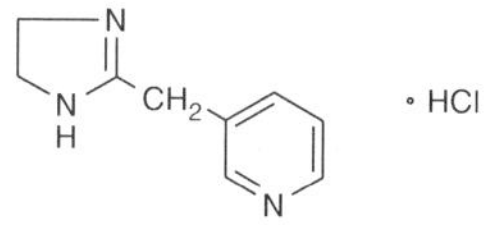
5 gm

S57,983-1
2,2'-ETHYLENEDIPYRIDINE
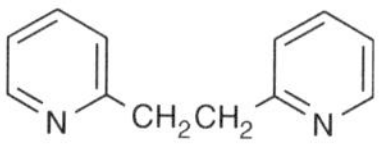
5 gm

S90,618-2
1,2-BIS(6-METHYL-2-PYRIDYL)ETHYLENE
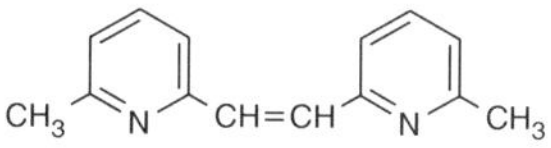
250 mg

S93,568-9
4-(HYDROXYIMINOMETHYL)-1-METHYLPYRIDINIUM IODIDE
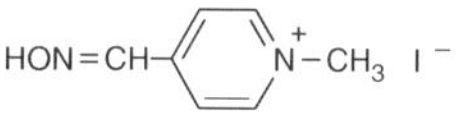
250 mg

S88,777-3
4-ACETYLPYRIDINE AZINE
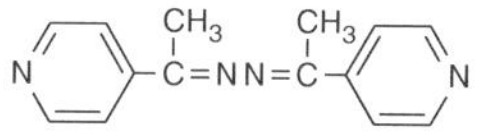
250 mg

S50,524-2
ALPHA-(4-NITROBENZYLIDENE)-N-PHENYL-2-PYRIDINEMETHYLAMINE
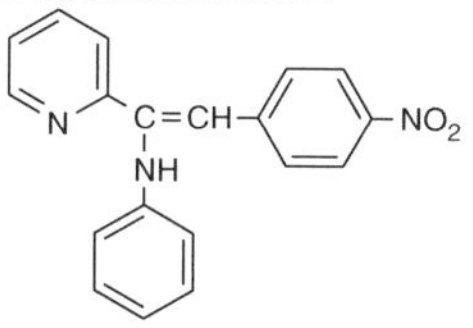
5 gm

S43,245-8
2-(1,2-DIPHENYLETHYLAMINO)PYRIDINE
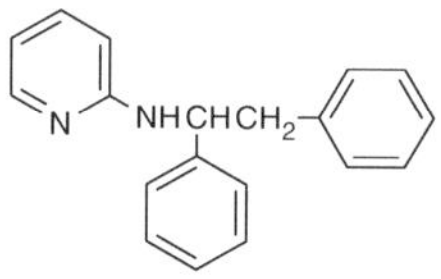
250 mg

S61,814-4
1-PHENACYL-2-PHENYLPYRIDINIUM BROMIDE
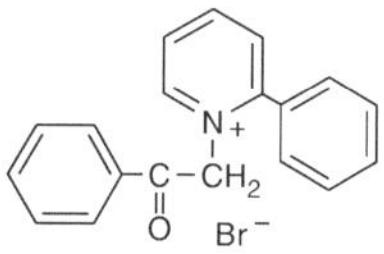
5 gm

S42,253-3
N-(1,2-DIPHENYLETHYL)PICOLINAMIDE
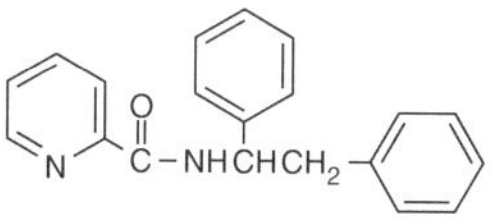
250 mg

S39,330-4
3-[2-(4-PYRIDYL)ETHYL]-4(3H)-QUINAZOLINONE
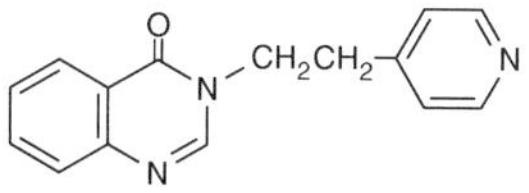
250 mg

S58,099-6
ALPHA-PHENYL-4-PYRIDINEBUTYRONITRILE
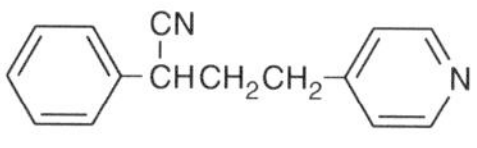
250 mg

S65,153-2
2,4-DIBENZYLPYRIDINE
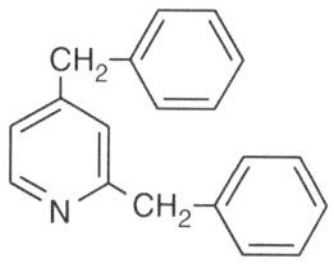
5 gm

S59,371-0
2-(ALPHA-BUTYLBENZYL)PYRIDINE
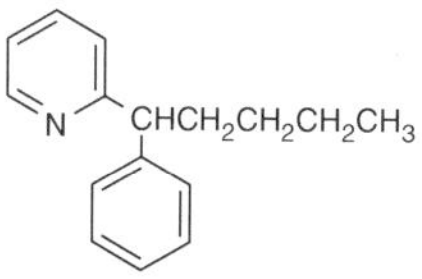
5 gm

S83,416-5
1-PHENYL-N,N'-DI-2-PYRIDINYLMETHANEDIAMINE
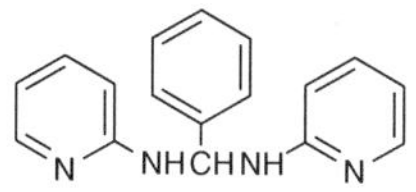
250 mg

S51,241-9
2-CYCLOHEXYLAMINOPYRIDINE
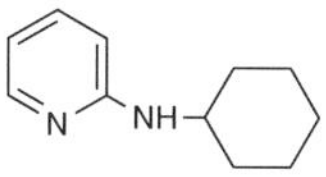
250 mg

S52,992-3
2,6-BIS(PHTHALIMIDOMETHYLAMINO)-PYRIDINE
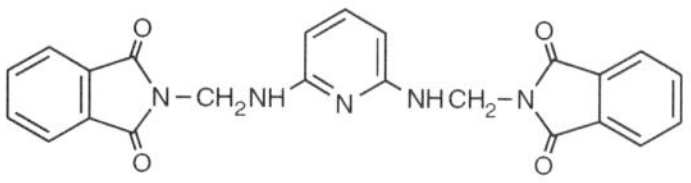
5 gm

S53,936-8
DIETHYL (6-METHYL-2-PYRIDYLAMINOMETHYLENE)MALONATE
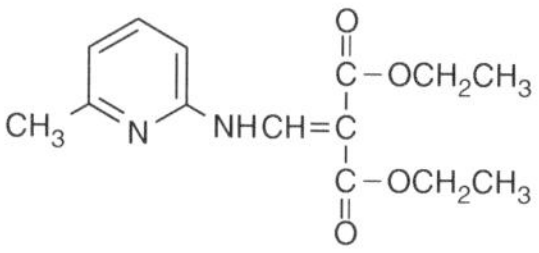
5 gm

S93,387-2
DIETHYL (6-METHYL-2-PYRIDYLAMINOMETHYLENE)MALONATE
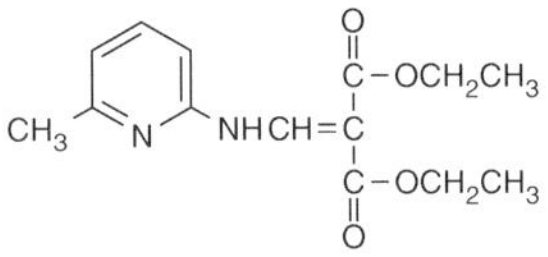
250 mg

S70,403-2
3-CHLORO-2-(3-METHYL-2-PYRIDYLAMINO)-1,4-NAPHTHOQUINONE
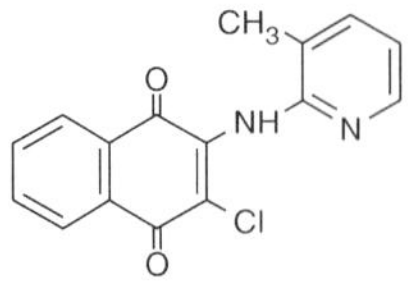
5 gm

S70,431-8
2-CHLORO-3-(4-METHYL-2-PYRIDYLAMINO)-1,4-NAPHTHOQUINONE
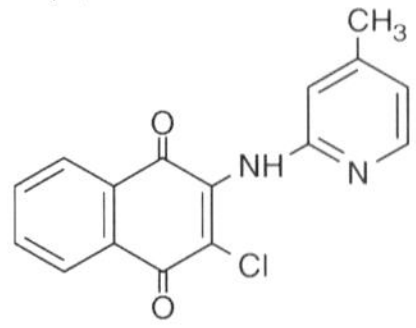
5 gm

S94,019-4
2-ANILINO-N-NITROSOPYRIDINE
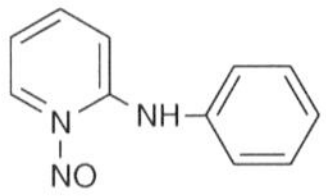
250 mg

S70,438-5
2-CHLORO-3-(6-METHYL-2-PYRIDYLAMINO)-1,4-NAPHTHOQUINONE
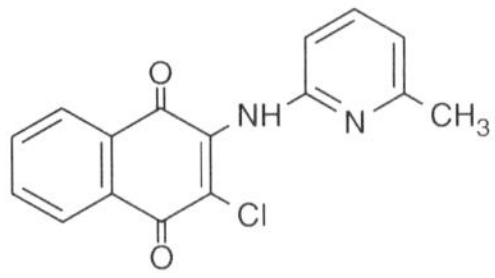
5 gm

S94,224-3
DIETHYL (2,6-DICHLORO-3-PYRIDYLAMINOMETHYLENE)MALONATE
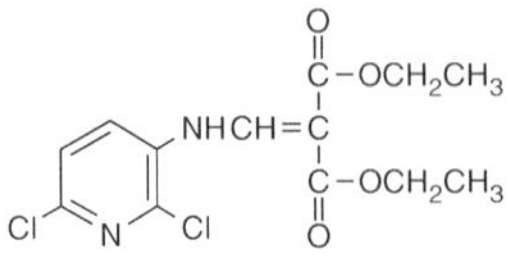
250 mg

S94,580-3
DIETHYL (2-PYRIDYLAMINOMETHYLENE)-MALONATE
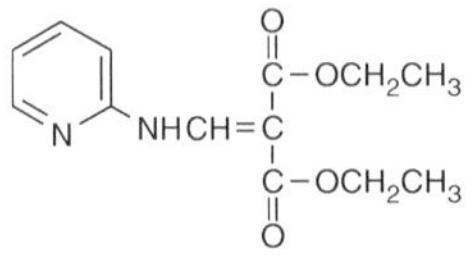
250 mg

S42,073-5
2,6-DIMETHYL-1-NICOTINOYLPIPERIDINE
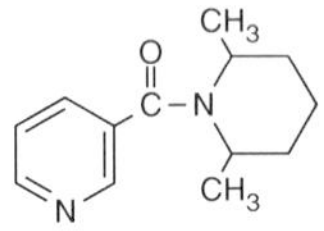
250 mg

S53,635-0
1-(2-PYRIDYL)-3-(3,4-XYLYL)-2-THIOUREA
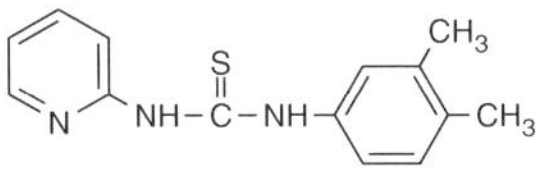
250 mg

S54,846-4
4-METHYL-2-(4-METHYLBENZYLAMINO)-PYRIDINE
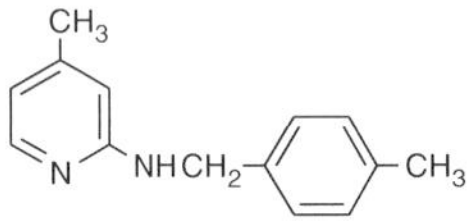
250 mg

S36,334-0
3-(4-METHYLAMINOBUTYL)PYRIDINE
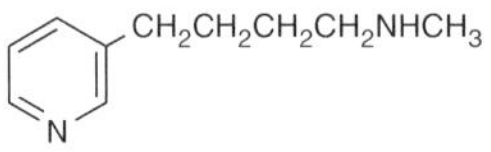
250 mg

S57,069-9
2,2'-IMINODI-3-PICOLINE
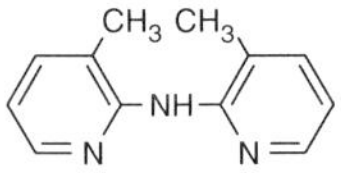
5 gm

S62,700-3
2-ALLYLAMINOPYRIDINE
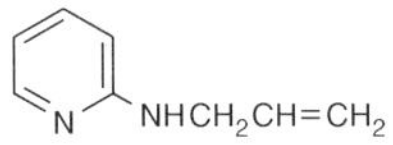
5 gm

S53,334-3
N-(2-PYRIDYLAMINOMETHYL)-PHTHALIMIDE
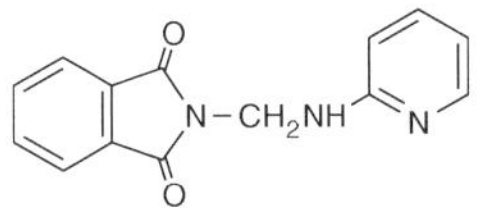
250 mg

S53,148-0
4-[4-(DIMETHYLAMINO)STYRYL]-1-ETHYLPYRIDINIUM IODIDE
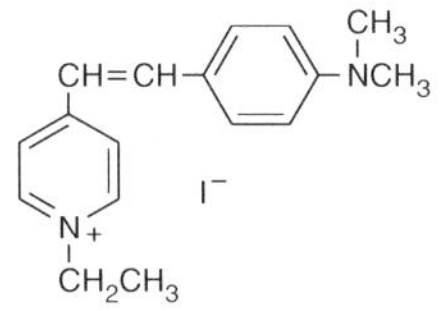
250 mg

S37,977-8
4-METHOXYCARBONYL-1-METHYLPYRIDINIUM IODIDE
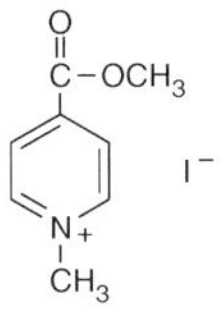
250 mg

S67,946-1
2,4-BIS(4-DIMETHYLAMINOSTYRYL)-1-METHYLPYRIDINIUM IODIDE
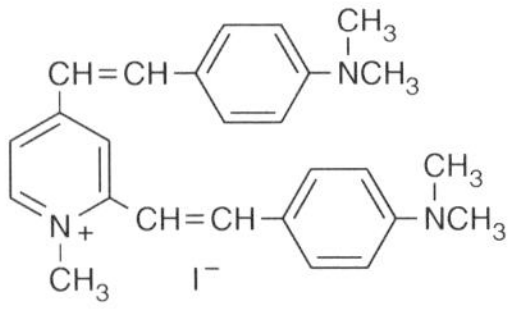
5 gm

S50,772-5
4-[4-(DIETHYLAMINO)STYRYL]-1-METHYLPYRIDINIUM IODIDE
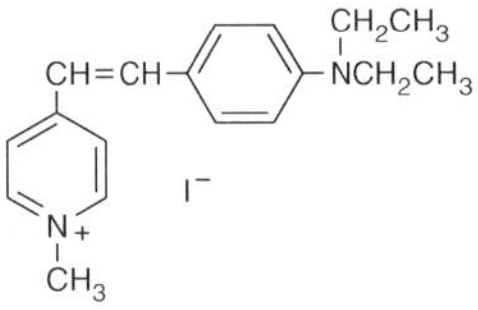
5 gm

S92,226-9
2-(4-[4-(DIMETHYLAMINO)PHENYL]-1,3-BUTADIEN-1-YL)-1-ETHYLPYRIDINIUM PERCHLORATE
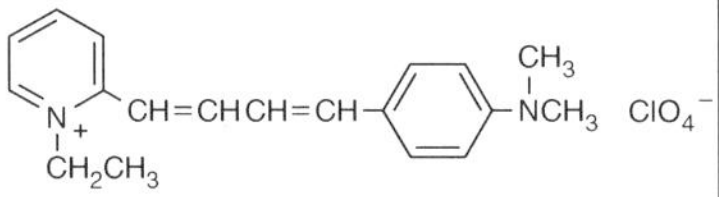
250 mg

S84,013-0
N,N-DIETHYL-4-(2-PYRIDYL)ANILINE
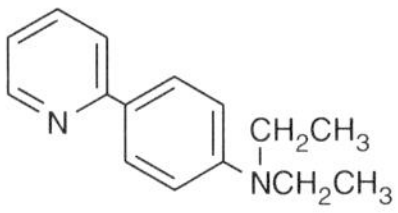
250 mg

S82,078-4
2,6-BIS[4-(DIMETHYLAMINO)PHENYL]-PYRIDINE
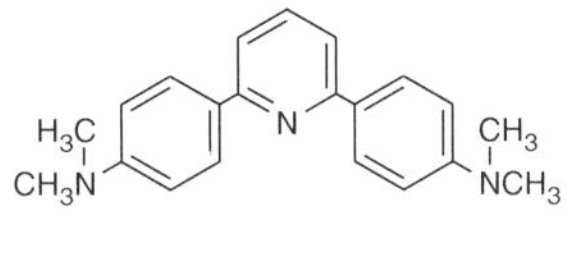
250 mg

S87,203-2
7-METHYL-5,8-QUINOLINEDIONE 5-[(3-METHYL-2-BENZOTHIAZOLINYLIDENE)-HYDRAZONE]
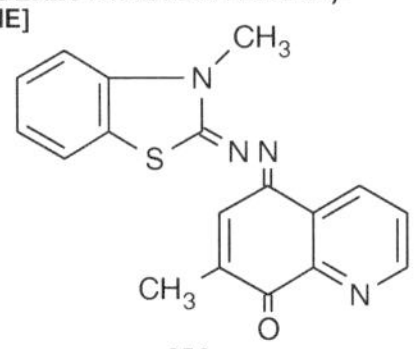
250 mg

S87,204-0
5-METHYL-7,8-QUINOLINEDIONE 7-[(3-METHYL-2-BENZOTHIAZOLINYLIDENE)-HYDRAZONE]
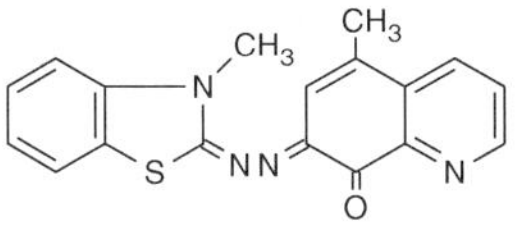
250 mg

S87,205-9
5,8-QUINOLINEDIONE 5-[(1,3-DIMETHYL-2-BENZIMIDAZOLINYLIDENE)-HYDRAZONE]
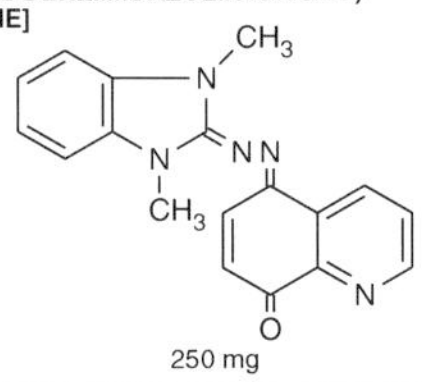
250 mg

S87,196-6
5,8-QUINOLINEDIONE 5-[(1,2-DIHYDRO-1-METHYL-2-QUINOLYLIDENE)HYDRAZONE]-
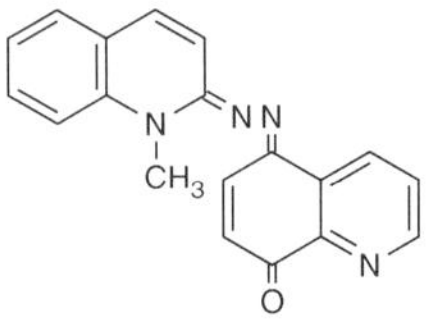
250 mg

S87,186-9
5,8-QUINOLINEDIONE 5-[(3-METHYL-4,5,6,7-TETRAHYDRO-2-BENZOTHIAZOLINYLIDENE)HYDRAZONE]

250 mg

S43,117-6
2-MORPHOLINO-5-NITROPYRIDINE
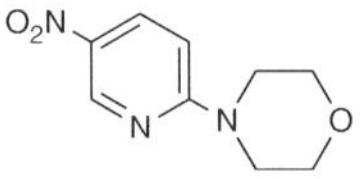
5 gm

S87,191-5
5,8-QUINOLINEDIONE 5-[(3-METHYL-2-BENZOTHIAZOLINYLIDENE)HYDRAZONE]
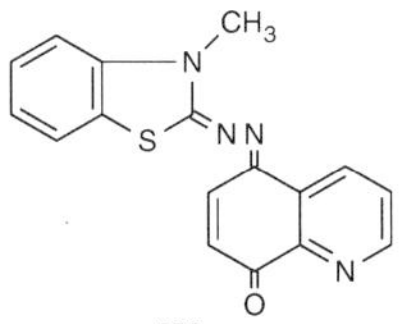
250 mg

S40,705-4
4-CARBOXY-1-METHYLPYRIDINIUM CHLORIDE
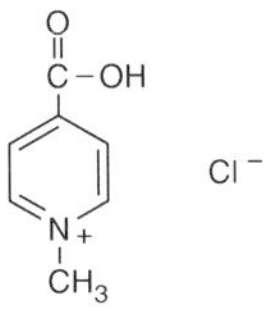
5 gm

S76,725-5
8-DIMETHYLAMINO-2-(4-PYRIDYL)-2-OCTENENITRILE
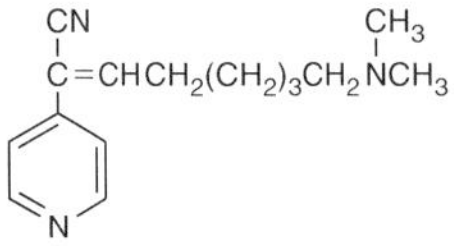
250 mg

S44,108-2
1-(2-PYRIDYL)PYRROLE
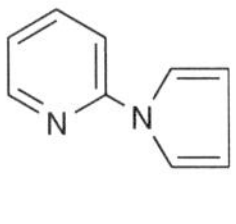
5 gm

S35,537-2
4-(2-PYRROLIDINOETHYL)PYRIDINE
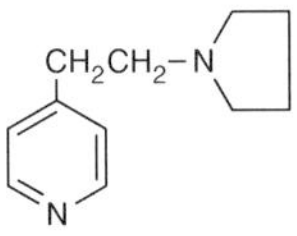
5 gm

S36,006-6
4-(1,5-DIAZABICYCLO[3.2.1]OCT-8-YL)-PYRIDINE

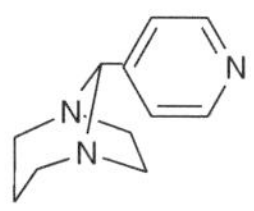

250 mg

S34,696-9
1,3-DIMETHYL-4,5-DI-(4-PYRIDYL)-IMIDAZOLINE

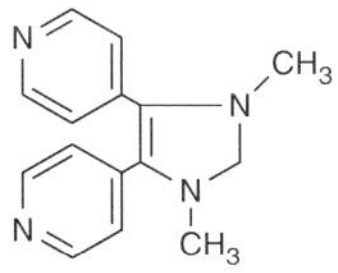

5 gm

S35,487-2
4-(2-PIPERIDINOETHYL)PYRIDINE

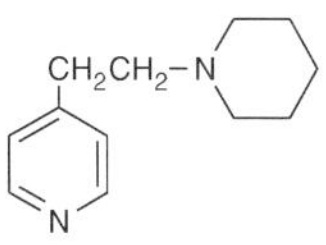

5 gm

S81,079-7
2-[4-(DIMETHYLAMINO)PHENYL]PYRIDINE

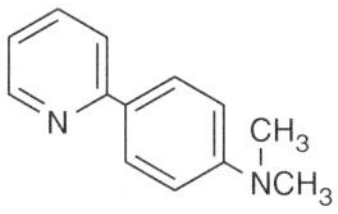

250 mg

S51,820-4
2-[4-(DIETHYLAMINO)STYRYL]-1-ETHYLPYRIDINIUM IODIDE

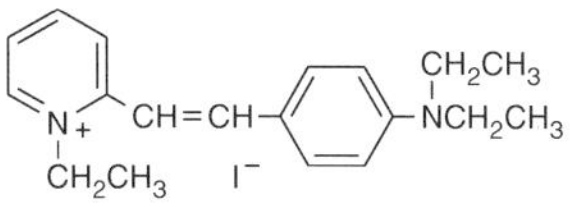

5 gm

S36,045-7
4-(2-DIETHYLAMINOETHYL)PYRIDINE

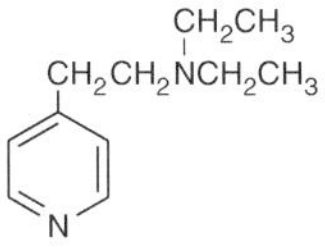

5 gm

S87,639-9
5,5'-METHYLENEBIS[2-(DIMETHYLAMINO)-PYRIDINE]

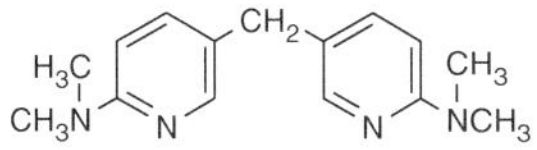

250 mg

S45,627-6
2-[2-(DIMETHYLAMINO)ETHYL]-PYRIDINE

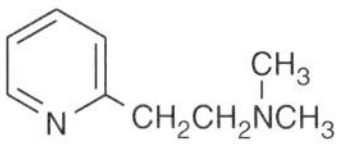

5 gm

S74,880-3
1-BENZYL-2-(PHENYLSULFONYLMETHYL)-PYRIDINIUM BROMIDE

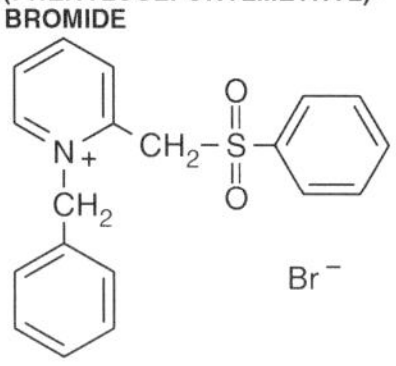

250 mg

S40,301-6
3-BENZYLCARBAMOYL-1-(DODECYLOXYCARBONYLMETHYL)-PYRIDINIUM CHLORIDE

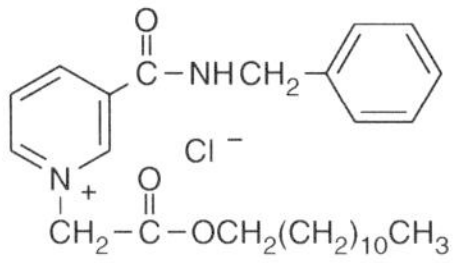

250 mg

S40,123-4
1-(BENZYLCARBAMOYLMETHYL)-3-CARBAMOYLPYRIDINIUM CHLORIDE

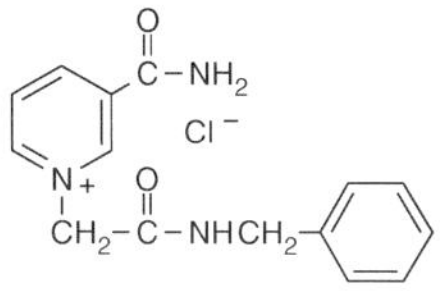

250 mg

S40,222-2
1-BENZYLCARBAMOYLMETHYL-3-(PHENYLCARBAMOYL)-PYRIDINIUM CHLORIDE

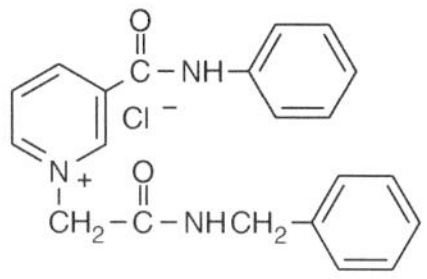

250 mg

S40,223-0
3-BENZYLCARBAMOYL-1-(PHENYLCARBAMOYLMETHYL)-PYRIDINIUM CHLORIDE

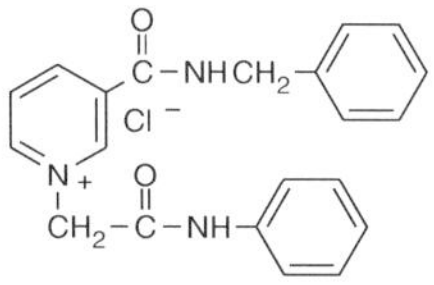

250 mg

S40,317-2
3-BENZYLCARBAMOYL-1-(BENZYLCARBAMOYLMETHYL)-PYRIDINIUM CHLORIDE

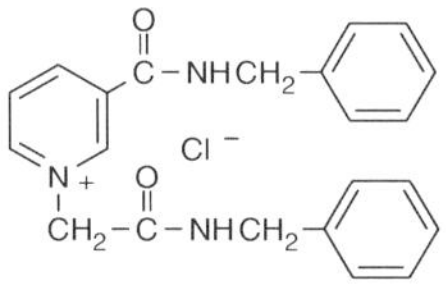

250 mg

S40,221-4
1-BENZYLCARBAMOYLMETHYL-3-(4-CHLOROPHENYLCARBAMOYL)-PYRIDINIUM CHLORIDE

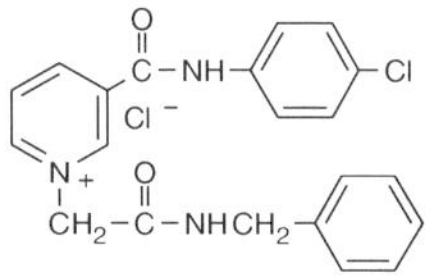

250 mg

S40,403-9
3-BENZYLCARBAMOYL-1-(4-CHLOROPHENYLCARBAMOYLMETHYL)-PYRIDINIUM CHLORIDE

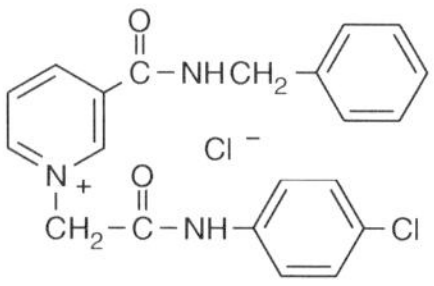

5 gm

S76,729-8
1-BENZYL-3-(PYRROLIDINYLSULFONYL)-PYRIDINIUM BROMIDE

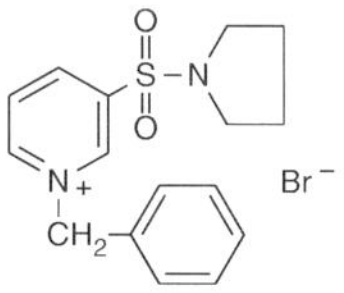

250 mg

S73,505-1
1-BENZYL-4-(2-IMIDAZOLIN-2-YL)-PYRIDINIUM BROMIDE

250 mg

S50,558-7
1-BENZYLPYRIDINIUM P-TOLUENESULFONATE

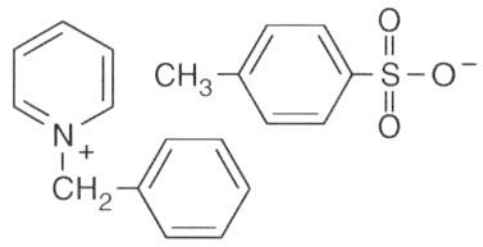

250 mg

S76,147-8
1-BENZYL-4-[4,5-DIHYDRO(1)-BENZOTHIEPINO(5,4-C)ISOXAZOL-3-YL]-PYRIDINIUM CHLORIDE

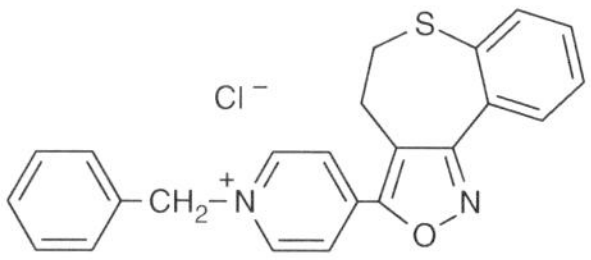

250 mg

S76,721-2
1-(2-METHYLBENZYL)-PYRIDINIUM BROMIDE

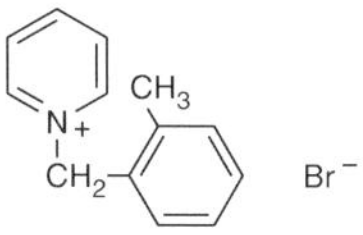

250 mg

S90,958-0
1-AMINO-2-(ETHOXYCARBONYL)-PYRIDINIUM PERCHLORATE

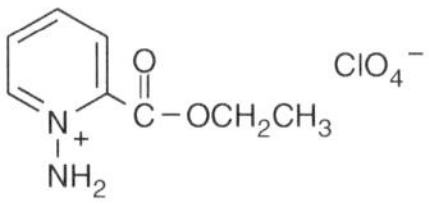

250 mg

S90,961-0
1-AMINO-2-CARBAMOYLPYRIDINIUM P-TOLUENESULFONATE

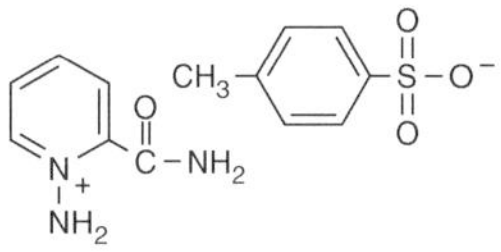

250 mg

S90,963-7
1-AMINOPYRIDINIUM P-TOLUENESULFONATE

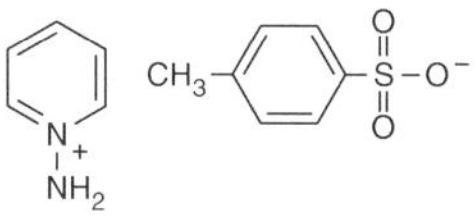

250 mg

S90,968-8
1-AMINO-2-CYANOPYRIDINIUM PERCHLORATE

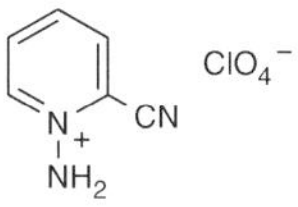

250 mg

S90,957-2
1-AMINO-2-CHLOROPYRIDINIUM P-TOLUENESULFONATE

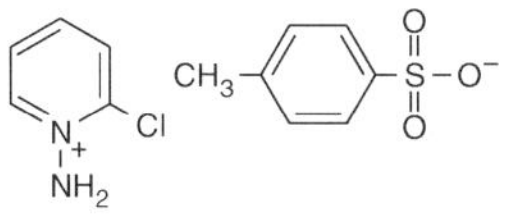

250 mg

S44,343-3
1-PHENACYLPYRIDINIUM BROMIDE

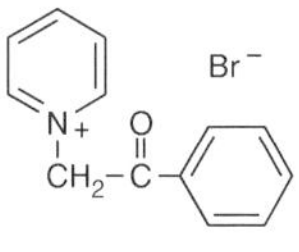

5 gm

S51,548-5
1-(4'-DODECYLPHENACYL)-PYRIDINIUM IODIDE

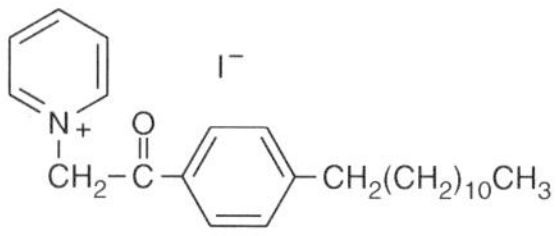

250 mg

S61,658-3
1-(2-OXOPROPYL)PYRIDINIUM BROMIDE

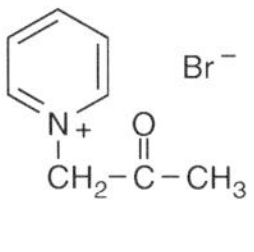

5 gm

S40,296-6
3-(4-CHLOROPHENYLCARBAMOYL)-1-(DODECYLOXYCARBONYLMETHYL)-PYRIDINIUM CHLORIDE

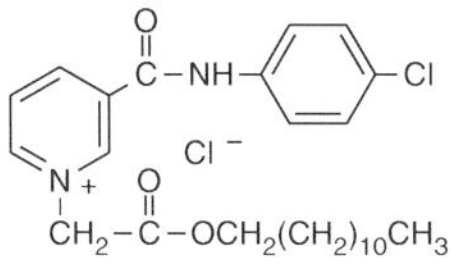

250 mg

S62,309-1
2,4-DIMETHYL-1-(2-METHOXYETHYL)-PYRIDINIUM P-TOLUENESULFONATE

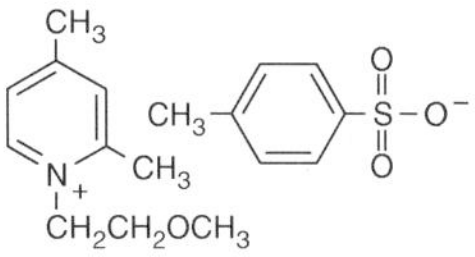

250 mg

S44,111-2
1-METHYLPYRIDINIUM 3-SULFONATE

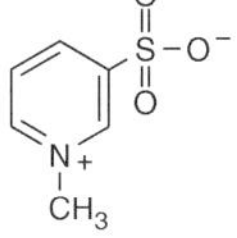

5 gm

S44,112-0
1-ALLYLPYRIDINIUM 3-SULFONATE

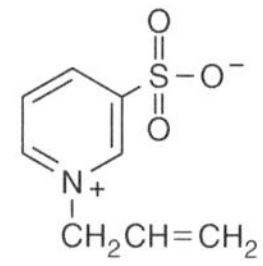

5 gm

S50,553-6
1-(4-NITROBENZYL)-PYRIDINIUM CHLORIDE

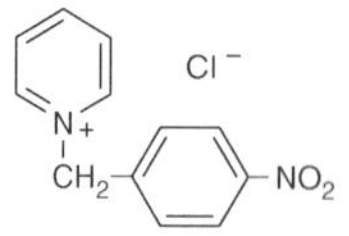

250 mg

S50,533-1
1-(4-NITROBENZYL)PYRIDINIUM BROMIDE

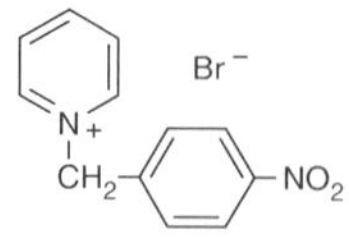

250 mg

S63,960-5
1-ETHYL-2-(3-NITROSTYRYL)-PYRIDINIUM IODIDE

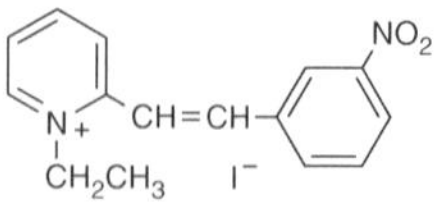

250 mg

S51,595-7
1-METHYL-4-(3-NITROSTYRYL)-PYRIDINIUM IODIDE

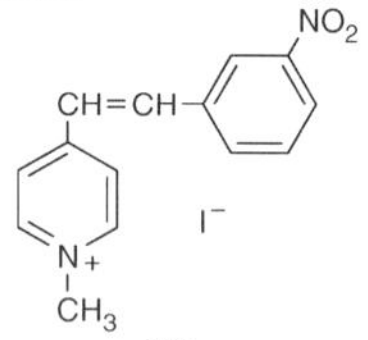

250 mg

S51,255-9
1-ETHYL-4-(3-NITROSTYRYL)-PYRIDINIUM IODIDE

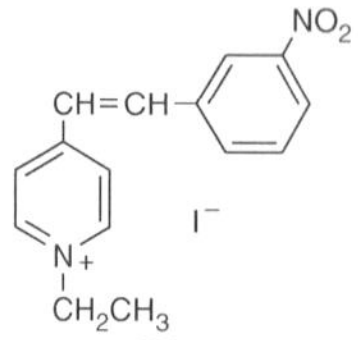

250 mg

S87,040-4
4-(1,3-DIOXO-2-INDANYLIDENEMETHYL)-1-METHYLPYRIDINIUM TETRAFLUOROBORATE

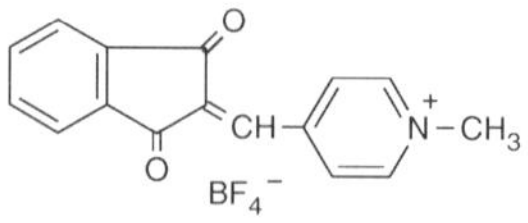

250 mg

S63,063-2
1-(4'-BROMOPHENACYL)-PYRIDINIUM BROMIDE

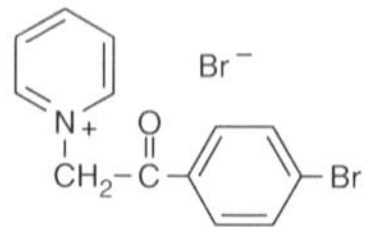

250 mg

S61,448-3
1-(4'-METHYLPHENACYL)-PYRIDINIUM BROMIDE

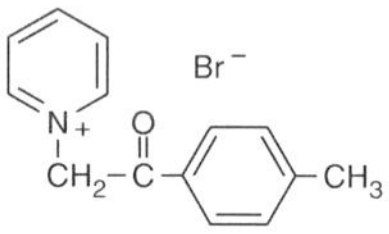

5 gm

S58,191-7
1-(2-ACETYLVINYL)-PYRIDINIUM CHLORIDE

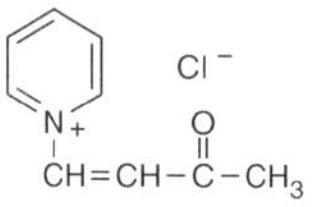

250 mg

S61,911-6
1-(3-CARBETHOXY-2-OXOPROPYL)-PYRIDINIUM CHLORIDE

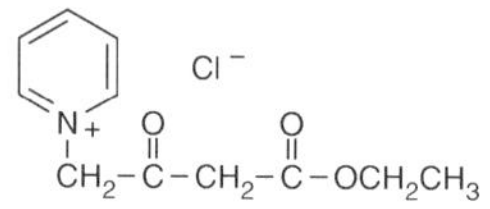

5 gm

S59,660-4
1-ACETONYLPYRIDINIUM CHLORIDE

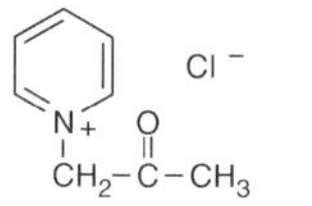

5 gm

S51,350-4
1-(4-CARBOXYBENZYL)-2-METHYLPYRIDINIUM BROMIDE

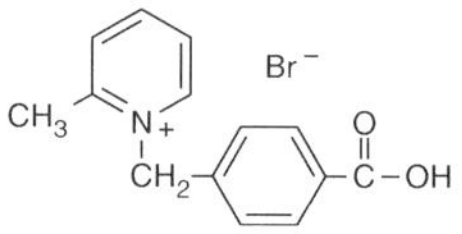

5 gm

S35,932-7
4-CARBETHOXY-1-CARBETHOXYMETHYLPYRIDINIUM CHLORIDE

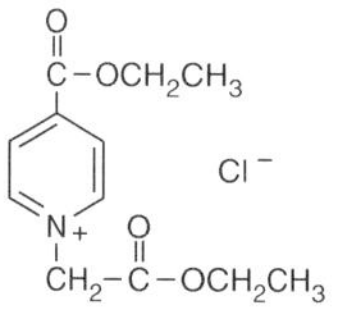

5 gm

S40,238-9
1-DODECYLOXYCARBONYLMETHYL-3-(METHYLCARBAMOYL)-PYRIDINIUM CHLORIDE

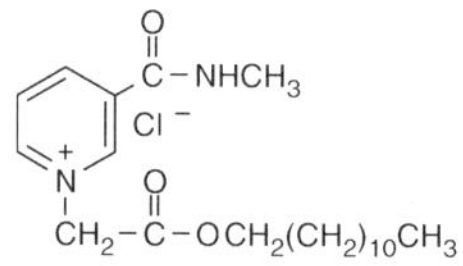

250 mg

S73,430-6
1-(ETHOXYCARBONYLALLYL)-2-(PHENYLSULFONYLMETHYL)-PYRIDINIUM BROMIDE

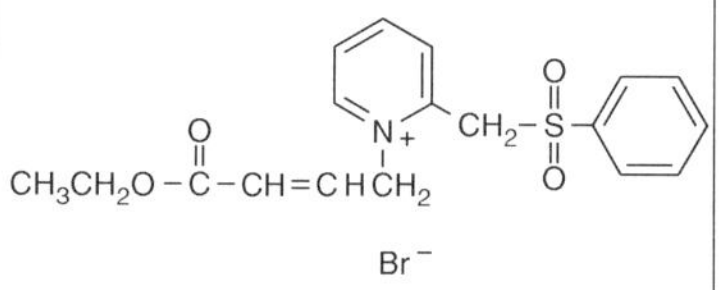

250 mg

S39,849-7
4-CARBAMOYL-1-DODECYLPYRIDINIUM BROMIDE

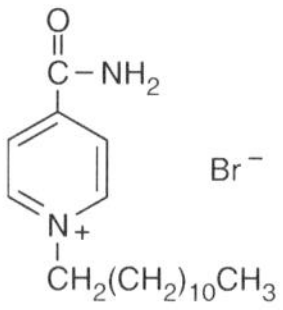

5 gm

S39,852-7
4-CARBAMOYL-1-TETRADECYLPYRIDINIUM BROMIDE

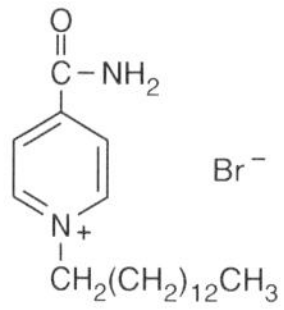

5 gm

S40,195-1
3-(4-CHLOROPHENYLCARBAMOYL)-1-(PHENYLCARBAMOYLMETHYL)-PYRIDINIUM CHLORIDE

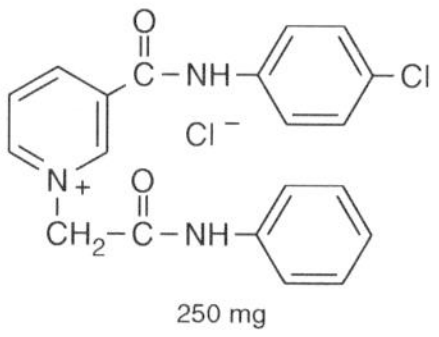

250 mg

S35,522-4
PYRIDINIUM-1-ACETATE COMPOUND WITH PYRIDINE HYDROCHLORIDE

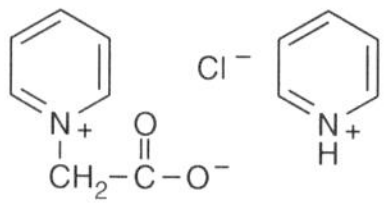

5 gm

S42,251-7
NICOTINIC ACID, 1,2-DIPHENYLETHYLAMMONIUM SALT

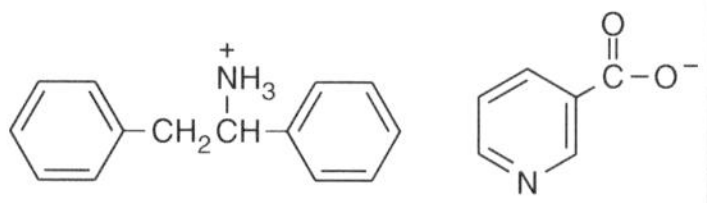

250 mg

S41,425-5
TRIGONELLINE HYDROCHLORIDE

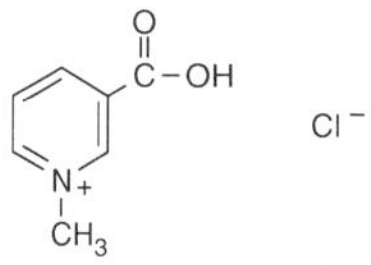

5 gm

S34,347-1
1-(CARBETHOXYMETHYL)-PYRIDINIUM CHLORIDE

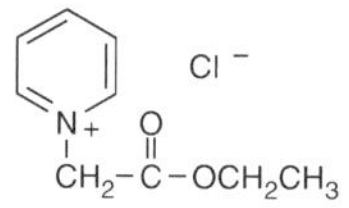

5 gm

S38,458-5
3-METHOXYCARBONYL-1-METHYLPYRIDINIUM P-TOLUENESULFONATE

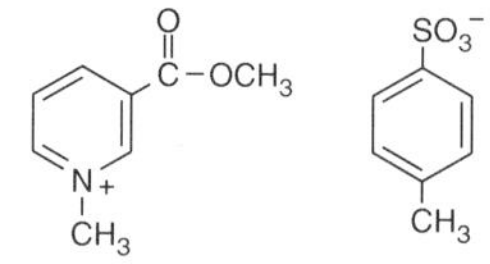

250 mg

S38,536-0
3-METHOXYCARBONYL-1-METHYLPYRIDINIUM IODIDE

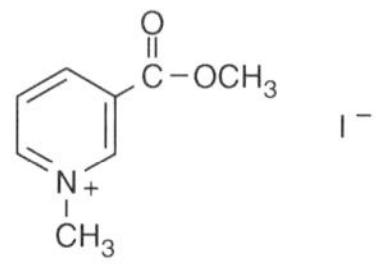

5 gm

S92,006-1
2-(HYDROXYMETHYL)-1-METHYLPYRIDINIUM IODIDE

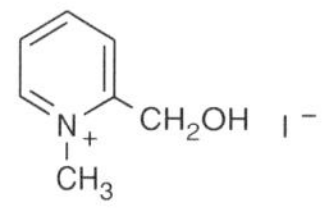

250 mg

S63,099-3
1-ALLYL-2-(4-ACETAMIDOSTYRYL)-PYRIDINIUM IODIDE

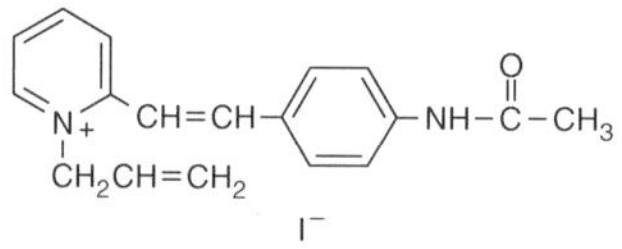

250 mg

S53,606-7
4-(4-ACETAMIDOSTYRYL)-1-ETHYLPYRIDINIUM IODIDE

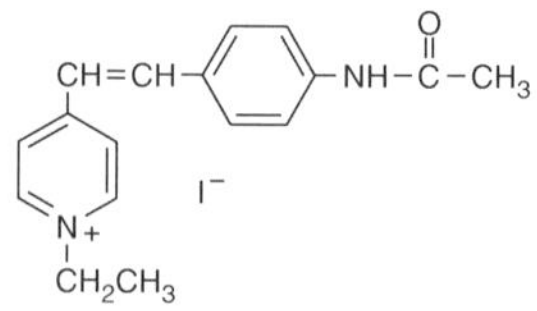

250 mg

S50,144-1
1-(CARBAMOYLMETHYL)-PYRIDINIUM CHLORIDE

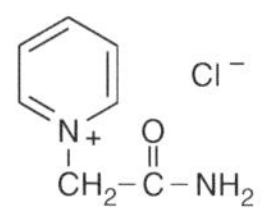

250 mg

S59,682-5
N-(CYANOMETHYL)PYRIDINIUM CHLORIDE

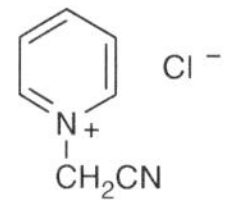

5 gm

S76,608-9
1-METHYL-2-(PHENYLSULFONYLMETHYL)-PYRIDINIUM IODIDE

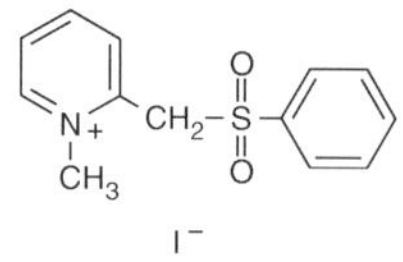

250 mg

S83,636-2
1,4-DIMETHYLPYRIDINIUM P-TOLUENESULFONATE

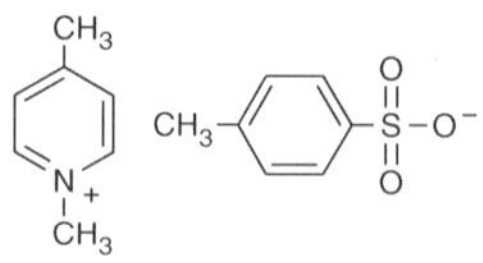

250 mg

S58,229-8
1-OCTADECYLPYRIDINIUM P-TOLUENESULFONATE

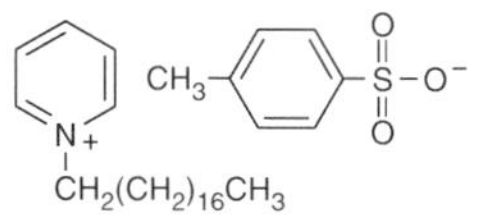

250 mg

S63,066-7
2-METHYL-1-(3-SULFOPROPYL)-PYRIDINIUM HYDROXIDE, INNER SALT

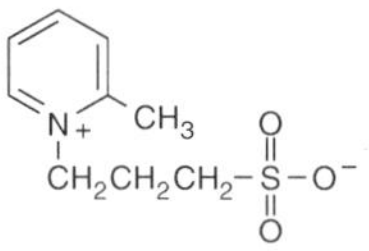

250 mg

S73,331-8
3-(MORPHOLINOSULFONYL)-1-(3-PHENYLALLYL)PYRIDINIUM BROMIDE

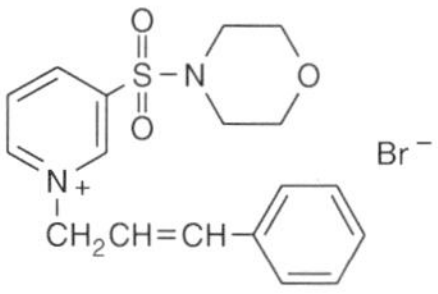

250 mg

S59,975-1
2-(2-THIENYLMETHYLENE)PICOLINE METHIODIDE

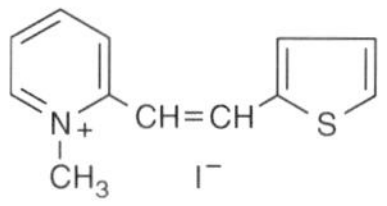

5 gm

S63,961-3
1-ETHYL-2-(2-THENYLIDENEMETHYL)-PYRIDINIUM IODIDE

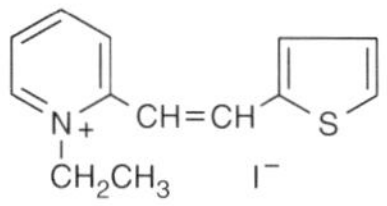

5 gm

S62,325-3
1-ETHYL-4-(2-THENYLIDENEMETHYL)-PYRIDINIUM IODIDE

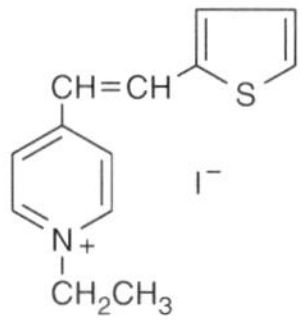

250 mg

S63,967-2
1-ETHYL-2-[3,4-(METHYLENEDIOXY)-STYRYL]PYRIDINIUM IODIDE

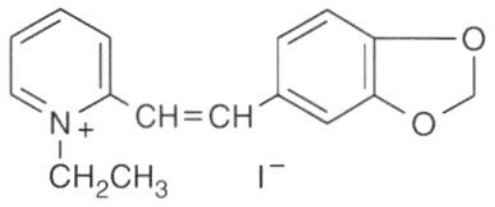

250 mg

S39,978-7
1-ALLYL-2-(3,4-METHYLENEDIOXYSTYRYL)-PYRIDINIUM IODIDE

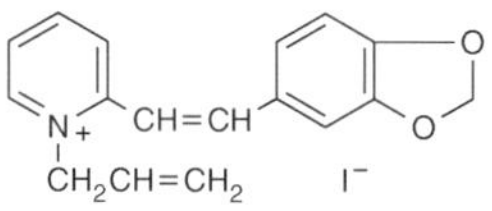

250 mg

S50,426-2
1-METHYL-4-[3,4-(METHYLENEDIOXY)-STYRYL]PYRIDINIUM IODIDE

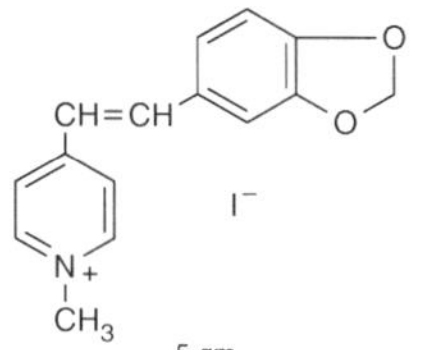

5 gm

S53,610-5
1-ETHYL-4-[3,4-(METHYLENEDIOXY)-STYRYL]PYRIDINIUM IODIDE

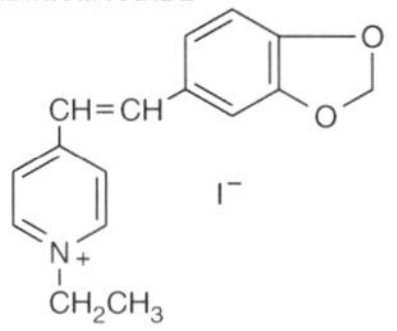

250 mg

S39,171-9
1-METHYL-2-[3,4-(METHYLENEDIOXY)-STYRYL]PYRIDINIUM IODIDE

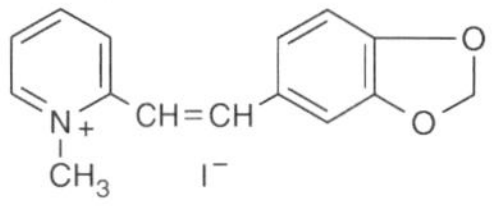

5 gm

S90,950-5
1-PHENYLPYRIDINIUM CHLORIDE

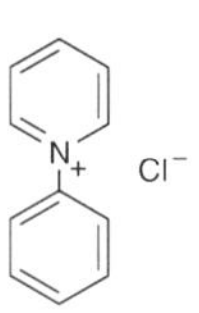

250 mg

S87,410-8
2,6-DIPHENYL-1-METHYLPYRIDINIUM CHLORIDE

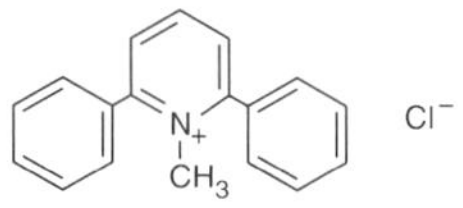

250 mg

S87,411-6
2,6-DIPHENYL-1-METHYLPYRIDINIUM TETRAFLUOROBORATE

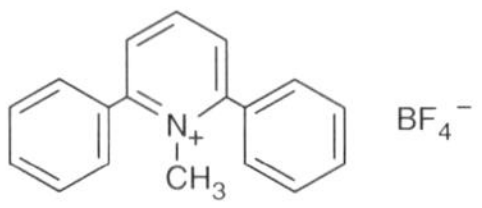

250 mg

S92,081-9
1,2,4,6-TETRAPHENYLPYRIDINIUM TETRAFLUOROBORATE

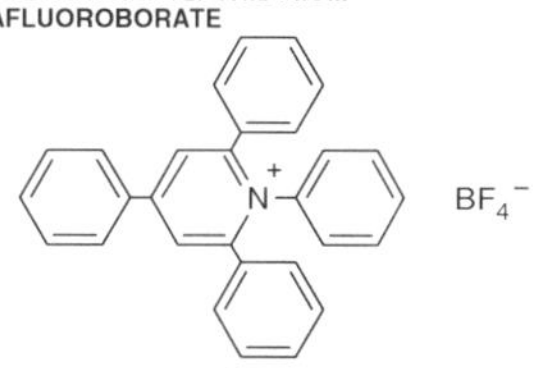

250 mg

S87,476-0
4,4'-BIS(2,6-DIPHENYL-1-METHYLPYRIDINIUM TETRAFLUOROBORATE)

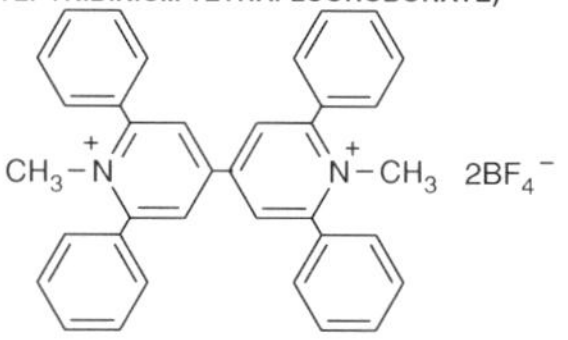

250 mg

S87,389-6
4,4'-BIS(2,6-DISTYRYL-1-METHYLPYRIDINIUM TETRAFLUOROBORATE)

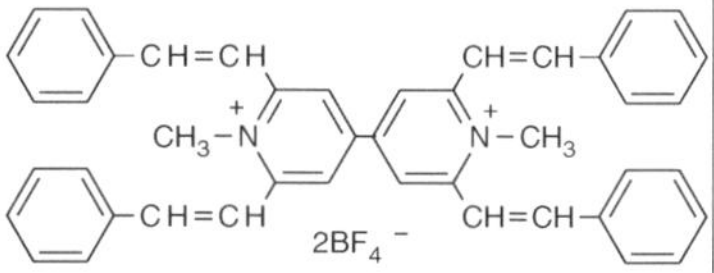

250 mg

S43,080-3
1-METHYL-2-STYRYLPYRIDINIUM IODIDE

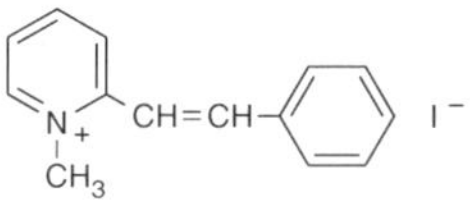

5 gm

S58,076-7
1-METHYL-3-STYRYLPYRIDINIUM IODIDE

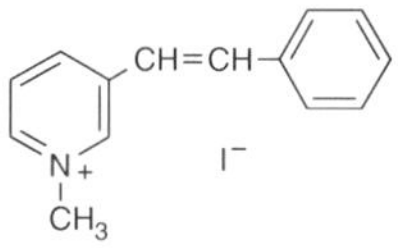

5 gm

S90,982-3
4-METHYL-1-PHENYLPYRIDINIUM PERCHLORATE

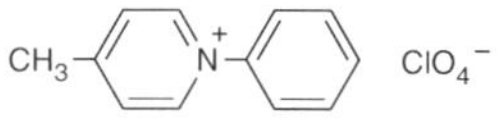

250 mg

S51,606-6
1-ETHYL-4-STYRYLPYRIDINIUM IODIDE

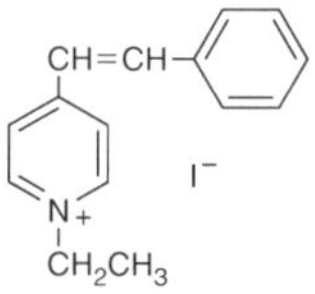

5 gm

S63,866-8
1-HEXADECYL-4-METHYLPYRIDINIUM BENZENESULFONATE

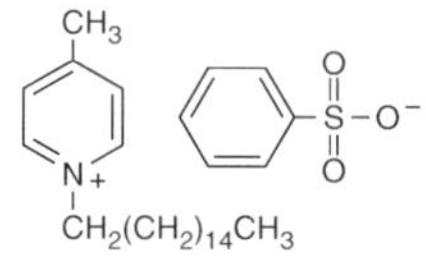

250 mg

S87,274-1
1-BENZHYDRYL-4-METHYLPYRIDINIUM BROMIDE

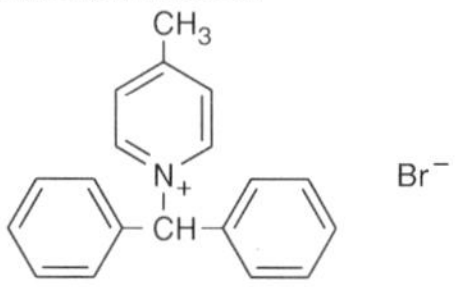

250 mg

S76,620-8
1-(DIPHENYLMETHOXY)-PYRIDINIUM BROMIDE

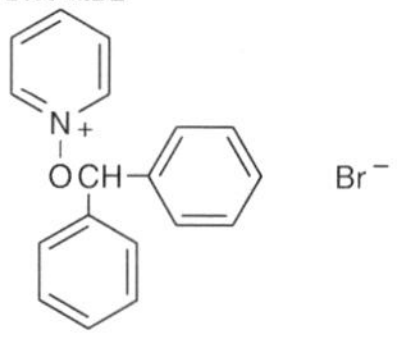

250 mg

S92,193-9
5,6-DIHYDRO-2,4-DIPHENYL-1-METHYLBENZO(H)QUINOLINIUM TETRAFLUOROBORATE

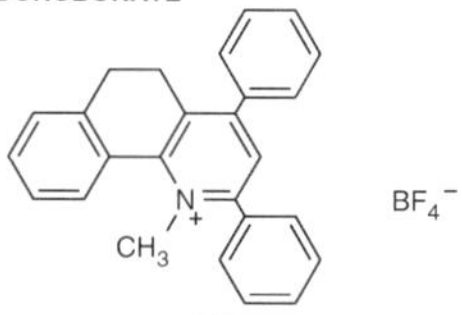

250 mg

S50,449-1
1-METHYL-2-(4-METHYLSTYRYL)-PYRIDINIUM IODIDE

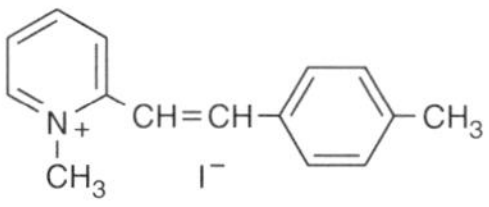

5 gm

S63,964-8
1-ETHYL-2-(4-METHYLSTYRYL)-PYRIDINIUM IODIDE

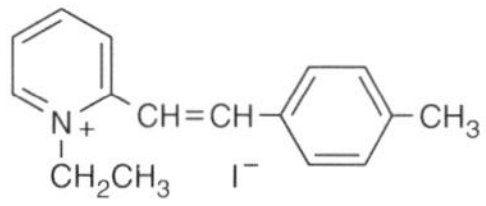

250 mg

S59,991-3
2-(4-ISOPROPYLSTYRYL)PYRIDINE METHIODIDE

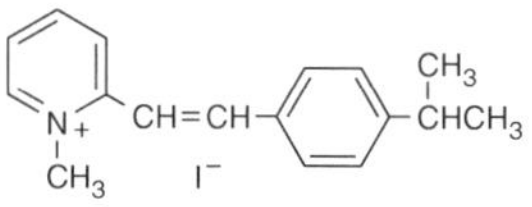

250 mg

S50,567-6
1-METHYL-4-(4-METHYLSTYRYL)-PYRIDINIUM IODIDE

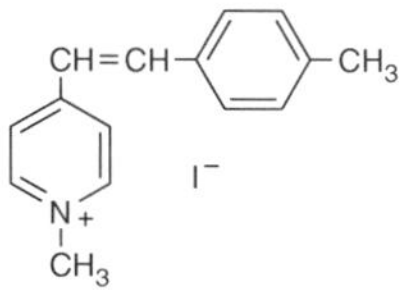

250 mg

S53,882-5
4-(4-CHLOROSTYRYL)-1-ETHYLPYRIDINIUM IODIDE

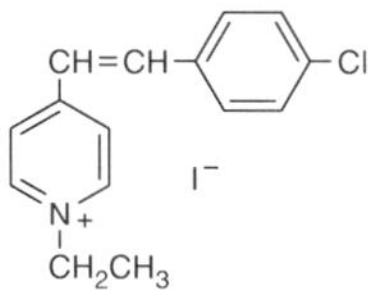

5 gm

S50,566-8
1-ETHYL-4-(4-METHYLSTYRYL)-PYRIDINIUM IODIDE

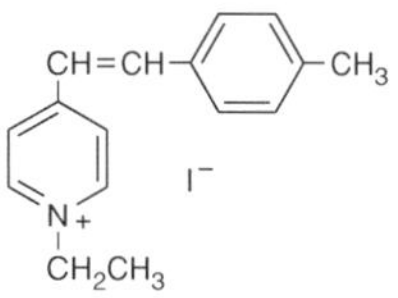

250 mg

S56,900-3
1-ETHYLPYRIDINIUM P-TOLUENESULFONATE

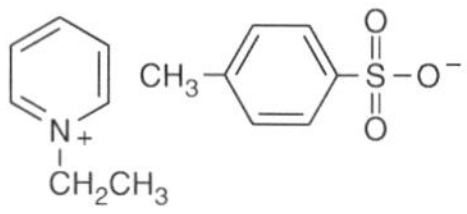

250 mg

S50,568-4
2-(2,6-DICHLOROSTYRYL)-1-METHYLPYRIDINIUM IODIDE

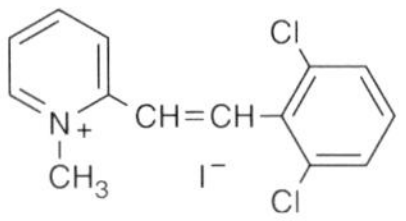

5 gm

S50,562-5
4-(2,6-DICHLOROSTYRYL)-1-METHYLPYRIDINIUM IODIDE

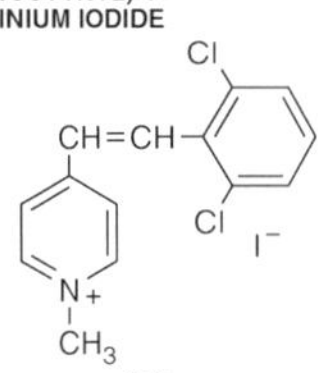

250 mg

S51,243-5
2-(3,4-DICHLOROSTYRYL)-1-METHYLPYRIDINIUM IODIDE

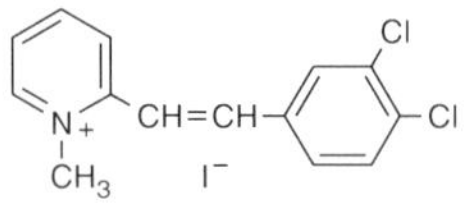

250 mg

S54,366-7
4-(3,4-DICHLOROSTYRYL)-1-METHYLPYRIDINIUM IODIDE

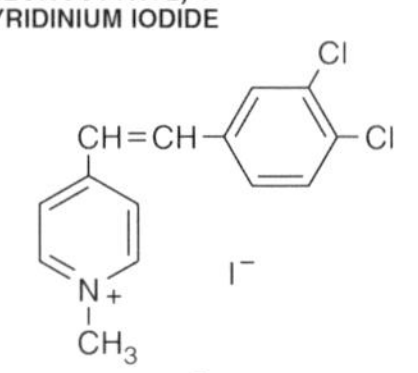

5 gm

S50,353-3
PYRIDINIUM TETRAFLUOROBORATE

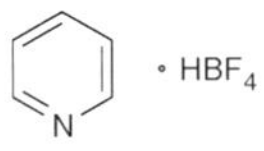

5 gm

S50,349-5
PYRIDINIUM HEXAFLUOROPHOSPHATE

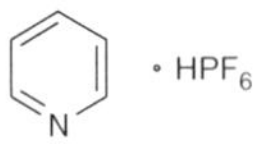

250 mg

S91,430-4
1-METHYLPYRIDINIUM IODIDE

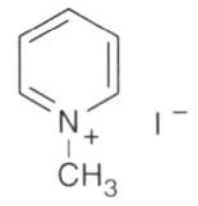

250 mg

S61,939-6
1,3-DIMETHYLPYRIDINIUM IODIDE

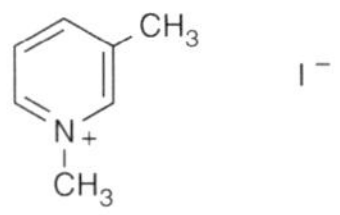

250 mg

S87,424-8
1,2,6-TRIMETHYLPYRIDINIUM IODIDE

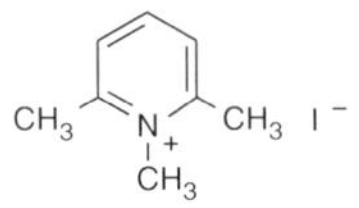

250 mg

S61,920-5
1,2,3-TRIMETHYLPYRIDINIUM IODIDE

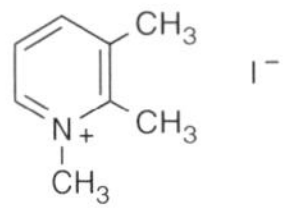

250 mg

S60,519-0
1,2-DIMETHYL-5-VINYLPYRIDINIUM METHOSULFATE

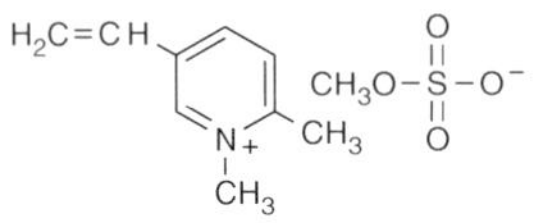

250 mg

S62,342-3
1,1'-HEXAMETHYLENEBIS(2,5-DIMETHYLPYRIDINIUM IODIDE)

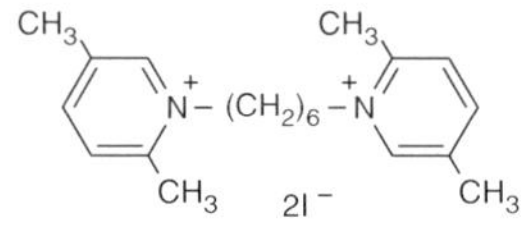

250 mg

S61,919-1
1,2,4-TRIMETHYLPYRIDINIUM IODIDE

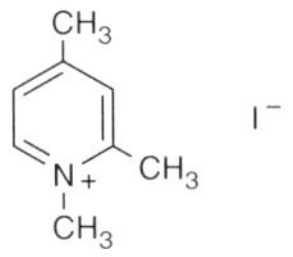

250 mg

S63,537-5
2,4-DIMETHYL-1-ETHYLPYRIDINIUM IODIDE

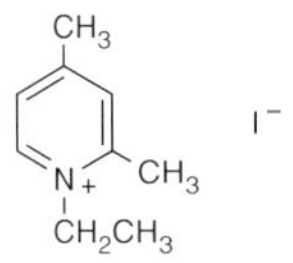

250 mg

S61,932-9
1,2-DIMETHYLPYRIDINIUM IODIDE

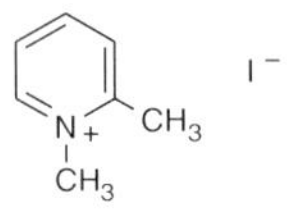

250 mg

S62,314-8
1-METHYL-2-[2-(1-NAPHTHYL)VINYL]-PYRIDINIUM IODIDE

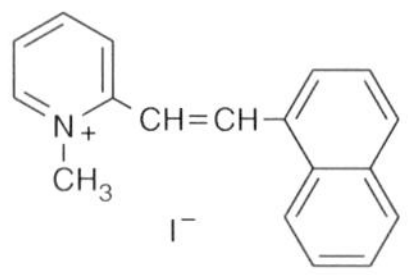

5 gm

S62,101-3
3,5-DIMETHYL-1-ETHYLPYRIDINIUM IODIDE

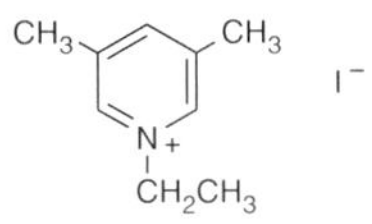

250 mg

S63,540-5
3,5-DIMETHYL-1-PROPYLPYRIDINIUM IODIDE

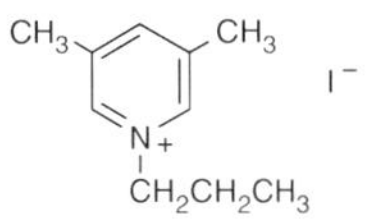

250 mg

S62,346-6
1,1'-PENTAMETHYLENEBIS(3,5-DIMETHYLPYRIDINIUM BROMIDE)

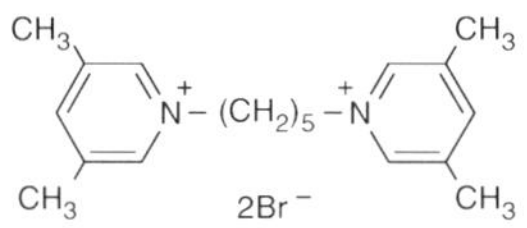

5 gm

S62,350-4
1,1'-HEXAMETHYLENEBIS(3,5-DIMETHYLPYRIDINIUM IODIDE)

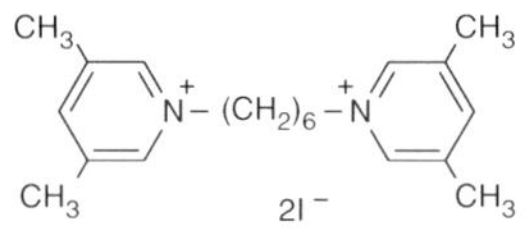

5 gm

S62,100-5
3,4-DIMETHYL-1-ETHYLPYRIDINIUM IODIDE

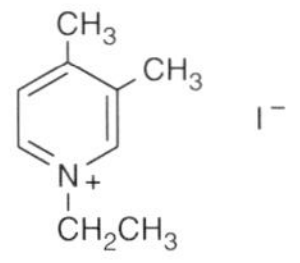

250 mg

S63,546-4
3,4-DIMETHYL-1-PROPYLPYRIDINIUM IODIDE

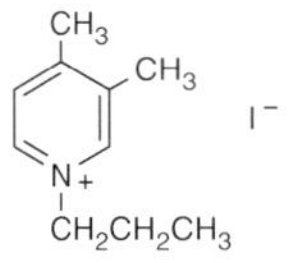

250 mg

S62,345-8
1,1'-PENTAMETHYLENEBIS(3,4-DIMETHYLPYRIDINIUM BROMIDE)

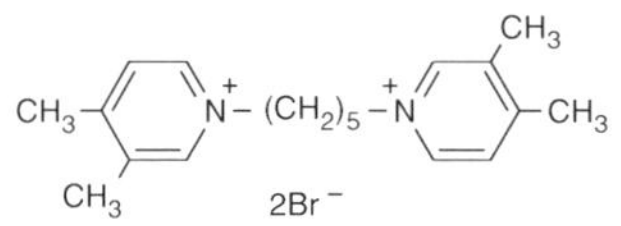

250 mg

S62,339-3
1,1'-HEXAMETHYLENEBIS(3,4-DIMETHYLPYRIDINIUM IODIDE)

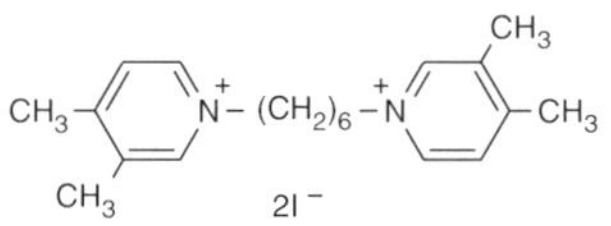

250 mg

S61,935-3
1-ETHYL-3-METHYLPYRIDINIUM IODIDE

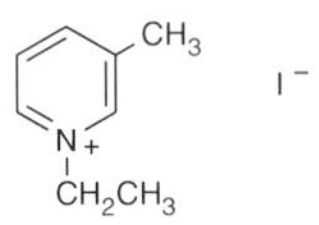

250 mg

S62,354-7
1,1'-HEXAMETHYLENEBIS(3-METHYLPYRIDINIUM IODIDE)

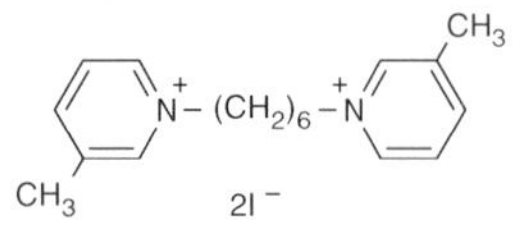

250 mg

S88,781-1
1-METHYL-4-PYRIDINIOPYRIDINIUM DIIODIDE

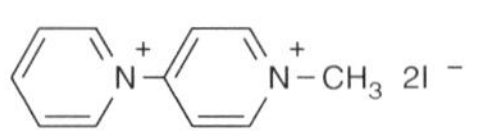

250 mg

S87,363-2
4,4'-VINYLENEBIS(1-METHYLPYRIDINIUM TETRAFLUOROBORATE)

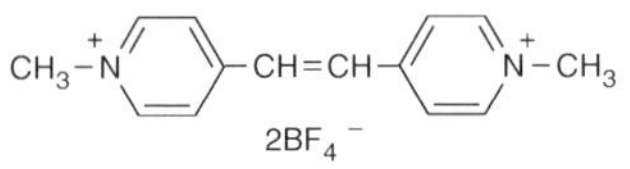

250 mg

S42,886-8
1-(4-PYRIDYL)PYRIDINIUM CHLORIDE

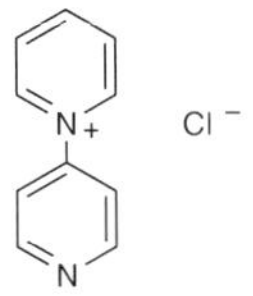

5 gm

S61,938-8
1-ETHYL-4-METHYLPYRIDINIUM IODIDE

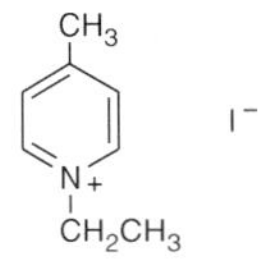

250 mg

S63,558-8
4-METHYL-1-PROPYLPYRIDINIUM IODIDE

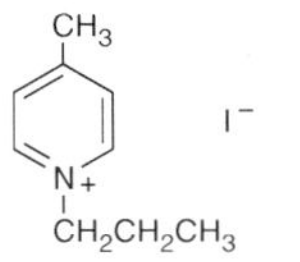

250 mg

S62,341-5
1,1'-HEXAMETHYLENEBIS(4-METHYLPYRIDINIUM IODIDE)

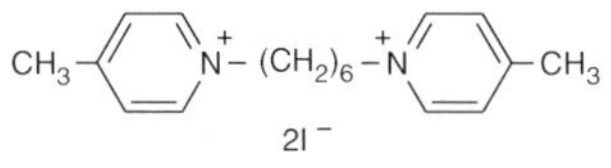

250 mg

S63,592-8
1,1'-DECAMETHYLENEBIS(4-METHYLPYRIDINIUM PERCHLORATE)
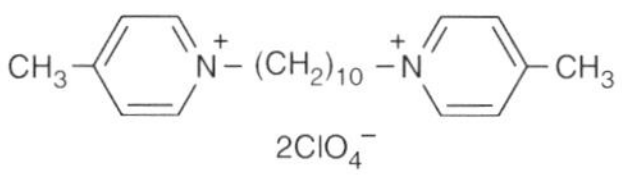
250 mg

S63,264-3
1-HEXADECYL-4-METHYLPYRIDINIUM DICHROMATE
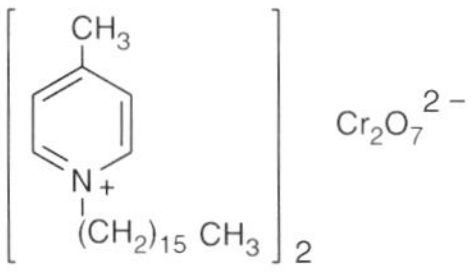
250 mg

S76,568-6
2-BROMO-1-ETHYLPYRIDINIUM TETRAFLUOROBORATE
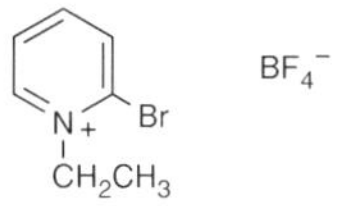
5 gm

S76,567-8
2-CHLORO-1-METHYLPYRIDINIUM TRIFLUOROMETHANESULFONATE
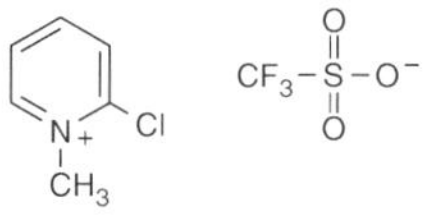
5 gm

S51,962-6
1-DODECYLPYRIDINIUM BROMIDE
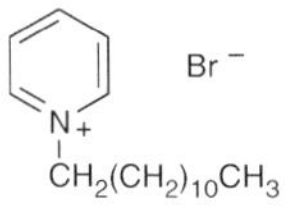
5 gm

S77,920-2
1-(1-NAPHTHYLMETHYL)-PYRIDINIUM BROMIDE
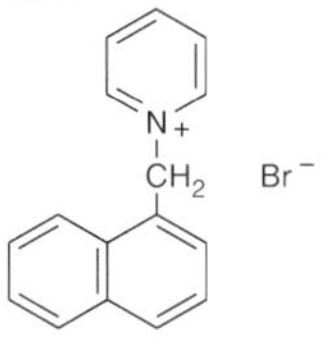
250 mg

S91,835-0
1-(9-ANTHRACENYLMETHYL)PYRIDINIUM IODIDE DIHYDRATE
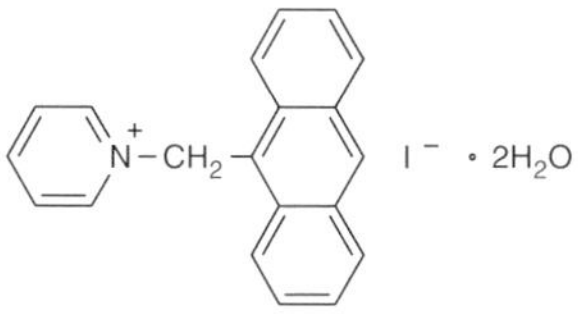
250 mg

S65,492-2
4,4'-METHYLENEBIS[N-(2-PYRIDYLOXYCARBONYL)ANILINE]
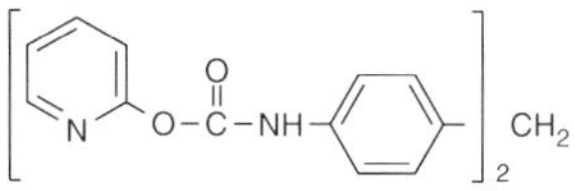
5 gm

S42,270-3
N-(2-CYCLOHEXYL-1-PHENYLETHYL)-NICOTINAMIDE
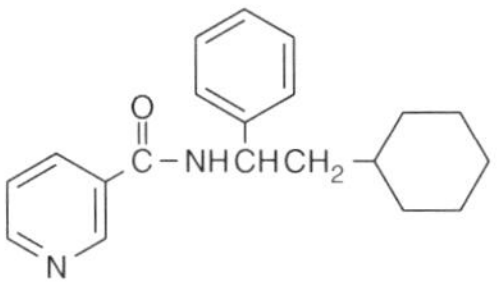
250 mg

S40,670-8
N-BENZYLNICOTINAMIDE
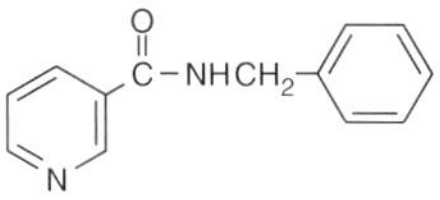
5 gm

S76,093-5
2-BENZYL-4,5-DIHYDRO-3-(4-PYRIDYL)-2H-BENZ(G)INDAZOLE HYDROBROMIDE
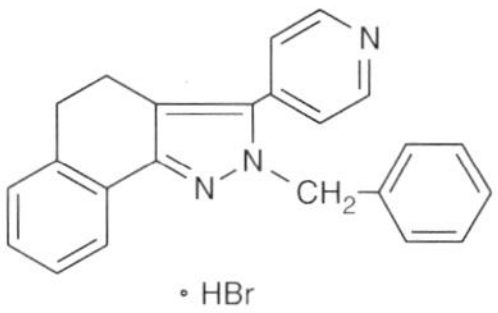
250 mg

S95,486-1
2,4,6-TRIPHENYLPYRIDINE
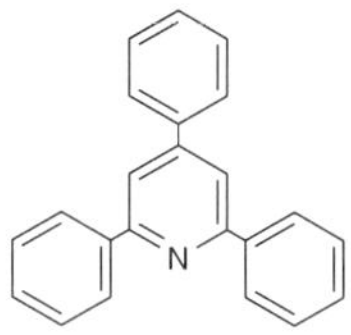
50 mg

S68,438-4
1-ISONICOTINOYL-4-(2-METHYL-6-NITROPHENYL)SEMICARBAZIDE
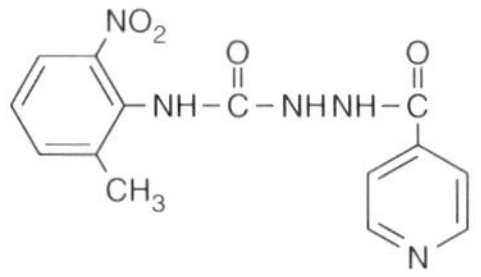
250 mg

S68,423-6
4-(3-ETHYLPHENYL)-1-(ISONICOTINOYL)-SEMICARBAZIDE
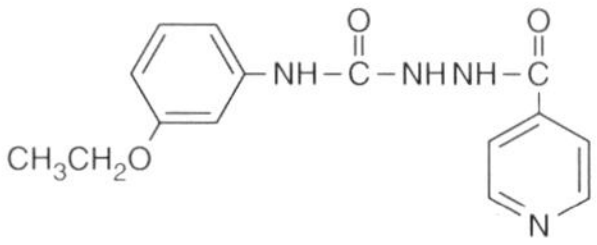
250 mg

S72,901-9
1-(4-METHYLBENZOYL)-2-(4-PYRIDYLCARBONYL)HYDRAZINE HYDROCHLORIDE
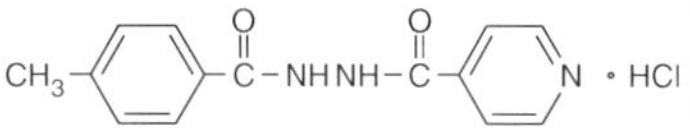
250 mg

S69,812-1
4-(2,4-DIMETHYLPHENYL)-1-(5-METHYL-2-PYRIDYLUREIDOACETYL)SEMICARBAZIDE
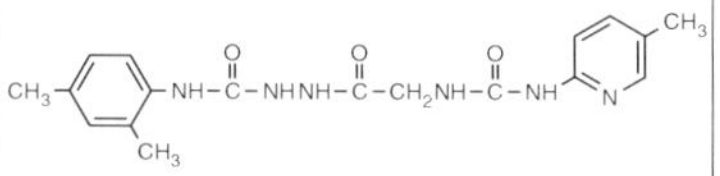
250 mg

S77,934-2
1-(5-METHYL-2-PYRIDYLUREIDOACETYL)-4-PHENYL-3-THIOSEMICARBAZIDE
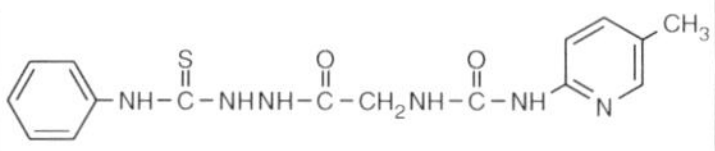
250 mg

S38,929-3
N-HEXADECYL-6-HYDRAZINO-N-METHYL-3-PYRIDINESULFONAMIDE
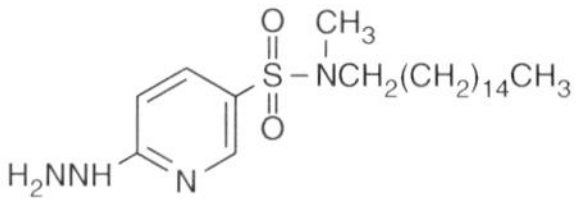
250 mg

S39,076-3
2-ACETOACETAMIDO-3-METHYLPYRIDINE
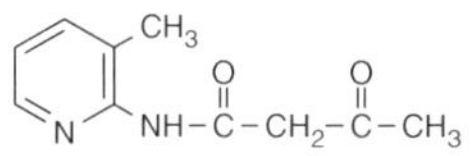
5 gm

S35,859-2
2,6-BIS(ACETOACETAMIDO)PYRIDINE
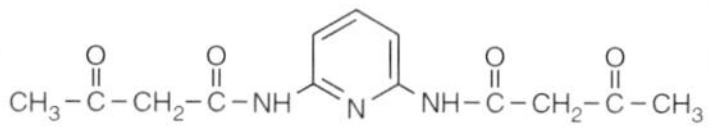
5 gm

S56,367-6
2-ACETOACETAMIDO-4-METHYLPYRIDINE
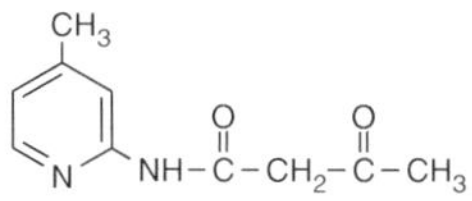
5 gm

S38,728-2
6-ACETAMIDO-3-PYRIDINESULFONIC ACID
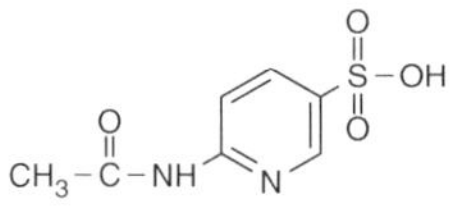
250 mg

S35,714-6
2-ACETAMIDOPYRIDINE
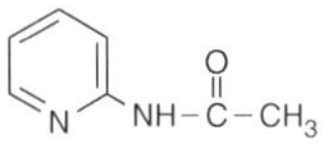
5 gm

S37,831-3
2,6-BIS(ACETAMIDO)PYRIDINE
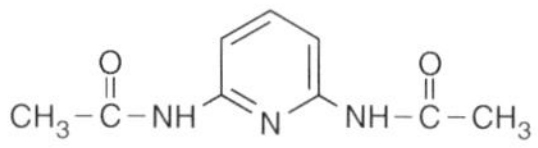
5 gm

S36,229-8
HEXAHYDRO-1-(2-PYRIDYL)-2H-AZEPIN-2-ONE
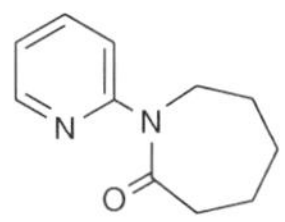
5 gm

S53,975-9
1-(2-PYRIDYL)-3-(O-TOLYL)-2-THIOUREA
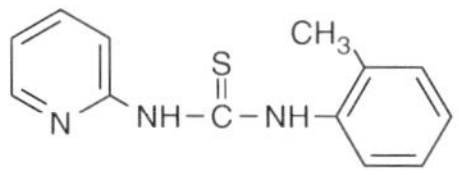
250 mg

S53,990-2
1-(4-CHLOROPHENYL)-3-(2-PYRIDYL)-2-THIOUREA
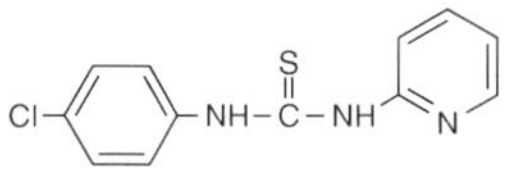
5 gm

S50,122-0
DIETHYL (5-METHYL-2-PYRIDYLAMINOMETHYLENE)MALONATE
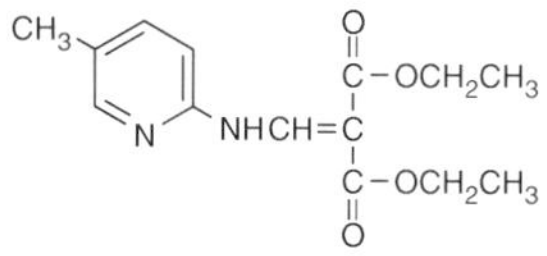
5 gm

S89,308-0
4,6-DIMETHYL-2-PYRIDINECARBOXALDEHYDE
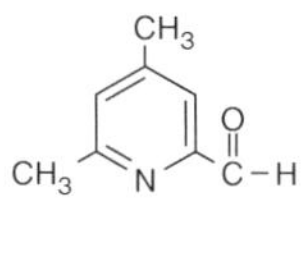
250 mg

S66,188-0
2-(2-PYRIDYLCARBONYL)BENZOIC ACID
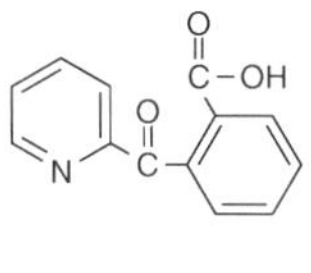
5 gm

S45,434-6
BENZYL 2-PYRIDYL KETONE
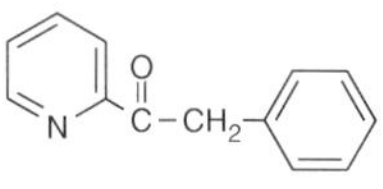
5 gm

S36,929-2
ISOBUTYL 6-METHYL-2-PYRIDYL KETONE
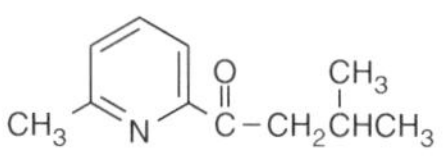
250 mg

S86,082-4
DIMETHYL 6-METHYL-2,4-PYRIDINEDICARBOXYLATE

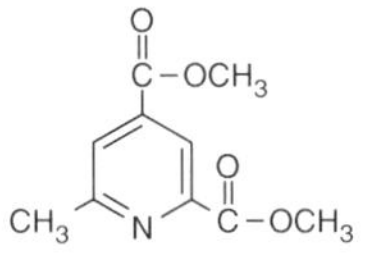

250 mg

S36,988-8
6-METHYLPICOLINIC ACID

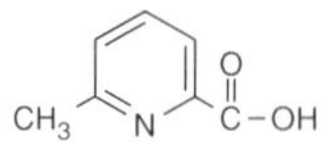

5 gm

S65,638-0
4,4'-METHYLENEBIS[N-(3-PYRIDYLOXYCARBONYL)ANILINE]

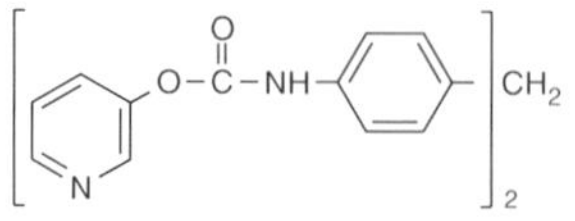

5 gm

S59,416-4
3-PYRIDYL N-PHENYLCARBAMATE

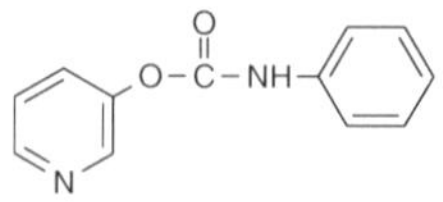

5 gm

S63,197-3
3-PYRIDYL N-(O-TOLYL)CARBAMATE

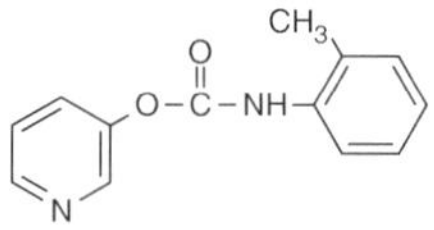

5 gm

S63,661-4
3-PYRIDYL N-(4-BROMOPHENYL)-CARBAMATE

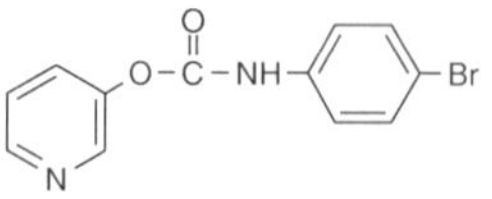

5 gm

S85,511-1
3-PYRIDYL N-(P-TOLYL)CARBAMATE

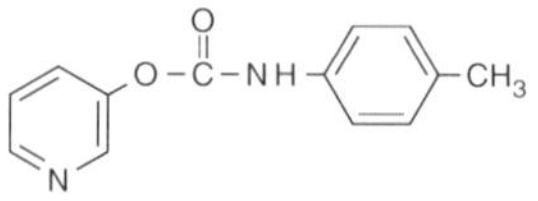

250 mg

S63,804-8
3-PYRIDYL N-(2,3-DICHLOROPHENYL)-CARBAMATE

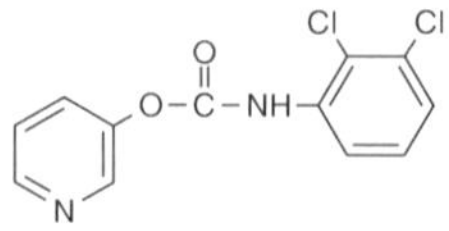

5 gm

S60,701-0
2-PYRIDYL N-(4-CHLORO-2-METHYLPHENYL)CARBAMATE

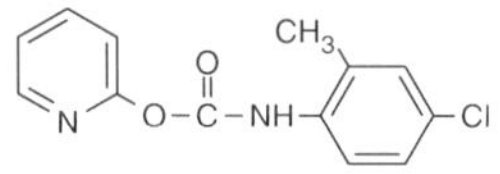

5 gm

S60,943-9
3-PYRIDYL N-(2,5-DICHLOROPHENYL)-CARBAMATE

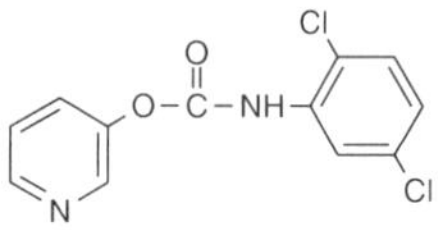

5 gm

S61,277-4
3-PYRIDYL N-(3,4-DICHLOROPHENYL)-CARBAMATE

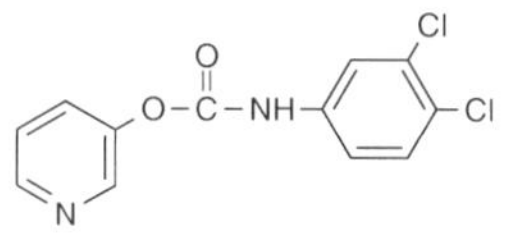

5 gm

S61,303-7
3-PYRIDYL N-(2,4-DICHLOROPHENYL)-CARBAMATE

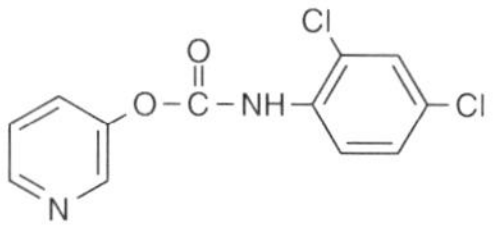

5 gm

S60,947-1
3-PYRIDYL N-(5-CHLORO-2-METHYLPHENYL)CARBAMATE

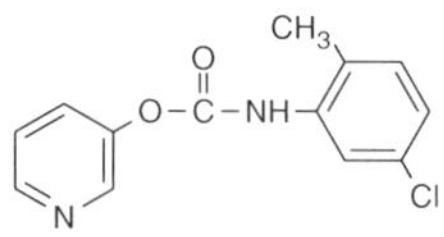

5 gm

S61,017-8
3-PYRIDYL N-(3-CHLORO-4-METHYLPHENYL)CARBAMATE

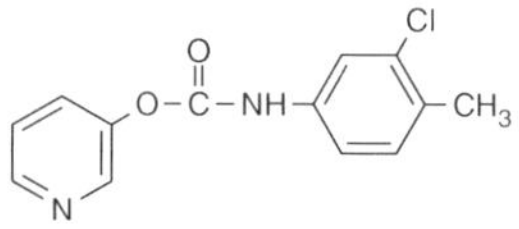

5 gm

S63,698-3
3-PYRIDYL N-(4-CHLORO-2-METHYLPHENYL)CARBAMATE

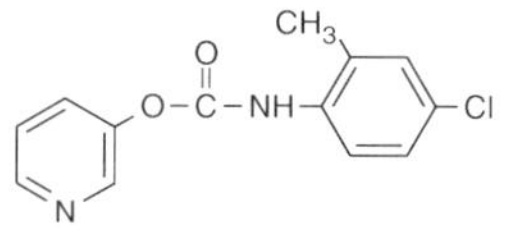

5 gm

S60,990-0
3-PYRIDYL N-(2,4-XYLYL)CARBAMATE

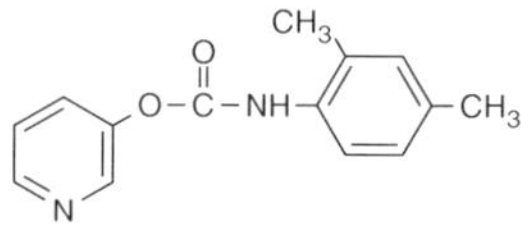

5 gm

S59,555-1
3-PYRIDYL N-(4-CHLOROPHENYL)-CARBAMATE

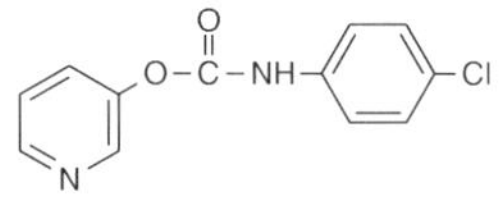

5 gm

S59,559-4
3-PYRIDYL N-(2-CHLOROPHENYL)-CARBAMATE

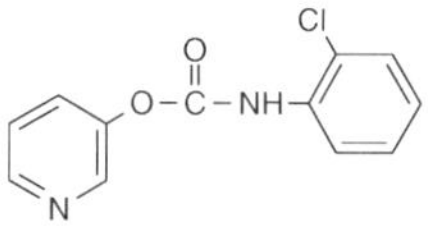

5 gm

S64,701-2
3-PYRIDYL N-(2,4,5-TRICHLOROPHENYL)-CARBAMATE

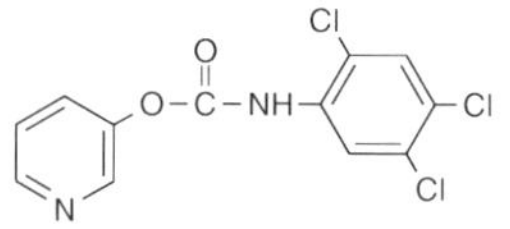

5 gm

S38,946-3
2-BROMO-3-PYRIDYL ACETATE

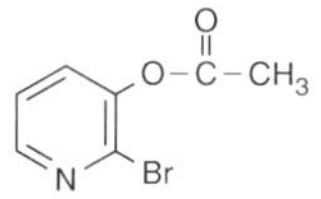

250 mg

S56,037-5
3-SULFOISONICOTINIC ACID, BARIUM SALT

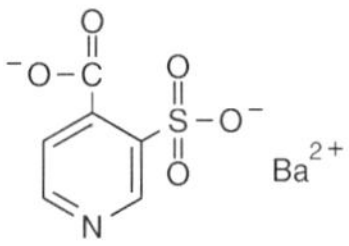

250 mg

S67,242-4
PHENYL 2-PYRIDYLMETHYL SULFONE

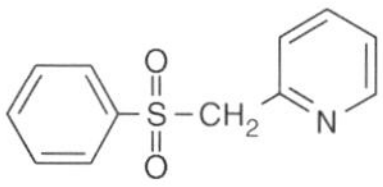

5 gm

S59,715-5
5-METHYL-3-PYRIDINESULFONIC ACID

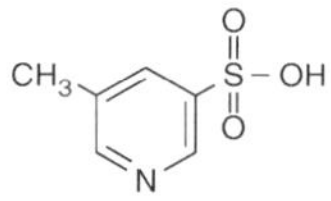

250 mg

S51,396-2
6-CHLORO-N-HEXADECYL-3-PYRIDINESULFONAMIDE

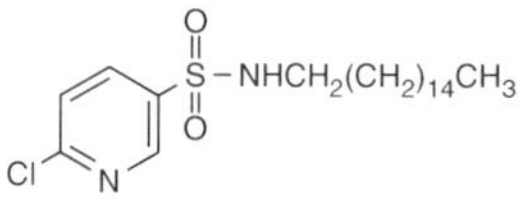

250 mg

S77,095-7
NICOTINOHYDROXAMIC ACID

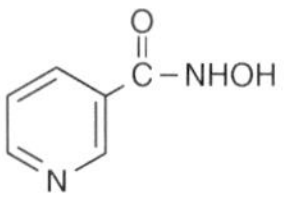

250 mg

S74,223-6
3-(4-PYRIDYL)ACRYLOHYDROXAMIC ACID HYDROCHLORIDE MONOHYDRATE

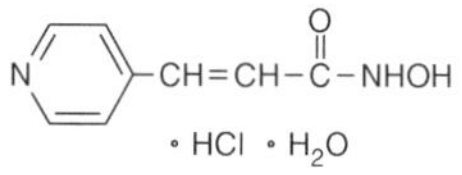

250 mg

S37,695-7
2-NITRAMINOPYRIDINE

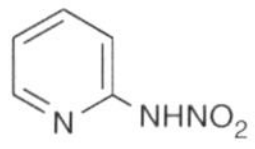

5 gm

S81,256-0
2-(3-NITROCINNAMOYL)PYRIDINE

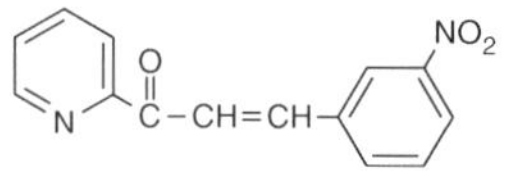

250 mg

S92,277-3
2-(ALPHA-CYANO-4-NITROBENZYLIDENE)-1,2-DIHYDRO-1-METHYLPYRIDINE

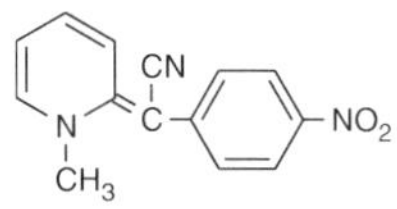

250 mg

S92,084-3
2,6-DIPHENYL-4-(4-NITRO-ALPHA,3,5-TRIPHENYLBENZYLIDENE)-4H-PYRAN

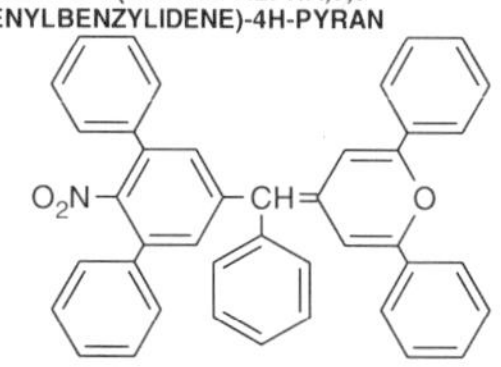

250 mg

S81,815-1
2-(4-NITROPHENYL)PYRIDINE

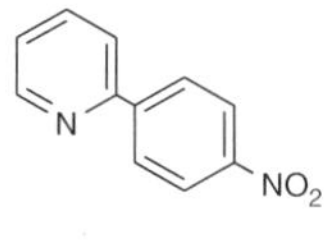

250 mg

S87,669-0
2-(4-NITROBENZYL)PYRIDINE

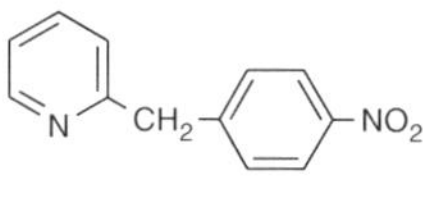

250 mg

S76,480-9
4-(4-NITROBENZYL)PYRIDINE PERCHLORATE
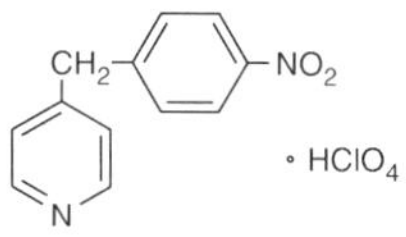
5 gm

S81,905-0
2,4-BIS(3-NITROSTYRYL)PYRIDINE
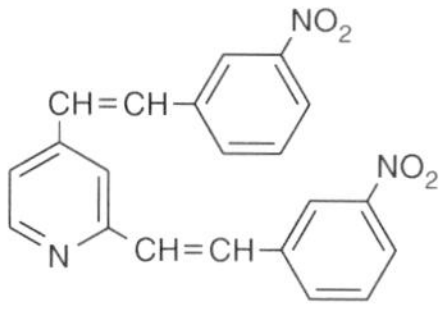
250 mg

S81,906-9
2,6-BIS(3-NITROSTYRYL)PYRIDINE
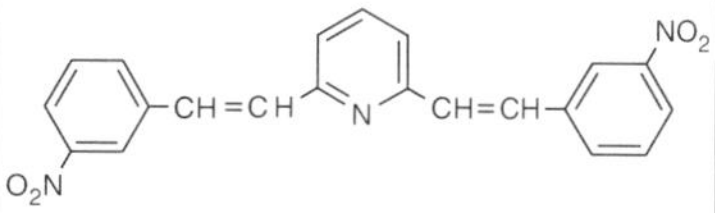
250 mg

S91,124-0
ETHYL 4-[3-(3-NITRO-2-PYRIDYL)-UREIDO]BENZOATE
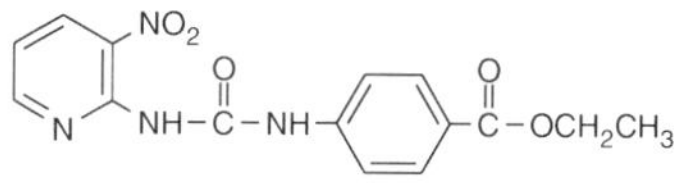
250 mg

S79,335-3
METHYL 6-NITRO-3-PYRIDINECARBOXYLATE
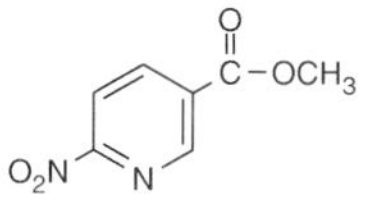
250 mg

S93,370-8
BENZOIC N2-(3-NITRO-2-PYRIDYL)-HYDRAZIDE
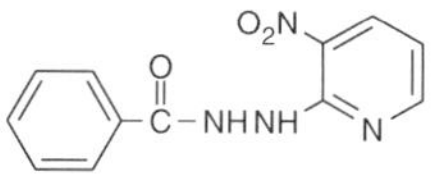
250 mg

S87,618-6
5-BROMO-2-NITROPYRIDINE
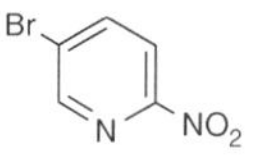
250 mg

S87,735-2
2-NITROPYRIDINE
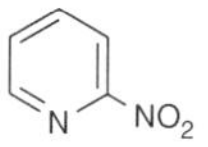
250 mg

S87,414-0
2,6-DIPHENYL-4-PYRIDINECARBOXALDEHYDE
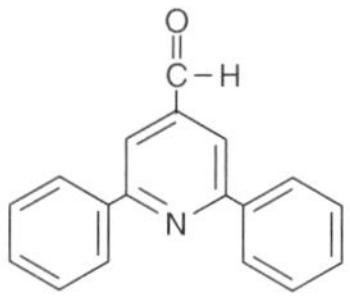
250 mg

S83,623-0
ETHYL 3-OXO-3-(2-PYRIDYL)PROPIONATE
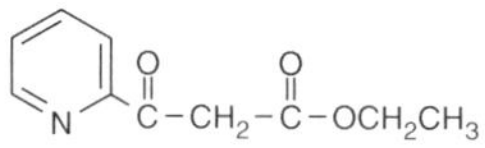
250 mg

S83,622-2
ETHYL 3-OXO-3-(4-PYRIDYL)PROPIONATE
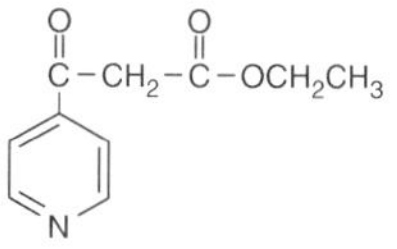
250 mg

S87,027-7
2-(4-PYRIDYLMETHYLENE)-1,3-INDANDIONE
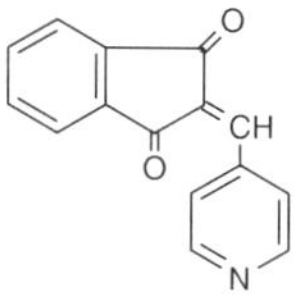
250 mg

S80,792-3
1-(3-ACETYLPHENYL)-3-(3-METHYL-2-PYRIDYL)UREA
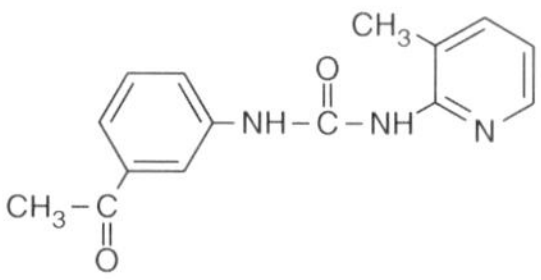
250 mg

S75,850-7
3,4-DIHYDRO-4,4-DIMETHYL-ALPHA,2-EPOXY-2-(4-PYRIDYLMETHYL)-1(2H)-NAPHTHALENONE
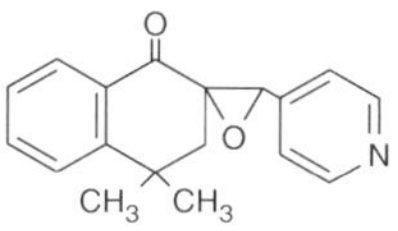
250 mg

S75,896-5
1,2,3,4,5,6,7,8-OCTAHYDRO-9-(4-PYRIDYL)-3,3,6,6-TETRAMETHYL-1,8-XANTHENEDIONE
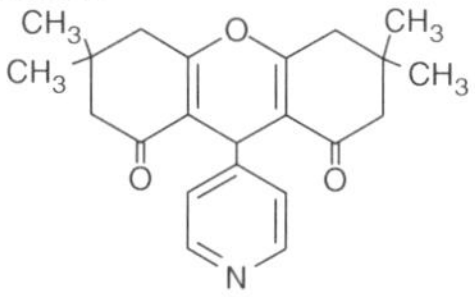
250 mg

S79,850-9
4-BIPHENYLYL 2-PYRIDYL KETONE
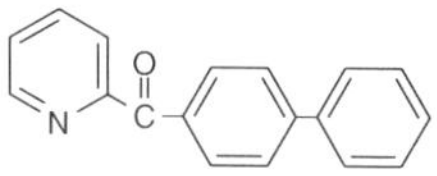
250 mg

S82,389-9
2,5-DICHLORO-4-PYRIDYL PHENYL KETONE
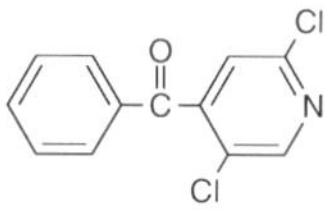
250 mg

S58,152-6
2,2-BIS[2-(2-PYRIDYL)ETHYL]-PROPIOPHENONE
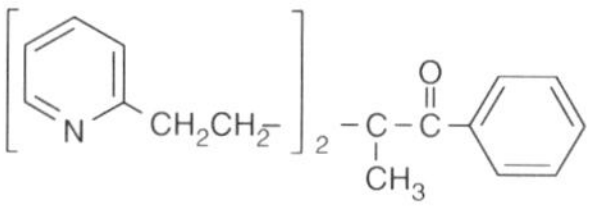
250 mg

S95,478-0
2-(6-METHYL-2-PYRIDYL)ACETOPHENONE HYDROCHLORIDE
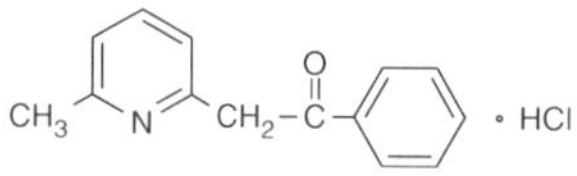
50 mg

S83,171-9
4-ACETYL-2,3,5,6-TETRACHLOROPYRIDINE
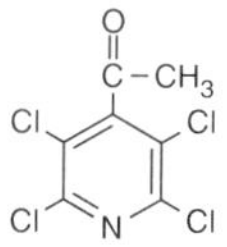
250 mg

S82,980-3
BIS(2,3,5,6-TETRACHLORO-4-PYRIDYL)-KETONE
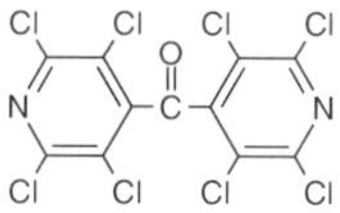
250 mg

S87,540-6
4-PYRIDYL 4-PYRIDYLMETHYL KETONE
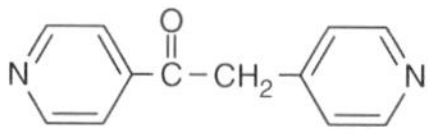
250 mg

S88,303-4
2-(4-PYRIDYLMETHYL)-1-TETRALONE
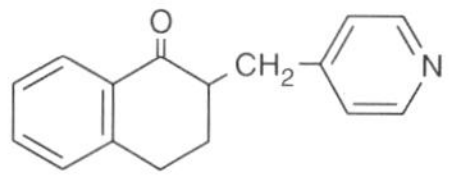
250 mg

S89,732-9
7,8-DIHYDRO-3,7,7-TRIMETHYL-5(6H)-QUINOLINONE HYDROCHLORIDE
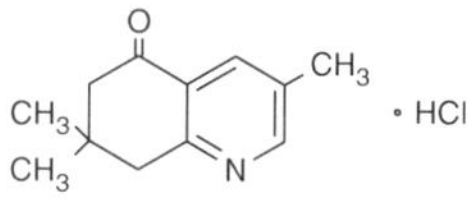
250 mg

S45,702-7
2-ACETOACETAMIDOPYRIDINE
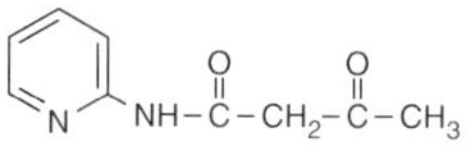
5 gm

S56,136-3
2-ACETOACETAMIDO-6-METHYLPYRIDINE
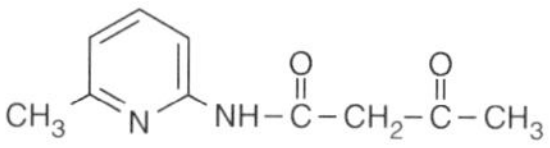
5 gm

S38,029-6
METHYL 2,4,6-TRIMETHYL-3-PYRIDYL KETONE
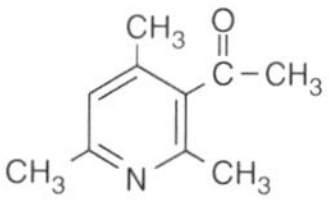
250 mg

S81,715-5
2-PYRIDINEACETONE
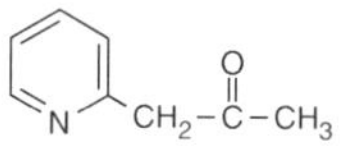
250 mg

S58,127-5
3-METHYL-5-(2-PYRIDYL)-2-PENTANONE
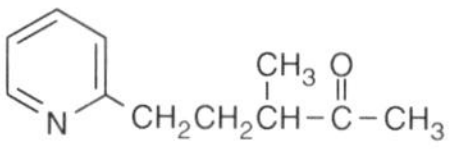
250 mg

S58,112-7
6-(2-PYRIDYL)-2,4,4-TRIMETHYL-3-HEXANONE
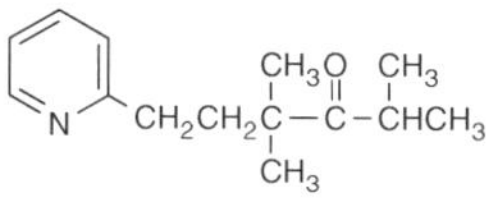
250 mg

S59,381-8
3,3-DIMETHYL-1,7-DI-2-PYRIDYL-4-HEPTANONE
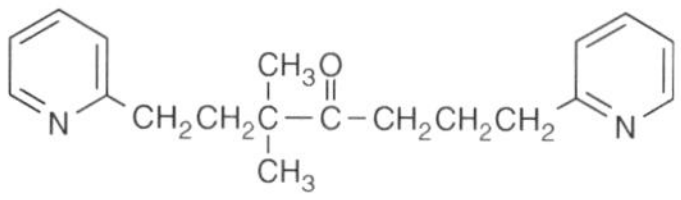
5 gm

S59,369-9
1,1-BIS[2-(2-PYRIDYL)ETHYL]-3,3-DIMETHYL-5-(2-PYRIDYL)-2-PENTANONE
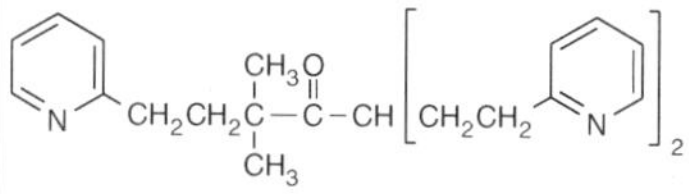
5 gm

S54,376-4
4-AZA-9-FLUORENONE

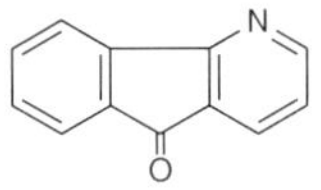

250 mg

S51,484-5
ETHYL BETA-OXO-3-PYRIDINEPROPIONATE

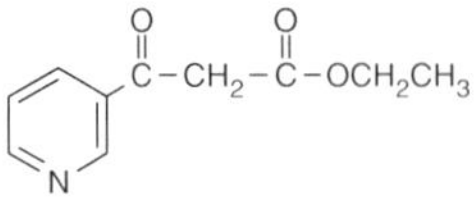

250 mg

S71,909-9
N-(4-METHYL-2-PYRIDYL)-PHTHALAMIC ACID

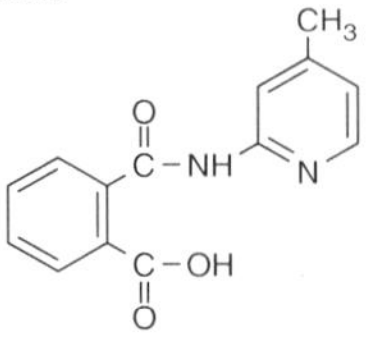

5 gm

S70,477-6
N-(2-PYRIDYL)PHTHALAMIC ACID

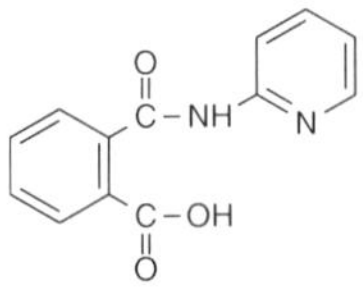

5 gm

S71,923-4
N-(6-METHYL-2-PYRIDYL)-PHTHALAMIC ACID

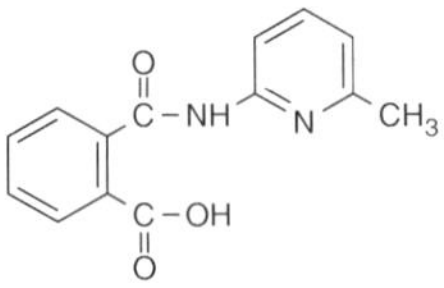

5 gm

S79,406-6
2,5-BIS[N-(4,6-DIMETHYL-2-PYRIDYL)-CARBOMYL]TEREPHTHALIC ACID MONOHYDRATE

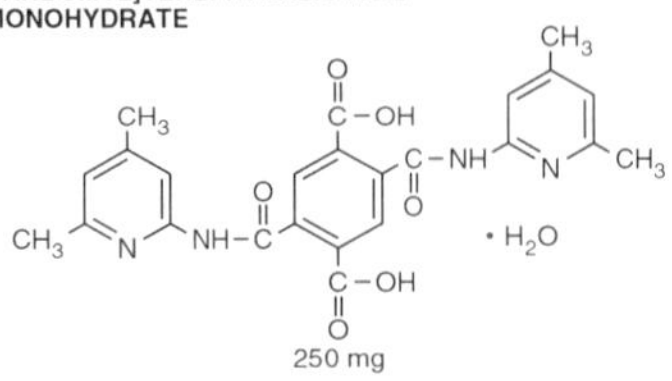

250 mg

S72,075-5
N-(3-METHYL-2-PYRIDYL)-PHTHALAMIC ACID

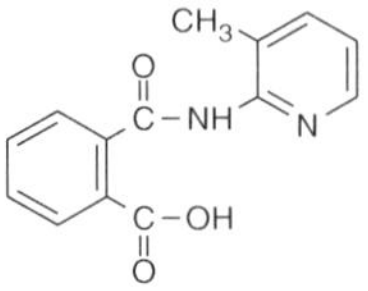

5 gm

S75,797-7
DIMETHYL 2,5-PYRIDINEDICARBOXYLATE (DIMETHYL ISOCINCHOMERONATE)

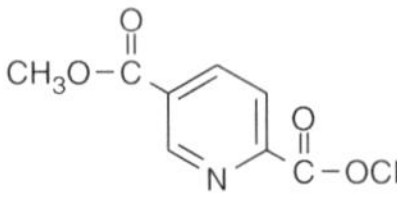

5 gm

S57,930-0
N-NICOTINOYLGLYCINE

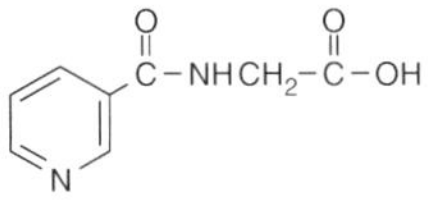

250 mg

S36,591-2
2-CARBAMYLNICOTINIC ACID

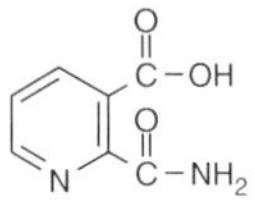

5 gm

S63,862-5
N-[2-(2-PYRIDYL)-5-BENZIMIDAZOLYL]-MALEAMIC ACID

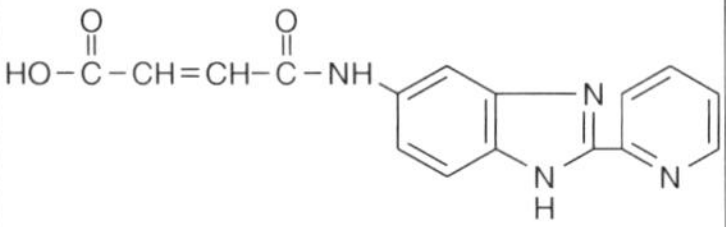

250 mg

S81,225-0
2,3,5,6-TETRACHLORO-4-PYRIDINECARBOXYLIC ACID

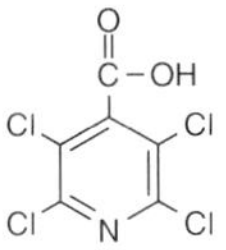

250 mg

S82,594-8
NICOTINIC ACID, OCTADECYLAMINE SALT

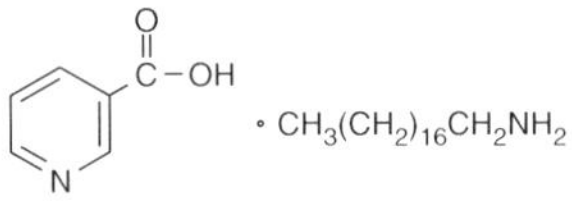

250 mg

S78,699-3
3-(4-PYRIDYL)ACRYLIC ACID

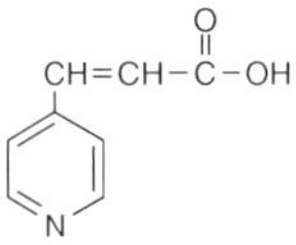

250 mg

S34,312-9
BUTYL PICOLINATE

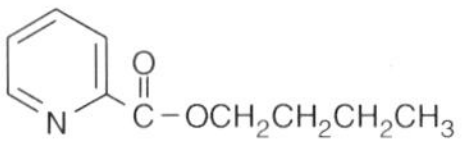

5 gm

S79,460-0
2-PYRIDYL N-[4-(ETHOXYCARBONYL)-PHENYL]CARBAMATE

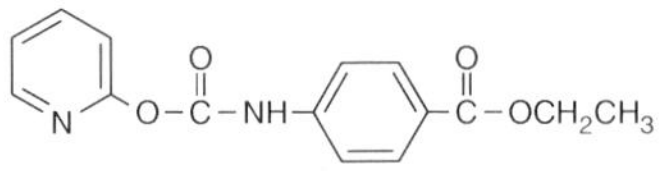

250 mg

S34,613-6
DIETHYL 4-PHENYL-1,4-DIHYDRO-2,6-LUTIDINE-3,5-DICARBOXYLATE

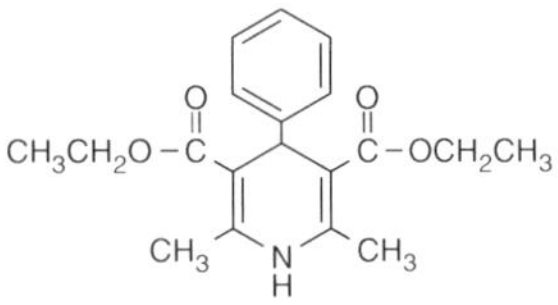

5 gm

S44,350-6
DIETHYL 2,4,6-TRIMETHYL-3,5-PYRIDINEDICARBOXYLATE

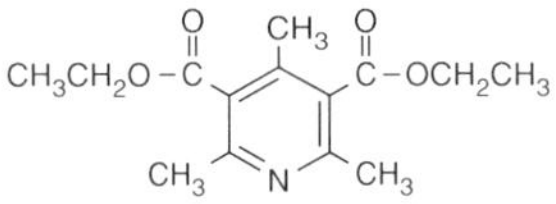

5 gm

S38,148-9
DIETHYL 3,5-PYRIDINEDICARBOXYLATE

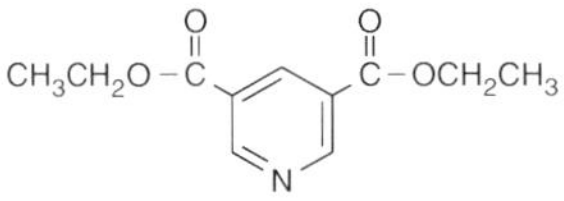

5 gm

S40,086-6
DIPROPYL 3,5-PYRIDINEDICARBOXYLATE

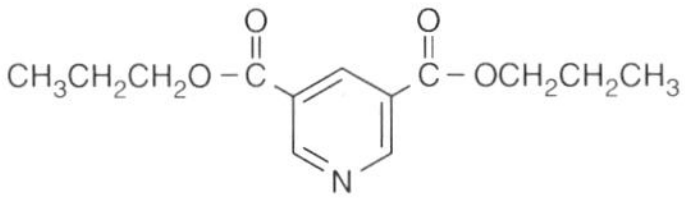

250 mg

S40,115-3
DIBUTYL 3,5-PYRIDINEDICARBOXYLATE

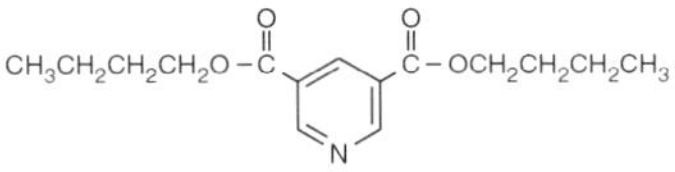

250 mg

S91,125-9
ETHYL 4-[3-(5-CHLORO-2-PYRIDYL)-UREIDO]BENZOATE

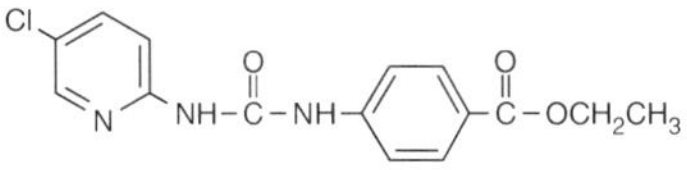

250 mg

S91,123-2
ETHYL 4-[3-(5-BROMO-2-PYRIDYL)-UREIDO]BENZOATE

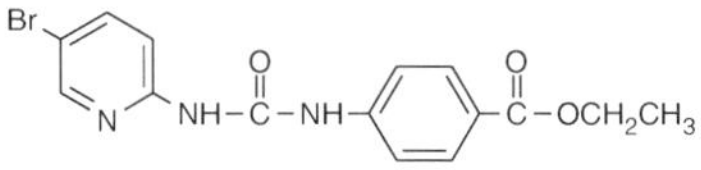

250 mg

S64,355-6
METHYL 6-CHLORONICOTINATE

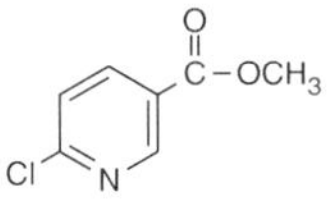

5 gm

S36,467-3
PROPYL NICOTINATE

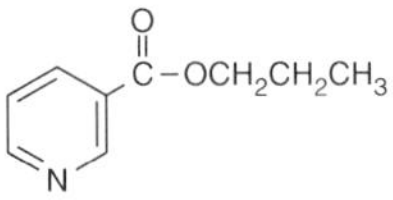

5 gm

S77,029-9
ETHYL 4-PYRIDINECARBOXYLATE PICRATE

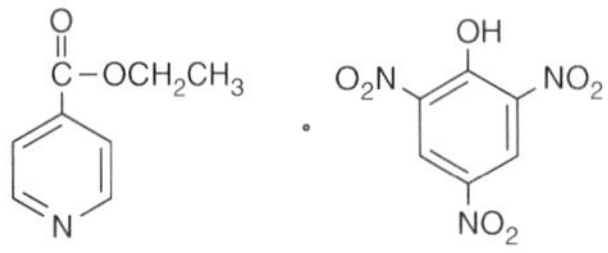

250 mg

S40,941-3
METHYL 4-PYRIDINEACRYLATE

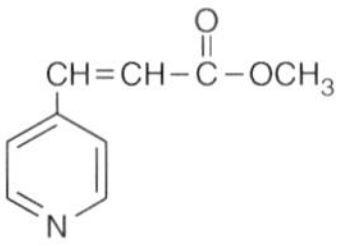

5 gm

S87,632-1
S-BENZYL 3-PYRIDINETHIOCARBOXYLATE

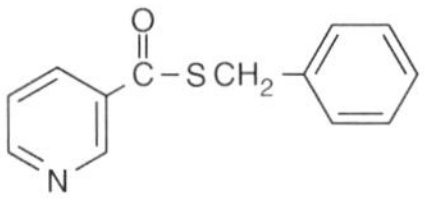

250 mg

S87,415-9
2,6-DIPHENYLISONICOTINOYL CHLORIDE

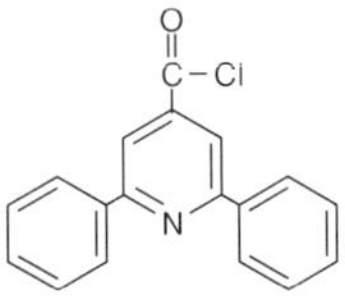

250 mg

S87,393-4
2,6-DIPHENYL-4-PYRIDINECARBOXYLIC ANHYDRIDE

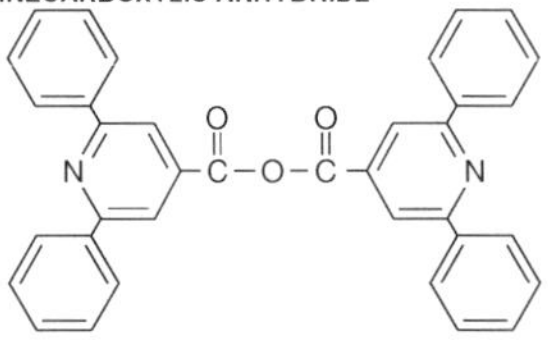

250 mg

S92,830-5
2-PHENYL-2-(2-PYRIDYL)ACETAMIDE

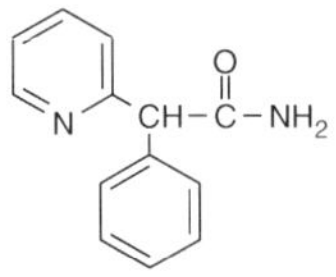

250 mg

S76,882-0
5-(4-PYRIDYLMETHYLENE)RHODANINE

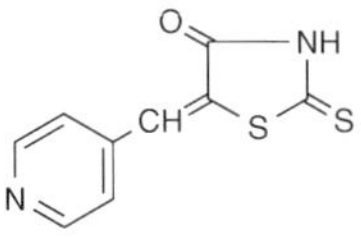

250 mg

S92,828-3
2-CHLORO-N-(2-PYRIDYL)BENZAMIDE

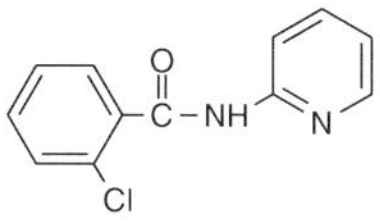

250 mg

S92,829-1
2'-CHLOROPICOLINANILIDE

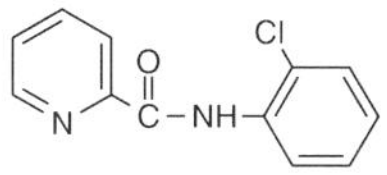

250 mg

S91,017-1
4-METHYL-N-(4-METHYL-2-PYRIDYL)-BENZAMIDE HYDROCHLORIDE

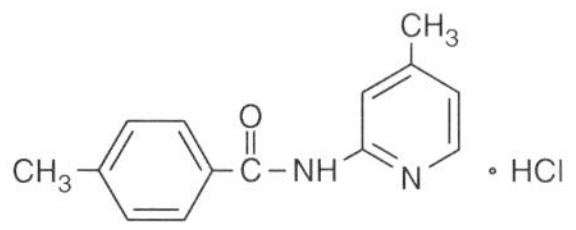

250 mg

S91,009-0
3-METHYL-N-(4-METHYL-2-PYRIDYL)-BENZAMIDE HYDROCHLORIDE

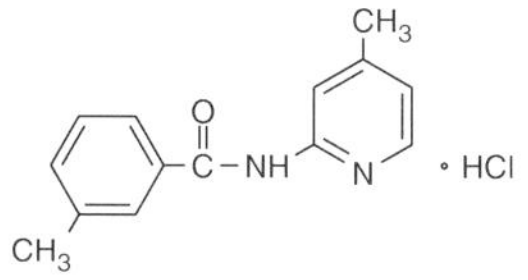

250 mg

S91,007-4
3-METHYL-N-(2-PYRIDYL)BENZAMIDE HYDROCHLORIDE

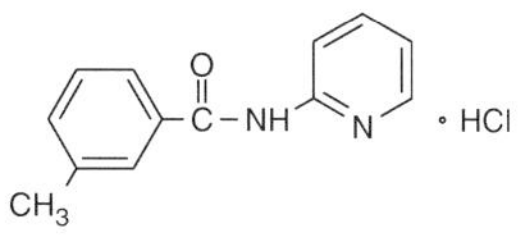

250 mg

S38,392-9
N-DODECYLNICOTINAMIDE

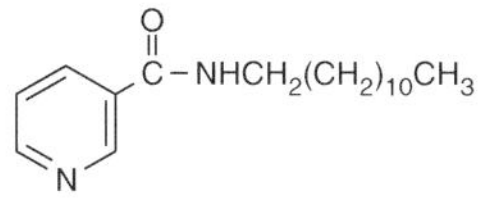

5 gm

S67,520-2
N2-(METHYLSULFONYL)ISONICOTINIC HYDRAZIDE HYDROCHLORIDE

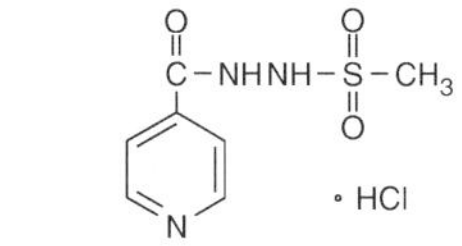

250 mg

S43,182-6
2,3-DIHYDRO-2-[2-(4-PYRIDYL)ETHYL]-1,4-PHTHALAZINEDIONE

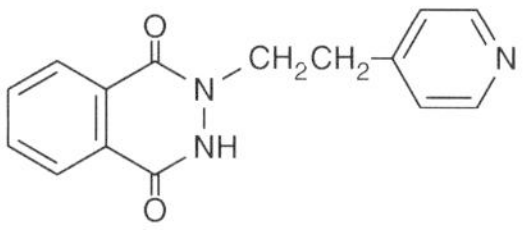

5 gm

S79,373-6
4-CHLORO-2,6-PYRIDINEDICARBOXYLIC HYDRAZIDE

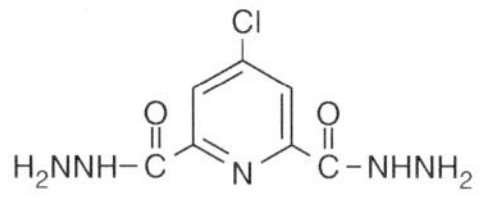

250 mg

S40,381-4
ISONICOTINIC ACID METHYLENEHYDRAZIDE

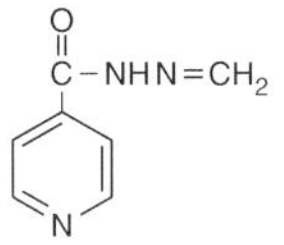

250 mg

S87,418-3
2-(4-PYRIDYLMETHYLENE)ACETOPHENONE

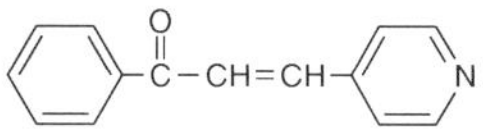

250 mg

S72,027-5
3-[N-(2-PYRIDYL)CARBAMOYL]-5-NORBORNENE-2-CARBOXYLIC ACID

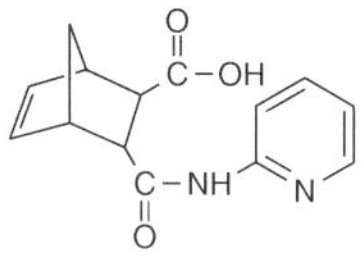

5 gm

S73,066-1
3-[N-(3-PYRIDYL)CARBAMOYL]-5-NORBORNENE-2-CARBOXYLIC ACID

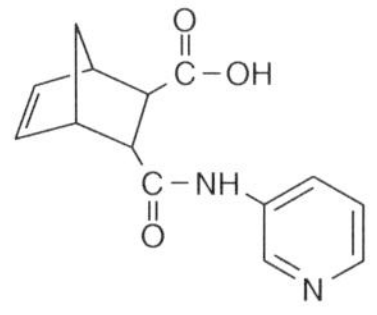

5 gm

S71,952-8
CIS-HEXAHYDRO-N-(4-METHYL-2-PYRIDYL)PHTHALAMIC ACID

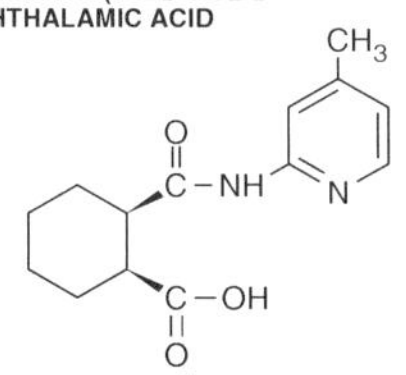

5 gm

S72,018-6
CIS-HEXAHYDRO-N-(2-PYRIDYL)-PHTHALAMIC ACID

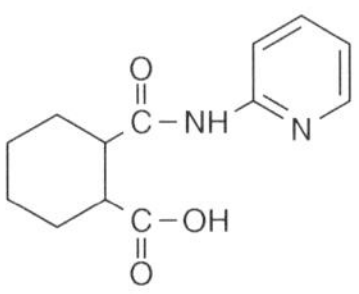

5 gm

S70,478-4
N-(3-PYRIDYL)-HEXAHYDROPHTHALAMIC ACID

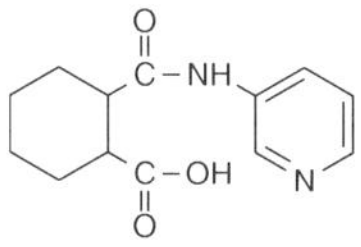

5 gm

S71,933-1
N-(2-PYRIDYL)-1,2,5,6-TETRAHYDROPHTHALAMIC ACID

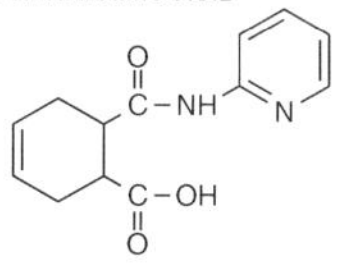

5 gm

S70,303-6
N-(2-PYRIDYL)SUCCINAMIC ACID

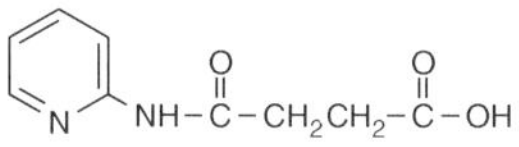

5 gm

S72,071-2
N-(3-METHYL-2-PYRIDYL)-GLUTARAMIC ACID

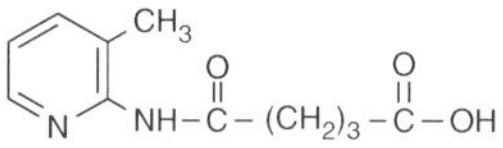

5 gm

S70,111-4
N-(4-METHYL-2-PYRIDYL)-SUCCINAMIC ACID

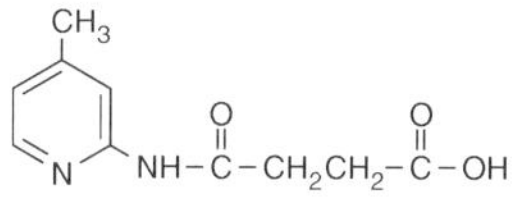

5 gm

S72,066-6
N-(4-METHYL-2-PYRIDYL)-GLUTARAMIC ACID

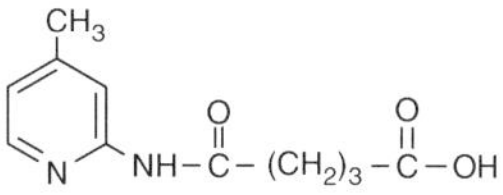

5 gm

S72,032-7
N-(2-PYRIDYL)GLUTARAMIC ACID

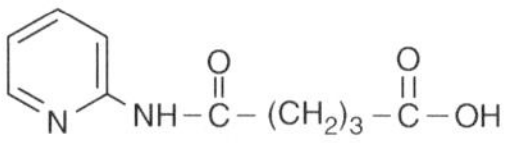

5 gm

S73,075-0
N-(6-METHYL-2-PYRIDYL)-GLUTARAMIC ACID

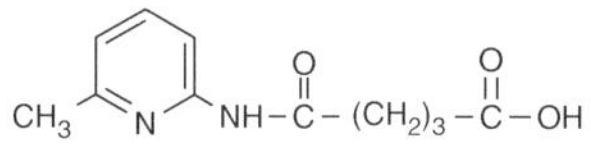

5 gm

S70,488-1
N-(3-PYRIDYL)SUCCINAMIC ACID

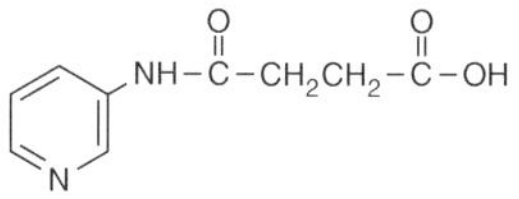

5 gm

S72,065-8
N-(3-PYRIDYLMETHYL)GLUTARAMIC ACID

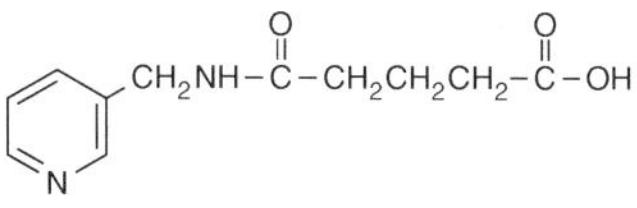

5 gm

S62,861-1
2-CARBOXY-2-PROPANESULFONIC ACID PYRIDINIUM SALT

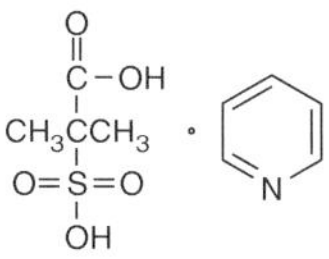

5 gm

S50,514-5
2-METHYL-2-SULFOOCTADECANOIC ACID PYRIDINE SALT

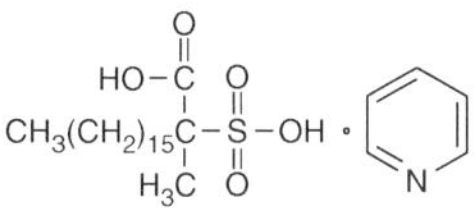

250 mg

S39,356-8
2-SULFODOCOSANOIC ACID, PYRIDINIUM SALT

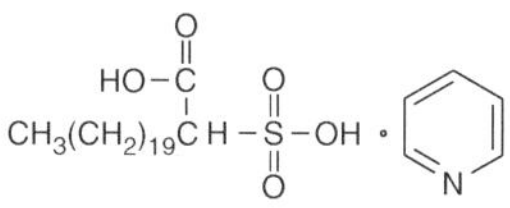

250 mg

S57,661-1
SULFOACETIC ACID, PYRIDINE COMPLEX

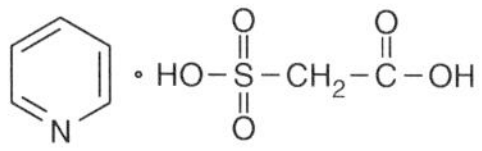

250 mg

S56,403-6
3-PYRIDINEACETIC ACID

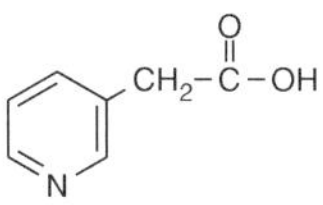

5 gm

S94,523-4
3,4,5-TRICHLORO-2-PYRIDINECARBOXYLIC ACID

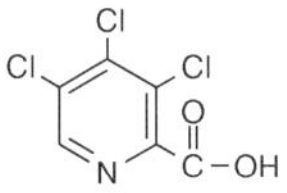

250 mg

S78,241-6
ALPHA-METHYL-2-PYRIDINEBUTYRIC ACID

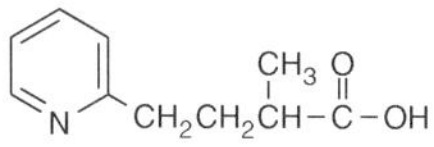

5 gm

S36,470-3
3-PYRIDYLACETIC ACID, SODIUM SALT

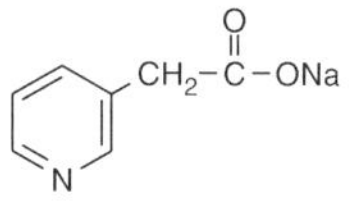

5 gm

S71,560-3
ETHYL 2-[3-(5-CHLORO-2-PYRIDYL)-UREIDO]ACETATE

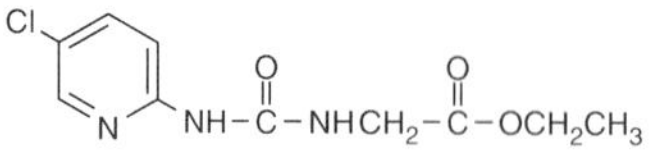

250 mg

S59,418-0
2-PYRIDYL N-PHENYLCARBAMATE

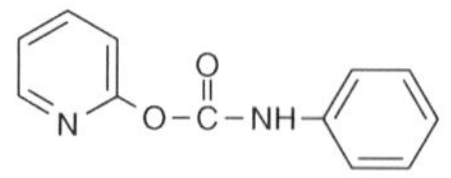

5 gm

S63,208-2
2-PYRIDYL N-(O-TOLYL)CARBAMATE

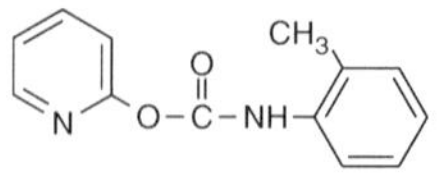

5 gm

S65,981-9
2-PYRIDYL N-(P-TOLYL)CARBAMATE

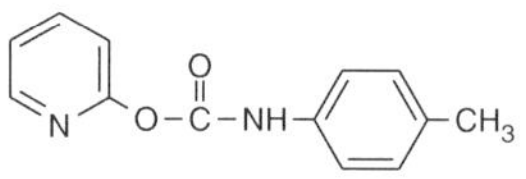

5 gm

S68,812-6
4-PYRIDYLMETHYL N-(2,6-DIMETHYLPHENYL)CARBAMATE

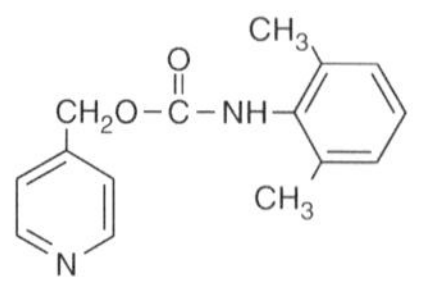

250 mg

S76,981-9
4-PYRIDYL N-(2,6-DIETHYLPHENYL)-CARBAMATE

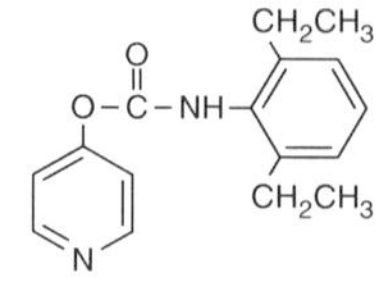

250 mg

S71,095-4
2-PYRIDYL N-(2,4-XYLYL)CARBAMATE

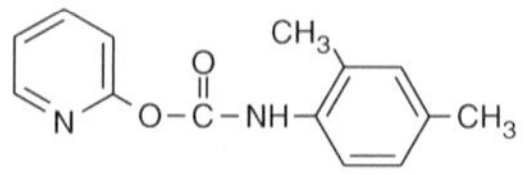

5 gm

S63,219-8
2-PYRIDYL N-(3,4-DICHLOROPHENYL)-CARBAMATE

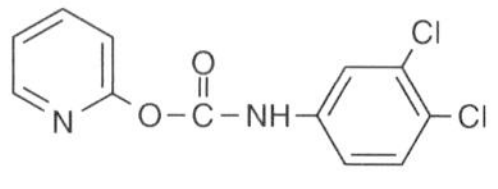

5 gm

S61,123-9
2-PYRIDYL N-(5-CHLORO-2-METHYLPHENYL)CARBAMATE

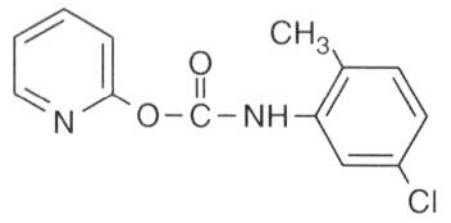

5 gm

S72,154-9
3-PYRIDYLMETHYL N-(2,5-DIFLUOROPHENYL)CARBAMATE

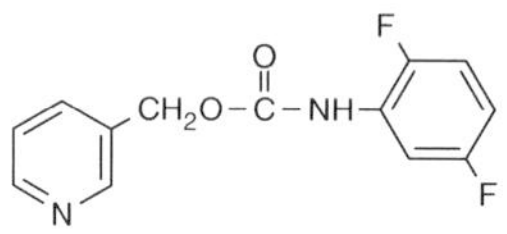

5 gm

S68,556-9
4-PYRIDYLMETHYL N-(2,4-DIMETHYLPHENYL)CARBAMATE

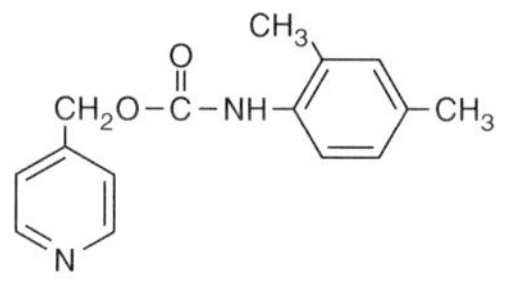

250 mg

S72,409-2
4-PYRIDYL N-(3,4-DICHLOROPHENYL)-CARBAMATE

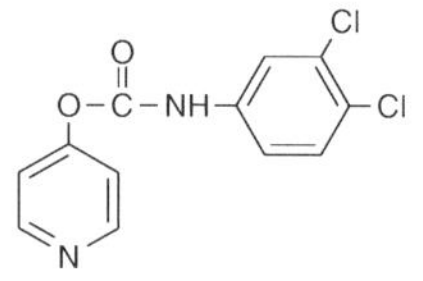

5 gm

S59,548-9
2-PYRIDYL N-(2-CHLOROPHENYL)-CARBAMATE

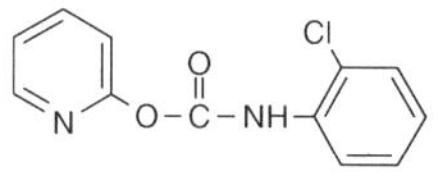

5 gm

S59,409-1
2-PYRIDYL N-(3-CHLOROPHENYL)-CARBAMATE

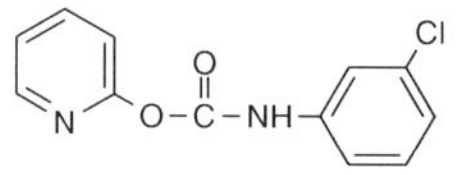

5 gm

S71,535-2
2-PYRIDYL N-(M-TOLYL)CARBAMATE

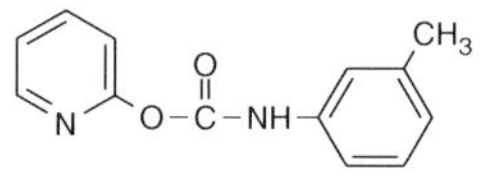

5 gm

S73,796-8
2-PYRIDYL N-(3-ETHYLPHENYL)-CARBAMATE

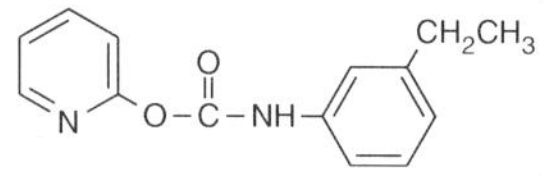

5 gm

S73,920-0
3-PYRIDYLMETHYL N-(3-ETHYLPHENYL)-CARBAMATE

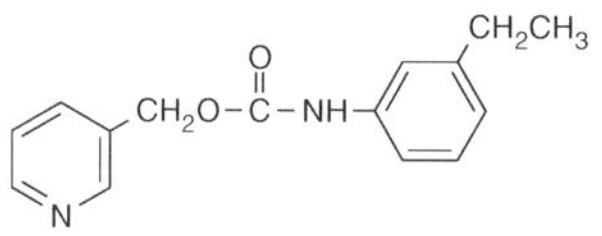

5 gm

S68,671-9
4-PYRIDYLMETHYL N-(3-ETHYLPHENYL)-CARBAMATE

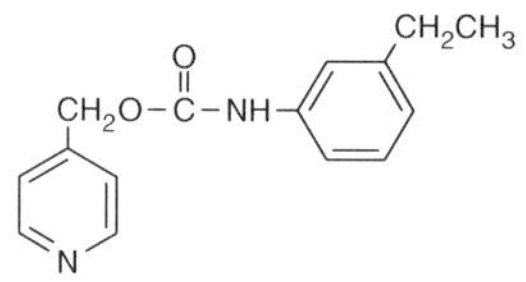

250 mg

S72,668-0
S-(2,3,5,6-TETRACHLORO-4-PYRIDYL) N-(ETHOXYCARBONYLMETHYL)-THIOCARBAMATE

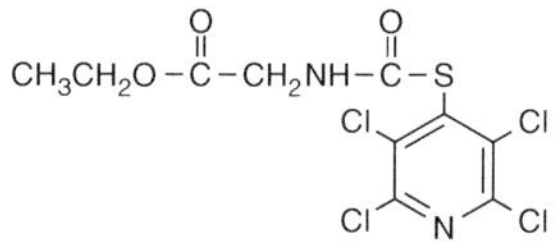

5 gm

S59,373-7
ETHYL ALPHA-PHENYL-2-PYRIDINEBUTYRATE

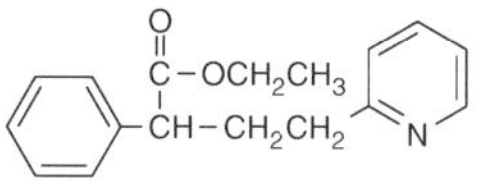

5 gm

S37,035-5
PHENYL NICOTINATE

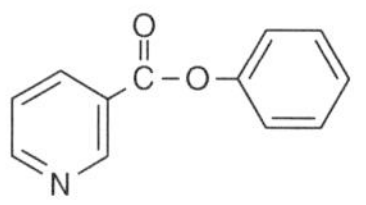

5 gm

S94,524-2
METHYL 3,4,5-TRICHLORO-2-PYRIDINECARBOXYLATE

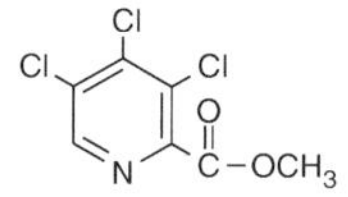

250 mg

S54,562-7
DIETHYL 2,6-PYRIDINEDICARBOXYLATE

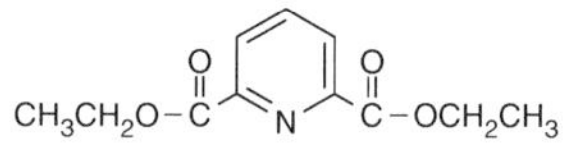

250 mg

S36,385-5
METHYL 3-PYRIDYLACETATE

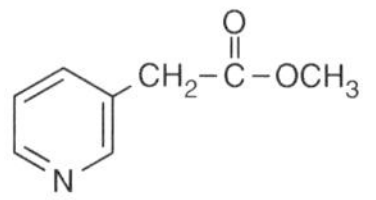

5 gm

S34,309-9
BUTYL NICOTINATE

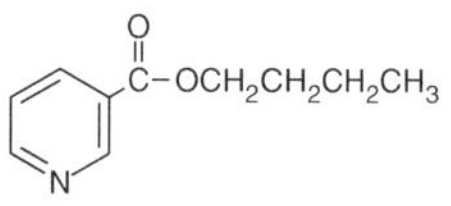

5 gm

S39,799-7
N3,N3-DIETHYL-3,5-PYRIDINEDICARBOXAMIDE

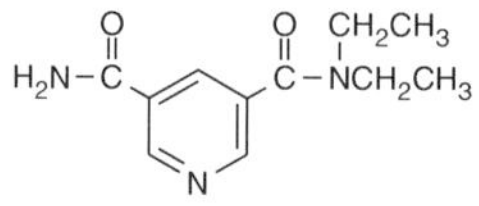

5 gm

S56,152-5
2-PYRIDINEACETAMIDE

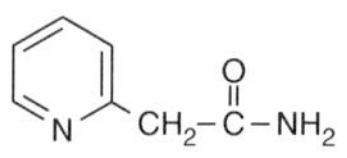

5 gm

S40,034-3
1-BUTYL-3-CARBAMOYLPYRIDINIUM BROMIDE

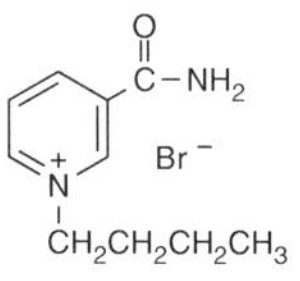

250 mg

S69,385-5
1-(5-METHYL-2-PYRIDYLUREIDOACETYL)-4-(1-NAPHTHYL)SEMICARBAZIDE

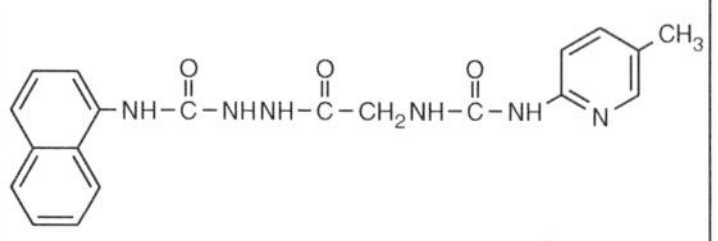

250 mg

S90,540-2
1,4,5,6,7,7-HEXACHLORO-N-(6-METHYL-2-PYRIDYL)-5-NORBORNENE-2,3-DICARBOXIMIDE

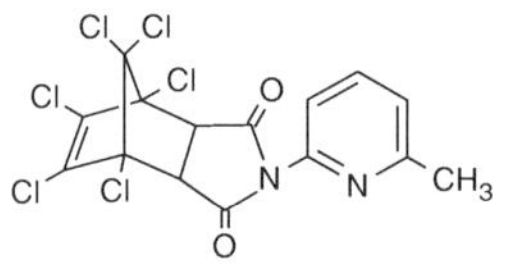

250 mg

S89,456-7
3,4-DICHLORO-5,6-DIMETHYL-2(1H)-PYRIDONE

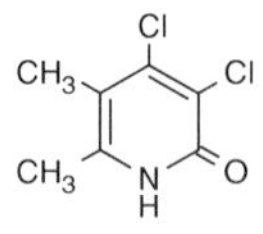

250 mg

S94,375-4
2,4-DICHLORO-N-(3-PYRIDYL)BENZAMIDE
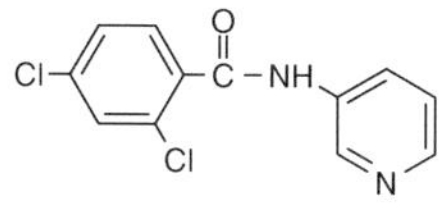
5 gm

S76,636-4
6-METHYL-2(1H)-PYRIDONE
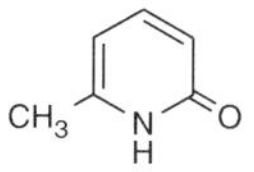
250 mg

S90,669-7
N-(3-METHYL-2-PYRIDYL)FORMAMIDE
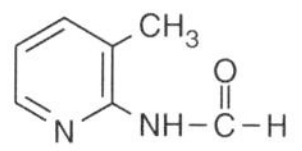
250 mg

S80,337-5
N-(5-METHYL-2-PYRIDYL)ACETAMIDE
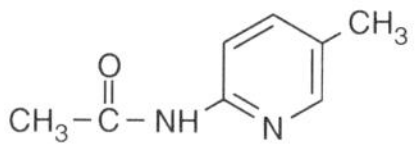
250 mg

S80,336-7
N-(6-METHYL-2-PYRIDYL)ACETAMIDE
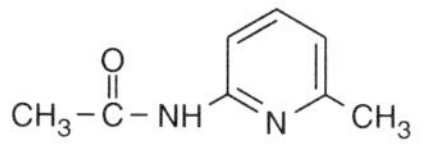
250 mg

S42,072-7
2,6-DIMETHYL-1-ISONICOTINOYLPIPERIDINE
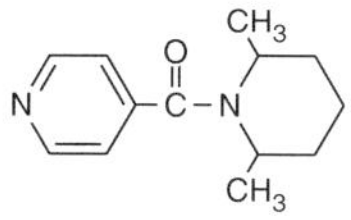
250 mg

S75,922-8
10,11-DIHYDRO-5H-DIBENZ(B,F)AZEPIN-5-YL 2-PYRIDYLMETHYL KETONE
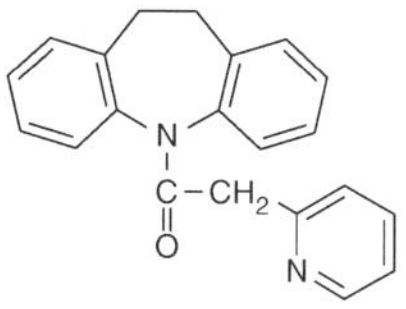
250 mg

S86,977-5
FORMIC N1-PHENYL-N2-(2-PYRIDYL)-HYDRAZIDE
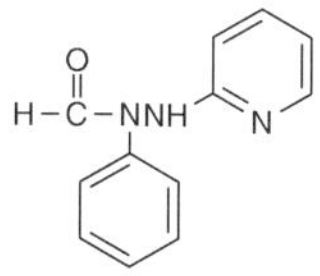
250 mg

S73,206-0
2-(3-PYRIDYL)ACETIC HYDRAZIDE
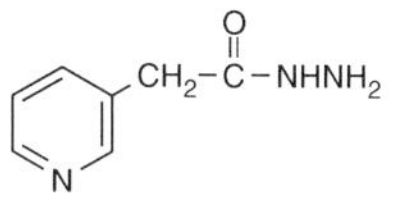
250 mg

S89,932-1
1-CYCLOHEXYL-3-(4-METHYL-2-PYRIDYL)UREA
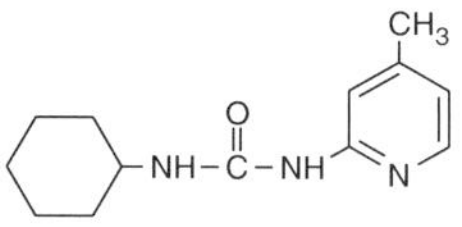
250 mg

S84,420-9
1-CYCLOHEXYL-3-(2-PYRIDYL)UREA
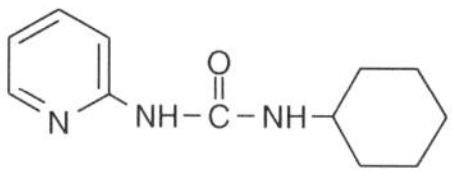
250 mg

S64,473-0
1-PHENYL-3-(2-PYRIDYL)UREA
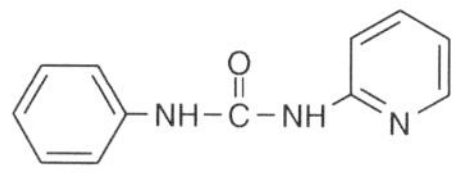
5 gm

S61,185-9
1-(6-METHYL-2-PYRIDYL)-3-PHENYLUREA
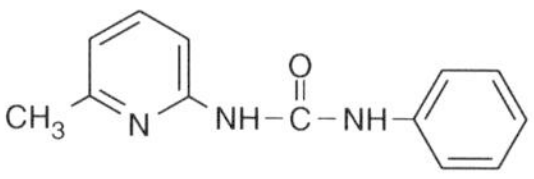
5 gm

S60,801-7
1-PHENYL-3-(3-PYRIDYLMETHYL)UREA
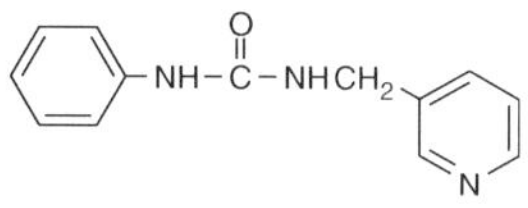
5 gm

S84,536-1
1-(2-ETHYLPHENYL)-3-(4-METHYL-2-PYRIDYL)UREA
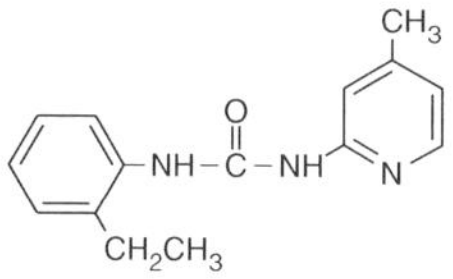
250 mg

S65,269-5
1-(2,6-DICHLOROPHENYL)-3-(2-PYRIDYL)UREA
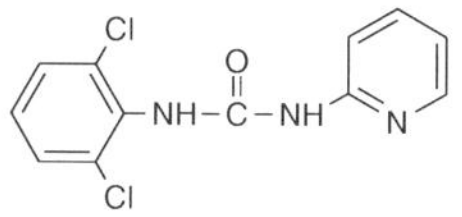
5 gm

S71,074-1
1-(3-PYRIDYLMETHYL)-3-(2,3-XYLYL)UREA
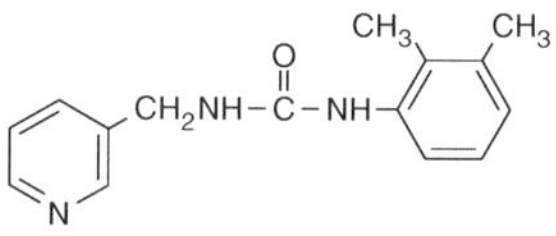
5 gm

S91,247-6
1-(2-PYRIDYL)-3-(3,4-XYLYL)UREA
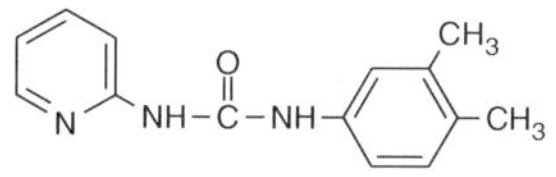
250 mg

S72,149-2
1-(2-PYRIDYLMETHYL)-3-(2,4-XYLYL)UREA
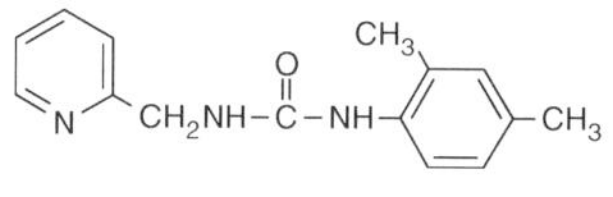
250 mg

S69,451-7
1-(3,5-DIMETHYLPHENYL)-3-(4-METHYL-2-PYRIDYL)UREA
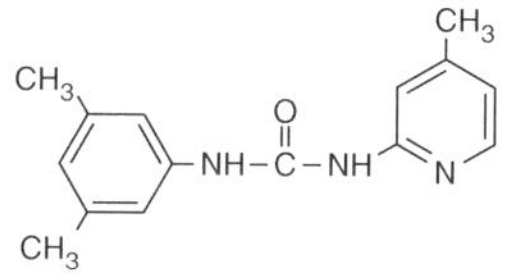
250 mg

S71,849-1
1-(5-CHLORO-2-PYRIDYL)-3-(3-ETHYLPHENYL)UREA
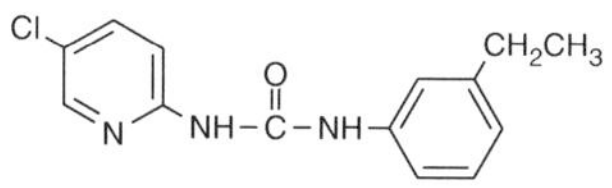
250 mg

S91,027-9
1-BUTYL-3-(4-METHYL-2-PYRIDYL)UREA
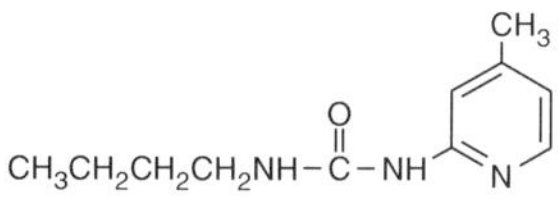
250 mg

S68,840-1
1-(3-ETHYLPHENYL)-3-(2-PYRIDYL)UREA
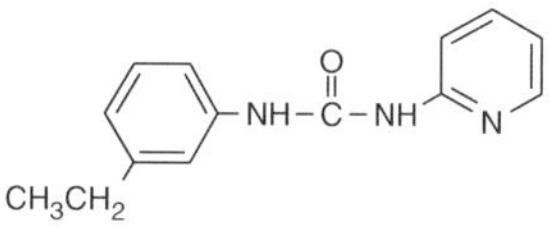
250 mg

S52,733-5
1-PHENYL-3-(2-PYRIDYL)-2-THIOUREA
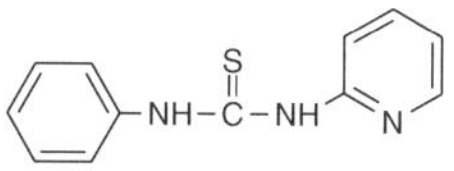
250 mg

S68,677-8
1-BUTYL-3-(2-PYRIDYL)-2-THIOUREA
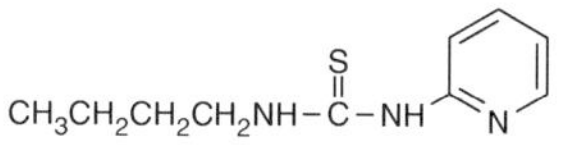
250 mg

S69,042-2
1-(3-METHYLPHENYL)-3-(2-PYRIDYL)-2-THIOUREA
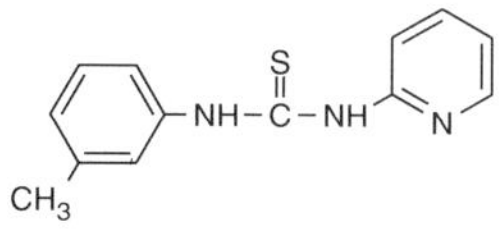
250 mg

S90,485-6
1-(6-METHYL-2-PYRIDYL)-3-(2-NAPHTHYL)-2-THIOUREA
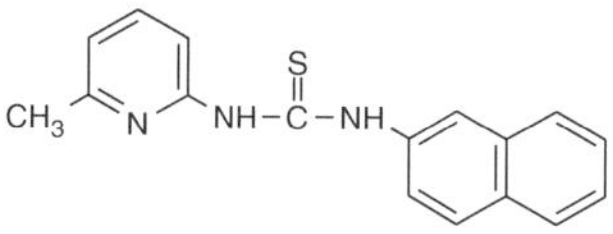
250 mg

S90,974-2
3-(4-CYANOPHENYL)-1-(2-PYRIDYL)-TRIAZENE
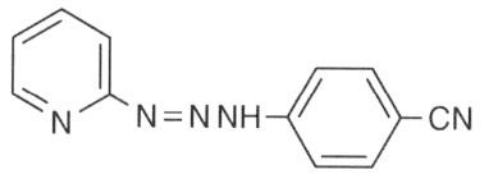
250 mg

S87,536-8
TRANS-2,3-BIS(4-PYRIDYL)-ACRYLONITRILE
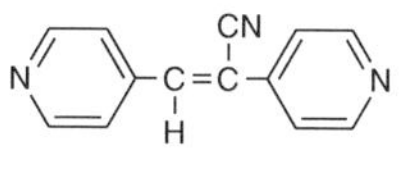
250 mg

S65,666-6
PHENYL 4-PYRIDYLMETHYL SULFONE
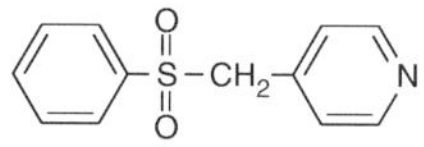
5 gm

S37,052-5
3-PYRIDINESULFONIC ACID, SODIUM SALT
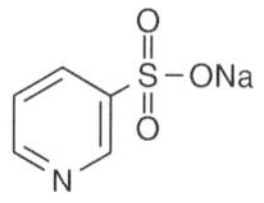
250 mg

S93,533-6
N-(2-PYRIDYL)-2-THIOPHENESULFONAMIDE
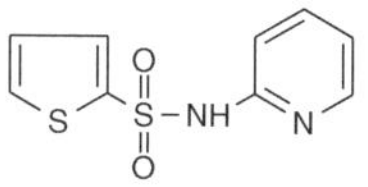
250 mg

S76,103-6
3-(MORPHOLINOSULFONYL)PYRIDINE

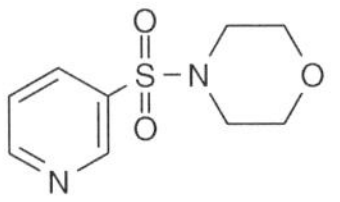

250 mg

S38,919-6
6-CHLORO-N-HEXADECYL-N-METHYL-3-PYRIDINESULFONAMIDE

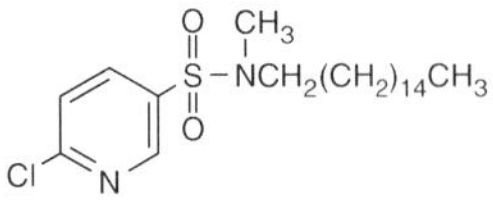

250 mg

S87,494-9
1,2,3-TRIS(4-PYRIDYL)CYCLOPROPANE

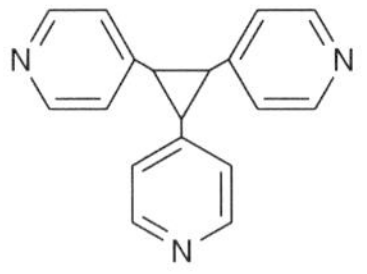

250 mg

S79,876-2
6-[2-(2-PYRIDYL)ETHYL]-1,4-DIOXASPIRO(4.5)DECANE

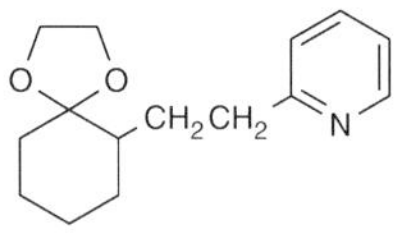

250 mg

S88,466-9
3-(2-METHYL-1,3-DIOXOLAN-2-YL)-PYRIDINE

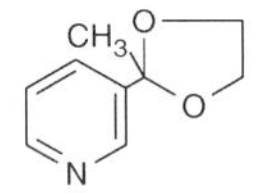

250 mg

S66,868-0
4-PYRIDINECARBOXALDEHYDE ETHYLENE KETAL

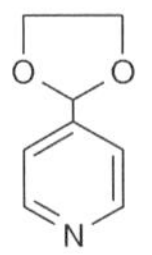

250 mg

S75,860-4
2-METHYL-3-(4-PYRIDYL)-3,3A,4,5-TETRAHYDRO-2H-BENZ(G)INDAZOLE

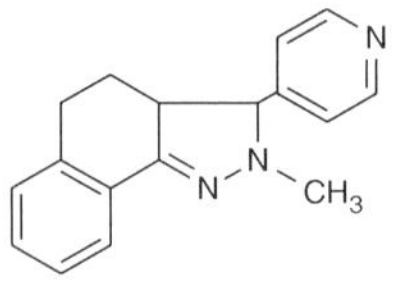

250 mg

S73,309-1
2,3,3A,4,5,6-HEXAHYDRO-3-(4-PYRIDYL)BENZO(6,7)CYCLOHEPTA(1,2-C)-PYRAZOLE

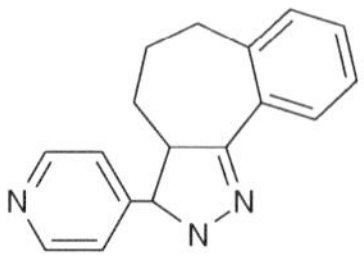

250 mg

S51,448-9
2-[4,5-DI-(2-PYRIDYL)-2H-IMIDAZOL-2-YLIDENEAMINO]BENZIMIDAZOLE

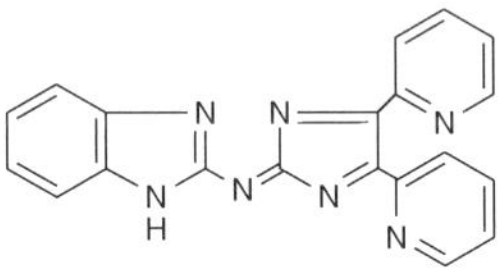

250 mg

S92,067-3
4-BENZYLIDENE-2,6-DIPHENYL-4H-PYRAN

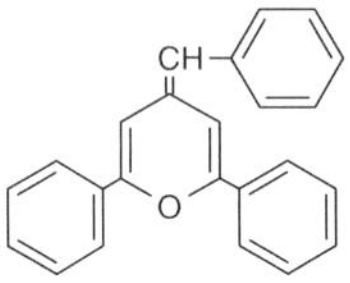

250 mg

S54,626-7
4-(TETRAHYDROPYRAN-4-YL)PYRIDINE

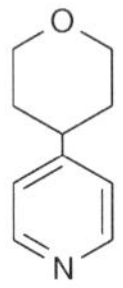

5 gm

S92,274-9
2,6-DI-TERT-BUTYLTHIOPYRYLIUM TETRAFLUOROBORATE

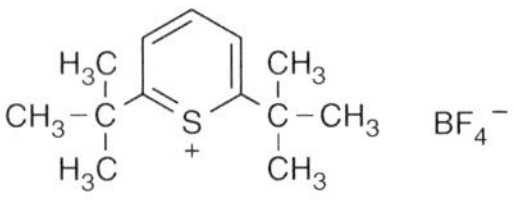

250 mg

S76,159-1
2-(4-PYRIDYL)-1,4,5,6-TETRAHYDROPYRIMIDINE

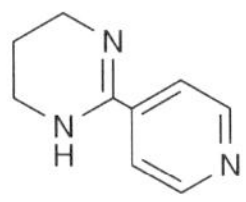

250 mg

S43,197-4
2,3-DIHYDRO-5,6-DI-(2-PYRIDYL)-2-METHYLPYRAZINE

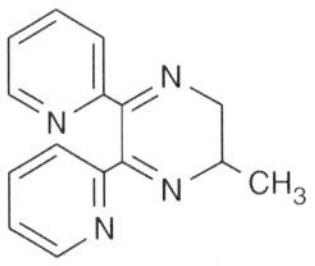

250 mg

S46,294-2
10H-PYRIDO(3,2-B)(1,4)BENZOTHIAZINE

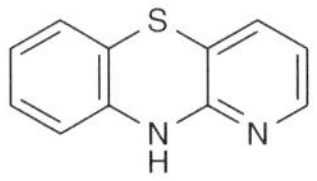

5 gm

S91,969-1
2,6-DIPHENYL-4-METHYLPHOSPHORIN

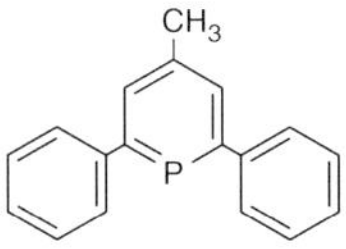

250 mg

S91,986-1
2,4,6-TRIPHENYLPHOSPHORIN

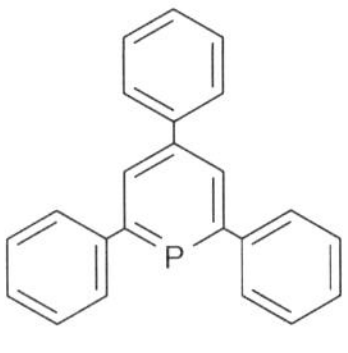

250 mg

S92,089-4
4-TERT-BUTYL-1-DIMETHOXY-2,6-DIPHENYLPHOSPHORIN

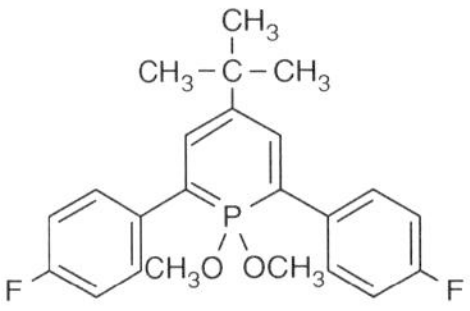

250 mg

S82,639-1
4-BENZYL-2,3,5,6-TETRACHLOROPYRIDINE

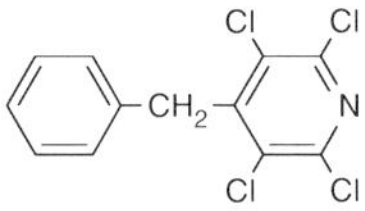

250 mg

S87,413-2
2,6-DIPHENYL-4-METHYLPYRIDINE

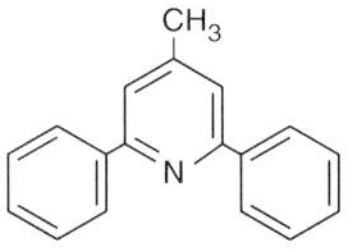

250 mg

S87,403-5
2,6-DIPHENYL-4,4'-DIPYRIDYL

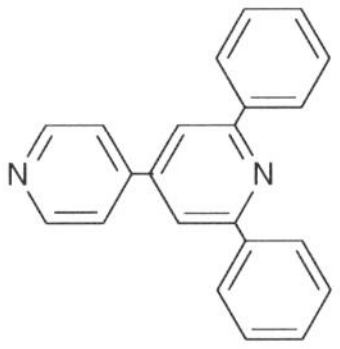

250 mg

S58,111-9
2-PHENYL-4-PICOLINE

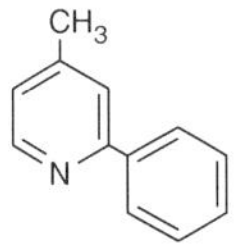

250 mg

S95,485-3
2,4-DISTYRYLPYRIDINE

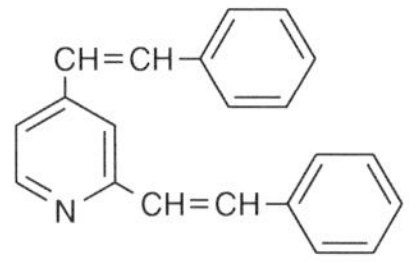

50 mg

S43,984-3
2-STILBAZOLE

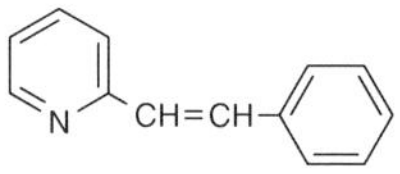

250 mg

S95,477-2
2-METHYL-6-STYRYLPYRIDINE HYDROCHLORIDE

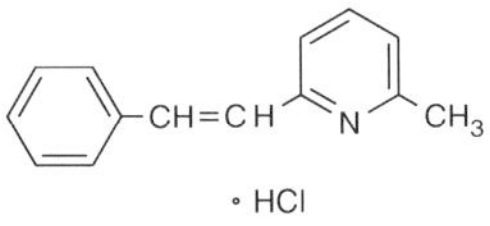

50 mg

S87,404-3
2,6-DISTYRYLPYRIDINE

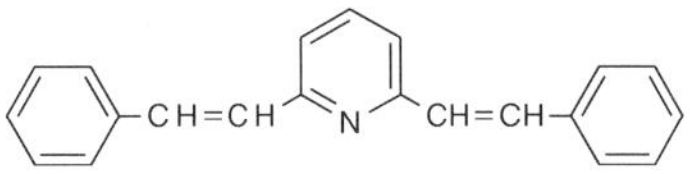

250 mg

S43,537-6
DIPHENYL-3-PYRIDYLMETHANE

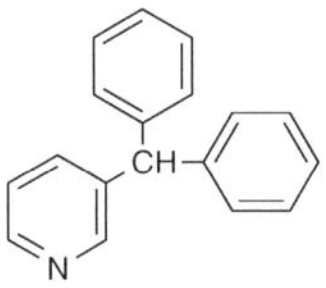

5 gm

S79,977-7
4-(2-PHENYLETHYL)PYRIDINE

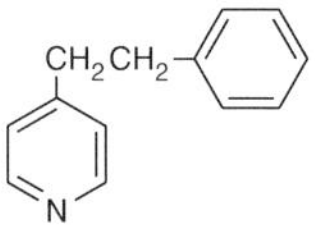

250 mg

S87,032-3
2-PHENYL-5,6,7,8-TETRAHYDROQUINOLINE HYDROCHLORIDE

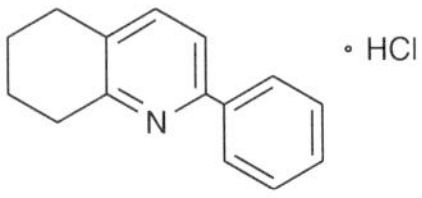

250 mg

S68,900-9
7-PHENYL-5,6,8,9-TETRAHYDRODIBENZ (C,H)ACRIDINE

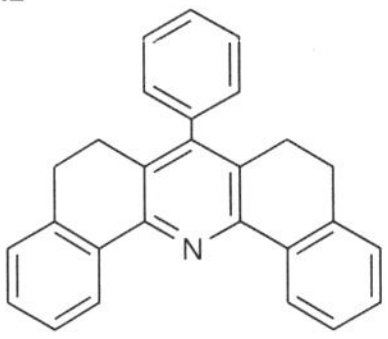

5 gm

S93,191-8
8-BENZYLIDENE-2,4-DIPHENYL-5,6,7,8-TETRAHYDROPHOSPHINOLINE

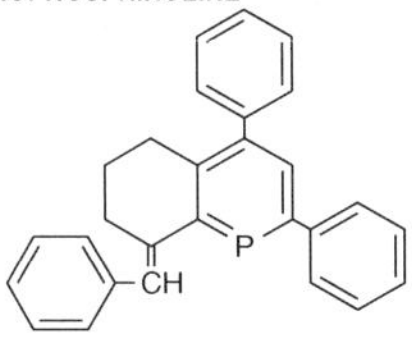

250 mg

S81,720-1
PYRIDINE PICRATE

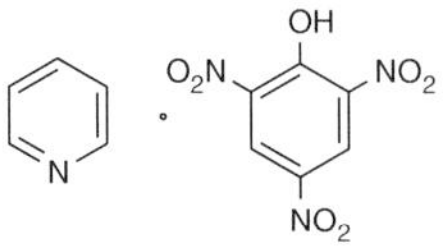

250 mg

S37,700-7
3-CHLORO-2,6-DIAMINOPYRIDINE

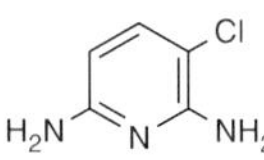

250 mg

S81,226-9
2,3,5,6-TETRACHLOROPYRIDINE

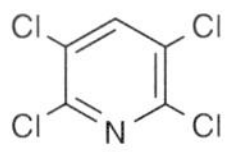

250 mg

S83,005-4
4-BROMO-2,3,5,6-TETRACHLOROPYRIDINE

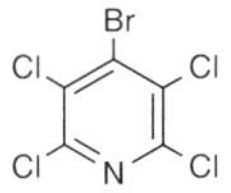

250 mg

S82,825-4
4-IODO-2,3,5,6-TETRACHLOROPYRIDINE

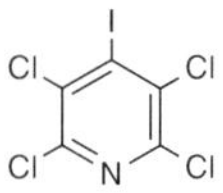

250 mg

S94,091-7
4-CHLORO-2,3,5,6-TETRABROMOPYRIDINE

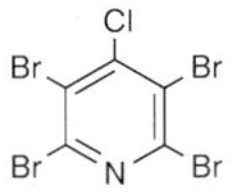

250 mg

S73,954-5
2,4-DICHLORO-3-METHYLPYRIDINE HYDROCHLORIDE

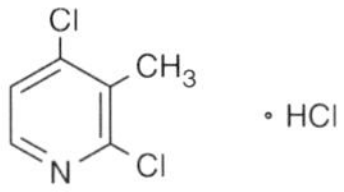

5 gm

S89,454-0
3-METHYL-2,4,5,6-TETRACHLOROPYRIDINE

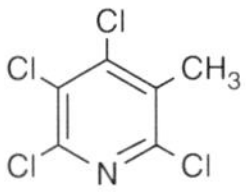

250 mg

S82,826-2
4-ALLYL-2,3,5,6-TETRACHLOROPYRIDINE

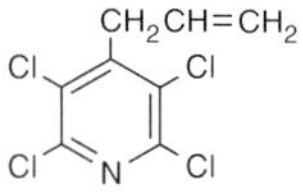

250 mg

S79,832-0
3,4,5,6-TETRACHLORO-N-[(2,3,5,6-TETRAFLUORO-4-PYRIDYL)THIO]-PHTHALIMIDE

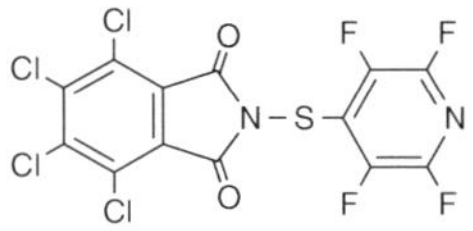

250 mg

S35,532-1
1-(2-PYRIDYL)-2-(3-PYRIDYL)-ETHYLENE

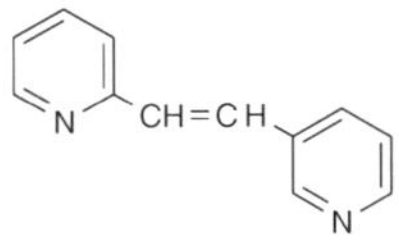

5 gm

S86,346-7
2,4,6-COLLIDINE HYDROCHLORIDE

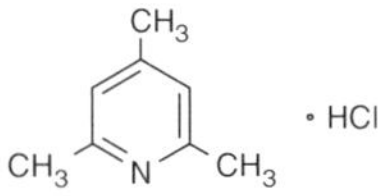

250 mg

S64,349-1
2,6-DI-TERT-BUTYL-4-METHYLPYRIDINE

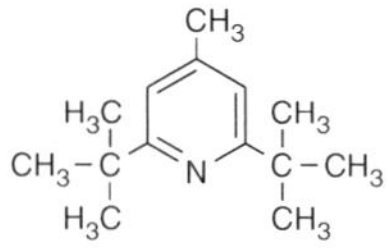

250 mg

S87,522-8
2,2',6,6'-TETRAMETHYL-4,4'-DIPYRIDYL

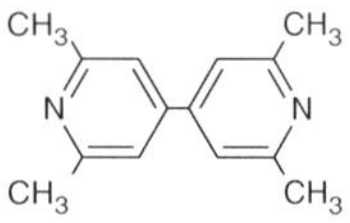

250 mg

S81,717-1
2-PICOLINE PICRATE

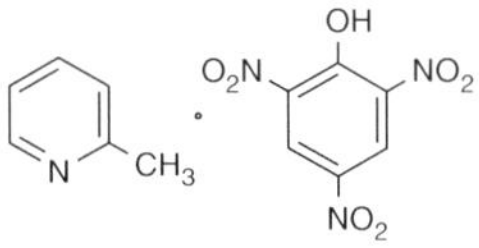

250 mg

S81,716-3
2-VINYLPYRIDINE PICRATE

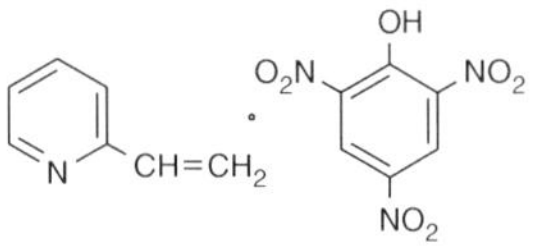

250 mg

S69,006-6
2-TERT-BUTYLPYRIDINE

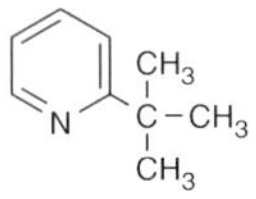

5 gm

S54,625-9
2-(1-ETHYLPROPYL)PYRIDINE

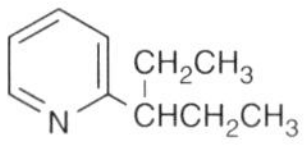

5 gm

S66,812-5
5-(2-PYRIDYL)2,7-NONADIENE

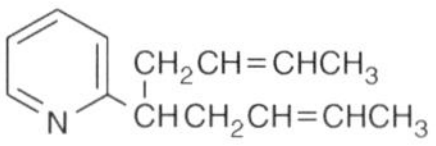

5 gm

S37,176-9
2-(1-BUTYLPENTYL)PYRIDINE

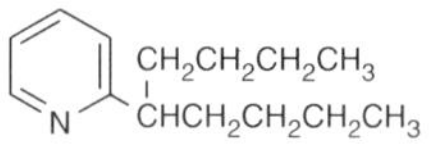

5 gm

S64,176-6
3-[2-(2-PYRIDYL)ETHYL]INDOLE

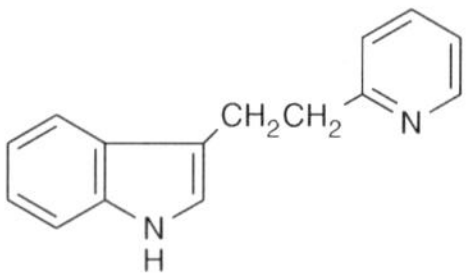

5 gm

S58,003-1
6-ETHYL-2-PICOLINE

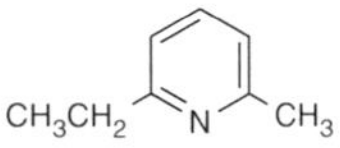

5 gm

S88,809-5
6,6'-DIMETHYL-2,2'-DIPYRIDYL

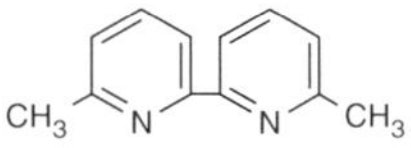

250 mg

S52,049-7
1-ALLYL-2-[2-(3-INDOLYL)VINYL]-PYRIDINIUM IODIDE

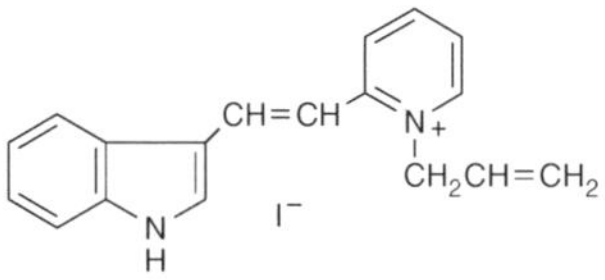

250 mg

S37,056-8
1-(3-PYRIDYL)-2-(4-PYRIDYL)-ETHYLENE

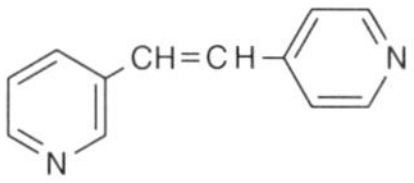

5 gm

S35,531-3
1-(3-PYRIDYL)-2-(4-PYRIDYL)ETHANE

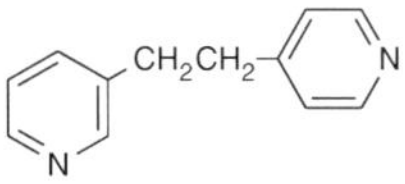

250 mg

S36,891-1
3-ETHYL-4-METHYLPYRIDINE

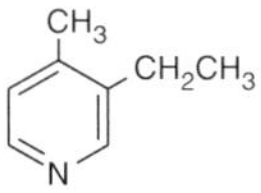

5 gm

S59,655-8
3-IODOPYRIDINE

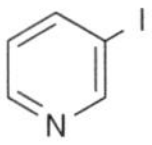

250 mg

S57,308-6
4-PROPYLPYRIDINE

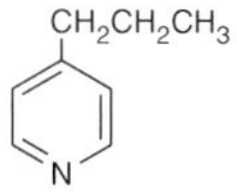

5 gm

S88,154-6
4-STYRYLPYRIDINE

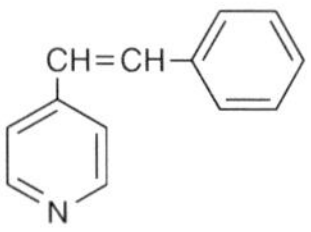

250 mg

S64,177-4
3-[2-(4-PYRIDYL)ETHYL]INDOLE

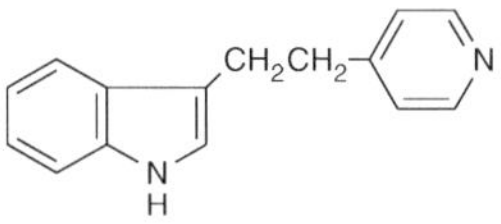

250 mg

S65,726-3
3,4-DIHYDRO-3'-(4-PYRIDYL)-1-OXOSPIRO[NAPHTHALENE-2(1H)-2'-OXIRANE]

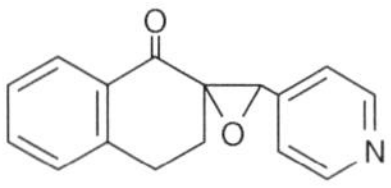

5 gm

S87,952-5
5-ACETYL-3,4-BIS(4-CHLOROPHENYL)-6-(1H)-PYRIDAZINONE

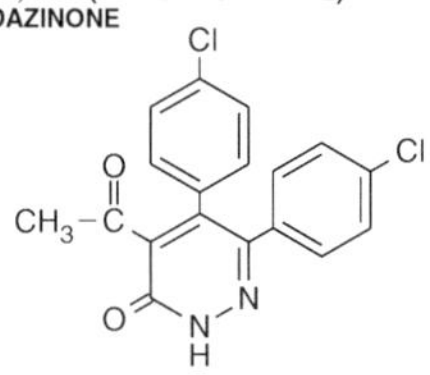

5 gm

S34,407-9
4-CHLORO-3,6-DIHYDROXYPYRIDAZINE

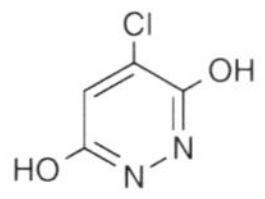

5 gm

S93,711-8
3-CHLORO-1,6-DIAMINOPYRIDAZINIUM TOSYLATE

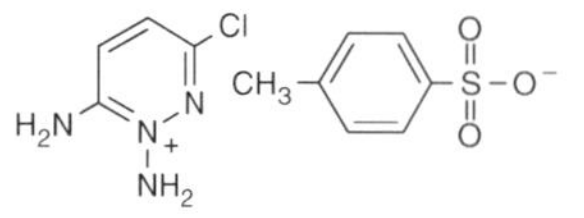

250 mg

S73,236-2
4-(DIALLYLAMINO)-3,6-DICHLOROPYRIDAZINE

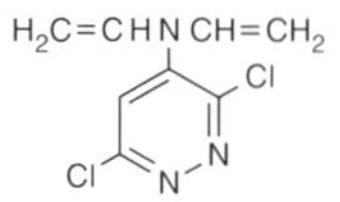

250 mg

S90,947-5
1,6-DIHYDRO-6-OXO-3-PYRIDAZINECARBOXYLIC ACID

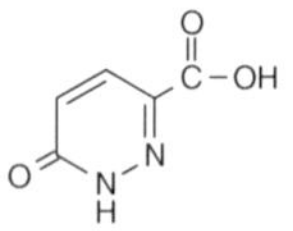

250 mg

S93,261-2
ETHYL 1,6-DIHYDRO-6-OXO-3-PYRIDAZINECARBOXYLATE

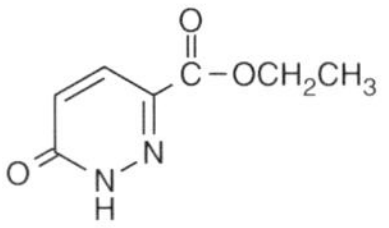

250 mg

S90,948-3
6-CHLORO-3-PYRIDAZINECARBOXYLIC ACID

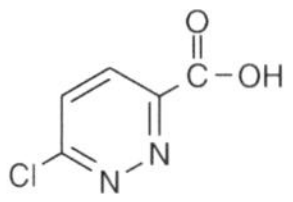

250 mg

S74,490-5
5-PHENYL-4-PYRIDAZINECARBOXYLIC ACID

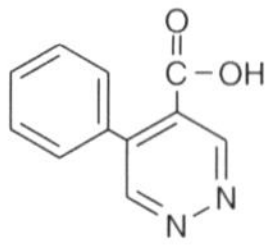

250 mg

S74,492-1
METHYL 5-PHENYL-4-PYRIDAZINECARBOXYLATE

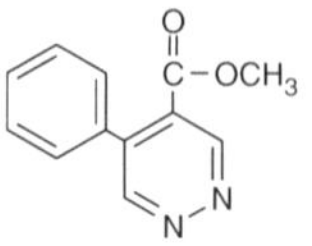

250 mg

S85,988-5
6-CHLORO-3(2H)-PYRIDAZINONE

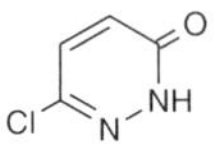

250 mg

S92,743-0
6-PHENYL-3(2H)-PYRIDAZINONE

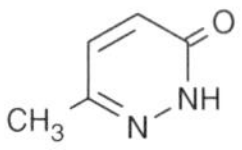

250 mg

S41,253-8
4,5-DIBROMO-3(2H)-PYRIDAZINONE

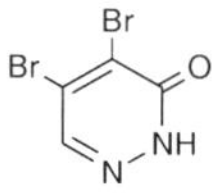

5 gm

S54,086-2
1,2-DIHYDRO-4-METHYL-3,6-PYRIDAZINEDIONE

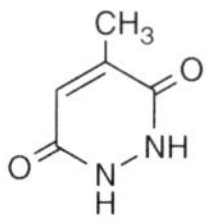

250 mg

S76,218-0
3-CHLORO-10H-PYRIDAZINO(4,3-B)(1,4)-BENZOTHIAZINE

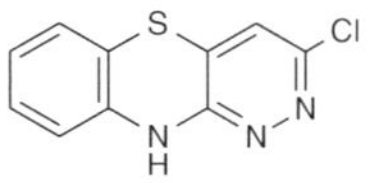

250 mg

S74,245-7
3-CHLORO-6-METHYL-4-PHENYLPYRIDAZINE

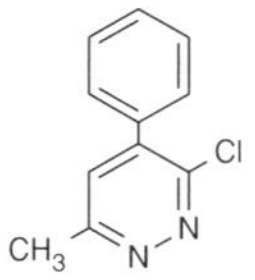

5 gm

S88,697-1
3,6-DIPHENYLPYRIDAZINE

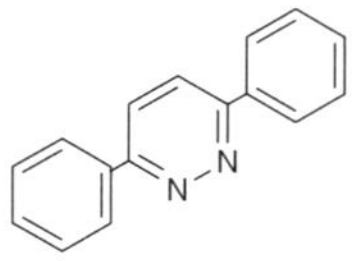

250 mg

S76,699-2
3-(3,6-DICHLORO-4-PYRIDAZINYL)-3-AZABICYCLO[3.2.2]NONANE

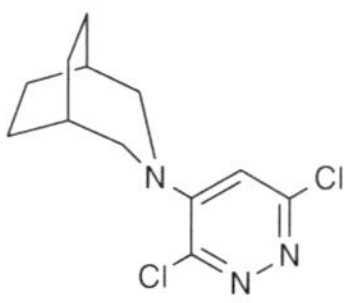

250 mg

S37,506-3
5-[BIS(2-CHLOROETHYL)AMINO]-URACIL

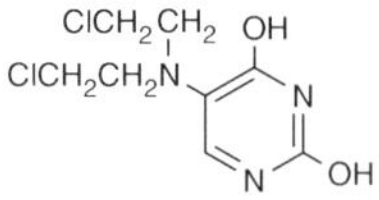

250 mg

S68,566-6
6-CHLOROMETHYL-4-HYDROXY-2-ISOPROPYLPYRIMIDINE

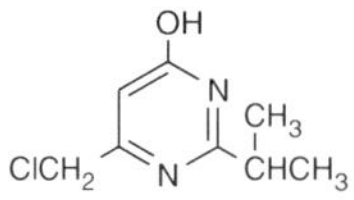

5 gm

S90,456-2
2-CHLORO-N-(2-PYRIMIDINYL)ACETAMIDE

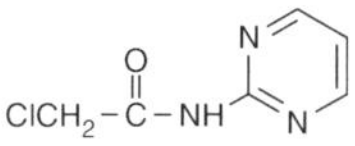

250 mg

S79,610-7
1-(4-CHLORO-3-NITROPHENYL)-3-[4-HYDROXY-6-(TRIFLUOROMETHYL)-2-PYRIMIDINYL]UREA

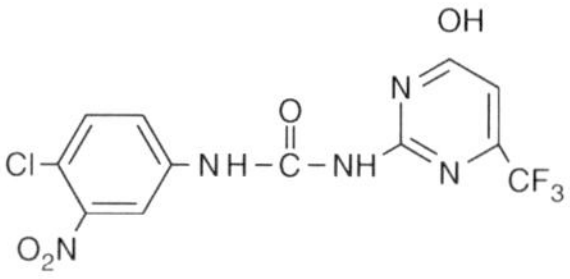

250 mg

S76,995-9
1-[2-CHLORO-5-(TRIFLUOROMETHYL)-PHENYL]-3-(2-PYRIMIDINYL)UREA

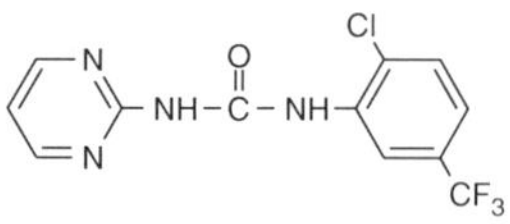

250 mg

S68,882-7
2-AMINO-6-HYDROXY-5-[(1-HYDROXY-2,2,2-TRICHLORO)ETHYL]-4-METHYLPYRIMIDINE

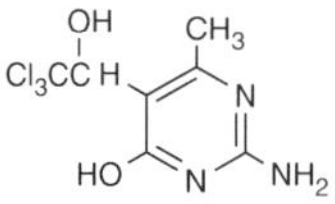

5 gm

S93,893-9
2-AMINO-4-(2-THIENYL)-6-(TRIFLUOROMETHYL)PYRIMIDINE

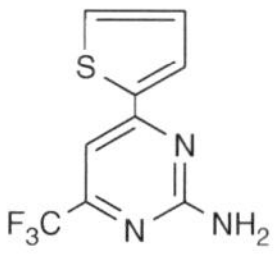

250 mg

S93,888-2
2-AMINO-4-METHYL-6-(TRIFLUOROMETHYL)PYRIMIDINE

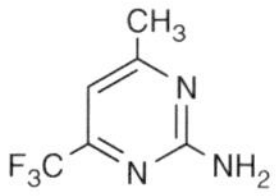

250 mg

S93,886-6
2-AMINO-4-PHENYL-6-(TRIFLUOROMETHYL)PYRIMIDINE

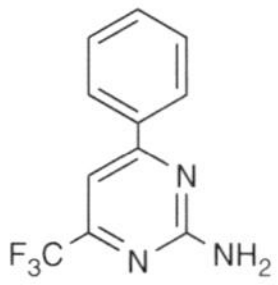

250 mg

S93,889-0
2-(ETHYLAMINO)-4-METHYL-6-(TRIFLUOROMETHYL)PYRIMIDINE

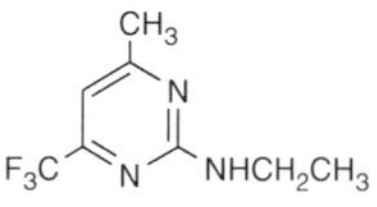

250 mg

S93,892-0
2-(DIMETHYLAMINO)-4-METHYL-6-(TRIFLUOROMETHYL)PYRIMIDINE

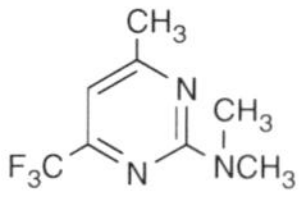

250 mg

S93,887-4
2-(DIMETHYLAMINO)-4-PHENYL-6-(TRIFLUOROMETHYL)PYRIMIDINE

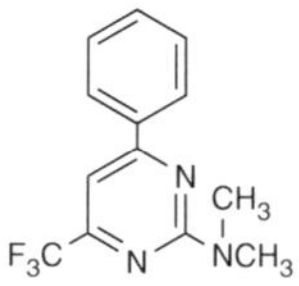

250 mg

S91,174-7
1-(5-CHLORO-2-METHOXYPHENYL)-3-(2-PYRIMIDINYL)UREA

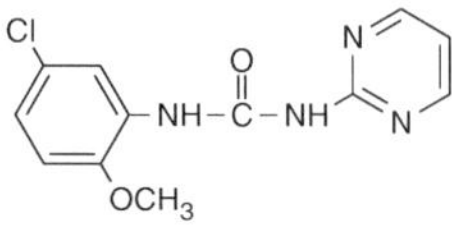

250 mg

S77,970-9
1-(6-CHLORO-4-PYRIMIDINYL)-3-(2,3-DICHLOROPHENYL)UREA

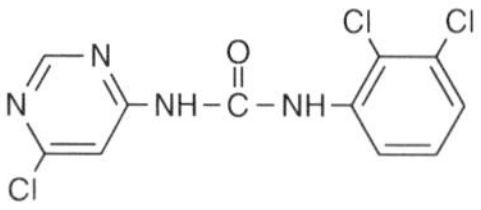

250 mg

S68,200-4
4,6-BIS(4-CHLOROPHENYL)-2-CHLOROPYRIMIDINE

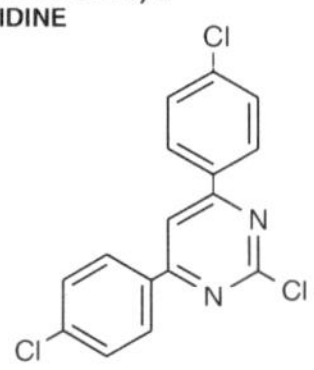

250 mg

S90,932-7
2-(4-CHLOROBENZOYL)PYRIMIDINE

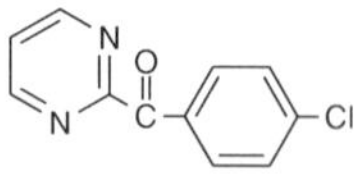

250 mg

S74,355-0
N-(2-PYRIMIDINYL)-3,4,5,6-TETRACHLOROPHTHALAMIC ACID

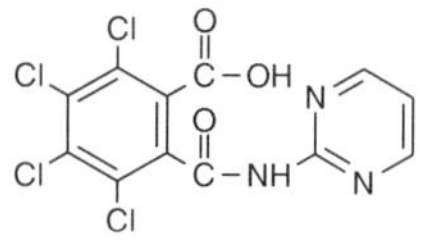

250 mg

S71,791-6
1-(3-BROMOPHENYL)-3-(2-PYRIMIDINYL)UREA

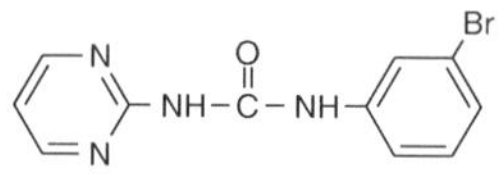

250 mg

S71,552-2
1-(2-FLUOROPHENYL)-3-(2-PYRIMIDINYL)UREA

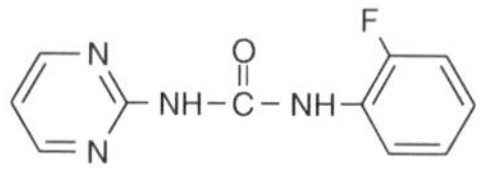

250 mg

S91,232-8
1-(4-FLUOROPHENYL)-3-(2-PYRIMIDINYL)UREA

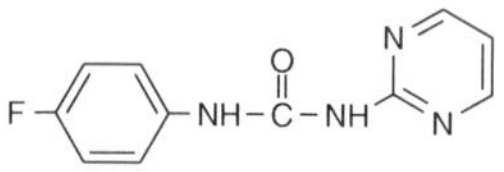

250 mg

S91,086-4
1-(2,3-DICHLOROPHENYL)-3-(2-PYRIMIDINYL)UREA

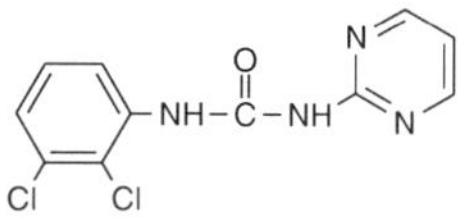

250 mg

S72,760-1
1-(2,5-DICHLOROPHENYL)-3-(2-PYRIMIDINYL)UREA

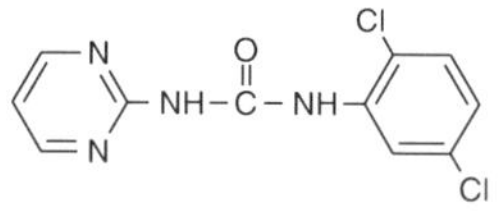

5 gm

S91,246-8
1-(3-CHLORO-4-METHYLPHENYL)-3-(2-PYRIMIDINYL)UREA

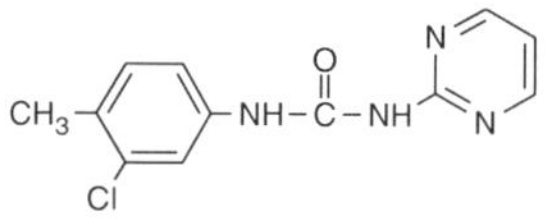

250 mg

S88,296-8
2,4-BIS(4-FLUOROPHENYL)-7-METHYL-5,6,7,8-TETRAHYDROQUINAZOLINE

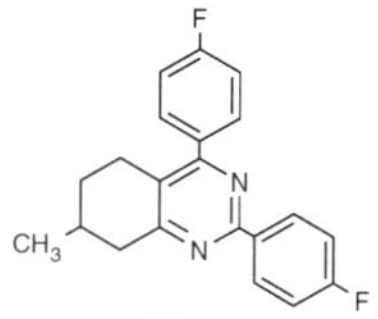

250 mg

S88,590-8
2-AMINO-6-METHYL-4,5-PYRIMIDINEDIOL

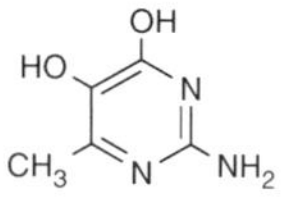

250 mg

S93,264-7
5-PHENYL-2-(2-PYRIDYL)-4,6-PYRIMIDINEDIOL

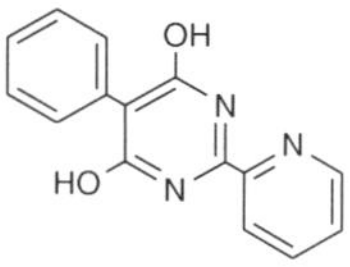

250 mg

S37,655-8
ETHYL 4-HYDROXY-2-[N-(2-HYDROXYETHYL)-N-METHYLAMINO]-5-PYRIMIDINECARBOXYLATE

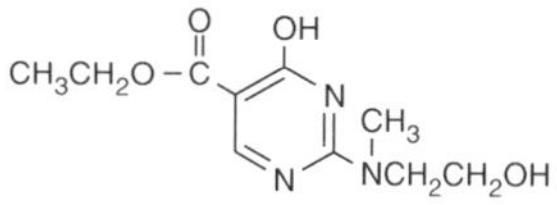

250 mg

S78,400-1
4,6-DIHYDROXY-2-MERCAPTO-5-NITROSOPYRIMIDINE

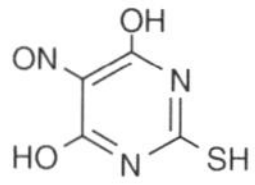

250 mg

S92,837-2
6-AMINO-5-(METHYLAMINO)-2,4-PYRIMIDINEDIOL

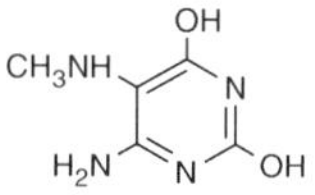

250 mg

S78,129-0
4-AMINO-2,6-DIHYDROXY-5-NITROPYRIMIDINE

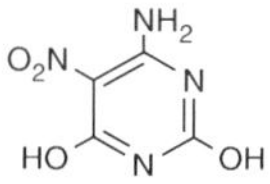

250 mg

S76,594-5
4,5-DIAMINO-2,6-DIHYDROXYPYRIMIDINE SULFATE

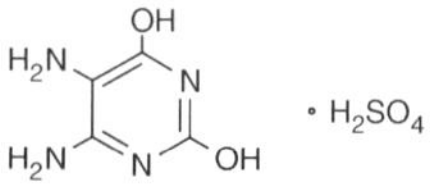

5 gm

S93,266-3
4-AMINO-5-BROMO-2,6-DIHYDROXYPYRIMIDINE

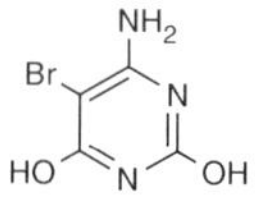

250 mg

S78,364-1
2-AMINO-4,6-DIHYDROXY-5-NITROSOPYRIMIDINE

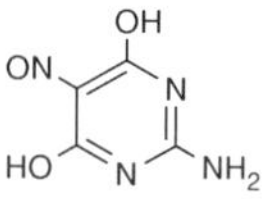

5 gm

S78,886-4
2,5-DIAMINO-4,6-DIHYDROXYPYRIMIDINE SULFATE

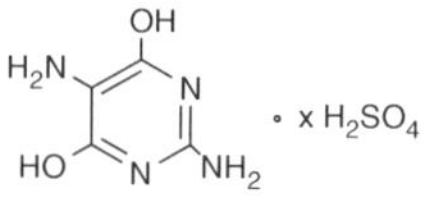

250 mg

S78,343-9
5-AMINO-2,4-DIHYDROXY-6-METHYLPYRIMIDINE

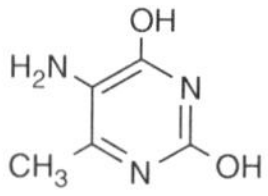

5 gm

S78,715-9
4-AMINO-2-HYDROXY-5-(HYDROXYMETHYL)-PYRIMIDINE

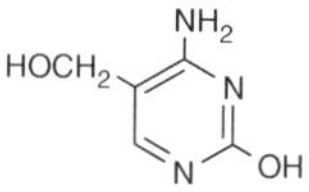

250 mg

S78,345-5
2-AMINO-4,6-DIHYDROXY-5-ETHYLPYRIMIDINE

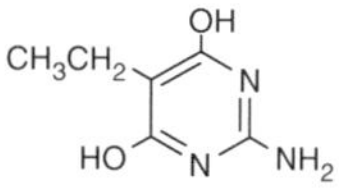

5 gm

S78,384-6
ETHYL 2,4-DIHYDROXY-5-NITRO-6-PYRIMIDINECARBOXYLATE

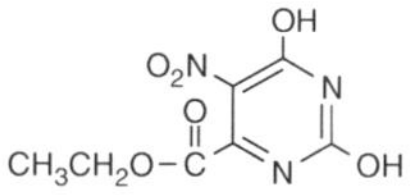

5 gm

S77,676-9
5-CHLORO-2,6-DIHYDROXY-4-PYRIMIDINECARBOXYLIC ACID

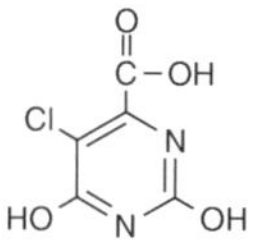

250 mg

S76,593-7
2,4-DIHYDROXYPYRIMIDINE

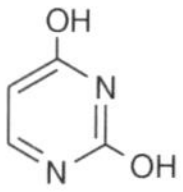

5 gm

S84,367-9
5-METHOXY-2-THIOURACIL

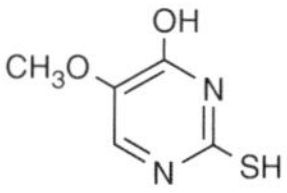

250 mg

S68,561-5
4-HYDROXY-2-ISOPROPYL-6-METHOXYMETHYLPYRIMIDINE

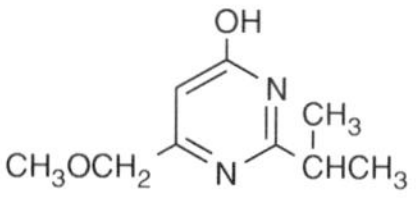

5 gm

S39,092-5
ETHYL 4-HYDROXY-2-(2-METHOXYETHYLAMINO)-5-PYRIMIDINECARBOXYLATE

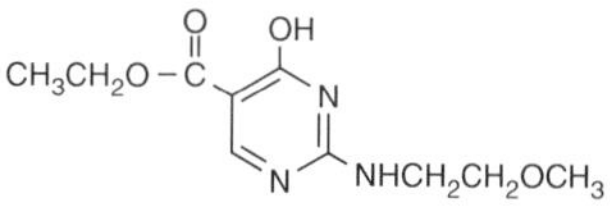

250 mg

S59,179-3
6-METHOXYMETHYL-2-METHYL-4-PYRIMIDINOL

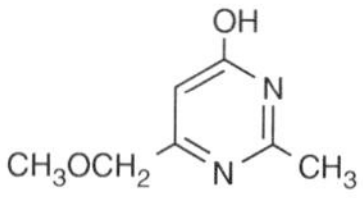

250 mg

S59,165-3
2-ISOPROPYL-6-PHENOXYMETHYL-4-PYRIMIDINOL

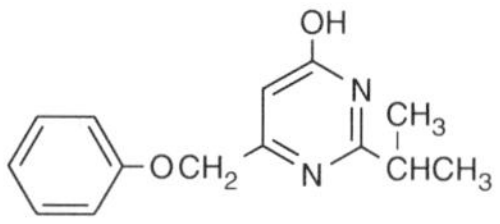

5 gm

S78,292-0
ETHYL 4-HYDROXY-2-MERCAPTO-5-PYRIMIDINECARBOXYLATE, SODIUM SALT

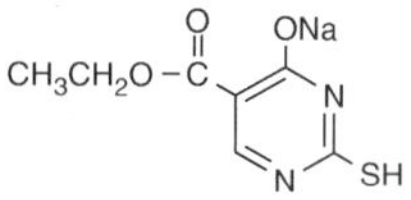

5 gm

S34,919-4
ETHYL 4-HYDROXY-2-MERCAPTOPYRIMIDINE-5-CARBOXYLATE

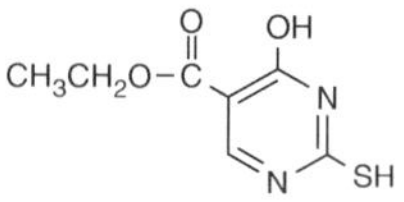

5 gm

S78,422-2
ETHYL 4-HYDROXY-2-(METHYLTHIO)-5-PYRIMIDINECARBOXYLATE, SODIUM SALT

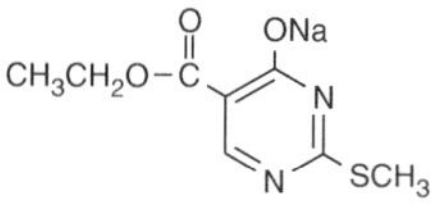

250 mg

S45,095-2
5-BROMO-2',3'-ISOPROPYLIDENEURIDINE
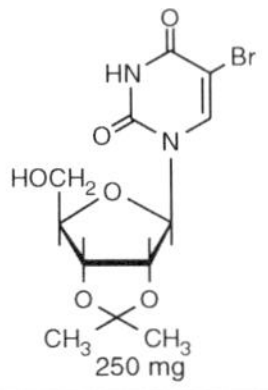
250 mg

S79,006-0
4-AMINO-2-HYDROXY-6-MERCAPTOPYRIMIDINE
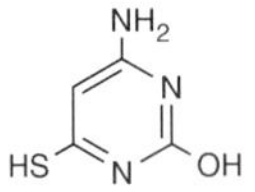
250 mg

S40,642-2
4-AMINO-6-HYDROXY-2-MERCAPTO-5-NITROSOPYRIMIDINE, AMMONIUM SALT
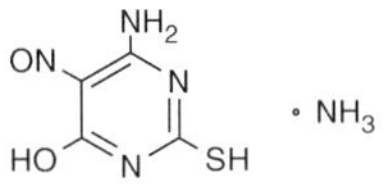
5 gm

S40,914-6
2-MERCAPTOOROTIC ACID
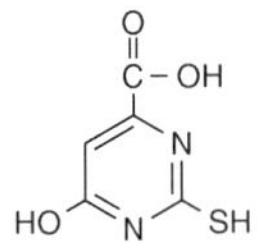
5 gm

S34,302-1
5-BUTYL-4-HYDROXY-2-MERCAPTO-6-METHYLPYRIMIDINE
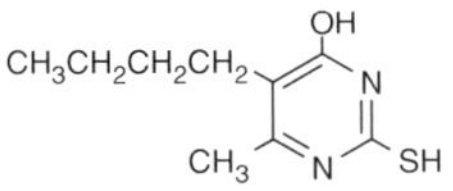
5 gm

S34,185-1
4-BENZYL-6-HYDROXY-2-MERCAPTOPYRIMIDINE
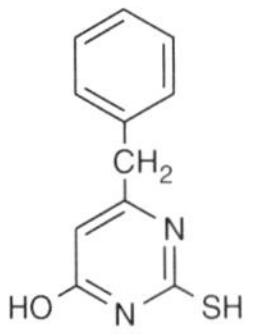
5 gm

S67,246-7
2-(CARBOXYMETHYLTHIO)-5,6-DIAMINO-4-HYDROXYPYRIMIDINE
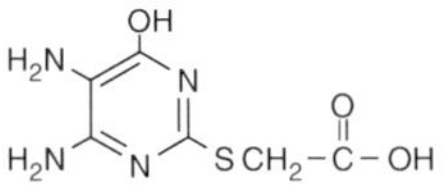
5 gm

S67,004-9
5,6-DIAMINO-4-HYDROXY-2-METHYLTHIOPYRIMIDINE
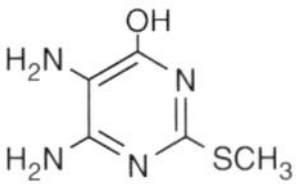
5 gm

S78,569-5
4-AMINO-2-(ETHYLTHIO)-5-(HYDROXYMETHYL)PYRIMIDINE
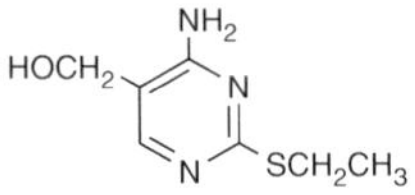
250 mg

S40,643-0
4-AMINO-6-HYDROXY-2-(METHYLTHIO)-PYRIMIDINE
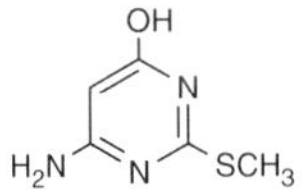
5 gm

S40,641-4
6-AMINO-2-(ETHYLTHIO)-4-PYRIMIDINOL
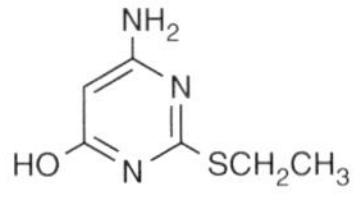
5 gm

S36,251-4
6-HYDROXY-2-(METHYLTHIO)-4-PYRIMIDINECARBOXYLIC ACID
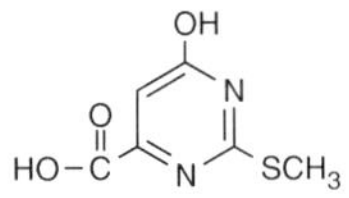
5 gm

S38,034-2
ETHYL 4-HYDROXY-2-[2-(METHYLTHIO)-ETHYLAMINO]-5-PYRIMIDINECARBOXYLATE
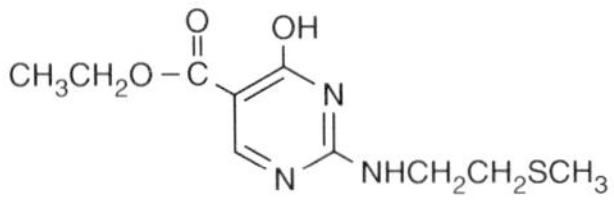
250 mg

S37,652-3
ETHYL 4-HYDROXY-2-METHYLTHIO-5-PYRIMIDINECARBOXYLATE
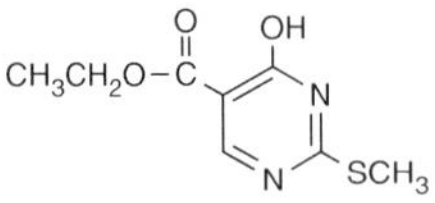
5 gm

S44,616-5
ETHYL 2-(ETHYLTHIO)-4-HYDROXY-5-PYRIMIDINECARBOXYLATE
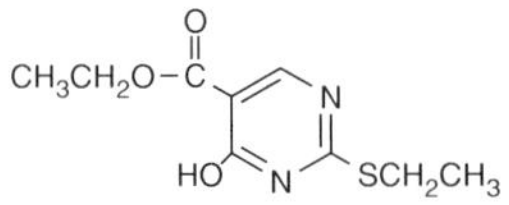
5 gm

S85,916-8
4-(METHYLTHIO)-2-PYRIMIDINOL, SODIUM SALT
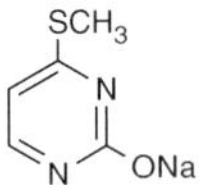
250 mg

S44,004-3
2-METHYLTHIO-4-PYRIMIDINOL
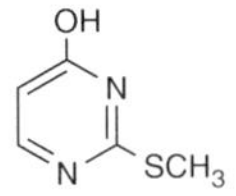
5 gm

S77,918-0
2-(ETHYLTHIO)-4-HYDROXY-5-METHYLPYRIMIDINE
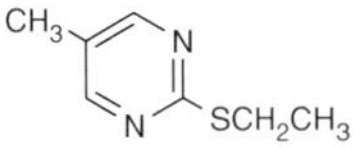
250 mg

S38,666-9
5,6-DIMETHYL-2-ETHYLTHIO-4-PYRIMIDINOL
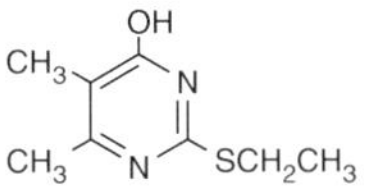
250 mg

S78,412-5
2-HYDROXY-4-METHYL-6-(METHYLTHIO)-PYRIMIDINE
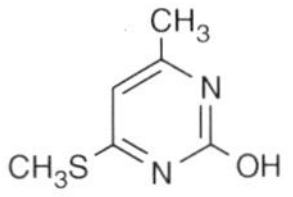
250 mg

S35,778-2
THIAMINE PROPYL DISULFIDE
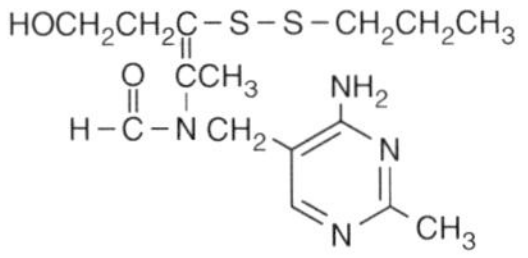
250 mg

S54,089-7
2,5,6-TRIAMINO-4-PYRIMIDINOL HYDROCHLORIDE
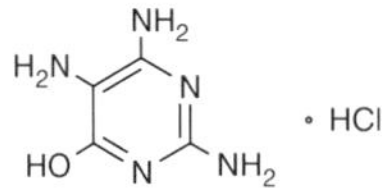
5 gm

S92,736-8
5-AMINO-4-HYDROXY-6-(METHYLAMINO)-PYRIMIDINE
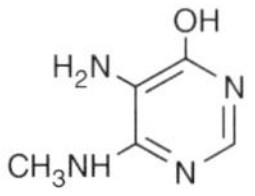
250 mg

S45,450-8
5,6-DIAMINO-2,4-DIHYDROXYPYRIMIDINE MONOHYDROCHLORIDE
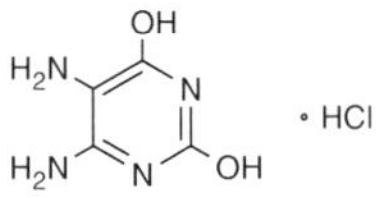
250 mg

S78,342-0
4-AMINO-5-FORMAMIDO-6-HYDROXYPYRIMIDINE
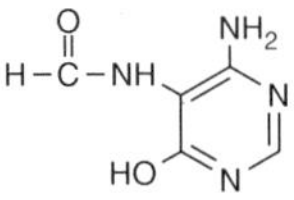
5 gm

S76,573-2
4,6-DIAMINO-2-HYDROXY-5-NITROSOPYRIMIDINE
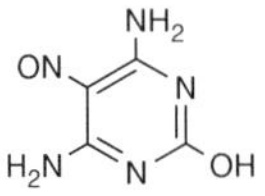
5 gm

S78,281-5
4,5-DIAMINO-2-HYDROXYPYRIMIDINE SULFATE
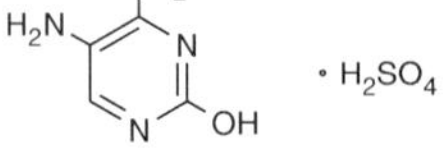
5 gm

S41,571-5
2,6-DIAMINO-5-NITROSO-4-PYRIMIDINOL
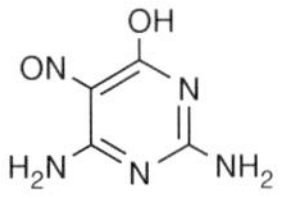
5 gm

S78,713-2
4,6-DIAMINO-2-HYDROXYPYRIMIDINE HYDROCHLORIDE
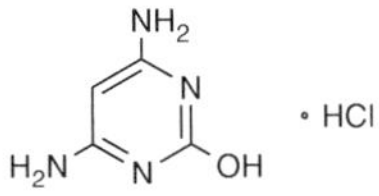
250 mg

S53,371-8
5-ACETAMIDO-6-AMINO-2-(2-BENZIMIDAZOLYLAMINO)-4-PYRIMIDINOL
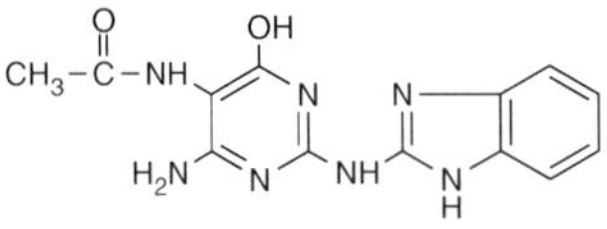
5 gm

S37,655-8
6-AMINO-2-(DIMETHYLAMINO)-4-HYDROXYPYRIMIDINE
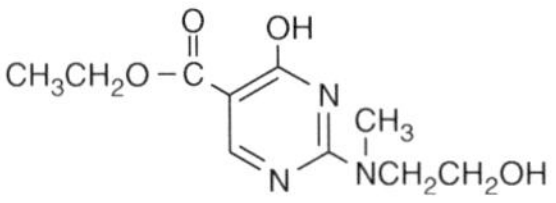
250 mg

S35,754-5
4-AMINO-2-CYANOAMINO-6-HYDROXYPYRIMIDINE
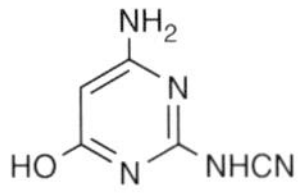
5 gm

S88,591-6
2,4-DIAMINO-6-METHYL-5-PYRIMIDINOL

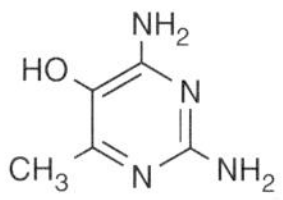

250 mg

S78,366-8
2-AMINO-6-HYDROXY-4-PYRIMIDINECARBOXYLIC ACID

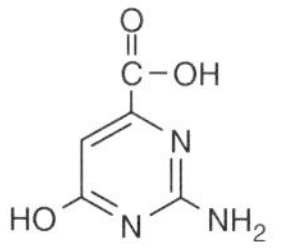

5 gm

S76,591-0
ETHYL 4-AMINO-2-HYDROXY-5-PYRIMIDINECARBOXYLATE

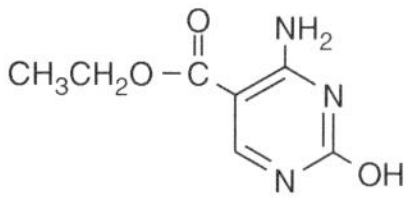

5 gm

S35,771-5
2-AMINO-4-HYDROXYPYRIMIDINE-6-CARBOXYLIC ACID

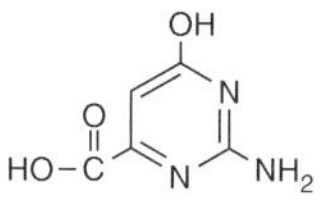

5 gm

S42,347-5
ETHYL 2-AMINO-4-HYDROXY-5-PYRIMIDINECARBOXYLATE

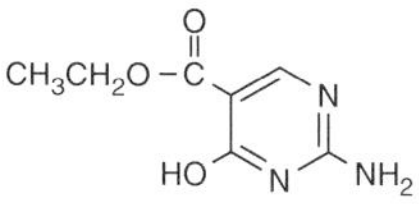

5 gm

S72,049-6
3-(4-AMINO-2-METHYL-5-PYRIMIDYL)-4,5-DIMETHYL-5-(2-HYDROXYETHYL)-2-THIAZOLIDINONE

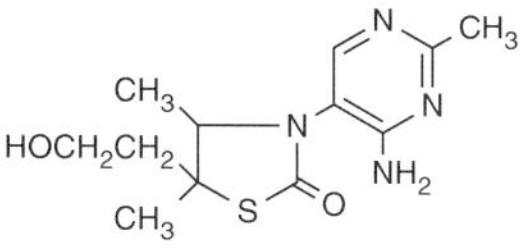

5 gm

S79,233-0
2-AMINO-6,7-DIHYDRO-4-HYDROXY-5H-CYCLOPENTAPYRIMIDINE

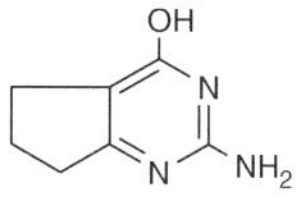

250 mg

S36,644-7
2-AMINO-4-PYRIMIDINOL

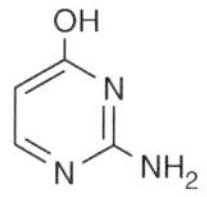

250 mg

S85,979-6
2-AMINO-4-HYDROXYPYRIMIDINE SULFATE

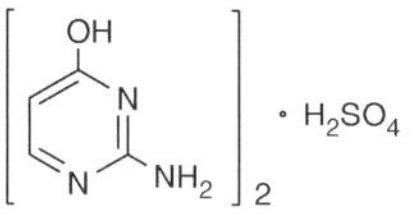

250 mg

S34,250-5
5-BROMO-2,4-DIAMINO-6-HYDROXYPYRIMIDINE

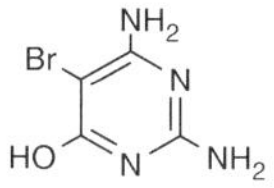

5 gm

S35,769-3
4-AMINO-6-HYDROXY-2-METHYLPYRIMIDINE

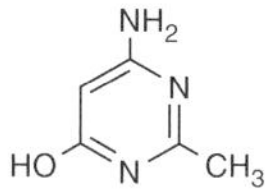

5 gm

S35,758-8
2-AMINO-4,6-DIHYDROXY-5-METHYLPYRIMIDINE

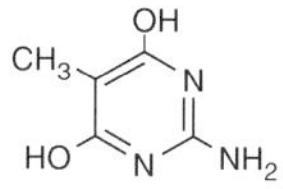

5 gm

S43,773-5
2-AMINO-5,6-DIMETHYL-4-PYRIMIDINOL

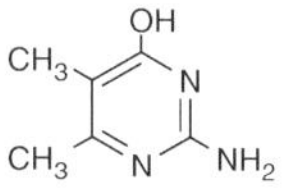

5 gm

S93,174-8
6-AMINO-2-(4-PYRIDYL)-4-PYRIMIDINOL

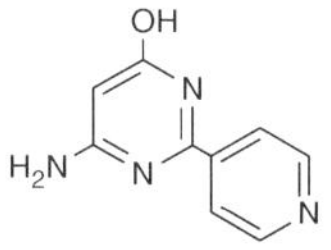

250 mg

S59,998-0
6-(DIETHYLAMINOMETHYL)-4-HYDROXY-2-METHYLPYRIMIDINE

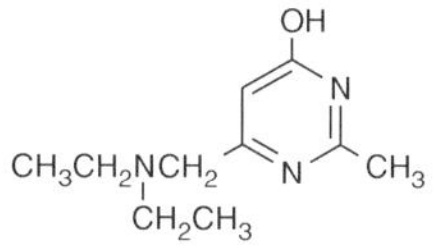

250 mg

S92,836-4
6-(METHYLAMINO)-5-NITRO-4-PYRIMIDINOL

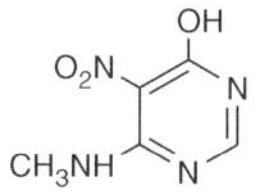

250 mg

S36,829-6
6-METHYL-5-NITROURACIL

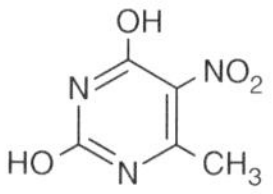

5 gm

S54,239-3
2-(5-NITRO-2-BENZIMIDAZOLYLAMINO)-6-PHENYL-4-PYRIMIDINOL

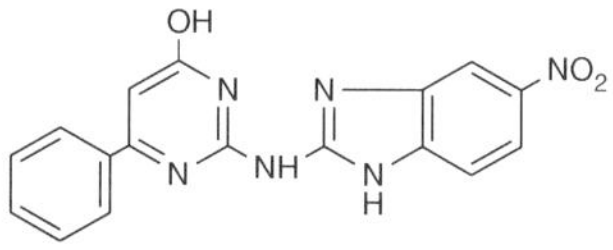

250 mg

S60,614-6
2-HEXADECYLAMINO-4-HYDROXY-6-METHYLPYRIMIDINE

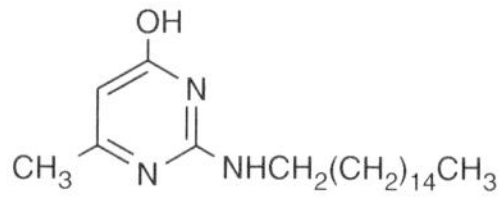

250 mg

S41,277-5
4-HYDROXY-2-(ISOPROPYLIDENEHYDRAZINO)-6-METHYLPYRIMIDINE

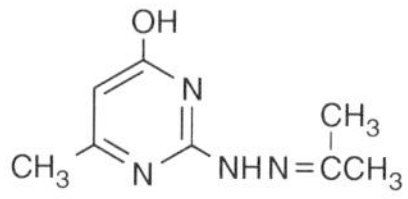

5 gm

S60,315-5
OROTIC ACID, SODIUM SALT

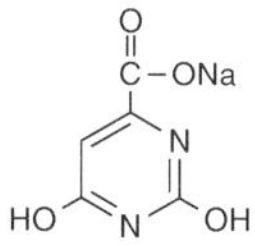

5 gm

S53,480-3
4-HYDROXY-5-PYRIMIDINECARBONITRILE

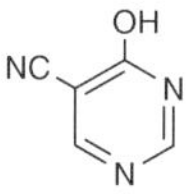

250 mg

S61,738-5
5-CYANO-2,6-DIHYDROXY-4-METHYLPYRIMIDINE

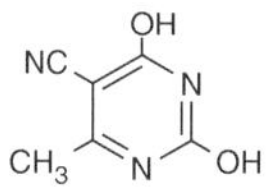

5 gm

S60,045-8
2-(3,3-DIMETHYLGUANIDINO)-6-HYDROXY-4-METHYLPYRIMIDINE

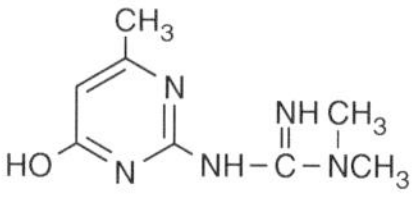

5 gm

S37,760-0
ETHYL 2-(1-ETHYLPROPYLAMINO)-4-HYDROXY-5-PYRIMIDINECARBOXYLATE

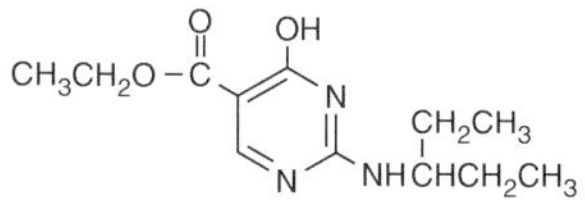

250 mg

S42,012-3
4-HYDROXY-2-PROPYLAMINO-5-PYRIMIDINECARBOXAMIDE

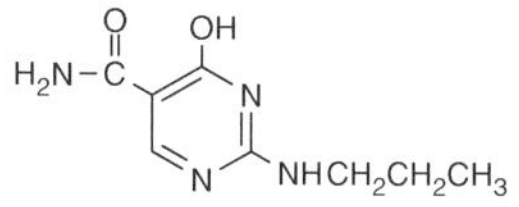

250 mg

S34,132-0
2-N-ANILINO-4-HYDROXY-6-METHYLPYRIMIDINE

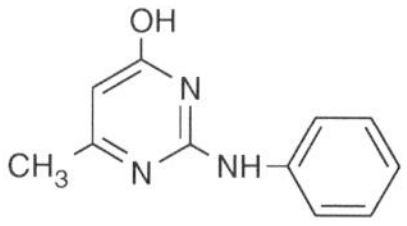

5 gm

S37,784-8
ETHYL 2-DIETHYLAMINO-4-HYDROXY-5-PYRIMIDINECARBOXYLATE

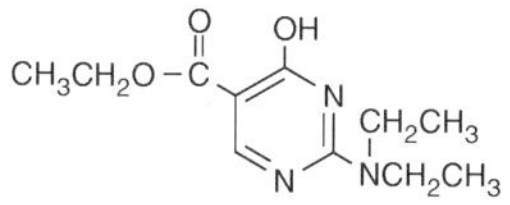

250 mg

S84,717-8
4-HYDROXY-6-METHYL-2-PIPERIDINOPYRIMIDINE

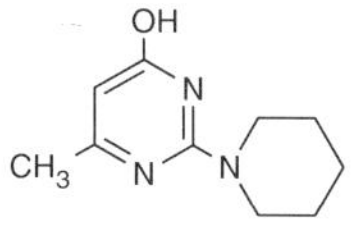

250 mg

S84,384-9
2-(DIMETHYLAMINO)-4-HYDROXY-6-METHYLPYRIMIDINE

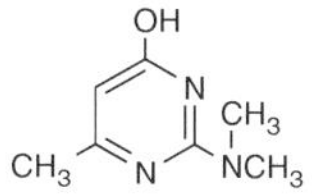

250 mg

S36,253-0
4-HYDROXY-6-METHYL-2-[2-(3-OXO-SEC-BUTYLIDENE)HYDRAZINO]PYRIMIDINE

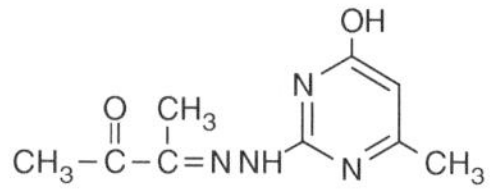

5 gm

S35,989-0
2-(CYCLOPENTYLIDENEHYDRAZINO)-4-HYDROXY-6-METHYLPYRIMIDINE

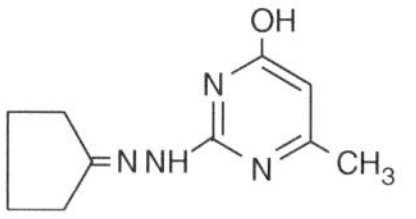

5 gm

S35,099-0
4-HYDROXY-6-METHYL-2-(2-METHYLCYCLOHEXYLIDENEHYDRAZINO)-PYRIMIDINE
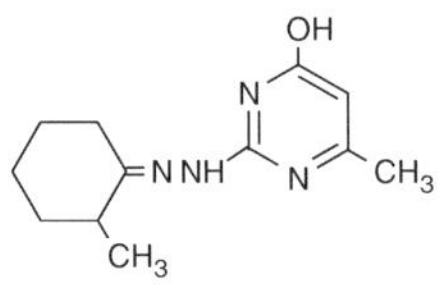
5 gm

S35,987-4
2-(CYCLOHEXYLIDENEHYDRAZINO)-4-HYDROXY-6-METHYLPYRIMIDINE
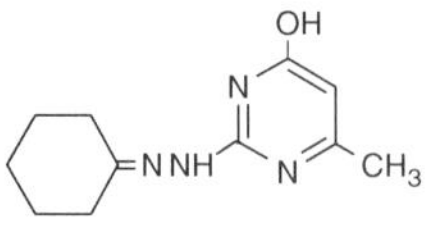
250 mg

S35,813-4
2-(SEC-BUTYLIDENEHYDRAZINO)-4-HYDROXY-6-METHYLPYRIMIDINE
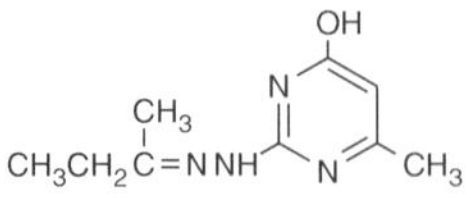
5 gm

S35,101-6
4-HYDROXY-6-METHYL-2-(2-PENTYLIDENEHYDRAZINO)PYRIMIDINE
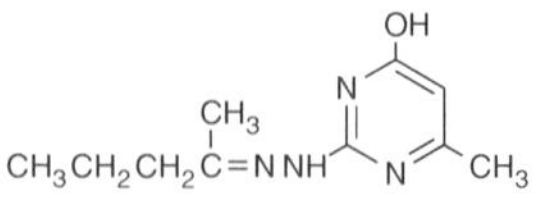
250 mg

S34,684-5
2-(1,3-DIMETHYLBUTYLIDENEHYDRAZINO)-4-HYDROXY-6-METHYLPYRIMIDINE
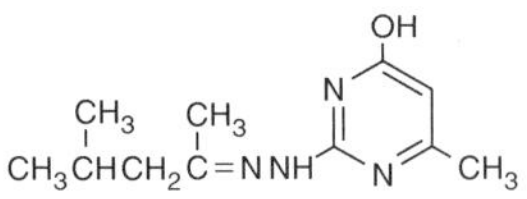
5 gm

S56,591-1
N-(4-HYDROXY-6-METHYL-2-PYRIMIDINYL)CARBAMONITRILE
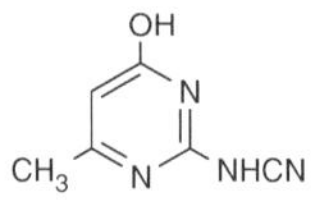
5 gm

S56,496-6
N-[4,6-DIMETHYL-5-(2-HYDROXYETHYL)-2-PYRIMIDINYL]CARBAMONITRILE
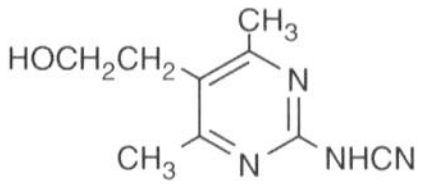
5 gm

S36,273-5
6-HYDROXY-4-PYRIMIDINECARBOXYLIC ACID
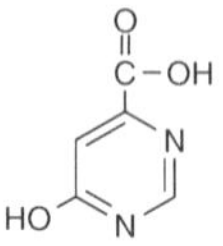
250 mg

S88,593-2
4,6-DIMETHYL-2-HYDROXY-5-PYRIMIDINYL SULFATE
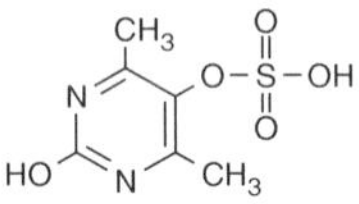
250 mg

S51,584-1
6-METHYL-2-NITRAMINO-4-PYRIMIDINOL
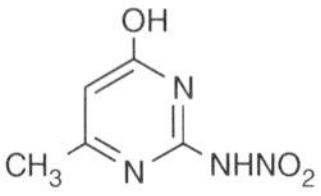
250 mg

S37,341-9
ETHYL 2-BENZYLTHIO-4-HYDROPYRIMIDINE-5-CARBOXYLATE
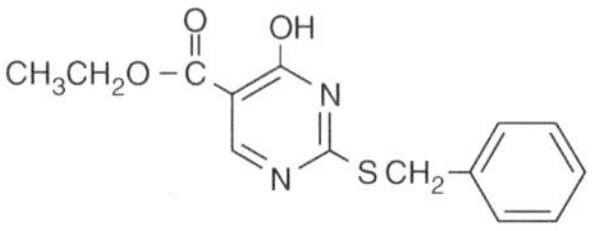
5 gm

S37,763-5
ETHYL 2-CYCLOPENTYLAMINO-4-HYDROXY-5-PYRIMIDINECARBOXYLATE
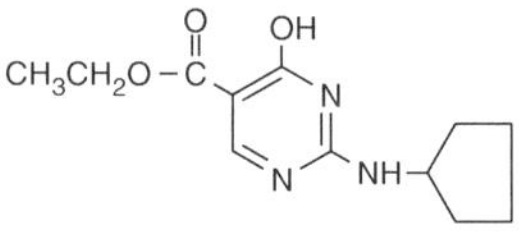
250 mg

S38,153-5
ETHYL 4-HYDROXY-2-PIPERIDINO-5-PYRIMIDINECARBOXYLATE
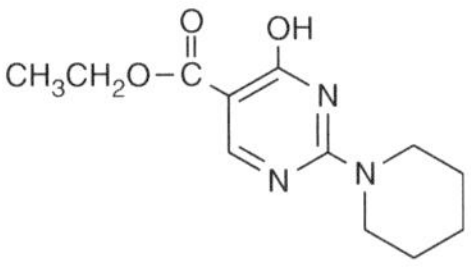
250 mg

S38,147-0
ETHYL 4-HYDROXY-2-ISOBUTYLAMINO-5-PYRIMIDINECARBOXYLATE
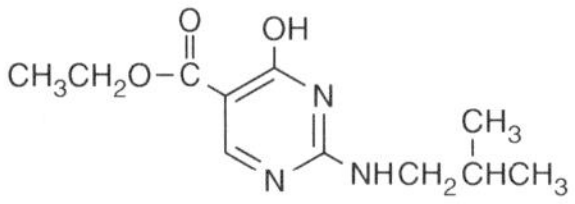
250 mg

S37,769-4
ETHYL 4-HYDROXY-2-NEOPENTYLAMINO-5-PYRIMIDINECARBOXYLATE
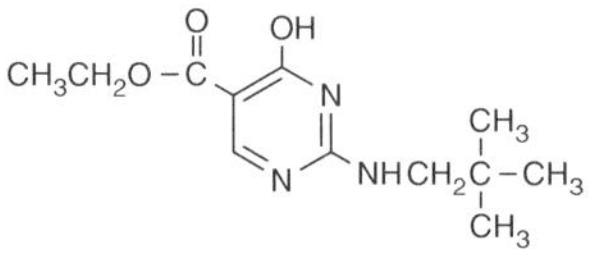
250 mg

S38,210-8
ETHYL 2-(4-CHLOROANILINO)-4-HYDROXY-5-PYRIMIDINECARBOXYLATE
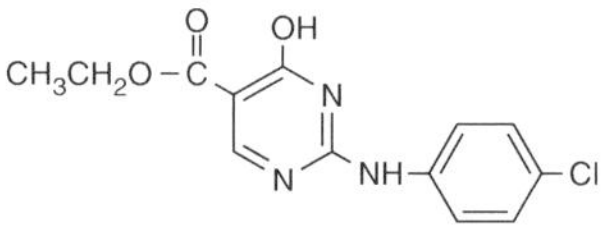
250 mg

S37,353-2
6-HYDROXY-4-METHYL-2-UREIDOPYRIMIDINE
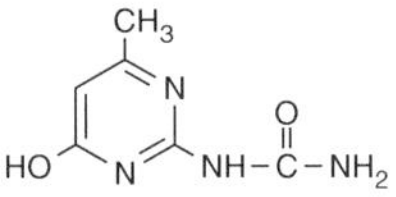
5 gm

S35,944-0
5-CHLORO-2,4-DIHYDROXY-6-METHYLPYRIMIDINE
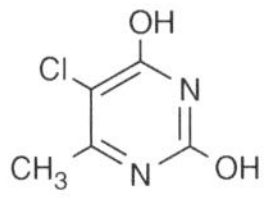
250 mg

S36,851-2
4,6-DIMETHYL-2-HYDROXYPYRIMIDINE SULFATE
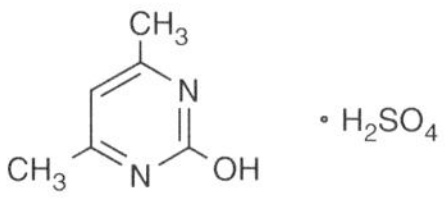
5 gm

S35,973-4
2-CYANOAMINO-4-HYDROXY-6-METHYLPYRIMIDINE
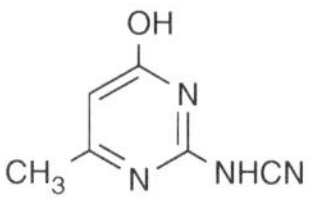
5 gm

S37,858-5
2-METHYL-5-METHYLSULFONYL-4-PYRIMIDINOL
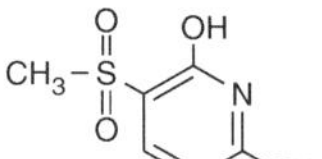
250 mg

S35,809-6
2-(2-BENZIMIDAZOLYLAMINO)-4-HYDROXY-6-METHYLPYRIMIDINE
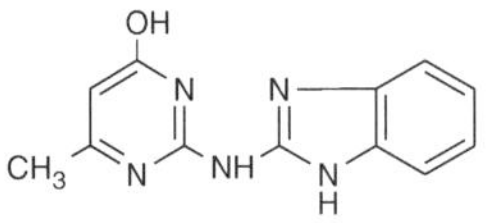
5 gm

S37,409-1
2-(5,6-DIMETHYL-2-BENZIMIDAZOLYLAMINO)-6-METHYL-4-PYRIMIDINOL
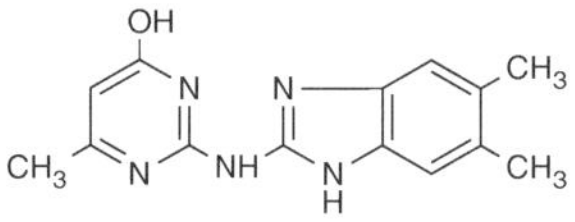
5 gm

S84,306-7
5-CHLORO-4-PYRIMIDINOL HYDROCHLORIDE
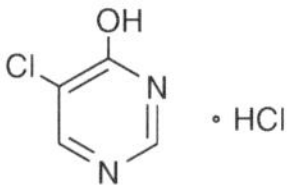
250 mg

S37,901-8
6-CHLORO-2-METHYL-4-PYRIMIDINOL
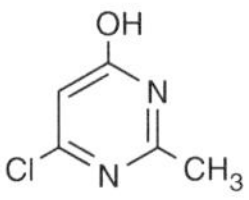
5 gm

S87,357-8
2,4,6-TRIPHENYL-1,3-OXAZIN-1-IUM PERCHLORATE
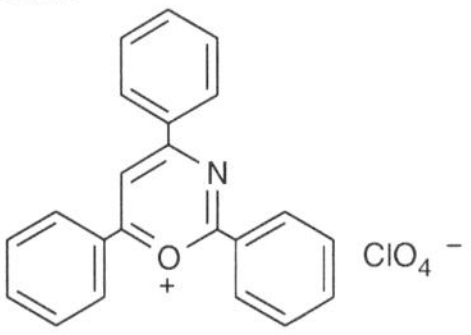
250 mg

S90,934-3
2,4,6-TRIMETHOXYPYRIMIDINE
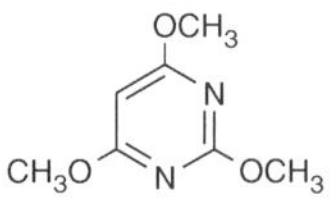
250 mg

S68,372-8
4,6-BIS(4-METHOXYPHENYL)PYRIMIDINE
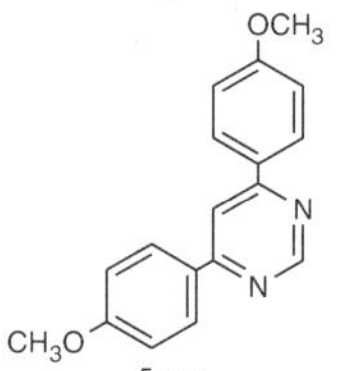
5 gm

S67,681-0
N4-(3,4,5-TRIMETHOXYBENZOYL)-SULFAMETHAZINE
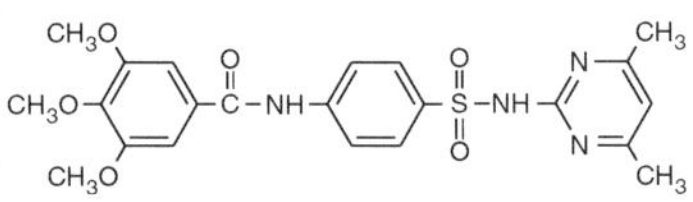
5 gm

S78,398-6
4-CHLORO-2,6-DIMETHOXYPYRIMIDINE
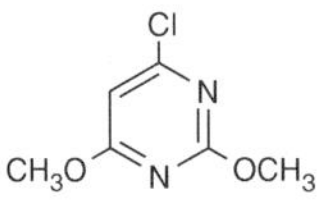
5 gm

S78,795-7
2,4-DIMETHOXY-5-IODOPYRIMIDINE
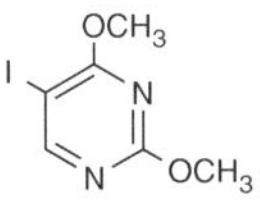
250 mg

S78,373-0
2,4-DIETHOXY-5-METHYLPYRIMIDINE
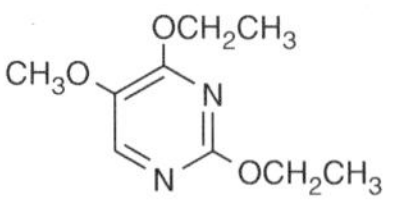
5 gm

S78,428-1
2,4-DIETHOXY-6-METHYLPYRIMIDINE

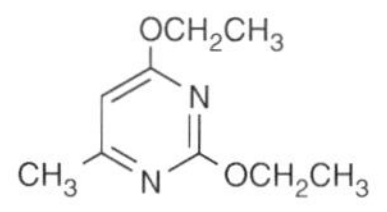

250 mg

S67,064-2
2-CHLORO-5-(3,4-DIMETHOXYPHENYL)-PYRIMIDINE

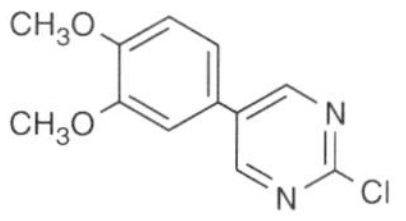

5 gm

S84,305-9
4-MERCAPTO-5-METHOXYPYRIMIDINE

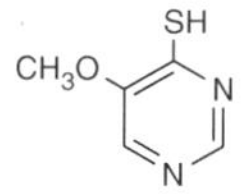

250 mg

S54,965-7
5,6-BIS(ACETAMIDO)URACIL

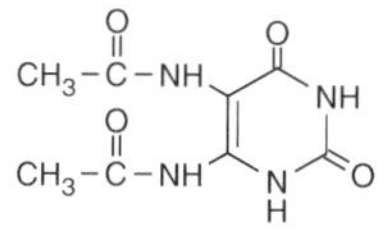

5 gm

S69,477-0
1-(2-METHOXY-5-NITROPHENYL)-3-(2-PYRIMIDINYL)UREA

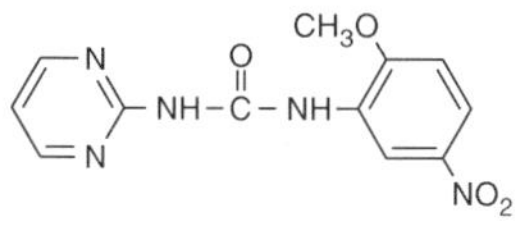

250 mg

S57,942-4
4,6-DICHLORO-5-METHOXYPYRIMIDINE

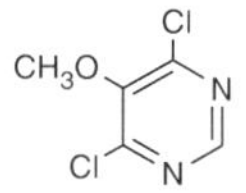

250 mg

S38,619-7
4-ACETYL-N-(2,6-DIMETHOXY-4-PYRIMIDINYL)BENZENESULFONAMIDE

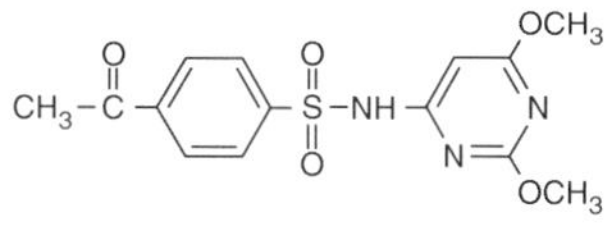

5 gm

S78,718-3
4-METHOXY-1-METHYL-2(1H)-PYRIMIDINONE

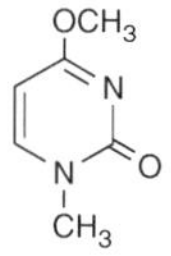

250 mg

S44,371-9
2,4-DIMETHOXYPYRIMIDINE

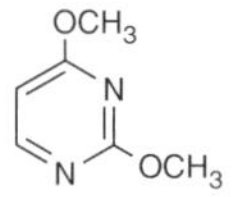

250 mg

S93,960-9
5-BROMO-2,4-DIMETHOXYPYRIMIDINE

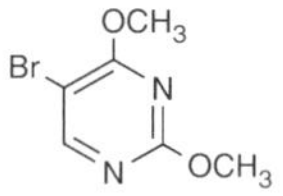

250 mg

S35,207-1
2-(4-METHOXYBENZYLAMINO)PYRIMIDINE

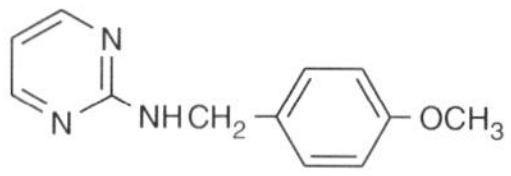

5 gm

S36,362-6
4-METHYL-6-MERCAPTO-2-METHYLTHIO-PYRIMIDINE

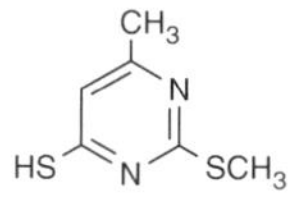

5 gm

S34,927-5
2-ETHYLTHIO-4-MERCAPTO-6-METHYLPYRIMIDINE

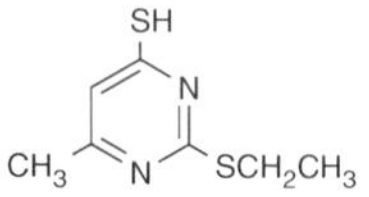

5 gm

S41,774-2
4,6-DIAMINO-2-MERCAPTO-5-NITROSOPYRIMIDINE

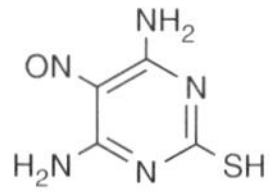

5 gm

S42,746-2
4-AMINO-5-METHYL-2-PYRIMIDINETHIOL

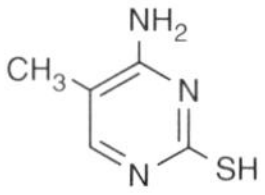

250 mg

S45,633-0
4,5-DIAMINO-2-MERCAPTOPYRIMIDINE

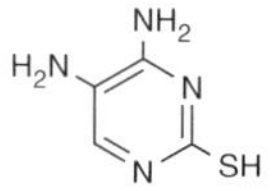

250 mg

S37,918-2
4-AMINO-2-MERCAPTO-5-PYRIMIDINECARBONITRILE

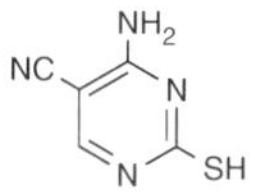

250 mg

S41,374-7
2-AMINO-4,6-PYRIMIDINEDITHIOL

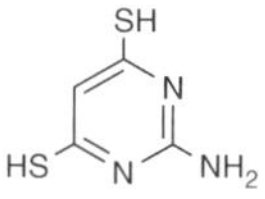

5 gm

S78,346-3
2-AMINO-4-(ETHYLAMINO)-6-MERCAPTOPYRIMIDINE

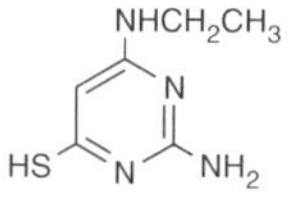

5 gm

S77,655-6
2,4-DIAMINO-6-MERCAPTOPYRIMIDINE

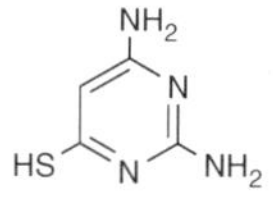

250 mg

S34,807-4
ETHYL 4-AMINO-2-MERCAPTOPYRIMIDINE-5-CARBOXYLATE

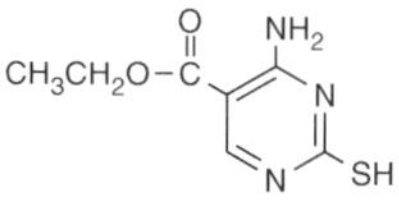

5 gm

S78,567-9
4-AMINO-2-MERCAPTO-6-METHYLPYRIMIDINE

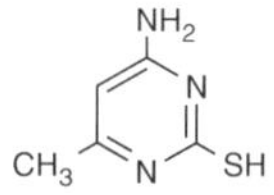

250 mg

S78,794-9
5-MERCAPTOURACIL

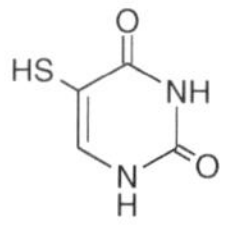

250 mg

S56,565-2
2-PYRIMIDINETHIOL HYDROCHLORIDE

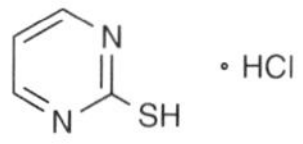

5 gm

S34,709-4
4,6-DIMETHYL-2-MERCAPTOPYRIMIDINE MONOSULFATE

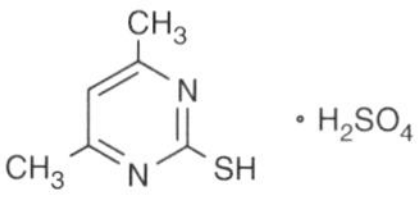

5 gm

S43,902-9
4,6-DIMETHYL-2-MERCAPTOPYRIMIDINE HYDROCHLORIDE MONOHYDRATE

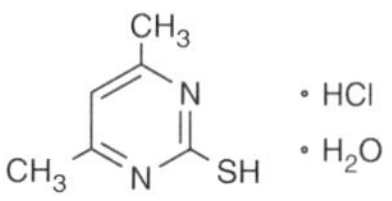

5 gm

S78,349-8
2,4-DIMERCAPTO-5-METHYLPYRIMIDINE

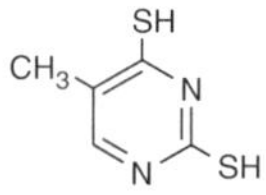

5 gm

S35,303-5
2-METHYLTHIO-4,5,6-TRIAMINOPYRIMIDINE

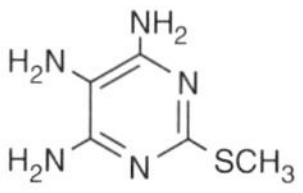

250 mg

S53,761-6
4-BENZYLTHIO-5,6-DIAMINOPYRIMIDINE

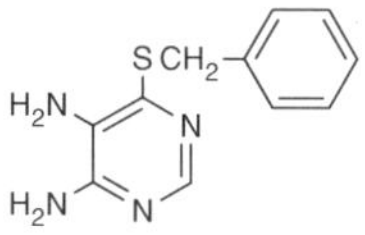

5 gm

S45,065-0
4,6-DIAMINO-2-METHYLMERCAPTOPYRIMIDINE

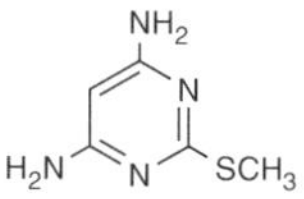

5 gm

S66,219-4
4,6-DIAMINO-2-METHYLTHIOPYRIMIDINE-5-SULFONIC ACID

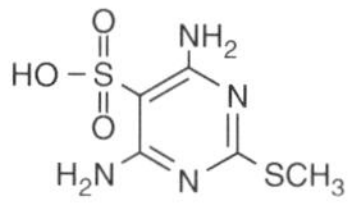

5 gm

S67,760-4
4-AMINO-5-CYANO-2-(METHYLTHIO)-PYRIMIDINE

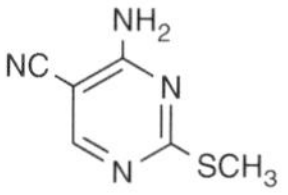

5 gm

S78,395-1
2,4-DIAMINO-6-(METHYLTHIO)-PYRIMIDINE
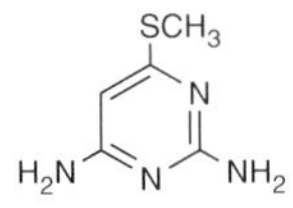
5 gm

S75,677-6
4-AMINO-2-(ETHYLTHIO)-5-PYRIMIDINECARBOXYLIC ACID
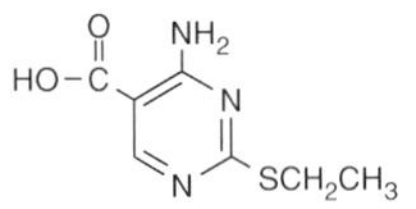
5 gm

S78,421-4
ETHYL 4-AMINO-2-(METHYLTHIO)-5-PYRIMIDINECARBOXYLATE
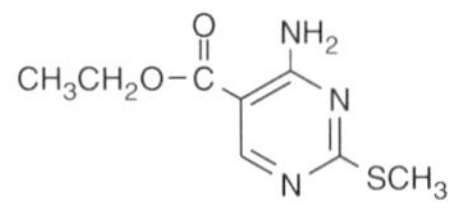
250 mg

S93,759-2
2-[4-CHLORO-2-(METHYLTHIO)-6-PYRIMIDYL]-2-(1-METHYL-2-BENZIMIDAZOLYL)ACETONITRILE
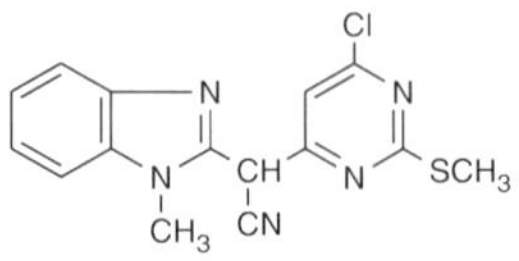
250 mg

S66,204-6
2-(CARBOXYMETHYLTHIO)-5-PHENYLPYRIMIDINE
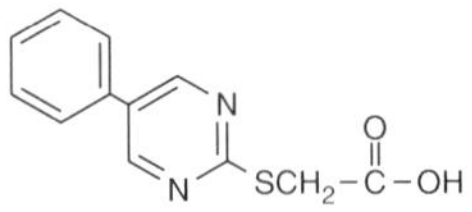
250 mg

S34,342-0
5-CARBETHOXY-2-ETHYLTHIO-6-THIOCYANOPYRIMIDINE
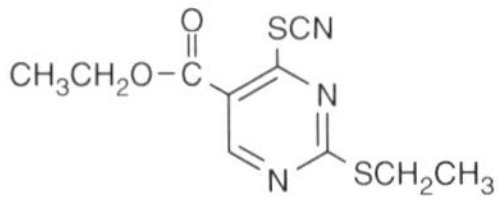
5 gm

S78,273-4
ETHYL 4-CHLORO-2-(ETHYLTHIO)-5-PYRIMIDINECARBOXYLATE
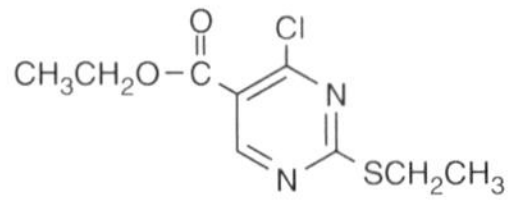
5 gm

S34,372-2
2-CARBOXYMETHYLMERCAPTOPYRIMIDINE
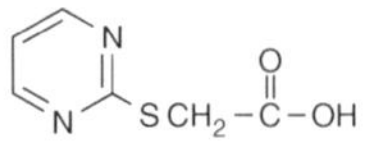
5 gm

S35,935-1
2-(CARBOXYMETHYLTHIO)-4-METHYLPYRIMIDINE
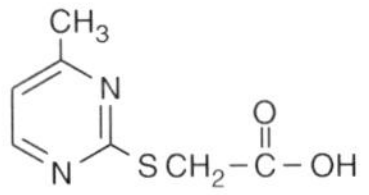
5 gm

S44,986-5
2-(4,6-DIMETHYLPYRIMIDIN-2-YLTHIO)ACETIC ACID
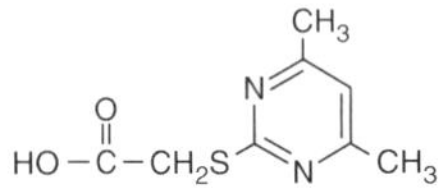
5 gm

S91,060-0
1-[4-(METHYLTHIO)PHENYL]-3-(2-PYRIMIDINYL)UREA
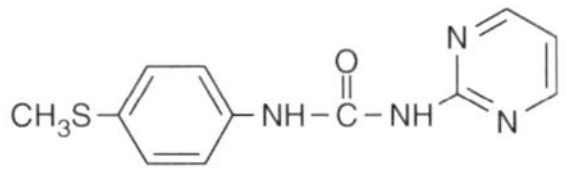
250 mg

S43,312-8
4-METHYL-2-(METHYLTHIO)PYRIMIDINE
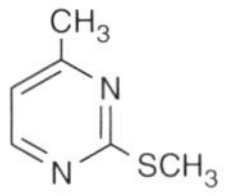
5 gm

S34,435-4
4-CHLORO-6-METHYL-2-METHYLTHIOPYRIMIDINE
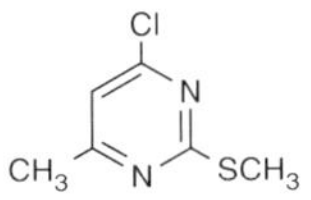
5 gm

S34,422-2
4-CHLORO-2-ETHYLTHIO-6-METHYLPYRIMIDINE
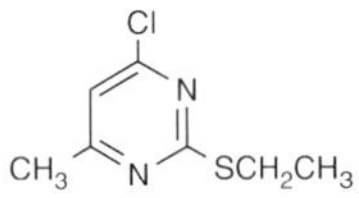
5 gm

S34,846-5
4,6-DIAMINO-5-NITROSOPYRIMIDINE
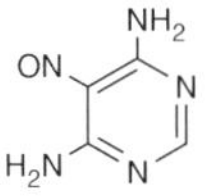
5 gm

S59,692-2
2-AMINO-4,6-DIACETAMIDO-5-NITROSOPYRIMIDINE
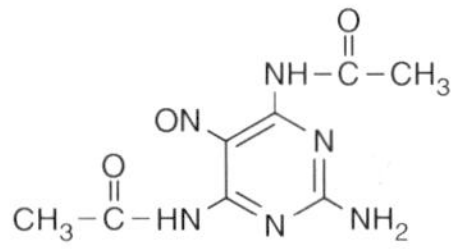
5 gm

S59,691-4
4,6-DIAMINO-2-METHYL-5-NITROSOPYRIMIDINE
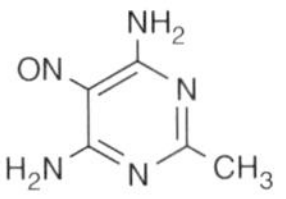
5 gm

S59,661-2
4,6-DIAMINO-2-PHENYL-5-NITROSOPYRIMIDINE
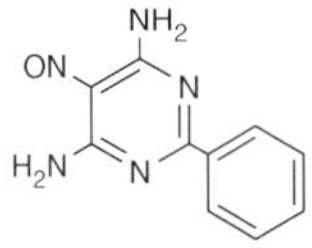
250 mg

S45,460-5
4,6-DIAMINO-5-NITROPYRIMIDINE
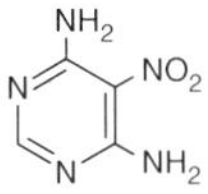
5 gm

S92,734-1
6-AMINO-5-(METHYLAMINO)-2-THIOURACIL
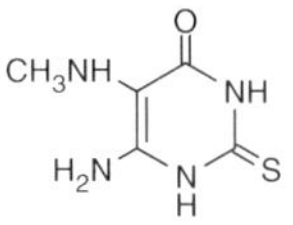
250 mg

S78,789-2
2,4-DIAMINO-5-NITROPYRIMIDINE
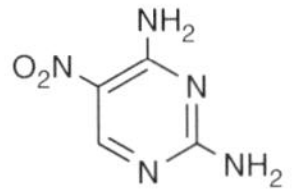
250 mg

S77,686-6
4-AMINO-2-CHLORO-6-METHYL-5-NITROPYRIMIDINE
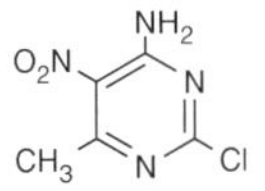
250 mg

S78,329-3
6-METHYL-2,4,5-TRIAMINOPYRIMIDINE SULFATE
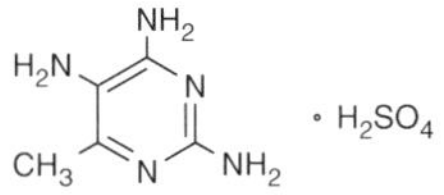
5 gm

S50,736-9
6-CHLORO-4,5-DIAMINOPYRIMIDINE
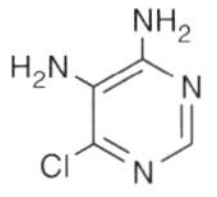
250 mg

S42,056-5
4-AMINO-2-CYCLOHEXYLAMINO-5-PYRIMIDINECARBONITRILE
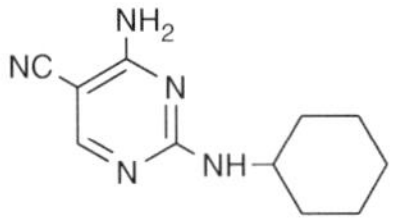
250 mg

S43,475-2
N4-(6-AMINOPYRIMIDIN-4-YL)-SULFANILAMIDE MONOHYDROCHLORIDE
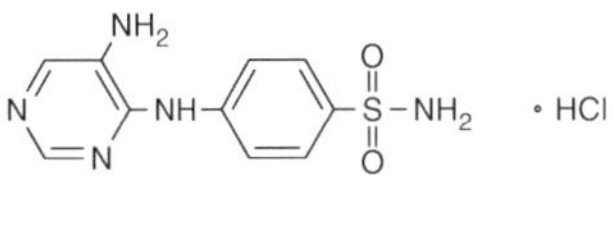
5 gm

S85,911-7
4,6-DIAMINOPYRIMIDINE
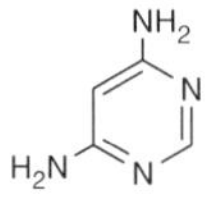
250 mg

S34,847-3
4,6-DIAMINOPYRIMIDINE-2-SULFINIC ACID
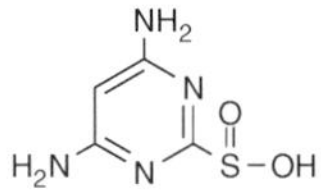
5 gm

S36,637-4
4-AMINO-2-METHYL-5-PYRIMIDINECARBONITRILE
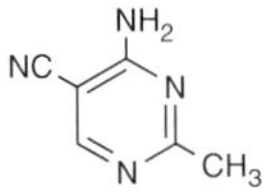
250 mg

S75,676-8
4,6-DIAMINO-2-HYDROXYPYRIMIDINE SULFATE
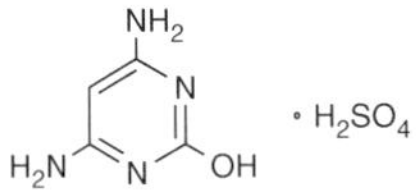
5 gm

S45,461-3
4,6-DIAMINOPYRIMIDINE HYDROCHLORIDE
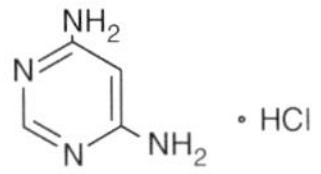
5 gm

S79,822-3
4,6-DIAMINO-2-METHYLPYRIMIDINE
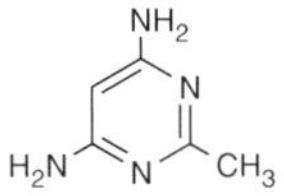
250 mg

S45,533-4
5-METHYLCYTOSINE

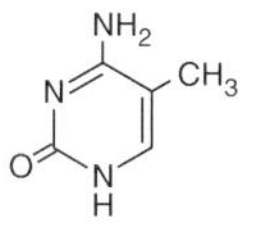

250 mg

S73,622-8
4-AMINO-5-CYANO-2,6-DIMETHYLPYRIMIDINE

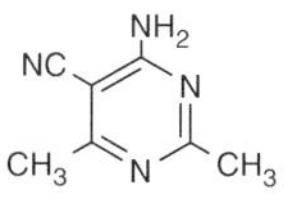

5 gm

S84,667-8
4-AMINO-5-PHENYLPYRIMIDINE

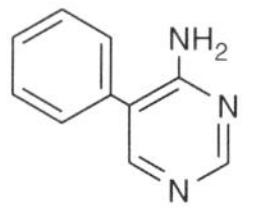

250 mg

S34,337-4
5-CARBETHOXYCYTOSINE

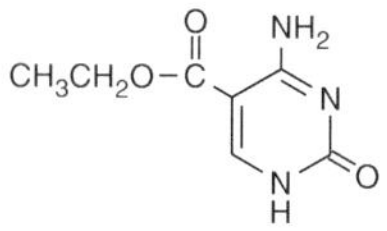

5 gm

S78,361-7
2-AMINO-6-CHLORO-4-(PROPYLAMINO)-PYRIMIDINE

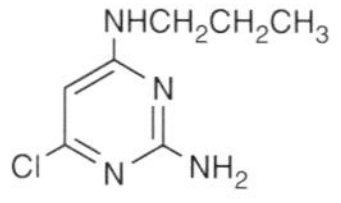

5 gm

S84,665-1
2-AMINO-4-PHENYLPYRIMIDINE

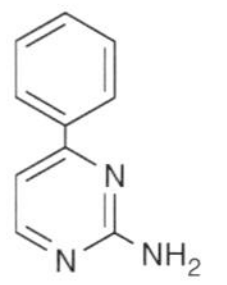

250 mg

S56,773-6
2-AMINO-5-CHLORO-4-PYRIMIDINECARBOXYLIC ACID

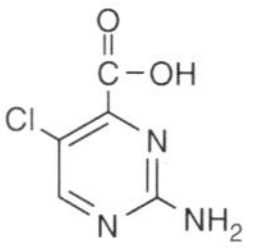

250 mg

S89,702-7
METHYL 2-AMINO-1,6-DIHYDRO-6-OXO-4-PYRIMIDINEACETATE

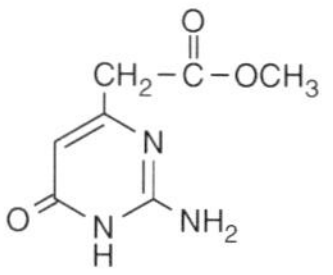

250 mg

S89,747-7
ETHYL 2-AMINO-1,6-DIHYDRO-6-OXO-4-PYRIMIDINEACETATE

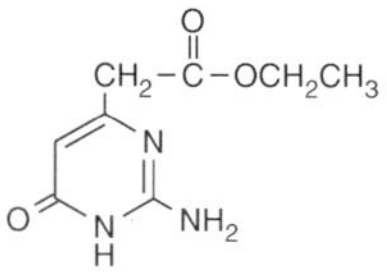

250 mg

S79,800-2
5-(AMINOMETHYL)URACIL

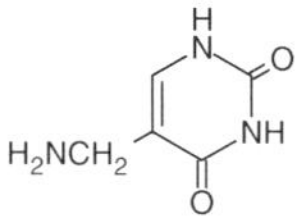

250 mg

S44,810-9
5-IODOCYTOSINE

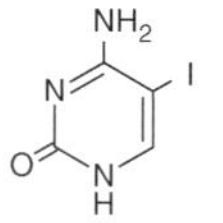

250 mg

S36,756-7
5-CYANO-2,4-DIAMINOPYRIMIDINE

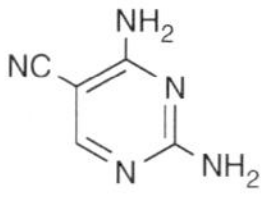

5 gm

S64,012-3
SULFAMERAZINE

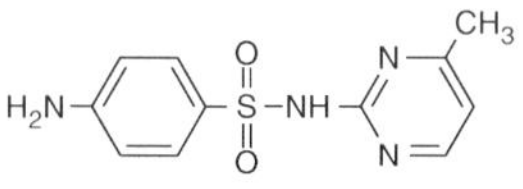

250 mg

S84,271-0
2-AMINO-5-CHLOROPYRIMIDINE

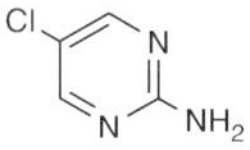

250 mg

S84,366-0
2-AMINO-5-BROMO-4-METHYLPYRIMIDINE

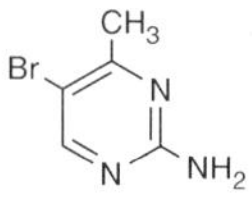

250 mg

S93,668-5
2-AMINO-4,6-DIMETHYLPYRIMIDINE HYDROBROMIDE

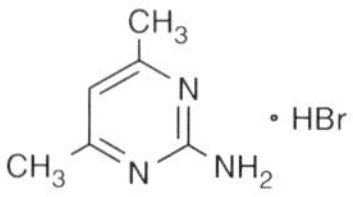

250 mg

S88,592-4
2-AMINO-5-BROMO-4,6-DIMETHYLPYRIMIDINE

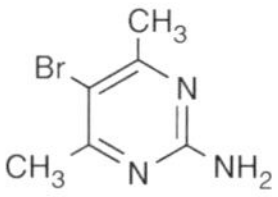

250 mg

S59,658-2
2,4,6-TRIACETAMIDO-5-NITROSOPYRIMIDINE

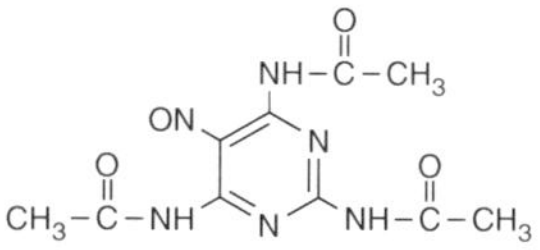

250 mg

S44,282-8
5-NITROBARBITURIC ACID HYDRATE

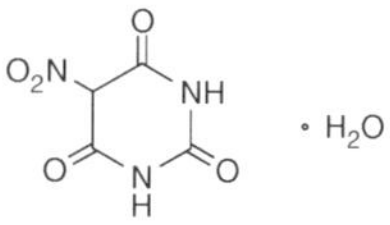

250 mg

S93,171-3
6-(METHYLAMINO)-5-NITRO-4(3H)-PYRIMIDONE

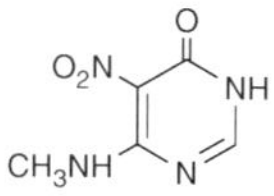

250 mg

S53,812-4
2-(4,6-DIMETHYL-2-PYRIMIDINYLAMINO)-6-METHYLBENZIMIDAZOLE

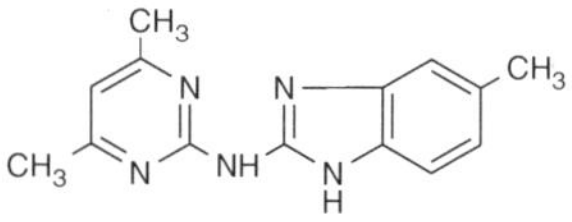

250 mg

S60,607-3
4-CHLORO-2-HEXADECYLAMINO-6-METHYLPYRIMIDINE

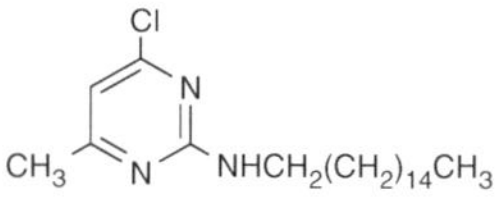

250 mg

S91,143-7
1-(4-METHYL-3-NITROPHENYL)-3-(2-PYRIMIDINYL)UREA

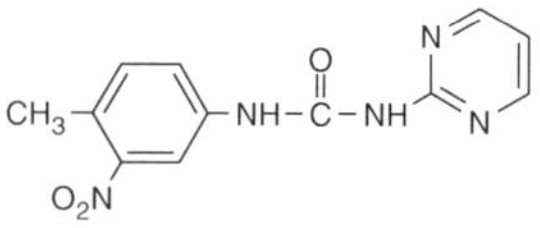

250 mg

S60,317-1
2-ACETOACETAMIDO-4-METHYLPYRIMIDINE

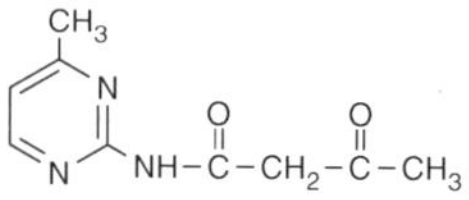

5 gm

S41,553-7
OROTIC ACID, POTASSIUM SALT

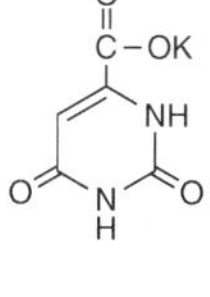

5 gm

S74,745-9
1-(2,6-DIMETHYL-4-PYRIMIDINYL)-3-PHENYLUREA

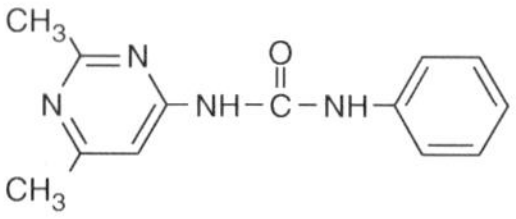

5 gm

S93,760-6
2-(2-CHLORO-4-METHYL-6-PYRIMIDYL)-2-(1-METHYL-2-BENZIMIDAZOLYL)-ACETONITRILE

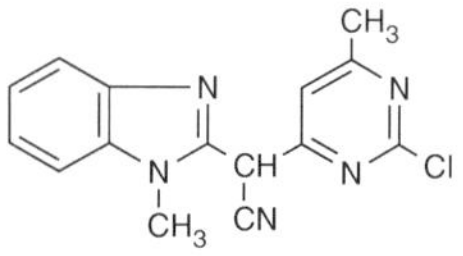

250 mg

S60,063-6
2-(3,3-DIMETHYLGUANIDINO)PYRIMIDINE

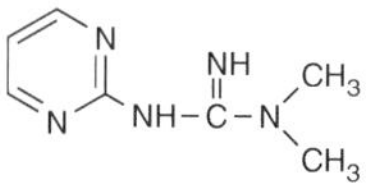

5 gm

S36,671-4
2-(4-METHYL-2-PYRIMIDINYLAMINO)-BENZIMIDAZOLE

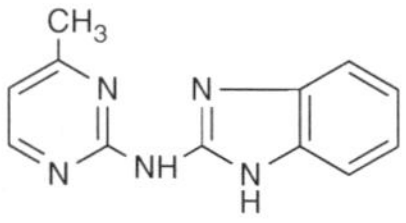

250 mg

S34,131-2
2-ANILINO-4-CHLORO-6-METHYLPYRIMIDINE

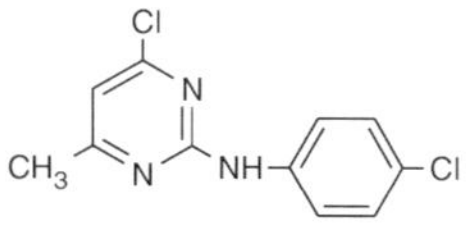

5 gm

S93,890-4
4,6-DIMETHYL-2-(DIMETHYLAMINO)-PYRIMIDINE

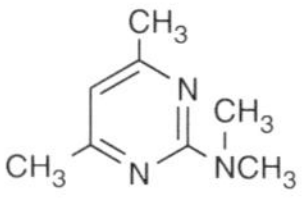

250 mg

S93,891-2
2-(DIMETHYLAMINO)-4,5,6-TRIMETHYLPYRIMIDINE

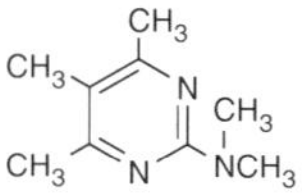

250 mg

S93,885-8
2-DIMETHYLAMINO-4-METHYL-6-PHENYLPYRIMIDINE

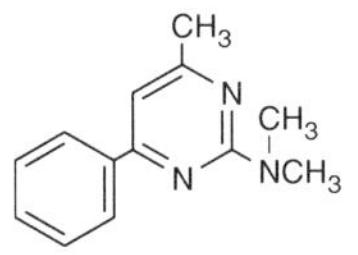

250 mg

S90,935-1
2,4-DIMETHYL-6-HYDRAZINOPYRIMIDINE

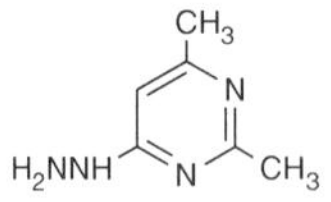

250 mg

S93,835-1
(4,6-DIPHENYL-2-PYRIMIDINYL)-HYDRAZINE

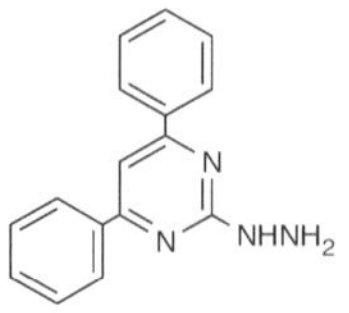

250 mg

S78,419-2
METHYL 5-NITROOROTATE HYDRATE

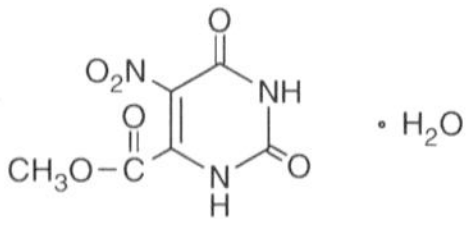

250 mg

S44,725-0
N-(4,6-DIMETHYL-2-PYRIMIDINYL)-4-NITROBENZENESULFONAMIDE

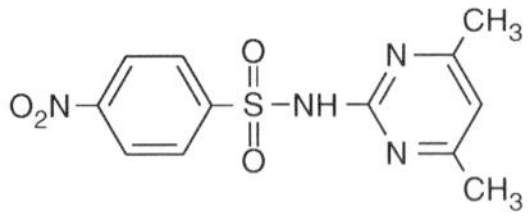

5 gm

S79,820-7
5-NITRO-2,4,6-TRICHLOROPYRIMIDINE

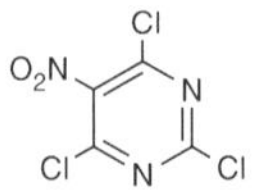

250 mg

S80,202-6
5-FORMYL-6-METHYLURACIL

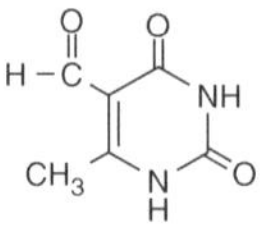

250 mg

S44,521-5
2-ACETOACETAMIDOPYRIMIDINE

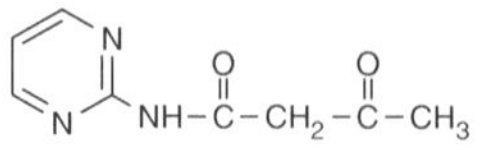

5 gm

S84,371-7
4,6-PYRIMIDINEDICARBOXYLIC ACID

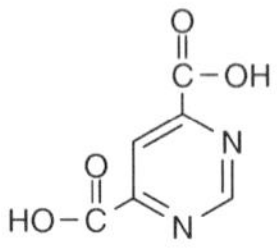

250 mg

S91,354-5
N-(2-PYRIMIDINYL)PHTHALAMIC ACID

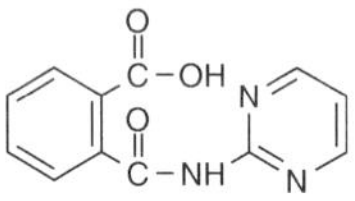

250 mg

S78,951-8
6-CARBETHOXY-2-THIOURACIL SODIUM SALT

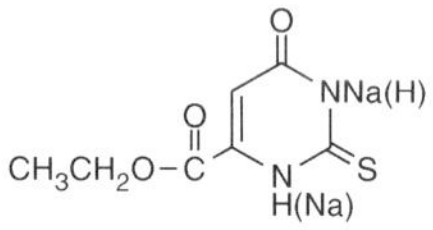

250 mg

S84,488-8
1-CINNAMOYL-3-(2-PYRIMIDINYL)-2-THIOUREA

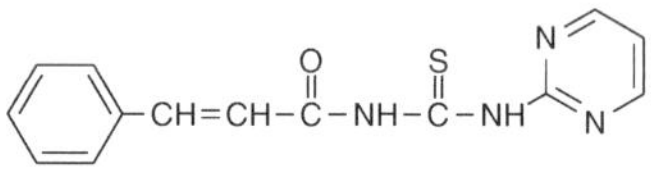

250 mg

S80,692-7
6-METHYL-2-(PHENYLSULFONYLMETHYL)-4(3H)-PYRIMIDINONE

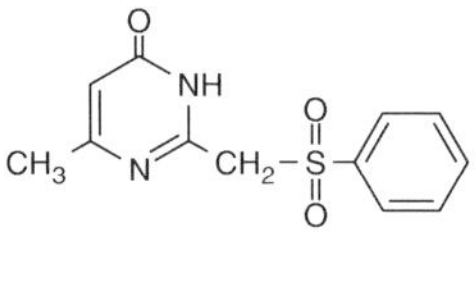

250 mg

S50,967-1
6-METHYL-4(3H)-PYRIMIDONE

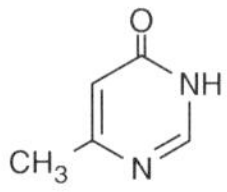

250 mg

S58,037-6
5-ETHYL-2-THIOBARBITURIC ACID

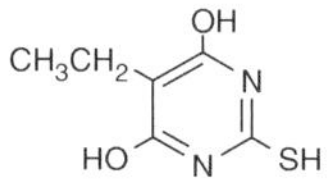

5 gm

S78,324-2
6-CHLOROURACIL

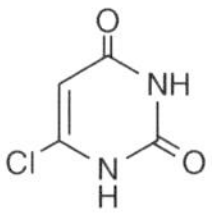

5 gm

S84,595-7
1-DODECANOYL-3-(2-PYRIMIDINYL)-2-THIOUREA

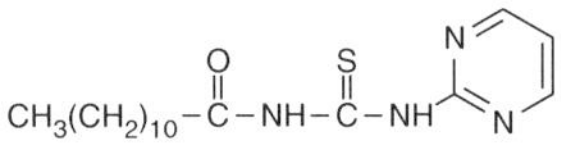

250 mg

S36,476-2
N-(2-PYRIMIDINYL)FORMAMIDE

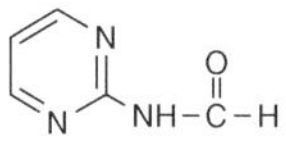

5 gm

S66,955-5
2-ACETAMIDOPYRIMIDINE

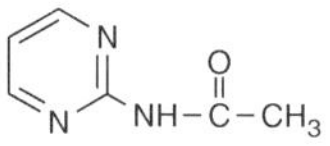

5 gm

S84,385-7
2-ACETAMIDO-4,6-DIMETHYLPYRIMIDINE

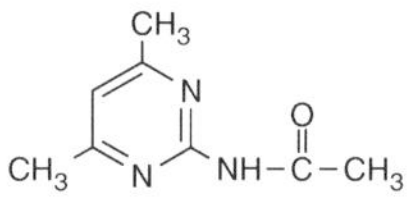

250 mg

S82,010-5
5-BUTYL-2-THIOBARBITURIC ACID

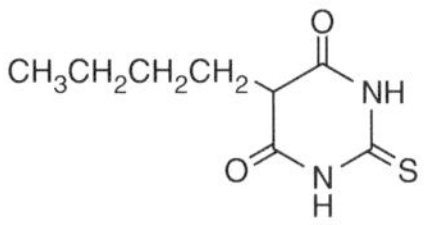

250 mg

S78,712-4
4-THIOURACIL

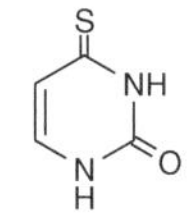

250 mg

S67,377-3
1-[4-(4-METHYL-2-PYRIMIDINYLSULFAMOYL)PHENYL]-3-(1-NAPHTHYL)UREA

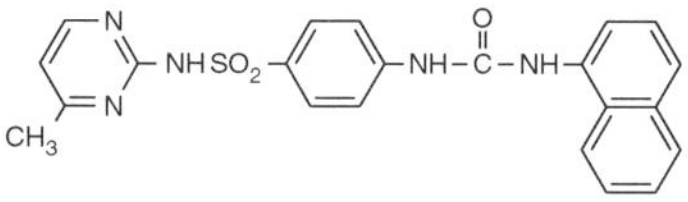

250 mg

S55,120-1
4-HYDROXYBENZOFURO(2,3-D)PYRIMID-2(1H)-ONE, SODIUM SALT

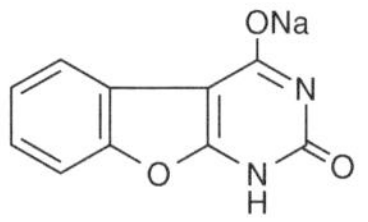

5 gm

S76,922-3
1-PHENYL-3-(2-PYRIMIDINYL)UREA

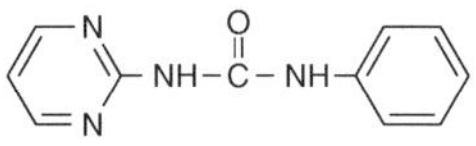

250 mg

S73,880-8
1-(4-BUTYLPHENYL)-3-(2-PYRIMIDINYL)UREA

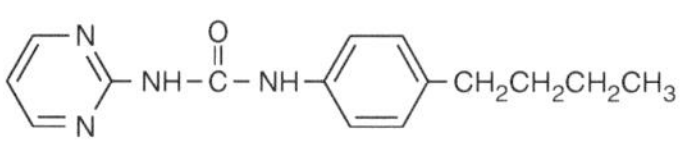

5 gm

S89,846-5
1-(2-PYRIMIDINYL)-3-(2,6-XYLYL)UREA

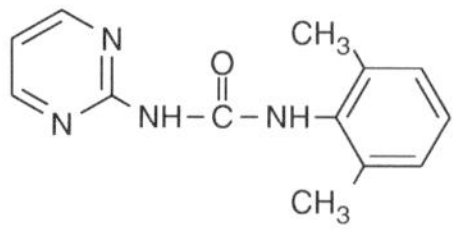

250 mg

S91,239-5
1-(2-PYRIMIDINYL)-3-(3,4-XYLYL)UREA

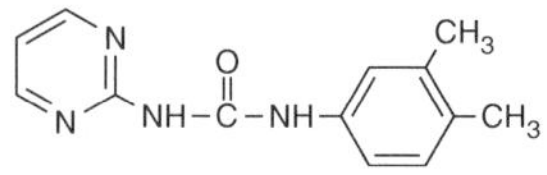

250 mg

S75,239-8
1-(1-NAPHTHYL)-3-(2-PYRIMIDINYL)UREA

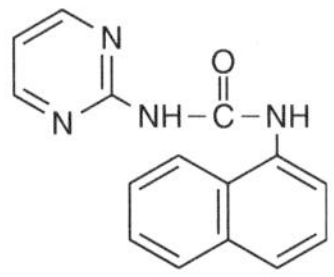

250 mg

S90,822-3
2-PYRIMIDINESULFONIC ACID, SODIUM SALT

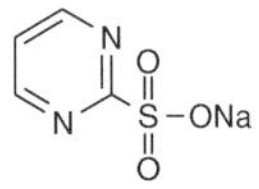

250 mg

S90,937-8
2-PYRIMIDINECARBONITRILE

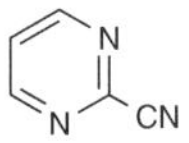

250 mg

S78,300-5
BIS(2-METHYL-1-AZIRIDINYL)-2-PYRIMIDINYLAMINOPHOSPHINE OXIDE

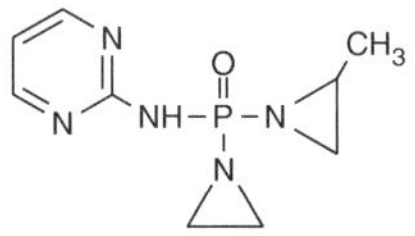

5 gm

S39,104-2
2-(4,6-DIMETHYL-2-PYRIMIDINYLAMINO)-BENZIMIDAZOLE

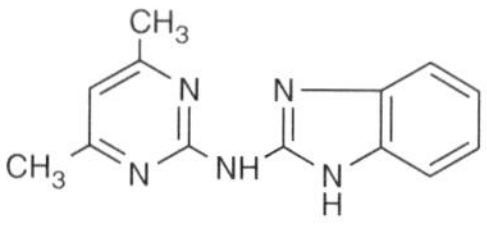

5 gm

S46,217-9
2-BROMOPYRIMIDINE

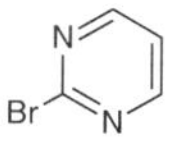

250 mg

S43,961-4
2,4-DICHLORO-5-METHYLPYRIMIDINE

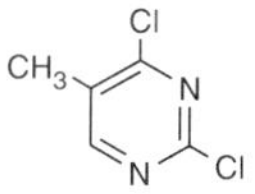

5 gm

S76,542-2
4,6-DICHLORO-2-METHYLPYRIMIDINE

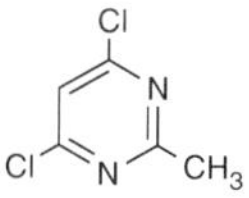

250 mg

S37,141-6
2-AMINO-4,6-DICHLORO-5-METHYLPYRIMIDINE

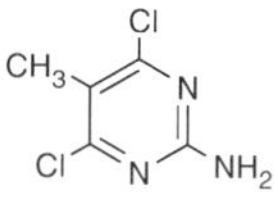

250 mg

S76,107-9
5,6-DIHYDROBENZO(H)QUINAZOLINE HYDROCHLORIDE

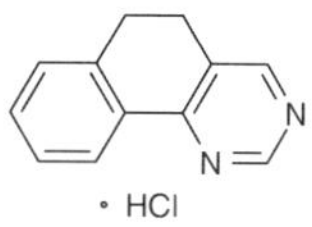

250 mg

S92,219-6
5,6-DIHYDRO-2,4,4,6-TETRAMETHYL-4H-1,3-OXAZINE

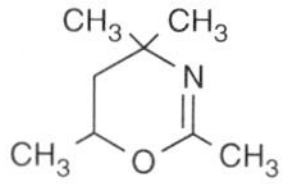

5 gm

S91,252-2
1-[2-CHLORO-5-(TRIFLUOROMETHYL)-PHENYL]-3-(2-PYRAZINYL)UREA

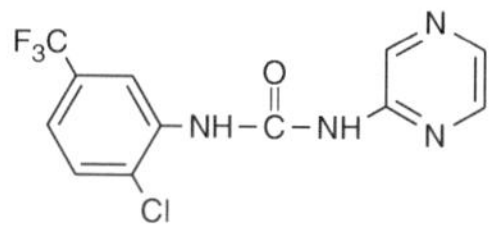

250 mg

S91,128-3
1-(5-CHLORO-2,4-DIMETHOXYPHENYL)-3-(2-PYRAZINYL)UREA

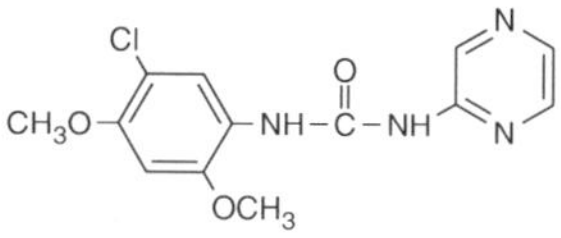

250 mg

S76,870-7
1-(5-CHLORO-2-METHOXYPHENYL)-3-(2-PYRAZINYL)UREA

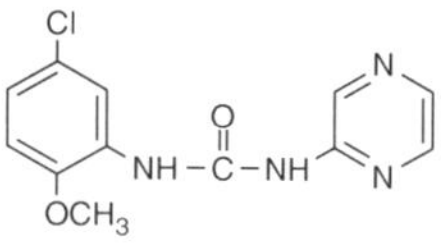

250 mg

S77,261-5
1-(4-CHLORO-2-METHOXYPHENYL)-3-(2-PYRAZINYL)UREA

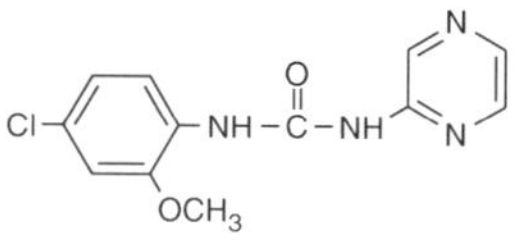

250 mg

S94,587-0
2-(4-CHLOROBENZYLAMINO)PYRAZINE

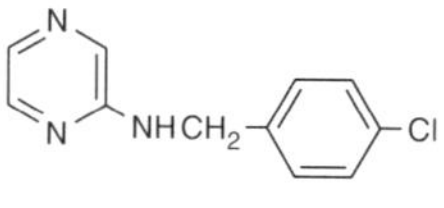

250 mg

S91,233-6
1-(4-FLUOROPHENYL)-3-(2-PYRAZINYL)UREA

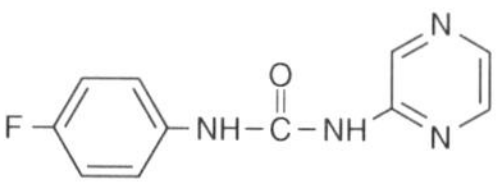

250 mg

S91,691-9
1-(4-IODOPHENYL)-3-(2-PYRAZINYL)UREA

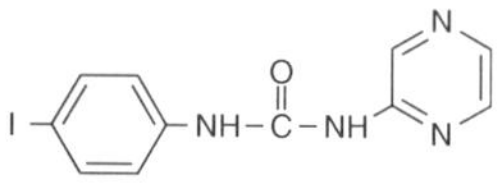

250 mg

S91,692-7
1-(4-BROMO-2,6-DIMETHYLPHENYL)-3-(2-PYRAZINYL)UREA

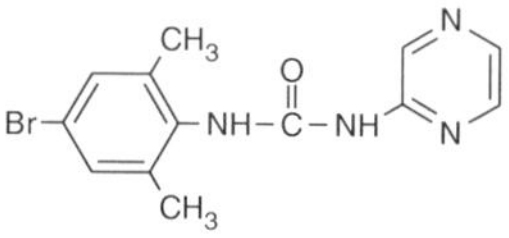

250 mg

S90,939-4
2-(4-CHLOROBENZOYL)PYRAZINE

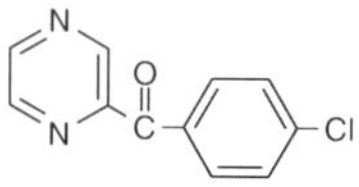

250 mg

S71,634-0
N-(2-PYRAZINYL)-3,4,5,6-TETRACHLOROPHTHALAMIC ACID

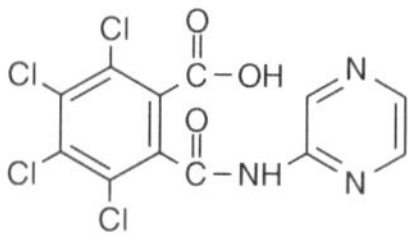

250 mg

S80,348-0
1,3-DIMETHYL-2,4-DIOXO-7-HYDROXY-1,2,3,4-TETRAHYDRO-6-PTERIDINECARBOXYLIC ACID

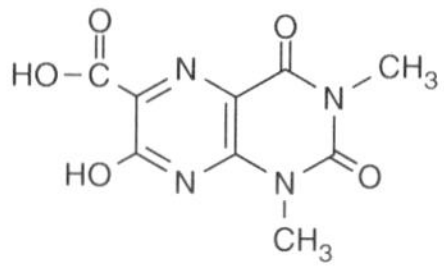

250 mg

S80,538-6
6-CARBAMOYL-1,3-DIETHYL-7-HYDROXY-2,4(1H,3H)-PTERIDINEDIONE

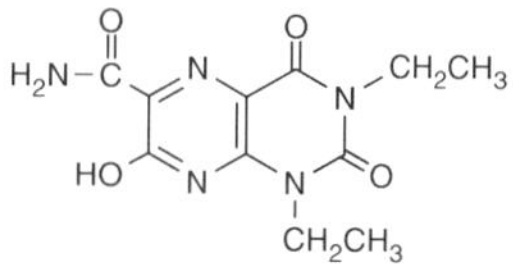

250 mg

S80,544-0
1,3-DIMETHYL-2,4-DIOXO-7-HYDROXY-1,2,3,4-TETRAHYDRO-6-PTERIDINECARBOXAMIDE

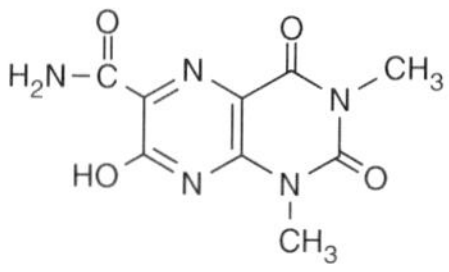

250 mg

S45,145-2
PYRAZINOL, SODIUM SALT

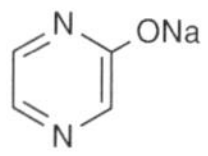

5 gm

S91,043-0
1-(2,5-DIMETHOXYPHENYL)-3-(2-PYRAZINYL)UREA

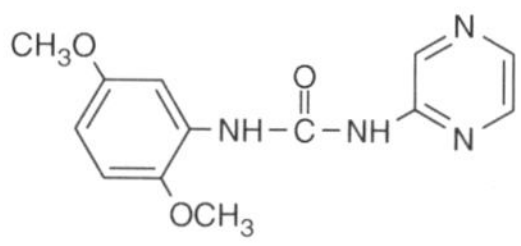

250 mg

S92,384-2
2,3,5,6-TETRAPHENYL-1,4-DIOXIN

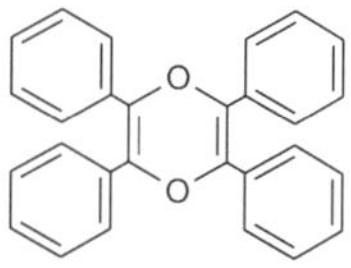

250 mg

S63,990-7
BENZYL 3-AMINOPYRAZINE-2-CARBOXYLATE

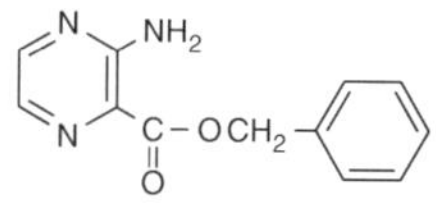

5 gm

S41,581-2
2-AMINO-6-CHLOROPYRAZINE

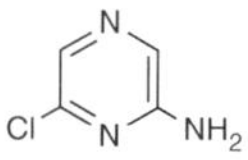

250 mg

S80,433-9
7-AMINO-1,3-DIMETHYL-2,4(1H,3H)-PTERIDINEDIONE

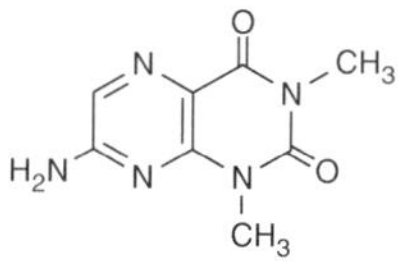

250 mg

S80,384-7
7-AMINO-1,3-DIETHYL-2,4(1H,3H)-PTERIDINEDIONE

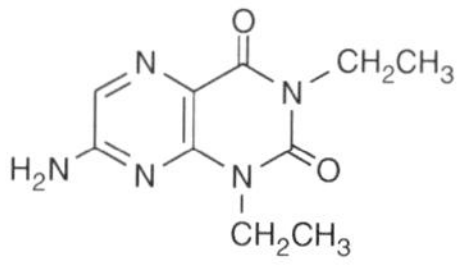

250 mg

S91,117-8
ETHYL 4-[3-(2-PYRAZINYL)UREIDO]-BENZOATE

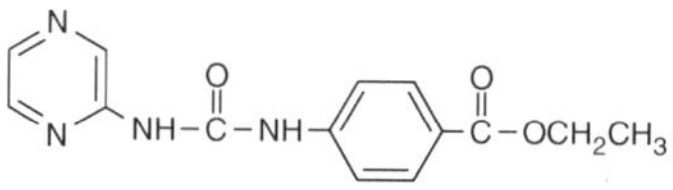

250 mg

S92,845-3
1-METHYL-2,4,6(1H,3H,5H)-PTERIDINETRIONE

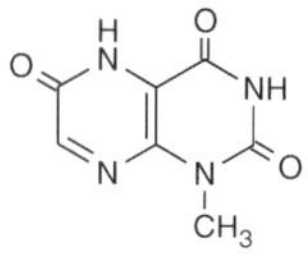

250 mg

S91,212-3
1-(2,6-DIETHYLPHENYL)-3-(2-PYRAZINYL)UREA

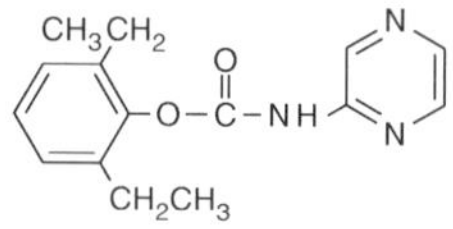

250 mg

S40,940-5
2-METHYL-3-PIPERIDINOPYRAZINE SULFATE

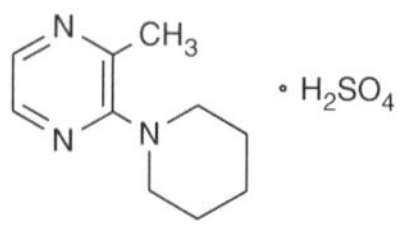

5 gm

S58,206-9
2-[2-(DIMETHYLAMINO)ETHYL]PYRAZINE

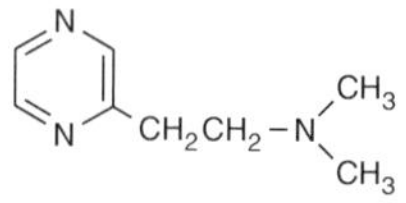

5 gm

S58,159-3
2-[2-(DIETHYLAMINO)ETHYL]PYRAZINE
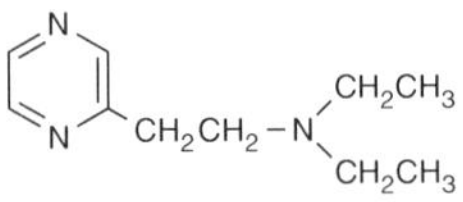
5 gm

S91,895-4
1-(ALPHA,ALPHA-DIMETHYL-4-NITROBENZYL)PYRAZINIUM CHLORIDE
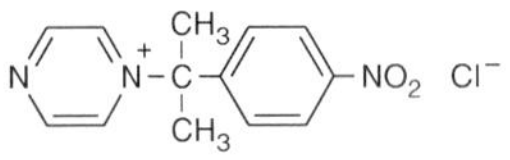
250 mg

S84,282-6
2,3-DIPHENYLPYRAZINE
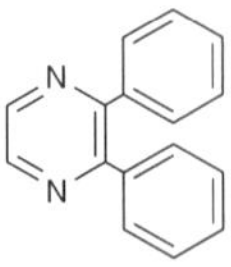
250 mg

S74,215-5
3,6-DIMETHYLPYRAZINE-2,5-DIYL BIS(PHENYLACETATE)
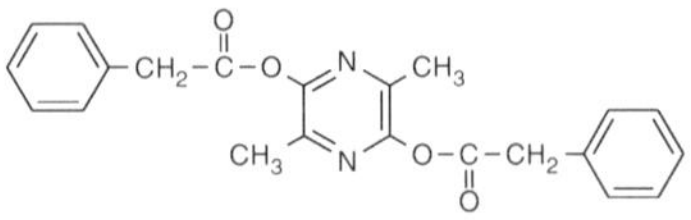
250 mg

S72,827-6
N-(2-PYRAZINYL)PHTHALAMIC ACID
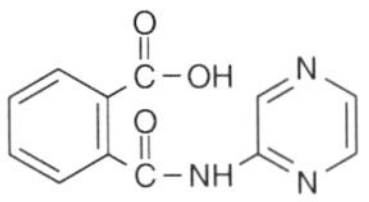
250 mg

S44,916-4
1,3-DIMETHYL-6,7-DIPHENYLLUMAZINE
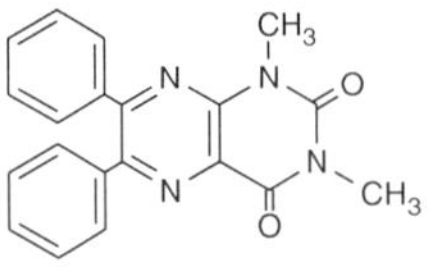
250 mg

S92,790-2
1,3-DIMETHYL-2,4,6(1H,3H,5H)-PTERIDINETRIONE
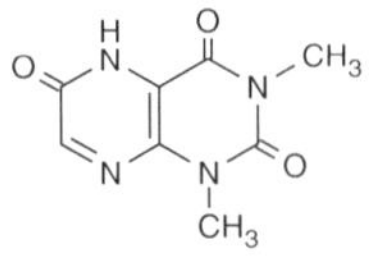
250 mg

S35,505-4
PYRAZINE, MERCURIC CHLORIDE COMPLEX
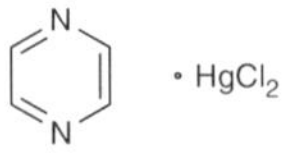
5 gm

S72,899-3
2,3,5,6-TETRAPHENYLPYRAZINE (AMARON)
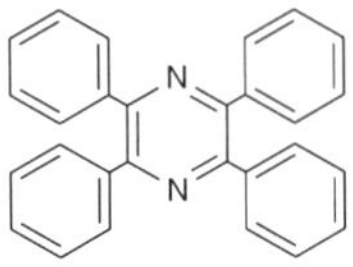
5 gm

S35,745-6
2-AMINO-4-ANILINO-6-(CHLOROMETHYL)-1,3,5-TRIAZINE
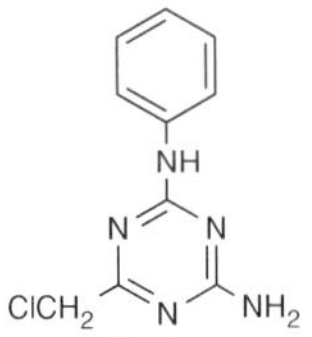
5 gm

S54,773-5
1,1'-[(6-AMINO-1,3,5-TRIAZINE-2,4-DIYL)DIIMINO]BIS(2,2,2-TRICHLOROETHANOL)
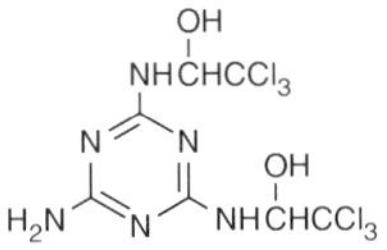
100 mg

S51,742-9
6-METHYL-2,4-(TRICHLOROMETHYL)-1,3,5-TRIAZINE
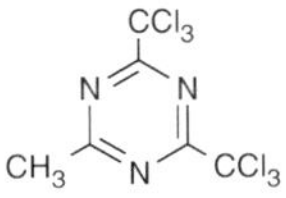
250 mg

S39,409-2
2,4,6-TRIS(TRICHLOROMETHYL)-1,3,5-TRIAZINE
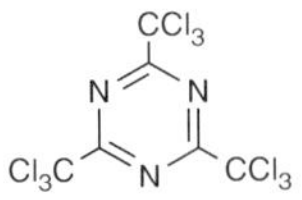
250 mg

S89,105-3
6-[2-(4-CHLORO-2-METHYLPHENOXY)-ACETAMIDO]-2,4-DIAMINO-1,3,5-TRIAZINE
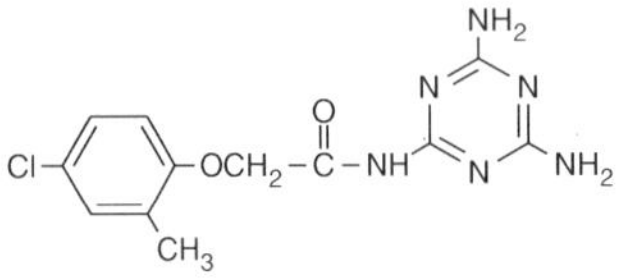
250 mg

S89,106-1
2,4-DIAMINO-6-[2-(2,4,5-TRICHLOROPHENOXY)ACETAMIDO]-1,3,5-TRIAZINE
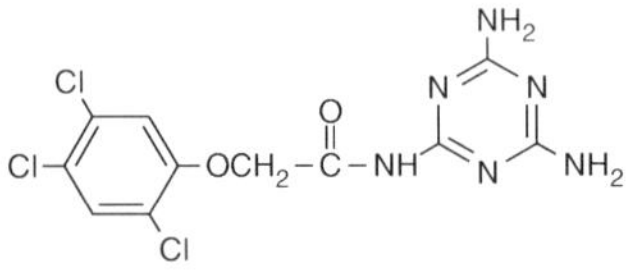
250 mg

S92,026-6
2-(4-BROMOPHENYL)PYRIDO(1,2-B)-(1,2,4)TRIAZIN-5-IUM PERCHLORATE
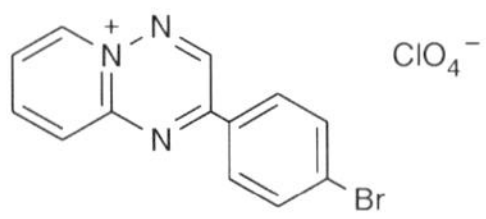
250 mg

S74,869-2
3-(4-BROMOPHENYL)-5,6-DIPHENYL-1,2,4-TRIAZINE
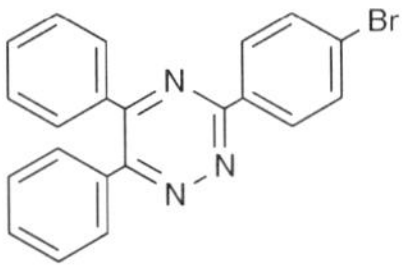
250 mg

S65,982-7
4,6-DIHYDROXY-1,3,5-TRIAZINE-2-ACETIC ACID O-ANISIDIDE
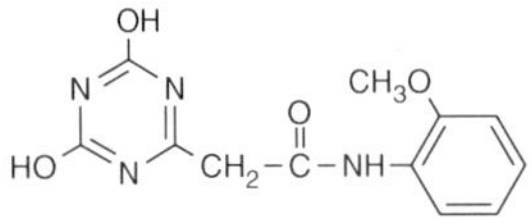
5 gm

S78,411-7
ETHYL 5-HYDROXY-3-MERCAPTO-1,2,4-TRIAZINE-6-CARBOXYLATE
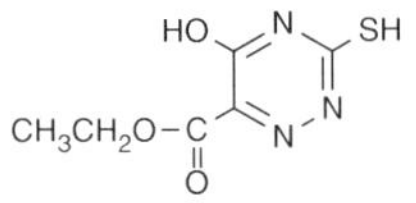
250 mg

S70,256-0
4,6-DIAMINO-1,3,5-TRIAZIN-2-OL
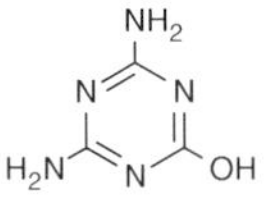
5 gm

S40,825-5
5,6-DIPHENYL-3-HYDROXY-1,2,4-TRIAZINE
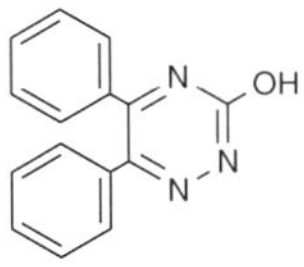
5 gm

S67,185-1
2,4-BIS(2-HYDROXYETHYLAMINO)-6-CHLORO-1,3,5-TRIAZINE
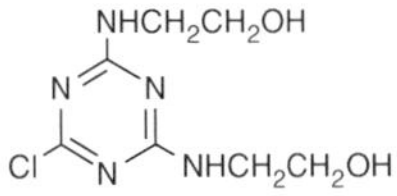
250 mg

S52,013-6
4-(2-BENZIMIDAZOLYLAMINO)-6-HYDROXY-1,3,5-TRIAZINE-2-ACETONITRILE
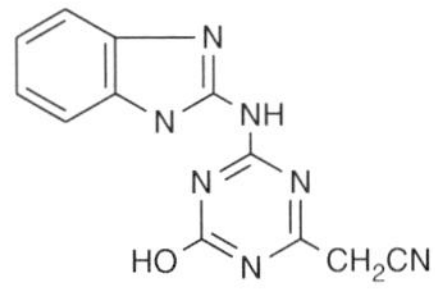
250 mg

S74,873-0
5,6-DIPHENYL-3-(4-HYDROXYPHENYL)-1,2,4-TRIAZINE
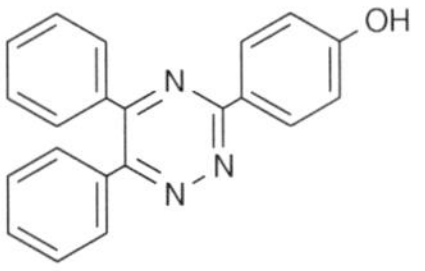
250 mg

S87,426-4
2,6-DIMETHYL-4-(METHYLTHIO)-PYRYLIUM IODIDE
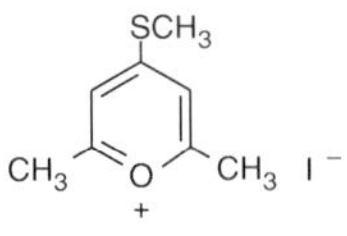
250 mg

S36,635-8
2-AMINO-4-CHLORO-6-METHOXY-S-TRIAZINE
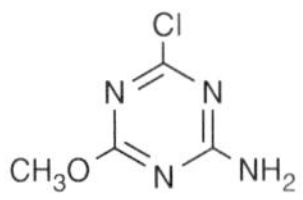
250 mg

S92,027-4
2-METHOXY(1,2,4)TRIAZINO(6,1-A)-ISOQUINOLIN-5-IUM PERCHLORATE
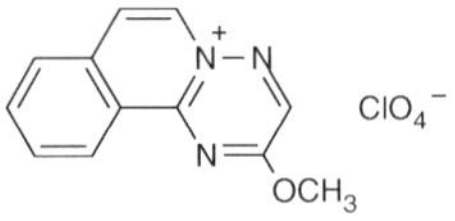
250 mg

S69,302-2
2-(4,6-DICHLORO-1,3,5-TRIAZIN-2-YL)-O-ACETANISIDIDE
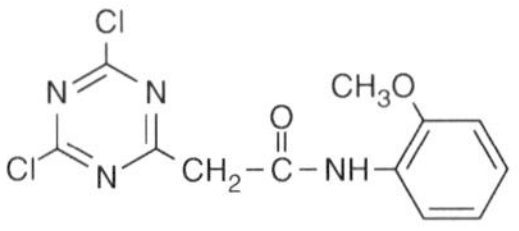
5 gm

S61,801-2
2-CHLORO-4,6-DIPHENOXY-1,3,5-TRIAZINE
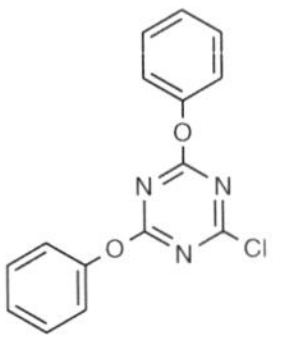
250 mg

S74,527-8
5,6-DIPHENYL-3-MERCAPTO-1,2,4-TRIAZINE
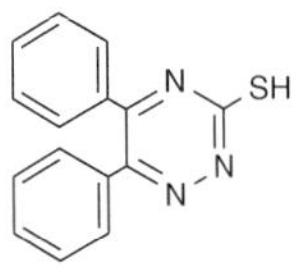
250 mg

S55,709-9
2,4-DIAMINO-6-(2,3-XYLYL)-1,3,5-TRIAZINE
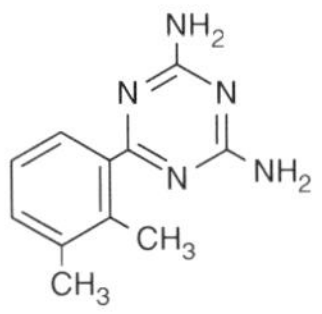
5 gm

S45,433-8
6-AZACYTOSINE
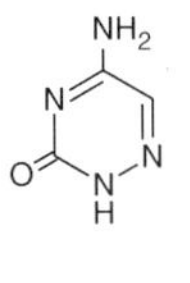
250 mg

S53,713-6
4-AMINO-6-PHENYL-S-TRIAZINE-2-CARBONITRILE

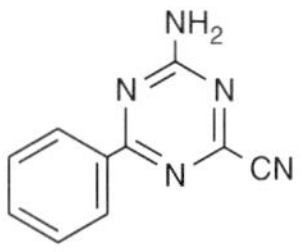

250 mg

S53,104-9
2,4-DIAMINO-6-(3-NITROPHENYL)-1,3,5-TRIAZINE

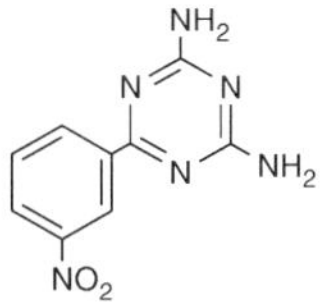

250 mg

S43,449-3
TRANS-1,2-CYCLOBUTANEDICARBOGUANAMINE

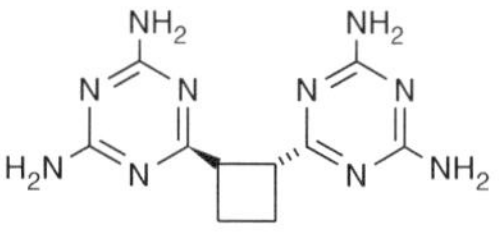

5 gm

S55,708-0
2,4-DIAMINO-6-(4-PYRIDYL)-1,3,5-TRIAZINE

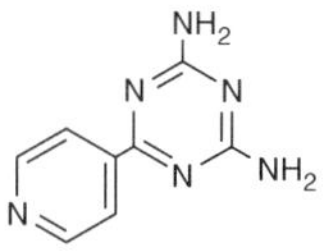

5 gm

S37,318-4
N(4-AMINO-S-TRIAZIN-2-YL)-SULFANILIC ACID

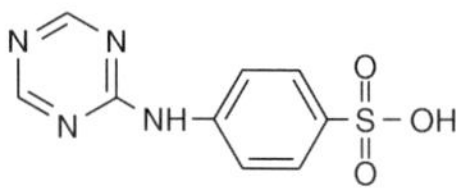

5 gm

S35,755-3
2-AMINO-4-CYCLOHEXYLAMINO-1,3,5-TRIAZINE

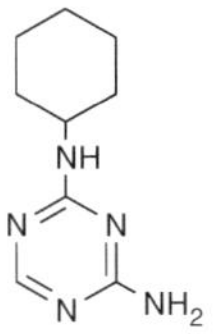

5 gm

S35,783-9
2-AMINO-4-PHENYLAMINO-1,3,5-TRIAZINE HYDROCHLORIDE

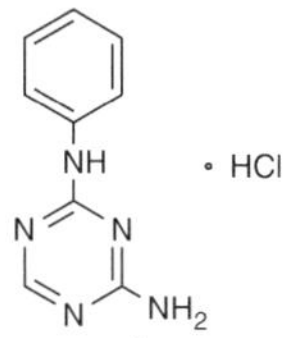

5 gm

S40,636-8
2-AMINO-4-(4-CHLOROANILINO)-1,3,5-TRIAZINE HYDROCHLORIDE

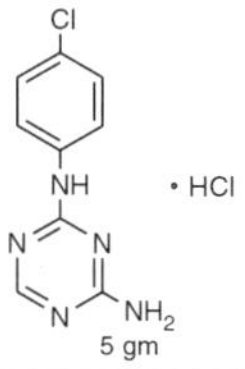

5 gm

S39,246-4
4-AMINO-2-DIMETHYLAMINO-1,3,5-TRIAZINE

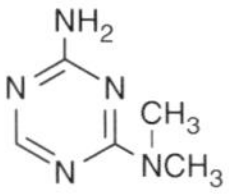

5 gm

S44,194-5
2,4-DIAMINO-6-PHENYL-1,3,5-TRIAZINE

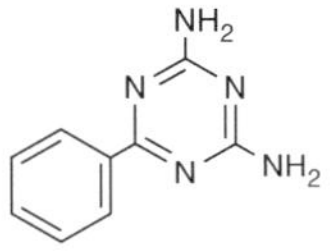

5 gm

S55,704-8
4,6-DIAMINO-GAMMA-OXO-1,3,5-TRIAZINE-2-BUTYRIC ACID

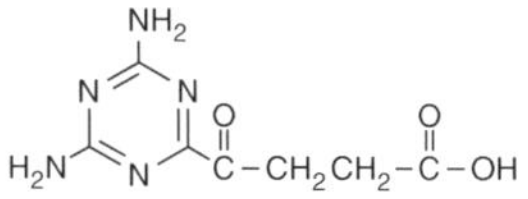

5 gm

S55,706-4
4,6-DIAMINO-GAMMA-OXO-1,3,5-TRIAZINE-2-BUTYRONITRILE

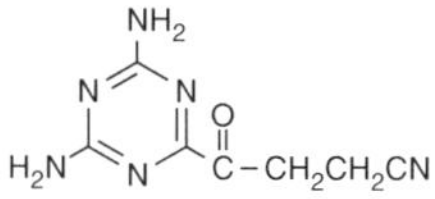

5 gm

S55,705-6
2,4-DIAMINO-6-ISOBUTYRYL-1,3,5-TRIAZINE

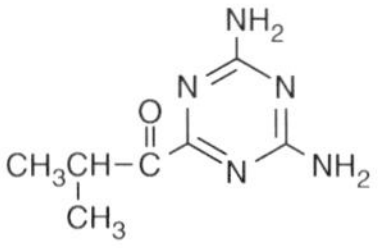

5 gm

S55,710-2
2-DECANOYL-4,6-DIAMINO-1,3,5-TRIAZINE

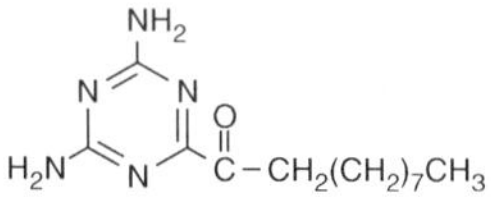

5 gm

S55,714-5
2,4-DIAMINO-6-PHENYLACETYL-1,3,5-TRIAZINE

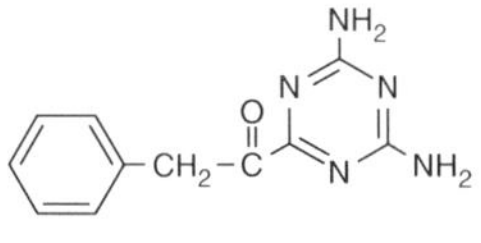

250 mg

S40,374-1
N1-(4,6-DIAMINO-1,3,5-TRIAZIN-2-YL)-SULFANILAMIDE MONOHYDRATE

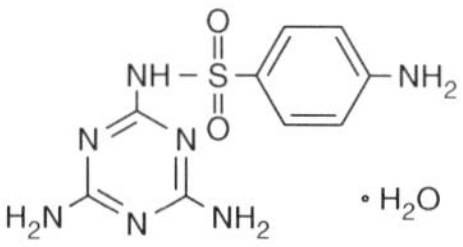

250 mg

S72,044-5
2-AMINO-1,3,5-TRIAZINE

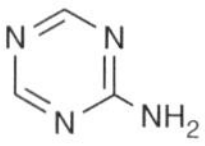

5 gm

S36,775-3
2,4-DIAMINO-1,3,5-TRIAZINE

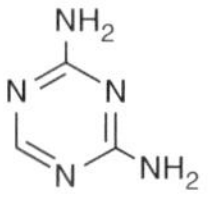

5 gm

S54,783-2
2,4-DIAMINO-6-METHYL-1,3,5-TRIAZINE

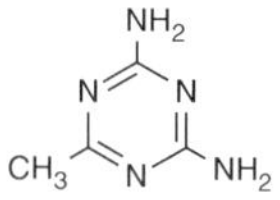

250 mg

S55,707-2
2-(2-CHLOROPHENYL)-4,6-DIAMINO-1,3,5-TRIAZINE

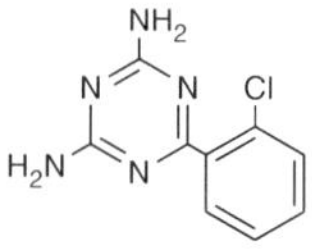

5 gm

S55,711-0
2,4-DIAMINO-6-(M-TOLYL)-1,3,5-TRIAZINE

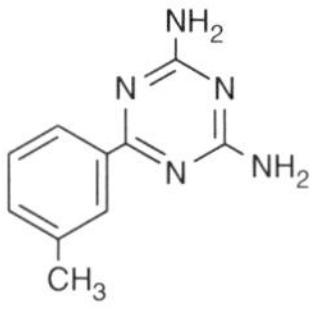

5 gm

S39,250-2
1-(4-CHLOROPHENYL)-4,6-DIAMINO-1,2-DIHYDRO-2,2-DIMETHYL-1,3,5-TRIAZINE HYDROCHLORIDE

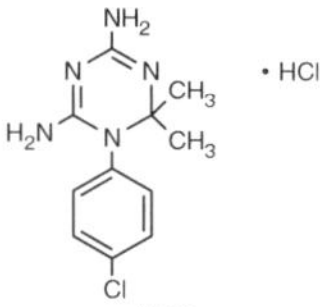

5 gm

S44,356-5
2-AMINO-4-MORPHOLINO-1,3,5-TRIAZINE

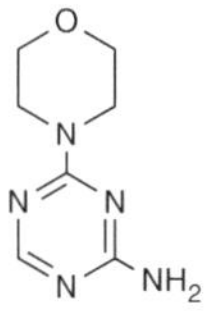

5 gm

S93,718-5
5-METHYL-3-(2-PYRIDYL)-1,2,4-TRIAZINE

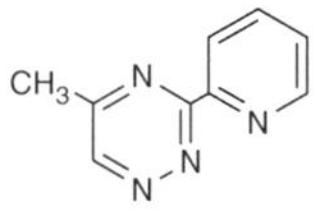

250 mg

S93,791-6
5-METHYL-3-(2-PYRIDYL)-1,2,4-TRIAZINE

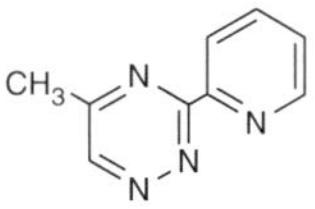

250 mg

S93,715-0
2,6-BIS(5,6-DIMETHYL-1,2,4-TRIAZIN-3-YL)PYRIDINE

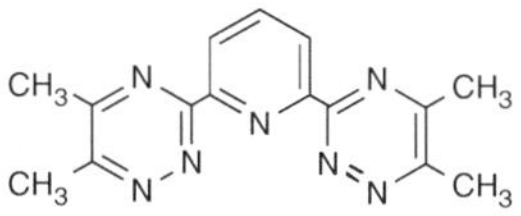

250 mg

S92,011-8
4,5-DIHYDRO(1,2,4)TRIAZOLO(3,4-C)-(1,2,4)BENZOTRIAZINE

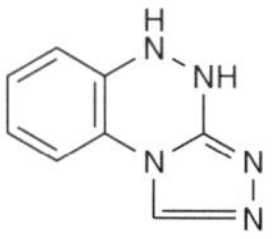

250 mg

S93,896-3
3-(3-NITROPHENYL)-5,6,7,8-TETRAHYDRO-1,2,4-BENZOTRIAZINE

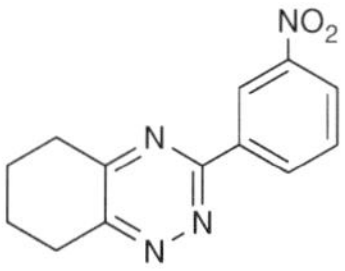

250 mg

S93,897-1
6,7-DIHYDRO-5(H)-5-METHYL-3-(3-NITROPHENYL)CYCLOPENTA(E)-1,2,4-TRIAZINE

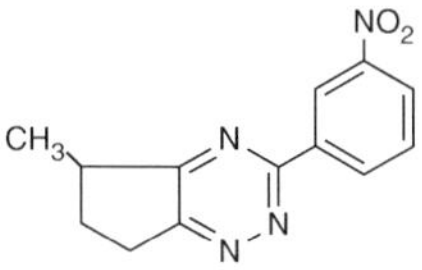

250 mg

S44,237-2
2,4-BIS(ETHYLAMINO)-6-CHLORO-1,3,5-TRIAZINE

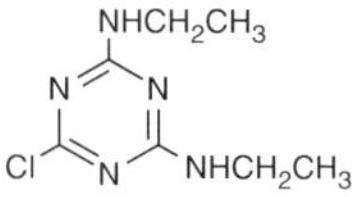

5 gm

S43,656-9
2-CHLORO-4-ETHYLAMINO-6-ISOPROPYLAMINO-S-TRIAZINE

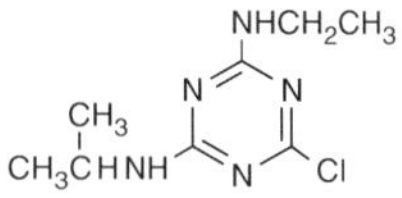

5 gm

S86,298-3
2,6,4-TRIANILINO-1,3,5-TRIAZINE

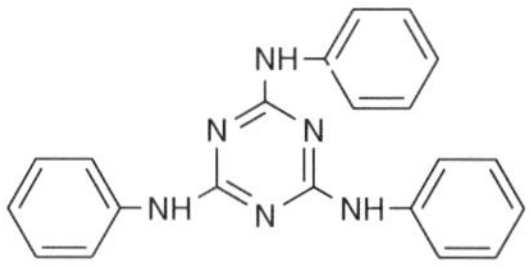

250 mg

S92,054-1
2-MORPHOLINO(1,2,4)TRIAZINO(6,1-A)-ISOQUINOLIN-5-IUM ETHANESULFONATE

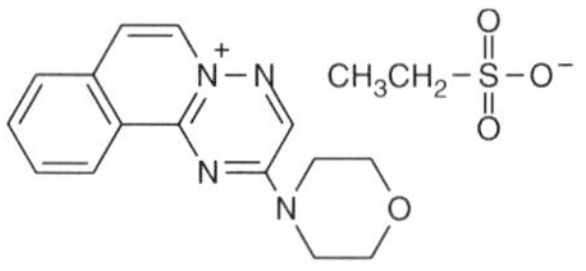

250 mg

S93,669-3
2,4-DICHLORO-6-(DIMETHYLAMINO)-1,3,5-TRIAZINE

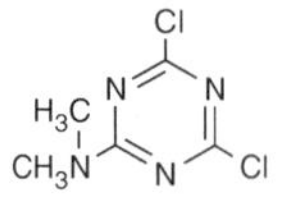

250 mg

S61,840-3
2,6-DICHLORO-4-(1,2,3,4-TETRAHYDRO-1-QUINOLYL)-1,3,5-TRIAZINE

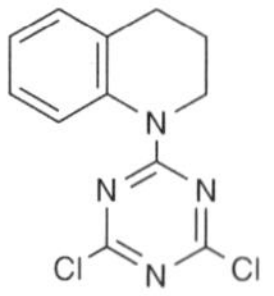

5 gm

S93,717-7
5-PHENYL-3-(2-PYRIDYL)-1,2,4-TRIAZINE

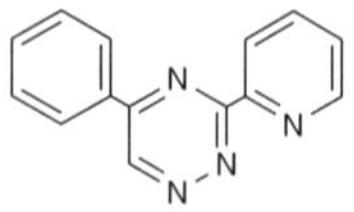

250 mg

S78,402-8
6-AZA-2-THIOURACIL-5-CARBOXYLIC ACID

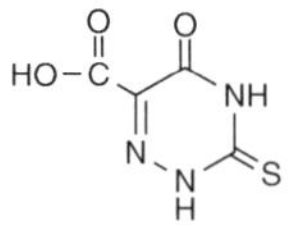

250 mg

S92,053-3
4,5-DIHYDRO-2,4,5-TRIPROPIONYL (1,2,4)TRIAZOLO(3,4-C)(1,2,4)-BENZOTRIAZIN-1(2H)ONE

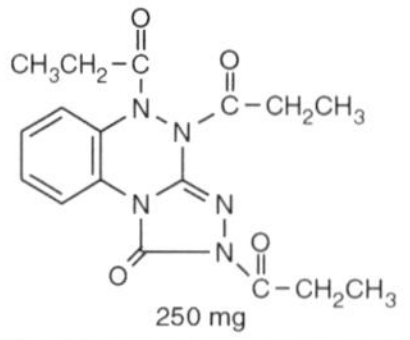

250 mg

S73,918-9
5-AZAURACIL

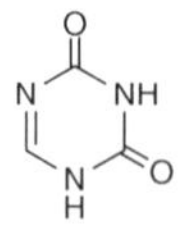

5 gm

S74,874-9
3-(4-ACETAMIDOPHENYL)-5,6-DIPHENYL-1,2,4-TRIAZINE

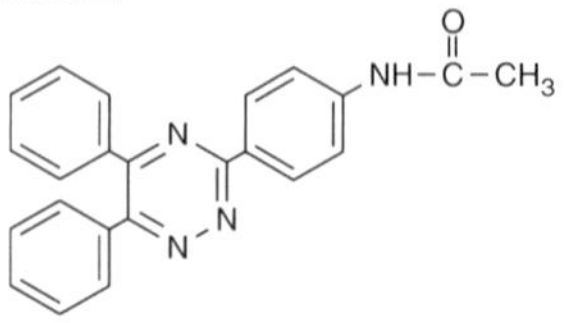

250 mg

S53,368-8
2,4-DICHLORO-6-(2-PYRROLYL)-1,3,5-TRIAZINE

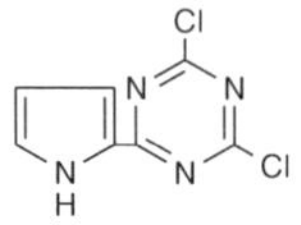

250 mg

S69,540-8
7-(4-METHYLPHENYL)-5,6,8,9-TETRAHYDRODIBENZO(C,H)XANTHYLIUM PERCHLORATE

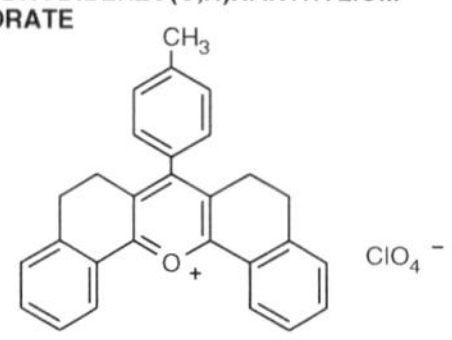

5 gm

S77,527-4
2,4,6-TRIPHENYLTHIOPYRYLIUM PERCHLORATE

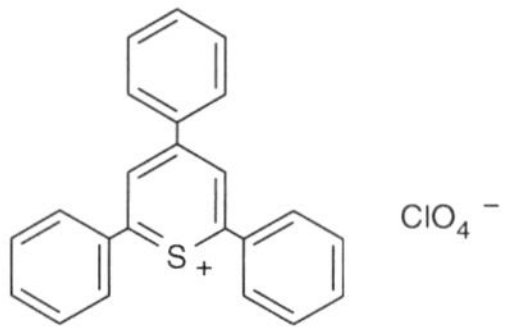

250 mg

S36,128-3
5,6-DIPHENYL-1,2,4-TRIAZINE

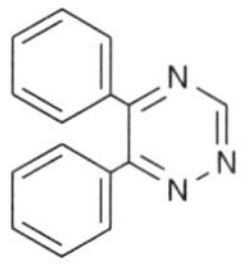

250 mg

S74,872-2
3,5,6-TRIPHENYL-1,2,4-TRIAZINE

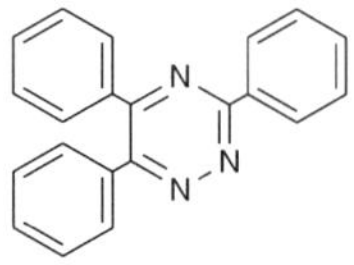

250 mg

S75,803-5
3,5,6-TRIPHENYL-1,2,4-TRIAZINE METHIODIDE

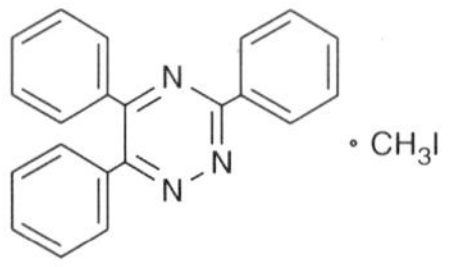

250 mg

S74,876-5
5,6-DIPHENYL-3-(4-PYRIDYL)-1,2,4-TRIAZINE

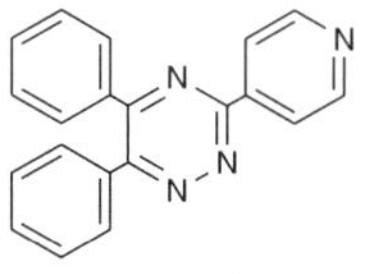

250 mg

S93,714-2
2,4-BIS(5,6-DIMETHYL-1,2,4-TRIAZIN-3-YL)PYRIDINE

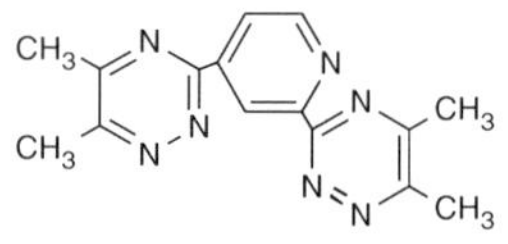

250 mg

S93,716-9
2,6-BIS(5,6-DIPHENYL-1,2,4-TRIAZIN-3-YL)PYRIDINE

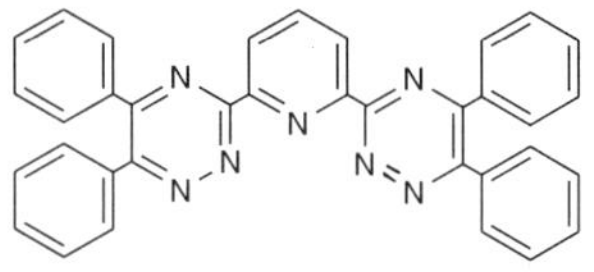

250 mg

S77,007-8
4-(2-BROMOACETYL)-2-(4-CHLOROPHENYL)-6-METHOXYQUINOLINE HYDROBROMIDE

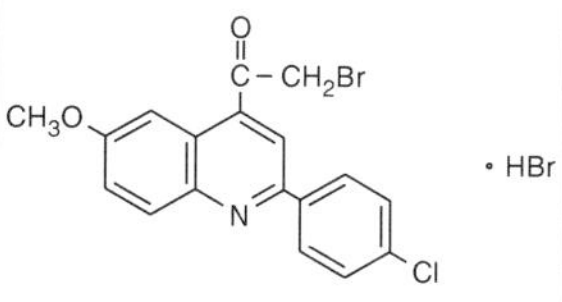

250 mg

S84,078-5
3-CHLORO-2-(CHLOROMETHYL)-6-METHOXY-4-(4-METHOXYANILINO)-QUINOLINE PICRATE

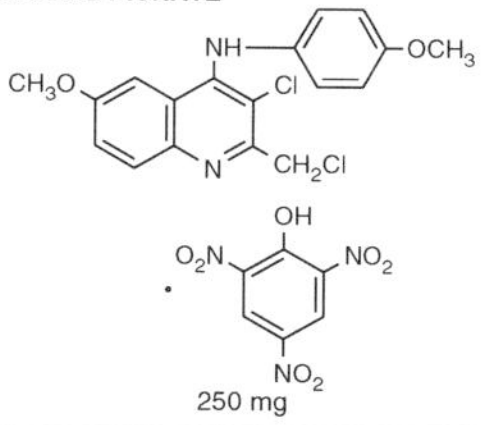

250 mg

S83,866-7
3-(2-BROMOACETYL)-6-METHOXY-2-PHENYLQUINOLINE HYDROBROMIDE

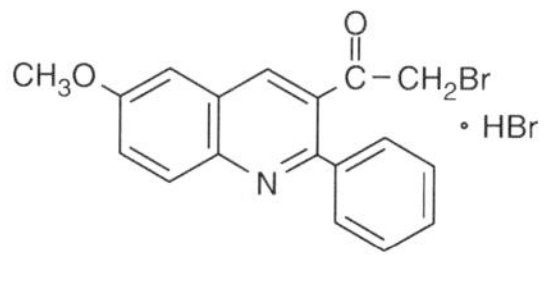

250 mg

S83,867-5
4-(2-BROMOACETYL)-6-METHOXY-2-PHENYLQUINOLINE HYDROBROMIDE

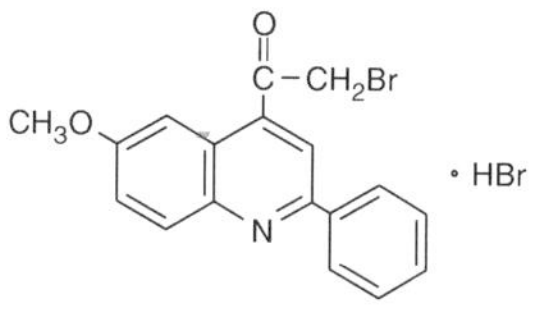

250 mg

S83,960-4
2-(2-CHLOROETHYLAMINO)QUINOLINE HYDROCHLORIDE

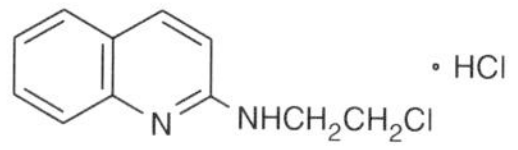

250 mg

S54,282-2
ALPHA-CHLOROQUINALDINE

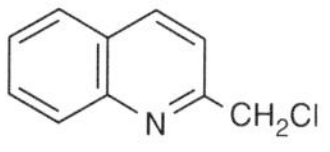

250 mg

S54,297-0
ALPHA,ALPHA-DIBROMOQUINALDINE

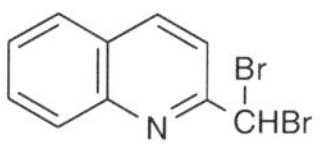

250 mg

S92,773-2
4-CHLORO-6,8-DIMETHOXY-2-(TRIFLUOROMETHYL)QUINOLINE

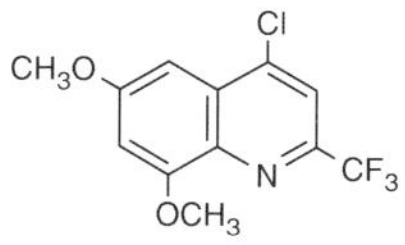

250 mg

S92,772-4
4-CHLORO-8-METHOXY-2-(TRIFLUOROMETHYL)QUINOLINE

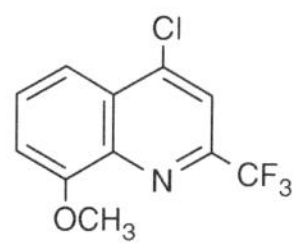

250 mg

S92,771-6
2-(TRIFLUOROMETHYL)-4,7-QUINOLINEDIOL

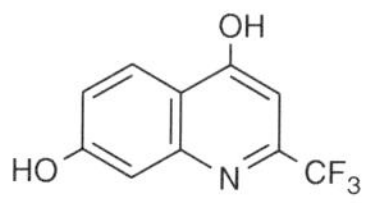

250 mg

S69,672-2
ALPHA-(TRICHLOROMETHYL)-2-QUINOLINEETHANOL

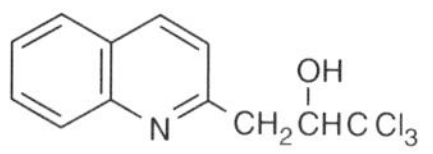

5 gm

S76,133-8
3-AMINO-4-[2,5-BIS(TRIFLUOROMETHYL)-PHENYL]-2(1H)-QUINOLINONE

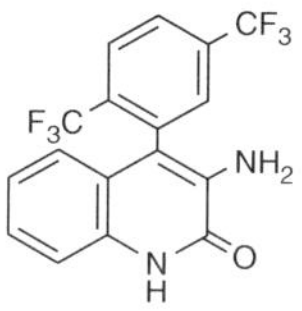

250 mg

S61,163-8
8-QUINOLYL N-[3-(TRIFLUOROMETHYL)-PHENYL]CARBAMATE

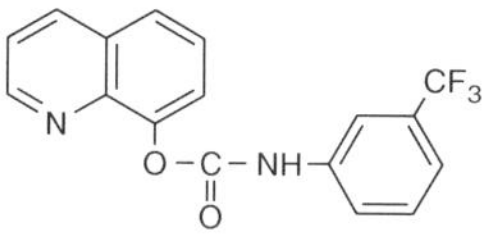

5 gm

S92,770-8
2-(TRIFLUOROMETHYL)-4,5,8-QUINOLINETRIOL

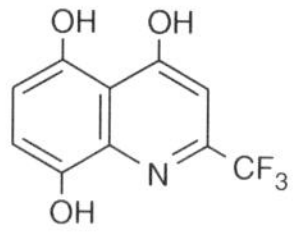

250 mg

S71,281-7
5-CHLORO-8-NAPHTHYL N-(3-TRIFLUOROMETHYLPHENYL)-CARBAMATE

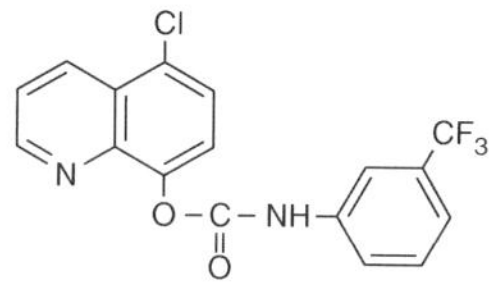

250 mg

S71,771-1
5,7-DICHLORO-8-QUINOLYL N-(3-TRIFLUOROMETHYLPHENYL)-CARBAMATE

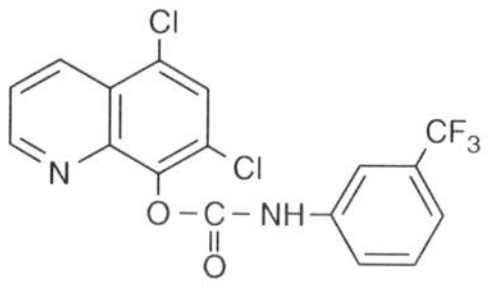

1 gm

S70,956-5
2-METHYL-8-QUINOLYL N-(3-TRIFLUOROMETHYLPHENYL)-CARBAMATE

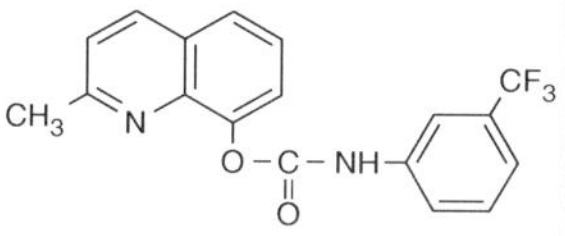

250 mg

S77,079-5
2-(4-CHLOROPHENYL)-ALPHA-(DIHEXYLAMINOMETHYL)-6,8-DIMETHYL-4-QUINOLINEMETHANOL HYDROCHLORIDE

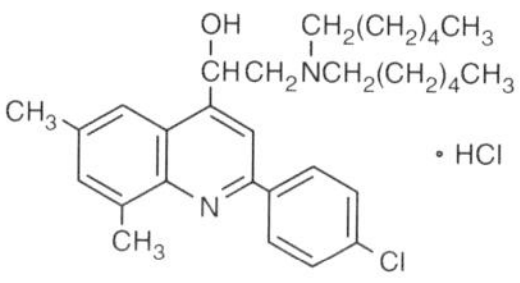

250 mg

S75,016-6
2-(8-QUINOLYLOXY)ETHYL N-(4-CHLORO-2-METHOXYPHENYL)CARBAMATE

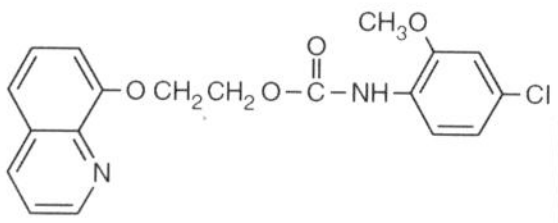

250 mg

S91,284-0
1-(5-CHLORO-2-METHOXYPHENYL)-3-(2-METHYL-4-QUINOLYL)UREA

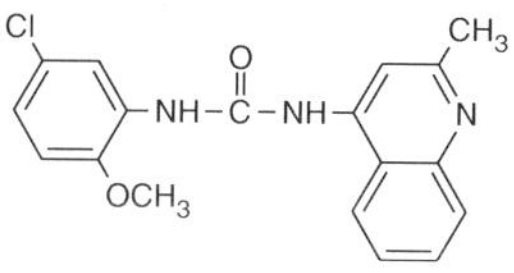

250 mg

S69,952-7
8-QUINOLYL N-(3-CHLORO-2-METHOXYPHENYL)CARBAMATE

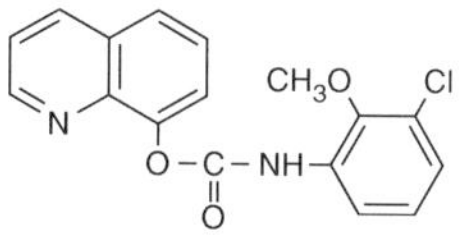

250 mg

S91,211-5
8-QUINOLYL N-(4-CHLORO-2-METHOXYPHENYL)CARBAMATE

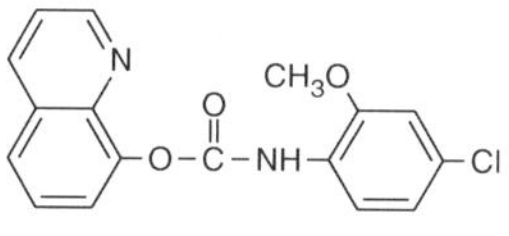

250 mg

S91,401-0
2-METHYL-8-QUINOLYL N-(4-CHLORO-2-METHOXYPHENYL)CARBAMATE

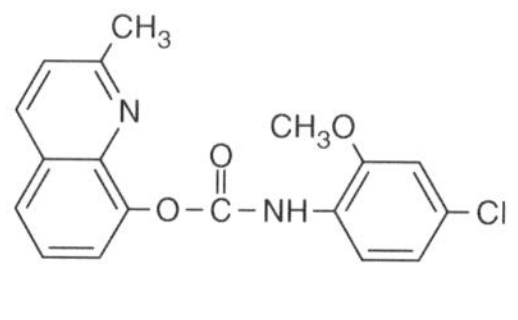

250 mg

S91,404-5
4-METHYL-2-QUINOLYL N-(4-CHLORO-2-METHOXYPHENYL)CARBAMATE

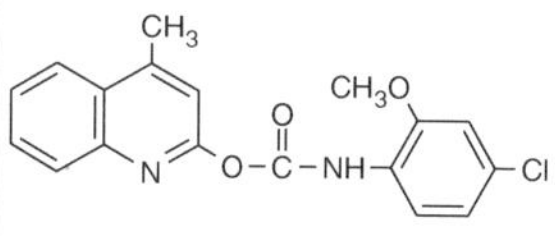

250 mg

S91,405-3
2-METHYL-4-QUINOLYL N-(4-CHLORO-2-METHOXYPHENYL)CARBAMATE

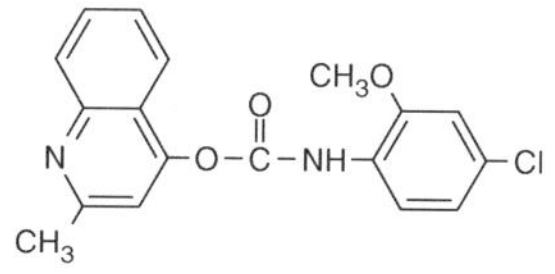

250 mg

S79,849-5
2-(4-CHLOROPHENYL)-6-METHOXY-4-(2-PYRIDYLCARBONYL)QUINOLINE

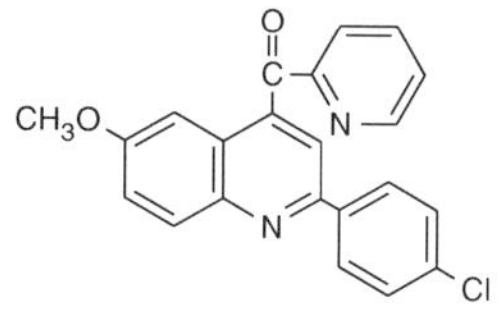

250 mg

S73,157-9
8-QUINOLYL N-(2-CHLORO-4-METHOXYPHENYL)CARBAMATE

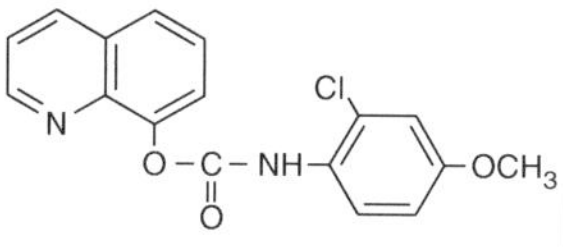

5 gm

S75,718-7
2-(4-CHLOROPHENYL)-4-(1,2-EPOXYETHYL)-6-METHOXYQUINOLINE

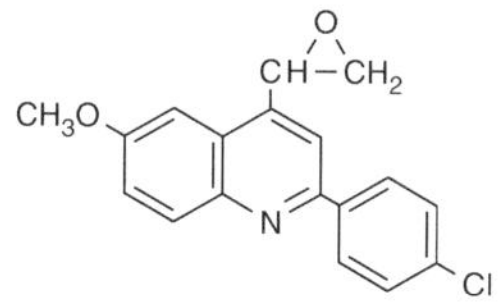

5 gm

S93,480-1
2-(2-FLUOROBENZOYL)-2-QUINOLYLACETONITRILE

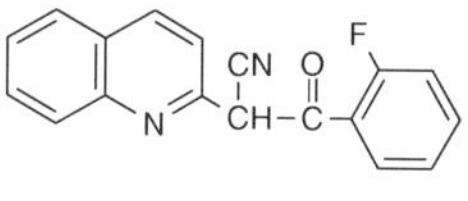

250 mg

S72,306-1
8-QUINOLYL N-(4-CHLORO-3-NITROPHENYL)CARBAMATE

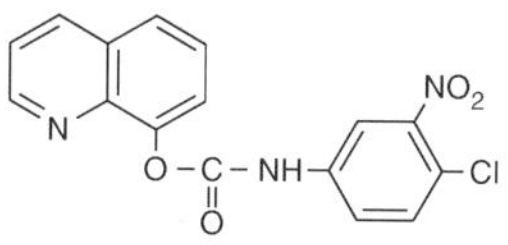

250 mg

S72,977-9
5-NITRO-8-QUINOLYL N-(4-FLUOROPHENYL)CARBAMATE

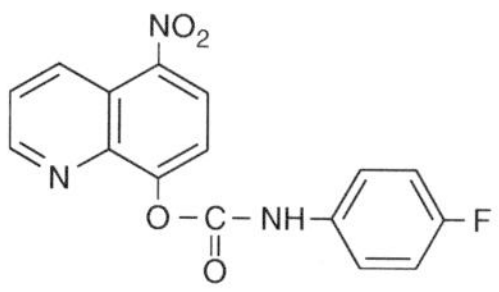

5 gm

S91,317-0
1-(3-CHLORO-4-METHYLPHENYL)-3-(2-METHYL-4-QUINOLYL)UREA

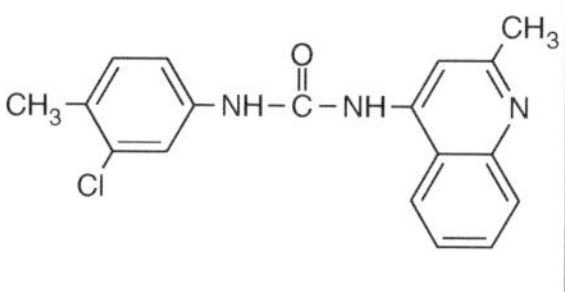

250 mg

S91,649-8
2-METHYL-8-QUINOLYL N-(2-CHLORO-4-NITROPHENYL)CARBAMATE

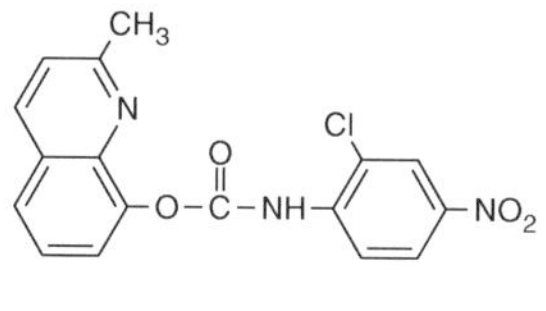

250 mg

S68,626-3
2-(4-CHLOROSTYRYL)-1-ETHYLQUINOLINIUM IODIDE

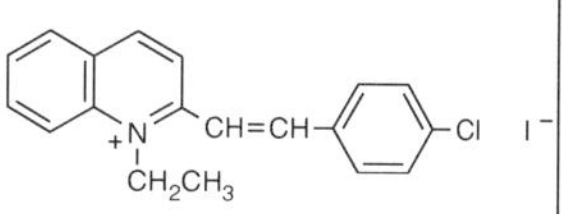

5 gm

S94,955-8
2-(4-BROMOPHENYL)-4-QUINOLINECARBOXYLIC ACID

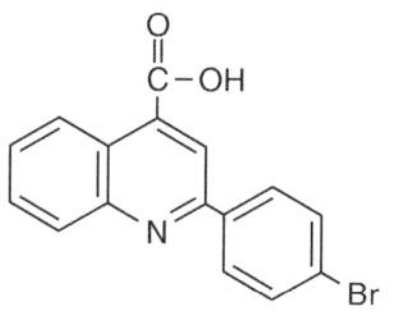

250 mg

S91,622-6
8-QUINOLYL N-(2,5-DIFLUOROPHENYL)-CARBAMATE

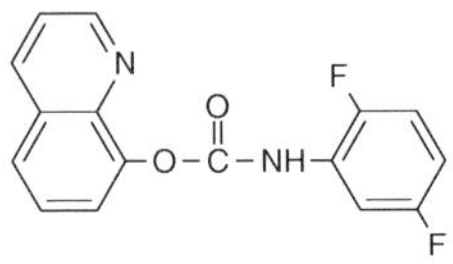

250 mg

S77,972-5
2-METHYL-8-QUINOLYL N-(2,5-DIFLUOROPHENYL)CARBAMATE

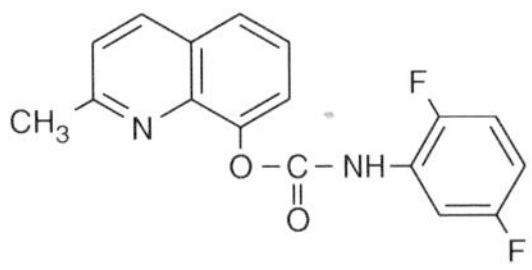

250 mg

S71,077-6
4-CHLORO-8-QUINOLYL N-(5-CHLORO-2-METHYLPHENYL)CARBAMATE

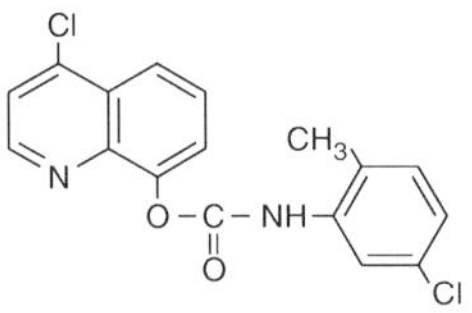

5 gm

S71,078-4
4-CHLORO-8-QUINOLYL N-(4-CHLORO-2-METHYLPHENYL)CARBAMATE

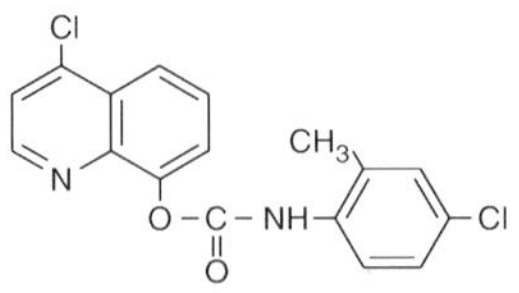

5 gm

S92,595-0
5-NITRO-8-QUINOLYL N-(3-CHLOROPHENYL)CARBAMATE

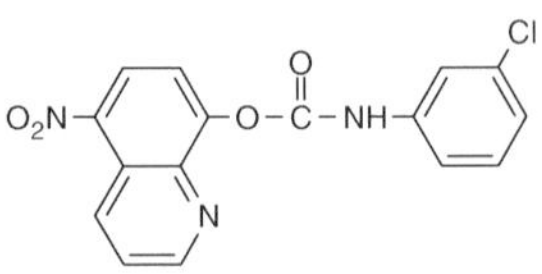

250 mg

S76,494-9
6-CHLORO-2-(4-CHLOROPHENYL)-4-QUINOLINECARBOXYLIC ACID

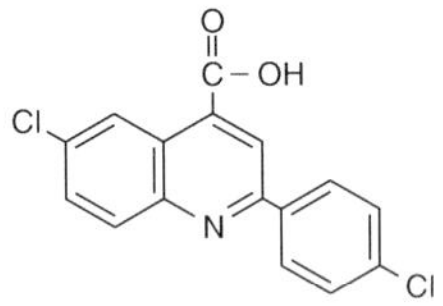

5 gm

S77,009-4
7-CHLORO-2-(4-CHLOROPHENYL)-6-METHYL-4-QUINOLINECARBOXYLIC ACID

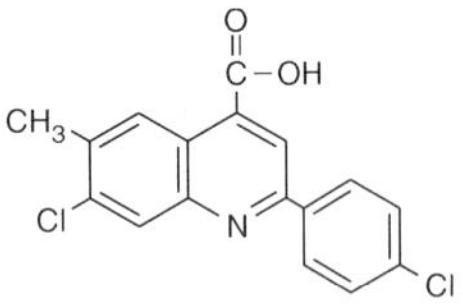

250 mg

S75,716-0
7-CHLORO-2-(4-CHLOROPHENYL)-6-METHYL-4-QUINOLINECARBONYL CHLORIDE

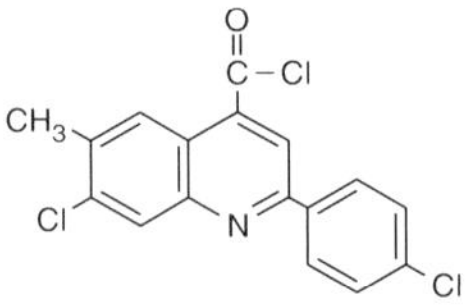

5 gm

S34,459-1
2-(P-CHLOROPHENYL)QUINOLINE-4-CARBOXYLIC ACID

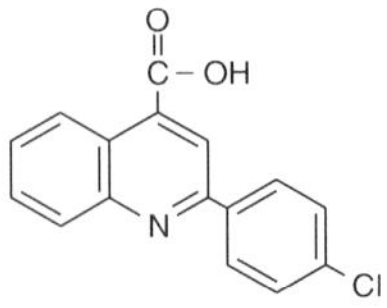

250 mg

S71,726-6
5,7-DICHLORO-8-QUINOLYL N-(4-CHLOROPHENYL)CARBAMATE

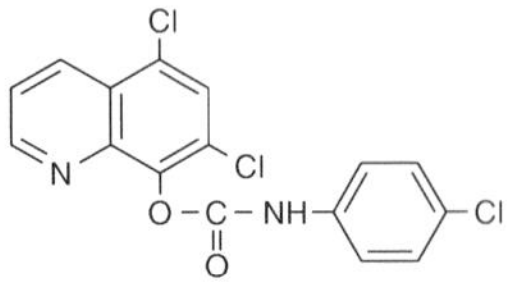

50 mg

S71,767-3
5-CHLORO-8-QUINOLYL N-(4-BROMOPHENYL)CARBAMATE

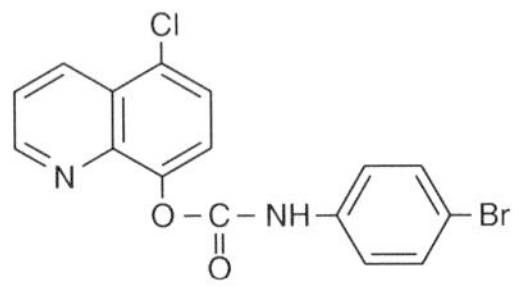

5 gm

S71,768-1
2-METHYL-8-QUINOLYL N-(4-FLUOROPHENYL)CARBAMATE

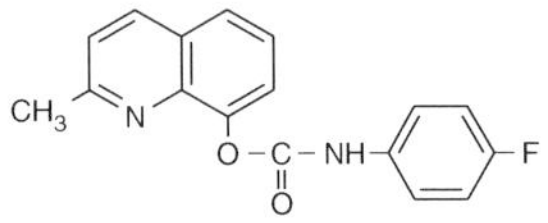

5 gm

S71,141-1
8-QUINOLYL N-(2,6-DICHLOROPHENYL)-CARBAMATE

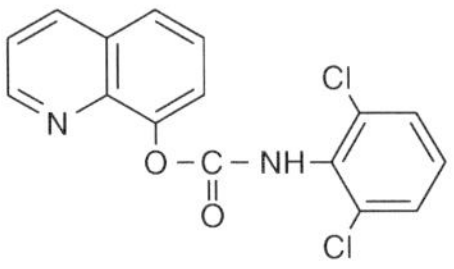

250 mg

S70,927-1
8-QUINOLYL N-(2-CHLORO-6-METHYLPHENYL)CARBAMATE

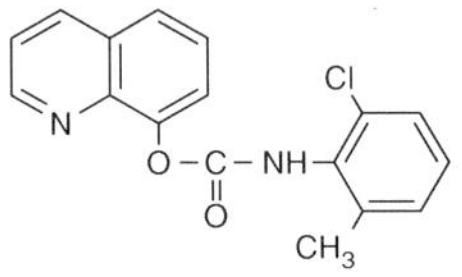

5 gm

S69,948-9
2,4-DICHLORO-8-QUINOLYL N-(3-CHLORO-2-METHYLPHENYL)-CARBAMATE

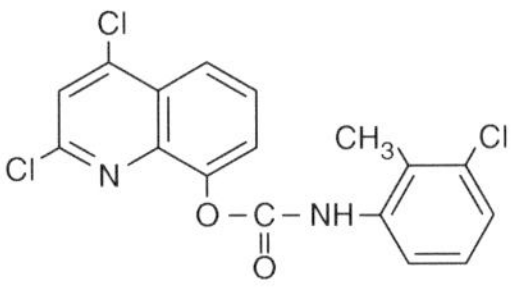

250 mg

S71,684-7
5,7-DICHLORO-8-QUINOLYL N-(3,4-DICHLOROPHENYL)CARBAMATE

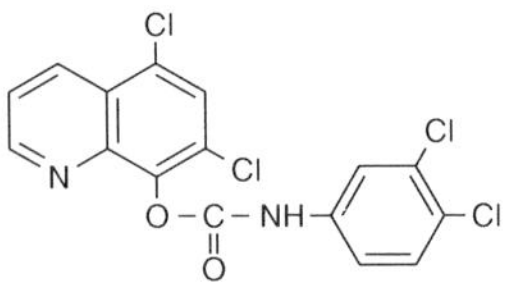

250 mg

S71,763-0
5,7-DICHLORO-8-QUINOLYL N-(2,4-DICHLOROPHENYL)CARBAMATE

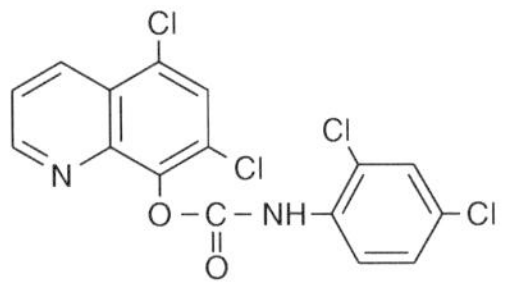

5 gm

S70,926-3
2-METHYL-8-QUINOLYL N-(3,5-DICHLOROPHENYL)CARBAMATE

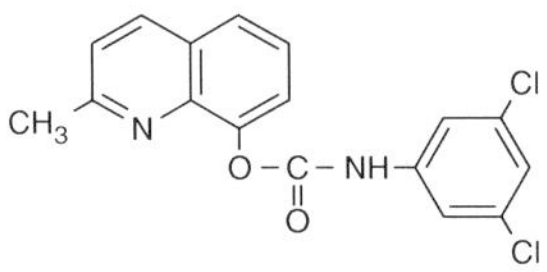

250 mg

S71,133-0
2-METHYL-8-QUINOLYL N-(2,4,5-TRICHLOROPHENYL)CARBAMATE

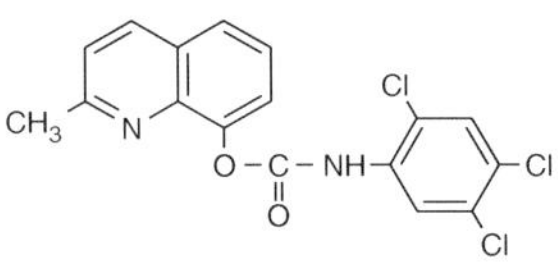

5 gm

S71,274-4
8-QUINOLINYL N-(3-BROMOPHENYL)-CARBAMATE

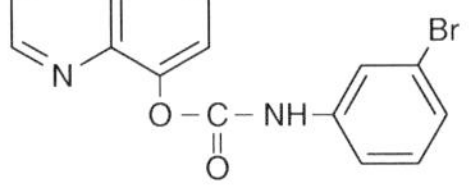

250 mg

S45,665-9
2,4-BIS[1-ETHYL-2(1H)-QUINOLYLIDENEMETHYL]-1,3-CYCLOBUTADIENEDIYLIUM-1,3-DIOLATE

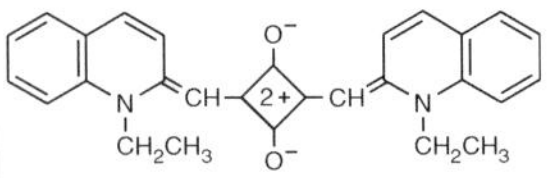

250 mg

S60,600-6
5,8-DIMETHOXY-4-PENTADECYL-2-HYDROXYQUINOLINE

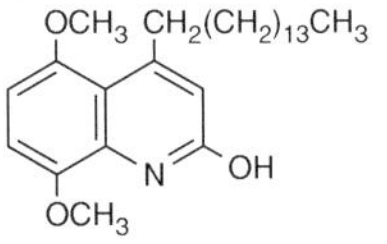

250 mg

S86,161-8
DIMETHYL 7-HYDROXY-8-METHOXY-2,4-QUINOLINEDICARBOXYLATE

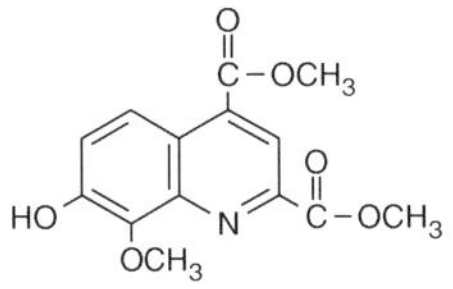

250 mg

S65,711-5
4-[N-(5-HYDROXYPENTYL)AMINO]-8-METHOXYQUINOLINE

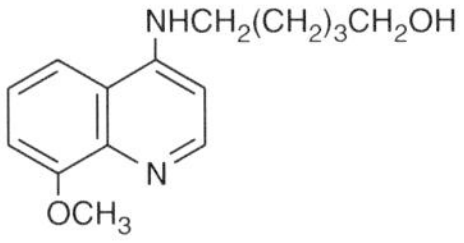

250 mg

S42,590-7
NIQUINE DIOXALATE

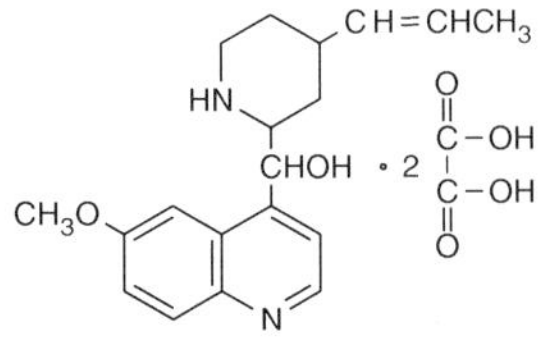

250 mg

S64,130-8
QUININE SALICYLATE

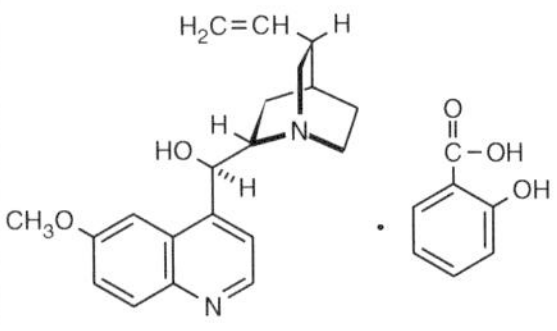

5 gm

S64,141-3
QUININE BISULFATE

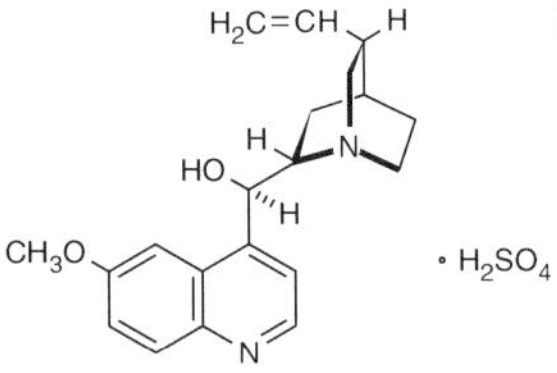

5 gm

S76,274-1
EPIQUINIDINE

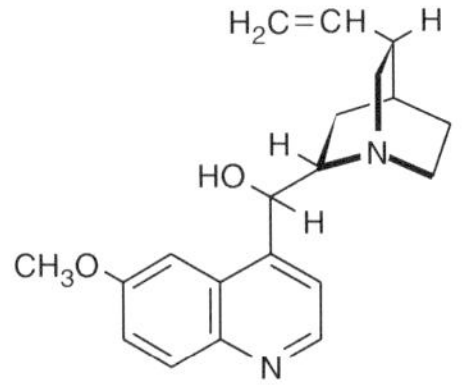

250 mg

S83,637-0
2-(4-HYDROXYPHENYL)-6-METHOXY-4-QUINOLINECARBOXYLIC ACID

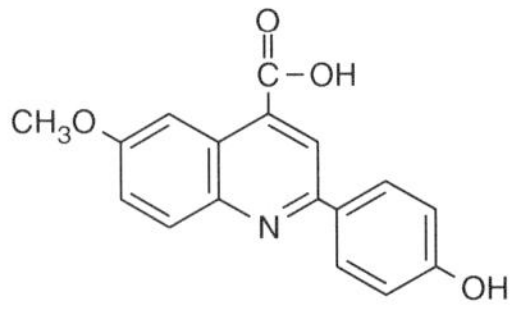

250 mg

S39,292-8
ETHYL 4-HYDROXY-7-METHOXY-3-QUINOLINECARBOXYLATE

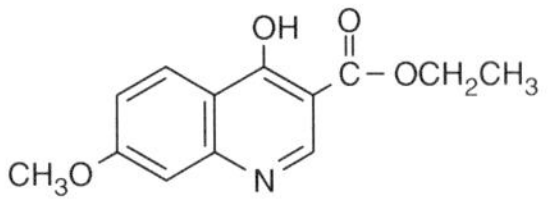

5 gm

S43,087-0
3-CARBOXY-4-HYDROXY-7-METHOXYQUINOLINE

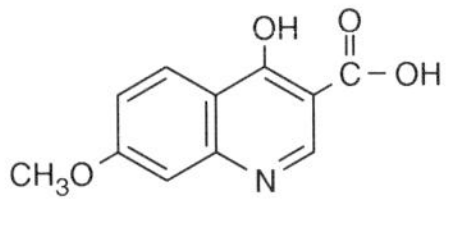

250 mg

S40,703-8
3-CARBOXY-4-HYDROXY-6-METHOXYQUINOLINE

5 gm

S37,114-9
ETHYL 4-HYDROXY-6-METHOXY-3-QUINOLINECARBOXYLATE

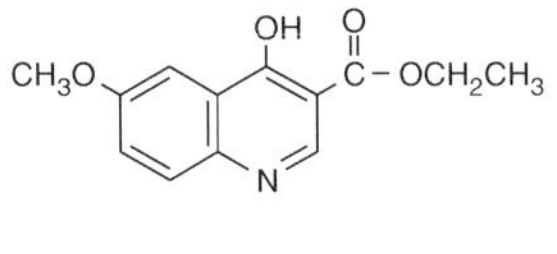

5 gm

S51,219-2
4-HYDROXY-2[4-(P-TOLYLOXY)PHENYL]-6-QUINOLINESULFONYL FLUORIDE

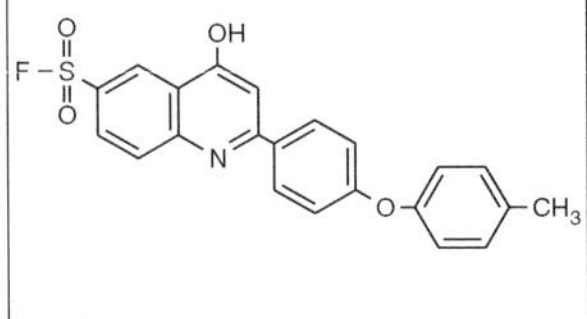

250 mg

S45,810-4
QUININE DIHYDROCHLORIDE MONOHYDRATE

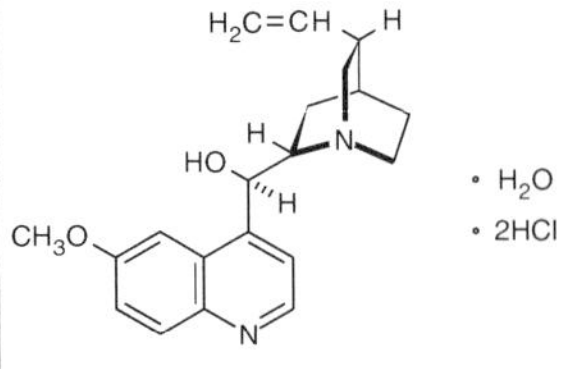

5 gm

S81,000-2
6-METHOXY-ALPHA-METHYL-2-PHENYL-4-QUINOLINEMETHANOL
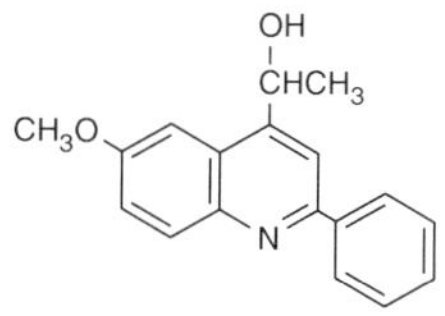
250 mg

S43,819-7
4-HYDROXY-6-METHOXYQUINOLINE
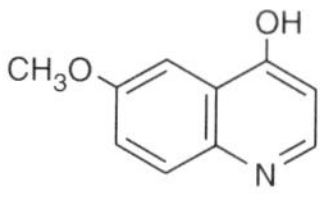
250 mg

S51,224-9
7-METHOXY-2-METHYL-4-QUINOLINOL
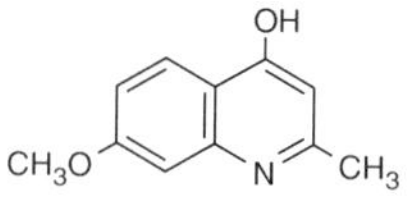
250 mg

S60,606-5
4-HYDROXY-6-METHOXY-2-PENTADECYLQUINOLINE
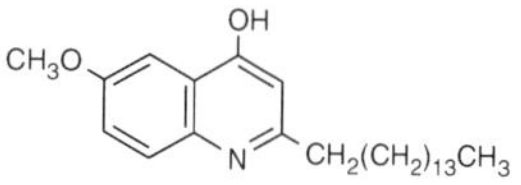
250 mg

S82,505-0
ALPHA,ALPHA-DIPHENYL-2-QUINOLINEMETHANOL
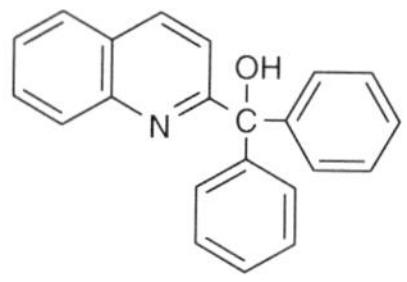
250 mg

S87,094-3
ALPHA,ALPHA-DIPHENYL-5-QUINOLINEMETHANOL
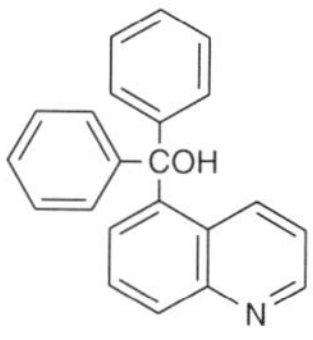
250 mg

S88,695-5
2-HYDROXY-3-PHENYL-4-QUINOLINECARBOXYLIC ACID
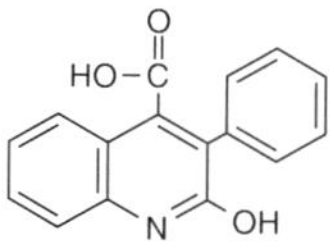
250 mg

S43,703-4
2-HYDROXYQUINOLINE-4-CARBOXYLIC ACID
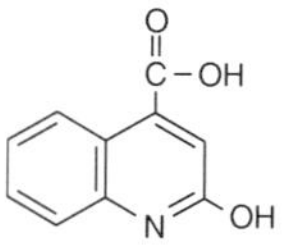
5 gm

S75,729-2
6,8-DICHLORO-2-HYDROXY-4-QUINOLINECARBOXYLIC ACID
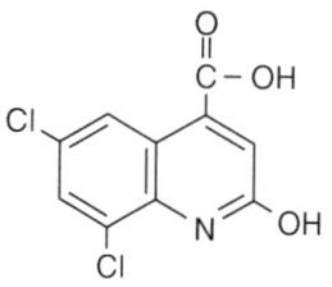
5 gm

S77,432-4
ETHYL 6-CHLORO-2-HYDROXY-4-QUINOLINECARBOXYLATE
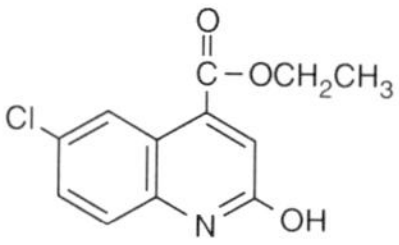
250 mg

S77,076-0
ETHYL 6,8-DICHLORO-2-HYDROXY-4-QUINOLINECARBOXYLATE
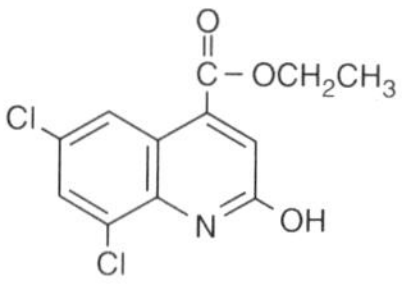
250 mg

S63,034-9
6-AMINO-2-(4-FLUOROSULFONYLPHENYL)-4-HYDROXYQUINOLINE
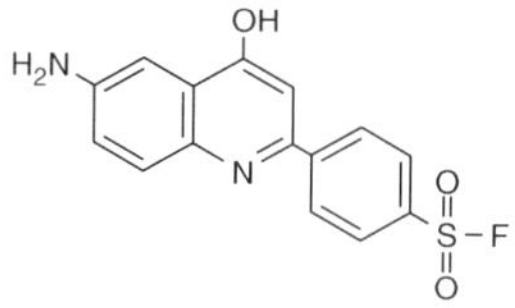
5 gm

S93,840-8
3-(2-HYDROXYBENZYLIDENEAMINO)-QUINOLINE
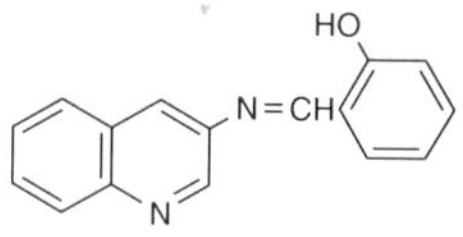
250 mg

S77,004-3
7-CHLORO-4-(4-HYDROXYANILINO)-QUINOLINE
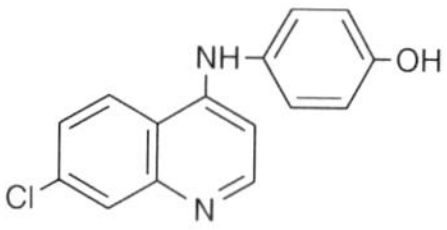
250 mg

S77,025-6
7-CHLORO-4-(2-HYDROXYETHYLAMINO)-QUINOLINE
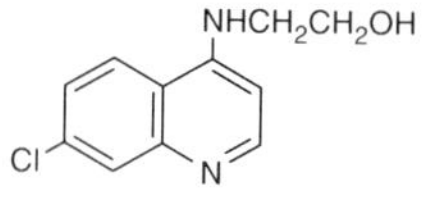
250 mg

S37,152-1
5,7-DINITRO-8-QUINOLINOL
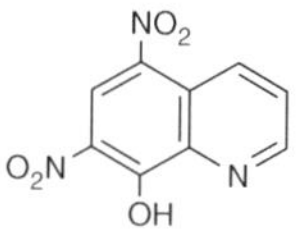
250 mg

S37,285-4
ETHYL 4-HYDROXY-7-NITRO-3-QUINOLINECARBOXYLATE
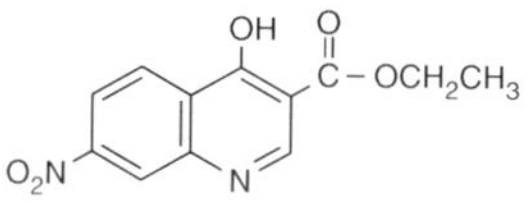
250 mg

S59,307-9
N-(2-HYDROXY-4-METHYL-7-QUINOLYL)-ACETOACETAMIDE
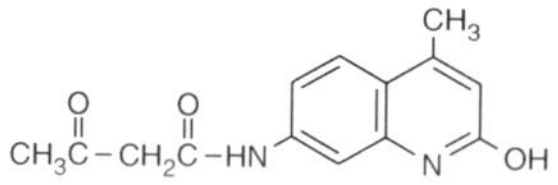
250 mg

S65,502-3
2-(4-FLUOROSULFONYLPHENYL)-4-HYDROXY-6-NITROQUINOLINE
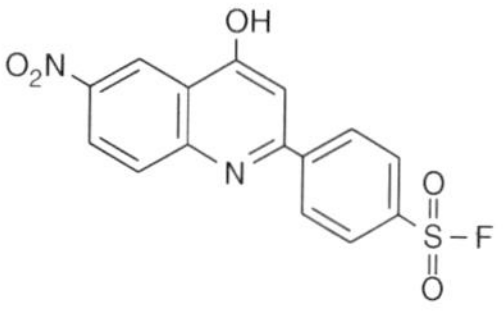
5 gm

S81,285-4
2-BENZYL-8-HYDROXYQUINOLINE
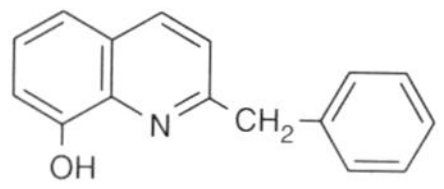
250 mg

S65,475-2
CINCHONINE SULFATE
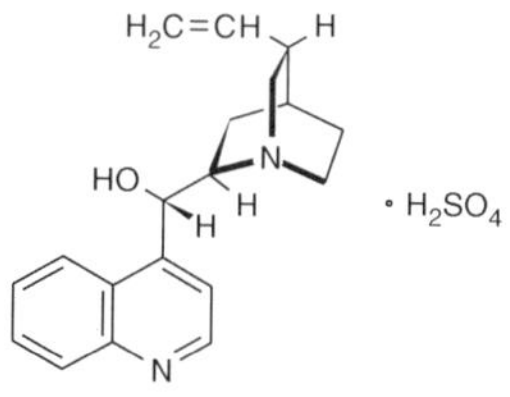
5 gm

S87,964-9
1-(2-HYDROXYETHYL)QUINOLINIUM P-TOLUENESULFONATE
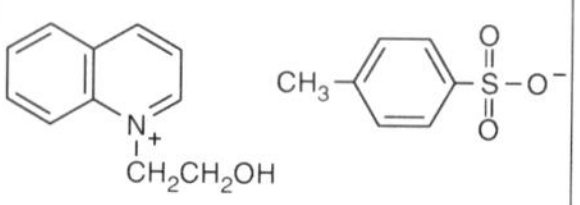
5 gm

S67,927-5
1-ETHYL-2-(2-HYDROXYSTYRYL)-QUINOLINIUM IODIDE
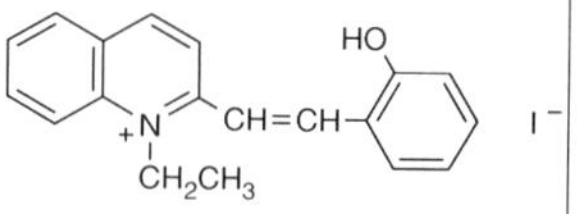
1 gm

S39,921-3
2-(2-HYDROXY-1-NAPHTHYLVINYL)-1-METHYLQUINOLINIUM IODIDE
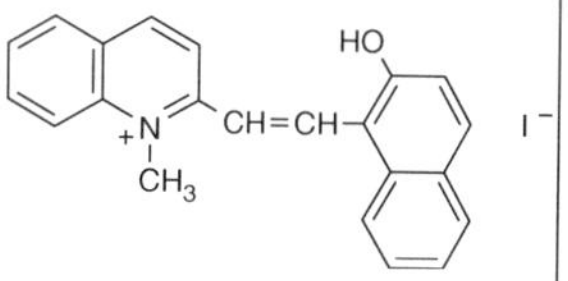
250 mg

S63,116-7
4-HYDROXY-2-PHENYL-6-QUINOLINESULFONIC ACID
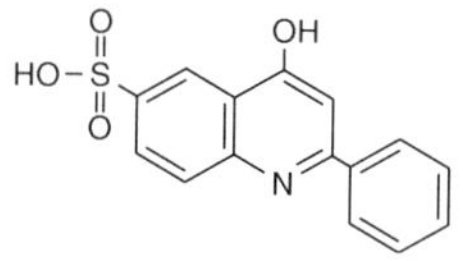
250 mg

S93,280-9
4-HYDROXY-2-PHENYLQUINOLINE
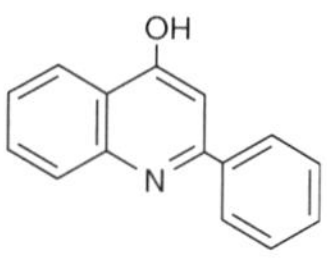
250 mg

S64,642-3
6,6'-BI(4-HYDROXY-2-PHENYLQUINOLINE)
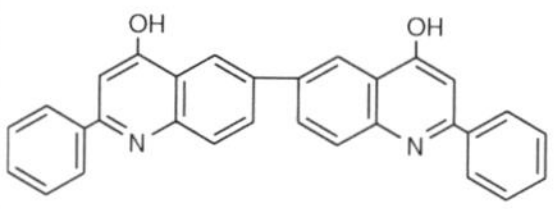
250 mg

S39,788-1
8-HYDROXY-5-QUINOLYL METHYL KETONE
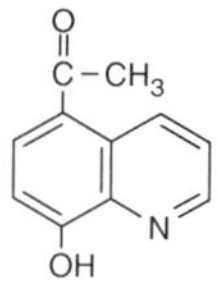
5 gm

S35,116-4
3-HYDROXY-2-PHENYLCINCHONINIC ACID
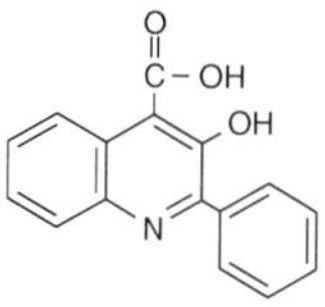
5 gm

S83,863-2
8-HYDROXYQUINOLINE-4-CARBOXYLIC ACID
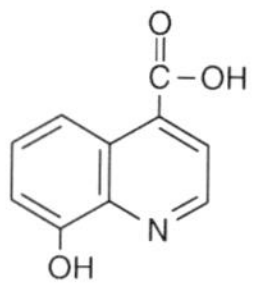
250 mg

S44,067-1
8-HYDROXYQUINOLINE, ZINC SALT
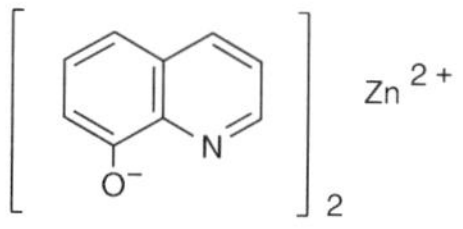
5 gm

S44,089-2
8-HYDROXYQUINOLINE BENZOATE
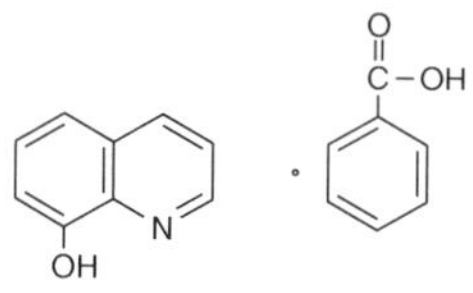
5 gm

S44,287-9
8-HYDROXYQUINOLINE, ZONE REFINED
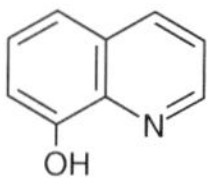
5 gm

S84,071-8
8-HYDROXYQUINOLINE PICRATE
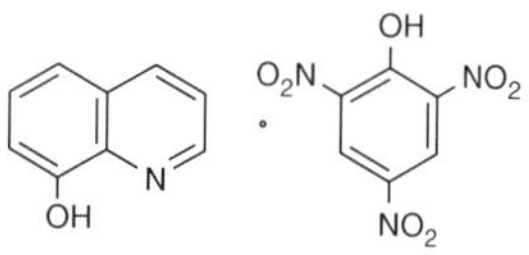
250 mg

S56,374-9
7-BROMO-5-CHLORO-8-QUINOLINOL
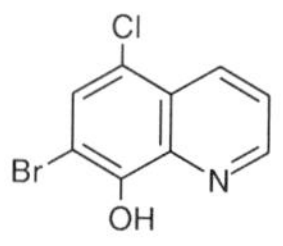
5 gm

S43,818-9
5,7-DIIODO-8-HYDROXY-2-METHYLQUINOLINE

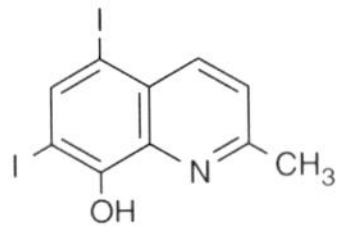

5 gm

S66,244-5
CINCHONINE SULFATE HEXAHYDRATE

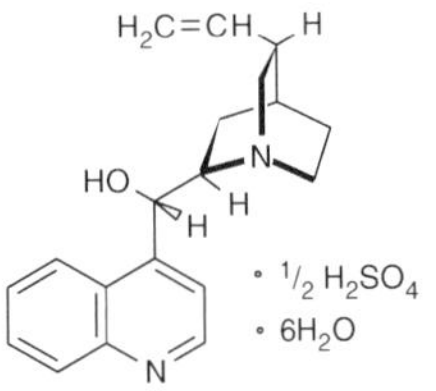

250 mg

S34,369-2
2-(3-CARBOXY-4-HYDROXYPHENYL)-QUINOLINE-4-CARBOXYLIC ACID

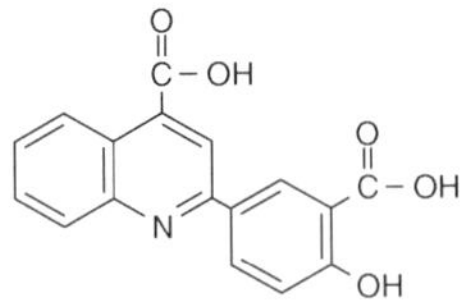

250 mg

S37,499-7
7-CHLORO-4-HYDROXY-2-PHENYLQUINOLINE-3-CARBOXYLIC ACID

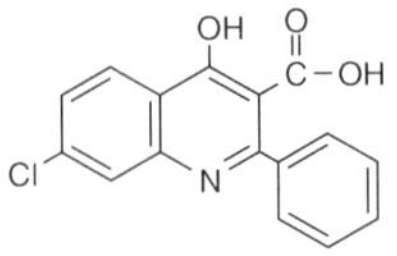

250 mg

S58,126-7
7,8-DIMETHYL-4-HYDROXY-3-QUINOLINECARBOXYLIC ACID

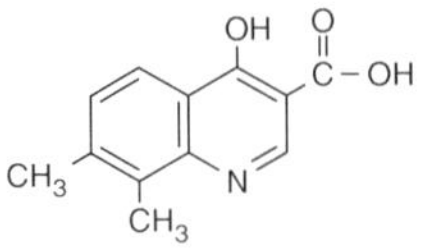

250 mg

S43,764-6
ETHYL 8-CHLORO-4-HYDROXY-3-QUINOLINECARBOXYLATE

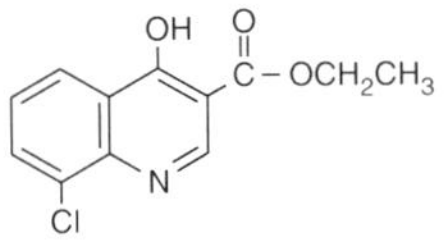

5 gm

S58,148-8
ETHYL 4-HYDROXY-6-METHYL-3-QUINOLINECARBOXYLATE

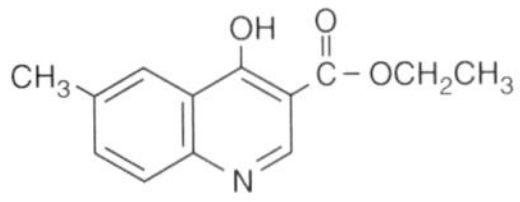

250 mg

S50,060-7
4-HYDROXY-2-PHENYL-3-QUINOLINECARBOXYLIC ACID

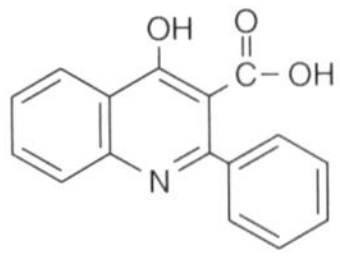

50 mg

S44,021-3
6-CHLORO-4-HYDROXYGUINOLINE-3-CARBOXYLIC ACID

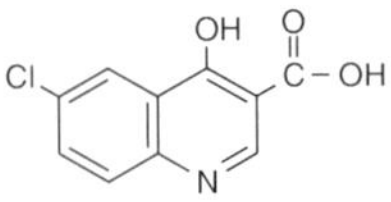

5 gm

S44,784-6
4-HYDROXY-2-PHENYL-6-QUINOLINESULFONYL FLUORIDE

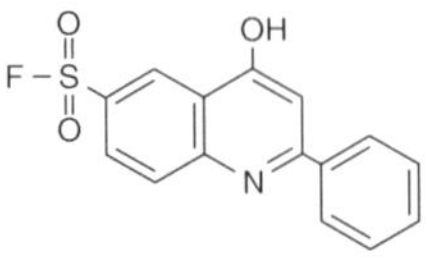

5 gm

S77,022-1
4-HYDROXY-3-PHENYLQUINOLINE

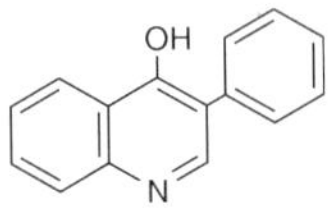

250 mg

S86,955-4
8-QUINOLINEMETHANOL HYDROCHLORIDE

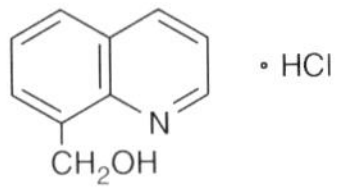

250 mg

S86,956-2
5-QUINOLINEMETHANOL

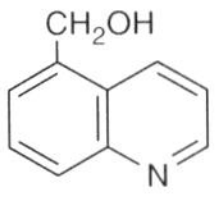

250 mg

S50,549-8
6-CHLORO-2-METHYL-4-QUINOLINOL

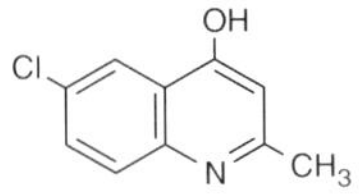

250 mg

S37,618-3
4,8-DIMETHYL-2-HYDROXYQUINOLINE

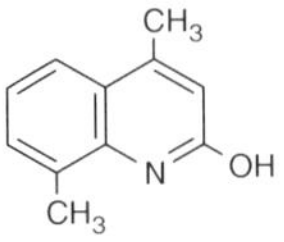

5 gm

S37,619-1
4,6-DIMETHYL-2-HYDROXYQUINOLINE

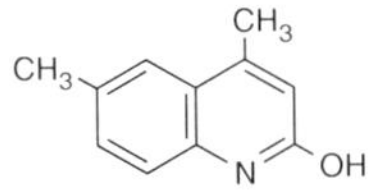

5 gm

S38,549-2
4,7-DIMETHYL-2-QUINOLINOL

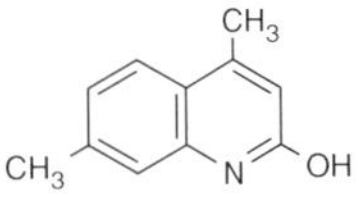

5 gm

S58,156-9
8-ETHYL-4-QUINOLINOL

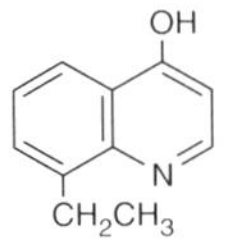

250 mg

S64,761-6
5-CHLORO-4-HYDROXY-8-METHYLQUINALDINE

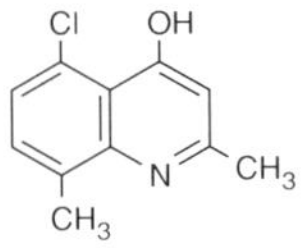

5 gm

S86,980-5
ALPHA,ALPHA-DIMETHYL-3-QUINOLINEMETHANOL

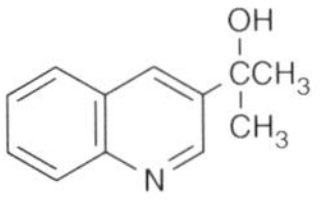

50 mg

S86,981-3
ALPHA,ALPHA-DIMETHYL-4-QUINOLINEMETHANOL

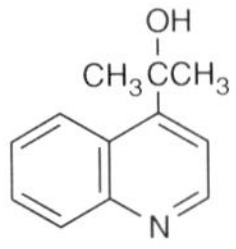

250 mg

S86,982-1
ALPHA,ALPHA-DIMETHYL-6-QUINOLINEMETHANOL PERCHLORATE

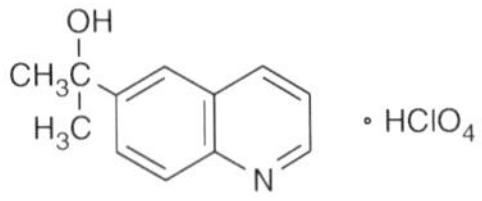

250 mg

S38,499-2
4,4'-DIMETHYL-2,2'-DIHYDROXY-6,6'-BIQUINOLINE

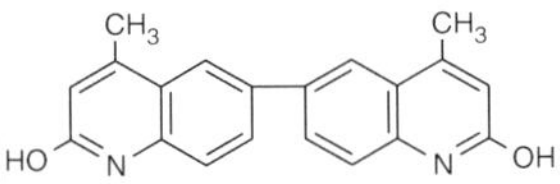

5 gm

S38,501-8
4,4',8,8'-TETRAMETHYL-2,2'-DIHYDROXY-6,6'-BIQUINOLINE

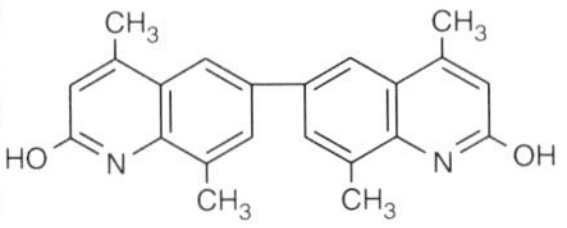

5 gm

S73,588-4
6,7,8,9,10,11,12,13,14,15-DECAHYDRO-16-HYDROXY-CYCLODODECA(B)QUINOLINE

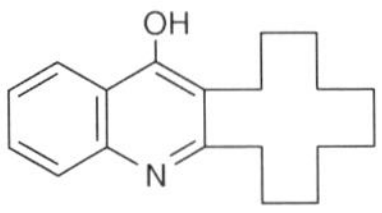

250 mg

S68,632-8
2-(4-BENZYLOXYSTYRYL)-1-ETHYLQUINOLINIUM IODIDE

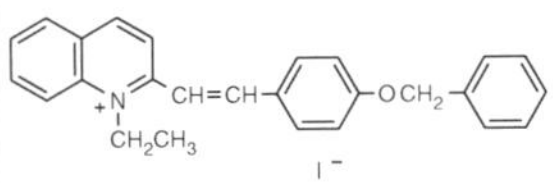

5 gm

S67,856-2
2-(3,4-DIMETHOXYSTYRYL)-1-ETHYLQUINOLINIUM IODIDE

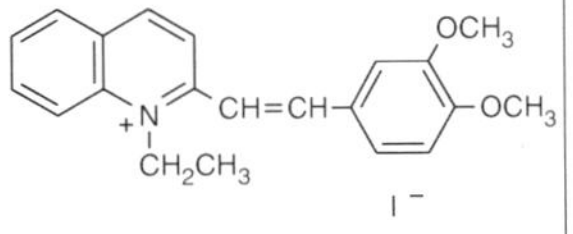

5 gm

S89,733-7
6,7-DIMETHOXY-4-METHYLQUINOLINE

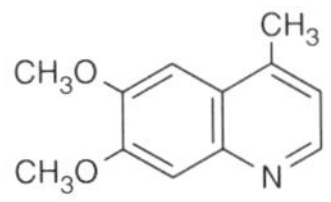

250 mg

S76,871-5
8-QUINOLYL N-(2,4-DIMETHOXYPHENYL)CARBAMATE

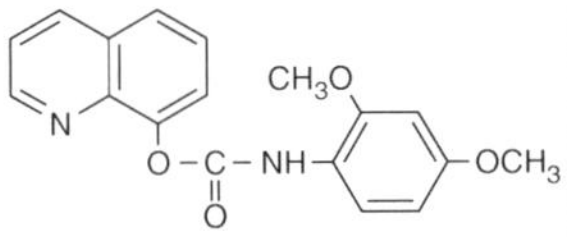

250 mg

S72,332-0
4-CHLORO-8-QUINOLYL N-(2,5-DIMETHOXYPHENYL)CARBAMATE

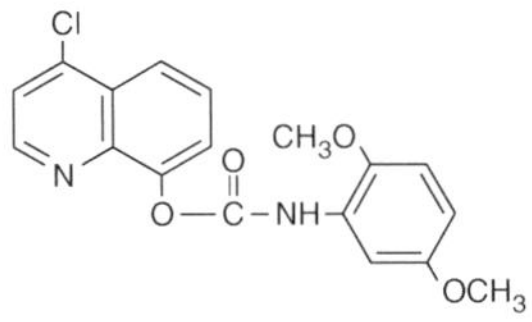

5 gm

S91,735-4
2,4-DICHLORO-8-QUINOLYL N-(2,5-DIMETHOXYPHENYL)CARBAMATE

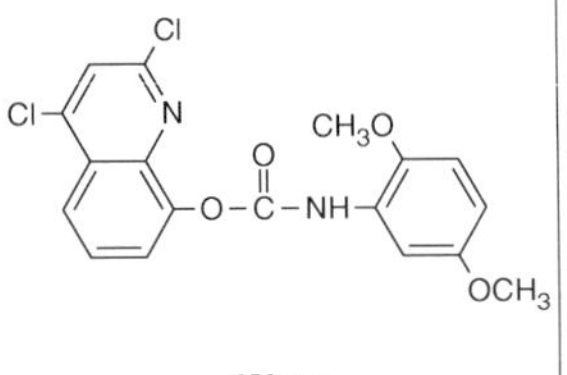

250 mg

S72,086-0
BIS(8-QUINOLYL) N,N'-(3,3'-DIMETHOXYBIPHENYL-4,4'-DIYL)-BISCARBAMATE

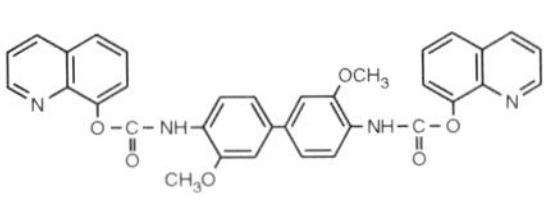

250 mg

S92,431-8
BIS(8-QUINOLYL)-N,N'-HEXAMETHYLENEBISCARBAMATE

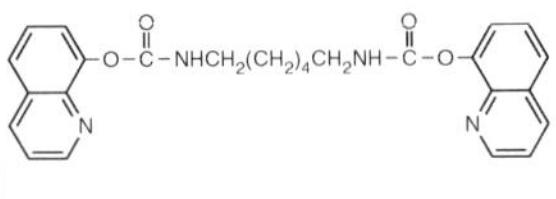

250 mg

S50,451-3
2-(3,4-DIMETHOXYSTYRYL)-1-METHYLQUINOLINIUM IODIDE

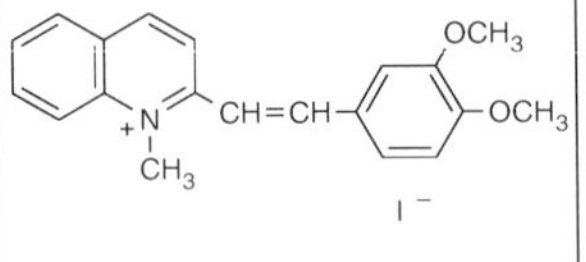

250 mg

S83,530-7
2-(8-QUINOLYLOXY)ACETIC ACID HYDROCHLORIDE

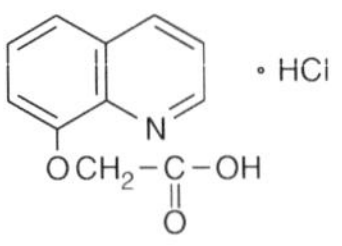

250 mg

S53,344-0
2-(3,4-DIMETHOXYPHENYL)-7-METHYLCINCHONINIC ACID

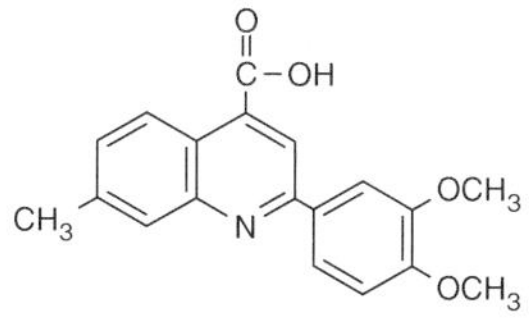

5 gm

S83,529-3
2-(5,7-DIIODO-8-QUINOLYLOXY)ACETIC ACID

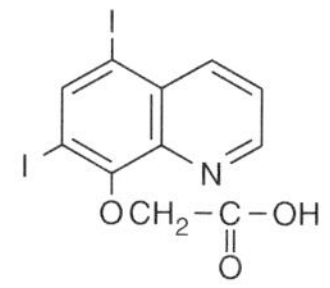

250 mg

S76,687-9
2-[10-CHLORO-5H-BENZO(6,7)-CYCLOHEPTA(1,2-B)QUINOLIN-8-YLOXY]ACETIC ACID

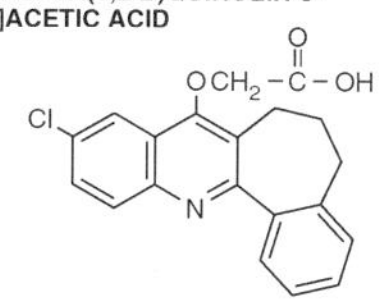

250 mg

S59,800-3
4-CHLORO-6-FLUOROSULFONYL-2-(4-ETHOXY-3-METHOXYPHENYL)QUINOLINE

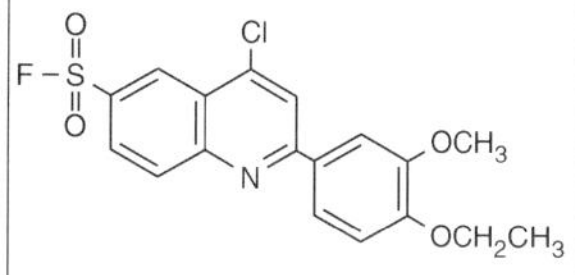

5 gm

S81,071-1
8-AMINO-6-METHOXYQUINOLINE HYDROCHLORIDE

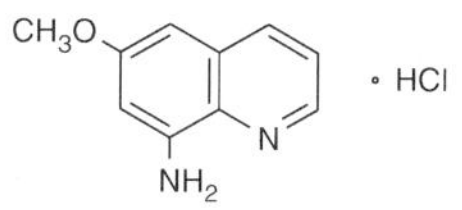

250 mg

S81,711-2
8-AMINO-6-METHOXYQUINOLINE PICRATE

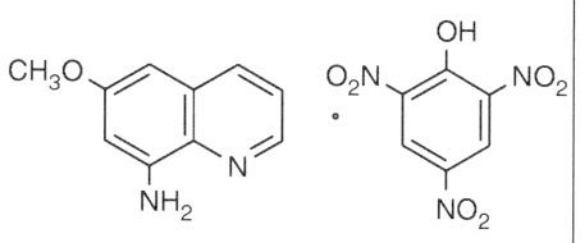

250 mg

S70,725-2
6-AMINO-4,5-DIMETHYL-8-METHOXYQUINOLINE

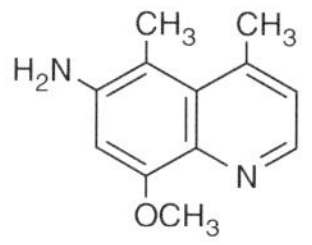

5 gm

S45,315-3
PLASMOCID DIHYDROIODIDE

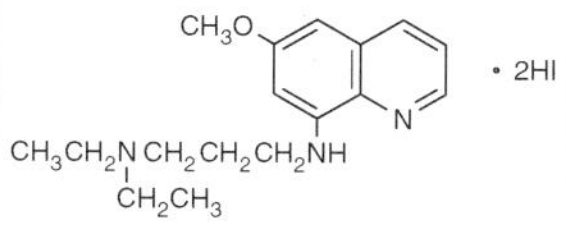

250 mg

S74,457-3
PAMAQUINE

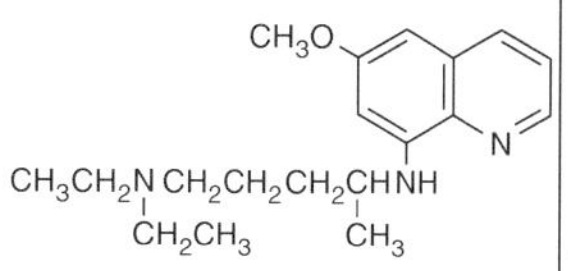

250 mg

S40,309-1
DIBUCAINE, N-(4-SULFANILYLPHENYL)-GLYCINE SALT

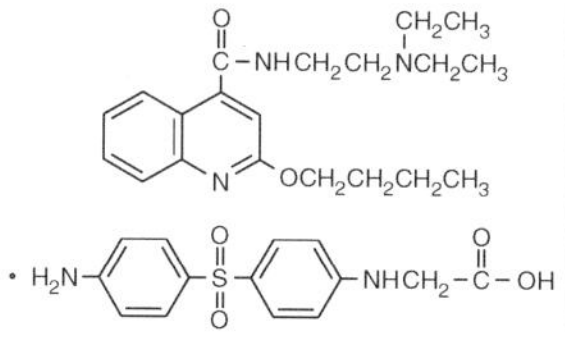

250 mg

S81,761-9
8-(ETHYLAMINO)-6-METHOXYQUINOLINE

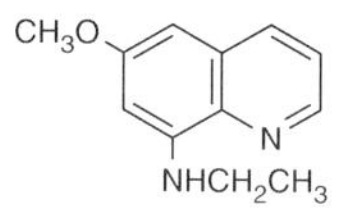

250 mg

S35,892-4
5-BROMO-6-METHOXY-8-NITROQUINOLINE

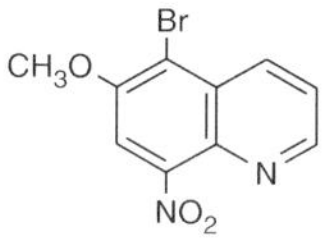

250 mg

S77,021-3
6-METHOXY-4-METHYL-8-NITROQUINOLINE

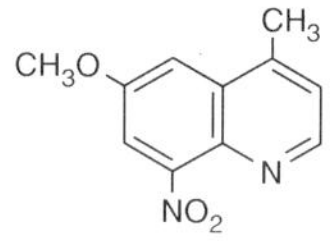

250 mg

S84,200-1
6-METHOXY-2-METHYL-8-NITROQUINOLINE

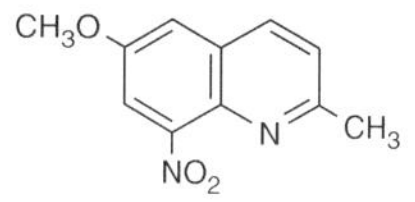

250 mg

S64,018-2
8-ACETAMIDO-6-METHOXYQUINOLINE

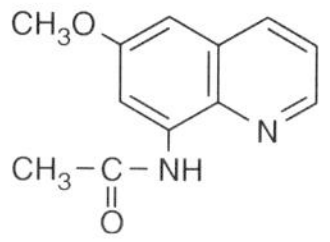

5 gm

S36,758-3
2-CYANO-6-METHOXYQUINOLINE

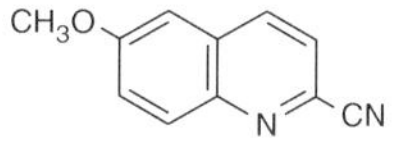

250 mg

S93,511-5
7-(3-DIMETHYLAMINOPROPYLAMINO)-9-METHOXYBENZ(C)ACRIDINE DIHYDROCHLORIDE

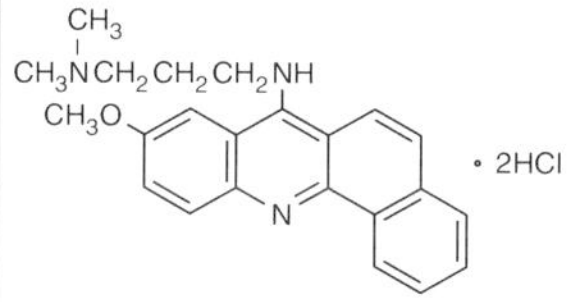

250 mg

S65,680-1
5-[N-(8-METHOXY-4-QUINOLYL)-AMINO]PENTYL NITRATE

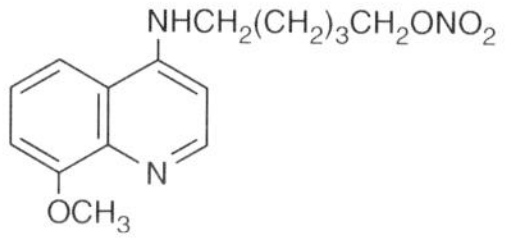

5 gm

S83,396-7
4-HYDRAZINO-6-METHOXY-2-METHYLQUINOLINE

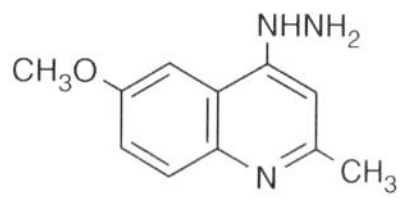

250 mg

S91,338-3
8-QUINOLYL N-(2-METHOXY-4-NITROPHENYL)CARBAMATE

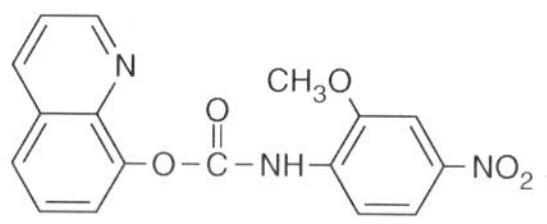

250 mg

S83,275-8
6-METHOXY-2-(3-NITROPHENYL)-4-QUINOLINECARBOXYLIC ACID

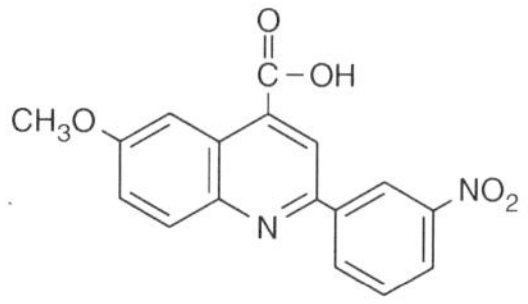

250 mg

S51,539-6
4-CHLORO-2-(4-METHOXYPHENYL)-6-NITROQUINOLINE

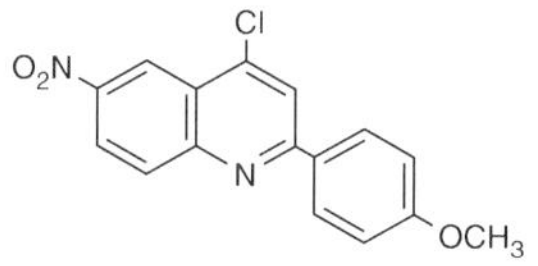

5 gm

S76,294-6
PENTAQUINE THIODISALICYLATE

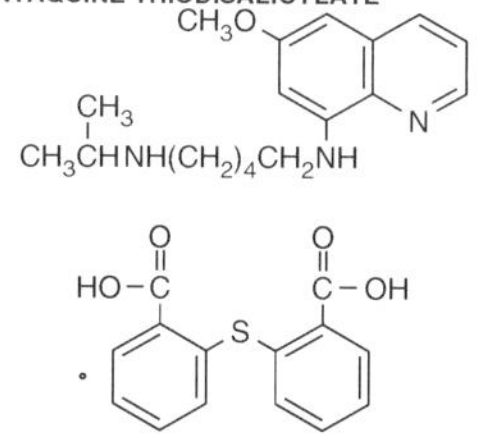

1 gm

S63,312-7
1-(2-METHOXYETHYL)QUINOLINIUM P-TOLUENESULFONATE

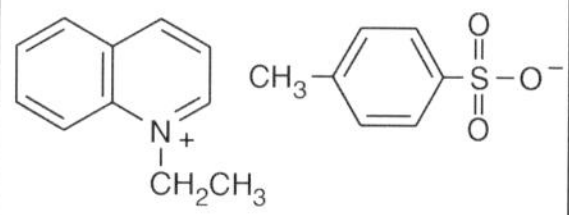

250 mg

S39,181-6
2-(4-METHOXYSTYRYL)-1-METHYLQUINOLINIUM IODIDE

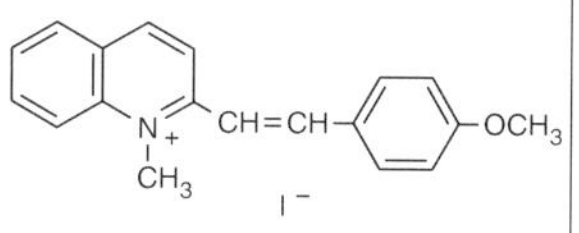

1 gm

S67,940-2
1-ETHYL-2-(4-METHOXYSTYRYL)-QUINOLINIUM IODIDE

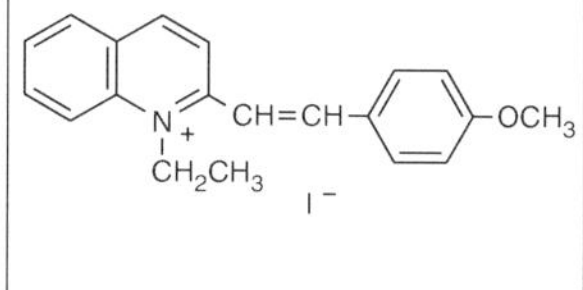

5 gm

S68,331-0
6-METHOXY-1-METHYLQUINOLINIUM IODIDE

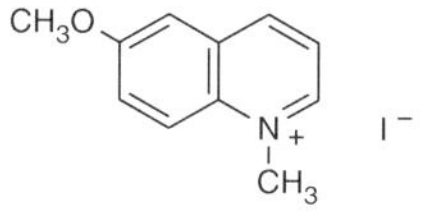

5 gm

S87,310-1
1,2-DIMETHYL-4-METHOXYQUINOLINIUM IODIDE

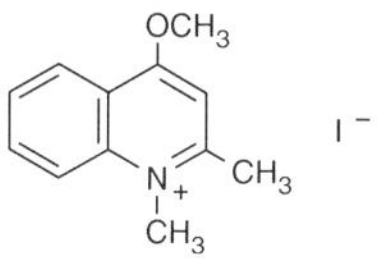

250 mg

S66,178-3
5-[N-(8-METHOXY-4-QUINOLYL)-AMINO]PENTYL NITRATE METHIODIDE

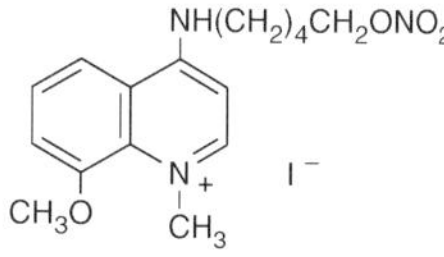

5 gm

S50,153-0
6-METHOXY-2-PHENYL-4-QUINOLINECARBOXYLIC ACID

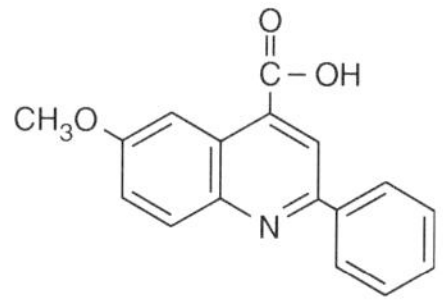

5 gm

S63,112-4
ETHYL 6-METHOXY-2-PHENYLQUINOLINE-4-CARBOXYLATE

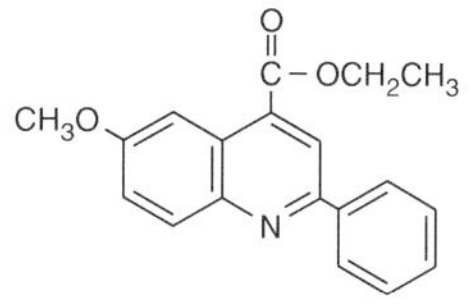

250 mg

S94,956-6
2-(4-METHOXYPHENYL)-4-QUINOLINECARBOXYLIC ACID

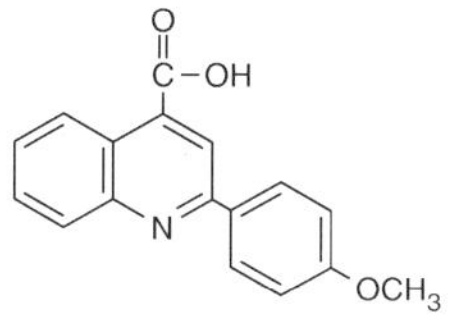

250 mg

S69,152-6
8-QUINOLYL N-(2-METHOXY-5-NITROPHENYL)CARBAMATE

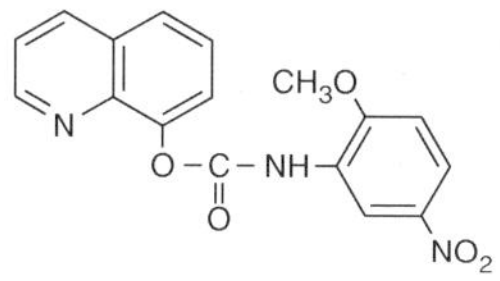

250 mg

S59,748-1
8-QUINOLYL N-(2-METHOXYPHENYL)-CARBAMATE

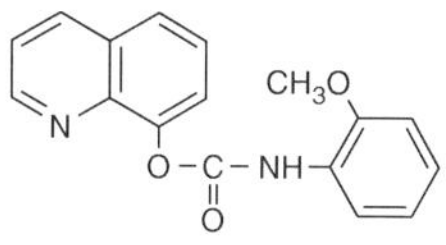

5 gm

S60,361-9
8-QUINOLYL N-(P-ANISYL)CARBAMATE

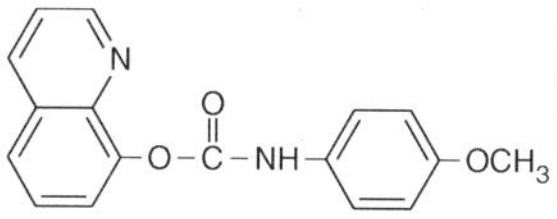

5 gm

S76,825-1
8-QUINOLINYL N-(4-ETHOXYPHENYL)-CARBAMATE
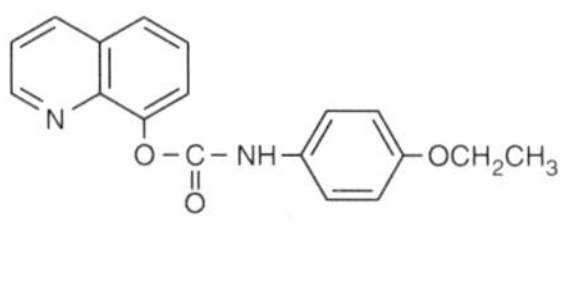
250 mg

S63,928-1
8-QUINOLYL N-(2-METHOXY-5-METHYLPHENYL)CARBAMATE
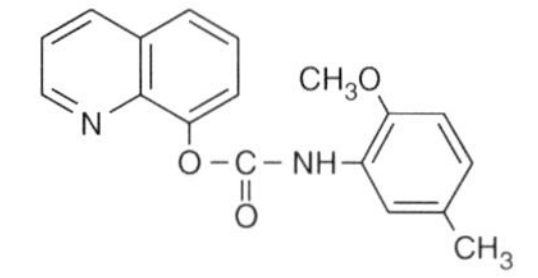
5 gm

S60,989-7
8-QUINOLYL N-(2,5-DIMETHOXYPHENYL)-CARBAMATE
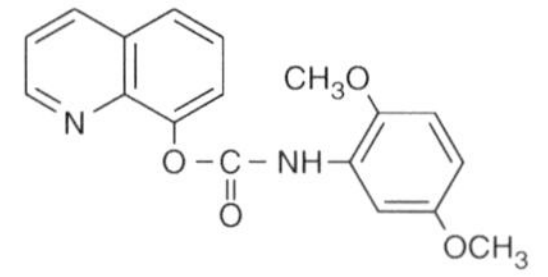
5 gm

S63,936-2
5-CHLORO-8-QUINOLYL N-(2-METHOXY-5-METHYLPHENYL)CARBAMATE
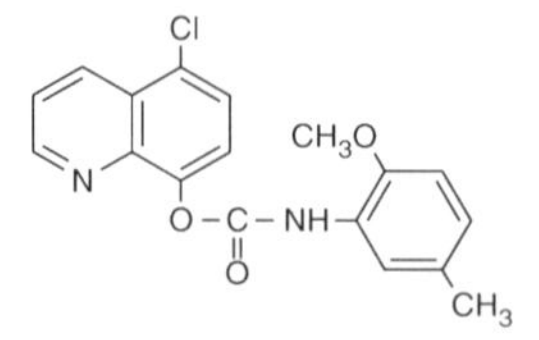
5 gm

S71,157-8
2-METHYL-8-QUINOLYL N-(2-METHOXY-5-METHYLPHENYL)CARBAMATE
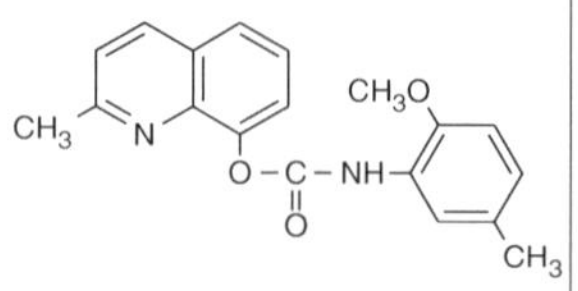
5 gm

S60,958-7
8-QUINOLYL N-(3-METHOXYPHENYL)-CARBAMATE
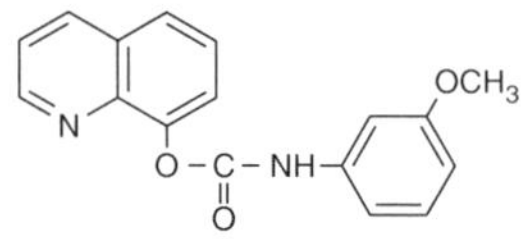
5 gm

S37,391-5
8-ETHOXY-5-QUINOLINESULFONIC ACID
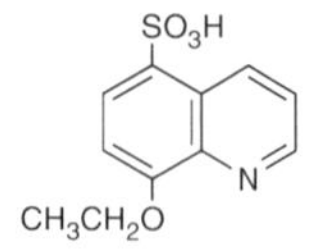
5 gm

S77,034-5
8-METHOXY-2-METHYLQUINOLINE
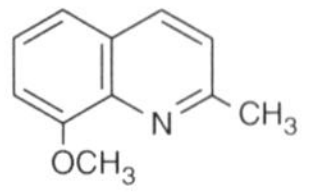
250 mg

S83,478-5
8-ETHOXYQUINOLINE
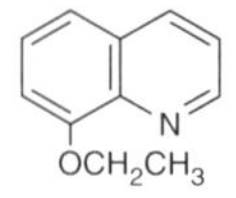
250 mg

S61,766-0
2-CHLORO-5,8-DIMETHOXY-4-(PENTADECYL)QUINOLINE
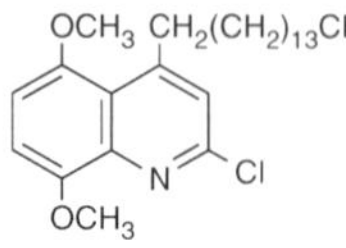
250 mg

S81,028-2
6-METHOXY-2-PHENYL-4-QUINOLINECARBOXALDEHYDE
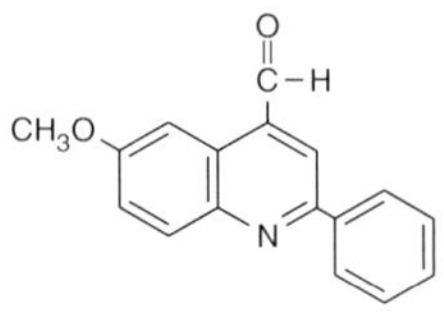
250 mg

S91,540-8
2-CHLORO-6-METHOXY-3-QUINOLINECARBOXALDEHYDE
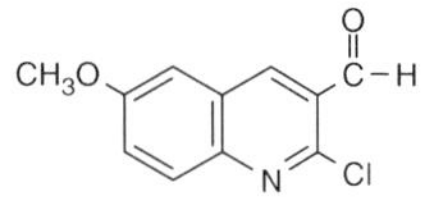
250 mg

S82,262-0
ETHYL 3-(6-METHOXY-2-PHENYL-4-QUINOLYL)-3-OXOPROPIONATE
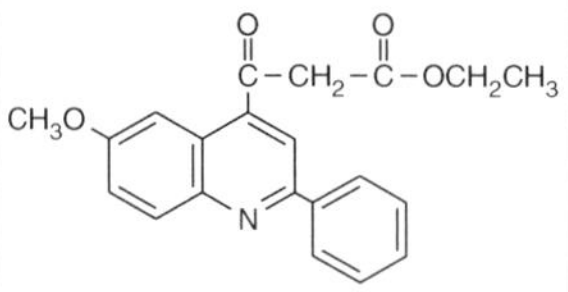
250 mg

S81,029-0
4-ACETYL-6-METHOXY-2-PHENYLQUINOLINE
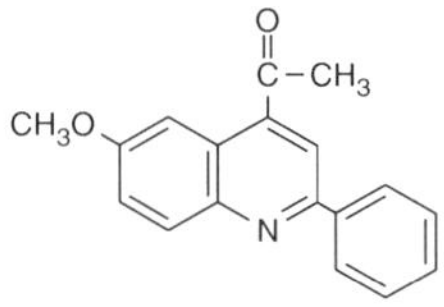
250 mg

S81,031-2
6-METHOXY-2-PHENYL-4-PROPIONYLQUINOLINE
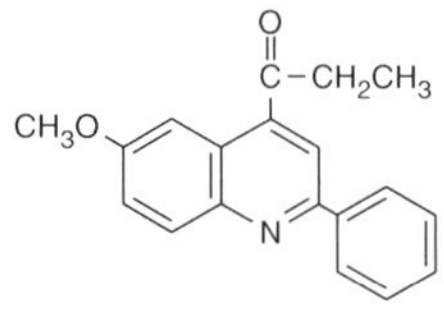
250 mg

S95,869-7
QUININONE
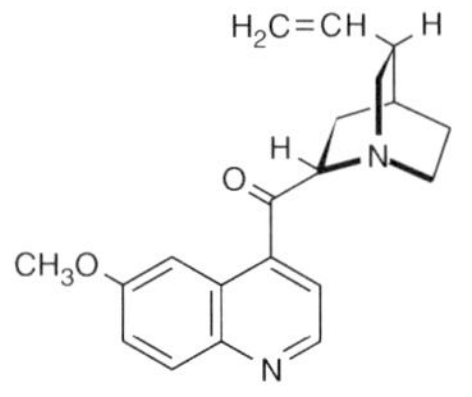
50 mg

S83,466-1
METHYL 6-METHOXY-4-QUINOLINECARBOXYLATE
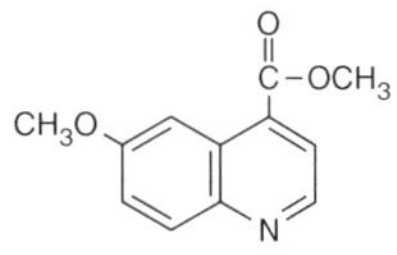
250 mg

S95,662-7
6-METHOXY-4-QUINOLINECARBOXYLIC ACID
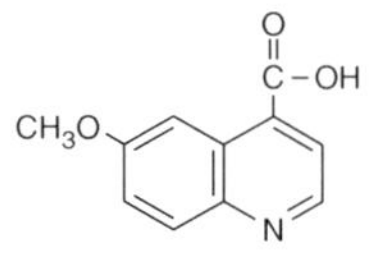
5 gm

S95,663-5
6-METHOXY-4-QUINOLINEACRYLIC ACID
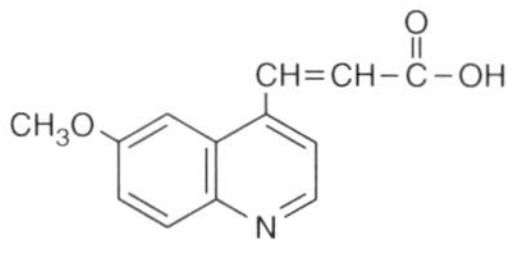
50 mg

S95,665-1
METHYL 6-METHOXY-4-QUINOLINEACRYLATE
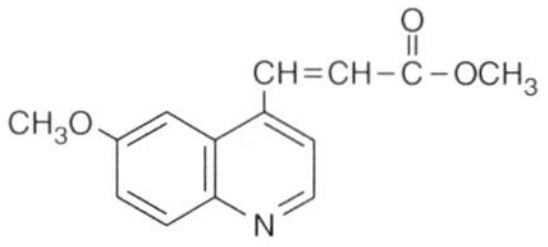
50 mg

S66,391-3
1-(4-CHLORO-7-METHOXY-2-QUINOLYL)-3-PHENYLUREA
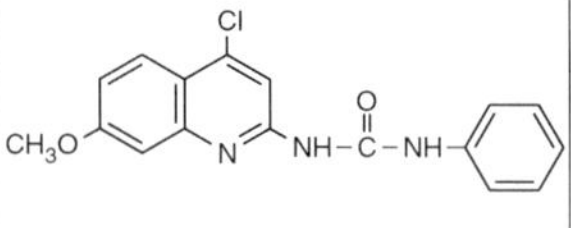
5 gm

S66,394-8
1-(4-CHLORO-7-METHOXY-2-QUINOLYL)-3-(P-TOLYL)UREA
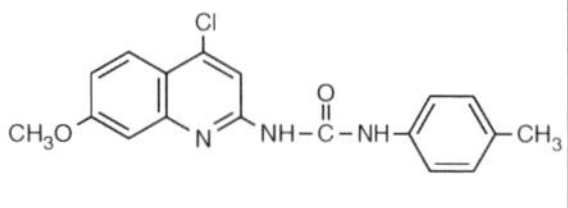
5 gm

S66,392-1
1-(P-ANISYL)-3-(4-CHLORO-7-METHOXY-2-QUINOLYL)UREA
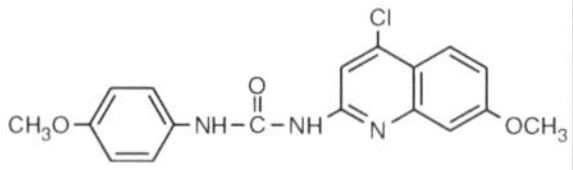
5 gm

S65,777-8
1-(4-CHLORO-7-METHOXY-2-QUINOLYL)-3-(2,6-XYLYL)UREA
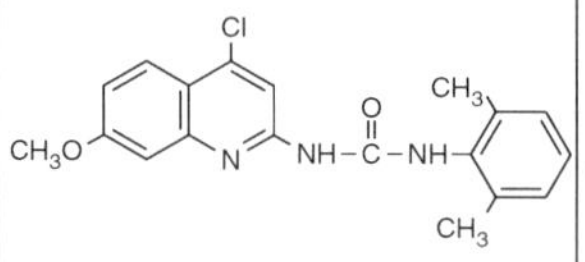
5 gm

S66,097-3
1-(4-CHLORO-7-METHOXY-2-QUINOLYL)-3-PHENYL-2-THIOUREA
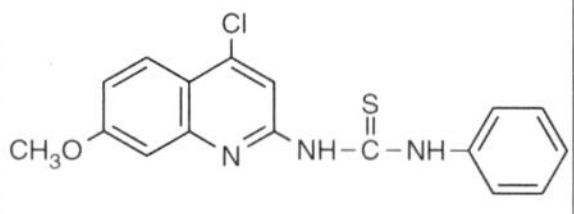
250 mg

S79,868-1
2-CHLORO-4-(4-METHOXYPHENYL)-3-PHENYLQUINOLINE
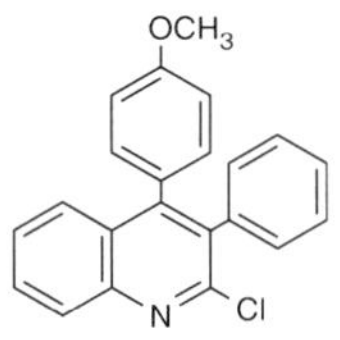
250 mg

S83,769-5
6-METHOXY-2-PHENYLQUINOLINE
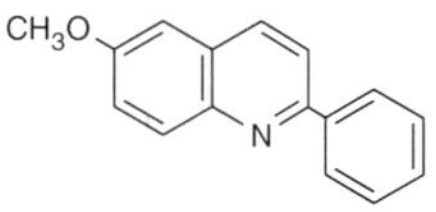
250 mg

S81,953-0
2-(2-METHOXYPHENYL)-QUINOLINE PICRATE
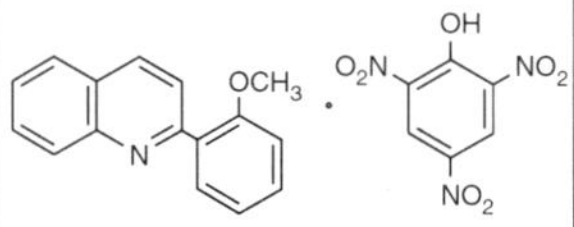
250 mg

S82,037-7
2-(4-METHOXYPHENYL)-7-METHYLQUINOLINE
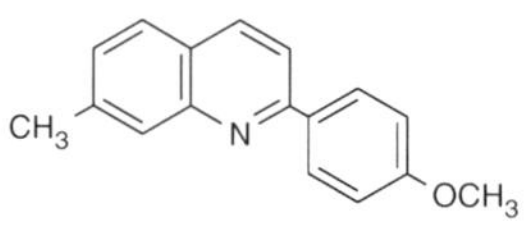
250 mg

S84,077-7
2-(4-METHOXYPHENYL)-6-METHYLQUINOLINE PICRATE
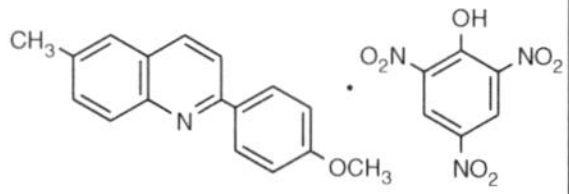
250 mg

S64,008-5
6-METHOXYQUINOLINE PICRATE
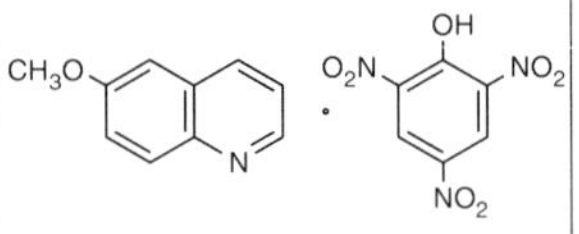
250 mg

S36,743-5
4-CHLORO-6-METHOXYQUINOLINE
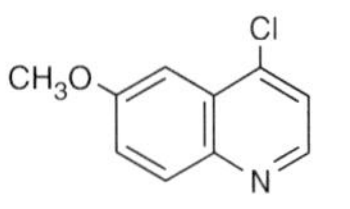
5 gm

S73,227-3
4-CHLORO-8-METHOXYQUINOLINE
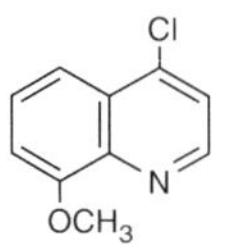
250 mg

S56,390-0
6-METHOXY-4-METHYLQUINOLINE HYDROCHLORIDE
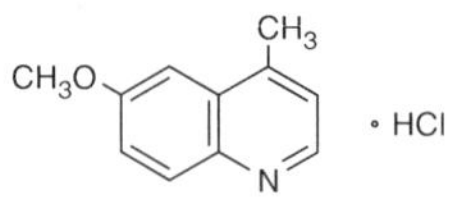
5 gm

S81,789-9
6-ETHOXYQUINOLINE HYDROCHLORIDE
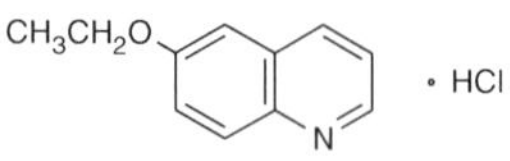
250 mg

S83,485-8
6-ETHOXYQUINOLINE
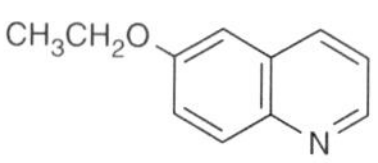
250 mg

S87,313-6
4-METHOXY-2-METHYLQUINOLINE
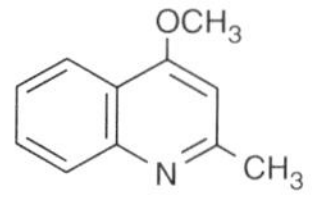
250 mg

S56,331-5
2-CHLORO-6-METHOXY-4-METHYLQUINOLINE
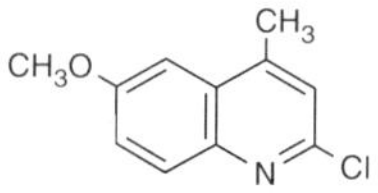
5 gm

S81,915-8
7-CHLORO-4-ETHOXYQUINOLINE
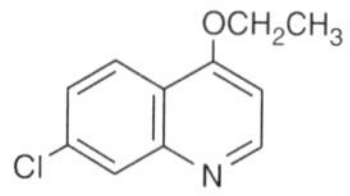
250 mg

S83,395-9
4-CHLORO-6-METHOXY-2-METHYLQUINOLINE
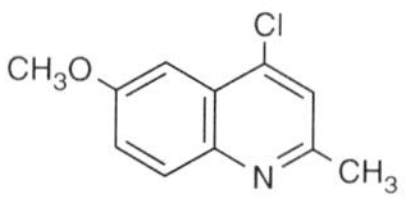
250 mg

S40,841-7
6-ETHOXYQUINALDINE
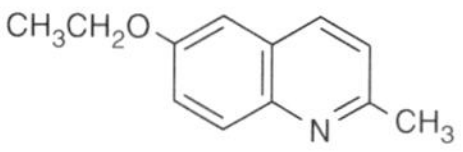
5 gm

S39,956-6
2,4-DIMETHYL-6-ETHOXYQUINOLINE
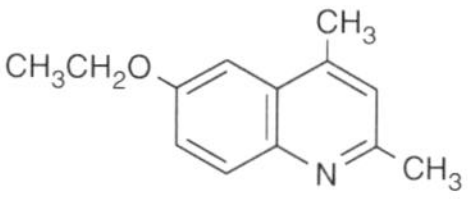
250 mg

S42,084-0
4,8-DIMETHYL-2-ETHOXYQUINOLINE
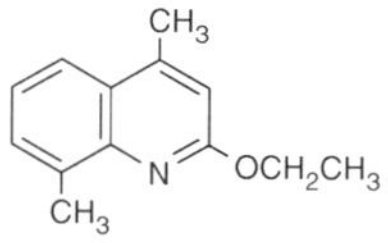
250 mg

S60,619-7
4-CHLORO-6-METHOXY-2-PENTADECYLQUINOLINE
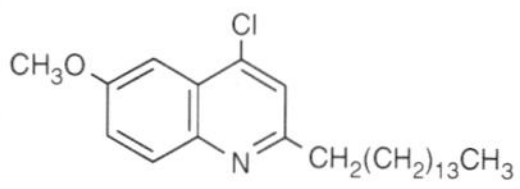
250 mg

S81,686-8
2-(BENZYLOXY)QUINOLINE
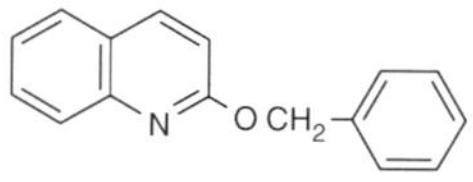
250 mg

S39,184-0
1-METHYL-2-[3,4-(METHYLENEDIOXY)-STYRYL]QUINOLINIUM IODIDE
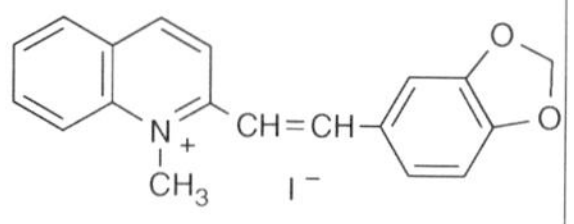
250 mg

S88,692-0
2-PROPYL-3,4-QUINOLINEDICARBOXYLIC ACID
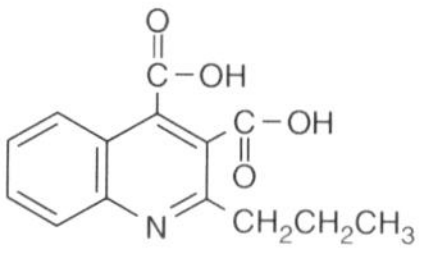
250 mg

S81,472-5
8-AMINO-6-(METHYLTHIO)-5-NITROQUINOLINE
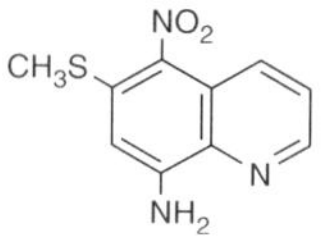
250 mg

S84,010-6
8-ACETAMIDO-5-AMINO-6-(METHYLTHIO)-QUINOLINE
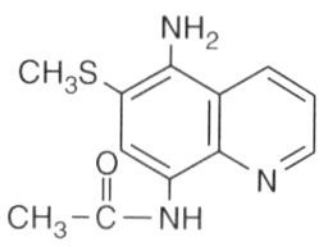
250 mg

S83,093-3
8-ACETAMIDO-5-AMINO-6-(DODECYLTHIO)-QUINOLINE
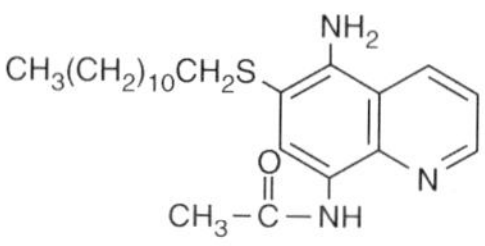
250 mg

S84,023-8
5-AMINO-6-(DODECYLTHIO)QUINOLINE
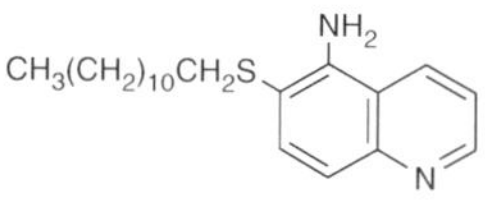
250 mg

S82,186-1
2-(4-AMINOPHENYLTHIO)-4-QUINOLINECARBOXYLIC ACID
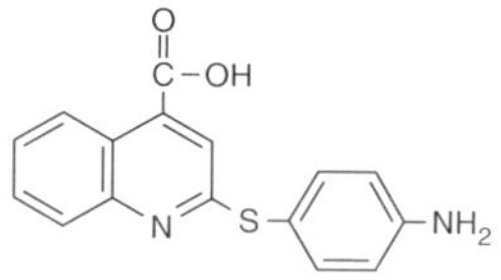
250 mg

S82,184-5
2-(4-AMINOPHENYLTHIO)QUINOLINE
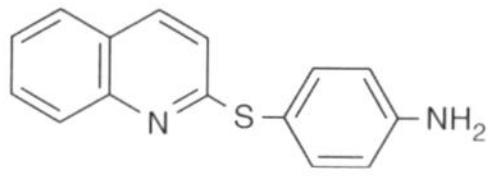
250 mg

S82,185-3
2-(4-AMINOPHENYLTHIO)QUINOLINE HYDROCHLORIDE
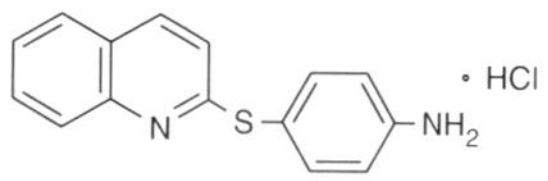
250 mg

S51,238-9
1-METHYL-2-METHYLTHIO-4-PHENYLQUINOLINIUM IODIDE
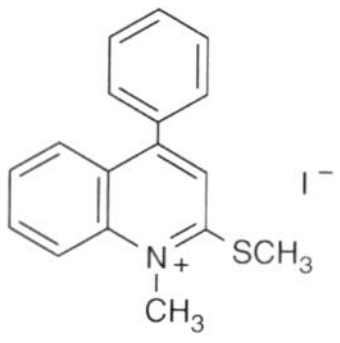
250 mg

S76,828-6
8-QUINOLINYL N-[4-(METHYLTHIO)-PHENYL]CARBAMATE
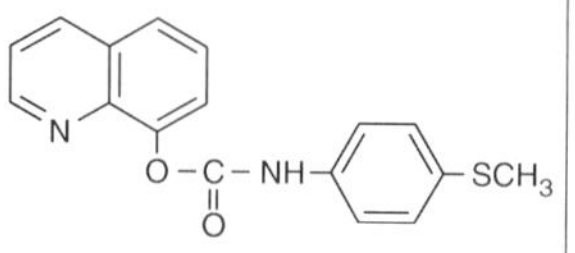
250 mg

S77,355-7
2-METHYL-8-QUINOLINYL N-[4-(METHYLTHIO)PHENYL]CARBAMATE
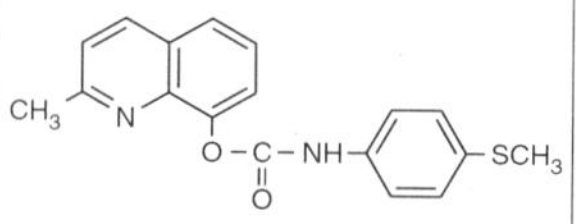
250 mg

S83,089-5
6-(DODECYLTHIO)-5-NITROQUINOLINE
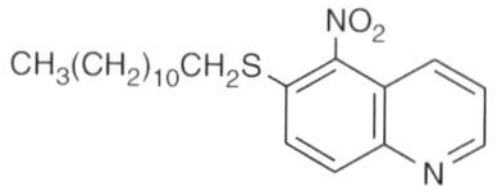
250 mg

S82,448-8
2-(2-QUINOLYLTHIO)BENZOTHIAZOLE
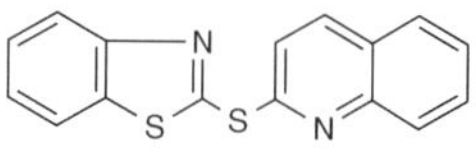
250 mg

S83,403-3
7-CHLORO-4-(4-METHYL-2-THIAZOLYLTHIO)QUINOLINE HYDROCHLORIDE
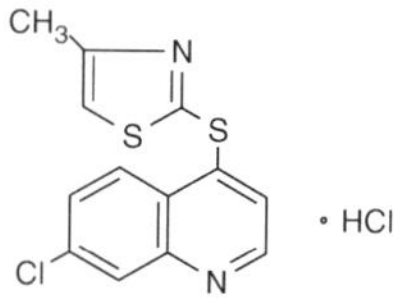
250 mg

S83,817-9
2-(4,5-DIMETHYL-2-THIAZOLYLTHIO)-QUINOLINE
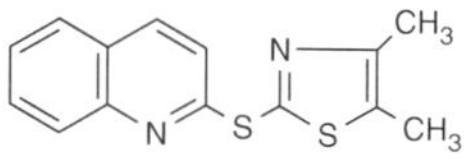
250 mg

S83,818-7
2-(PHENYLTHIO)QUINOLINE
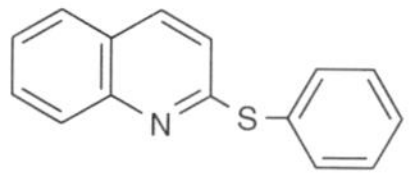
250 mg

S82,451-8
2-(BENZYLTHIO)QUINOLINE
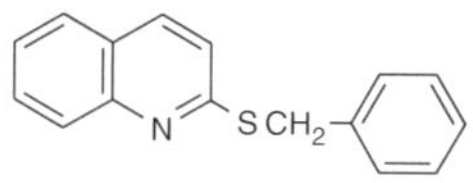
250 mg

S76,795-6
4-(TERT-BUTYLTHIO)-7-CHLOROQUINOLINE
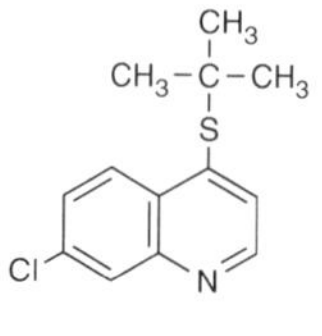
250 mg

S82,453-4
2-(OCTADECYLTHIO)QUINOLINE
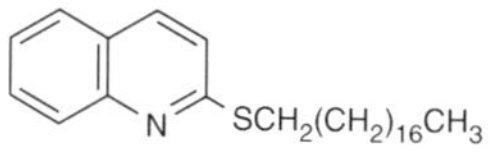
250 mg

S82,191-8
2-QUINOLYL DISULFIDE
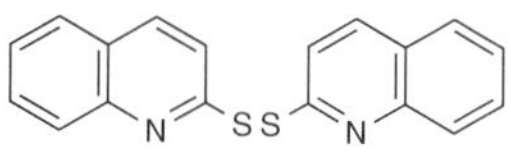
250 mg

S82,192-6
4-METHYL-2-QUINOLYL DISULFIDE
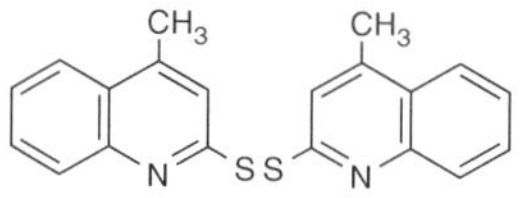
250 mg

S45,607-1
8-AMINOQUINALDINE
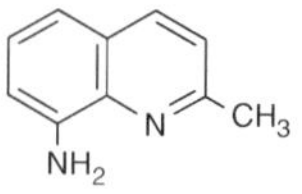
250 mg

S80,852-0
5-AMINO-1,2,3,4-TETRAHYDROACRIDINE
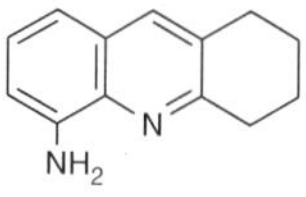
250 mg

S93,281-7
5-AMINO-9-METHYL-1,2,3,4-TETRAHYDROACRIDINE
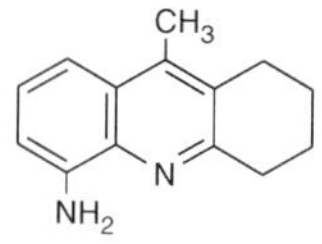
250 mg

S81,192-0
2-(2-AMINOPHENYL)-4-QUINOLINECARBOXYLIC ACID
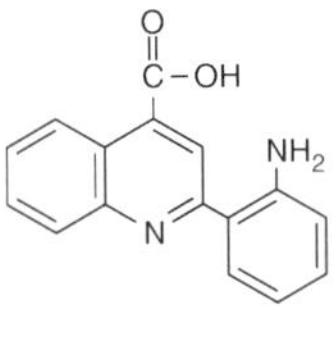
250 mg

S65,143-5
2-AMINO-3-CYANOQUINOLINE-4-CARBOXYLIC ACID, SODIUM SALT
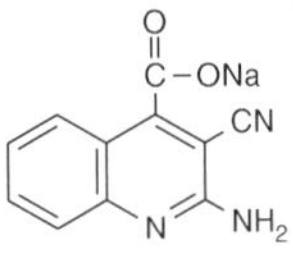
5 gm

S82,228-0
5-AMINOQUINOLINE HYDROCHLORIDE
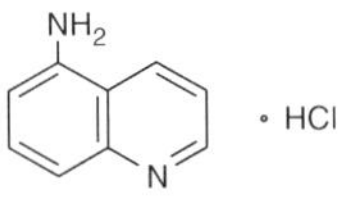
250 mg

S83,864-0
5-AMINO-6-METHYLQUINOLINE
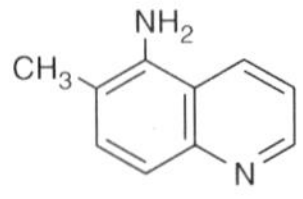
250 mg

S95,664-3
7-AMINO-5-ISOPROPYL-8-METHYLQUINOLINE HYDROCHLORIDE
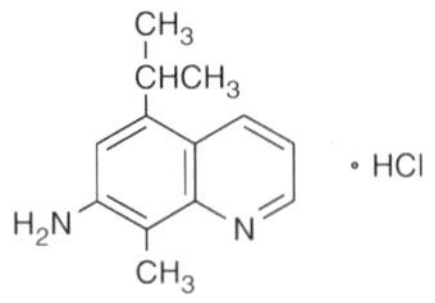
50 mg

S51,223-0
2-PHENYL-4-QUINOLINECARBOXAMIDE
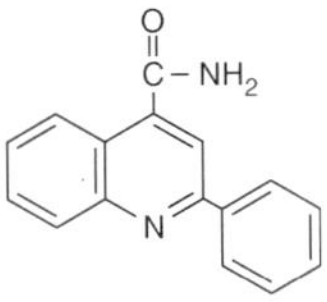
250 mg

S80,678-1
2-[2-(2-AMINOPHENYL)ETHYL]QUINOLINE
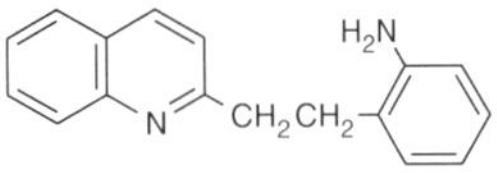
250 mg

S41,332-1
2-AMINOQUINOLINE
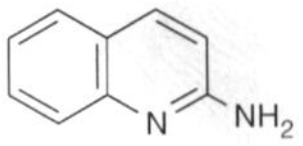
250 mg

S92,769-4
4-AMINO-2-QUINOLINECARBOXYLIC ACID
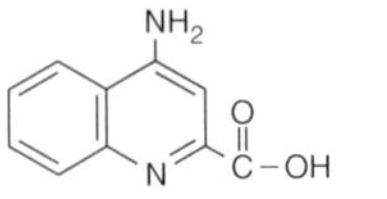
250 mg

S82,040-7
2-[4-(DIMETHYLAMINO)-3,5-DINITROPHENYL]-8-METHYL-5-NITROQUINOLINE
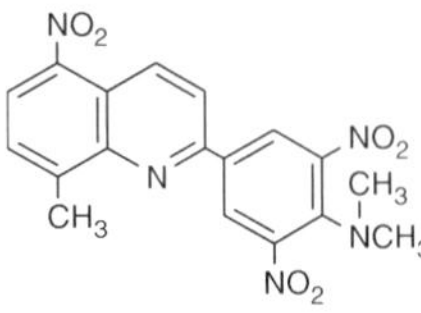
250 mg

S84,060-2
8-ANILINOQUINOLINE
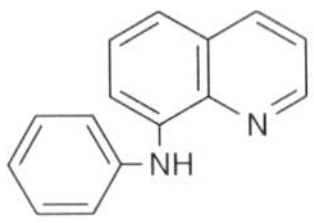
250 mg

S84,059-9
6-CHLORO-8-(2,5-DIMETHYL-1-PYRROLYL)QUINOLINE
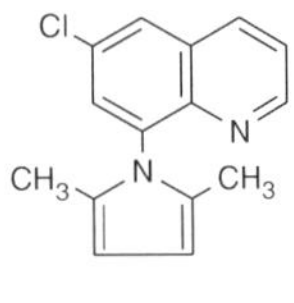
250 mg

S85,163-9
8-QUINOLYL N-(4-METHYL-2-NITROPHENYL)CARBAMATE
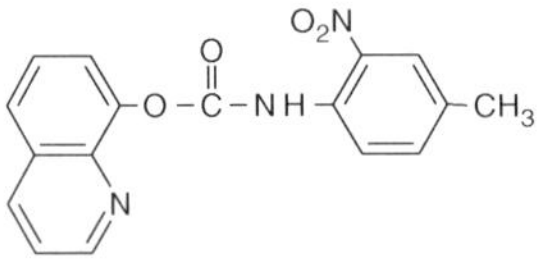
250 mg

S83,083-6
8-ACETAMIDO-6-CHLORO-5-NITROQUINOLINE
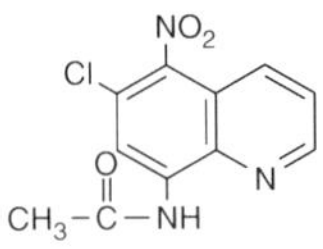
250 mg

S81,161-0
6-CHLORO-8-NITROQUINOLINE
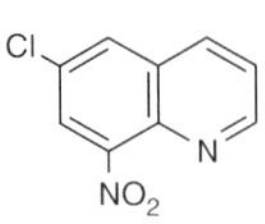
250 mg

S81,792-9
2-CHLORO-8-NITROQUINOLINE
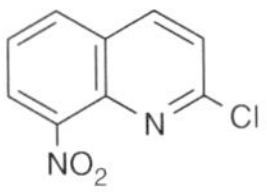
250 mg

S84,910-3
2-(DIBROMONITROMETHYL)QUINOLINE
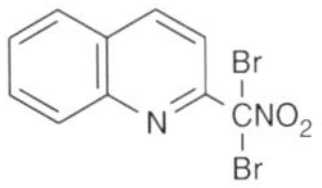
250 mg

S93,282-5
2-ETHYL-4-METHYL-8-NITROQUINOLINE
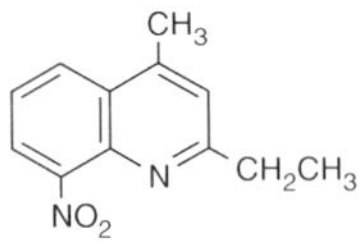
250 mg

S87,306-3
FORMIC N2-(3-QUINOLYL)HYDRAZIDE
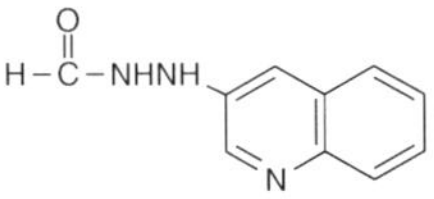
250 mg

S73,153-6
2-PHENYL-3-PHTHALIMIDOQUINOLINE-4-CARBOXYLIC ACID
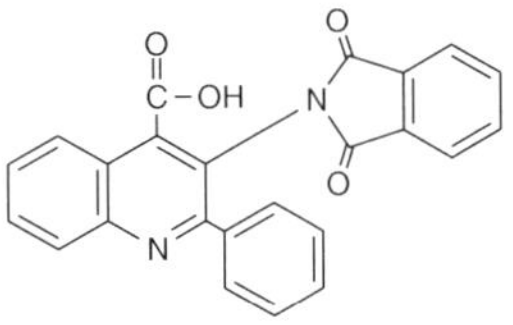
5 gm

S67,128-2
ETHYL 3-CARBOXYMETHYL-1,3-DIHYDRO-2H-PYRROLO(3,4-B)QUINOLINE-2-CARBOXYLATE
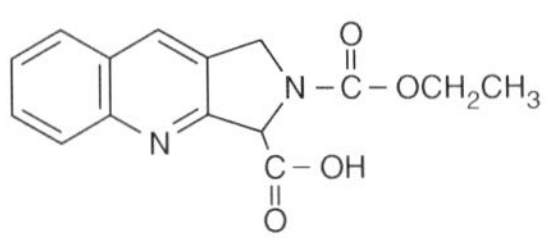
250 mg

S81,787-2
3-ACETAMIDOQUINOLINE
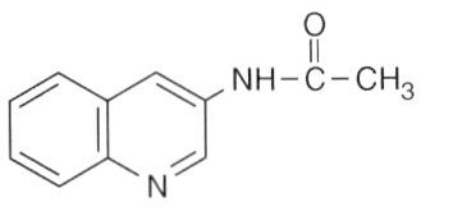
250 mg

S93,277-9
6,7'-BIQUINOLINE
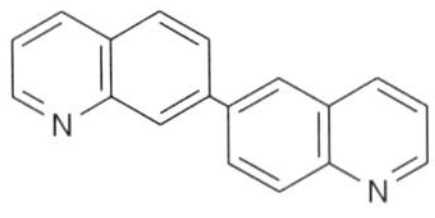
250 mg

S93,590-5
3-CHLORO-9-[4-(DIETHYLAMINO)-1-METHYLBUTYLAMINO]ACRIDINE DIHYDROCHLORIDE
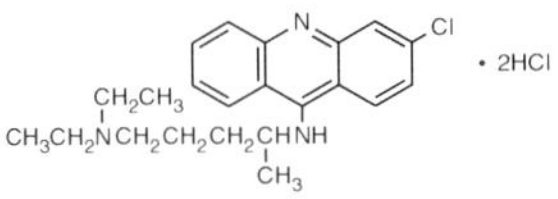
250 mg

S81,027-4
7-CHLORO-4-[3-(DIETHYLAMINO)-PROPYLAMINO]QUINOLINE
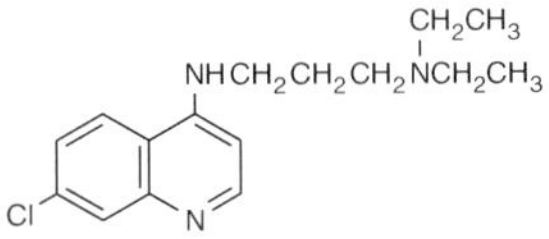
250 mg

S87,061-7
1-ETHYL-4-(PHENYLSULFONYLAZO)-QUINOLINIUM TETRAFLUOROBORATE
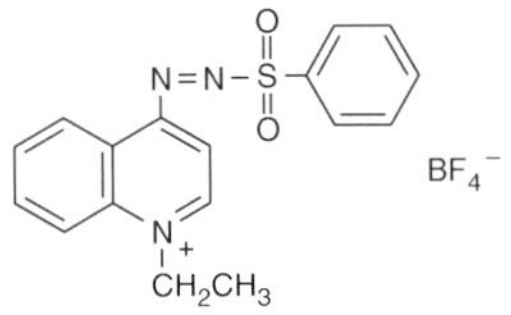
250 mg

S87,488-4
4,4'-AZOBIS(1-ETHYLQUINOLINIUM TETRAFLUOROBORATE)
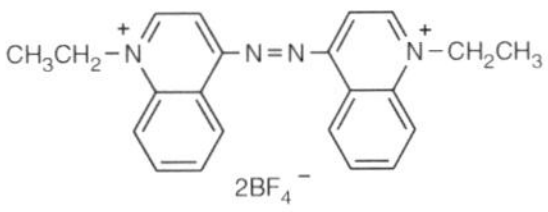
250 mg

S63,938-9
N,N'-BIS(8-QUINOLYLOXYCARBONYL)-4-METHYL-1,3-PHENYLENEDIAMINE
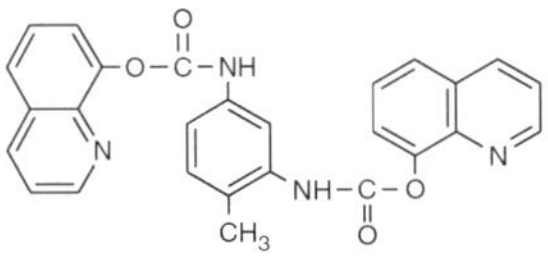
5 gm

S83,397-5
1,2-BIS(7-CHLORO-4-QUINOLYL)-HYDRAZINE
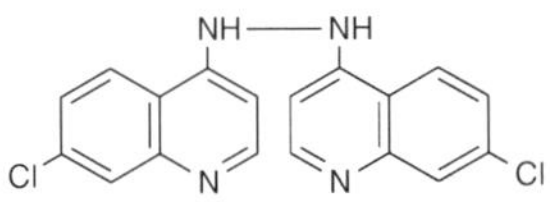
250 mg

S83,469-6
DIMETHYL 5-NITRO-2,4-QUINOLINEDICARBOXYLATE
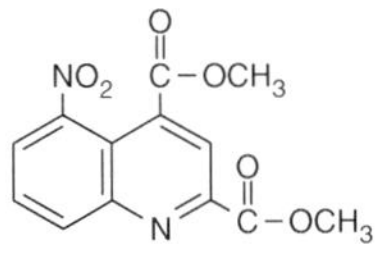
250 mg

S62,170-6
8-QUINOLYL N-(3-NITROPHENYL)-CARBAMATE
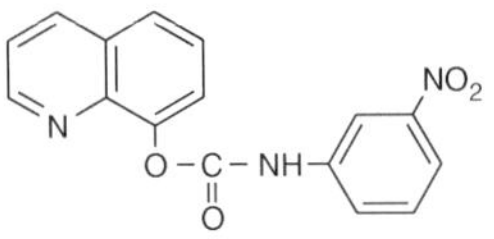
5 gm

S62,492-6
5-CHLORO-8-QUINOLYL N-(3-NITROPHENYL)CARBAMATE
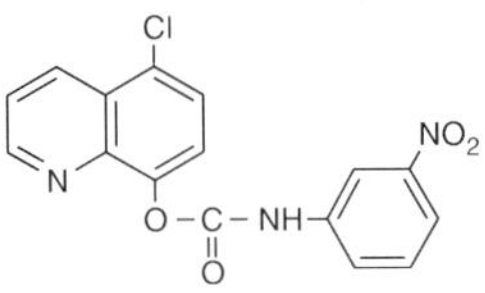
5 gm

S64,395-5
2-METHYL-8-QUINOLYL N-(3-NITROPHENYL)CARBAMATE
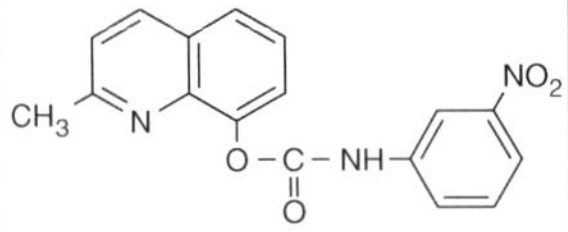
5 gm

S71,309-0
N,N'-(4-METHYL-1,3-PHENYLENE)BIS(2-METHYL-8-QUINOLYL CARBAMATE)
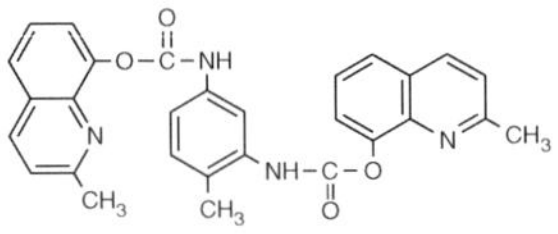
250 mg

S83,461-0
5-NITRO-4-QUINOLINECARBOXYLIC ACID
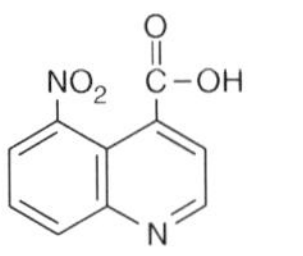
250 mg

S83,465-3
METHYL 2-METHYL-5-NITRO-4-QUINOLINECARBOXYLATE
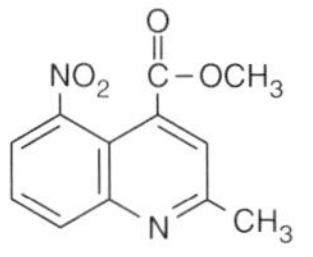
250 mg

S35,394-9
5-NITROQUINALDIC ACID
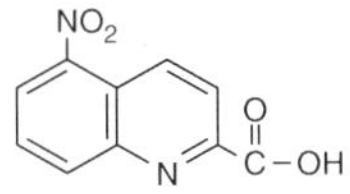
5 gm

S36,420-7
5-NITROQUINOLINE NITRATE
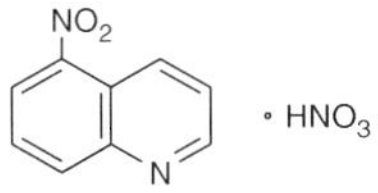
5 gm

S55,920-2
6-METHYL-5-NITROQUINOLINE
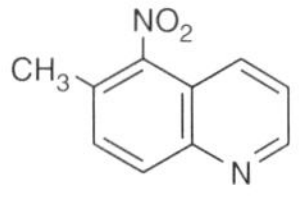
5 gm

S74,601-0
5-NITROQUINALDINE
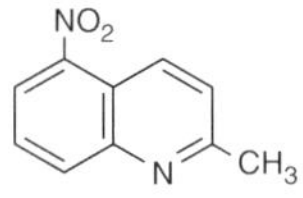
5 gm

S81,039-8
7-METHYL-4-NITROQUINOLINE
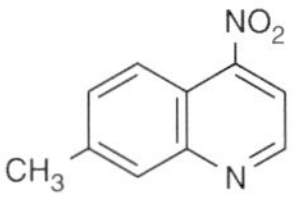
250 mg

S71.549-2
N,N'-(4-METHYL-1,3-PHENYLENE)BIS(5-CHLORO-8-QUINOLYL CARBAMATE)
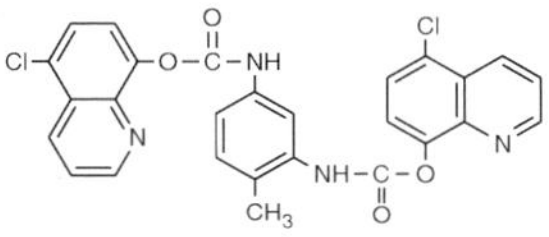
100 mg

S93,298-1
METHYL [5,6-DIHYDRO-5,5-DIMETHYLBENZ(C)ACRIDINE-7-YL] CARBAMATE
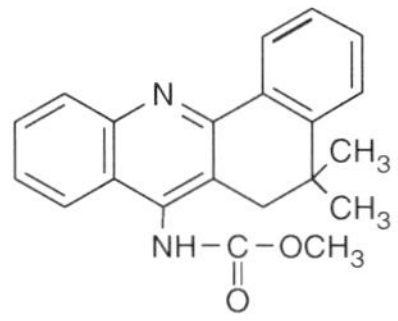
250 mg

S91,341-3
1-CYCLOHEXYL-3-(4-QUINOLYL)UREA
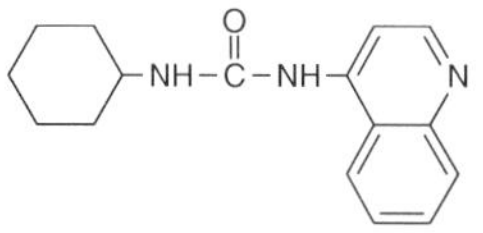
250 mg

S91,321-9
1-(2-METHYL-4-QUINOLYL)-3-(3,4-XYLYL)UREA
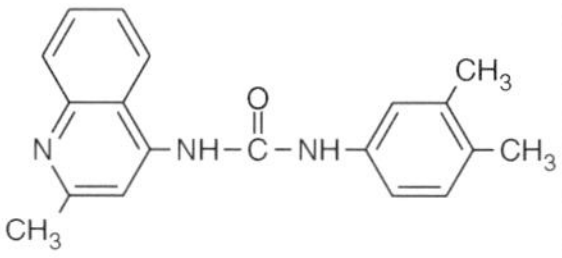
250 mg

S94,900-0
2-CHLORO-3-QUINOLINECARBOXALDEHYDE OXIME
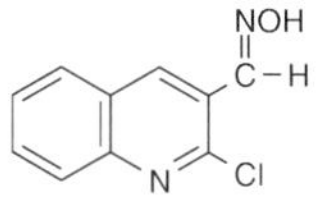
250 mg

S76,466-3
CHLOROQUINE
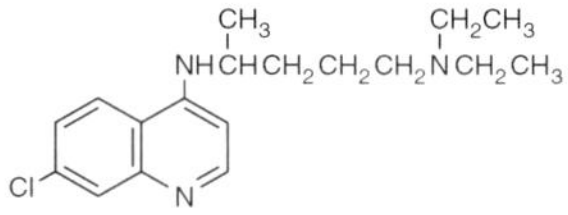
250 mg

S94,149-2
2-(4-NITROANILINO)QUINOLINE
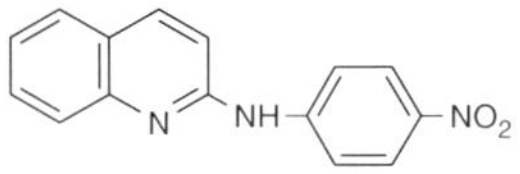
250 mg

S82,084-9
2-[3-(DIMETHYLAMINO)PHENYL]-8-METHYLQUINOLINE PICRATE
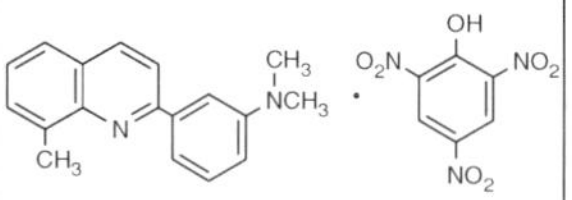
250 mg

S64,874-4
6-DIMETHYLAMINOQUINALDINE
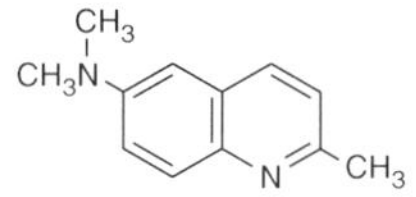
5 gm

S82,081-4
6-(DIETHYLAMINO)-5-METHYLQUINOLINE
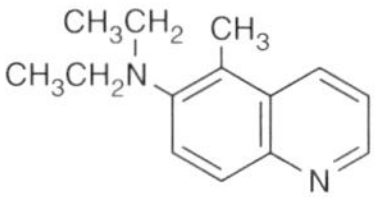
250 mg

S60,748-7
8-QUINOLYL N-(2-CHLORO-4-NITROPHENYL)CARBAMATE
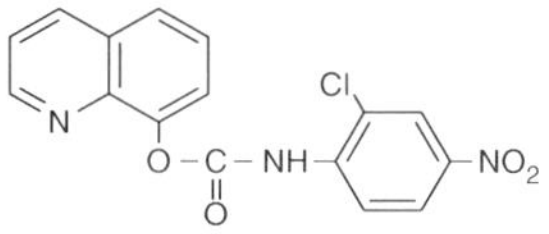
5 gm

S50,151-4
4-(4-CHLORO-6-NITRO-2-QUINOLYL)-BENZENESULFONYL FLUORIDE
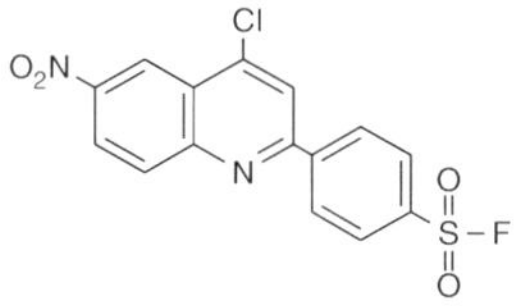
5 gm

S63,288-0
4-CHLORO-6-FLUOROSULFONYL-2-(3-NITROPHENYL)QUINOLINE
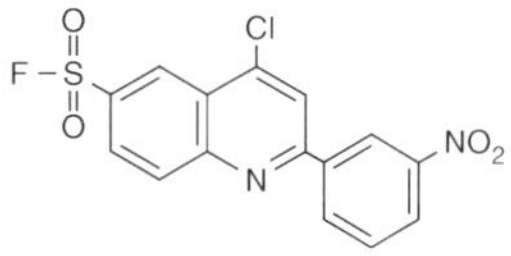
50 mg

S83,862-4
8-CHLORO-6-NITROQUINOLINE
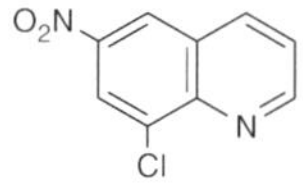
250 mg

S65,426-4
4-CHLORO-2-METHYL-6-NITROQUINOLINE
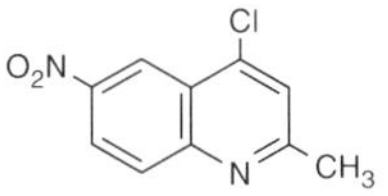
250 mg

S95,659-7
7-METHYL-6-NITROQUINOLINE
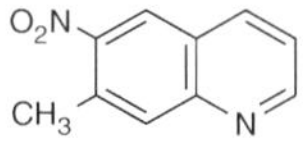
50 mg

S87,697-6
2-PHENYL-4-QUINOLINECARBOXYLIC HYDRAZIDE
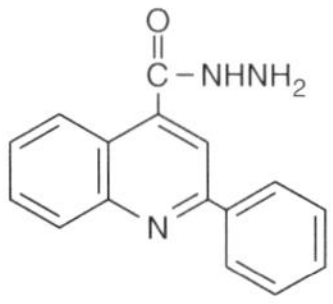
250 mg

S34,781-7
1,3-DI-6-QUINOLYLUREA
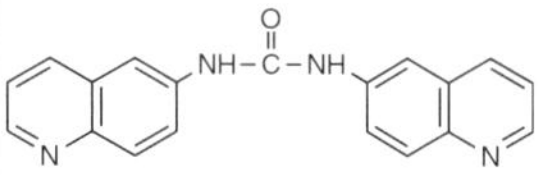
5 gm

S88,694-7
2-PHENYL-4-QUINOLINECARBONITRILE
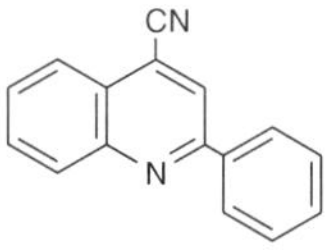
250 mg

S90,951-3
2-BENZOYLQUINOLINE
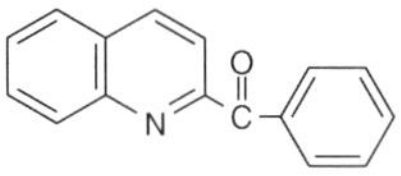
250 mg

S94,953-1
5,6-DIMETHYL-11H-INDOLO(3,2-C)-QUINOLINIUM P-TOSYLATE
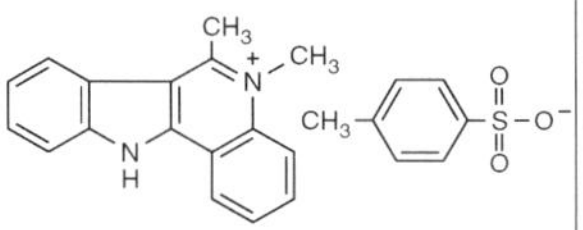
100 mg

S93,279-5
3-(CYCLOHEXYLAMINO)-2-PHENYLQUINOLINE PICRATE
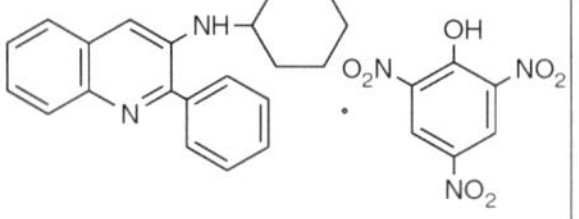
250 mg

S42,443-9
CINCHOTOXINE
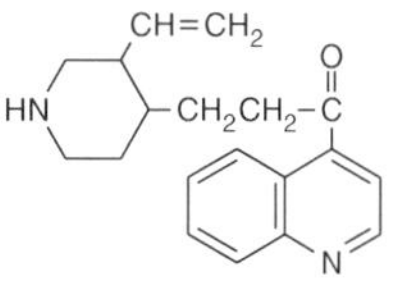
250 mg

S93,745-2
6-CYANOBENZIMIDAZO(1,2-A)-QUINOLIN-5(7H)-ONE
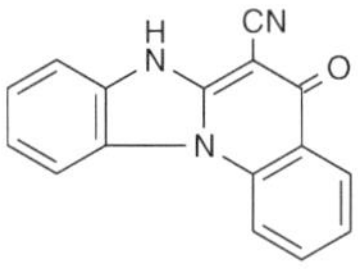
250 mg

S90,956-4
2,2'-IMINOBIS(QUINOLINE)
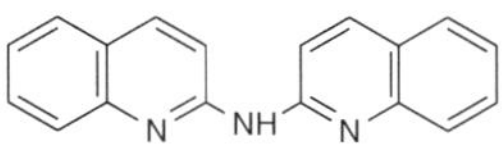
250 mg

S51,256-7
4-[4-(DIMETHYLAMINO)STYRYL]-1-METHYLQUINOLINIUM IODIDE
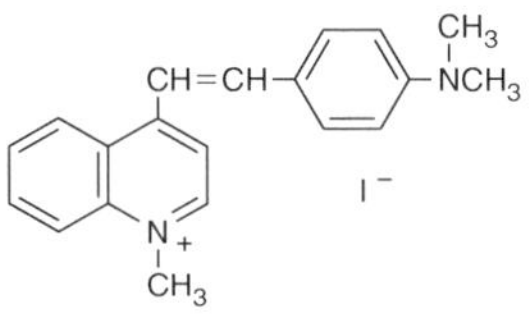
250 mg

S83,992-2
2-[4-(DIETHYLAMINO)PHENYL]QUINOLINE
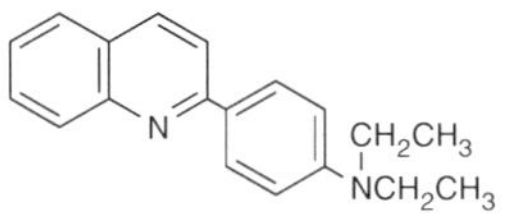
250 mg

S82,096-2
3-[4-(DIMETHYLAMINO)PHENYL]BENZO(F)-QUINOLINE
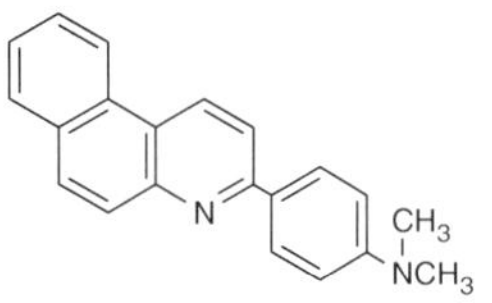
250 mg

S82,097-0
3,4-DIHYDRO-3-[4-(DIMETHYLAMINO)-PHENYL]BENZO(F)QUINOLINE
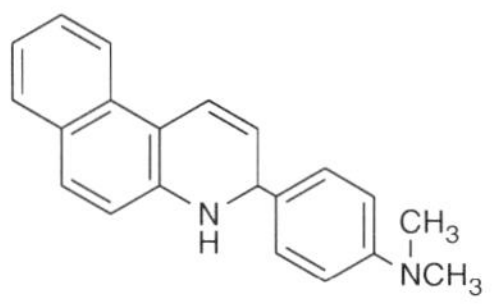
250 mg

S82,082-2
6-CHLORO-2-[4-(DIETHYLAMINO)PHENYL]-QUINOLINE
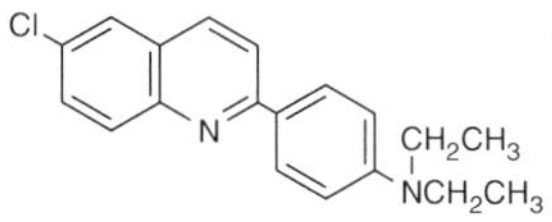
250 mg

S84,076-9
2-[4-(DIMETHYLAMINO)PHENYL]-QUINOLINE DIPICRATE
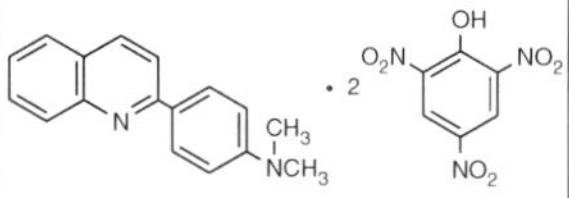
250 mg

S81,909-3
2-[4-(DIMETHYLAMINO)STYRYL]-QUINOLINE
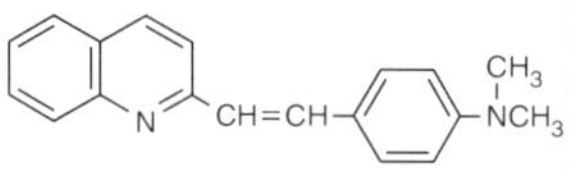
250 mg

S82,036-9
2-(DIMETHYLAMINO)QUINOLINE
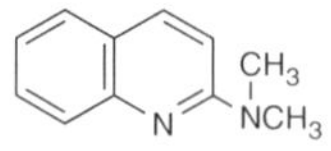
250 mg

S81,002-9
7-CHLORO-4-(DIMETHYLAMINO)-QUINOLINE DIPICRATE
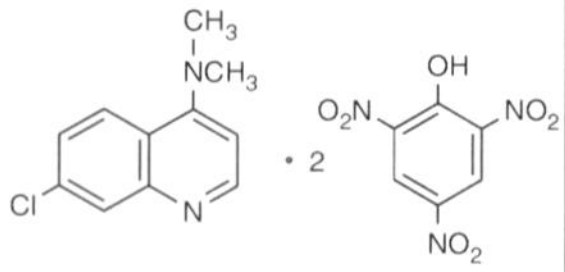
250 mg

S39,528-5
1-BENZYLQUINOLINIUM BROMIDE
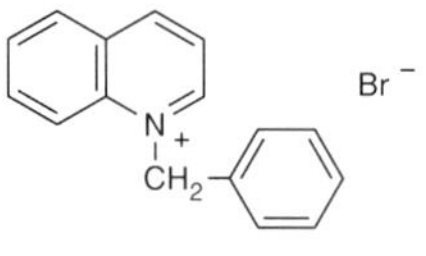
250 mg

S60,595-6
1-BENZYLQUINOLINIUM CHLORIDE
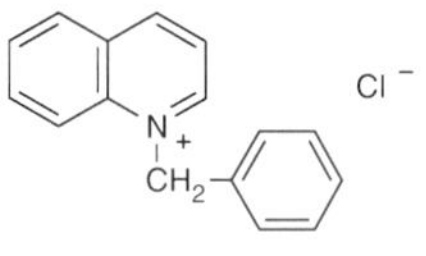
250 mg

S87,330-6
1-METHYL-2-(PHENYLSULFONYLAZO)-QUINOLINIUM TETRACHLOROZINCATE
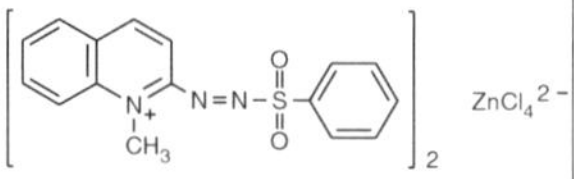
250 mg

S67,934-8
1-ETHYL-2-(3-NITROSTYRYL)-QUINOLINIUM IODIDE
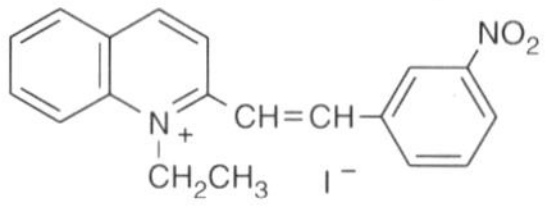
5 gm

S51,708-9
1-METHYL-2-(3-NITROSTYRYL)-QUINOLINIUM IODIDE
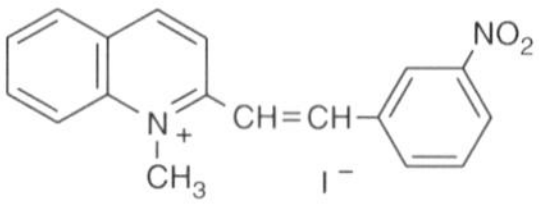
5 gm

S87,060-9
2-(2-BENZOYLHYDRAZINO)-1-METHYLQUINOLINIUM TETRAFLUOROBORATE
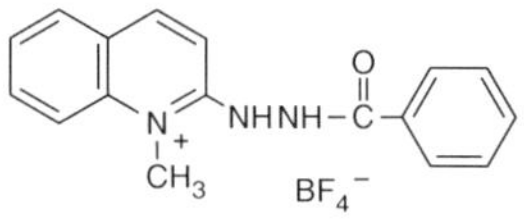
250 mg

S87,080-3
2-(2-BENZOYLHYDRAZINO)-1-ETHYLQUINOLINIUM TETRAFLUOROBORATE

250 mg

S51,253-2
4-(4-ACETAMIDOSTYRYL)-1-METHYLQUINOLINIUM IODIDE
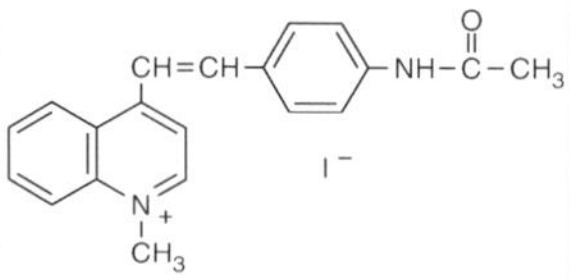
1 gm

S87,010-2
4-(2,2-DICYANOVINYL)-1-METHYLQUINOLINIUM TETRAFLUOROBORATE
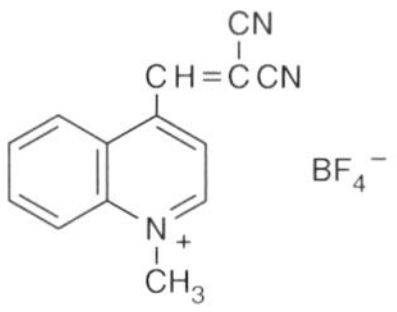
250 mg

S80,943-8
QUINOLINIUMMETHYL SULFATE INNER SALT
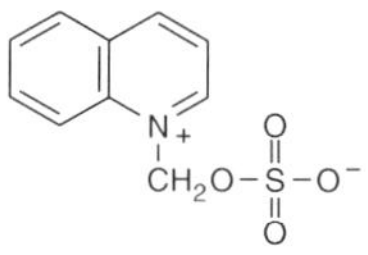
250 mg

S70,681-7
1-METHYLQUINOLINIUM METHYL SULFATE
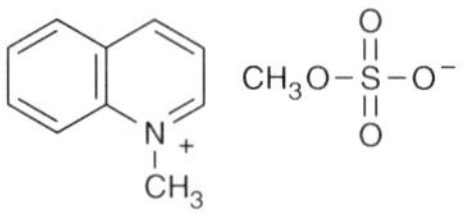
5 gm

S87,062-5
1-ETHYL-2-(PHENYLSULFONYLAZO)-QUINOLINIUM TETRAFLUOROBORATE
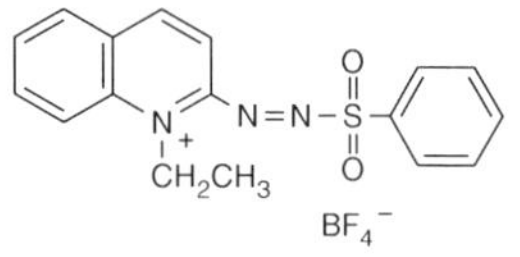
250 mg

S87,066-8
1-ETHYL-2-[2-(PHENYLSULFONYL)-HYDRAZINO]QUINOLINIUM TETRAFLUOROBORATE
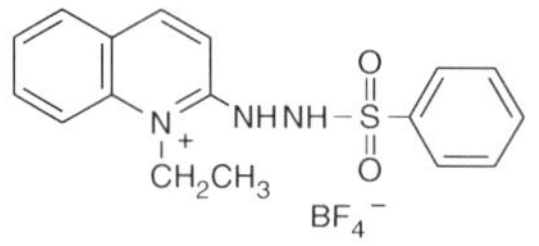
250 mg

S67,851-1
1-ETHYL-2-[3,4-(METHYLENEDIOXY)-STYRYL]QUINOLINIUM IODIDE
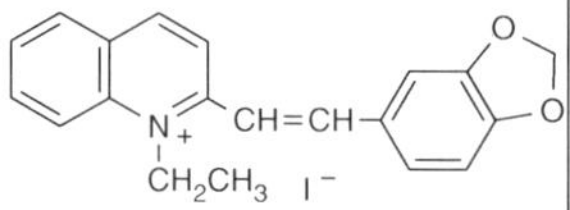
5 gm

S88,309-3
1-PHENYLQUINOLINIUM PERCHLORATE
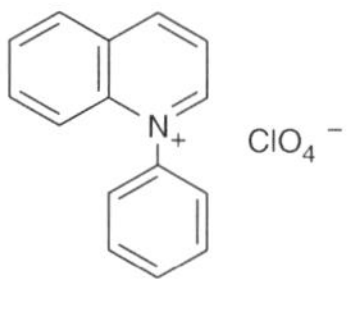
250 mg

S67,921-6
1-ETHYL-2-STYRYLQUINOLINIUM IODIDE
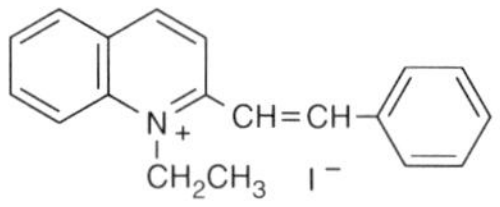
5 gm

S53,447-1
2-(4-FLUOROSTYRYL)-1-METHYLQUINOLINIUM IODIDE
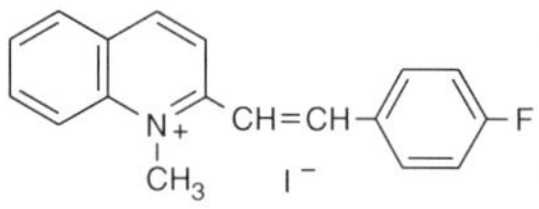
250 mg

S50,403-3
1-METHYL-2-(4-METHYLSTYRYL)-QUINOLINIUM IODIDE
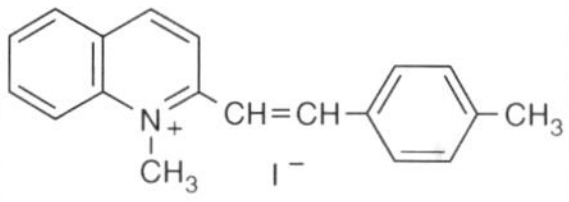
5 gm

S67,867-8
1-ETHYL-2-(4-METHYLSTYRYL)-QUINOLINIUM IODIDE
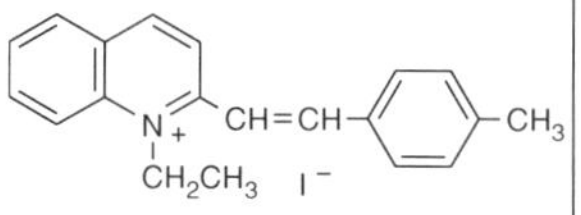
5 gm

S39,976-0
2-(2,6-DICHLOROSTYRYL)-1-METHYLQUINOLINIUM IODIDE
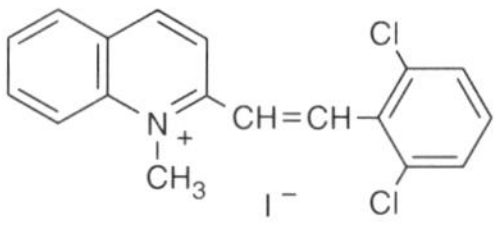
250 mg

S51,254-0
4-(2,6-DICHLOROSTYRYL)-1-METHYLQUINOLINIUM IODIDE
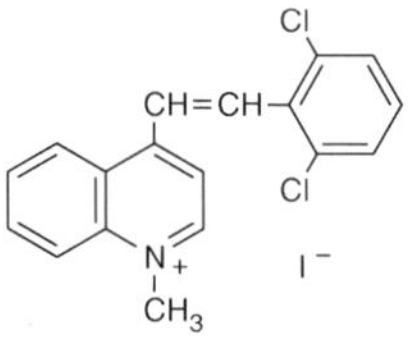
250 mg

S51,245-1
4-(3,4-DICHLOROSTYRYL)-1-METHYLQUINOLINIUM IODIDE
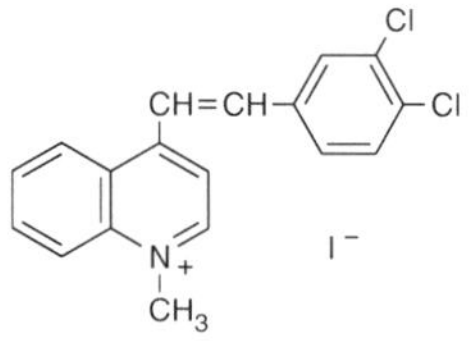
250 mg

S62,942-1
2-(3,4-DICHLOROSTYRYL)-1-METHYLQUINOLINIUM IODIDE
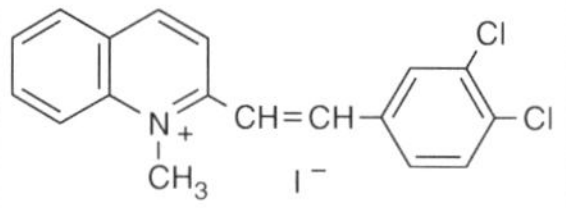
5 gm

S82,275-2
QUINOLINIUM BROMIDE PERBROMIDE
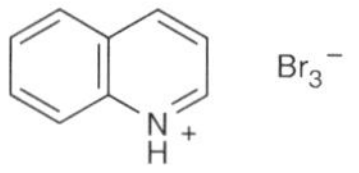
250 mg

S70,677-9
1-METHYLQUINOLINIUM IODIDE
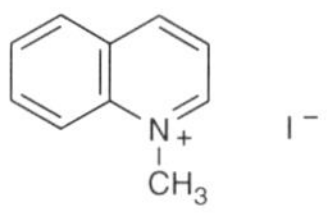
5 gm

S69,711-7
1-ETHYLQUINOLINIUM PERCHLORATE
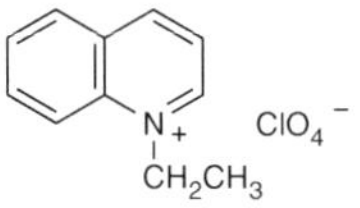
250 mg

S38,970-6
1-ETHYLQUINOLINIUM IODIDE
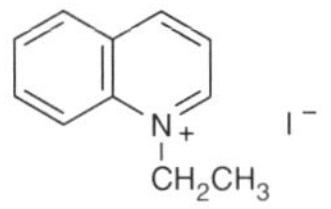
5 gm

S83,286-3
1-QUINALDINIUM IODIDE
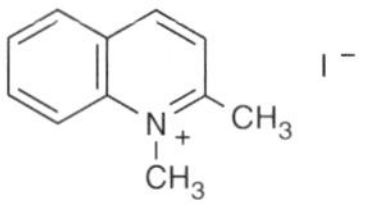
250 mg

S51,563-9
4-CHLORO-1,2-DIMETHYLQUINOLINIUM IODIDE
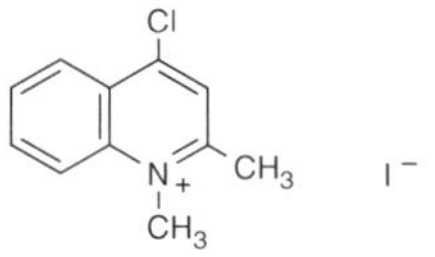
250 mg

S92,029-0
1-ETHYL-2-METHYLQUINOLINIUM TETRAFLUOROBORATE
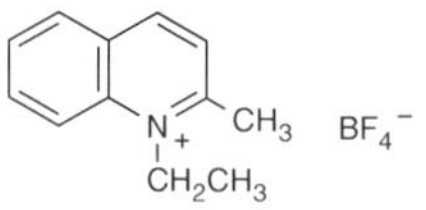
250 mg

S39,515-3
1,4-DIMETHYLQUINOLINIUM METHYLSULFATE
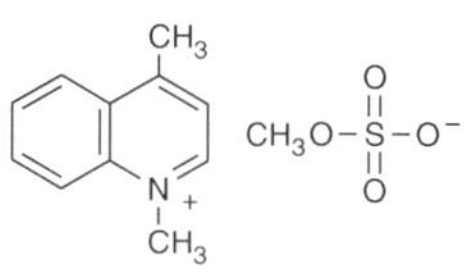
250 mg

S51,221-4
4-CHLORO-1,2-DIMETHYLQUINOLINIUM METHYL SULFATE
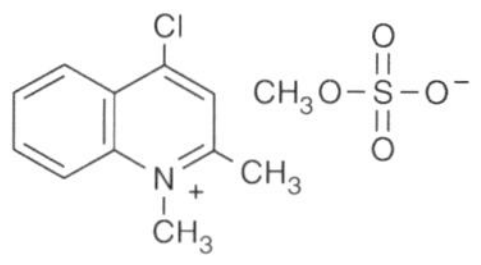
250 mg

S76,396-9
12-CHLORO-6,7,8,9,10,11-HEXAHYDRO-5-METHYLCYCLOOCTA(B)-QUINOLINIUM IODIDE
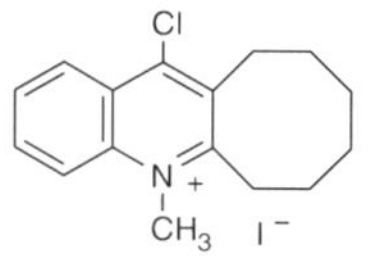
250 mg

S65,077-3
4,4'-METHYLENEBIS[N-(8-QUINOLYLOXYCARBONYL)ANILINE]
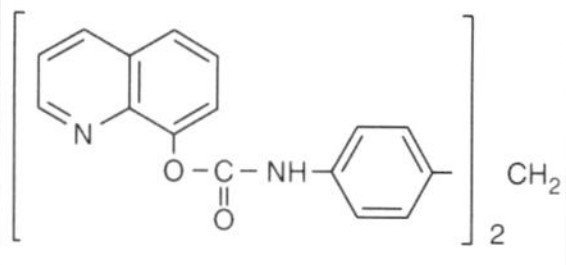
5 gm

S64,961-9
4,4'-METHYLENEBIS[N-(5-CHLORO-8-QUINOLYLOXYCARBONYL)ANILINE]
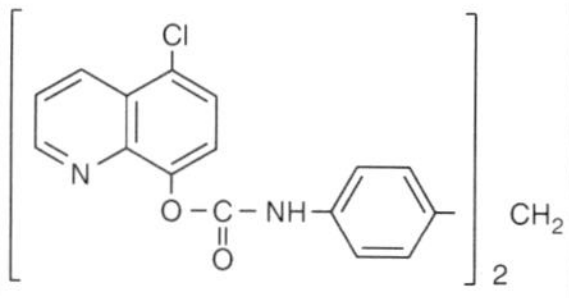
5 gm

S65,038-2
BIS(2-METHYL-8-QUINOLYL)-4,4'-METHYLENEBIS(N-PHENYLCARBAMATE)
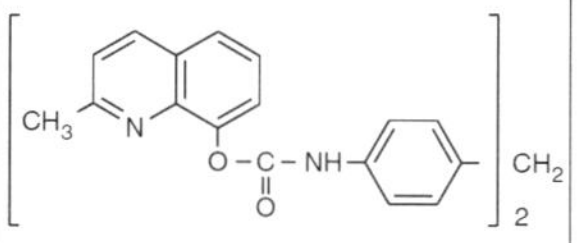
5 gm

S65,561-9
4-CHLORO-6-FLUOROSULFONYL-2-(4-NITROPHENYL)QUINOLINE
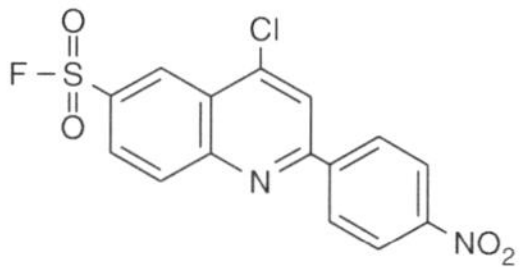
250 mg

S94,957-4
2-(2-NAPHTHYL)-4-QUINOLINECARBOXAMIDE
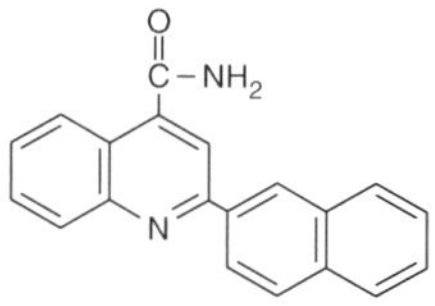
250 mg

S95,667-8
4-METHYL-2-PHENYLQUINOLINE (IN HEXANE)
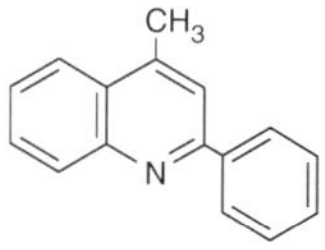
50 mg

S82,023-7
2-MESITYLQUINOLINE
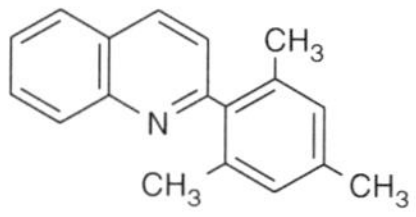
250 mg

S45,560-1
2-PYRIDINECARBOXALDEHYDE 2-QUINOLINYL-HYDRAZONE
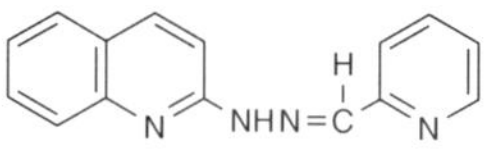
5 gm

S83,982-5
8-SULFO-2,4-QUINOLINEDICARBOXYLIC ACID
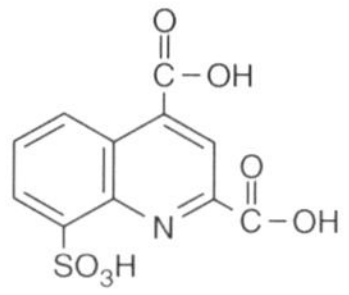
250 mg

S83,471-8
DIMETHYL 2,4-QUINOLINEDICARBOXYLATE
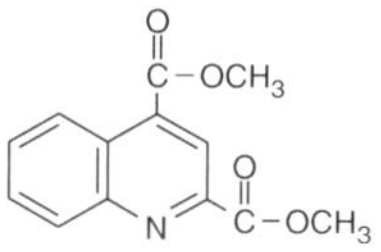
250 mg

S81,166-1
2-(2-FURYL)-4-QUINOLINECARBOXYLIC ACID
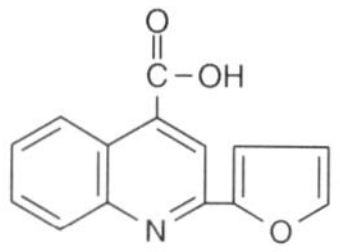
250 mg

S91,333-2
8-QUINOLYL N-[4-(ETHOXYCARBONYL)-PHENYL]CARBAMATE
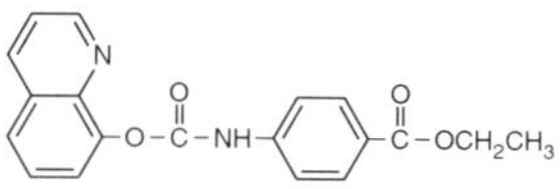
250 mg

S60,639-1
8-QUINOLYL N-CYCLOHEXYLCARBAMATE
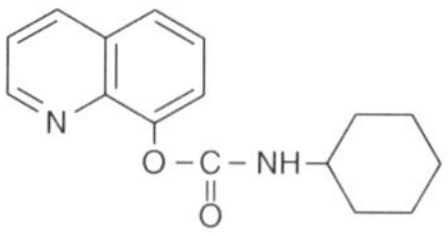
5 gm

S59,415-6
8-QUINOLYL N-PHENYLCARBAMATE
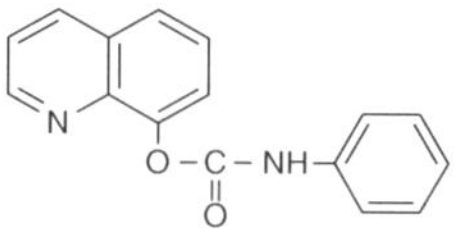
5 gm

S60,363-5
5-CHLORO-8-QUINOLYL N-PHENYLCARBAMATE
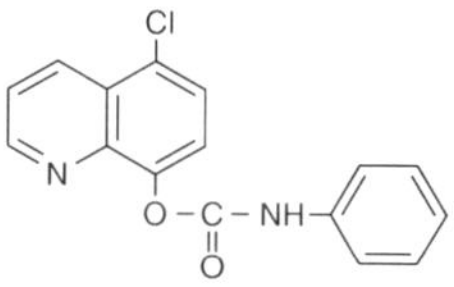
250 mg

S63,942-7
5,7-DICHLORO-8-QUINOLYL N-PHENYLCARBAMATE
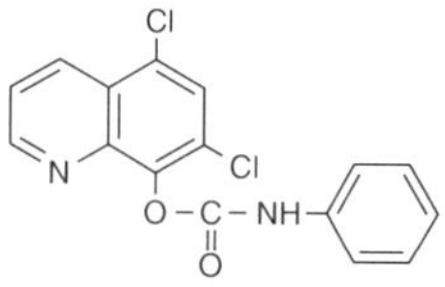
5 gm

S59,545-4
2-METHYL-8-QUINOLYL N-PHENYLCARBAMATE
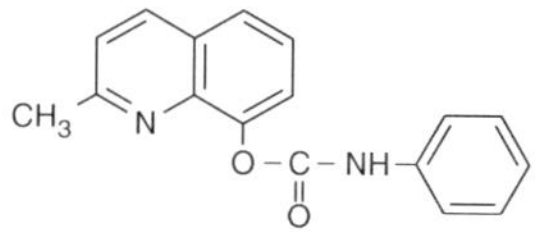
5 gm

S59,547-0
8-QUINOLYL N-(2-CHLOROPHENYL)-CARBAMATE
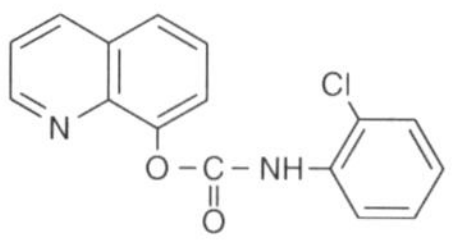
5 gm

S64,534-6
5-CHLORO-8-QUINOLYL N-(2-CHLOROPHENYL)CARBAMATE
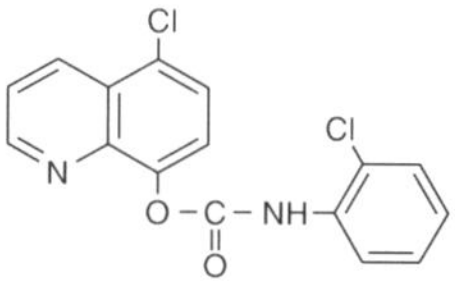
5 gm

S60,791-6
8-QUINOLYL N-(O-TOLYL)CARBAMATE
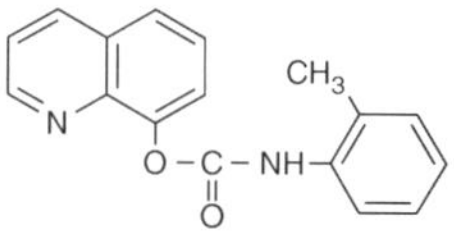
5 gm

S59,570-5
2-METHYL-8-QUINOLYL N-(2-CHLOROPHENYL)CARBAMATE
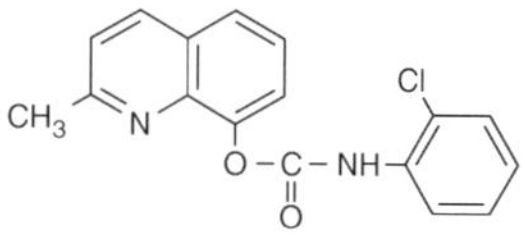
5 gm

S61,105-0
8-QUINOLYL N-(4-FLUOROPHENYL)-CARBAMATE
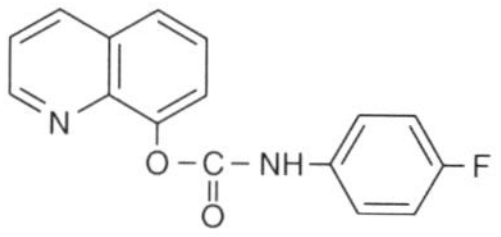
5 gm

S61,346-0
8-QUINOLYL N-(4-CHLOROPHENYL)-CARBAMATE
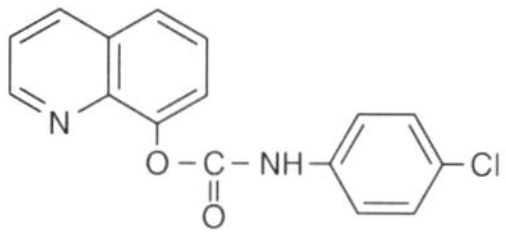
5 gm

S63,827-7
5-CHLORO-8-QUINOLYL N-(4-CHLOROPHENYL)CARBAMATE
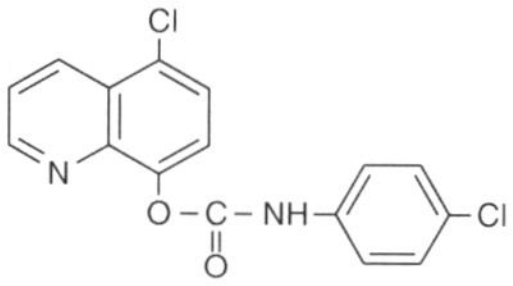
5 gm

S59,745-7
8-QUINOLYL N-(4-BROMOPHENYL)-CARBAMATE
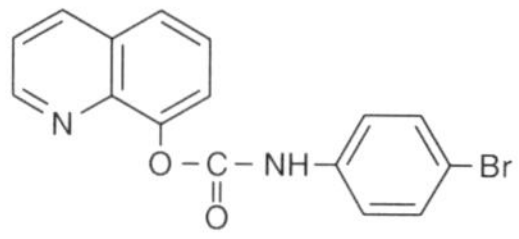
5 gm

S61,278-2
8-QUINOLYL N-(P-TOLYL)CARBAMATE
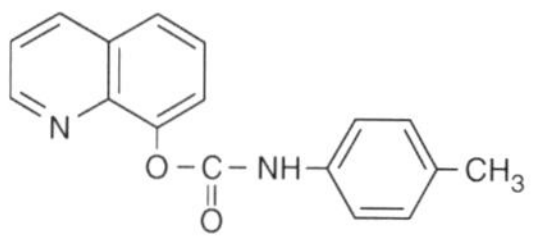
5 gm

S63,209-0
2-METHYL-8-QUINOLYL N-(4-CHLOROPHENYL)CARBAMATE
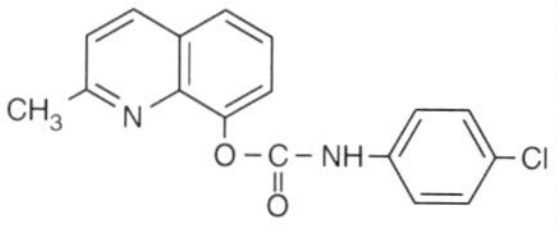
5 gm

S64,651-2
5-CHLORO-8-QUINOLYL N-(P-TOLYL)-CARBAMATE
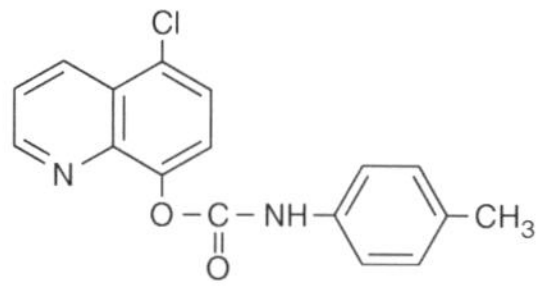
250 mg

S59,768-6
2-METHYL-8-QUINOLYL N-(4-BROMOPHENYL)CARBAMATE
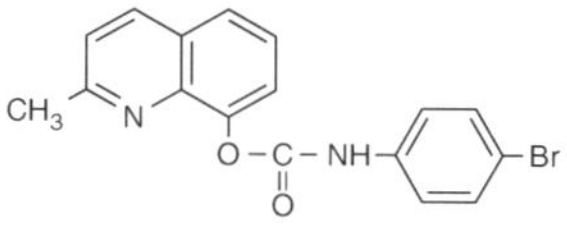
5 gm

S59,771-6
8-QUINOLYL N-(2,3-DICHLOROPHENYL)-CARBAMATE
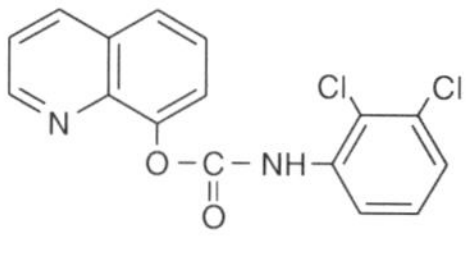
5 gm

S64,535-4
5-CHLORO-8-QUINOLYL N-(2,3-DICHLOROPHENYL)CARBAMATE
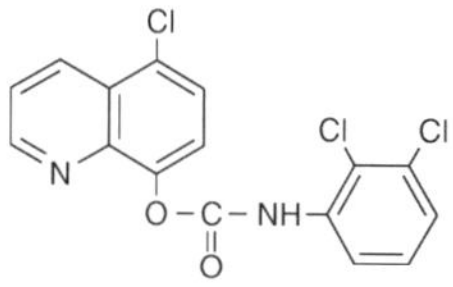
5 gm

S59,541-1
2-METHYL-8-QUINOLYL N-(2,6-DICHLOROPHENYL)CARBAMATE
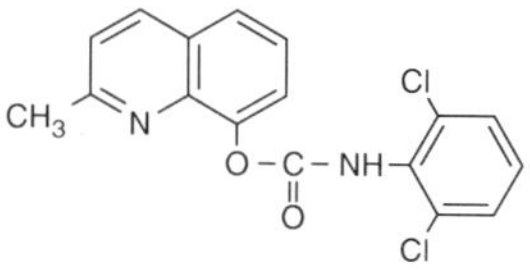
250 mg

S62,484-5
2-METHYL-8-QUINOLYL N-(2,3-DICHLOROPHENYL)CARBAMATE
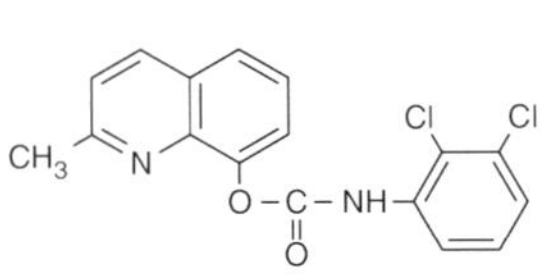
5 gm

S60,704-5
8-QUINOLYL N-(3,4-DICHLOROPHENYL)-CARBAMATE
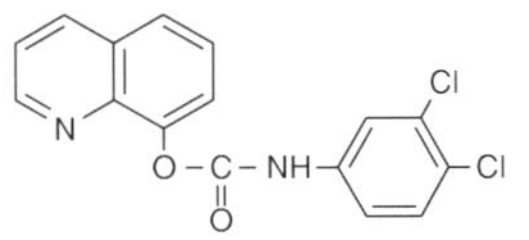
5 gm

S61,281-2
8-QUINOLYL N-(2,4-DICHLOROPHENYL)-CARBAMATE
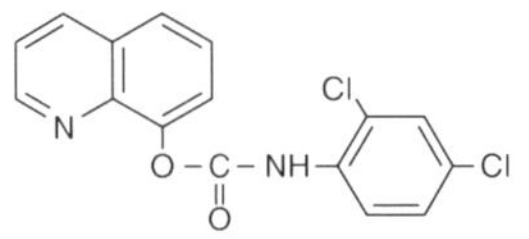
5 gm

S61,612-5
8-QUINOLYL N-(2,5-DICHLOROPHENYL)-CARBAMATE
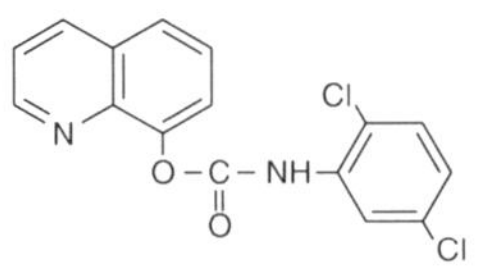
5 gm

S63,413-1
5-CHLORO-8-QUINOLYL N-(2,4-DICHLOROPHENYL)CARBAMATE
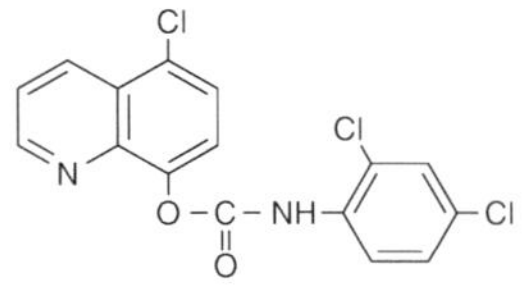
5 gm

S63,679-7
5-CHLORO-8-QUINOLYL N-(2,5-DICHLOROPHENYL)CARBAMATE
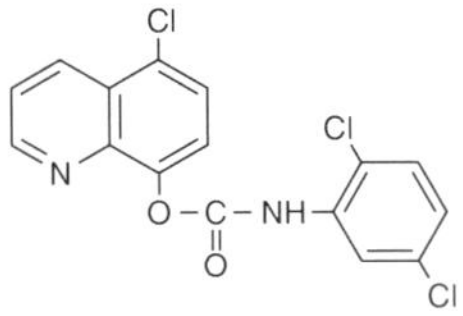
5 gm

S64,700-4
5-CHLORO-8-QUINOLYL N-(3,4-DICHLOROPHENYL)CARBAMATE
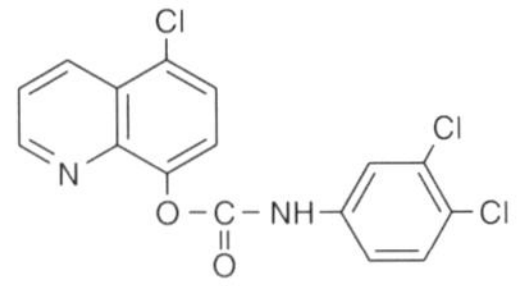
5 gm

S60,453-4
8-QUINOLYL N-(3-CHLORO-4-METHYLPHENYL)CARBAMATE
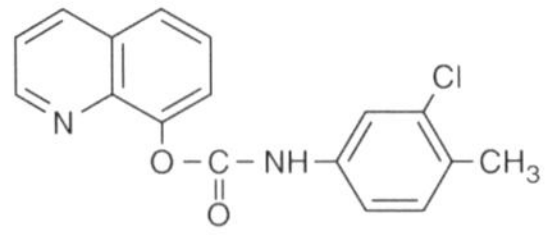
5 gm

S60,457-7
8-QUINOLYL N-(4-CHLORO-2-METHYLPHENYL)CARBAMATE
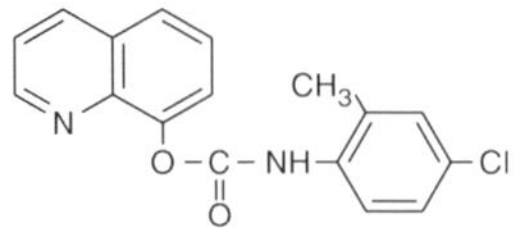
5 gm

S61,010-0
8-QUINOLYL N-(5-CHLORO-2-METHYLPHENYL)CARBAMATE
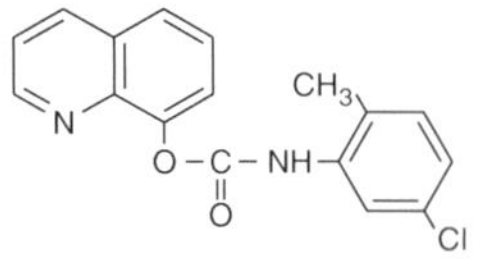
5 gm

S63,640-1
2-METHYL-8-QUINOLYL N-(2,4-DICHLOROPHENYL)CARBAMATE
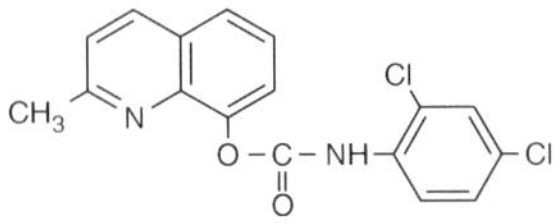
5 gm

S64,502-8
2-METHYL-8-QUINOLYL N-(3,4-DICHLOROPHENYL)CARBAMATE
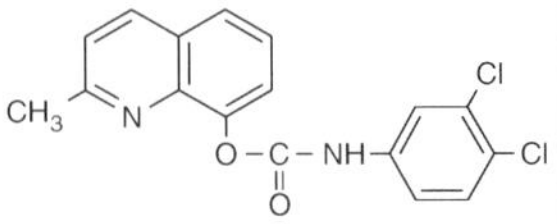
5 gm

S64,665-2
5-CHLORO-8-QUINOLYL N-(3-CHLORO-4-METHYLPHENYL)CARBAMATE
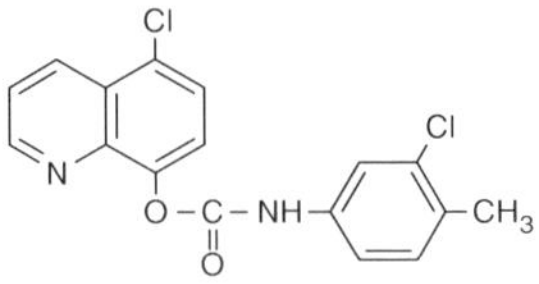
5 gm

S64,669-5
2-METHYL-8-QUINOLYL N-(2,5-DICHLOROPHENYL)CARBAMATE
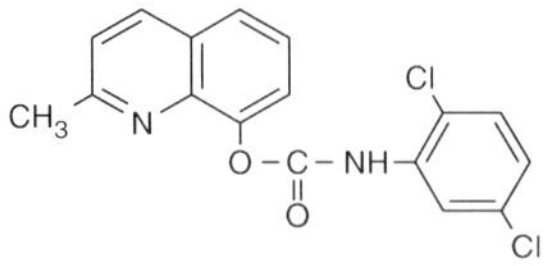
5 gm

S60,705-3
8-QUINOLYL N-(2,4-XYLYL)CARBAMATE
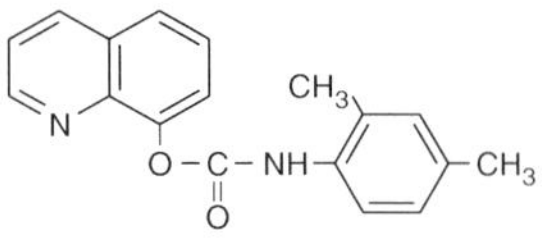
5 gm

S64,504-4
2-METHYL-8-QUINOLYL N-(4-CHLORO-2-METHYLPHENYL)CARBAMATE
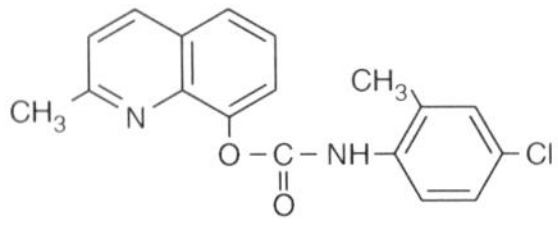
5 gm

S65,052-8
5-CHLORO-8-QUINOLYL N-(2,4-XYLYL)-CARBAMATE
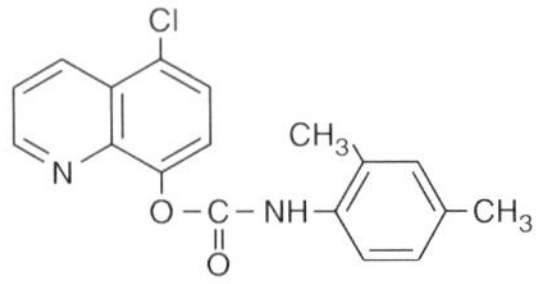
5 gm

S79,592-5
8-QUINOLINYL N-METHYLCARBAMATE
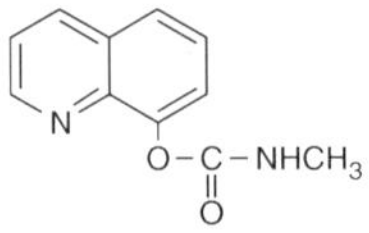
250 mg

S91,335-9
8-QUINOLYL N-ALLYLCARBAMATE
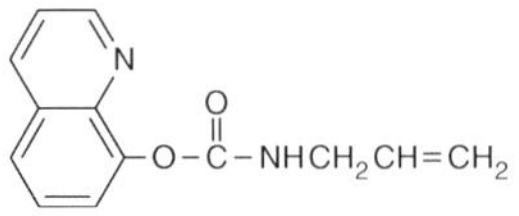
250 mg

S79,597-6
8-QUINOLINYL N-PROPYLCARBAMATE
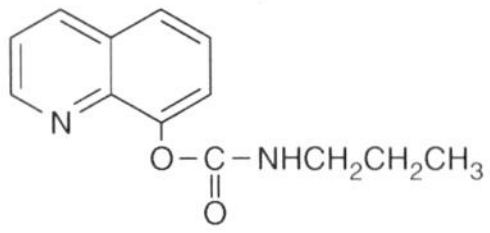
250 mg

S91,042-2
8-QUINOLYL N-ISOPROPYLCARBAMATE
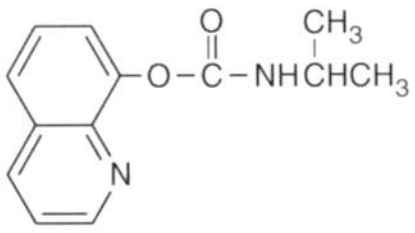
250 mg

S71,581-6
4-CHLORO-8-QUINOLYL N-ISOPROPYLCARBAMATE
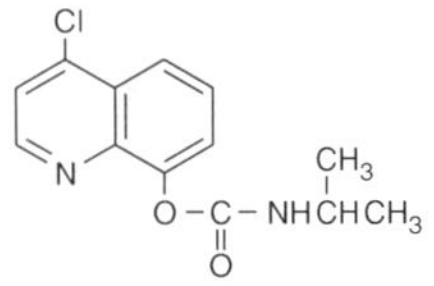
50 mg

S91,718-4
2,4-DIBROMO-8-QUINOLYL N-ISOPROPYLCARBAMATE
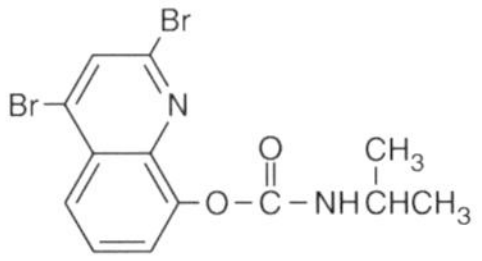
250 mg

S91,030-9
8-QUINOLYL N-BUTYLCARBAMATE
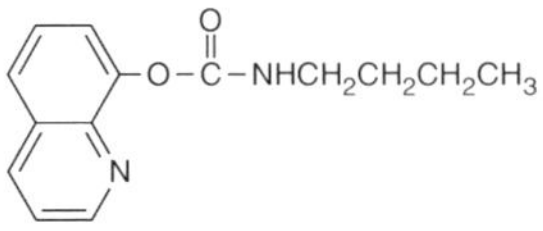
250 mg

S91,724-9
2,4-DICHLORO-8-QUINOLYL N-BUTYLCARBAMATE
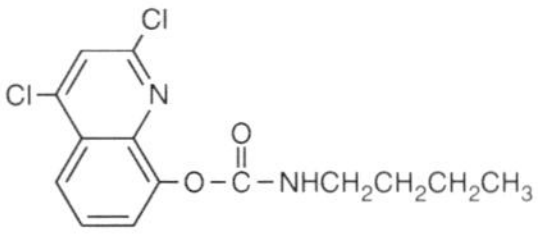
250 mg

S59,402-4
8-QUINOLYL N-(3-CHLOROPHENYL)-CARBAMATE
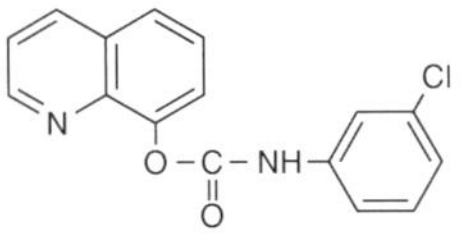
5 gm

S64,500-1
5-CHLORO-8-QUINOLYL N-(3-CHLOROPHENYL)CARBAMATE
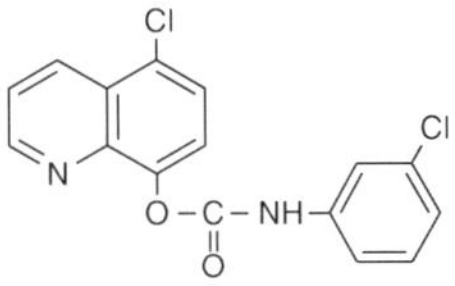
5 gm

S60,764-9
8-QUINOLYL N-(2,4,5-TRICHLOROPHENYL)CARBAMATE
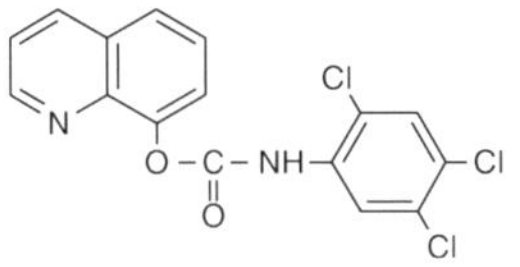
5 gm

S64,483-8
5,7-DICHLORO-8-QUINOLYL N-(3-CHLOROPHENYL)CARBAMATE
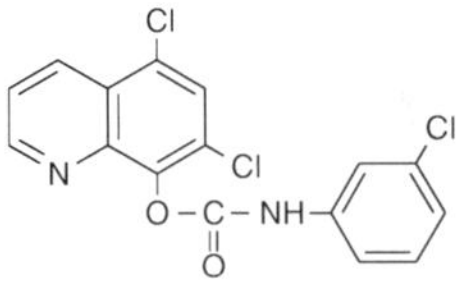
5 gm

S60,720-7
5-CHLORO-8-QUINOLYL N-(2,4,5-TRICHLOROPHENYL)CARBAMATE
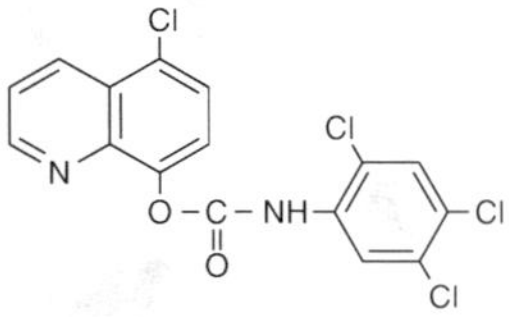
5 gm

S64,524-9
5,7-DICHLORO-8-QUINOLYL N-(2,4,5-TRICHLOROPHENYL)CARBAMATE
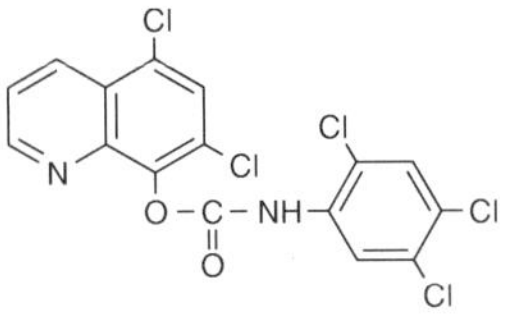
5 gm

S60,698-7
8-QUINOLYL N-(M-TOLYL)CARBAMATE
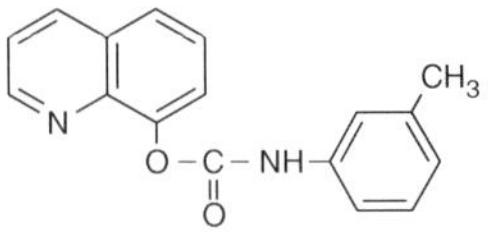
5 gm

S59,552-7
2-METHYL-8-QUINOLYL N-(3-CHLOROPHENYL)CARBAMATE
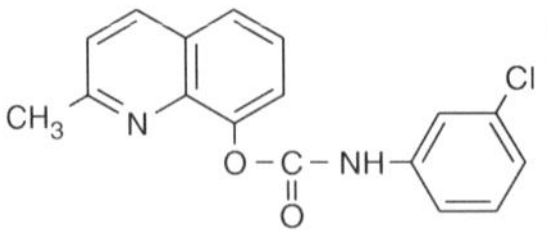
5 gm

S66,004-3
5-CHLORO-8-QUINOLYL N-(M-TOLYL)-CARBAMATE
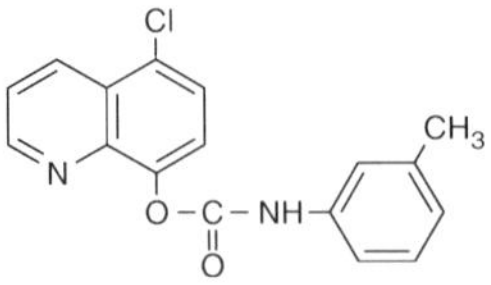
250 mg

S64,646-6
2-METHYL-8-QUINOLYL N-(3-METHOXYPHENYL)CARBAMATE
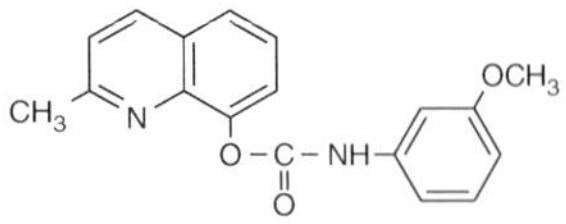
5 gm

S91,334-0
8-QUINOLYL N-DODECYLCARBAMATE

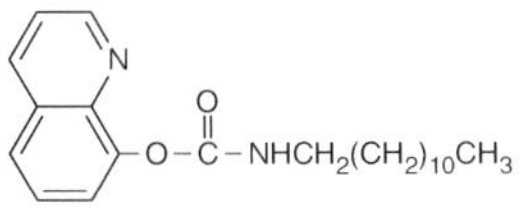

250 mg

S91,336-7
8-QUINOLYL N-OCTADECYLCARBAMATE

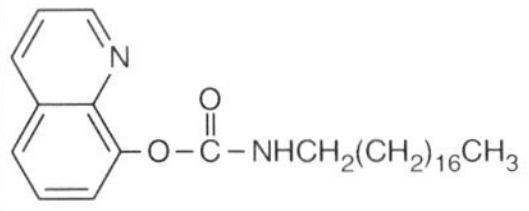

250 mg

S50,948-5
8-ACETOXY-2-PHENYL-4-QUINOLINECARBOXYLIC ACID

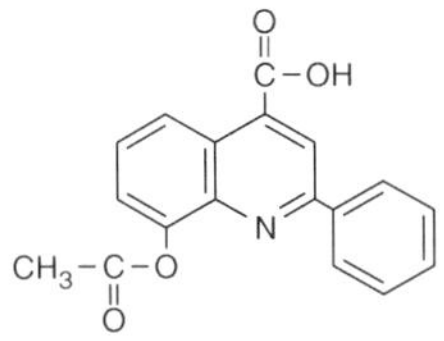

5 gm

S87,719-0
METHYL 2-QUINOLINECARBOXYLATE

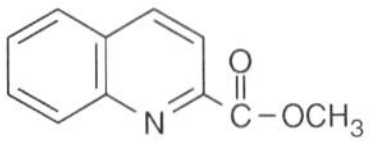

250 mg

S45,377-3
8-QUINOLINESULFONIC ACID

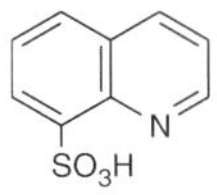

5 gm

S82,080-6
2-(2-THIENYL)QUINOLINE

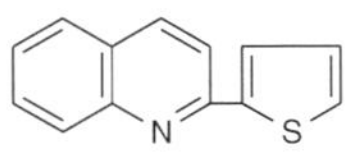

250 mg

S92,412-1
1-(4-METHYL-2-NITROPHENYL)-3-(2-METHYL-4-QUINOLYL)UREA

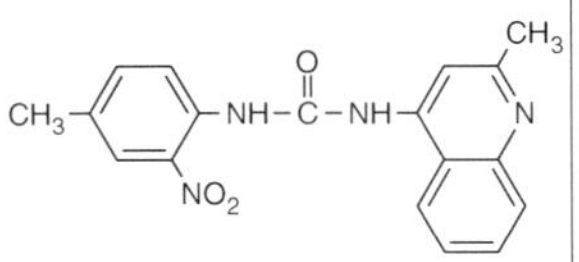

250 mg

S83,082-8
6-CHLORO-5-NITROQUINOLINE

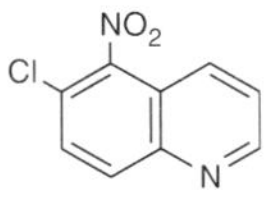

250 mg

S50,393-2
2-(3-NITROSTYRYL)QUINOLINE

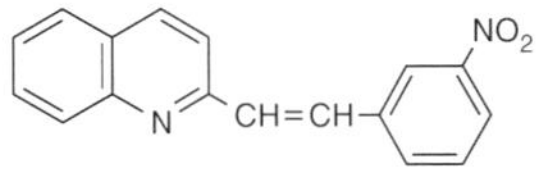

250 mg

S86,978-3
FORMIC N1-METHYL-N2-(2-QUINOLYL)-HYDRAZIDE

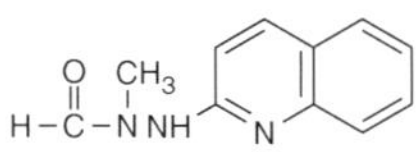

250 mg

S57,194-6
ETHYL 6-BROMO-DIHYDRO-DIOXO-4-METHYL-7H-DIBENZO(F,IJ)-ISOQUINOLINE-1-CARBOXYLATE

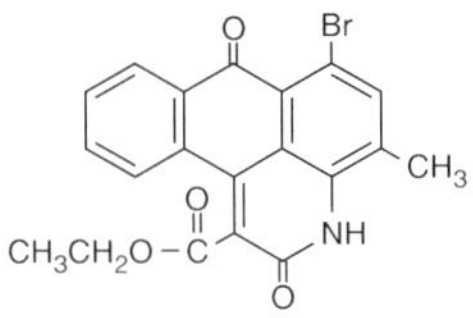

5 gm

S86,225-8
3-PYRIDYL 2-QUINOLYLMETHYL KETONE

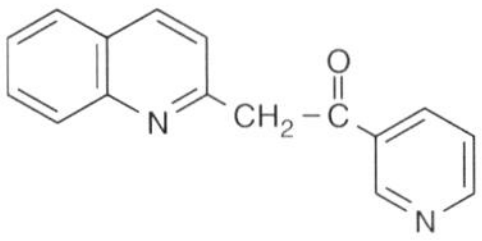

250 mg

S95,661-9
3-ACETYLQUINOLINE

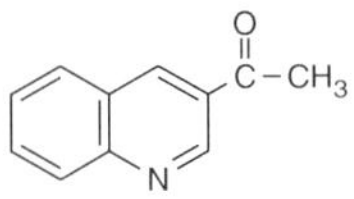

50 mg

S91,477-0
1,2-DIHYDRO-3-METHYL-2-OXO-4-QUINOLINECARBOXYLIC ACID

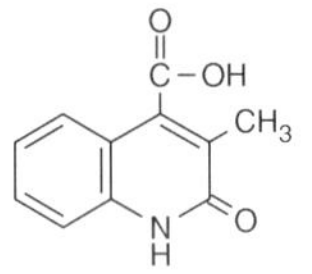

250 mg

S79,852-5
2-CYCLOHEXYL-4-QUINOLINECARBOXYLIC ACID

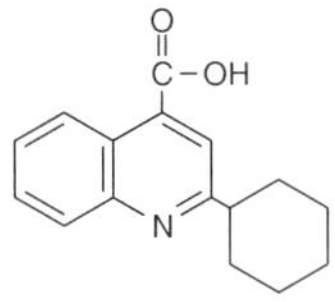

250 mg

S43,527-9
6,8-DICHLORO-2-PHENYL-4-QUINOLINE CARBOXYLIC ACID

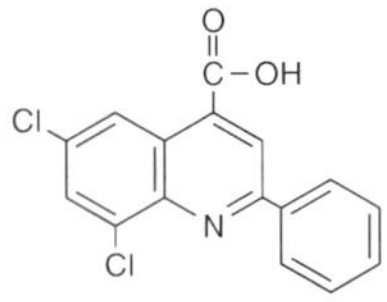

5 gm

S74,855-2
2-METHYL-3-PHTHALIMIDO-4-QUINOLINECARBOXYLIC ACID

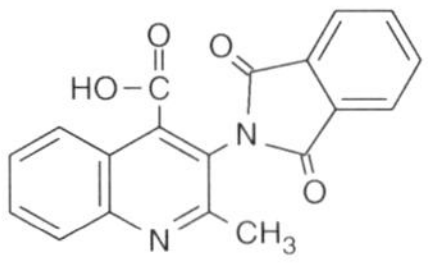

250 mg

S83,464-5
2-METHYL-4-QUINOLINECARBOXYLIC ACID

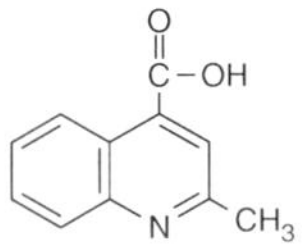

250 mg

S83,975-2
4-QUINOLINEACRYLIC ACID

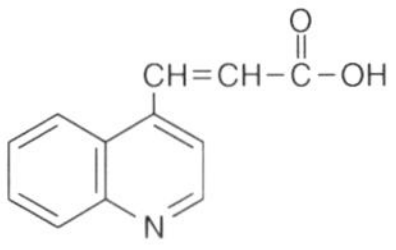

250 mg

S72,231-6
ETHYL 2-PHENYL-4-QUINOLINECARBOXYLATE

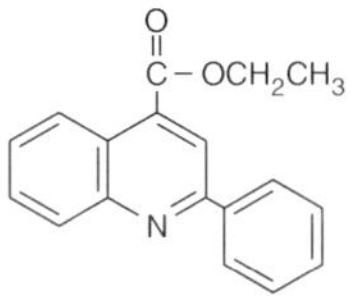

5 gm

S41,749-1
ETHYL 6,8-DICHLORO-2-PHENYL-4-QUINOLINECARBOXYLATE

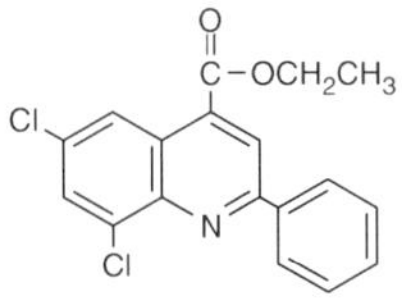

250 mg

S72,232-4
ETHYL 6-METHYL-2-PHENYLQUINOLINECARBOXYLATE

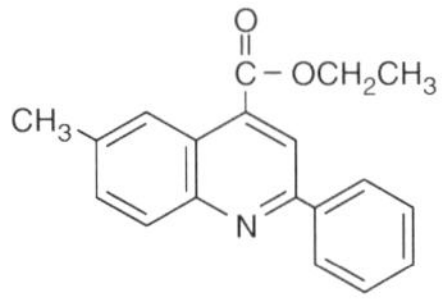

5 gm

S87,710-7
ETHYL 7-QUINOLINECARBOXYLATE

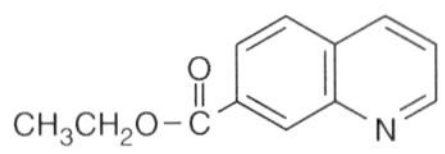

250 mg

S87,711-5
ETHYL 7-QUINOLINECARBOXYLATE HYDROCHLORIDE

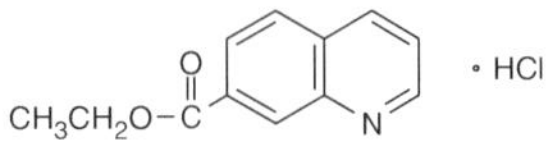

250 mg

S64,767-5
ETHYL 7-BROMO-2-METHYLQUINOLINE-4-CARBOXYLATE

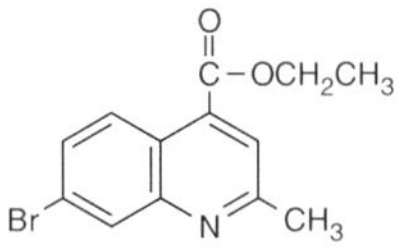

250 mg

S36,348-0
2-METHYLCINCHONINAMIDE

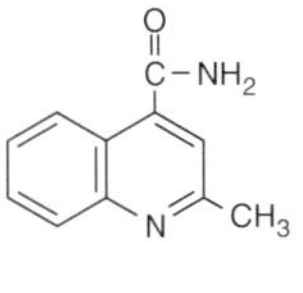

5 gm

S51,225-7
4-CYCLOHEXYL-2(1H)QUINOLONE

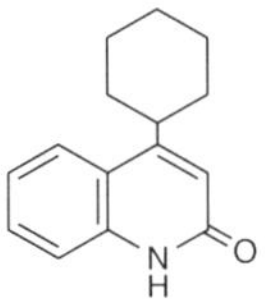

250 mg

S79,859-2
3,4-DIPHENYL-2(1H)QUINOLINONE

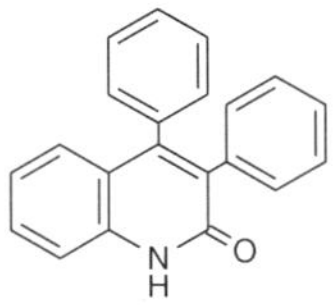

250 mg

S51,528-0
4-METHYLBENZO(H)QUINOLIN-2(1H)ONE

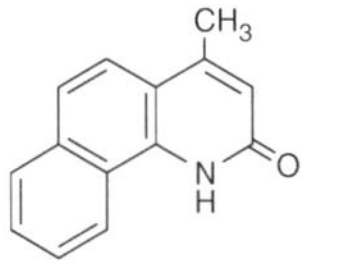

250 mg

S81,046-0
N,N-DIPROPYL-3-QUINOLINECARBOXAMIDE HYDROCHLORIDE

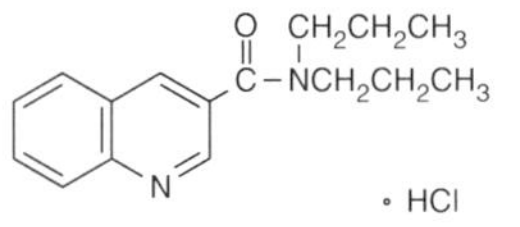

250 mg

S42,102-2
8-QUINOLYL 3-METHYLCROTONATE

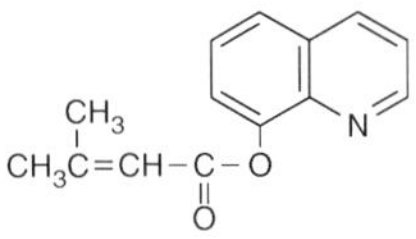

5 gm

S44,778-1
2-PIPERIDINOCINCHONINIC ACID

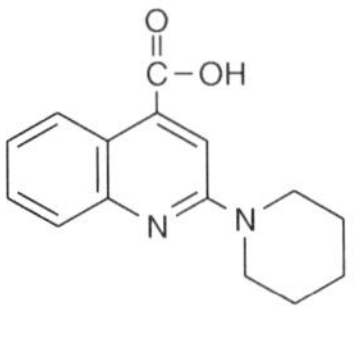

5 gm

S53,342-4
2-(4-CHLOROPHENYL)-7-METHYLCINCHONINIC ACID

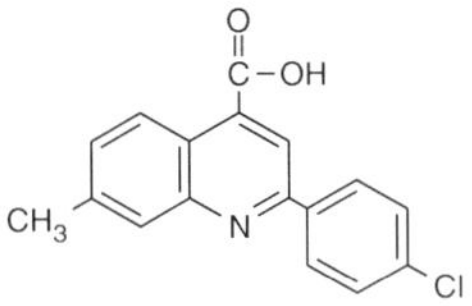

5 gm

S93,296-5
5,6-DIHYDROBENZ(C)ACRIDINE-7-CARBOXYLIC ACID

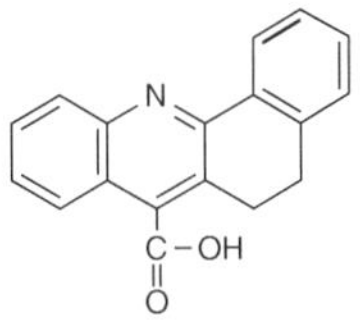

250 mg

S70,777-5
5-CHLORO-8-QUINOLYL N-(O-TOLYL)-CARBAMATE

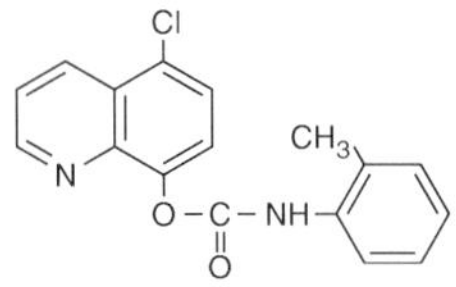

5 gm

S71,523-9
8-QUINOLYL N-(2,6-XYLYL)CARBAMATE
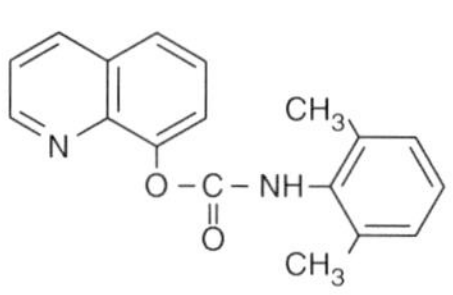
250 mg

S71,701-0
8-QUINOLYL N-(1-NAPHTHYL)CARBAMATE
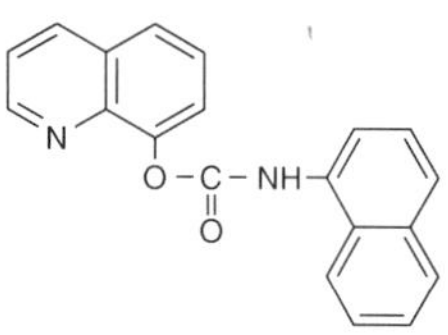
5 gm

S69,482-7
2,4-DIBROMO-8-QUINOLYL N-PROPYLCARBAMATE
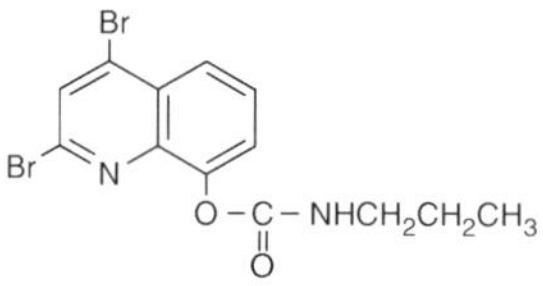
250 mg

S71,533-6
2-METHYL-8-QUINOLYL N-(M-TOLYL)-CARBAMATE
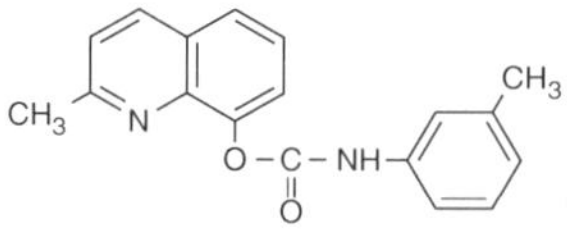
5 gm

S54,450-7
METHYL 4-(4-CHLORO-6-FLUOROSULFONYL-2-QUINOLYL)BENZOATE
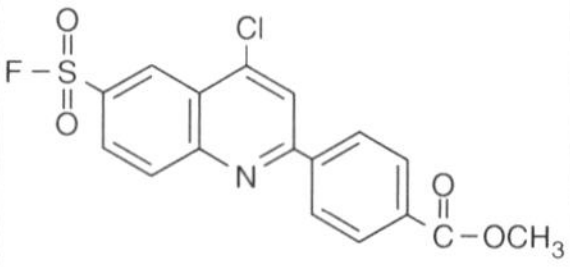
250 mg

S42,368-8
5,7-DICHLORO-8-QUINOLYL ACETATE
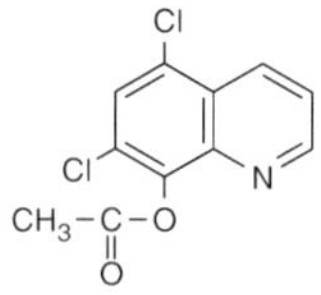
5 gm

S93,301-5
7-ACETOXY-5,6-DIHYDRO-5,5-DIMETHYLBENZ(C)ACRIDINE
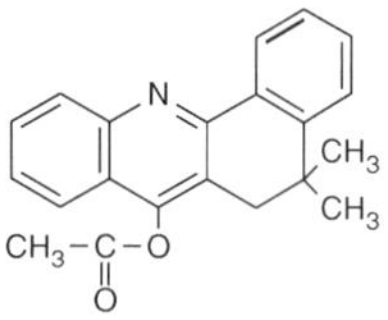
250 mg

S70,347-8
TERT-BUTYL 8-QUINOLINYL CARBONATE
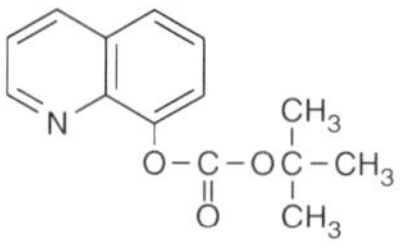
5 gm

S84,101-3
4-QUINOLINECARBONITRILE METHIODIDE
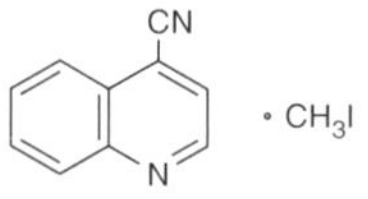
250 mg

S82,449-6
2-(BENZYLSULFONYL)QUINOLINE
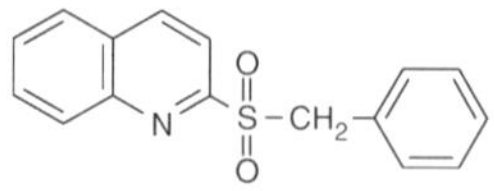
250 mg

S81,473-3
6-(METHYLSULFONYL)QUINOLINE
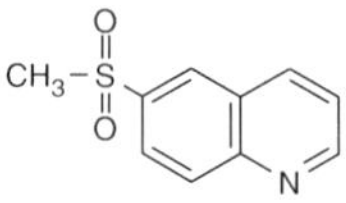
250 mg

S54,448-5
4-CHLORO-2-[4-(N,N-DIETHYLAMINOSULFONYL)PHENYL]-6-QUINOLINESULFONYL FLUORIDE
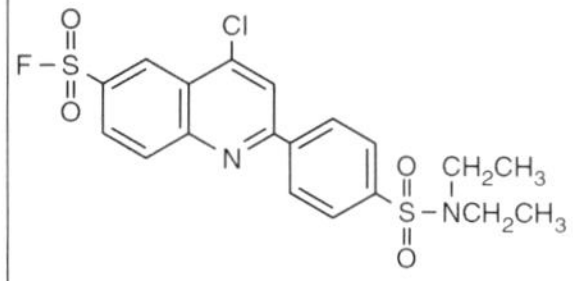
5 gm

S43,393-4
4-CHLORO-2-PHENYL-6-QUINOLINESULFONYL FLUORIDE
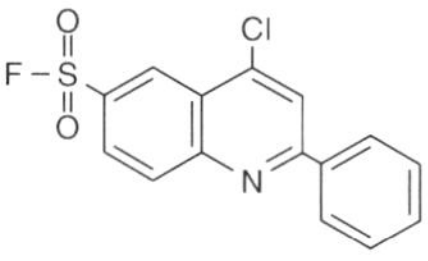
5 gm

S50,742-3
4-CHLORO-2-PHENYLQUINOLINE-4',6-DISULFONYL FLUORIDE
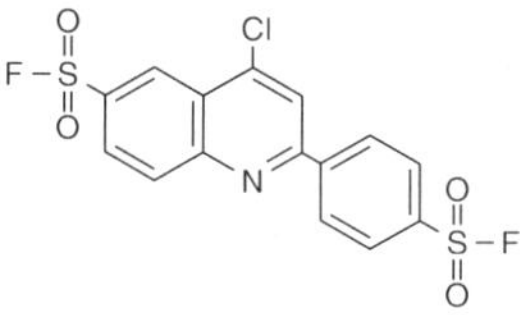
250 mg

S54,656-9
4-CHLORO-2-(P-TOLYL)-6-QUINOLINESULFONYL FLUORIDE
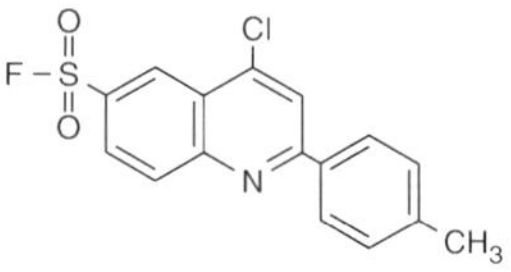
250 mg

S54,648-8
4-CHLORO-2-(M-TOLYL)-6-QUINOLINESULFONYL FLUORIDE
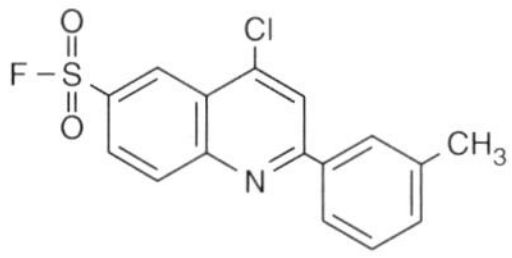
250 mg

S87,042-0
2-(P-TOSYLAZO)QUINOLINE
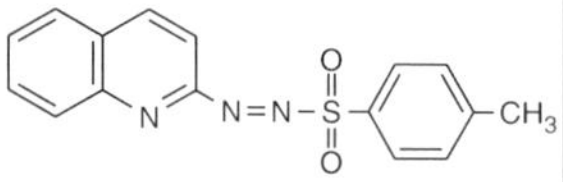
250 mg

S87,059-5
P-TOLUENESULFONIC N2-(2-QUINOLYL)-HYDRAZIDE
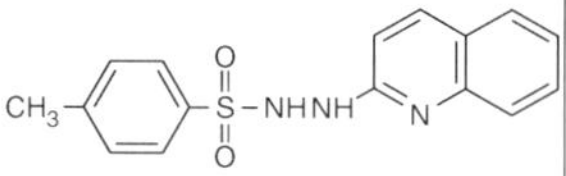
250 mg

S44,732-3
9-CHLORO-1,2,3,4-TETRAHYDROACRIDINE
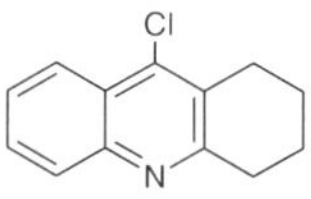
5 gm

S83,289-8
2-PIPERIDINOQUINOLINE
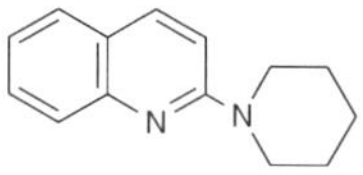
250 mg

S83,404-1
2-MORPHOLINOQUINOLINE
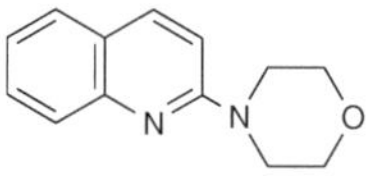
250 mg

S82,411-9
10-(2-QUINOLYL)PHENOTHIAZINE
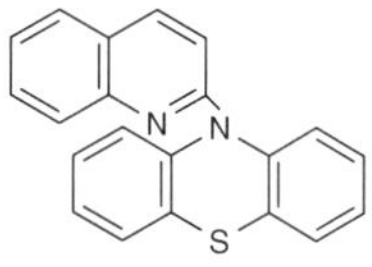
250 mg

S83,134-4
2-(4-BIPHENYLYL)QUINOLINE
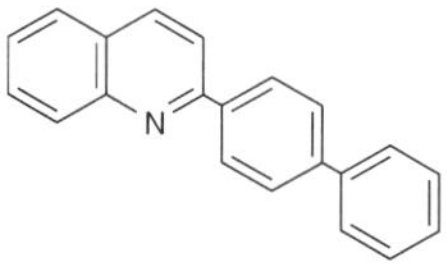
250 mg

S93,278-7
7-PHENYLQUINOLINE PICRATE
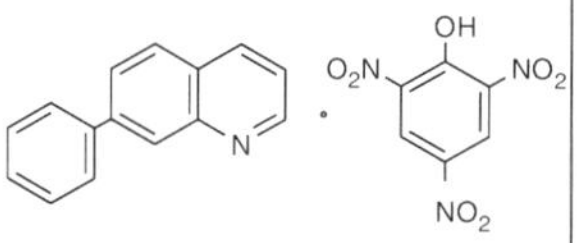
250 mg

S45,025-1
4-CHLORO-2-PHENYLQUINOLINE
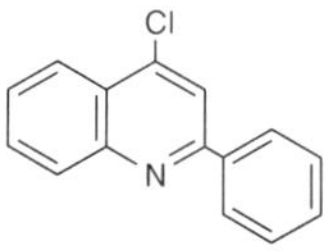
5 gm

S77,060-4
7-CHLORO-3-PHENYLQUINOLINE
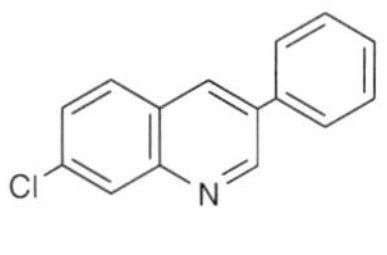
250 mg

S81,026-6
4,7-DICHLORO-2-PHENYLQUINOLINE
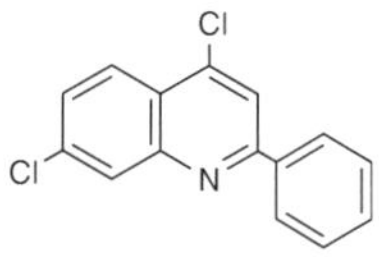
250 mg

S71,084-9
O-(8-QUINOLYL)-N-PHENYLTHIOCARBAMATE
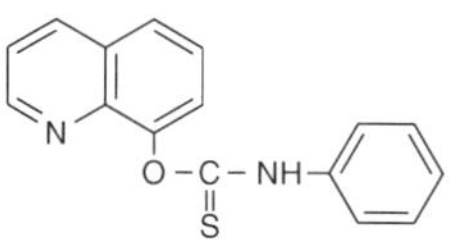
5 gm

S83,771-7
4-STYRYLQUINOLINE NITRATE
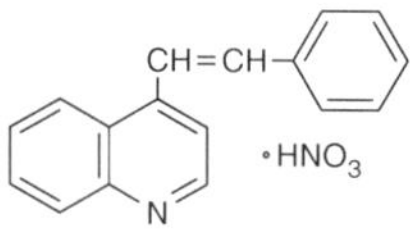
250 mg

S81,074-6
4-(PHENETHYL)QUINOLINE
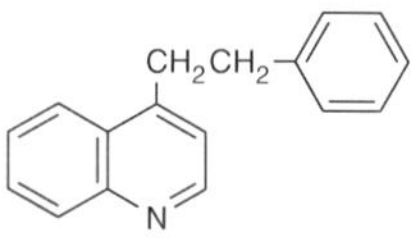
250 mg

S81,705-8
2-(PHENETHYL)QUINOLINE
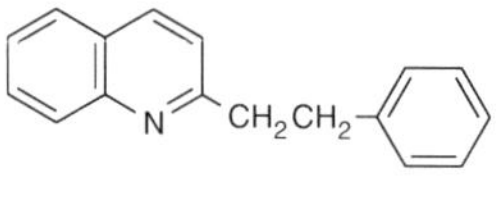
250 mg

S94,641-9
2-(4-PHENYL-1,3-BUTADIEN-1-YL)-QUINOLINE
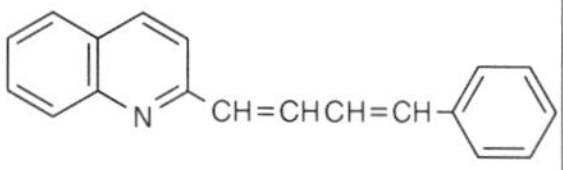
250 mg

S93,792-4
6-CHLORO-4-PHENYLQUINOLINE PICRATE
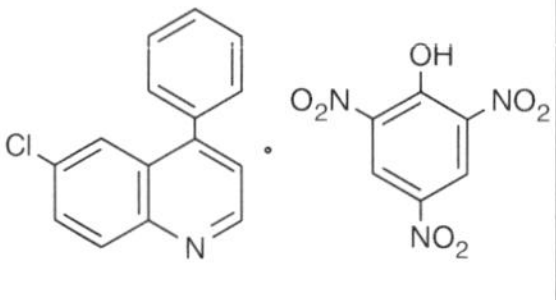
250 mg

S84,037-8
4,4'-BIS(2-QUINOLYL)BIPHENYL
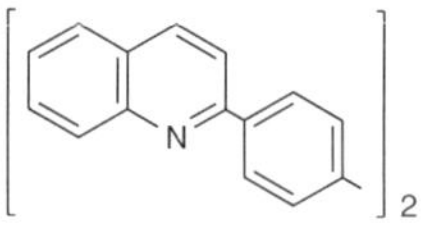
250 mg

S81,191-2
2-(P-TOLYL)QUINOLINE
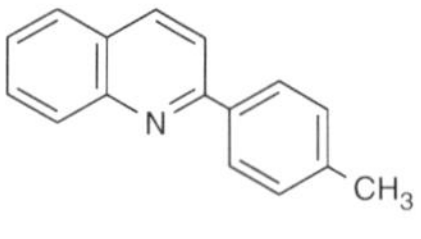
250 mg

S83,487-4
1,4-PHENYLENEBIS(2-QUINOLINE)

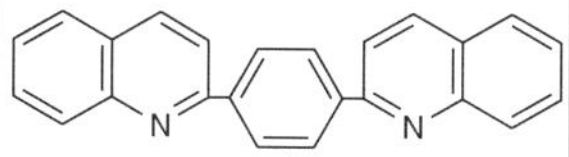

250 mg

S92,098-3
1-ETHYLQUINOLINIUM TETRAFLUOROBORATE

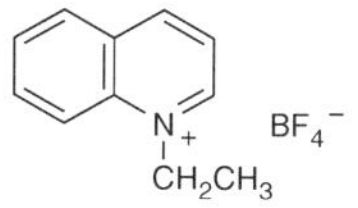

250 mg

S93,306-6
11-BROMO-11H-INDENO(1,2-B)QUINOLINE

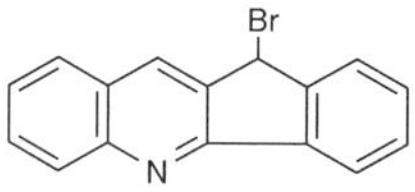

250 mg

S93,302-3
6H-INDENO(2,1-B)QUINOLINE

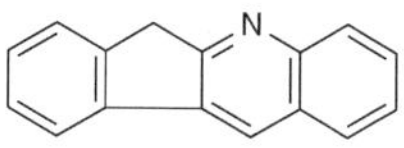

250 mg

S93,303-1
11H-INDENO(1,2-B)QUINOLINE

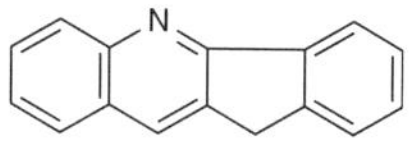

250 mg

S95,745-3
4-METHYL-2-(2-NAPHTHYL)BENZO(H)-QUINOLINE

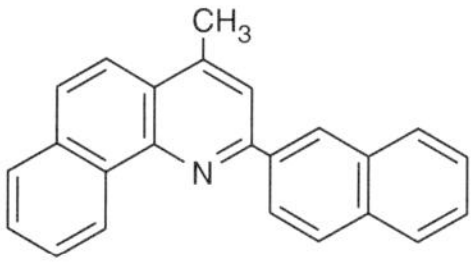

50 mg

S77,031-0
BENZO(F)QUINOLINE

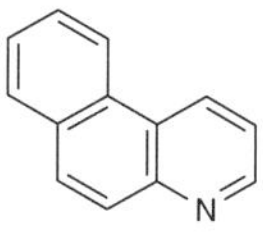

250 mg

S81,714-7
QUINOLINE PICRATE

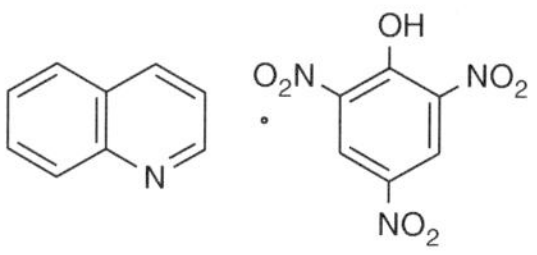

250 mg

S64,866-3
6-CHLOROQUINOLINE

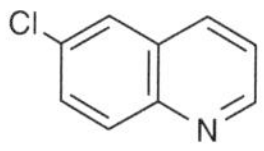

250 mg

S51,319-9
4,8-DICHLOROQUINOLINE

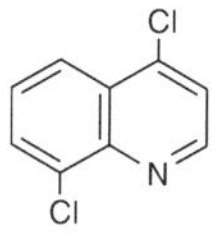

250 mg

S63,674-6
6-CHLOROLEPIDINE

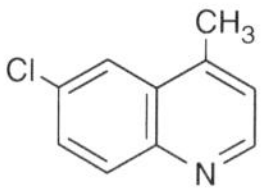

250 mg

S86,953-8
8-(BROMOMETHYL)QUINOLINE HYDROBROMIDE

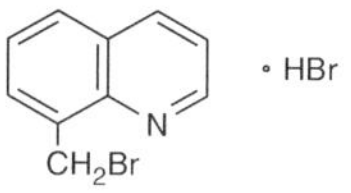

250 mg

S58,136-4
4-CHLORO-6,8-DIMETHYLQUINOLINE

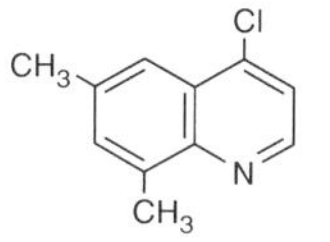

250 mg

S58,141-0
4-CHLORO-8-ETHYLQUINOLINE

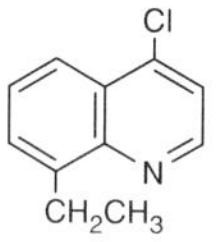

5 gm

S81,917-4
2-(1-PROPENYL)QUINOLINE

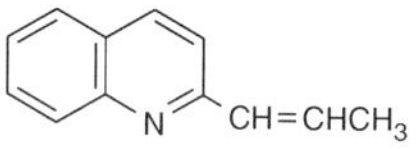

250 mg

S35,671-9
2,4,6-TRIMETHYLQUINOLINE SULFATE MONOHYDRATE

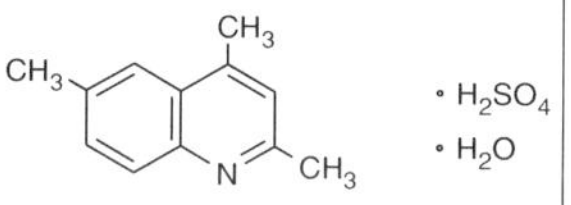

5 gm

S82,503-4
2-PROPYLQUINOLINE

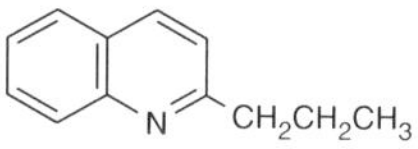

250 mg

S81,042-8
2-BUTYLQUINOLINE PICRATE

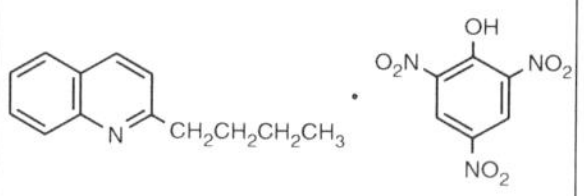

250 mg

S82,504-2
2-BUTYLQUINOLINE

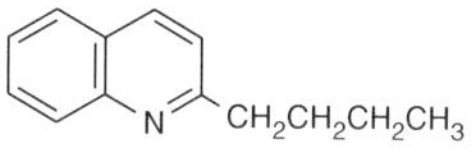

250 mg

S84,106-4
4-(1-PENTEN-1-YL)QUINOLINE

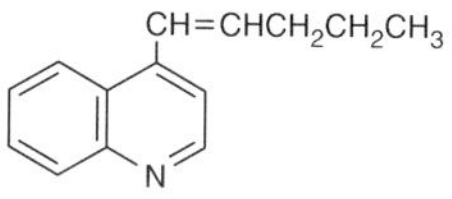

250 mg

S51,526-4
2-CHLORO-4-PENTADECYLQUINOLINE

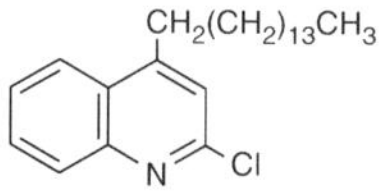

250 mg

S81,239-0
2-HEXADECYL-6-METHYLQUINOLINE

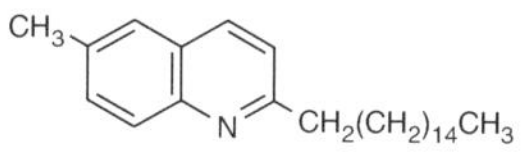

250 mg

S50,784-9
6,6'-METHYLENEBIS(4-CHLORO-2-PHENYLQUINOLINE)

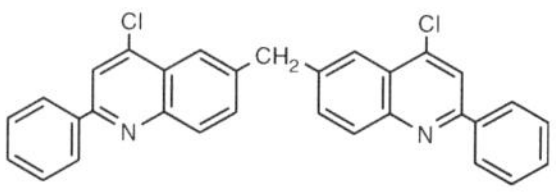

250 mg

S44,361-1
6-CHLORO-9-(3-DIETHYLAMINO-2-HYDROXYPROPYLAMINO)-2-METHOXYACRIDINE DIHYDROCHLORIDE

5 gm

S34,844-9
6,9-DIAMINO-2-ETHOXYACRIDINE HYDROCHLORIDE

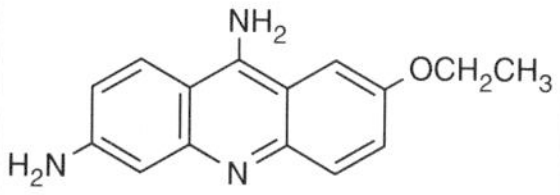

5 gm

S83,587-0
9-(DODECYLAMINO)-2-METHOXYACRIDINE

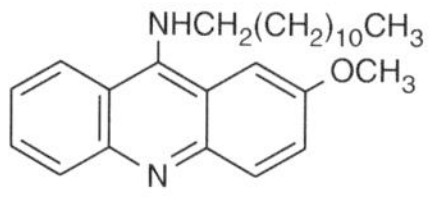

100 mg

S34,411-7
9-CHLORO-6,7-DIMETHOXY-3-NITROACRIDINE

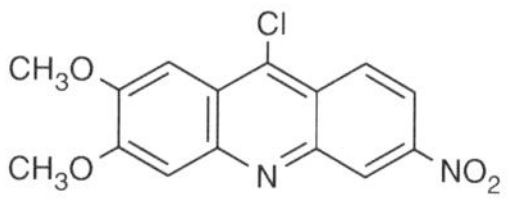

5 gm

S94,660-5
N-(6-CHLORO-2-METHOXY-9-ACRIDINYL)-GLYCINE ETHYL ESTER

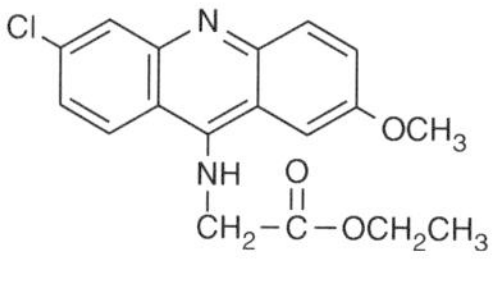

250 mg

S93,790-8
9-PHENOXYACRIDINE

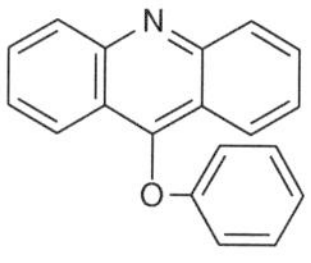

250 mg

S93,504-2
6,9-DICHLORO-2-METHOXY-N-METHYLACRIDINIUM IODIDE

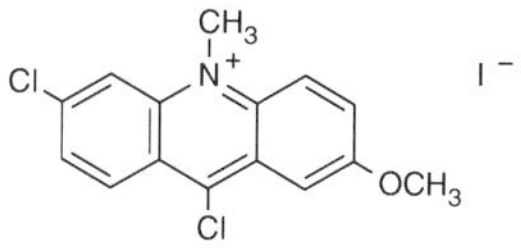

250 mg

S51,721-6
9-AMINOACRIDINE

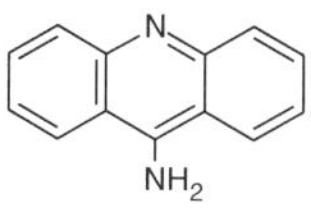

250 mg

S36,771-0
3,6-DIAMINOACRIDINE HEMISULFATE HEMIHYDRATE

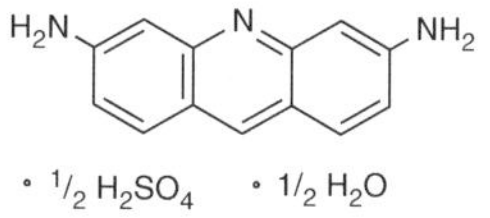

5 gm

S93,510-7
6-CHLORO-9-[4-(DIETHYLAMINO)-1-METHYLBUTYLAMINO]-2-METHYLACRIDINE DIHYDROCHLORIDE

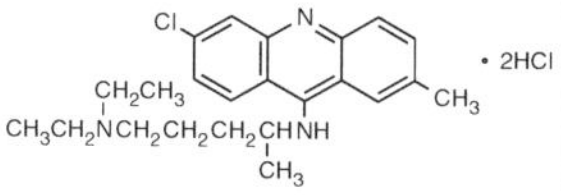

250 mg

S93,725-8
9-(3-DIETHYLAMINOPROPYLAMINO)-ACRIDINE DIHYDROCHLORIDE

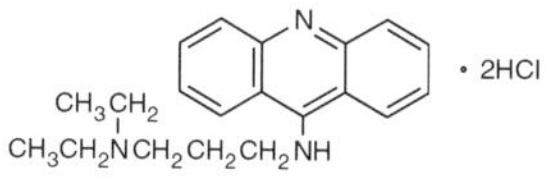

250 mg

S83,588-9
9-(OCTADECYLAMINO)ACRIDINE

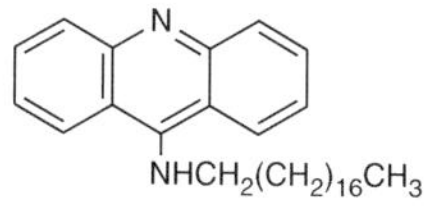

250 mg

S93,751-7
9-(3-DIMETHYLAMINOPROPYLAMINO)-2-FLUOROACRIDINE HYDROCHLORIDE

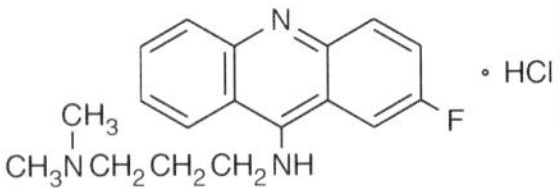

250 mg

S93,509-3
6-CHLORO-9-[4-(DIETHYLAMINO)-1-METHYLBUTYLAMINO]-2-NITROACRIDINE

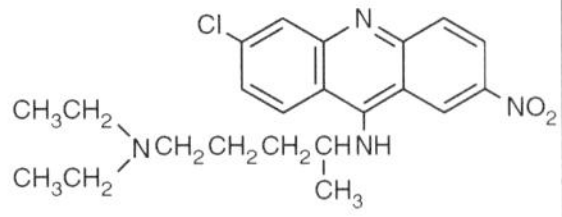

250 mg

S93,507-7
9-CHLORO-2-NITROACRIDINE

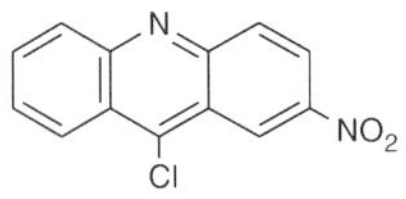

250 mg

S93,508-5
6,9-DICHLORO-2-NITROACRIDINE

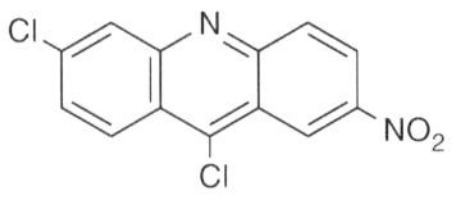

250 mg

S83,420-3
9-[4-(2,5-DIMETHYL-1-PYRROLYL)-PHENYL]ACRIDINE

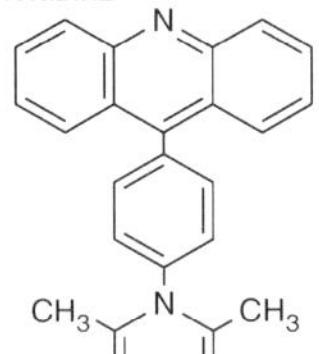

250 mg

S94,491-2
9-CYANO-10-METHYLACRIDINIUM METHYL SULFATE

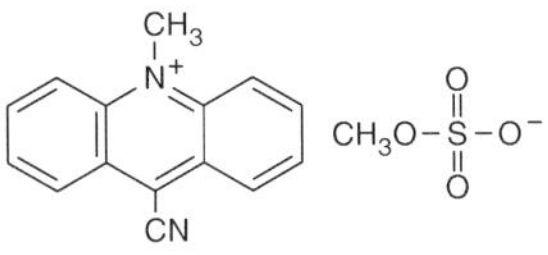

250 mg

S58,220-4
10-METHYLACRIDINIUM METHYL SULFATE

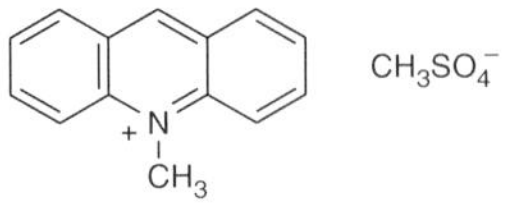

250 mg

S94,490-4
10-METHYL-9-PHENYLACRIDINIUM CHLORIDE

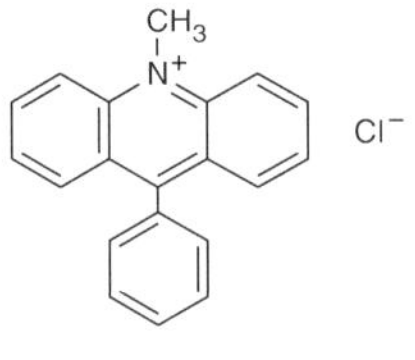

250 mg

S87,015-3
10-METHYLACRIDINE TETRAFLUOROBORATE

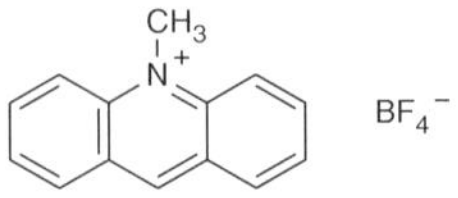

250 mg

S81,337-0
10-METHYLACRIDINIUM IODIDE

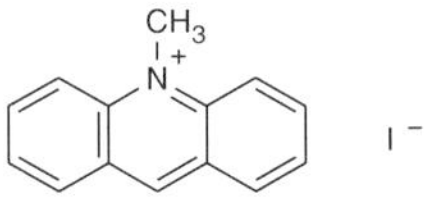

250 mg

S81,454-7
9-PHENYLACRIDINE

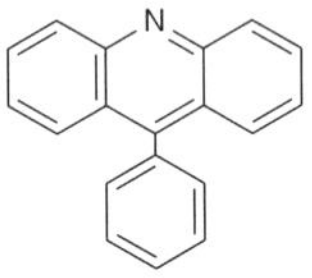

250 mg

S81,710-4
ACRIDINE PICRATE

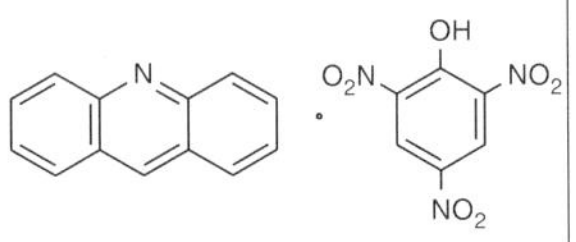

250 mg

S45,339-0
ACRIDINE HYDROCHLORIDE

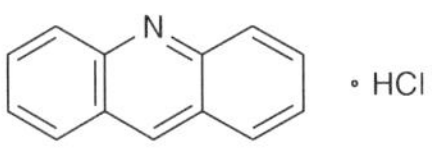

5 gm

S93,505-0
6,9-DICHLORO-2-METHYLACRIDINE

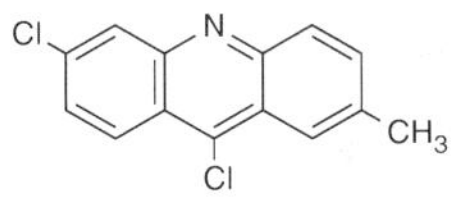

250 mg

S93,503-4
9-CHLORO-N-METHYLACRIDINIUM IODIDE

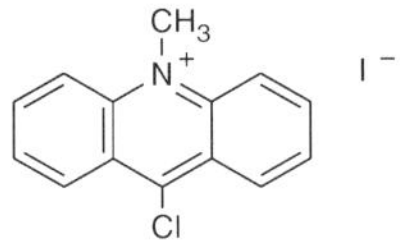

250 mg

S93,506-9
3,9-DICHLOROACRIDINE

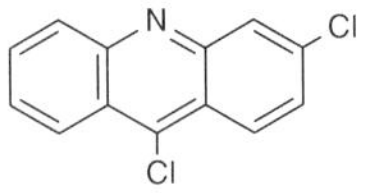

250 mg

S94,148-4
7-CHLOROBENZ(C)ACRIDINE

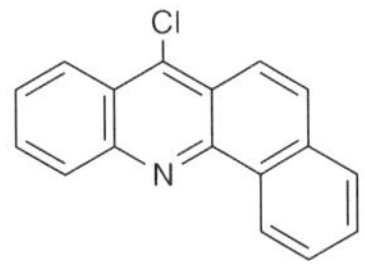

250 mg

S95,748-8
4-METHYL-2-(2-NAPHTHYL)BENZ(F)-ISOQUINOLINE

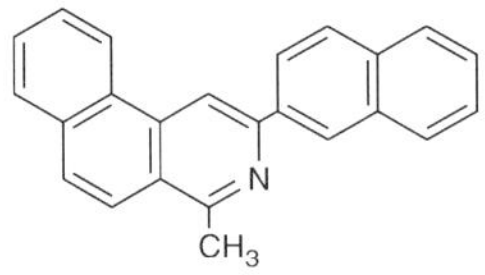

25 mg

S76,144-3
3-TERT-BUTYL-1-CHLORO-4-[3-(TRIFLUOROMETHYL)PHENYL]-ISOQUINOLINE

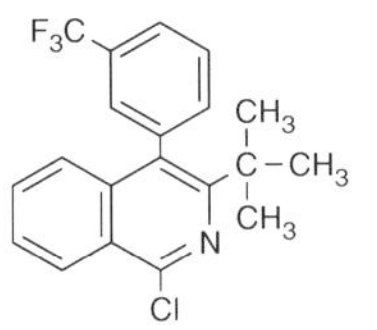

250 mg

S91,402-9
5-ISOQUINOLYL N-(4-CHLORO-2-METHOXYPHENYL)CARBAMATE

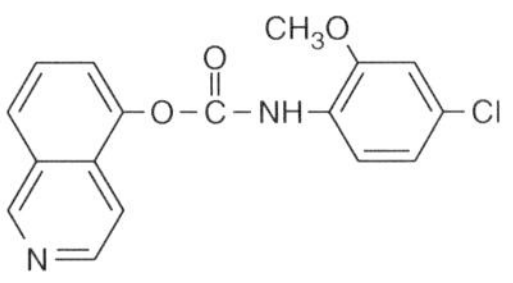

250 mg

S91,403-7
1-ISOQUINOLYL N-(4-CHLORO-2-METHOXYPHENYL)CARBAMATE

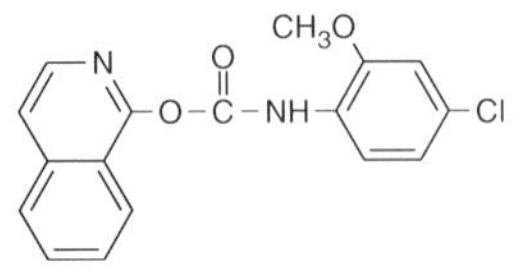

250 mg

S93,289-2
1-(4-BROMOPHENYL)-2-[1-(3-ISOQUINOLINYL)ETHYLIDENE]HYDRAZINE

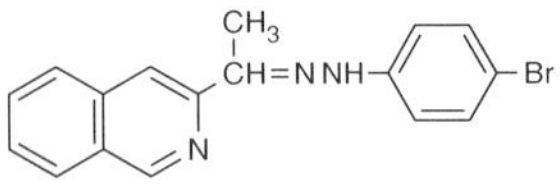

250 mg

S93,388-0
N-(1-BROMO-4-NITRO-3-ISOQUINOLYL)-BENZAMIDE

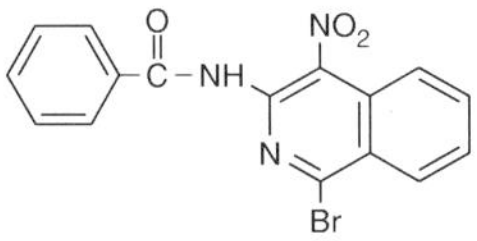

250 mg

S91,639-0
1-ISOQUINOLYL N-(4-FLUOROPHENYL)-CARBAMATE

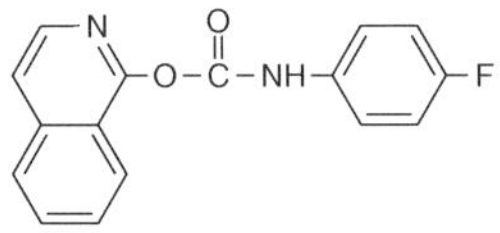

250 mg

S73,724-0
1-ISOQUINOLYL N-(2,5-DIFLUOROPHENYL)CARBAMATE

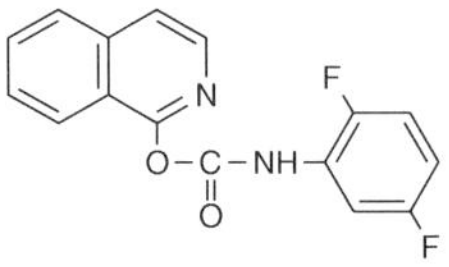

5 gm

S94,346-0
1,3-ISOQUINOLINEDIOL

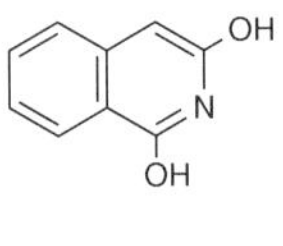

250 mg

S75,843-4
ALPHA-(4-METHOXYPHENYL)-1-ISOQUINOLINEMETHANOL

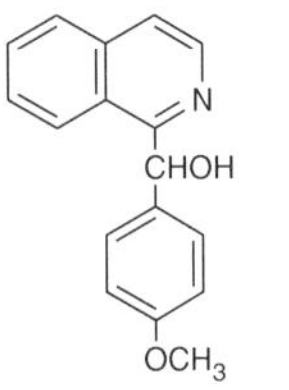

250 mg

S42,628-8
PROTOPAPAVERINE HYDROCHLORIDE

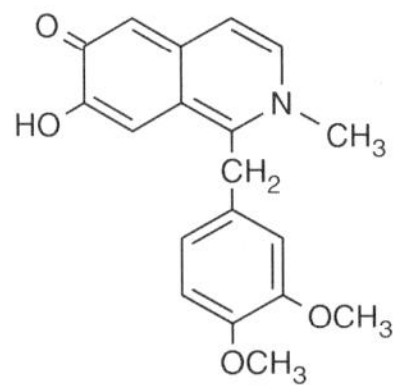

250 mg

S42,569-9
NORPAPAVERINE METHOCHLORIDE

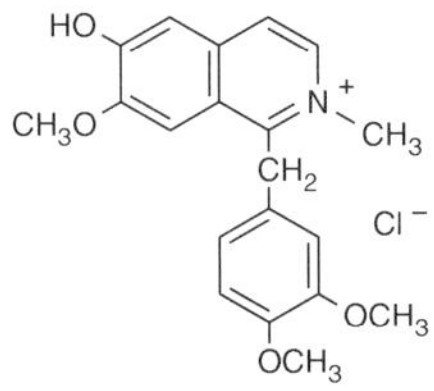

250 mg

S53,483-8
5,6,7-TRIMETHOXYISOCARBOSTYRIL

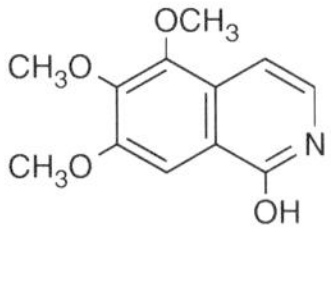

5 gm

S45,928-3
4-NITROISOCARBOSTYRIL

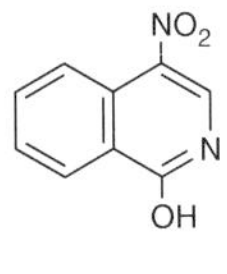

250 mg

S75,313-0
6-(3-HYDROXYPROPYLAMINO)-2-NITROPHENANTHRIDINE

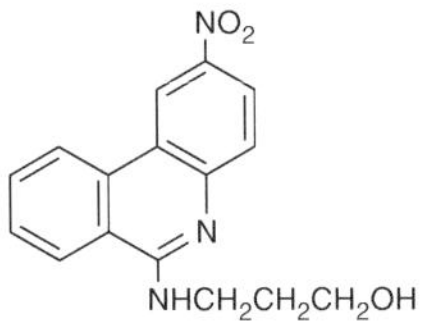

250 mg

S95,668-6
7-HYDROXY-8-METHYLISOQUINOLINE

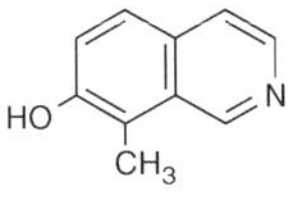

50 mg

S53,366-1
2-ETHYL-5-HYDROXYISOQUINOLINIUM BROMIDE

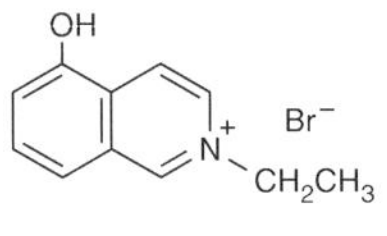

5 gm

S76,148-6
4-[10,11-DIHYDRO-5-HYDROXY-5H-DIBENZO(A,D)CYCLOHEPTEN-5-YL]-ISOQUINOLINE

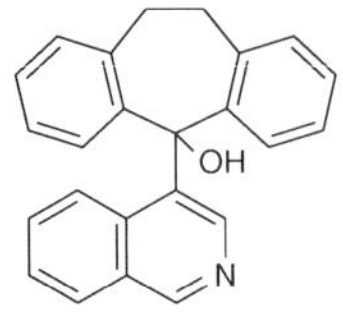

250 mg

S81,650-7
N,N-DIETHYL-BETA-(1-ISOQUINOLYLOXY)-PHENETHYLAMINE

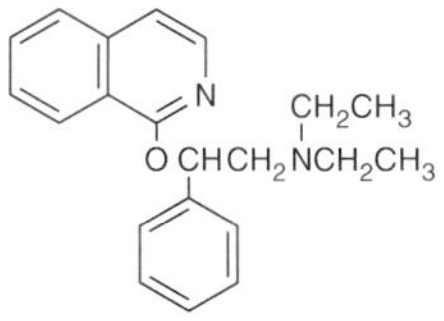

250 mg

S81,330-3
1-[2-(DIETHYLAMINO)-1-PHENYLETHOXY]-ISOQUINOLINE PICRATE

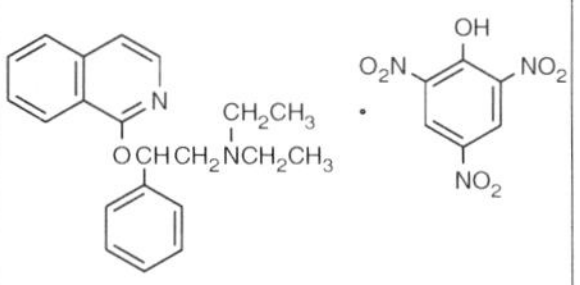

250 mg

S71,060-1
PAPAVERINE

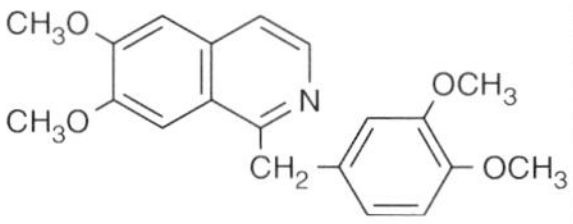

250 mg

S42,601-6
DL-NORLAUDANOSINE HYDROCHLORIDE

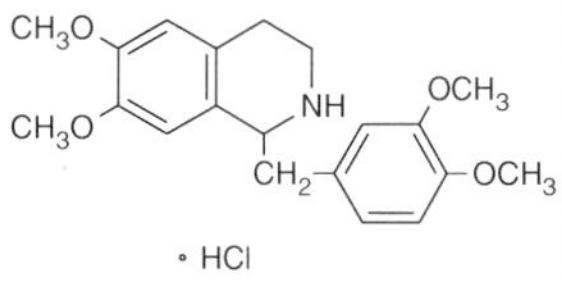

250 mg

S42,570-2
N-METHYLNORPAPAVERINE, BETAINE SALT

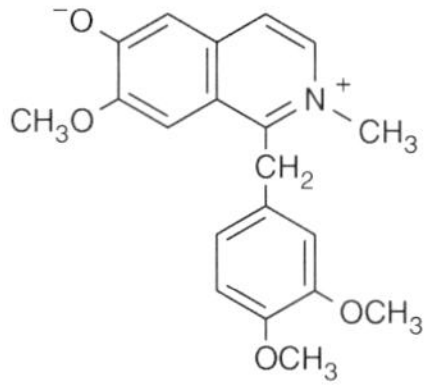

250 mg

S42,615-6
PAPAVERINE SULFONIC ACID (SULFONIC ACID UNSPECIFIED)

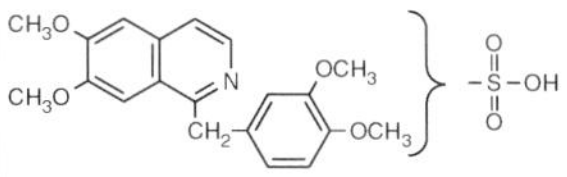

5 gm

S34,208-4
BERBERINIUM IODIDE

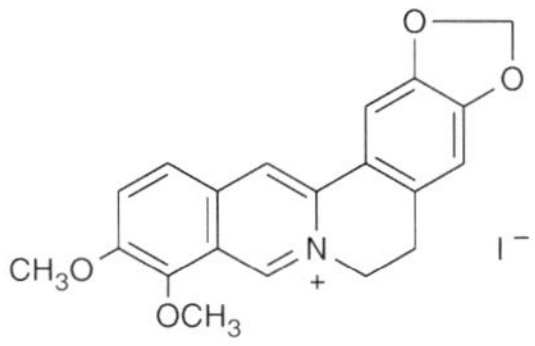

250 mg

S58,363-4
BERBERINE BISULFATE

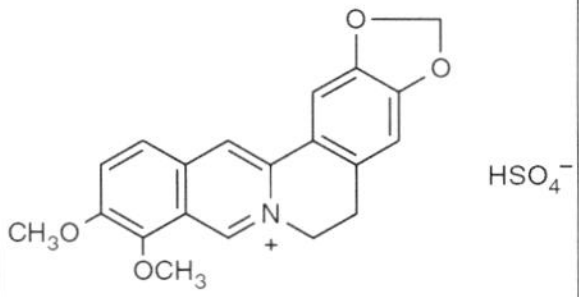

5 gm

S56,757-4
BERBERINE ACETONE ADDUCT

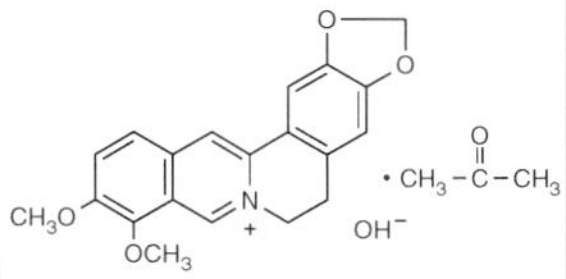

5 gm

S57,948-3
BERBERINE SULFATE

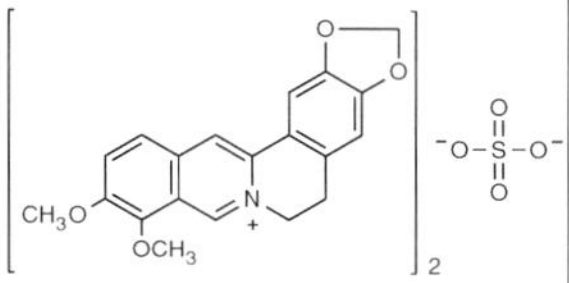

5 gm

S42,453-6
CORALYNE SULFOACETATE

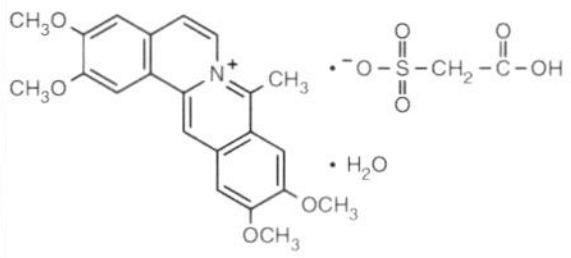

250 mg

S43,226-1
PSEUDOCORALYNE

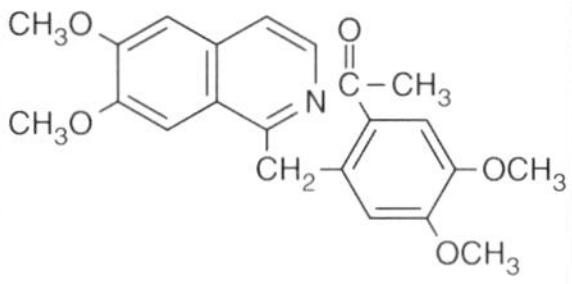

250 mg

S35,411-2
OCTAVERINE HYDROCHLORIDE

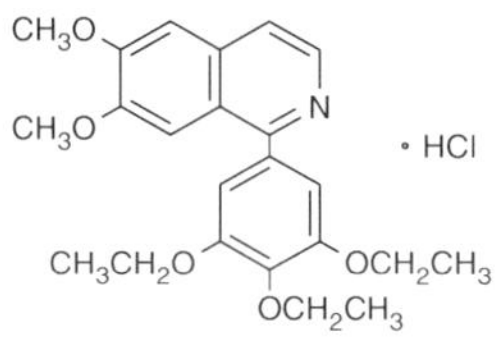

1 gm

S95,669-4
6,7-DIMETHOXY-1-METHYLISOQUINOLINE

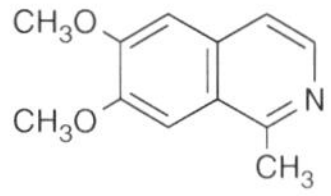

50 mg

S94,272-3
3-CHLORO-1-METHOXYISOQUINOLINE

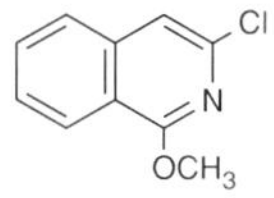

250 mg

S90,924-6
1-ISOQUINOLINETHIOL

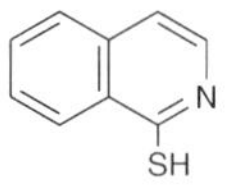

250 mg

S93,670-7
2,4-DIAMINO-6-(1-ISOQUINOLYL)-1,3,5-TRIAZINE

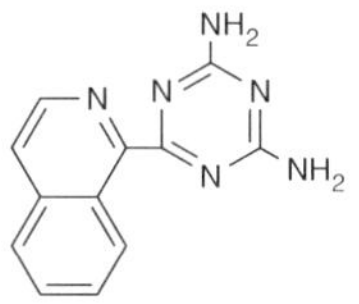

250 mg

S51,325-3
5-AMINOISOQUINOLINE DIHYDROCHLORIDE

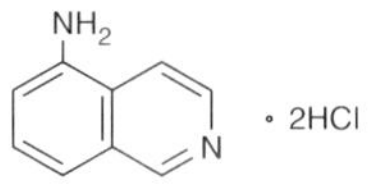

250 mg

S90,910-6
1,2-DIAMINOISOQUINOLINIUM P-TOLUENESULFONATE

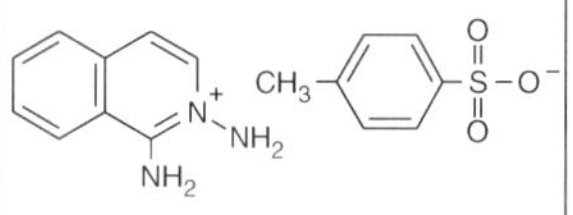

250 mg

S90,926-2
3-AMINOISOQUINOLINE

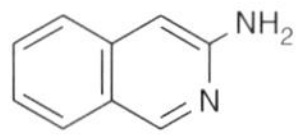

250 mg

S90,927-0
3-AMINO-1-BROMOISOQUINOLINE

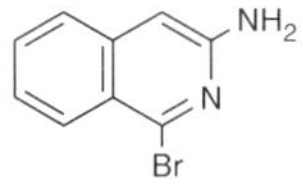

250 mg

S90,908-4
3-AMINO-4-METHYLISOQUINOLINE

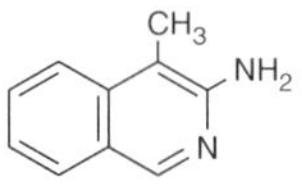

250 mg

S92,776-7
5-AMINO-3-METHYLISOQUINOLINE

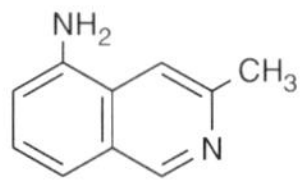

250 mg

S90,907-6
3-AMINO-1-BROMO-4-METHYLISOQUINOLINE

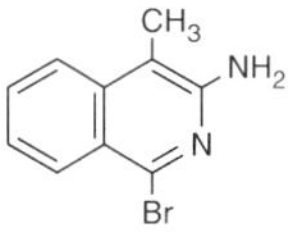

250 mg

S83,133-6
4-(2,5-DIMETHYL-1-PYRROLYL)-ISOQUINOLINE

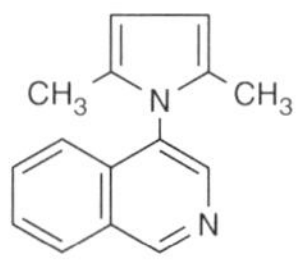

250 mg

S50,240-5
5-NITRO-2-PHENACYLISOQUINOLINIUM BROMIDE

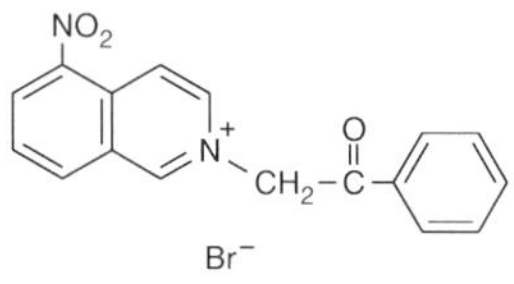

250 mg

S62,386-5
2,2'-TRIMETHYLENEBIS-(ISOQUINOLINIUM BROMIDE)

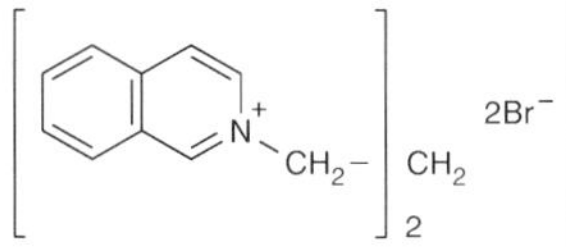

250 mg

S62,395-4
2,2'-TRIMETHYLENEBIS-(ISOQUINOLINIUM PERCHLORATE)

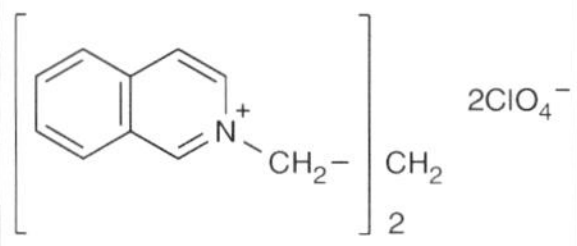

5 gm

S84,130-7
1-ISOQUINOLINECARBONITRILE

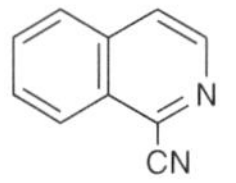

250 mg

S62,352-0
2,2'-TETRAMETHYLENEBIS-(ISOQUINOLINIUM BROMIDE)

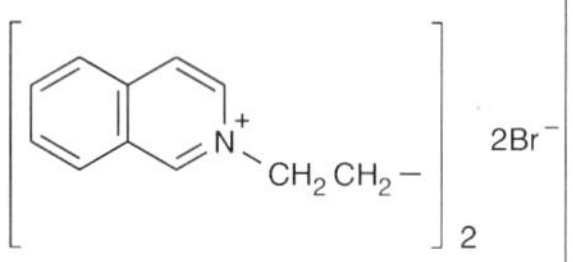

250 mg

S50,241-3
3-(5-ISOQUINOLYLCARBAMOYL)-5-NONADECENOIC ACID

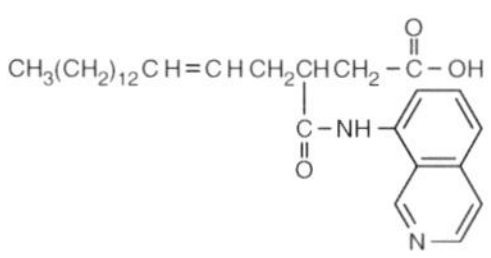

5 gm

S64,489-7
2-PHENACYLISOQUINOLINIUM BROMIDE

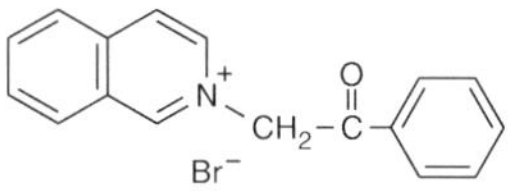

5 gm

S60,877-7
2-PHENETHYLISOQUINOLINIUM BROMIDE

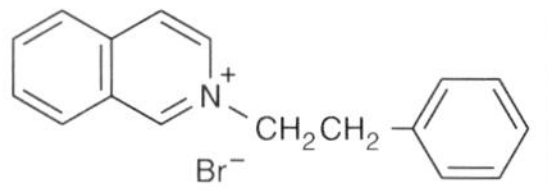

5 gm

S50,242-1
2-(4'-ACETAMIDOPHENACYL)-ISOQUINOLINIUM CHLORIDE

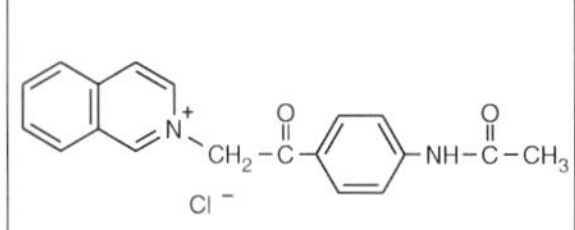

5 gm

S69,688-9
2-BENZYLISOQUINOLINIUM PERCHLORATE

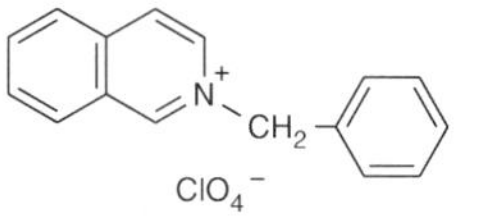

250 mg

S90,931-9
2-AMINO-1-(ETHOXYCARBONYL)-ISOQUINOLINIUM P-TOLUENESULFONATE
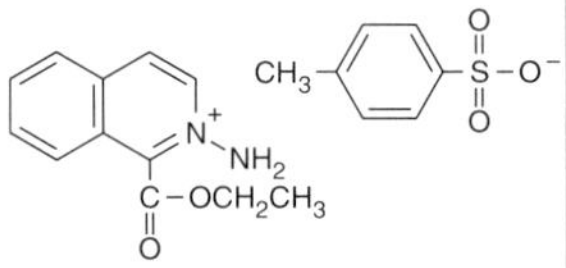
250 mg

S90,928-9
2-AMINO-1-CARBAMOYLISOQUINOLINIUM P-TOLUENESULFONATE
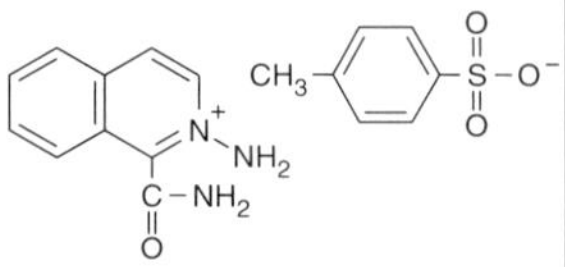
250 mg

S90,904-1
2-AMINO-1-CYANOISOQUINOLINIUM P-TOLUENESULFONATE
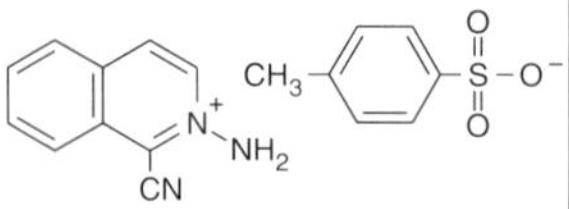
250 mg

S61,921-3
2-METHYLISOQUINOLINIUM IODIDE
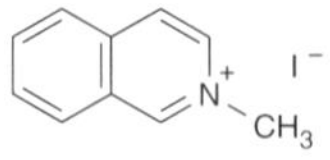
5 gm

S61,929-9
2-ETHYLISOQUINOLINIUM IODIDE
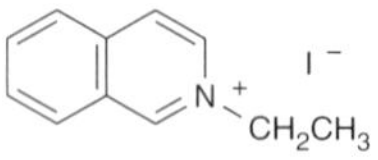
250 mg

S67,653-5
2,3-DIMETHYLISOQUINOLINIUM IODIDE
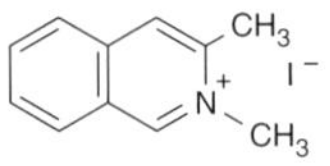
250 mg

S62,351-2
2,2'-HEXAMETHYLENEBIS-(ISOQUINOLINIUM IODIDE)
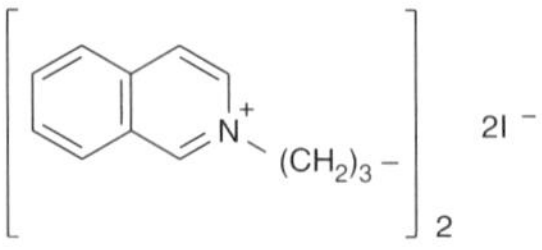
5 gm

S83,486-6
1-PHENYLISOQUINOLINE
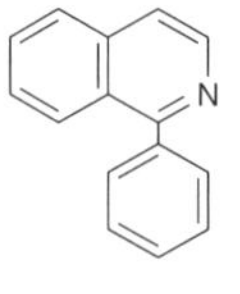
250 mg

S84,100-5
4-ISOQUINOLINECARBOXYLIC ACID
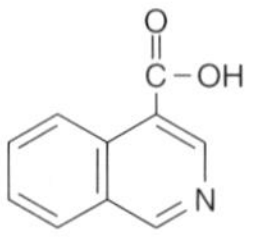
250 mg

S90,925-4
3-ISOQUINOLINECARBOXAMIDE
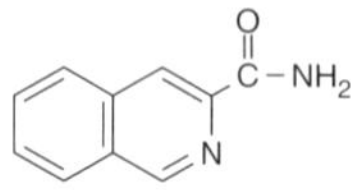
250 mg

S93,288-4
1-ISOQUINOLINECARBOXAMIDE
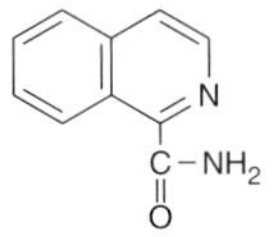
250 mg

S81,920-4
1-(4-BIPHENYLYL)ISOQUINOLINE
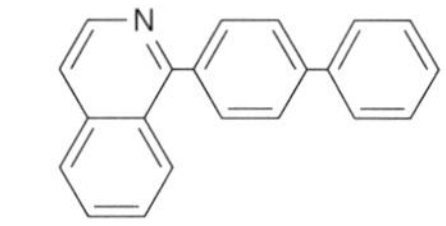
250 mg

S82,024-5
6-PHENYLPHENANTHRIDINE PICRATE
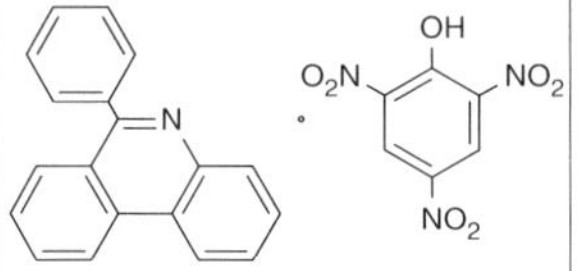
250 mg

S89,645-4
6-PHENYLPHENANTHRIDINE
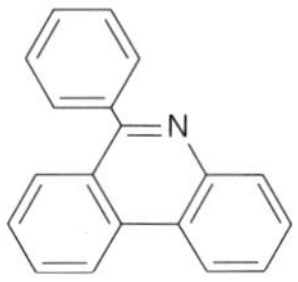
250 mg

S82,025-3
6-(P-TOLYL)PHENANTHRIDINE PICRATE
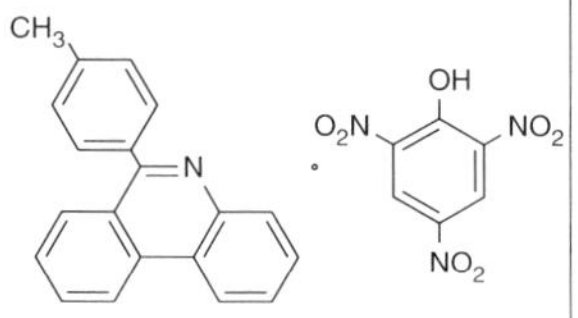
250 mg

S81,919-0
1-(P-TOLYL)ISOQUINOLINE
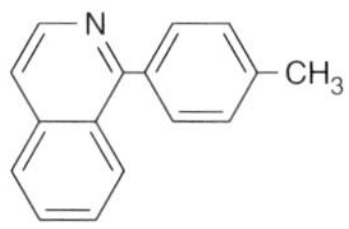
250 mg

S81,918-2
1-ALLYLISOQUINOLINE
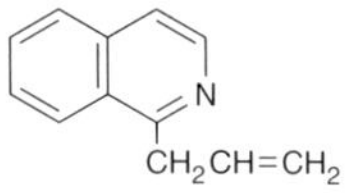
250 mg

S81,052-5
ISOQUINOLINE PICRATE
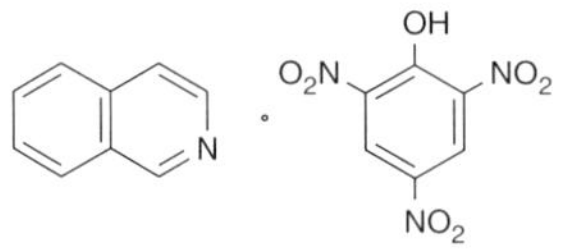
250 mg

S85,022-5
7-CHLOROISOQUINOLINE
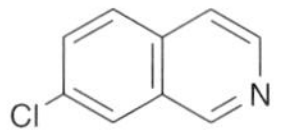
250 mg

S66,180-5
6-CHLORO-8-METHYLISOQUINOLINE
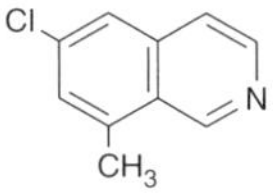
5 gm

S37,278-1
2-TRIBROMOMETHYLQUINOXALINE
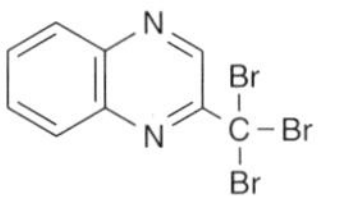
5 gm

S83,641-9
6-AMINO-2,3-BIS(2-CHLOROPHENYL)-QUINOXALINE
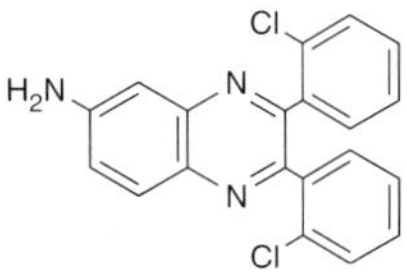
250 mg

S91,631-5
4-QUINAZOLINYL N-(3-CHLOROPHENYL)-CARBAMATE
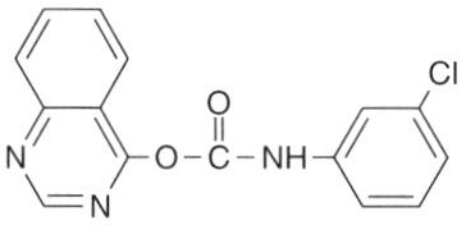
250 mg

S92,784-8
7-BROMO-2,3,6,8-TETRAMETHYLPYRIDO-(2,3-B)PYRAZINE
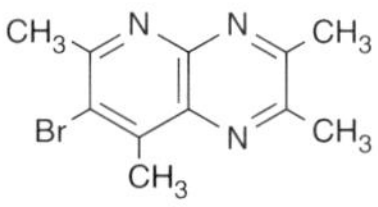
250 mg

S79,871-1
6-(3-CHLOROPHENETHYL)PHENANTHRIDINE
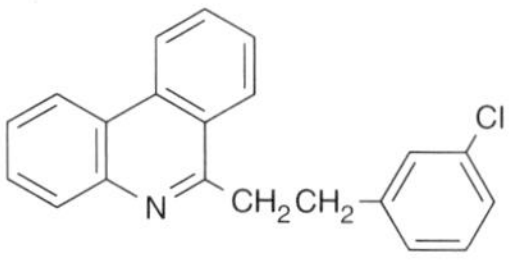
250 mg

S87,302-0
1,3,6,8-TETRAHYDROXY-2,7-NAPHTHYRIDINE
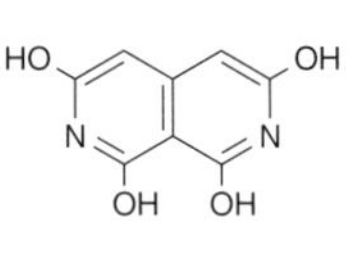
250 mg

S92,780-5
2,3-DIPHENYL-5,8-QUINOXALINEDIOL
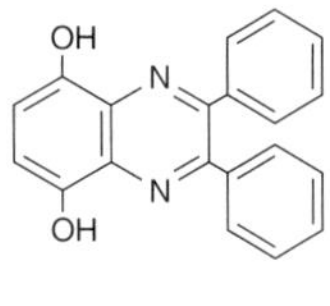
250 mg

S74,458-1
1,4-DIHYDROXYPHTHALAZINE
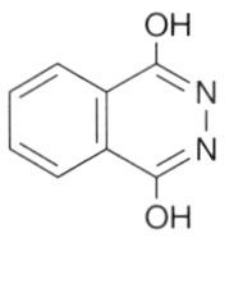
250 mg

S92,781-3
5,8-DIMETHOXY-3-METHYL-2-QUINOXALINOL
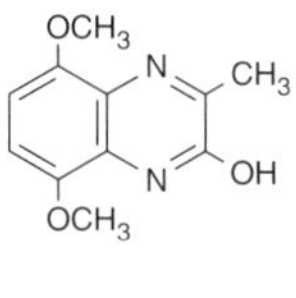
250 mg

S92,777-5
5,8-DIMETHOXY-2-ETHYL-3-HYDROXY-QUINOXALINE
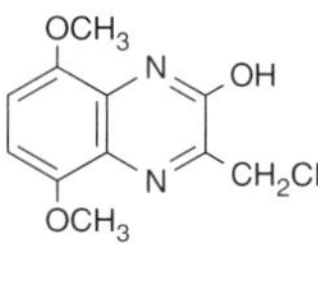
250 mg

S44,574-6
2,3-DIHYDROXY-1,4-PHENAZINEDIONE
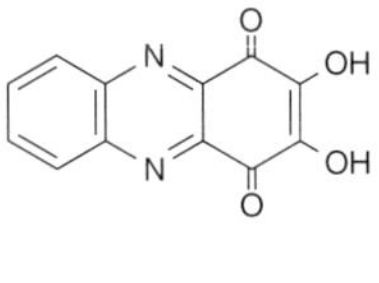
250 mg

S34,734-5
6,7-DIMETHYL-2,3-QUINOXALINEDIOL
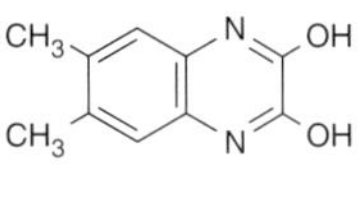
5 gm

S94,364-9
8-CHLORO-4-HYDROXYCINNOLINE-3-CARBOXYLIC ACID
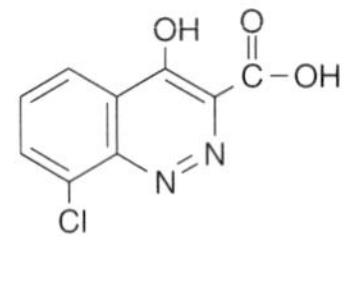
250 mg

S94,363-0
6-BROMO-4-HYDROXYCINNOLINE-3-CARBOXYLIC ACID
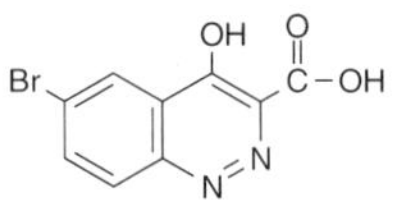
250 mg

S40,647-3
8-AMINO-7-METHYL-2-PHENAZINOL
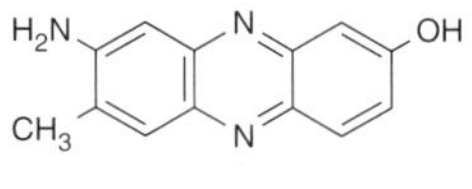
5 gm

S74,466-2
4-HYDROXY-3-METHYL-8-NITROCINNOLINE

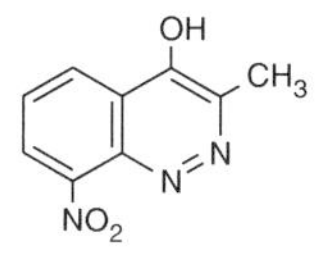

250 mg

S39,101-8
3-HYDROXY-2-QUINOXALINEACETIC ACID HYDRAZIDE

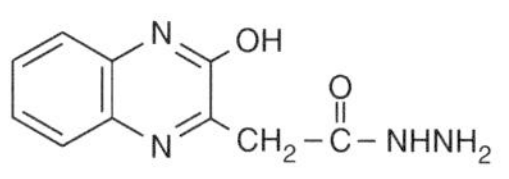

5 gm

S77,116-3
2-(2-NITRO-6-PHENANTHRIDINYLAMINO)-1-BUTANOL

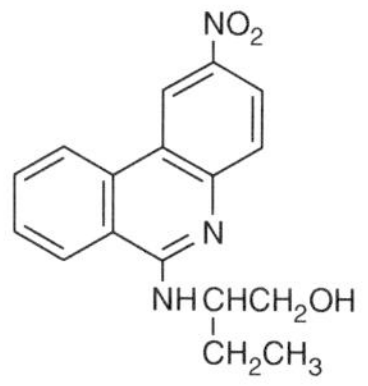

250 mg

S77,063-9
3,8-DICHLORO-6-(2-HYDROXYETHYLAMINO)PHENANTHRIDINE

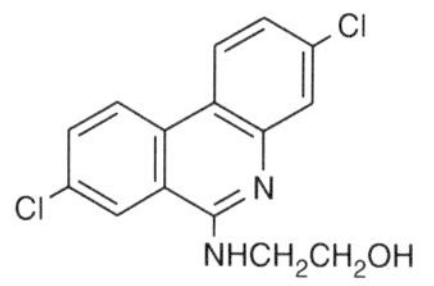

250 mg

S50,340-1
3-HYDROXY-2-QUINOXALINEACETIC ACID (5-INDANYLMETHYLENE)HYDRAZIDE

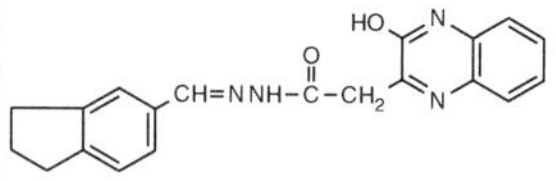

5 gm

S74,482-4
7-CHLORO-4-HYDROXYCINNOLINE

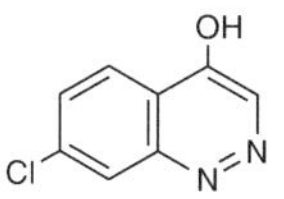

250 mg

S94,366-5
8-METHYL-4-HYDROXYCINNOLINE

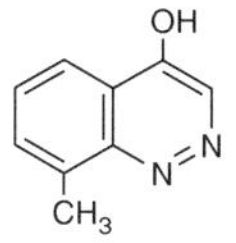

250 mg

S94,365-7
8-CHLORO-4-HYDROXYCINNOLINE

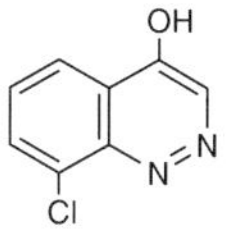

250 mg

S83,618-4
2-HYDROXY-3-METHYLQUINOXALINE

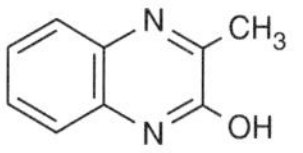

250 mg

S35,134-2
2-HYDROXY-3,6,7-TRIMETHYLQUINOXALINE

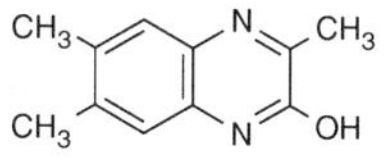

5 gm

S84,919-7
4-HYDROXY-1(2H)PHTHALAZINONE, SILVER SALT

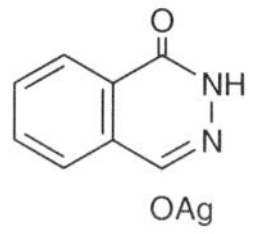

250 mg

S34,834-1
ETHYL 6,7-DIMETHYL-3-HYDROXY-2-QUINOXALINECARBOXYLATE

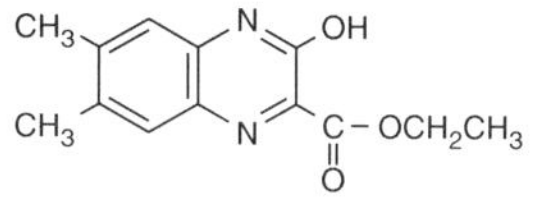

5 gm

S45,048-0
ETHYL 3-HYDROXY-2-QUINOXALINEACETATE

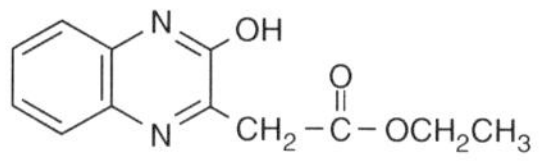

5 gm

S56,125-8
ETHYL 6,7-DIMETHYL-3-HYDROXY-2-QUINOXALINEACETATE

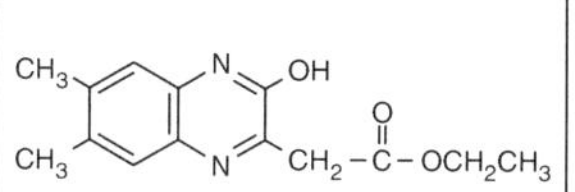

5 gm

S43,772-7
4-METHYL-1,8-NAPHTHYRIDINE-2,7-DIOL

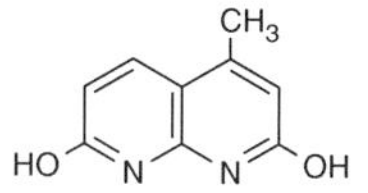

250 mg

S76,749-2
5-[(6,7,8-TRIMETHOXY-4-QUINAZOLINYL)AMINO]-1-PENTANOL NITRATE ESTER

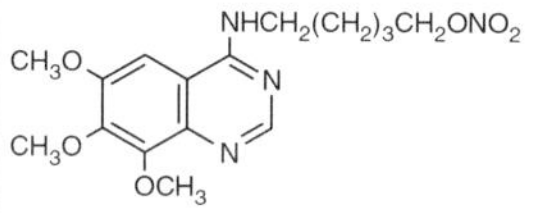

250 mg

S73,509-4
6,7-DIETHOXY-2,4(1H,3H)-QUINAZOLINEDIONE

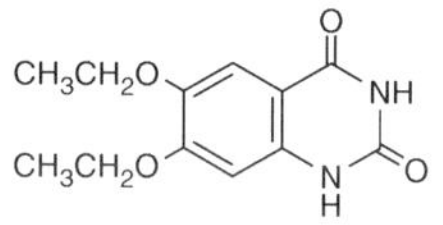

250 mg

S83,642-7
6-AMINO-2,3-BIS(4-METHOXYPHENYL)-QUINOXALINE

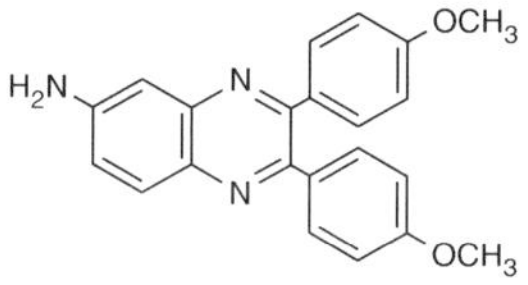

250 mg

S68,362-0
4-(2,3-DIHYDROXYPROPYLAMINO)-6,7,8-TRIMETHOXYQUINAZOLINE DINITRATE ESTER MALEATE

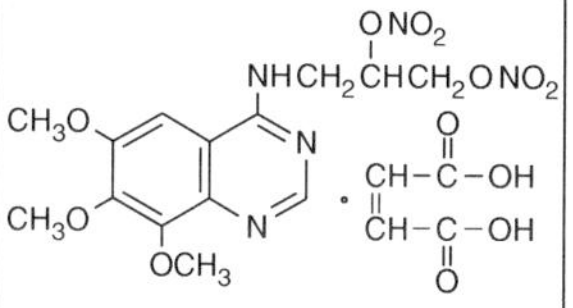

5 gm

S67,089-8
6-[N-(6,7,8-TRIMETHOXY-4-QUINAZOLINYL)AMINO]HEXYL NITRATE MALEATE

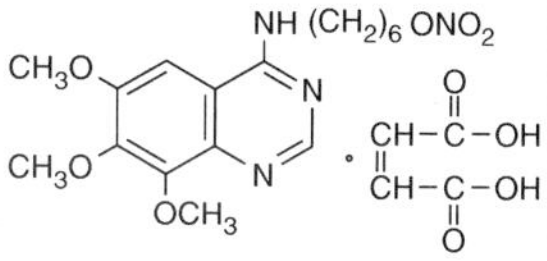

250 mg

S55,872-9
2,7-BIS[2-(DIETHYLAMINO)ETHOXY]-4-METHYL-1,8-NAPHTHYRIDINE HYDROCHLORIDE

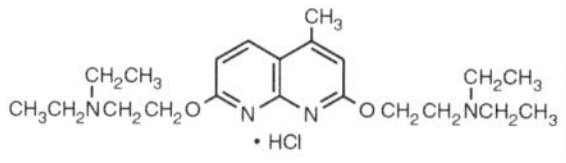

250 mg

S74,464-6
4-(4-METHOXYPHENYL)-3-METHYLCINNOLINE

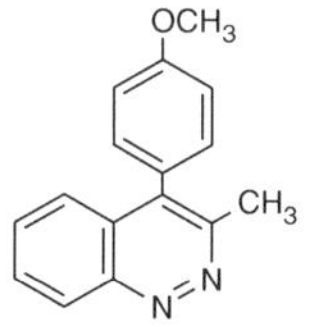

250 mg

S80,266-2
2-CHLORO-4-METHOXYQUINAZOLINE

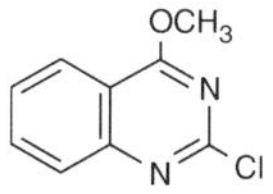

250 mg

S80,558-0
2-CHLORO-4-ETHOXYQUINAZOLINE

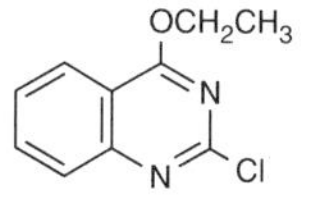

250 mg

S92,779-1
5,8-DIHYDRO-2,3-DIPHENYL-5,8-QUINOXALINEDIONE

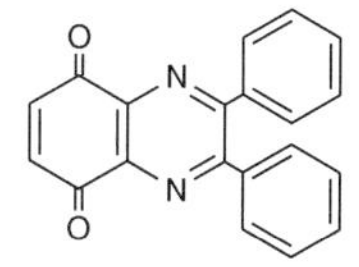

250 mg

S80,330-8
4-MERCAPTOQUINAZOLINE

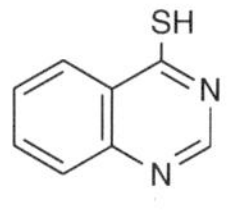

250 mg

S42,165-0
7-AMINO-2-[2-(DIETHYLAMINO)-ETHYLTHIO]-4-METHYL-1,8-NAPHTHYRIDINE

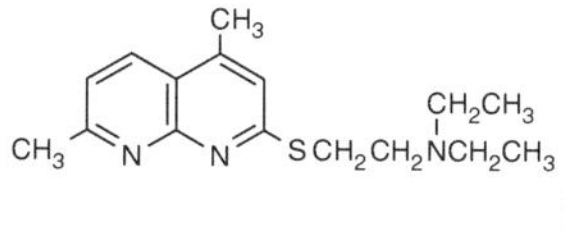

5 gm

S42,291-6
2,7-BIS[2-(DIETHYLAMINO)-ETHYLTHIO]-4-METHYL-1,8-NAPHTHYRIDINE

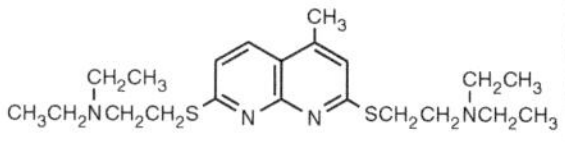

5 gm

S75,940-6
5-ISOPROPYL-6-(METHYLTHIO)-8-PHENYL-1,3-DIOXOLO(4,5-G)-QUINAZOLIN-5-IUM IODIDE

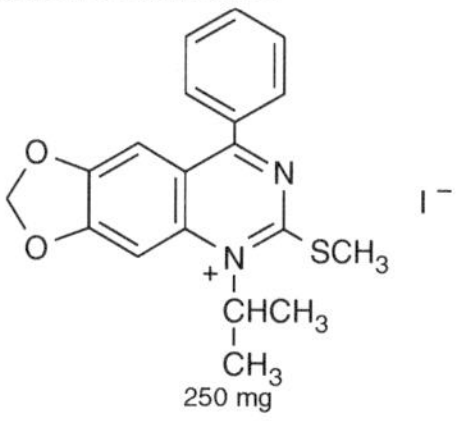

250 mg

S43,780-8
NEUTRAL VIOLET

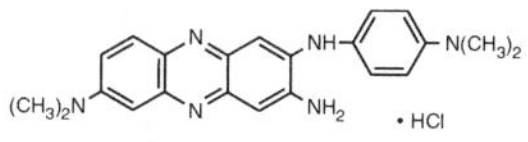

250 mg

S73,150-1
4-AMINO-6-NITROCINNOLINE METHOCHLORIDE

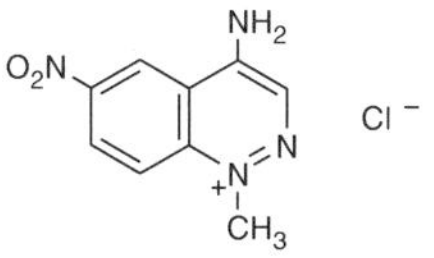

5 gm

S94,648-6
2,3-DIAMINOPHENAZINE

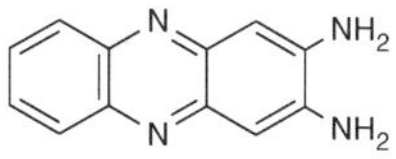

250 mg

S77,419-7
1-AMINO-3,8-DICHLORO-6(5H)-PHENANTHRIDINONE

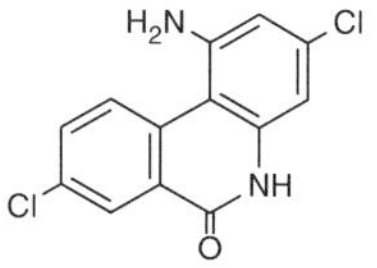

250 mg

S77,078-7
2,4-DIAMINO-6-NITROQUINAZOLINE ACETATE

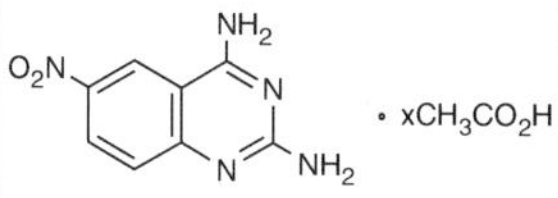

250 mg

S74,871-4
4-AMINO-3-METHYL-6-NITROCINNOLINE

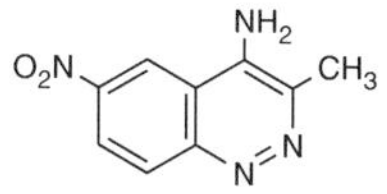

250 mg

S39,294-4
2,4-DIAMINO-6-NITROQUINAZOLINE
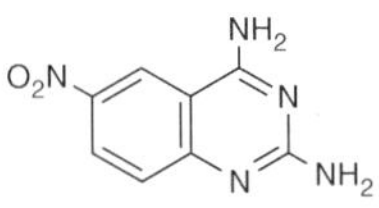
5 gm

S87,293-8
3-AMINO-1-BROMO-2,6-NAPHTHYRIDINE
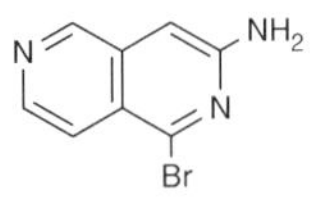
250 mg

S83,619-2
3-AMINO-2-QUINOXALINECARBOXYLIC ACID
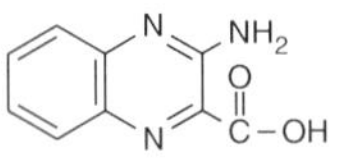
250 mg

S80,560-2
4-AMINOQUINAZOLINE
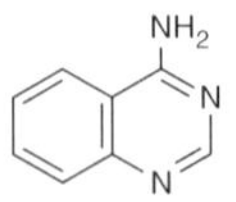
250 mg

S92,785-6
6-AMINO-2,3-DIMETHYLPYRIDO(2,3-B)-PYRAZINE
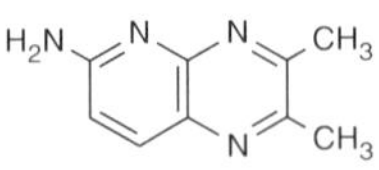
250 mg

S83,617-6
2-AMINOQUINOXALINE
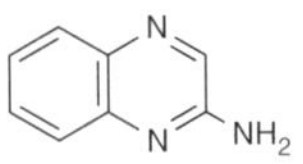
250 mg

S93,286-8
N-[2,3-BIS(2-PYRIDYL)QUINOXALIN-5-YL]ACETAMIDE
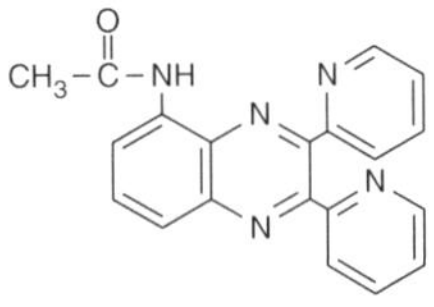
250 mg

S87,300-4
1,5-NAPHTHYRIDINE
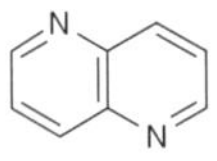
250 mg

S93,307-4
11H-INDENO(1,2-B)QUINOXALINE
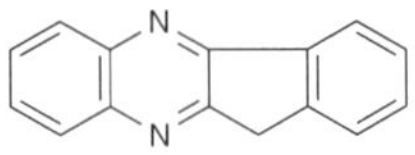
250 mg

S53,129-4
2,3-DIPHENYL-6-METHYLQUINOXALINE
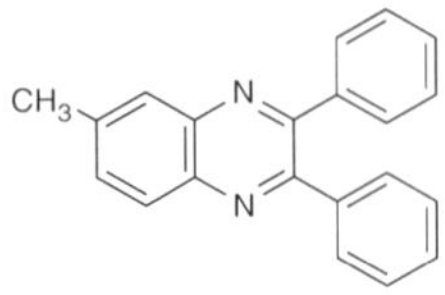
250 mg

S34,767-1
2,3-DIPHENYL-7-NITROQUINOXALINE
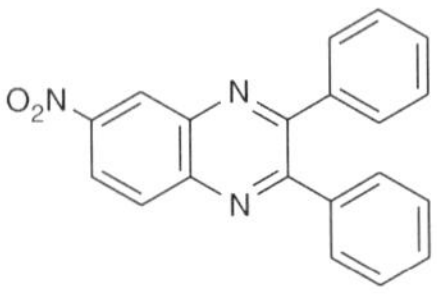
5 gm

S36,414-2
11-NITRODIBENZO(A,C)PHENAZINE
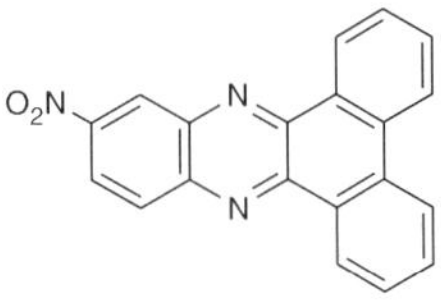
5 gm

S67,874-0
4,4'-(1,3-PROPANEDIYLDI-4,1-PIPERIDINEDIYL)BIS(7-NITRO-QUINAZOLINE) BIS(METHANESULFONATE)
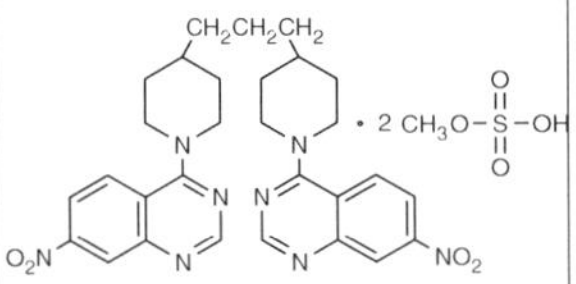
5 gm

S65,842-1
6-CHLORO-4-HYDRAZINOQUINAZOLINE
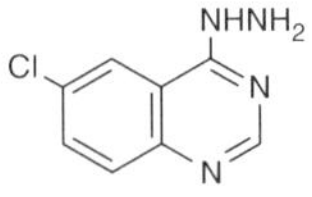
5 gm

S67,209-2
7-NITROQUINAZOLIN-4(3H)ONE
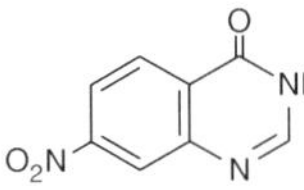
5 gm

S73,209-5
4-CHLORO-7-NITROQUINAZOLINE
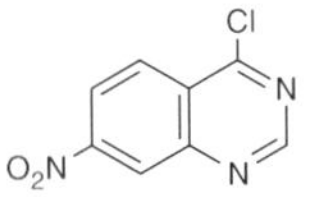
250 mg

S74,517-0
2-CHLORO-7-NITROQUINOXALINE
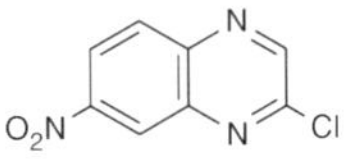
250 mg

S35,380-9
9-NITROACENAPHTHO(1,2-B)QUINOXALINE
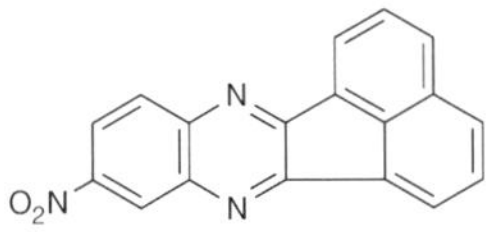
5 gm

S87,299-7
2,7-NAPHTHYRIDINE
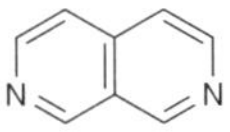
250 mg

S58,130-5
AZOCARMINE G (C.I. 50085)
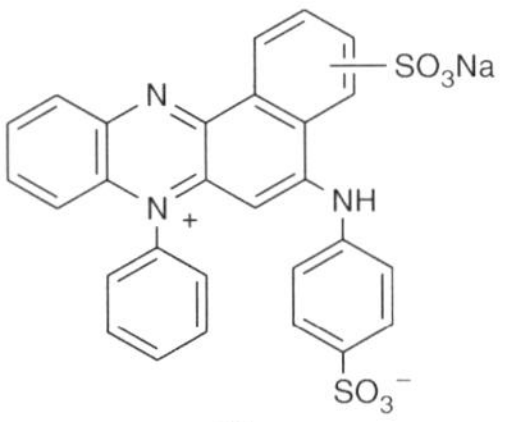
250 mg

S42,765-9
1,4-DIHYDRAZINOPHTHALAZINE SULFATE DIHYDRATE
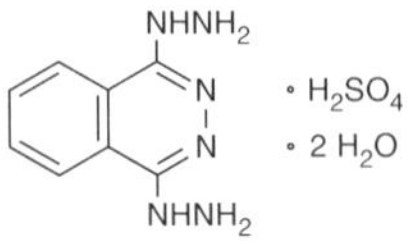
5 gm

S80,491-6
(1 OR 2 OR 3 OR 4)CHLORO-9-NITRO-6H-INDOLO(2,3-B)-QUINOXALINE-6-ACETIC ACID
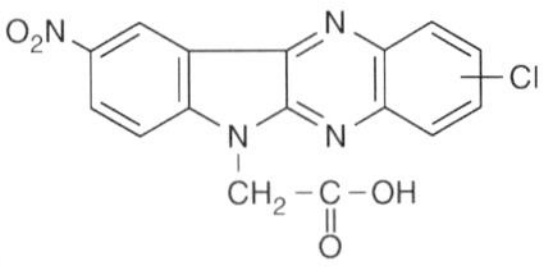
250 mg

S68,587-9
4,4'-(1,2-ETHANEDIYLDI-4,1-PIPERIDINEDIYL)BIS(7-CHLORO-QUINAZOLINE)
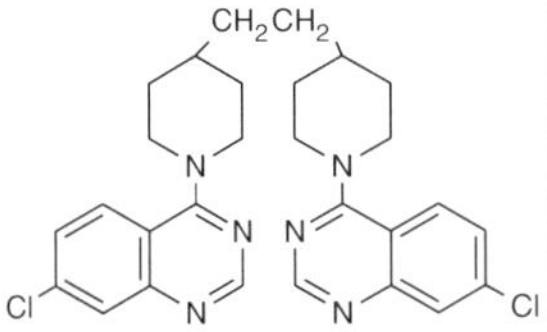
5 gm

S94,642-7
2-PHENYL-3-(3,4-XYLYL)QUINOXALINE
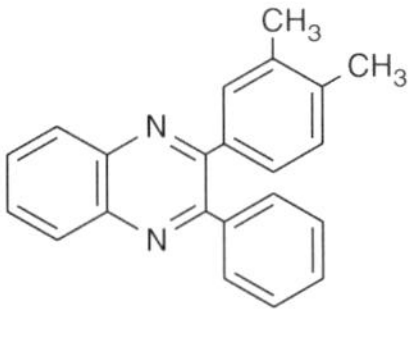
250 mg

S39,031-3
N-(5,7-DIMETHYL-1,8-NAPHTHYRIDIN-2-YL)SUCCINAMIC ACID
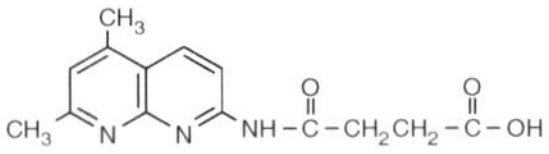
5 gm

S93,284-1
5-NITRO-3-(2-PYRIDYL)-2-QUINOXALINOL
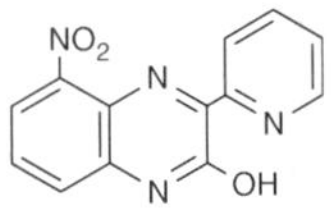
250 mg

S59,846-1
2-[4-(2-BENZOYLACETAMIDO)]PHENYL-4-(3H)QUINAZOLINONE
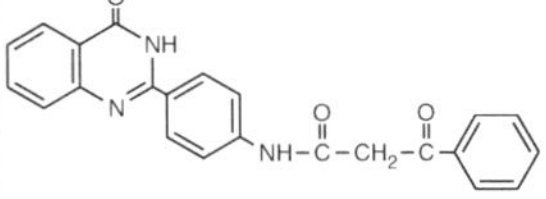
5 gm

S63,900-1
ETHYL 3,4-DIHYDRO-4-OXO-1-PHTHALAZINEACETATE
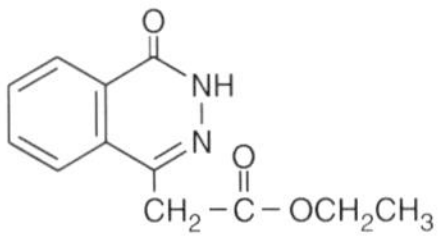
250 mg

S65,832-4
7-METHYL-4(3H)QUINAZOLINONE
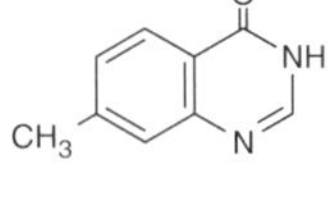
5 gm

S76,033-1
4-(4-CARBOXYPHENYL)PHTHALAZIN-1-(2H)ONE
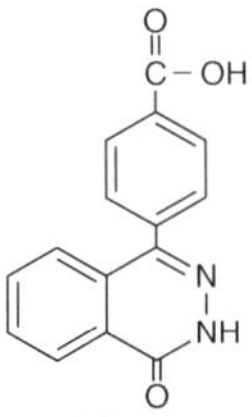
250 mg

S80,240-9
7-CHLORO-2,4(1H,3H)-QUINAZOLINEDIONE
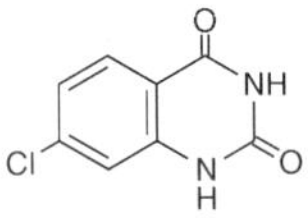
250 mg

S80,244-1
7-CHLORO-6-METHYL-2,4(1H,3H)-QUINAZOLINEDIONE
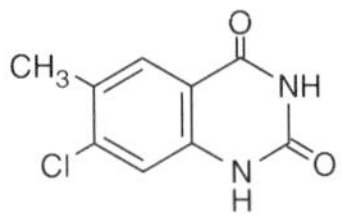
250 mg

S80,822-9
2-PHENYL-4(3H)QUINAZOLINONE
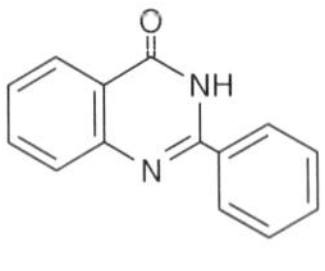
250 mg

S80,821-0
2-PHENYL-1,2,3,4-TETRAHYDRO-4-QUINAZOLINONE
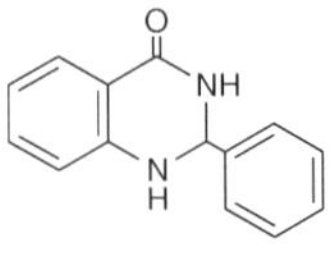
250 mg

S75,511-7
3-CHLORO-6(5H)PHENANTHRIDINONE
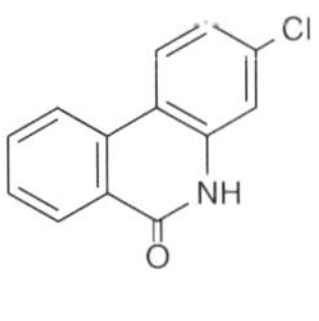
10 mg

S75,461-7
4-TERT-BUTYL-1(2H)PHTHALAZINONE
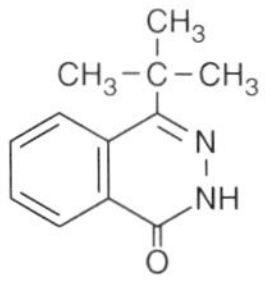
250 mg

S80,409-6
(6 OR 7 OR 8 OR 9)CHLORO-1,3-DIMETHYLBENZO(G)PTERIDINE-2,4 (1H,3H)-DIONE
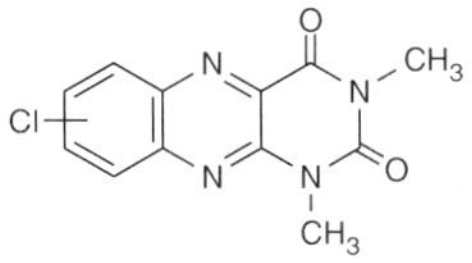
100 mg

S68,188-1
6-CHLORO-2,4(1H,3H)-QUINAZOLINEDIONE
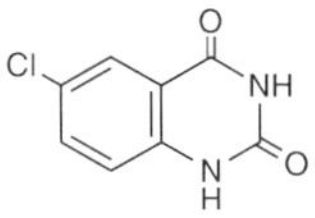
250 mg

S93,658-8
4-CHLORO-1(2H)PHTHALAZINONE
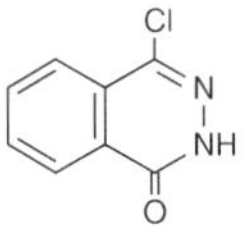
250 mg

S97,243-6
1,2-DIHYDRO-3,3-DIMETHYL-METHYL-3H-BENZO(A)PYRANO(2,3-C)-PHENAZINE
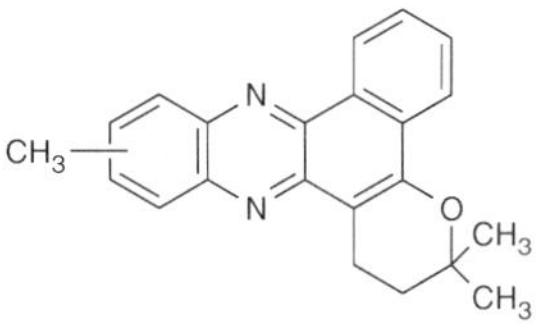
100 mg

S93,317-1
3-METHYL-4-PHENYLCINNOLINE
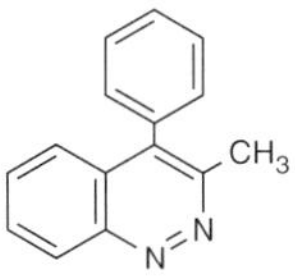
250 mg

S41,447-6
4-STYRYLCINNOLINE
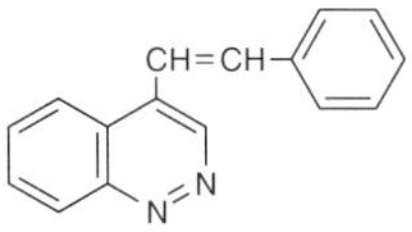
5 gm

S75,811-6
CINNOLINE PICRATE
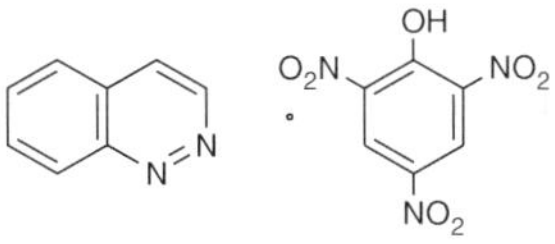
250 mg

S74,460-3
7-CHLORO-4-METHYLCINNOLINE
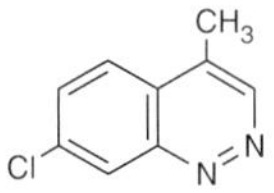
250 mg

S41,681-9
4-PHENYLQUINAZOLINE
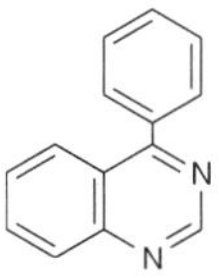
5 gm

S76,095-1
2,4-DIPHENYLBENZO(H)QUINAZOLINE
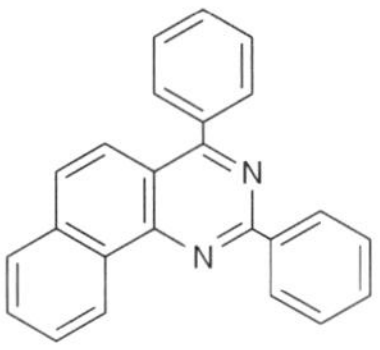
250 mg

S74,506-5
3,4-DIHYDRO-2-METHYLINDOLO(1,2-C)-QUINAZOLINE
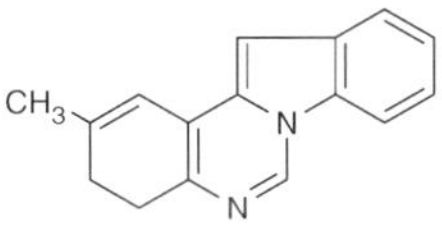
250 mg

S65,836-7
4,6-DICHLOROQUINAZOLINE
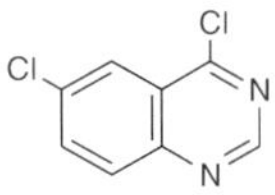
5 gm

S67,148-7
4,7-DICHLOROQUINAZOLINE
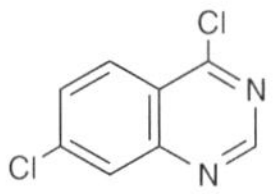
5 gm

S80,451-7
2,4-DICHLOROQUINAZOLINE
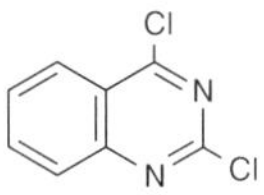
250 mg

S41,680-0
2-GUANIDINO-4-METHYLQUINAZOLINE HYDROCHLORIDE
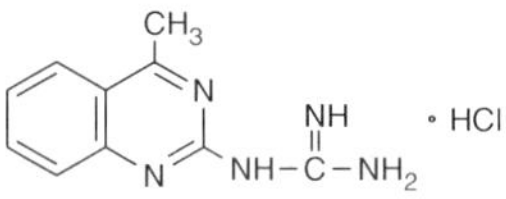
5 gm

S56,588-1
2-GUANIDINO-4-METHYLQUINAZOLINE
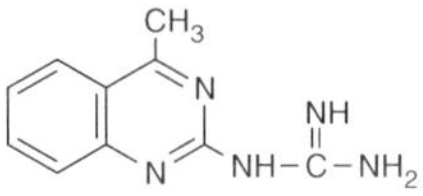
5 gm

S34,686-1
11,12-DIMETHYLDIBENZO(A,C)PHENAZINE
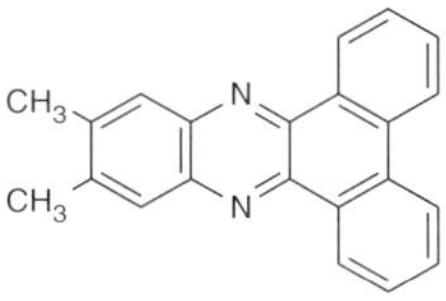
5 gm

S82,047-4
2-BROMOPHENAZINE
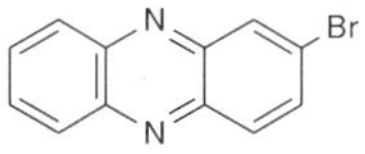
250 mg

S81,190-4
2-PHENYLBENZO(H)QUINOLINE
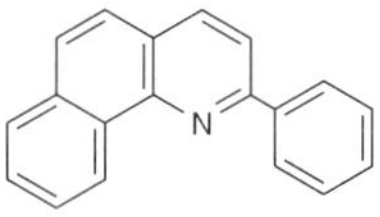
250 mg

S87,301-2
1,8-NAPHTHYRIDINE
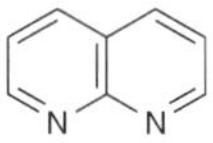
250 mg

S97,449-8
AR-F15 BENZO(A)NAPHTHO(2,1-C)-PHENAZINE
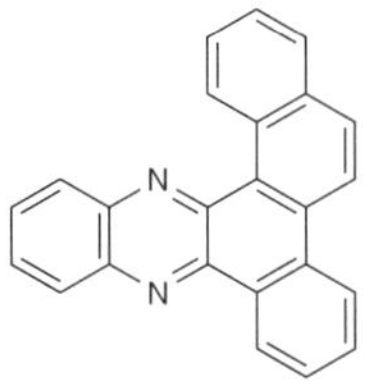
10 mg

S43,526-0
DIBENZO(A,C)PHENAZINE
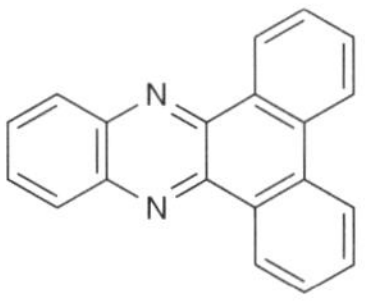
5 gm

S36,848-2
10,11-DIMETHYLDIBENZO(A,C)PHENAZINE
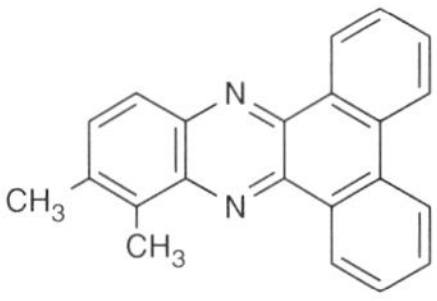
250 mg

S81,123-8
2-BROMO-6-METHYLPHENANTHRIDINE NITRATE
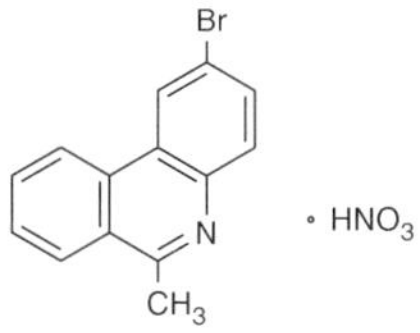
250 mg

S46,056-7
2-CHLORO-3-METHYLQUINOXALINE
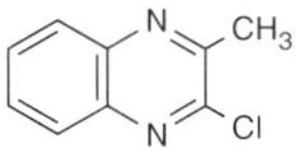
250 mg

S35,591-7
2,3,6,7-TETRAMETHYLQUINOXALINE
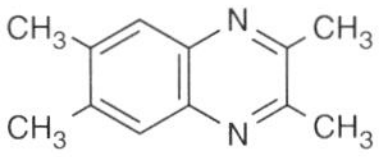
5 gm

S34,693-4
6,7-DIMETHYL-2,3-DIPHENYLQUINOXALINE
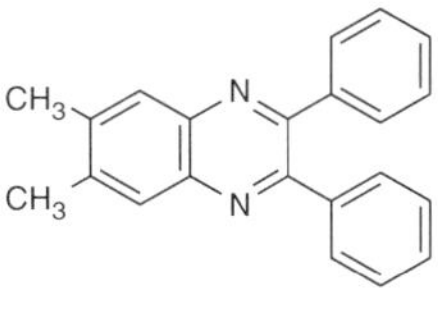
5 gm

S34,759-0
2,3-DIPHENYL-5,6-BENZOQUINOXALINE
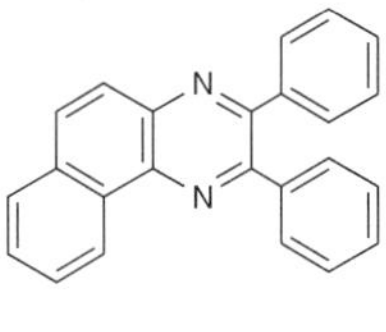
5 gm

S80,376-6
(8 OR 9 OR 10 OR 11)CHLOROBENZO(A)-PHENAZINE
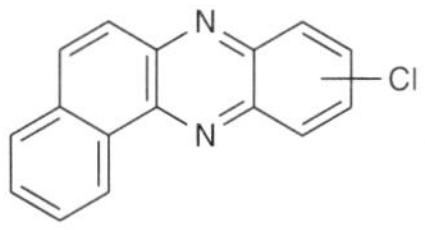
250 mg

S35,700-6
ACENAPHTHO(1,2-B)QUINOXALINE
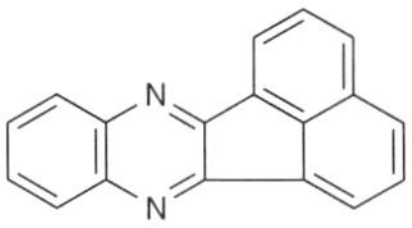
5 gm

S93,300-7
5,6-DIMETHYLBENZO(A)PHENAZINE
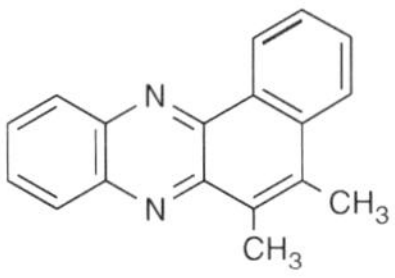
250 mg

S35,258-6
9-METHYLACENAPHTHO(1,2-B)-QUINOXALINE
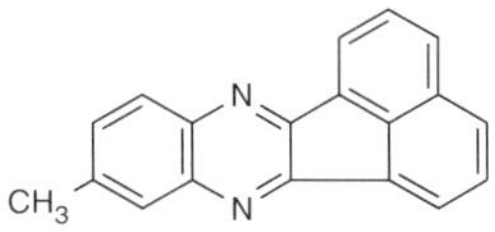
5 gm

S80,531-9
9,10-DIMETHYLACENAPHTHO(1,2-B)-QUINOXALINE
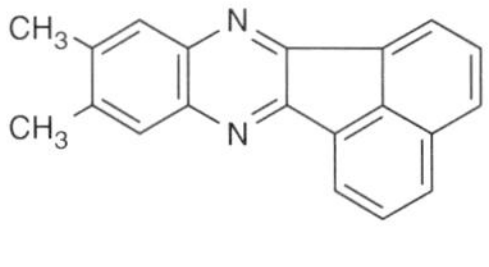
250 mg

S36,351-0
7-METHYL-1,2,3,4-DIBENZOPHENAZINE
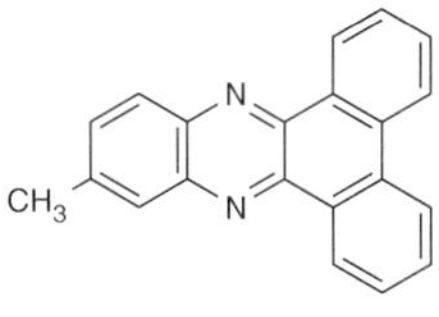
5 gm

S40,656-2
8,9-BENZOACENAPHTHO(1,2-B)-QUINOXALINE
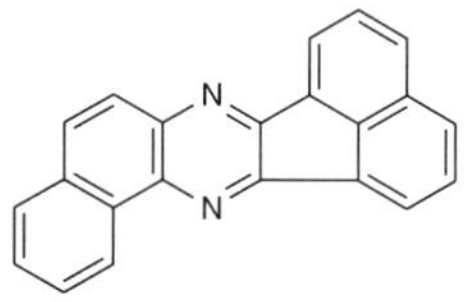
5 gm

S87,340-3
1,10-ETHANO-1,10-PHENANTHROLIN-1,10-DIIUM DIBROMIDE
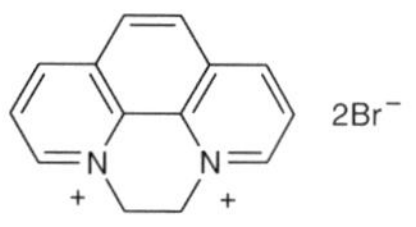
250 mg

S92,778-3
2-CHLORO-4-HYDROXY-4-PHENYL-4H-1,3-BENZOXAZINE
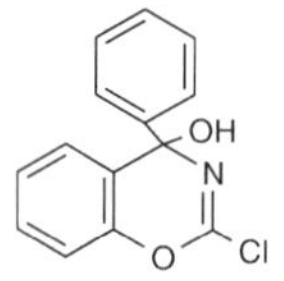
250 mg

S90,971-8
3-(4-CHLOROPHENYL)-3H-TETRAZOLO-(1,5-A)PYRIDIN-4-IUM BROMIDE
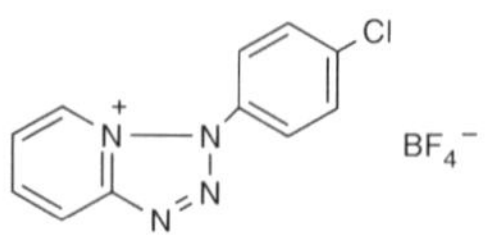
250 mg

S90,969-6
3-(4-CHLOROPHENYL)-6-METHYL-3H-TETRAZOLO(1,5-A)PYRIDIN-4-IUM BROMIDE
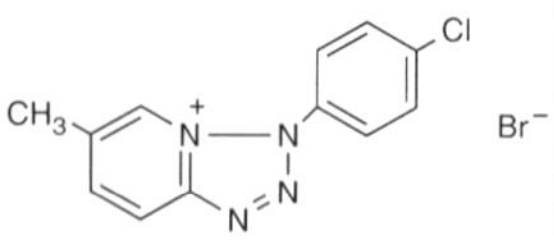
250 mg

S90,966-1
3-(4-CHLOROPHENYL)-7-METHYL-3H-TETRAZOLO(1,5-A)PYRIDIN-4-IUM TETRAFLUOROBORATE
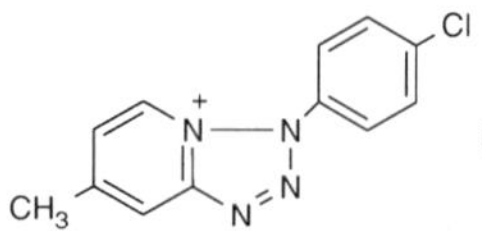
250 mg

S87,037-4
2-(4-CHLOROPHENYL)-1,2,4-TRIAZOLO-(3,4-A)ISOQUINOLINIUM TETRAFLUOROBORATE
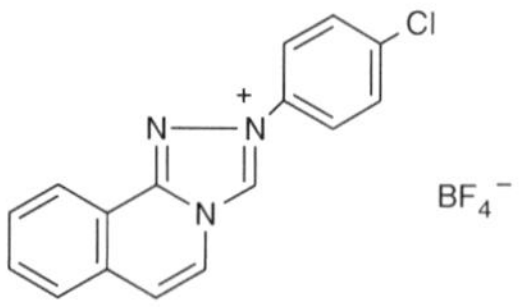
250 mg

S75,570-2
METHYL 2,4-DIHYDROXYPYRIMIDO(4,5-B)-QUINOLINE-5-CARBOXYLATE
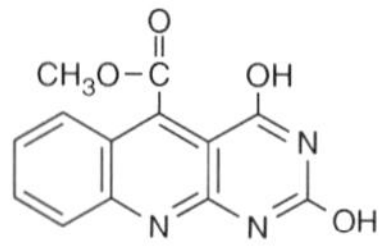
250 mg

S36,306-5
2-MERCAPTO-4,6,7-TRIHYDROXYPTERIDINE
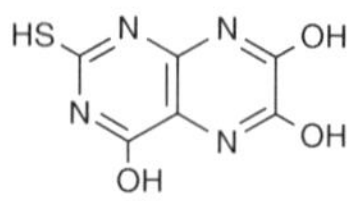
5 gm

S40,806-9
5,7-DIMETHYL-4-HYDROXY-2-MERCAPTOPYRIDO(2,3-D)PYRIMIDINE
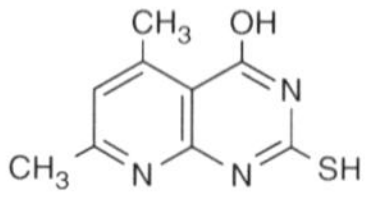
5 gm

S36,064-3
4,6-DIHYDROXY-2-MERCAPTO-7-METHYLPYRIMIDO(4,5-B)PYRAZINE
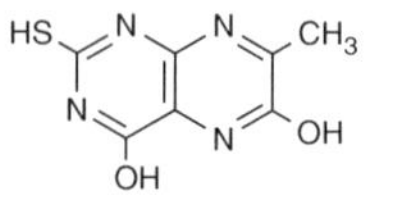
250 mg

S36,065-1
4,7-DIHYDROXY-2-MERCAPTO-6-METHYLPYRIMIDO(4,5-B)PYRAZINE
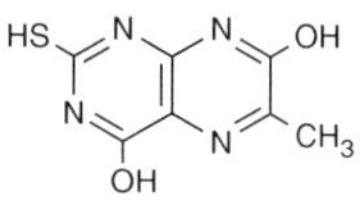
5 gm

S34,703-5
6,7-DIMETHYL-4-HYDROXY-2-MERCAPTOPTERIDINE
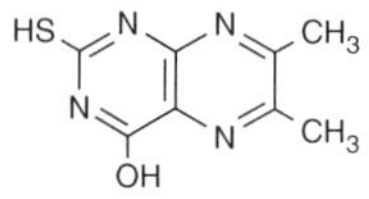
5 gm

S46,140-7
4-HYDROXY-2-MERCAPTOPTERIDINE
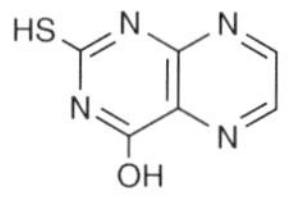
250 mg

S34,926-7
2-ETHYLTHIO-4-HYDROXYPTERIDINE
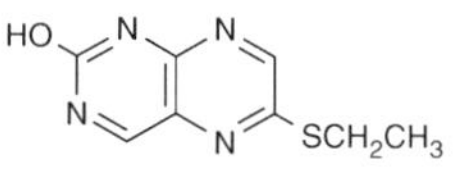
5 gm

S87,653-4
2-AMINO-4-HYDROXY-6-PTERIDINECARBOXALDEHYDE
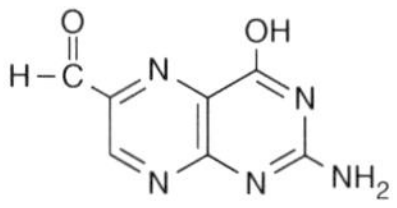
250 mg

S44,990-3
2-AMINO-6-METHYL-4,7-PTERIDINEDIOL
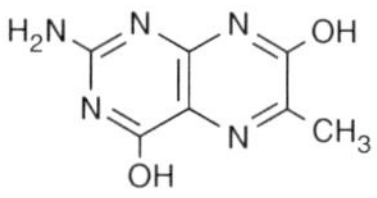
5 gm

S37,306-0
2-AMINO-6,7-DIPHENYL-4-HYDROXYPYRIMIDO-(4,5-B)PYRAZINE
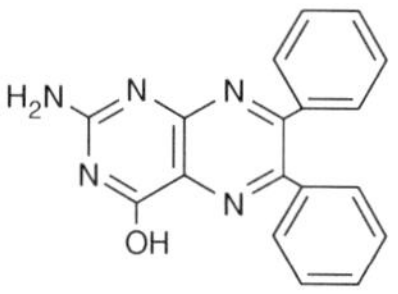
5 gm

S80,504-1
4-HYDROXY-6H-INDENO(2,1G)-PTERIDIN-6-ONE
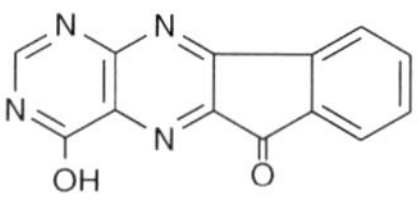
250 mg

S80,363-4
1,3-DIMETHYL-9-HYDROXYPYRIMIDO(4,5-G)PTERIDINE-2,4(1H,3H)DIONE
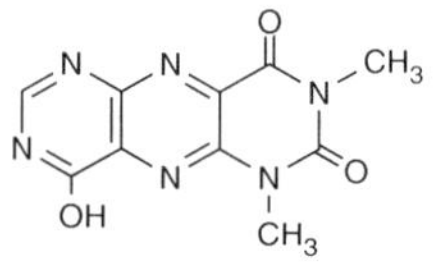
250 mg

S80,283-2
1,3-DIETHYL-2,4-DIOXO-7-HYDROXY-1,2,3,4-TETRAHYDRO-6-PTERIDINECARBOXYLIC ACID
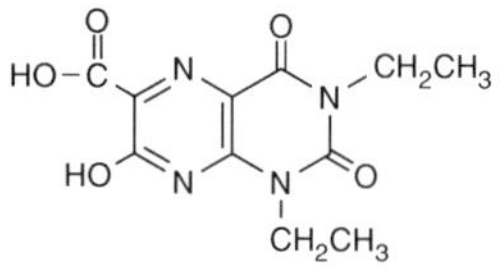
250 mg

S57,927-0
FOLIC ACID
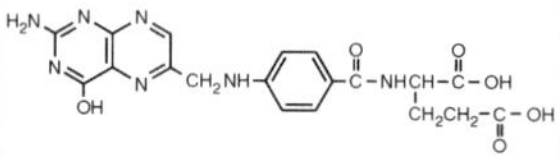
250 mg

S44,910-5
6,7-DIMETHYLLUMAZINE
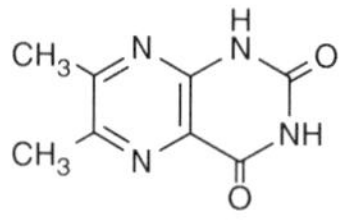
250 mg

S92,787-2
4-HYDROXYPTERIDINE
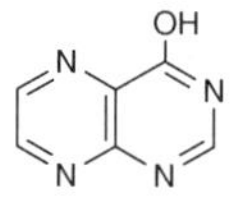
250 mg

S43,964-9
4-HYDROXY-6,7-DIMETHYLPTERIDINE
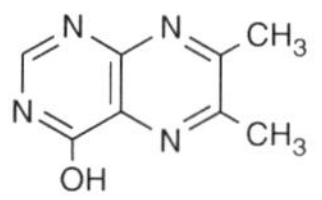
250 mg

S39,110-7
2,2'-[PYRIDO(2,3-B)PYRAZINE-2,3-DIYL]BIS(4-BROMOPHENOL)
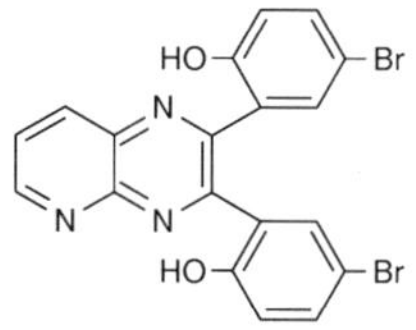
250 mg

S88,681-5
4-METHYL-1H-1,5-BENZODIAZEPIN-2-OL
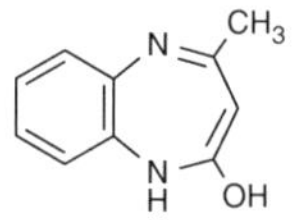
250 mg

S92,679-5
2-TERT-BUTYL-5,6-DIHYDRO-4-(4-METHOXYPHENYL)NAPHTHO(1,2-B)-PYRYLIUM TETRAFLUOROBORATE
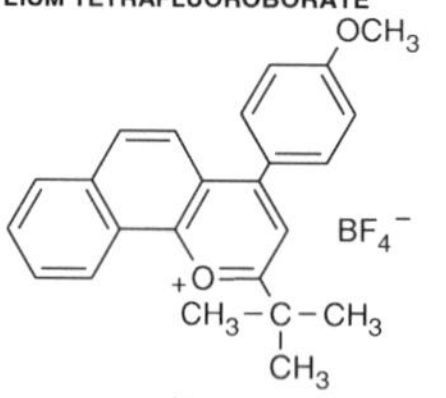
250 mg

S92,366-4
1,3-DIPHENYL-2-BENZOPYRYLIUM TETRAFLUOROBORATE
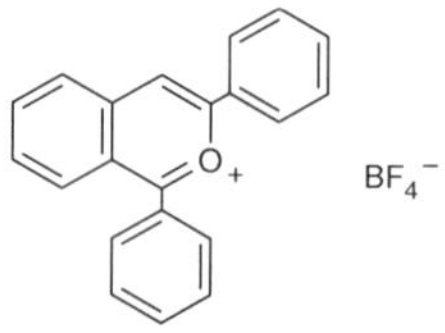
250 mg

S92,071-1
2,4-DIPHENYLNAPHTHO(1,2-B)PYRYLIUM TETRAFLUOROBORATE
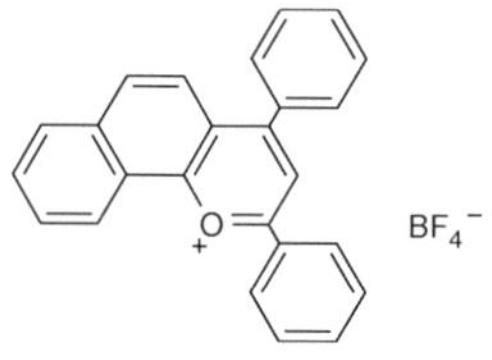
250 mg

S92,124-6
1,3-DIPHENYLNAPHTHO(2,1-B)PYRYLIUM TETRAFLUOROBORATE
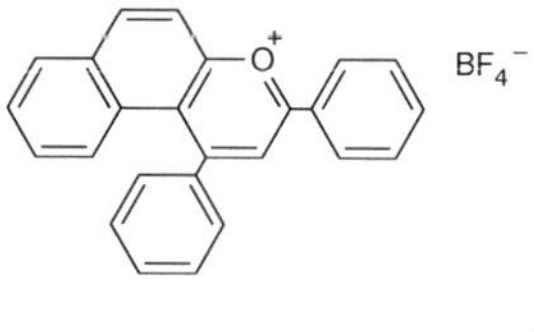
250 mg

S41,709-2
2-BUTOXY-7,10-DICHLOROPYRIDO(3,2-B)-QUINOLINE
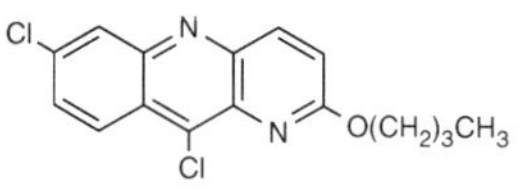
5 gm

S90,913-0
3-HYDRAZINO-7-METHOXY-1,2,4-BENZOTRIAZINE
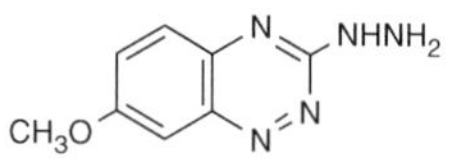
250 mg

S46,301-9
7,10-DICHLORO-2-METHOXYPYRIDO(3,2-B)QUINOLINE
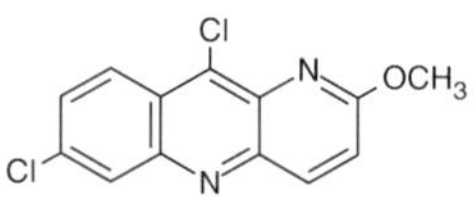
250 mg

S92,059-2
3-(METHYLTHIO)-2-PHENYL(1,2,4)-TRIAZINO(2,3-A)QUINOLIN-11-IUM PERCHLORATE
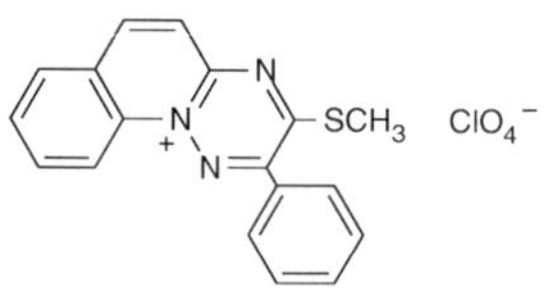
250 mg

S40,734-8
2,4-DIAMINO-6,7-DIMETHYLPTERIDINE

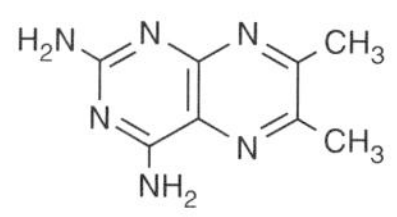

5 gm

S59,676-0
3-AMINO-7-NITRO-1,2,4-BENZOTRIAZINE-1-OXIDE

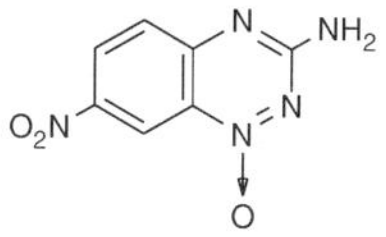

5 gm

S80,534-3
4-AMINO-6H-INDENO(2,1-G)-PTERIDIN-6-ONE

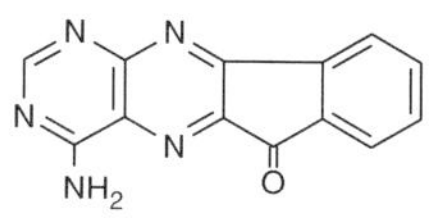

250 mg

S59,681-7
3-AMINO-7-FLUORO-1,2,4-BENZOTRIAZINE

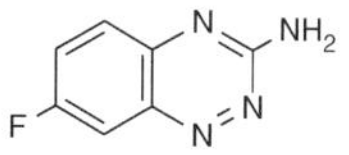

250 mg

S59,659-0
3-AMINO-7-CHLORO-1,2,4-BENZOTRIAZINE

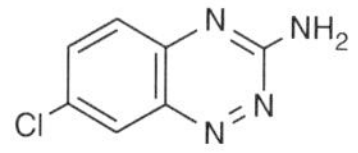

250 mg

S59,662-0
3-AMINO-7-METHYL-1,2,4-BENZOTRIAZINE

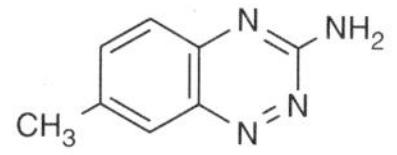

250 mg

S80,313-8
3-AMINOPYRIMIDO(4,5-E)-1,2,4-TRIAZINE-6,8(5H,7H)DIONE

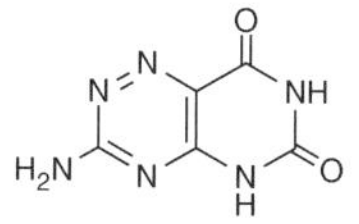

250 mg

S92,786-4
6-AMINOPYRIDO(2,3-B)PYRAZINE

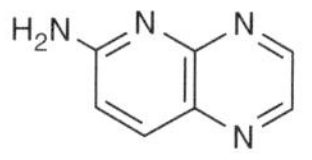

250 mg

S80,265-4
6-METHYLPYRIDO(3,2-D)PYRIMIDINE-2,4(1H,3H)DIONE

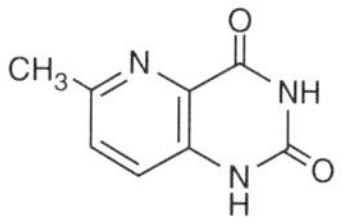

250 mg

S92,844-5
2,4,6(1H,3H,5H)PTERIDINETRIONE

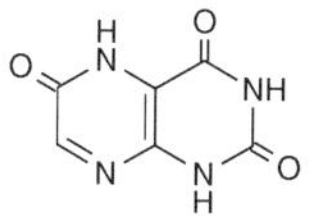

250 mg

S54,975-4
BENZO(B)NAPHTHO(2,3-F)(1,4)-DIAZOCINE

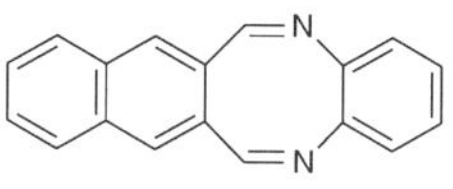

250 mg

S40,655-4
10-AZADIBENZO(A,C)PHENAZINE

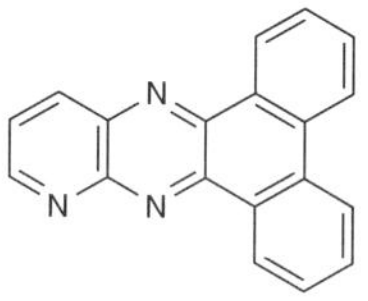

250 mg

S42,728-4
6,7-DIPHENYLDIBENZO(E,G)(1,4)-DIAZOCINE

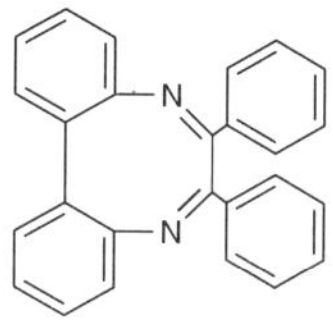

250 mg

S89,356-0
4-[3-(DIMETHYLAMINO)PROPYLAMINO]-1METHYL-1H-PYRAZOLO(3,4-B)-QUINOLINE TRIPHOSPHATE

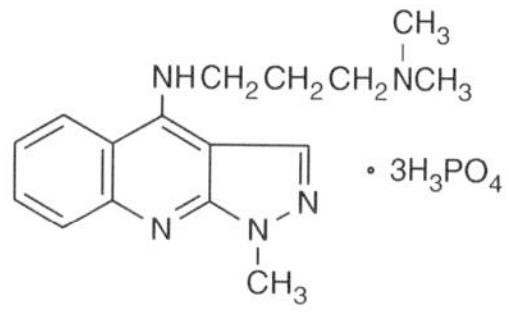

250 mg

S89,324-2
1,3-DIMETHYL-4-[3-(DIMETHYLAMINO)-PROPYLAMINO]-1H-PYRAZOLO(3,4-B)-QUINOLINE

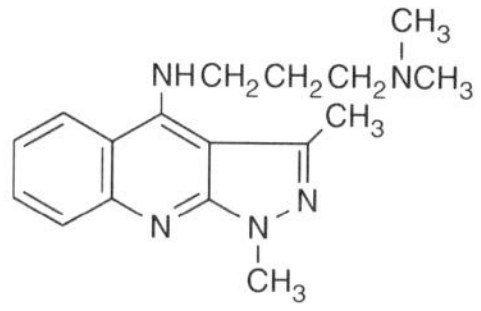

250 mg

S36,499-1
1,3,7,9-TETRAMETHYLPYRIMIDO(5,4-G)-PTERIDINE-2,4,6,8-(1H,3H,7H,9H)-TETRAONE

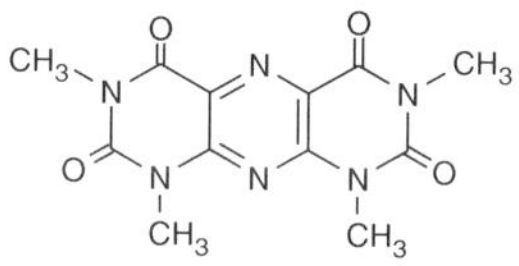

5 gm

S93,747-9
5-CYANO-9-FLUORO(1)BENZOTHIENO(2,3-C)QUINOLIZIN-6-ONE

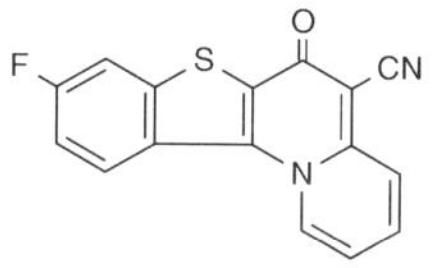

250 mg

S87,038-2
2-(4-NITROPHENYL)-1,2,4-TRIAZOLO-(3,4-A)ISOQUINOLINIUM TETRAFLUOROBORATE

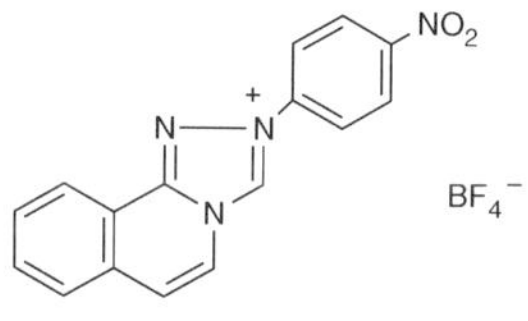

250 mg

S93,894-7
5,7-DI-TERT-BUTYL-3-(3-NITROPHENYL)-1,2,4-BENZOTRIAZINE

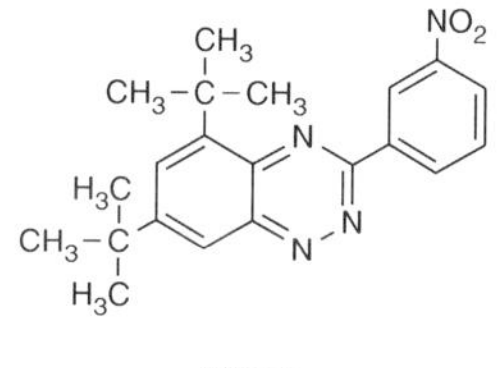

250 mg

S90,903-3
2-PHENYLPYRIDO(1,2-B)(1,2,4)-TRIAZIN-5-IUM PERCHLORATE

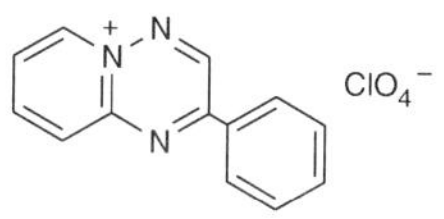

250 mg

S42,641-5
SEMPERVIRINE NITRATE

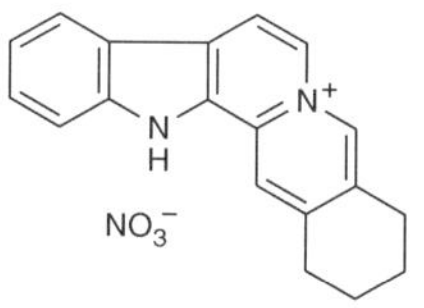

250 mg

S92,238-2
ACID VIOLET 9 (C.I. 45190)

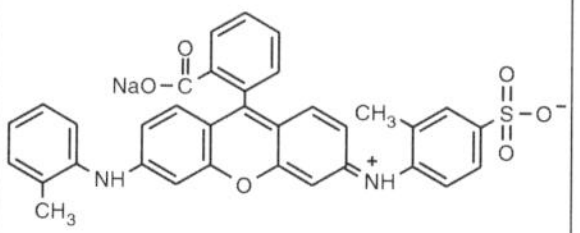

5 gm

S89,358-7
4-CHLORO-3-METHYL-1-PHENYL-1H-PYRROLO(2,3-B)QUINOLINE

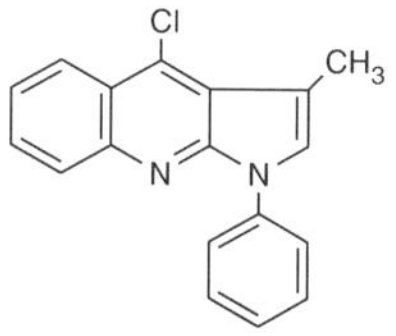

250 mg

S76,483-3
3-(DIMETHYLAMINO)-7-METHYL-1,2,4-BENZOTRIAZINE-1-OXIDE

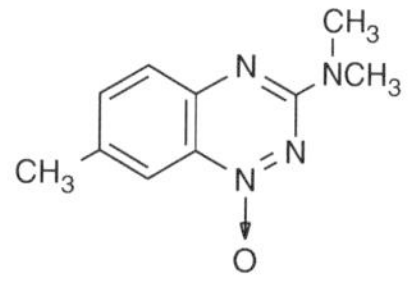

5 gm

S87,050-1
2-(P-TOLYL)-1,2,4-TRIAZOLO(3,4-A)-ISOQUINOLINIUM TETRAFLUOROBORATE

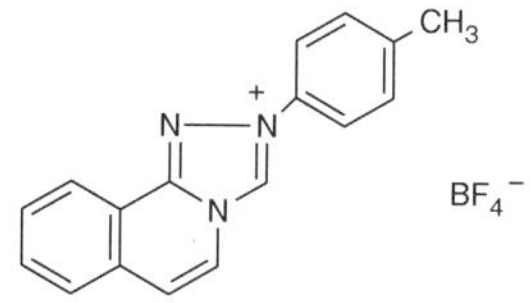

250 mg

S69,649-8
NAPHTHO(2,1-B)QUINOLIZINIUM BROMIDE

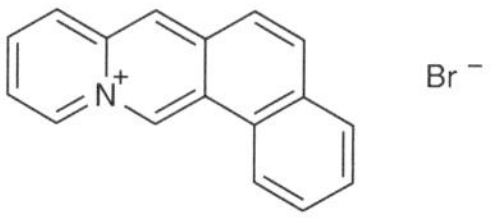

250 mg

S69,646-3
BENZO(B)QUINOLIZINIUM PERCHLORATE

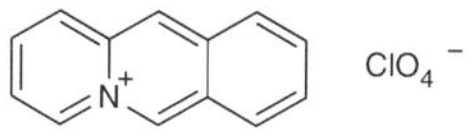

5 gm

S76,601-1
6-NITRO-1,2,3-BENZOTRIAZIN-4-(3H)ONE

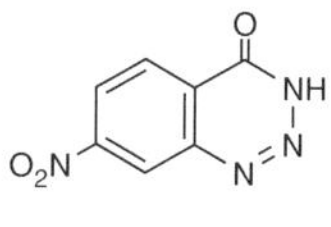

250 mg

S77,047-7
7,8-DIMETHYL-10-(FORMYLMETHYL)-BENZO(G)PTERIDINE-2,4(3H,10H)DIONE

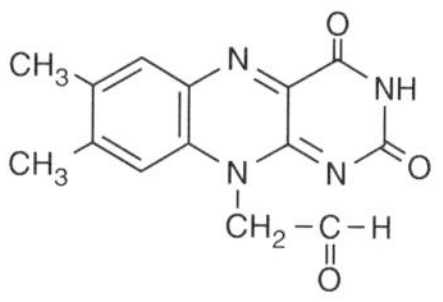

250 mg

S89,915-1
2-METHYL-7H-DIBENZ(F,IJ)-ISOQUINOLIN-7-ONE

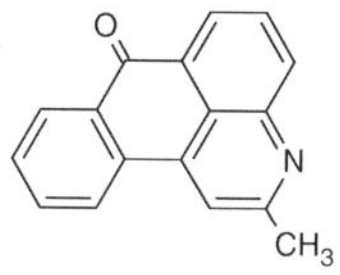

250 mg

S92,789-9
4,6,7(3H,5H,8H)PTERIDINETRIONE

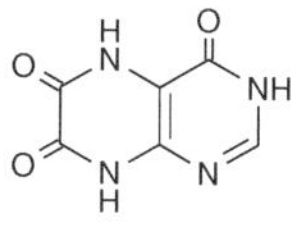

250 mg

S92,788-0
1,6,7-TRIMETHYL-2,4(1H,3H)-PTERIDINEDIONE

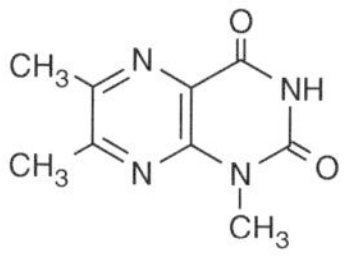

250 mg

S94,513-7
2-PHENYLPYRIMIDO(1,2-A)-BENZIMIDAZOL-4(3H)ONE

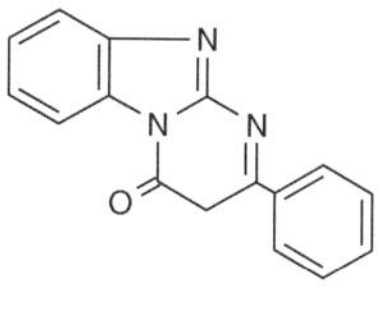

250 mg

S89,349-8
4,7-DICHLORO-1,3-DIMETHYL-1H-PYRAZOLO(3,4-B)QUINOLINE

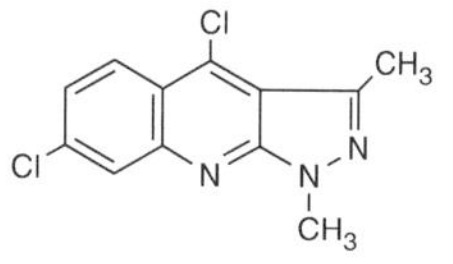

250 mg

S90,630-1
2-METHYL-2H-CYCLOPENTA(D)PYRIDAZINE

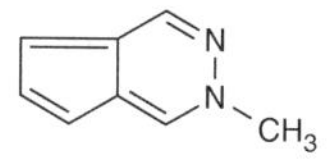

250 mg

S92,037-1
5,12-DIHYDROQUINOXALINO(2,3-B)-QUINOXALINE

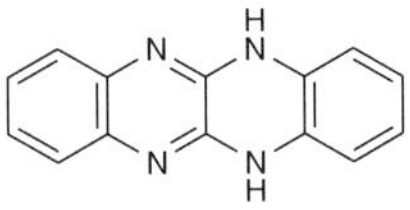

250 mg

S52,376-3
1H-NAPTHO(1,8-D,E)TRIAZINE

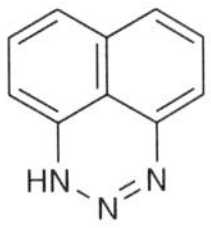

5 gm

S90,912-2
3-METHYLPYRIDO(3,2-E)-1,2,4-TRIAZINE

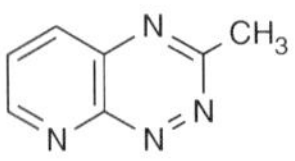

250 mg

S45,935-6
1,2,4-BENZOTRIAZINE

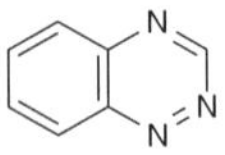

250 mg

S88,705-6
9-PHENYLPHENANTHRIDINE PICRATE

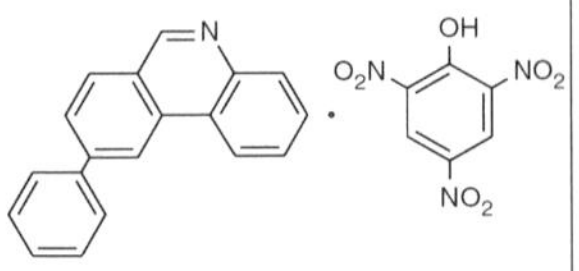

250 mg

S88,706-4
9-PHENYLPHENANTHRIDINE

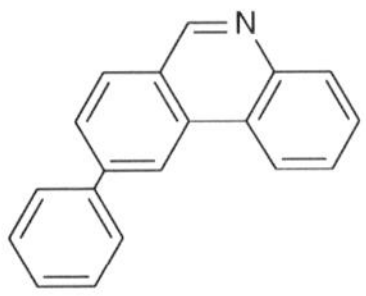

250 mg

S92,730-9
2-BROMO-4-NITROPYRIDINE N-OXIDE
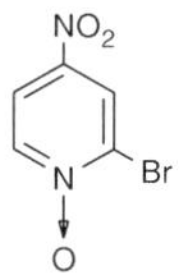

250 mg

S71,062-8
QUININE-N-OXIDE
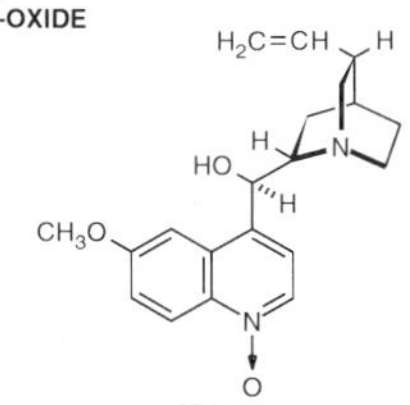

250 mg

S66,830-3
3-(ALPHA-HYDROXYBENZYL)-PYRIDINE-N-OXIDE
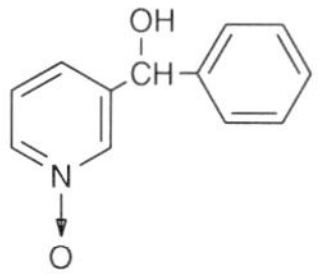

5 gm

S75,942-2
ALPHA-4-PYRIDYLBENZHYDROL-N-OXIDE
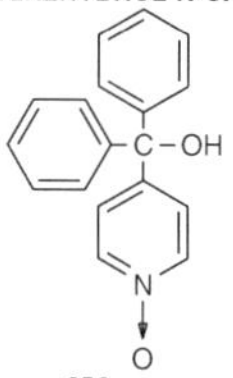

250 mg

S44,905-9
HOMATROPINE-N-OXIDE HYDROBROMIDE
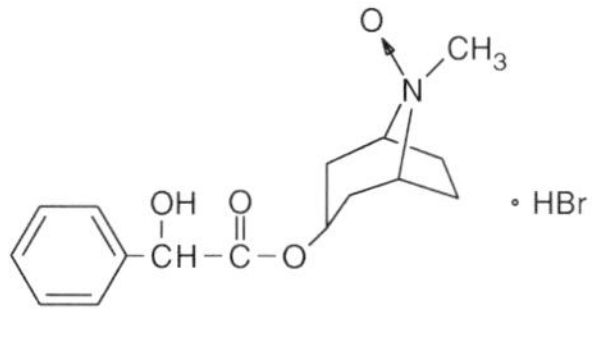

250 mg

S94,747-4
10-(3-DIMETHYLAMINOPROPYL)-10H-PHENOTHIAZINE-N,5-DIOXIDE PICRATE
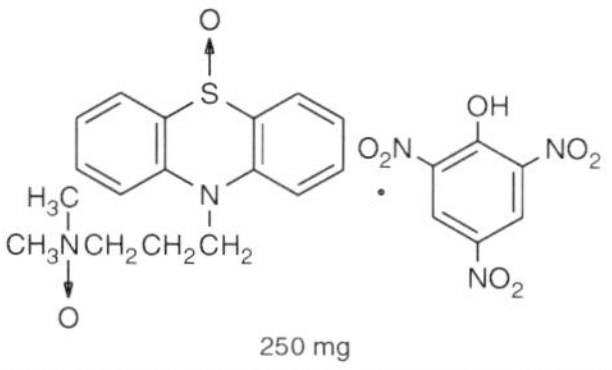

250 mg

S58,014-7
6-METHYL-3-PYRIDINOL-N-OXIDE
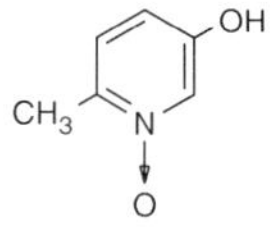

250 mg

S45,924-0
2-PYRIDYLCARBINOL N-OXIDE
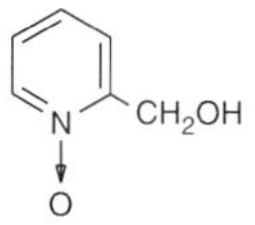

250 mg

S42,423-4
ATROPINE-N-OXIDE HYDROCHLORIDE
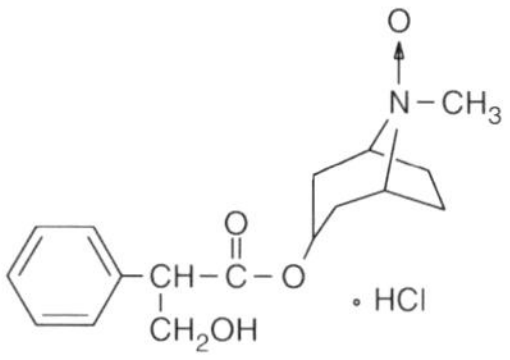

250 mg

S42,522-2
(-)-HYOSCYAMINE-N-OXIDE HYDROCHLORIDE
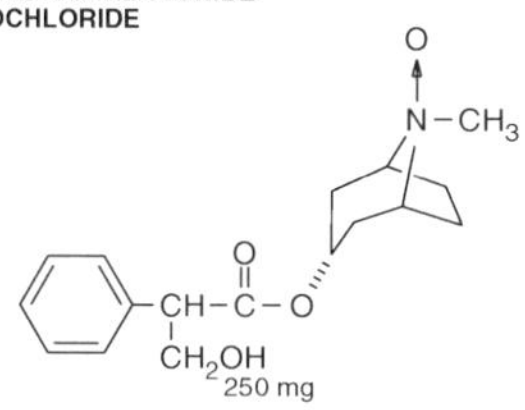

250 mg

S66,841-9
3-(1-HYDROXYETHYL)PYRIDINE-N-OXIDE
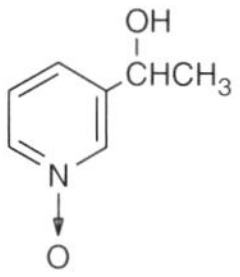

5 gm

S66,833-8
3-(1-HYDROXY-1-METHYLETHYL)-PYRIDINE-N-OXIDE
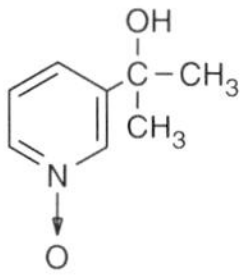

250 mg

S66,840-0
3-(1-HYDROXYBUTYL)PYRIDINE-N-OXIDE
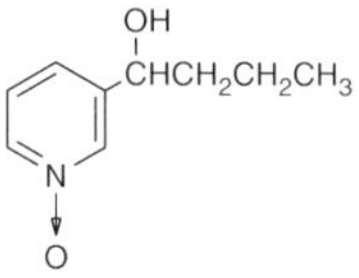

5 gm

S66,824-9
3-(1-HYDROXYPENTYL)PYRIDINE-N-OXIDE
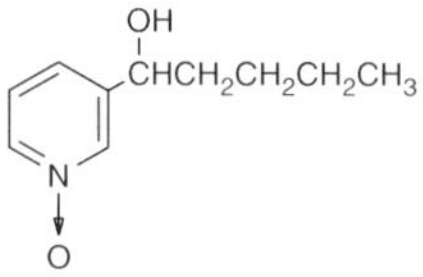

5 gm

S86,046-8
ADENOSINE-N1-OXIDE
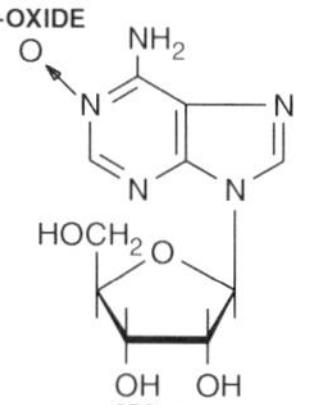

250 mg

S42,586-9
NARCOTINE-N-OXIDE HYDROCHLORIDE
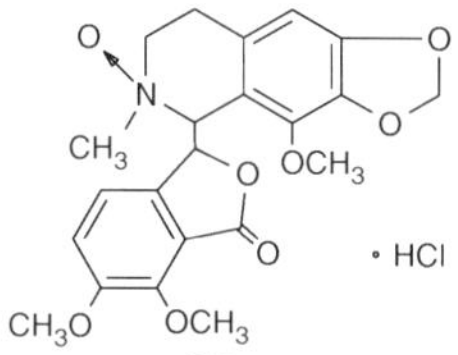

250 mg

S81,082-7
2-(4-METHOXYPHENYL)PYRIDINE-N-OXIDE
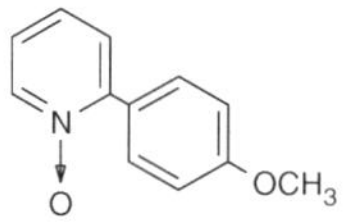

250 mg

S37,230-7
4-ETHOXYPYRIDINE-1-OXIDE
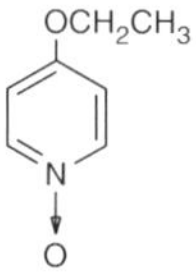

5 gm

S90,918-1
3-MERCAPTO-1,2,4-BENZOTRIAZINE N1-OXIDE
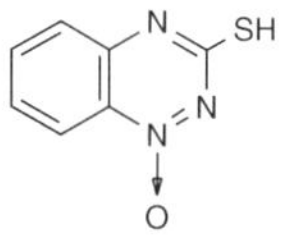

250 mg

S92,093-2
2-(1,2,4-BENZOTRIAZIN-3-YLTHIO)ACETIC ACIDE N1-OXIDE
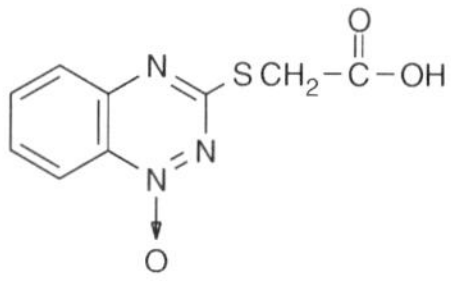

250 mg

S90,915-7
7-METHYL-3-(METHYLTHIO)-1,2,4-BENZOTRIAZINE N1-OXIDE
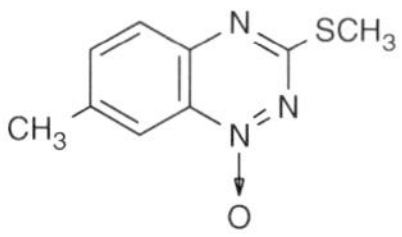

250 mg

S59,694-9
3-AMINO-1,2,4-BENZOTRIAZINE-1-OXIDE
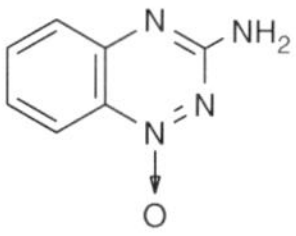

5 gm

S59,695-7
3-AMINO-7-METHYL-1,2,4-BENZOTRIAZINE-1-OXIDE
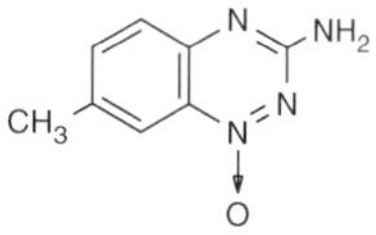

5 gm

S92,832-1
4-AMINO-3-NITROPYRIDINE N-OXIDE
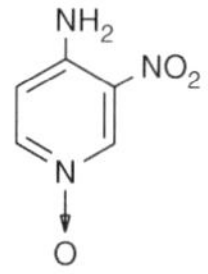

250 mg

S61,487-4
3-AMINO-7-CHLORO-1,2,4-BENZOTRIAZINE-1-OXIDE
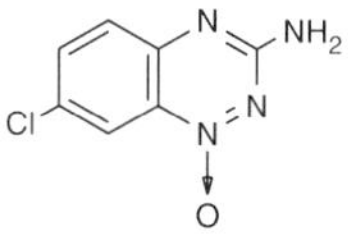

5 gm

S59,604-3
3-AMINO-7-BROMO-1,2,4-BENZOTRIAZINE 1-0XIDE
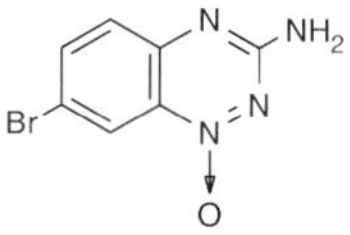

250 mg

S54,066-8
ADENINE N1-OXIDE
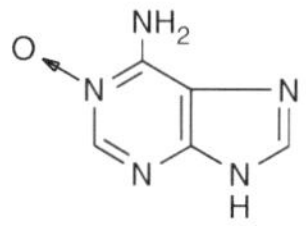

5 gm

S95,466-7
4-AMINO-2,3-LUTIDINE-N-OXIDE
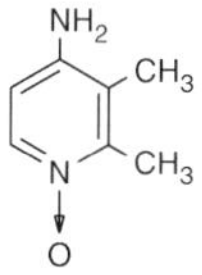

50 mg

S90,941-6
3-HYDRAZINO-4-NITROPYRIDINE N-OXIDE
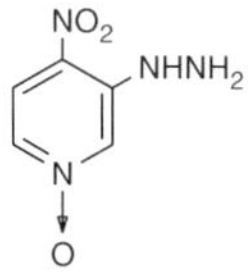

250 mg

S84,370-9
BENZO(1,2-C:3,4-C')BIS(1,2,5)-OXADIAZOLE 3,6-DIOXIDE
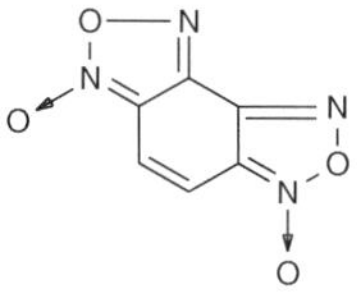

250 mg

S67,467-2
5,7-DIHYDRO-3-(2-NITROPHENYL)-5,5,7,7-TETRAMETHYLFURO(3,4-E)-1,2,4-TRIAZINE-4-OXIDE
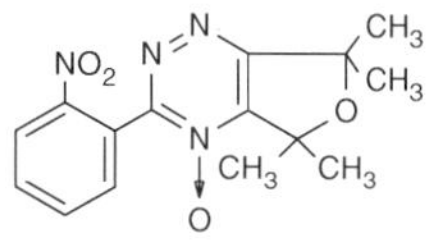

5 gm

S95,463-2
4-NITRO-2,3-LUTIDINE-N-OXIDE
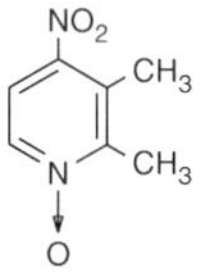

50 mg

S58,327-8
2,6-DIMETHYL-4-NITROPYRIDINE-N-OXIDE

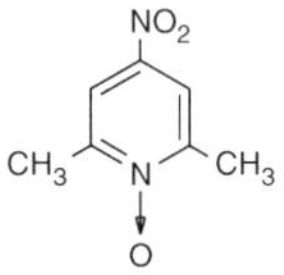

250 mg

S95,493-4
4'-NITRO-2,2'-BIPYRIDINE-N-OXIDE

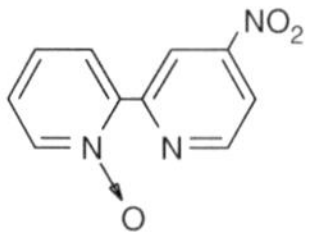

50 mg

S95,489-6
4,4'-DINITRO-2,2'-BIPYRIDINE-N-OXIDE

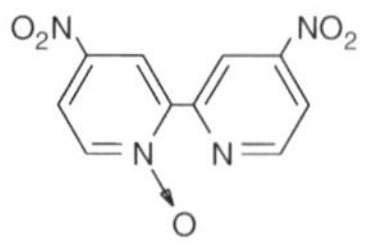

50 mg

S95,491-8
4,4'-DINITRO-2,2'-BIPYRIDINE-N,N'-DIOXIDE

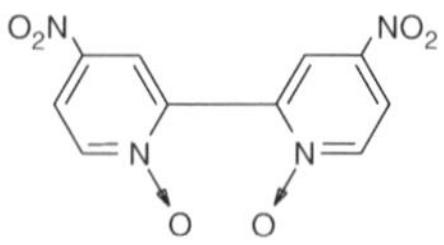

50 mg

S94,020-8
2-METHYL-4-NITROPYRIDINE-N-OXIDE

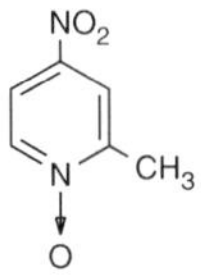

250 mg

S38,928-5
3-BROMO-4-NITROPYRIDINE 1-OXIDE

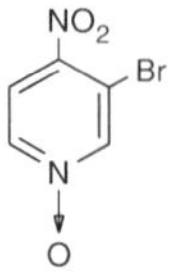

250 mg

S39,241-3
N,N-DIETHYLNICOTINAMIDE 1-OXIDE

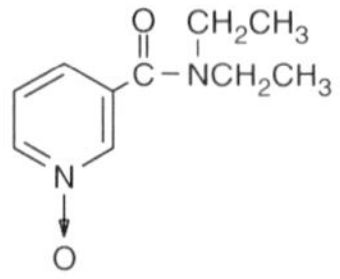

250 mg

S37,000-2
NICOTINIC HYDRAZIDE N-OXIDE

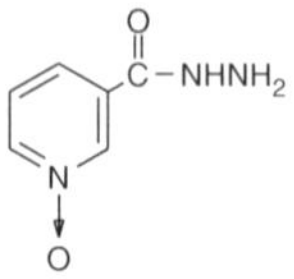

250 mg

S67,465-6
5,7-DIHYDRO-3-(3-NITROPHENYL)-5,5,7,7-TETRAMETHYLFURO(3,4-E)-1,2,4-TRIAZINE-4-OXIDE

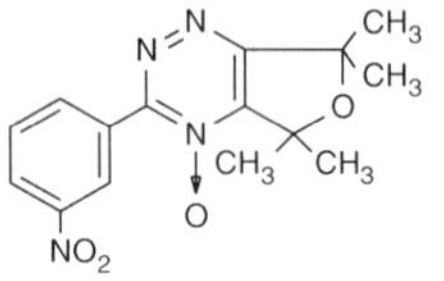

5 gm

S79,351-5
N,N,2,2-TETRAMETHYLPROPYLAMINE N-OXIDE

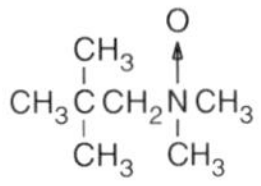

250 mg

S87,130-3
3-HYDRAZINOISOQUINOLINE N-OXIDE

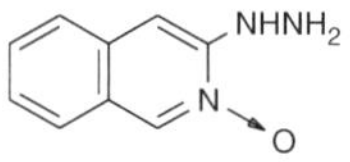

250 mg

S67,468-0
5,7-DIHYDRO-3-PHENYL-5,5,7,7-TETRAMETHYLFURO(3,4-E)-1,2,4-TRIAZINE-4-OXIDE

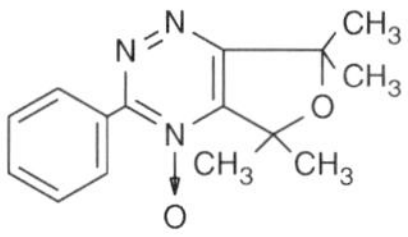

5 gm

S80,742-7
2-[2-(2-FORMYLCYCLOHEXYL)ETHYL]-2H-BENZOTRIAZOLE 1-OXIDE

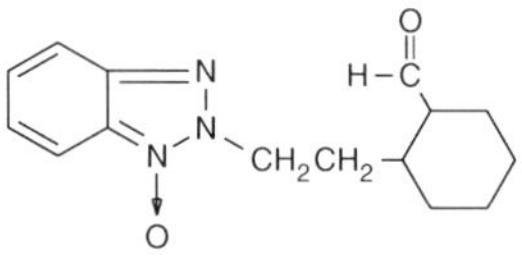

250 mg

S66,842-7
3-PYRIDYLACETALDEHYDE N-OXIDE

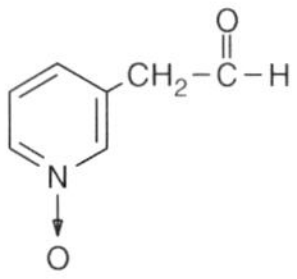

250 mg

S58,144-5
4-PHENACYLPYRIDINE-N-OXIDE MONOHYDRATE

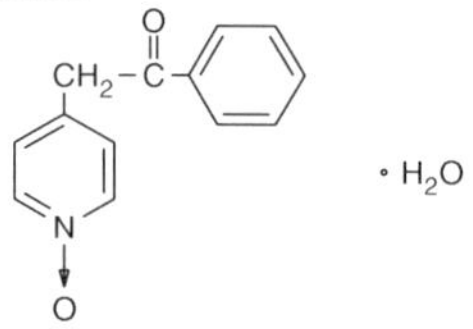

250 mg

S55,174-0
3,4-DITHENOYL-1,2,5-OXADIAZOLE N2-OXIDE

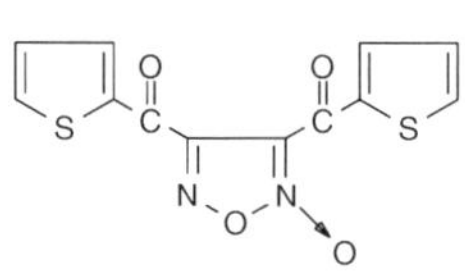

250 mg

S93,287-6
3-ISOQUINOLINECARBOXYLIC ACID N-OXIDE

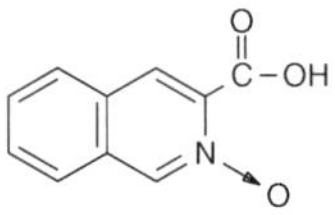

250 mg

S85,300-3
4-QUINOLINECARBOXYLIC ACID N-OXIDE

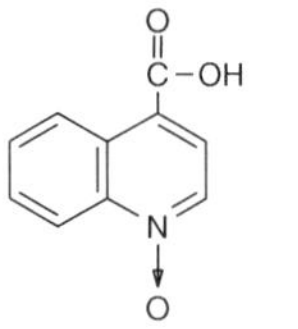

250 mg

S93,290-6
METHYL 3-ISOQUINOLINECARBOXYLATE N-OXIDE

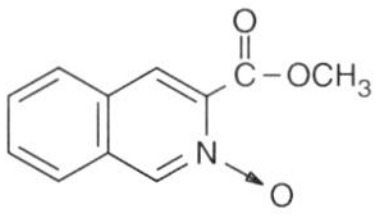

250 mg

S95,467-5
4-CYANO-2,3-LUTIDINE-N-OXIDE

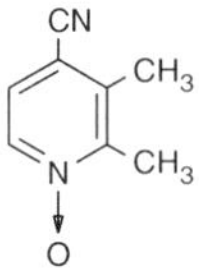

50 mg

S80,744-3
2,4,6-TRIMETHYLBENZONITRILE N-OXIDE

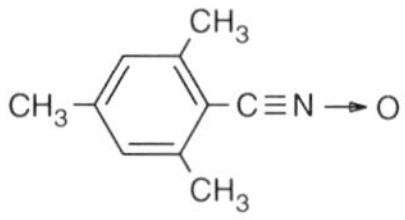

250 mg

S73,208-7
2-(METHYLSULFONYLMETHYL)-PYRIDINE-N-OXIDE

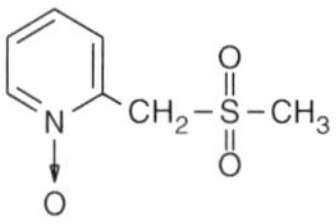

250 mg

S53,010-7
BENZO(C)CINNOLINE N-OXIDE

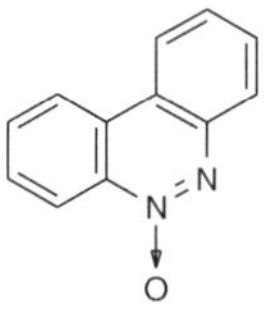

250 mg

S87,127-3
3-CHLOROPYRIDO(2,3-E)-1,2,4-TRIAZINE N'-OXIDE

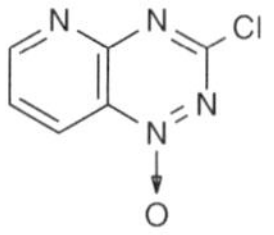

250 mg

S83,994-9
2,6-DIPHENYLPYRIDINE N-OXIDE

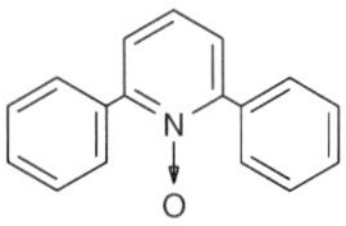

250 mg

S36,897-0
2-ETHYLPYRIDINE-N-OXIDE

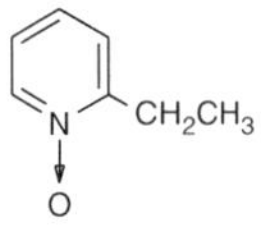

5 gm

S86,113-8
PENTACHLOROPYRIDINE N-OXIDE

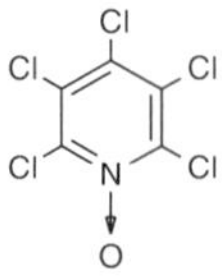

250 mg

S95,494-2
2,2'-BIPYRIDINE-N-OXIDE

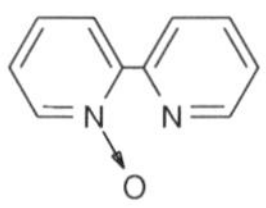

50 mg

S87,736-0
3,5-DIBROMOPYRIDINE N-OXIDE

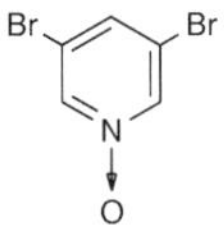

250 mg

S86,932-5
4-ETHYLPYRIDINE N-OXIDE

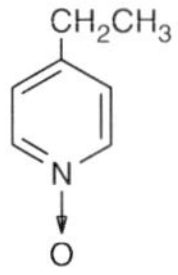

250 mg

S74,801-3
4-TERT-BUTYLPYRIDINE-N-OXIDE

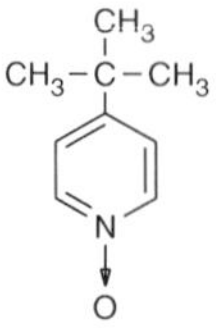

250 mg

S84,905-7
2,4-DIMETHYLPYRIDINE N-OXIDE

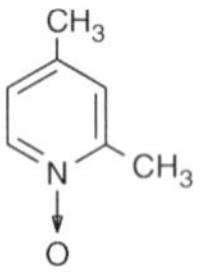

250 mg

S93,305-8
11H-INDENO(1,2-B)QUINOLINE N-OXIDE

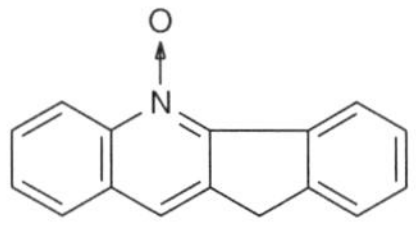

250 mg

S90,923-8
PHENANTHRIDINE N-OXIDE

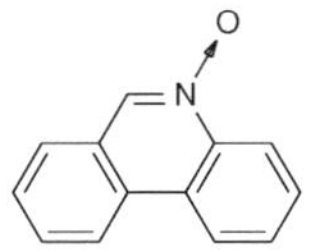

250 mg

S82,034-2
QUINOLINE N-OXIDE PICRATE

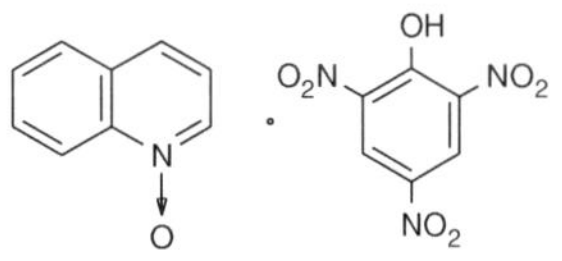

250 mg

S38,837-8 ISOBUTYRALDEHYDE OXIME — 5 gm

S42,913-9 BUTYRALDOXIME — 5 gm

S86,429-3 TRIMETHYLACETALDEHYDE OXIME — 250 mg

S35,017-6 GLUTARALDEHYDE DIOXIME — 5 gm

S38,833-5 4-METHYL-2-PENTANONE OXIME — 5 gm

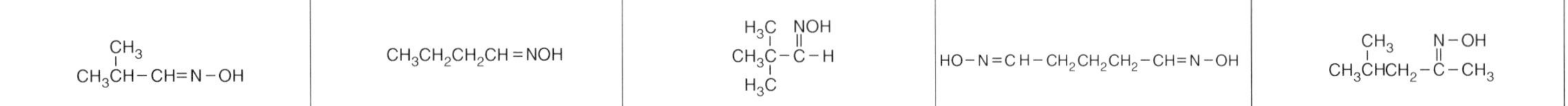

S66,112-0 3,3-DIMETHYL-2-BUTANONE OXIME — 250 mg

S55,055-8 3-FLUOROBENZALDEHYDE OXIME — 5 gm

S55,058-2 ACETOPHENONE OXIME — 5 gm

S38,832-7 5-METHYL-3-HEPTANONE OXIME — 5 gm

S64,830-2 2-OCTANONE OXIME — 5 gm

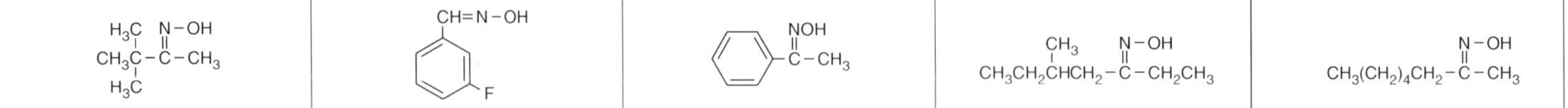

S91,000-7 EUCARVONE OXIME — 250 mg

S94,099-2 CITRAL OXIME — 5 gm

S93,196-9 CYCLODODECANONE OXIME — 250 mg

S67,264-5 OCTADECANOHYDROXAMIC ACID — 250 mg

S55,370-0 12-TRICOSANONE OXIME — 250 mg

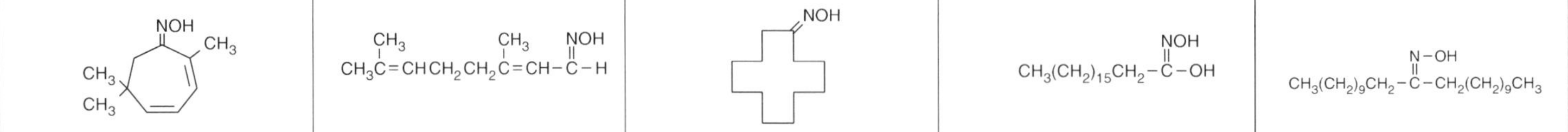

S73,211-7 3-(TRIFLUOROMETHYL)BENZALDOXIME — 250 mg

S66,595-9 O-[N-(2-CHLORO-5-TRIFLUOROMETHYLPHENYL)CARBAMOYL]-ACETONE OXIME — 5 gm

S81,899-2 ETHYL 2-HYDROXYIMINO-3-OXO-4,4,4-TRIFLUOROBUTYRATE — 250 mg

S66,694-7 O-(N-[2-(TRIFLUOROMETHYL)PHENYL]-CARBAMOYL)CYCLOHEXANONE OXIME — 5 gm

S66,576-2 O-[N-(2-CHLORO-5-TRIFLUOROMETHYLPHENYL)CARBAMOYL]-CYCLOHEXANONE OXIME — 5 gm

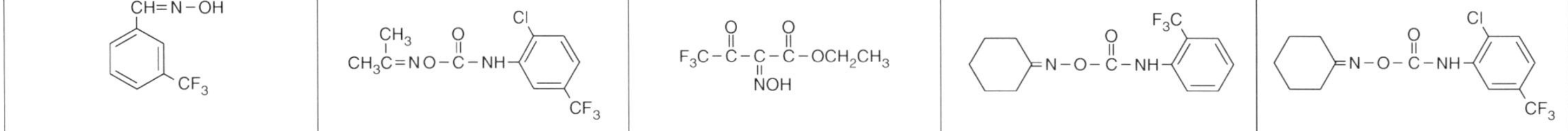

S71,746-0 O-(N-[3-(TRIFLUOROMETHYL)PHENYL]-CARBAMOYL)ACETONE OXIME — 5 gm

S61,174-3 O-(N-[3-(TRIFLUOROMETHYL)PHENYL]-CARBAMOYL)CYCLOHEXANONE OXIME — 5 gm

S90,704-9 SYN-3-IODOBENZALDEHYDE OXIME — 250 mg

S41,344-5 2-AMINO-5-CHLOROBENZOPHENONE OXIME (MIXTURE OF SYN AND ANTI ISOMERS) — 5 gm

S67,133-9 2-(HYDROXYIMINO)-N-(2-IODOPHENYL)-ACETAMIDE — 5 gm

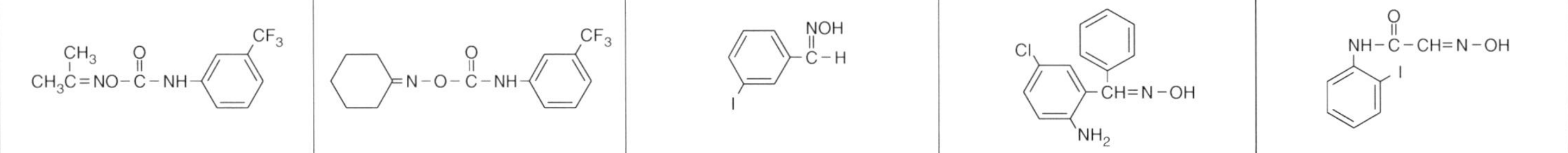

S87,903-7 4-CHLORO-3-NITROBENZALDEHYDE OXIME — 5 gm

S87,904-5 2-CHLORO-5-NITROBENZALDEHYDE OXIME — 5 gm

S76,712-3 2-(4-CHLOROPHENYL)-2,5,5-TRIMETHYLTETRAHYDROFURAN-3,4-DIONE 3-OXIME — 250 mg

S66,774-9 O-[N-(2-CHLORO-6-METHYLPHENYL)-CARBAMOYL]ACETONE OXIME — 5 gm

S71,346-5 O-[N-(2,6-DICHLOROPHENYL)CARBAMOYL]-ACETONE OXIME — 5 gm

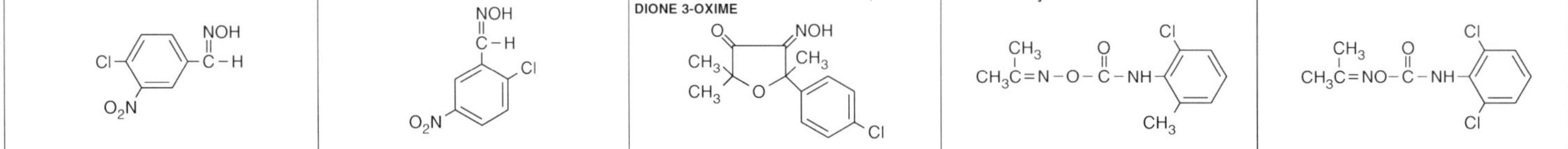

S71,121-7 O-[N-(3-BROMOPHENYL)CARBAMOYL]-CYCLOHEXANONE OXIME — 5 gm

S71,221-3 O-[N-(4-FLUOROPHENYL)CARBAMOYL]-CYCLOHEXANONE OXIME — 250 mg

S84,892-1 4-BROMOPHENYLAZO PHENYL KETONE OXIME — 250 mg

S75,908-2 ALPHA,4-DICHLOROBENZALDEHYDE OXIME — 250 mg

S90,714-6 SYN-4-IODOBENZALDEHYDE OXIME — 250 mg

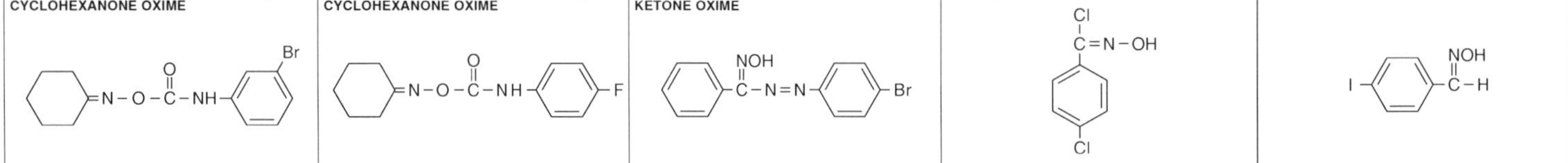

S53,957-0 VANILLIN OXIME — 5 gm

S86,087-5 2',4'-DIHYDROXY-3'-METHYL-5'-NITROACETOPHENONE OXIME — 250 mg

S81,168-8 DIMETHYL 3-HYDROXY-4-METHOXY-2,5-FURANDICARBOXYLATE — 250 mg

S60,038-5 ISOVANILLIN OXIME — 5 gm

S66,200-3 3-HYDROXY-3-PHENYLPROPYLHYDROXYLAMINE HEMIOXALATE — 5 gm

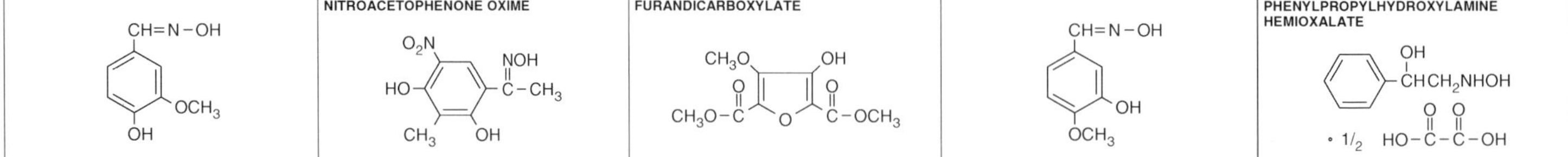

S75,948-1 1-[4-(BENZYLOXY)PHENYL]-4,4-DIMETHYL-1-HYDROXY-3-PENTANONE OXIME — 250 mg

S73,512-4 5-(4-HYDROXYIMINO-1-METHYL-3-PIPERIDYL)DIBENZOSUBEROL — 250 mg

S76,634-8 3-(1-HYDROXYCYCLOHEXYL)-1-METHYL-4-PIPERIDONE OXIME — 250 mg

S56,659-4 P-HYDROXYBENZOHYDROXAMIC ACID — 250 mg

S53,596-6 3-HYDROXY-3-[2-(HYDROXYIMINO)-PROPYL]-2-INDOLINONE — 5 gm

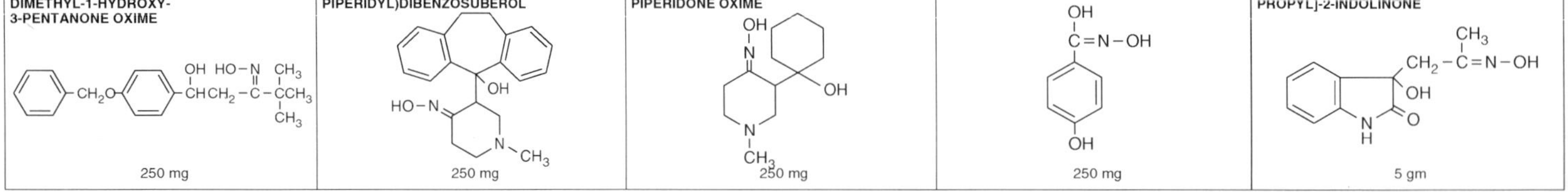

S38,693-6
4-HYDROXYBENZALDEHYDE OXIME
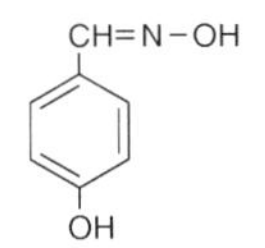
250 mg

S87,905-3
4,5-DIMETHOXY-2-NITROBENZALDEHYDE OXIME
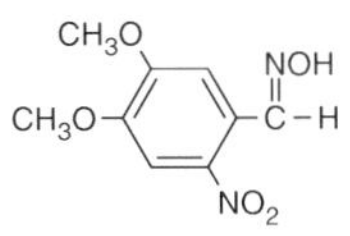
5 gm

S77,358-1
2,3-DIMETHOXY-SYN-BENZALDEHYDE OXIME
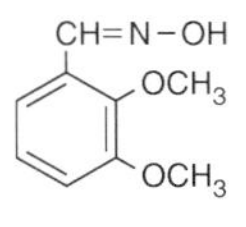
250 mg

S93,169-1
2,4,6-TRIMETHOXYBENZALDEHYDE OXIME
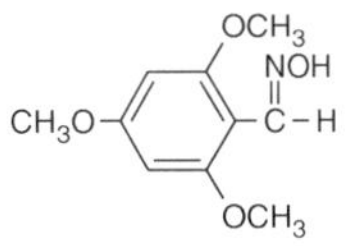
250 mg

S36,841-5
VERATRALDEHYDE OXIME
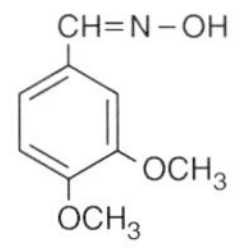
5 gm

S76,698-4
1-(2-ETHOXYPHENYL)-1,2-PROPANEDIONE 1-HYDRAZONE 2-OXIME
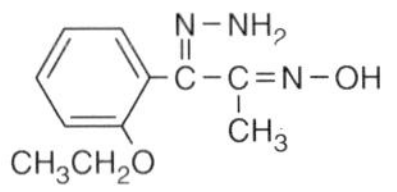
250 mg

S50,540-4
4-HEXADECYLOXY-3-[2-(HYDROXYIMINO)-ACETAMIDO]BENZOIC ACID
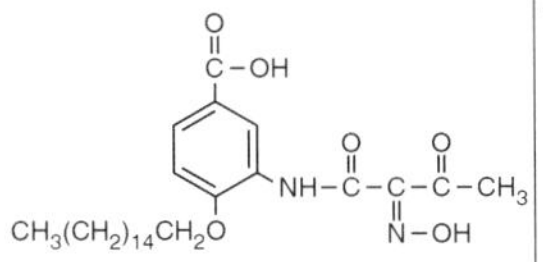
250 mg

S53,399-8
A-(DICHLORO-4-HYDROXYIMINO-2,5-CYCLOHEXADIENYLIDENE)-4-METHOXYPHENYLACETONITRILE
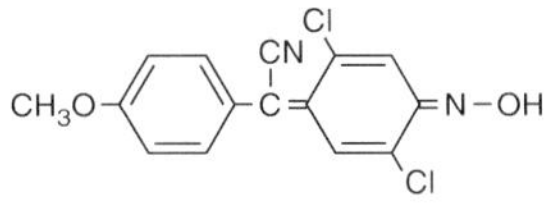
5 gm

S53,332-7
A-(4-HYDROXYIMINO-2,5-CYCLOHEXADIEN-1-YLIDENE)-4-METHOXYPHENYLACETONITRILE
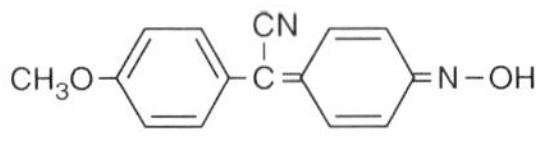
5 gm

S65,460-4
2'-(HEXADECYLOXY)ACETOPHENONE OXIME
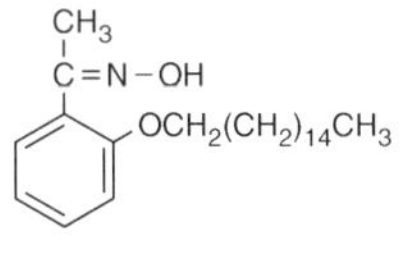
250 mg

S68,138-5
2-METHOXYACETOPHENONE, OXIME
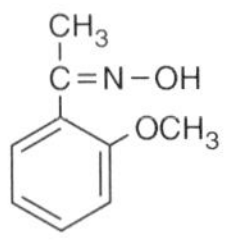
5 gm

S63,377-1
O-[N-(O-ANISYL)CARBAMOYL]-CYCLOHEXANONE OXIME
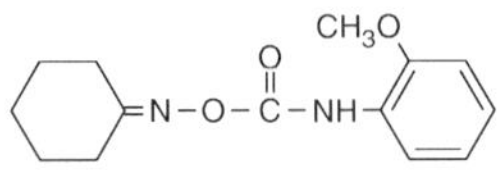
5 gm

S60,084-9
O-[N-(2-METHOXYPHENYL)CARBAMOYL]-ACETONE OXIME
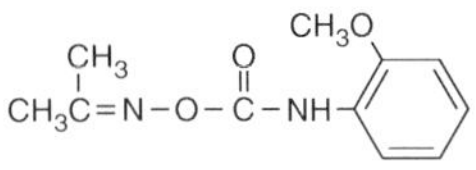
5 gm

S63,935-4
O-[N-(3-CHLORO-2-METHOXYPHENYL)-CARBAMOYL]ACETONE OXIME
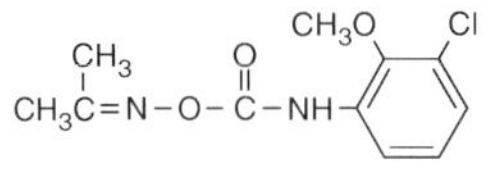
5 gm

S73,110-2
2-ISONITROSO-2'-METHOXY-3'-METHYLACETANILIDE
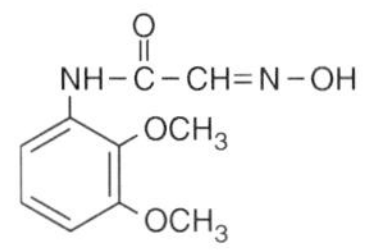
5 gm

S61,246-4
O-[N-(2,5-DIMETHOXYPHENYL)-CARBAMOYL]CYCLOHEXANONE OXIME
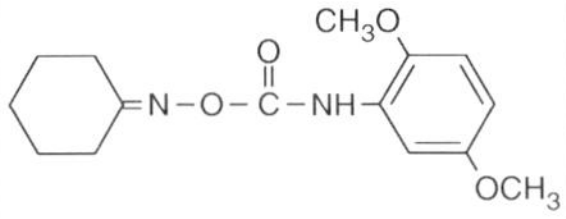
5 gm

S93,363-5
2-(4-METHOXYPHENYL)GLYOXAL 1-OXIME
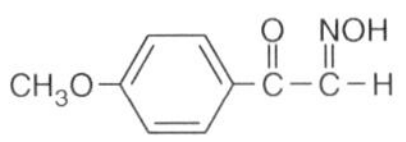
250 mg

S64,452-8
O-[N-(2-METHOXY-5-METHYLPHENYL)-CARBAMOYL]ACETONE OXIME
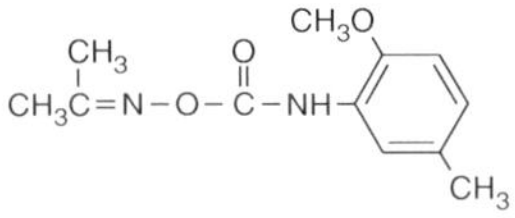
5 gm

S63,941-9
O-[N-(3-CHLORO-2-METHOXYPHENYL)-CARBAMOYL]CYCLOHEXANONE OXIME
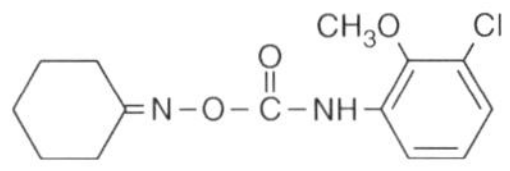
5 gm

S60,086-5
O-[N-(4-METHOXYPHENYL)CARBAMOYL]-ACETONE OXIME
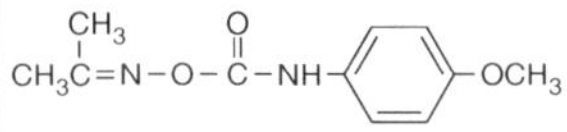
5 gm

S60,950-1
O-[N-(3-METHOXYPHENYL)CARBAMOYL]-ACETONE OXIME
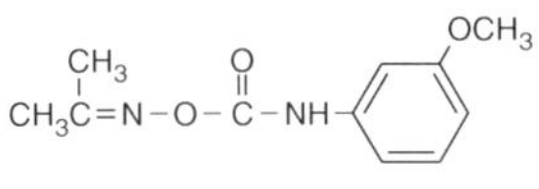
5 gm

S61,183-2
O-[N-(P-ANISYL)CARBAMOYL]-CYCLOHEXANONE OXIME
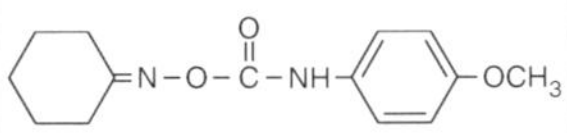
5 gm

S60,985-4
O-[N-(2,5-DIMETHOXYPHENYL)-CARBAMOYL]ACETONE OXIME
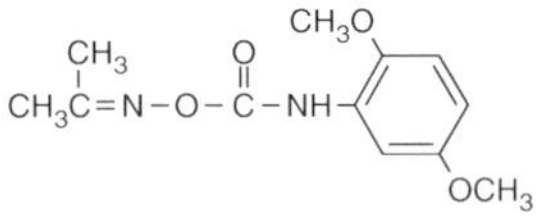
5 gm

S61,247-2
O-[N-(M-ANISYL)CARBAMOYL]-CYCLOHEXANONE OXIME
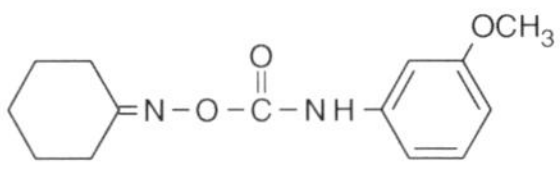
5 gm

S92,215-3
2-METHOXYBENZALDEHYDE OXIME
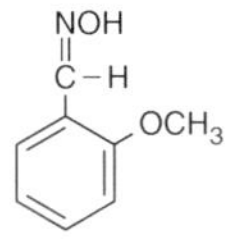
250 mg

S56,899-6
2'-(HEXADECYLOXY)-PROPIOPHENONE OXIME
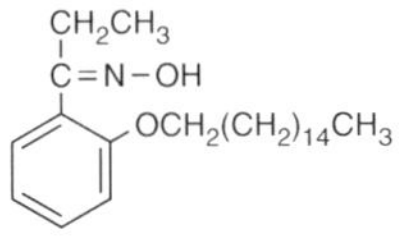
250 mg

S68,275-6
4'-METHOXYACETOPHENONE, OXIME
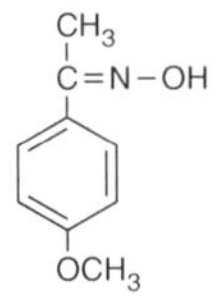
250 mg

S38,683-9
4-(HEXADECYLOXY)BENZALDEHYDE OXIME
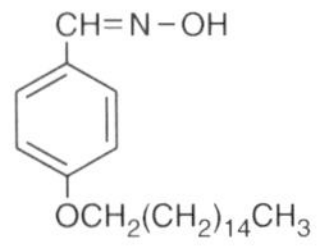
250 mg

S56,240-8
2'-HEXADECYLOXY-5'-METHYLACETOPHENONE OXIME
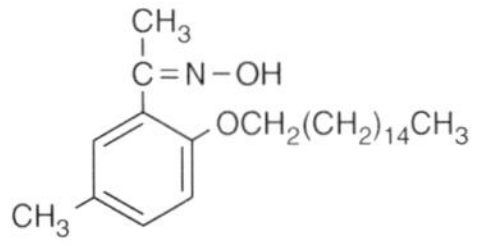
250 mg

S95,325-3
BENZYL 2-CHLORO-2-(HYDROXYIMINO)-ACETATE
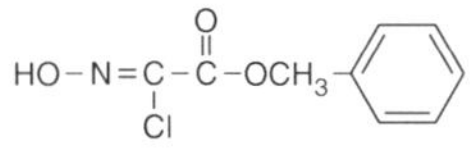
50 mg

S67,135-5
DIMETHYL 2-[N-(2,4,5,7-TETRANITROFLUOREN-9-YLIDENE)-AMINOXY]SUCCINATE
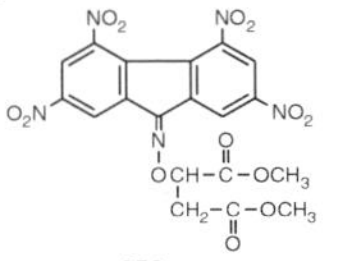
250 mg

S87,894-4
6-NITRO-1,3-BENZODIOXOLE-5-CARBOXALDEHYDE OXIME
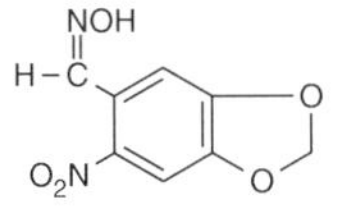
5 gm

S68,597-6
4-HYDROXYIMINO-2,2,5,5-TETRAMETHYLTETRAHYDROFURAN-3-ONE
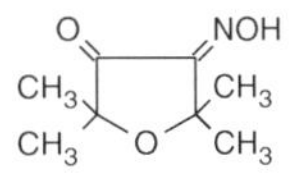
250 mg

S45,937-2
METHYL THIOACETOHYDROXAMATE
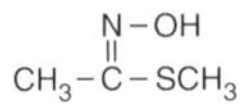
250 mg

S95,402-0
S-(4-NITROPHENYL)-BENZOTHIOHYDROXAMATE
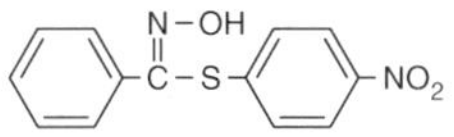
50 mg

S76,031-5
ETHYL 2-(3-HYDROXYIMINO-1-INDANYLTHIO)ACETATE

S95,382-2
S-TERT-BUTYL BENZOTHIOHYDROXAMATE

S88,497-9
2'-(METHYLTHIO)ACETOPHENONE OXIME

S78,796-5
4-AMINO-5-IMIDAZOLECARBOXAMIDOXIME DIHYDROCHLORIDE

S65,808-1
2-AMINOBENZOPHENONE OXIME

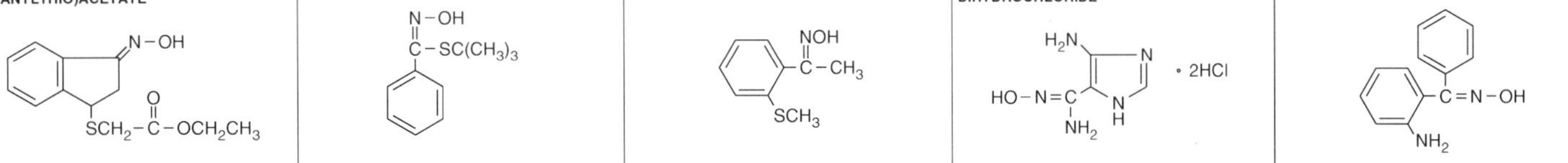

250 mg | 50 mg | 250 mg | 250 mg | 5 gm

S45,909-7
DIMETHYLGLYOXIME, DISODIUM SALT, OCTAHYDRATE

S54,956-8
3,4-DIHYDRO-1(2H)-PHENAZINONE OXIME

S45,701-9
DIACETYL MONOXIME 4-NITROPHENYLHYDRAZONE

S84,900-6
BENZIL OXIME HYDRAZONE

S73,532-9
3,3-DIMETHYL-2-HYDRAZONOBUTYROPHENONE OXIME

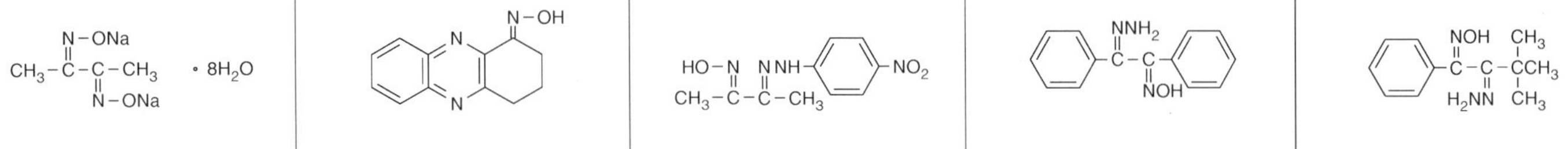

250 mg | 250 mg | 250 mg | 250 mg | 250 mg

S80,300-6 2-CYANO-2-HYDROXIMINOACETAMIDE — 250 mg

S66,266-6 ALPHA-HYDROXYIMINO-O-ACETOTOLUIDIDE — 250 mg

S54,500-7 METHYL 2-CYANO-2-(HYDROXYIMINO)-ACETATE — 5 gm

S53,946-5 2'-CHLORO-2-(HYDROXYIMINO)-ACETANILIDE — 5 gm

S54,524-4 2',5'-DICHLOROGLYOXYLANILIDE OXIME — 250 mg

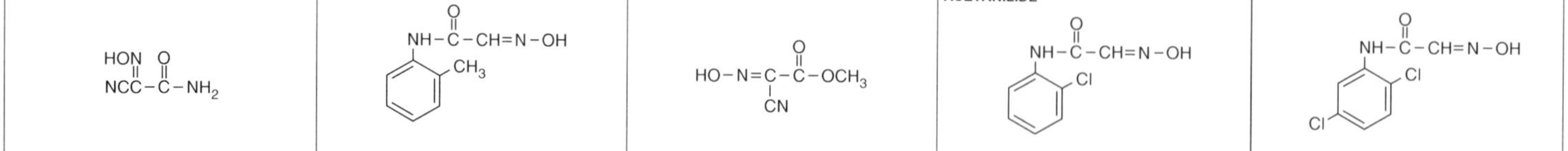

S54,520-1 2',4',5'-TRICHLOROGLYOXYLANILIDE OXIME — 250 mg

S92,388-5 3-(DIMETHYLAMINO)-2-(HYDROXYIMINO)-3-OXOPROPIONITRILE — 250 mg

S35,517-8 2-PYRIDINEALDOXIME O-PROPYL ETHER — 5 gm

S35,506-2 2-PYRIDINEALDOXIME O-BENZYL ETHER — 5 gm

S35,513-5 2-PYRIDINEALDOXIME O-PHENETHYL ETHER — 5 gm

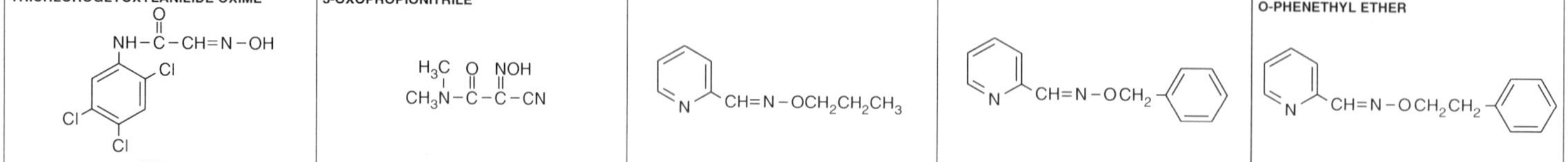

S84,671-6 PHENANTHRENEQUINONE DIOXIME — 250 mg

S74,267-8 1,3-DIPHTHALIMIDOACETONE OXIME — 5 gm

S38,266-3 MALONOHYDROXAMAMIDE — 250 mg

S45,385-4 INDOLE-3-CARBOXALDEHYDE OXIME — 250 mg

S94,967-1 2,2,6,6-TETRAMETHYL-4-PIPERIDONE OXIME HYDROCHLORIDE — 1 gm

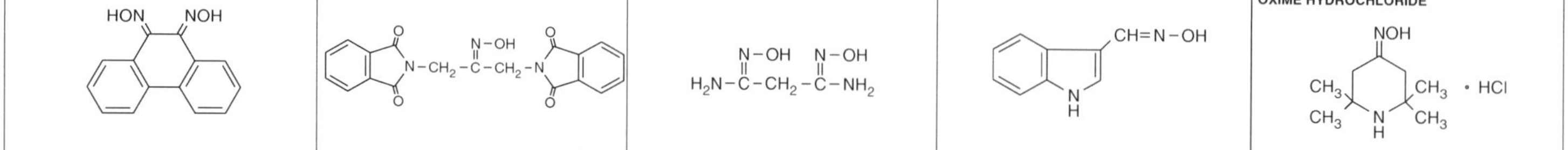

S45,805-8 1,1'-TRIMETHYLENEBIS[4-(HYDROXYIMINOMETHYL)-PYRIDINIUM BROMIDE] — 5 gm

S79,221-7 2,6-DINITRO-3-METHYLBENZALDEHYDE OXIME — 250 mg

S63,461-1 O-[N-(3-NITROPHENYL)CARBAMOYL]-ACETONE OXIME — 5 gm

S63,449-2 O-[N-(3-NITROPHENYL)CARBAMOYL]-CYCLOHEXANONE OXIME — 5 gm

S61,120-4 O-[N-(4-METHYL-3-NITROPHENYL)-CARBAMOYL]CYCLOHEXANONE OXIME — 5 gm

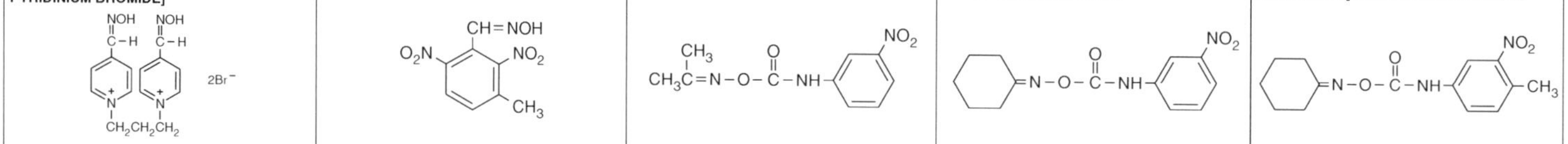

S60,976-5 O-[N-(4-METHYL-3-NITROPHENYL)-CARBAMOYL]ACETONE OXIME — 5 gm

S45,458-3 1,2,3-CYCLOHEXANETRIONE-1,3-DIOXIME — 5 gm

S76,670-4 3-(9-HYDROXYIMINO-6,7,8,9-TETRAHYDRO-5H-BENZOCYCLOHEPTEN-8-YL)-2-METHYL-1-ISOINDOLINONE — 250 mg

S63,949-4 N,N'-(4-METHYL-1,3-PHENYLENE)BIS(O-CARBAMOYLCYCLOHEXANONE OXIME) — 5 gm

S63,953-2 N,N'-(4-METHYL-1,3-PHENYLENE)BIS(O-CARBAMOYLACETONE OXIME) — 5 gm

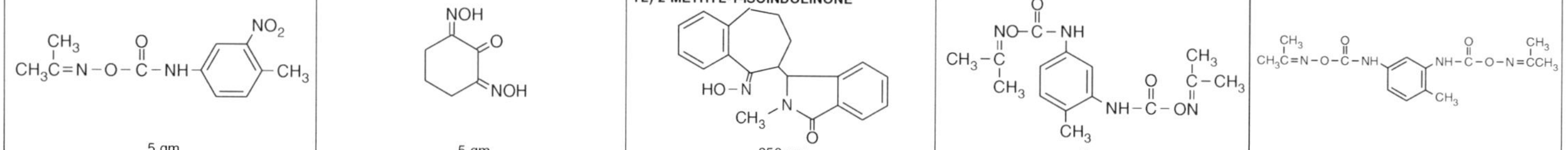

S79,231-4 3-ACETYL-4A,5,6,7,8,8A-HEXAHYDRO-2-METHYL-4H-1-BENZOPYRAN-4-ONE DIOXIME — 250 mg

S35,518-6 3-PYRIDINEALDOXIME O-PROPYL ETHER — 5 gm

S35,514-3 3-PYRIDINEALDOXIME O-PHENETHYL ETHER — 5 gm

S41,031-4 1,1'-TRIMETHYLENEBIS[4-(HYDROXYIMINOMETHYL)-PYRIDINIUM CHLORIDE] — 5 gm

S37,849-6 2,5-HEXANEDIONE DIOXIME — 250 mg

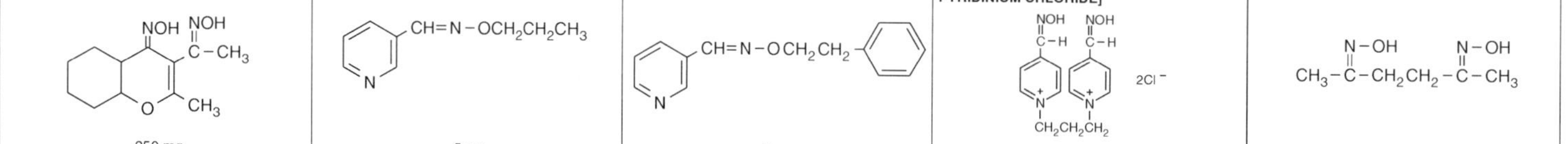

S91,948-9 1-(HYDROXYAMINO)-5-NITRONAPHTHALENE — 250 mg

S44,766-8 3-NITROBENZAMIDOXIME — 5 gm

S60,761-4 O-[N-(2-CHLORO-4-NITROPHENYL)-CARBAMOYL]CYCLOHEXANONE OXIME — 5 gm

S60,718-5 O-[N-(2-CHLORO-4-NITROPHENYL)-CARBAMOYL]ACETONE OXIME — 5 gm

S61,087-9 O-[N-(2-METHYL-4-NITROPHENYL)-CARBAMOYL]CYCLOHEXANONE OXIME — 5 gm

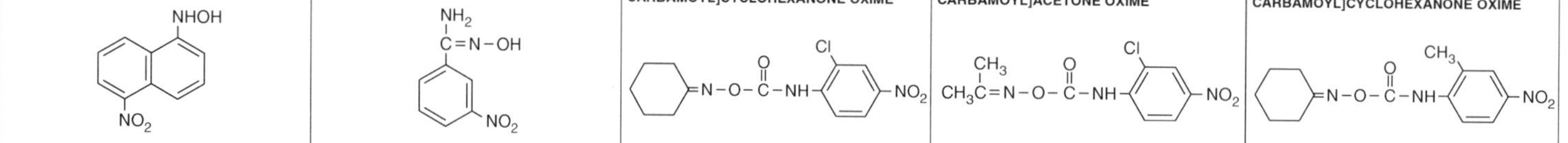

S61,127-1 O-[N-(2-METHYL-4-NITROPHENYL)-CARBAMOYL]ACETONE OXIME — 5 gm

S35,508-9 4-PYRIDINEALDOXIME O-BENZYL ETHER — 5 gm

S35,515-1 4-PYRIDINEALDOXIME O-PHENETHYL ETHER — 5 gm

S61,909-4 2-METHYL-2-PHENYLACETALDEHYDE OXIME — 5 gm

S95,536-1 2-HYDROXIMINO-3-PHENYL-1-INDANONE — 50 mg

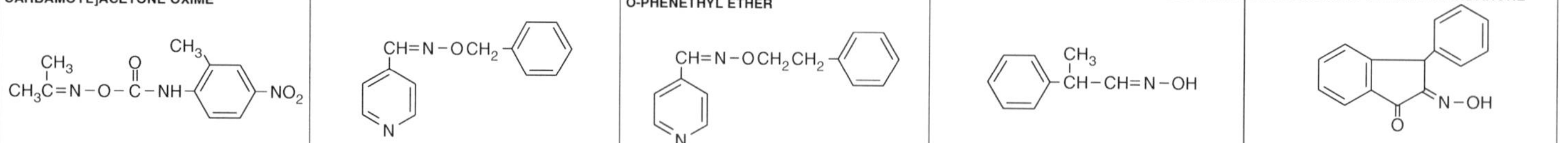

S35,086-9 4-(HYDROXYIMINO)-ALPHA-PHENYL-2,5-CYCLOHEXADIENE DELTA1,ALPHA ACETONITRILE — 5 gm

S86,355-6 2-PHENYLCYCLOHEXANONE OXIME — 250 mg

S68,132-6 BENZYLAMIDOXIME — 5 gm

S79,877-0 1,3-DIPHENYLACETONE OXIME — 250 mg

S64,786-1 4-DIMETHYLAMINOBENZALDEHYDE OXIME — 250 mg

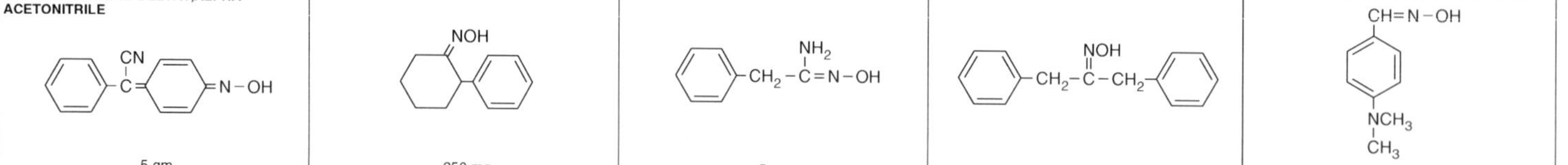

S89,377-3
3-METHYL-1-PHENYL-4,5-PYRAZOLIDINEDIONE 4-OXIME
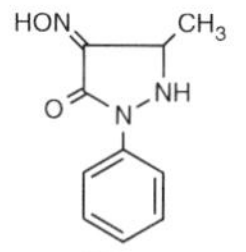
250 mg

S35,509-7
2-PYRIDINEALDOXIME DIMETHYL SULFATE
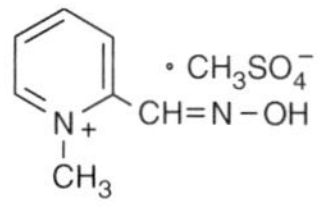
5 gm

S65,752-2
2-(ETHOXYCARBONYL)BENZIL MONOXIME
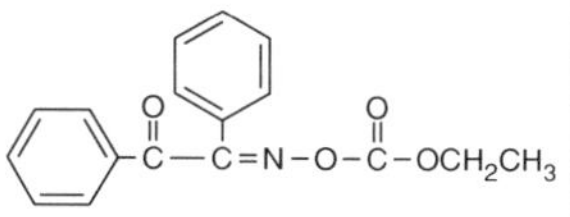
5 gm

S82,608-1
2-BENZOYLFURAN OXIME
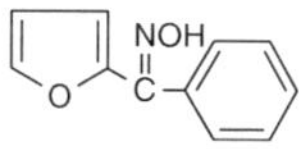
250 mg

S65,409-4
O-(MORPHOLINOCARBONYL)-BENZALDEHYDE OXIME
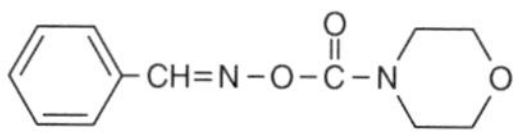
5 gm

S65,831-6
ACETOPHENONE OXIME
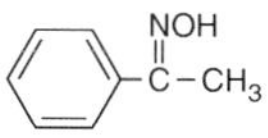
250 mg

S79,994-7
PROPIOPHENONE OXIME P-TOSYLATE
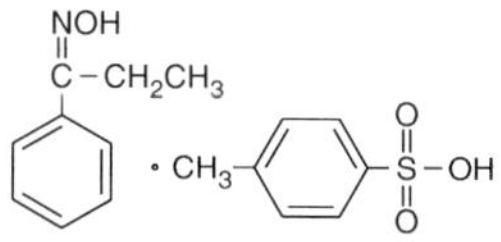
250 mg

S66,319-0
O-(2-METHYLALLYL)BENZOPHENONE OXIME
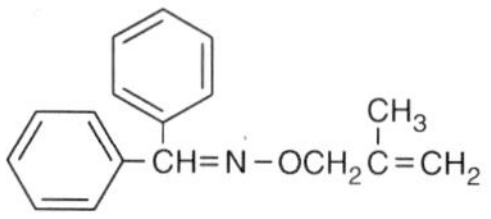
5 gm

S67,096-0
2,6-DIMETHYLBENZALDEHYDE OXIME
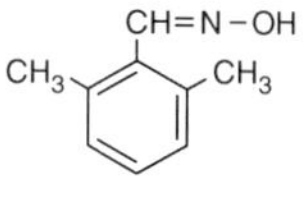
5 gm

S76,182-6
BENZOIC HYDRAZIDE OXIME
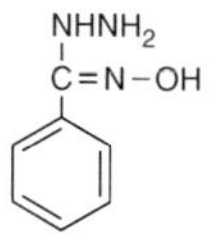
250 mg

S86,341-6
2,3-DIHYDRO-1,1-DIOXO-O-(2-NITROPHENYL)-4H-BENZOTHIOPYRAN-4-ONE OXIME
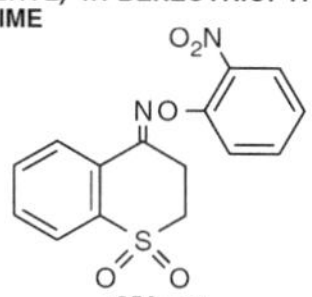
250 mg

S85,021-7
3-PHENYL-1,2-INDANDIONE 2-OXIME
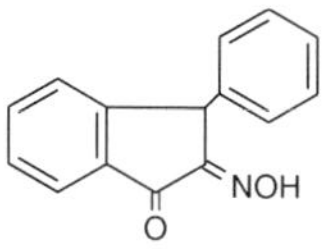
250 mg

S73,221-4
2,2,6,6-TETRAMETHYLTETRAHYDROPYRAN-3,4,5-TRIONE-3,5-DIOXIME
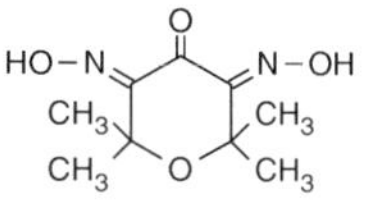
250 mg

S83,729-6
ETHYL 2-OXOBUTYRATE 2-OXIME
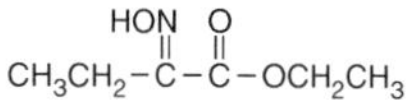
250 mg

S37,538-1
FURFURAL OXIME
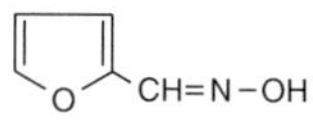
250 mg

S60,290-6
2-ACETYLFURAN OXIME
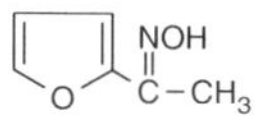
250 mg

S73,276-1
3,3-DIMETHYL-1-(PHENYLSULFONYL)-2-BUTANONE OXIME
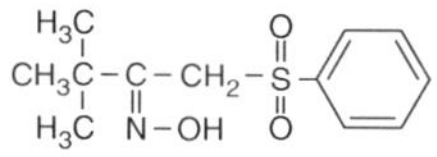
250 mg

S90,821-5
2-ACETYL-5-BROMOTHIOPHENE OXIME
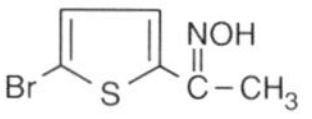
250 mg

S58,132-1
O-(3-OXOBUTYL)-N-PROPYLHYDROXYLAMINE OXALATE SALT
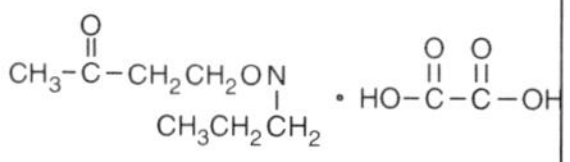
250 mg

S67,160-6
O-(4-NITROPHENYL)ACETONE OXIME
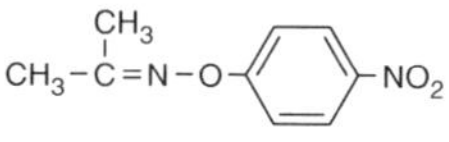
5 gm

S72,903-5
1-PHENYL-1,2,3-BUTANETRIONE 2-OXIME
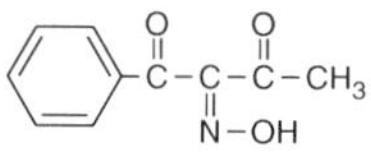
5 gm

S84,636-8
3,5-DIMETHYL-1,4-BENZOQUINONE 1-OXIME
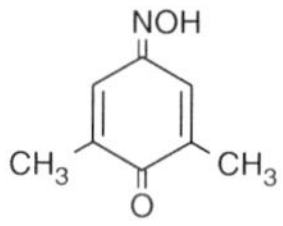
250 mg

S90,687-5
PHENANTHRENEQUINONE 9-OXIME
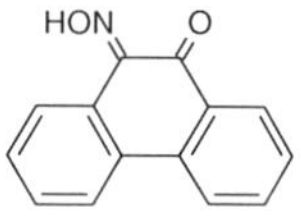
250 mg

S89,502-4
METHYL 3,5-BIS(HYDROXYIMINO)-1-(METHOXYCARBONYL)-4-OXO-1-CYCLOHEXANEACETATE
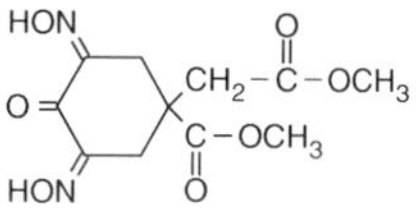
100 mg

S93,159-4
1-CYANO-3,3-DIMETHYL-1,2-BUTANEDIONE 1-OXIME
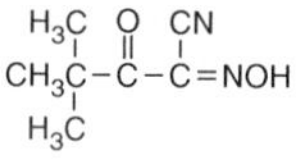
250 mg

S95,856-5
3-PHENYLACETYL-2-OXIMINOINDOLINE
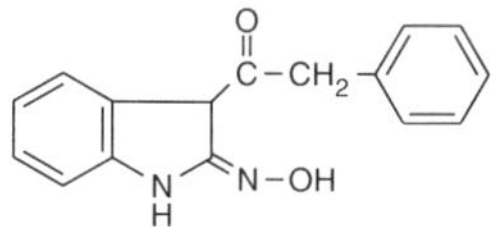
50 mg

S80,453-3
5-(HYDROXYIMINOAMINO)-6-(HYDROXYIMINOMETHYL)URACIL
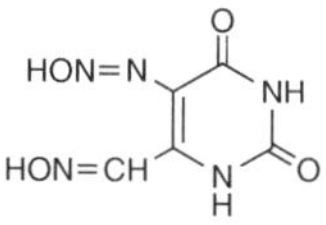
250 mg

S65,157-5
DIETHYL OXIMINOMALONATE
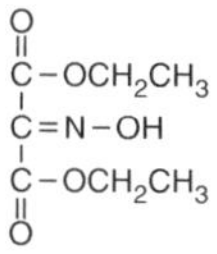
5 gm

S54,551-1
DIMETHYL (HYDROXYIMINO)MALONATE
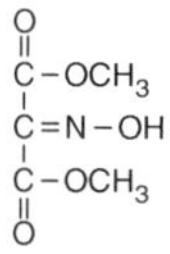
5 gm

S90,665-4
5-(HYDROXYIMINO)AZELAIC ACID
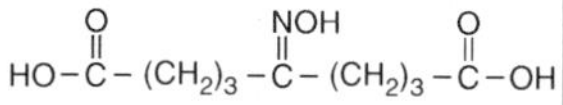
250 mg

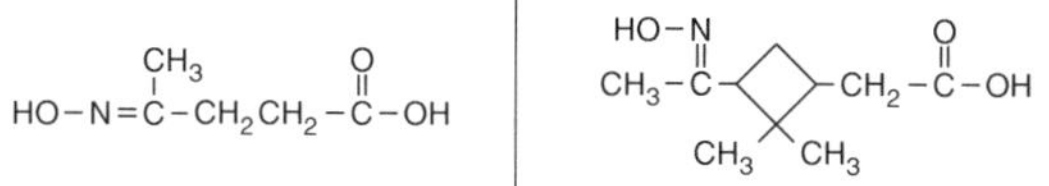

S35,420-1
4-OXIMINOVALERIC ACID
5 gm

S43,467-1
PINONIC ACID OXIME
5 gm

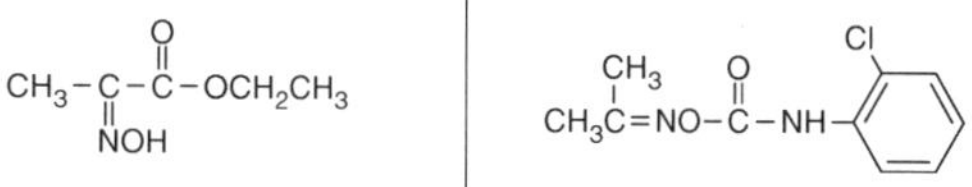

S94,297-9
ETHYL 2-HYDROXYIMINOPROPIONATE
250 mg

S59,644-2
O-[N-(2-CHLOROPHENYL)CARBAMOYL]-ACETONE OXIME
250 mg

S59,643-4
O-(4-BROMOPHENYLCARBAMOYL)-ACETONE OXIME
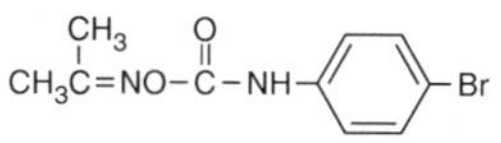
5 gm

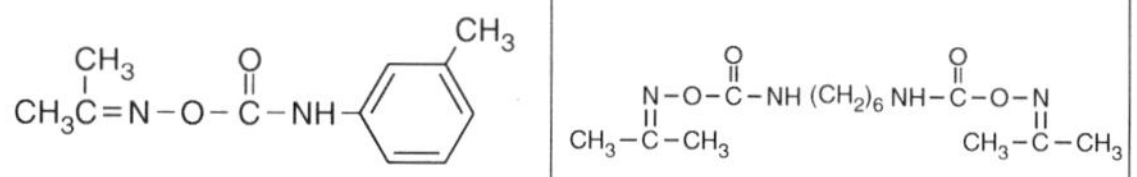

S60,638-3
O-[N-(M-TOLYL)CARBAMOYL]-ACETONE OXIME
5 gm

S63,840-4
N,N'-HEXAMETHYLENEBIS[O-(CARBAMOYL)-ACETONE OXIME]
5 gm

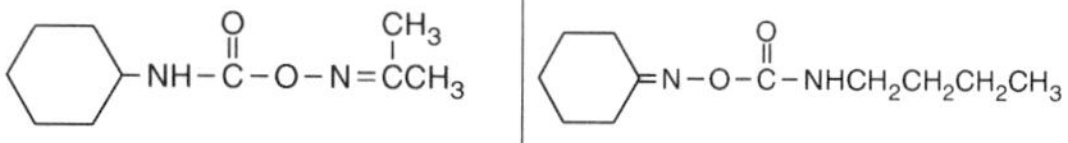

S62,496-9
O-[N-(CYCLOHEXYL)CARBAMOYL]-ACETONE OXIME
5 gm

S60,753-3
O-(BUTYLCARBAMOYL)-CYCLOHEXANONE OXIME
5 gm

S60,684-7
O-(N-PHENYLCARBAMOYL)-CYCLOHEXANONE OXIME
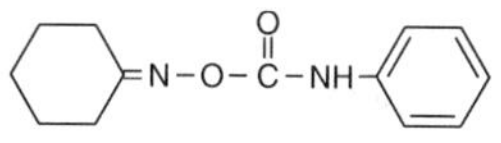
5 gm

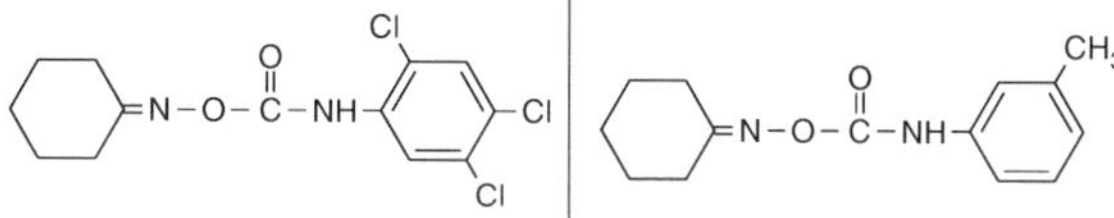

S60,733-9
O-[N-(2,4,5-TRICHLOROPHENYL)-CARBAMOYL]CYCLOHEXANONE OXIME
5 gm

S60,624-3
O-[N-(M-TOLYL)CARBAMOYL]-CYCLOHEXANONE OXIME
5 gm

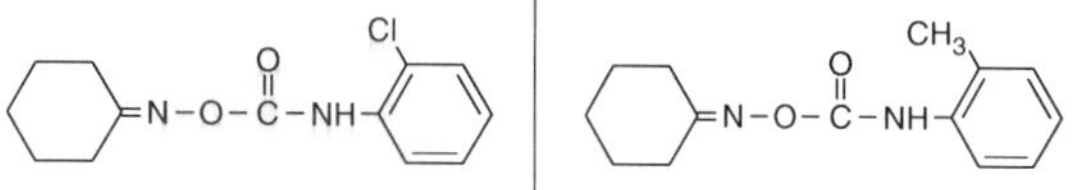

S63,386-0
O-[N-(2-CHLOROPHENYL)CARBAMOYL]-CYCLOHEXANONE OXIME
5 gm

S60,722-3
O-[N-(O-TOLYL)CARBAMOYL]-CYCLOHEXANONE OXIME
5 gm

S66,791-9
O-[N-(2-ETHYLPHENYL)CARBAMOYL]-CYCLOHEXANONE OXIME
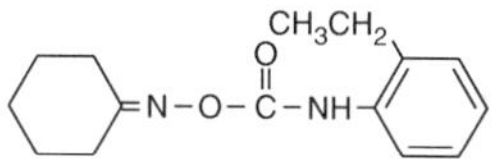
5 gm

S60,709-6
O-[N-(4-CHLOROPHENYL)CARBAMOYL]-CYCLOHEXANONE OXIME
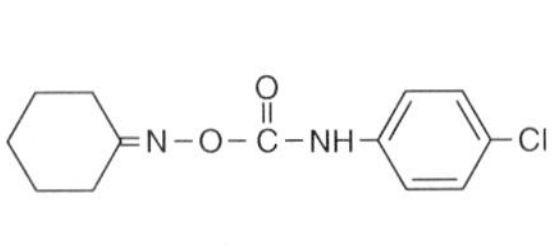
5 gm

S66,760-9
O-[N-(2-CHLORO-6-METHYLPHENYL)-CARBAMOYL]CYCLOHEXANONE OXIME
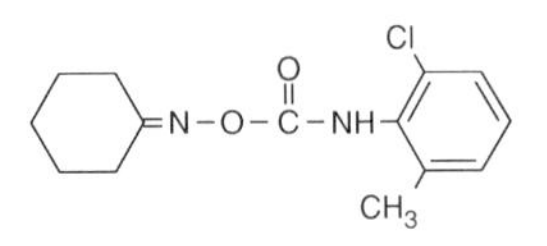
5 gm

S64,042-5
O-[N-(2,3-XYLYL)CARBAMOYL]-CYCLOHEXANONE OXIME
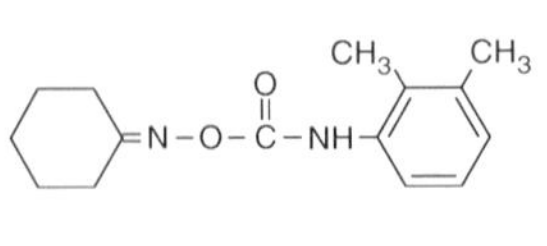
5 gm

S66,632-7
O-[N-(2,6-XYLYL)CARBAMOYL]-CYCLOHEXANONE OXIME
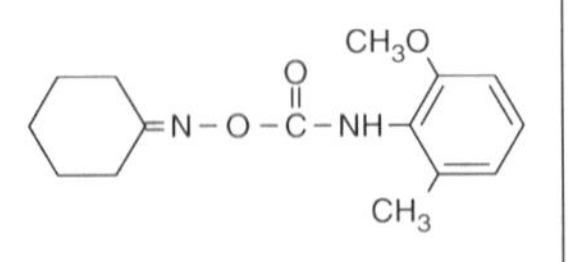
5 gm

S65,036-6
O-[N-(2-ETHYL-6-METHYLPHENYL)-CARBAMOYL]CYCLOHEXANONE OXIME
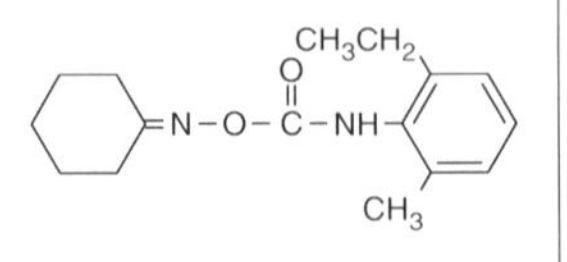
5 gm

S66,767-6
O-[N-(2,6-DIETHYLPHENYL)CARBAMOYL]-CYCLOHEXANONE OXIME
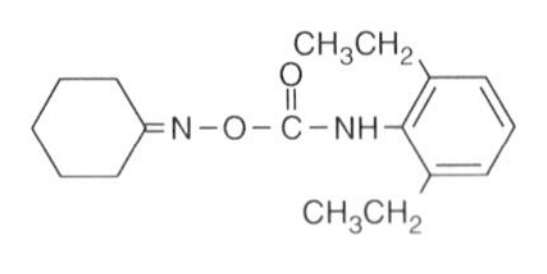
5 gm

S65,007-2
O-[N-(2,6-DIISOPROPYLPHENYL)-CARBAMOYL]CYCLOHEXANONE OXIME
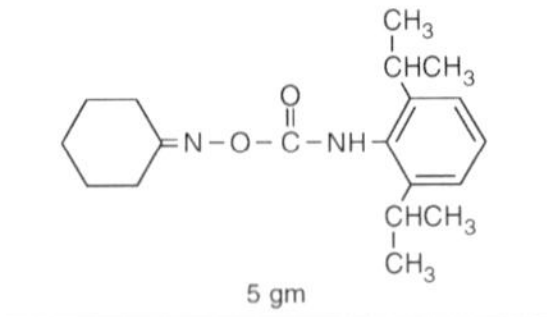
5 gm

S60,784-3
O-[N-(3,4-DICHLOROPHENYL)CARBAMOYL]-CYCLOHEXANONE OXIME
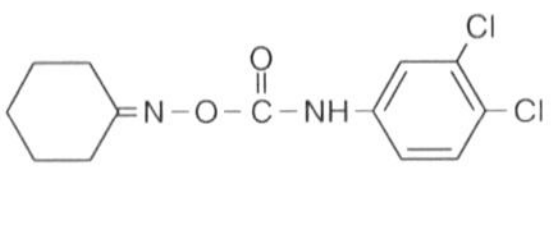
5 gm

S63,927-3
O-[N-(2,4-DIMETHYLPHENYL)CARBAMOYL]-CYCLOHEXANONE OXIME
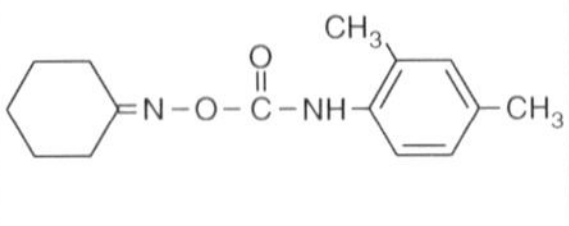
5 gm

S64,054-9
O-[N-(2,5-XYLYL)CARBAMOYL]-CYCLOHEXANONE OXIME
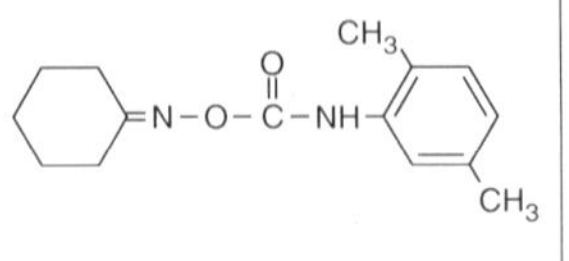
5 gm

S66,743-9
O-[N-(3,4-XYLYL)CARBAMOYL]-CYCLOHEXANONE OXIME
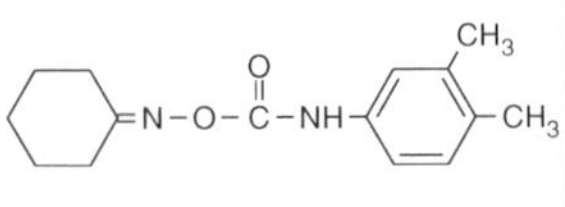
5 gm

S66,788-9
O-[N-(3,5-XYLYL)CARBAMOYL]-CYCLOHEXANONE OXIME
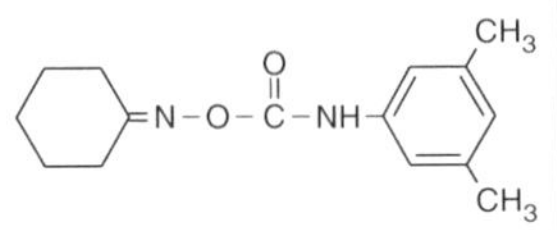
5 gm

S62,509-4
O-[N-(1-NAPHTHYL)CARBAMOYL]-CYCLOHEXANONE OXIME
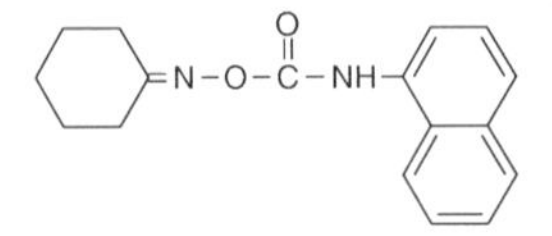
5 gm

S63,447-6
N,N'-(1,5-NAPHTHALENEDIYL)BIS[O-(CARBAMOYL)CYCLOHEXANONE OXIME]
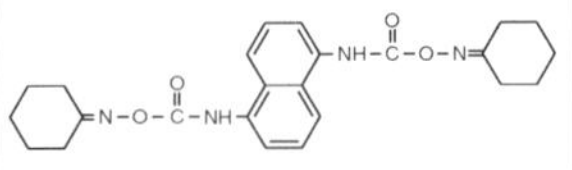
5 gm

S60,773-8
O-[N-(O-TOLYL)CARBAMOYL]-ACETONE OXIME
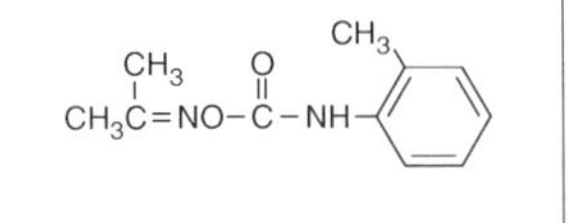
5 gm

S66,813-3
O-[N-(2-ETHYLPHENYL)CARBAMOYL]-ACETONE OXIME
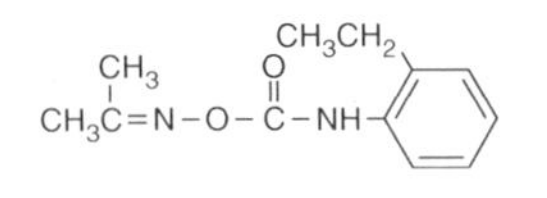
5 gm

S61,604-4
O-[N-(4-CHLOROPHENYL)CARBAMOYL]-ACETONE OXIME
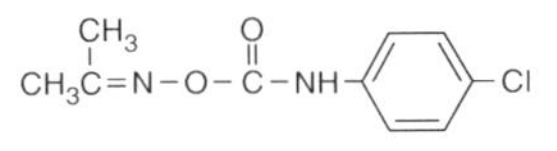
250 mg

S63,663-0
O-[N-(2,3-XYLYL)CARBAMOYL]-ACETONE OXIME
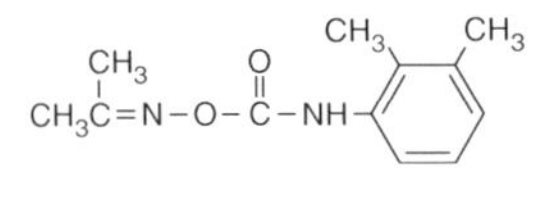
5 gm

S66,768-4
O-[N-(2,6-XYLYL)CARBAMOYL]-ACETONE OXIME
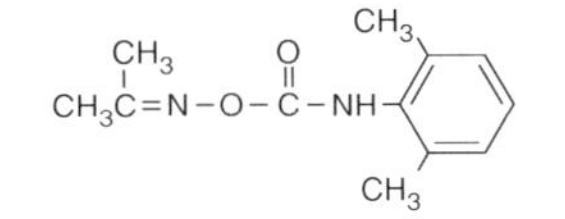
5 gm

S65,011-0
O-[N-(2-ETHYL-6-METHYLPHENYL)-CARBAMOYL]ACETONE OXIME
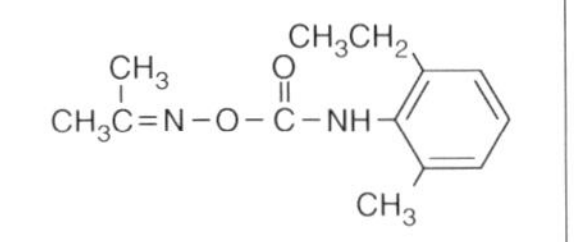
5 gm

S66,750-1
O-[N-(2,6-DIETHYLPHENYL)CARBAMOYL]-ACETONE OXIME
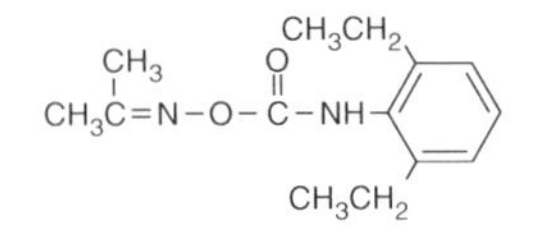
5 gm

S65,037-4
O-[N-(2,6-DIISOPROPYLPHENYL)-CARBAMOYL]ACETONE OXIME
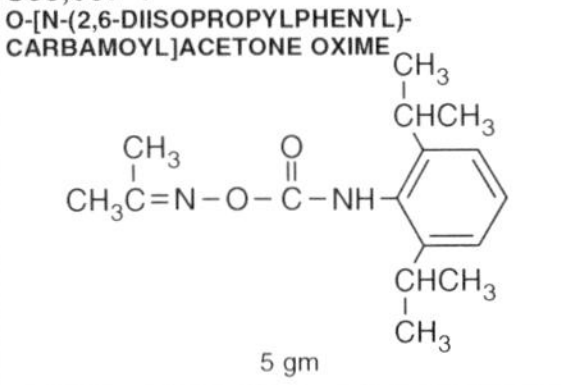
5 gm

S60,703-7
O-[N-(3,4-DICHLOROPHENYL)CARBAMOYL]-ACETONE OXIME
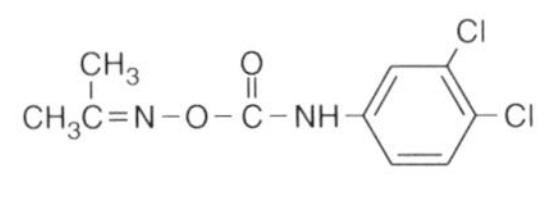
5 gm

S60,343-0
O-[N-(3-CHLORO-4-METHYLPHENYL)-CARBAMOYL]ACETONE OXIME
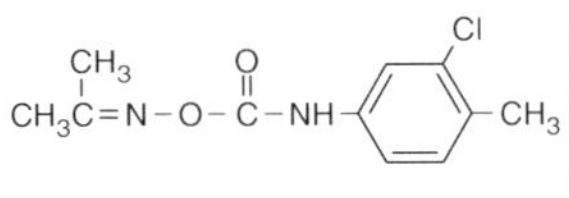
5 gm

S60,706-1
O-[N-(2,4-XYLYL)CARBAMOYL]-ACETONE OXIME
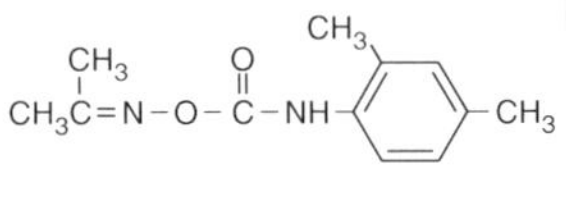
5 gm

S64,043-3
O-[N-(2,5-XYLYL)CARBAMOYL]-ACETONE OXIME
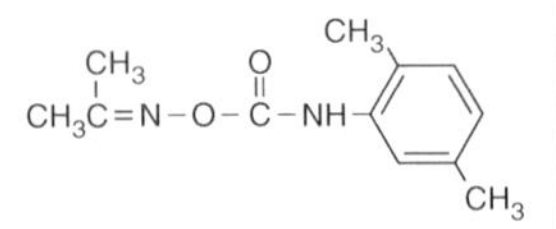
5 gm

S66,758-7
O-[N-(3,4-XYLYL)CARBAMOYL]-ACETONE OXIME
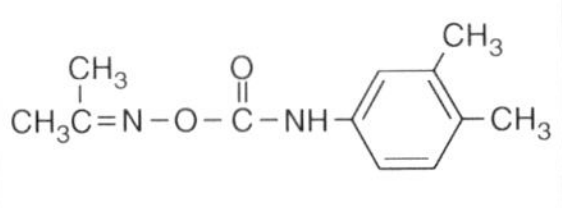
5 gm

S64,040-9
O-[N-(3,5-XYLYL)CARBAMOYL]-ACETONE OXIME
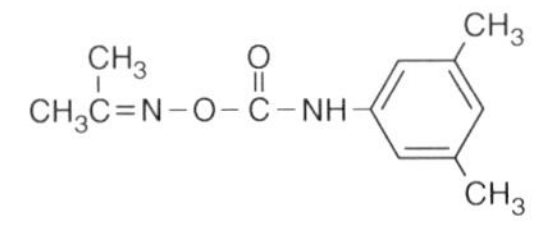
5 gm

S63,441-7
O-[N-(2,4,5-TRICHLOROPHENYL)-CARBAMOYL]ACETONE OXIME
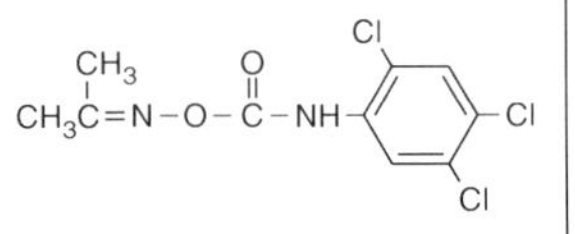
5 gm

S63,950-8
O-[N-(1-NAPHTHYL)CARBAMOYL]-ACETONE OXIME
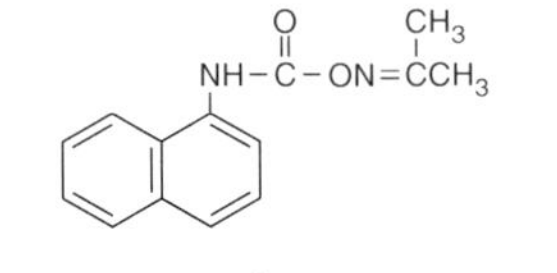
5 gm

S63,469-7
N,N'-(1,5-NAPHTHALENEDIYL)BIS[O-(CARBAMOYL)ACETONE OXIME]

S44,774-9
NORCAMPHOR OXIME
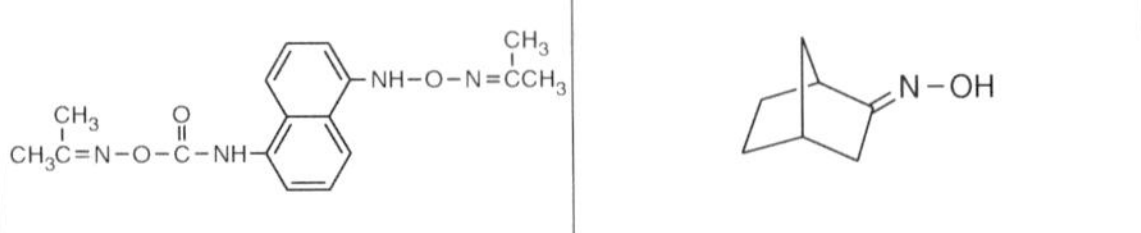
5 gm | 5 gm

S41,794-7
ETHYL 2-PYRIDYLGLYOXYLATE OXIME
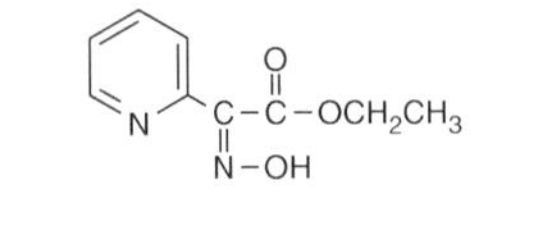
5 gm

S76,245-8
2,3,5,6,10,10A-HEXAHYDRO-1H,4H-ACENAPHTHO(1,8A-C)PYRROLE-1,3,10-TRIONE 10-OXIME
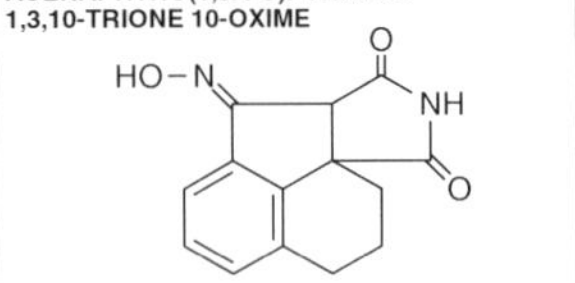
250 mg

S76,330-6
1-ACETYL-3-(HYDROXYIMINO)-N-METHYL-2-INDOLINECARBOXAMIDE
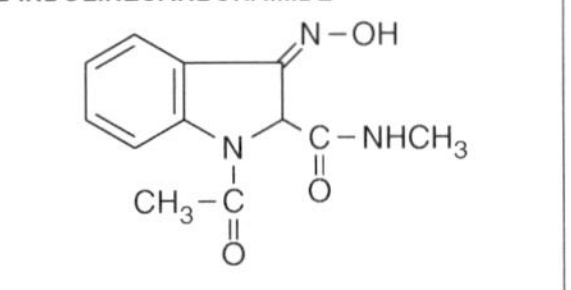
250 mg

S57,219-5
ALPHA-HYDROXYIMINO-P-ACETOTOLUIDIDE

S67,645-4
2-HYDROXIMINO-2',5'-ACETOXYLIDIDE
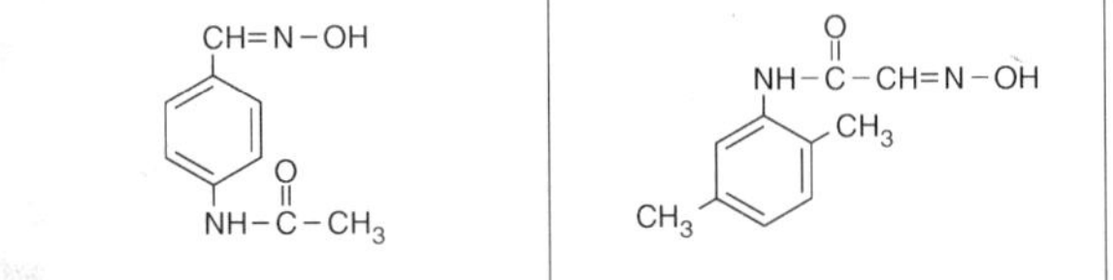
250 mg | 250 mg

S92,616-7
2-(HYDROXYIMINO)-3',4'-DIMETHYLACETANILIDE

S80,329-4
5-BROMOISATIN 3-OXIME
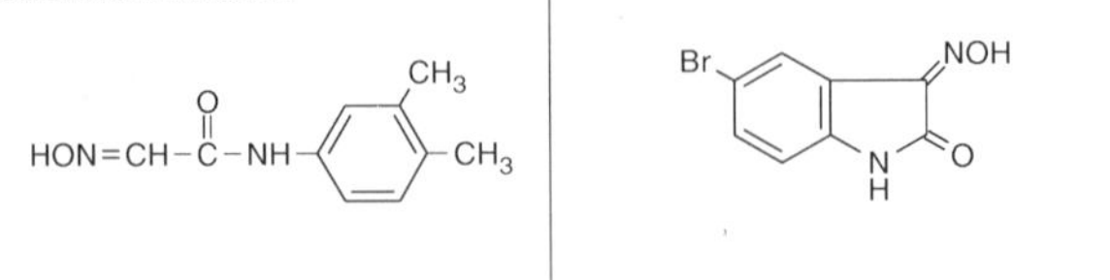
250 mg | 250 mg

S95,808-5
5,8,8-TRIMETHYL-6-AZABICYCLO[3.2.1]-OCTANE-4,7-DIONE 4-OXIME
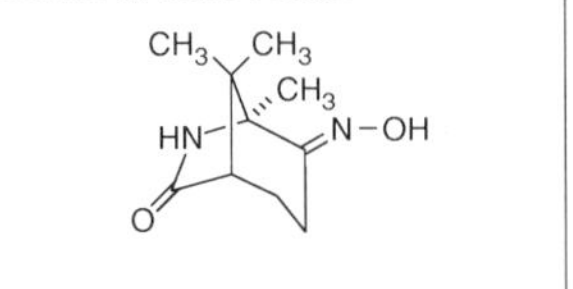
50 mg

S59,585-3
O-[N-(3-CHLOROPHENYL)CARBAMOYL]ACETONE OXIME

S60,956-0
O-[N-(CYCLOHEXYL)CARBAMOYL]-CYCLOHEXANONE OXIME
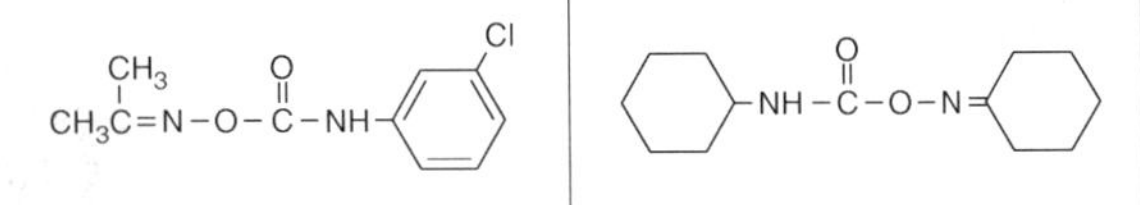
250 mg | 5 gm

S60,978-1
O-[N-(3-CHLOROPHENYL)CARBAMOYL]-CYCLOHEXANONE OXIME

S64,522-2
4,4'-METHYLENEBIS[O-(CARBAMOYL)-CYCLOHEXANONE OXIME]
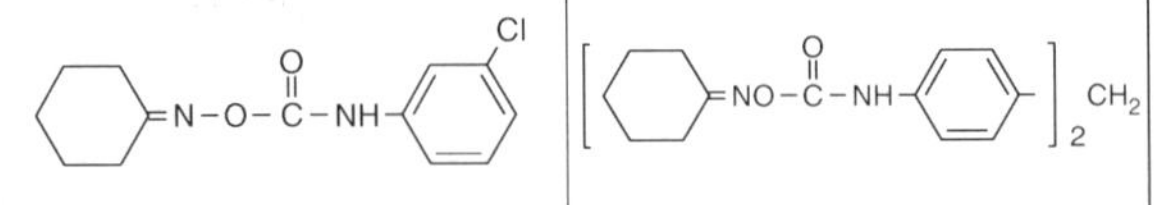
5 gm | 5 gm

S60,971-4
O-(4-BROMOPHENYLCARBAMOYL)-CYCLOHEXANONE OXIME
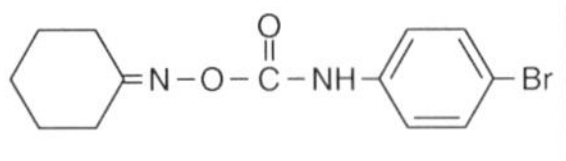
5 gm

S61,306-1
O-[N-(P-TOLYL)CARBAMOYL]-CYCLOHEXANONE OXIME

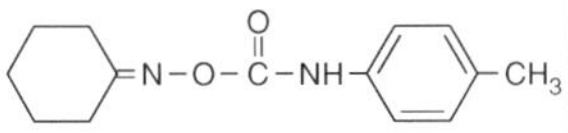

5 gm

S61,157-3
O-[N-(2,6-DICHLOROPHENYL)CARBAMOYL]-CYCLOHEXANONE OXIME

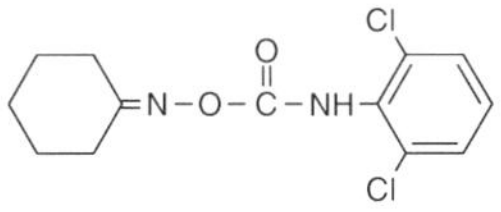

5 gm

S61,165-4
O-[N-(2,3-DICHLOROPHENYL)CARBAMOYL]-CYCLOHEXANONE OXIME

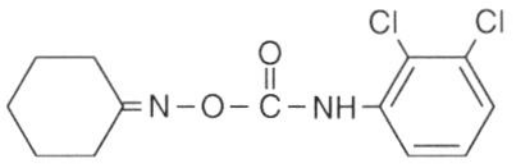

5 gm

S60,944-7
O-[N-(2,5-DICHLOROPHENYL)CARBAMOYL]-CYCLOHEXANONE OXIME

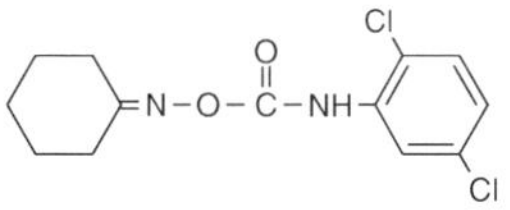

5 gm

S61,279-0
O-[N-(2,4-DICHLOROPHENYL)CARBAMOYL]-CYCLOHEXANONE OXIME

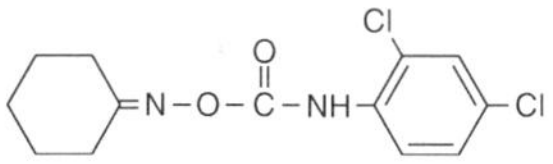

5 gm

S60,998-6
O-[N-(5-CHLORO-2-METHYLPHENYL)-CARBAMOYL]CYCLOHEXANONE OXIME

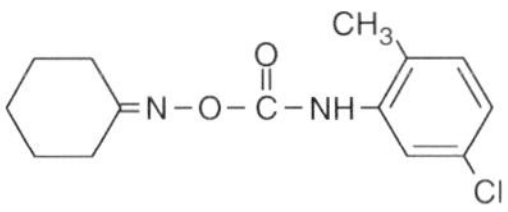

5 gm

S61,020-8
O-[N-(3-CHLORO-4-METHYLPHENYL)-CARBAMOYL]CYCLOHEXANONE OXIME

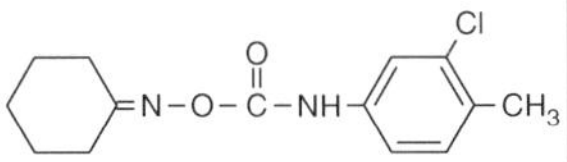

5 gm

S76,803-0
9-FLUORENYLIDENEAMINO N-CYCLOHEXYLCARBAMATE

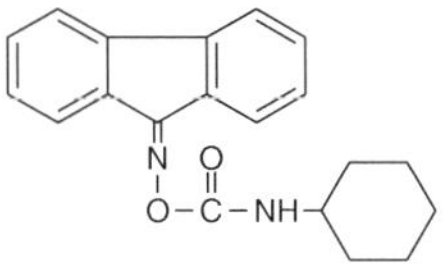

250 mg

S58,030-9
ACETONE O-(PHENYLCARBAMOYL)OXIME

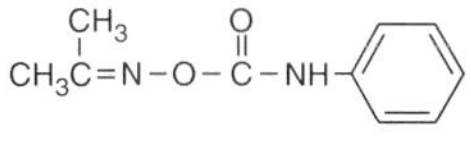

5 gm

S60,935-8
O-[N-(4-FLUOROPHENYL)CARBAMOYL]-ACETONE OXIME

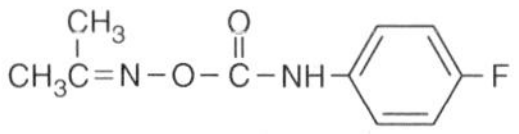

5 gm

S61,298-7
O-[N-(P-TOLYL)CARBAMOYL]-ACETONE OXIME

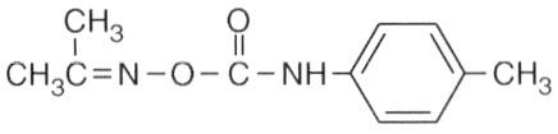

5 gm

S61,138-7
O-[N-(2,3-DICHLOROPHENYL)CARBAMOYL]-ACETONE OXIME

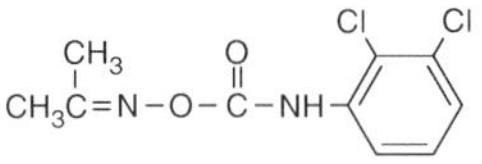

5 gm

S61,188-3
O-[N-(2,5-DICHLOROPHENYL)CARBAMOYL]-ACETONE OXIME

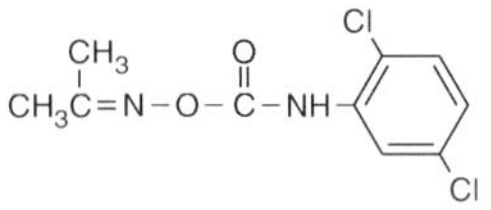

5 gm

S61,331-2
O-[N-(2,4-DICHLOROPHENYL)CARBAMOYL]-ACETONE OXIME

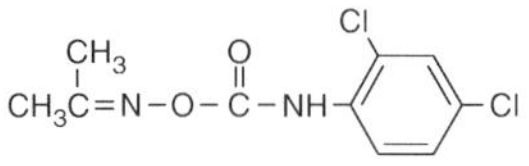

5 gm

S60,962-5
O-[N-(4-CHLORO-2-METHYLPHENYL)-CARBAMOYL]ACETONE OXIME

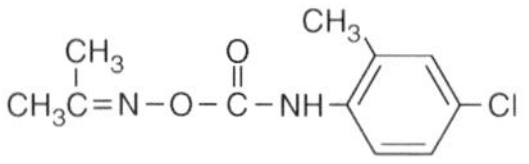

5 gm

S61,067-4
O-[N-(5-CHLORO-2-METHYLPHENYL)-CARBAMOYL]ACETONE OXIME

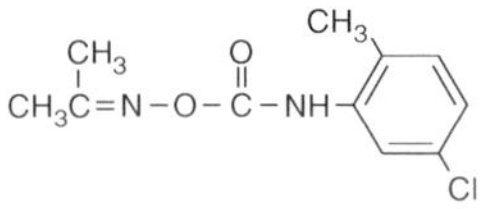

5 gm

S64,526-5
N,N'-(METHYLENEDI-4,1-PHENYLENE)-BIS[O-(CARBAMOYL)ACETONE OXIME]

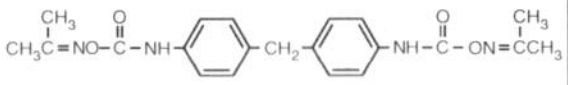

5 gm

S52,403-4
2,3-DICHLORO-4-HYDROXYIMINO-ALPHA-PHENYL-2,5-CYCLOHEXADIENE-DELTA1,ALPHA-ACETONITRILE

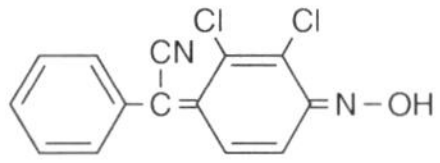

5 gm

S53,797-7
3-BROMO-6-CHLORO-4-HYDROXYIMINO-ALPHA-(P-TOLYL)-2,5-CYCLOHEXADIENEACETONITRILE

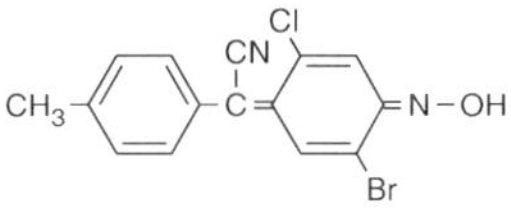

5 gm

S53,402-1
4-CHLORO-ALPHA-(3-CHLORO-4-HYDROXYIMINOCYCLOHEXADIEN-1-YLIDENE)PHENYLACETONITRILE

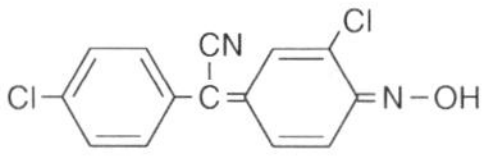

5 gm

S53,400-5
ALPHA-(3-BROMO-4-HYDROXYIMINO-2,5-CYCLOHEXADIEN-1-YLIDENE)-4-CHLOROPHENYLACETONITRILE

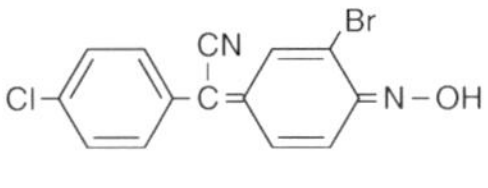

5 gm

S53,401-3
ALPHA-(5-BROMO-2-CHLORO-4-HYDROXY-IMINOCYCLOHEXADIENYLIDENE)-4-CHLOROPHENYLACETONITRILE

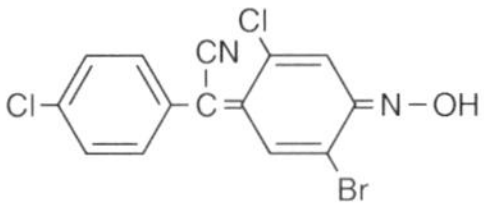

250 mg

S53,346-7
ALPHA-(2,5-DICHLORO-4-HYDROXYIMINOCYCLOHEXADIEN-1-YLIDENE)-1-NAPHTHALENEACETONITRILE

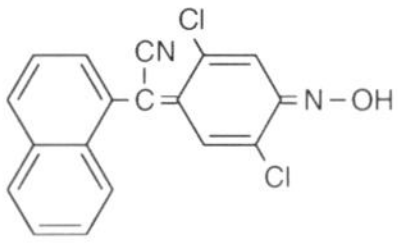

5 gm

S53,329-7
ALPHA-(5-BROMO-2-CHLORO-4-HYDROXY-IMINOCYCLOHEXADIENYLIDENE)-1-NAPHTHALENEACETONITRILE

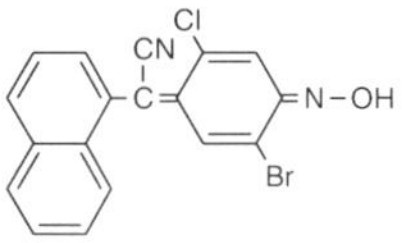

5 gm

S36,040-6
DICYCLOPROPYL KETOXIME

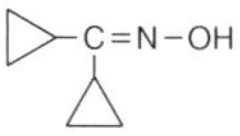

5 gm

S95,535-3
4',5',6',7'-TETRAHYDROSPIRO-(CYCLOHEXANE-1,1'-INDAN)-3'-OXIME

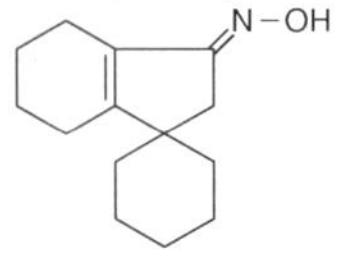

50 mg

S95,790-9
1,4,4A,4B,5,8,8A,9A-OCTAHYDRO-1,4:5,8-DIMETHANOFLUOREN-9-ONE OXIME

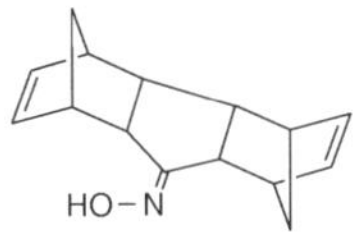

50 mg

S90,730-8
5-ACETYLINDAN OXIME

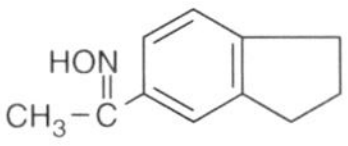

250 mg

S95,755-0
5,10-DIHYDROINDENO(2,1-A)INDENE-5,10-DIONE DIOXIME

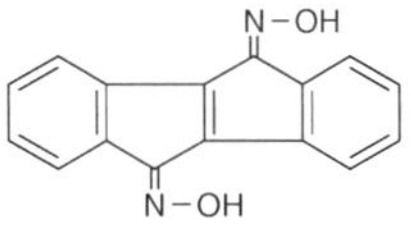

50 mg

S83,647-8
3-(2-FURYL)ACROLEIN OXIME

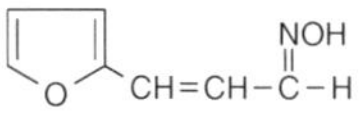

250 mg

S55,184-8
2-THIOPHENECARBOXALDEHYDE OXIME

250 mg

S75,591-5
2-ACETYLTHIOPHENE OXIME

250 mg

S75,573-7
2-ACETYL-5-METHYLTHIOPHENE OXIME

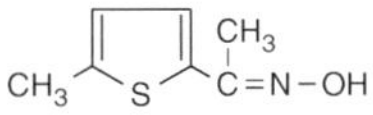

250 mg

S52,618-5
PIPERONAL OXIME

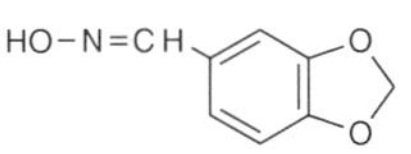

5 gm

S64,835-3
O-ETHYLCYCLOHEXANONE OXIME

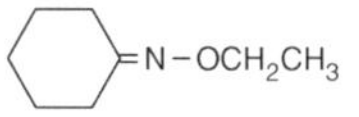

5 gm

S79,972-6
4-PROPYLCYCLOHEXANONE OXIME

250 mg

S79,268-3
3,5-DIMETHYL-2-CYCLOHEXEN-1-ONE OXIME

250 mg

S45,918-6
PERILLARTINE

250 mg

S90,839-8
6-ACETYLTETRALIN OXIME

250 mg

S65,925-8
THIOCHROMAN-4-ONE OXIME

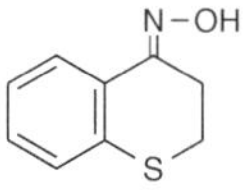

250 mg

S89,314-5
BENZALDEHYDE OXIME HYDROCHLORIDE

250 mg

S77,817-6
BENZAMIDE OXIME

250 mg

S56,620-9
CINNAMALDEHYDE OXIME

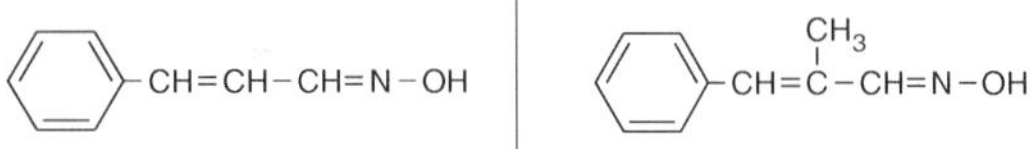

5 gm

S62,685-6
ALPHA-METHYLCINNAMALDOXIME

5 gm

S81,988-3
BENZOPHENONE O-METHYLOXIME

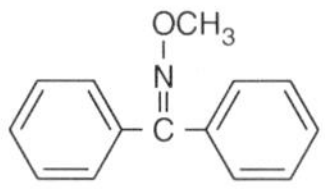

250 mg

S59,950-6
3-PHENYL-1-INDANONE OXIME

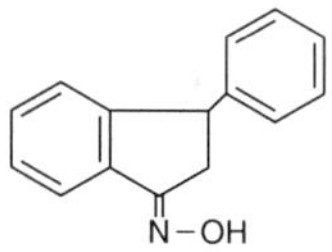

250 mg

S55,053-1
2-FLUOROBENZALDEHYDE OXIME

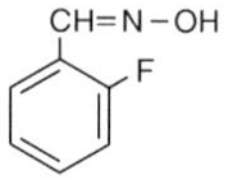

5 gm

S35,938-6
2-CHLOROBENZALDEHYDE OXIME (MIXTURE OF SYN AND ANTI)

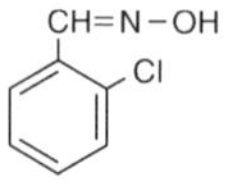

5 gm

S52,059-4
4-CHLOROBENZALDEHYDE OXIME

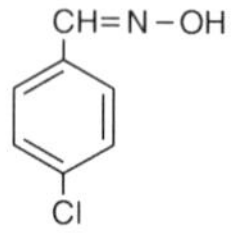

5 gm

S90,715-4
SYN-P-TOLUALDEHYDE OXIME

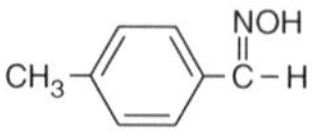

250 mg

S41,125-6
P-TOLUAMIDOXIME

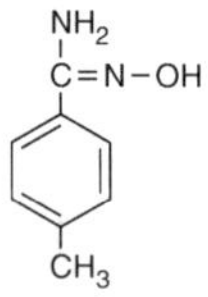

5 gm

S68,280-2
4'-METHYLACETOPHENONE OXIME

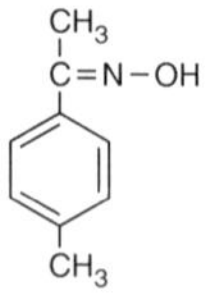

5 gm

S43,102-8
4-BIPHENYLCARBOXALDEHYDE OXIME

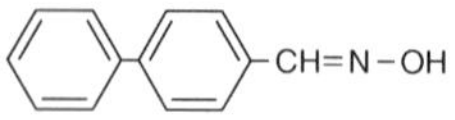

5 gm

S53,139-1
2,6-DICHLOROBENZALDEHYDE OXIME

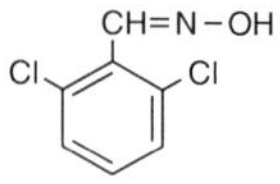

5 gm

S63,070-5
3,4-DICHLOROBENZAMIDE OXIME

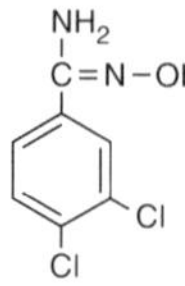

250 mg

S92,201-3
CIS-1-DECALONE OXIME

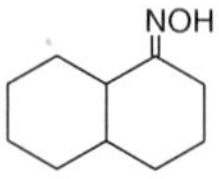

250 mg

S56,753-1
2-INDANONE OXIME

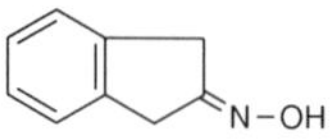

5 gm

S64,006-9
CHLOROPROPIONYLPHENANTHRENE OXIME (MIXED ISOMERS)

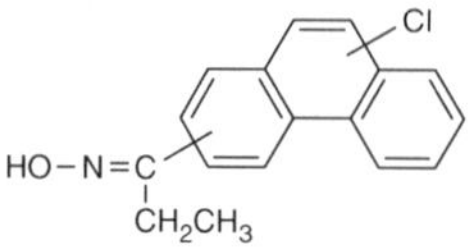

250 mg

S70,249-8
ALPHA-TETRALONE OXIME

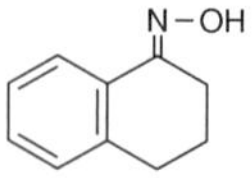

5 gm

S67,122-3
7-METHYL-1-TETRALONE OXIME

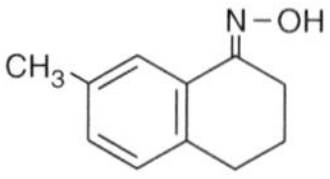

250 mg

S37,500-4
9-FLUORENONE OXIME

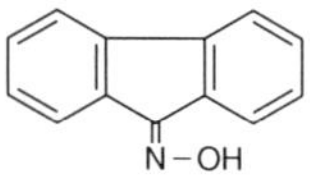

5 gm

S74,587-1
2,4,7-TRICHLORO-9-FLUORENONE OXIME

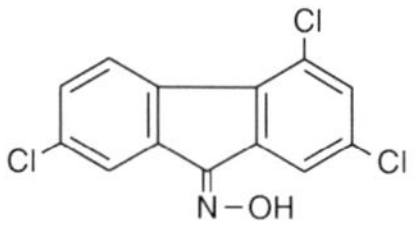

250 mg

S50,156-5
METHYL 3-PHENANTHRYL KETONE OXIME

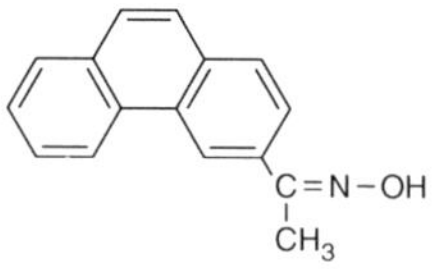

5 gm

S77,065-5
2-ACETYLPHENANTHRENE OXIME

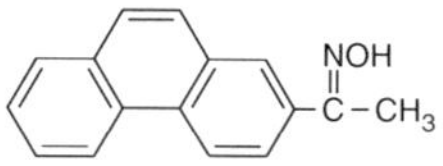

250 mg

S42,909-0
5H-DIBENZO(A,D)CYCLOHEPTEN-5-ONE OXIME

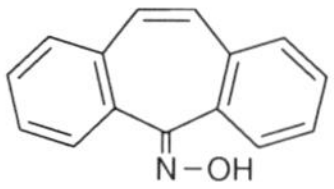

250 mg

S45,636-5
10,11-DIHYDRO-5H-DIBENZO(A,D)-CYCLOHEPTEN-5-ONE OXIME

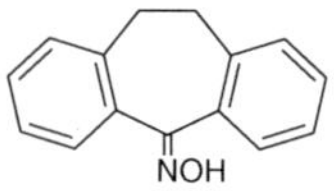

5 gm

S76,220-2
3,4-DIHYDRO-1-BENZOTHIEPIN-5(2H)-ONE OXIME

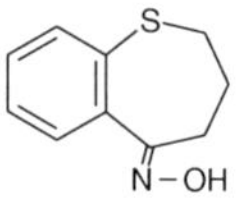

250 mg

S88,124-4
1-IODO-1-HEXYNE

$CH_3CH_2CH_2CH_2C\equiv C-I$

250 mg

S89,369-2
HEXACHLORO-1,5-HEXADIEN-3-YNE

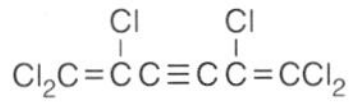

250 mg

S76,477-9
1,6-HEPTADIYNE

$HC\equiv CCH_2CH_2CH_2C\equiv CH$

5 gm

S94,320-7
5-METHYL-1-HEXYNE

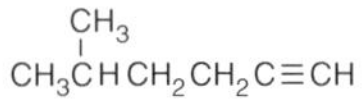

5 gm

S57,283-7
2-OCTYNE

$CH_3C\equiv CCH_2(CH_2)_3CH_3$

5 gm

S57,979-3
3-OCTYNE

$CH_3CH_2C\equiv CCH_2CH_2CH_2CH_3$

5 gm

S89,423-0
1,1,2,3,6,7,8,8-OCTACHLORO-1,7-OCTADIEN-4-YNE

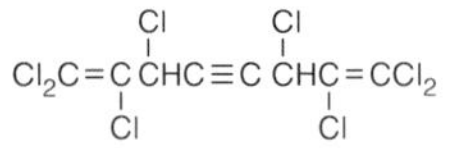

250 mg

S50,919-1
5-DECYNE

$CH_3CH_2CH_2CH_2C\equiv CCH_2CH_2CH_2CH_3$

5 gm

S88,444-8
5,6-DIMETHYL-5-DECEN-1,9-DIYNE

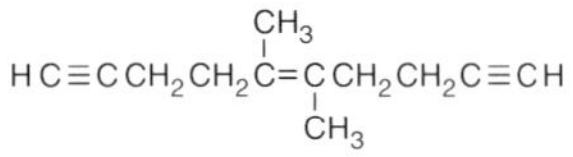

250 mg

S87,165-6
1,3,7,9-DECATETRAEN-5-YNE-1,10-DICARBOXALDEHYDE DIOXIME

250 mg

S66,820-6
N-BENZYL-N-(2-CHLOROETHYL)-PROPARGYLAMINE HYDROCHLORIDE

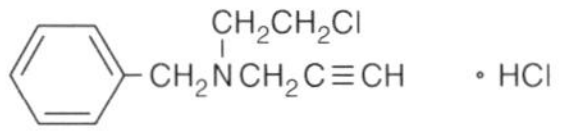

250 mg

S89,313-7
3-CHLORO-N-(2-PROPYN-1-YL)-PROPIONAMIDE

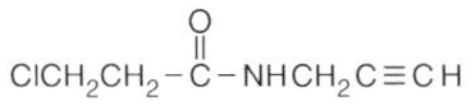

250 mg

S88,164-3
5-BROMO-1-PHENYL-1-PENTYNE

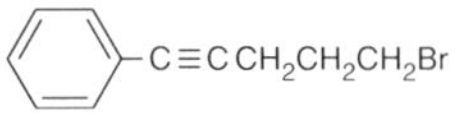

250 mg

S52,965-6
2,2,2-TRICHLOROETHYL 3-[N-(2-PROPYNYL)BENZYLAMINO]PROPIONATE

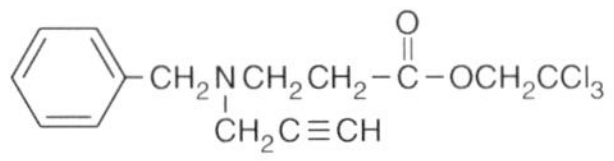

250 mg

S83,010-0
1,2-BIS(PENTAFLUOROPHENYL)ACETYLENE

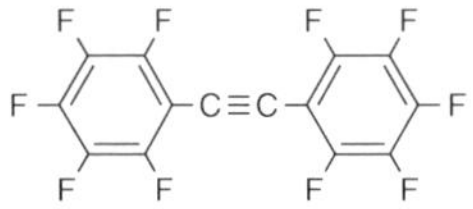

250 mg

S83,008-9
1,2-BIS(PENTACHLOROPHENYL)ACETYLENE

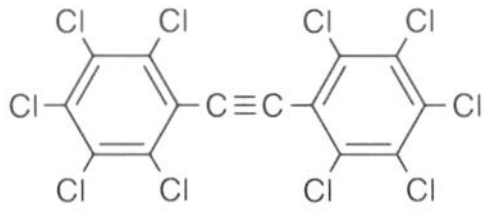

250 mg

S68,276-4
3-(3-CHLORO-4-PROPOXYPHENYL)-4,4-DIMETHYL-1-PENTYN-3-OL

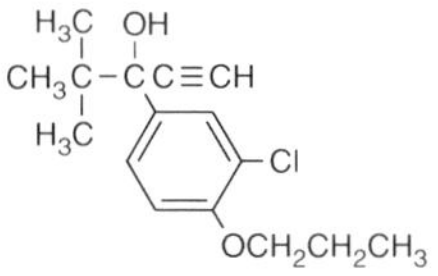

5 gm

S75,015-8
3-BUTYN-1-YL N-(2-CHLORO-4-NITROPHENYL)CARBAMATE

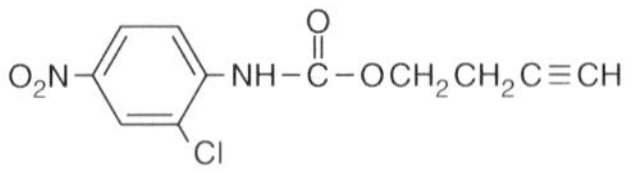

5 gm

S75,912-0
N-(2,4-DICHLOROBENZYL)-PROPARGYLAMINE

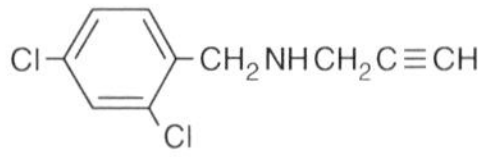

250 mg

S46,236-5
1-(4-CHLOROPHENYL)-3-(4-METHYL-1-PIPERAZINYL)-2-PROPYN-1-ONE

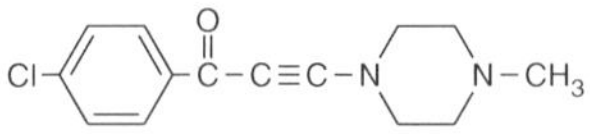

250 mg

S39,931-0
2-BUTYNE-1,4-DIOL BIS(3-CHLOROCARBANILATE)

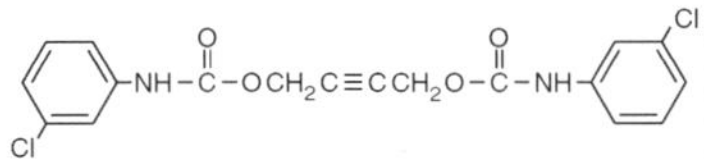

250 mg

S39,878-0
2-BUTYNE-1,4-DIOL BIS(2-CHLOROCARBANILATE)

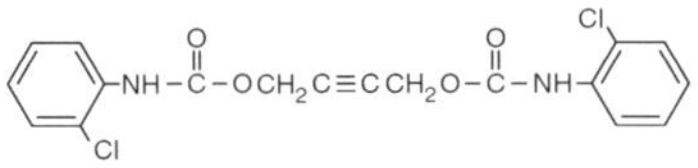

5 gm

S76,835-9
2-PROPYNYL N-(2,5-DICHLOROPHENYL)-CARBAMATE

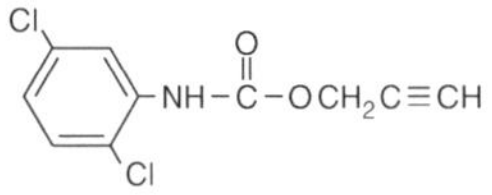

250 mg

S39,930-2
2-BUTYNE-1,4-DIOL BIS(3,4-DICHLOROCARBANILATE)

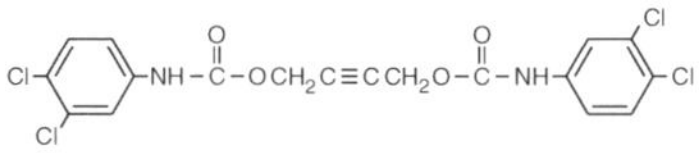

5 gm

S39,880-2
2-BUTYNE-1,4-DIOL BIS(2-METHYLCARBANILATE)

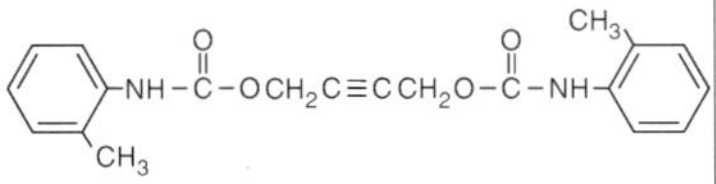

5 gm

S50,320-7
2-BUTYNE-1,4-DIOL DICARBANILATE

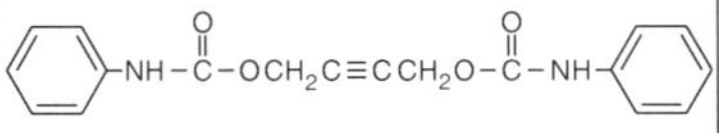

5 gm

S88,130-9
4-BROMO-2-METHYL-3-BUTYN-2-OL

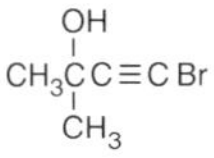

250 mg

S43,013-7
1-HEXYN-3-OL

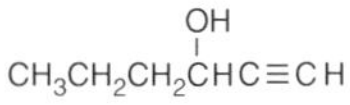

5 gm

S88,495-2
2-OCTYN-1-OL

$CH_3(CH_2)_3CH_2C\equiv CCH_2OH$

250 mg

S94,321-5
3-HYDROXY-3-METHYL-1-TRIMETHYLSILYL-1-BUTYNE

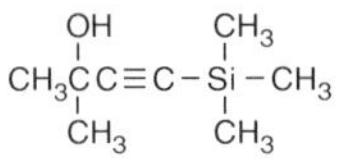

250 mg

S34,951-8
4-ETHYL-1-OCTYN-3-OL

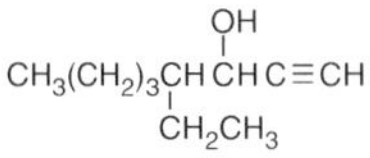

5 gm

S34,630-6
3,8-DIHYDROXY-3,8-DIMETHYL-4,6-DECADIYNE

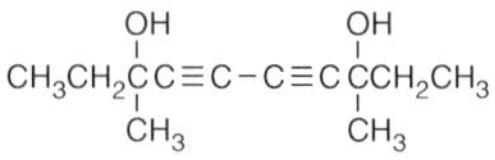

5 gm

S34,685-3
4,7-DIMETHYL-5-DECYNE-4,7-DIOL

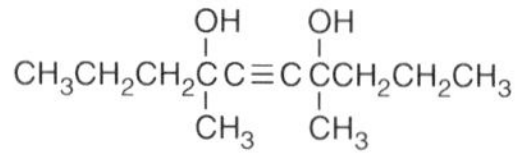

5 gm

S70,462-8
2,4-HEXADIYNE-1,6-DIOL

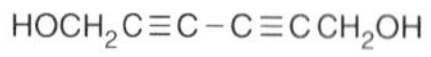

250 mg

S86,067-0
7,7-DIETHOXY-5-HEPTYNE-2,4-DIOL

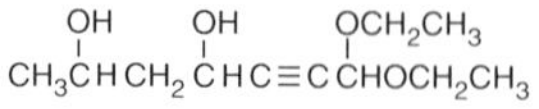

250 mg

S89,424-9
1,1,2,7,8,8-HEXACHLORO-1,7-OCTADIEN-4-YNE-3,6-DIOL

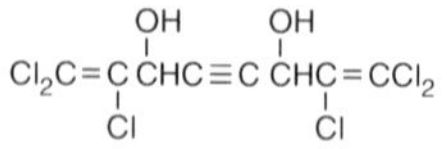

250 mg

S81,264-1
1,1,4,4-TETRAUNDECYL-2-BUTYNE-1,4-DIOL

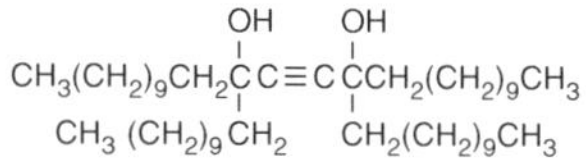

250 mg

S95,447-0
ALPHA,ALPHA'-ETHYNYLENEBIS-(CINNAMYL ALCOHOL)

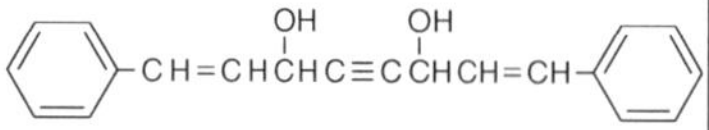

50 mg

S94,219-7
1-(1-HYDROXY-2,2,6-TRIMETHYLCYCLOHEXYL)-1-BUTYN-3-OL

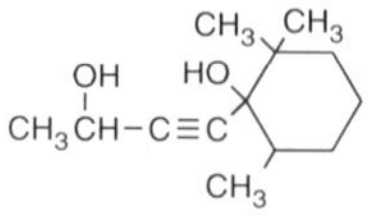

250 mg

S70,471-7
2-BROMO-1-PHENYL-1-HEXEN-4-YNE-3,6-DIOL

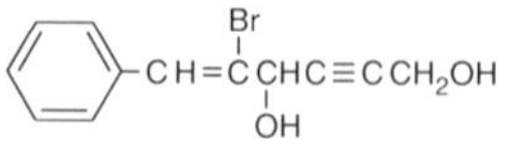

250 mg

S93,547-6
1,2-BIS(4-METHOXYPHENYL)-3-BUTYN-2-OL

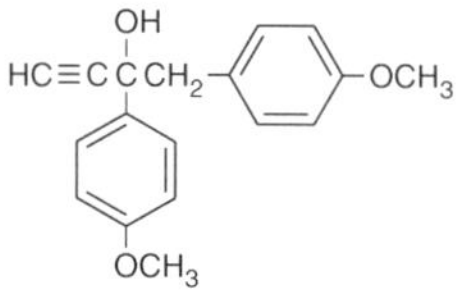

250 mg

S79,295-0
4,4-DIETHOXY-2-BUTYN-1-OL

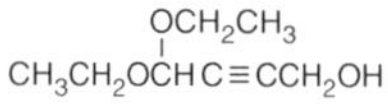

250 mg

S72,218-9
2-(3-HEXYN-1-YLOXY)ETHANOL

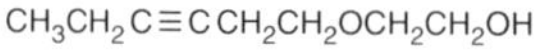

5 gm

S68,387-6
4,4-DIMETHYL-3-(4-PHENOXYPHENYL)-1-PENTYN-3-OL

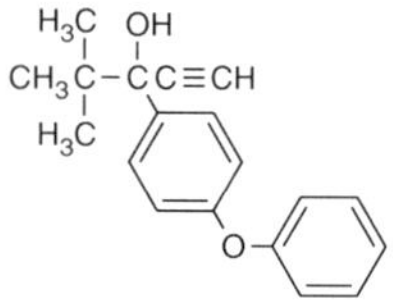

5 gm

S75,855-8
2,2-DIMETHYL-3-(6-METHOXY-2-NAPHTHYL)-4-HEXYN-3-OL

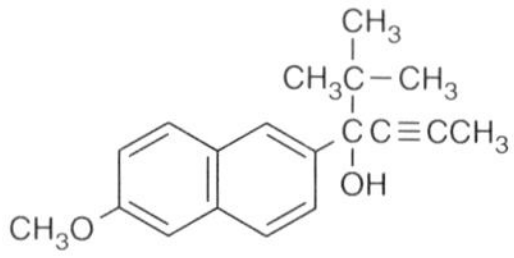

250 mg

S44,770-6
1,1,4,4-TETRAPHENYL-2-BUTYNE-1,4-DIOL

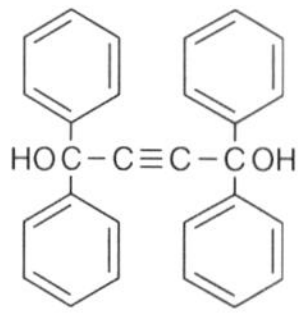

5 gm

S79,332-9
5-HYDROXY-2-(1-OCTYNYL)-3-OXO-1-CYCLOPENTENE-1-HEPTANOIC ACID

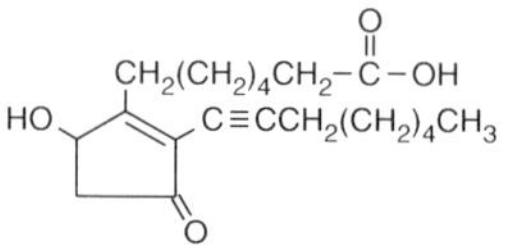

250 mg

S95,002-5
METHYL 4-HYDROXYTETROLATE

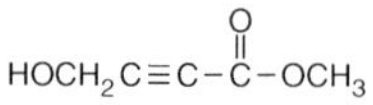

50 mg

S39,782-2
4-[N-(2-HYDROXYETHYL)-N-ISOPROPYLAMINO]-2-BUTYN-1-OL

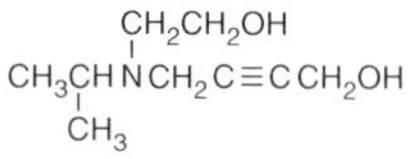

5 gm

S39,783-0
2-[N-ISOPROPYL-N-(4-METHYLPENT-4-EN-2-YN-1-YL)AMINO]ETHANOL

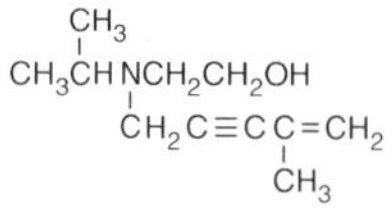

5 gm

S39,785-7
5-[N-BUTYL-N-(2-HYDROXYETHYL)-AMINO]-2,6-DIMETHYL-3-HEPTYN-2-OL

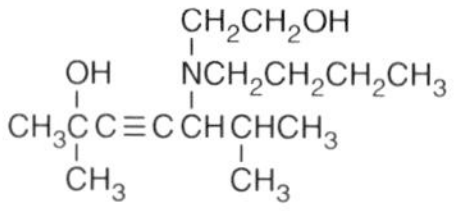

5 gm

S70,476-8
TRANS-2-CHLORO-1-ETHYNYLCYCLOHEXANOL

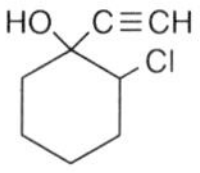

250 mg

S66,293-3
1-(1-HYDROXYCYCLOHEXYL)PROPYN-3-OL

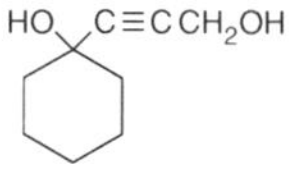

5 gm

S34,968-2
1,1'-ETHYNYLENE-BISCYCLOHEXANOL

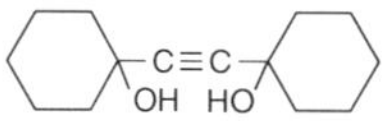

5 gm

S57,688-3
1,1'-BUTADIYNYLENEDICYCLOHEXANOL

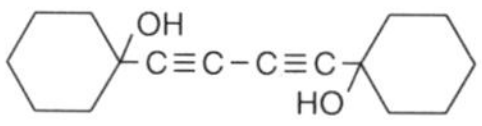

5 gm

S68,108-3
4-(4-CYCLOHEXYLPHENYL)-3-BUTYN-1-OL

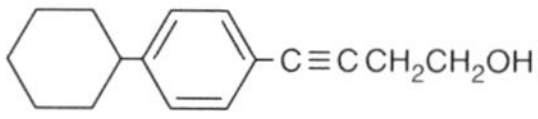

5 gm

S68,258-6
3-(4-CYCLOHEXYLPHENYL)-4,4-DIMETHYL-1-PENTYN-3-OL

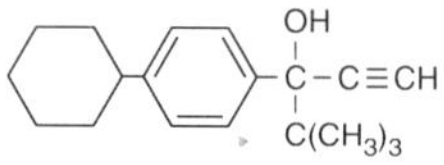

250 mg

S46,089-3
1-PHENYLPROPARGYL ALCOHOL

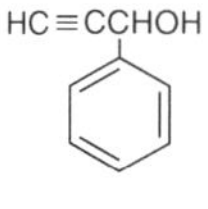

5 gm

S93,341-4
2-METHYL-4-PHENYL-3-BUTYN-2-OL

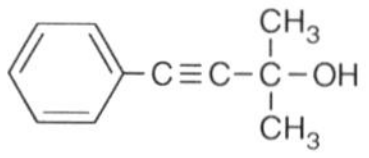

250 mg

S93,546-8
1-PHENYL-3-PENTYN-2-OL

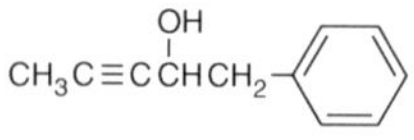

250 mg

S93,797-5
1-PHENYL-2,4-HEXADIYN-1-OL

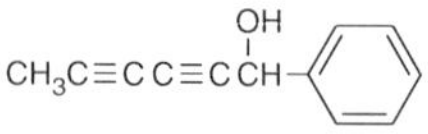

250 mg

S77,090-6
6-PHENYL-5-HEXYN-3-OL

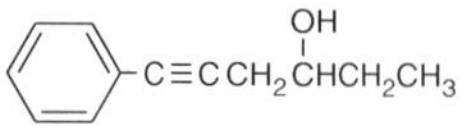

250 mg

S77,089-2
7-PHENYL-6-HEPTYN-3-OL

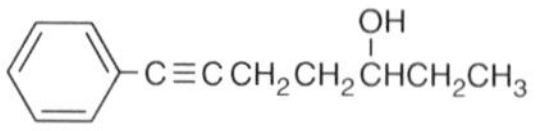

250 mg

S92,050-9
1,5-DIPHENYL-1,4-PENTADIYN-3-OL

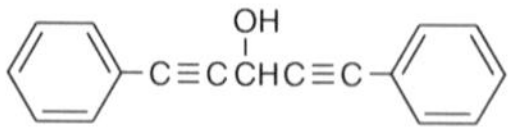

250 mg

S68,318-3
4-[4-(2-METHYLPROPYL)PHENYL]-3-BUTYN-1-OL

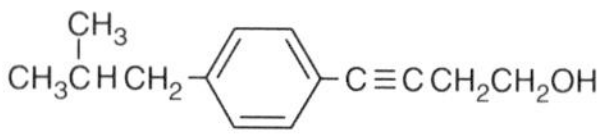

5 gm

S68,196-2
4,4-DIMETHYL-3-(2-NAPHTHYL)-1-PENTYN-3-OL

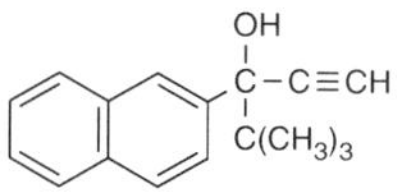

250 mg

S88,465-0
(3-METHOXYPHENYL)ACETYLENE

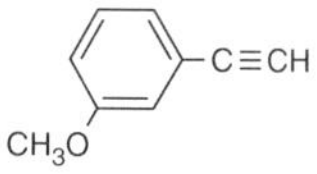

250 mg

S68,157-1
5,5-DIMETHYL-4-HYDROXY-4-(4-PHENOXYPHENYL)-2-HEXYNOIC ACID

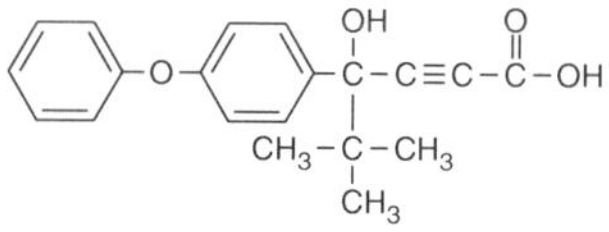

250 mg

S91,652-8
3-BUTYN-1-YL N-(2,5-DIMETHOXYPHENYL)CARBAMATE

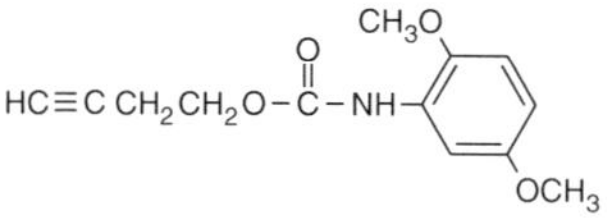

250 mg

S89,787-6
3-BUTYN-1-YL N-(2,5-DIMETHOXYPHENYL)CARBAMATE

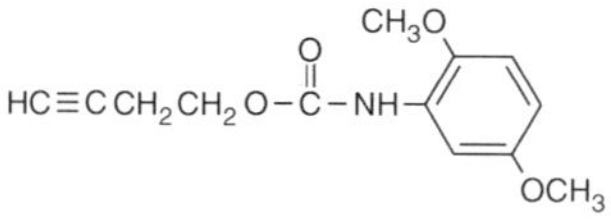

250 mg

S95,658-9
BIS(3,4-DIHYDRO-6-METHOXY-1-NAPHTHYL)ACETYLENE

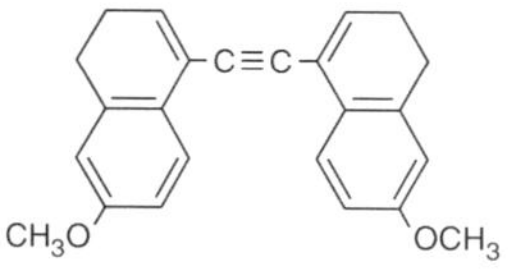

50 mg

S95,427-6
1,2-BIS(4-METHOXYPHENYL)ACETYLENE

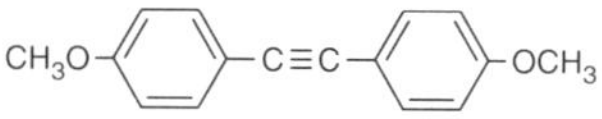

50 mg

S93,795-9
1,2-BIS(4-METHOXYPHENYL)-2-ETHOXY-3-BUTYNE

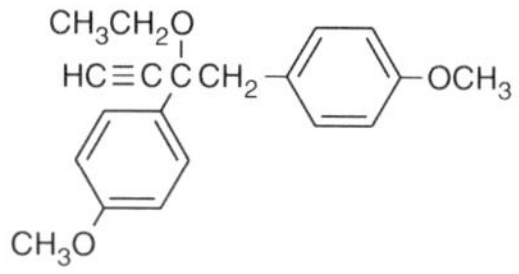

250 mg

S93,796-7
1,2,5,6-TETRAKIS(4-METHOXYPHENYL)-3-HEXYNE

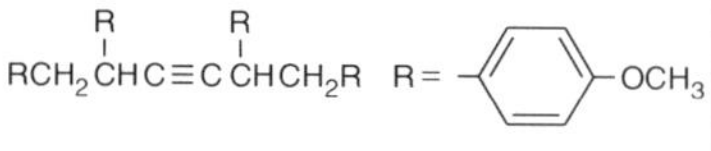

250 mg

S86,153-7
TERT-BUTYL-4-(2-TETRAHYDROPYRANYLOXY)-2-BUTYNOATE

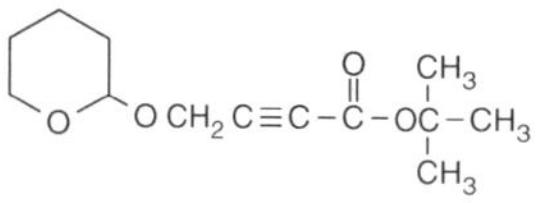

250 mg

S39,879-9
2-BUTYNE-1,4-DIOL BIS(2-METHOXYCARBANILATE)

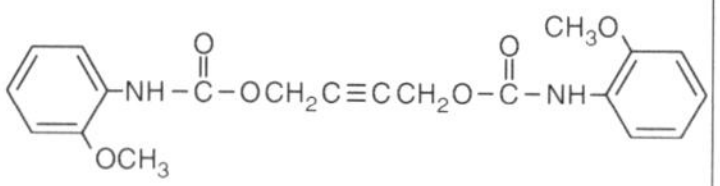

5 gm

S88,144-9
1-(4-METHOXYPHENYL)-2-PHENYLACETYLENE

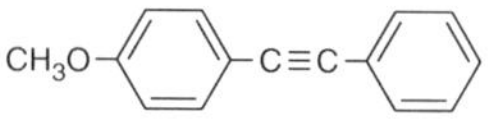

250 mg

S60,807-6
1,4-BIS(4-CHLOROPHENOXY)-2-BUTYNE

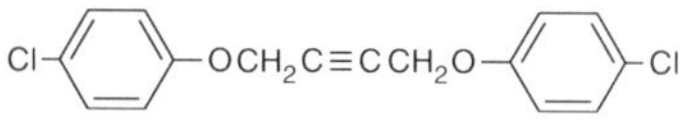

250 mg

S62,549-3
1,4-BIS(4-CHLORO-3-METHYLPHENOXY)-2-BUTYNE

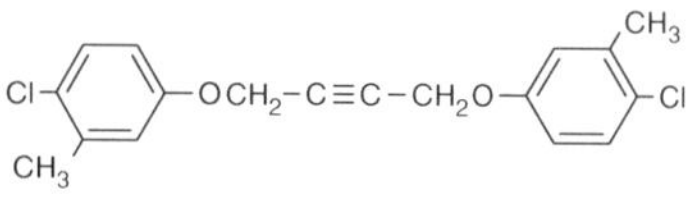

5 gm

S86,116-2
4-(2-PROPYN-1-YLOXY)-2,3,5,6-TETRACHLOROPYRIDINE

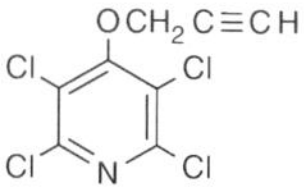

250 mg

S67,065-0
6-METHOXY-2-NAPHTHYLACETYLENE

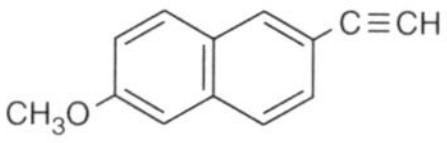

5 gm

S61,877-2
ACETALDEHYDE DIPROPARGYL ACETAL

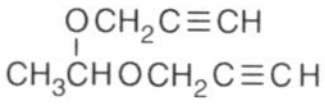

250 mg

S86,165-0
7,7-DIETHOXY-5-HEPTYNE-2,4-DIOL DIACETATE

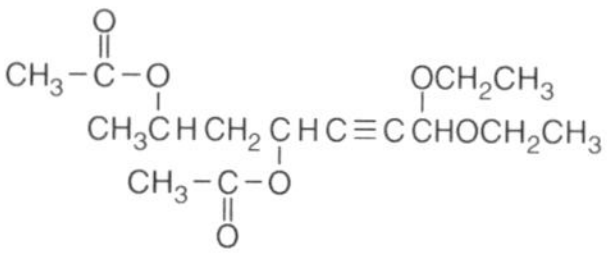

250 mg

S89,102-9
1,1,4,4-TETRAETHOXY-2-BUTYNE

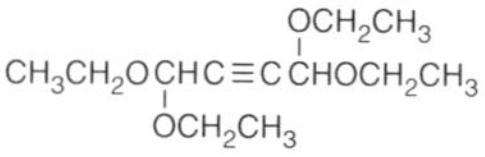

250 mg

S77,826-5
2-METHYL-3-BUTYN-2-YL 2-TETRAHYDROPYRANYL ETHER

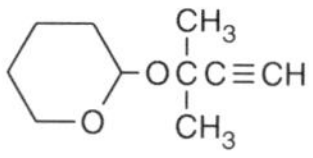

250 mg

S94,929-9
2-(2-PROPYNYL)MALONIC ACID

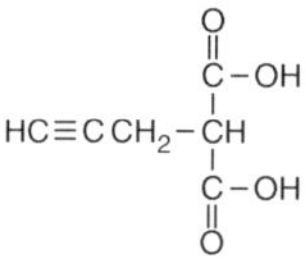

250 mg

S95,000-9
ACETYLENEDICARBOXAMIDE

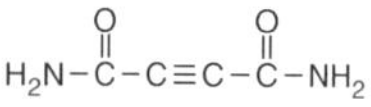

50 mg

S56,788-4
2-BUTYNE-1,4-DIOL DIACETATE

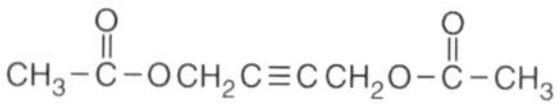

5 gm

S79,993-9
1,2-BIS(ETHYLTHIO)ETHYNE

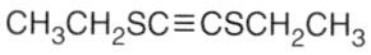

250 mg

S80,897-0
4-NITROPHENYL 1-PROPYNYL SULFIDE

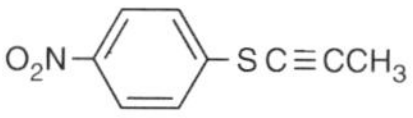

250 mg

S62,030-0
PHENYL PROPARGYL SULFIDE

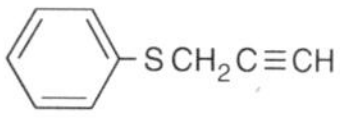

250 mg

S80,748-6
1,2-BIS(PHENYLTHIO)ETHYNE

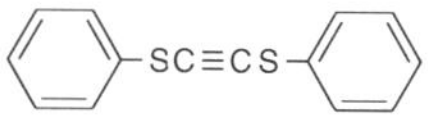

250 mg

S77,833-8
1-BUTYN-1-YL P-TOLYL SULFIDE

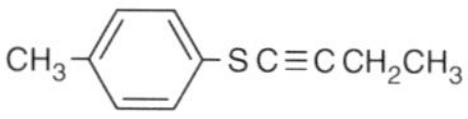

250 mg

S80,747-8
1,2-BIS(P-TOLYLTHIO)ETHYNE

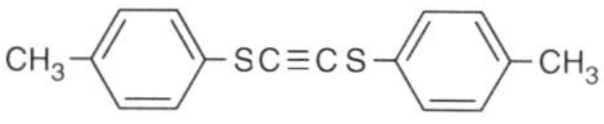

250 mg

S86,117-0
4-(2-PROPYN-1-YLTHIO)-2,3,5,6-TETRACHLOROPYRIDINE

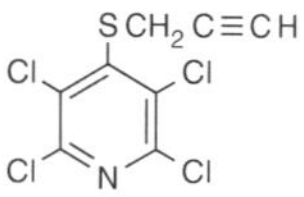

250 mg

S61,657-5
DIMETHYLPROPARGYLSULFONIUM BROMIDE

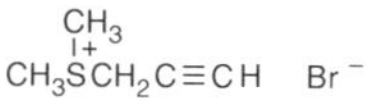

250 mg

S88,165-1
(5-PHENYL-4-PENTYN-1-YL)AMINE

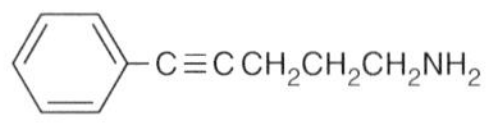

250 mg

S87,365-9
1,2-BIS(DIETHYLAMINO)ETHYNE

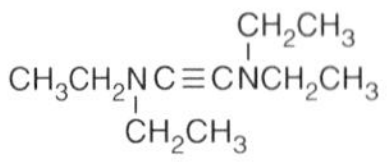

250 mg

S38,842-4
N,N'-DIETHYL-2-BUTYNE-1,4-DIAMINE

$CH_3CH_2NHCH_2C{\equiv}CCH_2NHCH_2CH_3$

5 gm

S44,587-8
N,N,N',N'-TETRAMETHYL-2-BUTYNE-1,4-DIAMINE

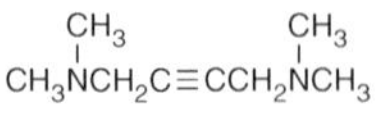

250 mg

S38,857-2
N,N,N',N'-TETRAETHYL-2-BUTYNE-1,4-DIAMINE

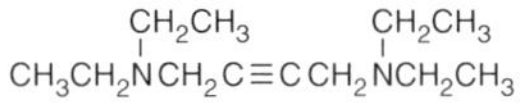

5 gm

S76,750-6
N-(2-PROPYNYL)BENZYLAMINE HYDROCHLORIDE

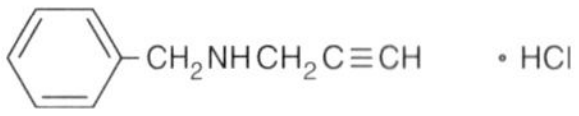

250 mg

S73,451-9
N-(1-INDANYL)PROPARGYLAMINE

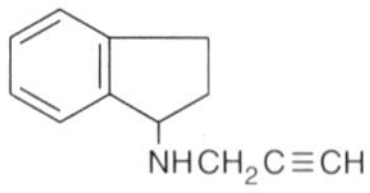

250 mg

S88,134-1
BENZYLMETHYL(5-PHENYL-4-PENTYN-1-YL)AMINE

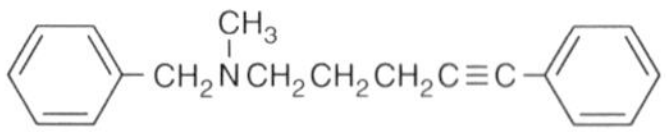

250 mg

S34,220-3
1,6-BIS(DIETHYLAMINO)-2-HEXYNE

5 gm

S36,129-1
1,4-DIPHTHALIMIDO-2-BUTYNE

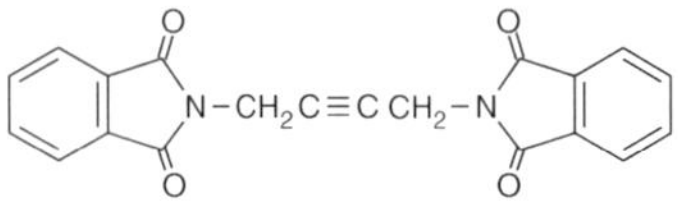

5 gm

S34,241-6
1,6-BIS(PYRROLIDINO)-2-HEXYNE

5 gm

S35,003-6
6-DIETHYLAMINO-1-PYRROLIDINO-2-HEXYNE

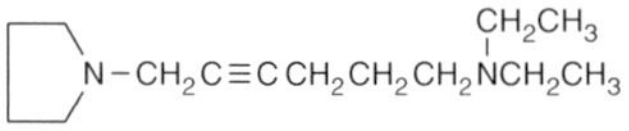

5 gm

S89,354-4
1-(1-PIPERIDINOCYCLOHEXYL)-1-PROPYNE

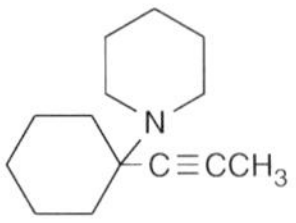

250 mg

S34,598-9
6-DIETHYLAMINO-1-PIPERIDINO-2-HEXYNE

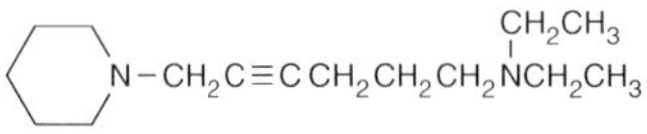

5 gm

S34,597-0
6-DIETHYLAMINO-1-MORPHOLINO-2-HEXYNE

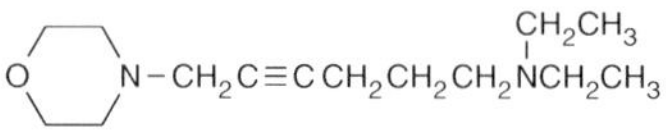

5 gm

S34,744-2
1,6-DIMORPHOLINO-2,4-HEXADIYNE

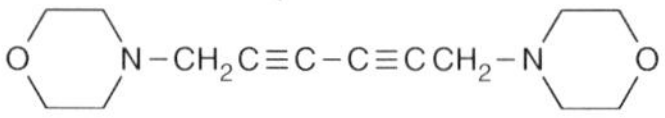

5 gm

S34,235-1
1,6-BIS(MORPHOLINO)-2-HEXYNE

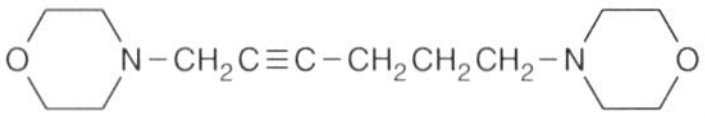

5 gm

S80,898-9
4-NITROPHENYL 1-PROPYNYL SULFONE

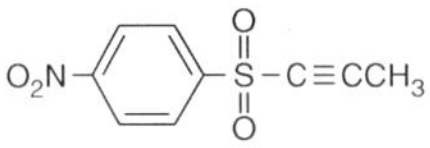

250 mg

S67,749-3
2-METHYL-3-BUTYN-2-YL 4-NITROBENZOATE

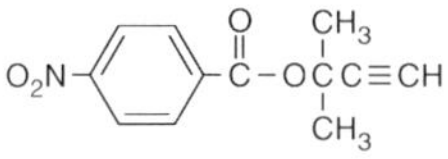

250 mg

S58,669-2
3-PENTYN-1-AL

250 mg

S88,490-1
5-UNDECYN-2-ONE

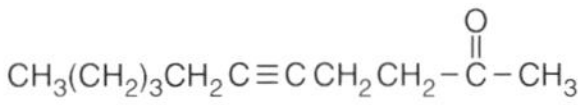

250 mg

S73,121-8
DIBENZOYLACETYLENE

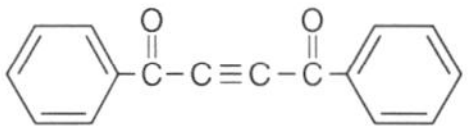

5 gm

S59,176-9
2,5-DIMETHYL-3-HEXYNE-2,5-DIYL BIS(3-OXOBUTYRATE)

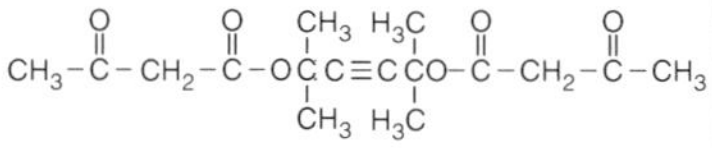

5 gm

S63,158-2
PHENYLPROPIOLIC ACID, SODIUM SALT

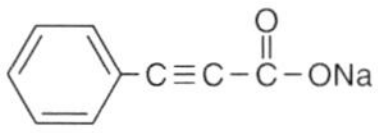

5 gm

S87,764-6
METHYL 1-OCTEN-3-YNE-1-CARBOXYLATE

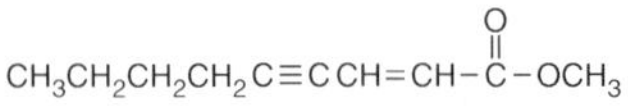

250 mg

S88,212-7
METHYL 2-NONEN-4-YNE-2-CARBOXYLATE

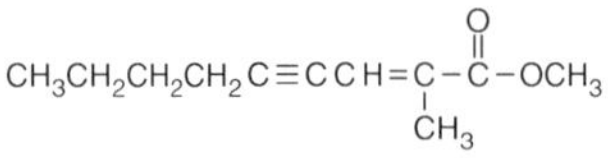

250 mg

S88,030-2
METHYL 5-CYCLOHEXYL-2-PENTEN-4-YNE-2-CARBOXYLATE

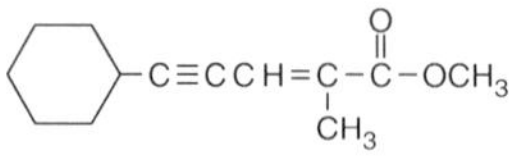

250 mg

S88,032-9
METHYL 5-PHENYL-2-PENTEN-4-YNE-2-CARBOXYLATE

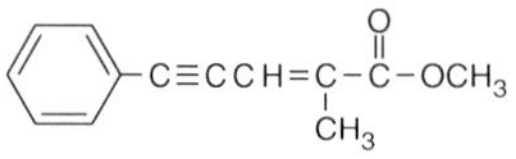

250 mg

S76,359-4
(1,1-DIMETHYL-2-PROPYN-1-YL)-MALONIC ACID

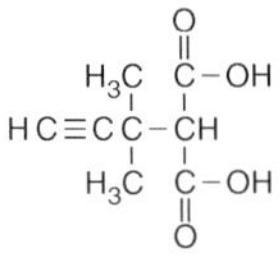

250 mg

S80,475-4
DIETHYL 2-(2-PROPYNYL)MALONATE

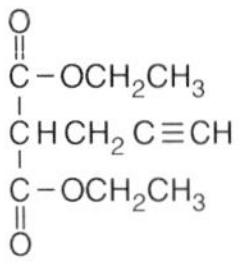

250 mg

S76,467-1
DIETHYL 2-PHENYL-2-(2-PROPYNYL)-MALONATE

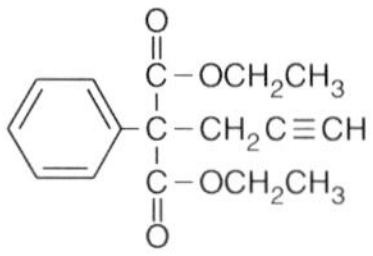

5 gm

S70,469-5
3-HEXYNE-1,1,6,6-TETRACARBOXYLIC ACID

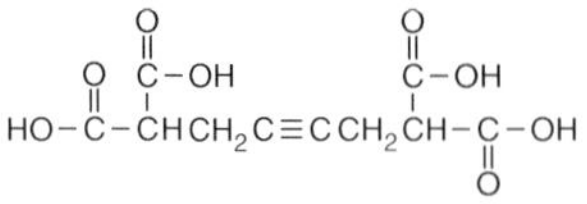

250 mg

S70,459-8
2,7-NONADIYNEDIAMIDE
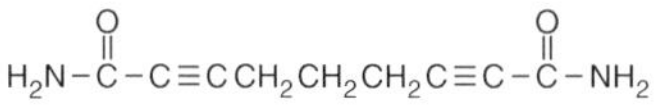
250 mg

S70,460-1
2,8-DECADIYNEDIAMIDE
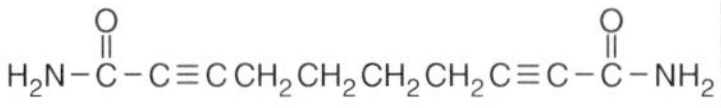
5 gm

S35,402-3
ALPHA-NONYNOIC ACID
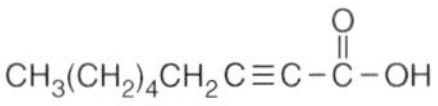
5 gm

S41,410-7
ETHYL 2-OCTYNOATE
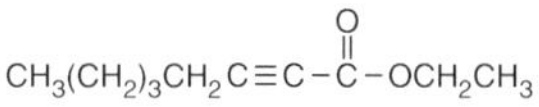
5 gm

S88,448-0
METHYL 4-DECYNOATE
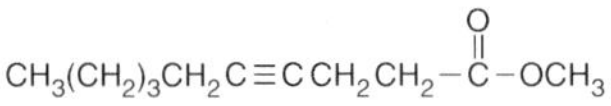
250 mg

S37,047-9
PROPYNYL N-PHENYLCARBAMATE
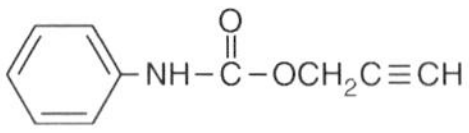
250 mg

S88,633-5
3-PHENYLPROPIOLAMIDE
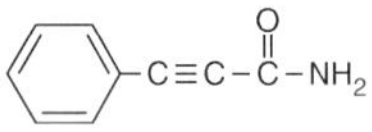
250 mg

S94,652-4
N-(2-BUTYNYL)SUCCINIMIDE
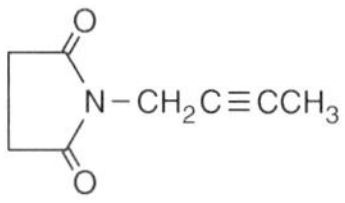
100 mg

S86,146-4
3,3-DIMETHYL-1-(2-PROPYN-1-YL)-2-INDOLINONE
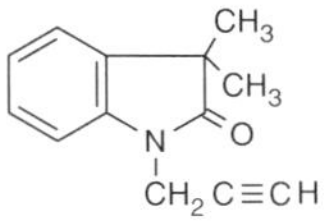
250 mg

S37,046-0
PROPYNYL N-(3-CHLOROPHENYL)-CARBAMATE
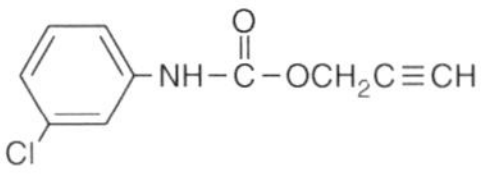
250 mg

S77,809-5
ISOPROPYL 1-PROPYNYL SULFONE
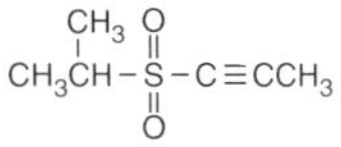
250 mg

S77,816-8
ETHYL 3-METHYL-1-BUTYNYL SULFONE
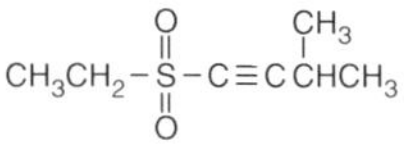
250 mg

S77,828-1
2-PHENYLETHYL 1-PROPYNYL SULFONE
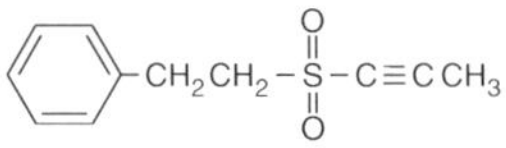
250 mg

S80,913-6
2-METHYL-4-PHENYL-1-BUTEN-3-YN-1-YL PHENYL SULFONE
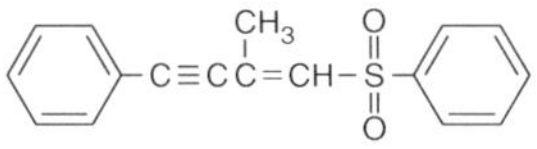
250 mg

S77,847-8
2-PHENYLETHYNYL P-TOLYL SULFONE
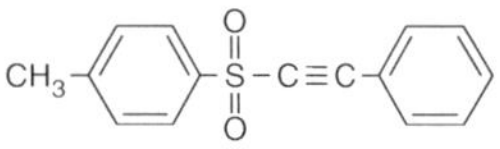
250 mg

S80,912-8
4-PHENYL-1-BUTEN-3-YN-1-YL P-TOLYL SULFONE
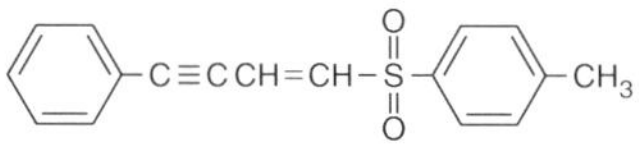
250 mg

S80,914-4
2,4-DIPHENYL-1-BUTEN-3-YN-1-YL P-TOLYL SULFONE
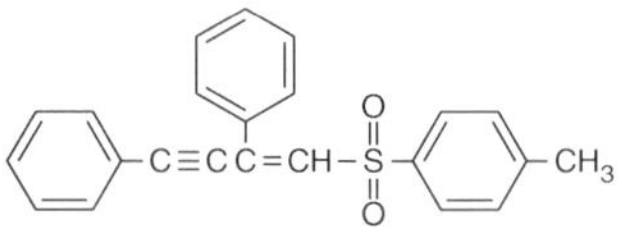
250 mg

S80,900-4
1-PROPYNYL P-TOLYL SULFONE
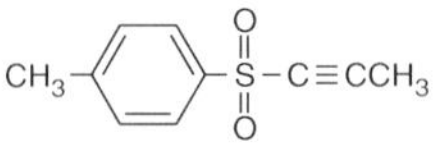
250 mg

S86,589-3
1,4-BIS(P-TOSYL)-2-BUTYNE
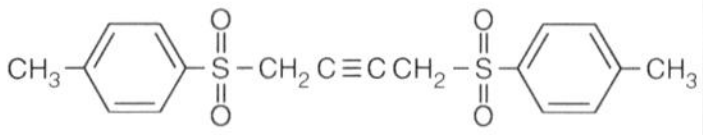
250 mg

S70,463-6
3,5-OCTADIYNE-1,8-DIYL DI-P-TOLUENESULFONATE
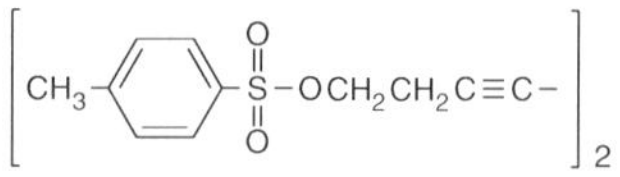
250 mg

S87,091-9
1,2-BIS(PHENYLETHYNYL)BENZENE
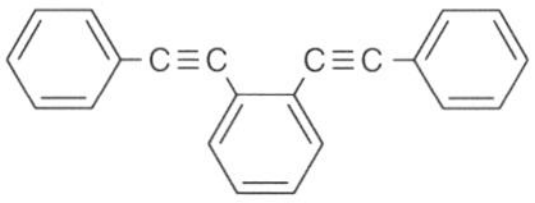
250 mg

S87,092-7
1,4-BIS(PHENYLETHYNYL)BENZENE

250 mg

S93,773-8
4-ETHYNYLBIPHENYL
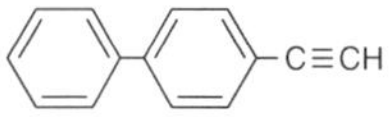
250 mg

S42,980-5
6-PHENYL-2-HEXYNE
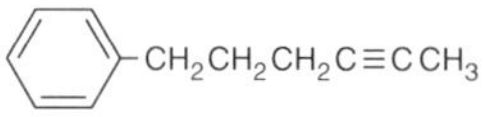
250 mg

S68,309-4
[4-(2-METHYLPROPYL)PHENYL]ACETYLENE
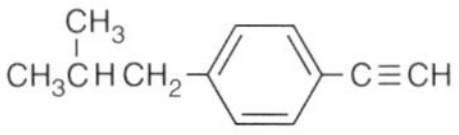
5 gm

S93,785-1
1-PHTHALAMIDO-2-BUTYNE
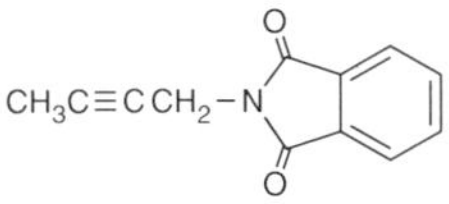
250 mg

S45,743-4
BIGUANIDE
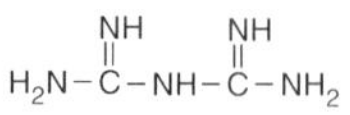
250 mg

S88,599-1
TERT-BUTYLETHYLIDENEAMINE
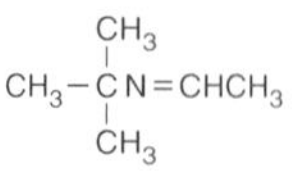
250 mg

S80,616-1
N-ISOPROPYL-N-(3-METHYL-2-BUTEN-1-YLIDENE)AMINE
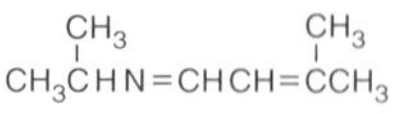
250 mg

S88,606-8
TERT-BUTYLCYCLOHEPTYLIDENEAMINE
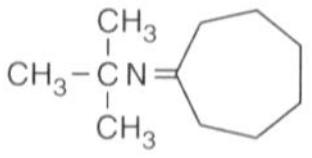
250 mg

S88,607-6
TERT-BUTYLCYCLOOCTYLIDENEAMINE
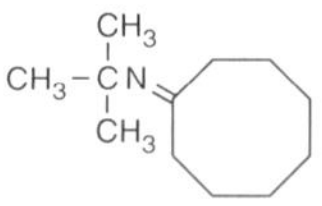
250 mg

S62,399-7
N-(CINNAMYLIDENE)-M-TOLUIDINE
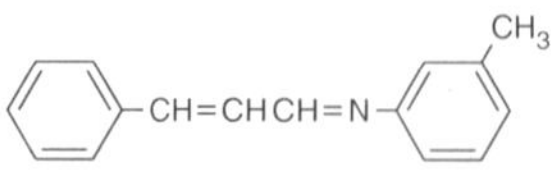
250 mg

S54,527-9
OCTYL ALDEHYDE AZINE

250 mg

S77343-3
N-(2-CHLOROETHYLIDENE)-2,4-DINITROANILINE
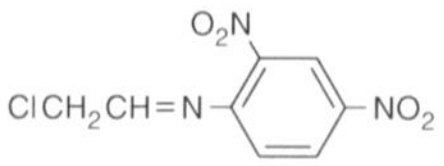
250 mg

S72,021-6
ALPHA-[2-CHLORO-5-(TRIFLUORO-METHYL)PHENYLIMINO]-6-METHOXY-O-CRESOL
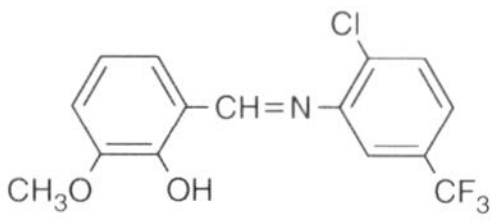
5 gm

S87,916-9
N,N'-BIS[2-CHLORO-5-(TRIFLUOROMETHYL)PHENYL]-1,4-DIAZABUTADIENE
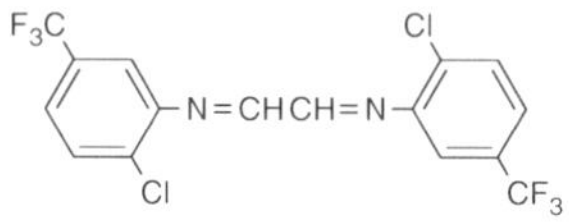
5 gm

S87,911-8
N-(4-CHLORO-3-NITROBENZYLIDENE)-3-(TRIFLUOROMETHYL)ANILINE
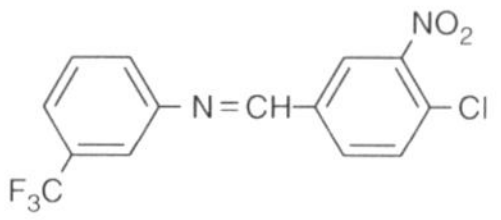
5 gm

S87,685-2
1-[4-BROMO-4-(TRIFLUOROMETHYL)-DIPHENYLMETHYLENEAMINO]GUANIDINE P-TOLUENESULFONATE
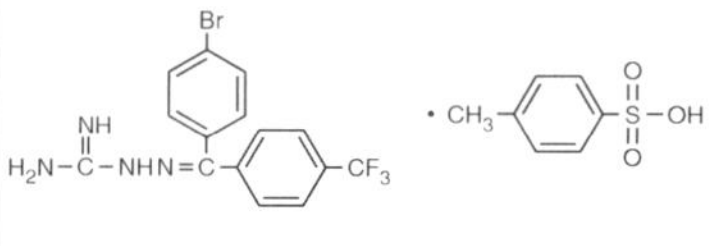
250 mg

S53,282-7
ALPHA-[3-(TRIFLUOROMETHYL)-PHENYLIMINO]-O-CRESOL
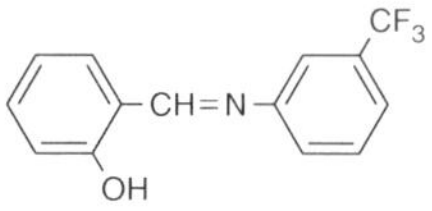
5 gm

S68,912-2
2-METHOXY-ALPHA-[3-(TRIFLUOROMETHYL)PHENYLIMINO]-O-CRESOL
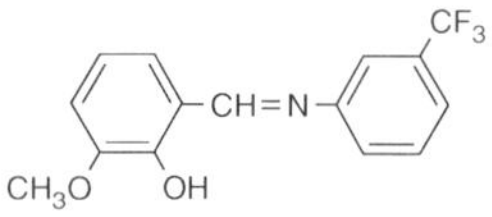
5 gm

S54,304-7
N-(3-ETHOXY-4-HYDROXYBENZYLIDENE)-3-(TRIFLUOROMETHYL)ANILINE
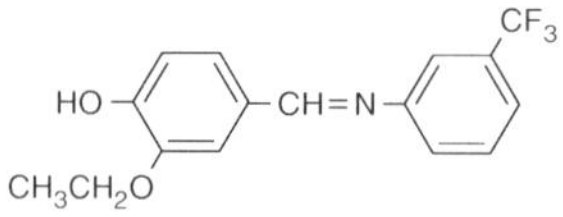
5 gm

S87,901-0
2-METHOXY-4-NITRO-6-[3-(TRIFLUOROMETHYL)PHENYLIMINO-METHYL]PHENOL
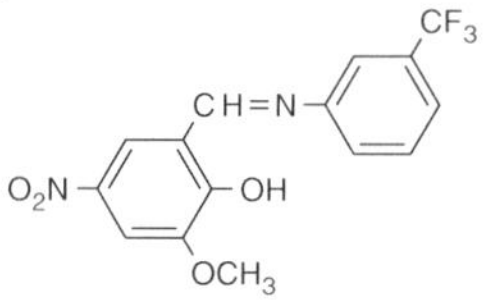
5 gm

S54,370-5
2-METHOXY-ALPHA-[3-(TRIFLUOROMETHYL)PHENYLIMINO]-P-CRESOL
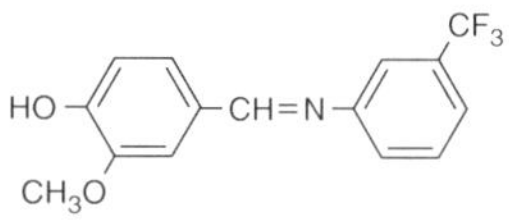
5 gm

S58,395-2
ALPHA-[2-(TRIFLUOROMETHYL)-PHENYLIMINO]-O-CRESOL
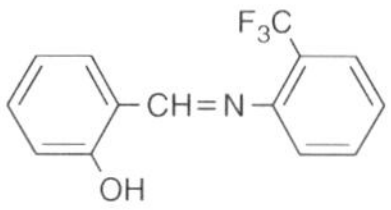
5 gm

S58,394-4
ALPHA-[2-CHLORO-5-(TRIFLUORO-METHYL)PHENYLIMINO]-O-CRESOL
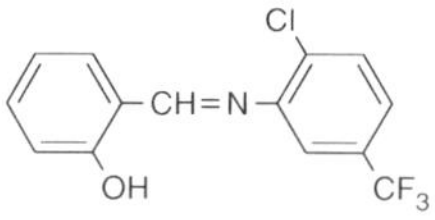
5 gm

S64,933-3
ALPHA-[3,5-BIS(TRIFLUOROMETHYL)-PHENYLIMINO]-O-CRESOL
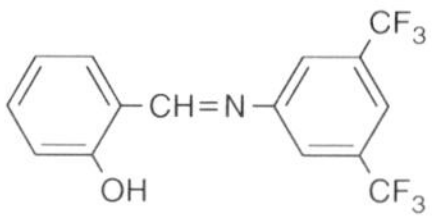
5 gm

S53,318-1
1-(ALPHA,ALPHA,ALPHA-TRIFLUORO-M-TOLYLIMINOMETHYL)-2-NAPHTHOL
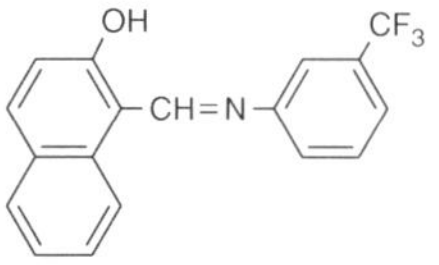
5 gm

S63,427-1
N-(4-BENZYLOXYBENZYLIDENE)-3-TRIFLUOROMETHYLANILINE
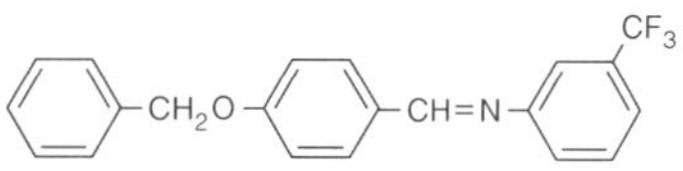
5 gm

S53,284-3
VERATRYLIDENE-3-(TRIFLUOROMETHYL)ANILINE
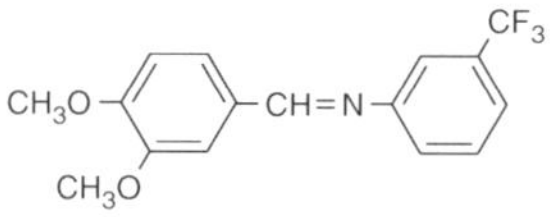
250 mg

S62,536-1
N-(9-ANTHRACENYLMETHYLENE)-5-CHLORO-2,4-DIMETHOXYANILINE
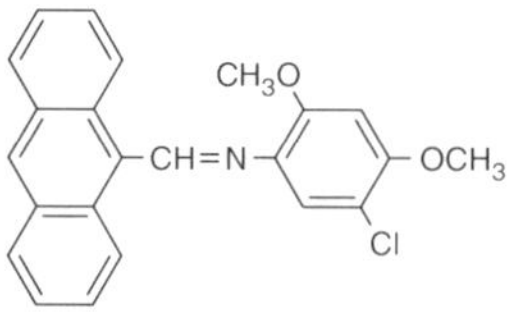
5 gm

S62,627-9
N-(4-BENZYLOXYBENZYLIDENE)-2-(TRIFLUOROMETHYL)ANILINE
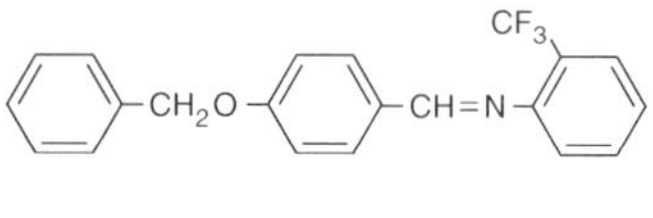
5 gm

S53,287-8
N-(4-METHOXYBENZYLIDENE)-3-(TRIFLUOROMETHYL)ANILINE
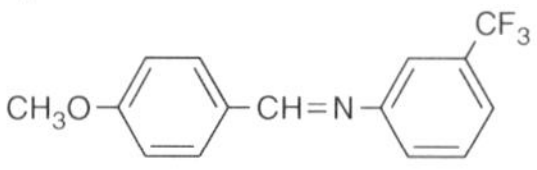
5 gm

S53,311-4
N-PIPERONYLIDENE-3-(TRIFLUOROMETHYL)ANILINE
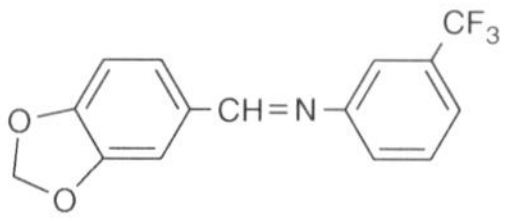
5 gm

S54,134-6
N,N-DIETHYL-ALPHA-[3-(TRIFLUOROMETHYL)PHENYLIMINO]-P-TOLUIDINE
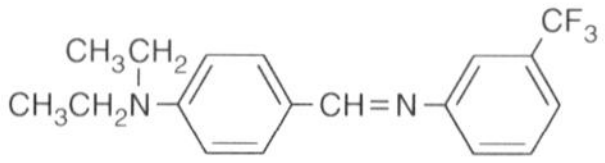
250 mg

S53,306-8
N-[4-(DIMETHYLAMINO)BENZYLIDENE]-3-(TRIFLUOROMETHYL)ANILINE
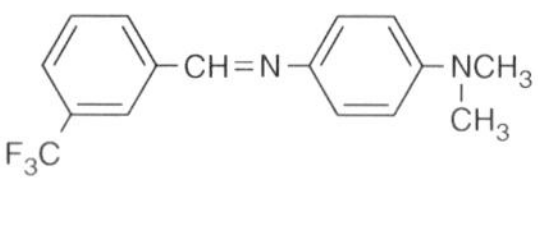
5 gm

S53,307-6
N-[2-CHLORO-4-(DIMETHYLAMINO)-BENZYLIDENE]-3-(TRIFLUOROMETHYL)-ANILINE
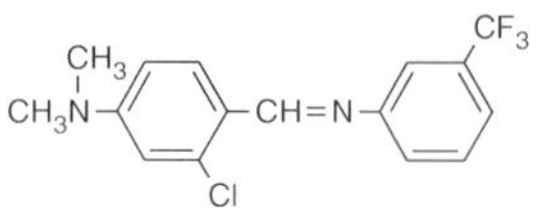
5 gm

S62,469-1
N-(9-ANTHRACENYLMETHYLENE)-2-(TRIFLUOROMETHYL)ANILINE
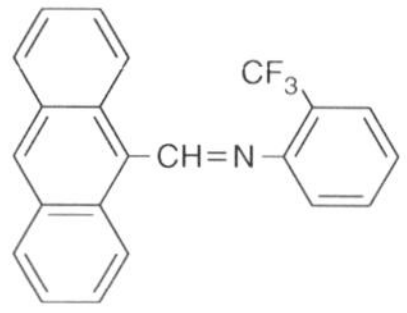
5 gm

S62,533-7
N-(9-ANTHRACENYLMETHYLENE)-3-(TRIFLUOROMETHYL)ANILINE
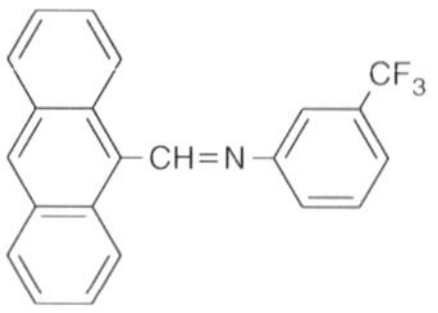
5 gm

S77,164-3
1-[ALPHA-METHYL-3-(TRIFLUORO-METHYL)BENZYLIDENEAMINO]GUANIDINE NITRATE

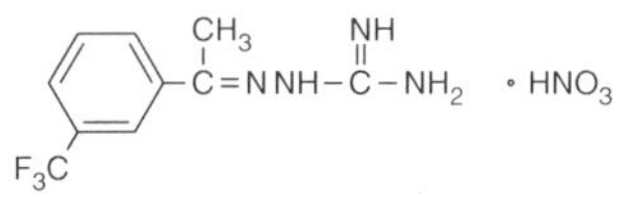

250 mg

S77,189-9
1-[ALPHA-PENTYL-3-(TRIFLUORO-METHYL)BENZYLIDENEAMINO]GUANIDINE NITRATE

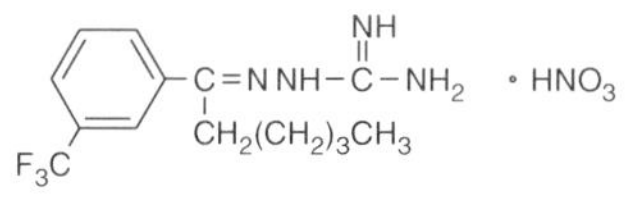

250 mg

S38,313-9
4-NITROBENZALDEHYDE (2,2,2-TRIFLUOROETHYL)HYDRAZONE

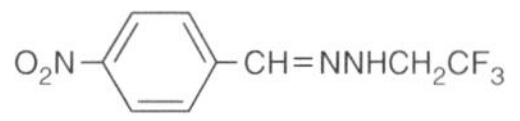

250 mg

S77,146-5
3-(GUANIDINOIMINO)-4,4,4-TRIFLUOROBUTYROPHENONE NITRATE

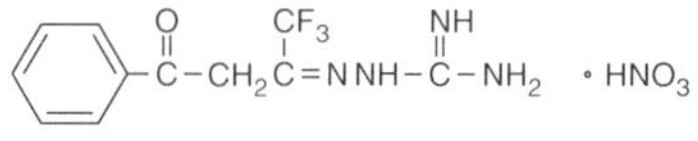

250 mg

S79,640-9
2-[3-(TRIFLUOROMETHYL)-PHENYLUREIDO]ACETIC FURFURYLIDENEHYDRAZIDE

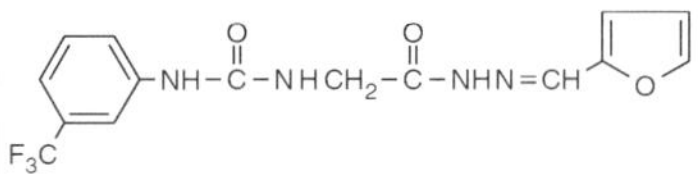

250 mg

S76,295-4
1-(3-TRIFLUOROMETHYLCYCLOHEXYLIDENE)-3-THIOSEMICARBAZIDE

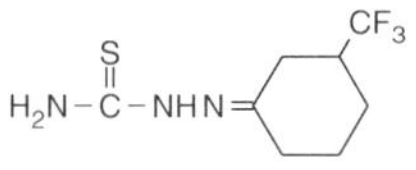

250 mg

S59,978-6
N-(2-THIENYLMETHYLENE)-2-(TRIFLUOROMETHYL)ANILINE

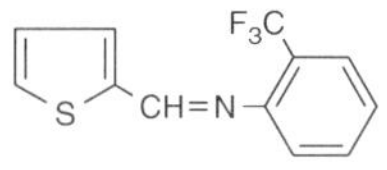

5 gm

S54,301-2
N-(3-NITROBENZYLIDENE)-3-(TRIFLUOROMETHYL)ANILINE

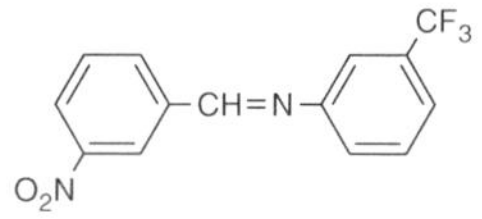

5 gm

S54,303-9
N-(4-NITROBENZYLIDENE)-3-(TRIFLUOROMETHYL)ANILINE

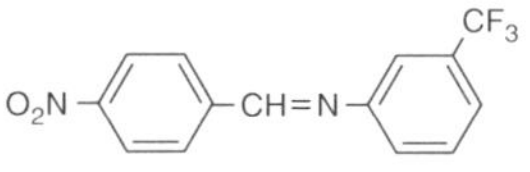

5 gm

S53,316-5
ALPHA-[3-(TRIFLUOROMETHYL)-PHENYLIMINO]-P-ACETOTOLUIDIDE

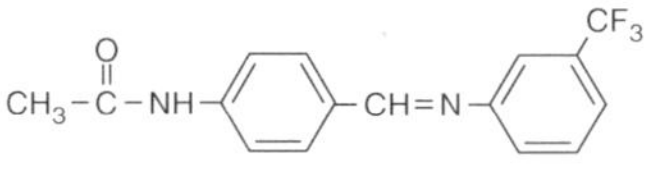

5 gm

S58,432-0
N,N'-(1,4-PHENYLENEDIMETHYLIDYNE)-BIS[(3-TRIFLUOROMETHYL)ANILINE]

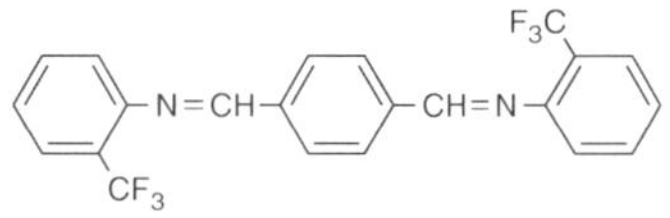

250 mg

S53,313-0
N-(4-FLUOROBENZYLIDENE)-3-(TRIFLUOROMETHYL)ANILINE

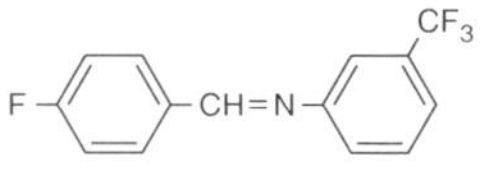

5 gm

S53,309-2
N-(4-CHLOROBENZYLIDENE)-3-(TRIFLUOROMETHYL)ANILINE

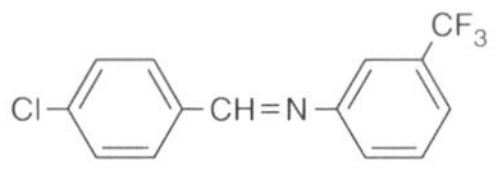

5 gm

S53,283-5
N-(4-METHYLBENZYLIDENE)-3-(TRIFLUOROMETHYL)ANILINE

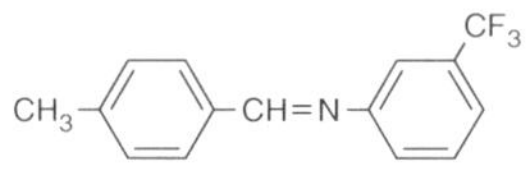

5 gm

S58,433-9
N,N'-(1,4-PHENYLENEDIMETHYLIDYNE)-BIS[(3-TRIFLUOROMETHYL)ANILINE]

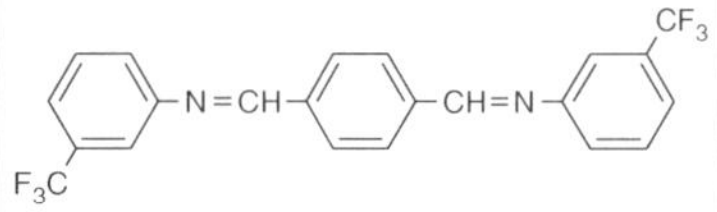

250 mg

S58,431-2
N,N'-(1,4-PHENYLENEDIMETHYLIDYNE)-BIS(6-CHLORO-3-TRIFLUOROMETHYL)ANILINE

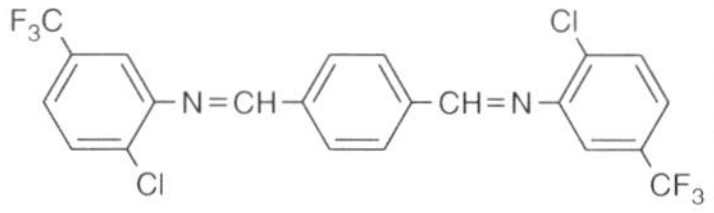

5 gm

S53,310-6
N-(2,6-DICHLOROBENZYLIDENE)-3-(TRIFLUOROMETHYL)ANILINE

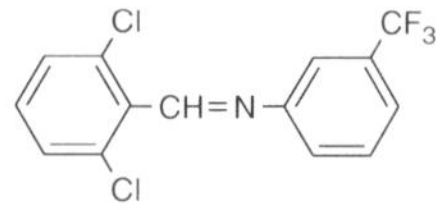

5 gm

S72,006-2
ALPHA-(5-CHLORO-2-HYDROXYPHENYLIMINO)-O-CRESOL

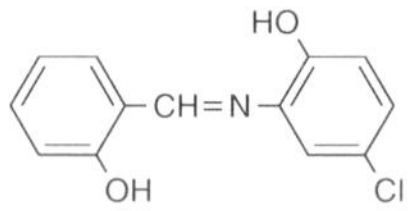

5 gm

S69,107-0
ALPHA-(5-CHLORO-2,4-DIMETHOXYPHENYLIMINO)-6-METHOXY-O-CRESOL

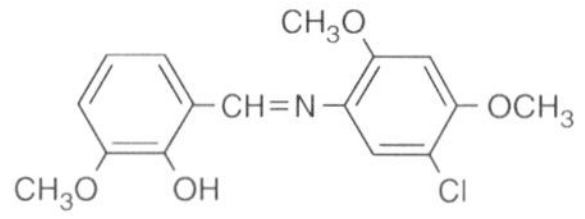

5 gm

S68,913-0
ALPHA-(5-CHLORO-2-METHOXYPHENYLIMINO)-6-METHOXY-O-CRESOL

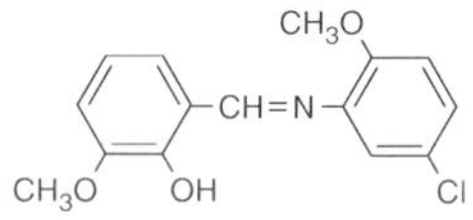

5 gm

S71,735-5
ALPHA-(2-CHLORO-5-NITROPHENYL-IMINO)-6-METHOXY-O-CRESOL

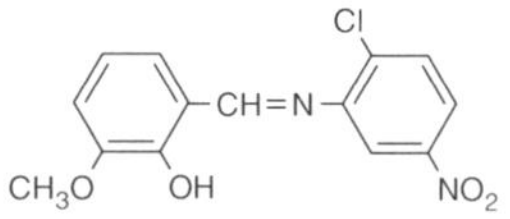

5 gm

S92,292-7
2-(4-BROMO-3-METHYLPHENYLIMINOMETHYL)-6-METHOXYPHENOL

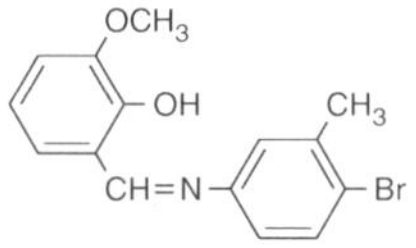

250 mg

S87,912-6
2-(3-CHLORO-4-NITROPHENYLIMINOMETHYL)-6-METHOXY-4-NITROPHENOL

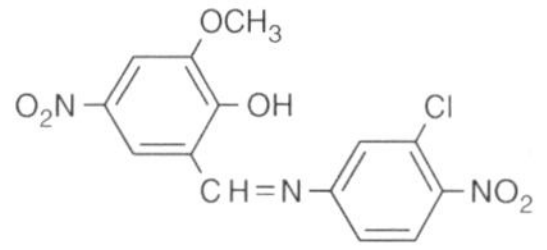

5 gm

S67,307-2
2-CHLORO-N-(3-ETHOXY-4-HYDROXYBENZYLIDENE)ANILINE

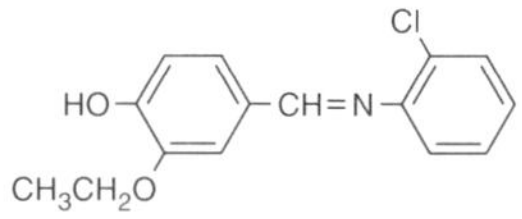

250 mg

S68,902-5
ALPHA-(4-FLUOROPHENYLIMINO)-2-METHOXY-O-CRESOL

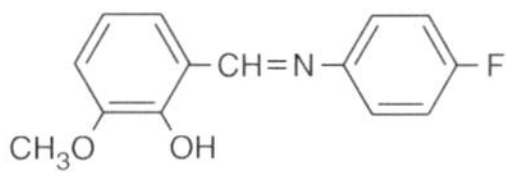

5 gm

S69,103-8
ALPHA-(4-CHLOROPHENYLIMINO)-6-METHOXY-O-CRESOL

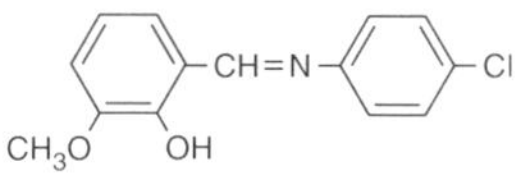

5 gm

S69,110-0
ALPHA-(4-BROMOPHENYLIMINO)-6-METHOXY-O-CRESOL

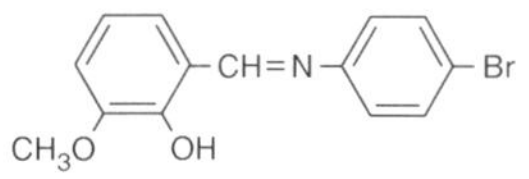

5 gm

S68,896-7
ALPHA-(3-CHLOROPHENYLIMINO)-6-METHOXY-O-CRESOL

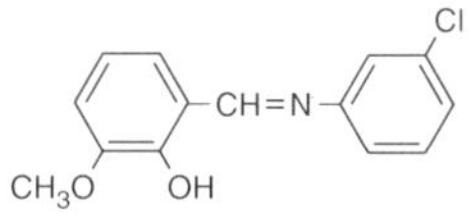

5 gm

S68,914-9
ALPHA-(2,4-DICHLOROPHENYLIMINO)-6-METHOXY-O-CRESOL

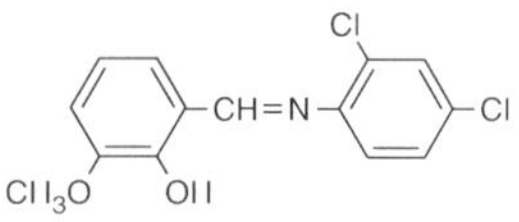

5 gm

S68,926-2
ALPHA-(3,4-DICHLOROPHENYLIMINO)-6-METHOXY-O-CRESOL

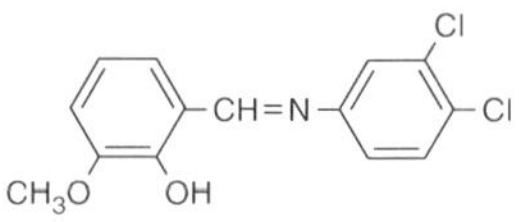

5 gm

S71,732-0
ALPHA-(2,5-DICHLOROPHENYLIMINO)-6-METHOXY-O-CRESOL

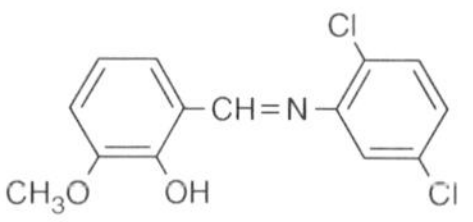

5 gm

S92,271-4
2-(FURFURYLIMINOMETHYL)-6-METHOXY-4-NITROPHENOL

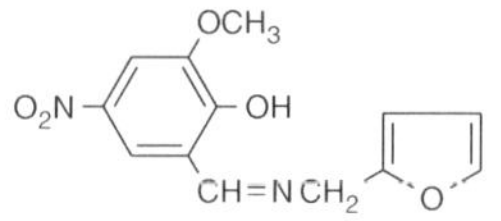

250 mg

S71,740-1
2-CHLORO-N-(2-HYDROXY-3-METHOXY-BENZYLIDENE)ANILINE

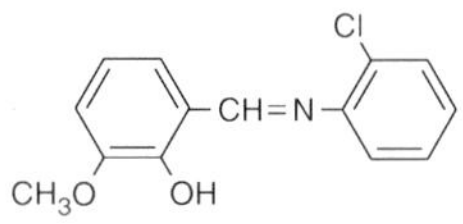

5 gm

S69,872-5
N-(2-HYDROXY-3-METHOXYBENZYLIDENE)-4-IODOANILINE

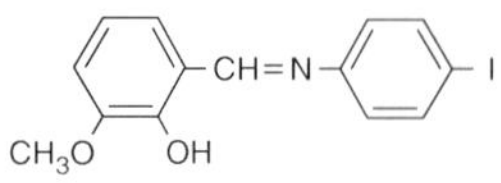

5 gm

S69,533-5
N-(2-HYDROXY-3-METHOXYBENZYLIDENE)-2,4,5-TRICHLOROANILINE

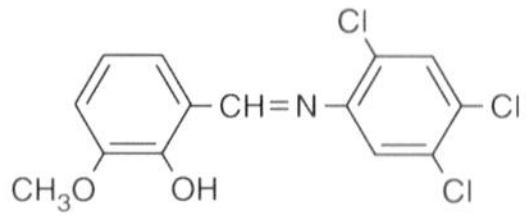

5 gm

S69,884-9
3-CHLORO-2-METHYL-N-(2-HYDROXY-3-METHOXYBENZYLIDENE)ANILINE

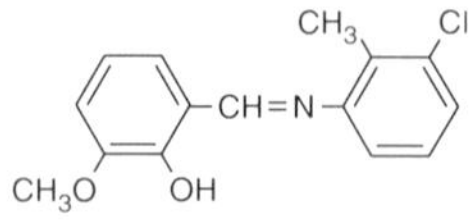

5 gm

S87,810-3
4-BROMO-2-CHLORO-N-(2-HYDROXY-3-METHOXYBENZYLIDENE)ANILINE

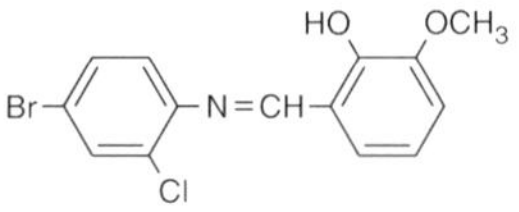

5 gm

S69,531-9
5-CHLORO-2-METHYL-N-(2-HYDROXY-3-METHOXYBENZYLIDENE)ANILINE

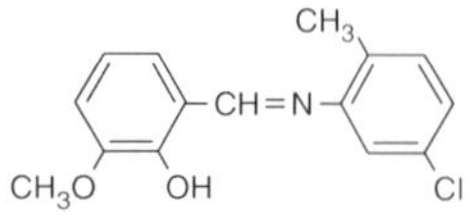

5 gm

S69,882-2
3-CHLORO-4-METHYL-N-(2-HYDROXY-3-METHOXYBENZYLIDENE)ANILINE

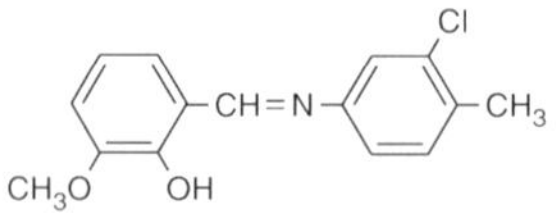

5 gm

S69,885-7
4-CHLORO-2-METHYL-N-(2-HYDROXY-3-METHOXYBENZYLIDENE)ANILINE

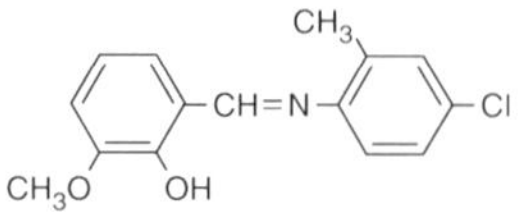

5 gm

S87,813-8
4-BROMO-N-(2-HYDROXY-3-METHOXYBENZYLIDENE)-2-METHYLANILINE

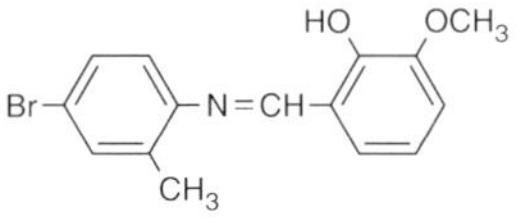

5 gm

S92,306-0
2-(4-BROMO-3-METHYLPHENYLIMINOMETHYL)-6-METHOXY-4-NITROPHENOL

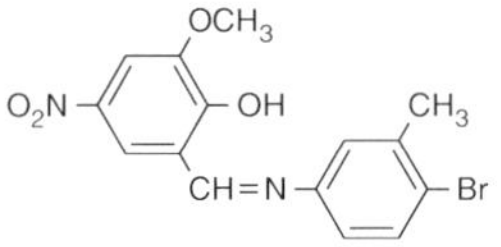

250 mg

S72,003-8
5-CHLORO-2-(O-VANILLYLIDENEAMINO)-BENZOPHENONE

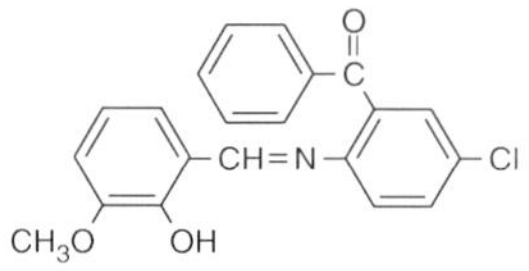

5 gm

S79,405-8
4-(5-HYDROXY-1-NAPHTHYLIMINO-METHYL)-2-METHOXYPHENYL N-(3-CHLOROPHENYL)CARBAMATE

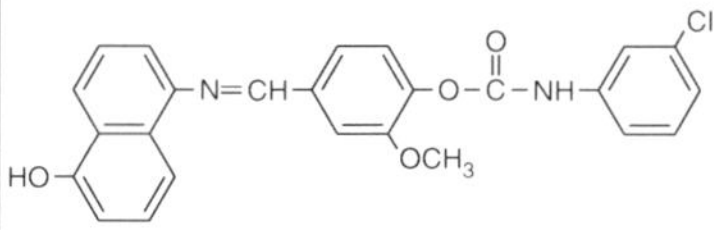

250 mg

S68,648-4
1-(5-CHLORO-2-METHOXYPHENYLIMINOMETHYL)-2-NAPHTHOL

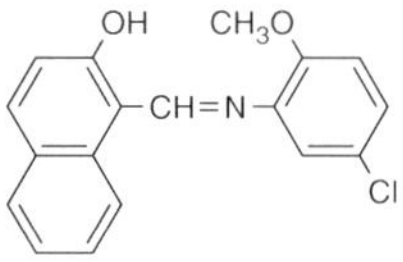

5 gm

S70,311-7
4-(4-CHLOROBENZYLIDENEAMINO)-SALICYLIC ACID

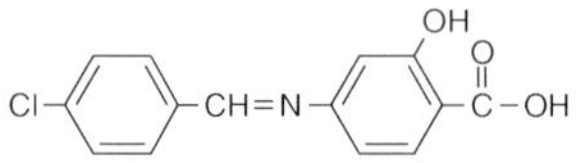

5 gm

S69,560-2
ALPHA-(4-CHLORO-3-NITROPHENYL-IMINO)-O-CRESOL

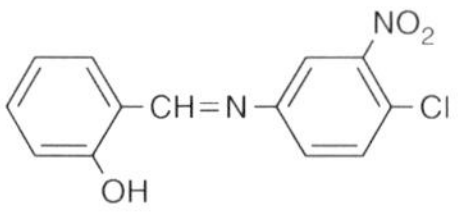

5 gm

S68,517-8
4-CHLORO-3-NITRO-N-(4-HYDROXY-BENZYLIDENE)ANILINE

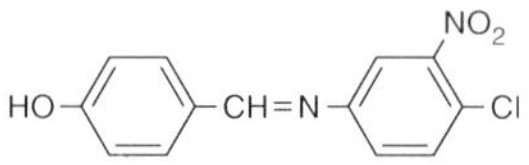

5 gm

S87,868-5
N-(4-CHLORO-3-NITROBENZYLIDENE)-2-HYDROXYANILINE

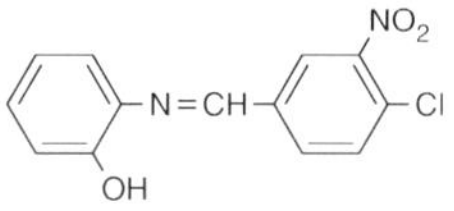

5 gm

S74,678-9
ALPHA-(2,4-DICHLOROPHENYLIMINO)-5-(DIETHYLAMINO)-O-CRESOL

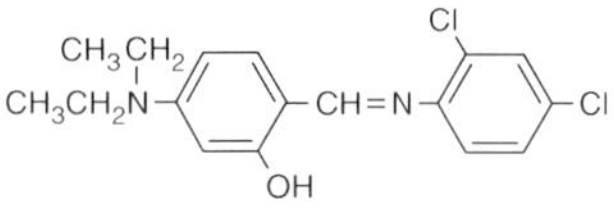

5 gm

S92,855-0
1-BROMO-3-(2-HYDROXY-3,4,5,6-TETRAHYDROBENZYLIDENEAMINO)-4-METHYLISOQUINOLINIUM PERCHLORATE

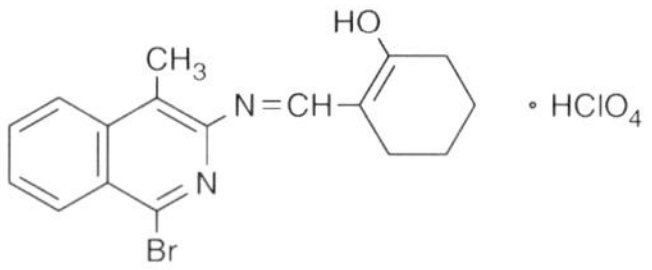

250 mg

S77,198-8
1-[(3-HYDROXY-2,4,6-TRICHLOROBENZYLIDENE)AMINO]-GUANIDINE ACETATE

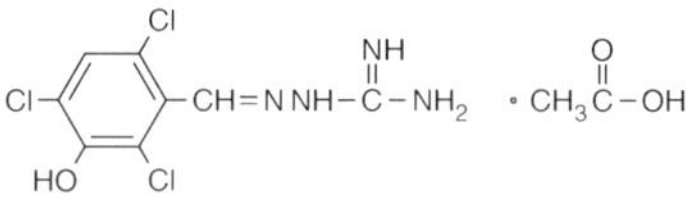

250 mg

S53,667-9
3,5-DIBROMOSALICYLALDEHYDE THIOSEMICARBAZONE

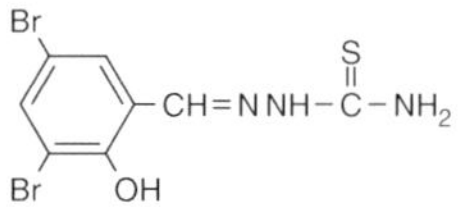

250 mg

S73,076-9
N-(2,4-DICHLOROBENZYLIDENE)-2-HYDROXYANILINE

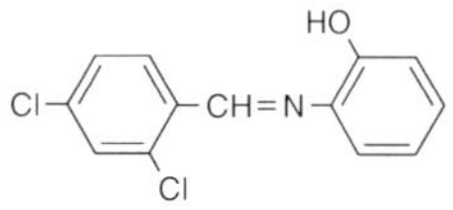

5 gm

S75,327-0
N-(2,4-DICHLOROBENZYLIDENE)-ETHANOLAMINE

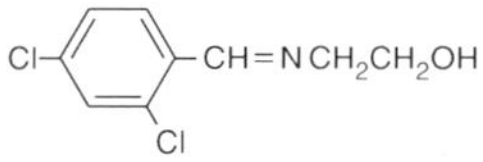

5 gm

S73,083-1
4-CHLORO-2-(2,4-DICHLOROBENZYLIDENEAMINO)PHENOL

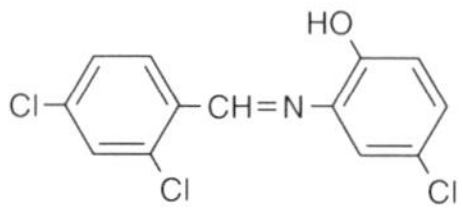

5 gm

S92,507-1
2-[3-(2-CHLORO-4-NITROPHENYL)-UREIDO]ACETIC (2-HYDROXYBENZYLIDENE)HYDRAZIDE

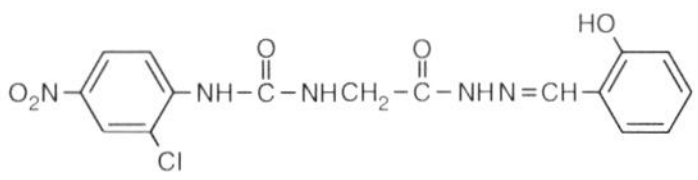

250 mg

S58,981-0
3-FLUORO-N-(2-HYDROXYBENZYLIDENE)-ANILINE

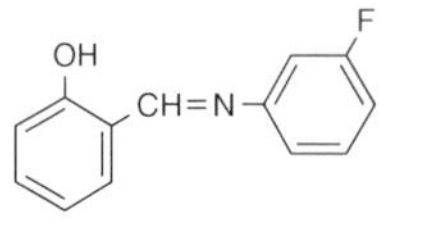

250 mg

S87,925-8
3-CHLORO-N-(2-HYDROXYBENZYLIDENE)-ANILINE

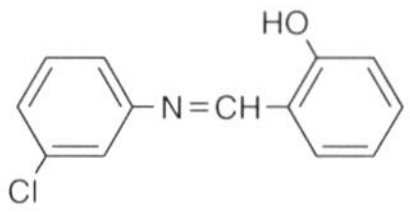

5 gm

S87,928-2
2-CHLORO-N-(2-HYDROXYBENZYLIDENE)-ANILINE

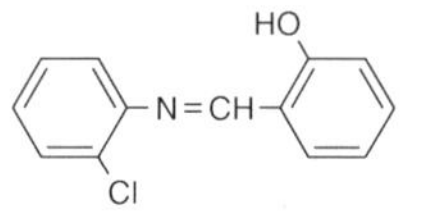

5 gm

S59,646-9
N-(2-HYDROXYBENZYLIDENE)-2-IODOANILINE

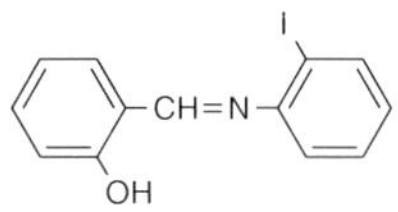

250 mg

S58,914-4
N-(2-HYDROXYBENZYLIDENE)-4-FLUOROANILINE

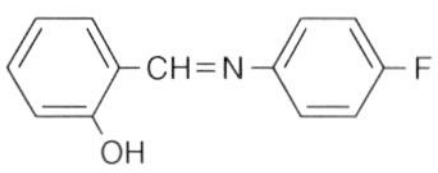

5 gm

S87,806-5
4-BROMO-2-CHLORO-N-(2-HYDROXYBENZYLIDENE)ANILINE

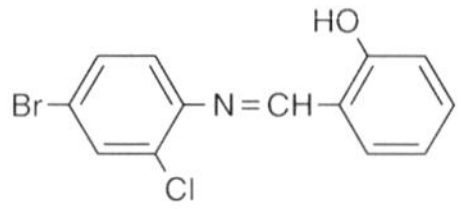

5 gm

S59,390-7
ALPHA-(4-BROMO-3-METHYLPHENYL-IMINO)-O-CRESOL

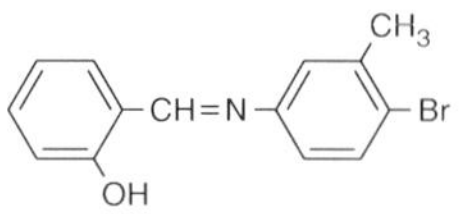

250 mg

S68,454-6
ALPHA-(4-FLUOROPHENYLIMINO)-P-CRESOL
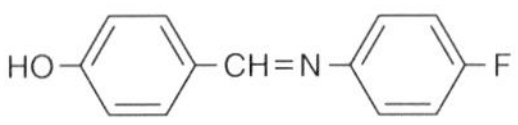
5 gm

S68,453-8
ALPHA-(4-CHLOROPHENYLIMINO)-P-CRESOL
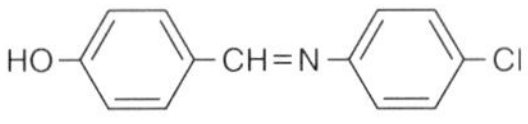
5 gm

S68,457-0
ALPHA-(3-CHLOROPHENYLIMINO)-P-CRESOL
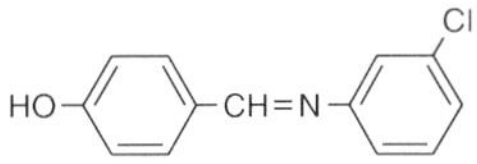
5 gm

S68,448-1
ALPHA-(4-BROMOPHENYLIMINO)-P-CRESOL
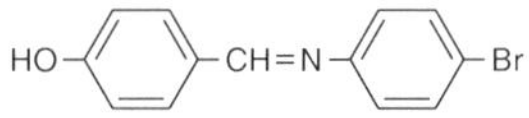
5 gm

S68,450-3
ALPHA-(3-CHLORO-4-METHYLPHENYLIMINO)-P-CRESOL
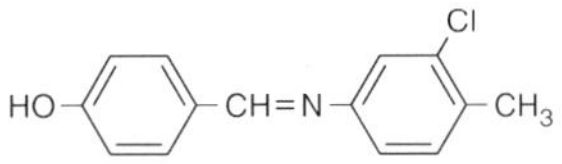
5 gm

S68,456-2
ALPHA-(4-CHLORO-2-METHYLPHENYLIMINO)-P-CRESOL
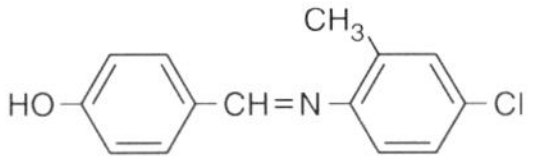
5 gm

S68,481-3
ALPHA-(3-CHLORO-6-METHYLPHENYLIMINO)-P-CRESOL
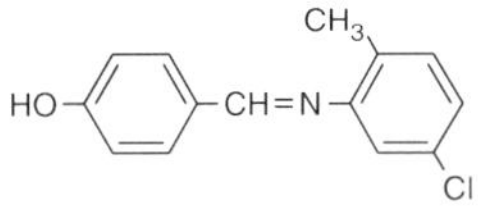
5 gm

S77,182-1
1-(3-BROMO-ALPHA-HEPTYL-4-HYDROXY-2-PHENYLBENZYLIDENEAMINO)-GUANIDINE HYDROCHLORIDE
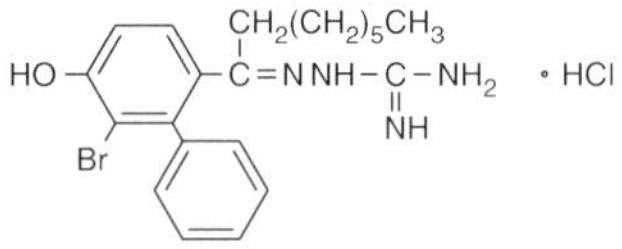
250 mg

S79,093-1
N-(3,4-DICHLOROBENZYLIDENE)-ETHANOLAMINE
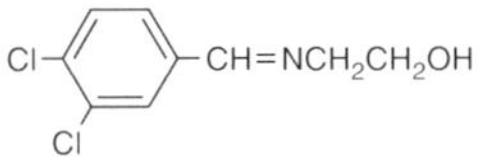
250 mg

S73,047-5
2-(2,4-DICHLOROBENZYLIDENEAMINO)-P-CRESOL
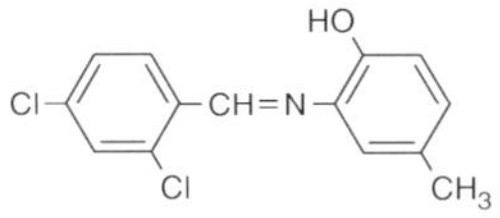
5 gm

S69,870-9
1-(4-CHLORO-3-NITROPHENYLIMINOMETHYL)-2-NAPHTHOL
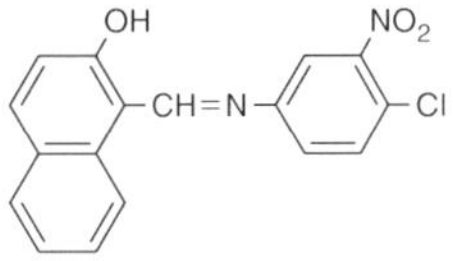
5 gm

S69,881-4
1-(2-CHLORO-5-NITROPHENYLIMINOMETHYL)-2-NAPHTHOL
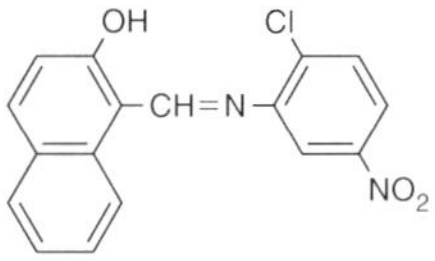
5 gm

S72,005-4
5-CHLORO-2-(2-HYDROXYBENZYLIDENEAMINO)-BENZOPHENONE
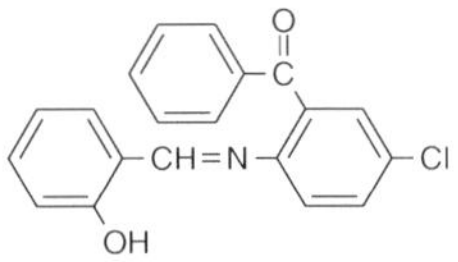
5 gm

S71,972-2
5-CHLORO-2-(2-HYDROXY-1-NAPHTHYLMETHYLENEAMINO)BENZOPHENONE
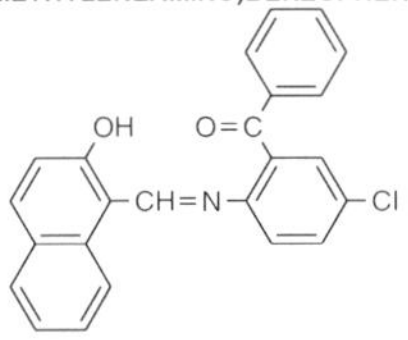
5 gm

S68,630-1
1-(2-FLUOROPHENYLIMINOMETHYL)-2-NAPHTHOL
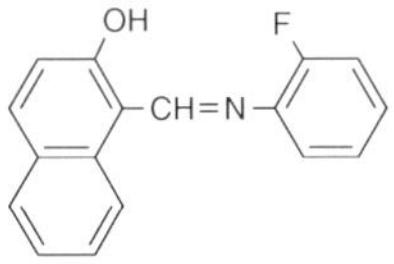
5 gm

S68,664-6
1-(3,4-DICHLOROPHENYLIMINOMETHYL)-2-NAPHTHOL
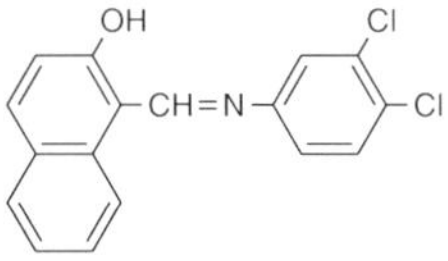
5 gm

S66,991-1
N-(4-BENZYLOXYBENZYLIDENE)-3-CHLOROANILINE
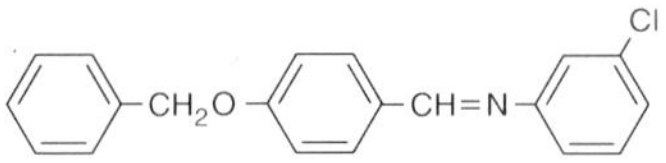
5 gm

S66,988-1
N-(4-BENZYLOXYBENZYLIDENE)-4-CHLORO-2-METHYLANILINE
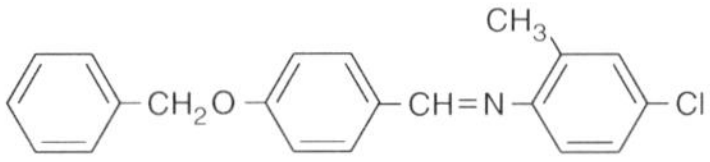
5 gm

S77,158-9
1-[2-(4-CHLOROBENZYLOXY)-BENZYLIDENEAMINO]GUANIDINE NITRATE
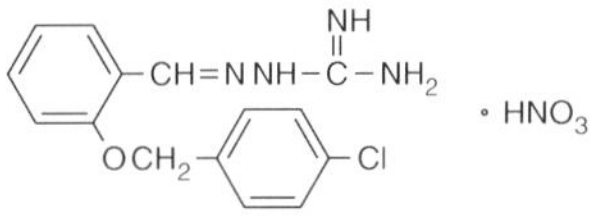
250 mg

S77,155-4
1-[2-(2,4-DICHLOROBENZYLOXY)-BENZYLIDENEAMINO]GUANIDINE NITRATE
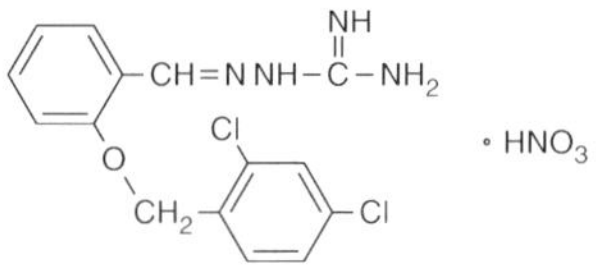
250 mg

S77,156-2
1-[2-(3,4-DICHLOROBENZYLOXY)-BENZYLIDENEAMINO]GUANIDINE NITRATE
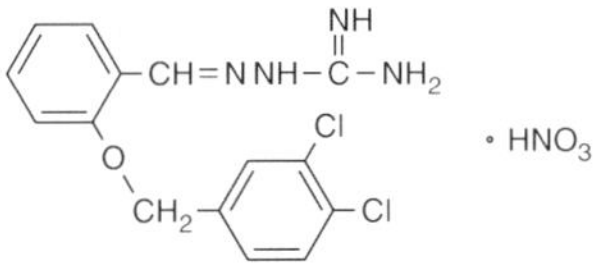
250 mg

S77,157-0
1-[4-(3,4-DICHLOROBENZYLOXY)-BENZYLIDENEAMINO]GUANIDINE ACETATE
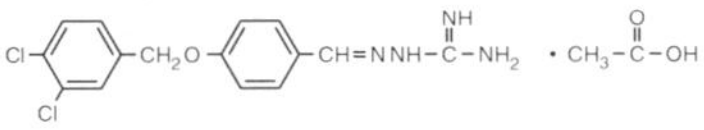
250 mg

S77,154-6
1-[5-BROMO-2-(4-CHLOROBENZYLOXY)-BENZYLIDENEAMINO]GUANIDINE HYDROCHLORIDE
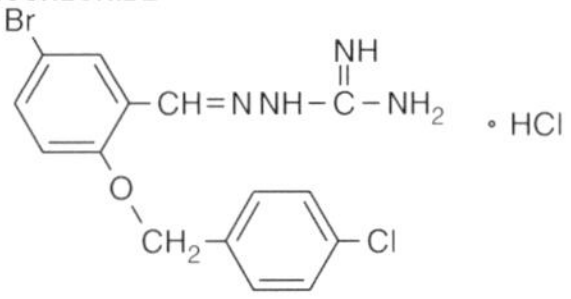
250 mg

S77,177-5
1-[4-(3,4-DICHLOROBENZYLOXY)-ALPHA-HEPTYLBENZYLIDENEAMINO]-GUANIDINE NITRATE
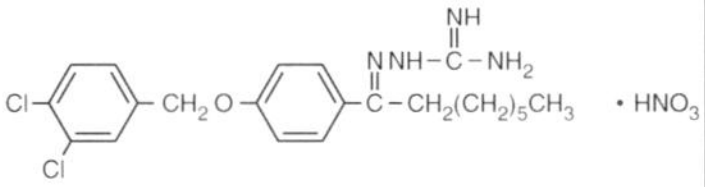
250 mg

S77,153-8
1-[3-(3,4-DICHLOROBENZYLOXY)-BENZYLIDENEAMINO]GUANIDINE HYDROCHLORIDE
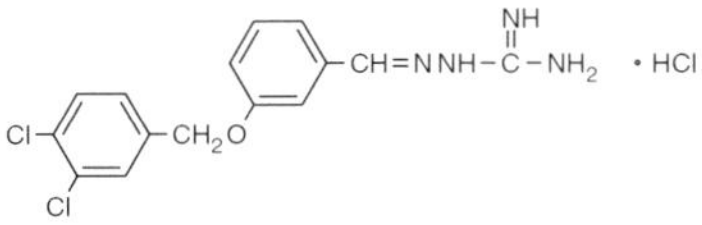
250 mg

S73,026-2
5-CHLORO-N-(2,4-DICHLOROBENZYLIDENE)-2,4-DIMETHOXYANILINE
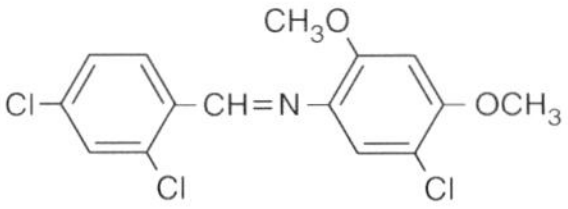
5 gm

S67,647-0
2,2'-OXYDIACETIC ACID BIS[(4-CHLOROBENZYLIDENE)HYDRAZIDE]
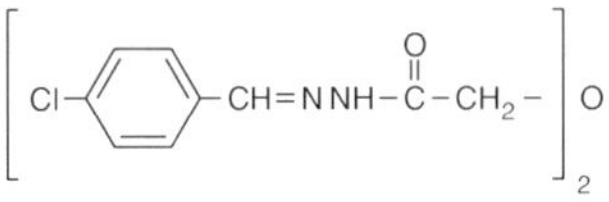
5 gm

S72,430-0
2,2'-OXYBIS[ACETIC ACID (2,4-DICHLOROBENZYLIDENE)HYDRAZIDE]
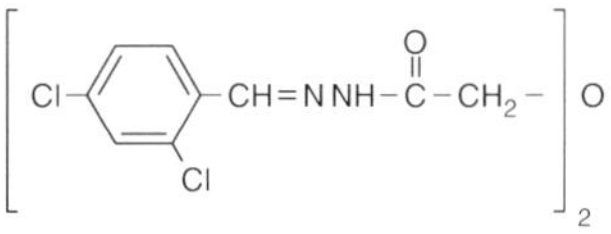
5 gm

S92,506-3
2-PHENOXYACETIC (4-CHLOROBENZYLIDENE)HYDRAZIDE
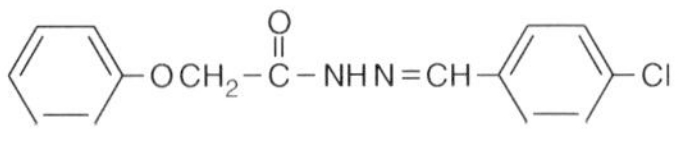
250 mg

S79,407-4
2METHOXY4[4(N-METHYLACETAMIDO)-PHENYLIMINOMETHYL]PHENYL N-(3-CHLOROPHENYL)CARBAMATE
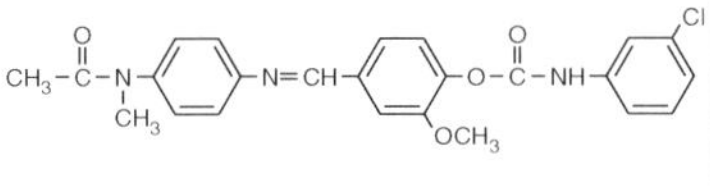
250 mg

S79,410-4
4-(3-CHLOROPHENYLIMINOMETHYL)-2-METHOXYPHENYL N-(3-CHLOROPHENYL)-CARBAMATE
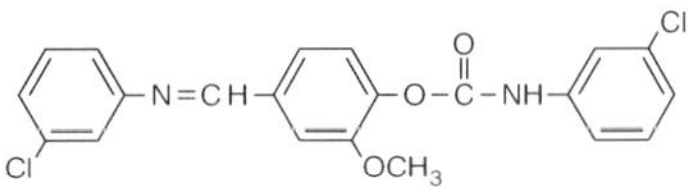
250 mg

S90,167-9
N-(2,4-DICHLOROBENZYLIDENE)-2-METHOXY-5-NITROANILINE
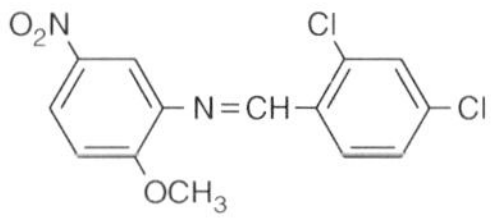
250 mg

S58,440-1
N,N'-(1,4-PHENYLENEDIMETHYLIDYNE)-BIS(3-CHLORO-O-ANISIDINE)
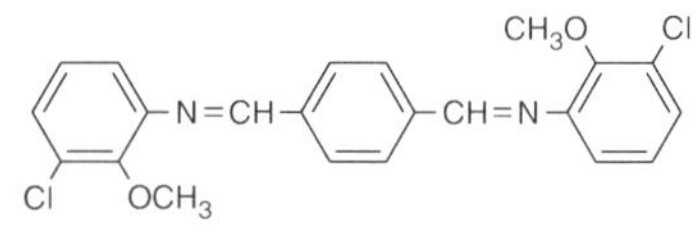
5 gm

S67,773-6
N-(4-BENZYLOXYBENZYLIDENE)-4-CHLORO-3-NITROANILINE
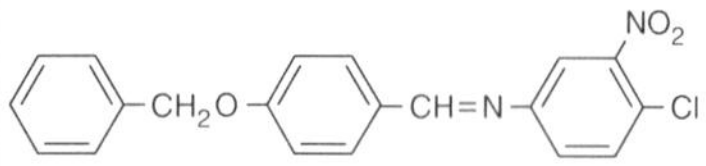
5 gm

S92,501-2
4-METHOXYBENZOIC (2,6-DICHLOROBENZYLIDENE)HYDRAZIDE
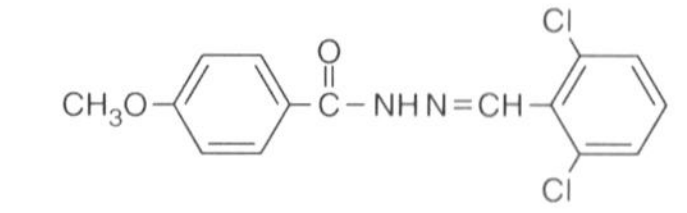
250 mg

S67,329-3
2-CHLORO-N-(2-ETHOXYBENZYLIDENE)-ANILINE
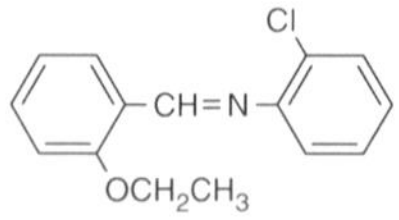
250 mg

S67,309-9
2,5-DICHLORO-N-(2-ETHOXYBENZYLIDENE)ANILINE
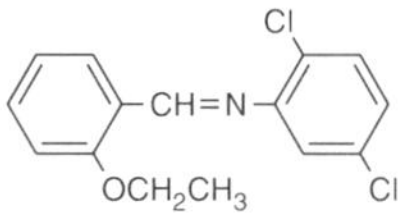
250 mg

S72,302-9
1-(2,4-DICHLOROBENZYLIDENE)-4-(2-METHOXY-5-METHYLPHENYL)-3-THIOSEMICARBAZIDE
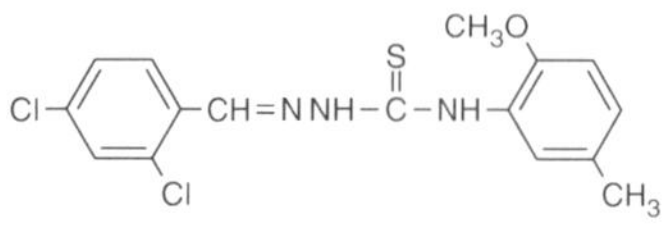
250 mg

S77,190-2
1-[(3-IODO-4-METHOXYBENZYLIDENE)-AMINO]GUANIDINE NITRATE
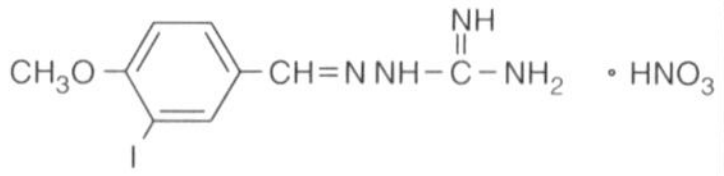
250 mg

S68,487-2
N-(4-FLUOROBENZYLIDENE)-4-PHENOXYANILINE
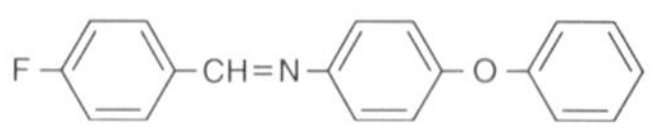
5 gm

S91,422-3
N-[ALPHA-(4-CHLOROPHENOXY)-BENZYLIDENE]ANILINE
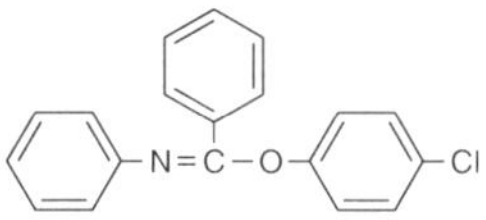
250 mg

S69,808-3
N-(2,4-DICHLOROBENZYLIDENE)-4-PHENOXYANILINE
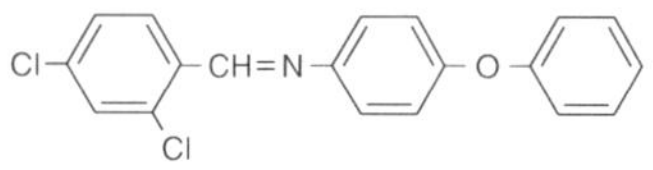
5 gm

S70,096-7
4-FLUORO-N-(4-METHOXYBENZYLIDENE)-ANILINE
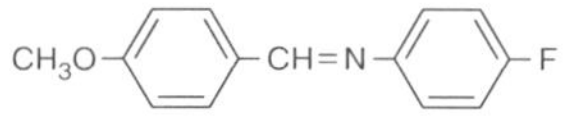
5 gm

S83,964-7
3-CHLORO-N-(4-METHOXYBENZYLIDENE)-ANILINE
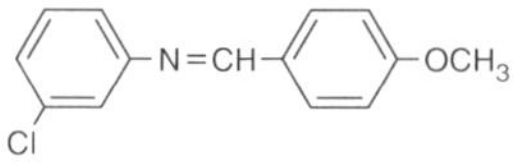
250 mg

S70,125-4
4-IODO-N-(4-METHOXYBENZYLIDENE)-ANILINE
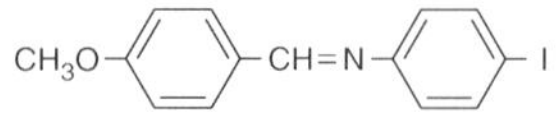
5 gm

S90,168-7
N-(2,4-DICHLOROBENZYLIDENE)-4-METHOXYANILINE
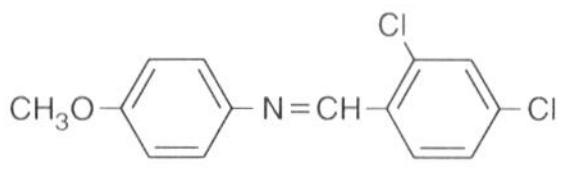
250 mg

S73,357-1
2-(4-CHLOROBENZOYL)-1-[ALPHA-(METHYLTHIO)PHENYLIMINOMETHYL]-PIPERIDINE HYDRIODIDE
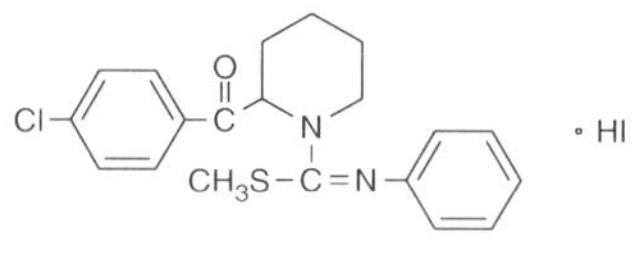
250 mg

S67,853-8
4,4'-DITHIOBIS[N-(4-CHLOROBENZYLIDENE)ANILINE]
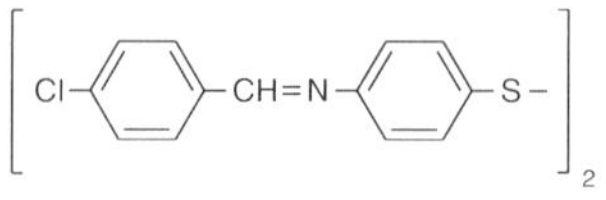
5 gm

S74,808-0
4-CHLOROPHENYL 2-PYRROLYL KETONE HYDRAZONE
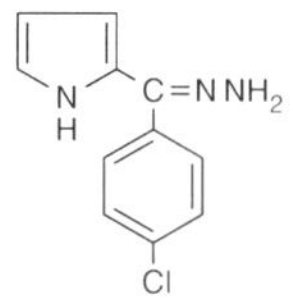
250 mg

S90,220-9
2-(3-FURFURYLUREIDO)ACETIC (2-FLUOROBENZYLIDENE)HYDRAZIDE
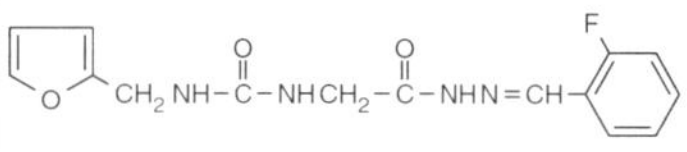
250 mg

S90,221-7
2-(3-FURFURYLUREIDO)ACETIC (4-FLUOROBENZYLIDENE)HYDRAZIDE
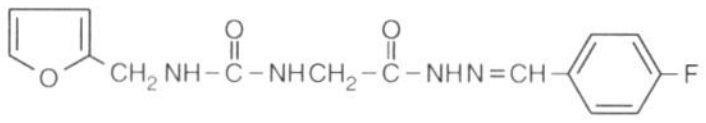
250 mg

S68,725-1
N-(3,4-DICHLOROPHENYL)-N'-(FURFURYLIDENEAMINO)OXAMIDE
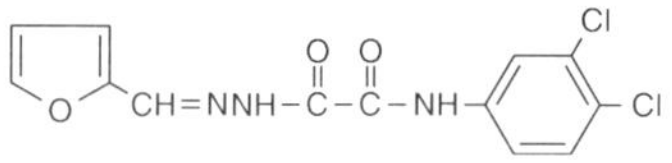
250 mg

S92,473-3
OXALIC BIS(4-CHLOROBENZYLIDENE)-HYDRAZIDE
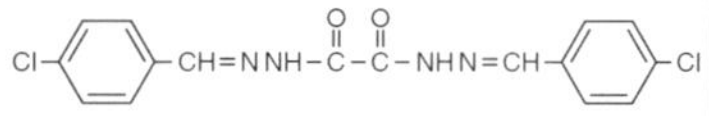
250 mg

S93,850-5
1-BENZYL-4-(4-CHLOROPHENYLHYDRAZONO)-3-METHYL-2-PYRAZOLIN-5-ONE
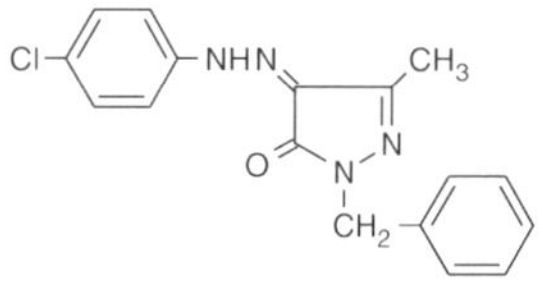
250 mg

S70,297-8
N,N'-BIS(2,4-DICHLOROBENZYLIDENE)-ETHYLENEDIAMINE
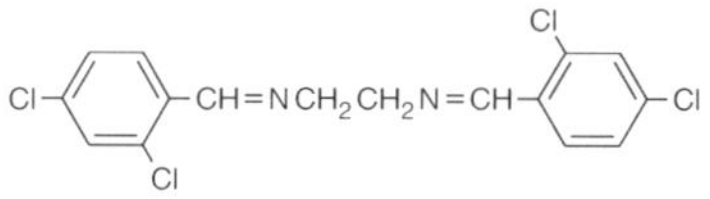
5 gm

S76,198-2
N-(4-CHLOROBENZYLIDENE)-2-MORPHOLINOETHYLAMINE
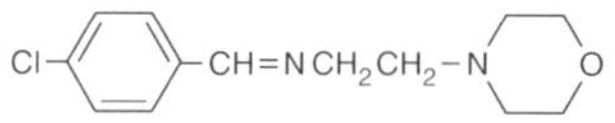
250 mg

S71,915-3
N-(2,4-DICHLOROBENZYLIDENE)-3-NITROANILINE
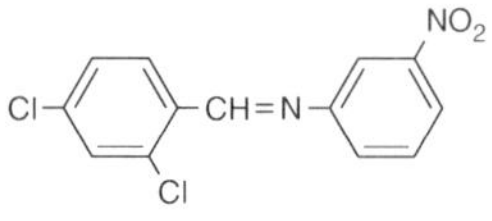
5 gm

S73,081-5
4-CHLORO-N-(2,4-DICHLOROBENZYLIDENE)-3-NITROANILINE
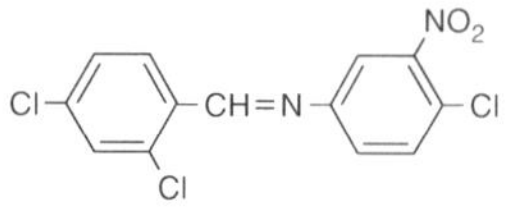
5 gm

S73,080-7
N-(2,4-DICHLOROBENZYLIDENE)-3-NITRO-P-TOLUIDINE
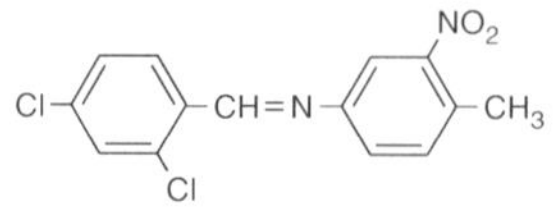
5 gm

S87,896-0
4-FLUORO-N-[4,5-(METHYLENEDIOXY)-2-NITROBENZYLIDENE]ANILINE
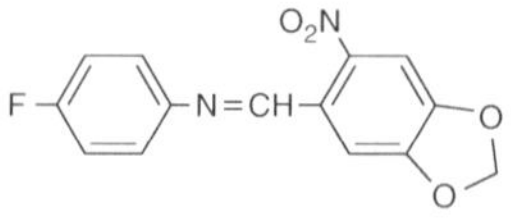
5 gm

S68,925-4
N-(4-CHLOROBENZYLIDENE)-N'-PHENYL-1,4-PHENYLENEDIAMINE
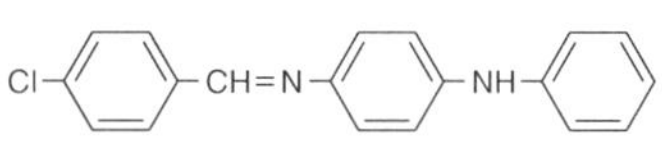
5 gm

S68,901-7
N-(2,4-DICHLOROBENZYLIDENE)-N'-PHENYL-1,4-PHENYLENEDIAMINE
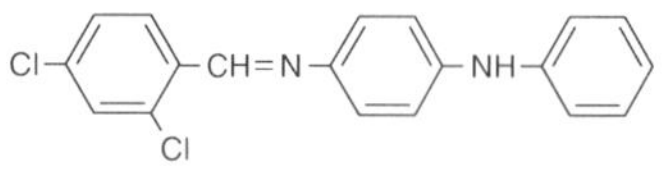
5 gm

S71,913-7
N-(2,4-DICHLOROBENZYLIDENE)-4-NITROANILINE
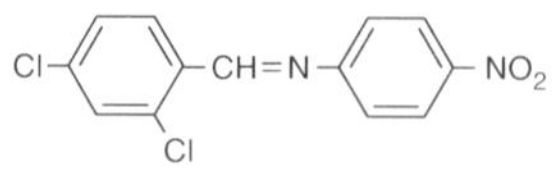
5 gm

S87,906-1
N-(4-CHLORO-3-NITROBENZYLIDENE)-4-FLUOROANILINE
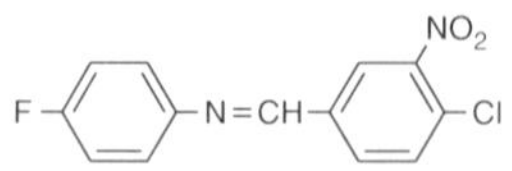
5 gm

S87,914-2
N-(4-CHLORO-3-NITROBENZYLIDENE)-4-ETHYLANILINE
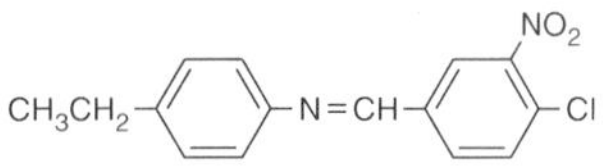
5 gm

S73,028-9
N-(2,4-DICHLOROBENZYLIDENE)-2-METHYL-4-NITROANILINE
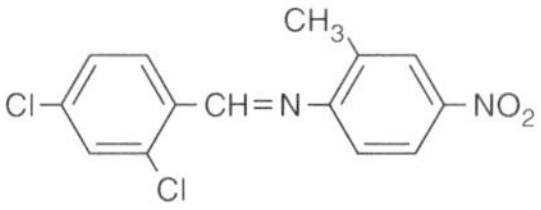
5 gm

S92,487-3
N-METHYLANTHRANILIC (2,4-DICHLOROBENZYLIDENE)HYDRAZIDE
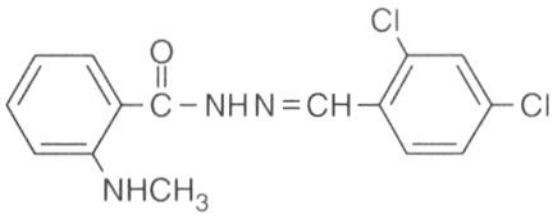
250 mg

S92,490-3
P-TOLUIC [2-CHLORO-4-(DIMETHYLAMINO)BENZYLIDENE]-HYDRAZIDE
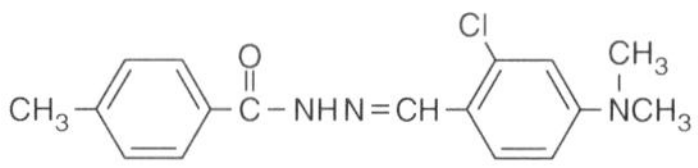
250 mg

S92,509-8
ACETIC (2-CHLORO-4-DIMETHYLAMINOBENZYLIDENE)HYDRAZIDE
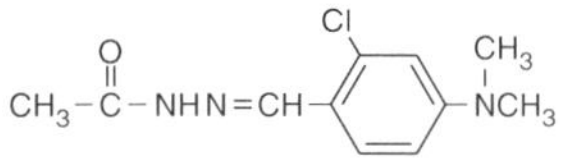
250 mg

S79,680-8
2-(3-FURFURYLUREIDO)ACETIC [2-CHLORO-4-(DIMETHYLAMINO)-BENZYLIDENE]HYDRAZIDE
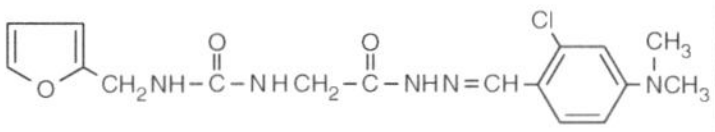
250 mg

S68,653-0
4,4'-METHYLENEBIS(N-[2-CHLORO-4-(DIMETHYLAMINO)BENZYLIDENE]ANILINE)
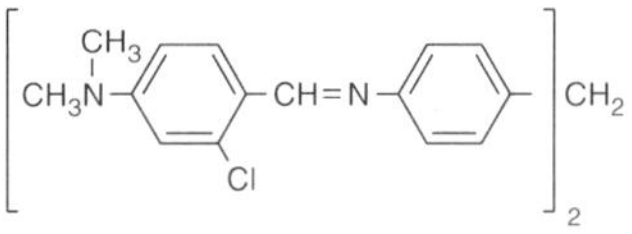
5 gm

S70,300-1
N-[4-(DIMETHYLAMINO)BENZYLIDENE]-4-FLUOROANILINE
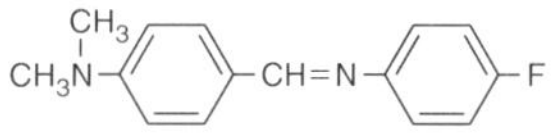
5 gm

S71,947-1
N,N-DIMETHYL-ALPHA-(4-IODOPHENYLIMINO)-P-TOLUIDINE
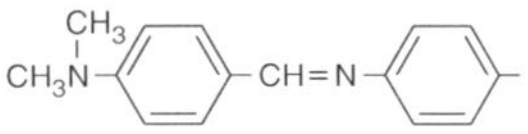
5 gm

S60,423-2
N-(BENZYLIDENE)-2,5-DICHLOROANILINE
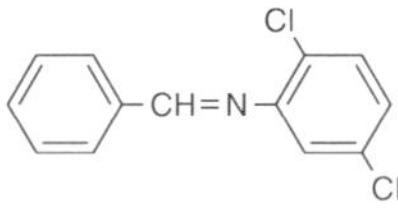
250 mg

S39,618-4
4-CHLORO-N-(2-CHLOROBENZYLIDENE)-ANILINE
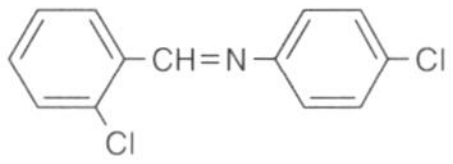
5 gm

S92,494-6
P-TOLUIC (2-CHLOROBENZYLIDENE)-HYDRAZIDE
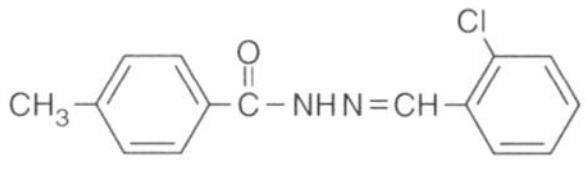
250 mg

S67,674-8
SEBACIC ACID BIS(2-CHLOROBENZYLIDENEHYDRAZIDE)
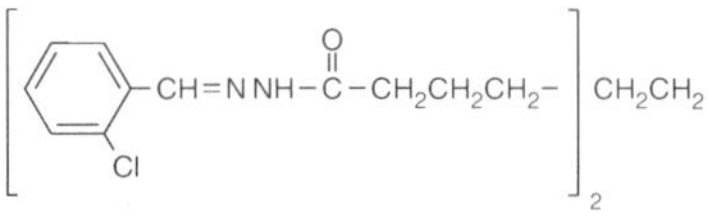
5 gm

S92,445-8
METHYL (2-CHLOROBENZYLIDENE)-CARBAZATE
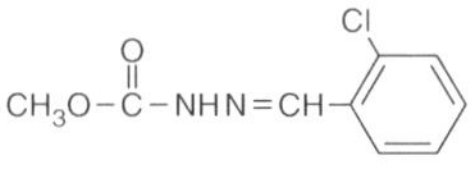
250 mg

S92,510-1
ACETIC (3-FLUOROBENZYLIDENE)-HYDRAZIDE
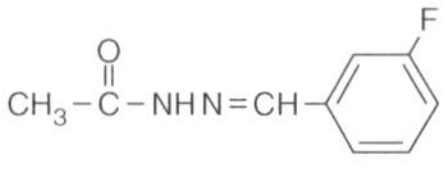
250 mg

S76,209-1
N-(2,6-DICHLOROBENZYLIDENEIMINO)-GUANIDINE
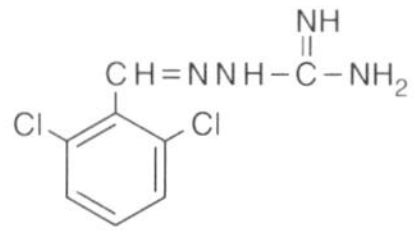
250 mg

S75,328-9
N-(4-CHLOROBENZYLIDENE)ETHANOLAMINE
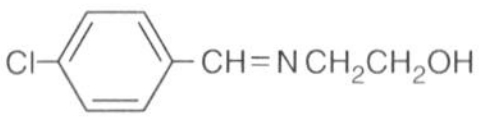
5 gm

S77,094-9
3-(2-CHLOROBENZYLIDENEAMINO)-6H-DIBENZO(B,D)PYRAN-6-ONE
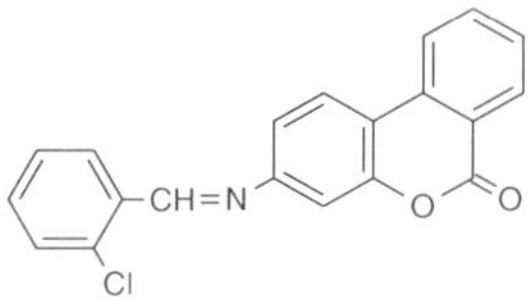
250 mg

S81,840-2
N-BENZYLIDENE-4-IODOANILINE
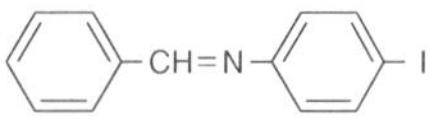
250 mg

S68,666-2
4,4'-METHYLENEBIS[N-(2-CHLOROBENZYLIDENE)ANILINE]
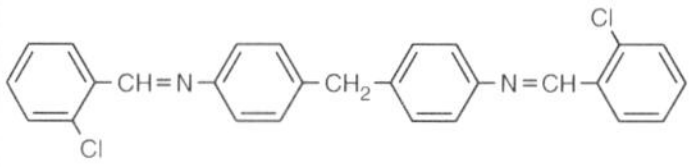
5 gm

S68,163-6
4-CHLOROBENZAMIDOXIME
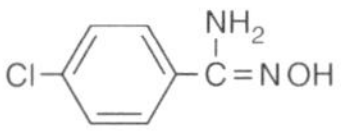
250 mg

S93,451-8
4-(3-BROMOBENZYLIDENEAMINO)-4H-1,2,4-TRIAZOLE
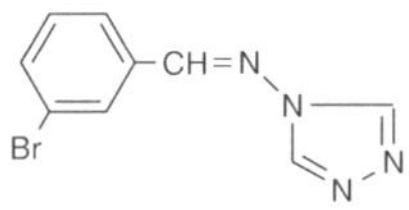
250 mg

S90,285-3
2,4-DICHLOROBENZOIC (FURFURYLIDENE)HYDRAZIDE
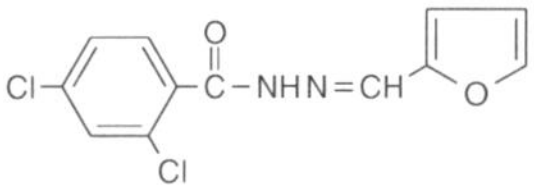
250 mg

S90,288-8
3,4-DICHLOROBENZOIC (FURFURYLIDENE)HYDRAZIDE
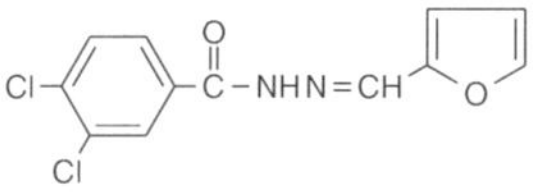
250 mg

S88,611-4
1-(4-BROMO-ALPHA-NITROBENZYLIDENE)-2-PHENYLHYDRAZINE
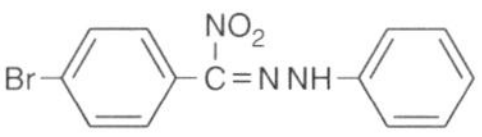
250 mg

S77,173-2
8-(2,6-DICHLOROBENZYLIDENE-HYDRAZINO)XANTHINE
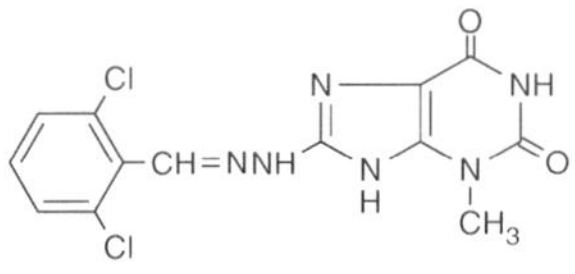
250 mg

S73,096-3
ISONICOTINIC 3,4-DICHLOROBENZYLIDENEHYDRAZIDE
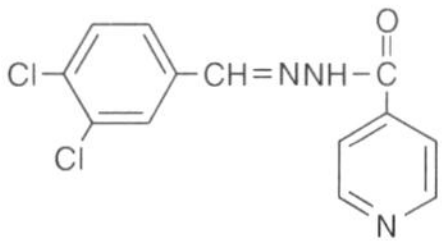
5 gm

S68,741-3
N-(3-CHLOROPHENYL)-N'-(ALPHA-METHYLCINNAMYLIDENEAMINO)OXAMIDE
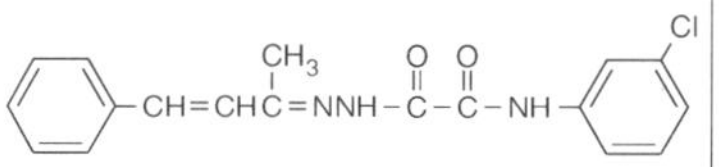
250 mg

S71,631-6
SEBACIC ACID BIS(3,4-DICHLOROBENZYLIDENEHYDRAZIDE)
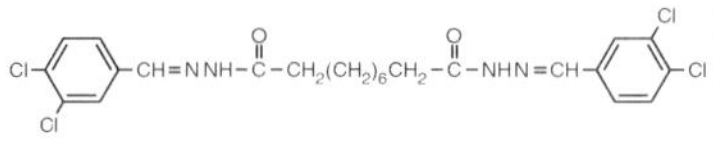
250 mg

S93,154-3
2-(4-CHLOROPHENYLHYDRAZONO)-2-CYANOACETAMIDE
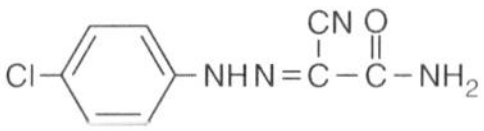
250 mg

S92,449-0
1-(4-METHYL-2-PENTYLIDENE)-4-(2-CHLOROPHENYL)SEMICARBAZIDE
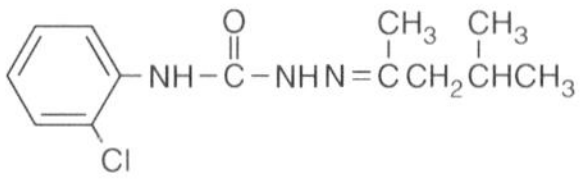
250 mg

S92,447-4
1-(4-METHYL-2-PENTYLIDENE)-4-(3-CHLORO-2-METHYLPHENYL)SEMICARBAZIDE
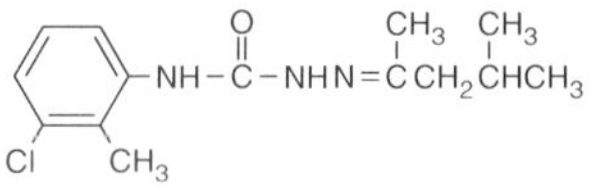
250 mg

S69,452-5
4-(3-CHLOROPHENYL)-1-(FLUOREN-9-YLIDENE)SEMICARBAZIDE
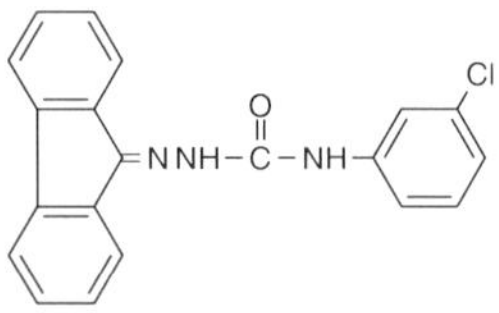
250 mg

S72,346-0
2-(FURFURYLUREIDO)ACETIC ACID(2-CHLOROBENZYLIDENE)HYDRAZIDE
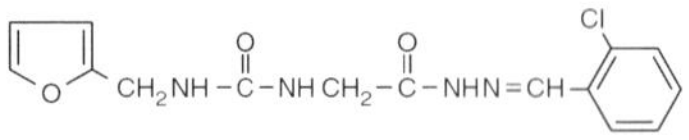
250 mg

S69,433-9
2-[3-(2-THIAZOLYL)UREIDO]ACETIC ACID N2-(2,6-DICHLOROBENZYLIDENE)-HYDRAZIDE
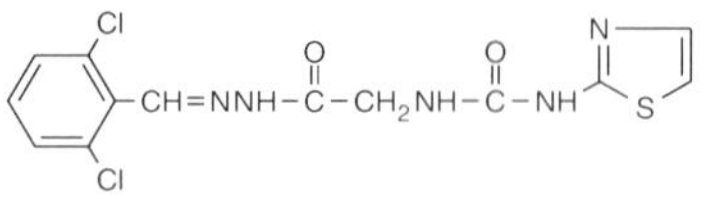
250 mg

S92,550-0
10-UNDECENOIC (4-CHLORO-ALPHA-PHENYLBENZYLIDENE)HYDRAZIDE
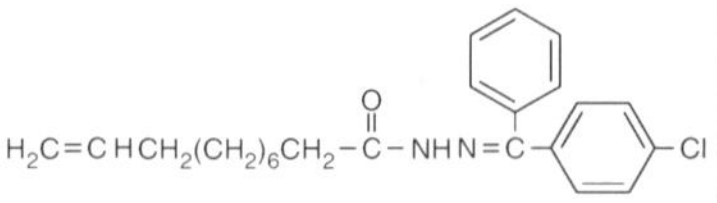
250 mg

S92,545-4
10-UNDECENOIC (4-CHLOROBENZYLIDENE)-HYDRAZIDE
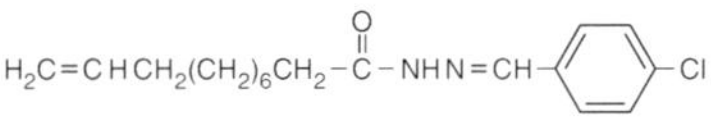
250 mg

S91,346-4
OCTADECANOIC (4-CHLOROBENZYLIDENE)-HYDRAZIDE
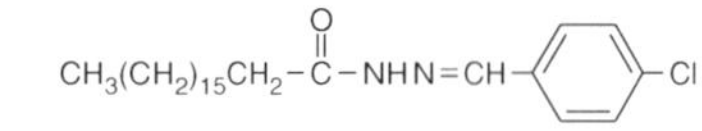
250 mg

S74,837-4
3-(4-CHLORO-2-FLUOROBENZYLIDENEAMINO)-2-METHYL-2-THIOPSEUDOUREA HYDRIODIDE
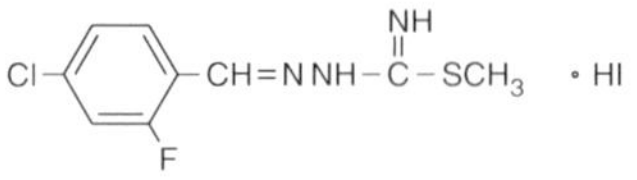
250 mg

S77,195-3
1-[(4-CHLORO-ALPHA-CYCLOPROPYLBENZYLIDENE)AMINO]-GUANIDINE NITRATE
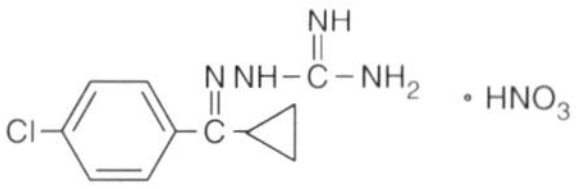
250 mg

S76,371-3
2,6-DICHLOROBENZALDEHYDE 2-IMIDAZOLIN-2-YLHYDRAZONE HYDROCHLORIDE
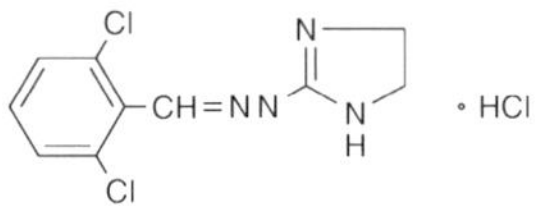
250 mg

S69,550-5
4-CYCLOHEXYL-1-(2,4-DICHLOROBENZYLIDENE)-3-THIOSEMICARBAZIDE
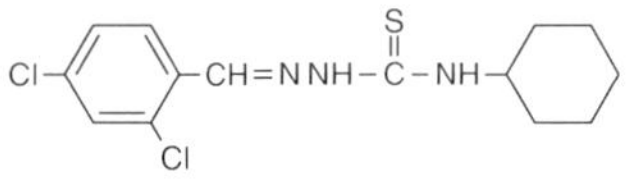
250 mg

S53,654-7
2-CHLOROBENZALDEHYDE THIOSEMICARBAZONE
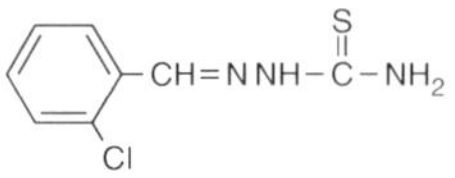
5 gm

S77,158-3
N-(2-CHLOROBENZYLIDENEAMINO)-ACETAMIDINE HYDRIODIDE
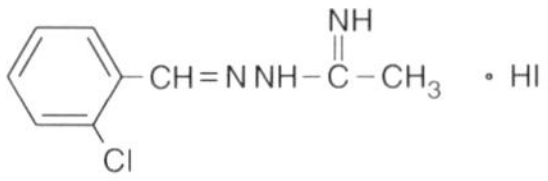
250 mg

S77,196-1
1-[(4-BROMOBENZYLIDENE)AMINO]-GUANIDINE ACETATE
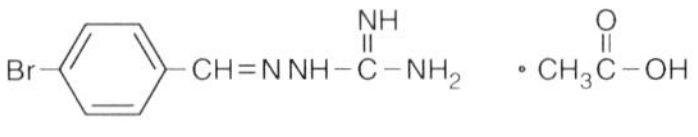
250 mg

S93,441-0
4-(4-FLUOROBENZYLIDENEAMINO)-4H-1,2,4-TRIAZOLE
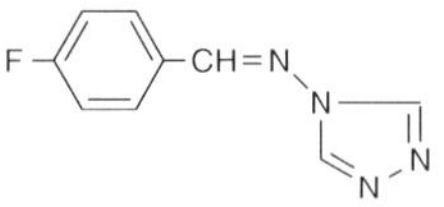
250 mg

S93,440-2
4-(4-BROMOBENZYLIDENEAMINO)-4H-1,2,4-TRIAZOLE
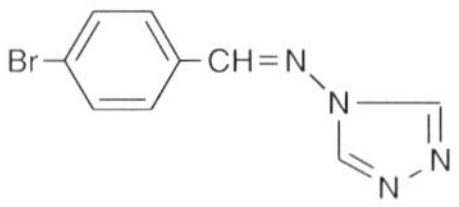
250 mg

S77,138-4
1-(4-CHLORO-ALPHA-PENTYLBENZYLIDENEAMINO)-GUANIDINE NITRATE
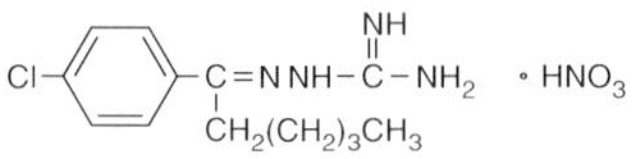
250 mg

S53,663-6
2,6-DICHLOROBENZALDEHYDE THIOSEMICARBAZONE
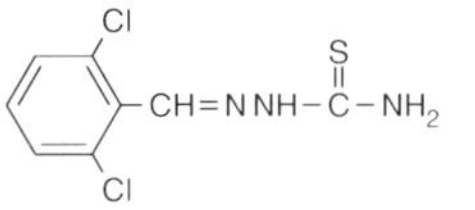
250 mg

S77,160-0
2,6-DICHLOROBENZALDEHYDE 3-PYRIDYLHYDRAZONE NITRATE
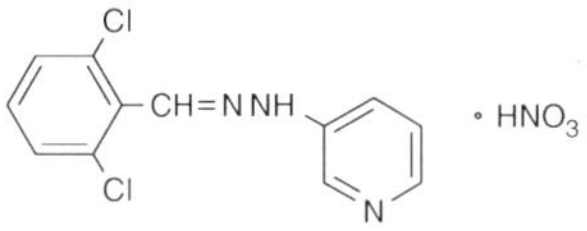
250 mg

S77,163-5
1-[(3,4-DICHLOROBENZYLIDENE)AMINO]-GUANIDINE ACETATE
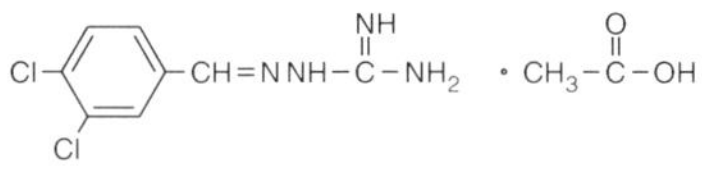
250 mg

S77,191-0
1-[(2,5-DIBROMO-ALPHA-METHYLBENZYLIDENE)AMINO]-GUANIDINE NITRATE
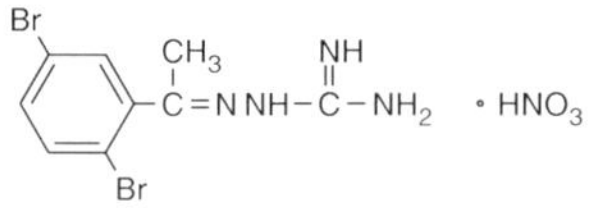
250 mg

S77,142-2
1-(2,4-DICHLORO-ALPHA-METHYLCINNAMYLIDENEAMINO)-GUANIDINE NITRATE
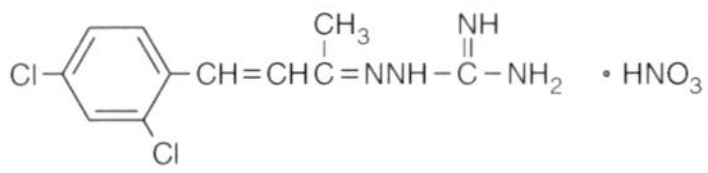
250 mg

S76,886-3
3-PYRIDYLMETHYLENEAMINO N-(2-CHLORO-4-NITROPHENYL)CARBAMATE
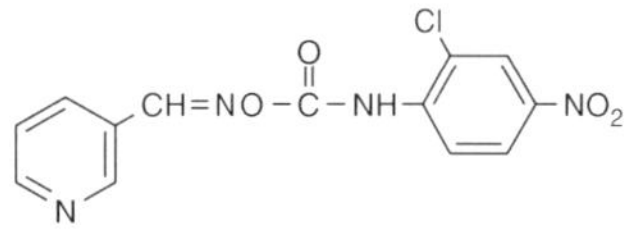
250 mg

S91,809-1
2-(4-CHLOROPHENYLAZO)-2-NITROPROPANE
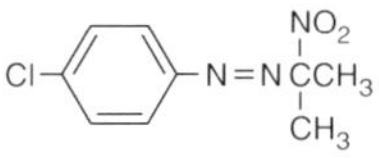
250 mg

S71,973-0
4'-(4-CHLOROBENZYLIDENEAMINO)-ACETOPHENONE
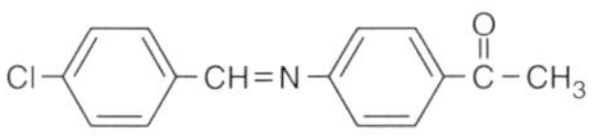
5 gm

S71,968-4
4'-(2,4-DICHLOROBENZYLIDENEAMINO)-ACETOPHENONE
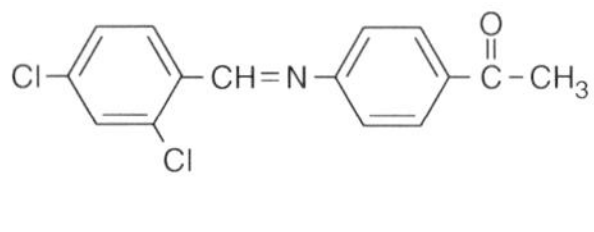
5 gm

S84,510-8
3'-(2,4-DICHLOROBENZYLIDENEAMINO)-ACETOPHENONE
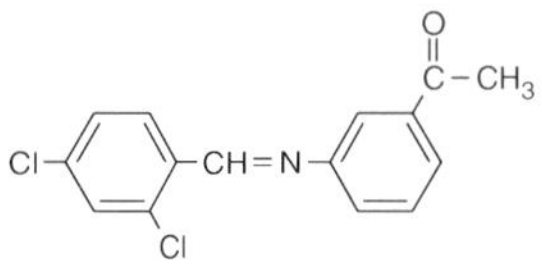
250 mg

S72,180-8
4-(2,4-DICHLOROBENZYLIDENEAMINO)-BENZOIC ACID
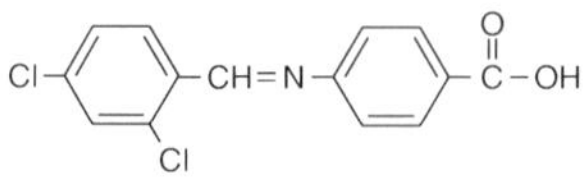
5 gm

S73,027-0
3-(2,4-DICHLOROBENZYLIDENEAMINO)-BENZOIC ACID
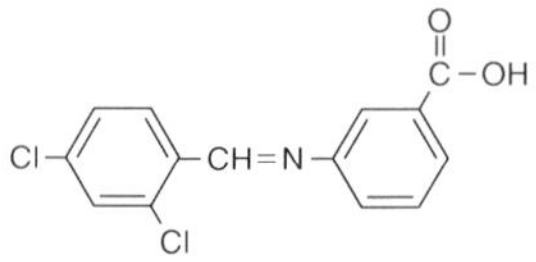
5 gm

S73,082-3
ETHYL 4-(2,4-DICHLOROBENZYLIDENEAMINO)BENZOATE
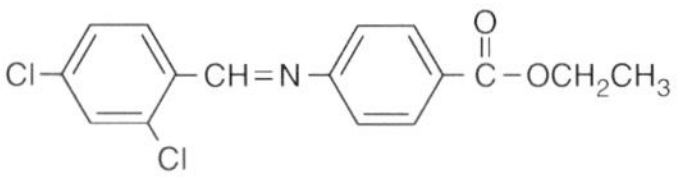
5 gm

S89,972-0
N-[3-(4-CHLOROPHENYL)-4-PHENYL-4-THIAZOLIN-2-YLIDENE]CINNAMAMIDE
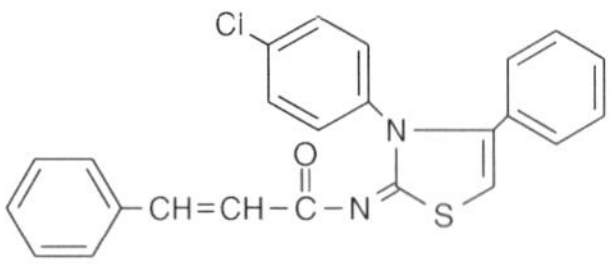
250 mg

S90,171-7
4-(4-FLUOROPHENYLIMINOMETHYL)-PHENYL N-METHYLCARBAMATE
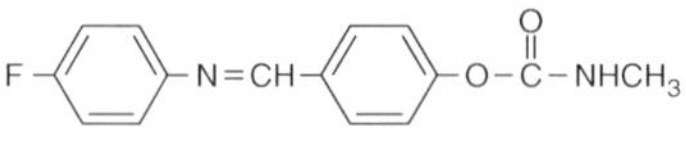
250 mg

S90,170-9
4-(3-CHLOROPHENYLIMINOMETHYL)-PHENYL N-METHYLCARBAMATE
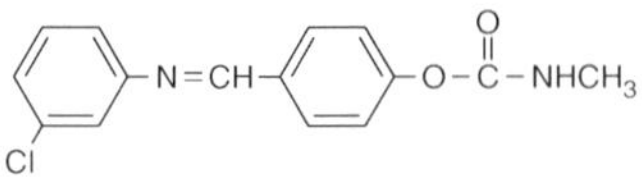
250 mg

S73,737-2
4-PYRIDYLMETHYLENEAMINO N-(4-CHLORO-2-METHYLPHENYL)-CARBAMATE

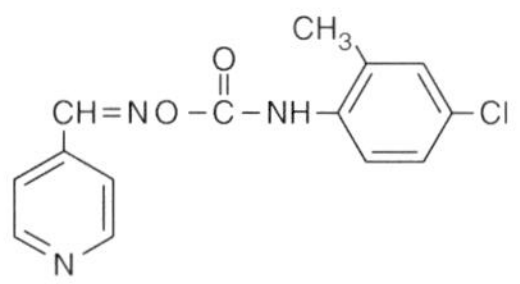

5 gm

S76,915-0
4'-(4-FLUOROPHENYLIMINOMETHYL)-ACETANILIDE

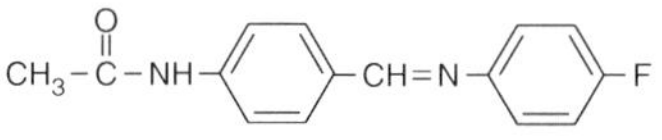

250 mg

S90,169-5
4'-(3-CHLOROPHENYLIMINOMETHYL)-ACETANILIDE

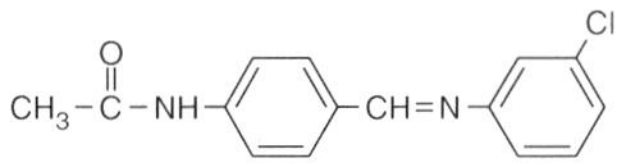

250 mg

S71,981-1
4-(4-CHLOROBENZYLIDENEAMINO)-ANTIPYRINE

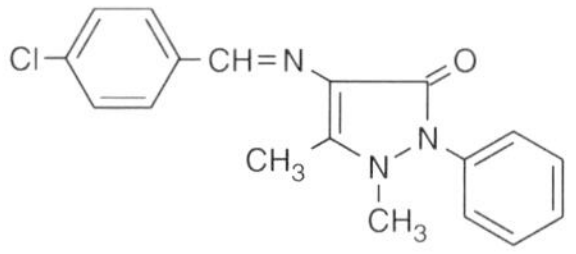

5 gm

S71,980-3
4-(2,4-DICHLOROBENZYLIDENEAMINO)-ANTIPYRINE

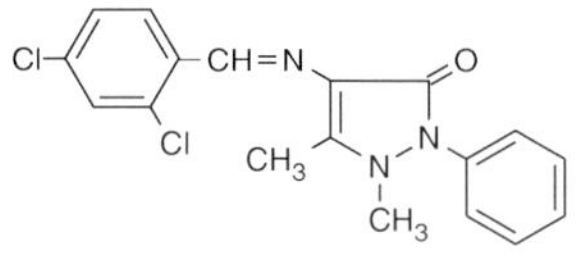

5 gm

S76,810-3
(2-PYRIDYL)METHYLENEAMINO N-(2,5-DICHLOROPHENYL)CARBAMATE

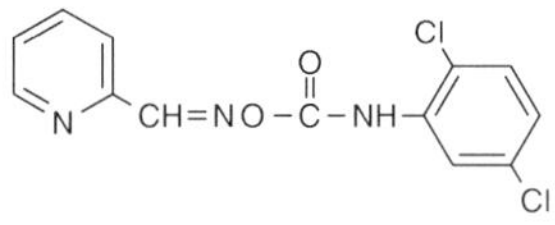

250 mg

S92,564-0
TRIS(6-CHLORO-3-METHYL-2-BENZOTHIAZOLINYLIDENEAMINO)-CARBENIUM TETRAFLUOROBORATE

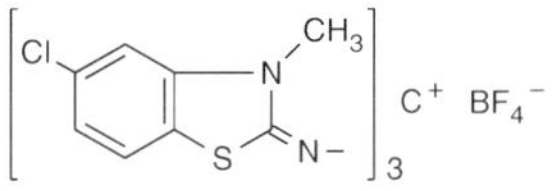

250 mg

S82,619-7
N-(3-BROMOBENZYLIDENE)-ANILINE PICRATE

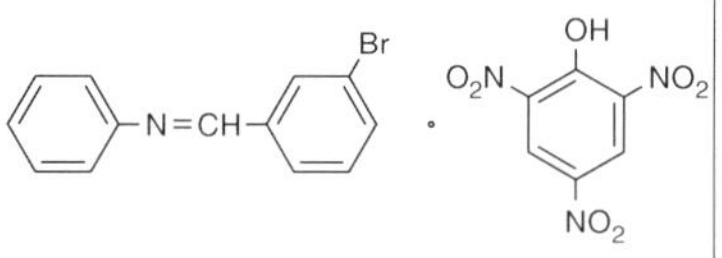

250 mg

S89,631-4
N-BENZYLIDENE-3-BROMOANILINE

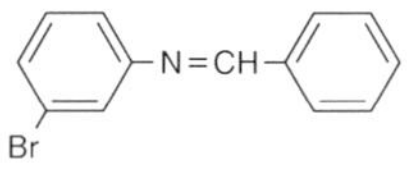

250 mg

S90,076-1
3-CHLORO-N-CINNAMYLIDENEANILINE

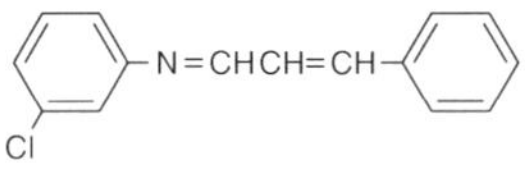

250 mg

S88,035-3
N-(2-BROMOBENZYLIDENE)ANILINE

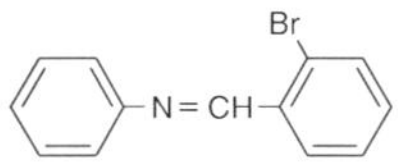

250 mg

S83,488-2
N-(4-BROMOBENZYLIDENE)ANILINE

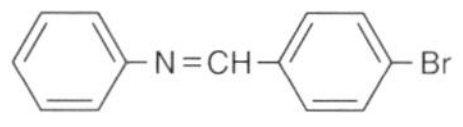

250 mg

S75,001-8
N-CINNAMYLIDENE-4-FLUOROANILINE

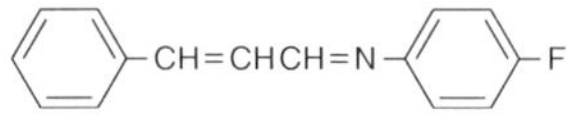

5 gm

S67,864-3
N-(4-ISOPROPYLBENZYLIDENE)-4-FLUOROANILINE

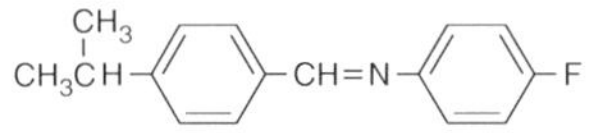

5 gm

S67,865-1
N-(4-ISOPROPYLBENZYLIDENE)-4-CHLOROANILINE

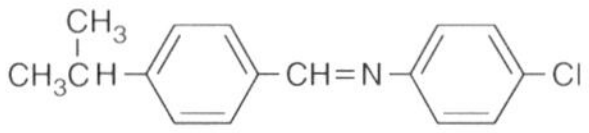

5 gm

S67,871-6
N-(4-ISOPROPYLBENZYLIDENE)-4-BROMOANILINE

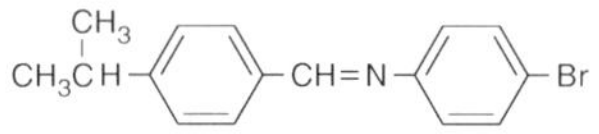

5 gm

S73,046-7
N-(2,4-DICHLOROBENZYLIDENE)-4-FLUOROANILINE

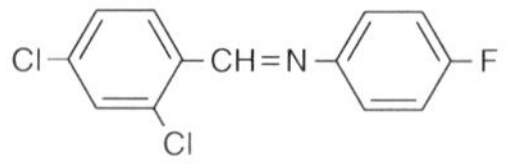

5 gm

S73,025-4
N-(2,4-DICHLOROBENZYLIDENE)-4-IODOANILINE

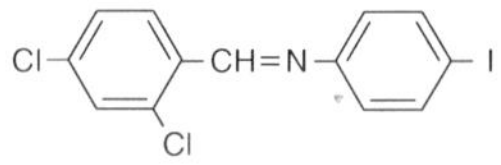

5 gm

S74,636-3
N,N'-TEREPHTHALYLIDENEBIS(4-BROMO-2-CHLOROANILINE)

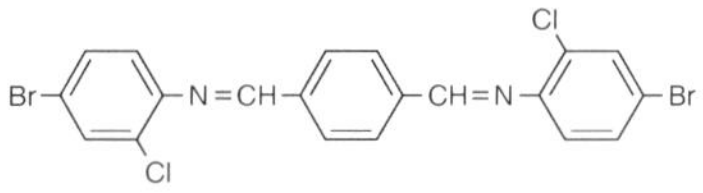

5 gm

S68,515-1
4,4'-METHYLENEBIS[N-(2,4-DICHLOROBENZYLIDENE)ANILINE]

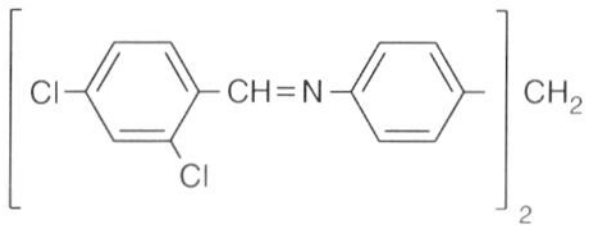

5 gm

S68,623-9
4,4'-METHYLENEBIS[N-(4-FLUOROBENZYLIDENE)ANILINE]

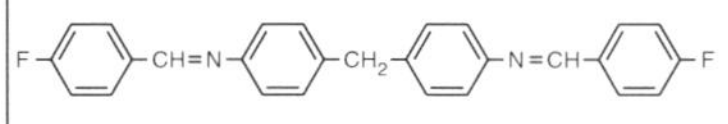

5 gm

S68,669-7
4,4'-METHYLENEBIS[N-(4-CHLOROBENZYLIDENE)ANILINE]

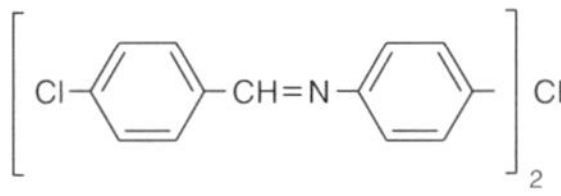

5 gm

S73,077-7
N-(2,4-DICHLOROBENZYLIDENE)-2,5-XYLIDINE

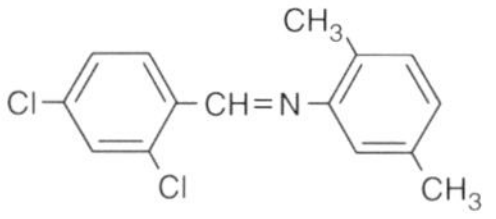

5 gm

S73,079-3
N-(2,4-DICHLOROBENZYLIDENE)-3,4-XYLIDINE

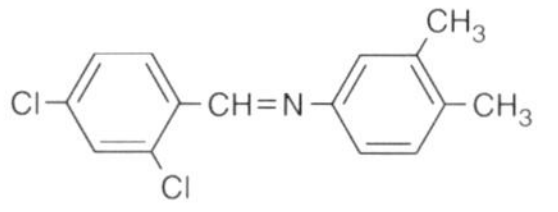

5 gm

S73,078-5
N-(2,4-DICHLOROBENZYLIDENE)-3,5-XYLIDINE

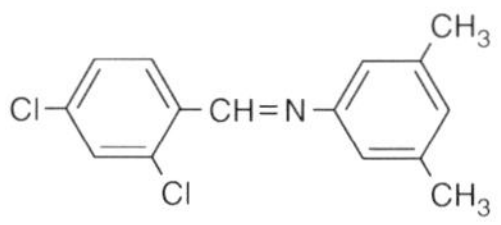

5 gm

S90,174-1
N-(2,4-DICHLOROBENZYLIDENE)-2,4,5-TRIMETHYLANILINE

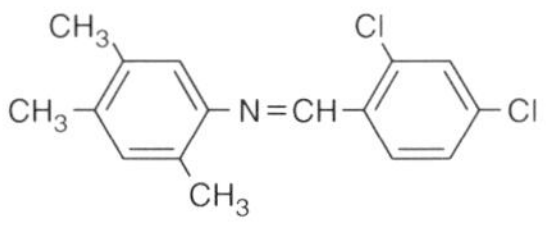

250 mg

S90,176-8
(2,4-DICHLOROBENZYLIDENE)(1-NAPHTHYL)AMINE

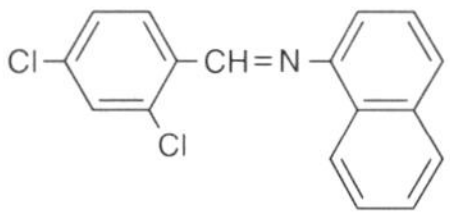

250 mg

S76,217-2
4-ETHYL-2-IMINO-4-HEPTANOL

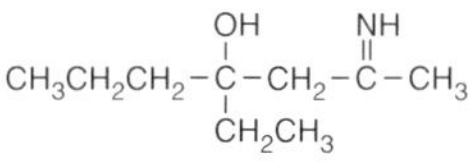

250 mg

S67,497-4
2,2'-OXYDIACETIC ACID VANILLYLIDENEHYDRAZIDE

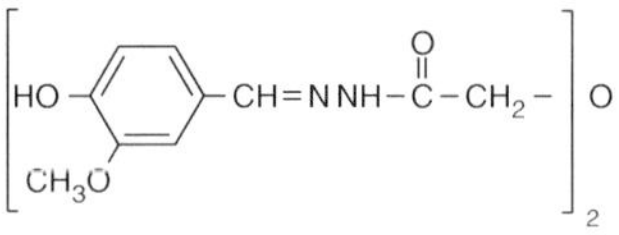

250 mg

S71,611-1
2,2'-OXYDIACETIC ACID BIS[(3-ETHOXY-4-HYDROXYBENZYLIDENE)-HYDRAZIDE]

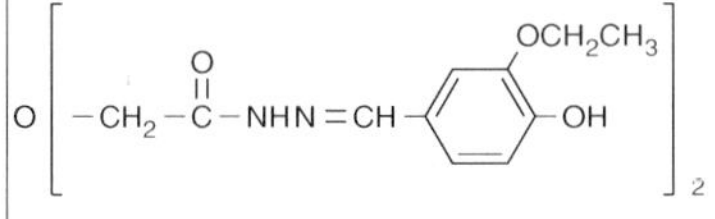

5 gm

S71,627-8
SEBACIC ACID BIS(3-HYDROXY-4-METHOXYBENZYLIDENEHYDRAZIDE)

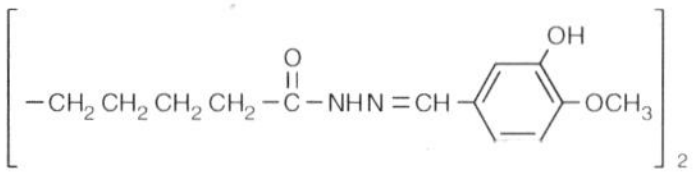

50 mg

S68,521-6
4,4'-METHYLENEBIS[N-(4-HYDROXY-3-METHOXYBENZYLIDENE)ANILINE]

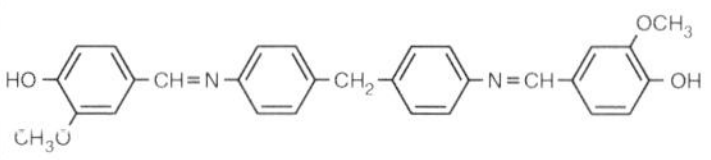

5 gm

S92,455-5
GLUCONIC (3-ETHOXY-4-HYDROXYBENZYLIDENE)HYDRAZIDE

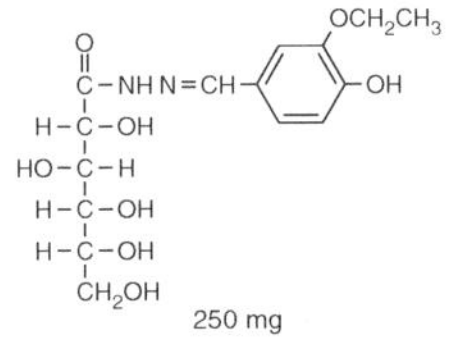

250 mg

S69,886-5
ALPHA-(2-HYDROXYPHENYLIMINO)-6-METHOXY-O-CRESOL

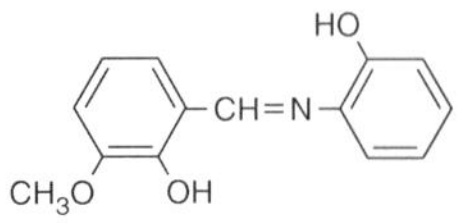

5 gm

S69,883-0
ALPHA-(2-HYDROXY-5-METHYLPHENYLIMINO)-6-METHOXY-O-CRESOL

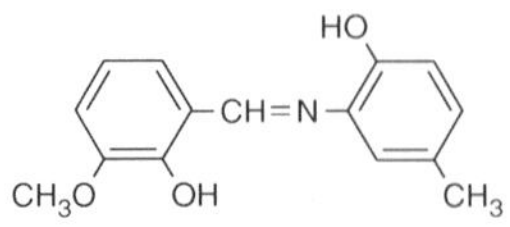

5 gm

S70,322-2
ALPHA-(4-CARBOXY-3-HYDROXYPHENYLIMINO)-6-METHOXY-O-CRESOL

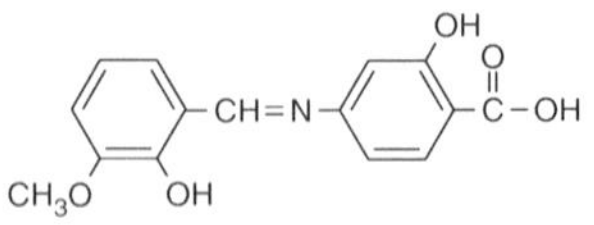

5 gm

S90,683-2
N-(4-METHOXYBENZYLIDENE)-ALPHA-GLUCOSAMINE

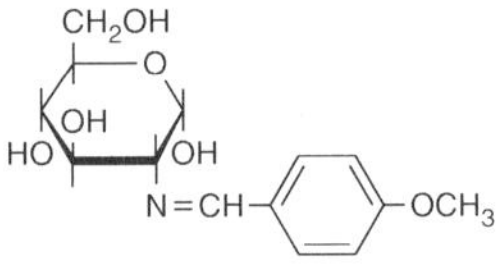

250 mg

S92,280-3
2-HYDROXY-N-(2-HYDROXY-3-METHOXY-5-NITROBENZYLIDENE)ANILINE

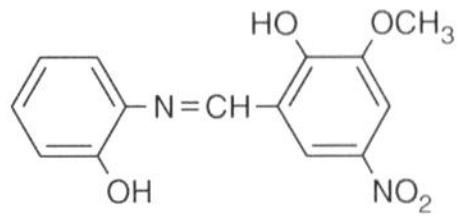

250 mg

S63,783-1
4,4'-BIS(2-HYDROXYBENZYLIDENEAMINO) DIPHENYL ETHER

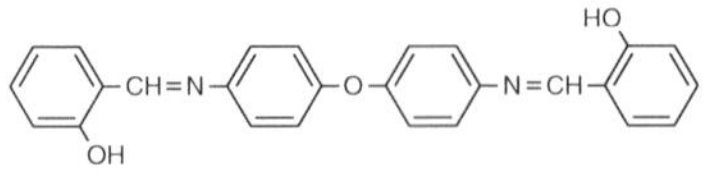

5 gm

S71,615-4
BENZILIC ACID N-(2-HYDROXYBENZYLIDENE)HYDRAZIDE

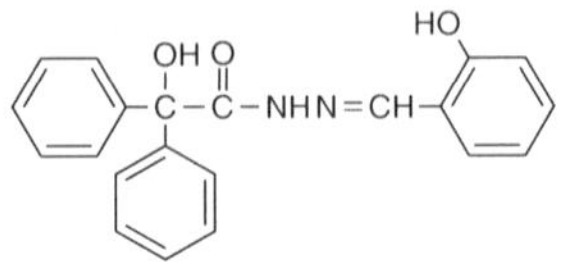

5 gm

S77,266-6
1-(2,3-DIHYDROXYBENZYLIDENE)-SEMICARBAZIDE

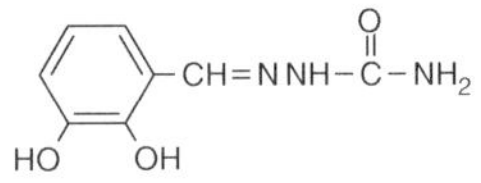

250 mg

S93,712-6
1-(2-PHENYLTRIAZOLYL)-MESO-ERYTHRITOL

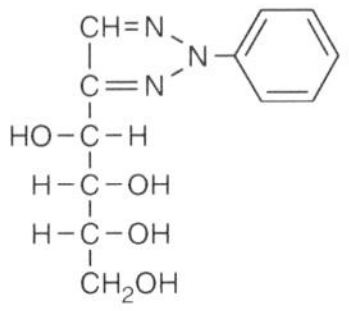

250 mg

S90,410-4
2,4-DIHYDROXYBENZOIC (FURFURYLIDENE)HYDRAZIDE

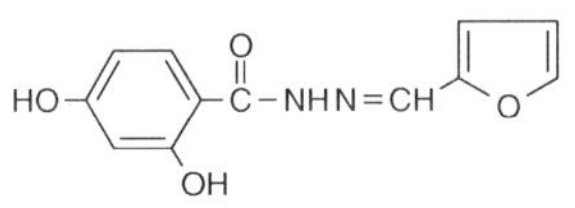

250 mg

S92,479-2
SALICYLIC (2-HYDROXYBENZYLIDENE)-HYDRAZIDE

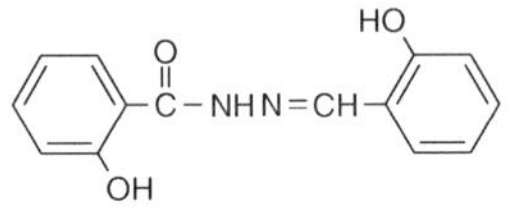

250 mg

S70,328-1
N,N'-TEREPHTHALYLIDENEBIS(4-AMINOSALICYLIC ACID)

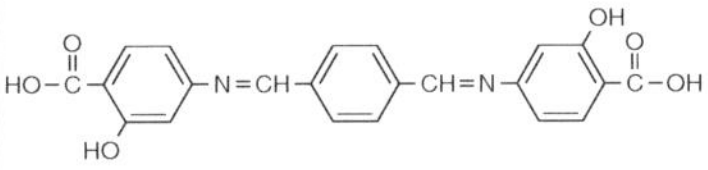

5 gm

S70,283-8
4-(2-HYDROXYBENZYLIDENEAMINO)-SALICYLIC ACID

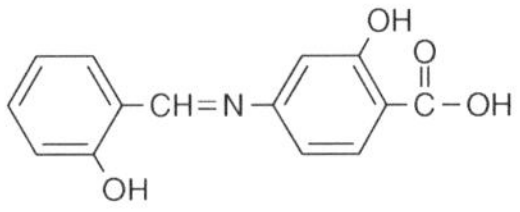

5 gm

S70,405-9
4-(4-HYDROXYBENZYLIDENEAMINO)-SALICYLIC ACID

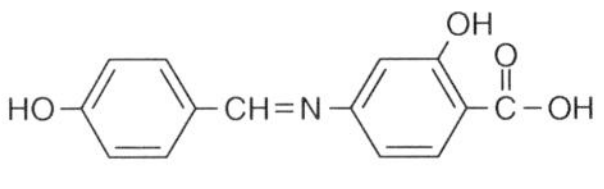

5 gm

S63,856-0
2',4',6'-TRIHYDROXYACETOPHENONE IMINE HYDROCHLORIDE

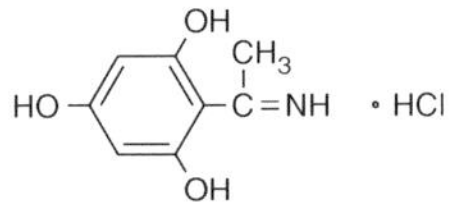

5 gm

S69,015-5
4,4'-DITHIOBIS[N-(2-HYDROXY-1-NAPHTHYLMETHYLENE)ANILINE]

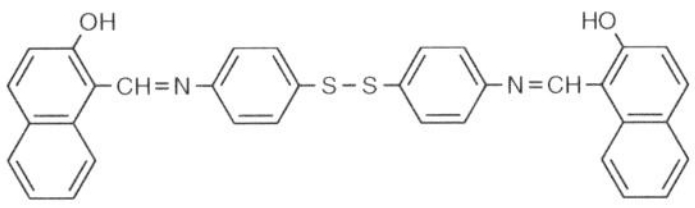

5 gm

S87,965-7
N,N'-BIS(2-HYDROXY-5-METHYLPHENYL)-1,4-DIAZABUTADIENE

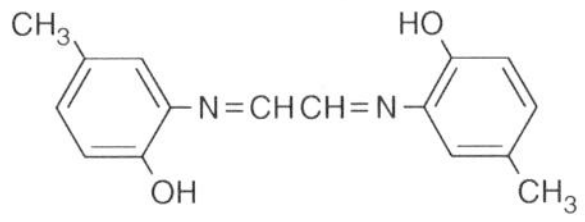

5 gm

S64,324-6
N,N'-BIS(2-HYDROXYBENZYLIDENE)-1,4-PHENYLENEDIAMINE

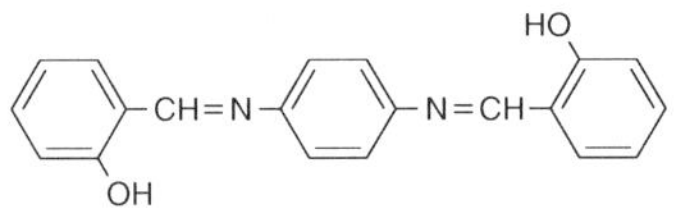

5 gm

S64,357-2
N,N'-BIS(4-HYDROXYBENZYLIDENE)-1,4-PHENYLENEDIAMINE

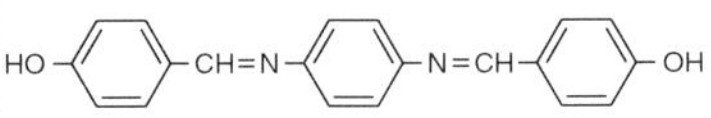

5 gm

S68,569-0
4,4'-METHYLENEBIS[N-(2-HYDROXY-1-NAPHTHYLMETHYLENE)ANILINE]

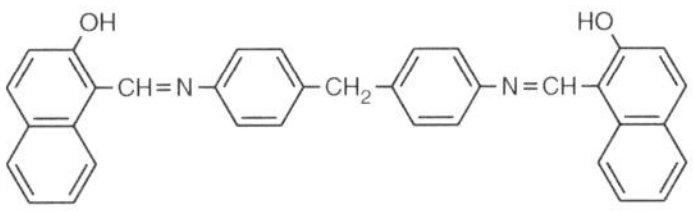

5 gm

S92,497-0
4-HYDROXYBENZOIC (2,4-DIHYDROXYBENZYLIDENE)HYDRAZIDE

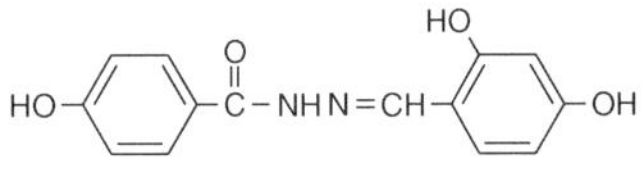

250 mg

S92,554-3
ADIPIC BIS(2-HYDROXYBENZYLIDENEHYDRAZIDE)

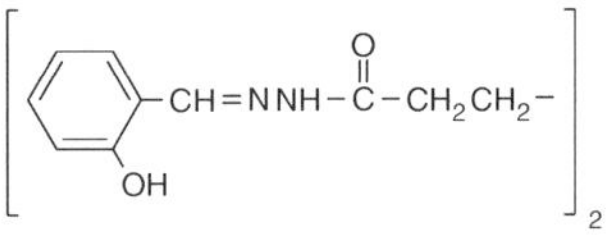

250 mg

S92,557-8
SEBACIC BIS(4-HYDROXYBENZYLIDENE)-HYDRAZIDE

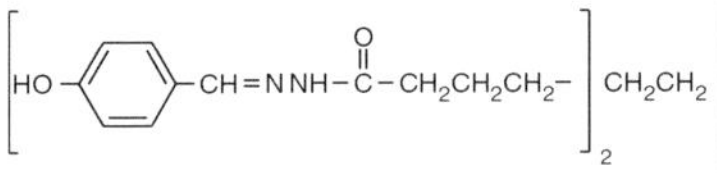

250 mg

S64,887-6
ALPHA-(2-HYDROXYETHYLIMINO)-ALPHA-METHYL-O-CRESOL

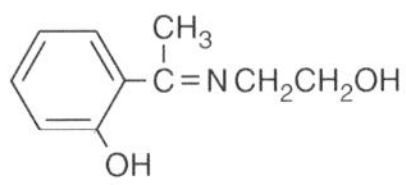

250 mg

S64,298-3
N-(5-CHLORO-2-HYDROXYBENZYLIDENE)-2-HYDROXYANILINE

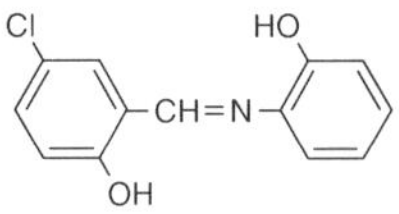

250 mg

S64,297-5
N-(5-BROMO-2-HYDROXYBENZYLIDENE)-2-HYDROXYANILINE

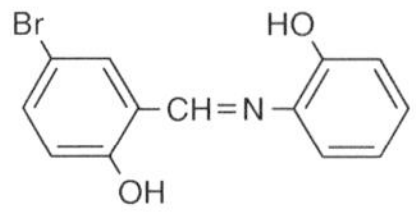

250 mg

S69,101-1
ALPHA-(2-HYDROXYETHYLIMINO)-P-CRESOL

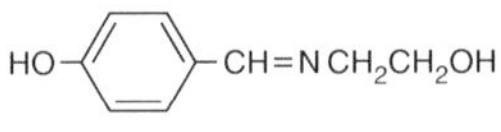

5 gm

S63,425-5
N,N'-(1,4-PHENYLENEDIMETHYLIDYNE)-BIS(2-HYDROXY-5-METHYLANILINE)

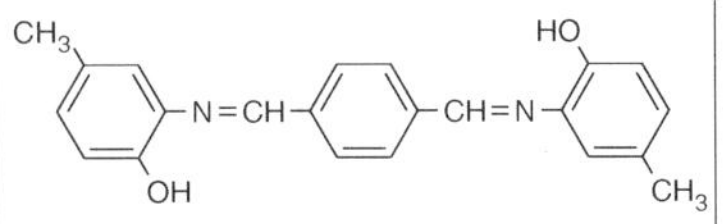

5 gm

S63,780-7
4-(2-HYDROXYBENZYLIDENEAMINO) PHENYL SULFONE

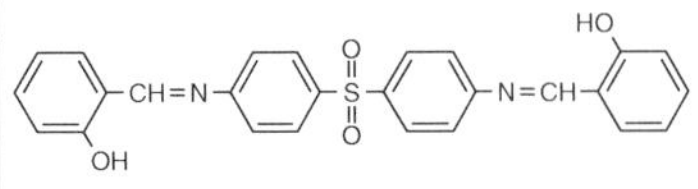

5 gm

S65,314-4
4,4'-BIS(4-HYDROXYBENZYLIDENE-AMINO)DIPHENYL SULFONE

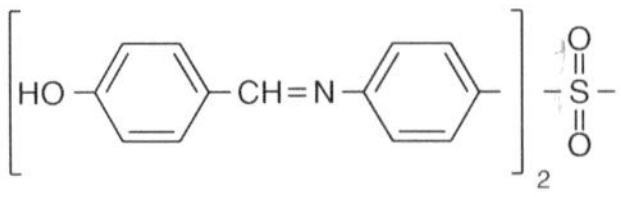

250 mg

S41,913-3
2-(2-HYDROXYBENZYLIDENEAMINO)PHENOL

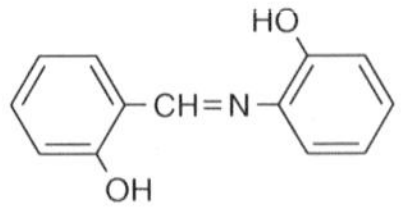

5 gm

S58,389-8
ALPHA-(3-HYDROXYPHENYLIMINO)-O-CRESOL

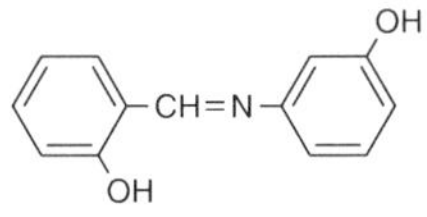

5 gm

S58,387-1
ALPHA-(4-HYDROXYPHENYLIMINO)-O-CRESOL

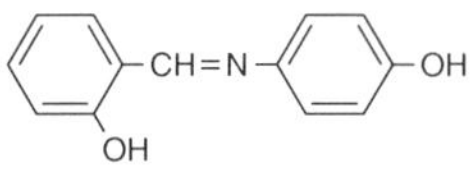

5 gm

S68,461-9
4,4'-METHYLENEBIS[N-(2-HYDROXYBENZYLIDENE)ANILINE]

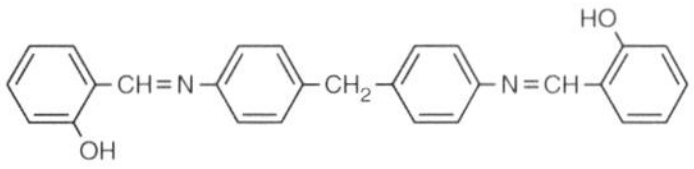

5 gm

S68,455-4
ALPHA-(4-HYDROXYPHENYLIMINO)-P-CRESOL

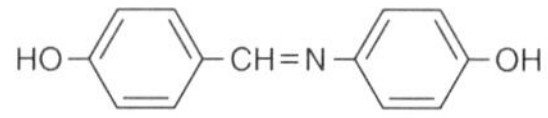

5 gm

S58,941-1
N,N'-(1,4-PHENYLENEDIMETHYLIDYNE)-BIS(3-AMINOPHENOL)

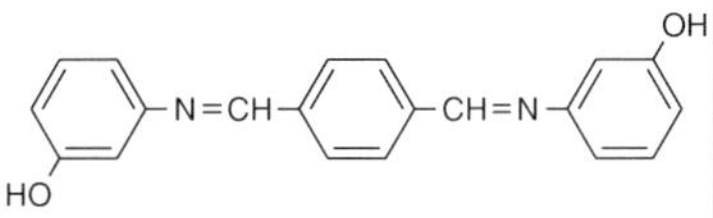

5 gm

S71,630-8
SEBACIC ACID BIS(2-HYDROXY-1-NAPHTHYLMETHYLENEHYDRAZIDE)

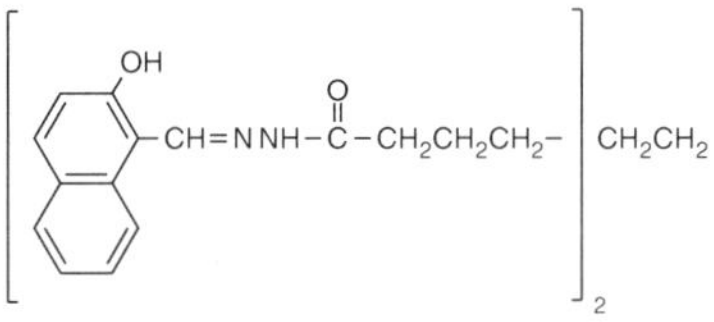

250 mg

S93,439-9
4-(BENZYLIDENEAMINO)-3,5-BIS-(HYDROXYMETHYL)-4H-1,2,4-TRIAZOLE

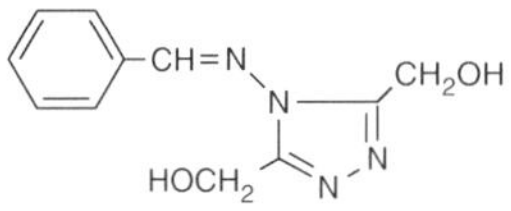

250 mg

S93,448-8
4-(2,4-DIHYDROXYBENZYLIDENEAMINO)-4H-1,2,4-TRIAZOLE

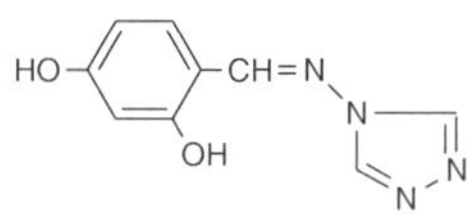

250 mg

S36,558-0
N,N'-TEREPHTHALYLIDENEBIS(2-AMINOPHENOL)

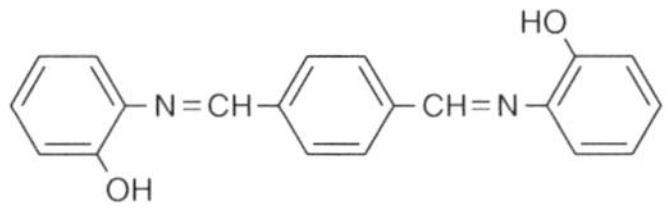

5 gm

S69,278-6
ALPHA-(2-HYDROXY-5-METHYLPHENYLIMINO)-O-CRESOL

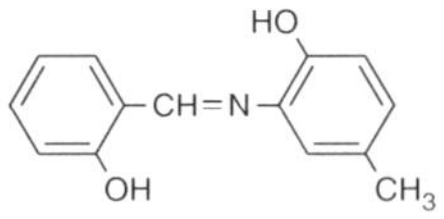

5 gm

S77,307-7
1-[(2-HYDROXYETHYL)IMINOMETHYL]-2-NAPHTHOL

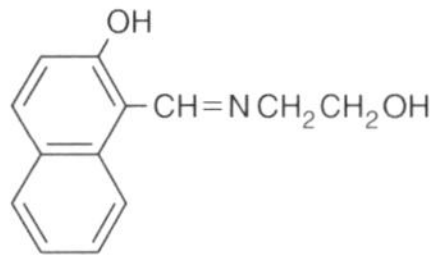

250 mg

S78,027-8
4-HYDROXYBENZALDEHYDE AZINE

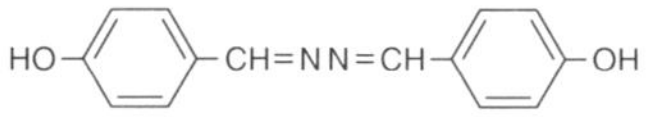

250 mg

S36,559-9
N,N'-TEREPHTHALYLIDENEBIS(4-AMINOPHENOL)

5 gm

S69,871-7
1-(4-HYDROXYPHENYLIMINOMETHYL)-2-NAPHTHOL

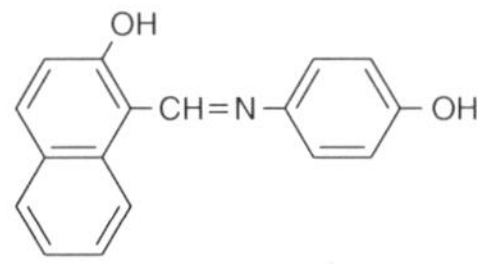

5 gm

S53,791-8
5-BROMOSALICYLALDEHYDE AZINE

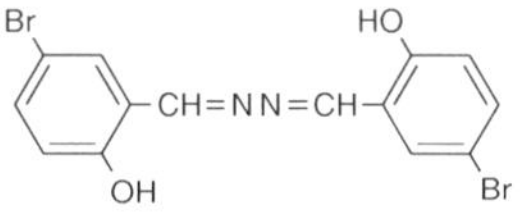

5 gm

S68,629-8
1-(2-HYDROXY-5-METHYLPHENYLIMINOMETHYL)-2-NAPHTHOL

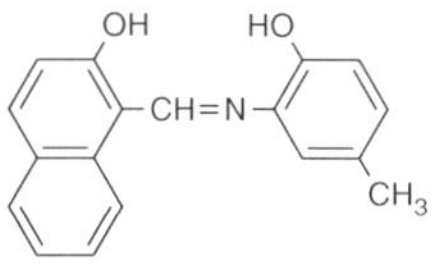

5 gm

S54,155-9
3,5-DIBROMOSALICYLALDEHYDE AZINE

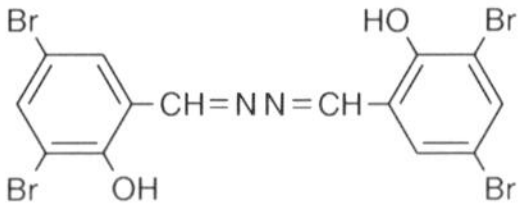

250 mg

S45,906-2
2,2'-DIHYDROXY-1,1'-NAPHTHALAZINE

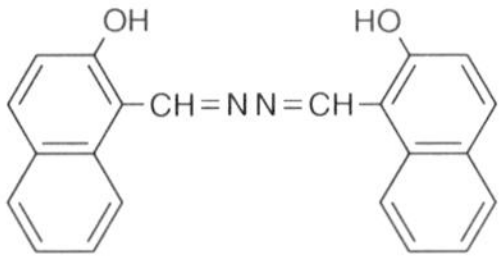

250 mg

S77,238-0
2,7'-DIHYDROXY-1,1'-NAPHTHALAZINE

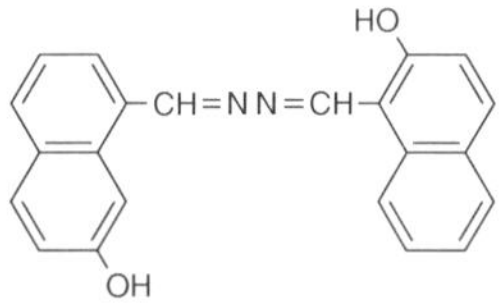

250 mg

S66,990-3
N-(4-BENZYLOXYBENZYLIDENE)-4-HYDROXYANILINE

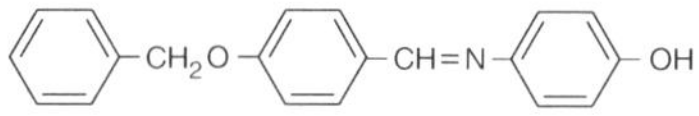

5 gm

S76,709-3
ALPHA-[2-(3,4-DIMETHOXYPHENYL)-ETHYLIMINO]-P-CRESOL

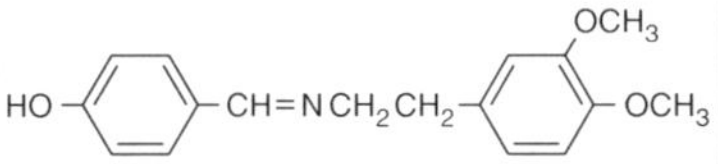

250 mg

S69,104-6
ALPHA-(2-ETHOXYPHENYLIMINO)-6-METHOXY-O-CRESOL

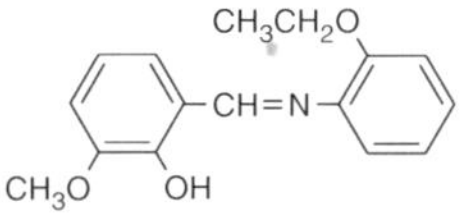

5 gm

S68,920-3
ALPHA-(4-ETHOXYPHENYLIMINO)-6-METHOXY-O-CRESOL

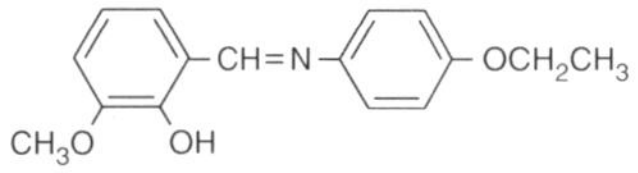

5 gm

S87,917-7
2-(4-ETHOXYPHENYLIMINOMETHYL)-6-METHOXY-4-NITROPHENOL

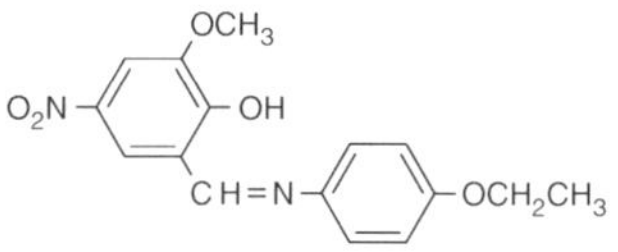

5 gm

S70,124-6
6-METHOXY-ALPHA-(4-PHENOXYPHENYLIMINO)-O-CRESOL

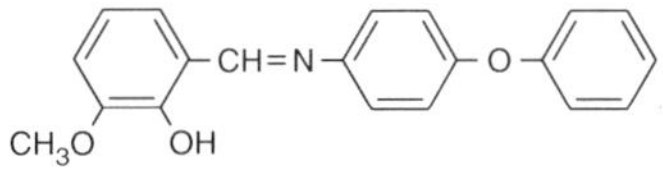

5 gm

S72,619-2
2-(2,5-DIMETHOXYBENZYLIDENEAMINO)-ETHANOL

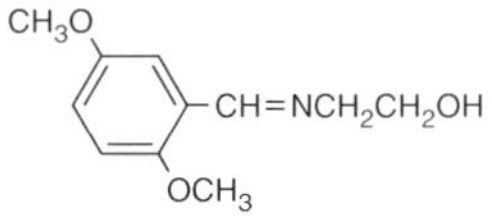

5 gm

S68,622-0
4,4'-METHYLENEBIS[N-(3-ETHOXY-4-HYDROXYBENZYLIDENE)ANILINE]

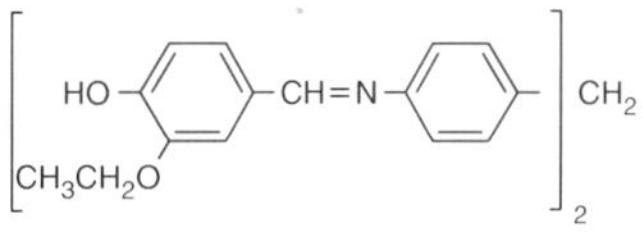

5 gm

S87,867-7
N-(4-HYDROXY-3-METHOXY-5-NITROBENZYLIDENE)-4-ETHOXYANILINE

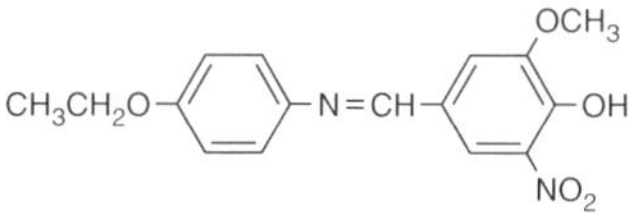

5 gm

S40,501-9
N-VANILLYLIDENEBENZIDINE

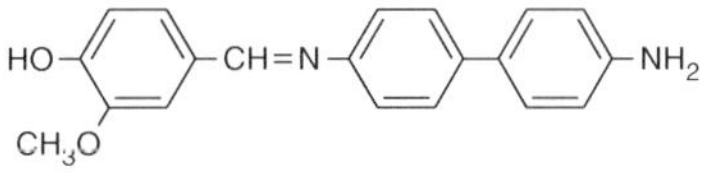

250 mg

S74,787-4
3-HYDROXY-4-METHOXYBENZALDEHYDE (2,4-DINITROPHENYL)METHYLHYDRAZONE

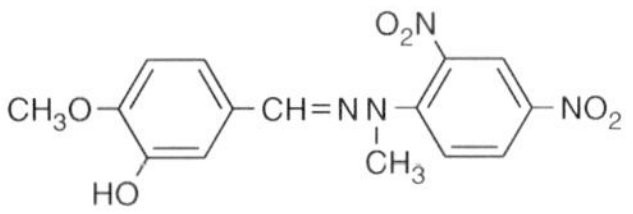

5 gm

S68,540-2
N,N'-DIVANILLYLIDENEETHYLENEDIAMINE

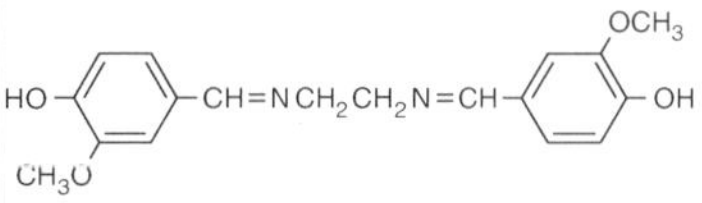

5 gm

S69,873-3
6-METHOXY-ALPHA-(3-NITROPHENYLIMINO)-O-CRESOL

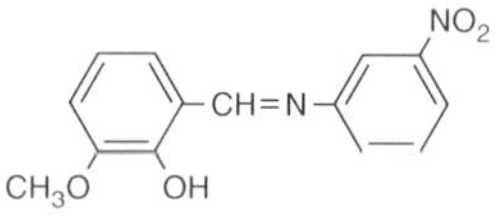

5 gm

S69,887-3
6-METHOXY-ALPHA-(2-METHYL-5-NITROPHENYLIMINO)-O-CRESOL

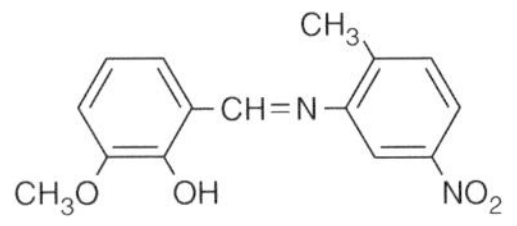

5 gm

S69,876-8
6-METHOXY-ALPHA-(2-METHOXY-5-NITROPHENYLIMINO)-O-CRESOL

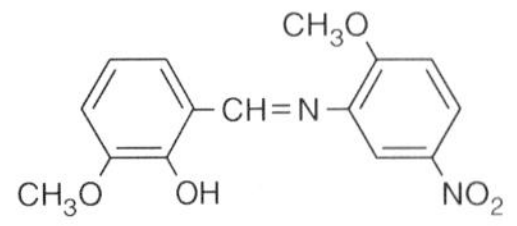

5 gm

S87,918-5
2-(4-BUTYLPHENYLIMINOMETHYL)-6-METHOXY-4-NITROPHENOL

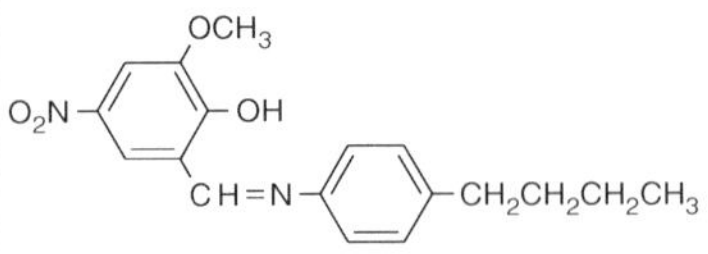

5 gm

S69,874-1
6-METHOXY-ALPHA-(4-NITROPHENYLIMINO)-O-CRESOL

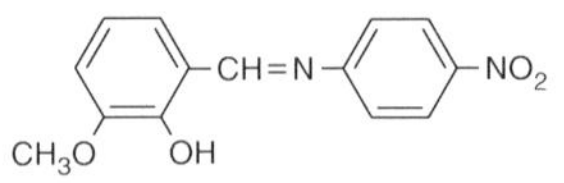

5 gm

S64,597-4
ALPHA-(3,4-DIMETHOXYPHENETHYLIMINO)-ALPHA-METHYL-O-CRESOL

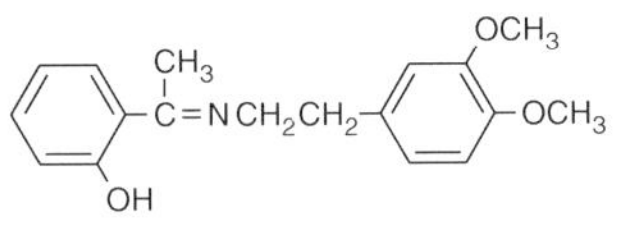

5 gm

S92,304-4
2-METHOXY-4-NITRO-6-(PHENETHYLIMINOMETHYL)PHENOL

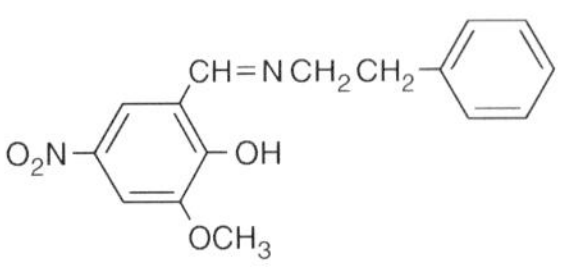

250 mg

S69,880-6
ALPHA-(4-ANILINOPHENYLIMINO)-6-METHOXY-O-CRESOL

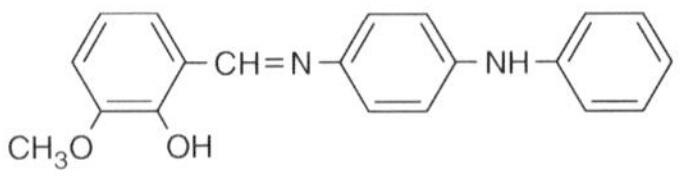

5 gm

S67,390-0
4,4-DIPHENYL-1-(3-ETHOXY-4-HYDROXYBENZYLIDENE)SEMICARBAZIDE

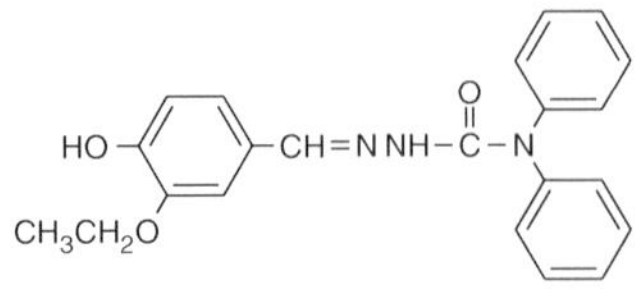

250 mg

S69,108-9
ALPHA-(4-CHLORO-3-NITROPHENYLIMINO)-6-METHOXY-O-CRESOL

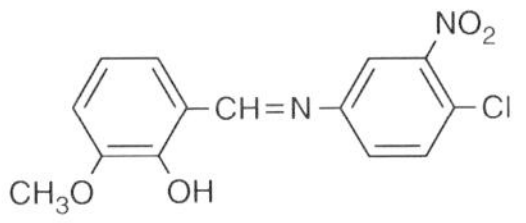

5 gm

S68,929-7$
6-METHOXY-ALPHA-(4-METHYL-3-NITROPHENYLIMINO)-O-CRESOL

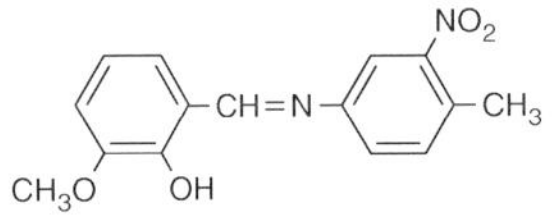

5 gm

S68,915-7
6-METHOXY-ALPHA-(2-METHOXY-4-NITROPHENYLIMINO)-O-CRESOL

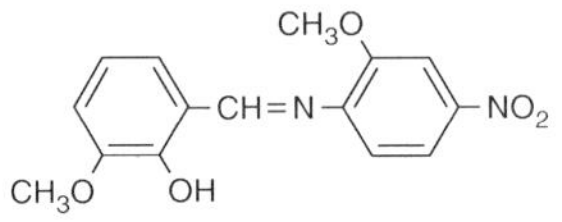

5 gm

S68,910-6
ALPHA-(4-CARBOXYPHENYLIMINO)-6-METHOXY-O-CRESOL

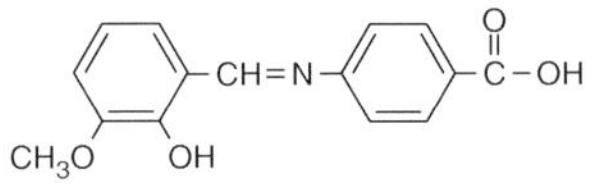

5 gm

S69,109-7
ALPHA-(3-CARBOXYPHENYLIMINO)-6-METHOXY-O-CRESOL

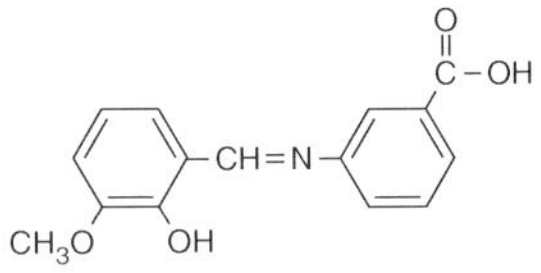

5 gm

S69,297-2
ALPHA-[4-(ETHOXYCARBONYL)-PHENYLIMINO]-6-METHOXY-O-CRESOL

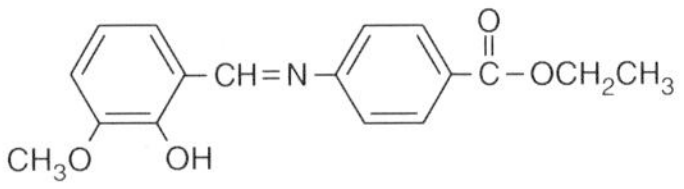

5 gm

S68,911-4
4,4'-SULFONYLBIS[N-(2-HYDROXY-3-METHOXYBENZYLIDENE)ANILINE]

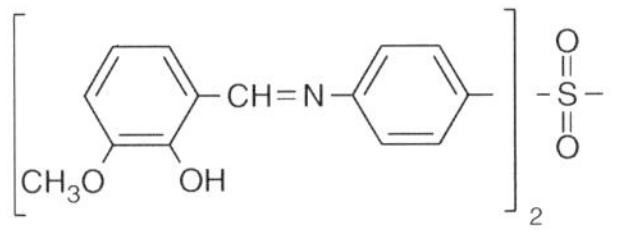

5 gm

S70,128-9
6-METHOXY-ALPHA-(2-THIAZOLYLIMINO)-O-CRESOL

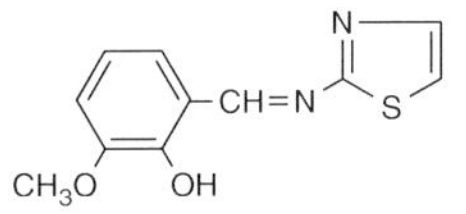

5 gm

S71,964-1
ALPHA-BENZYLIMINO-6-METHOXY-O-CRESOL

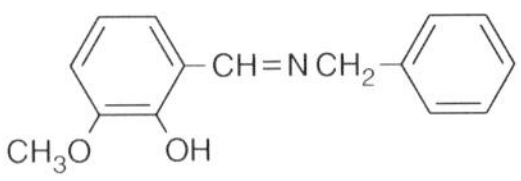

5 gm

S68,927-0
6-METHOXY-ALPHA-(4-METHYLPHENYLIMINO)-O-CRESOL

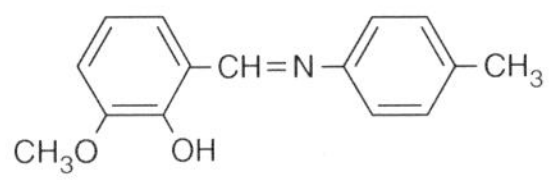

5 gm

S68,907-6
ALPHA-(4-ETHYLPHENYLIMINO)-6-METHOXY-O-CRESOL

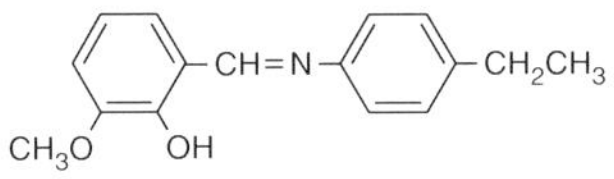

5 gm

S39,604-4
ALPHA-(4-CHLOROPHENYLIMINO)-2-METHOXY-P-CRESOL

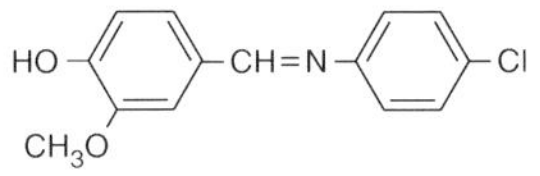

5 gm

S52,267-8
2-METHOXY-ALPHA-(4-METHOXYPHENYLIMINO)-P-CRESOL

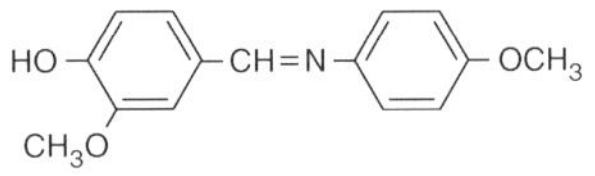

5 gm

S69,106-2
ALPHA-(3,4-DIMETHYLPHENYLIMINO)-6-METHOXY-O-CRESOL

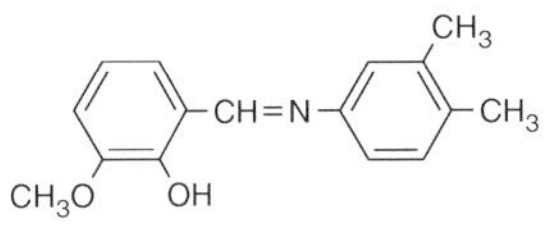

5 gm

S69,100-3
6-METHOXY-ALPHA-(4-METHYL-2-PYRIDYLIMINO)-O-CRESOL

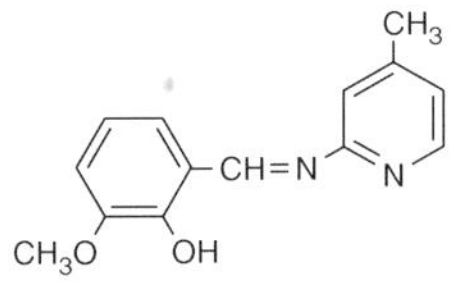

5 gm

S68,899-1
6-METHOXY-ALPHA-(2-PYRIDYLIMINO)-O-CRESOL

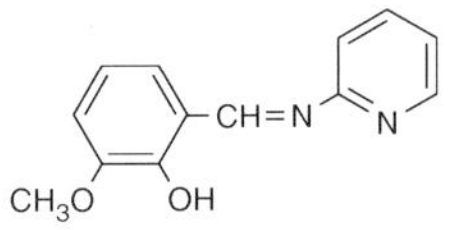

5 gm

S53,017-4
N-VANILLYLIDENE-2,5-XYLIDINE

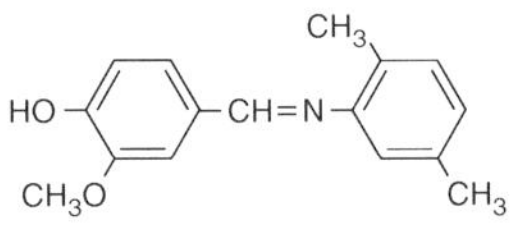

5 gm

S53,019-0
N-VANILLYLIDENE-2,4-XYLIDINE

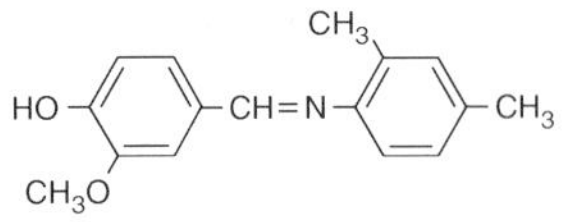

5 gm

S72,558-7
4-METHYLBENZOIC ACID (4-HYDROXY-3-METHOXYBENZYLIDENE)HYDRAZIDE

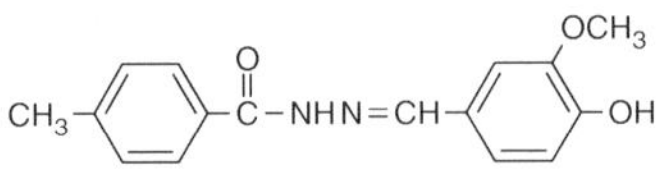

250 mg

S91,287-5
ADIPIC BIS[(4-HYDROXY-3-METHOXYBENZYLIDENE)HYDRAZIDE]

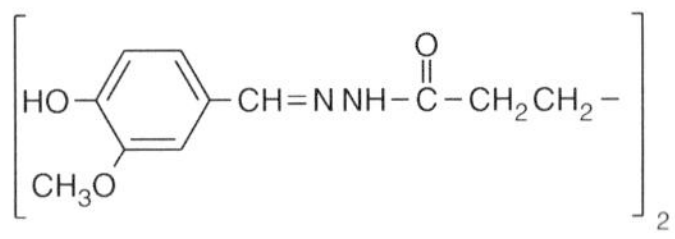

250 mg

S69,215-8
2-(FURFURYLUREIDO)ACETIC ACID N2-(4-HYDROXY-3-METHOXYBENZYLIDENE)-HYDRAZIDE

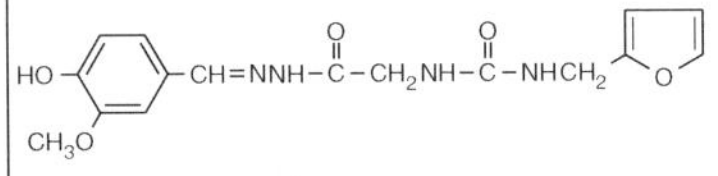

250 mg

S72,089-5
2,2'-OXYBIS[ACETIC ACID (2-HYDROXY-1-NAPHTHYLMETHYLENE)HYDRAZIDE]

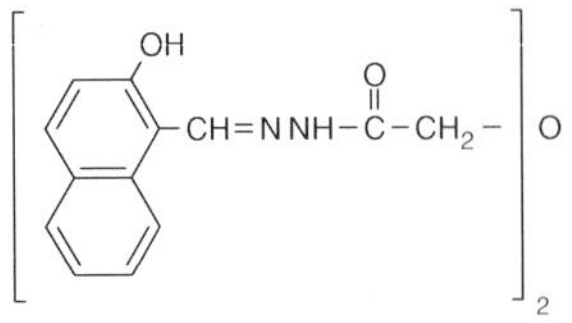

250 mg

S69,154-2
4-(CYCLOHEXYL)-1-(3-ETHOXY-2-HYDROXYBENZYLIDENE)-3-THIOSEMICARBAZIDE

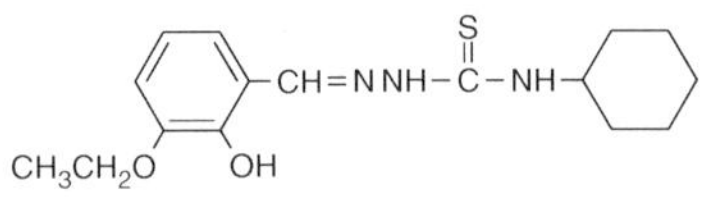

5 gm

S53,980-5
N-(3-ETHOXY-4-HYDROXYBENZYLIDENEAMINO)-PHTHALIMIDE

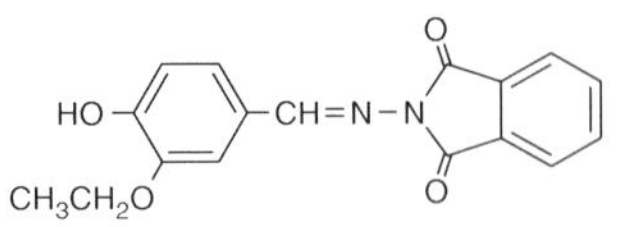

5 gm

S59,809-7
2-(2-HYDROXYETHOXY)BENZALDEHYDE, 3-METHYL-2-BENZOTHIAZOLINONE AZINE

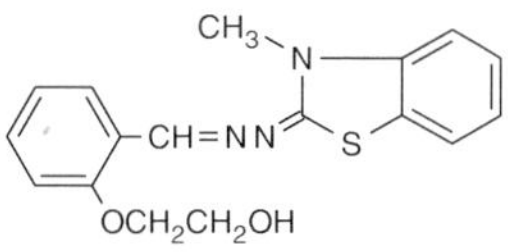

5 gm

S92,281-1
2-METHOXY-4-NITRO-6-(3-PYRIDYLMETHYLIMINOMETHYL)PHENOL

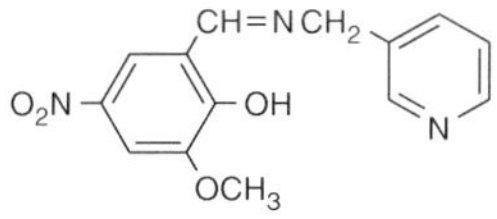

250 mg

S71,974-9
4'-(2-HYDROXY-3-METHOXYBENZYLIDENEAMINO) ACETOPHENONE

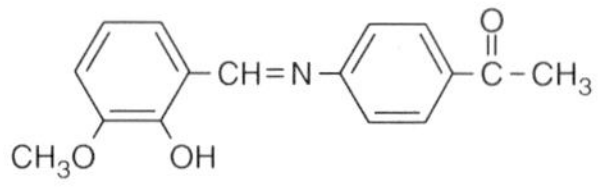

5 gm

S71,978-1
3'-(2-HYDROXY-3-METHOXYBENZYLIDENEAMINO) ACETOPHENONE

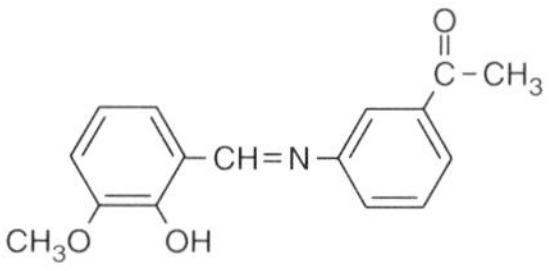

5 gm

S73,006-8
4-(2-HYDROXY-3-METHOXYBENZYLIDENEAMINO)ANTIPYRINE

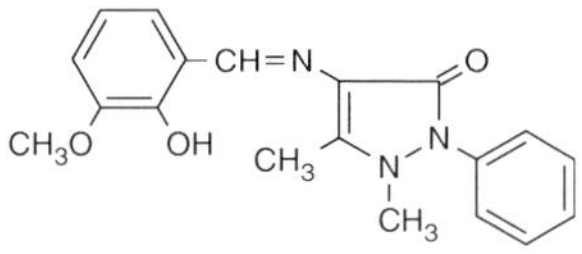

5 gm

S72,000-3
2,3-DIMETHYL-4-(3-ETHOXY-4-HYDROXYBENZYLIDENEAMINO)-1-PHENYL-3-PYRAZOLIN-5-ONE

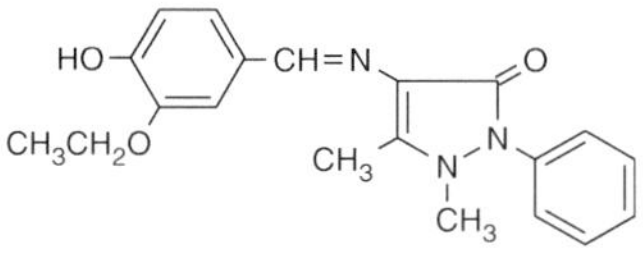

5 gm

S72,012-7
5-(O-VANILLYLIDENEAMINO)-1H-TETRAZOLE

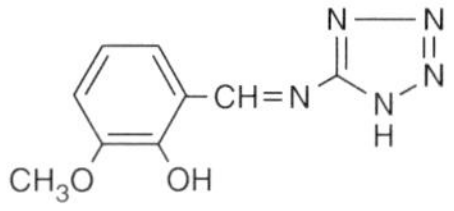

5 gm

S70,127-0
6-METHOXY-ALPHA-(2,4,6-TRIMETHYLPHENYLIMINO)-O-CRESOL

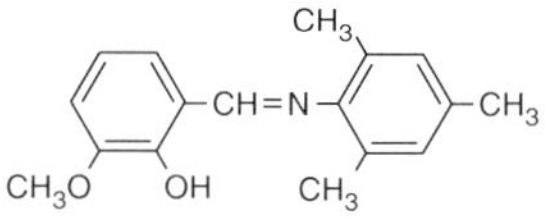

5 gm

S68,917-3
6-METHOXY-ALPHA-(3-METHYL-2-PYRIDYLIMINO)-O-CRESOL

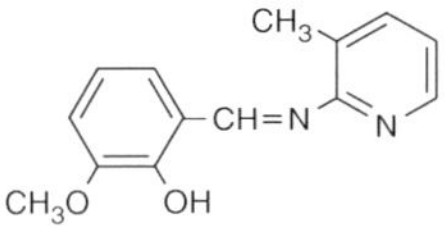

250 mg

S36,547-5
2-(N-VANILLYLIDENEAMINO)FLUORENE

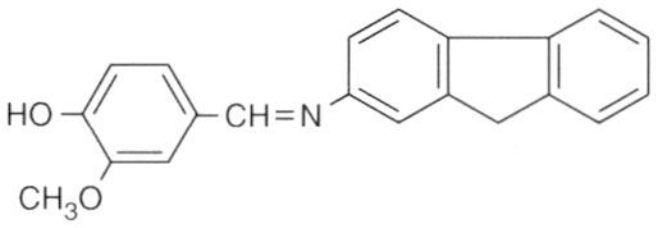

250 mg

S92,481-4
2-METHOXYBENZOIC (2-HYDROXYBENZYLIDENE)HYDRAZIDE

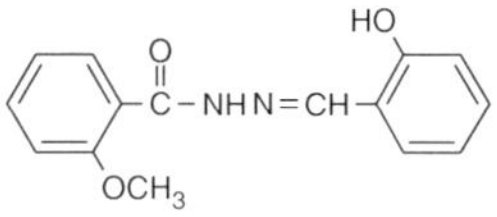

250 mg

S90,227-6
2-(3-FURFURYLUREIDO)ACETIC (3-HYDROXY-4-METHOXYBENZYLIDENE)-HYDRAZIDE

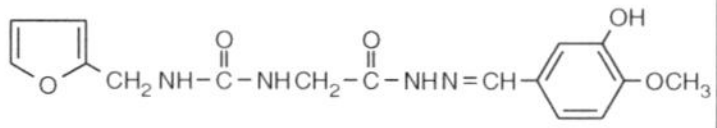

250 mg

S58,223-9
3-CARBOXY-4-HYDROXYBENZENESULFONIC ACID [2-(4-HEXADECYLOXY)PHENYL]-1-METHYL-6-SULFO-4(1H)QUINOLYLIDENE HYDRAZIDE

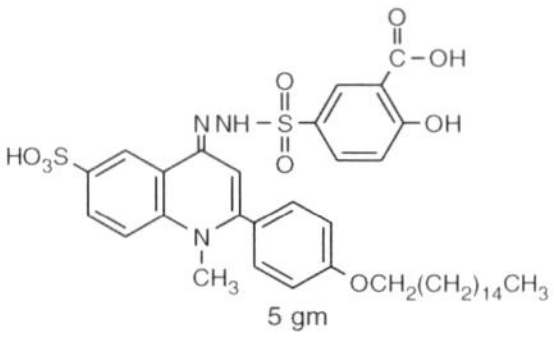

5 gm

S62,607-4
N-(9-ANTHRACENYLMETHYLENE)-2-METHOXY-5-NITROANILINE

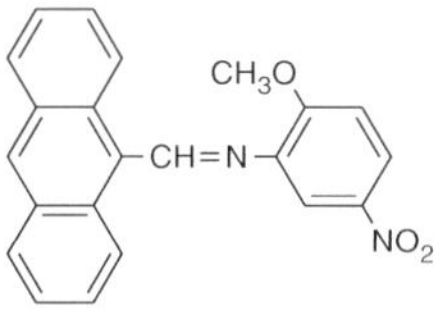

5 gm

S58,400-2
ALPHA-(2-METHOXY-5-NITROPHENYLIMINO)-O-CRESOL

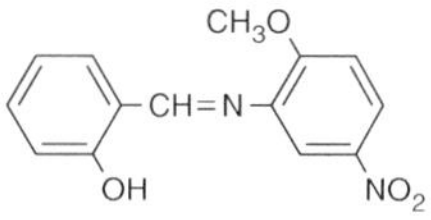

250 mg

S58,401-0
ALPHA-(2-METHOXY-4-NITROPHENYLIMINO)-O-CRESOL

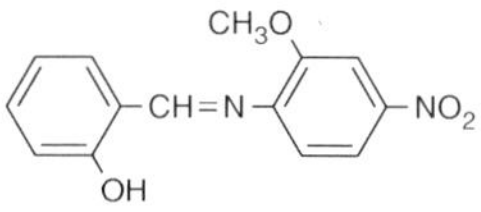

5 gm

S92,486-5
N-METHYLANTHRANILIC (4-HYDROXY-3-METHOXYBENZYLIDENE)HYDRAZIDE

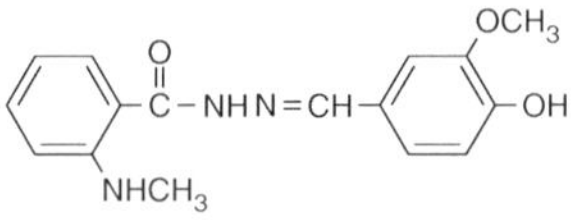

250 mg

S58,405-3
ALPHA-(4-ETHOXYPHENYLIMINO)-O-CRESOL

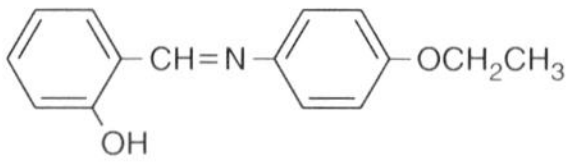

5 gm

S92,495-4
4-HYDROXYBENZOIC (4-METHOXYBENZYLIDENE)HYDRAZIDE

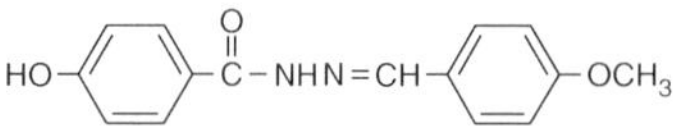

250 mg

S58,913-6
ALPHA-(2-METHOXYPHENYLIMINO)-O-CRESOL

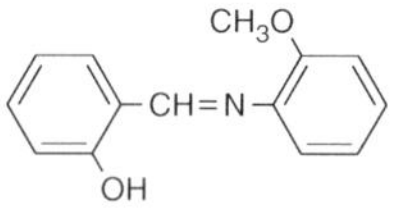

5 gm

S58,658-7
ALPHA-(2,5-DIMETHOXYPHENYLIMINO)-O-CRESOL

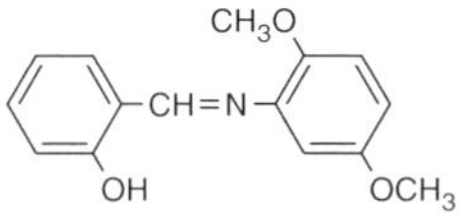

5 gm

S50,521-8
ALPHA-(5-CHLORO-2,4-DIMETHOXYPHENYLIMINO)-O-CRESOL

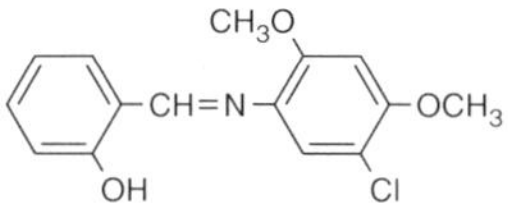

5 gm

S62,086-6
ALPHA-(4-PHENOXYPHENYLIMINO)-O-CRESOL

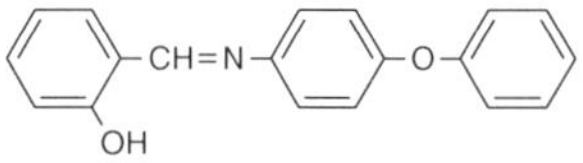

250 mg

S53,648-2
4,6-DICHLORO-ALPHA-(4-PHENOXYPHENYLIMINO)-O-CRESOL

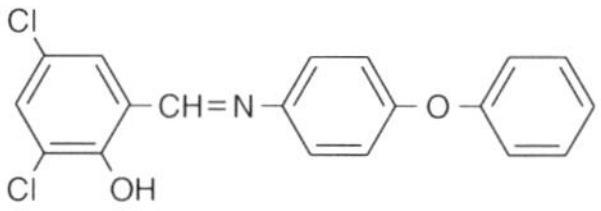

5 gm

S59,388-5
ALPHA-(3-METHOXYPHENYLIMINO)-O-CRESOL

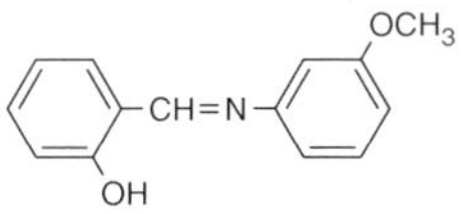

5 gm

S58,912-8
ALPHA-(4-METHOXYPHENYLIMINO)-O-CRESOL

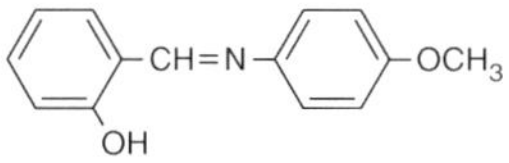

5 gm

S59,387-7
ALPHA-(4-METHOXY-2-METHYLPHENYLIMINO)-O-CRESOL

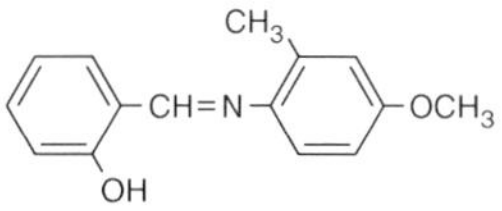

5 gm

S68,451-1
ALPHA-(4-METHOXYPHENYLIMINO)-P-CRESOL

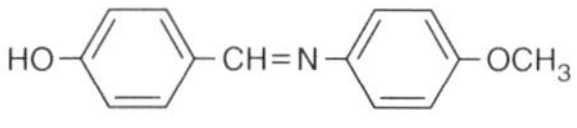

5 gm

S52,573-1
ALPHA-(4-IODOPHENYLIMINO)-2-METHOXY-P-CRESOL

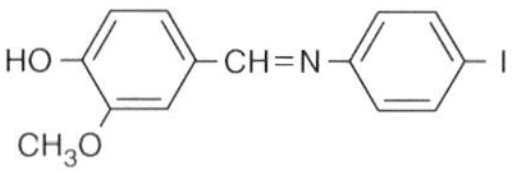

5 gm

S62,612-0
N-(9-ANTHRACENYLMETHYLENE)-4-METHOXY-2-METHYLANILINE

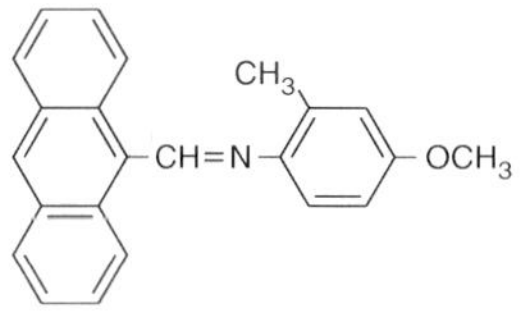

5 gm

S51,773-9
1,2-DIHYDRO-4,6-DIMETHYL-1-(3-METHOXYSALICYLIDENEAMINO)-2-OXONICOTINONITRILE

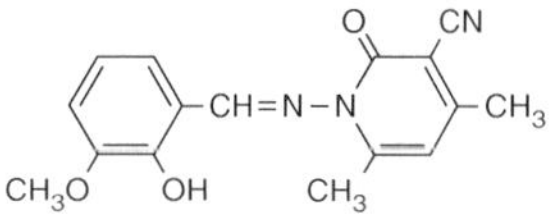

250 mg

S90,413-9
NICOTINIC (4-HYDROXY-3-METHOXYBENZYLIDENE)HYDRAZIDE

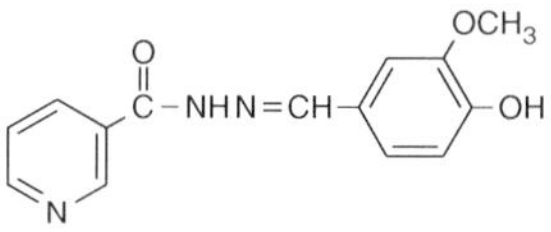

250 mg

S52,467-0
N-(VANILLYLIDENEAMINO)PHTHALIMIDE
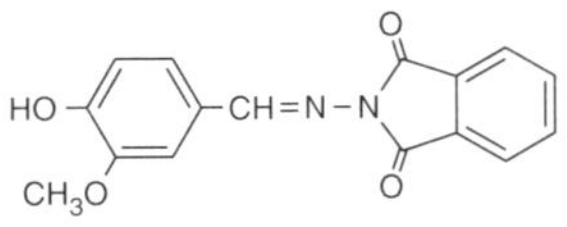
250 mg

S51,455-1
1,2-DIHYDRO-4,6-DIMETHYL-2-OXO-1-(VANILLYLIDENEAMINO)NICOTINONITRILE
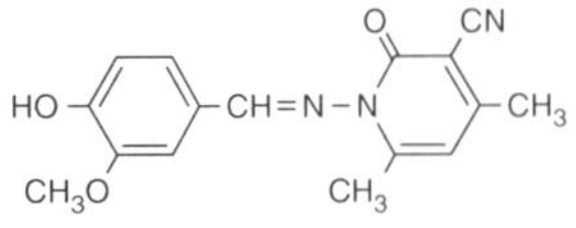
5 gm

S92,547-0
10-UNDECENOIC (4-HYDROXY-3-METHOXYBENZYLIDENE)HYDRAZIDE
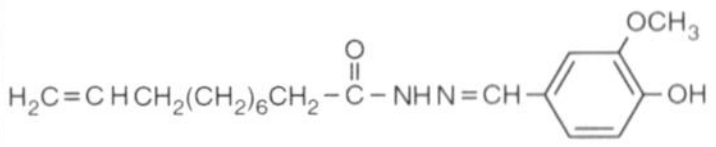
250 mg

S69,558-0
1-(2-ETHOXYPHENYLIMINOMETHYL)-2-NAPHTHOL
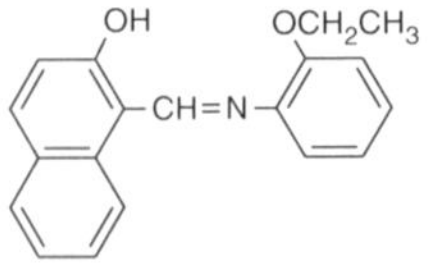
5 gm

S68,514-3
ALPHA-(2-METHOXY-5-METHYLPHENYLIMINO)-P-CRESOL
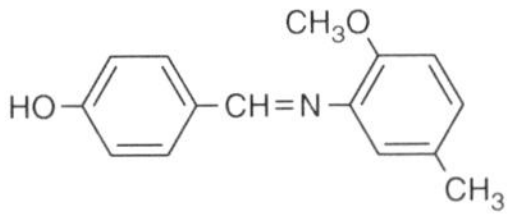
5 gm

S68,627-1
1-(2-METHOXY-5-METHYLPHENYLIMINOMETHYL)-2-NAPHTHOL
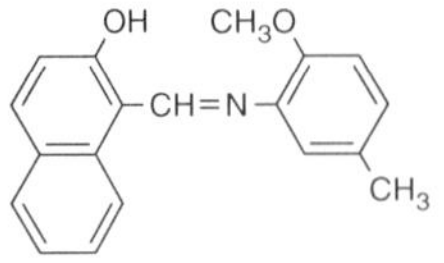
5 gm

S87,902-9
3'-(2-HYDROXY-3-METHOXY-5-NITROBENZYLIDENEAMINO)ACETOPHENONE
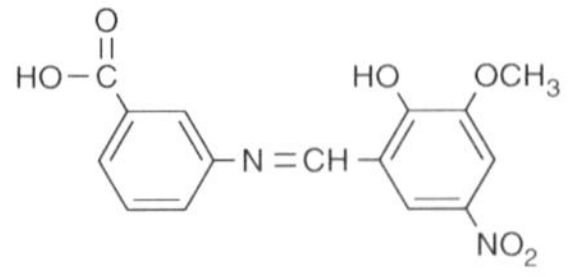
5 gm

S92,256-0
N-(2-HYDROXY-3-METHOXY-5-NITROBENZYLIDENE)-BETA-ALANINE
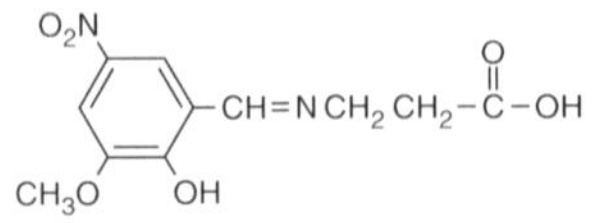
250 mg

S87,897-9
2-METHOXY-6-(3-METHYL-2-PYRIDYLIMINOMETHYL)-4-NITROPHENOL
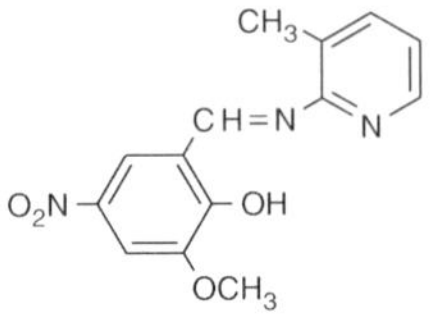
5 gm

S69,296-4
1-(2-METHOXY-5-NITROPHENYLIMINOMETHYL)-2-NAPHTHOL
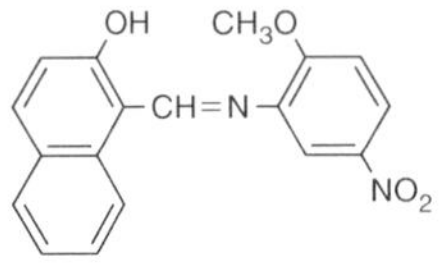
5 gm

S68,663-8
ALPHA-(4-PHENOXYPHENYLIMINO)-P-CRESOL
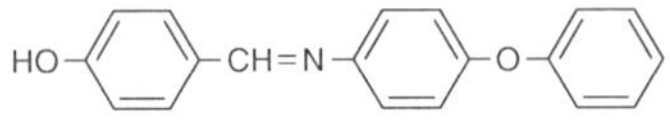
5 gm

S69,888-1
1-(4-PHENOXYPHENYLIMINOMETHYL)-2-NAPHTHOL
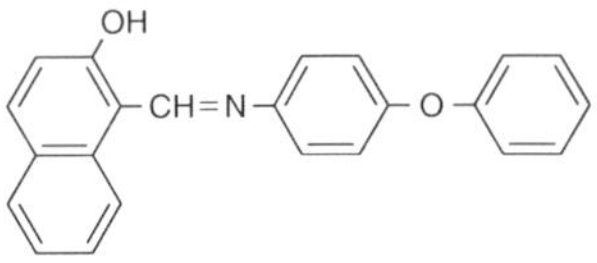
5 gm

S68,665-4
1-(2-METHOXYPHENYLIMINOMETHYL)-2-NAPHTHOL
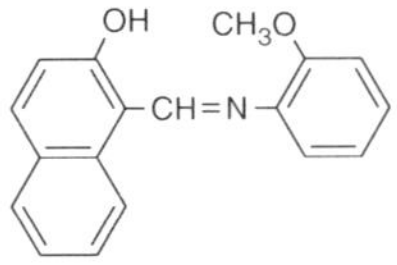
5 gm

S68,492-9
ALPHA-(4-ETHOXYPHENYLIMINO)-P-CRESOL
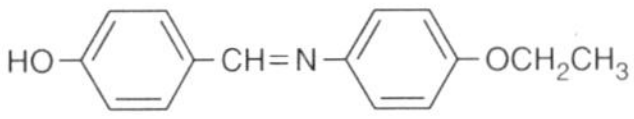
5 gm

S52,339-9
1-(4-METHOXYPHENYLIMINOMETHYL)-2-NAPHTHOL
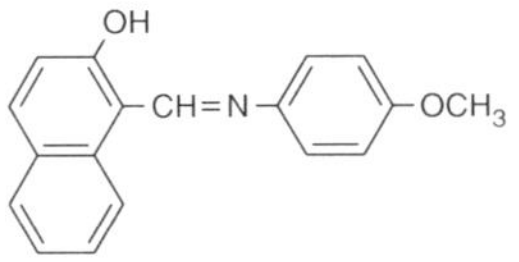
5 gm

S68,649-2
1-(4-ETHOXYPHENYLIMINOMETHYL)-2-NAPHTHOL
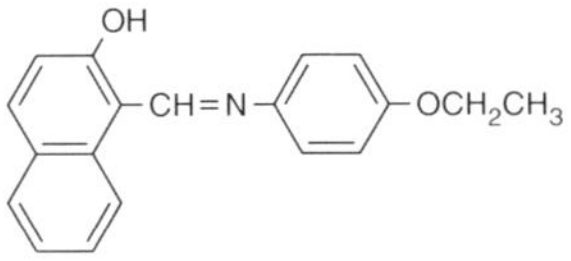
5 gm

S64,592-3
ALPHA-BENZYLIMINO-O-CRESOL
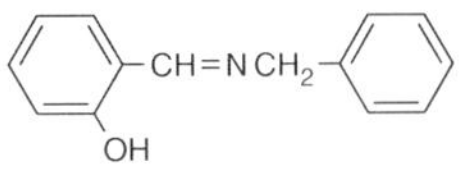
5 gm

S68,923-8
6-METHOXY-ALPHA-(6-METHYL-2-PYRIDYLIMINO)-O-CRESOL
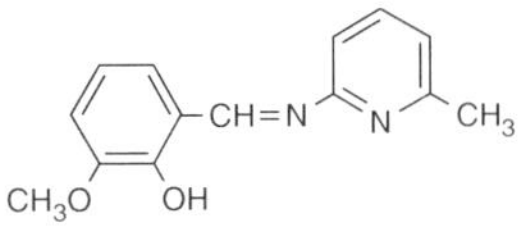
5 gm

S80,957-8
4-HYDROXY-3-[1-(PHENYLIMINO)ETHYL]-2(5H)-FURANONE
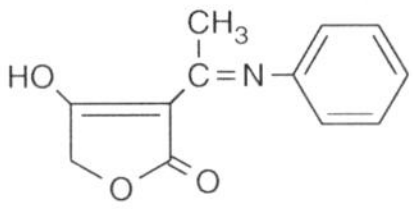
250 mg

S92,480-6
SALICYLIC (4-DIMETHYLAMINO-BENZYLIDENE)HYDRAZIDE
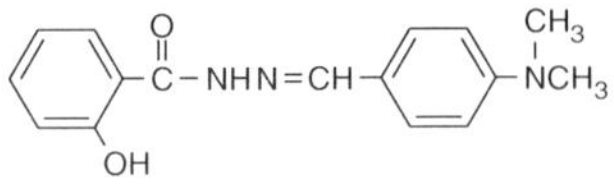
250 mg

S71,950-1
4-[4-(DIMETHYLAMINO)-BENZYLIDENEAMINO]SALICYLIC ACID
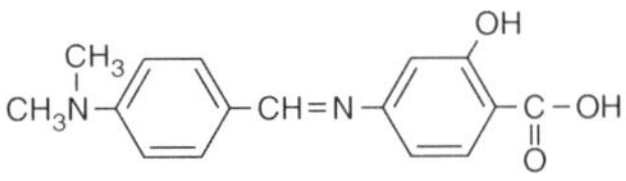
5 gm

S71,949-8
4-[4-(DIETHYLAMINO)-BENZYLIDENEAMINO]SALICYLIC ACID
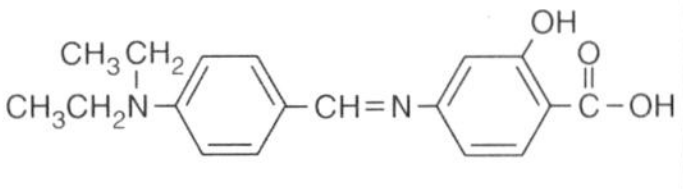
5 gm

S92,537-3
SALICYLIC (FURFURYLIDENE)HYDRAZIDE
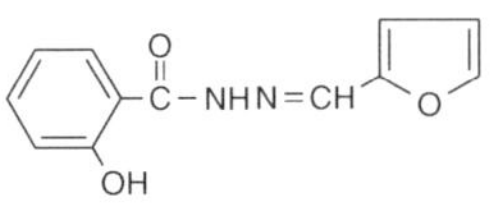
250 mg

S74,726-2
SALICYLIC (1-METHYLOCTYLIDENE)-HYDRAZIDE
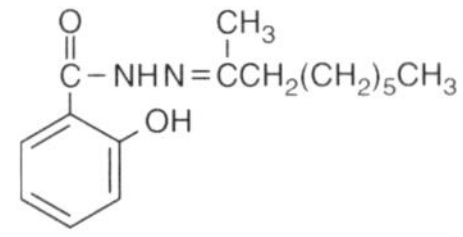
250 mg

S70,321-4
4-(2-THIENYLMETHYLENEAMINO)-SALICYLIC ACID
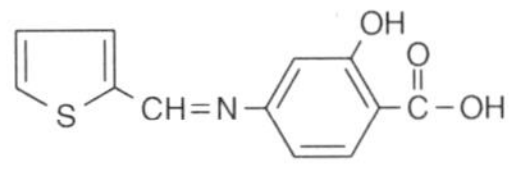
250 mg

S70,310-9
4-[3,4-(METHYLENEDIOXY)-BENZYLIDENEAMINO]SALICYLIC ACID
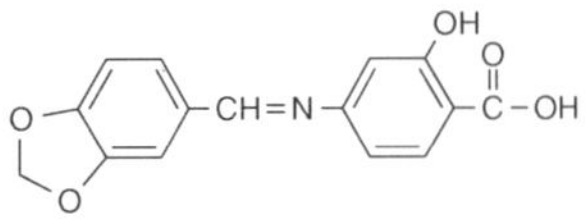
5 gm

S70,306-0
4-(4-METHYLBENZYLIDENEAMINO)-SALICYLIC ACID
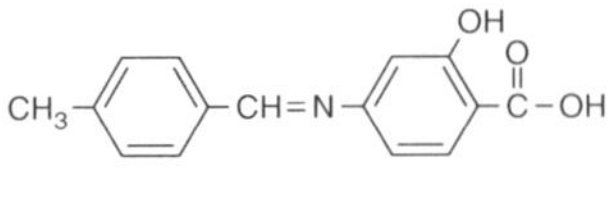
5 gm

S70,409-1
2-(SALICYLIDENEAMINO)THIOPHENOL
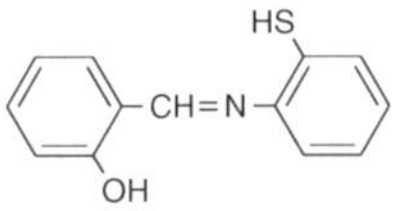
5 gm

S59,389-3
ALPHA-[3-(METHYLTHIO)PHENYLIMINO]-O-CRESOL
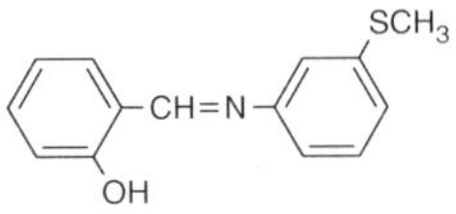
5 gm

S38,617-0
ALPHA-[4-(METHYLTHIO)PHENYLIMINO]-O-CRESOL
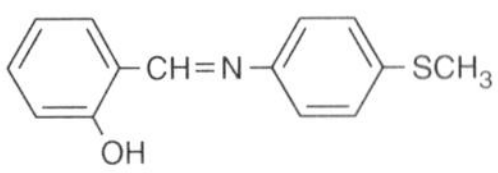
5 gm

S50,716-4
1-[4-(METHYLTHIO)PHENYLIMINOMETHYL]-2-NAPHTHOL
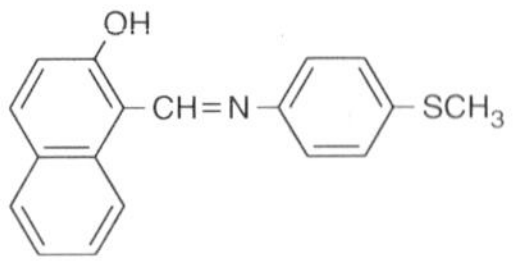
5 gm

S67,232-7
N,N'-BIS(2-HYDROXYBENZYLIDENE)-4-AMINOPHENYL DISULFIDE
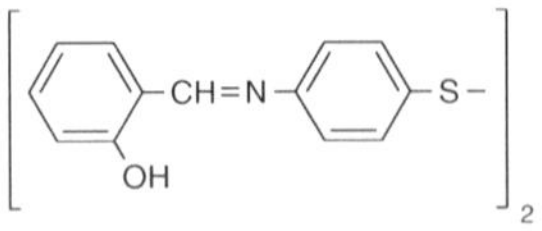
5 gm

S63,888-9
3-CARBOXY-4-HYDROXYBENZENESULFONIC ACID [6-AMINO-1-METHYL-2-PHENYL-4(1H)-QUINOLYLIDENE]HYDRAZIDE
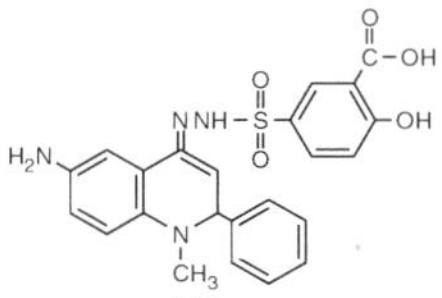
250 mg

S53,192-8
4,6-DICHLORO-ALPHA-[2-(4-NITROANILINO)ETHYLIMINO]-O-CRESOL
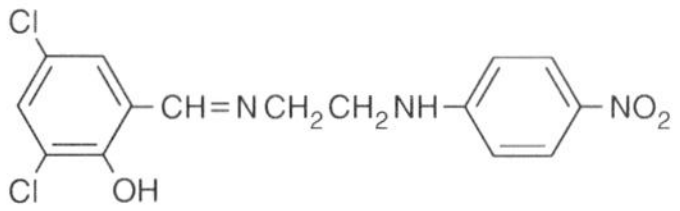
5 gm

S35,879-7
N,N-BIS(2-SALICYLIDENEAMINOETHYL)-ANILINE
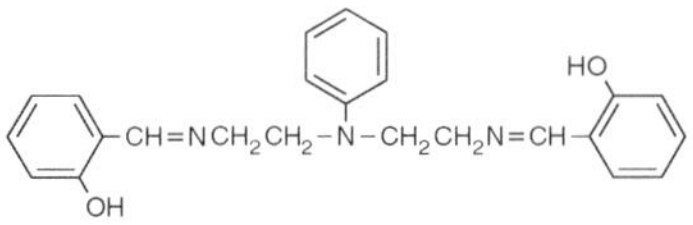
5 gm

S90,222-5
2-(3-FURFURYLUREIDO)ACETIC (3-HYDROXYBENZYLIDENE)HYDRAZIDE
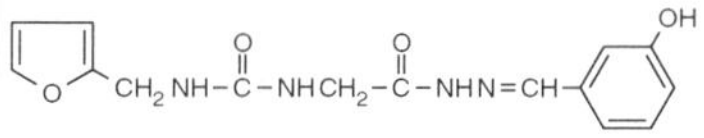
250 mg

S38,863-7
ALPHA,ALPHA'-(1-METHYLETHYLENE-DIIMINO)DI-O-CRESOL
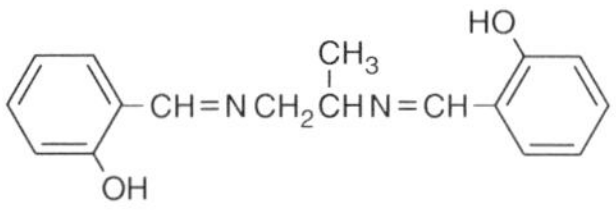
5 gm

S87,872-3
N,N'-BIS(4-HYDROXYPHENYL)-1,4-DIAZABUTADIENE
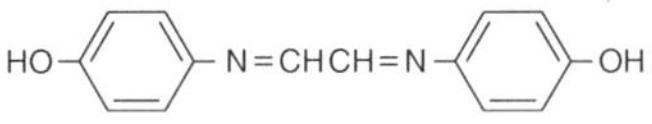
5 gm

S50,700-8
ALPHA,ALPHA'-(ETHYLENEDINITRILO)-BIS(1-METHYL-2-NAPHTHOL)
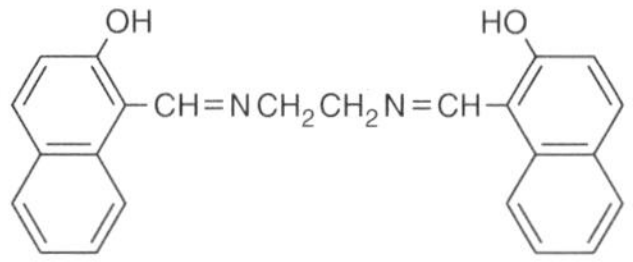
5 gm

S63,788-2
ALPHA-(2-CHLORO-5-NITROPHENYL-IMINO)-O-CRESOL
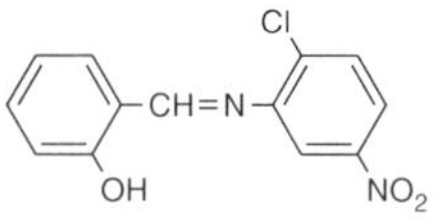
5 gm

S58,397-9
ALPHA-(3-NITRO-P-TOLYLIMINO)-O-CRESOL
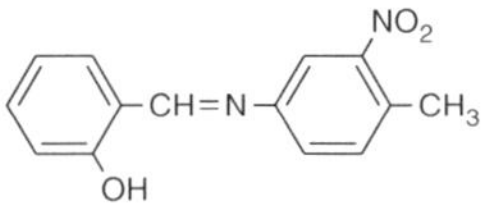
250 mg

S51,307-5
ALPHA-(3-NITROPHENYLIMINO)-O-CRESOL
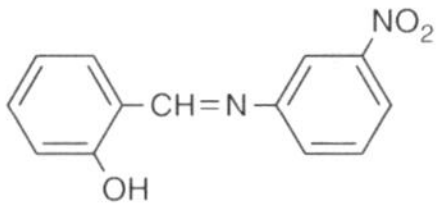
5 gm

S58,398-7
ALPHA-(5-NITRO-O-TOLYLIMINO)-O-CRESOL
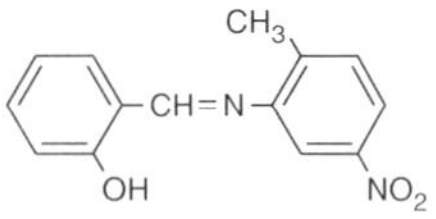
5 gm

S64,889-2
ALPHA-METHYL-ALPHA-(3-PYRIDYLMETHYLIMINO)-O-CRESOL
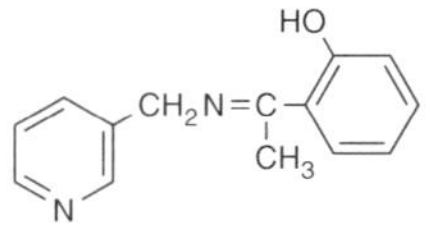
5 gm

S40,959-6
5-NITRO-2-FURALDEHYDE (4-HYDROXY-6-METHYLPYRIMIDIN-2-YL)-HYDRAZONE
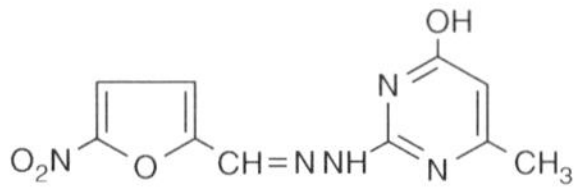
5 gm

S52,419-0
P-TOLUENESULFONIC ACID (2-HYDROXY-1-NAPHTHYLMETHYLENE)-HYDRAZIDE
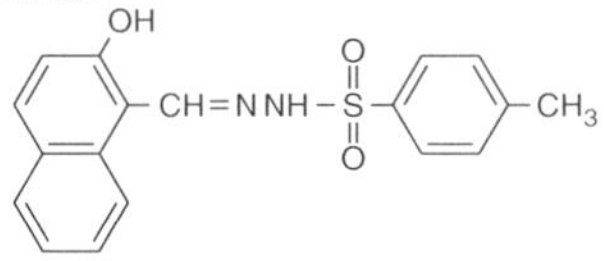
5 gm

S35,558-5
N,N'-TEREPHTHALYLIDENEBIS(2-HYDROXY-5-NITROANILINE)
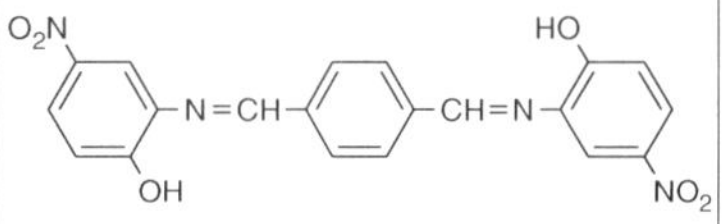
5 gm

S36,280-8
2-(3-INDOLYLMETHYLENEAMINO)ETHANOL
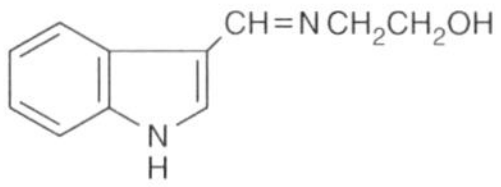
5 gm

S50,952-3
1-(3-NITROPHENYLIMINOMETHYL)-2-NAPHTHOL
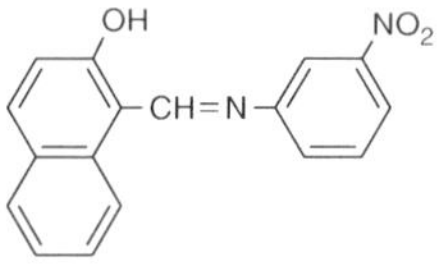
5 gm

S73,963-4
2-[1-(3-MORPHOLINOPROPYLIMINO)ETHYL]-PHENOL
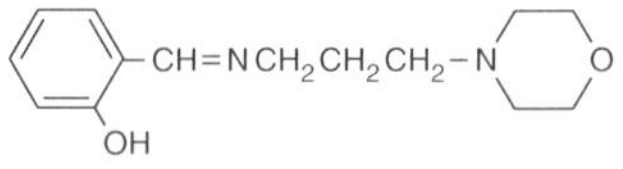
5 gm

S68,916-5
N-(2-HYDROXYBENZYLIDENE)-N'-PHENYL-1,4-PHENYLENEDIAMINE
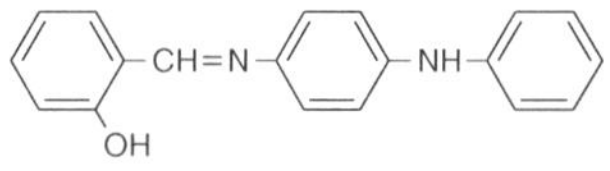
5 gm

S87,435-3
2,6-DI-TERT-BUTYL-4-[4-(DIMETHYLAMINO)PHENYLIMINOMETHYL]-PHENOL
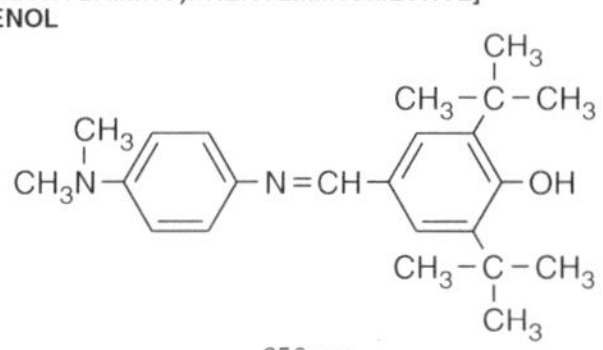
250 mg

S65,608-9
4-[BIS(2-HYDROXYETHYL)AMINO]-N-(4-NITROBENZYLIDENE)ANILINE
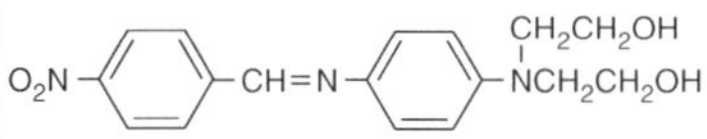
250 mg

S90,075-3
2-(3-NITROBENZYLIDENEAMINO)PHENOL
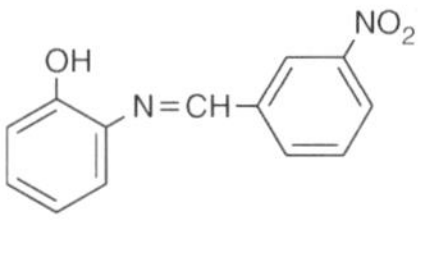
250 mg

S63,779-3
ALPHA-(2-CHLORO-4-NITROPHENYL-IMINO)-O-CRESOL
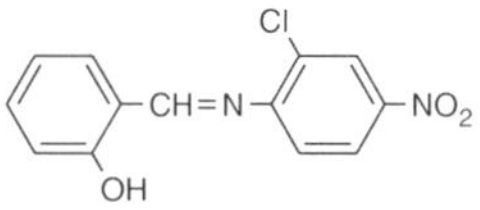
5 gm

S58,399-5
ALPHA-(4-NITRO-O-TOLYLIMINO)-O-CRESOL
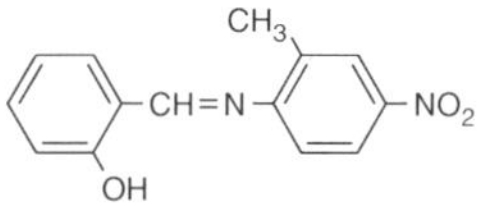
5 gm

S62,615-5
N-(9-ANTHRACENYLMETHYLENE)-2-METHYL-4-NITROANILINE
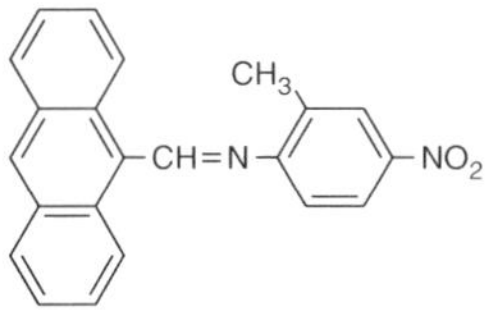
5 gm

S69,277-8
ALPHA-(4-ACETAMIDOPHENYLIMINO)-O-CRESOL
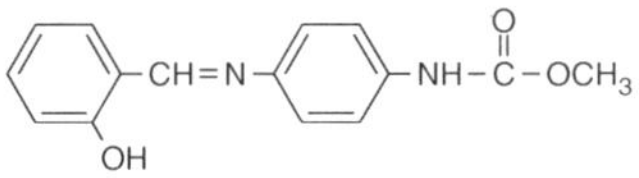
5 gm

S64,326-2
N,N'-BIS(2-HYDROXY-1-NAPHTHYLMETHYLENE)-1,4-PHENYLENEDIAMINE
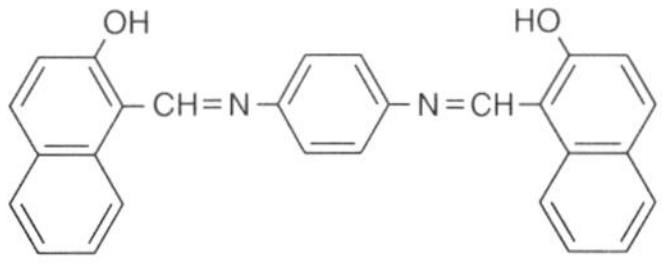
5 gm

S42,356-4
ISONICOTINIC ACID SALICYLIDENEHYDRAZIDE
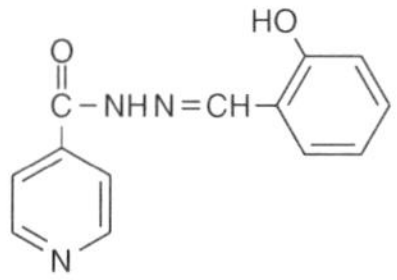
5 gm

S75,007-7
4-(3-NITROBENZYLIDENEAMINO)-PHENETHYL ALCOHOL
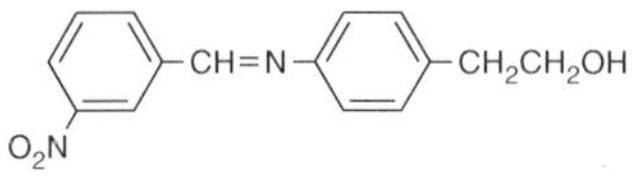
5 gm

S73,352-0
6-BENZYL-4-HYDROXY-3-(ALPHA-IMINOBENZYL)PYRIDIN-2(1H)-ONE
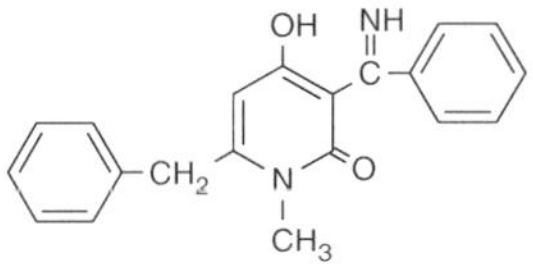
250 mg

S71,739-8
ALPHA-(PHENETHYLIMINO)-O-CRESOL
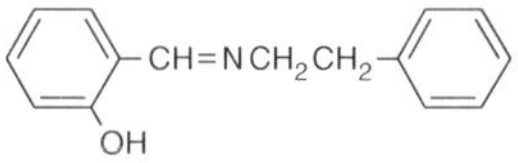
5 gm

S92,485-7
N-METHYLANTHRANILIC (2-HYDROXYBENZYLIDENE)HYDRAZIDE
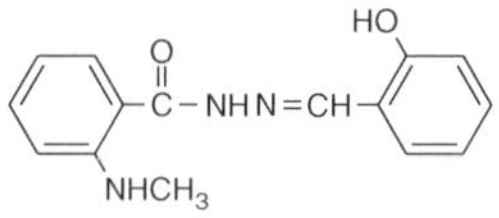
250 mg

S69,869-5
1-(4-ANILINOPHENYLIMINOMETHYL)-2-NAPHTHOL

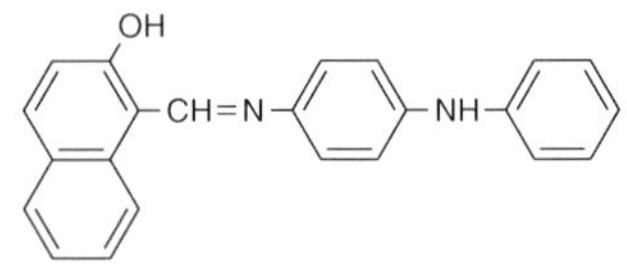

5 gm

S55,949-0
ALPHA-(4-METHYL-2-PYRIDYLAMINO)-O-CRESOL

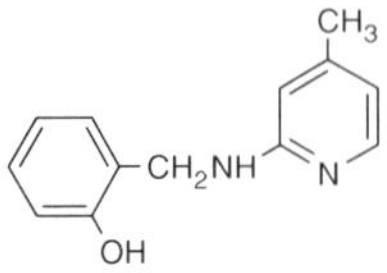

250 mg

S50,763-6
2-[4-(DIMETHYLAMINO)-BENZYLIDENEAMINO]PHENOL

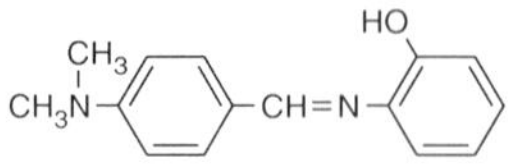

5 gm

S63,969-9
4-[4-(DIMETHYLAMINO)-BENZYLIDENEAMINO]PHENOL

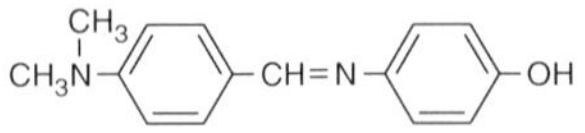

5 gm

S87,320-9
2,6-DIMETHYL-4'-(DIMETHYLAMINO)-4-HYDROXYBENZOPHENONE IMINE

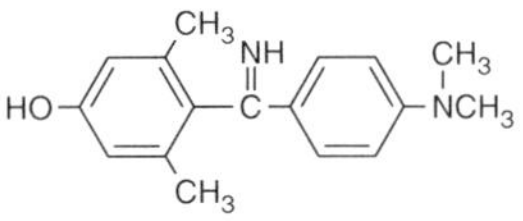

250 mg

S92,496-2
4-HYDROXYBENZOIC (4-DIMETHYLAMINOBENZYLIDENE)-HYDRAZIDE

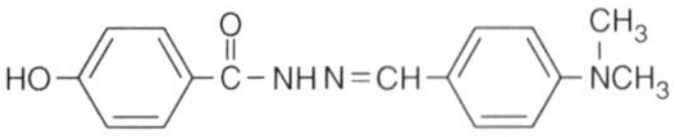

250 mg

S34,661-6
2-[4-(DIMETHYLAMINO)-BENZYLIDENEHYDRAZINO]-4-HYDROXY-6-METHYLPYRIMIDINE

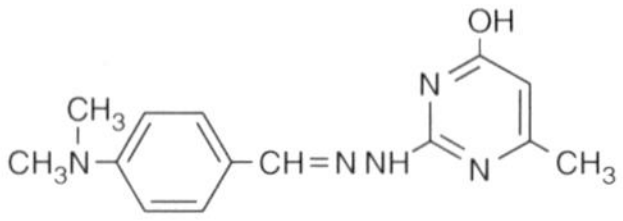

5 gm

S60,823-8
1-[2-HYDROXY-5-(1,2,3,4-TETRAHYDROQUINOL-1-YL)-2,4-PENTADIEN-1-YLIDENE]-1,2,3,4-TETRAHYDROQUINOLINIUM PERCHLORATE

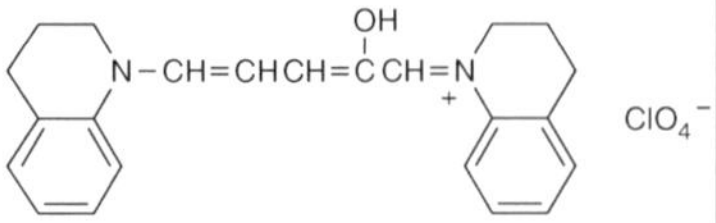

250 mg

S35,732-4
ADRENOCHROME MONOSEMICARBAZONE

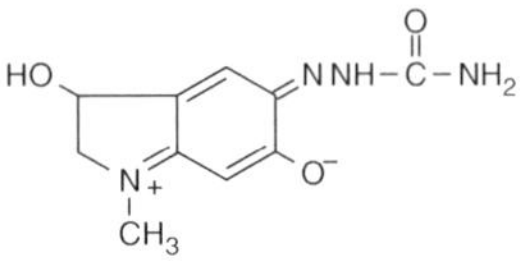

5 gm

S64,185-5
ALPHA-(4-BUTYLPHENYLIMINO)-O-CRESOL

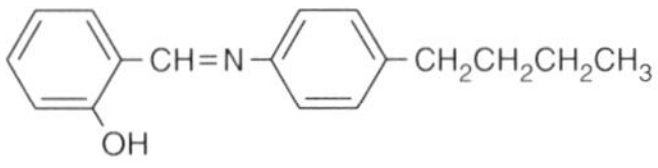

5 gm

S63,184-1
ALPHA-(4-CYANOPHENYLIMINO)-O-CRESOL

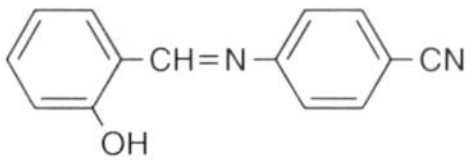

250 mg

S61,587-0
ALPHA-(O-TOLYLIMINO)-O-CRESOL

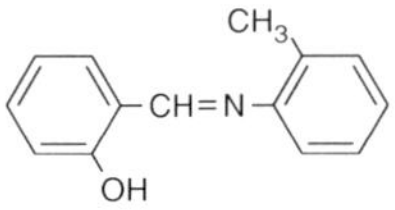

250 mg

S61,933-7
ALPHA-(6-CHLORO-2-METHYLPHENYLIMINO)-O-CRESOL

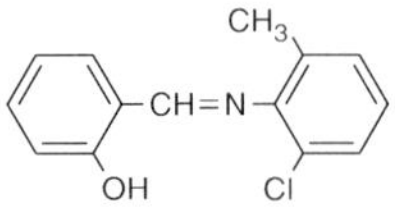

250 mg

S61,942-6
ALPHA-(3,5-XYLYLIMINO)-O-CRESOL

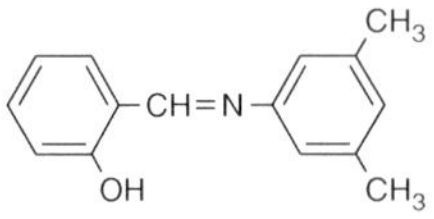

250 mg

S92,498-9
4-HYDROXYBENZOIC (4-NITROBENZYLIDENE)HYDRAZIDE

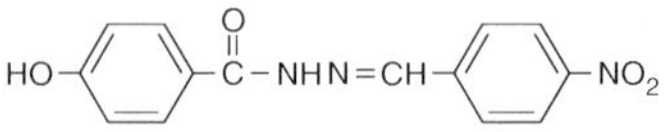

250 mg

S92,493-8
P-TOLUIC (2-HYDROXYBENZYLIDENE)-HYDRAZIDE

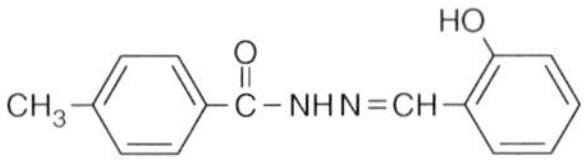

250 mg

S92,544-6
10-UNDECENOIC (2-HYDROXYBENZYLIDENE)HYDRAZIDE

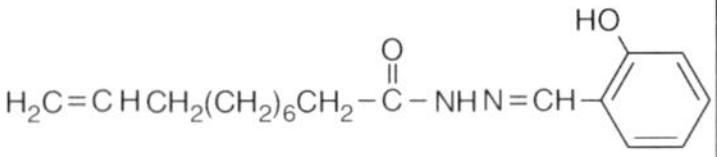

250 mg

S54,864-2
P-TOLUENESULFONIC ACID (5-CHLOROSALICYLIDENE)HYDRAZIDE

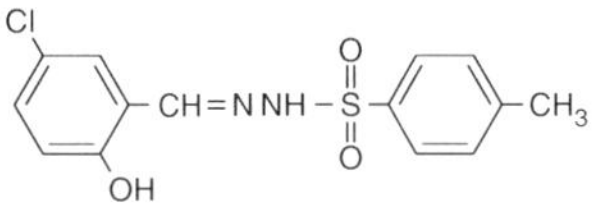

250 mg

S39,358-4
2-NITRO-ALPHA-PHENYLIMINO-P-CRESOL

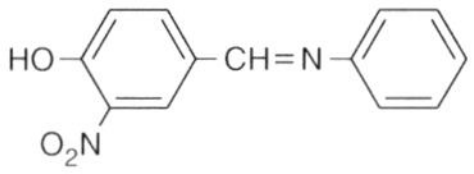

250 mg

S62,466-7
N-(9-ANTHRACENYLMETHYLENE)-2-HYDROXYANILINE

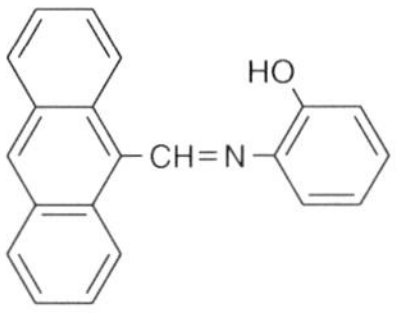

250 mg

S62,613-9
N,N'-(1,4-PHENYLENEDIMETHYLIDYNE)-BIS(2-HYDROXY-4-METHYLANILINE)

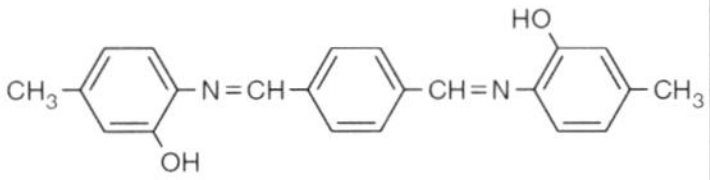

5 gm

S94,088-7
4-CHLORO-2-(2,6-DICHLOROBENZYLIDENEAMINO)PHENOL

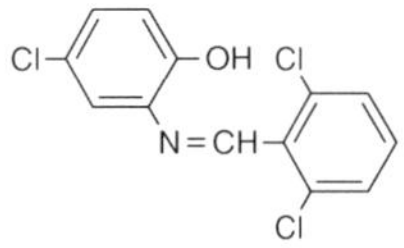

250 mg

S68,459-7
ALPHA-(3-NITROPHENYLIMINO)-P-CRESOL

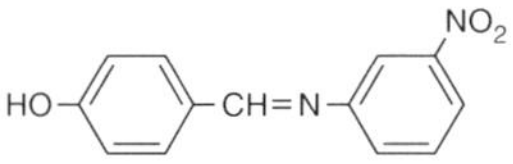

5 gm

S63,086-1
1-(4-NITROPHENYLIMINOMETHYL)-2-NAPHTHOL

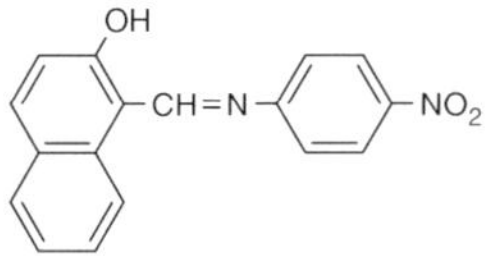

250 mg

S63,513-8
ALPHA-(3-BENZOYLPHENYLIMINO)-O-CRESOL

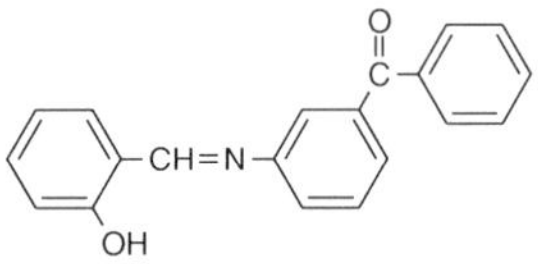

250 mg

S58,921-7
ETHYL 4-(2-HYDROXYBENZYLIDENE-AMINO)-1-BENZOATE

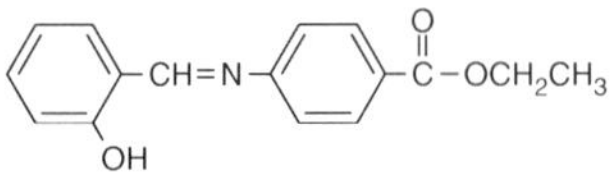

5 gm

S63,782-3
4-(2-HYDROXY-1-NAPHTHYLMETHYLENEAMINO)BENZOIC ACID

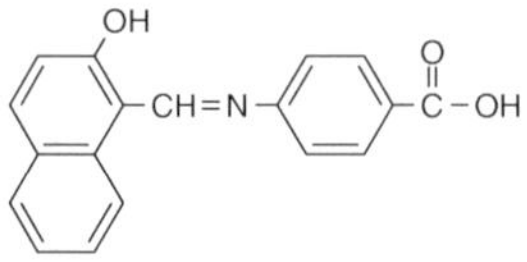

5 gm

S63,781-5
3-(2-HYDROXY-1-NAPHTHYLMETHYLENEAMINO)BENZOIC ACID

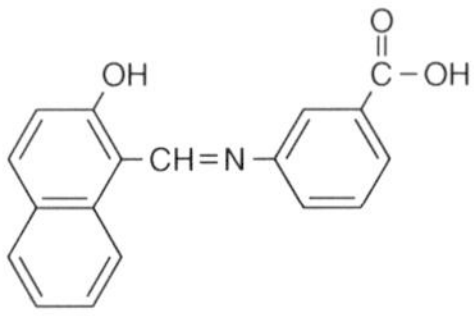

5 gm

S63,957-5
4-(ETHOXYCARBONYL)-N-(2-HYDROXY-1-NAPHTHYLMETHYLENE)ANILINE

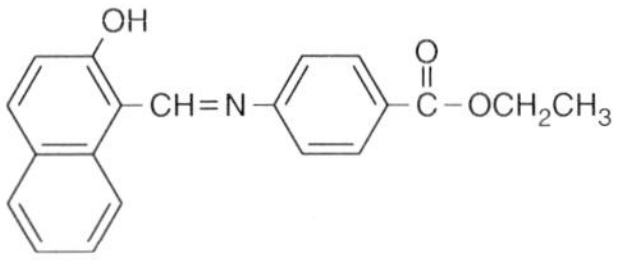

5 gm

S65,316-0
4,4'-BIS[N-(2-HYDROXY-1-NAPHTHYLMETHYLENE)AMINO]-DIPHENYL SULFONE

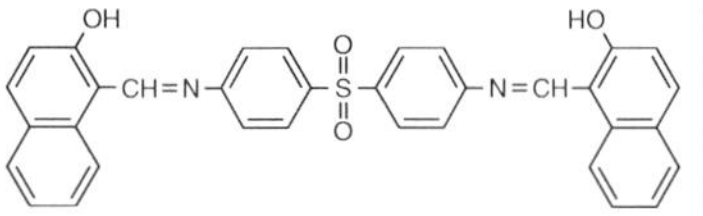

5 gm

S50,135-2
4-HYDROXY-ALPHA-PHENYLIMINO-M-TOLUENESULFONIC ACID

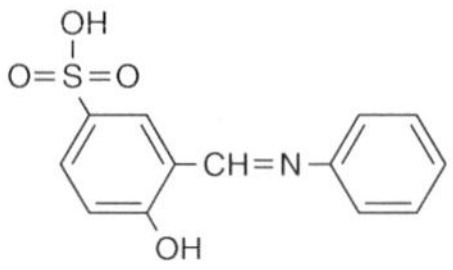

250 mg

S62,675-9
4-BENZYLIDENEAMINO-5-HYDROXY-2,7-NAPHTHALENEDISULFONIC ACID MONOSODIUM SALT

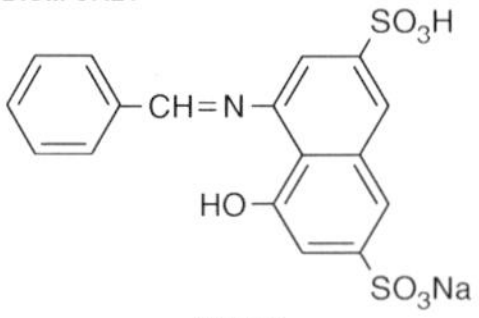

250 mg

S63,778-5
ALPHA-(2-BENZOTHIAZOLYLIMINO)-O-CRESOL

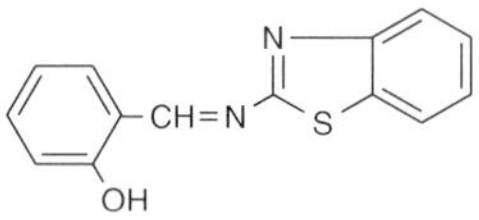

5 gm

S64,186-3
1-(2-BENZOTHIAZOLYLIMINOMETHYL)-2-NAPHTHOL

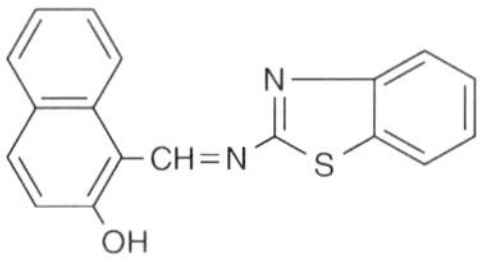

5 gm

S38,883-1
ALPHA-(PHENYLIMINO)-O-CRESOL

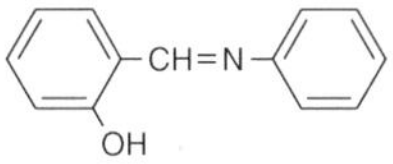

5 gm

S64,602-4
ALPHA-BENZYLIMINO-ALPHA-METHYL-O-CRESOL

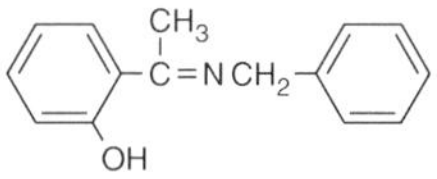

5 gm

S59,064-9
4-(BENZYLIDENEAMINO)PHENOL

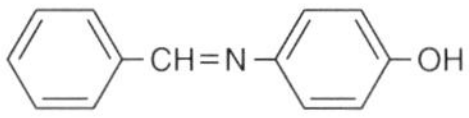

250 mg

S68,458-9
ALPHA-(PHENYLIMINO)-P-CRESOL

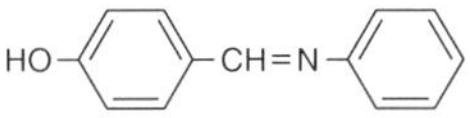

5 gm

S72,059-3
ALPHA-METHYL-ALPHA-(PROPYLIMINO)-O-CRESOL

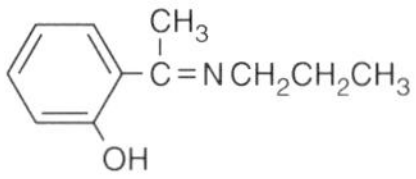

5 gm

S73,010-6
2-(1-BUTYLIMINOETHYL)PHENOL

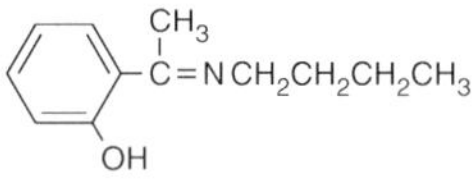

5 gm

S58,385-5
ALPHA-(2-FLUOROPHENYLIMINO)-O-CRESOL

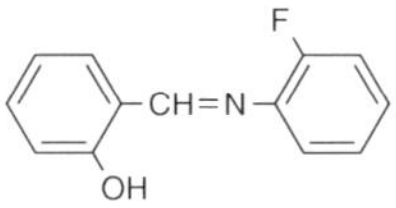

5 gm

S50,581-1
ALPHA-(2-CHLOROPHENYLIMINO)-O-CRESOL

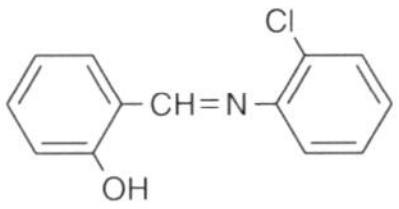

5 gm

S51,670-8
ALPHA-(3-CHLOROPHENYLIMINO)-O-CRESOL

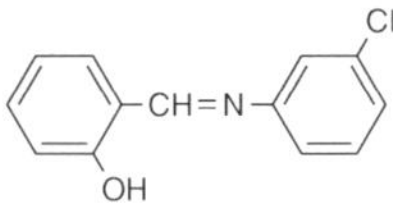

5 gm

S58,384-7
ALPHA-(3-BROMOPHENYLIMINO)-O-CRESOL

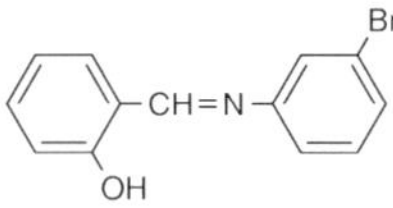

5 gm

S50,522-6
ALPHA-(M-TOLYLIMINO)-O-CRESOL

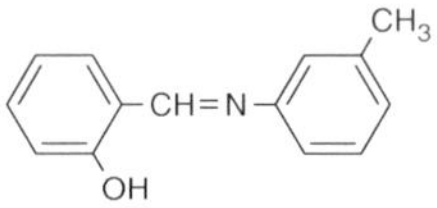

5 gm

S52,411-5
2-(4-FLUOROBENZYLIDENEAMINO)PHENOL

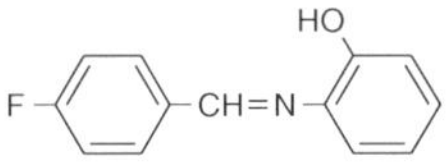

5 gm

S50,718-0
ALPHA-(4-CHLOROPHENYLIMINO)-O-CRESOL

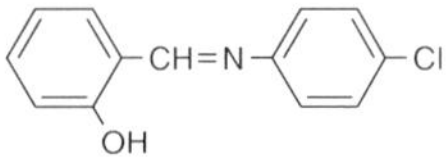

5 gm

S41,240-6
ALPHA-(4-BROMOPHENYLIMINO)-O-CRESOL

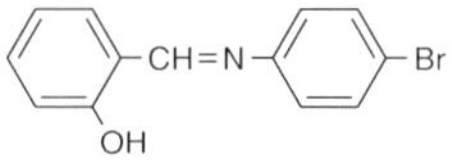

5 gm

S50,523-4
ALPHA-(P-TOLYLIMINO)-O-CRESOL

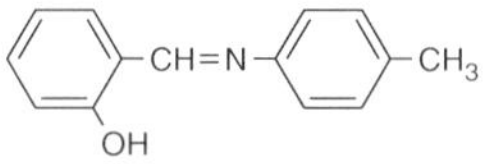

5 gm

S58,404-5
ALPHA-(4-ETHYLPHENYLIMINO)-O-CRESOL

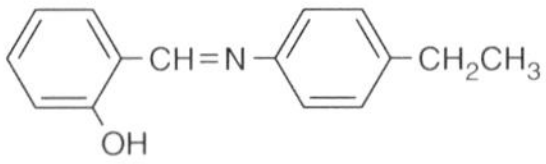

5 gm

S51,305-9
ALPHA-(2,3-DICHLOROPHENYLIMINO)-O-CRESOL

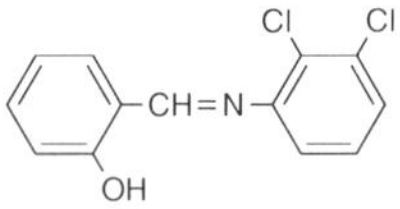

5 gm

S39,914-0
ALPHA-(3-CHLORO-O-TOLYLIMINO)-O-CRESOL

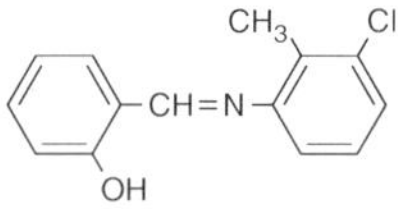

5 gm

S58,915-2
ALPHA-(2,3-DIMETHYLPHENYLIMINO)-O-CRESOL

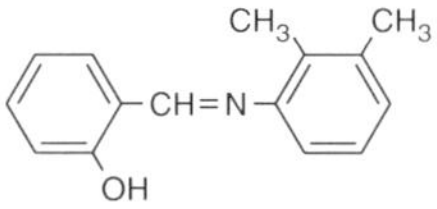

5 gm

S51,306-7
ALPHA-(2,4-DICHLOROPHENYLIMINO)-O-CRESOL

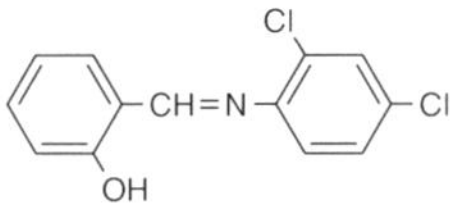

5 gm

S51,713-5
ALPHA-(2,5-DICHLOROPHENYLIMINO)-O-CRESOL

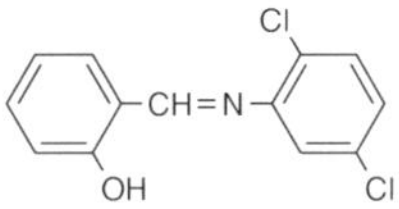

5 gm

S58,380-4
ALPHA-(3,4-DICHLOROPHENYLIMINO)-O-CRESOL

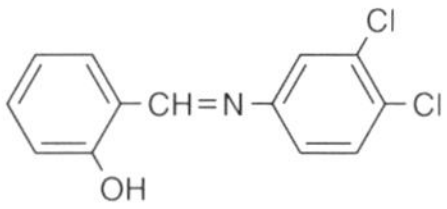

5 gm

S38,504-2
ALPHA-(5-CHLORO-O-TOLYLIMINO)-O-CRESOL

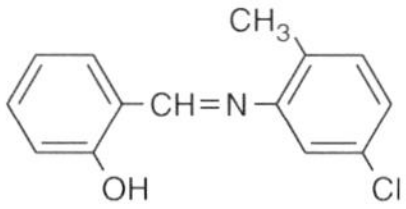

5 gm

S39,617-6
ALPHA-(3-CHLORO-P-TOLYLIMINO)-O-CRESOL

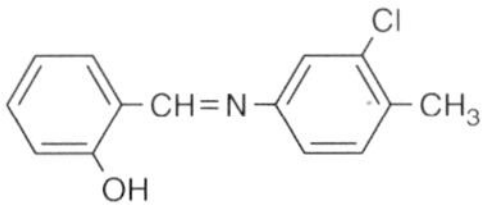

5 gm

S58,396-0
ALPHA-(2-BROMO-P-TOLYLIMINO)-O-CRESOL

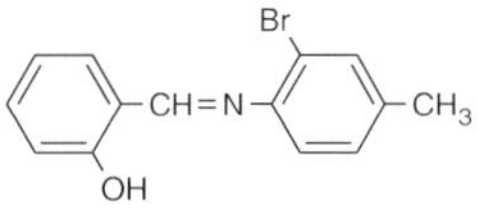

5 gm

S38,502-6
ALPHA-(3,4-XYLYLIMINO)-O-CRESOL

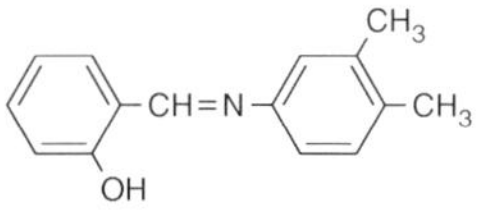

5 gm

S59,082-7
ALPHA-(2,4-XYLYLIMINO)-O-CRESOL

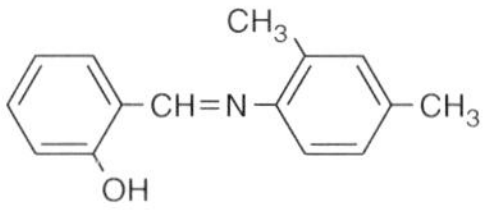

5 gm

S59,083-5
ALPHA-(2,5-XYLYLIMINO)-O-CRESOL

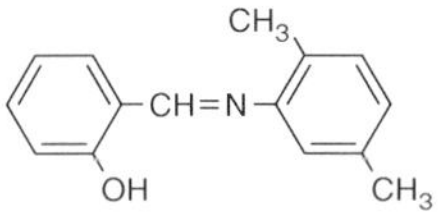

5 gm

S58,909-8
ALPHA-(2,4,6-TRIMETHYLPHENYLIMINO)-O-CRESOL

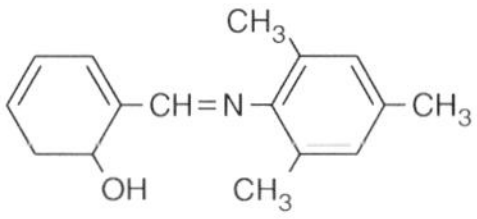

5 gm

S60,117-9
ALPHA-(5-CHLORO-2-PYRIDYLIMINO)-O-CRESOL COPPER(II) SALT

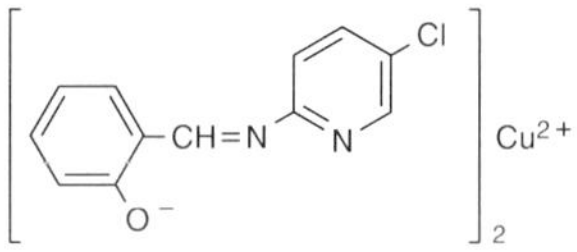

250 mg

S60,128-4
ALPHA-(3-METHYL-2-PYRIDYLIMINO)-O-CRESOL COPPER(II) SALT
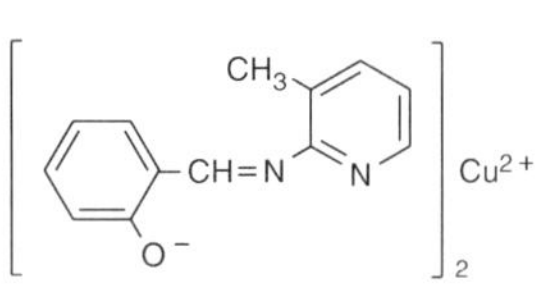
250 mg

S60,498-4
ALPHA-(4-METHYL-2-PYRIDYLIMINO)-O-CRESOL COPPER(II) SALT
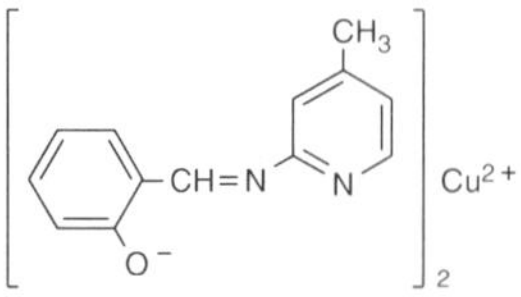
5 gm

S60,490-9
ALPHA-(2-PYRIDYLIMINO)-O-CRESOL COPPER(II) SALT
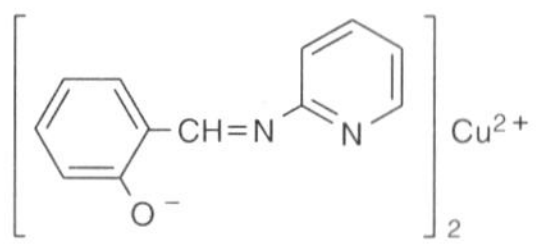
250 mg

S60,499-2
ALPHA-(6-METHYL-2-PYRIDYLIMINO)-O-CRESOL COPPER(II) SALT
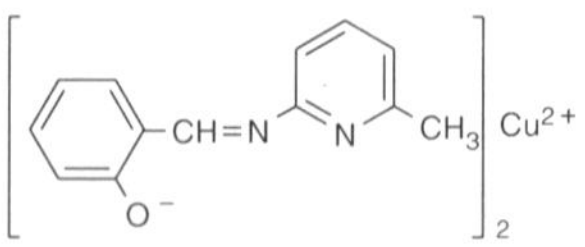
5 gm

S76,415-9
ALPHA-(3-PYRIDYLMETHYLIMINO)-O-CRESOL
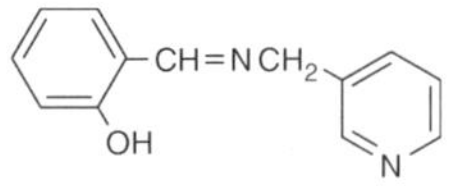
5 gm

S62,617-1
N-(9-ANTHRACENYLMETHYLENE)-2-ETHYLANILINE
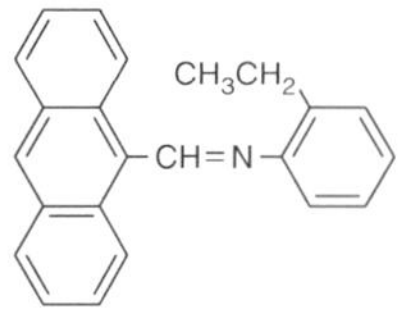
5 gm

S68,480-5
ALPHA-(4-TOLYLIMINO)-P-CRESOL
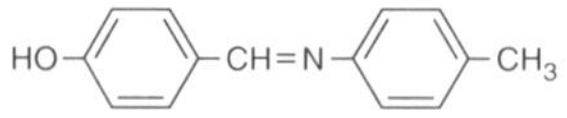
5 gm

S43,392-6
ALPHA-(4-BROMOPHENYLIMINO)-4,6-DIBROMO-O-CRESOL
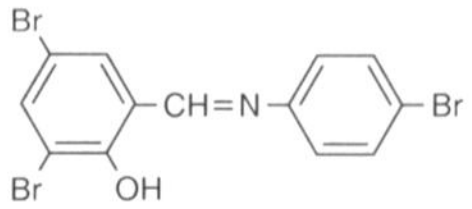
5 gm

S68,909-2
4,4'-METHYLENEBIS[N-(4-HYDROXYBENZYLIDENE)ANILINE]
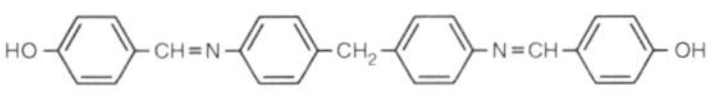
5 gm

S63,959-1
4-ETHYL-N-(2-HYDROXY-1-NAPHTHYLMETHYLENE)ANILINE
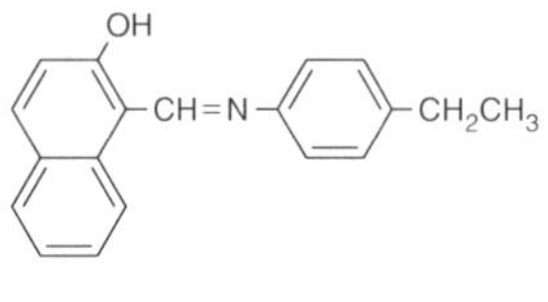
5 gm

S62,475-6
N-(9-ANTHRACENYLMETHYLENE)-4-HYDROXYANILINE
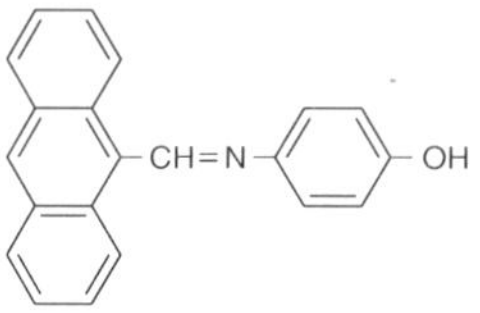
5 gm

S62,614-7
N-(9-ANTHRACENYLMETHYLENE)-3-HYDROXYANILINE
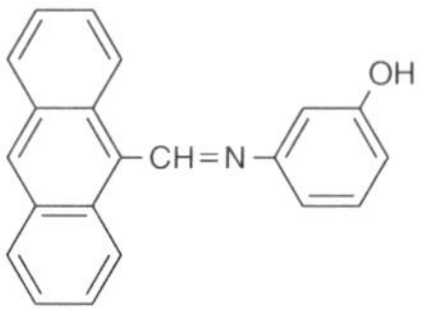
5 gm

S36,250-6
3-(4-HYDROXY-6-METHYL-2-PYRIMIDINYLHYDRAZONO)-2-INDOLINONE
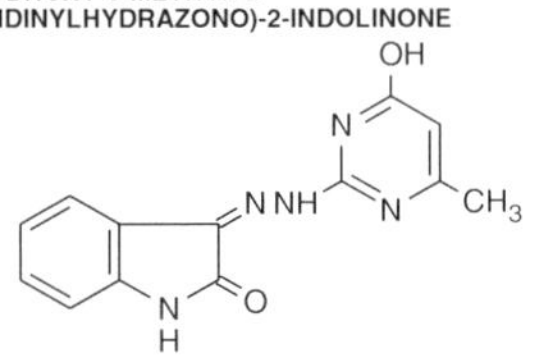
250 mg

S92,514-4
3-HYDROXYBUTYRIC (FURFURYLIDENE)-HYDRAZIDE MONOHYDRATE
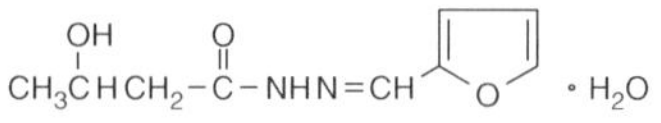
250 mg

S92,466-0
SALICYALDEHYDE 2,6-DINITRO-4-METHYLPHENYLHYDRAZONE
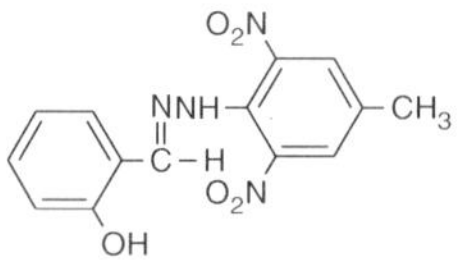
250 mg

S85,451-4
4-(3-FURFURYLUREIDO)BENZOIC (2-HYDROXYBENZYLIDENE)HYDRAZIDE
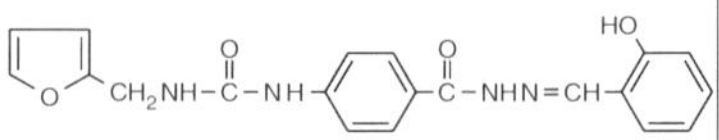
250 mg

S92,527-6
SALICYLIC (2-BUTEN-1-YLIDENE)-HYDRAZIDE
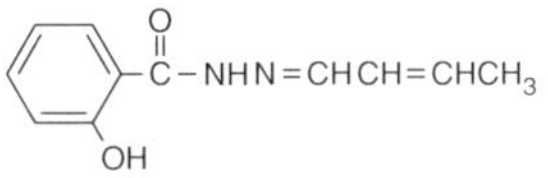
250 mg

S92,552-7
4-PYRIDINECARBOXYLIC (4-HYDROXYBENZYLIDENE)HYDRAZIDE
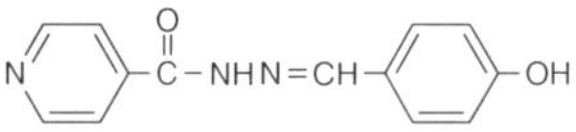
250 mg

S52,461-1
N-(5-BROMOSALICYLIDENEAMINO)-PHTHALIMIDE
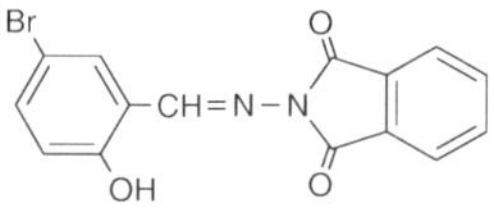
5 gm

S52,470-0
N-(3,5-DICHLOROSALICYLIDENEAMINO)-PHTHALIMIDE
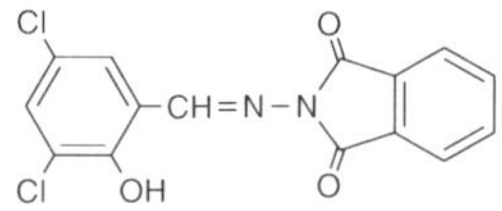
5 gm

S52,458-1
N-(3,5-DIBROMOSALICYLIDENEAMINO)-PHTHALIMIDE
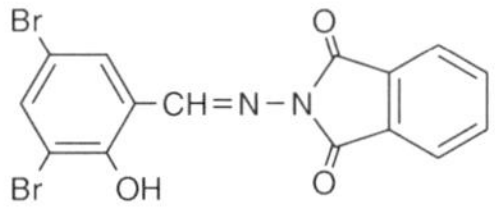
250 mg

S41,166-3
1,2-DIHYDRO-4,6-DIMETHYL-2-OXO-1-(SALICYLIDENEAMINO)NICOTINONITRILE
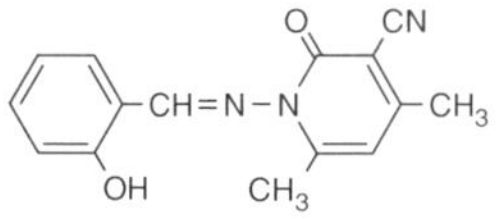
5 gm

S68,874-6
2-(FURFURYLUREIDO)ACETIC ACID (2-HYDROXYBENZYLIDENE)HYDRAZIDE
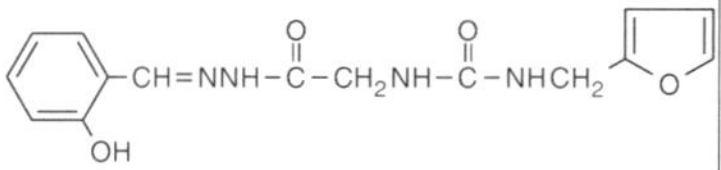
250 mg

S92,535-7
SALICYLIC (CYCLOPENTYLIDENE)-HYDRAZIDE
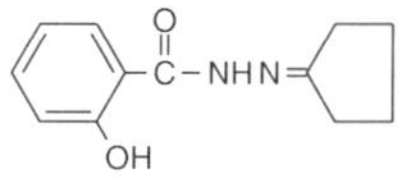
250 mg

S92,534-9
SALICYLIC (CYCLOHEXYLIDENE)-HYDRAZIDE
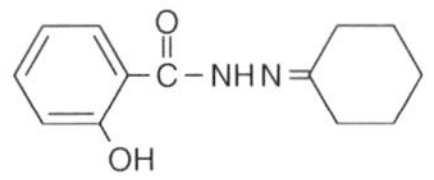
250 mg

S92,515-2
SALICYLIC (ALPHA-PROPYL-BENZYLIDENE)HYDRAZIDE
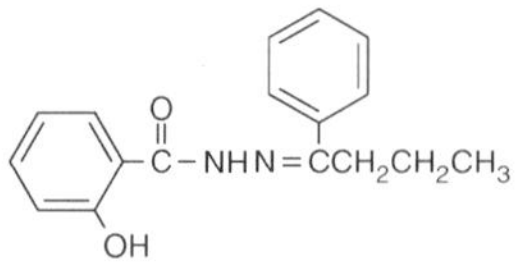
250 mg

S68,594-1
4,4-DIPHENYL-1-(2-HYDROXYBENZYLIDENE)SEMICARBAZIDE
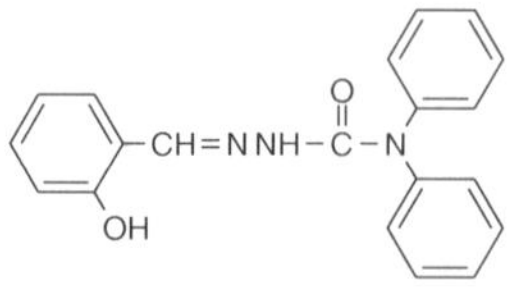
250 mg

S86,369-6
3-HYDROXYVALEROPHENONE (P-TOSYL)-HYDRAZONE
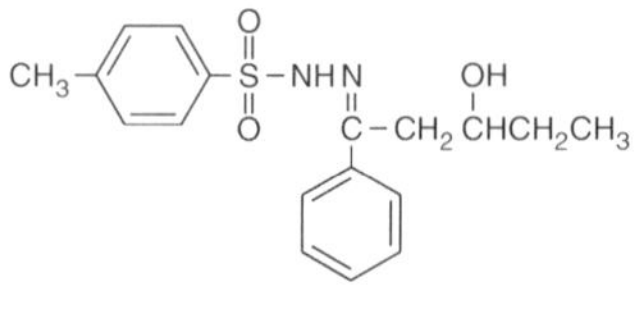
250 mg

S86,518-4
4-HYDROXY-2-HEXANONE (P-TOSYL)-HYDRAZONE
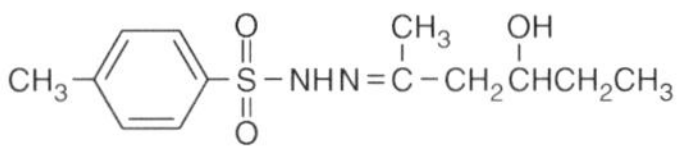
250 mg

S35,010-9
2-(FURFURYLIDENEHYDRAZINO)-4-HYDROXY-6-METHYLPYRIMIDINE
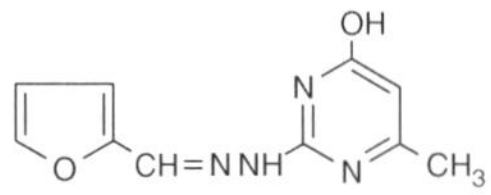
5 gm

S76,049-8
2-ACETYL-3-HYDROXYBENZO(B)-THIOPHENE HYDRAZONE HYDRATE
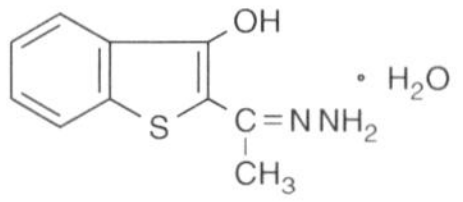
250 mg

S35,089-3
4-HYDROXY-2-(ISOBUTYLIDENEHYDRAZINO)-6-METHYLPYRIMIDINE
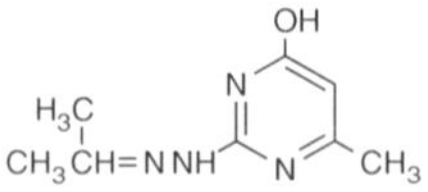
5 gm

S35,923-8
2-(BUTYLIDENEHYDRAZINO)-4-HYDROXY-6-METHYLPYRIMIDINE
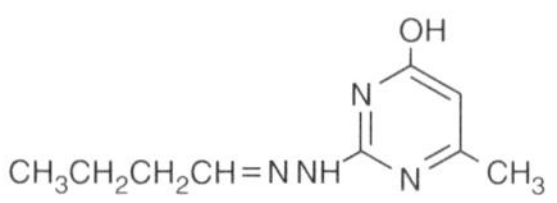
250 mg

S35,085-0
4-HYDROXY-2-(4-HYDROXYBENZYLIDENEHYDRAZINO)-6-METHYLPYRIMIDINE
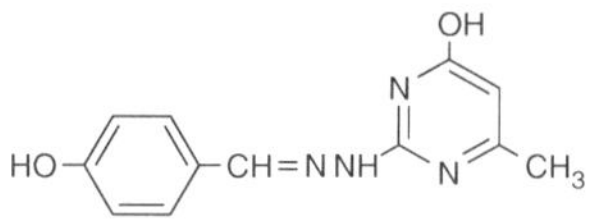
5 gm

S36,248-4
4-HYDROXY-2-(2-HYDROXYBENZYLIDENEHYDRAZINO)-6-METHYLPYRIMIDINE
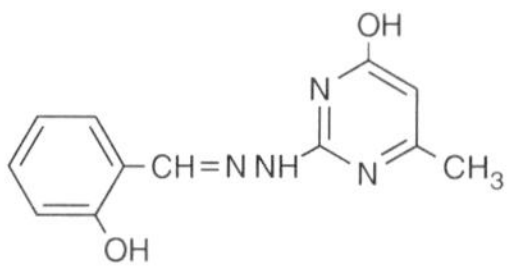
250 mg

S35,971-8
2-(CINNAMYLIDENEHYDRAZINO)-4-HYDROXY-6-METHYLPYRIMIDINE
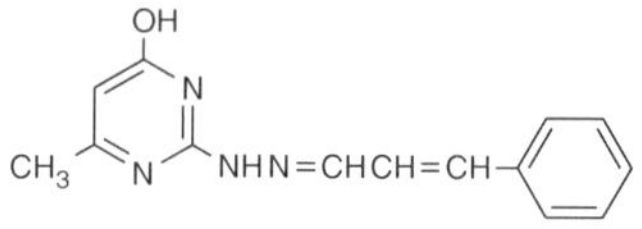
5 gm

S50,668-0
5-INDANCARBOXALDEHYDE (4-HYDROXY-6-METHYL-2-PYRIMIDINYL)HYDRAZONE
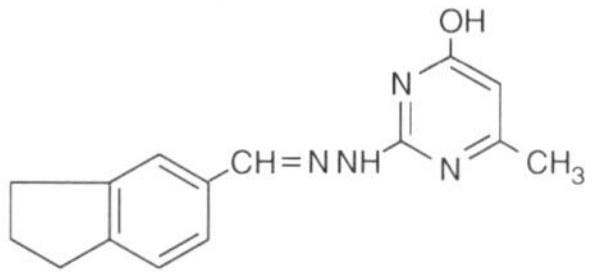
250 mg

S69,198-4
1-(2-HYDROXYBENZYLIDENE)-4-PHENYL-3-THIOSEMICARBAZIDE
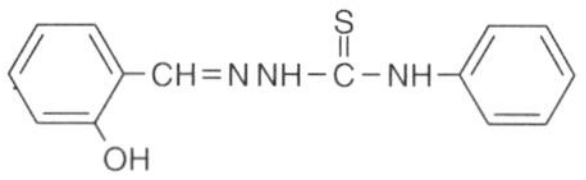
250 mg

S50,699-0
1-PHENYLIMINOMETHYL-2-NAPHTHOL
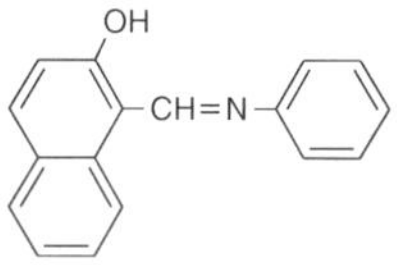
5 gm

S93,837-8
4-(2-HYDROXYBENZYLIDENEAMINO)-4H-1,2,4-TRIAZOLE
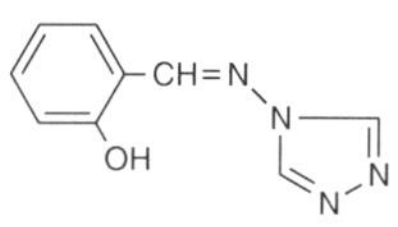
250 mg

S77,141-4
1-(ALPHA-BUTYL-4-HYDROXYBENZYLIDENEAMINO)-GUANIDINE NITRATE
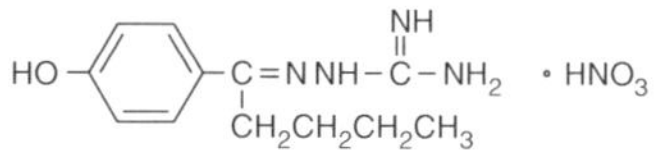
250 mg

S50,698-2
1-(2,5-DICHLOROPHENYLIMINOMETHYL)-2-NAPHTHOL
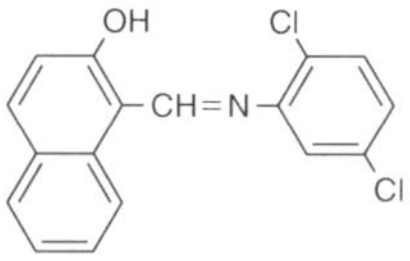
5 gm

S39,026-7
ALPHA-(2-PYRIDYLIMINO)-O-CRESOL
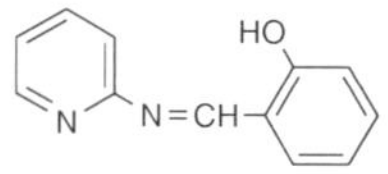
250 mg

S79,096-6
N-(3-NITROBENZYLIDENE)ETHANOLAMINE
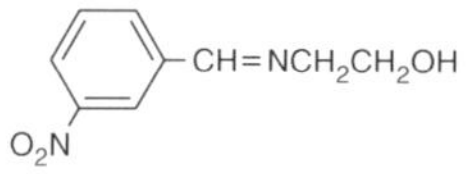
250 mg

S39,918-3
2-(PIPERONYLIDENEAMINO)PHENOL
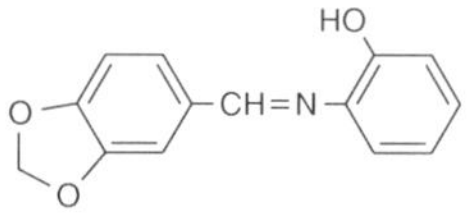
5 gm

S51,783-6
2-(5,6-DIMETHYL-2-BENZIMIDAZOLYLIMINOMETHYL)-4-METHYL-5-IMIDAZOLE
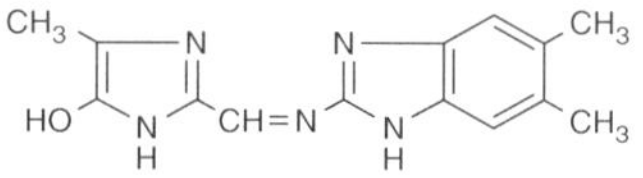
250 mg

S66,114-7
METHYL 1-HYDROXYCYCLOHEXANE-CARBOXIMIDATE HYDROCHLORIDE
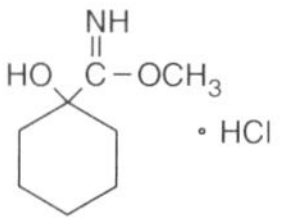
5 gm

S51,669-4
1-(2-HYDROXYPHENYLIMINOMETHYL)-2-NAPHTHOL
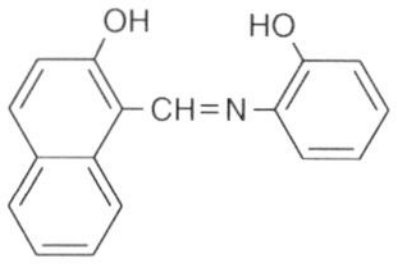
5 gm

S51,471-3
1-(5-CHLORO-2,4-DIMETHOXYPHENYLIMINOMETHYL)-2-NAPHTHOL
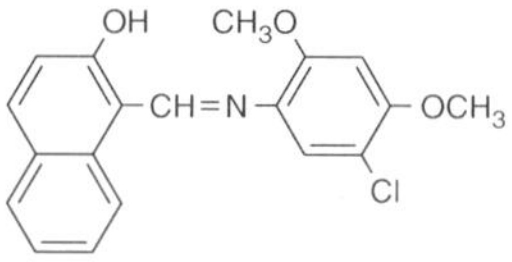
5 gm

S58,653-6
4-(2-THIENYLMETHYLENEAMINO)PHENOL
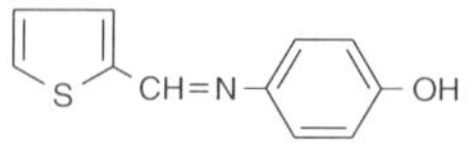
5 gm

S68,650-6
1-(4-METHYL-3-NITROPHENYLIMINOMETHYL)-2-NAPHTHOL
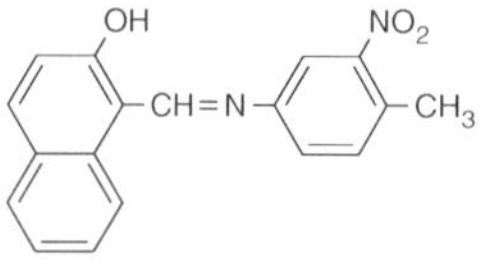
5 gm

S68,661-1
1-(2-METHYL-5-NITROPHENYLIMINOMETHYL)-2-NAPHTHOL
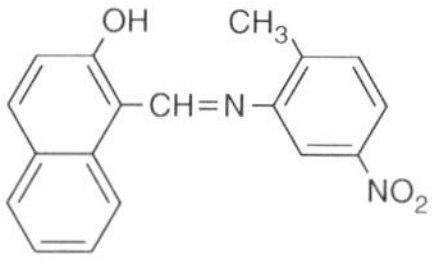
5 gm

S71,736-3
ALPHA-(3-ACETYLPHENYLIMINO)-O-CRESOL
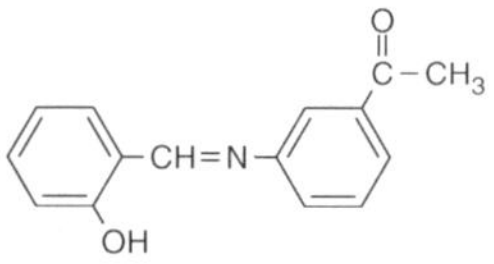
5 gm

S72,004-6
4'-(2-HYDROXYBENZYLIDENEAMINO)-ACETOPHENONE
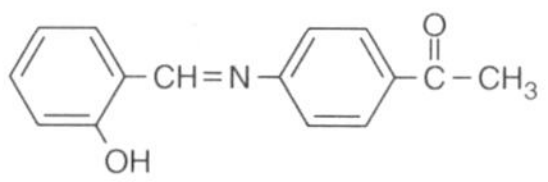
5 gm

S71,970-6
4'-(2-HYDROXY-1-NAPHTHYLMETHYLENEAMINO)ACETOPHENONE
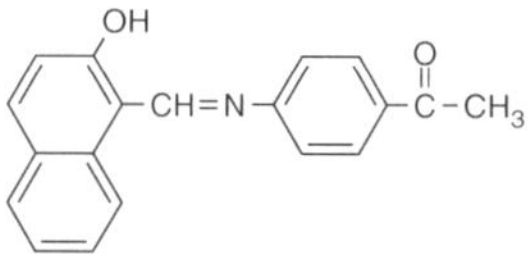
5 gm

S53,233-9
BENZIL, AZINE WITH 2-HYDROXY-1-NAPHTHALDEHYDE
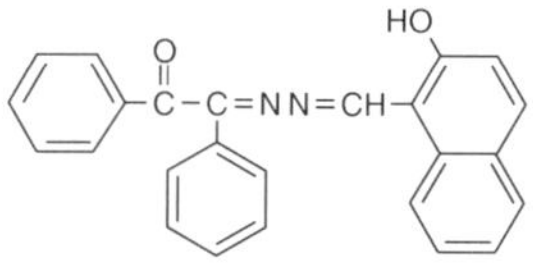
5 gm

S43,257-1
2,6-DICHLOROINDO-M-CRESOL, SODIUM SALT
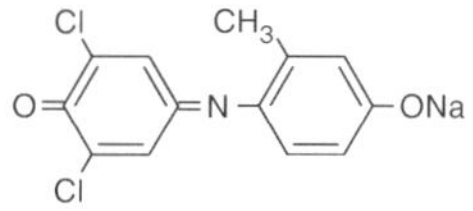
5 gm

S63,973-7
2-(2-HYDROXY-1-NAPHTHYLMETHYLENEAMINO)BENZOIC ACID
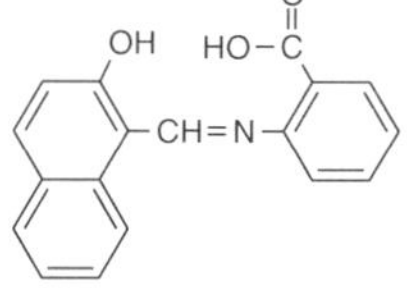
5 gm

S70,288-9
4-(2-HYDROXY-1-NAPHTHYLMETHYLENEAMINO)-SALICYLIC ACID
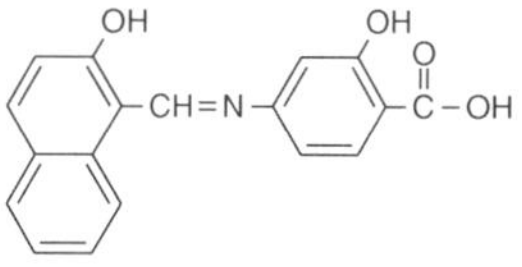
5 gm

S68,520-8
ETHYL 4-(4-HYDROXYBENZYLIDENE-AMINO)BENZOATE
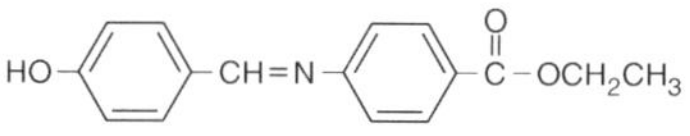
5 gm

S51,452-7
5-(5-INDANYLMETHYLENEAMINO)-SALICYLIC ACID
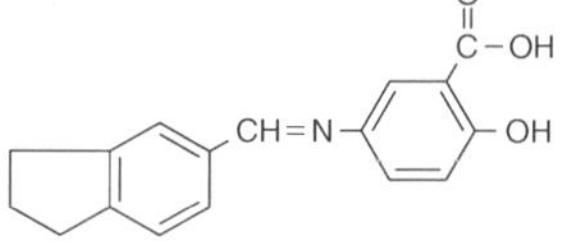
250 mg

S36,352-9
METHYL N-(3,7-DIMETHYL-7-HYDROXYOCTYLIDENE)ANTHRANILATE
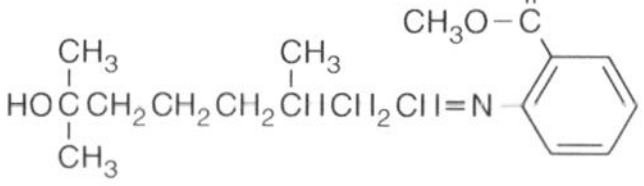
5 gm

S75,950-3
4-HYDROXY-3-(ALPHA-IMINOBENZYL)-1-METHYL-2,6(1H,5H)-PYRIDINEDIONE
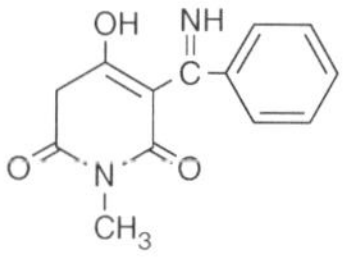
250 mg

S71,995-1
2,3-DIMETHYL-4-(2-HYDROXYBENZYLIDENEAMINO)-1-PHENYL-3-PYRAZOLIN-5-ONE
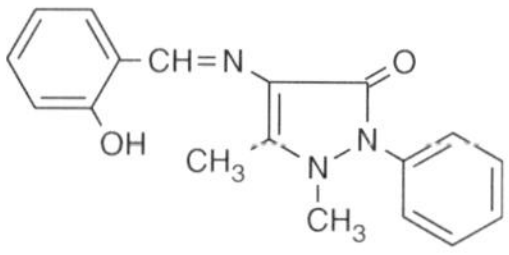
5 gm

S71,997-8
2,3-DIMETHYL-4-(4-HYDROXYBENZYLIDENEAMINO)-1-PHENYL-3-PYRAZOLIN-5-ONE

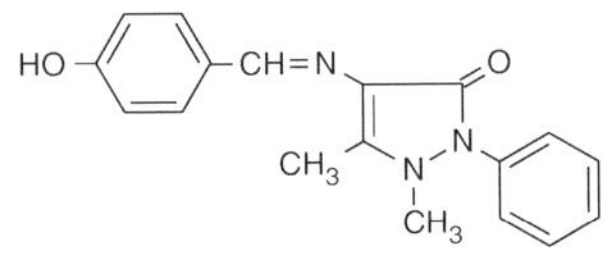

5 gm

S73,003-3
4-(2-HYDROXY-1-NAPHTHYLMETHYLENEAMINO)ANTIPYRINE

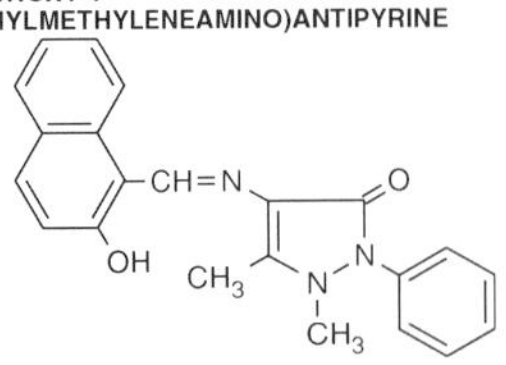

5 gm

S63,972-9
2-(2-HYDROXY-1-NAPHTHYLMETHYLENEAMINO)THIAZOLE

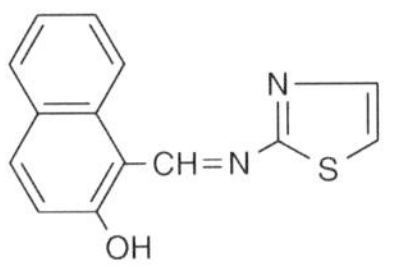

5 gm

S65,750-6
ALPHA-[4-(6-METHYL-2-BENZOTHIAZOLYL)PHENYLIMINO]-O-CRESOL

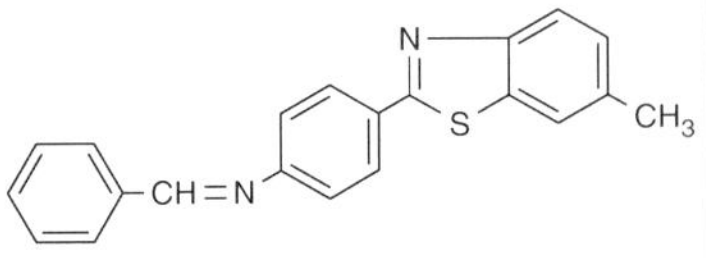

250 mg

S52,606-1
ALPHA,ALPHA-[1,4-CYCLOHEXYLENEBIS-(METHYLIMINO)]-DI-O-CRESOL

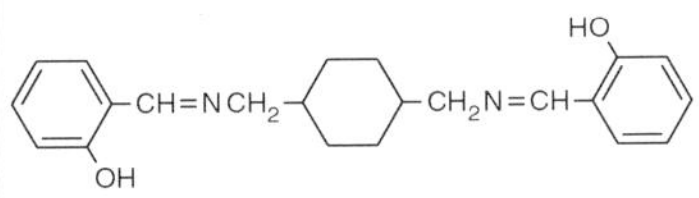

5 gm

S52,335-6
ALPHA,ALPHA'-(1,4-CYCLOHEXYLENEBIS-METHYLIMINO)BIS(4,6-DICHLORO-O-CRESOL)

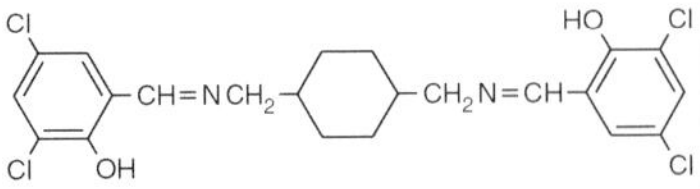

5 gm

S71,975-7
2-[1-(CYCLOHEXYLIMINO)ETHYL]PHENOL

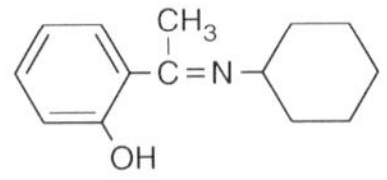

5 gm

S54,196-6
4-BROMO-ALPHA-(4-PIPERIDINOPHENYLIMINO)-O-CRESOL

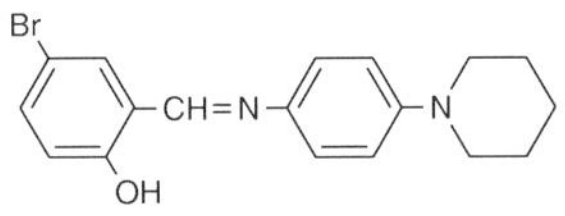

5 gm

S78,116-9
3-(CINNAMYLIDENEAMINO)PHENOL

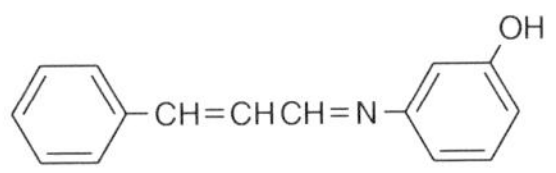

250 mg

S71,967-6
5-(2-HYDROXYBENZYLIDENEAMINO)-1H-TETRAZOLE

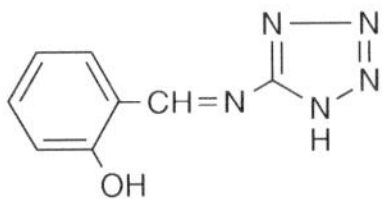

5 gm

S71,976-5
2-[1-(DECYLIMINO)ETHYL]PHENOL

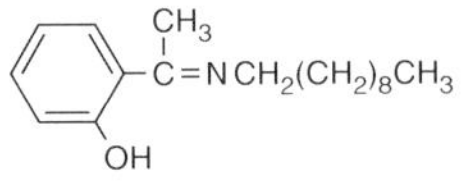

5 gm

S73,958-8
N,N'-BIS(2-HYDROXYBENZYLIDENE)-M-XYLYLENEDIAMINE

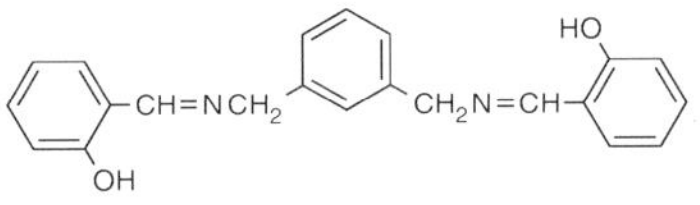

5 gm

S61,984-1
ALPHA-(3,5-DICHLORO-2-PYRIDYL-IMINO)-O-CRESOL COPPER(II) SALT

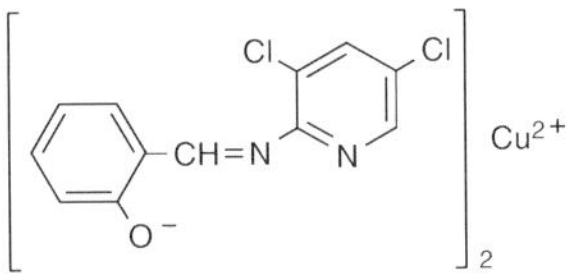

250 mg

S58,392-8
ALPHA-(4-METHYL-2-PYRIDYLIMINO)-O-CRESOL

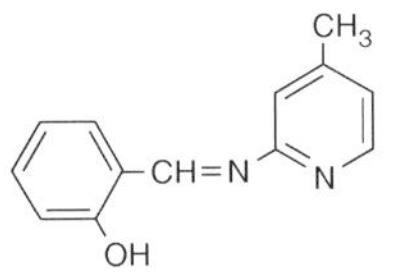

250 mg

S58,390-1
ALPHA-(6-METHYL-2-PYRIDYLIMINO)-O-CRESOL

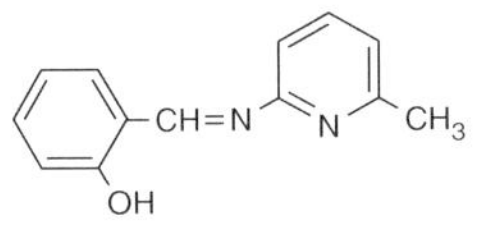

5 gm

S50,711-3
1-(2-CHLOROPHENYLIMINOMETHYL)-2-NAPHTHOL

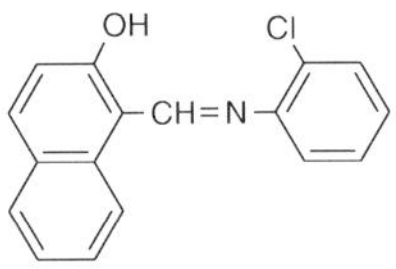

5 gm

S50,714-8
1-(O-TOLYLIMINOMETHYL)-2-NAPHTHOL

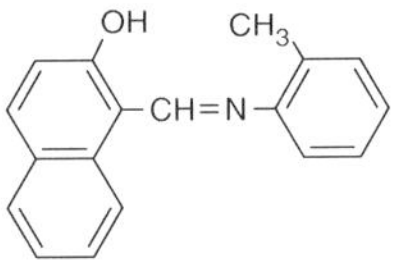

5 gm

S68,513-5
ALPHA-(4-ETHYLPHENYLIMINO)-P-CRESOL

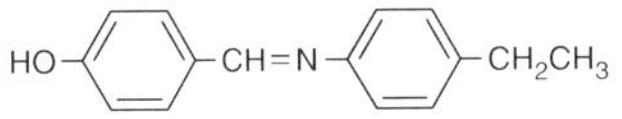

5 gm

S67,855-4
4-(4-ISOPROPYLBENZYLIDENEAMINO)-PHENOL

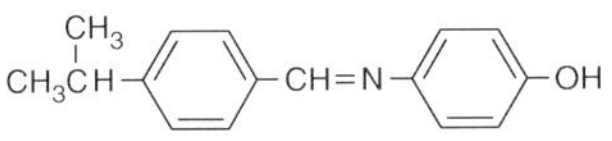

5 gm

S68,491-0
ALPHA-(3,4-XYLYLIMINO)-P-CRESOL

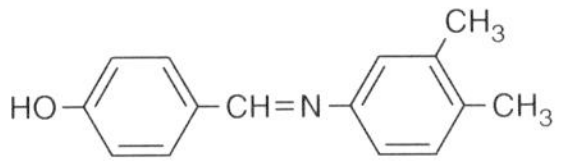

5 gm

S68,489-9
ALPHA-(2,4,6-TRIMETHYLPHENYLIMINO)-P-CRESOL

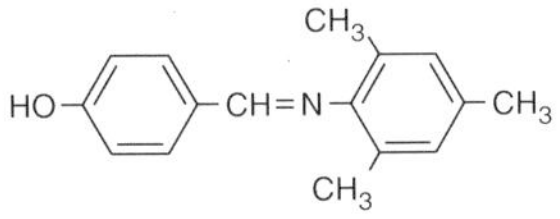

5 gm

S50,715-6
1-(4-CHLOROPHENYLIMINOMETHYL)-2-NAPHTHOL

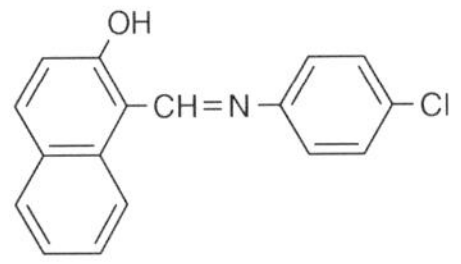

5 gm

S51,257-5
1-(4-BROMOPHENYLIMINOMETHYL)-2-NAPHTHOL

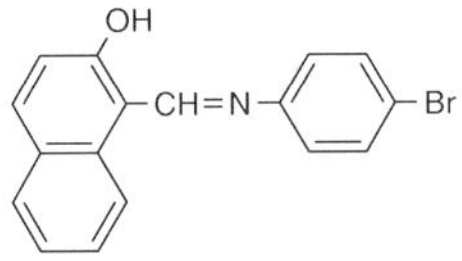

5 gm

S52,261-9
1-(4-IODOPHENYLIMINOMETHYL)-2-NAPHTHOL

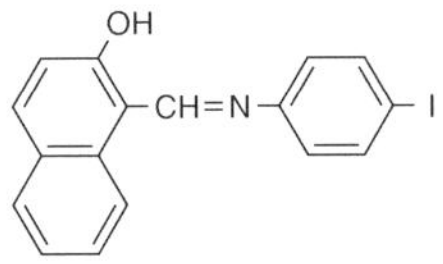

5 gm

S50,701-6
1-(P-TOLYLIMINOMETHYL)-2-NAPHTHOL

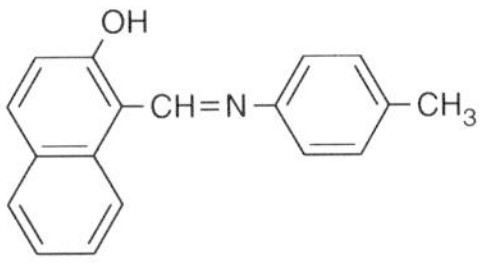

5 gm

S64,182-0
1-(4-BUTYLPHENYLIMINOMETHYL)-2-NAPHTHOL

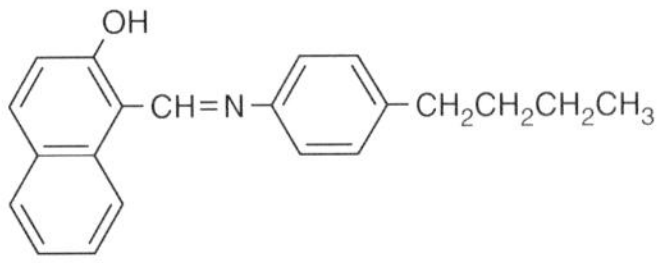

5 gm

S50,719-9
1-(2,3-DICHLOROPHENYLIMINOMETHYL)-2-NAPHTHOL

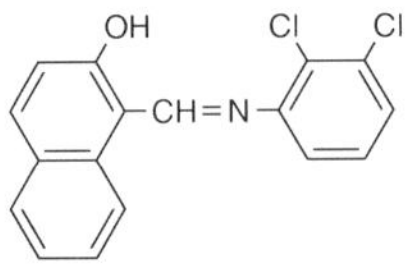

5 gm

S50,704-0
1-(3-CHLORO-O-TOLYLIMINOMETHYL)-2-NAPHTHOL

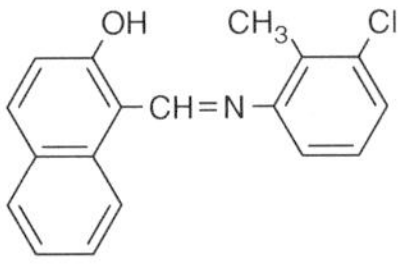

5 gm

S50,702-4
1-(2,4-DICHLOROPHENYLIMINOMETHYL)-2-NAPHTHOL

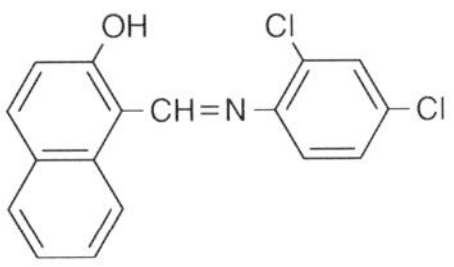

5 gm

S50,697-4
1-(3-CHLORO-P-TOLYLIMINOMETHYL)-2-NAPHTHOL

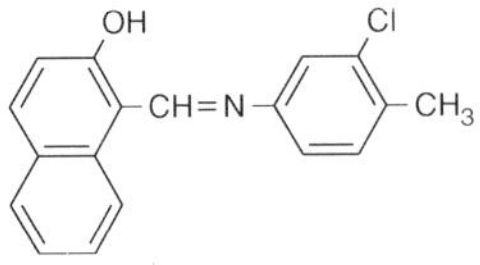

5 gm

S50,717-2
1-(5-CHLORO-O-TOLYLIMINOMETHYL)-2-NAPHTHOL

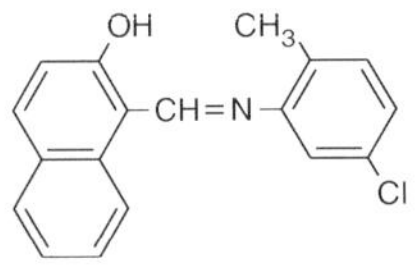

5 gm

S51,673-2
1-(4-CHLORO-O-TOLYLIMINO-METHYL)2-NAPHTHOL

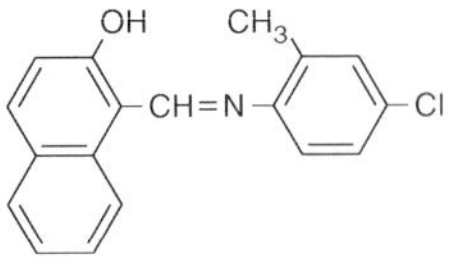

5 gm

S50,703-2
1-(3,4-XYLYLIMINOMETHYL)-2-NAPHTHOL

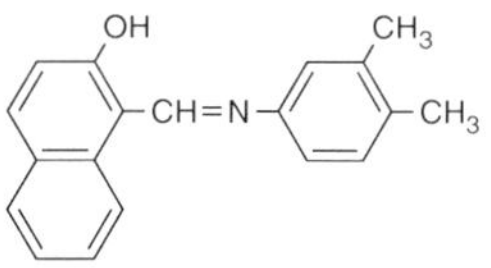

5 gm

S52,365-8
1-(2,4-XYLYLIMINOMETHYL)-2-NAPHTHOL

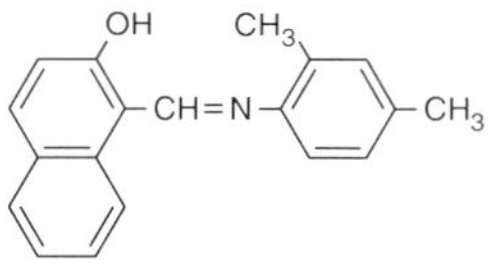

5 gm

S52,397-6
1-(2,5-XYLYLIMINOMETHYL)-2-NAPHTHOL

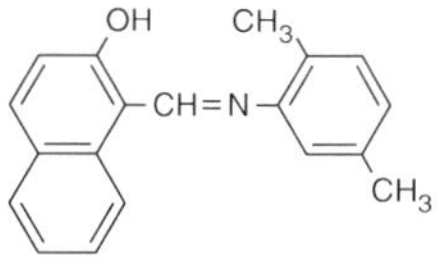

5 gm

S70,317-6
1-(2,4,6-TRIMETHYLPHENYLIMINO-METHYL)2-NAPHTHOL

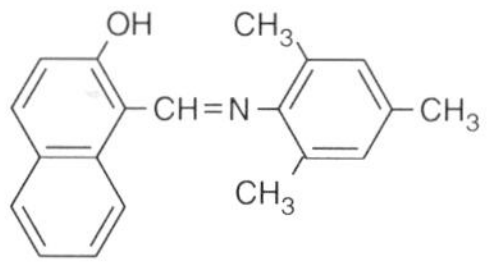

5 gm

S53,219-3
ISATIN, AZINE WITH 3,5-DICHLOROSALICYLALDEHYDE

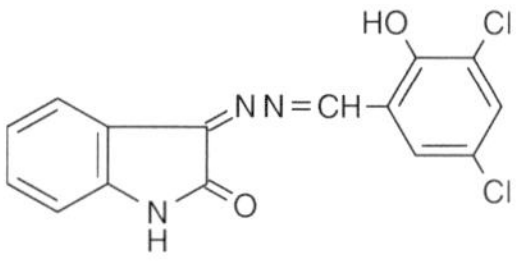

5 gm

S62,632-5
1-(4-METHYL-2-PYRIDYLIMINOMETHYL)-2-NAPHTHOL

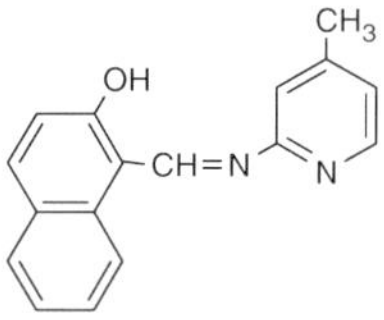

250 mg

S63,432-8
2-(2-HYDROXY-1-NAPHTHYLMETHYLENEAMINO)-6-METHYLPYRIDINE

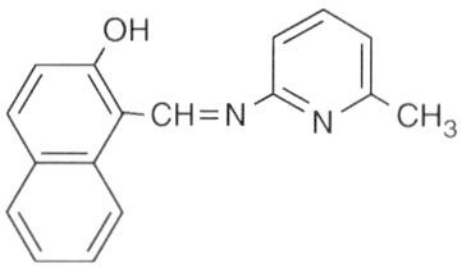

5 gm

S50,720-2
1-(2-PYRIDYLIMINOMETHYL)-2-NAPHTHOL

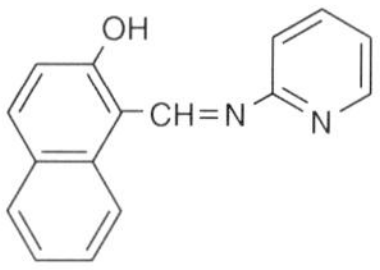

5 gm

S71,971-4
1-(3-PYRIDYLMETHYLIMINOMETHYL)-2-NAPHTHOL

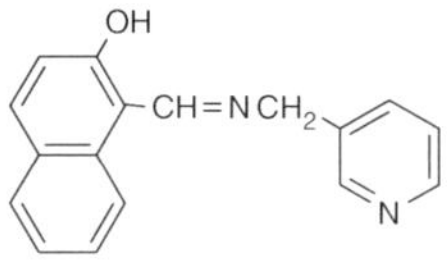

5 gm

S63,433-6
2-(2-HYDROXY-1-NAPHTHYLMETHYLENEAMINO)-3-METHYLPYRIDINE

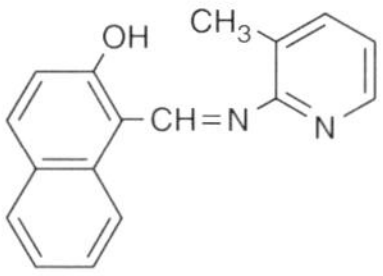

5 gm

S51,100-5
1-(3-CHLOROPHENYLIMINOMETHYL)-2-NAPHTHOL

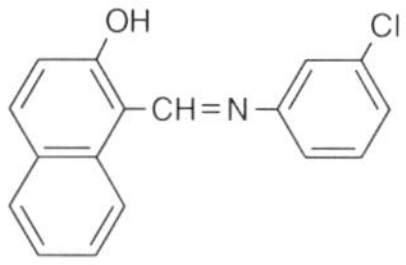

5 gm

S50,712-1
1-(M-TOLYLIMINOMETHYL)-2-NAPHTHOL

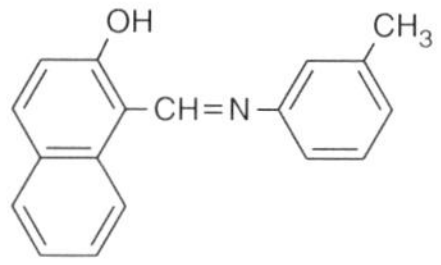

5 gm

S59,157-2
4-(1-NAPHTHYLIMINO)-2-BUTANOL

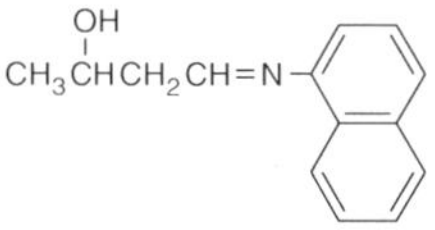

250 mg

S51,609-0
1-(1-NAPHTHYLIMINOMETHYL)-2-NAPHTHOL

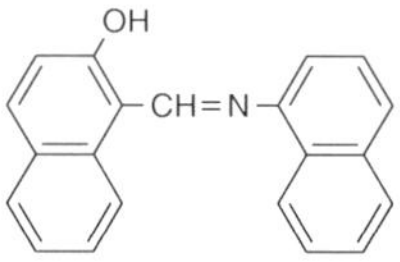

5 gm

S86,053-0
TERT-BUTYL(3-METHOXYBENZYLIDENE)-AMINE

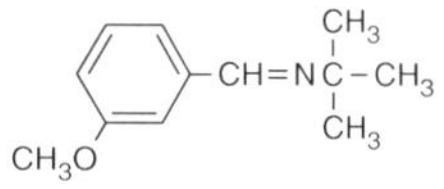

250 mg

S62,590-6
N-(CINNAMYLIDENE)-M-ANISIDINE

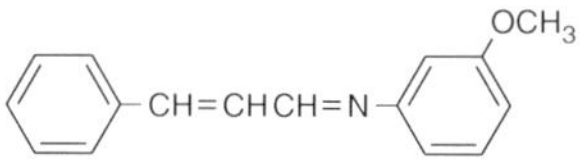

250 mg

S53,800-0
3-ETHOXYBENZALDEHYDE AZINE

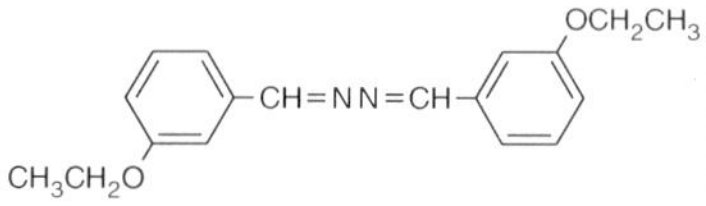

250 mg

S77,179-1
1-[3,4-BIS(BENZYLOXY)-BENZYLIDENEAMINO]GUANIDINE ACETATE

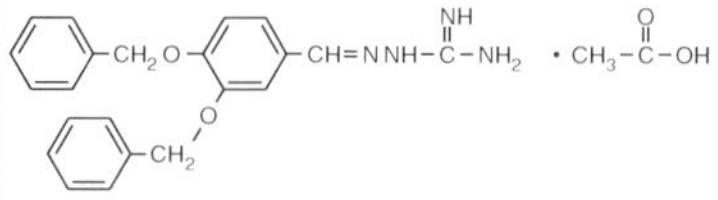

250 mg

S88,138-4
[3-(BENZYLOXY)-4-METHOXYBENZYLIDENE]METHYLAMINE

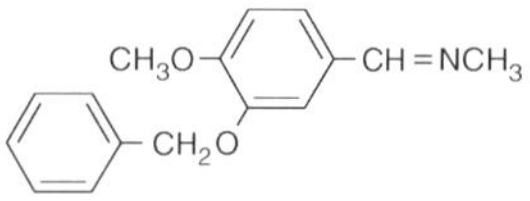

250 mg

S77,184-8
1-[(4-BENZYLOXY-3-ETHOXYBENZYLIDENE)AMINO]-GUANIDINE NITRATE

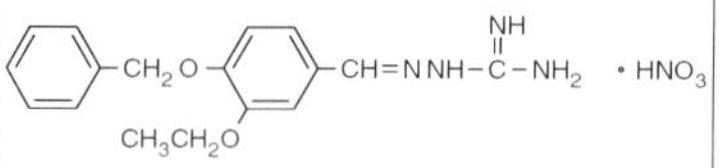

250 mg

S88,229-1
(3-BENZYLOXY-4-METHOXYBENZYLIDENE)-TERT-BUTYLAMINE

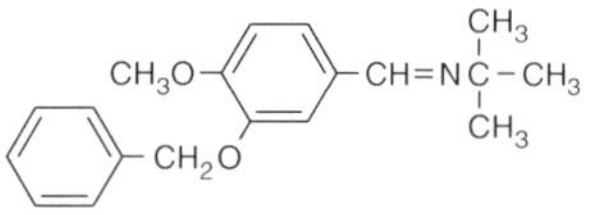

250 mg

S53,804-3
4-BENZYLOXY-3-METHOXYBENZALDEHYDE AZINE

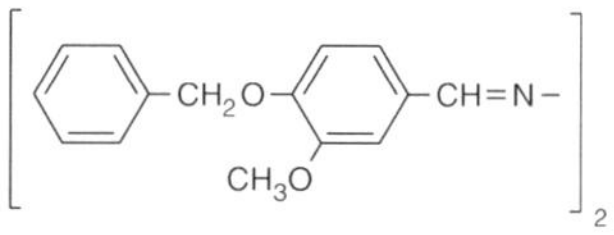

5 gm

S62,746-1
N-(4-BENZYLOXYBENZYLIDENE)-3-NITROANILINE

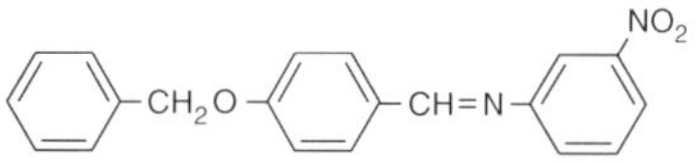

5 gm

S63,422-0
N-(4-BENZYLOXYBENZYLIDENE)-2-METHYL-5-NITROANILINE

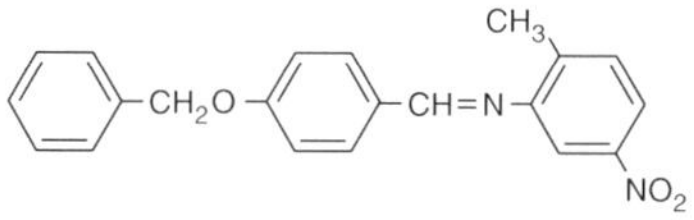

5 gm

S63,423-9
N-(4-BENZYLOXYBENZYLIDENE)-4-METHYL-3-NITROANILINE

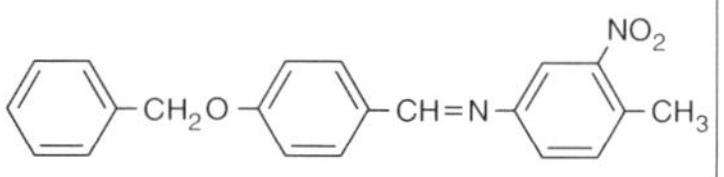

5 gm

S65,489-2
N,N'-BIS(4-BENZYLOXYBENZYLIDENE)-1,4-PHENYLENEDIAMINE

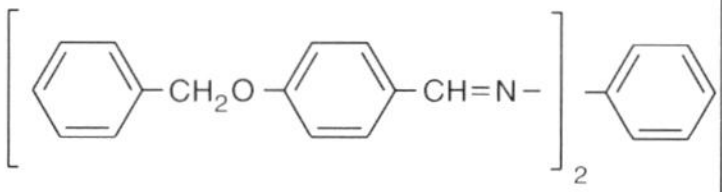

5 gm

S62,748-8
N-(4-BENZYLOXYBENZYLIDENE)-5-CHLORO-2,4-DIMETHOXYANILINE

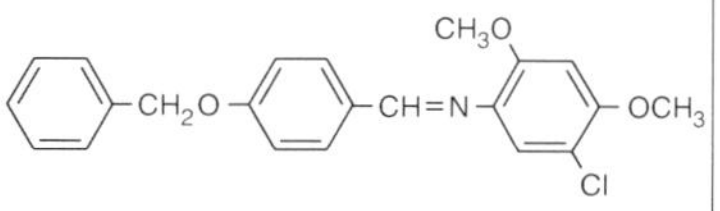

5 gm

S63,612-6
4-(4-BENZYLOXYBENZYLIDENEAMINO)-BENZOIC ACID

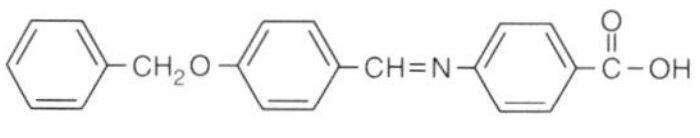

5 gm

S63,618-5
N-(4-BENZYLOXYBENZYLIDENE)-4-ETHYLANILINE

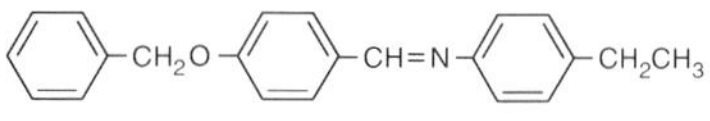

5 gm

S63,430-1
N-(4-BENZYLOXYBENZYLIDENE)-4-(ETHOXYCARBONYL)ANILINE

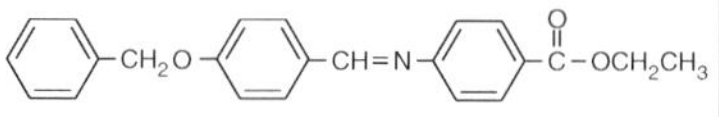

5 gm

S65,491-4
4,4'-SULFONYLBIS[N-(4-BENZYLOXYBENZYLIDENE)ANILINE]

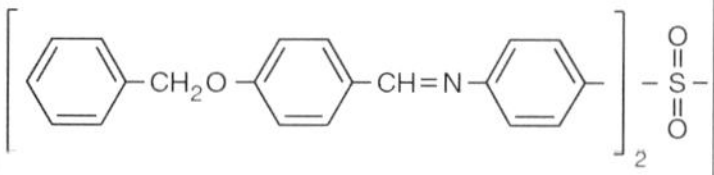

250 mg

S62,701-1
N-(4-BENZYLOXYBENZYLIDENE)ANILINE

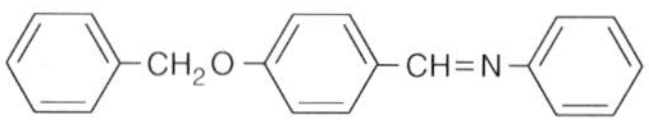

5 gm

S63,616-9
N-(4-BENZYLOXYBENZYLIDENE)-4-FLUOROANILINE
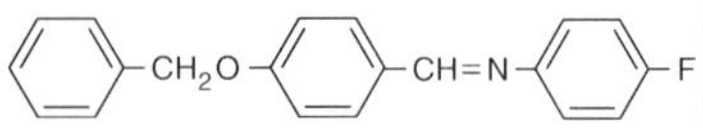
5 gm

S63,429-8
N-(4-BENZYLOXYBENZYLIDENE)-4-CHLOROANILINE
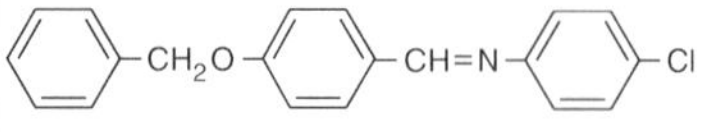
5 gm

S63,426-3
N-(4-BENZYLOXYBENZYLIDENE)-4-BROMOANILINE
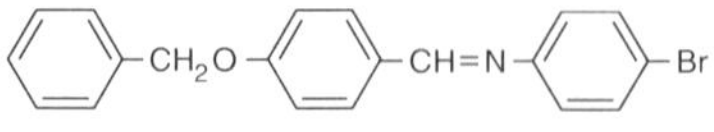
5 gm

S62,770-4
N-(4-BENZYLOXYBENZYLIDENE)-P-TOLUIDINE
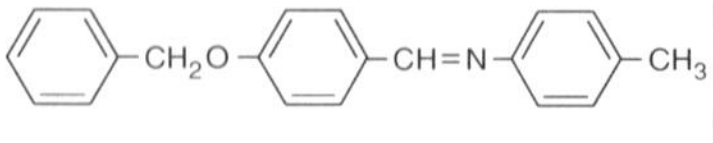
5 gm

S62,721-6
N-(4-BENZYLOXYBENZYLIDENE)-P-ANISIDINE
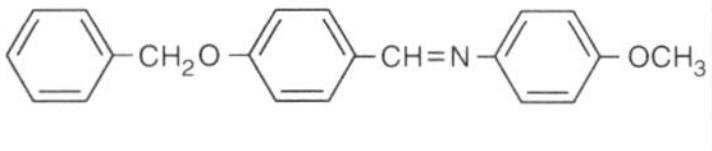
5 gm

S63,776-9
N-(4-BENZYLOXYBENZYLIDENE)-3,4-DICHLOROANILINE
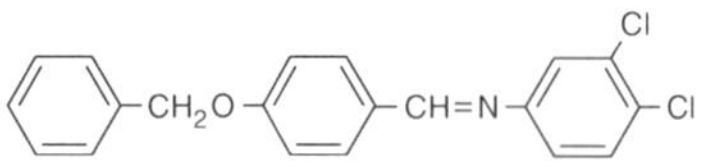
5 gm

S64,113-8
N-(4-BENZYLOXYBENZYLIDENE)-2,4-DICHLOROANILINE
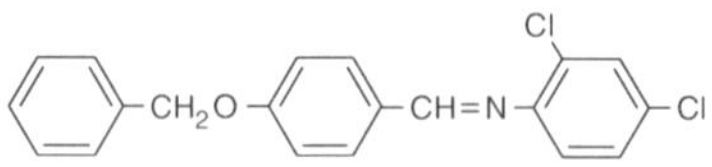
250 mg

S62,771-2
N-(4-BENZYLOXYBENZYLIDENE)-3,4-XYLIDINE
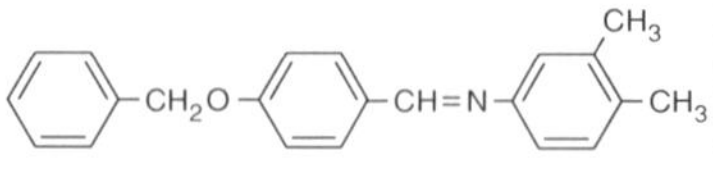
5 gm

S77,168-6
1-[(4-BENZYLOXY-ALPHA-ETHYL)-BENZYLIDENEAMINO]GUANIDINE NITRATE
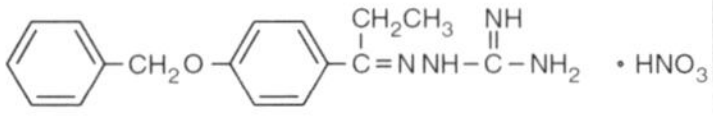
250 mg

S77,169-4
1-[(4-BENZYLOXY-ALPHA-ETHYL)-BENZYLIDENEAMINO]GUANIDINE ACETATE
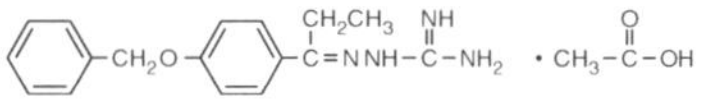
250 mg

S63,787-4
N-(4-BENZYLOXYBENZYLIDENE)-P-PHENETIDINE
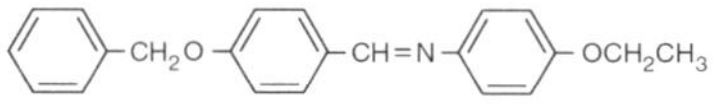
5 gm

S90,223-3
2-(3-FURFURYLUREIDO)ACETIC (3,4-DIMETHOXYBENZYLIDENE)HYDRAZIDE
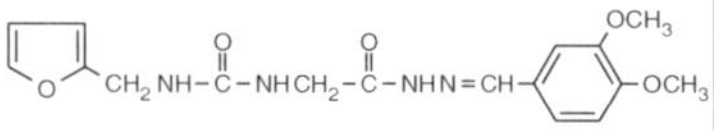
250 mg

S73,002-5
4-(3,4-DIMETHOXYBENZYLIDENEAMINO)-ANTIPYRINE
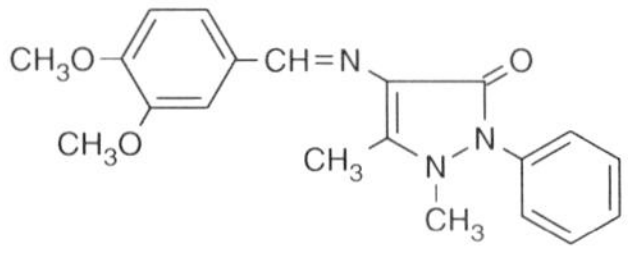
5 gm

S69,658-7
N-(3,4-DIMETHOXYBENZYLIDENE)-N'-PHENYL-1,4-PHENYLENEDIAMINE
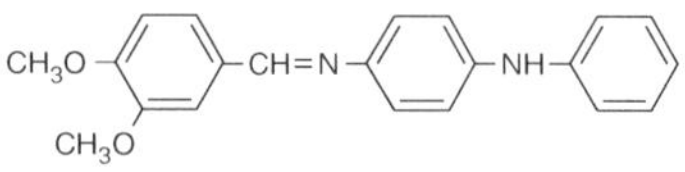
5 gm

S67,473-7
1-(2,3,4-TRIMETHOXYBENZYLIDENE)-3-THIOSEMICARBAZIDE
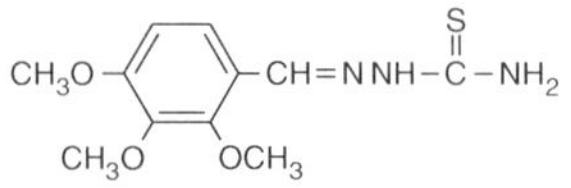
250 mg

S95,569-8
ETHYL (2-[2-(3,4-DIMETHOXYPHENYL)-INDOL-3-YL]ETHYLIMINO)ACETATE
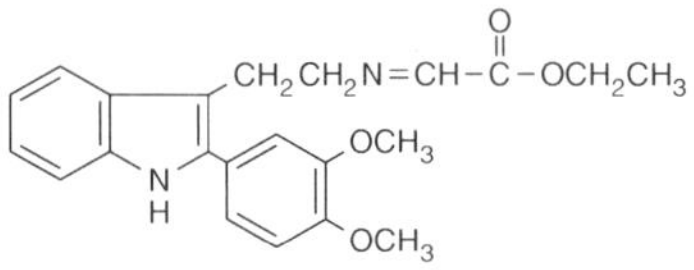
50 mg

S67,380-3
O-(3,4-DIMETHOXYBENZYLIDENEAMINO)-N-CYCLOHEXYLTHIOCARBAMATE
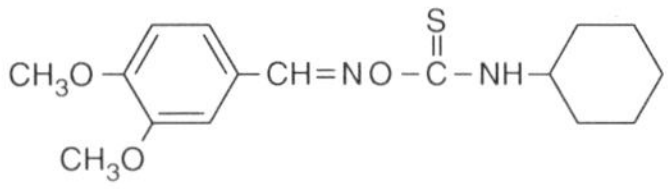
250 mg

S76,711-5
2-METHOXY-4-(METHYLIMINOMETHYL)-PHENETOLE
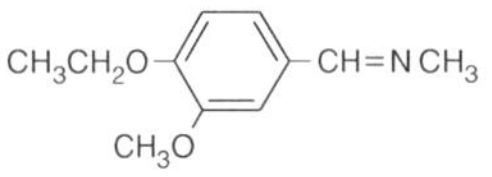
250 mg

S69,093-7
2,3-DIMETHOXYBENZALDEHYDE SEMICARBAZONE
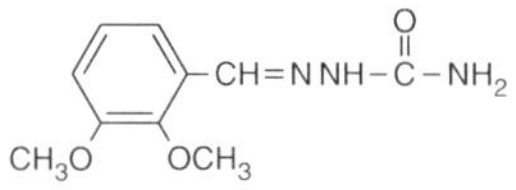
250 mg

S86,198-7
1-(4-METHOXYBENZYLIDENEAMINO)-3-PHENOXY-4-(P-TOLYL)-2-AZETIDINONE
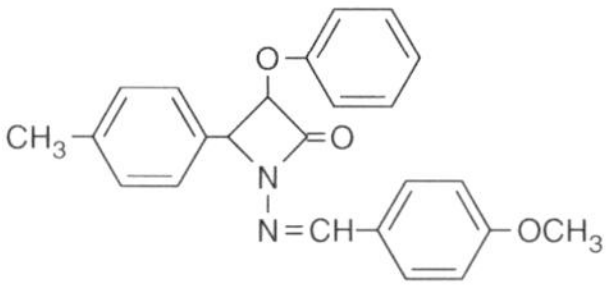
250 mg

S69,194-1
4-CYCLOHEXYL-1-(3,4-DIMETHOXYBENZYLIDENE)-3-THIOSEMICARBAZIDE
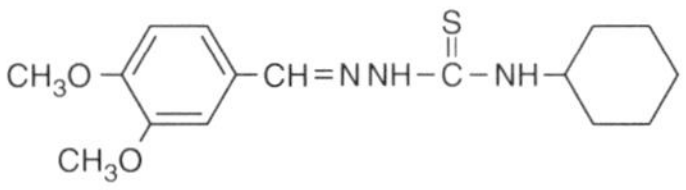
250 mg

S69,437-1
1-(3,4-DIMETHOXYBENZYLIDENE)-4-(3-MORPHOLINOPROPYL)-3-THIOSEMICARBAZIDE
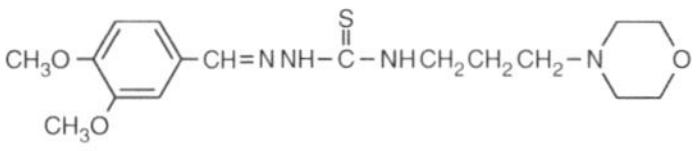
250 mg

S70,118-1
N-(3,4-DIMETHOXYBENZYLIDENE)-4-PHENOXYANILINE
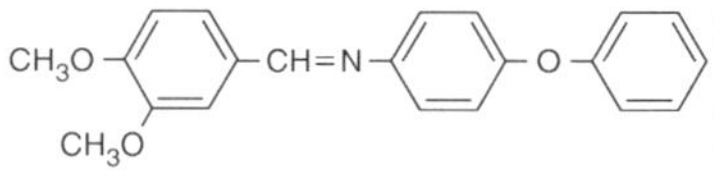
5 gm

S88,231-3
(3,4-DIMETHOXYBENZYLIDENE)-(DIPHENYLMETHYL)AMINE
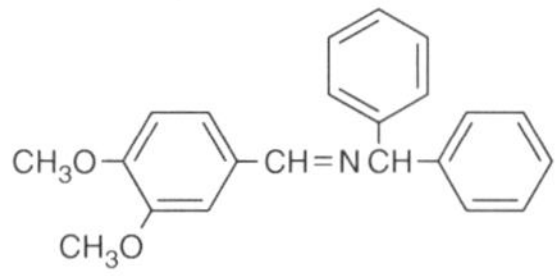
250 mg

S88,065-5
(3,4-DIMETHOXYBENZYLIDENE)METHYLAMINE
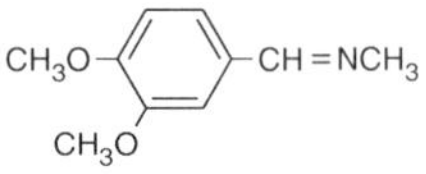
250 mg

S74,706-8
2,5-DIMETHOXYBENZALDEHYDE (2,4-DINITROPHENYL)METHYLHYDRAZONE
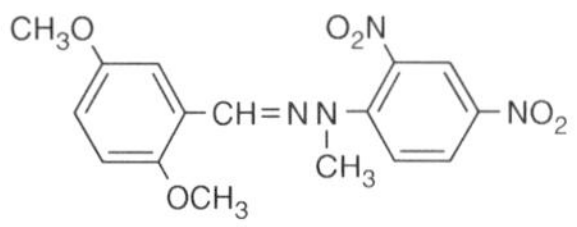
5 gm

S71,628-6
SEBACIC ACID BIS(2,5-DIMETHOXYBENZYLIDENEHYDRAZIDE)
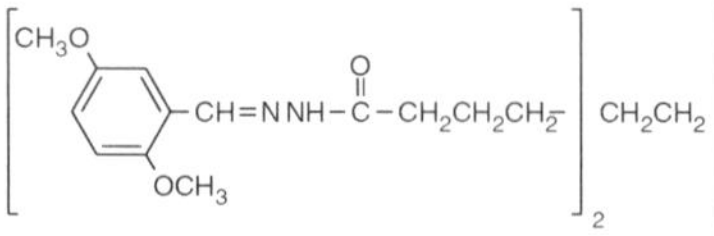
250 mg

S88,066-3
(2,5-DIMETHOXYBENZYLIDENE)METHYLAMINE
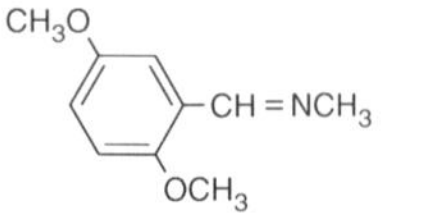
250 mg

S92,482-2
ANTHRANILIC (4-ETHOXY-3-METHOXYBENZYLIDENE)HYDRAZIDE
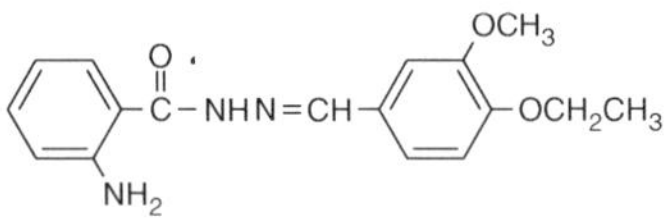
250 mg

S87,892-8
N,N'-BIS(4-PHENOXYPHENYL)-1,4-DIAZABUTADIENE
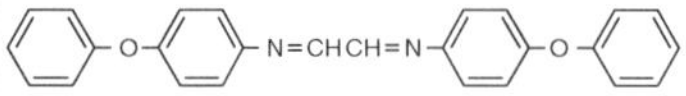
5 gm

S87,882-0
N,N'-BIS(4-ETHOXYPHENYL)-1,4-DIAZABUTADIENE
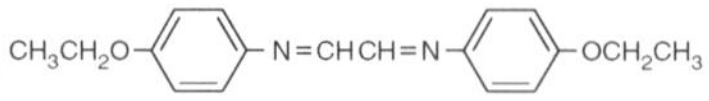
5 gm

S85,472-7
4-ETHOXY-3-METHOXYBENZALDEHYDE (4-NITROPHENYL)HYDRAZONE
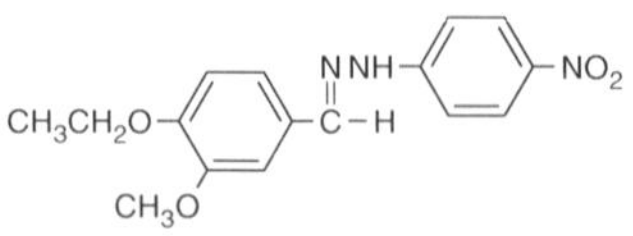
250 mg

S88,755-2
4-METHOXY-N-[5-(4-METHOXYANILINO)-2,4-PENTADIEN-1-YLIDENE]ANILINE PERCHLORATE

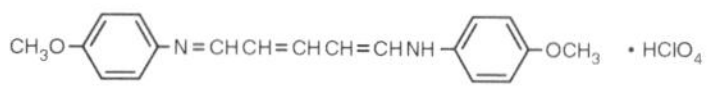

250 mg

S91,972-1
2,6-DIMETHOXY-4-[4-(DIMETHYLAMINO)-PHENYLIMINO]-2,5-CYCLOHEXADIEN-1-ONE

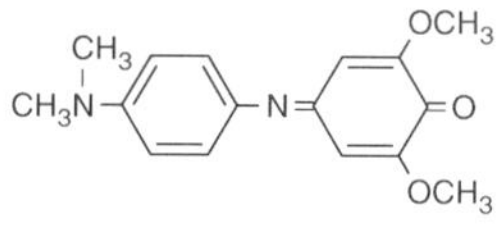

250 mg

S92,505-5
4-DIMETHYLAMINOBENZOIC (3,4-DIMETHOXYBENZYLIDENE)HYDRAZIDE

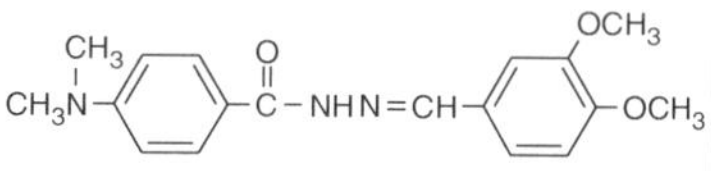

250 mg

S92,500-4
4-METHOXYBENZOIC (4-METHOXYBENZYLIDENE)HYDRAZIDE

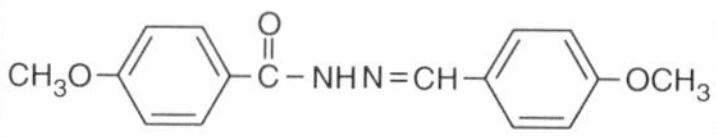

250 mg

S92,502-0
4-METHOXYBENZOIC (2,3-DIMETHOXYBENZYLIDENE)HYDRAZIDE

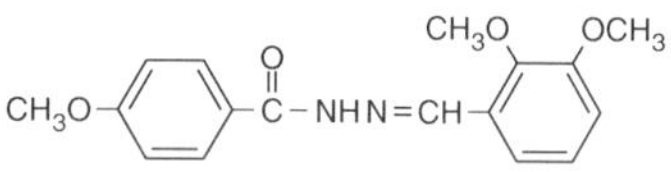

250 mg

S67,636-5
2,2'-OXYDIACETIC BIS[(4-METHOXYBENZYLIDENE)HYDRAZIDE]

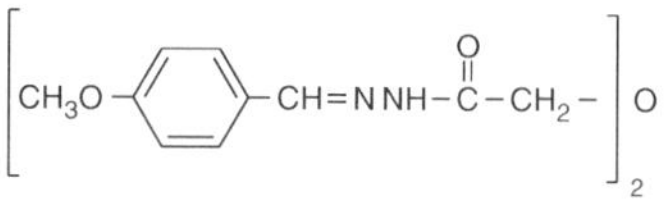

5 gm

S68,482-1
4,4'-METHYLENEBIS[N-(4-METHOXYBENZYLIDENE)ANILINE]

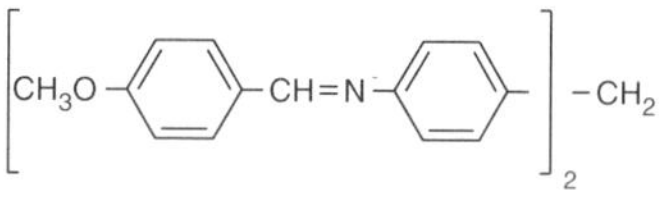

5 gm

S68,316-7
N-(3,4,5-TRIMETHOXYBENZYLIDENE)-METHYLAMINE

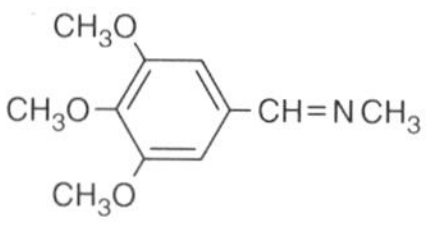

5 gm

S85,450-6
4-(3-FURFURYLUREIDO)BENZOIC (3,4-DIMETHOXYBENZYLIDENE)HYDRAZIDE

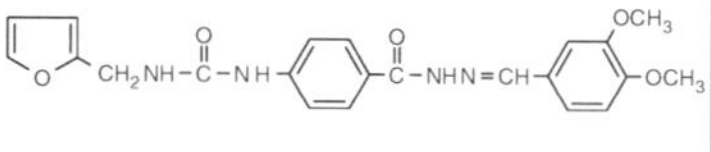

250 mg

S73,799-2
2,2'-OXYDIACETIC BIS[(2-METHOXYBENZYLIDENE)HYDRAZIDE]

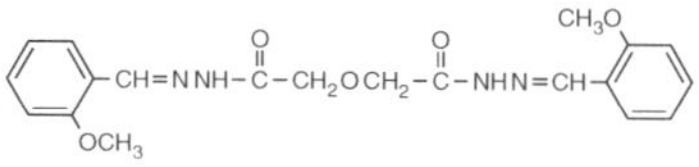

5 gm

S70,121-1
N,N'-TEREPHTHALYLIDENEBIS(4-PHENOXYANILINE)

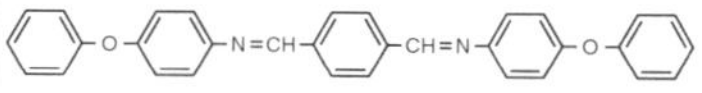

5 gm

S93,833-5
BIS[2-(4-METHOXYBENZYLIDENEAMINO)-PHENYL]DITELLURIDE

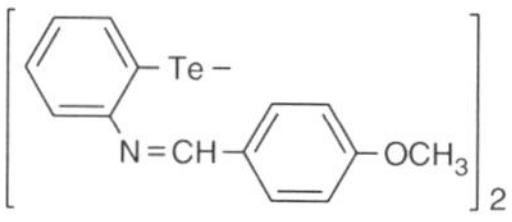

250 mg

S63,601-0
2-NITRO-N-PIPERONYLIDENEANILINE

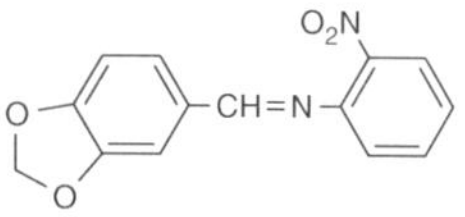

250 mg

S68,557-7
N,N'-DIVERATRYLIDENEETHYLENEDIAMINE

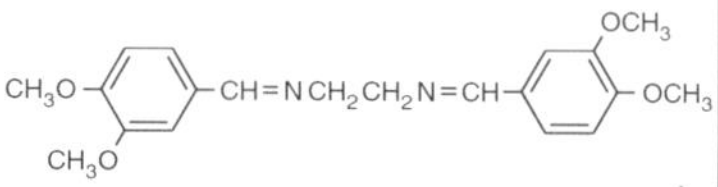

250 mg

S58,225-5
1-HEXADECANESULFONIC ACID [2-(3,4-DIMETHOXYPHENYL)-1-METHYL-6-SULFO-4(1H)-QUINOLYLIDENE]HYDRAZIDE

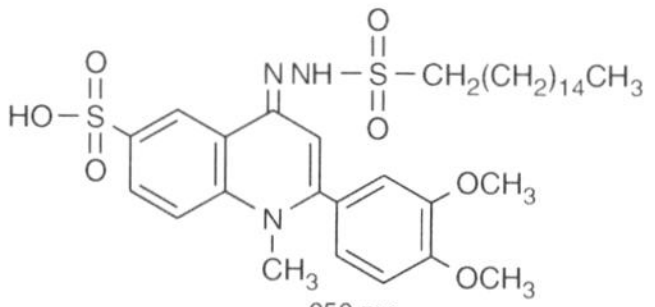

250 mg

S64,259-2
N,N'-BIS(3,4-DIMETHOXYBENZYLIDENE)-1,3-PROPANEDIAMINE

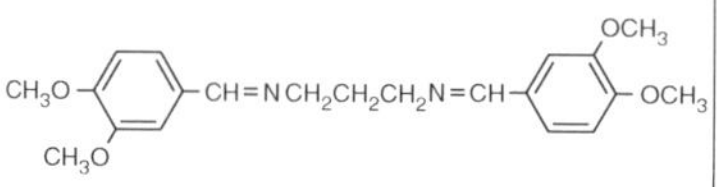

250 mg

S71,608-1
2,2'-OXYDIACETIC ACID BIS[(3-NITROBENZYLIDENE)HYDRAZIDE]

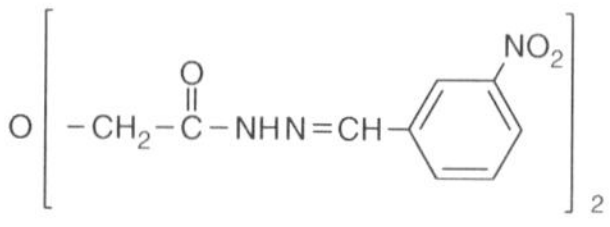

5 gm

S64365-3
N,N'-BIS(3,4-DIMETHOXYBENZYLIDENE)-1,4-PHENYLENEDIAMINE

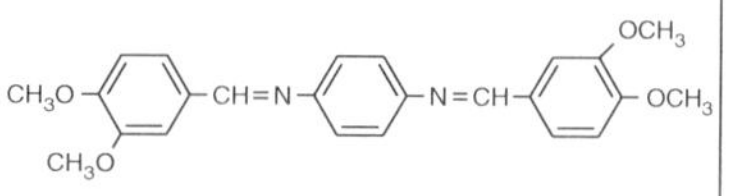

5 gm

S54,951-7
N-(4-CHLOROBENZYLIDENE)-3,4-DIMETHOXYPHENETHYLAMINE

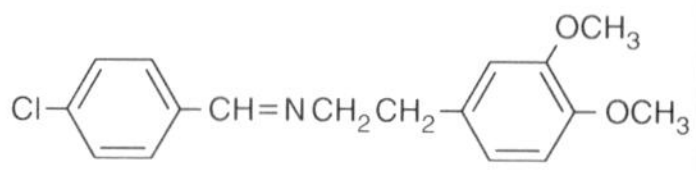

5 gm

S61,875-6
N-[4-(DIMETHYLAMINO)BENZYLIDENE]-3-METHOXY-4-(2-PHENOXYETHOXY)ANILINE

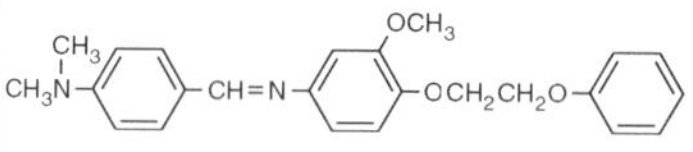

250 mg

S71,606-5
2,2'-OXYDIACETIC ACID BIS[(4-ACETAMIDOBENZYLIDENE)HYDRAZIDE]

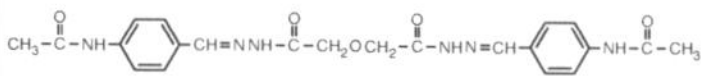

250 mg

S63,571-5
2,5-DIMETHOXY-N-(PIPERONYLIDENE)ANILINE

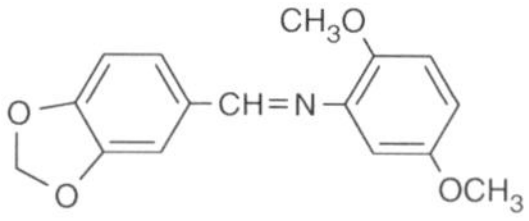

250 mg

S64,111-1
2,5-DIMETHOXY-N-(VERATRYLIDENE)ANILINE

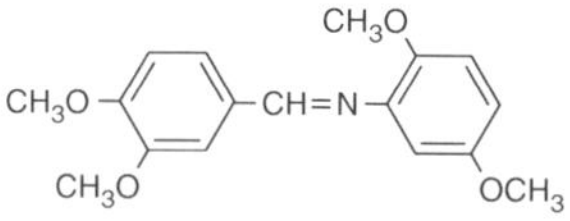

250 mg

S63,180-9
N-PIPERONYLIDENE-P-PHENETIDINE

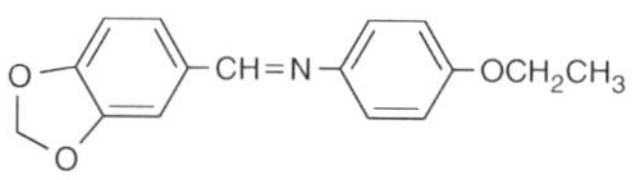

250 mg

S63,518-9
3,4-DICHLORO-N-PIPERONYLIDENEANILINE

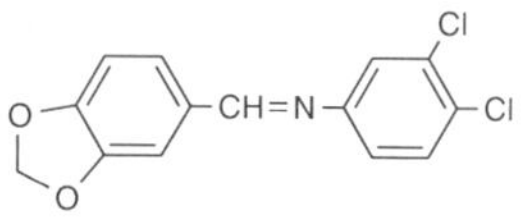

5 gm

S63,570-7
N-PIPERONYLIDENE-3,5-XYLIDINE

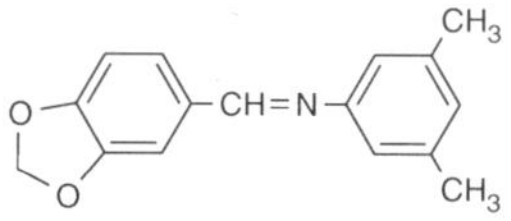

250 mg

S63,516-2
3-CHLORO-N-PIPERONYLIDENEANILINE

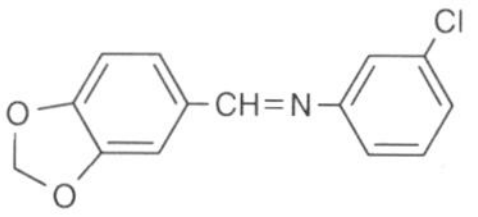

250 mg

S61,941-8
2-CHLORO-N-(3,4-DIMETHOXYBENZYLIDENE)ANILINE

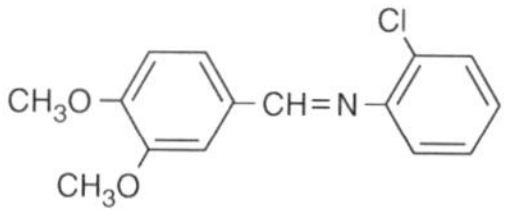

250 mg

S61,930-2
N-(3,4-DIMETHOXYBENZYLIDENE)-O-TOLUIDINE

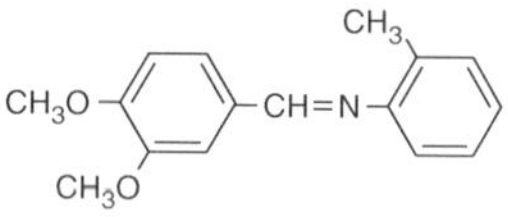

250 mg

S61,581-1
4-CHLORO-N-(3,4-DIMETHOXYBENZYLIDENE)ANILINE

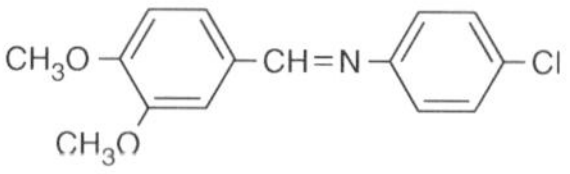

5 gm

S59,476-8
N-(3,4-DIMETHOXYBENZYLIDENE)-4-BROMOANILINE

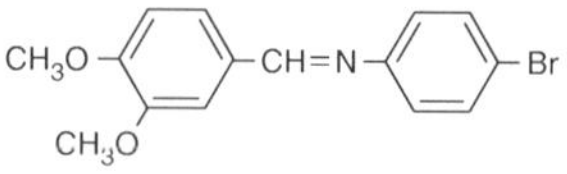

250 mg

S61,586-2
N-(3,4-DIMETHOXYBENZYLIDENE)-P-TOLUIDINE

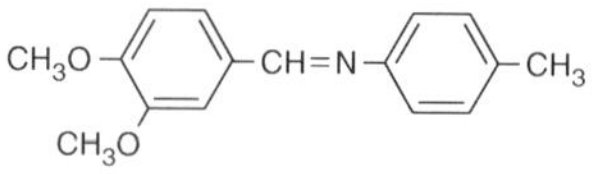

250 mg

S63,565-0
2,3-DICHLORO-N-(VERATRYLIDENE)ANILINE
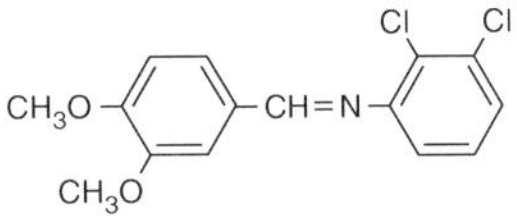
250 mg

S62,743-7
N-VERATRYLIDENE-2,3-XYLIDINE
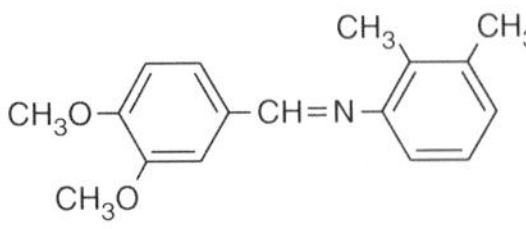
250 mg

S61,585-4
N-(3,4-DIMETHOXYBENZYLIDENE)ANILINE
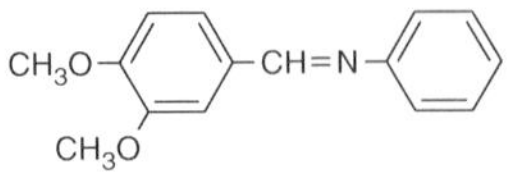
250 mg

S61,584-6
3-CHLORO-N-(3,4-DIMETHOXYBENZYLIDENE)ANILINE
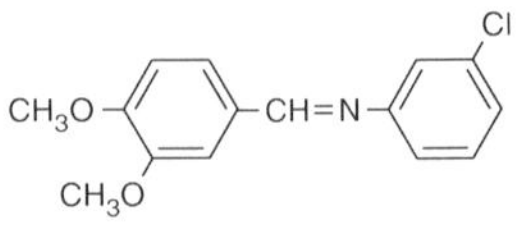
250 mg

S62,696-1
2,5-DICHLORO-N-(VERATRYLIDENE)ANILINE
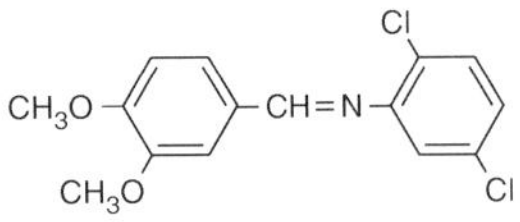
250 mg

S62,768-2
3,4-DICHLORO-N-(VERATRYLIDENE)ANILINE
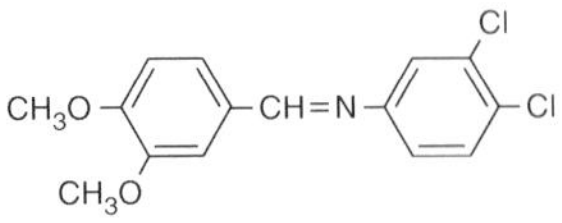
250 mg

S63,573-1
N-VERATRYLIDENE-M-TOLUIDINE
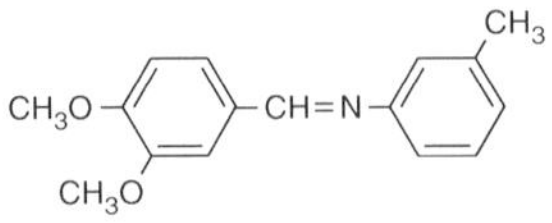
250 mg

S62,598-1
N-VERATRYLIDENE-M-ANISIDINE
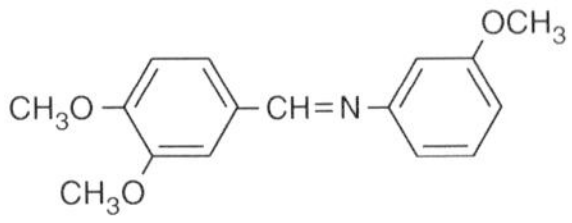
250 mg

S62,694-5
3-CHLORO-4-METHYL-N-(VERATRYLIDENE)ANILINE
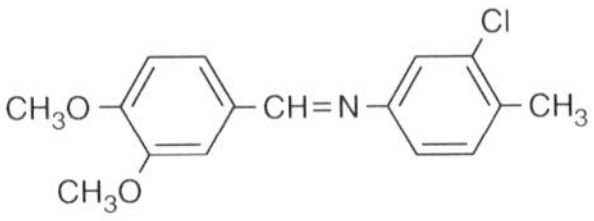
250 mg

S62,601-5
N-VERATRYLIDENE-3,4-XYLIDINE
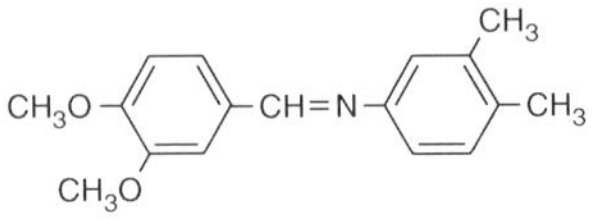
250 mg

S73,801-8
2,2'-OXYDIACETIC BIS(CINNAMYLIDENEHYDRAZIDE)
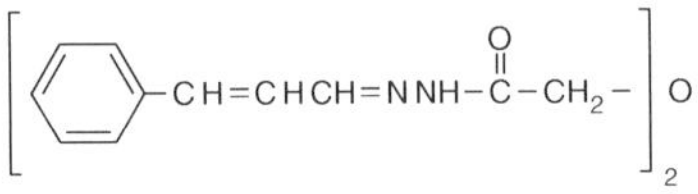
5 gm

S53,373-4
3,4,5-TRIMETHOXYBENZALDEHYDE N,N-HEXAMETHYLENEHYDRAZONE
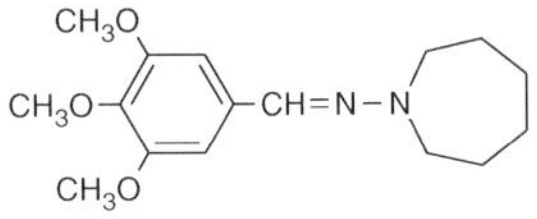
5 gm

S52,016-0
6-PHENYL-4-(3,4,5-TRIMETHOXYBENZYLIDENEAMINO)-1,2,4-TRIAZINE-3,5(2H,4H)-DIONE
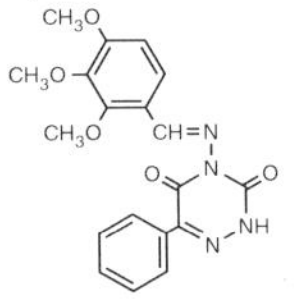
250 mg

S52,021-7
6-METHYL-4-VERATRYLIDENEAMINO-1,2,4-TRIAZINE-3,5(2H,4H)-DIONE
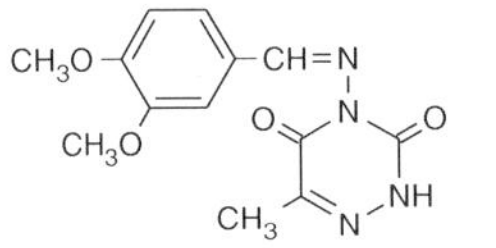
250 mg

S51,454-3
1,2-DIHYDRO-4,6-DIMETHYL-2-OXO-1-(VERATRYLIDENEAMINO)NICOTINONITRILE
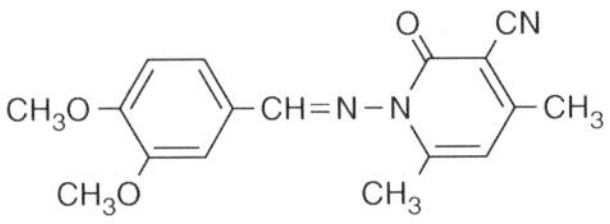
250 mg

S53,374-2
NICOTINIC ACID (3,4,5-TRIMETHOXYBENZYLIDENE)HYDRAZIDE
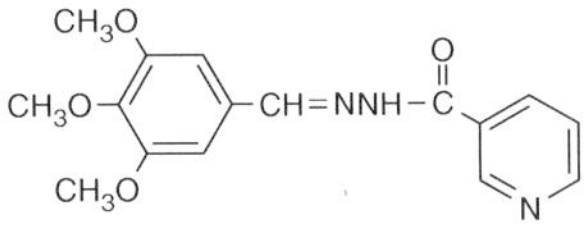
5 gm

S52,469-7
N-(VERATRYLIDENEAMINO)PHTHALIMIDE
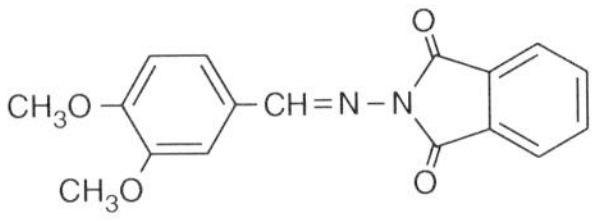
5 gm

S53,228-2
BENZIL, AZINE WITH 3,4,5-TRIMETHOXYBENZALDEHYDE
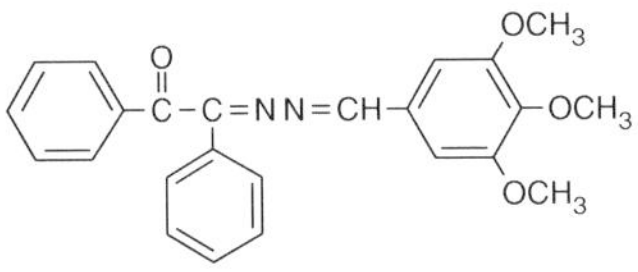
5 gm

S53,989-9
N-VERATRYLIDENEANTHRANILIC ACID
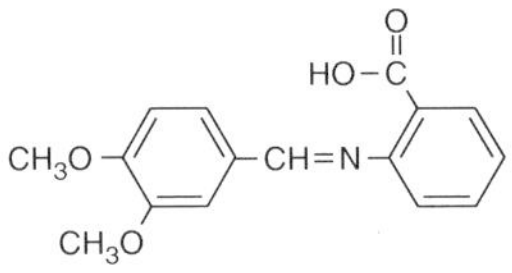
250 mg

S53,959-7
4-(VERATRYLIDENEAMINO)BENZOIC ACID
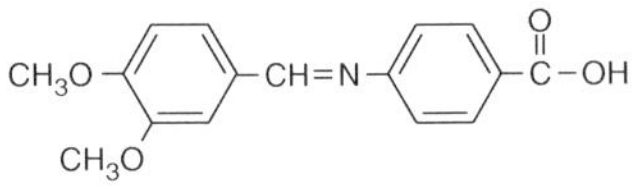
5 gm

S53,811-6
PICOLINIC ACID (3,4,5-TRIMETHOXYBENZYLIDENE)HYDRAZIDE
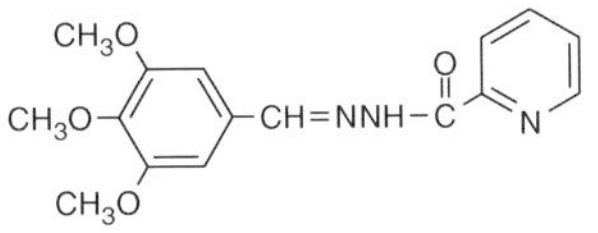
5 gm

S54,976-2
4-(2-METHOXYPHENOXY)-N-(4-METHYLBENZYLIDENE)ANILINE
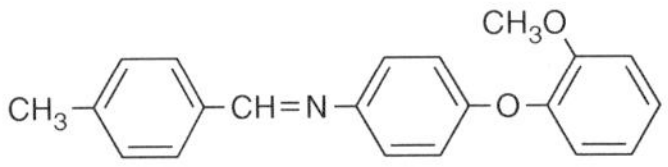
250 mg

S53,012-3
N-VERATRYLIDENE-4-FLUOROANILINE
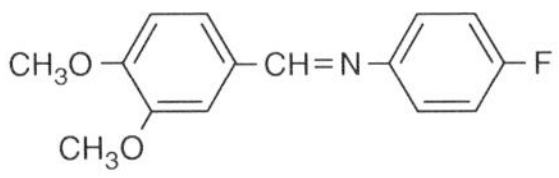
250 mg

S53,878-7
4-IODO-N-VERATRYLIDENEANILINE
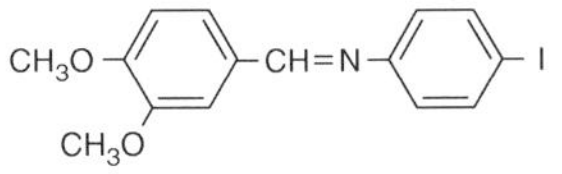
5 gm

S53,020-4
N-VERATRYLIDENE-P-ANISIDINE
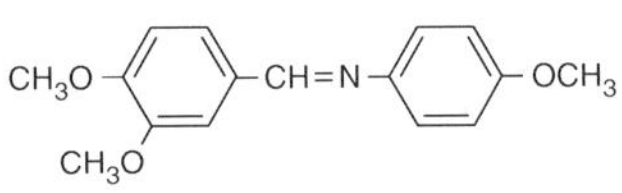
5 gm

S53,016-6
N-VERATRYLIDENE-2,4-XYLIDINE
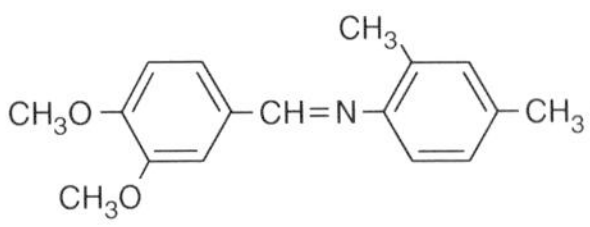
5 gm

S53,140-5
N-VERATRYLIDENE-2,5-XYLIDINE
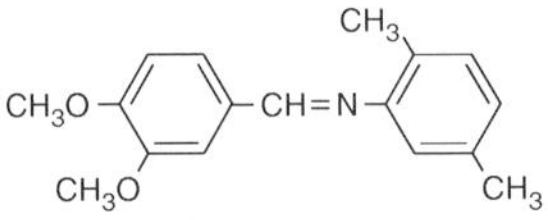
5 gm

S52,986-9
VERATRALDEHYDE AZINE
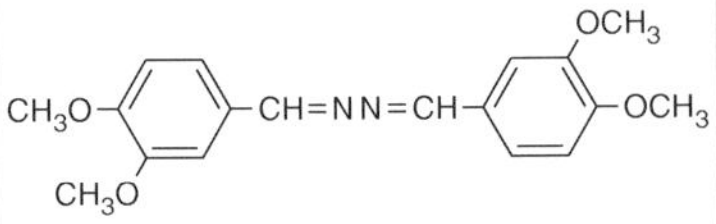
5 gm

S92,521-7
2-METHOXYBENZOIC (3-INDOLYLMETHYLENE)HYDRAZIDE
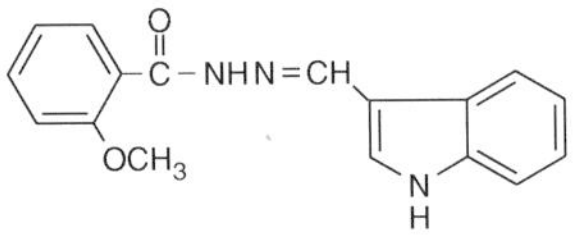
250 mg

S53,218-5
ISATIN, AZINE WITH 5-BROMO-O-ANISALDEHYDE
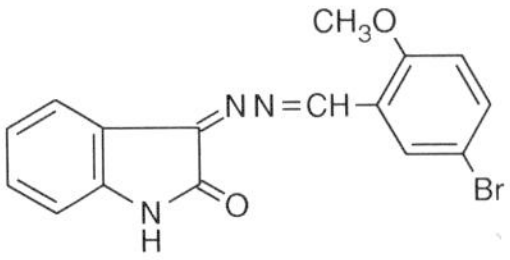
5 gm

S51,267-2
HEXADECYL[2-METHOXY-ALPHA-(METHYLTHIO)BENZYLIDENE]-METHYLAMMONIUM IODIDE
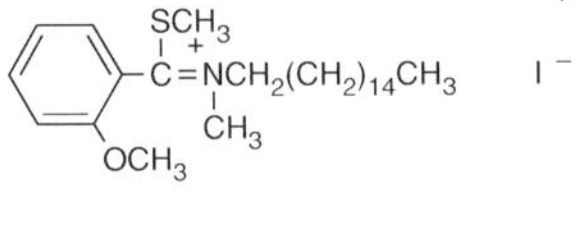
250 mg

S61,945-0
N-(4-METHOXYBENZYLIDENE)-3-(METHYLTHIO)ANILINE
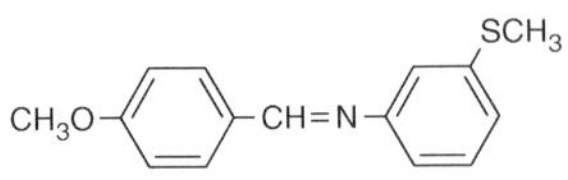
250 mg

S94,425-4
BASIC RED 3 (C.I. 45210)
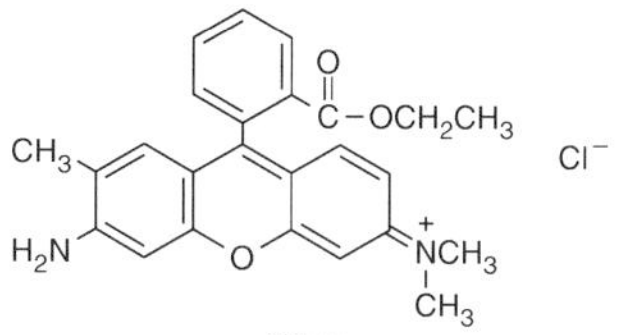
250 mg

S71,422-4
N,N-DIETHYL-N'-(4-METHOXYBENZYLIDENE)ETHYLENEDIAMINE
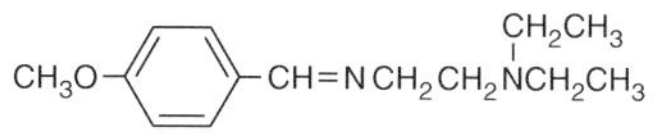
5 gm

S64,283-5
N,N'-BIS(2-METHOXYBENZYLIDENE)-ETHYLENEDIAMINE
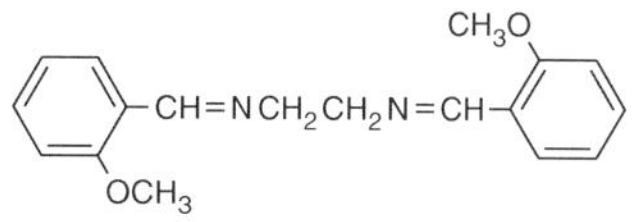
250 mg

S54,917-7
3,4,5-TRIMETHOXYBENZALDEHYDE 2-NITROPHENYLHYDRAZONE
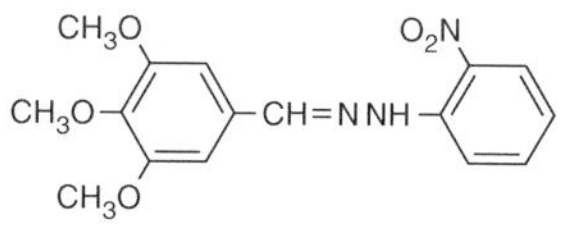
5 gm

S67,357-9
N-(4-ETHOXYPHENYL)-N'-(ALPHA-METHYLCINNAMYLIDENEAMINO)OXAMIDE
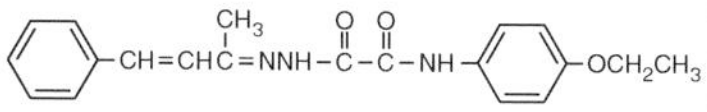
5 gm

S38,882-3
N,N'-BIS(4-METHOXYBENZYLIDENE)-ETHYLENEDIAMINE
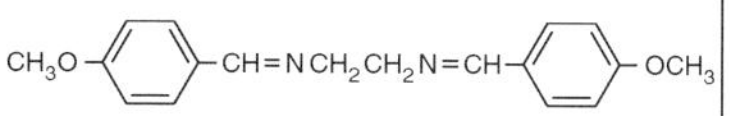
5 gm

S34,242-4
N,N'-BIS(3,4,5-TRIMETHOXYBENZYLIDENE)-ETHYLENEDIAMINE
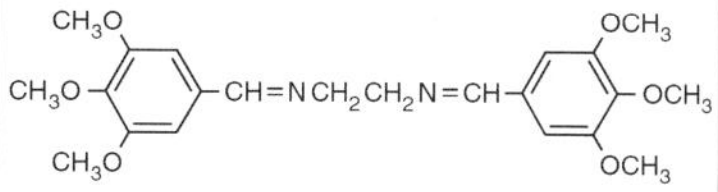
5 gm

S88,769-2
2,4-BIS[3-(4-METHOXYANILINO)-PROPYLAMINO]-3-OXO-1-CYCLOBUTENYLIUM-1-OLATE
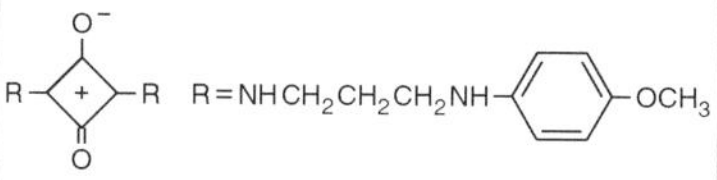
250 mg

S64,356-4
N-[4-(DIMETHYLAMINO)BENZYLIDENE]-2-METHOXY-5-NITROANILINE
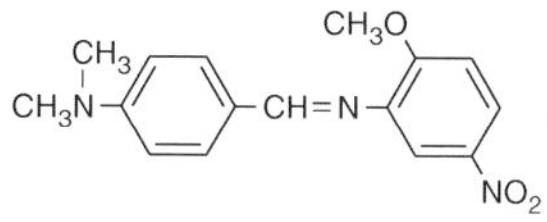
5 gm

S92,483-0
ANTHRANILIC (4-METHOXYBENZYLIDENE)-HYDRAZIDE
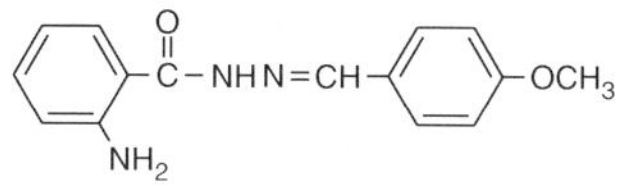
250 mg

S64,082-4
N-(2-NITROBENZYLIDENE)-O-ANISIDINE
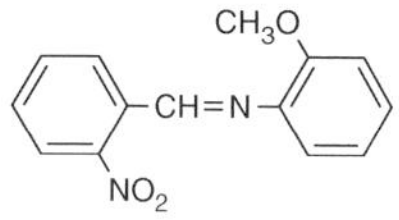
250 mg

S64,097-2
2,5-DIMETHOXY-N-(2-NITROBENZYLIDENE)ANILINE
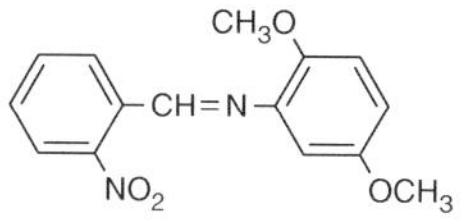
250 mg

S65,572-4
ETHYL ALPHA-CYANO-3-METHOXY-BETA-(4-NITROPHENYLIMINO)HYDROCINNAMATE
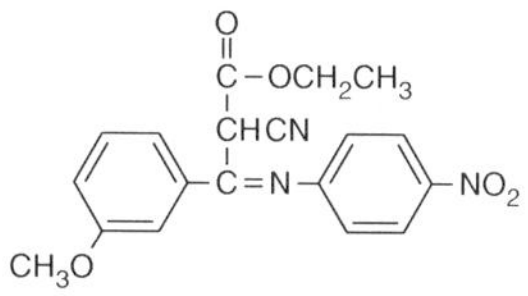
250 mg

S64,083-2
N-(2-NITROBENZYLIDENE)-M-ANISIDINE
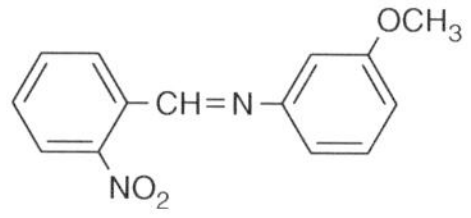
5 gm

S64,096-4
N-(2-NITROBENZYLIDENE)-P-PHENETIDINE
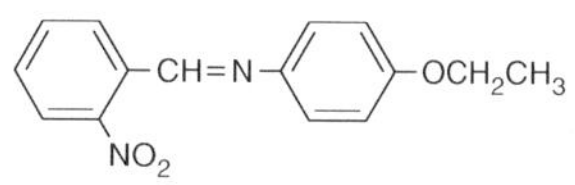
250 mg

S59,471-7
N-(4-METHOXYBENZYLIDENE)-3-NITROANILINE
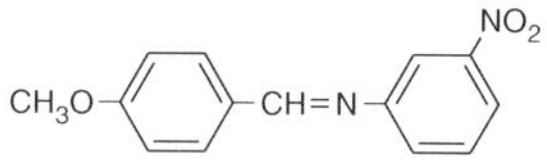
250 mg

S54,912-6
CYANOACETIC ACID (3,4,5-TRIMETHOXYBENZYLIDENE)HYDRAZIDE
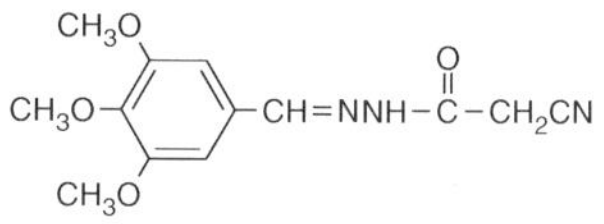
5 gm

S60,210-8
5-NITRO-N-(2-THIENYLMETHYLENE)-O-ANISIDINE
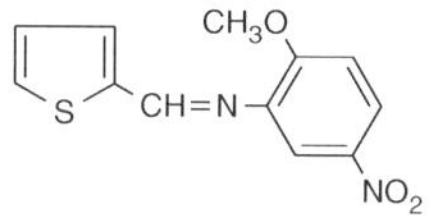
250 mg

S63,621-5
2-METHOXY-5-NITRO-N-PIPERONYLIDENEANILINE
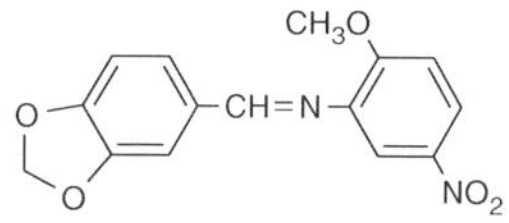
5 gm

S58,444-4
N,N'-(1,4-PHENYLENEDIMETHYLIDYNE)-BIS(5-NITRO-O-ANISIDINE)
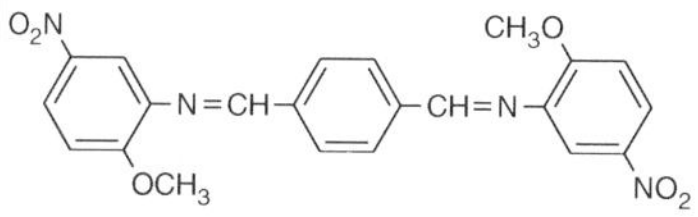
250 mg

S70,276-5
N-(4-METHOXYBENZYLIDENE)-2-METHYL-5-NITROANILINE
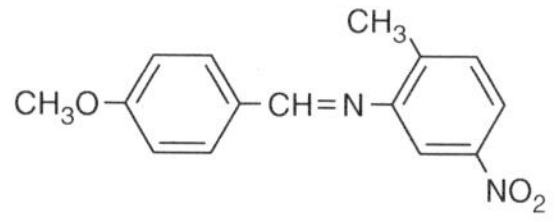
5 gm

S52,012-8
4,5-DIPHENYL-2-(3,4,5-TRIMETHOXYBENZYLIDENEAMINO)-3-FURANCARBONITRILE
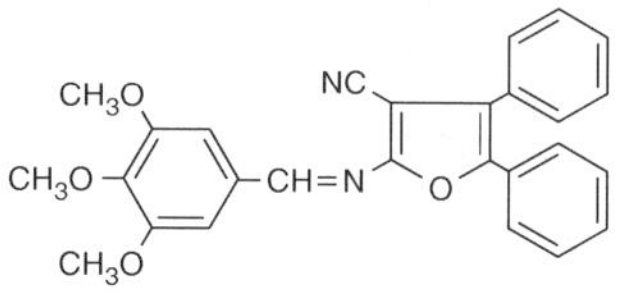
250 mg

S68,906-8
N-(4-METHOXYBENZYLIDENE)-N'-PHENYL-1,4-PHENYLENEDIAMINE
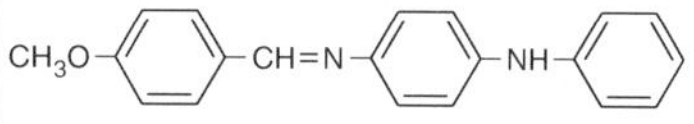
5 gm

S69,877-6
N-(4-BENZYLOXYBENZYLIDENE)-N'-PHENYL-1,4-PHENYLENEDIAMINE
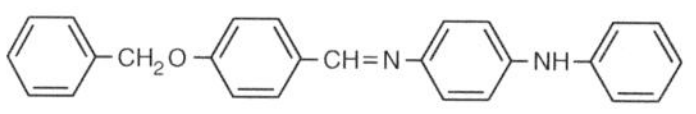
5 gm

S64,362-9
N-[4-(DIMETHYLAMINO)BENZYLIDENE]-2-METHOXY-4-NITROANILINE
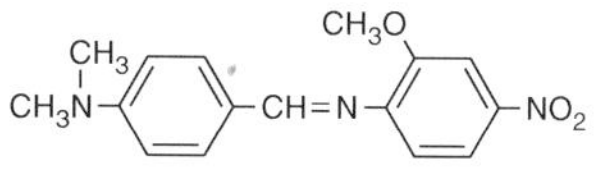
5 gm

S83,472-6
4-METHOXY-N-(3-NITROBENZYLIDENE)-ANILINE
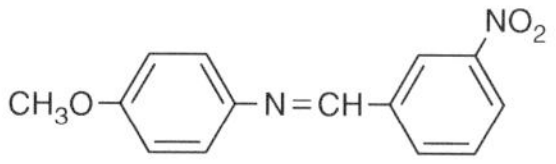
250 mg

S63,563-4
N-(2-CHLORO-5-NITROBENZYLIDENE)-O-ANISIDINE
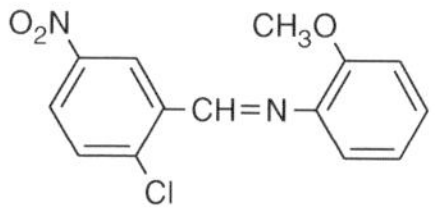
250 mg

S60,040-7
N,N'-TEREPHTHALYLIDENEBIS(2-METHOXY-4-NITROANILINE)
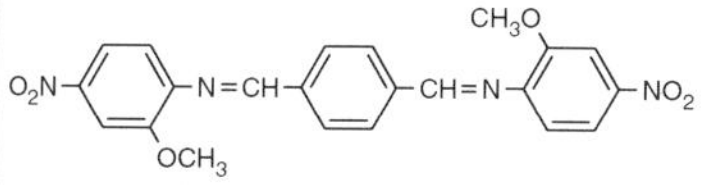
5 gm

S64,084-0
5-CHLORO-N-(2-CHLORO-5-NITROBENZYLIDENE)-2,4-DIMETHOXYANILINE
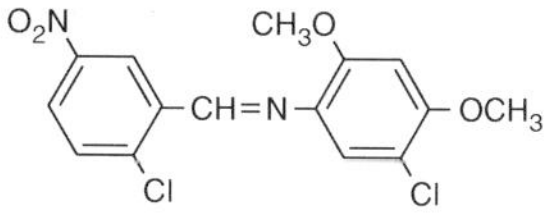
250 mg

S64,086-7
N-(2-CHLORO-5-NITROBENZYLIDENE)-O-PHENETIDINE
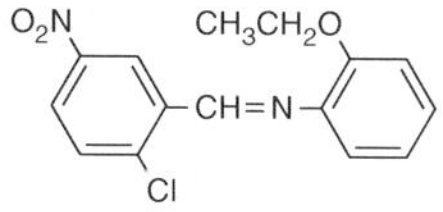
250 mg

S61,925-6
N-(4-METHOXYBENZYLIDENE)-4-NITROANILINE
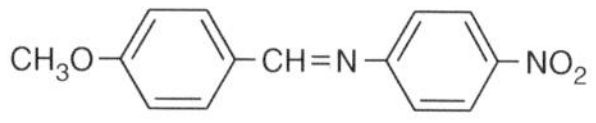
250 mg

S62,724-0
N-(2-CHLORO-5-NITROBENZYLIDENE)-P-ANISIDINE
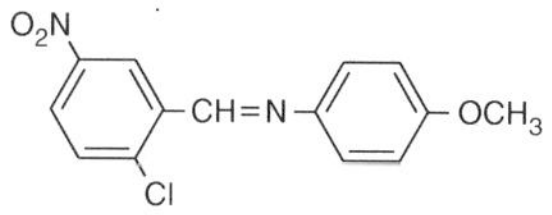
250 mg

S64,329-7
N,N'-BIS(4-METHOXYBENZYLIDENE)-1,4-PHENYLENEDIAMINE

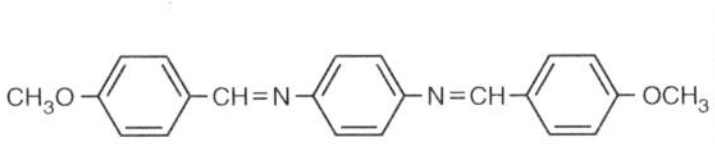

5 gm

S53,534-6
N,N'-BIS(4-NONYLOXYBENZYLIDENE)-1,4-PHENYLENEDIAMINE

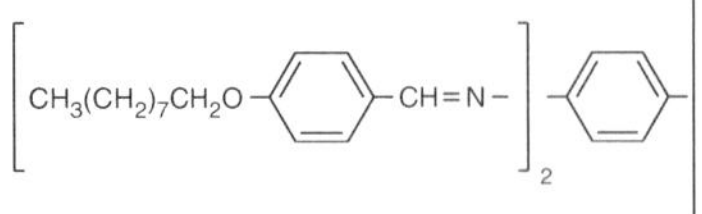

5 gm

S53,535-4
BIS(4-N-DECYLOXYBENZYLIDENE)-1,4-PHENYLENEDIAMINE

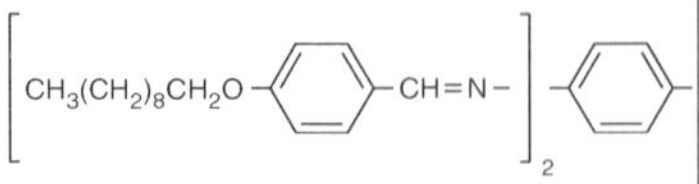

250 mg

S53,536-2
N,N'-BIS(4-DODECYLOXYBENZYLIDENE)-1,4-PHENYLENEDIAMINE

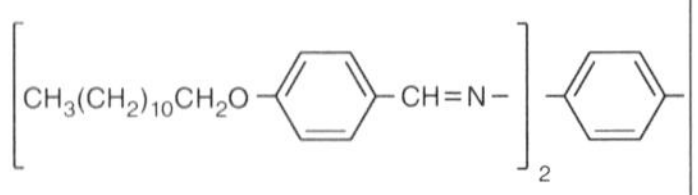

5 gm

S53,539-7
BIS(4-METHOXYBENZYLIDENE)-2-CHLORO-1,4-PHENYLENEDIAMINE

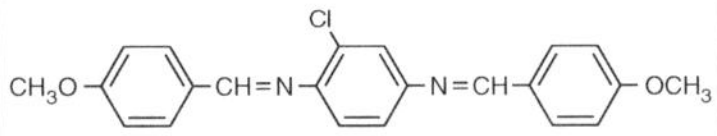

250 mg

S53,540-0
N,N'-BIS(4-PENTYLOXYBENZYLIDENE)-2-CHLORO-1,4-PHENYLENEDIAMINE

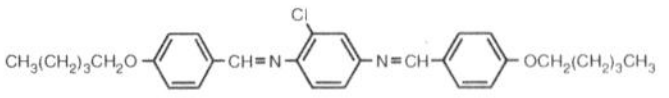

5 gm

S53,541-9
N,N'-BIS(4-HEXYLOXYBENZYLIDENE)-2-CHLORO-1,4-PHENYLENEDIAMINE

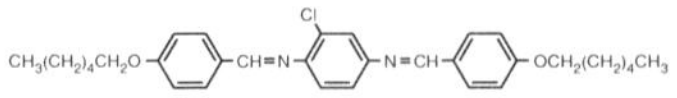

5 gm

S53,542-7
N,N'-BIS(4-OCTYLOXYBENZYLIDENE)-2-CHLORO-1,4-PHENYLENEDIAMINE

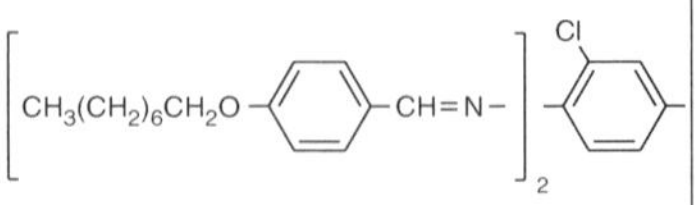

5 gm

S53,544-3
BIS(4-N-DECYLOXYBENZYLIDENE)-2-CHLORO-1,4-PHENYLENEDIAMINE

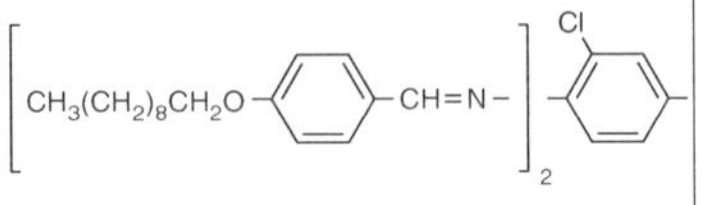

250 mg

S51,468-3
5-CHLORO-2,4-DIMETHOXY-N-(3-NITROBENZYLIDENE)ANILINE

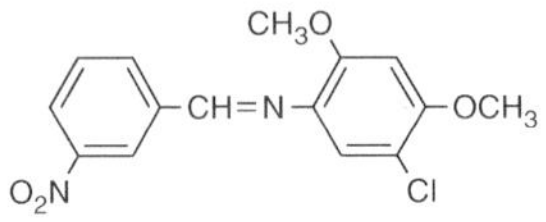

5 gm

S75,002-6
4-ETHOXY-N-(3-NITROBENZYLIDENE)-ANILINE

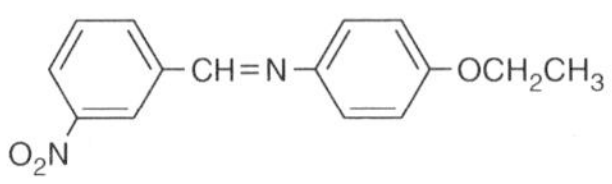

5 gm

S73,936-7
N-(4-ETHOXYPHENYL)-N'-(4-NITROPHENYL)ACETAMIDINE

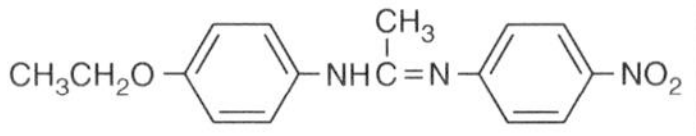

5 gm

S52,270-8
N-(3-NITROBENZYLIDENE)-P-ANISIDINE

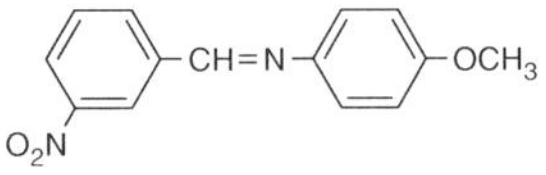

5 gm

S92,484-9
N-METHYLANTHRANILIC (4-METHOXYBENZYLIDENE)HYDRAZIDE

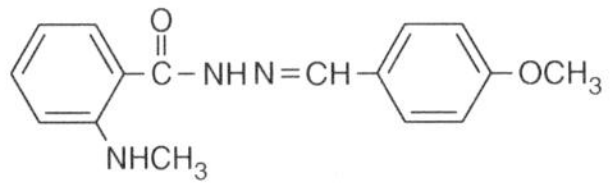

250 mg

S61,869-1
ETHYL 11-[4-(4-[BIS(2-CHLOROETHYL)AMINO]BENZYLIDENEAMINO)-PHENOXY]UNDECANOATE

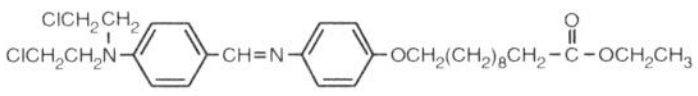

250 mg

S58,332-4
3-METHYL-2-BENZOTHIAZOLINONE, AZINE WITH 4'-HEXADECYLOXY-3'-SULFOACETOPHENONE, SODIUM SALT

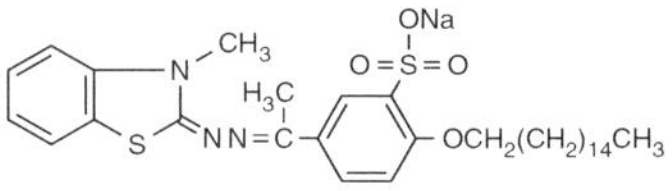

250 mg

S61,802-0
N-[4-(DIMETHYLAMINO)BENZYLIDENE]-P-ANISIDINE

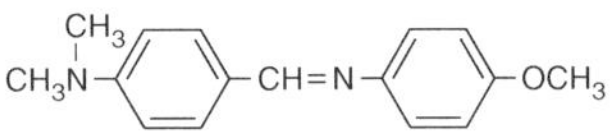

250 mg

S61,828-4
N-[4-(DIMETHYLAMINO)BENZYLIDENE]-2-METHYL-P-ANISIDINE

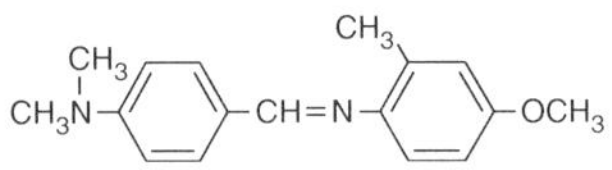

250 mg

S63,975-3
N-[4-(DIMETHYLAMINO)BENZYLIDENE]-P-PHENETIDINE

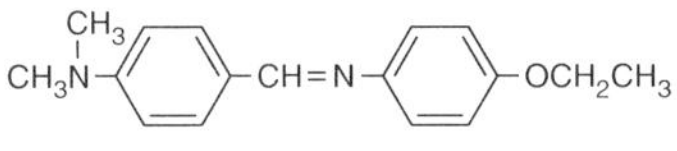

5 gm

S52,589-8
ALPHA-(5-CHLORO-2,4-DIMETHOXYPHENYLIMINO)-N,N-DIMETHYL-P-TOLUIDINE

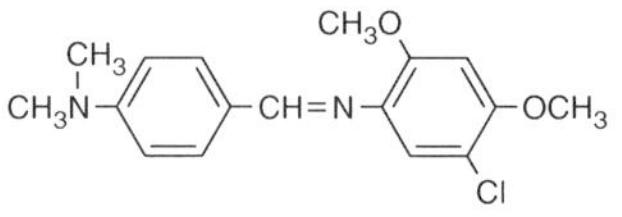

250 mg

S61,831-4
N-[4-(DIMETHYLAMINO)BENZYLIDENE]-4-HEXADECYLOXYANILINE

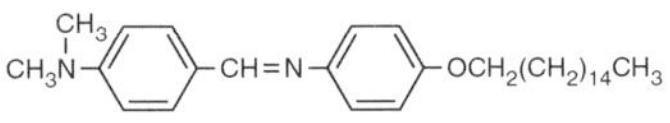

5 gm

S59,231-5
3-METHYL-2-BENZOTHIAZOLINONE, AZINE WITH 4'[2-(METHACRYLOYLOXY)ETHOXY]-ACETOPHENONE

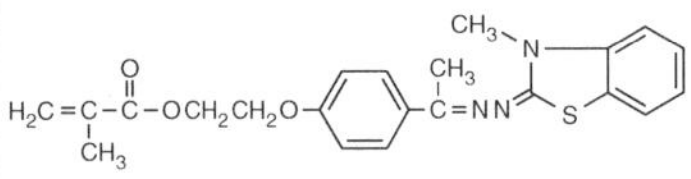

250 mg

S87,491-4
6-ETHOXY-3-ETHYL-2-[2-(PHENYL-IMINO)ETHYLIDENE]BENZOTHIAZOLINE

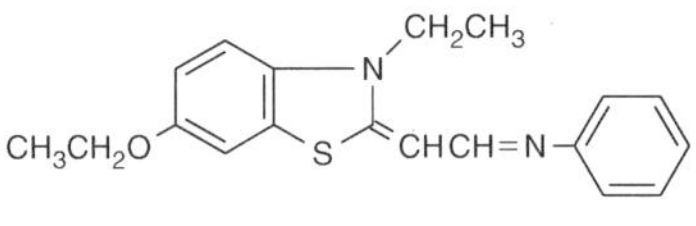

250 mg

S70,117-3
N-(4-DIMETHYLAMINOBENZYLIDENE)-4-PHENOXYANILINE

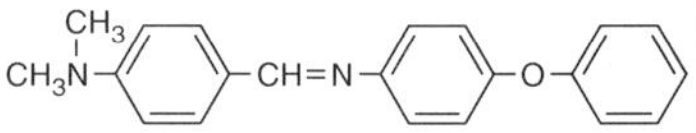

5 gm

S70,110-6
N-(4-DIETHYLAMINOBENZYLIDENE)-4-PHENOXYANILINE

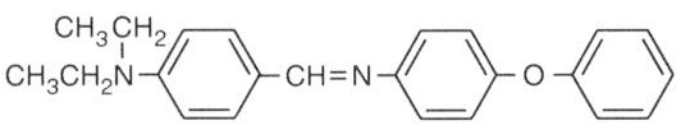

5 gm

S59,292-7
1-(2-CHLORO-ALPHA-METHOXYBENZYLIDENE)PIPERIDINIUM METHYL SULFATE

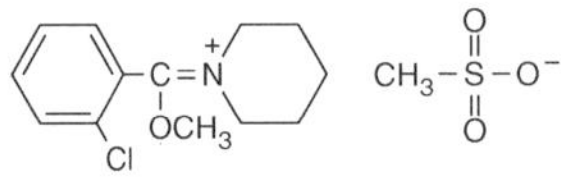

5 gm

S51,589-2
(GAMMA-CHLORO-4-METHOXYCINNAMYLIDENE) DIMETHYLAMMONIUM PERCHLORATE

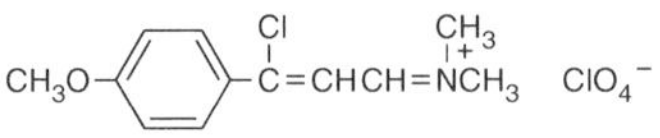

5 gm

S68,324-8
N-(3-METHOXYBENZYLIDENE)METHYLAMINE

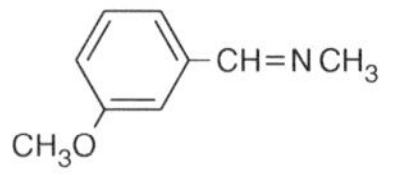

5 gm

S62,699-6
N-BENZYLIDENE-M-ANISIDINE

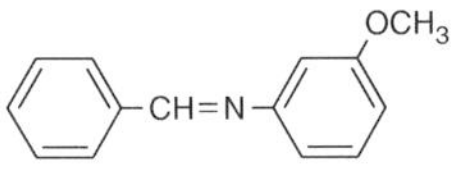

250 mg

S63,188-4
3-CHLORO-N-(3-METHOXYBENZYLIDENE)-ANILINE

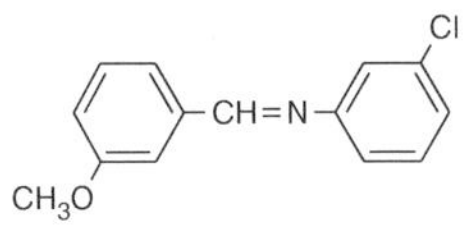

250 mg

S62,586-8
N-(3-METHOXYBENZYLIDENE)-M-TOLUIDINE

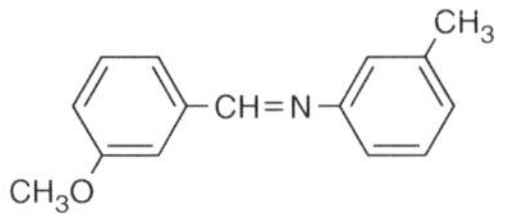

250 mg

S62,588-4
N-(3-METHOXYBENZYLIDENE)-M-ANISIDINE

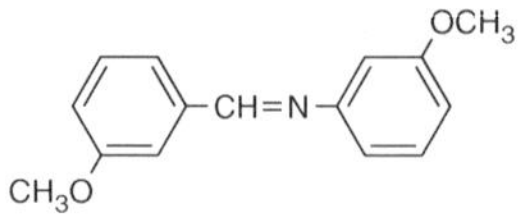

250 mg

S92,503-9
4-METHOXYBENZOIC (3-NITROBENZYLIDENE)HYDRAZIDE

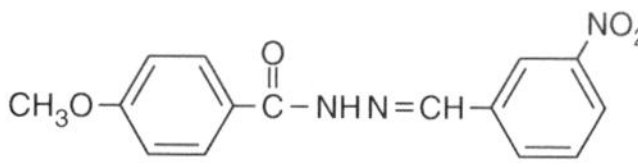

250 mg

S92,499-7
4-METHOXYBENZOIC BENZYLIDENEHYDRAZIDE

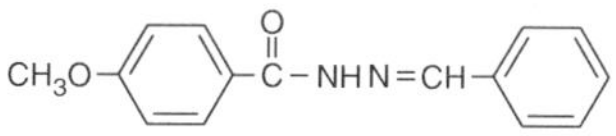

250 mg

S92,516-0
4-METHOXYBENZOIC (ALPHA-PROPYLBENZYLIDENE)HYDRAZIDE

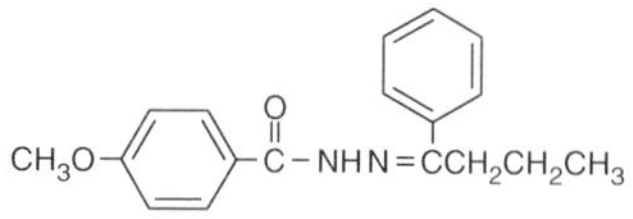

250 mg

S92,541-1
DODECANOIC (2-METHOXYBENZYLIDENE)-HYDRAZIDE

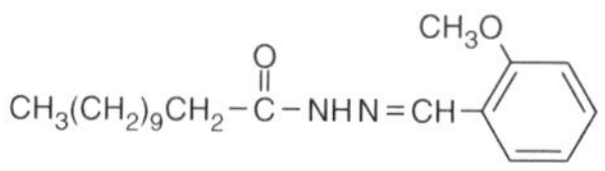

250 mg

S63,189-2
4-CYANO-N-(PIPERONYLIDENE)ANILINE

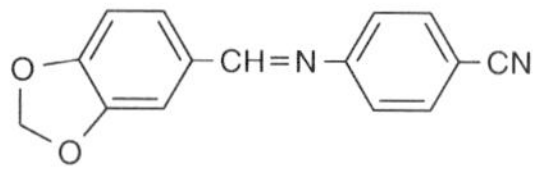

250 mg

S63,593-6
N-PIPERONYLIDENE-O-ANISIDINE

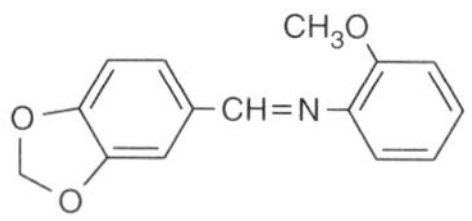

250 mg

S63,193-0
N-PIPERONYLIDENE-2,4-XYLIDINE

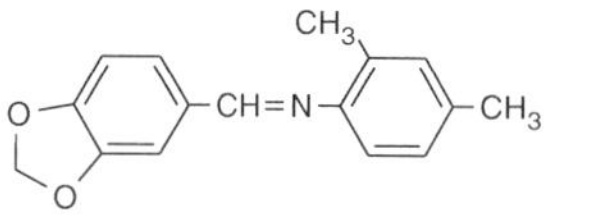

250 mg

S63,191-4
N-PIPERONYLIDENE-M-ANISIDINE

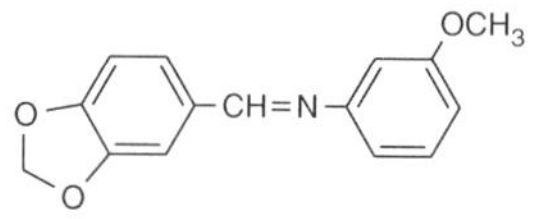

250 mg

S60,928-5
N-BENZYLIDENE-O-ANISIDINE

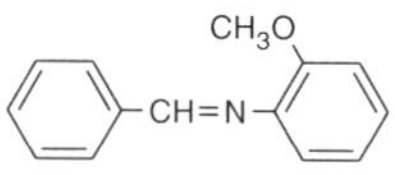

250 mg

S63,181-7
N-(3-CHLOROBENZYLIDENE)-O-ANISIDINE

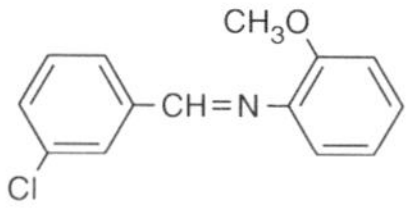

250 mg

S62,587-6
N-(3-METHOXYBENZYLIDENE)-O-ANISIDINE

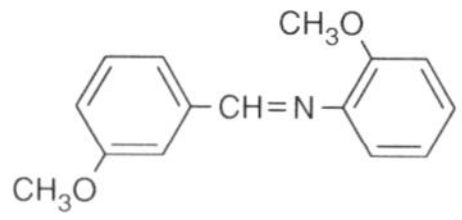

250 mg

S61,928-0
N-(4-CHLOROBENZYLIDENE)-O-ANISIDINE

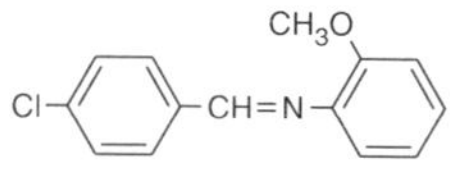

250 mg

S60,228-0
N-(4-METHYLBENZYLIDENE)-O-ANISIDINE

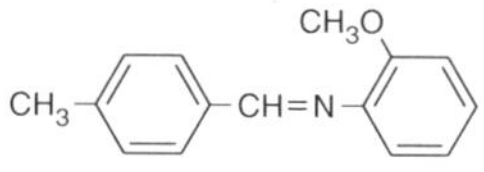

250 mg

S60,216-7
N-(4-METHOXYBENZYLIDENE)-O-ANISIDINE

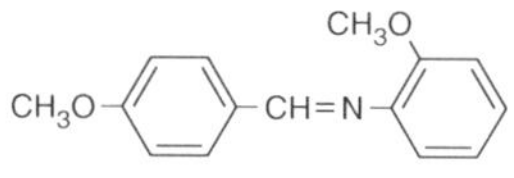

250 mg

S64,945-7
N-(4-CHLOROBENZYLIDENE)-O-PHENETIDINE

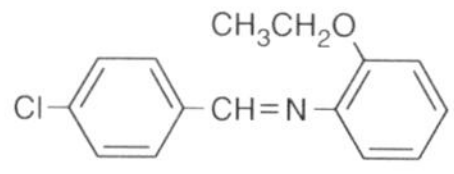

250 mg

S67,295-5
N-(4-METHYLBENZYLIDENE)-O-PHENETIDINE

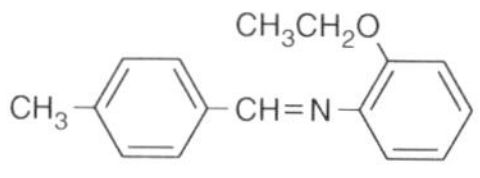

250 mg

S59,645-0
N,N'-(1,4-PHENYLENEDIMETHYLIDYNE)-DI-O-PHENETIDINE

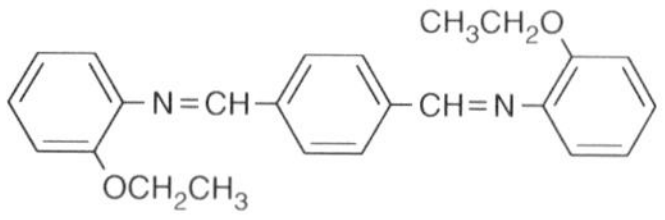

5 gm

S63,535-9
N-(3,4-DICHLOROBENZYLIDENE)-O-ANISIDINE

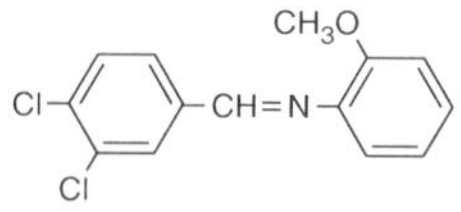

250 mg

S63,561-8
N-(2,4-DICHLOROBENZYLIDENE)-O-ANISIDINE

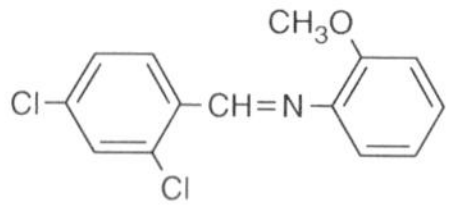

250 mg

S64,089-1
N-(2-CHLOROBENZYLIDENE)-2,5-DIMETHOXYANILINE

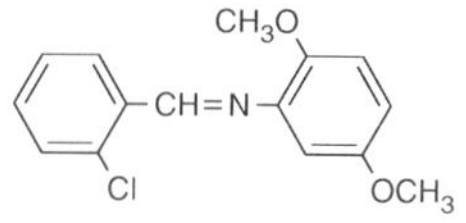

250 mg

S64,104-9
N-(2,5-DIMETHOXYBENZYLIDENE)-O-ANISIDINE

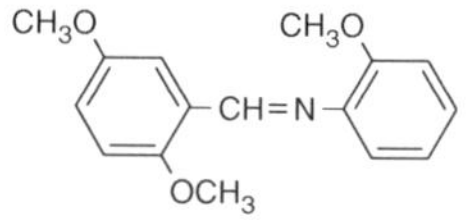

250 mg

S63,568-5
N-(4-CHLOROBENZYLIDENE)-2,5-DIMETHOXYANILINE

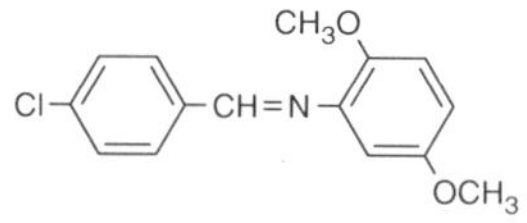

250 mg

S64,101-4
2,5-DIMETHOXY-N-(4-METHYLBENZYLIDENE)ANILINE

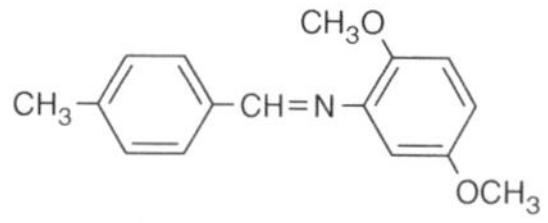

250 mg

S64,103-0
2,5-DIMETHOXY-N-(4-METHOXYBENZYLIDENE)ANILINE

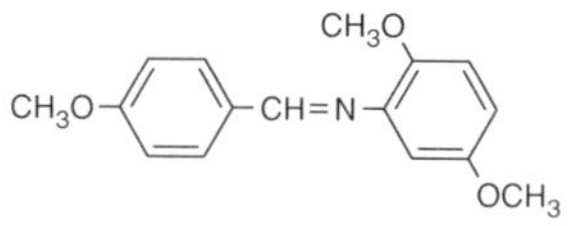

250 mg

S58,917-9
N,N'-(1,4-PHENYLENEDIMETHYLIDYNE)-BIS(5-METHYL-O-ANISIDINE)

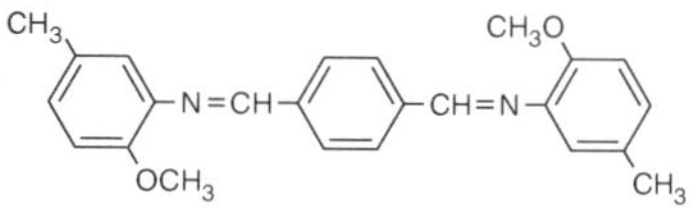

5 gm

S58,903-9
N,N'-(1,4-PHENYLENEDIMETHYLIDYNE)-BIS(2,4-DIMETHOXYANILINE)

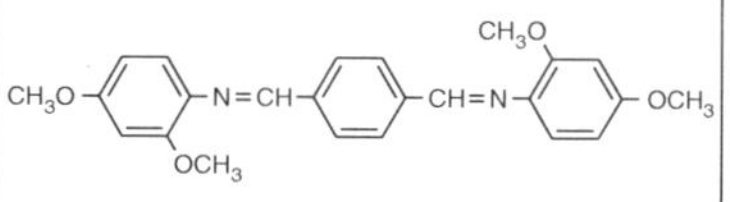

5 gm

S64,098-0
N-BENZYLIDENE-2,5-DIMETHOXYANILINE

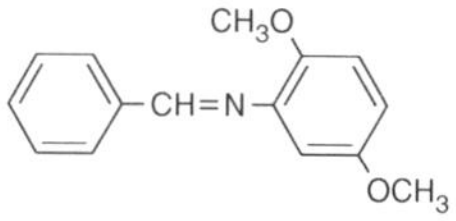

250 mg

S63,525-1
N-(3,4-DICHLOROBENZYLIDENE)-2,5-DIMETHOXYANILINE

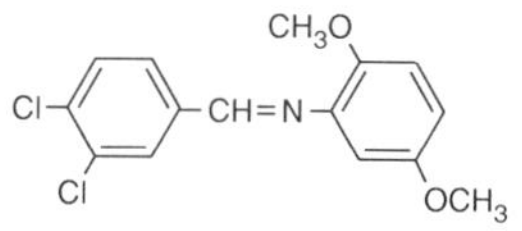

250 mg

S64,110-3
2,5-DIMETHOXY-N-(2,5-DIMETHOXYBENZYLIDENE)ANILINE

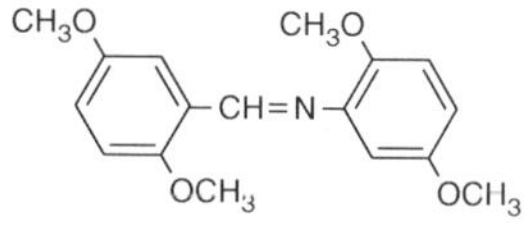

250 mg

S62,546-9
N-(9-ANTHRACENYLMETHYLENE)-2-METHOXY-5-METHYLANILINE

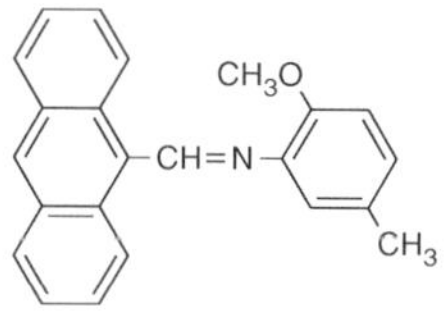

5 gm

S53,502-8
4'-(4-PROPOXYBENZYLIDENEAMINO)-PROPIOPHENONE

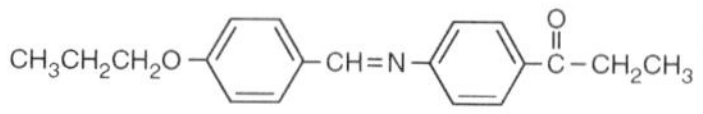

250 mg

S58,778-8
3-METHYL-2-BENZOTHIAZOLINONE, AZINE WITH 4-[2-(METHACRYLOXYLOXY) ETHOXY]-M-ANISALDEHYDE

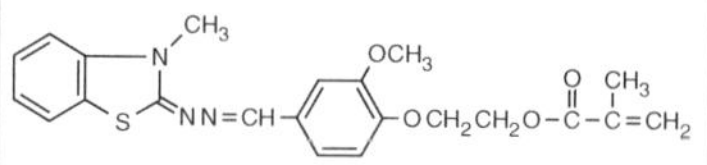

250 mg

S63,590-1
N-PIPERONYLIDENE-P-ANISIDINE

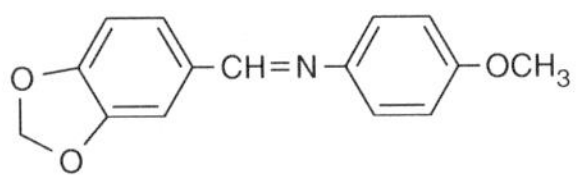

250 mg

S34,191-6
BENZYLIMINOBIS(4-METHOXYPHENYL)-METHANE

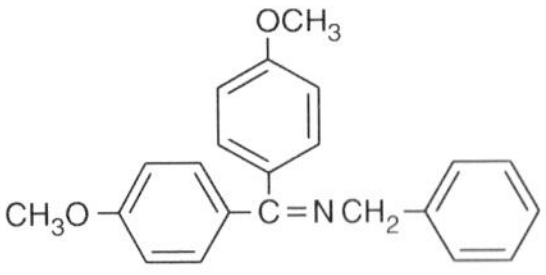

250 mg

S95,572-8
3-[2-(BENZYLIDENEAMINO)ETHYL]-2-(4-METHOXYPHENYL)INDOLE

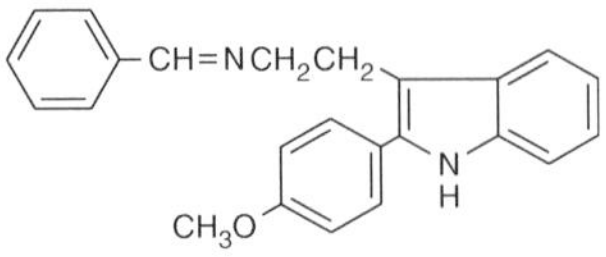

50 mg

S63,508-1
N-(2-METHOXYBENZYLIDENE)ANILINE

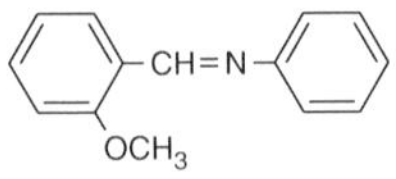

250 mg

S60,221-3
N-(2-CHLOROBENZYLIDENE)-M-ANISIDINE

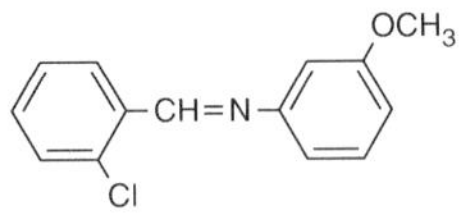

5 gm

S62,577-9
N-(3-METHOXYBENZYLIDENE)-2-CHLOROANILINE

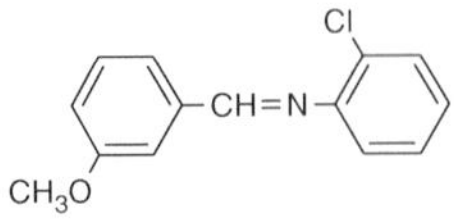

250 mg

S63,536-7
3-CHLORO-N-(2-METHOXYBENZYLIDENE)-ANILINE

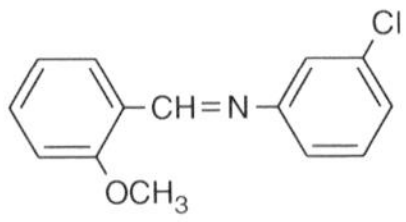

250 mg

S62,584-1
N-(3-METHOXYBENZYLIDENE)-O-TOLUIDINE

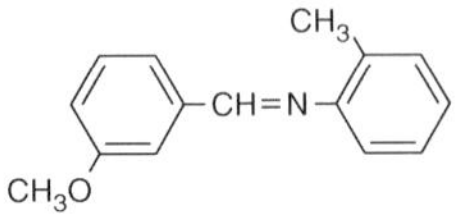

250 mg

S67,301-3
N-(2-ETHOXYBENZYLIDENE)ANILINE

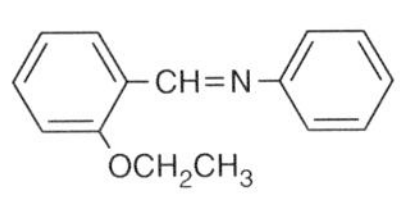

250 mg

S67,312-9
N-(2-ETHOXYBENZYLIDENE)-M-TOLUIDINE

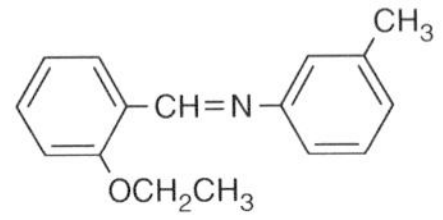

250 mg

S62,576-0
N-(2-CHLOROBENZYLIDENE)-P-ANISIDINE

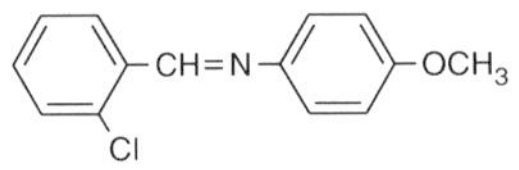

250 mg

S63,562-6
4-CHLORO-N-(2-METHOXYBENZYLIDENE)-ANILINE

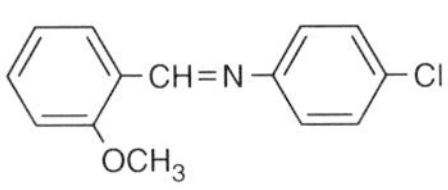

250 mg

S60,230-2
N-(4-METHOXYBENZYLIDENE)-O-TOLUIDINE

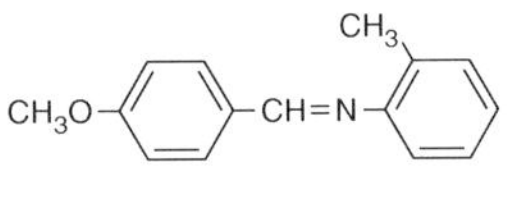

250 mg

S63,569-3
N-(2-METHOXYBENZYLIDENE)-P-TOLUIDINE

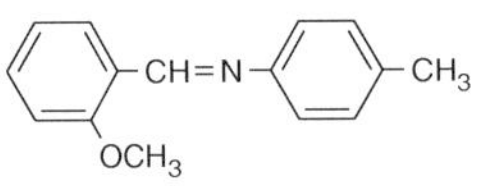

250 mg

S63,534-0
N-(2-METHOXYBENZYLIDENE)-P-ANISIDINE

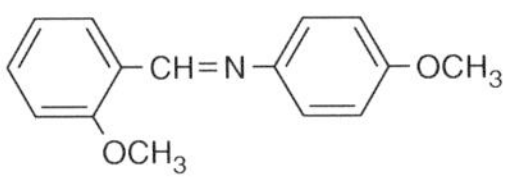

250 mg

S60,418-6
N-(2-CHLOROBENZYLIDENE)-P-PHENETIDINE

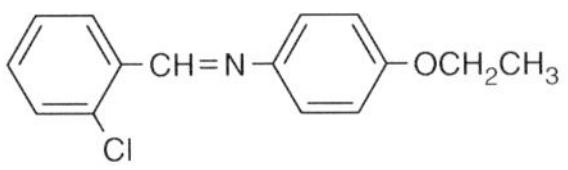

250 mg

S63,584-7
N-(2-METHOXYBENZYLIDENE)-P-PHENETIDINE

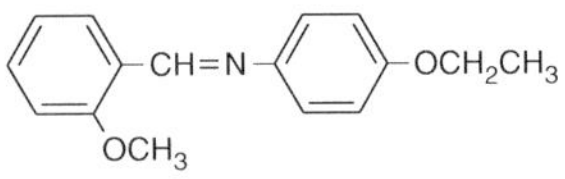

250 mg

S62,710-0
N-(2-CHLOROBENZYLIDENE)-4-PHENOXYANILINE

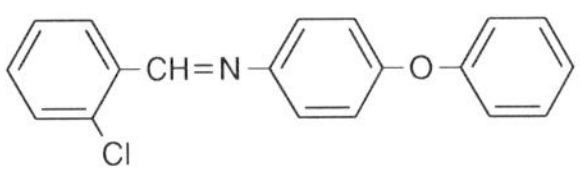

250 mg

S62,622-8
N-(4-BENZYLOXYBENZYLIDENE)-2-FLUOROANILINE

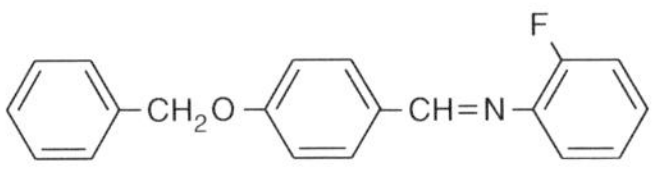

5 gm

S63,533-2
2,3-DICHLORO-N-(2-METHOXYBENZYLIDENE)ANILINE

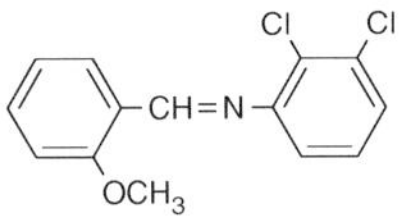

250 mg

S67,324-2
N-(2-ETHOXYBENZYLIDENE)-2,3-XYLIDINE

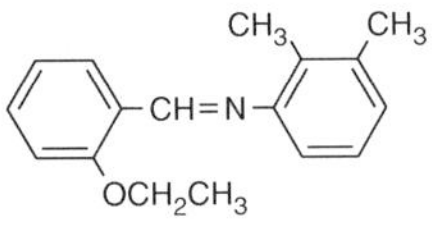

5 gm

S67,326-9
N-(2-ETHOXYBENZYLIDENE)-2,6-XYLIDINE

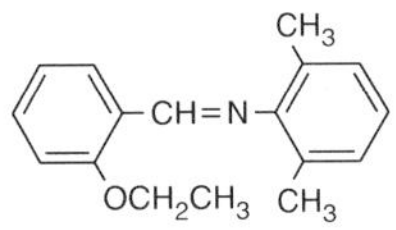

250 mg

S64,088-3
2-CHLORO-N-(2,5-DIMETHOXYBENZYLIDENE)ANILINE

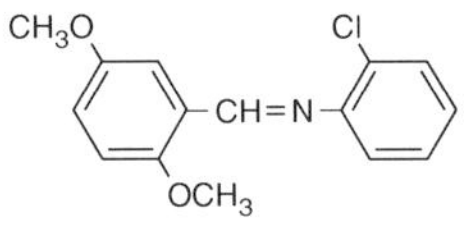

250 mg

S64,946-5
N-(2,4-DICHLOROBENZYLIDENE)-O-PHENETIDINE

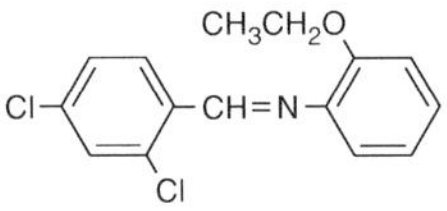

250 mg

S67,325-0
N-(2-ETHOXYBENZYLIDENE)-2,5-XYLIDINE

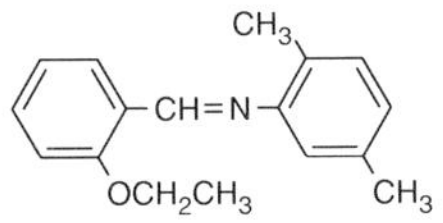

250 mg

S63,572-3
N-(2-METHOXYBENZYLIDENE)-3,5-XYLIDINE

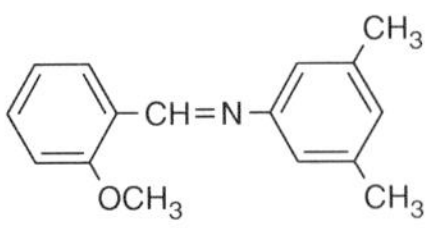

250 mg

S62,104-8
N-(9-ANTHRACENYLMETHYLENE)-2-ETHOXYANILINE

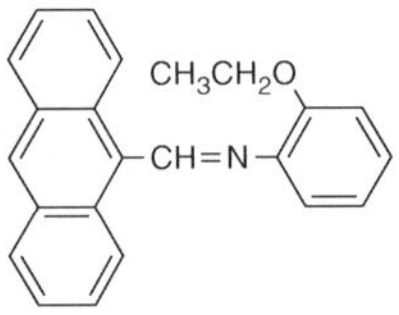

5 gm

S59,477-6
N-(4-METHOXYBENZYLIDENE)ANILINE

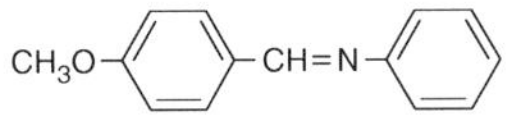

250 mg

S39,336-3
4-CHLORO-N-(4-METHOXYBENZYLIDENE)-ANILINE

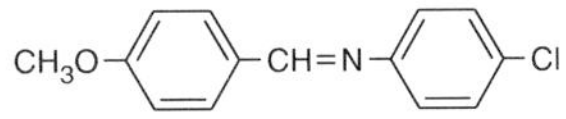

5 gm

S62,578-7
4-CHLORO-N-(3-METHOXYBENZYLIDENE)-ANILINE

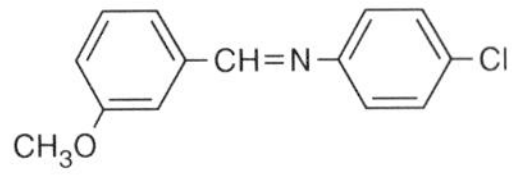

5 gm

S59,438-5
N-(4-METHOXYBENZYLIDENE)-4-BROMOANILINE

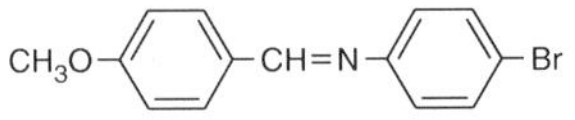

250 mg

S63,186-8
4-BROMO-N-(3-METHOXYBENZYLIDENE)-ANILINE

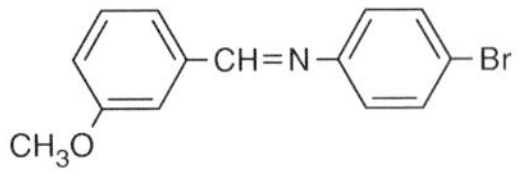

250 mg

S59,465-2
N-(4-METHOXYBENZYLIDENE)-P-TOLUIDINE

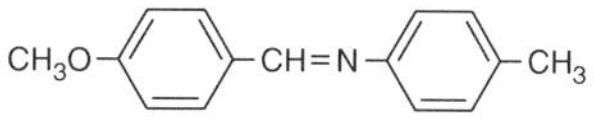

250 mg

S60,410-0
N-BENZYLIDENE-P-PHENETIDINE

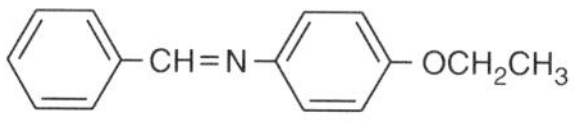

250 mg

S60,412-7
N-(4-METHYLBENZYLIDENE)-M-ANISIDINE

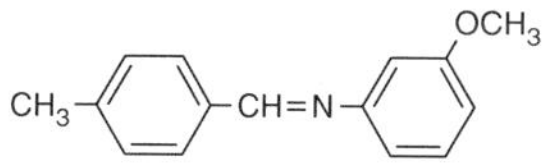

250 mg

S63,234-1
N-(4-METHYLBENZYLIDENE)-P-ANISIDINE

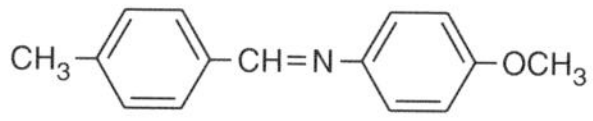

250 mg

S60,225-6
N-(4-METHOXYBENZYLIDENE)-M-ANISIDINE

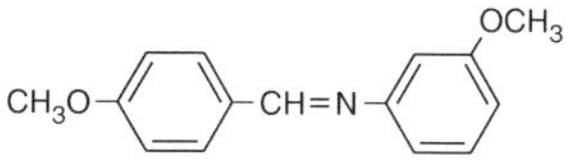

250 mg

S61,573-0
N-(4-METHOXYBENZYLIDENE)-P-ANISIDINE

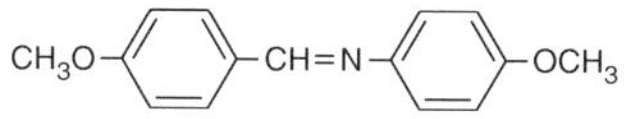

250 mg

S63,567-7
N-(4-CHLOROBENZYLIDENE)-P-PHENETIDINE

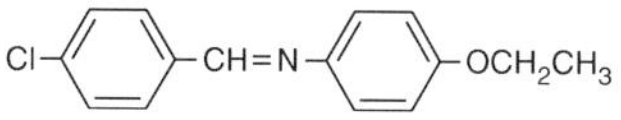

250 mg

S60,420-8
N-(4-METHYLBENZYLIDENE)-P-PHENETIDINE

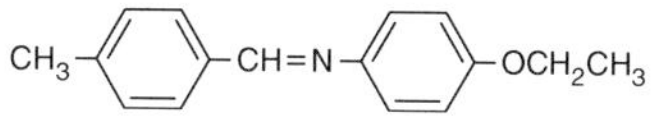

250 mg

S60,417-8
N-(4-METHOXYBENZYLIDENE)-P-PHENETIDINE

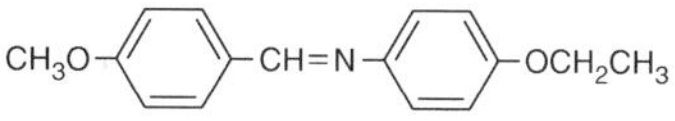

250 mg

S63,582-0
N-(4-CHLOROBENZYLIDENE)-4-PHENOXYANILINE

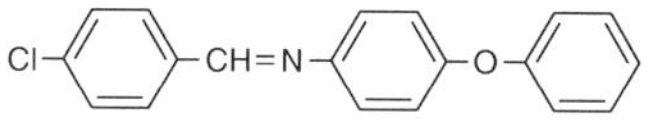

250 mg

S62,620-1
N-(4-BENZYLOXYBENZYLIDENE)-3-BROMOANILINE

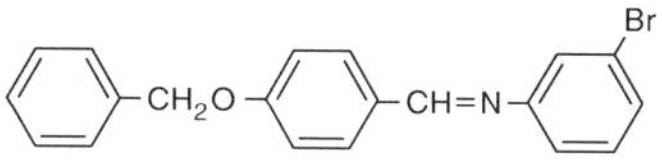

5 gm

S62,623-6
N-(4-BENZYLOXYBENZYLIDENE)-4-IODOANILINE

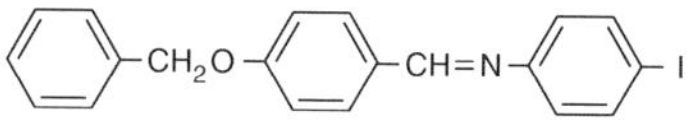

5 gm

S53,528-1
N,N'-BIS(4-PENTYLOXYBENZYLIDENE)-1,4-PHENYLENEDIAMINE

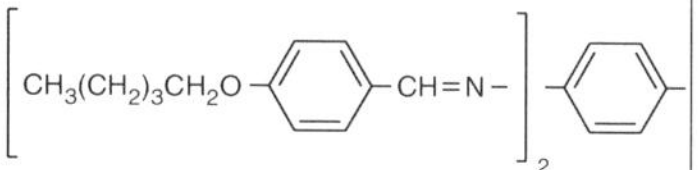

250 mg

S53,531-1
N,N'-BIS(4-HEPTYLOXYBENZYLIDENE)-1,4-PHENYLENEDIAMINE

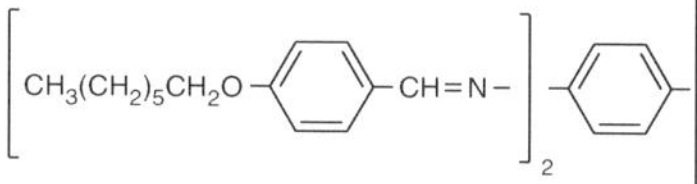

250 mg

S53,533-8
N,N'-BIS(4-OCTYLOXYBENZYLIDENE)-1,4-PHENYLENEDIAMINE

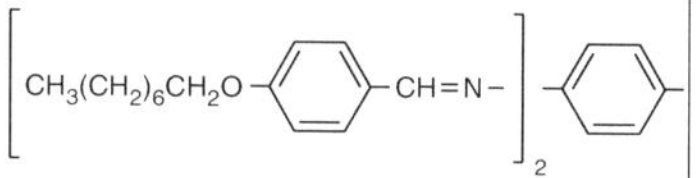

5 gm

S63,510-3
2,3-DICHLORO-N-(4-METHOXYBENZYLIDENE)ANILINE

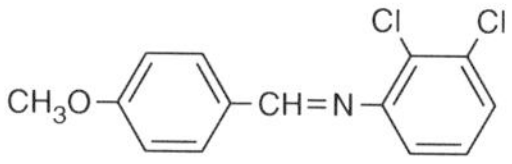

250 mg

S61,569-2
N-(4-METHOXYBENZYLIDENE)-2,3-XYLIDINE

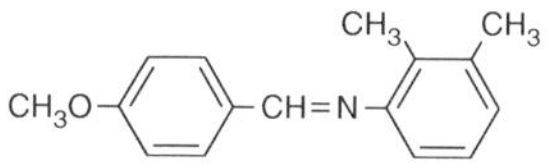

250 mg

S60,415-1
N-(4-METHOXYBENZYLIDENE)-2,6-DIETHYLANILINE

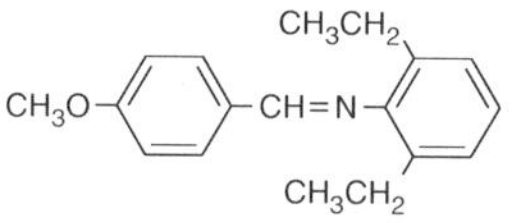

250 mg

S60,231-0
N-(4-METHOXYBENZYLIDENE)-2,5-DICHLOROANILINE

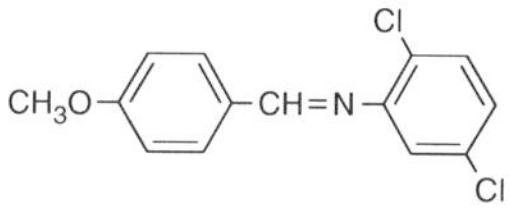

250 mg

S60,234-5
N-(4-METHOXYBENZYLIDENE)-2,4-DICHLOROANILINE

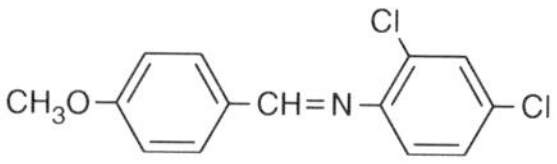

250 mg

S60,413-5
N-(4-METHOXYBENZYLIDENE)-3,4-DICHLOROANILINE

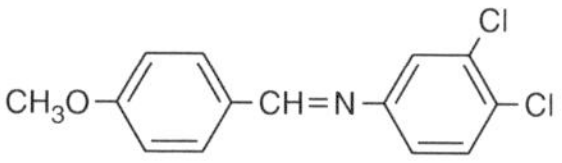

250 mg

S62,848-4
N-(3,4-DICHLOROBENZYLIDENE)-P-ANISIDINE

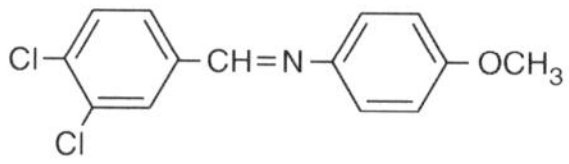

250 mg

S61,915-9
3-CHLORO-N-(4-METHOXYBENZYLIDENE)-6-METHYLANILINE

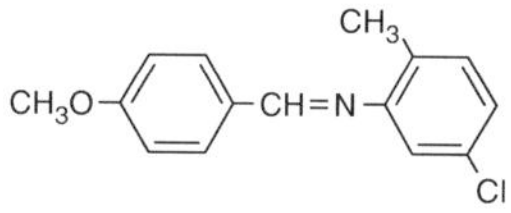

250 mg

S64,090-5
4-CHLORO-N-(2,5-DIMETHOXYBENZYLIDENE)ANILINE

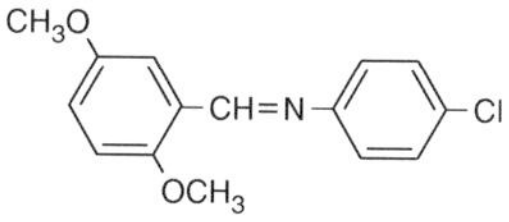

250 mg

S63,566-9
N-(2,4-DICHLOROBENZYLIDENE)-P-PHENETIDINE

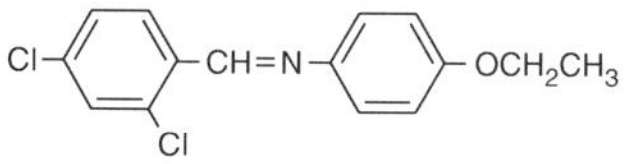

250 mg

S60,243-4
N-(4-METHOXYBENZYLIDENE)-2,5-XYLIDINE

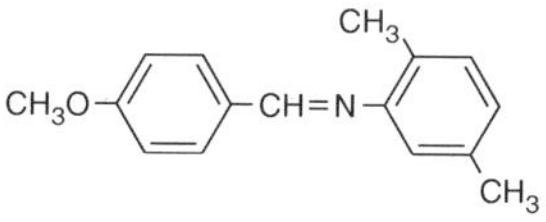

250 mg

S62,592-2
N-(4-METHOXYBENZYLIDENE)-3,4-XYLIDINE

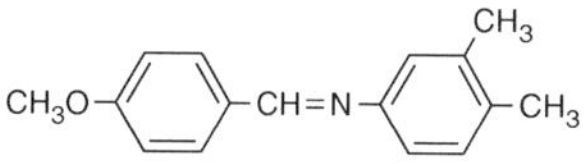

250 mg

S64,102-2
N-(2,5-DIMETHOXYBENZYLIDENE)-P-TOLUIDINE

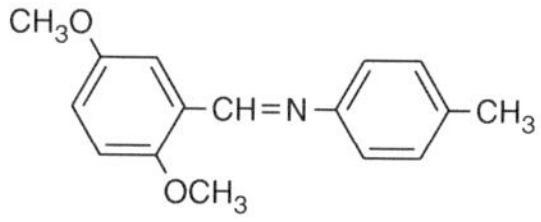

250 mg

S64,108-1
N-(2,5-DIMETHOXYBENZYLIDENE)-P-PHENETIDINE

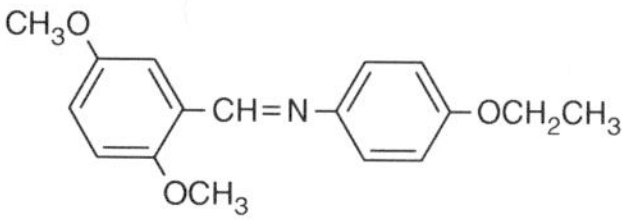

250 mg

S62,619-8
N-(4-BENZYLOXYBENZYLIDENE)-2-BROMO-4-METHYLANILINE

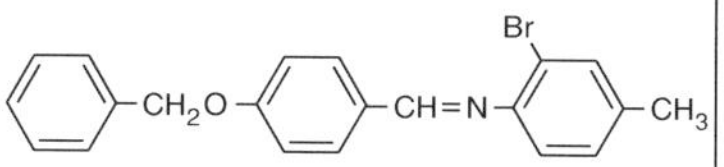

5 gm

S60,211-6
N,N'-TEREPHTHALYLIDENEBIS(2-METHYL-P-ANISIDINE)

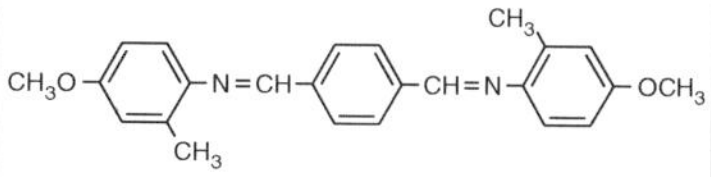

5 gm

S62,593-0
N-(4-METHOXYBENZYLIDENE)-3,5-XYLIDINE

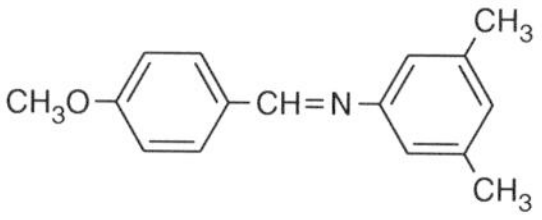

250 mg

S60,127-6
N-(9-ANTHRACENYLMETHYLENE)-P-ANISIDINE
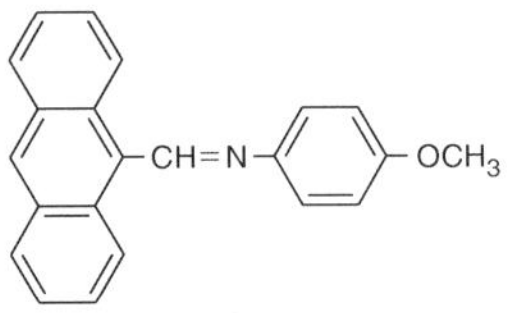
5 gm

S62,105-6
N-(9-ANTHRACENYLMETHYLENE)-4-ETHOXYANILINE
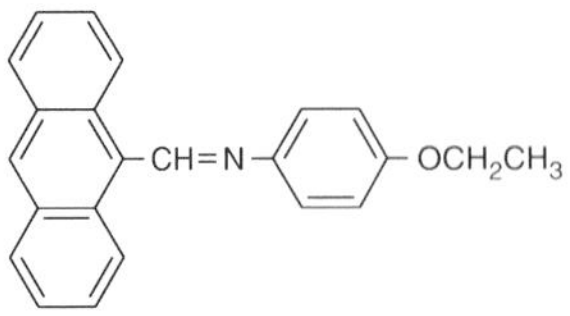
5 gm

S63,194-9
N-(3-METHOXYBENZYLIDENE)-2,3-XYLIDINE
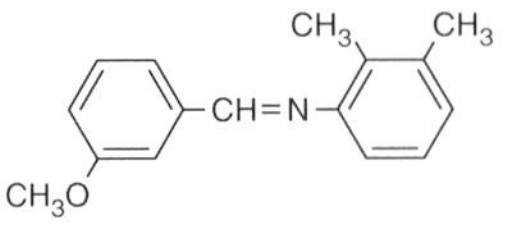
250 mg

S63,559-6
N-(3,4-DICHLOROBENZYLIDENE)-M-ANISIDINE
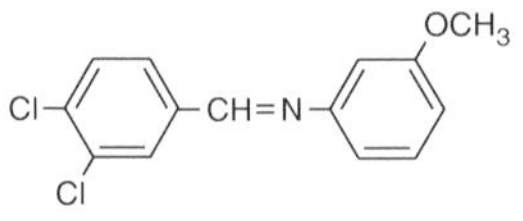
250 mg

S63,583-9
N-(2,4-DICHLOROBENZYLIDENE)-M-ANISIDINE
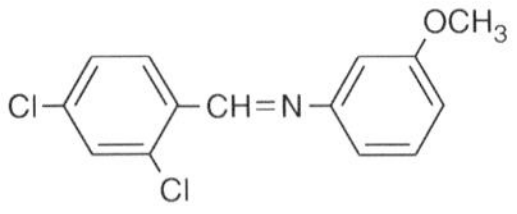
250 mg

S64,099-9
N-(2,5-DIMETHOXYBENZYLIDENE)ANILINE
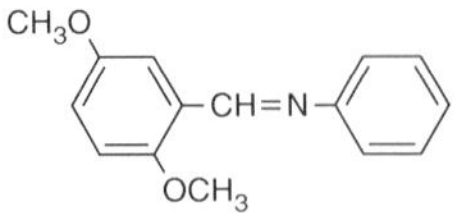
250 mg

S64,087-5
2,5-DICHLORO-N-(2,5-DIMETHOXYBENZYLIDENE)ANILINE
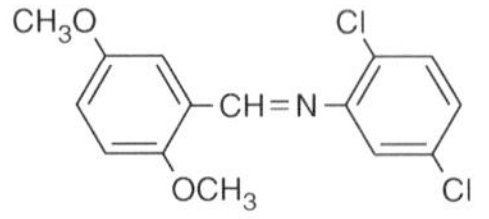
250 mg

S63,195-7
N-(3-METHOXYBENZYLIDENE)-2,4-XYLIDINE
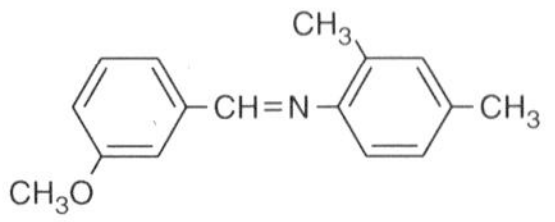
250 mg

S63,196-5
N-(3-METHOXYBENZYLIDENE)-2,5-XYLIDINE
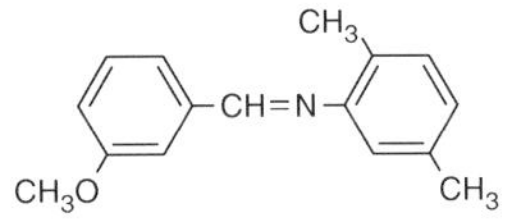
250 mg

S64,100-6
N-(2,5-DIMETHOXYBENZYLIDENE)-M-TOLUIDINE
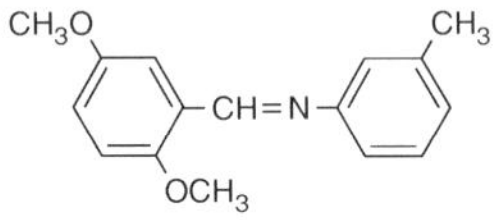
250 mg

S64,107-3
N-(2,5-DIMETHOXYBENZYLIDENE)-3,4-XYLIDINE
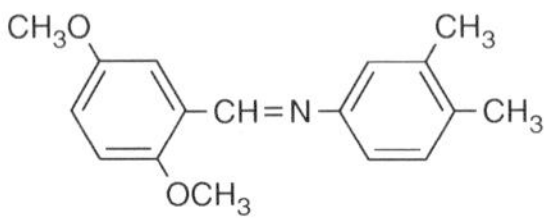
250 mg

S66,321-2
ALPHA-BUTYLIMINO-2-METHYLALLYLOXYMESITYLENE
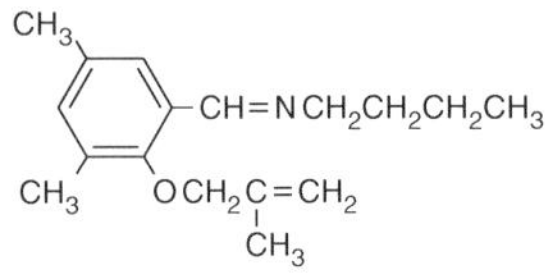
5 gm

S60,119-5
N-(9-ANTHRACENYLMETHYLENE)-M-ANISIDINE
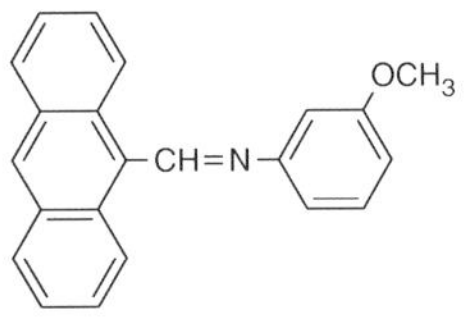
5 gm

S93,449-6
4-(3-METHOXYBENZYLIDENEAMINO)-4H-1,2,4-TRIAZOLE
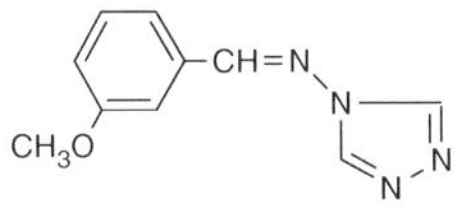
250 mg

S77,193-7
1-[3-(ISOPENTYLOXY)-BENZYLIDENEAMINO]GUANIDINE NITRATE
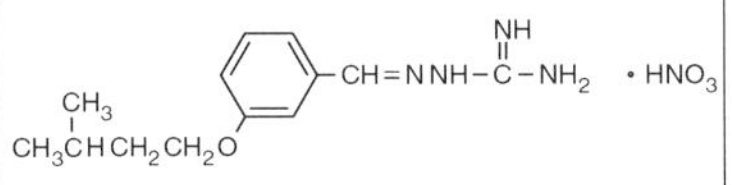
250 mg

S69,159-3
4-(2-METHOXY-5-METHYLPHENYL)-1-(3-NITROBENZYLIDENE)-3-THIOSEMICARBAZIDE
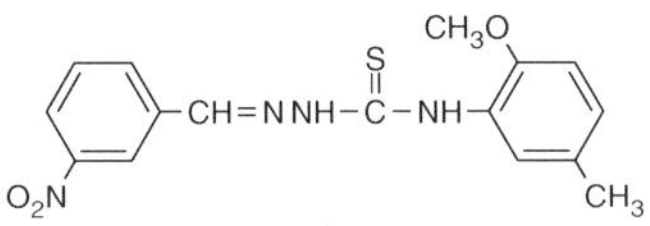
250 mg

S69,383-9
4-(2-METHOXY-5-METHYLPHENYL)-1-(1-NAPHTHYLMETHYLENE)-3-THIOSEMICARBAZIDE
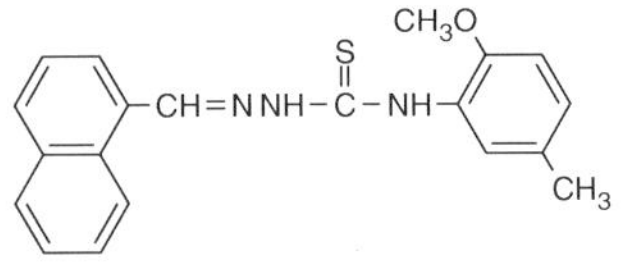
250 mg

S54,918-5
3,4,5-TRIMETHOXYBENZALDEHYDE 4-NITROPHENYLHYDRAZONE
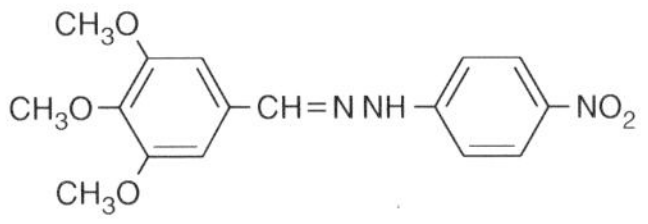
5 gm

S51,805-0
1,2-DIHYDRO-1-(2,3-DIMETHOXYBENZYLIDENEAMINO)-4,6-DIMETHYL-2-OXONICOTINONITRILE
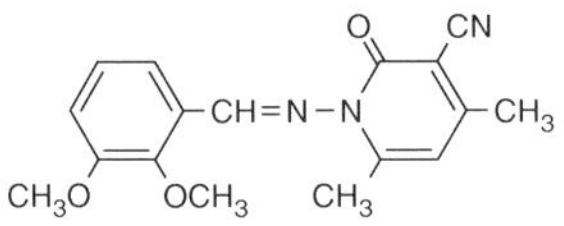
250 mg

S92,531-4
2-METHOXYBENZOIC CINNAMYLIDENEHYDRAZIDE
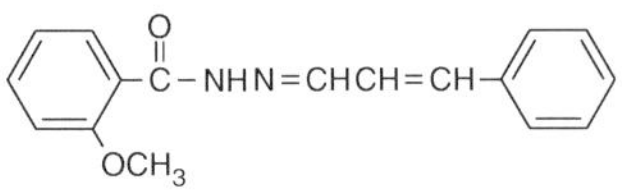
250 mg

S92,551-9
CINNAMIC (4-METHOXYBENZYLIDENE)-HYDRAZIDE
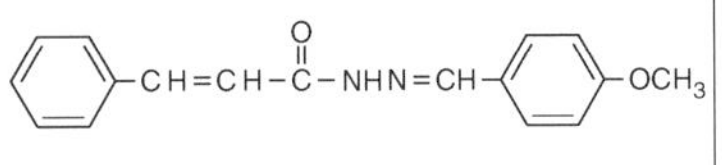
250 mg

S92,478-4
BENZOIC (4-HYDROXY-3-METHOXYBENZYLIDENE)HYDRAZIDE
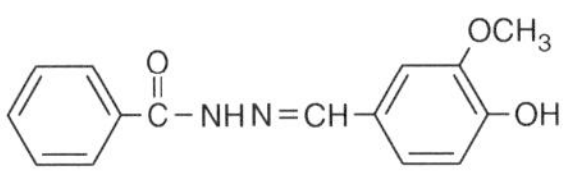
250 mg

S92,528-4
4-METHOXYBENZOIC (2-BUTEN-1-YLIDENE)HYDRAZIDE
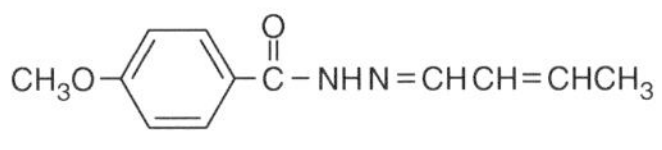
250 mg

S94,872-1
N-(2-THIAZOLYL)OXAMIC (4-METHOXYBENZYLIDENE)HYDRAZIDE
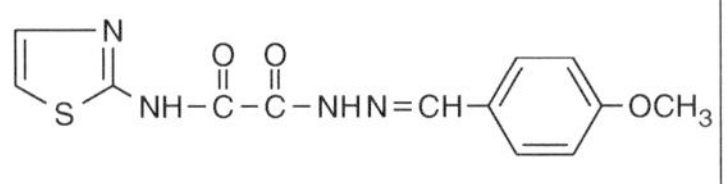
250 mg

S92,443-1
METHYL N-(4-ACETAMIDOBENZYLIDENE)-CARBAZATE
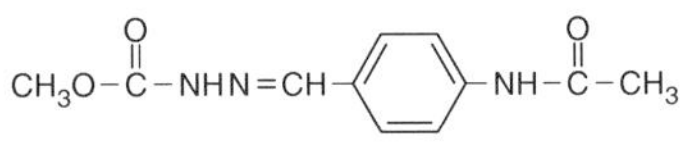
250 mg

S92,448-2
1-(4-METHYL-2-PENTYLIDENE)-4-(4-ETHOXYPHENYL)SEMICARBAZIDE
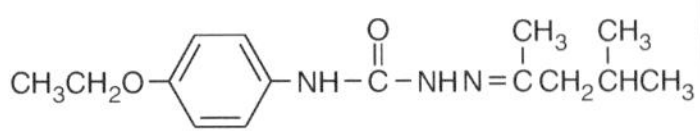
250 mg

S54,820-0
2-THIOPHENECARBOXYLIC ACID (3,4,5-TRIMETHOXYBENZYLIDENE)HYDRAZIDE
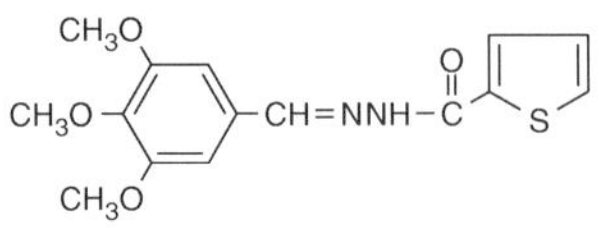
5 gm

S95,654-6
(+/-)-2-ISOPROPYL-5-METHYLCYCLOHEXYL N-DECAHYDRONAPHTH-2-YLIDENECARBAZATE
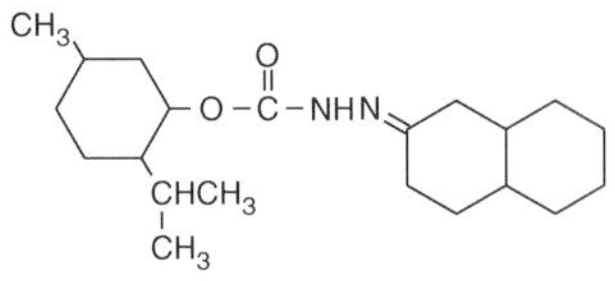
50 mg

S76,710-7
3-(4-METHOXYBENZYLIDENEAMINO)-2-METHYL-2-THIOPSEUDOUREA
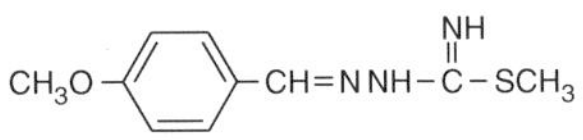
250 mg

S74,363-1
2-METHOXYBENZALDEHYDE N-PHENYL-N-(3-SULFOLANYL)HYDRAZONE
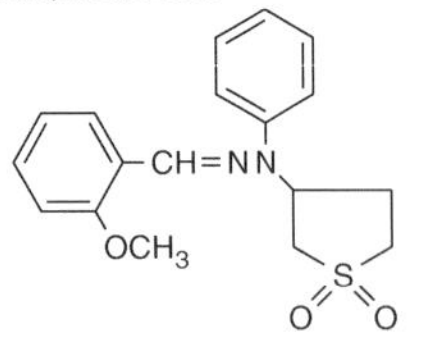
250 mg

S54,940-1
P-TOLUENESULFONIC ACID (4-ETHOXYBENZYLIDENE)HYDRAZIDE
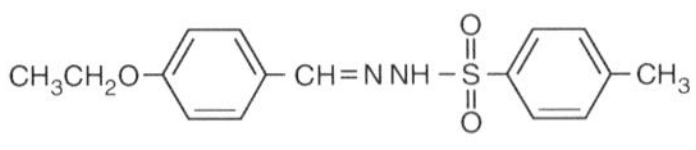
250 mg

S54,514-7
P-TOLUENESULFONIC ACID (5-BROMO-2-METHOXYBENZYLIDENE)HYDRAZIDE
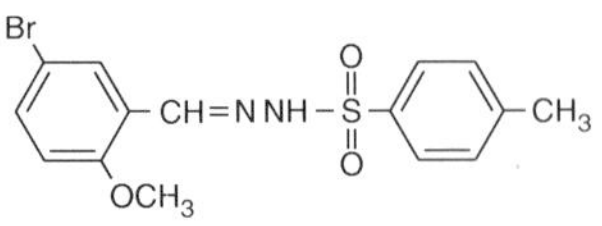
250 mg

S54,519-8
P-TOLUENESULFONIC ACID (5-BROMO-2-ETHOXYBENZYLIDENE)HYDRAZIDE

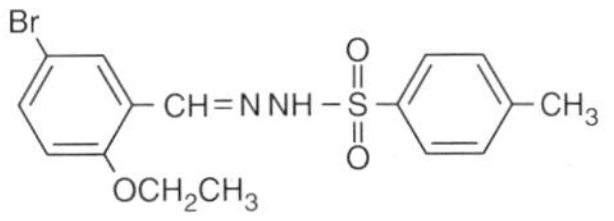

250 mg

S54,606-2
P-TOLUENESULFONIC ACID (3,5-DIBROMO-2-ETHOXYBENZYLIDENE)-HYDRAZIDE

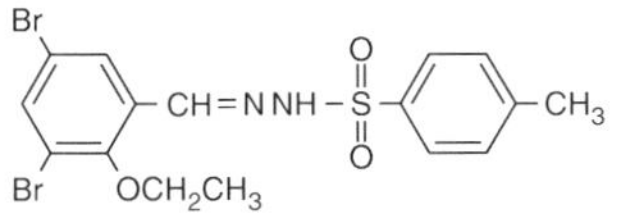

250 mg

S54,817-0
1-(3,4,5-TRIMETHOXYBENZYLIDENE-AMINO)PIPERIDINE

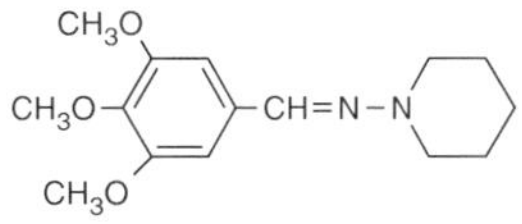

5 gm

S54,823-5
1-(2-METHOXY-1-NAPHTHYLMETHYLENEAMINO)PIPERIDINE

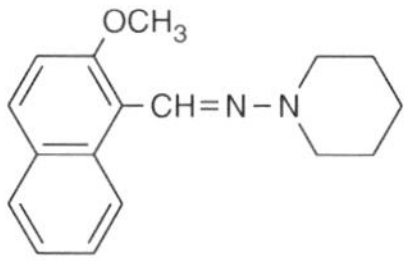

5 gm

S77,185-6
1-[ALPHA-METHYL-3-(2-PHENYLETHOXY)-BENZYLIDENEAMINO]GUANIDINE NITRATE

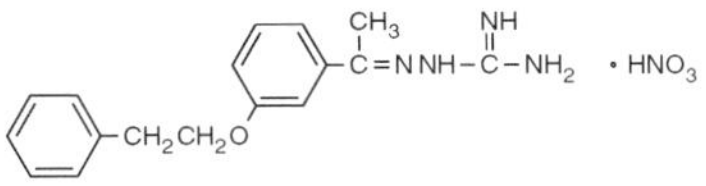

250 mg

S77,188-0
1-[4-(PHENOXY)BENZYLIDENEAMINO]-GUANIDINE NITRATE

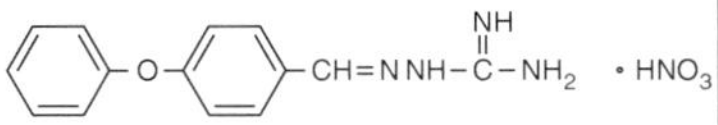

250 mg

S80,099-6
1-(4-BUTOXYBENZYLIDENE)-4-PHENYL-3-THIOSEMICARBAZIDE

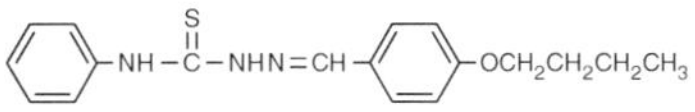

250 mg

S77,192-9
1(2-METHOXYBENZYLIDENEAMINO)-GUANIDIUM NITRATE

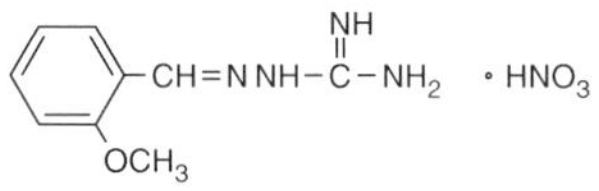

250 mg

S77,134-1
1-[1-(2-METHOXYBENZYL)-ETHYLIDENEAMINO]GUANIDIUM NITRATE

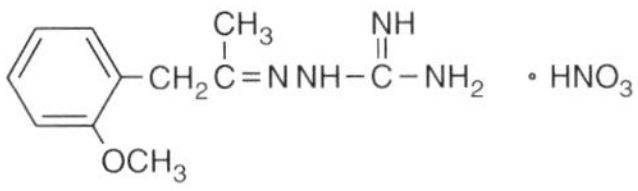

250 mg

S93,442-9
4-(4-METHOXYBENZYLIDENEAMINO)-4H-1,2,4-TRIAZOLE

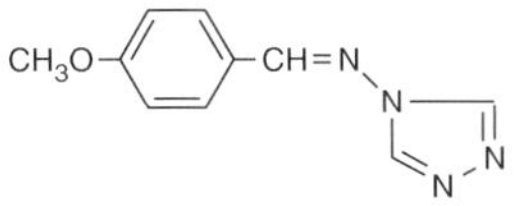

250 mg

S77,178-3
1-[(4-DECYLOXYBENZYLIDENE)AMINO]-GUANIDIUM NITRATE

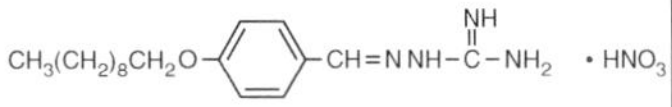

250 mg

S54,920-7
4-ALLYL-1-(3,4,5-TRIMETHOXYBENZYLIDENE)-3-THIOSEMICARBAZIDE

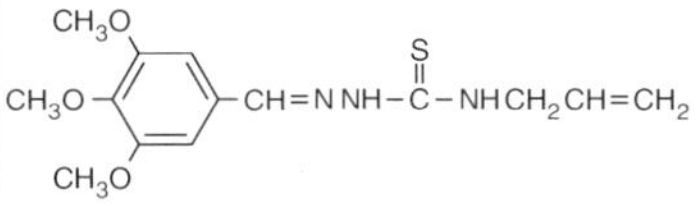

5 gm

S58,991-8
2,5-DIMETHOXY-N-(2-THIENYLMETHYLENE)ANILINE

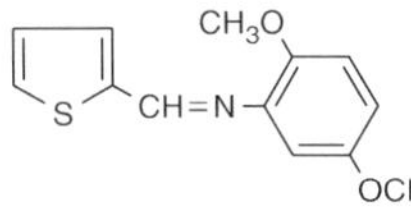

250 mg

S58,976-4
5-CHLORO-2,4-DIMETHOXY-N-(2-THIENYLMETHYLENE)ANILINE

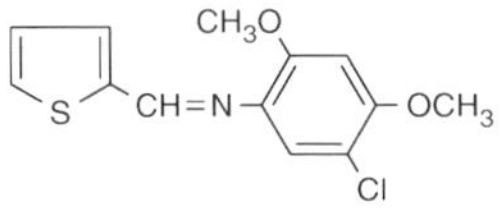

5 gm

S34,612-8
DIETHYL 2-(2-METHOXYPHENYLIMINO)-SUCCINATE

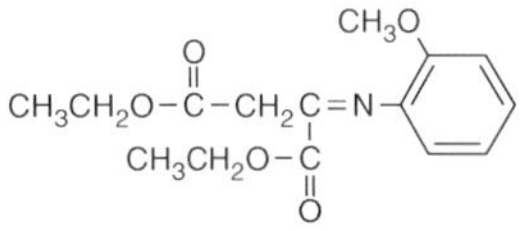

5 gm

S68,655-7
2-PYRIDYLMETHYLENEAMINO N-(2-METHOXYPHENYL)CARBAMATE

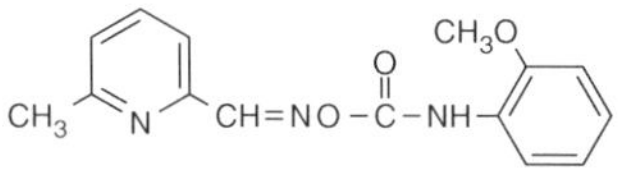

5 gm

S60,926-9
N-FURFURYLIDENE-M-ANISIDINE

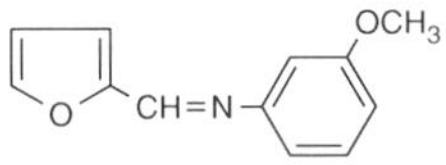

250 mg

S61,558-7
N-(FURFURYLIDENE)-P-ANISIDINE

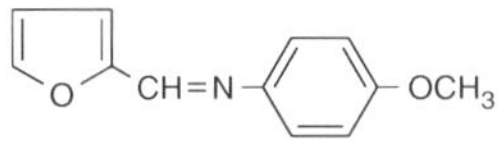

250 mg

S60,930-7
N-FURFURYLIDENE-P-PHENETIDINE

250 mg

S60,923-4
N-(CINNAMYLIDENE)-O-ANISIDINE

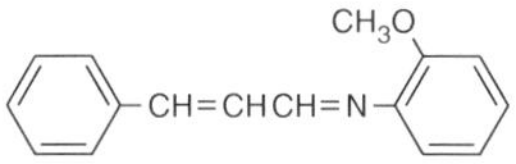

250 mg

S58,447-9
N,N'-(1,4-PHENYLENEDIMETHYLIDYNE)-DI-O-ANISIDINE

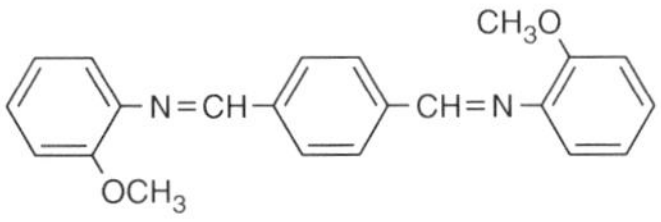

250 mg

S60,124-1
N-(9-ANTHRACENYLMETHYLENE)-O-ANISIDINE

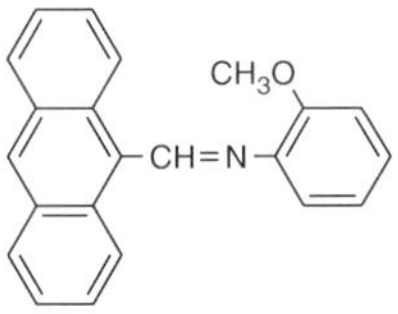

5 gm

S58,453-3
N,N'-(1,4-PHENYLENEDIMETHYLIDYNE)-BIS(5-CHLORO-2,4-DIMETHOXYANILINE)

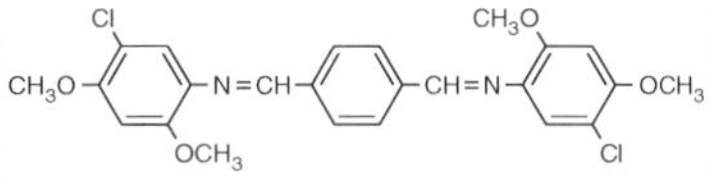

5 gm

S62,616-3
4-(2-THIENYLMETHYLENEAMINO)PHENYL ETHER

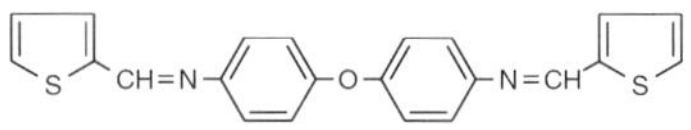

5 gm

S56,820-1
ETHYL 3-ETHOXY-3-(3-NITROPHENYLIMINO)PROPIONATE

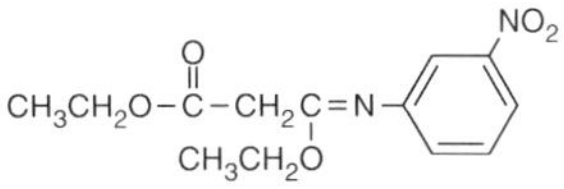

250 mg

S74,926-5
2,5-DINITRO-9-(4-METHOXYPHENYLIMINO)FLUORENE

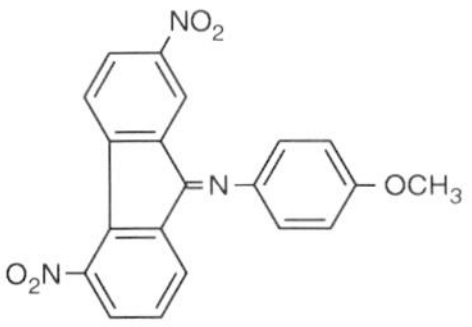

250 mg

S56,546-6
4'-(4-BUTOXYBENZYLIDENEAMINO)-PROPIOPHENONE

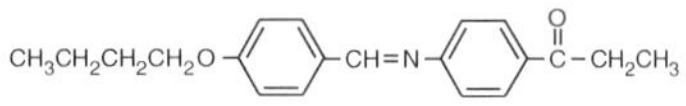

5 gm

S62,127-7
ETHYL 3-(2-CARBOXYPHENYLIMINO)-3-ETHOXYPROPIONATE

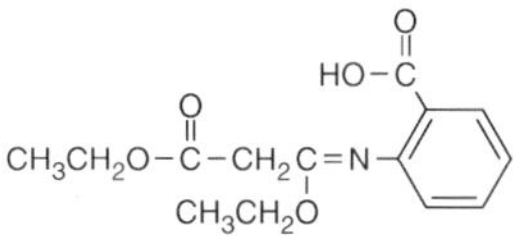

250 mg

S35,802-9
4-(ANISYLIDENEAMINO)CINNAMIC ACID

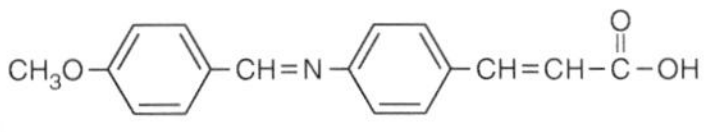

5 gm

S53,705-5
N-(4-METHOXYBENZYLIDENE)-ANTHRANILIC ACID

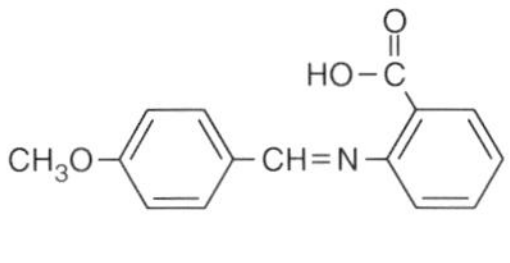

5 gm

S53,650-4
4-(4-METHOXYBENZYLIDENEAMINO)-BENZOIC ACID

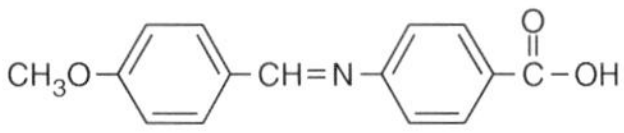

5 gm

S69,838-5
BIS[2-METHOXY-6-(4-METHYL-2-PYRIDINYLIMINOMETHYL)PHENOLATO] COPPER

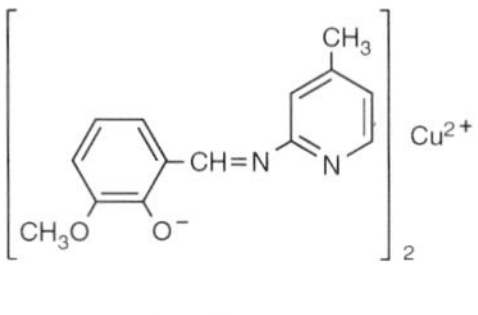

5 gm

S88,060-4
N-(4-METHOXYBENZYLIDENE)GLYCINE METHYL ESTER

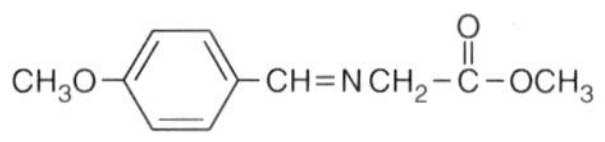

250 mg

S65,540-6
ETHYL 3-(2,6-DICHLOROPHENYLIMINO)-3-METHOXYPROPIONATE

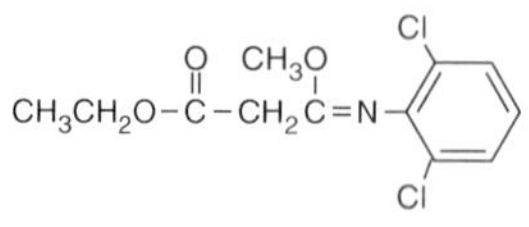

5 gm

S70,116-5
N-(4-ACETAMIDOBENZYLIDENE)-4-PHENOXYANILINE

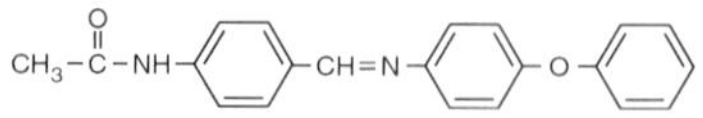

5 gm

S53,782-9
3-BENZYLOXYBENZALDEHYDE AZINE

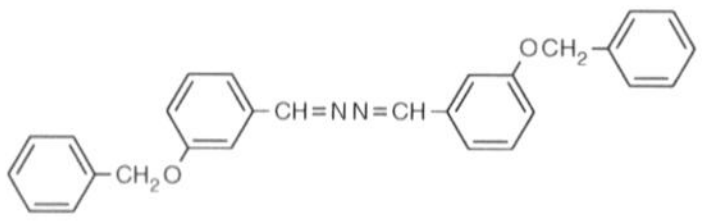

5 gm

S71993-5
2,3-DIMETHYL-4-(2-METHOXYBENZYLIDENEAMINO)-1-PHENYL-3-PYRAZOLIN-5-ONE

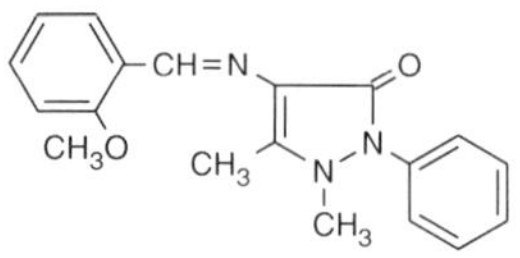

5 gm

S71,992-7
2,3-DIMETHYL-4-(4-METHOXYBENZYLIDENEAMINO)-1-PHENYL-3-PYRAZOLIN-5-ONE

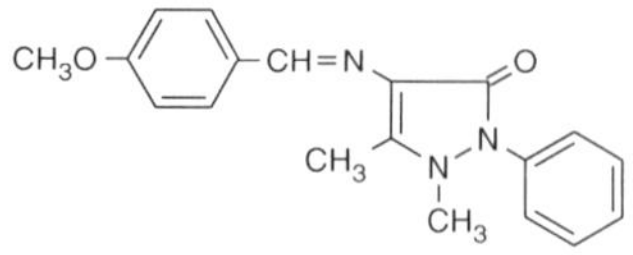

5 gm

S76,404-3
4-(4-BENZYLOXYBENZYLIDENEAMINO)-2,3-DIMETHYL-1-PHENYL-3-PYRAZOLIN-5-ONE

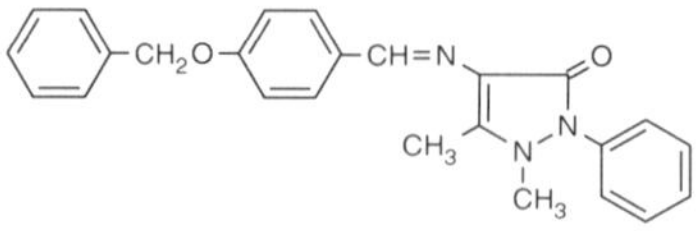

5 gm

S68,494-5
3-PYRIDYLMETHYLENEAMINO N-(2-METHOXYPHENYL)CARBAMATE

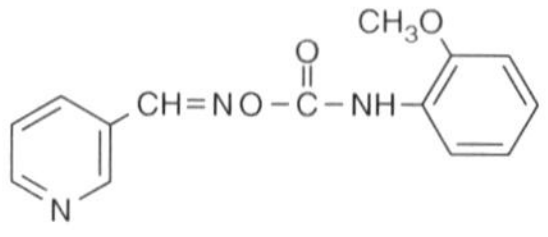

250 mg

S45,063-4
N-(4-CYANOBENZYLIDENE)-P-ANISIDINE

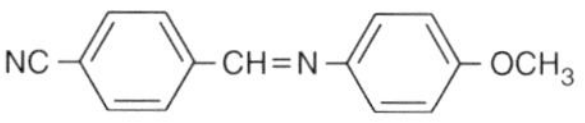

250 mg

S65,243-1
4-[N-(2-ETHOXYBENZYLIDENE)-AMINO]PHENYL SULFONE

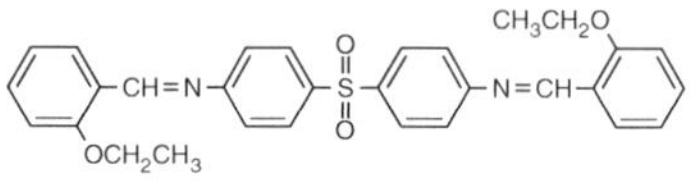

250 mg

S65,242-3
4-[N-(4-METHOXYBENZYLIDENE)-AMINO]PHENYL SULFONE

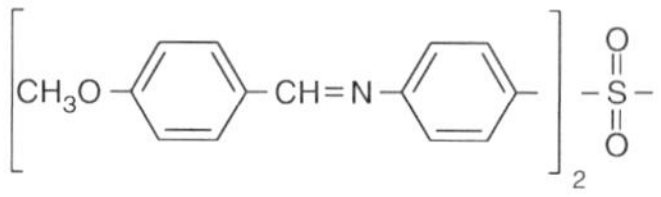

5 gm

S68,488-0
4-PHENOXY-N-(2-THIENYLMETHYLENE)-ANILINE

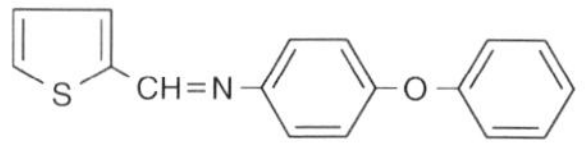

5 gm

S69,810-5
N-(3,4-METHYLENEDIOXYBENZYLIDENE)-4-PHENOXYANILINE

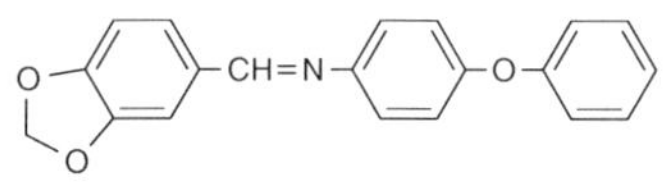

5 gm

S89,739-6
5,5-DIMETHYL-3-METHOXY-2-CYCLOHEXEN-1-IMINE METHOSULFATE

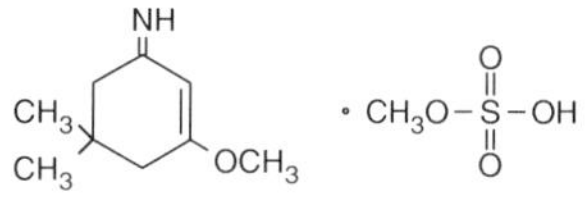

250 mg

S68,624-7
N-BENZYLIDENE-4-PHENOXYANILINE

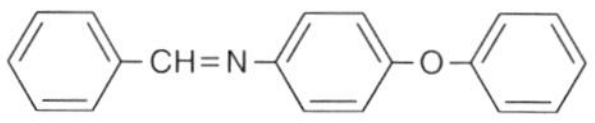

5 gm

S68,625-5
N-(4-METHYLBENZYLIDENE)-4-PHENOXYANILINE

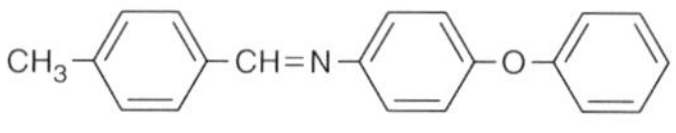

5 gm

S68,486-4
N-(4-METHOXYBENZYLIDENE)-4-PHENOXYANILINE

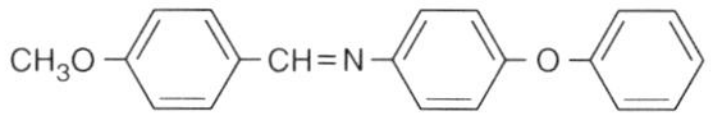

5 gm

S52,396-8
N-(4-FLUOROBENZYLIDENE)-P-ANISIDINE

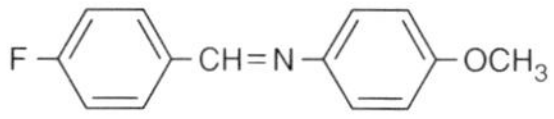

5 gm

S53,021-2
N-(4-CHLOROBENZYLIDENE)-P-ANISIDINE

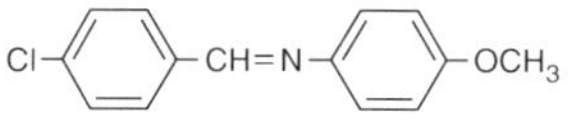

5 gm

S52,344-5
5-CHLORO-2,4-DIMETHOXY-N-(4-FLUOROBENZYLIDENE)ANILINE

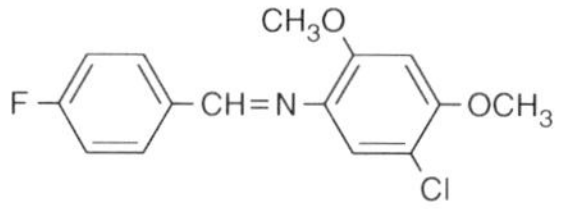

5 gm

S62,398-9
N-(CINNAMYLIDENE)-P-ANISIDINE

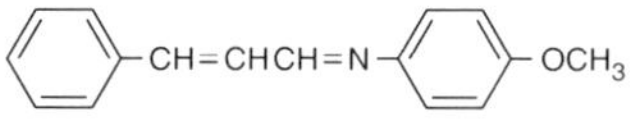

250 mg

S88,101-5
(4-METHOXYBENZYL)(4-METHYLBENZYLIDENE)AMINE

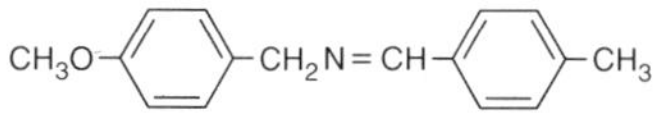

250 mg

S53,004-2
P-ANISALDEHYDE AZINE

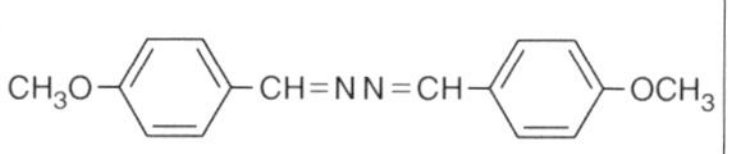

250 mg

S59,586-1
N-CINNAMYLIDENE-P-PHENETIDINE

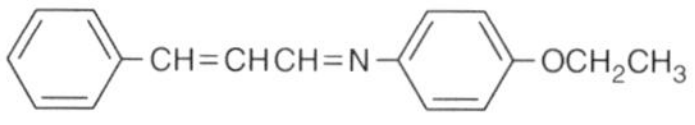

250 mg

S67,870-8
N-(4-ISOPROPYLBENZYLIDENE)-P-ANISIDINE

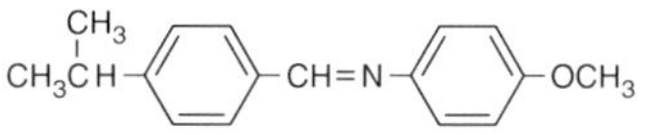

5 gm

S67,863-5
N-(4-ISOPROPYLBENZYLIDENE)-P-PHENETIDINE

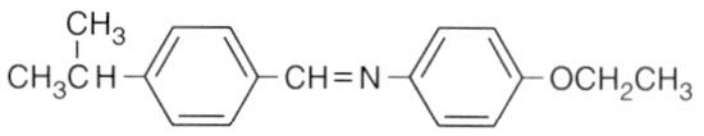

5 gm

S62,602-3
N-(CINNAMYLIDENE)-4-PHENOXYANILINE

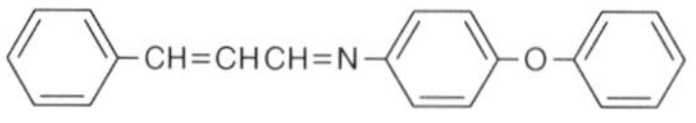

250 mg

S35,556-9
N,N'-TEREPHTHALYLIDENEDI-BIS-(M-ANISIDINE)

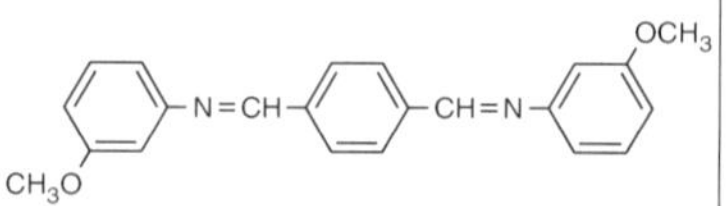

5 gm

S35,557-7
N,N'-TEREPHTHALYLIDENEDI-P-ANISIDINE

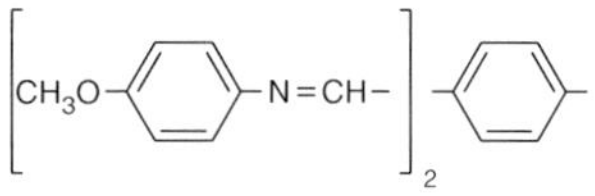

5 gm

S58,419-3
N,N'-(1,4-PHENYLENEDIMETHYLIDYNE)-DI-P-PHENETIDINE

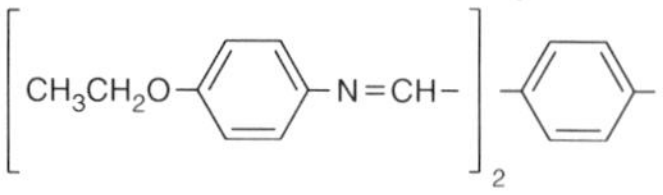

5 gm

S34,137-1
2-(N-P-ANISYLIDENEAMINO)FLUORENE

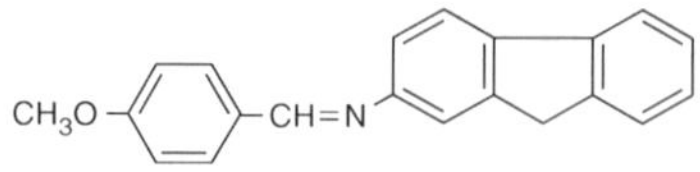

250 mg

S54,513-9
5-BROMO-O-ANISALDEHYDE AZINE

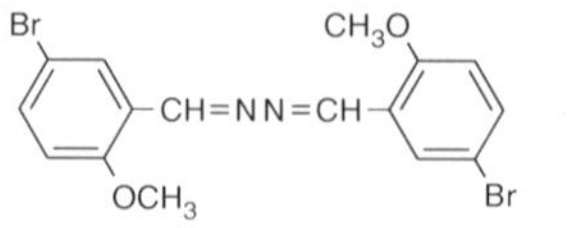

250 mg

S35,347-7
6-METHYL-2-PYRIDINEALDOXIME O-BENZYL ETHER

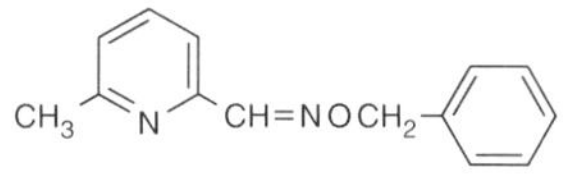

5 gm

S64,066-2
N-(2-CHLORO-5-NITROBENZYLIDENE)-2,5-DICHLOROANILINE

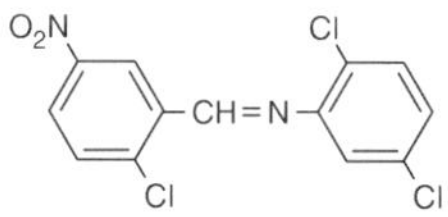

250 mg

S61,368-1
N-(9-ANTHRACENYLMETHYLENE)-2,3-DICHLOROANILINE

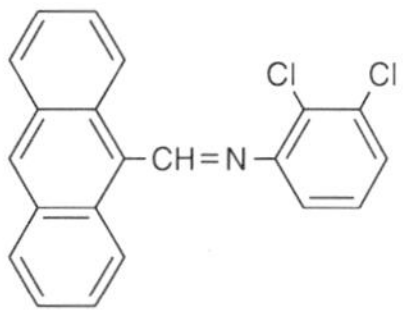

5 gm

S61,366-5
N-(9-ANTHRACENYLMETHYLENE)-3,4-DICHLOROANILINE

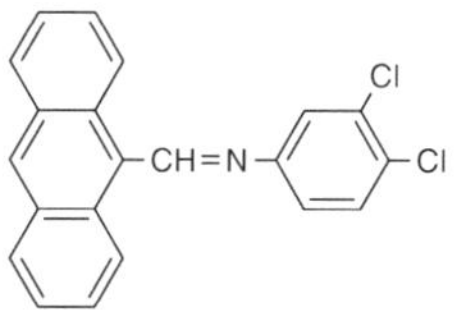

5 gm

S61,365-7
N-(9-ANTHRACENYLMETHYLENE)-3,5-DICHLOROANILINE

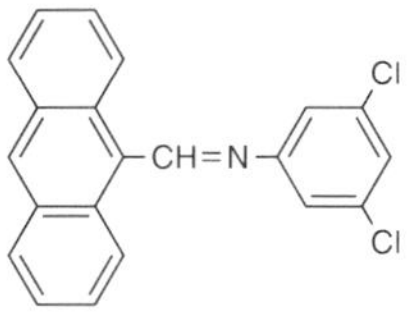

250 mg

S63,890-0
4-[4-(DIMETHYLAMINO)PHENYLIMINO]-5-[3-(FLUOROSULFONYL)BENZAMIDO]-2-(FURAMIDO)-2,5-CYCLOHEXADIEN-1-ONE

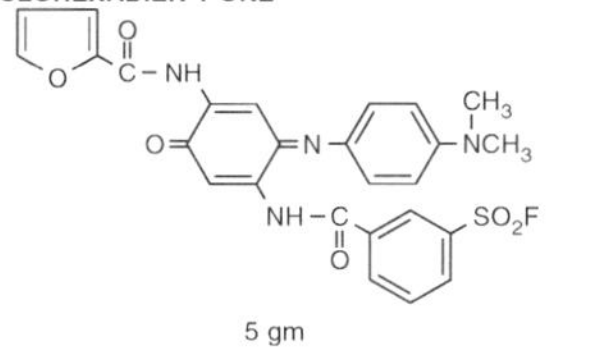

5 gm

S63,958-3
2-METHYL-5-NITRO-N-(PIPERONYLIDENE)ANILINE

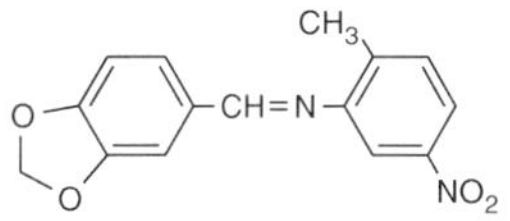

5 gm

S63,974-5
2-METHYL-4-NITRO-N-(PIPERONYLIDENE)ANILINE

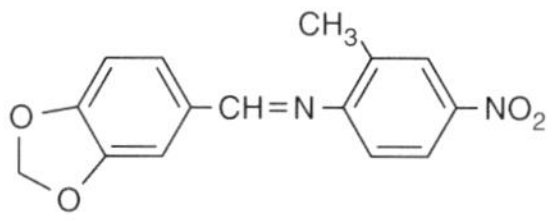

5 gm

S63,970-2
4-(ETHOXYCARBONYL)-N-(PIPERONYLIDENE)ANILINE

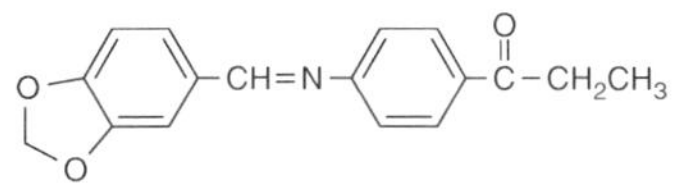

5 gm

S62,093-9
N-PIPERONYLIDENE-2,3-XYLIDINE

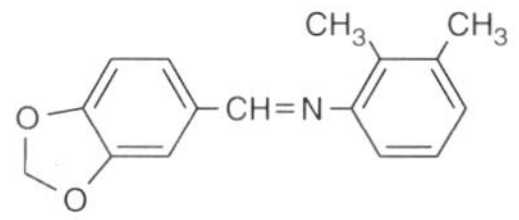

250 mg

S62,095-5
N-PIPERONYLIDENE-2,5-XYLIDINE

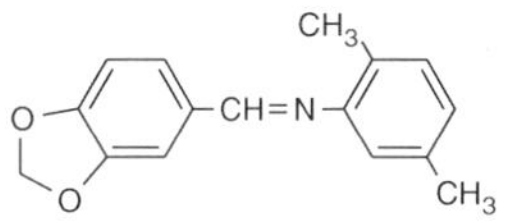

250 mg

S51,451-9
1,2-DIHYDRO-4,6-DIMETHYL-1-(ALPHA-METHYLTRIMETHOXYBENZYLIDENEAMINO)-2-OXONICOTINONITRILE

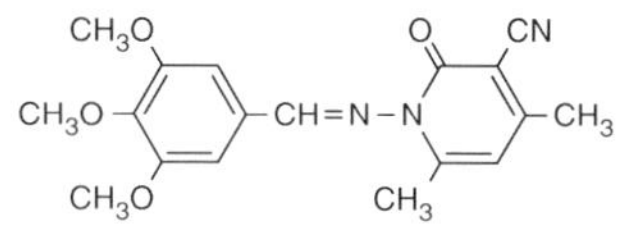

250 mg

S92,540-3
TETRAHYDROFUROIC (2-THENYLIDENE)-HYDRAZIDE

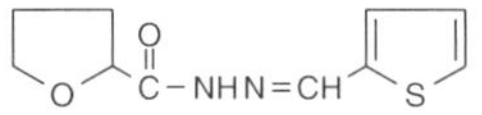

250 mg

S50,630-3
ADIPIC ACID (3,4,5-TRIMETHOXYBENZYLIDENE)HYDRAZIDE

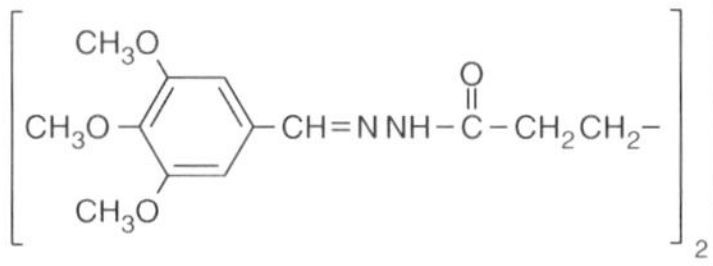

250 mg

S50,631-1
PIMELIC ACID (3,4,5-TRIMETHOXYBENZYLIDENE)HYDRAZIDE

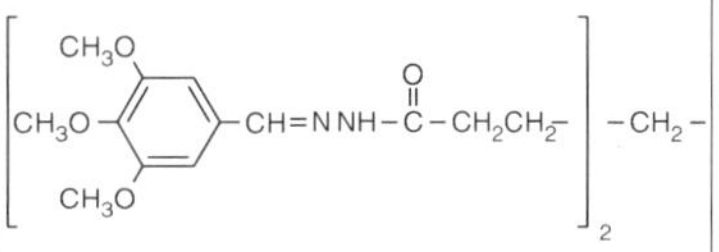

250 mg

S50,666-4
AZELAIC ACID (3,4,5-TRIMETHOXYBENZYLIDENE)HYDRAZIDE

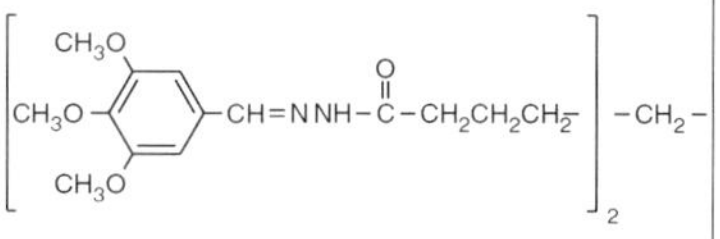

250 mg

S50,667-2
DODECANEDIOIC ACID (3,4,5-TRIMETHOXYBENZYLIDENE)HYDRAZIDE

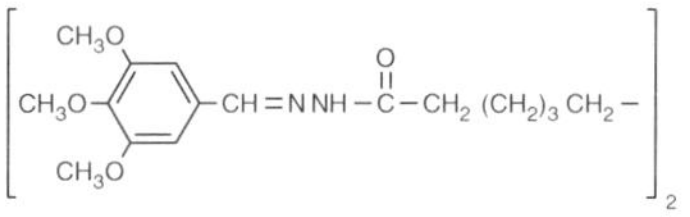

250 mg

S50,987-6
1,4-BIS(3,4,5-TRIMETHOXYBENZYLIDENE)AMINO-PIPERAZINE

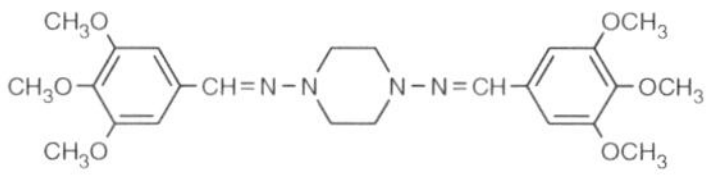

250 mg

S51,460-8
N1-(2-THIAZOLYL)-N4-(3,4,5-TRIMETHOXYBENZYLIDENE)SULFANILAMIDE

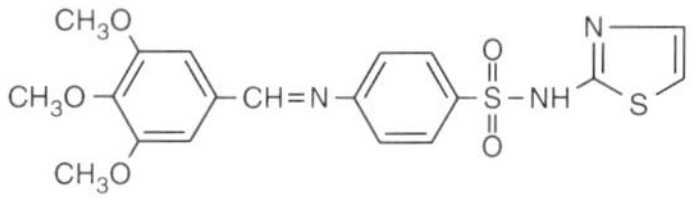

250 mg

S92,538-1
N-METHYLANTHRANILIC FURFURYLIDENEHYDRAZIDE

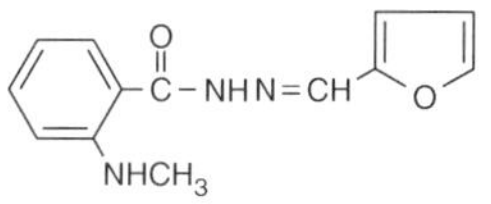

250 mg

S79,874-6
2-ACETYLPHENYL N-PHENYLBENZIMIDATE

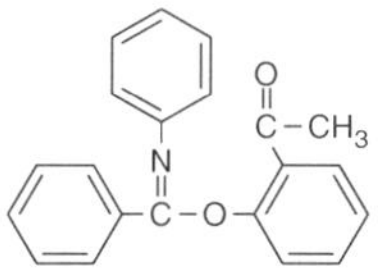

250 mg

S88,687-4
N-(2,4-DINITROBENZYLIDENE)-2-MERCAPTOANILINE

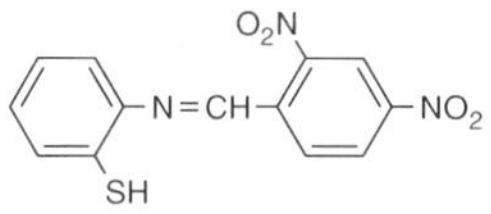

250 mg

S61,936-1
N-(CINNAMYLIDENE)-3-(METHYLTHIO)-ANILINE

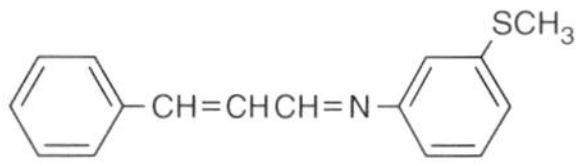

250 mg

S50,765-2
N,N-DIMETHYL-ALPHA-[4-(METHYLTHIO)-PHENYLIMINO]-P-TOLUIDINE

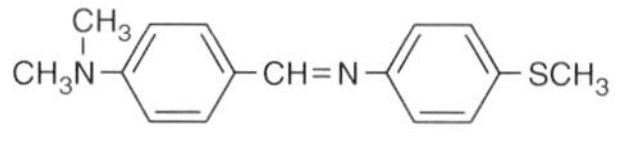

5 gm

S57,796-0
1-[ALPHA-(CARBOXYMETHYLTHIO)-3-CHLOROBENZYLIDENE]-PIPERIDINIUM BROMIDE

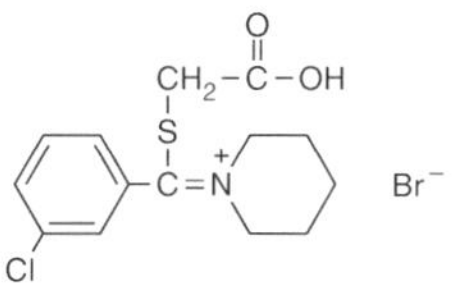

5 gm

S58,273-5
4-[ALPHA-(CARBOXYMETHYLTHIO)-2-CHLOROBENZYLIDENE]-MORPHOLINIUM BROMIDE

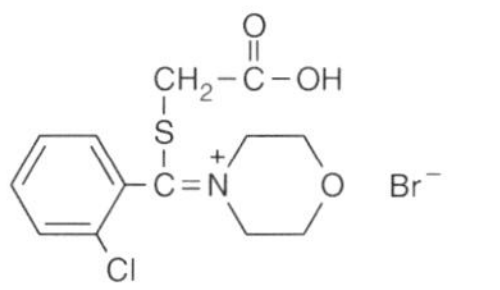

250 mg

S63,530-8
N-(4-CHLOROBENZYLIDENE)-3-(METHYLTHIO)ANILINE

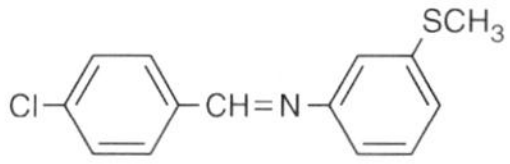

250 mg

S56,885-6
4-NITROBENZALDEHYDE 2-(OCTADECYLTHIO)PHENYLHYDRAZONE

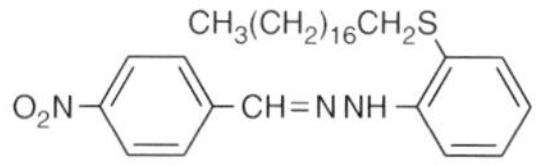

250 mg

S40,856-5
ETHYL N-PHENYLTHIOBUTYRIMIDATE

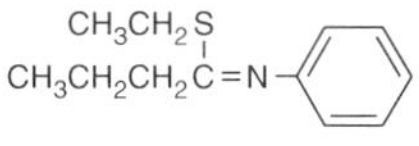

5 gm

S35,180-6
6-ISOTHIOUREIDO-4-METHYL-2-(METHYLTHIO)PYRIMIDINE HYDROCHLORIDE

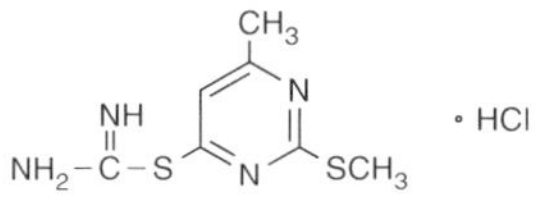

5 gm

S39,912-4
4-METHYLTHIO-N-PIPERONYLIDENEANILINE

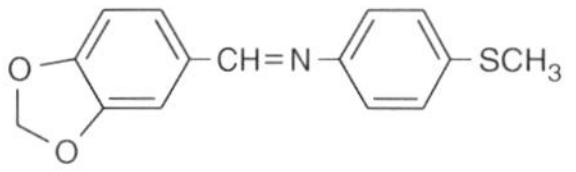

5 gm

S50,330-4
N-CINNAMYLIDENE-4-(METHYLTHIO)-ANILINE

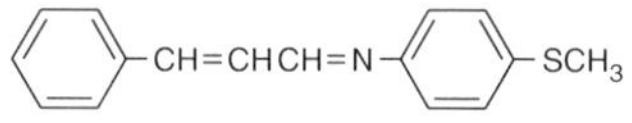

5 gm

S69,014-7
4,4'-DITHIOBIS[N-(4-ACETAMIDOBENZYLIDENE)ANILINE]

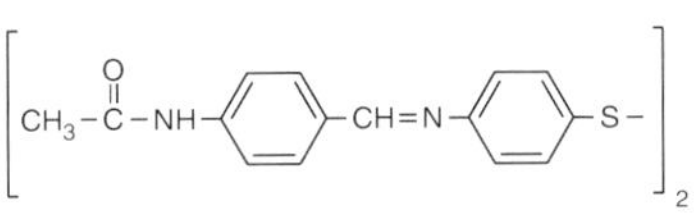

5 gm

S67,850-3
4,4'-DITHIOBIS[N-(4-DIMETHYLAMINOBENZYLIDENE)ANILINE]

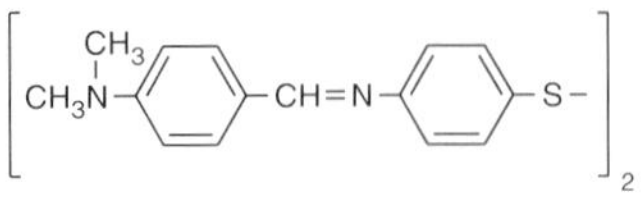

5 gm

S67,231-9
4-(PIPERONYLIDENEAMINO)PHENYL DISULFIDE

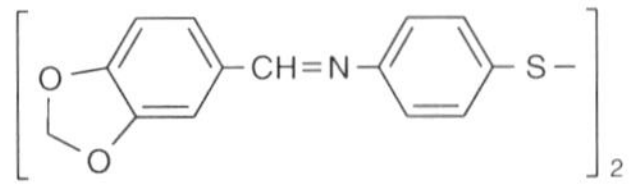

5 gm

S67,233-5
4-(BENZYLIDENEAMINO)PHENYL DISULFIDE

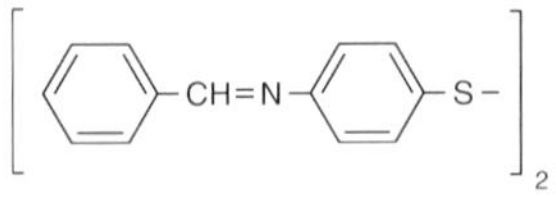

5 gm

S67,859-7
4,4'-DITHIOBIS[N-(4-METHYLBENZYLIDENE)ANILINE]

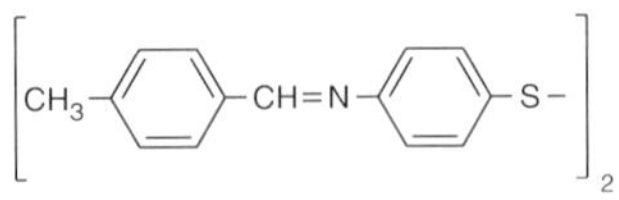

5 gm

S67,858-9
4,4'-DITHIOBIS[N-(4-ISOPROPYLBENZYLIDENE)ANILINE]

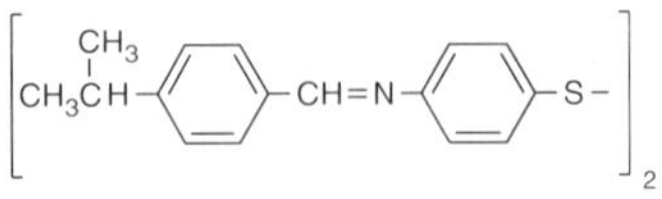

5 gm

S67,854-6
4,4'-DITHIOBIS(N-CINNAMYLIDENEANILINE)

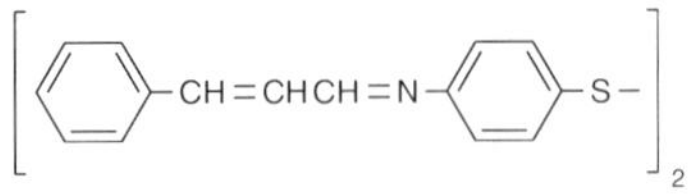

5 gm

S61,640-0
4-(4,6-DIAMINO-M-TOLYL)IMINO-2,5-(CYCLOHEXADIEN-1-YLIDENE)DI-METHYLAMMONIUM CHLORIDE

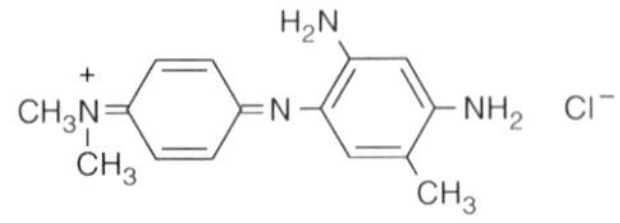

250 mg

S44,692-0
N-(2,4-DIAMINO-5-METHYLPHENYL)-1,4-BENZOQUINONEIMINE TRIHYDRATE

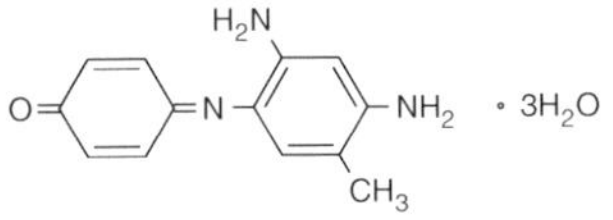

5 gm

S65,560-0
3-AMINOBENZENESULFONIC [6-(FLUORO-SULFONYL)-1-METHYL-2-PHENYL-4(1H)-QUINOLYLIDENE]HYDRAZIDE

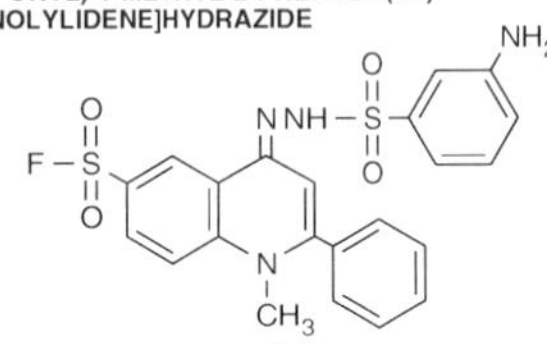

5 gm

S63,134-5
1-HEXADECANESULFONIC ACID 2-(4-AMINOPHENYL)-1-METHYL-4(1H)-QUINOLYLIDENE HYDRAZIDE

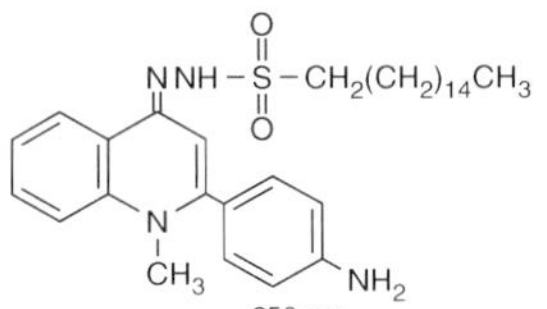

250 mg

S65,611-9
3'-AMINOACETOPHENONE AZINE WITH 3-METHYL-2-BENZOTHIAZOLINONE

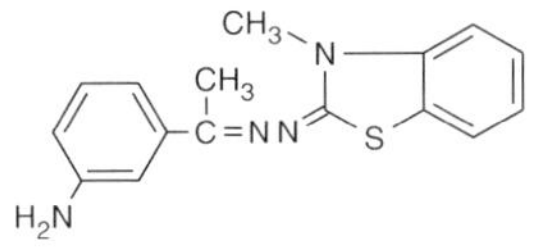

250 mg

S65,930-4
3-[(3-[(3-AMINOBENZYLIDENE)AMINO]-BENZYLIDENE)AMINO]BENZALDEHYDE

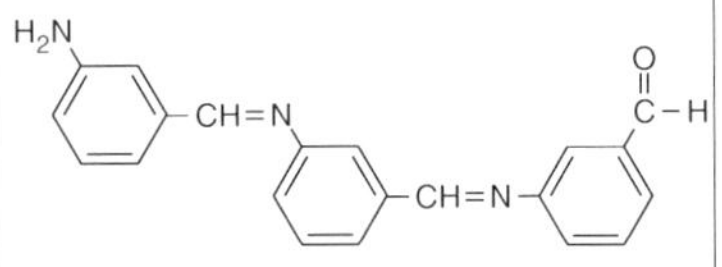

250 mg

S75,508-7
2-AMINO-9-FLUORENONE HYDRAZONE

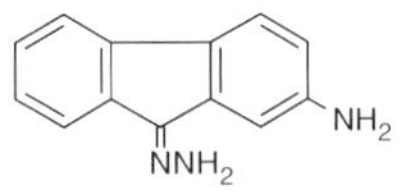

100 mg

S94,430-0
BASIC DYE (C.I. 49400)

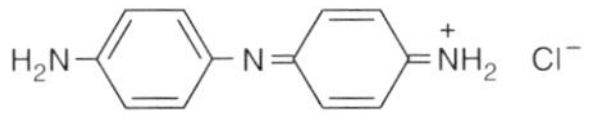

250 mg

S92,898-4
N1-(2-NITROBENZYLIDENE)-5-NITRO-1,2-PHENYLENEDIAMINE

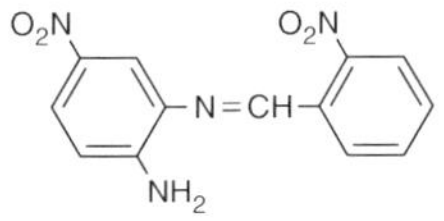

250 mg

S92,530-6
4-AMINOBENZOIC CINNAMYLIDENEHYDRAZIDE

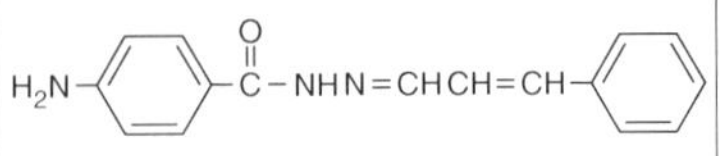

250 mg

S91,286-7
4-AMINOBENZOIC [1-(2-NAPHTHYL)-ETHYLIDENE]HYDRAZIDE

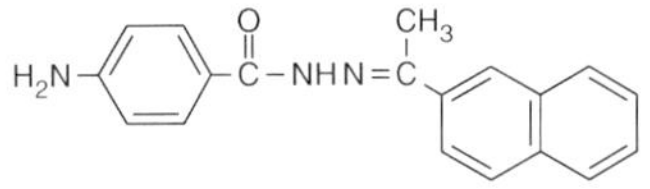

250 mg

S40,633-3
4-AMINOBENZALDEHYDE THIOSEMICARBAZONE

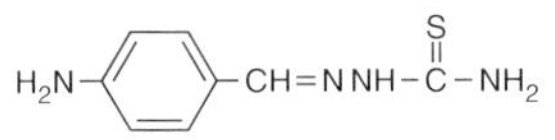

5 gm

S92,504-7
4-AMINOBENZOIC (1-NAPHTHYLMETHYLENE)HYDRAZIDE

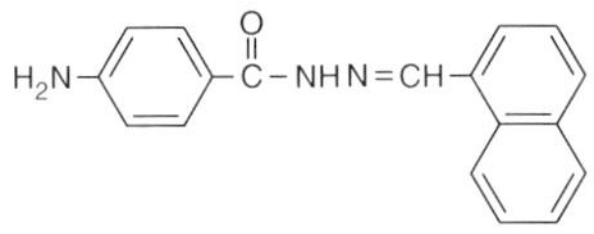

250 mg

S53,655-5
PIPERONAL THIOSEMICARBAZONE

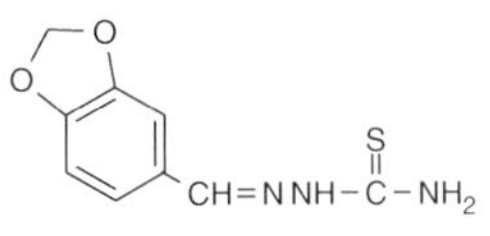

250 mg

S56,585-7
RHODAMINE SULFATE

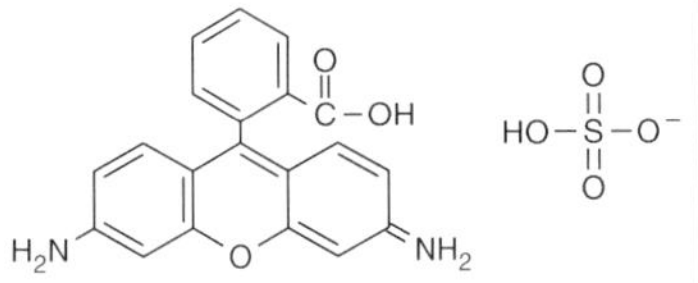

5 gm

S80,499-1
6-AMINO-1,3-DIMETHYL-5-(METHYLIMINOMETHYL)URACIL

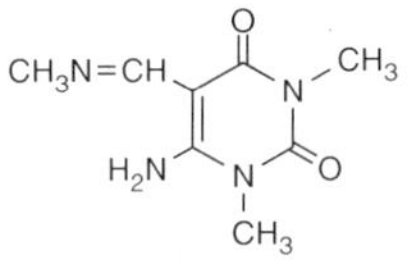

250 mg

S80,378-2
6-AMINO-1,3-DIETHYL-5-(METHYLIMINOMETHYL)URACIL

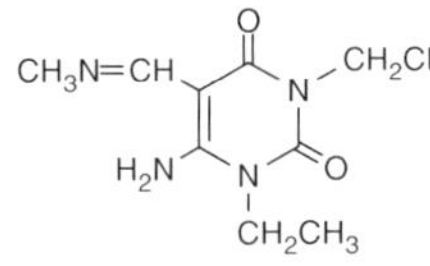

250 mg

S54,076-5
N,N'-BIS(1-ISOPROPYL-2-METHYLPROPYLIDENE)ETHYLENEDIAMINE

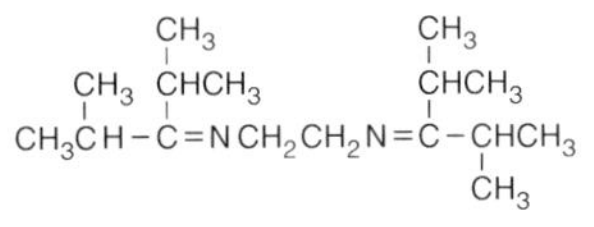

5 gm

S64,815-9
2-CHLOROBENZALDEHYDE 2,4-DINITROPHENYLHYDRAZONE

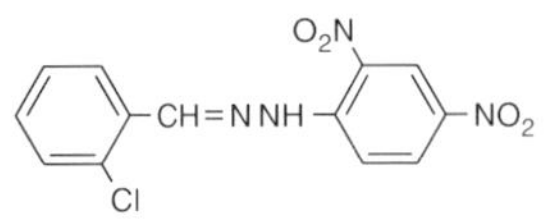

250 mg

S64,802-7
4-NITROBENZALDEHYDE 2,4-DINITROPHENYLHYDRAZONE

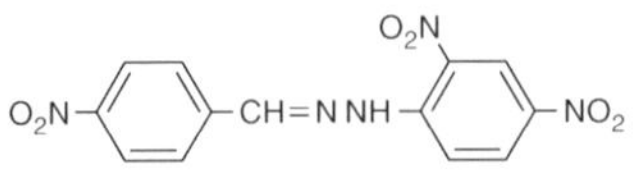

5 gm

S64,804-3
HEPTALDEHYDE 2,4-DINITROPHENYLHYDRAZONE

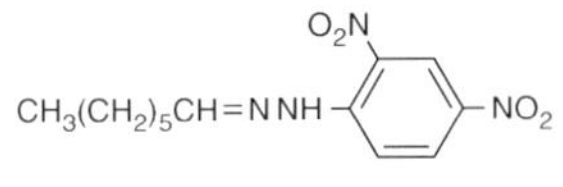

5 gm

S64,807-8
OCTYL ALDEHYDE 2,4-DINITROPHENYLHYDRAZONE

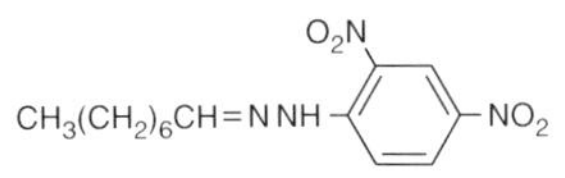

250 mg

S74,786-6
3-NITROBENZALDEHYDE (2,4-DINITROPHENYL)METHYLHYDRAZONE

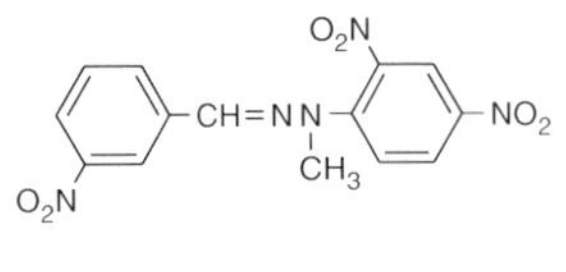

5 gm

S64,622-9
4-BROMOBENZALDEHYDE, 2,4-DINITROPHENYLHYDRAZONE

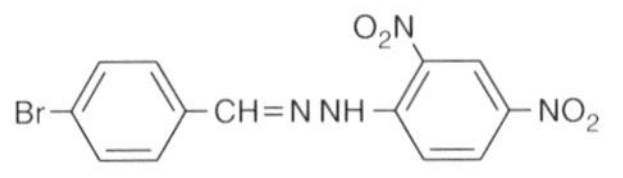

5 gm

S64,641-5
P-TOLUALDEHYDE, 2,4-DINITROPHENYLHYDRAZONE

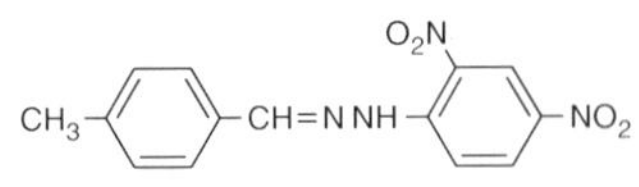

250 mg

S51,204-4
5-INDANCARBOXALDEHYDE 4-(2,4-DINITROPHENYL)SEMICARBAZONE

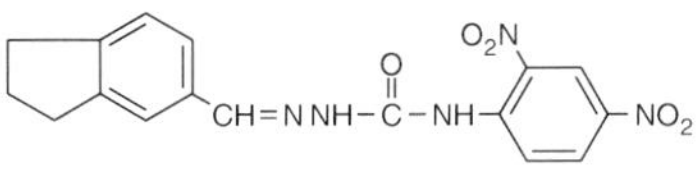

250 mg

S74,771-8
N-CINNAMYLIDENE-2,4-DINITROANILINE

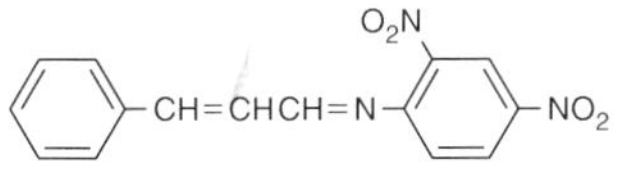

5 gm

S91,532-7
2-(1-METHYL-5-NITRO-2-IMIDAZOLYL)-ACETALDEHYDE SEMICARBAZONE

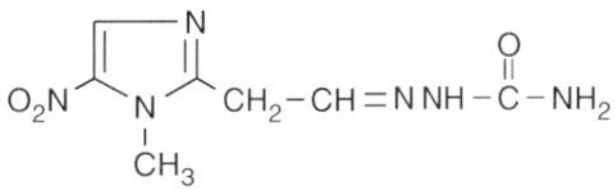

250 mg

S86,201-0
2-PIPERIDINO-N-(2-PIPERIDINOBENZYLIDENE)ANILINE

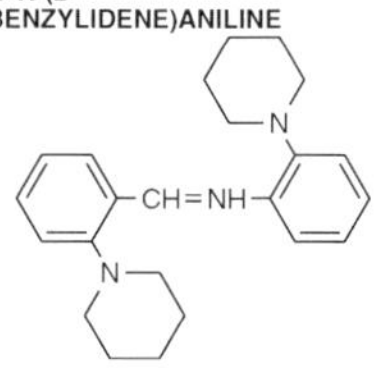

250 mg

S69,747-8
3-(2,4-DINITROPHENYLHYDRAZONOMETHYL)-CAMPHOR

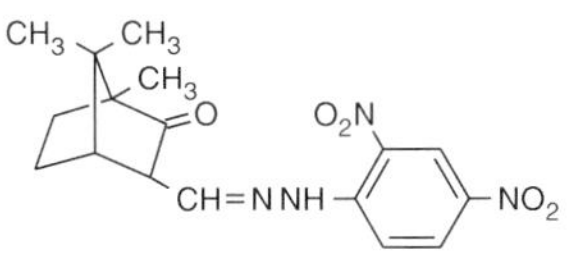

250 mg

S90,230-6
2-[3-(3-NITROPHENYL)UREIDO]ACETIC (2-THIENYLMETHYLENE)HYDRAZIDE

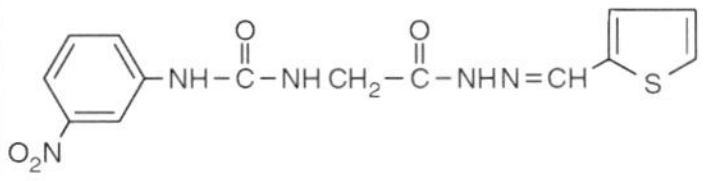

250 mg

S42,920-1
2-PYRIDINEALDAZINE

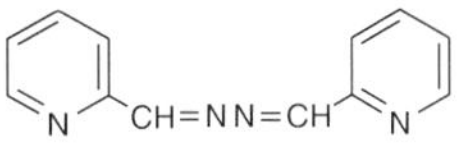

5 gm

S87,893-6
N,N'-BIS(4-ANILINOPHENYL)-1,4-DIAZABUTADIENE

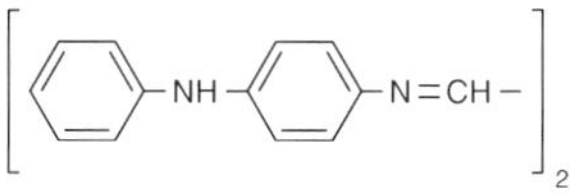

5 gm

S69,298-0
2-(4-DIMETHYLAMINOPHENYLIMINOMETHYL)-1-METHYLBENZO(F)QUINOLINIUM IODIDE

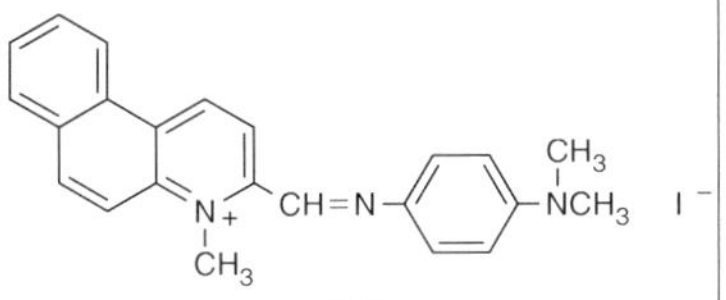

5 gm

S51,106-4
N,N'-BIS(3-NITROBENZYLIDENE)-ETHYLENEDIAMINE

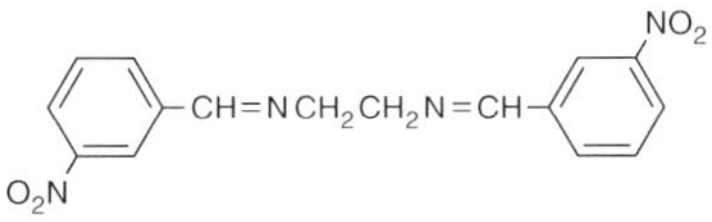

5 gm

S41,708-4
N-(2-BENZYLIDENEAMINO-1,2-DIPHENYLETHYL)BENZAMIDE

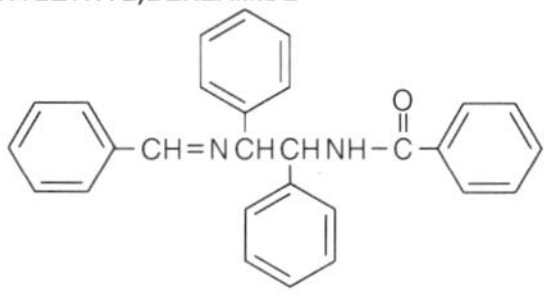

250 mg

S90,226-8
2-(3-FURFURYLUREIDO)ACETIC (2-PHENYLETHYLIDENE)HYDRAZIDE

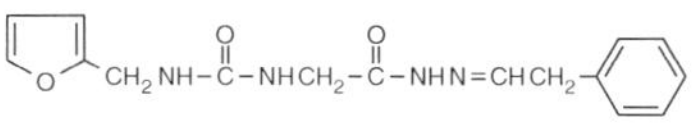

250 mg

S90,224-1
2-(3-FURFURYLUREIDO)ACETIC [4-(DIMETHYLAMINO)BENZYLIDENE]-HYDRAZIDE

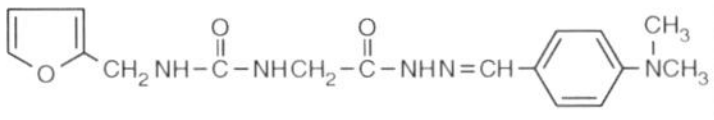

250 mg

S50,177-8
N,N'-BIS(4-DIMETHYLAMINOBENZYLIDENE)-ETHYLENEDIAMINE

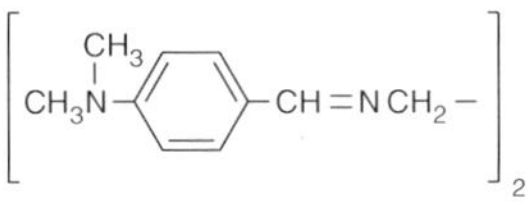

5 gm

S87,315-2
1-METHYL-3-PIPERIDINONE (4-NITROBENZYLIDENE)HYDRAZONE

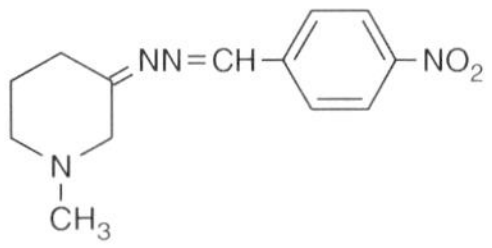

250 mg

S87,496-5
3-METHYL-2-BENZOTHIAZOLINONE (1,2-DIHYDRO-2-IMINO-1-NAPHTHYLIDENE)HYDRAZONE HYRIODIDE

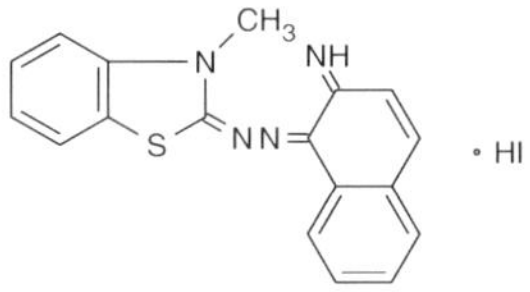

250 mg

S73,001-7
4-[4-(DIETHYLAMINO)-BENZYLIDENEAMINO]ANTIPYRINE

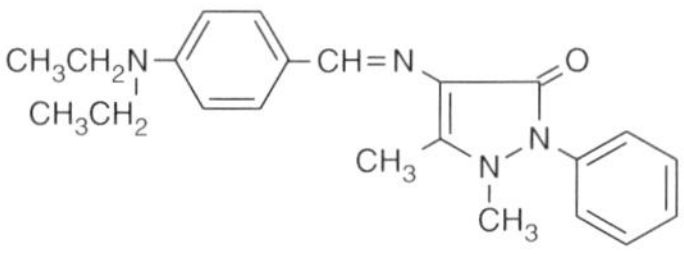

5 gm

S67,928-3
N,N'-DIPIPERONYLIDENE-ETHYLENEDIAMINE

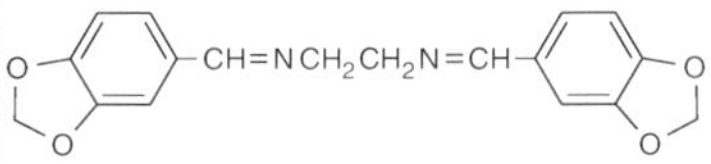

250 mg

S86,931-7
GLYOXAL BIS(DIMETHYLHYDRAZONE)

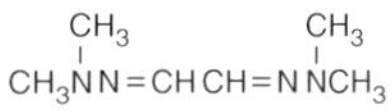

250 mg

S92,476-8
OXALIC BIS(FURFURYLIDENE)HYDRAZIDE

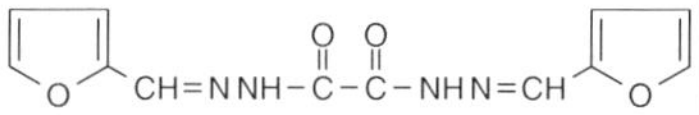

250 mg

S90,298-5
2-(3-BENZYLUREIDO)ACETIC (FURFURYLIDENE)HYDRAZIDE

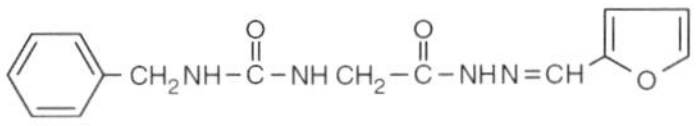

250 mg

S89,977-1
TETRAHYDRO-2-FUROIC (2-PYRIDYLMETHYLENE)HYDRAZIDE

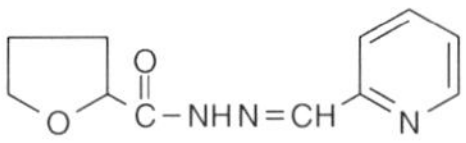

250 mg

S53,817-5
PICOLINIC ACID (4-NITRO-BENZYLIDENE)HYDRAZIDE

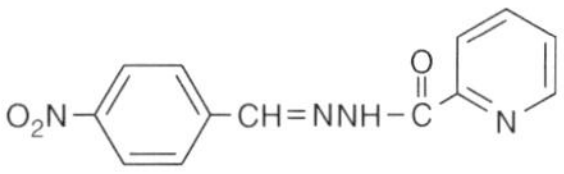

5 gm

S52,015-2
6-PHENYL-4-(2-PYRIDYLMETHYLENEAMINO)-1,2,4-TRIAZINE-3,5(2H,4H)-DIONE

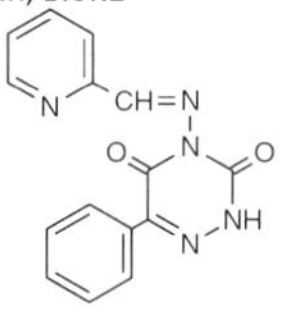

250 mg

S51,086-6
1,2-DIHYDRO-4,6-DIMETHYL-2-OXO-1-(2-PYRIDYLMETHYLENEAMINO)-NICOTINONITRILE

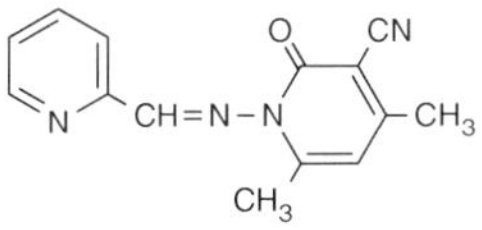

250 mg

S53,814-0
OXALIC ACID BIS[(2-PYRIDYLMETHYLENE)HYDRAZIDE]

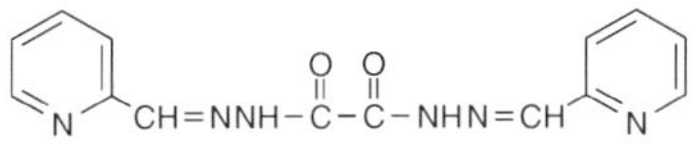

5 gm

S92,474-1
OXALIC BIS(1-NAPHTHYLMETHYLENE)-HYDRAZIDE

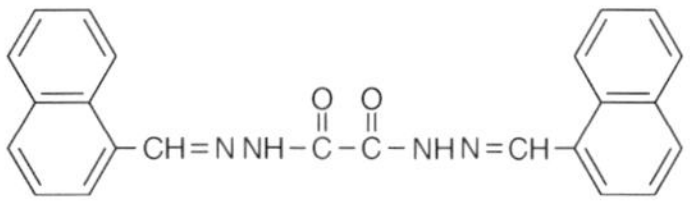

250 mg

S65,615-1
FORMIC ACID (2-PYRROLYLMETHYLENE)-HYDRAZIDE

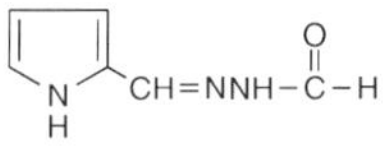

250 mg

53,818-3
GLUTARIC ACID BIS[(2-PYRIDYLMETHYLENE)HYDRAZIDE]

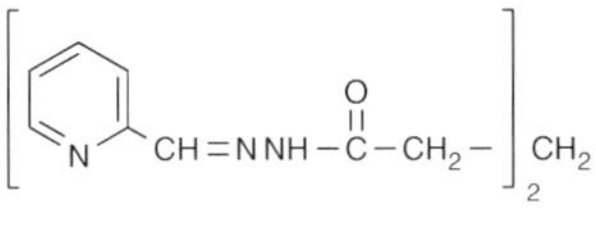

5 gm

S93,847-5
3-METHYL-1-PHENYL-4-(4-TOLYLHYDRAZONO)-2-PYRAZOLIN-5-ONE

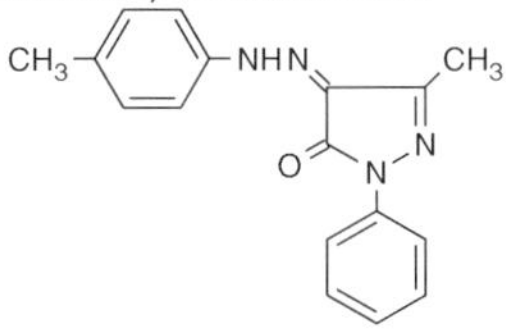

250 mg

S51,772-0
1,4-BIS(2-PYRIDYLMETHYLENEAMINO)-PIPERAZINE

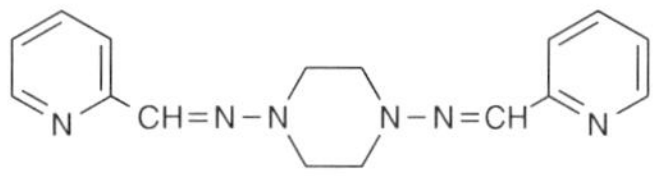

250 mg

S37,756-2
2-PYRIDINECARBOXALDEHYDE PHENYLHYDRAZONE

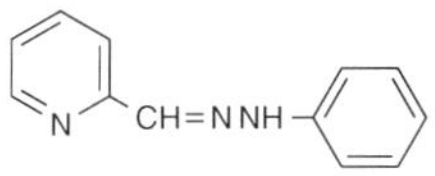

250 mg

S35,978-5
1-CYANO-N,N'-DIPHENYLFORMAMIDINE

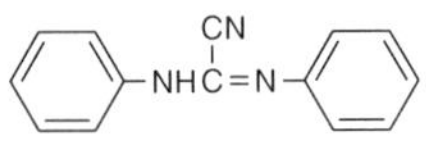

5 gm

S84,938-3
N-TERT-BUTYL-2-ACI-NITROCYCLOHEPTANIMINIUM HYDROXIDE, INNER SALT

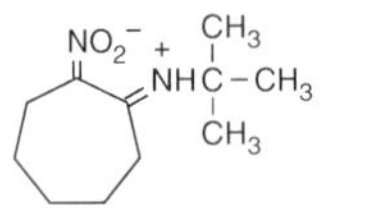

250 mg

S84,937-5
N-TERT-BUTYL-2-ACI-NITROCYCLOHEXANIMINIUM HYDROXIDE, INNER SALT

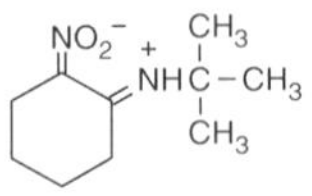

250 mg

S50,174-3
ALPHA,ALPHA'-(ETHYLENEDINITRILO)-BIS(P-ACETOTOLUIDIDE)

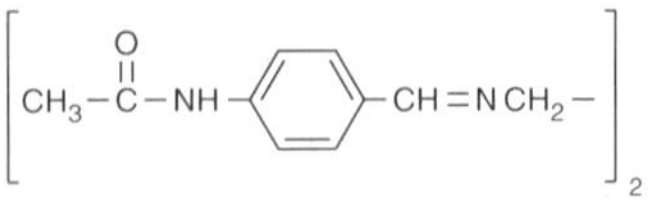

5 gm

S73,000-9
4-(PIPERONYLIDENEAMINO)ANTIPYRINE

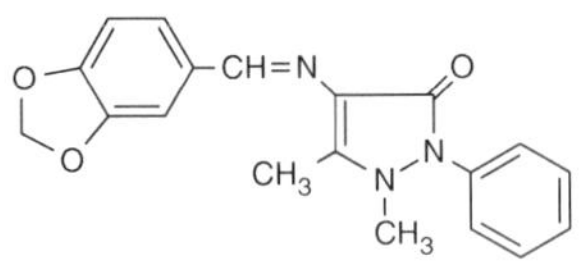

5 gm

S70,286-2
N,N'-BIS(2-THENYLIDENE)-ETHYLENEDIAMINE

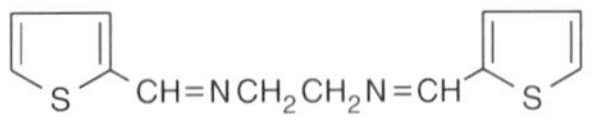

5 gm

S38,861-0
N,N'-DIBENZYLIDENEETHYLENEDIAMINE

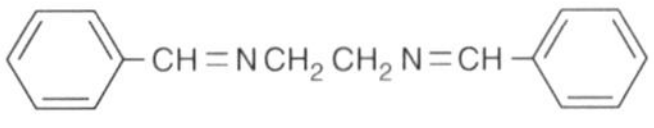

5 gm

S89,626-8
2,3-BIS(TRIPHENYLMETHYLAZO)-NAPHTHALENE

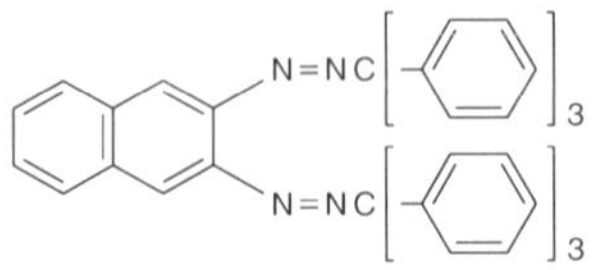

250 mg

S52,348-8
N,N'-BIS(4-FLUOROBENZYLIDENE)-ETHYLENEDIAMINE

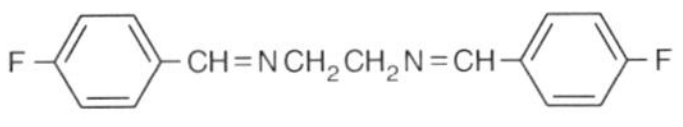

5 gm

S70,298-6
N,N'-BIS(4-METHYLBENZYLIDENE)-ETHYLENEDIAMINE

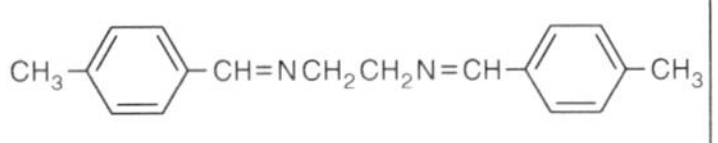

5 gm

S55,469-3
N,N'-BIS(4-CHLOROBENZYLIDENE)-4,5-DIMETHYL-1,2-PHENYLENEDIAMINE

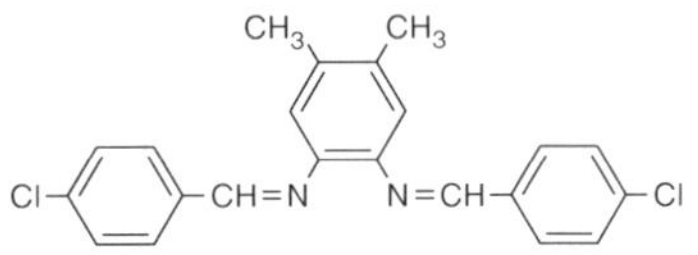

5 gm

S54,834-0
N,N'-BIS(3,4-DICHLOROBENZYLIDENE)-ETHYLENEDIAMINE

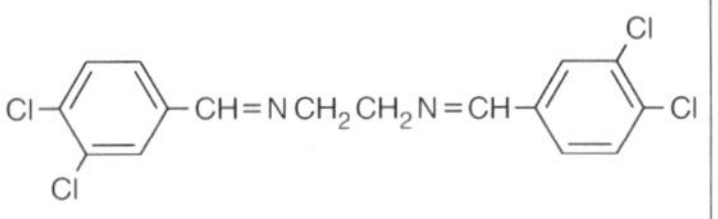

5 gm

S87,881-2
N,N'-BIS(2,5-DIMETHYLPHENYL)-1,4-DIAZABUTADIENE

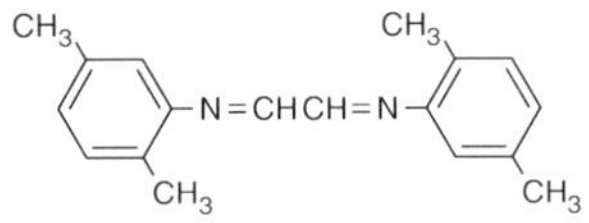

5 gm

S35,510-0
2-PYRIDINEALDOXIME O-ISOAMYL ETHER

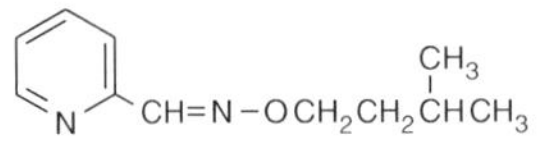

5 gm

S35,350-7
6-METHYL-2-PYRIDINEALDOXIME O-PROPYL ETHER

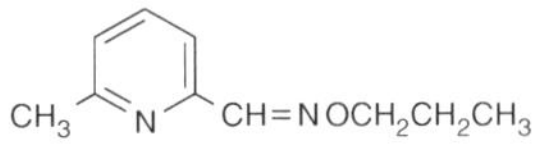

250 mg

S35,348-5
6-METHYL-2-PYRIDINEALDOXIME O-ISOAMYL ETHER

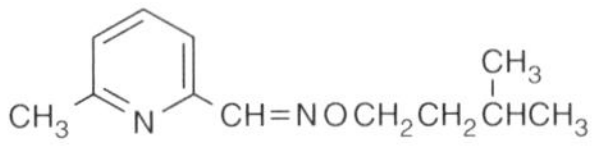

5 gm

S36,384-7
6-METHYL-2-PYRIDINEALDOXIME O-4-CHLOROBENZYL ETHER

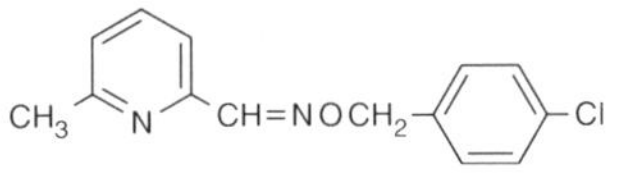

250 mg

S35,349-3
6-METHYL-2-PYRIDINEALDOXIME O-PHENETHYL ETHER

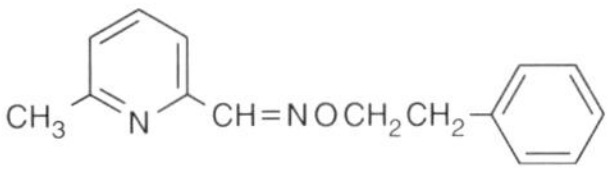

5 gm

S50,344-4
BENZIL, AZINE WITH 5-INDANCARBOXALDEHYDE

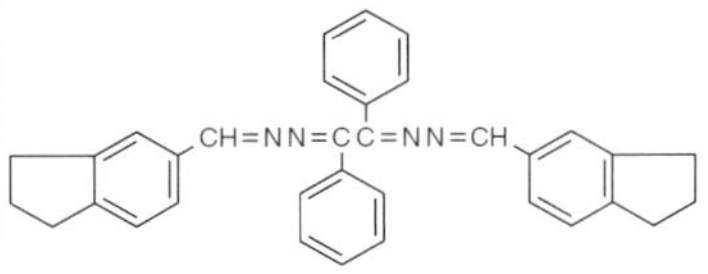

250 mg

S41,403-4
N,N'-BIS(ISOCOUMARAN-5-YLIDENE)-ETHYLENEDIAMINE

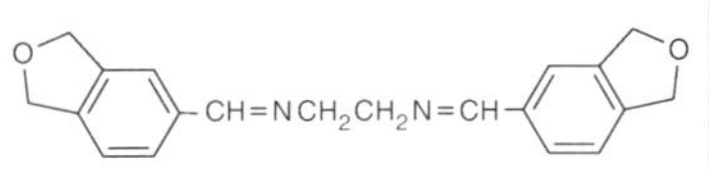

5 gm

S73,522-1
DIMETHYL([2-(DIMETHYLAMINO-METHYLENE)2-(3-INDOLYL)ETHYLIDENE]-AMMONIUM)TETRAFLUOROBORATE

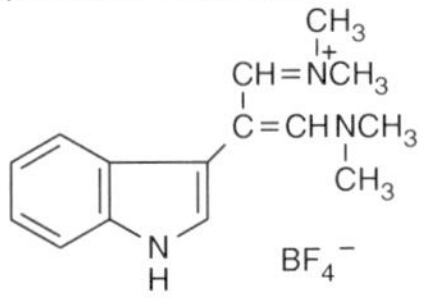

250 mg

S92,085-1
2-ACETAMIDO-4-[4-(DIMETHYLAMINO)-PHENYLIMINO]-2,5-CYCLOHEXADIEN-1-ONE

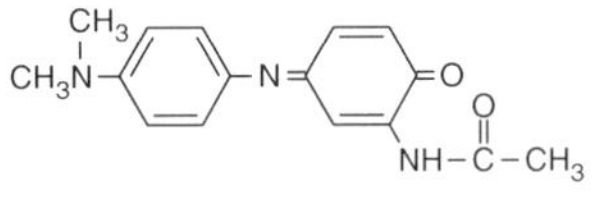

250 mg

S61,251-0
ALPHA-(2-CHLORO-6-NITROPHENYL)-N-(4-DIMETHYLAMINOPHENYL)NITRONE

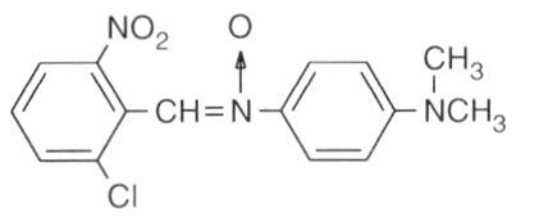

5 gm

S51,259-1
3-NITRO-N-(3-NITROBENZYLIDENE)-ANILINE

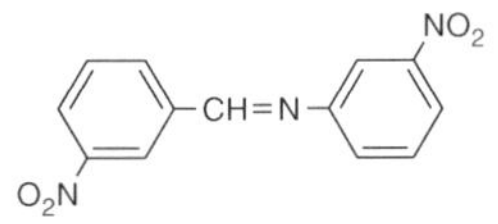

5 gm

S87,084-6
2-(2-ANILINOVINYL)-1-ETHYLQUINOLINIUM PERCHLORATE

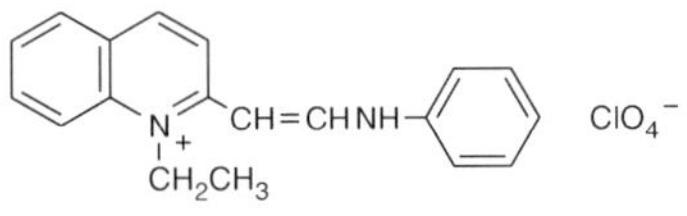

250 mg

S93,454-2
4-(3-INDOLYLMETHYLENEAMINO)-4H-1,2,4-TRIAZOLE

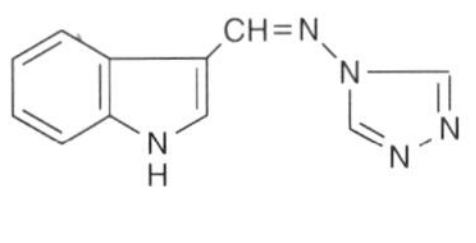

250 mg

S87,814-6
N-[3-(PHENYLIMINO)PROPENYL]ANILINE

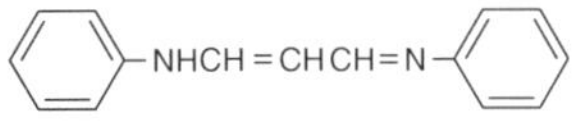

5 gm

S87,838-3
N-[2-BROMO-3-(PHENYLIMINO)-PROPENYL]ANILINE

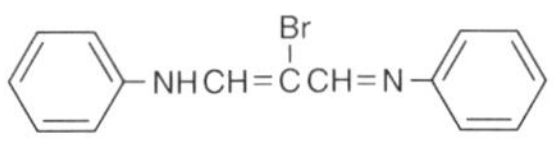

5 gm

S62,208-7
METHYL[3-(N-METHYLANILINO)-ALLYLYLIDENE]PHENYLAMMONIUM IODIDE

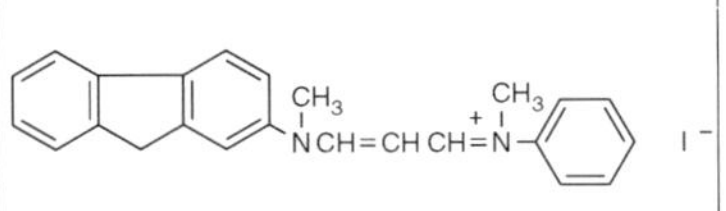

5 gm

S63,870-6
ACETIC ACID (1-METHYL-2-PHENYL-3-INDOLYLMETHYLENE)HYDRAZIDE

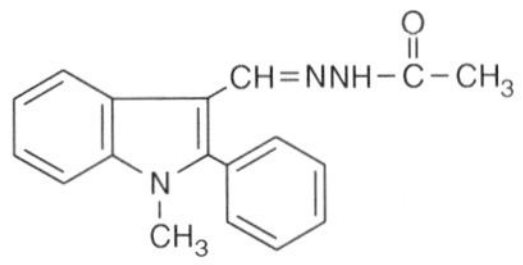

250 mg

S58,218-2
1-HEXADECANESULFONIC ACID [1-METHYL-2-(P-NITROPHENYL)-6-SULFO-4(1H)-QUINOLYLIDENE]HYDRAZIDE

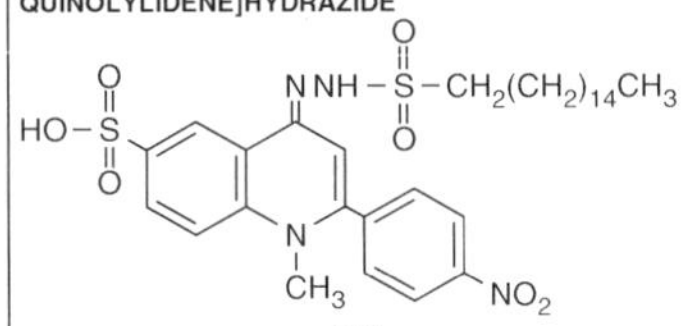

250 mg

S51,311-3
N,N-DIMETHYL-ALPHA-(3-NITROPHENYLIMINO)-P-TOLUIDINE

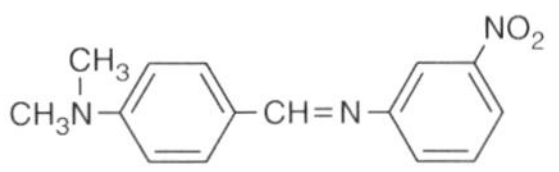

5 gm

S63,965-6
N-[4-(DIMETHYLAMINO)BENZYLIDENE]-2-METHYL-5-NITROANILINE
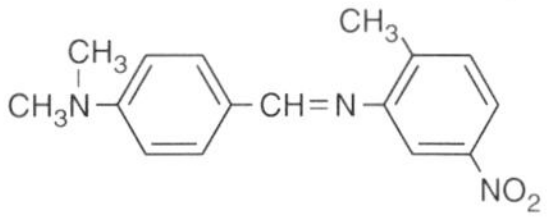
5 gm

S64,327-0
N-[4-(DIMETHYLAMINO)BENZYLIDENE]-4-METHYL-3-NITROANILINE
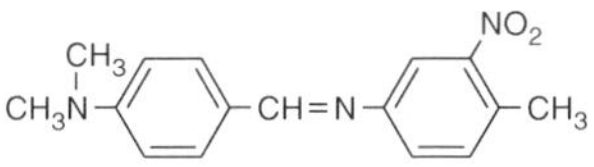
5 gm

S58,176-3
5-CARBOXY-2-(HEXADECYLOXY)BENZENE-SULFONIC ACID [1-METHYL-6-SULFO-2-TOLYL-4(1H)-QUINOLYLIDENE]HYDRAZIDE
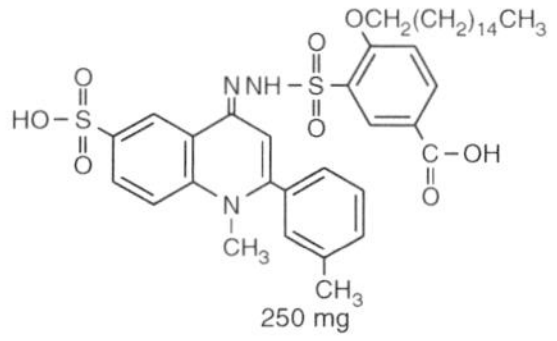
250 mg

S62,209-5
N,N-BIS(2-CYANOETHYL)-ALPHA-PHENYLIMINO-P-TOLUIDINE
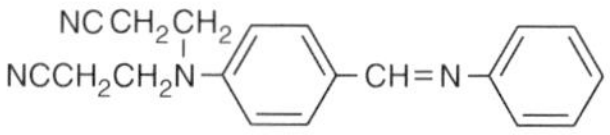
250 mg

S57,887-8
1-HEXADECANESULFONIC ACID (1-METHYL-2-[P-(METHYLSULFONYL)PHENYL]-6-SULFO-4(1H)-QUINOLYLIDENE)HYDRAZIDE
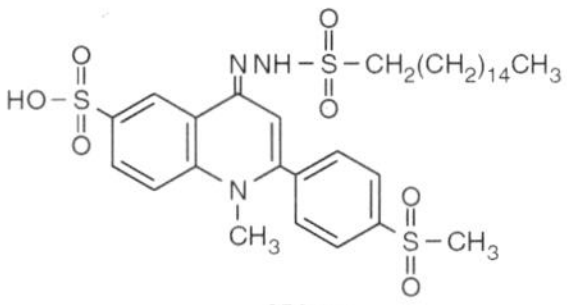
250 mg

S58,162-3
1-HEXADECANESULFONIC ACID [2-(P-CYCLOHEXYLPHENYL)-1-METHYL-6-SULFO-4(1H)-QUINOLYLIDENE]HYDRAZIDE
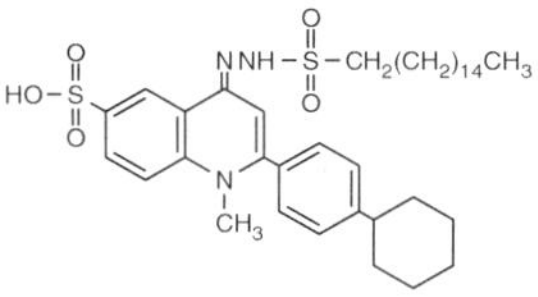
250 mg

S57,857-6
1-HEXADECANESULFONIC ACID [1-METHYL-6-SULFO-2-(O-TOLYL)-4(1H)-QUINOLYLIDENE]HYDRAZIDE
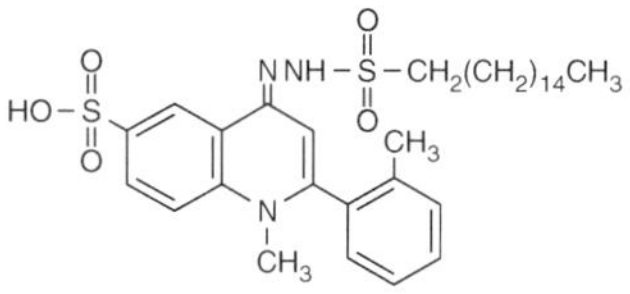
5 gm

S57,881-9
1-HEXADECANESULFONIC ACID [2-(4-BIPHENYL)-1-METHYL-6-SULFO-4(1H)-QUINOLYLIDENE]HYDRAZIDE
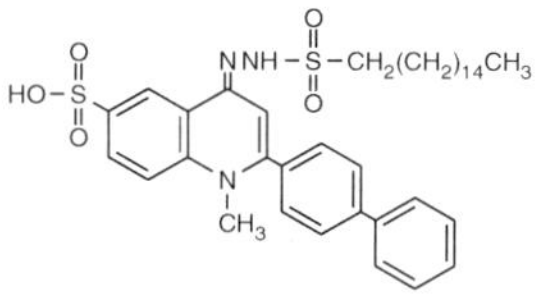
5 gm

S57,859-2
1-HEXADECANESULFONIC ACID [2-(4-FLUOROPHENYL)-1-METHYL-6-SULFO-4(1H)-QUINOLYLIDENE]HYDRAZIDE
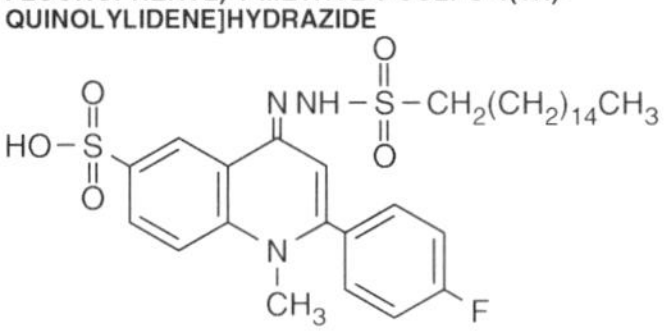
5 gm

S86,516-8
1-METHYL-4-PIPERIDINONE (P-TOSYL)-HYDRAZONE
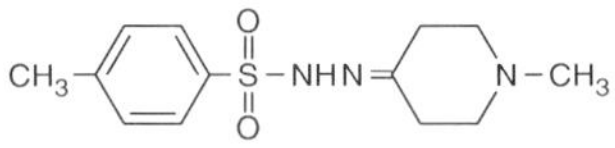
250 mg

S87,278-4
BENZENESULFONIC (9,10-DIHYDRO-10-METHYL-9-ACRIDINYLIDENE)HYDRAZIDE
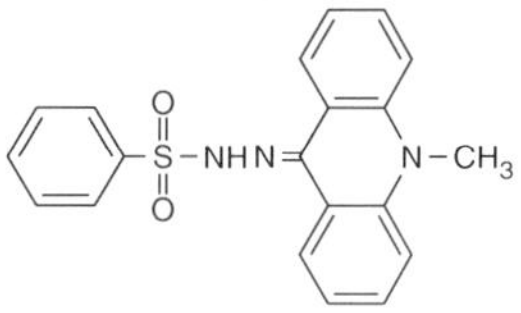
250 mg

S87,303-9
1,2,6-TRIMETHYL-4(1H)-PYRIDONE HYDRAZONE DIHYDROCHLORIDE
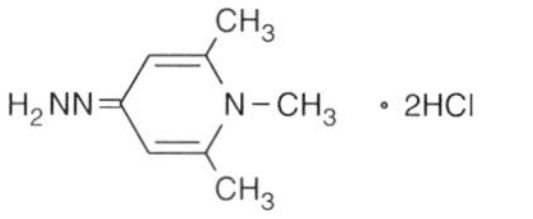
250 mg

S56,961-5
1-HEXADECANESULFONIC ACID [1-METHYL-2-PHENYL-4(1H)-QUINOLYLIDENE]HYDRAZIDE HYDROCHLORIDE
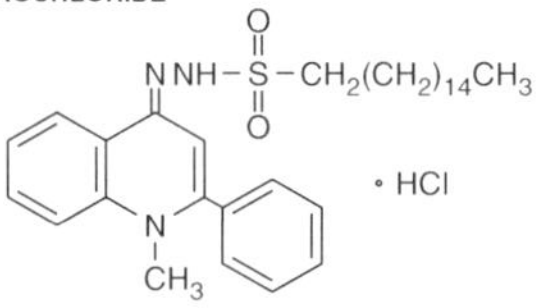
5 gm

S63,238-4
2-NITROBENZALDEHYDE PHENYLHYDRAZONE
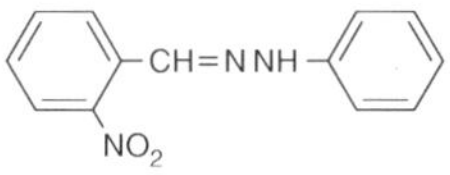
250 mg

S93,421-6
N'-(4-NITROPHENYL)-N-(4-TOLYL)-BENZAMIDINE
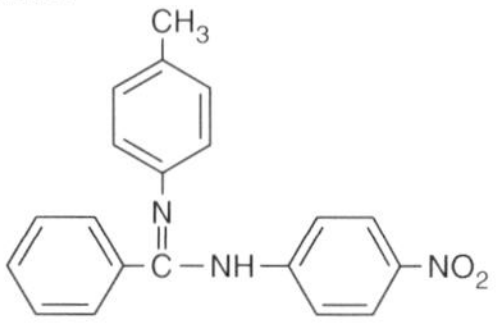
250 mg

S64,071-9
3-CHLORO-N-(2-NITROBENZYLIDENE)-ANILINE
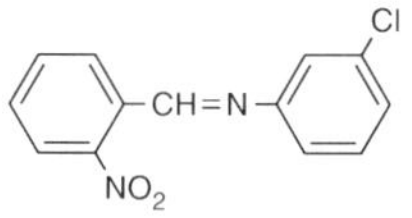
250 mg

S64,072-7
2-CHLORO-N-(2-NITROBENZYLIDENE)-ANILINE
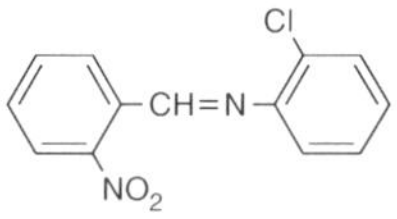
250 mg

S64,065-4
N-(2-NITROBENZYLIDENE)-2,4,5-TRICHLOROANILINE
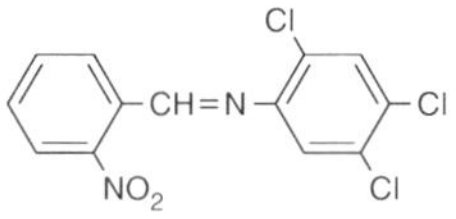
250 mg

S64,070-0
4-CHLORO-N-(2-NITROBENZYLIDENE)-ANILINE
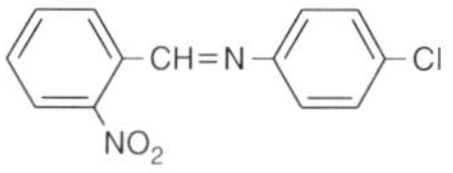
250 mg

S64,081-6
N-(2-NITROBENZYLIDENE)-P-TOLUIDINE
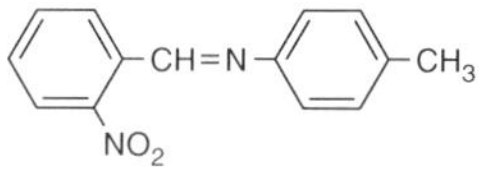
250 mg

S64,069-7
2,6-DICHLORO-N-(2-NITRO-BENZYLIDENE)ANILINE
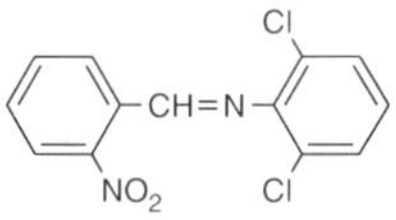
250 mg

S64,078-6
6-CHLORO-N-(2-NITROBENZYLIDENE)-O-TOLUIDINE
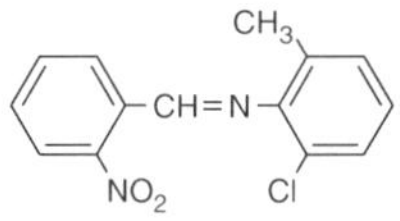
250 mg

S64,092-1
N-(2-NITROBENZYLIDENE)-2,6-XYLIDINE
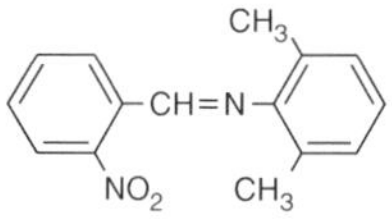
250 mg

S64,106-5
2,6-DIETHYL-N-(2-NITROBENZYLIDENE)-ANILINE
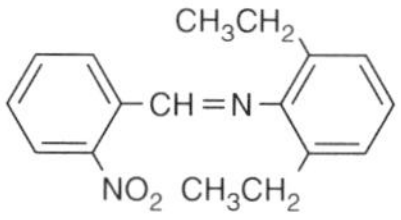
250 mg

S64,068-9
2,5-DICHLORO-N-(2-NITRO-BENZYLIDENE)ANILINE
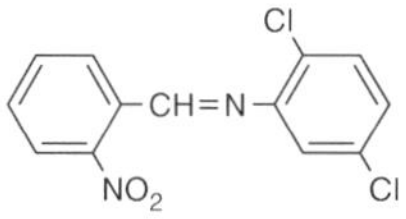
250 mg

S64,077-8
3-CHLORO-N-(2-NITROBENZYLIDENE)-P-TOLUIDINE
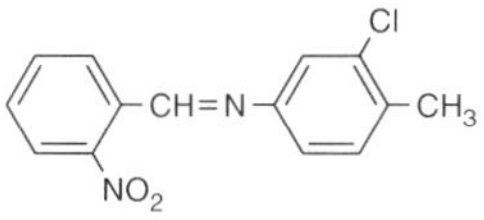
250 mg

S64,091-3
N-(2-NITROBENZYLIDENE)-3,4-XYLIDINE
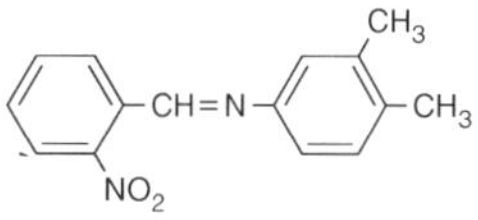
250 mg

S64,095-6
N-(2-NITROBENZYLIDENE)-2,5-XYLIDINE
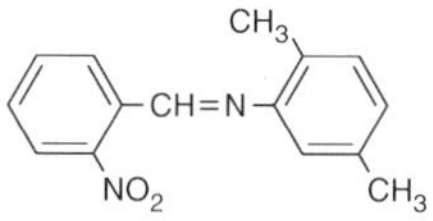
250 mg

S64,094-8
N-(2-NITROBENZYLIDENE)-3,5-XYLIDINE
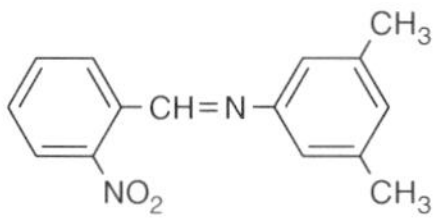
250 mg

S60,238-8
N-(4-METHYLBENZYLIDENE)-3-NITROANILINE
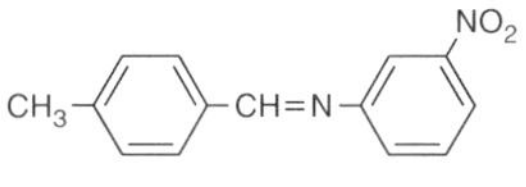
250 mg

S62,457-8
N,N'-TEREPHTHALYLIDENEBIS(4-CHLORO-3-NITROANILINE)
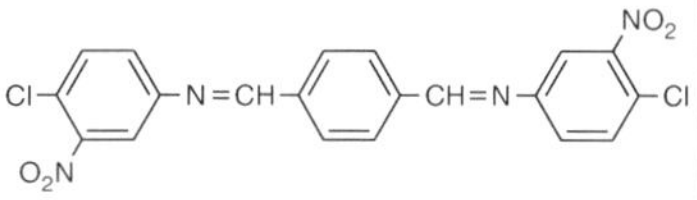
5 gm

S62,464-0
N-(9-ANTHRACENYLMETHYLENE)-4-METHYL-3-NITROANILINE
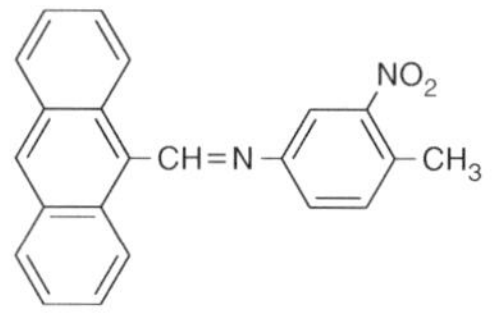
5 gm

S62,476-4
N-(9-ANTHRACENYLMETHYLENE)-2-METHYL-5-NITROANILINE
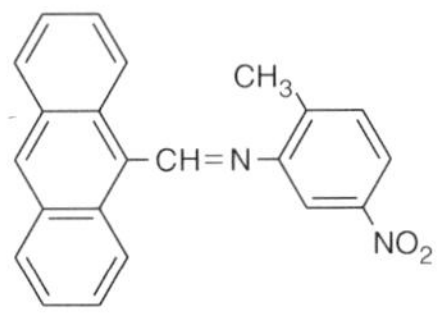
5 gm

S63,764-5
1-HEXADECANESULFONIC ACID [1-METHYL-2-PHENYL-5-SULFO-4(1H)-QUINOLYLIDENE]-HYDRAZIDE
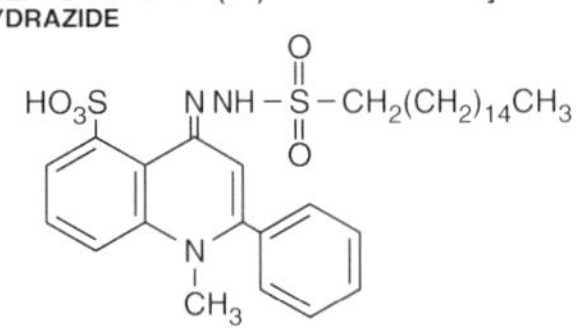
250 mg

S61,794-6
METHANESULFONIC ACID [1-METHYL-2-PHENYL-6-SULFO-4(1H)-QUINOLYLIDENE]-HYDRAZIDE
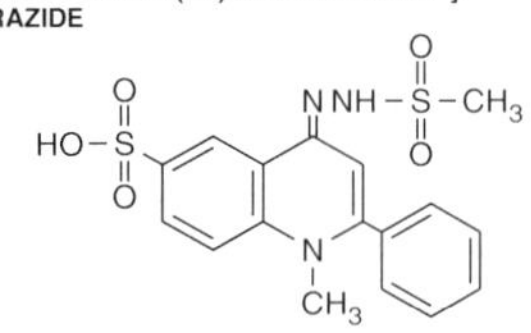
250 mg

S64,267-3
N,N'-BIS(4-CHLOROBENZYLIDENE)-ETHYLENEDIAMINE
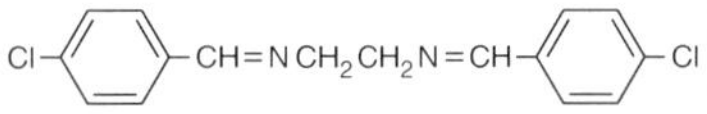
5 gm

S91,520-3
2-NITROBENZALDEHYDE SEMICARBAZONE
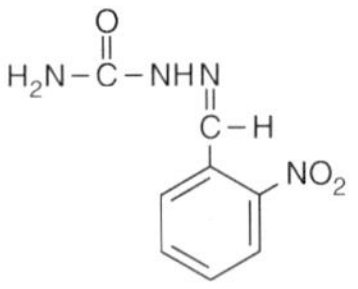
250 mg

S53,906-6
2-NITROBENZALDEHYDE THIOSEMICARBAZONE
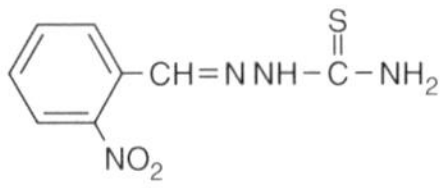
5 gm

S37,612-4
1-(BENZYLIDENEIMINO)-2,4-DIMETHYL-5-CYANO-6-PYRIDONE
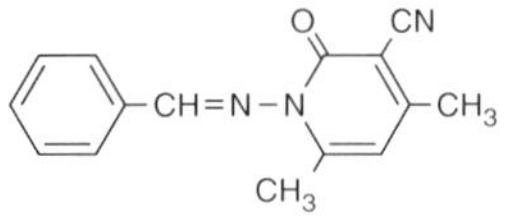
5 gm

S51,209-5
N-(5-INDANYLMETHYLENE)ANTHRANILIC ACID (5-INDANYLMETHYLENE)HYDRAZIDE
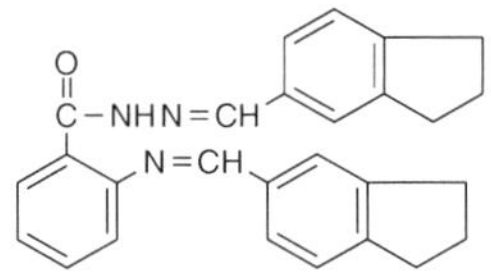
250 mg

S50,040-2
CYANOACETIC ACID 5-INDANYLIDENEHYDRAZIDE
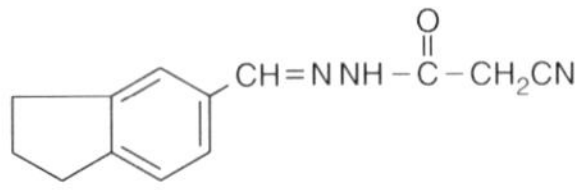
250 mg

S78,018-9
DECANOIC (3-PYRIDYLMETHYLENE)-HYDRAZIDE
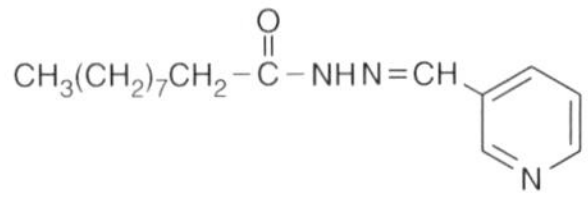
250 mg

S50,665-6
NICOTINIC ACID (5-INDANYL-METHYLENE)HYDRAZIDE
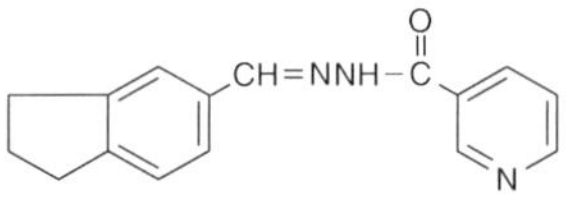
250 mg

S57,870-3
1-HEXADECANESULFONIC ACID[1-METHYL-2-(2-FLUOROPHENYL)-6-SULFO-4(1H)-QUINOLYLIDENE]HYDRAZIDE
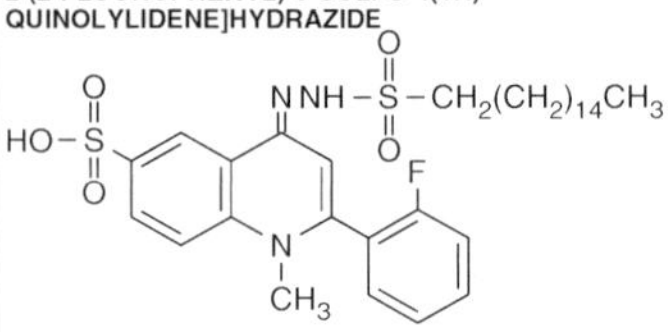
250 mg

S86,597-4
2,4-PENTANEDIONE BIS[(P-TOSYL)-HYDRAZONE]
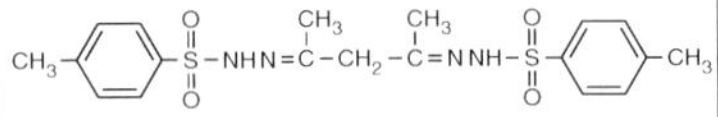
250 mg

S89,570-9
3-PYRROLECARBOXALDEHYDE PHENYLHYDRAZONE
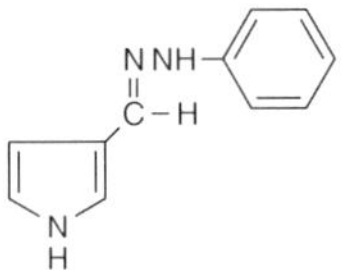
250 mg

S80,079-1
1-(3-PYRIDYLMETHYLENE)-4-PHENYL-3-THIOSEMICARBAZIDE
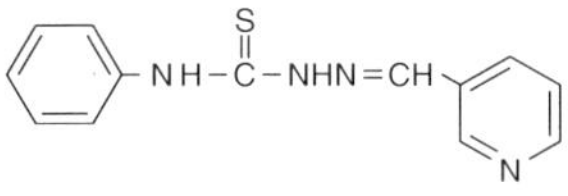
250 mg

S77,180-5
1-(3-PYRIDYLMETHYLENEAMINO)-GUANIDINE DINITRATE HYDRATE
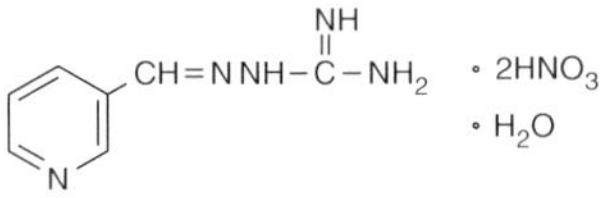
250 mg

S35,520-8
3-PYRIDINECARBOXALDEHYDE THIOSEMICARBAZONE
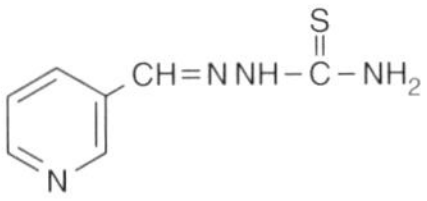
5 gm

S58,645-5
3-NITRO-N-(2-THIENYLMETHYLENE)-ANILINE
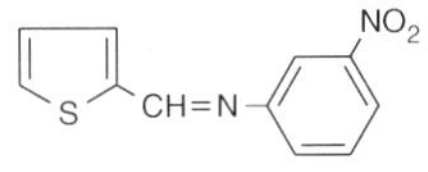
5 gm

S58,657-9
5-NITRO-N-(2-THIENYLMETHYLENE)-O-TOLUIDINE
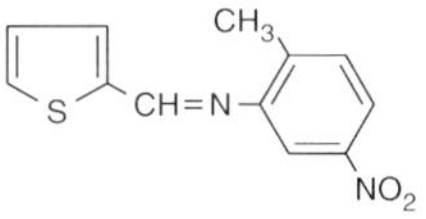
5 gm

S58,975-6
3-NITRO-N-(2-THIENYLMETHYLENE)-P-TOLUIDINE
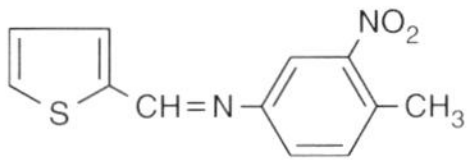
250 mg

S80,797-4
4-(3-NITROPHENYLIMINOMETHYL)PHENYL N-METHYLCARBAMATE
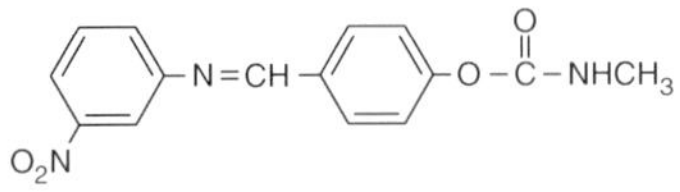
250 mg

S90,172-5
4'-(3-NITROPHENYLIMINOMETHYL)-ACETANILIDE
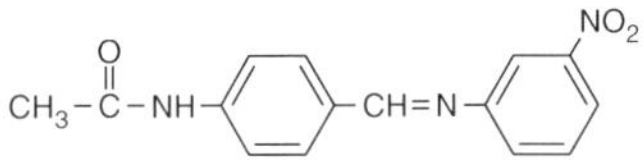
250 mg

S37,106-8
N-BENZYLIDENE-3-NITROANILINE
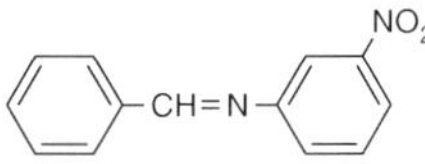
5 gm

S60,543-3
ETHYL 2-[N,N'-BIS(3-NITROPHENYL)-AMIDINO]ACETATE
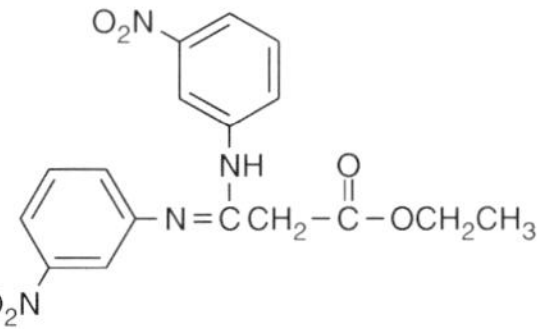
250 mg

S52,017-9
4,5-DIPHENYL-2-(4-NITROBENZYLIDENEAMINO)-3-FURANCARBONITRILE
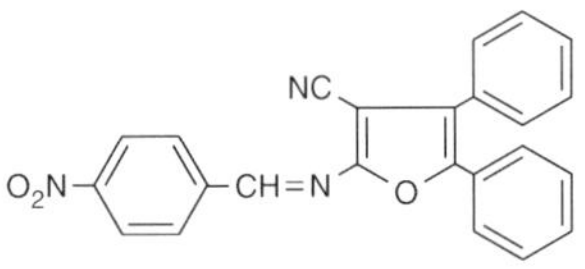
250 mg

S62,448-9
4-CHLORO-N-(2-THENYLIDENE)-3-NITROANILINE
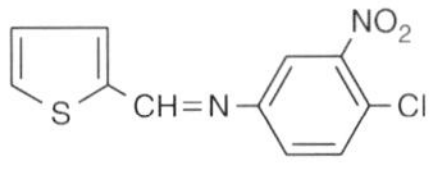
5 gm

S39,924-8
3-NITRO-N-PIPERONYLIDENEANILINE
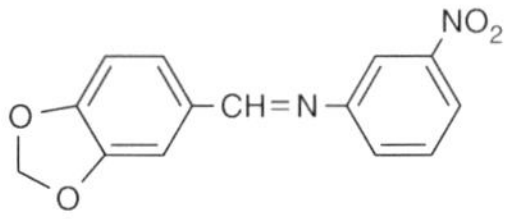
5 gm

S58,442-8
N,N'-(1,4-PHENYLENEDIMETHYLIDYNE)-BIS(5-NITRO-O-TOLUIDINE)
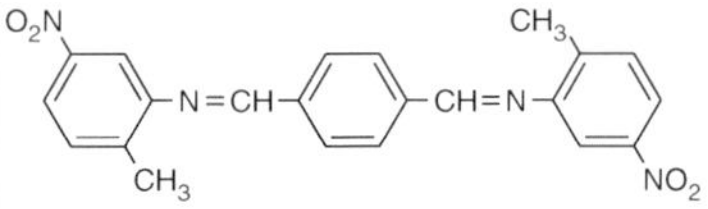
5 gm

S58,443-6
N,N'-(1,4-PHENYLENEDIMETHYLIDYNE)-BIS(3-NITRO-P-TOLUIDINE)
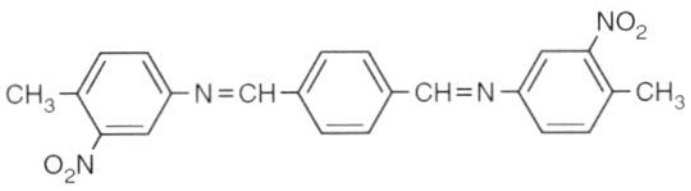
250 mg

S61,952-3
N-(9-ANTHRACENYLMETHYLENE)-3-NITROANILINE
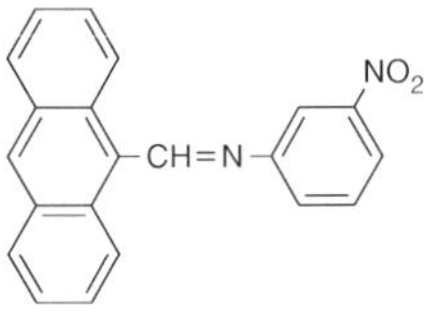
5 gm

S93,180-2
2-(CYCLOHEXYLIDENEAMINO)-1-CYCLOHEXENE-1-CARBOXAMIDE
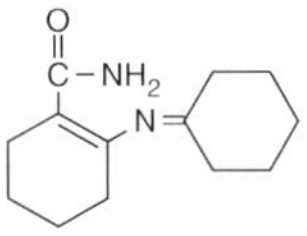
250 mg

S76,125-7
5-[3-(MORPHOLINO)PROPYLIMINO-METHYL]-1,3-BENZODIOXOLE
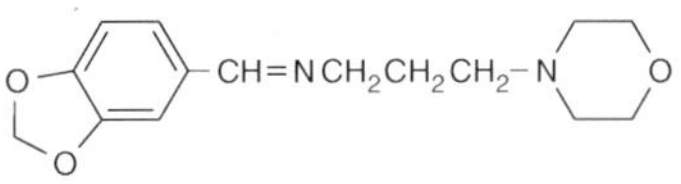
250 mg

S34,496-6
3-(N-CYCLOHEXYLFORMIMIDOYL)INDOLE
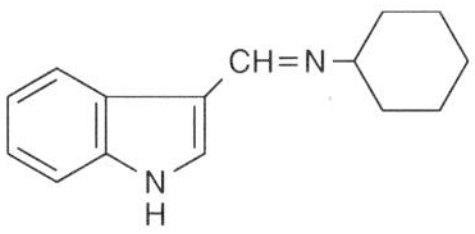
5 gm

S35,511-9
3-PYRIDINEALDOXIME O-ISOAMYL ETHER
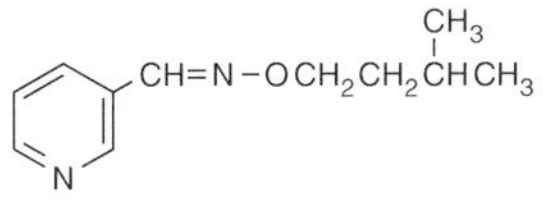
5 gm

S37,050-9
3-PYRIDINEALDAZINE
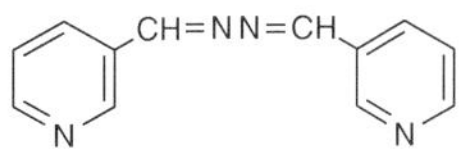
250 mg

S68,893-2
N-BENZYLIDENE-N'-PHENYL-1,4-PHENYLENEDIAMINE
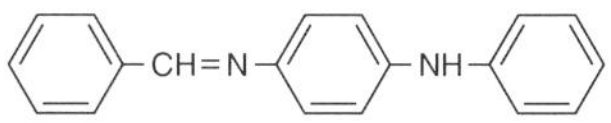
5 gm

S68,895-9
N-(4-METHYLBENZYLIDENE)-N'-PHENYL-1,4-PHENYLENEDIAMINE
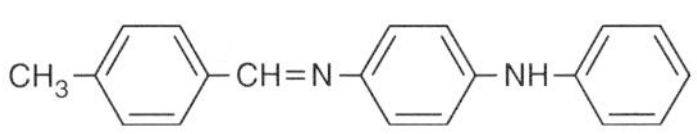
5 gm

S69,879-2
N-(3-NITROBENZYLIDENE)-N'-PHENYL-1,4-PHENYLENEDIAMINE
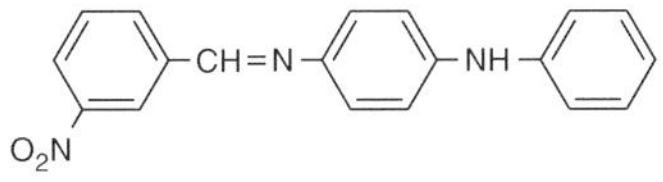
5 gm

S68,904-1
N-PHENYL-N'-PIPERONYLIDENE-1,4-PHENYLENEDIAMINE
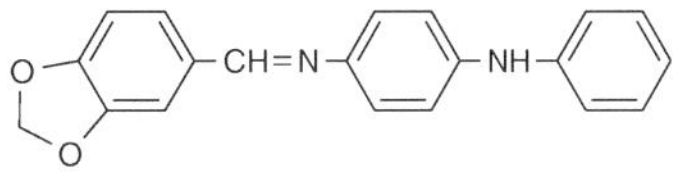
250 mg

S68,918-1
N-CINNAMYLIDENE-N'-PHENYL-1,4-PHENYLENEDIAMINE
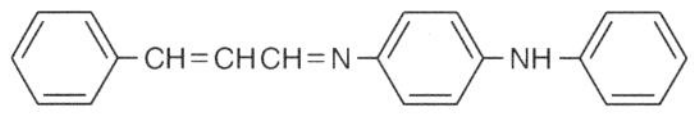
5 gm

S69,272-7
N-(4-ISOPROPYLBENZYLIDENE)-N'-PHENYL-1,4-PHENYLENEDIAMINE
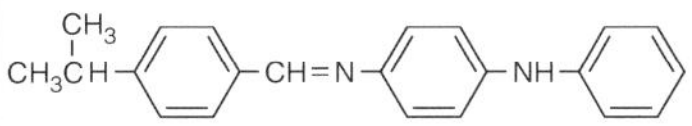
5 gm

S53,176-6
N-(4-NITROBENZYLIDENE)-N'-(4-NITROPHENYL)ETHYLENEDIAMINE
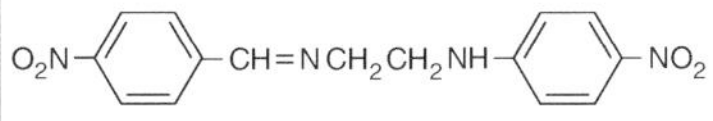
5 gm

S69,657-9
N-PHENYL-N'-(2-THIENYLMETHYLENE)-1,4-PHENYLENEDIAMINE
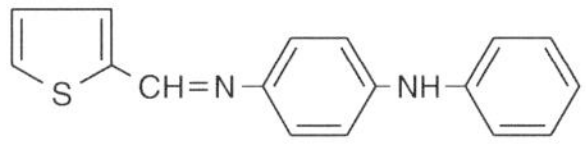
5 gm

S41,960-5
BINDSCHEDLERS GREEN, INDICATOR GRADE
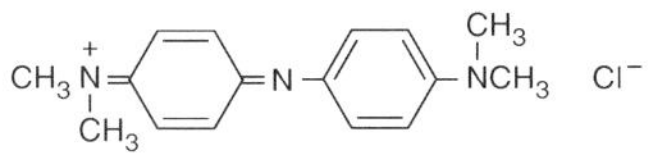
5 gm

S92,099-1
2,6-DIBROMO-4-[4-(DIMETHYLAMINO)-PHENYLIMINO]-2,5-CYCLOHEXADIEN-1-ONE
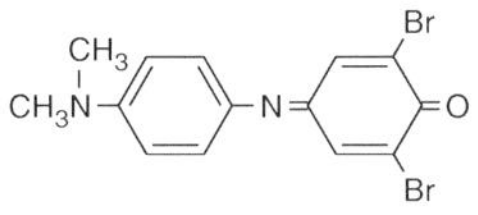
250 mg

S92,046-0
4-[4-(DIMETHYLAMINO)PHENYLIMINO]-2-METHYL-2,5-CYCLOHEXADIEN-1-ONE
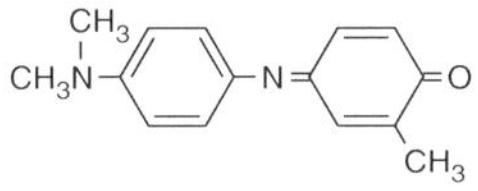
250 mg

S91,971-3
2,5-DIMETHYL-4-[4-(DIMETHYLAMINO)-PHENYLIMINO]-2,5-CYCLOHEXADIEN-1-ONE
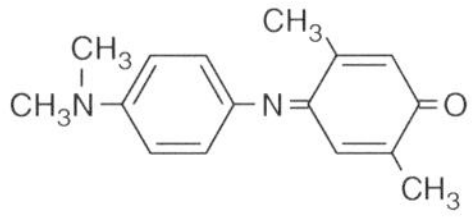
250 mg

S92,058-4
4-[4-(DIMETHYLAMINO)PHENYLIMINO]-1(4H)-NAPHTHALENONE
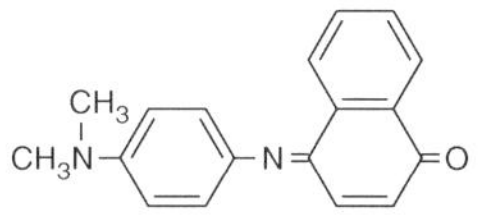
250 mg

S92,123-8
N'-BENZYLIDENE-N,N-DIMETHYL-P-PHENYLENEDIAMINE
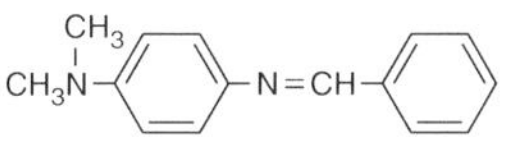
250 mg

S87,502-3
N-CINNAMYLIDENE-N',N'-DIMETHYL-1,4-PHENYLENEDIAMINE
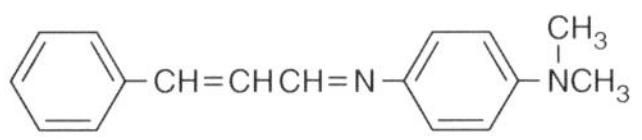
250 mg

S89,498-2
N-CINNAMYLIDENE-N',N'-DIETHYL-1,4-PHENYLENEDIAMINE
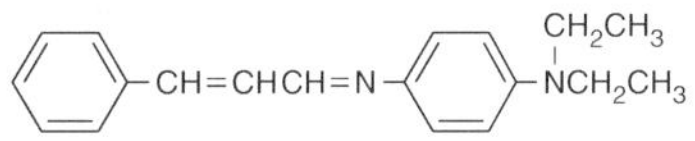
250 mg

S87,443-4
1,4-DIHYDRO-4-[4-(DIMETHYLAMINO)-PHENYLIMINO]-1-ETHYLQUINOLINE HYDROCHLORIDE
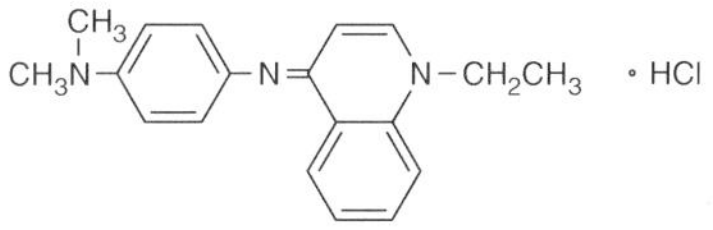
250 mg

S50,767-9
N,N-DIMETHYL-ALPHA-(4-NITROPHENYLIMINO)-P-TOLUIDINE
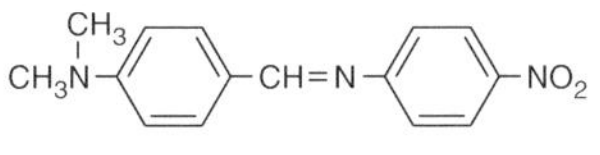
5 gm

S64,555-9
N-[4-(DIMETHYLAMINO)BENZYLIDENE]-2-METHYL-4-NITROANILINE
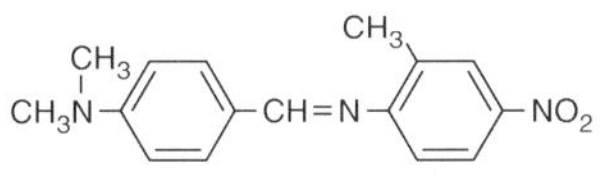
5 gm

S61,980-9
N,N-DIBENZYL-N',N'-DI-TERT-BUTYL-1,4-DIAMINO-2-BUTENE
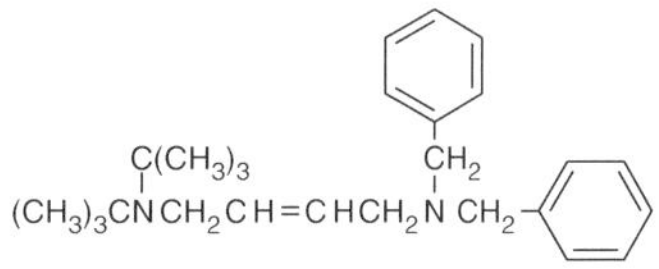
250 mg

S64,363-7
N,N'-BIS(4-DIMETHYLAMINOBENZYLIDENE)-1,4-PHENYLENEDIAMINE
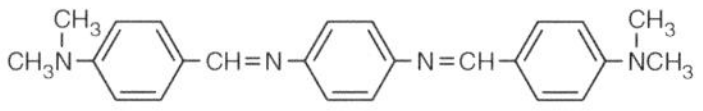
5 gm

S63,276-7
2-BENZOYL-2-(4-DIMETHYLAMINOPHENYLIMINO)-4'-(FLUOROSULFONYL)ACETANILIDE
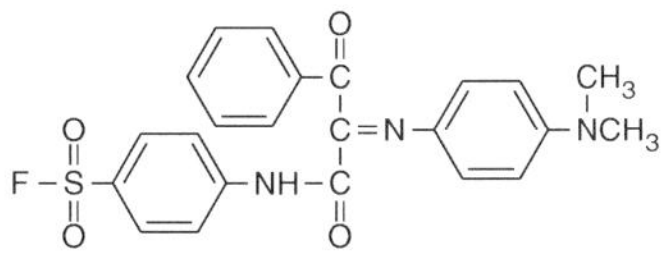
250 mg

S90,352-3
1-[2-(3-NITROBENZYLIDENE)-HYDRAZINOCARBONYLMETHYL]-PYRIDINIUM CHLORIDE
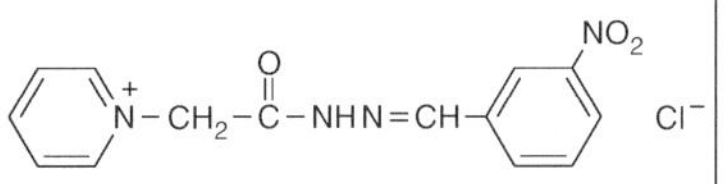
250 mg

S88,759-5
1-(4-IMINO-2,5-CYCLOHEXADIEN-1-YLIDENE)PIPERIDINIUM DIPERCHLORATE
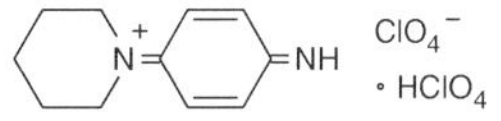
250 mg

S87,510-4
[4-(BENZYLIDENEAMINO)PHENYL]-TRIMETHYLAMMONIUM IODIDE
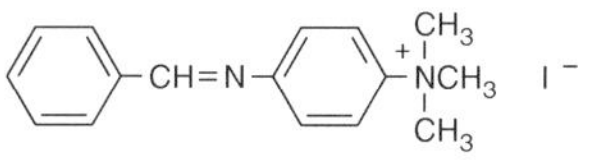
250 mg

S62,289-3
BENZALDEHYDE 4-NITROPHENYLHYDRAZONE
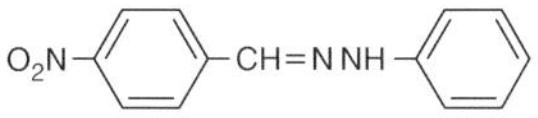
250 mg

S39,609-5
N,N'-BIS(4-NITROPHENYL)-BENZAMIDINE
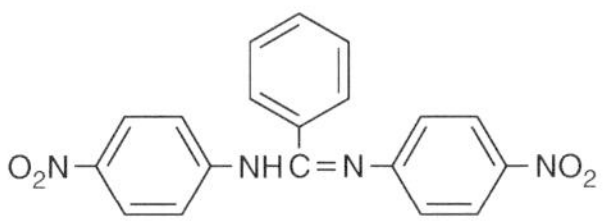
5 gm

S59,478-4
N-(4-METHYLBENZYLIDENE)-4-NITROANILINE
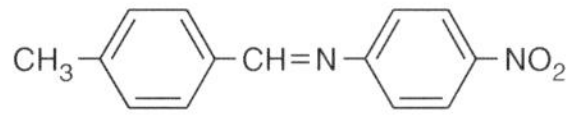
250 mg

S64,331-9
N,N'-BIS(3-NITROBENZYLIDENE)-1,4-PHENYLENEDIAMINE
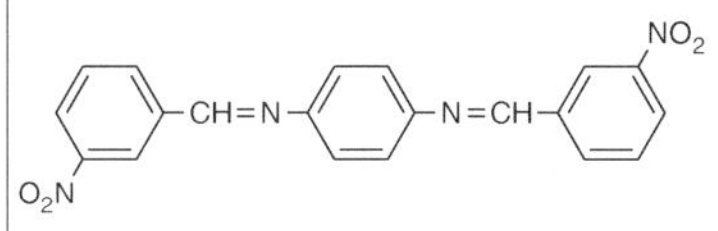
5 gm

S62,973-1
4-CHLORO-N-(2-CHLORO-5-NITROBENZYLIDENE)ANILINE

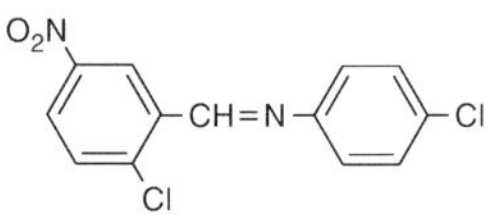

250 mg

S62,755-0
N-(2-CHLORO-5-NITROBENZYLIDENE)-P-TOLUIDINE

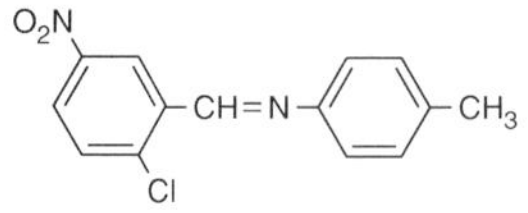

250 mg

S63,500-6
N,N'-(1,4-PHENYLENEDIMETHYLIDYNE)-BIS(2-CHLORO-4-NITROANILINE)

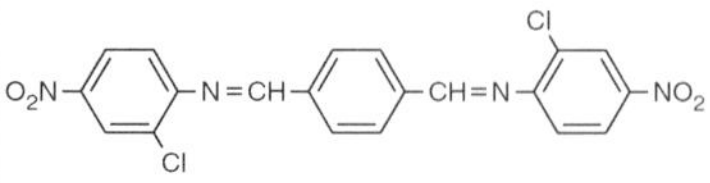

250 mg

S59,965-4
N,N'-TEREPHTHALYLIDENEBIS(2-METHYL-4-NITROANILINE)

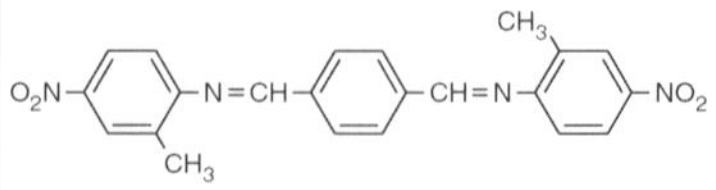

250 mg

S62,463-2
N-(9-ANTHRACENYLMETHYLENE)-4-NITROANILINE

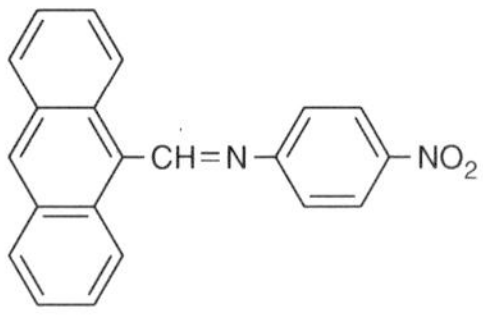

5 gm

S62,692-9
N-(2-CHLORO-5-NITROBENZYLIDENE)-2,3-XYLIDINE

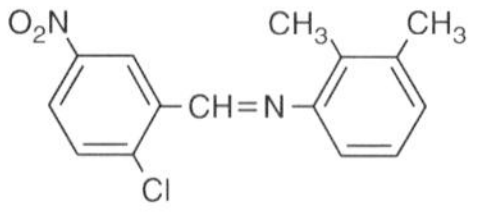

250 mg

S62,835-2
N-(2-CHLORO-5-NITROBENZYLIDENE)-ANILINE

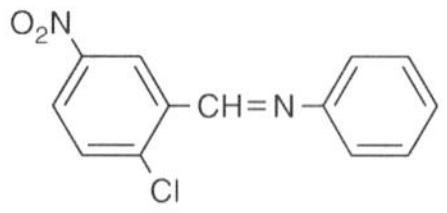

250 mg

S62,840-9
3-CHLORO-N-(2-CHLORO-5-NITROBENZYLIDENE)ANILINE

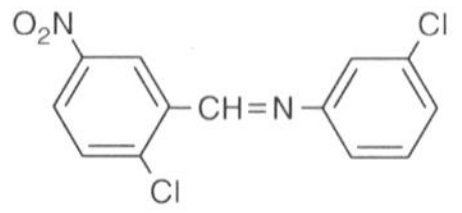

250 mg

S62,725-9
N-(2-CHLORO-5-NITROBENZYLIDENE)-3,4-DICHLOROANILINE

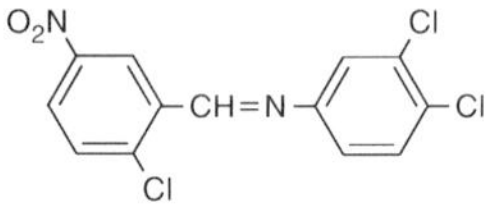

250 mg

S64,067-0
N-(2-CHLORO-5-NITROBENZYLIDENE)-2,4-DICHLOROANILINE

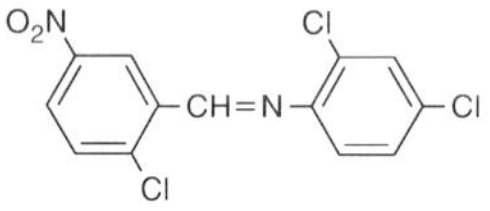

250 mg

S62,715-1
3-BROMO-N-(2-CHLORO-5-NITROBENZYLIDENE)ANILINE

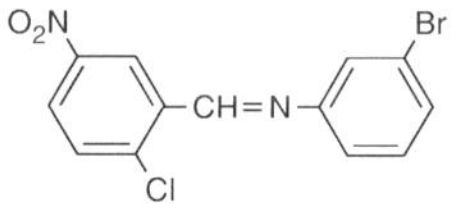

5 gm

S64,064-6
N-(2-CHLORO-5-NITROBENZYLIDENE)-2,4,5-TRICHLOROANILINE

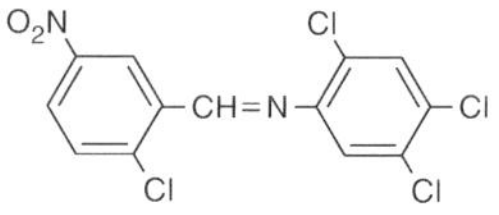

250 mg

S62,896-4
N-(2-CHLORO-5-NITROBENZYLIDENE)-M-TOLUIDINE

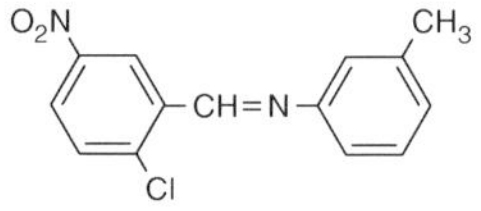

5 gm

S62,765-8
5-CHLORO-N-(2-CHLORO-5-NITROBENZYLIDENE)-2-METHYLANILINE

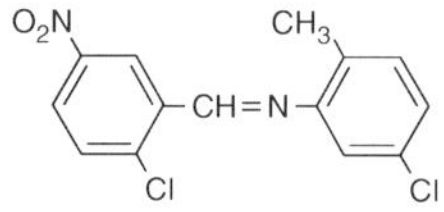

250 mg

S64,075-1
3-CHLORO-N-(2-CHLORO-5-NITROBENZYLIDENE)-4-METHYLANILINE

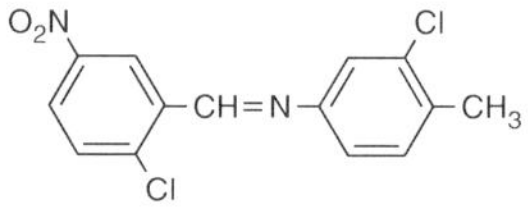

250 mg

S62,740-2
N-(2-CHLORO-5-NITROBENZYLIDENE)-3,4-XYLIDINE

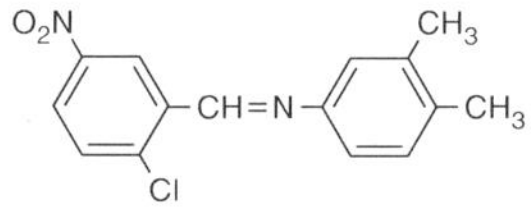

250 mg

S62,744-5
N-(2-CHLORO-5-NITROBENZYLIDENE)-3,5-XYLIDINE

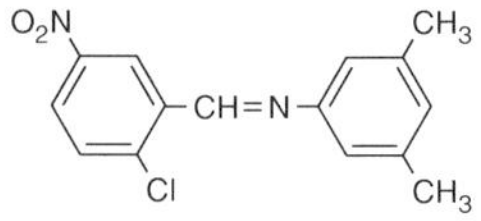

250 mg

S64,321-1
N,N'-BIS(PIPERONYLIDENE)-1,4-PHENYLENEDIAMINE

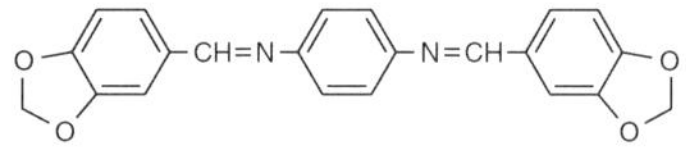

5 gm

S64,360-2
N,N'-BIS(2-CHLOROBENZYLIDENE)-1,4-PHENYLENEDIAMINE

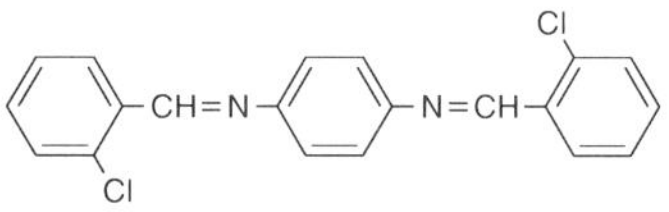

5 gm

S64,361-0
N,N'-DIBENZYLIDENE-1,4-PHENYLENEDIAMINE

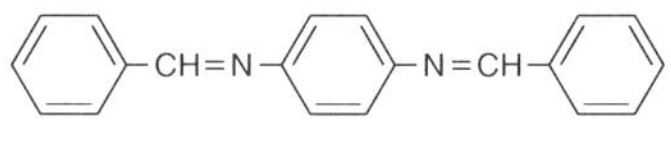

5 gm

S64,325-4
N,N'-BIS(4-CHLOROBENZYLIDENE)-1,4-PHENYLENEDIAMINE

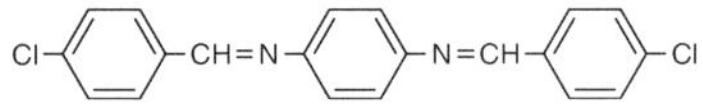

5 gm

S64,320-3
N,N'-BIS(4-METHYLBENZYLIDENE)-1,4-PHENYLENEDIAMINE

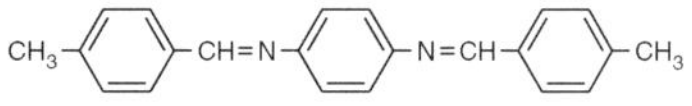

250 mg

S64,323-8
N,N'-BIS(CINNAMYLIDENE)-1,4-PHENYLENEDIAMINE

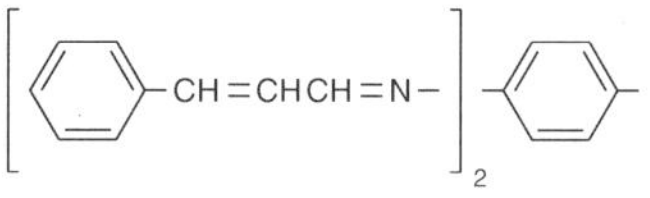

250 mg

S64,332-7
N,N'-BIS(4-ISOPROPYLBENZYLIDENE)-1,4-PHENYLENEDIAMINE

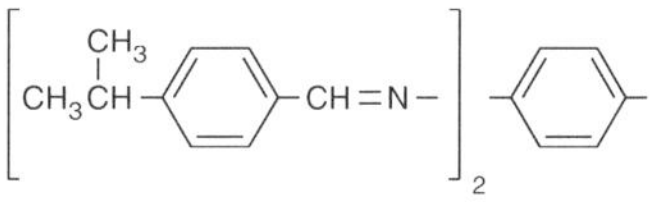

5 gm

S64,358-0
N,N'-BIS(2,4-DICHLOROBENZYLIDENE)-1,4-PHENYLENEDIAMINE

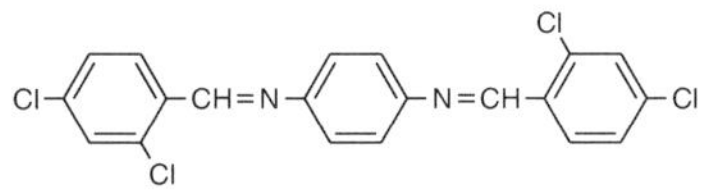

5 gm

S53,538-9
BIS(4-BROMOBENZYLIDENE)-2-CHLORO-1,4-PHENYLENEDIAMINE

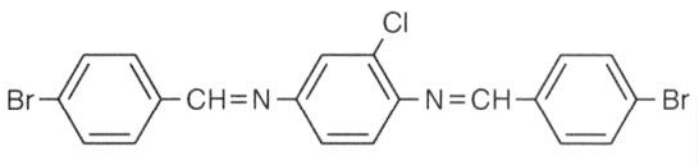

250 mg

S64,330-0
N,N'-BIS(9-ANTHRACENYLMETHYLENE)-1,4-PHENYLENEDIAMINE

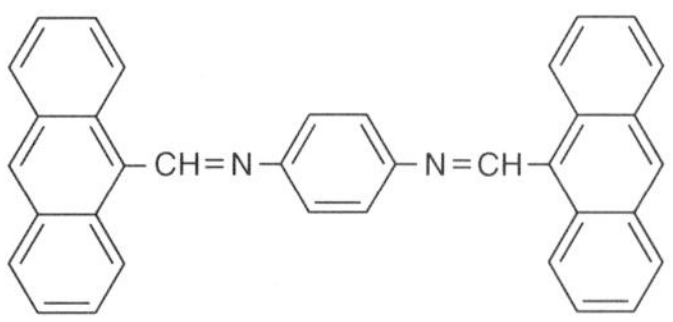

5 gm

S90,081-8
3-NITROBENZALDEHYDE N-PHENYL-N-(3-SULFOLANYL)HYDRAZONE

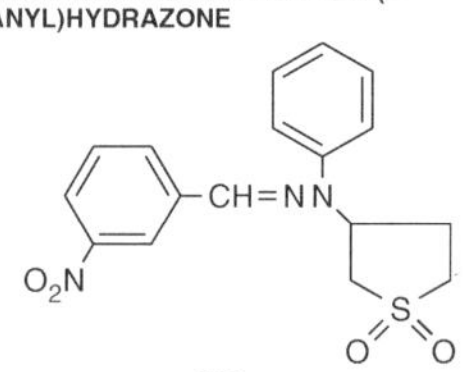

250 mg

S91,291-3
4-PYRIDINECARBOXYLIC (3-NITROBENZYLIDENE)HYDRAZIDE

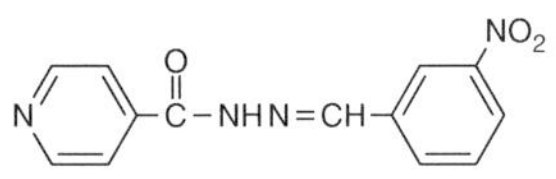

250 mg

S91,545-9
3-NITROBENZALDEHYDE SEMICARBAZONE

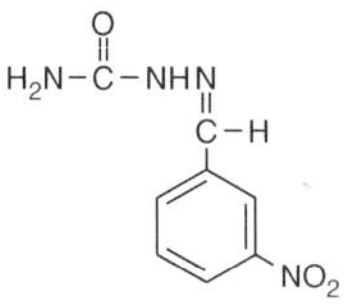

250 mg

S91,954-3
3-NITROBENZALDEHYDE PHENYLHYDRAZONE

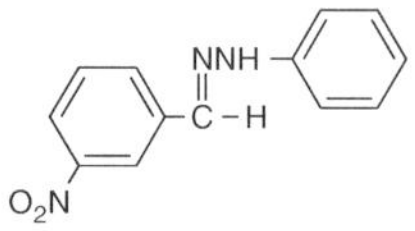

250 mg

S93,502-6
4-(3-NITROBENZYLIDENEAMINO)-4H-1,2,4-TRIAZOLE

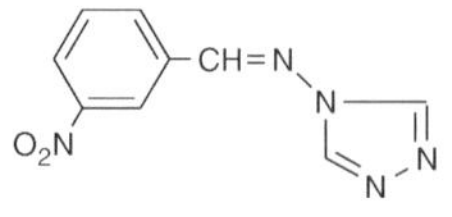

250 mg

S90,388-4
HIPPURIC (4-PYRIDYLMETHYLENE)-HYDRAZIDE
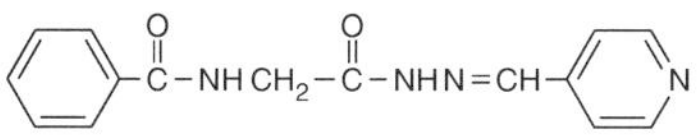
250 mg

S89,566-0
4-INDOLECARBOXALDEHYDE SEMICARBAZONE
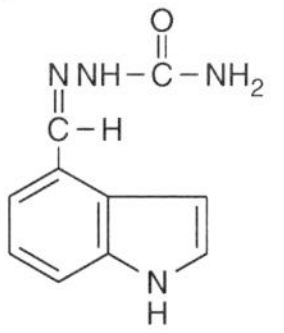
250 mg

S89,545-8
3-INDOLEACETALDEHYDE SEMICARBAZONE
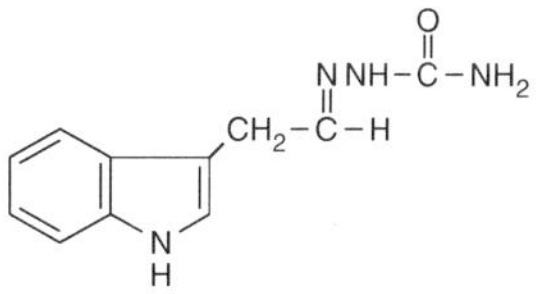
250 mg

S89,546-6
6,7-DIHYDRO-3-INDOLEACETALDEHYDE SEMICARBAZONE
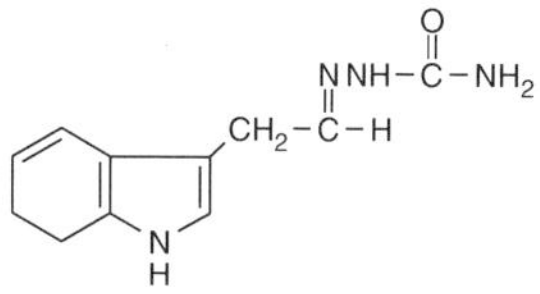
250 mg

S54,232-6
ISONICOTINIC ACID (3,4,5-TRIMETHOXYBENZYLIDENE)HYDRAZIDE
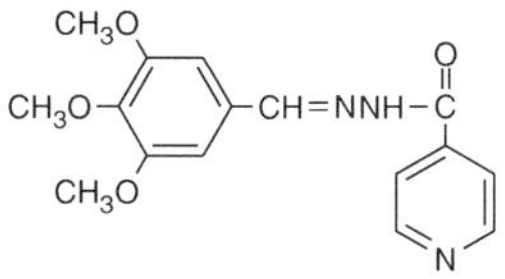
250 mg

S51,672-4
2-(3-NITROBENZYLIDENEAMINO)PYRIDINE
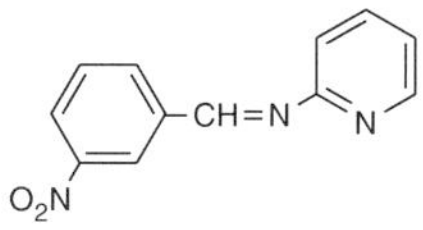
250 mg

S58,373-1
4-NITRO-N-(2-THIENYLMETHYLENE)-ANILINE
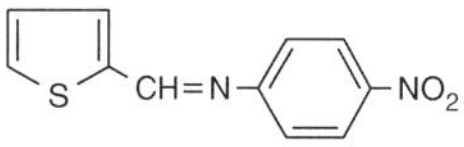
250 mg

S58,978-0
4-NITRO-N-(2-THIENYLMETHYLENE)-O-TOLUIDINE
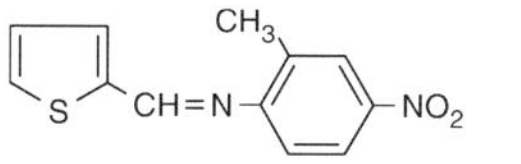
5 gm

S64,364-5
N,N'-BIS(2-THENYLIDENE)-1,4-PHENYLENEDIAMINE
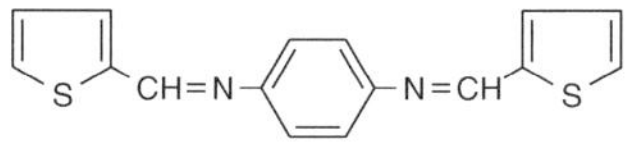
5 gm

S71,990-0
2,3-DIMETHYL-4-(3-NITROBENZYLIDENEAMINO)-1-PHENYL-3-PYRAZOLIN-5-ONE
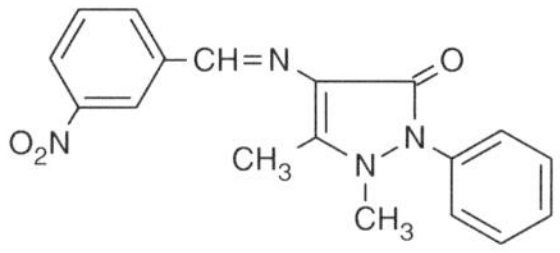
5 gm

S92,270-6
2-(4-NITROPHENYLIMINO)-1,3-BENZODITHIOLE
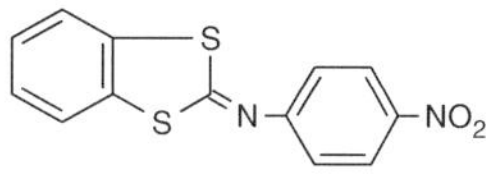
250 mg

S90,074-5
N-(3-NITROBENZYLIDENE)ANILINE
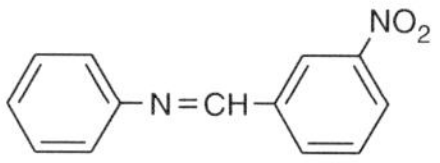
250 mg

S51,661-9
3-CHLORO-N-(3-NITROBENZYLIDENE)-ANILINE
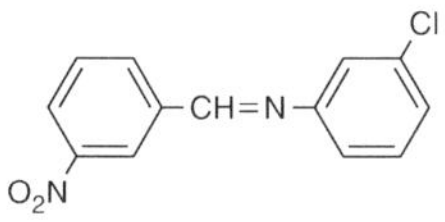
5 gm

S51,096-3
N-(3-NITROBENZYLIDENE)-M-TOLUIDINE
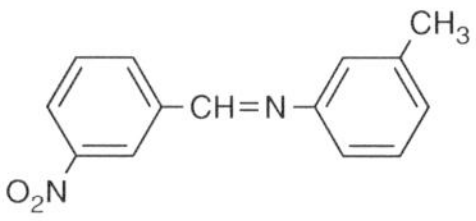
5 gm

S53,801-9
3-NITROBENZALDEHYDE AZINE
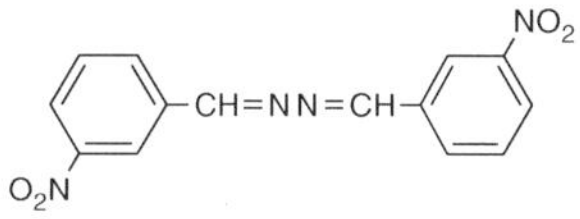
5 gm

S65,582-1
ETHYL 3-ETHOXY-3-(4-NITROPHENYLIMINO)PROPIONATE
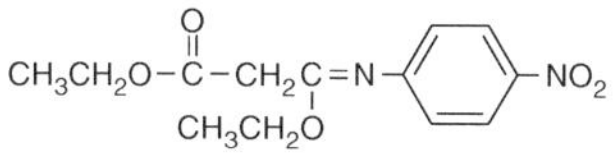
5 gm

S65,241-5
4-[N-(3-NITROBENZYLIDENE)-AMINO]PHENYL SULFONE
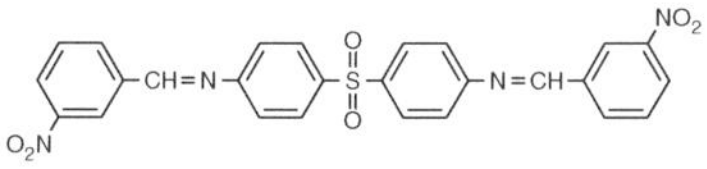
250 mg

S39,904-3
4-NITRO-N-PIPERONYLIDENEANILINE
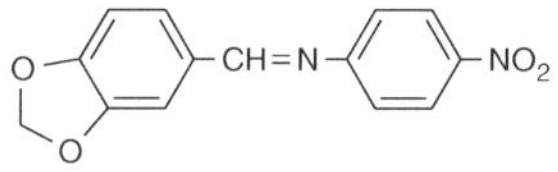
5 gm

S51,104-8
2-CHLORO-N-(3-NITROBENZYLIDENE)-ANILINE
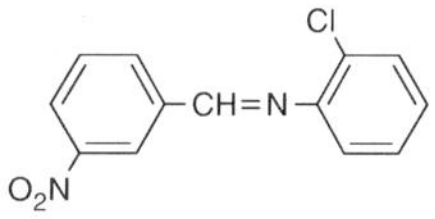
5 gm

S51,105-6
N-(3-NITROBENZYLIDENE)-O-TOLUIDINE
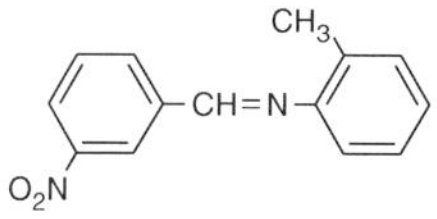
5 gm

S36,555-6
N-BENZYLIDENE-4-NITROANILINE
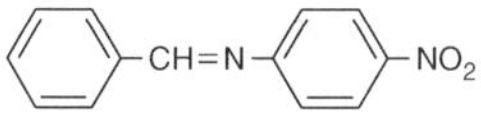
5 gm

S51,092-0
4-CHLORO-N-(3-NITROBENZYLIDENE)-ANILINE
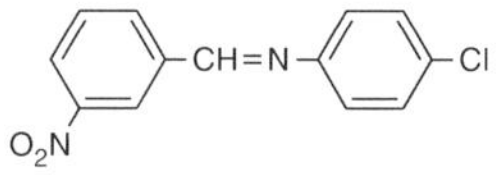
5 gm

S51,094-7
N-(3-NITROBENZYLIDENE)-P-TOLUIDINE
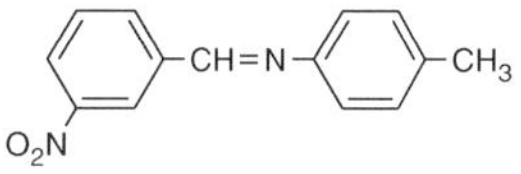
5 gm

S75,440-4
4-ETHYL-N-(2-NITRO-9-FLUORENYLIDENE)ANILINE
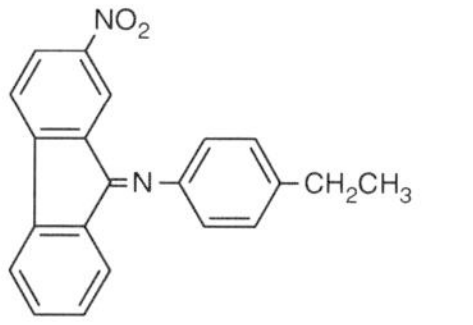
250 mg

S51,099-8
2,3-DICHLORO-N-(3-NITROBENZYLIDENE)-ANILINE
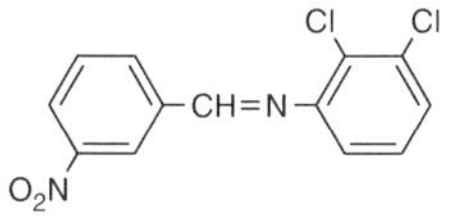
5 gm

S51,093-9
2,4-DICHLORO-N-(3-NITRO-BENZYLIDENE)ANILINE
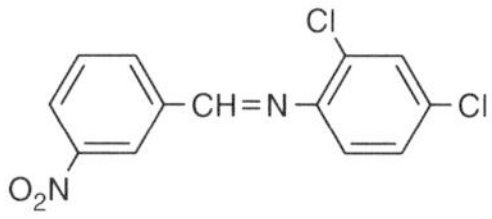
5 gm

S51,101-3
2,5-DICHLORO-N-(3-NITROBENZYLIDENE)-ANILINE
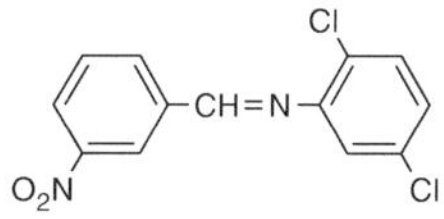
5 gm

S51,097-1
5-CHLORO-N-(3-NITROBENZYLIDENE)-O-TOLUIDINE
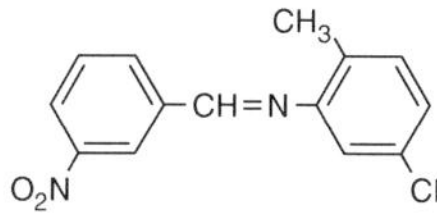
5 gm

S51,707-0
N-(3-NITROBENZYLIDENE)-3,4-XYLIDINE
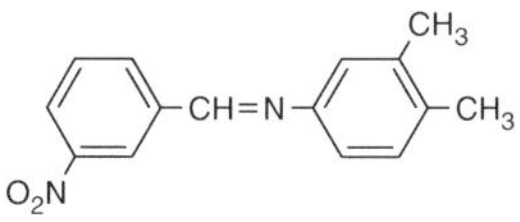
5 gm

S52,269-4
N-(3-NITROBENZYLIDENE)-2,4-XYLIDINE
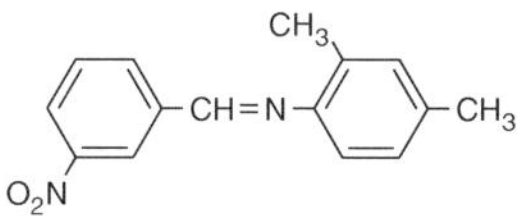
5 gm

S51,608-2
N-(3-NITROBENZYLIDENE)-1-NAPHTHYLAMINE
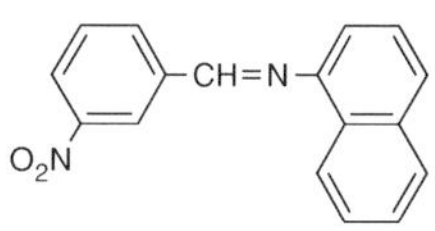
5 gm

S50,375-4
ISONICOTINIC ACID BENZYLIDENEHYDRAZIDE
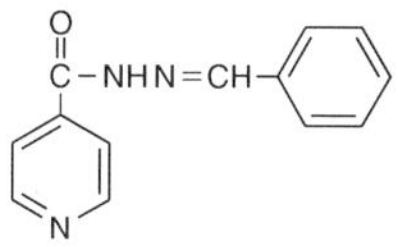
5 gm

S45,102-9
N-ALPHA-ACETYL-L-ARGININE DIHYDRATE
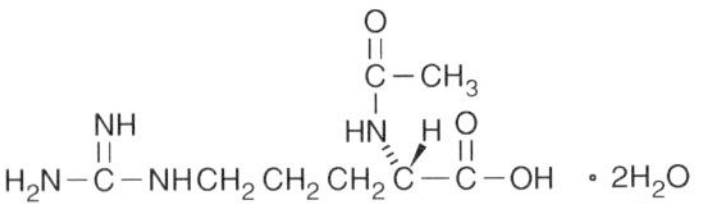
5 gm

S72,593-5
4-(4-ACETAMIDOPHENYLIMINO)-2,3-DICHLORO-2-BUTENOIC ACID
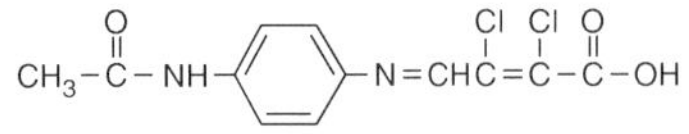
5 gm

S75,567-2
4'-(9-FLUORENYLIDENEAMINO)-ACETANILIDE
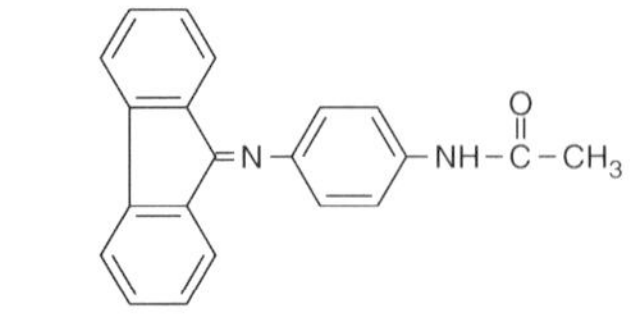
250 mg

S50,624-9
SUCCINIC ACID BIS(3,4,5-TRIMETHOXYBENZYLIDENEHYDRAZIDE)
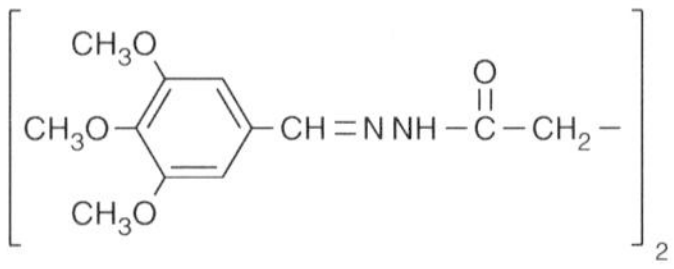
250 mg

S35,521-6
4-PYRIDINECARBOXALDEHYDE THIOSEMICARBAZONE
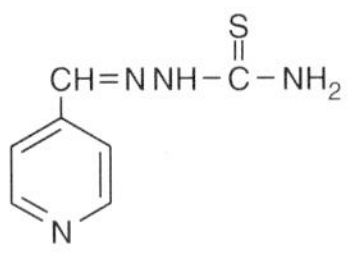
5 gm

S88,263-1
2,2'-BIS(BENZYLIDENEAMINO)BIPHENYL
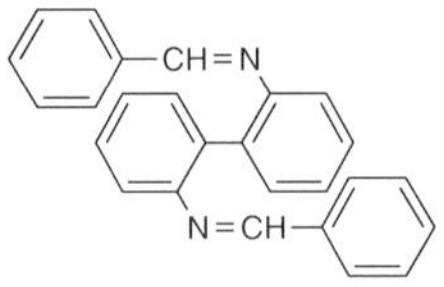
250 mg

S51,132-3
ISONICOTINALDEHYDE O-(4-CHLOROBENZYL)OXIME
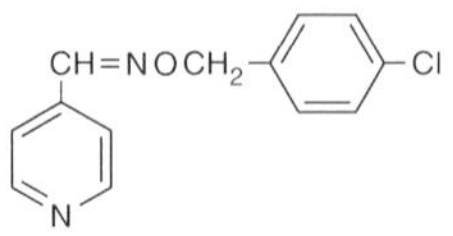
250 mg

S35,512-7
4-PYRIDINEALDOXIME O-ISOAMYL ETHER
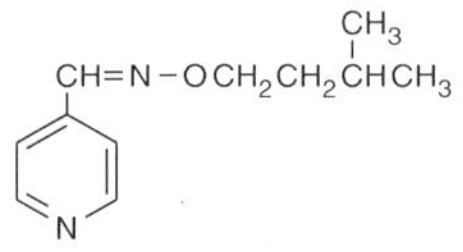
5 gm

S92,524-1
N-METHYLANTHRANILIC (2-PHENYLETHYLIDENE)HYDRAZIDE
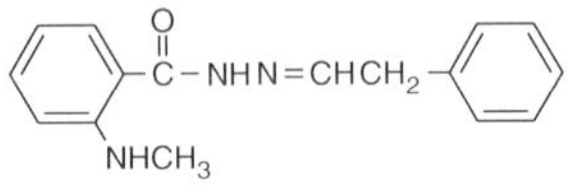
250 mg

S87,961-4
N-BENZYLIDENE-ALPHA-PHENYLPHENETHYLAMINE
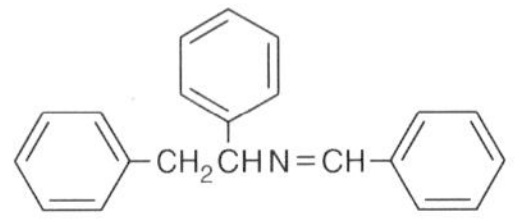
5 gm

S92,450-4
4-PHENYL-1-(2-PHENYLETHYLIDENE)-3-THIOSEMICARBAZIDE
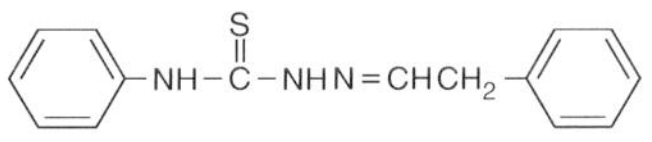
250 mg

S92,472-5
OXALIC BIS(2-PHENYLETHYLIDENE)-HYDRAZIDE
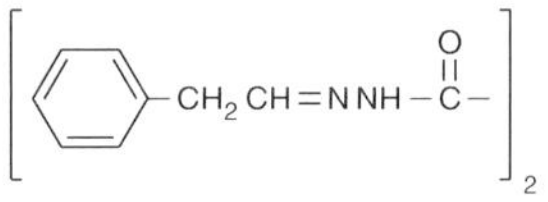
250 mg

S73,173-0
ADIPIC BIS[(2-PHENYLETHYLIDENE)-HYDRAZIDE]
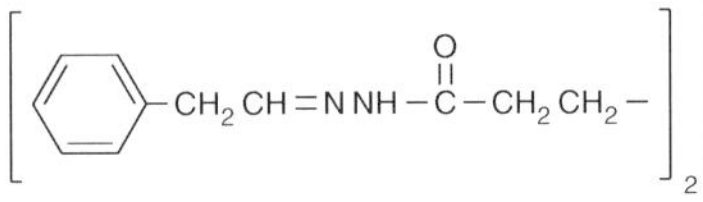
5 gm

S86,425-0
DEOXYBENZOIN (PHENYLSULFONYL)-HYDRAZONE
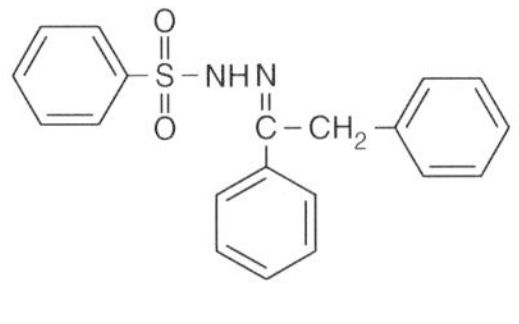
250 mg

S53,664-4
PHENYLACETALDEHYDE THIOSEMICARBAZONE
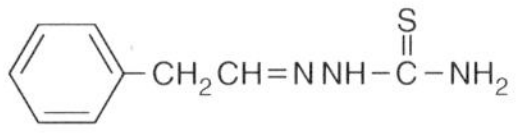
250 mg

S35,974-2
4-(ALPHA-CYANOBENZYL)PIPERIDINE HYDROCHLORIDE
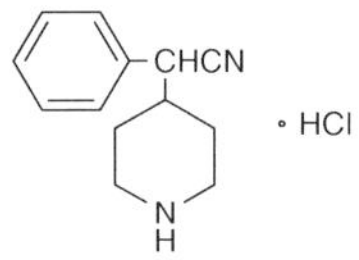
250 mg

S69,878-4
N-[4-(DIETHYLAMINO)BENZYLIDENE]-N'-PHENYL-1,4-PHENYLENEDIAMINE
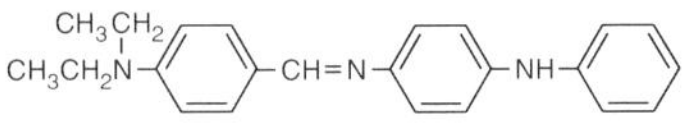
5 gm

S87,350-0
4-(2-ANILINOVINYL)-1-ETHYLQUINOLINIUM P-TOLUENESULFONATE
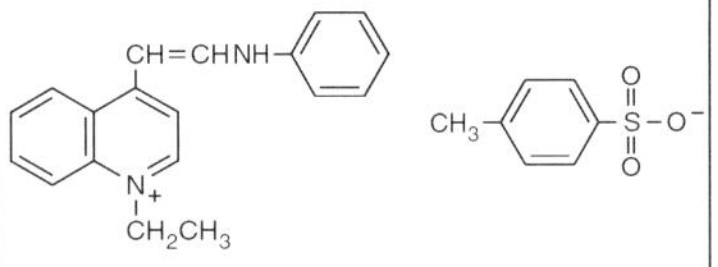
250 mg

S93,431-3
4-(BENZYLAMINOMETHYLENEAMINO)-2-(ETHOXYCARBONYL)-4H-1,2,4-TRIAZOL-2-IUM CHLORIDE
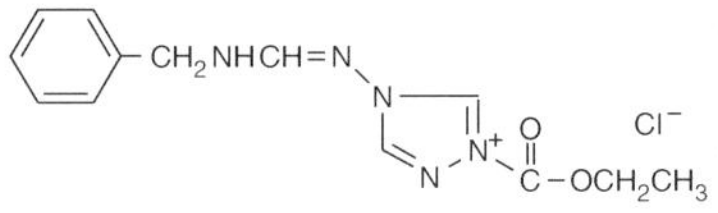
250 mg

S69,949-7
N,N'-TEREPHTHALYLIDENEBIS(4-ANILINOANILINE)
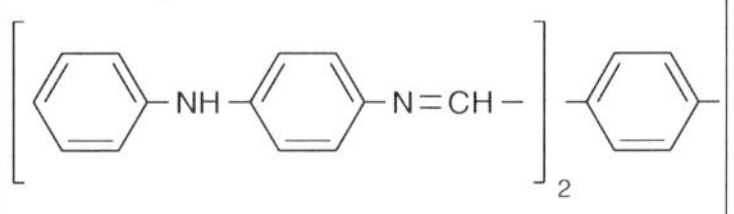
5 gm

S54,439-6
N-[2-BROMO-3-(PHENYLIMINO)-PROPENYL]ANILINE HYDROBROMIDE
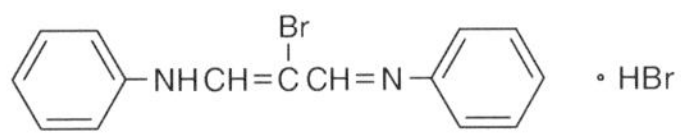
250 mg

S92,529-2
N-METHYLANTHRANILIC CINNAMYLIDENEHYDRAZIDE
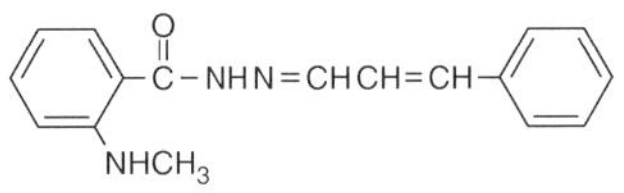
250 mg

S94,424-6
BASIC RED 8 (C.I. 45150)
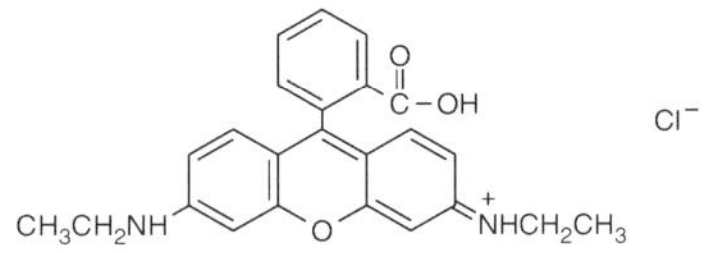
250 mg

S94,420-3
BASIC DYE (C.I. 45000)
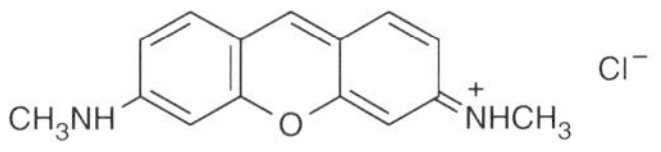
250 mg

S53,331-9
ISATIN, AZINE WITH 2-CHLOROBENZALDEHYDE
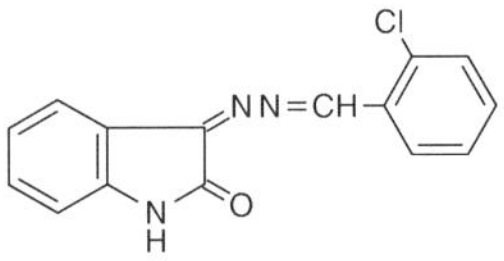
5 gm

S94,417-3
BASIC YELLOW 3 (C.I. 41005)
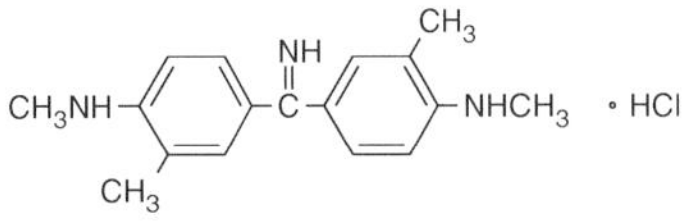
250 mg

S94,418-1
BASIC DYE (C.I. 44520)
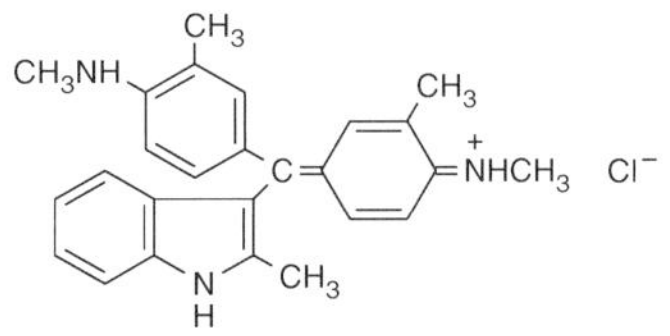
250 mg

S41,212-0
N,N'-DI-1-NAPHTHYLFORMAMIDINE
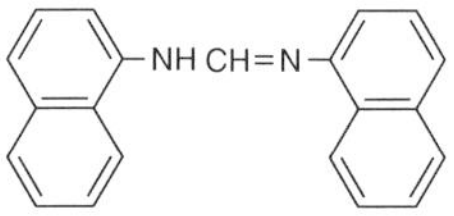
5 gm

S58,010-4
METHYL GREEN (C.I. 42585)
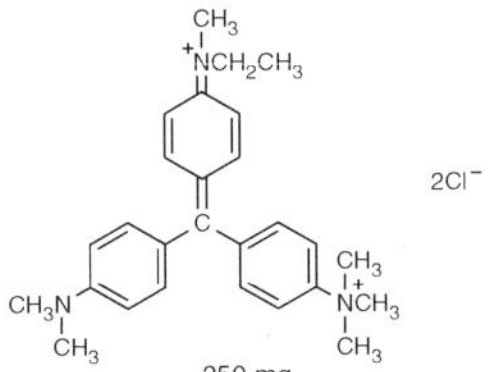
250 mg

S87,272-5
3-METHYL-2-BENZOTHIAZOLINONE [4-(DIMETHYLAMINO)BENZYLIDENE]-HYDRAZONE
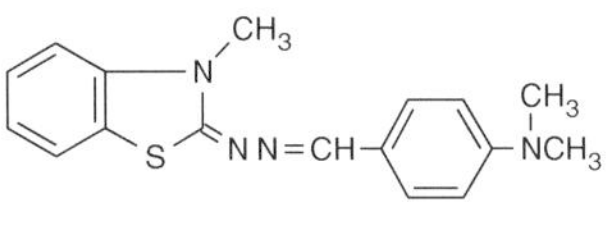
250 mg

S67,441-9
4-(3-MORPHOLINOPROPYL)-1-(4-NITROBENZYLIDENE)THIOSEMICARBAZIDE
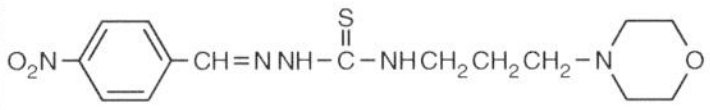
250 mg

S92,512-8
HIPPURIC N-(4-DIMETHYLAMINO-BENZYLIDENE)HYDRAZIDE
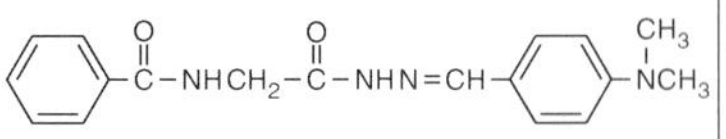
250 mg

S63,962-1
N-[4-(DIMETHYLAMINO)BENZYLIDENE]-4-(ETHOXYCARBONYL)ANILINE
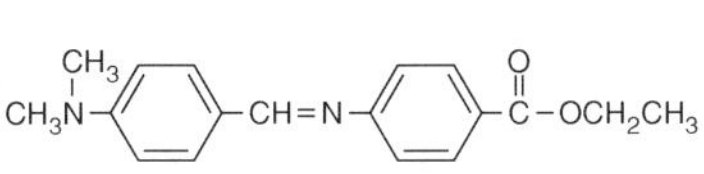
5 gm

S51,646-5
ALPHA-(2-CHLOROPHENYLIMINO)-N,N-DIMETHYL-P-TOLUIDINE
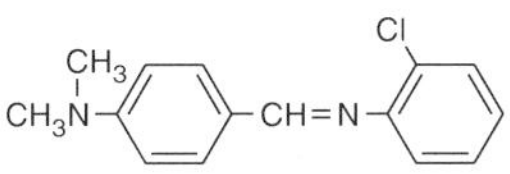
5 gm

S51,420-9
N,N-DIMETHYL-ALPHA-(O-TOLYLIMINO)-P-TOLUIDINE
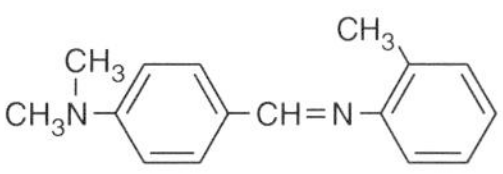
5 gm

S50,764-4
N,N-DIMETHYL-ALPHA-PHENYLIMINO-P-TOLUIDINE
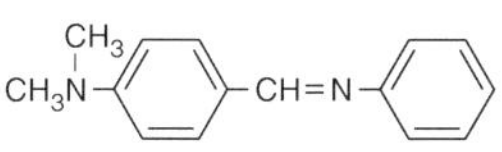
5 gm

S51,198-6
ALPHA-(3-CHLOROPHENYLIMINO)-N,N-DIMETHYL-P-TOLUIDINE
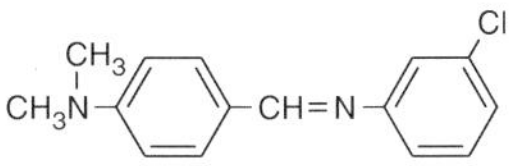
5 gm

S51,310-5
ALPHA-(4-CHLOROPHENYLIMINO)-N,N-DIMETHYL-P-TOLUIDINE
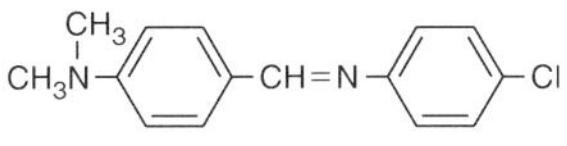
5 gm

S51,421-7
N,N-DIMETHYL-ALPHA-(M-TOLYLIMINO)-P-TOLUIDINE
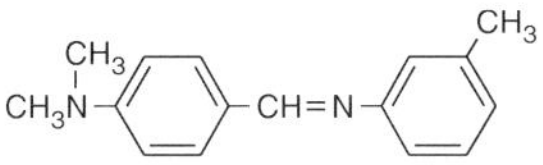
5 gm

S51,422-5
N,N-DIMETHYL-ALPHA-(P-TOLYLIMINO)-P-TOLUIDINE
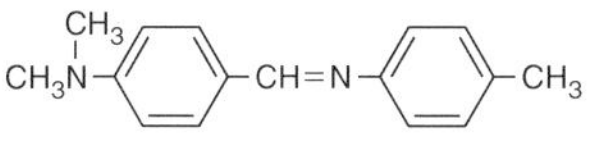
5 gm

S65,585-6
N-[4-(DIETHYLAMINO)BENZYLIDENE]-ANILINE
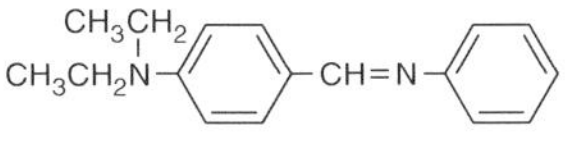
250 mg

S61,865-9
N-ETHYL-N-[4-(PHENYLIMINOMETHYL)-PHENYL]AMINOACETONITRILE
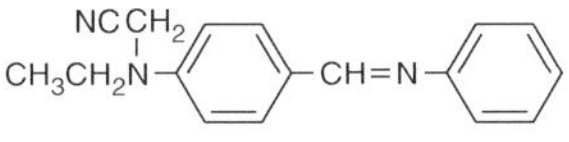
250 mg

S51,414-4
ALPHA-(2,3-DICHLOROPHENYLIMINO)-N,N-DIMETHYL-P-TOLUIDINE
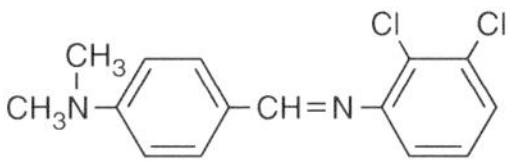
5 gm

S50,766-0
ALPHA-(3-CHLORO-O-TOLYLIMINO)-N,N-DIMETHYL-P-TOLUIDINE
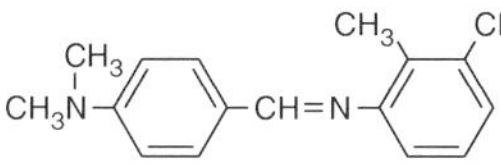
5 gm

S61,819-5
2,6-DIETHYL-N-[4-(DIMETHYLAMINO)-BENZYLIDENE]ANILINE
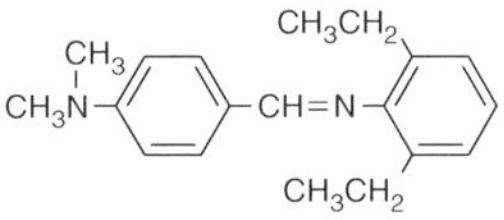
250 mg

S51,295-8
ALPHA-(2,5-DICHLOROPHENYLIMINO)-N,N-DIMETHYL-P-TOLUIDINE
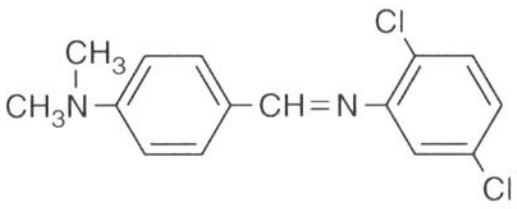
5 gm

S51,309-1
ALPHA-(2,4-DICHLOROPHENYLIMINO)-N,N-DIMETHYL-P-TOLUIDINE
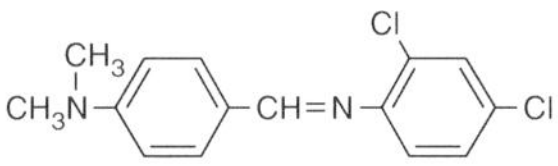
5 gm

S63,609-6
ALPHA-(3,4-DICHLOROPHENYLIMINO)-N,N-DIMETHYL-P-TOLUIDINE
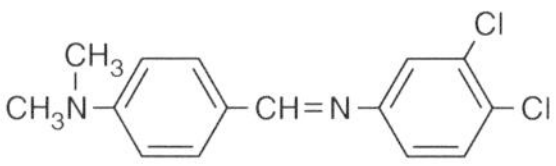
5 gm

S50,768-7
ALPHA-(3-CHLORO-P-TOLYLIMINO)-N,N-DIMETHYL-P-TOLUIDINE
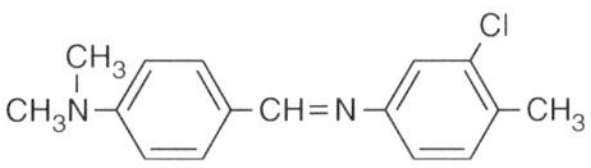
5 gm

S50,769-5
ALPHA-(4-CHLORO-O-TOLYLIMINO)-N,N-DIMETHYL-P-TOLUIDINE
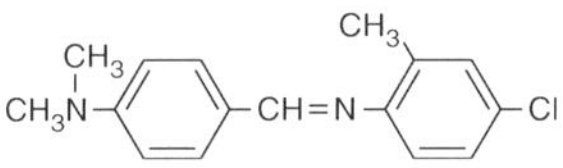
5 gm

S51,315-6
ALPHA-(5-CHLORO-O-TOLYLIMINO)-N,N-DIMETHYL-P-TOLUIDINE
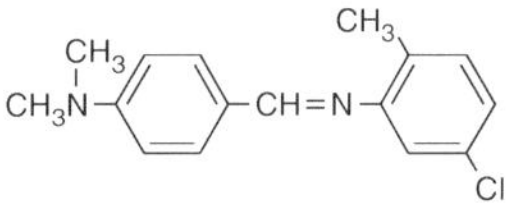
5 gm

S64,328-9
4-BROMO-N-[4-(DIMETHYLAMINO)-BENZYLIDENE]-3-METHYLANILINE
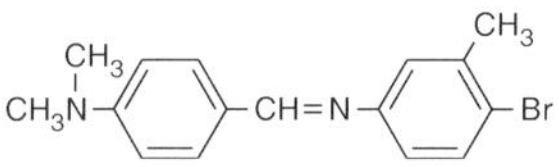
250 mg

S52,268-6
N,N-DIMETHYL-ALPHA-(2,4-XYLYLIMINO)-P-TOLUIDINE
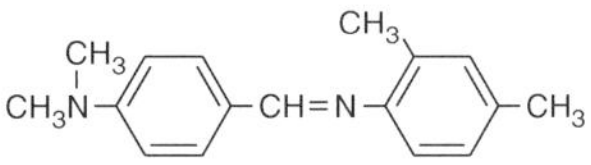
5 gm

S52,433-6
N-[4-(DIMETHYLAMINO)BENZYLIDENE]-2,5-XYLIDINE
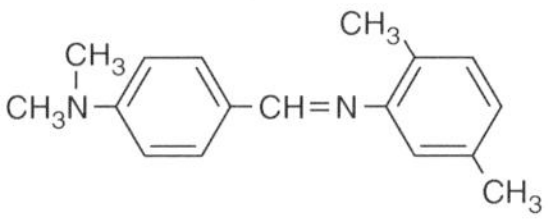
5 gm

S62,275-3
N-(4-DIMETHYLAMINOBENZYLIDENE)-4-AMINOBIPHENYL
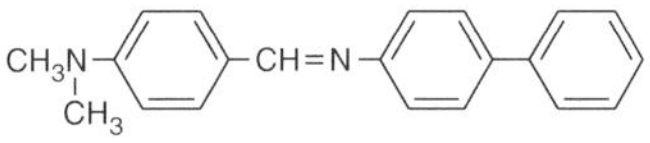
250 mg

S69,362-6
N,N'-BIS[4-(DIMETHYLAMINO)-BENZYLIDENE]HYDRAZINE
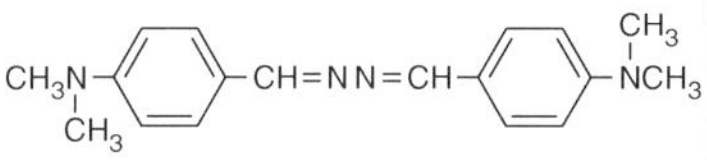
250 mg

S87,497-3
3-METHYL-2-BENZOTHIAZOLINONE (1,2-DIHYDRO-2-OXO-1-NAPHTHYLIDENE)HYDRAZONE
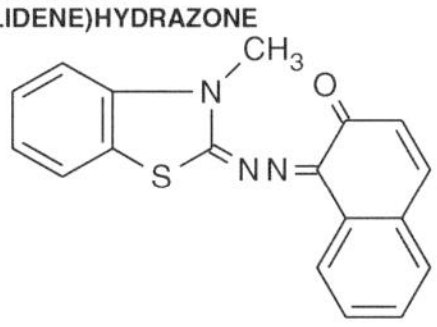
250 mg

S87,511-2
3-METHYL-2-BENZOTHIAZOLINONE [2,3-DIHYDRO-2-OXOBENZO(B)-THIOPHEN-3-YLIDENE]HYDRAZONE
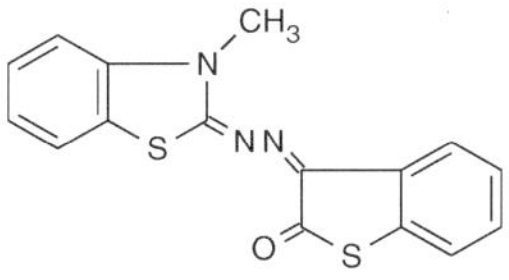
250 mg

S87,317-9
2,3-DIHYDRO-3-METHYL-2-BENZOXAZOLONE (4-NITROBENZYLIDENE)-HYDRAZONE
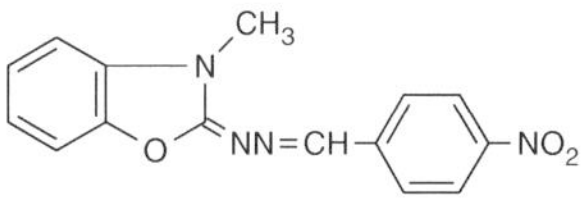
250 mg

S87,270-9
3-METHYL-2-BENZOTHIAZOLINONE (4-NITROBENZYLIDENE)HYDRAZONE
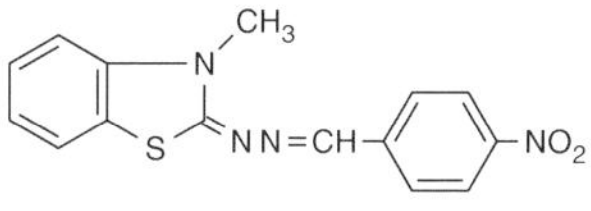
250 mg

S79,690-5
4-(4-METHYL-1-PIPERAZINYLCARBOXAMIDO)BENZOIC (3-NITROBENZYLIDENE)HYDRAZIDE
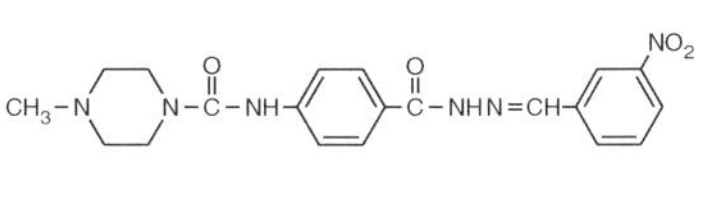
250 mg

S87,486-8
3-METHYL-2-BENZOTHIAZOLINONE [1,2-DIHYDRO-3-(N-PHENYLCARBAMOYL)-2-OXO-1-NAPHTHYLIDENE]HYDRAZONE
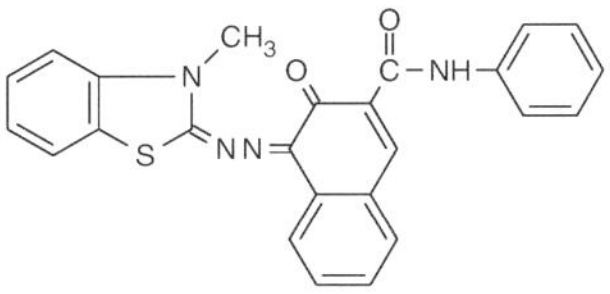
250 mg

S51,206-0
1,2-DIHYDRO-4,6-DIMETHYL-1-[4-(DIMETHYLAMINO)BENZYLIDENEAMINO]-2-OXONICOTINONITRILE
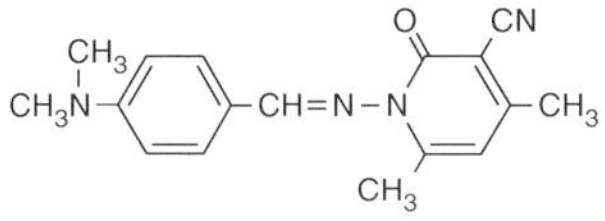
250 mg

S73,701-1
4-PYRIDINECARBOXYLIC 4-(DIMETHYLAMINO)-BENZYLIDENEHYDRAZIDE
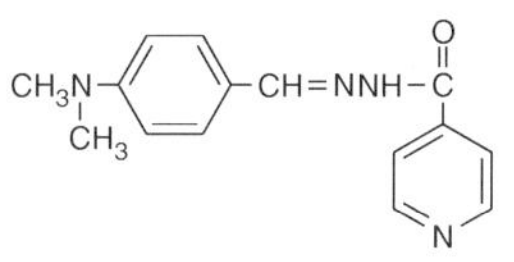
5 gm

S52,023-3
4-[4-(DIMETHYLAMINO)-BENZYLIDENEAMINO]-6-METHYL-1,2,4-TRIAZINE-3,5(2H,4H)-DIONE
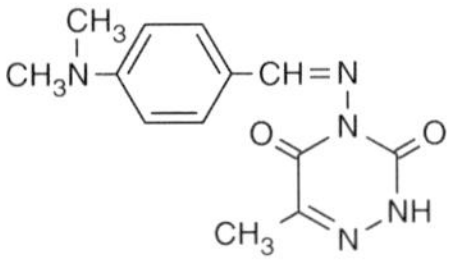
250 mg

S67,837-6
1-METHYL-2-PYRROLECARBOXALDEHYDE SEMICARBAZONE
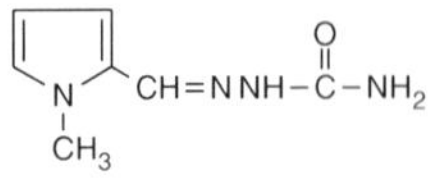
250 mg

S63,881-1
ACETIC ACID, (9-ETHYL-2-CARBAZOLYLMETHYLENE)HYDRAZIDE
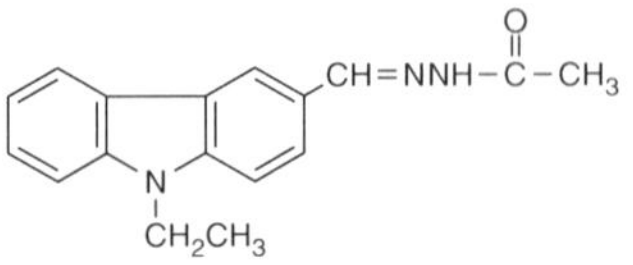
5 gm

S87,379-9
3-METHYL-2-BENZOTHIAZOLINONE (1H-PHENALEN-1-YLIDENE)HYDRAZONE
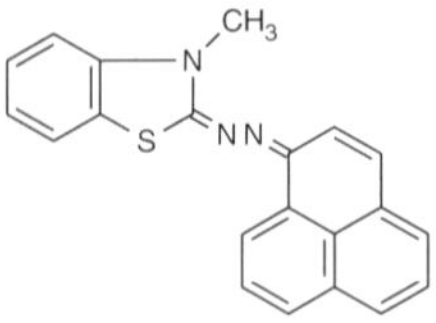
250 mg

S69,211-5
4-CYCLOHEXYL-1-(4-DIMETHYLAMINOBENZYLIDENE)-3-THIOSEMICARBAZIDE
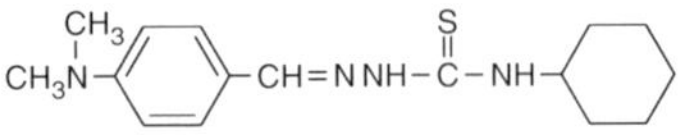
250 mg

S36,111-9
N,N-DIMETHYL-N'-PHENYLFORMAMIDINE
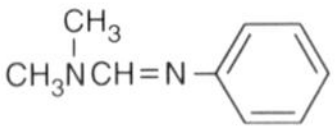
5 gm

S51,821-2
2-[P-(DIMETHYLAMINO)-BENZYLIDENEAMINO]PYRIDINE
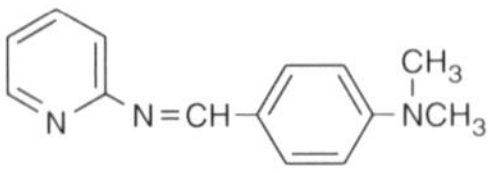
250 mg

S68,654-9
4,4'-METHYLENEBIS(N-[4-(DIMETHYLAMINO)BENZYLIDENE]ANILINE)
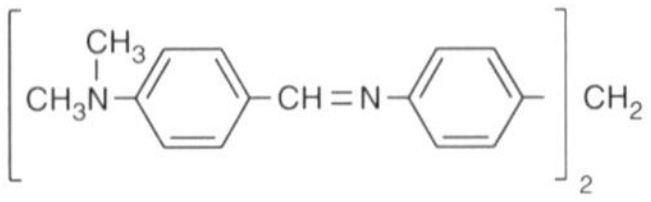
5 gm

S68,652-2
4,4'-METHYLENEBIS(N-[4-(DIETHYLAMINO)BENZYLIDENE]ANILINE)
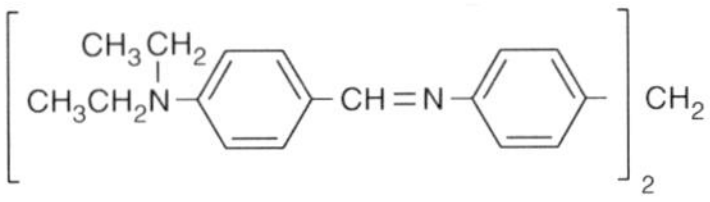
5 gm

S39,058-5
4-(4-DIETHYLAMINO-O-TOLYLIMINO)-1,4-DIHYDRO-4'-METHYLSULFONYL-1-OXONAPHTHANILIDE
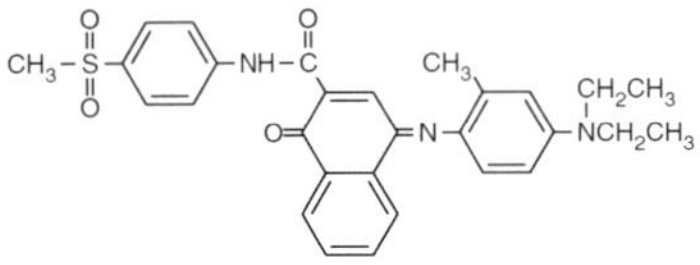
250 mg

S94,421-1
BASIC RED 11 (C.I. 45050)
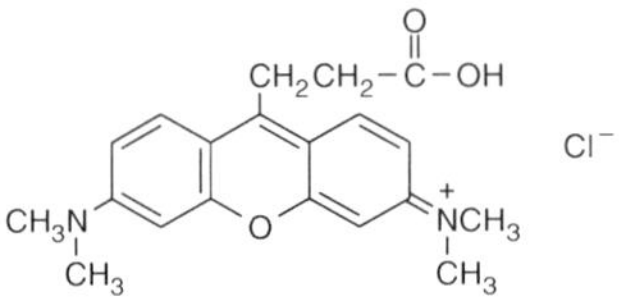
250 mg

S50,640-0
2,3-DIHYDRO-2-(ETHOXYCARBONYLHYDRAZINYLYLIDENE)-3-METHYLBENZOTHIAZOLE
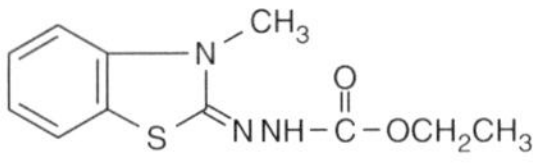
250 mg

S72,051-8
2,3-DIMETHYL-4-[4-(DIMETHYLAMINO)-BENZYLIDENEAMINO]-1-PHENYL-3-PYRAZOLIN-5-ONE
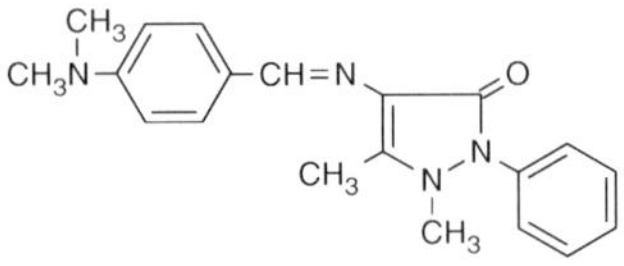
5 gm

S50,628-1
3-METHYL-2-BENZOTHIAZOLINONE, AZINE WITH 5-INDANCARBOXALDEHYDE
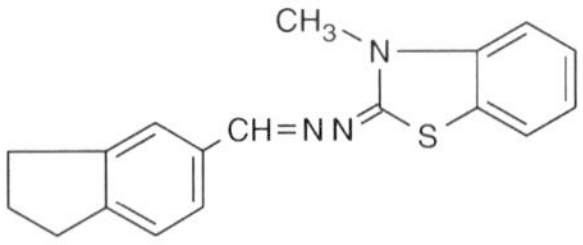
250 mg

S65,248-2
4-[4-(DIMETHYLAMINO)-BENZYLIDENEAMINO]PHENYL SULFONE
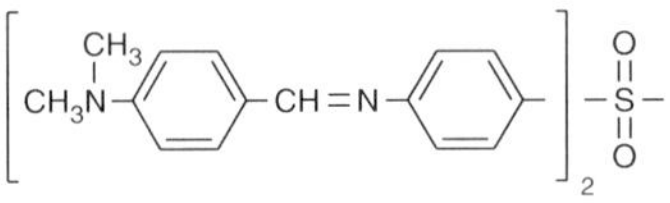
5 gm

S87,447-7
3-ETHYL-2-[2-(PHENYLIMINO)-ETHYLIDENE]BENZOTHIAZOLINE
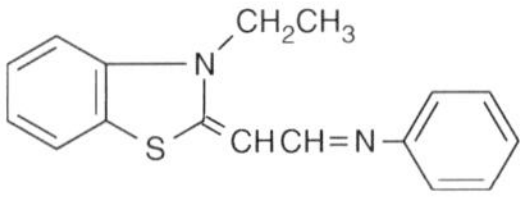
250 mg

S71,946-3
N,N-DIMETHYL-ALPHA-(2-ETHYLPHENYLIMINO)-P-TOLUIDINE
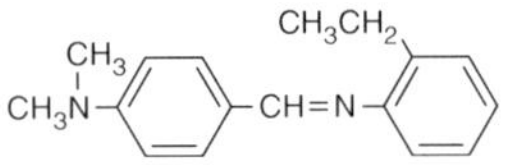
5 gm

S41,055-1
ALPHA-(4-BROMOPHENYLIMINO)-N,N-DIMETHYL-P-TOLUIDINE
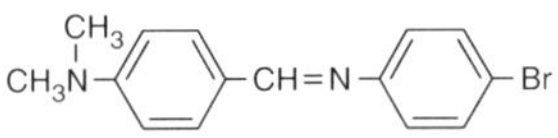
5 gm

S70,304-4
N-[4-(DIMETHYLAMINO)BENZYLIDENE]-4-ETHYLANILINE
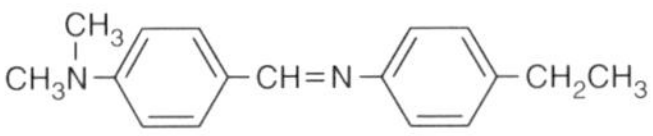
5 gm

S50,762-8
N,N-DIMETHYL-ALPHA-(3,4-XYLYLIMINO)-P-TOLUIDINE
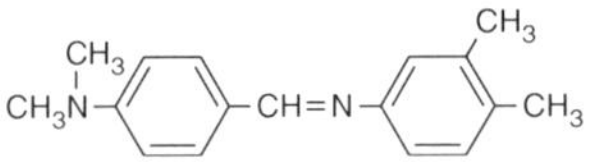
5 gm

S71,945-5
N,N-DIMETHYL-ALPHA-(2,4,6-TRIMETHYLPHENYLIMINO)-P-TOLUIDINE
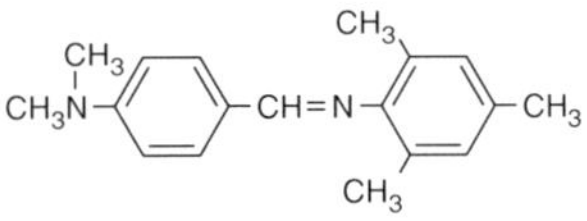
5 gm

S51,610-4
N,N-DIMETHYL-ALPHA-(1-NAPHTHYLIMINO)-P-TOLUIDINE
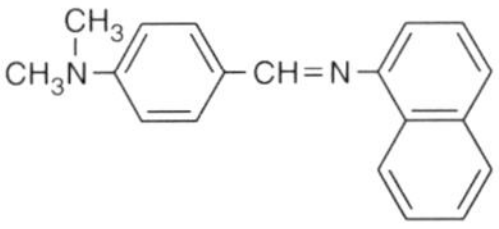
5 gm

S34,659-4
2-[4-(DIMETHYLAMINO)-BENZYLIDENEAMINO]FLUORENE
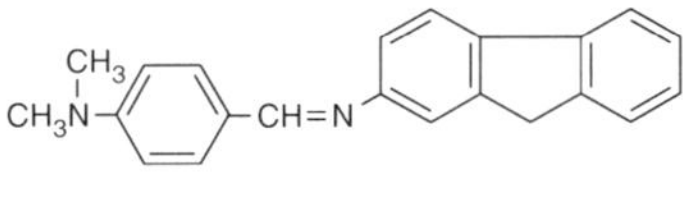
5 gm

S53,680-6
2-CHLORO-4-(DIMETHYLAMINO)-BENZALDEHYDE AZINE
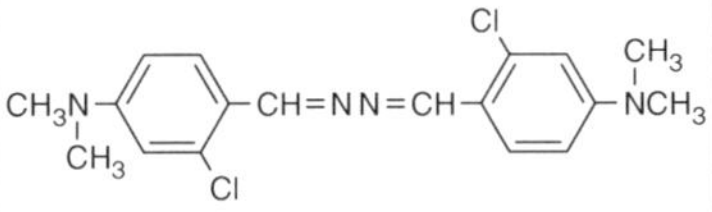
250 mg

S63,877-3
(BETA-CHLOROCINNAMYLIDENE)-DIMETHYLAMMONIUM PERCHLORATE
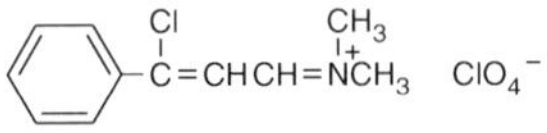
250 mg

S51,581-7
(GAMMA,4-DICHLOROCINNAMYLIDENE)-DIMETHYLAMMONIUM PERCHLORATE
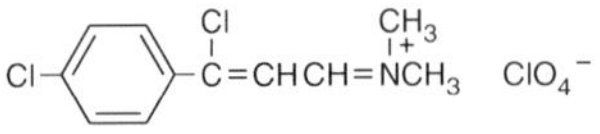
250 mg

S58,959-4
N,N'-(1,4-PHENYLENEDIMETHYLIDYNE)-BIS(ETHYL 4-AMINOBENZOATE)
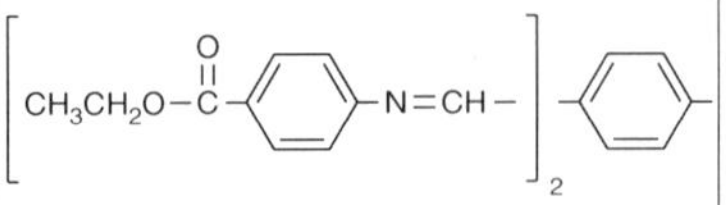
5 gm

S62,331-8
N,N'-TEREPHTHALYLIDENEDIANTHRANILIC ACID
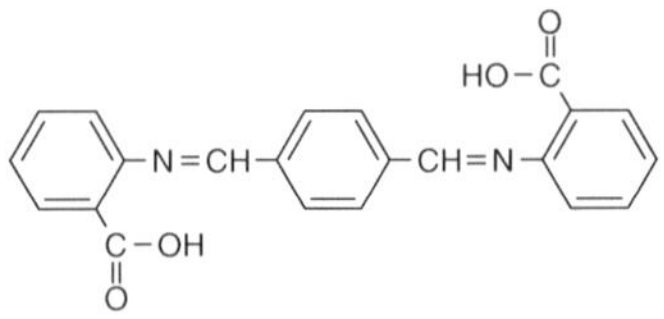
5 gm

S62,407-1
N,N'-(1,4-PHENYLENEDIMETHYLIDYNE)-BIS(4-CARBOXYANILINE)
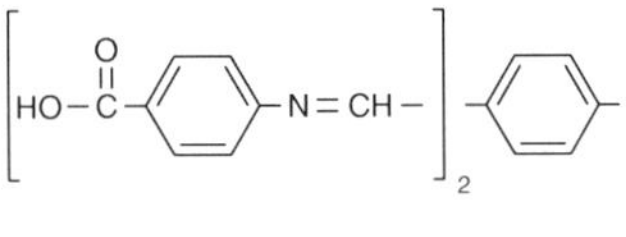
5 gm

S62,453-5
N,N'-TEREPHTHALYLIDENEBIS(3-CARBOXYANILINE)
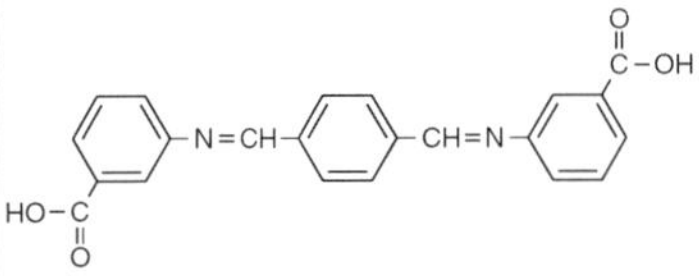
250 mg

S82,209-4
N-(5-NITROFURFURYLIDENE)ANILINE
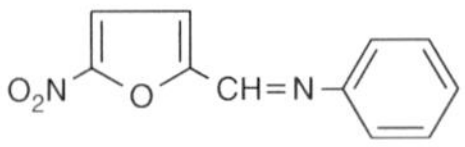
250 mg

S63,233-3
N-BENZYLIDENE-3-CHLOROANILINE
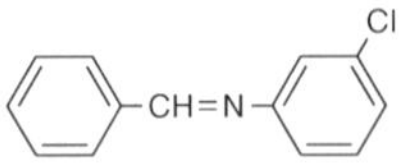
5 gm

S63,187-6
N-(3-CHLOROBENZYLIDENE)-M-TOLUIDINE

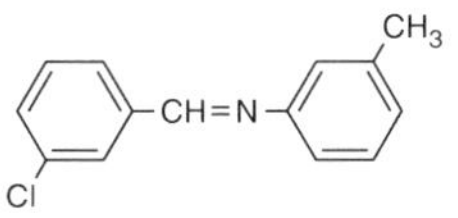

250 mg

S67,672-1
SEBACIC ACID BIS(3-NITROBENZYLIDENE)HYDRAZIDE

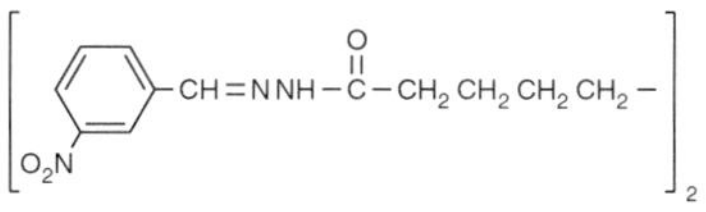

5 gm

S92,543-8
DECANOIC (3-NITROBENZYLIDENE)-HYDRAZIDE

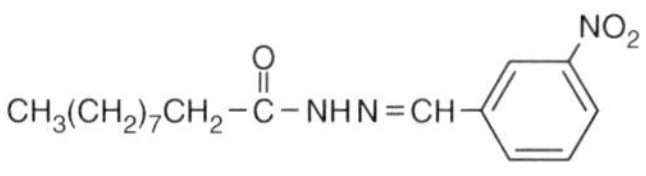

250 mg

S67,439-7
4-CYCLOHEXYL-1-(3-NITRO-BENZYLIDENE)-3-THIOSEMICARBAZIDE

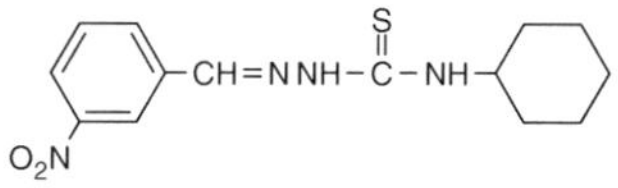

250 mg

S92,488-1
P-TOLUIC (4-ACETAMIDOBENZYLIDENE)-HYDRAZIDE

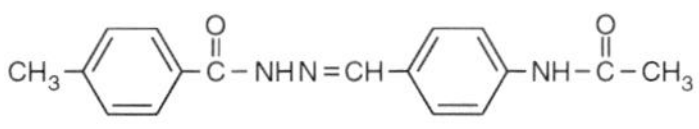

250 mg

S92,517-9
BENZOIC (ALPHA-PROPYLBENZYLIDENE)-HYDRAZIDE

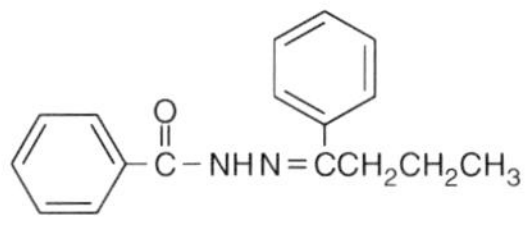

250 mg

S92,446-6
METHYL (4-METHYLBENZYLIDENE)-CARBAZATE

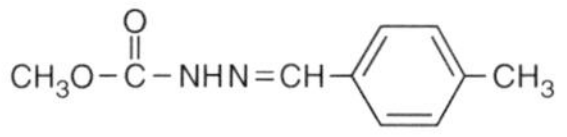

250 mg

S67,657-8
3-ACETAMIDOACETOPHENONE SEMICARBAZONE

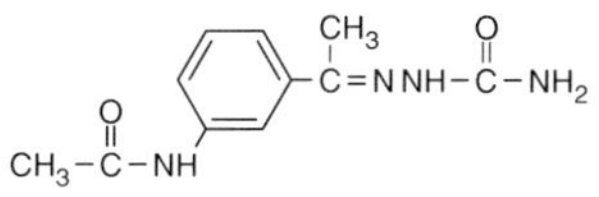

5 gm

S76,845-6
4,4-DIBENZYL-1-(3-PYRIDYL-METHYLENE)SEMICARBAZIDE

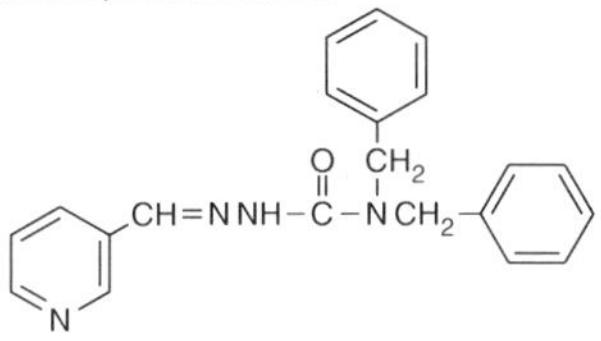

250 mg

S53,668-7
4-ISOPROPYLBENZALDEHYDE THIOSEMICARBAZONE

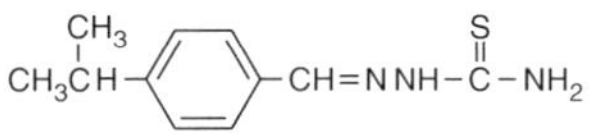

5 gm

S68,579-8
ALPHA,ALPHA-BIS(BENZYLIDENIMINO)-TOLUENE

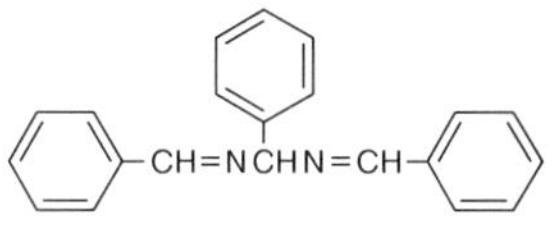

5 gm

S62,318-0
N-(2-THIENYLMETHYLENE)-ANTHRANILIC ACID

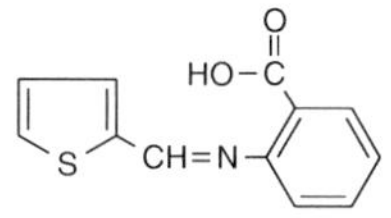

5 gm

S62,319-9
N-(2-THIENYLMETHYLENE)-2,4,5-TRICHLOROANILINE

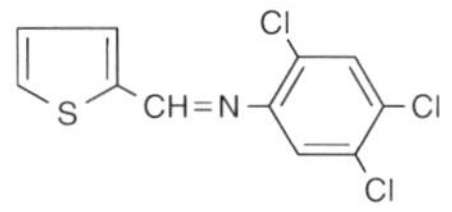

5 gm

S52,384-4
N-(4-NITROBENZYLIDENE)-4-TRITYLANILINE

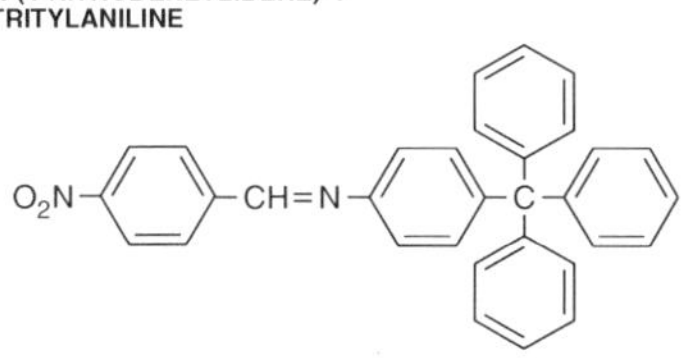

5 gm

S59,066-5
4-(BENZYLIDENEAMINO)BENZOIC ACID

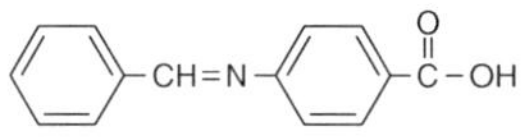

250 mg

S66,993-8
4-CARBOXY-N-(4-ISOPROPYLBENZYLIDENE)ANILINE

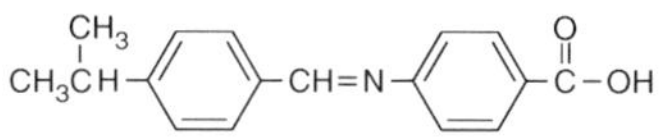

5 gm

S62,542-6
ETHYL 4-(9-ANTHRACENYLMETHYLENEAMINO)-BENZOATE

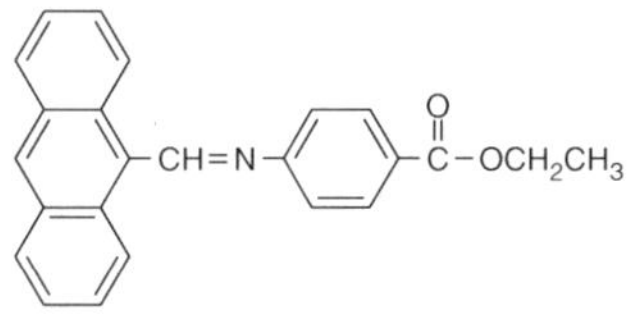

250 mg

S65,484-1
4,4'-SULFONYLBIS[N-(4-ACETAMIDOBENZYLIDENE)ANILINE]

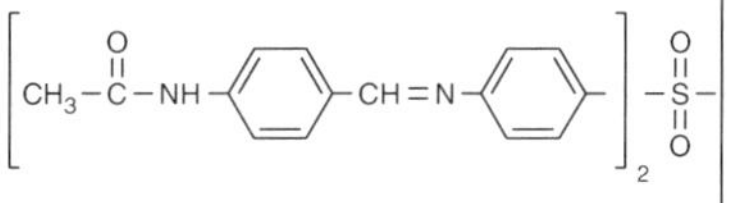

5 gm

S80,427-4
3-(BENZYLIMINO)-2-INDOLINONE

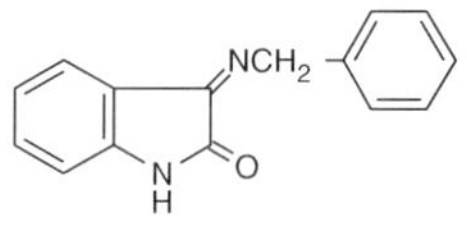

250 mg

S63,602-9
N-(2-CHLOROBENZYLIDENE)-4-CYANOANILINE

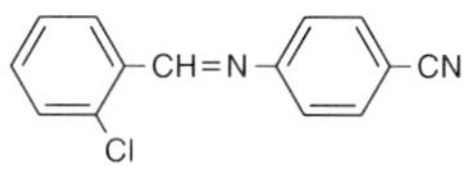

250 mg

S65,313-6
4-(PIPERONYLIDENEAMINO)PHENYL SULFONE

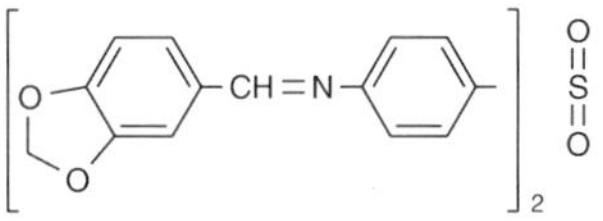

5 gm

S65,319-5
4,4'-BIS(4-CHLOROBENZYLIDENEAMINO)-DIPHENYL SULFONE

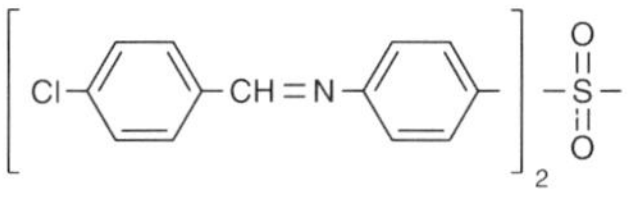

5 gm

S65,315-2
4-(4-METHYLBENZYLIDENEAMINO)PHENYL SULFONE

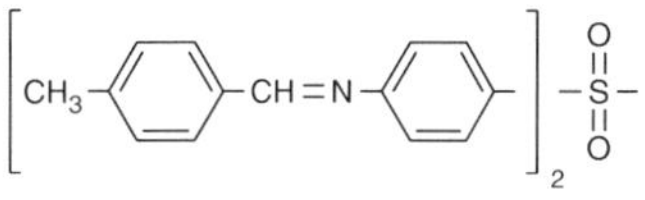

5 gm

S60,479-8
N-(9-ANTHRACENYLMETHYLENE)-P-TOLUIDINE

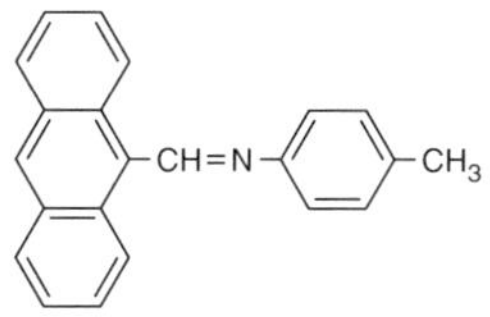

5 gm

S63,968-0
4-ETHYL-N-(PIPERONYLIDENE)ANILINE

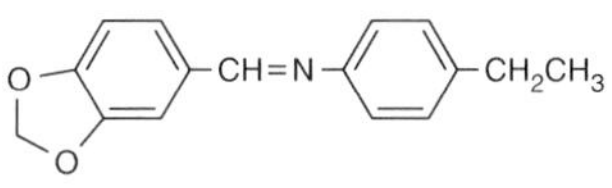

5 gm

S37,663-9
N-BENZYLIDENEPROPYLAMINE

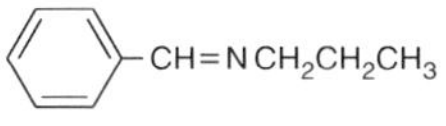

5 gm

S37,682-5
N-BENZYLIDENEISOPROPYLAMINE

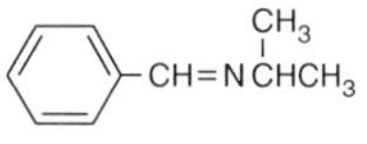

5 gm

S61,654-0
N,ALPHA-DIPHENYLNITRONE

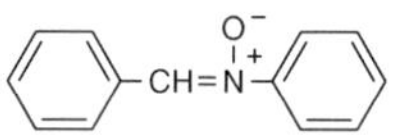

5 gm

S37,481-4
ACETOPHENONE AZINE

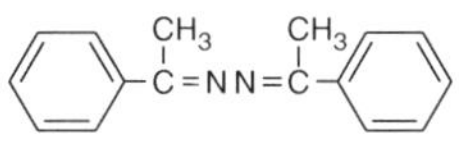

5 gm

S52,436-0
N-BENZYLIDENECYCLODODECYLAMINE

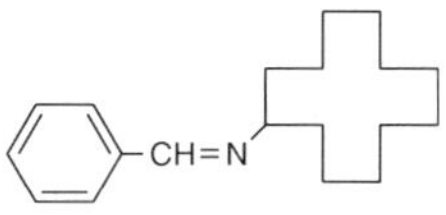

5 gm

S68,667-0
4,4'-METHYLENEBIS(N-BENZYLIDENEANILINE)

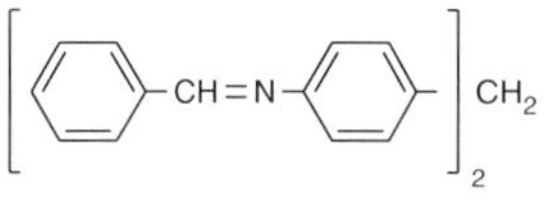

5 gm

S84,114-5
N-BENZYLIDENE-2-NAPHTHYLAMINE

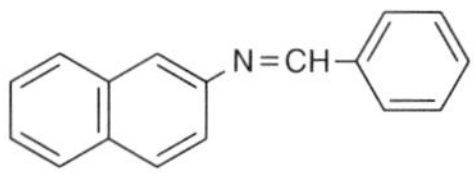

250 mg

S34,189-4
2-(BENZYLIDENEAMINO)FLUORENE

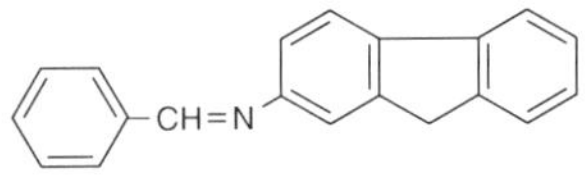

5 gm

S60,130-6
9-(PHENYLIMINOMETHYL)ANTHRACENE

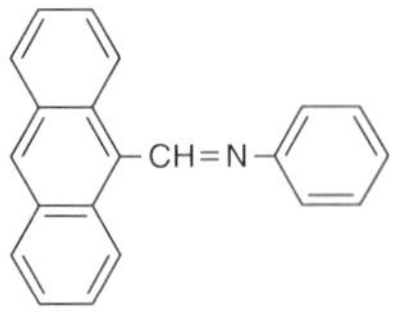

5 gm

S60,220-5
N-(2-CHLOROBENZYLIDENE)ANILINE

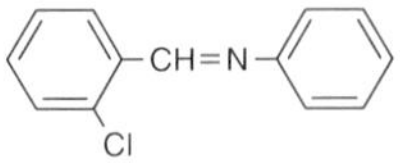

250 mg

S60,925-0
N-BENZYLIDENE-2-CHLOROANILINE

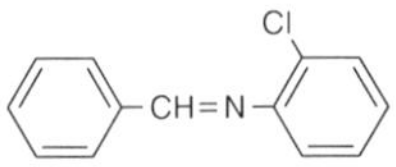

250 mg

S62,564-7
3-CHLORO-N-(2-CHLOROBENZYLIDENE)-ANILINE

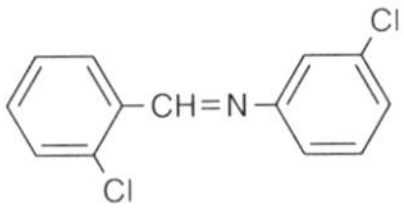

250 mg

S63,183-3
2-CHLORO-N-(3-CHLOROBENZYLIDENE)-ANILINE

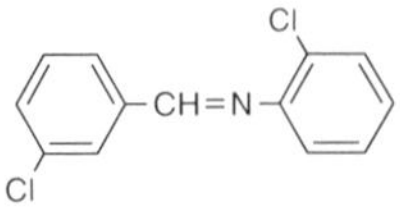

250 mg

S60,929-3
N-BENZYLIDENE-O-TOLUIDINE

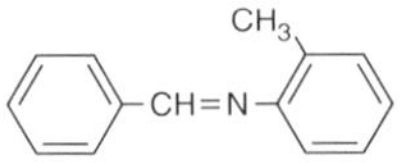

5 gm

S60,226-4
N-(2-CHLOROBENZYLIDENE)-O-TOLUIDINE

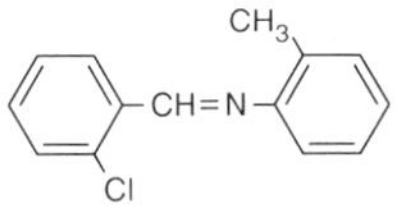

250 mg

S62,574-4
N-(2-CHLOROBENZYLIDENE)-M-TOLUIDINE

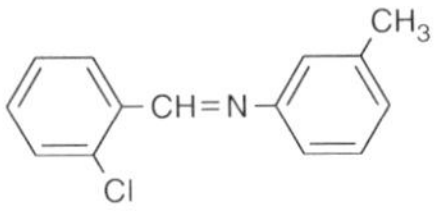

250 mg

S63,519-7
2-CHLORO-N-(4-CHLOROBENZYLIDENE)-ANILINE

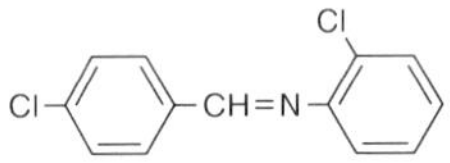

250 mg

S62,563-9
4-BROMO-N-(2-CHLOROBENZYLIDENE)-ANILINE

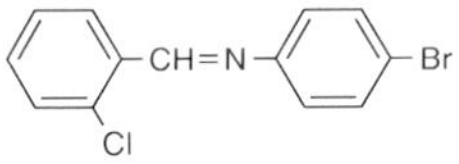

250 mg

S61,583-8
N-(4-CHLOROBENZYLIDENE)-O-TOLUIDINE

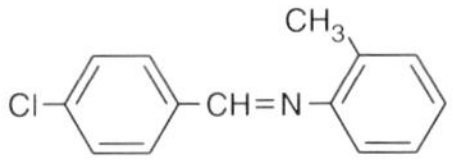

5 gm

S62,572-8
N-(4-METHYLBENZYLIDENE)-2-CHLOROANILINE

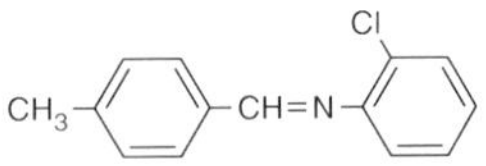

250 mg

S62,575-2
N-(2-CHLOROBENZYLIDENE)-P-TOLUIDINE

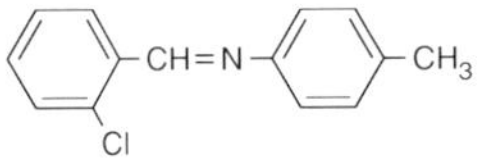

250 mg

S62,583-3
N-(4-METHYLBENZYLIDENE)-O-TOLUIDINE

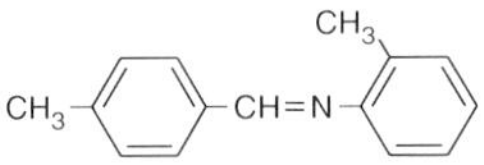

250 mg

S59,989-1
N,N'-TEREPHTHALYLIDENEBIS(2-BROMOANILINE)

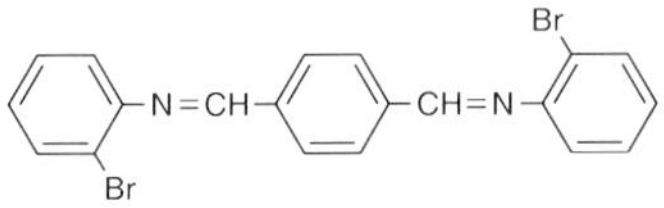

250 mg

S59,955-7
N,N'-TEREPHTHALYLIDENEBIS(2-ETHYLANILINE)

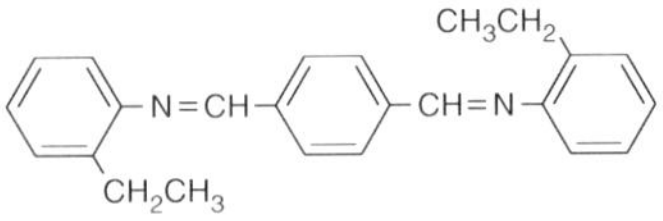

5 gm

S63,550-2
N-(2-CHLOROBENZYLIDENE)-2,3-DICHLOROANILINE

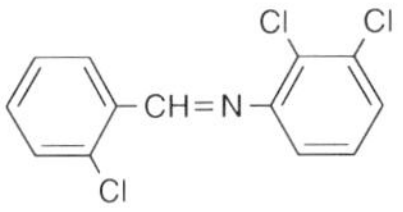

5 gm

S60,435-6
N-(2-CHLOROBENZYLIDENE)-6-CHLORO-2-METHYLANILINE

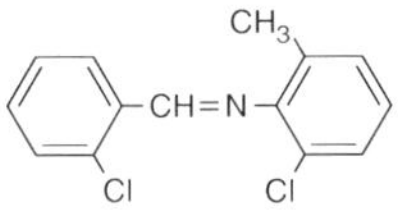

5 gm

S60,215-9
N-(2-CHLOROBENZYLIDENE)-2,3-XYLIDINE

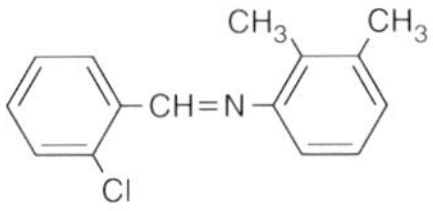

250 mg

S60,441-0
N-(2-CHLOROBENZYLIDENE)-2,6-DIETHYLANILINE

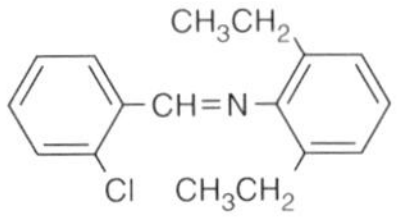

250 mg

S62,096-3
N-(2-CHLOROBENZYLIDENE)-3,4-DICHLOROANILINE

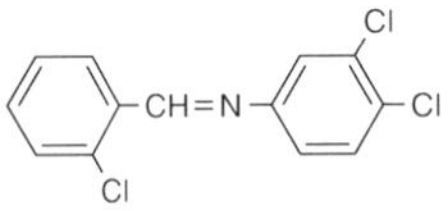

5 gm

S62,561-2
N-(2-CHLOROBENZYLIDENE)-2,4-DICHLOROANILINE

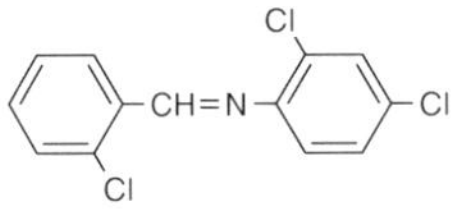

250 mg

S62,562-0
N-(2-CHLOROBENZYLIDENE)-2,5-DICHLOROANILINE

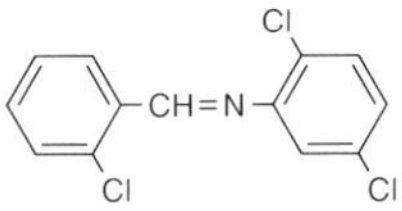

250 mg

S62,965-0
2-CHLORO-N-(3,4-DICHLOROBENZYLIDENE)ANILINE

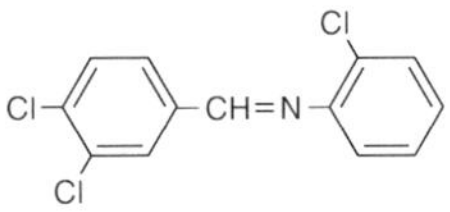

250 mg

S63,548-0
2-CHLORO-N-(2,4-DICHLOROBENZYLIDENE)ANILINE

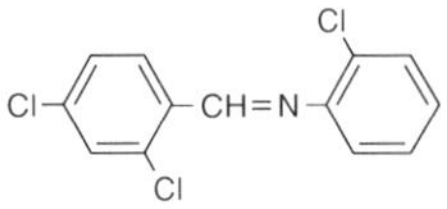

250 mg

S62,565-5
5-CHLORO-N-(2-CHLOROBENZYLIDENE)-2-METHYLANILINE

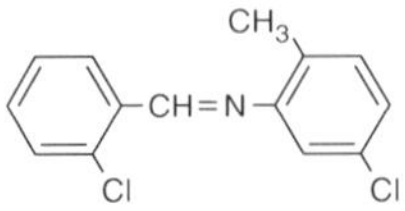

250 mg

S62,603-1
3-CHLORO-N-(2-CHLOROBENZYLIDENE)-4-METHYLANILINE

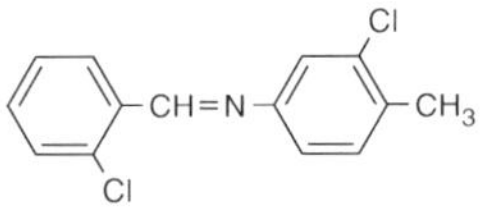

5 gm

S63,507-3
N-(3,4-DICHLOROBENZYLIDENE)-O-TOLUIDINE

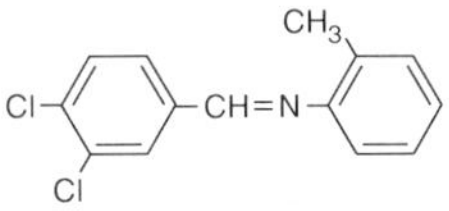

250 mg

S63,517-0
N-(2,4-DICHLOROBENZYLIDENE)-O-TOLUIDINE

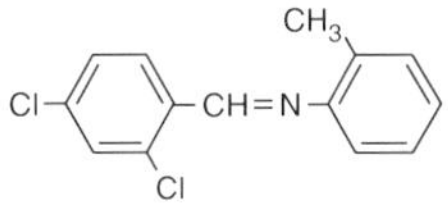

250 mg

S60,436-4
N-(2-CHLOROBENZYLIDENE)-2,5-XYLIDINE

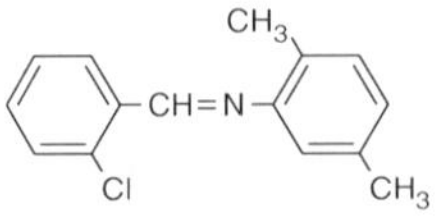

5 gm

S61,576-5
N-(2-CHLOROBENZYLIDENE)-2,4-XYLIDINE

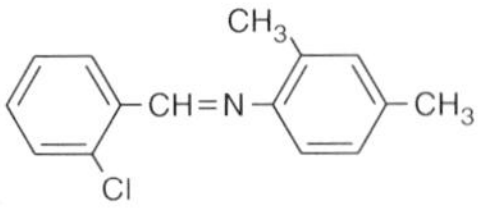

250 mg

S63,527-8
N-(2-CHLOROBENZYLIDENE)-3,5-XYLIDINE
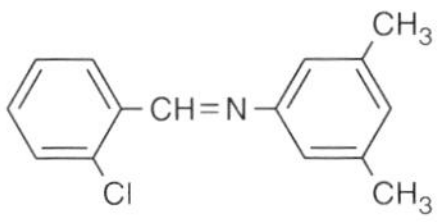
250 mg

S34,403-6
2-(2-CHLOROBENZYLIDENEAMINO)-FLUORENE
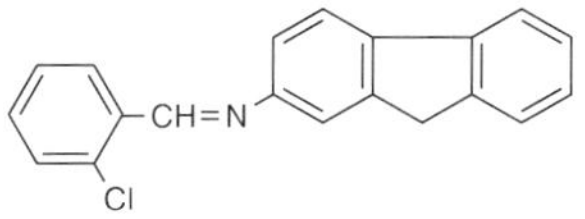
5 gm

S62,106-4
N-(9-ANTHRACENYLMETHYLENE)-2-CHLOROANILINE
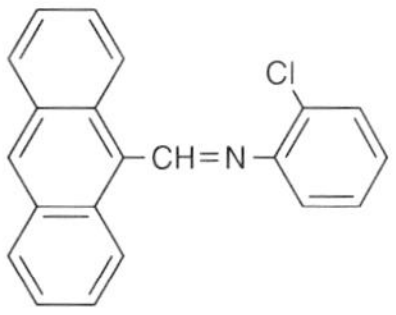
5 gm

S61,580-3
N-(4-CHLOROBENZYLIDENE)ANILINE
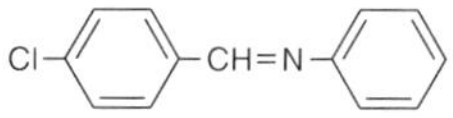
5 gm

S61,578-1
3-CHLORO-N-(4-CHLOROBENZYLIDENE)-ANILINE
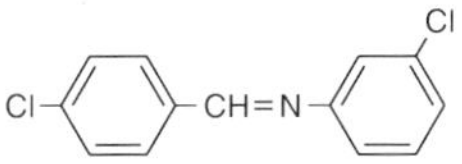
5 gm

S61,590-0
4-CHLORO-N-(4-CHLOROBENZYLIDENE)-ANILINE
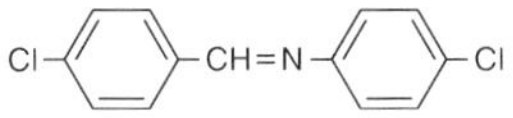
5 gm

S59,210-2
N-BENZYLIDENE-4-BROMOANILINE
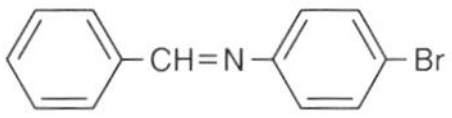
250 mg

S63,522-7
4-BROMO-N-(4-CHLOROBENZYLIDENE)-ANILINE
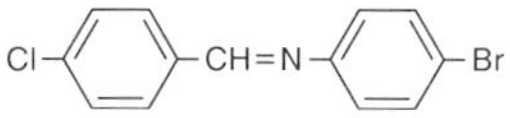
250 mg

S59,442-3
N-(4-METHYLBENZYLIDENE)-4-CHLOROANILINE
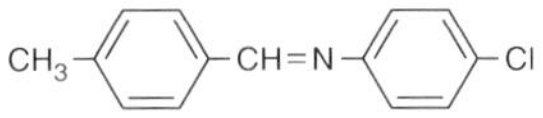
250 mg

S61,588-9
N-(4-CHLOROBENZYLIDENE)-P-TOLUIDINE
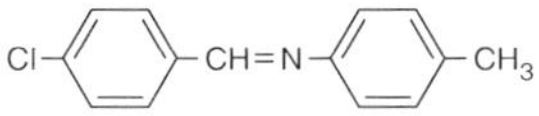
5 gm

S62,573-6
N-(4-METHYLBENZYLIDENE)-3-CHLOROANILINE
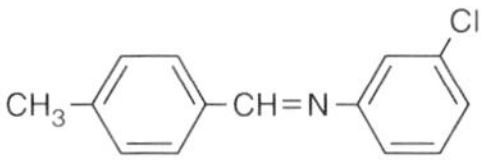
250 mg

S63,532-4
N-(4-CHLOROBENZYLIDENE)-M-TOLUIDINE
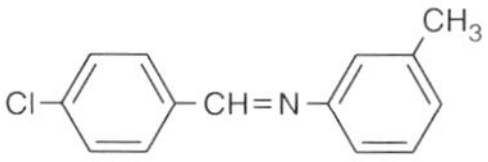
250 mg

S59,446-6
N-(4-METHYLBENZYLIDENE)-4-BROMOANILINE
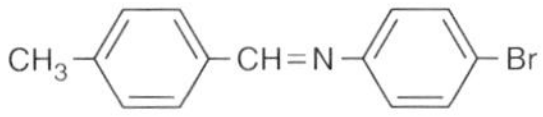
250 mg

S59,435-0
N-(4-METHYLBENZYLIDENE)-P-TOLUIDINE
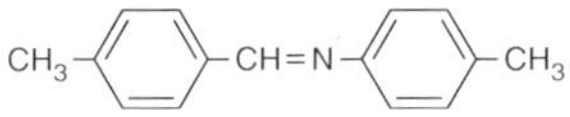
250 mg

S66,994-6
N-(4-ISOPROPYLBENZYLIDENE)-P-TOLUIDINE
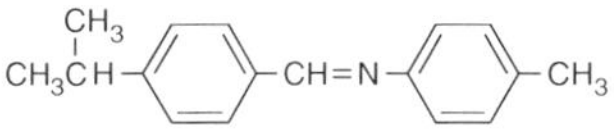
5 gm

S53,525-7
N,N'-BIS(4-FLUOROBENZYLIDENE)-1,4-PHENYLENEDIAMINE
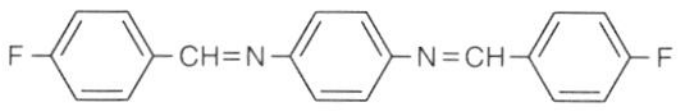
5 gm

S58,907-1
N,N'-(1,4-PHENYLENEDIMETHYLIDYNE)-BIS(4-FLUOROANILINE)

5 gm

S59,964-6
N,N'-TEREPHTHALYLIDENEBIS(3-FLUOROANILINE)
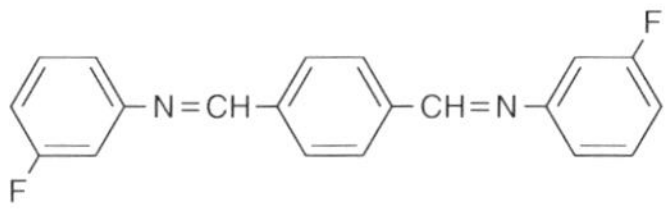
5 gm

S53,527-3
N,N'-BIS(4-BROMOBENZYLIDENE)-1,4-PHENYLENEDIAMINE
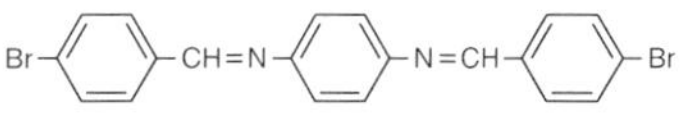
5 gm

S59,958-1
N,N'-TEREPHTHALYLIDENEBIS(3-BROMOANILINE)
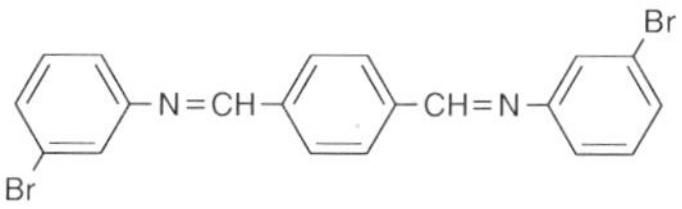
5 gm

S62,332-6
N,N'-TEREPHTHALYLIDENEBIS(2,4,5-TRICHLOROANILINE)
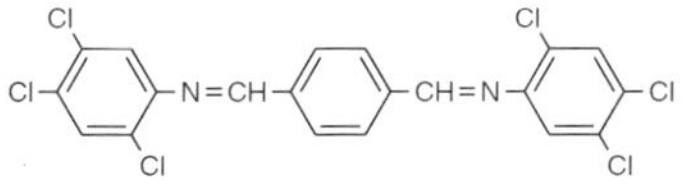
5 gm

S44,754-4
N,N'-TEREPHTHALYLIDENEDI-P-TOLUIDINE
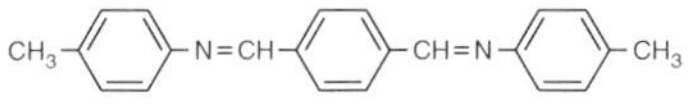
5 gm

S58,966-7
N,N'-(1,4-PHENYLENEDIMETHYLIDYNE)-DI-M-TOLUIDINE
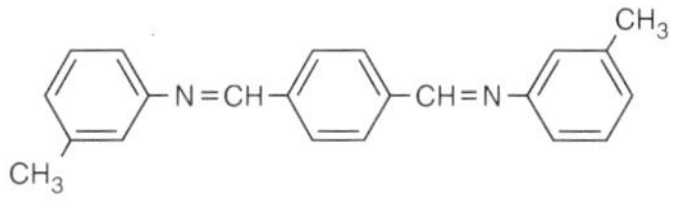
5 gm

S58,953-5
N,N'-(1,4-PHENYLENEDIMETHYLIDYNE)-BIS(4-ETHYLANILINE)
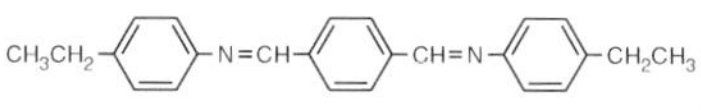
5 gm

S58,970-5
N,N'-(1,4-PHENYLENEDIMETHYLIDYNE)-BIS(2,4,5-TRIMETHYLANILINE)
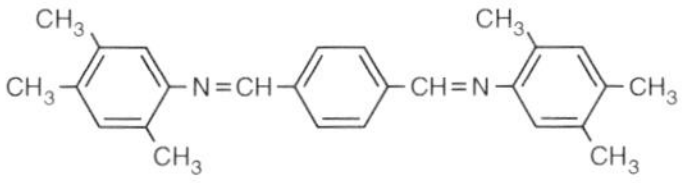
5 gm

S62,538-8
N,N'-(1,4-PHENYLENEDIMETHYLIDYNE)-BIS(4-BUTYLANILINE)
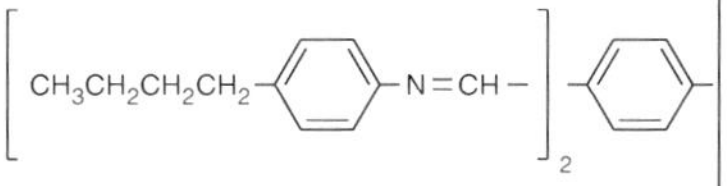
250 mg

S63,555-3
N-(4-CHLOROBENZYLIDENE)-2,3-DICHLOROANILINE
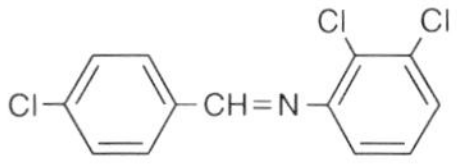
250 mg

S63,553-7
2,3-DICHLORO-N-(4-METHYLBENZYLIDENE)ANILINE
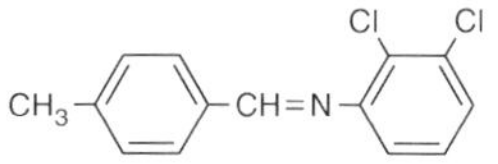
250 mg

S61,577-3
N-(4-CHLOROBENZYLIDENE)-2,3-XYLIDINE
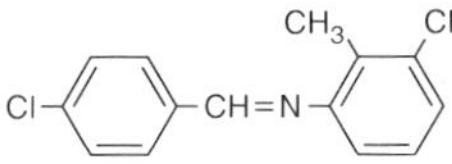
250 mg

S60,411-9
N-(4-METHYLBENZYLIDENE)-2,3-XYLIDINE
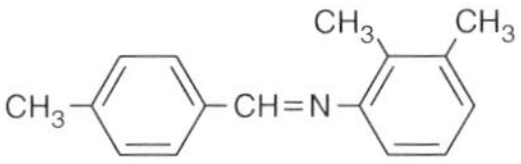
250 mg

S58,952-7
N,N'-(1,4-PHENYLENEDIMETHYLIDYNE)-BIS(6-CHLORO-O-TOLUIDINE)
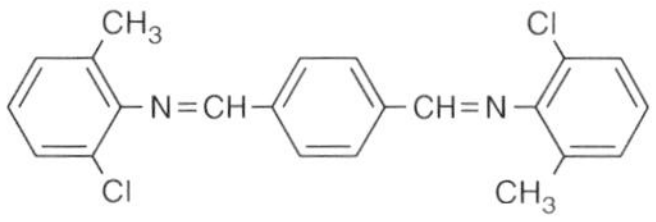
5 gm

S61,591-9
N-(4-CHLOROBENZYLIDENE)-3,4-DICHLOROANILINE
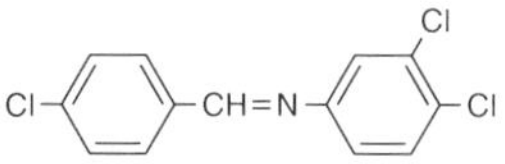
5 gm

S61,592-7
N-(4-CHLOROBENZYLIDENE)-2,5-DICHLOROANILINE

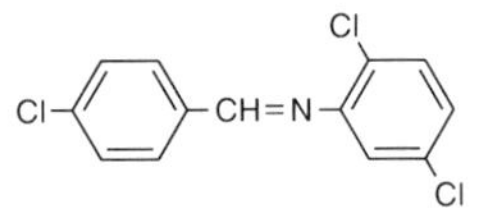

5 gm

S61,593-5
N-(4-CHLOROBENZYLIDENE)-2,4-DICHLOROANILINE

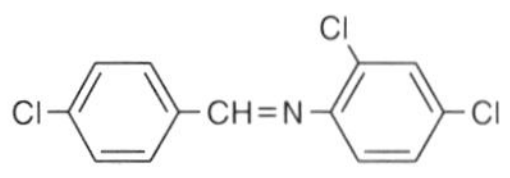

5 gm

S62,949-9
4-CHLORO-N-(3,4-DICHLOROBENZYLIDENE)ANILINE

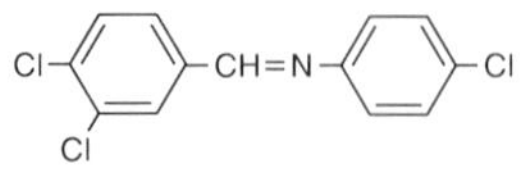

5 gm

S63,549-9
4-CHLORO-N-(2,4-DICHLOROBENZYLIDENE)ANILINE

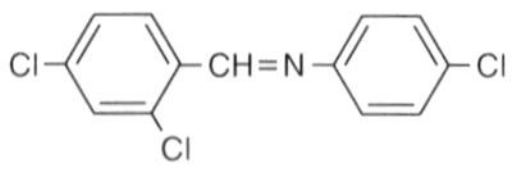

250 mg

S62,976-6
4-BROMO-N-(3,4-DICHLOROBENZYLIDENE)-ANILINE

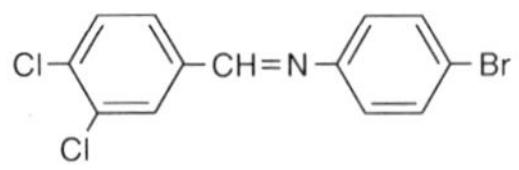

250 mg

S63,523-5
4-BROMO-N-(2,4-DICHLOROBENZYLIDENE)-ANILINE

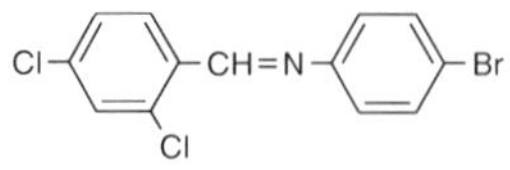

250 mg

S60,438-0
N-(4-METHYLBENZYLIDENE)-2,4-DICHLOROANILINE

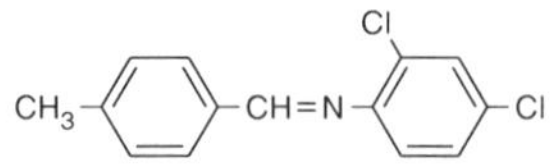

250 mg

S61,589-7
N-(4-CHLOROBENZYLIDENE)-3-CHLORO-4-METHYLANILINE

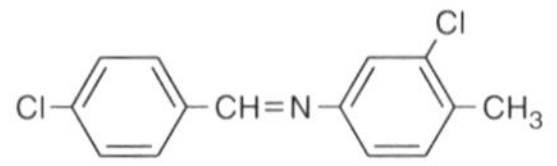

5 gm

S61,927-2
N-(4-CHLOROBENZYLIDENE)-5-CHLORO-2-METHYLANILINE

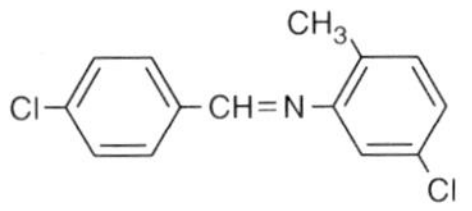

5 gm

S62,567-1
2,5-DICHLORO-N-(4-METHYLBENZYLIDENE)ANILINE

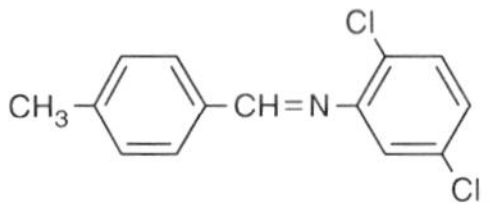

250 mg

S62,569-8
3,4-DICHLORO-N-(4-METHYLBENZYLIDENE)ANILINE

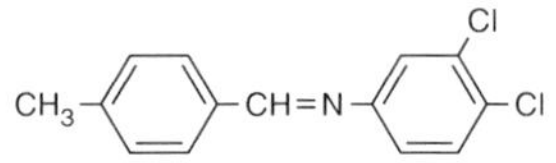

250 mg

S63,511-1
N-(2,4-DICHLOROBENZYLIDENE)-P-TOLUIDINE

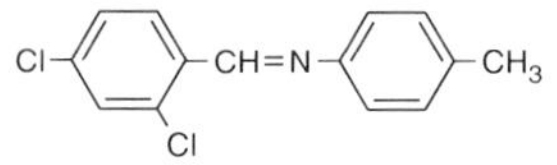

250 mg

S63,556-1
N-(3,4-DICHLOROBENZYLIDENE)-P-TOLUIDINE

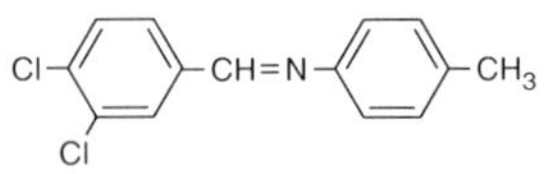

250 mg

S61,918-3
N-(4-CHLOROBENZYLIDENE)-3,4-XYLIDINE

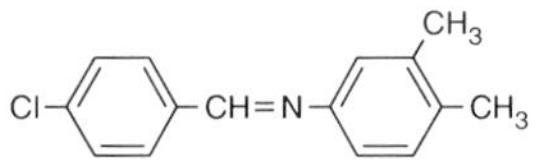

250 mg

S62,581-7
5-CHLORO-2-METHYL-N-(4-METHYLBENZYLIDENE)ANILINE

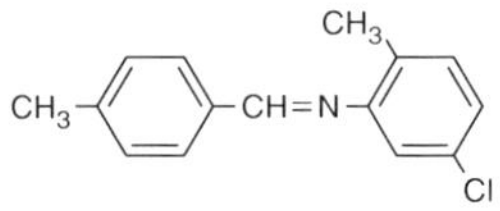

250 mg

S62,606-6
3-CHLORO-4-METHYL-N-(4-METHYLBENZYLIDENE)ANILINE

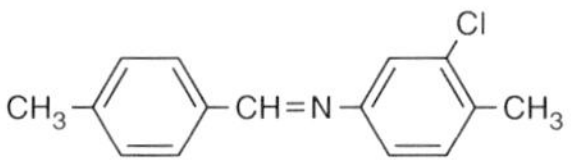

250 mg

S59,440-7
N-(4-METHYLBENZYLIDENE)-3,4-DIMETHYLANILINE

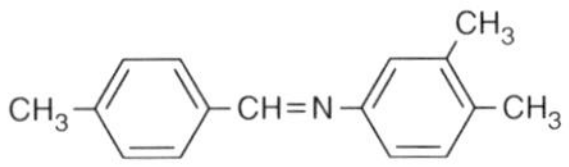

250 mg

S60,229-9
N-(4-METHYLBENZYLIDENE)-2,4-XYLIDINE

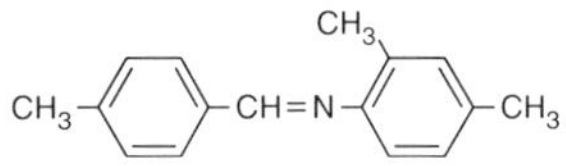

250 mg

S60,422-4
N-(4-METHYLBENZYLIDENE)-2,5-XYLIDINE

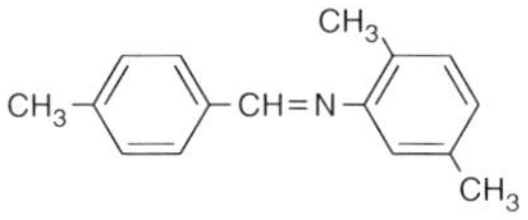

250 mg

S58,951-9
N,N'-(1,4-PHENYLENEDIMETHYLIDYNE)-BIS(2,5-DICHLOROANILINE)

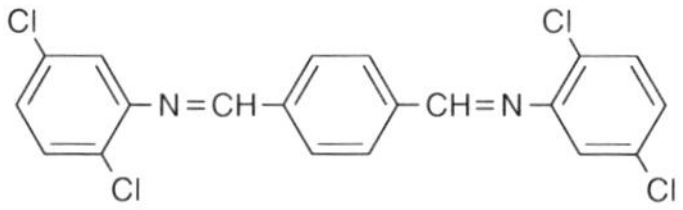

5 gm

S58,936-5
N,N'-(1,4-PHENYLENEDIMETHYLIDYNE)-BIS(5-CHLORO-O-TOLUIDINE)

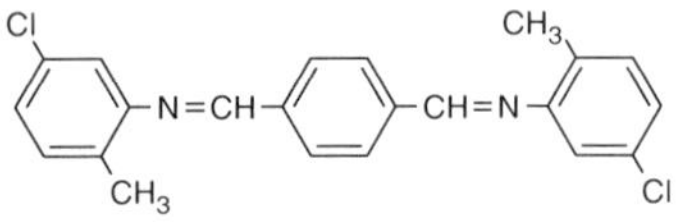

5 gm

S58,922-5
N,N'-(1,4-PHENYLENEDIMETHYLIDYNE)-BIS(4-BROMO-O-TOLUIDINE)

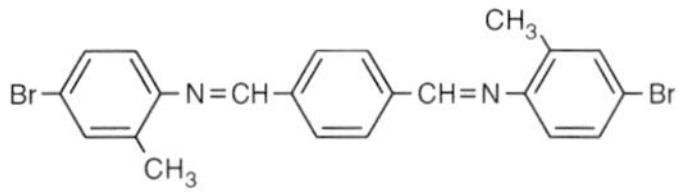

5 gm

S59,956-5
N,N'-TEREPHTHALYLIDENEBIS(4-BROMO-3-METHYLANILINE)

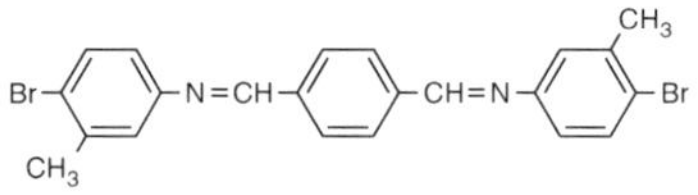

5 gm

S59,961-1
N,N'-TEREPHTHALYLIDENEBIS(2-BROMO-4-METHYLANILINE)

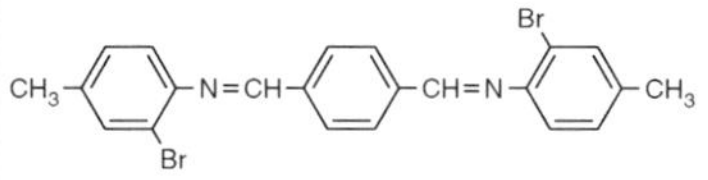

250 mg

S63,521-9
N-(4-CHLOROBENZYLIDENE)-3,5-XYLIDINE

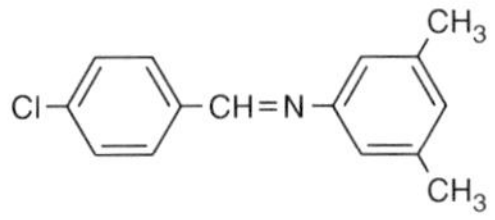

250 mg

S58,958-6
N,N'-(1,4-PHENYLENEDIMETHYLIDYNE)-BIS(3,5-DICHLOROANILINE)

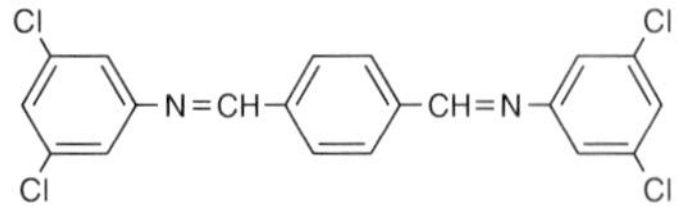

5 gm

S59,960-3
N,N'-TEREPHTHALYLIDENEBIS(3,5-DIMETHYLANILINE)

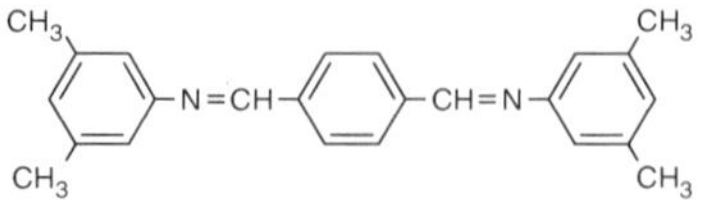

250 mg

S59,963-8
N,N'-TEREPHTHALYLIDENEBIS(2,4,6-TRIMETHYLANILINE)

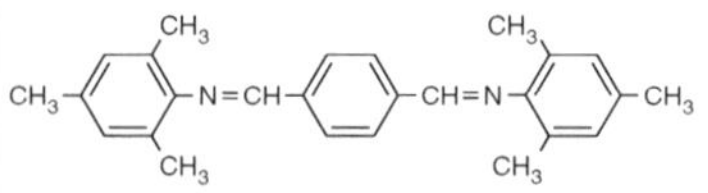

5 gm

S58,927-6
N,N'-(1,4-PHENYLENEDIMETHYLIDYNE)-BIS(2-AMINOPYRIDINE)

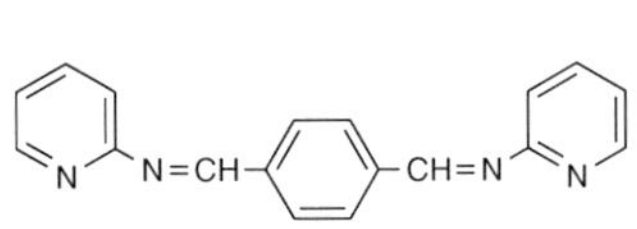

5 gm

S68,668-9
4,4'-METHYLENEBIS[N-(4-METHYLBENZYLIDENE)ANILINE]

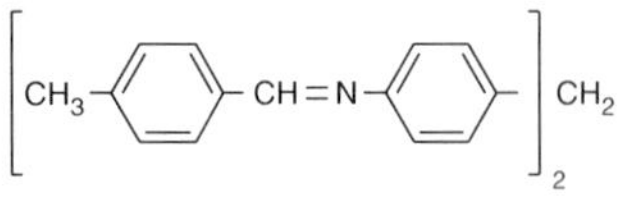

5 gm

S34,404-4
2-(4-CHLOROBENZYLIDENEAMINO)-FLUORENE

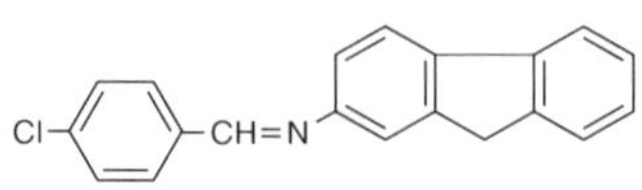

5 gm

S62,474-8
N-(9-ANTHRACENYLMETHYLENE)-4-FLUOROANILINE

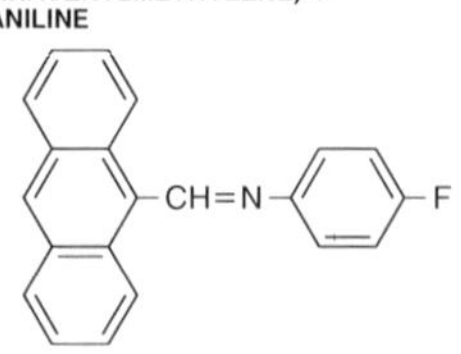

5 gm

S62,111-0
N-(9-ANTHRACENYLMETHYLENE)-4-CHLOROANILINE
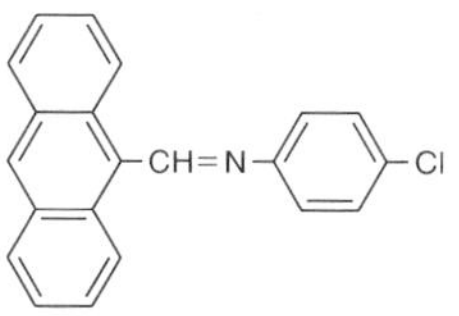
5 gm

S62,470-5
N-(9-ANTHRACENYLMETHYLENE)-4-BROMOANILINE
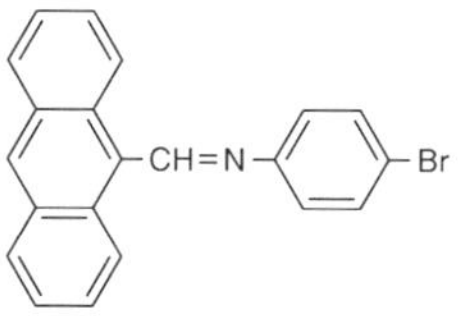
5 gm

S62,465-9
N-(9-ANTHRACENYLMETHYLENE)-4-ETHYLANILINE
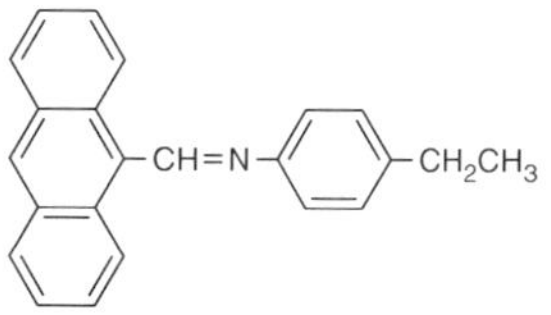
5 gm

S63,520-0
N-BENZYLIDENE-2,3-DICHLOROANILINE
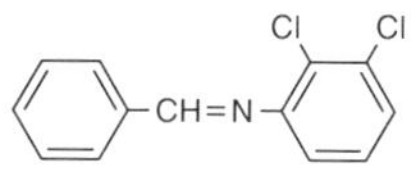
250 mg

S60,222-1
N-BENZYLIDENE-2-CHLORO-6-METHYLANILINE
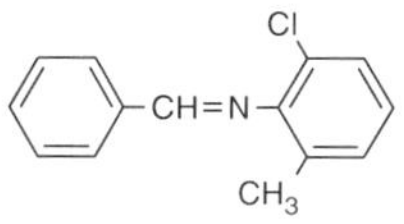
250 mg

S60,432-1
N-BENZYLIDENE-2,3-XYLIDINE
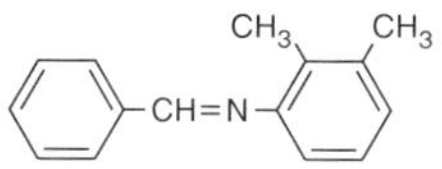
5 gm

S60,414-3
N-BENZYLIDENE-2,6-DIETHYLANILINE
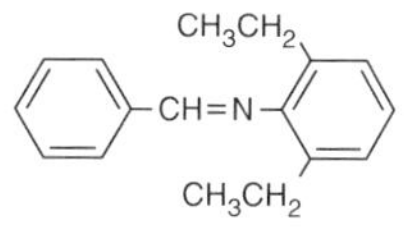
5 gm

S63,547-2
2,3-DICHLORO-N-(3,4-DICHLOROBENZYLIDENE)ANILINE
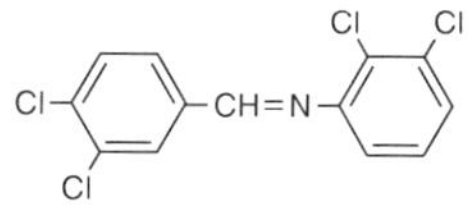
250 mg

S63,552-9
2-CHLORO-N-(3,4-DICHLOROBENZYLIDENE)-6-METHYLANILINE
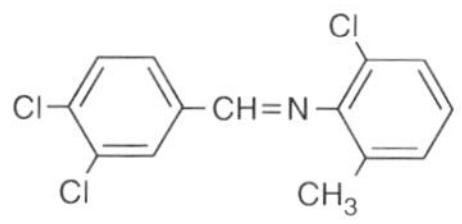
250 mg

S62,884-0
N-(3,4-DICHLOROBENZYLIDENE)-2,3-XYLIDINE
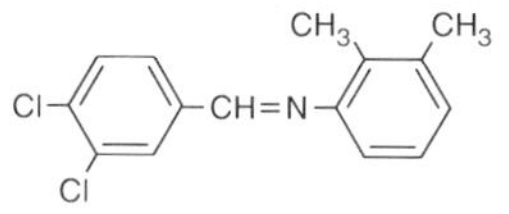
250 mg

S63,564-2
N-(2,4-DICHLOROBENZYLIDENE)-2,3-XYLIDINE
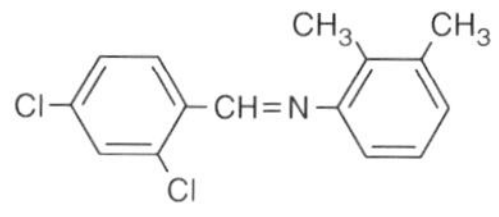
250 mg

S60,485-2
N-(9-ANTHRACENYLMETHYLENE)-2,3-XYLIDINE
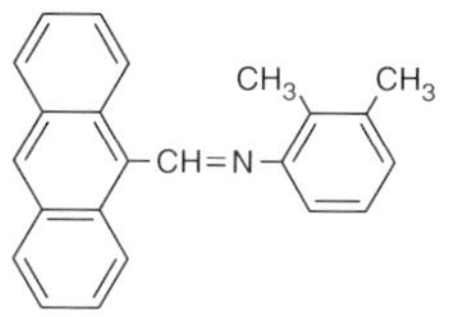
5 gm

S60,217-5
N-BENZYLIDENE-2,4-DICHLOROANILINE
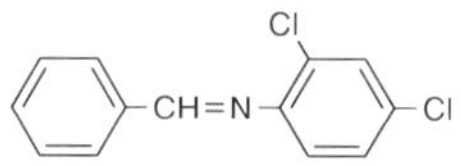
250 mg

S60,219-1
N-BENZYLIDENE-3,4-DICHLOROANILINE
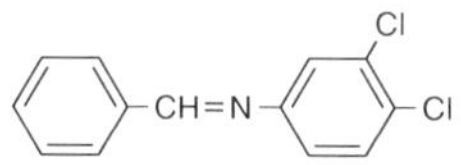
250 mg

S62,903-0
N-(3,4-DICHLOROBENZYLIDENE)ANILINE
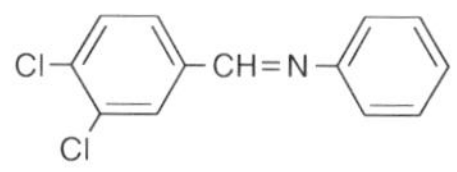
250 mg

S63,598-7
N-(2,4-DICHLOROBENZYLIDENE)ANILINE
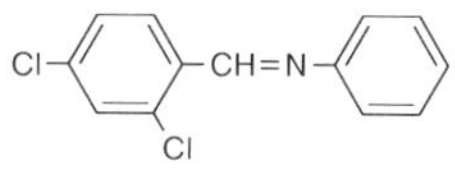
250 mg

S62,843-3
3-CHLORO-N-(3,4-DICHLOROBENZYLIDENE)ANILINE
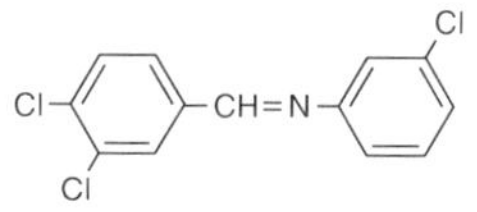
250 mg

S62,709-7
2,5-DICHLORO-N-(3,4-DICHLOROBENZYLIDENE)ANILINE
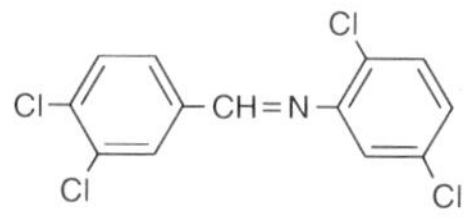
250 mg

S62,930-8
2,4-DICHLORO-N-(3,4-DICHLOROBENZYLIDENE)ANILINE
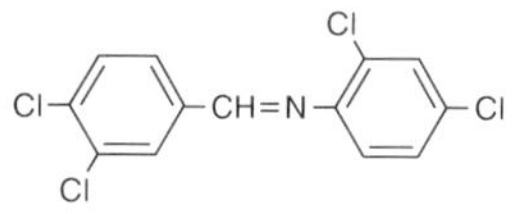
250 mg

S63,524-3
2,4-DICHLORO-N-(2,4-DICHLOROBENZYLIDENE)ANILINE
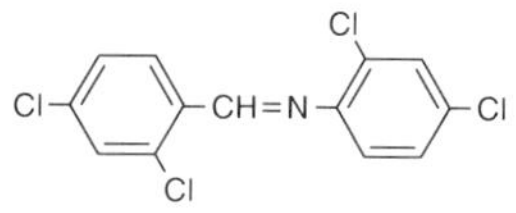
250 mg

S63,554-5
2,5-DICHLORO-N-(2,4-DICHLOROBENZYLIDENE)ANILINE
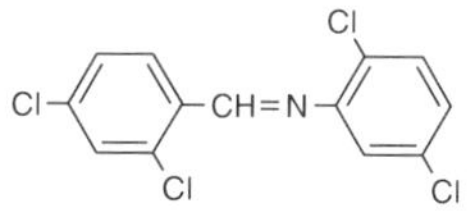
250 mg

S63,604-5
3,4-DICHLORO-N-(2,4-DICHLOROBENZYLIDENE)ANILINE
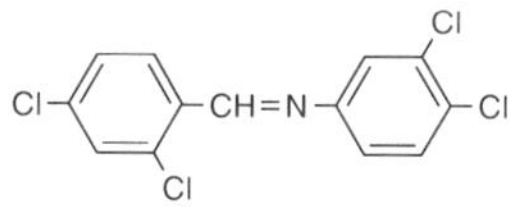
250 mg

S63,545-6
N-(3,4-DICHLOROBENZYLIDENE)-2,4,5-TRICHLOROANILINE
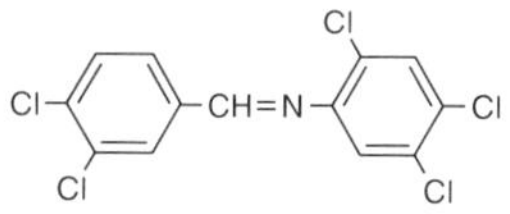
250 mg

S60,931-5
N-BENZYLIDENE-3-CHLORO-4-METHYLANILINE
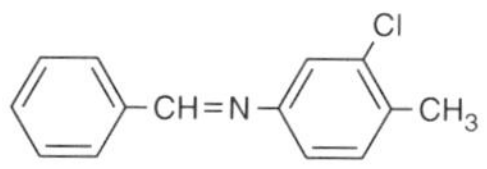
250 mg

S62,854-9
N-(3,4-DICHLOROBENZYLIDENE)-M-TOLUIDINE
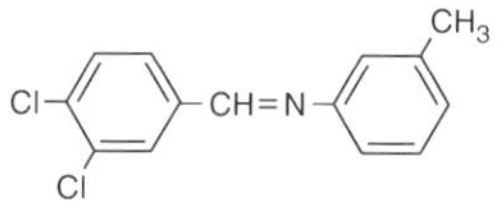
250 mg

S63,587-1
N-(2,4-DICHLOROBENZYLIDENE)-M-TOLUIDINE
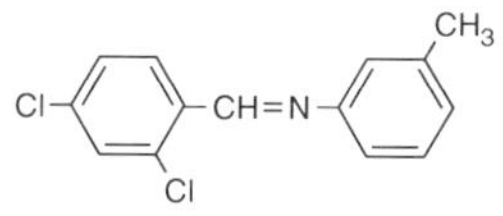
250 mg

S62,882-4
5-CHLORO-N-(3,4-DICHLOROBENZYLIDENE)-2-METHYLANILINE
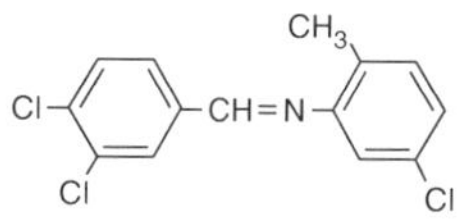
250 mg

S63,515-4
3-CHLORO-N-(2,4-DICHLOROBENZYLIDENE)-4-METHYLANILINE
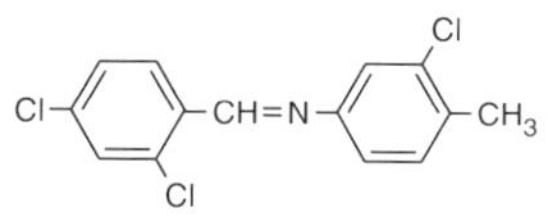
250 mg

S59,202-1
N-BENZYLIDENE-2,5-XYLIDINE
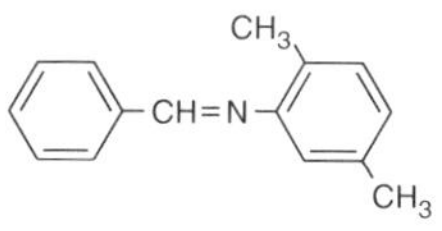
250 mg

S60,218-3
N-BENZYLIDENE-2,4-XYLIDINE
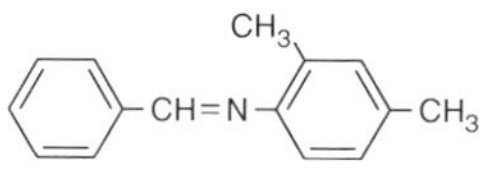
250 mg

S63,528-6
N-(2,4-DICHLOROBENZYLIDENE)-2,4-XYLIDINE
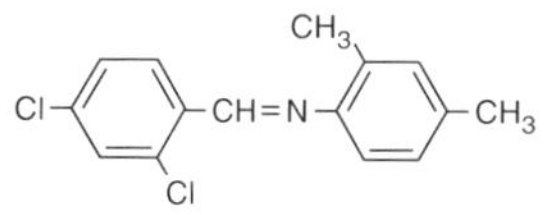
250 mg

S63,585-5
N-(3,4-DICHLOROBENZYLIDENE)-2,5-XYLIDINE
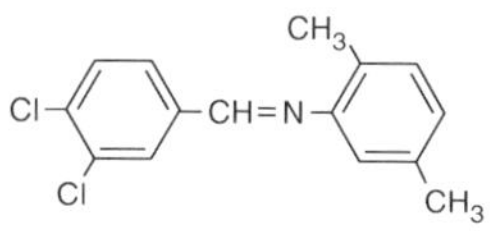
250 mg

S63,603-7
N-(3,4-DICHLOROBENZYLIDENE)-3,5-XYLIDINE

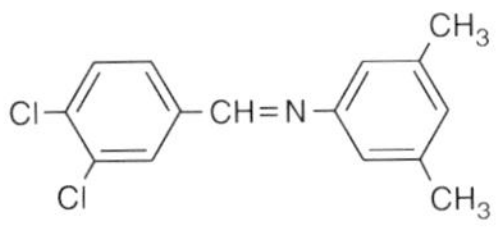

250 mg

S62,112-9
N-(9-ANTHRACENYLMETHYLENE)-2,5-DICHLOROANILINE

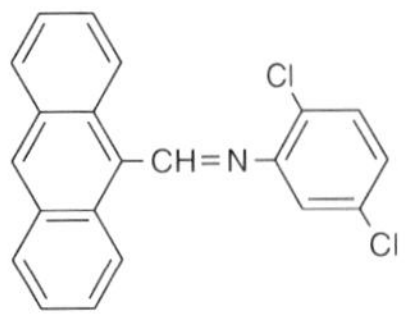

5 gm

S62,125-0
N-(9-ANTHRACENYLMETHYLENE)-2,4-DICHLOROANILINE

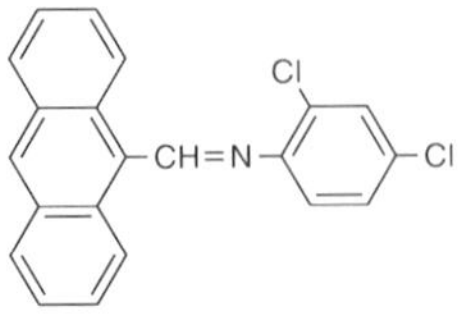

5 gm

S62,462-4
N-(9-ANTHRACENYLMETHYLENE)-5-CHLORO-2-METHYLANILINE

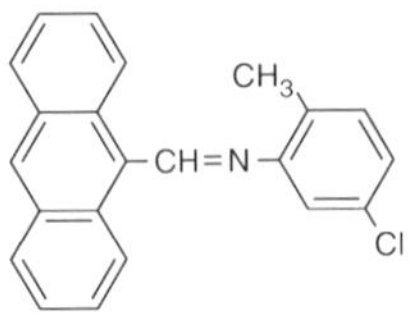

5 gm

S62,467-5
N-(9-ANTHRACENYLMETHYLENE)-3-CHLORO-4-METHYLANILINE

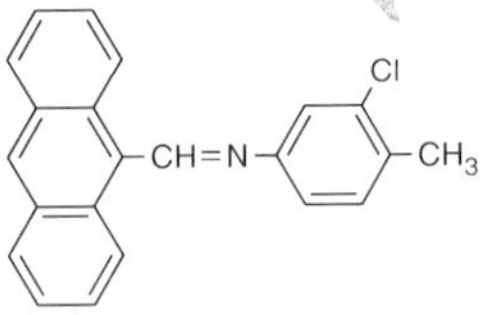

5 gm

S62,477-2
N-(9-ANTHRACENYLMETHYLENE)-4-CHLORO-2-METHYLANILINE

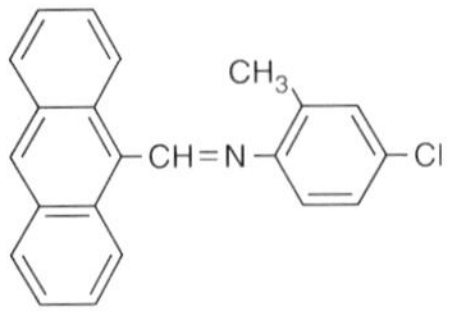

5 gm

S62,472-1
N-(9-ANTHRACENYLMETHYLENE)-4-BROMO-2-METHYLANILINE

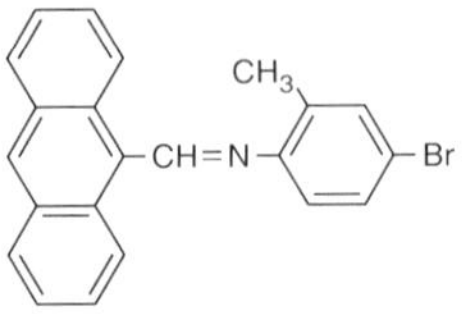

5 gm

S63,611-8
N-(9-ANTHRACENYLMETHYLENE)-4-BROMO-3-METHYLANILINE

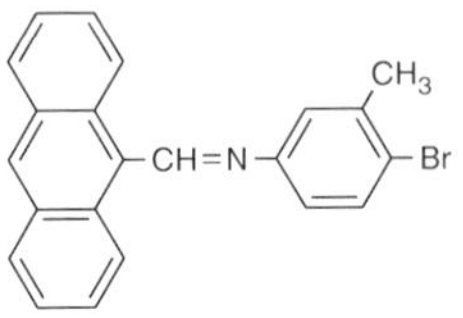

5 gm

S60,477-1
N-(9-ANTHRACENYLMETHYLENE)-3,4-XYLIDINE

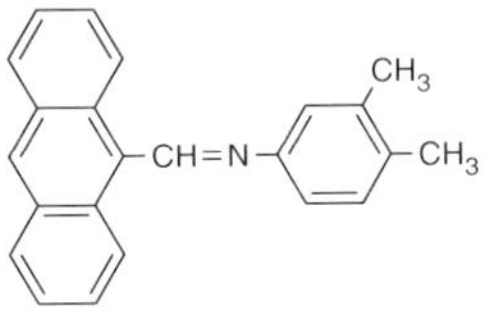

250 mg

S61,379-7
N-(9-ANTHRACENYLMETHYLENE)-2,5-XYLIDINE

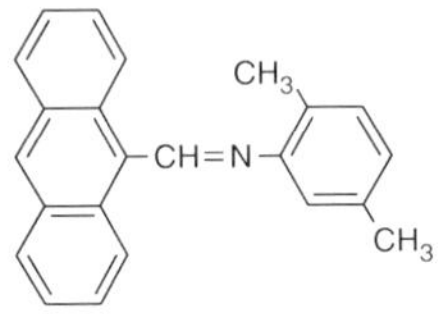

5 gm

S60,123-3
N-(9-ANTHRACENYLMETHYLENE)-3,5-XYLIDINE

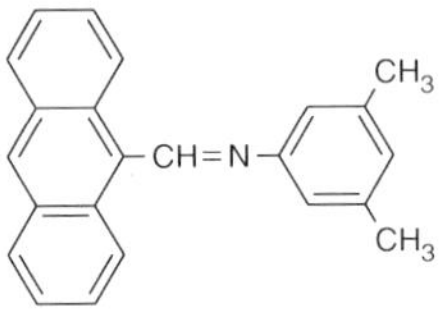

5 gm

S62,624-4
N-(9-ANTHRACENYLMETHYLENE)-2,4,6-TRIMETHYLANILINE

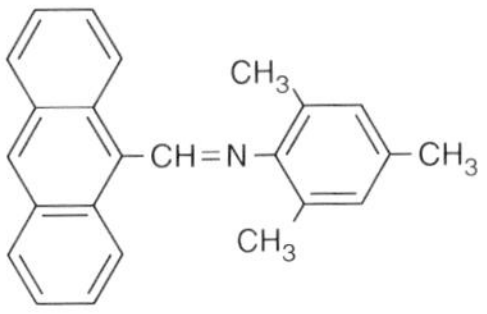

5 gm

S62,107-2
N-(9-ANTHRACENYLMETHYLENE)-3-CHLOROANILINE

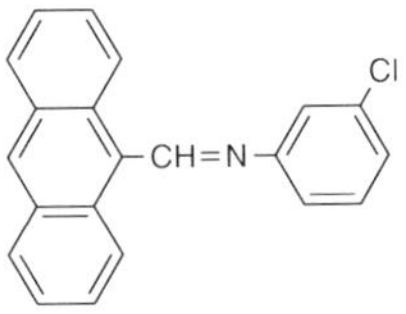

5 gm

S62,323-7
N-(9-ANTHRACENYLMETHYLENE)-3-BROMOANILINE

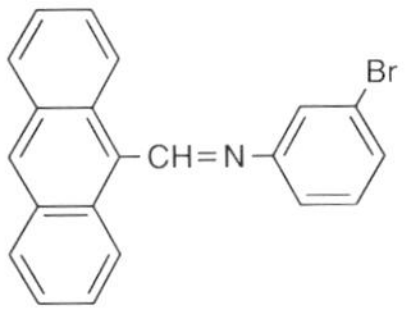

5 gm

S60,120-9
N-(9-ANTHRACENYLMETHYLENE)-M-TOLUIDINE

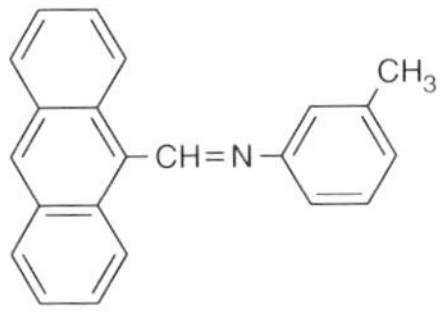

250 mg

S88,609-2
1-(ALPHA-NITROFURFURYLIDENE)-2-PHENYLHYDRAZINE

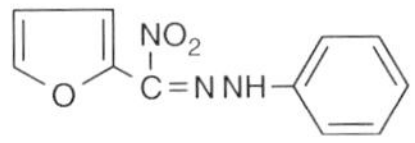

250 mg

S88,610-6
2-THIOPHENECARBOXALDEHYDE PHENYLHYDRAZONE

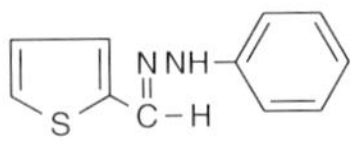

250 mg

S69,382-0
N,N'-BIS(PIPERONYLIDENE)HYDRAZINE

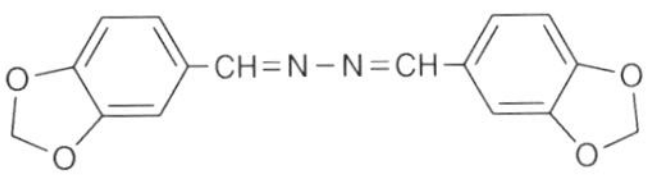

5 gm

S94,292-8
2,3-BUTANEDIONE BIS(PHENYLHYDRAZONE)

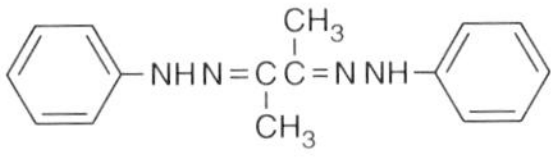

250 mg

S67,168-1
N-(CINNAMYLIDENE)PHENYLHYDRAZONE

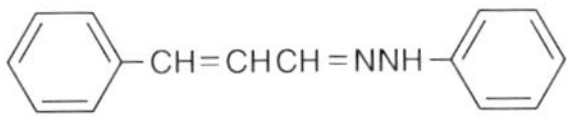

250 mg

S93,438-0
4-(BENZYLIDENEAMINO)-4H-1,2,4-TRIAZOLE

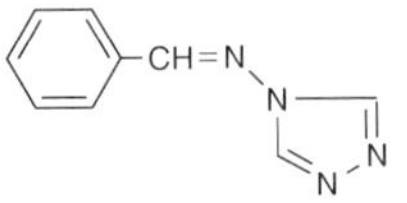

250 mg

S54,818-9
4-(3,4,5-TRIMETHOXYBENZYLIDENEAMINO)-4H-1,2,4-TRIAZOLE

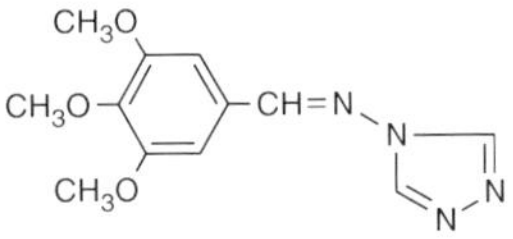

250 mg

S77,136-8
1-(3-METHYL-ALPHA-PENTYLBENZYLIDENEAMINO)-GUANIDINE NITRATE

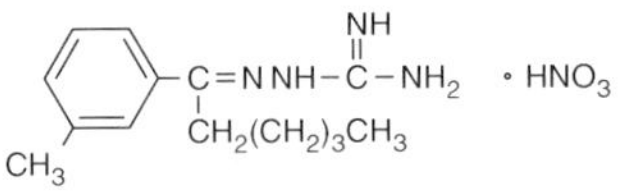

250 mg

S77,139-2
1-(ALPHA-HEXYL-3-METHYLBENZYLIDENEAMINO)-GUANIDINE NITRATE

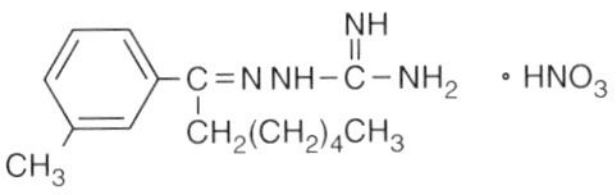

250 mg

S50,664-8
3-(5-INDANYLMETHYLENEAMINO)-2-METHYL-4(3H)-QUINAZOLONE

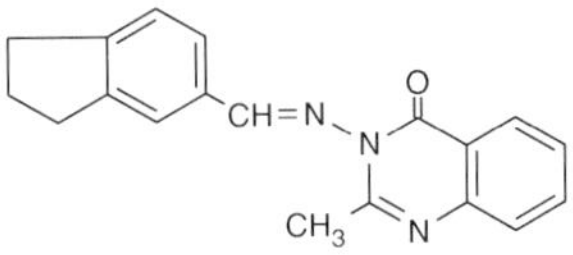

250 mg

S51,458-6
5-INDANCARBOXALDEHYDE 2-PYRIDYLHYDRAZONE

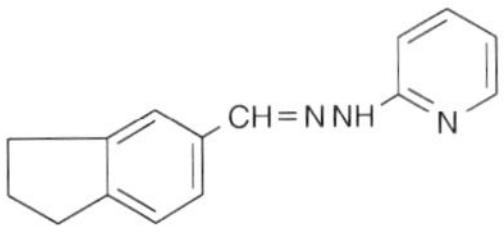

250 mg

S50,610-9
4-(5-INDANYLMETHYLENEAMINO)-4H-1,2,4-TRIAZOLE

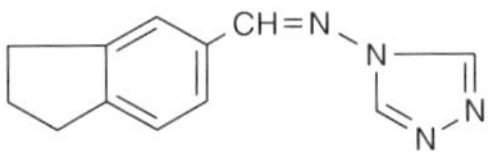

250 mg

S90,218-7
2-(3-FURFURYLUREIDO)ACETIC (2-THIENYLMETHYLENE)HYDRAZIDE

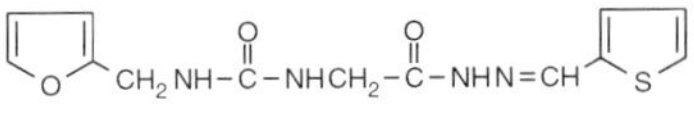

250 mg

S85,449-2
4-(3-FURFURYLUREIDO)BENZOIC (4-ACETAMIDOBENZYLIDENE)HYDRAZIDE

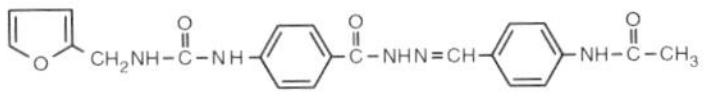

250 mg

S90,391-4
HIPPURIC (FURFURYLIDENE)HYDRAZIDE

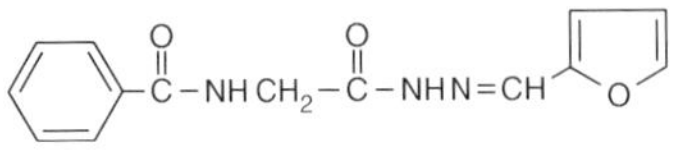

250 mg

S90,383-3
ADIPIC BIS(FURFURYLIDENEHYDRAZIDE)

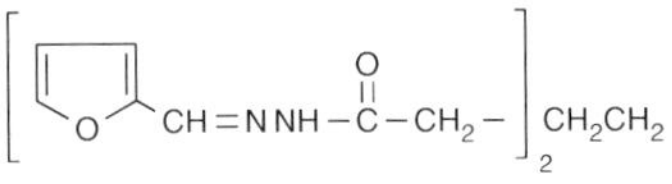

250 mg

S92,549-7
10-UNDECENOIC (FURFURYLIDENE)-HYDRAZIDE

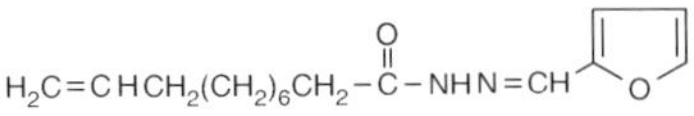

250 mg

S51,784-4
1,2-DIHYDRO-4,6-DIMETHYL-2-OXO-1-(2-THIENYLMETHYLENEAMINO)-NICOTINONITRILE

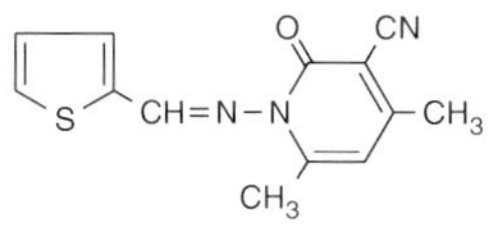

250 mg

S86,994-5
BENZENESULFONIC N2-BENZYLIDENE-N1-NITROHYDRAZIDE

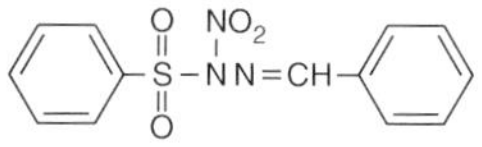

250 mg

S85,452-2
4-(3-FURFURYLUREIDO)BENZOIC (3-NITROBENZYLIDENE)HYDRAZIDE

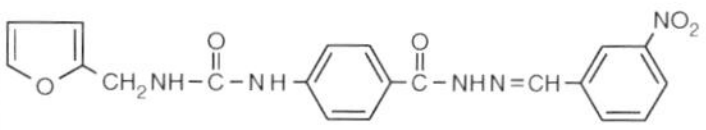

250 mg

S92,477-6
BENZOIC (3-NITROBENZYLIDENE)-HYDRAZIDE

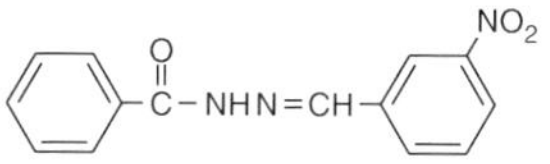

250 mg

S92,522-5
3-NITROBENZOIC (3-PYRIDYLMETHYLENE)-HYDRAZIDE

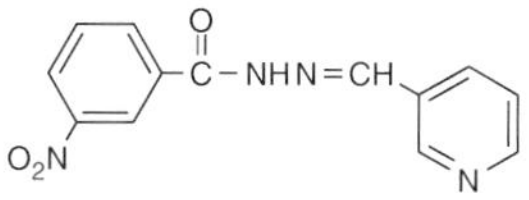

250 mg

S91,546-7
2-NITROCINNAMALDEHYDE SEMICARBAZONE

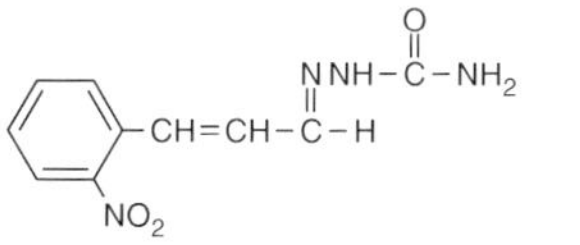

250 mg

S91,534-3
4-NITROBENZALDEHYDE SEMICARBAZONE

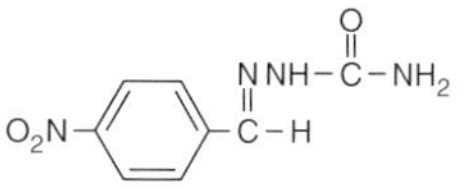

250 mg

S80,081-3
1-(3-NITROBENZYLIDENE)-4-PHENYL-3-THIOSEMICARBAZIDE

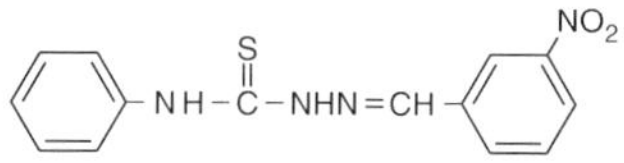

250 mg

S93,444-5
4-(4-NITROBENZYLIDENEAMINO)-4H-1,2,4-TRIAZOLE

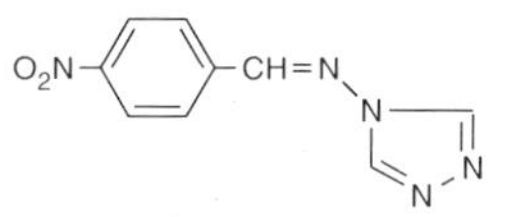

5 gm

S37,613-2
1-(5-NITROFURFURYLIDENE-2-IMINO)-2,4-DIMETHYL-5-CYANO-6-PYRIDONE

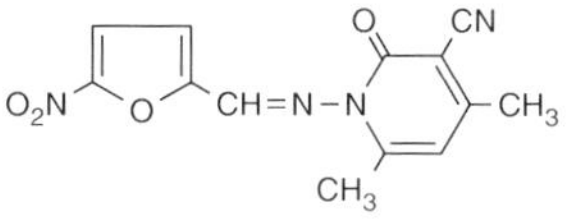

5 gm

S51,089-0
1,2-DIHYDRO-4,6-DIMETHYL-1-(2-NITROBENZYLIDENEAMINO)-2-OXONICOTINONITRILE

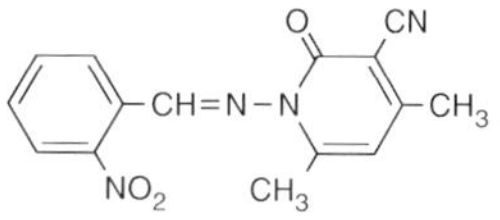

250 mg

S52,472-7
N-(4-NITROBENZYLIDENEAMINO)-PHTHALIMIDE

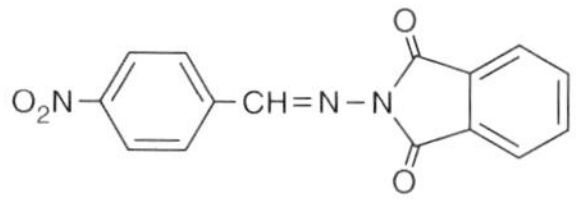

5 gm

S52,717-3
N-(2-CHLORO-5-NITROBENZYLIDENEAMINO)PHTHALIMIDE

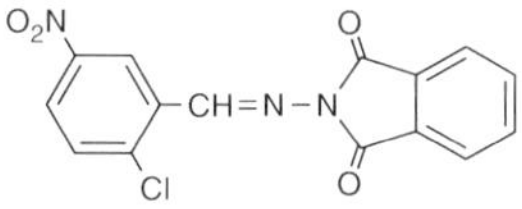

250 mg

S40,898-0
ISATIN 3-(4-NITROPHENYLHYDRAZONE)

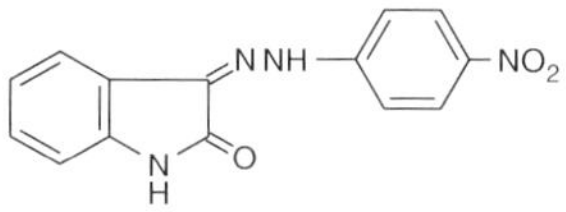

5 gm

S52,594-4
1,2-DIHYDRO-4,6-DIMETHYL-1-(3-NITROBENZYLIDENEAMINO)-2-OXONICOTINONITRILE

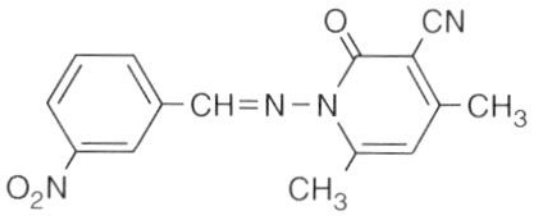

250 mg

S95,791-7
TRICYCLO[5.2.1.02,6]DEC-8-EN-10-ONE 2,4-DINITROPHENYLHYDRAZONE

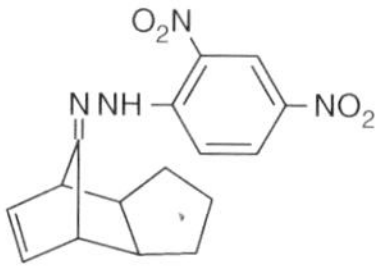

50 mg

S95,844-1
1-ACETYLISATIN 3-HYDRAZONE

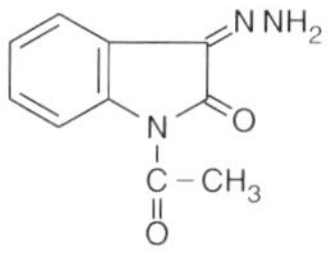

50 mg

S51,785-2
1-CINNAMYLIDENEAMINO-1,2-DIHYDRO-4,6-DIMETHYL-2-OXONICOTINONITRILE

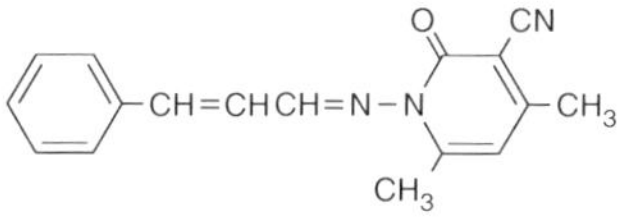

250 mg

S36,478-9
PYRUVIC ALDEHYDE 1-PHENYLHYDRAZONE

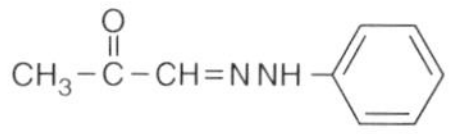

5 gm

S92,513-6
HIPPURIC N-(4-METHYLBENZYLIDENE)-HYDRAZIDE

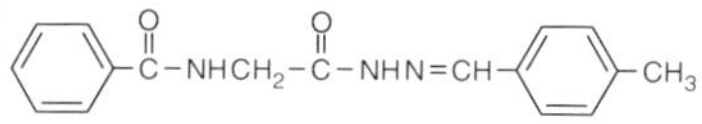

250 mg

S37,110-6
1,2-DIHYDRO-4,6-DIMETHYL-1-FURFURYLIDENEAMINO-2-OXONICOTINONITRILE

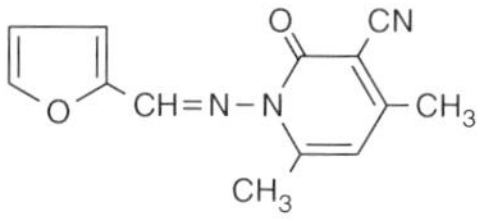

250 mg

S51,087-4
1-CYCLOHEXYLIDENEAMINO-1,2-DIHYDRO-4,6-DIMETHYL-2-OXONICOTINONITRILE

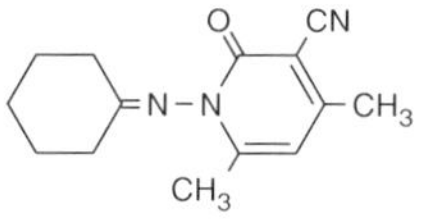

250 mg

S93,381-3
BENZOIC (4-CYANOBENZYLIDENE)-HYDRAZIDE

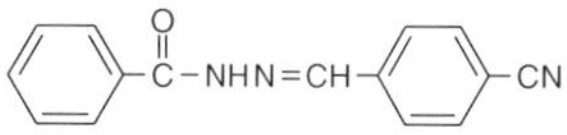

250 mg

S51,091-2
N,N'-(P-PHENYLENEDIMETHYLIDYNE)-BIS(1-AMINO-1,2-DIHYDRO-4,6-DIMETHYL-2-OXONICOTINONITRILE)

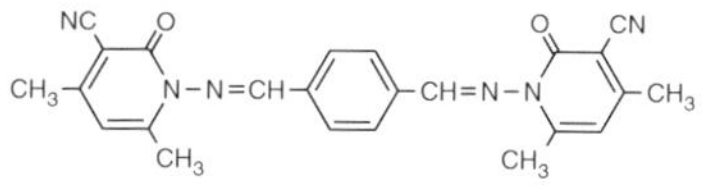

250 mg

S51,207-9
1,2-DIHYDRO-4,6-DIMETHYL-1-(5-INDANYLMETHYLENEAMINO)-2-OXONICOTINONITRILE

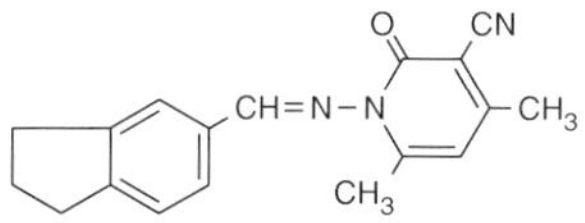

250 mg

S75,070-0
BENZOIC (3-PYRIDYLMETHYLENE)-HYDRAZIDE

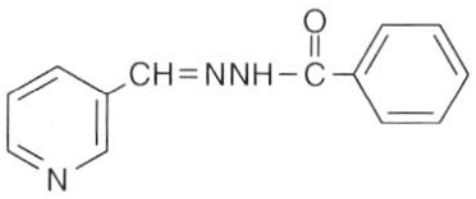

5 gm

S92,532-2
BENZOIC (4-PYRIDYLMETHYLENE)-HYDRAZIDE

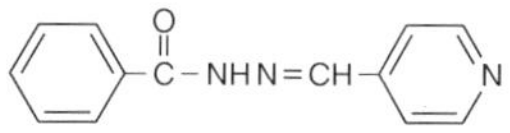

250 mg

S73,180-3
N,N'-TEREPHTHALYLIDENEBIS-(NICOTINIC HYDRAZIDE)

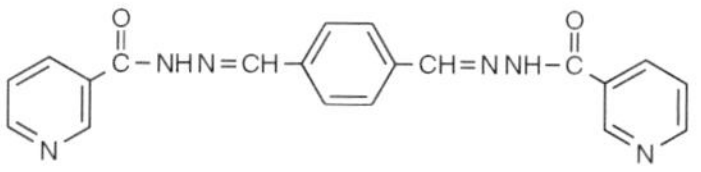

250 mg

S92,533-0
P-TOLUIC (4-PYRIDYLMETHYLENE)-HYDRAZIDE

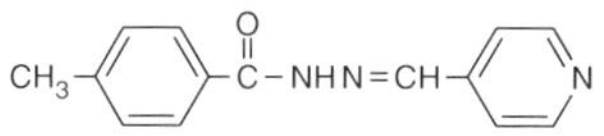

250 mg

S92,553-5
4-PYRIDINECARBOXYLIC (5-NORBORNEN-2-YLMETHYLENE)HYDRAZIDE

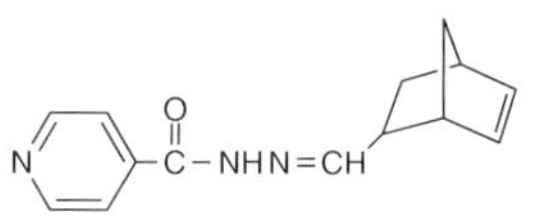

250 mg

S50,028-3
ISONICOTINIC ACID 5-INDANYLIDENEHYDRAZIDE

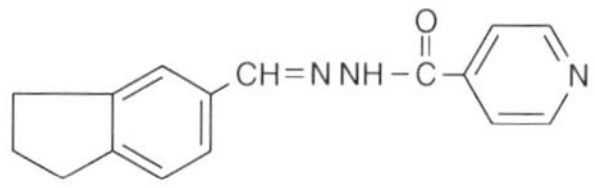

250 mg

S94,874-8
N-(2-THIAZOLYL)OXAMIC FURFURYLIDENEHYDRAZIDE

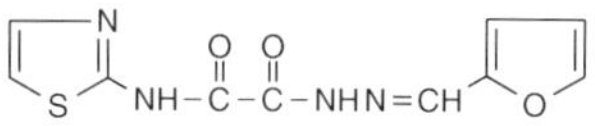

250 mg

S92,555-1
SEBACIC BIS(ISOBUTYLIDENE)HYDRAZIDE
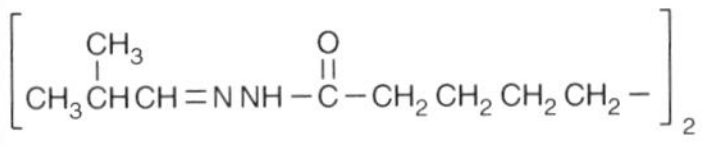
250 mg

S72,595-1
SEBACIC BIS(CYCLOPENTYLIDENEHYDRAZIDE)
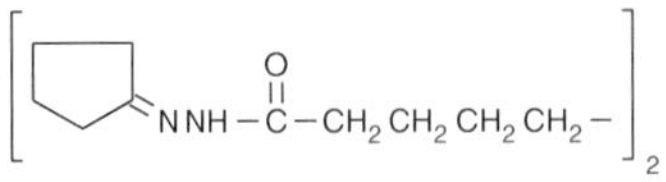
5 gm

S73,185-4
ADIPIC BIS(BENZYLIDENEHYDRAZIDE)
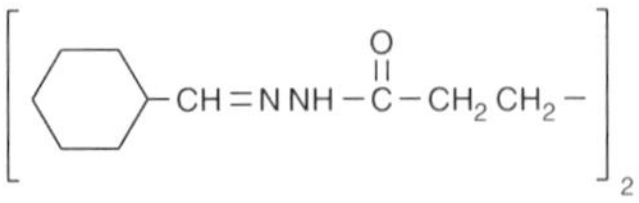
5 gm

S91,288-3
ADIPIC BIS(BENZYLIDENEHYDRAZIDE)
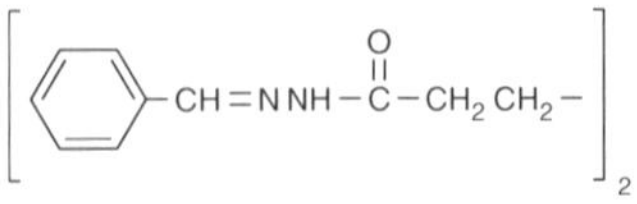
250 mg

S73,661-9
ADIPIC BIS(CINNAMYLIDENEHYDRAZIDE)
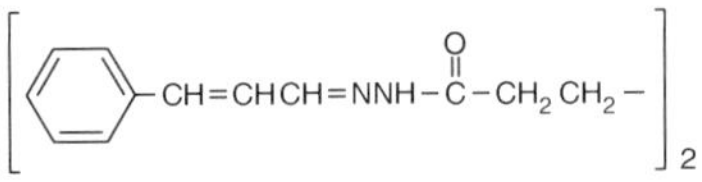
5 gm

S76,929-0
ADIPIC BIS(ALPHA,4-DIMETHYLBENZYLIDENE)HYDRAZIDE
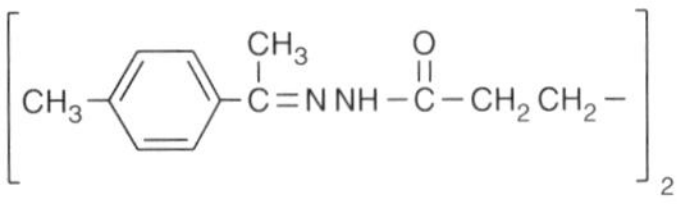
250 mg

S77,187-2
1-[3-(BENZOYLOXY)-ALPHA-METHYLBENZYLIDENEAMINO]GUANIDINE HYDROCHLORIDE
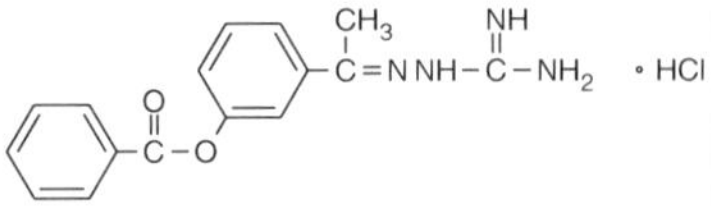
250 mg

S77,147-3
ISOPROPYL 6-(GUANIDINOIMINO)-6-PHENYLHEXANOATE NITRATE
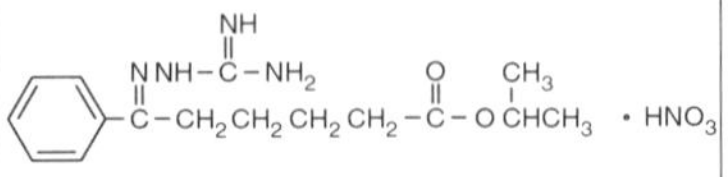
250 mg

S52,019-5
4-BENZYLIDENEAMINO-6-METHYL-1,2,4-TRIAZINE-3,5(2H,4H)-DIONE
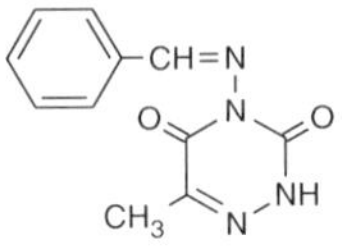
250 mg

S52,022-5
4-(4-CHLOROBENZYLIDENEAMINO)-6-METHYL-1,2,4-TRIAZINE-3,5(2H,4H)-DIONE
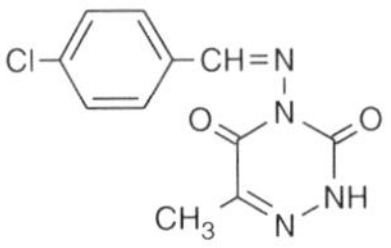
5 gm

S94,535-8
3-METHYL-1H-PYRAZOLE-5-CARBOXALDEHYDE SEMICARBAZIDE
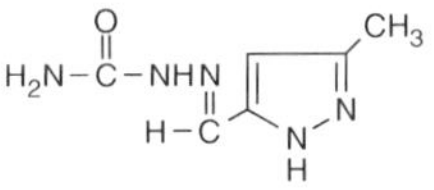
50 mg

S95,362-8
2'-METHYL-1',4',5',6'-(AND 3',4',5',6')TETRAHYDROACETOPHENONE SEMICARBAZONE
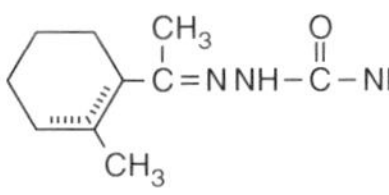
50 mg

S95,312-1
1-(2-CYCLOHEXEN-1-YLIDENE)-SEMICARBAZIDE
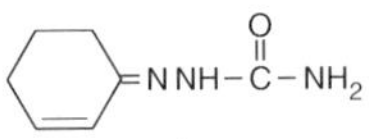
50 mg

S95,360-1
2'-METHYL-1',4',5',6'-TETRAHYDROACETOPHENONE SEMICARBAZONE
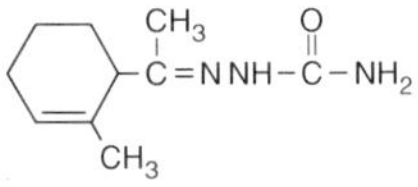
50 mg

S95,711-9
1,8A-DIMETHYL-2,3,4,4A,4B,5,8,8A-OCTAHYDROPHENANTHREN-2-ONE SEMICARBAZONE
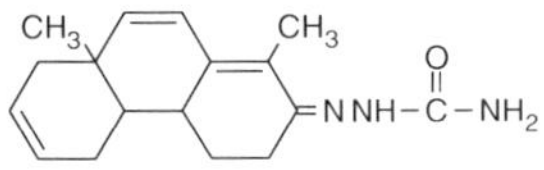
50 mg

S51,656-2
CYCLOOCTANONE SEMICARBAZONE
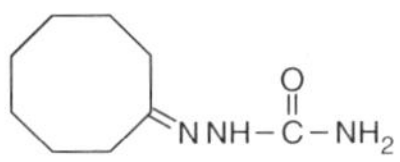
5 gm

S81,445-8
5-BUTYL-2-FURALDEHYDE SEMICARBAZONE
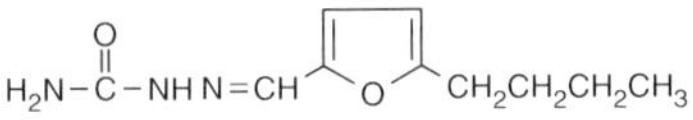
250 mg

S51,940-5
BENZALDEHYDE SEMICARBAZONE
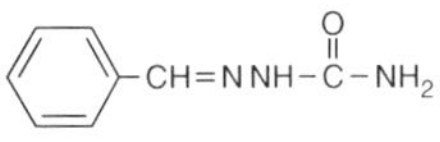
5 gm

S52,197-3
1-(1,2,3,4,5,6,7,8-OCTAHYDRO-3-PHENANTHRYL)-1-PROPANONE SEMICARBAZONE
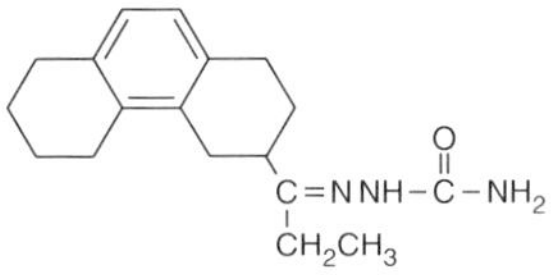
250 mg

S52,457-3
N-(2-PHENYLPROPYLIDENEAMINO)-PHTHALIMIDE
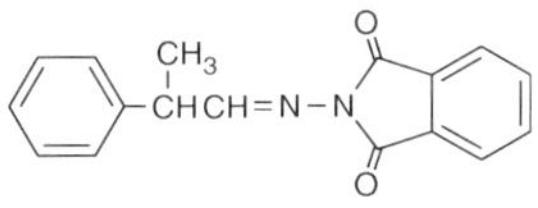
5 gm

S52,468-9
N-(2,4-DICHLOROBENZYLIDENEAMINO)-PHTHALIMIDE
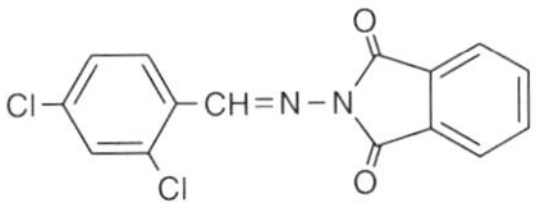
5 gm

S52,473-5
N-(4-METHYLBENZYLIDENEAMINO)-PHTHALIMIDE
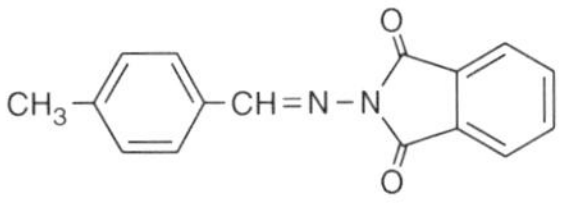
250 mg

S70,419-9
3-ACETAMIDOBENZALDEHYDE THIOSEMICARBAZONE
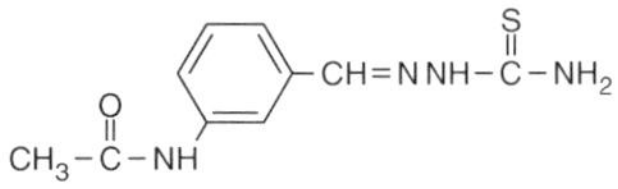
5 gm

S92,546-2
10-UNDECENOIC (4-ACETAMIDOBENZYLIDENE)HYDRAZIDE
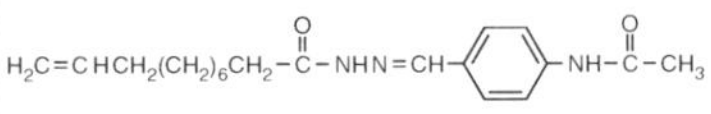
250 mg

S80,091-0
1-(4-ACETAMIDOBENZYLIDENE)-4-PHENYL-3-THIOSEMICARRBAZIDE
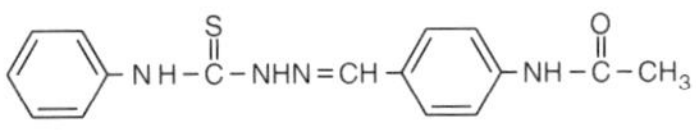
250 mg

S53,649-0
4-ACETAMIDOBENZALDEHYDE THIOSEMICARBAZONE
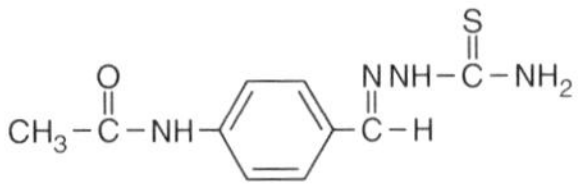
250 mg

S95,843-3
ISATIN 3-HYDRAZONE
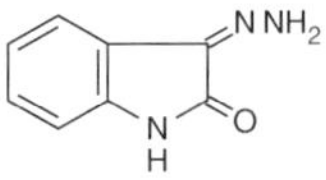
50 mg

S40,353-9
ISONICOTINALDEHYDE THIOSEMICARBAZONE
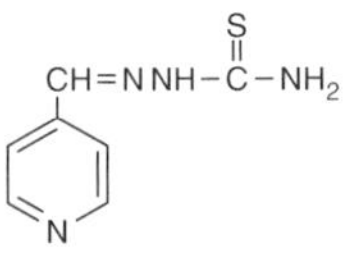
5 gm

S92,526-8
DODECANOIC (2-BUTEN-1-YLIDENE)-HYDRAZIDE
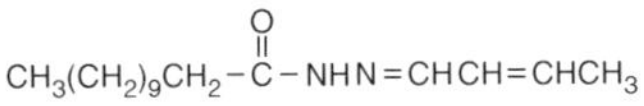
250 mg

S53,909-0
3-(4-FLUOROBENZYLIDENEAMINO)-RHODANINE
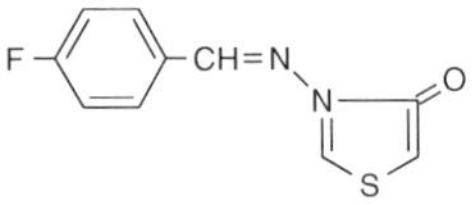
250 mg

S77,867-2
2-[3-(5-METHYL-2-PYRIDYL)-UREIDO]ACETIC (PIPERONYLMETHYLENE)-HYDRAZIDE
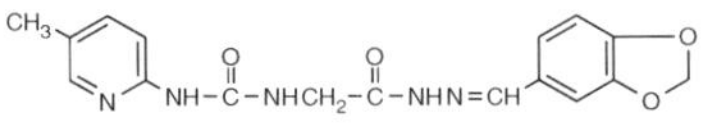
250 mg

S77,931-8
2-[3-(5-METHYL-2-PYRIDYL)-UREIDO]ACETIC BENZYLIDENEHYDRAZIDE
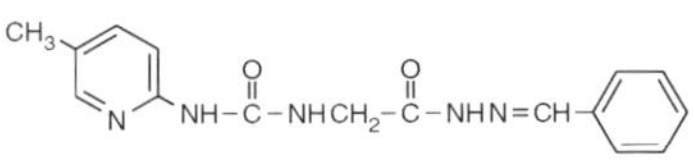
250 mg

S51,447-0
PICOLINIC ACID (5-INDANYLMETHYLENE)-HYDRAZIDE
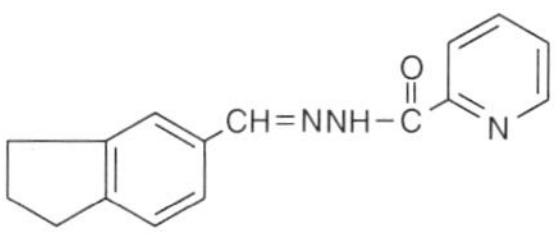
250 mg

S90,407-4
2-CYANOACETIC (ALPHA-PROPYLBENZYLIDENE)HYDRAZIDE
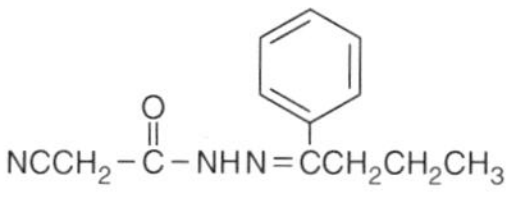
250 mg

S51,090-4
1,2-DIHYDRO-4,6-DIMETHYL-1-(3-INDOLYLMETHYLENEAMINO)-2-OXONICOTINONITRILE
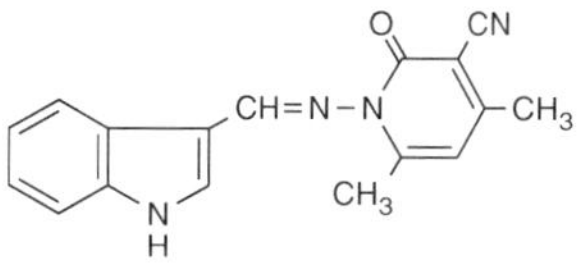
250 mg

S92,548-9
10-UNDECENOIC (THENYLIDENE)-HYDRAZIDE
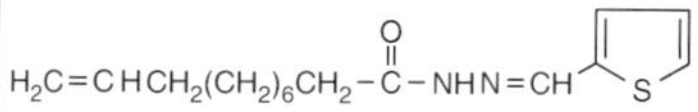
250 mg

S50,625-7
5-INDANCARBOXALDEHYDE 4-PHENYLSEMICARBAZONE
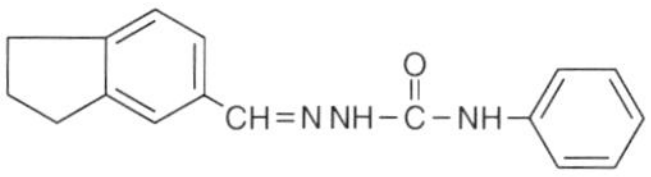
250 mg

S52,729-7
N-(3,4-DICHLOROBENZYLIDENEAMINO)-PHTHALIMIDE
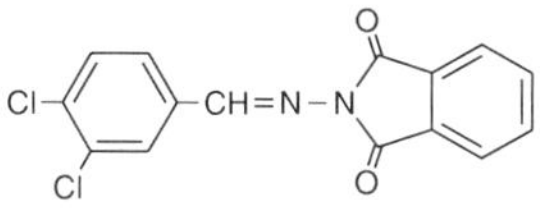
250 mg

S54,241-5
ADIPIC ACID BIS[(2-PYRIDYLMETHYLENE)HYDRAZIDE]
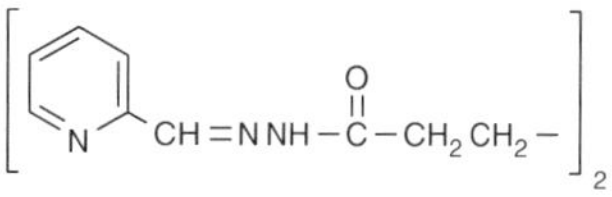
250 mg

S92,520-9
DECANOIC (3-INDOLYLMETHYLENE)-HYDRAZIDE
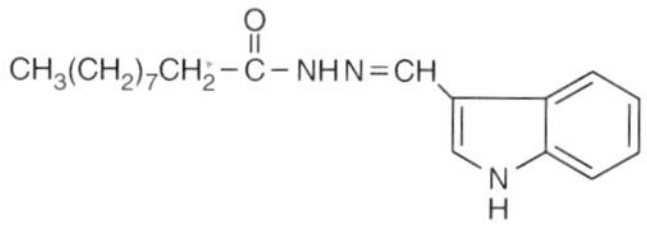
250 mg

S95,337-7
4-(ALPHA-METHYLBENZYLIDENEAMINO)-3-THIOSEMICARBAZIDE
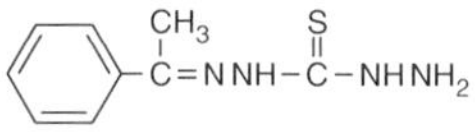
50 mg

S35,141-5
INDOLE-3-CARBOXALDEHYDE THIOSEMICARBAZONE
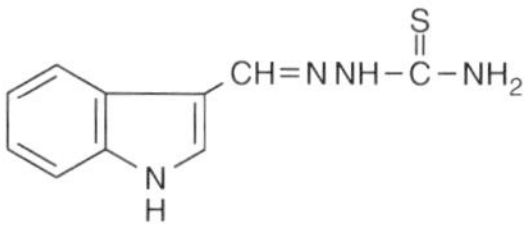
5 gm

S50,977-9
5-INDANCARBOXALDEHYDE 4-METHYL-3-THIOSEMICARBAZONE
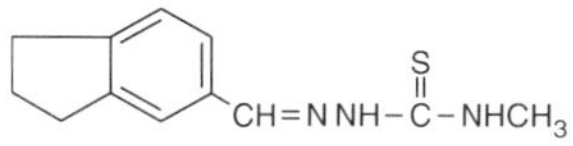
250 mg

S51,088-2
N4-(2-PYRIDYLMETHYLENE)-N1-(2-THIAZOLYL)SULFANILAMIDE
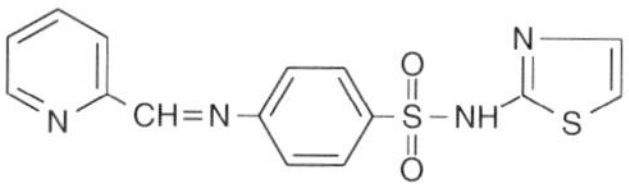
250 mg

S73,511-6
S-METHYL-1-(4-PHENYLBENZYLIDENE) ISOTHIOSEMICARBAZIDE
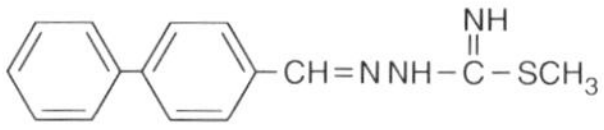
250 mg

S86,531-1
2-NORBORNANONE (P-TOSYL)HYDRAZONE
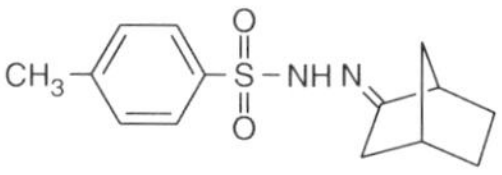
250 mg

S86,357-2
CYCLOHEXANONE (P-TOSYL)HYDRAZONE
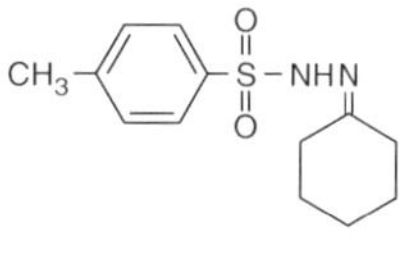
250 mg

S86,533-8
2-METHYLCYCLOHEXANONE (P-TOSYL)-HYDRAZONE
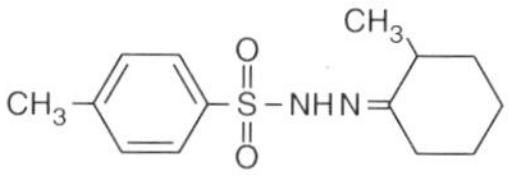
250 mg

S86,572-9
2-SEC-BUTYLCYCLOHEXANONE (P-TOSYL)-HYDRAZONE
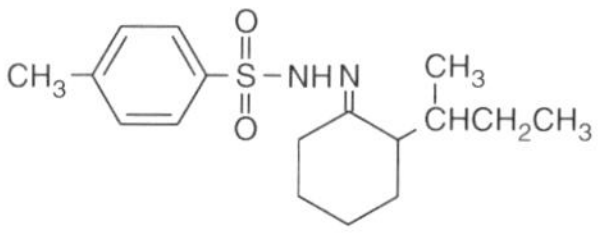
250 mg

S88,220-8
P-TOLUENESULFONIC (3-METHYL-2-CYCLOHEXEN-1-YLIDENE)HYDRAZIDE
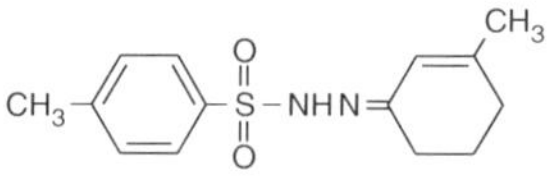
250 mg

S87,265-2
MESITYLENESULFONIC BENZYLIDENEHYDRAZIDE
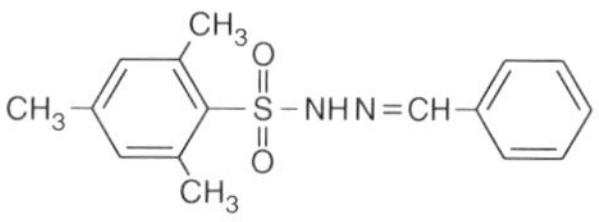
250 mg

S52,393-3
P-TOLUENESULFONIC ACID (4-CHLOROBENZYLIDENE)HYDRAZIDE
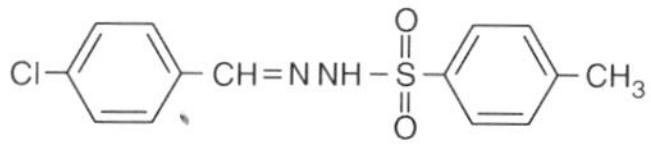
5 gm

S92,465-2
P-TOLUALDEHYDE P-TOSYLHYDRAZONE
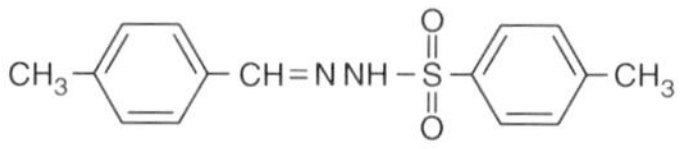
250 mg

S51,647-3
P-TOLUENESULFONIC ACID CYCLOOCTYLIDENEHYDRAZIDE
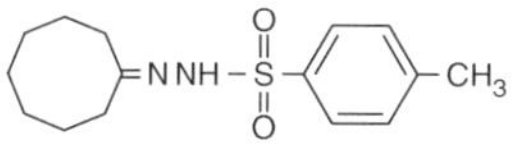
250 mg

S52,202-3
NICOTINALDEHYDE (P-TOLYLSULFONYL)HYDRAZONE
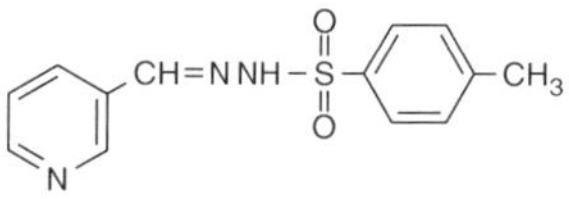
250 mg

S86,396-3
ANTHRONE (P-TOSYL)HYDRAZONE
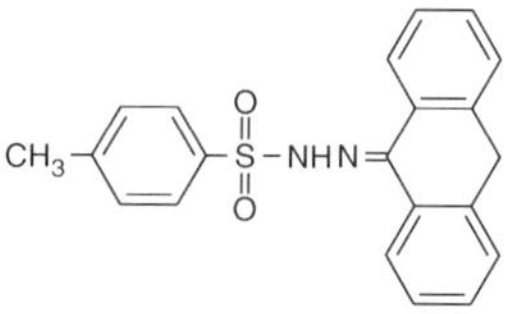
250 mg

S85,299-6
9-ANTHRALDEHYDE (P-TOSYL)HYDRAZONE
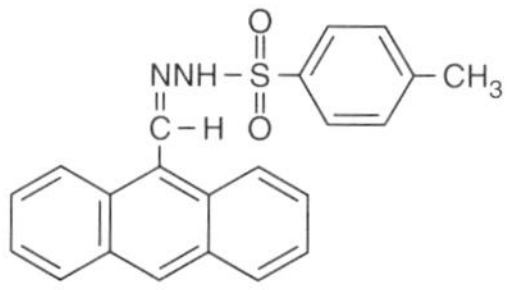
250 mg

S87,250-4
MESITYLENESULFONIC ETHYLIDENEHYDRAZIDE
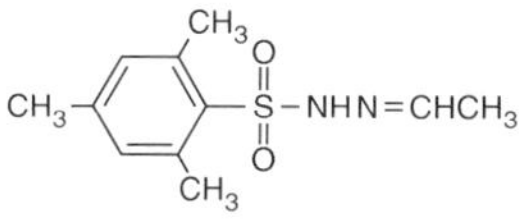
250 mg

S81,435-0
2,5-DIMETHYL-3-FURALDEHYDE HYDRAZONE
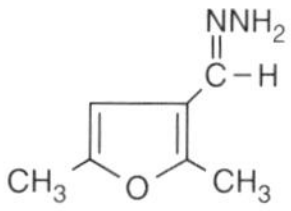
250 mg

S52,256-2
4-FURFURYLIDENEAMINO-4H-1,2,4-TRIAZOLE
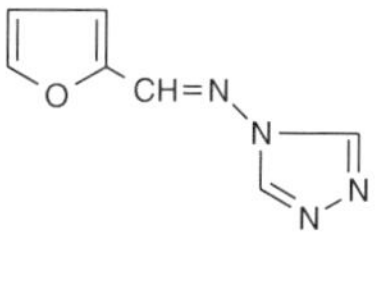
5 gm

S80,078-3
1-FURFURYLIDENE-4-PHENYL-3-THIOSEMICARBAZIDE
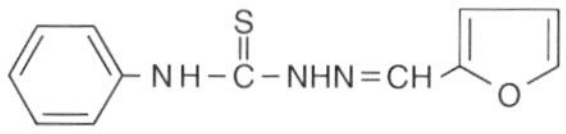
250 mg

S93,455-0
4-(2-THENYLIDENEAMINO)-4H-1,2,4-TRIAZOLE
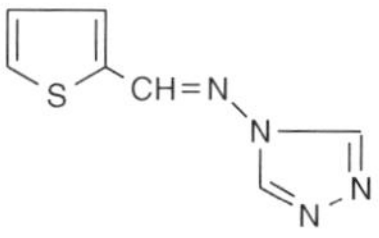
250 mg

S77,170-8
1-[ALPHA-(2-THIENYL)-BENZYLIDENEAMINO]GUANIDINE NITRATE
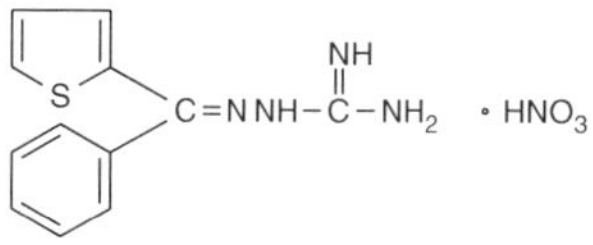
250 mg

S77,285-2
1-(1,3-BENZODIOXOL-6-YLMETHYLENE)-4-CYCLOHEXYL-3-THIOSEMICARBAZIDE
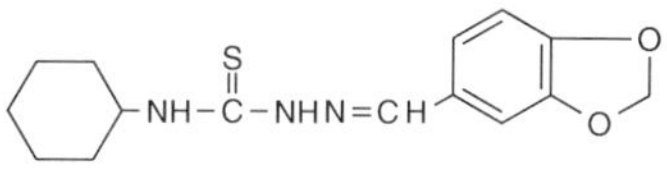
250 mg

S69,221-2
4-(3-MORPHOLINOPROPYL)-1-(PIPERONYLIDENE)-3-THIOSEMICARBAZIDE
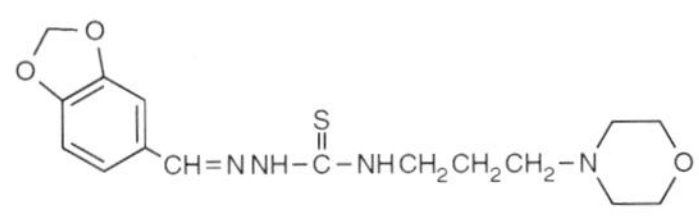
250 mg

S76,685-2
DECANAL (2-IMIDAZOLIN-2-YL)-HYDRAZONE HYDRIODIDE
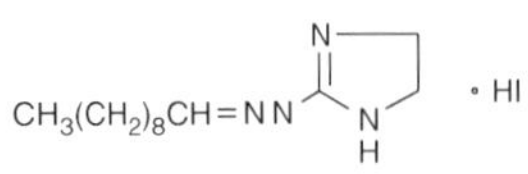
250 mg

S93,501-8
2-PYRIDYLCARBOXALDEHYDE (2-BENZOTHIAZOLYL)HYDRAZONE
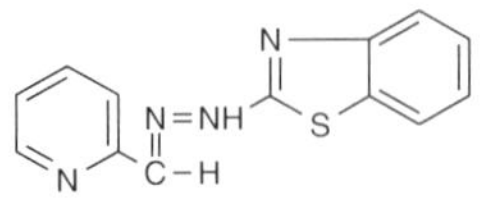
250 mg

S93,437-2
4-(CYCLOHEXYLIDENEAMINO)-4H-1,2,4-TRIAZOLE
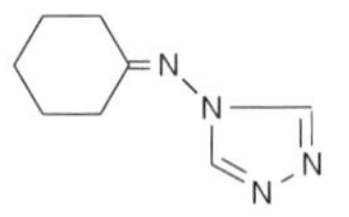
250 mg

S69,126-7
4-CYCLOHEXYL-1-(1-NAPHTHYLMETHYLENE)-3-THIOSEMICARBAZIDE
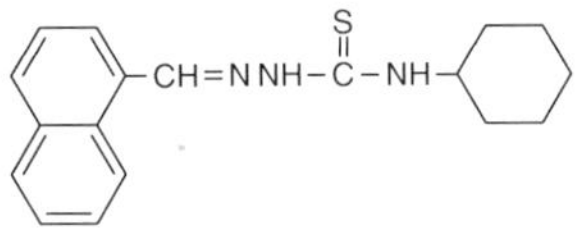
250 mg

S77,159-7
1-[ALPHA-(2-PIPERIDINOETHYL)-BENZYLIDENEAMINO]GUANIDINE DIHYDROCHLORIDE
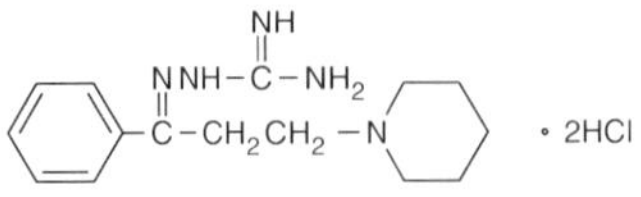
250 mg

S50,613-3
1-(5-INDANYLMETHYLENEAMINO)-PIPERIDINE
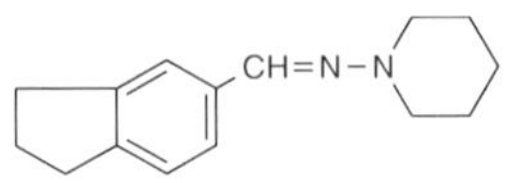
250 mg

S76,272-5
4-(BENZYLIDENEAMINO)-3-THIOSEMICARBAZIDE
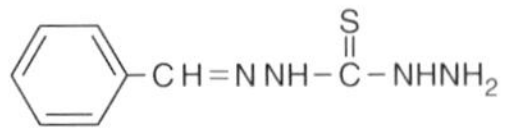
250 mg

S77,175-9
1-(1-BENZYLETHYLIDENEAMINO)-GUANIDINE NITRATE
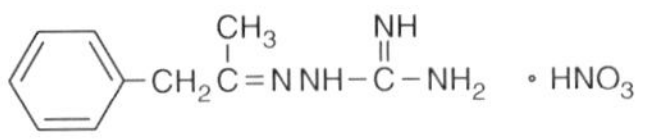
250 mg

S82,588-3
BENZALDEHYDE PHENYLHYDRAZONE
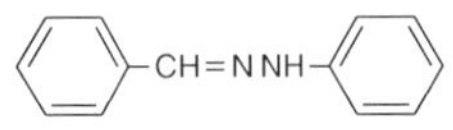
250 mg

S83,784-9
BENZALDEHYDE 2-METHYL-2-PHENYLHYDRAZONE
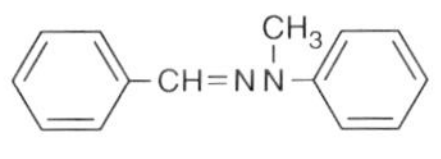
250 mg

S77,143-0
1-(BETA-PENTYLCINNAMYLIDENEAMINO)-GUANIDINE
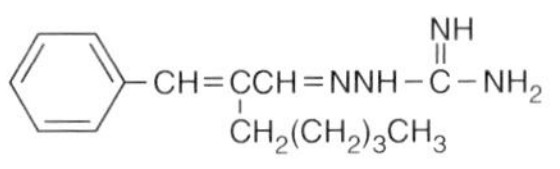
250 mg

S77,186-4
1-(ALPHA-PHENETHYLBENZYLIDENEAMINO)-GUANIDINE NITRATE
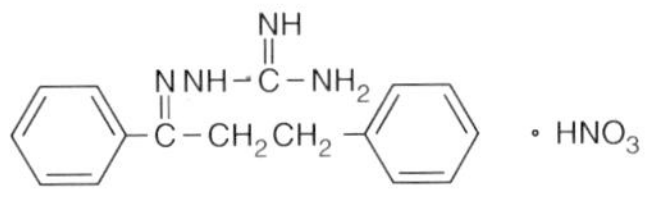
250 mg

S77,151-1
1-[4-(3-BIPHENYLYL)-ALPHA-METHYLBENZYLIDENEAMINO]-GUANIDINE NITRATE
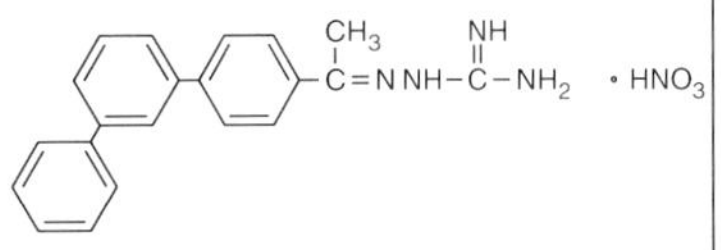
250 mg

S69,673-0
2-ACETYLPYRIDINE PHENYLHYDRAZONE
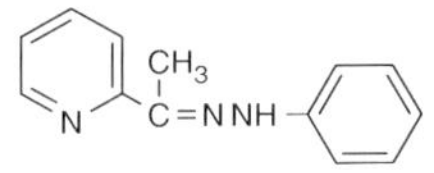
250 mg

S59,953-0
BENZOPHENONE THIOSEMICARBAZIDE
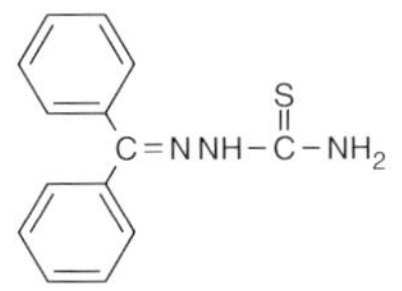
250 mg

S71,031-8
[(DIPHENYLMETHYLENE)HYDRAZONO]-TRIPHENYLPHOSPHORANE
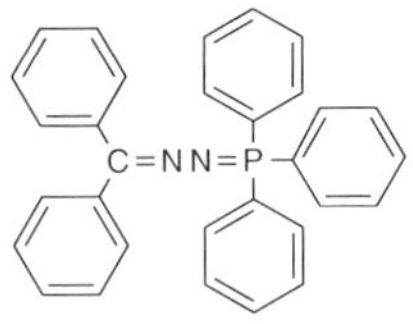
5 gm

S92,467-9
2'-ACETONAPHTHONE PHENYLHYDRAZONE
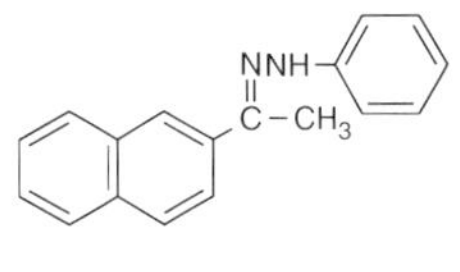
250 mg

S77,140-6
1-(ALPHA-HEXYL-2-METHYLBENZYLIDENEAMINO)-GUANIDINE NITRATE
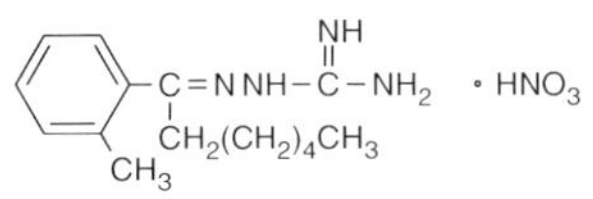
250 mg

S93,443-7
4-(4-METHYLBENZYLIDENEAMINO)-4H-1,2,4-TRIAZOLE
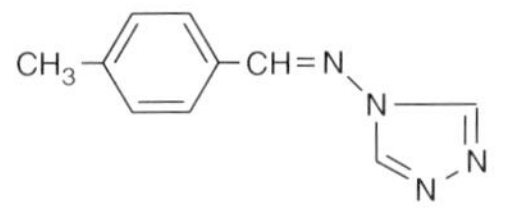
250 mg

S53,826-4
P-TOLYLACETALDEHYDE THIOSEMICARBAZONE
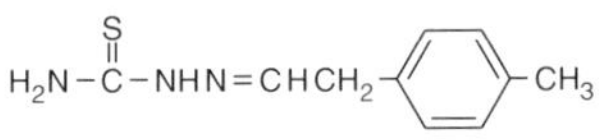
5 gm

S93,445-3
4-(4-ISOPROPYLBENZYLIDENEAMINO)-4H-1,2,4-TRIAZOLE
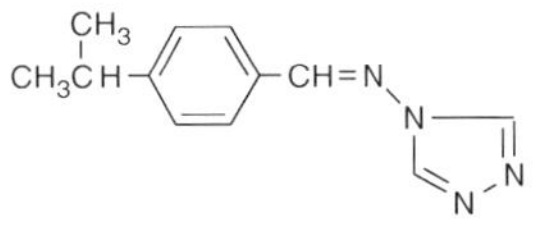
250 mg

S77,144-9
1-(ALPHA,4-DIMETHYLCINNAMYLIDENEAMINO)-GUANIDINE NITRATE
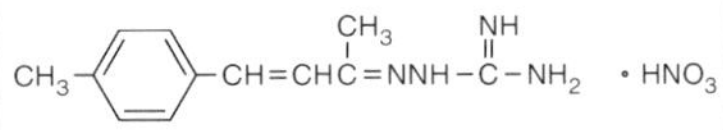
250 mg

S55,471-5
4-CHLOROBENZALDEHYDE PHENYLHYDRAZONE
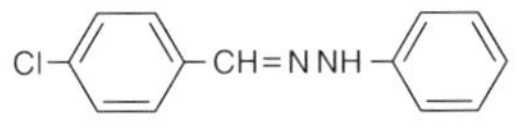
250 mg

S77,137-6
1-(4-METHYL-ALPHA-PENTYLBENZYLIDENEAMINO) GUANIDINE NITRATE
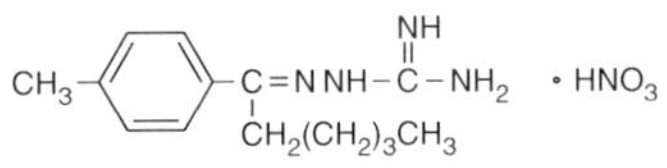
250 mg

S93,446-1
4-(4-TERT-BUTYLBENZYLIDENEAMINO)-4H-1,2,4-TRIAZOLE
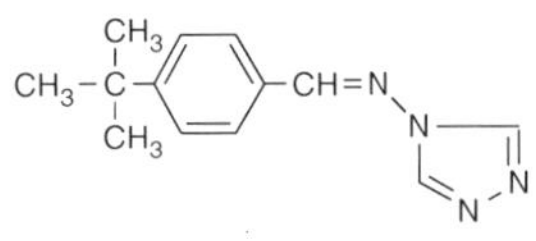
250 mg

S86,672-5
2-ADAMANTANONE DIMETHYLHYDRAZONE
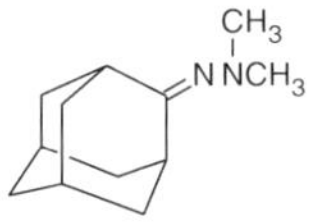
250 mg

S51,453-5
5-INDANCARBOXALDEHYDE HEXAMETHYLENEHYDRAZONE
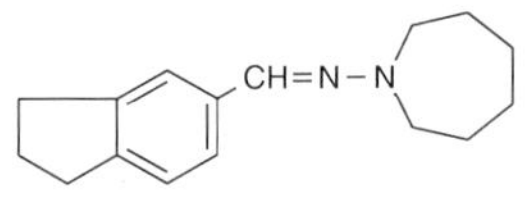
250 mg

S78,097-9
1-[1-(2-NAPHTHYL)ETHYLIDENE]-3-THIOSEMICARBAZIDE
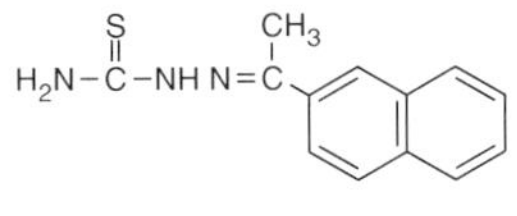
250 mg

S93,453-4
4-(2-NAPHTHYLMETHYLENEAMINO)-4H-1,2,4-TRIAZOLE
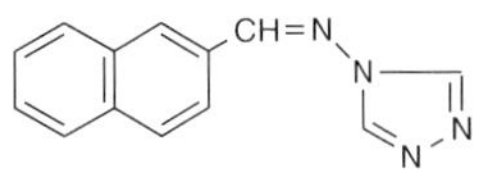
250 mg

S93,452-6
4-(9-ANTHRACENYLMETHYLENEAMINO)-4H-1,2,4-TRIAZOLE
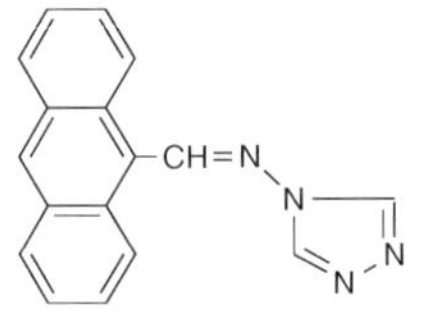
250 mg

S50,660-5
5-INDANCARBOXALDEHYDE, AZINE WITH BENZIL
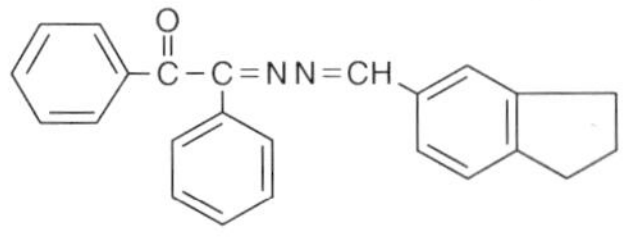
250 mg

S61,561-7
4-CHLORO-N-(FURFURYLIDENE)ANILINE

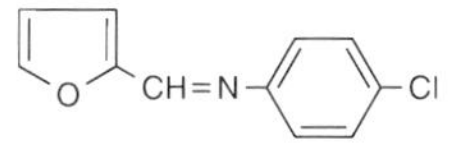

5 gm

S62,901-2
2,5-DICHLORO-N-(FURFURYLIDENE)-ANILINE

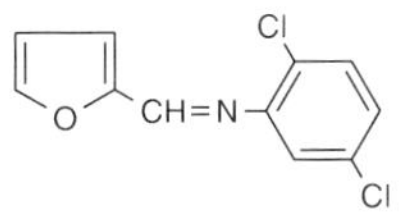

250 mg

S60,431-3
N-(FURFURYLIDENE)-2,5-XYLIDINE

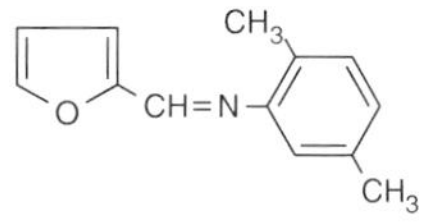

250 mg

S61,556-0
N-(FURFURYLIDENE)-2,4-XYLIDINE

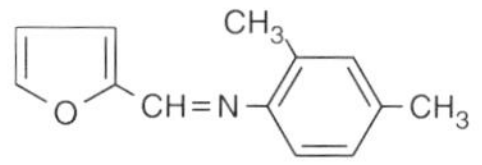

250 mg

S62,087-4
N-(FURFURYLIDENE)-3,5-XYLIDINE

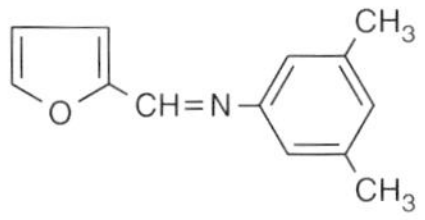

250 mg

S62,089-0
N-(FURFURYLIDENE)-2,3-XYLIDINE

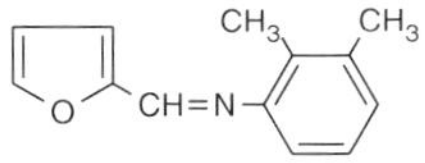

250 mg

S62,092-0
N-(FURFURYLIDENE)-3,4-XYLIDINE

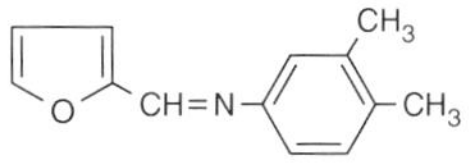

250 mg

S51,787-9
2-[4,5-DI-(2-FURYL)-2-IMIDAZOLYLMETHYLENEAMINO]-BENZIMIDAZOLE

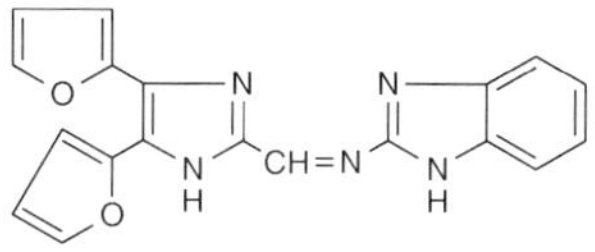

250 mg

S59,997-2
ETHYL 4-(2-THIENYLMETHYLENEAMINO)-BENZOATE

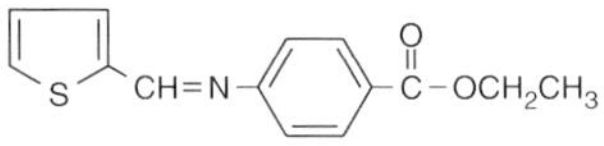

5 gm

S50,611-7
2-THIOPHENECARBOXYLIC ACID (5-INDANYLMETHYLENE)HYDRAZIDE

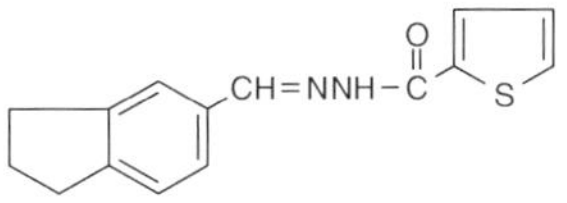

250 mg

S58,647-1
4-FLUORO-N-(2-THIENYLMETHYLENE)-ANILINE

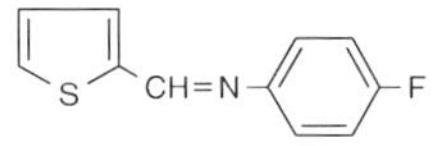

250 mg

S58,650-1
4-CHLORO-N-(2-THIENYLMETHYLENE)-ANILINE

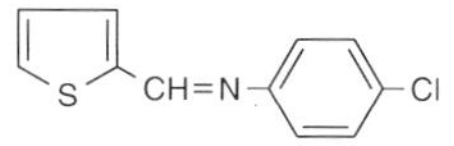

5 gm

S58,646-3
3,4-DICHLORO-N-(2-THIENYLMETHYLENE)ANILINE

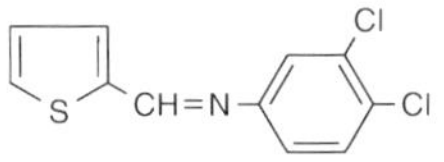

5 gm

S58,652-8
2,4-DICHLORO-N-(2-THIENYLMETHYLENE)-ANILINE

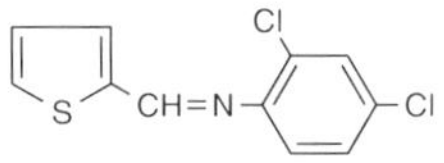

5 gm

S64,060-3
2,5-DICHLORO-N-(2-THIENYLMETHYLENE)-ANILINE

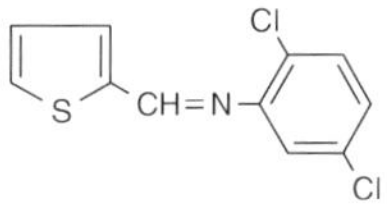

250 mg

S59,603-5
4-BROMO-N-(2-THIENYLMETHYLENE)-ANILINE

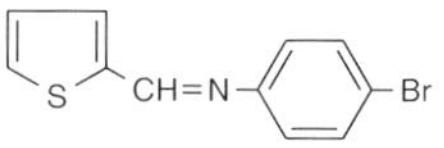

250 mg

S58,986-1
4-IODO-N-(2-THIENYLMETHYLENE)-ANILINE

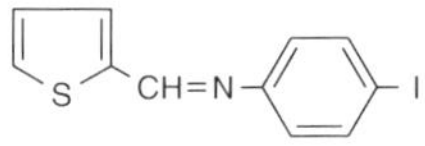

250 mg

S58,378-2
N-(2-THIENYLMETHYLENE)-P-TOLUIDINE

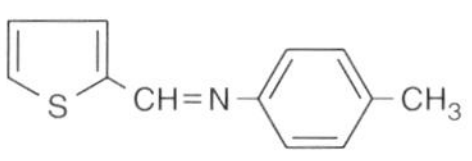

5 gm

S58,654-4
5-CHLORO-N-(2-THIENYLMETHYLENE)-O-TOLUIDINE

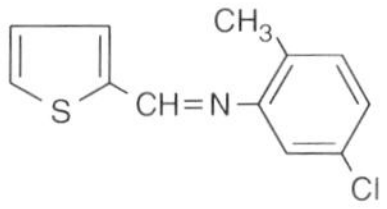

5 gm

S58,923-3
4-BROMO-N-(2-THIENYLMETHYLENE)-O-TOLUIDINE

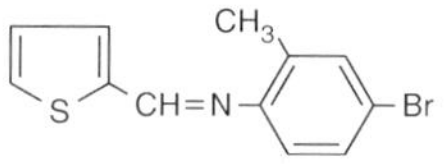

5 gm

S59,602-7
2-BROMO-N-(2-THIENYLMETHYLENE)-P-TOLUIDINE

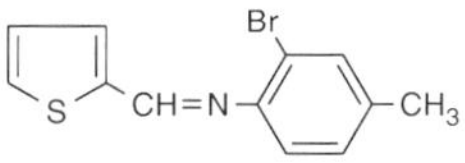

250 mg

S58,649-8
N-(2-THIENYLMETHYLENE)-3,4-XYLIDINE

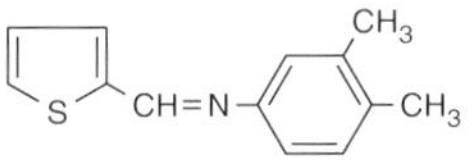

5 gm

S58,988-8
N-(2-THIENYLMETHYLENE)-2,4,6-TRIMETHYLANILINE

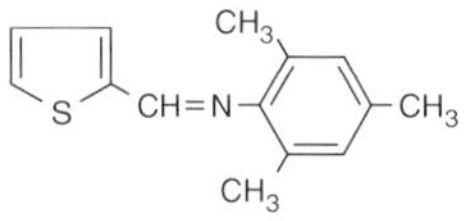

250 mg

S80,839-3
1-CINNAMOYL-3-(DIAZOACETYL)-4-PHENYL-2-PYRAZOLINE

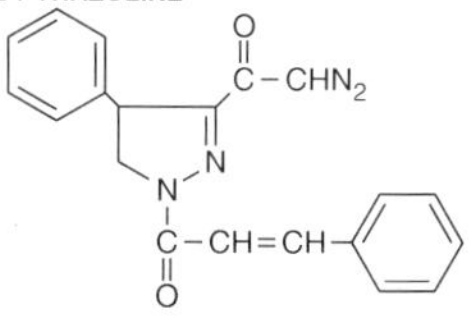

250 mg

S92,565-9
TRIS(6-NITRO-2-BENZOTHIAZOLINYLIDENEAMINO) CARBENIUM TETRAFLUOROBORATE

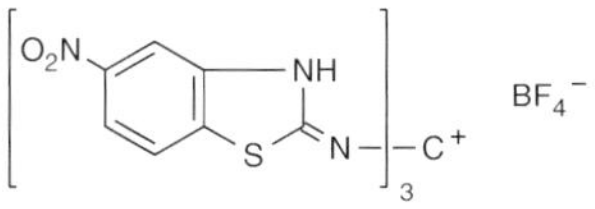

250 mg

S88,613-0
N-(4-NITROBENZYLIDENE)ANILINE

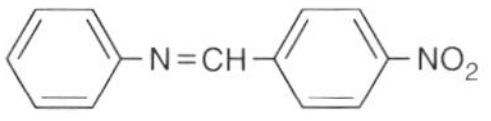

250 mg

S88,219-4
4-METHYL-N-(4-NITROBENZYLIDENE)-ANILINE

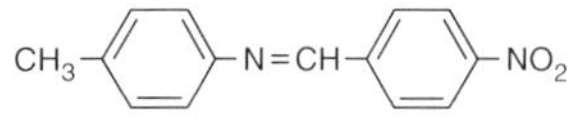

250 mg

S86,223-1
4-NITROBENZALDEHYDE AZINE

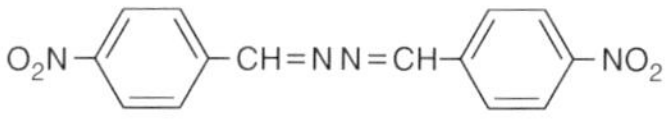

250 mg

S53,231-2
BENZIL, AZINE WITH 4-NITROBENZALDEHYDE

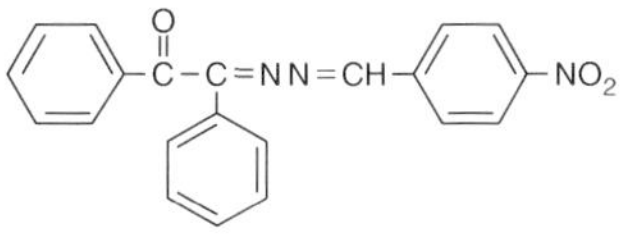

250 mg

S532231
ISATIN, AZINE WITH 4-NITROBENZALDEHYDE

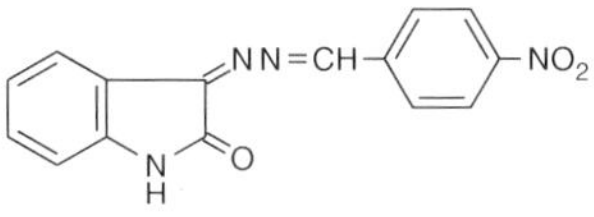

5 gm

S95,797-6
TRICYCLO[5.2.1.0(2,6)]DECAN-8-ONE AZINE

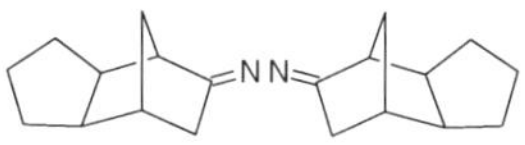

50 mg

S77,425-1
2-(BENZYLIDENEAMINO)-7-NITROFLUORENE

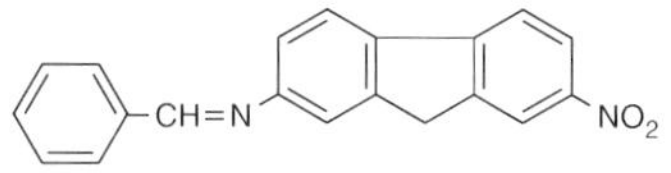

250 mg

S52,258-9
4-FLUORO-N-(3-NITROBENZYLIDENE)-ANILINE

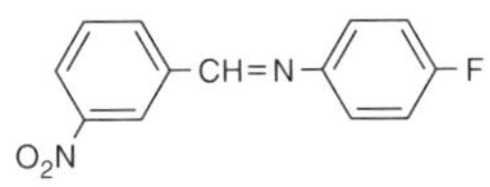

5 gm

S52,341-0
N-(4-FLUOROBENZYLIDENE)-4-NITROANILINE

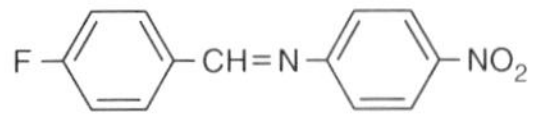

250 mg

S52,380-1
N-(4-FLUOROBENZYLIDENE)-3-NITROANILINE

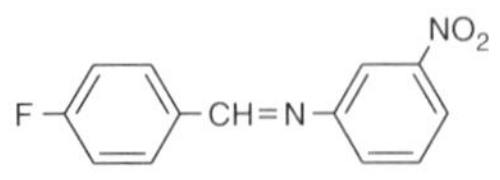

5 gm

S52,265-1
4-IODO-N-(3-NITROBENZYLIDENE)-ANILINE

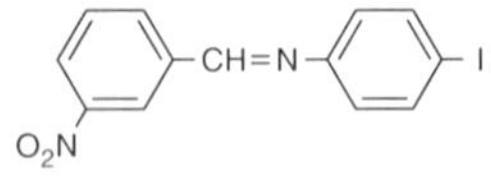

5 gm

S52,398-4
N-(3-NITROBENZYLIDENE)-2,5-XYLIDINE

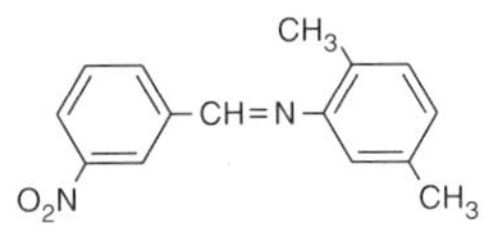

5 gm

S90,178-4
4-(3-ACETYLPHENYLIMINOMETHYL)-PHENYL N-PHENYLCARBAMATE

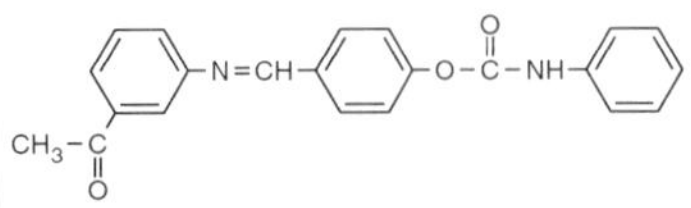

250 mg

S71,969-2
4'-(4-METHYLBENZYLIDENEAMINO)-ACETOPHENONE

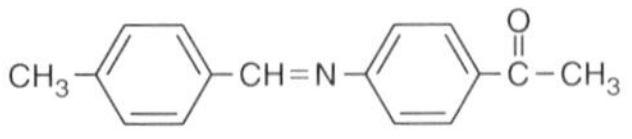

5 gm

S72,001-1
3'-(2-HYDROXY-1-NAPHTHYLMETHYLENEAMINO)ACETOPHENONE

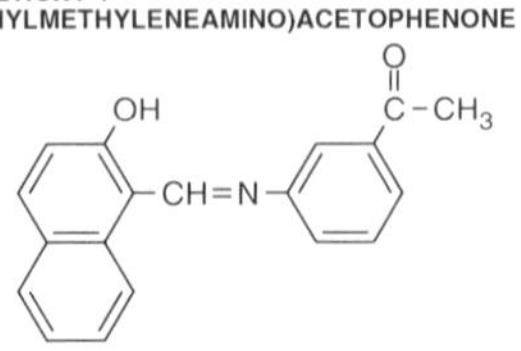

5 gm

S73,004-1
4-(PIPERONYLIDENEAMINO)ACETOPHENONE

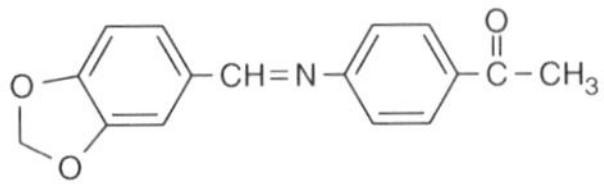

5 gm

S53,251-7
4-PHENYLIMINO-2-PENTANONE

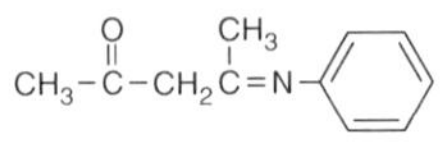

5 gm

S53,214-2
4-(4-CHLOROPHENYLIMINO)-2-PENTANONE

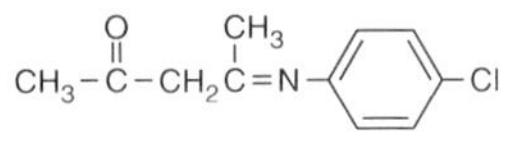

5 gm

S53,229-0
4-(P-TOLYLIMINO)-2-PENTANONE

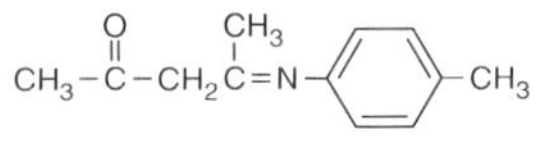

250 mg

S53,252-5
BENZIL, AZINE WITH 4-CHLOROBENZALDEHYDE

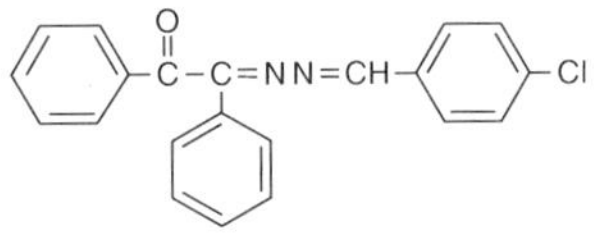

250 mg

S53,230-4
4-(2,4-DICHLOROPHENYLIMINO)-2-PENTANONE

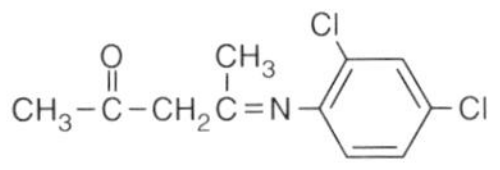

250 mg

S87,638-0
4-(2-NAPHTHYLIMINO)-2-BUTANONE

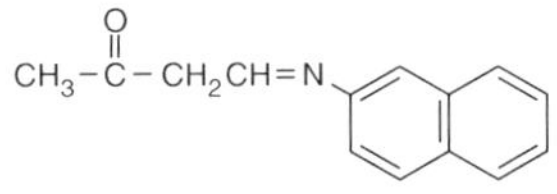

250 mg

S69,260-3
4-(2-CARBAMOYLPHENYLIMINO)-2,3-DICHLORO-2-BUTENOIC ACID

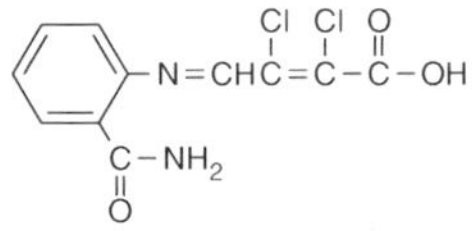

250 mg

S90,101-6
N-(4-ACETAMIDOBENZYLIDENE)-ANTHRANILIC ACID

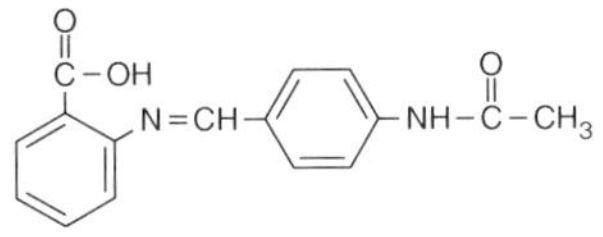

250 mg

S62,404-7
4-(2-THIENYLMETHYLENEAMINO)-BENZOIC ACID

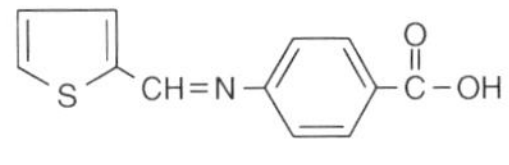

5 gm

S62,447-0
3-CARBOXY-N-(2-THENYLIDENE)ANILINE

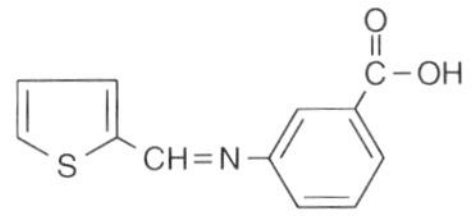

5 gm

S50,654-0
N-(5-INDANYLMETHYLENE)-ANTHRANILIC ACID

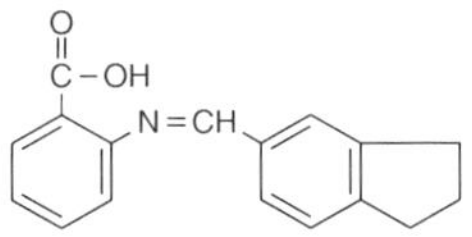

250 mg

S50,652-4
4-(5-INDANYLMETHYLENEAMINO)-BENZOIC ACID

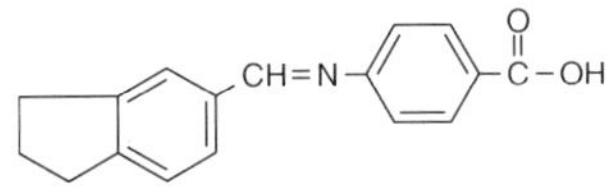

250 mg

S50,979-5
4-SULFOBENZOIC ACID (5-INDANYLMETHYLENE)HYDRAZIDE

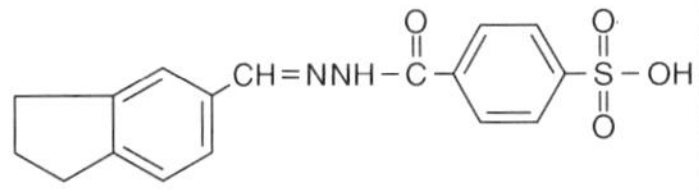

250 mg

S39,832-2
ISONICOTINIC ACID (PROPIONYLHYDRAZONOETHYLIDENE)-HYDRAZIDE

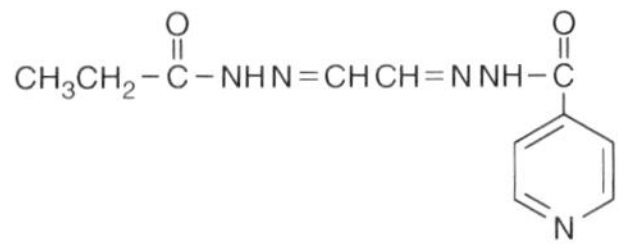

5 gm

S50,622-2
MALONIC ACID (3,4,5-TRIMETHOXYBENZYLIDENE)HYDRAZIDE

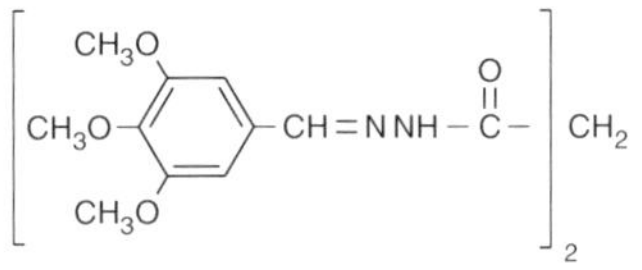

250 mg

S93,985-4
ALPHA-[(ETHOXYCARBONYLMETHOXY)-IMINO]-3-PYRAZOLEACETIC ACID

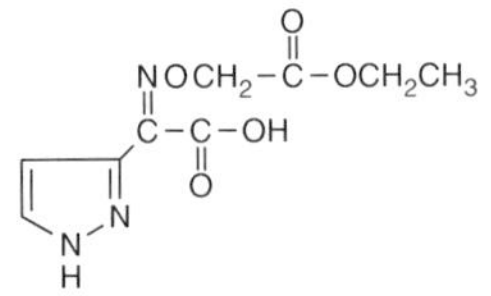

5 gm

S53,960-0
4-(2,6-DICHLOROBENZYLIDENEAMINO)-BENZOIC ACID

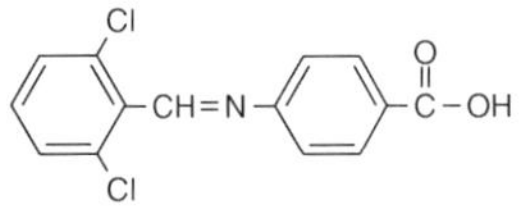

5 gm

S90,179-2
4-[4-(ETHOXYCARBONYL)-PHENYLIMINOMETHYL]PHENYL N-PHENYLCARBAMATE

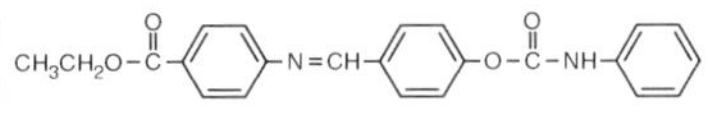

250 mg

S90,177-6
4-(2,4,5-TRIMETHYLPHENYLIMINOMETHYL)PHENYL N-METHYLCARBAMATE

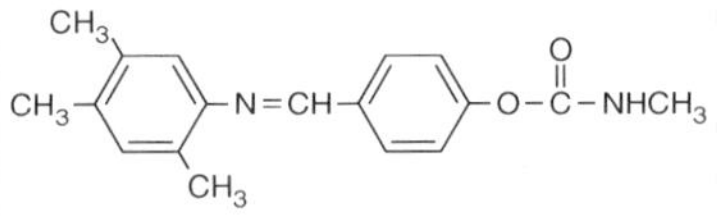

250 mg

S71,994-3
4-(4-ACETAMIDOBENZYLIDENEAMINO)-2,3-DIMETHYL-1-PHENYL-3-PYRAZOLIN-5-ONE

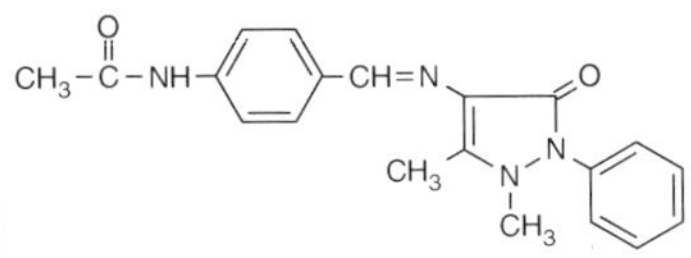

5 gm

S53,266-5
3-HEXADECYL-2-PHENYLIMINO-4-THIAZOLIDINONE

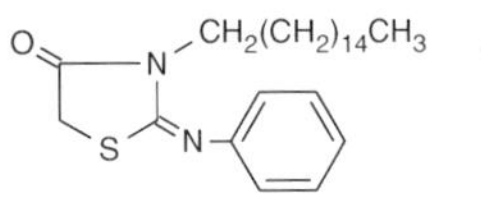

250 mg

S53,396-3
4-ACETAMIDOBENZALDEHYDE AZINE

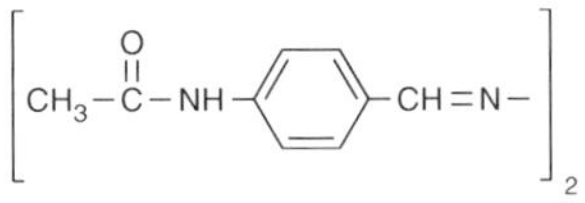

5 gm

S53,222-3
ISATIN, AZINE WITH 4-CHLOROBENZALDEHYDE

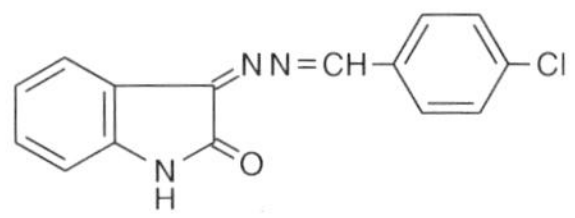

5 gm

S57,755-3
N-(3-HEXADECYL-4-OXO-2-THIAZOLIDINYLIDENE) SULFANILYL FLUORIDE

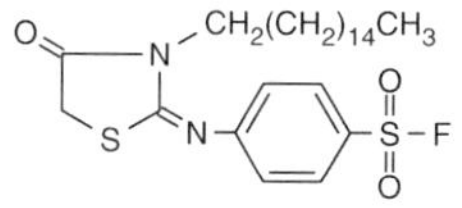

5 gm

S74,404-2
3-PHENYL-1-(M-TOLYL)-5-(M-TOLYLCARBAMOYLIMINO)-4-THIOHYDANTOIN

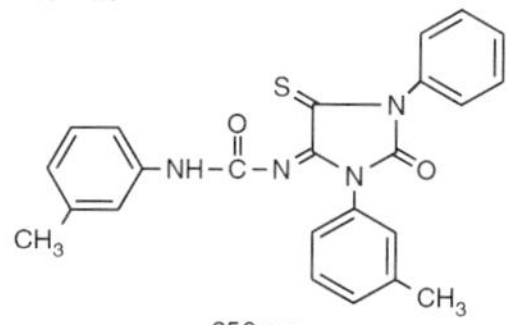

250 mg

S71,998-6
4-CINNAMYLIDENEAMINO-2,3-DIMETHYL-1-PHENYL-3-PYRAZOLIN-5-ONE

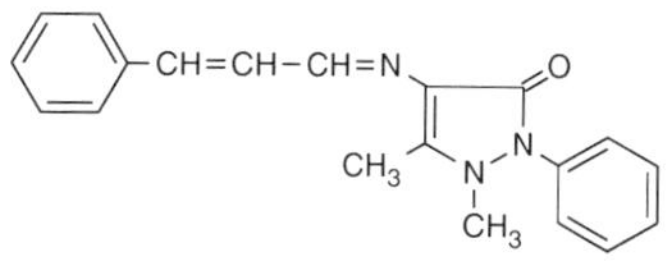

5 gm

S72,052-6
2,3-DIMETHYL-4-(4-METHYLBENZYLIDENEAMINO)-1-PHENYL-3-PYRAZOLIN-5-ONE

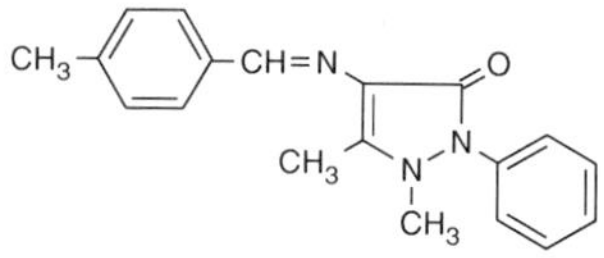

5 gm

S71,991-9
2,3-DIMETHYL-4-(4-ISOPROPYLBENZYLIDENEAMINO)-1-PHENYL-3-PYRAZOLIN-5-ONE

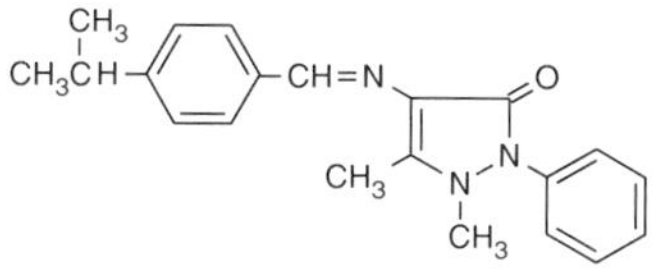

5 gm

S71,999-4
2,3-DIMETHYL-4-(1-NAPHTHYLMETHYLENEAMINO)-1-PHENYL-3-PYRAZOLIN-5-ONE

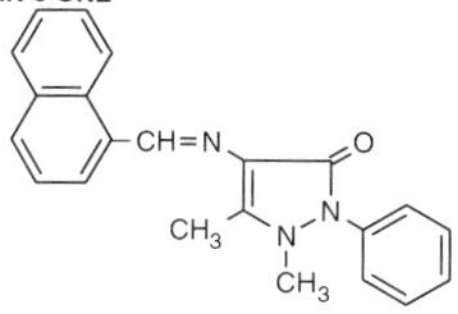

5 gm

S53,807-8
SUBERIC ACID BIS(2-PYRIDYLMETHYLENEHYDRAZIDE)

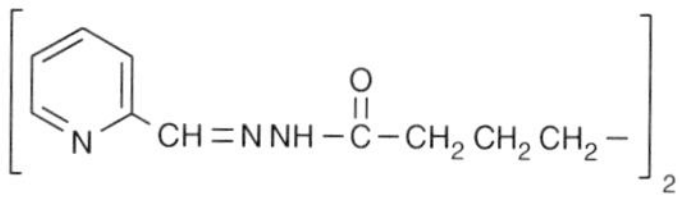

250 mg

S50,179-4
GLUTARIC ACID (5-INDANYLMETHYLENE)-HYDRAZIDE

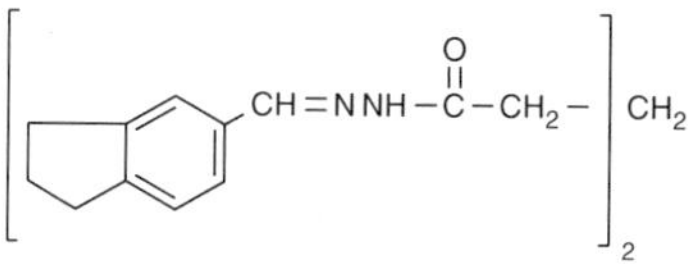

250 mg

S50,614-1
ADIPIC ACID (5-INDANYLMETHYLENE)-HYDRAZIDE

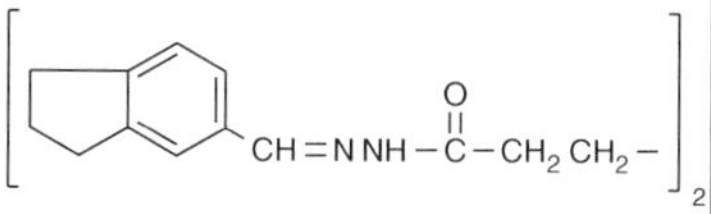

250 mg

S50,616-8
SUBERIC ACID BIS(5-INDANYLMETHYLENEHYDRAZIDE)

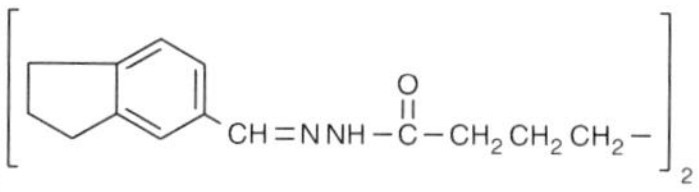

250 mg

S50,617-6
SEBACIC ACID BIS(5-INDANYLMETHYLENEHYDRAZIDE)

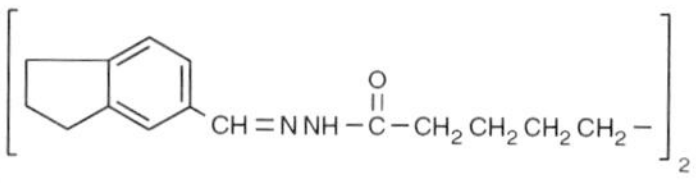

250 mg

S50,618-4
DODECANEDIOIC ACID (5-INDANYLMETHYLENE)HYDRAZIDE

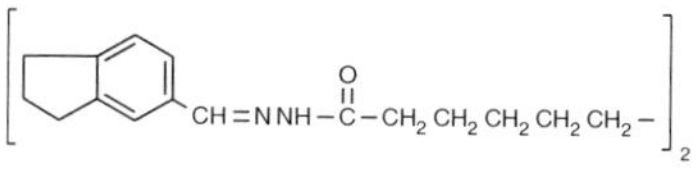

250 mg

S94,574-9
2,3-DIHYDRO-2-IMINO-3-METHYLTHIAZOLE HYDROIODIDE

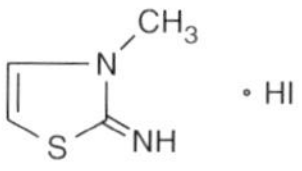

5 gm

S63,673-8
N,N'-(1,4-PHENYLENEDIMETHYLIDYNE)-BIS(2-AMINOTHIAZOLE)

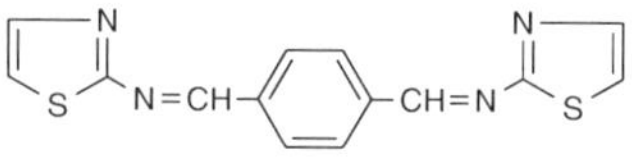

250 mg

S62,455-1
P-[N-(2-THENYLIDENE)-AMINO]PHENYL SULFONE

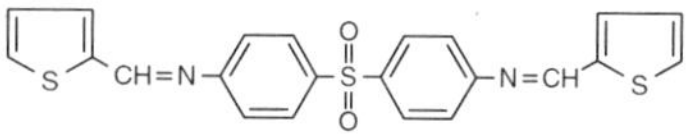

5 gm

S65,249-0
4-(2-CHLOROBENZYLIDENEAMINO)PHENYL SULFONE

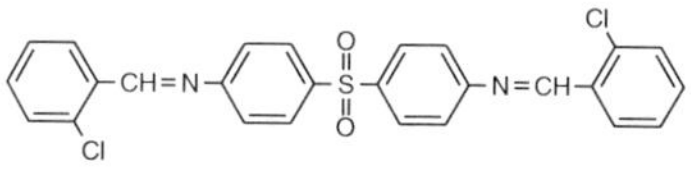

5 gm

S65,251-2
4-(CINNAMYLIDENEAMINO)PHENYL SULFONE

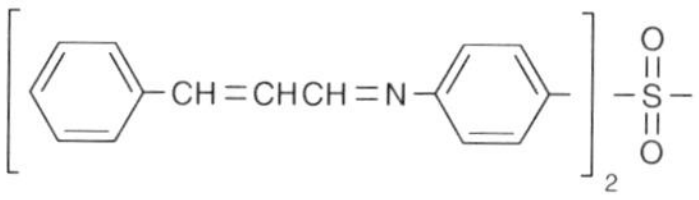

5 gm

S65,247-4
4-[N-(4-ISOPROPYLBENZYLIDENE)-AMINO]PHENYL SULFONE

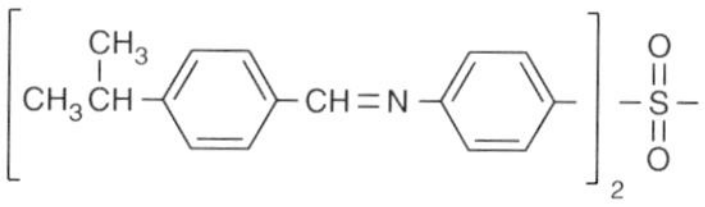

5 gm

S65,250-4
4-[N-(2,4-DICHLOROBENZYLIDENE)-AMINO]PHENYL SULFONE

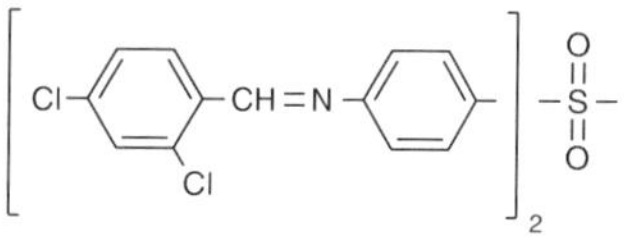

5 gm

S51,446-2
4,4'-SULFONYLBIS[N-(5-INDANYLMETHYLENE)ANILINE]

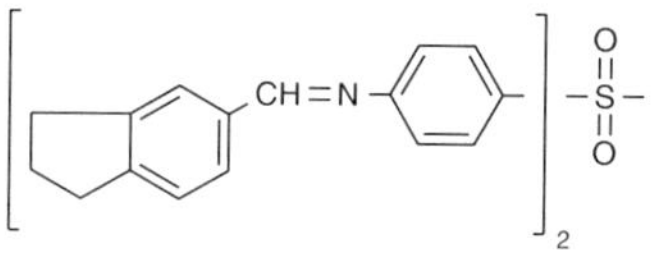

250 mg

S51,450-0
N4-(5-INDANYLMETHYLENE)-SULFANILAMIDE

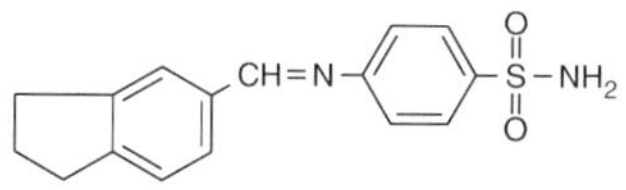

250 mg

S94,754-7
DICYCLOPROPYL KETONE P-TOSYLHYDRAZONE

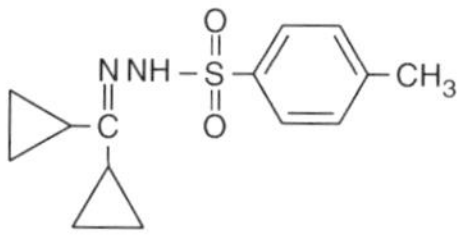

250 mg

S94,599-4
2,4,6-CYCLOHEPTATRIEN-1-ONE P-TOSYLHYDRAZONE HYDROCHLORIDE

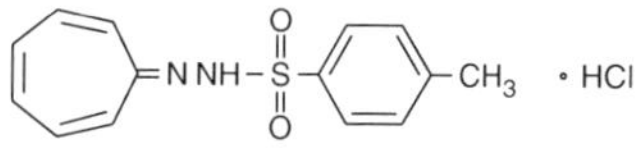

250 mg

S52,505-7
P-TOLUENESULFONIC ACID 2,3-BUTANEDIYLIDENEDIHYDRAZIDE

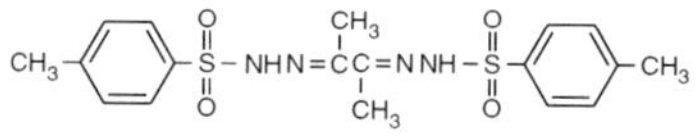

5 gm

S64,575-3
N-(2,2-DIMETHYLCYCLOPENTYLIDENE)-ANILINE

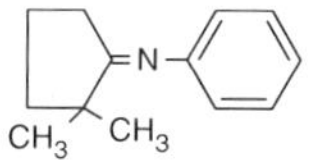

250 mg

S94,654-0
4,5-DIHYDRO-3,3-DIPHENYL-5-METHYL-2(3H)FURANIMINE HYDROCHLORIDE

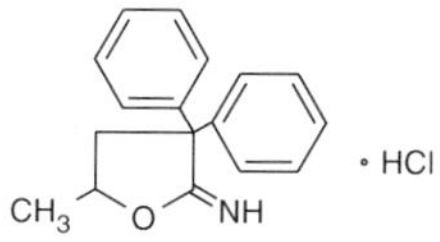

250 mg

S64,600-8
2-FURALDEHYDE THIOSEMICARBAZONE

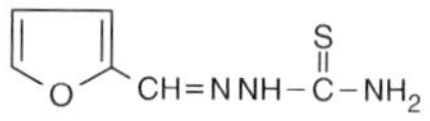

5 gm

S53,130-8
2-FURALDEHYDE AZINE

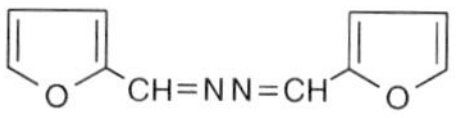

5 gm

S60,434-8
N-(FURFURYLIDENE)-2-CHLOROANILINE

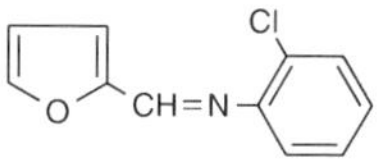

250 mg

S63,541-3
2,3-DICHLORO-N-FURFURYLIDENE ANILINE

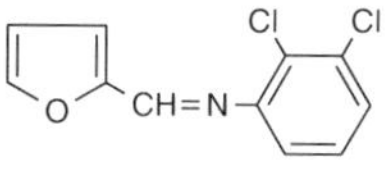

250 mg

S63,542-1
N-FURFURYLIDENE-M-TOLUIDINE

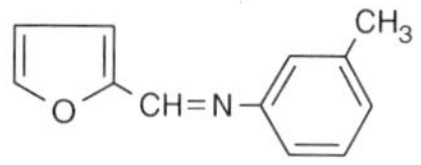

250 mg

S36,910-1
HYDROFURAMIDE

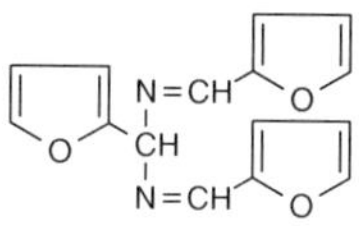

5 gm

S73,959-6
N,N'-(P-PHENYLENEDIMETHYLIDYNE)BIS-(FURFURYLAMINE)

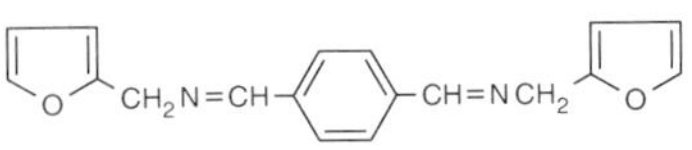

5 gm

S92,714-7
2-THIOPHENECARBOXALDEHYDE AZINE

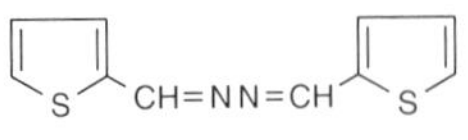

250 mg

S59,977-8
4-BROMO-N-(2-THIENYLMETHYLENE)-M-TOLUIDINE

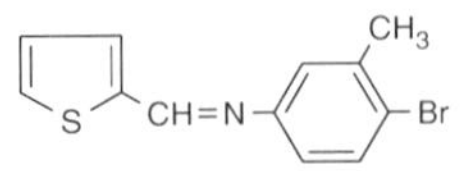

5 gm

S68,485-6
4,4'-METHYLENEBIS[N-(2-THENYLIDENE)ANILINE]

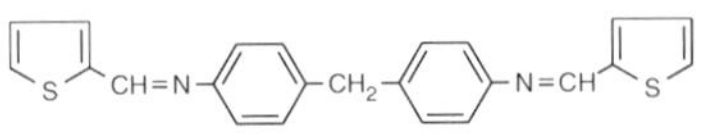

5 gm

S94.429-7
BASIC YELLOW 12 (C.I. 48065)

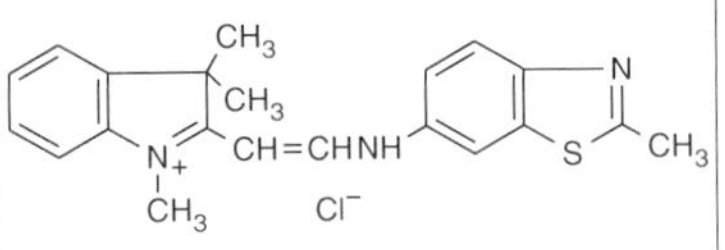

250 mg

S50,964-7
N-PIPERONYLIDENEANILINE

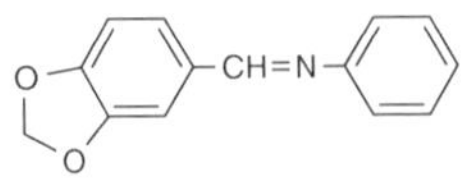

5 gm

S39,917-5
2-CHLORO-N-PIPERONYLIDENEANILINE

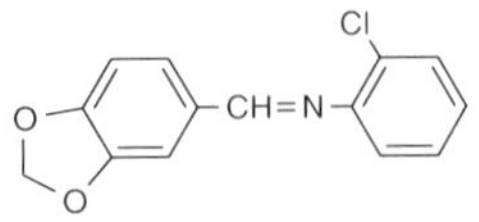

5 gm

S39,907-8
N-PIPERONYLIDENE-O-TOLUIDINE

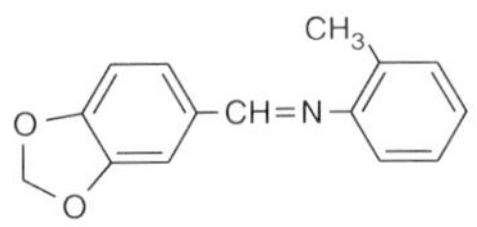

5 gm

S69,811-3
4,4'-METHYLENEBIS(N-[3,4-(METHYLENEDIOXY)BENZYLIDENE] ANILINE)

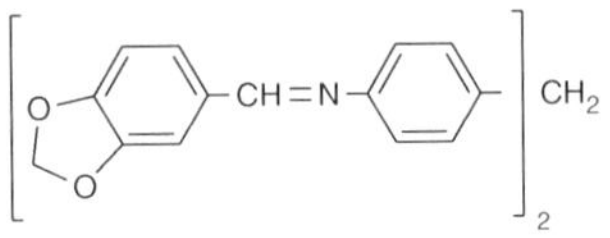

5 gm

S51,258-3
4-BROMO-N-PIPERONYLIDENEANILINE

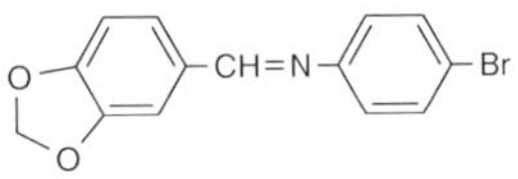

5 gm

S39,916-7
N-PIPERONYLIDENE-P-TOLUIDINE

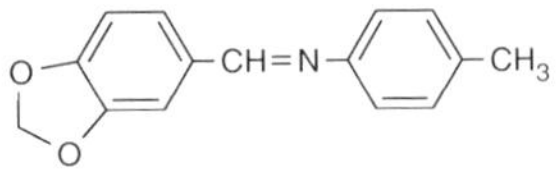

5 gm

S39,881-0
2,3-DICHLORO-N-PIPERONYLIDENEANILINE

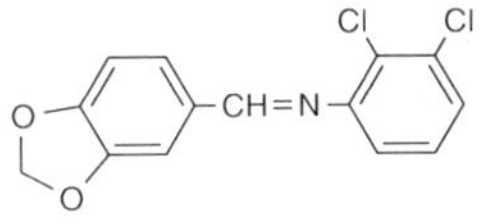

5 gm

S39,929-9
3-CHLORO-N-PIPERONYLIDENE-O-TOLUIDINE

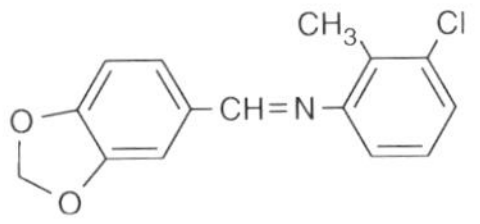

5 gm

S39,913-2
2,5-DICHLORO-N-PIPERONYLIDENEANILINE

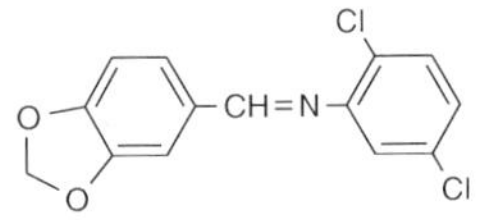

5 gm

S39,926-4
2,4-DICHLORO-N-PIPERONYLIDENEANILINE

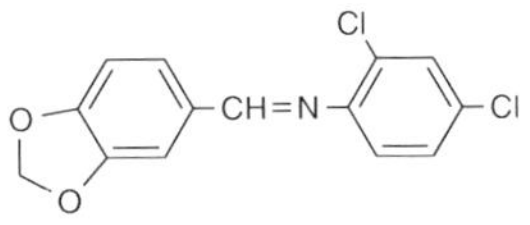

5 gm

S39,911-6
4-CHLORO-N-PIPERONYLIDENE-O-TOLUIDINE

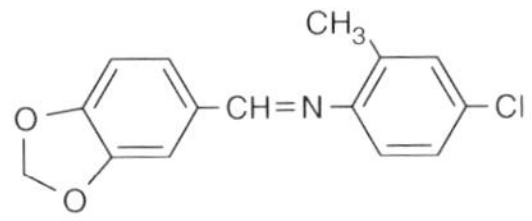

5 gm

S51,308-3
5-CHLORO-N-PIPERONYLIDENE-O-TOLUIDINE

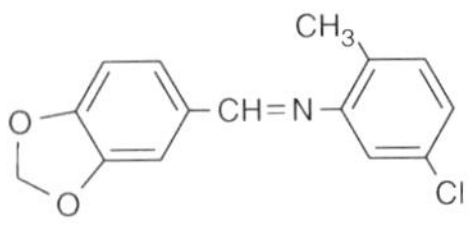

5 gm

S51,714-3
3-CHLORO-N-PIPERONYLIDENE-P-TOLUIDINE

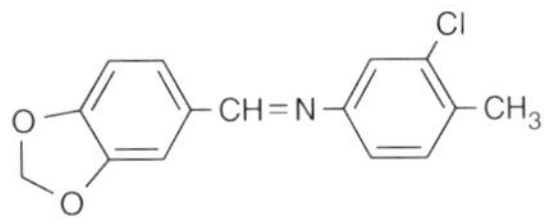

5 gm

S51,712-7
N-PIPERONYLIDENE-3,4-XYLIDINE

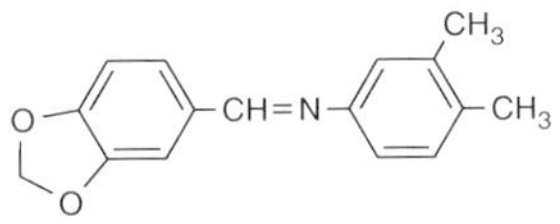

5 gm

S39,905-1
N-PIPERONYLIDENE-1-NAPHTHYLAMINE

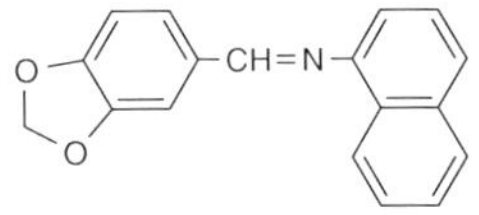

5 gm

S39,928-0
N-PIPERONYLIDENE-M-TOLUIDINE

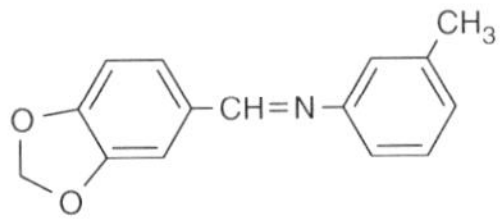

5 gm

S92,591-8
TRIS(4,5-DIPHENYL-1,3-DITHIOLAN-2-YLIDENEAMINO)CARBENIUM TETRAFLUOROBORATE

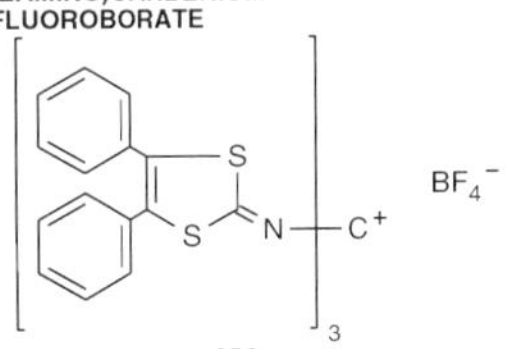

250 mg

S92,811-9
TRIS[4-(P-TOLYL)-1,3-DITHIOLAN-2-YLIDENEAMINO]CARBENIUM TETRAFLUOROBORATE

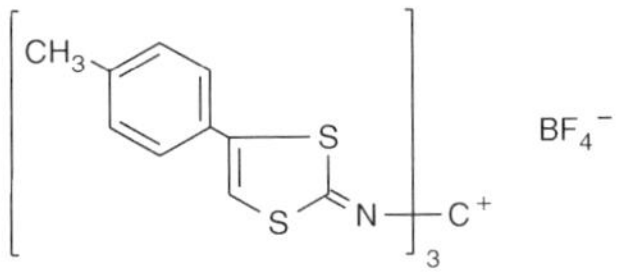

250 mg

TERT-BUTYLCYCLOHEXYLIDENEAMINE

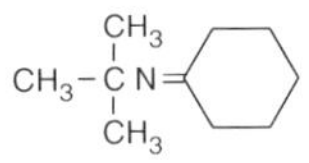

250 mg

S89,652-7
[2-(CYCLOHEXYLIMINO)ETHYL]-TRIPHENYLPHOSPHONIUM CHLORIDE

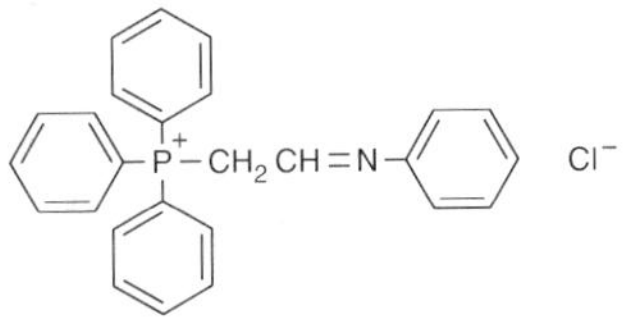

250 mg

S90,173-3
N-(2,4-DICHLOROBENZYLIDENE)-2,3-DIMETHYLANILINE

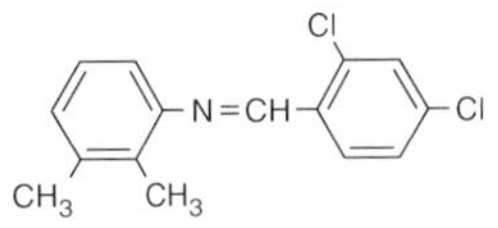

250 mg

S38,791-6
BENZALDEHYDE N-METHYLOXIME

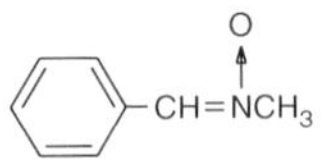

250 mg

S38,132-2
N-BENZYLIDENEBUTYLAMINE

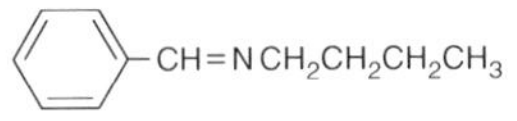

5 gm

S62,258-3
N,N'-DIPHENYLACETAMIDINE

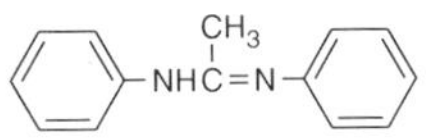

5 gm

S59,067-3
N-CINNAMYLIDENEANILINE

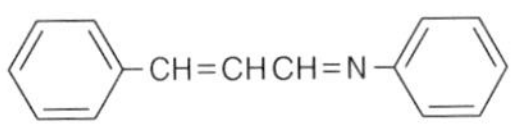

5 gm

S34,470-2
CINNAMALDEHYDE AZINE

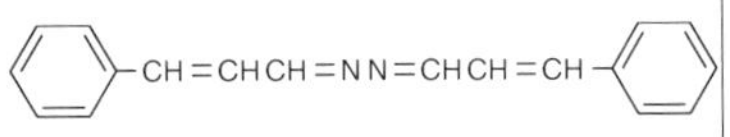

5 gm

S89,650-0
[2-(PHENYLIMINO)ETHYL]-TRIPHENYLPHOSPHONIUM BROMIDE

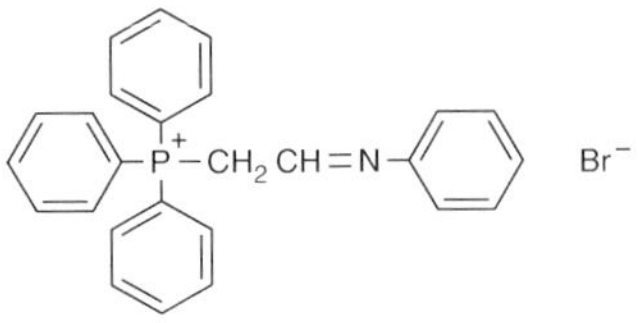

250 mg

S89,635-7
N-BENZYLIDENE-4-ETHYLANILINE
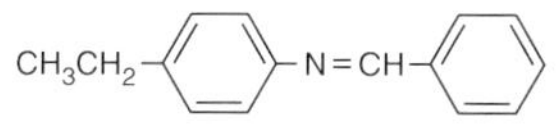
250 mg

S39,337-1
4-CHLORO-N-CINNAMYLIDENEANILINE
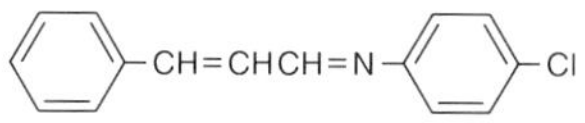
5 gm

S83,967-1
N-BENZYLIDENE-4-(2-PYRIDYL)ANILINE
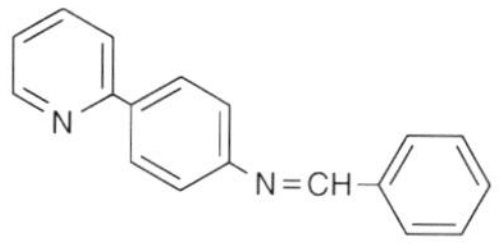
250 mg

S68,651-4
4,4'-METHYLENEBIS(N-CINNAMYLIDENEANILINE)
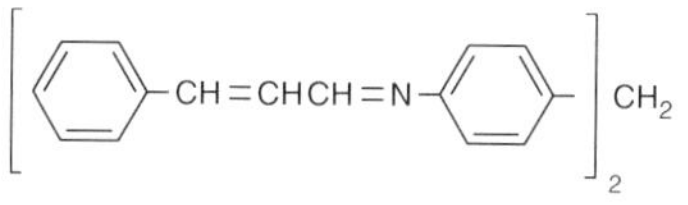
5 gm

S88,641-6
BENZOPHENONE IMINE HYDROCHLORIDE
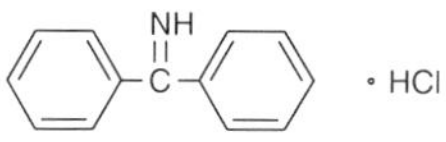
250 mg

S71,029-6
N-(DIPHENYLMETHYLENE)ANILINE
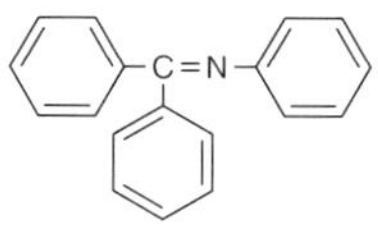
5 gm

S88,675-0
2,2-DIPHENYL-3-METHYLBUTYROPHENONE IMINE HYDROCHLORIDE
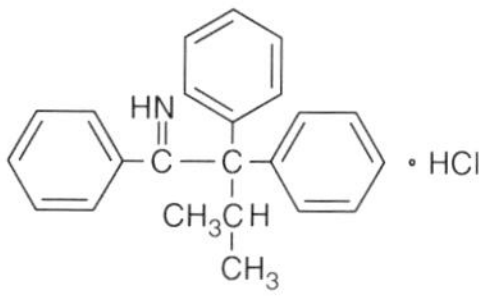
250 mg

S51,550-7
BENZOPHENONE AZINE
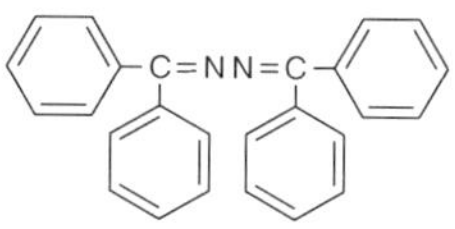
250 mg

S88,265-8
2,2'-BIS(PHENYLIMINOMETHYL)BIPHENYL
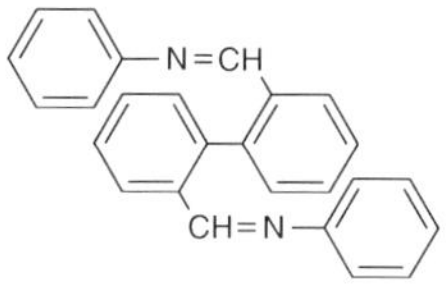
250 mg

S50,626-5
5-INDANCARBOXALDEHYDE 4-PHENYL-3-THIOSEMICARBAZONE
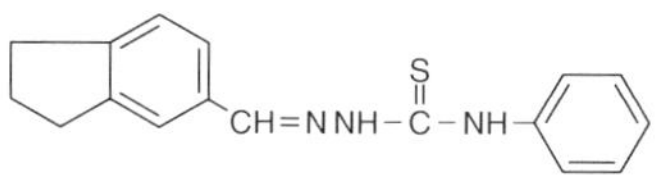
250 mg

S74,919-2
9-(PHENYLIMINO)FLUORENE
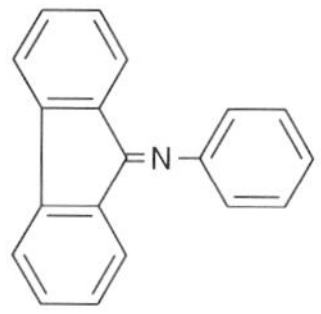
250 mg

S34,471-0
2-(CINNAMYLIDENEAMINO)FLUORENE
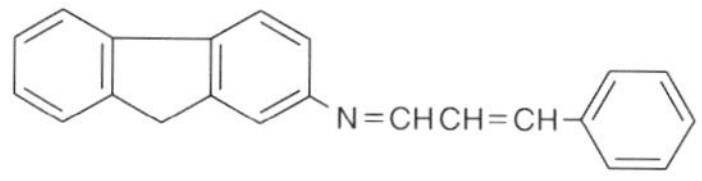
250 mg

S61,557-9
2-CHLORO-N-(2-CHLOROBENZYLIDENE)-ANILINE
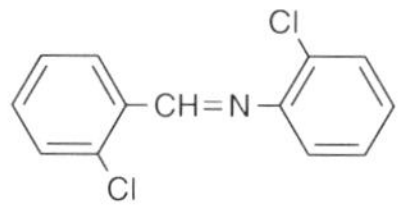
5 gm

S51,590-6
2-CHLOROBENZALDEHYDE AZINE
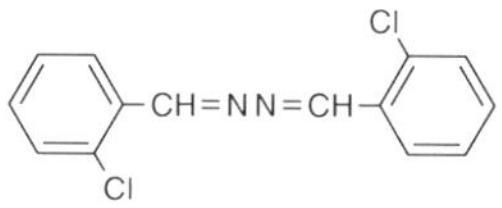
5 gm

S60,924-2
2-CHLORO-N-CINNAMYLIDENEANILINE
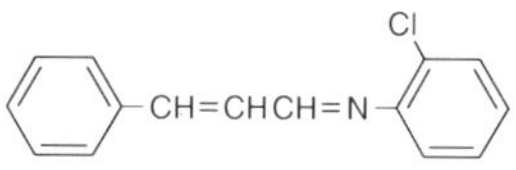
250 mg

S58,427-4
N,N'-(1,4-PHENYLENEDIMETHYLIDYNE)-BIS(2-CHLOROANILINE)
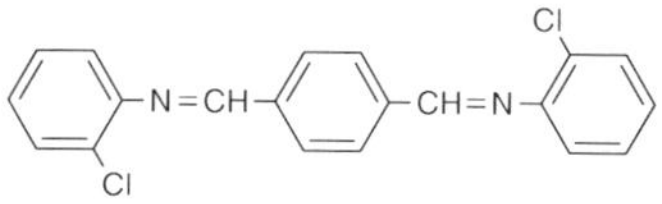
5 gm

S58,446-0
N,N'-(1,4-PHENYLENEDIMETHYLIDYNE)-DI-O-TOLUIDINE
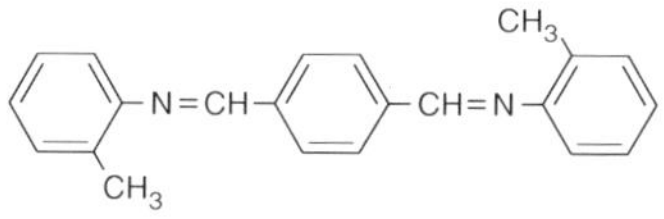
5 gm

S69,546-7
2-ETHYL-N-(1-NAPHTHYLMETHYLENE)-ANILINE
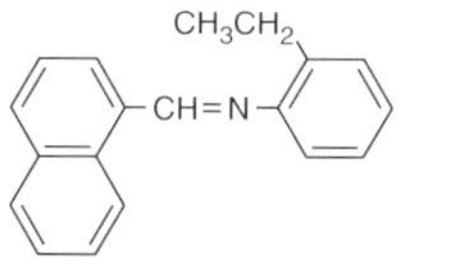
5 gm

S61,610-9
N-(9-ANTHRACENYLMETHYLENE)-O-TOLUIDINE
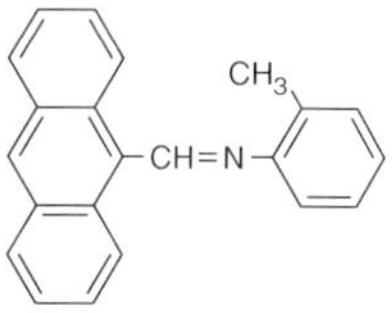
5 gm

S45,985-2
3-ETHYL-1-P-TOLYLTRIAZENE
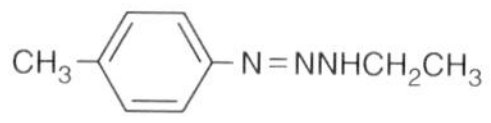
250 mg

S44,468-5
3-BUTYL-1-P-TOLYLTRIAZENE
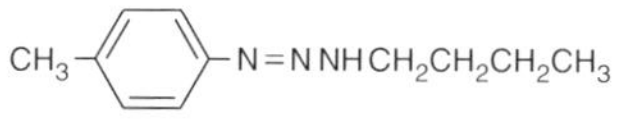
5 gm

S52,399-2
N-(4-FLUOROBENZYLIDENE)ANILINE
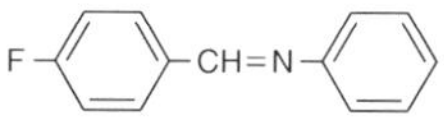
5 gm

S36,554-8
N-BENZYLIDENE-4-CHLOROANILINE
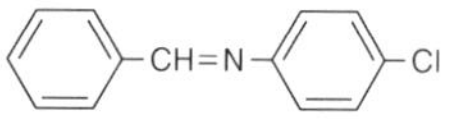
5 gm

S52,345-3
4-FLUORO-N-(4-FLUOROBENZYLIDENE)-ANILINE
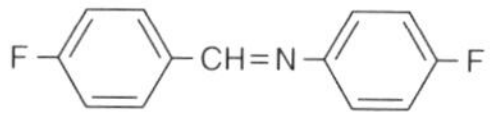
5 gm

S52,415-8
4-CHLORO-N-(4-FLUOROBENZYLIDENE)-ANILINE
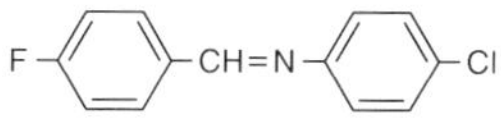
5 gm

S54,305-5
3-CHLORO-N-(4-FLUOROBENZYLIDENE)-ANILINE
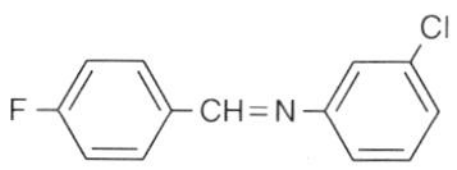
250 mg

S53,288-6
4-BROMO-N-(4-FLUOROBENZYLIDENE)-ANILINE
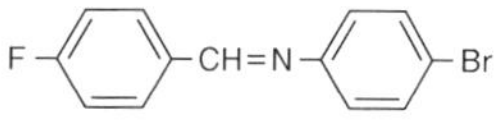
5 gm

S52,349-6
N-(4-FLUOROBENZYLIDENE)-4-IODOANILINE
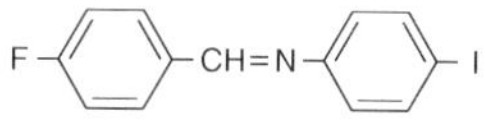
5 gm

S52,444-1
N-(4-CHLOROBENZYLIDENE)-4-IODOANILINE
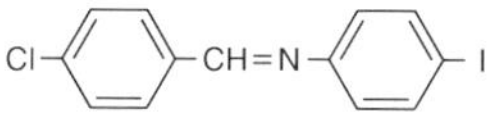
5 gm

S52,377-1
N-(4-FLUOROBENZYLIDENE)-P-TOLUIDINE
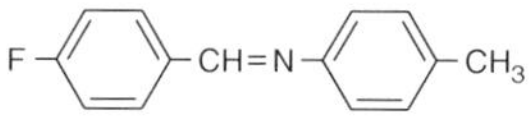
5 gm

S54,568-6
4-CHLOROBENZALDEHYDE AZINE
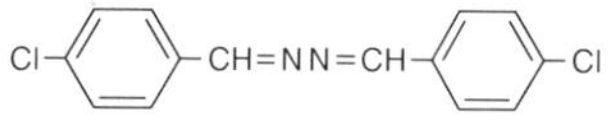
5 gm

S62,579-5
4-BROMO-N-(CINNAMYLIDENE)ANILINE
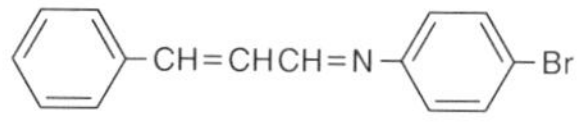
5 gm

S61,926-4
N-(CINNAMYLIDENE)-P-TOLUIDINE

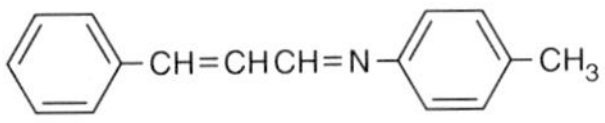

250 mg

S54,263-6
P-TOLUALDEHYDE AZINE

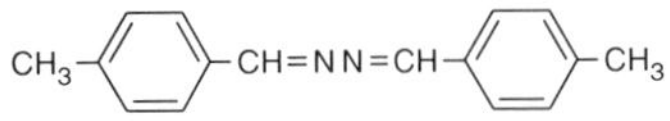

250 mg

S54,630-5
1,5-DI-P-TOLYL-3-ETHYLPENTAAZADIENE

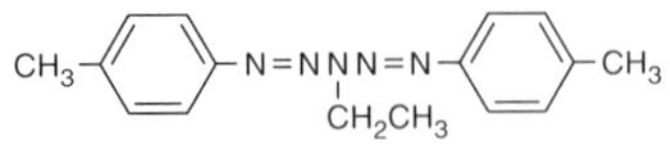

250 mg

S86,202-9
P-TOLUALDEHYDE AZINE BORON TRIFLUORIDE COMPLEX

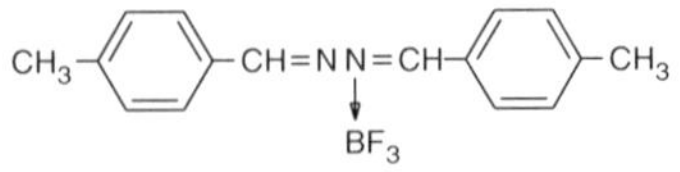

250 mg

S37,501-2
4'-CHLOROACETOPHENONE AZINE

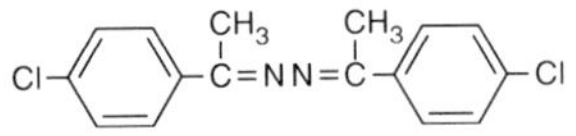

250 mg

S74,034-9
4'-METHYLACETOPHENONE AZINE

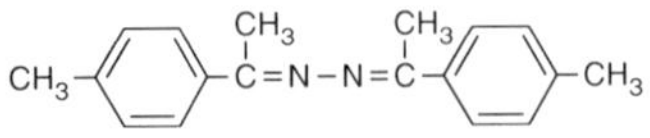

250 mg

S41,059-4
N,N'-TEREPHTHALYLIDENEDIANILINE

5 gm

S58,425-8
N,N'-(1,4-PHENYLENEDIMETHYLIDYNE)-BIS(4-CHLOROANILINE)

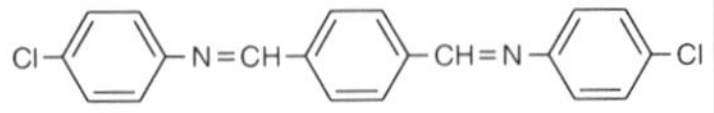

5 gm

S58,426-6
N,N'-(1,4-PHENYLENEDIMETHYLIDYNE)-BIS(3-CHLOROANILINE)

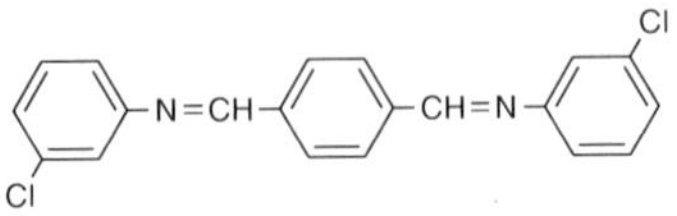

250 mg

S58,423-1
N,N'-(1,4-PHENYLENEDIMETHYLIDYNE)-BIS(4-BROMOANILINE)

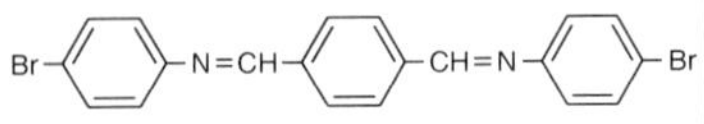

250 mg

S58,428-2
N,N'-(1,4-PHENYLENEDIMETHYLIDYNE)-BIS(4-IODOANILINE)

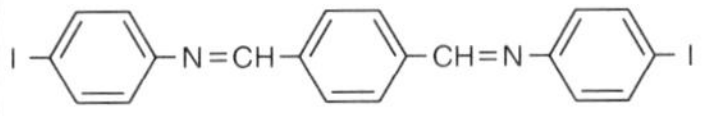

5 gm

S52,346-1
2,3-DICHLORO-N-(4-FLUOROBENZYLIDENE)ANILINE

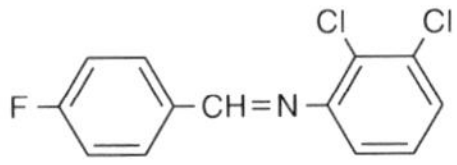

5 gm

S52,413-1
3-CHLORO-N-(4-FLUOROBENZYLIDENE)-O-TOLUIDINE

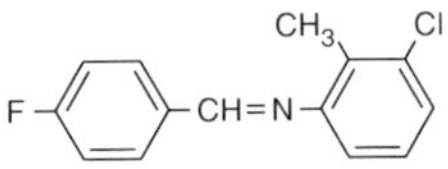

5 gm

S58,421-5
N,N'-(1,4-PHENYLENEDIMETHYLIDYNE)-BIS(2,3-DICHLOROANILINE)

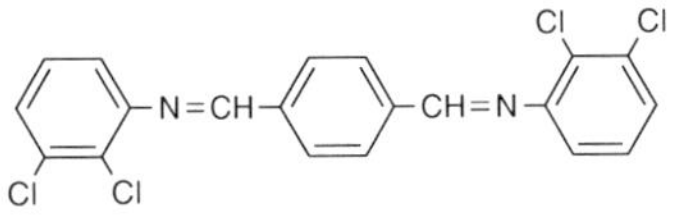

5 gm

S58,439-8
N,N'-(1,4-PHENYLENEDIMETHYLIDYNE)-BIS(3-CHLORO-O-TOLUIDINE)

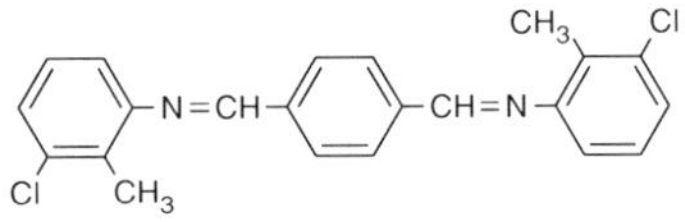

250 mg

S58,456-8
N,N'-(1,4-PHENYLENEDIMETHYLIDYNE)-DI-2,3-XYLIDINE

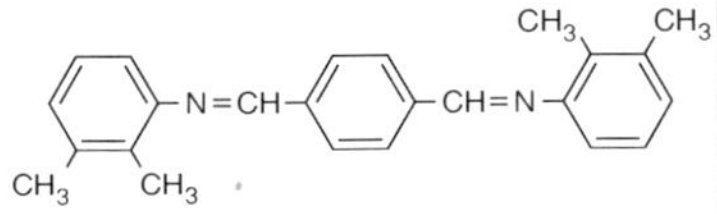

250 mg

S58,459-2
N,N'-(1,4-PHENYLENEDIMETHYLIDYNE)-DI-2,6-XYLIDINE

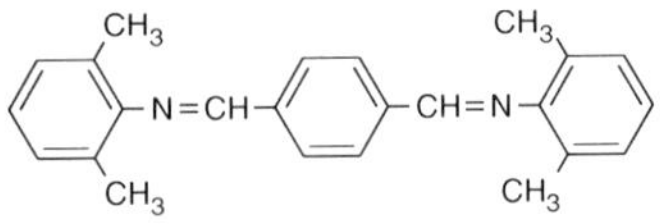

5 gm

S52,394-1
2,5-DICHLORO-N-(4-FLUOROBENZYLIDENE)ANILINE

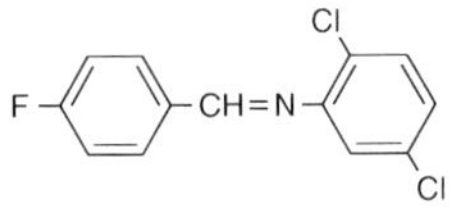

5 gm

S52,382-8
5-CHLORO-N-(4-FLUOROBENZYLIDENE)-O-TOLUIDINE

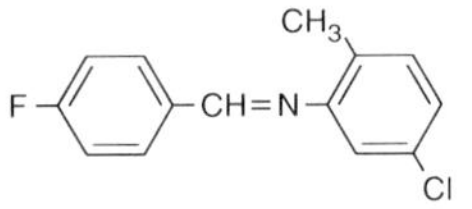

5 gm

S52,412-3
3-CHLORO-N-(4-FLUOROBENZYLIDENE)-P-TOLUIDINE

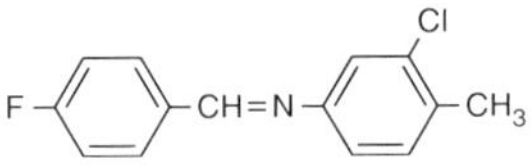

5 gm

S53,880-9
4-CHLORO-N-(4-FLUOROBENZYLIDENE)-O-TOLUIDINE

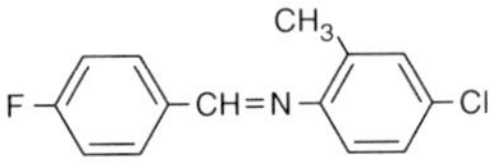

5 gm

S52,340-2
N-(4-FLUOROBENZYLIDENE)-3,4-XYLIDINE

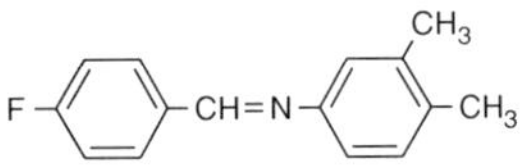

5 gm

S52,362-3
N-(4-FLUOROBENZYLIDENE)-2,5-XYLIDINE

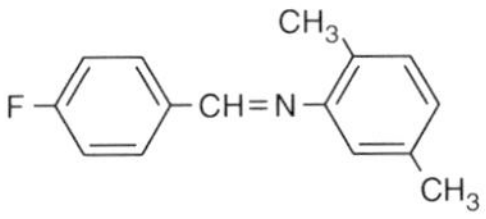

5 gm

S52,580-4
N-(4-FLUOROBENZYLIDENE)-2,4-XYLIDINE

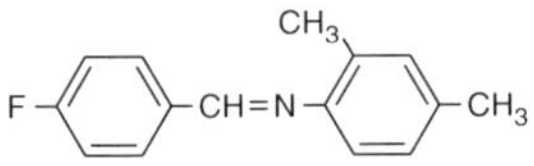

5 gm

S53,018-2
N-(4-CHLOROBENZYLIDENE)-2,4-XYLIDINE

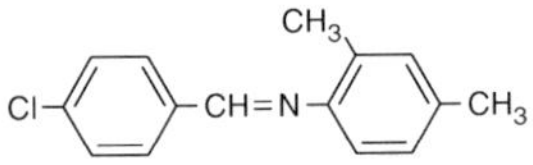

5 gm

S53,022-0
N-(4-CHLOROBENZYLIDENE)-2,5-XYLIDINE

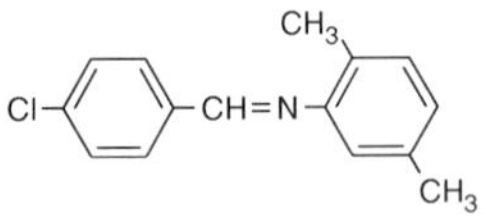

5 gm

S58,420-7
N,N'-(1,4-PHENYLENEDIMETHYLIDYNE)-BIS(2,4-DICHLOROANILINE)

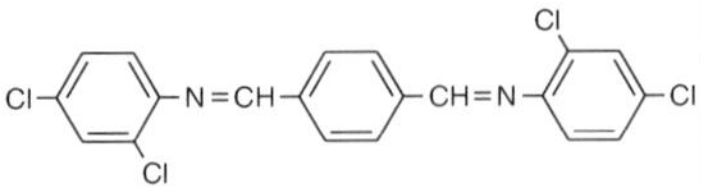

5 gm

S58,422-3
N,N'-(1,4-PHENYLENEDIMETHYLIDYNE)-BIS(3,4-DICHLOROANILINE)

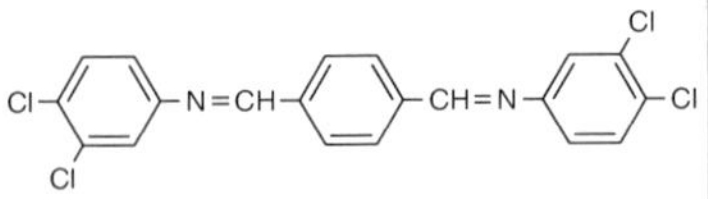

5 gm

S58,436-3
N,N'-(1,4-PHENYLENEDIMETHYLIDYNE)-BIS(4-CHLORO-O-TOLUIDINE)

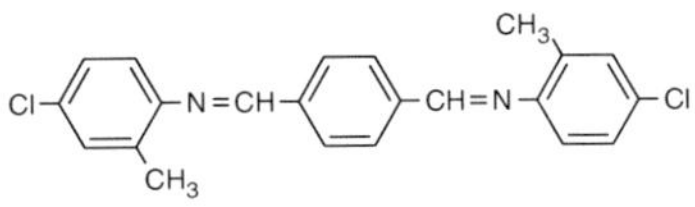

5 gm

S58,437-1
N,N'-(1,4-PHENYLENEDIMETHYLIDYNE)-BIS(3-CHLORO-P-TOLUIDINE)

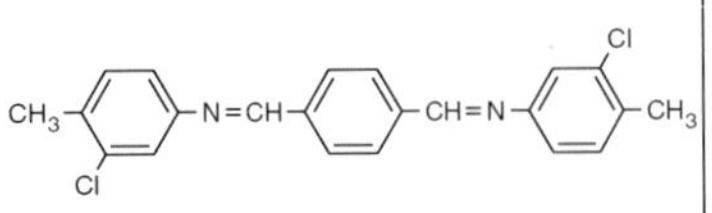

250 mg

S58,454-1
N,N'-(1,4-PHENYLENEDIMETHYLIDYNE)-DI-3,4-XYLIDINE

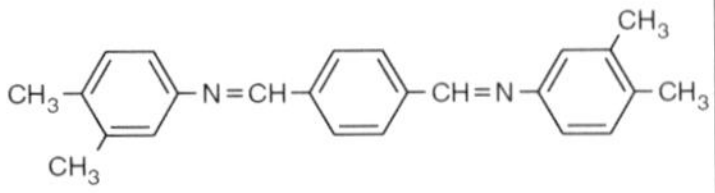

250 mg

S58,458-4
N,N'-(1,4-PHENYLENEDIMETHYLIDYNE)-DI-2,4-XYLIDINE

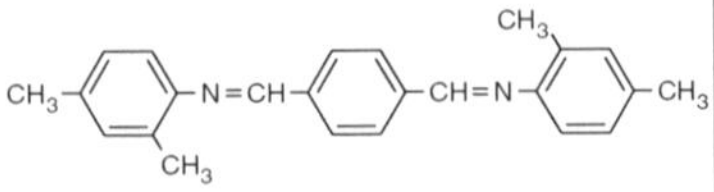

250 mg

S54,444-2
N-BENZYLIDENE-4-BIPHENYLAMINE

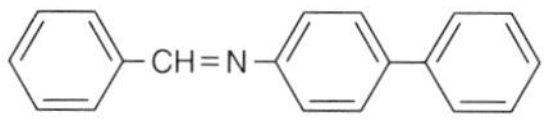

250 mg

S52,379-8
N-(4-FLUOROBENZYLIDENE)-1-NAPHTHYLAMINE

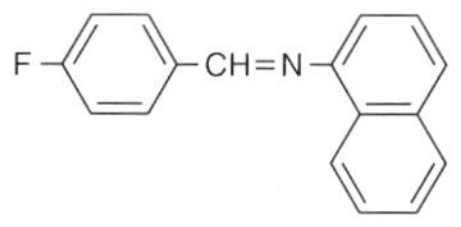

5 gm

S92,693-0
9-[4-(TRIFLUOROMETHYL)PHENYLIMINO]-FLUORENE

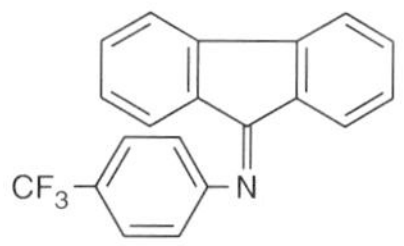

250 mg

S53,799-3
2,6-DICHLOROBENZALDEHYDE AZINE

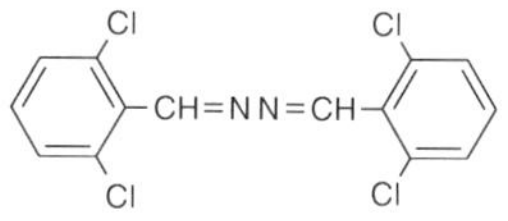

5 gm

S63,599-5
N-(CINNAMYLIDENE)-2,3-DICHLOROANILINE

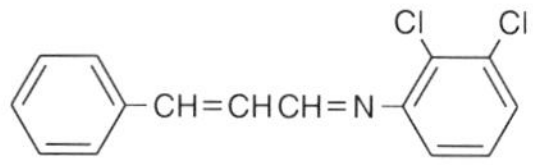

250 mg

S61,931-0
N-(CINNAMYLIDENE)-2,3-XYLIDINE

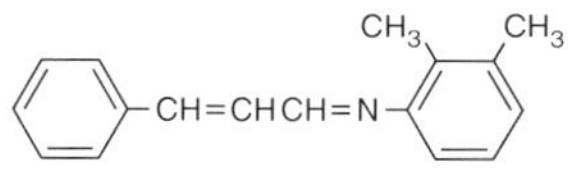

250 mg

S61,609-5
N-(9-ANTHRACENYLMETHYLENE)-2,6-XYLIDINE

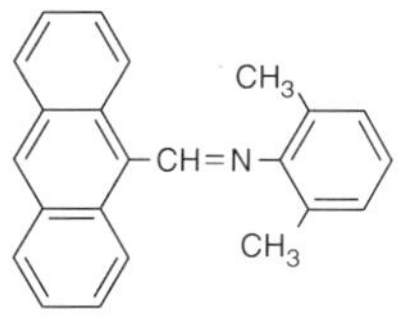

5 gm

S53,178-2
2,4-DICHLOROBENZALDEHYDE AZINE

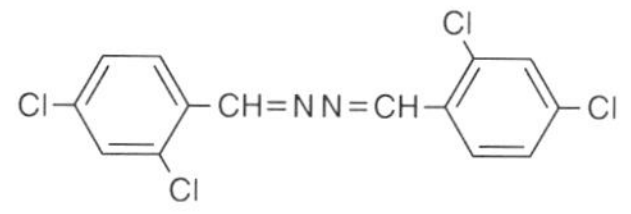

5 gm

S54,886-3
3,4-DICHLOROBENZALDEHYDE AZINE

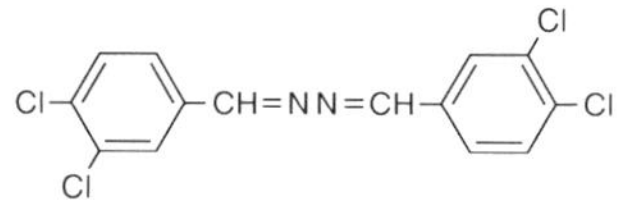

250 mg

S62,388-1
N-(CINNAMYLIDENE)-3,4-DICHLOROANILINE

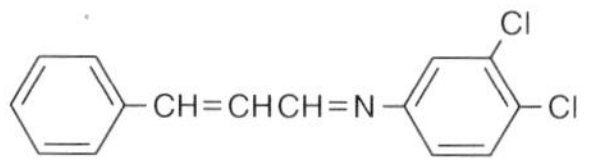

250 mg

S62,394-6
N-(CINNAMYLIDENE)-2,4-DICHLOROANILINE

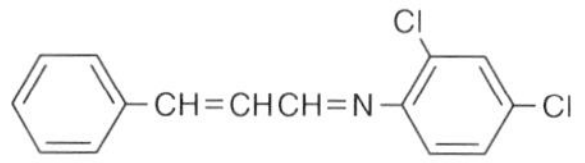

250 mg

S62,589-2
3-CHLORO-N-(CINNAMYLIDENE)-4-METHYLANILINE

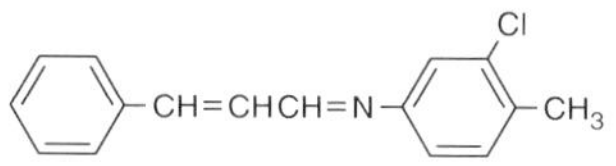

250 mg

S61,922-1
N-(CINNAMYLIDENE)-2,4-XYLIDINE

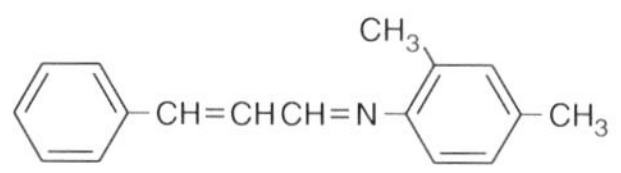

250 mg

S61,943-4
N-(CINNAMYLIDENE)-3,4-XYLIDINE

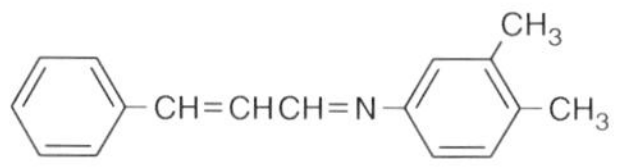

250 mg

S61,608-7
N-(9-ANTHRACENYLMETHYLENE)-2,4-XYLIDINE

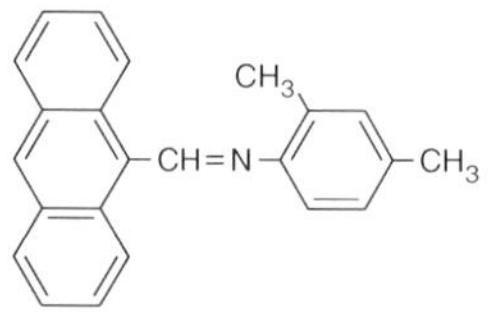

5 gm

S61,944-2
N-(CINNAMYLIDENE)-3,5-XYLIDINE

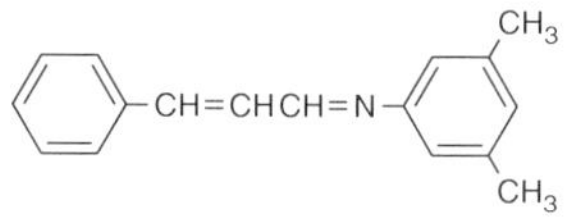

250 mg

S34,970-4
N-2-FLUORENYL-3-PYRIDINEALDIMINE

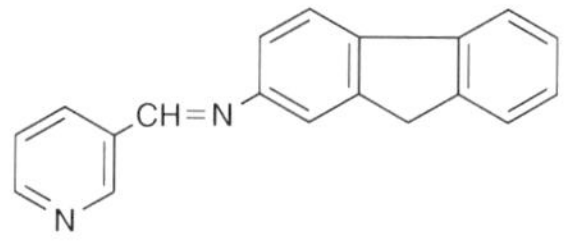

250 mg

S41,000-4
4-PYRIDINEALDAZINE

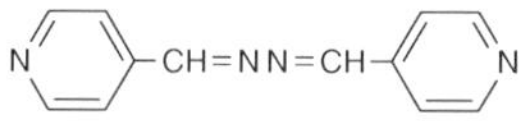

5 gm

S34,971-2
N-2-FLUORENYL-4-PYRIDINEALDIMINE

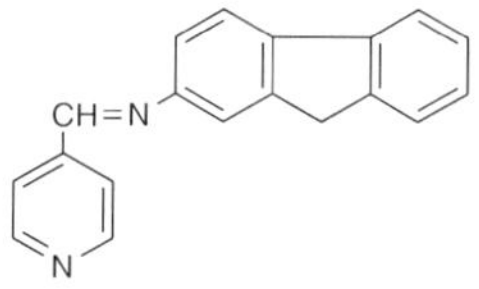

250 mg

S50,345-2
BENZOPHENONE AZINE WITH 5-INDANCARBOXALDEHYDE

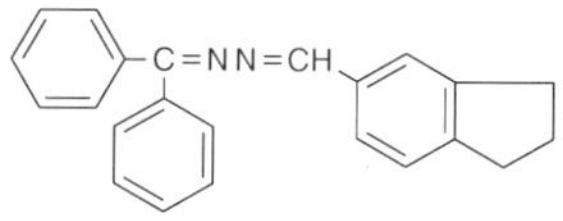

250 mg

S95,798-4
CAMPHOR AZINE

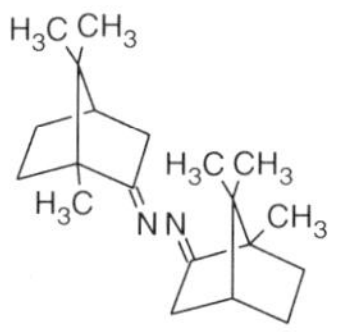

50 mg

S76,422-1
N,N'-(TEREPHTHALYLIDENE)-DODECAMETHYLENEDIAMINE

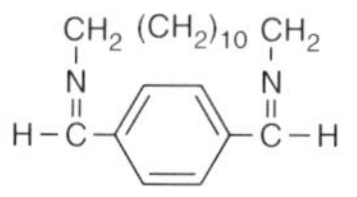

5 gm

S50,018-6
5-INDANCARBOXALDEHYDE AZINE

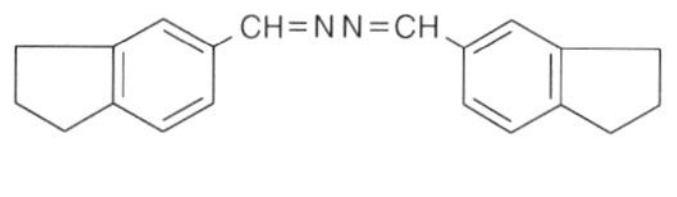

250 mg

S95,657-0
1'-ACETONAPHTHONE AZINE

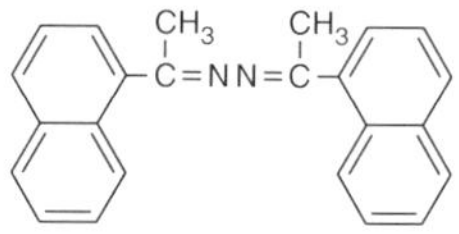

50 mg

S34,972-0
N-2-FLUORENYL-4-QUINOLINEALDIMINE

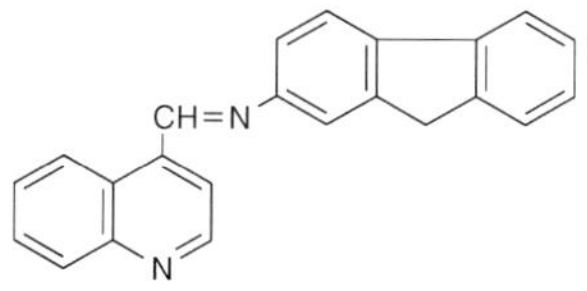

5 gm

S75,505-2
9-IMINOFLUORENE

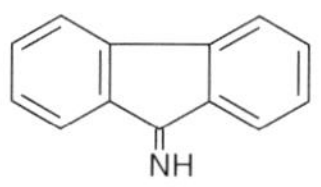

250 mg

S35,139-3
INDOLE-3-CARBOXALDEHYDE AZINE

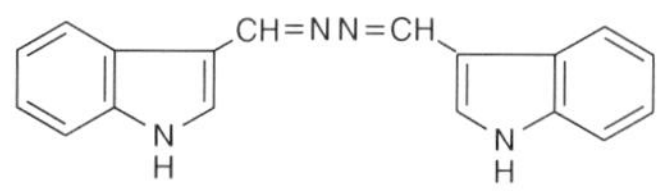

5 gm

S92,491-1
P-TOLUIC [4-(DIMETHYLAMINO)-BENZYLIDENE]HYDRAZIDE

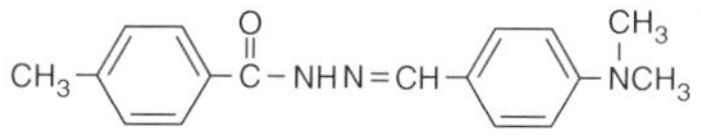

250 mg

S57,884-3
1-HEXADECANESULFONIC [1-(2-CHLORO-ETHYL)-2-PHENYL-6-SULFO-4(1H)-QUINOLYLIDENE]HYDRAZIDE

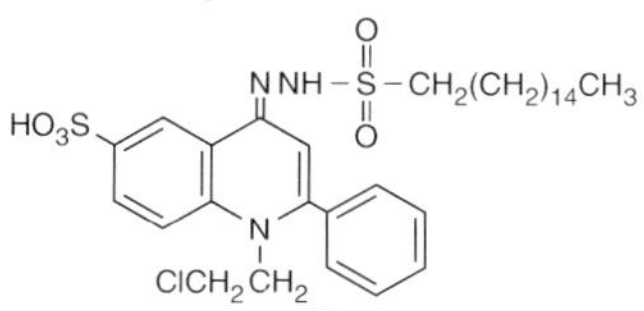

250 mg

S45,431-1
CYSSOR I

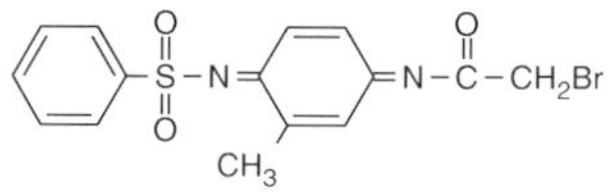

250 mg

S77,588-6
2-(2-BROMOETHYL)BENZALDEHYDE (P-TOSYL)HYDRAZONE

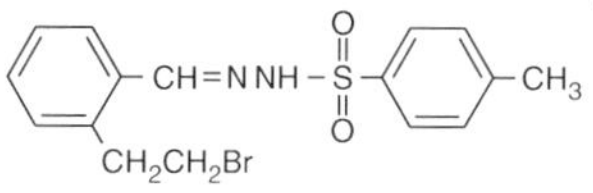

250 mg

S64,843-4
2-BROMO-4'-(PHENYLAZO)ACETOPHENONE

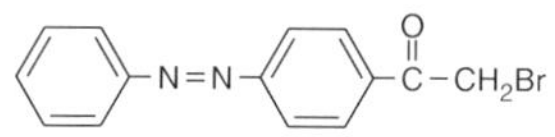

250 mg

S73,895-6
2-CHLORO-5'-METHYL-2'-(P-TOLYLAZO)-ACETANILIDE

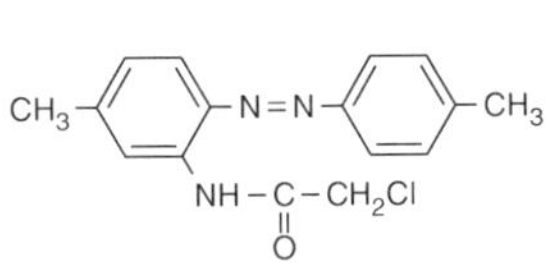

5 gm

S67,624-1
4-[2-CHLORO-5-(TRIFLUOROMETHYL)-PHENYLAZO]-3-HYDROXY-2-NAPHTHOIC ACID

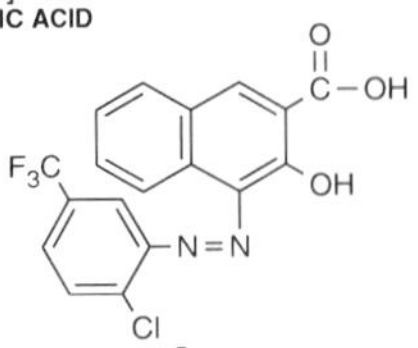

5 gm

S85,205-8
2,6-DIHYDROXY-3-[3-(TRIFLUOROMETHYL)PHENYLAZO]-BENZOIC ACID

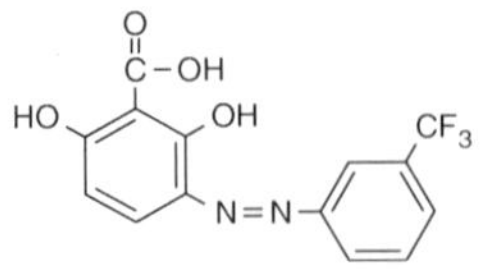

250 mg

S69,852-0
5'-METHYL-2'-(4-TOLYLAZO)-2,2,2-TRICHLOROACETANILIDE

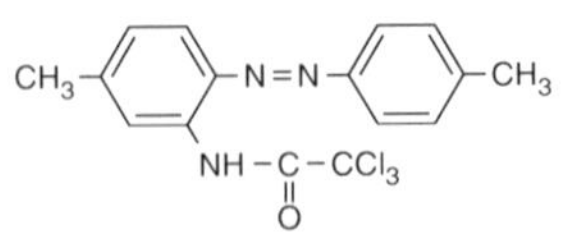

250 mg

S81,358-3
4-(DIMETHYLAMINO)-4'-NITRO-2'-(TRIFLUOROMETHYL)-1,1'-AZOBENZENE

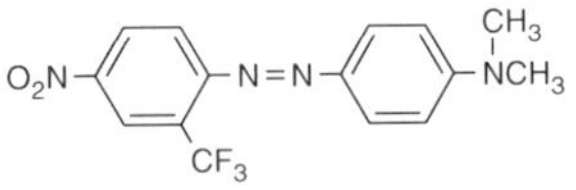

250 mg

S93,067-9
N,N-DIMETHYL-4-[4-(TRIFLUOROMETHYL)-PHENYLAZO]ANILINE

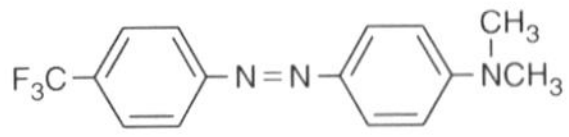

250 mg

S93,072-5
N-ETHYL-N-METHYL-4-[4-(TRIFLUOROMETHYL)PHENYLAZO]ANILINE

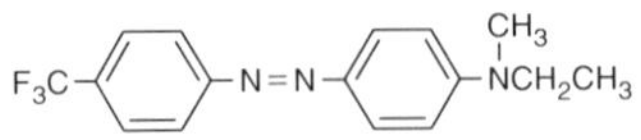

250 mg

S93,019-9
2-(3-METHYL-4-[4-(TRIFLUOROMETHYL)-PHENYLAZO]PHENYL)ISOINDOLINE

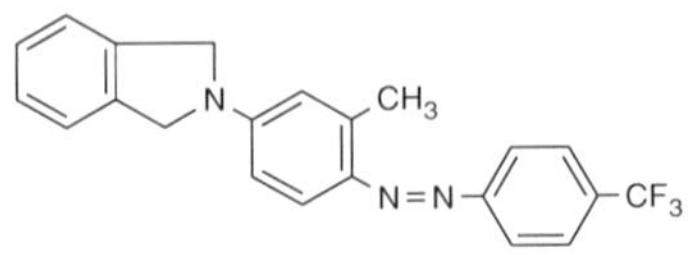

250 mg

S93,033-4
3-METHYL-1-(4-[4-(TRIFLUOROMETHYL)-PHENYLAZO]PHENYL)PIPERIDINE

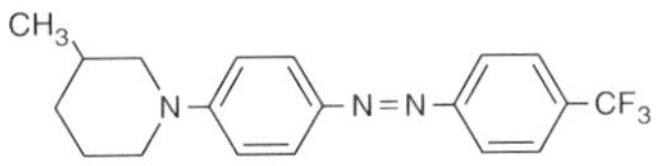

250 mg

S93,039-3
4-METHYL-1-(4-[4-(TRIFLUOROMETHYL)-PHENYLAZO]PHENYL)PIPERIDINE

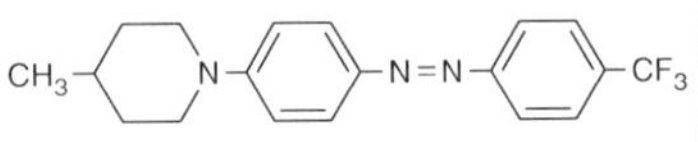

250 mg

S93,032-6
1-(4-[4-(TRIFLUOROMETHYL)PHENYLAZO]-PHENYL)HEXAHYDRO-1H-AZEPINE

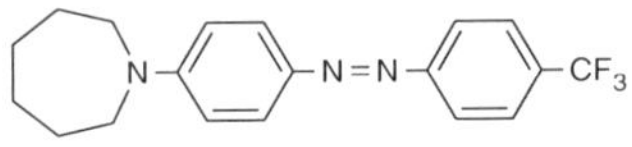

250 mg

S84,726-7
1-[5-METHYL-2-(P-TOLYLAZO)PHENYL]-3-[2-(TRIFLUOROMETHYL)PHENYL]UREA

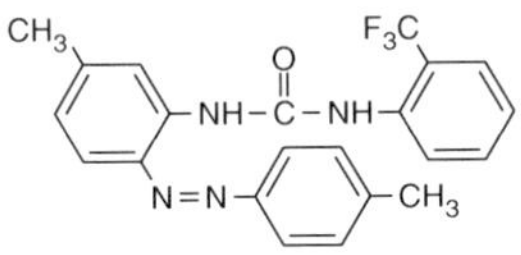

250 mg

S64,922-8
3,3',5,5'-TETRAKIS(TRIFLUOROMETHYL)-AZOXYBENZENE

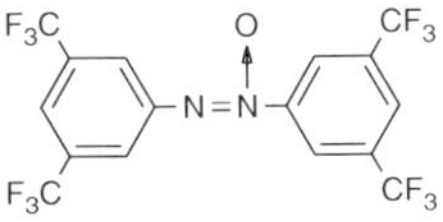

250 mg

S92,309-5
4'-[BIS(2-HYDROXYETHYL)AMINO]-2'-CHLORO-4-NITROAZOBENZENE

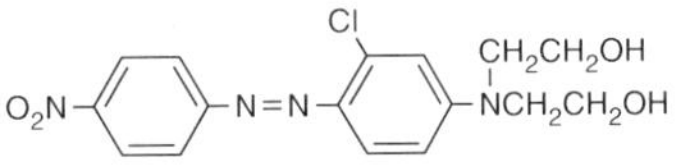

5 gm

S92,303-6
4'-[BIS(2-HYDROXYETHYL)AMINO]-2,6-DICHLORO-4-NITROAZOBENZENE

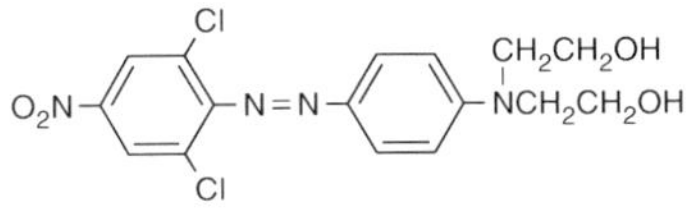

1 gm

S92,330-3
4'-[BIS(2-HYDROXYETHYL)AMINO]-2,6-DICHLORO-2'-METHYL-4-NITROAZOBENZENE

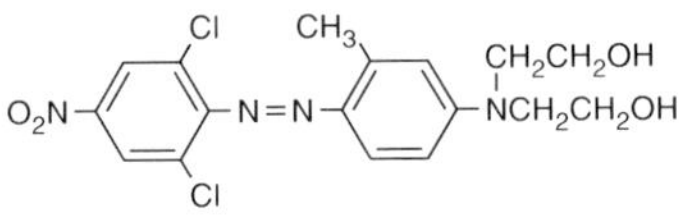

250 mg

S45,603-9
SODIUM FAST PURPLE

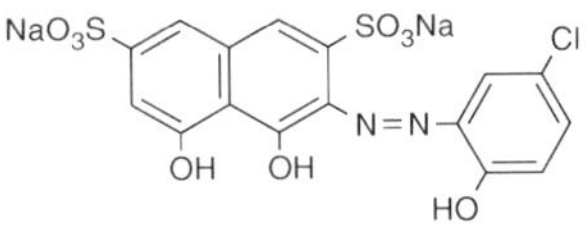

5 gm

S81,952-2
3,3'-DICHLORO-4,4'-DIHYDROXYAZOBENZENE

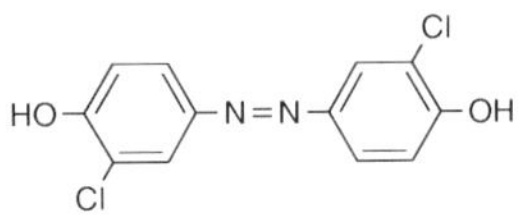

250 mg

S93,095-4
N-[4-(BROMOPHENYLAZO)PHENYL]-ETHANOLAMINE

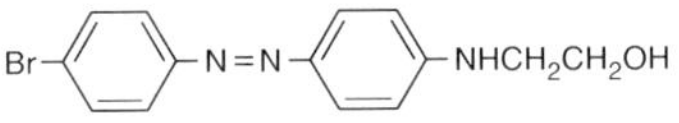

250 mg

S59,620-5
4-(4-NITROPHENYLAZO)-2,6-DICHLOROPHENOL

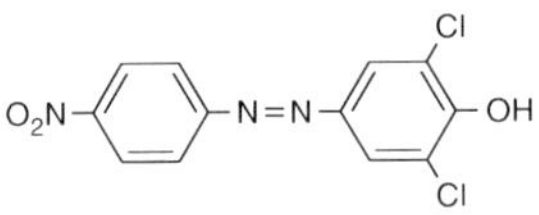

5 gm

S92,308-7
2-[4-(2-CHLORO-4-NITROPHENYLAZO)-N-ETHYLANILINO]-1-ETHANOL

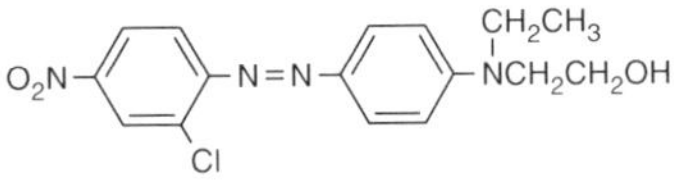

250 mg

S92,368-0
2'-(2-CHLORO-4,6-DINITROPHENYLAZO)-5'-[(2-CYANOETHYL)(2-HYDROXYETHYL)-AMINO]-4'-METHOXYACETANILIDE

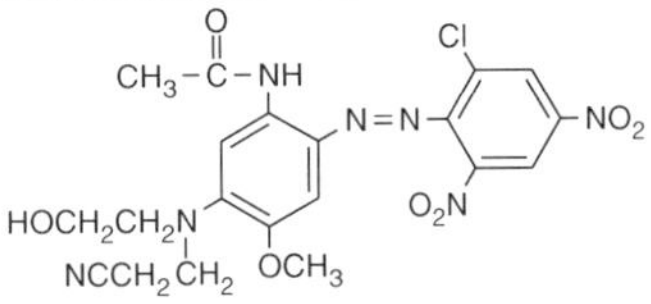

250 mg

S88,778-1
2-(4-BROMOBENZYL)ISOTHIOURONIUM 8-(4-HYDROXY-1-NAPHTHYLAZO)-2-NAPHTHALENESULFONATE

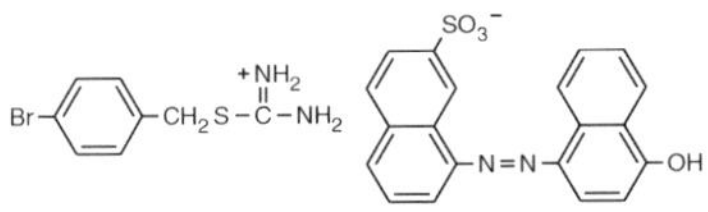

250 mg

S81,447-4
1-[2-(CHLOROPHENYL)AZOMETHYL]-2-NAPHTHOL

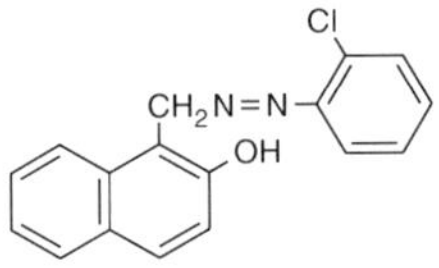

250 mg

S83,480-7
4-(4-BROMOPHENYLAZO)PHENOL

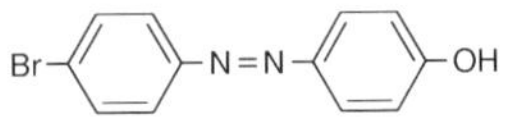

250 mg

S72,850-0
1-(5-CHLORO-2-METHOXYPHENYL)-3-[5-METHYL-2-(4-TOLYLAZO)PHENYL]UREA

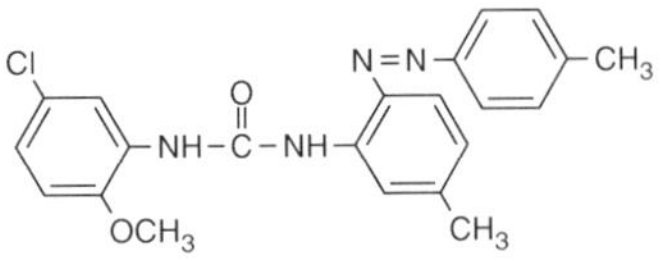

250 mg

S92,936-0
2-[4-(4-BROMOPHENYLAZO)-2-METHOXYPHENYL]ISOINDOLINE

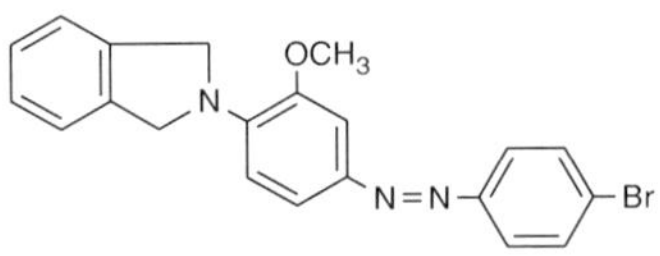

250 mg

S93,009-1
2-[4-(4-BROMOPHENYLAZO)-3-METHOXYPHENYL]ISOINDOLINE

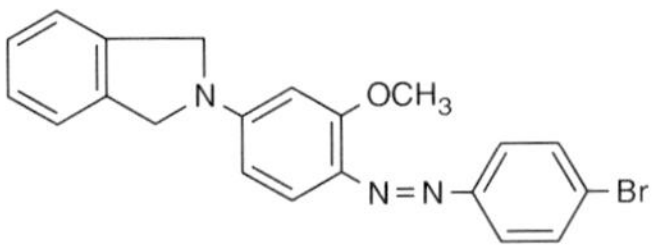

250 mg

S94,200-6
4-(4-CHLORO-2-METHOXYPHENYLAZO)-MORPHOLINE

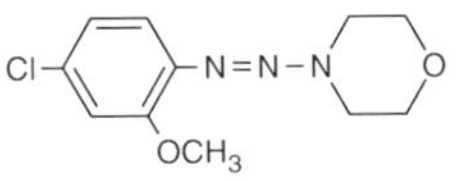

250 mg

S93,157-8
3-AMINO-3-MERCAPTO-2-(4-CHLOROPHENYLAZO)ACRYLONITRILE

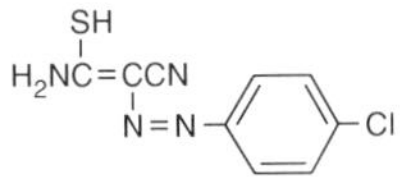

250 mg

S93,846-7
5-AMINO-4-(4-BROMOPHENYLAZO)-3-METHYL-1-PHENYL-1H-PYRAZOLE

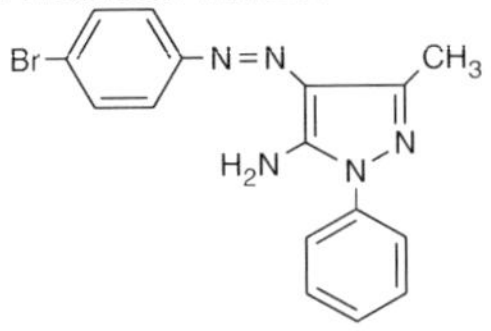

250 mg

S93,257-4
2-AMINO-3-(4-CHLOROPHENYLAZO)-6-METHYLPYRIDINE

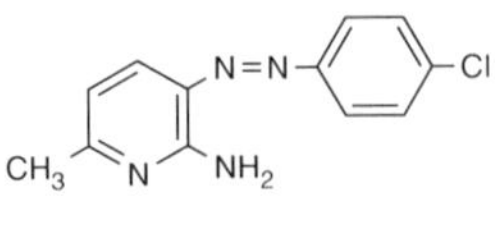

250 mg

S84,808-5
1-(2-FLUORO-5-NITROPHENYL)-3-[5-METHYL-2-(P-TOLYLAZO)PHENYL]UREA

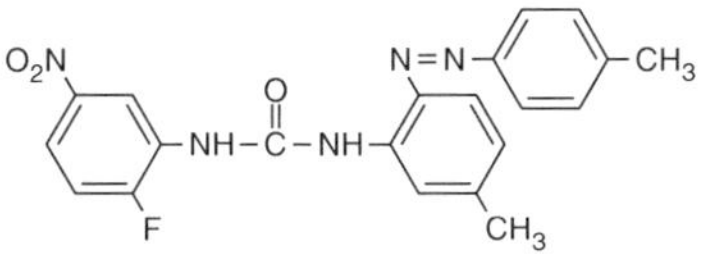

250 mg

S85,495-6
5'-METHYL-2'-(P-TOLYLAZO)-3,4,5,6-TETRABROMOPHTHALANILIC ACID

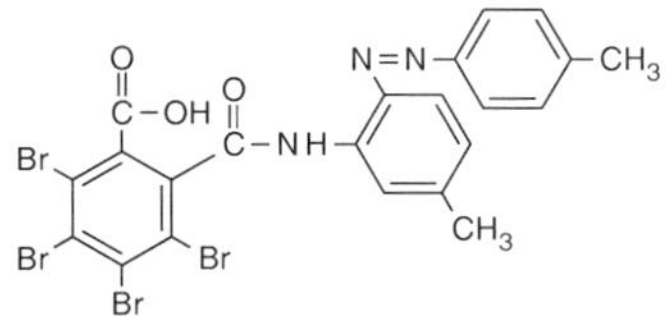

250 mg

S67,582-2
3,4-DICHLORO-2'-(4-METHYLPHENYLAZO)-BENZANILIDE

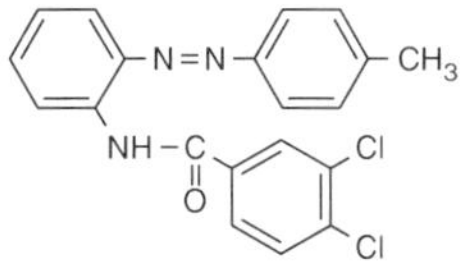

5 gm

S67,628-4
2-CHLORO-5'-METHYL-2'-(4-METHYLPHENYLAZO)BENZANILIDE

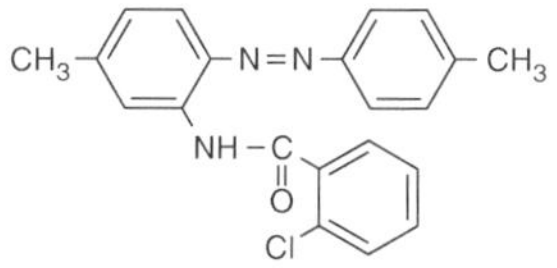

5 gm

S75,052-2
2,4-DICHLORO-5'-METHYL-2'-(4-TOLYLAZO)BENZANILIDE

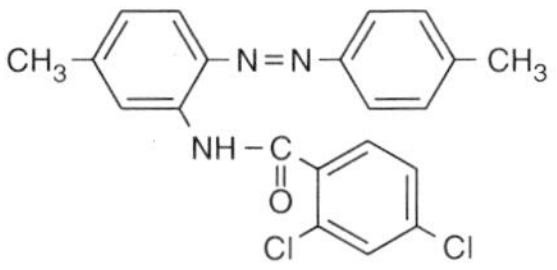

5 gm

S80,189-5
1-(3-FLUOROPHENYL)-3-[5-METHYL-2-(4-METHYLCYCLOHEXYLAZO)PHENYL]UREA

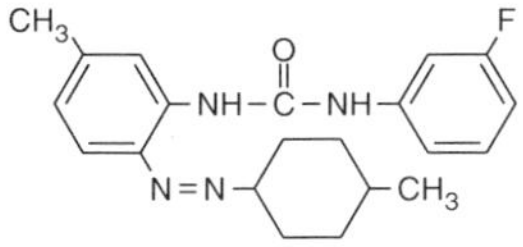

250 mg

S80,126-7
1-(2-FLUOROPHENYL)-3-[5-METHYL-2-(PHENYLAZO)PHENYL]UREA

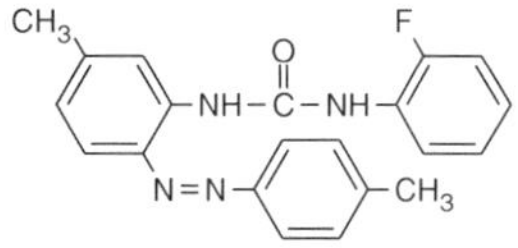

250 mg

S78,065-0
1-(3-FLUOROPHENYL)-3-[5-METHYL-2-(4-TOLYLAZO)PHENYL]UREA

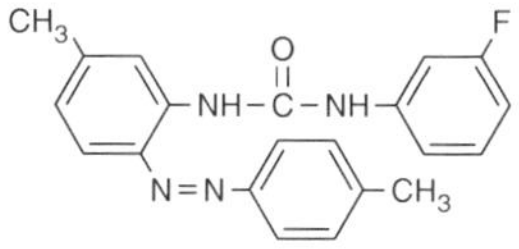

250 mg

S67,498-2
1-(5-CHLORO-2-METHYLPHENYL)-3-[5-METHYL-2-(4-METHYLPHENYLAZO)-PHENYL]UREA

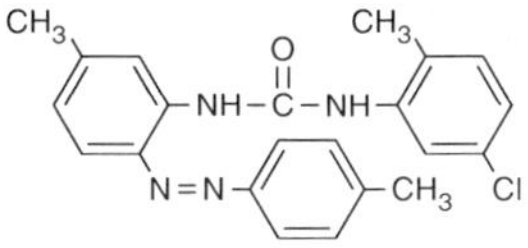

5 gm

S84,801-8
1-(2-FLUORO-5-NITROPHENYL)-3-[4-(PHENYLAZO)PHENYL]UREA

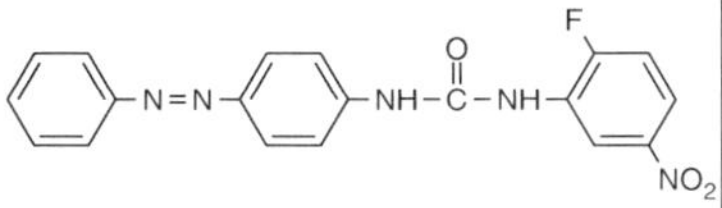

250 mg

S94,201-4
4-FLUORO-3-NITROAZOBENZENE

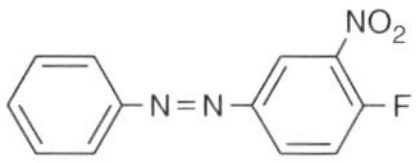

250 mg

S92,934-4
N-[4-(2-CHLORO-4,6-DINITROPHENYLAZO)-3-METHYLPHENYL]-DIETHANOLAMINE

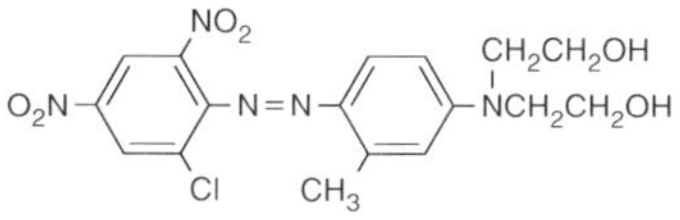

250 mg

S93,086-5
N,N-DIETHYL-4-(2-FLUOROPHENYLAZO)-ANILINE

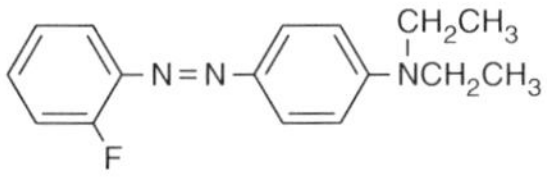

250 mg

S93,084-9
4-(2-CHLOROPHENYLAZO)-N,N-DIETHYLANILINE

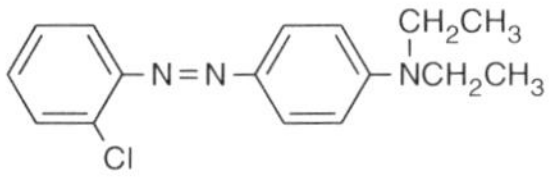

250 mg

S93,085-7
4-(2-BROMOPHENYLAZO)-N,N-DIETHYLANILINE

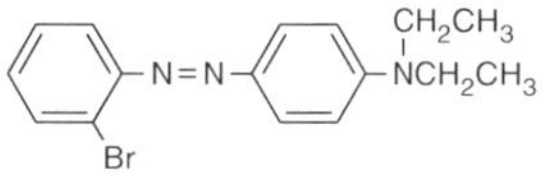

250 mg

S88,289-5
N,N-DIMETHYLAMINO-4-(4-FLUOROPHENYLAZO)ANILINE

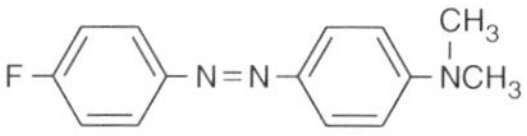

250 mg

S93,065-2
4-(4-BROMOPHENYLAZO)-N,N-DIMETHYLANILINE

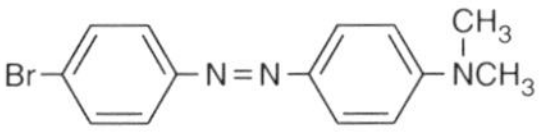

250 mg

S93,074-1
4-(4-BROMOPHENYLAZO)-N-ETHYL-N-METHYLANILINE

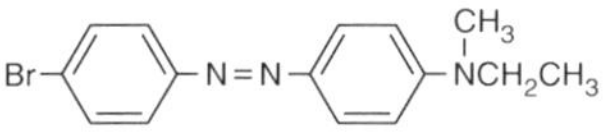

250 mg

S92,930-1
N,N-DIETHYL-4-(4-FLUOROPHENYLAZO)-ANILINE

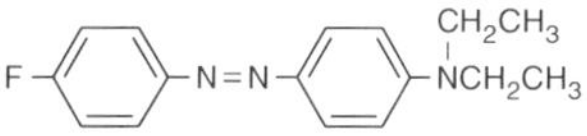

250 mg

S93,080-6
4-(3-CHLOROPHENYLAZO)-N,N-DIETHYLANILINE

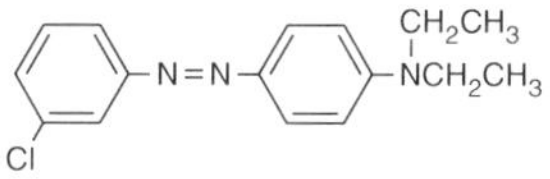

250 mg

S93,093-8
4-(4-CHLOROPHENYLAZO)-N,N-DIETHYLANILINE

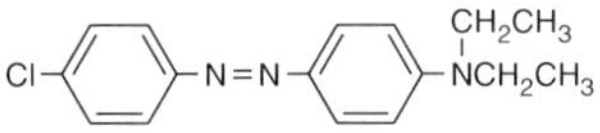

250 mg

S93,079-2
4-(3-BROMOPHENYLAZO)-N,N-DIETHYLANILINE

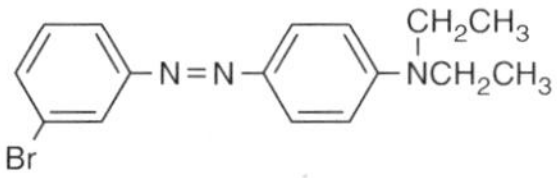

250 mg

S93,088-1
4-(4-BROMOPHENYLAZO)-N,N-DIETHYLANILINE

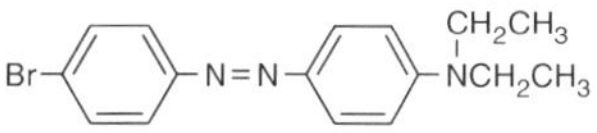

250 mg

S93,078-4
4-(2,3-DICHLOROPHENYLAZO)-N,N-DIETHYLANILINE

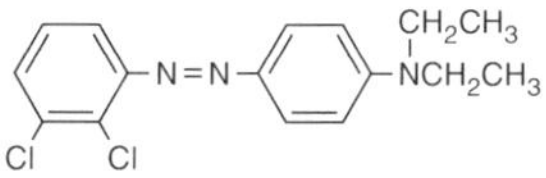

250 mg

S93,082-2
4-(2,6-DICHLOROPHENYLAZO)-N,N-DIETHYLANILINE

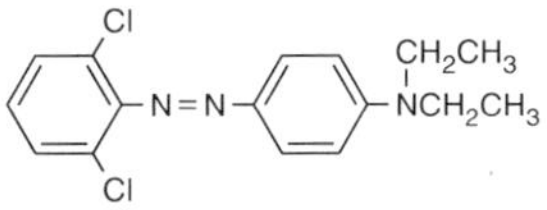

250 mg

S93,081-4
4-(2,6-DIBROMOPHENYLAZO)-N,N-DIETHYLANILINE

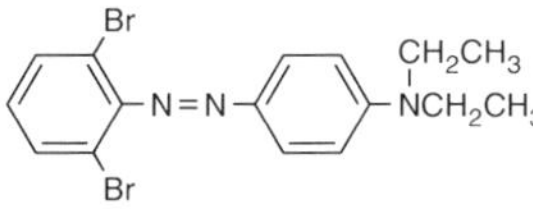

250 mg

S93,083-0
4-(2-BROMO-6-CHLORO-4-FLUOROPHENYLAZO)-N,N-DIETHYLANILINE

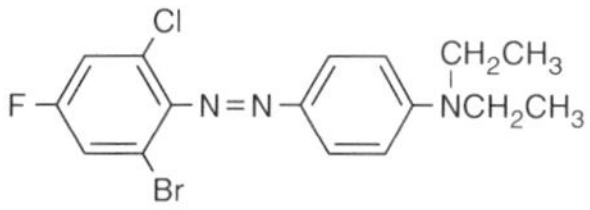

250 mg

S92,327-3
5'-[BIS(2-ACETOXYETHYL)AMINO]-2'-(2-BROMO-4,6-DINITROPHENYLAZO)-4'-ETHOXYACETANILIDE

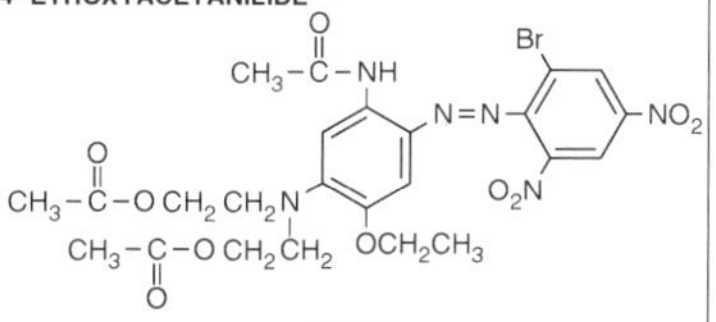

250 mg

S93,087-3
N,N-DIETHYL-4-(2,4-DIFLUOROPHENYLAZO)ANILINE

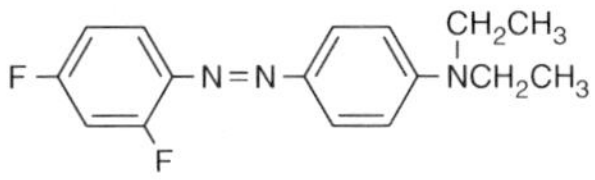

250 mg

S80,773-7
4-(3-CHLOROPHENYLAZO)-N,N-DIMETHYLANILINE

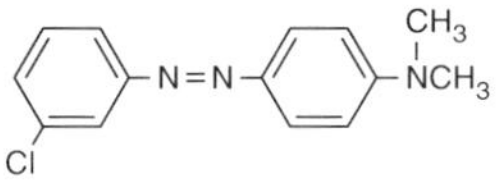

250 mg

S92,926-3
4-(4-BROMOPHENYLAZO)-3-METHYLANILINE TETRAFLUOROBORATE

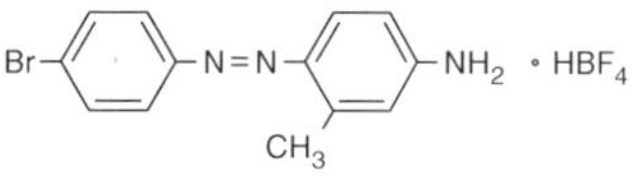

250 mg

S75,812-4
4-ACETAMIDO-2,6-DIFLUORO-4'-NITROAZOBENZENE

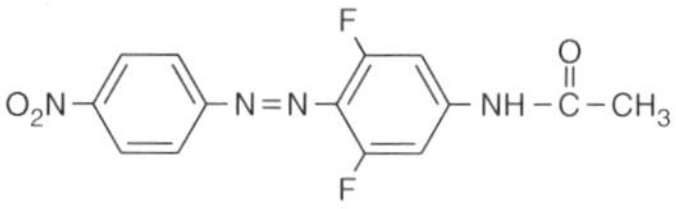

250 mg

S93,003-2
2-[2-CHLORO-4-(4-NITROPHENYLAZO)-PHENYL]ISOINDOLINE

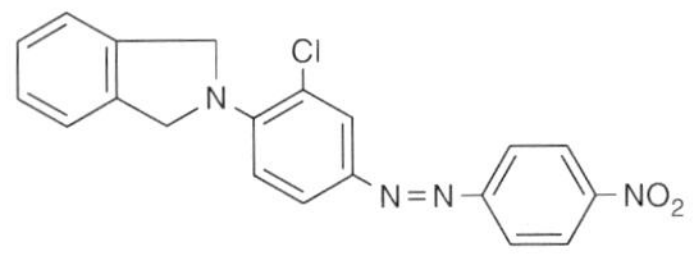

250 mg

S85,496-4
4-CHLORO-4'-(PHENYLAZO)BENZANILIDE

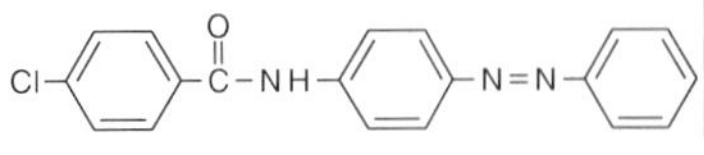

250 mg

S93,041-5
1-[4-(4-BROMOPHENYLAZO)PHENYL]-AZIRIDINE

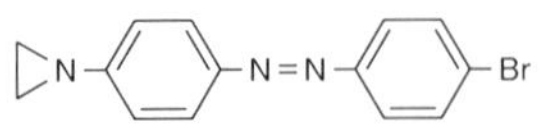

250 mg

S93,016-4
2-[4-(4-BROMOPHENYLAZO)-3-METHYLPHENYL]ISOINDOLINE

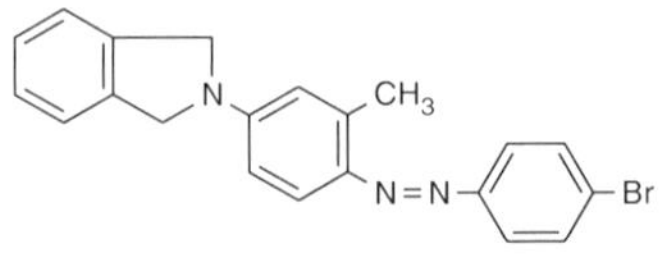

250 mg

S93,006-7
2-[4-(4-BROMOPHENYLAZO)-3,5-DIMETHYLPHENYL]ISOINDOLINE

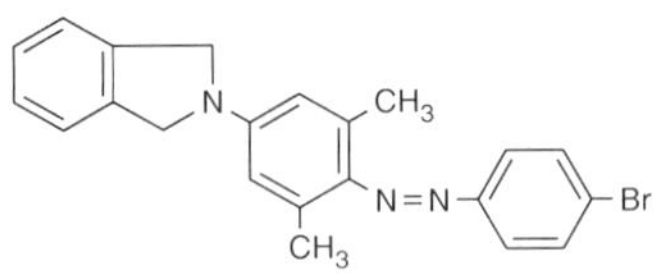

250 mg

S93,023-7
2-[4-(4-BROMOPHENYLAZO)PHENYL]-ISOINDOLINE

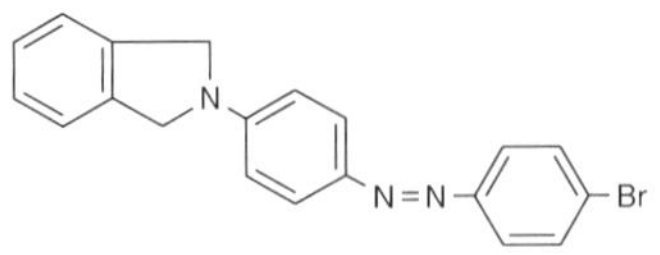

250 mg

S93,035-0
1-[4-(4-BROMOPHENYLAZO)PHENYL]-3-METHYLPIPERIDINE

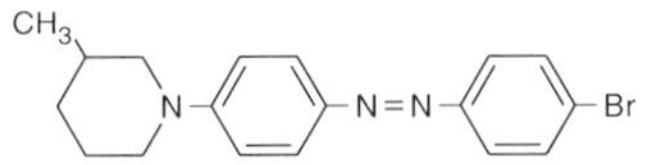

250 mg

S93,036-9
1-[4-(4-BROMOPHENYLAZO)PHENYL]-4-METHYLPIPERIDINE

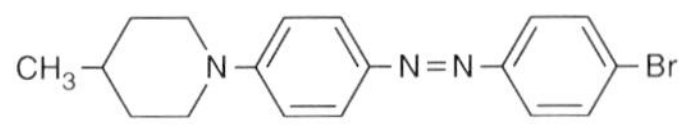

250 mg

S93,309-0
N-BENZYLIDENE-N-[4-(4-CHLOROPHENYLAZO)-1-NAPHTHYL]AMINE TETRAFLUOROBORATE

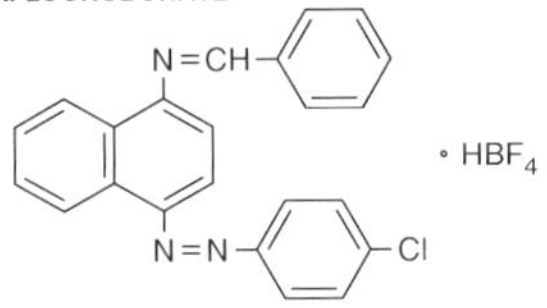

250 mg

S93,028-8
1-(4-[4-(BROMOPHENYLAZO)PHENYL])-HEXAHYDRO-1H-AZEPINE

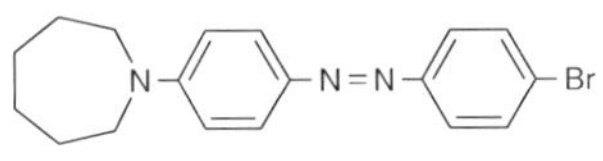

250 mg

S93,066-0
4-(4-BROMOPHENYLAZO)-N-ISOPROPYL-N-METHYLANILINE

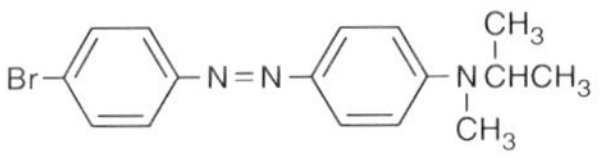

250 mg

S92,932-8
4-(DIETHYLAMINO)-3'-FLUOROAZOBENZENE

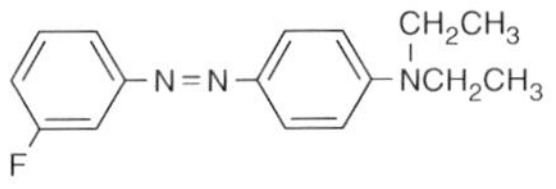

250 mg

S93,077-6
4-(3,5-DICHLOROPHENYLAZO)-N,N-DIETHYLANILINE

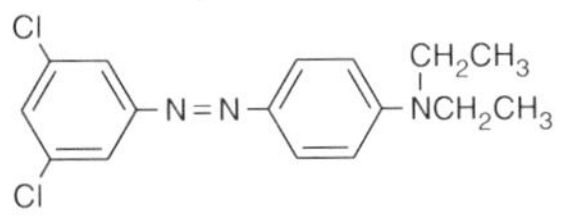

250 mg

S37,236-6
3,3'-DICHLOROAZOXYBENZENE

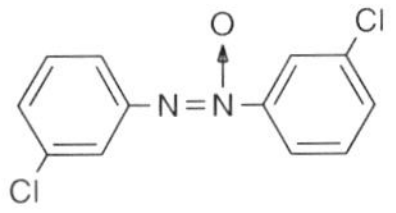

5 gm

S71,633-2
N-[3-METHYL-5-(4-METHYLPHENYLAZO)-PHENYL]-3,4,5,6-TETRABROMOPHTHALAMIC ACID

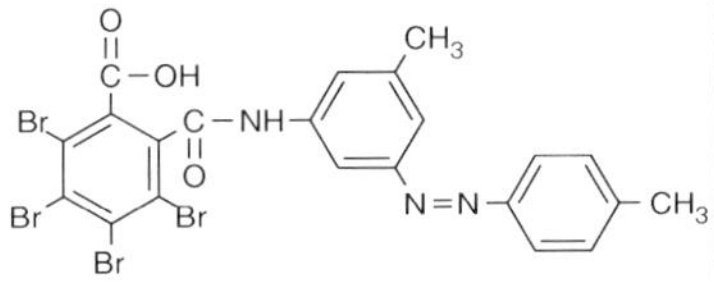

100 mg

S71,869-6
3,4-DICHLORO-4'-(PHENYLAZO)-BENZANILIDE

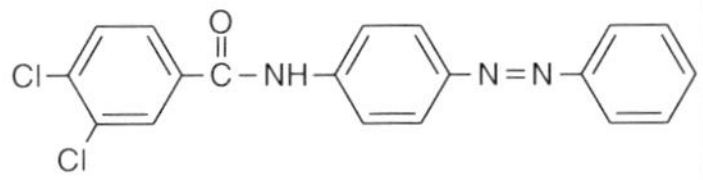

250 mg

S71,864-5
4-CHLORO-5'-METHYL-2'-(4-METHYLPHENYLAZO)BENZANILIDE

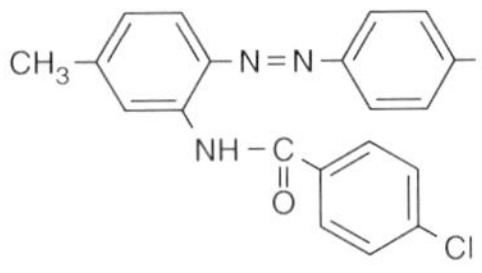

5 gm

S74,921-4
N,N-DIACETYL-2,6-DICHLORO-4-(PHENYLAZO)ANILINE

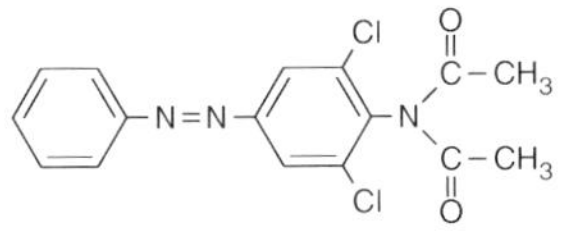

250 mg

S73,175-7
1-(2-FLUOROPHENYL)-3-[5-METHYL-2-(4-TOLYLAZO)PHENYL]UREA

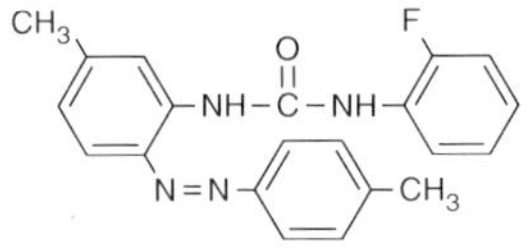

5 gm

S92,942-5
2-[4-(4-BROMOPHENYLAZO)-2,5-DIMETHYLPHENYL]ISOINDOLINE

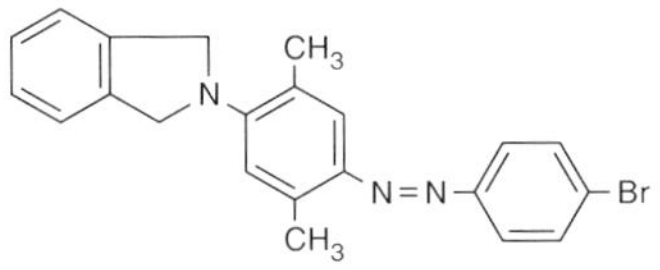

250 mg

S93,052-0
4-[4-(4-BROMOPHENYLAZO)PHENYL]-2,6-DIMETHYLMORPHOLINE

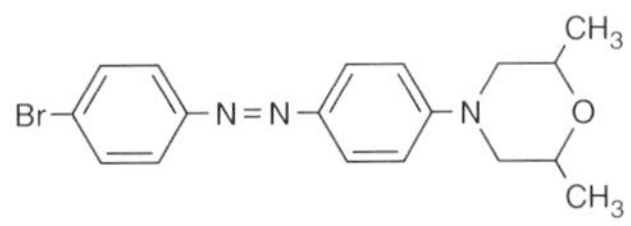

250 mg

S75,166-9
4-CHLOROAZOBENZENE

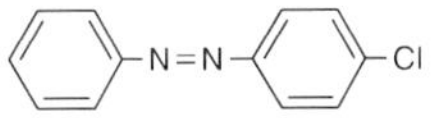

5 gm

S92,918-2
4-BROMOAZOBENZENE

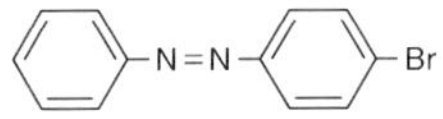

250 mg

S45,374-9
4,4'-DICHLOROAZOBENZENE

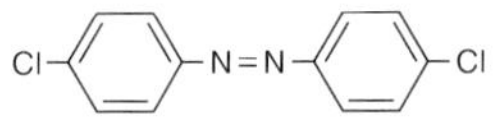

250 mg

S88,052-3
4,4'-DIBROMOAZOBENZENE

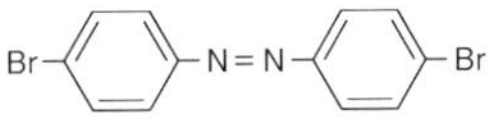

250 mg

S82,583-2
4,4'-DIBROMOAZOXYBENZENE

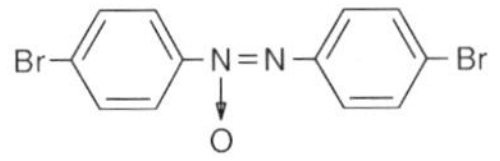

250 mg

S92,397-4
2-(4-CHLOROPHENYLAZO)NAPHTHALENE

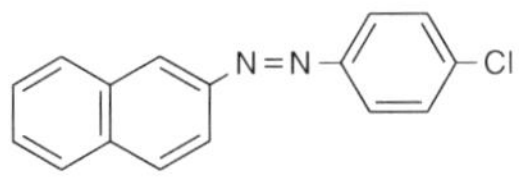

250 mg

S89,467-2
1,1-DICHLORO-2-(2,4-DICHLOROPHENYLAZO)ETHENE

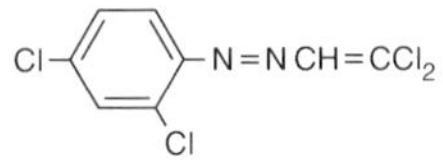

250 mg

S94,413-0
DISPERSE BLACK 4 (C.I. 11220)

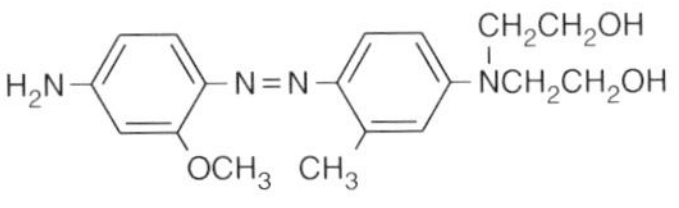

250 mg

S94,408-4
DISPERSE RED 58 (C.I. 11135)

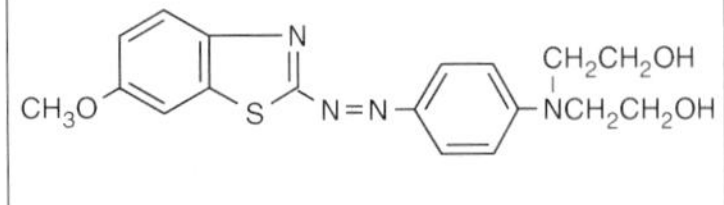

250 mg

S92,242-0
DIRECT BLUE 10 (C.I. 24340)

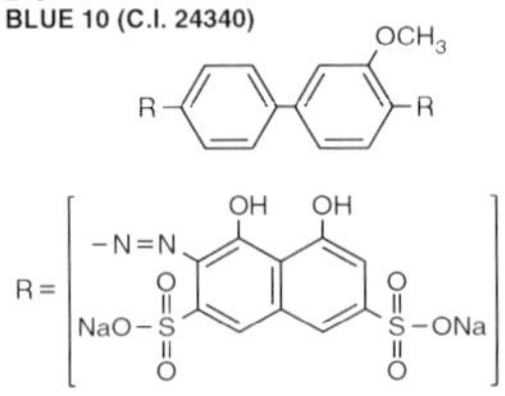

5 gm

S44,843-5
4-(4-NITROPHENYLAZO)CATECHOL

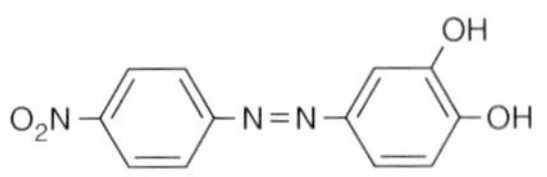

5 gm

S76,520-1
4,5-DIHYDROXY-3-(4-SULFO-1-NAPHTHYLAZO)-2,7-NAPHTHALENEDISULFONIC ACID TRISODIUM SALT

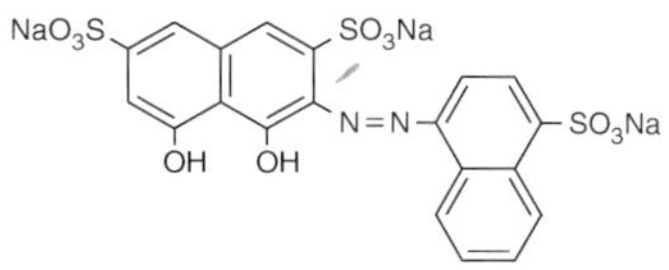

5 gm

S92,324-9
4-AMINO-4'-[BIS(2-HYDROXYETHYL)-AMINO]AZOBENZENE

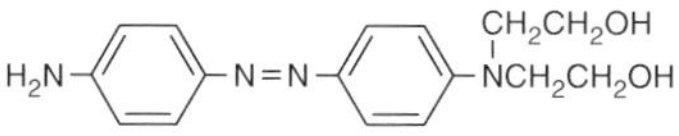

250 mg

S92,316-8
4'-[BIS(2-HYDROXYETHYL)AMINO]-4-NITROAZOBENZENE

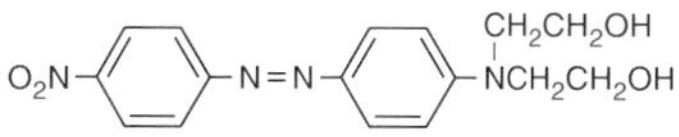

250 mg

S94,414-9
DISPERSE RED 16 (C.I. 11225)

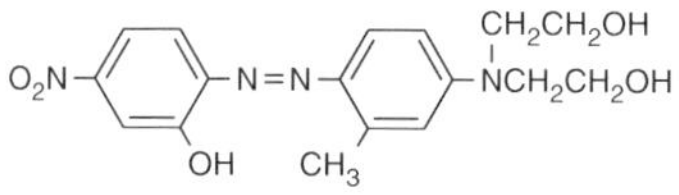

250 mg

S44,100-7
4-(4-NITROPHENYLAZO)ORCINOL

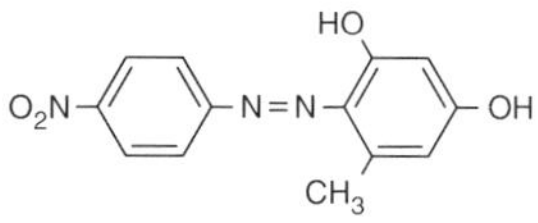

5 gm

S92,307-9
4'-[BIS(2-HYDROXYETHYL)AMINO]-2,4-DINITROAZOBENZENE

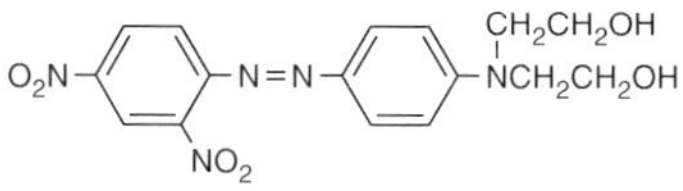

250 mg

S92,339-7
N,N-[BIS(2-HYDROXYETHYL)-3-METHYL]-4-(4-NITROPHENYLAZO)ANILINE

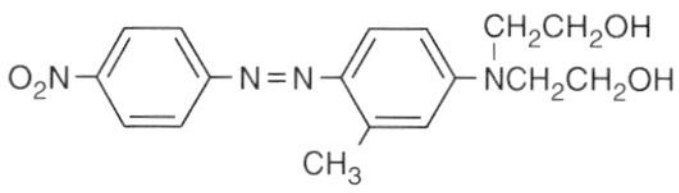

250 mg

S92,331-1
4'-[BIS(2-HYDROXYETHYL)AMINO]-2-CHLORO-2'-METHYL-4-NITROAZOBENZENE

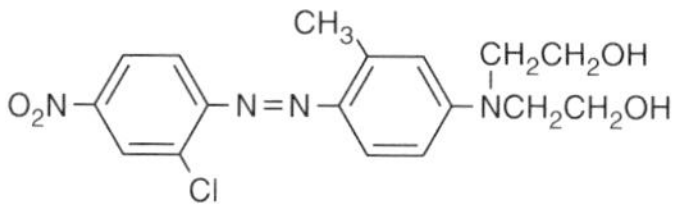

250 mg

S41,667-3
3-(4-DIETHYLAMINO-2-HYDROXYPHENYLAZO)-4-HYDROXYBENZENESULFONIC ACID

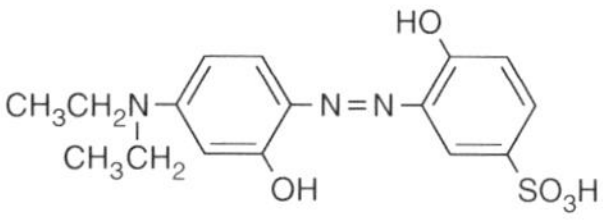

5 gm

S55,883-4
4-(2,5-DICHLOROPHENYLAZO)-RESORCINOL

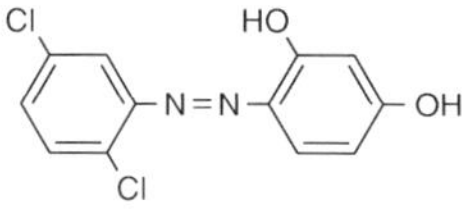

250 mg

S46,006-0
FAST SULPHON BLACK F

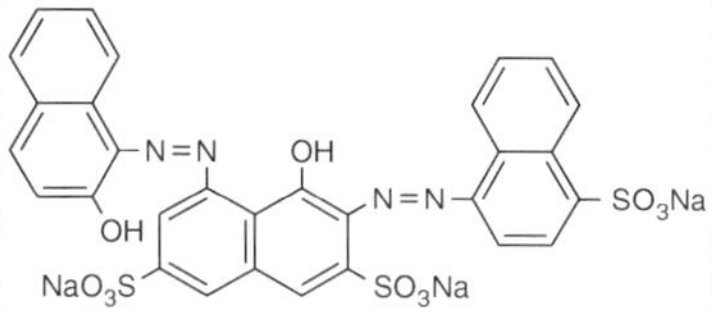

5 gm

S81,951-4
4,4'-DIHYDROXYAZOBENZENE

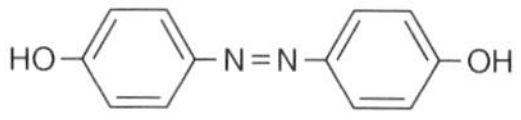

250 mg

S81,950-6
4,4'-DIHYDROXYAZOXYBENZENE

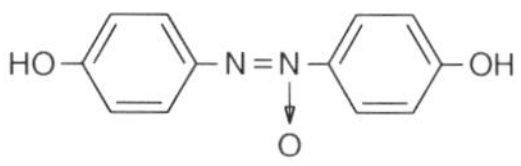

250 mg

S92,239-0
DIRECT BLUE 1 (C.I. 24410)

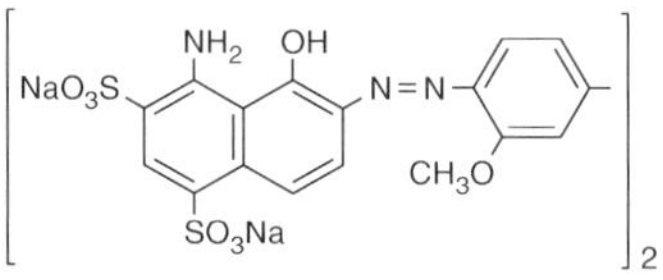

250 mg

S55,699-8
5-METHOXY-2-(2-THIAZOLYLAZO)PHENOL

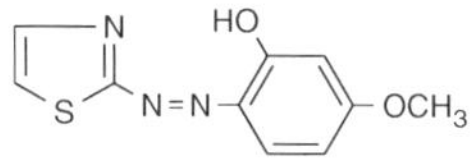

250 mg

S45,312-9
CHLORAZOL AZURINE

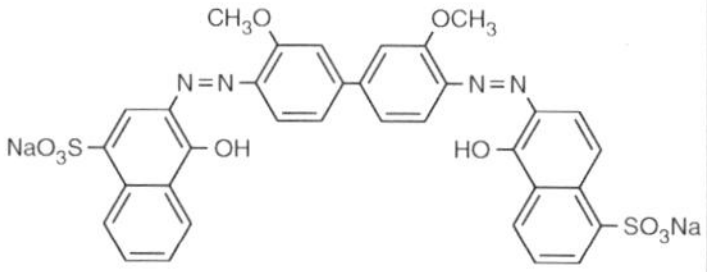

5 gm

S92,305-2
2-(4-METHOXYPHENYLAZO)-4-METHYLPHENOL

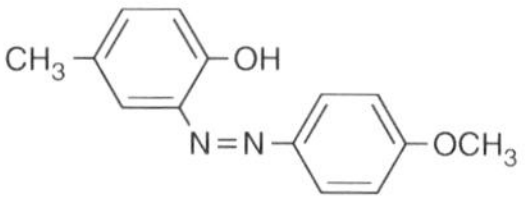

250 mg

S92,949-2
3-HYDROXY-4-(2-METHOXYPHENYLAZO)-2-NAPHTHANILIDE

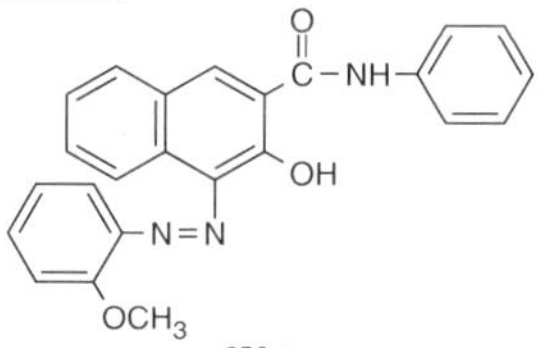

250 mg

S82,485-2
2-[4-(2-HYDROXY-1-NAPHTHYLAZO)-PHENOXY]ACETIC ACID

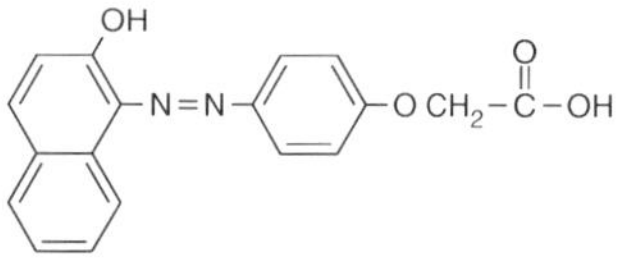

250 mg

S55,701-3
4-METHOXY-2-(2-THIAZOLYLAZO)PHENOL

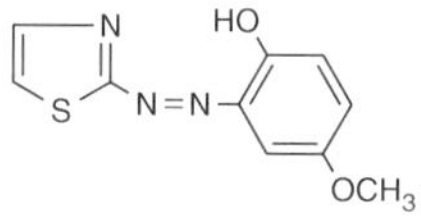

5 gm

S92,372-9
1-[4-(BENZYLOXY)PHENYLAZO]-2-NAPHTHOL

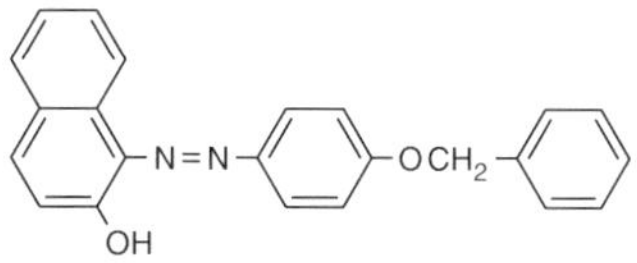

250 mg

S64,824-8
4-(4-DECYLOXYPHENYLAZO)PHENOL

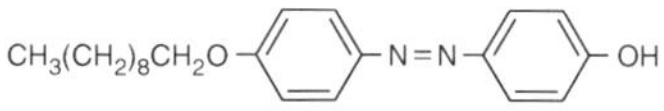

250 mg

S60,599-9
5-(2,5-DICHLOROPHENYLAZO)-4-METHYLSALICYLIC ACID

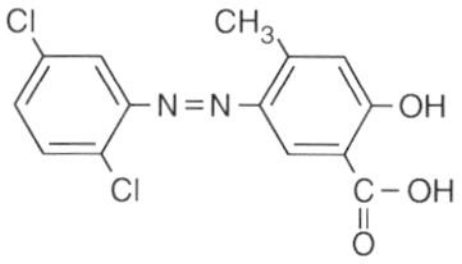

250 mg

S62,288-5
ETHYL 5-PHENYLAZOSALICYLATE

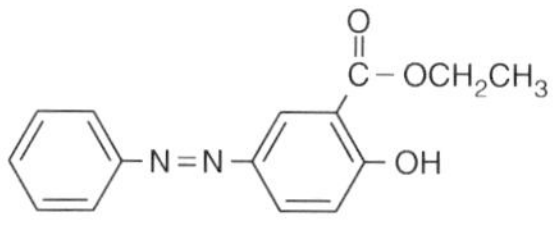

250 mg

S87,022-6
5,5-DIMETHYL-3-HYDROXY-2-(PHENYLAZO)-2-CYCLOHEXEN-1-ONE

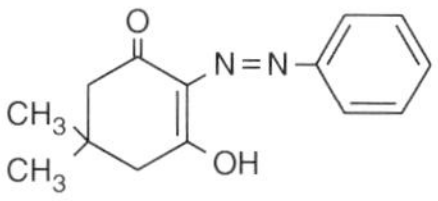

250 mg

S41,178-7
6-HEXYL-4-(2-THIAZOLYLAZO)-RESORCINOL

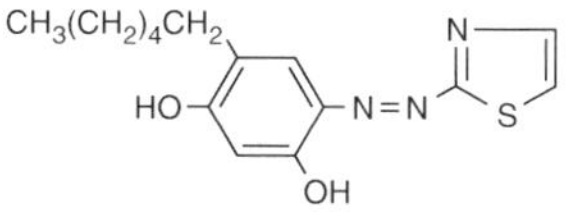

5 gm

S45,322-6
MORDANT BROWN 12

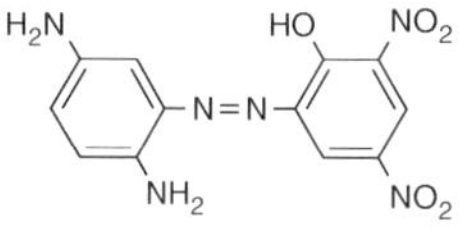

5 gm

S38,138-1
4-(2,6-DIAMINO-3-PYRIDYLAZO)PHENOL

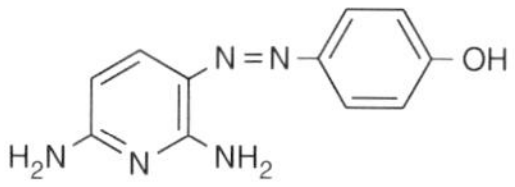

5 gm

S45,255-6
5-AMINO-4-PHENYLAZO-3-PYRAZOLOL

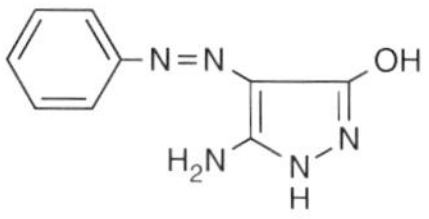

5 gm

S56,237-8
2-(2,4-DIAMINOPHENYLAZO)PHENOL

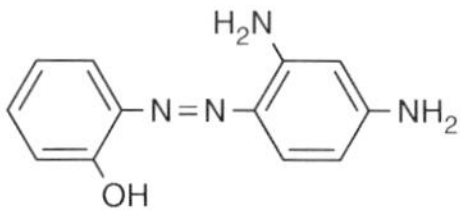

5 gm

S44,737-4
GALLION TETRAHYDRATE

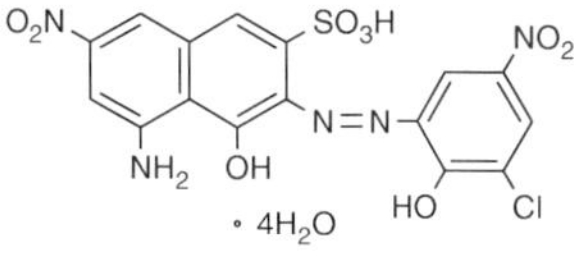

5 gm

S44,735-8
OWENS BLUE

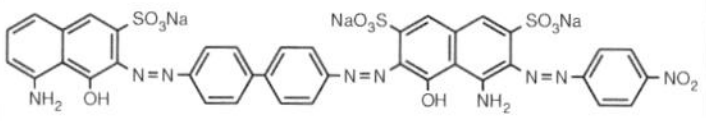

5 gm

S92,241-2
DIRECT DYE (C.I. 31793)
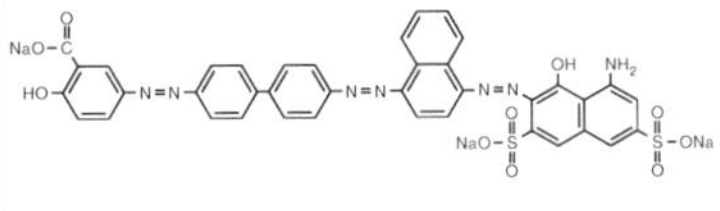
5 gm

S92,237-4
DIRECT GREEN 1 (C.I. 30280)
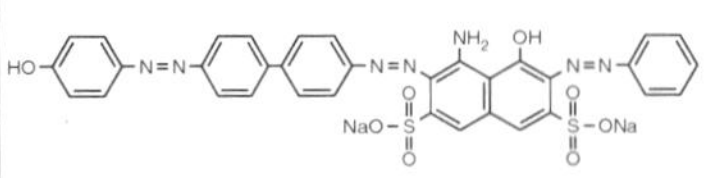
250 mg

S60,567-0
2-(2,4-DINITROPHENYLAZO)-P-CRESOL
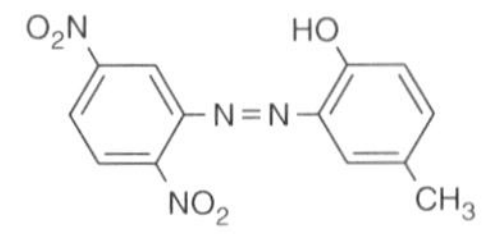
250 mg

S94,485-8
4-(2,4-DINITROPHENYLAZO)-1-NAPHTHOL
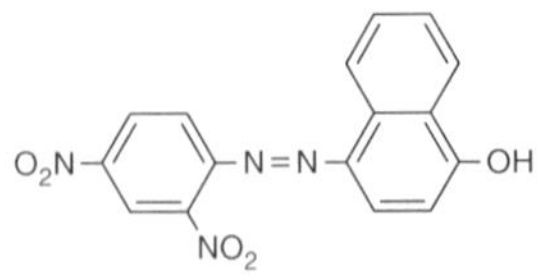
250 mg

S94,411-4
DISPERSE VIOLET 13 (C.I. 11195)
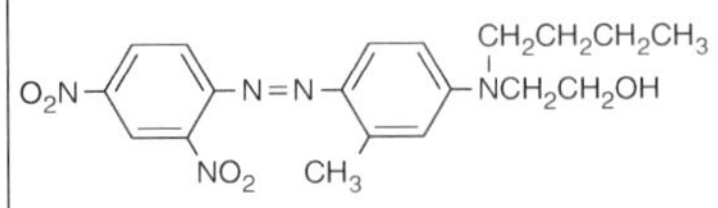
250 mg

S92,317-6
2-[4-(2,4-DINITROPHENYLAZO)-N-ETHYLANILINO]-1-ETHANOL
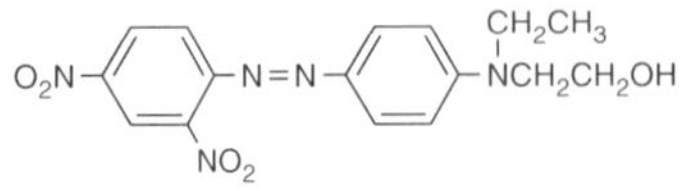
250 mg

S79,830-4
3,5-DIMETHYL-4-(2-NITROPHENYLAZO)-PHENOL
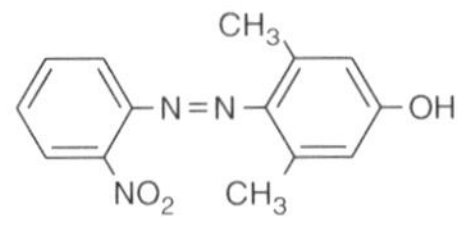
250 mg

S88,595-9
6-METHYL-4-PHENYL-5-(PHENYLAZO)-2-PYRIMIDINOL
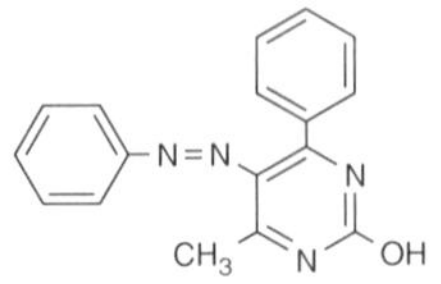
250 mg

S94,407-6
DISPERSE VIOLET 12 (C.I. 11120)
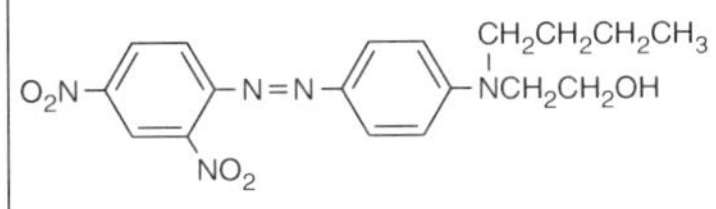
250 mg

S94,409-2
DISPERSE DYE (C.I. 11180)
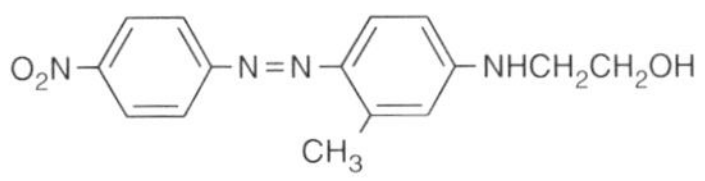
250 mg

S80,494-0
4-(DIETHYLAMINO)-2-HYDROXYAZOBENZENE
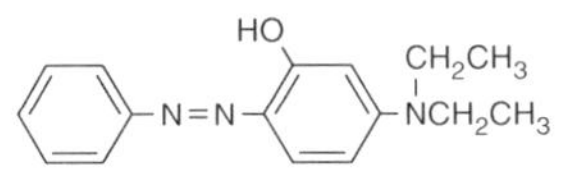
250 mg

S60,539-5
N,N-BIS(2-HYDROXYETHYL)-4-(2,5-DICHLOROPHENYLAZO)ANILINE
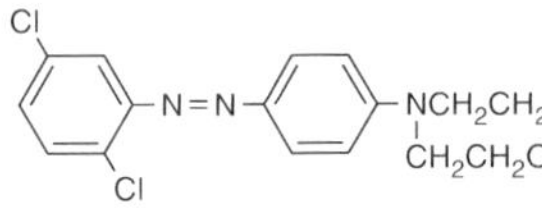
250 mg

S94,410-6
DISPERSE RED 32 (C.I. 11190)
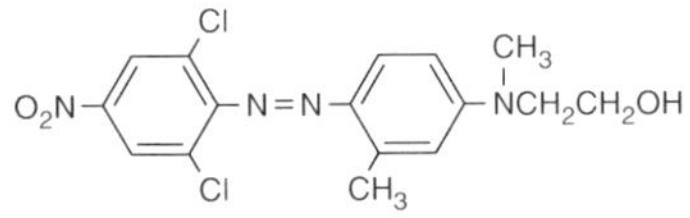
250 mg

S87,434-5
[8-(4-HYDROXY-1-NAPHTHYLAZO)-2-NAPHTHYL]TRIMETHYLAMMONIUM ACETATE
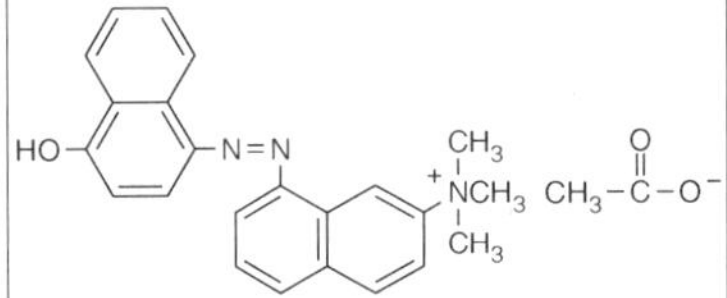
250 mg

S87,326-8
6-(4-HYDROXY-1-NAPHTHYLAZO)-1-METHYLQUINOLINIUM CHLORIDE
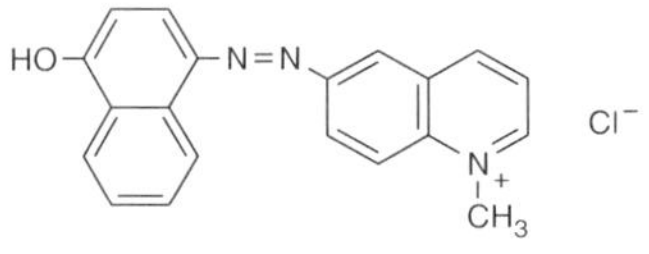
250 mg

S92,952-2
6-(4-ACETYLPHENYLAZO)-1,2,3,4-TETRAHYDRO-2,2,4,7-TETRAMETHYL-1-QUINOLINEETHANOL
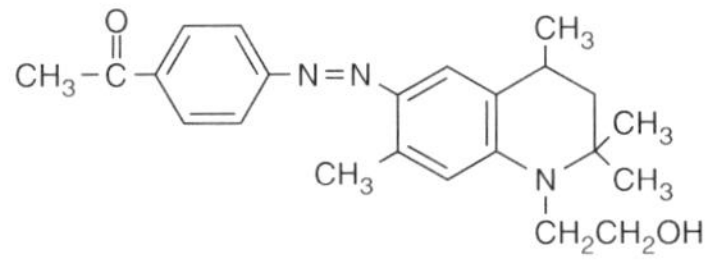
250 mg

S92,922-0
4-(4-NITROPHENYLAZO)PHENOL
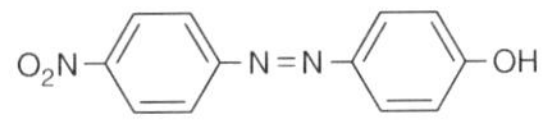
250 mg

S80,866-0
5-HYDROXY-2-(4-NITROPHENYLAZO)-PYRIDINE
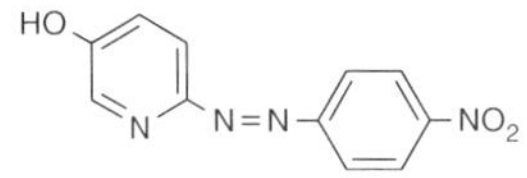
250 mg

S94,415-7
DISPERSE DYE (C.I. 11230)
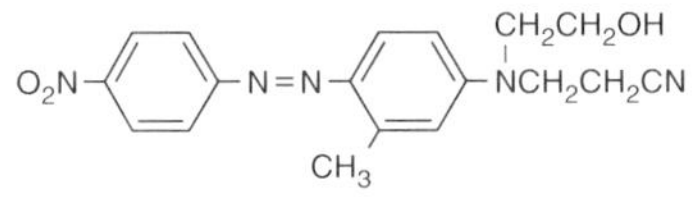
250 mg

S94,406-8
DISPERSE ORANGE 5 (C.I. 11100)
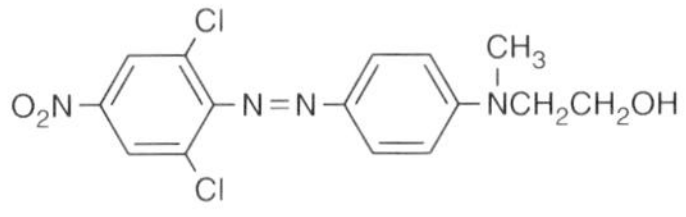
250 mg

S57,608-5
3,6-BIS(4-BROMO-2-SULFOPHENYLAZO)-4,5-DIHYDROXY-2,7-NAPHTHALENEDISULFONIC ACID
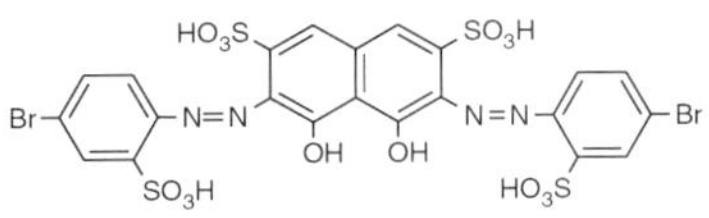
5 gm

S81,364-8
1-(4-NITROPHENYLAZO)-6-(TRIMETHYLSILYL)-2-NAPHTHOL
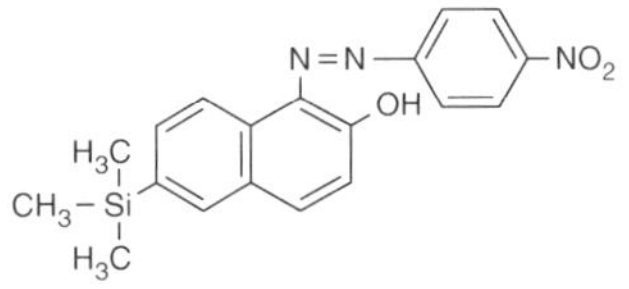
250 mg

S57,610-7
3-HYDROXY-4-(4-NITROPHENYLAZO)-2,7-NAPHTHALENEDISULFONIC ACID, DISODIUM SALT
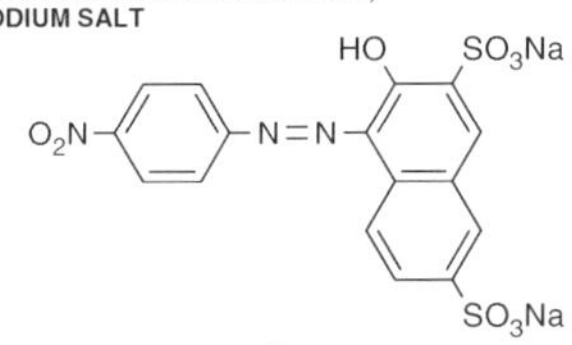
5 gm

S57,609-3
3,6-BIS(2-HYDROXY-5-NITROPHENYLAZO)-4,5-DIHYDROXY-2,7-NAPHTHALENEDISULFONIC ACID
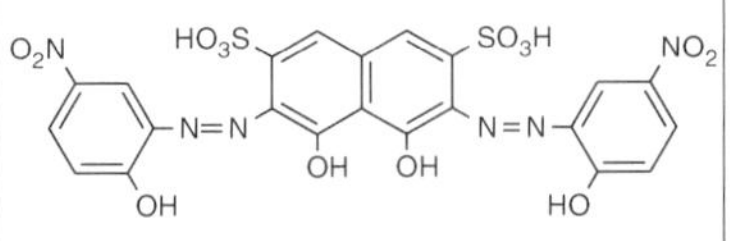
5 gm

S81,365-6
4-[2-HYDROXY-6-(TRIMETHYLSILYL)-1-NAPHTHYLAZO]BENZOIC ACID
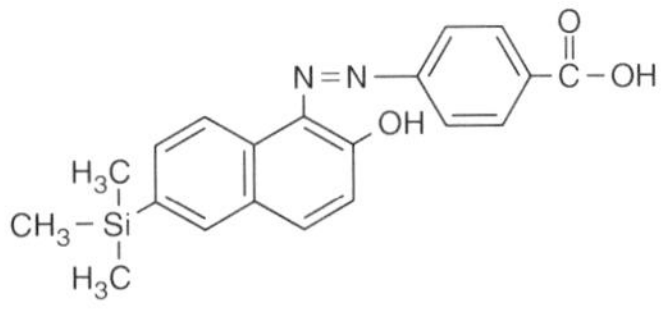
250 mg

S36,277-8
4-(2-HYDROXY-4,5-DIMETHYLPHENYLAZO)-BENZOIC ACID
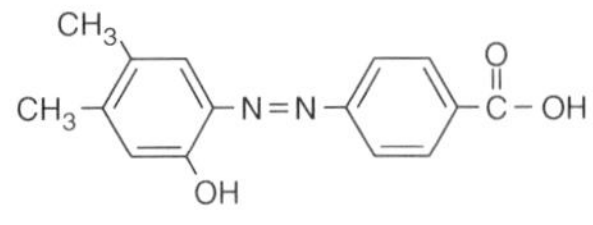
5 gm

S58,766-4
4-HYDROXY-3-(PHENYLAZO)-1-NAPHTHALENESULFONIC ACID, SODIUM SALT
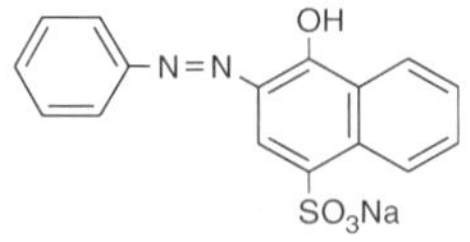
250 mg

S44,907-5
ORANGE G
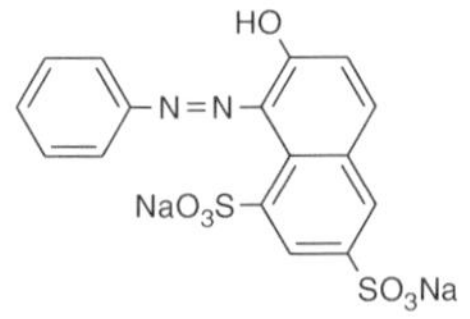
5 gm

S65,932-0
SCARLET R
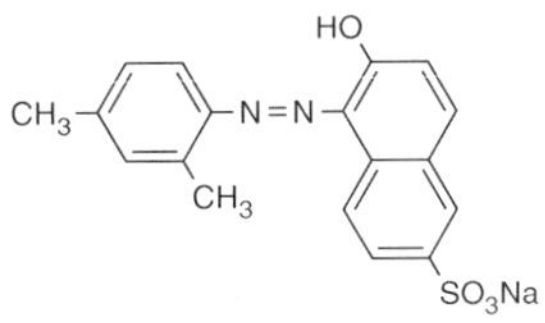
250 mg

S59,115-7
6-HYDROXY-5-(4-SULFO-1-NAPHTHYLAZO)-1,3-NAPHTHALENEDISULFONIC ACID, TRISODIUM SALT
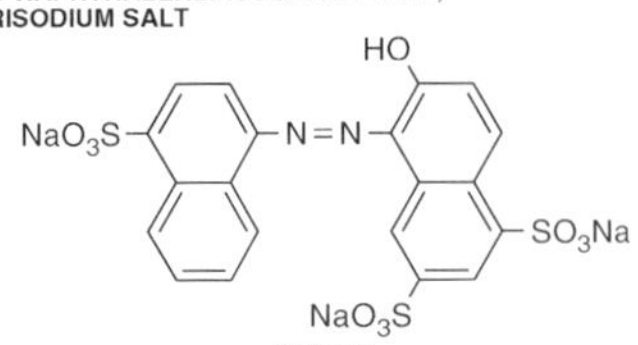
250 mg

S60,102-0
SCARLET R (C.I. 16150)
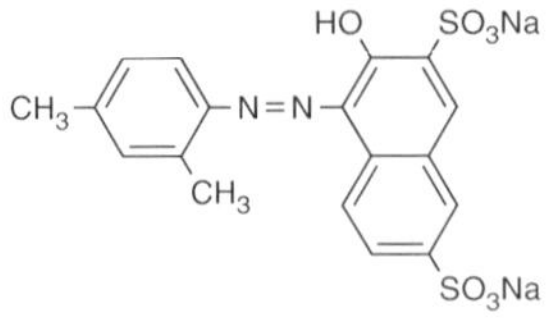
250 mg

S43,140-0
6-METHYL-4-(2-THIAZOLYLAZO)-RESORCINOL
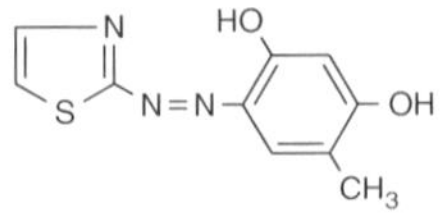
250 mg

S45,927-5
1-(2-THIAZOLYLAZO)-2-NAPHTHOL

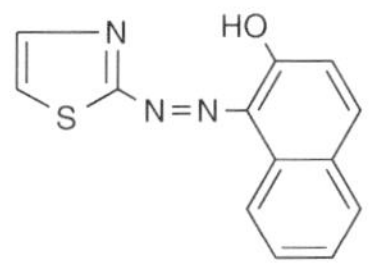

250 mg

S41,563-4
6-PHENYLAZO-3,4-XYLENOL

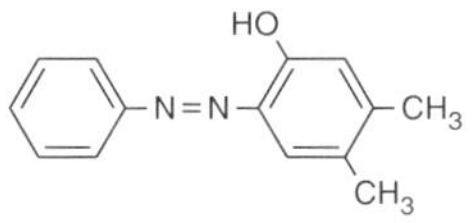

5 gm

S77,434-0
3-HYDROXY-2-(PHENYLAZO)FLUORENE

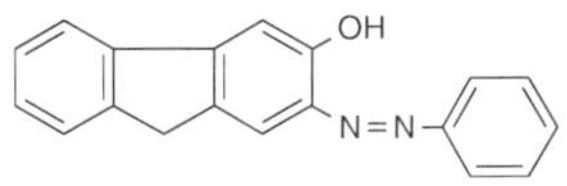

250 mg

S43,444-2
1-(2-HYDROXYPHENYLAZO)-2-NAPHTHOL

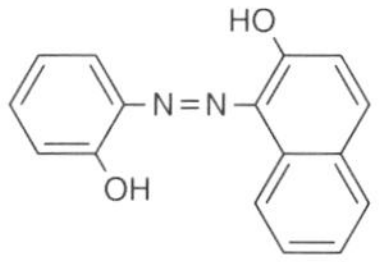

5 gm

S82,329-5
1-(1-CHLORO-2-NAPHTHYLAZO)-2-NAPHTHOL

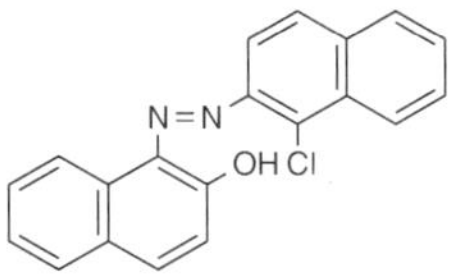

250 mg

S35,131-8
2-HYDROXY-4-SULFO-1-NAPHTHALENEDIAZONIUM CHLORIDE

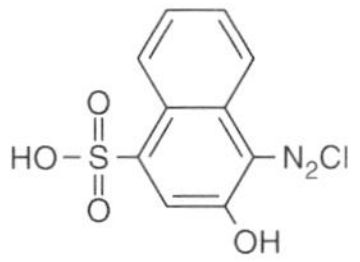

5 gm

S81,446-6
1-[(4-NITROPHENYL)AZOMETHYL]-2-NAPHTHOL

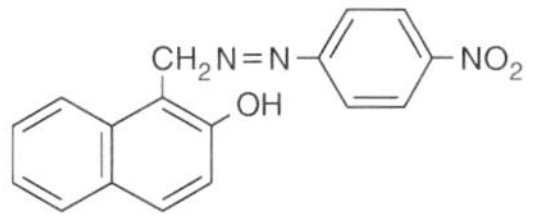

250 mg

S81,448-2
1-[(2,4-DINITROPHENYL)AZOMETHYL]-2-NAPHTHOL

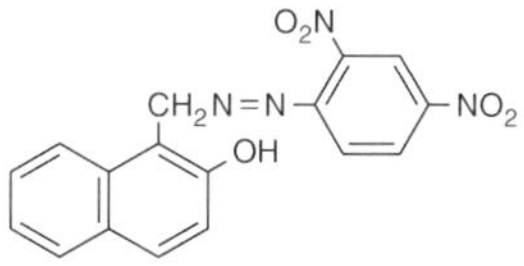

250 mg

S50,519-6
PHENYL 1-HYDROXY-4-PHENYLAZO-2-NAPHTHOATE

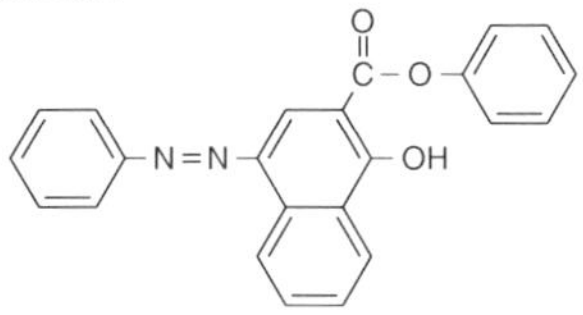

250 mg

S92,711-2
4'-(4-HYDROXYPHENYLAZO)ACETOPHENONE

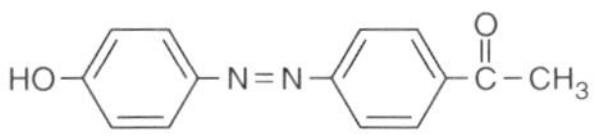

250 mg

S59,112-2
5-(6-SULFO-2-NAPHTHYLAZO)SALICYLIC ACID, DISODIUM SALT

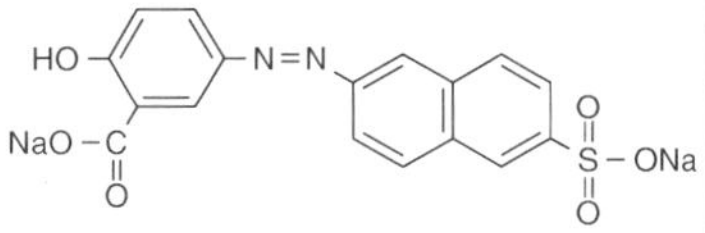

250 mg

S45,327-7
5-(PHENYLAZO)SALICYLIC ACID

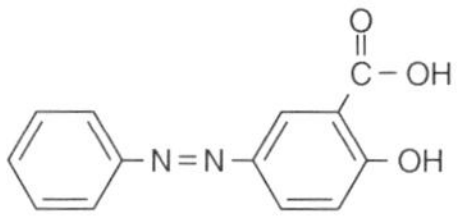

5 gm

S92,232-3
METHYL 4-(2-HYDROXY-5-METHYLPHENYLAZO)BENZOATE

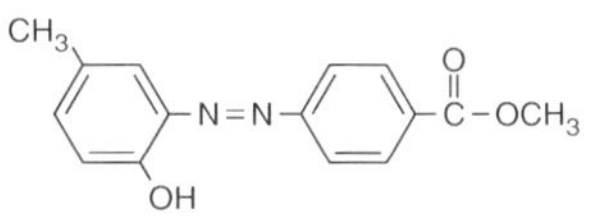

250 mg

S43,779-4
LIGNIN PINK

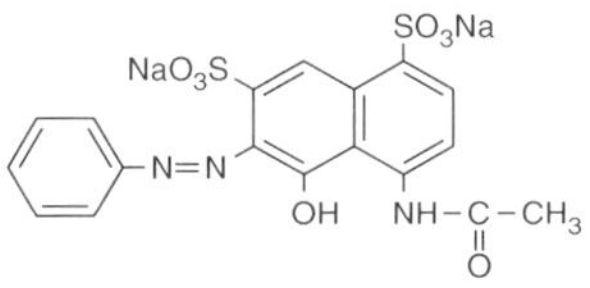

250 mg

S63,263-5
4-(2,5-DICHLOROPHENYLAZO)-4'-FLUOROSULFONYL-1-HYDROXY-2-NAPHTHANILIDE

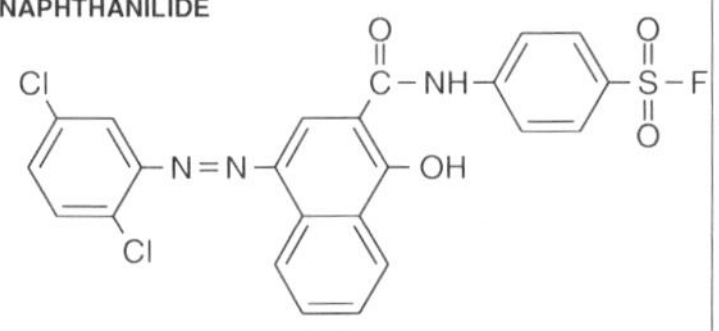

250 mg

S92,240-4
DIRECT RED 26 (C.I. 29190)

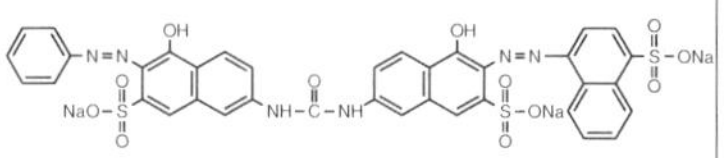

5 gm

S45,033-2
4,5-DIHYDROXY-3-(2-THIAZOLYLAZO)-2,7-NAPHTHALENEDISULFONIC ACID, SODIUM SALT

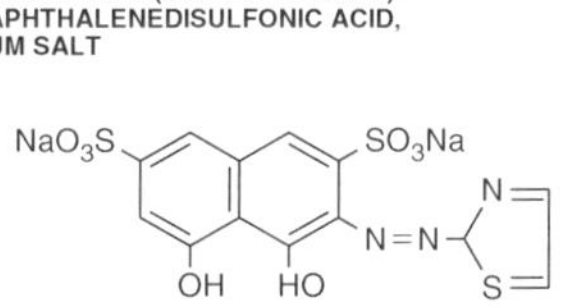

5 gm

S92,360-5
ERIOCHROME RED B (C.I. 18760)

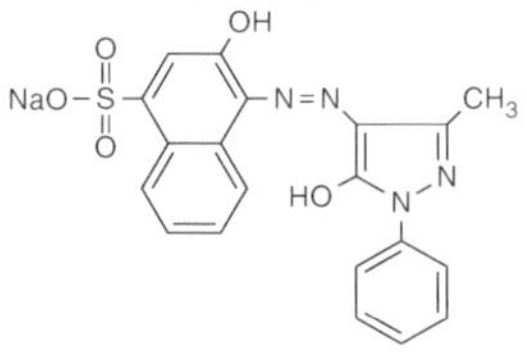

250 mg

S57,043-5
THIAZINE RED R

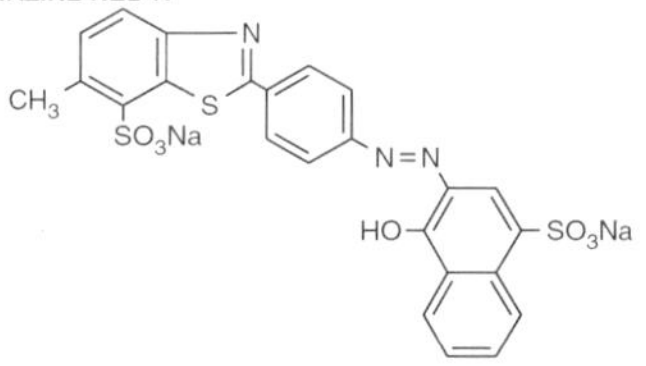

5 gm

S92,112-2
4-(5-HYDROXY-1-PHENYL-1,2,3-TRIAZOL-4-YLAZO)BENZENESULFONIC ACID, SODIUM SALT

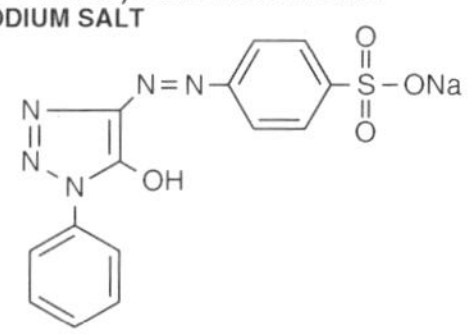

250 mg

S35,064-8
4-(4-HYDROXYPHENYLAZO)-BENZENESULFONIC ACID, DISODIUM SALT DIHYDRATE

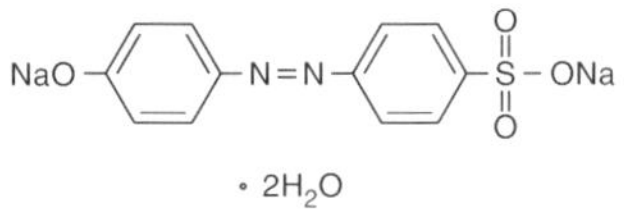

5 gm

S92,923-9
4-(4-HYDROXYPHENYLAZO)-BENZENESULFONIC ACID, SODIUM SALT

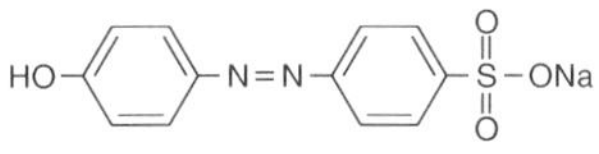

250 mg

S89,479-6
4-(4-HYDROXY-2-METHYLPHENYLAZO)-BENZENESULFONIC ACID, SODIUM SALT

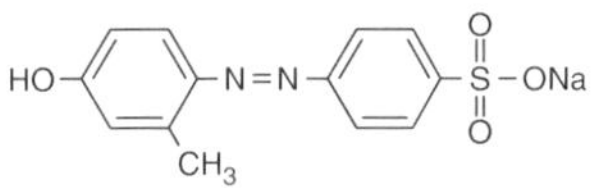

250 mg

S92,040-1
4-(3-METHYL-4-HYDROXYPHENYLAZO)-BENZENESULFONIC ACID

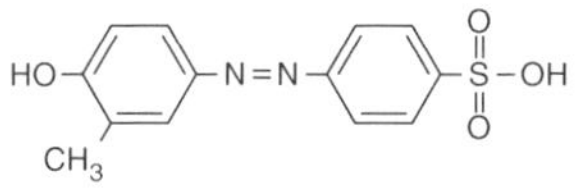

250 mg

S92,039-8
4-(2,6-DIMETHYL-4-HYDROXYPHENYLAZO)-BENZENESULFONIC ACID

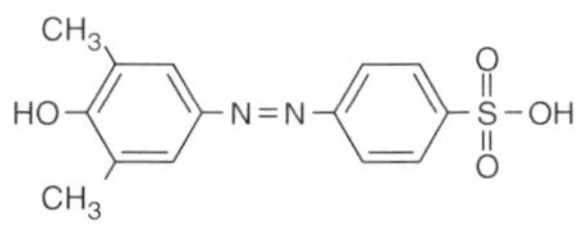

250 mg

S87,337-3
8-HYDROXY-5-(4-SULFOPHENYLAZO)-2-NAPHTHALENESULFONIC ACID

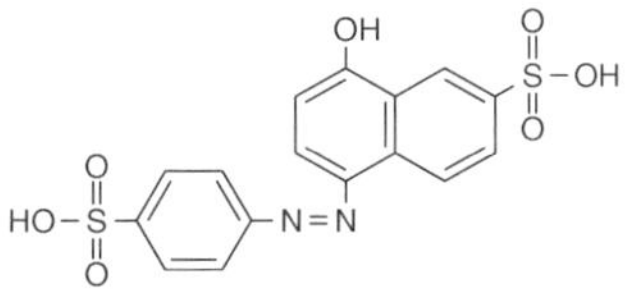

250 mg

S94,486-6
4-(4-HYDROXY-1-NAPHTHYLAZO)-BENZENESULFONIC ACID

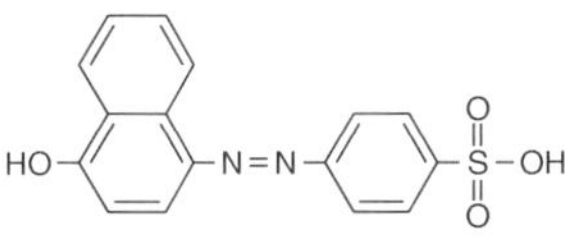

250 mg

S60,267-1
ORANGE I (C.I. 14600)

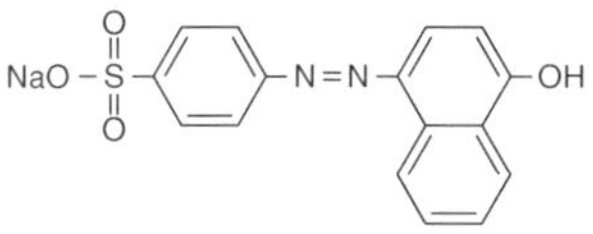

5 gm

S40,780-1
4-(2,4-DIHYDROXYPHENYLAZO)-3-HYDROXY-1-NAPHTHALENESULFONIC ACID

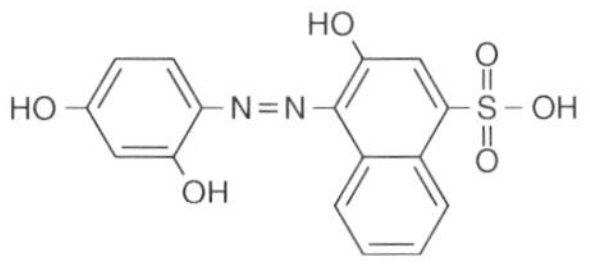

5 gm

S87,441-8
8-HYDROXY-5-(1-NAPHTHYLAZO)-2-NAPHTHALENESULFONIC ACID, SODIUM SALT

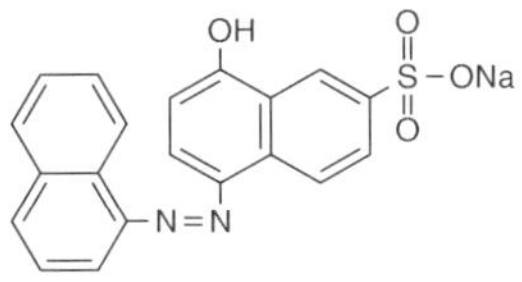

250 mg

S43,283-0
3,6-BIS(4-SULFO-1-NAPHTHYLAZO)-4,5-DIHYDROXY-2,7-NAPHTHALENE DISULFONIC ACID, SODIUM SALT

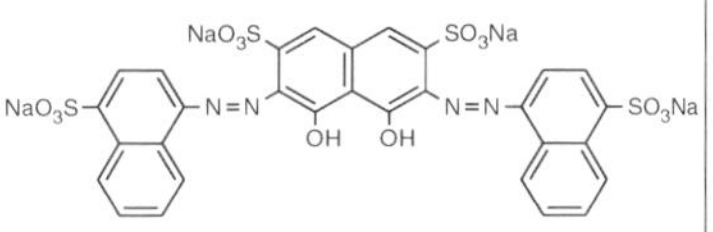

5 gm

S94,898-5
1-(1-METHYL-2-BENZIMIDAZOLYLAZO)-2-NAPTHOL

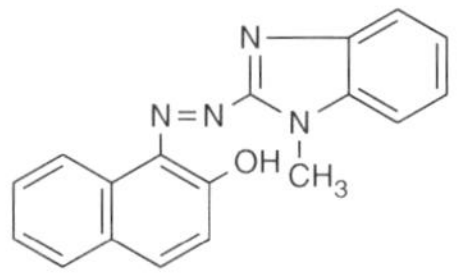

250 mg

S92,713-9
2-(PHENYLAZO)PHENOL

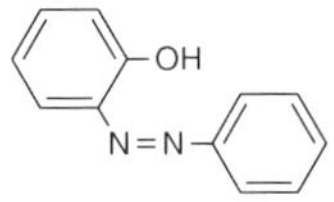

250 mg

S92,712-0
4-METHYL-2-(PHENYLAZO)PHENOL

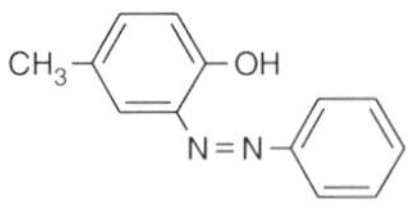

250 mg

S92,924-7
2-METHYL-4-(PHENYLAZO)PHENOL

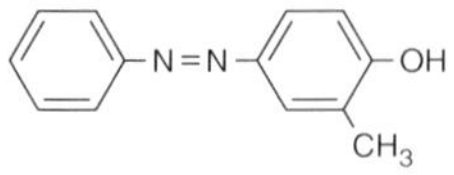

250 mg

S92,925-5
3-METHYL-4-(PHENYLAZO)PHENOL

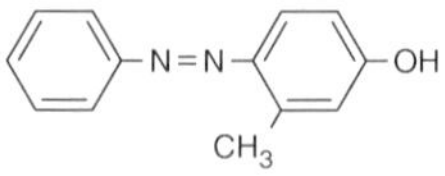

250 mg

S59,617-5
4-(PHENYLAZO)-2,6-DICHLOROPHENOL

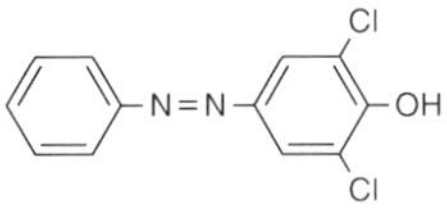

5 gm

S84,689-9
4-(PHENYLAZO)-1-NAPHTHOL, POTASSIUM SALT

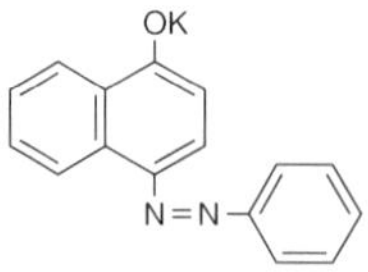

250 mg

S94,483-1
4-(PHENYLAZO)-1-NAPHTHOL

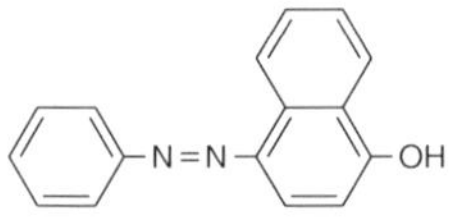

100 mg

S92,929-8
4-(4-TERT-BUTYLPHENYLAZO)PHENOL

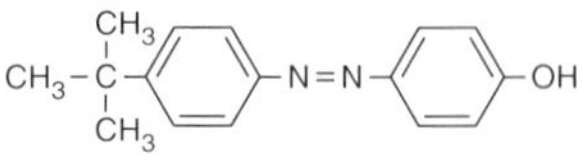

250 mg

S81,369-9
1,1'-(1,1'-BIPHENYL-4,4'-DIYLBISAZO)BIS[6-(TRIMETHYLSILYL)-2-NAPHTHOL]

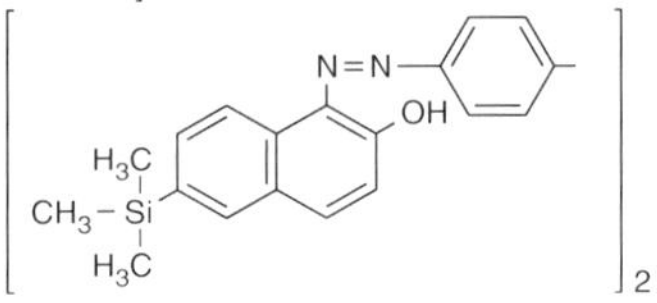

250 mg

S92,948-4
2'-(2,6-DICYANO-4-NITROPHENYLAZO)-5-(N-[2-(2-ETHOXYETHOXY)ETHYL]-N-ETHYLAMINO)ACETANILIDE

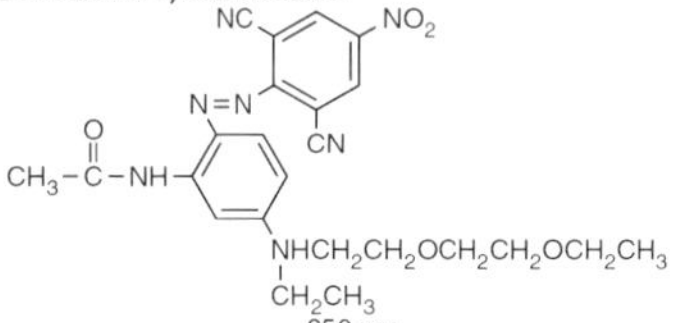

250 mg

S92,370-2
5,6-DIHYDRO-7-(4-METHOXYPHENYLAZO)-3-(4-METHOXYPHENYLTHIO)-4H-1,2-BENZODITHIOLE

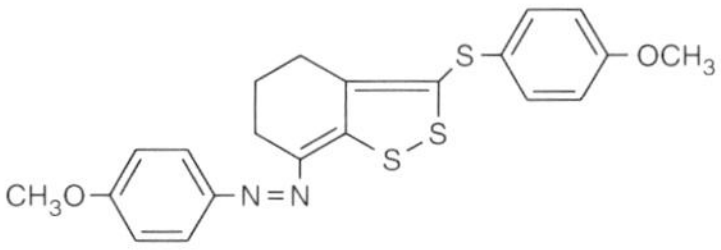

250 mg

S93,308-2
N-(4-METHOXYBENZYLIDENE)-N-[4-(4-METHOXYPHENYLAZO)-1-NAPHTHYL]AMINE

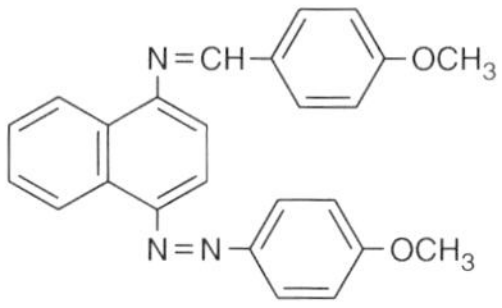

250 mg

S94,202-2
4-DECYLOXYPHENYLDIAZONIUM CHLORIDE, ZINC CHLORIDE DOUBLE SALT

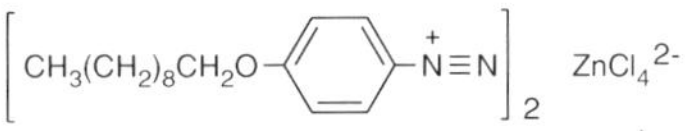

250 mg

S93,004-0
2-[3,5-DIMETHOXY-4-(4-NITROPHENYLAZO)PHENYL]ISOINDOLINE

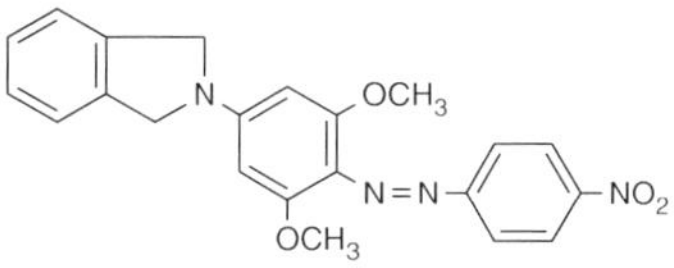

250 mg

S92,944-1
2-[3-METHOXY-4-(4-METHOXYPHENYLAZO)-PHENYL]ISOINDOLINE

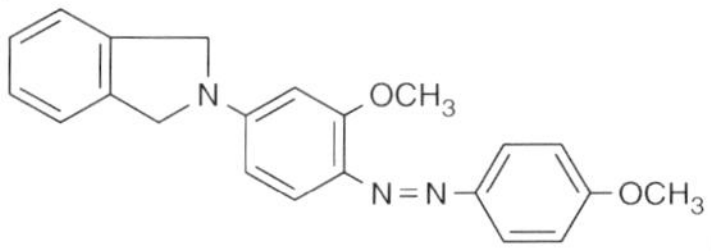

250 mg

S90,981-5
4,4'-DIMETHOXYAZOBENZENE N-OXIDE

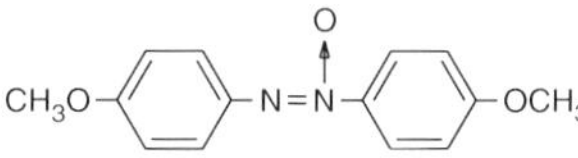

250 mg

S93,158-6
3-AMINO-3-MERCAPTO-2-(4-METHOXYPHENYLAZO)ACRYLONITRILE

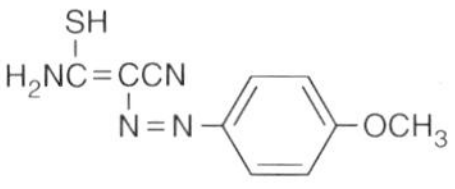

250 mg

S34,286-6
2'-BUTOXY-2,6-DIAMINO-3,5'-AZOPYRIDINE

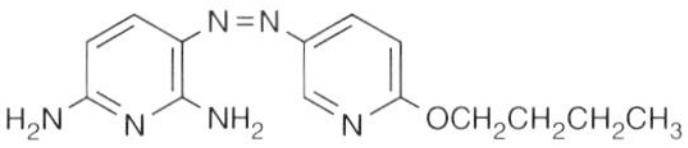

5 gm

S92,284-6
4'-AMINO-2-METHOXY-3'-METHYL-4-NITROAZOBENZENE

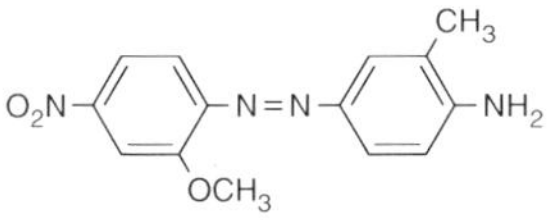

250 mg

S80,510-6
4-AMINO-3-METHOXYAZOBENZENE

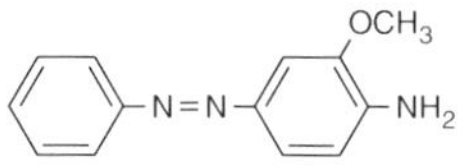

250 mg

S94,416-5
DISPERSE RED 31 (C.I. 11250)

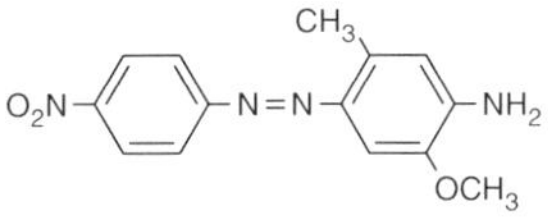

250 mg

S92,927-1
4,4'-DIAMINO-5-METHOXY-2-METHYLAZOBENZENE

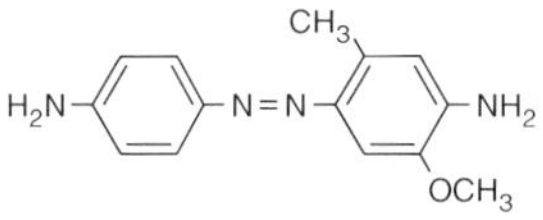

250 mg

S42,206-1
2,6-DIAMINO-3-[4-(2-DIETHYLAMINOETHOXY)PHENYLAZO]-PYRIDINE

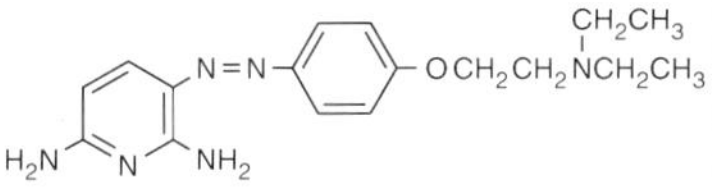

5 gm

S42,239-8
2,6-DIAMINO-3-[4-(2-DIETHYLAMINOETHOXY)PHENYLAZO]-PYRIDINE METHANESULFONATE

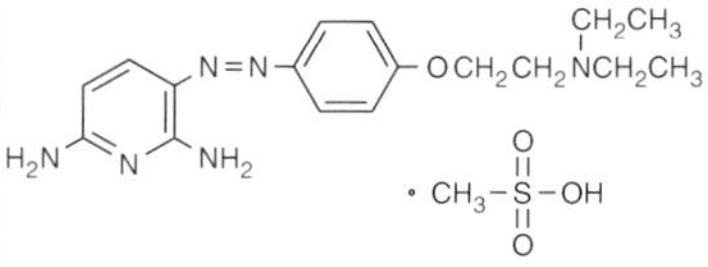

5 gm

S79,831-2
2,6-DINITRO-4'-ETHOXYAZOBENZENE

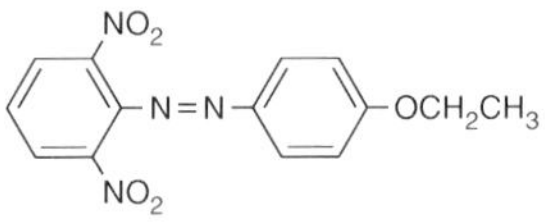

250 mg

S93,311-2
N-(4-METHOXYBENZYLIDENE)-N-[1-(4-METHOXYPHENYLAZO)-2-NAPHTHYL]-AMINE TETRAFLUOROBORATE

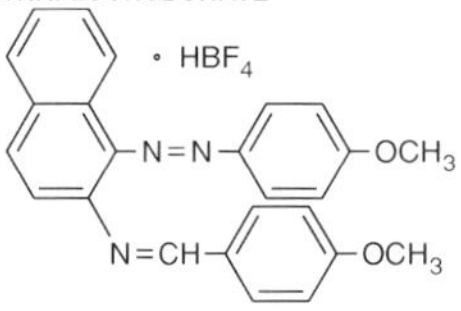

250 mg

S93,049-0
3-METHOXY-4'-(4-NITROPHENYLAZO)-DIPHENYLAMINE

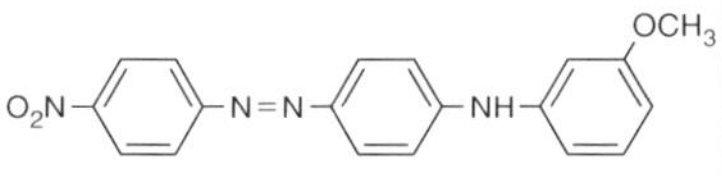

250 mg

S92,295-1
2-METHOXY-4'-(DIMETHYLAMINO)-4-NITROAZOBENZENE

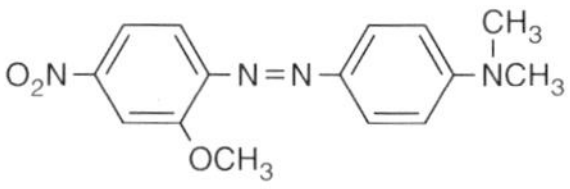

250 mg

S88,286-0
4-(DIMETHYLAMINO)-3'-METHOXYAZOBENZENE

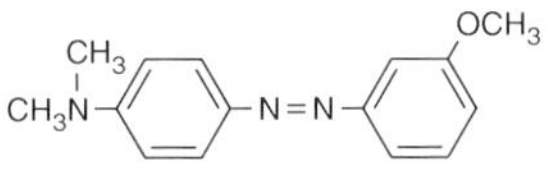

250 mg

S88,284-4
4-(DIMETHYLAMINO)-4'-ETHOXYAZOBENZENE

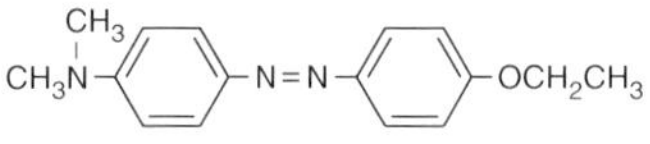

250 mg

S88,285-2
4-(DIMETHYLAMINO)-3'-ETHOXYAZOBENZENE

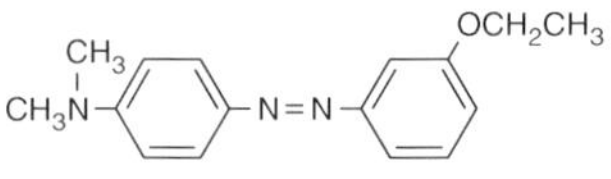

250 mg

S92,937-9
4-[4-(2-ISOINDOLINYL)-3-METHOXYPHENYLAZO]BENZONITRILE

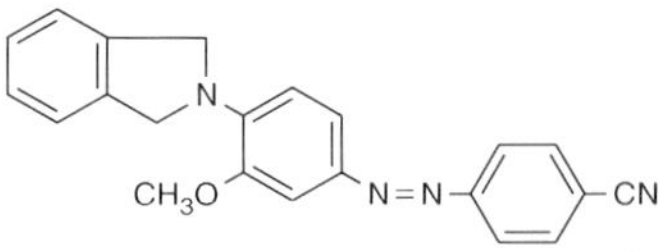

250 mg

S80,524-6
4-(DIMETHYLAMINO)-4'-METHOXYAZOBENZENE
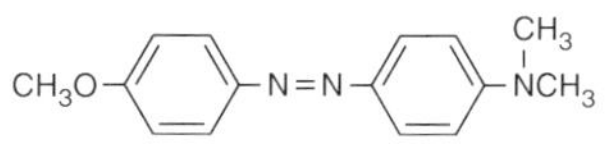
250 mg

S92,921-2
4-METHOXY-4'-NITROAZOBENZENE
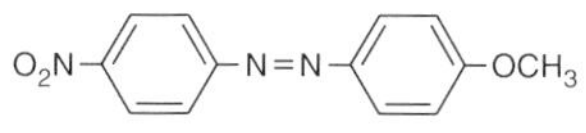
250 mg

S93,014-8
2-[3-METHOXY-4-(4-NITROPHENYLAZO)-PHENYL]ISOINDOLINE
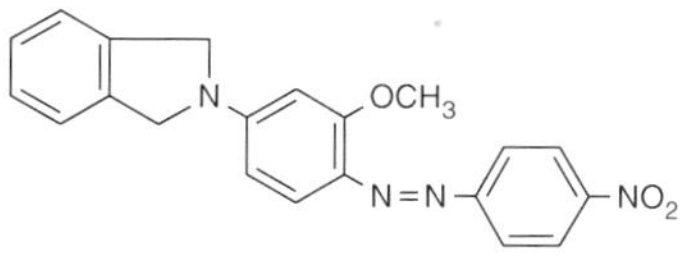
250 mg

S93,008-3
2-[3-METHOXY-4-(PHENYLAZO)PHENYL]-ISOINDOLINE
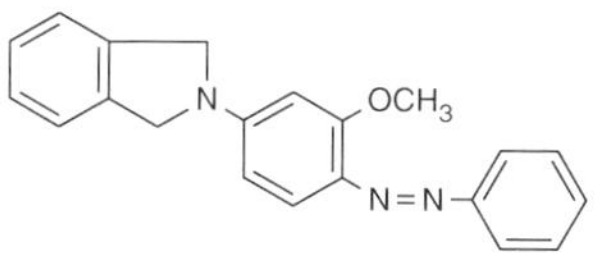
250 mg

S93,011-3
2-[3-METHOXY-4-(P-TOLYLAZO)PHENYL]-ISOINDOLINE
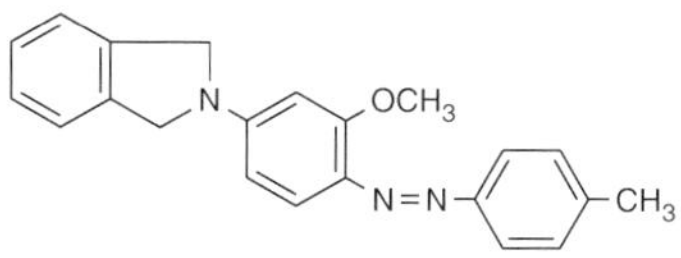
250 mg

S93,017-2
2-[4-(4-METHOXYPHENYLAZO)-3-METHYLPHENYL]ISOINDOLINE
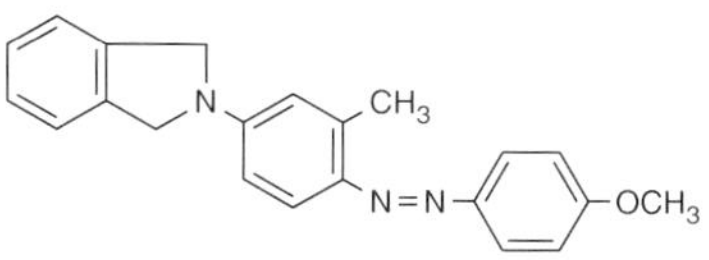
250 mg

S40,738-0
4-DIAZO-4'-METHOXYDIPHENYLAMINE CHLORIDE
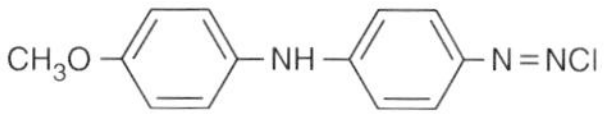
5 gm

S88,287-9
4-(DIMETHYLAMINO)-2-METHOXYAZOBENZENE
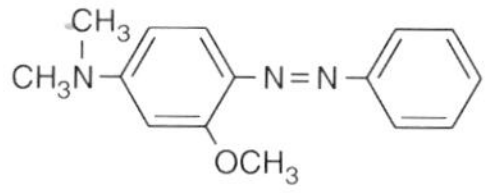
250 mg

S59,972-7
4,4'-BIS(4-ETHOXYPHENYLAZO)-3,3'-DISULFOSTILBENE DISODIUM SALT
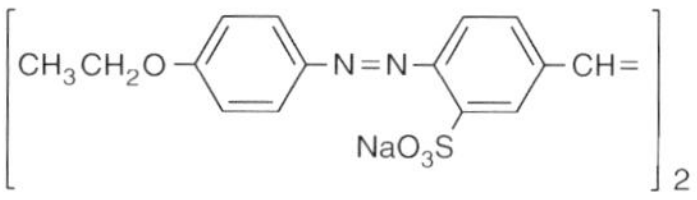
250 mg

S93,045-8
2-METHOXY-1-(4-NITROPHENYLAZO)-NAPHTHALENE
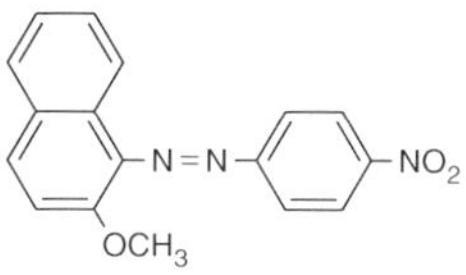
250 mg

S93,001-6
4'-[4-(2-ISOINDOLINYL)-3-METHOXYPHENYLAZO]ACETOPHENONE
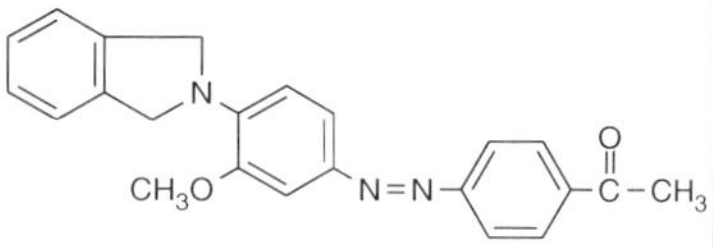
250 mg

S94,482-3
4-(4-METHOXYPHENYLAZO)-BENZENESULFONIC ACID, SODIUM SALT
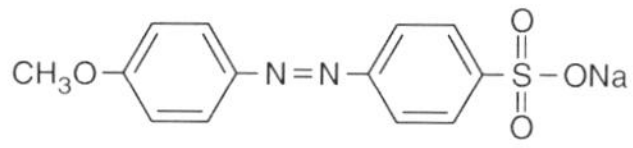
50 mg

S93,195-0
4-(DECYLOXY)AZOBENZENE
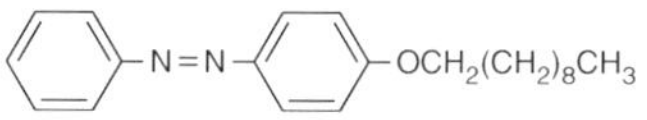
250 mg

S94,924-8
4-METHOXY-4'-PHENYLAZOBENZENE
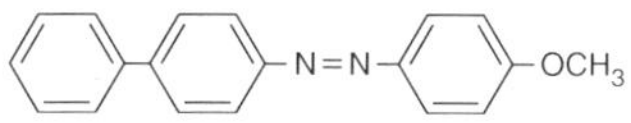
250 mg

S94,487-4
1-METHOXY-4-(PHENYLAZO)NAPHTHALENE
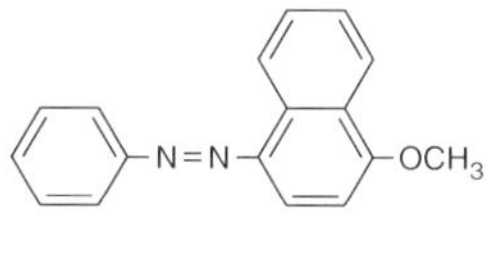
5 gm

S94,488-2
1-ETHOXY-4-(PHENYLAZO)NAPHTHALENE
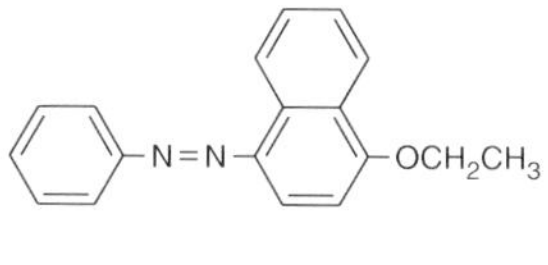
250 mg

S53,438-2
2,2'-AZOXYANISOLE
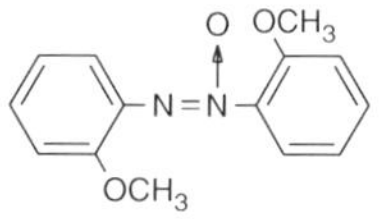
5 gm

S41,883-8
4-METHOXYAZOBENZENE
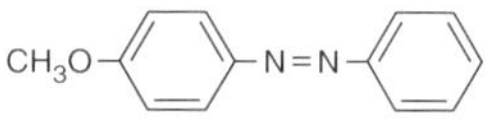
5 gm

S39,234-0
4,4'-AZOXYDIPHENETOLE
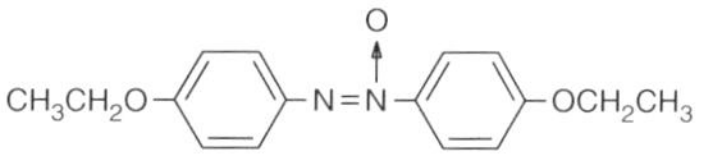
5 gm

S62,296-6
3,4'-DIMETHYL-4-ETHOXYAZOBENZENE
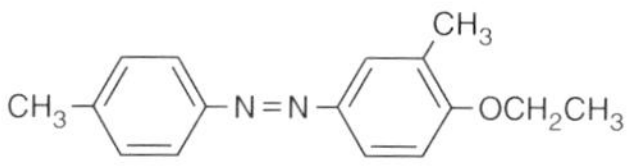
250 mg

S79,834-7
4-(4-METHOXYPHENYLAZO)-2,3,5,6-TETRAFLUOROPYRIDINE

250 mg

S71,575-1
1-[4-(METHYLTHIO)PHENYL]-3-[2-(P-TOLYLAZO)PHENYL]UREA
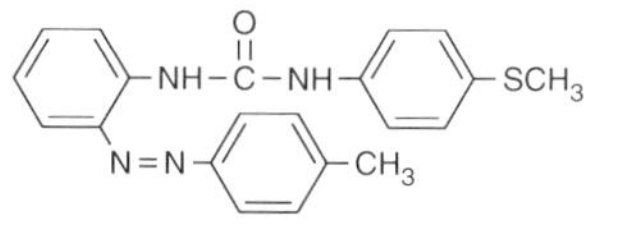
250 mg

S79,797-9
1-[4-(METHYLTHIO)PHENYL]-3-[5-METHYL-2-(P-TOLYLAZO)PHENYL]UREA
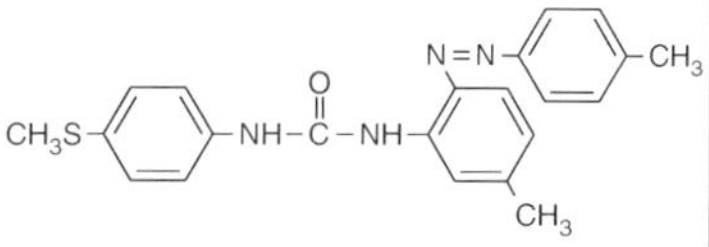
250 mg

S79,423-6
1-[4-(METHYLTHIO)PHENYL]-3-[4-(PHENYLAZO)PHENYL]UREA
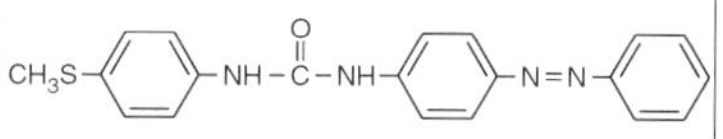
250 mg

S68,344-2
2,4-DIAMINOAZOBENZENE HYDROCHLORIDE
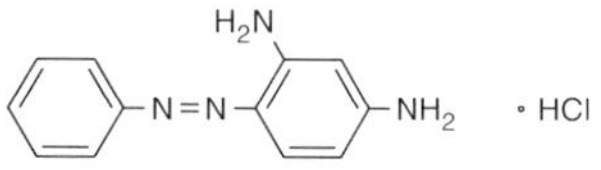
5 gm

S75,423-4
4-AMINO-3-NITROAZOBENZENE
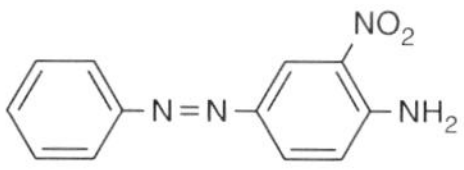
250 mg

S46,048-6
BISMARCK BROWN B
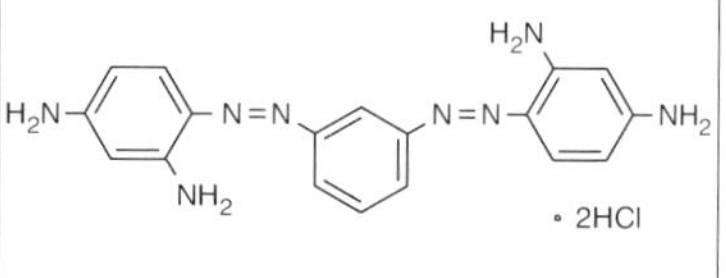
5 gm

S93,845-9
5-AMINO-3-METHYL-4-(4-NITROPHENYLAZO)-1-PHENYL-1H-PYRAZOLE
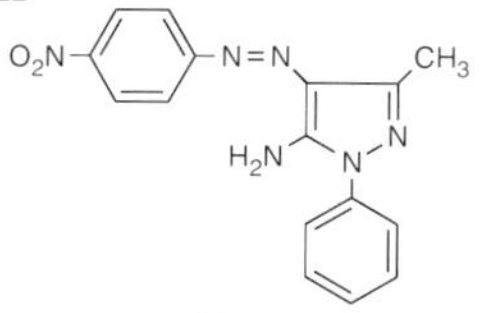
250 mg

S56,062-6
2,6-DIAMINO-3-(PHENYLAZO)PYRIDINE HYDROCHLORIDE
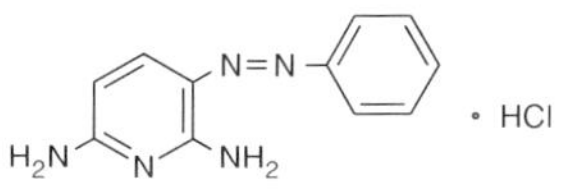
5 gm

S62,779-8
2,6-DIAMINO-3-PHENYLAZOPYRIDINE ETHANEDISULFONATE
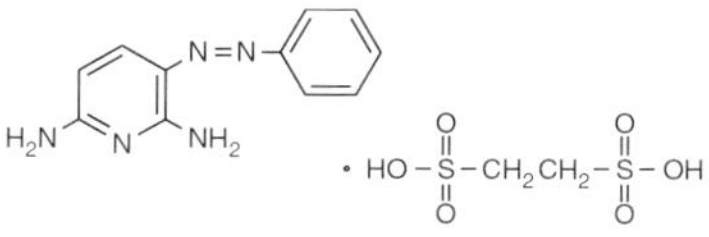
250 mg

S80,254-9
2,6-DIAMINO-3-(PHENYLAZO)PYRIDINE
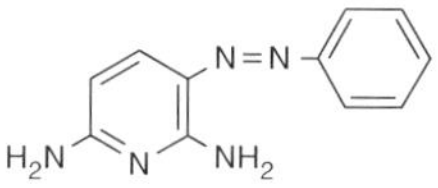
250 mg

S93,844-0
5-AMINO-3-METHYL-4-(PHENYLAZO)-1-PHENYL-1H-PYRAZOLE
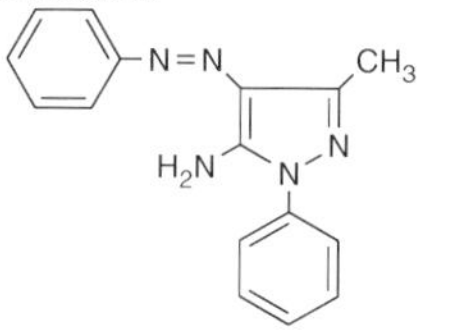
250 mg

S63,681-9
2,6-DIAMINO-3-(PHENYLAZO)PYRIDINE 2-HYDROXYETHANESULFONATE
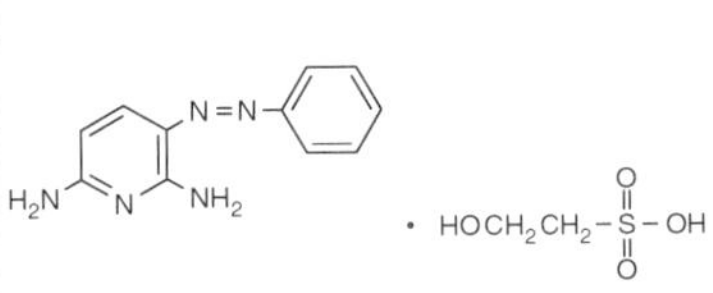
250 mg

S87,133-8
1-(PHENYLAZO)-2-NAPHTHALENEAMINE
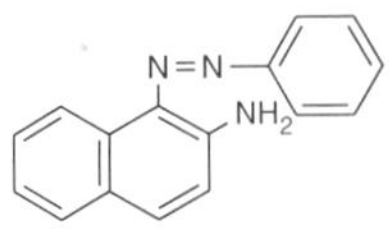
250 mg

S95,413-6
2-(P-TOLYLAZO)-P-TOLUIDINE
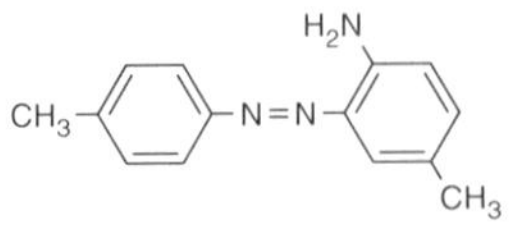
50 mg

S92,933-6
4-(DIETHYLAMINO)-4'-(DIMETHYLAMINO)-2-(UREIDO)AZOBENZENE
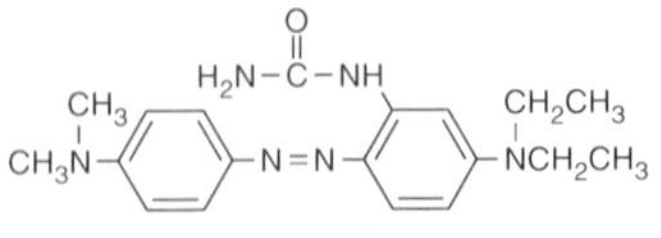
250 mg

S34,084-7
4-(4-AMINOPHENYLAZO)-BENZENESULFONIC ACID, SODIUM SALT
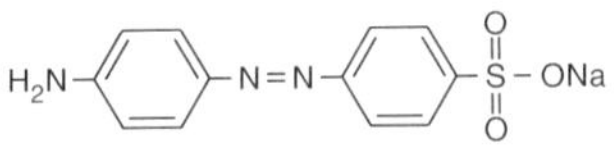
250 mg

S92,091-6
4-AMINO-3,5-DIMETHYLAZOBENZENE
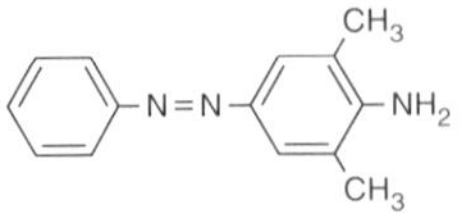
250 mg

S80,528-9
4-(M-TOLYLAZO)-M-TOLUIDINE
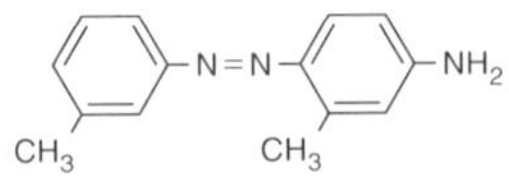
250 mg

S82,910-2
4-(4-AMINO-1-NAPHTHYLAZO)-BENZENESULFONIC ACID, THALLIUM SALT
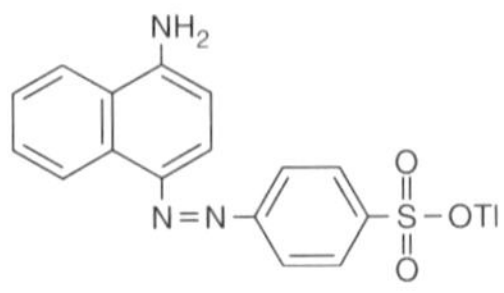
250 mg

S87,081-1
2-(4-AMINO-1-NAPHTHYLAZO)-3-METHYL-4,5,6,7-TETRAHYDROBENZOTHIAZOLIUM CHLORIDE
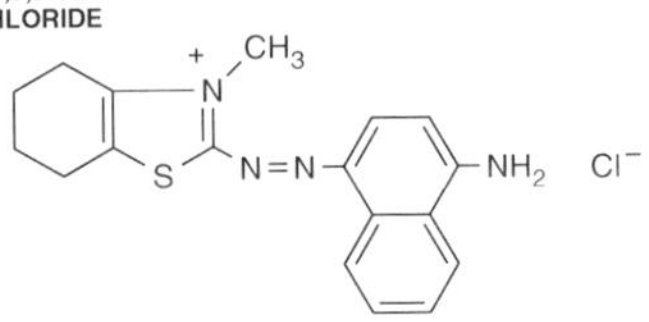
250 mg

S92,703-1
4-AMINO-2-METHYL-4'-NITROAZOBENZENE
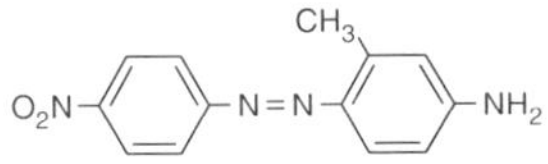
250 mg

S92,701-5
2-AMINO-1-(4-NITROPHENYLAZO)-ANTHRACENE
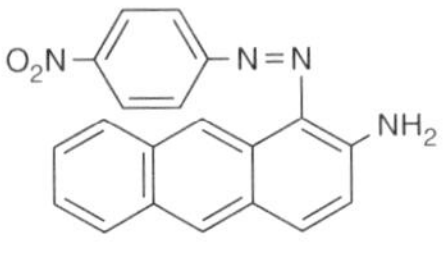
250 mg

S56,762-0
BENZOPURPURIN B
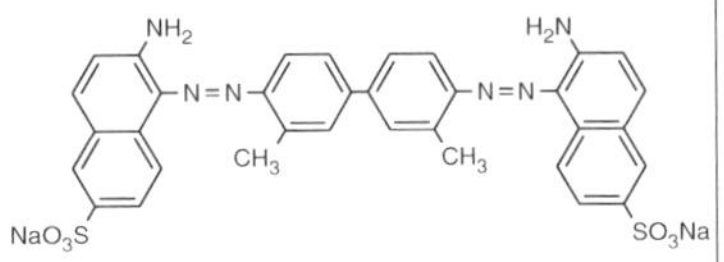
5 gm

S92,706-6
4-AMINO-2,6-DIMETHYLAZOBENZENE
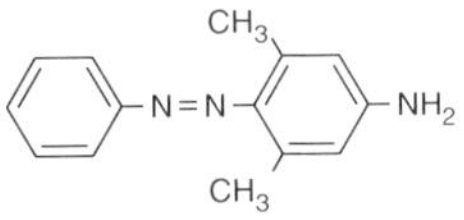
250 mg

S92,314-1
4'-(DIETHYLAMINO)-2,4-DINITROAZOBENZENE
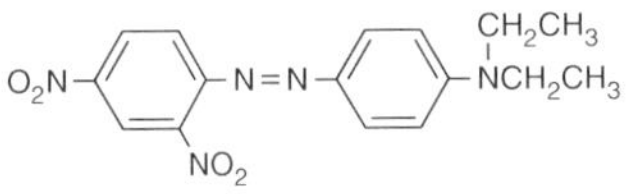
250 mg

S50,202-2
3-[2-(2,5-DICHLOROPHENYLAZO)-5-(DIETHYLAMINO)PHENYLCARBAMOYL]-PROPIONIC ACID
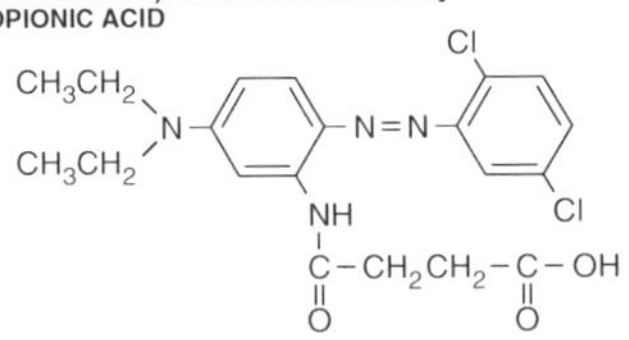
250 mg

S73,149-8
4,4'-DIACETAMIDO-3,3'-DINITROAZOXYBENZENE
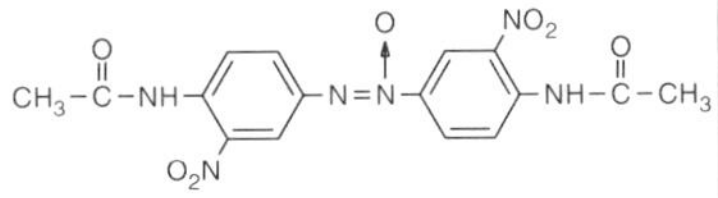
5 gm

S71,795-9
1-(2-METHYL-5-NITROPHENYL)-3-[5-METHYL-2-(4-TOLYLAZO)PHENYL]UREA
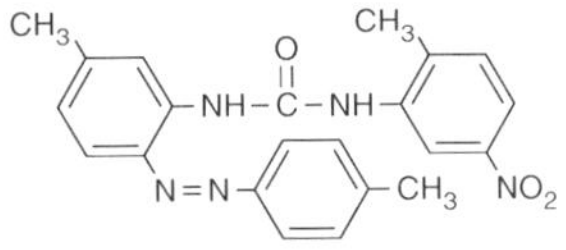
250 mg

S94,412-2
DISPERSE VIOLET 24 (C.I. 11200)
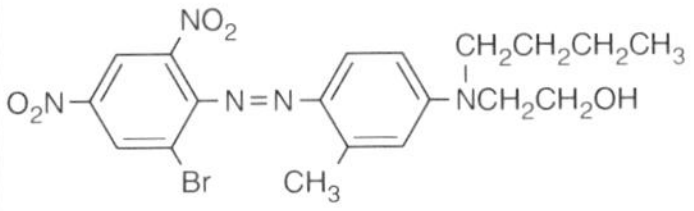
250 mg

S50,555-2
N-[4-(2,5-DICHLOROPHENYLAZO)-PHENYL]-N-ETHYLGLYCINONITRILE
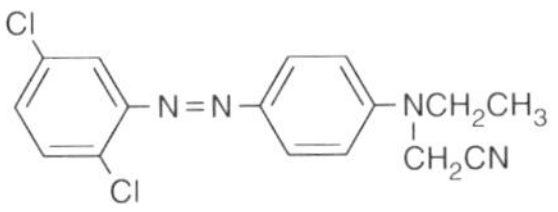
250 mg

S87,240-7
1,1'-AZOBIS[2-(NITROMETHYL)-PIPERIDINE]
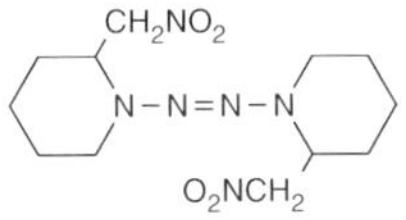
250 mg

S85,480-8
5'-METHYL-2'-[(P-TOLYL)AZO]-1-NAPHTHALENE CARBOXANILIDE
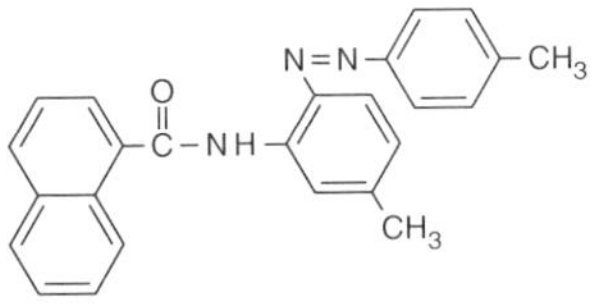
250 mg

S51,079-3
3-METHYL-2-PHENYL-4-PHENYLAZO-3-PYRAZOLIN-5-ONE
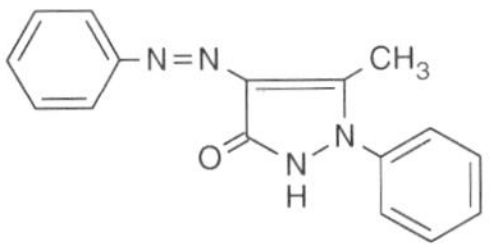
250 mg

S34,953-4
ETHYL 2-(PHENYLAZO)CYANOACETATE
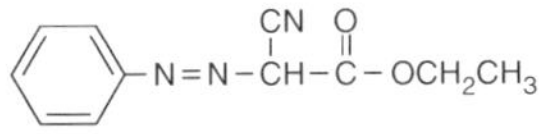
5 gm

S68,721-9
5'-METHYL-2'-(4-TOLYLAZO)-2-BUTENANILIDE
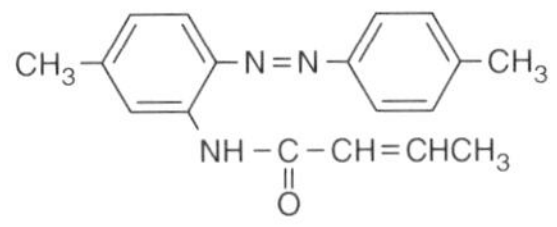
250 mg

S73,896-4
5'-METHYL-2'-(P-TOLYLAZO)-BUTYRANILIDE
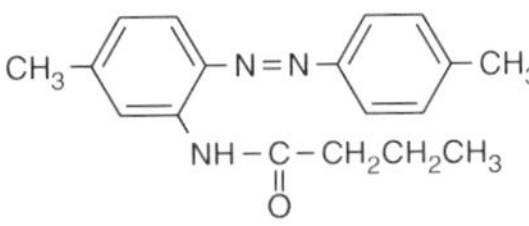
5 gm

S93,396-1
4-(4-ARSONOPHENYLAZO)-3-METHYL-5-PYRAZOLONE
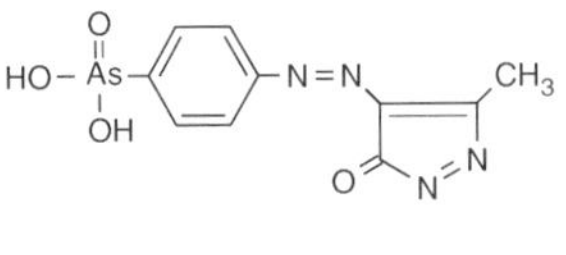
250 mg

S84,742-9
1-[5-METHYL-2-(P-TOLYLAZO)PHENYL]-3-(2,5-XYLYL)UREA
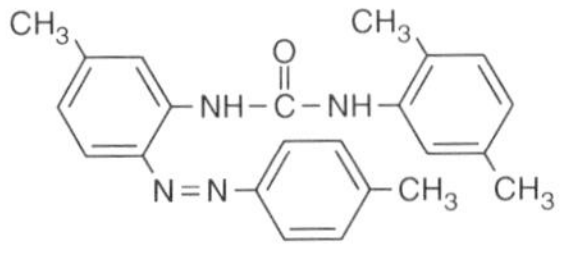
250 mg

S93,843-2
5-CHLORO-3-METHYL-1-PHENYL-4-(4-TOLYLAZO)-1H-PYRAZOLE
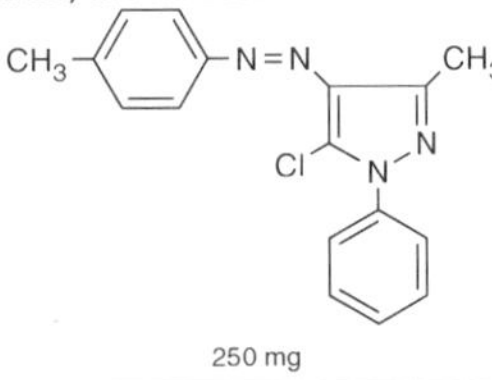
250 mg

S93,310-4
N-BENZYLIDENE-1-[(4-METHOXYPHENYLAZO)-2-NAPHTHYL]AMINE TETRAFLUOROBORATE
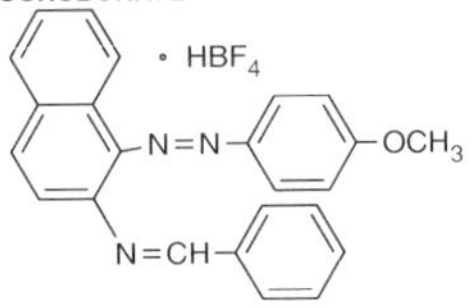
250 mg

S88,288-7
4-(DIMETHYLAMINO)-3'-NITROAZOBENZENE
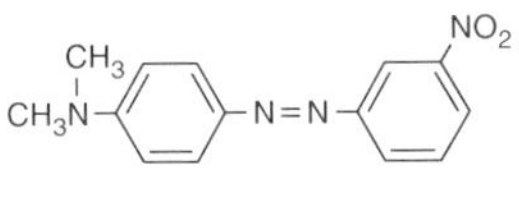
250 mg

S87,475-2
4,4'-AZOBIS(2,6-DIPHENYL-1-METHYLPYRIDINIUM TETRAFLUOROBORATE)
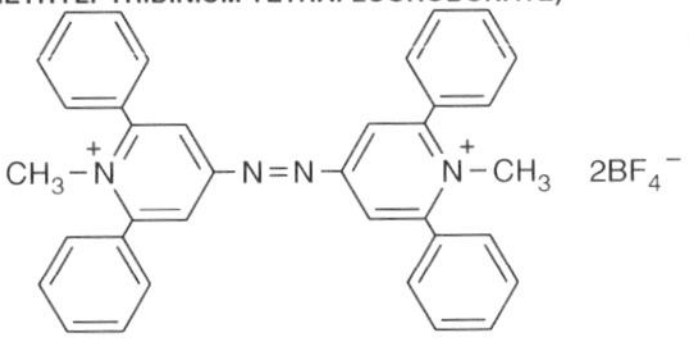
250 mg

S87,474-4
4,4'-AZOBIS(2,6-DIPHENYL-1-METHYLPYRIDINIUM PERCHLORATE)
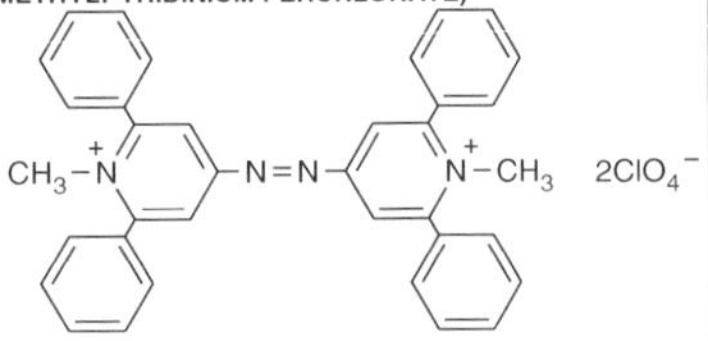
250 mg

S44,674-2
4,4'-AZOBIS(PYRIDINE-1-OXIDE)

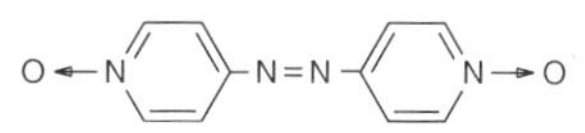

250 mg

S92,919-0
3-NITROAZOBENZENE

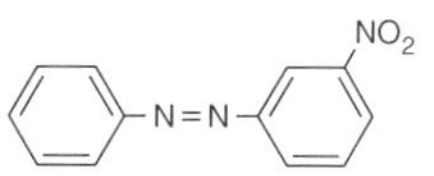

250 mg

S92,920-4
4-METHOXY-3'-NITROAZOBENZENE

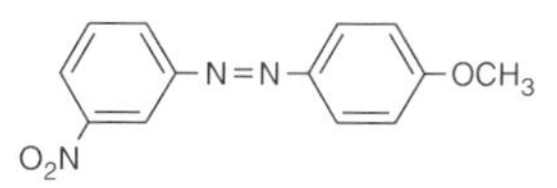

250 mg

S72,233-2
4-(4-QUINOLYLAZO)-BENZENESULFONIC ACID

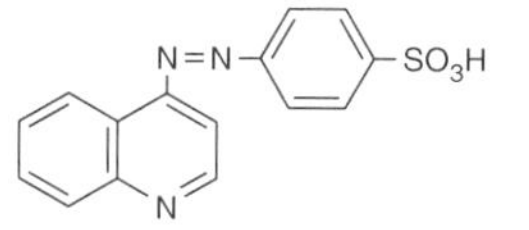

5 gm

S87,544-9
4,4'-AZOBIS(PYRIDINE)

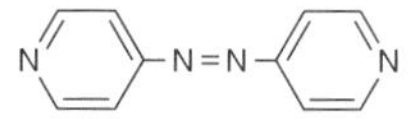

250 mg

S93,063-6
N-ETHYL-N-ISOPROPYL-4-(4-NITROPHENYLAZO)ANILINE

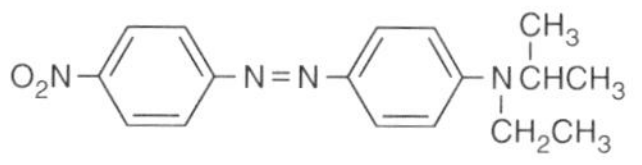

250 mg

S56,765-5
4-[4-(BENZYLAMINO)PHENYLAZO]-BENZENESULFONIC ACID, POTASSIUM SALT

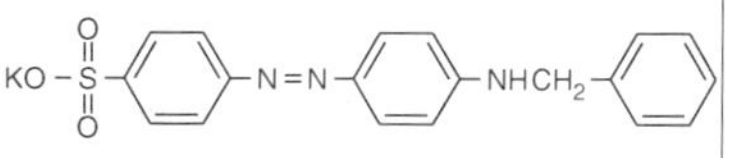

5 gm

S93,048-2
3-METHYL-4'-(4-NITROPHENYLAZO)-DIPHENYLAMINE

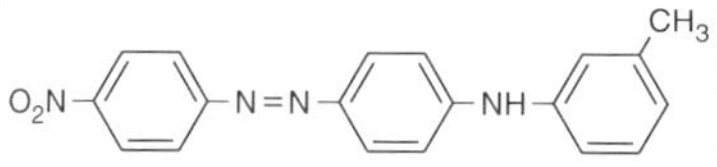

250 mg

S93,050-4
4-(P-TOLYLAZO)DIPHENYLAMINE

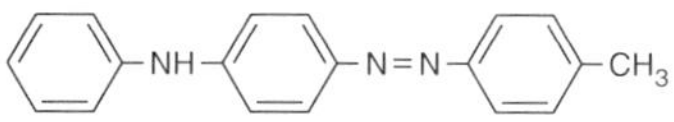

250 mg

S80,445-2
3-METHYL-4'-(METHYLAMINO)AZOBENZENE

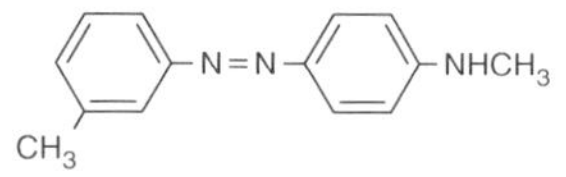

250 mg

S53,156-1
4-(P-CHLOROPHENYLAZO)-N,N-DIMETHYLANILINE

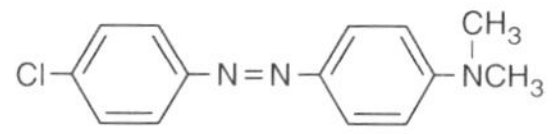

250 mg

S82,621-9
4-(DIMETHYLAMINO)-4'-NITROAZOBENZENE

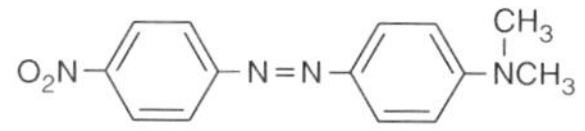

250 mg

S93,076-8
N-ETHYL-N-METHYL-4-(4-NITROPHENYLAZO)ANILINE

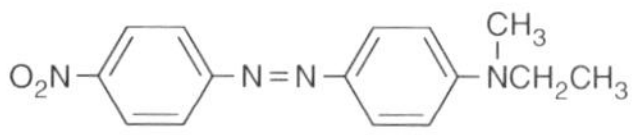

250 mg

S93,091-1
N,N-DIETHYL-4-(4-NITROPHENYLAZO)-ANILINE

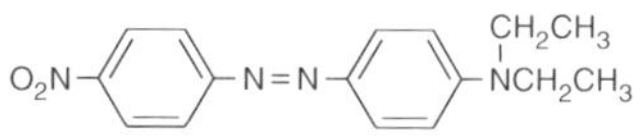

250 mg

S91,966-7
4'-(DIMETHYLAMINO)-3-METHYL-4-NITROAZOBENZENE

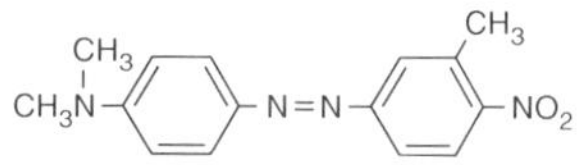

250 mg

S92,705-8
4-(DIMETHYLAMINO)-2-METHYL-4'-NITROAZOBENZENE

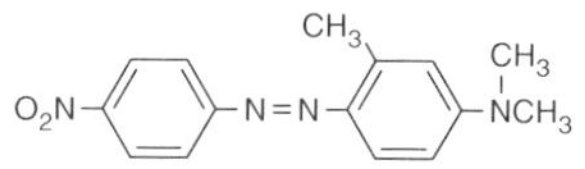

250 mg

S92,707-4
4'-(DIMETHYLAMINO)-2-METHYL-4-NITROAZOBENZENE

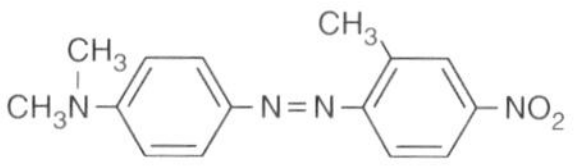

250 mg

S92,708-2
4-(DIMETHYLAMINO)-3-METHYL-4'-NITROAZOBENZENE

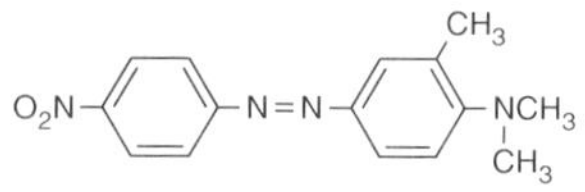

250 mg

S92,709-0
4-(DIMETHYLAMINO)-2,6-DIMETHYL-4'-NITROAZOBENZENE

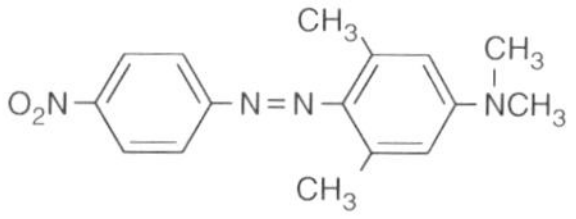

250 mg

S92,321-4
4'-(DIMETHYLAMINO)-2',3-DIMETHYL-4-NITROAZOBENZENE

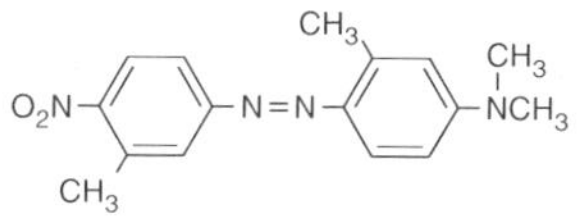

250 mg

S91,975-6
4-(DIMETHYLAMINO)-4'-NITRO-2,2',6-TRIMETHYLAZOBENZENE

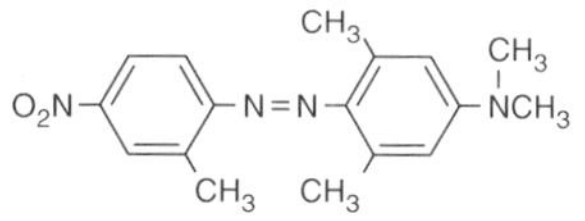

250 mg

S92,118-1
4-(DIMETHYLAMINO)-4'-NITRO-2,3',6-TRIMETHYLAZOBENZENE

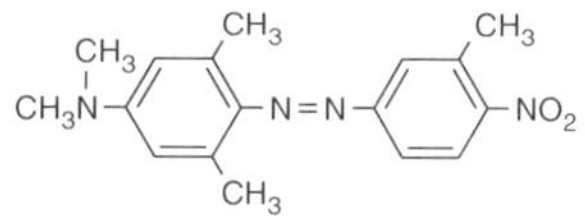

250 mg

S92,120-3
4'-(DIMETHYLAMINO)-4-NITRO-2,2',5,6'-TETRAMETHYLAZOBENZENE

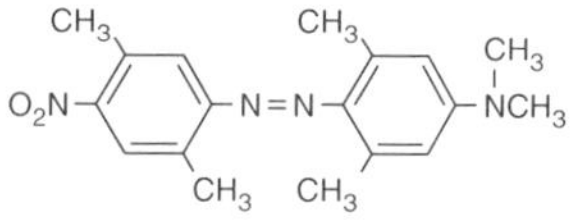

250 mg

S93,094-6
3-(N-[4-(2-CYANO-4-NITROPHENYLAZO)-PHENYL]-N-ETHYLAMINO)PROPIONITRILE

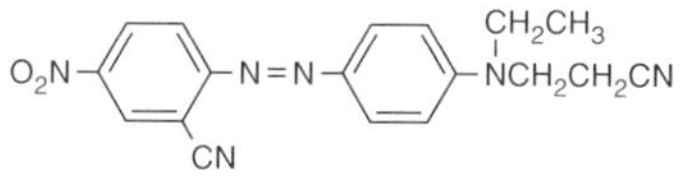

250 mg

S87,074-9
N,N-DIMETHYL-4-(4-NITRO-1-NAPHTHYLAZO)ANILINE

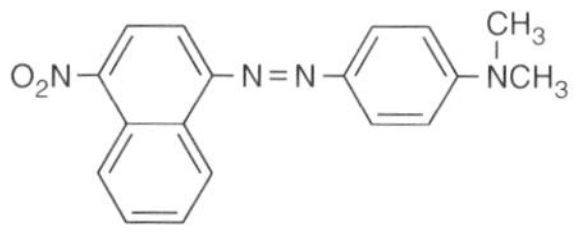

250 mg

S93,069-5
4'-[4-(DIMETHYLAMINO)PHENYLAZO]-ACETOPHENONE

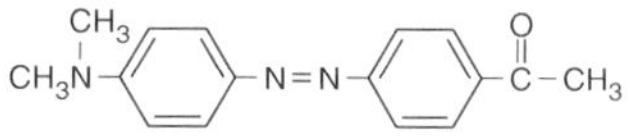

250 mg

S93,070-9
4'-[4-(N-ETHYL-N-METHYLAMINO)-PHENYLAZO]ACETOPHENONE

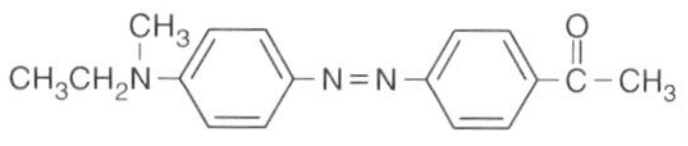

250 mg

S93,090-3
4'-[4-(DIETHYLAMINO)PHENYLAZO]-ACETOPHENONE

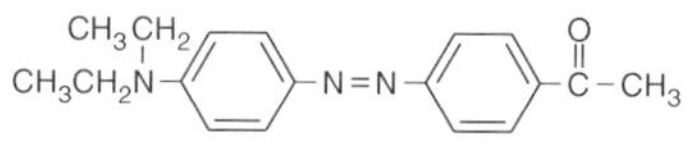

250 mg

S63,150-7
4-[4-(DIMETHYLAMINO)PHENYLAZO]-BENZOIC ACID, SODIUM SALT

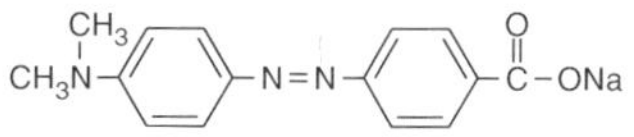

250 mg

S93,068-7
4-[4-(DIMETHYLAMINO)PHENYLAZO]-BENZONITRILE

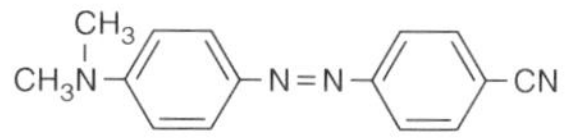

250 mg

S93,071-7
4-[4-(N-ETHYL-N-METHYLAMINO)-PHENYLAZO]BENZONITRILE

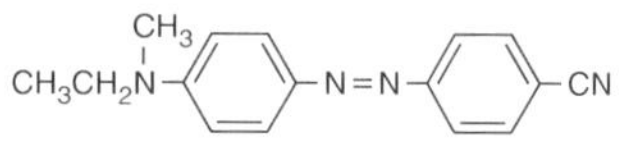

250 mg

S93,064-4
4-[4-(N-METHYL-N-ISOPROPYLAMINO)-PHENYLAZO]BENZONITRILE

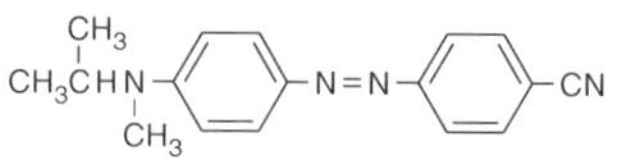

250 mg

S88,282-8
4-(DIMETHYLAMINO)-4'(PHENYLAZO)-AZOBENZENE

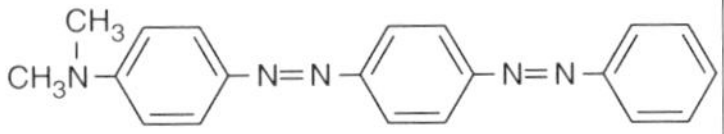

250 mg

S93,073-3
N-ETHYL-N-METHYL-4-(P-TOLYLAZO)-ANILINE

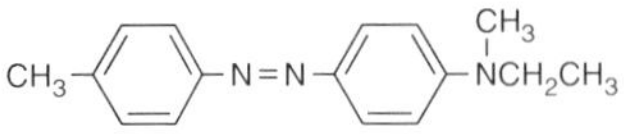

250 mg

S88,283-6
4-(DIMETHYLAMINO)-4'-ETHYL-2-METHYLAZOBENZENE

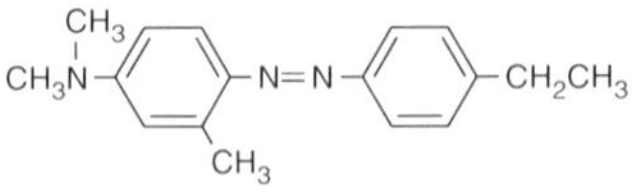

250 mg

S87,073-0
N,N-DIMETHYL-4-(4-NITROPHENYLAZO)-1-NAPHTHYLAMINE

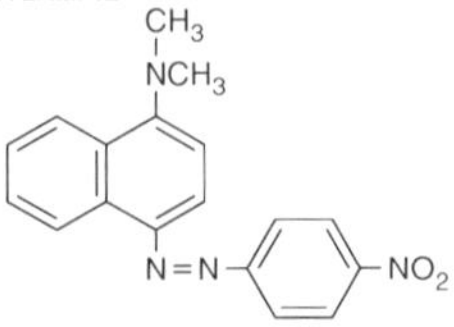

250 mg

S87,087-0
1-METHYL-6-(4-NITROPHENYLAZO)-1,2,3,4-TETRAHYDROBENZO(H)QUINOLINE

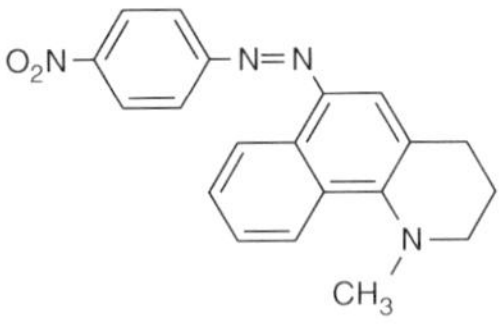

250 mg

S87,046-3
1-METHYL-6-(4-NITROPHENYLAZO)-1,2,3,4-TETRAHYDROQUINOLINE

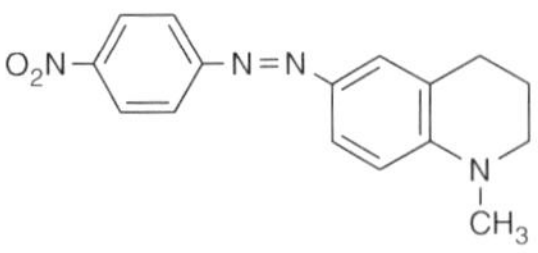

250 mg

S92,941-7
2-[3,5-DIMETHYL-4-(4-NITROPHENYLAZO)PHENYL]ISOINDOLINE

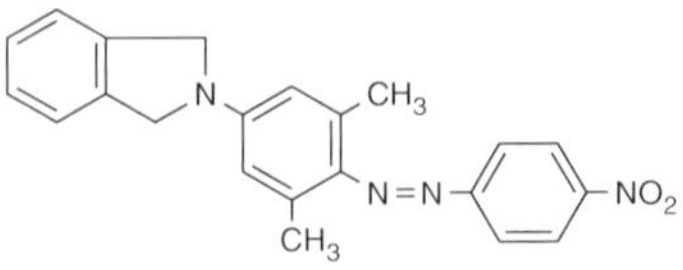

250 mg

S92,938-7
2-[2-METHYL-4-(4-NITROPHENYLAZO)-PHENYL]ISOINDOLINE

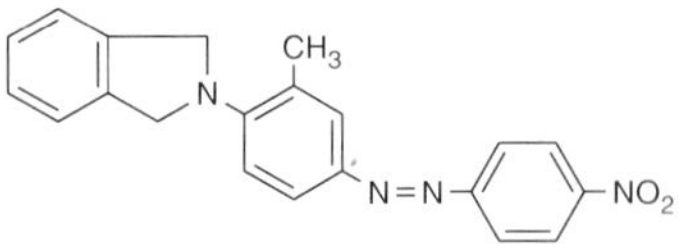

250 mg

S92,946-8
4'-[4-(2-ISOINDOLINYL)PHENYLAZO]-ACETOPHENONE

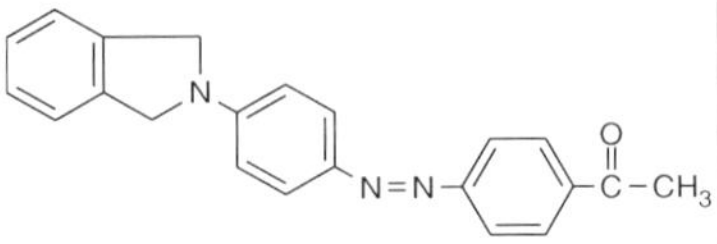

250 mg

S93,054-7
4'-[4-(2,6-DIMETHYLMORPHOLINO)-PHENYLAZO]ACETOPHENONE

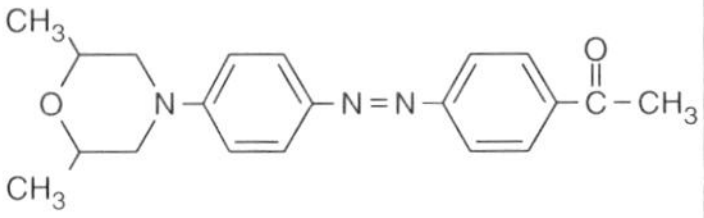

250 mg

S92,947-6
4-[4-(2-ISOINDOLINYL)PHENYLAZO]-BENZONITRILE

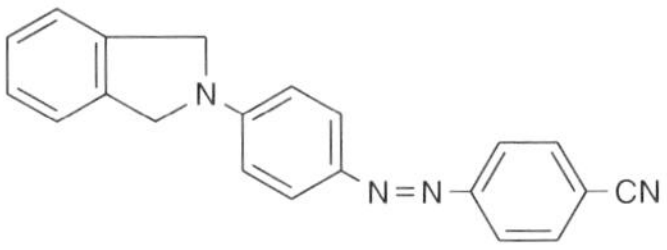

250 mg

S52,745-9
4-[4-(DIETHYLAMINO)PHENYLAZO]-BENZENESULFONIC ACID, SODIUM SALT

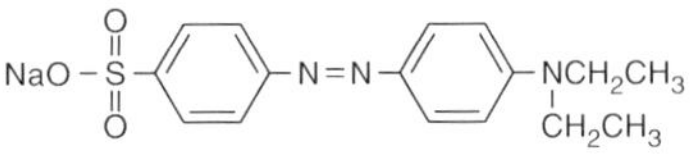

5 gm

S50,997-3
4-[N-BUTYL-4-(2,5-DICHLOROPHENYLAZO)ANILINO]-1-BUTANESULFONIC ACID, SODIUM SALT

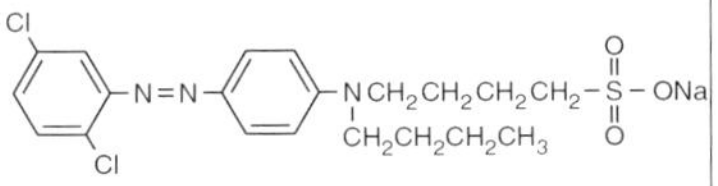

5 gm

S80,520-3
4-(DIMETHYLAMINO)AZOBENZENE N-OXIDE

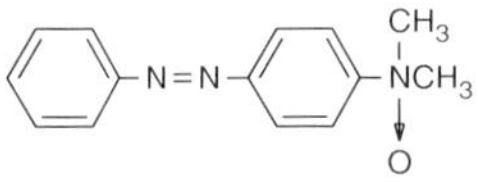

250 mg

S80,518-1
4'-(DIMETHYLAMINO)-AZOXYBENZENE N-OXIDE

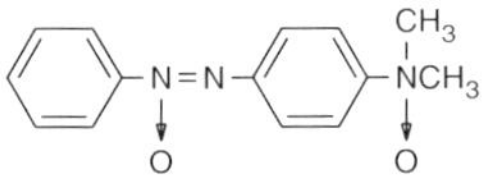

250 mg

S80,527-0
4-(N-ETHYL-N-METHYLAMINO)AZOBENZENE

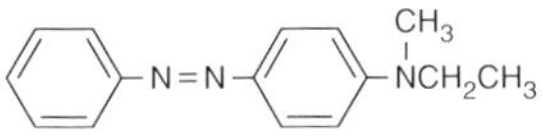

250 mg

S44,151-1
4-DIMETHYLAMINO-2'-METHYLAZOBENZENE

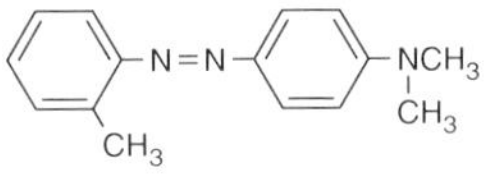

5 gm

S80,566-1
4-DIMETHYLAMINO-3'-METHYLAZOBENZENE

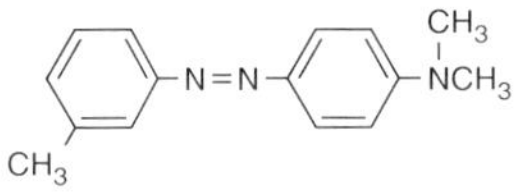

250 mg

S80,500-9
4-(DIMETHYLAMINO)-3'-METHYLAZOBENZENE N-OXIDE

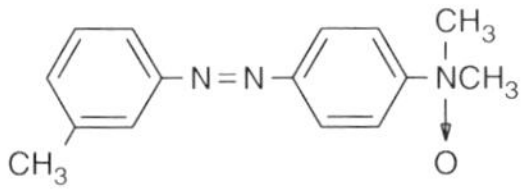

250 mg

S80,479-7
4-DIMETHYLAMINO-4'-ETHYLAZOBENZENE

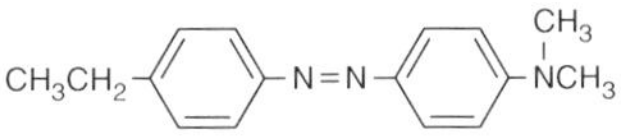

250 mg

S80,501-7
4-(DIETHYLAMINO)-4'-ETHYLAZOBENZENE

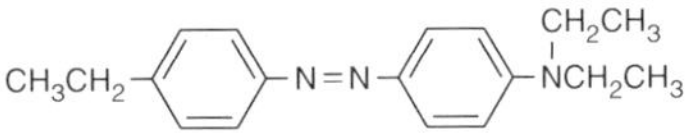

250 mg

S80,526-2
3',4'-DIMETHYL-4-(DIMETHYLAMINO)-AZOBENZENE

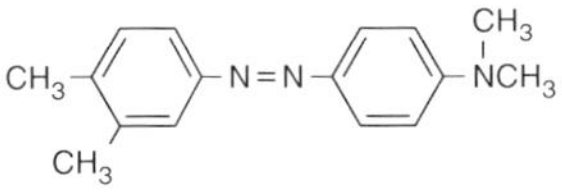

250 mg

S51,365-2
N-[4-(2,5-DICHLOROPHENYLAZO)-PHENYL]-N-METHYLHEXADECYLAMINE

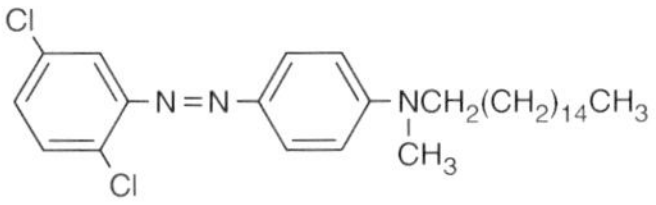

250 mg

S51,146-3
N,N-DIMETHYL-4-(2-NAPHTHYLAZO)-ANILINE

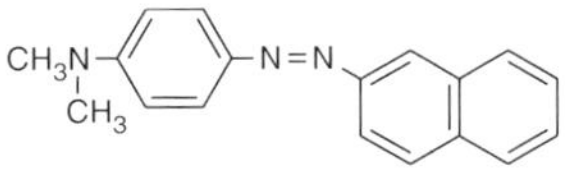

5 gm

S80,525-4
2,3'-DIMETHYL-4-(DIMETHYLAMINO)-AZOBENZENE

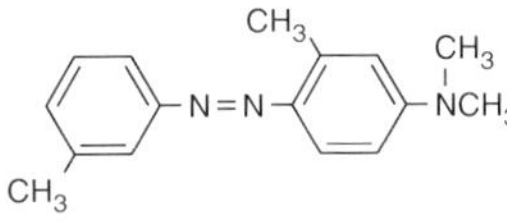

250 mg

S50,745-8
N,N-DIBUTYL-4-(2,5-DICHLOROPHENYLAZO)-3,5-XYLIDINE

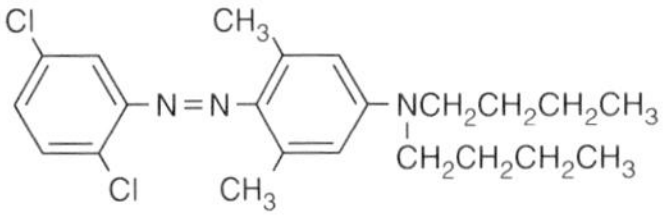

250 mg

S92,174-2
4-(DIMETHYLAMINO)-2',5'-DIMETHYL-4'-NITROAZOBENZENE

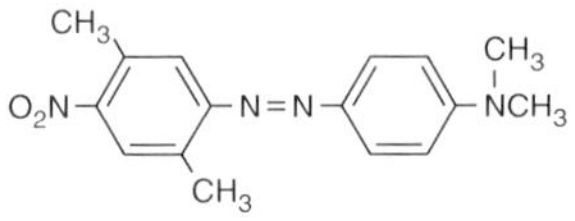

250 mg

S93,022-9
2-[4-(4-NITROPHENYLAZO)PHENYL]-ISOINDOLINE

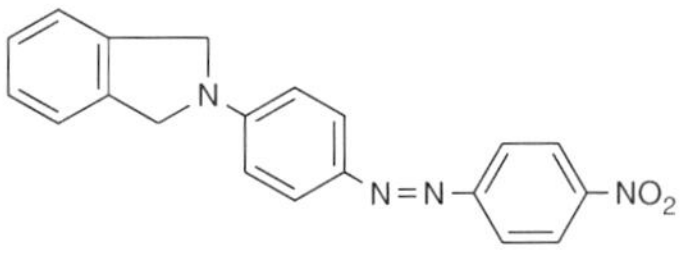

250 mg

S93,024-5
1-[4-(4-NITROPHENYLAZO)PHENYL]-OCTAHYDROAZOCINE

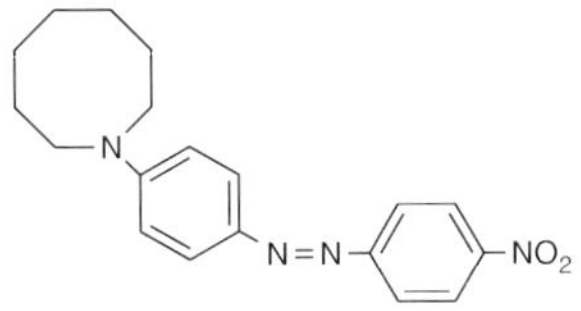

250 mg

S87,028-5
2-(4-NITROPHENYLAZO)QUINOLINE

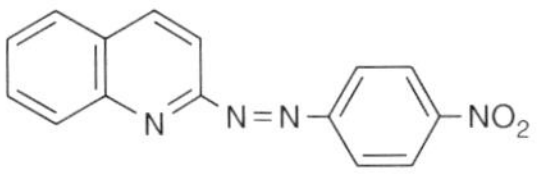

250 mg

S92,003-7
1-(HYDROXYIMINOAMINO)-4-NITROBENZENE, SODIUM SALT

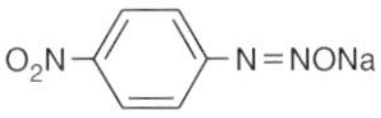

250 mg

S43,639-9
4-(4-NITROPHENYLAZO)-BENZOYL CHLORIDE

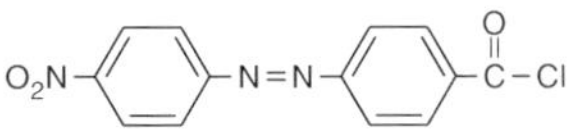

250 mg

S93,056-3
N-[1-ETHYL-6-(4-NITROPHENYLAZO)-1,2,3,4-TETRAHYDRO-2,2,4-TRIMETHYL-7-QUINOLYL]PROPIONAMIDE

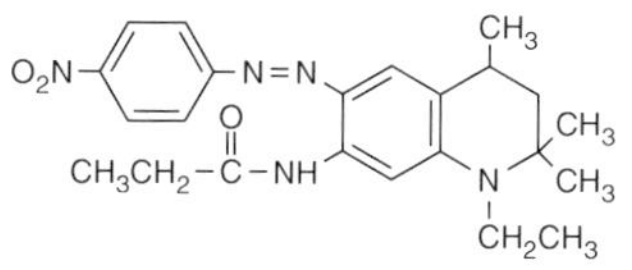

250 mg

S93,038-5
4-METHYL-1-[4-(4-NITROPHENYLAZO)-PHENYL]PIPERIDINE

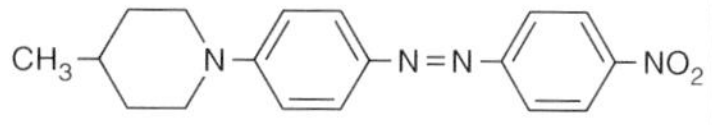

250 mg

S93,042-3
4'-[4-(1-AZIRIDINYL)PHENYLAZO]-ACETOPHENONE

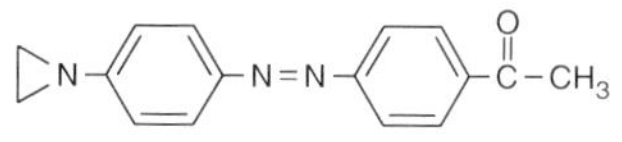

250 mg

S93,005-9
4'-[2,6-DIMETHYL-4-(2-ISOINDOLINYL)-PHENYLAZO]ACETOPHENONE

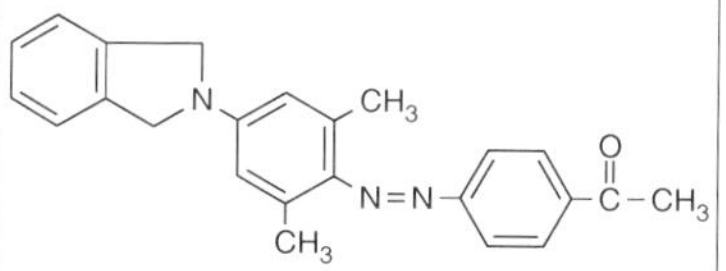

250 mg

S93,025-3
4'-[4-(OCTAHYDRO-1-AZOCINYL)-PHENYLAZO]ACETOPHENONE

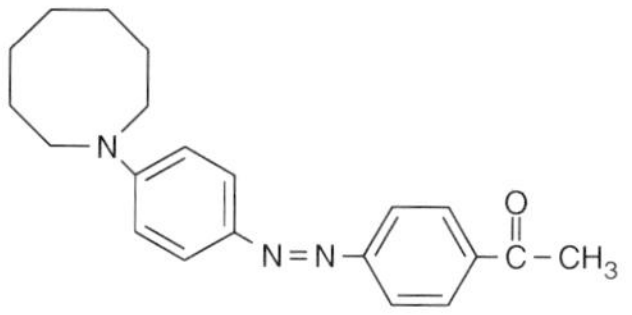

250 mg

S85,492-1
4'-(PHENYLAZO)CINNAMANILIDE

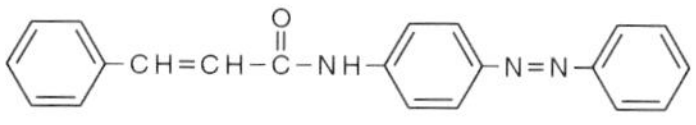

250 mg

S34,531-8
4-DIACETAMIDO-2',3-DIMETHYLAZOBENZENE

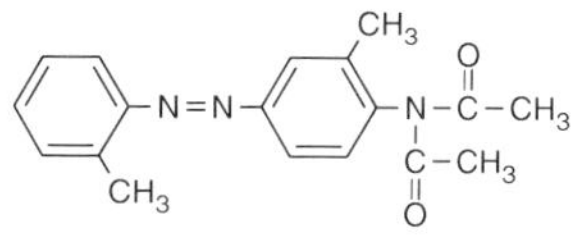

5 gm

S92,940-9
4-[2,6-DIMETHYL-4-(2-ISOINDOLINYL)-PHENYLAZO]BENZONITRILE

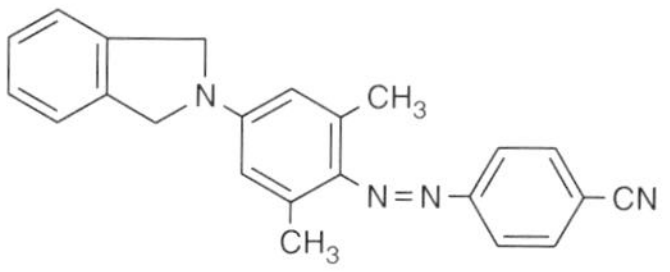

250 mg

S93,040-7
4-[4-(4-METHYLPIPERIDINO)PHENYLAZO]-BENZONITRILE

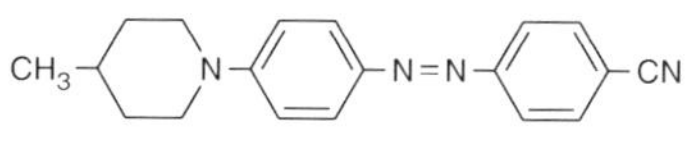

250 mg

S93,031-8
4-[4-(HEXAHYDRO-1H-AZEPIN-1-YL)-PHENYLAZO]BENZONITRILE

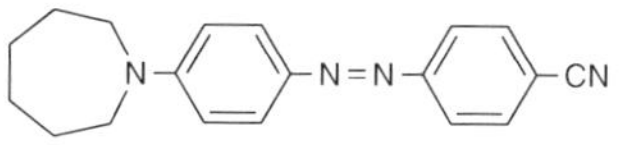

250 mg

S89,501-6
2-[4-(PHENYLAZO)PHENYL]-1-HYDRAZINESULFONIC ACID, 4-METHYLANILINE SALT

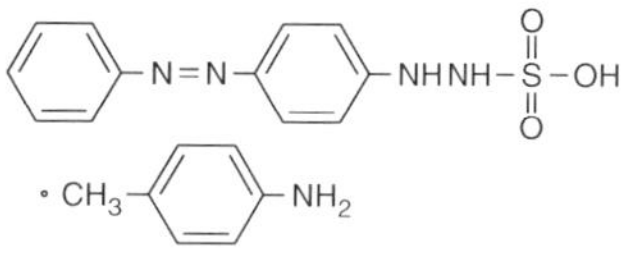

250 mg

S35,446-5
2-[4-(PHENYLAZO)PHENYL]-1-HYDRAZINESULFONIC ACID

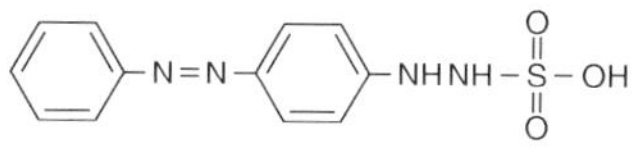

1 gm

S93,007-5
2-[3,5-DIMETHYL-4-(PHENYLAZO)-PHENYL]ISOINDOLINE

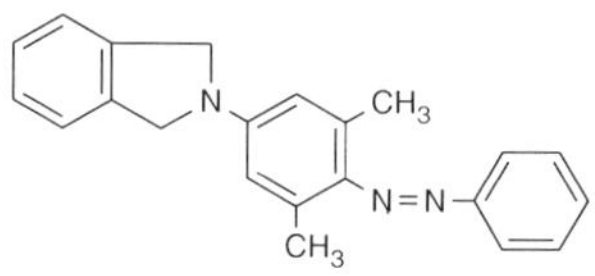

250 mg

S93,015-6
2-[3-METHYL-4-(P-TOLYLAZO)PHENYL]-ISOINDOLINE

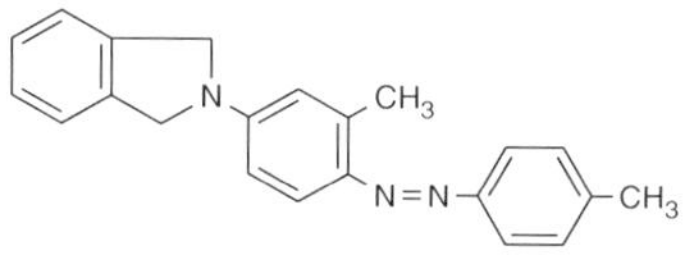

250 mg

S93,020-2
2-[4-(P-TOLYLAZO)PHENYL]ISOINDOLINE

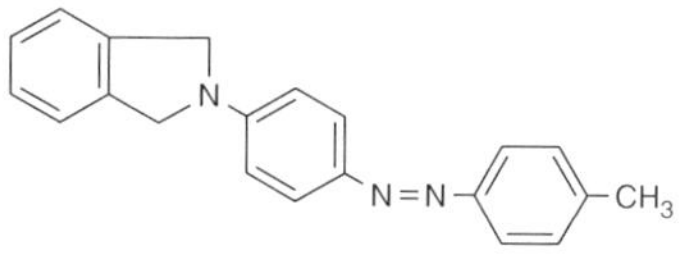

250 mg

S93,037-7
4-METHYL-1-[4-(PHENYLAZO)PHENYL]-PIPERIDINE

250 mg

S93,026-1
HEXAHYDRO-1-[4-(PHENYLAZO)PHENYL]-1H-AZEPINE

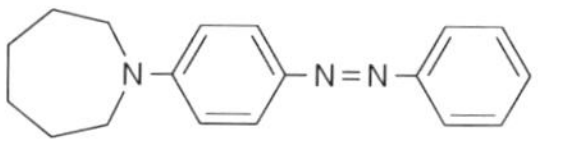

250 mg

S93,029-6
HEXAHYDRO-1-[4-(P-TOLYLAZO)PHENYL]-1H-AZEPINE

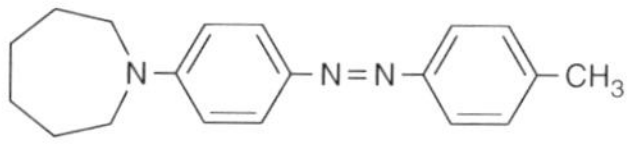

250 mg

S93,043-1
9-(4-TOLYLAZO)-2,3,5,6-TETRAHYDRO-1H,7H-BENZO(IJ)QUINOLIZINE

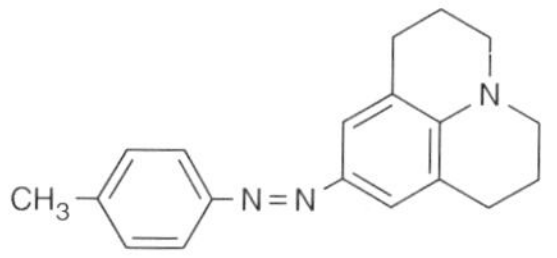

250 mg

S92,950-6
6-(4-NITROPHENYLAZO)-1,2,3,4-TETRAHYDRO-2,2,4,7-TETRAMETHYLQUINOLINE

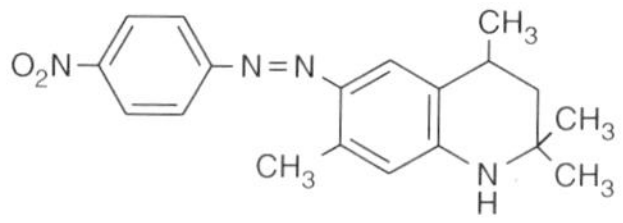

250 mg

S92,951-4
6-(4-ACETYLPHENYLAZO)-1-ETHYL-1,2,3,4-TETRAHYDRO-2,2,4-TRIMETHYLQUINOLINE

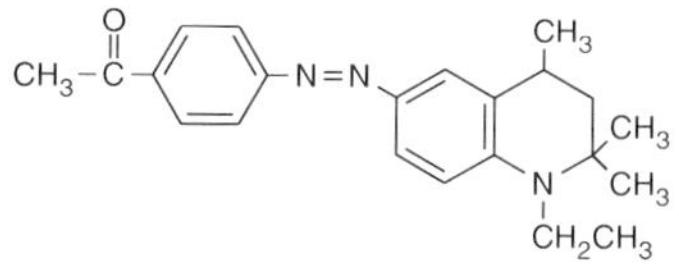

250 mg

S87,508-2
3-METHYL-2-[(3-METHYL-2-BENZOTHIAZOLINYLIDENE)HYDRAZONOMETHYLAZO]-BENZOTHIAZOLIUM PERCHLORATE

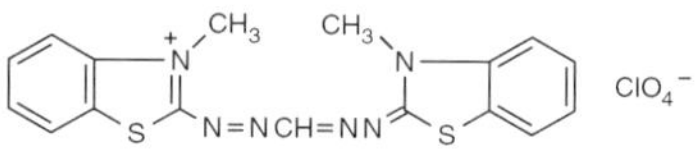

250 mg

S92,333-8
4'-DIETHYLAMINO-2-METHOXY-4-NITROAZOBENZENE

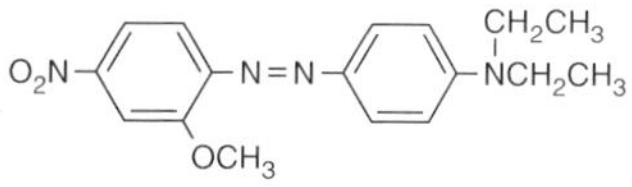

250 mg

S92,710-4
4'-(DIMETHYLAMINO)-4-NITRO-2,2',3,5,6-PENTAMETHYLAZOBENZENE

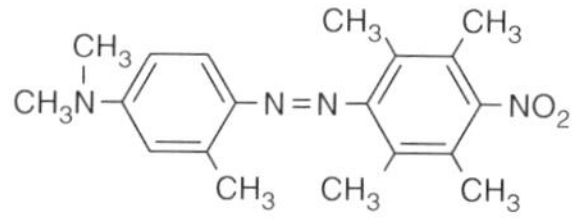

250 mg

S82,212-4
1-PHENYL-2-(PHENYLAZO)PYRROLE

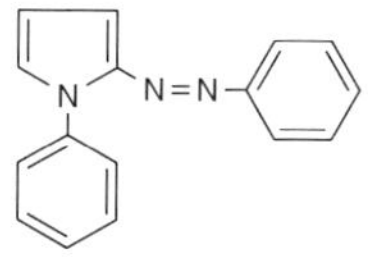

250 mg

S93,053-9
4-[4-(2,6-DIMETHYLMORPHOLINO)-PHENYLAZO]BENZONITRILE

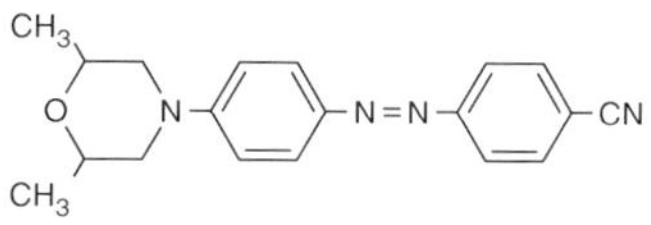

250 mg

S80,263-8
4-(DIMETHYLAMINO)-4'-METHYLAZOBENZENE

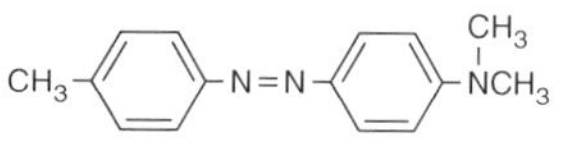

250 mg

S52,769-6
N,N-DIETHYL-4-(P-TOLYLAZO)-ANILINE

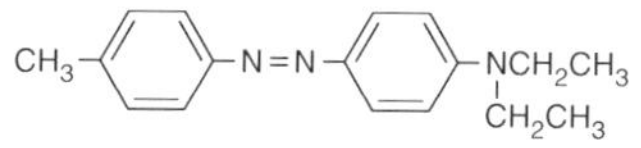

5 gm

S87,088-9
2-(4-ANILINOPHENYLAZO)-3-METHYL-4,5,6,7-TETRAHYDROBENZOTHIAZOLIUM TETRAFLUOROBORATE

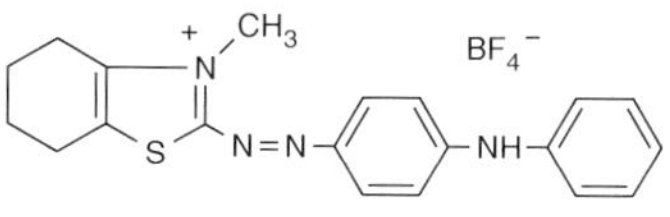

250 mg

S59,886-0
5-(2-CARBOXYPHENYL)-3-METHYL-1-PHENYLFORMAZAN

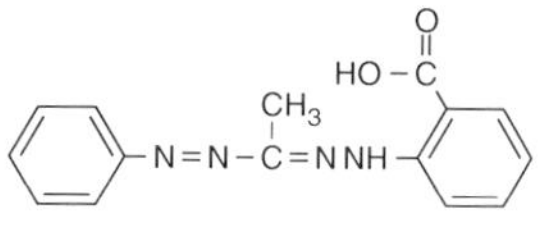

250 mg

S61,809-8
3-ETHYL-5-[4-(FLUOROSULFONYL)-PHENYL]-1-PHENYLFORMAZAN

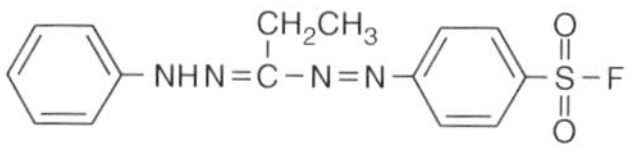

250 mg

S86,453-6
4-TERT-BUTYLCYCLOHEXANONE (P-TOSYL)-HYDRAZONE

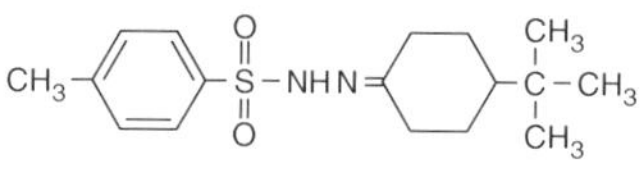

250 mg

S71,014-8
TRIMETHYLACETALDEHYDE(P-TOSYL)-HYDRAZONE
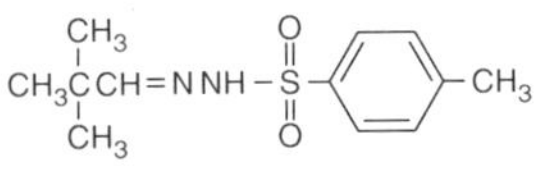
5 gm

S86,517-6
3,3-DIMETHYL-2-BUTANONE (P-TOSYL)-HYDRAZONE
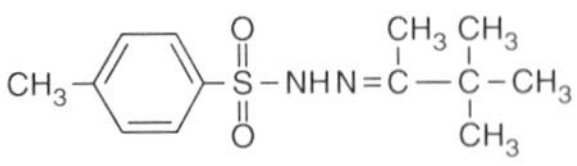
250 mg

S44,174-0
TRIPHENYLFORMAZAN
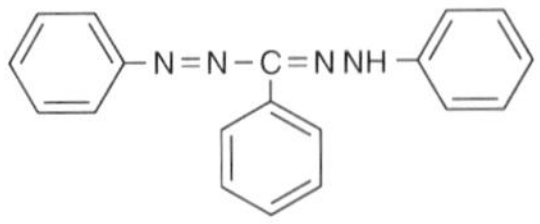
5 gm

S35,444-9
ETHYL 2-(PHENYLAZO)ACETOACETATE
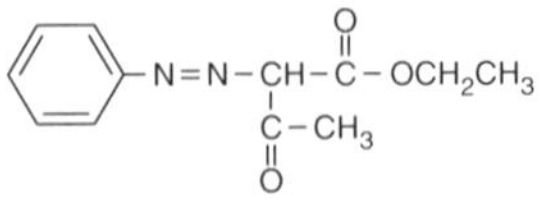
5 gm

S35,444-9
3-(PHENYLAZO)-2,4-PENTANEDIONE
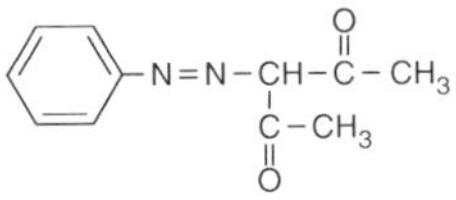
5 gm

S53,352-1
3,3'-AZOXYTOLUENE
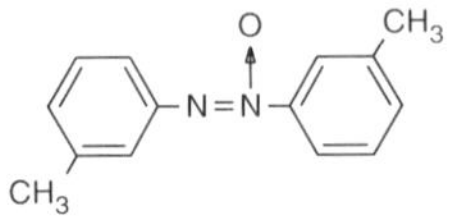
5 gm

S74,715-7
2-(2-NITROSO-1-NAPHTHYLAZO)BENZOIC ACID, SODIUM SALT
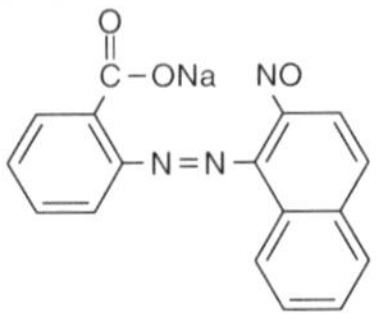
5 gm

S92,818-6
3-METHYL-4-(4-NITROPHENYLAZO)-1-PHENYL-1-PYRAZOLIN-5-ONE
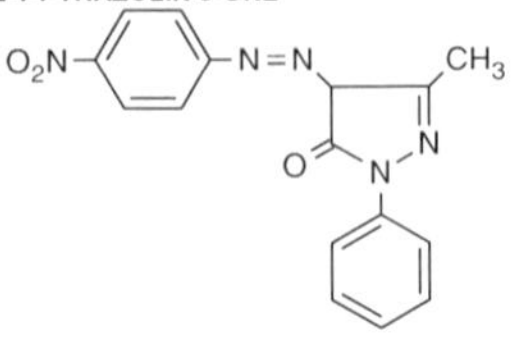
250 mg

S93,034-2
3-METHYL-1-[4-(4-NITROPHENYLAZO)-PHENYL]PIPERIDINE
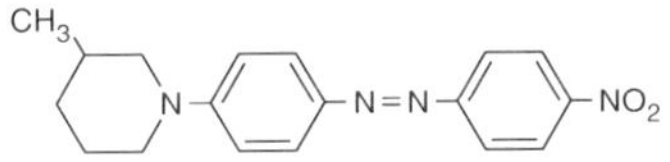
250 mg

S92,943-3
4'-[4-(2-ISOINDOLINYL)-2-METHYLPHENYLAZO]ACETOPHENONE
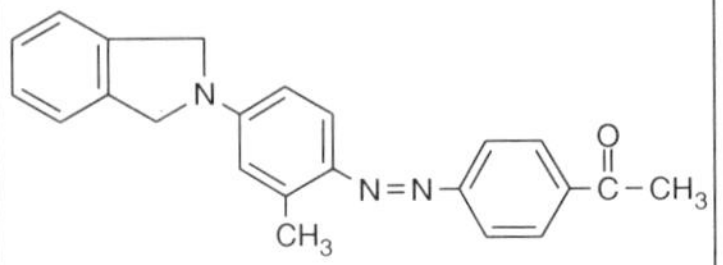
250 mg

S74,606-1
3'-(PHENYLAZO)ACETOPHENONE
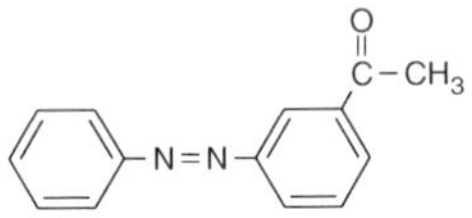
5 gm

S84,326-1
4'-(PHENYLAZO)ACETOPHENONE
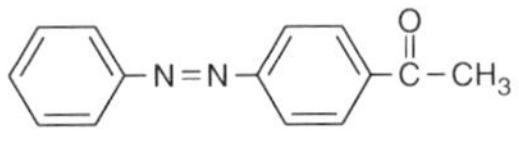
250 mg

S37,598-5
4-ACETOACETAMIDOAZOBENZENE
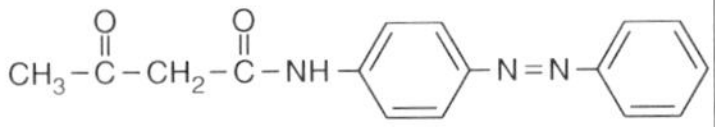
5 gm

S34,901-1
4,4'-DICARBOXYAZOXYBENZENE
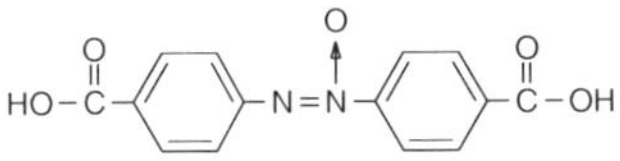
250 mg

S82,179-9
4-(3-BORONO-4-HYDROXYPHENYLAZO)-BENZOIC ACID
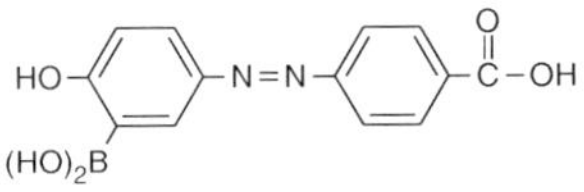
250 mg

S91,979-9
DIETHYL 4,4'-AZODIBENZOATE
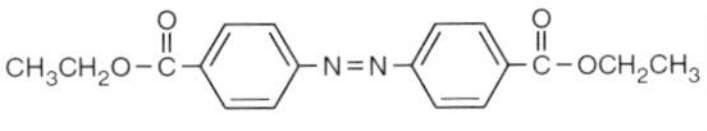
250 mg

S92,337-0
METHYL 4-(PHENYLAZO)BENZOATE
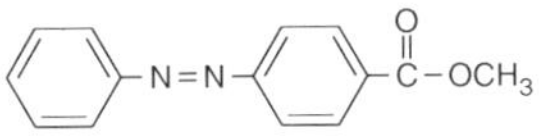
250 mg

S80,926-8
4-(PHENYLAZO)BENZOYL CHLORIDE
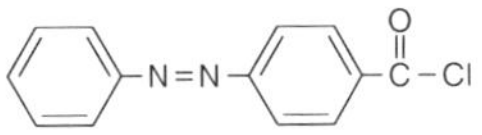
250 mg

S67,580-6
N-[5-METHYL-2-(4-METHYLPHENYLAZO)-PHENYL]CINNAMAMIDE
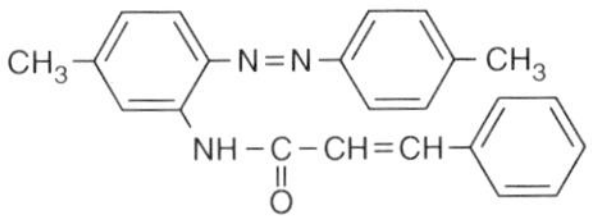
5 gm

S89,577-6
1-(PHENYLAZO)-1H-INDAZOL-3(2H)-ONE
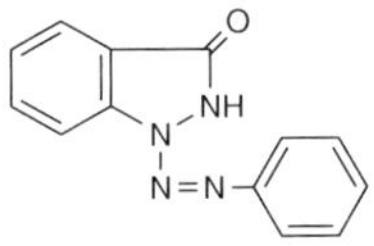
250 mg

S79,649-2
4-(4-TOLYLAZO)PHENYL N-METHYLCARBAMATE
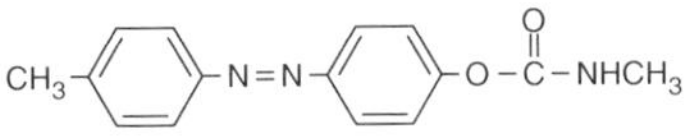
250 mg

S92,115-7
ETHYL 4-(PHENYLAZO)PHENYL CARBONATE
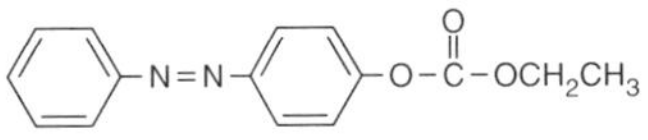
250 mg

S66,453-7
PHENYLAZOFORMAMIDE
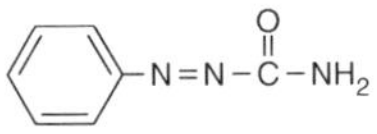
250 mg

S92,935-2
4'-(PHENYLAZO)ACETANILIDE
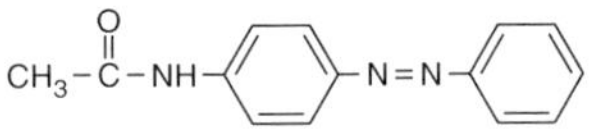
250 mg

S69,580-7
4'-(PHENYLAZO)PROPIONANILIDE
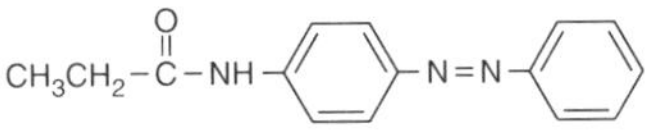
250 mg

S50,229-4
3-PENTADECYL-2-PHENYL-4-PHENYLAZO-2-PYRAZOLIN-5-ONE
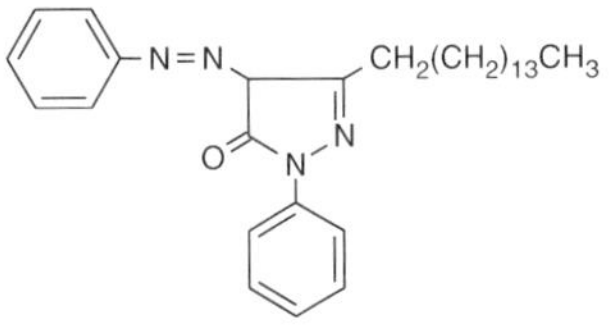
5 gm

S93,000-8
4-[4-(2-ISOINDOLINYL)-3-METHYLPHENYLAZO]BENZONITRILE
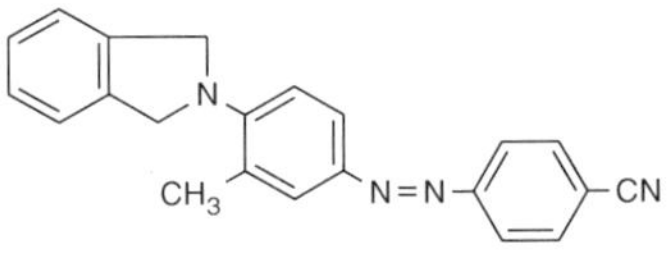
250 mg

S93,018-0
4-[4-(2-ISOINDOLINYL)-2-METHYLPHENYLAZO]BENZONITRILE
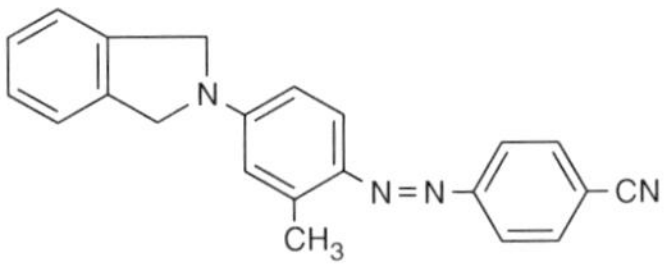
250 mg

S65,547-3
4-(PHENYLAZO)-BENZENESULFONYL CHLORIDE
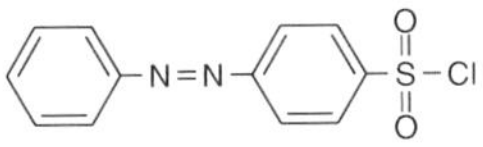
250 mg

S50,149-2
4-(PHENYLAZO)BENZENESULFONIC ACID HYDRAZIDE
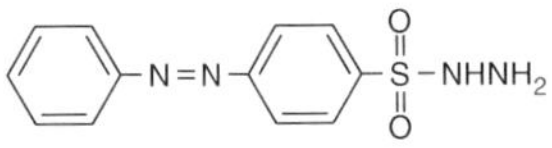
250 mg

S81,483-0
3,3'-AZOXYDIBENZOFURAN
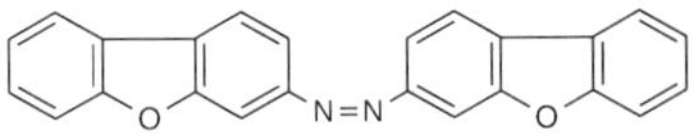
250 mg

S94,475-0
5-PHENYLAZO-1,3-BENZODIOXOLE
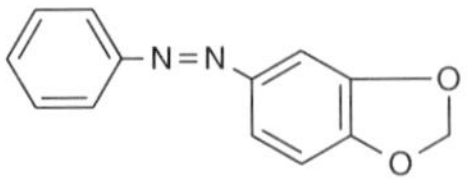
250 mg

S92,364-8
4-PHENYL-2-[ALPHA-(PHENYLAZO)-BENZYLIDENE]-1,3-DITHIOLE
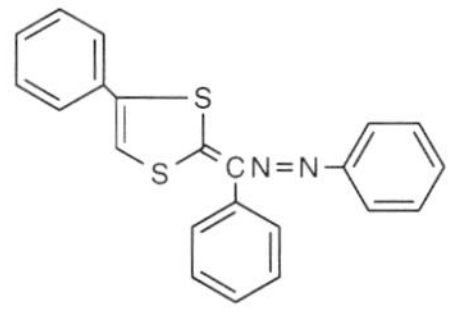
250 mg

S92,107-6
4-METHYLAZOBENZENE
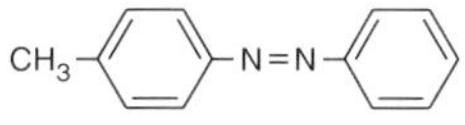
5 gm

S90,943-2
2-PHENYLAZOQUINOLINE
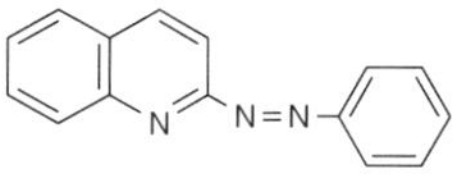
250 mg

S53,438-2
2,2'-DICHLOROAZOXYBENZENE
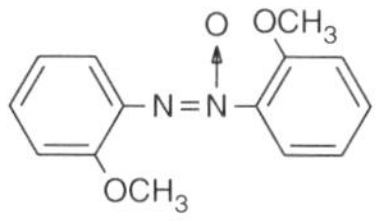
5 gm

S44,671-8
4,4'-DICHLOROAZOXYBENZENE
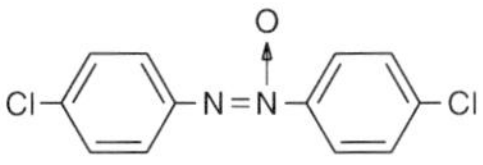
5 gm

S81,774-0
4,4'-DIMETHYLAZOBENZENE
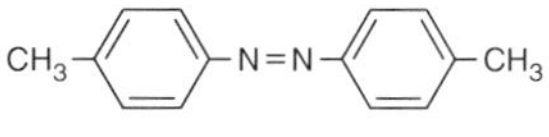
250 mg

S53,439-0
4,4'-AZOXYTOLUENE
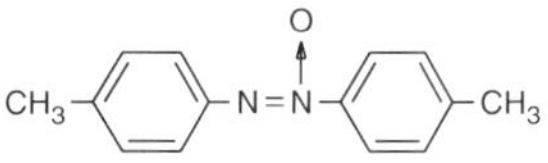
5 gm

S53,357-2
4,4'-DIETHYLAZOXYBENZENE
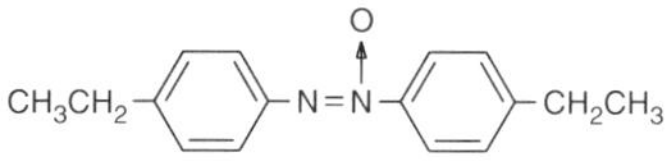
5 gm

S70,653-1
4,4'-DIMETHYL-2-(4-TOLYL)AZOBENZENE
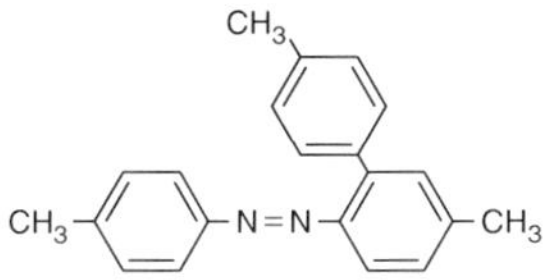
5 gm

S92,702-3
2,2',6,6'-TETRAMETHYLAZOBENZENE
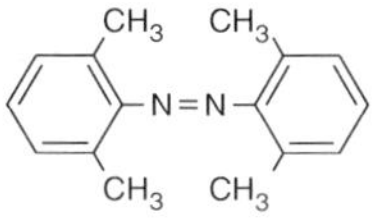
250 mg

S92,320-6
4-(PHENYLAZO)-2,3,5,6-TETRAMETHYLPHENOL
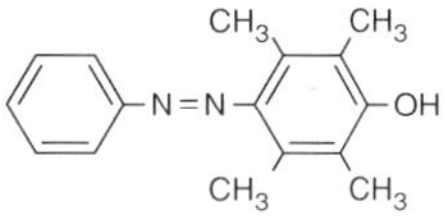
250 mg

S80,517-3
2,2'-DIAZONAPHTHALENE
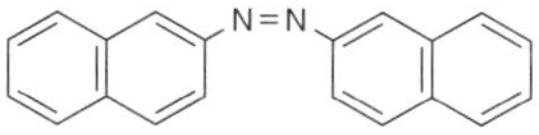
250 mg

S81,910-7
3,3'-BIS[2-(2-QUINOLYL)VINYL]-AZOBENZENE
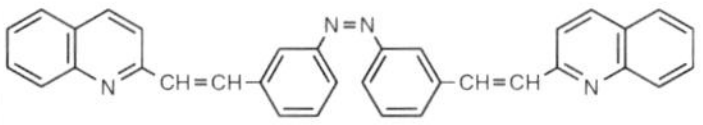
250 mg

S76,510-4
THYMOLPHTHALEIN MONOPHOSPHATE 2-AMINO-2-ETHYL-1,3-PROPANEDIOL SALT
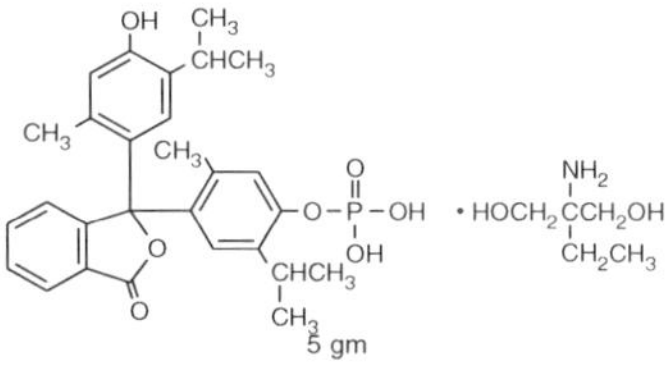
5 gm

S45,644-6
BROMOXYLENOL BLUE, WATER SOLUBLE
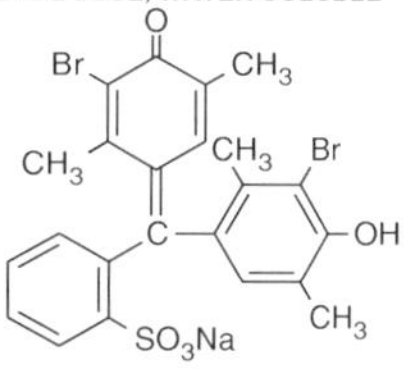
250 mg

S94,448-3
MORDANT DYE (C.I. 60875)
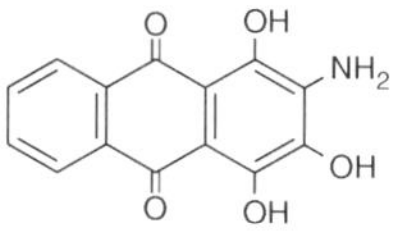
250 mg

S94,457-2
DISPERSE BLUE 26 (C.I. 63305)
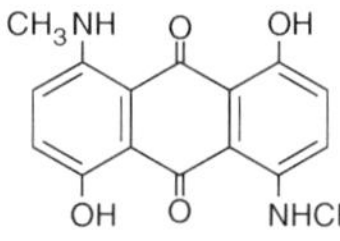
250 mg

S94,454-8
DISPERSE BLUE 23 (C.I. 61545)
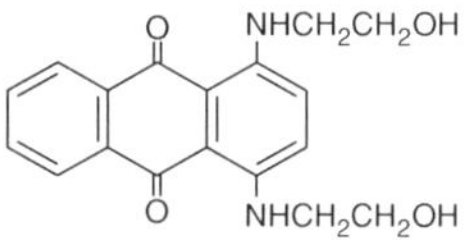
250 mg

S94,447-5
DISPERSE RED 4 (C.I. 60755)
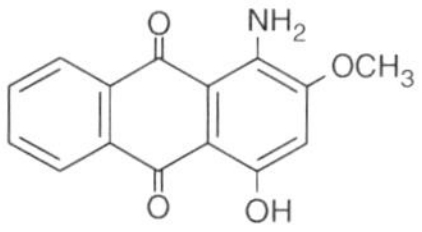
250 mg

S94,445-9
PIGMENT RED 89 (C.I. 60745)
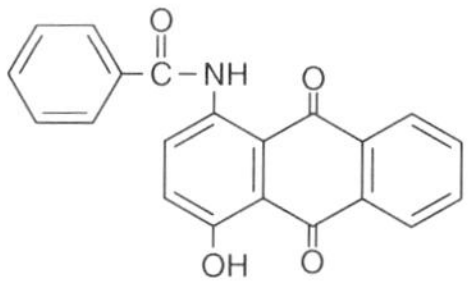
250 mg

S94,452-1
DISPERSE DYE (C.I. 61540)
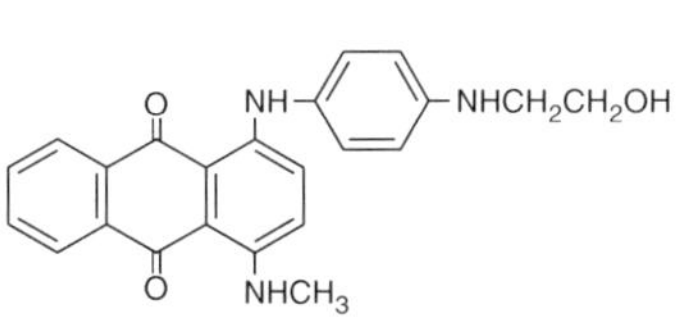
250 mg

S94,444-0
ACID VIOLET 43 (C.I. 60730)
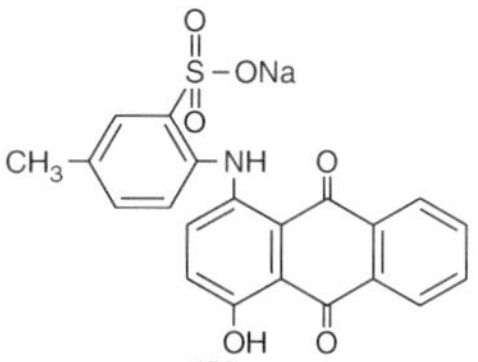
250 mg

S46,250-0
SARCOSINE CRESOL RED
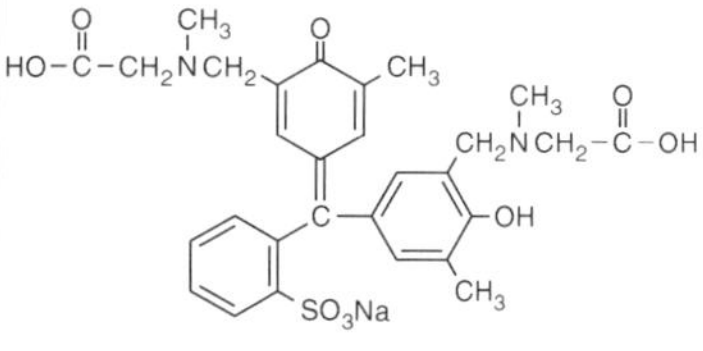
5 gm

S44,749-8
METHYL XYLENOL BLUE
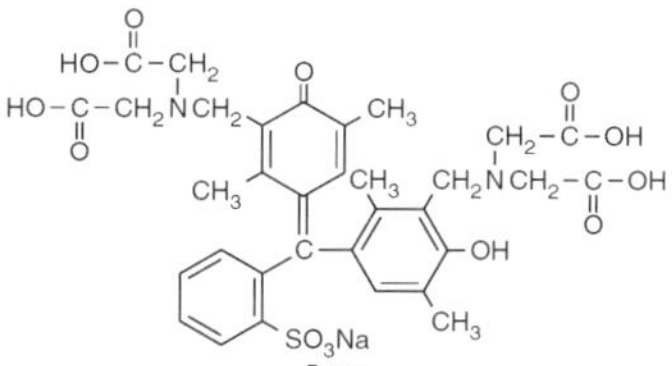
5 gm

S42,793-4
GLYCINETHYMOL BLUE
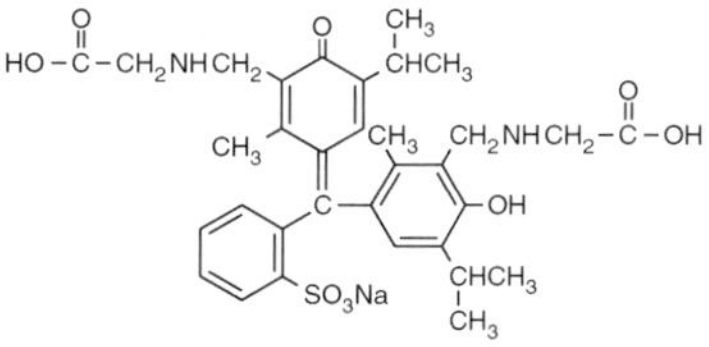
250 mg

S45,342-0
MORDANT BLUE (C.I. 43820)
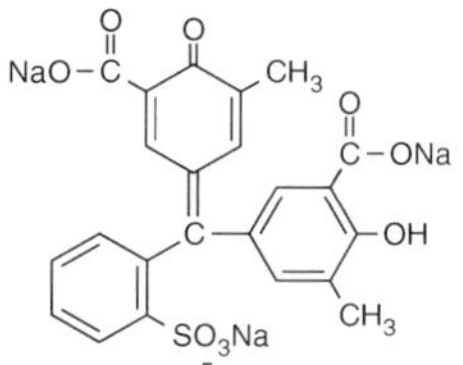
5 gm

S94,455-6
DISPERSE RED 11 (C.I. 62015)
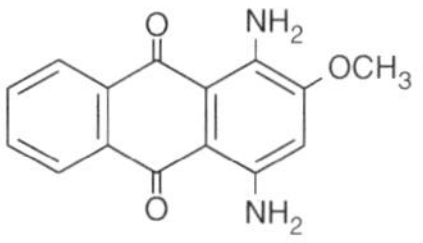
250 mg

S94,451-3
DISPERSE DYE (C.I. 61115)
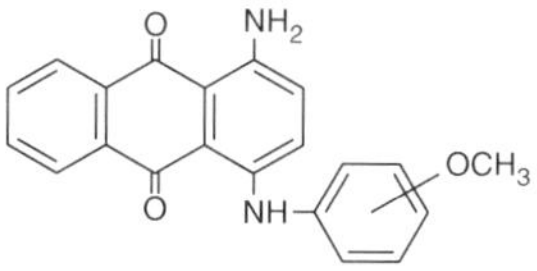
250 mg

S94,433-5
DISPERSE GREEN 1 (C.I. 56060)
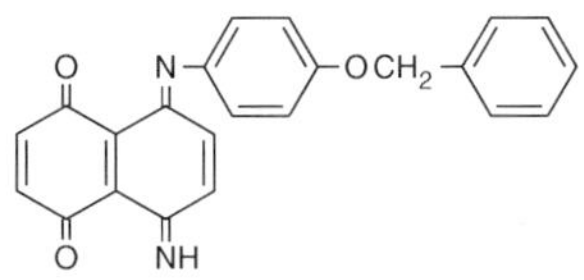
250 mg

S94,446-7
VAT DYE (C.I. 60750)
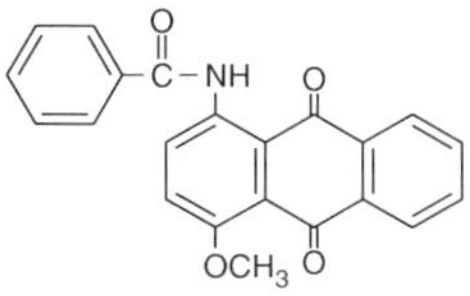
250 mg

S94,449-1
DISPERSE VIOLET 4 (C.I. 61105)
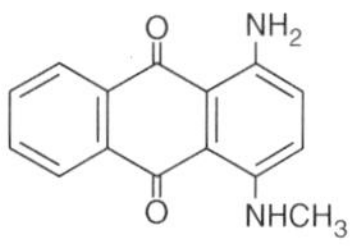
250 mg

S94,450-5
DISPERSE BLUE 19 (C.I. 61110)
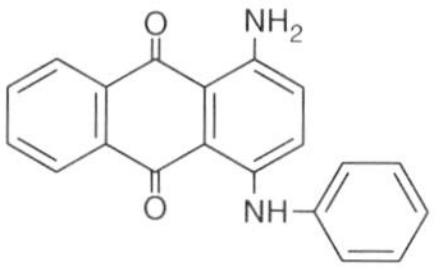
250 mg

S94,456-4
DISPERSE BLUE 5 (C.I. 62035)

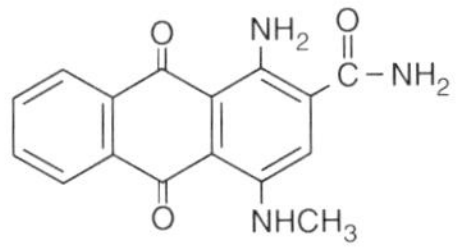

250 mg

S94,404-1
ACID BROWN 13 (C.I. 10410)

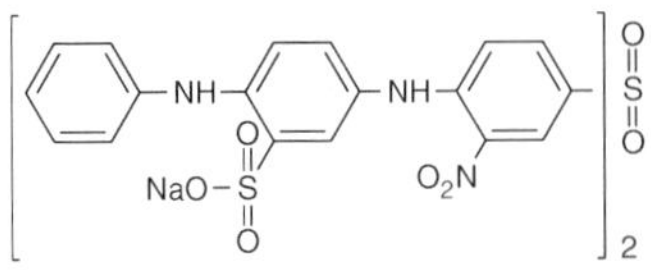

250 mg

S94,403-3
ACID ORANGE 3 (C.I. 10385)

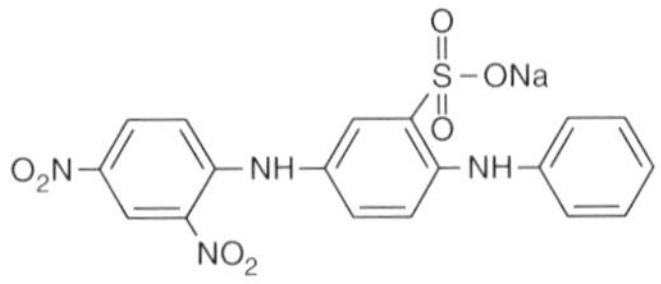

250 mg

S94,489-0
VAT RED 39 (C.I. 67900)

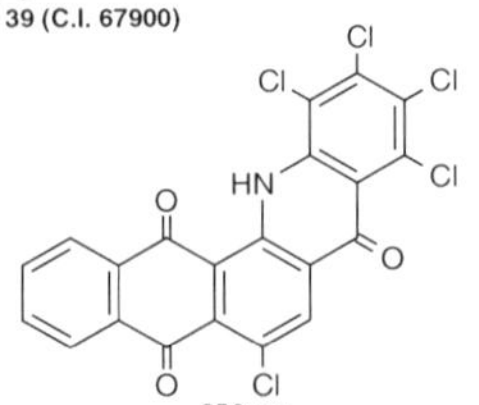

250 mg

S94,434-3
SOLUBILISED VAT GREEN 21 (C.I. 59051)

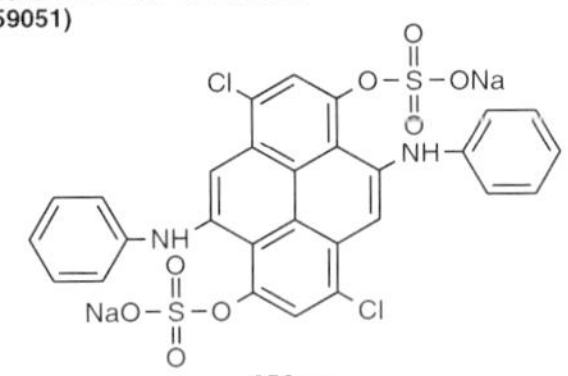

250 mg

S94,443-2
VAT DYE (C.I. 60515)

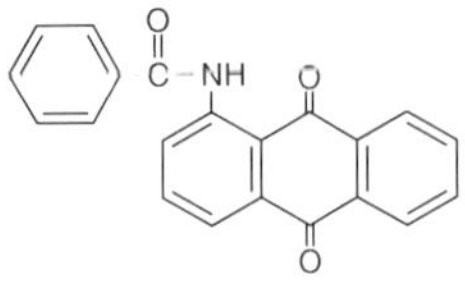

50 mg

S94,401-7
MORDANT GREEN 4 (C.I. 10005)

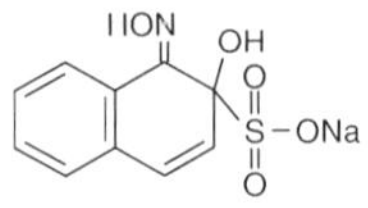

250 mg

S95,324-5
2'-BROMO-2-DIAZO-3'-HYDROXY-4'-METHOXYACETOPHENONE

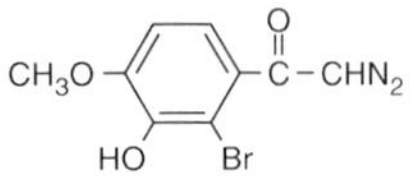

50 mg

S53,227-4
BENZIL, AZINE WITH 3,5-DICHLOROSALICYLALDEHYDE

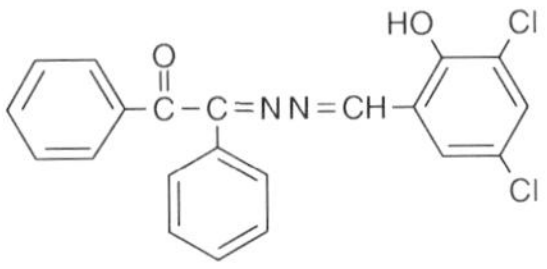

5 gm

S46,187-3
FAST DARK BLUE R SALT (C.I. 37195)

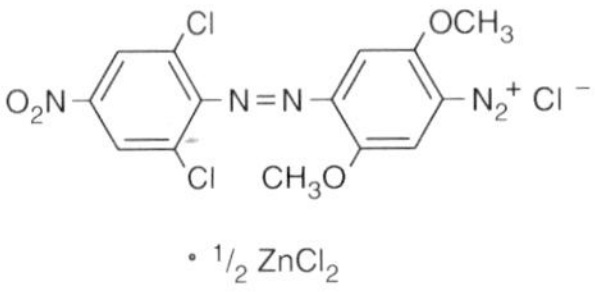

5 gm

S77,074-4
2-(4-CHLOROPHENYL)-4-(2-DIAZOACETYL)-6,8-DICHLORO-3-METHYLQUINOLINE

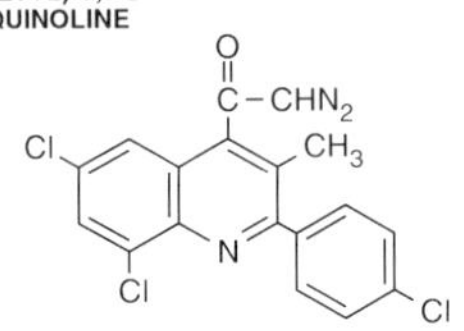

250 mg

S53,974-0
3-ETHOXY-4-HYDROXYBENZALDEHYDE AZINE

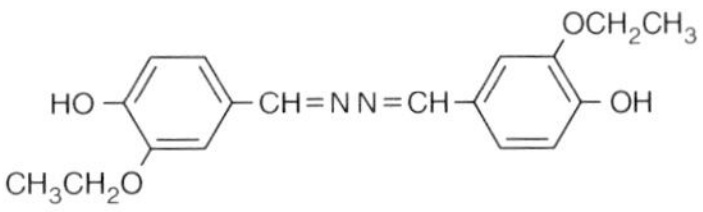

5 gm

S94,402-5
MORDANT DYE (C.I. 10015)

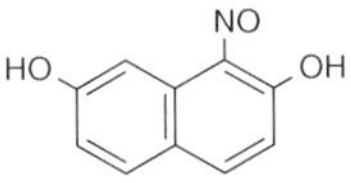

50 mg

S34,855-4
4-DIAZO-N-ETHYL-N-(2-HYDROXYETHYL)-ANILINE, ZINC CHLORIDE DOUBLE SALT

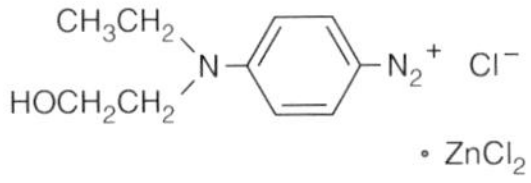

5 gm

S92,378-8
MORDANT DYE -ALIZARIN BLUE R (C.I. 52060)

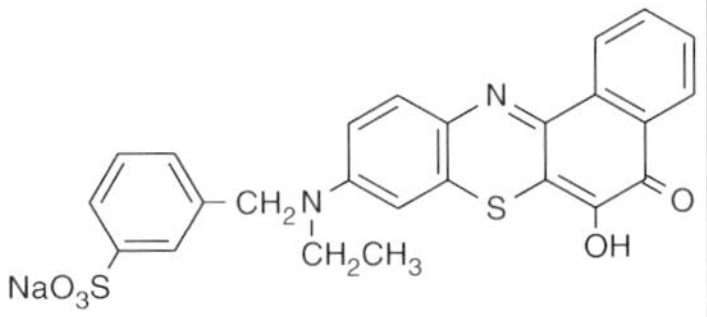

250 mg

S74,045-4
2-HYDROXY-1-NAPHTHALENEDIAZONIUM CHLORIDE

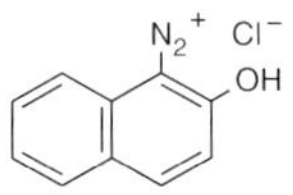

250 mg

S80,810-5
2-ACETYL-5-(DIAZOACETYL)-3-HYDROXY-5-METHYL-4-PHENYL-3-PYRAZOLINE

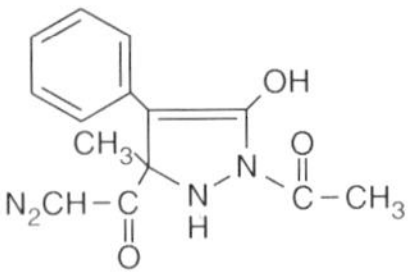

250 mg

S95,453-5
1-(3,4-DIBENZYLOXYPHENYL)-3,3-DIMETHYLTRIAZENE

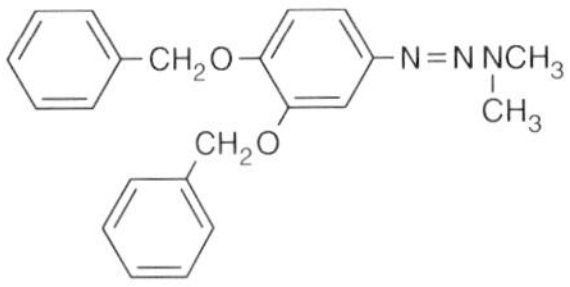

50 mg

S80,811-3
1-DIAZO-3-PHENOXY-4-PHENYL-3-BUTEN-2-ONE

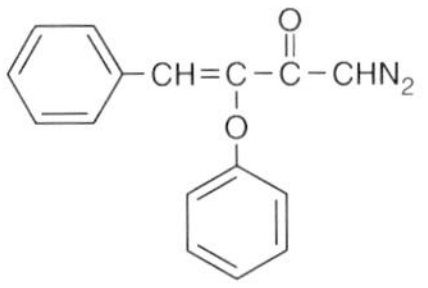

250 mg

S94,442-4
VAT GREEN 1 (C.I. 59825)

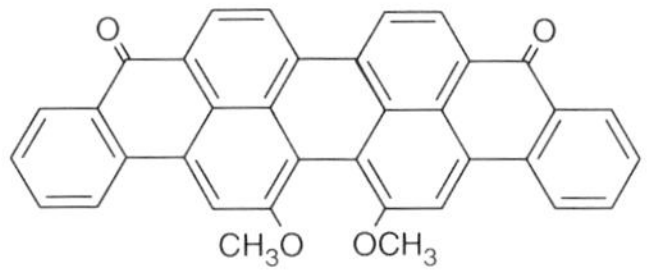

250 mg

S34,590-3
2,5-DIETHOXY-4-(P-TOLYLTHIO)-DIAZOBENZENE, ZINC CHLORIDE DOUBLE SALT

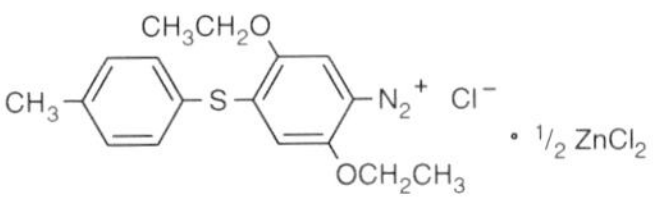

5 gm

S57,956-4
4-(2,6-DICHLORO-4-NITROPHENYLAZO)-3,5-DIMETHOXYBENZENEDIAZONIUM CHLORIDE, ZINC CHLORIDE SALT

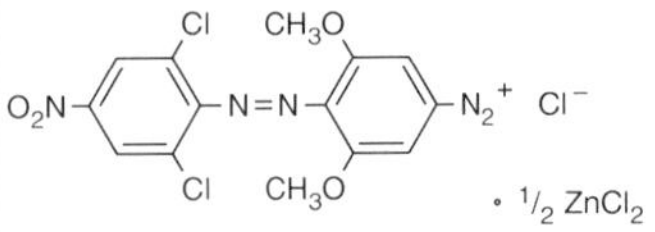

5 gm

S34,587-3
2,5-DIETHOXY-4-(2-PHENYLACETAMIDO)-BENZENEDIAZONIUM CHLORIDE, ZINC CHLORIDE SALT

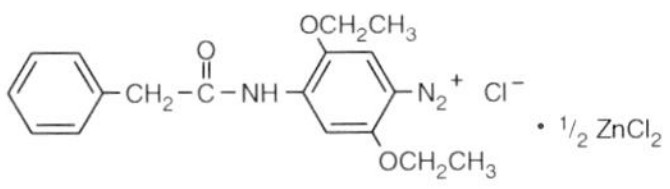

5 gm

S34,651-9
2,5-DIMETHOXY-4-(2-PHENYLACETAMIDO)-BENZENEDIAZONIUM CHLORIDE, ZINC CHLORIDE SALT

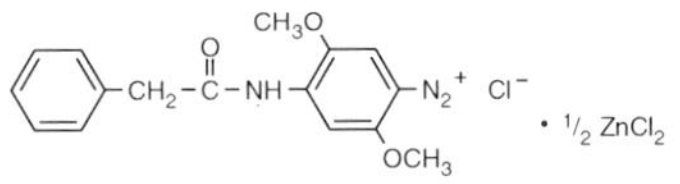

5 gm

S94,423-8
BASIC DYE (C.I. 45105)

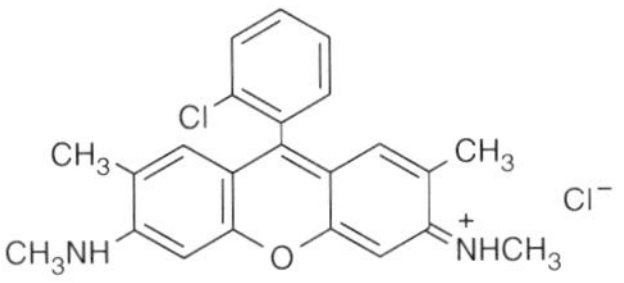

250 mg

S46,222-5
THIOCARBAMYL NITRO BLUE TETRAZOLIUM CHLORIDE

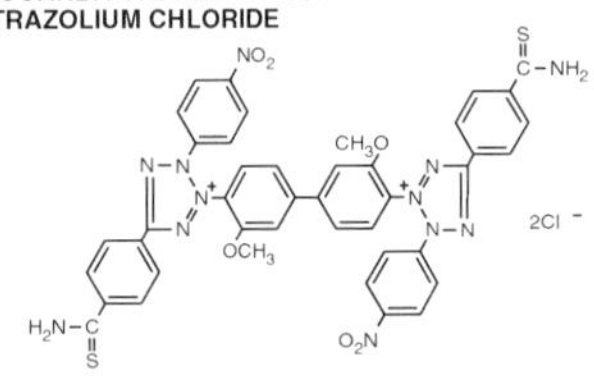

250 mg

S53,799-3
4-ETHOXYBENZALDEHYDE AZINE

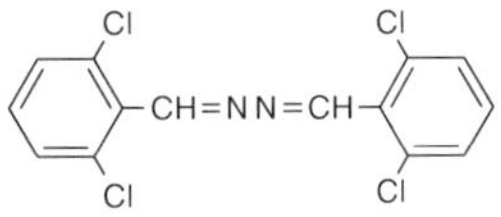

250 mg

S94,426-2
BASIC ORANGE 4-11(C.I. 46035)

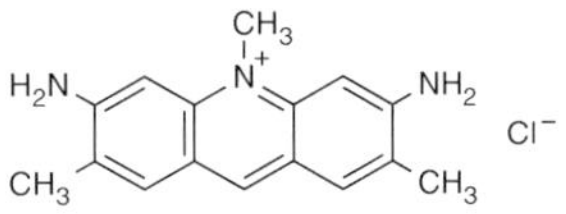

250 mg

S94,427-0
BASIC YELLOW 9 (C.I. 46040)

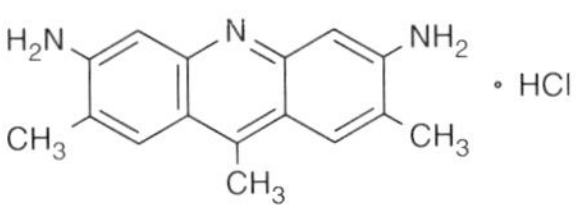

250 mg

S94,428-9
BASIC ORANGE 15 (C.I. 46045)

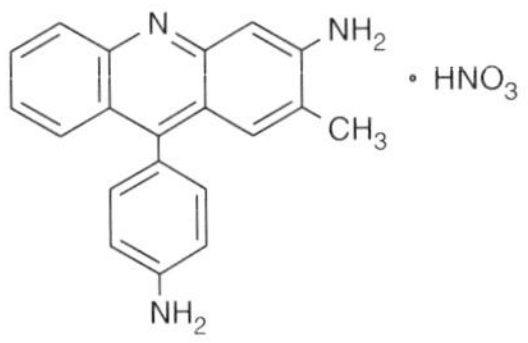

50 mg

S44,117-1
5-DIAZOURACIL MONOHYDRATE

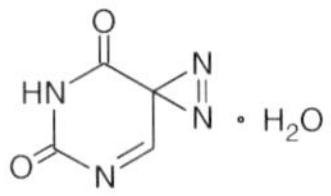

5 gm

S88,768-4
2,4-BIS(2-CYANOETHYLAMINO)-3-OXO-1-CYCLOBUTENYLIUM-1-OLATE

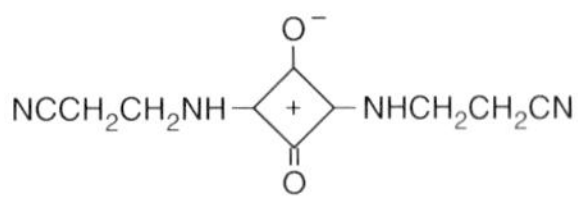

250 mg

S88,767-6
2,4-BIS(BUTYLAMINO)-3-OXO-1-CYCLOBUTENYLIUM-1-OLATE

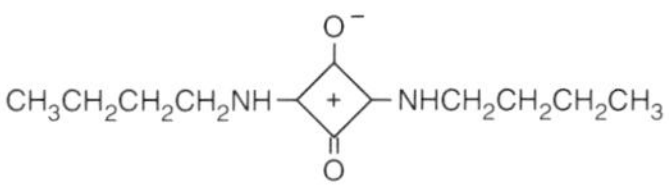

250 mg

S88,770-6
2,4-BIS(3-BENZAMIDOPROPYLAMINO)-3-OXO-1-CYCLOBUTENYLIUM-1-OLATE

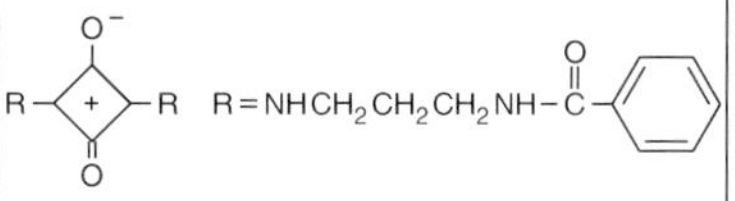

250 mg

S36,108-9
3,3-DIMETHYL-1-(3-NITROPHENYL)-TRIAZENE

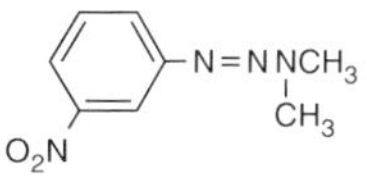

5 gm

S94,400-9
MORDANT DYE (C.I. 10000)

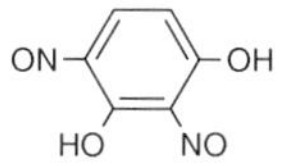

250 mg

S36,464-9
4-(PROPYLAMINO)BENZENEDIAZONIUM CHLORIDE, ZINC CHLORIDE DOUBLE SALT

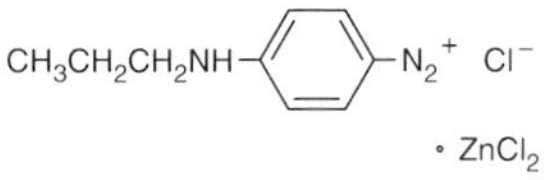

5 gm

S94,431-9
BASIC RED 10 (C.I. 50250)

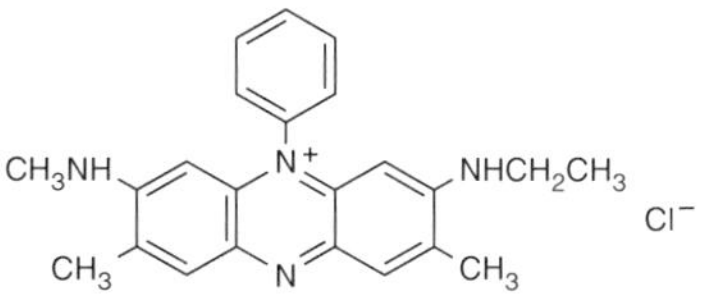

250 mg

S88,750-1
3,4-BIS[3-METHYL-2(3H)-BENZOTHIAZOLYLIDENEMETHYL]-1,3-CYCLOBUTADIENEDIYLIUM-1,2-DIOLATE

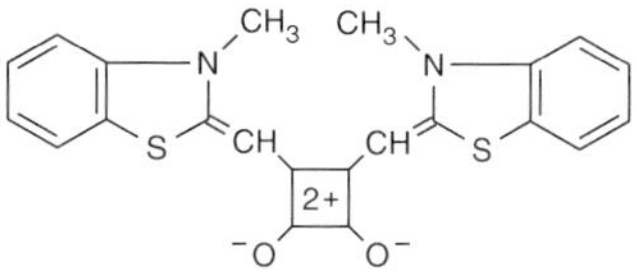

250 mg

S88,749-8
2,4-BIS[3-METHYL-2(3H)-BENZOTHIAZOLYLIDENEMETHYL]-1,3-CYCLOBUTADIENEDIYLIUM-1,3-DIOLATE

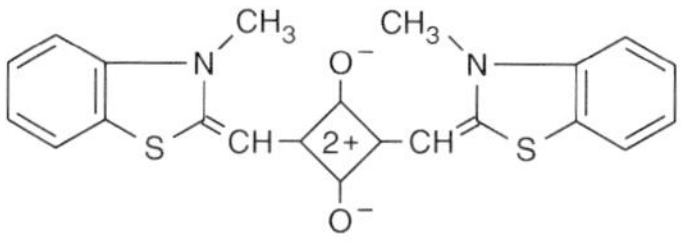

250 mg

S88,752-8
3,4-BIS[1-PHENYL-2(1H)-QUINOLINYLIDENEMETHYL]-1,3-CYCLOBUTADIENEDIYLIUM-1,2-DIOLATE

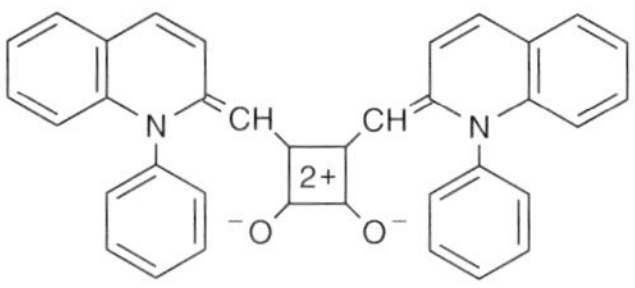

250 mg

S88,753-6
3,4-BIS[1-ETHYL-2(1H)-QUINOLINYLIDENEMETHYL]-1,3-CYCLOBUTADIENEDIYLIUM-1,2-DIOLATE

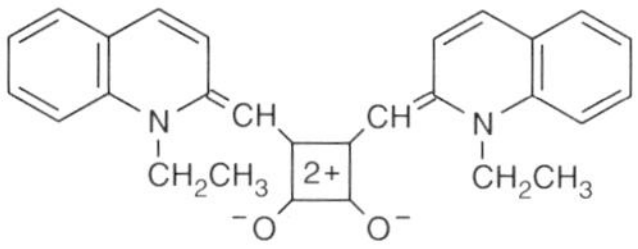

250 mg

S45,604-7
CRESYL FAST VIOLET

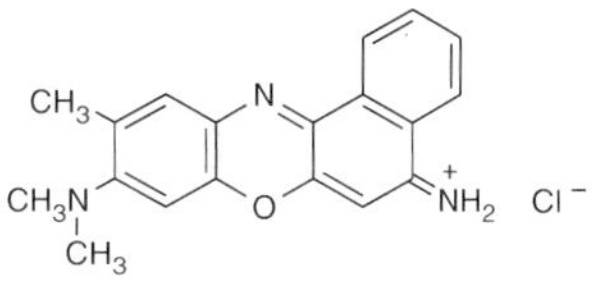

5 gm

S46,066-4
4-DIETHYLAMINO-2-METHYLBENZENEDIAZONIUM CHLORIDE, ZINC CHLORIDE DOUBLE SALT

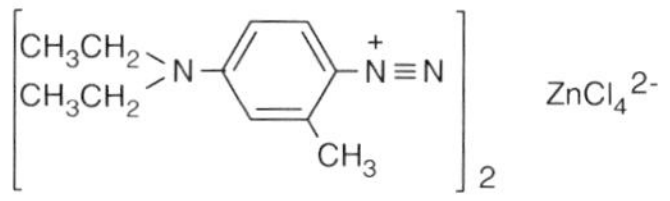

5 gm

S80,837-7
3-(DIAZOACETYL)-4,5-DIHYDRO-3-ETHYL-4-PHENYL-3H-PYRAZOLE

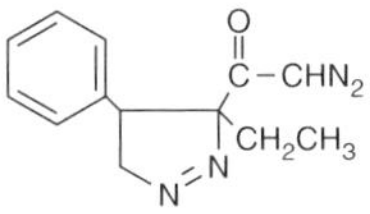

250 mg

S97,500-1
AR-H 2 1-PYRENYLDIAZONIUM TETRAFLUOROBORATE

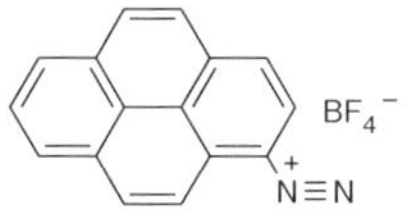

10 mg

S80,373-1
2-DIAZOFLUORENE CHLORIDE

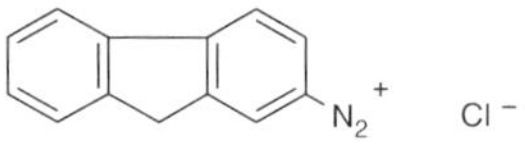

250 mg

S94,439-4
VAT ORANGE 3 (C.I. 59300)

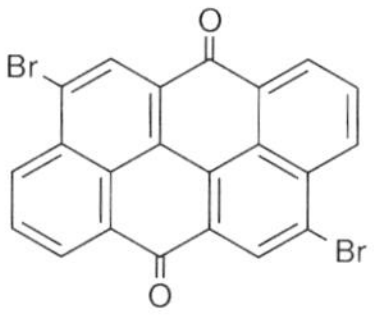

250 mg

S94,440-8
VAT DYE (C.I. 59310)

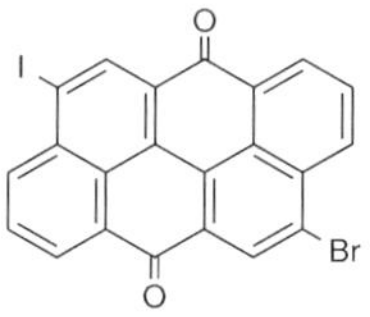

250 mg

S94,435-1
VAT YELLOW 4 (C.I. 59100)

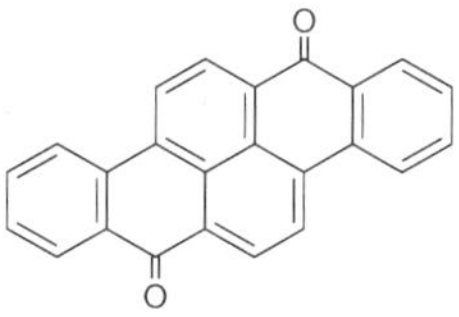

250 mg

S94,437-8
VAT ORANGE 1 (C.I. 59105)

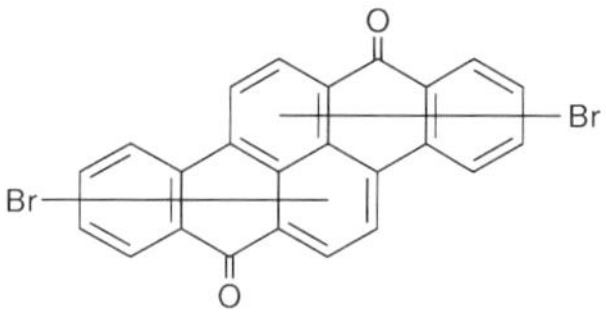

250 mg

S94,441-6
VAT ORANGE 9 (C.I. 59700)

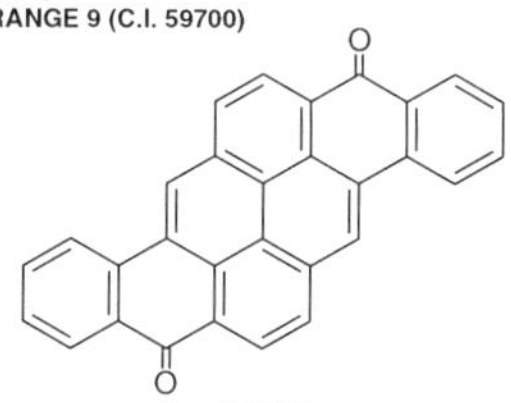

250 mg

S61,897-7
1-(2-CARBOXYPHENYL)-3,3-DIMETHYLTRIAZENE

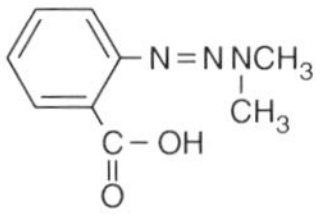

5 gm

S80,484-3
4-(3,3-DIMETHYL-1-TRIAZENYL)-BENZAMIDE

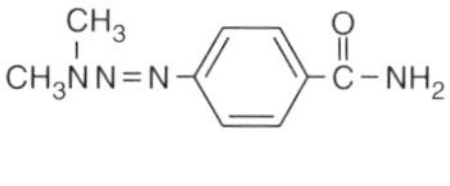

250 mg

S82,858-0
1,4-DIOXA-2,3,5,6-TETRASILACYCLOHEXANE
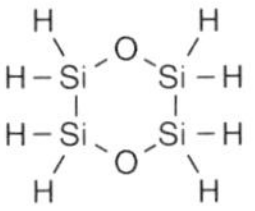
250 mg

S58,485-1
(1-CHLOROETHYL)TRICHLOROSILANE
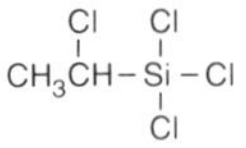
250 mg

S56,532-6
3-(DICHLOROMETHYLSILYL)PROPYL ISOCYANATE
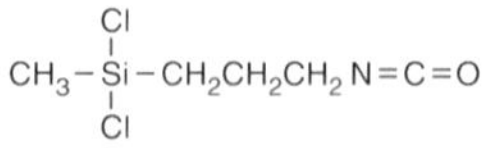
5 gm

S81,551-9
(3-CHLORO-1,1,2,2,3,3-HEXAFLUOROPROPYL)DIMETHYLSILANE
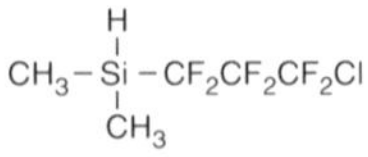
250 mg

S56,525-3
2,2-DIMETHYL-1-OXA-2-SILACYCLOHEXANE
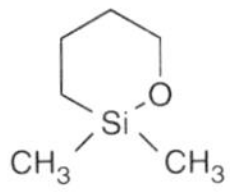
250 mg

S58,462-2
DICHLOROMETHYLPENTYLSILANE
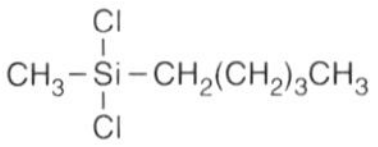
250 mg

S83,056-9
2-(DIMETHYLSILYL)-1,1,3,3-TETRAMETHYLTRISILANE
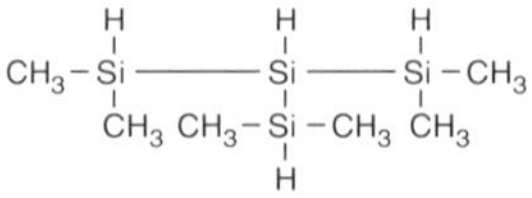
250 mg

S83,711-3
BIS(METHYLTHIO)DIETHOXYSILANE
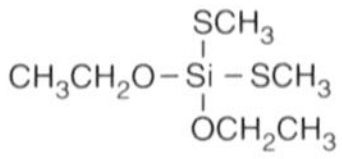
250 mg

S81,554-3
(3-CHLORO-1,1,2,2,3,3-HEXAFLUOROPROPYL)TRIMETHYLSILANE
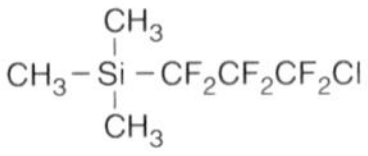
250 mg

S58,504-1
(DICHLOROPHENYL)TRICHLOROSILANE (MIXTURE OF ISOMERS)
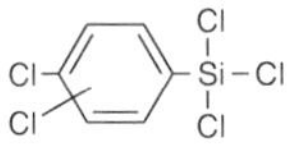
5 gm

S58,498-3
1,1-DIMETHYL-1-SILACYCLOHEXANE
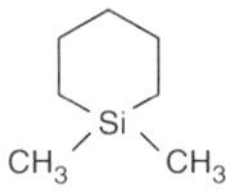
5 gm

S81,423-7
BUTYLTRIMETHYLSILANE
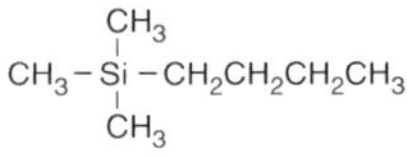
250 mg

S86,616-4
ISOPROPYL TRIMETHYLSILYL KETONE
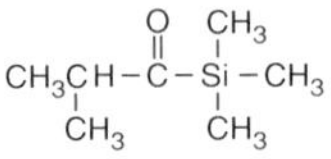
250 mg

S58,591-2
METHYLTRIS(DIMETHYLAMINO)SILANE
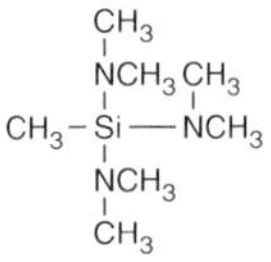
5 gm

S58,610-2
DIACETOXYMETHYLVINYLSILANE
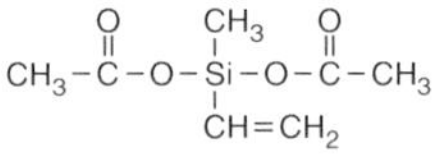
5 gm

S83,218-9
1,1,1,2,3,3,3-HEPTAMETHYL-2-HYDROXYTRISILANE
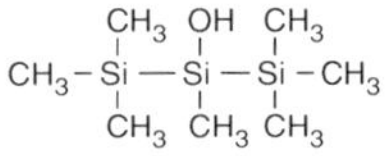
250 mg

S83,057-7
2-(DIMETHYLSILYL)-1,1,2,3,3-PENTAMETHYLTRISILANE
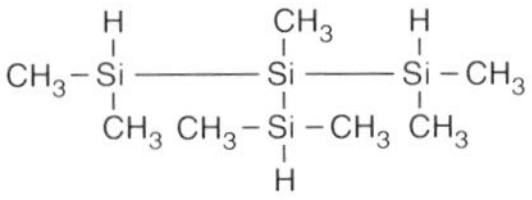
250 mg

S83,969-8
1,2-DIVINYL-1,1,2,2-TETRAMETHYLDISILANE
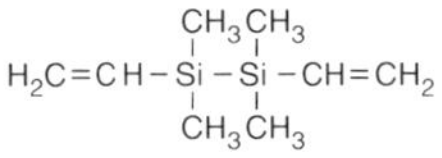
250 mg

S58,507-6
DIETHOXYDIETHYLSILANE
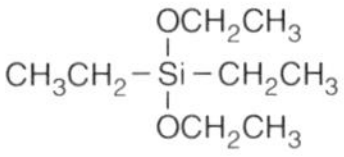
250 mg

S83,357-6
OCTAMETHYLTRISILANE
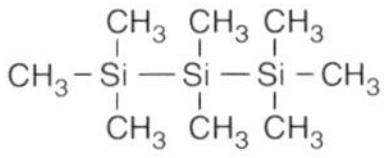
250 mg

S83,058-5
2-(DIMETHYLSILYL)-2-ETHYL-1,1,3,3-TETRAMETHYLTRISILANE
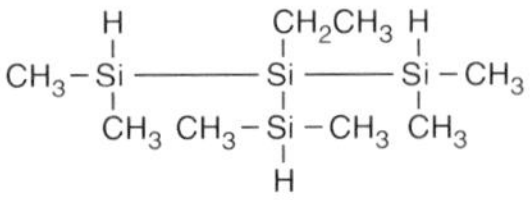
250 mg

S83,963-9
TETRAETHYL ORTHOTHIOSILICATE
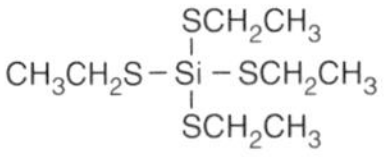
250 mg

S58,551-3
2,2,4,4,6,6,8,8-OCTAMETHYLCYCLOTETRASILAZANE
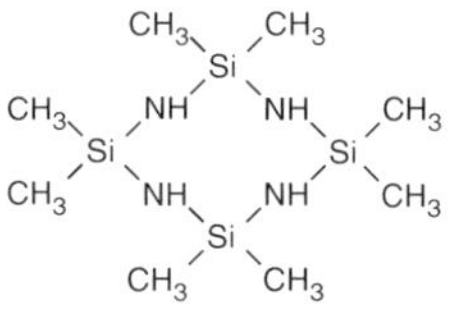
5 gm

S88,372-7
3,3-DIMETHYL-1-(TRIMETHYLSILYL)-1-BUTENE
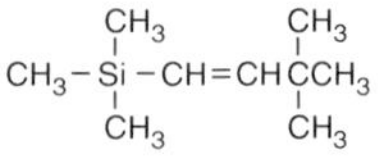
250 mg

S58,562-9
PROPYLTRIETHOXYSILANE
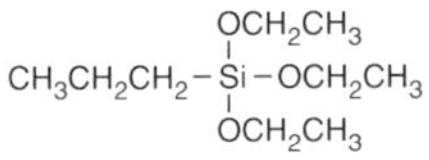
250 mg

S82,868-8
BIS(TRIMETHYLSILYLMETHYL)-METHYLSILANE
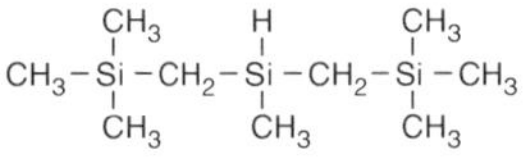
250 mg

S87,587-2
N-TRIMETHYLSILYL-3-(TRIMETHYLSILYLOXY)PROPYLAMINE
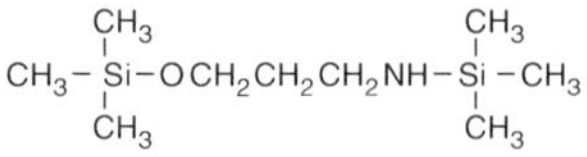
250 mg

S56,531-8
2,4,6-TRIMETHYL-2,4,6-TRIVINYLCYCLOTRISILOXANE
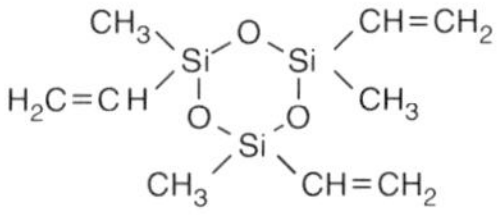
250 mg

S81,562-4
DIMETHYLOCTYLSILANE
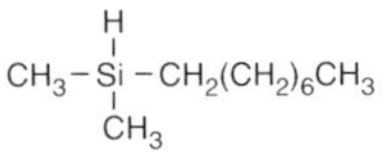
250 mg

S40,538-8
DIALLYLDIETHOXYSILANE
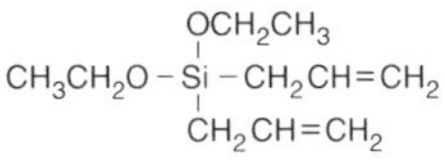
5 gm

S82,635-9
1,4-BIS(TRIMETHYLSILYL)BUTANE
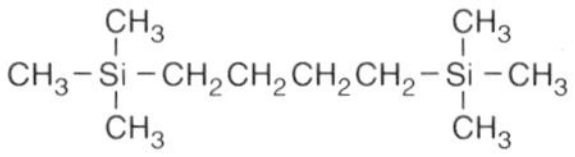
250 mg

S83,259-6
TRIMETHYL[4-(TRIMETHYLSILYL)BUTOXY]-SILANE
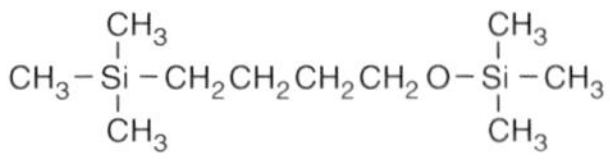
250 mg

S86,652-0
2-ISOPROPYL-2-(TRIMETHYLSILYL)-1,3-DITHIANE

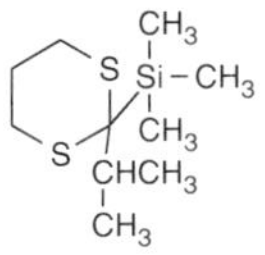

250 mg

S83,060-7
DECAMETHYLTETRASILANE

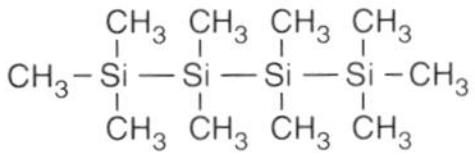

250 mg

S81,571-3
PERFLUOROHEXAMETHYLENEBIS-(DIMETHYLSILANE)

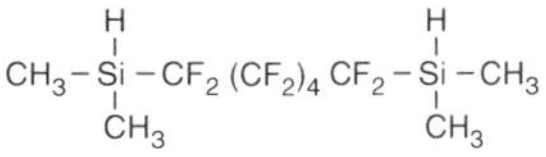

250 mg

S81,572-1
DIMETHYL(PERFLUOROOCTYL)SILANE

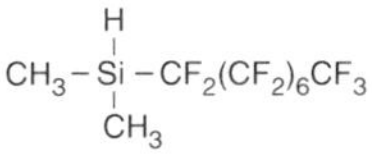

250 mg

S43,909-6
ETHOXYTRIALLYLSILANE

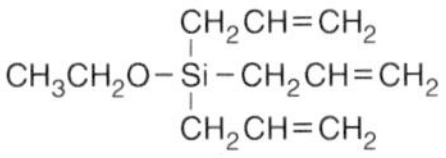

250 mg

S58,587-4
ETHOXYTRIPROPYLSILANE

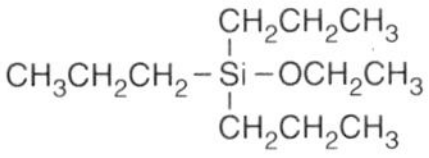

250 mg

S41,534-0
TRIALLYLOXYVINYLSILANE

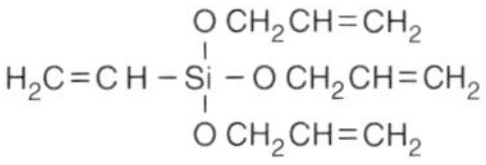

5 gm

S36,526-2
TRIISOPROPOXYVINYLSILANE

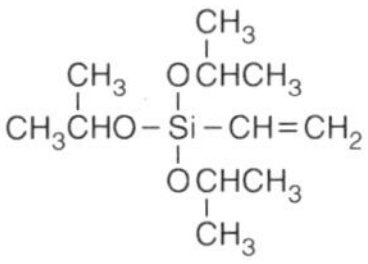

5 gm

S57,633-6
TRIETHOXYPENTYLSILANE

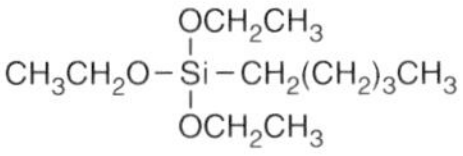

5 gm

S83,061-5
2-(DIMETHYLSILYL)-1,1,1-TRIETHYL-2,3,3-TRIMETHYLTRISILANE

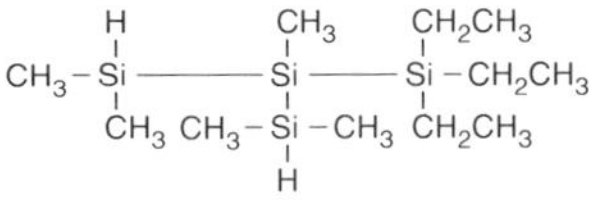

250 mg

S82,869-6
2-ACETYL-1,1,1,3,3,3-HEXAMETHYL-2-(TRIMETHYLSILYL)TRISILANE

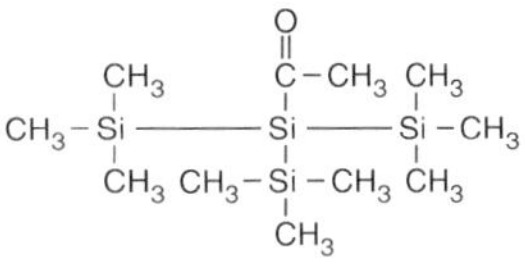

250 mg

S81,574-8
(PERFLUOROOCTYL)TRIMETHYLSILANE

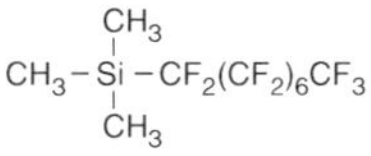

250 mg

S83,696-6
TETRAALLYLSILANE

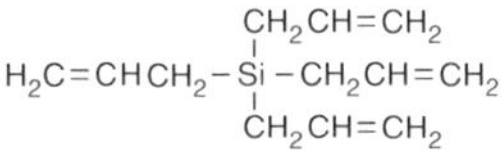

250 mg

S83,070-4
HEXAETHYLDISILANE

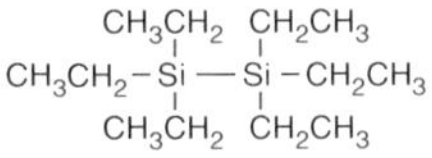

250 mg

S38,782-7
TETRAPROPYLGERMANE

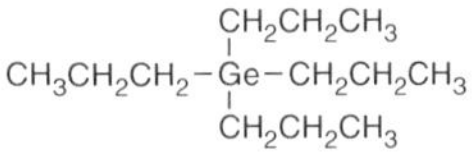

250 mg

S66,714-5
TRIBUTYLGERMANE

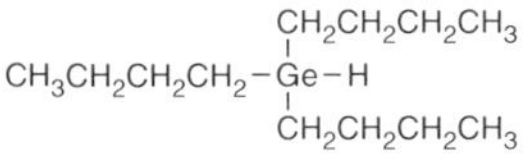

5 gm

S58,530-0
HEXYLTRIETHOXYSILANE

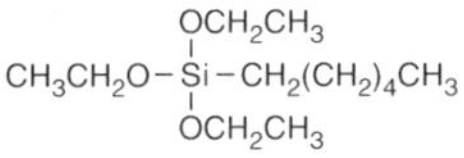

5 gm

S58,564-5
TETRAISOPROPYL ORTHOSILICATE

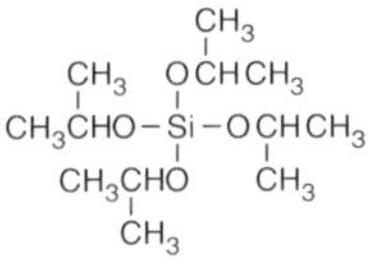

5 gm

S38,785-1
CHLOROTRIBUTYLGERMANE

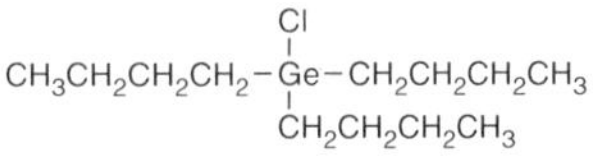

250 mg

S83,970-1
1,4-DIVINYL-1,1,2,2,3,3,4,4-OCTAMETHYLTETRASILANE

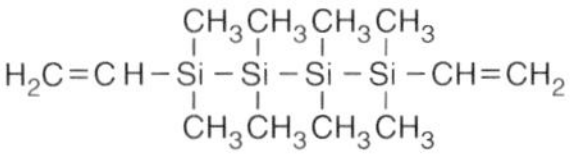

250 mg

S39,531-5
HEXAETHYLDIGERMANE

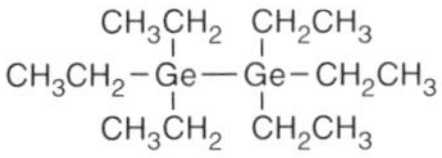

250 mg

S83,062-3
2,2-BIS(DIMETHYLSILYL)-1,1-DIMETHYL-3,3,3-TRIETHYLTRISILANE

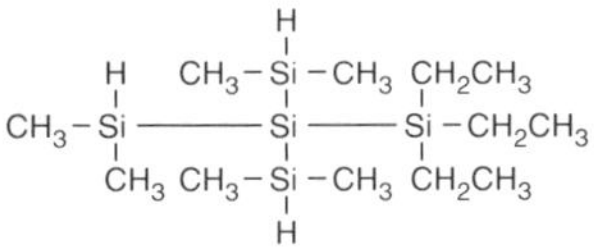

250 mg

S82,301-5
1,1,2,2,3,3,4,4,5,5,6,6-DODECAMETHYLHEXASILANE

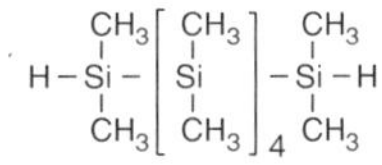

250 mg

S81,575-6
PERFLUOROHEXAMETHYLENE-1,6-BIS-(TRIMETHYLSILANE)

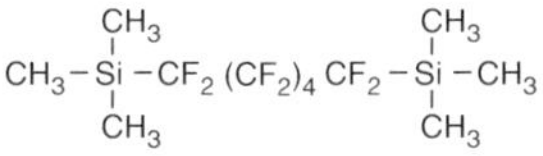

250 mg

S58,538-6
DICHLORODODECYLMETHYLSILANE

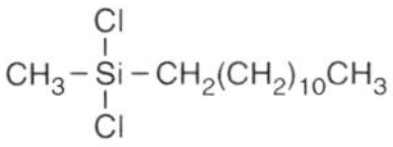

5 gm

S87,603-8
3-(TERT-BUTYLDIMETHYLSILYLOXY)-1-IODO-1-OCTENE

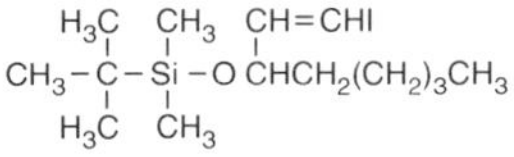

250 mg

S82,360-0
2,2,3,6,7,7-HEXAMETHYL-3,6-BIS-(TRIMETHYLSILYL)-4,5-DIOXA-2,3,6,7-TETRASILAOCTANE

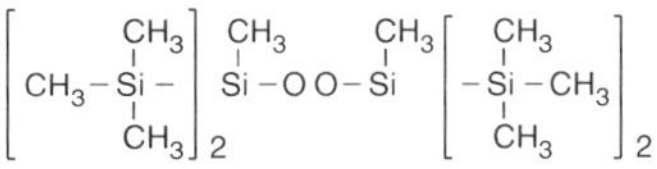

250 mg

S58,581-5
TRIPENTYLOXYSILANE

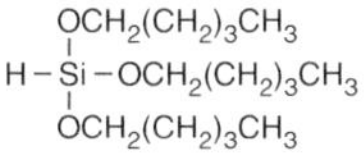

250 mg

S82,422-4
1,2-PROPADIENE-1,3-DIYLIDENETETRAKIS(TRIMETHYLSILANE)

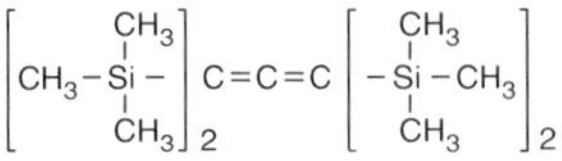

250 mg

S81,471-7
TETRABUTYLSILANE

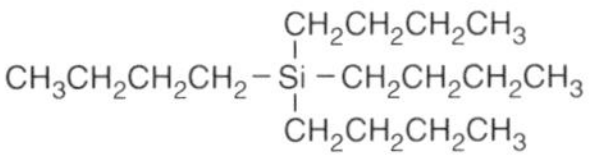

250 mg

S82,423-2
TETRAKIS[(TRIMETHYLSILYL)METHYL]-SILANE

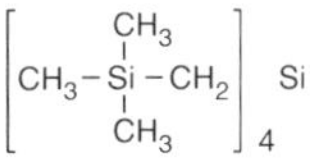

250 mg

S83,207-3
DODECAMETHYL-1-SILACYCLOHEXANE

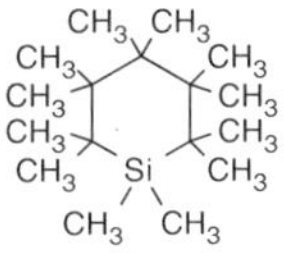

250 mg

S88,483-9
2,5,6,9-TETRAMETHYL-1-(TRIMETHYLSILYLOXY)-1,5,9-DECATRIENE
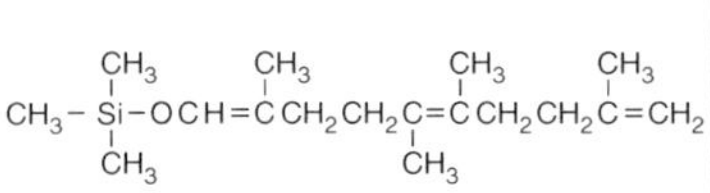
250 mg

S58,539-4
DIETHOXYDODECYLMETHYLSILANE
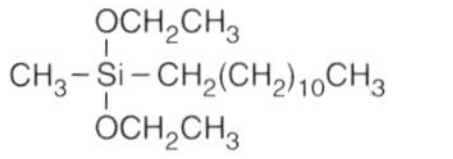
250 mg

S83,560-9
TRIHEXYLGERMANE
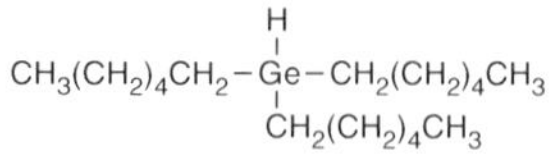
250 mg

S82,632-4
1,2,4,5-HEXATETRAENE-1,6-DIYLIDENETETRAKIS(TRIMETHYLSILANE)
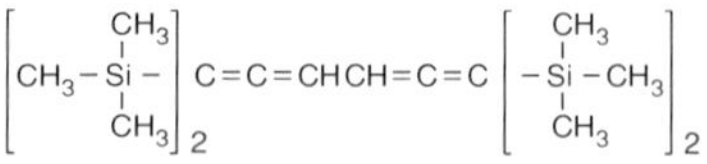
250 mg

S82,420-8
2,5-CYCLOHEXADIENE-1,2,4,5-TETRAYLTETRAKIS(TRIMETHYLSILANE)
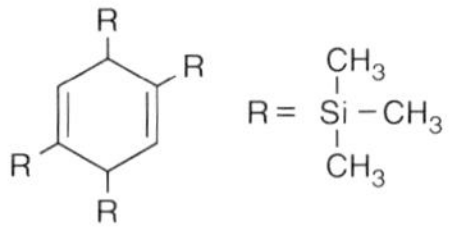
250 mg

S83,222-7
1,1,1,4,4,4-HEXAMETHYL-2,2,3,3-TETRAKIS(TRIMETHYLSILYL)TETRASILANE
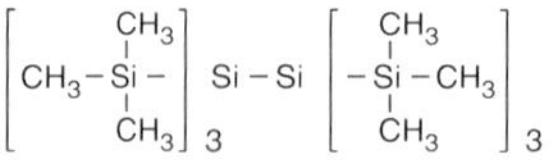
250 mg

S58,583-1
METHOXYTRIPHENYLSILANE
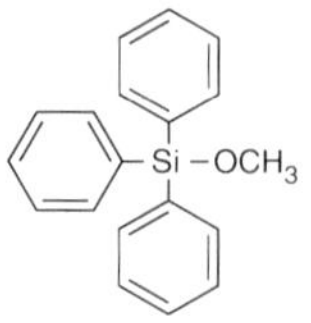
250 mg

S58,540-8
DICHLOROMETHYLOCTADECYLSILANE
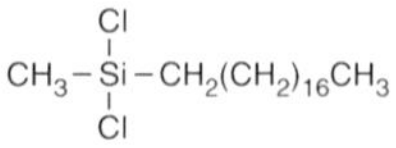
5 gm

S58,582-3
ETHOXYTRIPHENYLSILANE
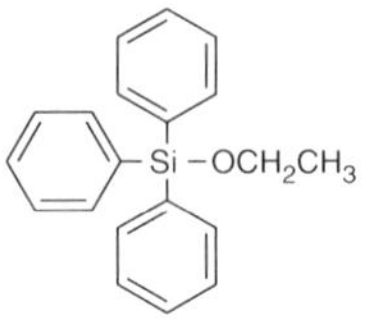
250 mg

S83,147-6
BUTYLHEXADECYLSILANE
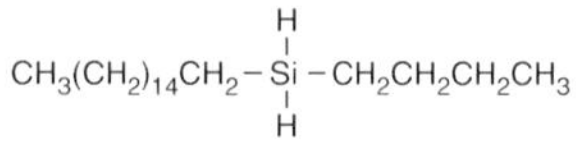
250 mg

S83,778-4
DIDECYLSILANE
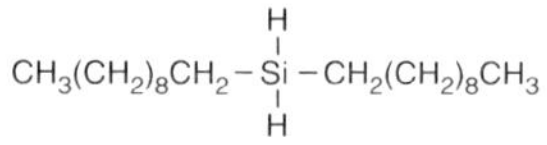
250 mg

S38,784-3
TETRAPENTYLGERMANE
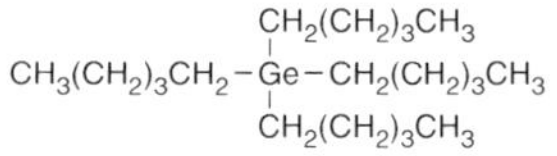
250 mg

S36,485-1
TETRAAMYL ORTHOSILICATE
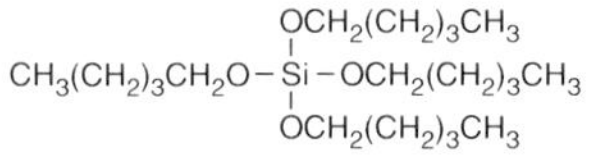
5 gm

S82,252-3
1,1,1,4,4,4-HEXAKIS(TRIMETHYLSILYL)-2-BUTYNE
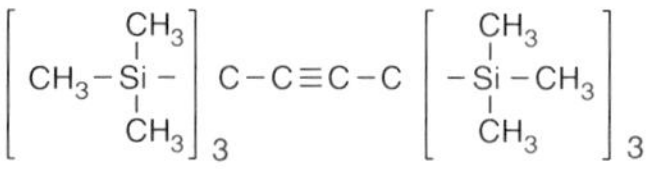
250 mg

S83,223-5
2,2,5,5-TETRAKIS(TRIMETHYLSILYL)-3,3,4,4-TETRAMETHYLHEXASILANE
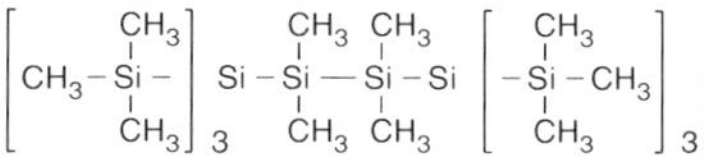
250 mg

S83,554-4
ALLYLDIDECYLSILANE
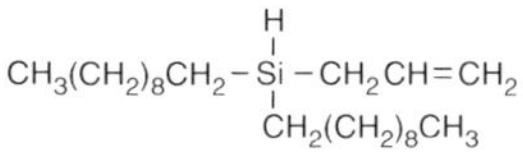
250 mg

S58,541-6
DIETHOXYMETHYLOCTADECYLSILANE
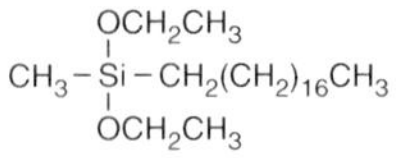
5 gm

S83,151-4
TRIOCTYLSILANE
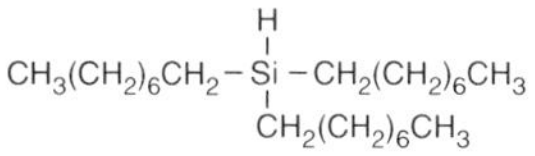
250 mg

S38,886-6
TETRAHEXYLGERMANE
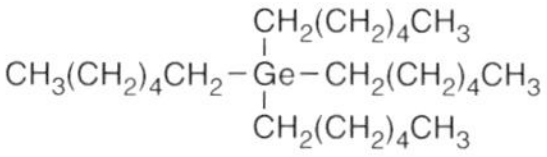
5 gm

S83,150-6
OCTADECYLTRIETHYLGERMANE
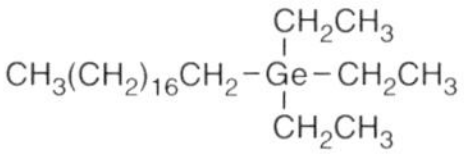
250 mg

S36,493-2
TETRAHEXYL ORTHOSILICATE
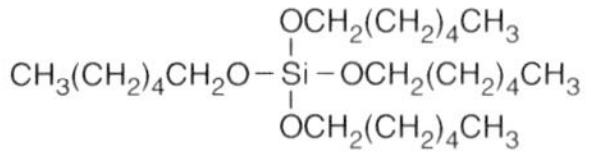
5 gm

S60,261-2
TETRAKIS(2-ETHYL-1-BUTYL)-ORTHOSILICATE
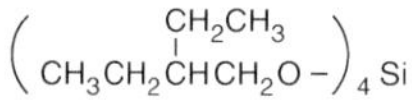
5 gm

S39,532-3
TRIBUTYLGERMYL OXIDE
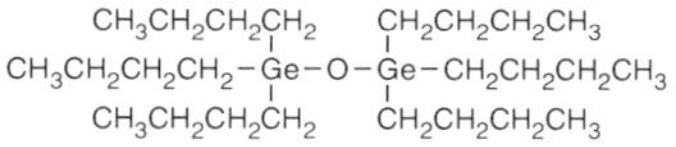
250 mg

S82,634-0
1,2,4,5-HEXATETRAENE-3,4-DIYL-1,6-DIYLIDENEHEXAKIS(TRIMETHYLSILANE)
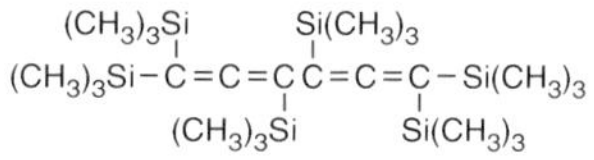
250 mg

S82,638-3
1,1,1,8,8,8-HEXAKIS(TRIMETHYLSILYL)-1,8-DISILAOCTANE
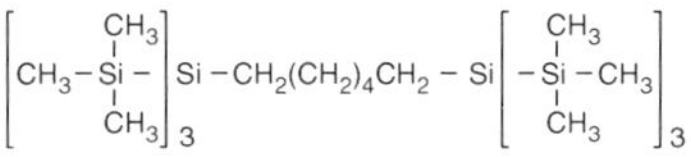
250 mg

S83,224-3
1,1,1,2,3,3,3-HEPTAMETHYL-2-[2,3,4,5-TETRAKIS(TRIMETHYLSILYL)-PHENYL]TRISILANE
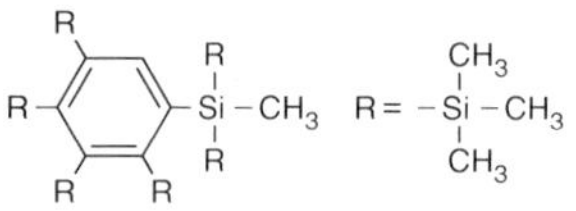
250 mg

S82,460-7
1,2,3,4-TETRAKIS(TRIMETHYLSILYL)-5-[TRIS(TRIMETHYLSILYL)SILYL]BENZENE
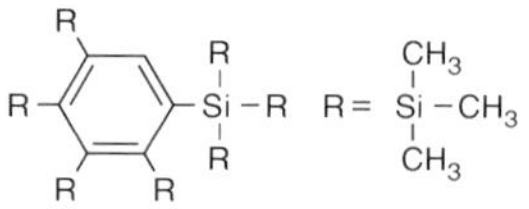
250 mg

S81,661-2
TETRAKIS(M-TOLYL)SILANE
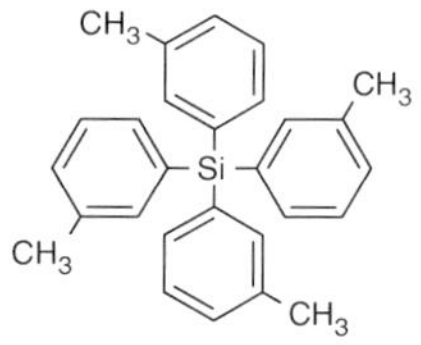
250 mg

S54,984-3
TETRATOLYL ORTHOSILICATE (MIXTURE OF ISOMERS)
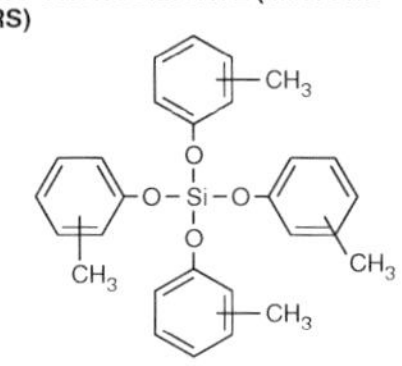
5 gm

S83,431-9
TRIS(DECYL)SILANE
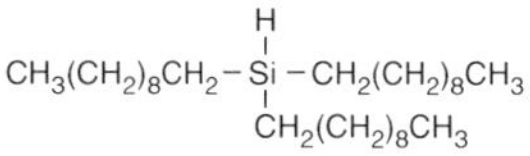
250 mg

S83,800-4
TETRAKIS(2-ETHYLHEXYL)SILANE
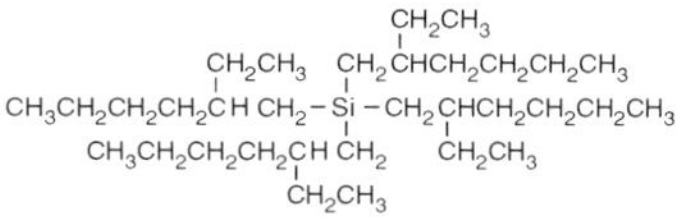
250 mg

S83,801-2
TETRAOCTYLSILANE
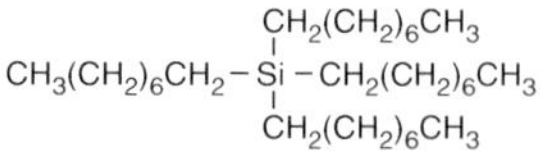
250 mg

S36,500-9
TETRAOCTYL ORTHOSILICATE

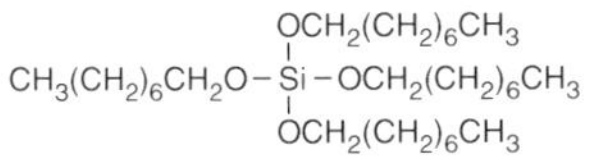

5 gm

S83,433-5
DODECYLTRIS(M-TOLYL)SILANE

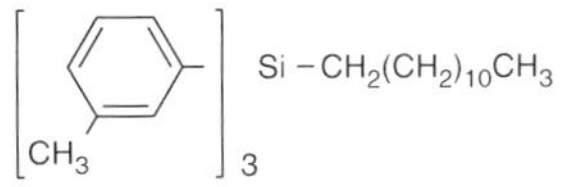

250 mg

S83,154-9
DIOCTADECYLSILANE

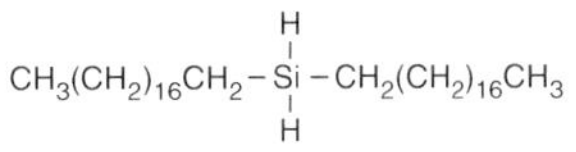

250 mg

S83,779-2
TRIDODECYLSILANE

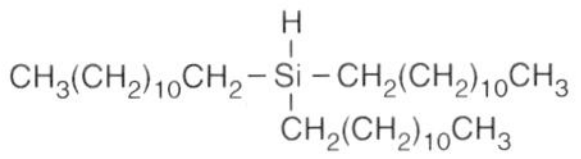

250 mg

S56,512-1
HEXAPHENYLCYCLOTRISILOXANE

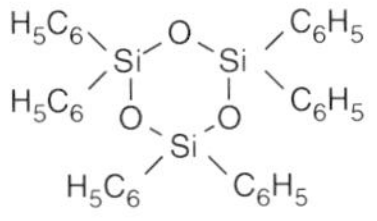

250 mg

S82,497-6
DICHLORODIOCTADECYLSILANE

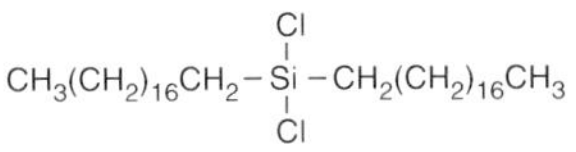

250 mg

S39,545-5
HEXAPHENYLDIGERMANE

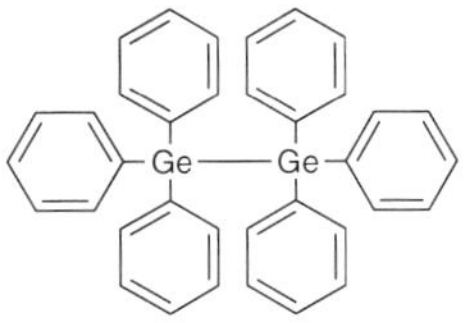

250 mg

S83,442-4
BIS(M-TOLYL)DIDODECYLSILANE

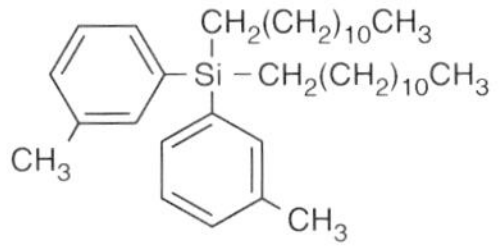

250 mg

S87,567-8
BIS(DIMETHYLHEXADECYLSILYLMETHYL)-MERCURY

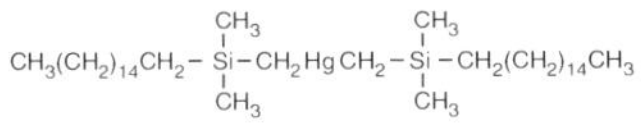

250 mg

S82,761-4
1,4-BIS(TRIPHENYLSILYL)-2,3,4,5,6,7-HEXAHYDRO-1H-1,4-DIAZEPINE

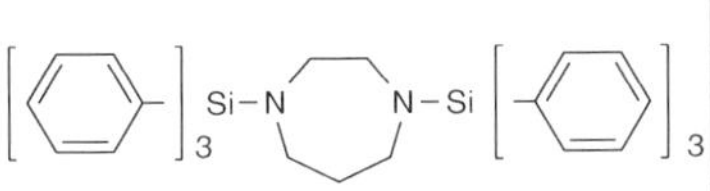

250 mg

S83,446-7
(3-TERT-BUTYLPHENYL)-TRIDODECYLSILANE

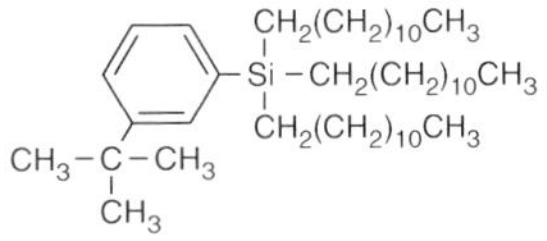

250 mg

S82,507-7
TETRADODECYLSILANE

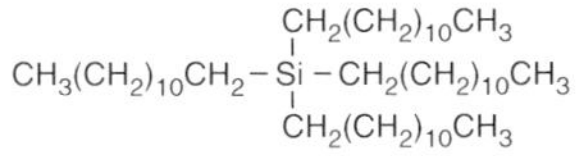

250 mg

S82,781-9
TRIS(HEXADECYL)SILANE

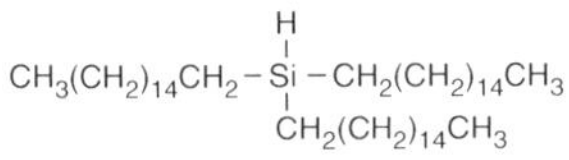

250 mg

S83,302-9
TETRADODECYLGERMANE

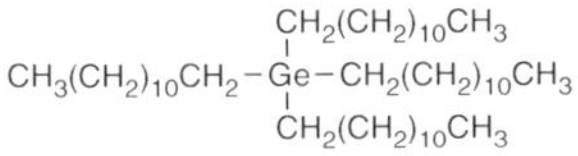

250 mg

S83,455-6
BUTYLTRIHEXADECYLSILANE

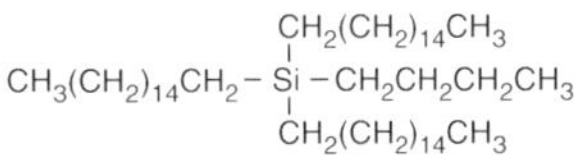

250 mg

S83,309-6
3,3,6,6-TETRABENZYL-1,2,4,5-TETRAPHENYL-1,2,4,5-TETRAAZA-3,6-DISILACYCLOHEXANE

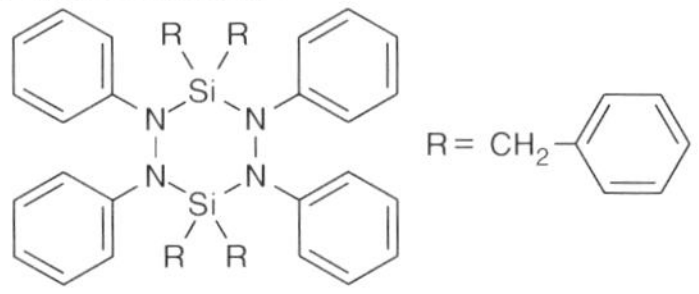

250 mg

S83,511-0
1,2-BIS[TRIS(DECYL)SILYL]ETHANE

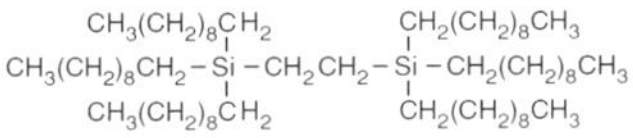

250 mg

S83,185-9
1,1,2,2,3,3,4,4,5,5-DECAPHENYL-6,6-DIMETHYLCYCLOHEXASILANE

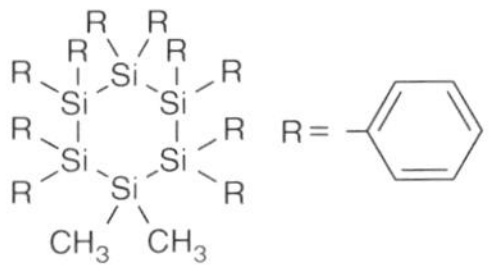

250 mg

S82,784-3
TETRA(HEXADECYL)GERMANE

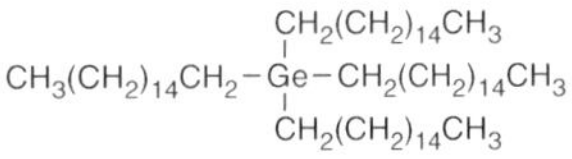

250 mg

S83,242-1
1,3-PHENYLENEBIS(TRISDECYLSILANE)

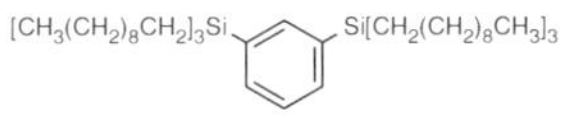

250 mg

S82,454-2
TETRAOCTADECYLSILANE

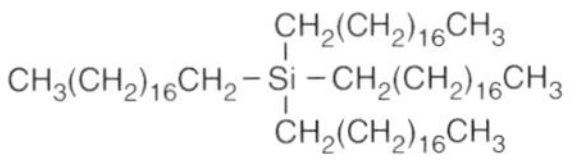

250 mg

S83,512-9
1-(TRIHEXADECYLSILYL)-2-(TRIOCTADECYLSILYL)ETHANE

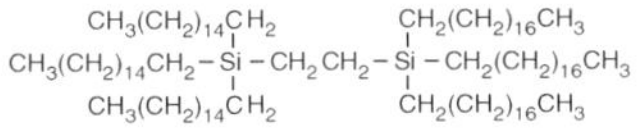

250 mg

S58,501-7
(1,2-DICHLOROETHYL)TRICHLOROSILANE

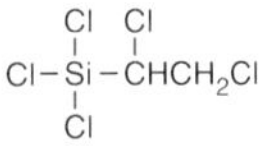

5 gm

S58,500-9
(1,2-DIBROMOETHYL)TRICHLOROSILANE

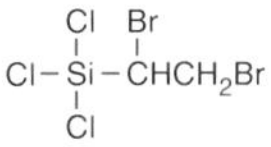

5 gm

S58,489-4
(CHLOROMETHYL)DIETHOXYMETHYLSILANE

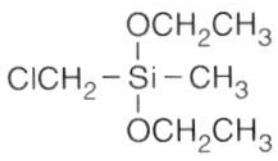

5 gm

S88,456-1
4-CHLORO-1-(TRIMETHYLSILYL)-1-BUTYNE

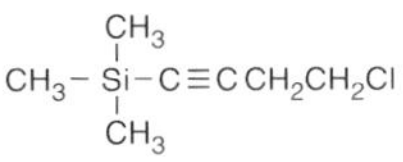

250 mg

S83,037-2
(2-BROMOETHYL)TRIPHENYLSILANE

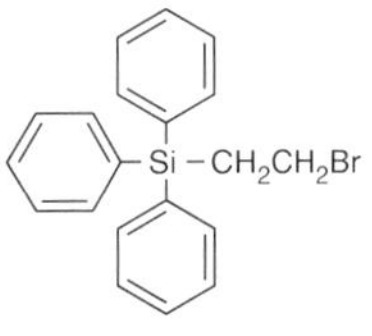

250 mg

S83,119-0
(CHLOROMETHYL)DIMETHYL-(PENTAFLUOROPHENYL)SILANE

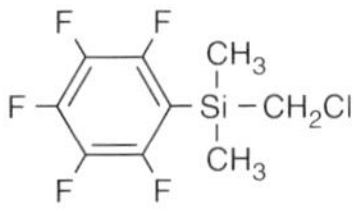

250 mg

S82,371-6
(CHLOROMETHYL)DIMETHYL-(PENTACHLOROPHENYL)SILANE

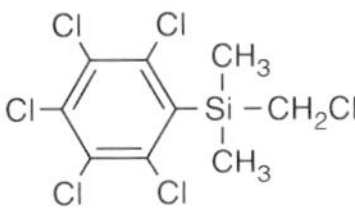

250 mg

S82,440-2
(2,3,5,6-TETRAFLUORO-1,4-PHENYLENE)-BIS[(CHLOROMETHYL)DIMETHYLSILANE]

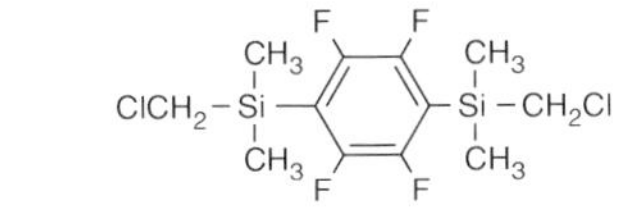

250 mg

S82,994-3
1-(2-CHLOROETHOXY)-3-(TRIPHENYLSILYL)-2-PROPANOL

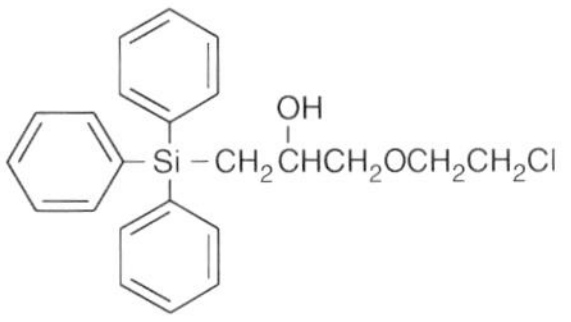

250 mg

S83,496-3
(CHLOROMETHYL)DIPHENYLMETHYLSILANE

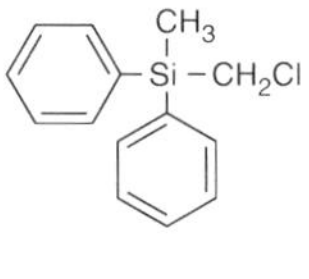

250 mg

S82,347-3
(CHLOROMETHYL)TRIPHENYLGERMANE
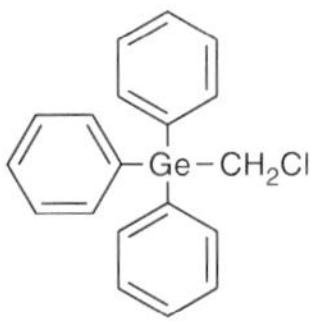
250 mg

S83,023-2
(3-BROMOPROPYL)TRIPHENYLSILANE
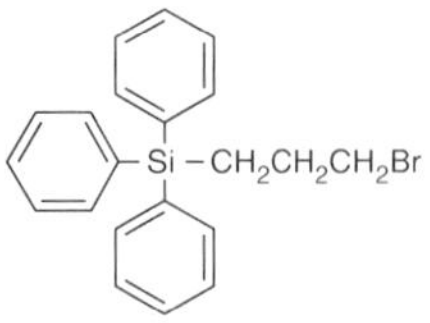
250 mg

S81,978-6
(3-BROMOPROPYL)TRIPHENYLSILANE
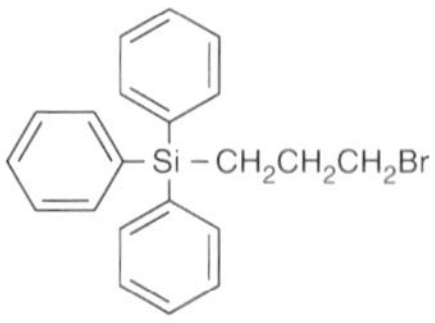
250 mg

S83,717-2
[4-(BROMOMETHYL)PHENYL]-TRIPHENYLSILANE
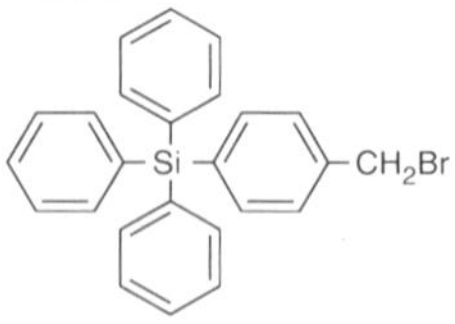
250 mg

S57,066-4
CHLORO(DICHLOROMETHYL)-DIMETHYLSILANE
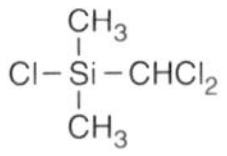
5 gm

S46,304-3
(DICHLOROMETHYL)TRIMETHYLSILANE
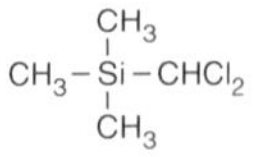
5 gm

S81,304-4
(DICHLOROMETHYL)TRIPHENYLSILANE
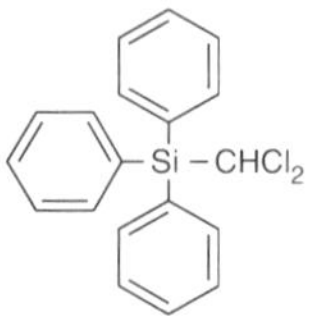
250 mg

S58,577-7
N-(HEXAFLUOROISOPROPYLIDENE)-TRIMETHYLSILYLAMINE
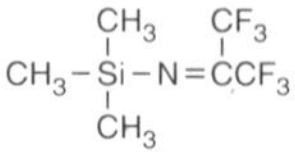
5 gm

S77,581-9
BIS(1,1,2,2,3,3,3-HEPTAFLUOROPROPYL)DIMETHYLSILANE
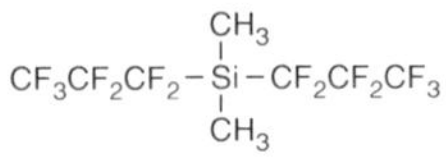
250 mg

S86,641-5
3-(TRIFLUOROMETHYL)PHENYL TRIMETHYLSILYL ETHER
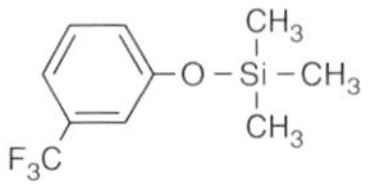
250 mg

S77,596-7
BIS(3,3,4,4,5,5,5-HEPTAFLUOROPENTYL)DIETHOXYSILANE
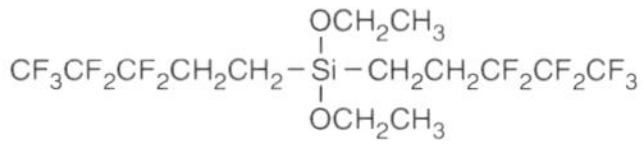
250 mg

S83,251-0
TETRAKIS[3-(TRIFLUOROMETHYL)PHENYL]-SILANE
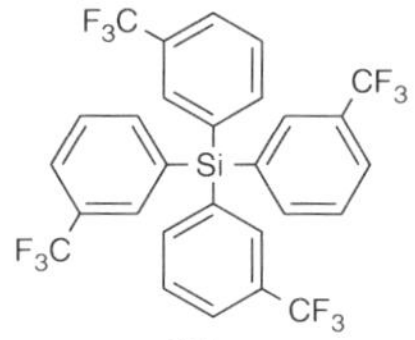
250 mg

S83,884-5
[3-(TRIFLUOROMETHYL)PHENYL]-TRIPHENYLSILANE
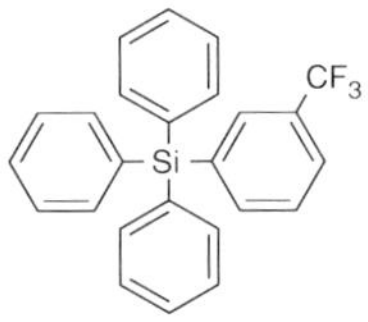
250 mg

S84,123-4
BIS[3-(TRIFLUOROMETHYL)PHENYL]-DIPHENYLSILANE
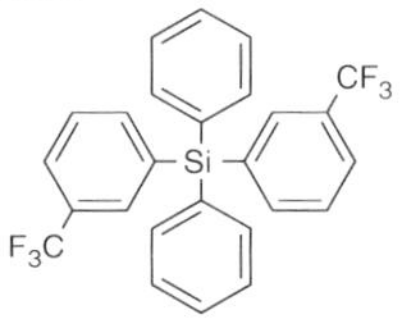
250 mg

S81,555-1
DIMETHYL(2,3,5,6-TETRAFLUOROPHENYL)-SILANE
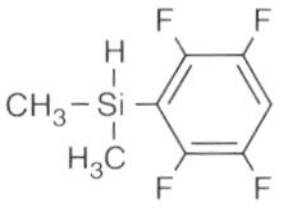
250 mg

S83,996-5
DIMETHYL(PENTACHLOROPHENYL)SILANE
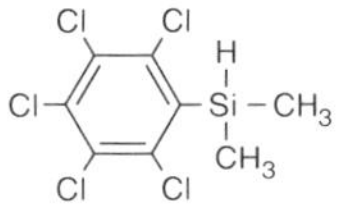
250 mg

S86,632-6
3-CHLOROPHENYL TRIMETHYLSILYL ETHER
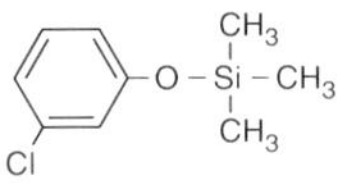
250 mg

S81,559-4
(2,3,5,6-TETRAFLUOROPHENYL)-TRIMETHYLSILANE
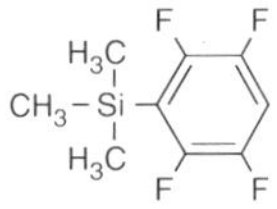
250 mg

S82,552-2
(2,3,4,5-TETRACHLOROPHENYL)-TRIMETHYLSILANE
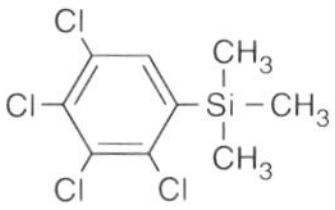
250 mg

S82,866-1
(2,3,5,6-TETRACHLOROPHENYL)-TRIMETHYLSILANE
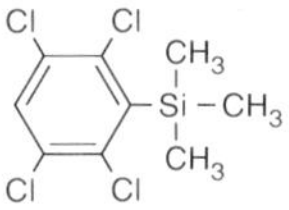
250 mg

S81,221-8
(PENTACHLOROPHENYL)TRIMETHYLSILANE
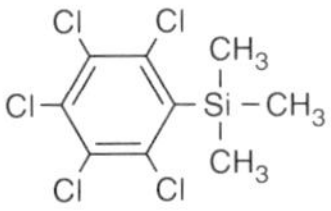
250 mg

S82,914-5
2,4,6-TRICHLORO-1,3-PHENYLENEBIS-(DIMETHYLSILANE)
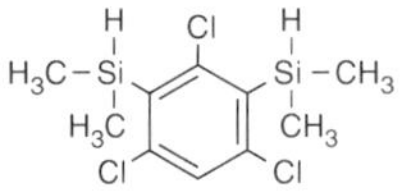
250 mg

S82,441-0
(2,3,5,6-TETRACHLORO-1,4-PHENYLENE)-BIS(DIMETHYLSILANE)
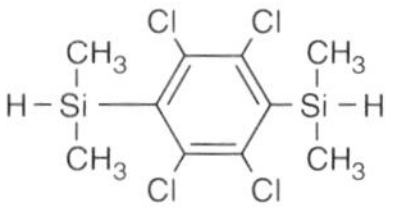
250 mg

S82,544-1
1-(PENTACHLOROPHENYL)-1,1,3,3,3-PENTAMETHYLDISILOXANE
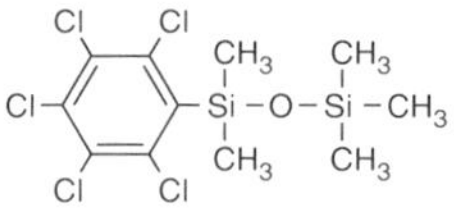
250 mg

S82,541-7
BUTYLDIMETHYL(2,3,4,5-TETRAFLUOROPHENYL)SILANE
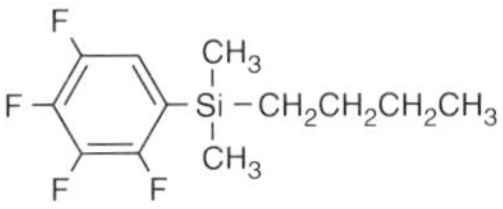
250 mg

S82,542-5
BUTYLDIMETHYL(2,3,5,6-TETRAFLUOROPHENYL)SILANE
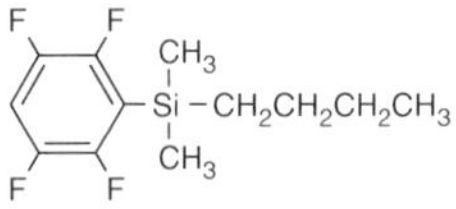
250 mg

S82,543-3
BUTYLDIMETHYL(PENTAFLUOROPHENYL)-SILANE
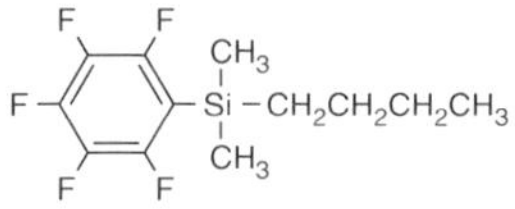
250 mg

S82,384-8
1,4-BIS(TRIMETHYLSILYL)-2,3,5,6-TETRAFLUOROBENZENE
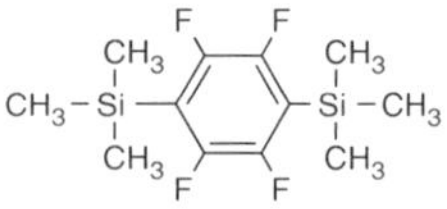
250 mg

S82,556-5
(2,3,5-TRICHLORO-1,4-PHENYLENE)BIS-(TRIMETHYLSILANE)
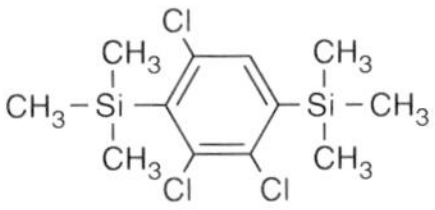
250 mg

S82,442-9
(2,3,5,6-TETRACHLORO-1,4-PHENYLENE)-BIS(TRIMETHYLSILANE)
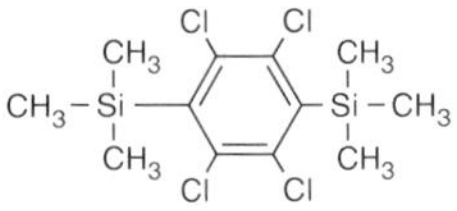
250 mg

S82,540-9
(3,4,5,6-TETRACHLORO-1,2-PHENYLENE)-BIS(TRIMETHYLSILANE)
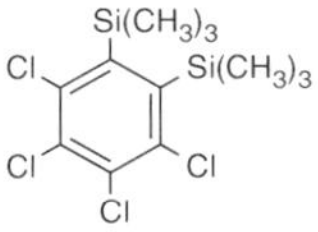
250 mg

S82,300-7
BIS(PENTACHLOROPHENYL)-DICHLOROSILANE
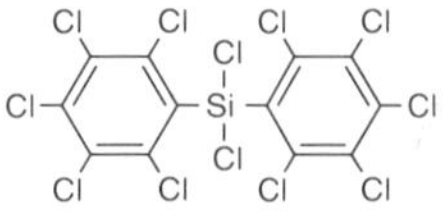
250 mg

S82,370-8
BIS(PENTAFLUOROPHENYL)METHYLSILANE

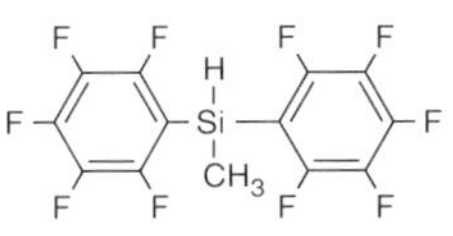

250 mg

S81,215-3
BIS(PENTACHLOROPHENYL)METHYLSILANE

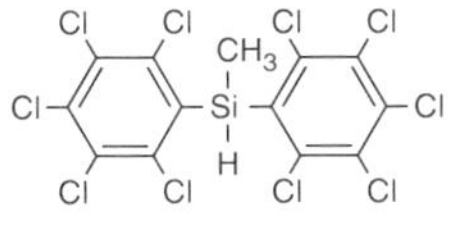

250 mg

S82,430-5
(3-[DIMETHYL(PENTAFLUOROPHENYL)-SILYL]PROPYL)TRIMETHYLSILANE

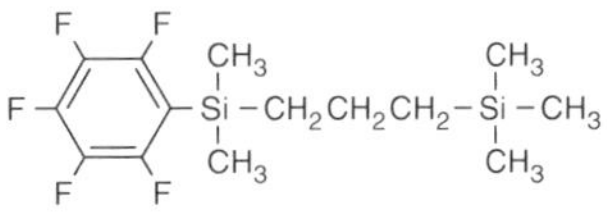

250 mg

S82,548-4
BIS(2,3,5,6-TETRAFLUOROPHENYL)-DIMETHYLSILANE

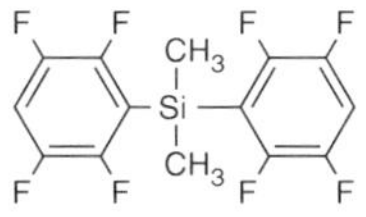

250 mg

S82,915-3
BIS(2,3,4,5-TETRAFLUOROPHENYL)-DIMETHYLSILANE

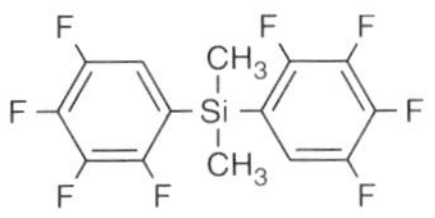

250 mg

S83,539-0
BIS(PENTAFLUOROPHENYL)-DIMETHYLSILANE

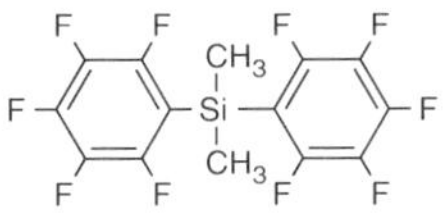

250 mg

S82,549-2
BIS(2,3,5,6-TETRACHLOROPHENYL)-DIMETHYLSILANE

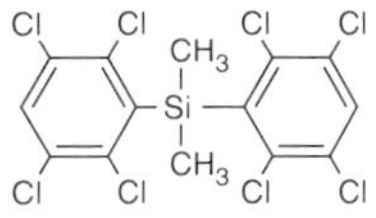

250 mg

S82,560-3
BIS(2,3,4,5-TETRACHLOROPHENYL)-DIMETHYLSILANE

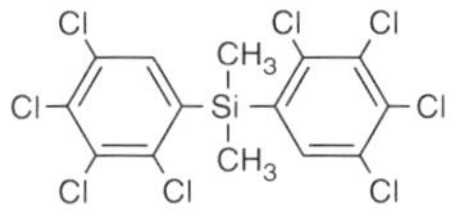

250 mg

S82,376-7
BIS(PENTACHLOROPHENYL)-DIMETHYLSILANE

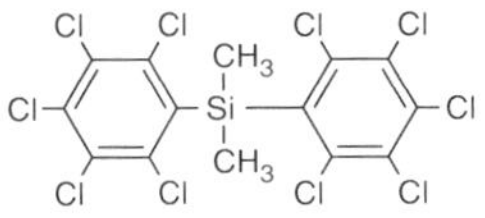

250 mg

S81,636-1
1,3-DIFLUORO-2,4,6-TRIS-(TRIMETHYLSILYL)BENZENE

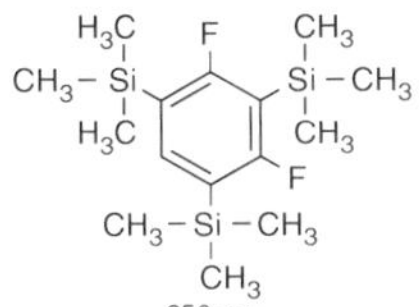

250 mg

S81,635-3
1,3,5-TRIFLUORO-2,4,6-TRIS-(TRIMETHYLSILYL)BENZENE

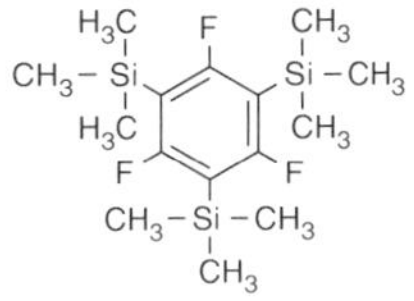

250 mg

S82,437-2
1,3-BIS(PENTAFLUOROPHENYL)-1,1,3,3-TETRAMETHYLDISILOXANE

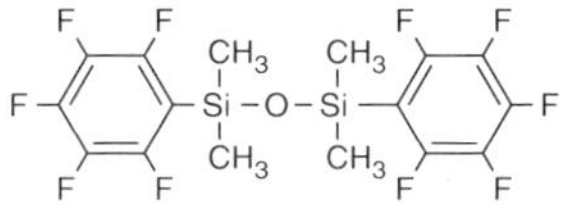

250 mg

S82,546-8
1,1'-(2,3,5,6-TETRACHLORO-1,4-PHENYLENE)BIS(1,1,3,3,3-PENTAMETHYLDISILOXANE)

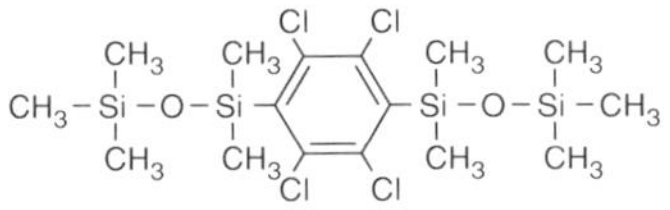

250 mg

S83,127-1
1,2-BIS(PENTACHLOROPHENYL)-1,1,2,2-TETRAMETHYLDISILANE

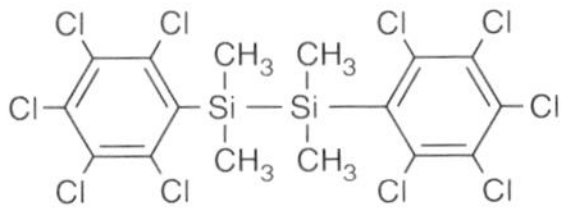

250 mg

S82,916-1
1,3-BIS(PENTACHLOROPHENYL)-1,1,3,3-TETRAMETHYLDISILOXANE

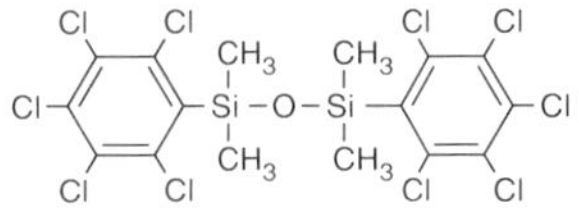

250 mg

S82,436-4
DIMETHYL(PENTAFLUOROPHENYL)[3-(PENTAFLUOROPHENYL)PROPYL]SILANE

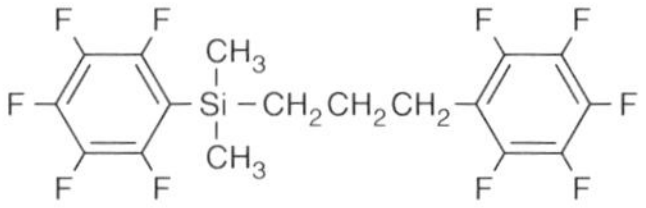

250 mg

S82,443-7
DIMETHYL[3-(PENTACHLOROPHENYL)-PROPYL](PENTAFLUOROPHENYL)SILANE

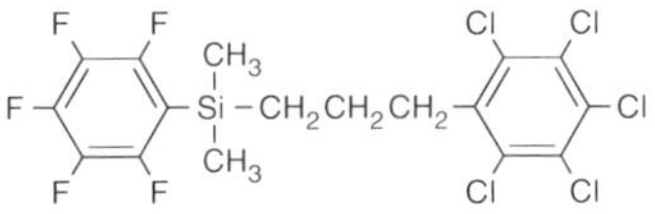

250 mg

S82,659-6
1,3-BIS(PERFLUOROPHENYL)-1,1,2,2,3,3-HEXAMETHYLTRISILANE

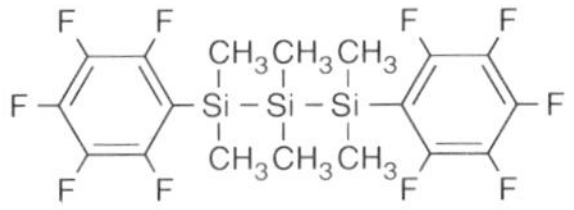

250 mg

S82,438-0
TRIS(PENTAFLUOROPHENYL)SILANE

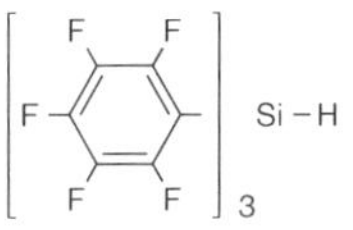

250 mg

S83,129-8
1,3-BIS(PENTACHLOROPHENYL)-1,1,2,2,3,3-HEXAMETHYLTRISILANE

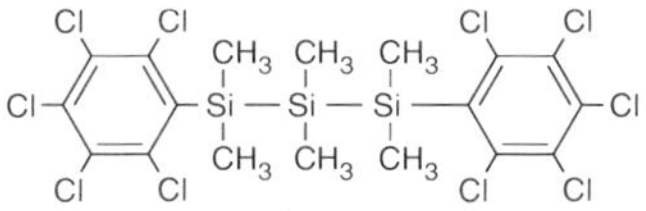

250 mg

S82,918-8
1,5-BIS(PENTACHLOROPHENYL)-1,1,3,3,5,5-HEXAMETHYLTRISILOXANE

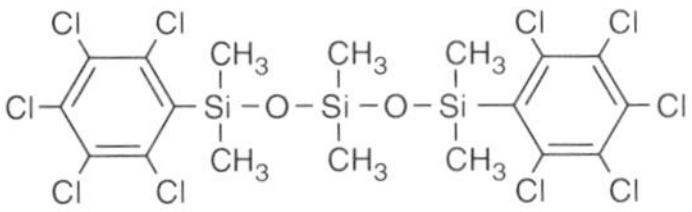

250 mg

S81,213-7
TRIS(PENTACHLOROPHENYL)SILANE

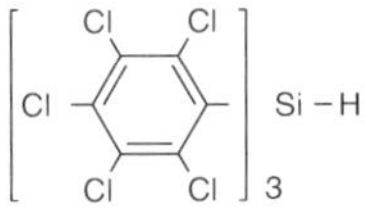

250 mg

S83,423-8
(3-BROMOPHENYL)TRIMETHYLSILANE

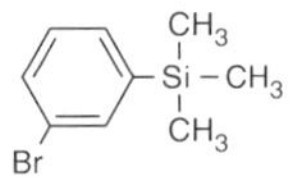

250 mg

S83,541-2
METHYLTRIS(PENTAFLUOROPHENYL)SILANE

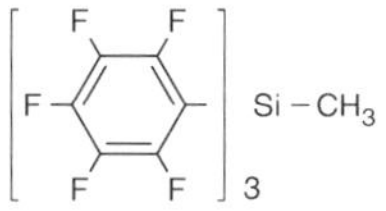

250 mg

S81,639-6
METHYLTRIS(PENTACHLOROPHENYL)SILANE

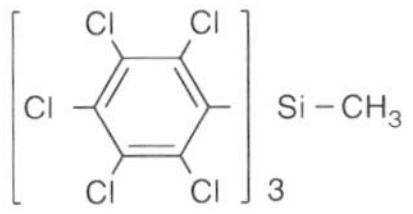

250 mg

S83,132-8
1,4-BIS(PENTACHLOROPHENYL)-1,1,2,2,3,3,4,4-OCTAMETHYLTETRASILANE

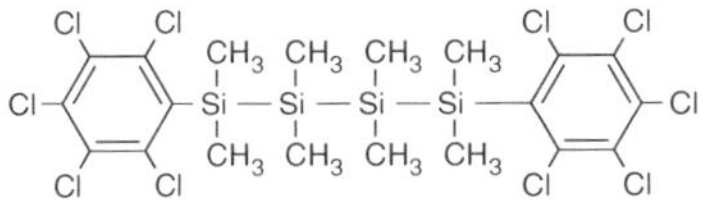

250 mg

S82,316-3
TETRAKIS(3-FLUOROPHENYL)SILANE

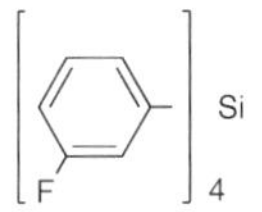

250 mg

S81,213-7
TETRAKIS(3-CHLOROPHENYL)SILANE

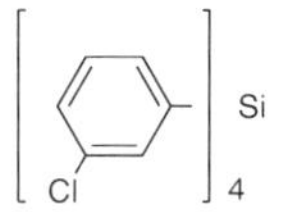

250 mg

S83,543-9
TETRAKIS(PENTAFLUOROPHENYL)SILANE

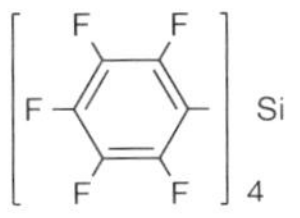

250 mg

S83,504-8
DODECYLTRIS(3-FLUOROPHENYL)SILANE

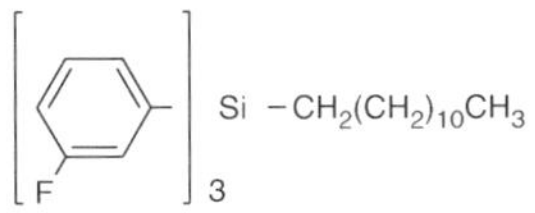

250 mg

S83,502-1
DODECYLTRIS(3-CHLOROPHENYL)SILANE

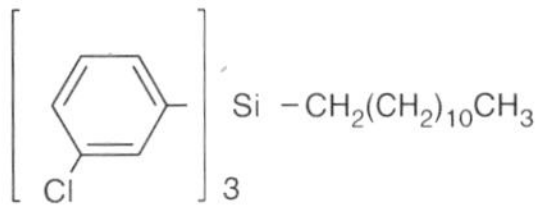

250 mg

S82,731-2
HEXADECYLTRIS(3-CHLOROPHENYL)SILANE

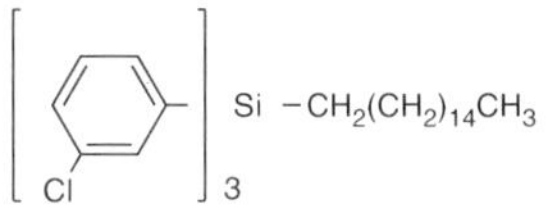

250 mg

S83,438-6
BIS(3-FLUOROPHENYL)DIDODECYLSILANE

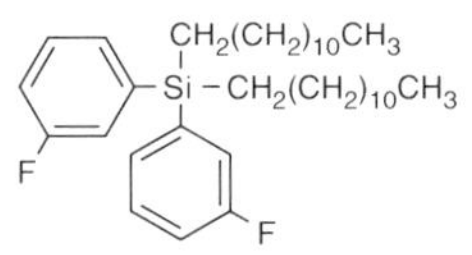

250 mg

S82,314-7
OCTADECYLTRIS(3-FLUOROPHENYL)SILANE

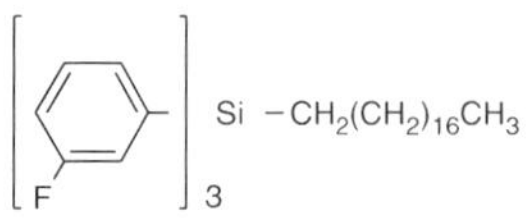

250 mg

S83,437-8
BIS(3-CHLOROPHENYL)DIDODECYLSILANE

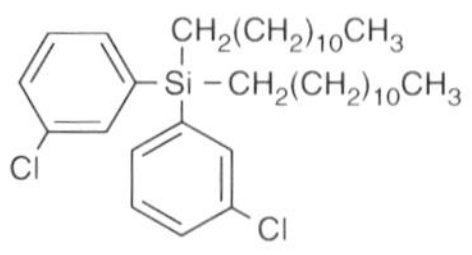

250 mg

S82,728-2
OCTADECYLTRIS(3-CHLOROPHENYL)SILANE

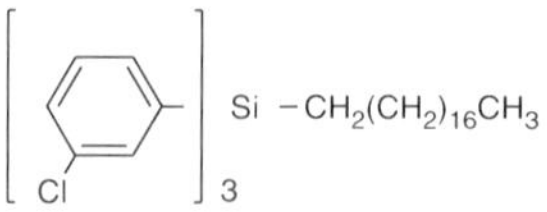

250 mg

S82,930-7
HEXADECYLTRIS(3-FLUOROBENZYL)SILANE

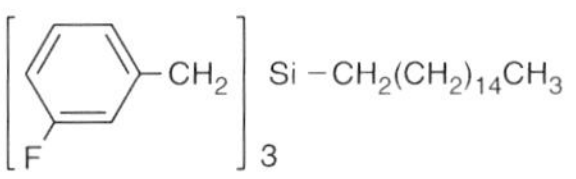

250 mg

S83,074-7
BIS(3-FLUOROBENZYL)DIDODECYLSILANE

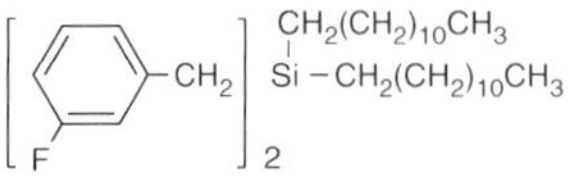

250 mg

S82,932-3
1,1,1,3,3,3-HEXAKIS(3-FLUOROBENZYL)-DISILOXANE

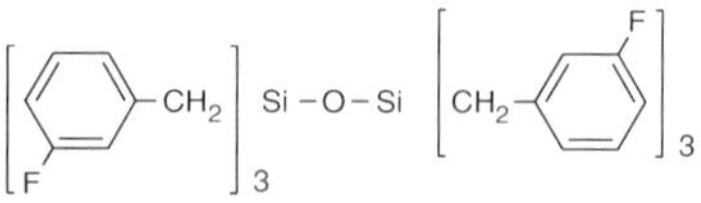

250 mg

S82,765-7
[5-CHLORO-2-(PHENOXY)PHENYL]-TRIPHENYLSILANE

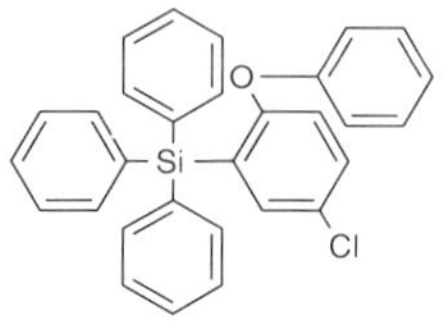

250 mg

S82,428-3
DIMETHYL(PENTAFLUOROPHENYL)(1-PHENYLETHOXY)SILANE

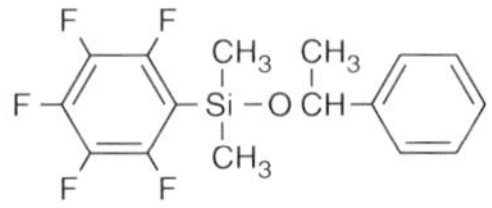

250 mg

S82,385-6
[(PENTACHLOROPHENYL)PHENYLMETHOXY]TRIMETHYLSILANE

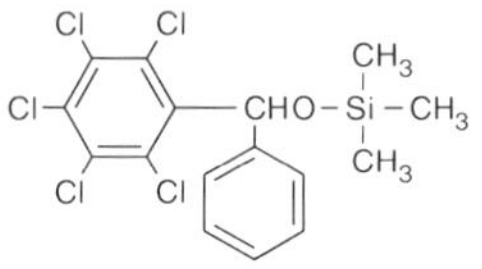

250 mg

S82,429-1
DIMETHYL(DIPHENYLMETHOXY)-(PENTAFLUOROPHENYL)SILANE

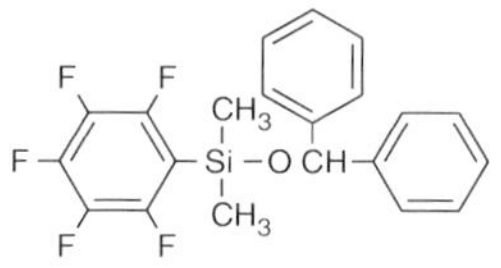

250 mg

S82,378-3
N,N-BIS(TRIMETHYLSILYL)-2,3,4,5,6-PENTAFLUOROANILINE

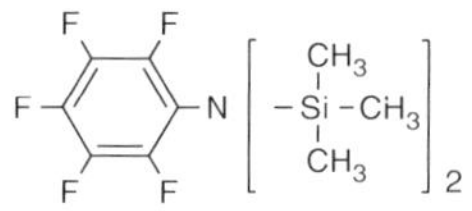

250 mg

S82,379-1
N,N-BIS(DIMETHYLPHENYLSILYL)-2,3,4,5,6-PENTAFLUOROANILINE

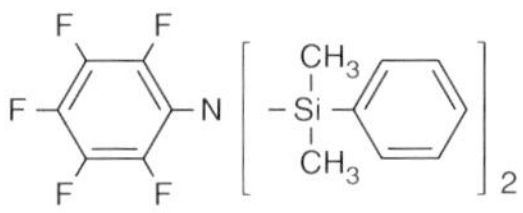

250 mg

S82,380-5
N,N-BIS[(DIPHENYL)METHYLSILYL]-2,3,4,5,6-PENTAFLUOROANILINE

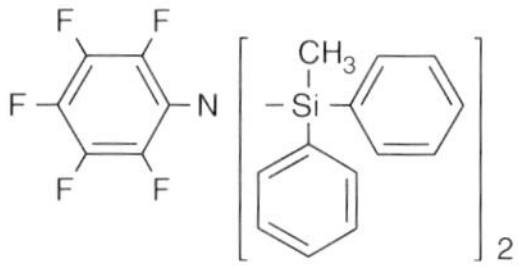

250 mg

S83,007-0
2',3',5',6'-TETRACHLORO-4'-(TRIMETHYLSILYL)ACETOPHENONE

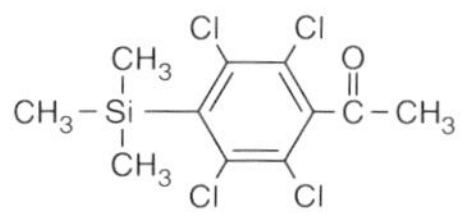

250 mg

S81,642-6
METHYL(PENTAFLUOROPHENYL)-PHENYLSILANE

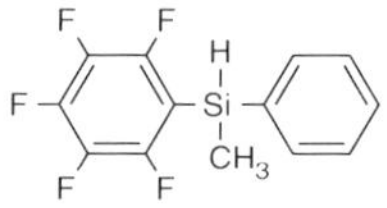

250 mg

S83,174-3
DIMETHYL(PENTAFLUOROPHENYL)-PHENYLSILANE

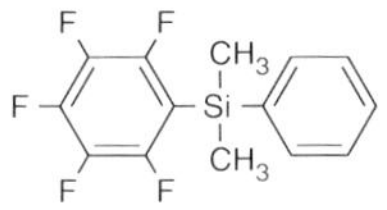

250 mg

S82,386-4
DIMETHYL(PENTACHLOROPHENYL)-PHENYLSILANE

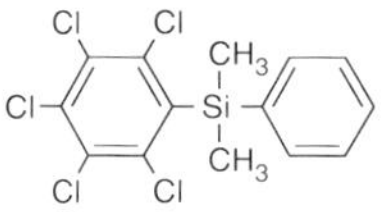

250 mg

S82,431-3
DIMETHYL(PENTAFLUOROPHENYL)-STYRYLSILANE

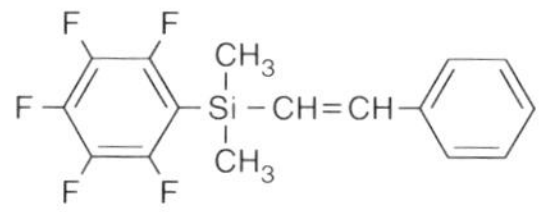

250 mg

S82,367-8
DIPHENYL(PENTACHLOROPHENYL)SILANE

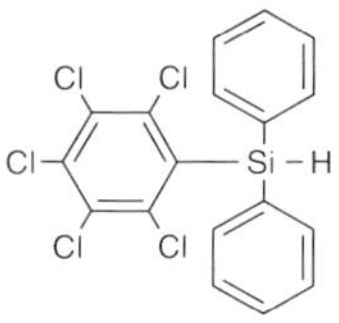

250 mg

S83,542-0
DIPHENYLMETHYL(PENTAFLUOROPHENYL)-SILANE

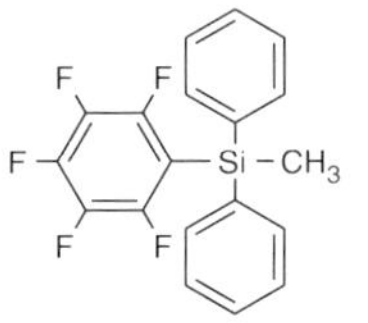

250 mg

S82,387-2
DIPHENYLMETHYL(PENTACHLOROPHENYL)-SILANE

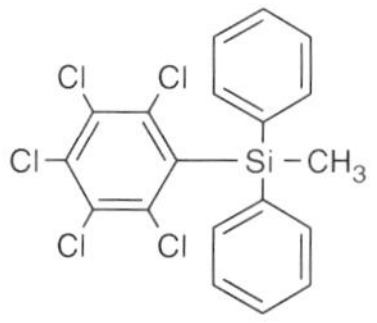

250 mg

S82,561-1
BIS(2,3,4,5-TETRACHLOROPHENYL)-METHYLPHENYLSILANE

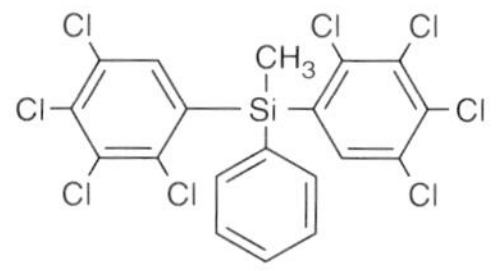

250 mg

S83,130-1
BIS(PENTACHLOROPHENYL)-METHYLPHENYLSILANE

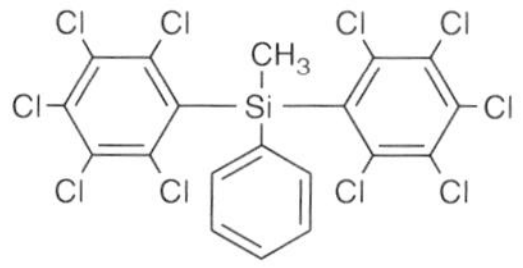

250 mg

S82,434-8
BIS(PENTAFLUOROPHENYL)METHYL-(BETA-STYRYL)SILANE

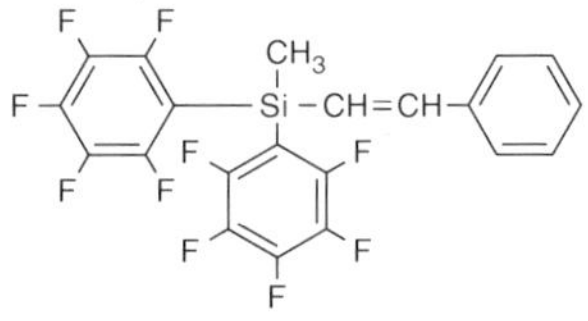

250 mg

S82,435-6
BIS(PENTAFLUOROPHENYL)METHYL(ALPHA-STYRYL)SILANE

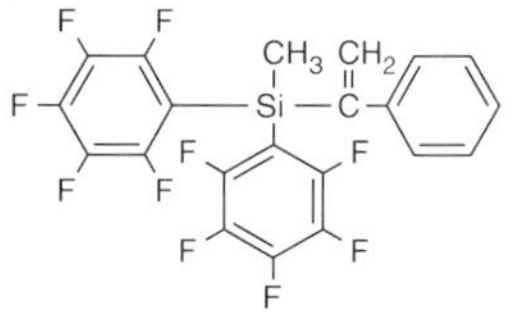

250 mg

S82,563-8
(2,3,6-TRICHLORO-1,4-PHENYLENE)BIS-(DIMETHYLPHENYLSILANE)

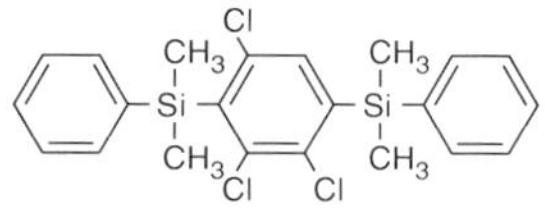

250 mg

S83,262-6
2,3,5,6-TETRACHLORO-1,4-PHENYLENEBIS(DIMETHYLPHENYLSILANE)

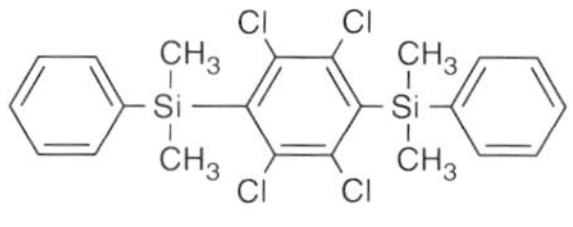

250 mg

S82,432-1
1,3-DIPHENYL-1-PROPENYLOXY(DIMETHYL)(PENTAFLUOROPHENYL)SILANE

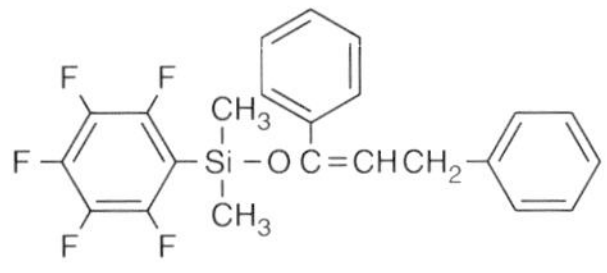

250 mg

S83,499-8
(3-CHLOROPHENYL)TRIPHENYLSILANE

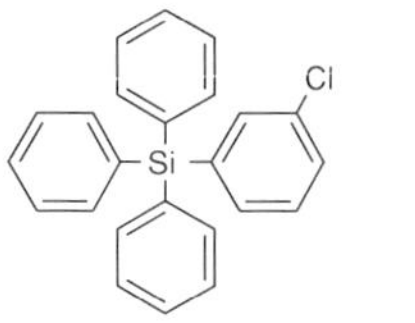

250 mg

S84,030-0
BIS(3-FLUOROPHENYL)DIPHENYLSILANE

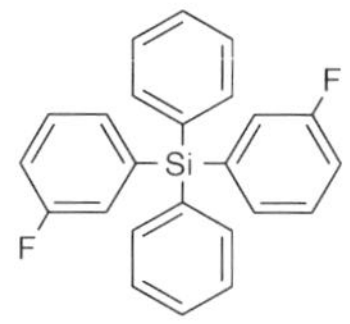

250 mg

S82,315-5
PHENYLTRIS(3-FLUOROPHENYL)SILANE

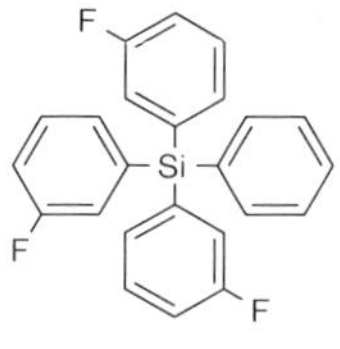

250 mg

S84,031-9
BIS(3-CHLOROPHENYL)DIPHENYLSILANE
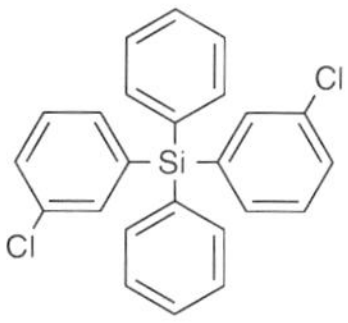
250 mg

S82,568-9
(2,3,4,5-TETRAFLUOROPHENYL)-TRIPHENYLSILANE
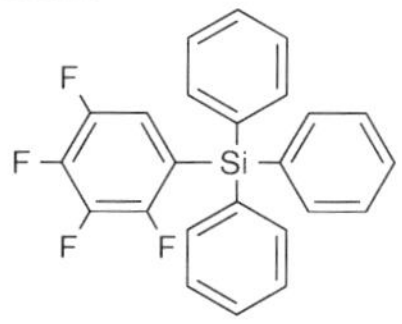
250 mg

S83,124-7
(PENTAFLUOROPHENYL)TRIPHENYLGERMANE
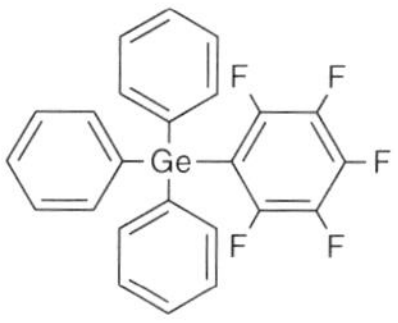
250 mg

S82,566-2
(2,3,4,5-TETRACHLOROPHENYL)-TRIPHENYLSILANE
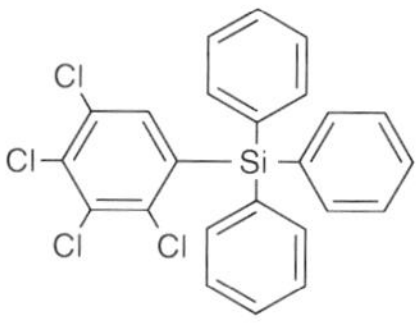
250 mg

S82,388-0
(PENTACHLOROPHENYL)TRIPHENYLSILANE
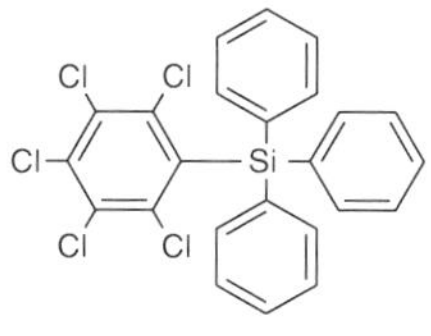
250 mg

S81,425-3
BIS(PENTAFLUOROPHENYL)DIPHENYSILANE
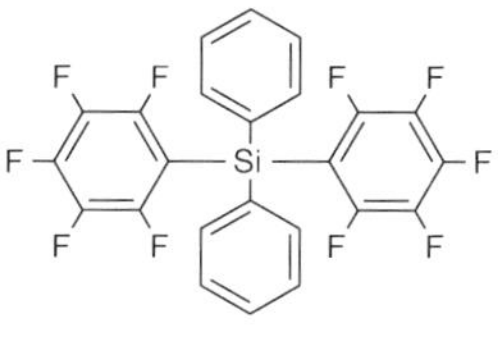
250 mg

S82,565-4
(2,3,4,5-TETRACHLOROPHENYL)-TRIPHENYLGERMANE
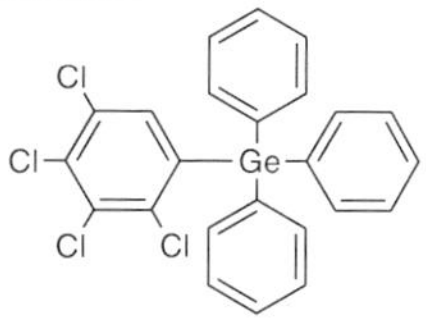
250 mg

S83,211-1
(PENTACHLOROPHENYL)TRIPHENYLGERMANE
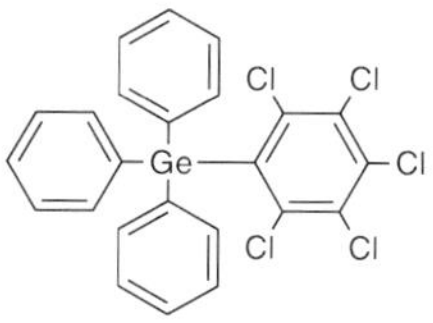
250 mg

S83,123-9
BIS(PENTAFLUOROPHENYL)-DIPHENYLGERMANE
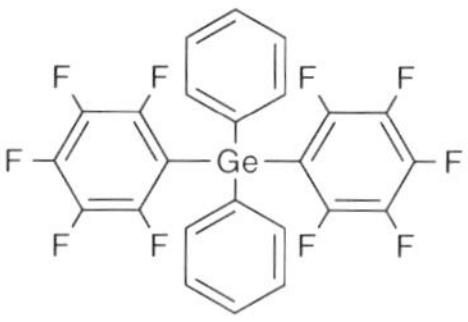
250 mg

S82,564-6
BIS(2,3,4,5-TETRACHLOROPHENYL)-DIPHENYLSILANE
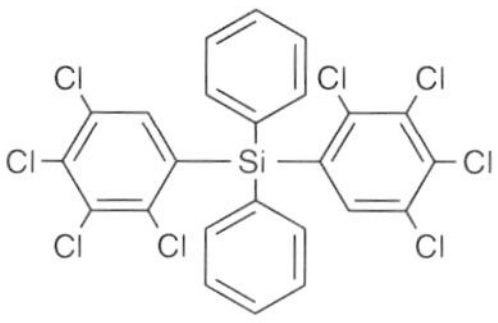
250 mg

S83,122-0
BIS(PENTACHLOROPHENYL)-DIPHENYLGERMANE
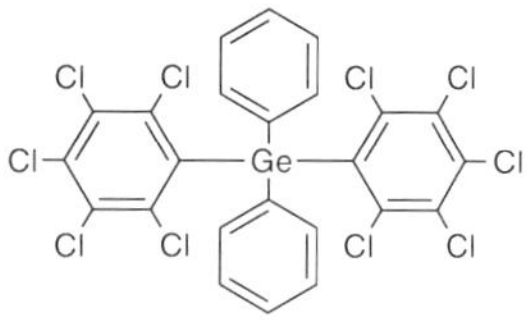
250 mg

S82,439-9
TRANS-STYRYLTRIS(PENTAFLUOROPHENYL)-SILANE
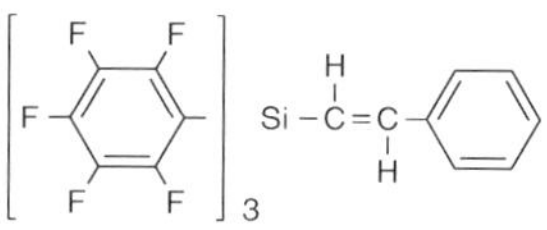
250 mg

S82,661-8
2,3,5,6-TETRAFLUORO-1-(TRIMETHYLSILYL)-4-(TRIPHENYLSILYL)-BENZENE
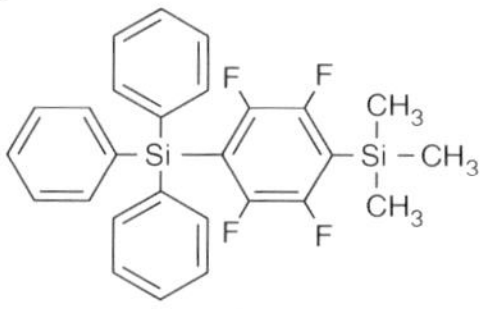
250 mg

S82,662-6
1,4-BIS(DIPHENYLSILYL)-2,3,5,6-TETRAFLUOROBENZENE
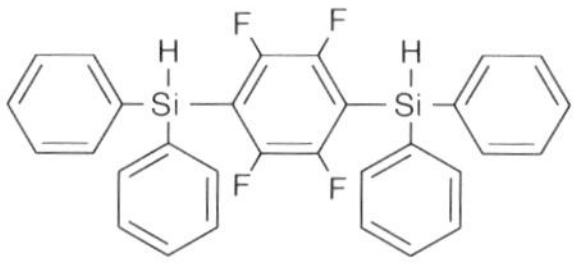
250 mg

S82,571-9
(2,3,6-TRICHLORO-1,4-PHENYLENE)BIS-(DIPHENYLMETHYLSILANE)
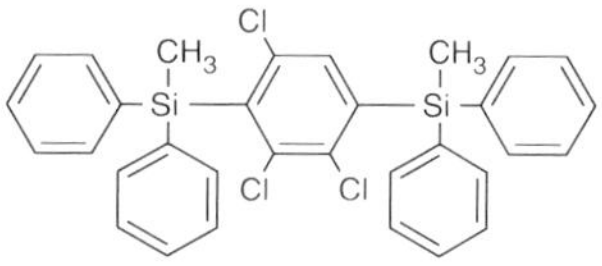
250 mg

S83,263-4
2,3,5,6-TETRACHLORO-1,4-PHENYLENEBIS(DIPHENYLMETHYLSILANE)
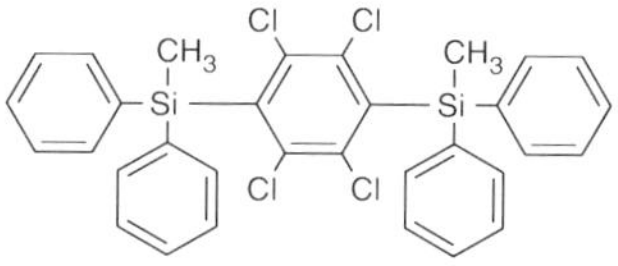
250 mg

S82,547-6
(2,3,5,6-TETRAFLUORO-1,4-PHENYLENE)-BIS(TRIPHENYLSILANE)
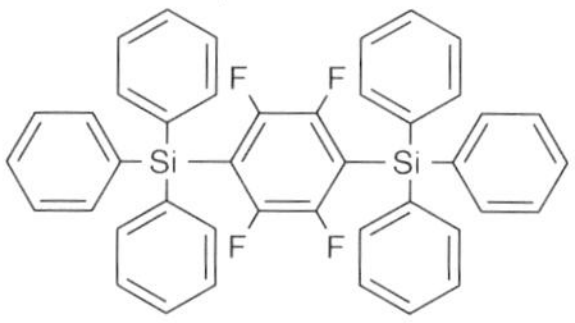
250 mg

S82,742-8
(2-CHLOROPHENYL)DIPHENYLSILANE
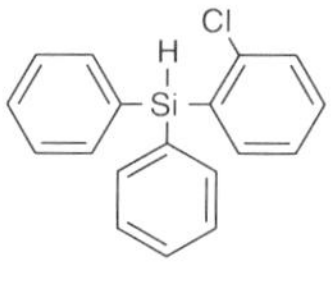
250 mg

S82,744-4
[3-(2-CHLOROPHENYL)PROPYL]-DIPHENYLSILANE
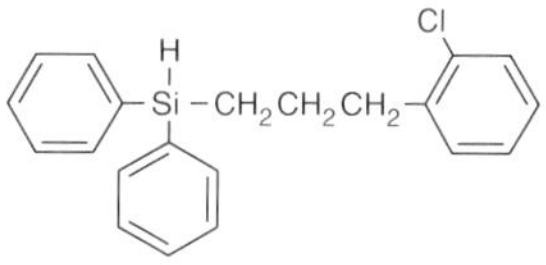
250 mg

S83,024-0
(2-CHLOROPHENETHYL)DIBENZYLSILANE
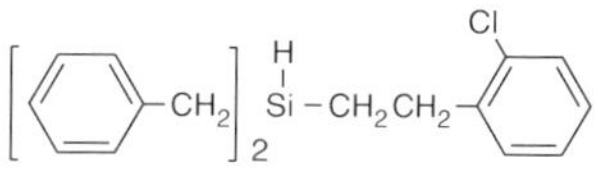
250 mg

S83,044-5
(2-CHLOROBENZYL)TRIPHENYLSILANE
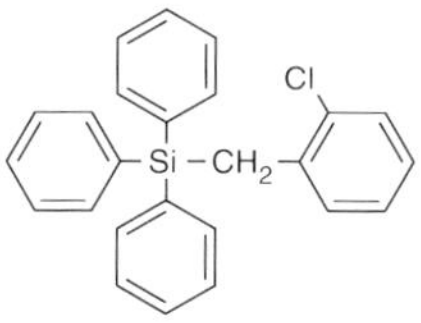
250 mg

S82,309-0
2-BROMO-2'-(TRIPHENYLSILYL)BIPHENYL
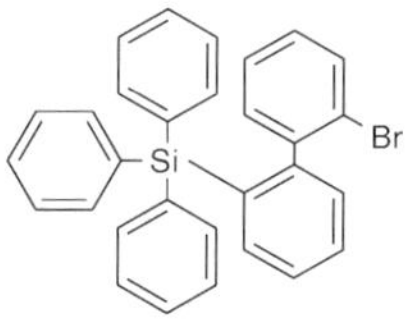
250 mg

S82,988-9
(4-CHLOROPHENYL)PHENYLSILANE
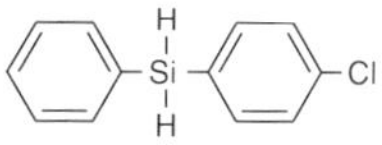
250 mg

S82,666-9
(4-BROMOPHENYL)METHYLPHENYLSILANE
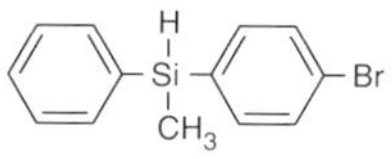
250 mg

S82,667-7
(4-CHLOROBENZYL)METHYLPHENYLSILANE
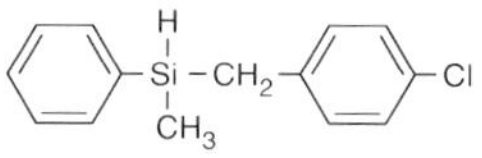
250 mg

S82,716-9
(4-CHLOROPHENYL)DIPHENYLSILANE
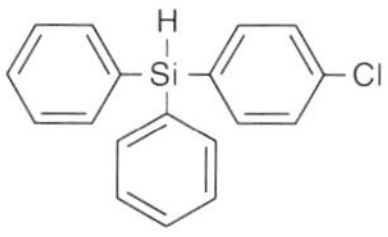
250 mg

S83,883-7
BIS(4-CHLOROPHENYL)DIPHENYLSILANE
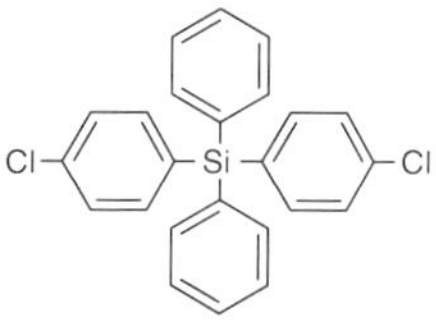
250 mg

S82,746-0
(4-BROMOPHENYL)TRIPHENYLSILANE
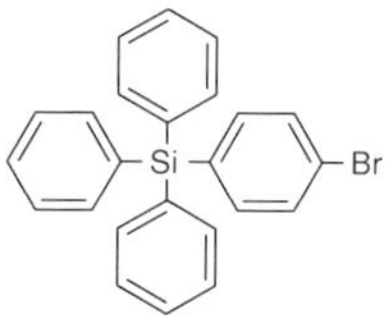
250 mg

S83,246-4
(4-BROMOPHENYL)TRIPHENYLGERMANE
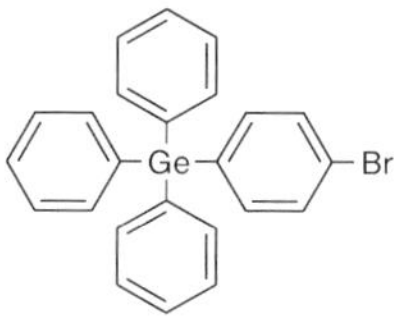
250 mg

S83,238-3
(4-FLUOROBENZYL)TRIPHENYLSILANE
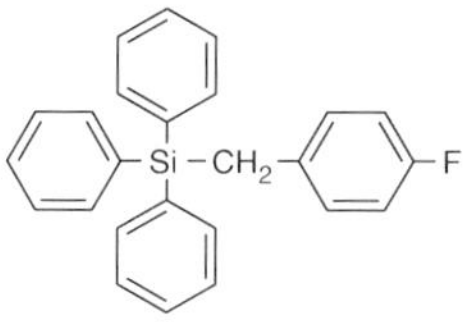
250 mg

S83,180-8
(4-CHLOROBENZYL)TRIPHENYLSILANE
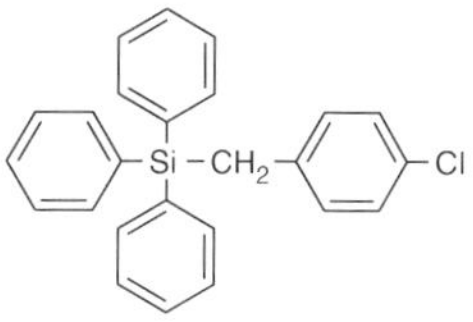
250 mg

S83,239-1
PHENYLTRIS(4-FLUOROBENZYL)SILANE
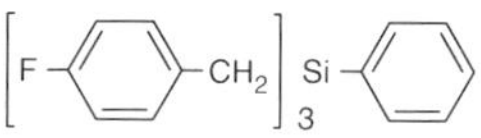
250 mg

S82,749-5
(4-BROMOPHENYL)TRIBENZYLSILANE

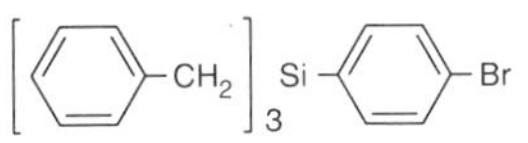

250 mg

S82,754-1
(4-BROMO-4'-BIPHENYLYL)-TRIBENZYLSILANE

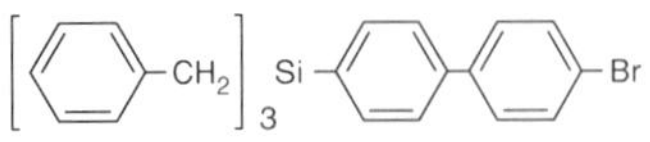

250 mg

S82,569-7
(2,6-DICHLOROPHENYL)-TRIPHENYLGERMANE

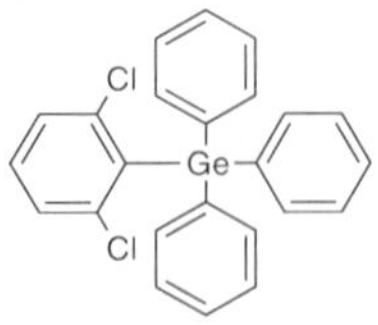

250 mg

S82,984-6
[(PENTAFLUOROPHENYL)PHENYLMETHOXY]-TRIMETHYLSILANE

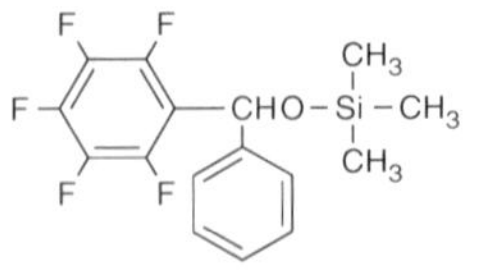

250 mg

S81,561-6
(2-FLUOROPHENYL)TRIMETHYLSILANE

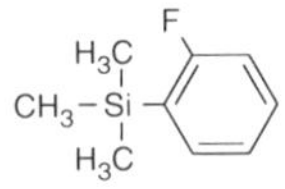

250 mg

S83,028-3
BIS(2-CHLOROBENZYL)SILANE

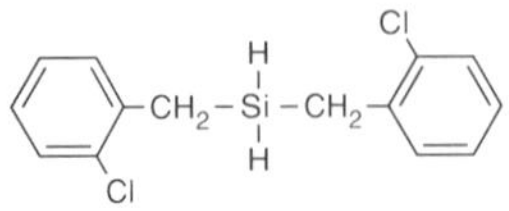

250 mg

S83,040-2
TRIS(2-CHLOROBENZYL)SILANE

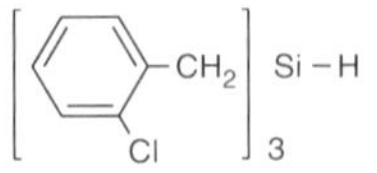

250 mg

S82,931-5
BIS(2-FLUOROBENZYL)DIDODECYLSILANE

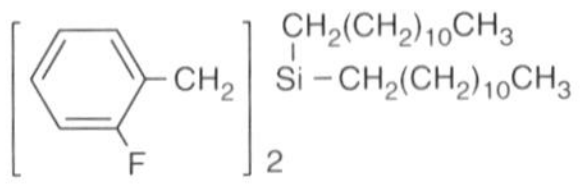

250 mg

S86,631-8
4-CHLOROPHENYL TRIMETHYLSILYL ETHER

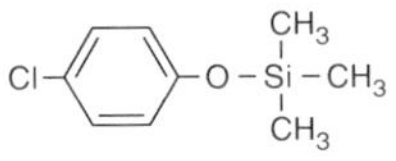

250 mg

S83,823-3
(4-BROMOPHENYL)TRIMETHYLSILANE

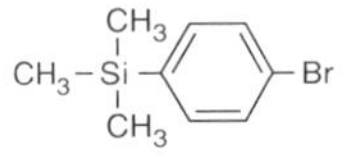

250 mg

S83,244-8
TRIS(4-BROMOPHENYL)SILANE

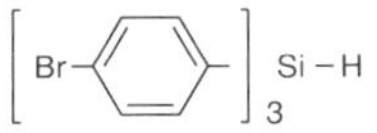

250 mg

S83,208-1
1,4-BIS(4-BROMOPHENYL)-1,1,2,2,3,3,4,4-OCTAMETHYLTETRASILANE

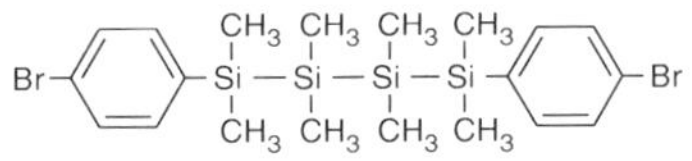

250 mg

S82,730-4
HEXADECYLTRIS(4-CHLOROPHENYL)SILANE

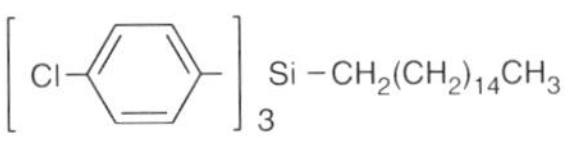

250 mg

S82,521-2
BIS(4-CHLOROPHENYL)DIDODECYLSILANE

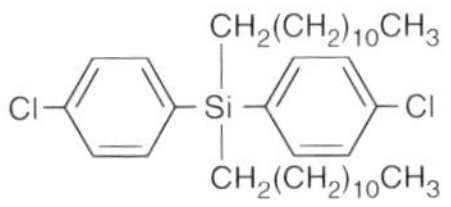

250 mg

S82,729-0
OCTADECYLTRIS(4-CHLOROPHENYL)SILANE

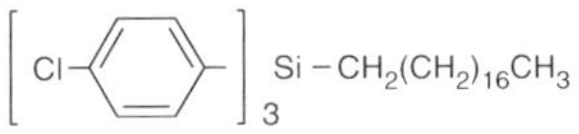

250 mg

S83,456-4
(4-CHLOROPHENYL)TRIHEXADECYLSILANE

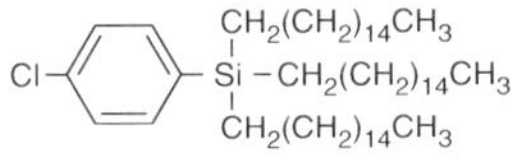

250 mg

S81,560-8
(2,6-DIFLUOROPHENYL)TRIMETHYLSILANE

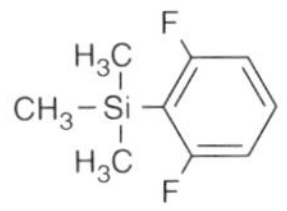

250 mg

S82,555-7
(2,6-DICHLOROPHENYL)TRIMETHYLSILANE

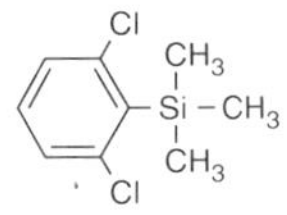

250 mg

S81,576-4
1,3-BIS(TRIMETHYLSILYL)-2-FLUOROBENZENE

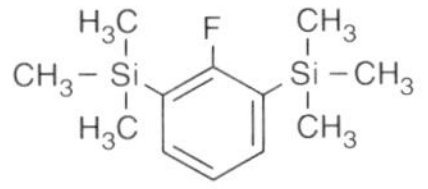

250 mg

S82,551-4
DIMETHYL(2,3,6-TRICHLOROPHENYL)-SILANE

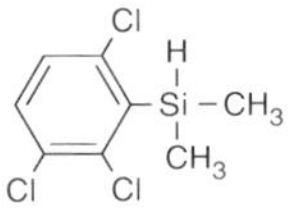

250 mg

S82,554-9
(2,3,6-TRICHLOROPHENYL)-TRIMETHYLSILANE

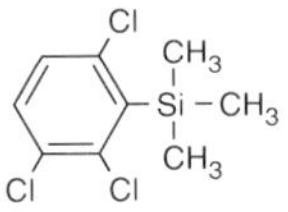

250 mg

S82,550-6
DIMETHYL(2,4,6-TRICHLOROPHENYL)-SILANE

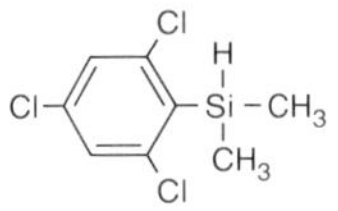

250 mg

S81,557-8
(2,4,6-TRIFLUOROPHENYL)-TRIMETHYLSILANE

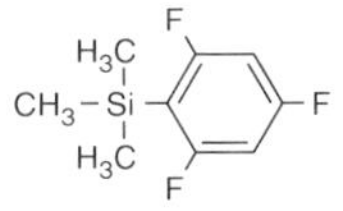

250 mg

S82,553-0
(3,4,5-TRICHLOROPHENYL)-TRIMETHYLSILANE

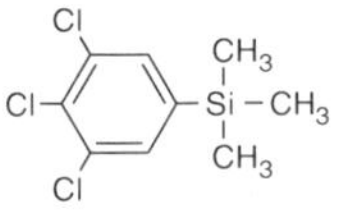

250 mg

S83,426-2
3-BROMO-3'-(TRIBUTYLSILYL)BIPHENYL

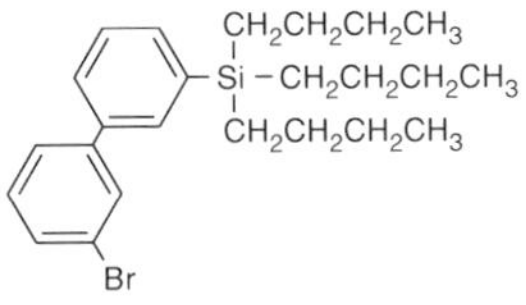

250 mg

S82,787-8
4-(TRIMETHYLSILYL)-1-BUTANOL

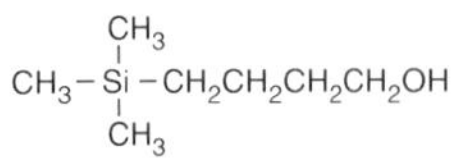

250 mg

S83,001-1
1,4-BIS(TRIPHENYLSILYL)-2,3-BUTANEDIOL

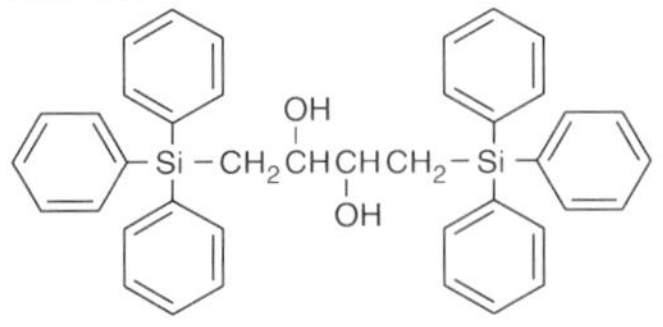

250 mg

S81,624-8
3-(TRIPHENYLSILYLOXY)-1,2-PROPANEDIOL

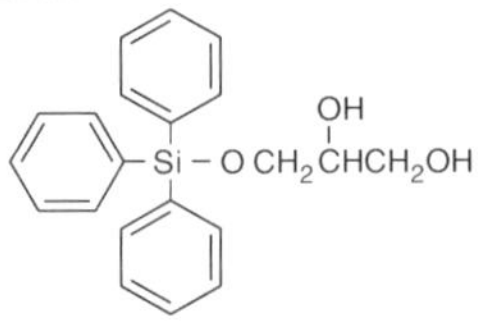

250 mg

S76,297-0
BIS(3-HYDROXYPROPYL)DIMETHYLSILANE

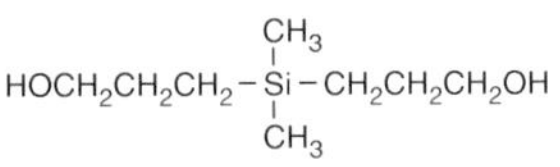

250 mg

S82,964-1
1,1,2,2,3,3,4,4-OCTAPHENYL-1,4-TETRASILANEDIOL

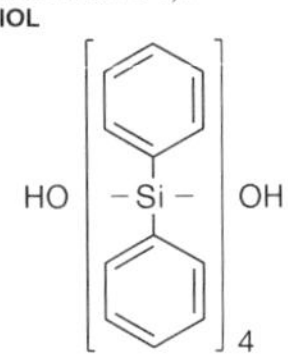

250 mg

S82,997-8
1-ALLYLOXY-3-(TRIPHENYLSILYL)-2-PROPANOL

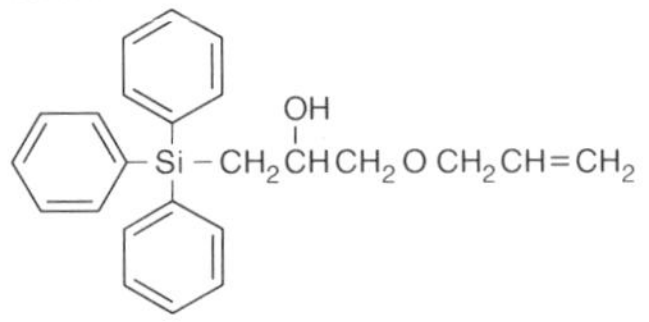

250 mg

S82,176-4
1-(TRIPHENYLSILYL)-3-[2-(TRIPHENYLSILYL)ETHOXY]-2-PROPANOL

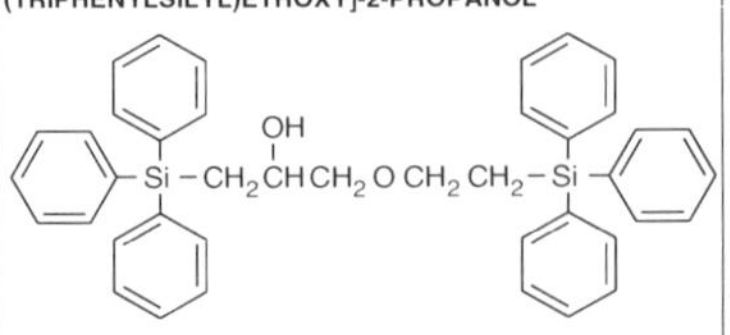

250 mg

S82,992-7
DIPHENYL(2-HYDROXY-2-PHENYLETHYL)-METHYLSILANE
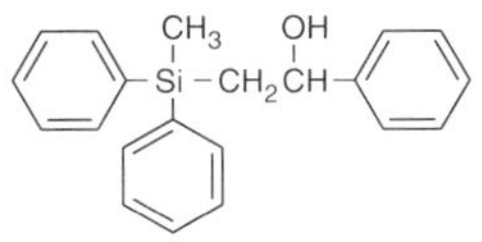
250 mg

S82,166-7
1-PHENYL-2-(TRIPHENYLSILYL)-1-ETHANOL
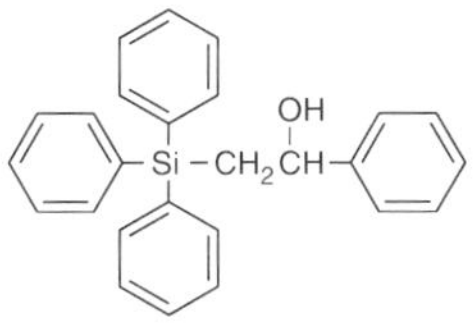
250 mg

S83,160-3
1-PHENYL-2-(TRIPHENYLGERMYL)ETHANOL
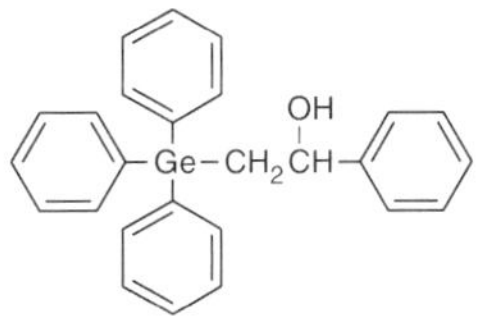
250 mg

S83,718-0
4-(TRIPHENYLSILYL)BENZYL ALCOHOL
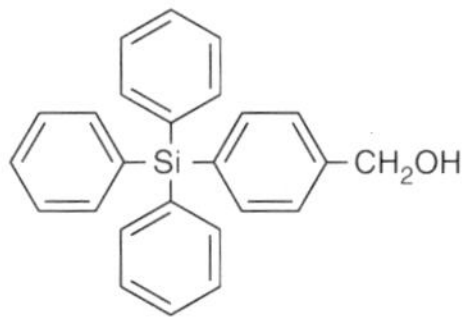
250 mg

S81,325-7
4-(TRIMETHYLSILYL)BENZHYDROL
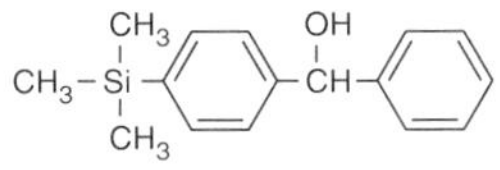
250 mg

S81,187-4
ALPHA[4-(TRIMETHYLSILYL)PHENYL]-BENZHYDROL
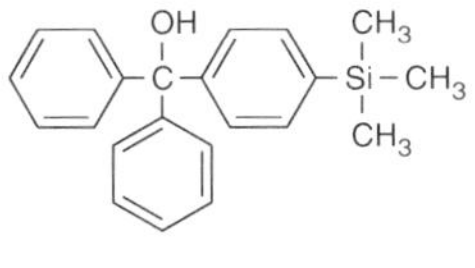
250 mg

S82,354-6
ALPHA-(TRIPHENYLGERMYL)BENZHYDROL
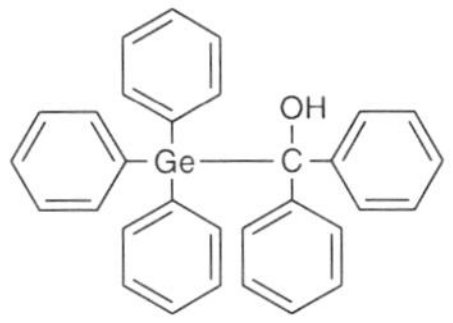
250 mg

S86,054-9
4-[2,3-EPOXY-1-HYDROXY-4-(TRIMETHYLSILYLOXY)BUTYL]-GLUTARIMIDE
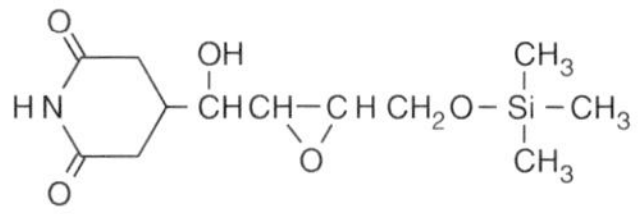
250 mg

S84,095-5
(2-HYDROXYETHOXY)TRIPHENYLSILANE
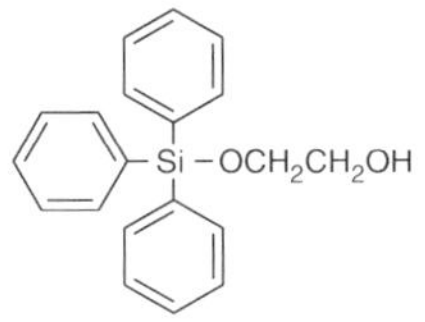
250 mg

S82,747-9
1-(TRIPHENYLSILYL)CYCLOHEXANOL
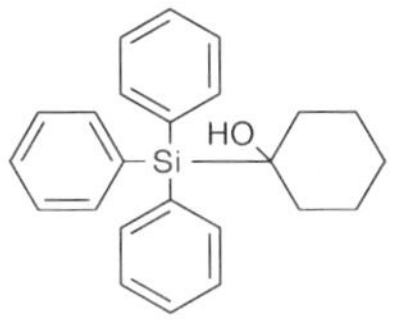
250 mg

S83,231-6
(DIMETHYLPHENYLSILYL)METHANOL
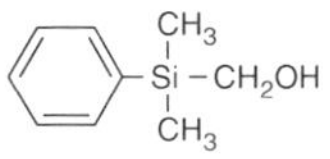
250 mg

S88,366-2
1-PHENYL-3-(TRIMETHYLSILYL)-2-PROPANOL
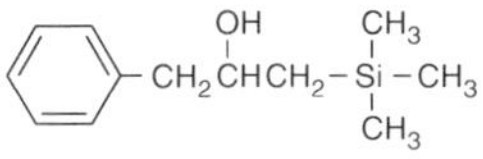
250 mg

S82,841-6
2-(DIPHENYLMETHYLSILYL)-2-PROPANOL
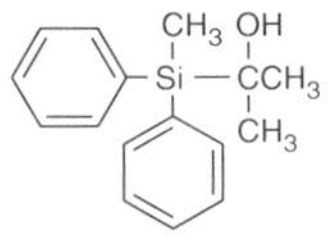
250 mg

S82,989-7
DIPHENYL(4-HYDROXYBUTYL)-METHYLSILANE
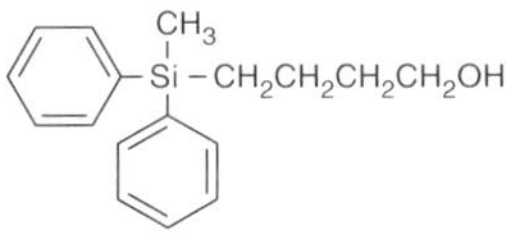
250 mg

S82,735-5
1-(TRIPHENYLSILYL)ETHANOL
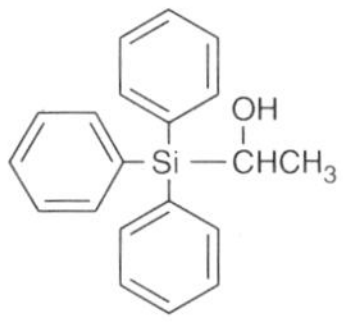
250 mg

S82,348-1
2-(TRIPHENYLGERMYL)-1-ETHANOL
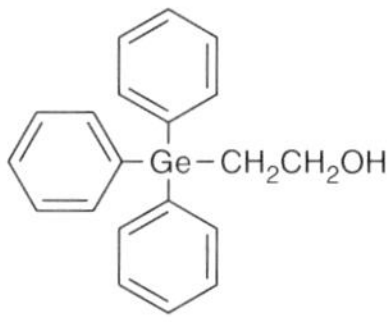
250 mg

S81,979-4
2-(TRIPHENYLSILYL)-2-PROPANOL
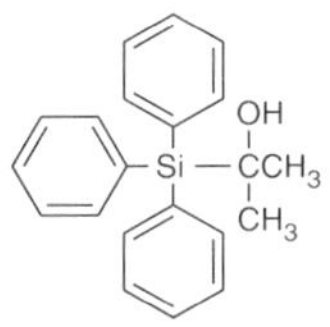
250 mg

S82,160-8
1-(TRIPHENYLSILYL)-2-PROPANOL
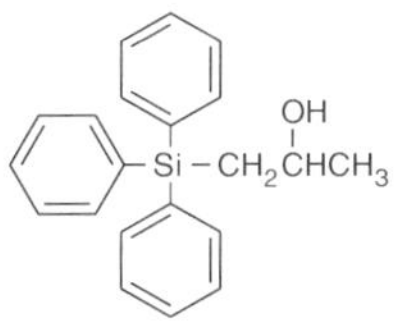
250 mg

S82,499-2
3-(TRIPHENYLSILYL)-1-PROPANOL
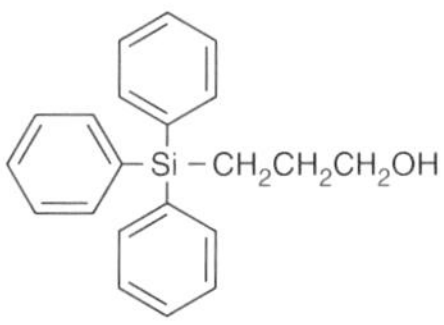
250 mg

S82,162-4
4-(TRIPHENYLSILYL)-1-BUTANOL
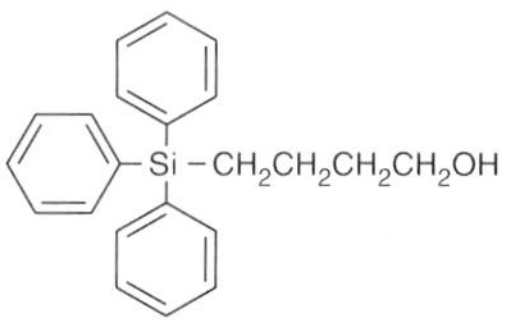
250 mg

S83,042-9
(4-HYDROXYBUTYL)TRIPHENYLGERMANE
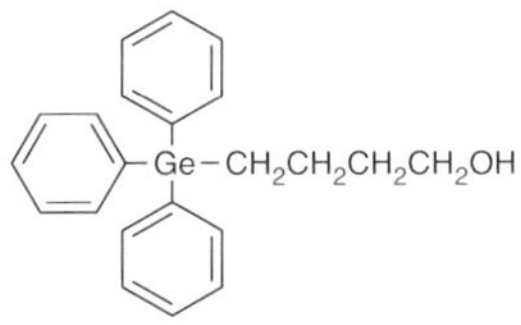
250 mg

S82,167-5
1-PHENYL-3-(TRIPHENYLSILYL)-2-PROPANOL
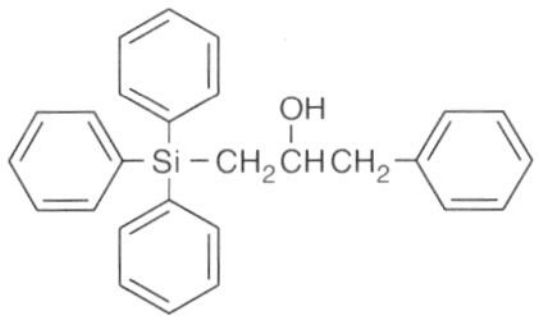
250 mg

S82,846-7
3-(TRIPHENYLSILYL)-3-OCTADECANOL
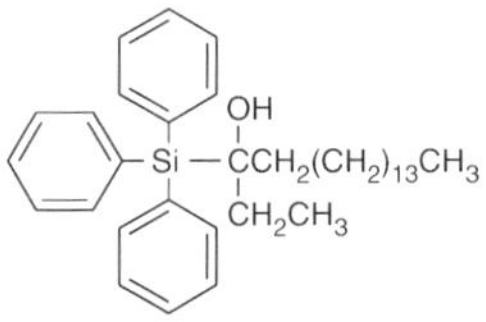
250 mg

S83,106-9
1,1-BIS(TRIPHENYLSILYL)ETHANOL
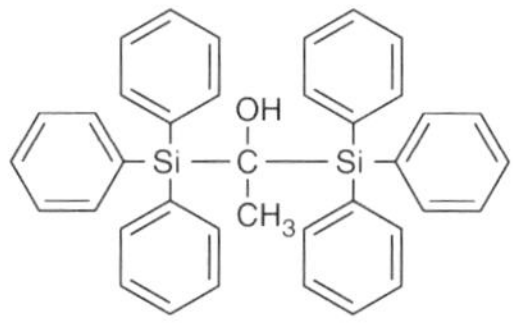
250 mg

S82,175-6
1,3-BIS(TRIPHENYLSILYL)-2-PROPANOL
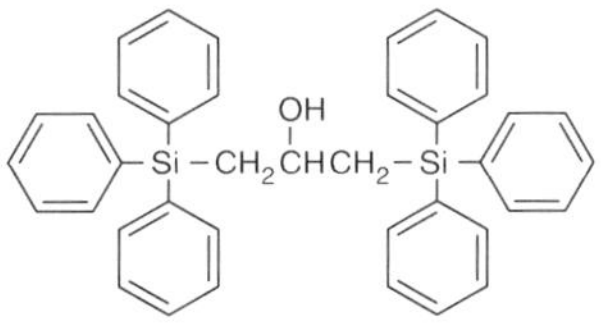
250 mg

S83,829-2
4-(TRIPHENYLSILYL)PHENOL
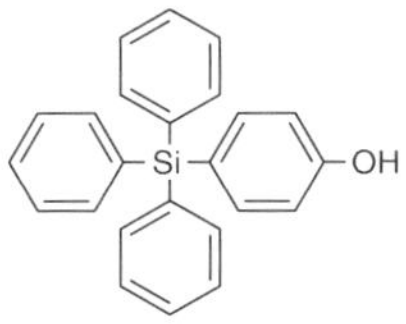
250 mg

S81,367-2
6-(TRIPHENYLSILYL)-2-NAPHTHOL
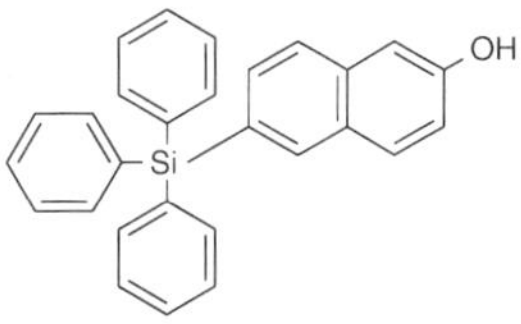
250 mg

S75,736-5
4-(TRIMETHYLSILYL)PHENOL
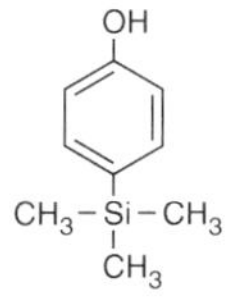
5 gm

S81,326-5
4-(TRIMETHYLSILYL)PHENETHYL ALCOHOL
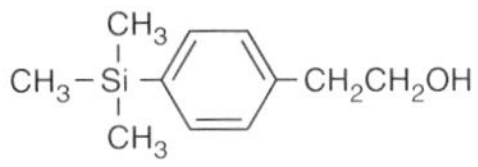
250 mg

S88,427-8
4-(3-METHOXYPHENYL)-1-(TRIMETHYLSILYL)-1-BUTYNE
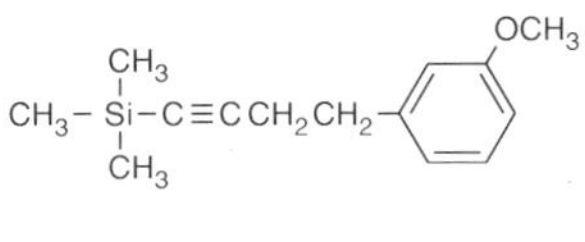
250 mg

S87,978-9
4-(TERT-BUTYLDIPHENYLSILYLOXY)-2-METHOXY-6-(TRITYLOXY)-TETRAHYDROPYRAN
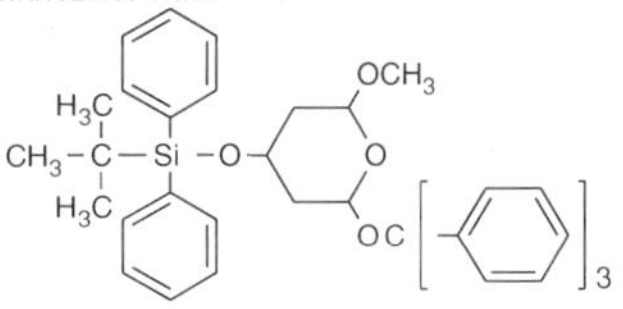
5 gm

S82,785-1
[4-(BENZYLOXY)PHENYL]-TRIMETHYLSILANE
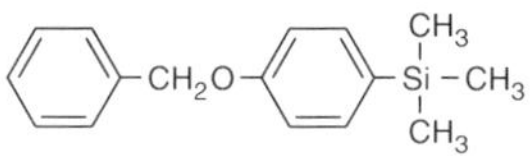
250 mg

S82,768-1
BIS[2-(2-PHENOXYETHOXY)PHENYL]-DIPHENYLSILANE
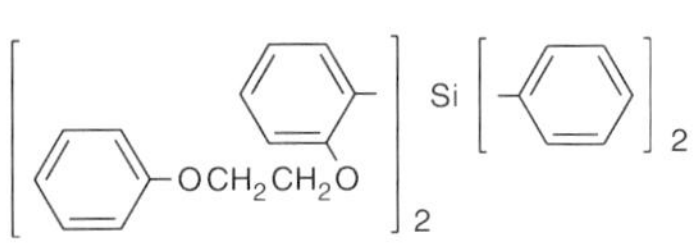
250 mg

S78,208-4
1,3-BIS(2,3-EPOXY-3-PROPOXYPROPYL)-1,1,3,3-TETRAMETHYLDISILOXANE
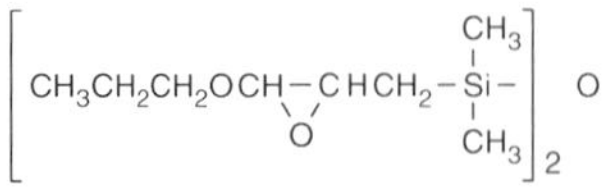
5 gm

S82,631-6
DIMETHYLDIPHENOXYSILANE
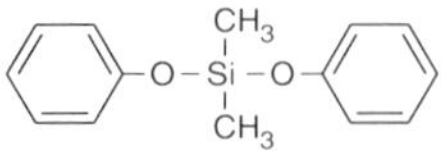
250 mg

S83,048-8
(4-METHOXYPHENYL)(4-PHENOXYPHENYL)-PHENYLSILANE
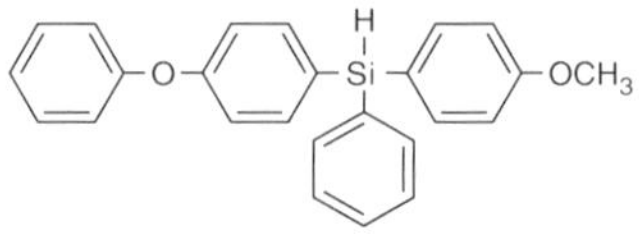
250 mg

S81,662-0
ETHOXYTRIS(2-METHOXYPHENYL)SILANE
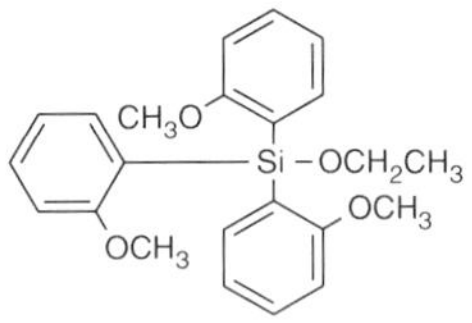
250 mg

S83,139-5
BIS(4-METHOXYPHENYL)SILANE
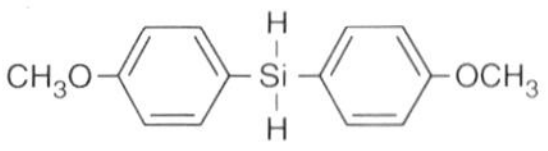
250 mg

S40,947-2
METHYLTRIS(2-METHOXYETHOXY)SILANE
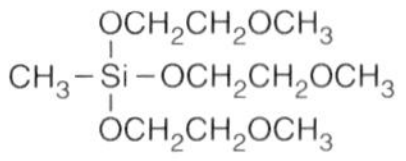
5 gm

S58,565-3
TETRAKIS(2-METHOXYETHYL)-ORTHOSILICATE
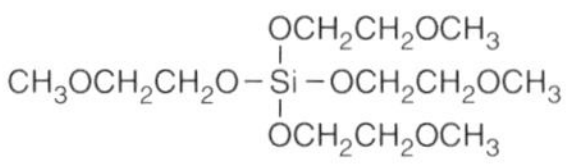
5 gm

S58,521-1
DIETHOXY(3-GLYCIDYLOXYPROPYL)-METHYLSILANE
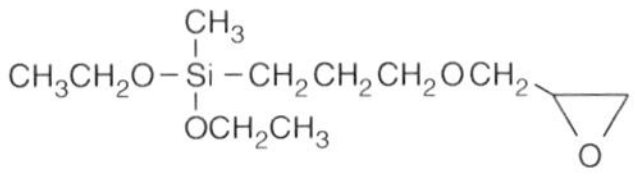
5 gm

S58,558-0
PHENYLTRIS(2-METHOXYETHOXY)SILANE
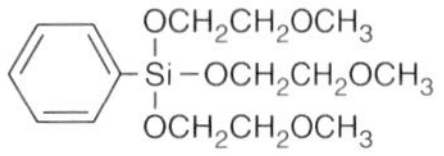
5 gm

S86,671-7
3-METHOXY-1,2,3,6-TETRAHYDRO-5-(TRIMETHYLSILYLOXY)-PHTHALIC ANHYDRIDE
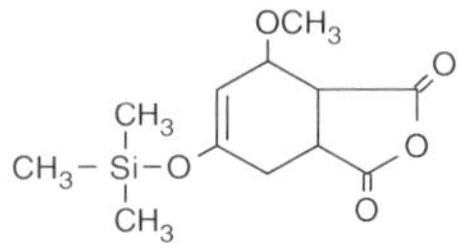
250 mg

S82,203-5
(4-METHOXYPHENYL)TRIS[4-(DIMETHYLAMINO)PHENYL]SILANE
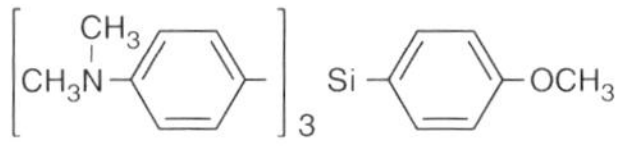
250 mg

S82,878-5
(4-METHOXYBENZOYL)TRIPHENYLSILANE
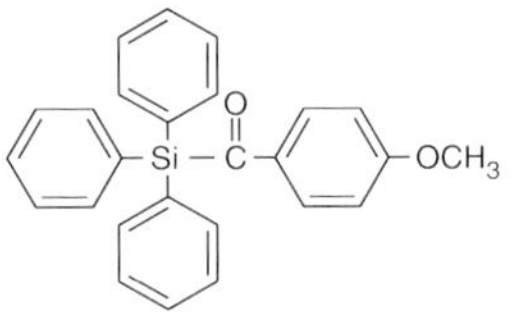
250 mg

S82,756-8
(3-PHENOXYPHENYL)TRIBENZYLSILANE
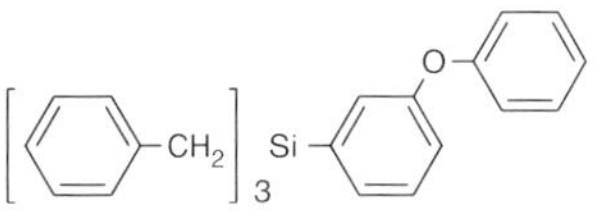
250 mg

S82,774-6
(2-PHENOXYPHENYL)TRIPHENYLSILANE
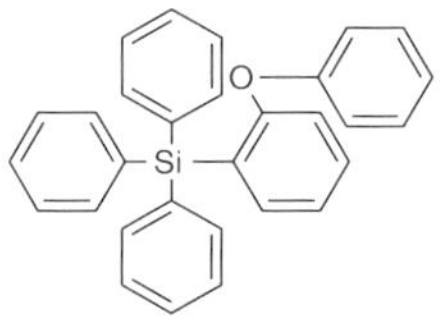
250 mg

S82,919-6
2-(TRIPHENYLSILYL)PHENYL ETHER
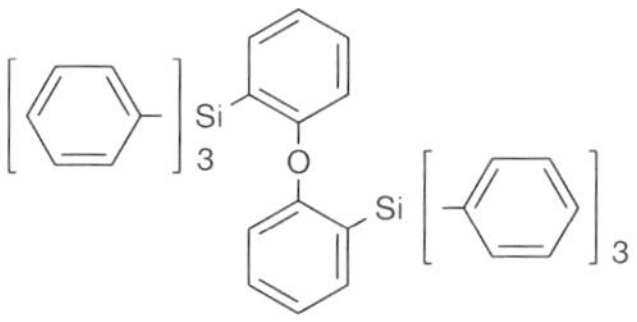
250 mg

S83,448-3
BIS(2-PHENOXYPHENYL)DIDODECYLSILANE
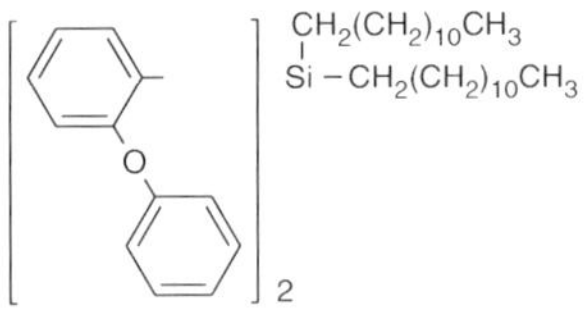
250 mg

S83,046-1
(4-PHENOXYPHENYL)(PHENYL)(O-TOLYL)-SILANE
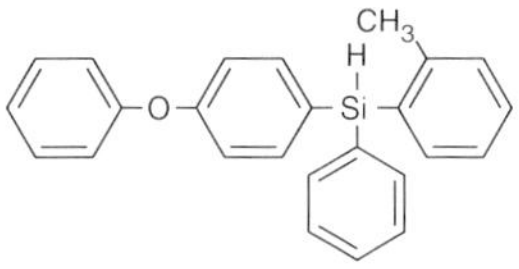
250 mg

S83,051-8
(4-PHENOXYPHENYL)PHENYL(O-TOLYL)(P-TOLYL)SILANE
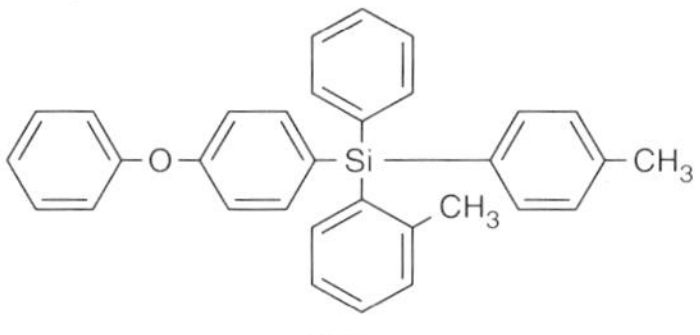
250 mg

S83,032-1
(4-PHENOXYPHENYL)PHENYLSILANE
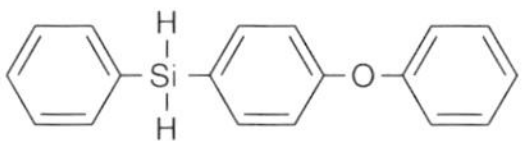
250 mg

S82,678-2
DIPHENYL(4-METHOXYPHENYL)SILANE
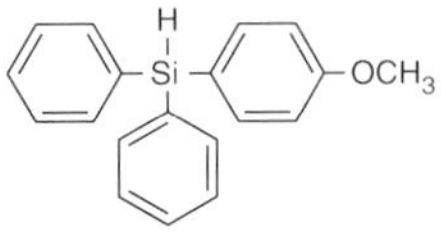
250 mg

S82,679-0
METHYL(4-PHENOXYPHENYL)PHENYLSILANE
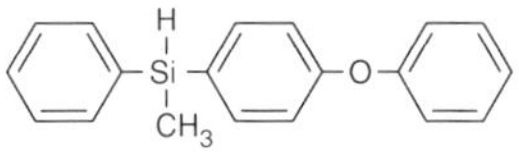
250 mg

S82,692-8
DIPHENYL(4-PHENOXYPHENYL)SILANE
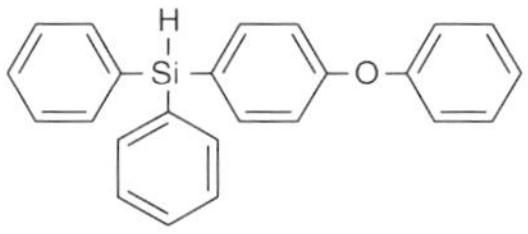
250 mg

S81,341-9
(4-METHOXYPHENYL)TRIPHENYLSILANE
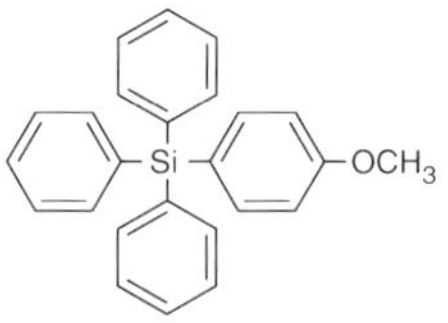
250 mg

S81,466-0
(4-METHOXYPHENYL)TRIBENZYLSILANE
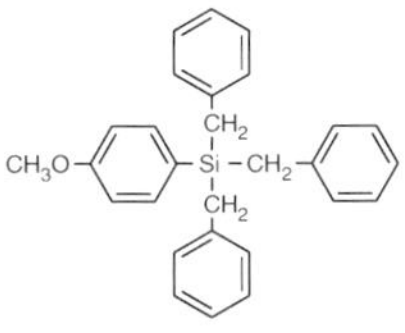
250 mg

S83,255-3
4-(TRIPHENYLSILYL)PHENYL ETHER
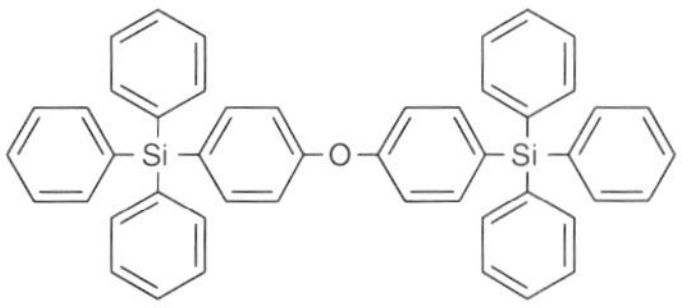
250 mg

S83,449-1
BIS(4-PHENOXYPHENYL)DIDODECYLSILANE
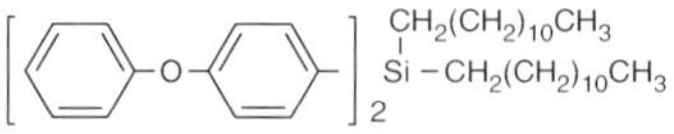
250 mg

S82,515-8
(4-PHENOXYPHENYL)TRIHEXADECYLSILANE
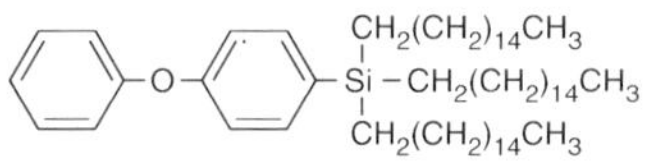
250 mg

S81,873-9
2-METHOXY-1-(TRIPHENYLSILYL)-NAPHTHALENE
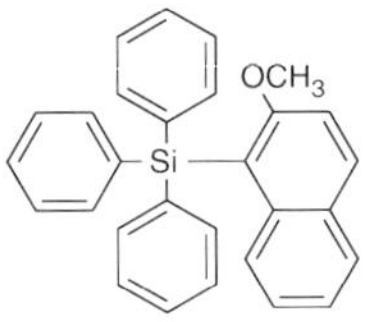
250 mg

S81,872-0
2-METHOXY-3-(TRIPHENYLSILYL)-NAPHTHALENE
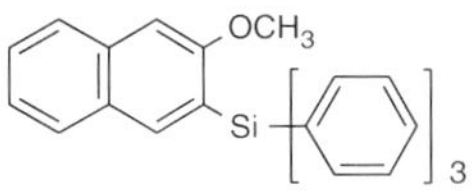
250 mg

S82,516-6
2-(TRIMETHYLSILYL)PHENYL ETHER
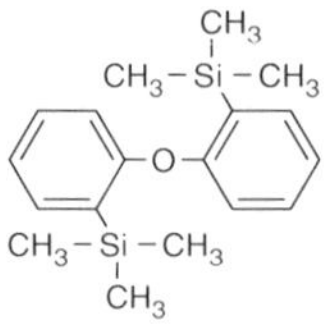
250 mg

S83,563-3
CHLOROTRIS(2-METHOXYPHENYL)SILANE
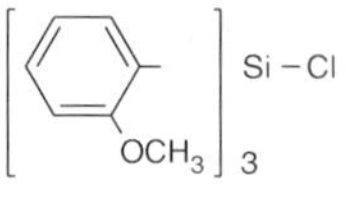
250 mg

S81,294-3
TETRAKIS(2-METHOXYPHENYL)SILANE

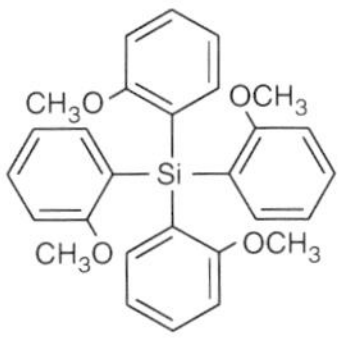

250 mg

S81,295-1
(4-METHOXYPHENYL)TRI(O-TOLYL)SILANE

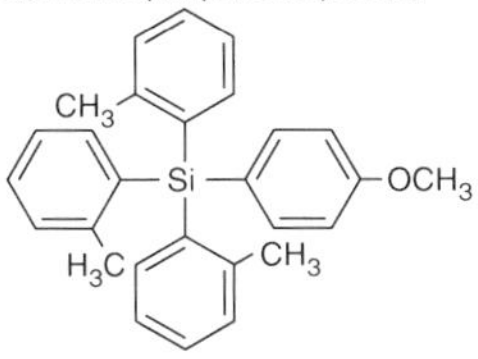

250 mg

S81,018-5
(4-METHOXYPHENYL)TRIETHYLSILANE

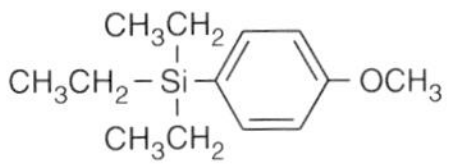

250 mg

S82,394-5
4-(TRIMETHYLSILYL)PHENYL ETHER

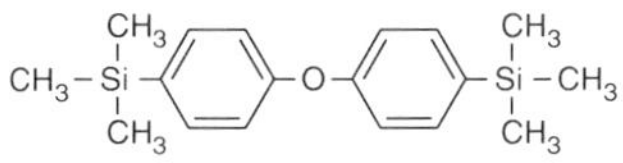

250 mg

S88,392-1
1-PHENYL-2-(TRIMETHYLSILYL)ETHYL N-PHENYLCARBAMATE

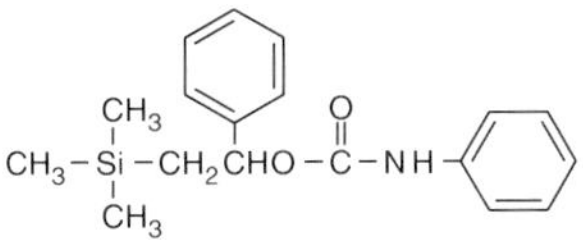

250 mg

S82,844-0
DIMETHYL(DIPHENYLMETHOXY)-PHENYLSILANE

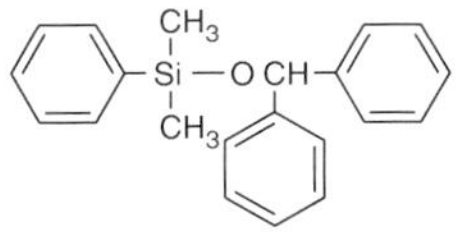

250 mg

S82,426-7
1,3-DIPHENYL-1-PROPENYLOXY-(DIMETHYL)(TRIMETHYLSILYLMETHYL)-SILANE

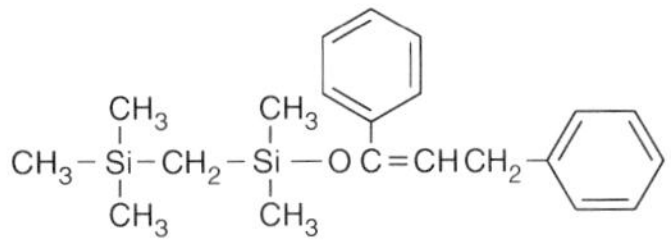

250 mg

S81,955-7
(BENZYLOXY)TRIPHENYLSILANE

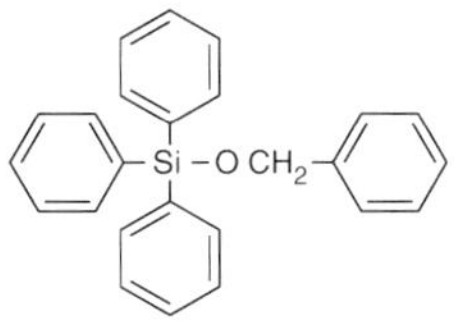

250 mg

S82,985-4
DIMETHYLPHENYL[PHENYL(2,3,5,6-TETRACHLORO-4-PYRIDYL)METHOXY]-SILANE

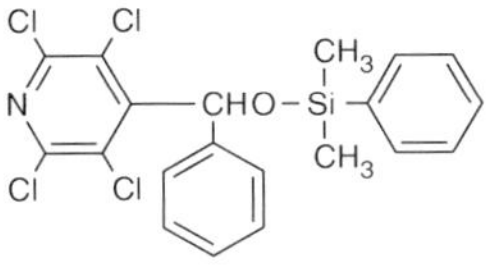

250 mg

S81,982-4
(DIPHENYLMETHOXY)TRIPHENYLSILANE

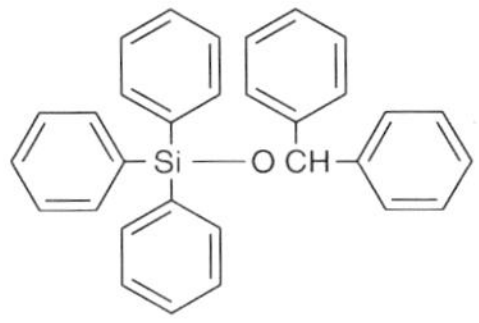

250 mg

S82,986-2
DIPHENYL(DIPHENYLMETHOXY)-(DIPHENYLMETHYL)SILANE

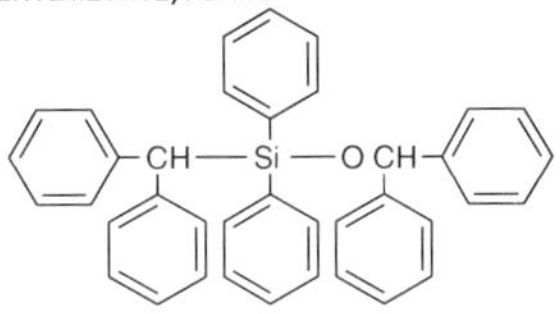

250 mg

S86,490-0
1,3-BIS(TRIMETHYLSILYLOXY)PROPANE

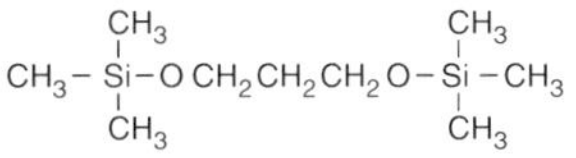

250 mg

S87,977-0
3,5-BIS(TERT-BUTYLDIPHENYLSILYLOXY)-1-(TRIMETHYLSILYLOXY)-1-CYCLOHEXENE

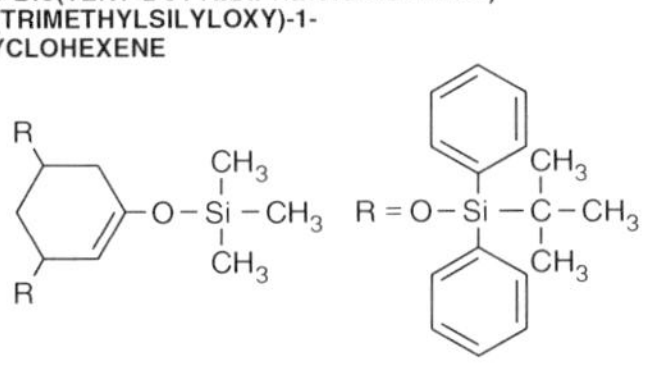

5 gm

S81,252-8
TETRAKIS(PHENYLTHIOMETHYL)SILANE

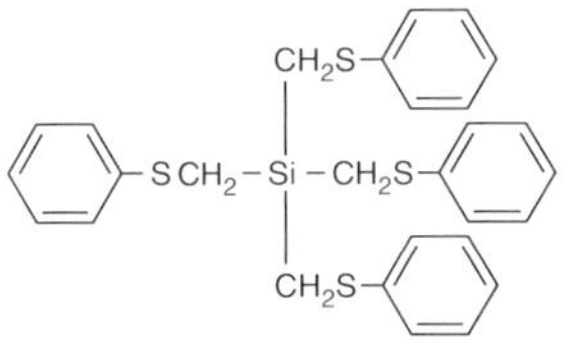

250 mg

S81,238-2
TETRAKIS[(P-TOLYL)THIOMETHYL]SILANE

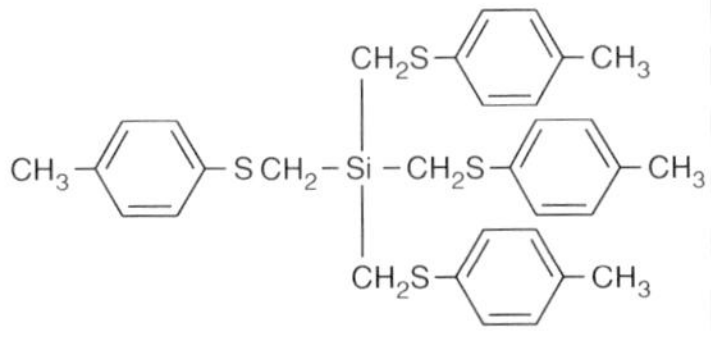

250 mg

S56,516-4
3-(DIMETHOXYMETHYLSILYL)-PROPYLAMINE

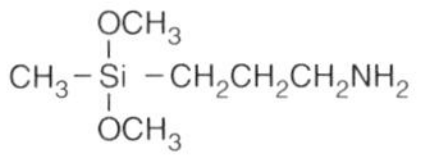

5 gm

S56,535-0
4-(DIMETHYLMETHOXYSILYL)BUTYLAMINE

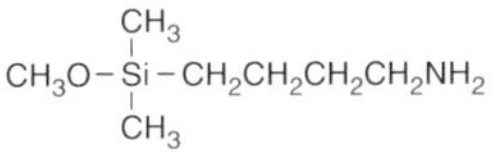

1 gm

S41,131-0
N-[3-(DIMETHOXYMETHYLSILYL)-ISOBUTYL]ETHYLENEDIAMINE

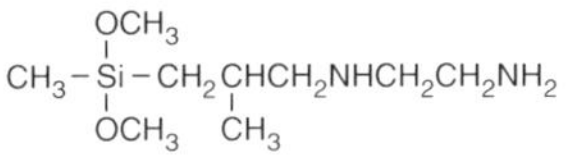

5 gm

S82,719-3
1-(TRIPHENYLSILYL)ETHYLAMINE

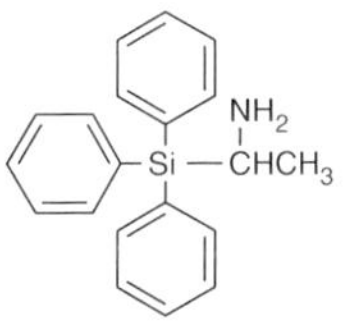

250 mg

S86,638-5
1,3-BIS(TRIMETHYLSILYLAMINO)PROPANE

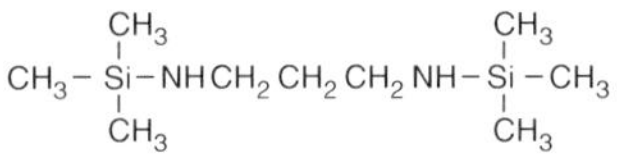

250 mg

S80,619-6
3-(DECYLDIMETHYLSILYL)-N-[1-PHENYL-2-(M-TOLYL)ETHYL]PROPIONAMIDE

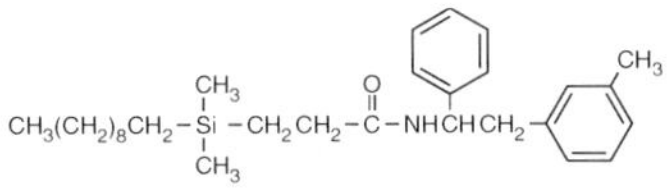

250 mg

S88,291-7
3-(DECYLDIMETHYLSILYL)-N-[1-PHENYL-2-(O-TOLYL)ETHYL]PROPIONAMIDE

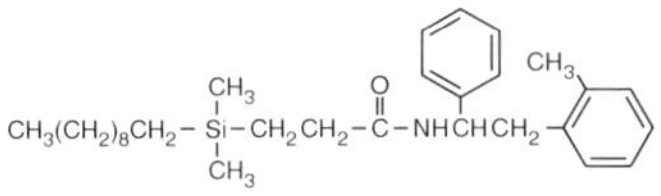

250 mg

S87,980-0
3-(DECYLDIMETHYLSILYL)-N-[1-PHENYL-2-(P-TOLYL)ETHYL]PROPIONAMIDE

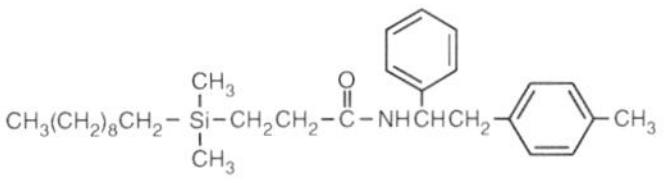

5 gm

S83,719-9
4-(TRIPHENYLSILYL)BENZYL CYANIDE

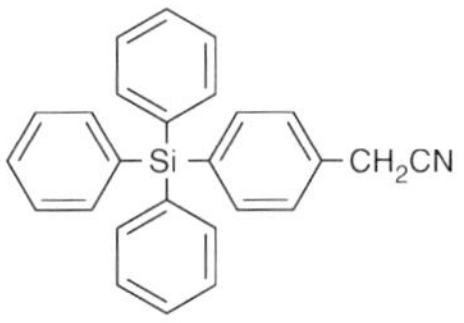

250 mg

S88,389-1
PHENYL[1-PHENYL-2-(TRIMETHYLSILYL)-ETHYL]AMINE

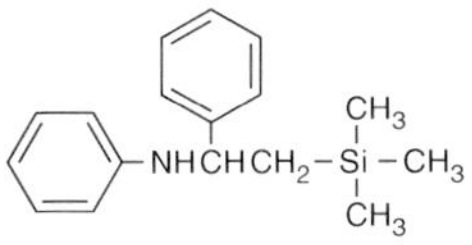

250 mg

S83,878-0
N-[DIPHENYL(TRIPHENYLSILYL)METHYL]-ANILINE

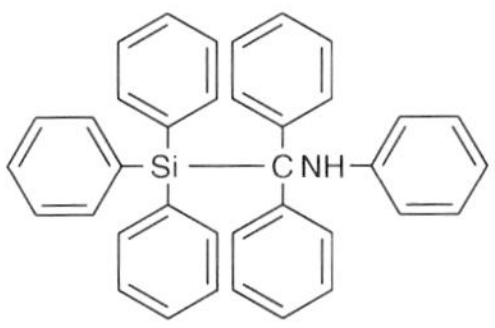

250 mg

S83,947-7
N,N-DIMETHYL-3-(TRIMETHYLSILYL)-ANILINE

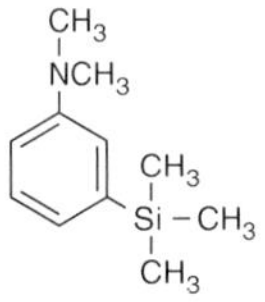

250 mg

S82,205-1
1,1,1,3,3,3-HEXAKIS[4-(DIMETHYLAMINO)PHENYL]DISILOXANE

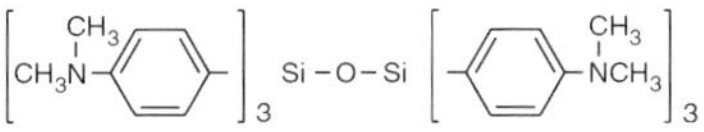

250 mg

S83,393-2
N,N-DIMETHYL-3-(DIPHENYLSILYL)-ANILINE

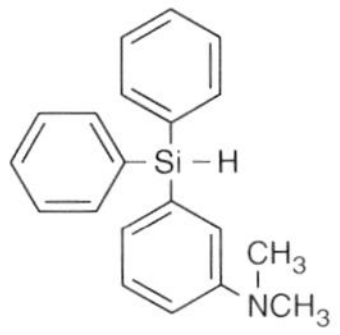

250 mg

S81,366-4
N,N-DIMETHYL-3-(TRIPHENYLSILYL)-ANILINE

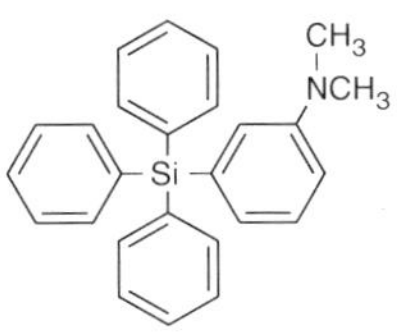

250 mg

S81,793-7
N,N-DIMETHYL-2-(TRIPHENYLSILYL)-ANILINE

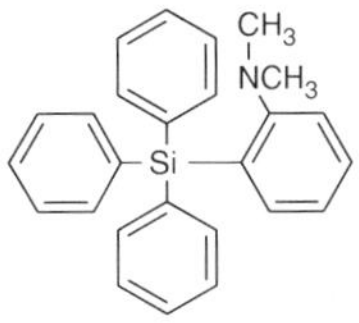

250 mg

S83,390-8
N,N-DIMETHYL-4-(TRIMETHYLSILYL)-ANILINE

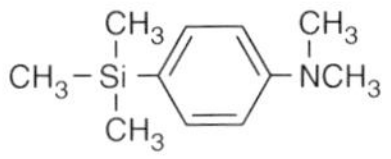

250 mg

S83,348-7
1,2-BIS[4-(DIMETHYLAMINO)PHENYL]-1,1,2,2-TETRAMETHYLDISILANE
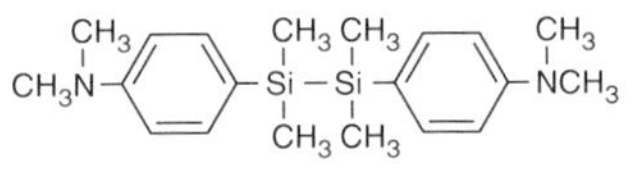
250 mg

S82,204-3
TETRAKIS[4-(DIMETHYLAMINO)PHENYL]-SILANE
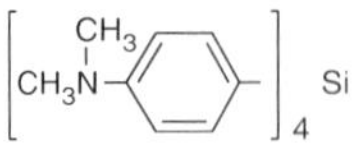
250 mg

S82,721-5
N,N-DIMETHYL-ALPHA-(TRIPHENYLSILYL)-BENZYLAMINE
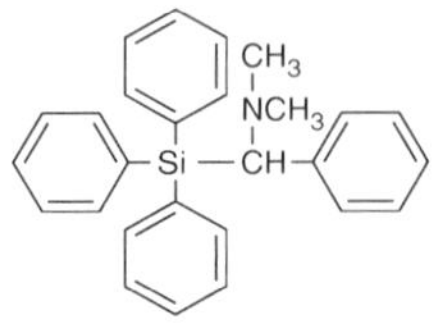
250 mg

S82,723-1
N,N,ALPHA-TRIMETHYL-ALPHA-(TRIPHENYLSILYL)BENZYLAMINE
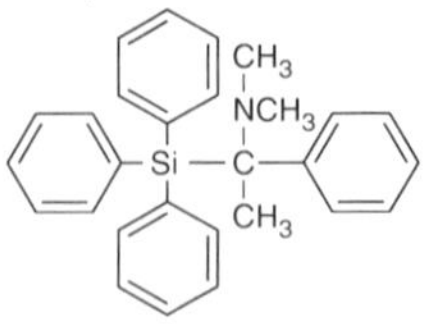
250 mg

S82,726-6
N,N-DIPHENYL-ALPHA-(TRIPHENYLSILYL)-BENZYLAMINE
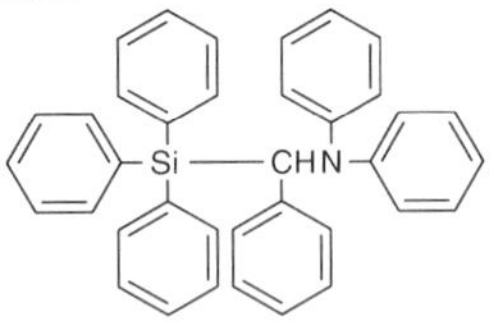
250 mg

S82,722-3
N,N,4-TRIMETHYL-ALPHA-(TRIPHENYLSILYL)BENZYLAMINE
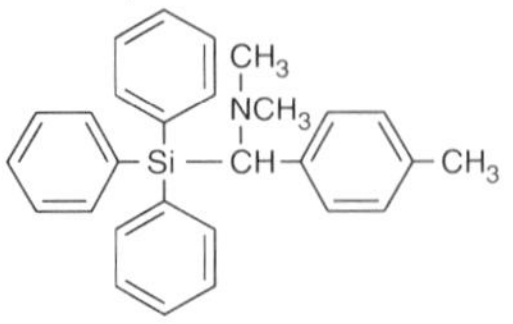
250 mg

S81,174-2
9-ETHYL-3-(TRIPHENYLSILYL)CARBAZOLE
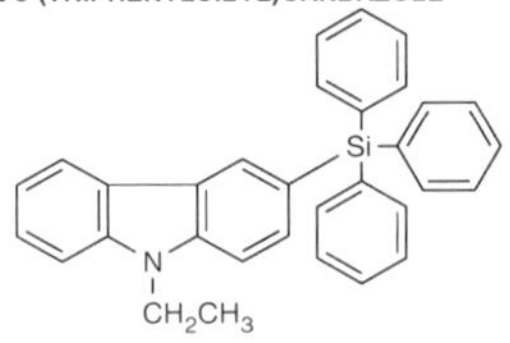
250 mg

S81,175-0
9-ETHYL-1-(TRIPHENYLSILYL)CARBAZOLE
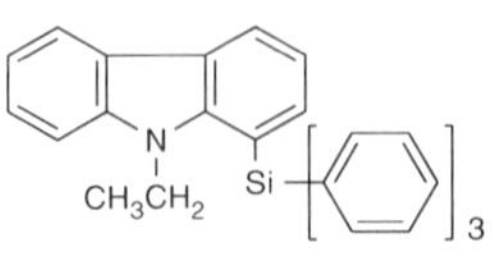
250 mg

S83,799-7
3-CHLORO-9-ETHYL-6-(TRIPHENYLSILYL)-CARBAZOLE
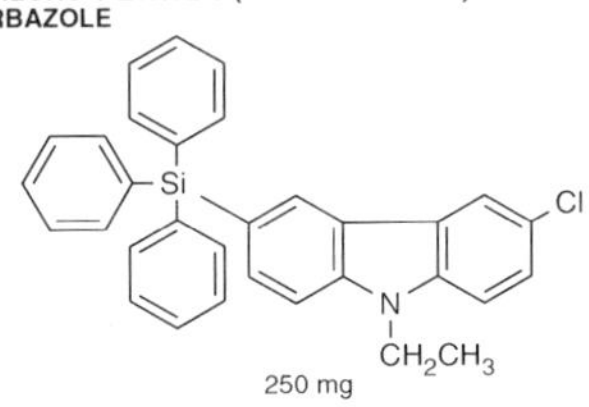
250 mg

S83,804-7
BIS(9-ETHYL-3-CARBAZOLYL)-DIPHENYLSILANE
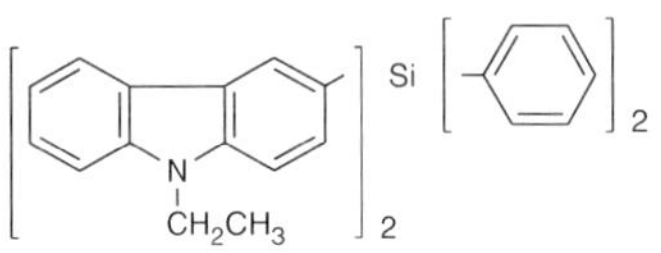
250 mg

S83,806-3
PHENYLTRIS(9-ETHYL-3-CARBAZOLYL)-SILANE
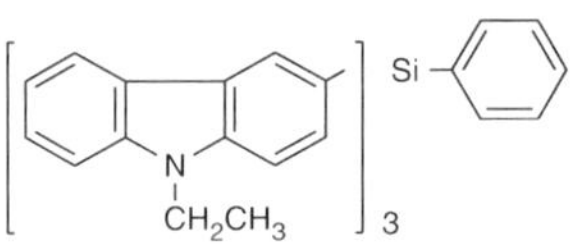
250 mg

S82,718-5
N,N-DIMETHYL-1-(TRIPHENYLSILYL)-ETHYLAMINE
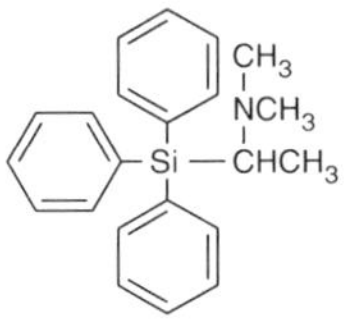
250 mg

S82,845-9
9,10-DIHYDRO-10-METHYL-9-(TRIPHENYLSILYL)ACRIDINE
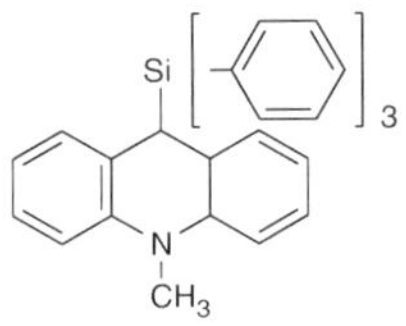
250 mg

S82,298-1
5,10-DIHYDRO-10,10-DIPHENYL-5-ETHYLPHENAZAGERMANE
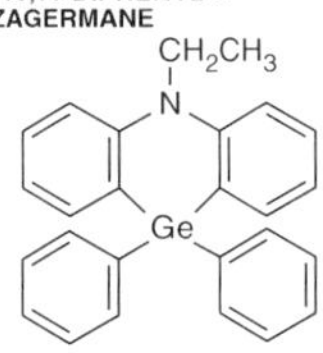
250 mg

S82,292-2
5,10-DIHYDRO-10,10-DIPHENYL-2,5,8-TRIMETHYLPHENAZASILINE
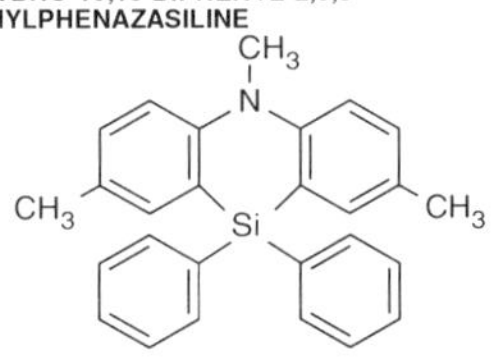
250 mg

S82,294-9
5,10-DIHYDRO-5-ETHYL-2,10,10-TRIPHENYLPHENAZASILINE
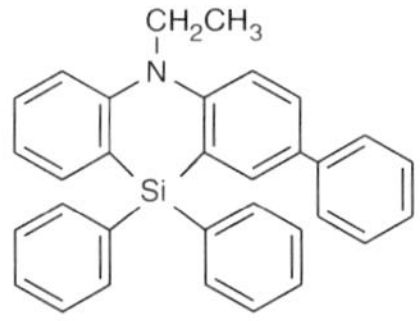
250 mg

S83,352-5
1,6-BIS[4-(DIMETHYLAMINO)PHENYL]-DODECAMETHYLHEXASILANE
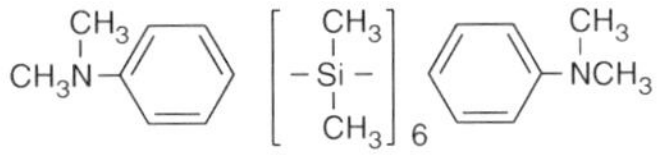
250 mg

S58,597-1
1-VINYL-2,8,9-TRIOXA-5-AZA-1-SILABICYCLO[3.3.3]UNDECANE
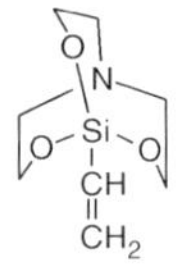
5 gm

S82,296-5
10,10-DIBENZYL-5,10-DIHYDRO-2,5,8-TRIMETHYLPHENAZASILINE
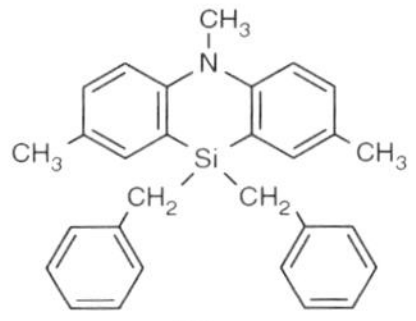
250 mg

S81,983-2
N-(DIPHENYLMETHYL)-N-(TRIPHENYLSILYL)ANILINE
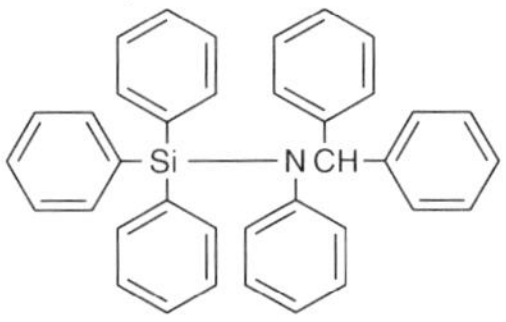
250 mg

S83,555-2
TRIS[4-(DIMETHYLAMINO)PHENYL]-SILANOL
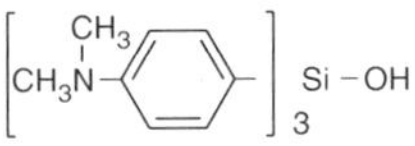
250 mg

S83,303-7
1,2-BIS(TRIMETHYLSILYL)-1,2-DIPHENYLHYDRAZINE
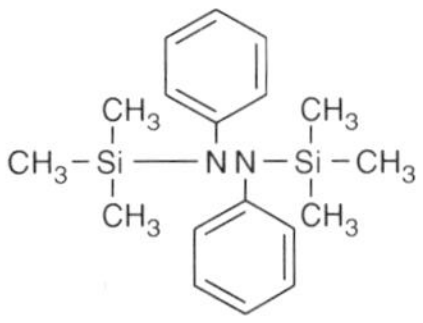
250 mg

S82,882-3
1,2-BIS(TRIETHYLSILYL)-1,2-DIPHENYLHYDRAZINE
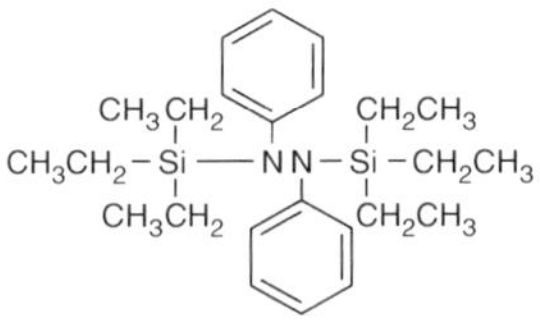
250 mg

S82,498-4
1,2-DIPHENYL-1-[(DIPHENYL)-METHYLSILYL]HYDRAZINE
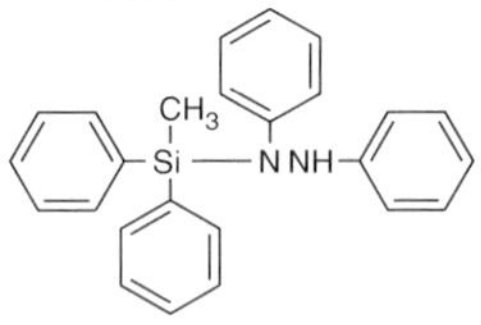
250 mg

S83,304-5
1,2-BIS(DIMETHYLPHENYLSILYL)-1,2-DIPHENYLHYDRAZINE
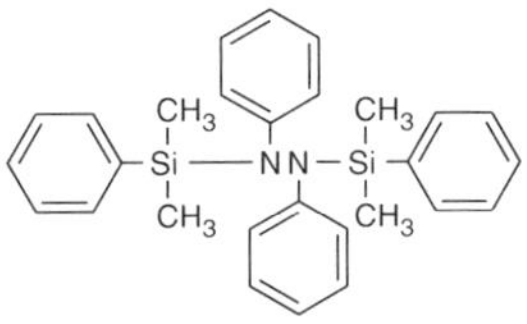
250 mg

S82,883-1
1,2-DIPHENYL-1-(TRIPHENYLGERMYL)-HYDRAZINE
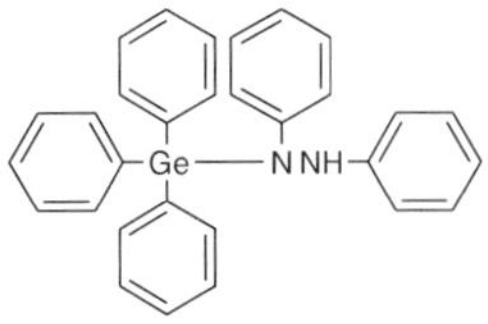
250 mg

S83,306-1
1,2-DIPHENYL-1-(TRIPHENYLGERMYL)-HYDRAZINE
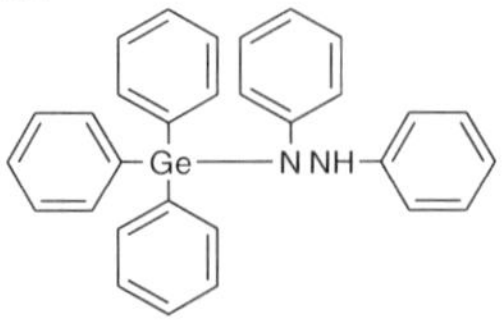
250 mg

S83,308-8
1,2-BIS(DIPHENYLMETHYLSILYL)-1,2-DIPHENYLHYDRAZINE
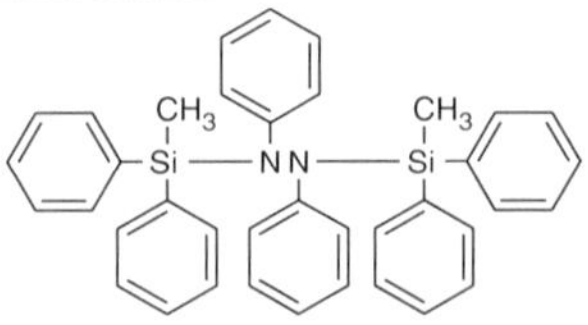
250 mg

S81,747-3
1,2-DIPHENYL-1-(TRIPHENYLSILYL)-HYDRAZINE
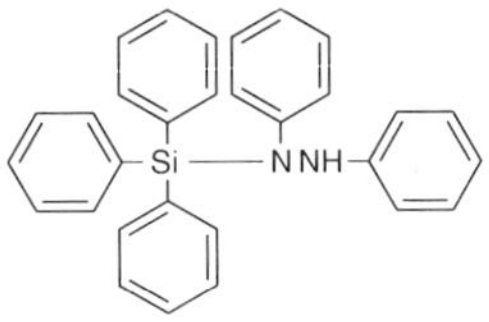
250 mg

S86,503-6
N-NITROSO-N-(TRIMETHYLSILYLMETHYL)-P-TOLUENESULFONAMIDE
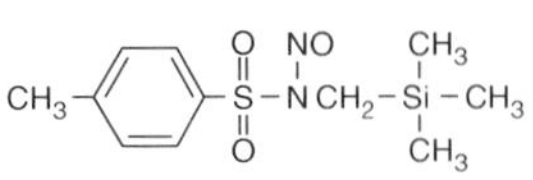
250 mg

S86,633-4
3-NITROPHENYL TRIMETHYLSILYL ETHER
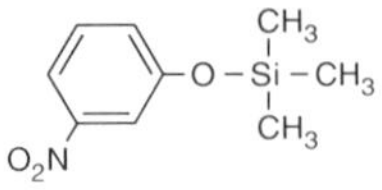
250 mg

S81,356-7
(4-NITROPHENYL)TRIMETHYLSILANE
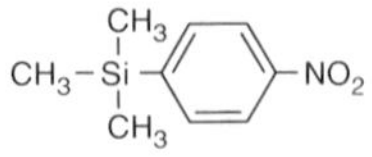
250 mg

S86,508-7
OCTYL TRIMETHYLSILYL KETONE

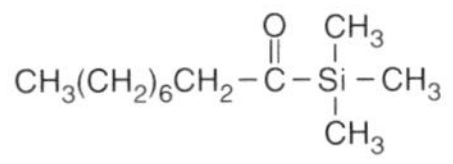

250 mg

S84,354-7
4-METHYL-3-[1-(TRIMETHYLSILYL)-VINYL]-2-CYCLOHEXEN-1-ONE

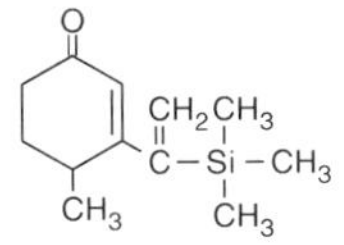

250 mg

S83,989-2
4,4'-BIS(TRIMETHYLSILYL)-BENZOPHENONE

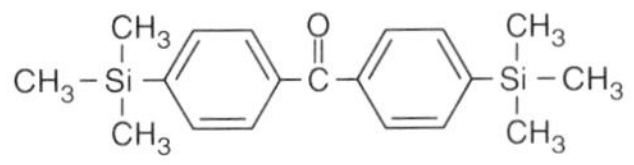

250 mg

S76,462-0
4-(TRIMETHYLSILYL)CYCLOHEXANONE

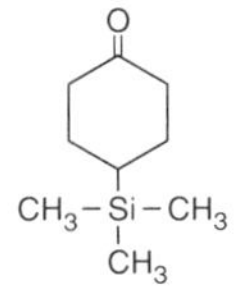

5 gm

S70,446-6
METHYL 1-(TRIMETHYLSILYLOXY)-CYCLOHEX-1-EN-4(5)-YL KETONE

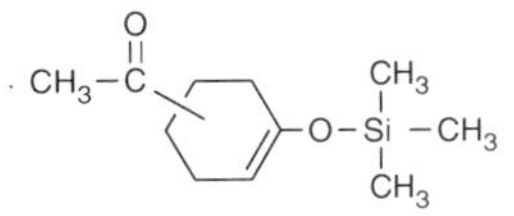

5 gm

S83,716-4
4-(TRIPHENYLSILYL)BENZOIC ACID

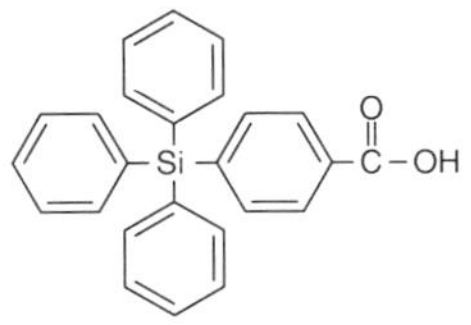

250 mg

S81,327-3
4-(TRIMETHYLSILYL)BENZOIC ACID

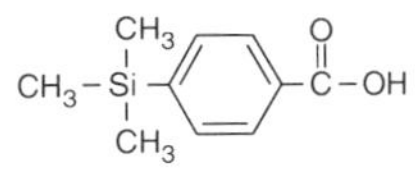

250 mg

S82,928-5
S-(TRIPHENYLSILYL)THIOBENZOATE

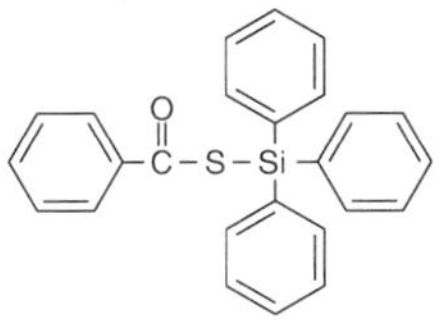

250 mg

S82,873-4
2-BENZOYL-1,1,1,3,3,3-HEXAMETHYL-2-(TRIMETHYLSILYL)TRISILANE

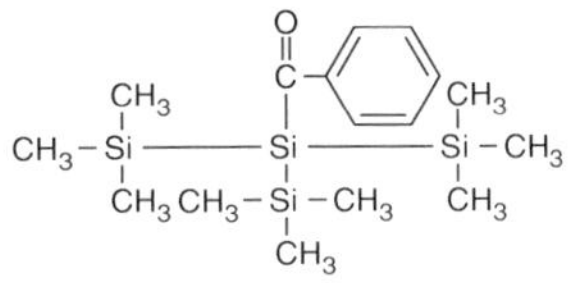

250 mg

S82,875-0
BENZOYLDIPHENYLMETHYLSILANE

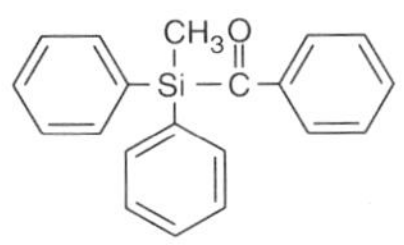

250 mg

S82,877-7
BENZOYLTRIPHENYLSILANE

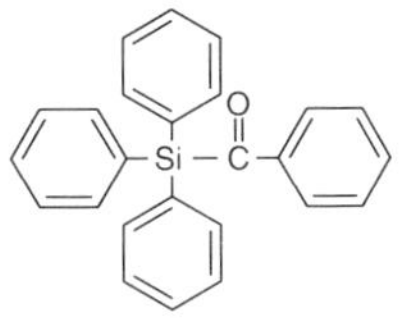

250 mg

S83,102-6
1,3-BIS(2,2-DICARBOXYPROPYL)-1,1,3,3-TETRAMETHYLDISILOXANE

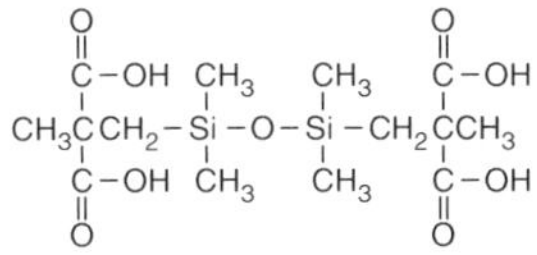

250 mg

S76,287-3
1-CARBOXY-4-(TRIMETHYLSILYL)-CYCLOHEXANEACETIC ACID

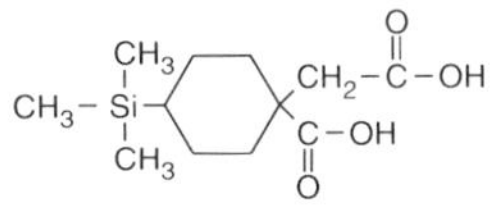

250 mg

S82,225-6
TRICHLOROGERMANECARBOXYLIC ACID

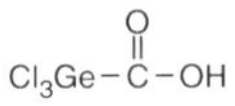

250 mg

S43,975-4
TRIACETOXYVINYLSILANE

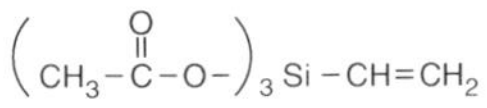

5 gm

S82,457-7
1,1,1,2,3,3,3-HEPTAMETHYL-2-TRISILANECARBOXYLIC ACID

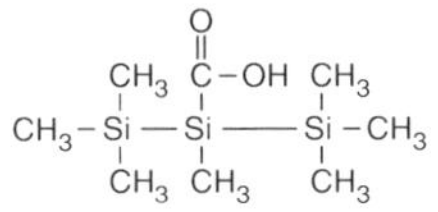

250 mg

S51,634-1
PROPYLTRIACETOXYSILANE

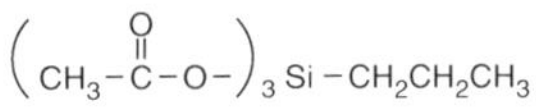

5 gm

S51,678-3
DIACETOXYDIISOPROPOXYSILANE

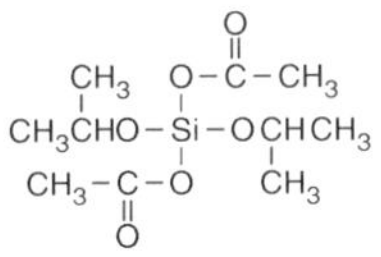

5 gm

S82,358-9
1,1,1,3,3,3-HEXAMETHYL-2-(TRIMETHYLSILYL)-2-TRISILANECARBOXYLIC ACID

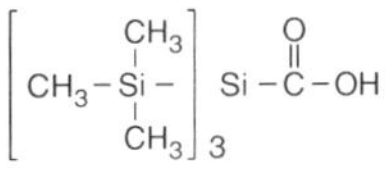

250 mg

S77,126-0
3,3'-(DIMETHYLSILYLENE)-BISPROPIONIC ACID

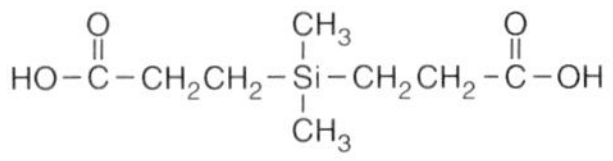

250 mg

S51,638-4
PHENYLTRIACETOXYSILANE

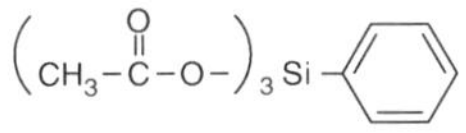

5 gm

S83,236-7
TRIPHENYLSILYLCARBOXYLIC ACID

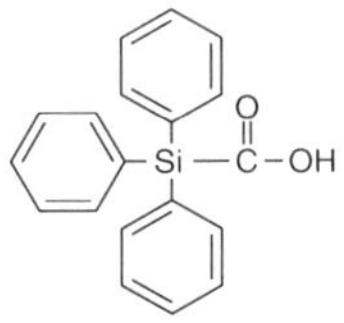

250 mg

S82,346-5
TRIPHENYLGERMANECARBOXYLIC ACID

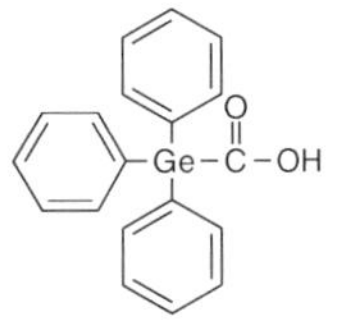

250 mg

S83,041-0
3-(TRIPHENYLSILYL)PROPIONIC ACID

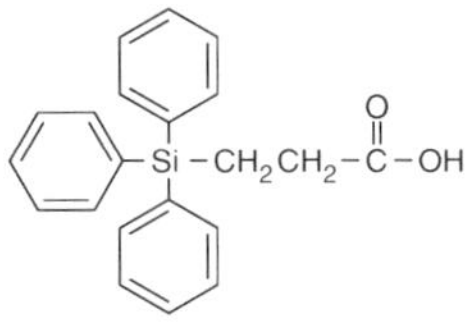

250 mg

S77,131-7
DIETHYL 3,3'-(DIMETHYLSILYLENE)-DIPROPIONATE

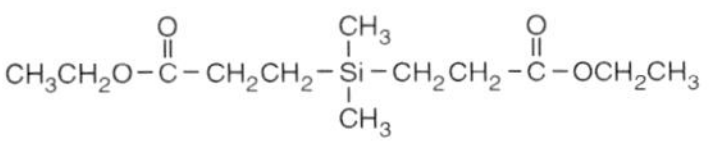

250 mg

S88,175-9
ETHYL 2-(TRIMETHYLSILYL)-1-CYCLOPROPANE CARBOXYLATE

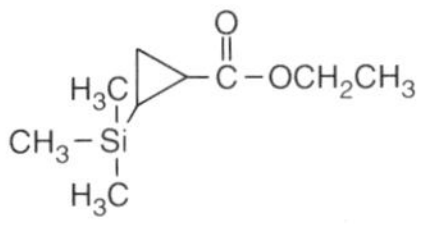

250 mg

S83,235-9
METHYL (METHYLDIPHENYLSILYL)-CARBOXYLATE

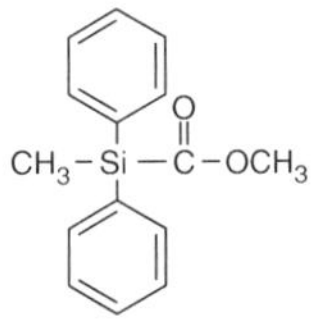

250 mg

S80,606-4
ETHYL 2-(2,2-DIPHENYL-6-PROPENYL-1,3-DIOXA-2-SILACYCLOHEX-4-YL)-ACETATE

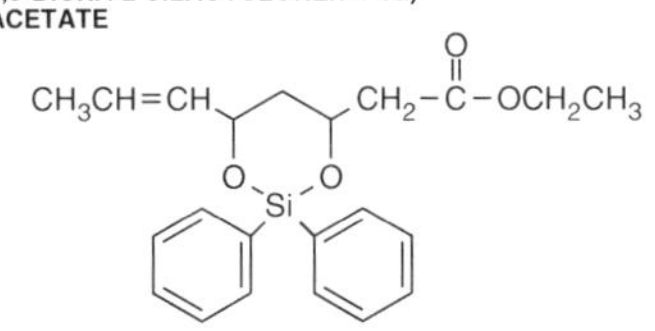

250 mg

S83,219-7
TRIS(TRIMETHYLSILYL)SILANOL

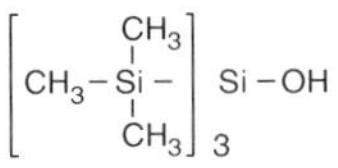

250 mg

S81,303-6
TRI(O-TOLYL)SILANOL

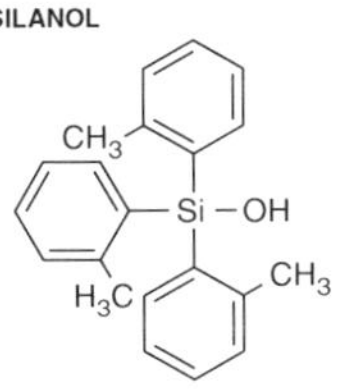

250 mg

S83,564-1
TRIS(2-METHOXYPHENYL)SILANOL

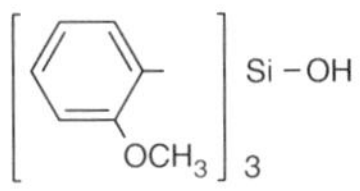

250 mg

S83,643-5
TRIS(2,4,6-TRIMETHYLPHENYL)SILANOL

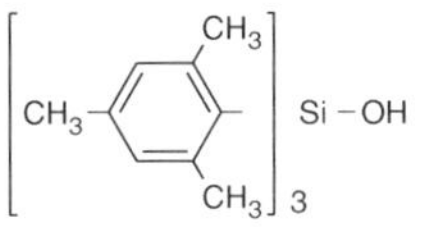

250 mg

S83,566-8
TRIS(4-BIPHENYLYL)SILANOL

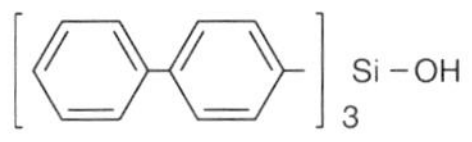

250 mg

S83,721-0
TRIS(1-NAPHTHYL)SILANOL

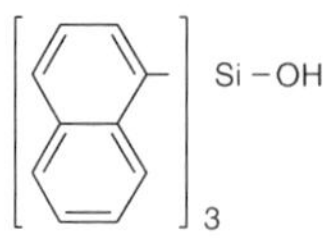

250 mg

S82,879-3
METHYL (PHENYLSILYL)DITHIOFORMATE

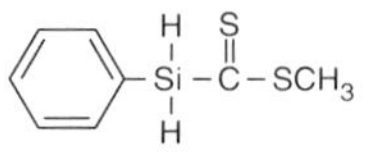

250 mg

S58,560-2
4,5-DIHYDRO-1-PHENYL-4-TRIMETHYLSILYL-1H-TETRAZOL-5-ONE

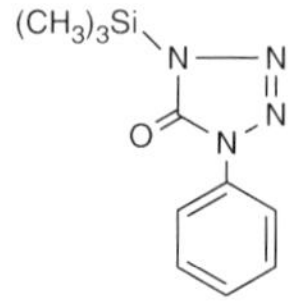

5 gm

S45,540-7
O,O'-BIS(TRIMETHYLSILYL)THYMINE

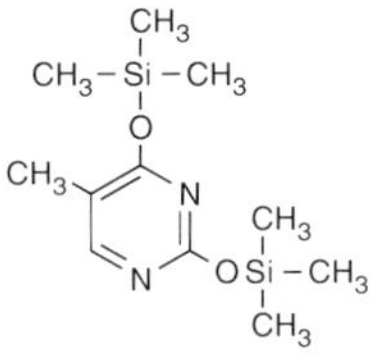

250 mg

S56,542-3
3-(DICHLOROMETHYLSILYL)-PROPIONITRILE

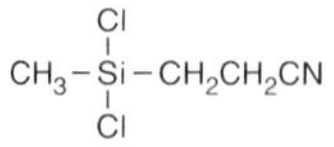

5 gm

S56,543-1
4-(DICHLOROMETHYLSILYL)-BUTYRONITRILE

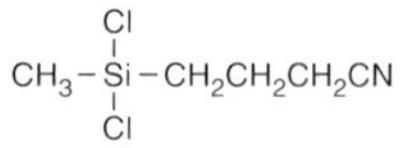

5 gm

S56,526-1
4-(TRIMETHOXYSILYL)BUTYRONITRILE

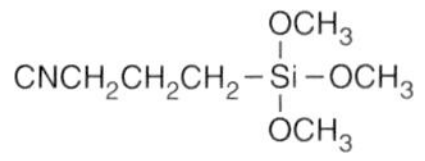

5 gm

S36,616-1
(2-CYANOETHYL)DIETHOXYMETHYLSILANE

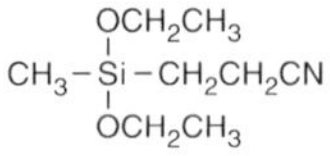

5 gm

S86,642-3
3-(TRIMETHYLSILYLOXY)BENZONITRILE

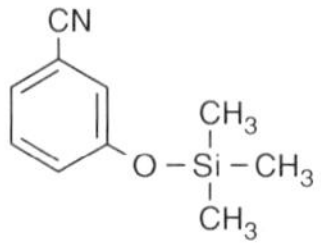

250 mg

S56,652-7
5,5,7,7-TETRAMETHYL-6-OXA-5,7-DISILAUNDECANEDINITRILE

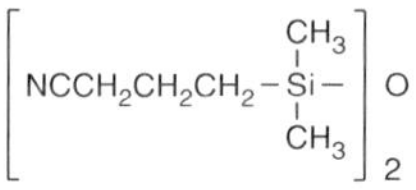

5 gm

S82,575-1
TRIPHENYLSILYL ISOTHIOCYANATE

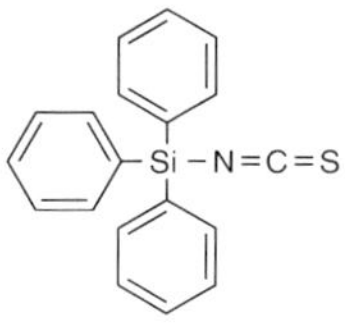

250 mg

S83,586-2
P-TOLYL TRIPHENYLSILYLMETHYL SULFONE

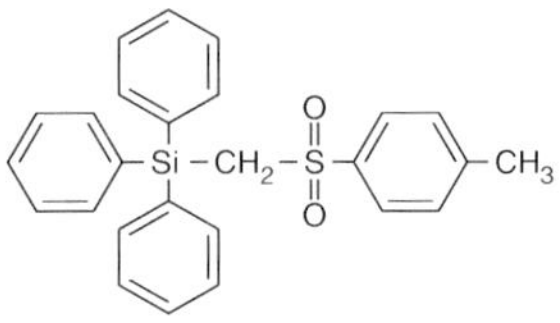

250 mg

S82,764-9
10,10-DIPHENYL-10H-PHENOTHIASILIN-5,5-DIOXIDE

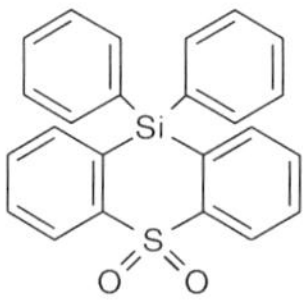

250 mg

S86,505-2
N-(TRIMETHYLSILYLMETHYL)-P-TOLUENESULFONAMIDE

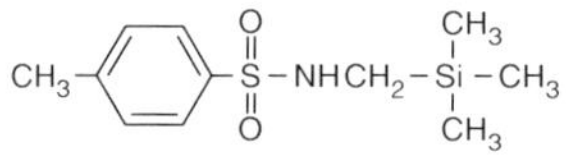

250 mg

S58,467-3
5-NORBORNEN-2-YLTRICHLOROSILANE

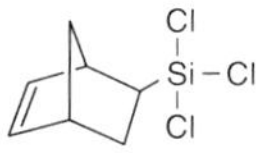

5 gm

S58,468-1
5-NORBORNEN-2-YLTRIETHOXYSILANE

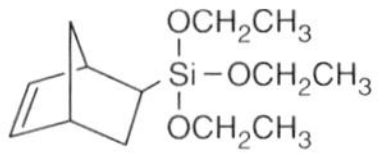

5 gm

S82,313-9
CYCLOPENTYLTRIPHENYLSILANE

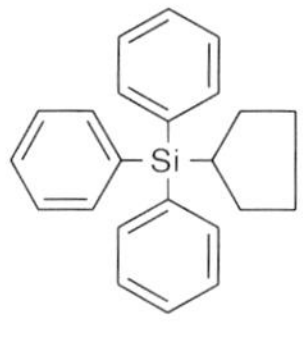

250 mg

S58,470-3
2-NORBORNYLTRICHLOROSILANE

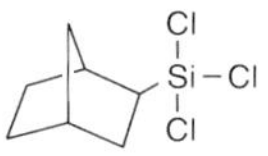

5 gm

S84,353-9
2,3-DIMETHYL-1-(TRIMETHYLSILYLOXY)-1-CYCLOPENTENE

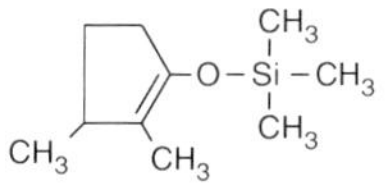

250 mg

S83,699-0
2-(TRIPHENYLSILYL)DIBENZOFURAN

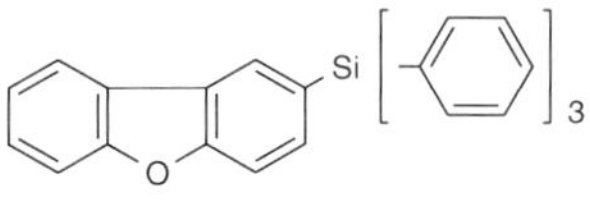

250 mg

S83,700-8
4-(TRIPHENYLSILYL)DIBENZOFURAN

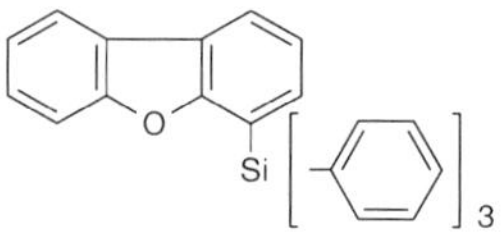

250 mg

S81,118-1
2-DIBENZOFURANSULFINIC ACID

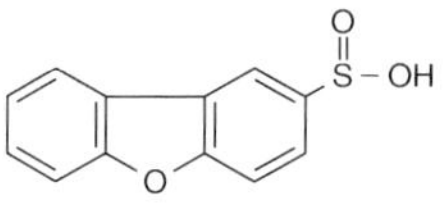

250 mg

S81,224-2
2-(DIMETHYLSILYL)-3,4,5-TRICHLOROTHIOPHENE

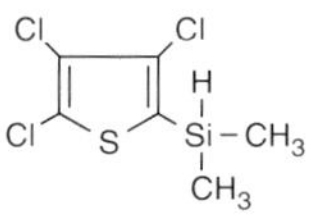

250 mg

S82,557-3
(3,4,5-TRICHLORO-2-THIENYL)-TRIMETHYLSILANE

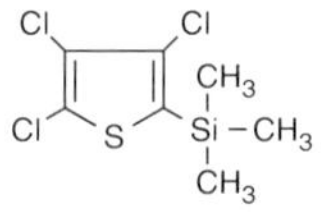

250 mg

S81,432-6
2,5-BIS(DIMETHYLSILYL)-3,4-DICHLOROTHIOPHENE

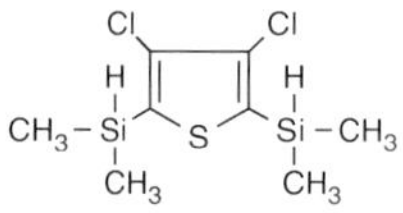

250 mg

S81,222-6
BIS(3,4,5-TRICHLORO-2-THIENYL)-METHYLSILANE

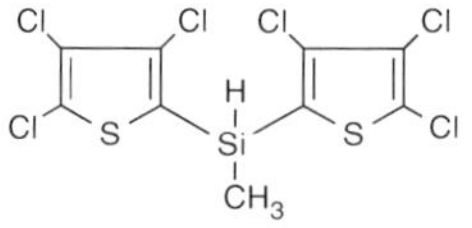

250 mg

S83,977-9
2,5-BIS(TRIMETHYLSILYL)-3,4-DICHLOROTHIOPHENE

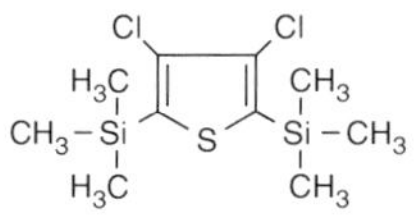

250 mg

S81,219-6
BIS(3,4,5-TRICHLORO-2-THIENYL)-DIMETHYLSILANE

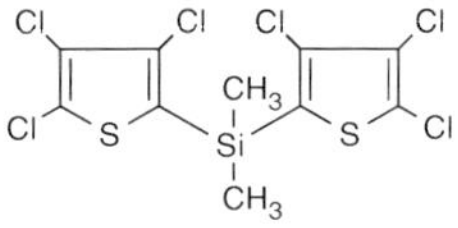

250 mg

S81,218-8
TRIS(3,4,5-TRICHLORO-2-THIENYL)-SILANE

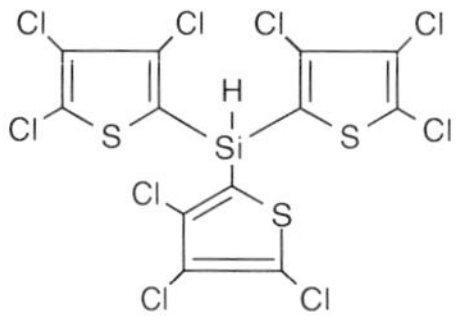

250 mg

S81,216-1
METHYLTRIS(3,4,5-TRICHLORO-2-THIENYL)SILANE

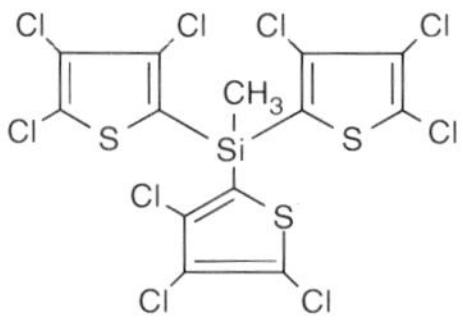

250 mg

S81,214-5
TETRAKIS(3,4,5-TRICHLORO-2-THIENYL)-SILANE

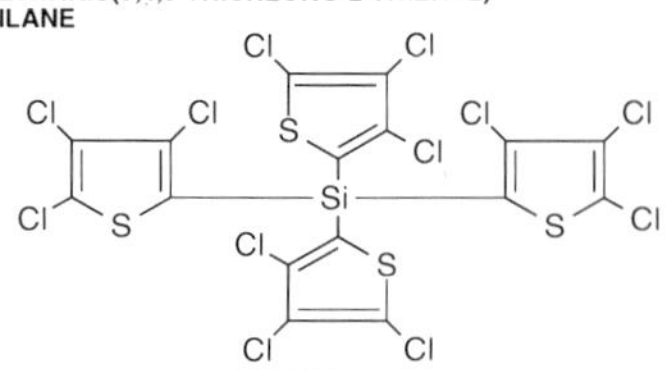

250 mg

S83,164-6
2-(DIMETHYLPHENYLSILYL)-3,4,5-TRICHLOROTHIOPHENE

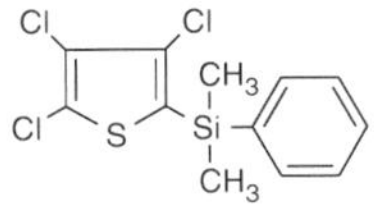

250 mg

S81,344-3
BIS(3,4,5-TRICHLORO-2-THIENYL)-DIPHENYLSILANE

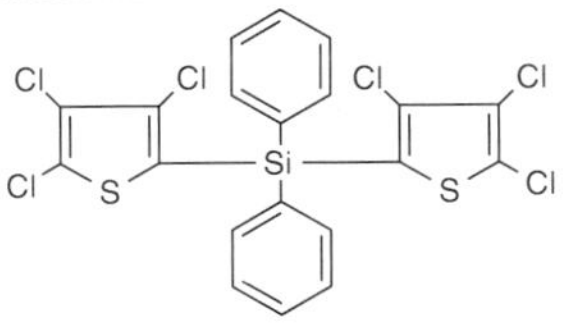

250 mg

S81,468-7
(2-THIENYL)TRIPHENYLSILANE

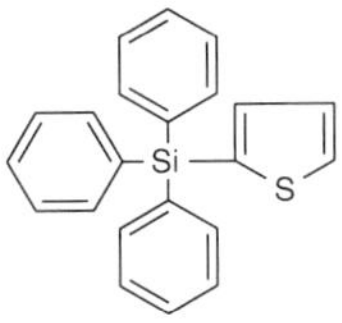

250 mg

S81,212-9
(3,4,5-TRICHLORO-2-THIENYL)-TRIPHENYLSILANE

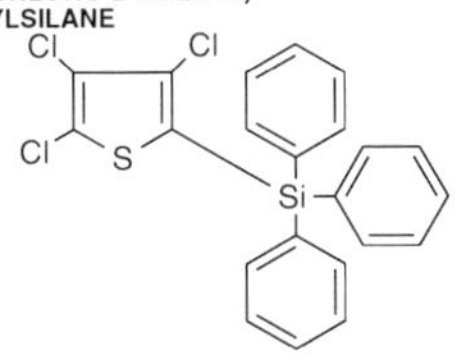

250 mg

S81,176-9
2-(TRIPHENYLSILYL)DIBENZOTHIOPHENE

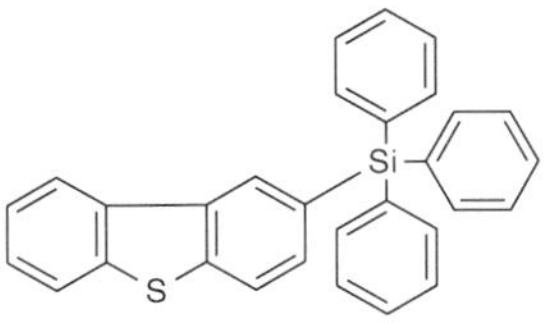

250 mg

S83,701-6
4-(TRIPHENYLSILYL)DIBENZOTHIOPHENE

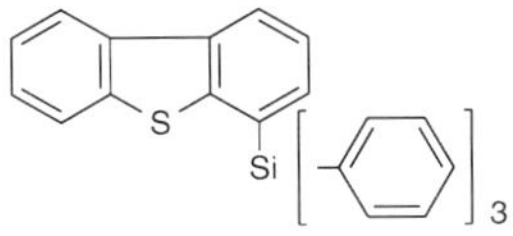

250 mg

S83,697-4
2-(TRIPHENYLSILYL)BENZOTHIAZOLE

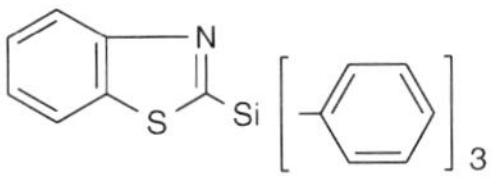

250 mg

S58,496-7
(CYCLOHEXYLMETHYL)TRICHLOROSILANE

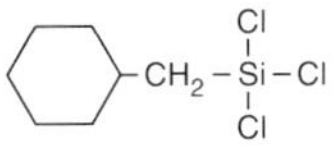

5 gm

S83,371-1
TETRACYCLOHEXYLSILANE

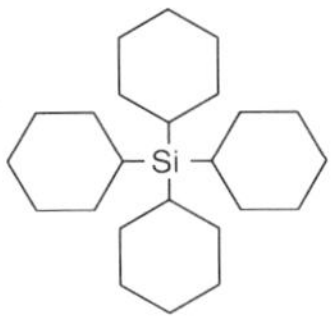

250 mg

S83,562-5
TETRAKIS(2-CYCLOHEXYLETHYL)SILANE

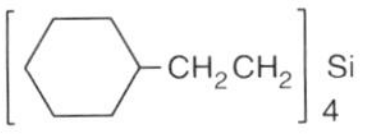

250 mg

S83,803-9
DODECYLTRIS(2-CYCLOHEXYLETHYL)-SILANE

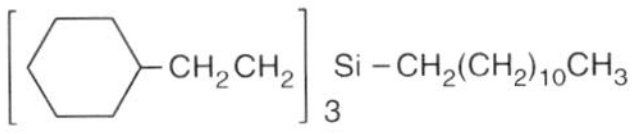

250 mg

S83,805-5
OCTADECYLTRIS(2-CYCLOHEXYLETHYL)-SILANE

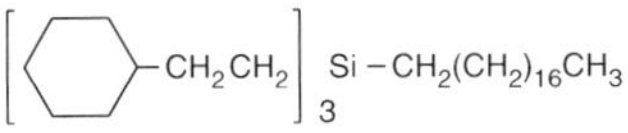

250 mg

S82,508-5
CYCLOHEXYLTRIDODECYLSILANE

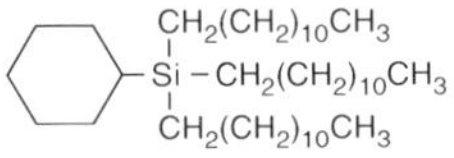

250 mg

S82,935-8
CYCLOHEXYLDIPHENYLSILANE

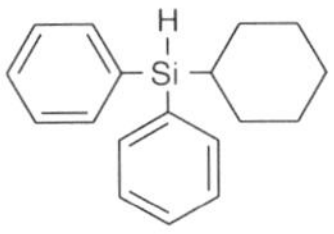

250 mg

S82,398-8
CYCLOHEXYLTRIPHENYLSILANE

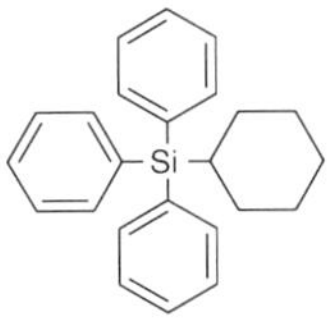

250 mg

S83,787-3
(2-CYCLOHEXYLETHYL)TRIPHENYLSILANE

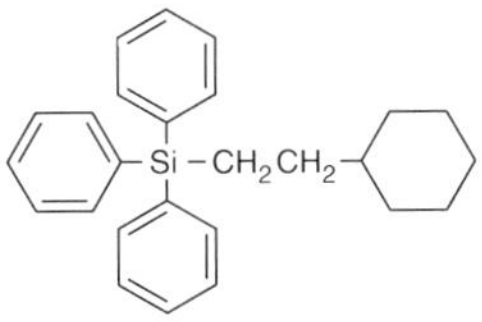

250 mg

S82,523-9
CYCLOHEXYLTRIBENZYLSILANE

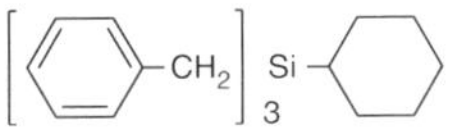

250 mg

S83,791-1
BIS(2-CYCLOHEXYLETHYL)-DIPHENYLSILANE

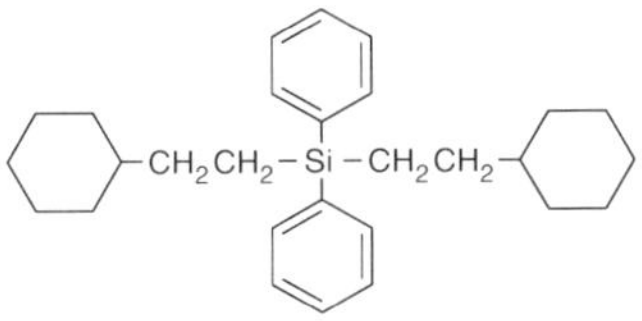

250 mg

S83,796-2
PHENYLTRIS(2-CYCLOHEXYLETHYL)SILANE

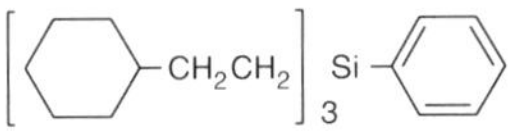

250 mg

S86,649-0
2-METHYL-1-CYCLOHEXEN-1-YL TRIMETHYLSILYL ETHER

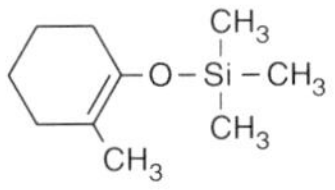

250 mg

S86,679-2
TERT-BUTYLDIMETHYLSILYL 1-CYCLOHEXEN-1-YL ETHER

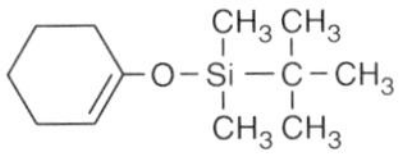

250 mg

S88,484-7
TERT-BUTYLDIMETHYLSILYL 2-(1,5,5-TRIMETHYL-2-CYCLOHEXEN-1-YL)ACETATE

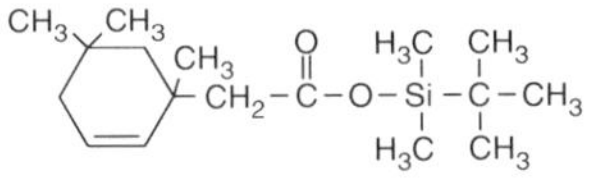

250 mg

S82,359-7
3,4,5,6-TETRAKIS(TRIMETHYLSILYL)-1-[TRIS(TRIMETHYLSILYL)SILYL]-1-CYCLOHEXENE

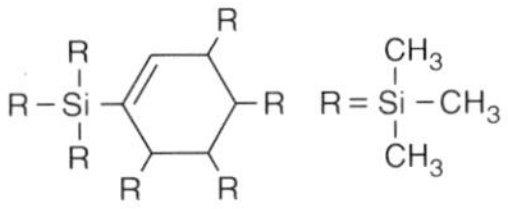

250 mg

S82,933-1
(1-CYCLOHEXEN-3-YL)DIPHENYLSILANE

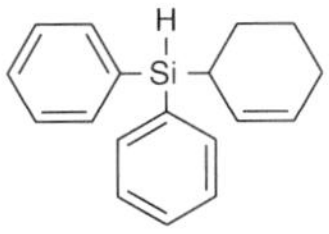

250 mg

S82,170-5
(1-CYCLOHEXENYL)TRIPHENYLSILANE

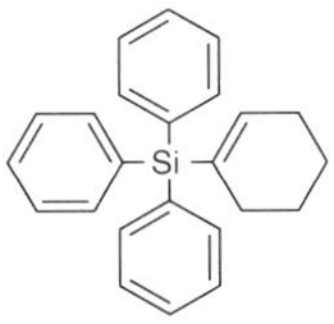

250 mg

S82,937-4
(1-CYCLOHEXEN-3-YL)TRIPHENYLSILANE

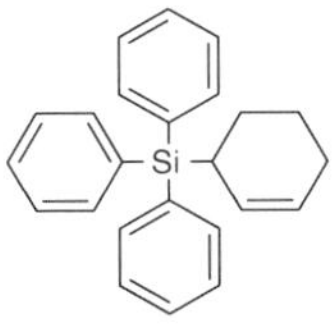

250 mg

S82,857-2
9,9-DIPHENYL-4-(TRIPHENYLSILYL)-XANTHENE

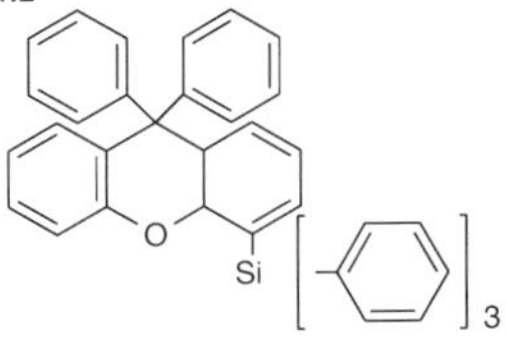

250 mg

S82,854-8
9-(TRIPHENYLSILYL)XANTHENE

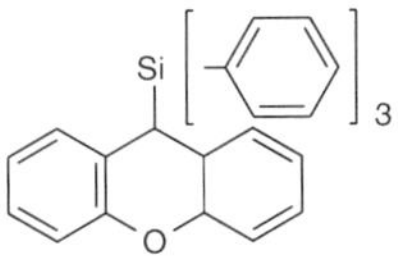

250 mg

S82,855-6
BIS(9-XANTHENYL)DIPHENYLSILANE

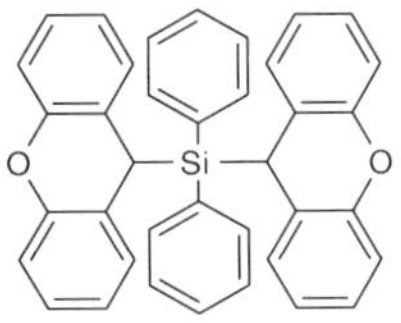

250 mg

S82,848-3
9-(TRIMETHYLSILYL)XANTHENE

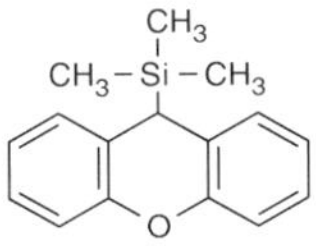

250 mg

S82,852-1
BIS(9-XANTHENYL)DIMETHYLSILANE

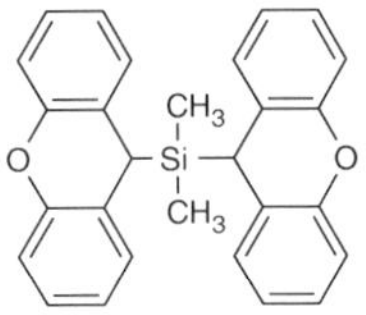

250 mg

S82,957-9
1-(TRIPHENYLSILYL)PIPERIDINE

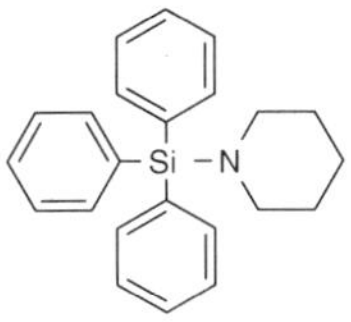

250 mg

S82,847-5
1,4-PIPERAZINEDIYLBIS-(TRIPHENYLSILANE)

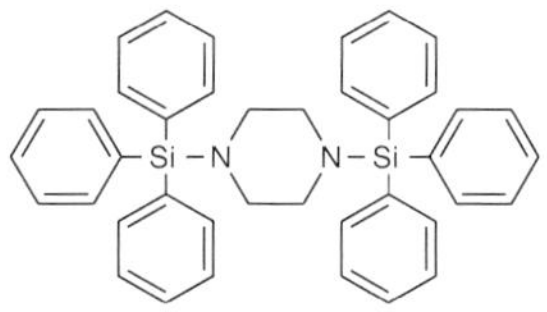

250 mg

S94,026-7
2,2,6,6-TETRAMETHYL-4-(3-METHYLBENZYL)-1-OXA-4-AZA-2,6-DISILACYCLOHEXANE HYDROCHLORIDE

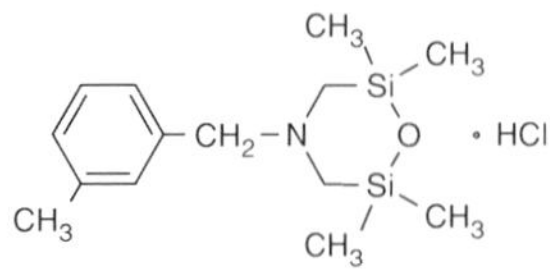

5 gm

S82,745-2
MORPHOLINOTRIPHENYLSILANE

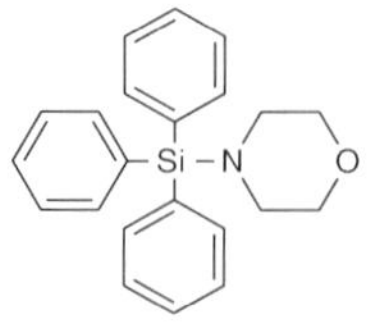

250 mg

S83,356-8
BENZYLSILANE

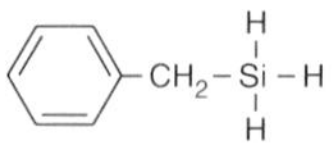

250 mg

S83,358-4
(3-PHENYLPROPYL)SILANE

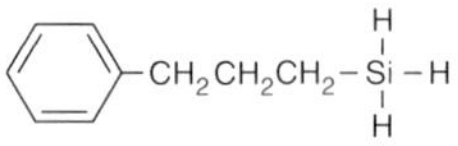

250 mg

S41,621-5
BENZYLCHLORODIMETHYLSILANE

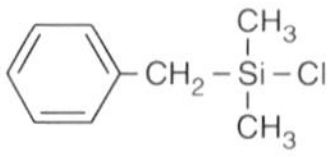

5 gm

S42,820-5
PHENYLTRIMETHOXYSILANE

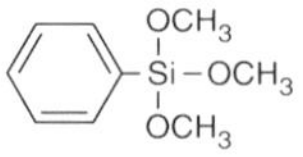

5 gm

S58,544-0
DICHLOROMETHYLPHENETHYLSILANE

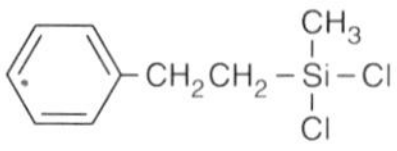

5 gm

S83,140-9
TERT-BUTYLPHENYLSILANE

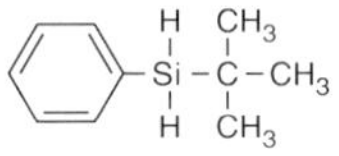

250 mg

S83,346-0
1-PHENYL-1,1,2,2-TETRAMETHYLDISILANE

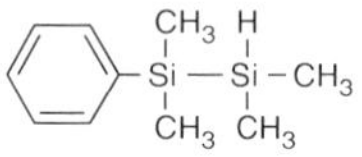

250 mg

S83,025-9
1-PHENYLSILACYCLOHEXANE

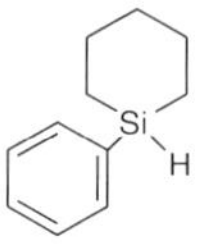

250 mg

S82,987-0
DIMETHYL(3-PHENYLPROPYL)SILANE

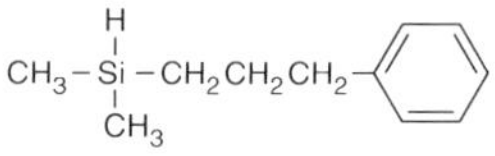

250 mg

S82,739-8
BENZYLDIETHYLSILANOL

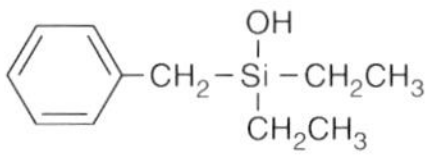

250 mg

S82,463-1
1,1,2,2,2-PENTAMETHYL-1-PHENYLDISILANE

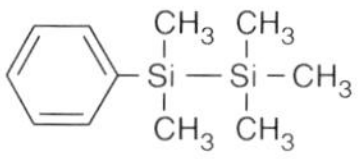

250 mg

S58,604-8
DIACETOXYMETHYLPHENYLSILANE

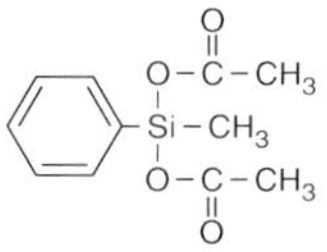

250 mg

S59,982-4
DIFLUORODIPHENYLSILANE

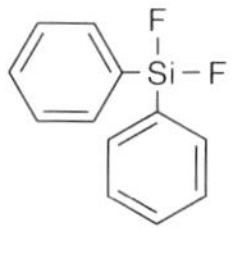

250 mg

S82,362-7
1,1,1,3,3,3-HEXAMETHYL-2-PHENYLTRISILANE

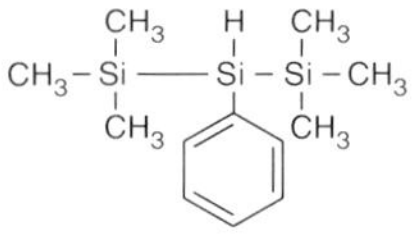

250 mg

S83,360-6
BENZYLPHENYLSILANE

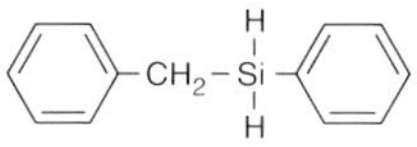

250 mg

S88,435-9
4-PHENYL-1-(TRIMETHYLSILYL)-1-BUTYNE

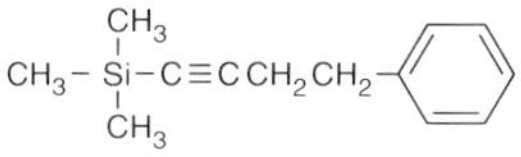

250 mg

S58,513-0
DIPHENYLMETHYLSILYL AZIDE

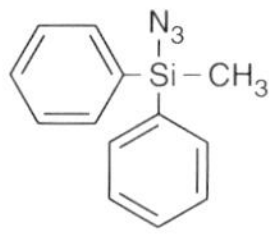

5 gm

S58,465-7
BENZYLTRIETHOXYSILANE

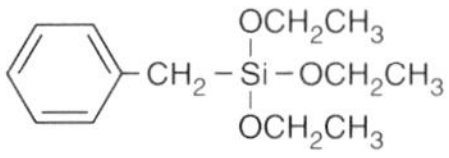

250 mg

S83,220-0
1,1,1,2,3,3,3-HEPTAMETHYL-2-PHENYLTRISILANE

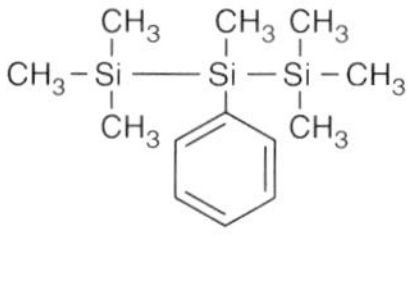

250 mg

S83,063-1
2,2-BIS(DIMETHYLSILYL)-1-PHENYL-1,3,3-TRIMETHYLTRISILANE

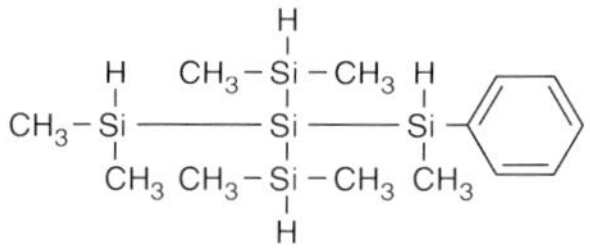

250 mg

S82,462-3
DIMETHYLDIPHENYLSILANE

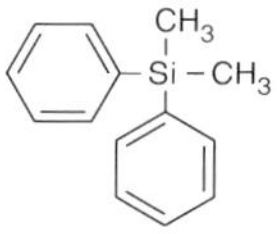

250 mg

S82,668-5
BENZYLMETHYLPHENYLSILANE

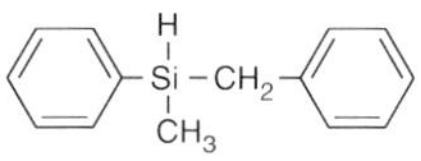

250 mg

S82,669-3
METHYLPHENYL(M-TOLYL)SILANE

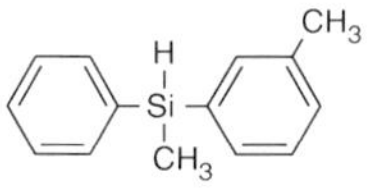

250 mg

S83,021-6
DIPHENYLETHYLSILANE

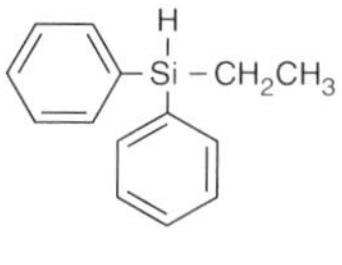

250 mg

S83,361-4
DIBENZYLSILANE

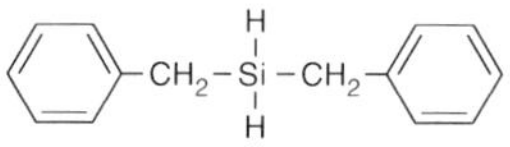

250 mg

S83,984-1
DIPHENYLSILYL ACETATE

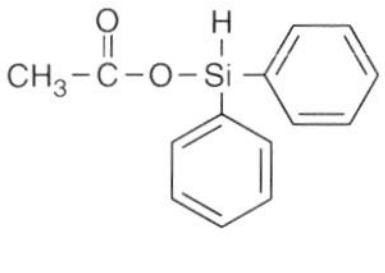

250 mg

S58,508-4
CHLORO(DIMETHYLAMINO)DIPHENYLSILANE

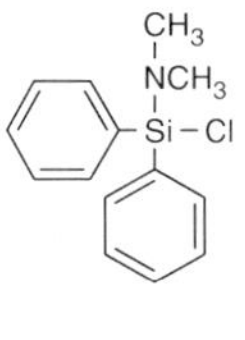

5 gm

S56,538-5
1,3-DICHLORO-1,3-DIMETHYL-1,3-DIPHENYLDISILOXANE

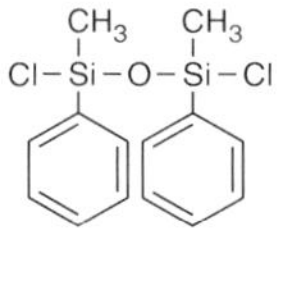

250 mg

S83,029-1
ALLYLDIPHENYLSILANE

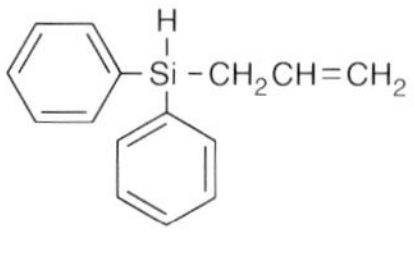

250 mg

S82,671-5
METHYL(3-METHYLBENZYL)PHENYLSILANE

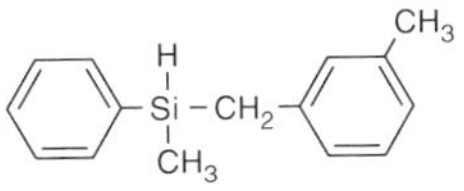

250 mg

S82,673-1
METHYL(PHENETHYL)PHENYLSILANE

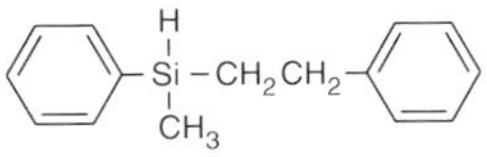

250 mg

S83,786-5
PHENYLTRIALLYLSILANE

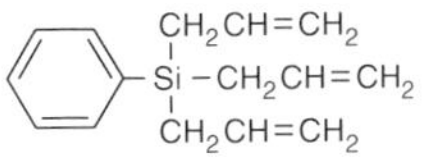

250 mg

S82,872-6
ACETYLDIPHENYLMETHYLSILANE

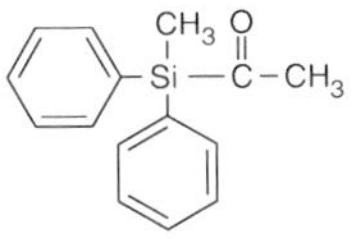

250 mg

S58,514-9
DIPHENYLETHOXYMETHYLSILANE

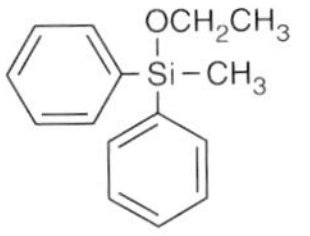

5 gm

S83,347-9
2,2-DIPHENYL-1,1,1-TRIMETHYLDISILANE

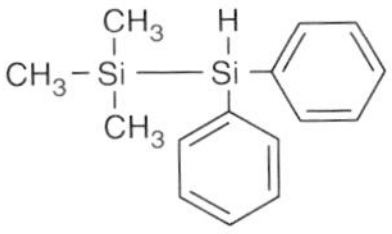

250 mg

S83,221-9
1,1,1,3,3,3-HEXAMETHYL-2-PHENYL-2-(TRIMETHYLSILYL)TRISILANE

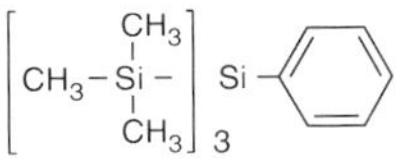

250 mg

S83,030-5
BUTYLDIPHENYLSILANE

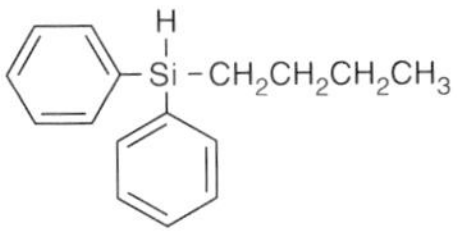

250 mg

S83,103-4
DIBENZYLDIMETHYLSILANE

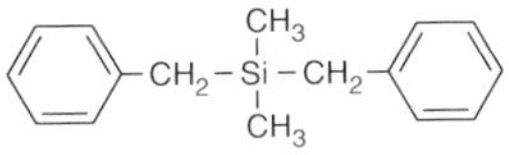

250 mg

S83,363-0
BENZYL(3-PHENYLPROPYL)SILANE

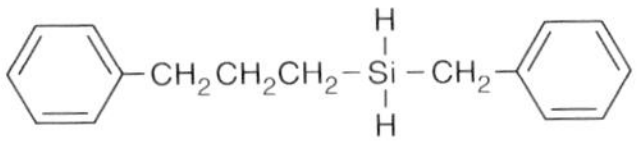

250 mg

S83,031-3
DECYLPHENYLSILANE

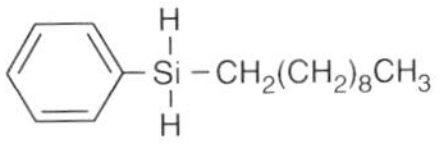

250 mg

S82,217-5
(DIPHENYLMETHOXY)TRIMETHYLSILANE

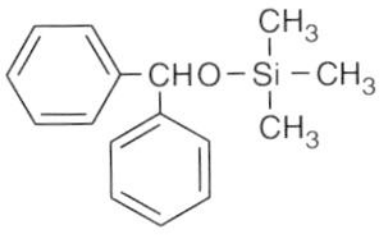

250 mg

S82,363-5
1,2-DIPHENYL-1,1,2,2-TETRAMETHYLDISILANE

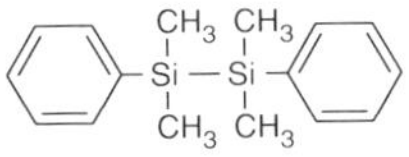

250 mg

S82,424-0
[(DIPHENYLSILYL)METHYL]-TRIMETHYLSILANE

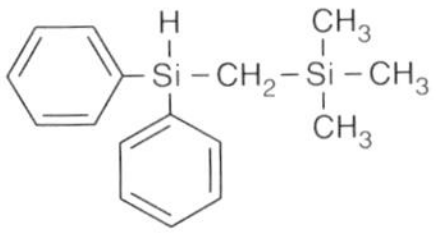

250 mg

S45,865-1
DIETHOXYDIPHENYLSILANE

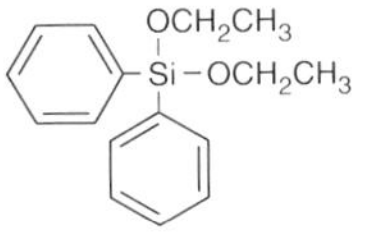

5 gm

S83,104-2
BIS(ALPHA,ALPHA-DIBROMOBENZYL)-DIMETHYLSILANE

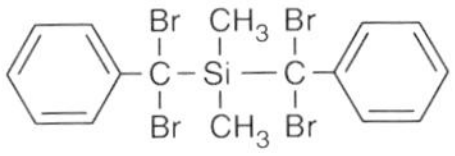

250 mg

S82,216-7
BUTYLDIPHENYLMETHYLSILANE

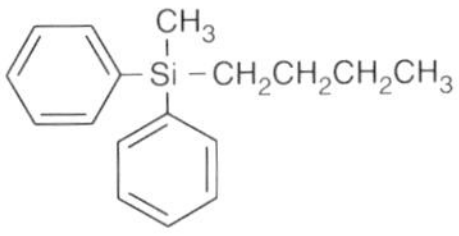

250 mg

S82,676-6
(4-TERT-BUTYLPHENYL)-METHYLPHENYLSILANE

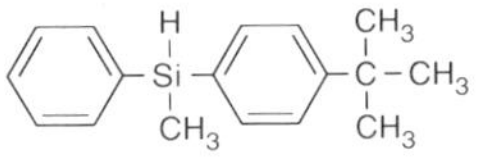

250 mg

S83,066-6
1-PHENYL-2,2,3,3-TETRAKIS-(DIMETHYLSILYL)-1,4,4-TRIMETHYLTETRASILANE

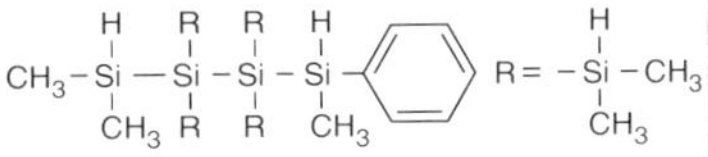

250 mg

S58,512-2
DIALLYLDIPHENYLSILANE

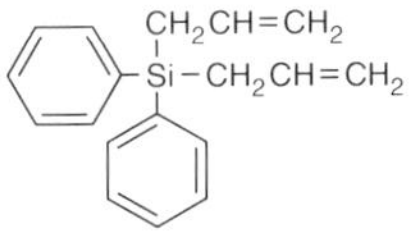

250 mg

S83,364-9
BIS(3-PHENYLPROPYL)SILANE

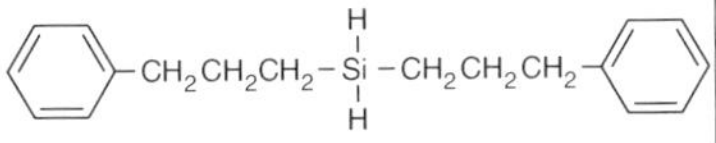

250 mg

S83,034-8
DODECYLPHENYLSILANE

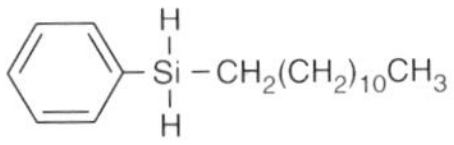

250 mg

S83,425-4
PHENYLTRIBUTYLSILANE

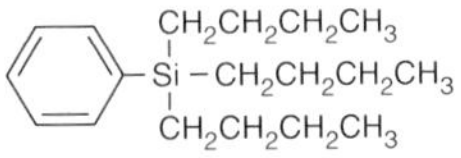

250 mg

S82,734-7
1,2-DIBENZYL-1,1,2,2-TETRAMETHYLDISILANE

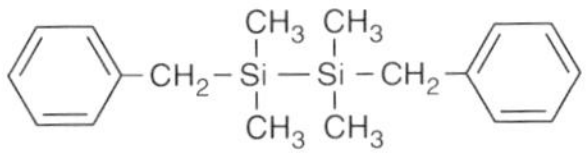

250 mg

S58,585-8
TRIPHENYLSILYL AZIDE

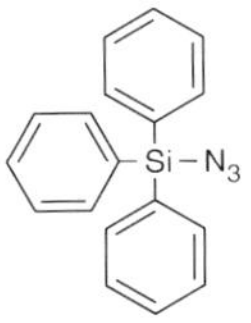

250 mg

S82,345-7
TRIPHENYLGERMANE

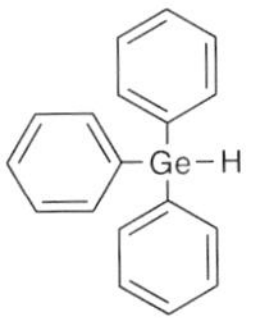

250 mg

S84,043-2
1,3-DIBENZYL-1,1,3,3-TETRAMETHYLDISILOXANE

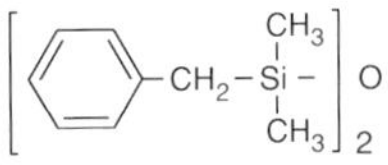

250 mg

S82,290-6
TRIPHENYLGERMANOL

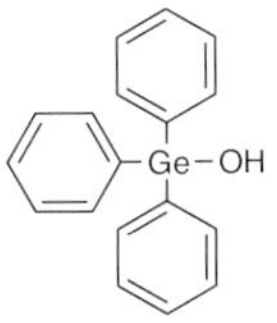

250 mg

S82,458-5
2,2-DIPHENYL-1,1,1,3,3,3-HEXAMETHYLTRISILANE

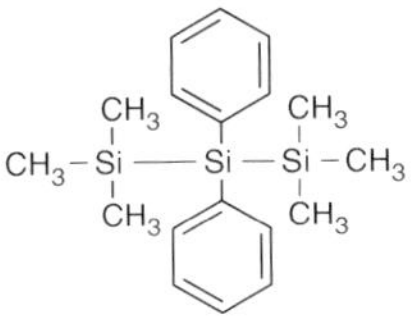

250 mg

S81,402-4
METHYLTRIPHENYLSILANE

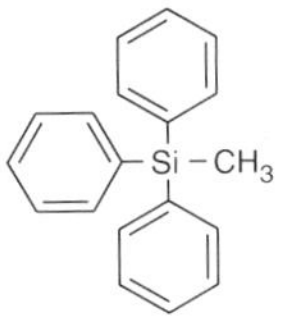

250 mg

S82,681-2
DIPHENYL(M-TOLYL)SILANE

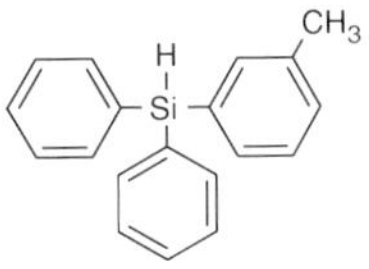

250 mg

S83,035-6
BENZYLDIPHENYLSILANE

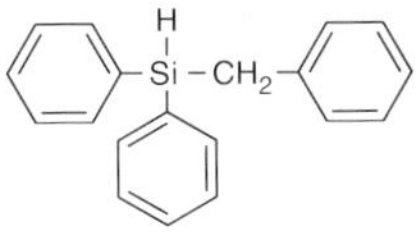

250 mg

S82,577-8
TRIPHENYLSILYL ISOCYANATE
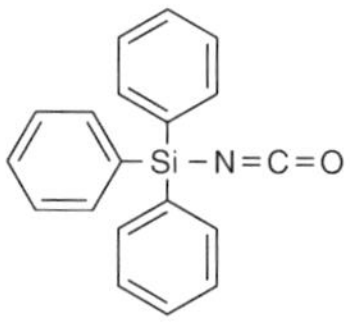
250 mg

S82,880-7
(METHYLTHIO)TRIPHENYLSILANE
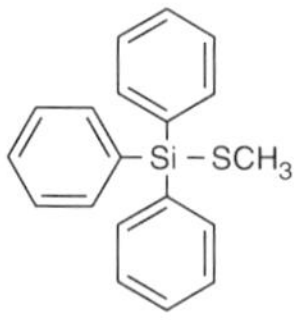
250 mg

S82,399-6
(METHYLSULFONYL)TRIPHENYLSILANE
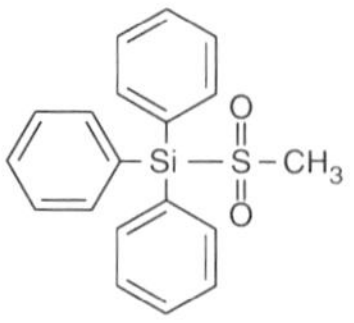
250 mg

S83,067-4
2,2-BIS(DIMETHYLSILYL)-1,1-DIPHENYL-1,3,3-TRIMETHYLTRISILANE
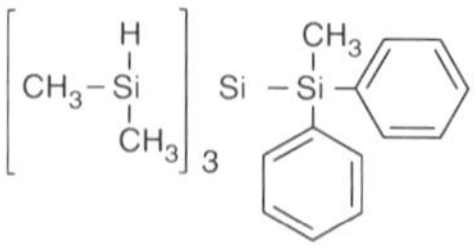
250 mg

S82,684-7
DIPHENYLSTYRYLSILANE
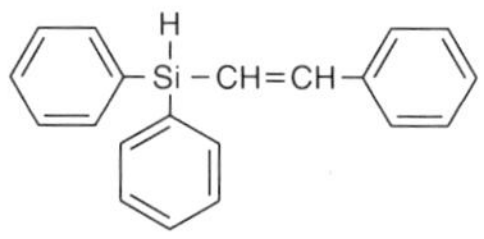
250 mg

S81,120-3
ETHYLTRIPHENYLSILANE
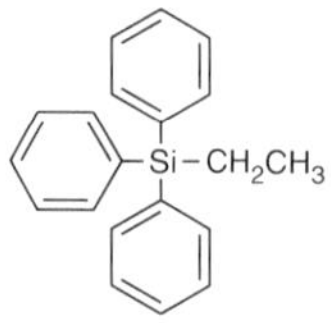
250 mg

S83,365-7
DIBENZYLPHENYLSILANE
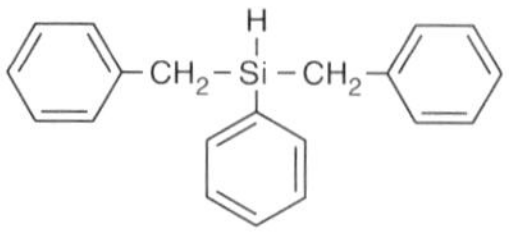
250 mg

S82,646-4
ACETYLTRIPHENYLSILANE
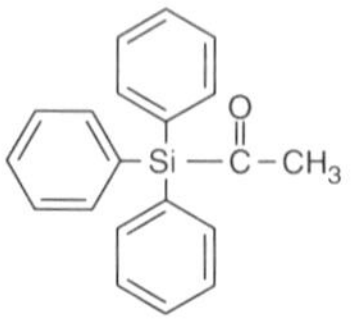
250 mg

S83,366-5
1,1-DIMETHYL-1,2,2-TRIPHENYLDISILANE
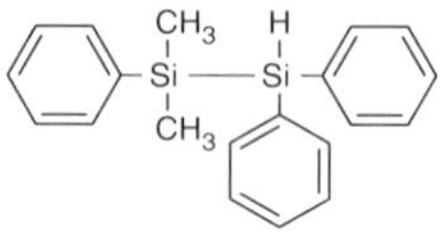
250 mg

S83,860-8
(1-CHLOROETHYL)TRIPHENYLSILANE
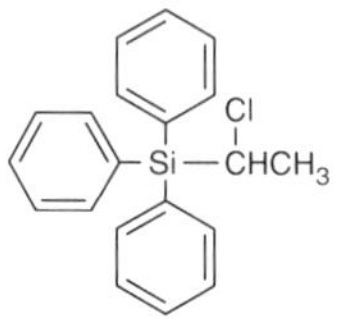
250 mg

S82,736-3
1,2-DIPHENYL-1,1,2,2-TETRAETHYLDISILANE
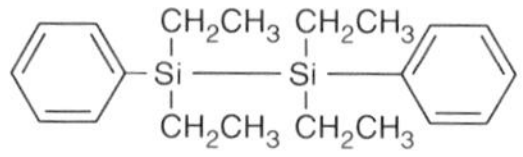
250 mg

S58,600-5
TRIPHENOXYVINYLSILANE
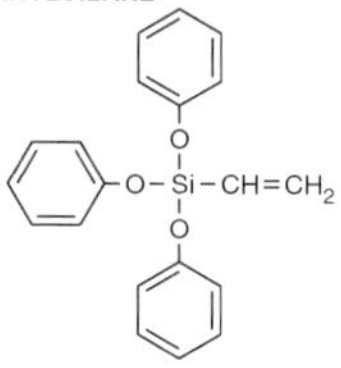
5 gm

S82,874-2
ACETYLTRIPHENYLGERMANE
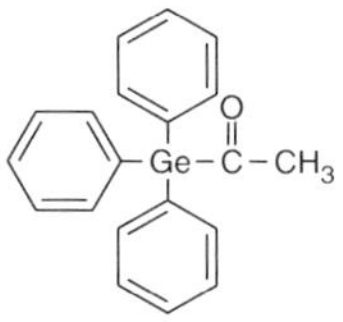
250 mg

S82,876-9
BIS(TRIMETHYLSILYLMETHYL)-DIPHENYLSILANE
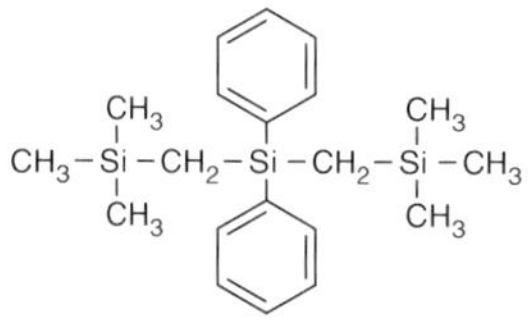
250 mg

S82,169-1
1-PROPYNYLTRIPHENYLSILANE
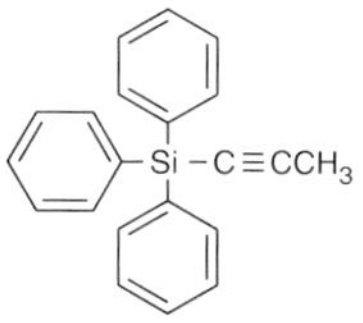
250 mg

S82,991-9
(1-PROPEN-1-YL)TRIPHENYLSILANE
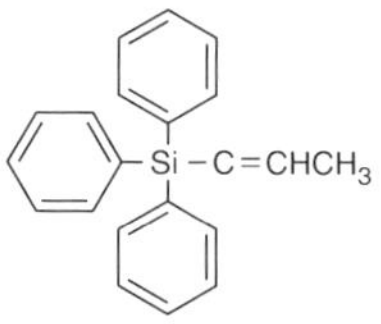
250 mg

S83,071-2
ISOPROPYLTRIPHENYLSILANE
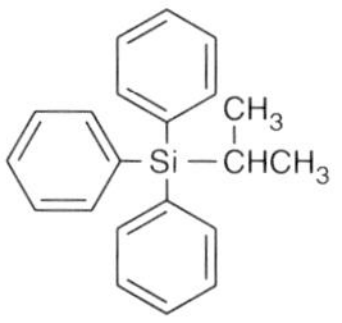
250 mg

S83,105-0
DIPHENYL(3-PHENYLPROPYL)SILANE
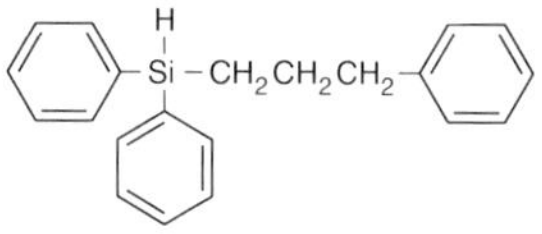
250 mg

S82,218-3
(ALLYLOXY)TRIPHENYLSILANE
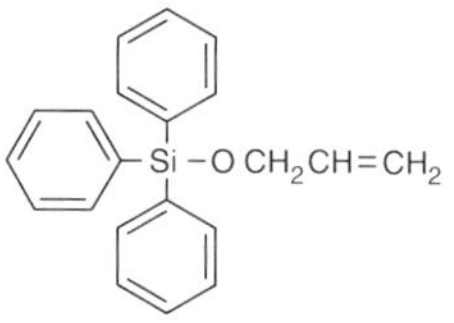
250 mg

S82,222-1
1,1,1-TRIMETHYL-2,2,2-TRIPHENYLDISILANE
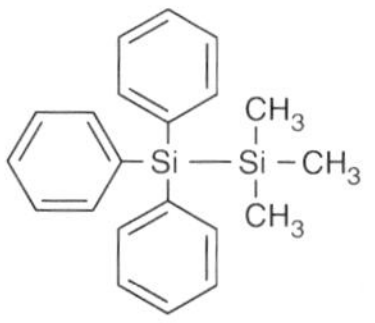
250 mg

S83,350-9
1,1,2-TRIMETHYL-1,2,2-TRIPHENYLDISILANE
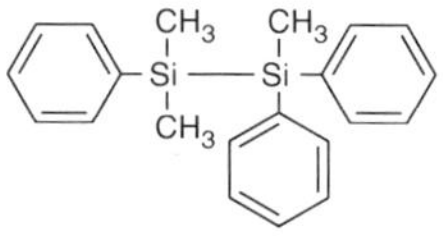
250 mg

S83,158-1
ALLYLTRIPHENYLGERMANE
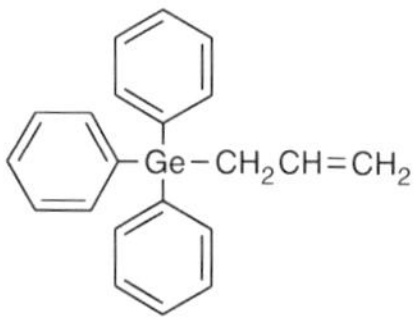
250 mg

S82,350-3
PROPYLTRIPHENYLGERMANE
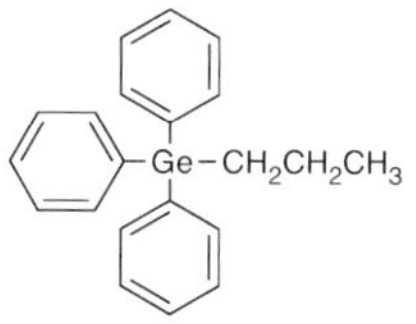
250 mg

S83,681-8
1,1,1-TRIMETHYL-3,3,3-TRIPHENYLDISILOXANE
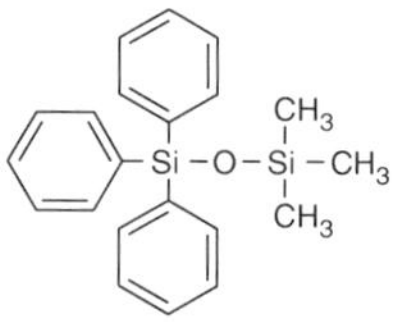
250 mg

S82,993-5
(1-BUTYN-1-YL)TRIPHENYLSILANE
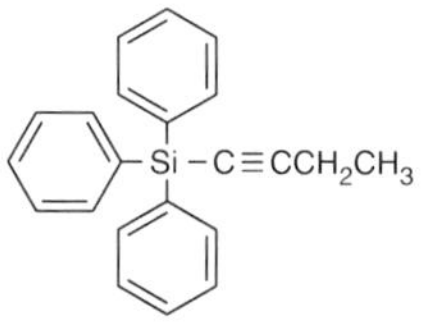
250 mg

S82,219-1
(2-BUTEN-1-YL)TRIPHENYLSILANE
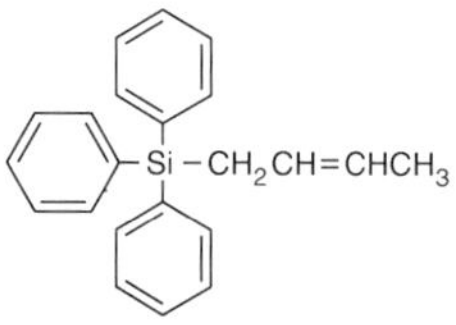
250 mg

S82,163-2
BUTYLTRIPHENYLSILANE
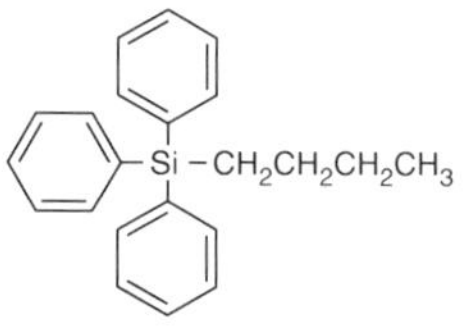
250 mg

S82,737-1
ISOBUTYLTRIPHENYLSILANE
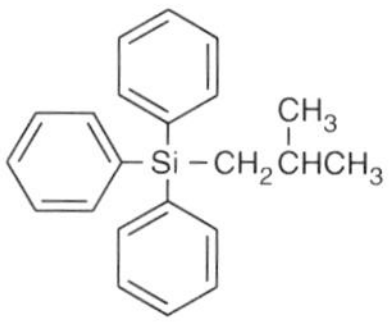
250 mg

S83,043-7
DECYLDIPHENYLSILANE
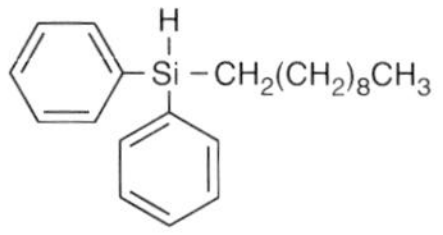
250 mg

S82,220-5
(2-BUTEN-1-YLOXY)TRIPHENYLSILANE
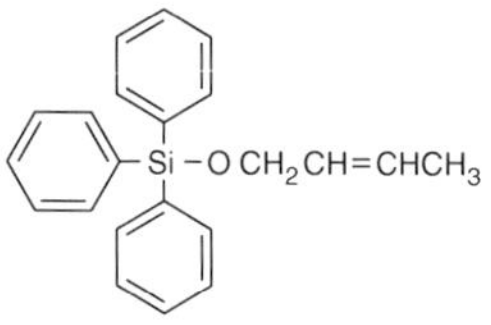
250 mg

S83,148-4
HEXADECYLPHENYLSILANE
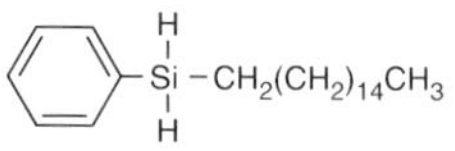
250 mg

S82,177-2
(TRIMETHYLSILYL)(TRIPHENYLSILYL)-METHANE
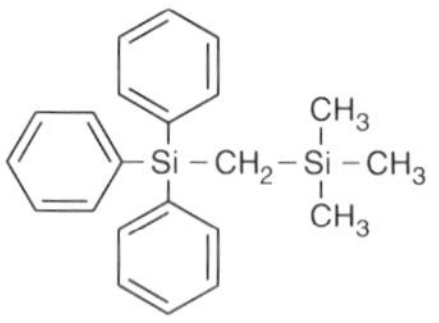
250 mg

S82,226-4
BUTYLTRIPHENYLGERMANE

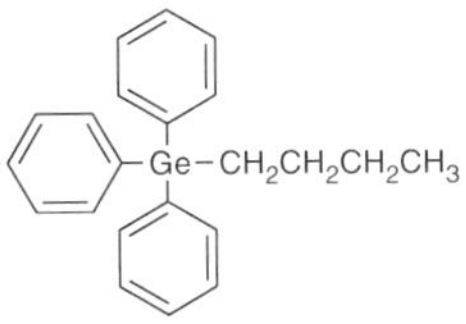

250 mg

S83,256-1
1,1'-OXYBIS(1-PHENYLSILACYCLOHEXANE)

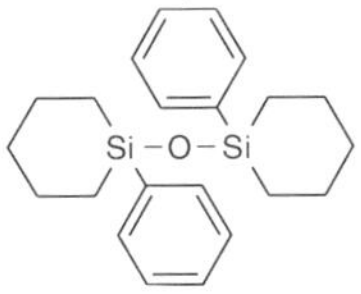

250 mg

S82,361-9
2,2-BIS(TRIMETHYLSILYL)-1,1-DIPHENYL-1,3,3,3-TETRAMETHYLTRISILANE

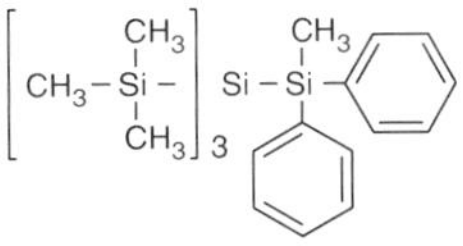

250 mg

S83,351-7
1,1,2,2,3,3,4,4,5,5-DECAMETHYL-1,5-DIPHENYLPENTASILANE

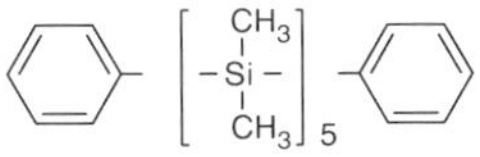

250 mg

S83,882-9
PENTYLTRIPHENYLSILANE

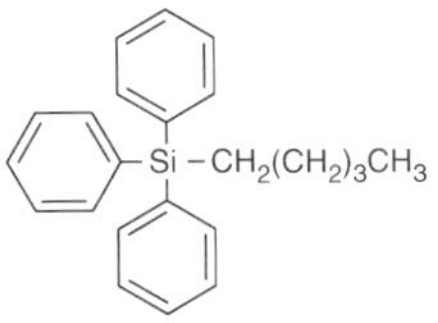

250 mg

S83,149-2
BENZYLHEXADECYLSILANE

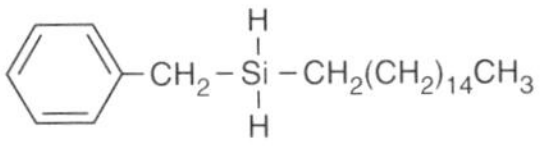

250 mg

S83,245-6
(4-PENTEN-1-YL)TRIPHENYLGERMANE

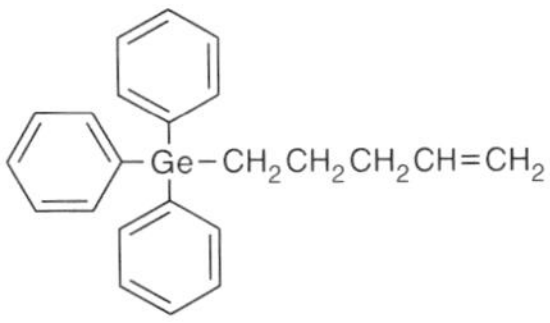

250 mg

S75,203-7
HEXYLTRIPHENYLSILANE

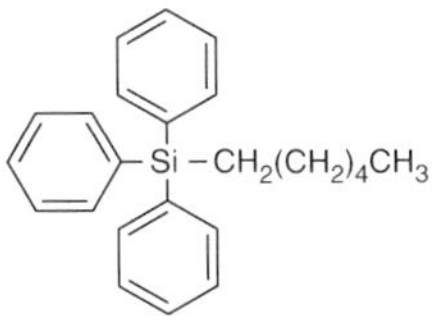

250 mg

S81,627-2
TRIS(PHENETHYL)SILANE

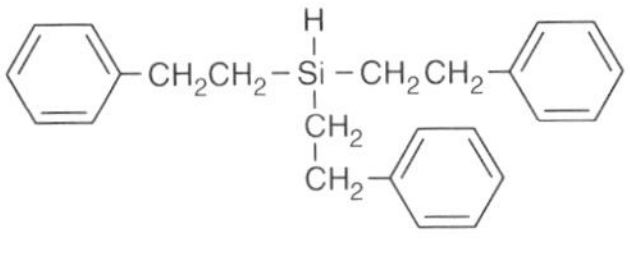

250 mg

S81,299-4
(PHENOXY)TRIPHENYLSILANE

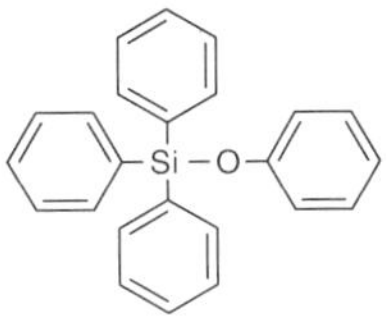

250 mg

S82,465-8
1,1,2,2-TETRAPHENYLDISILANE

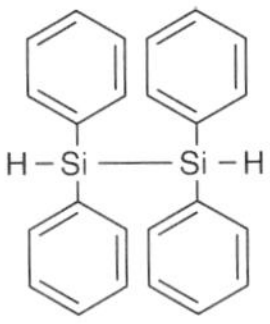

250 mg

S82,241-8
1,1,1-TRIETHYL-2,2,2-TRIPHENYLDISILANE

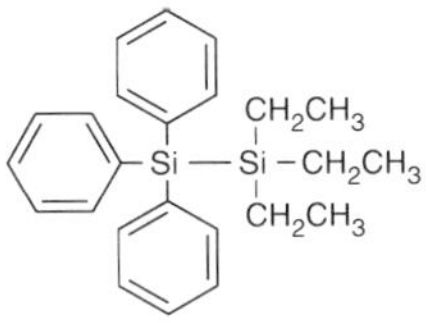

250 mg

S58,566-1
TETRAPHENYL ORTHOSILICATE

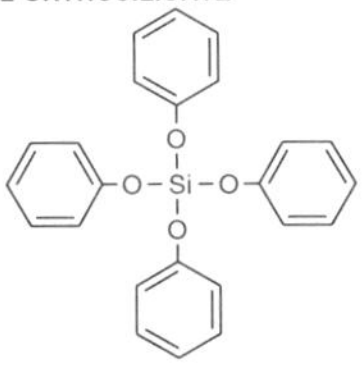

5 gm

S56,518-0
TETRAPHENYL-1,3-DISILOXANEDIOL

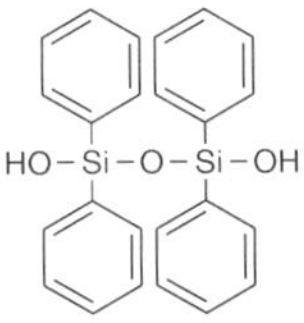

5 gm

S82,355-4
TRIETHYL(TRIPHENYLGERMYL)SILANE

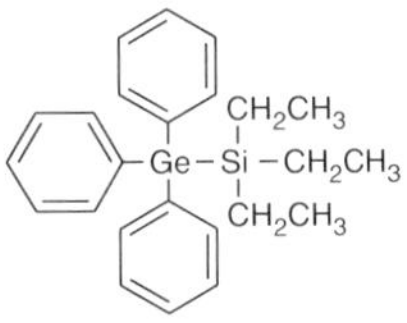

250 mg

S81,980-8
BENZYLTRIPHENYLSILANE

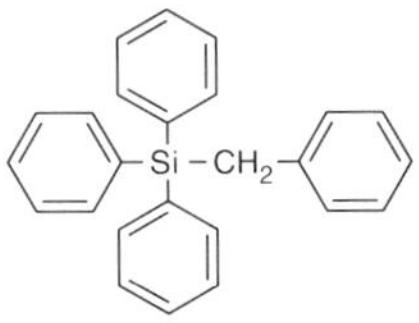

250 mg

S83,247-2
(M-TOLYL)TRIPHENYLSILANE

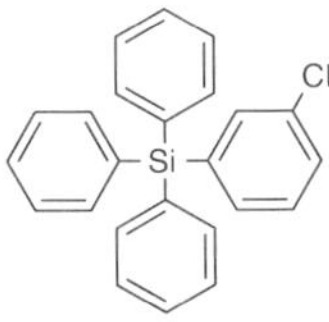

250 mg

S82,536-0
BENZYLTRIPHENYLGERMANE

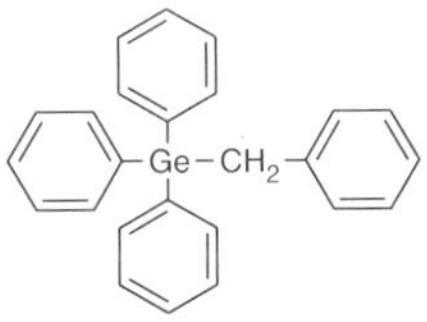

250 mg

S82,732-0
(ALPHA-BROMOBENZYL)TRIPHENYLSILANE

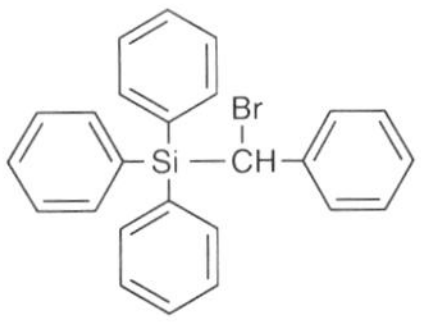

250 mg

S82,364-3
1,1,1,2,3,3,3-HEPTAMETHYL-2-(TRIPHENYLSILYL)TRISILANE

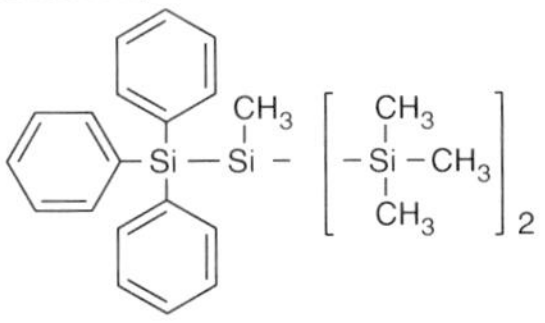

250 mg

S82,938-2
1-PHENYL-2-(TRIPHENYLSILYL)-ACETYLENE

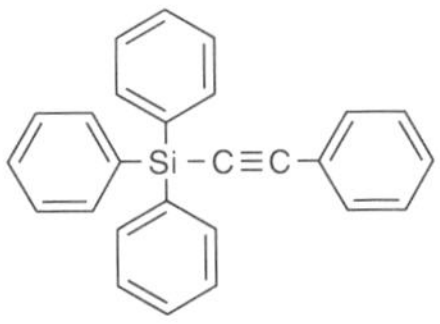

250 mg

S82,748-7
STYRYLTRIPHENYLSILANE

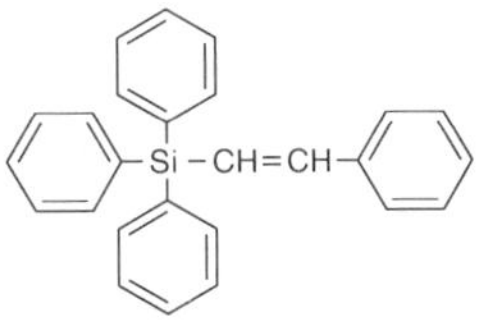

250 mg

S82,775-4
PHENETHYLTRIPHENYLSILANE

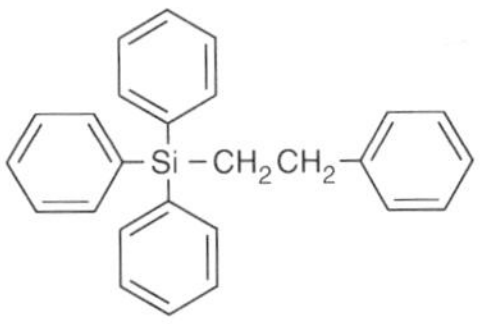

250 mg

S83,248-0
DIPHENYLDI(M-TOLYL)SILANE

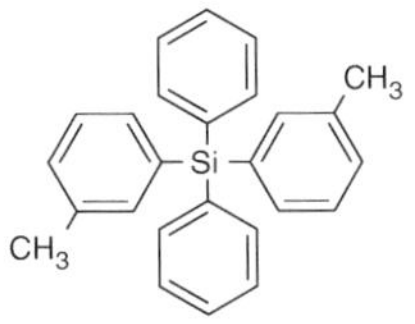

250 mg

S83,072-0
OCTYLTRIPHENYLSILANE

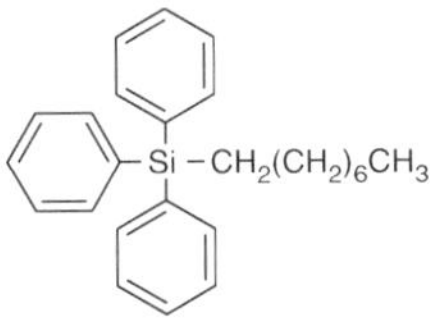

250 mg

S83,788-1
(2-ETHYLHEXYL)TRIPHENYLSILANE

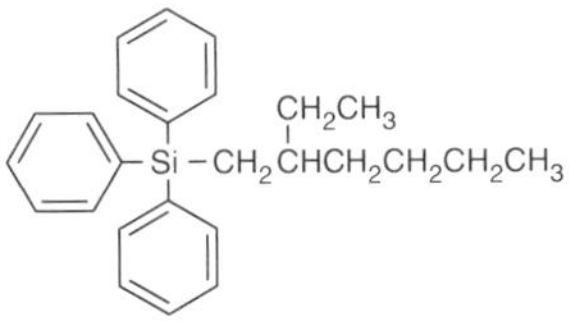

250 mg

S83,790-3
(1-METHYLHEPTYL)TRIPHENYLSILANE

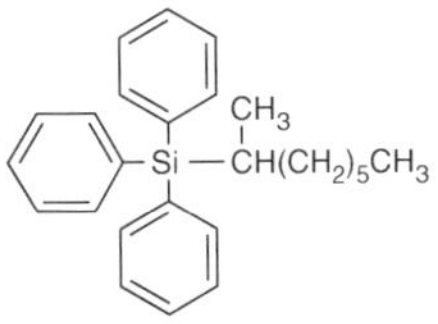

250 mg

S82,165-9
(2-PHENYLETHOXY)TRIPHENYLSILANE

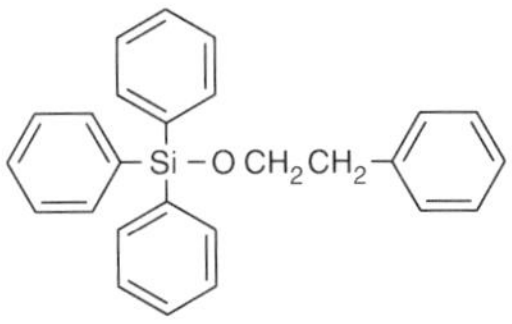

250 mg

S82,171-3
1,2-DIMETHYL-1,1,2,2-TETRAPHENYLDISILANE

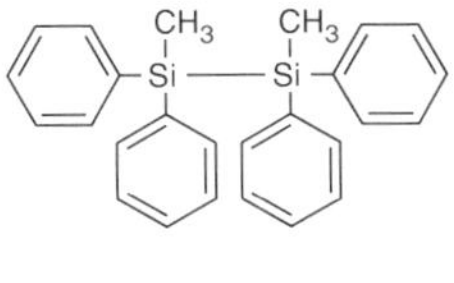

250 mg

S82,242-6
(PHENETHYL)TRIPHENYLGERMANE

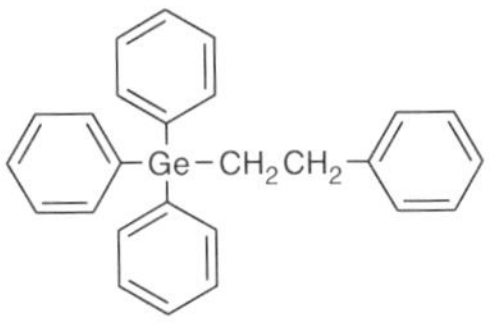

250 mg

S82,351-1
OCTYLTRIPHENYLGERMANE

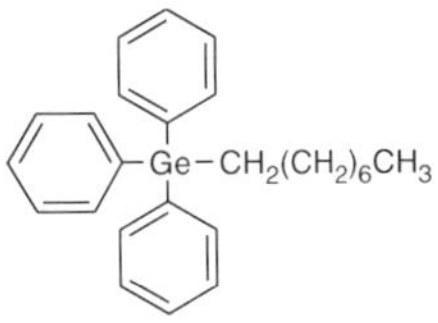

250 mg

S82,776-2
PHENYLTRIBENZYLSILANE

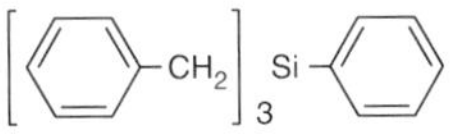

250 mg

S83,249-9
PHENYLTRI(M-TOLYL)SILANE
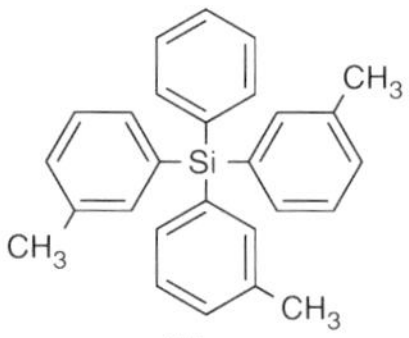
250 mg

S82,365-1
2,2-BIS(TRIMETHYLSILYL)-1,1,1-TRIMETHYL-3,3,3-TRIPHENYLTRISILANE
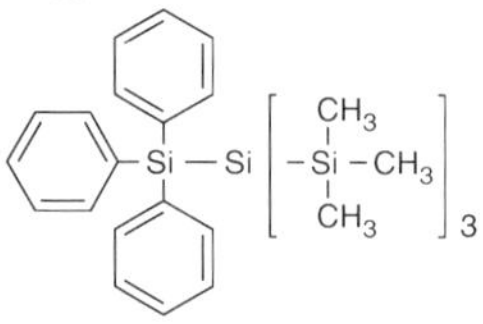
250 mg

S82,939-0
BIS(PHENYLETHYNYL)DIPHENYLSILANE
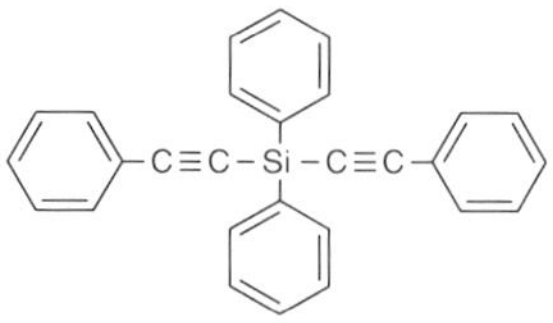
250 mg

S82,396-1
DIPHENETHYLDIPHENYLSILANE
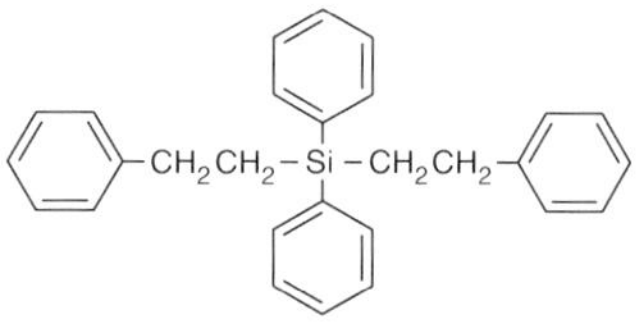
250 mg

S82,751-7
DIBENZYLDI(M-TOLYL)SILANE
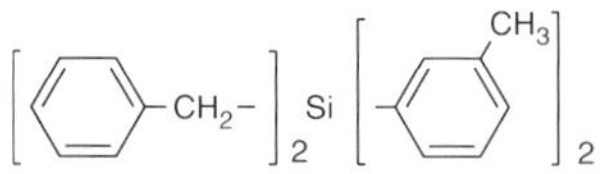
250 mg

S82,243-4
DECYLTRIPHENYLSILANE
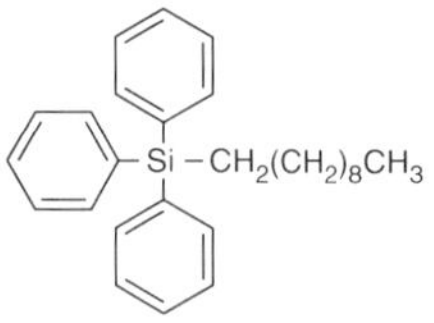
250 mg

S83,152-2
DIPHENYLHEXADECYLSILANE
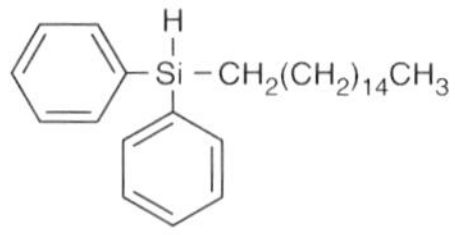
250 mg

S83,793-8
BIS(2-ETHYLHEXYL)DIPHENYLSILANE
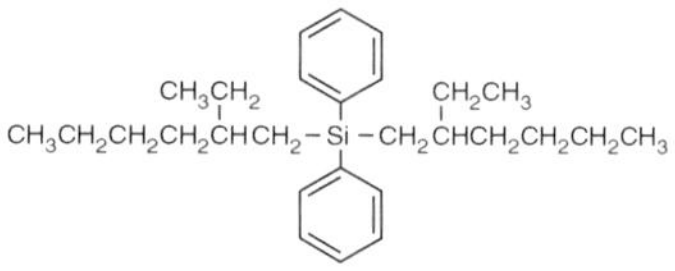
250 mg

S83,794-6
BIS(1-METHYLHEPTYL)DIPHENYLSILANE
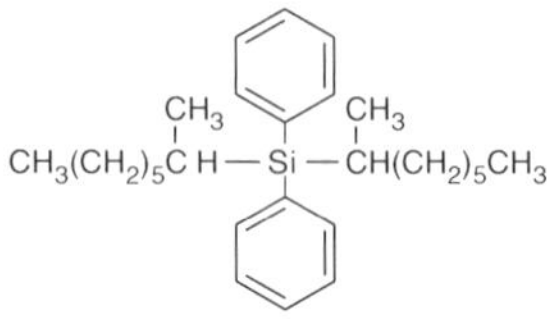
250 mg

S82,725-8
1,1-DIETHYL-1,2,2,2-TETRAPHENYLDISILANE
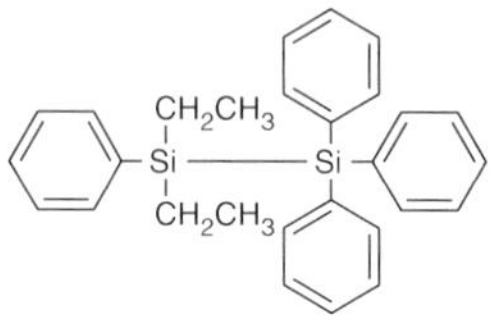
250 mg

S82,777-0
TETRABENZYLGERMANE
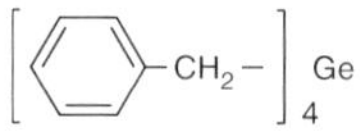
250 mg

S83,305-3
3,3,6,6-TETRAMETHYL-1,2,4,5-TETRAPHENYL-1,2,4,5-TETRAAZA-3,6-DISILACYCLOHEXANE
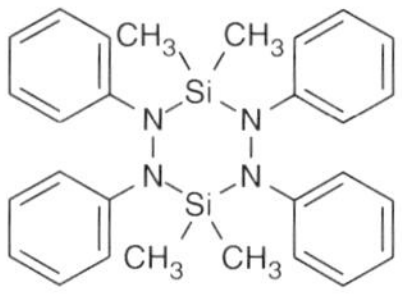
250 mg

S83,225-1
1,6-BIS(2-PHENYL-1-ETHYNYL)-1,1,2,2,3,3,4,4,5,5,6,6-DODECAMETHYLHEXASILANE
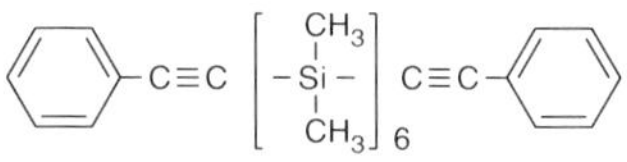
250 mg

S83,227-8
1,6-DISTYRYL-1,1,2,2,3,3,4,4,5,5,6,6-DODECAMETHYLHEXASILANE
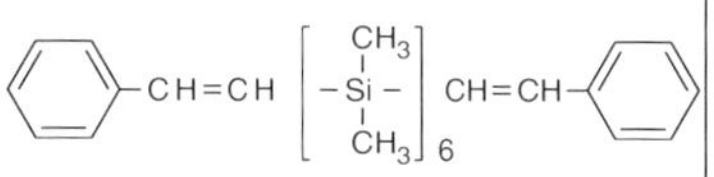
250 mg

S82,289-2
PENTAMETHYLENEBIS(DIPHENYLSILANE)
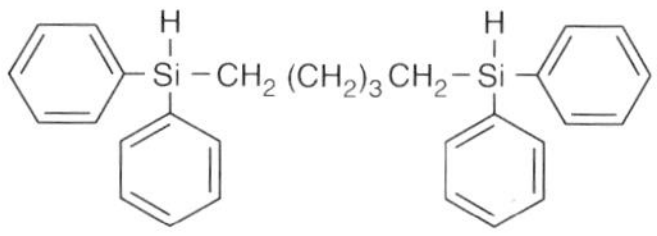
250 mg

S82,779-7
PHENYLTRIPHENETHYLSILANE
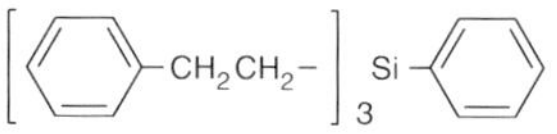
250 mg

S83,429-7
BIS(3-PHENYLPROPYL)DIPHENYLSILANE
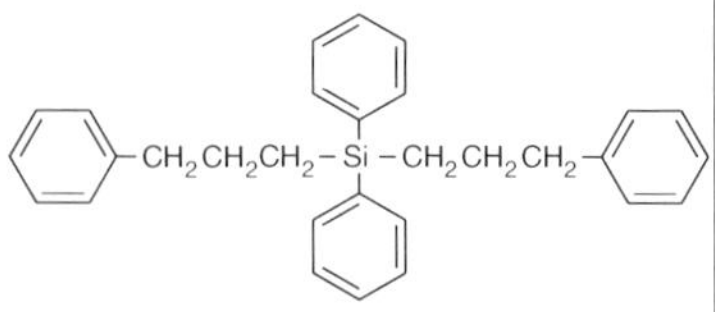
250 mg

S82,245-0
DODECYLTRIPHENYLSILANE
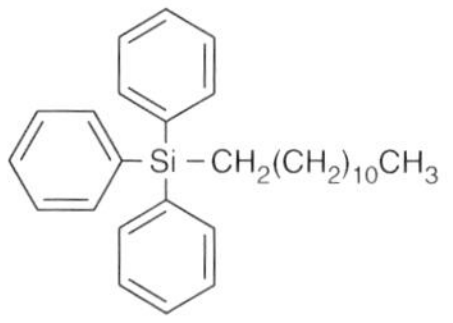
250 mg

S83,108-5
1,1,1,2,2-PENTAPHENYLDISILANE
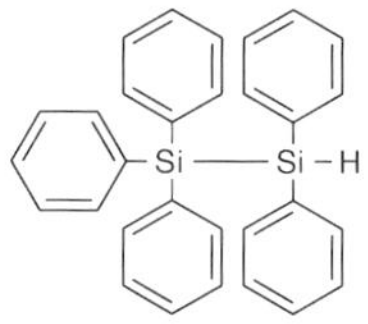
250 mg

S83,797-0
PHENYLTRIS(2-ETHYLHEXYL)SILANE
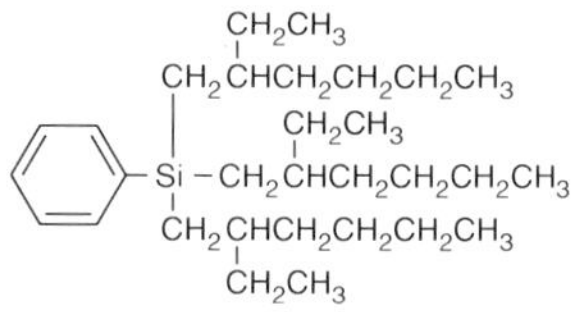
250 mg

S83,798-9
PHENYLTRIOCTYLSILANE
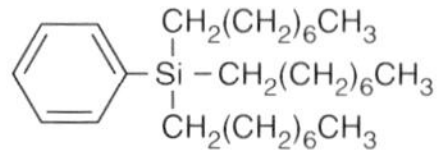
250 mg

S81,957-3
1-CHLORO-1,1,2,2,2-PENTAPHENYLDISILANE
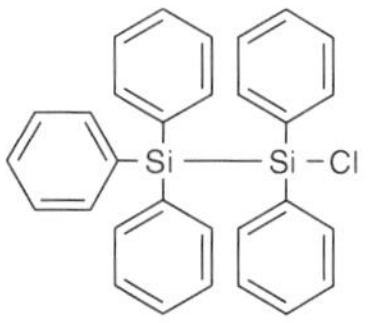
250 mg

S82,859-9
1,2-DIPROPOXY-1,1,2,2-TETRAPHENYLDISILANE
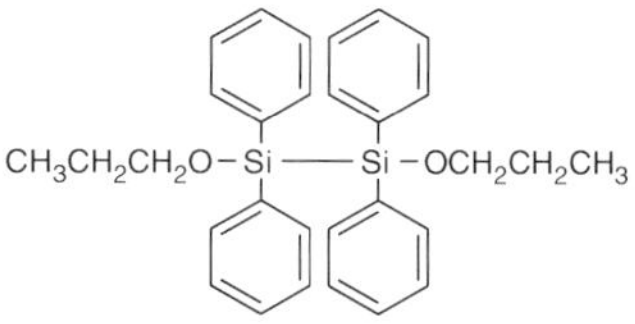
250 mg

S82,221-3
(DIPHENYLMETHYL)TRIPHENYLSILANE
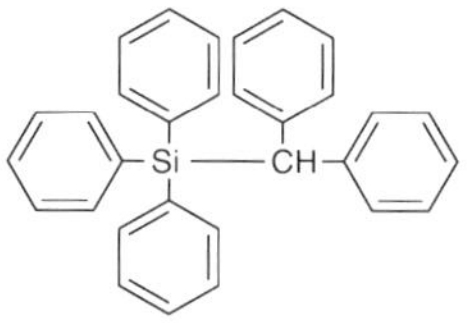
250 mg

S82,466-6
1-METHYL-1,1,2,2,2-PENTAPHENYLDISILANE
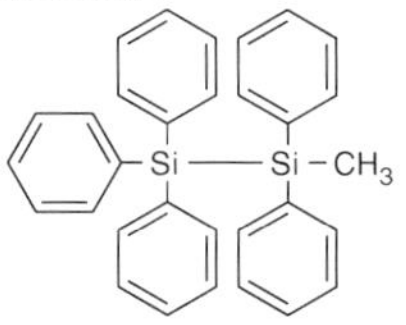
250 mg

S84,038-6
1-METHOXY-1,1,2,2,2-PENTAPHENYLDISILANE
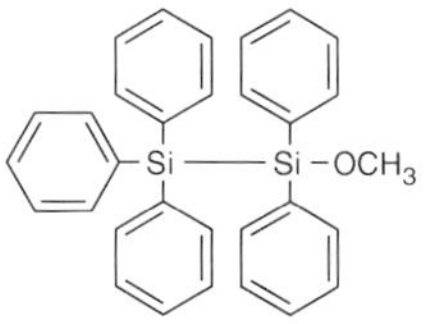
250 mg

S81,746-5
(1,2-DIPHENYLETHYL)TRIPHENYLSILANE
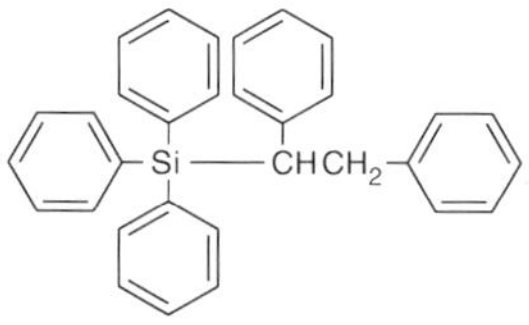
250 mg

S82,943-9
(2,2-DIPHENYLETHYL)TRIPHENYLSILANE
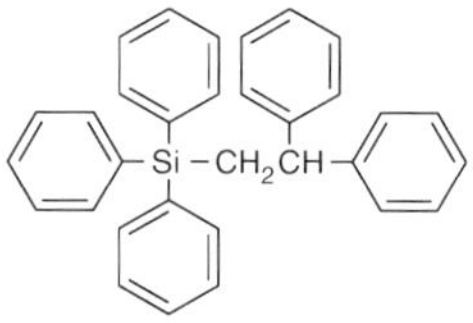
250 mg

S82,944-7
(1,1-DIPHENYLETHYL)TRIPHENYLSILANE
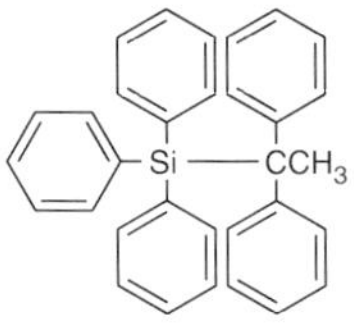
250 mg

S82,524-7
TETRA(PHENETHYL)SILANE
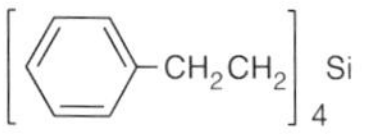
250 mg

S83,161-1
(2,2-DIPHENYLETHYL)TRIPHENYLGERMANE
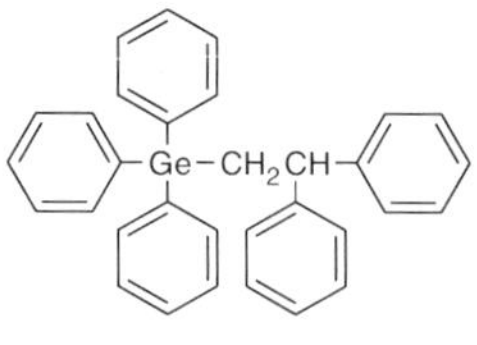
250 mg

S82,885-8
3,3,6,6-TETRAETHYL-1,2,4,5-TETRAPHENYL-1,2,4,5-TETRAAZA-3,6-DISILACYCLOHEXANE
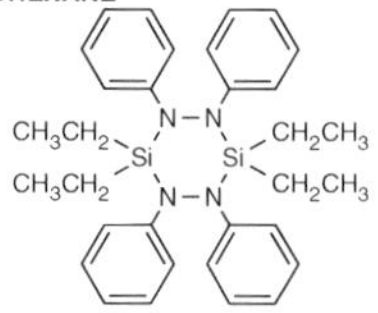
250 mg

S82,468-2
TETRAKIS(DIMETHYLPHENYLSILYL)SILANE

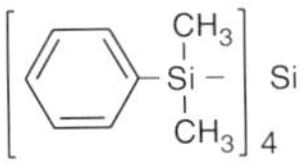

250 mg

S82,780-0
PHENYLTRIS(3-PHENYLPROPYL)SILANE

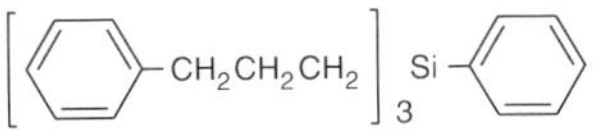

250 mg

S82,425-9
(1,3-DIPHENYL-1-PROPENYLOXY)-TRIPHENYLSILANE

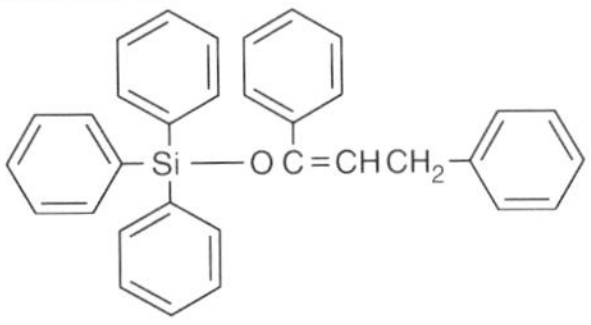

250 mg

S83,432-7
DODECYLTRIBENZYLSILANE

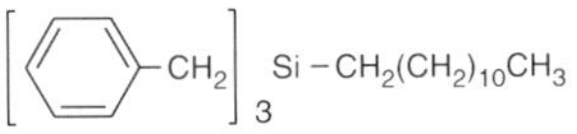

250 mg

S82,459-3
1,1,1,2,2-PENTAPHENYL-3,3,3-TRIMETHYLTRISILANE

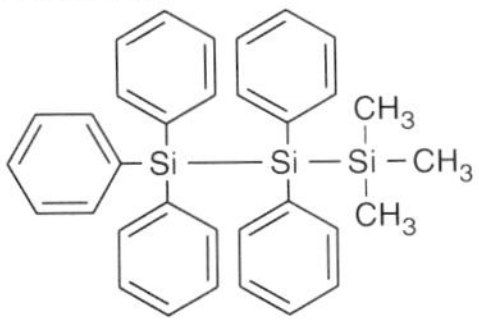

250 mg

S83,430-0
PHENYLTRIS[3-(TRIMETHYLSILYL)-PHENYL]SILANE

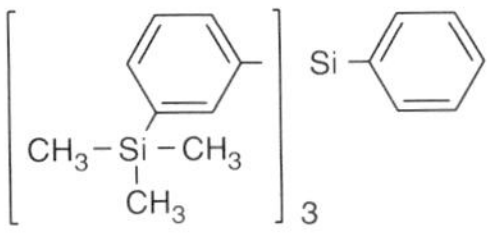

250 mg

S83,434-3
BENZYLTRIS(3-PHENYLPROPYL)SILANE

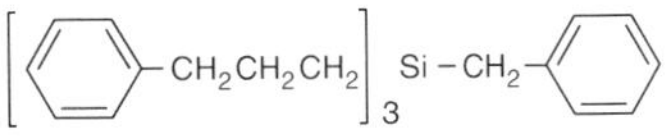

250 mg

S83,398-3
HEXADECYLTRIPHENYLSILANE

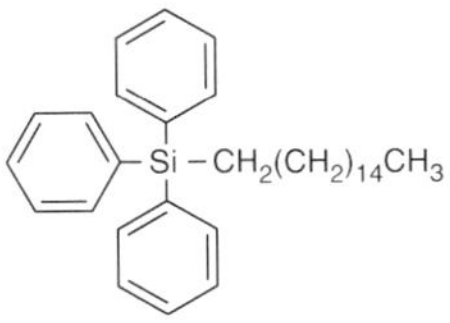

250 mg

S82,224-8
1-BUTYL-1,1,2,2,2-PENTAPHENYLDISILANE

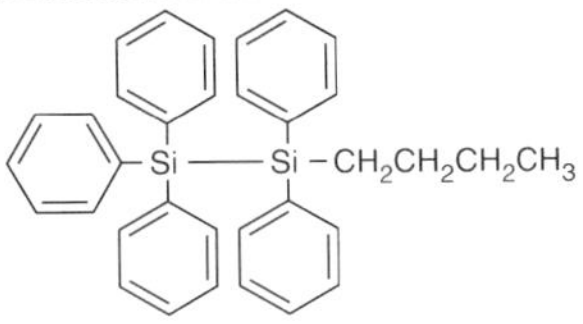

250 mg

S83,399-1
HEPTADECYLTRIPHENYLSILANE

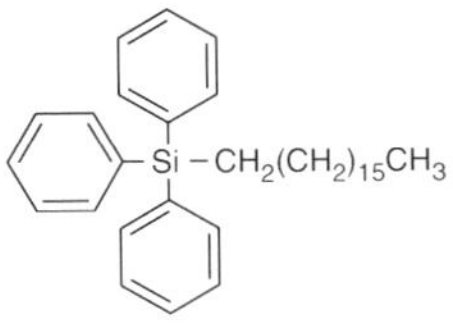

250 mg

S83,435-1
TETRAKIS(3-PHENYLPROPYL)SILANE

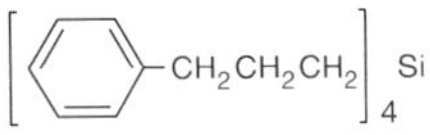

250 mg

S82,246-9
OCTADECYLTRIPHENYLSILANE

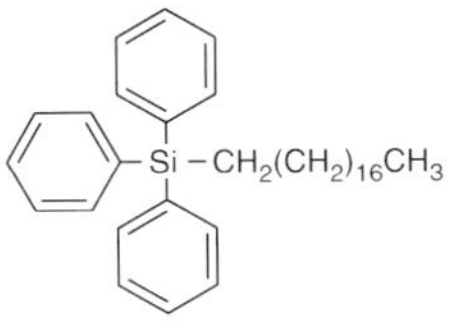

250 mg

S81,426-1
HEXAPHENYLDISILANE

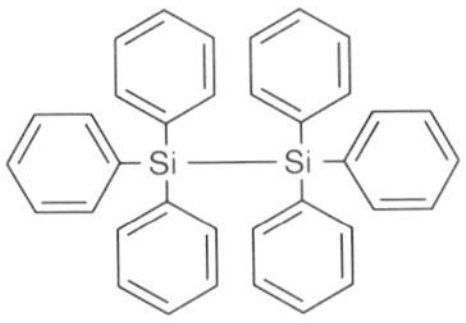

250 mg

S54,707-7
DIDODECYLDIPHENYLSILANE

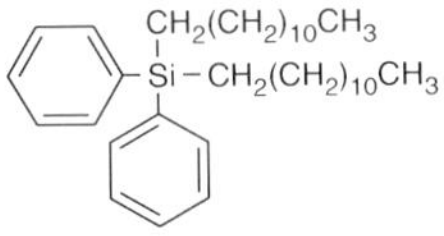

5 gm

S81,532-2
1,1,1,3,3,3-HEXAPHENYLDISILOXANE

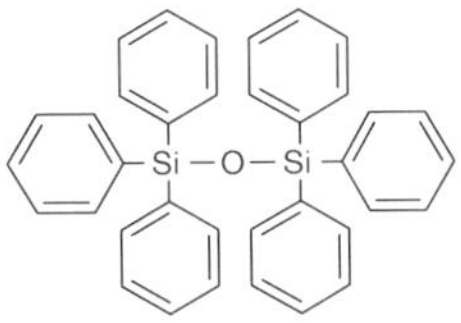

250 mg

S83,228-6
1,1,2,2,3,3-HEXAPHENYLTRISILANE

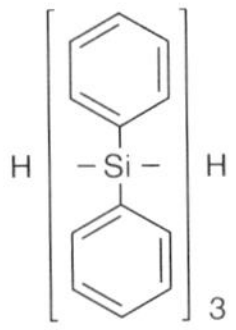

250 mg

S82,353-8
OCTADECYLTRIPHENYLGERMANE

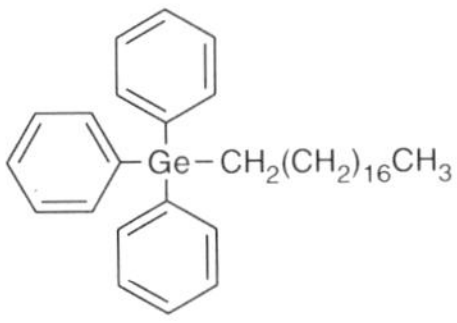

250 mg

S82,886-6
HEXAPHENYLDIGERMOXANE

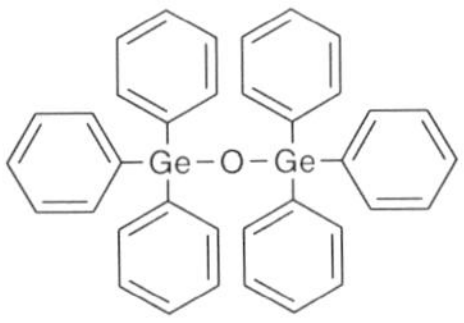

250 mg

S83,183-2
TRIPHENYL(TRIPHENYLMETHYL)SILANE

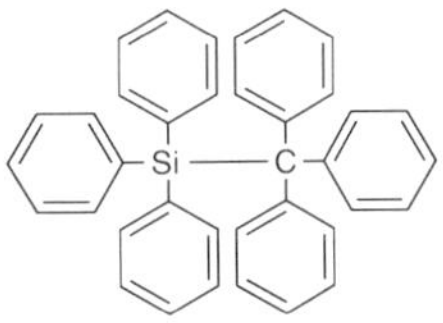

250 mg

S83,508-0
HEXADECYLTRIBENZYLSILANE

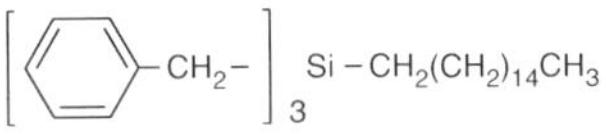

250 mg

S82,995-1
METHYLENEBIS(TRIPHENYLSILANE)

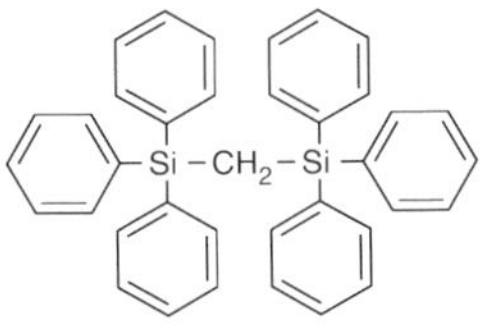

250 mg

S82,946-3
TRIPHENYL(1,2,2-TRIPHENYLETHYL)-SILANE

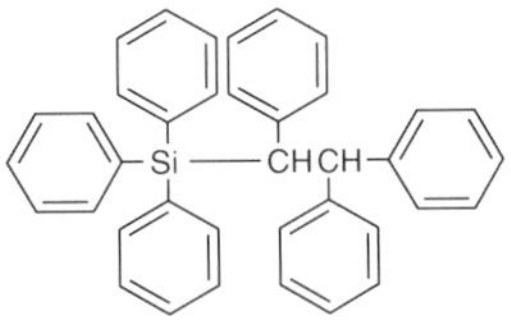

250 mg

S83,374-6
BIS(DIPHENYLMETHYL)DIPHENYLSILANE

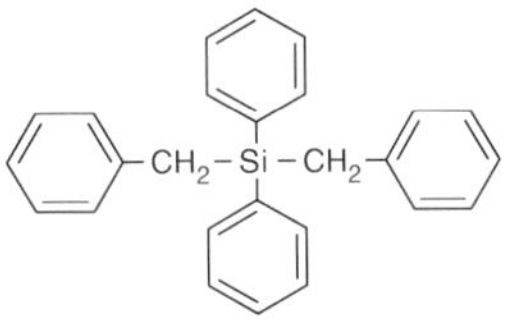

250 mg

S82,172-1
1,2-BIS(TRIPHENYLSILYL)ACETYLENE

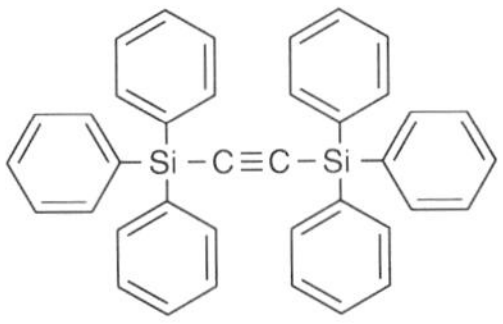

250 mg

S82,174-8
1,2-BIS(TRIPHENYLSILYL)ETHANE

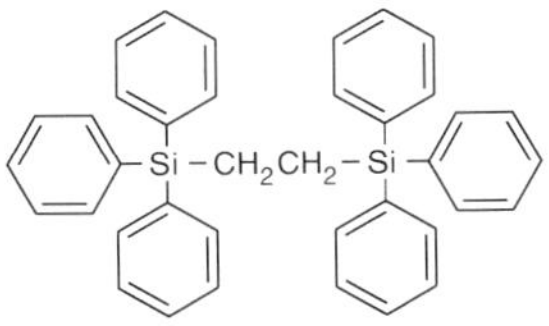

250 mg

S82,887-4
3,6-DIMETHYL-1,2,3,4,5,6-HEXAPHENYL-1,2,4,5-TETRAAZA-3,6-DISILACYCLOHEXANE

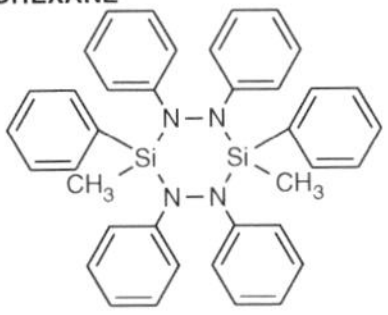

250 mg

S83,299-5
OCTADECYLTRIBENZYLSILANE

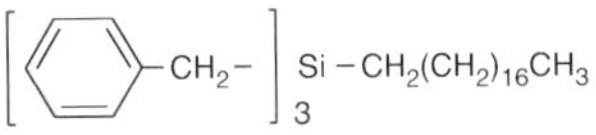

250 mg

S82,998-6
1,2-BIS(TRIPHENYLSILYL)-1-PROPENE

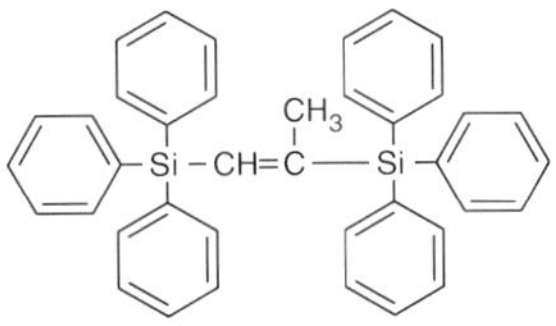

250 mg

S82,999-4
1,3-BIS(TRIPHENYLSILYL)-1-PROPENE

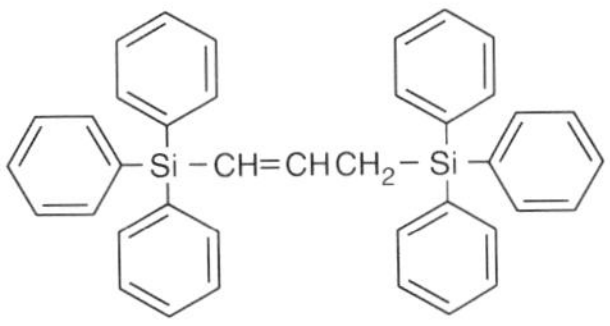

250 mg

S82,758-4
1,3-TRIMETHYLENEBIS-(TRIPHENYLSILANE)

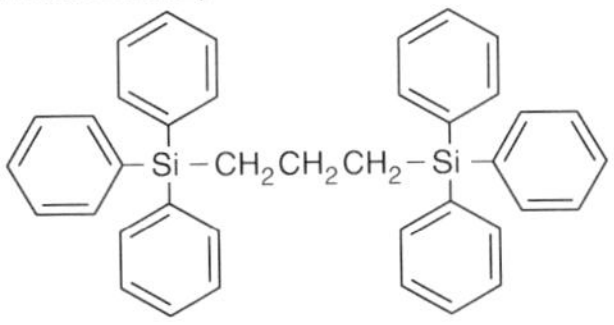

250 mg

S83,194-8
1,2-PROPANEDIYLBIS(TRIPHENYLSILANE)

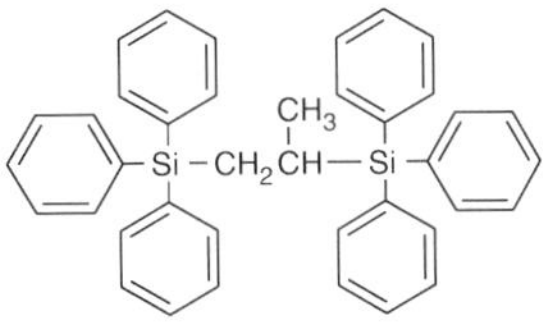

250 mg

S83,556-0
1,1,1-TRIPHENYL-3,3,3-TRIS(M-TOLYL)-DISILOXANE

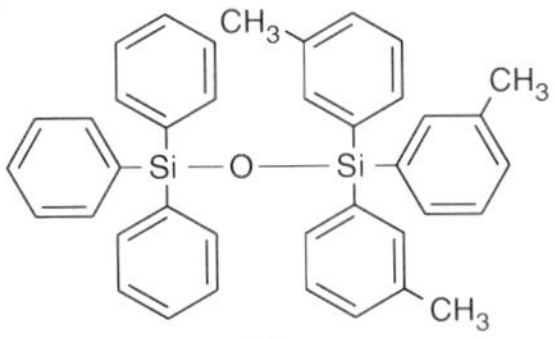

250 mg

S83,704-0
TRIPHENYL[3-(TRIPHENYLGERMYL)-PROPYL]SILANE

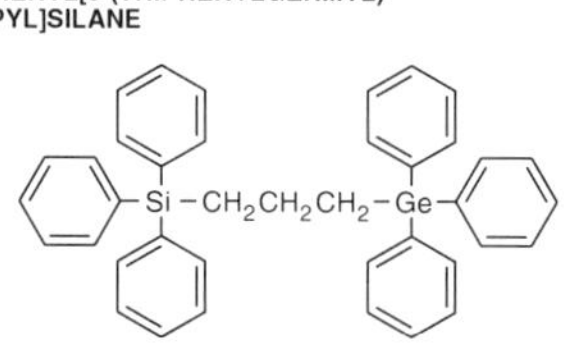

250 mg

S82,356-2
TRIMETHYLENEBIS(TRIPHENYLGERMANE)

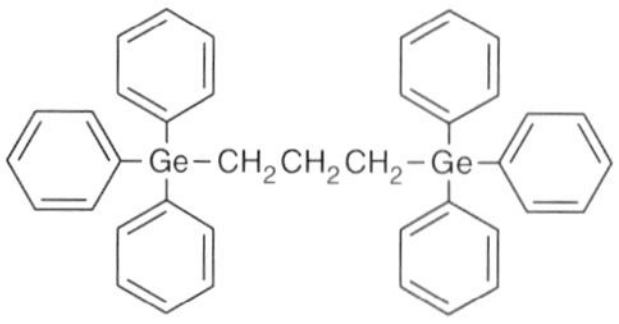

250 mg

S83,000-3
1,4-BIS(TRIPHENYLSILYL)-1,3-BUTADIYNE

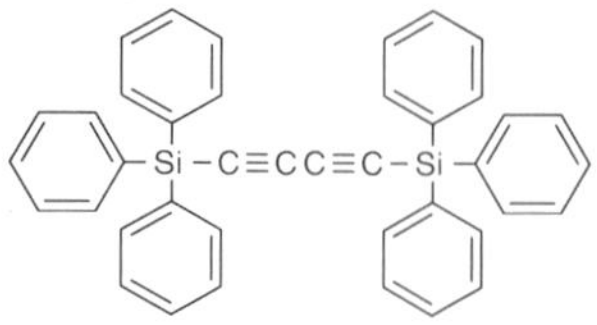

250 mg

S81,986-7
TETRAMETHYLENEBIS(TRIPHENYLSILANE)

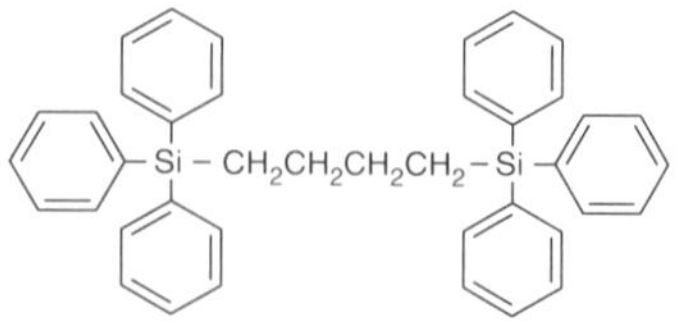

250 mg

S83,192-1
1,2-BUTANEDIYLBIS(TRIPHENYLSILANE)

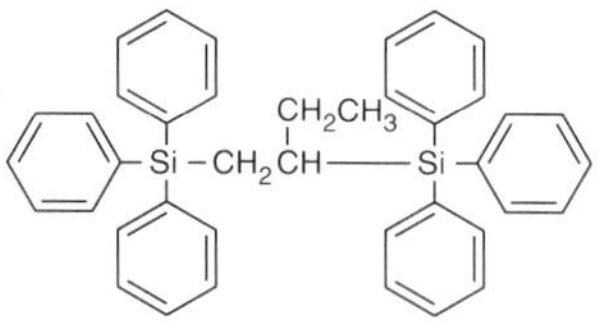

250 mg

S83,443-2
DIPHENYLDITETRADECYLSILANE

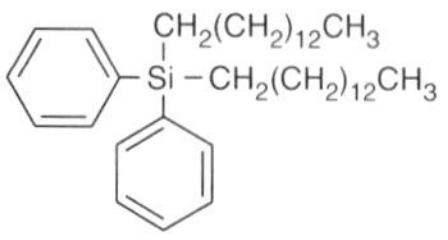

250 mg

S83,186-7
1,1,2,2,3,3-HEXAPHENYL-4,4,5,5-TETRAMETHYLCYCLOPENTASILANE

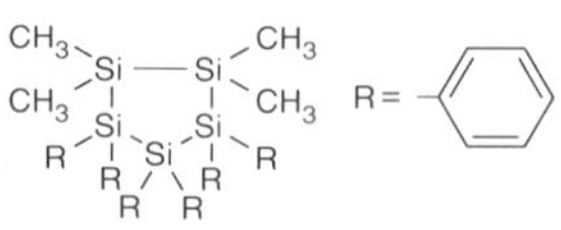

250 mg

S82,312-0
PENTAMETHYLENEBIS(TRIPHENYLSILANE)

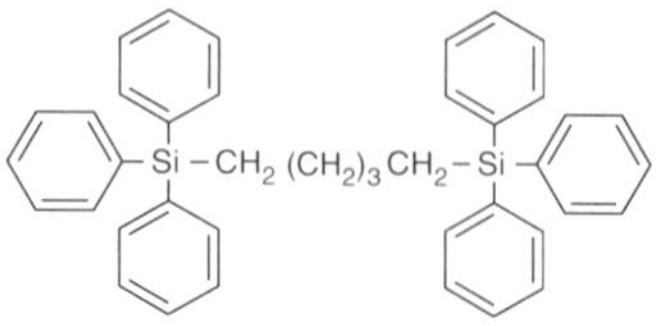

250 mg

S81,985-9
1,3-BIS(TRIPHENYLSILYL)BENZENE

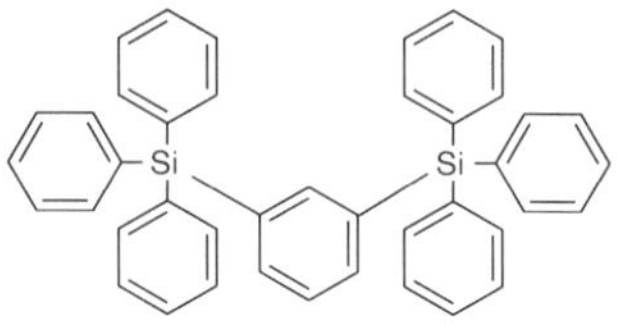

250 mg

S82,455-0
HEXABENZYLDISILANE

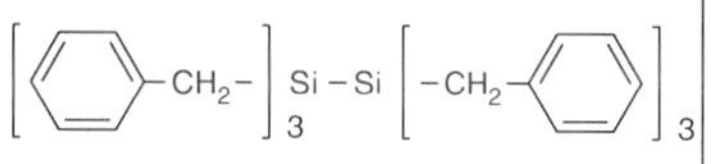

250 mg

S82,510-7
PHENYLTRIDODECYLSILANE

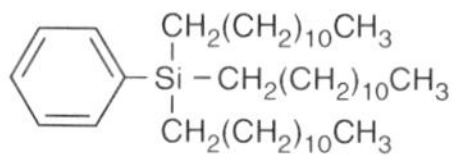

250 mg

S83,510-2
HEXADECYLTRIS(3-PHENYLPROPYL)SILANE

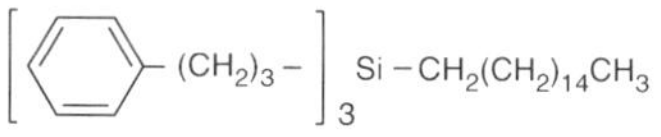

250 mg

S82,311-2
ETHYLENEBIS(TRIBENZYLSILANE)

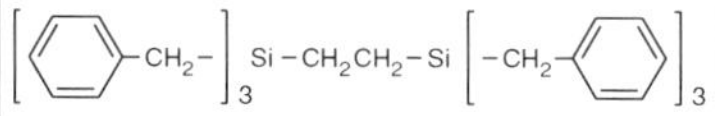

250 mg

S83,300-2
DIHEXADECYLDIPHENYLSILANE

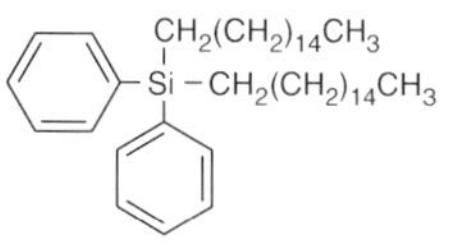

250 mg

S82,395-3
DIOCTADECYLDIPHENYLSILANE

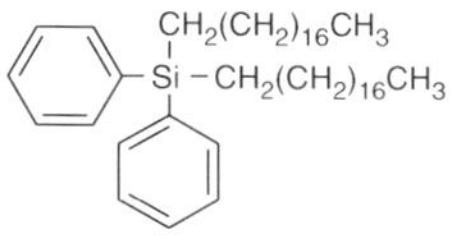

250 mg

S83,301-0
PHENYLTRITETRADECYLSILANE

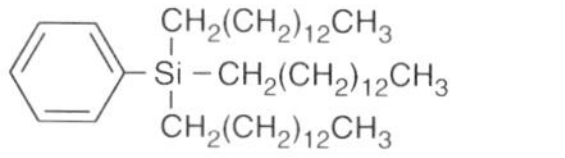

250 mg

S82,247-7
OCTAPHENYLTRISILANE

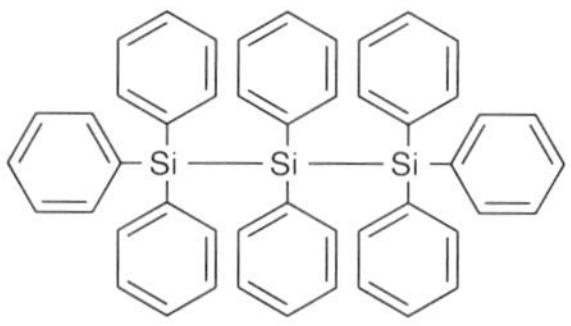

250 mg

S82,963-3
OCTAPHENYLCYCLOTETRASILANE

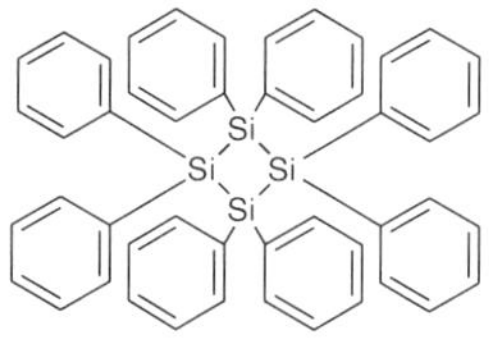

250 mg

S82,302-3
1,1,2,2,3,3,4,4-OCTAPHENYLTETRASILANE

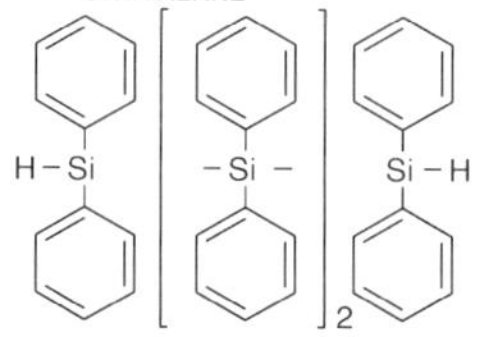

250 mg

S83,155-7
DIOCTADECYLDIPHENYLGERMANE

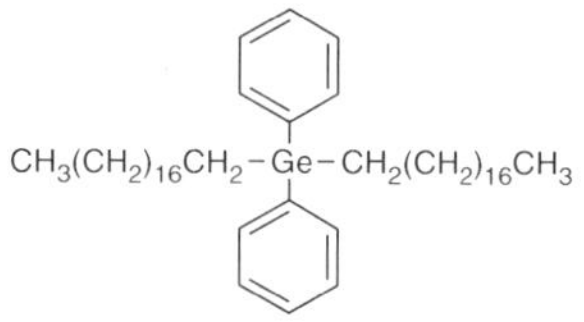

250 mg

S82,527-1
2,2,3,3,5,5,6,6-OCTAPHENYL-1,4-DIOXA-2,3,5,6-TETRASILACYCLOHEXANE

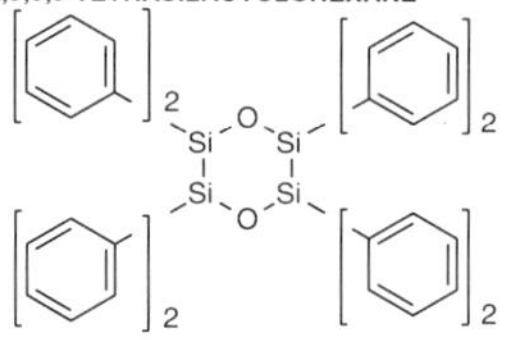

250 mg

S83,137-9
2,2,4,4,5,5,6,6-OCTAPHENYL-1,3-DIOXA-2,4,5,6-TETRASILACYCLOHEXANE

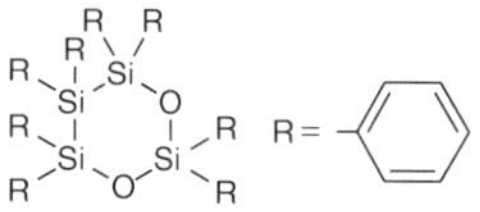

250 mg

S83,138-7
1-CHLORO-1,1,2,2,3,3,4,4-OCTAPHENYLTETRASILANE

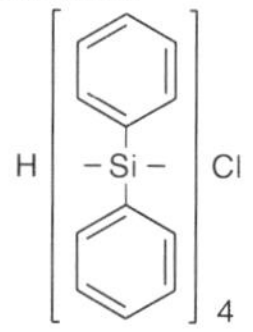

250 mg

S82,861-0
2,2,4,4,6,6,7,7-OCTAPHENYL-1,3,5-TRIOXA-2,4,6,7-TETRASILACYCLOHEPTANE

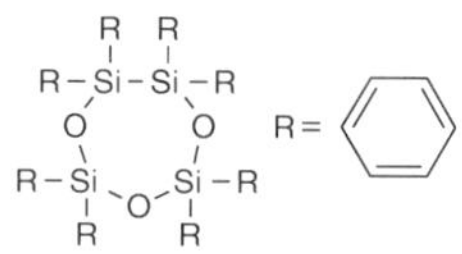

250 mg

S82,303-1
1,4-DICHLORO-1,1,2,2,3,3,4,4-OCTAPHENYLTETRASILANE

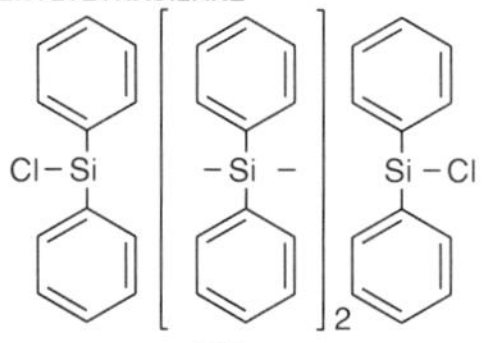

250 mg

S83,264-2
1-BROMO-1,1,2,2,3,3,4,4-OCTAPHENYLTETRASILANE

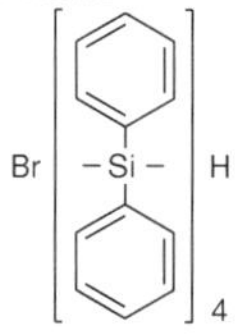

250 mg

S82,860-2
1,4-DIBROMO-1,1,2,2,3,3,4,4-OCTAPHENYLTETRASILANE

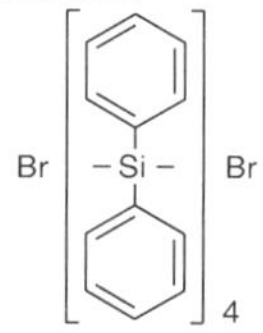

250 mg

S84,041-6
1-CHLORO-4-METHYL-1,1,2,2,3,3,4,4-OCTAPHENYLTETRASILANE

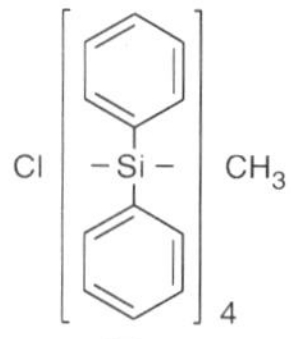

250 mg

S83,451-3
DIBENZYLDIOCTADECYLSILANE

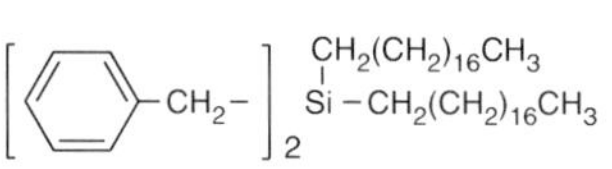

250 mg

S84,042-4
1,4-DIMETHYL-1,1,2,2,3,3,4,4-OCTAPHENYLTETRASILANE

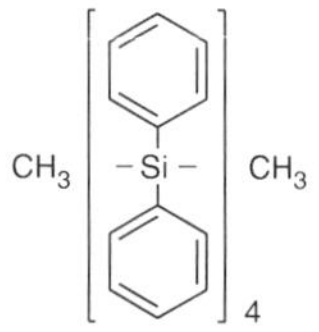

250 mg

S82,862-9
1,1,2,2,3,3,4,4-OCTAPHENYL-1-PROPOXYTETRASILANE

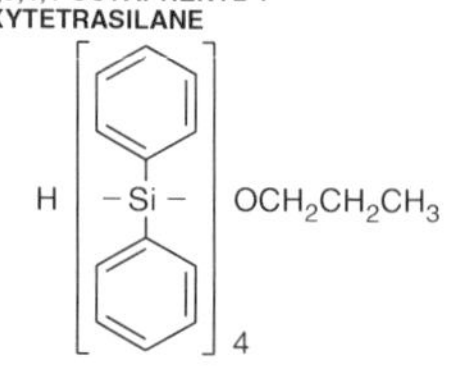

250 mg

S83,454-8
BIS(3-PHENYLPROPYL)-DIOCTADECYLSILANE

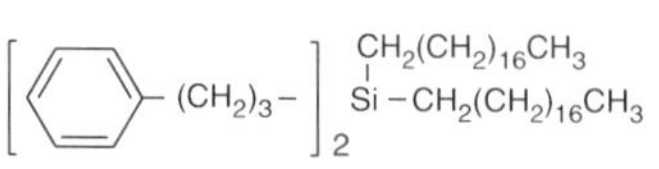

250 mg

S82,511-5
PHENYLTRIHEXADECYLSILANE

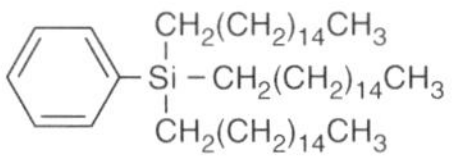

250 mg

S83,229-4
1,1,1,6,6,6-HEXAMETHYL-2,2,3,3,4,4,5,5-OCTAPHENYLHEXASILANE

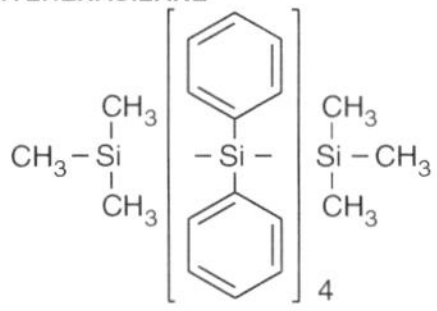

250 mg

S83,003-8
1,3,3-TRIS(TRIPHENYLSILYL)ALLENE

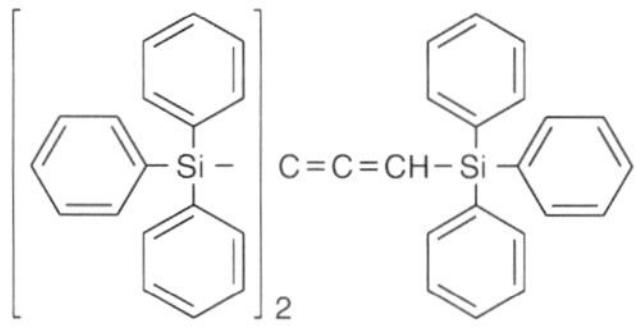

250 mg

S83,004-6
1,3,3-TRIS(TRIPHENYLSILYL)-1-PROPENE

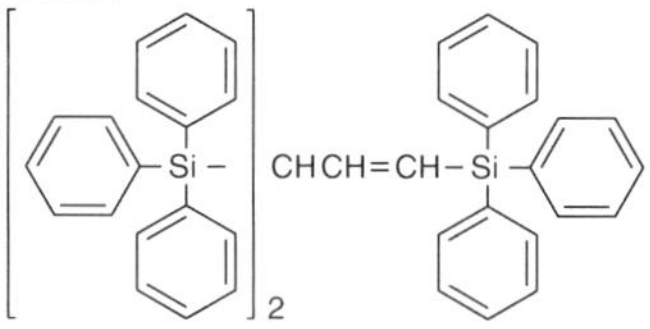

250 mg

S83,376-2
DECAPHENYLTETRASILANE

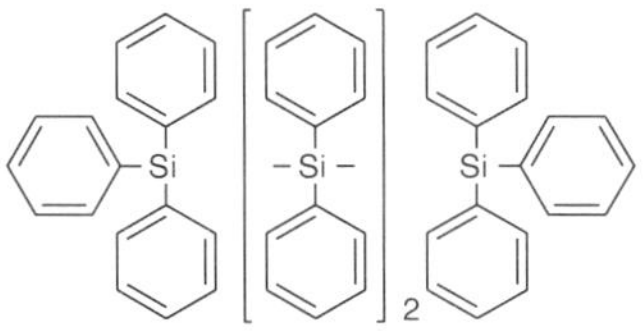

250 mg

S83,230-8
1,1,2,2,3,3,4,4,5,5-DECAPHENYLPENTASILANE

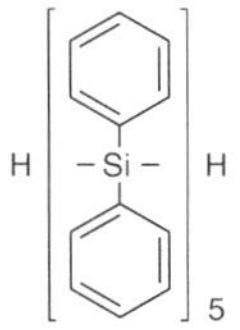

250 mg

S83,353-3
DECAPHENYL-1,5-PENTASILANEDIOL

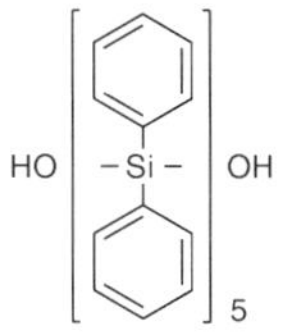

250 mg

S82,512-3
BENZYLTRIOCTADECYLSILANE

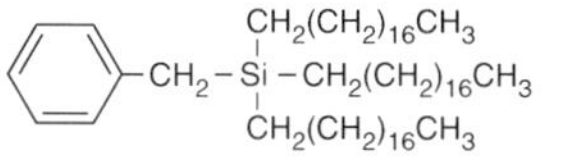

250 mg

S83,459-9
1,1,1-TRIHEXADECYL-2,2,2-TRIPHENYLDISILANE

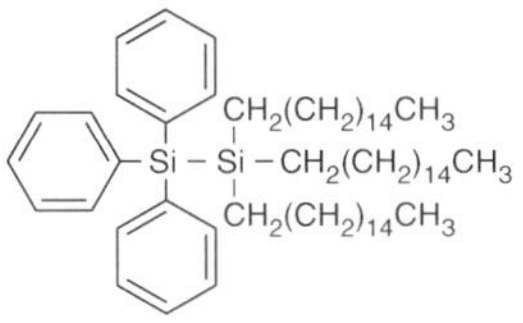

250 mg

S82,863-7
1,1,2,2,3,3,4,4,5,5-DECAPHENYL-1,5-DIPROPOXYPENTASILANE

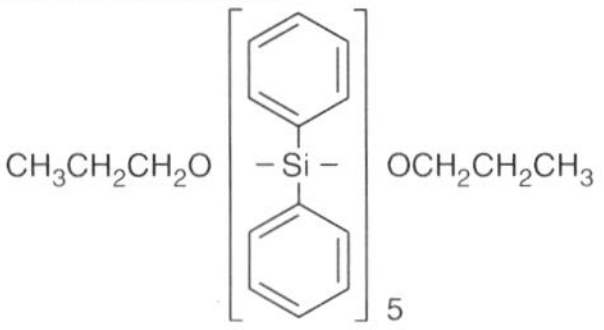

250 mg

S83,380-0
1-METHYLUNDECAPHENYLPENTASILANE

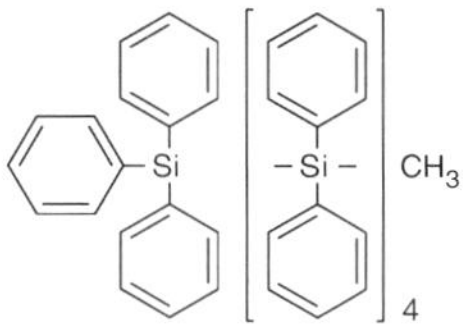

250 mg

S83,972-8
1,1,2,2,3,3,4,4,5,5-DECAPHENYL-1,5-DIBUTYLPENTASILANE

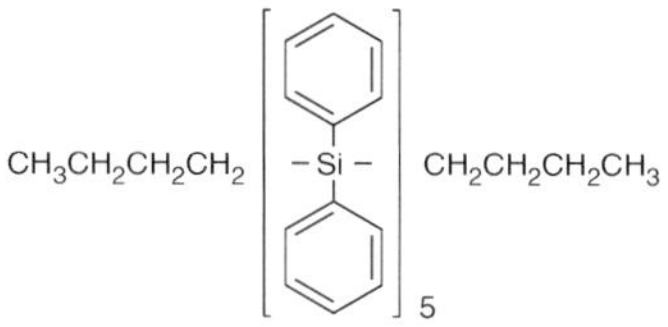

250 mg

S82,305-8
1,1,2,2,3,3,4,4,5,5,6,6-DODECAPHENYLHEXASILANE

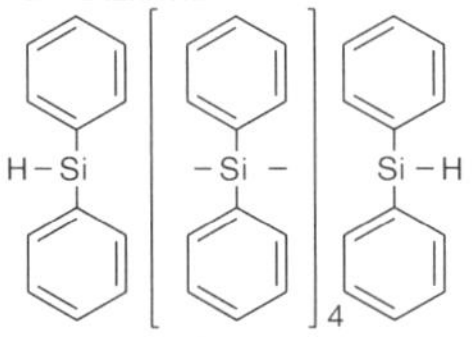

250 mg

S83,184-0
1,6-DIMETHYL-1,1,2,2,3,3,4,4,5,5,6,6-DODECAPHENYLHEXASILANE

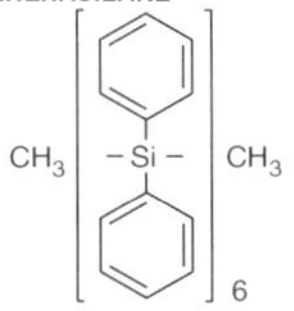

250 mg

S82,526-3
1,2,2,3,3,4,4,5,5,6,6,7-DODECAPHENYL-1,1,7,7-TETRAMETHYLHEPTASILANE

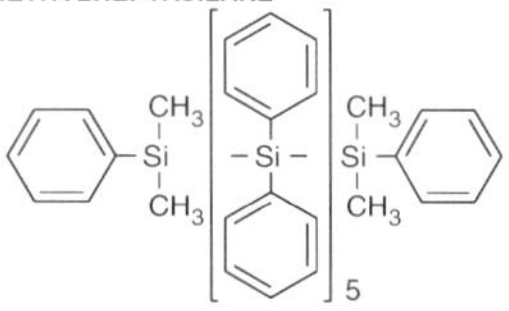

250 mg

S83,217-0
1,4-BIS(TRIPHENYLGERMYL)-1,1,2,2,3,3,4,4-OCTAPHENYLTETRASILANE

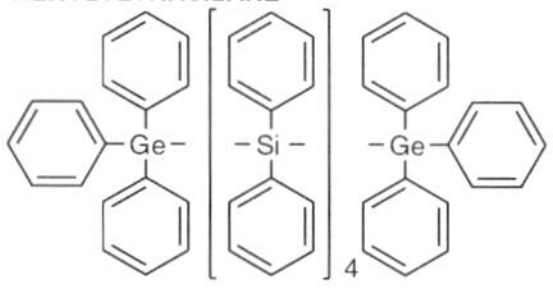

250 mg

S83,354-1
1,1,2,2,3,3,4,4,5,5,6,6,7,7-TETRADECAPHENYLHEPTASILANE

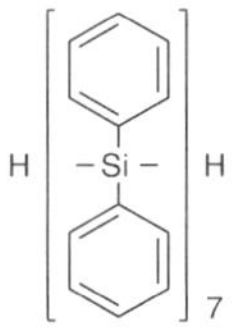

250 mg

S82,864-5
HEXADECAPHENYLHEPTASILANE

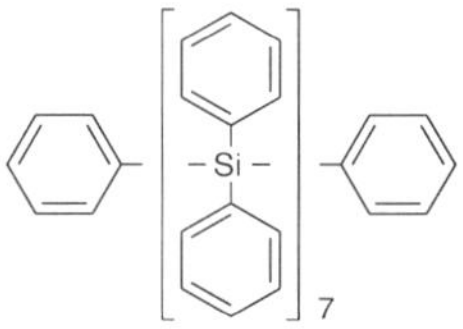

250 mg

S83,027-5
PHENYL(O-TOLYL)SILANE

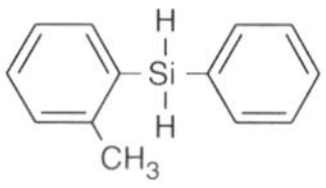

250 mg

S83,026-7
(2-CHLOROBENZYL)PHENYLSILANE

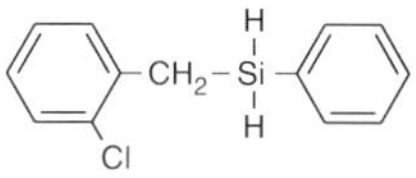

250 mg

S82,670-7
METHYL(2-METHYLBENZYL)PHENYLSILANE

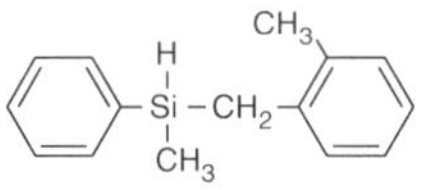

250 mg

S82,159-4
2-(DIPHENYLMETHYLSILYL)PHENOL

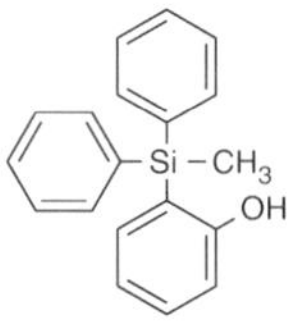

250 mg

S83,038-0
DIPHENYLMETHYL(O-TOLYL)SILANE

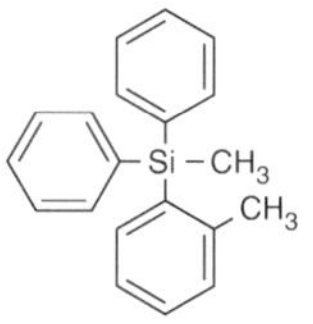

250 mg

S83,039-9
PHENYL(O-TOLYL)(M-TOLYL)SILANE

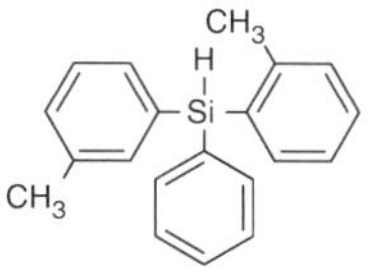

250 mg

S82,164-0
2-(TRIPHENYLSILYL)PHENOL

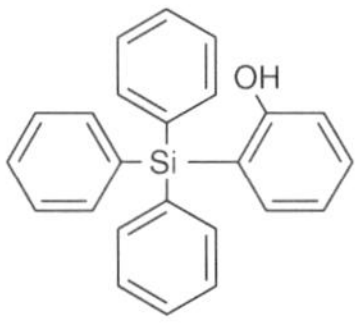

250 mg

S81,298-6
(O-TOLYL)TRIPHENYLSILANE

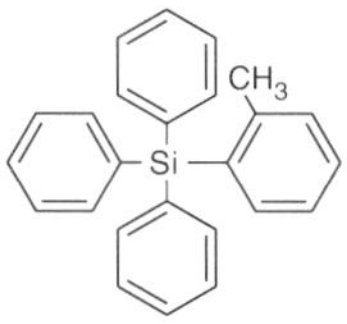

250 mg

S81,660-4
DIPHENYLDI(O-TOLYL)SILANE

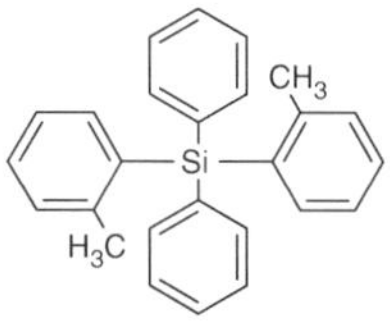

250 mg

S81,293-5
PHENYLTRI(O-TOLYL)SILANE

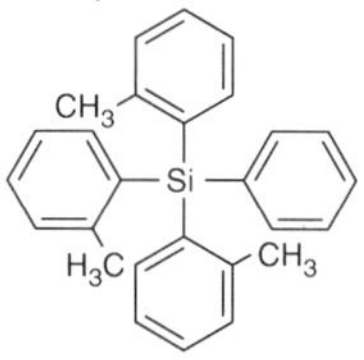

250 mg

S82,693-6
DIPHENYLPHENETHYL(O-TOLYL)SILANE

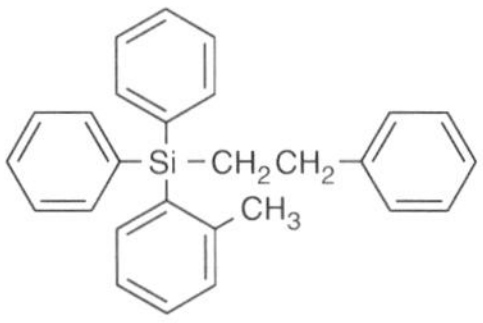

250 mg

S82,773-8
OCTADECYLTRIS(2-BIPHENYLYL)SILANE

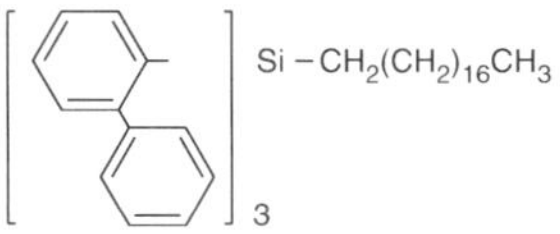

250 mg

S83,049-6
PHENYL(O-TOLYL)(M-TOLYL)(P-TOLYL)-SILANE

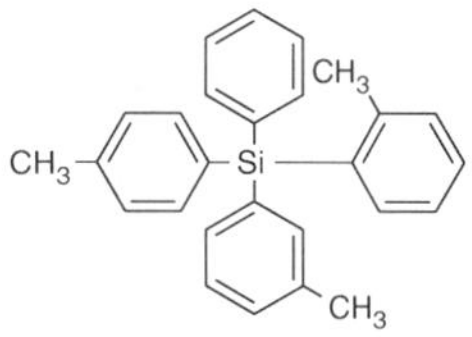

250 mg

S82,694-4
DIPHENYL(O-TOLYL)(4-TRIMETHYLSILYL)-PHENYLSILANE

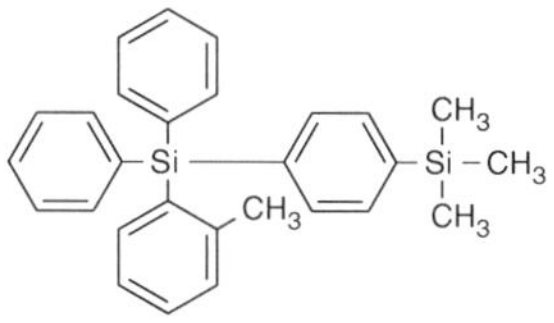

250 mg

S82,696-0
(4-TERT-BUTYLPHENYL)DIPHENYL(O-TOLYL)SILANE

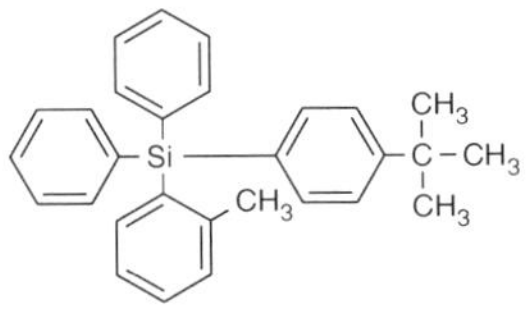

250 mg

S83,777-6
(2-BIPHENYLYL)TRIETHYLSILANE

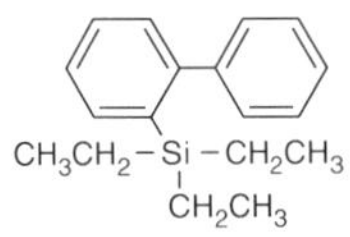

250 mg

S83,143-3
(2-BIPHENYLYL)TRIETHYLGERMANE

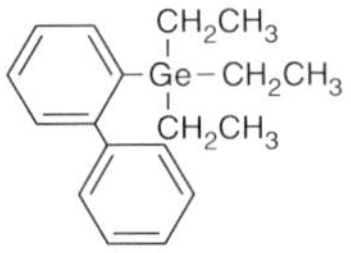

250 mg

S82,936-6
(2-BIPHENYLYL)METHYLPHENYLSILANE

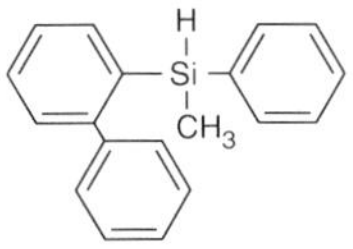

250 mg

S81,036-3
TRIS(2-BIPHENYLYL)SILANE

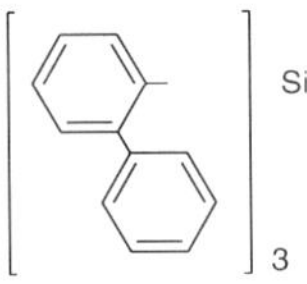

250 mg

S81,338-9
CHLOROTRI(2-BIPHENYLYL)SILANE

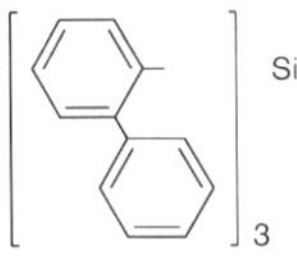

250 mg

S83,444-0
(2-BIPHENYLYL)TRIS(DECYL)SILANE

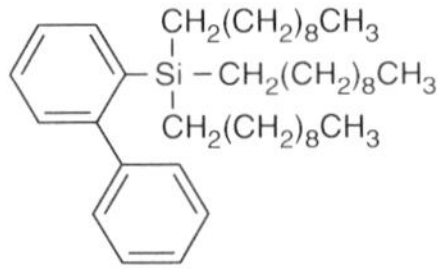

250 mg

S82,771-1
DODECYLTRIS(2-BIPHENYLYL)SILANE

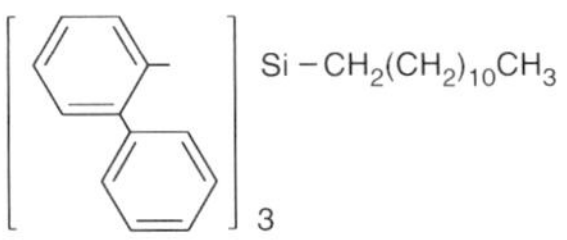

250 mg

S83,450-5
BIS(2-BIPHENYLYL)DIDODECYLSILANE

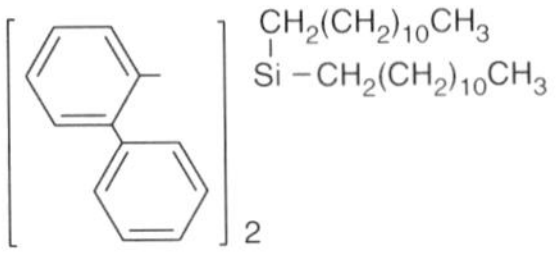

250 mg

S82,518-2
(2-BIPHENYLYL)TRIDODECYLSILANE

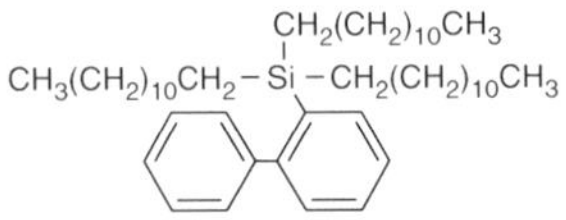

250 mg

S83,458-0
(2-BIPHENYLYL)TRIHEXADECYLSILANE

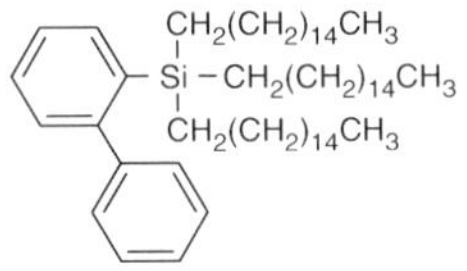

250 mg

S83,319-3
9-(2-BIPHENYLYL)-9-METHYL-9H-9-SILAFLUORENE

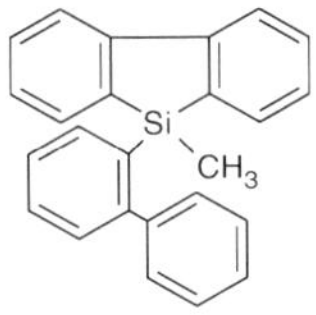

250 mg

S84,026-2
1-PHENYL-1,2,3,4-TETRAHYDRO-1-(O-TOLYL)-1-SILANAPHTHALENE

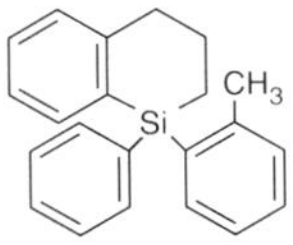

250 mg

S82,672-3
METHYL(4-METHYLBENZYL)PHENYLSILANE

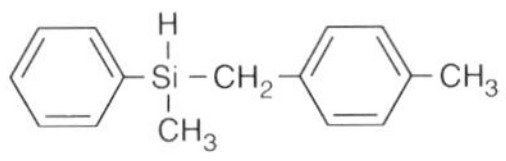

250 mg

S82,677-4
METHYLPHENYL[4-(TRIMETHYLSILYLMETHYL)PHENYL]SILANE

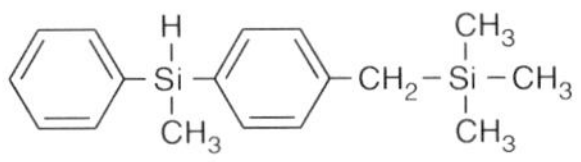

250 mg

S82,680-4
DIPHENYL(P-TOLYL)SILANE

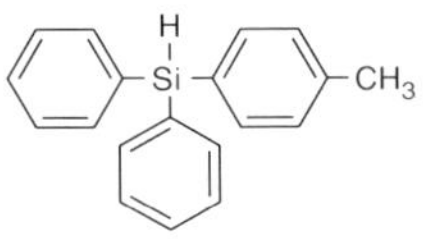

250 mg

S82,686-3
[4-(DIPHENYLSILYL)PHENYL]-TRIMETHYLSILANE

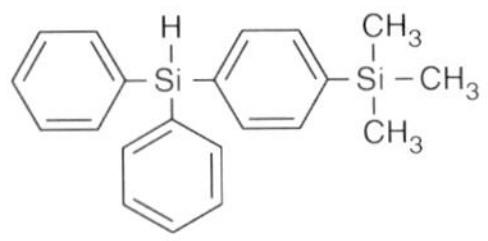

250 mg

S82,689-8
(4-TERT-BUTYLPHENYL)DIPHENYLSILANE

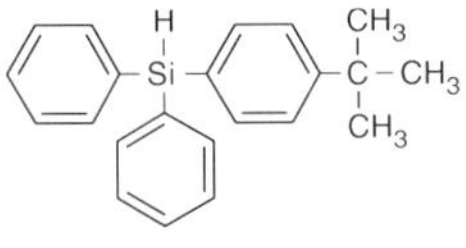

250 mg

S82,589-1
(P-TOLYL)TRIPHENYLSILANE

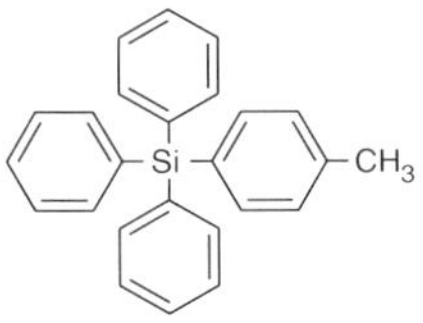

250 mg

S83,720-2
PHENYLTRI(P-TOLYL)SILANE

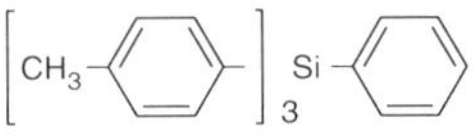

250 mg

S81,177-7
[4-(TRIMETHYLSILYL)PHENYL]-TRIPHENYLSILANE

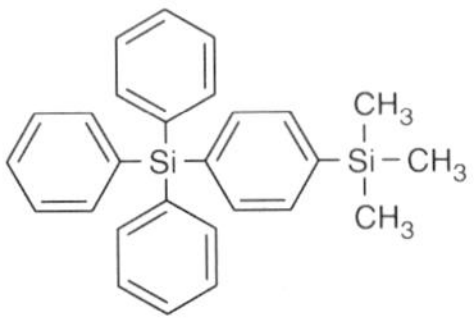

250 mg

S83,795-4
BIS[4-(TRIMETHYLSILYL)PHENYL]-DIPHENYLSILANE

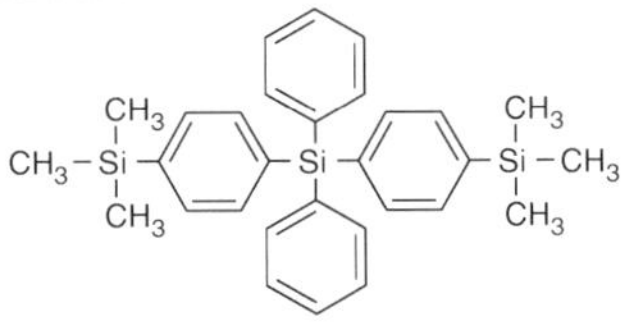

250 mg

S83,252-9
PHENYLTRIS[4-(TRIMETHYLSILYL)-PHENYL]SILANE

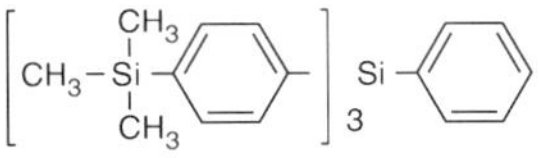

250 mg

S81,743-0
1,1,1,2,2-PENTAPHENYL-2-(P-TOLYL)-DISILANE

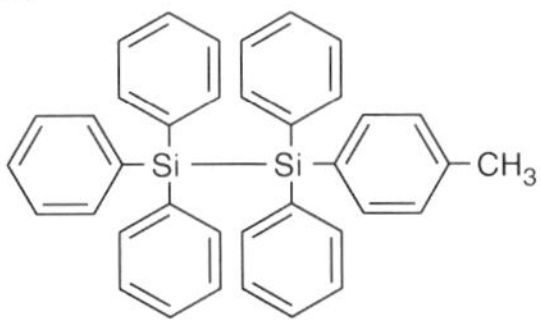

250 mg

S81,741-4
1,1-DI(P-TOLYL)-1,2,2,2-TETRAPHENYLDISILANE

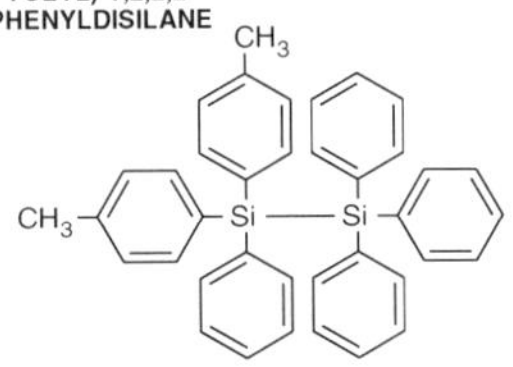

250 mg

S81,742-2
1,2-DI(P-TOLYL)-1,1,2,2-TETRAPHENYLDISILANE

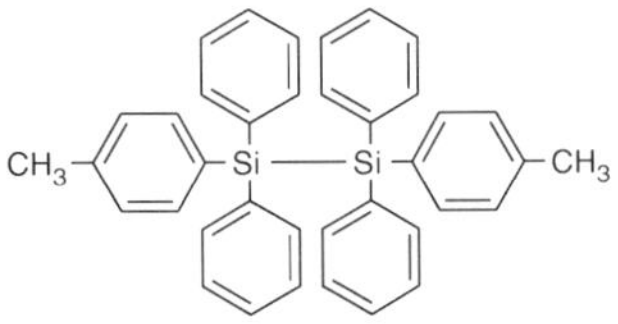

250 mg

S81,739-2
1,2-DIPHENYL-1,1,2,2-TETRA(P-TOLYL)-DISILANE

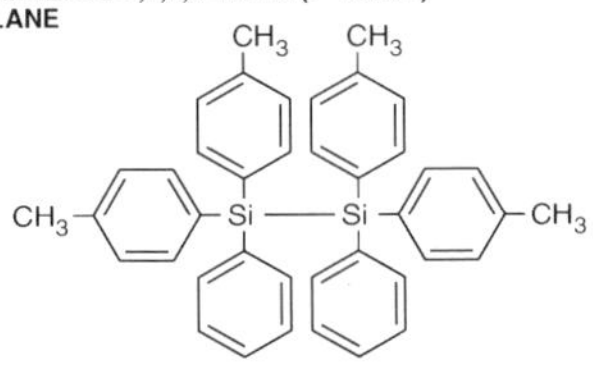

250 mg

S81,740-6
1,1-DIPHENYL-1,2,2,2-TETRA(P-TOLYL)-DISILANE

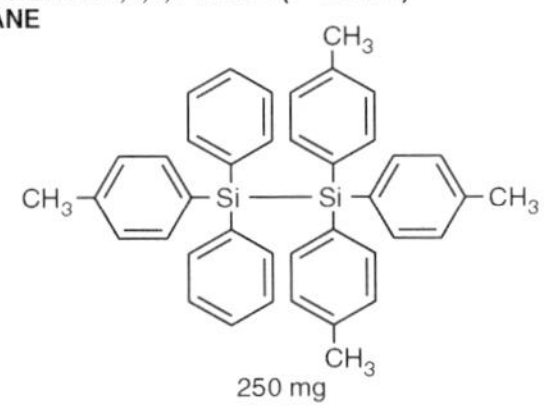

250 mg

S82,759-2
1-[METHYLPHENYL(P-TOLYL)SILYL]-4-(TRIBENZYLSILYL)BENZENE

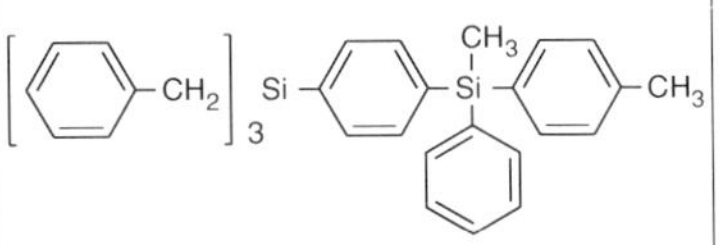

250 mg

S81,037-1
1,4-PHENYLENEBIS(TRIPHENYLSILANE)

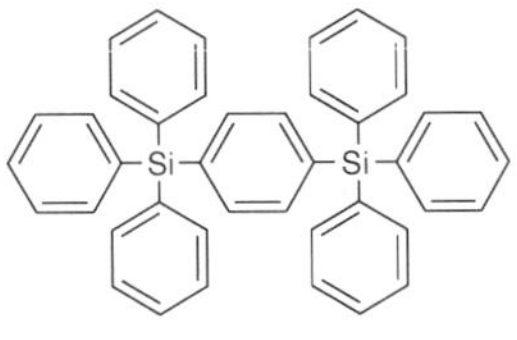

250 mg

S82,760-6
1,4-BIS(TRIBENZYLSILYL)BENZENE

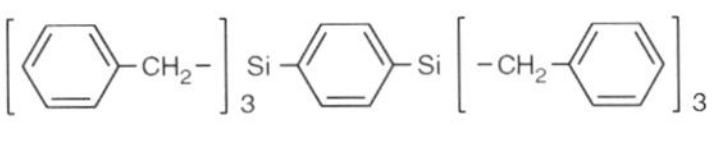

250 mg

S83,375-4
METHYLTRIS(P-TERPHENYL-4-YL)SILANE

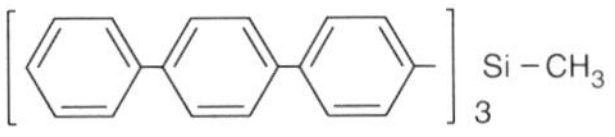

250 mg

S82,727-4
PHENYLTRIS(P-TERPHENYLYL)SILANE
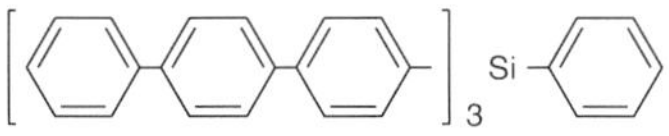
250 mg

S81,744-9
[DI(P-TOLYL)METHOXY]TRIPHENYLSILANE
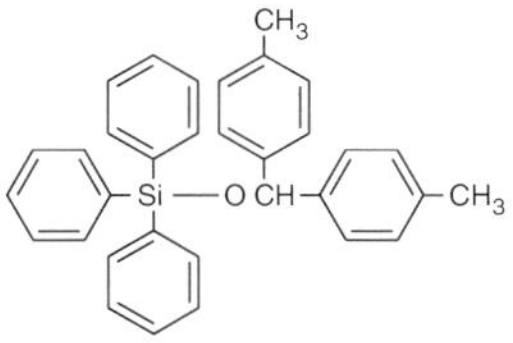
250 mg

S82,682-0
(4-BIPHENYLYL)METHYLPHENYLSILANE
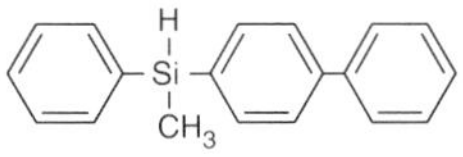
250 mg

S82,690-1
(4-BIPHENYLYL)DIPHENYLSILANE
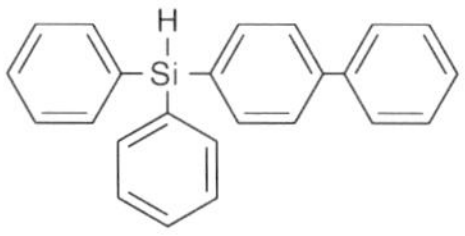
250 mg

S82,467-4
4,4'-BIS(DIMETHYLPHENYLSILYL)-BIPHENYL
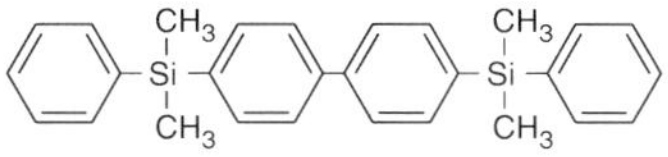
250 mg

S57,177-6
4-BIPHENYLYLTRIPHENYLSILANE
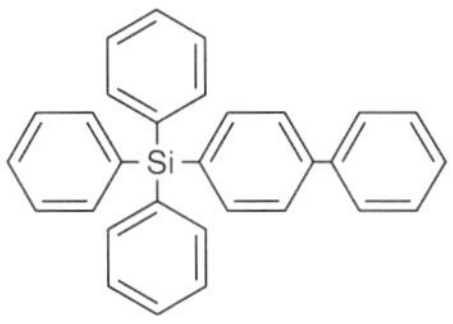
5 gm

S82,757-6
4-BIPHENYLYLTRIBENZYLSILANE
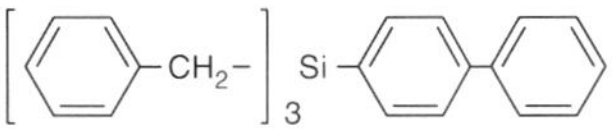
250 mg

S82,307-4
BIS(4-BIPHENYLYL)DIPHENYLSILANE
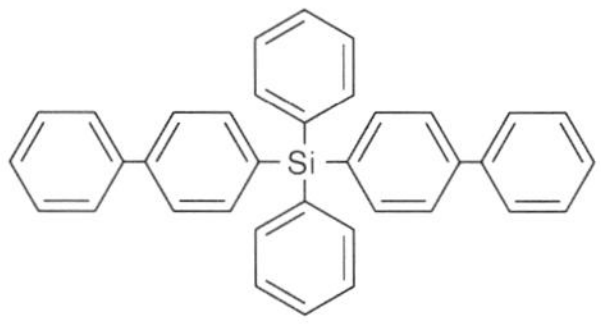
250 mg

S83,373-8
BIS(P-TERPHENYL-4-YL)DIMETHYLSILANE
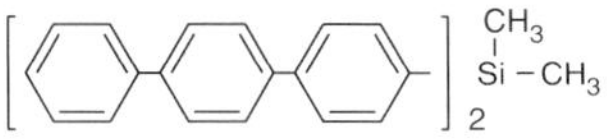
250 mg

S82,308-2
TETRAKIS(4-BIPHENYLYL)SILANE
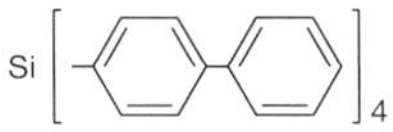
250 mg

S83,447-5
4,4'-BIS(TRIPHENYLSILYL)BIPHENYL
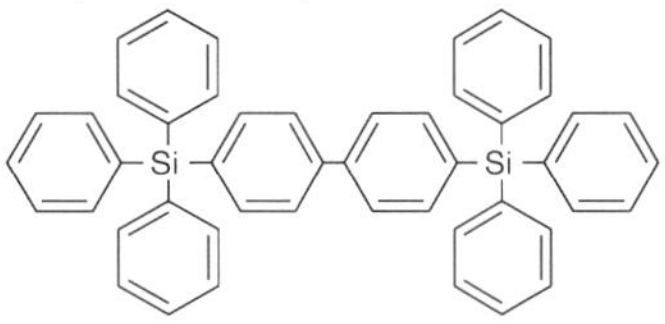
250 mg

S83,053-4
BIS(P-TERPHENYLYL)DIBENZYLSILANE
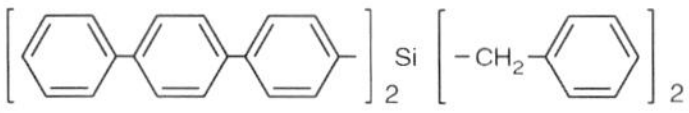
250 mg

S82,783-5
4,4'-BIS(TRIBENZYLSILYL)BIPHENYL
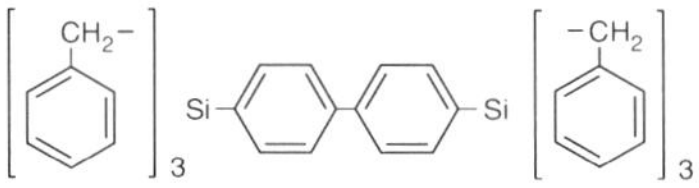
250 mg

S82,519-0
(4-BIPHENYLYL)TRIHEXADECYLSILANE
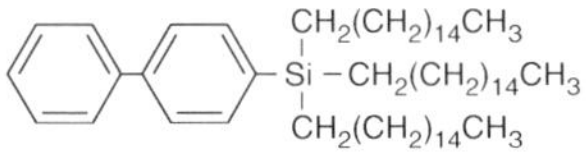
250 mg

S83,055-0
BENZYLTRIS(P-TERPHENYLYL)SILANE
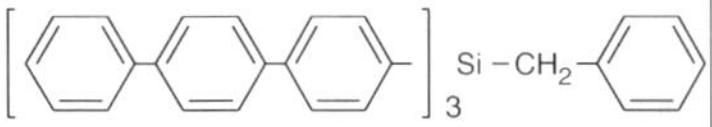
250 mg

S82,695-2
DIPHENYL(1-NAPHTHYL)(P-TOLYL)SILANE
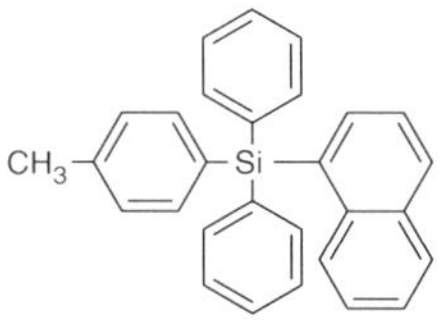
250 mg

S82,392-9
4-(DIMETHYLPHENYLSILYL)-2,3,5,6-TETRACHLOROPYRIDINE
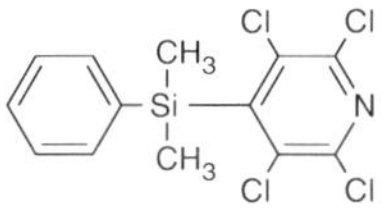
250 mg

S81,431-8
4-[(DIPHENYL)METHYLSILYL]-2,3,5,6-TETRACHLOROPYRIDINE
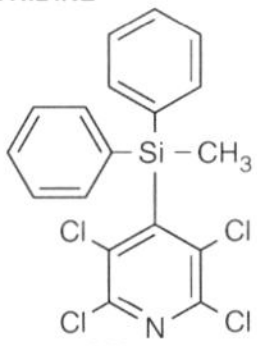
250 mg

S83,045-3
DIPHENYL(DIPHENYLMETHYL)SILANE
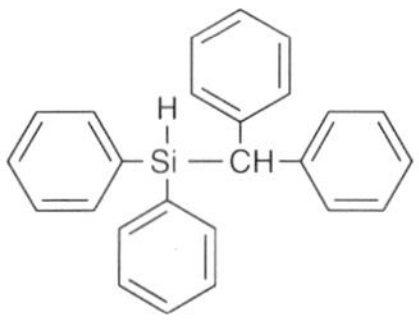
250 mg

S82,762-2
CHLORODIPHENYL(DIPHENYLMETHYL)-SILANE
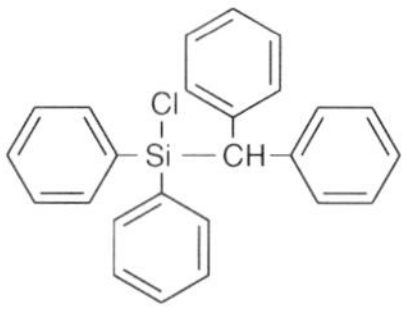
250 mg

S82,945-5
BIS(DIPHENYLMETHOXY)PHENYLSILANE
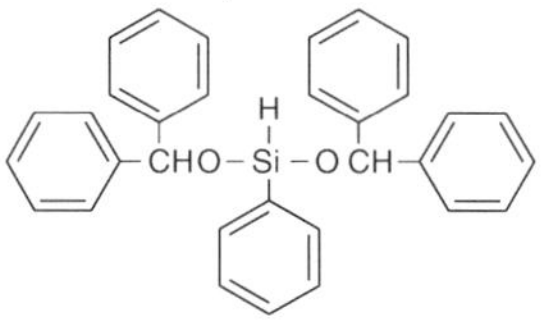
250 mg

S82,927-7
(3-BIPHENYLYL)TRIMETHYLSILANE
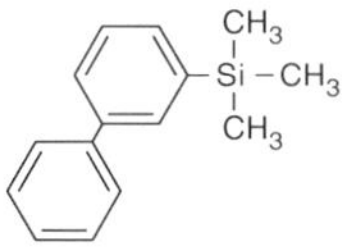
250 mg

S82,683-9
(3-BIPHENYLYL)METHYLPHENYLSILANE
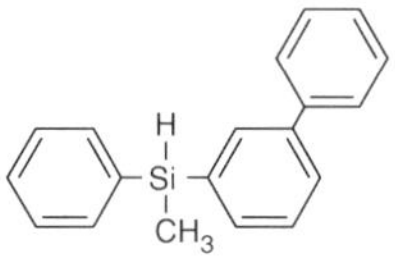
250 mg

S82,929-3
(3-BIPHENYLYL)TRIPHENYLSILANE
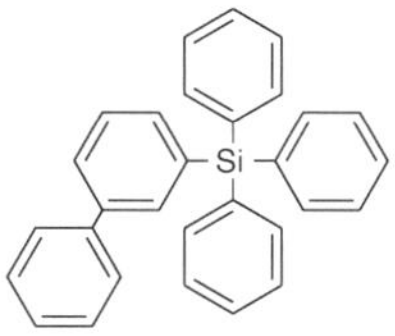
250 mg

S83,240-5
TRIS(3-BIPHENYLYL)SILANE
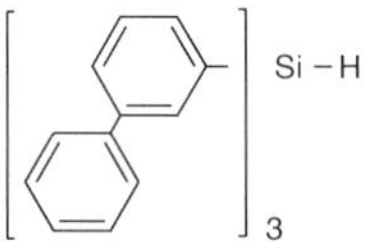
250 mg

S83,241-3
(3-BIPHENYLYL)TRIS(3-PHENYLPROPYL)-SILANE
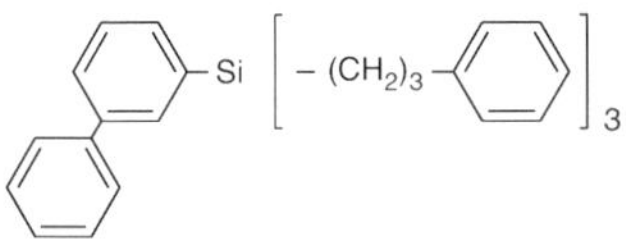
250 mg

S82,310-4
3,3'-BIS(TRIPHENYLSILYL)BIPHENYL
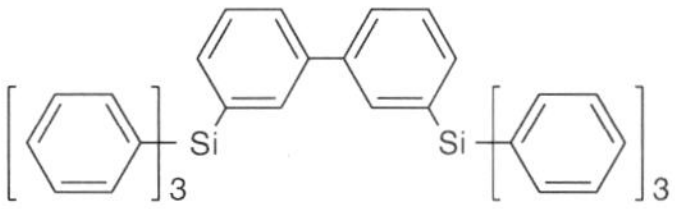
250 mg

S83,441-6
BIS(3-BIPHENYLYL)DIDODECYLSILANE
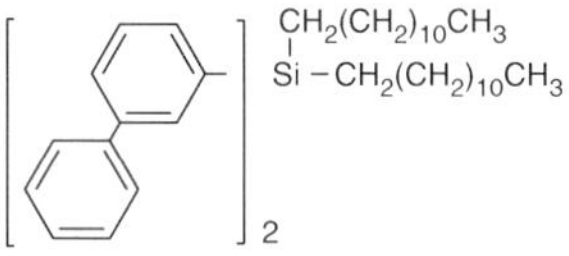
250 mg

S82,782-7
3,3'-BIS(TRIBENZYLSILYL)BIPHENYL
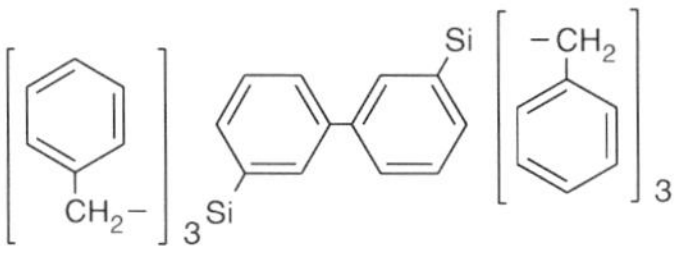
250 mg

S83,885-3
1-(TRIPHENYLSILYL)INDENE
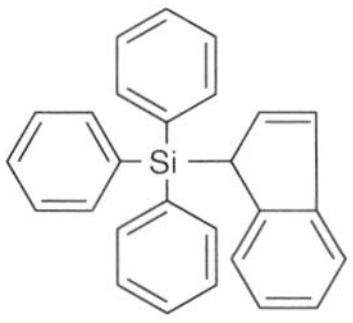
250 mg

S83,886-1
1,1-BIS(TRIPHENYLSILYL)INDENE
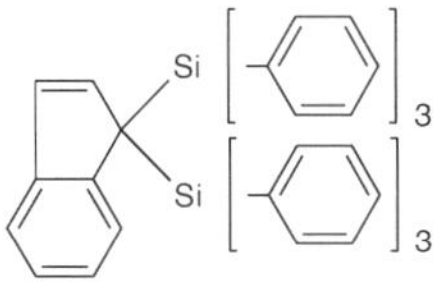
250 mg

S82,675-8
METHYL(1-NAPHTHYL)PHENYLSILANE
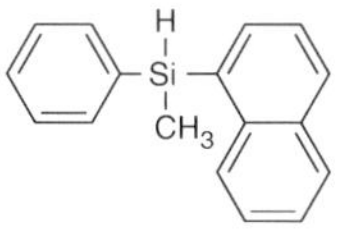
250 mg

S82,687-1
DIPHENYL(1-NAPHTHYL)SILANE

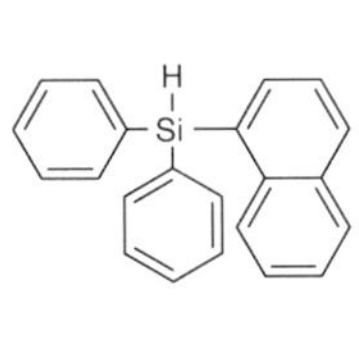

250 mg

S82,168-3
(2-NAPHTHYL)TRIPHENYLSILANE

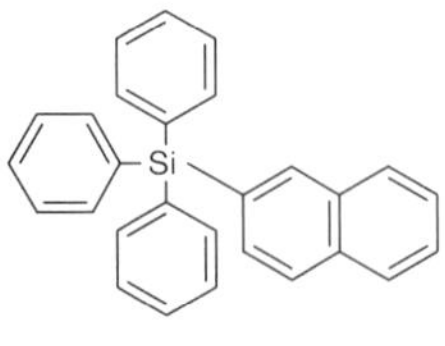

250 mg

S82,763-0
DIPHENYL(9-FLUORENYL)SILANE

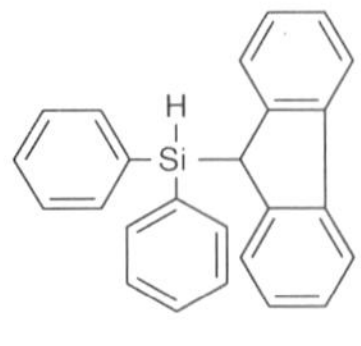

250 mg

S83,253-7
BIS(9-FLUORENYL)DIPHENYLSILANE

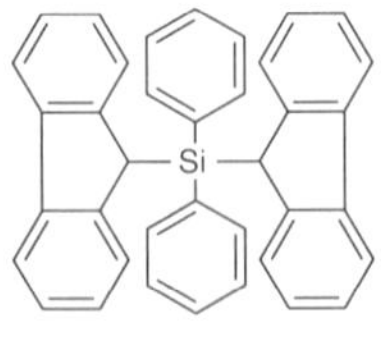

250 mg

S83,254-5
PHENYLTRIS(9-FLUORENYL)SILANE

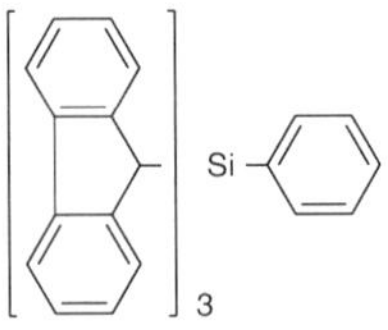

250 mg

S82,753-3
(9,10-DIHYDRO-9-ANTHRACENYL)-TRIPHENYLSILANE

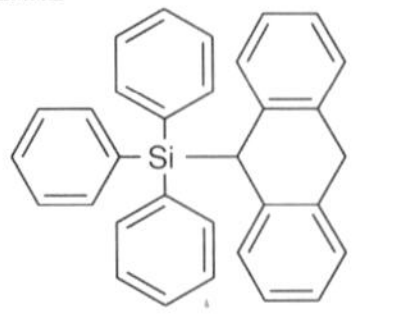

250 mg

S82,850-5
10-PHENYL-10H-PHENOXASILIN

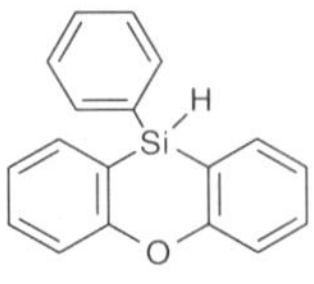

250 mg

S82,766-5
10,10-DIPHENYL-10H-PHENOXASILIN

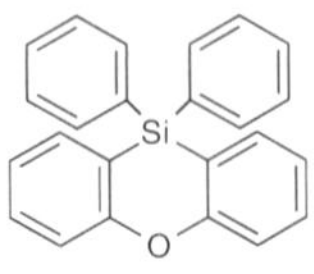

250 mg

S82,942-0
10,10-DIPHENYL-10H-PHENOTHIASILIN

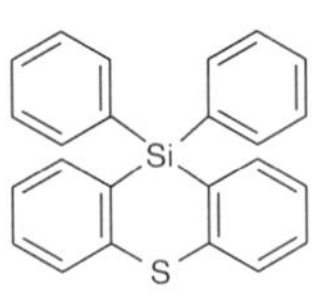

250 mg

S83,427-0
10,10-DIBENZYL-10H-PHENOXASILIN

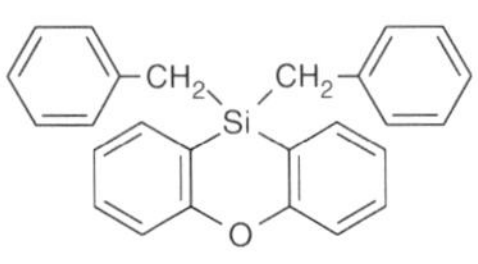

250 mg

S83,505-6
9-DODECYL-9-PHENYL-9H-9-SILAFLUORENE

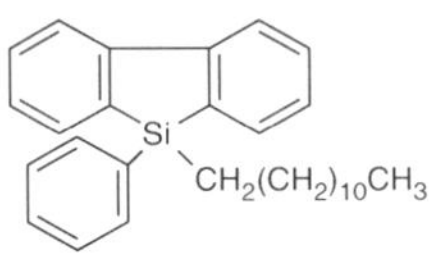

250 mg

S82,856-4
10,10'-BIS(10-PHENYL-10H-PHENOXASILIN)

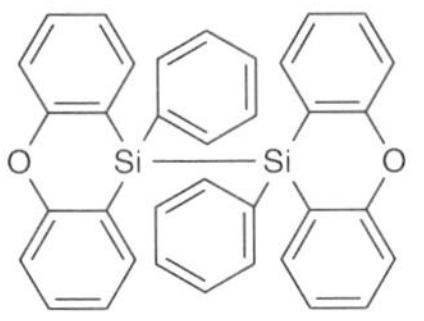

250 mg

S83,509-9
9-[(DIBENZYLMETHYLSILYL)METHYL]-9-DODECYL-9H-9-SILAFLUORENE

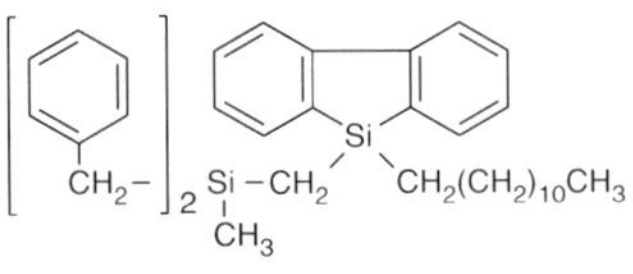

250 mg

S83,774-1
(O-TOLYL)TRIMETHYLSILANE

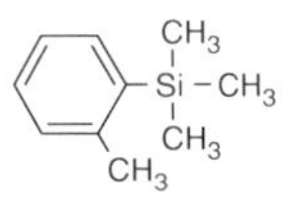

250 mg

S83,261-8
1,2-PHENYLENEBIS(TRIMETHYLSILANE)

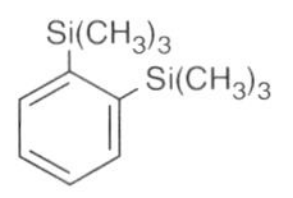

250 mg

S81,665-5
DI(O-TOLYL)ETHOXYSILANE

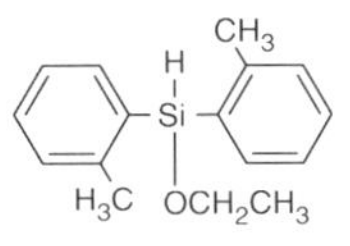

250 mg

S81,302-8
TRI(O-TOLYL)SILANE

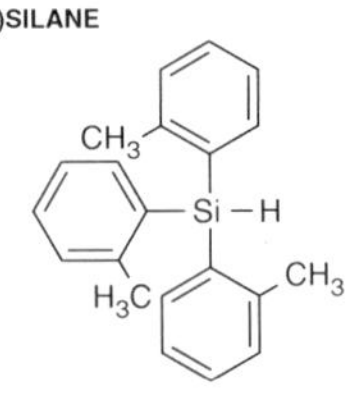

250 mg

S81,663-9
CHLOROTRI(O-TOLYL)SILANE

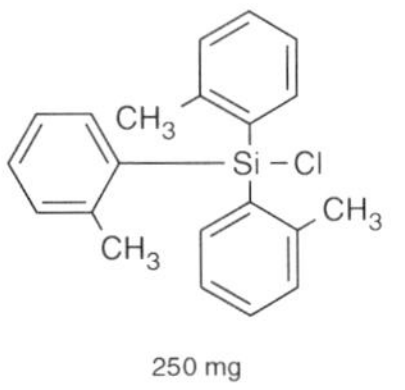

250 mg

S81,300-1
ETHOXYTRI(O-TOLYL)SILANE

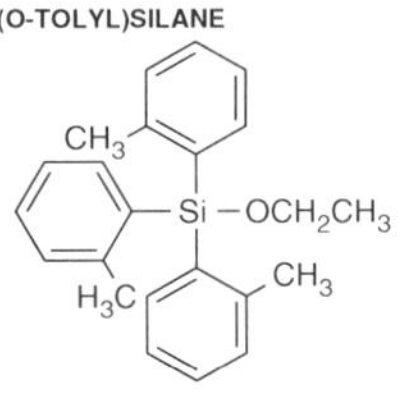

250 mg

S81,297-8
(O-TOLYLOXY)TRI(O-TOLYL)SILANE

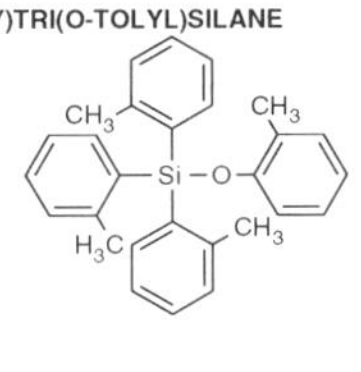

250 mg

S81,465-2
1,1,2,2-TETRA(O-TOLYL)-1,2-DISILANEDIOL

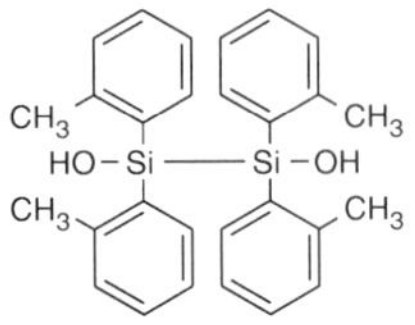

250 mg

S81,658-2
TRI(O-TOLYL)SILYL STEARATE

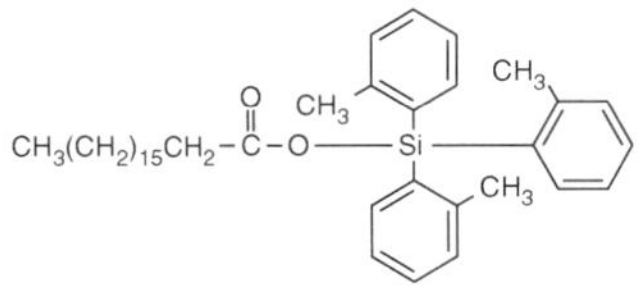

250 mg

S83,054-2
BIS(2-BIPHENYLYL)DIBENZYLSILANE

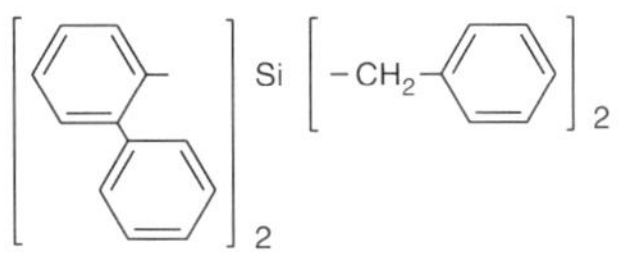

250 mg

S81,604-3
1,4-BIS(TRIMETHYLSILYL)BENZENE

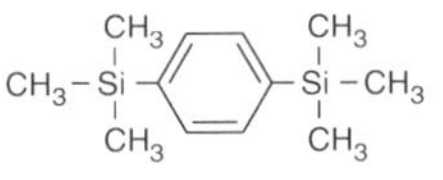

250 mg

S81,017-7
(P-TOLYL)TRIETHYLSILANE

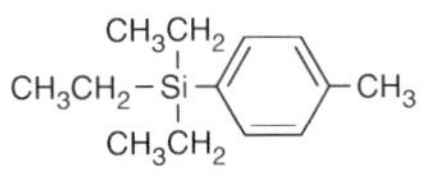

250 mg

S83,141-7
DI(P-TOLYL)SILANE

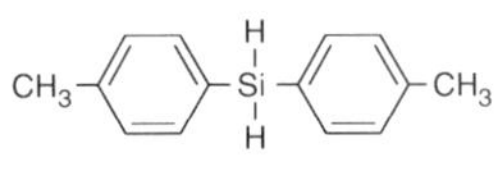

250 mg

S81,322-2
1,2-BIS[4-(TRIMETHYLSILYL)PHENYL]-ETHANE

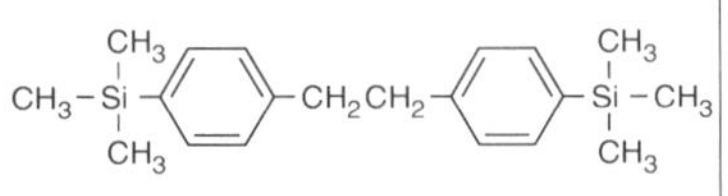

250 mg

S83,349-5
1,2-BIS[4-(TRIMETHYLSILYL)PHENYL]-1,1,2,2-TETRAMETHYLDISILANE

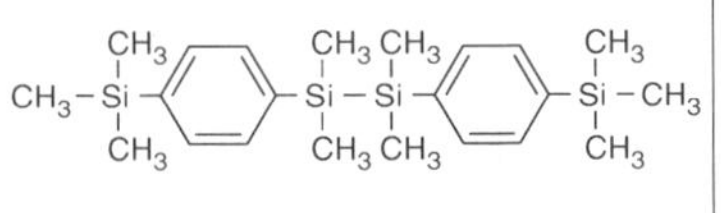

250 mg

S83,370-3
4,4''-BIS(TRIMETHYLSILYL)-P-TERPHENYL

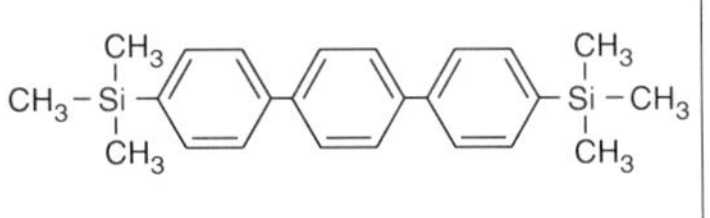

250 mg

S83,250-2
TRIS[4-(TRIMETHYLSILYL)PHENYL]-SILANE

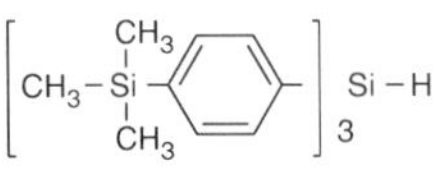

250 mg

S81,653-1
TETRAKIS(P-TOLYL)SILANE

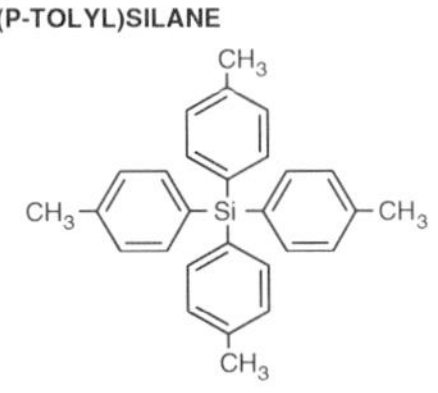

250 mg

S83,702-4
TETRAKIS[4-(TRIMETHYLSILYL)PHENYL]-SILANE

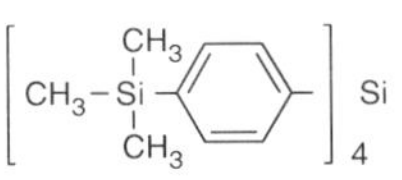

250 mg

S83,703-2
TETRAKIS[4-(TRIMETHYLSILYL)PHENYL]-GERMANE

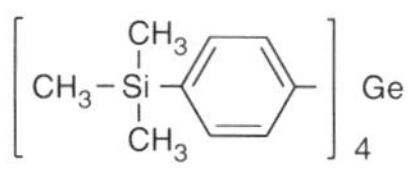

250 mg

S83,069-0
HEXAKIS(P-TOLYL)DISILANE

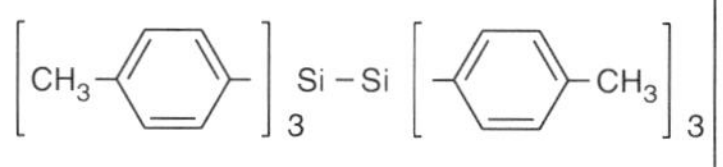

250 mg

S83,445-9
BIS[4-(TRIMETHYLSILYL)PHENYL]-DIDODECYLSILANE

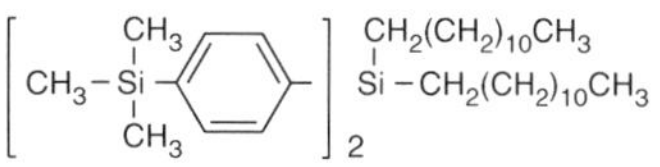

250 mg

S83,452-1
BIS(P-TOLYL)DIOCTADECYLSILANE

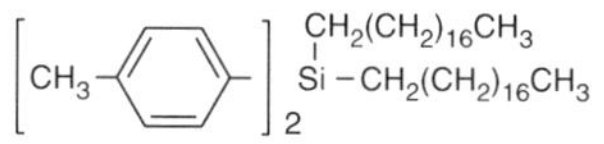

250 mg

S83,457-2
TRIHEXADECYL[4-(TRIMETHYLSILYL)-PHENYL]SILANE

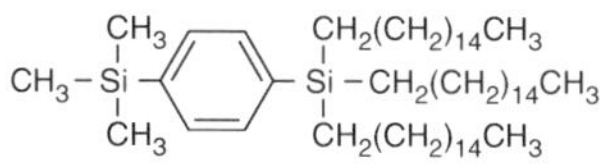

250 mg

S81,323-0
4,4'-BIS(TRIMETHYLSILYL)BIPHENYL

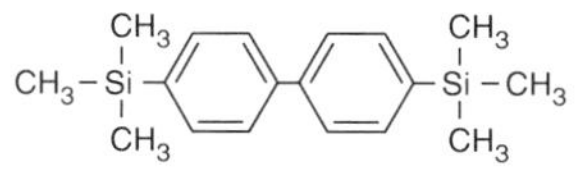

250 mg

S83,723-7
(P-TOLYL)TRIS(1-NAPHTHYL)SILANE

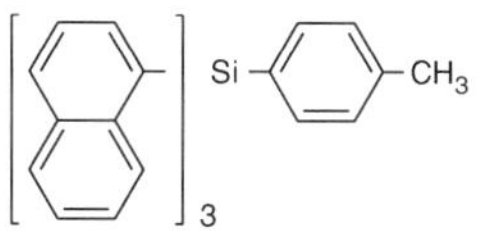

250 mg

S82,558-1
1,3,5-BENZENETRIYLTRIS-(TRIMETHYLSILANE)

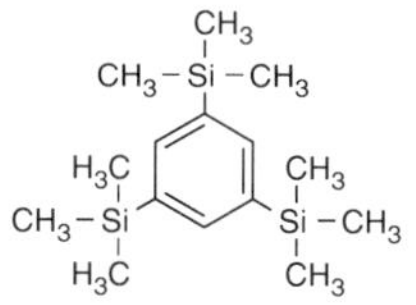

250 mg

S82,421-6
1,2,4,5-BENZENETETRAYLTETRAKIS-(DIMETHYLSILANE)

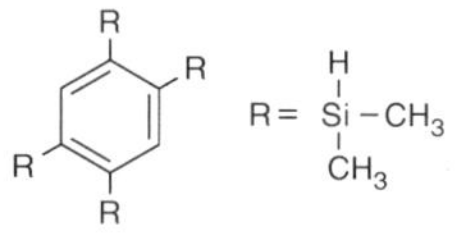

250 mg

S82,977-3
2,3,5,6-TETRACHLORO-4-(TRIMETHYLSILYL)PYRIDINE

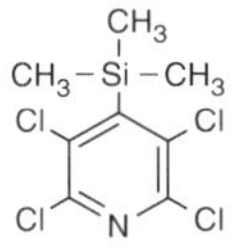

250 mg

S83,538-2
BIS(2,3,5,6-TETRACHLORO-4-PYRIDYL)-METHYLSILANE

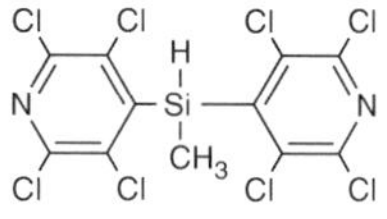

250 mg

S83,424-6
3-BROMO-3'-(TRIMETHYLSILYL)BIPHENYL

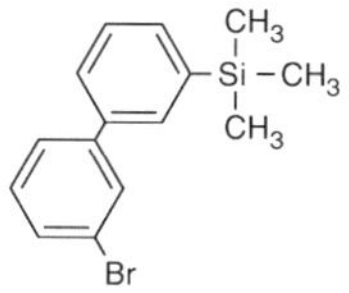

250 mg

S83,440-8
3,3'-BIS(TRIBUTYLSILYL)BIPHENYL

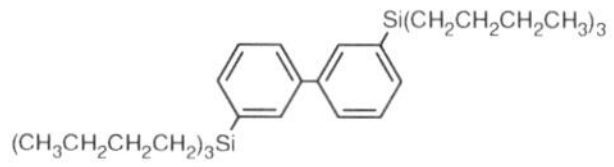

250 mg

S83,318-5
9,9'-SPIROBI(9H-9-SILAFLUORENE)

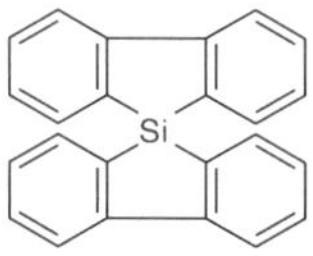

250 mg

S83,776-8
(1-NAPHTHYL)TRIMETHYLSILANE

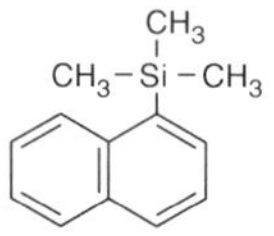

250 mg

S83,714-8
(1-NAPHTHYL)TRIETHOXYSILANE

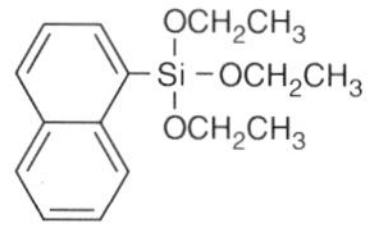

250 mg

S83,715-6
BIS(1-NAPHTHYL)DIETHOXYSILANE

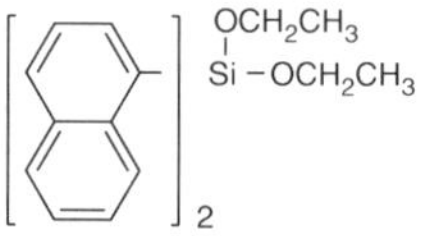

250 mg

S83,722-9
TRIS(1-NAPHTHYL)SILANE

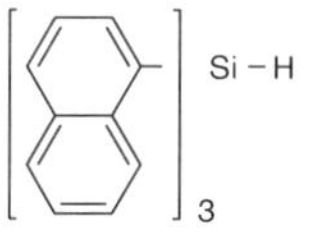

250 mg

S83,394-0
CHLOROTRIS(1-NAPHTHYL)SILANE

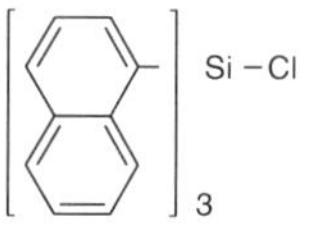

250 mg

S83,146-8
1,4-DIHYDRO-1,1,4-TRIS-(TRIMETHYLSILYL)NAPHTHALENE

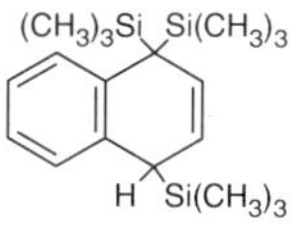

250 mg

S83,724-5
TETRAKIS(2-NAPHTHYL)SILANE

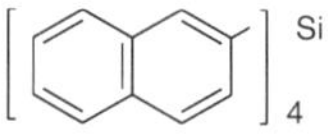

250 mg

S81,470-9
(9-FLUORENYL)TRIMETHYLSILANE

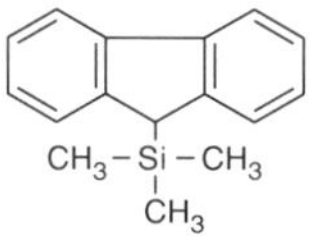

250 mg

S83,142-5
9-METHYL-9-(TRIMETHYLSILYL)FLUORENE

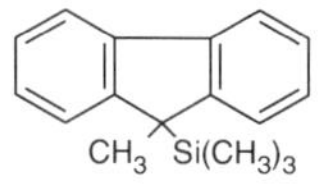

250 mg

S83,144-1
9,9-BIS(TRIMETHYLSILYL)FLUORENE

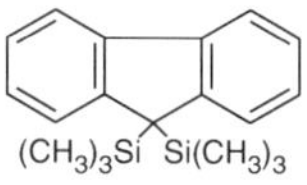

250 mg

S58,580-7
N-(TRIMETHYLSILYL)PHTHALIMIDE

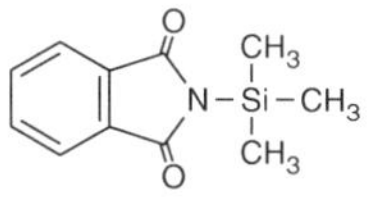

250 mg

S83,359-2
9-CHLORO-9-METHYL-9H-9-SILAFLUORENE

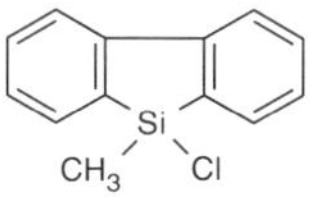

250 mg

S83,315-0
9,9-DIMETHYL-9H-9-SILAFLUORENE

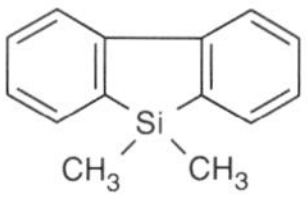

250 mg

S83,234-0
10,10-DIMETHYL-10H-PHENOXASILIN

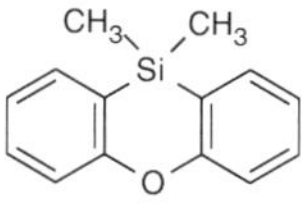

250 mg

S83,379-7
9,9'-DIMETHYL-9,9'-BI(9H-9-SILAFLUORENE)

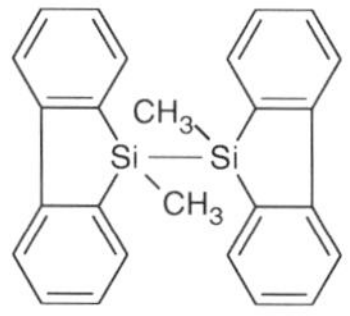

250 mg

S83,507-2
9,9-DIDODECYL-9H-9-SILAFLUORENE

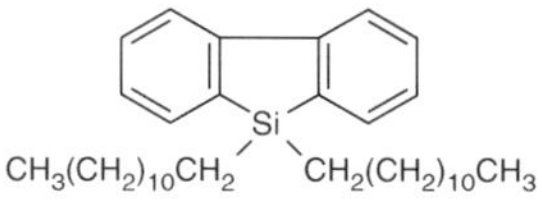

250 mg

S83,439-4
10,10-DIDODECYL-10H-PHENOXASILIN

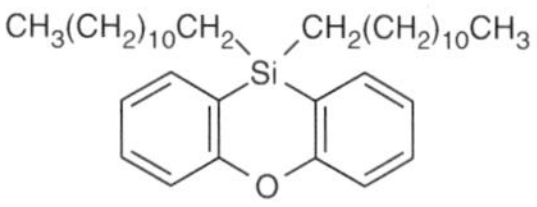

250 mg

S59,936-0
2,4,6-TRICHLOROBORAZINE
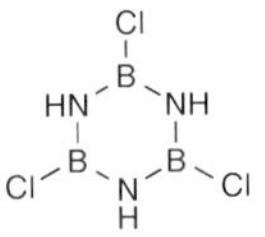
5 gm

S34,306-4
BUTYL METABORATE
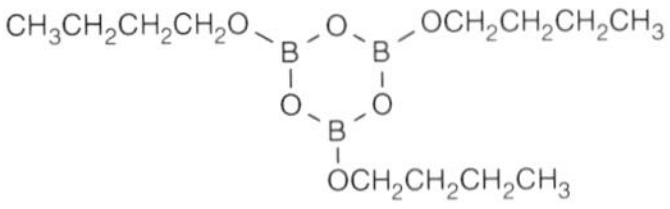
5 gm

S56,654-3
1,3,5-TRIETHYLBORAZINE
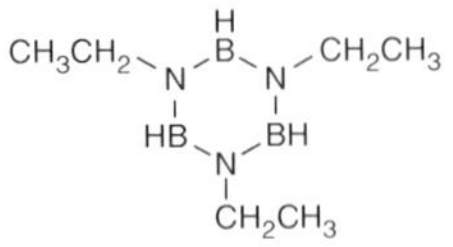
5 gm

S86,227-4
HEXAMETHYLBORAZINE
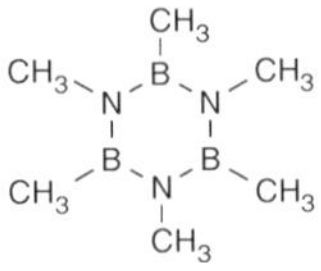
250 mg

S43,434-5
METHYL POLYBORATE, TECH.
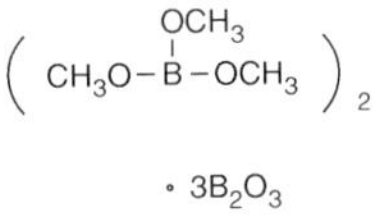
5 gm

S45,832-5
4,4,6-TRIMETHYL-2-VINYL-1,3,2-DIOXABORINANE
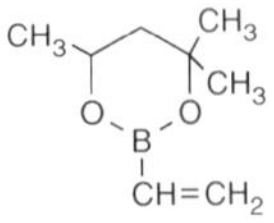
5 gm

S35,571-2
SODIUM TETRAETHOXYBORON
$Na^+ [B(OCH_2CH_3)_4]^-$
5 gm

S37,085-1
TRIALLYL BORATE
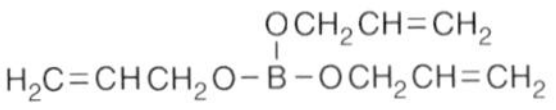
5 gm

S45,023-5
TRIISOPROPOXYBOROXINE
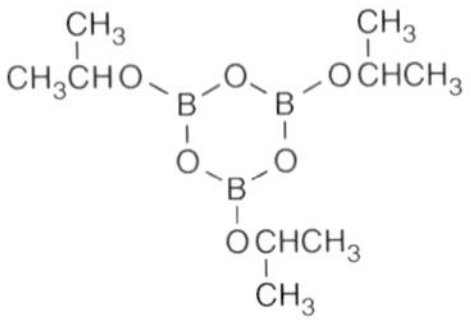
5 gm

S51,285-0
2-BUTOXY-4,4,6-TRIMETHYL-1,3,2-DIOXABORINANE
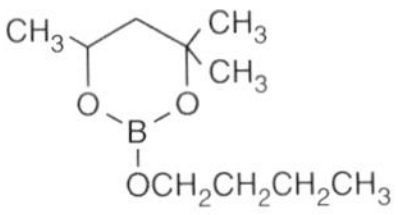
5 gm

S37,452-0
TRIMETHALLYL BORATE
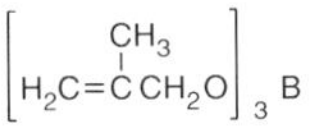
5 gm

S83,156-5
TETRADECANEBORONIC ACID
$CH_3(CH_2)_{12}CH_2B(OH)_2$
250 mg

S51,290-7
TRIPENTYL BORATE
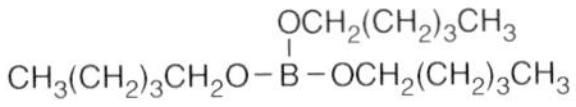
5 gm

S83,157-3
HEXADECANEBORONIC ACID
$CH_3(CH_2)_{14}CH_2B(OH)_2$
250 mg

S51,291-5
TRIHEXYL BORATE
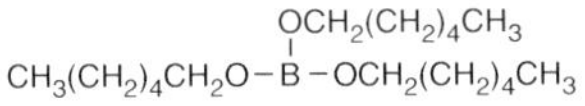
5 gm

S51,226-5
2,2'-(2-METHYL-2,4-PENTANEDIYLDIOXY)BIS(4,4,6-TRIMETHYL-1,3,2-DIOXABORINANE)
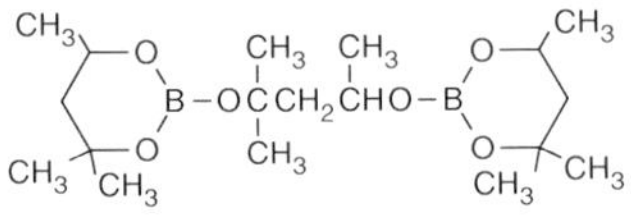
5 gm

S43,432-9
ISOPROPYL POLYBORATE
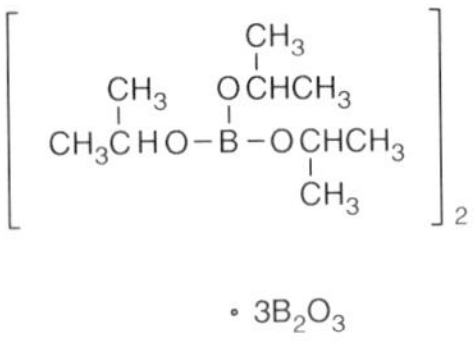
5 gm

S51,292-3
TRITOLYL BORATE (MIXTURE OF M- AND P-)
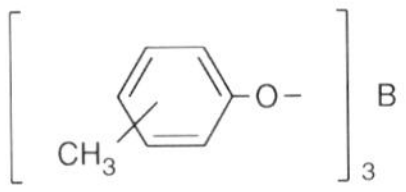
5 gm

S51,299-0
TRIS(1-ISOPROPYL-2-METHYLPROPYL)-BORATE
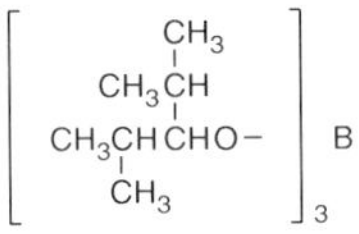
5 gm

S37,578-0
TRIS(2-ETHYLHEXYL) BORATE
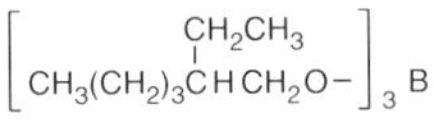
5 gm

S37,579-9
2-OCTYL BORATE
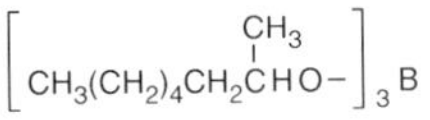
5 gm

S51,297-4
TRIOCTYL BORATE
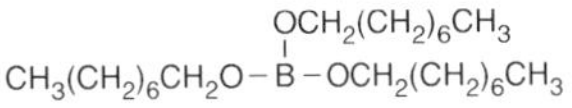
5 gm

S51,300-8
TRIS(1-ISOBUTYL-3-METHYLBUTYL) BORATE
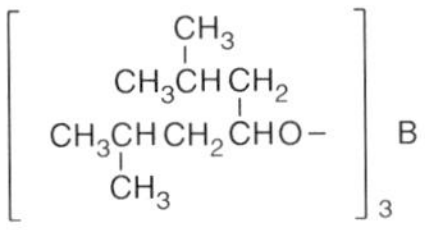
250 mg

S51,706-2
TRIS(3,3,5-TRIMETHYLHEXYL) BORATE
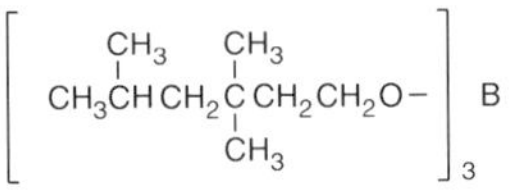
5 gm

S37,582-9
TRIS(DECYL) BORATE
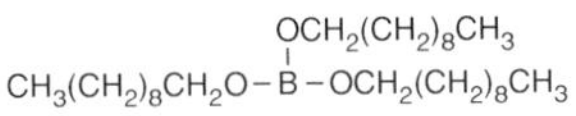
5 gm

S51,298-2
TRIHEXADECYL BORATE
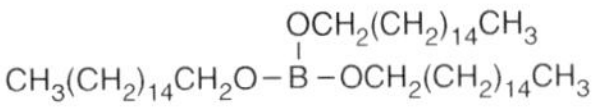
5 gm

S51,304-0
TRIOCTADECYL BORATE
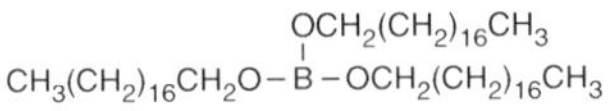
5 gm

S83,269-3
TRIS(2,2,2-TRIFLUOROETHYL) BORATE
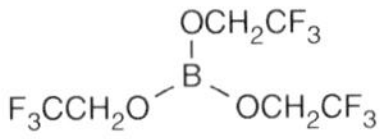
250 mg

S81,131-9
2-HYDROXY-5-[3-(TRIFLUOROMETHYL)PHENZYLAZO]BENZENEBORONIC ACID
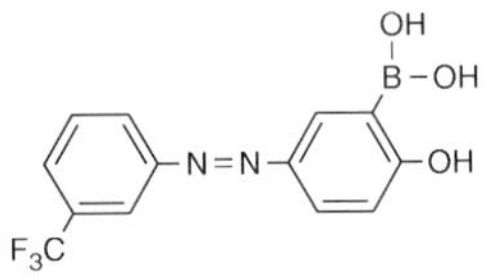
250 mg

S94,171-9
3-CHLOROPHENYLBORONIC ACID
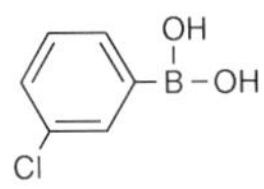
250 mg

S82,790-8
5-BROMO-2-METHOXYBENZENEBORONIC ACID
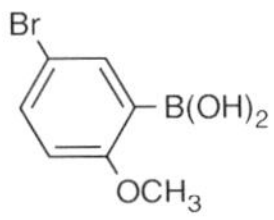
250 mg

S94,177-8
4-CHLOROPHENYLBORONIC ACID
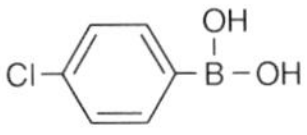
250 mg

S82,182-9
2,4,6-TRIS[5-(4-BROMOPHENYLAZO)-2-HYDROXYPHENYL]BOROXIN
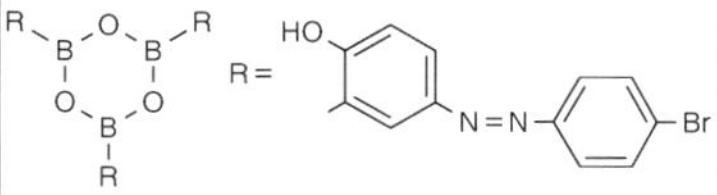
250 mg

S82,789-4
5-BROMO-2-HYDROXYBENZENEBORONIC ACID
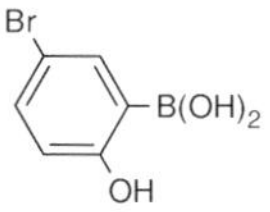
250 mg

S51,335-0
PENTAERYTHRITOL BORATE
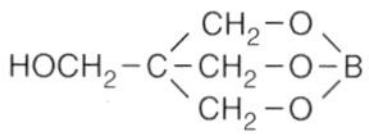
250 mg

S39,662-1
TRIS(HYDROXYBENZYL) BORATE
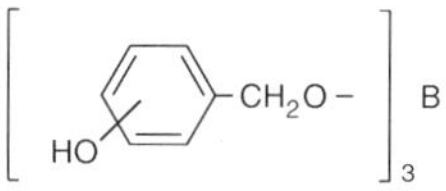
5 gm

S83,953-1
1-(3-BORONO-4-HYDROXYPHENYLAZO)-ANTHRAQUINONE
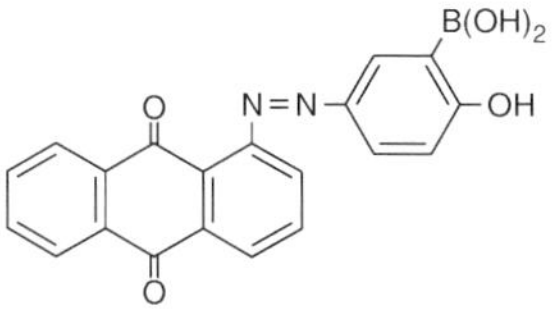
250 mg

S82,178-0
6-HYDROXY-2-NAPHTHALENEBORONIC ACID
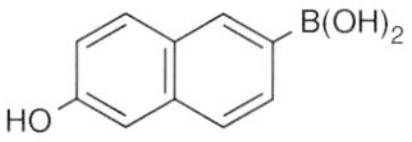
250 mg

S94,169-7
3-METHOXYPHENYLBORONIC ACID
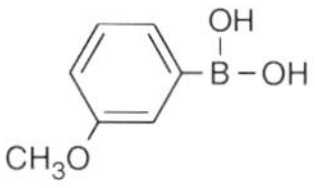
250 mg

S82,665-0
4-(PHENOXY)BENZENEBORONIC ACID
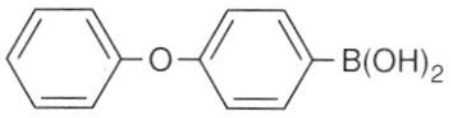
250 mg

S94,174-3
4-METHOXYPHENYLBORONIC ACID
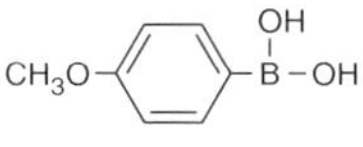
250 mg

S87,593-7
[2-(2-BENZYLTHIOETHYL)-6,6-DIMETHYLBICYCLO[3.1.1]HEPT-3-YL]-BROMOBORANE
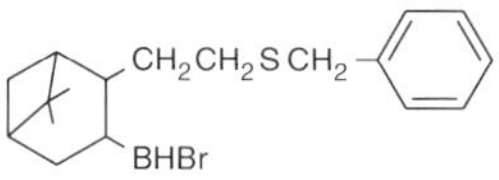
250 mg

S82,791-6
4-(METHYLTHIO)BENZENEBORONIC ACID
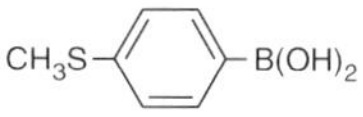
250 mg

S45,078-2
2-PHENYL-1,3,2-BENZODIAZABOROLE
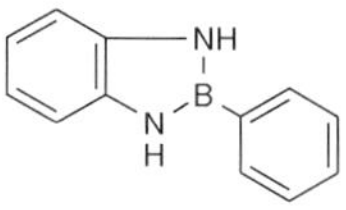
5 gm

S43,433-7
2-(2-DIMETHYLAMINOETHOXY)-4-METHYL-1,3,2-DIOXABORINANE
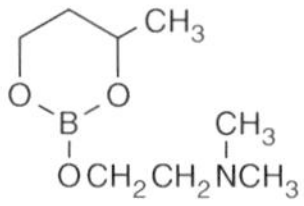
5 gm

S94,167-0
2-NITROPHENYLBORONIC ACID
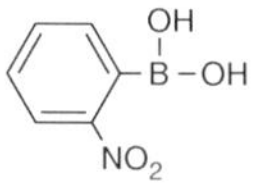
250 mg

S94,175-1
4-NITROPHENYLBORONIC ACID
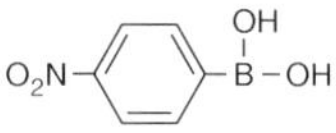
250 mg

S82,183-7
2,4,6-TRIS[5-(2-NITROPHENYLAZO)-2-HYDROXYPHENYL]BOROXIN
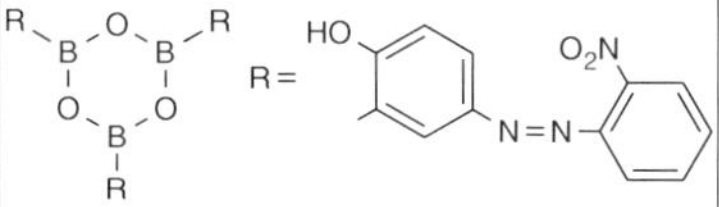
250 mg

S94,168-9
3-FORMYLPHENYLBORONIC ACID
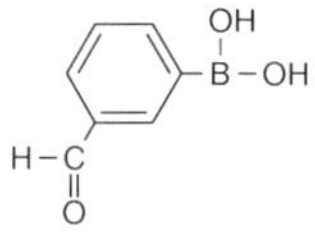
250 mg

S94,172-7
4-FORMYLPHENYLBORONIC ACID
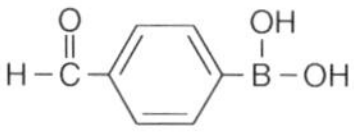
250 mg

S81,091-6
4-DIBENZOFURANBORONIC ACID
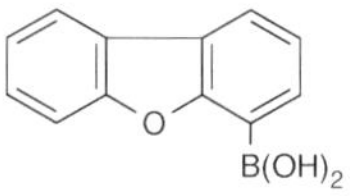
250 mg

S51,296-6
TRIS(1-ETHYNYLCYCLOHEXYL) BORATE
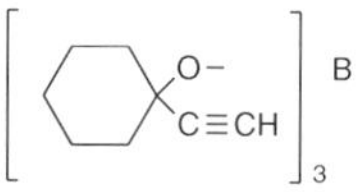
5 gm

S51,301-6
(+/-)-TRIMENTHYL BORATE
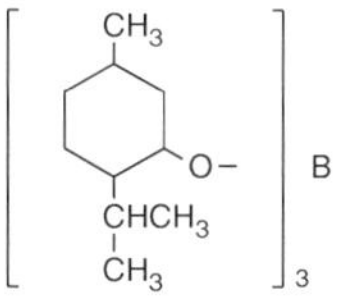
5 gm

S51,302-4
(-)-TRIMENTHYL BORATE
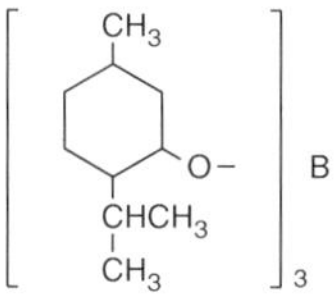
5 gm

S51,303-2
TRIS(2-CYCLOHEXYLCYCLOHEXYL) BORATE
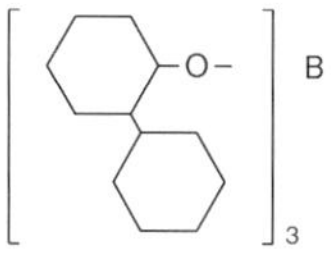
5 gm

S39,664-8
TRIPHENYL BORATE
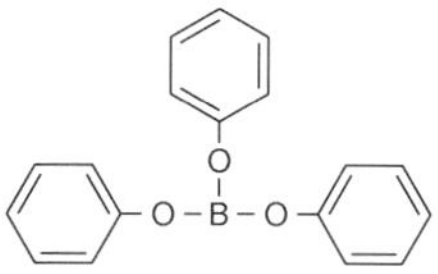
5 gm

S82,180-2
2,4,6-TRIPHENYLBOROXIN
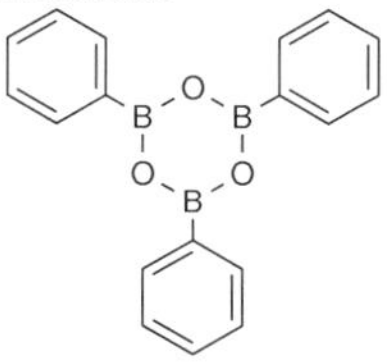
250 mg

S37,241-2
TRIBENZYL BORATE
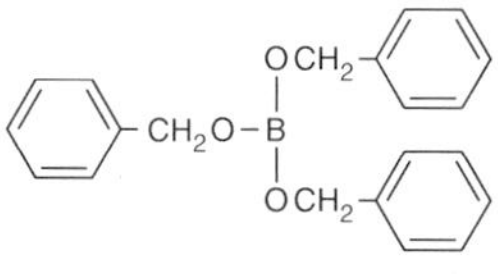
5 gm

S81,013-4
2-BIPHENYLYLBORONIC ACID HYDRATE
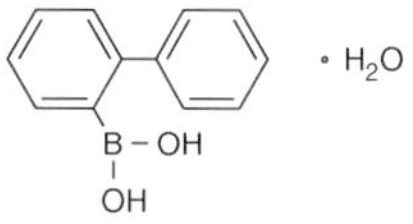
250 mg

S44,862-1
TRI-O-TOLYL BORATE
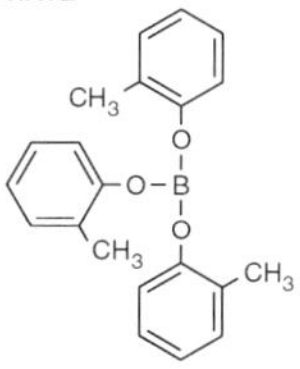
5 gm

S82,717-7
4-(TRIMETHYLSILYL)-BENZENEBORONIC ACID
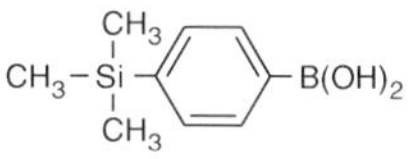
250 mg

S34,890-2
2,6-DI-TERT-BUTYLPHENYL DIBUTYL BORATE
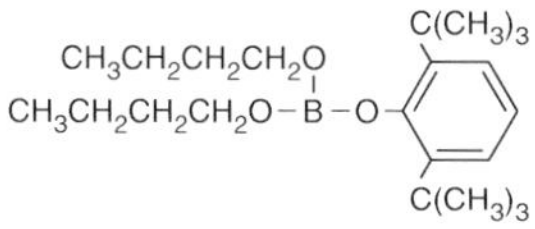
5 gm

S40,950-2
1-NAPHTHALENEBORONIC ACID
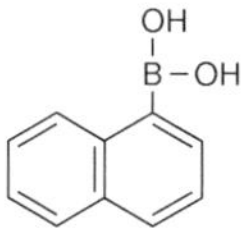
250 mg

S82,281-7
1-THIANTHRENEBORONIC ACID
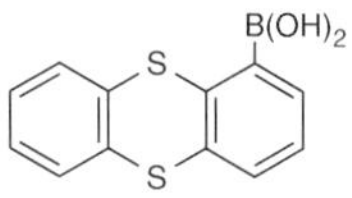
250 mg

S93,208-6
M-TOLUENEARSONIC ACID

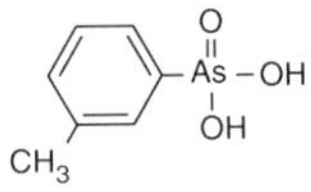

250 mg

S80,970-5
1,4-DIPHENYL-1,2,3,4,5,6-HEXAHYDRO-1,4-AZARSENINE

250 mg

S34,390-0
N-CHLOROACETYLARSANILIC ACID

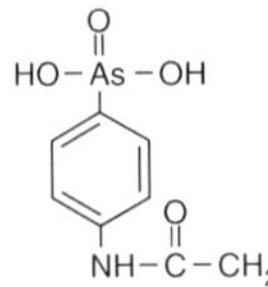

5 gm

S72,733-4
3-CHLOROBENZENEARSONIC ACID

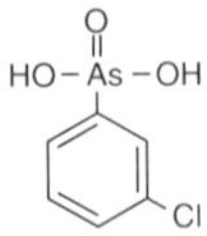

5 gm

S85,149-3
4-[3-(2,3-DICHLOROPHENYL)UREIDO]-BENZENEARSONIC ACID

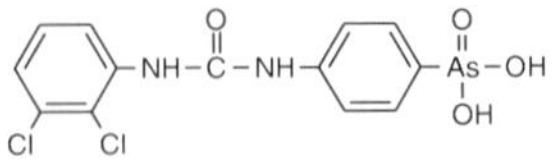

250 mg

S74,186-8
(2-BROMOPHENYL)DIPHENYLARSINE

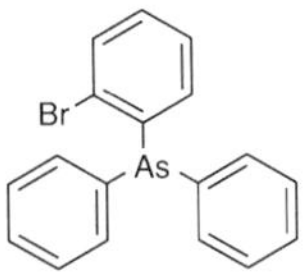

250 mg

S80,959-4
2-BROMOPHENYLARSINE OXIDE

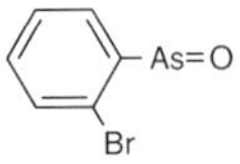

250 mg

S74,188-4
2-IODOBENZENEARSONIC ACID

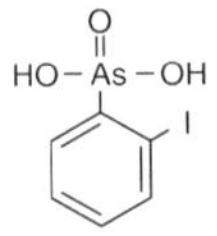

250 mg

S83,861-6
(4-BROMOPHENYL)DIMETHYLARSINE

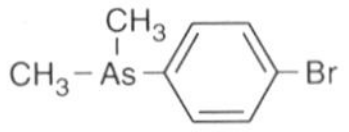

250 mg

S83,807-1
3-(DIMETHYLARSINO)PHENOL METHIODIDE

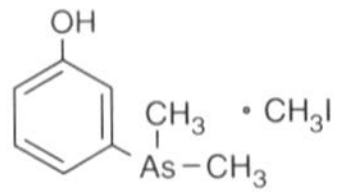

250 mg

S93,203-5
4-(2,3-DIHYDROXYPROPOXY)-BENZENEARSONIC ACID

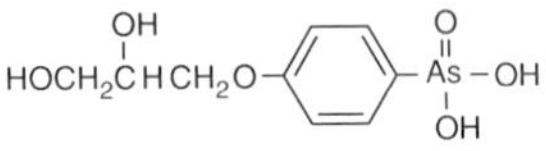

250 mg

S93,404-6
2,4-DIHYDROXYBENZENEARSONIC ACID

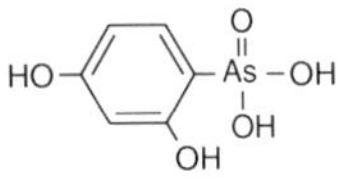

250 mg

S93,201-9
4-[2-(2-HYDROXYETHOXY)ETHOXY]-BENZENEARSONIC ACID

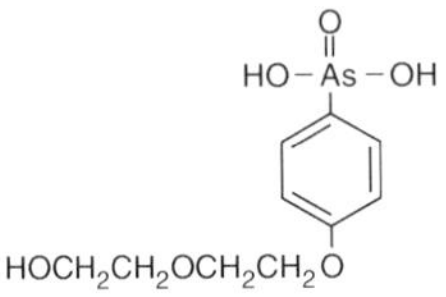

250 mg

S93,204-3
4-(2,3-DIHYDROXYPROPOXY)-3-NITROBENZENEARSONIC ACID, DISODIUM SALT

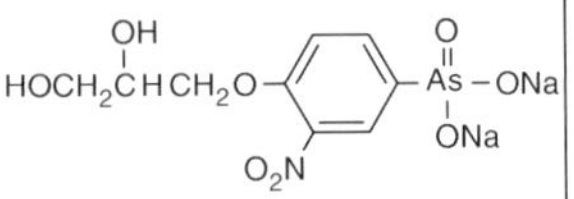

250 mg

S93,399-6
4-(2-HYDROXYPROPOXY)-3-NITROBENZENEARSONIC ACID

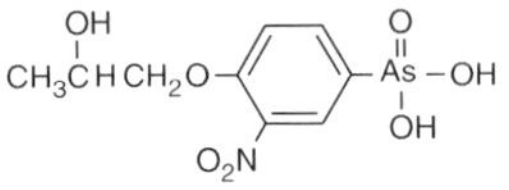

250 mg

S93,403-8
BENZYL N-[4-ARSONO-3-(2-HYDROXYETHOXY)PHENYL]CARBAMATE

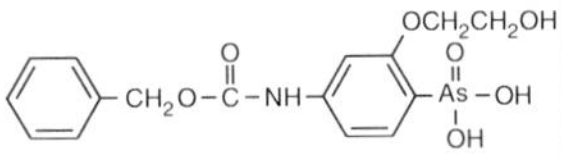

250 mg

S93,202-7
4-(2-HYDROXYPROPOXY)-BENZENEARSONIC ACID

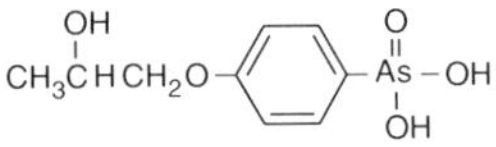

250 mg

S93,413-5
4-(2-HYDROXYETHYLTHIO)-BENZENEARSONIC ACID

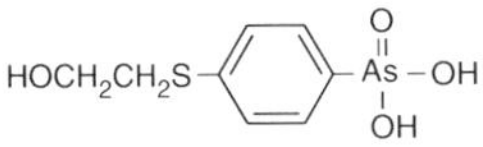

250 mg

S36,640-4
3-AMINO-4-HYDROXYBENZENEARSONIC ACID

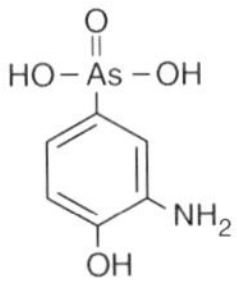

5 gm

S36,641-2
4-AMINO-2-HYDROXYBENZENEARSONIC ACID

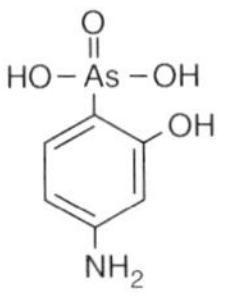

5 gm

S93,313-9
N-(2-HYDROXYACETYL)-M-ARSANILIC ACID

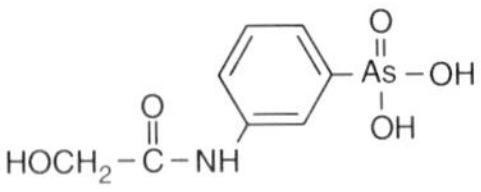

250 mg

S93,213-2
4-(BENZYLAMINO)-2-HYDROXYBENZENEARSONIC ACID

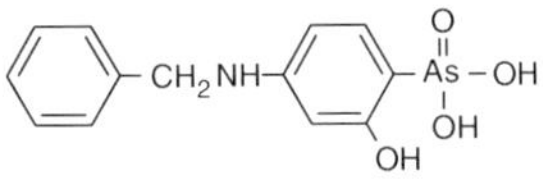

250 mg

S36,912-8
N-GLYCOLOYLARSANILIC ACID, SODIUM SALT

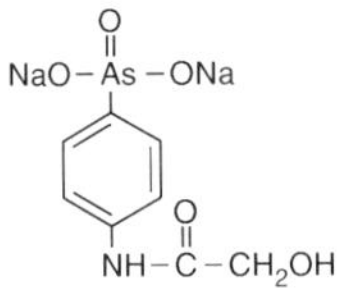

5 gm

S45,093-6
ACETARSONE

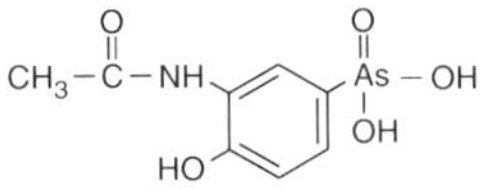

5 gm

S93,401-1
ETHYL N-(4-ARSONO-3-HYDROXYPHENYL)-CARBAMATE

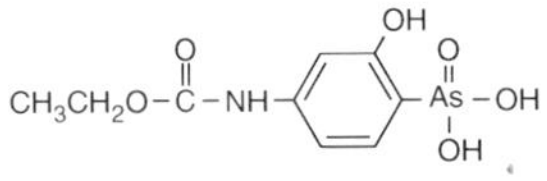

250 mg

S74,272-4
2-HYDROXYBENZENEARSONIC ACID

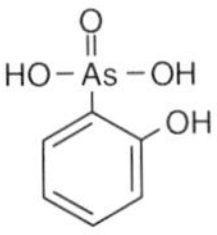

5 gm

S45,511-3
4-HYDROXYBENZENEARSONIC ACID

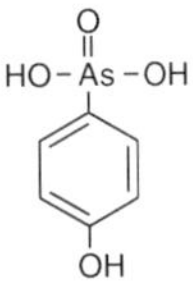

250 mg

S83,809-8
4-(DIMETHYLARSINO)PHENOL METHIODIDE

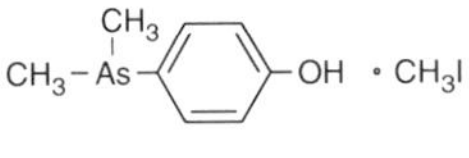

250 mg

S93,400-3
4,4'-(ETHYLENEDIOXY)BIS(3-AMINOBENZENEARSONIC ACID)

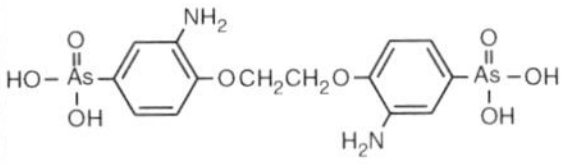

250 mg

S84,870-0
4-[3-(2,5-DIMETHOXYPHENYL)UREIDO]-PHENYLARSONIC ACID

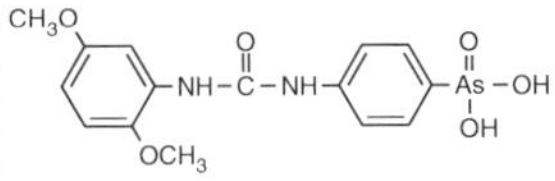

250 mg

S74,551-0
8-METHOXY-2,3,5,6-TETRAHYDRO-ARSENINO[3,2,1-IJ]-ARSINOLINE-1,7-DIONE

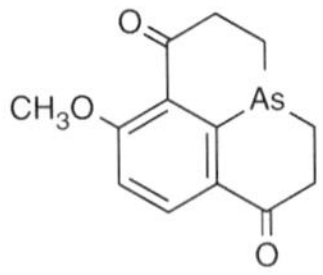

250 mg

S93,408-9
4-BUTOXY-3-NITROBENZENEARSONIC ACID

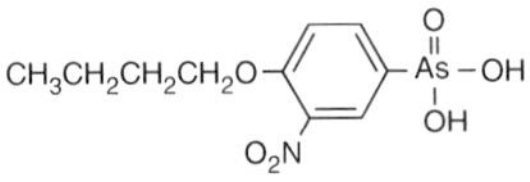

250 mg

S93,418-6
4-(ETHOXYCARBONYLOXY)-3-NITROBENZENEARSONIC ACID

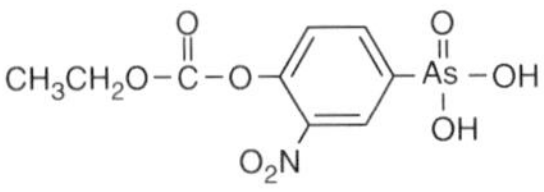

250 mg

S93,407-0
4-METHOXY-2-NITROBENZENEARSONIC ACID

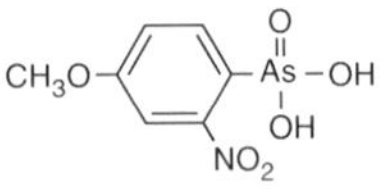

250 mg

S74,520-0
3,3'-(3-METHOXYPHENYLARSINO)-DIPROPIONIC ACID

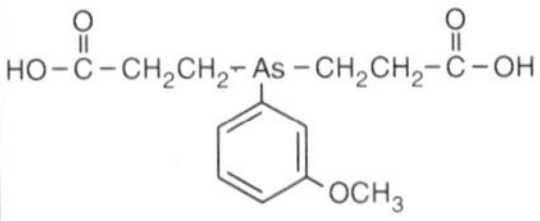

250 mg

S93,419-4
3-(METHOXYCARBONYLOXY)-BENZENEARSONIC ACID

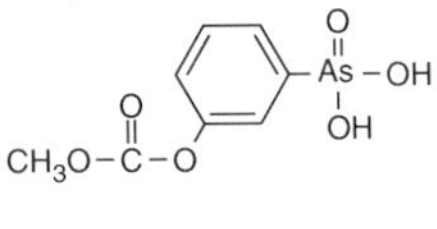

250 mg

S93,409-7
4-(ISOPROPOXYCARBONYLOXY)-BENZENEARSONIC ACID

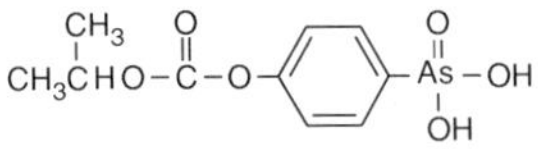

250 mg

S93,410-0
4-(4-TOLYLOXY)BENZENEARSONIC ACID

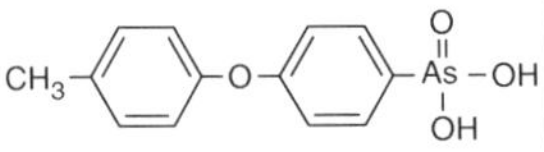

250 mg

S93,214-0
4-(2-FURAMIDO)BENZENEARSONIC ACID

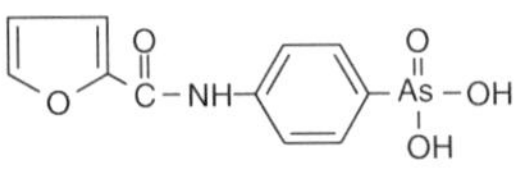

250 mg

S93,411-9
4-(4-NITROPHENYLTHIO)-BENZENEARSONIC ACID

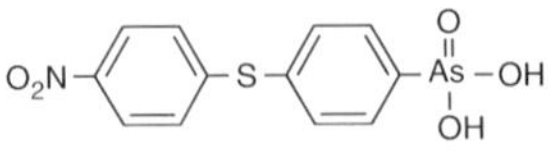

250 mg

S83,569-2
4-AMINO-3,5-DINITROBENZENEARSONIC ACID
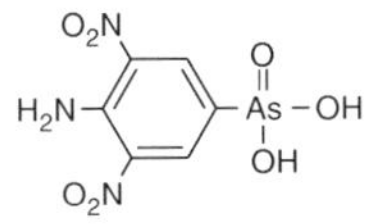
250 mg

S93,412-7
4-(3-AMINOPROPYLAMINO)-BENZENEARSONIC ACID
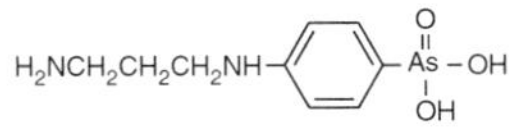
250 mg

S74,577-4
3,3'-(4-AMINOPHENYLARSINO)-DIPROPIONITRILE
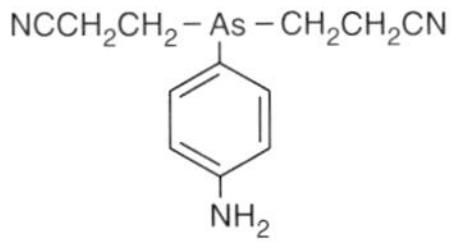
250 mg

S93,211-6
3,4-DIAMINOBENZENEARSONIC ACID
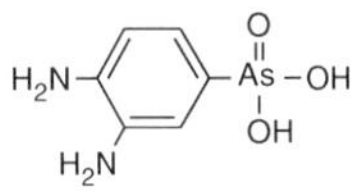
250 mg

S93,398-8
4-(4-ARSONO-2-NITROPHENYLAZO)-3-METHYL-5-PYRAZOLONE
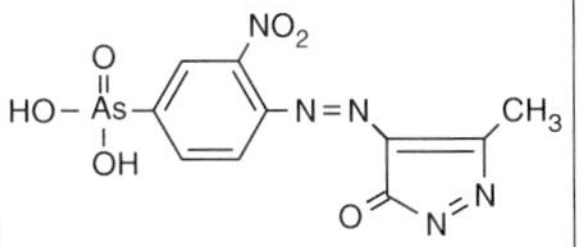
250 mg

S93,420-8
4-(DIPHENYLAMINO)-2-NITROBENZENEARSONIC ACID
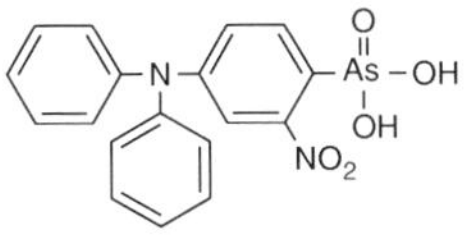
250 mg

S93,405-4
4-(ARSONOPHENYL)GLYCINE
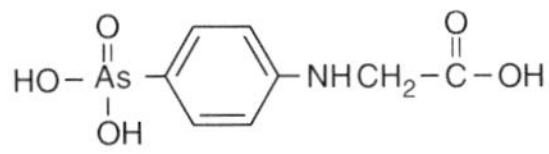
250 mg

S93,200-0
4-(ETHYLAMINO)BENZENEARSONIC ACID
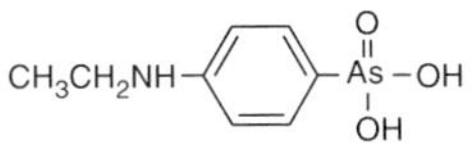
250 mg

S81,869-0
3-CARBAZOLEARSONIC ACID
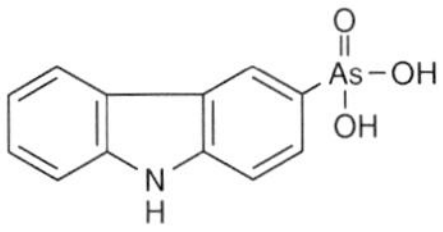
250 mg

S93,212-4
2-(DIPHENYLAMINO)-BENZENEARSONIC ACID
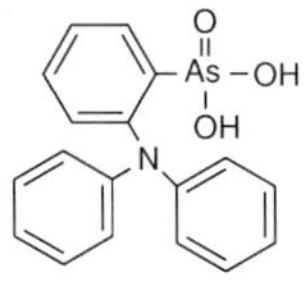
250 mg

S81,868-2
9-ETHYL-3-CARBAZOLEARSONIC ACID
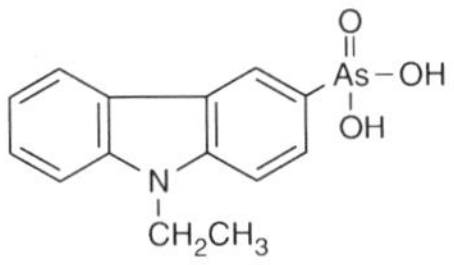
250 mg

S93,209-4
4-(SEMICARBAZONOMETHYL)-BENZENEARSONIC ACID
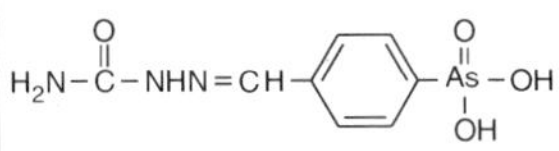
250 mg

S43,554-6
4-NITROBENZENEARSONIC ACID
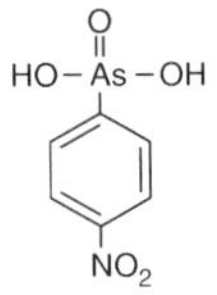
5 gm

S93,207-8
(3-NITROPHENYL)METHANEARSONIC ACID
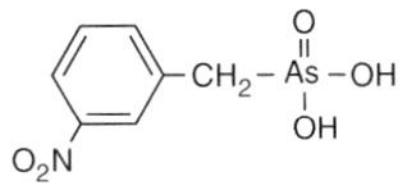
250 mg

S93,417-8
4-(BUTOXYCARBONYLOXY)-3-NITROBENZENEARSONIC ACID
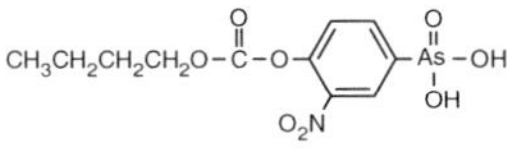
250 mg

S65,877-4
2-NITROBENZENEARSONIC ACID
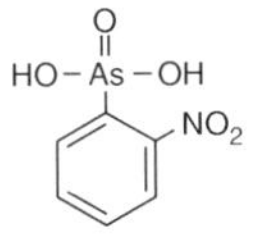
5 gm

S34,443-5
4-CHLORO-3-NITROPHENYLARSONIC ACID
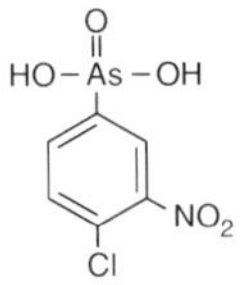
5 gm

S93,205-1
4-METHYL-3-NITROBENZENEARSONIC ACID
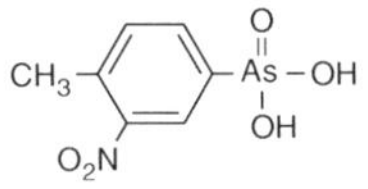
250 mg

S93,215-9
4-NITRO-1-NAPHTHALENEARSONIC ACID
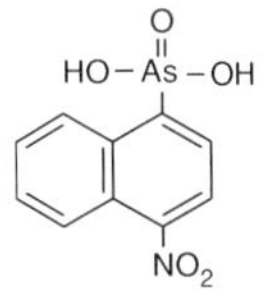
250 mg

S72,712-1
4-(IODOMETHYLARSINO)BENZOIC ACID
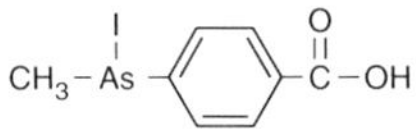
5 gm

S83,191-3
4-(DIMETHYLARSINO)BENZOIC ACID
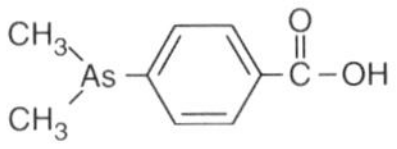
250 mg

S74,625-8
10-CHLORO-5,10-DIHYDRO-1-PHENARSAZINECARBOXYLIC ACID
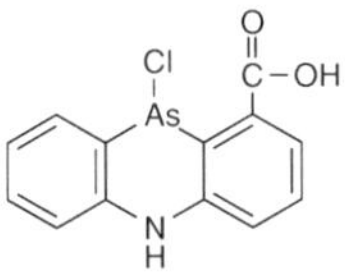
5 gm

S93,216-7
3-ARSONOPHTHALIC ANHYDRIDE
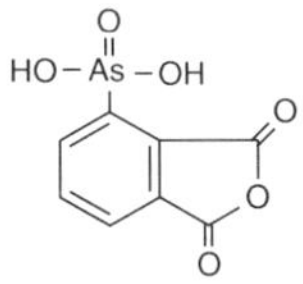
250 mg

S64,557-5
4,4'-ARSENOBENZAMIDE
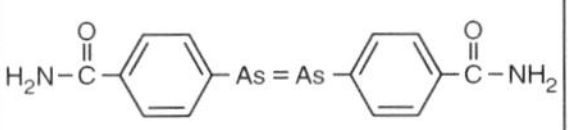
5 gm

S81,871-2
ARSONOACETIC ACID
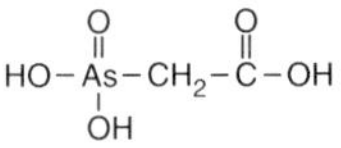
250 mg

S74,179-5
2,2'-(PHENYLARSINO)DIACETIC ACID
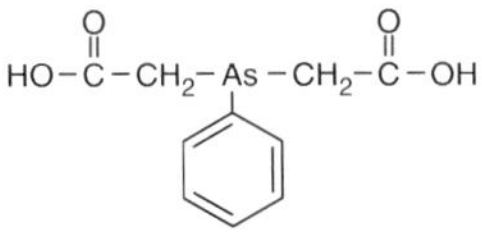
250 mg

S41,852-8
ARSONOACETIC ACID, DISODIUM SALT MONOHYDRATE
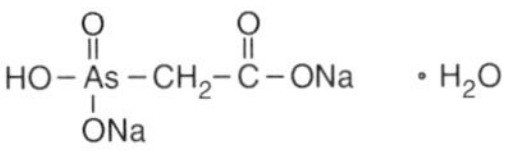
5 gm

S74,540-5
[CHLORO(2-NAPHTHYL)-ARSINO]ACETIC ACID
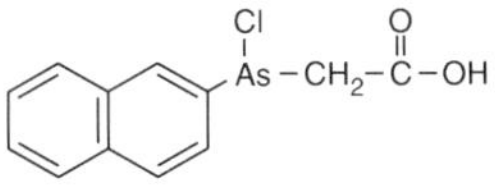
250 mg

S83,812-8
3-(DIMETHYLARSINO)PHENYL N-METHYLCARBAMATE METHIODIDE
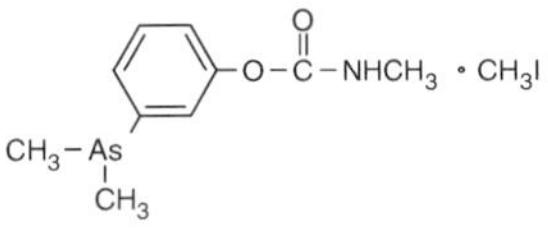
250 mg

S57,048-6
N-CARBAMOYLARSANILIC ACID
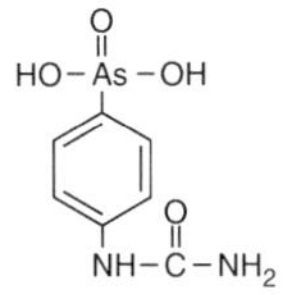
5 gm

S72,717-2
(3-CYANOPROPYL)DIMETHYL(2-PHENYLETHYL)ARSONIUM BROMIDE
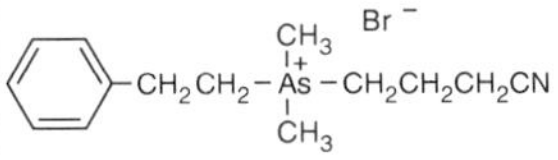
5 gm

S72,735-0
2-THIOCYANATOBENZENEARSONIC ACID
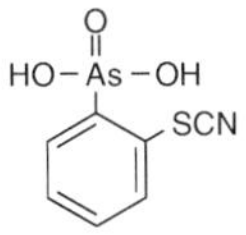
5 gm

S93,415-1
4-TOLUENEARSONIC ACID
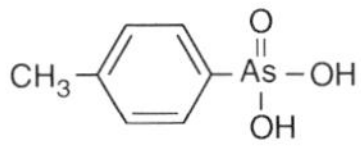
250 mg

S72,741-5
2-HYDROXY-2-PHENYL-1,2,3,4-TETRAHYDROISOARSINOLINIUM NITRATE
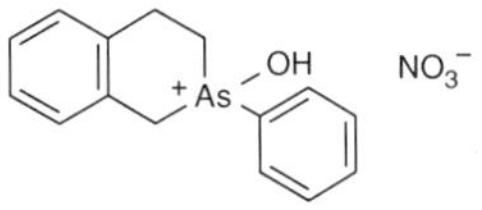
5 gm

S83,188-3
TRIS(PHENYLTHIO)ARSINE
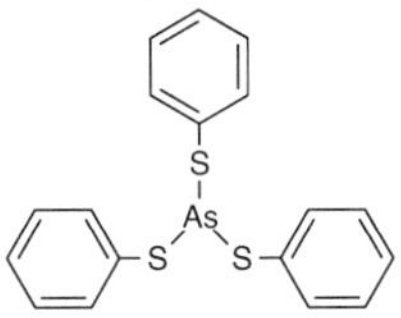
250 mg

S45,925-9
(METHYL)TRIPHENYLARSONIUM IODIDE
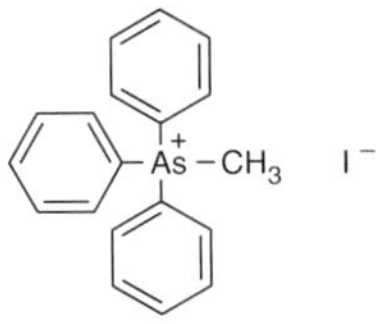
250 mg

S54,344-6
(2-PROPYNYL)-TRIPHENYLARSONIUM BROMIDE
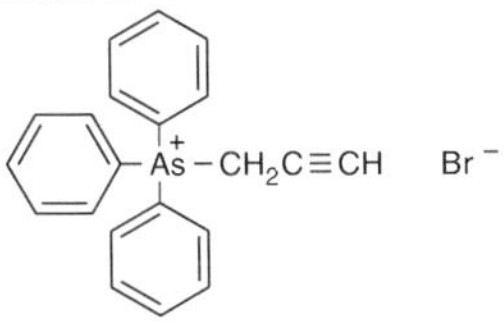
250 mg

S53,643-1
TETRAPHENYLARSONIUM BROMIDE
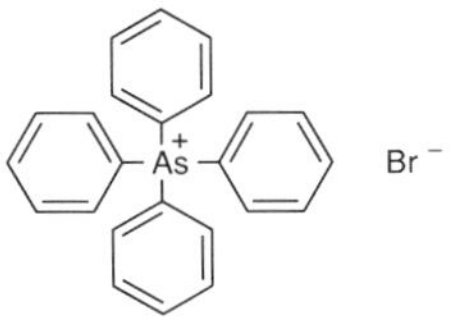
5 gm

S44,750-1
METHYLENEBIS(DIPHENYLARSINE)
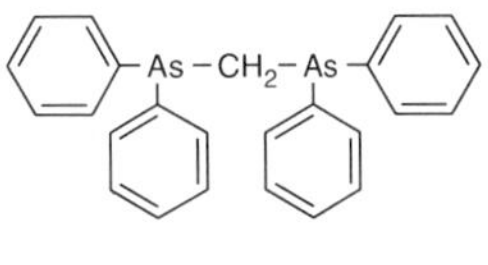
5 gm

S58,719-2
BENZYLTRIPHENYLARSONIUM BROMIDE
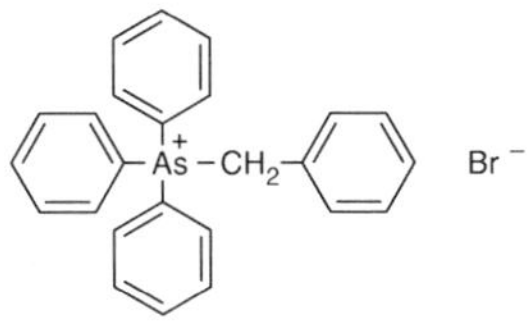
250 mg

S72,720-2
ETHYLENEBIS-(BUTYLMETHYLPHENYLARSONIUM)-DIPICRATE
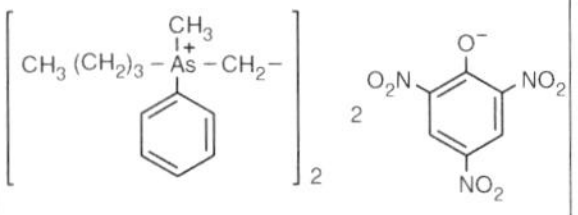
5 gm

S72,715-6
(4-BIPHENYLYL)(4-CHLOROPHENYL)-PHENYLARSINE OXIDE
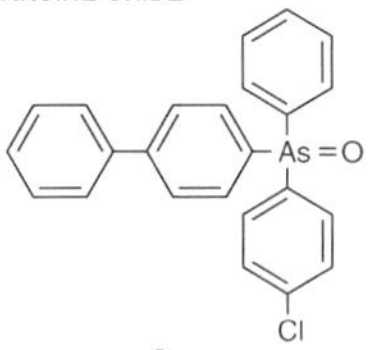
5 gm

S80,979-9
2,3-DIHYDRO-2-HYDROXY-2-PHENYL-1H-ISOARSINDOLIUM NITRATE
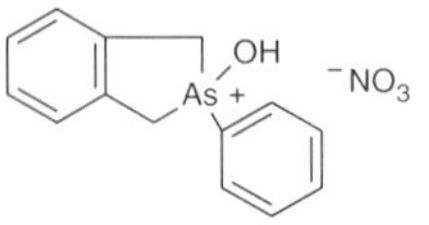
250 mg

S84,262-1
5-CHLORO-2,3-DIHYDRO-2-HYDROXY-2-PHENYL-1H-ISOARSINDOLIUM NITRATE
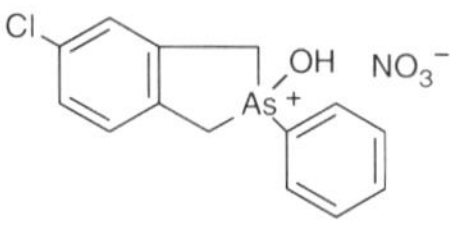
250 mg

S72,723-7
6,7-DIHYDRO-6-PHENYL-5H-DIBENZ(C,E)-ARSEPIN
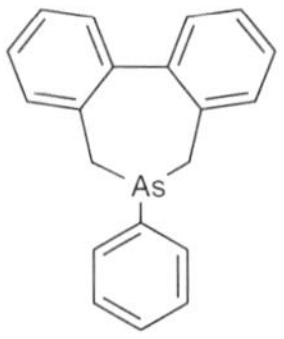
5 gm

S74,537-5
10-BENZYL-5,10-DIHYDRO-10-PROPYL-1-PHENARSAZINIUM BROMIDE
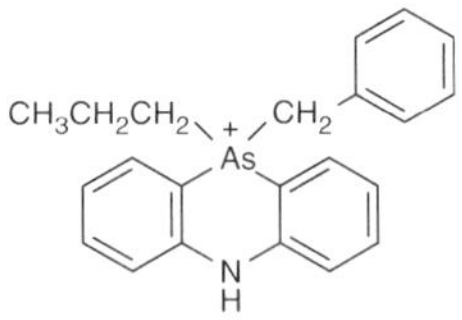
250 mg

S44,697-1
1,2-PHENYLENEDIARSONIC ACID
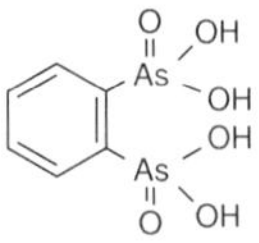
5 gm

S93,406-2
2-(2-METHYLALLYL)BENZENZE-ARSONIC ACID
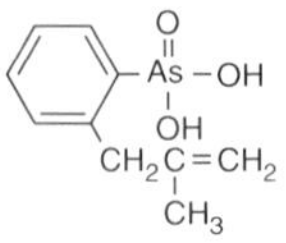
250 mg

S74,176-0
5,10-DIHYDRO-5,7,10-TRIMETHYL-5,10-(O-XYLYLENE)ARSANTHRENE-5,10-DIIUM DIIODODE
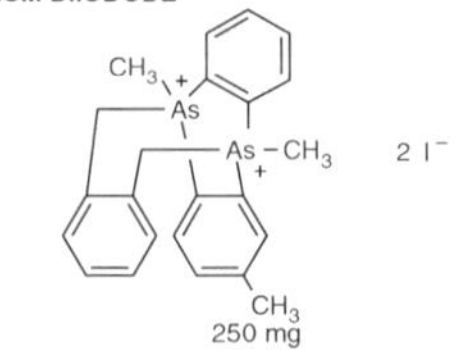
250 mg

S40,706-2
4-CHLOROBENZENEARSONIC ACID
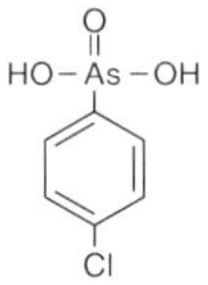
5 gm

S93,210-8
4-[1-(HYDROXYIMINO)ETHYL]-BENZENEARSONIC ACID
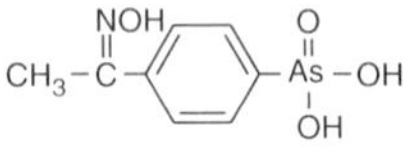
250 mg

S74,183-3
4,4'-BIS(DICHLOROARSINO)AZOBENZENE
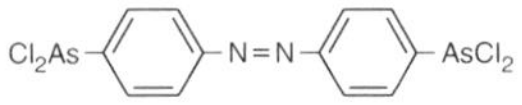
250 mg

S81,999-9
TRIS(2-PYRIDYL)ARSINE
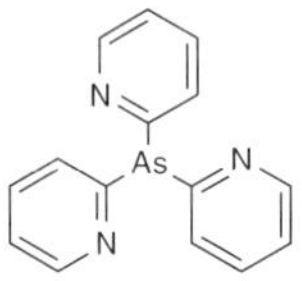
250 mg

S72,700-8
TRIS(1-NAPHTHYL)ARSINE
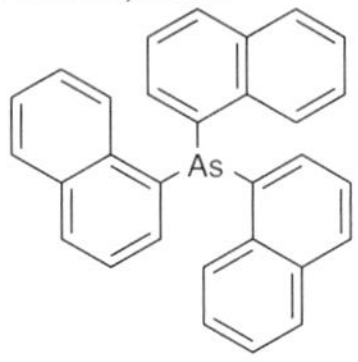
5 gm

S81,997-2
DICHLORO(8-QUINOLYL)ARSINE HYDROCHLORIDE
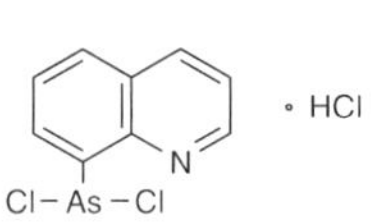
250 mg

S81,870-4
8-METHYL-5-QUINOLINEARSONIC ACID
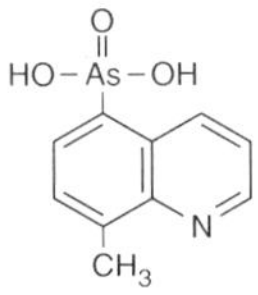
250 mg

S81,998-0
DICHLORO(8-METHYL-5-QUINOLYL)ARSINE HYDROCHLORIDE
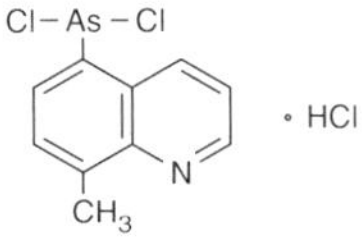
250 mg

S81,867-4
5,5'-ARSENOBIS(8-METHYLQUINOLINE)-DIHYDROCHLORIDE
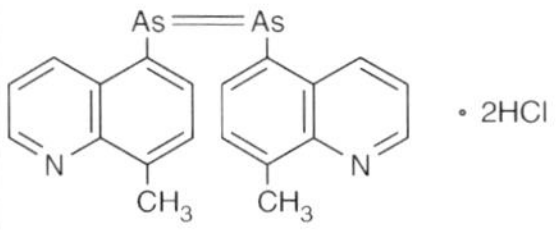
250 mg

S74,576-6
2-IODOISOARSINDOLINE
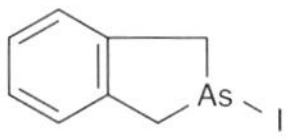
250 mg

S54,603-8
2,2'-OXYBIS(1,3,2-BENZODIOXARSOLE)
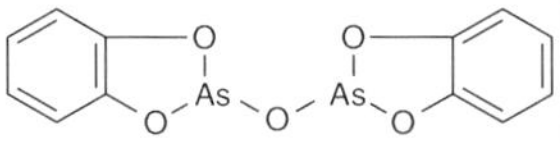
250 mg

S74,539-1
10-CHLORO-5,10-DIHYDRO-2-METHYLPHENARSAZINE
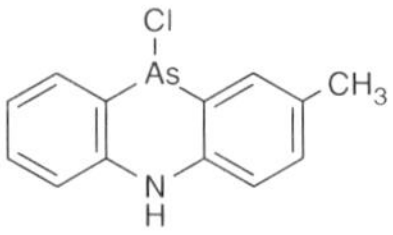
250 mg

S74,271-6
5,10-DIHYDRO-10-ETHYLPHENARSAZINE
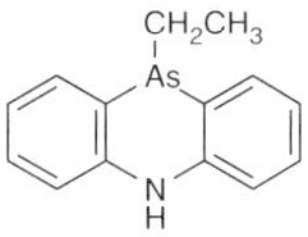
5 gm

S74,243-0
7-CHLORO-7,12-DIHYDROBENZO(C)-PHENARSAZINE
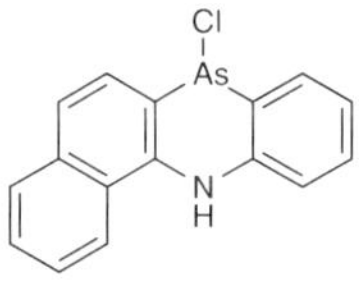
5 gm

S74,237-6
12-CHLORO-7,12-DIHYDRO-10-METHYLBENZO(A)PHENARSAZINE
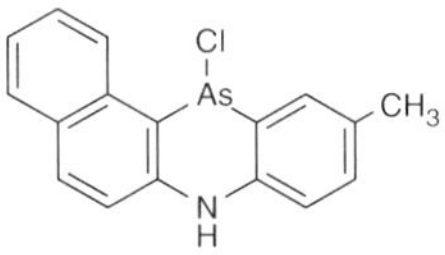
5 gm

S72,734-2
5,6,7,8-TETRAHYDRO-5,5,8,8-TETRAMETHYLDIBENZO(E,G)(1,4)-DIARSOCINIUM DIBROMIDE
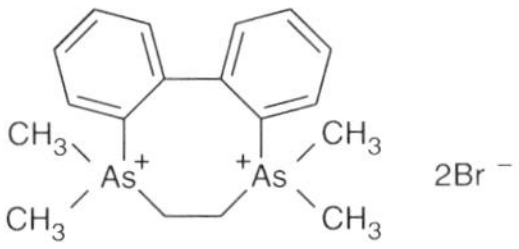
5 gm

S74,195-7
6,7,8,9-TETRAHYDRO-5,5,9,9-TETRAMETHYL-5H-DIBENZO(F,H)(1,5)-DIARSONINIUM DIIODIDE
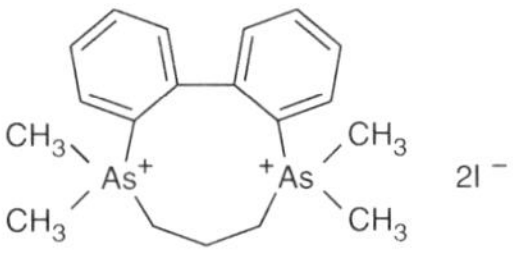
250 mg

S74,196-5
5,6,15,16-TETRAHYDRO-6,6,15,15-TETRAMETHYL-TRIBENZO(B,D,H)(1,6)-DIARSECINIUM DIBROMIDE
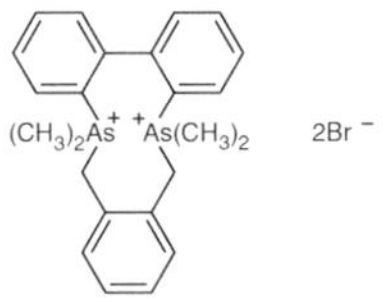
250 mg

S74,169-8
2-CHLORO-1,3,2-DITHIASTIBOLANE
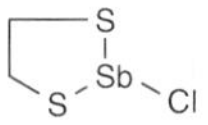
250 mg

S44,959-8
PENTACHLOROALLYL HEXACHLOROANTIMONATE
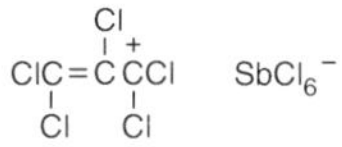
5 gm

S81,672-8
DIBROMOTRIBUTYLANTIMONY
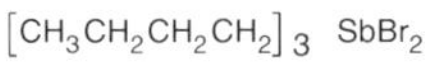
250 mg

S81,831-3
TRIS[3-(TRIFLUOROMETHYL)PHENYL]-ANTIMONY
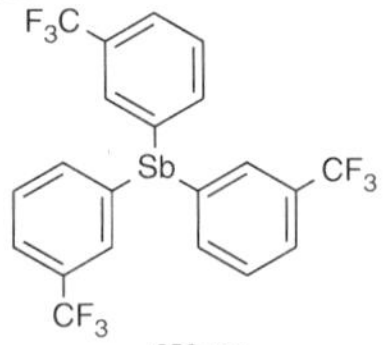
250 mg

S81,832-1
DIPHENYL[4-(TRIFLUOROMETHYL)PHENYL]-ANTIMONY
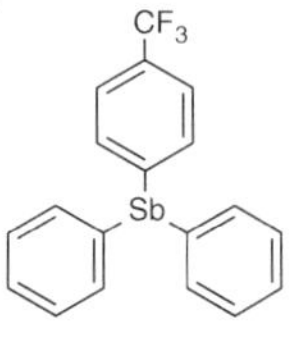
250 mg

S81,762-7
TRIS(4-BROMOPHENYL)ANTIMONY
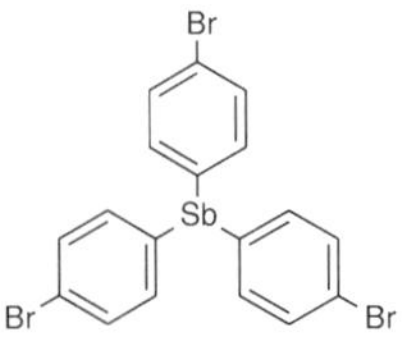
250 mg

S81,842-9
DICHLOROTRIS(4-BROMOPHENYL)-ANTIMONY
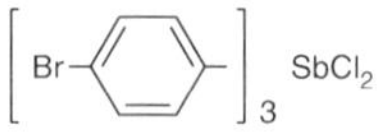
250 mg

S81,822-4
TRIS[4-(DIETHYLAMINO)PHENYL]-ANTIMONY
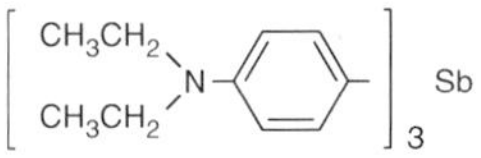
250 mg

S81,306-0
TRIS[4-(DIMETHYLAMINO)PHENYL]-ANTIMONY
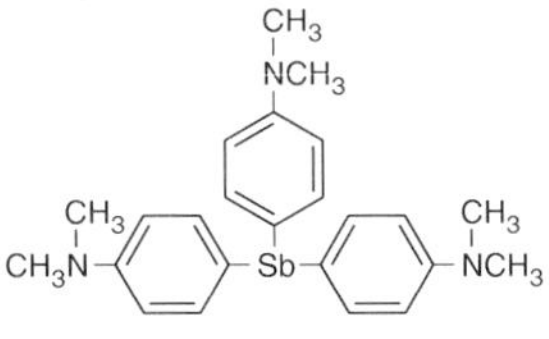
250 mg

S88,738-2
(TERT-BUTOXYIMINO)DIMETHYLAMMONIUM HEXACHLOROANTIMONATE
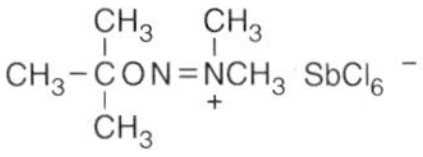
250 mg

S88,740-4
(TERT-BUTOXYIMINO)(2,2-DIMETHYLPROPYL)METHYLAMMONIUM HEXACHLOROANTIMONATE
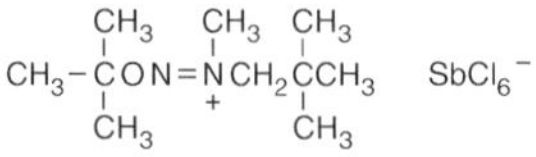
250 mg

S44,777-3
TRIPHENYLANTIMONY DIACETATE
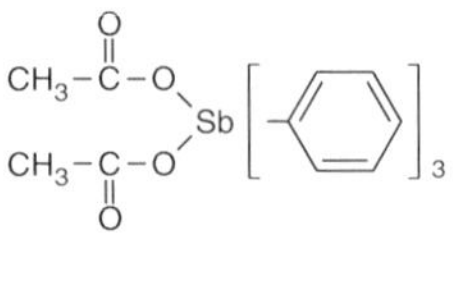
5 gm

S81,821-6
TRIS(4-DIBENZOFURYL)ANTIMONY
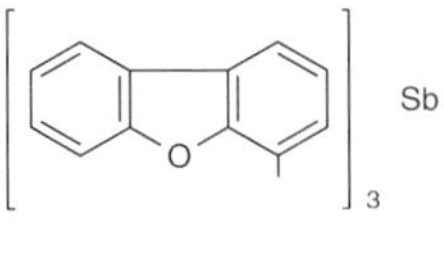
250 mg

S81,834-8
(2,5-DIMETHYL-3-PYRROLYL)-DIPHENYLANTIMONY
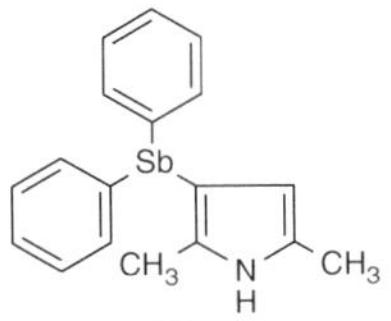
250 mg

S72,714-8
PHENYLSTIBINE OXIDE
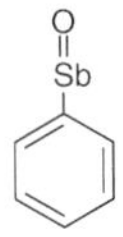
5 gm

S44,957-1
DICHLOROPHENYLCARBENIUM HEXACHLOROANTIMONATE
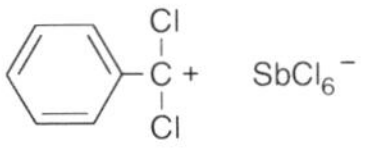
5 gm

S61,705-9
TRIPHENYLANTIMONY SULFIDE
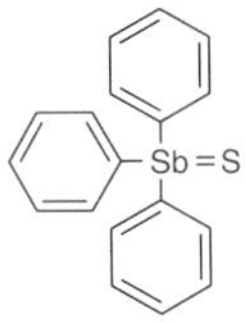
5 gm

S81,845-3
CHLOROHYDROXYTRIPHENYLANTIMONY
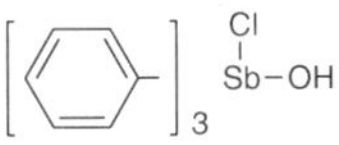
250 mg

S94,058-5
TRIPHENYLANTIMONY DIBROMIDE
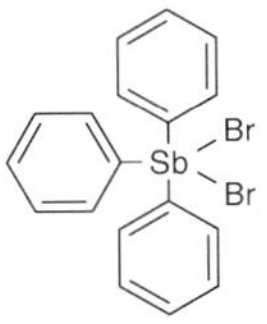
250 mg

S81,764-3
DICHLOROTRIBENZYLANTIMONY
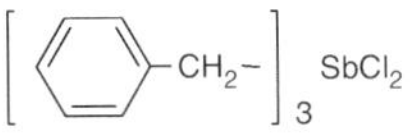
250 mg

S81,767-8
TRIS(O-TOLYL)ANTIMONY
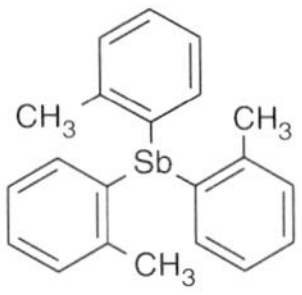
250 mg

S81,765-1
TRIS(P-TOLYL)ANTIMONY
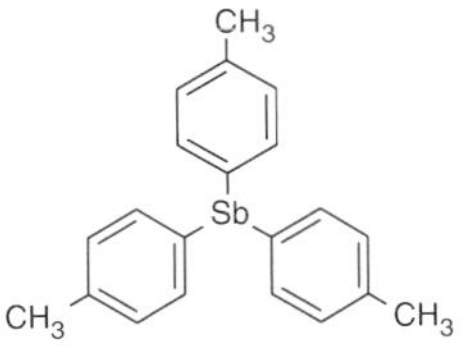
250 mg

S81,856-9
TRIS(2-PYRIDYL)ANTIMONY
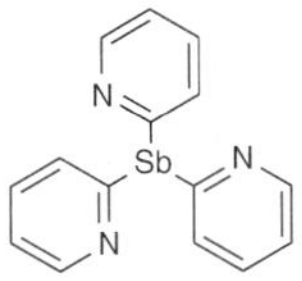
250 mg

S81,820-8
TRIS(1-NAPHTHYL)ANTIMONY
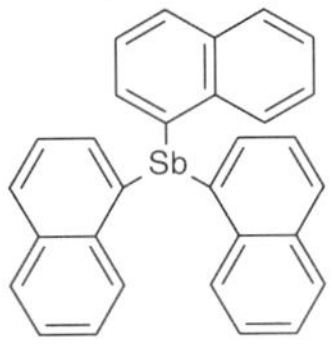
250 mg

S57,253-5
IODOTRIMETHYLTIN
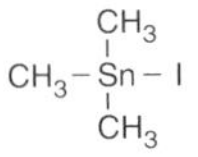
5 gm

S50,359-2
DIBUTYLDIFLUOROTIN
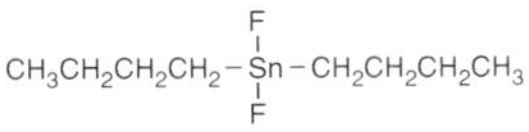
5 gm

S39,573-0
DIISOPENTYLOXOTIN
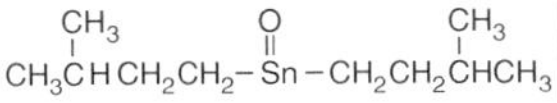
250 mg

S39,585-4
DIPENTYLOXOTIN
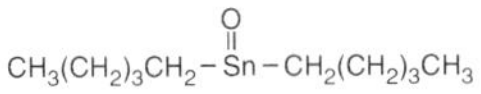
250 mg

S54,723-9
DIBUTYLDIVINYLTIN
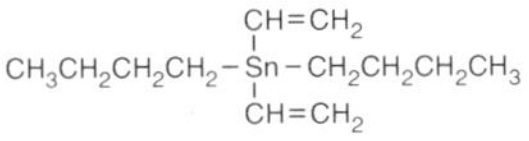
5 gm

S39,562-5
DIETHYLDIISOBUTYLTIN
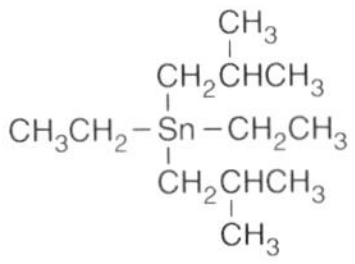
250 mg

S55,207-0
TETRAISOPROPYLTIN
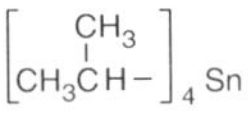
250 mg

S55,210-0
TETRAPROPYLTIN
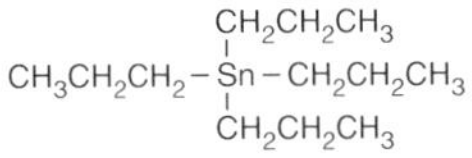
250 mg

S39,574-9
DIHEXYLOXOTIN
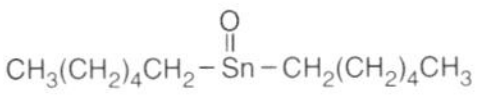
250 mg

S83,206-5
TETRAKIS(TRIMETHYLSILYL)TIN
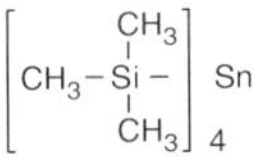
250 mg

S82,949-8
TETRAISOBUTYLTIN
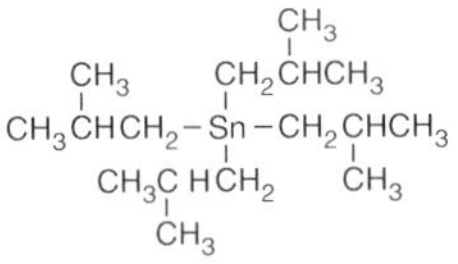
250 mg

S82,950-1
TETRA(SEC-BUTYL)TIN
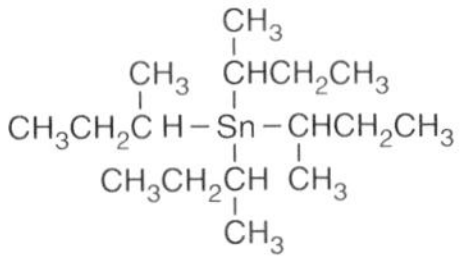
250 mg

S51,261-3
DIOCTYLOXOTIN
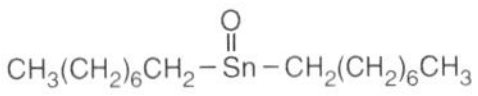
5 gm

S34,246-7
BIS(TRIPROPYLTIN) OXIDE
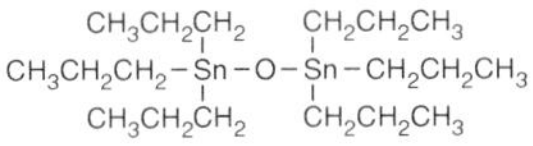
5 gm

S74,189-2
TETRACHLOROBIS(TRIPROPYL-PHOSPHINE)TIN(IV)
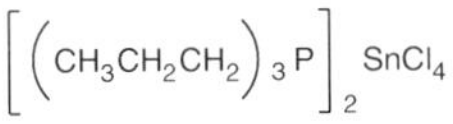
250 mg

S57,116-4
DICHLORODIDODECYLTIN
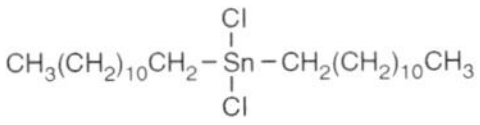
5 gm

S81,941-7
TETRAKIS(M-TOLYL)TIN
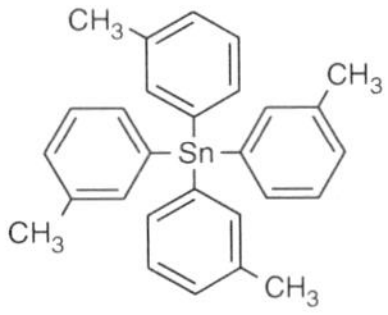
250 mg

S39,561-7
TETRAHEPTYLTIN
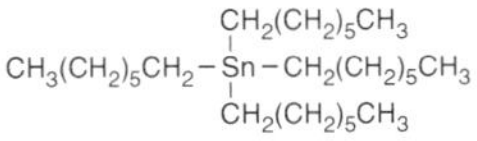
250 mg

S35,594-1
TETRAOCTYLTIN
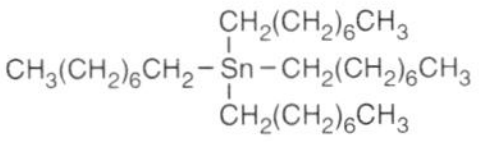
5 gm

S54,725-5
BIS(LAUROYLOXY)DIOCTYLTIN
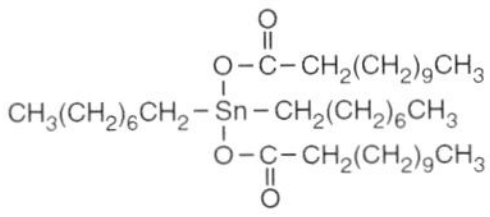
5 gm

S83,521-8
CHLOROTRITETRADECYLTIN
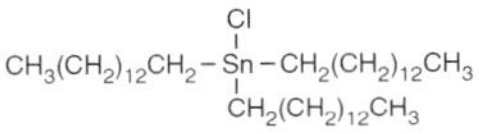
250 mg

S35,583-6
TETRALAURYLTIN
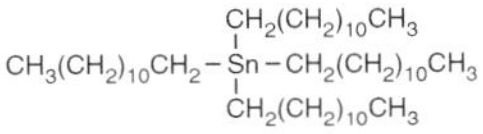
5 gm

S83,524-2
TETRAOCTADECYLTIN
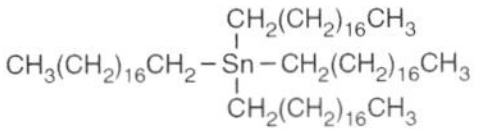
250 mg

S86,693-8
(IODOMETHYL)TRIBUTYLTIN
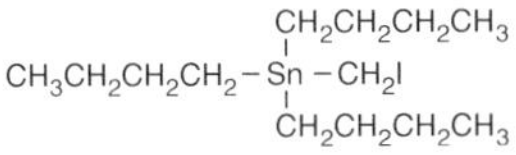
250 mg

S35,639-5
TRIBUTYLTIN CHLOROACETATE
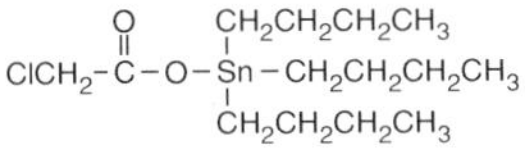
5 gm

S86,715-2
4,5-DIETHYL-6-(IODOMETHYL)-2-(TRIBUTYLSTANNYL)-1,3-DIOXANE
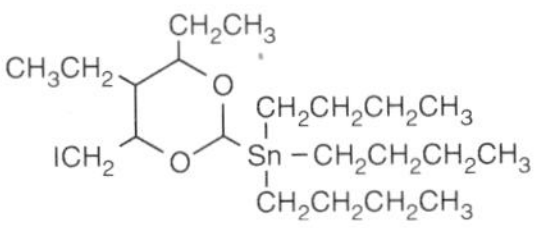
250 mg

S37,087-8
TRIBUTYLTIN TRICHLOROACETATE
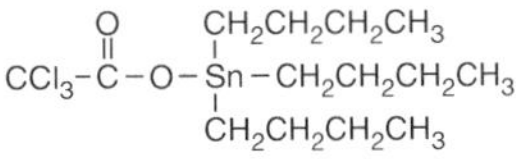
5 gm

S82,427-5
(PENTAFLUOROPHENYL)TRIMETHYLTIN
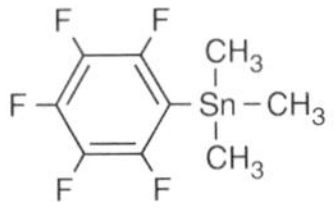
250 mg

S82,978-1
(PENTACHLOROPHENYL)TRIMETHYLTIN
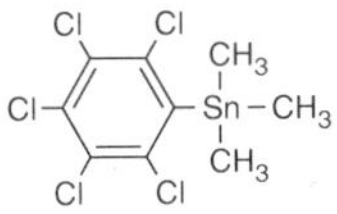
250 mg

S83,121-2
TETRAKIS(PENTAFLUOROPHENYL)TIN
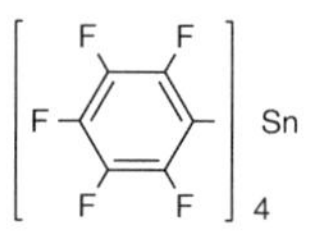
250 mg

S84,027-0
(4-METHOXY-2,3,5,6-TETRAFLUOROPHENYL)TRIPHENYLTIN
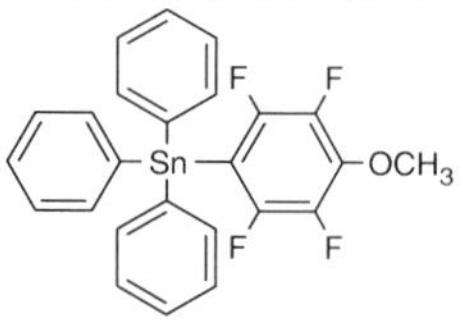
250 mg

S84,039-4
4-BROMO-4'-(DIMETHYLAMINO)-2'-(TRIPHENYLSTANNYL)AZOBENZENE
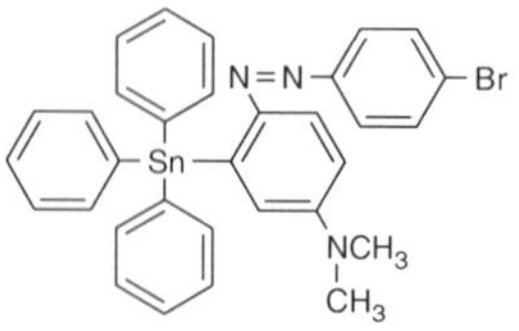
250 mg

S82,567-0
(2,3,4,5-TETRACHLOROPHENYL)-TRIPHENYLTIN
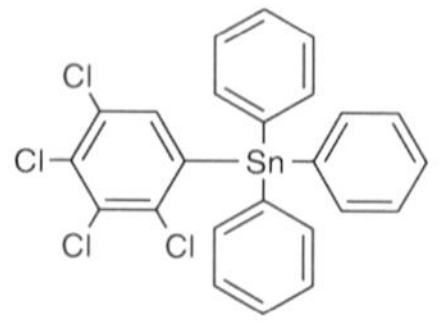
250 mg

S81,768-6
PHENYLTRIS(2,3,4,5,6-PENTAFLUOROPHENYL)TIN

250 mg

S83,210-3
BIS(PENTACHLOROPHENYL)DIPHENYLTIN
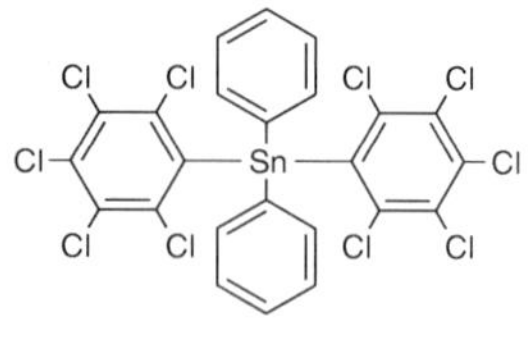
250 mg

S82,750-9
[3-(2-CHLOROPHENYL)PROPYL]-TRIPHENYLTIN
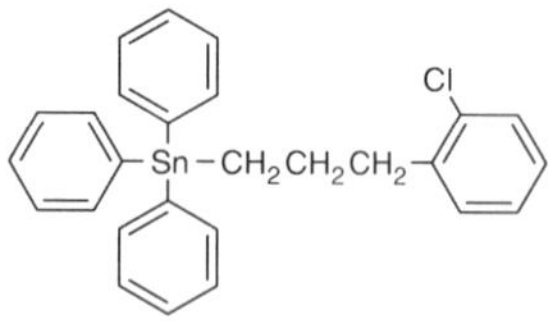
250 mg

S81,777-5
(4-CHLOROPHENYL)TRIPHENYLTIN
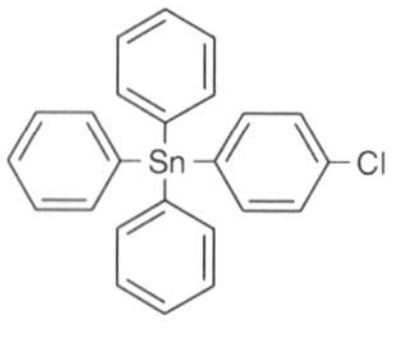
250 mg

S81,771-6
(4-BROMOPHENYL)TRIPHENYLTIN
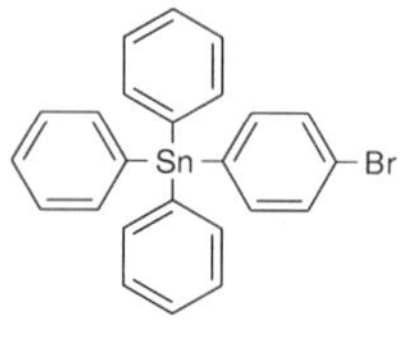
250 mg

S82,570-0
(2,6-DICHLOROPHENYL)TRIPHENYLTIN
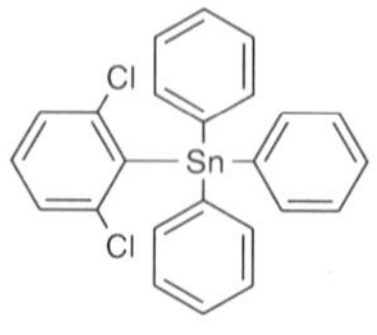
250 mg

S83,693-1
TETRAKIS(4-CHLOROPHENYL)TIN
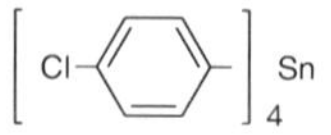
250 mg

S81,509-8
2,2'-(DIPHENYLSTANNYLENE)DIPHENOL
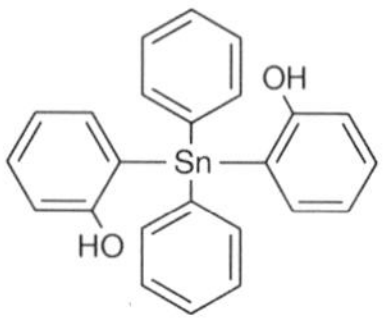
250 mg

S81,507-1
2-(TRIPHENYLSTANNYL)BENZYL ALCOHOL
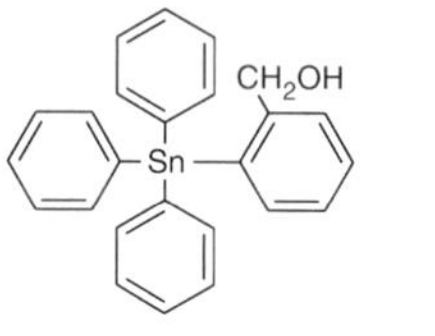
250 mg

S81,506-3
4-(TRIPHENYLSTANNYL)BENZYL ALCOHOL
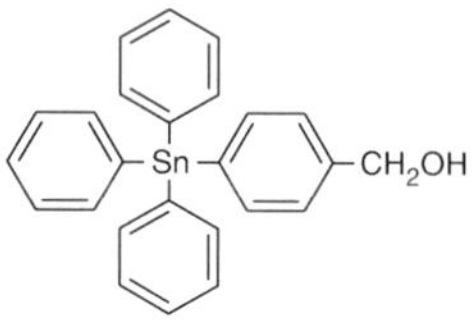
250 mg

S36,518-1
TRIBUTYLTIN SALICYLATE
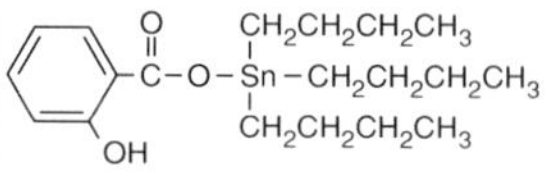
5 gm

S81,512-8
2-(TRIPHENYLSTANNYL)ETHANOL
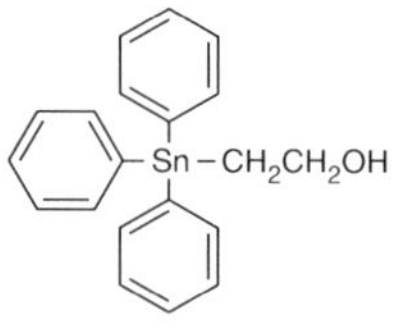
250 mg

S82,479-8
3-(TRIPHENYLSTANNYL)PHENOL
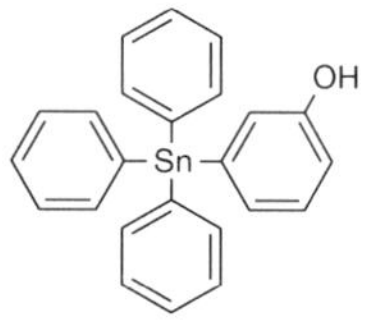
250 mg

S81,510-1
2-(TRIPHENYLSTANNYL)PHENOL
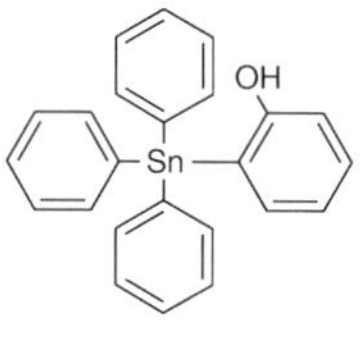
250 mg

S81,940-9
6-(TRIPHENYLSTANNYL)-2-NAPHTHOL
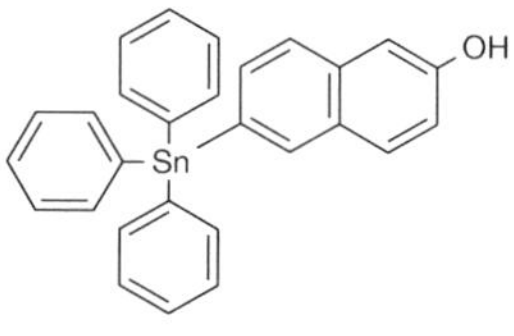
250 mg

S82,483-6
TETRAKIS(3-HYDROXYPHENYL)TIN
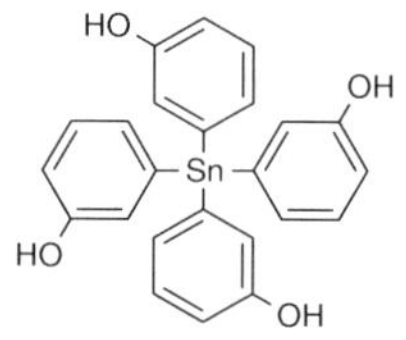
250 mg

S82,480-1
3,3'-(DIPHENYLSTANNYLENE)DIPHENOL
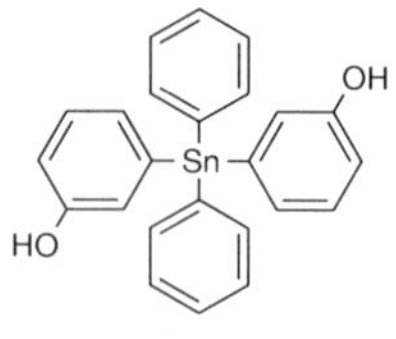
250 mg

S82,482-8
3,3',3''-(PHENYLSTANNYLIDYNE)-TRIPHENOL
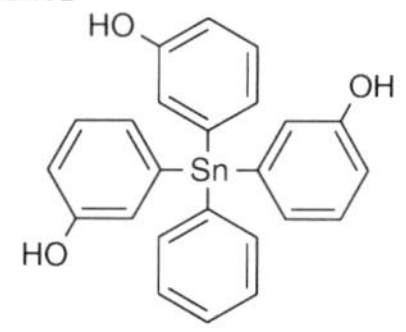
250 mg

S86,714-4
BENZYL (TRIBUTYLSTANNYL)METHYL ETHER
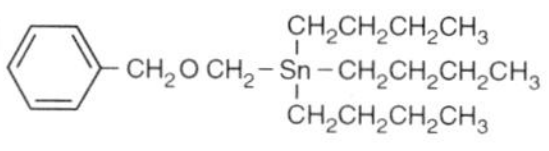
250 mg

S81,501-2
[2-(METHOXYMETHYL)PHENYL]-TRIPHENYLTIN
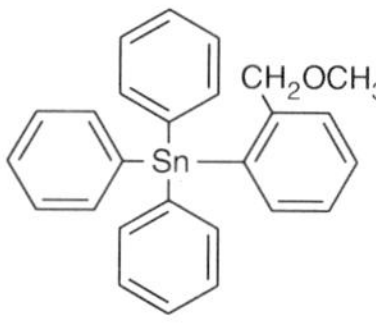
250 mg

S87,585-6
(DIETHOXYMETHYL)TRIBUTYLTIN
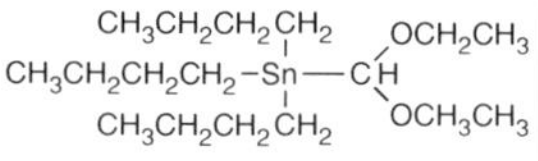
250 mg

S82,477-1
ETHYL 2-[3-(TRIPHENYLSTANNYL)-PHENOXY]ACETATE
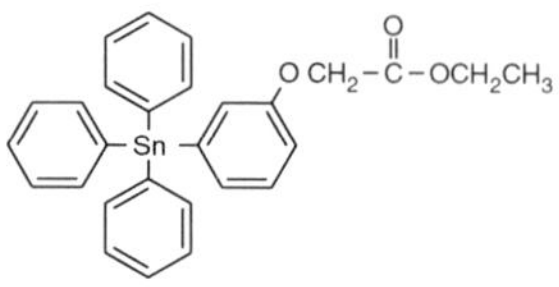
250 mg

S86,712-8
(2-METHOXY-2-CYCLOHEXEN-1-YL)-TRIBUTYLTIN
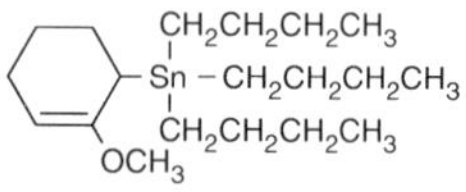
250 mg

S88,397-2
(3-METHOXYBENZYL)TRIPHENYLTIN
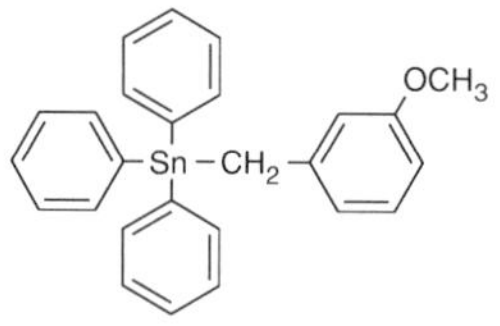
250 mg

S81,933-6
(4-METHOXYPHENYL)TRIPHENYLTIN
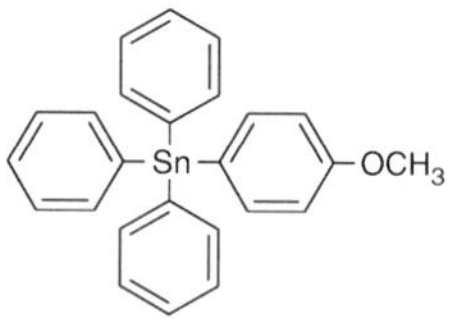
250 mg

S81,608-6
4-(TRIPHENYLSTANNYL)ANILINE
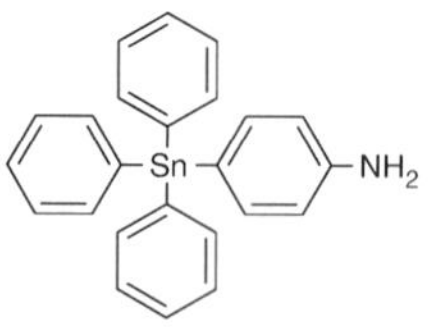
250 mg

S81,502-0
N,N-DIMETHYL-2-(4-NITROPHENYLAZO)-4-(TRIPHENYLSTANNYL)ANILINE
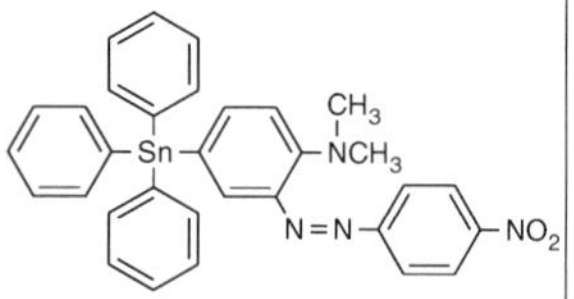
250 mg

S34,245-9
BIS(TRIPHENYLTIN) ETHYLENEBIS-(DITHIOCARBAMATE)
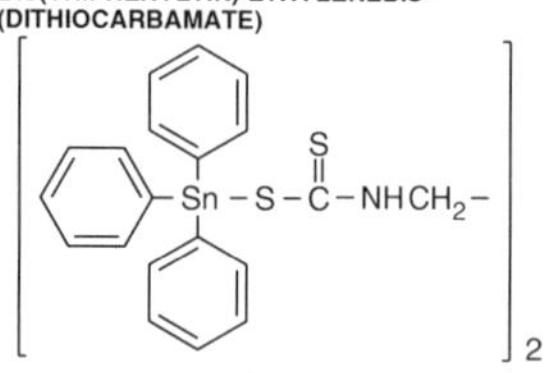
5 gm

S81,944-1
4-(DIMETHYLAMINO)-4'-NITRO-2-(TRIPHENYLSTANNYL)AZOBENZENE
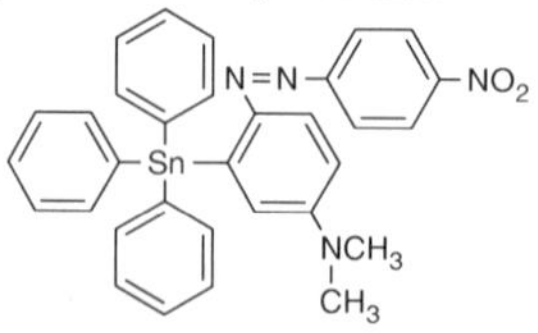
250 mg

S81,946-8
4-([4-(DIMETHYLAMINO)-2-(TRIPHENYLSTANNYL)PHENYL]-IMINOMETHYL)BENZOIC ACID
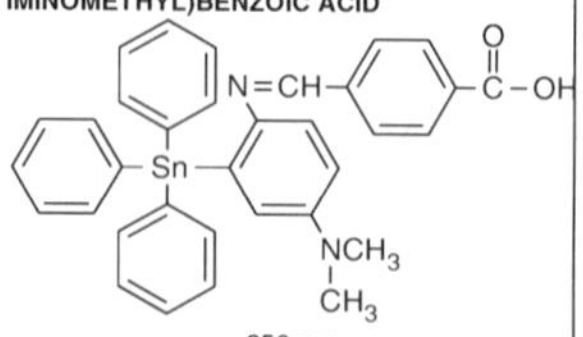
250 mg

S84,035-1
N,N-DIMETHYL-3-(TRIPHENYLSTANNYL)-ANILINE
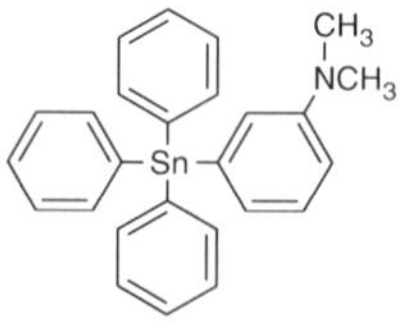
250 mg

S81,938-7
N,N-DIMETHYL-2-(TRIPHENYLSTANNYL)-ANILINE
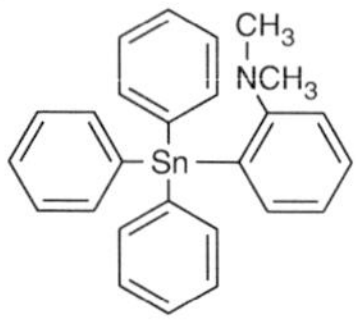
250 mg

S81,503-9
N,N-DIMETHYL-4-(TRIPHENYLSTANNYL)-ANILINE
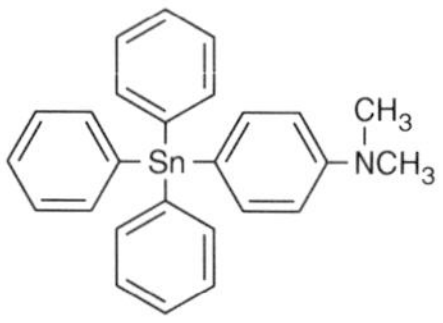
250 mg

S83,876-4
TRIMETHYL[4-(TRIPHENYLSTANNYL)-PHENYL]AMMONIUM METHOSULFATE
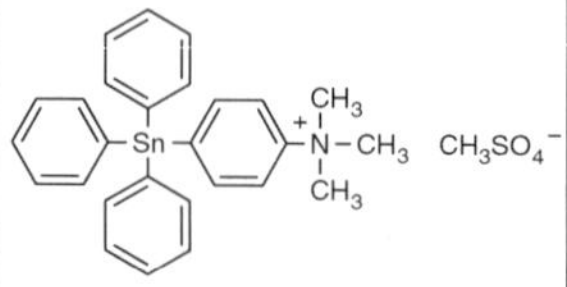
250 mg

S35,638-7
TRIBUTYLTIN 8-HYDROXYQUINOLATE
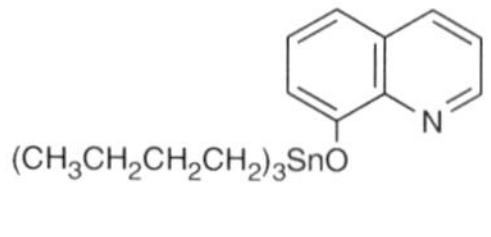
5 gm

S83,909-4
BIS(3-NITROPHENYL)TIN OXIDE
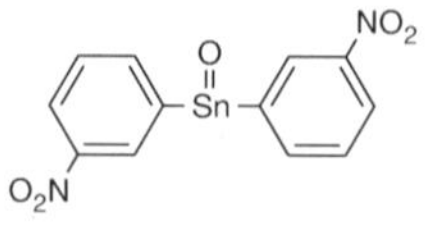
250 mg

S83,908-6
BIS(4-NITROPHENYL)TIN OXIDE
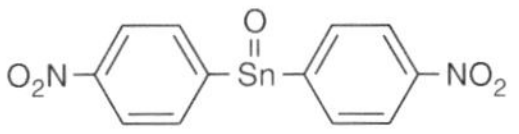
250 mg

S83,495-5
2-(TRIPHENYLSTANNYL)ACETOPHENONE
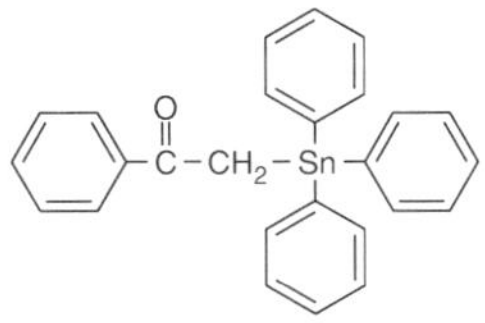
250 mg

S35,673-5
TRIPHENYLTIN 4-ACETAMIDOBENZOATE
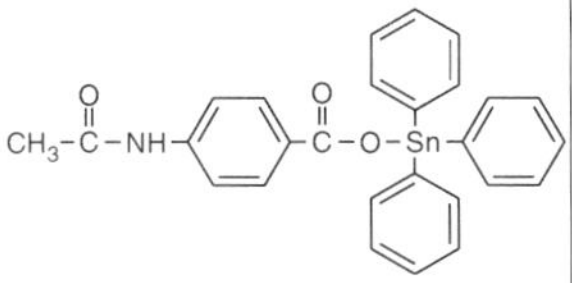
5 gm

S35,634-4
TRIBUTYLTIN 4-ACETAMIDOBENZOATE
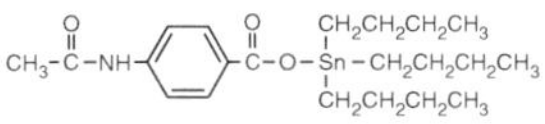
5 gm

S81,505-5
4-(TRIPHENYLSTANNYL)BENZOIC ACID
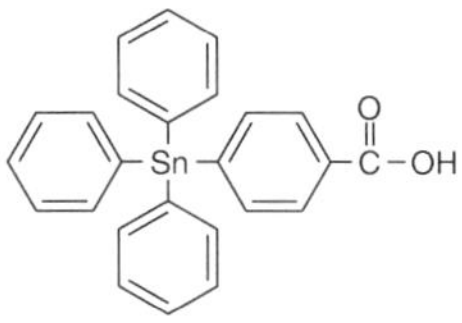
250 mg

S60,079-2
BIS(TRIBUTYLSTANNYL)-METHYLENESUCCINATE
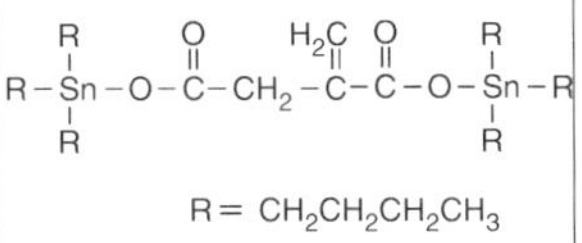
5 gm

S81,504-7
METHYL 2-(CHLORODIPHENYLSTANNYL)-BENZOATE
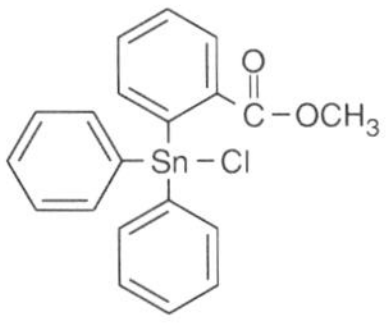
250 mg

S39,581-1
BIS(TRIPHENYLSTANNYL) SUCCINATE
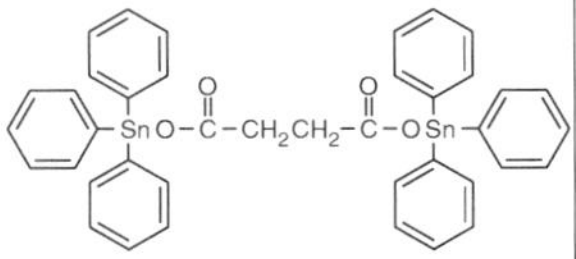
250 mg

S39,570-6
ACETOXYBUTYLDIETHYLTIN
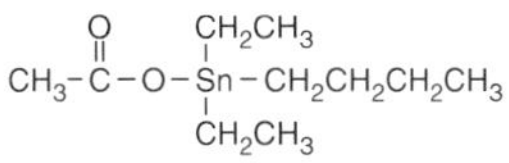
250 mg

S39,568-4
ACETOXYDIETHYLHEXYLTIN
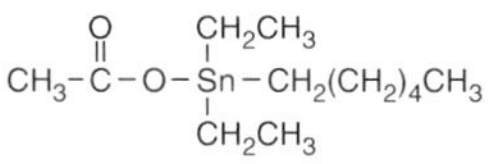
250 mg

S39,569-2
ACETOXYDIETHYLOCTYLTIN
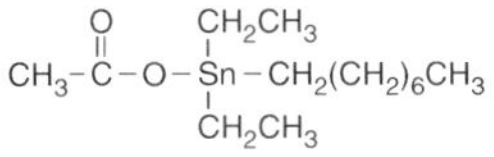
250 mg

S35,641-7
TRIBUTYLTIN PROPIONATE
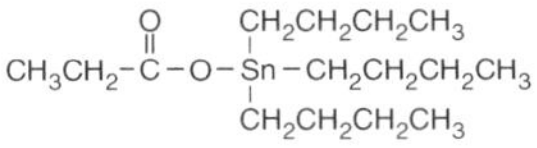
5 gm

S35,637-9
TRIBUTYLTIN 2-ETHYLHEXANOATE
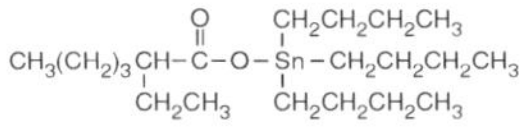
5 gm

S39,564-1
DIACETOXYDIISOOCTYLTIN
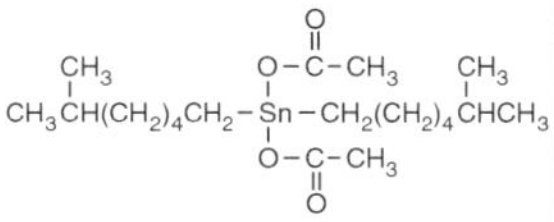
250 mg

S35,646-8
TRIBUTYLTIN 10-UNDECYLENATE
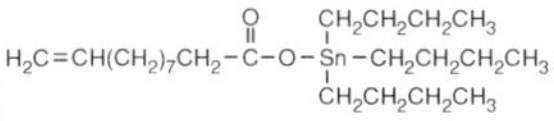
5 gm

S37,137-8
LAUROYLOXYTRIBUTYLTIN
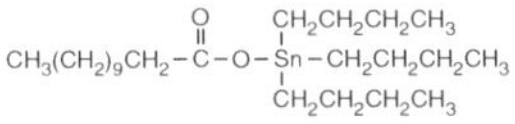
5 gm

S39,577-3
DIETHYLDILAUROYLOXYTIN
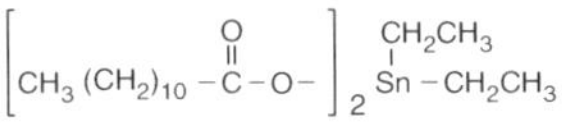
250 mg

S57,627-1
TRIBUTYL(OLEOYLOXY)STANNANE
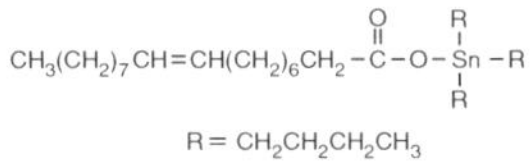
5 gm

S35,643-3
TRIBUTYLTIN STEARATE
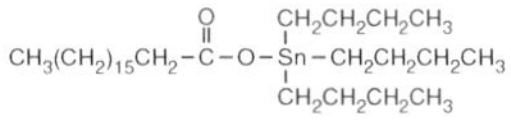
5 gm

S39,560-9
BIS(2-ETHYLHEXANOYLOXY)DIOCTYLTIN
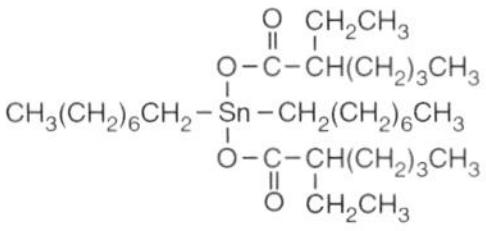
250 mg

S39,575-7
BIS(2-ETHYLHEXANOYLOXY)BIS(2-ETHYLHEXYL)TIN
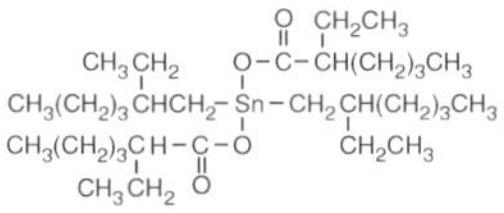
250 mg

S39,589-7
BIS(2-ETHYLHEXANOYLOXY)BIS-(3,5,5-TRIMETHYLHEXYL)TIN
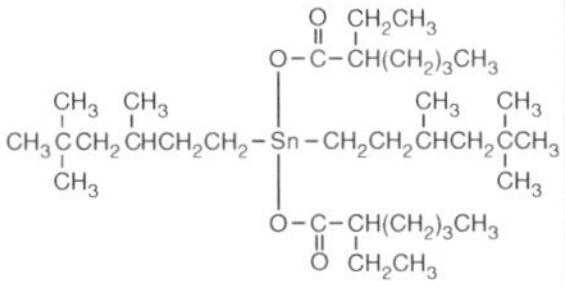
250 mg

S35,642-5
BIS(TRIBUTYLTIN) SEBACATE
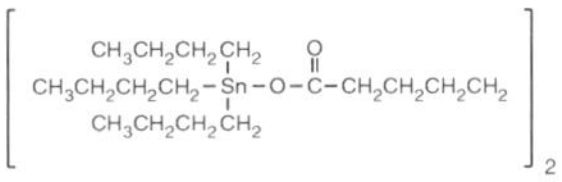
5 gm

S34,891-0
DIBUTYLTIN DIMYRISTATE
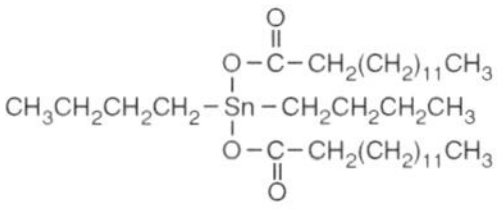
5 gm

S39,559-5
DIHEXYLDILAUROYLOXYTIN
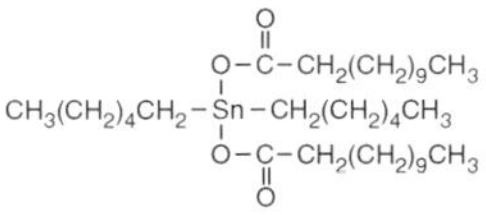
250 mg

S36,023-6
DIBUTYLTIN DIOLEATE
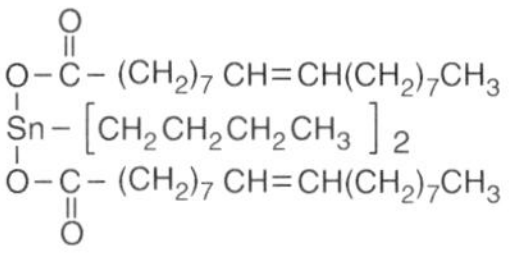
5 gm

S36,793-1
DIBUTYLTIN DISTEARATE
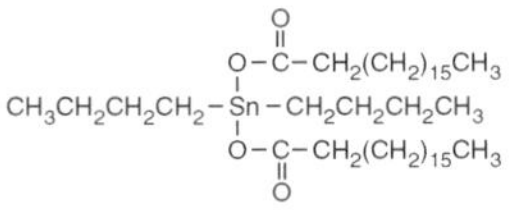
5 gm

S35,624-7
TRIBENZYLTIN ACETATE
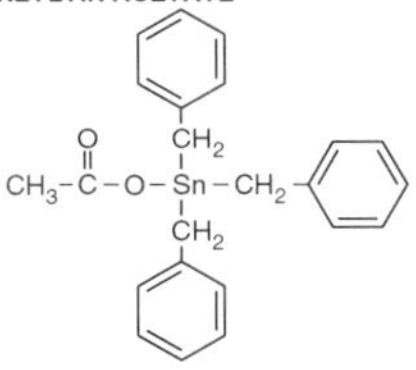
5 gm

S35,627-1
TRIBENZYLTIN LAURATE
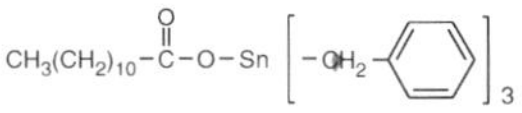
5 gm

S82,018-0
ETHYL 2-(TRIPHENYLSTANNYL)ACETATE
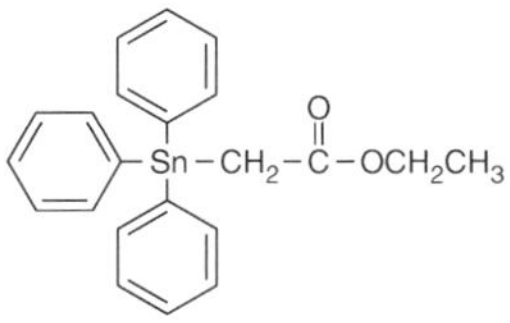
250 mg

S82,956-0
ETHYL 2-(TRIPHENYLSTANNYL)-PROPIONATE
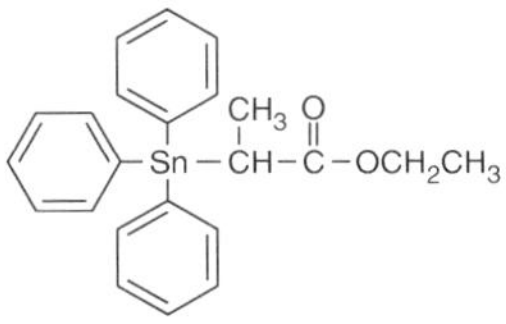
250 mg

S37,118-1
TRIPHENYLTIN N,N-DIMETHYLDITHIOCARBAMATE
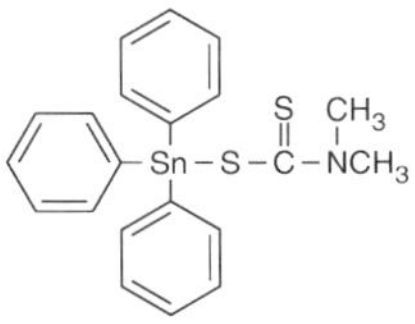
5 gm

S61,711-3
TRIBUTYLTIN ISOTHIOCYANATE
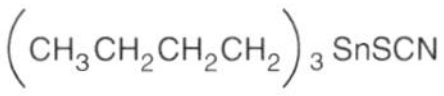
250 mg

S61,709-1
TRIPHENYLTIN ISOTHIOCYANATE
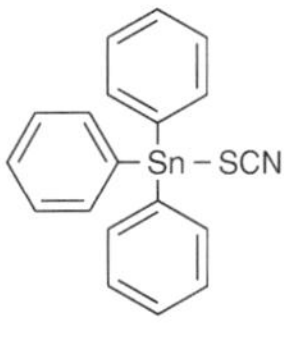
5 gm

S82,475-5
BIS[2-(PHENYLSULFONYL)PHENYL]-DIPHENYLTIN
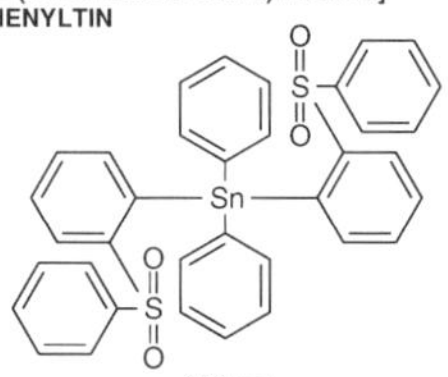
250 mg

S39,572-2
TRIBUTYLTIN 2-NAPHTHALENESULFONATE
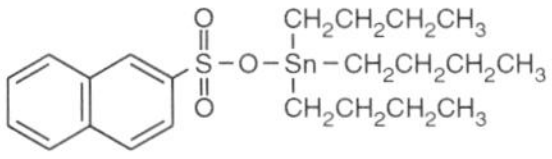
250 mg

S39,588-9
BIS(TRIBUTYLSTANNYL) SULFATE
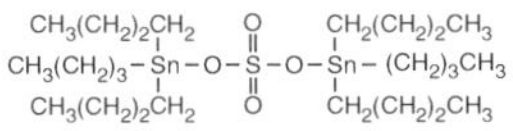
250 mg

S39,584-6
N-TRIETHYLSTANNYL-P-TOLUENESULFONAMIDE
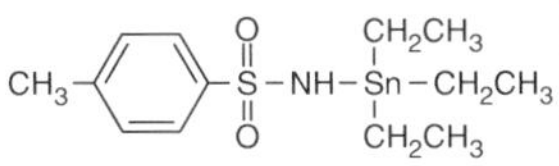
250 mg

S39,586-2
DIOCTYLDI-P-TOLUENESULFONAMIDOTIN
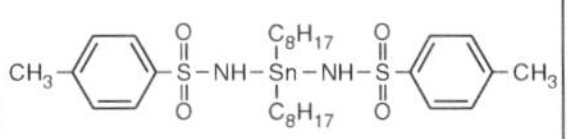
250 mg

S82,471-2
TETRAKIS(2,4-CYCLOPENTADIEN-1-YL)-TIN
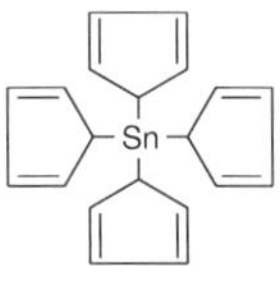
250 mg

S82,470-4
PHENYLTRIS(2,4-CYCLOPENTADIEN-1-YL)TIN

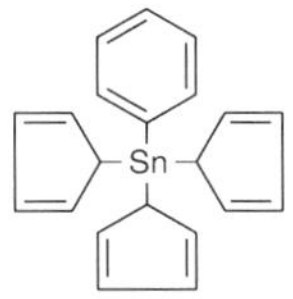

250 mg

S82,469-0
BIS(2,4-CYCLOPENTADIEN-1-YL)-DIPHENYLTIN

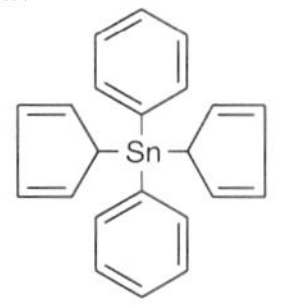

250 mg

S82,476-3
(2-FURYL)TRIPHENYLTIN

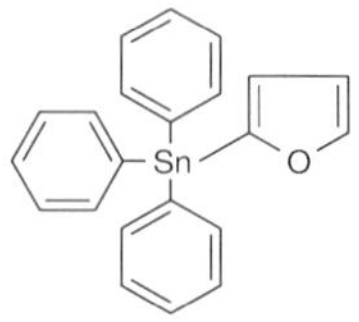

250 mg

S39,583-8
1-(TRIPHENYLTIN)IMIDAZOLE

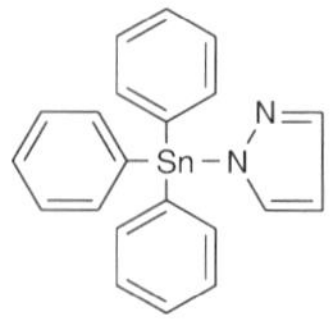

250 mg

S83,494-7
2-(TRIPHENYLSTANNYL)BENZOTHIAZOLE

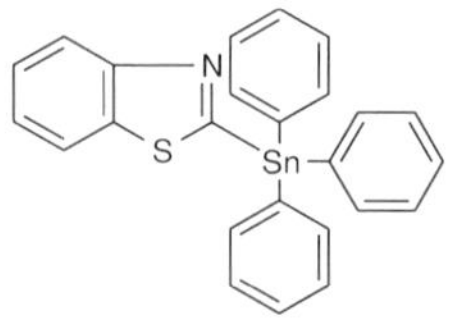

250 mg

S76,531-7
TETRACYCLOHEXYLTIN

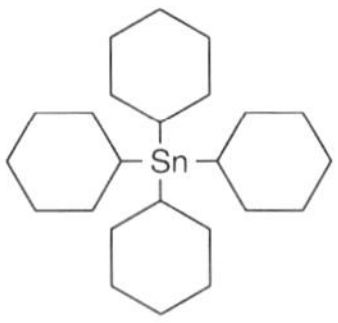

5 gm

S86,709-8
(2-CYCLOHEXEN-1-YL)TRIBUTYLTIN

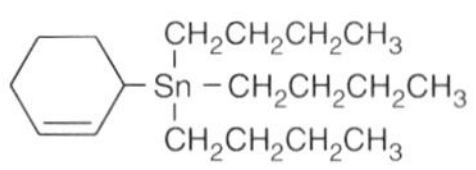

250 mg

S86,702-0
5,6-DIHYDRO-2-(TRIBUTYLSTANNYL)-4H-PYRAN

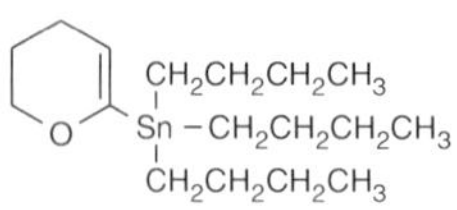

250 mg

S82,501-8
DIPHENYLTIN

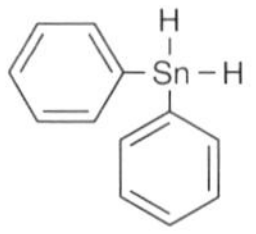

250 mg

S83,910-8
DIPHENYLTIN OXIDE

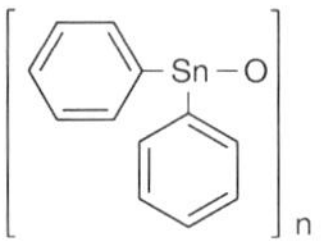

250 mg

S81,601-9
DIIODODIPHENYLTIN

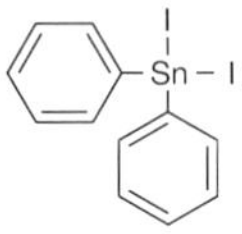

250 mg

S83,911-6
BROMOTRIPHENYLTIN

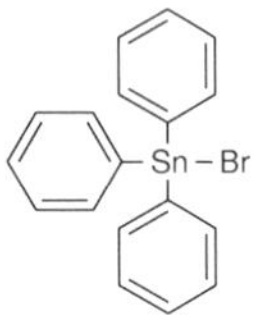

250 mg

S81,772-4
IODOTRIPHENYLTIN

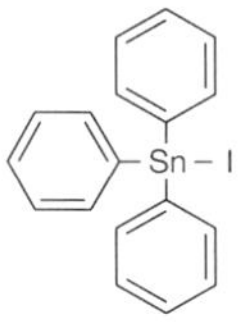

250 mg

S81,932-8
ETHYLTRIPHENYLTIN

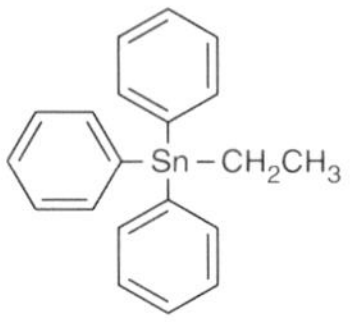

250 mg

S54,724-7
DIBUTYLDIPHENYLTIN

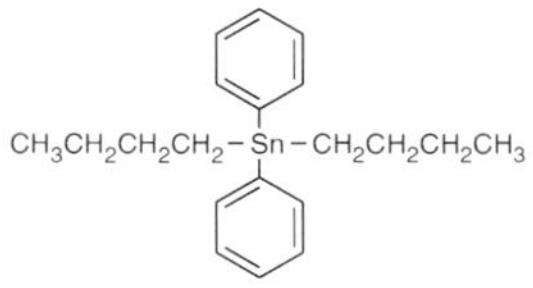

5 gm

S82,952-8
DIISOBUTYLDIPHENYLTIN

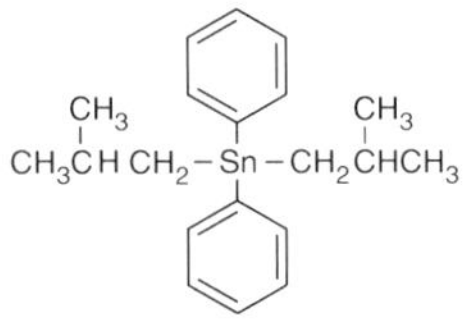

250 mg

S35,626-3
TRIBENZYLTIN HYDROXIDE

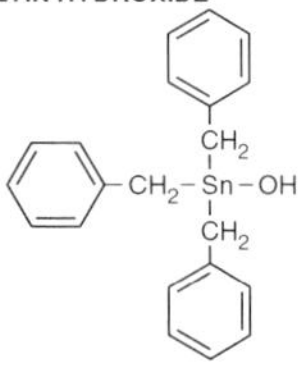

5 gm

S81,778-3
CHLOROTRIBENZYLTIN

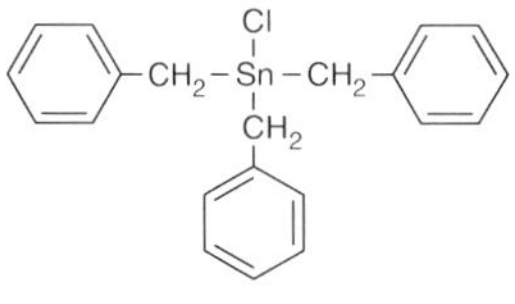

250 mg

S82,953-6
TERT-BUTYLTRIPHENYLTIN

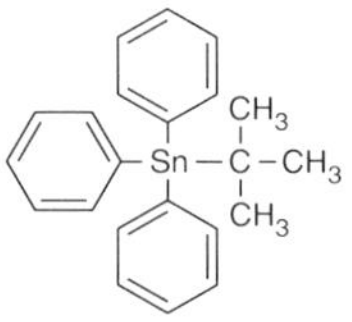

250 mg

S82,954-4
ISOBUTYLTRIPHENYLTIN

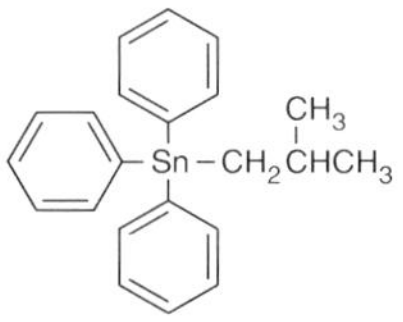

250 mg

S82,955-2
BUTYLTRIPHENYLTIN

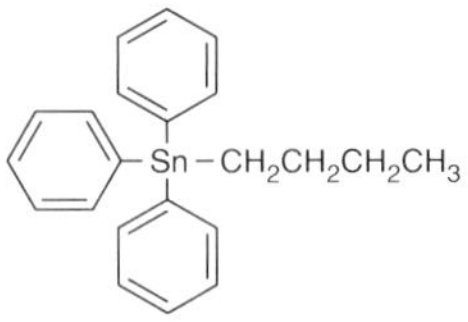

250 mg

S83,277-4
(4-PENTEN-1-YL)TRIPHENYLTIN

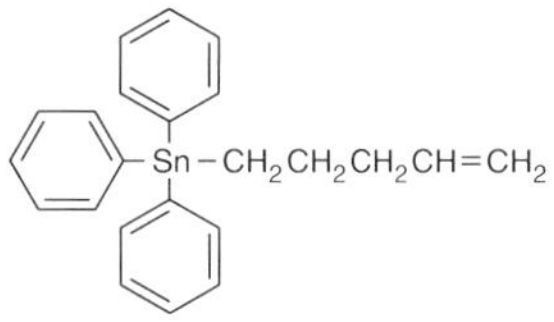

250 mg

S83,428-9
DIBENZYLDIPHENYLTIN

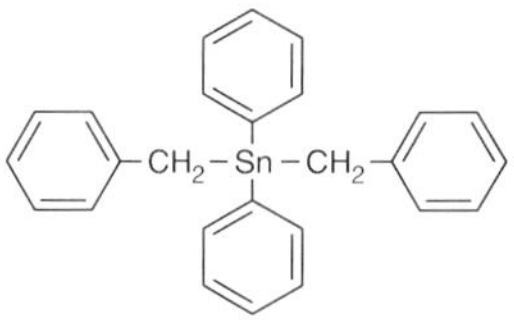

250 mg

S88,552-5
(3-METHYLBENZYL)TRIPHENYLTIN

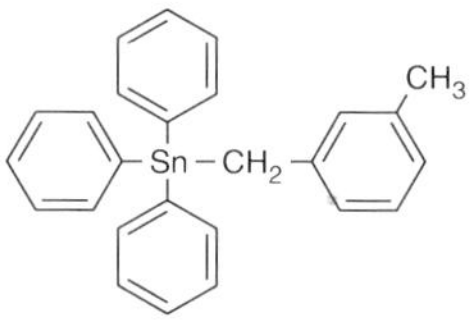

250 mg

S83,214-6
OCTYLTRIPHENYLTIN

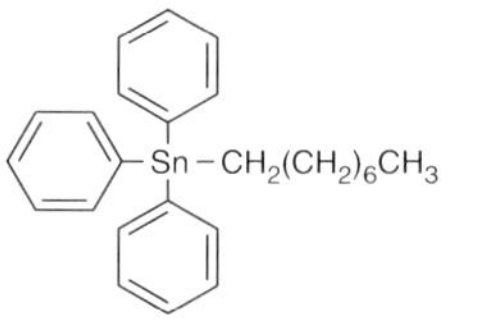

250 mg

S82,778-9
DIPHENETHYLDIPHENYLTIN

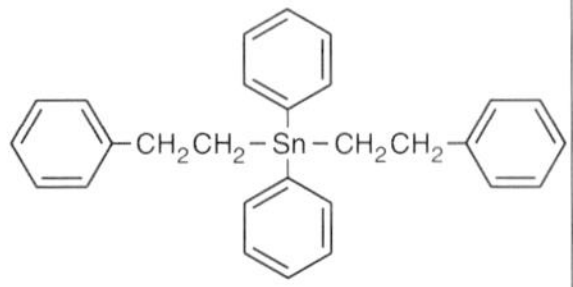

250 mg

S85,291-0
BROMO[(4-BROMO-1,2,3,4-TETRAPHENYL)-1,3-BUTADIENYL]DIMETHYLTIN

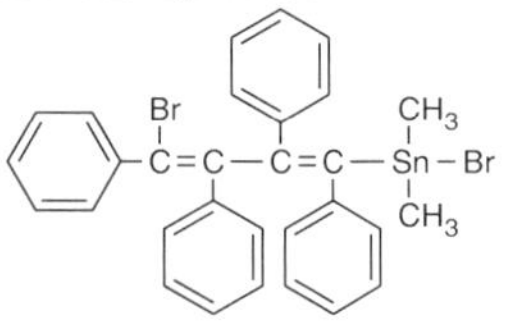

250 mg

S82,456-9
OCTADECYLTRIPHENYLTIN

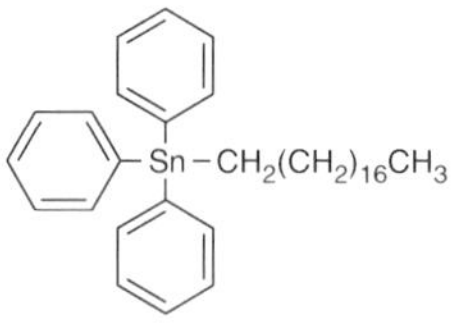

250 mg

S83,213-8
TRIPHENYL(TRIPHENYLSTANNYL)GERMANE

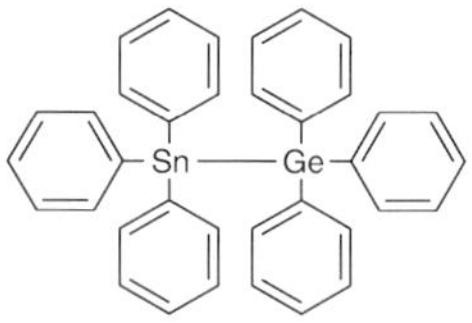

250 mg

S57,232-2
HEXAPHENYLDITIN

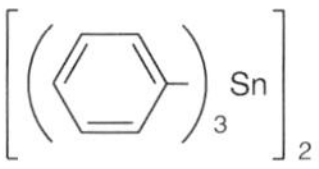

250 mg

S39,567-6
TRIPHENYLTIN SULFIDE

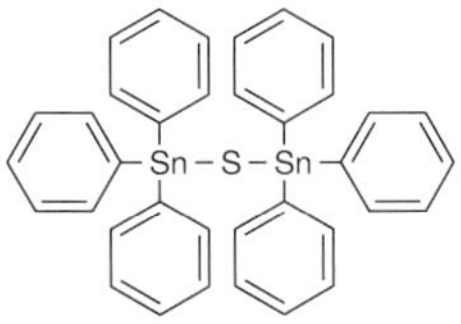

250 mg

S81,513-6
TRIPHENYL(TRIPHENYLMETHYL)TIN

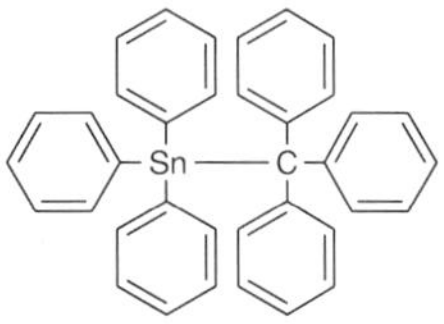

250 mg

S83,215-4
BIS(TRIPHENYLSTANNYL)DIPHENYLSILANE

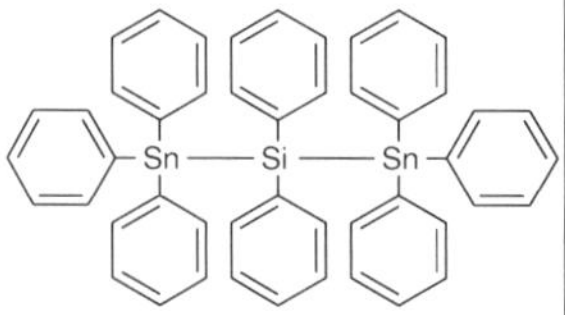

250 mg

S82,227-2
TETRAKIS(TRIPHENYLSTANNYL)TIN

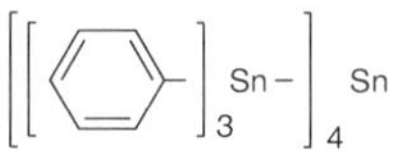

250 mg

S83,216-2
TETRAKIS(TRIPHENYLSILYL)TIN

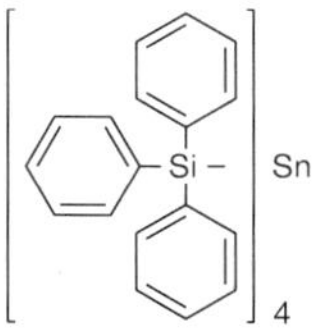

250 mg

S83,988-4
TETRAKIS(TRIPHENYLSTANNYL)SILANE

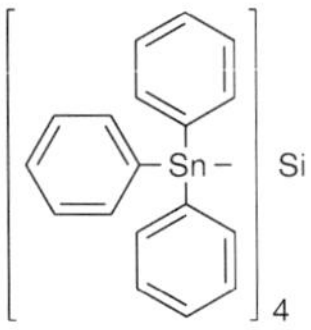

250 mg

S81,934-4
(O-TOLYL)TRIPHENYLTIN

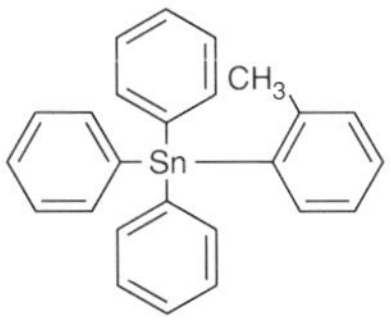

250 mg

S81,776-7
(P-TOLYL)TRIPHENYLTIN

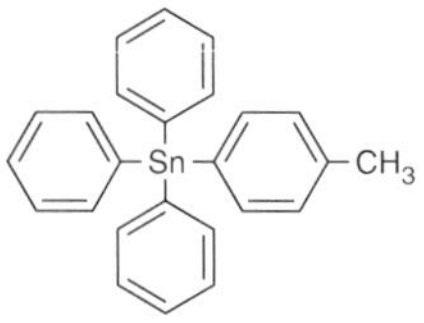

250 mg

S83,153-0
4-(TRIPHENYLSTANNYL)-P-TERPHENYL

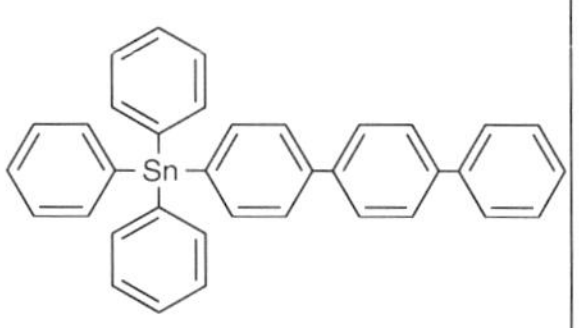

250 mg

S81,935-2
TRIPHENYL(2,6-XYLYL)TIN

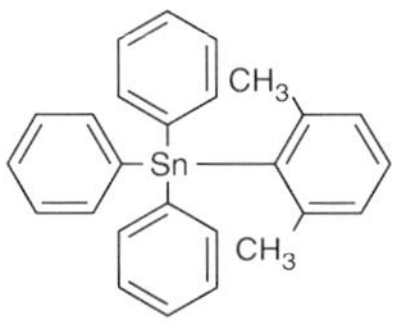

250 mg

S81,936-0
TRIPHENYL(2,4-XYLYL)TIN
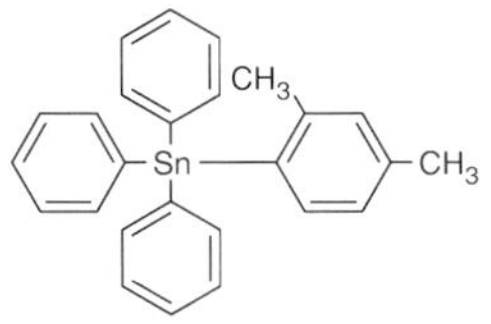
250 mg

S81,937-9
TRIPHENYL(2,5-XYLYL)TIN
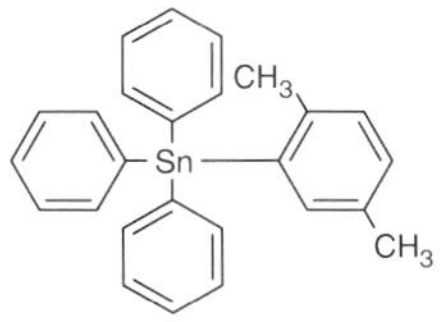
250 mg

S81,939-5
MESITYLTRIPHENYLTIN
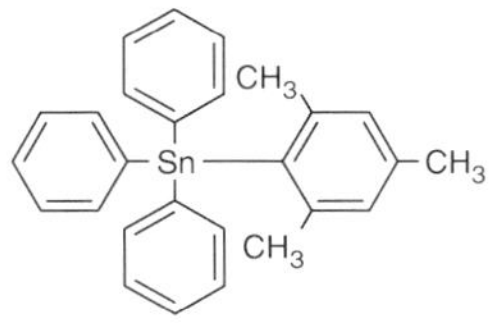
250 mg

S83,493-9
(3-PYRIDYL)TRIPHENYLTIN
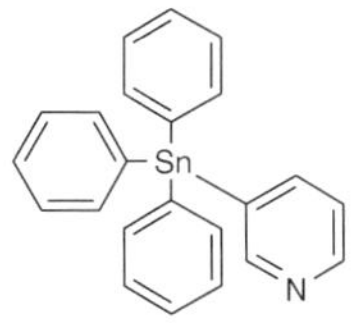
250 mg

S82,474-7
BIS(1-INDENYL)DIPHENYLTIN
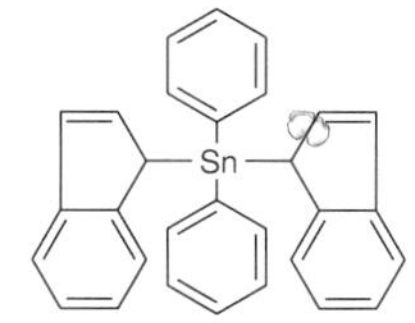
250 mg

S83,073-9
(1-NAPHTHYL)TRIPHENYLTIN
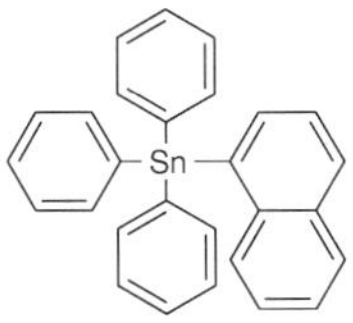
250 mg

S82,472-0
10,10-DIPHENYL-10H-PHENOXASTANNIN
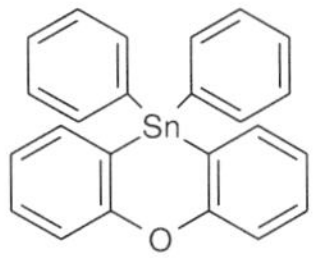
250 mg

S81,942-5
TETRAKIS(O-TOLYL)TIN
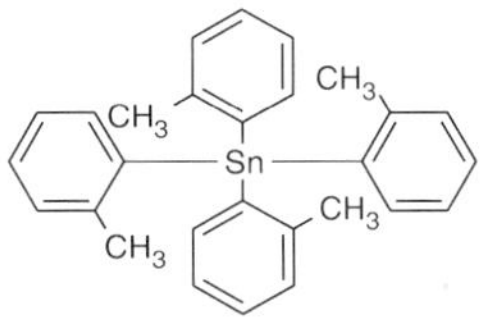
250 mg

S81,943-3
HEXA(O-TOLYL)DITIN
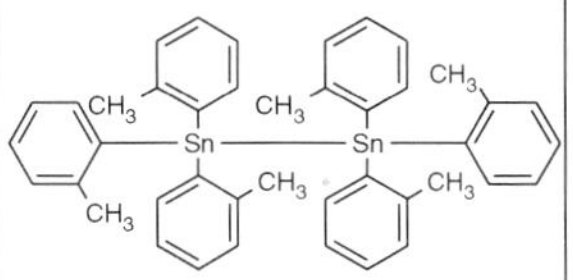
250 mg

S35,640-9
TRIBUTYLTIN 2-PHENYLPHENOXIDE
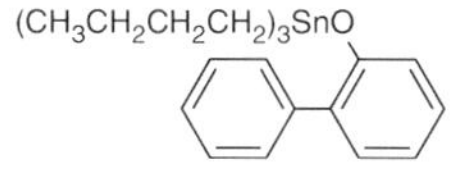
5 gm

S83,392-4
TETRAKIS(P-TOLYL)TIN
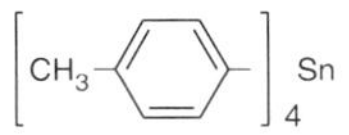
250 mg

S35,636-0
TRIBUTYLTIN 2,4-DICHLOROPHENOXIDE
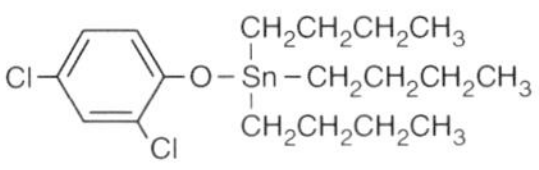
5 gm

S35,635-2
TRIBUTYLTIN 4-CHLORO-3-METHYLPHENOXIDE
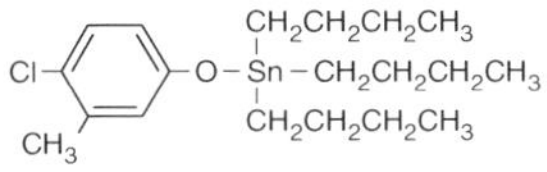
5 gm

S35,644-1
TRIBUTYLTIN 2,4,5-TRICHLOROPHENOXIDE
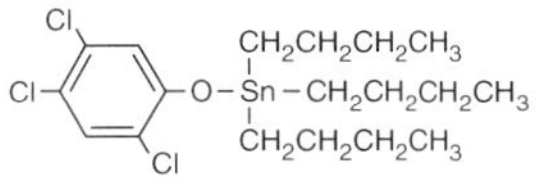
5 gm

S82,473-9
10,10'-SPIROBI(10H-PHENOXASTANNIN)
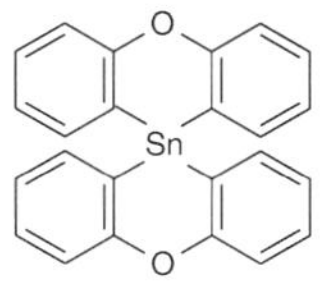
250 mg

S39,571-4
DIBROMODI-1-NAPHTHYLTIN
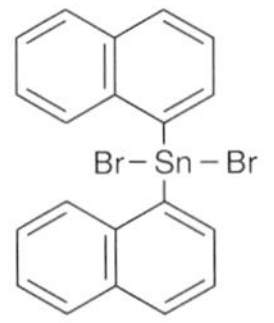
250 mg

S54,999-1
BIS(METHYLTHIO)LEAD

$CH_3S-Pb-SCH_3$

5 gm

S55,000-0
BIS(ETHYLTHIO)LEAD

$CH_3CH_2S-Pb-SCH_2CH_3$

5 gm

S79,840-1
DICHLORODIETHYLLEAD
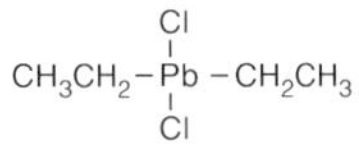
250 mg

S81,320-6
DIETHYLLEAD DIBROMIDE
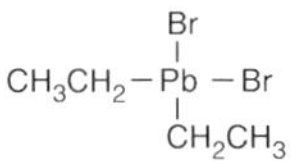
250 mg

S81,158-0
TRIETHYLLEAD HYDROXIDE
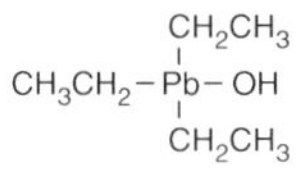
250 mg

S79,839-8
CHLOROTRIETHYLLEAD
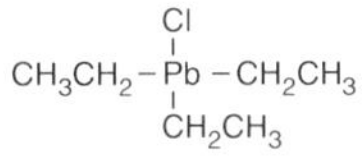
250 mg

S81,518-7
(3-BUTENYL)DICHLOROETHYLLEAD
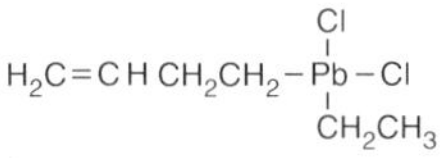
250 mg

S81,319-2
BROMOTRIETHYLLEAD
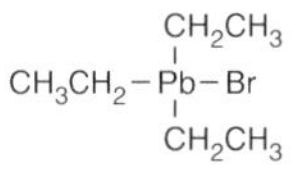
250 mg

S81,524-1
(3-BUTENYL)TRIETHYLLEAD
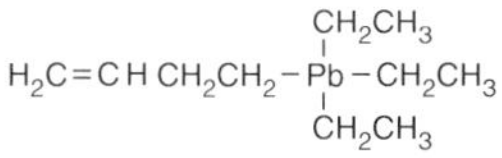
250 mg

S79,837-1
N-(TRIETHYLPLUMBYL)-HEXAHYDROPHTHALIMIDE
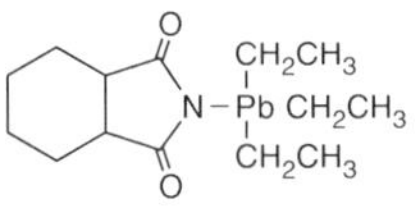
250 mg

S39,449-1
BROMOTRIPHENYLLEAD
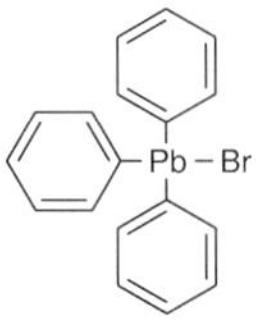
250 mg

S81,972-7
TETRAKIS(M-TOLYL)LEAD
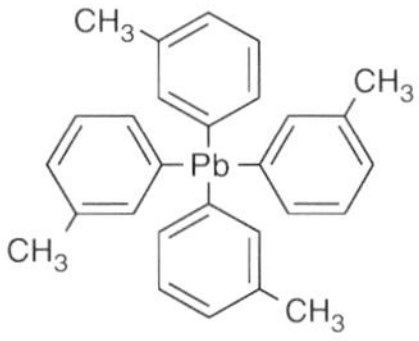
250 mg

S38,294-9
BIS(HEXADECYLTHIO)LEAD(II)

$CH_3(CH_2)_{14}CH_2S-Pb-SCH_2(CH_2)_{14}CH_3$

250 mg

S83,518-8
CHLOROTRIDODECYLLEAD
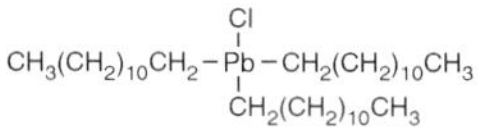
250 mg

S83,519-6
TRIDODECYLPLUMBYL ACETATE
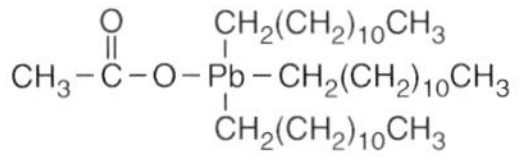
250 mg

S83,522-6
CHLOROTRIOCTADECYLLEAD
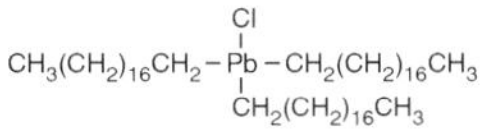
250 mg

S83,523-4
TETRAHEXADECYLLEAD
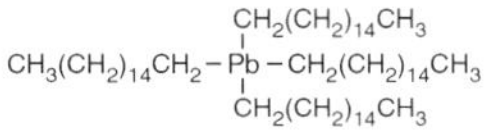
250 mg

S83,821-7
DIPHENYLPLUMBYLENE BIS(2-BROMOACETATE)
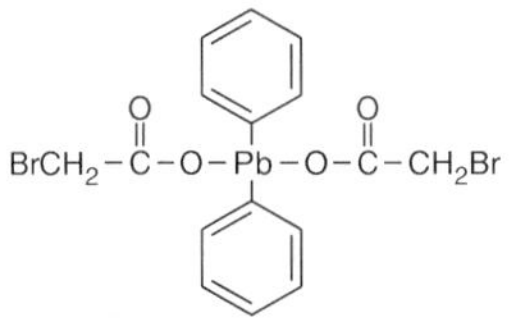
250 mg

S82,648-0
(4-METHOXY-2,3,5,6-TETRAFLUOROPHENYL)TRIPHENYLLEAD
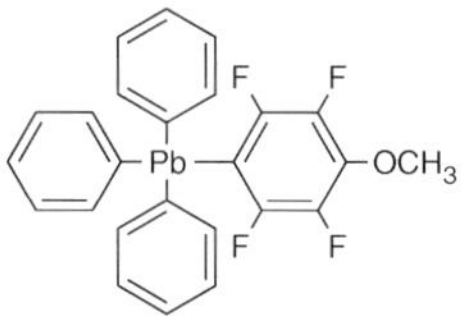
250 mg

S81,859-3
TRIETHYLPLUMBYL 3-BROMOBENZOATE
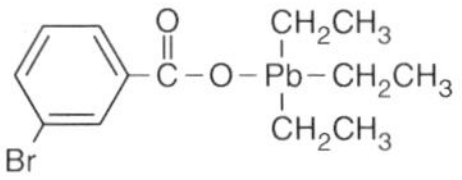
250 mg

S81,857-7
TRIETHYLPLUMBYL 3-IODOBENZOATE
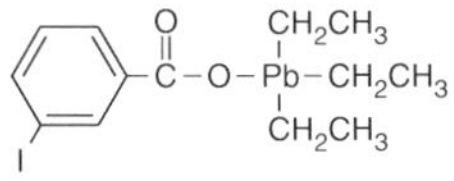
250 mg

S81,865-8
DIETHYLPLUMBYL BIS(3-BROMOBENZOATE)
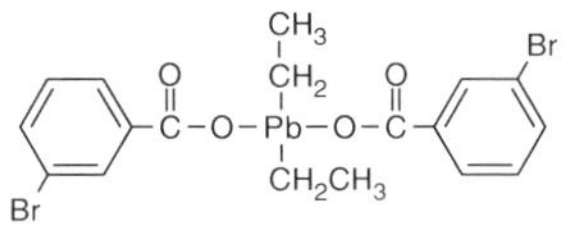
250 mg

S81,858-5
TRIETHYLPLUMBYL 2-IODOBENZOATE
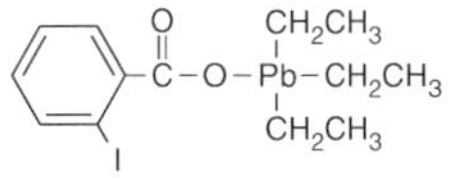
250 mg

S81,523-3
(3-BROMOPHENYL)TRIPHENYLLEAD
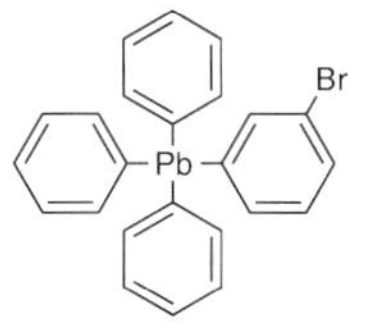
250 mg

S83,126-3
(PENTAFLUOROPHENYL)TRIPHENYLLEAD

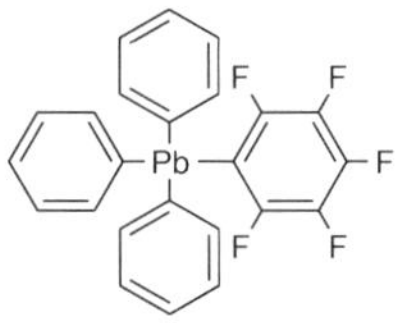

250 mg

S83,125-5
(PENTACHLOROPHENYL)TRIPHENYLLEAD

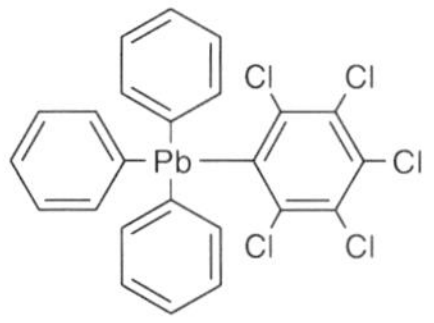

250 mg

S81,829-1
(4-CHLOROPHENYL)TRIPHENYLLEAD

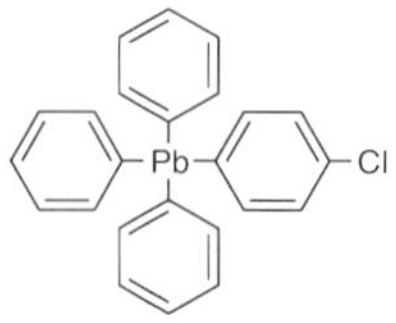

250 mg

S83,695-8
BIS(4-CHLOROPHENYL)DIPHENYLLEAD

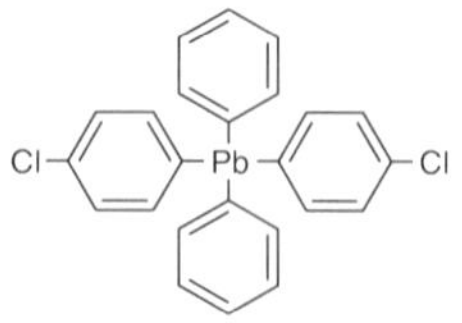

250 mg

S81,403-2
(4-BROMOPHENYL)TRIPHENYLLEAD

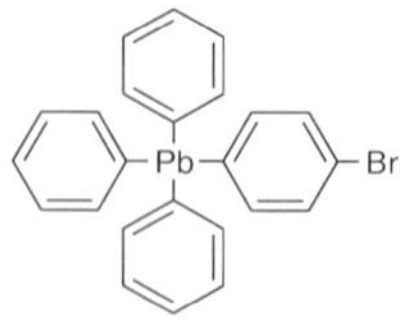

250 mg

S81,622-1
(4-BROMOBENZYL)TRIPHENYLLEAD

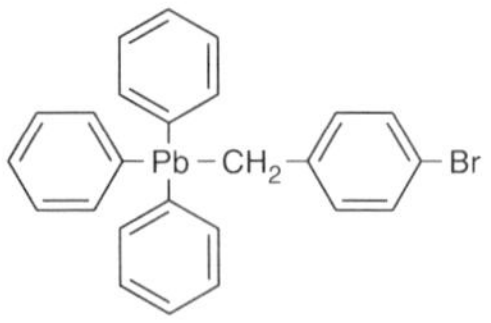

250 mg

S81,498-9
(4-BROMOPHENYL)TRIETHYLLEAD

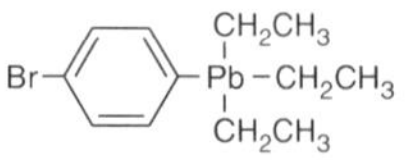

250 mg

S81,861-5
BIS(4-CHLOROPHENYL)DICHLOROLEAD

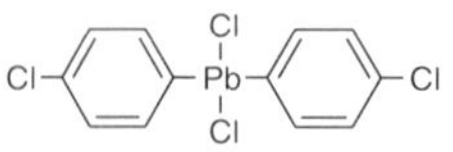

250 mg

S81,835-6
CHLOROTRIS(4-CHLOROPHENYL)LEAD

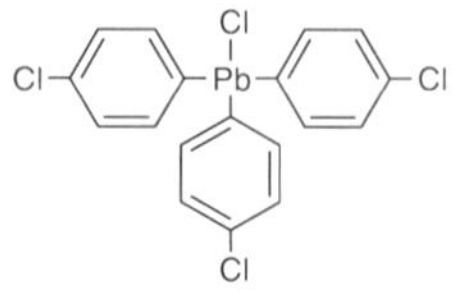

250 mg

S81,830-5
TETRAKIS(4-CHLOROPHENYL)LEAD

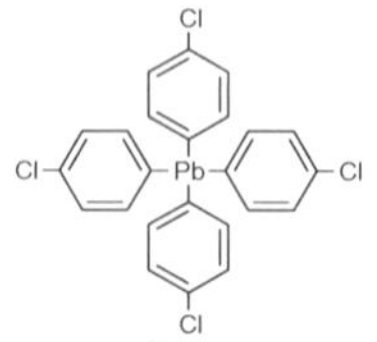

250 mg

S81,408-3
[3-(HYDROXYMETHYL)PHENYL]-TRIPHENYLLEAD

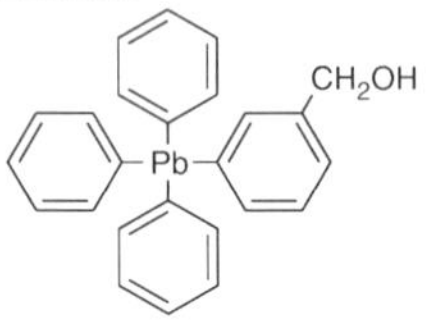

250 mg

S81,409-1
[2-(HYDROXYMETHYL)PHENYL]-TRIPHENYLLEAD

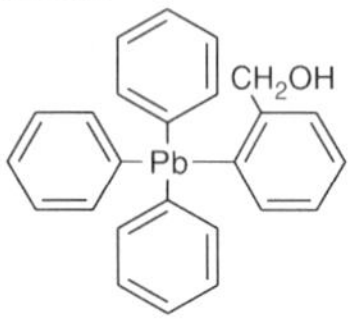

250 mg

S81,497-0
[2-(HYDROXYMETHYL)PHENYL]-TRIPHENYLLEAD

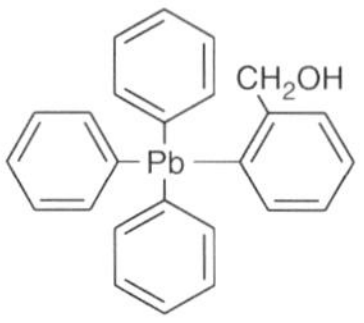

250 mg

S81,410-5
[4-(HYDROXYMETHYL)PHENYL]-TRIPHENYLLEAD

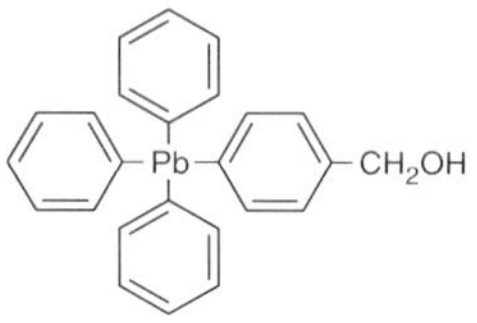

250 mg

S81,411-3
[4-(1-HYDROXYETHYL)PHENYL]-TRIPHENYLLEAD

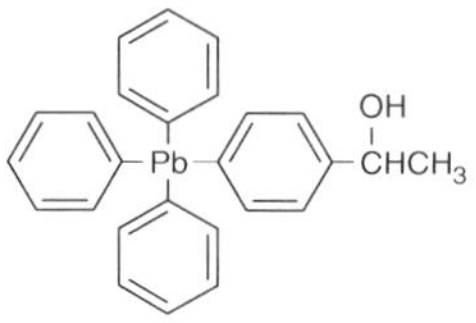

250 mg

S81,828-3
(2-HYDROXYPHENYL)TRIPHENYLLEAD

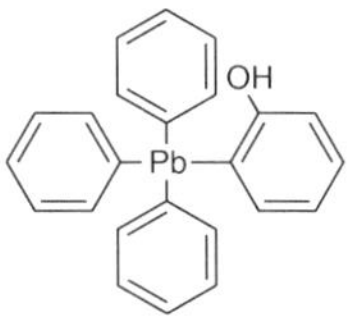

250 mg

S81,412-1
[4-(2-HYDROXYETHYL)PHENYL]-TRIPHENYLLEAD

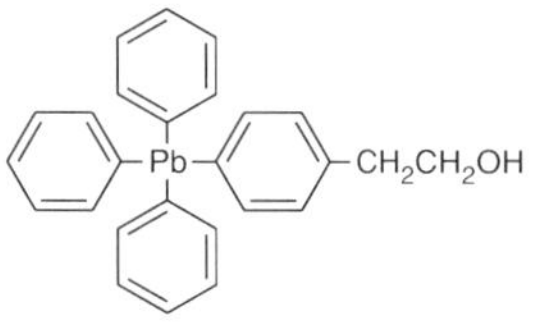

250 mg

S81,612-4
DI(2-FURYL)BIS(4-METHOXYPHENYL)LEAD

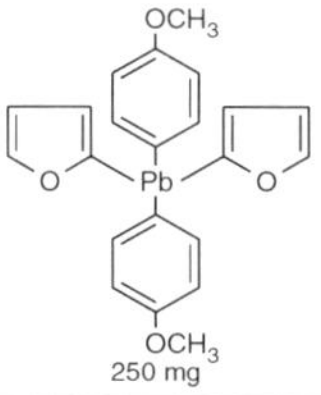

250 mg

S81,974-3
TRIS(2-ETHOXYPHENYL)LEAD

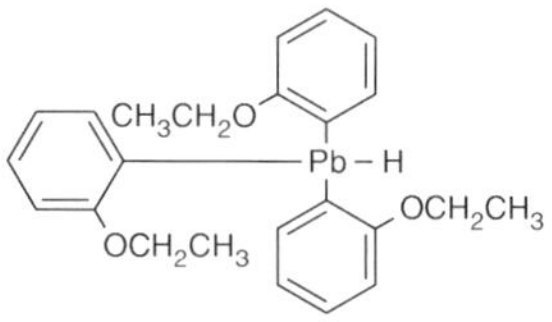

250 mg

S81,971-9
TETRAKIS(2-METHOXYPHENYL)LEAD

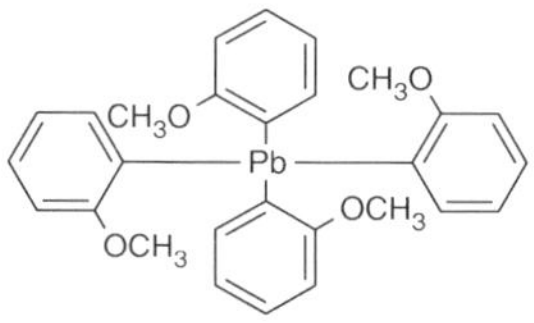

250 mg

S81,311-7
CHLOROTRIS(4-METHOXYPHENYL)LEAD

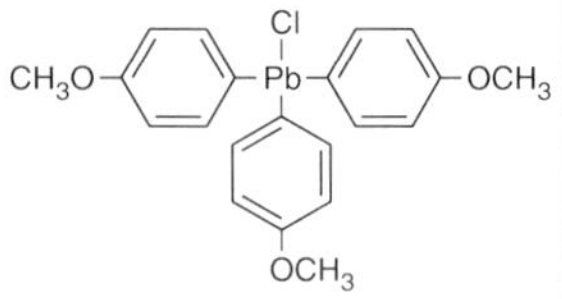

250 mg

S81,975-1
TRIS(4-ETHOXYPHENYL)LEAD

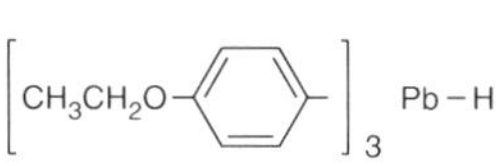

250 mg

S81,309-5
TETRAKIS(4-METHOXYPHENYL)LEAD

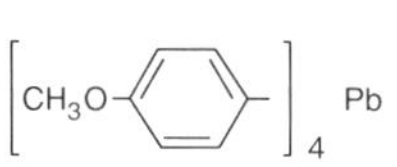

250 mg

S81,977-8
TETRAKIS(4-ETHOXYPHENYL)LEAD

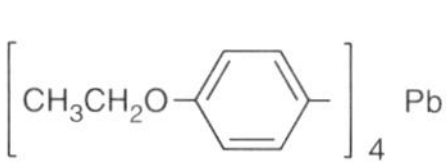

250 mg

S81,407-5
(2-METHOXYPHENYL)TRIPHENYLLEAD

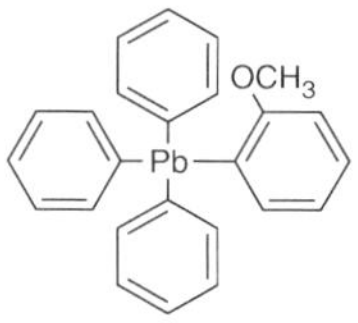

250 mg

S81,826-7
(4-METHOXYPHENYL)TRIPHENYLLEAD

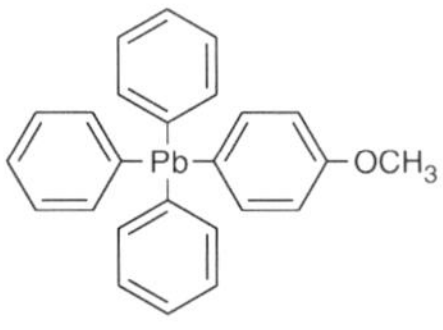

250 mg

S83,907-8
2,4-DINITROBENZENESULFONIC ACID, LEAD SALT

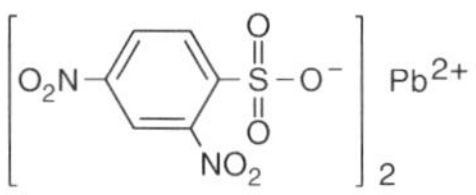

250 mg

S81,863-1
DIETHYLPLUMBYL BIS[2-(BUTYLAMINO)-BENZOATE]

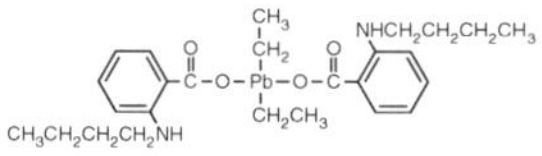

250 mg

S83,781-4
N-[3-(CHLORODIPHENYLPLUMBYL)-PROPYL]DIETHYLAMINE HYDROCHLORIDE

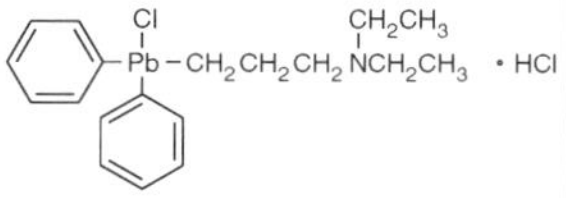

250 mg

S81,500-4
N-[3-(TRIPHENYLPLUMBYL)PROPYL]-DIETHYLAMINE METHIODIDE

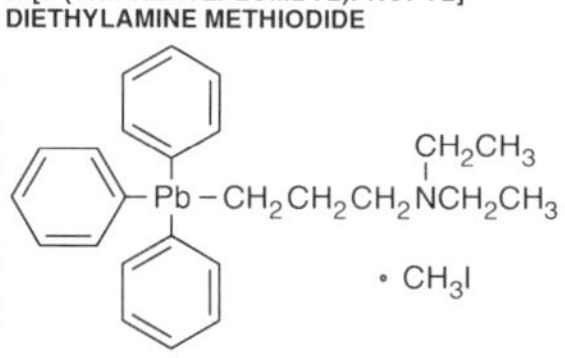

250 mg

S83,782-2
N-[3-(TRIPHENYLPLUMBYL)PROPYL]-DIETHYLAMINE

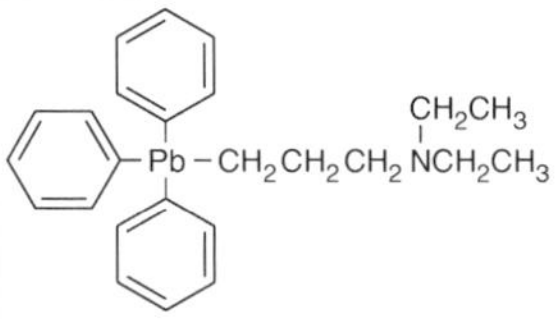

250 mg

S81,520-9
N,N-DIMETHYL-3-(TRIPHENYLPLUMBYL)-ANILINE

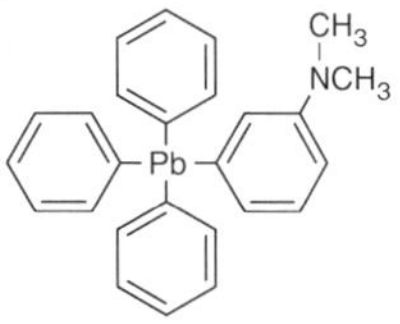

250 mg

S81,621-3
N,N-DIMETHYL-4-(TRIPHENYLPLUMBYL)-ANILINE

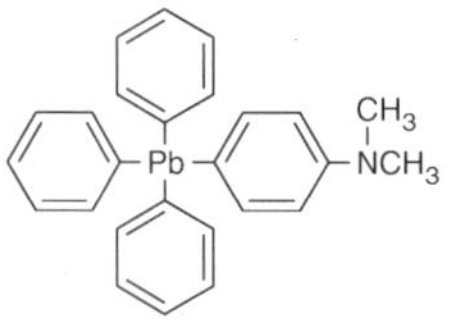

250 mg

S81,824-0
N,N-DIMETHYL-4-(TRIPHENYLPLUMBYL)-ANILINE

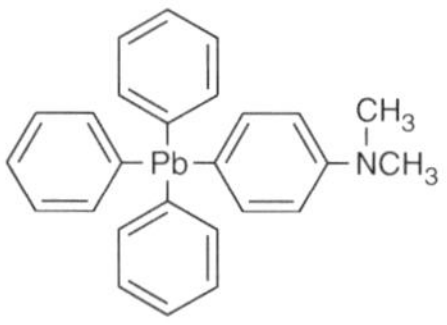

250 mg

S81,823-2
N,N-DIETHYL-4-(TRIPHENYLPLUMBYL)-ANILINE

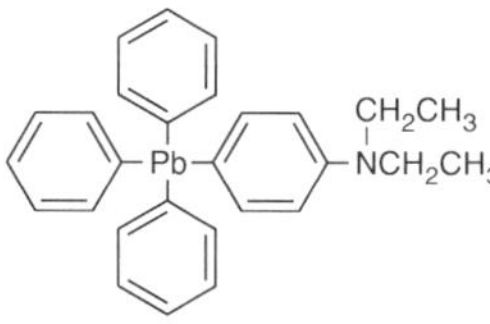

250 mg

S81,522-5
BIS[4-(DIMETHYLAMINO)PHENYL]-DIPHENYLLEAD

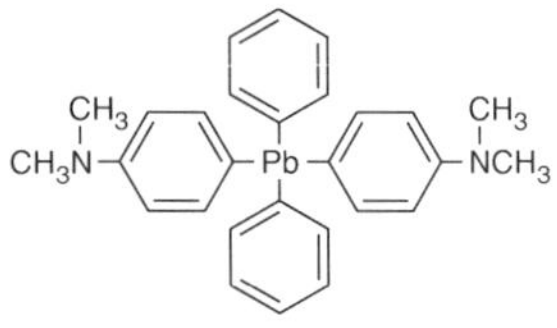

250 mg

S81,530-6
TETRAKIS[4-(DIMETHYLAMINO)-PHENYL]LEAD

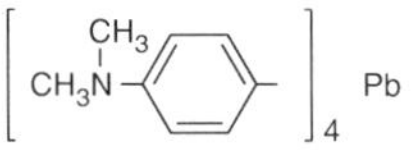

250 mg

S81,848-8
TRIETHYLPLUMBYL 9-ETHYL-3-CARBAZOLECARBOXYLATE

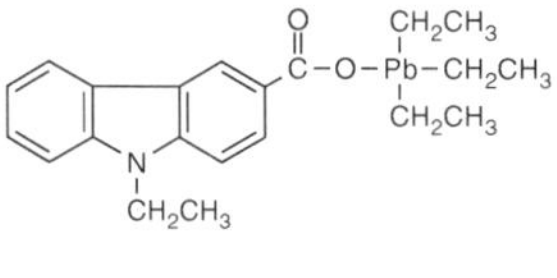

250 mg

S83,783-0
DIETHYLMETHYL[3-(TRIPHENYLPLUMBLY)-PROPYL]AMMONIUM METHOSULFATE

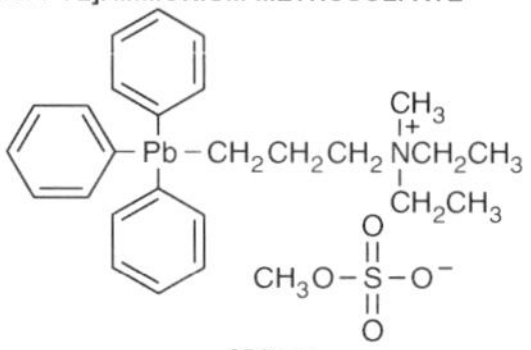

250 mg

S81,849-6
TRIETHYLPLUMBYL 4-(2-NAPHTHYL)-4-OXOBUTYRATE

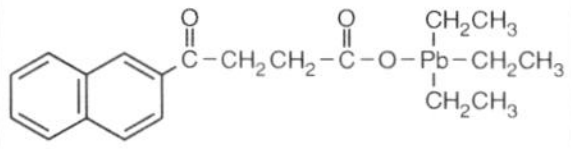

250 mg

S81,866-6
TRIETHYLPLUMBYL (+)-10-CAMPHORSULFONATE
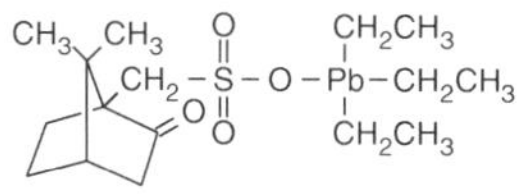
250 mg

S41,725-4
TRIPROPYLPLUMBYL ACETATE
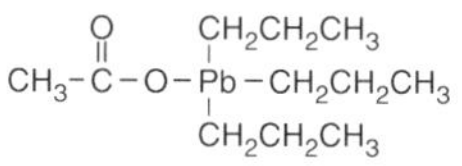
5 gm

S83,705-9
TRIPHENYLSTANNYL BENZOATE
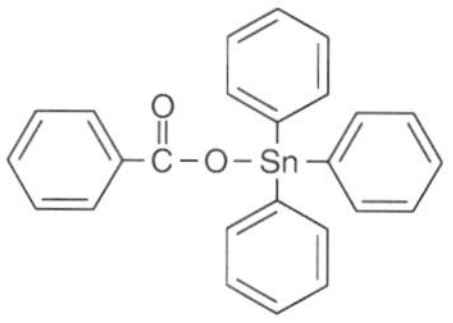
250 mg

S81,851-8
TRIETHYLPLUMBYL 9-FLUORENECARBOXYLATE
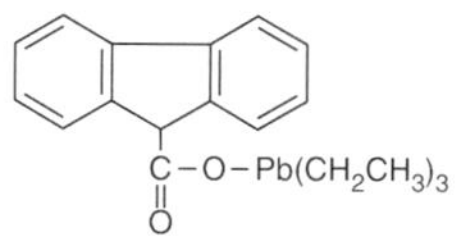
250 mg

S83,948-5
4-OXO-4H-3,2,1-BENZOXATHIAPLUMBIN
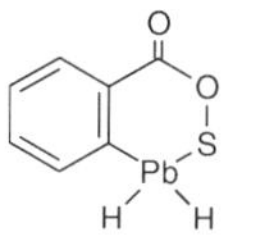
250 mg

S83,949-3
1,1-DIETHYL-3-OXO-3H-2,1-BENZOXAPLUMBOLE
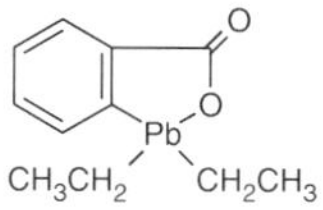
250 mg

S81,846-1
BIS(TRIETHYLPLUMBYL) (+)-CAMPHORATE
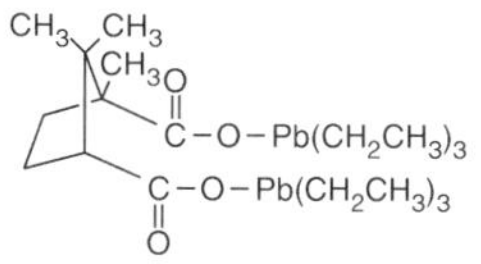
250 mg

S39,547-1
TRIMETHYLPLUMBYL 2-ETHYLHEXANOATE
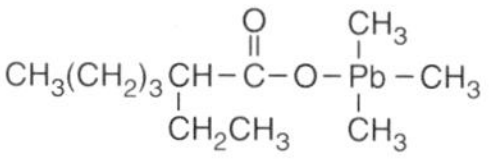
250 mg

S43,201-6
DIBUTYLPLUMBYLENE DIACETATE
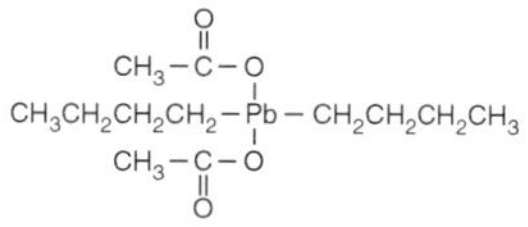
5 gm

S39,558-7
TRIETHYLPLUMBYL 2-ETHYLHEXANOATE
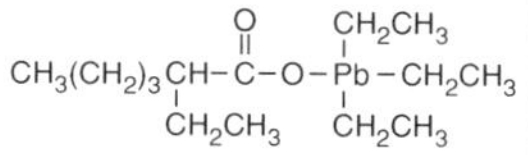
250 mg

S42,792-6
TRIBUTYLPLUMBYL ACETATE
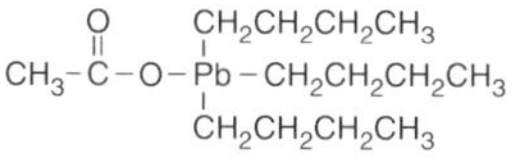
250 mg

S39,542-0
TRIPENTYLPLUMBYL ACETATE
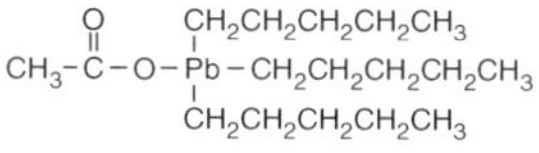
250 mg

S44,066-3
DIPHENYLPLUMBYLENE DIACETATE (CONTAINS ACETIC ACID OF CRYSTALLIZATION)
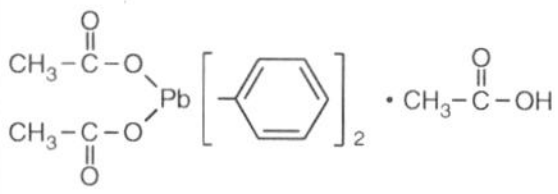
5 gm

S39,554-4
BIS(2-ETHYLHEXANOYLOXY)-DIPHENYLLEAD
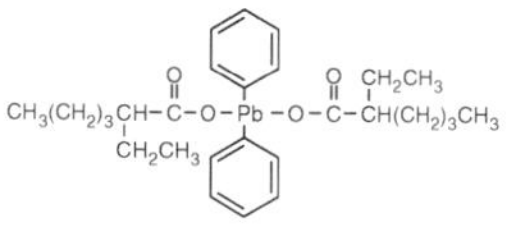
250 mg

S39,557-9
P-TOLYLTRIACETOXYLEAD
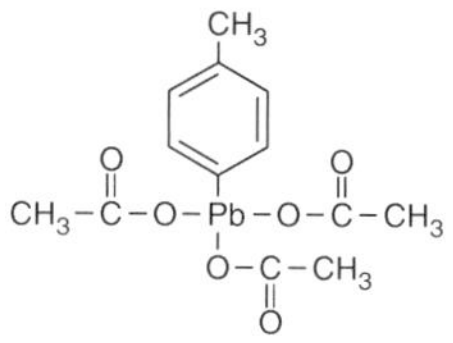
250 mg

S79,843-6
METHYL (TRIETHYLPLUMBYL)FORMATE
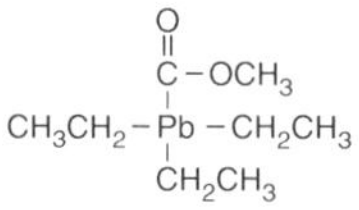
250 mg

S83,877-2
BIS(TRIPHENYLPLUMBYL) CARBONATE
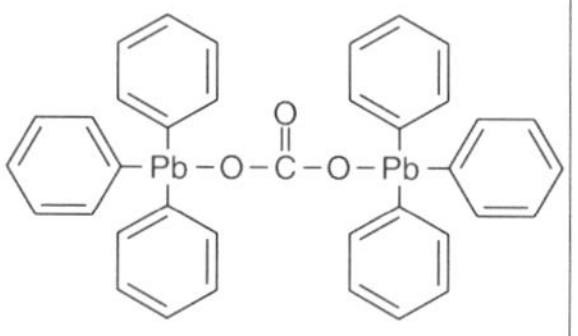
250 mg

S41,728-9
S-TRIPHENYLPLUMBYL THIOACETATE
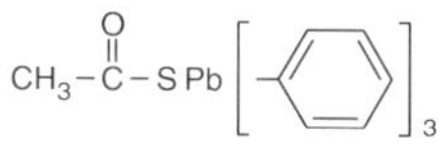
5 gm

S84,121-8
1-(TRIPHENYLPLUMBYL)SUCCINIMIDE
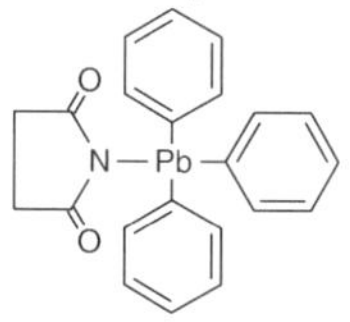
250 mg

S79,842-8
CYANOTRIETHYLLEAD
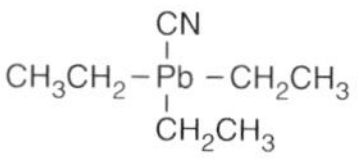
250 mg

S81,853-4
S-(TRIETHYLPLUMBYL)-P-TOLUENETHIOSULFONATE
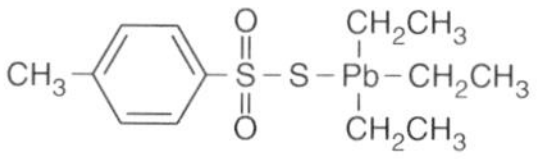
250 mg

S81,313-3
BIS(2-FURYL)DIPHENYLLEAD
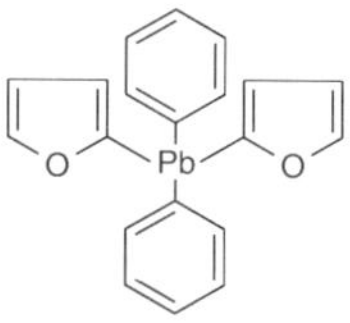
250 mg

S81,310-9
(2-FURYL)TRIPHENYLLEAD
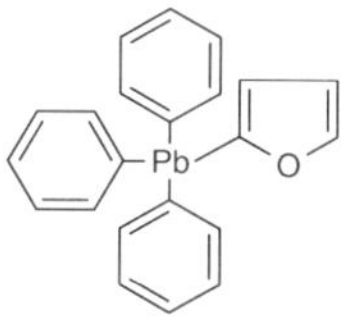
250 mg

S81,585-3
BIS(2-THIENYL)DICHLOROLEAD
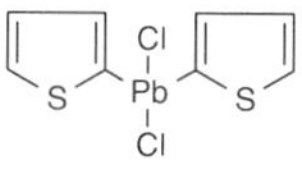
250 mg

S81,616-7
TETRAKIS(2-THIENYL)LEAD
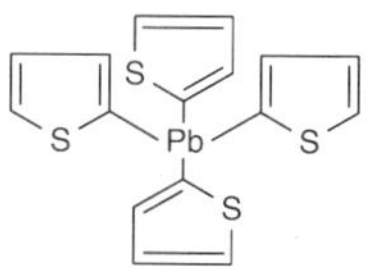
250 mg

S81,312-5
BIS(2-THIENYL)DIPHENYLLEAD
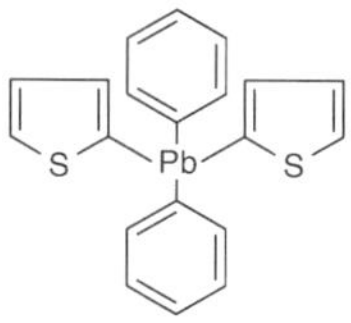
250 mg

S81,588-8
TRIS(1-PYRROLYL)LEAD
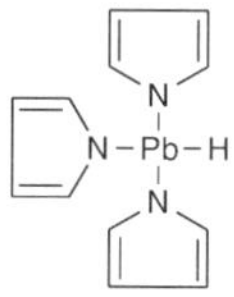
250 mg

S81,613-2
DIPHENYLDI(1-PYRROLYL)LEAD
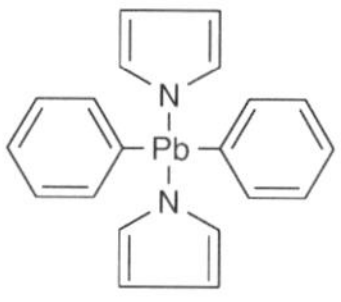
250 mg

S81,860-7
TRIETHYLPLUMBYL CYCLOHEXANESULFINATE
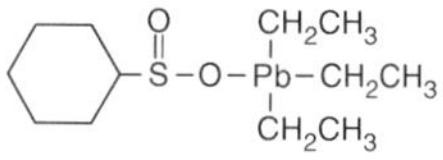
250 mg

S81,968-9
TRICYCLOHEXYLLEAD
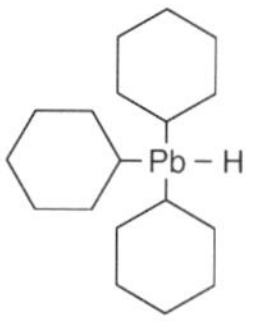
250 mg

S82,325-2
CHLOROTRICYCLOHEXYLLEAD
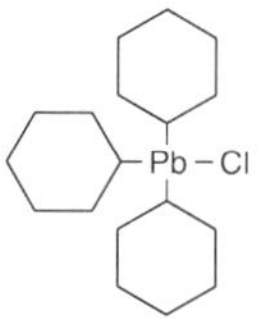
250 mg

S82,324-4
IODOTRICYCLOHEXYLLEAD
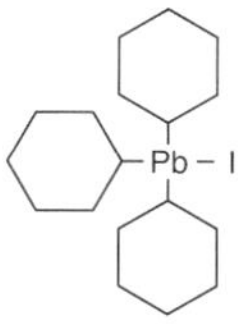
250 mg

S72,743-1
DICHLORODIPHENYLLEAD
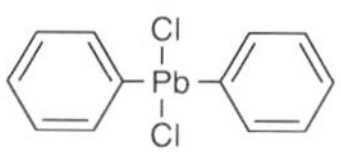
5 gm

S81,157-2
DIPHENYLLEAD DIBROMIDE
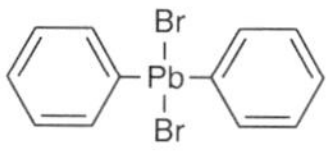
250 mg

S81,156-4
DIIODODIPHENYLLEAD
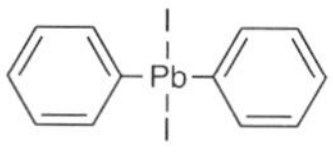
250 mg

S83,871-3
PHENETHYLTRIETHYLLEAD
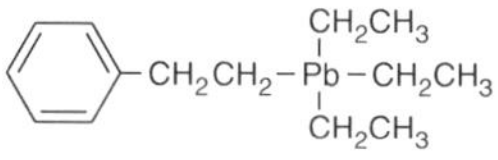
250 mg

S81,855-0
TRIETHYLPLUMBYL 3-PHENYLPROPIONATE
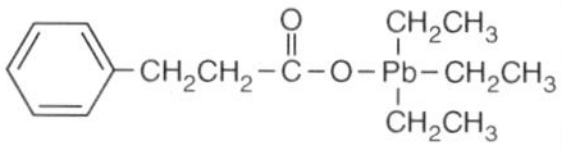
250 mg

S83,822-5
DIETHYLDIPHENYLLEAD
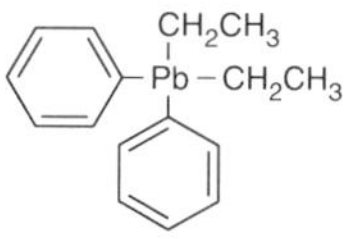
250 mg

S81,615-9
TERT-BUTYLCHLORODIPHENYLLEAD
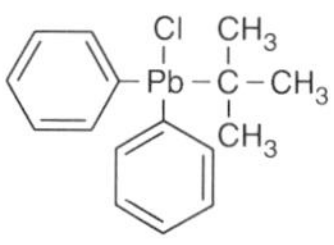
250 mg

S81,517-9
CHLORODIPHENYL(4-PENTENYL)LEAD
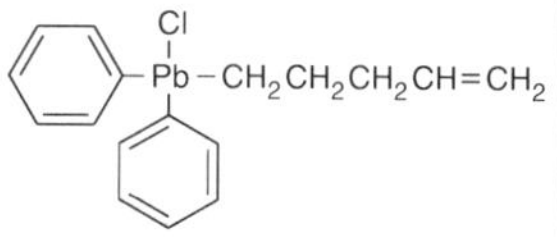
250 mg

S81,314-1
TRIPHENYLLEAD

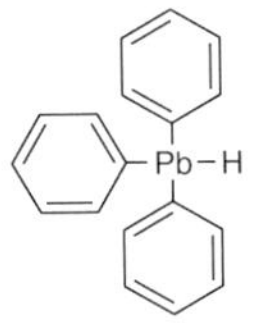

250 mg

S83,879-9
FLUOROTRIPHENYLLEAD

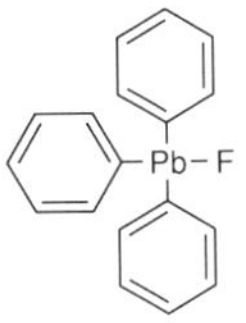

250 mg

S81,614-0
CHLOROTRIPHENYLLEAD

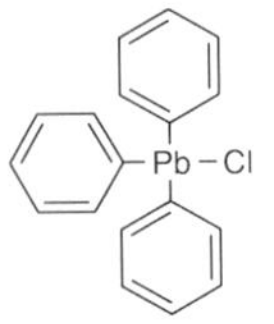

250 mg

S39,553-6
IODOTRIPHENYLLEAD

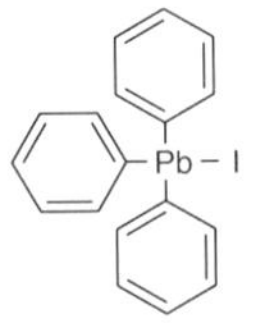

250 mg

S83,987-6
METHYLTRIPHENYLLEAD

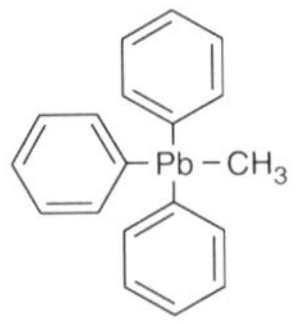

250 mg

S41,726-2
(METHYLTHIO)TRIPHENYLLEAD

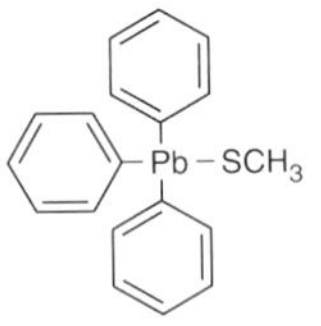

5 gm

S81,289-7
BENZYLCHLORODIPHENYLLEAD

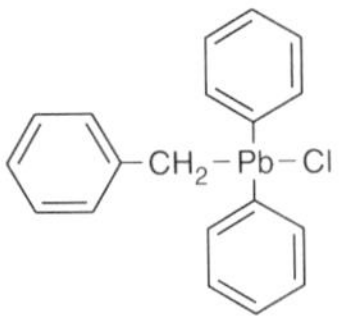

250 mg

S81,516-0
ETHYLTRIPHENYLLEAD

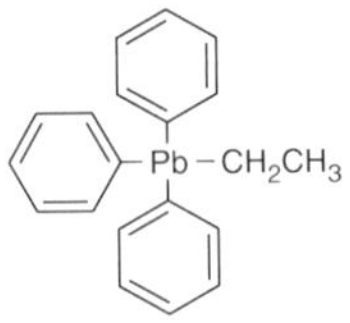

250 mg

S39,552-8
(ETHYLTHIO)TRIPHENYLLEAD

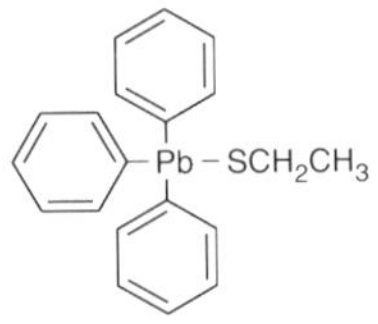

250 mg

S81,515-2
PROPYLTRIPHENYLLEAD

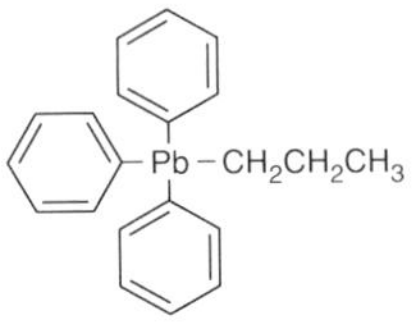

250 mg

S39,543-9
(PROPYLTHIO)TRIPHENYLLEAD

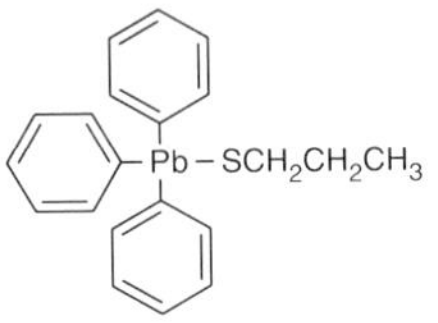

250 mg

S81,611-6
TERT-BUTYLTRIPHENYLLEAD

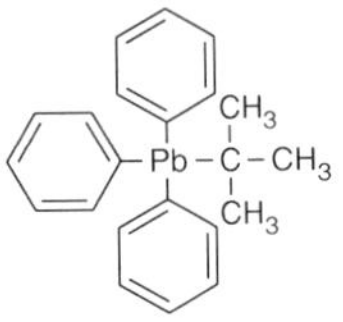

250 mg

S81,991-3
BUTYLTRIPHENYLLEAD

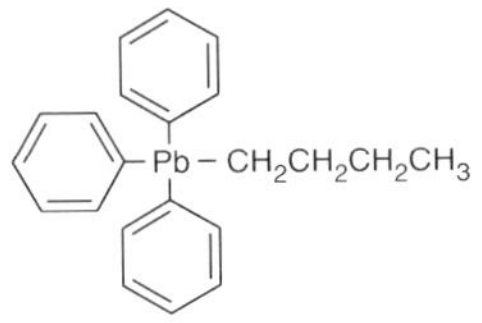

250 mg

S81,992-1
ISOBUTYLTRIPHENYLLEAD

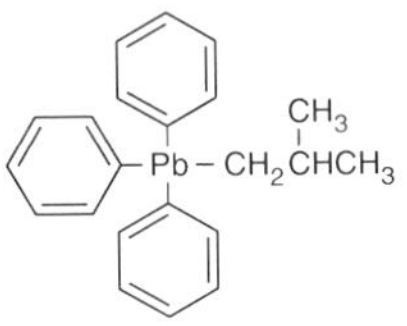

250 mg

S41,727-0
(THIOPHENOXY)TRIPHENYLLEAD

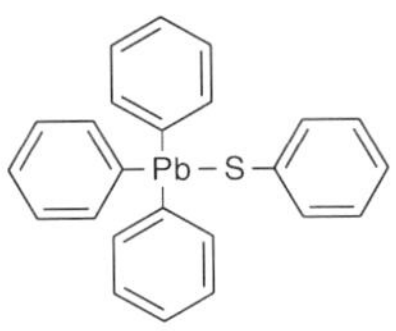

5 gm

S39,537-4
BENZYLTRIPHENYLLEAD

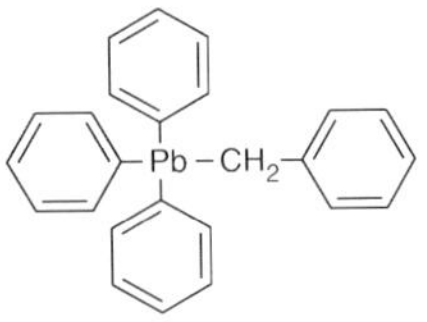

250 mg

S81,519-5
1-PHENYL-2-(TRIPHENYLPLUMBYL)-ACETYLENE

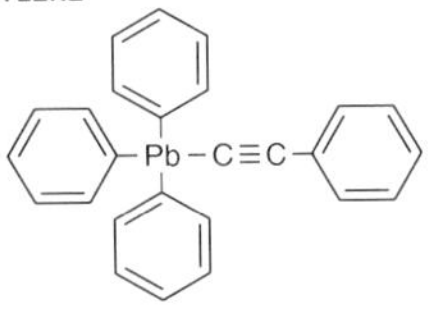

250 mg

S81,825-9
(PHENETHYL)TRIPHENYLLEAD

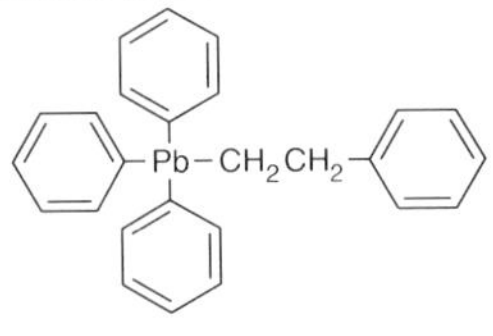

250 mg

S82,143-8
TRIPHENYLPLUMBYL LAURATE

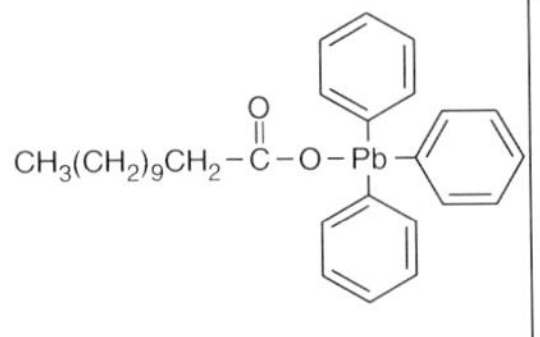

250 mg

S81,619-1
TRIPHENYLPLUMBYL TRIPHENYLSILYL OXIDE

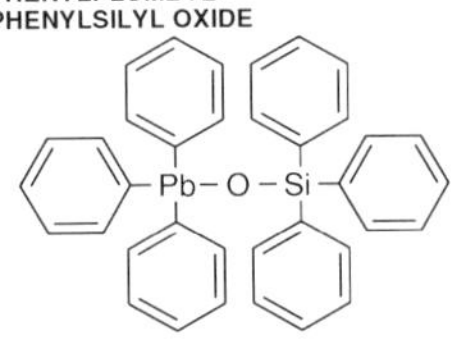

250 mg

S41,754-8
HEXAPHENYLDILEAD

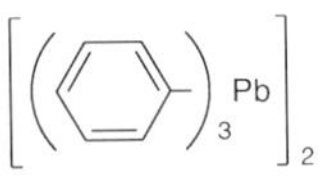

5 gm

S81,521-7
TRIPHENYL(TRIPHENYLMETHYL)LEAD

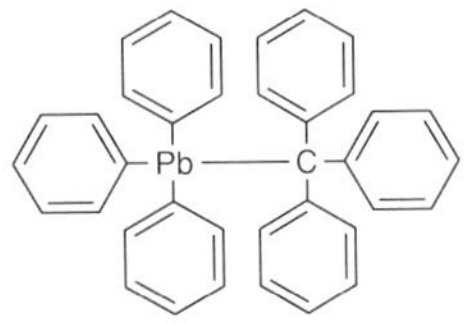

250 mg

S81,620-5
1,3-BIS(TRIPHENYLPLUMBYL)PROPANE

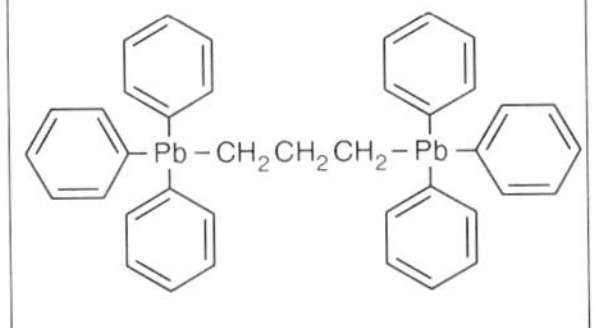

250 mg

S81,499-7
1,4-BIS(TRIPHENYLPLUMBYL)BUTANE

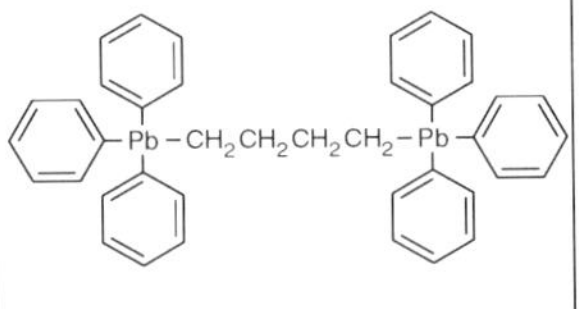

250 mg

S81,827-5
(P-TOLYL)TRIPHENYLLEAD

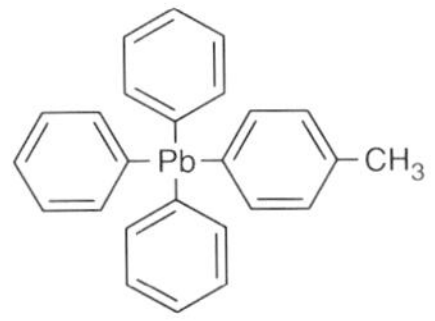

250 mg

S81,970-0
DIPHENYLDI(P-TOLYL)LEAD

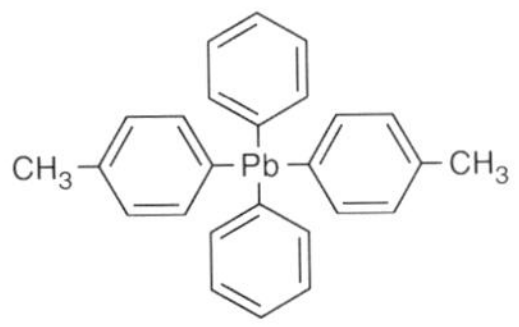

250 mg

S81,514-4
BENZYLTRI(P-TOLYL)LEAD

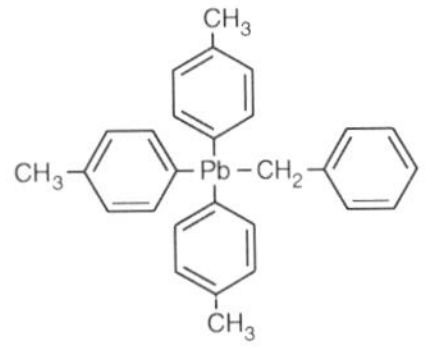

250 mg

S81,496-2
1,4-BIS(TRIPHENYLPLUMBYL)BENZENE

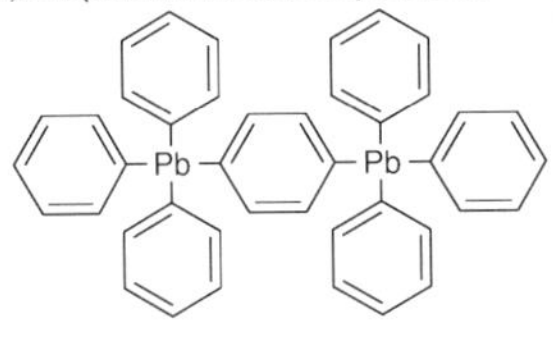

250 mg

S81,404-0
(4-BIPHENYLYL)TRIPHENYLLEAD

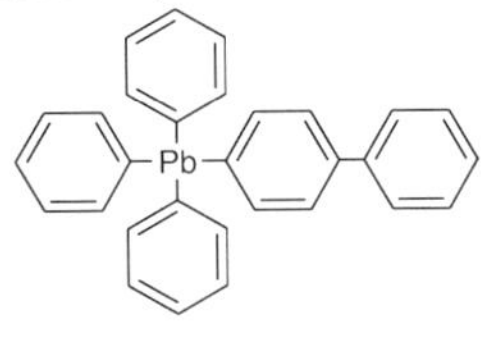

250 mg

S81,610-8
TRIETHYL(TRIPHENYLMETHYL)LEAD

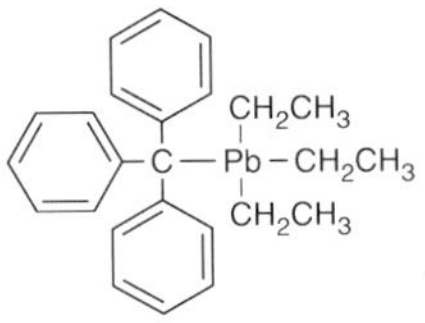

250 mg

S81,288-9
(1-NAPHTHYL)TRIPHENYLLEAD

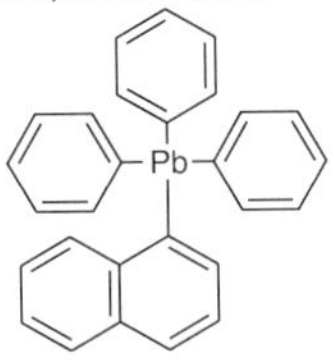

250 mg

S81,287-0
DI(1-NAPHTHYL)DIPHENYLLEAD

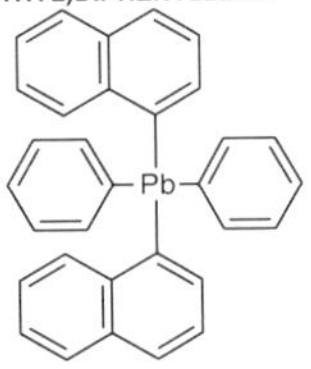

250 mg

S81,969-7
TRI(O-TOLYL)LEAD

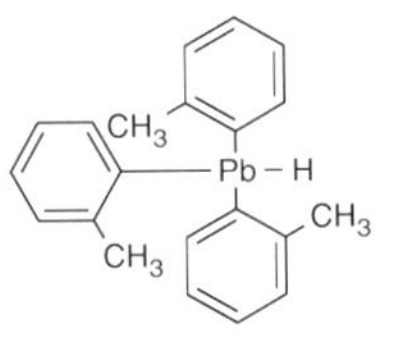

250 mg

S81,533-0
TRI(P-TOLYL)LEAD

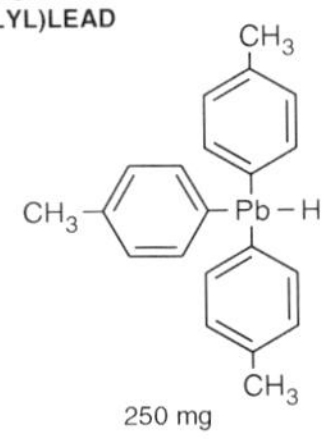

250 mg

S82,326-0
TRIMESITYLLEAD

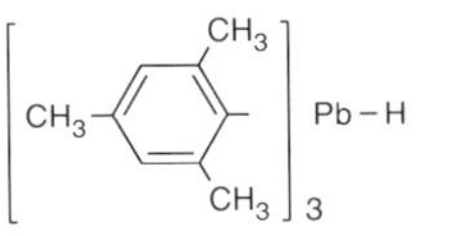

250 mg

S82,327-9
IODOTRIS(MESITYL)LEAD

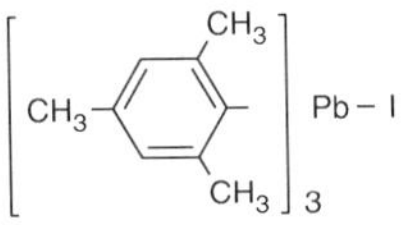

250 mg

S81,973-5
TRIS(1-NAPHTHYL)LEAD

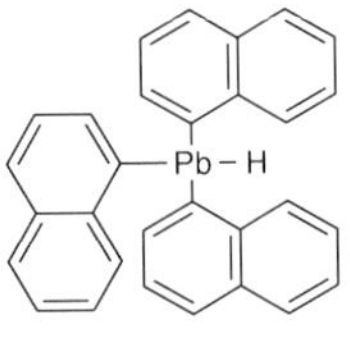

250 mg

S81,852-6
TRIETHYLPLUMBYL 4-METHYL-2-QUINOLINECARBOXYLATE

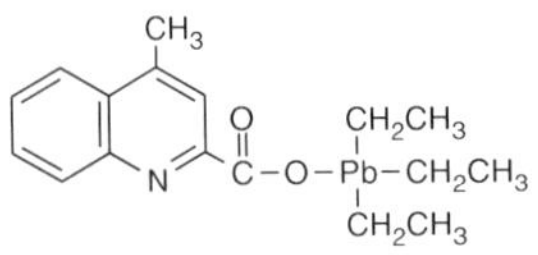

250 mg

S88,189-9
VINYLMERCURIC CHLORIDE

$H_2C{=}CHHgCl$

250 mg

S65,954-1
ETHYLMERCURIC CHLORIDE

CH_3CH_2HgCl

5 gm

S82,234-5
BIS(METHYLTHIO)MERCURY
$CH_3SHgSCH_3$
250 mg

S82,961-7
BUTYLMERCURIC CHLORIDE
$CH_3CH_2CH_2CH_2HgCl$
250 mg

S89,562-8
BIS(2-CHLOROETHYNYL)MERCURY
$ClC{\equiv}CHgC{\equiv}CCl$
250 mg

S81,451-2
BUTYLMERCURI ACETATE
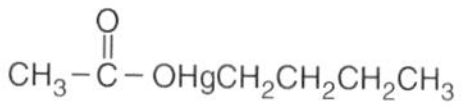
250 mg

S57,251-9
CHLOROHEXYLMERCURY
$CH_3(CH_2)_4CH_2HgCl$
5 gm

S88,180-5
(2,2-DIMETHYLPROPYL)MERCURI ACETATE
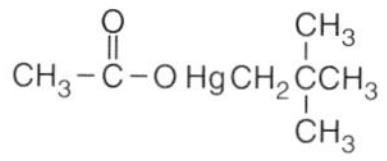
250 mg

S57,250-0
CHLOROHEPTYLMERCURY
$CH_3(CH_2)_5CH_2HgCl$
5 gm

S57,255-1
CHLOROOCTYLMERCURY
$CH_3(CH_2)_6CH_2HgCl$
5 gm

S57,258-6
CHLORODECYLMERCURY
$CH_3(CH_2)_8CH_2HgCl$
5 gm

S57,261-6
CHLORODODECYLMERCURY
$CH_3(CH_2)_{10}CH_2HgCl$
5 gm

S74,178-7
MERCURY(II) (N-HEXYLSULFIDE)
$CH_3(CH_2)_4CH_2S-Hg-SCH_2(CH_2)_4CH_3$
250 mg

S57,252-7
CHLOROTETRADECYLMERCURY
$CH_3(CH_2)_{12}CH_2HgCl$
5 gm

S74,181-7
MERCURY(II) (N-HEPTYLSULFIDE)
$CH_3(CH_2)_5CH_2S-Hg-SCH_2(CH_2)_5CH_3$
250 mg

S83,513-7
BROMOTETRADECYLMERCURY
$CH_3(CH_2)_{12}CH_2HgBr$
250 mg

S74,180-9
MERCURY(II) (N-OCTYLSULFIDE)
$CH_3(CH_2)_6CH_2S-Hg-SCH_2(CH_2)_6CH_3$
250 mg

S83,514-5
BROMOHEXADECYLMERCURY
$CH_3(CH_2)_{14}CH_2HgBr$
250 mg

S57,262-4
CHLOROOCTADECYLMERCURY
$CH_3(CH_2)_{16}CH_2HgCl$
5 gm

S83,930-2
OCTADECYLMERCURY(II) NITRATE
$CH_3(CH_2)_{16}CH_2Hg^+ \; NO_3^-$
250 mg

S83,515-3
BROMOOCTADECYLMERCURY
$CH_3(CH_2)_{16}CH_2HgBr$
250 mg

S83,516-1
DIDODECYLMERCURY
$CH_3(CH_2)_{10}CH_2HgCH_2(CH_2)_{10}CH_3$
250 mg

S83,933-7
DIOCTADECYLMERCURY
$CH_3(CH_2)_{16}CH_2HgCH_2(CH_2)_{16}CH_3$
250 mg

S61,684-2
BIS(TRICHLOROMETHYL)MERCURY(II)
$Cl_3C-Hg-CCl_3$
5 gm

S85,196-5
PHENYL TRICHLOROMETHYLMERCURY
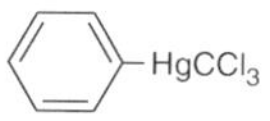
250 mg

S91,508-4
PHENYL(TRIBROMOMETHYL)MERCURY
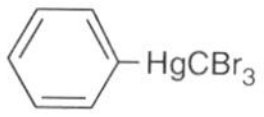
250 mg

S88,091-4
2-AMINO-5-IODOPHENYLMERCURI ACETATE
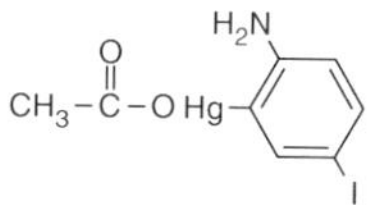
250 mg

S88,187-2
4-FLUORO-3-NITROPHENYLMERCURIC CHLORIDE
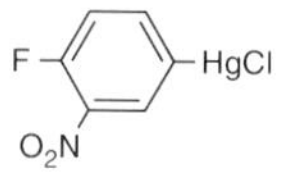
250 mg

S79,237-3
BIS(2-FLUOROPHENYL)MERCURY
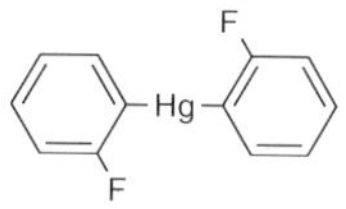
250 mg

S81,286-2
BIS(4-CHLOROPHENYL)MERCURY
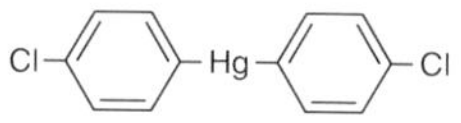
250 mg

S83,709-1
BIS(4-BROMOPHENYL)MERCURY
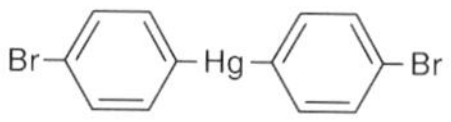
250 mg

S86,695-4
ALPHA-(CHLOROMERCURIMETHYL)-4-HYDROXYBENZHYDROL
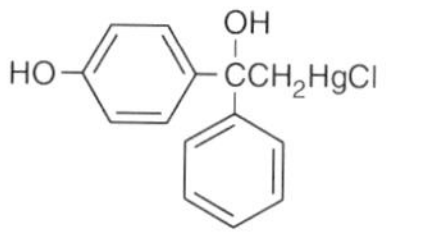
250 mg

S46,114-8
CHLORANILIC ACID, MERCURY SALT
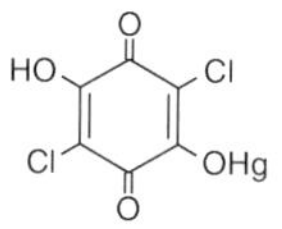
5 gm

S87,589-9
2-HYDROXY-2-(4-METHOXYPHENYL)-2-PHENYLETHYLMERCURI ACETATE
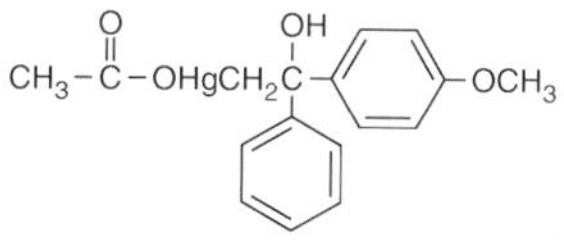
250 mg

S86,401-3
BETA-(CHLOROMERCURI)-SEC-PHENETHYLALCOHOL
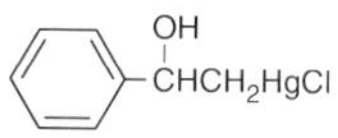
250 mg

S86,699-7
(2,2-DIPHENYL-2-HYDROXYETHYL)-MERCURI ACETATE
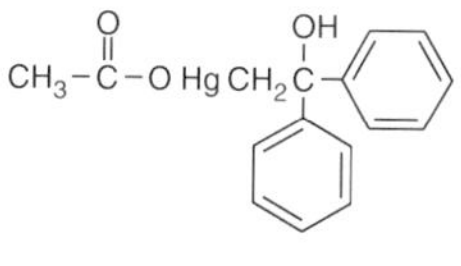
250 mg

S55,390-5
2-CHLOROMERCURI-4,6-DINITROPHENOL
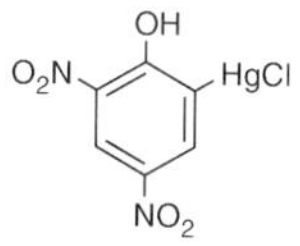
250 mg

S61,742-3
PHENYLMERCURIC SALICYLATE
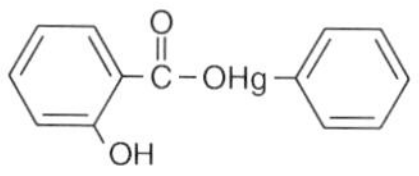
5 gm

S86,520-6
DIHYDRO-5,11-DIMETHYL-11-HYDROXY-5,10-METHANO(5H)DIBENZO(AD)-CYCLOHEPTEN-12-YL MERCURI ACETATE
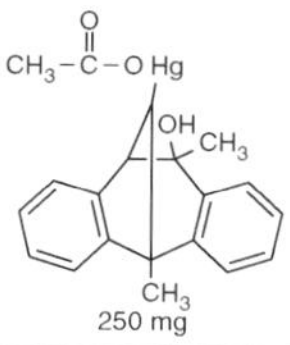
250 mg

S87,579-1
2-HYDROXY-2-PHENYLPROPYLMERCURI ACETATE
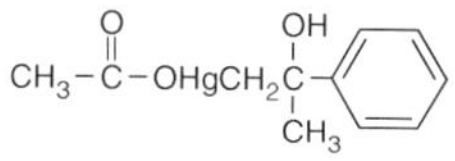
250 mg

S88,123-6
2-(CHLOROMERCURI)PHENOL
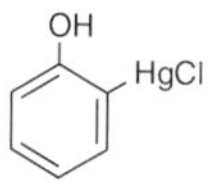
250 mg

S87,556-2
1-[4-(CHLOROMERCURI)PHENYLAZO]-2-NAPHTHOL
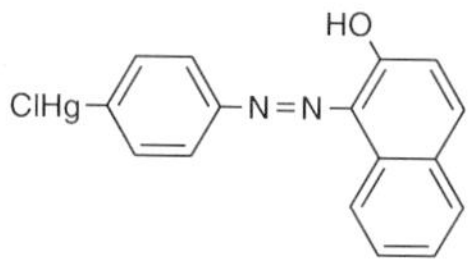
250 mg

S56,444-3
2-(CHLOROMERCURI)ETHYL ETHYL ETHER

5 gm

S88,090-6
2,4-DIMETHOXYPHENYLMERCURIC CHLORIDE

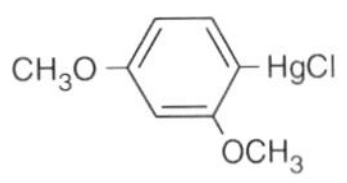

250 mg

S39,423-8
MESO-1,4-BIS(ACETOXYMERCURI)-2,3-DIETHOXYBUTANE

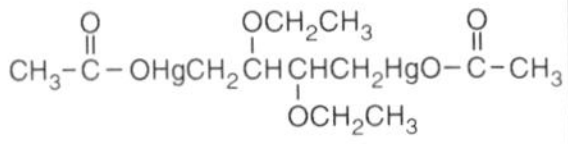

250 mg

S81,237-4
2-(CHLOROMERCURI)-5-(METHOXYMETHYL)FURAN

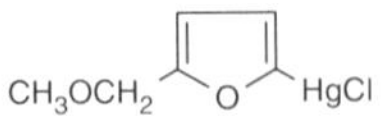

250 mg

S86,596-6
12-(CHLOROMERCURI)DIHYDRO-5,11-DIMETHYL-11-METHOXY-5,10-METHANO(5H)DIBENZO(AD)CYCLOHEPTENE

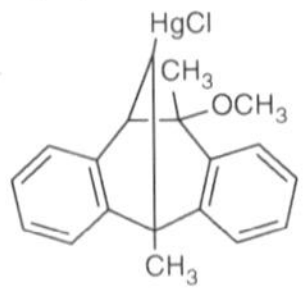

250 mg

S82,624-3
2-(CHLOROMERCURI)ANISOLE

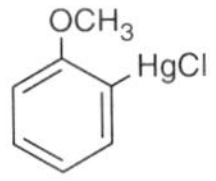

250 mg

S88,119-8
4-(CHLOROMERCURI)ANISOLE

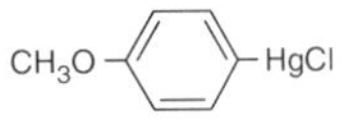

250 mg

S88,178-3
(4-METHOXYPHENYL)MERCURI ACETATE

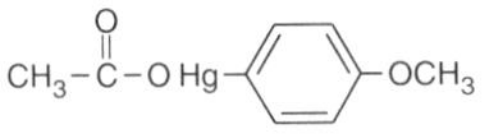

250 mg

S87,568-6
11-(CHLOROMERCURI)-9,10-DIHYDRO-12-METHOXY-9,10-ETHANOANTHRACENE

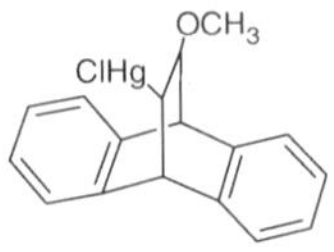

250 mg

S81,085-1
DIMETHYL 3,4-BIS(CHLOROMERCURI)-2,5-FURANDICARBOXYLATE

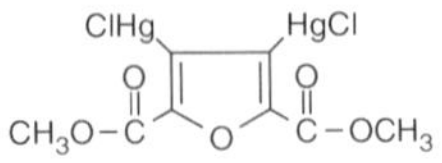

250 mg

S84,044-0
METHYL 5-BROMO-4-(CHLOROMERCURI)-2-FURANCARBOXYLATE

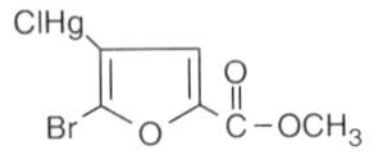

250 mg

S84,057-2
METHYL 4-(CHLOROMERCURI)-5-PHENYL-2-FURANCARBOXYLATE

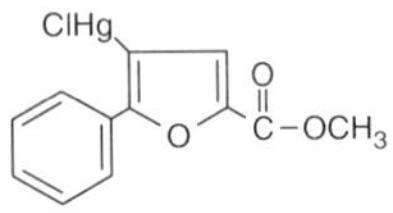

250 mg

S85,914-1
6-BENZAMIDO-7-(CHLOROMERCURI)PURINE

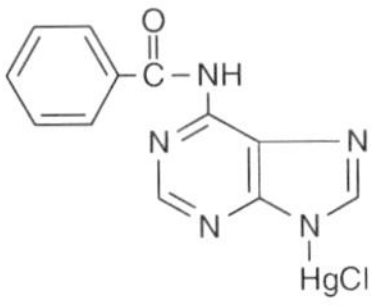

250 mg

S88,171-6
4-(DIMETHYLAMINO)-PHENYLMERCURI ACETATE

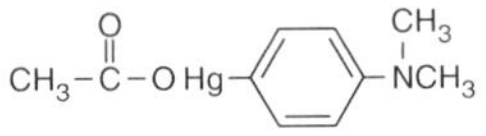

250 mg

S81,045-2
(9-ETHYL-3-CARBAZOLYL)-MERCURI ACETATE

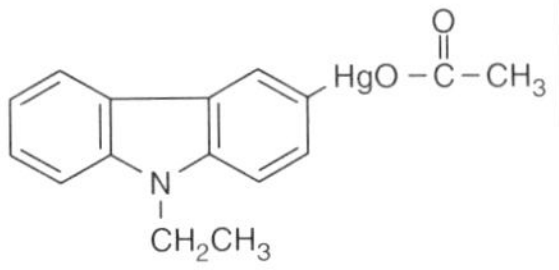

250 mg

S95,308-3
1-BENZYLIDENEAMINO-1H-TETRAZOL-5-YLMERCURI ACETATE

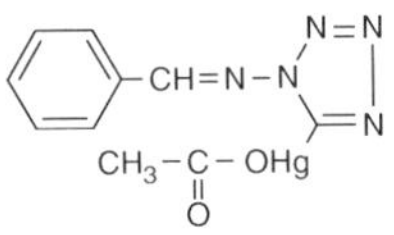

50 mg

S34,129-0
ANHYDRO-2-HYDROXYMERCURI-3-NITROBENZOIC ACID

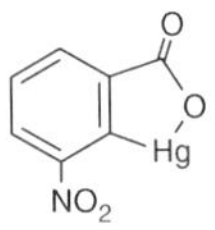

5 gm

S88,185-6
3-(CHLOROMERCURI)BENZALDEHYDE

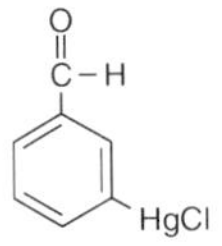

250 mg

S88,182-1
3-(CHLOROMERCURI)BENZOIC ACID

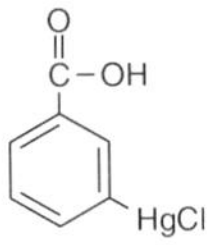

250 mg

S56,051-0
4-(HYDROXYMERCURI)BENZOIC ACID, POTASSIUM SALT

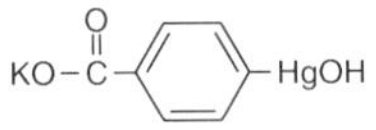

5 gm

S82,003-2
5-CHLORO-4-(CHLOROMERCURI)-2-FUROIC ACID

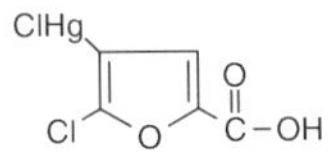

250 mg

S82,005-9
METHYL 3,4-BIS[(ACETOXY)MERCURI]-5-BROMO-2-FUROATE

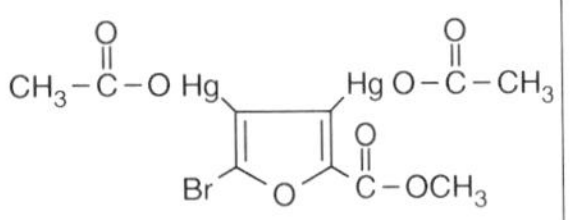

250 mg

S81,331-1
METHYL 5-CHLORO-4-(CHLOROMERCURI)-2-FUROATE

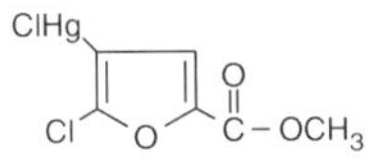

250 mg

S88,200-3
METHYL 4-(ACETOXYMERCURI)BENZOATE

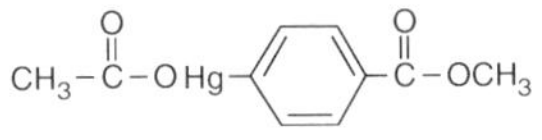

250 mg

S88,191-0
2-(CHLOROMERCURI)ACETIC ACID

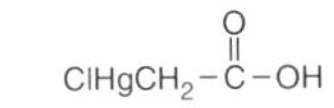

250 mg

S35,459-7
PHENYLMERCURIC STEARATE

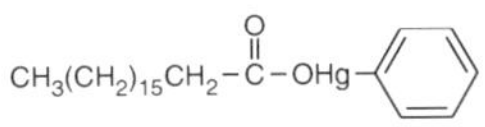

250 mg

S52,246-5
BIS(PHENYLMERCURI) PHTHALATE

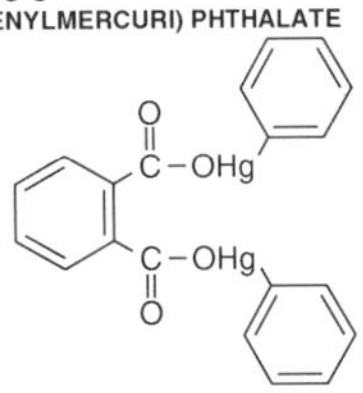

5 gm

S89,093-6
3-ACETOXY-4-(ACETOXYMERCURI)-3-HEXENE

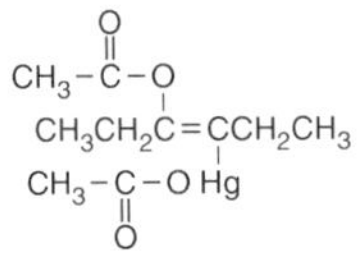

250 mg

S86,710-1
12-(CHLOROMERCURIMETHYL)-10,11-DIHYDRO-5,10-METHANO(5H)DIBENZO(AD)-CYCLOHEPTEN-11-YL ACETATE

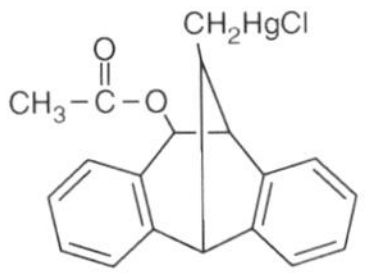

250 mg

S86,576-1
12-(CHLOROMERCURI)-9,10-DIHYDRO-9,10-ETHANOANTHRACEN-11-YL ACETATE

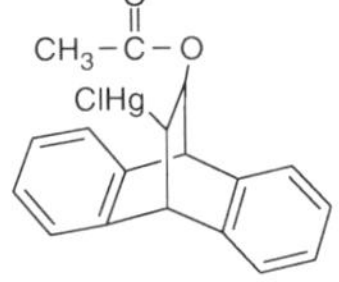

250 mg

S86,599-0
12-(CHLOROMERCURI)-9,10-DIHYDRO-9,10-DIMETHYL-9,10-ETHANOANTHRACEN-11-YL ACETATE

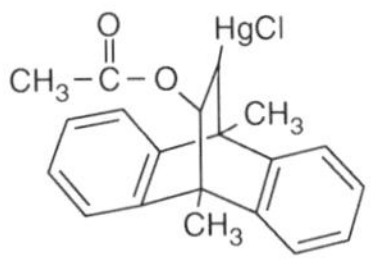

250 mg

S85,930-3
S-[1-(CHLOROMERCURI)-4-OXO-1,2,3,4-TETRAHYDRO-2-PYRIMIDINYL]-THIOACETATE

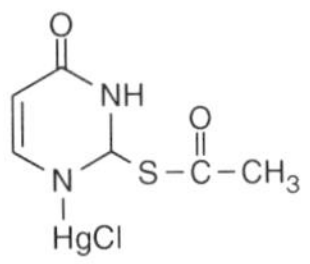

250 mg

S36,224-7
1,4,5,6,7,7-HEXACHLORO-N-(PHENYLMERCURI)BICYCLO[2.2.1]HEPT-5-ENE-2,3-DICARBOXIMIDE

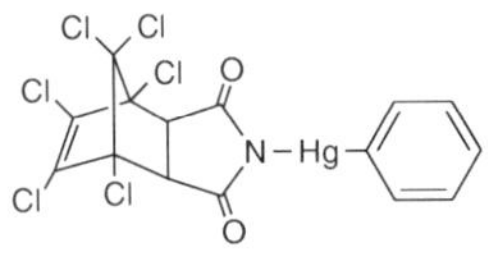

250 mg

S36,950-0
1,1'-MERCURIDI-2-PYRROLIDINONE

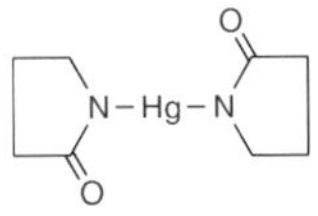

5 gm

S88,115-5
4'-(CHLOROMERCURI)ACETANILIDE

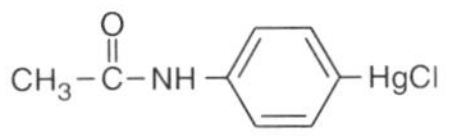

250 mg

S83,931-0
OCTADECYLMERCURY(II)CYANIDE

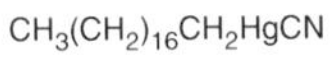

250 mg

S81,581-0
2-(CYANOMERCURI)FURAN

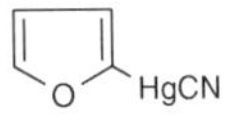

250 mg

S86,473-0
2-(BROMOMERCURI)TRICYCLO[2.2.1.0 (2,6)]HEPTANE

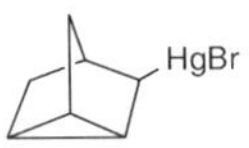

250 mg

S81,317-6
2-(CHLOROMERCURI)FURAN

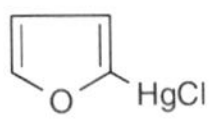

250 mg

S81,414-8
3-(CHLOROMERCURI)FURAN

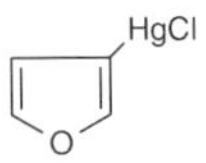

250 mg

S81,449-0
2-CHLORO-5-(CHLOROMERCURI)FURAN

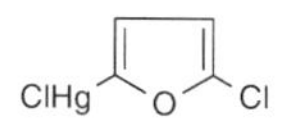

250 mg

S82,000-8
5-(CHLOROMERCURI)-2,3-DIBROMOFURAN

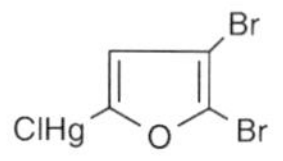

250 mg

S81,318-4
2,5-BIS(CHLOROMERCURI)FURAN

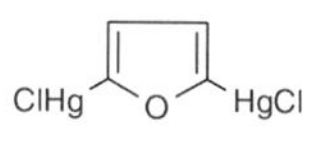

250 mg

S83,847-0
2,3,4,5-TETRAKIS-(CHLOROMERCURI)FURAN

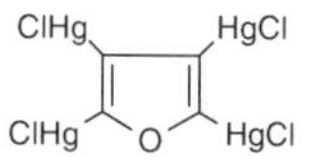

250 mg

S83,819-5
BIS(2-FURYL)MERCURY

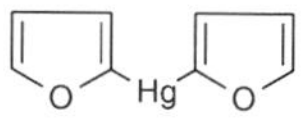

250 mg

S81,154-8
3-(CHLOROMERCURI)DIBENZOFURAN

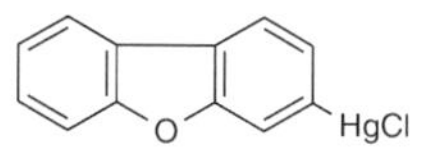

250 mg

S83,905-1
2,8-BIS(CHLOROMERCURI)DIBENZOFURAN

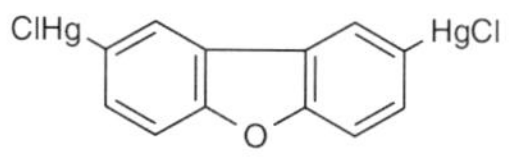

250 mg

S83,946-9
(4-DIBENZOFURYLMERCURI) ACETATE

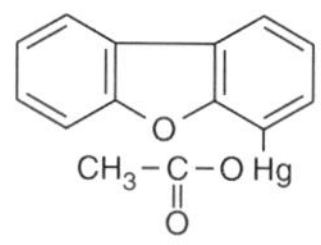

250 mg

S81,580-2
2-(CHLOROMERCURI)THIOPHENE

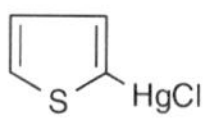

250 mg

S81,579-9
2,5-BIS(CHLOROMERCURI)THIOPHENE

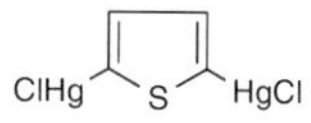

250 mg

S93,537-9
5,5'-BIS(CHLOROMERCURI)-2,2'-BITHIOPHENE

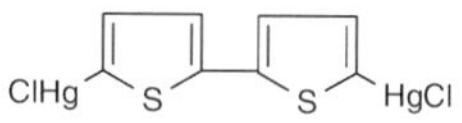

250 mg

S87,208-3
BIS[2-(2-BENZOXAZOLYL)ETHYNYL]-MERCURY

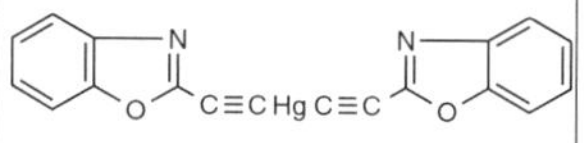

250 mg

S36,613-7
PHENYLMERCURIC BORATE

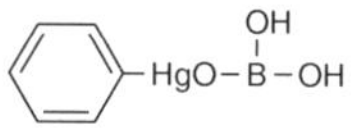

5 gm

S58,067-8
(BROMOMERCURI)BENZENE

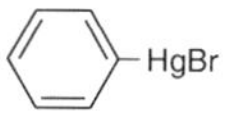

250 mg

S85,002-0
PHENYLMERCURIC IODIDE

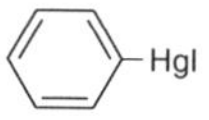

250 mg

S81,089-4
BENZYLMERCURIC CHLORIDE

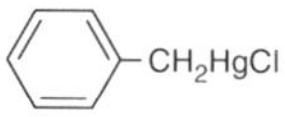

250 mg

S83,870-5
BENZYLMERCURIC IODIDE

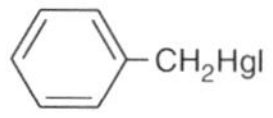

250 mg

S88,112-0
BENZYLMERCURI ACETATE

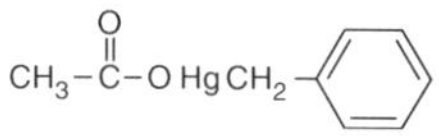

250 mg

S88,203-8
2-METHYL-2-PHENYLPROPYLMERCURIC CHLORIDE

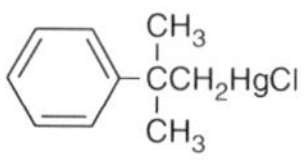

250 mg

S88,044-2
2-METHYL-2-PHENYLPROPYLMERCURI ACETATE

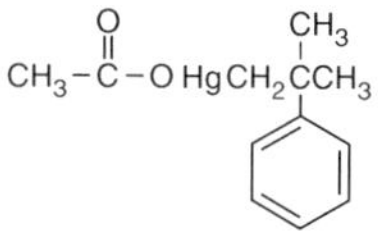

250 mg

S81,629-9
DIBENZYLMERCURY

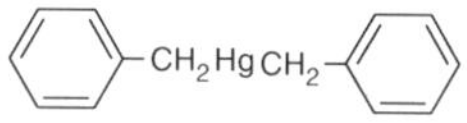

250 mg

S70,673-6
P-TOLYLMERCURIC CHLORIDE

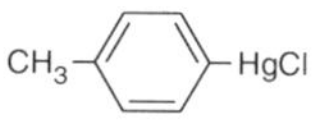

5 gm

S81,290-0
DI(P-TOLYL)MERCURY

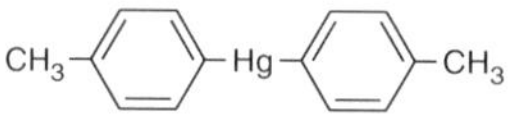

250 mg

S82,118-7
P-TOLYLMERCURI HEXADECANOATE

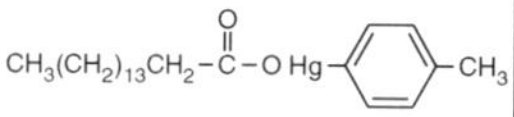

250 mg

S88,114-7
3,4-XYLYLMERCURIC CHLORIDE

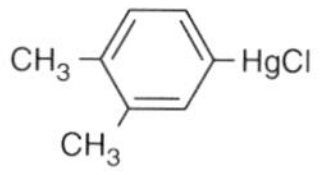

250 mg

S88,077-9
MESITYLMERCURIC CHLORIDE

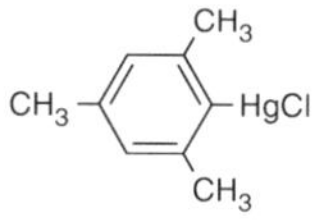

250 mg

S88,142-2
2,4,6-(TRIISOPROPYL)-PHENYLMERCURIC CHLORIDE

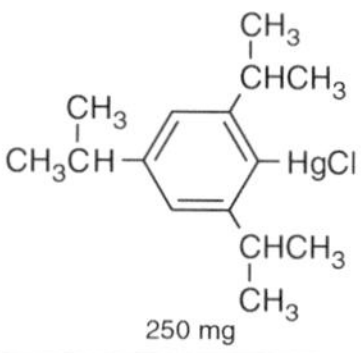

250 mg

S88,137-6
2,4,6-(TRIISOPROPYL)-PHENYLMERCURI ACETATE

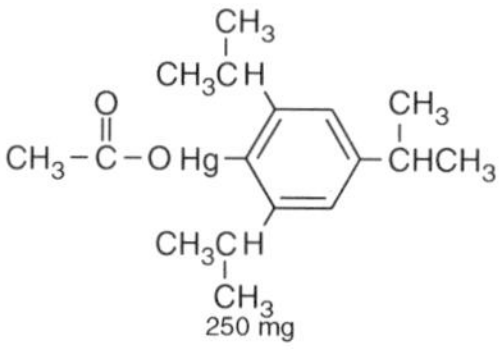

250 mg

S81,681-7
DIMESITYLMERCURY

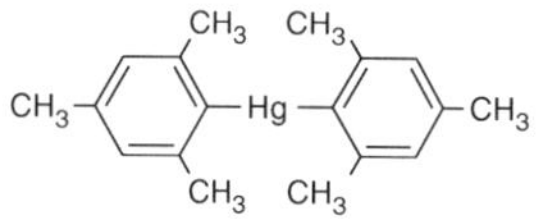

250 mg

S87,563-5
9,10-DIHYDRO-9,10-ETHANOANTHRACENE-11,12-DICARBOXYLIC ACID, MERCURY(II) SALT

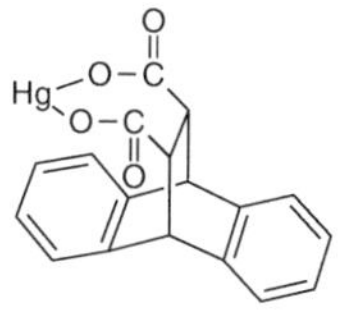

250 mg

S82,940-4
DIBENZO(B,E)(1,4)DIMERCURIN

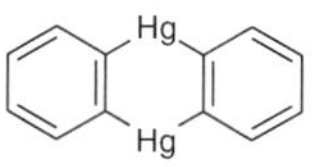

250 mg

S60,276-0
TRIMETHYLTIN IODIDE AMMONIA COMPLEX

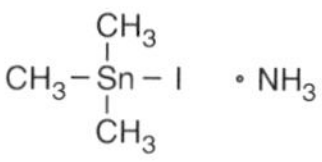

5 gm

S89,927-5
2-(1-HYDROXYETHYLIDENE)-1-CYCLOPENTANONE, COPPER(II) COMPLEX

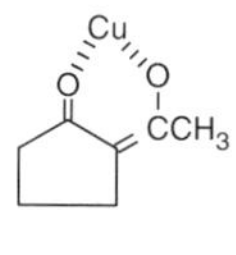

250 mg

S64,485-4
ETHYL 2-[4-(PENTYLOXY)BENZOYL]-ACETATE, COPPER(II) DERIVATIVE

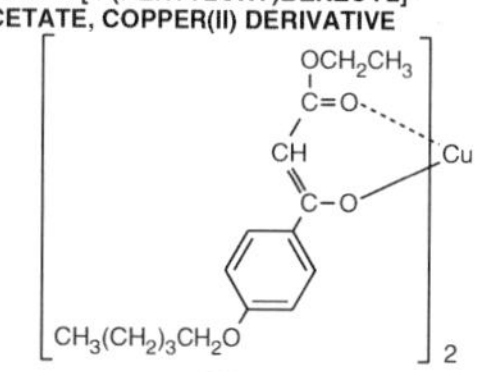

250 mg

S94,351-7
5-(FERROCENYLMETHYL)-2,2,5-TRIMETHYL-1,3-DIOXANE-4,6-DIONE

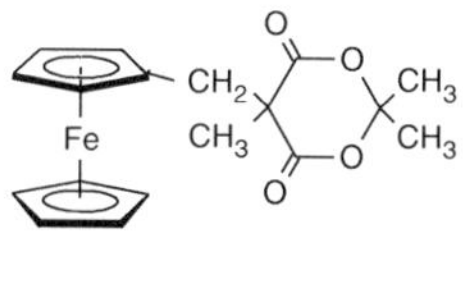

250 mg

S88,775-7
TRIS(ETHYLENEDIAMINE)NICKEL(II) CHLORIDE DIHYDRATE

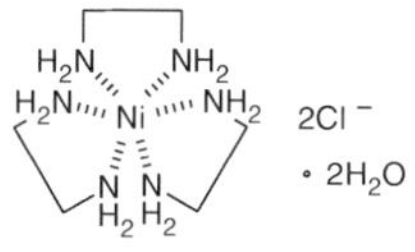

250 mg

S88,774-9
(5,7,7,12,14,14-HEXAMETHYL-1,4,8,11-TETRAAZACYCLOTETRADECA-4,11-DIENE-N1,N4,N8,N11)NICKEL(II) BIS(PERCHLORATE)

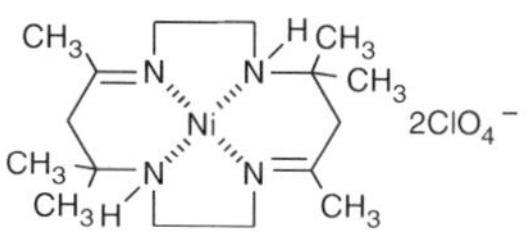

250 mg

S88,773-0
(5,7,7,12,14,14-HEXAMETHYL-1,4,8,11-TETRAAZACYCLOTETRADECANE-N1,N4,N8,N11)NICKEL(II) BIS(PERCHLORATE)

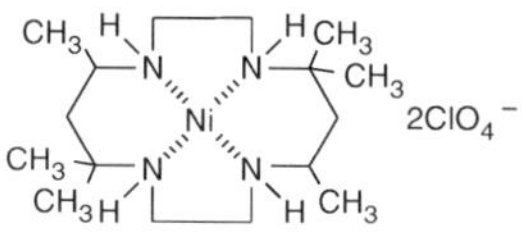

250 mg

S88,764-1
DICHLORO[N-(2-PYRIDYLMETHYLENE)-ANILINE-N,N']COPPER(II)

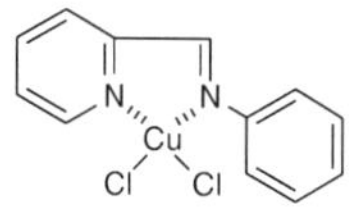

250 mg

S88,745-5
FERROCENECARBOXALDEHYDE (FERROCENYLMETHYL)HYDRAZONE

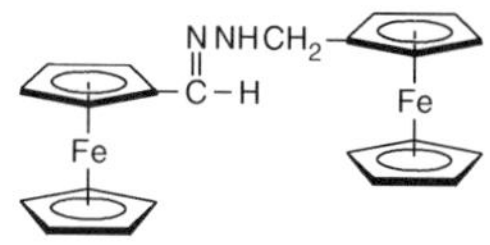

250 mg

S39,232-4
TRIMETHYLTIN NITRATE MONOHYDRATE
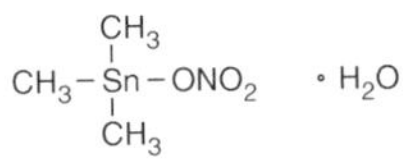
5 gm

S55,666-1
N,N-DIPHENYL-1-FERROCENECARBOXAMIDE
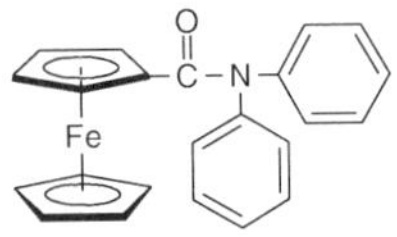
250 mg

S46,074-5
1,1'-BIS(CHLOROMERCURI) FERROCENE
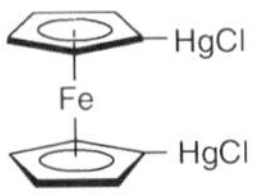
5 gm

S88,746-3
1,1''-BIFERROCENE
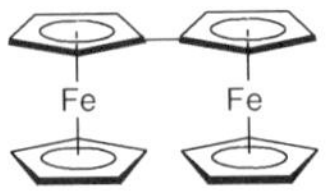
250 mg

S88,744-7
(FERROCENYLMETHYL)-TRIPHENYLPHOSPHONIUM IODIDE
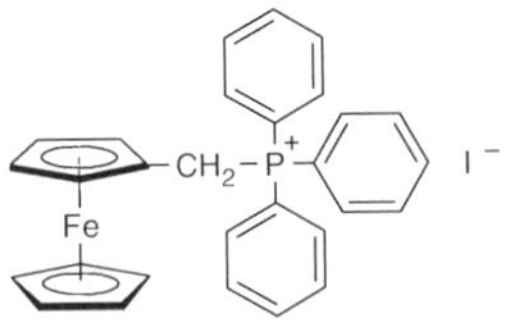
250 mg

S88,756-0
2-BENZOTHIAZOLYLACETYLENE COPPER(I) SALT
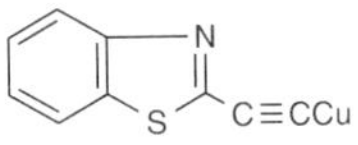
250 mg

S88,763-3
(2,12-DIMETHYL-3,7,11,17-TETRAAZA-BICYCLO[11.3.1]HEPTADECA-1(17),2,11,13,15-PENTAENE-N3,N7,N11,N17)NICKEL(II) ZNCL4
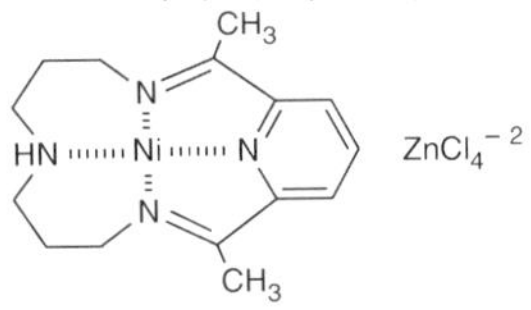
250 mg

S53,435-8
M-TOLUIDINE CHROMIUM TRICARBONYL
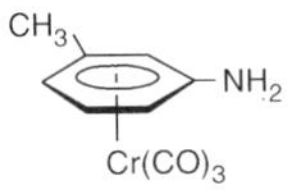
5 gm

S53,433-1
P-TOLUIDINE CHROMIUM TRICARBONYL
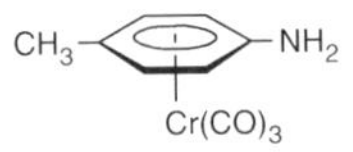
5 gm

S54,222-9
2,6-XYLIDINE CHROMIUM TRICARBONYL
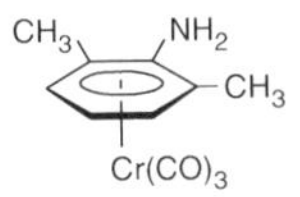
5 gm

S53,358-0
N,N-DIMETHYLANILINE CHROMIUM TRICARBONYL
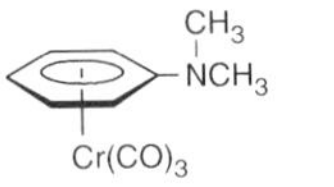
250 mg

S81,628-0
TEREPHTHALIC ACID, COPPER(II) SALT
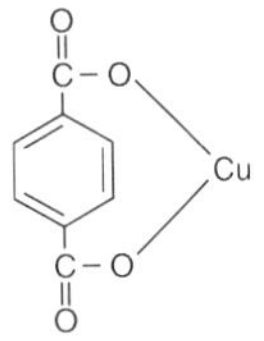
250 mg

S43,223-7
BIS(CYCLOPENTADIENYLCHROMIUM TRICARBONYL) MERCURY
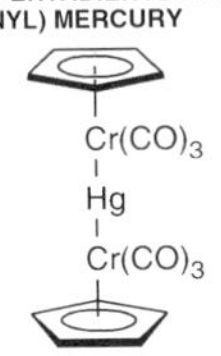
250 mg

S61,710-5
BIS[DICARBONYL (METHYLCYCLOPENTADIENYL)IRON]
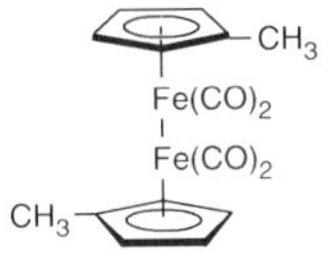
5 gm

S53,450-1
O-XYLENE CHROMIUM TRICARBONYL
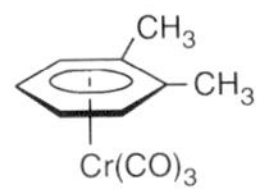
5 gm

S53,449-8
P-XYLENE CHROMIUM TRICARBONYL
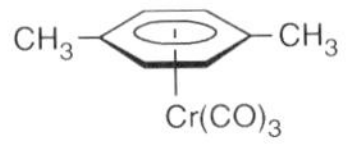
5 gm

S46,055-9
SILVER FLUOROSULFATE
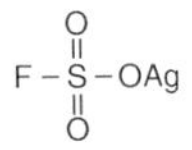
5 gm

S94,160-3
SODIUM HEXAFLUOROFERRATE(III)
Na_3FeF_6
5 gm

S74,541-3
CHLOROPENTAMMINECHROMIUM(III) CHLORIDE
$[Cr(NH_3)_5Cl]Cl_2$
250 mg

S83,706-7
OSMIUM TETRACHLORIDE
$OsCl_4$
250 mg

S74,187-6
HEXAAMMINECHROMIUM(III) NITRATE
$[Cr(NH_3)_6]\ (NO_3)_3$
250 mg

S45,931-3
GALLIUM PERCHLORATE HEXAHYDRATE
$Ga(ClO_4)_3 \cdot 6H_2O$
250 mg

S46,176-8
PRASEODYMIUM SULFATE OCTAHYDRATE
$Pr_2(SO_4)_3 \cdot 8H_2O$
5 gm

S46,108-3
EUROPIUM(III) SULFATE OCTAHYDRATE, 99.99%
$Eu_2(SO_4)_3 \cdot 8H_2O$
5 gm

S46,023-0
HOLMIUM SULFATE OCTAHYDRATE
$Ho_2(SO_4)_3 \cdot 8H_2O$
5 gm

S46,107-5
ERBIUM SULFATE OCTAHYDRATE, 99.99%
$Er_2(SO_4)_3 \cdot 8H_2O$
5 gm

S46,269-1
YTTERBIUM(III) SULFATE OCTAHYDRATE
$Yb_2(SO_4)_3 \cdot 8H_2O$
250 mg

S76,598-8
HEXABROMOPLATINIC ACID (PLATINIC BROMIDE)
H_2PtBr_6
250 mg

S81,376-1
COPPER MERCURIC IODIDE
Cu_2HgI_4
250 mg

S66,849-4
ALUMINUM TRIMETHOXIDE
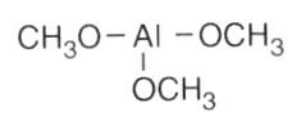
250 mg

S35,741-3
ALOXALIC ACID TETRAHYDRATE
$H_3[Al(C_2O_4)_2(OH)_2] \cdot 4H_2O$
5 gm

S72,738-5
TRANS-POTASSIUM DIAQUODIOXALATOCHROMATE(III) TRIHYDRATE
$K[Cr(H_2O)_2(C_2O_4)_2] \cdot 3H_2O$
5 gm

S83,707-5
BROMODIETHYLTHALLIUM
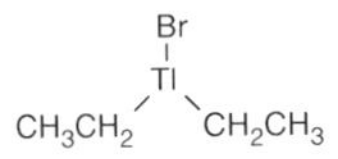
250 mg

S83,708-3
DIETHYLIODOTHALLIUM
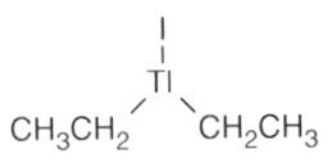
250 mg

S81,854-2
DIETHYLLEAD SELENITE
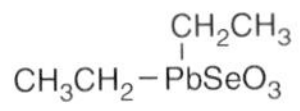
250 mg

S72,740-7
(TRIETHYLENETETRAMINE)CADMIUM(II) IODIDE
$[Cd(NH_2CH_2CH_2NHCH_2)_2]\ I_2$
5 gm

S43,343-8
POTASSIUM HEXATHIOCYANATOCHROMATE(III)
$K_3Cr(SCN)_6$
5 gm

S74,184-1
BIS(1,2,3-PROPANETRIYLTRIAMINE)-COBALT(III) IODIDE
$\{Co[(NH_2)_3C_3H_5]\}\ I_3$
250 mg

S72,742-3
BIS(1,3-DIAMINO-2-PROPANOL)NICKEL(II) THIOCYANATE
$\{Ni[(H_2NCH_2)_2CHOH]_2\}\ (SCN)_2$
5 gm

S72,736-9
BIS(N,N-DIETHYLETHYLENEDIAMINE)-NICKEL(II) THIOCYANATE
$\{Ni[(CH_3CH_2)_2NCH_2CH_2NH_2]_2\}\ (SCN)_2$
5 gm

S74,276-7
TRIETHYLPHOSPHINE-SILVER IODIDE COMPLEX

$[(CH_3CH_2)_3P \cdot AgI]_4$

5 gm

S81,667-1
TRIPHENYLBISMUTH BIS(2-CHLOROACETATE)

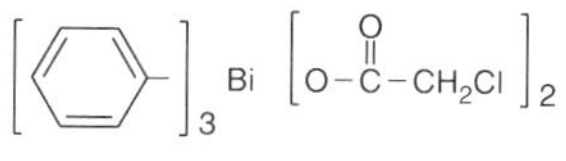

250 mg

S66,431-6
TRIS(1,1,1,5,5,6,6,7,7,7-DECAFLUORO-2,4-HEPTANEDIONATO)-PRASEODYMIUM DIHYDRATE

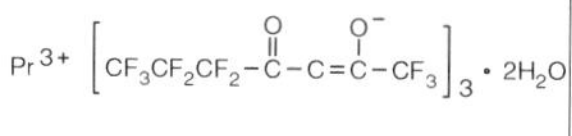

250 mg

S66,433-2
TRIS(1,1,1,5,5,6,6,7,7,7-DECAFLUORO-2,4-HEPTANEDIONATO)-EUROPIUM DIHYDRATE

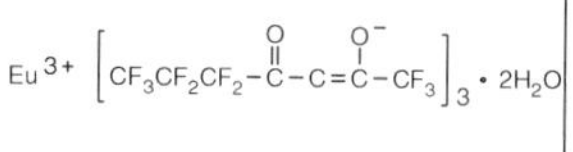

250 mg

S66,439-1
TRIS(1,1,1,5,5,6,6,7,7,7-DECAFLUORO-2,4-HEPTANEDIONATO)-YTTERBIUM DIHYDRATE

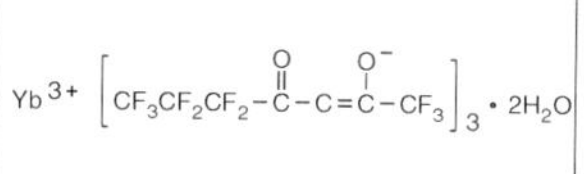

250 mg

S84,001-7
DIETHYLTHALLIUM TRIFLUOROACETATE

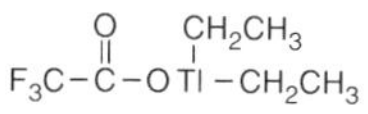

250 mg

S81,280-3
BIS(PHENYLTHIO)(4-CHLOROPHENYL)-BISMUTHINE

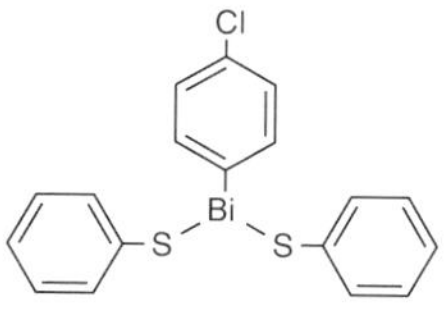

250 mg

S81,680-9
BIS(4-CHLOROPHENYL)(O-TOLYL)-BISMUTHINE

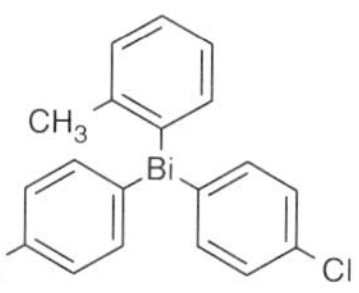

250 mg

S83,980-9
(4-CHLOROPHENYL)DIMETHOXYBISMUTHINE

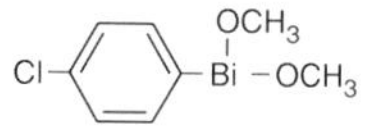

250 mg

S81,281-1
TRIS(4-FLUOROPHENYL)BISMUTHINE

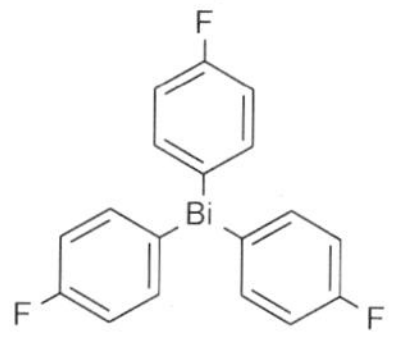

250 mg

S81,284-6
TRIS(4-CHLOROPHENYL)BISMUTHINE

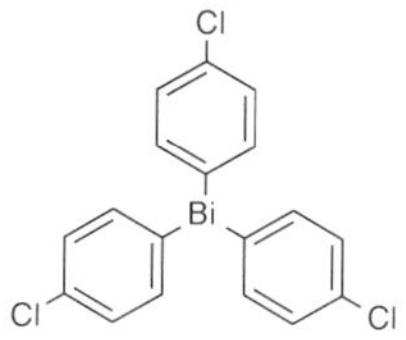

250 mg

S81,283-8
DICHLOROTRIS(4-CHLOROPHENYL)BISMUTH

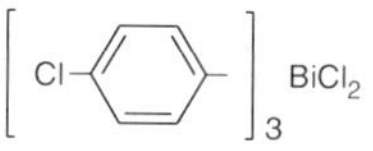

250 mg

S81,684-1
TRIS(4-BROMOPHENYL)BISMUTHINE

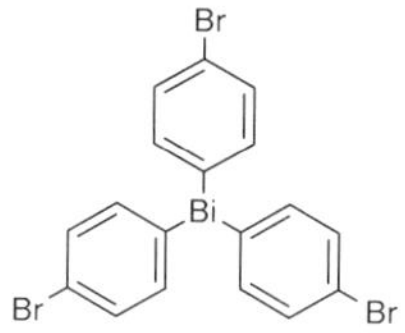

250 mg

S81,275-7
BIS(4-CHLOROPHENYL)-1-NAPHTHYLBISMUTHINE

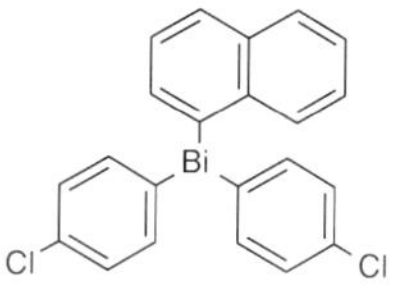

250 mg

S88,840-0
3,4-DIHYDRO-6-METHYL-4-PHENYL-2H-1-BENZOSELENIN-4-OL

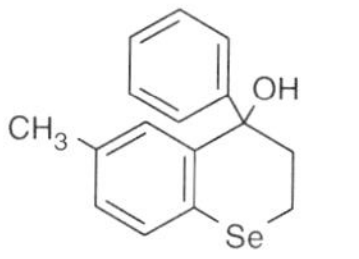

250 mg

S43,142-7
CHLORANILIC ACID, LANTHANUM SALT

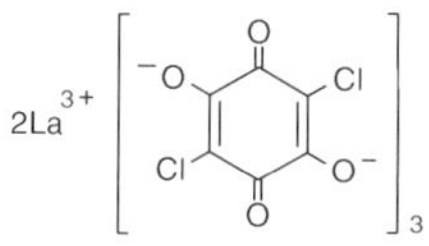

5 gm

S43,186-9
RHODIZONIC ACID, BARIUM SALT

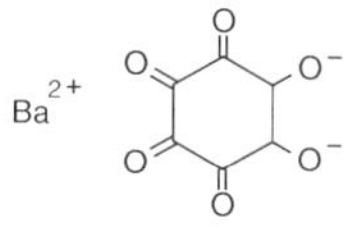

5 gm

S76,500-7
BISMUTH(III) GALLATE, BASIC

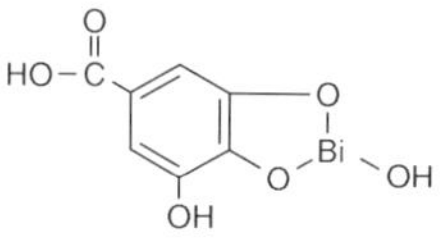

5 gm

S81,273-0
TRIPHENYLBISMUTH DISALICYLATE

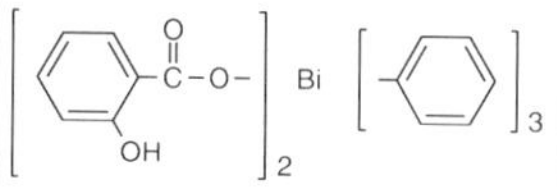

250 mg

S81,678-7
TRIS(2-METHOXYPHENYL)BISMUTHINE

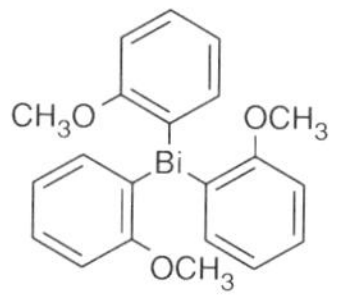

250 mg

S94,939-6
(4-METHOXYPHENYL)TRIBROMOTELLURIUM

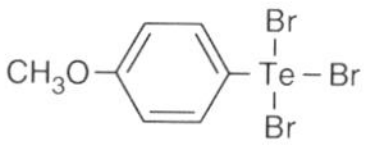

250 mg

S94,943-4
(4-METHOXYPHENYL)TRIIODOTELLURIUM

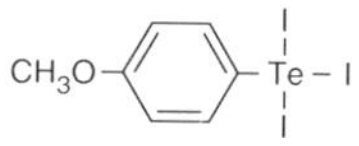

250 mg

S94,941-8
(4-ETHOXYPHENYL)TRIBROMOTELLURIUM

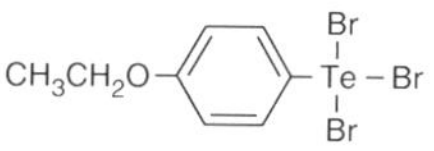

250 mg

S81,276-5
TRIS(4-METHOXYPHENYL)BISMUTHINE

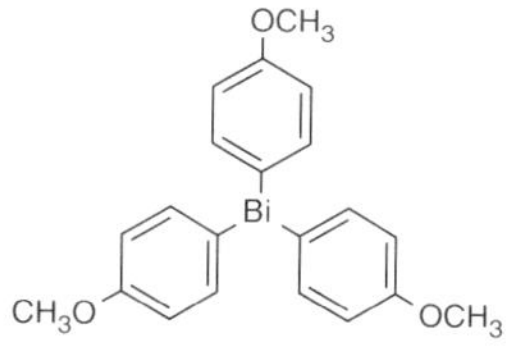

250 mg

S95,080-7
GAMMA-OXOFERROCENEBUTYRIC ACID

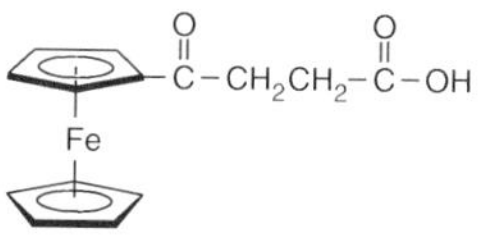

50 mg

S81,343-5
2-BENZOTHIAZOLETHIOL, SILVER(I) SALT

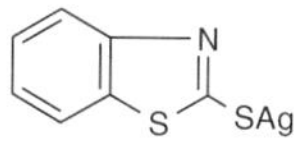

250 mg

S41,017-9
2,2'-THIOBIS(4-TERT-OCTYLPHENOLATO)-N-BUTYLAMINE NICKEL(II)

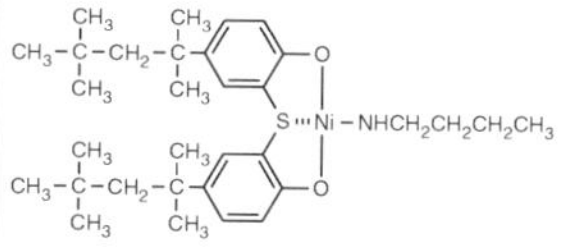

5 gm

S36,539-4
TRIS(ETHYLENEDIAMINE)COBALT(III) SULFATE

$Co(H_2NCH_2CH_2NH_2)_3SO_4$

5 gm

S36,540-8
TRIS(ETHYLENEDIAMINE)COPPER(II) SULFATE

$Cu(H_2NCH_2CH_2NH_2)_3SO_4$

5 gm

S36,538-6
TRIS(ETHYLENEDIAMINE)CHROMIUM(III) SULFATE

$[Cr(H_2NCH_2CH_2NH_2)_3]_2(SO_4)_3$

5 gm

S36,541-6
TRIS(ETHYLENEDIAMINE)IRON(II) SULFATE

$[Fe(H_2NCH_2CH_2NH_2)_3]_2(SO_4)_3$

5 gm

S36,038-4
DL-CIS-ALPHA-DICHLOROTRIETHYLENETETRAAMINE-COBALT(III) CHLORIDE

$[Co(H_2NCH_2CH_2NHCH_2CH_2NHCH_2CH_2NH_2)Cl_2]Cl$

250 mg

S36,039-2
TRANS-DICHLOROTRIETHYLENETETRAAMINE-COBALT(III) CHLORIDE

$[Co(H_2NCH_2CH_2NHCH_2CH_2NHCH_2CH_2NH_2)Cl_2]Cl$

250 mg

S81,272-2
TRIPHENYLBISMUTH BIS(4-AMINOBENZOATE)

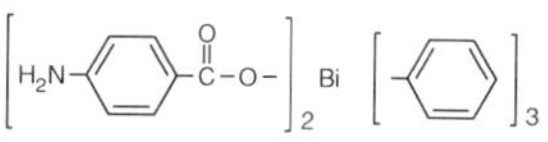

250 mg

S54,821-9
PICOLINIC ACID (FERROCENYLMETHYLENE)HYDRAZIDE

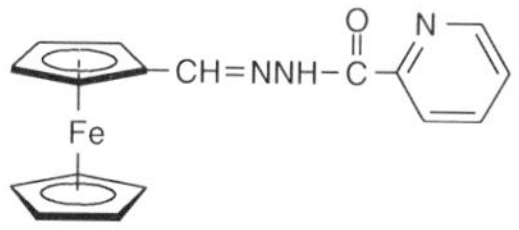

5 gm

S54,822-7
ISONICOTINIC ACID (FERROCENYLMETHYLENE)HYDRAZIDE

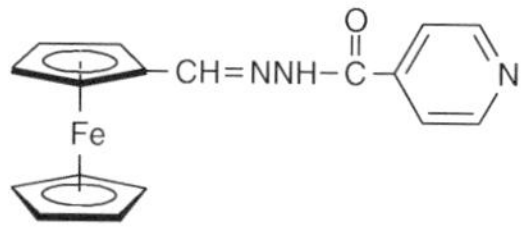

5 gm

S38,876-9
N-FERROCENYLMETHYLENE-4-(PHENYLAZO)ANILINE

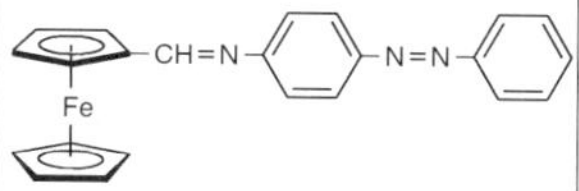

5 gm

S38,878-5
N,N'-BIS(FERROCENYLMETHYLENE)-1,4-PHENYLENEDIAMINE

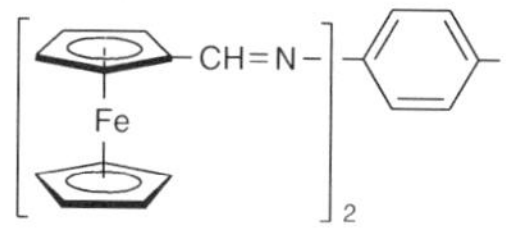

5 gm

S52,596-0
1,2-DIHYDRO-2-FERROCENYL-4(3H)-QUINAZOLINONE

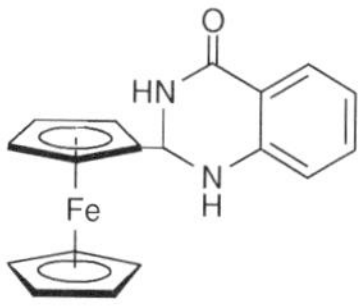

250 mg

S65,369-1
(R)-1-(DIMETHYLAMINO)-ETHYLFERROCENE TARTRATE

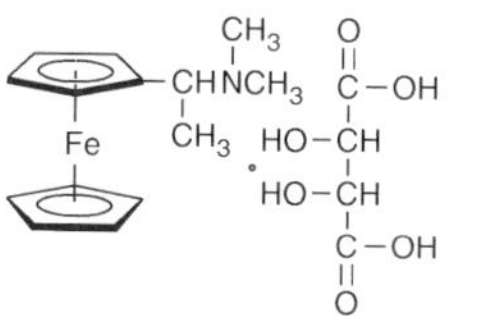

250 mg

S65,375-6
(S)-1-(DIMETHYLAMINO)-ETHYLFERROCENE TARTRATE

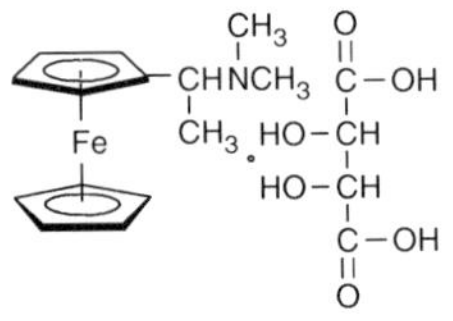

250 mg

S82,622-7
BIS[2-(DIMETHYLAMINO)PHENYL]-BROMOTHALLIUM(III)

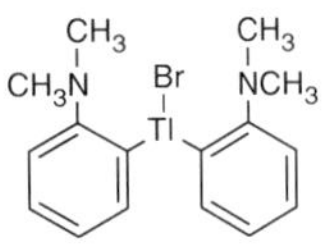

250 mg

S54,827-8
TERT-BUTYL 3-(FERROCENYLMETHYLENE)-CARBAZATE

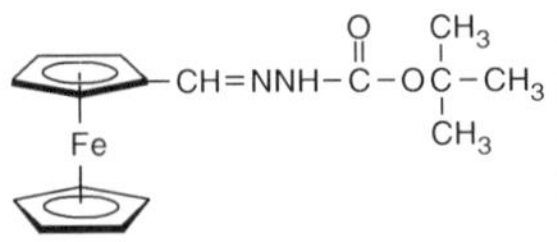

5 gm

S54,819-7
3-(FERROCENYLMETHYLENE)-CARBAZIC ACID

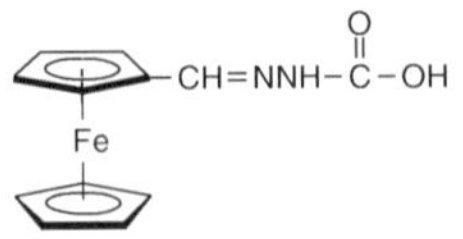

5 gm

S67,945-3
SALICYLALDEHYDE CHROMIUM COMPLEX

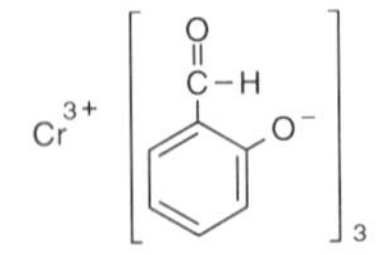

5 gm

S44,740-4
1,1'-DIBENZOYLFERROCENE

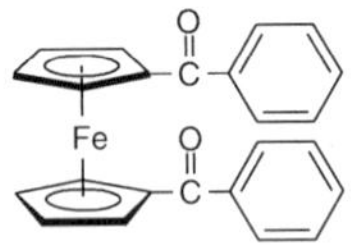

5 gm

S81,676-0
BIS[2-(METHOXYCARBONYL)PHENYLTHIO]-PHENYLBISMUTH

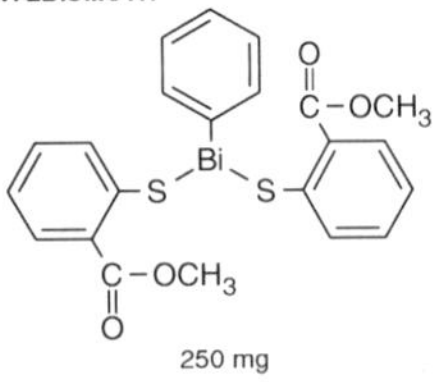

250 mg

S89,311-0
N-[2-(METHYLSELENO)ETHYL]BENZAMIDE

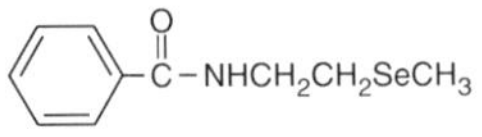

250 mg

S88,897-4
2-BENZAMIDOETHYL SELENIDE

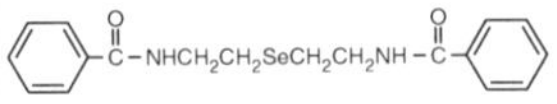

250 mg

S95,081-5
ETHYL BETA-OXOFERROCENEPROPIONATE

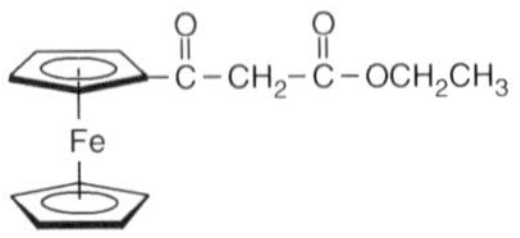

50 mg

S36,619-6
CESIUM TETROXALATE DIHYDRATE

$CsH_3(C_2O_4)_2 \cdot 2H_2O$

5 gm

S84,982-0
4-(METHYLSELENO)BUTYRIC ACID

$CH_3SeCH_2CH_2CH_2-C(=O)-OH$

250 mg

S46,283-7
MANGANESE(II) 4-CYCLOHEXANEBUTYRATE

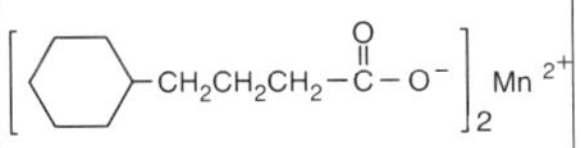

250 mg

S58,662-5
METHYL ACETOACETATE, COPPER(II) DERIVATIVE

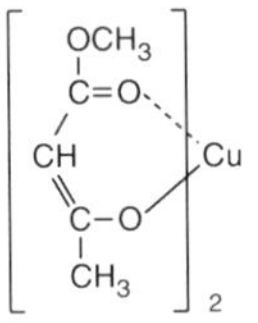

5 gm

S44,933-4
N-PYRUVYLIDENEGLYCINATO(AQUO)-COPPER(II) DIHYDRATE

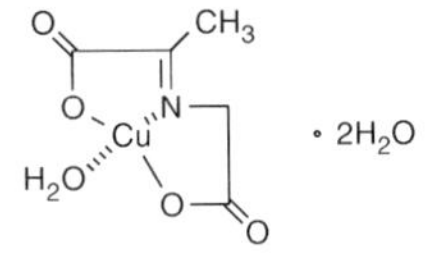

5 gm

S94,944-2
3-(PHENYLTELLURO)PROPIONIC ACID

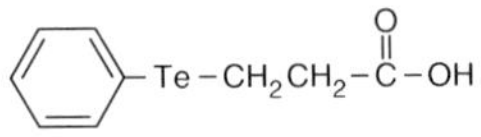

250 mg

S94,947-7
3-(PHENYLTELLURO)PROPIONIC ACID TE,TE-DICHLORIDE

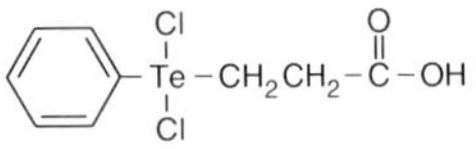

250 mg

S94,948-5
3-(PHENYLTELLURO)PROPIONIC ACID TE,TE-DIIODIDE

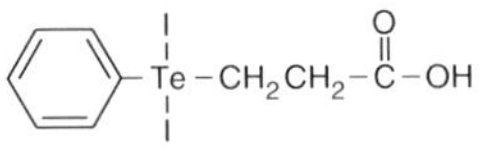

250 mg

S94,945-0
3-(1-NAPHTHYLTELLURO)PROPIONIC ACID

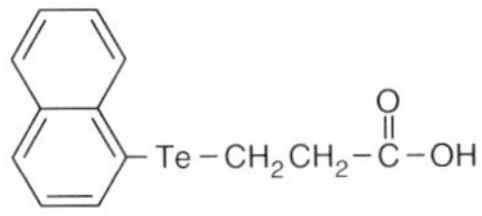

250 mg

S94,946-9
3-(1-NAPHTHYLTELLURO)PROPIONIC ACID TE,TE-DIBROMIDE

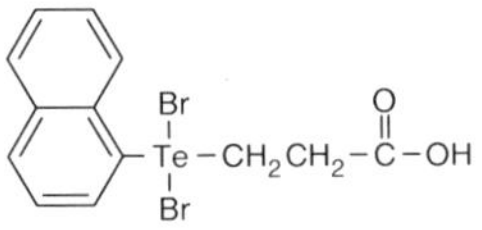

250 mg

S89,097-9
1,1'-DISELENOBIS(3-PHTHALIMIDOPROPANE)

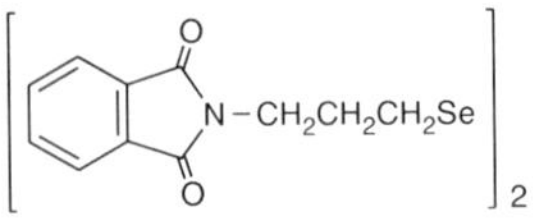

250 mg

S95,078-5
N-METHYLFERROCENECARBOXAMIDE

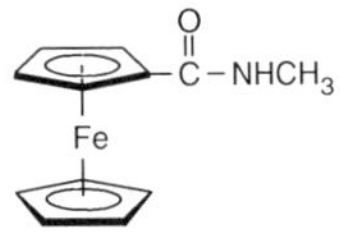

50 mg

S71,094-6
1,1-DIMETHYLUREA, LITHIUM ADDUCT

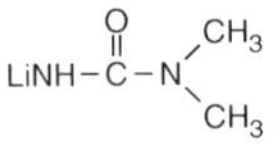

5 gm

S54,828-6
4-(FERROCENYLMETHYLENE)-IMIDAZOLIDINE-2-THIONE

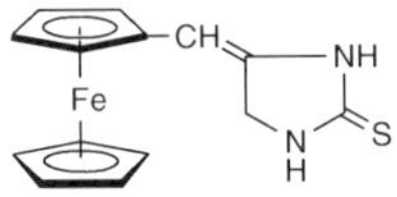

5 gm

S77,544-4
MANGANESE(III) METHANESULFONATE

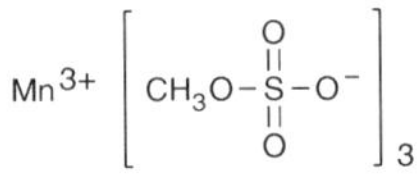

250 mg

S95,789-5
5-METHYL-9-(PHENYLSELENENYL)-2-OXATRICYCLO[4.2.1.0(4,8)]NONAN-3-ONE

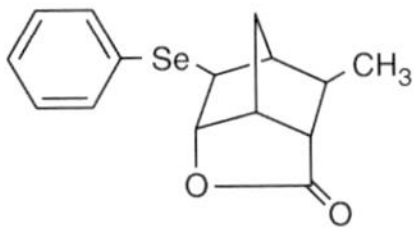

50 mg

S40,593-0
COBALTOCENE (10% SOLUTION IN BENZENE)

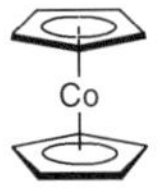

5 gm

S44,342-5
COBALTOCENE (7.5% SOLUTION IN DIETHYLBENZENE)

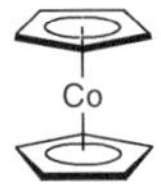

5 gm

S95,077-7
METHYL FERROCENECARBOXYLATE

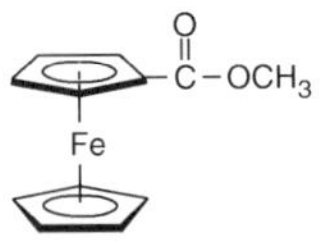

50 mg

S62,285-0
P-TOLYL SELENIDE

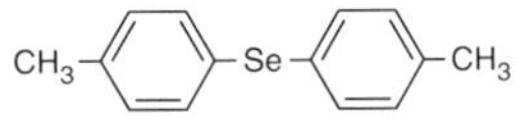

250 mg

S38,877-7
N-FERROCENYLMETHYLENE-P-TOLUIDINE

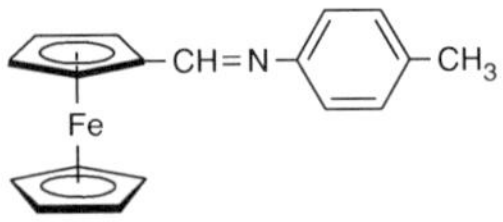

5 gm

S38,873-4
1,4-BIS(2-FERROCENYLVINYL)BENZENE

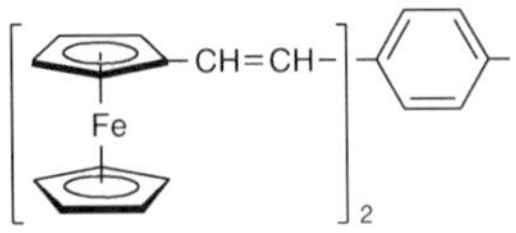

5 gm

S39,090-9
N-FERROCENYLMETHYLENE-2-NAPHTHYLAMINE

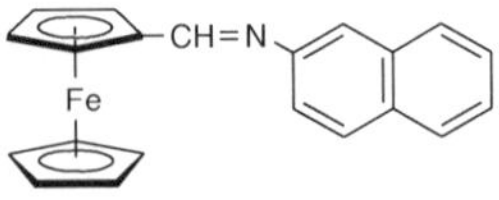

5 gm

S94,936-1
(PHENYL)TRIBROMOTELLURIUM

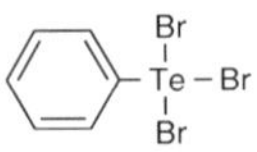

250 mg

S72,198-0
(+)-TRANS-DICHLORO(ETHYLENE)(ALPHA-METHYLPHENETHYLAMINE)PLATINUM(II)

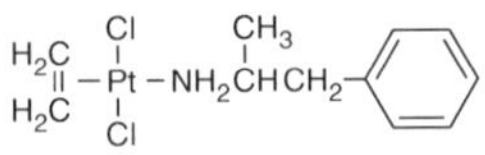

5 gm

S83,929-9
BROMODIPHENYLTHALLIUM

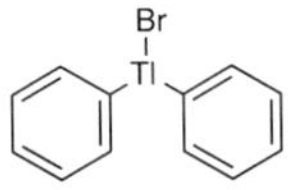

250 mg

S81,674-4
TRIPHENYLBISMUTH DIHYDROXIDE

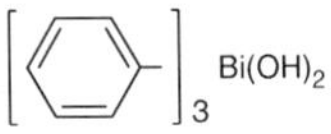

250 mg

S81,669-8
DIFLUOROTRIPHENYLBISMUTH

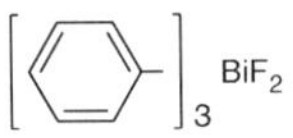

250 mg

S81,278-1
BIS(PHENYLTHIO)PHENYLBISMUTHINE

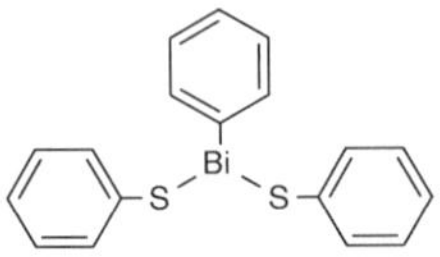

250 mg

S81,626-4
TRIPHENYLBISMUTH DINITRATE

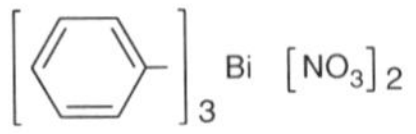

250 mg

S81,670-1
DIBROMOTRIPHENYLBISMUTH

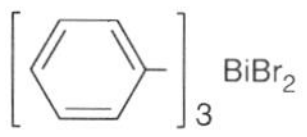

250 mg

S81,666-3
TRIPHENYLBISMUTH DIACETATE

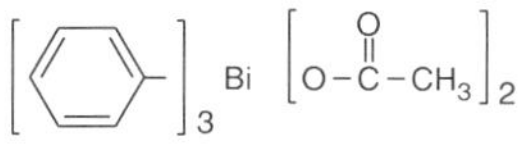

250 mg

S90,705-7
HEXAKIS(BENZYLSELENOMETHYL)BENZENE

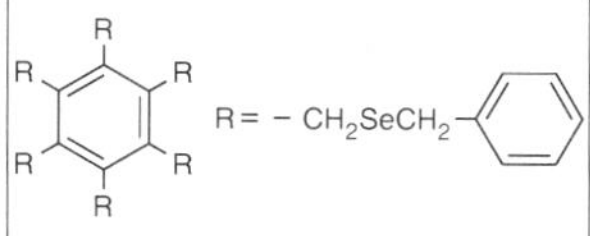

250 mg

S86,719-5
CHLORO(1,2,3,4-TETRAPHENYLCYCLOBUTADIENE)-MU-CHLOROPALLADIUM(II) DIMER

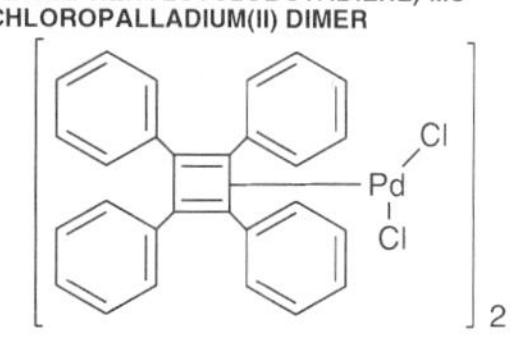

250 mg

S81,673-6
DICHLORODIPHENYL(P-TOLYL)BISMUTH

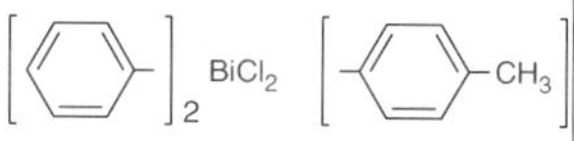

250 mg

S44,598-3
CHLORODIPHENYLCARBENIUM HEXACHLOROANTIMONATE

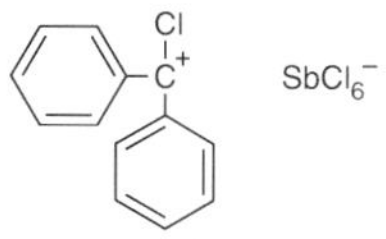

5 gm

S94,942-6
(O-TOLYL)TRICHLOROTELLURIUM

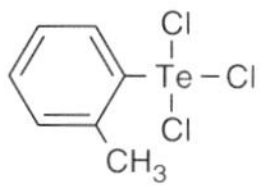

250 mg

S94,938-8
(O-TOLYL)TRIBROMOTELLURIUM

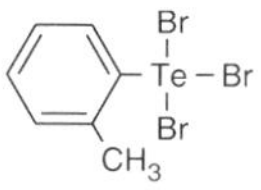

250 mg

S81,277-3
DICHLORITRI(O-TOLYL)BISMUTH

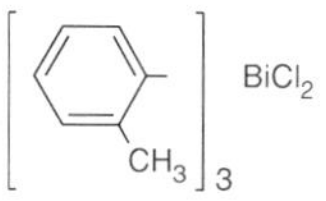

250 mg

S53,044-1
[(4-BROMOPHENYL)-THIO]COPPER(I)

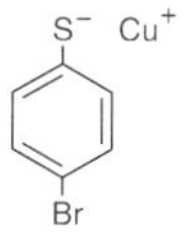

250 mg

S81,274-9
TRI(MESITYL)BISMUTHINE

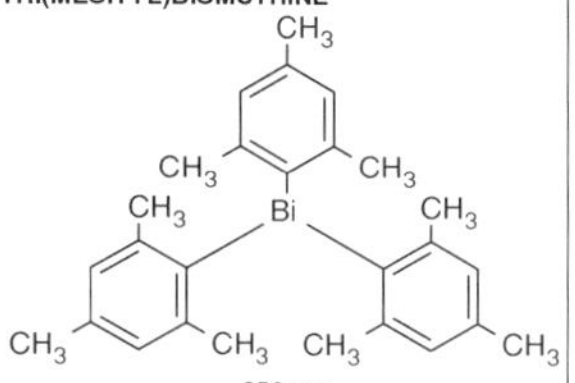

250 mg

S81,677-9
DIMESITYL(1-NAPHTHYL)BISMUTHINE

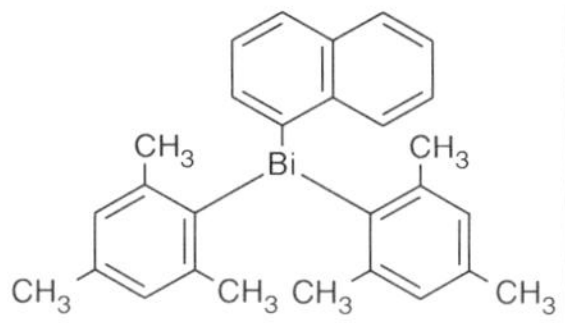

250 mg

S94,934-5
(1-NAPHTHYL)TRICHLOROTELLURIUM

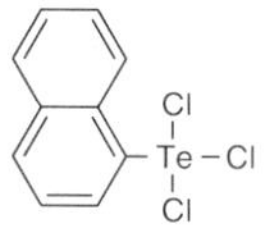

250 mg

S94,935-3
(1-NAPHTHYL)TRIBROMOTELLURIUM

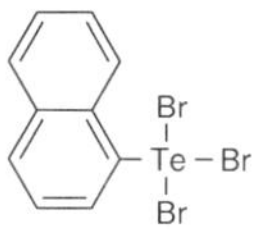

250 mg

S94,949-3
1-NAPHTHYL DITELLURIDE

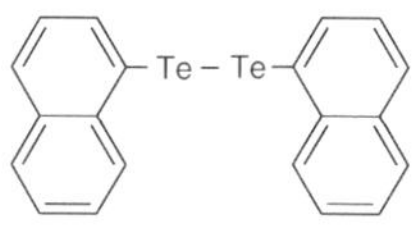

250 mg

S44,005-1
BENZOSELENAZOLE

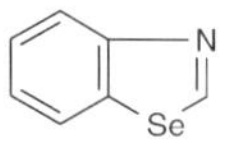

250 mg

Molecular Formula Index

All compounds are separately referenced in the Molecular Formula Index in ascending order of molecular formula. For your convenience, the page where the structure can be located, is next to the chemical name.

The arrangement used in this index is based on the general formula "$C_xH_yElement_z$" where the order of elements is alphabetical:

* Inorganics precede carbon-containing compounds.

* Carbon-containing inorganics such as Ag_2CO_3 will be found with C listed first, i.e., CAg_2O_3.

* Inorganics containing hydrogen will have hydrogen listed first.

* Organics lacking hydrogen such as C_6N_4 are listed before any C_6H's.

* Water of hydration is not included.

* Salts of amines, except quaternary salts, are listed under the formulas of the free amines.

* Simple salts of organic acids are listed under the formulas of the free acids.

$AgFO_3S$
Silver fluorosulfate 1140
Cl_3GaO_{12}
Gallium perchlorate hexahydrate 1140
Cl_4Os
Osmium tetrachloride 1140
Cu_2HgI_4
Copper mercuric iodide 1140
$Er_2O_{12}S_3$
Erbium sulfate octahydrate, 1140
$Eu_2O_{12}S_3$
Europium(III) sulfate octahydrate, 1140
F_6FeNa_3
Sodium hexafluoroferrate(III) 1140
$Ho_2O_{12}S_3$
Holmium sulfate octahydrate 1140
$O_{12}Pr_2S_3$
Praseodymium sulfate octahydrate. 1140
$O_{12}S_3Yb_2$
Ytterbium(III) sulfate octahydrate 1140
HO_3STl
Thallium(*I*) bisulfite 157
H_2PtBr_6
Hexabromoplatinic acid (platinic bromide) 1140
$H_3B_3Cl_3N_3$
2,4,6-Trichloroborazine 1124
H_3NO_3S
Sulfamic acid, cesium salt 157
H_3N_5O
1-Amino-4,5-dihydro-1*H*-tetrazol-5-one 870
$H_4N_4S_4$
1,3,5,7,2,4,6,8-Tetrathiatetrazocine 36
$H_8O_2Si_4$
1,4-Dioxa-2,3,5,6-tetrasilacyclohexane 1096
$H_{15}Cl_3CrN_5$
Chloropentamminechromium(III) chloride 1140
$H_{18}CrN_9O_9$
Hexaamminechromium(III) nitrate 1140
$CHCl_3GeO_2$
Trichlorogermanecarboxylic acid 1109
$CHCl_4OP$
(Dichloromethyl)phosphonic dichloride 833
$CH_2Cl_3O_3P$
Trichloromethylphosphonic acid 159
CH_2N_4S
5-Amino-1,2,3,4-thiatriazole 879
CH_2O_2
Formic acid, octadecylamine salt 69
CH_3BrO_3S
Bromomethanesulfonic acid sodium salt 155
CH_3ClN_2
Chloroformamidine hydrochloride 136
CH_3NO_4S
Carbamoylsulfonic acid, potassium salt 156
$CH_3N_3O_3$
Nitrourea 128
CH_4O_2S
Methanesulfinic acid, lithium salt 156
CH_4O_3S
Ammonium methanesulfonate 156
Manganese(III) methanesulfonate 1142
CH_5NO
N-Methylhydroxylamine oxalate 29
CH_5N_3
Guanidine *p*-toluenesulfonate 795
CH_6NO_3P
(Aminomethyl)-phosphonic acid, iron (II) salt 159
CH_6N_2
Methylhydrazine hydrochloride 30
Methylhydrazine sulfate 30
$CH_6N_2O_2S$
Methanesulfonic acid hydrazide 829
CH_6N_4
Aminoguanidine sulfate monohydrate 138
CH_8N_6
Triaminoguanidine hydrochloride 138
C_2Cl_6Hg
Bis(trichloromethyl)mercury (II) 1137
$C_2Cl_6O_2S$
Trichloromethyl sulfone 152
$C_2HBr_2Cl_3$
1,2-Dibromo-1,1,2-trichloroethane 5
$C_2HBr_2N_3$
3,5-Dibromo-4*H*-1,2,4-triazole 869
$C_2HF_3N_4$
5-Trifluoromethyl-2*H*-1,2,3,4-tetrazole 870
$C_2HF_3O_2$
Barium trifluoroacetate 69
Trifluoroacetic acid, thallium salt 69
$C_2H_2ClN_3$
3-Chloro-1*H*-1,2,4-triazole 869
$C_2H_2Cl_4N_2$
N',2,2,2-Tetrachloroacetamidine 136
$C_2H_2N_2O_4S$
4,5-Dihydroxy-2,1,3-thiadiazole-2,2-dioxide, disodium salt dihydrate 157
$C_2H_2O_3$
Glyoxylic acid, barium salt 69
$C_2H_3Br_2Cl_3Si$
(1,2-Dibromoethyl)trichlorosilane 1099
C_2H_3ClHg
Vinylmercuric chloride 1136
$C_2H_3ClHgO_2$
2-(Chloromercuri)acetic acid 1138
$C_2H_3ClO_2$
2-Chloroacetic acid, dodecylamine salt 69
$C_2H_3Cl_5Si$
(1,2-Dichloroethyl)trichlorosilane 1099
$C_2H_3NO_3$
Oxamic acid, ammonium salt 110
$C_2H_3N_3OS$
5-Amino-1,3,4-thiadiazol-2-ol 878
$C_2H_3N_5O$
2-(Formylamino)-1*H*-tetrazole 870
C_2H_4BrCl
1-Bromo-1-chloroethane 4
$C_2H_4ClS_2Sb$
2-Chloro-1,3,2-dithiastibolane 1128
$C_2H_4Cl_2S$
2,2-Dichloro-1-ethanethiol 24
$C_2H_4Cl_4Si$
(1-Chloroethyl)trichlorosilane 1096
$C_2H_4F_2O$
2,2-Difluoroethanol 15
$C_2H_4N_2O_2$
Formylurea 128
$C_2H_4N_2O_4$
1,1-Dinitroethane (30% solution in ethylene chloride) 46
$C_2H_4N_4$
4-Amino-1,2,4-triazole hydrochloride 869
$C_2H_4N_4O_2$
4-Amino-4*H*-1,2,4-triazole-3,5-diol 869
Hexahydro-1,2,4,5-tetrazine-3,6-dione 136
$C_2H_4N_4O_4$
1-Nitrobiuret 127
$C_2H_4N_4S$
3,5-Diamino-1,2,4-thiadiazole 878
3,5-Diamino-1,2,4-thiadiazole *p*-toluenesulfonate 878
5-(Methylthio)-1*H*-tetrazole 870
$C_2H_4O_2$
Berylium acetate 70
$C_2H_4O_2S$
Mercaptoacetic acid, calcium salt 69
$C_2H_4O_5S$
Sulfoacetic acid, pyridine complex 963
$C_2H_5AsO_5$
Arsonoacetic acid 1127
Arsonoacetic acid, disodium salt monohydrate 1127
C_2H_5ClHg
Ethylmercuric chloride 1136
$C_2H_5ClO_4S_2$
(Methylsulfonyl)methanesulfonyl chloride 156
$C_2H_5NO_2$
Glycine, cupric salt 72
$C_2H_5N_3O$
1-Formylguanidine 138
$C_2H_5N_3OS$
1-Formylthiosemicarbazide 141
$C_2H_5O_3P$
Ethylene phosphite 158
$C_2H_6HgS_2$
Bis(methylthio)mercury 1137
$C_2H_6N_2O$
O-Methylisourea hydrochloride 136
$C_2H_6N_2O_2$
1-Methoxyurea 128
$C_2H_6N_2S$
2-Methyl-2-thiopseudourea hydriodide 138
2-Methyl-2-thiopseudourea *p*-toluenesulfonate 749
$C_2H_6N_4O$
1-Amidinourea benzenesulfonate 795
Amidinourea sulfate 138
Amidinourea *p*-toluenesulfonate 795
Guanylurea 138
Guanylurea phosphate 137
$C_2H_6N_4OS$
2-Thiobiurea 140
$C_2H_6N_4O_2$
Allophanic acid hydrazide 128
Biurea 127
$C_2H_6N_4S$
1-Amidino-2-thiourea oxalate 142
$C_2H_6N_4S_2$
Guanidinodithiocarbamic acid 140
$C_2H_6O_3S$
Ethanesulfonic acid, morpholine adduct 156
$C_2H_6O_6S_2$
1,2-Ethanedisulfonic acid, disodium salt 156
$C_2H_6PbS_2$
Bis-(methylthio)-lead 1133
$C_2H_7NO_3S_2$
2-Aminoethylthiosulfonic acid 155
C_2H_7NS
2-Aminoethanethiol bitartrate 35
$C_2H_7N_3$
1-Methylguanidine sulfate 136
$C_2H_7N_3S$
Methyl hydrazinocarboximidothioate tosylate salt 137
$C_2H_7N_5$
Biguanide 1020
$C_2H_7O_4P$
Bis-(hydroxymethyl)-phosphinic acid 158
$C_2H_8N_2$
Ethylenediamine dihydroiodide 28
Ethylenediamine *p*-toluenesulfonate 793
$C_2H_8N_6O_2$
4,4'-Bisemicarbazide 126
$C_2H_8O_7P_2$
Dimethyl acid pyrophosphate 160
$C_2H_{12}B_{10}S_2$
o-Carborane-1,2-dithiol 209
m-Carborane-1,7-dithiol 209
$C_3Cl_2F_6$
1,2-Dichlorohexafluoropropane 5
1,3-Dichlorohexafluoropropane 5
$C_3Cl_3F_5$
1,1,2,2,3-Pentafluorotrichloropropane 5
$C_3Cl_4F_4$
1,1,1,3-Tetrachlorotetrafluoropropane 5
1,1,3,3-Tetrachlorotetrafluoropropane 5
C_3Cl_4O
Trichloroacryloyl chloride 100
$C_3Cl_4O_3$
4,4,5,5-Tetrachloro-1,3-dioxolan-2-one 97
$C_3Cl_5F_3$
1,1,1,3,3-Pentachlorotrifluoropropane 5
C_3Cl_5N
Pentachloropropionitrile 144
C_3Cl_6
Hexachlorocyclopropane 6
$C_3Cl_6F_2$
2,2-Difluorohexachloropropane 5
C_3Cl_6O
Pentachloropropionyl chloride 100
$C_3Cl_{11}Sb$
Pentachloroallyl hexachloroantimonate 1128
C_3N_3O
(Hydroxyimino)malononitrile, potassium salt 150
$C_3HBrCl_2N_2$
2-Bromo-4,5-dichloroimidazole 868
$C_3HBr_3N_2$
3,4,5-Tribromo-1*H*-pyrazole 863
$C_3HCl_2N_3O_3$
Dichloro-1,3,5-triazine-2,4,6(1*H*,3*H*,5*H*)-trione, potassium salt 135
C_3HCl_3O
Trichloroacrolein 60
$C_3HCl_3O_2$
Trichloroacrylic acid 62
$C_3H_2BrCl_3$
3-Bromo-1,1,2-trichloro-1-propene 5
$C_3H_2BrN_3O_2$
5-Bromo-4-nitroimidazole 868
6-Bromo-1,2,4-triazine-3,5-(2*H*,4*H*)-dione 134
$C_3H_2Br_2N_2$
4,5-Dibromoimidazole 868
3,4-Dibromo-1*H*-pyrazole 863
3,5-Dibromo-1*H*-pyrazole 863
$C_3H_2Br_2N_2O_2$
1,3-Dibromohydantoin 135
$C_3H_2Cl_2F_2$
1,2-Dichloro-3,3-difluoro-1-propene 5
$C_3H_2Cl_3F_3O$
1,1,1-Trichloro-3,3,3-trifluoro-2-propanol 15
$C_3H_2Cl_4O$
1,1,3,3-Tetrachloroacetone 50
$C_3H_2Cl_5NO$
2,2,3,3,3-Pentachloropropionamide 101
$C_3H_2Cl_6O$
1,1,1,3,3,3-Hexachloro-2-propanol 15
$C_3H_2N_2O_2S$
Thioparabanic acid 141
$C_3H_2N_4O_4$
3,4-Dinitro-1*H*-pyrazole 861
$C_3H_3BrN_2$
4-Bromoimidazole 868
5-Bromoimidazole 868
$C_3H_3ClN_2$
4-Chloro-1*H*-pyrazole 863
$C_3H_3ClN_2S$
2-Amino-5-chlorothiazole hydrochloride 876
$C_3H_3Cl_2F_3S$
Chloromethyl 2-chloro-1,1,2-trifluoroethyl sulfide 24
$C_3H_3Cl_3$
1,2,3-Trichloropropene 5
$C_3H_3Cl_5$
1,1,2,3,3-Pentachloropropane 4
$C_3H_3Cl_5O$
2,2,3,3,3-Pentachloro-1-propanol 15
$C_3H_3FN_2O_2S_2$
2-Amino-5-thiazolesulfonyl fluoride 876
$C_3H_3F_4I$
2,2,3,3-Tetrafluoropropyl iodide 4
$C_3H_3IN_2$
2-Iodoimidazole 868
$C_3H_3NS_2$
2-Mercaptothiazole 878
$C_3H_3N_3O$
3-Amino-5-pyrazolone 861
$C_3H_3N_3O_2$
5-Azauracil 980
2-Cyano-2-hydroximinoacetamide 1010
2-Cyano-2-nitrosoacetamide 145
5-Nitroimidazole 867
1-Nitro-1*H*-pyrazole 862
3-Nitro-1*H*-pyrazole 862
4-Nitro-1*H*-pyrazole 862
$C_3H_4Br_2Cl_2$
1,2-Dibromo-2,3-dichloropropane 4
$C_3H_4Br_2N_2O_2$
Dibromomalonic diamide 111
$C_3H_4Cl_2O_2$
2,3-Dichloropropionic acid 61
$C_3H_4Cl_3NO$
2,2,3-Trichloropropionamide 101

$C_3H_4Cl_3NO_2$
Chloral formamide 103
$C_3H_4F_6N_2$
1,1,1,3,3,3-Hexafluoro-2,2-propanediamine 32
$C_3H_4N_2$
Imidazole perfluorobutyrate salt 868
Imidazole perfluoropropionic acid salt 868
$C_3H_4N_2OS$
4-Imino-2-thiazolidinone 133
$C_3H_4N_4$
2-Amino-1,3,5-triazine 979
$C_3H_4N_4O$
6-Azacytosine 978
$C_3H_4N_4O_2$
5-Amino-4-nitroso-4-pyrazolin-3-one 860
C_3H_5BrO
3-Bromo-2-propen-1-ol 15
$C_3H_5Br_2NO$
2,3-Dibromopropionamide 101
$C_3H_5ClO_2$
2-Chloroethyl formate 81
C_3H_5ClS
2-Chloroallyl mercaptan 24
$C_3H_5Cl_2NO_2$
1,1-Dichloro-1-nitropropane 46
1,2-Dichloro-3-nitropropane 45
$C_3H_5Cl_3$
1,1,2-Trichloropropane 4
C_3H_5FO
Propionyl fluoride 100
$C_3H_5F_3O$
1,1,1-Trifluoro-2-propanol 15
$C_3H_5F_3O_3$
2,2,2-Trifluoroethyl mesylate 156
$C_3H_5NO_3$
Malonamic acid 101
$C_3H_5NO_4$
2-Nitropropionic acid 67
$C_3H_5N_3$
2-Aminoimidazole hydrogen sulfate 866
$C_3H_5N_3S$
2-Amino-5-methyl-1,3,4-thiadiazole hydrochloride 879
$C_3H_5N_3S_2$
2-Mercapto-5-methylamino-1,3,4-thiadiazole 878
$C_3H_5N_5$
2,4-Diamino-1,3,5-triazine 979
$C_3H_5N_5O$
4,6-Diamino-1,3,5-triazin-2-ol 978
$C_3H_5O_6P$
Phosphoenolpyruvic acid, silver barium salt, dihydrate 835
C_3H_6BrClO
2-Bromo-2-chloro-1-propanol 15
C_3H_6ClNO
2-Chloro-*N*-methylacetamide 102
$C_3H_6ClNO_2$
1-Chloro-1-nitropropane 46
2-Chloro-2-nitropropane 46
$C_3H_6ClN_3O$
(Chloroacetyl)guanidine hydrochloride 136
$C_3H_6Cl_2O_2$
2,2-Dichloro-1,3-propanediol 16
$C_3H_6N_2$
N-Methylglycinonitrile sulfate 145
3-Pyrazoline hydrochloride 41
$C_3H_6N_2O$
2-Amino-2-oxazoline picrate 724
$C_3H_6N_2O_2$
3-Amino-2-oxazolidinone 134
3-Amino-2-oxazolidinone sulfate 125
Formimidoylglycine 137
N,N'-Methylenebis(formamide) 111
$C_3H_6N_2O_2S$
N-(Thiocarbamoyl)-glycine 141
$C_3H_6N_2O_3$
Hydantoic acid 127
Methyl *N*-carbamoylcarbamate 127
$C_3H_6N_2O_5$
2,2-Dinitropropanol (30% solution in ethylene chloride) 45
$C_3H_6N_2O_6S_2$
1-Amino-2-cyano-1,1-ethanedisulfonic acid, disodium salt trihydrate 155
$C_3H_6N_4O_3$
Triuret 127
$C_3H_6N_4S$
3-Amino-5-(methylamino)-1,2,4-thiadiazole 878
3-Amino-5-(methylamino)-1,2,4-thiadiazole *p*-toluenesulfonate 878
1-Methyl-5-(methylthio)-1*H*-tetrazole 870
$C_3H_6OS_2$
O-Ethylxanthic acid, thallium salt 74
$C_3H_6O_2$
Propionic acid, octadecylamine salt 69
$C_3H_6O_2S$
2-Thioglyceraldehyde 60
Trimethylene sulfone 154
$C_3H_6O_3$
L-(-)-Glyceraldehyde 60
L-(+)-Lactic acid, calcium salt 69
L-(+)-Lactic acid lithium salt 69
$C_3H_6O_4S$
Acetyl methanesulfonate 156
$C_3H_6O_5S$
Acetoxymethanesulfonic acid, potassium salt 156
3-Sulfopropionic acid, diammonium salt 155
$C_3H_6O_6S_3$
1,3,5-Trithiane-1,1,3,3,5,5-hexaoxide 154
$C_3H_7BrO_2S$
Bromomethyl ethyl sulfone 152
$C_3H_7BrO_3S$
3-Bromo-1-propanesulfonic acid, sodium salt 155
$C_3H_7Cl_2OPS$
Propyl phosphorodichloridothionate 160
$C_3H_7Cl_3Si$
Chloro(dichloromethyl)dimethylsilane 1100
$C_3H_7IO_3S$
3-Iodo-1-propanesulfonic acid, potassium salt 155
$C_3H_7IO_4S$
3-Iodopropyl sulfate, sodium salt 157
C_3H_7NOS
Methyl thioacetohydroxamate 1009
$C_3H_7NO_2$
N-(2-Hydroxyethyl)formamide 104
Methoxyacetamide 101
Propionohydroxamic acid 105
$C_3H_7NO_3$
3-(Aminooxy)propionic acid hydrochloride 71
Ethyl *N*-hydroxycarbamate, sodium salt 115
$C_3H_7NO_3S$
2-(Methylsulfonyl)acetamide 153
$C_3H_7NO_5S$
DL-Cysteic acid 155
$C_3H_7N_3$
2-Amino-2-imidazoline hydrobromide 38
3-Amino-3-pyrazoline dihydrochloride 38
$C_3H_7N_3OS$
1-Formyl-4-methyl-3-thiosemicarbazide 140
$C_3H_7N_3O_2$
2-Aminomalonamide 110
$C_3H_8Cl_2NPS$
N-Isopropylthiophosphoramidic dichloride 160
$C_3H_8NO_6P$
DL-Serine phosphate hydrate 161
$C_3H_8N_2O$
1,1-Dimethylurea, lithium derivative 1142
$C_3H_8N_2O_2$
1-Ethoxyurea 128
$C_3H_8N_2O_3$
2-Hydroxyethyl carbazate 644
$C_3H_8N_2S$
2-Ethyl-2-thiopseudourea *p*-toluenesulfonate 749
$C_3H_8N_2S_2$
(2-Aminoethyl)-dithiocarbamic acid 139
$C_3H_8N_4O_2$
Malonic dihydrazide 112
Malonohydroxamamide 1010
$C_3H_8N_4S$
1-Amidino-3-methyl-2-thiourea *p*-toluenesulfonate 795
$C_3H_8O_4S$
1-Hydroxy-1-methyl-1-ethanesulfonic acid sodium salt 155
$C_3H_8O_4SS$
Acetone sodium bisulfite addition compound 155
$C_3H_8O_4S_2$
Bis(methylsulfonyl)methane 154
$C_3H_8S_2$
1,2-Propanedithiol 24
$C_3H_9AlO_3$
Aluminum trimethoxide 1140
$C_3H_9BF_4S$
Trimethylsulfonium tetrafluoroborate 27
C_3H_9ISn
Iodotrimethyltin 1129
Trimethyltin iodide ammonia complex 1139
C_3H_9N
Trimethylamine formate 29
$C_3H_9NO_2S$
N,N-Dimethylmethanesulfonamide 157
$C_3H_9NO_3Sn$
Trimethyltin nitrate monohydrate 1140
$C_3H_9N_3O_3S$
4-Methyl-1-(methylsulfonyl)-semicarbazide 157
1-(2-Sulfamoylethyl)urea 157
$C_3H_9O_3P$
Tris-(hydroxymethyl)-phosphine 158
$C_3H_9O_5P$
2-Methoxyethyl phosphate, barium salt 161
$C_3H_9O_6P$
Glycerol-1-phosphate, disodium salt 161
$C_4Br_2I_2S$
2,5-Dibromo-3,4-diiodothiophene 859
$C_4Br_2O_3$
2,3-Dibromomaleic anhydride 98
$C_4Cl_2F_6$
1,2-Dichloro-1,2,3,3,4,4-hexafluorocyclobutane 6
$C_4Cl_2F_8$
2,3-Dichlorooctafluorobutane 5
C_4Cl_2Hg
Bis(2-chloroethynyl)mercury 1137
$C_4Cl_3F_7$
Heptafluoro-2,2,3-trichlorobutane 5
C_4Cl_3IS
2-Iodo-3,4,5-trichlorothiophene 859
$C_4Cl_3N_3O_2$
5-Nitro-2,4,6-trichloropyrimidine 976
$C_4Cl_4Hg_4O$
2,3,4,5-Tetrakis(chloromercuri)furan 1139
C_4Cl_4O
Perchloro-2-cyclobuten-1-one 53
C_4I_4S
Tetraiodothiophene 859
C_4HBr_2ClHg
5-(Chloromercuri)-2,3-dibromofuran 1139
C_4HBr_3S
2,3,5-Tribromothiophene 859
$C_4HCl_7O_2$
Heptachloro-1,4-dioxane 23
C_4HCl_9O
Pentachloroethyl 1,1,2,2-tetrachloroethyl ether 17
$C_4H_2BrNO_2S$
2-Bromo-5-nitrothiophene 856
$C_4H_2Br_2N_2O$
4,5-Dibromo-3(2*H*)-pyridazinone 968
$C_4H_2ClF_3O$
4,4,4-Trifluorocrotonyl chloride 100
$C_4H_2Cl_2HgO$
2-Chloro-5-(chloromercuri)furan 1139
$C_4H_2Cl_2Hg_2O$
2,5-Bis(chloromercuri)furan 1139
$C_4H_2Cl_2Hg_2S$
2,5-Bis(chloromercuri)thiophene 1139
$C_4H_2Cl_4N_4O_2$
N,N',N'',N'''-Tetrachloroglycoluril 136
$C_4H_2Cl_4O_2$
cis-3,4,4,4-Tetrachlorocrotonic acid 62
$C_4H_2Cl_6O$
1,1,1,3,4,4-Hexachloro-3-buten-2-ol 15
$C_4H_2Cl_8S_2$
1,2,2,2-Tetrachloroethyl disulfide 24
$C_4H_2N_2O_5$
2,5-Dinitrofuran 852
$C_4H_2N_2S$
4-Cyanoisothiazole 874
$C_4H_2N_4O_2$
5-Diazouracil monohydrate 1094
$C_4H_3BrN_2$
2-Bromopyrimidine 977
$C_4H_3BrO_4$
Bromomaleic acid 62
$C_4H_3Br_2F_3O$
2-Bromo-4,4,4-trifluorobutyryl bromide 100
C_4H_3ClHgO
2-(Chloromercuri)furan 1138
3-(Chloromercuri)furan 1138
C_4H_3ClHgS
2-(Chloromercuri)thiophene 1139
$C_4H_3ClN_2O$
6-Chloro-3(2*H*)-pyridazinone 968
5-Chloro-4-pyrimidinol hydrochloride 972
$C_4H_3ClN_2O_2$
4-Chloro-3,6-dihydroxypyridazine 968
6-Chlorouracil 976
$C_4H_3ClO_4$
Chlorofumaric acid 62
$C_4H_3Cl_5O_2$
Methyl pentachloropropionate 82
$C_4H_3CsO_8$
Cesium tetroxalate 1142
$C_4H_3N_3O_3S$
6-Aza-2-thiouracil-5-carboxylic acid 980
4,6-Dihydroxy-2-mercapto-5-nitrosopyrimidine 969
$C_4H_3N_3O_4$
4-Nitro-5-pyrazolecarboxylic acid 862
$C_4H_3N_3O_4S$
2-Amino-3,5-dinitrothiophene 856
$C_4H_3N_3O_5$
5-Nitrobarbituric acid-.hydrate 975
$C_4H_4BrF_3O$
4,4,4-Trifluorobutyryl bromide 100
$C_4H_4BrN_3O_2$
4-Amino-5-bromo-2,6-dihydroxypyrimidine 969
$C_4H_4Br_2ClF_3$
2-Chloro-1,4-dibromo-1,1,2-trifluorobutane 4
$C_4H_4Br_2N_2$
3,4-Dibromo-5-methyl-1*H*-pyrazole 864
$C_4H_4Br_2N_2O$
4,4-Dibromo-3-methyl-2-pyrazolin-5-one 134
$C_4H_4Br_2O_4$
α,α'-Dibromosuccinic acid 62
$C_4H_4Br_4N_2O_4$
1,4-Dinitro-1,1,4,4-tetrabromobutane 46
$C_4H_4ClF_3O$
4,4,4-Trifluorobutyryl chloride 100
$C_4H_4ClN_3$
2-Amino-6-chloropyrazine 977
2-Amino-5-chloropyrimidine 975
$C_4H_4Cl_2N_2O_2$
1,2-Bis(dichloroacetyl)hydrazine 111
$C_4H_4Cl_4N_2O_4$
1,4-Dinitro-1,1,4,4-tetrachlorobutane 46
$C_4H_4Cl_4O_2S$
2,5-Dihydro-3,3,4,4-tetrachlorothiophene-1,1-dioxide 154
$C_4H_4Cl_6$
1,1,2,3,4,4-Hexachlorobutane 4
$C_4H_4CrKO_{10}$
trans-Potassium diaquodioxalatochromate(III) trihydrate 1140
$C_4H_4F_3NO_3$
N-(Trifluoroacetyl)glycine 117

$C_4H_4F_3N_3S$
4-Methyl-5-(trifluoromethyl)-4*H*-1,2,4-triazoline-3(2*H*)-thione 869
$C_4H_4IN_3O$
5-Iodocytosine 975
$C_4H_4N_2$
Pyrazine, mercuric chloride complex 978
$C_4H_4N_2O$
Pyrazinol, sodium salt 977
3(2*H*)-Pyridazinone 136
$C_4H_4N_2OS$
4-Thiouracil 976
$C_4H_4N_2O_2$
Acetylenedicarboxamide 1017
2-Cyano-3-hydroxyacrylamide, sodium salt 144
2,4-Dihydroxypyrimidine 969
1*H*-Pyrazole-3-carboxylic acid 863
$C_4H_4N_2O_2S$
5-Mercaptouracil 973
2-Thiobarbituric acid, sodium derivative 141
$C_4H_4N_2O_3$
Methyl 2-cyano-2-(hydroxyimino)-acetate 1010
$C_4H_4N_2O_3S$
2-Pyrimidinesulfonic acid, sodium salt 976
$C_4H_4N_2S$
2-Pyrimidinethiol hydrochloride 973
$C_4H_4N_2S_2$
Ethylene dithiocyanate 151
$C_4H_4N_4$
5-Amino-4-pyrazolecarbonitrile 860
$C_4H_4N_4O_2S$
4-Amino-6-hydroxy-2-mercapto-5-nitrosopyrimidine, ammonium salt 970
$C_4H_4N_4O_3$
2-Amino-4,6-dihydroxy-5-nitrosopyrimidine 969
$C_4H_4N_4O_4$
4-Amino-2,6-dihydroxy-5-nitropyrimidine 969
$C_4H_4O_2$
1,2-Cyclobutanedione 57
$C_4H_4O_3S_2$
2-Thiophenesulfonic acid, sodium salt monohydrate 859
$C_4H_5AlO_{10}$
Aloxalic acid 1140
$C_4H_5BrN_2O_2$
5-Bromo-5,6-dihydrouracil 135
$C_4H_5BrN_2S$
5-Amino-4-bromo-3-methylisothiazole 874
$C_4H_5BrN_4O$
5-Bromo-2,4-diamino-6-hydroxypyrimidine 971
$C_4H_5BrO_2$
2-Bromo-2-butenoic acid 62
3-Bromo-2-methylacrylic acid 62
Methyl 3-bromoacrylate 83
$C_4H_5BrO_2S$
2-Bromo-4,5-dihydrothiophene 1,1-dioxide 154
$C_4H_5BrO_3S$
2-Bromo-4-hydroxy-4,5-dihydrothiophene 1,1-dioxide 152
$C_4H_5Br_3O_2$
2,3,4-Tribromobutyric acid 62
$C_4H_5ClN_4$
6-Chloro-4,5-diaminopyrimidine 974
$C_4H_5ClN_4O$
2-Amino-4-chloro-6-methoxy-*s*-triazine 978
$C_4H_5ClO_2$
trans-3-Chlorocrotonic acid 62
$C_4H_5ClO_2S$
3-Chloro-2,5-dihydrothiophene-1,1-dioxide 154
$C_4H_5ClO_3$
(Chloromethyl)ethylene carbonate 89
$C_4H_5Cl_3O_3$
3-Hydroxy-4,4,4-trichlorobutyric acid 62
$C_4H_5F_3O_2$
4,4,4-Trifluorobutyric acid 62
$C_4H_5F_3O_3$
3-Hydroxy-4,4,4-trifluorobutyric acid 62
$C_4H_5F_3O_3S$
Imidazole trifluoromethanesulfonate salt 864
C_4H_5NO
3-Methylisoxazole 870
$C_4H_5NO_3$
Maleamic acid 121
Morpholine-2,3-dione 134
$C_4H_5N_3O$
2-Amino-4-hydroxypyrimidine sulfate 971
2-Amino-4-pyrimidinol 971
$C_4H_5N_3OS$
4-Amino-2-hydroxy-6-mercaptopyrimidine 970
$C_4H_5N_3O_2$
1-Methyl-5-azauracil 135
1-Methyl-4-nitroimidazole 867
5-Methyl-4-nitroimidazole 868
3-Methyl-4-nitro-1*H*-pyrazole 862
$C_4H_5N_3O_3$
5-Aminobarbituric acid 132
$C_4H_5N_3S_2$
2-Amino-4,6-pyrimidinedithiol 973
$C_4H_5N_5O$
4,6-Diamino-5-nitrosopyrimidine 974
$C_4H_5N_5OS$
4,6-Diamino-2-mercapto-5-nitrosopyrimidine 973
$C_4H_5N_5O_2$
4,6-Diamino-2-hydroxy-5-nitrosopyrimidine 970
2,4-Diamino-5-nitropyrimidine 974
4,6-Diamino-5-nitropyrimidine 974
2,6-Diamino-5-nitroso-4-pyrimidinol 970
$C_4H_6BrF_3O_2S$
2-Bromopropyl trifluoromethyl sulfone 152
$C_4H_6Br_2Cl_2$
2,3-Dibromo-1,4-dichlorobutane 4
$C_4H_6Br_2O_2$
2,3-Dibromobutyric acid 62
2,3-Dibromo-2-methylpropionic acid 62
Methyl 2,3-dibromopropionate 81
$C_4H_6Br_2O_2S$
2,3-Dibromotetrahydrothiophene-1,1-dioxide 154
$C_4H_6Br_4$
1,2,3,4-Tetrabromobutane 4
$C_4H_6ClF_3O$
2-Chloro-1,1,2-trifluoroethyl ethyl ether 17
$C_4H_6ClNO_3$
N-(Chloroacetyl)glycine 117
$C_4H_6Cl_2O$
3,4-Dichlorotetrahydrofuran 22
$C_4H_6Cl_2OS$
1-(Acetylmercapto)-2,2-dichloroethane 82
$C_4H_6Cl_2O_2$
Ethyl dichloroacetate 82
$C_4H_6Cl_3NO$
2-(Trichloromethyl)-oxazolidine 36
$C_4H_6Cl_3NO_2$
2,2,2-Trichloroethyl *N*-methylcarbamate 114
Tris(chloromethyl)nitromethane 45
$C_4H_6Cl_3NO_3$
Methyl *N*-(1-hydroxy-2,2,2-trichloroethyl)carbamate 113
$C_4H_6F_4O$
3,3,4,4-Tetrafluoro-2-butanol 15
$C_4H_6N_2O$
N-(Cyanomethyl)-acetamide 145
2-Cyano-*N*-methylacetamide 146
3,4-Dimethylfurazan 38
4-(Hydroxymethyl)imidazole picrate 866
3-Methyl-2-pyrazolin-5-one 862
$C_4H_6N_2O_2$
Maleamide 111
5-Methylhydantoin 135
2,3-Piperazinedione 134
$C_4H_6N_2O_3$
5-Hydroxy-5-methylhydantoin 131
$C_4H_6N_2O_4$
L-Aspartic acid hydantoin 868
$C_4H_6N_2S$
2-Amino-4-methylthiazole hydrochloride 876
2,3-Dihydro-2-imino-3-methylthiazole hydroiodide 1075
2,5-Dimethyl-1,3,4-thiadiazole, 879
$C_4H_6N_4$
4,6-Diaminopyrimidine 974
4,6-Diaminopyrimidine hydrochloride 974
6,7-Dihydro-5*H*-pyrrolotetrazole 870
$C_4H_6N_4O$
3-Acetamido-1*H*-1,2,4-triazole 870
4-Amino-5-pyrazolecarboxamide hemisulfate 861
4,6-Diamino-2-hydroxypyrimidine hydrochloride 970
4,5-Diamino-2-hydroxypyrimidine sulfate 970
4,6-Diamino-2-hydroxypyrimidine sulfate 974
$C_4H_6N_4OS$
1-(4-Oxo-2-thiazolin-2-yl)-guanidine 138
$C_4H_6N_4O_2$
5,6-Diamino-2,4-dihydroxypyrimidine monohydrochloride 970
2,5-Diamino-4,6-dihydroxypyrimidine sulfate 969
4,5-Diamino-2,6-dihydroxypyrimidine sulfate 969
5,6-Diaminouracil 132
$C_4H_6N_4O_2S$
4,6-Diaminopyrimidine-2-sulfinic acid 974
$C_4H_6N_4S$
2,4-Diamino-6-mercaptopyrimidine 973
4,5-Diamino-2-mercaptopyrimidine 973
$C_4H_6O_3$
1,3-Dioxan-2-one 97
β-Hydroxy-γ-butyrolactone 95
$C_4H_6O_3S$
3-Oxosulfolane 153
$C_4H_6O_4$
Glycerine carbonate 95
Succinic acid, diammonium salt 70
$C_4H_6O_5$
Malic acid 63
Malic acid, diammonium salt 69
$C_4H_6O_5S_2$
2,3-Dihydrothiophene-1,1-dioxide-3-sulfonic acid, sodium salt 156
$C_4H_6O_6$
Tartaric acid, monopotassim salt 69
C_4H_7BrO
3-Bromo-2-butanone 50
3-Bromo-3-buten-1-ol 16
3-Bromo-2-methyl-2-propen-1-ol 16
$C_4H_7BrO_3S$
3-Bromo-4-hydroxytetrahydrothiophene-1,1-dioxide 152
$C_4H_7Br_3$
2-Methyl-1,2,3-tribromopropane 4
$C_4H_7Br_3O$
2-Methyl-1,1,1-tribromo-2-propanol 15
$C_4H_7ClN_2O$
5-Chloromethyl-2-iminooxazolidine 136
C_4H_7ClOS
Propyl chlorothiolformate 100
$C_4H_7ClO_2$
2-Chlorobutyric acid 62
3-Chlorobutyric acid 62
$C_4H_7ClO_2S$
3-Chlorotetrahydrothiophene-1,1-dioxide 154
$C_4H_7Cl_2NSi$
3-(Dichloromethylsilyl)-propionitrile 1110
$C_4H_7Cl_3$
2-Methyl-1,2,3-trichloropropane 4
$C_4H_7Cl_3O_2$
2,2,3-Trichloro-1,1-butanediol 16
$C_4H_7F_3O$
2-Methyl-1,1,1-trifluoro-2-propanol 15
$C_4H_7IO_2S$
2-Iodo-1-propenyl methyl sulfone 154
C_4H_7NO
3-Cyano-1-propanol 144
Cyclopropanecarboxamide 102
C_4H_7NOS
L-Homocysteine thiolactone hydrochloride 96
Thioacetylacetamide hydrochloride 101
$C_4H_7NO_2$
Allyl carbamate 115
$C_4H_7NO_2S_2$
3-(Thiocarbamylthio)propionic acid 141
$C_4H_7NO_3S$
3-Cyanopropanesulfonic acid potassium salt 156
$C_4H_7NO_4$
Aspartic acid, disodium salt 71
L-Aspartic acid, potassium magnesium salt 71
$C_4H_7NS_2$
3-Methyl-2-thiazolidinethione 139
5-Methyl-2-thiazolidinethione 139
Tetrahydro-2*H*-1,3-thiazine-2-thione 143
$C_4H_7N_3O$
1-(2-Cyanoethyl)urea 146
$C_4H_7N_3O_2$
Ethyl 2-(1*H*-triazirin-1-yl)acetate 90
$C_4H_7N_5$
2-Allyl-5-amino-2*H*-tetrazole 870
2,4-Diamino-6-methyl-1,3,5-triazine 979
$C_4H_7N_5O$
4-Amino-5-imidazolecarboxamidoxime dihydrochloride 1009
2,5,6-Triamino-4-pyrimidinol hydrochloride 970
$C_4H_8Br_2O_2S$
2-Bromoethyl sulfone 152
$C_4H_8ClO_2S$
trans-3-Amino-4-chlorotetrahydrothiophene-1,1-dioxide hydrochloride 153
$C_4H_8Cl_2$
1,2-Dichlorobutane 4
2,2-Dichlorobutane 4
$C_4H_8Cl_2O_2S$
Bis(2-chloroethyl)sulfone 152
$C_4H_8F_4N_2$
(1-Methyl-2,2,3,3-tetrafluoropropyl)-hydrazine 32
$C_4H_8I_2O$
2-Iodoethyl ether 17
$C_4H_8N_2$
2-Methyl-2-imidazoline tartrate 41
$C_4H_8N_2O_2$
1-Acetyl-3-methylurea 128
Dimethylglyoxime, disodium salt, octahydrate 1009
N,N'-Dimethyloxamide 110
N,N'-Ethylenebis(formamide) 110
2-Methylmalonamide 104
$C_4H_8N_2O_2S$
3-Isothioureidopropionic acid 137
2,2'-Thiobisacetamide 110
$C_4H_8N_2O_3$
Glycylglycine hydrochloride monohydrate 119
$C_4H_8N_2O_4$
1,4-Dinitrobutane 46
2,2-Dinitrobutane 46
N-Hydroxy-L-asparagine 121
Nα-Hydroxy-D-asparagine 121
$C_4H_8N_2O_4S$
N,N'-Diacetylsulfamide 157
$C_4H_8N_2S_2$
N,N'-Dimethyldithiooxamide 139
$C_4H_8N_4O_2$
Maleic acid dihydrazide 111
$C_4H_8OS_2$
1,3-Dithiane-1-oxide 154
$C_4H_8O_2$
Aldol 60
Butyric acid, octadecylamine salt 69
Lithium butyrate 70
2-Methylglycidol 20
$C_4H_8O_2S$
2-(Ethylthio)acetic acid 63
2-Mercaptoethyl acetate 86

$C_4H_8O_3$
m-Dioxan-5-ol 21
(-)-3-Hydroxybutyric acid, quinine salt 63
2-Methoxypropionic acid 63
2,5-Tetrahydrofurandiol 21
$C_4H_8O_3S$
1,3-Butylene glycol sulfite 155
2-Hydroxyethyl mercaptoacetate 86
Tetrahydrothiophene-3-ol-1,1-dioxide 152
$C_4H_8O_4S$
cis-3,4-Dihydroxytetrahydrothiophene-1,1-dioxide 152
trans-3,4-Dihydroxytetrahydrothiophene-1,1-dioxide 152
$C_4H_8O_5S$
2-Carboxy-2-propanesulfonic acid pyridinium salt 963
$C_4H_8O_6S_2$
trans-4-Hydroxytetrahydrothiophene-1,1-dioxide-3-sulfonic acid, sodium salt 155
$C_4H_8O_8S_2$
S,S'-Methylenebis(sulfonylacetic acid) 153
$C_4H_9BO_2$
Butyl metaborate 1124
$C_4H_9BrO_3S$
4-Bromo-1-butanesulfonic acid, sodium salt 155
C_4H_9ClHg
Butylmercuric chloride 1137
C_4H_9ClHgO
2-(Chloromercuri)-ethyl ethyl ether 1138
C_4H_9ClO
3-Chloro-2-butanol 16
1-Chloro-2-methyl-2-propanol 15
$C_4H_9ClO_2$
3-Chloro-2-hydroxypropyl methyl ether 19
$C_4H_9ClO_3S$
4-Chloro-1-butanesulfonic acid, sodium salt 155
$C_4H_9Cl_2OPS$
Butyl phosphorodichloridothionate 160
C_4H_9NO
Butyraldoxime 1008
Isobutyraldehyde oxime 1008
N-Isopropylformamide 105
Morpholine hydrochloride 42
1-Pyrrolidinol 41
$C_4H_9NO_2$
Alanine methyl ester hydrochloride 89
Isopropyl carbamate 115
2-Methyl-1-nitropropane 46
2-Nitrobutane 46
$C_4H_9NO_2S$
Tetrahydro-3-thiophenamine 1,1-dioxide 153
$C_4H_9NO_3$
Butyl nitrate 45
O-(2-Hydroxyethyl)-glycolamide 101
2-Hydroxyethyl *N*-methylcarbamate 115
DL-2-Methylserine 71
2-Nitro-1-butanol 45
$C_4H_9NO_4$
2-Methyl-2-nitro-1,3-propanediol 45
$C_4H_9NO_4S_2$
3-Sulfolanesulfonamide 157
C_4H_9NS
N,N-Dimethylthioacetamide 141
$C_4H_9NS_2$
N-(Bis-(methylthio)-methylene)-methylamine hydriodide 139
S-Methyl *N,N*-dimethyldithiocarbamate 140
$C_4H_9N_3$
1-Allylguanidine sulfate 136
2-Amino-1-methyl-2-imidazoline hydrobromide 137
2-Amino-1,4,5,6-tetrahydropyrimidine hydrobromide 137
2-Amino-1,4,5,6-tetrahydropyrimidine *p*-toluenesulfonate 793
2-(Methylamino)-2-imidazoline hydriodide 137
$C_4H_9N_3O$
Acetone semicarbazone 126
$C_4H_9N_3O_2$
N-Methyl-2-ureidoacetamide 126
$C_4H_9N_3O_2S$
Ethyl 3-thiocarbamoylcarbazate 141
$C_4H_9N_3S$
Acetone thiosemicarbazone 140
5-Methylhexahydro-1,3,5-triazine-2-thione 140
$C_4H_{10}BF_4NO$
Morpholinium tetrafluoroborate 44
$C_4H_{10}BrTl$
Bromodiethylthallium 1140
$C_4H_{10}Br_2Pb$
Diethyllead dibromide 1133
$C_4H_{10}Cl_2NPS$
N,N-Diethylthiophosphoramidic dichloride 160
$C_4H_{10}Cl_2Pb$
Dichlorodiethyllead 1133
$C_4H_{10}Cl_2Si$
(Dichloromethyl)trimethylsilane 1100
$C_4H_{10}ITl$
Diethyliodothallium 1140
$C_4H_{10}NO_6P$
DL-Threonine phosphate hydrate 161
$C_4H_{10}N_2O$
L-2-Aminobutyramide 119
L-2-Aminobutyramide hydrochloride 121
Butyric acid hydrazide 112
2-(Ethylamino)acetamide hydrochloride 121
Isobutyric acid hydrazide 112
N-Nitrosodiethylamine 45
1,1,3-Trimethylurea 128
$C_4H_{10}N_2O_2$
N,N-Dimethyl-2-nitroethylamine 45
3-Hydroxybutyric acid hydrazide 111
2-Hydroxy-2-methylpropionic hydrazide 111
DL-Threoninamide hydrochloride 118
$C_4H_{10}N_2O_2S$
trans-3,4-Diaminotetrahydrothiophene-1,1-dioxide dihydrochloride 153
$C_4H_{10}N_4O_2$
1,2-Diureidoethane 126
1-Nitro-3-propylguanidine 138
$C_4H_{10}N_4S_2$
2,2'-Ethylenebis(2-thiopseudourea hydrobromide) 138
$C_4H_{10}OS$
1-Methoxy-2-(methylthio)ethane 25
$C_4H_{10}O_3S$
2,2'-Sulfinyldiethanol 152
$C_4H_{10}O_4S$
Butyl sulfate sodium salt 157
4-Hydroxy-1-butanesulfonic acid, sodium salt 155
$C_4H_{10}O_8S_2$
Succinaldehyde sodium bisulfite addition compound 155
$C_4H_{10}PbS_2$
Bis-(ethylthio)-lead 1133
$C_4H_{11}I_2N$
(Iodomethyl)trimethylammonium iodide 42
$C_4H_{11}N$
Diethylamine picrate 277
$C_4H_{11}NO$
N,N-Diethylhydroxylamine hemioxalate 35
2-(Dimethylamino)-ethanol bitartrate 34
2-(Dimethylamino)-ethanol hydrochloride 34
2-Ethoxyethylamine 35
$C_4H_{11}NOS$
2-((2-Aminoethyl)-thio)-ethanol 35
$C_4H_{11}NO_4S$
2-Amino-2-methyl-1-propanol hydrogen sulfate 157
$C_4H_{11}N_3$
1-Isopropylguanidine sulfate 136
1-Propylguanidine sulfate 136
$C_4H_{11}N_3O$
N-(3-Hydroxypropyl)guanidine nitrate 136
$C_4H_{11}N_3S$
4-Isopropyl-3-thiosemicarbazide 138
$C_4H_{11}O_3P$
Dimethyl ethylphosphonate 158
$C_4H_{11}O_4P$
tert-Butyl phosphate, dicyclohexylamine salt 160
$C_4H_{12}BF_4N$
Butylammonium tetrafluoroborate 43
$C_4H_{12}B_{10}O_4$
m-Carborane-1,7-dicarboxylic acid 346
$C_4H_{12}Br_3N$
Tetramethylammonium tribromide 43
$C_4H_{12}ClN$
Tetramethylammonium chloride (50% aqueous solution) 43
$C_4H_{12}F_6NP$
Butylammonium hexafluorophosphate 43
$C_4H_{12}N_2$
1,2-Diethylhydrazine 28
$C_4H_{12}N_2O$
2-Hydroxybutylhydrazine 34
$C_4H_{12}O_7P_2$
Diethyl acid pyrophosphate 160
$C_4H_{12}P_2S_2$
Tetramethyldiphosphine disulfide 160
$C_4H_{14}O_{10}P_2$
1,4-Dioxane diphosphate 161
$C_4H_{16}B_{10}O_2$
1,2-Bis(hydroxymethyl)-*o*-carborane 199
$C_4H_{20}B_3N$
Tetramethylammonium octahydrotriborate 43
C_5BrCl_4N
4-Bromo-2,3,5,6-tetrachloropyridine 967
$C_5BrCl_5O_2$
5-(Bromodichloromethyl)-3,4,5-trichloro-2(5*H*)-furanone 96
$C_5Br_2Cl_4O$
2,5-Dibromo-2,3,4,5-tetrachloro-3-cyclopenten-1-one 53
C_5Br_4ClN
4-Chloro-2,3,5,6-tetrabromopyridine 967
$C_5Cl_2F_6$
1,2-Dichlorohexafluorocyclopentene 6
C_5Cl_4IN
4-Iodo-2,3,5,6-tetrachloropyridine 967
$C_5Cl_4O_2$
2,2,4,5-Tetrachlorocyclopentene-1,3-dione 53
3,4,5,6-Tetrachloro-2*H*-pyran-2-one 96
C_5Cl_5N
1,2,3,4,4-Pentachloro-1,3-butadiene-1-carbonitrile 150
C_5Cl_5NO
Pentachloropyridine *N*-oxide 1006
C_5Cl_6O
Hexachloro-2-cyclopenten-1-one 53
C_5Cl_8
Octachlorocyclopentene 6
Octachloro-1,3-pentadiene 5
$C_5HBrCl_2O_3$
5-Bromo-3,4-dichloro-2-furoic acid 848
$C_5HBrCl_4O_2$
4-Bromo-1,2,3,4-tetrachloro-1,3-butadiene-1-carboxylic acid 62
$C_5HBr_3O_3$
3,4,5-Tribromo-2-furoic acid 848
$C_5HCl_3N_4$
2,6,8-Trichloropurine, ammonium salt 924
$C_5HCl_3O_2S$
3,4,5-Trichloro-2-thiophenecarboxylic acid 858
3,4,5-Trichloro-2-thiophenecarboxylic acid, sodium salt 857
C_5HCl_4N
2,3,5,6-Tetrachloropyridine 967
C_5HCl_5O
2,3,4,4,5-Pentachloro-2-cyclopenten-1-one 53
$C_5HCl_5O_2$
Pentachloro-*cis*-2,4-pentadienoic acid 62
$C_5H_2Br_2O_2$
2,2-Dibromo-4-cyclopentene-1,3-dione 53
$C_5H_2Br_2O_3$
3,5-Dibromo-2-furoic acid 848
4,5-Dibromo-2-furoic acid 848
$C_5H_2ClNO_4$
5-Nitro-2-furoyl chloride 847
$C_5H_2Cl_2HgO_3$
5-Chloro-4-(chloromercuri)-2-furoic acid 1138
$C_5H_2Cl_2O_2S$
2,5-Dichloro-3-thiophenecarboxylic acid 858
3,4-Dichloro-2-thiophenecarboxylic acid 858
$C_5H_2Cl_2O_3$
3,4-Dichloro-2-furoic acid 848
$C_5H_2Cl_5NO$
1,2,3,4,4-Pentachloro-1,3-butadiene-1-carboxamide 101
$C_5H_3BrCl_2O_2$
1-Bromo-4,4-dichloro-1,3-butadiene-1-carboxylic acid 62
$C_5H_3BrN_2O_2$
5-Bromo-2-nitropyridine 961
$C_5H_3BrN_2O_3$
2-Bromo-4-nitropyridine *N*-oxide 1005
3-Bromo-4-nitropyridine *N*-oxide 1006
5-Bromo-3-nitro-2-pyridinol 932
$C_5H_3BrO_2$
5-Bromo-2-furaldehyde 847
$C_5H_3BrO_3$
3-Bromo-2-furoic acid 848
4-Bromo-2-furoic acid 848
$C_5H_3Br_2NO$
3,5-Dibromopyridine *N*-oxide 1006
3,5-Dibromo-2-pyridinol 934
3,5-Dibromo-2-pyridone, sodium salt 134
$C_5H_3ClN_2O_2$
6-Chloro-3-pyridazinecarboxylic acid 968
$C_5H_3ClN_2O_3$
5-(Chlorocarbonyl)uracil 134
$C_5H_3ClN_2O_4$
5-Chloro-2,6-dihydroxy-4-pyrimidinecarboxylic acid 969
$C_5H_3ClO_3$
5-Chloro-2-furoic acid 848
$C_5H_3Cl_3O_4$
2,3,4-Trichloroglutaconic acid 62
$C_5H_3Cl_7O_2$
Methyl heptachlorobutyrate 82
C_5H_3HgNO
2-(Cyanomercuri)furan 1138
$C_5H_3IO_2$
5-Iodo-2-furaldehyde 847
$C_5H_3I_3S$
3-Methyl-2,4,5-triiodothiophene 859
$C_5H_3NO_4S$
5-Nitro-2-thiophenecarboxylic acid 856
$C_5H_3N_3$
2-Pyrimidinecarbonitrile 976
$C_5H_3N_3O$
4-Hydroxy-5-pyrimidinecarbonitrile 971
$C_5H_4BrNO_2$
5-Bromo-2-furoamide 851
$C_5H_4BrN_5O$
8-Bromoguanine 924
8-Bromoguanine hydrobromide 924
$C_5H_4Br_2O_3$
3,4-Dibromo-5-methoxy-2(5*H*)-furanone 96
$C_5H_4ClN_3O_2$
2-Amino-3-chloro-5-nitropyridine 941
2-Amino-5-chloro-3-nitropyridine 941
2-Amino-5-chloro-4-pyrimidinecarboxylic acid 975
$C_5H_4Cl_2N_2$
2,4-Dichloro-5-methylpyrimidine 977
4,6-Dichloro-2-methylpyrimidine 977
$C_5H_4Cl_2N_2O$
4,6-Dichloro-5-methoxypyrimidine 973

$C_5H_4Cl_2O_2$
3,4-Dichloro-5-methyl-2(5*H*)-furanone 96
$C_5H_4Cl_2O_3$
3,4-Dichloro-5-methoxy-2(5*H*)-furanone 95
$C_5H_4Cl_2S$
2,5-Dichloro-3-methylthiophene 859
C_5H_4IN
3-Iodopyridine 967
C_5H_4INO
2-Iodo-3-pyridinol 934
$C_5H_4N_2O_2$
2-Nitropyridine 961
$C_5H_4N_2O_2S$
2-Mercapto-5-nitropyridine 940
$C_5H_4N_2O_3$
1,6-Dihydro-6-oxo-3-pyridazinecarboxylic acid 968
2,5-Dioxo-3-pyrroline-1-carboxamide 125
4-Hydroxy-3-nitropyridine 932
6-Hydroxy-4-pyrimidinecarboxylic acid 972
3-Nitro-4(1*H*)-pyridone 944
$C_5H_4N_2O_3S$
2-Mercaptoorotic acid 970
$C_5H_4N_2O_4$
Orotic acid, potassium salt 975
Orotic acid, sodium salt 971
3,5-Pyrazoledicarboxylic acid, monopotassium salt 863
$C_5H_4N_4O$
2-Hydroxypurine monohydrate 923
$C_5H_4N_4OS$
6-Hydroxy-2-mercaptopurine 920
6-Hydroxy-8-mercaptopurine monohydrate 920
4-Hydroxy-6-mercaptopyrazolo(3,4-d)-pyrimidine 920
$C_5H_4N_4OS_2$
2,8-Dimercapto-6-hydroxypurine 920
6,8-Dimercapto-2-hydroxypurine 920
$C_5H_4N_4O_2$
5,7-Dihydroxy-*s*-triazolo-(1,5-a)-pyrimidine 918
$C_5H_4N_4O_2S$
2-Mercapto-6,8-purinediol 921
$C_5H_4N_4O_4$
2-Amino-3,5-dinitropyridine 941
4-Amino-3,5-dinitropyridine 941
$C_5H_4N_4S$
4-Amino-2-mercapto-5-pyrimidinecarbonitrile 973
$C_5H_4N_4S_2$
2,6-Dimercaptopurine 922
6,8-Dimercaptopurine 922
$C_5H_4N_4S_3$
2,6,8-Trimercaptopurine 922
$C_5H_4N_6O_2$
3-Aminopyrimido(4,5-e)-1,2,4-triazine-6,8(5*H*,7*H*)-dione 1003
C_5H_4OS
4*H*-Pyran-4-thione 59
4*H*-Thiopyran-4-one 59
$C_5H_4O_3$
2-Furoic acid, dodecylamine salt 848
$C_5H_4S_2$
4*H*-Thiopyran-4-thione 59
$C_5H_5BrN_2O_2$
5-Bromo-1-methyluracil 134
4-Bromo-1*H*-pyrazole-1-acetic acid 863
$C_5H_5ClN_2O$
6-Chloro-2-methyl-4-pyrimidinol 972
$C_5H_5ClN_2O_2$
5-Chloro-2,4-dihydroxy-6-methylpyrimidine 972
$C_5H_5ClN_4O_2$
4-Amino-2-chloro-6-methyl-5-nitropyrimidine 974
N-(5-Chloro-2-thiazolyl)oxamic hydrazide 876
$C_5H_5Cl_2N_3$
2-Amino-4,6-dichloro-5-methylpyrimidine 977
$C_5H_5Cl_3N_2OS$
N-(3-Thiazolin-2-yl)-2,2,2-trichloroacetamide 103
$C_5H_5F_3N_2O_2$
Imidazole trifluoroacetate salt 868
$C_5H_5F_3O_2$
3-(Trifluoromethyl)-2-butenoic acid 62
$C_5H_5IN_2$
2-Amino-5-iodopyridine 942
C_5H_5IS
2-Iodo-5-methylthiophene 859
C_5H_5N
Pyridine picrate 966
C_5H_5NOS
2-Thiophenecarboxaldehyde oxime 1013
$C_5H_5NO_2$
2,6-Dihydroxypyridine hemisulfate 931
Furfural oxime 1011
$C_5H_5NO_2S$
3-Methyl-4-isothiazolecarboxylic acid 874
3-Methyl-2-nitrothiophene 857
$C_5H_5NO_2S_2$
2,3-Dihydro-3-thiocyanatothiophene-1,1-dioxide 153
$C_5H_5NO_3S$
3-Pyridinesulfonic acid, sodium salt 965
$C_5H_5NO_4S$
2,4-Dihydroxythiazole-5-acetic acid 875
$C_5H_5N_3O$
N-(2-Pyrimidinyl)-formamide 976
$C_5H_5N_3O_2$
4-Amino-3-nitropyridine 941
2-Nitraminopyridine 960
4-Nitraminopyridine 949
$C_5H_5N_3O_3$
2-Amino-4-hydroxypyrimidine-6-carboxylic acid 971
2-Amino-6-hydroxy-4-pyrimidinecarboxylic acid 971
4-Amino-3-nitropyridine *N*-oxide 1005
$C_5H_5N_3O_3S$
2-Acetamido-5-nitrothiazole 877
$C_5H_5N_3O_4$
6-Methyl-5-nitrouracil 971
$C_5H_5N_5$
Adenine phosphate 923
5-Cyano-2,4-diaminopyrimidine 975
$C_5H_5N_5O$
Adenine N^1-oxide 1005
4-Amino-2-cyanoamino-6-hydroxypyrimidine 970
6-Amino-2-hydroxypurine sulfate 921
2-Amino-6-purinol 921
7-Amino-*s*-triazolo(1,5-a)pyrimidin-5(4*H*)-one 919
Guanine hydrochloride 924
Guanine sulfate 924
5-Hydroxy-7-methyl-1,2,3,8-tetraazaindolizine 918
5-Methyl-*s*-triazolo(4,3-a)-*s*-triazin-7-ol 918
$C_5H_5N_5O_2$
2,8-Dihydroxyadenine 920
$C_5H_5N_5O_4$
5-(Hydroxyiminoamino)-6-(hydroxyiminomethyl)uracil 1011
$C_5H_5N_5S$
6-Amino-2-mercaptopurine sulfate 922
$C_5H_6BF_4N$
Pyridinium tetrafluoroborate 958
$C_5H_6BrNO_2$
N-Bromoglutarimide 126
$C_5H_6BrN_3$
2-Amino-5-bromo-4-methylpyrimidine 975
$C_5H_6BrN_3O_2$
6-Amino-5-bromo-3-methyluracil 133
$C_5H_6Br_2O_3$
3,5-Dibromolevulinic acid 63
$C_5H_6Br_2O_4$
Dimethyl 2,2-dibromomalonate 83
$C_5H_6ClNO_3$
2-(2-Chloroacetamido)acrylic acid 117
$C_5H_6ClN_3$
3-Chloro-2,6-diaminopyridine 967
$C_5H_6Cl_2N_4$
2,4-Dichloro-6-(dimethylamino)-1,3,5-triazine 980
$C_5H_6Cl_2O_4$
3,3-Dichloroglutaric acid 62
$C_5H_6Cl_4O_3$
1,5-Dihydroxy-2,2,4,4-tetrachloro-3-pentanone 50
$C_5H_6Cl_6N_2O_3$
1,3-Bis-(1-hydroxy-2,2,2-trichloroethyl)-urea 126
$C_5H_6F_6NP$
Pyridinium hexafluorophosphate 958
$C_5H_6N_2$
2-Aminopyridine hydrochloride 942
2-Aminopyridine picrate 942
$C_5H_6N_2O$
2-Cyano-*N*-vinylacetamide 146
3-Methyl-1*H*-pyrazole-5-carboxaldehyde 862
6-Methyl-4(3*H*)-pyrimidone 976
$C_5H_6N_2OS$
2-Acetamidothiazole 877
4-Mercapto-5-methoxypyrimidine 973
2-Methylthio-4-pyrimidinol 970
4-(Methylthio)-2-pyrimidinol, sodium salt 970
$C_5H_6N_2O_2$
1,2-Dihydro-4-methyl-3,6-pyridazinedione 968
1-Methyl-4-pyrazolecarboxylic acid 863
3-Methyl-5-pyrazolecarboxylic acid 863
4-Methyl-1*H*-pyrazole-5-carboxylic acid 863
$C_5H_6N_2O_2S$
5-Methoxy-2-thiouracil 969
$C_5H_6N_2O_2S_2$
N-Acetamidorhodanine hydrate 141
$C_5H_6N_2O_3S$
6-Amino-3-pyridinesulfonic acid 942
4-Hydroxy-2-imino-4-thiazoline-5-acetic acid 136
$C_5H_6N_2O_4$
N-Carbamoylmaleamic acid 122
(*R*,*S*)-Hydroorotic acid 131
$C_5H_6N_2S_2$
2,4-Dimercapto-5-methylpyrimidine 973
$C_5H_6N_4O_2$
4-Amino-5-formamido-6-hydroxypyrimidine 970
2-Hydrazino-3-nitropyridine 944
$C_5H_6N_4O_3$
3-Hydrazino-4-nitropyridine *N*-oxide 1005
6-(Methylamino)-5-nitro-4-pyrimidinol 971
6-(Methylamino)-5-nitro-4(3*H*)-pyrimidone 975
6-Methyl-2-nitramino-4-pyrimidinol 972
$C_5H_6N_4O_4$
6-Amino-1-methyl-5-nitrouracil 132
$C_5H_6N_6O$
3-Methyl-8-azaguanine 924
$C_5H_6N_6S$
2,8-Diamino-6-mercaptopurine 922
C_5H_6O
3-Pentyn-1-al 1018
$C_5H_6O_3$
Methyl 4-hydroxytetrolate 1016
3-Methyl-2,3,4,5-tetrahydro-2,4-furandione 853
4-Oxo-2-pentenoic acid 63
$C_5H_6O_6$
1,1,2-Ethanetricarboxylic acid 67
$C_5H_7BrN_2O_2$
5-Bromo-5,6-dihydro-1-methyluracil 135
5-Bromo-5,6-dihydro-5-methyluracil 131
3-Bromo-5,5-dimethylhydantoin 135
C_5H_7BrO
4-Bromo-2-methyl-3-butyn-2-ol 1015
$C_5H_7BrO_2$
3-(Bromomethyl)crotonic acid 62
Methyl 3-bromo-2-methylacrylate 83
$C_5H_7Br_2ClO_2$
2,3-Dibromopropyl chloroacetate 81
$C_5H_7ClF_6Si$
(3-Chloro-1,1,2,2,3,3-hexafluoropropyl)dimethylsilane 1096
$C_5H_7ClN_2$
5-Chloromethyl-4-methylimidazole hydrochloride 864
$C_5H_7ClN_2OS$
2-Amino-3-(2-chloroacetyl)-4-thiazoline 876
$C_5H_7Cl_2NO_4$
N-Dichloroacetyl-DL-serine, sodium salt 117
$C_5H_7Cl_3O$
2,2-Bis(chloromethyl)propionyl chloride 100
3,3-Dichloropivaloyl chloride 100
$C_5H_7CuNO_5$
N-Pyruvylideneglycinato(aquo)-copper(II) dihydrate 1142
$C_5H_7NO_2$
Methyl 3-cyanopropionate 149
$C_5H_7NO_2S$
5,5-Dimethyl-2-thioxo-4-oxazolidinone 140
3-Ethyl-2,4-thiazolidinedione 745
$C_5H_7NO_3$
N-Methylmaleamic acid 122
$C_5H_7NO_5$
Dimethyl (hydroxyimino)-malonate 1011
C_5H_7NS
2,4-Dimethylthiazole 878
$C_5H_7N_2O_3$
Methyl 5-hydroxy-1-methyl-1*H*-1,2,3-triazole-4-carboxylate 869
$C_5H_7N_3O$
4-Amino-6-hydroxy-2-methylpyrimidine 971
3,5-Dimethyl-4-nitroso-1*H*-pyrazole 861
5-Methylcytosine 975
5-Methylcytosine hydrochloride 133
3-Methyl-1-pyrazolecarboxamide 863
$C_5H_7N_3OS$
4-Amino-6-hydroxy-2-(methylthio)-pyrimidine 970
$C_5H_7N_3O_2$
2-Amino-4,6-dihydroxy-5-methylpyrimidine 971
5-Amino-2,4-dihydroxy-6-methylpyrimidine 969
4-Amino-2-hydroxy-5-(hydroxymethyl)pyrimidine 969
2-Amino-6-methyl-4,5-pyrimidinediol 969
5-(Aminomethyl)uracil 975
3-(Dimethylamino)-2-(hydroxyimino)-3-oxopropionitrile 1010
1,2-Dimethyl-5-nitroimidazole 867
3,5-Dimethyl-4-nitro-1*H*-pyrazole 862
$C_5H_7N_3O_3$
Methyl 3-(*N*-methylcarbamoyl)-3-diazirinecarboxylate 105
1-Methyl-5-nitro-2-imidazolemethanol 865
5-Methyl-4-nitro-1*H*-pyrazole-3-methanol 860
2-(5-Nitro-1-imidazolyl)-1-ethanol 865
$C_5H_7N_3S$
4-Amino-2-mercapto-6-methylpyrimidine 973
4-Amino-5-methyl-2-pyrimidinethiol 973
4-Amino-3-methyl-2(3*H*)-pyrimidinethione 248
3-Methyl-1-pyrazolethiocarboxamide 864
$C_5H_7N_5O$
4,6-Diamino-2-methyl-5-nitrosopyrimidine 974
C_5H_8BrN
3-Bromo-2,2-dimethylpropionitrile 144
$C_5H_8BrNO_3$
DL-2-Bromopropionylglycine 123
$C_5H_8Br_2O_2$
Methyl 2,3-dibromoisobutyrate 83
Methyl 2,3-dibromo-2-methylpropionate 81
$C_5H_8Br_2O_2S$
trans-3,4-Dibromo-3-methyltetrahydrothiophene-1,1-dioxide 154
C_5H_8ClNO
N-Allyl-2-chloroacetamide 102
$C_5H_8ClNO_3$
Chloroacetyl-DL-alanine 117
N-Chloroacetyl-L-alanine 117

$C_5H_8ClNO_4$
Chloroacetyl-DL-serine 102
$C_5H_8Cl_2O_2$
cis-3,3-Dichloro-1,2-cyclopropanedimethanol 12
Ethyl 2,3-dichloropropionate 81
$C_5H_8Cl_2O_3$
2-Chloroethyl carbonate 82
$C_5H_8Cl_3NO_2S$
Methyl *N*-(1-(methylthio)-2,2,2-trichloroethyl)carbamate 113
$C_5H_8Cl_3NO_3$
Ethyl *N*-(1-hydroxy-2,2,2-trichloroethyl)carbamate 113
Methyl *N*-(1-methoxy-2,2,2-trichloroethyl)carbamate 113
$C_5H_8F_3NO_2$
4,4,4-Trifluoro-DL-valine 71
$C_5H_8F_4O$
2-Methyl-3,3,4,4-tetrafluoro-2-butanol 15
$C_5H_8I_4$
Pentaerythrityl tetraiodide 4
$C_5H_8N_2$
1-(2-Cyanoethyl)aziridine 147
$C_5H_8N_2O$
3-(Dimethylamino)-3-oxopropionitrile 150
3,5-Dimethyl-4-hydroxypyrazole 860
3,4-Dimethyl-5-pyrazolone 863
3-Methyl-1*H*-pyrazole-5-methanol 860
$C_5H_8N_2OS$
5,6-Dihydro-1-methyl-2-thiouracil 141
$C_5H_8N_2OS_2$
3-(Dimethylamino)rhodanine 140
S-(2-Thiazolin-2-yl) *N*-methylthiocarbamate 688
$C_5H_8N_2O_2$
2-Cyano-*N*-(2-hydroxyethyl)acetamide 144
5,6-Dihydro-5-methyluracil 133
5,6-Dihydro-6-methyluracil 133
2-Methyl-2-nitrobutyronitrile 147
5-Methyl-2,3-piperazinedione 134
Methyl 2-pyrazoline-3-carboxylate 90
L-Pyroglutamic acid amide 125
$C_5H_8N_2O_4$
1,1-Dinitrocyclopentane 46
$C_5H_8N_2O_6$
2,2-Dinitropropyl acetate 89
$C_5H_8N_2S$
2-Amino-4,5-dimethylthiazole hydrobromide 876
$C_5H_8N_4$
5-Amino-4-cyano-1-methyl-2-pyrazoline 145
4,6-Diamino-2-methylpyrimidine 974
3,4,5-Triaminopyridine hydrochloride 942
$C_5H_8N_4O$
5-Amino-4-hydroxy-6-(methylamino)pyrimidine 970
4-Amino-3-methyl-5-pyrazolecarboxamide sulfate 861
5-Amino-1-methyl-4-pyrazolecarboxamide sulfate 861
3-Cyanopropionaldehyde semicarbazone 147
2,4-Diamino-6-methyl-5-pyrimidinol 971
$C_5H_8N_4OS$
6-Amino-5-(methylamino)-2-thiouracil 974
5,6-Diamino-4-hydroxy-2-methylthiopyrimidine 970
5,6-Diamino-1-methyl-2-thiouracil 133
1-Methyl-3-(5-methyl-1,3,4-thiadiazol-2-yl)urea 879
$C_5H_8N_4O_2$
6-Amino-5-(methylamino)-2,4-pyrimidinediol 969
5,6-Diamino-1-methyluracil 132
$C_5H_8N_4O_3S_2$
4,6-Diamino-2-methylthiopyrimidine-5-sulfonic acid 973
$C_5H_8N_4S$
4,6-Diamino-2-methylmercaptopyrimidine 973
2,4-Diamino-6-(methylthio)-pyrimidine 974
C_5H_8O
3-Methyl-3-buten-2-one 50
4-Pentenal 60
$C_5H_8O_2$
4-Cyclopentene-1,3-diol 16
4-Hydroxy-3-methyl-3-buten-2-one, sodium salt 50
$C_5H_8O_2S$
4,5-Dihydro-3-methylthiophene 1,1-dioxide 154
$C_5H_8O_3$
Tetrahydro-2-furoic acid 68
$C_5H_8O_3S_2$
Ethylxanthogenacetic acid 94
$C_5H_8O_4$
Monomethyl methylmalonate 95
Propyl oxalate 75
$C_5H_8O_4S_2$
Methylenebis-(thioglycollic acid) 63
$C_5H_8O_5$
DL-Citramalic acid 63
L-2-Hydroxyglutaric acid disodium salt 69
$C_5H_9BO_4$
Pentaerythritol borate 1125
$C_5H_9BrO_2$
Methyl 2-bromo-2-methylpropionate 83
C_5H_9BrS
Dimethylpropargylsulfonium bromide 1017
$C_5H_9Br_2NO_3$
N,*N*-Bis-(hydroxymethyl)-2,3-dibromopropionamide 106
$C_5H_9Br_3O$
3-Bromo-2,2-bis(bromomethyl)-1-propanol 15
$C_5H_9ClF_2O$
1-Chloro-1,1-difluoro-2-pentanol 15
$C_5H_9ClO_2$
3-Chloro-5-hydroxy-2-pentanone 50
2-(Chloromethyl)-1,3-dioxane 21
Methyl 3-chlorobutyrate 83
Propyl chloroacetate 81
$C_5H_9Cl_2NOSi$
3-(Dichloromethylsilyl)-propyl isocyanate 1096
$C_5H_9Cl_2NSi$
4-(Dichloromethylsilyl)-butyronitrile. 1110
$C_5H_9Cl_3N_2O_2$
Methyl *N*-(1-methylamino)-2,2,2-trichloroethyl)carbamate 114
$C_5H_9Cl_3O$
2,2,2-Tris-(chloromethyl)-ethanol 15
$C_5H_9IN_2$
1,3-Dimethylimidazolium iodide 867
C_5H_9NO
Acetylacetone imide 58
Cyclobutanecarboxamide 101
Ethyl methyl ketone cyanohydrin 144
C_5H_9NOS
3-(2-Hydroxyethylthio)propionitrile 144
3-Methoxypropyl isothiocyanate 151
$C_5H_9NOS_2$
4-Morpholinecarbodithioic acid, sodium salt 142
$C_5H_9NO_2$
2-Allyl-L-glycine 72
1-Aza-4,6-dioxabicyclo(3.3.0)octane. 38
N-Hydroxymethylmethacrylamide 104
1-(Hydroxymethyl)-2-pyrrolidinone 124
3-Hydroxy-2-piperidone 124
Methallyl carbamate 115
Methyl 3-amino-2-butenoate 87
N-Methyldiacetamide 104
$C_5H_9NO_2S_2$
2-(Dimethylthiocarbamoylthio)-acetic acid 139
N-(Dithioethoxycarbonyl)-glycine 140
$C_5H_9NO_3$
N-Acetyl-DL-alanine 122
N-Acetyl-β-alanine 124
N-Acetylglycine methyl ester 124
N-Acetylsarcosine 124
Acetylurethan 105
Ethyl 2-hydroxyiminopropionate 1011
N-Formyl-L-2-aminobutyric acid 122
4-Hydroxymethyl-4-methyl-2-oxazolidinone 131
3-(Methylcarbamoyl)-propionic acid. 124
5-Nitro-2-pentanone 48
4-Oximinovaleric acid 1011
$C_5H_9NO_4$
4,4-Bis(hydroxymethyl)-2-oxazolidinone 131
Ethyl 3-nitropropionate 73
Glutamic acid calcium salt trihydrate 71
Glutamic acid magnesium salt dihydrate 71
L-Glutamic acid, monopotassium salt 71
N-Methyl-DL-aspartic acid 72
$C_5H_9NO_4S$
3-(Nitromethyl)-tetrahydrothiophene-1,1-dioxide 153
$C_5H_9NO_5$
2-(Methylaminomethyl)-tartronic acid 71
C_5H_9NS
n-Butylthiocyanate 151
Isobutyl isothiocyanate 151
$C_5H_9N_3$
5-Amino-1,3-dimethyl-1*H*-pyrazole 861
$C_5H_9N_3O$
1-Acetyl-3-amino-2-pyrazoline 108
4-Amino-1,6-dihydro-2-hydroxy-1-methylpyrimidine 36
4-Amino-5,6-dihydro-1-methyl-2(1*H*)-pyrimidinone 132
1-(2-Cyanoethyl)-1-methylurea 146
$C_5H_9N_3S$
Crotonaldehyde thiosemicarbazone 140
4,6-Dimethylhexahydro-*s*-triazine-2-thione 142
$C_5H_9N_5$
4-Amino-2-dimethylamino-1,3,5-triazine 979
6-Methyl-2,4,5-triaminopyrimidine sulfate 974
$C_5H_9N_5S$
2-Methylthio-4,5,6-triaminopyrimidine 973
$C_5H_{10}BrNO_3S$
4-(Bromomethylsulfonyl)-morpholine 157
$C_5H_{10}Br_2$
2,4-Dibromo-2-methylbutane 4
$C_5H_{10}Cl_2$
1,3-Dichloro-3-methylbutane 4
$C_5H_{10}N_2$
tert-Butylcyanamide 147
3-(Ethylamino)-propionitrile 146
$C_5H_{10}N_2O$
Acetic acid isopropylidenehydrazide 112
3-Methyl-2-piperazinone 124
1-Nitrosopiperidine 45
$C_5H_{10}N_2OS$
1-(2-Hydroxyethyl)-2-imidazolidinethione 139
$C_5H_{10}N_2O_2$
N-Acetyl-DL-alaninamide 124
N,*N'*-Dimethylmalonamide 111
Glutaraldehyde dioxime 1008
N,*N'*-Methylenediacetamide 111
4-Morpholinecarboxamide 130
Nα-Acetyl-*N*-methylglycinamide 121
$C_5H_{10}N_2O_3$
N-Acetyl-DL-serinamide 118
N-Acetyl-L-serinamide 118
D-Alanylglycine 119
β-Alanylglycine 118
L-Alanylglycine 119
Ethyl hydantoate 126
D(-)-Glutamine 71
DL-Glutamine 118
Glycyl-DL-alanine 119
Glycyl-β-alanine 118
Glycyl-L-alanine 119
$C_5H_{10}N_2O_3S$
N-Methyl-*N*-nitroso-3-tetrahydrothiophenamine 1,1-dioxide 153
$C_5H_{10}N_2O_4$
2,2-Dimethyl-1,3-dinitropropane 46
Glycyl-DL-serine 118
Glycyl-L-serine 117
N,*N'*-Methylenebis(methyl carbamate) 115
$C_5H_{10}N_2S$
Hexahydro-1-methyl-2-pyrimidinethione 140
2-(2-Methylallyl)-2-thiopseudourea hydrochloride 138
$C_5H_{10}N_2S_2$
3,5-Dimethyltetrahydro-1,3,5-thiadiazine-2-thione 140
$C_5H_{10}N_4O$
1-(1-Cyano-1-methylethyl)-semicarbazide 149
$C_5H_{10}O$
3-Methyl-3-buten-2-ol 11
$C_5H_{10}OS$
(Ethylthio)acetone 48
Tetrahydrothiopyran-3-ol 25
$C_5H_{10}O_2$
1,3-Dioxepane 22
Ethylene glycol monoallyl ether 19
4-Hydroxy-3-methyl-2-butanone 50
2-Methoxytetrahydrofuran 22
3-Methylbutyric acid, strontium salt 70
$C_5H_{10}O_2S$
Isopropyl mercaptoacetate 86
3-Mercaptopropyl acetate 86
2-Methyltetrahydrothiophene 1,1-dioxide 154
2-Methylthioethyl acetate 86
Pentamethylene sulfone 155
$C_5H_{10}O_2Se$
4-(Methylseleno)butyric acid 1142
$C_5H_{10}O_3$
2,2-Dimethyl-3-hydroxypropionic acid 63
3-Ethoxypropionic acid 63
2-Hydroxy-2-(hydroxymethyl)-tetrahydrofuran 21
2-Hydroxyvaleric acid 63
3-Methoxybutyric acid 63
4-Methoxybutyric acid 63
$C_5H_{10}O_4S$
1-(Ethoxycarbonyl)-1-ethanesulfonic acid, potassium salt 156
1-Hydroxy-1-cyclopentanesulfonic acid, sodium salt 155
$C_5H_{10}O_4S_2$
2,2-Dimethyl-1,3-dithiolane-1,1,3,3-tetroxide 155
$C_5H_{10}O_5$
DL-Arabinose 16
$C_5H_{10}O_7S_2$
3,3-Oxetanebis(methanesulfonic acid) disodium salt 156
$C_5H_{10}S$
Allyl ethyl sulfide 25
Methyl 2-methylallyl sulfide 25
$C_5H_{11}BrO$
2-Bromopropyl ethyl ether 17
$C_5H_{11}BrO_3S$
2,2-Dimethylpropyl bromosulfonate. 156
$C_5H_{11}Cl$
1-Chloro-3-methylbutane 4
2-Chloropentane 4
$C_5H_{11}ClO_2$
1-Chloro-3-ethoxy-2-propanol 19
$C_5H_{11}ClO_2S$
tert-Butyl chloromethyl sulfone 152
$C_5H_{11}I$
2-Iodo-2-methylbutane 4
$C_5H_{11}NO$
N-Butylformamide 105
N-*sec*-Butylformamide 106
4,4-Dimethyloxazolidine 40
3-Hydroxypiperidine 37
N-Isopropylacetamide 105
Isovaleramide 101
(*S*)-(+)-2-Methylbutyramide 101
Trimethylacetaldehyde oxime 1008

$C_5H_{11}NO_2$
N,N-Dimethyl-L-alanine 72
N,N-Dimethylformamide ethylene acetal 35
5-Hydroxyvaleramide 101
Isobutyl carbamate 117
N-(3-Methoxypropyl)-formamide 104

$C_5H_{11}NO_2S$
S-Ethyl-L-cysteine 71
trans-4-Methyltetrahydro-3-thiophenamine 1,1-dioxide hydrochloride 153
Penicillamine hydrochloride 71

$C_5H_{11}NO_3$
Ethyl *N*-(2-hydroxyethyl)-carbamate 115
N-(2-(2-Hydroxyethoxy)-ethyl)-formamide 103
N-(2-Hydroxyethyl)-*β*-alanine 71
DL-*allo*-Threonine methyl ester 87
DL-Threonine methyl ester hydrochloride 87

$C_5H_{11}NO_3S$
4-(Methylsulfonyl)morpholine 157

$C_5H_{11}NO_4$
2-Ethyl-2-nitro-1,3-propanediol 45

$C_5H_{11}NO_4S$
D-Methionine sulfone 777
DL-Methionine sulfone 153

$C_5H_{11}NS_2$
Diethyldithiocarbamic acid, lead salt, moist 70

$C_5H_{11}N_3$
2-Amino-1-methyl-1,4,5,6-tetrahydropyrimidine hydrobromide 137
1-Methyl-2-pyrrolidinone hydrazone dihydrochloride 137

$C_5H_{11}N_3O$
1-(2-Aminoethyl)-2-imidazolidone 133

$C_5H_{11}N_3O_3$
L-Glutamic acid δ-hydrazide 111

$C_5H_{11}N_3S$
Isobutyraldehyde thiosemicarbazone 140

$C_5H_{11}O_4PS_2$
Methyl (dimethoxyphosphinothioylthio)acetate 160

$C_5H_{12}BrNO_2S$
N,N-Diethylbromomethanesulfonamide 157

$C_5H_{12}ClN$
N-(2-Chloroethyl)isopropylamine hydrochloride 32

$C_5H_{12}ClNO_2$
(Carboxymethyl)-trimethylammonium chloride 67

$C_5H_{12}Cl_3NO$
N,N-Bis-(2-chloroethyl)-*N*-hydroxy-*N*-methylammonium chloride 42

$C_5H_{12}N_2O$
sec-Butylurea 128
tert-Butylurea 128
1,1-Diethylurea, sodium derivative 128
1-Isobutylurea 128
3,3-Oxetanebis(methylamine) dihydrobromide 35
Valeric acid hydrazide 112
D-Valinamide acetate 101

$C_5H_{12}N_2OS$
1-(3-Methoxypropyl)-2-thiourea 139

$C_5H_{12}N_2O_2$
DL-Ornithine hydrobromide 71
DL-Ornithine monohydrochloride 71

$C_5H_{12}N_2O_3$
1,1-Bis-(2-hydroxyethyl)-urea 126

$C_5H_{12}N_2S$
2-Butyl-2-thiopseudourea hydrobromide 137
2-Butyl-2-thiopseudourea hydrochloride 138
2-Butyl-2-thiopseudourea *p*-toluenesulfonate 795
1-Butyl-2-thiourea 142
1-Methyl-3-propyl-2-thiourea 142

$C_5H_{12}N_4O_2$
Ethylmalonic acid, dihydrazide 112
Glutaric acid, dihydrazide 112

$C_5H_{12}N_4S$
1-Amidino-3-propyl-2-thiourea 142
1-Amidino-3-propyl-2-thiourea *p*-toluenesulfonate 795

$C_5H_{12}OS$
2-Hydroxyethyl isopropyl sulfide 24
2-Hydroxyethyl propyl sulfide 24

$C_5H_{12}OS_2$
Ethyl (ethylsulfinylmethyl) sulfide 153

$C_5H_{12}O_2$
2-Methyl-1,4-butanediol 10
DL-2-Methyl-2,3-butanediol 10

$C_5H_{12}O_4S$
Trimethylacetaldehyde sodium bisulfite addition compound 155

$C_5H_{12}S$
Butyl methyl sulfide 25
Isobutyl methyl sulfide 25

$C_5H_{13}N$
Neopentylamine hydrochloride 28

$C_5H_{13}NO$
4-Methoxybutylamine 35

$C_5H_{13}NO_2$
1-(2-Hydroxyethylamino)-2-propanol 33
3-(2-Hydroxyethylamino)-1-propanol 33

$C_5H_{13}NO_2S$
N,N-Diethylmethanesulfonamide 157

$C_5H_{13}NO_5S$
N,N-Bis(2-hydroxyethyl)-aminomethanesulfonic acid, sodium salt 155
Formyltrimethylammonium methyl sulfate 157

$C_5H_{13}N_3$
1-Butylguanidine sulfate 137
1,1-Diethylguanidine hydrochloride 136
1,3-Diethylguanidine picrate 265
1,1-Diethylguanidine sulfate 137

$C_5H_{14}NO_3P$
Diethyl methylamidophosphate 159

$C_5H_{14}N_2$
N[1],*N*[1]-Dimethyl-1,2-propanediamine 29
N,N'-Dimethyl-1,3-propanediamine 30
N[2],*N*[2]-Dimethyl-1,2-propanediamine 29
N-Ethyl-*N'*-methylethylenediamine 30
1-*tert*-Pentylhydrazine hydrochloride 30

$C_5H_{14}N_2O$
1-(2-Aminoethylamino)-2-propanol 33
1,3-Diamino-*N*-(2-hydroxyethyl)-propane 33
2-(2-(Methylamino)-ethylamino)-ethanol 33

$C_6Br_2F_4$
1,3-Dibromotetrafluorobenzene 170

$C_6Br_5NO_2$
1-Nitro-2,3,4,5,6-pentabromobenzene 262

$C_6Cl_4N_2$
3,4,5,6-Tetrachloropicolinonitrile 945

$C_6Cl_4N_2O_4$
1,3-Dinitro-2,4,5,6-tetrachlorobenzene 261

$C_6Cl_4O_2S$
3,4-Dichloro-2,5-thiophenedicarbonyl chloride 858

C_6Cl_6
3,4-Bis(dichloromethylene)-1,2-dichloro-1-cyclobutene 6
5-(Dichloromethylene)-1,2,3,4-tetrachloro-1,3-cyclopentadiene 6
Hexachloro-1,5-hexadien-3-yne 1015

C_6Cl_8
3-(Dichloromethylene)-1,2,4,4,5,5-hexachloro-1-cyclopentene 6
Perchloro-1,3,5-hexatriene 5

$C_6Cl_9N_3$
2,4,6-Tris(trichloromethyl)-1,3,5-triazine 978

C_6Cl_{10}
Perchloro-1,5-hexadiene 5

$C_6CrK_3N_6S_6$
Potassium hexathiocyanatochromate(III) 1140

C_6HBrF_4
2,3,4,6-Tetrafluorobromobenzene 170

$C_6HBr_3Cl_2$
2,4-Dichloro-1,3,5-tribromobenzene 170

$C_6HCl_2HgO_4$
Chloranilic acid, mercury salt 1137

$C_6HCl_3N_2$
3,4,5-Trichloro-2-pyridinecarbonitrile 945

$C_6HCl_3N_2O_4$
1,3-Dinitro-2,4,5-trichlorobenzene 261

$C_6HCl_3O_2$
Trichloro-1,4-benzoquinone 53

C_6HCl_4I
1-Iodo-2,3,5,6-tetrachlorobenzene 170

$C_6HCl_4NO_2$
2,3,5,6-Tetrachloro-4-pyridinecarboxylic acid 962

C_6HCl_5S
Pentachlorothiophenol 209

C_6HCl_6N
3,4,5-Trichloro-2-(trichloromethyl)-pyridine 926

$C_6H_2BrCl_2I$
5-Bromo-1,3-dichloro-2-iodobenzene 171

$C_6H_2BrCl_3$
1-Bromo-2,3,5-trichlorobenzene 171
1-Bromo-2,4,5-trichlorobenzene 171
2-Bromo-1,3,5-trichlorobenzene 171
5-Bromo-1,2,3-trichlorobenzene 171

$C_6H_2BrFN_2O_4$
3-Bromo-4,6-dinitrofluorobenzene 276

$C_6H_2BrINO_2$
1-Bromo-2,3-diiodo-5-nitrobenzene 171

$C_6H_2BrI_2NO_2$
2-Bromo-1,3-diiodo-5-nitrobenzene 262

$C_6H_2Br_2O_2$
2,5-Dibromo-1,4-benzoquinone 315

$C_6H_2Br_3I$
1-Iodo-2,4,6-tribromobenzene 171

$C_6H_2Br_3NO_2$
1,3,5-Tribromo-2-nitrobenzene 262

$C_6H_2Br_4$
1,2,3,5-Tetrabromobenzene 171

$C_6H_2Br_4O_2$
Tetrabromohydroquinone 193

$C_6H_2Br_5N$
2,3,4,5,6-Pentabromoaniline 215

$C_6H_2ClN_2O_7S$
4-Chloro-2,6-dinitrobenzenesulfonic acid 789

$C_6H_2Cl_2N_2O_4$
1,3-Dichloro-4,6-dinitrobenzene 276

$C_6H_2Cl_2N_2S$
5,6-Dichloro-1,2,3-benzothiadiazole 919

$C_6H_2Cl_2O_2$
2,5-Dichloro-1,4-benzoquinone 315

$C_6H_2Cl_2O_4$
Chloranilic acid, barium salt 53
Chloranilic acid, disodium salt dihydrate 51
Chloranilic acid, lanthanum salt 1141

$C_6H_2Cl_3NO_2$
1,3,5-Trichloro-2-nitrobenzene 279
3,4,5-Trichloro-2-pyridinecarboxylic acid 963

$C_6H_2Cl_4O$
2,3,4,6-Tetrachlorophenol 190
2,3,4,6-Tetrachlorophenol, sodium salt monohydrate 190

$C_6H_2Cl_4O_2S$
2,3,4,-Trichlorobenzenesulfonyl chloride 801

$C_6H_2Cl_7NO$
Heptachloro-2,4-pentadiene-1-carboxamide 102

$C_6H_2N_4O_4$
Benzo(1,2-c:3,4-c')bis(1,2,5)-oxadiazole 3,6-dioxide 1005

$C_6H_2O_6$
Rhodizonic acid, barium salt 1141

$C_6H_3BrClNO_2$
1-Bromo-4-chloro-2-nitrobenzene 279

$C_6H_3BrCl_2$
1-Bromo-2,5-dichlorobenzene 171

$C_6H_3BrCl_2O$
4-Bromo-2,6-dichlorophenol 192

$C_6H_3BrN_2O_4$
1-Bromo-2,4-dinitrobenzene 275
1-Bromo-3,5-dinitrobenzene 261

$C_6H_3BrO_5$
5-Bromo-2,4-furandicarboxylic acid 848

$C_6H_3Br_2Cl$
1-Chloro-2,5-dibromobenzene 171

$C_6H_3Br_2ClO$
2-Chloro-4,6-dibromophenol 190

$C_6H_3Br_2Cl_2N$
2,4-Dibromo-3,6-dichloroaniline 215

$C_6H_3Br_2NO_2$
1,3-Dibromo-5-nitrobenzene 262

$C_6H_3Br_2NO_3$
2,4-Dibromo-6-nitrophenol 264
3,6-Dibromo-2-nitrophenol 260

$C_6H_3Br_3O$
2,4,6-Tribromophenol, sodium salt 190

$C_6H_3ClFHgNO_2$
4-Fluoro-3-nitrophenylmercuric chloride 1137

$C_6H_3ClFNO_4S$
4-Chloro-3-nitrobenzenesulfonyl fluoride 803

$C_6H_3ClHgN_2O_5$
2-Chloromercuri-4,6-dinitrophenol 1137

$C_6H_3ClINO_2$
1-Chloro-2-iodo-5-nitrobenzene 262

$C_6H_3ClN_2O_5$
2-Chloro-4,6-dinitrophenol, moist 264
4-Chloro-2,6-dinitrophenol, moist 259

$C_6H_3ClN_2O_7S$
4-Chloro-3,5-dinitrobenzenesulfonic acid 789
4-Chloro-3,5-dinitrobenzenesulfonic acid, potassium salt 793

$C_6H_3ClN_4O$
3-Chloropyrido(2,3-e)-1,2,4-triazine *N'*-oxide 1006

$C_6H_3Cl_2FO_4S_2$
2-Chloro-5-fluorosulfonylbenzenesulfonyl chloride 804

$C_6H_3Cl_2I$
1,2-Dichloro-4-iodobenzene 171
1,3-Dichloro-2-iodobenzene 171

$C_6H_3Cl_2IO$
2,4-Dichloro-6-iodophenol 192

$C_6H_3Cl_3N_2O_2$
3-Nitro-2,4,6-trichloroaniline 260

$C_6H_3Cl_3N_4$
6-(Trichloromethyl)purine 920

$C_6H_3Cl_3OS$
2-Acetyl-3,4,5-trichlorothiophene 857

$C_6H_3Cl_3O_3S$
2,4,5-Trichlorobenzenesulfonic acid, potassium salt 791
2,4,5-Trichlorobenzenesulfonic acid, sodium salt 796

$C_6H_3Cl_3S$
2,4,5-Trichlorothiophenol 209

$C_6H_3Cl_4N$
2-Chloro-3-(trichloromethyl)pyridine 926
2-Chloro-5-(trichloromethyl)pyridine 926
3-Methyl-2,4,5,6-tetrachloropyridine 967

$C_6H_3Cl_5$
5-Methyl-1,2,3,4,5-pentachloro-1,3-cyclopentadiene 6

$C_6H_3Cl_5N_2$
4-Amino-3,5-dichloro-2-(trichloromethyl)pyridine 925

$C_6H_3Cl_5O_3$
Methyl 3-oxo-2,2,4,5,5-pentachloro-4-pentenoate 92

$C_6H_3Cl_5Si$
(Dichlorophenyl)trichlorosilane (mixture of isomers) 1096

$C_6H_3Cl_6N_3$
α2,α2,α2,α4,α4,α4-Hexachloro-2,4,6-trimethyl-1,3,5-triazine 978

$C_6H_3FINO_2$
1-Fluoro-2-iodo-5-nitrobenzene 262

$C_6H_3IN_2O_4$
3,5-Dinitroiodobenzene 276

$C_6H_3I_2NO_2$
3,5-Diiodo-1-nitrobenzene 262

$C_6H_3N_3O_5$
2-(5-Nitro-2-furyl)-1,3,4-oxadiazol-2-in-5-one 873

$C_6H_3N_3O_8$
2,4,6-Trinitroresorcinol, wet 262

C_6H_4AsBrO
2-Bromophenylarsine oxide 1126

$C_6H_4BrClHgO_3$
Methyl 5-bromo-4-(chloromercuri)-2-furancarboxylate 1138

$C_6H_4BrClN_2O_2$
2-Bromo-6-chloro-4-nitroaniline 260

$C_6H_4BrCl_2N$
4-Bromo-2,6-dichloroaniline 216

C_6H_4BrCuS
((4-Bromophenyl)-thio)-copper(I) 1143

$C_6H_4BrNO_5S$
4-Bromo-3-nitrobenzenesulfonic acid, thallium salt 791

$C_6H_4BrO_2P$
1,2-Phenylene phosphorobromidite 833

$C_6H_4Br_2$
1,4-Dibromobenzene, zone refined 169

$C_6H_4Br_2ClN$
2-Chloro-4,6-dibromoaniline 216

$C_6H_4Br_2O_3$
Methyl 4,5-dibromo-2-furoate 853

$C_6H_4Br_2O_4$
4,5-Dibromo-2-(hydroxymethyl)-3-furoic acid 845

C_6H_4ClF
p-Chlorofluorobenzene 169

$C_6H_4ClFO_4S_2$
o-Fluorosulfonylbenzenesulfonyl chloride 804

$C_6H_4ClNO_2$
2-Chloro-4-nitrosophenol, moist 257

$C_6H_4ClNO_3$
2-Chloro-4-nitrophenol 265
2-Chloro-6-nitrophenol 264

$C_6H_4ClNO_4S$
2-Chloro-5-nitrobenzenesulfinic acid 786
4-Chloro-3-nitrobenzenesulfinic acid, barium salt 786
2-Chloro-5-nitrobenzenesulfinic acid, sodium salt 786

$C_6H_4ClNO_5S$
3-Chloro-4-nitrobenzenesulfonic acid 789
2-Chloro-3-nitrobenzenesulfonic acid, potassium salt 791
2-Chloro-5-nitrobenzenesulfonic acid, sodium salt 794
4-Chloro-3-nitrobenzenesulfonic acid, sodium salt 791

$C_6H_4ClN_3O_2S$
4-Chlorobenzenesulfonyl azide 810

$C_6H_4Cl_2HgO_3$
Methyl 5-chloro-4-(chloromercuri)-2-furoate 1138

$C_6H_4Cl_2N_2O_2$
2,4-Dichloro-5-nitroaniline 260
2,5-Dichloro-4-nitroaniline 260

$C_6H_4Cl_2N_4S$
6,8-Dichloro-2-(methylthio)purine 924

$C_6H_4Cl_2O$
3,4-Dichlorophenol sodium salt 190

$C_6H_4Cl_2OS$
2,5-Dichloro-3-thienyl methyl ketone 857

$C_6H_4Cl_2O_2$
2,5-Dichlorohydroquinone 193

$C_6H_4Cl_2O_3S$
3,4-Dichlorobenzenesulfonic acid 790

$C_6H_4Cl_2O_4S$
3,5-Dichloro-4-hydroxybenzenesulfonic acid 787

$C_6H_4Cl_3N$
2-(Trichloromethyl)pyridine hydrochloride 926

$C_6H_4Cl_3NO_4S_2$
4-Amino-6-chloro-1,3-benzenedisulfonyl chloride 801

$C_6H_4Cl_5NO$
N-Methyl-1,2,3,4,4-pentachloro-1,3-butadiene-1-carboxamide 105

$C_6H_4Cl_6O_2$
Methyl 2,2,3,4,5,5-hexachloro-4-pentenoate 83

$C_6H_4Cl_8$
1,2,3,4,5,5,6,6-Octachloro-1-hexene 5

$C_6H_4FNO_4S$
3-Nitrobenzenesulfonyl fluoride 803

$C_6H_4N_2O_3S$
4-Nitro-*N*-sulfinylaniline 771

$C_6H_4N_2O_4$
Mordant Dye (C.I. 10000) 1095
4,6-Pyrimidinedicarboxylic acid 976

$C_6H_4N_2O_5$
2,3-Dinitrophenol 264

$C_6H_4N_2O_6$
2,6-Dinitrohydroquinone 262

$C_6H_4N_2O_7S$
2,4-Dinitrobenzenesulfonic acid, lead salt 1134
2,5-Dinitrobenzenesulfonic acid, sodium salt 793

$C_6H_4N_2S_2$
2,2′-Bis(thiazole) dihydrochloride 879

$C_6H_4N_2Se$
2,1,3-Benzoselenadiazole 213

$C_6H_4N_4O$
4-Hydroxypteridine 1002

$C_6H_4N_4OS$
4-Hydroxy-2-mercaptopteridine 1002

$C_6H_4N_4O_2$
4-Nitrobenzotriazole 909

$C_6H_4N_4O_3$
5-Amino-4-nitrobenzofurazan 917
2,4,6(1*H*,3*H*,5*H*)-Pteridinetrione 1003
4,6,7(3*H*,5*H*,8*H*)-Pteridinetrione 1003

$C_6H_4N_4O_3S$
2-Mercapto-4,6,7-trihydroxypteridine 1002

$C_6H_4O_4S$
2,4-Thiophenedicarboxylic acid 857

$C_6H_4O_5$
2,4-Furandicarboxylic acid 853
3,4-Furandicarboxylic acid 849

$C_6H_4O_6$
2,3,5,6-Tetrahydroxy-1,4-benzoquinone 52

$C_6H_5AsClNO_5$
4-Chloro-3-nitrophenylarsonic acid 1127

$C_6H_5AsNO_4$
2-Nitrobenzenearsonic acid 1127

$C_6H_5BrCl_2O_2$
Methyl 1-bromo-4,4-dichloro-1,3-butadiene-1-carboxylate 83

$C_6H_5BrFNO_2S$
3-Bromosulfanilyl fluoride 801

C_6H_5BrHg
(Bromomercuri)benzene 1139

$C_6H_5BrN_2O_2$
2-Bromo-4-nitroaniline 260

$C_6H_5BrO_2$
Bromohydroquinone 193

$C_6H_5BrO_3$
Methyl 4-bromo-2-furoate 848
Methyl 5-bromo-2-furoate 848

$C_6H_5BrO_3S$
4-Bromobenzenesulfonic acid, sodium salt 791

$C_6H_5Br_2N$
3,5-Dibromoaniline 216

$C_6H_5Br_2NO$
3,4-Dibromo-5-methyl-2-pyrrolecarboxaldehyde 842

$C_6H_5Br_2NO_3S$
3,5-Dibromosulfanilic acid 787

$C_6H_5Br_3N_2$
1-Allyl-3,4,5-tribromopyrazole 864

$C_6H_5Br_3Te$
(Phenyl)tribromotellurium 1142

C_6H_5ClHgO
2-(Chloromercuri)phenol 1137

C_6H_5ClIN
2-Chloro-4-iodoaniline 215

$C_6H_5ClN_2O_2$
3-Chloro-4-nitroaniline 260

$C_6H_5ClN_2O_4S$
2-Chloro-5-nitrobenzenesulfonamide 822
4-Chloro-3-nitrobenzenesulfonamide 822

C_6H_5ClOS
5-Chloro-2-thienyl methyl ketone 857

$C_6H_5ClO_2$
4-Chlorocatechol 192

$C_6H_5ClO_2S$
4-Chlorobenzenesulfinic acid sodium salt 786

$C_6H_5ClO_3$
Methyl 5-chloro-2-furoate 853

$C_6H_5ClO_3S$
4-Chlorobenzenesulfonic acid, sodium salt 795

$C_6H_5Cl_2N$
2,4-Dichloroaniline hydrochloride 215
2,4-Dichloro-3-methylpyridine hydrochloride 967

$C_6H_5Cl_2NOS$
2′,5′-Dichloro-3-thiopheneacetamide 858

$C_6H_5Cl_2NO_2S$
N,*N*-Dichlorobenzenesulfonamide 827

$C_6H_5Cl_2NO_3S$
2,5-Dichlorosulfanilic acid 787

$C_6H_5Cl_2OPS$
Phenyl phosphorodichloridothionate 835

$C_6H_5Cl_3Te$
Phenyltrichlorotellurium 214

$C_6H_5Cl_5O_2$
Ethyl pentachloro-3-butenoate 83

$C_6H_5Cl_5O_3$
Methyl 3-hydroxy-2,2,4,5,5-pentachloro-4-pentenoate 84

$C_6H_5FN_2O$
5-Fluoro-3-pyridinecarboxamide 949

$C_6H_5FN_2O_2$
5-Fluoro-2-nitroaniline 260

$C_6H_5F_3N_2OS$
2-Acetamido-4-(trifluoromethyl)-thiazole 874

$C_6H_5F_8NO$
4-Ethyl-octafluoromorpholine 41

C_6H_5HgI
Phenylmercuric iodide 1139

$C_6H_5IN_2O_2$
3-Iodo-5-nitroaniline 260

$C_6H_5I_2NO$
3,5-Diiodo-4-methyl-2-pyrrolecarboxaldehyde 839

$C_6H_5NO_2$
Nicotinic acid, 1,2-diphenylethylammonium salt 956
Nicotinic acid, octadecylamine salt 962

$C_6H_5NO_3$
4-Nitrophenol, potassium salt 265
2-Nitrophenol, sodium salt 264
2-(2-Nitrovinyl)furan 852

$C_6H_5NO_3S$
2-Formamido-3-thiophenecarboxylic acid 858

$C_6H_5NO_5$
Methyl 5-nitro-2-furoate 847
2-Methyl-5-nitro-3-furoic acid 852

$C_6H_5NO_5S$
4-Nitrobenzenesulfonic acid, ammonium salt 794
4-Nitrobenzenesulfonic acid, sodium salt 794
3-Sulfoisonicotinic acid, barium salt 960

$C_6H_5NO_6S$
2-Hydroxy-5-nitrobenzenesulfonic acid, disodium salt 791
4-Hydroxy-3-nitrobenzenesulfonic acid, disodium salt 791

$C_6H_5NO_9S_2$
4-Hydroxy-5-nitro-1,3-benzenedisulfonic acid, dipotassium salt 791

$C_6H_5N_3O_2$
5-Cyano-2,6-dihydroxy-4-methylpyrimidine 971

$C_6H_5N_3O_3$
1-(Hydroxyiminoamino)-4-nitrobenzene, sodium salt 1090

$C_6H_5N_3O_6$
Methyl 5-nitroorotate hydrate 976

$C_6H_5N_3S$
4-Aza-2-mercaptobenzimidazole 918

$C_6H_5N_5$
5-Amino-4-cyano-3-cyanomethylpyrazole 860

$C_6H_5N_5O_2$
4-Nitro-5-(1*H*-pyrazol-1-yl)-1*H*-pyrazole 861

C_6H_5OSb
Phenylstibine oxide 1129

$C_6H_6AsClO_3$
3-Chlorobenzenearsonic acid 1126
4-Chlorobenzenearsonic acid 1128

$C_6H_6AsIO_3$
2-Iodobenzenearsonic acid 1126

$C_6H_6AsNO_5$
4-Nitrobenzenearsonic acid 1127

$C_6H_6AsN_3O_7$
4-Amino-3,5-dinitrophenylarsonic acid 1127

$C_6H_6BBrO_3$
5-Bromo-2-hydroxybenzeneboronic acid 1125

$C_6H_6BClO_2$
3-Chlorophenylboronic acid 1124
4-Chlorophenylboronic acid 1124

$C_6H_6BF_9O_3$
Tris(2,2,2-trifluoroethyl) borate 1124

$C_6H_6BNO_4$
2-Nitrophenylboronic acid 1125
4-Nitrophenylboronic acid 1125

C_6H_6BrN
4-Bromoaniline hydrochloride 215

C_6H_6BrNO
2-Bromo-6-methoxypyridine 940

C_6H_6BrNOS
2-Acetyl-5-bromothiophene oxime 1011

$C_6H_6BrN_3O_2$
2-Amino-5-bromo-6-methyl-3-nitropyridine 941
2-Amino-6-bromo-3-methyl-5-nitropyridine 942

$C_6H_6BrO_3P$
(2-Bromophenyl)phosphonic acid 833
(4-Bromophenyl)phosphonic acid 833

$C_6H_6Br_2Cl_2O$
2,3-Dibromo-5,5-dichloro-4-methyl-2,4-pentadien-1-ol 16

$C_6H_6Br_2N_2O_2S$
2,6-Dibromosulfanilamide 807
3,5-Dibromosulfanilamide 807

C_6H_6ClN
3-Chloroaniline hydrochloride 215

$C_6H_6ClNO_3S$
2-Chloroaniline-4-sulfonic acid 787
4-Chloroaniline-3-sulfonic acid 787

C_6H_6ClNS
5-Chloro-2-mercaptoaniline hydrochloride 215

$C_6H_6ClN_3O$
2-Chloro-*N*-(2-pyrimidinyl)-acetamide 968

$C_6H_6ClN_3O_2$
4-Chloro-2-nitro-1-phenylhydrazine 262

$C_6H_6ClN_3O_4$
1-Amino-2-cyanopyridinium perchlorate 955

$C_6H_6ClO_3P$
(2-Chlorophenyl)phosphonic acid 833
(3-Chlorophenyl)phosphonic acid 833
(4-Chlorophenyl)phosphonic acid 833

$C_6H_6Cl_2N_2$
4-Amino-3,5-dichloro-2-methylpyridine 942
2,3-Dichlorophenylhydrazine hydrochloride 234

$C_6H_6Cl_2N_2O_2$
Ethyl *N*-(1-cyano-2,2-dichlorovinyl)-carbamate 144

$C_6H_6Cl_2N_2O_2S$
2,5-Dichlorobenzenesulfonic hydrazide 810
3,5-Dichlorosulfanilamide 817

$C_6H_6Cl_2O_3$
Methyl 5,5-dichloro-3-oxo-4-pentenoate 92

$C_6H_6Cl_4O$
2,2,6,6-Tetrachlorocyclohexanone 55

$C_6H_6Cl_4O_2$
5,5-Dimethoxy-1,2,3,4-tetrachloro-1,3-cyclopentadiene ... 18
C_6H_6FN
4-Fluoroaniline hydrochloride ... 215
$C_6H_6FNO_2S$
4-Fluorobenzenesulfonamide ... 827
$C_6H_6FO_3P$
(2-Fluorophenyl)phosphonic acid ... 833
(3-Fluorophenyl)phosphonic acid ... 833
$C_6H_6F_3NO_4$
Ethyl 2-hydroxyimino-3-oxo-4,4,4-trifluorobutyrate ... 1008
$C_6H_6F_3NO_5$
N-(Trifluoroacetyl)-DL-aspartic acid ... 117
$C_6H_6F_3N_3$
2-Amino-4-methyl-6-(trifluoromethyl)pyrimidine ... 968
$C_6H_6INO_2S$
4-Iodobenzenesulfonamide ... 809
$C_6H_6IO_3P$
(2-Iodophenyl)phosphonic acid ... 833
$C_6H_6NO_5P$
(2-Nitrophenyl)phosphonic acid ... 835
(3-Nitrophenyl)phosphonic acid ... 835
(4-Nitrophenyl)phosphonic acid ... 835
$C_6H_6N_2$
2-Hexenedinitrile ... 150
$C_6H_6N_2O$
Picolinamide ... 944
$C_6H_6N_2O_2$
3-Aminopicolinic acid ... 942
Nicotinohydroxamic acid ... 960
$C_6H_6N_2O_2S$
2-Carboxymethylmercaptopyrimidine ... 974
$C_6H_6N_2O_3$
5-Formyl-6-methyluracil ... 976
6-Hydroxy-2-methyl-3-nitropyridine ... 932
2-Methoxy-3-nitropyridine ... 937
2-Methyl-4-nitropyridine-*N*-oxide ... 1006
6-Methyl-2-nitro-3-pyridinol ... 933
$C_6H_6N_2O_3S$
5-Acetyl-2-amino-3-nitrothiophene ... 856
6-Hydroxy-2-(methylthio)-4-pyrimidinecarboxylic acid ... 970
$C_6H_6N_2O_4$
5-Acetylbarbituric acid ... 131
1,3-Dimethyl-hexahydro-2,4,5,6-pyrimidinetetraone ... 134
$C_6H_6N_4$
4-Amino-2-methyl-5-pyrimidinecarbonitrile ... 974
2,2′-Bi-1*H*-imidazole ... 866
5-Methyl-*s*-triazolo(1,5-a)pyrimidine ... 920
Nitrilotriacetonitrile ... 145
$C_6H_6N_4O$
2-Cyanoamino-4-hydroxy-6-methylpyrimidine ... 972
2-Hydroxy-6-methylpurine ... 921
6-Hydroxy-2-methylpurine ... 921
N-(4-Hydroxy-6-methyl-2-pyrimidinyl)-carbamonitrile ... 972
$C_6H_6N_4OS$
2-(Methylthio)hypoxanthine ... 923
3-Methyl-2-thioxanthine ... 868
$C_6H_6N_4O_2$
3-Aminopyrazolo(4,5-c)pyridine-4,6-diol ... 918
$C_6H_6N_4O_2S$
6,8-Dihydroxy-2-methylmercaptopurine ... 920
2,6-Dihydroxy-8-(methylthio)purine ... 920
$C_6H_6N_4O_4$
2-Amino-3,5-dinitro-6-methylpyridine ... 941
2,3-Dinitro-*p*-phenylenediamine ... 270
$C_6H_6N_4S$
4-Amino-5-cyano-2-(methylthio)-pyrimidine ... 973
6-Mercapto-9-methylpurine ... 922
4-(Methylthio)-1*H*-pyrazolo(3,4-d)-pyrimidine ... 919
$C_6H_6N_4S_2$
2,6-Dimercapto-7-methylpurine ... 922
2-(Methylthio)-6-thiopurine ... 922
$C_6H_6N_4S_3$
7-Methyl-2,6,8-trimercaptopurine ... 922
C_6H_6OS
4-Methyl-2-thiophenecarboxaldehyde ... 857
$C_6H_6O_2$
2,4-Hexadiyne-1,6-diol ... 1016
$C_6H_6O_2S$
Benzenesulfinic acid nitrate ... 786
4-Methyl-2-thiophenecarboxylic acid ... 858
$C_6H_6O_3$
cis-1,2-Cyclobutanedicarboxylic anhydride ... 98
2-Methyl-3-furoic acid ... 853
5-Methyl-2-furoic acid ... 848
$C_6H_6O_3S$
Benzenesulfonic acid, thallium salt. 795
$C_6H_6O_4$
4-Acetyldihydro-2,3-furandione ... 96
2-(2-Propynyl)malonic acid ... 1017
$C_6H_6O_4S$
2-(Acetylthio)succinic anhydride ... 98
4-Hydroxybenzenesulfonic acid, potassium salt ... 792
$C_6H_6O_5$
Methyl 4-carboxy-3-furoate ... 849
Methyl 2,5-dihydro-4-hydroxy-2-oxo-3-furancarboxylate ... 95
$C_6H_6O_5S$
Hydroquinonesulfonic acid, sodium salt ... 791
$C_6H_6O_6S_2$
1,2-Benzenedisulfonic acid, disodium salt ... 795
C_6H_6S
Thiophenol, thallium salt ... 209
$C_6H_7AsO_4$
2-Hydroxybenzenearsonic acid ... 1126
4-Hydroxybenzenearsonic acid ... 1126
$C_6H_7AsO_5$
2,4-Dihydroxybenzenearsonic acid ... 1126
$C_6H_7BHgO_3$
Phenylmercuric borate ... 1139
$C_6H_7BrN_2$
6-Amino-3-bromo-2-methylpyridine ... 942
$C_6H_7BrN_2O_2$
5-Bromo-2,4-dimethoxypyrimidine ... 973
$C_6H_7BrN_2O_2S$
4-Bromobenzenesulfonic hydrazide. 810
$C_6H_7BrO_4$
2-Bromoethyl maleate ... 94
$C_6H_7ClHgN_2O_2S$
S-(1-(Chloromercuri)-4-oxo-1,2,3,4-tetrahydro-2-pyrimidinyl) thioacetate ... 1138
$C_6H_7ClHgO_2$
2-(Chloromercuri)-5-(methoxymethyl)furan ... 1138
$C_6H_7ClN_2$
4-Chloro-1,3-phenylenediamine ... 220
5-Chloro-1,3-phenylenediamine ... 220
$C_6H_7ClN_2OS$
2-Acetamido-4-(chloromethyl)-thiazole ... 874
$C_6H_7ClN_2O_2$
4-Chloro-2,6-dimethoxypyrimidine ... 972
5-Chloro-1,3-dimethyluracil ... 134
6-Chloro-1,3-dimethyluracil ... 134
$C_6H_7ClN_2O_2S$
4-Chlorobenzenesulfonic hydrazide. 810
3-(2-Chloroethyl)-1-methyl-5-thioxohydantoin ... 139
$C_6H_7ClN_2O_3S$
2-Chlorophenylhydrazine-4-sulfonic acid ... 789
$C_6H_7ClN_2S$
4-Chloro-6-methyl-2-methylthiopyrimidine ... 974
$C_6H_7Cl_2N_3$
3-(2-Chloro-2-cyanoethyl)-imidazolium chloride ... 867
$C_6H_7Cl_3O$
2,6,6-Trichloro-1,2-epoxycyclohexane ... 20
$C_6H_7Cl_3O_2$
Ethyl 4,4,4-trichloro-2-butenoate ... 82
$C_6H_7Cl_3SSi$
Dimethyl(3,4,5-trichloro-2-thienyl)-silane ... 1110
$C_6H_7Cl_4NO_3S$
N-(4-Chloro-3-sulfolanyl)-2,2,2-trichloroacetamide ... 152
$C_6H_7F_3O$
4-Methyl-5,5,5-trifluoro-3-penten-2-one ... 50
$C_6H_7F_3O_2$
Cyclopropylmethyl 2,2,2-trifluoroacetate ... 82
$C_6H_7F_4N_3O$
3-Amino-1-(2,2,3,3-tetrafluoropropyl)-2-pyrazolin-5-one 130
$C_6H_7IN_2O_2$
2,4-Dimethoxy-5-iodopyrimidine ... 972
$C_6H_7IN_2O_2S$
4-Iodobenzenesulfonic acid hydrazide ... 810
C_6H_7N
Aniline hydrobromide ... 219
Aniline picrate ... 271
2-Furoic acid, aniline salt ... 848
2-Picoline picrate ... 967
C_6H_7NO
1-Acetylpyrrole ... 843
2-Aminophenol hydrochloride ... 216
6-Methyl-2(1*H*)-pyridone ... 965
3-Methyl-2-pyrrolecarboxaldehyde ... 840
2-Oxo-1-cyclopentanecarbonitrile ... 148
C_6H_7NOS
2-Acetylthiophene oxime ... 1013
2-(Aminoacetyl)thiophene hydrochloride ... 856
N-(2-Thienyl)acetamide ... 858
N-(3-Thienyl)acetamide ... 858
$C_6H_7NO_2$
2-Acetylfuran oxime ... 1011
4-Ethylidene-2-methyl-2-oxazolin-5-one ... 97
N-Furfurylformamide ... 851
6-Methyl-3-pyridinol-*N*-oxide ... 1005
4-Methyl-3-pyrrolecarboxylic acid ... 840
2-Pyridylcarbinol *N*-oxide ... 1005
$C_6H_7NO_2S$
Methyl 2-amino-3-thiophenecarboxylate ... 856
$C_6H_7NO_3$
2-Cyano-3-ethoxyacrylic acid ... 144
2-(2-Furyl)glycine ... 850
$C_6H_7NO_3S$
Metanilic acid, sodium salt ... 793
5-Methyl-3-pyridinesulfonic acid ... 960
1-Methylpyridinium 3-sulfonate ... 955
Sulfanilic acid, thallium(I) salt ... 793
$C_6H_7NO_3S_2$
3-Rhodaninepropionic acid ... 141
$C_6H_7NO_4$
2,6-Dioxo-4-piperidinecarboxylic acid ... 125
Ethyl 5-oxo-2-isoxazoline-4-carboxylate ... 97
$C_6H_7NO_4S$
2-Amino-6-hydroxybenzenesulfonic acid ... 787
5-Amino-2-hydroxybenzenesulfonic acid ... 787
4-Pyridylhydroxymethanesulfonic acid ... 934
$C_6H_7N_3O$
2-Acetamidopyrimidine ... 976
2-Cyano-*N*-(2-cyanoethyl)acetamide. 146
Formic acid (2-pyrrolylmethylene)-hydrazide ... 1053
$C_6H_7N_3OS$
2-Furaldehyde thiosemicarbazone ... 1075
$C_6H_7N_3O_2$
6-Amino-2-methyl-3-nitropyridine ... 942
2-Methylamino-5-nitropyridine ... 943
Nicotinic hydrazide *N*-oxide ... 1006
4-Nitro-1,3-phenylenediamine ... 270
4-Nitro-1,2-phenylenediamine sulfate ... 270
$C_6H_7N_3O_3S$
Ethyl 5-hydroxy-3-mercapto-1,2,4-triazine-6-carboxylate ... 978
$C_6H_7N_3O_4$
1,3-Dimethyl-5-nitrouracil monohydrate ... 133
$C_6H_7N_3S$
1-(3-Pyridyl)-2-thiourea ... 943
$C_6H_7N_5$
5,6-Diamino-1*H*-benzotriazole dihydrochloride ... 908
6,7-Diamino-1*H*-benzotriazole dihydrochloride ... 908
$C_6H_7N_5O$
6-Amino-2-hydroxy-7-methylpurine ... 921
$C_6H_7N_5O_2$
4,6-Dimethyl-1*H*-triazolo(4,5-d)-pyrimidine-5,7(4*H*,6*H*)-dione ... 870
$C_6H_7N_5S$
2-Amino-1,7-dihydro-1-methyl-6*H*-purine-6-thione ... 924
$C_6H_7O_2P$
Phenylphosphinic acid, sodium salt ... 835
$C_6H_7O_3P$
Phenylphosphonic acid, dithallium salt ... 835
$C_6H_8AsNO_4$
3-Amino-4-hydroxybenzenearsonic acid ... 1126
4-Amino-2-hydroxybenzenearsonic acid ... 1126
$C_6H_8As_2O_6$
1,2-Phenylenediarsonic acid ... 1128
C_6H_8BrF
1-Bromo-2-fluoro-1-cyclohexene ... 6
$C_6H_8BrF_3O_2$
Ethyl 3-bromo-4,4,4-trifluorobutyrate ... 82
C_6H_8BrNO
3-Hydroxy-1-methylpyridinium bromide ... 933
$C_6H_8BrN_3$
2-Amino-5-bromo-4,6-dimethylpyrimidine ... 975
5-Bromo-2,3-diamino-6-methylpyridine ... 927
$C_6H_8Br_2N_2O_2$
1,3-Dibromo-5-ethyl-5-methylhydantoin ... 135
5-(Dibromomethyl)-5-ethylhydantoin ... 130
$C_6H_8Br_2O_2$
4,5-Dibromo-2-hexenoic acid ... 62
$C_6H_8Br_2O_4$
2,5-Dibromoadipic acid ... 62
$C_6H_8Br_4$
2,3-Bis-(bromomethyl)-1,4-dibromo-2-butene ... 5
$C_6H_8Br_4N_2O_4$
1,6-Dinitro-1,1,6,6-tetrabromohexane ... 46
C_6H_8ClN
1-Chloro-1-cyclopentanecarbonitrile ... 150
C_6H_8ClNO
3-Chloro-*N*-(2-propyn-1-yl)-propionamide ... 1015
$C_6H_8ClNO_5$
N-(Chloroacetyl)-L-aspartic acid ... 117
$C_6H_8ClN_3OS$
1-(2-Chloroethyl)-5-imino-3-methyl-4-thiohydantoin ... 139
$C_6H_8Cl_2N_2$
1-(2,3-Dichloropropyl)imidazole ... 864
$C_6H_8Cl_2O$
1,1-Dichloro-3,4-epoxy-4-methyl-1-pentene ... 20
$C_6H_8Cl_2O_2$
α,α′-Dimethylsuccinyl chloride ... 100
$C_6H_8Cl_2O_3$
3,3-Dichloro-2,2-dihydroxycyclohexanone ... 51
$C_6H_8Cl_2O_4$
3,4-Dichloroadipic acid ... 62
$C_6H_8Cl_2O_5$
Diglycol chloroformate ... 100
$C_6H_8Cl_3NO_2$
N-(2,2,2-Trichloroacetyl)morpholine. 107
$C_6H_8Cl_3NO_4$
Methyl *N*-(1-acetoxy-2,2,2-trichloroethyl)carbamate ... 114
C_6H_8IN
1-Methylpyridinium iodide ... 958

$C_6H_8N_2$
1-Allylpyrazole ... 864
3-(Aminomethyl)pyridine dihydrochloride ... 942
2-Imino-1-cyclopentanecarbonitrile .. 150
1,4-Phenylenediamine oxalate ... 221
1,4-Phenylenediamine sulfate ... 221

$C_6H_8N_2O$
4-Acetyl-1-methyl-1*H*-pyrazole ... 863
2,4-Diaminophenol ... 221
2,5-Diaminophenol dihydrochloride . 221
4,6-Dimethyl-2-hydroxypyrimidine sulfate ... 972
1,3-Dimethyl-1*H*-pyrazole-4-carboxaldehyde ... 863
1,5-Dimethyl-1*H*-pyrazole-4-carboxyaldehyde ... 863

$C_6H_8N_2OS$
2-Amino-5-acetyl-4-methylthiazole hydrochloride ... 877
3,6-Dimethyl-2-thiouracil ... 141
2-Hydroxy-4-methyl-6-(methylthio)-pyrimidine ... 970
N-(3-Methyl-5-isothiazolyl)-acetamide ... 874
3-Methyl-2-(methylthio)-4(3*H*)-pyrimidinone ... 136

$C_6H_8N_2O_2$
3,4-Bis(methylamino)-3-cyclobutene-1,2-dione ... 57
2,4-Dimethoxypyrimidine ... 973
1,3-Dimethyl-1*H*-pyrazole-5-carboxylic acid ... 863
3,5-Dimethyl-1*H*-pyrazole-4-carboxylic acid ... 863
1,3-Dimethyluracil ... 135
Ethyl 1*H*-pyrazole-3-carboxylate ... 863
1-Ethyluracil ... 135
3-Hydroxy-2-(methylamino)-4-(methylimino)-2-cyclobuten-1-one ... 57
4-Methoxy-1-methyl-2(1*H*)-pyrimidinone ... 973
Methyl 4-methyl-1*H*-pyrazole-3-carboxylate ... 863

$C_6H_8N_2O_2S$
5-Ethyl-2-thiobarbituric acid ... 976
3,3′-Sulfonyldipropionitrile ... 153

$C_6H_8N_2O_3$
1,2,3-Cyclohexanetrione-1,3-dioxime ... 1010
5-Hydroxymethyl-6-methyluracil ... 131

$C_6H_8N_2O_3S$
4-Hydrazinobenzenesulfonic acid ... 789
4-Hydroxymetanilamide ... 811
2-Methyl-5-methylsulfonyl-4-pyrimidinol ... 972

$C_6H_8N_2O_4$
Methyl maleurate ... 127

$C_6H_8N_2O_4S_2$
1,3-Benzenedisulfonamide ... 827
N,N′-Bis(carboxymethyl)-dithiooxamide ... 141

$C_6H_8N_2O_5S$
4,6-Dimethyl-2-hydroxy-5-pyrimidinyl sulfate ... 972

$C_6H_8N_2O_6$
2,2-Dinitropropyl acrylate ... 89

$C_6H_8N_2S$
2-Amino-5,6-dihydro-4*H*-cyclopentathiazole ... 876
2-Amino-5,6-dihydro-4*H*-cyclopentathiazole hydrochloride ... 876
2-Cyanoethyl thioether ... 145
4,6-Dimethyl-2-mercaptopyrimidine hydrochloride monohydrate ... 973
4,6-Dimethyl-2-mercaptopyrimidine monosulfate ... 973
4-Methyl-2-(methylthio)pyrimidine ... 974

$C_6H_8N_2S_2$
4-Methyl-6-mercapto-2-methylthio)-pyrimidine ... 973
Tetramethylene dithiocyanate ... 151

$C_6H_8N_4O$
3-Amino-1-(2-cyanoethyl)-2-pyrazolin-5-one ... 145
5-Amino-4-cyano-1-(2-hydroxyethyl)-pyrazole ... 860

$C_6H_8N_4O_2$
6-Hydroxy-4-methyl-2-ureidopyrimidine ... 972

$C_6H_8N_4O_2S$
2,5-Diacetamido-1,3,4-thiadiazole ... 879
3,5-Diacetamido-1,2,4-thiadiazole ... 879

$C_6H_8N_4O_3$
5-Amino-4-carbamyl-3-pyrazoleacetic acid ... 861
6-Amino-1,3-dimethyl-5-nitrosouracil ... 132
6-Amino-1-ethyl-5-nitrosouracil ... 132

$C_6H_8N_4O_3S$
2-(Carboxymethylthio)-5,6-diamino-4-hydroxypyrimidine ... 970

$C_6H_8N_4O_4$
6-Amino-1-ethyl-5-nitrouracil ... 132
Dimethyl 1,4-dihdyro-1,2,4,5-tetrazine-3,6-dicarboxylate ... 90
(2-(2,4-Dioxo-5-imidazolidinyl)-acetyl)urea ... 135

$C_6H_8N_4O_4S$
4-Amino-3-nitrobenzenesulfonyl hydrazide ... 816

C_6H_8O
3-Cyclohexen-1-one ... 56

$C_6H_8O_2$
Allyl acrylate ... 79
1-Cyclopentene-1-carboxylic acid ... 66
2-Hydroxy-3-methyl-2-cyclopenten-1-one hydrate ... 52
2-Methyl-3-furanmethanol ... 845

$C_6H_8O_3$
3,4-Dihydroxy-2-methyl-2-cyclopenten-1-one ... 51
3,4-Furandimethanol ... 845
6-Hydroxytetrahydropyran-2-carboxylic acid lactone ... 97
3-Methyl-4-oxo-2-pentenoic acid ... 64

$C_6H_8O_3S_2$
2,5-Dimethyl-3-thiophenesulfonic acid sodium salt ... 859

$C_6H_8O_4$
4-Methyl itaconate ... 95
Monoethyl maleate, thallium salt ... 95

$C_6H_8O_7$
Citric acid, tin(II)salt ... 69

$C_6H_9AsN_2O_3$
3,4-Diaminobenzenearsonic acid ... 1127

$C_6H_9BrN_2O$
4-Bromo-3,5-dimethyl-1-pyrazolemethanol ... 860

$C_6H_9Br_2NO$
3,3-Dibromo-4,5,6,7-tetrahydro-1*H*-azepin-2(3*H*)-one ... 125

$C_6H_9ClF_6Si$
(3-Chloro-1,1,2,2,3,3-hexafluoropropyl)trimethylsilane ... 1096

$C_6H_9ClN_2O_4$
Chloroacetyl-DL-asparagine ... 117
N-(Chloroacetyl)glycylglycine ... 117

$C_6H_9ClO_4S$
3-Acetoxy-4-chlorotetrahydrothiophene 1,1-dioxide ... 153

$C_6H_9Cl_3N_2O_3S$
trans-*N*-(4-Amino-3-sulfolanyl)-2,2,2-trichloroacetamide hydrochloride ... 152

$C_6H_9Cl_3O_2$
tert.-Butyl trichloroacetate ... 82
1-Chloromethyl-2,3-dichloropropyl acetate ... 82
2,3,4-Trichlorobutyl acetate ... 82

$C_6H_9Cl_3O_3$
Ethyl 3-hydroxy-4,4,4-trichlorobutyrate ... 84
2-(Trichloromethyl)-1,3,6-trioxocane. 21

$C_6H_9Cl_4NO_3$
Methyl *N*-(1-(2-chloroethoxy)-2,2,2-trichloroethyl)carbamate ... 112

$C_6H_9F_3O$
4-Methyl-5,5,5-trifluoro-2-pentanone. 50

$C_6H_9F_3O_2$
Ethyl 4,4,4-trifluorobutyrate ... 82

$C_6H_9F_3O_3$
Ethyl 3-hydroxy-4,4,4-trifluorobutyrate ... 84

$C_6H_9F_6NSi$
N-(Hexafluoroisopropylidene)-trimethylsilylamine ... 1100

C_6H_9I
1-Iodo-1-hexyne ... 1015

C_6H_9NO
3-Amino-2-cyclohexen-1-one hydrobromide ... 57
N-Methylfurfurylamine ... 850

C_6H_9NOS
2-(Isothiocyanatomethyl)-tetrahydrofuran ... 151

$C_6H_9NO_2$
Adipimide ... 125
1-Cyano-1-methylethyl acetate ... 147
2,2-Dimethylsuccinimide ... 125
Isopropyl cyanoacetate ... 147

$C_6H_9NO_2S$
Methyl 2-thiocyanatobutyrate ... 151
4-Thioxo-2,2,3-trimethyl-5-oxazolidinone ... 140

$C_6H_9NO_3$
5,6-Dimethylmorpholine-2,3-dione ... 134
N-Ethylmaleamic acid ... 122
1-Methyl-5-oxo-3-pyrrolidinecarboxylic acid ... 125

$C_6H_9NO_3S$
3-Acetamido-2,3-dihydrothiophene 1,1-dioxide ... 153

$C_6H_9NO_3S_2$
S-(Ethoxycarbonylmethyl) 1,1-dithiooxamate ... 94

$C_6H_9NO_5$
N-Acetyl-DL-aspartic acid ... 122
N-Acetyl-L-aspartic acid ... 123

$C_6H_9NS_3$
Hexahydro-1,3,5-trithia-6b-azacyclopenta(cd)pentalene ... 41

$C_6H_9N_2P$
Bis(2-cyanoethyl)phosphine ... 158

$C_6H_9N_3$
2-Amino-4,6-dimethylpyrimidine hydrobromide ... 975
2-Amino-5-methyl-3-nitropyridine ... 941
3,3′-Iminodipropionitrile hydrochloride ... 146
1-Methyl-4(1*H*)-pyridone hydrazone dihydrochloride ... 39
1,2,3-Triaminobenzene dihydrochloride ... 255
1,2,4-Triaminobenzene dihydrochloride ... 221

$C_6H_9N_3O$
2-Amino-5,6-dimethyl-4-pyrimidinol .. 971
1-Amino-4,6-dimethylpyrimidin-2(1*H*)-one ... 133
Cyanoacetic acid isopropylidenehydrazide ... 147
3,4-Diamino-5-methyl-2(5*H*)-pyridone ... 942

$C_6H_9N_3OS$
Acetic N^2-(4-methyl-2-thiazolyl)-hydrazide hydrochloride ... 877
6-Amino-2-(ethylthio)-4-pyrimidinol ... 970

$C_6H_9N_3O_2$
2-Amino-4,6-dihydroxy-5-ethylpyrimidine ... 969
6-Amino-1-ethyluracil ... 132
4-Ethyl-1-methyl-3-nitro-1*H*-pyrazole. 862
DL-Histidine dihydrochloride ... 866
DL-Histidine hydrochloride hydrate .. 866
4-Nitro-1,3,5-trimethyl-1*H*-pyrazole ... 861

$C_6H_9N_3O_4S_2$
4-Sulfamoylbenzenesulfonic acid hydrazide ... 827

$C_6H_9N_5O$
3-Methyl-1*H*-pyrazole-5-carboxaldehyde semicarbazide ... 1070

$C_6H_{10}BrNO_3$
2-Bromopropionyl-DL-alanine ... 123

$C_6H_{10}BrNO_4$
2-Bromopropionyl-DL-serine ... 118

$C_6H_{10}Br_2$
1,4-Dibromo-2,3-dimethyl-2-butene ... 5
2,6-Dibromo-1-hexene ... 5
2,6-Dibromo-2-hexene ... 5

$C_6H_{10}Br_2O_2$
Ethyl 3-bromo-2-(bromomethyl)-propionate ... 82
Ethyl 2,3-dibromobutyrate ... 83
Ethyl 2,3-dibromoisobutyrate ... 82

$C_6H_{10}ClOPS$
2-Chloro-2,3-dihydro-3,3,5-trimethyl-1,2-oxaphosphole-2-thione ... 160

$C_6H_{10}Cl_2O_2$
3-Chloro-1-(chloromethyl)-propyl acetate ... 82
Methyl 3,3-dichloro-2,2-dimethylpropionate ... 82

$C_6H_{10}Cl_3NO_2$
Ethyl 3-hydroxy-4,4,4-trichlorobutyrimidate hydrochloride ... 136

$C_6H_{10}Cl_3NO_2S$
Ethyl *N*-(1-(methylthio)-2,2,2-trichloroethyl)carbamate ... 114
Methyl *N*-(1-(ethylthio)-2,2,2-trichloroethyl)carbamate ... 114

$C_6H_{10}Cl_3NO_3$
Ethyl *N*-(1-methoxy-2,2,2-trichloroethyl)carbamate ... 113
Isopropyl *N*-(1-hydroxy-2,2,2-trichloroethyl)carbamate ... 113
Methyl *N*-(1-ethoxy-2,2,2-trichloroethyl)carbamate ... 113
Propyl *N*-(1-hydroxy-2,2,2-trichloroethyl)carbamate ... 113

$C_6H_{10}Cl_4S_2$
Bis(chloromethyl)methyl disulfide ... 24

$C_6H_{10}F_3O_2Tl$
Diethylthallium trifluoroacetate ... 1141

$C_6H_{10}N_2$
4-Amino-1,4-dihydro-1-methylpyridine dihydrochloride ... 38
1-Pyrrolidineacetonitrile ... 145

$C_6H_{10}N_2O$
3-Cyano-2-methyl-3,4,5,6-tetrahydro-2*H*-1,2-oxazine ... 146
3,5-Dimethyl-1*H*-pyrazole-4-methanol ... 860
4-Morpholineacetonitrile ... 145

$C_6H_{10}N_2O_2$
1-Cyano-3,3-dimethyl-1,2-butanedione 1-oxime ... 1011
5,6-Dihydro-1-ethyluracil ... 135
Ethyl *N*-cyano-*N*-methylaminoacetate ... 147
5-Ethyl-5-methylhydantoin ... 135
4-(2-Hydroxyethyl)-3-methyl-2-pyrazolin-5-one ... 860
5-Isopropylhydantoin ... 135
4-Methyl-4-nitrovaleronitrile ... 147
6-Aci-nitro-1-cyclohexene-1-amine ... 271

$C_6H_{10}N_2O_3$
5,5-Dimethyl-1-hydroxymethylhydantoin ... 131
4,4-Dimethyl-2-(*aci*-nitromethyl)-2-oxazolin-3-ium hydroxide, inner salt ... 45

$C_6H_{10}N_2O_4$
N-Acetylglycylglycine ... 123
1,1-Dinitrocyclohexane ... 47
*N*α-Acetyl-D-asparagine hydrate ... 122
*N*α-Acetyl-DL-asparagine ... 122

$C_6H_{10}N_2O_5$
L-α-Aspartylglycine hemiacetate ... 119
Glycyl-D-aspartic acid ... 119
Glycyl-L-aspartic acid ... 119

$C_6H_{10}N_2O_6$
2,2-Dinitropropyl propionate ... 89

$C_6H_{10}N_2S_4$
1,4-Piperazinedicarbodithioic acid, disodium salt hexahydrate ... 142

$C_6H_{10}N_4$
2,4-Dimethyl-6-hydrazinopyrimidine. 976

$C_6H_{10}N_4O$
6-Amino-2-(dimethylamino)-4-hydroxypyrimidine ... 970
L-Histidinamide dihydrochloride ... 866

$C_6H_{10}N_4O_2$
1-(2-Aminoethyl)-2-methyl-5-nitroimidazole dihydrochloride monohydrate ... 866
5-Amino-1-(2-hydroxyethyl)-4-pyrazolecarboxamide ... 860

$C_6H_{10}N_4O_2$ (Cont'd)
6-Amino-5-(methylamino)-1-methyluracil … 131
1,3-Dimethyl-6-hydrazinouracil … 133
$C_6H_{10}N_4O_8$
1,1,6,6-Tetranitrohexane … 46
$C_6H_{10}N_4O_{10}$
2,2,5,5-Tetranitro-1,6-hexanediol … 45
$C_6H_{10}N_4S$
2-Amino-4-(ethylamino)-6-mercaptopyrimidine … 973
$C_6H_{10}O$
cis-Bicyclo(3.1.0)hexan-2-ol brucine salt … 13
1-Hexyn-3-ol … 1015
3-Methoxy-1-cyclopentene … 18
$C_6H_{10}O_2$
3-Acetyl-2,2-dimethyloxirane … 59
2-Buten-1-yl acetate … 80
3-Butenyl acetate … 80
1-Methylallyl acetate … 79
Methyl cyclobutanecarboxylate … 77
$C_6H_{10}O_2S$
2,5-Dihydro-3,4-dimethylthiophene-1,1-dioxide … 154
2,4-Dimethyl-3-sulfolene … 154
Isopropyl 1-propynyl sulfone … 1019
$C_6H_{10}O_2S_4$
O,O-Diethyl dithiobis(thioformate) … 74
$C_6H_{10}O_3$
1-(1,3-Dioxolan-2-yl)-2-propanone … 58
3-Ethoxy-2-butenoic acid … 63
Ethyl 2,3-epoxybutyrate … 89
1-Hydroxy-2-butanone acetate … 92
Methyl 2-tetrahydrofuran-carboxylate … 90
4-Oxohexanoic acid … 64
Tetrahydropyran-4-carboxylic acid … 68
3,3,3-Trimethylpyruvic acid … 64
$C_6H_{10}O_3S$
3-(Acetylmercapto)-butyric acid … 95
$C_6H_{10}O_4$
Adipic acid, disilver salt … 69
2,4-*O*-Ethylidene-D-erythrose … 58
Isopropylmalonic acid … 61
$C_6H_{10}O_4S$
3-Acetoxytetrahydrothiophene-1,1-dioxide … 153
$C_6H_{10}O_4S_2$
2,2′-(Ethylenedithio)-diacetic acid … 63
S,S′-Ethylidenebis(thioglycolic acid) … 63
$C_6H_{10}O_4Se_2$
3,3′-Diselenodipropionic acid … 213
$C_6H_{10}O_5$
Dimethyl malate … 84
2-(Methoxymethyl)succinic acid … 63
$C_6H_{10}O_6$
Ethylene glycol bis-(methyl carbonate) … 74
D-Mannonic γ-lactone … 95
$C_6H_{10}O_7$
D-(+)-Glucuronic acid … 67
Glucuronic acid, sodium salt … 69
$C_6H_{10}O_8$
D-Glucaric acid, calcium salt … 69
$C_6H_{10}S_2$
1,2-Bis(ethylthio)ethyne … 1017
$C_6H_{11}Br$
1-Bromo-1-hexene … 5
1-Bromo-3-hexene … 5
$C_6H_{11}BrO_2$
2-Bromo-3-methylvaleric acid … 62
$C_6H_{11}Br_3$
2,2-Bis(bromomethyl)-1-bromobutane … 4
$C_6H_{11}ClO$
1-Chloro-5-methoxy-2-pentene … 17
2-Ethylbutyryl chloride … 100
$C_6H_{11}ClOS$
3-Chloro-2-methyl-1-propanethiol acetate … 82
$C_6H_{11}ClO_2$
1-Allyloxy-3-chloro-2-propanol … 19
Butyl chloroacetate … 82
4-Chlorobutyl acetate … 82
4-(Chloromethyl)-2,2-dimethyl-1,3-dioxolane … 21
Ethyl 2-chlorobutyrate … 83
Ethyl 2-chloro-2-methylpropionate … 62
Isopropyl 3-chloropropionate … 82
$C_6H_{11}ClO_3$
Ethyl 2-chloro-2-ethoxyacetate … 85
$C_6H_{11}Cl_2N_3O_2$
1,5-Bis(2-chloroethyl)biuret … 126
$C_6H_{11}Cl_2O_3P$
Bis(2-chloroethyl) vinyl-phosphonate … 159
$C_6H_{11}Cl_3N_2O_2$
Methyl *N*-(1-(dimethylamino)-2,2,2-trichloroethyl)carbamate … 114
Methyl *N*-(1-(ethylamino)-2,2,2-trichloroethyl)carbamate … 114
$C_6H_{11}FO$
2-Fluoro-1-cyclohexanol … 13
$C_6H_{11}FO_2$
Ethyl 2-fluorobutyrate … 83
$C_6H_{11}IO_2$
Methyl 5-iodovalerate … 82
$C_6H_{11}N$
7-Azabicyclo(2.2.1)heptane hydrochloride … 41
$C_6H_{11}NO$
2-Aminomethyl-3,4-dihydro-2*H*-pyran … 35
4-Amino-3-methyl-3-penten-2-one … 57
1-Butyrylaziridine … 109
ε-Caprolactam, zone refined … 125
N-Ethylcyclopropanecarboxamide … 106
1-Isobutyrylaziridine … 109
3-Isopropoxypropionitrile … 144
2,4,4-Trimethyl-1-oxazoline … 41
$C_6H_{11}NOS$
4-(Thioacetyl)morpholine … 746
$C_6H_{11}NO_2$
α-Amino-γ-methyl-γ-valerolactone hydrochloride … 89
1-Aza-4,6-dioxa-5-methylbicyclo(3.3.0)-octane … 38
3,4-Dimethyl-4-hydroxy-2-pyrrolidinone … 124
2,4-Dimethyl-2-oxazoline-4-methanol … 37
Ethyl *N*-allylcarbamate … 117
Methyl 3-(methylamino)crotonate … 88
D-Valine methyl ester … 89
$C_6H_{11}NO_2S$
4-Carboxy--3,3-dimethylthiazolidinium hydroxide, inner salt … 69
2,2-Dimethyl-4-thiazolidinecarboxylic acid … 68
$C_6H_{11}NO_2S_2$
3-(*N*-Ethylthiocarbamoylthio)-propionic acid … 141
$C_6H_{11}NO_3$
L-2-Acetamidobutyric acid … 123
4-Acetamidobutyric acid … 124
N-(2-Acetoxyethyl)acetamide … 105
N-Acetyl-L-alanine methyl ester … 94
Cyclohexyl nitrate … 45
Ethyl 2-oxobutyrate 2-oxime … 1011
N-Formyl-DL-valine … 123
N-Formyl-L-valine … 123
$C_6H_{11}NO_3S$
N-Acetyl-*S*-methylcysteine … 118
N-Formyl-DL-methionine … 118
D-*N*-Formylpenicillamine … 104
$C_6H_{11}NO_4$
N-Acetyl-DL-threonine … 118
DL-Aspartic acid dimethyl ester hydrochloride … 87
Ethyl *N*-(ethoxycarbonyl)carbamate … 116
Ethyl 2-methyl-2-nitropropionate … 73
Ethyl 4-nitrobutyrate … 73
2-Nitrobutyl acetate … 73
$C_6H_{11}NS$
Pentyl isothiocyanate … 151
$C_6H_{11}NS_2$
Cyclopentamethylenedithio-carbamic acid, cadmium salt … 142
Cyclopentamethylenedithio-carbamic acid, sodium salt dihydrate … 140
$C_6H_{11}N_3O_2$
3-(2-Imidazolin-2-ylamino)-propionic acid … 137
$C_6H_{11}N_3O_3$
Levulinic acid semicarbazone … 127
$C_6H_{11}N_3O_4$
Glycyl-D-asparagine … 119
Glycylglycylglycine … 119
$C_6H_{11}N_3S_2$
2-Amino-5-(butylthio)-1,3,4-thiadiazole … 879
$C_6H_{11}O_3P$
Diallyl phosphite … 158
C_6H_{12}
4-Methyl-2-pentene (mixture of *cis* and *trans*) … 1
$C_6H_{12}ClNO$
N-*tert*-Butyl-2-chloroacetamide … 102
$C_6H_{12}Cl_2Pb$
(3-Butenyl)dichloroethyllead … 1133
$C_6H_{12}HgO_2$
Butylmercuri acetate … 1137
$C_6H_{12}INO$
3,4,4-Trimethyl-2-oxazolinium iodide … 44
$C_6H_{12}N_2$
4-Dimethylaminobutyronitrile … 147
2-(Isobutylamino)acetonitrile … 145
$C_6H_{12}N_2O$
1-Allyl-3-ethylurea … 128
Cyclopentylurea … 129
1-Formyl-4-methylpiperazine … 109
1-Pyrrolidineacetamide … 101
$C_6H_{12}N_2OS$
1-(2-Hydroxyethyl)-3,4,5,6-tetrahydropyrimidine-2(1*H*)-thione … 139
$C_6H_{12}N_2O_2$
Diethyl oxalimidate dihydrochloride … 136
N^2,N^2-Dimethylacetoacethydrazide … 111
N,N′-Dimethylsuccinamide … 111
1,2-Dipropionylhydrazine … 112
N,N′-Ethylenebisacetamide … 110
N,N′-Ethylidenebis(acetamide) … 111
2,5-Hexanedione dioxime … 1010
3-Methylglutaramide … 111
N,N,N′,N′-Tetramethyloxalamide … 110
$C_6H_{12}N_2O_2S$
5-(2-Thiopseudoureido)pentanoic acid hydrochloride … 137
$C_6H_{12}N_2O_2S_2$
N,N′-Bis(2-hydroxyethyl)-dithiooxamide … 139
$C_6H_{12}N_2O_3$
β-Alanyl-L-alanine … 118
L-Alanyl-β-alanine … 120
Ethyl 3-ureidopropionate … 127
Glycylglycine ethyl ester hydrochloride … 119
N-(2-Hydroxyethyl)-*N,N′*-ethylenebisformamide … 110
Sarcosylalanine … 121
$C_6H_{12}N_2O_3S$
S-(Ethylcarbamoyl)-L-cysteine … 119
$C_6H_{12}N_2O_4$
DL-Alanyl-DL-serine … 119
β-Alanyl-L-serine … 117
N,N′-Bis-(2-hydroxyethyl)-oxamide … 110
1,3-Dinitro-2-ethyl-2-methylpropane … 46
Sarcosyl-L-serine … 118
$C_6H_{12}N_2S$
1-Propyl-2-imidazolidinethione … 747
$C_6H_{12}N_3P$
1,3,5-Triaza-7-phosphaadamantane … 158
$C_6H_{12}N_4$
1-(2,3-Diaminopropyl)imidazole trihydrochloride … 866
$C_6H_{12}O$
2-Cyclopropyl-2-propanol … 13
5-Hexen-2-ol … 11
5-Hexen-3-ol … 11
2-Methyl-2-penten-1-ol … 11
2-Methyl-4-penten-2-ol … 11
4-Methyl-1-penten-3-ol … 11
2-Methyltetrahydropyran … 23
3-Methylvaleraldehyde … 60
$C_6H_{12}OS$
2,6-Dimethyl-1,4-thioxane … 27
S-Ethyl 2-methylthiopropionate … 87
$C_6H_{12}O_2$
sec.-Butyl acetate … 73
1,3-Cyclohexanediol (80% aqu. soln.) … 14
1,4-Dimethoxy-2-butene … 18
2,4-Dimethyl-1,3-dioxane … 23
Hexanoic acid, octadecylamine salt … 69
3-Hexene-2,5-diol … 11
2-Hydroxymethyl-2-methylbutyraldehyde … 60
Isoamyl formate … 73
3-Methyl-1,2-cyclopentanediol (mixture of isomers) … 12
2-Oxepanol … 21
2-Propyl-1,3-dioxolane … 22
2,2,4-Trimethyl-1,3-dioxolane … 22
$C_6H_{12}O_2S$
Methyl 3-(ethylthio)-propionate … 86
$C_6H_{12}O_3$
2,2-Bis(hydroxymethyl)-tetrahydrofuran … 21
5-Hydroxymethyl-5-methyl-1,3-dioxane … 21
2-Methoxy-5-tetrahydropyranol … 21
$C_6H_{12}O_4$
Methyl 2,2-bis(hydroxymethyl)-propionate … 85
$C_6H_{12}O_4S$
trans-3,4-Dimethoxytetrahydrothiophene-1,1-dioxide … 152
3,3-Dimethyl-2-oxo-1-butanesulfonic acid hydrate … 155
1-Hydroxy-1-cyclohexanesulfonic acid, sodium salt … 155
Tetrahydrofurfuryl methanesulfonate … 156
$C_6H_{12}O_5$
1,5-Anhydroglucitol … 21
2,5-Dimethoxytetrahydrofuran-3,4-*cis*-diol … 21
1-*O*-Methyl-β-D-xyloside … 16
$C_6H_{12}O_6$
D-Galtose … 50
Methyl arabonate … 84
$C_6H_{12}O_7$
Gluconic acid, manganese (II) salt … 69
Gluconic acid, strontium salt … 69
$C_6H_{12}S$
Allyl propyl sulfide … 25
$C_6H_{12}S_2$
2-Ethyl-2-methyl-1,3-dithiolane … 27
$C_6H_{12}S_3$
Trithioacetaldehyde … 24
$C_6H_{13}BrO_2S$
(Carbethoxymethyl)-dimethylsulfonium bromide … 87
$C_6H_{13}Cl$
1-Chloro-3,3-dimethylbutane … 4
$C_6H_{13}ClHg$
Chlorohexylmercury … 1137
$C_6H_{13}ClO_2$
1-Chloro-3-isopropoxy-2-propanol … 19
$C_6H_{13}I_2N$
1-(Iodomethyl)-1-methylpyrrolidinium iodide … 42
$C_6H_{13}N$
tert-Butylethylideneamine … 1020
Cyclohexylamine diborate monohydrate … 31
Cyclohexylamine hydrobromide … 31
Cyclohexylamine hydrochloride … 31
Cyclohexylamine tetraborate tetrahydrate … 31
Cyclohexylamine triborate hydrate … 31
1,2-Dimethylpyrrolidine … 40
2,5-Dimethylpyrrolidine hydrobromide … 41
N-Methylcyclopentylamine … 31
N,N,2-Trimethylpropenylamine … 32
$C_6H_{13}NO$
3,3-Dimethyl-2-butanone oxime … 1008
N,N-Dimethyl-2-(vinyloxy)-ethylamine … 34
2-Ethylbutyramide … 101
Ethyl 2-methylpropionimidate hydrochloride … 136

$C_6H_{13}NO$ (Cont'd)
1-Isopropyl-3-azetidinol hydrochloride 36
4-Methyl-2-pentanone oxime 1008
N-Methyltetrahydrofurfurylamine 35
1-Pyrrolidineethanol 36

$C_6H_{13}NO_2$
N-Ethyl-2-hydroxy-2-methylpropionamide 104
Ethyl *N*-propylcarbamate 117
Hexanohydroxamic acid 102
Methyl *N*-butylcarbamate 106
2-Methyl-2-(methylaminomethyl)-1,3-dioxolane 39

$C_6H_{13}NO_2S$
N,*N*-Dimethyltetrahydro-3-thiophenamine 1,1-dioxide 153
D-Methionine methyl ester 87
α-Methyl-DL-methionine 71

$C_6H_{13}NO_3$
3-Methoxy-DL-valine 71
4-Nitro-3-hexanol 45
DL-Serine isopropyl ester hydrochloride 87

$C_6H_{13}NO_3S$
Cyclohexylsulfamic acid 157
Cyclohexylsulfamic acid, calcium salt dihydrate 157
4-(Ethylsulfonyl)morpholine 157

$C_6H_{13}NO_4$
2-Isopropyl-2-nitro-1,3-propanediol 45
2-Nitro-2-propyl-1,3-propanediol 45
N-(Tris-(hydroxymethyl)-methyl)-acetamide 103

$C_6H_{13}NO_4S$
D-Ethionine sulfone 153

$C_6H_{13}NO_6$
D-Gluconamide 101

$C_6H_{13}NO_7$
1-Deoxy-1-nitro-D-mannitol 45

$C_6H_{13}NS_2$
Thialdine 36

$C_6H_{13}N_3$
Galegine sulfate 136
1-Piperidinecarboxamidine sulfate 138

$C_6H_{13}N_3O$
2-(1,4,5,6-Tetrahydro-2-pyrimidinylamino)-1-ethanol 136

$C_6H_{13}N_3OS$
1-(2-Hydroxypropyl)-4-thioxohexahydro-1,3,5-triazine 139

$C_6H_{13}N_3O_2$
L-Alanyl-β-alaninamide hydrochloride 120
5-Guanidinovaleric acid hemihydrate 138
Methyl 4-guanidinobutyrate hydrochloride 138

$C_6H_{13}N_3O_3$
L-Citrulline monohydrochloride 126

$C_6H_{13}N_5O_3$
1-Isoamyl-3-nitro-1-nitrosoguanidine 138

$C_6H_{13}O_5P$
trans-1,2-Cyclohexanediol monophosphate 161

$C_6H_{13}O_9P$
Fructose-1-phosphate, barium salt, trihydrate, cryst 161
Fructose-1-phosphate, barium salt heptahydrate 161
myo-Inositol 1-(dihydrogen phosphate), barium salt 161

$C_6H_{13}O_{10}P$
Gluconic acid 6-(dihydrogen phosphate), barium salt 161

$C_6H_{14}BrNO_2S$
(3-Amino-3-carboxypropyl)-dimethylsulfonium bromide 71

$C_6H_{14}ClNO_2S$
DL-Methionine methylsulfonium chloride 71

$C_6H_{14}ClO_2PS$
Dipropyl chlorothiophosphate 160

$C_6H_{14}Cl_2Si$
Dichloromethylpentylsilane 1096

$C_6H_{14}INO_2S$
Methionine methiodide 71

$C_6H_{14}NO_4P$
Diethyl carbamoylmethylphosphonate 159

$C_6H_{14}N_2$
Cyclohexylhydrazine oxalate 31
1,4-Diaminocyclohexane, mixture of *cis* and *trans* 30
N,*N'*-Dimethyl-2-butene-1,4-diamine 32

$C_6H_{14}N_2O$
1-(*sec*-Butyl)-3-methylurea 128
N-(3-Dimethylaminopropyl)-formamide 104
1-(1,1-Dimethylpropyl)urea 128
1-Ethyl-3-propylurea 128
Hexanoic hydrazide 112
1-Isobutyl-3-methylurea 128
DL-Leucinamide hydrochloride 120
2-Methylpentanoic hydrazide 112
Trimethylacetic N^2-methylhydrazide 112

$C_6H_{14}N_2O_2$
Diisopropylnitramine 46
1-Ethyl-3-(3-hydroxypropyl)urea 126
1-Ethyl-3-(2-methoxyethyl)urea 126

$C_6H_{14}N_2O_2S$
Hexahydro-2-(methylsulfonylmethyl)pyrimidine hydrochloride 155

$C_6H_{14}N_2O_6$
Gluconic acid hydrazide 111

$C_6H_{14}N_2S_2$
N-(3-(Dimethylamino)-propyl)-dithiocarbamic acid, ammonium salt 139

$C_6H_{14}N_4$
Glyoxal bis(dimethylhydrazone) 1053

$C_6H_{14}N_4O_2$
1,6-Diethylbiurea 127
1,2-Dimethylenebis(3-methyl-1-urea) 126

$C_6H_{14}N_4S_2$
Tetramethylenebis(isothiuronium bromide) 142

$C_6H_{14}O$
2,3-Dimethyl-1-butanol 8

$C_6H_{14}OS$
2-(Butylthio)-ethanol 24
Propyl sulfoxide 154

$C_6H_{14}OS_2$
1,3-Bis-(methylthio)-2-methoxypropane 25

$C_6H_{14}OSi$
2,2-Dimethyl-1-oxa-2-silacyclohexane 1096

$C_6H_{14}O_2$
1,1-Diethyl-1,2-ethanediol 11
2,2-Dimethyl-1,3-butanediol 10
5-Methoxy-2-pentanol 20
Pinacol hexahydrate 10

$C_6H_{14}O_2S$
Butyl ethyl sulfone 154
Isopropyl sulfone 154

$C_6H_{14}O_3$
3-(Hydroxymethyl)-2,4-pentanediol 16

$C_6H_{14}O_3S$
3,3-Dimethyl-1-butanesulfonic acid, sodium salt hydrate 156

$C_6H_{14}O_4$
1,2,3,6-Hexanetetrol 16

$C_6H_{14}O_4S$
Hexyl sulfate, sodium salt 157

$C_6H_{14}O_6$
L-Mannitol 16
l-Rhamnitol 16

$C_6H_{14}O_6S_3$
3,3'-Thiobis(1-propanesulfonic acid, sodium salt) 155

$C_6H_{14}S$
Amyl methyl sulfide 25
2,3-Dimethyl-2-butanethiol 24
Isoamyl methyl sulfide 25
Isopropyl propyl sulfide 25

$C_6H_{14}S_2$
3,6-Dithiaoctane 25

$C_6H_{15}BF_4N_2O$
tert-Butyl(methoxyimino)-methylammonium tetrafluoroborate 43

$C_6H_{15}BF_4S$
Triethylsulfonium tetrafluoroborate 27

$C_6H_{15}BrPb$
Bromotriethyllead 1133

$C_6H_{15}ClO_2Si$
(Chloromethyl)-diethoxymethylsilane 1099

$C_6H_{15}ClO_3S$
Tris-(2-hydroxyethyl)-sulfonium chloride 25

$C_6H_{15}ClPb$
Chlorotriethyllead 1133

$C_6H_{15}Cl_6N_2OSb$
(*tert*-Butoxyimino)-dimethylammonium hexachloroantimonate 1128

$C_6H_{15}N$
Diisopropylamine hydrochloride 28
sec.-Hexylamine 28
N-Isopropylpropylamine 28
Methyl-*tert*-pentylamine hydrochloride 28
Triethylamine methanesulfonate 29

$C_6H_{15}NO$
2-(Butylamino)-ethanol 33
4-Dimethylamino-1-butanol 33
2-Dimethylamino-2-methyl-1-propanol 33
1-Ethylamino-2-methyl-2-propanol hydrochloride 33
N-(Methoxymethyl)-diethylamine 35

$C_6H_{15}NO_2$
2-(Aminomethyl)-2-ethyl-1,3-propanediol 32
3-(2-Methoxyethoxy)-propylamine 34

$C_6H_{15}NO_4$
2-Hydroxyethyltrimethyl-ammonium bicarbonate (50% aqueous solution) 42

$C_6H_{15}NO_5$
1-Amino-1-deoxy-D-dulcitol 32

$C_6H_{15}OP$
Triethylphosphine oxide hydrate 158

$C_6H_{15}O_3PbSe$
Diethyllead selenite 1140

$C_6H_{16}INO$
(3-Hydroxypropyl)-trimethylammonium iodide 42

$C_6H_{16}NO_3P$
Diethyl ethylamidophosphate 159
Diisopropyl phosphoramidate 159

$C_6H_{16}N_2$
N-Butylethylenediamine 29
N-(*sec*-Butyl)-ethylenediamine 29
N,*N'*-Diisopropylhydrazine monohydrochloride 28
N,*N*,*N'*,*N'*-Tetramethylethylenediamine dihydrochloride 30
N,*N'*,2-Trimethyl-1,3-propanediamine 30

$C_6H_{16}N_2O$
*N*1-(2-Hydroxyethyl)-2-methyl-1,2-propanediamine 33

$C_6H_{16}N_2O_2$
N,*N*-Bis(2-hydroxyethyl)-ethylenediamine dihydrochloride 32
N,*N*-Bis(2-hydroxyethyl)-ethylenediamine hydrochloride 32

$C_6H_{16}OPb$
Triethyllead hydroxide 1133

$C_6H_{16}O_2S_2Si$
Bis(methylthio)diethoxysilane 1096

$C_6H_{17}NO_2Si$
3-(Dimethoxymethylsilyl)-propylamine 1107

$C_6H_{17}N_3$
Bis(2-aminopropyl)-amine 29
1,1-Dimethyldiethylenetriamine 29

$C_6H_{18}B_3N_3$
Hexamethylborazine 1124
1,3,5-Triethylborazine 1124

$C_6H_{18}B_8O_{15}$
Methyl polyborate 1124

$C_6H_{18}CdI_2N_4$
(Triethylenetetramine)cadmium(II) iodide 1140

$C_6H_{18}Cl_3CoN_4$
DL-*cis*-α-Dichlorotriethylene-tetraaminecobalt(II) chloride 1141
trans-Dichlorotriethylene-tetraaminecobalt(II) chloride 1141

$C_6H_{20}B_{10}O_2$
1,2-Bis(2-hydroxyethyl)-*o*-carborane 199

$C_6H_{22}CoI_3$
Bis(1,2,3-propanetriyltriammine)-cobalt(III) iodide 1140

$C_6H_{22}Si_4$
2-(Dimethylsilyl)-1,1,3,3-tetramethyltrisilane 1096

$C_6H_{24}Cl_2N_6Ni$
Tris(ethylenediamine)nickel(II) chloride dihydrate 1139

$C_6H_{24}CoN_6O_4S$
Tris(ethylenediamine)cobalt(III) sulfate 1141

$C_6H_{24}CuN_6O_4S$
Tris(ethylenediamine)copper(II) sulfate 1141

$C_6H_{24}N_6NiO_4S$
Tris(ethylenediamine)nickel(II) sulfate 157

C_7Cl_6O
Pentachlorobenzoyl chloride 389

C_7Cl_8
Octachloro-1,3,5-cycloheptatriene 5

C_7F_5N
Pentafluorobenzonitrile 760

$C_7HCl_2N_3O_4$
2,4-Dichloro-3,5-dinitrobenzonitrile 759

$C_7HCl_5O_2$
2,3,4,5,6-Pentachlorobenzoic acid 337

C_7HCl_7O
1,3,4,5,6,6,7-Heptachlorobicyclo(3.2.0)hept-3-en-2-one 53
1,2,3,4,5,6,6-Heptachloro-1,3,5-hexatriene-1-carboxaldehyde 60

$C_7HF_5O_2$
Pentafluorobenzoic acid, barium salt 350

C_7HF_7
α,α,2,3,4,5,6-Heptafluorotoluene 170

$C_7H_2ClI_3O$
3,4,5-Triiodobenzoyl chloride 389

$C_7H_2Cl_2N_2O_6$
2,6-Dichloro-3,5-dinitrobenzoic acid 337

$C_7H_2Cl_4N_2$
2-Chloro-3-cyano-5-(trichloromethyl)pyridine 926

$C_7H_2Cl_4O_2$
4-Hydroxy-2,3,5,6-tetrachlorobenzaldehyde 322
2,3,5,6-Tetrachlorobenzoic acid 337

$C_7H_2Cl_5NO$
4-(Chloroacetyl)-2,3,5,6-tetrachloropyridine 925

$C_7H_2Cl_6S$
Chloromethyl pentachlorophenyl sulfide 209

$C_7H_2F_3N_3O_7$
3-Trifluoromethyl-2,4,6-trinitrophenol 259

C_7H_3BrClN
2-Bromo-4-chlorobenzonitrile 760

$C_7H_3BrN_4O_4$
3-Bromo-1,5-dinitro-1*H*-indazole 900

$C_7H_3BrO_3$
4-Bromocatechol cyclic carbonate 387

$C_7H_3Br_3O_2$
2-Hydroxy-3,5,7-tribromo-2,4,6-cycloheptatriene-1-one 310
Methyltribromo-*p*-benzoquinone 56
3,5,7-Tribromotropolone sodium salt 53

C_7H_3ClFNO
3-Chloro-4-fluorophenyl isocyanate 768

$C_7H_3ClN_2O_2$
4-Chloro-2-nitrobenzonitrile 759

$C_7H_3ClN_2O_3$
4-Chloro-2-nitrophenyl isocyanate 769

$C_7H_3ClN_2O_6$
5-Chloro-2,4-dinitrobenzoic acid 337

$C_7H_3Cl_2FO_2$
2,3-Dichloro-6-fluorobenzoic acid 337

$C_7H_3Cl_2N$
2,4-Dichlorobenzonitrile 767
2,5-Dichlorobenzonitrile 760
$C_7H_3Cl_2NO_3$
2,6-Dichloro-3-nitrobenzaldehyde 322
$C_7H_3Cl_2NO_6S$
4-Chloro-3-(chlorosulfonyl)-5-nitrobenzoic acid 801
$C_7H_3Cl_4NO$
4-Acetyl-2,3,5,6-tetrachloropyridine 961
$C_7H_3Cl_5$
2,4-Dichlorobenzotrichloride 170
3,4-Dichlorobenzotrichloride 170
$C_7H_3HgNO_4$
Anhydro-2-hydroxymercuri-3-nitrobenzoic acid 1138
$C_7H_3I_2NO$
3,5-Diiodo-4-hydroxybenzonitrile 761
$C_7H_3I_3O_2$
3,4,5-Triiodobenzoic acid 337
$C_7H_3I_3O_3$
3-Hydroxy-2,4,6-triiodobenzoic acid 336
$C_7H_4BrClO_2$
2-Bromo-5-chlorobenzoic acid 346
C_7H_4BrNOS
7-Bromothieno(3,2-c)pyrid-4(5*H*)-one 897
$C_7H_4BrNO_4$
5-Bromo-6-nitro-1,3-benzodioxole 262
3-Bromo-5-nitrobenzoic acid 337
5-Bromo-2-nitrobenzoic acid 337
$C_7H_4BrN_3O_2$
3-Bromo-7-nitro-1*H*-indazole 900
$C_7H_4Br_2N_2$
3,5-Dibromoindazole 900
$C_7H_4Br_2O$
3,5-Dibromobenzaldehyde 322
$C_7H_4Br_2O_2$
5,6-Dibromo-1,3-benzodioxole 184
3,4-Dibromobenzoic acid 337
3,5-Dibromobenzoic acid 337
$C_7H_4Br_2O_3$
2,3-Dibromo-4,5-dihydroxybenzaldehyde 321
$C_7H_4Br_2O_4$
3,5-Dibromo-2,4-dihydroxybenzoic acid 336
$C_7H_4Br_3NO_2$
3-Amino-2,4,6-tribromobenzoic acid 337
$C_7H_4ClFO_2$
2-Chloro-5-fluorobenzoic acid 337
$C_7H_4ClF_3O_2S$
4-Chlorophenyl trifluoromethyl sulfone 773
$C_7H_4ClN_3O_2$
2-Amino-3-chloro-5-nitrobenzonitrile 758
N-(5-Chloro-2-nitrophenyl)-carbamonitrile 754
$C_7H_4ClN_3O_5$
4-Chloro-3,5-dinitrobenzamide 469
$C_7H_4Cl_2F_2O_2S$
2,5-Dichlorophenyl difluoromethyl sulfone 772
$C_7H_4Cl_2F_2S$
3,4-Dichlorophenyldifluoromethyl-sulfide 209
$C_7H_4Cl_2N_2$
2-Amino-3,5-dichlorobenzonitrile 759
2,5-Dichlorobenzimidazole 907
$C_7H_4Cl_2N_4$
2,4-Dichloro-6-(2-pyrrolyl)-1,3,5-triazine 980
$C_7H_4Cl_2OS$
S-(4-Chlorophenyl) chlorothiolformate 389
1,3-Dichloro-5,6-dihydro-4*H*-cyclopenta(c)thiophen-4-one 857
$C_7H_4Cl_3F$
4-Fluoro-α,α,α-trichlorotoluene 170
$C_7H_4Cl_3NO$
2′,4′,5′-Trichloroformanilide 440
$C_7H_4Cl_3NO_2$
Methyl 3,4,5-trichloro-2-pyridinecarboxylate 964
$C_7H_4Cl_3NO_2S$
4-Nitrophenyl trichloromethyl sulfide 259
$C_7H_4Cl_4$
3-Chlorobenzotrichloride 170
α,α,2,6-Tetrachlorotoluene 170
$C_7H_4Cl_4S$
Methyl 2,3,5,6-tetrachlorophenyl sulfide 210
$C_7H_4FNO_6S$
3-Fluorosulfonyl-5-nitrobenzoic acid 803
$C_7H_4F_2O_6S_2$
3,5-Bis(fluorosulfonyl)benzoic acid 804
$C_7H_4F_3N_3O_4$
2,6-Dinitro-4-trifluoromethylaniline 259
$C_7H_4HgO_2$
2-(Hydroxymercuri)benzoic acid cyclic anhydride 351
$C_7H_4I_2O_2$
3,5-Diiodo-4-hydroxybenzaldehyde 322
$C_7H_4N_2O_2S$
Nitrobenzothiazole (mixture of isomers) 914
5-Nitrobenzothiazole 916
$C_7H_4N_2O_4$
6-Nitro-2(3*H*)-benzoxazolone 910
$C_7H_4N_4O_3$
6-Nitro-1,2,3-benzotriazin-4(3*H*)-one 1003
$C_7H_4N_4O_4$
2-Amino-3,5-dinitrobenzonitrile 763
5,6-Dinitrobenzimidazole 904
5,7-Dinitrobenzimidazole 904
2,7-Dinitro-2*H*-indazole 900
$C_7H_4N_4O_5$
5,6-Dinitro-2-hydroxybenzimidazole 902
5,7-Dinitro-1*H*-indazol-3(2*H*)-one 900
$C_7H_4N_6$
Tetrazolo(5,1-c)(1,2,4)benzotriazine 919
$C_7H_4O_6$
Chelidonic acid, monosodium salt, dihydrate 69
2-Oxo-2*H*-pyran-4,6-dicarboxylic acid 95
$C_7H_5BiO_6$
Bismuth(III) gallate, basic 1141
$C_7H_5BrClNO$
4′-Bromo-2′-chloroformanilide 409
$C_7H_5BrI_2$
1-Bromo-2,5-diiodo-4-methylbenzene 171
$C_7H_5BrN_2$
2-Amino-5-bromobenzonitrile 759
2-Bromobenzimidazole 907
5-Bromobenzimidazole 907
3-Bromo-1*H*-indazole 900
$C_7H_5BrN_2O_3$
2-Bromo-3-nitrobenzamide 470
$C_7H_5BrN_2O_4$
2-Amino-5-bromo-4-nitrobenzoic acid 337
$C_7H_5BrN_2S$
2-Amino-6-bromobenzothiazole 914
$C_7H_5BrN_4$
5-Bromo-1-phenyl-1*H*-tetrazole 870
$C_7H_5BrN_4O$
3-Amino-7-bromo-1,2,4-benzotriazine 1-oxide 1005
$C_7H_5BrO_2$
2-Bromo-3-hydroxybenzaldehyde 322
2-Bromo-6-methyl-1,4-benzoquinone 54
$C_7H_5BrO_4$
5-Bromo-3,4-dihydroxybenzoic acid 336
$C_7H_5BrO_5S$
2-Bromo-5-sulfobenzoic acid 790
$C_7H_5Br_2Cl$
3-Chloro-α,α-dibromotoluene 169
4-Chloro-2,6-dibromotoluene 171
$C_7H_5Br_2I$
1,3-Dibromo-2-iodo-5-methylbenzene 171
$C_7H_5Br_2NO_3$
1,2-Dibromo-5-methoxy-3-nitrobenzene 260
1,4-Dibromo-2-methoxy-3-nitrobenzene 260
$C_7H_5Br_3$
α,α,3-Tribromotoluene 169
2,4,5-Tribromotoluene 171
$C_7H_5Br_3Hg$
Phenyl(tribromomethyl)mercury 1137
$C_7H_5Br_3N_2S$
1-(2,4,6-Tribromophenyl)-2-thiourea 733
$C_7H_5Br_3O$
2,4,6-Tribromo-*m*-cresol 190
$C_7H_5Br_3O_2$
5-Methyl-2,4,6-tribromoresorcinol 190
$C_7H_5ClF_2O_2S$
4-Chlorophenyl difluoromethyl sulfone 772
C_7H_5ClHgO
3-(Chloromercuri)benzaldehyde 1138
$C_7H_5ClHgO_2$
3-(Chloromercuri)benzoic acid 1138
$C_7H_5ClN_2$
2-Chlorobenzimidazole 905
4-Chloro-2-cyanoaniline 758
$C_7H_5ClN_2O$
5-Chloro-2-hydroxybenzimidazole 902
6-Chloro-2-hydroxybenzimidazole 902
$C_7H_5ClN_2O_3$
2-Chloro-5-nitrobenzaldehyde oxime 1008
4-Chloro-3-nitrobenzaldehyde oxime 1008
2-Chloro-5-nitrobenzamide 470
4-Chloro-3-nitrobenzamide 470
2′-Chloro-4′-nitroformanilide 428
4-Chloro-2-nitroformanilide 424
4′-Chloro-3′-nitroformanilide 404
$C_7H_5ClN_2O_4$
6-Chloro-2,3-dinitrotoluene 261
$C_7H_5ClN_2O_6S$
3,5-Dinitro-*p*-toluenesulfonyl chloride 803
$C_7H_5ClN_2S$
2-Amino-4-chlorobenzothiazole hydrobromide 914
2-Amino-6-chlorobenzothiazole hydrochloride 914
$C_7H_5ClN_4$
3-Amino-7-chloro-1,2,4-benzotriazine 1003
$C_7H_5ClN_4O$
3-Amino-7-chloro-1,2,4-benzotriazine-1-oxide 1005
C_7H_5ClOS
Phenyl chlorothiolformate 389
$C_7H_5ClO_2$
3-Chloro-4-hydroxybenzaldehyde 322
$C_7H_5ClO_4S$
3-(Chlorosulfonyl)-benzoic acid 804
$C_7H_5Cl_2N$
3,5-Dichloroanthranilic acid 337
$C_7H_5Cl_2NO$
α,4-Dichlorobenzaldehyde oxime 1008
2,6-Dichlorobenzaldehyde oxime 1014
2′,3′-Dichloroformanilide 439
2′,4′-Dichloroformanilide 439
2′,5′-Dichloroformanilide 439
3′,4′-Dichloroformanilide 439
$C_7H_5Cl_2NO_2$
3-Amino-2,5-dichlorobenzoic acid 337
4-Amino-3,5-dichlorobenzoic acid 337
3-Amino-2,5-dichlorobenzoic acid, sodium salt 350
α,α-Dichloro-3-nitrotoluene 258
2,4-Dichloro-5-nitrotoluene 262
2,5-Dichloro-4-nitrotoluene 262
2,6-Dichloro-3-nitrotoluene 262
$C_7H_5Cl_2NO_4S$
4-(*N*,*N*-Dichlorosulfamyl)-benzoic acid, sodium salt 823
$C_7H_5Cl_2NS_2$
Triethylammonium *N*-(3,4-dichlorophenyl)dithiocarbamate 730
$C_7H_5Cl_3$
2,3,4-Trichlorotoluene 171
2,3,6-Trichlorotoluene 171
$C_7H_5Cl_3Hg$
Phenyl trichloromethyl mercury 1137
$C_7H_5Cl_3N_2O$
N-(2-Pyridyl)-2,2,2-trichloroacetamide 926
N-(3-Pyridyl)-2,2,2-trichloroacetamide 925
1-(2,4,6-Trichlorophenyl)urea 602
(2,4,5-Trichlorophenyl)urea 546
$C_7H_5Cl_3N_2O_2$
Methyl 4-(trichlorovinyl)-3-pyrazolecarboxylate 862
$C_7H_5Cl_3O_3$
Ethyl 3,4,5-trichloro-2-furoate 848
$C_7H_5Cl_3O_3S$
2,4,6-Trichlorophenyl methanesulfonate 797
$C_7H_5Cl_3S$
2-Allyl-3,4,5-trichlorothiophene 859
Methyl 2,4,5-trichlorophenyl sulfide. 210
$C_7H_5Cl_8Sb$
Dichlorophenylcarbenium hexachloro-antimonate 1129
$C_7H_5FN_4$
3-Amino-7-fluoro-1,2,4-benzotriazine 1003
$C_7H_5FN_4O$
N-(Azidocarbonyl)-2-fluoroaniline 769
N-(Azidocarbonyl)-4-fluoroaniline 769
$C_7H_5FO_4S$
3-(Fluorosulfonyl)benzoic acid, copper(II) salt 804
$C_7H_5F_2NO$
2′,4′-Difluoroformanilide 409
3′,5′-Difluoroformanilide 410
$C_7H_5F_3N_4O_3$
2,2,2-Trifluoroacetic N^2-(3-nitro-2-pyridyl)hydrazide 925
$C_7H_5F_3O$
2,2,2-Trifluoroacetophenone 288
$C_7H_5F_3O_2S$
Phenyl trifluoromethyl sulfone 772
$C_7H_5IO_2$
4-Iodobenzoic acid, potassium salt 350
$C_7H_5I_2NO_3$
4-Amino-3,5-diiodosalicylic acid 336
1,4-Dihydro-3,5-diiodo-4-oxo-1-pyridineacetic acid, diethylamine salt 67
C_7H_5NOS
4-Hydroxyphenyl thiocyanate monohydrate 768
$C_7H_5NO_2S_2$
3-(2-Furyl)-rhodanine 854
$C_7H_5NO_3S$
1,2-Benzisothiazol-3(2*H*)-one-1,1-dioxide diethylamine salt 824
1,2-Benzisothiazol-3(2*H*)-one,1,1-dioxide tribenzylamine salt 824
Saccharin, silver salt 912
$C_7H_5NO_3S_3$
2-Mercaptobenzothiazole-5-sulfonic acid, sodium salt 913
$C_7H_5NO_4$
2-Hydroxy-3-nitrobenzaldehyde 322
2-Hydroxy-3-nitrobenzaldehyde, sodium salt 322
2-Hydroxy-5-nitrobenzaldehyde, sodium salt 322
2-Nitrobenzoic acid, sodium salt 351
4-Nitrophenyl formate 379
$C_7H_5NO_5$
2-Hydroxy-6-nitrobenzoic acid 338
4-Nitrosalicylic acid 338
$C_7H_5NO_6S_3$
Benzothiazole-2,5-disulfonic acid disodium salt 917
C_7H_5NS
Benzothiazole picrate 917
2-Thiopheneacrylonitrile 858
$C_7H_5NS_2$
2-Benzothiazolethiol, silver(I) salt 1141
C_7H_5NSe
Benzoselenazole 1143
$C_7H_5N_3$
1,2,4-Benzotriazine 1004
$C_7H_5N_3O$
6-Nitrobenzimidazole 904
6-Nitrobenzimidazole nitrate 904
$C_7H_5N_3OS$
3-Mercapto-1,2,4-benzotriazine N^1-oxide 1005
$C_7H_5N_3O_2$
6-Nitrobenzimidazole, sodium salt 905
5-Nitrobenzimidazole sulfate 904
3-Nitro-imidazo(1,2-a)pyridine 906

$C_7H_5N_3O_2$ (Cont'd)
7-Nitroindazole 900
$C_7H_5N_3O_2S$
2-Amino-5-nitrobenzothiazole 914
$C_7H_5N_3O_3$
2-Hydroxy-5-nitrobenzimidazole 902
6-Nitro-3-indazolinone 900
$C_7H_5N_3O_6$
4-Amino-3,5-dinitrobenzoic acid 342
3,5-Dinitroanthranilic acid 342
$C_7H_5N_5O_2$
2-Amino-4-hydroxy-6-pteridinecarboxaldehyde 1002
$C_7H_5N_5O_3$
3-Amino-7-nitro-1,2,4-benzotriazine-1-oxide 1003
$C_7H_6AsNO_3S$
2-Thiocyanatobenzenearsonic acid 1127
C_7H_6BrCl
5-Bromo-2-chlorotoluene 171
C_7H_6BrClO
6-Bromo-4-chloro-*o*-cresol 190
6-Bromo-4-chloro-*m*-cresol 190
C_7H_6BrNO
3-Bromobenzamide 477
4′-Bromoformanilide 438
$C_7H_6BrNO_2$
2-Amino-5-bromobenzoic acid, potassium salt 350
5-Bromo-2-hydroxybenzamide 446
4-Bromo-α-nitrotoluene 257
4-Bromo-3-nitrotoluene 279
2-Bromo-3-pyridyl acetate 960
$C_7H_6BrNO_3$
2-Bromo-1-methoxy-4-nitrobenzene 260
$C_7H_6BrNO_5S$
2-Bromo-3-nitro-α-toluenesulfonic acid sodium salt 794
$C_7H_6Br_2N_2O_2$
2-(Dibromonitromethyl)-3-methylpyridine 945
2-(Dibromonitromethyl)-6-methylpyridine 945
$C_7H_6Br_2O$
2,6-Dibromoanisole 177
3,5-Dibromo-2-methylphenol 190
$C_7H_6Br_2O_3$
Ethyl 3,4-dibromo-2-furoate 848
Ethyl 3,5-dibromo-2-furoate 848
Ethyl 4,5-dibromo-2-furoate 848
C_7H_6ClF
5-Chloro-2-fluorotoluene 171
$C_7H_6ClF_2NO_2S$
3-Amino-4-chlorophenyl difluoromethyl sulfone 772
C_7H_6ClI
2-Chlorobenzyl iodide 169
2-Chloro-6-iodotoluene 171
C_7H_6ClNO
2-Chlorobenzaldehyde oxime (mixture of *syn* and *anti*) 1014
4-Chlorobenzaldehyde oxime 1014
3-Chlorobenzamide 477
2′-Chloroformanilide 408
3′-Chloroformanilide 437
4-Chloroformanilide 438
$C_7H_6ClNO_2$
5-Amino-2-chlorobenzoic acid sulfate 337
6-Chloro-4-nitroso-*o*-cresol, moist 257
2-Chloro-5-nitrotoluene 262
5-Chloro-2-nitrotoluene 262
Methyl 6-chloronicotinate 962
$C_7H_6ClNO_3$
2-Chloro-6-nitroanisole 260
3-Chloro-4-nitroanisole 260
4-Chloro-2-nitroanisole 269
2-Chloro-6-nitrobenzyl alcohol 265
$C_7H_6ClNO_4$
5-Amino-3-chloro-2,4-dihydroxybenzoic acid 336
$C_7H_6ClNO_4S$
4-Chloro-3-nitrophenyl methyl sulfone 773
$C_7H_6ClNO_5S$
4-Chloro-5-nitro-*o*-toluenesulfonic acid 789
4-Chloro-5-nitro-*m*-toluenesulfonic acid, sodium salt 794
$C_7H_6ClN_3O$
2-(Chloromethyl)-2,3-dihydro-*s*-triazolo(4,3-a)pyridin-3-one 504
$C_7H_6Cl_2N_2O$
3,5-Dichloroanthranilamide 450
3,4-Dichlorobenzamide oxime 1014
2,4-Dichlorobenzoic acid hydrazide 503
1-(2,3-Dichlorophenyl)urea 511
1-(3,4-Dichlorophenyl)urea 543
1-(3,5-Dichlorophenyl)urea 545
(2,4-Dichlorophenyl)urea 543
$C_7H_6Cl_2N_2O_2$
1-(2,4-Dichlorophenyl)-3-hydroxyurea 549
1-(3,4-Dichlorophenyl)-3-hydroxyurea 529
$C_7H_6Cl_2N_2S$
1-(2,4-Dichlorophenyl)-2-thiourea 749
1-(3,5-Dichlorophenyl)-2-thiourea 733
$C_7H_6Cl_2O$
2,4-Dichloro-6-methylphenol 192
$C_7H_6Cl_2O_2$
2,3-Dichloro-4-methoxyphenol 191
3,5-Dichlorosalicyl alcohol 200
$C_7H_6Cl_2O_2S$
Chloromethyl 4-chlorophenyl sulfone 771
3,4-Dichlorophenyl methyl sulfone 786
$C_7H_6Cl_2O_3$
Ethyl 3,4-dichloro-2-furoate 848
Furfuryl 2,2-dichloroacetate 843
$C_7H_6Cl_2O_3S$
3,5-Dichloro-*p*-toluenesulfonic acid, sodium salt 796
$C_7H_6Cl_2S$
2,5-Dichlorophenyl methyl sulfide 213
3,4-Dichlorothioanisole 210
$C_7H_6Cl_3NO$
1,2-Epoxy-2,6,6-trichlorocyclohexanecarbonitrile 147
1,2,2-Trichloro-6-oxo-1-cyclohexanecarbonitrile 148
$C_7H_6Cl_3NO_2$
N-Furfuryl-2,2,2-trichloroacetamide 843
$C_7H_6Cl_3NO_3$
N-(1-Hydroxy-2,2,2-trichloroethyl)-2-furamide 843
$C_7H_6Cl_4O_3$
Ethyl 3-oxo-2,2,4,5-tetrachloro-4-pentenoate 92
C_7H_6FI
4-Fluoro-2-iodotoluene 171
C_7H_6FNO
2-Fluorobenzaldehyde oxime 1014
3-Fluorobenzaldehyde oxime 1008
4′-Fluoroformanilide 438
$C_7H_6FNO_4S$
4-Fluoro-3-nitrophenyl methyl sulfone 780
4-Methyl-3-nitrobenzenesulfonyl fluoride 803
$C_7H_6FNS_2$
4-Fluorodithiocarbanilic acid, ammonium salt 749
$C_7H_6F_2N_2O$
(2,4-Difluorophenyl)urea 543
$C_7H_6F_3N$
α,α,α-Trifluoro-*o*-toluidine hydrochloride 215
α,α,α-Trifluoro-*m*-toluidine hydrochloride 215
$C_7H_6F_3NO_2S$
α,α,α-Trifluoro-*p*-toluenesulfonamide 805
$C_7H_6F_3NO_2S_2$
4-(Trifluoromethylthio)-benzenesulfonamide 805
$C_7H_6F_3NO_4S_2$
4-(Trifluoromethylsulfonyl)-benzenesulfonamide 805
$C_7H_6HgO_3$
4-(Hydroxymercuri)benzoic acid, potassium salt 1138
C_7H_6INO
syn-3-Iodobenzaldehyde oxime 1008
syn-4-Iodobenzaldehyde oxime 1008
4′-Iodoformanilide 438
$C_7H_6INO_2$
α-Iodo-4-nitrotoluene 258
2-Iodo-3-nitrotoluene 262
2-Iodo-4-nitrotoluene 279
4-Iodo-3-nitrotoluene 279
$C_7H_6INO_3$
2-Iodo-4-nitroanisole 269
2-Iodo-5-nitroanisole 269
4-Iodo-3-nitroanisole 269
$C_7H_6I_2$
3,4-Diiodotoluene 169
$C_7H_6I_2N_2O_2$
2,6-Diiodo-3-methyl-4-nitroaniline 260
2,5-Diiodo-4-methyl-3-(2-nitrovinyl)-pyrrole 842
$C_7H_6I_2O$
2,6-Diiodo-4-methylphenol 190
$C_7H_6N_2$
3-Aminobenzonitrile hydrochloride 763
$C_7H_6N_2OS$
2-Amino-4-hydroxybenzothiazole 912
$C_7H_6N_2O_2$
3-((2-Cyanoethyl)amino)-3-cyclobutene-1,2-dione 146
$C_7H_6N_2O_3$
2-Carbamylnicotinic acid 962
5-(2-Furyl)hydantoin 854
2′-Nitroformanilide 424
3′-Nitroformanilide 426
4′-Nitroformanilide 427
$C_7H_6N_2O_3S$
2-Benzimidazolesulfonic acid 905
$C_7H_6N_2O_4$
3-Amino-5-nitrobenzoic acid 342
1-Methoxy-3-nitro-2-nitrosobenzene 267
Methyl 6-nitro-3-pyridinecarboxylate 961
3-Nitroanthranilic acid 342
4-Nitrobenzohydroxamic acid 472
$C_7H_6N_2O_5$
3-Amino-5-nitrosalicylic acid monohydrate 339
2,4-Dinitroanisole 267
2,6-Dinitro-*m*-cresol 264
3,5-Dinitro-*o*-cresol 264
3,5-Dinitro-*o*-cresol, sodium salt 279
$C_7H_6N_2O_6$
4,6-Dinitro-2-methylresorcinol 262
$C_7H_6N_2Se$
5-Methyl-2,1,3-benzoselenadiazole 919
$C_7H_6N_4$
6-Aminopyrido(2,3-b)pyrazine 1003
3-Methylpyrido(3,2-e)-1,2,4-triazine 1004
$C_7H_6N_4O$
3-Amino-1,2,4-benzotriazine-1-oxide 1005
N-(Azidocarbonyl)aniline 769
4-Furfurylideneamino-4*H*-1,2,4-triazole 1071
5-Phenoxy-1*H*-tetrazole 870
$C_7H_6N_4O_2S$
4,6-Dihydroxy-2-mercapto-7-methylpyrimido(4,5-b)pyrazine 1002
4,7-Dihydroxy-2-mercapto-6-methylpyrimido(4,5-b)pyrazine 1002
2-Hydrazino-6-nitrobenzothiazole 915
$C_7H_6N_4O_3$
1-Methyl-2,4,6(1*H*,3*H*,5*H*)-pteridinetrione 977
$C_7H_6N_4O_5$
3,5-Dinitrobenzhydrazide 499
$C_7H_6N_4O_6$
2,4,6-Trinitro-*m*-toluidine 270
$C_7H_6N_4S$
4-(2-Thenylideneamino)-4*H*,1,2,4-triazole 1071
C_7H_6OS
2-Thiopheneacrolein 857
$C_7H_6O_2$
Benzoic acid, copper (II) salt dihydrate 351
Benzoic acid, magnesium salt 351
Lithium benzoate 351
$C_7H_6O_2PbS$
4-Oxo-4*H*-3,2,1-benzoxathiaplumbin 1135
$C_7H_6O_3$
2-Furanacrylic acid, mercury(II) salt 853
2-Furanacrylic acid, potassium salt 853
2-Furanacrylic acid, sodium salt 849
4-Hydroxy-1,3-benzodioxole 195
Salicylic acid, octadecylamine salt 350
Salicylic acid, potassium salt 350
$C_7H_6O_3S$
2-Acetyl-3-thiophenecarboxylic acid 857
α-Hydroxy-*o*-toluenesulfonic acid γ-sultone 799
$C_7H_6O_4$
2,6-Dihydroxybenzoic acid, sodium salt 350
$C_7H_6O_5$
3-Carboxy-2-furanacetic acid 849
2,4,6-Trihydroxybenzoic acid, potassium salt 350
$C_7H_6O_5S$
2-Carboxybenzenesulfonic acid, ammonium salt 794
4-Carboxybenzenesulfonic acid, sodium salt 795
3-Sulfobenzoic acid 790
3-Sulfobenzoic acid, disodium salt 794
$C_7H_7BO_3$
3-Formylphenylboronic acid 1125
4-Formylphenylboronic acid 1125
$C_7H_7BrN_2O$
2-Bromobenzoic acid hydrazide 494
4-Bromophenylurea 589
$C_7H_7BrN_2O_2$
5-Bromo-4-methyl-2-nitroaniline 260
$C_7H_7BrN_2O_3$
3-Bromo-4-methoxy-6-nitroaniline 260
$C_7H_7BrN_4O_2$
8-Bromotheophylline 868
C_7H_7BrO
4-Bromo-2-methylphenol 190
C_7H_7BrOS
4-Bromophenyl methyl sulfoxide 771
$C_7H_7BrO_2$
2-Bromo-5-methylhydroquinone 193
$C_7H_7BrO_2S$
4-Bromophenyl methyl sulfone 785
$C_7H_7BrO_3$
Ethyl 5-bromo-2-furoate 848
$C_7H_7Br_3OTe$
(4-Methoxyphenyl)-tribromotellurium 1141
$C_7H_7Br_3Te$
(*o*-Tolyl)tribromotellurium 1143
$C_7H_7ClFN_3S$
4-(3-Chloro-4-fluorophenyl)-3-thiosemicarbazide 750
$C_7H_7ClF_2N_2O_2S$
3-Chloro-4-hydrazinophenyl difluoromethyl sulfone 772
$C_7H_7ClF_3NO_3S$
2-Chloro-1-methylpyridinium trifluoromethanesulfonate 959
C_7H_7ClHg
Benzylmercuric chloride 1139
p-Tolylmercuric chloride 1139
C_7H_7ClHgO
2-(Chloromercuri)anisole 1138
4-(Chloromercuri)anisole 1138
$C_7H_7ClN_2$
4-Chlorobenzaldehyde hydrazone 234
N-(Cyanomethyl)pyridinium chloride 956
$C_7H_7ClN_2O$
2-(2-Chloroacetamido)-pyridine 925
4-Chlorobenzamidoxime 1025
3-Chlorophenylurea 580
4-Chlorophenylurea 579
$C_7H_7ClN_2O_2$
4-Chloro-2-hydrazinobenzoic acid hemi-hydrochloride 343
4-Chloro-6-nitro-*m*-toluidine 260
1-(2-Chlorophenyl)-3-hydroxyurea 549
1-(3-Chlorophenyl)-3-hydroxyurea 574
1-(4-Chlorophenyl)-3-hydroxyurea 589
Methyl 4-(1-chlorovinyl)-3-pyrazolecarboxylate 860
$C_7H_7ClN_2O_3$
5-Chloro-2-methoxy-4-nitroaniline 266
$C_7H_7ClN_2S$
1-(3-Chlorophenyl)-2-thiourea 732
1-(4-Chlorophenyl)-2-thiourea 748

$C_7H_7ClN_4O$
6-Chloro-9-methoxymethylpurine...... 921
C_7H_7ClO
2-Chloro-6-methylphenol.................... 192
$C_7H_7ClO_2$
2-Chloro-5-methoxyphenol................. 191
5-Chlorosalicyl alcohol...................... 200
$C_7H_7ClO_2S$
4-Chlorophenyl methyl sulfone......... 785
o-Toluenesulfonyl chloride............... 804
$C_7H_7ClO_3$
Furfuryl 2-chloroacetate................... 843
Methyl 5-chloromethyl-2-furancarboxylate............................. 843
$C_7H_7ClO_3S$
3-Chloro-4-methoxybenzenesulfinic acid, sodium salt.. 786
C_7H_7ClS
4-Chlorophenyl methyl sulfide.......... 212
$C_7H_7Cl_2N$
2,3-Dichloro-*N*-methylaniline............ 216
$C_7H_7Cl_2NO$
4-Amino-2,6-dichloro-3-methylphenol....................................... 215
3,4-Dichloro-5,6-dimethyl-2(1*H*)-pyridone... 964
$C_7H_7Cl_2NO_2S$
N-(3,5-Dichlorophenyl)-methanesulfonamide........................ 810
$C_7H_7Cl_2N_3$
1-(3,4-Dichlorophenyl)guanidine carbonate... 727
$C_7H_7Cl_2OPS$
p-Tolyl phosphorodichloridothionate.......... 836
$C_7H_7Cl_3OTe$
(4-Methoxyphenyl)-trichlorotellurium............................. 211
$C_7H_7Cl_3Te$
(*o*-Tolyl)trichlorotellurium................. 1143
$C_7H_7Cl_4NO$
1-Hydroxy-2,2,6,6-tetrachlorocyclo-hexanecarbonitrile........................... 144
$C_7H_7Cl_5O_2$
Ethyl 2,2,3,4,5-pentachloro-4-pentenoate.. 83
$C_7H_7FN_2O$
(2-Fluorophenyl)urea........................ 536
4-(Fluorophenyl)urea......................... 589
$C_7H_7F_2NO_2S$
2′,5′-Difluoromethanesulfonanilide.. 810
$C_7H_7F_2NO_3S$
(3,5-Difluoroanilino)-methanesulfonic acid, sodium salt. 791
$C_7H_7F_2N_3O_4S$
4-Difluoromethylsulfonyl-2-nitrophenylhydrazine....................... 772
C_7H_7HgI
Benzyliodomercury............................ 1139
$C_7H_7IN_2$
4-Cyano-1-methylpyridinium iodide.. 952
$C_7H_7IN_2O$
2-Iodobenzoic acid hydrazide........... 503
$C_7H_7IN_4O_2$
8-Iodotheophylline............................. 924
$C_7H_7I_3OTe$
(4-Methoxyphenyl)triiodotellurium..... 1141
C_7H_7N
2-Vinylpyridine picrate...................... 967
C_7H_7NO
Benzaldehyde oxime hydrochloride. 1013
$C_7H_7NO_2$
4-Aminobenzoic acid hydrochloride. 342
4-Aminobenzoic acid, potassium salt................................... 351
Benzohydroxamic acid, potassium salt................................... 472
3-(2-Furyl)acrolein oxime.................. 1013
4-Hydroxybenzaldehyde oxime......... 1009
2′-Hydroxyformanilide........................ 413
4′-Hydroxyformanilide........................ 413
2-Methyl-4-nitrosophenol.................. 257
3-Methyl-4-nitrosophenol.................. 257
6-Methylpicolinic acid....................... 960
3-Pyridineacetic acid......................... 963
3-Pyridylacetaldehyde *N*-oxide......... 1006
3-Pyridylacetic acid, sodium salt...... 964
$C_7H_7NO_2S$
Methyl 3-nitrophenyl sulfide............. 270
3-Nitrobenzyl mercaptan.................. 270
4-Nitrobenzyl mercaptan.................. 269
$C_7H_7NO_3$
4-Aminosalicylic acid, potassium salt................................... 350
3,5-Dihydroxybenzamide.................. 461
p-Hydroxybenzohydroxamic acid..... 1008
4-Nitro-*m*-cresol, sodium salt........... 265
$C_7H_7NO_3S$
2-Acetamido-3-thiophenecarboxylic acid................ 858
$C_7H_7NO_4$
4-Methoxy-3-nitrophenol.................. 263
2-Methyl-4-nitroresorcinol................ 262
$C_7H_7NO_4S$
Methyl 3-nitrophenyl sulfone........... 779
N-Phenylcarbamoylsulfonic acid, potassium salt................................... 795
$C_7H_7NO_5$
2,4-Dimethyl-5-nitro-3-furoic acid...... 849
Ethyl 5-nitro-2-furoate...................... 847
$C_7H_7NO_5S$
4-Methylsulfonyl-2-nitrophenol......... 775
2-Nitrophenyl methanesulfonate...... 798
5-Nitro-*o*-toluenesulfonic acid.......... 789
$C_7H_7NO_6S$
5-Amino-2-hydroxy-3-sulfobenzoic acid................................ 787
C_7H_7NS
2,3-Dihydrobenzothiazole.................. 250
$C_7H_7N_3$
7-Aminoindazole................................ 900
$C_7H_7N_3O$
2-Hydrazinobenzoxazole.................... 910
Isonicotinic acid methylenehydrazide......................... 963
Phenylazoformamide......................... 1092
$C_7H_7N_3O_2$
3,4-Bis(formamido)pyridine.............. 945
1,2-Dihydro-3,6-dioxo-1-pyridazinepropionitrile...................... 146
3,4-Dihydroxy-6-methylpyrazolo(3,4-b)-pyridine.. 918
3,6-Dihydroxy-4-methyl-1-pyrazolo(3,4-b)pyridine........................ 899
$C_7H_7N_3O_3$
2-Amino-5-nitrobenzamide................ 468
3-Nitrobenzamidoxime....................... 1010
3-Nitrobenzoic acid hydrazide.......... 499
(3-Nitrophenyl)urea........................... 563
$C_7H_7N_3O_4$
2,3-Dinitro-6-methylaniline................ 271
2,4-Dinitro-*N*-methylaniline................ 273
4,5-Dinitro-2-methylaniline................ 270
$C_7H_7N_3O_5$
4-Methoxy-2,3-dinitroaniline.............. 266
$C_7H_7N_3O_6$
Ethyl 2,4-dihydroxy-5-nitro-6-pyrimidinecarboxylate..................... 969
$C_7H_7N_3O_6S$
2,6-Dinitro-4-(methylsulfonyl)aniline. 776
$C_7H_7N_3S$
7-Deaza-6-(methylthio)purine............ 919
4-Thiocyanatophenylhydrazine hydrochloride.................................... 768
$C_7H_7N_3S_2$
2,6-Pyridinedithiocarboxamide......... 941
$C_7H_7N_4$
5-Amino-3-(2-pyridyl)-4*H*-1,2,4-triazole.. 942
$C_7H_7N_5O_2$
2-Acetamido-6-hydroxypurine hemihydrate.................................... 924
2-Amino-6-methyl-4,7-pteridinediol... 1002
$C_7H_7N_5O_4$
8-Nitrotheophylline............................ 924
$C_7H_7N_5O_5$
1-(2,4-Dinitrophenyl)semicarbazide.. 561
$C_7H_7O_5P$
(3-Carboxyphenyl)phosphonic acid.. 835
4-Phosphonobenzoic acid................. 835
C_7H_8
1,6-Heptadiyne................................. 1015
Tetracyclo(2.2.1.0(2,6).0(3,5))-heptane... 2
$C_7H_8AsNO_5$
4-Methyl-3-nitrobenzenearsonic acid................ 1127
(3-Nitrophenyl)-methanearsonic acid........................ 1127
$C_7H_8AsNO_6$
4-Methoxy-2-nitrobenzenearsonic acid................ 1126
$C_7H_8BBrO_3$
5-Bromo-2-methoxybenzeneboronic acid.......... 1124
C_7H_8BrN
2-Bromobenzylamine......................... 236
2-Bromo-4-methylaniline hydrobromide..................................... 216
2-Bromo-4-methylaniline hydrochloride..................................... 216
C_7H_8BrNO
3-Bromo-4-methoxyaniline hydrochloride..................................... 215
$C_7H_8BrNO_2S$
N-(4-Bromophenyl)-methanesulfonamide........................ 809
$C_7H_8BrN_3O_3S$
N-(5-Bromo-2-thiazolyl)-*N*′-(2-hydroxyethyl)oxamide...................... 875
$C_7H_8ClFN_2O$
1-(Carbamoylmethyl)-3-fluoropyridinium chloride................ 944
C_7H_8ClN
4-Chloro-*m*-toluidine.......................... 215
5-Chloro-*o*-toluidine hydrochloride... 215
C_7H_8ClNO
5-Chloro-*o*-anisidine hydrochloride.. 215
2-Chloro-6-ethoxypyridine................ 940
$C_7H_8ClNO_2$
4-Carboxy-1-methylpyridinium chloride... 954
Trigonelline hydrochloride............... 956
$C_7H_8ClNO_2S$
2-Chloro-4-(methylsulfonyl)-aniline... 773
2-Chloro-5-(methylsulfonyl)aniline.... 773
3-Chloro-*p*-toluenesulfonamide........ 829
$C_7H_8ClNO_3S$
2-Amino-5-chloro-*p*-toluenesulfonic acid......................... 787
4-Amino-5-chloro-*m*-toluenesulfonic acid......................... 787
6-Amino-4-chloro-*m*-toluenesulfonic acid......................... 787
5′-Chloro-2′-hydroxymethanesulfonanilide........ 806
C_7H_8ClNS
3-Chloro-4-(methylthio)aniline.......... 215
$C_7H_8ClN_3$
1-(4-Chlorophenyl)guanidine carbonate... 727
$C_7H_8ClN_3S$
4-(3-Chlorophenyl)-3-thiosemicarbazide............................ 727
$C_7H_8ClN_5O_2$
4-Chloro-2,6-pyridinedicarboxylic hydrazide.. 963
$C_7H_8Cl_2N_2$
1-(2,6-Dichlorobenzyl)hydrazine....... 234
$C_7H_8Cl_3NO_2$
1,2-Epoxy-2,6,6-trichlorocyclohexanecarboxamide.. 109
$C_7H_8Cl_3N_3$
2-Amino-6-hydroxy-5-((1-hydroxy-2,2,2-trichloro)ethyl)-4-methylpyrimidine............................ 968
$C_7H_8Cl_6N_6O_2$
1,1′-(6-Amino-1,3,5-triazine-2,4-diyl)-diimino)bis(2,2,2-trichloroethanol)... 978
$C_7H_8Cl_6O_3$
2,6-Dihydroxy-1,1,1,7,7,7-hexachloro-4-heptanone.................. 50
C_7H_8FN
4-Fluoro-2-methylaniline.................. 215
4-Fluoro-3-methylaniline.................. 215
$C_7H_8FNO_2S$
3-Amino-4-methylbenzenesulfonyl fluoride... 803
$C_7H_8FNO_4S_2$
4-(Methylsulfonyl)-metanilyl fluoride... 803
$C_7H_8FN_3$
1-(4-Fluorophenyl)guanidine carbonate... 727
$C_7H_8F_3NO_3S$
Anilinium trifluoromethanesulfonate.............. 790
$C_7H_8F_8O$
2-Methyl-3,3,4,4,5,5,6,6-octafluoro-2-hexanol.. 15
C_7H_8IN
5-Iodo-2-methylaniline...................... 216
4-Iodo-*o*-toluidine.............................. 216
$C_7H_8INO_2S$
2-Iodo-*p*-toluenesulfonamide........... 810
$C_7H_8NO_5P$
(4-Nitrobenzyl)phosphonic acid........ 835
$C_7H_8N_2$
Benzaldehyde hydrazone.................. 235
$C_7H_8N_2O$
2-Acetamidopyridine......................... 959
Benzamide oxime............................. 1013
4-Hydroxybenzamidine hydrochloride..................................... 724
α-Methylamino-2-furanacetonitrile hydrochloride..................................... 850
N-Methyl-4-nitrosoaniline.................. 257
N-(3-Methyl-2-pyridyl)formamide....... 965
2-Pyridineacetamide......................... 964
$C_7H_8N_2O_2$
Acetic acid furfurylidenehydrazide.. 851
N-(2-Cyanoethyl)-succinimide........... 146
N-(Hydroxymethyl)-isonicotinamide. 933
1-Hydroxy-3-phenylurea.................... 575
3-Hydroxyphenylurea........................ 548
Methyl 2-aminonicotinate.................. 942
3-Methyl-4-nitroaniline...................... 271
3-Methyl-2-(nitromethyl)pyridine picrate... 944
o-Nitrobenzylamine hydrochloride... 271
3-Nitro-2,6-lutidine............................ 944
$C_7H_8N_2O_2S$
2-(Carboxymethylthio)-4-methylpyrimidine............................. 974
2-(2,5-Dihydroxyphenyl)-2-thiopseudourea hydrochloride........ 733
N-(2-Thiazolyl)acetoacetamide......... 877
$C_7H_8N_2O_3$
2,4-Dihydroxybenzoic hydrazide....... 494
2,6-Dimethyl-4-nitropyridine-*N*-oxide.. 1006
Ethyl 1,6-dihydro-6-oxo-3-pyridazinecarboxylate..................... 968
4-Nitro-2,3-lutidine-*N*-oxide............... 1005
$C_7H_8N_2O_3S$
6-Carbethoxy-2-thiouracil sodium salt.. 976
Ethyl 4-hydroxy-2-mercaptopyrimidine-5-carboxylate.. 969
Ethyl 4-hydroxy-2-mercapto-5-pyrimidinecarboxylate, sodium salt.. 969
1-(Phenylsulfonyl)urea...................... 825
N-(2-Thiazolyl)succinamic acid......... 877
$C_7H_8N_2O_3S_2$
3-Diacetylaminorhodanine................ 141
$C_7H_8N_2O_4$
1,3-Dimethyl-6-uracilcarboxylic acid......................... 134
Ethyl 2-carbamoyl-2-cyanoacetate, sodium salt.. 146
$C_7H_8N_2O_4P$
(Nicotinamidomethyl)phosphonic acid, iron(II) salt.............................. 949
$C_7H_8N_2O_4S$
6-Acetamido-3-pyridinesulfonic acid........................ 959
4-(Hydrazinosulfonyl)benzoic acid... 823
2-(Methylsulfonyl)-4-nitroaniline....... 776
4′-Nitromethanesulfonanilide........... 819
$C_7H_8N_2O_5S$
5-Hydrazinosulfonylsalicylic acid..... 811
2′-Hydroxy-5′-nitromethanesulfonanilide............... 811
$C_7H_8N_2O_6S$
5-Amino-4-methoxy-2-nitrobenzenesulfonic acid................ 788
$C_7H_8N_2S$
6-Methylpicolinic thionamide............ 941

$C_7H_8N_2S_2$
3-Phenyldithiocarbazic acid 747
Phenylhydrazinocarbodithioic acid, ammonium salt 749
$C_7H_8N_2S_3$
6-Ethyl-2-methylthiazolo(5,4-d)-thiazole-5(6*H*)-thione 879
$C_7H_8N_4$
4-Amino-5-cyano-2,6-dimethylpyrimidine 975
1,1′-Methylenebis(imidazole) 867
1,1′-Methylenebis(imidazole) dihydrochloride 867
$C_7H_8N_4O$
4-Amino-1-(2-cyanoethyl)-2(1*H*)-pyrimidinone 145
2-(Azidocarbonyl)-3,5-dimethylpyrrole 769
5,6-Diamino-2-hydroxybenzimidazole dihydrochloride 902
3-Hydroxy-5,6-dimethyl-1*H*-pyrazolo(3,4-b)pyrazine 921
2-Pyridinecarboxaldehyde semicarbazone 944
3-Pyridinecarboxaldehyde semicarbazone 948
4-Pyridinecarboxaldehyde semicarbazone 952
$C_7H_8N_4OS$
3,8-Dimethyl-2-thioxanthine 868
$C_7H_8N_4O_2$
3-Nitrobenzenecarboximidic hydrazide 725
1-(4-Nitrophenyl)guanidine hydrochloride 725
1-(2-Nitrophenyl)guanidine nitrate 725
Theobromine 867
$C_7H_8N_4O_2S$
1,3-Dimethyl-8-mercaptoxanthine 922
$C_7H_8N_4O_3S$
N-(2-Thiazolyl)oxamic N^2-acetylhydrazide 876
$C_7H_8N_4O_4$
1-(2,4-Dinitrophenyl)-1-methylhydrazine 273
8-Hydroxy-3-(2-hydroxyethyl)-xanthine 920
$C_7H_8N_4O_4S$
1-(4-Nitrophenylsulfonyl)guanidine 822
$C_7H_8N_4S$
6-Ethylthiopurine 922
Isonicotinaldehyde thiosemicarbazone 1070
2-Pyridinecarboxaldehyde thiosemicarbazone 944
3-Pyridinecarboxaldehyde thiosemicarbazone 1056
4-Pyridinecarboxaldehyde thiosemicarbazone 1060
$C_7H_8N_4S_2$
2,8-Di(methylthio)purine 923
$C_7H_8N_6O$
4,6-Diamino-γ-oxo-1,3,5-triazine-2-butyronitrile 979
$C_7H_8N_6OS$
6-(Carboxymethylthio)purine hydrazide 922
C_7H_8O
Nortricyclenone-3 57
C_7H_8OS
2,6-Dimethyl-4*H*-pyran-4-thione 59
2-Mercaptobenzyl alcohol 209
Methyl 5-methyl-2-thienyl ketone 857
$C_7H_8O_2$
2,2-Dimethyl-4-cyclopenten-1,3-dione 53
1-(2-Furyl)-1-propanone 853
Guaiacol picrate 191
3-Methylcatechol 192
3,3a,6,6a-Tetrahydro-2*H*-cyclopenta(b)furan-2-one 97
$C_7H_8O_2S$
Methyl phenyl sulfone 783
p-Toluenesulfinic acid 787
$C_7H_8O_3$
1,3-Cyclopentanedicarboxylic anhydride 98
3,6-Dimethyl-4-hydroxy-2*H*-pyran-2-one 95
2,4,6-Trihydroxytoluene 193
$C_7H_8O_3S$
1,4,4a,6,7,7a-Hexahydro-3*H*-furo(3,4-c)thiopyran-1,7-dione 96
Lithium *p*-toluenesulfonate 795
Methyl benzenesulfonate 800
Phenyl methanesulfonate 800
α-Toluenesulfonic acid, ethylamine salt 795
p-Toluenesulfonic acid, thallium(I) salt 795
α-Toluenesulfonic acid, triethylamine salt 795
$C_7H_8O_4$
1-Cyclopentene-1,2-dicarboxylic acid 197
Ethyl 2-hydroxy-3-furancarboxylate 845
$C_7H_8O_4S$
α-Hydroxy-α-toluenesulfonic acid, sodium salt 791
2-Hydroxy-α-toluenesulfonic acid, sodium salt 792
$C_7H_8O_5$
Ethyl 2,5-dihydro-4-hydroxy-2-oxo-3-furoate 95
$C_7H_8O_5S$
α,2-Dihydroxy-α-toluenesulfonic acid, sodium salt 791
$C_7H_8O_6$
3-Butene-1,2,3-tricarboxylic acid 61
$C_7H_8O_6S_2$
4-Methyl-1,2-benzenedisulfonic acid, disodium salt 796
C_7H_8S
p-Thiocresol, thallium salt 209
$C_7H_8S_2$
Methyl phenyl disulfide 213
C_7H_8Se
Methyl phenyl selenide 214
$C_7H_9AsN_2O_4$
N-Carbamoylarsanilic acid 1127
$C_7H_9AsO_3$
m-Toluenearsonic acid 1126
4-Toluenearsonic acid 1127
$C_7H_9BBrF_4N$
2-Bromo-1-ethylpyridiniumtetrafluoroborate 959
$C_7H_9BO_2S$
4-(Methylthio)benzeneboronic acid 1125
$C_7H_9BO_3$
3-Methoxyphenylboronic acid 1125
4-Methoxyphenylboronic acid 1125
C_7H_9Br
3-Bromotricyclo[2.2.1.$0^{2,6}$]heptane 6
C_7H_9BrHg
2-(Bromomercuri)-tricyclo(2.2.1.0(2,6))heptane 1138
$C_7H_9BrO_2$
3-Bromo-3-buten-1-yl acrylate 79
$C_7H_9ClN_2$
1-(2-Chlorobenzyl)hydrazine dihydrochloride 235
5-Chloro-*o*-tolylhydrazine 235
$C_7H_9ClN_2O$
1-(Carbamoylmethyl)-pyridinium chloride 956
$C_7H_9ClN_2O_2$
2-Carbamoyl-3-hydroxy-1-methylpyridinium chloride 933
Ethyl 4-chloro-5-methyl-3-pyrazolecarboxylate 863
$C_7H_9ClN_2O_2S$
2-Chloro-4-(methylsulfonyl)-phenylhydrazine 778
3-(3-Chloropropyl)-1-methyl-5-thioxohydantoin 138
$C_7H_9ClN_2S$
4-Chloro-2-ethylthio-6-methylpyrimidine 974
C_7H_9ClO
3-Chloro-2-methyl-2-cyclohexen-1-one 54
$C_7H_9ClO_4$
Dimethyl 2-chloro-1-propene-1,3-dicarboxylate 83
$C_7H_9Cl_3SSi$
(3,4,5-Trichloro-2-thienyl)-trimethylsilane 1110
$C_7H_9Cl_4NO_2$
1-Hydroxy-2,2,6,6-tetrachloro-cyclohexanecarboxamide 101
$C_7H_9F_3O$
3-(Trifluoromethyl)cyclohexanone 51
$C_7H_9F_3O_3$
Tetrahydrofurfuryl 2,2,2-trifluoroacetate 89
$C_7H_9F_3O_4$
Ethyl 2-methoxy-3-oxo-4,4,4-trifluorobutyrate 91
$C_7H_9IN_2O$
3-Carbamoyl-1-methylpyridinium iodide 947
4-(Hydroxyiminomethyl)-1-methylpyridinium iodide 953
C_7H_9N
Benzylamine hydrobromide 241
p-Toluidine, zone refined 219
C_7H_9NO
4-Amino-*m*-cresol hydrochloride 217
3-Amino-4-methylphenol 217
5-Amino-2-methylphenol 217
p-Anisidine hydrochloride 218
3-(Cyanomethyl)-1-cyclopentanone 148
2,4-Dimethylpyridine *N*-oxide 1006
3,4-Dimethyl-2-pyrrolecarboxaldehyde 841
2-Ethoxypyridine 939
2-Ethylpyridine-*N*-oxide 1006
4-Ethylpyridine *N*-oxide 1006
1-Ethyl-2-pyridone 692
2-(Methoxymethyl)-pyridine 939
2-Methyl-3-pyridinemethanol 934
6-Methyl-2-pyridinemethanol 934
2-Oxocyclohexanecarbonitrile 148
C_7H_9NOS
2-Acetyl-5-methylthiophene oxime 1013
2-Amino-5-(methylthio)phenol 216
2-(2-Pyridylthio)ethanol 932
$C_7H_9NOS_2$
5-Isobutylidenerhodanine 140
$C_7H_9NO_2$
2-Cyano-2-hexenoic acid 148
5-(Dimethylamino)-2-furaldehyde 847
2,5-Dimethyl-3-furamide 849
4-Ethoxypyridine-*N*-oxide 1005
Ethyl 2-cyano-1-cyclopropanecarboxylate 149
3-(1-Hydroxyethyl)pyridine-*N*-oxide 1005
Methyl 2-cyano-3-methylcrotonate 148
3,4-Pyridinedimethanol 931
1-(2-Pyridyl)-1,2-ethanediol 932
$C_7H_9NO_2S$
4-Aminophenyl methyl sulfone 777
Methanesulfonanilide 827
$C_7H_9NO_3$
N-Allylmaleamic acid 122
3-Hydroxy-2,6-pyridinedimethanol hydrochloride 934
Methyl 2-cyano-3-ethoxyacrylate 144
$C_7H_9NO_3S$
3-Amino-4-methylbenzenesulfonic acid, cesium salt 793
4-Amino-*m*-toluenesulfonic acid 788
4-Amino-α-toluenesulfonic acid, sodium salt 793
Anilinomethanesulfonic acid, sodium salt 793
4-Methoxybenzenesulfonamide 815
4-Methylmetanilic acid 788
2-(Methylsulfonylmethyl)pyridine-*N*-oxide 1006
3-(2-Thienyl)-L-serine monohydrate 855
$C_7H_9NO_3S_2$
Ethyl 3-rhodanineacetate 141
$C_7H_9NO_4$
3-Succinimidopropionic acid 125
$C_7H_9NO_4S$
4-Methoxymetanilic acid 787
C_7H_9NS
4-Methylmercaptoaniline hydrochloride 218
$C_7H_9N_2O_4P$
(Nicotinamidomethyl)phosphonic acid, disodium salt monohydrate 949
(Nicotinamidomethyl)phosphonic acid, disodium salt 949
$C_7H_9N_3$
Phenylguanidine 727
$C_7H_9N_3O$
2-Aminobenzoyl hydrazide 690
2-Amino-6,7-dihydro-4-hydroxy-5*H*-cyclopentapyrimidine 971
N-(5-Amino-2-pyridyl)acetamide 942
Benzoic hydrazide oxime 1011
2-(3-Pyridyl)acetic hydrazide 965
$C_7H_9N_3OS$
6-Methyl-1,2,3,5-tetrahydro-5-thioxoimidazo(1,2-c)pyrimidin-7(6*H*)-one 141
$C_7H_9N_3O_2$
1-Allyl-6-aminouracil 132
2-Ethylamino-5-nitropyridine 943
*N*1-Methyl-4-nitro-1,2-phenylenediamine 271
$C_7H_9N_3O_2S$
4-Amino-2-(ethylthio)-5-pyrimidinecarboxylic acid 974
Ethyl 4-amino-2-mercaptopyrimidine-5-carboxylate 973
$C_7H_9N_3O_3$
6-Amino-1,3-dimethyl-5-formyluracil 132
5-Carbethoxycytosine 975
Ethyl 2-amino-4-hydroxy-5-pyrimidinecarboxylate 971
Ethyl 4-amino-2-hydroxy-5-pyrimidinecarboxylate 971
5-Hydroxy-4-methyl-4-cyclopentene-1,3-dione 1-semicarbazone 126
Methyl 2-amino-1,6-dihydro-6-oxo-4-pyrimidineacetate 975
3-Methyl-1,2,4-cyclopentanetrione 1-semicarbazone 574
2-(5-Nitro-2-pyridylamino)-ethanol 933
$C_7H_9N_3O_3S$
1-Sulfanilylurea 817
$C_7H_9N_3O_4$
5-Nitro-1,3,6-trimethyluracil 134
$C_7H_9N_3O_4S$
Isonicotinic acid 2-(sulfomethyl)-hydrazide, calcium salt dihydrate 953
4-(Methylsulfonyl)-2-nitrophenylhydrazine 777
$C_7H_9N_5$
1-(3-Pyridylmethyleneamino)-guanidine dinitrate hydrate 1056
$C_7H_9N_5O$
6-(2-Hydroxyethylamino)purine 921
9-Methyl-2-(methylamino)-6(1*H*)-purinone 923
$C_7H_9N_5O_2$
8-Aminotheophylline 923
8-Aminotheophylline hydrochloride 923
$C_7H_9N_5O_3$
4,6-Diamino-γ-oxo-1,3,5-triazine-2-butyric acid 979
$C_7H_9N_5O_5$
1,3-Dimethyl-6-(formylhydrazino)-5-nitrouracil 133
$C_7H_9O_3P$
Benzyl phosphite, ammonium salt 835
α-Hydroxybenzylphosphinic acid 834
α-Hydroxybenzylphosphinic acid, sodium salt 833
(*o*-Tolyl)phosphonic acid 836
(*m*-Tolyl)phosphonic acid 833
$C_7H_9O_4P$
α-Hydroxy-α-phenylmethanephosphonic acid sodium salt 834
(2-Methoxyphenyl)phosphonic acid 834
(3-Methoxyphenyl)phosphonic acid 834
(4-Methoxyphenyl)phosphonic acid 834
$C_7H_9SiCl_3$
5-Norbornen-2-yltrichlorosilane 1110
$C_7H_{10}Br_2$
7,7-Dibromobicyclo(4.1.0)heptane 6
$C_7H_{10}Br_2N_2O_2$
1,3-Dibromo-5-isopropyl-5-methylhydantoin 135

$C_7H_{10}ClN$
1-Chlorocyclohexanecarbonitrile 150

$C_7H_{10}ClN_3O_3S$
N^2-(Methylsulfonyl)isonicotinic hydrazide hydrochloride 963

$C_7H_{10}Cl_2$
2,2-Dichlorobicyclo(4.1,0)heptane 6

$C_7H_{10}Cl_2O_2$
3-Chlorocyclohexyl chloroformate 100

$C_7H_{10}Cl_3NO$
1-(2,2,2-Trichloroacetyl)piperidine 107

$C_7H_{10}Cl_3NO_4$
Ethyl *N*-(1-acetoxy-2,2,2-trichloroethyl)carbamate 114

$C_7H_{10}CuO_2$
2-(1-Hydroxyethylidene)-1-cyclopentanone, copper(II) complex 1139

$C_7H_{10}F_3O_2Tl$
1,1,1-Trifluoro-2,4-pentanedione, dimethylthallium salt 50

$C_7H_{10}IN$
1,2-Dimethylpyridinium iodide 958
1,3-Dimethylpyridinium iodide 958

$C_7H_{10}INO$
2-(Hydroxymethyl)-1-methylpyridinium iodide 956
4-Methoxy-1-methylpyridinium iodide 939

$C_7H_{10}N_2$
4-Amino-2,3-lutidine 942
Benzylhydrazine oxalate 235
Diallylcyanamide 147
3,4-Diaminotoluene dihydrochloride 221
2,5-Diaminotoluene (2-methyl-*o*-phenylenediamine) dihydrochloride 221
3-(Methylamino)aniline sulfate 221
1-Methyl-1-phenylhydrazine sulfate dihydrate 235

$C_7H_{10}N_2O$
2-(2-Aminoethoxy)pyridine 937
4-Amino-2,3-lutidine-*N*-oxide 1005
1-Cyanoacetylpyrrolidine 146
2,4-Diamino-6-methylphenol dihydrochloride 221
2,5-Dimethyl-3-furaldehyde hydrazone 1071
4-(1-Pyrazolyl)-2-butanone 861
1,3,5-Trimethyl-1*H*-pyrazole-4-carboxaldehyde 862
1,4,6-Trimethyl-2(1*H*)-pyrimidinone hydrochloride 133

$C_7H_{10}N_2OS$
N-(3,4-Dimethyl-4-thiazolin-2-ylidene)acetamide 138
2-(Ethylthio)-4-hydroxy-5-methylpyrimidine 970

$C_7H_{10}N_2O_2$
4-Cyanoacetylmorpholine 146
1,7-Diazaspiro(4.4)nonan-2,6-dione 125
Ethyl 1*H*-pyrazole-1-acetate 862
2-(2-Hydroxyethyl)-6-methyl-3(2*H*)-pyridazinone 124
6-Methoxymethyl-2-methyl-4-pyrimidinol 969
Methyl 2-cyano-3-(dimethylamino)-acrylate 146
5,5-Tetramethylenehydantoin 135
1,3,5-Trimethyl-1*H*-pyrazole-4-carboxylic acid 863
1,3,5-Trimethyluracil 134

$C_7H_{10}N_2O_2S$
2-Carbethoxythio-1-methylimidazole 868

$C_7H_{10}N_2O_3$
2,4,6-Trimethoxypyrimidine 972
3-Ureidomethylene-2,4-pentanedione 127

$C_7H_{10}N_2O_3S$
4-Oxo-2-thioxohexahydropyrimidine-3-propionic acid 140

$C_7H_{10}N_2O_5S_2$
4-Sulfamoylanilinomethane-sulfonic acid, sodium salt 819

$C_7H_{10}N_2S$
2-Amino-4,5,6,7-tetrahydrobenzothiazole 876

$C_7H_{10}N_2S_2$
2-Ethylthio-4-mercapto-6-methylpyrimidine 973

$C_7H_{10}N_4O$
Acetic N^2-(3-amino-4-pyridyl)-hydrazide hydrochloride 941
3-Amino-1-(2-cyanopropyl)-2-pyrazolidin-5-one 145
1-Methyl-2-pyrrolecarboxaldehyde semicarbazone 1062

$C_7H_{10}N_4O_2S$
4-Amino-5-(formylamino)-1-methyl-2-(methylthio)-6(1*H*)-pyrimidinone 131

$C_7H_{10}N_4O_3$
6-Amino-1,3-dimethyl-5-(formylamino)uracil 132
6-Amino-1-ethyl-3-methyl-5-nitrosouracil 132

$C_7H_{10}N_4S_2$
6-Isothioureido-4-methyl-2-(methylthio)pyrimidine hydrochloride 1051

$C_7H_{10}N_6$
2-Amino-6-(dimethylamino)purine 923
6-Amino-2-(dimethylamino)purine 923
2,6-Diamino-7-ethyl-7*H*-purine 923

$C_7H_{10}N_6O_3$
2-(1-Methyl-5-nitro-2-imidazolyl)-acetaldehyde semicarbazone 1053

$C_7H_{10}O$
4-Methyl-1,3-pentadiene-1-carboxaldehyde 60

$C_7H_{10}OS$
1-(2-Thienyl)-2-propanol 855

$C_7H_{10}O_2$
Allyl 3-butenoate 80
Allyl crotonate 79
1-Cyclopenten-1-yl acetate 81
1-Cyclopentenylacetic acid 67
Cyclopentylideneacetic acid 66
3,5-Dihydroxytricyclo(2.2.1.0(2,6))-heptane 12
α,α-Dimethylfurfuryl alcohol 845
2-Methoxy-2-cyclohexen-1-one 52

$C_7H_{10}O_2S$
7-Thiabicyclo(4.2.0)oct-1-ene 7,7-dioxide 154

$C_7H_{10}O_3$
Allyl 2-hydroxy-3-butenoate 84
2,5-Dimethyl-2-hydroxymethyl-3(2*H*)-furanone 58
2,4,6-Heptatrione 51
4-Hydroxy-3-methoxy-2-methyl-2-cyclopenten-1-one 51
4-Hydroxy-3-penten-2-one acetate 92
7-Oxabicyclo(4.1.0)heptane-1-carboxylic acid, sodium salt 69
3-Oxo-1-cyclopentaneacetic acid 65
2-Tetrahydrofuryl acrylate 90

$C_7H_{10}O_4$
1,1-Cyclopentanedicarboxylic acid 66
5,5-Dihydroxy-3,4-epoxy-3-methylhexanoic acid δ-lactone 95
Ethyl 2-hydroxy-4-oxo-2-pentenoate, sodium salt 91
Methyl 3-acetoxycrotonate 78
Terebic acid 96

$C_7H_{10}O_5S_2$
1-(2-Furylsulfonyl)-2-(methylsulfonyl)ethane 854

$C_7H_{10}O_6$
1,2,4-Butanetricarboxylic acid 67

$C_7H_{10}Si$
Benzylsilane 1112

$C_7H_{11}BrN_2O_2$
1-Bromo-3,5-dimethyl-5-ethylhydantoin 135

$C_7H_{11}BrN_2O_4$
2-Bromopropionyl-DL-asparagine 121
2-Bromopropionyl-DL-glycylglycine 123

$C_7H_{11}BrO$
Allyl 3-bromo-2-buten-1-yl ether 17
2-Bromoallyl methallyl ether 17

$C_7H_{11}BrO_2$
1-Bromocyclohexane-carboxylic acid 67
Methyl 3-bromo-1-cyclopentanecarboxylate 84

$C_7H_{11}ClN_4$
2-Amino-6-chloro-4-(propylamino)-pyrimidine 975

$C_7H_{11}ClO$
3-Chloro-2-methyl-1-cyclohexanone 55

$C_7H_{11}ClO_3S$
3-(2-Chloroallyloxy)-tetrahydrothiophene 1,1-dioxide 152

$C_7H_{11}Cl_3O$
1-(Trichloromethyl)-cyclohexanol 15

$C_7H_{11}Cl_3O_2$
1-(Trichloromethyl)-butyl acetate 82

$C_7H_{11}Cl_3O_4$
2-(Trichloromethyl)-1,3-dioxane-5,5-dimethanol 21

$C_7H_{11}Cl_3Si$
2-Norbornyltrichlorosilane 1110

$C_7H_{11}Cl_4NO_2$
1,2,2,2-Tetrachloroethyl *N*,*N*-diethylcarbamate 114

$C_7H_{11}Cl_4NO_3$
Ethyl *N*-(1-(2-chloroethoxy)-2,2,2-trichloroethyl)carbamate 112

$C_7H_{11}Cl_4O_5P$
Diethyl (1,2,2,2-tetrachloroethoxycarbonyl)-phosphonate 159

$C_7H_{11}IN_2OS$
5-Acetamido-2,3-dimethylisothiazolium iodide 874

$C_7H_{11}NO$
5-*tert*-Butylisoxazole 870
N,*N*-Diallylformamide 107
Dicyclopropyl ketoxime 1013
Norcamphor oxime 1012
1,4,5-Trimethyl-3-pyrrolin-2-one 125

$C_7H_{11}NOS$
2-(Furfurylamino)-ethanethiol 850

$C_7H_{11}NO_2$
1-Cyclopentyl-2-nitroethene 47
Ethyl 2-cyano-2-methylpropionate 149

$C_7H_{11}NO_2S$
Ethyl 2-thiocyanatobutyrate 151
Methyl 3-methyl-2-thiocyanatobutyrate 151

$C_7H_{11}NO_3$
N-Acetyl-2-allyl-DL-glycine 123
N-Acetyl-DL-proline 124
5,5-Dimethyl-3-ethyl-2,4-oxazolidinedione 134
N-Propylmaleamic acid 122

$C_7H_{11}NO_3S$
S-(2-Carboxyethyl) *N*-allylthiocarbamate 116

$C_7H_{11}NO_4$
Dimethyl 3-aminoglutaconate 87

$C_7H_{11}NO_4S$
5,5-Dimethyl-2,4-thiazolidinedicarboxylic acid 68

$C_7H_{11}NO_5$
N-Acetyl-DL-glutamic acid 122
Diethyl oximinomalonate 1011
Ethyl 2-acetamidomalonate 122

$C_7H_{11}N_3$
2-Amino-3-aminomethyl-6-methylpyridine dihydrochloride 942

$C_7H_{11}N_3O$
1-(2-Cyclohexen-1-ylidene)-semicarbazide 1070
2-(Dimethylamino)-4-hydroxy-6-methylpyrimidine 971
2-(4,5-Dimethyl-1-imidazolyl)-acetamide 867
1-(2-Imidazolin-2-yl)-2-pyrrolidinone 136
1-(2-Imidazolin-2-yl)-2-pyrrolidinone picrate 710
8-Methyl-2,3,7,8-tetrahydroimidazo(1,2-a)pyrimidin-5(6*H*)-one 133
8-Methyl-2,3,7,8-tetrahydroimidazo(1,2-a)pyrimidin-5(6*H*)-one hydriodide 133
1-Methyl-1,2,3,7-tetrahydroimidazo(1,2-a)pyrimidin-5(6*H*)-one hydrobromide 133

$C_7H_{11}N_3OS$
4-Amino-2-(ethylthio)-5-(hydroxymethyl)pyrimidine 970

$C_7H_{11}N_3O_2$
6-Amino-3-ethyl-1-methyluracil 132
5-Amino-1,3,6-trimethyluracil 133
5-*tert*-Butyl-3-nitro-1*H*-pyrazole 862
3-Cyclohexyl-5-hydroxy-1,2,3,4-oxatriazol-3-ium, inner salt 874
3,5-Diamino-2,6-dimethoxypyridine dihydrochloride 936
1-Nitro-3-*tert*-butyl-1*H*-pyrazole 862

$C_7H_{11}N_3S$
Acetone (4-methyl-2-thiazolyl)-hydrazone hydrochloride 877

$C_7H_{11}N_5$
2-(3,3-Dimethylguanidino)-pyrimidine 975
4-Phenylsemicarbazide hydrazone 726

$C_7H_{11}N_5O$
2-Amino-4-morpholino-1,3,5-triazine 979
2,4-Diamino-6-isobutyryl-1,3,5-triazine 979

C_7H_{12}
1-Ethyl-1-cyclopentene 3
3-Ethylcyclopentene 3
Ethylidenecyclopentane 3
1,5-Heptadiene 2
5-Methyl-1-hexyne 1015

$C_7H_{12}BrNO_3$
DL-Bromoisovalerylglycine 123

$C_7H_{12}Br_2O_2$
Methyl 2,6-dibromocaproate 82

$C_7H_{12}Br_4$
1,2,6,7-Tetrabromoheptane 4

$C_7H_{12}Br_4O_2$
2,3,5,6-Tetrabromo-1,7-heptanediol 16

$C_7H_{12}ClNO$
4-(2-Chloroallyl)morpholine 42

$C_7H_{12}ClNO_3$
Chloroacetyl-DL-norvaline 117
Chloroacetyl-DL-valine 117
2-Chloroacetyl-L-valine 117
4-Methyl-4-nitrohexanoyl chloride 100

$C_7H_{12}ClN_5$
2,4-Bis(ethylamino)-6-chloro-1,3,5-triazine 979

$C_7H_{12}ClN_5O_2$
2,4-Bis(2-hydroxyethylamino)-6-chloro-1,3,5-triazine 978

$C_7H_{12}Cl_2O$
1,7-Dichloro-4-heptanone 50

$C_7H_{12}Cl_3NO_3$
Butyl *N*-(1-hydroxy-2,2,2-trichloroethyl)carbamate 113
Ethyl *N*-(1-ethoxy-2,2,2-trichloroethyl)carbamate 113
Methyl *N*-(1-propoxy-2,2,2-trichloroethyl)carbamate 113
Propyl *N*-(1-methoxy-2,2,2-trichloroethyl)carbamate 113

$C_7H_{12}Cl_3NO_4$
Methyl *N*-(1-(2-methoxyethoxy)-2,2,2-trichloroethyl)carbamate 113

$C_7H_{12}FNO_2S$
N-(1,1-Dimethyl-2-oxopropylthio)-*N*-methylcarbamoyl fluoride 107

$C_7H_{12}INS$
2,4-Dimethyl-3-ethylthiazolium iodide 876

$C_7H_{12}N_2$
5-*tert*-Butyl-1*H*-pyrazole 864
1-Piperidineacetonitrile 145
1-Pyrrolidinepropionitrile 146

$C_7H_{12}N_2O$
5,6-Dihydro-4,6,6-trimethyl-2(1*H*)-pyrimidinone 136
1-Hydroxyamino-1-cyclohexanecarbonitrile 147

$C_7H_{12}N_2O_2$
N-Acetyl-DL-prolinamide 121
1-(2-Ethylcrotonoyl)urea 127
5-Isopropyl-5-methylhydantoin 135
4-Methyl-4-nitrohexanenitrile 147
2-Methyl-1,4-piperazinedicarboxaldehyde 111

$C_7H_{12}N_2O_3$
1,3-Diacetamidoacetone 111
N-Glycyl-L-proline 124
$C_7H_{12}N_2O_4$
N-Acetyl-L-glutamine 105
N-Acetylglycylglycine methyl ester 124
2,2-(Diacetamido)propionic acid 121
$C_7H_{12}N_2S$
2-Amino-4-*tert*-butylthiazole 876
1,3-Diallyl-2-thiourea 142
3,4-Dihydro-4,4,6-trimethyl-2-pyrimidinethiol 38
4,5-Dihydro-4,4,6-trimethyl-2-pyrimidinethiol 38
$C_7H_{12}N_4$
2,3,5,6,7,8-Hexahydro-8-methylimidazo(1,2-a)pyrimidin-5-imine hydriodide 137
$C_7H_{12}N_4O_2$
6-Amino-*N*[1],*N*[3]-dimethyl-5-(methylamino)uracil 132
$C_7H_{12}N_4S$
4-Cyclohexyl-1*H*-tetrazole-5(4*H*)-thione 143
$C_7H_{12}N_8$
1,1-Bis(1*H*-1,2,4-triazol-4-ylamino)-propane 870
$C_7H_{12}O$
Dicyclopropyl carbinol 13
1,5-Heptadien-4-ol 11
2-Methyl-1,5-hexadien-3-ol 11
8-Oxabicyclo(5.1.0)octane 21
$C_7H_{12}O_2$
2-Buten-1-yl propionate 80
2,3-Dimethyl-4-pentenoic acid 61
1,4-Dioxaspiro(4.4)nonane 22
Ethyl 4-pentenoate 80
2-Methyl-1-butenyl acetate (*cis,trans* mixture) 78
3-Methyl-3-buten-1-yl acetate 80
Methyl cyclopentanecarboxylate 77
4-Methyl-2-propenyl-1,3-dioxolane 22
2,7-Norbornanediol 12
4-Penten-1-yl acetate 80
$C_7H_{12}O_2S$
Ethyl 3-methyl-1-butynyl sulfone 1019
$C_7H_{12}O_2S_2$
2-Methyl-1,3-dithiolane-2-propionic acid 68
$C_7H_{12}O_3$
β,*β*-Dimethyl-*α*-hydroxy-*γ*-valerolactone 95
3,3-Dimethyl-4-oxovaleric acid 63
6,8-Dioxa-1-(hydroxymethyl)-bicyclo(3.2.1)octane 21
4,5-Epoxypentyl acetate 91
Ethyl 4-methoxycrotonate 85
cis-2-Hydroxy-1-cyclohexanecarboxylic acid 64
Methyl 2,2-dimethylacetoacetate 93
Methyl 5-oxohexanoate 75
4-Methyl-5-oxohexanoic acid 64
5-Methyl-4-oxohexanoic acid 64
5-Methyl-2-propyl-1,3-dioxolan-4-one 97
Tetrahydrofurfuryl acetate 90
$C_7H_{12}O_3S$
3-Allyloxytetrahydrothiophene 1,1-dioxide 152
O,O-Diethyl thiomalonate 76
2-(Tetrahydropyran-2-ylthio)acetic acid potassium salt 69
$C_7H_{12}O_4$
Diethyl malonate thallium(I) salt 75
2,3-Dimethylglutaric acid 61
2-Methoxy-4-oxo-2,5,5-trimethyl-1,3-dioxolane 96
$C_7H_{12}O_4S$
4-Methyl-3-sulfolanyl acetate 153
$C_7H_{12}O_4S_2$
S,S'-Methylenebis(3-mercaptopropionic acid) 63
$C_7H_{12}O_4Si$
Diacetoxymethylvinylsilane 1096
$C_7H_{12}O_7$
Methyl 2-keto-L-gulonate 91
$C_7H_{13}BrO_2$
3-Bromo-2,2-diethylpropionic acid 62
Ethyl 2-bromo-3-methylbutyrate 83
$C_7H_{13}Cl$
Cycloheptyl chloride 5
$C_7H_{13}ClO$
3-Chloromethyl-2,4,6-trimethylphenol 190
2-(3-Chloropropyl)tetrahydrofuran 21
$C_7H_{13}ClO_2$
Isoamyl chloroacetate 82
$C_7H_{13}ClSi$
4-Chloro-1-(trimethylsilyl)-1-butyne 1099
$C_7H_{13}Cl_3N_2O_2$
Ethyl *N*-(1-(dimethylamino)-2,2,2-trichloroethyl)carbamate 114
Ethyl *N*-(1-(ethylamino)-2,2,2-trichloroethyl)carbamate 114
Methyl *N*-(1-(isopropylamino)-2,2,2-trichloroethyl)carbamate 114
Methyl *N*-(1-(propylamino)-2,2,2-trichloroethyl)carbamate 114
$C_7H_{13}Cl_3Si$
(Cyclohexylmethyl)trichlorosilane 1111
$C_7H_{13}FO_3$
n-Butyl 3-fluorolactate 84
$C_7H_{13}I$
6-Iodo-1-heptene 5
$C_7H_{13}IO_2$
Ethyl 5-iodovalerate 82
$C_7H_{13}N$
N-Ethyl-1,2,5,6-tetrahydropyridine 41
Hexyl isocyanide 150
$C_7H_{13}NO$
N-(*tert*-Butyl)acrylamide 105
4,4-Dimethyl-2-ethyl-2-oxazoline 41
N-Formylhexamethyleneimine 109
N-Propylcyclopropanecarboxamide 106
3,3,5-Trimethyl-2-pyrrolidinone 125
$C_7H_{13}NO_2$
trans-2-Amino-1-cyclohexanecarboxylic acid 72
1-Aza-3,5-dimethyl-4,6-dioxabicyclo-(3.3.0)octane 41
Cyclohexyl carbamate 116
5-Ethyl-1-aza-4,6-dioxabicyclo(3.3.0)octane 41
Ethyl 3-methylaminocrotonate 88
Ethyl 1-pyrrolidinecarboxylate 115
4-Glycidylmorpholine 40
1-Nitrocycloheptane 46
4-Piperidone ethyl hemiketal 41
$C_7H_{13}NO_2S$
N-Allyltetrahydro-3-thiophenamine 1,1-dioxide 153
$C_7H_{13}NO_2S$
Ethyl 2-thiazolidineacetate 91
$C_7H_{13}NO_2S_2$
(*N,N*-Diethylthiocarbamoylthio)-acetic acid 139
2-(Methoxycarbonyl)ethyl *N,N*-dimethyldithiocarbamate 140
$C_7H_{13}NO_3$
3-Acetamido-3-methylbutyric acid 105
N-Acetyl-DL-norvaline 123
N-Acetyl-D-valine 123
4,4-Dimethyl-5-nitro-2-pentanone 48
Ethyl 2,*N*-dimethylmalonamate 101
Ethyl morpholinocarboxylate 110
1-(Nitromethyl)-cyclohexanol 45
$C_7H_{13}NO_4$
Dimethyl 3-aminoglutarate 87
N-(Ethoxycarbonyl)-glycine ethyl ester 116
Ethyl 2,2-dimethyl-3-nitropropionate 73
4-Ethylglutamic acid hydrochloride 72
Ethyl 2-nitrovalerate 73
2-Isopropyl-5-nitro-1,3-dioxane 47
2-Nitropentyl acetate 73
$C_7H_{13}NO_5$
2,2-Dimethyl-5-nitro-1,3-dioxane-5-methanol 45
$C_7H_{13}NS_2$
N-Cyclohexyldithiocarbamic acid, ammonium salt 741
$C_7H_{13}N_3$
1,3,4,6,7,8-Hexahydro-2*H*-pyrimido(1,2-a)pyrimidine picrate 281
$C_7H_{13}N_3O$
Cyclohexanone semicarbazone 128
N-(Hexahydro-1-methyl-2-pyrimidinylidene)acetamide 137
$C_7H_{13}N_3O_3$
1-(2-Oxo-3-oxazolidinyl)-3-propylurea 127
$C_7H_{13}N_3O_4$
DL-Alanyl-DL-asparagine monohydrate 118
DL-Alanylglycylglycine 119
β-Alanylglycylglycine hemihydrate 119
Nα-Glycyl-L-glutamine 119
Glycylglycyl-L-alanine 119
Sarcosylglycylglycine 121
$C_7H_{13}N_3S$
1-Amino-1,4-dihydro-4,4,6-trimethyl-2-pyrimidinethiol 38
1-Cyclohexanone thiosemicarbazone 140
$C_7H_{13}N_5O_5$
Nα-Formylnitro-L-arginine 137
C_7H_{14}
cis-4,4-Dimethyl-2-pentene 1
2,3-Dimethyl-2-pentene 1
2,4-Dimethyl-1-pentene 1
4,4-Dimethyl-1-pentene 1
4,4-Dimethyl-2-pentene (mixture of *cis* and *trans*) 1
3-Ethyl-2-pentene 1
2-Methyl-2-hexene 1
$C_7H_{14}BrNO$
3-Bromo-2,2,*N*,*N*-tetramethylpropionamide 106
$C_7H_{14}ClN$
3-Chloro-1-ethylpiperidine hydrochloride 40
$C_7H_{14}ClNO$
2-Chloro-*N*-pentylacetamide 102
$C_7H_{14}HgO_2$
(2,2-Dimethylpropyl)mercuri acetate 1137
$C_7H_{14}INOS$
4-(1-Methylthioethylidene)-morpholinium iodide 42
$C_7H_{14}N_2$
3-(Butylamino)-propionitrile 146
3-Diethylaminopropionitrile 146
5,7-Dimethyl-2,3,6,7-tetrahydro-1*H*-1,4-diazepine perchlorate 39
$C_7H_{14}N_2O$
N-Allyl-2-(dimethylamino)-acetamide 104
Cyclohexylurea 126
1-Cyclopentyl-1-methylurea 129
1-Ethyl-3-(2-methylallyl)urea 128
Isobutyric acid isopropylidenehydrazide 112
N-Methyl-1-piperidinecarboxamide 130
$C_7H_{14}N_2OS$
O-Cyclohexyl thiocarbazate 141
$C_7H_{14}N_2O_2$
4-Amino-1-methyl-4-piperidinecarboxylic acid 67
Isobutyl(2-nitro-1-propen-1-yl)amine 45
$C_7H_{14}N_2O_2S$
N-Acetyl-DL-methioninamide 118
$C_7H_{14}N_2O_3$
Nα-Acetyl-DL-ornithine 119
Nα-Acetyl-L-ornithine 118
D-2-Aminobutyryl-D-alanine 123
L-2-Aminobutyryl-L-alanine 119
Glycyl-DL-valine 104
Glycyl-L-valine 119
L-Valylglycine 119
$C_7H_{14}N_2O_3S$
Glycyl-L-methionine 118
L-Methionylglycine 118
$C_7H_{14}N_2O_4$
L-Alanyl-L-threonine 118
L-2-Aminobutyryl-L-serine 117
2,2-Diethyl-1,3-dinitropropane 46
2,5-Dimethoxytetrahydro-2-furoic acid hydrazide (mixture of *cis* and *trans*) 111
2,3-Dimethyl-2,3-dinitropentane 45
N,N'-Methylenebis-(ethyl carbamate) 115
$C_7H_{14}N_2O_4S$
DL-Cystathionine 71
$C_7H_{14}N_2O_4S_2$
Djenkolic acid dihydrochloride 71
$C_7H_{14}N_2O_5$
2,3-Dimethyl-2,3-dinitro-1-methoxybutane 45
$C_7H_{14}N_2O_6$
Glucose ureide 126
$C_7H_{14}N_2S$
2-Cyclohexyl-2-thiopseudourea hydrobromide 138
2,2,4,4-Tetramethylimidazolidine-5-thione 139
$C_7H_{14}N_2S_2$
3,5-Diethyl-3,4,5,6-tetrahydro-2*H*-1,3,5-thiadiazine-2-thione 139
$C_7H_{14}N_4O_3$
Glycylglycyl-L-alaninamide acetate hemihydrate 121
$C_7H_{14}N_4O_4S$
2-(3-Sulfolanyl)malonic acid dihydrazide 153
$C_7H_{14}O$
Cyclohexyl methyl ether 22
1-Cyclopentylethanol 13
2-Cyclopentylethanol 13
2,3-Dimethyl-4-penten-2-ol 11
1-Ethylcyclopentanol 13
2-Ethylisovaleraldehyde 60
1-Hepten-3-ol 11
1-Hepten-4-ol 11
2-Hepten-4-ol 11
2-Methyl-3-hexanone 48
3-Methyl-2-hexanone 48
4-Methyl-3-hexanone 48
3-Methyl-1-hexen-3-ol 11
3-Methyl-5-hexen-3-ol 11
$C_7H_{14}OS$
2-Tetrahydrofuranpropanethiol 24
$C_7H_{14}OS_2$
Hexylxanthic acid, potassium salt 74
$C_7H_{14}O_2$
tert-Amyl acetate 73
2-*tert*-Butyl-1,3-dioxolane 22
2-Ethoxytetrahydropyran 22
Ethyl tetrahydrofurfuryl ether 22
2-Isopropyl-1,3-dioxane 23
2-Methoxycyclohexanol 20
1-Methyl-*trans*,-1,2-cyclohexanediol 12
Methyl 3,3-dimethylbutyrate 73
4-Methyl-2-propyl-1,3-dioxolane 22
$C_7H_{14}O_2S$
Isopentyl mercaptoacetate 86
1-Methoxy-4-((methylthio)methoxy)-2-butene 25
$C_7H_{14}O_3$
2,2-Bis-(hydroxymethyl)-tetrahydropyran 21
2,4-Dimethyl-1,3-dioxane-2-methanol (mixture of isomers) 21
$C_7H_{14}O_4$
Ethyl 3-hydroxy-4-methoxybutyrate 84
2-(2-Methoxyethoxy)ethyl acetate 85
Methyl diethoxyacetate 85
$C_7H_{14}O_5$
Pentaerythritol monoacetate 84
$C_7H_{14}O_5S$
Methyl-*β*-D-thiogalactoside 24
$C_7H_{14}O_6$
Methyl-*α*-D-galactopyranoside hydrate 16
$C_7H_{14}O_8$
Heptagluconic acid, calcium salt 69
$C_7H_{14}S$
Allyl *sec*-butyl sulfide 25
$C_7H_{14}S_2$
2-Isopropyl-1,3-dithiane 24
$C_7H_{15}Cl$
4-Chloroheptane 4
$C_7H_{15}ClHg$
Chloroheptylmercury 1137
$C_7H_{15}ClN_2O$
DL-Carnitinenitrile chloride 144
$C_7H_{15}N$
2,6-Dimethylpiperidine hydrobromide 40

$C_7H_{15}N$ (Cont'd)
2,6-Dimethylpiperidine hydrochloride 40
4,4-Dimethylpiperidine hydrochloride 41
$C_7H_{15}NO$
N,N-Diethylallylamine-*N*-oxide picrate 264
N,N-Diisopropylformamide 107
N,N-Dimethylvaleramide 107
1-(Ethoxymethyl)pyrrolidine 37
1-Ethyl-2-pyrrolidinemethanol 36
1-Ethyl-3-pyrrolidinemethanol 37
N-Glycidyldiethylamine 35
Methyl 3-piperidyl carbinol 37
1-Propylamino-3-buten-2-ol 33
1-Pyrrolidino-2-propanol 36
$C_7H_{15}NO_2$
L-Alanine *tert*-butyl ester dibenzenesulfimide 817
L-Alanine *tert*-butyl ester hydrochloride 87
N-Allyl-2,2′-iminodiethanol 34
DL-2-Aminoheptanoic acid 72
N-Butylurethane 116
2-(Dimethylamino)isopropyl acetate 88
5-(Dimethylamino)-valeric acid hydrochloride 67
N,N-Dimethylvaline 72
Ethyl 3-(methylamino)-butyrate 88
4-(2-Hydroxypropyl)morpholine 37
Methyl 6-aminohexanoate hydrochloride 87
N-Methyl-L-isoleucine 72
1-Nitroheptane 46
O-(3-Oxobutyl)-*N*-propyl-hydroxylamine oxalate salt 1011
$C_7H_{15}NO_3$
Ethyl *N*-(hydroxy-*tert.*-butyl)-carbamate 115
Ethyl 1-(hydroxymethyl)-propylcarbamate 115
$C_7H_{15}NO_3S$
cis-3-Hydroxy-4-(isopropylamino)-tetrahydrothiophene-1,1-dioxide 152
trans-3-Hydroxy-4-(isopropylamino)-tetrahydrothiophene-1,1-dioxide 152
$C_7H_{15}NO_3Si$
4-(Trimethoxysilyl)-butyronitrile 1110
$C_7H_{15}NO_7$
N-Hydroxymethylgluconamide 103
$C_7H_{15}NPb$
Cyanotriethyllead 1135
$C_7H_{15}N_3O$
4-Carbamoyl-4-methylaminopiperidine 108
1-(4-Methyl-2-pentylidene)-semicarbazide 127
$C_7H_{15}N_3OS$
4-(1,1-Dimethyl-3-oxobutyl)-3-thiosemicarbazide 140
α-Ethyl-α-methyl-4-thioxohexahydro-1,3,5-triazine-1-methanol 139
1-(3-Methoxypropyl)-4-thioxohexahydro-1,3,5-triazine 139
2-(2-Morpholinoethyl)-2-thiopseudourea dihydrochloride 137
$C_7H_{15}N_3O_3$
*N*6-Carbamoyl-DL-lysine 126
Ethyl 3-amino-3-ureidobutyrate 126
$C_7H_{15}N_3S$
1-(*tert*-Butyl)-4-thioxohexahydro-1,3,5-triazine 140
4-Cyclohexyl-3-thiosemicarbazide 142
$C_7H_{15}N_5O_3$
1-Hexyl-3-nitro-1-nitrosoguanidine 138
$C_7H_{15}O_3P$
Diethyl allylphosphonate 158
$C_7H_{15}O_4P$
Diethyl propionylphosphonate 158
C_7H_{16}
3-Ethylpentane 1
$C_7H_{16}BrNO$
(3-Oxobutyl)trimethylammonium bromide 57
$C_7H_{16}ClN$
3-Diethylaminopropyl chloride hydrochloride 32
$C_7H_{16}ClO_3P$
Dipropyl chloromethylphosphonate 159
$C_7H_{16}IN$
(Cyclopropylmethyl)-trimethylammonium iodide 44
1,1-Dimethylpiperidinium iodide 44
$C_7H_{16}N_2$
3-Amino-*N*-ethylpiperidine 38
4-(Dimethylamino)-piperidine 39
4-(Dimethylamino)-piperidine dihydrochloride 39
1,4-Dimethylhomopiperazine 39
N-Propyl-1,4-diamino-2-butene 32
1-Propylpiperazine dihydrochloride 39
1,2,4-Trimethylpiperazine 41
$C_7H_{16}N_2O$
1-Butyl-3-ethylurea 128
1-(1-Ethylbutyl)-urea 128
$C_7H_{16}N_2OS$
2-Mercapto-4,6,6-trimethylhexahydropyrimidin-4-ol 36
$C_7H_{16}N_2S$
1,3-Dipropyl-2-thiourea 142
$C_7H_{16}N_4$
4,7,10-Triazadecanenitrile trihydrochloride 145
$C_7H_{16}N_4O_2$
Pimelic acid dihydrazide 112
$C_7H_{16}N_4S_2$
S,S′-Pentamethylene-diisothiuronium dichloride 137
$C_7H_{16}O$
2,2-Dimethyl-1-pentanol 8
2,4-Dimethyl-1-pentanol 8
2,4-Dimethyl-2-pentanol 9
3-Ethyl-2-pentanol 8
4-Heptanol 8
2-Isopropyl-1-butanol 8
(2*R*,2*S*,4*S*)-(+)-4-Methyl-2-hexanol 8
3-Methyl-2-hexanol 8
3-Methyl-3-hexanol 9
4-Methyl-3-hexanol 8
5-Methyl-3-hexanol 8
$C_7H_{16}OS$
Amyl 2-hydroxyethyl sulfide 24
6-(Methylthio)-1-hexanol 24
$C_7H_{16}OSi$
Isopropyl trimethylsilyl ketone 1096
$C_7H_{16}O_2$
1-Butoxy-2-propanol 19
Dipropoxymethane 17
1-Ethoxy-3-pentanol 19
1,4-Heptanediol 10
6-Methoxy-2-hexanol 19
$C_7H_{16}O_2S$
Ethyl pentyl sulfone 154
2-(Methylthio)-acetaldehyde diethyl acetal 25
$C_7H_{16}O_3$
1,3-Diethoxy-2-propanol 19
$C_7H_{16}O_3S$
3-(Butylsulfonyl)propanol 152
$C_7H_{16}O_4$
Diethylene glycol mono-2-hydroxypropyl ether 19
2-Ethoxymethyl-2-hydroxymethyl-1,3-propanediol 19
$C_7H_{16}O_7$
α-Glucoheptitol 16
$C_7H_{16}S$
Butyl isopropyl sulfide 25
tert-Butyl isopropyl sulfide 26
Butyl propyl sulfide 25
sec-Butyl propyl sulfide 25
Hexyl methyl sulfide 25
Isobutyl isopropyl sulfide 25
Isobutyl propyl sulfide 25
$C_7H_{16}S_2$
1,5-Bis(methylthio)pentane 26
$C_7H_{16}S_3$
Tris(ethylthio)methane 26
$C_7H_{16}Si$
1,1-Dimethyl-1-silacyclohexane 1096
$C_7H_{17}N$
1,3-Dimethylpentylamine 28
1,4-Dimethylpentylamine 28
N-Isopropylbutylamine 28
1-Isopropyl-2-methylpropylamine 28
1-Methylhexylamine sulfate 28
$C_7H_{17}NO$
2-Amino-2-methyl-3-hexanol 33
2-Butylamino-1-propanol 33
5-Dimethylamino-1-pentanol 34
3-Isobutoxypropylamine 35
2-(Isopentylamino)-ethanol 33
1-Isopropylamino-2-methyl-2-propanol 33
2-(2-Methylbutylamino)-ethanol 33
2-(Pentylamino)-ethanol 33
N,N,2,2-Tetramethylpropylamine *N*-oxide 1006
$C_7H_{17}NO_2$
N-Isopropyl-2,2′-iminodiethanol 34
Methylaminoacetaldehyde diethyl acetal 35
$C_7H_{17}N_3O$
4-Butyl-1,1-dimethylsemicarbazide 128
$C_7H_{17}N_3O_2$
1-(2,2-Diethoxyethyl)guanidine 137
$C_7H_{17}N_3S$
4-Butyl-1,1-dimethyl-3-thiosemicarbazide 141
$C_7H_{18}BrN$
Dimethylethylpropylammonium bromide 43
$C_7H_{18}IN$
N-Butyltrimethylammonium iodide 43
$C_7H_{18}NO_3P$
Diethyl isopropylamidophosphate 160
$C_7H_{18}N_2$
2,2-Diethyl-1,3-propanediamine 29
Heptylhydrazine oxalate 30
N,N,N′,N′-Tetramethyl-1,2-propanediamine 30
$C_7H_{18}N_4$
1-(2-(Diethylamino)-ethyl)-guanidine sulfate 137
$C_7H_{18}OSi$
4-(Trimethylsilyl)-1-butanol 1104
$C_7H_{18}Si$
Butyltrimethylsilane 1096
$C_7H_{19}NOSi$
4-(Dimethylmethoxysilyl)-butylamine 1107
$C_7H_{19}N_3$
1,4,7-Trimethyldiethylenetriamine 30
$C_7H_{20}N_4$
1,4,8,11-Tetraazaundecane tetrahydrochloride 29
$C_7H_{21}N_3Si$
Methyltris(dimethylamino)silane 1096
$C_7H_{22}OSi_3$
1,1,1,2,3,3,3-Heptamethyl-2-hydroxytrisilane 1096
$C_7H_{24}Si_4$
2-(Dimethylsilyl)-1,1,2,3,3-pentamethyltrisilane 1096
C_8BrF_{15}
1-Bromopentadecafluoro-cyclooctane 5
$C_8Cl_6O_2$
2,3,5,6-Tetrachloroterephthaloyl chloride 389
$C_8Cl_6S_2$
3,4,5,3′,4′,5′-Hexachloro-2,2′-bithiophene 859
C_8Cl_8
Perchlorobenzocyclobutene 173
C_8Cl_{10}
Decachloro-1,3a,6,6a-tetrahydropentalene 6
$C_8F_{15}I$
1-Iodopentadecafluorocyclooctane 5
$C_8HF_4NO_2$
Tetrafluorophthalimide 723
C_8HF_{17}
1,1,2,2,3,3,4,4,5,5,6,6,7,7,8,8,8-Heptadecafluorooctane 4
$C_8H_2ClF_5O$
Pentafluorophenylacetyl chloride 389
$C_8H_2Cl_8$
1,1,2,3,6,7,8,8-Octachloro-1,7-octadien-4-yne 1015
α,α,α′,α′-2,3,5,6-Octachloro-*p*-xylene 170
$C_8H_3Br_2Cl_4NO$
N-(4-Chloro-2,6-dibromophenyl)-2,2,2-trichloroacetamide 396
$C_8H_3Br_2F_6N$
2,6-Dibromo-3,5-bis(trifluoromethyl)aniline 215
$C_8H_3Br_2NO_2$
5,7-Dibromoisatin 692
$C_8H_3Br_3Cl_3NO$
N-(2,4,6-Tribromophenyl)-2,2,2-trichloroacetamide 396
$C_8H_3ClO_3$
3-Chlorophthalic anhydride 388
$C_8H_3Cl_2F_4NO$
2,2-Dichloro-2′,3′,5′,6′-tetrafluoroacetanilide 394
$C_8H_3Cl_4NO$
4-(2-Propyn-1-yloxy)-2,3,5,6-tetrachloropyridine 1017
$C_8H_3Cl_4NS$
4-(2-Propyn-1-ylthio)-2,3,5,6-tetrachloropyridine 1017
$C_8H_3Cl_6NO$
2,2,2,2′,4′,5′-Hexachloroacetanilide 396
$C_8H_3Cl_7$
1,4-Bis(trichloromethyl)-2-chlorobenzene 170
4-Chloro-1,3-bis(trichloromethyl)-benzene 170
$C_8H_3F_3N_4O_4$
N-Cyano-2,6-dinitro-4-(trifluoromethyl)aniline 751
$C_8H_3F_7$
α,α,α,α′,α′,α′,5-Heptafluoro-*m*-xylene 170
$C_8H_3N_5O_4$
1,6-Dinitro-1*H*-indazole-3-carbonitrile 900
$C_8H_4BrF_3N_2S$
2-Amino-6-bromo-4-(trifluoromethyl)benzothiazole 910
$C_8H_4Br_2Cl_2FNO$
2′,6′-Dibromo-2,2-dichloro-4′-fluoroacetanilide 394
$C_8H_4ClFN_2O$
N-(3-Chloro-4-fluorophenyl)-cyanoformamide 759
$C_8H_4ClNO_2$
5-Chloroisatin 706
6-Chloroisatin 706
$C_8H_4ClNO_3$
6-Chloroisatoic anhydride 709
$C_8H_4ClN_3O_2$
4-Chloro-7-nitroquinazoline 1000
2-Chloro-7-nitroquinoxaline 1000
$C_8H_4Cl_2F_3NO_4S$
4-Chloro-3-nitrophenyl 2-chloro-1,1,2-trifluoroethyl sulfone 772
$C_8H_4Cl_2Hg_2S_2$
5,5′-Bis(chloromercuri)-2,2′-bithiophene 1139
$C_8H_4Cl_2N_2$
2,4-Dichloroquinazoline 1001
4,6-Dichloroquinazoline 1001
4,7-Dichloroquinazoline 1001
$C_8H_4Cl_2N_2O$
N-(2,3-Dichlorophenyl)-cyanoformamide 759
N-(2,5-Dichlorophenyl)-cyanoformamide 759
$C_8H_4Cl_2N_2S$
N-(2,3-Dichlorophenyl)-cyanothioformamide 759
N-(2,5-Dichlorophenyl)-cyanothioformamide 759
N-(2,6-Dichlorophenyl)-cyanothioformamide 759
$C_8H_4Cl_2OS$
4,7-Dichlorobenzo(b)thiophen-3(2*H*)-one 298
$C_8H_4Cl_2O_4$
2,5-Dichloroterephthalic acid 337
1,4-Phenylene bis(chloroformate) 389
$C_8H_4Cl_3FN_2O_3$
4′-Fluoro-3′-nitro-2,2,2-trichloroacetanilide 395

$C_8H_4Cl_3F_3S$
2-Chloro-1,1,2-trifluoroethyl 3,4-dichlorophenyl sulfide 209
$C_8H_4Cl_3NO$
2,4,5-Trichlorophenoxyacetonitrile 751
$C_8H_4Cl_3NO_3S_2$
2-(Trichloromethylthio)-1,2-benzisothiazol-3(2*H*)-one-1,1-dioxide 805
$C_8H_4Cl_3N_3O_5$
2′,4′-Dinitro-2,2,2-trichloroacetanilide 396
$C_8H_4Cl_4N_2$
1,1-Dichloro-2-(2,4-dichlorophenylazo)ethene 1082
$C_8H_4Cl_4O$
2,2,2′,4′-Tetrachloroacetophenone 288
2′,3′,5′,6′-Tetrachloroacetophenone 289
$C_8H_4Cl_4O_2$
4-(Trichloromethoxy)benzoyl chloride 389
2-(2,4,5-Trichlorophenoxy)acetyl chloride 389
$C_8H_4Cl_4O_2S$
2-Chloro-5-(trichloromethylthio)-benzoic acid 336
$C_8H_4Cl_5NO$
N-(2,5-Dichlorophenyl)-2,2,2-trichloroacetamide 395
N-(2,6-Dichlorophenyl)-2,2,2-trichloroacetamide 395
2,2,2,3′,4′-Pentachloroacetanilide 395
$C_8H_4Cl_6$
1,3-Bis-(trichloromethyl)-benzene 170
α,α′,2,4,5,6-Hexachloro-*m*-xylene 170
$C_8H_4Cl_6O_2$
1,1,2,7,8,8-Hexachloro-1,7-octadien-4-yne-3,6-diol 1016
$C_8H_4F_2N_2O$
N-(2,5-Difluorophenyl)-cyanoformamide 759
$C_8H_4F_3NO_4$
3-Nitro-5-(trifluoromethyl)-benzoic acid 336
$C_8H_4F_3NS$
2-(Trifluoromethyl)benzothiazole 911
$C_8H_4F_4O$
2′,3′,5′,6′-Tetrafluoroacetophenone 289
m-Trifluoromethylbenzoyl fluoride 389
$C_8H_4F_5NO_2$
2-(Pentafluorophenoxy)acetamide 399
C_8H_4INO
2-Iodobenzoyl cyanide 759
$C_8H_4I_2S_2$
5′,5′-Diiodo-2,2′-bithiophene 859
$C_8H_4N_2O_3$
4-Nitrobenzoyl cyanide 765
$C_8H_4N_2O_5$
5-Nitroisatoic anhydride 703
6-Nitroisatoic anhydride 703
$C_8H_4N_2S$
(2-Thenylidene)malononitrile 858
$C_8H_4N_6O_3$
2-Azido-5-(4-nitrophenyl)-1,3,4-oxadiazole 769
$C_8H_4O_7$
Tetrahydrofuran-2,3,4,5-tetracarboxylic dianhydride 98
$C_8H_4O_9$
2,3,4,5-Furantetracarboxylic acid 848
2,3,4,5-Furantetracarboxylic acid, monosodium salt 848
$C_8H_5AsO_6$
3-Arsonophthalic anhydride 1127
$C_8H_5BrCl_3NO$
4′-Bromo-2,2,2-trichloroacetanilide 395
$C_8H_5BrCl_6$
(-)-5-*endo*-Bromomethyl-1,2,3,4,7,7-hexachloro-2-norbornene 6
$C_8H_5BrN_2O$
N-(4-Bromophenyl)cyanoformamide 760
$C_8H_5BrN_2O_2$
5-Bromoisatin 3-oxime 1012
6-Bromo-1,4(2*H*,3*H*)-phthalazinedione 502
$C_8H_5BrO_2$
3-Bromophthalide 386
$C_8H_5BrO_3$
6-Bromo-1,3-benzodioxole-5-carboxaldehyde 324
$C_8H_5BrO_4$
6-Bromo-1,3-benzodioxole-5-carboxylic acid 341
2-Bromoisophthalic acid 337
C_8H_5BrS
3-Bromo-1-benzothiophene 899
$C_8H_5Br_2NS$
3-(Dibromomethyl)-1,2-benzisothiazole 910
$C_8H_5Br_3ClNO$
2-Bromo-*N*-(2-chloro-4,6-dibromophenyl)acetamide 391
2-Bromo-*N*-(4-chloro-2,6-dibromophenyl)acetamide 391
$C_8H_5Br_3FNO$
N-(2,6-Dibromo-4-fluorophenyl)-2-bromoacetamide 391
$C_8H_5Br_3O$
2,2,4′-Tribromoacetophenone 288
$C_8H_5ClF_3NO$
6′-Chloro-*α,α,α*-trifluoro-*m*-formotoluidide 398
$C_8H_5ClF_3NO_3$
2-Chloro-1,1,2-trifluoroethyl 4-nitrophenyl ether 258
$C_8H_5ClN_2O$
7-Chloro-4-hydroxycinnoline 999
8-Chloro-4-hydroxycinnoline 999
N-(2-Chlorophenyl)cyanoformamide 760
N-(4-Chlorophenyl)cyanoformamide 760
4-Chloro-1(2*H*)-phthalazinone 1001
$C_8H_5ClN_2O_2$
2-Chloro-4-nitrophenylacetonitrile 755
6-Chloro-2,4(1*H*,3*H*)-quinazolinedione 1001
7-Chloro-2,4(1*H*,3*H*)-quinazolinedione 1000
$C_8H_5ClN_2O_2S$
5-Chloro-2-methyl-4-nitrobenzothiazole 914
$C_8H_5ClN_2O_6$
Methyl 2-chloro-3,5-dinitrobenzoate 372
Methyl 4-chloro-3,5-dinitrobenzoate 373
$C_8H_5ClN_2S$
N-(3-Chlorophenyl)-cyanothioformamide 760
$C_8H_5ClO_2$
5-Chloro-3(2*H*)-benzofuranone 289
$C_8H_5ClO_3$
6-Chloropiperonal 324
$C_8H_5ClO_4$
2-Chloroisophthalic acid 338
$C_8H_5ClO_6S$
3,5-Dicarboxybenzenesulfonyl chloride 804
$C_8H_5Cl_2F_2NO$
2,2-Dichloro-2′,5′-difluoroacetanilide 394
$C_8H_5Cl_2F_3S$
4-Chlorophenyl 2-chloro-1,1,2-trifluoroethyl sulfide 209
$C_8H_5Cl_2NO$
2,4-Dichlorophenoxyacetonitrile 751
$C_8H_5Cl_2NO_2$
2,6-Dichloro-*β*-nitrostyrene 258
$C_8H_5Cl_2NS$
2,6-Dichlorobenzyl thiocyanate 768
4,5-Dichloro-2-methylbenzothiazole 917
4,6-Dichloro-2-methylbenzothiazole 917
$C_8H_5Cl_2N_3O$
1-(2,2-Dichloroacetyl)-1*H*-benzotriazole 908
$C_8H_5Cl_3FNO$
4′-Fluoro-2,2,2-trichloroacetanilide 395
$C_8H_5Cl_3INO$
N-(4-Iodophenyl)-2,2,2-trichloroacetamide 395
$C_8H_5Cl_3N_2$
2-(Trichloromethyl)-benzimidazole 901
$C_8H_5Cl_3N_2O_2$
2′,4′,5′-Trichloroglyoxylanilide oxime 1010
$C_8H_5Cl_3N_2O_3$
2′-Nitro-2,4′,6′-trichloroacetanilide 390
$C_8H_5Cl_3O$
2,2,4′-Trichloroacetophenone 288
$C_8H_5Cl_4N$
4-Allyl-2,3,5,6-tetrachloropyridine 967
$C_8H_5Cl_4NO$
2,2,2,3′-Tetrachloroacetanilide 395
2,2,2,4′-Tetrachloroacetanilide 395
2,2,3′,4′-Tetrachloroacetanilide 394
2,2,3′,5′-Tetrachloroacetanilide 394
2,2′,4′,5′-Tetrachloroacetanilide 393
2,2′,4′,6′-Tetrachloroacetanilide 393
$C_8H_5Cl_5O_3$
Methyl 2-oxo-3,4,5,6,6-pentachloro-3,5-hexadiene-1-carboxylate 92
$C_8H_5FN_2O$
N-(2-Fluorophenyl)cyanoformamide 760
N-(3-Fluorophenyl)cyanoformamide 760
$C_8H_5FN_2O_2$
5-Fluoro-2-nitrophenylacetonitrile 755
$C_8H_5FN_2S$
N-(4-Fluorophenyl)-cyanothioformamide 760
$C_8H_5FO_6S$
3,5-Dicarboxybenzenesulfonyl fluoride 804
4-(Fluorosulfonyl)-phthalic acid 804
$C_8H_5F_3N_2$
2-(Trifluoromethyl)benzimidazole 901
$C_8H_5F_3N_2O_5$
2,6-Dinitro-4-(trifluoromethyl)-phenyl methyl ether 259
$C_8H_5F_3N_2S$
2-Amino-4-(trifluoromethyl)-benzothiazole 910
2,3-Dihydro-2-thioxo-6-trifluoromethylbenzimidazole 901
$C_8H_5F_3O_2$
Phenyl 2,2,2-trifluoroacetate 376
$C_8H_5F_3O_3$
1-(2-Furyl)-4,4,4-trifluoro-1,3-butanedione 843
$C_8H_5F_4NO$
2′,3′,4′,5′-Tetrafluoroacetanilide 407
$C_8H_5F_5O$
2-(Pentafluorophenyl)ethanol 197
$C_8H_5IO_4$
2-Iodoisophthalic acid 338
$C_8H_5NO_2S_2$
5-Furfurylidenerhodanine 854
$C_8H_5NO_6$
2-Nitroisophthalic acid 344
4-Nitroisophthalic acid 343
3-Nitrophthalic acid, barium salt 351
$C_8H_5N_3$
4-Cyano-3-(cyanomethyl)pyridine 953
$C_8H_5N_3O_3$
N-(3-Nitrophenyl)cyanoformamide 764
7-Nitroquinazolin-4(3*H*)-one 1000
$C_8H_5N_3S$
2-Amino-6-benzothiazolyl thiocyanate 914
$C_8H_6BCl_2F_4N_3O$
4-Acetamido-3,5-dichlorobenzenediazonium tetrafluoroborate 431
C_8H_6BrClO
4′-Bromo-2-chloroacetophenone 287
$C_8H_6BrCl_2NO$
4′-Bromo-2,2′-chloroacetanilide 391
$C_8H_6BrCl_2NO_2$
2-Bromo-4,6-dichlorophenyl *N*-methylcarbamate 640
$C_8H_6BrCl_3O$
1-(4-Bromophenyl)-2,2,2-trichloroethanol 197
$C_8H_6BrFO_2$
Methyl 2-bromo-5-fluorobenzoate 376
$C_8H_6BrFO_3S$
2-Bromo-3′-fluorosulfonylacetophenone 801
$C_8H_6BrI_3O$
β-Bromo-2,4,6-triiodophenetole 176
C_8H_6BrN
2-Bromo-4-methylbenzonitrile 760
$C_8H_6BrNO_2$
trans-*β*-Bromo-4-nitrostyrene 278
$C_8H_6BrNO_3$
2′-Bromo-5′-nitroacetophenone 289
$C_8H_6BrNO_4$
Methyl 2-bromo-3-nitrobenzoate 374
Methyl 5-bromo-2-nitrobenzoate 366
$C_8H_6BrNO_5$
5-Bromo-2-nitrovanillin 321
$C_8H_6BrN_3$
3-Amino-1-bromo-2,6-naphthyridine 1000
$C_8H_6Br_2$
1,2-Dibromobenzocyclobutene 173
$C_8H_6Br_2Cl_2$
α,α′-Dibromo-2,5-dichloro-*p*-xylene 170
$C_8H_6Br_2F_3NO_3S$
2,2,2-Trifluoroethyl 4-amino-3,5-dibromobenzenesulfonate 796
$C_8H_6Br_2O_2$
2,3-Dibromo-5,6-dimethyl-1,4-benzoquinone 54
2,5-Dibromo-3,6-dimethyl-1,4-benzoquinone 51
2,6-Dibromo-3,5-dimethyl-1,4-benzoquinone 54
$C_8H_6Br_2O_3$
3,5-Dibromo-4-methoxybenzoic acid 337
$C_8H_6Br_2O_4$
2,6-Dibromo-3,5-dimethoxy-*p*-benzoquinone 53
$C_8H_6Br_4$
α,α,α′,α′-Tetrabromo-*o*-xylene 169
α,α,α′,α′-Tetrabromo-*p*-xylene 169
$C_8H_6ClFN_2O_3$
2-Chloro-4′-fluoro-3′-nitroacetanilide 390
5′-Chloro-4′-fluoro-2′-nitroacetanilide 425
$C_8H_6ClF_2NO$
2-Chloro-2′,5′-difluoroacetanilide 391
$C_8H_6ClF_3O$
1-(4-Chlorophenyl)-2,2,2-trifluoroethanol 197
C_8H_6ClN
2-(Chloromethyl)benzonitrile 758
C_8H_6ClNO
2-Chloro-6-methoxybenzonitrile 758
4-Chlorophenoxyacetonitrile 751
3-Chloro-*p*-tolyl isocyanate 769
4-Chloro-*o*-tolyl isocyanate 769
5-Chloro-*o*-tolyl isocyanate 769
C_8H_6ClNOS
2-Chloro-5-methoxybenzothiazole 913
$C_8H_6ClNO_2$
4-Chloro-2-methoxyphenyl isocyanate 769
5-Chloro-2-methoxyphenyl isocyanate 769
$C_8H_6ClNO_3$
2′-Chloro-5′-nitroacetophenone 288
$C_8H_6ClNO_4$
4′-Chloro-2-hydroxyacetophenone, nitrate ester 299
Methyl 2-chloro-4-nitrobenzoate 374
Methyl 4-chloro-3-nitrobenzoate 366
C_8H_6ClNS
5-Chloro-2-methylbenzothiazole 917
2-(4-Chlorophenyl)acetonitrile 754
3-Chloro-*p*-tolyl isothiocyanate 768
$C_8H_6ClN_3$
5-Chloro-1-phenyl-1*H*-1,2,3-triazole 870
$C_8H_6ClN_3O$
1-(4-Chlorophenyl)-3-cyanourea 752
1-(4-Chlorophenyl)-1,2,4-triazol-3-ol 869
$C_8H_6ClN_3O_2$
2-(Chloromethyl)-5-nitrobenzimidazole 901
$C_8H_6ClN_3O_4$
N-(2-Chloroethylidene)-2,4-dinitroaniline 1020
$C_8H_6Cl_2Hg_2O_5$
Dimethyl 3,4-bis(chloromercuri)-2,5-furandicarboxylate 1138
$C_8H_6Cl_2N_2O_2$
2′,5′-Dichloroglyoxylanilide oxime 1010
1-(3,4-Dichlorophenyl)-3-formylurea 533
$C_8H_6Cl_2N_2O_3$
2,4′-Chloro-3′-nitroacetanilide 390
2,2′-Dichloro-4′-nitroacetanilide 390
2,4′-Dichloro-2′-nitroacetanilide 392
2,4′-Dichloro-3′-nitroacetanilide 390

$C_8H_6Cl_2N_2O_3$ (Cont'd)
2′,5′-Dichloro-4′-nitroacetanilide ... 428
2′,6′-Dichloro-4′-nitroacetanilide ... 428
4′,5′-Dichloro-2′-nitroacetanilide ... 425
$C_8H_6Cl_2N_2S$
5,6-Dichloro-1-methyl-2-benzimidazolethiol ... 906
$C_8H_6Cl_2O$
2,4-Dichlorostyrene oxide ... 189
$C_8H_6Cl_2O_2$
2,6-Dichloro-3,5-dimethyl-1,4-benzoquinone ... 318
3′,5′-Dichloro-4′-hydroxyacetophenone ... 288
2,6-Dichloro-3-methylbenzoic acid ... 346
2-(2,4-Dichlorophenoxy)-acetaldehyde ... 321
Methyl 3,4-dichlorobenzoate ... 376
Methyl 3,5-dichlorobenzoate ... 367
$C_8H_6Cl_2O_2S$
2,5-Diacetyl-3,4-dichlorothiophene ... 857
2,5-Dichloro-α-methyl-3-thiopheneacrylic acid ... 858
$C_8H_6Cl_2O_4$
2,6-Dichloro-3,5-dimethoxy-1,4-benzoquinone ... 52
$C_8H_6Cl_2PbS_2$
Bis(2-thienyl)dichlorolead ... 1135
$C_8H_6Cl_3NO$
2,2,2-Trichloroacetanilide ... 397
2,2′,3′-Trichloroacetanilide ... 390
2,2,3′-Trichloroacetanilide ... 394
2,2′,4′-Trichloroacetanilide ... 393
2,2′,5′-Trichloroacetanilide ... 393
2,3′,4′-Trichloroacetanilide ... 391
2′,3′,4′-Trichloroacetanilide ... 440
2,3′,5′-Trichloroacetanilide ... 391
2′,4′,5′-Trichloroacetanilide ... 441
2′,4′,6′-Trichloroacetanilide ... 410
$C_8H_6Cl_3NO_2$
2′-Hydroxy-2,3′,5′-trichloroacetanilide ... 390
Methyl *N*-(2,4,5-trichlorophenyl)-carbamate ... 687
$C_8H_6Cl_4$
α,α,α′,α′-Tetrachloro-*p*-xylene ... 169
2,3,5,6-Tetrachloro-*p*-xylene ... 170
$C_8H_6Cl_4OS$
1-((4-Chlorophenyl)thio)-2,2,2-trichloroethanol ... 209
$C_8H_6Cl_4O_2$
2,3,5,6-Tetrachloro-*p*-xylene-α,α′-diol ... 198
$C_8H_6F_2N_2O_3$
2′,4′-Difluoro-6′-nitroacetanilide ... 403
$C_8H_6F_3NO$
α,α,α-Trifluoro-*o*-formotoluidide ... 398
α,α,α-Trifluoro-*m*-formotoluidide ... 397
3-(Trifluoromethyl)benzaldoxime ... 1008
2-(Trifluoromethyl)benzamide ... 446
3-(Trifluoromethyl)benzamide ... 446
$C_8H_6F_3NO_2$
2-(Trifluoromethyl)-benzohydroxamic acid ... 446
$C_8H_6F_3NO_2S$
4-Methylthio-3-nitrobenzotrifluoride. 259
$C_8H_6F_3NO_3$
4-(Trifluoromethoxy)-benzohydroxamic acid ... 446
$C_8H_6F_3N_3O_4$
2,6-Dinitro-*N*-methyl-4-(trifluoromethyl)aniline ... 259
$C_8H_6F_3N_3O_6S$
2′,6′-Dinitro-4′-(trifluoromethyl)-methanesulfonanilide ... 805
$C_8H_6F_{12}O$
3,3,4,4,5,5,6,6,7,7,8,8-Dodecafluoro-2-octanol ... 15
$C_8H_6F_{14}Si$
Bis(1,1,2,2,3,3,3-heptafluoropropyl)-dimethylsilane ... 1100
$C_8H_6HgO_2$
Bis(2-furyl)mercury ... 1139
$C_8H_6I_2O_5$
Dimethyl 3,4-diiodo-2,5-furandicarboxylate ... 848
$C_8H_6I_3NO_2$
(2,4,6-Triiodophenoxy)-acetamide ... 415
$C_8H_6N_2$
Cinnoline picrate ... 1001
1,5-Naphthyridine ... 1000
1,8-Naphthyridine ... 1001
2,7-Naphthyridine ... 1000
$C_8H_6N_2O$
Benzoylcyanamide ... 757
2-Cyanobenzamide ... 767
N-Phenylcyanoformamide ... 767
$C_8H_6N_2OS$
2-Cyano-3-(2-thienyl)acrylamide ... 856
2-Formamidobenzothiazole ... 916
$C_8H_6N_2O_2$
5-Aminoisatin ... 701
1,4-Dihydroxyphthalazine ... 998
4-Hydroxy-1(2*H*)-phthalazinone, silver salt ... 999
2-Nitro-*p*-tolunitrile ... 764
2-(1,3,4-Oxadiazol-2-yl)-phenol ... 872
2-Phenyl-1,3,4-oxadiazol-2-in-5-one ... 873
2*H*-Pyrido(1,2-a)pyrimidine-2,4-(3*H*)-dione ... 700
$C_8H_6N_2O_2S$
2-Aminobenzothiazole-5-carboxylic acid ... 914
2-Methyl-6-nitrobenzothiazole ... 916
4-Nitrobenzyl thiocyanate ... 768
$C_8H_6N_2O_3$
2-Methyl-6-nitrobenzoxazole ... 910
6-Nitro-*o*-anisonitrile ... 762
2-Nitro-*p*-tolyl isocyanate ... 769
$C_8H_6N_2O_4$
3,4-Dihydro-7-nitro-2*H*-1,4-benzoxazin-3-one ... 703
1,3,6,8-Tetrahydroxy-2,7-naphthyridine ... 998
$C_8H_6N_2O_5$
2-Carbamoyl-6-nitrobenzoic acid ... 470
N-Formyl-4-nitroanthranilic acid ... 426
6-Nitro-1,3-benzodioxole-5-carboxaldehyde oxime ... 1009
$C_8H_6N_2O_6$
2,5-Dimethyl-3,6-dinitro-1,4-benzoquinone ... 53
6,8-Dinitro-1,3-benzodioxane ... 269
2-Hydroxy-4′-nitroacetophenone, nitrate ester ... 293
$C_8H_6N_2S$
4-Mercaptoquinazoline ... 999
4-Phenyl-1,2,3-thiadiazole ... 879
$C_8H_6N_2S_2$
3-Phenyl-1,2,4-thiadiazole-5(4*H*)-thione ... 749
$C_8H_6N_2S_3$
5-Mercapto-3-phenyl-1,3,4-thiadiazole-2-thione, potassium salt ... 747
$C_8H_6N_4$
2-Benzimidazolecarbamonitrile ... 905
1,2-Dicyano-1-phenylhydrazine ... 756
$C_8H_6N_4OS$
5-Nitrosoamino-3-phenyl-1,2,4-thiadiazole ... 879
$C_8H_6N_4O_2S$
6-Nitro-2-benzothiazolinone methylenehydrazone ... 915
$C_8H_6N_4O_6S$
N-Cyano-2,6-dinitro-4-(methylsulfonyl)aniline ... 777
$C_8H_6N_4O_8$
Alloxantin ... 124
$C_8H_6O_2S$
Thionaphthene-1,1-dioxide ... 898
$C_8H_6O_2S_2$
2,2′-Bithiophene-5-carboxylic acid ... 858
Dithioterephthalic acid ... 347
$C_8H_6O_3$
6-Hydroxy-3(2*H*)-benzofuranone ... 290
$C_8H_6O_3S$
2,3-Dihydrobenzo(b)thiophen-3-one 1,1-dioxide ... 781
4-Oxo-4-(2-thienyl)-2-butenoic acid ... 857
$C_8H_6O_4$
2-Hydroxybenzoylformic acid, sodium salt ... 350
Phthalic acid copper(II), salt ... 350
Terephthalic acid, copper(II) salt ... 1140
$C_8H_6O_4S$
2-Benzofuransulfonic acid ... 896
$C_8H_6O_6$
3,6-Dihydroxyphthalic acid ... 338
$C_8H_6O_7S$
Sulfophthalic acid, mixture of isomers ... 789
4-Sulfophthalic acid, sodium salt ... 789
4-Sulfophthalic acid, trisodium salt dihydrate ... 794
$C_8H_7BrClNO$
2′-Bromo-4′-chloroacetanilide ... 409
3′-Bromo-2-chloroacetanilide ... 390
4′-Bromo-2-chloroacetanilide ... 393
$C_8H_7BrClNO_3$
2-(2-Bromoethoxy)-5-chloro-1-nitrobenzene ... 258
C_8H_7BrINO
4′-Bromo-2-iodoacetanilide ... 390
$C_8H_7BrN_2O_3$
4′-Bromo-2′-nitroacetanilide ... 424
$C_8H_7BrO_2$
3-Bromo-4-methoxybenzaldehyde ... 322
2-(Bromomethyl)benzoic acid ... 336
Methyl 3-bromobenzoate ... 367
Methyl 4-bromobenzoate ... 375
$C_8H_7BrO_3$
3-Bromo-*p*-anisic acid ... 341
2-Bromo-4-hydroxy-3-methoxybenzaldehyde ... 321
3-Bromo-2-hydroxy-5-methoxybenzaldehyde ... 322
5-Bromo-2-methoxybenzoic acid ... 336
5-Bromo-3-methylsalicylic acid ... 337
2-(4-Bromophenoxy)acetic acid ... 325
4-Bromophenylglyoxal hydrate ... 288
Methyl 5-bromosalicylate ... 368
$C_8H_7Br_2NO$
2′,4′-Dibromoacetanilide ... 409
$C_8H_7Br_2NO_2$
1-(2,2-Dibromoethyl)-3-nitrobenzene. 259
2,4-Dibromophenyl *N*-methylcarbamate ... 637
$C_8H_7Br_2N_3OS$
3,5-Dibromosalicylaldehyde thiosemicarbazone ... 1022
$C_8H_7Br_3$
1-Bromo-4-(1,2-dibromoethyl)-benzene ... 169
C_8H_7ClFNO
2-Chloro-4′-fluoroacetanilide ... 390
3′-Chloro-4′-fluoroacetanilide ... 439
$C_8H_7ClF_3N_3O_4S$
4-(2-Chloro-1,1,2-trifluoroethylsulfonyl)-2-nitrophenylhydrazine ... 772
C_8H_7ClINO
2-Chloro-4′-iodoacetanilide ... 390
$C_8H_7ClN_2$
5-Chloro-2-methylbenzimidazole ... 907
2-(Chloromethyl)benzimidazole hydrochloride ... 901
$C_8H_7ClN_2O_2$
2′-Chloro-2-(hydroxyimino)-acetanilide ... 1010
1-(4-Chlorophenyl)-3-formylurea ... 533
$C_8H_7ClN_2O_3$
5′-Amino-4′-chloro-2′-nitroacetophenone ... 288
2-Chloro-2′-nitroacetanilide ... 392
2-Chloro-3′-nitroacetanilide ... 392
5′-Chloro-2′-nitroacetanilide ... 405
$C_8H_7ClN_2O_7S$
2-Chloroethyl 2,4-dinitrobenzenesulfonate ... 796
$C_8H_7ClN_4$
6-Chloro-4-hydrazinoquinazoline ... 1000
$C_8H_7ClN_4O_3S$
5-(4-Chloro-3-nitrophenyl)-2-thiobiuret ... 730
$C_8H_7ClO_2$
4′-Chloro-3′-hydroxyacetophenone ... 288
(+/-)-α-Chlorophenylacetic acid ... 332
Methyl 2-chlorobenzoate ... 375
$C_8H_7ClO_2S$
2-(4-Chlorophenylthio)acetic acid ... 330
$C_8H_7ClO_3$
5-Acetyl-3-chloro-6-methyl-2*H*-pyran-2-one ... 96
2-Chloro-*p*-anisic acid ... 341
4-Chloromandelic acid ... 326
Methyl 5-chlorosalicylate ... 368
$C_8H_7ClO_5S$
3-(Chlorosulfonyl)-*p*-anisic acid ... 802
$C_8H_7Cl_2NO$
2,2′-Dichloroacetanilide ... 393
2,3′-Dichloroacetanilide ... 393
2,4′-Dichloroacetanilide ... 393
2′,5′-Dichloroacetanilide ... 439
3′,4′-Dichloroacetanilide ... 439
$C_8H_7Cl_2NO_2$
1,3-Dichloro-6-ethyl-4-nitrobenzene .. 262
2,3-Dichlorophenyl *N*-methylcarbamate ... 635
2,4-Dichlorophenyl *N*-methylcarbamate ... 637
3,4-Dichlorophenyl *N*-methylcarbamate ... 637
Methyl *N*-(2,3-dichlorophenyl)-carbamate ... 683
Methyl *N*-(2,4-dichlorophenyl)-carbamate ... 685
Methyl *N*-(2,5-dichlorophenyl)-carbamate ... 685
Methyl *N*-(2,6-dichlorophenyl)-carbamate ... 683
Methyl *N*-(3,4-dichlorophenyl)-carbamate ... 685
$C_8H_7Cl_2NO_3$
Methyl *N*-(3,4-dichlorophenyl)-*N*-hydroxycarbamate ... 628
$C_8H_7Cl_2NO_4$
1,5-Dichloro-2,4-dimethoxy-3-nitrobenzene ... 265
$C_8H_7Cl_2N_3S$
2,4-Dichlorobenzaldehyde thiosemicarbazone ... 730
2,6-Dichlorobenzaldehyde thiosemicarbazone ... 1026
$C_8H_7Cl_3N_2O$
N-(3-Methyl-2-pyridyl)-2,2,2-trichloroacetamide ... 926
N-(4-Methyl-2-pyridyl)-2,2,2-trichloroacetamide ... 926
1-Methyl-3-(2,4,5-trichlorophenyl)urea ... 546
N-(3-Pyridylmethyl)-2,2,2-trichloroacetamide ... 926
$C_8H_7Cl_3N_2O_3S$
3′-Sulfamoyl-2,2,2-trichloroacetanilide ... 805
$C_8H_7Cl_3N_2S$
1-Methyl-3-(2,4,5-trichlorophenyl)-2-thiourea ... 733
$C_8H_7Cl_3N_4O$
1-((3-Hydroxy-2,4,6-trichlorobenzylidene)amino)-guanidine acetate ... 1022
$C_8H_7Cl_3N_4OS$
1-(2,4,5-Trichlorophenyl)-5-thiobiurea ... 730
$C_8H_7Cl_3O$
1-(3-Chlorophenyl)-2,2-dichloroethanol ... 197
3,5-Dimethyl-2,4,6-trichlorophenol ... 190
$C_8H_7Cl_3O_3S$
2,2,2-Trichloroethyl benzenesulfonate ... 796
$C_8H_7Cl_5Si$
Dimethyl(pentachlorophenyl)silane. 1100
$C_8H_7FN_2O_3$
2′-Fluoro-4′-nitroacetanilide ... 404
$C_8H_7FO_2$
Methyl 2-fluorobenzoate ... 375
$C_8H_7FO_3$
4-Fluoromandelic acid ... 326
$C_8H_7FO_4S$
Methyl 2-(fluorosulfonyl)benzoate ... 804
$C_8H_7FO_5S$
3-(Fluorosulfonyl)-*p*-anisic acid ... 802
5-Fluorosulfonyl-*o*-anisic acid ... 802
2-(4-(Fluorosulfonyl)phenoxy)-acetic acid ... 802
$C_8H_7F_2NO$
2′,4′-Difluoroacetanilide ... 409

$C_8H_7F_2NO$ (Cont'd)
2′,5′-Difluoroacetanilide.................... 409
3′,4′-Difluoroacetanilide.................... 439
$C_8H_7F_2NO_2$
1-(1,1-Difluoroethyl)-4-nitrobenzene.. 278
$C_8H_7F_3N_2O_2$
1-Hydroxy-3-(3-trifluoromethylphenyl)urea.............. 509
N-Methyl-4-nitro-2-(trifluoromethyl)-aniline.. 259
$C_8H_7F_3N_2O_4S$
2′-Nitro-4′-(trifluoromethyl)-methanesulfonanilide...................... 805
$C_8H_7F_3N_2S$
1-(3-(Trifluoromethyl)phenyl)-2-thiourea.. 728
$C_8H_7HgNO_5$
4-Hydroxy-3-nitrophenylmercuri acetate... 350
$C_8H_7IN_2O_2$
2-(Hydroxyimino)-*N*-(2-iodophenyl)-acetamide... 1008
$C_8H_7IN_2O_3$
3′-Iodo-5′-nitroacetanilide................. 404
C_8H_7IO
2′-Iodoacetophenone........................ 289
$C_8H_7IO_2$
2-(4-Iodophenyl)acetic acid............... 325
$C_8H_7IO_3$
Methyl 5-iodosalicylate..................... 366
$C_8H_7IS_3$
(3*H*-1,2-Benzodithiol-3-ylidene)-methylsulfonium iodide.................. 372
$C_8H_7I_3O_2$
2-(2,4,6-Triiodophenoxy)-ethanol....... 202
$C_8H_7I_3O_4S$
2-(2,4,6-Triiodophenoxy)-ethanesulfonic acid.......................... 788
C_8H_7NOS
4-Hydroxy-*m*-tolyl thiocyanate.......... 768
2-Methoxybenzothiazole.................... 917
3-Methyl-2-benzothiazolinone........... 743
6-Methyl-2-benzoxazolethiol............. 909
2-(Methylthio)benzoxazole................ 909
4-Oxo-4-(2-thienyl)butyronitrile.......... 857
$C_8H_7NO_2$
6-Aminophthalide.............................. 384
2*H*-1,4-Benzoxazin-3(4*H*)-one............ 690
3-(4-Pyridyl)acrylic acid..................... 962
$C_8H_7NO_2S_2$
3-Furfurylrhodanine.......................... 852
$C_8H_7NO_3$
α-Cyanofurfuryl acetate.................... 851
4-Formamidobenzoic acid.................. 433
N-Formylanthranilic acid.................... 433
2-(4-Nitrophenyl)ethenol.................... 265
4-(2-Nitrovinyl)phenol........................ 257
Oxanilic acid...................................... 431
Phthalamic acid................................. 475
Piperonal oxime................................ 1013
$C_8H_7NO_3S$
2-Cyano-α-toluenesulfonic acid, sodium salt.................................... 795
S-(4-Nitrophenyl) thioacetate............ 379
$C_8H_7NO_3S_3$
2-Methylthio-5-benzothiazolesulfonic acid............. 914
$C_8H_7NO_4$
2-Methoxy-3-nitrobenzaldehyde........ 323
4-Methoxy-2-nitrobenzaldehyde........ 323
4-Methoxy-3-nitrobenzaldehyde........ 323
6-Nitro-1,3-benzodioxan.................... 279
6-Nitro-1,4-benzodioxan.................... 268
$C_8H_7NO_4S$
4-Methylthio-3-nitrobenzoic acid...... 341
2-(4-Nitrophenylthio)acetic acid........ 329
$C_8H_7NO_5$
3-Hydroxy-4-methoxy-2-nitrobenzaldehyde............................ 322
4-Hydroxy-3-methoxy-2-nitrobenzaldehyde............................ 322
3-Hydroxy-4-methoxy-6-nitrobenzaldehyde, sodium salt...... 322
5-Hydroxy-2-methyl-4-nitrobenzoic acid.............................. 338
4-Hydroxy-2-nitro-*m*-anisaldehyde.... 322
Methyl 3-hydroxy-4-nitrobenzoate.... 368
Methyl 4-hydroxy-3-nitrobenzoate.... 368
Methyl 4-hydroxy-3-nitrobenzoate, sodium salt.................................... 368
Methyl 5-nitrosalicylate..................... 368
2-(2-Nitrophenoxy)acetic acid........... 328
6-Nitrovanillin.................................... 322
$C_8H_7NO_6S$
4-Methylsulfonyl-3-nitrobenzoic acid.............................. 779
$C_8H_7NO_7S$
α-Carboxy-3-nitro-*p*-toluenesulfonic acid sodium salt.... 794
3-(Methylsulfonyloxy)-4-nitrobenzoic acid.............................. 799
C_8H_7NS
2-(Phenylthio)acetonitrile.................. 754
$C_8H_7NS_2$
2-Methyl-1,2-benzisothiazole-3(2*H*)-thione.. 746
$C_8H_7N_3$
4-Aminoquinazoline.......................... 1000
2-Aminoquinoxaline.......................... 1000
3-Phenyl-1*H*-1,2,4-triazole................. 869
$C_8H_7N_3O$
2-Amino-5-phenyl-1,3,4-oxadiazole... 873
4-Cyano-*N*-methyl-*N*-nitrosoaniline.. 765
1-Cyano-3-phenylurea....................... 758
Isatin 3-hydrazone............................ 1070
1-Methyl-4-cyanoformylpyridinium oximate... 944
1-Phenyl-1,2,4-triazol-3-ol.................. 869
$C_8H_7N_3OS$
5-Anilino-3-hydroxy-1,2,4-thiadiazole... 878
$C_8H_7N_3O_2$
5-Methyl-6-nitrobenzimidazole.......... 905
6-Methylpyrido(3,2-d)pyrimidine-2,4(1*H*,3*H*)-dione............................... 1003
4-Phenyl-1,2,4-triazolidine-3,5-dione. 870
$C_8H_7N_3O_2S$
6-Amino-2-methyl-7-nitrobenzothiazole........................... 914
2-Imino-3-methyl-6-nitrobenzothiazoline......................... 744
$C_8H_7N_3O_4$
5,7-Dinitroindoline............................. 274
1-(3-Nitrobenzoyl)urea....................... 567
$C_8H_7N_3O_5$
2′,4′-Dinitroacetanilide...................... 423
2,6-Dinitro-3-methylbenzaldehyde oxime... 1010
$C_8H_7N_3O_6$
(2,4-Dinitrophenyl)glycine.................. 348
$C_8H_7N_3S$
3-Amino-5-phenyl-1,2,4-thiadiazole... 878
2-Benzothiazolecarboxamidine......... 914
3-Phenyl-1*H*-1,2,4-triazole-5(4*H*)-thione.. 743
$C_8H_7N_5$
4,5-Dihydro(1,2,4)triazolo(3,4-c)-(1,2,4)benzotriazine........................... 979
$C_8H_7N_5O$
5-(2-Hydroxybenzylideneamino)-1*H*-tetrazole.. 1038
$C_8H_7N_5O_2$
1-(Azidocarbonyl)-2-benzoylhydrazine............................. 769
2,4-Diamino-6-nitroquinazoline......... 1000
2,4-Diamino-6-nitroquinazoline acetate... 999
C_8H_8AsI
2-Iodoisoarsindoline.......................... 1128
$C_8H_8AsIO_2$
4-(Iodomethylarsino)benzoic acid..... 1127
C_8H_8BrClO
2-Bromoethyl 4-chlorophenyl ether.. 176
$C_8H_8BrFO_2S$
4-(1-Bromoethyl)benzenesulfonyl fluoride... 804
C_8H_8BrNO
2′-Bromoacetanilide........................... 408
4′-Bromo-*o*-formotoluidide................ 439
$C_8H_8BrNO_2$
1-Bromo-2,4-dimethyl-3-nitrobenzene...................................... 262
N-(2-Bromophenyl)glycine................. 348
4-Bromophenyl *N*-methylcarbamate 634
Methyl 5-amino-2-bromobenzoate.... 366
Methyl *N*-(4-bromophenyl)-carbamate.. 634
$C_8H_8Br_2$
1-Bromo-2-(2-bromoethyl)benzene.... 169
4,5-Dibromo-*o*-xylene......................... 169
4,6-Dibromo-*m*-xylene........................ 169
$C_8H_8Br_2O$
2-Bromoethyl 4-bromphenyl ether.... 176
2,4-Dibromo-5-methylanisole............ 182
$C_8H_8Br_2O_2$
1,2-Dibromo-4,5-dimethoxybenzene. 177
1,4-Dibromo-2,5-dimethoxybenzene. 182
C_8H_8ClF
1-(2-Chloroethyl)-4-fluorobenzene..... 170
$C_8H_8ClHgNO$
4′-(Chloromercuri)acetanilide............ 1138
$C_8H_8ClIO_2$
5-Chloro-2,4-dimethoxyiodobenzene.................... 182
C_8H_8ClNO
3′-Chloro-*o*-formotoluidide................. 439
4′-Chloro-*o*-formotoluidide................. 439
5′-Chloro-*o*-formotoluidide................. 439
6′-Chloro-*o*-formotoluidide................. 439
2-Chloro-*N*-methylbenzamide........... 457
3-Chloro-4-methylbenzamide............ 456
3′-Chloro-4′-methylformanilide.......... 409
C_8H_8ClNOS
S-(4-Chlorophenyl) *N*-methylthiocarbamate.................... 641
2-(4-Chlorophenylthio)acetamide...... 421
$C_8H_8ClNO_2$
1-(1-Chloroethyl)-4-nitrobenzene....... 278
1-Chloro-2-ethyl-5-nitrobenzene........ 262
2-Chloro-2′-hydroxyacetanilide.......... 391
3′-Chloro-2′-methoxyformanilide....... 417
5′-Chloro-2′-methoxyformanilide....... 401
2-Chloro-5-nitro-*p*-xylene................... 280
Methyl *N*-(2-chlorophenyl)-carbamate.. 681
Methyl *N*-(3-chlorophenyl)-carbamate.. 677
Methyl *N*-(4-chlorophenyl)-carbamate.. 682
3-Methyl-4-nitrobenzyl chloride........ 258
$C_8H_8ClNO_2S$
2-Chloroethyl 4-nitrophenyl sulfide.. 258
$C_8H_8ClNO_3$
Methyl *N*-(3-chlorophenyl)-*N*-hydroxycarbamate........................... 627
$C_8H_8ClNO_3S$
2-(4-Chlorophenylsulfonyl)-acetamide.. 782
$C_8H_8ClNO_4$
4-Chloro-2,5-dimethoxynitrobenzene.................... 268
5-Chloro-2,4-dimethoxynitrobenzene.................... 268
$C_8H_8ClNO_4S$
3-Acetamido-4-chlorobenzenesulfonic acid, potassium salt................................. 795
2-Chloro-5-(methylsulfamoyl)-benzoic acid..................................... 823
3-Methylthiazolo(3,2-a)pyridinium perchlorate....................................... 919
$C_8H_8ClNO_6S$
2,5-Dimethoxy-4-nitrobenzenesulfonyl chloride......... 802
$C_8H_8ClNS_2$
3-Chloro-2-methyldithiocarbanilic acid, ammonium salt........................ 750
$C_8H_8ClN_3O$
3-Carbamoyl-1-(cyanomethyl)-pyridinium chloride.......................... 947
$C_8H_8ClN_3OS$
5-(3-Chlorophenyl)-2-thiobiuret.......... 732
$C_8H_8ClN_3O_3$
Acetic 2-(4-chloro-2-nitrophenyl)-hydrazide.. 494
1-(2-Chloro-4-nitrophenyl)-3-methylurea...................................... 568
$C_8H_8ClN_3S$
2-Chlorobenzaldehyde thiosemicarbazone........................... 1026
$C_8H_8ClN_5$
5-Chloro-2-guanidinobenzimidazole. 904
$C_8H_8Cl_2Hg_2O_3$
Ethyl 4,5-bis(chloromercuri)-2-methyl-3-furoate............................... 853
$C_8H_8Cl_2N_2O$
4′-Amino-2′,6′-dichloroacetanilide.... 403
1-(2,3-Dichlorophenyl)-3-methylurea. 592
1-(2,4-Dichlorophenyl)-3-methylurea. 597
1-(2,5-Dichlorophenyl)-3-methylurea. 597
1-(2,6-Dichlorophenyl)-3-methylurea. 540
1-(3,4-Dichlorophenyl)-3-methylurea. 597
1-(3,5-Dichlorophenyl)-3-methylurea. 545
$C_8H_8Cl_2N_2O_2$
4,5-Dichloro-*N*-ethyl-2-nitroaniline.... 274
$C_8H_8Cl_2N_2O_3$
2-(4,5-Dichloro-2-nitroanilino)-ethanol.. 259
$C_8H_8Cl_2N_2O_3S$
1-(3,4-Dichlorophenyl)-3-(methylsulfonyl)urea......................... 809
$C_8H_8Cl_2N_4$
N-(2,6-Dichlorobenzylideneamino)-guanidine.. 1025
1-((3,4-Dichlorobenzylidene)amino)-guanidine acetate............................ 1026
$C_8H_8Cl_2N_4O$
1-((3,4-Dichlorobenzoyl)amino)-guanidine acetate............................ 724
$C_8H_8Cl_2N_4OS$
1-(3,4-Dichlorophenyl)-5-thiobiurea... 730
$C_8H_8Cl_2N_4O_2$
8-Chloro-7-(chloromethyl)-1,3-dimethylxanthine.............................. 920
$C_8H_8Cl_2O$
2,4-Dichloro-3,5-dimethylphenol....... 190
2,5-Dichloro-α-methylbenzyl alcohol... 199
4-Dichloromethyl-4-methyl-2,5-cyclohexadien-1-one....................... 51
3,4-Dichlorophenyl methyl carbinol.. 198
$C_8H_8Cl_2O_2$
2-(2,4-Dichlorophenoxy)-ethanol....... 202
2-(2,6-Dichlorophenoxy)ethanol........ 202
$C_8H_8Cl_2O_2S$
2-Chloroethyl 4-chlorophenyl sulfone... 785
$C_8H_8Cl_2O_4S_2$
2,5-Dichlorophenyl methylsulfonylmethyl sulfone......... 786
3,4-Dichlorophenyl methylsulfonylmethyl sulfone......... 774
$C_8H_8Cl_2S$
2-Chloroethyl 4-chlorophenyl sulfide... 209
$C_8H_8Cl_2S_2$
2,5-Dichlorophenyl methylthiomethyl sulfide................ 214
3,4-Dichlorophenyl methylthiomethyl sulfide................ 214
$C_8H_8Cl_3NO$
1-(2-Pyridyl)-3,3,3-trichloro-2-propanol.. 925
$C_8H_8Cl_4N_2$
2,3,5,6-Tetrachloro-*p*-xylylenediamine................................ 236
$C_8H_8Cl_6N_2O_2$
1,4-Bis(trichloroacetyl)piperazine..... 110
C_8H_8FN
6-Fluoroindoline................................ 250
C_8H_8FNO
4-Fluoro-*N*-methylbenzamide........... 458
$C_8H_8FNO_2$
4-Fluorophenyl *N*-methylcarbamate. 634
$C_8H_8FNO_3S$
N-Acetylsulfanilyl fluoride................ 804
$C_8H_8FNO_4S$
2,4-Dimethyl-5-nitrobenzenesulfonyl fluoride.......... 803
$C_8H_8FN_3OS$
5-(4-Fluorophenyl)-2-thiobiuret.......... 732
$C_8H_8FN_3O_2S_2$
5-Fluorosulfonyl-3-methyl-2-benzothiazolinone hydrazone hydrochloride................................... 803
$C_8H_8F_2N_2O$
4′-Amino-3′,5′-difluoroacetanilide..... 403
$C_8H_8F_2N_4OS$
4-(2,4-Difluorophenyl)-1-(thiocarbamoyl)semicarbazide......... 730

$C_8H_8F_3N_3$
1-(3-(Trifluoromethyl)phenyl)-guanidine carbonate 723
$C_8H_8F_3N_3O_2$
3-(Acrylamido)-1-(2,2,2-trifluoroethyl)-2-pyrazolin-5-one 130
$C_8H_8F_4Si$
Dimethyl(2,3,5,6-tetrafluorophenyl)-silane 1100
$C_8H_8HgINO_2$
2-Amino-5-iodophenylmercuri acetate 1137
C_8H_8INO
2′-Iodoacetanilide 408
4′-Iodoacetanilide 408
$C_8H_8INO_2$
Methyl 4-amino-3-iodobenzoate 366
$C_8H_8INO_4$
1,2-Dimethoxy-4-iodo-5-nitrobenzene 260
$C_8H_8IN_3O$
4-(Cyano-(hydroximino)-methyl)-1-methylpyridinium iodide 943
$C_8H_8I_2O_2$
1,2-Diiodo-4,5-dimethoxybenzene 177
$C_8H_8N_2$
2-Aminobenzyl cyanide 754
2-Methylbenzimidazole hydrochloride 905
2-Methyl-2*H*-cyclopenta(d)-pyridazine 1003
$C_8H_8N_2O$
2-Benzimidazolemethanol 902
4-Cyano-2,3-lutidine-*N*-oxide 1006
1,2-Dihydro-4(3*H*)cinnolinone hydrochloride 290
$C_8H_8N_2OS$
4-Amino-3-methoxyphenyl thiocyanate hydrochloride 768
$C_8H_8N_2O_2$
Benzoylurea 577
5-Cyano-2,6-dihydroxy-3,4-dimethylpyridine 933
1-Formyl-1-phenylurea 579
Phthalic diamide 470
3-(4-Pyridyl)acrylohydroxamic acid hydrochloride monohydrate 960
Terephthalamide 477
$C_8H_8N_2O_3$
2′-Amino-4′-nitroacetophenone 294
2′-Amino-5′-nitroacetophenone 294
5′-Amino-2′-nitroacetophenone 294
2-Amino-4′-nitroacetophenone hydrochloride 294
1,3-Benzodioxol-5-ylurea 581
3-Methyl-4-nitrobenzamide 472
N-Nicotinoylglycine 962
2′-Nitro-*p*-formotoluidide 432
3′-Nitro-*p*-formotoluidide 426
4′-Nitro-*o*-formotoluidide 428
5′-Nitro-*o*-formotoluidide 426
4-Ureidobenzoic acid 576
$C_8H_8N_2O_3S_2$
2-Methylthiobenzimidazole-sulfonic acid 903
$C_8H_8N_2O_4$
2-Acetamido-4-nitrophenol, sodium salt 412
1,1′-Bisuccinimide(*N*,N′-bisuccinimidyl) 125
4,6-Dinitro-*m*-xylene 276
5-(*N*-(Hydroxymethyl)carbamoyl)-nicotinic acid, sodium salt 933
2′-Hydroxy-3′-nitroacetanilide 412
2′-Methoxy-4′-nitroformanilide 415
2′-Methoxy-5′-nitroformanilide 415
Methyl 4-nitroanthranilate 372
N-Methyl-4-nitrobenzohydroxamic acid 472
Methyl *N*-(4-nitrophenyl)carbamate 670
$C_8H_8N_2O_5$
2,4-Dinitrophenetole 267
2,4-Dinitrophenyl ethyl ether 268
$C_8H_8N_2O_6$
1,4-Dimethoxy-2,3-dinitrobenzene 267
4,5-Dinitroveratrole 266
$C_8H_8N_2O_6S$
4-Acetamido-3-nitrobenzenesulfonic acid, sodium salt 793
2-(Methylsulfonamido)-4-nitrobenzoic acid 818
$C_8H_8N_2S$
5-Amino-3-methyl-1,2-benzisothiazole 914
6-Amino-2-methylbenzothiazole 914
$C_8H_8N_2S_2$
6-Amino-2-(methylthio)-benzothiazole 913
$C_8H_8N_2Se$
5,6-Dimethyl-2,1,3-benzoselenadiazole 213
$C_8H_8N_4$
3-Amino-7-methyl-1,2,4-benzotriazine 1003
2-(2-Aminophenyl)-1*H*-1,3,4-triazole 869
3-Amino-5-phenyl-1*H*-1,2,4-triazole 869
2-Guanylbenzimidazole 725
5-Methyl-1-phenyl-1*H*-tetrazole 870
$C_8H_8N_4O$
3-Amino-7-methyl-1,2,4-benzotriazine-1-oxide 1005
N-(Azidocarbonyl)-4-methylaniline 770
1-(2-Benzoxazolyl)-guanidine 724
4-Hydroxy-6,7-dimethylpteridine 1002
$C_8H_8N_4OS$
4,5-Dihydro-4-(hydroxymethyl)-1-phenyl-1*H*-tetrazole-5-thione 733
6,7-Dimethyl-4-hydroxy-2-mercaptopteridine 1002
2-Ethylthio-4-hydroxypteridine 1002
$C_8H_8N_4O_2$
6,7-Dimethyllumazine 1002
$C_8H_8N_4O_2S$
2-Nitrobenzaldehyde thiosemicarbazone 1056
$C_8H_8N_4O_3$
1,3-Dimethyl-2,4,6(1*H*,3*H*,5*H*)-pteridinetrione 978
Ethyl 7-hydroxy-*s*-triazolo(1,5-a)-pyrimidine-6-carboxylate 917
2-Nitrobenzaldehyde semicarbazone 1056
3-Nitrobenzaldehyde semicarbazone 1058
4-Nitrobenzaldehyde semicarbazone 1069
$C_8H_8N_4O_3S$
4-(*p*-Tosyl)-1*H*-tetrazol-5(4*H*)-one 826
$C_8H_8N_4O_4$
1-(4-Nitrobenzoyl)semicarbazide 575
$C_8H_8N_4O_5$
Acetic 2-(2,4-dinitrophenyl)-hydrazide 500
$C_8H_8N_4S$
3-Amino-5-anilino-1,2,4-thiadiazole 878
5-Amino-3-anilino-1,2,4-thiadiazole 878
5-Amino-3-anilino-1,2,4-thiadiazole hydrochloride 878
2-Amino-5-(3-pyridylmethyl)-1,3,4-thiadiazole 942
(2-Benzothiazolyl)-guanidine 917
$C_8H_8N_6$
2,4-Diamino-6-(4-pyridyl)-1,3,5-triazine 979
$C_8H_8N_6O_2$
2-Guanidino-5-nitrobenzimidazole 904
$C_8H_8N_8$
5-Methyl-7-(1*H*-1,2,4-triazol-5-ylamino)-*s*-triazolo(1,5-a)pyrimidine. 919
C_8H_8O
2,3-Dihydrobenzofuran picrate 184
C_8H_8OS
S-Methyl thiobenzoate 376
$C_8H_8O_2$
Benzyl formate 356
2,5-Dimethylbenzoquinone 54
Furfuralacetone 853
2-Phenylacetic acid, octadecylamine salt 351
Phenylacetic acid, piperazine salt 351
1-Phthalanol 206
o-Toluic acid, zone refined 345
$C_8H_8O_2S$
α-Mercapto-*p*-toluic acid 341
3-(Methylthio)benzoic acid 341
α-Methyl-2-thiopheneacrylic acid 858
$C_8H_8O_3$
Allyl furoate 848
3-Cyclohexene-1,2-dicarboxylic anhydride 99
2,5-Diacetylfuran 847
2,4-Dihydroxy-3-methylbenzaldehyde 322
2,4-Dihydroxy-6-methylbenzaldehyde 322
2,5-Dimethyl-3-hydroxy-1,4-benzoquinone 52
2-Methoxy-5-methyl-1,4-benzoquinone 52
6-Methoxysalicylaldehyde 338
α-Methyl-2-furanacrylic acid 853
$C_8H_8O_3S$
5-(Methylthio)-salicylic acid 338
β-Styrenesulfonic acid, ammonium salt 795
3-(2-Thenoyl)-propionic acid 857
$C_8H_8O_4$
2,4-Dihydroxy-3-methylbenzoic acid. 338
4,5-Dihydroxy-2-methylbenzoic acid. 338
2,6-Dimethoxy-*p*-benzoquinone 52
4,6-Dimethyl-2-oxo-2*H*-pyran-5-carboxylic acid 940
3,6-Endoxohexahydrophthalic anhydride 99
3-Hydroxymandelic acid 325
2-(2-Hydroxyphenoxy)acetic acid 326
2-(4-Hydroxyphenoxy)acetic acid 326
Isodehydracetic acid 96
Methyl 2,6-dihydroxybenzoate 367
γ-Oxo-2-furanbutyric acid 847
Salicylic acid compound with formaldehyde 338
2,2′,4′-Trihydroxyacetophenone 291
$C_8H_8O_4S$
2-(Phenylsulfonyl)-acetic acid 781
$C_8H_8O_5$
2,3-Diacetylsuccinic anhydride 98
Dimethyl 2,5-furandicarboxylate 853
2,5-Dimethyl-3,4-furandicarboxylic acid 853
3,5-Dimethyl-2,4-furandicarboxylic acid 848
Methyl 2,4,6-trihydroxybenzoate 367
7-Oxabicyclo(2.2.1)hept-5-ene-2,3-dicarboxylic acid 67
$C_8H_8O_5S$
2-(4-Sulfophenyl)acetic acid, barium salt 790
α-Sulfophenylacetic acid, pyridine salt 795
α-Sulfophenylacetic acid, sodium salt 795
$C_8H_8O_5S_2$
S-Benzyloxycarbonylthiosulfonic acid sodium salt 793
$C_8H_8O_7$
3,4-Dimethoxy-2,5-furandicarboxylic acid 845
Dimethyl 3,4-dihydroxy-2,5-furandicarboxylate, disodium salt 845
$C_8H_9AsClNO_4$
N-Chloroacetylarsanilic acid 1126
$C_8H_9AsO_6$
3-(Methoxycarbonyloxy)-benzenearsonic acid 1126
$C_8H_9BrN_2O$
1-(3-Bromophenyl)-3-methylurea 534
1-(4-Bromophenyl)-3-methylurea 538
$C_8H_9BrN_2O_2$
3-Bromo-2,4-dimethyl-6-nitroaniline. 260
$C_8H_9BrN_2S$
1-(3-Bromophenyl)-3-methyl-2-thiourea 732
$C_8H_9BrN_4$
1-((4-Bromobenzylidene)amino)-guanidine acetate 1026
$C_8H_9BrN_4O_2$
8-Bromocaffeine 924
$C_8H_9BrN_4S$
1-Amidino-3-(4-bromophenyl)-2-thiourea 730
C_8H_9BrO
4-Bromo-2,5-dimethylphenol 190
2-Bromo-4-methylanisole 177
2-Bromophenethyl alcohol 198
2-Bromophenyl methyl carbinol 207
3-Bromophenyl methyl carbinol 199
$C_8H_9BrO_2$
3-Bromo-4-hydroxyphenethyl alcohol 198
6-Bromo-4-methoxy-*m*-cresol 191
5-Bromo-2-methoxy-4-methylphenol. 190
$C_8H_9BrO_2S$
2-Bromoethyl phenyl sulfone 772
Bromomethyl *p*-tolyl sulfone 772
$C_8H_9BrO_3$
3-Bromo-4-hydroxy-5-methoxybenzyl alcohol 197
$C_8H_9BrO_3S$
2-Bromoethyl benzenesulfonate 796
C_8H_9BrS
2-Bromobenzyl methyl sulfide 210
$C_8H_9Br_2N$
2,6-Dibromo-4-ethylaniline 216
$C_8H_9Br_3OTe$
(4-Ethoxyphenyl)tribromotellurium 1141
C_8H_9Cl
(+/-)-(1-Chloroethyl)benzene 169
C_8H_9ClHg
3,4-Xylylmercuric chloride 1139
C_8H_9ClHgO
β-(Chloromercuri)-*sec*-phenethylalcohol 1137
$C_8H_9ClHgO_2$
2,4-Dimethoxyphenylmercuric chloride 1138
$C_8H_9ClN_2O$
4′-Amino-2′-chloroacetanilide 403
5-Chloro-2-(methylamino)-benzamide 450
4-Chloro-2-methylphenylurea 597
4-Chlorophenylacetic hydrazide 494
1-(2-Chlorophenyl)-3-methylurea 587
1-(3-Chlorophenyl)-3-methylurea 580
1-(4-Chlorophenyl)-3-methylurea 589
1-(3-Chloro-*p*-tolyl)-urea 577
1-(5-Chloro-*o*-tolyl)urea 597
1-(6-Chloro-*o*-tolyl)urea 540
$C_8H_9ClN_2O_2$
2-Chloro-*N*,*N*-dimethyl-4-nitroaniline 261
4-Chloro-*N*-ethyl-2-nitroaniline 261
5-Chloro-4-ethyl-2-nitroaniline 260
1-(3-Chloro-4-methylphenyl)-3-hydroxyurea 529
Methyl 4-chloro-3-hydrazinobenzoate 373
$C_8H_9ClN_2O_3S$
1-(4-Chlorophenyl)-3-(methylsulfonyl)urea 808
2-Chloro-4′-sulfamoylacetanilide 805
$C_8H_9ClN_2O_4S$
4-Chloro-*N*,*N*-dimethyl-3-nitrobenzenesulfonamide 807
4-Dimethylamino-3-nitrobenzenesulfonyl chloride 803
$C_8H_9ClN_4OS$
1-(2-Chlorophenyl)-5-thiobiurea 730
1-(3-Chlorophenyl)-5-thiobiurea 730
$C_8H_9ClN_4O_2$
8-Chlorocaffeine 924
7-(Chloromethyl)-1,3-dimethylxanthine 920
C_8H_9ClO
3-Chloro-4-methylanisole 182
4-Chloro-3-methylanisole 182
β-Chlorophenetole 176
3-Chlorophenyl methyl carbinol 198
$C_8H_9ClO_2$
2-(3-Chloro-4-hydroxyphenyl)ethyl alcohol 198
5-Chloro-2-methoxybenzyl alcohol 201
2-Chloro-5-norbornene-2-carboxylic acid 66
2-(*p*-Chlorophenoxy)-ethanol 201
$C_8H_9ClO_2S$
2-Chloroethyl phenyl sulfone 772
2,5-Dimethylbenzenesulfonyl chloride 804

$C_8H_9ClO_3$
3-Chloro-4-hydroxy-5-methoxybenzyl alcohol ... 199
$C_8H_9ClO_4S$
2,5-Dimethoxybenzenesulfonyl chloride ... 802
$C_8H_9ClO_4S_2$
2,4-Bis-(methylsulfonyl)-1-chlorobenzene ... 786
(4-Chlorophenylsulfonyl)-(methylsulfonyl)methane ... 785
5-Methylsulfonyl-*o*-toluenesulfonyl chloride ... 804
C_8H_9ClS
4-Chlorobenzyl methyl sulfide ... 212
2-Chloroethyl phenyl sulfide ... 209
$C_8H_9ClS_2$
2,4-Bis-(methylthio)-1-chlorobenzene ... 213
1-Chloro-4-(methylthiomethylthio)-benzene ... 212
$C_8H_9Cl_2N$
3,5-Dichloro-*N*,*N*-dimethylaniline ... 216
$C_8H_9Cl_2NO$
2,6-Dichloro-4-dimethylaminophenol hydrochloride ... 217
$C_8H_9Cl_2NO_2S$
3,4-Dichloro-*N*,*N*-dimethylbenzenesulfonamide ... 829
$C_8H_9Cl_3N_2S$
2-(2,4-Dichlorobenzyl)-2-thiopseudourea hydrochloride ... 738
$C_8H_9Cl_3OTe$
(4-Ethoxyphenyl)-trichlorotellurium ... 211
$C_8H_9Cl_3Si$
Dimethyl(2,3,6-trichlorophenyl)-silane ... 1104
Dimethyl(2,4,6-trichlorophenyl)-silane ... 1104
$C_8H_9Cl_4NSi$
2,3,5,6-Tetrachloro-4-(trimethylsilyl)pyridine ... 1123
$C_8H_9FN_2O$
1-(4-Fluorophenyl)-3-methylurea ... 538
$C_8H_9FN_2O_3$
5-Ethoxy-4-fluoro-2-nitroaniline ... 260
$C_8H_9FN_2S$
1-(3-Fluorophenyl)-3-methyl-2-thiourea ... 732
C_8H_9FO
2-Fluorophenetole ... 180
4-Fluorophenetole ... 181
3-Fluorophenyl methyl carbinol ... 198
$C_8H_9F_2NO_3S$
5-(Difluoromethylsulfonyl)-*o*-anisidine ... 772
C_8H_9I
2-Iodo-*m*-xylene ... 169
C_8H_9IO
2,6-Dimethyl-4-iodophenol ... 190
$C_8H_9IO_2$
1,2-Dimethoxy-4-iodobenzene ... 177
6-*exo*-Iodo-5-*endo*-hydroxynorbornane-2-*endo*-carboxylic acid lactone ... 96
C_8H_9N
endo-5-Norbornene-2-carbonitrile ... 150
2-Phenylaziridine ... 252
C_8H_9NO
Acetophenone oxime ... 1011
2-Allyloxypyridine ... 939
2′-Aminoacetophenone hydrochloride ... 294
Benzaldehyde *N*-methyloxime ... 1076
4,6-Dimethyl-2-pyridinecarboxaldehyde ... 959
o-Formotoluidide ... 438
m-Formotoluidide ... 437
p-Formotoluidide ... 438
2-Phenylacetamide ... 429
2-Pyridineacetone ... 961
syn-*p*-Tolualdehyde oxime ... 1014
C_8H_9NOS
2-Mercaptoacetanilide ... 420
2-Mercapto-*N*-methylbenzamide ... 467
o-Methyl thiocarbanilate ... 745
$C_8H_9NO_2$
N-(2-Butynyl)succinimide ... 1019
3,5-Dimethyl-1,4-benzoquinone 1-oxime ... 1011
2,6-Dimethyl-4-nitrosophenol ... 257
3,4-Dimethyl-2,5-pyrroledicarboxaldehyde ... 840
2-Ethylnitrobenzene ... 279
Ethyl 4-pyridinecarboxylate picrate ... 962
o-Formanisidide ... 417
m-Formanisidide ... 418
p-Formanisidide ... 419
4′-Hydroxyacetanilide, sodium salt ... 413
N-(Hydroxymethyl)-benzamide ... 462
2-(2-Hydroxyphenyl)acetamide ... 413
Mandelamide ... 411
2-Methoxybenzaldehyde oxime ... 1009
2-Methoxybenzamide ... 467
4′-Methoxyformanilide ... 417
Methyl 3-aminobenzoate ... 372
Methyl *N*-phenylcarbamate ... 680
Methyl 3-pyridylacetate ... 964
(1-Nitroethyl)benzene ... 258
2-Nitro-*p*-xylene ... 279
4-Pyridinecarboxaldehyde ethylene ketal ... 966
4,5,6,7-Tetrahydroisatin ... 134
$C_8H_9NO_2S$
2-(4-Aminophenylthio)acetic acid ... 329
Methyl *N*-(4-mercaptophenyl)-carbamate ... 666
$C_8H_9NO_3$
2-(3-Aminophenoxy)acetic acid ... 328
2-(4-Aminophenoxy)acetic acid hydrochloride ... 327
3,5-Dimethyl-4-formyl-2-pyrrolecarboxylic acid ... 839
3,5-Dimethyl-2-nitrophenol ... 264
Ethyl 5-formyl-2-pyrrolecarboxylate ... 839
N-(Hydroxymethyl)-salicylamide ... 463
Isovanillin oxime ... 1008
Methyl 4-aminosalicylate ... 368
Methyl 1,6-dihydro-1-methyl-6-oxo-3-pyridinecarboxylate ... 134
Methyl 4-nitrobenzyl ether ... 265
3-Nitrophenetole ... 268
10-Oxa-4-azatricyclo[5.2.1.0^{2,6}]-decane-3,5-dione ... 126
2′,4′,6′-Trihydroxyacetophenone imine hydrochloride ... 1028
Vanillin oxime ... 1008
$C_8H_9NO_3S$
4-Acetamidobenzenesulfinic acid ... 786
4-Acetamidobenzenesulfinic acid, sodium salt ... 786
1-Allylpyridinium 3-sulfonate ... 956
Indoline-2-sulfonic acid ... 790
Methyl 5-acetyl-2-amino-3-thiophenecarboxylate ... 856
$C_8H_9NO_4$
1,3-Dimethoxy-4-nitrobenzene ... 269
1,4-Dimethoxy-3-nitrobenzene ... 268
3,4-Dimethyl-2,5-pyrroledicarboxylic acid ... 840
2-Methoxy-4-methyl-6-nitrophenol ... 263
2-Nitro-*p*-xylene-α,α'-diol ... 263
$C_8H_9NO_4S$
2-Acetamidobenzenesulfonic acid, barium salt ... 794
3-Acetamidobenzenesulfonic acid, sodium salt dihydrate ... 795
2-(4-Aminophenylsulfonyl)-acetic acid ... 777
Ethyl 4-nitrophenyl sulfone ... 779
4-(Methylsulfonamido)benzoic acid ... 823
$C_8H_9NO_5$
3,6-Dihydroxy-4-(methoxyimino)-2-methyl-2,5-cyclohexadien-1-one *N*-oxide ... 57
$C_8H_9NO_5S$
Ethyl 3-nitrobenzenesulfonate ... 799
4-Methoxy-3-nitrophenyl methyl sulfone ... 779
6-Nitro-*m*-tolyl methanesulfonate ... 798
5-Nitro-2,3-xylenesulfonic acid, sodium salt ... 794
5-Nitro-2,4-xylenesulfonic acid, sodium salt ... 794
N-(Sulfomethyl)-anthranilic acid ... 789
$C_8H_9NO_6$
2,3-Diacetoxysuccinimide ... 125
C_8H_9NS
2-Phenylthioacetamide ... 746
C_8H_9NSe
1-Phenyl-1,4-azaselenetane ... 213
$C_8H_9N_3$
6-Amino-2,4-dimethyl-3-pyridinecarbonitrile ... 942
2-Amino-1-methylbenzimidazole ... 904
3-(4,5-Dihydro-2-imidazolyl)pyridine ... 951
$C_8H_9N_3O$
2-Acetamido-3-methyl-5-nitropyridine ... 944
1-Amino-1,2-dihydro-4,6-dimethyl-2-oxonicotinonitrile ... 146
Benzaldehyde semicarbazone ... 1070
1-Benzoylguanidine ... 726
1-Benzoylguanidine hydrochloride ... 726
4,6-Dimethyl-3-hydroxypyrazolo-(3,4-b)-pyridine ... 918
$C_8H_9N_3OS$
1-Benzoyl-3-thiosemicarbazide ... 745
1-Formyl-4-phenyl-3-thiosemicarbazide ... 744
1-Phenyl-2-thiobiuret ... 741
$C_8H_9N_3O_2$
2-Acetoacetamidopyrimidine ... 976
1-Benzoylsemicarbazide ... 577
1-Phenylbiuret ... 570
$C_8H_9N_3O_2S$
1-Methyl-3-(4-nitrophenyl)-2-thiourea ... 740
2-(4-Nitrobenzyl)-2-thiopseudourea hydrobromide ... 743
2-(4-Nitrobenzyl)-2-thiopseudourea hydrochloride ... 744
$C_8H_9N_3O_2S_2$
2-(3-Methyl-2-benzothiazolinylidene)-1-hydrazinesulfonic acid ... 820
$C_8H_9N_3O_3$
6-Acetamido-2-methyl-3-nitropyridine ... 952
Acetic acid 2-(4-nitrophenyl)-hydrazide ... 499
4′-Amino-2′-nitroacetanilide ... 422
1-(2,3-Dihydroxybenzylidene)-semicarbazide ... 1028
1-Methyl-3-(3-nitrophenyl)urea ... 564
$C_8H_9N_3O_4$
2,4-Dimethyl-3,6-dinitroaniline ... 270
N,*N*-Dimethyl-3,4-dinitroaniline ... 273
2,6-Dinitro-*N*-ethylaniline ... 274
$C_8H_9N_3O_5$
2,4-Dinitro-*N*-(2-hydroxyethyl)-aniline ... 263
$C_8H_9N_3O_6S$
2,6-Dinitro-*N*-methyl-4-(methylsulfonyl)aniline ... 777
$C_8H_9N_3S$
4,5-Diamino-2-methylbenzothiazole ... 914
3-Methyl-2-benzothiazolinone hydrazone ... 743
$C_8H_9N_3S_2$
1-Phenyl-2,4-dithiobiuret ... 741
$C_8H_9N_3Se$
2,3-Dihydro-3-methyl-2-benzoselenazolone hydrazone ... 915
2,3-Dihydro-3-methyl-2-benzoselenazolone hydrazone *p*-toluenesulfonate ... 915
$C_8H_9N_5$
3-(3-Pyridylmethylamino)-1,2,4-triazole ... 948
$C_8H_9N_5O$
3-Hydrazino-7-methoxy-1,2,4-benzotriazine ... 1002
$C_8H_9N_5O_2$
7-Amino-1,3-dimethyl-2,4(1*H*,3*H*)-pteridinedione ... 977
3,5-Dimethyl-4′-nitro-1,5′-bi-1*H*-pyrazole ... 861
$C_8H_9N_5O_5$
1-(2,4-Dinitrophenyl)-4-methylsemicarbazide ... 561
$C_8H_9NaO_5S$
p-Anisaldehyde sodium bisulfite addition compound ... 791
$C_8H_9O_3P$
Phenyl ethylene phosphite ... 833
$C_8H_{10}AsBr$
(4-Bromophenyl)dimethylarsine ... 1126
$C_8H_{10}AsNO_4$
4-(1-(Hydroxyimino)ethyl)-benzenearsonic acid ... 1128
$C_8H_{10}AsNO_5$
Acetarsone ... 1126
4-(Arsonophenyl)glycine ... 1127
N-Glycoloylarsanilic acid, sodium salt ... 1126
N-(2-Hydroxyacetyl)-*m*-arsanilic acid ... 1126
$C_8H_{10}AsN_3O_4$
4-(Semicarbazonomethyl)-benzenearsonic acid ... 1127
$C_8H_{10}BiClO_2$
(4-Chlorophenyl)-dimethoxybismuthine ... 1141
$C_8H_{10}BrCl_3$
2-Bromo-3-(trichloromethyl)-norbornane ... 6
$C_8H_{10}BrN$
2-Bromo-4,5-dimethylaniline ... 216
2-Bromo-4,6-dimethylaniline ... 216
2-Bromo-*N*,*N*-dimethylaniline picrate ... 216
N-(2-Bromoethyl)aniline ... 215
N-(2-Bromoethyl)-aniline hydrobromide ... 215
4-Bromo-α-methylbenzylamine ... 241
$C_8H_{10}BrNO$
1-(2-Oxopropyl)pyridinium bromide ... 955
$C_8H_{10}Br_2O_2$
2,2-Dibromo-5,5-dimethyl-1,3-cyclohexanedione ... 55
$C_8H_{10}ClN$
3-Chloro-*N*,*N*-dimethylaniline ... 220
3-Chloro-*N*-ethylaniline ... 216
4-Chloro-*N*-ethylaniline ... 216
2-(3-Chloropropyl)pyridine ... 925
$C_8H_{10}ClNO$
1-Acetonylpyridinium chloride ... 956
2-Chloro-5-norbornene-2-carboxamide ... 102
Phenacylammonium chloride ... 295
$C_8H_{10}ClNO_2$
5-Chloro-2,4-dimethoxyaniline ... 215
$C_8H_{10}ClNO_2S$
5′-Chloro-2′-methylmethanesulfonanilide ... 810
N-(2-Chlorophenyl)-*N*-methylmethanesulfonamide ... 809
$C_8H_{10}ClNO_3S$
2-Amino-5-chloro-4-ethylbenzenesulfonic acid ... 787
$C_8H_{10}ClN_3$
4-Diazo-*N*,*N*-dimethylaniline, zinc chloride double salt ... 769
$C_8H_{10}ClN_3O_2$
1-Allyl-6-(chloromethylamino)uracil ... 130
3-Carbamoyl-1-(carbamoylmethyl)-pyridinium chloride ... 947
$C_8H_{10}ClN_3S$
4-(3-Chloro-*o*-tolyl)-3-thiosemicarbazide ... 750
4-(4-Chloro-*o*-tolyl)-3-thiosemicarbazide ... 750
$C_8H_{10}ClN_5$
1-(4-Chlorophenyl)biguanide hydrate ... 724
$C_8H_{10}Cl_2N_2O_2S$
2,5-Dichloro-*N*,*N*-dimethylsulfanilamide ... 807
3,5-Dichloro-*N*,*N*-dimethylsulfanilamide ... 807
$C_8H_{10}Cl_2N_2S$
2-(2-Chlorobenzyl)-2-thiopseudourea hydrochloride ... 743
2-(3-Chlorobenzyl)-2-thiopseudourea hydrochloride ... 749
$C_8H_{10}Cl_2N_4O$
1-(2,6-Dichloro-α-hydroxybenzylamino)guanidine

$C_8H_{10}Cl_2N_4O$ (Cont'd)
hydrochloride 723
$C_8H_{10}Cl_2O_2$
4-(2,2-Dichlorovinyl)-4,5-dihydro-5,5-dimethyl-2(3*H*)-furanone 97
$C_8H_{10}Cl_6N_2O_5$
Dimethyl *N*,*N'*-(1,1'-oxybis(2,2,2-trichloroethyl))biscarbamate 113
$C_8H_{10}FN$
4-Fluoro-α-methylbenzylamine 241
$C_8H_{10}FNO_3S$
3-Amino-4-ethoxybenzenesulfonyl fluoride 802
$C_8H_{10}F_3N_3$
2-(Dimethylamino)-4-methyl-6-(trifluoromethyl)pyrimidine 968
2-(Ethylamino)-4-methyl-6-(trifluoromethyl)pyrimidine 968
$C_8H_{10}INO_2$
3-Methoxycarbonyl-1-methylpyridinium iodide 956
4-Methoxycarbonyl-1-methylpyridinium iodide 954
$C_8H_{10}N_2$
2-Allylaminopyridine 954
6-Aminoindoline dihydrochloride 248
1,4-Cyclohexanedicarbonitrile 150
2-(2-Pyrrolyl)-1-pyrroline 841
1,2,3,4-Tetrahydroquinoxaline 254
$C_8H_{10}N_2O$
1-Acetyl-1-phenylhydrazine 501
3'-Aminoacetanilide 422
Benzylamidoxime 1010
4-(Cyanomethyl)-3-ethyl-5-methylisoxazole 871
N-Ethylnicotinamide 949
N-Glycylaniline hydrochloride 104
(3-Hydroxy-1-methyl-2-butenyl)-malononitrile 144
N^2-Methylanthranilamide 469
1-Methyl-3-phenylurea 585
N-(5-Methyl-2-pyridyl)acetamide 965
N-(6-Methyl-2-pyridyl)acetamide 965
DL-2-Phenylglycinamide 422
p-Toluamidoxime 1014
o-Toluic acid hydrazide 503
o-Tolylurea 587
m-Tolylurea 580
p-Tolylurea 589
$C_8H_{10}N_2OS$
4-Methoxythiobenzoic acid hydrazide 737
1-(4-(Methylthio)phenyl)urea 561
$C_8H_{10}N_2O_2$
p-Anisic acid hydrazide 499
o-Anisic acid hydrazide hydrochloride 498
3,5-Diamino-4-methylbenzoic acid 342
2,4-Dimethyl-5-nitroaniline 271
2,5-Dimethyl-4-nitroaniline 271
2,6-Dimethyl-3-nitroaniline 271
3,5-Dimethyl-4-nitroaniline 271
N,*N*-Dimethyl-*p*-nitroaniline 276
2-Ethyl-5-nitroaniline 271
N-Ethyl-4-nitroaniline 277
4-Ethyl-2-nitroaniline 270
Ethyl 4-pyridylcarbamate 949
6-Hydrazino-*m*-toluic acid hydrochloride 343
2-(2-Hydroxyphenyl)acethydrazide 497
2-Methoxyphenylurea 554
3-Methoxyphenylurea 557
4-Methoxyphenylurea 559
Methyl 2-hydrazinobenzoate hydrochloride 373
4-(1-Methyl-1-nitroethyl)pyridine 952
$C_8H_{10}N_2O_2S$
2-Acetoacetamido-4-methylthiazole 877
2-(4,6-Dimethylpyrimidin-2-ylthio)-acetic acid 974
$C_8H_{10}N_2O_3$
4-Amino-3-nitrophenetole 266
1-(*o*-Anisyl)-3-hydroxyurea 553
1-(*p*-Anisyl)-3-hydroxyurea 553
2-Ethoxy-4-nitroaniline 266
4-Methoxy-3-methyl-6-nitroaniline 266
2-(4-Nitroanilino)-1-ethanol 264
1-(3-Oxobutyl)-3,6(1*H*,2*H*)-pyridazinedione 134
3-(3-Pyridyl)-DL-serine 933
3-(4-Pyridyl)-DL-serine 933
$C_8H_{10}N_2O_3S$
Ethyl 4-hydroxy-2-methylthio-5-pyrimidinecarboxylate 970
Ethyl 4-hydroxy-2-(methylthio)-5-pyrimidinecarboxylate, sodium salt 969
1-(Methylsulfonyl)-3-phenylurea 825
N-(2-Thiazolyl)glutaramic acid 877
$C_8H_{10}N_2O_4$
2-(4-Amino-3-nitrophenoxy)-1-ethanol 263
4,5-Dimethoxy-2-nitroaniline 265
$C_8H_{10}N_2O_4S$
2'-Methyl-4'-nitromethanesulfonanilide 819
4'-Methyl-2'-nitromethanesulfonanilide 818
4'-Methyl-3'-nitromethanesulfonanilide 818
$C_8H_{10}N_2O_5S$
N,*N*-Dimethyl-3-nitrosulfanilic acid 789
N-Ethyl-3-nitrosulfanilic acid, potassium salt 793
2-(4-(Hydrazinosulfonyl)phenoxy)-acetic acid 813
2'-Methoxy-4'-nitromethanesulfonanilide 814
2'-Methoxy-5'-nitromethanesulfonanilide 813
4'-Methoxy-2'-nitromethanesulfonanilide 813
N-(4-Methoxy-3-nitrophenyl)-methanesulfonamide 813
$C_8H_{10}N_2S$
1-Benzyl-2-thiourea 747
2-Methyl-3-phenyl-2-thiopseudourea hydriodide 739
1-Methyl-1-phenyl-2-thiourea 742
1-Methyl-3-phenyl-2-thiourea 747
$C_8H_{10}N_2S_2$
N-(2-(2-Pyridyl)-ethyl)-dithiocarbamic acid, ammonium salt 951
$C_8H_{10}N_4$
1,4-(Diamidino)benzene dihydrochloride 727
1,1'-Dimethyl-3,3'-bi-1*H*-pyrazole 861
$C_8H_{10}N_4O$
6-Propoxypurine 921
$C_8H_{10}N_4O_2$
3,3-Dimethyl-1-(3-nitrophenyl)-triazene 1095
Isophthalic acid dihydrazide 503
$C_8H_{10}N_4O_2S$
4-Methyl-1-(4-nitrophenyl)-3-thiosemicarbazide 740
$C_8H_{10}N_4O_3$
1-Allyl-6-amino-3-methyl-5-nitrosouracil 132
1-Ethyl-8-hydroxy-3-methylxanthine 921
3-Ethyl-8-hydroxy-1-methylxanthine 921
$C_8H_{10}N_4O_4$
5,6-Bis-(acetamido)-uracil 973
$C_8H_{10}N_4O_5S$
Ethyl 2-(3-(5-nitro-2-thiazolyl)-ureido)acetate 877
$C_8H_{10}N_4S$
4-Aminobenzaldehyde thiosemicarbazone 1052
4-(Benzylideneamino)-3-thiosemicarbazide 1072
5,7-Dimethyl-2-(methylthio)-1,2,4-triazolo(1,5-a)pyrimidine 923
6-*n*-Propylmercaptopurine 922
$C_8H_{10}N_4S_2$
1,1'-(1,4-Phenylene)bis(2-thiourea) 741
$C_8H_{10}N_4S_3$
2,6,8-Tris(methylthio)purine 923
$C_8H_{10}N_6$
2,4-Diamino-6,7-dimethylpteridine 1003
1,4-Dihydrazinophthalazine sulfate dihydrate 1000
$C_8H_{10}N_6O_2$
1-(4-Nitrophenyl)biguanide 725
$C_8H_{10}N_6O_3$
2-Amino-4,6-diacetamido-5-nitrosopyrimidine 974
$C_8H_{10}O$
Benzyl methyl ether 187
3,5-Dimethylphenol, zone refined 192
$C_8H_{10}OS$
Benzyl methyl sulfoxide 771
α-Cyclopropyl-2-thiophenemethanol 855
4-Methylthio-*o*-cresol 210
4-Methylthio-*m*-cresol 210
1-(2-Thienyl)-1-butanone 856
$C_8H_{10}O_2$
Acetaldehyde dipropargyl acetal 1017
Bicyclo(2.2.1)hept-5-ene-2-carboxylic acid 67
exo-Bicyclo(2.2.1)hept-5-ene-2-carboxylic acid 67
2-Butyrylfuran 847
2,5-Dimethylresorcinol 193
5,6-Epoxy-4,7-methano-1-oxaspiro-(2,5)-octane 20
4-(Methoxymethyl)phenol 190
2,4,6-Octatrienoic acid 61
7-Oxabicyclo(4.2.1)non-4-en-8-one 97
$C_8H_{10}O_2S$
Benzyl methyl sulfone 783
Ethyl phenyl sulfone 783
Methyl *p*-tolyl sulfone 785
$C_8H_{10}O_2S_2$
2,3,4,6,7,8a-Hexahydro-5-hydroxy-thiopyrano(2,3-b)thiopyran-4-one 58
S-Methyl *p*-toluenethiosulfonate 801
$C_8H_{10}O_3$
1,2-Cyclohexanedicarboxylic anhydride 98
2,5-Dimethoxyphenol 191
3,5-Dimethoxyphenol, sodium salt 191
2-(2-Furyl)ethyl acetate 849
$C_8H_{10}O_3S$
Ethyl benzenesulfonate 800
4-Methoxyphenyl methyl sulfone 775
(*p*-Tolylsulfonyl)-methanol 775
2,5-Xylenesulfonic acid dihydrate 790
$C_8H_{10}O_4$
2-Butyne-1,4-diol diacetate 1017
(2-Cyclopenten-1-yl)-malonic acid 66
4,5-Dimethoxy-3-cyclohexene-1,2-dione 52
2,5-Dimethoxyresorcinol 193
cis,*cis*-2,5-Dimethylmuconic acid 61
(1,1-Dimethyl-2-propyn-1-yl)-malonic acid 1018
mono-Methyl 2-methylmuconate 95
$C_8H_{10}O_5$
Ethyl α,2-dioxo-3-tetrahydrofuranacetate 95
$C_8H_{10}O_5S$
2-Methoxybenzaldehyde sodium bisulfite addition compound 791
$C_8H_{10}O_7$
2-Hydroxy-2-(4-oxotetrahydropyran-3-yl)malonic acid dipotassium salt 69
$C_8H_{10}O_8S$
2,2'-Thiodisuccinic acid 63
$C_8H_{10}O_8S_2$
α,α'-Dihydroxy-*p*-xylene-α,α'-disulfonic acid, disodium salt 792
$C_8H_{10}S$
2-Methylbenzyl mercaptan 209
3-Methylbenzyl mercaptan 209
p-Xylene-α-thiol 209
$C_8H_{10}S_2Si$
Methyl (phenylsilyl)dithioformate 1110
$C_8H_{11}AsO$
3-(Dimethylarsino)phenol methiodide 1126
4-(Dimethylarsino)phenol methiodide 1126
$C_8H_{11}AsO_4S$
4-(2-Hydroxyethylthio)-benzenearsonic acid 1126
$C_8H_{11}BF_4O$
2,4,6-Trimethylpyrylium tetrafluoroborate 935
$C_8H_{11}Br$
endo-2-(Bromomethyl)-5-norbornene 6
$C_8H_{11}BrO_2$
2-Bromo-5,5-dimethyl-1,3-cyclohexanedione 55
$C_8H_{11}ClN_2$
2-Chloro-4-dimethylaminoaniline hydrochloride 220
$C_8H_{11}ClN_2O$
1-(2-Carbamoylethyl)pyridinium chloride 948
6-Chloromethyl-4-hydroxy-2-isopropylpyrimidine 968
$C_8H_{11}ClN_2O_2$
3-Carbamoyl-1-(2-hydroxyethyl)-pyridinium chloride 933
3-Carbamoyl-1-(methoxymethyl)-pyridinium chloride 938
$C_8H_{11}ClN_2O_2S$
4-Chloro-*N*-methyl-*N*-sulfamoylbenzylamine 821
$C_8H_{11}ClN_2O_4S_2$
3-Chloro-4-hydrazinophenyl methylsulfonylmethyl sulfone 778
$C_8H_{11}ClN_2O_6$
1-Amino-2-(ethoxycarbonyl)-pyridinium perchlorate 955
$C_8H_{11}ClO$
3-Chloro-5,5-dimethyl-2-cyclohexen-1-one 54
trans-2-Chloro-1-ethynylcyclohexanol 1016
$C_8H_{11}Cl_2N_3O_2$
5-(Bis-(2-chloroethyl)-amino)-uracil 968
$C_8H_{11}FN_2$
N-(4-Fluorophenyl)-ethylenediamine 236
$C_8H_{11}IOS$
2,6-Dimethyl-4-(methylthio)-pyrylium iodide 978
$C_8H_{11}IS_2$
2,6-Dimethyl-4-(methylthio)-thiopyrylium iodide 211
$C_8H_{11}N$
2,4,6-Collidine hydrochloride 967
3-Ethyl-4-methylpyridine 967
6-Ethyl-2-picoline 967
N-Methyl-*o*-toluidine 220
N-Methyl-*p*-toluidine 219
4-Propylpyridine 967
$C_8H_{11}NO$
5-Amino-2-methylbenzyl alcohol 216
3-Azatricyclo[4.2.1.$0^{2,5}$]nonan-4-one 135
3,3-Dimethyl-2-oxo-1-cyclopentanecarbonitrile 148
2-(2-Methoxyethyl)pyridine 939
4-(2-Methoxyethyl)pyridine 940
3-Methoxy-4-methylaniline 218
4-Methoxy-2-methylaniline hydrochloride 219
2-Methoxymethyl-6-methylpyridine 939
p-Phenetidine hydrochloride 218
2-(4-Pyridyl)-2-propanol 934
$C_8H_{11}NO_2$
4-Aminophenyl 2-hydroxyethyl ether 216
1-Cyanocyclopentyl acetate 149
Ethyl 2-cyano-2-pentenoate 148
Ethyl 2-methyl-3-pyrrolecarboxylate 840
Ethyl 4-methyl-3-pyrrolecarboxylate 841
4-Ethyl-3-methyl-2-pyrrolecarboxylic acid 841
3-(1-Hydroxy-1-methylethyl)-pyridine-*N*-oxide 1005
3-Hydroxy-3-phenylpropylhydroxylamine hemioxalate 1008
3,3'-Iminobis(methacrolein) 60
$C_8H_{11}NO_2S$
Ethyl 2,5-dimethyl-4-thiazolecarboxylate 877
2,5-Xylenesulfonamide 829
$C_8H_{11}NO_2S_2$
5-(Ethoxymethylene)-3-ethylrhodanine 139
4'-(Methylthio)-methanesulfonanilide 816

$C_8H_{11}NO_3$
3,5-Dimethoxy-4-hydroxyaniline hydrochloride 216
Ethyl-4-hydroxy-2-methyl-3-pyrrolecarboxylate 838
L-Noradrenalin L-bitartrate 237
L-Noradrenaline 236
Pyridoxine sulfate 933

$C_8H_{11}NO_3S$
5-Amino-2,4-xylenesulfonic acid 788
N,4-Dimethylmetanilic acid 789
5,6-Dimethylmetanilic acid 788
N-Phenyltaurine 789
N-(*o*-Tolyl)aminomethanesulfonic acid, sodium salt monohydrate 793

$C_8H_{11}NO_4S$
2-Methoxyanilinomethanesulfonic acid, sodium salt dihydrate 792

$C_8H_{11}NO_4S_2$
2,4-Bis(methylsulfonyl)aniline 776

$C_8H_{11}NO_5$
2-Acetamido-2-allylmalonic acid 121

$C_8H_{11}NO_5S$
2-(Methylsulfonylamino)phenyl methanesulfonate 822

$C_8H_{11}NO_6S_2$
mono-2-(4-Aminophenylsulfonyl)-ethyl sulfate 788

$C_8H_{11}N_3$
1-Benzylguanidine nitrate 726
1-Methyl-1-phenylguanidine hydrochloride 727
1-(*p*-Tolyl)guanidine carbonate 727
1-(*p*-Tolyl)-guanidine nitrate 727
1-(*p*-Tolyl)guanidine *p*-toluenesulfonate 725

$C_8H_{11}N_3O$
2-Acetamido-4,6-dimethylpyrimidine 976
Bicyclo(3.2.0)hept-2-en-6-one semicarbazone 127
2′,4′-Diaminoacetanilide 422
1-Ethyl-3-(4-pyridyl)urea 951
1-(4-Methoxyphenyl)guanidine carbonate 724
2-(Methylamino)benzoic acid hydrazide 500
4-Methyl-1-phenylsemicarbazide 578

$C_8H_{11}N_3OS$
4-(4-Methoxyphenyl)-3-thiosemicarbazide 738

$C_8H_{11}N_3O_2$
N^1-Ethyl-4-nitro-1,2-phenylenediamine 270

$C_8H_{11}N_3O_2S$
Ethyl 4-amino-2-(methylthio)-5-pyrimidinecarboxylate 974
1-(*p*-Tosyl)guanidine 828

$C_8H_{11}N_3O_2S_2$
4-Methyl-1-(phenylsulfonyl)-3-thiosemicarbazide 750

$C_8H_{11}N_3O_3$
2-(2-Amino-4-nitroanilino)-1-ethanol 263
Ethyl 2-amino-1,6-dihydro-6-oxo-4-pyrimidineacetate 975
2-(*N*-Methyl-*N*-(5-nitro-2-pyridyl)-amino)-ethanol 932
*N*α-Acetyl-DL-histidine hydrate 868

$C_8H_{11}N_3O_3S$
1-Methyl-3-(3-sulfamoylphenyl)urea 825
1-Methyl-3-(4-sulfamoylphenyl)urea 825

$C_8H_{11}N_3O_4S$
4-(Methylamino)-*N*-methyl-3-nitrobenzenesulfonamide 817

$C_8H_{11}N_3O_5S$
*N*1-(2-Hydroxyethyl)-3-nitrosulfanilamide 811

$C_8H_{11}N_3S$
4-Benzyl-3-thiosemicarbazide 743
4-Methyl-1-phenyl-3-thiosemicarbazide 743
1-Methyl-3-(2-pyridylmethyl)-2-thiourea 945

$C_8H_{11}N_5O$
2-(Dimethylamino)-6-hydroxy-9-methylpurine 921

$C_8H_{11}N_5O_2$
Glyoxal *N*-cyanoacetyl-*N*′-propionyldihydrazone 146

$C_8H_{11}N_5O_3S$
N-(Amidinoamidino)sulfanilic acid 790

$C_8H_{11}O_3P$
(2-Ethylphenyl)phosphonic acid 836
(4-Ethylphenyl)phosphonic acid 836
(2,3-Xylyl)phosphonic acid 837
(2,4-Xylyl)phosphonic acid 837
(2,5-Xylyl)phosphonic acid 837
(2,6-Xylyl)phosphonic acid 837
3,4-Xylylphosphonic acid 837
(3,5-Xylyl)phosphonic acid 837

$C_8H_{11}O_4P$
(2-Ethoxyphenyl)phosphonic acid 834
4-Ethoxyphenylphosphonic acid 834

C_8H_{12}
1,3-Cyclooctadiene 2
5-Methyl-1,3,6-heptatriene 2
3-Methyl-2-norbornene 2

$C_8H_{12}AsNO_3$
4-(Ethylamino)benzenearsonic acid 1127

$C_8H_{12}BrNO$
N-Allyl-*N*-(2-bromoallyl)acetamide 107

$C_8H_{12}Br_2O_4$
Diethyl 2-bromo-2-(bromomethyl)-malonate 81
Dimethyl 2,5-dibromoadipate 83

$C_8H_{12}Br_4$
1,2-Dibromo-4-(1,2-dibromoethyl)-cyclohexane 6

$C_8H_{12}ClNO$
2-Chloro-*N*,*N*-diallylacetamide 106

$C_8H_{12}ClNO_2S$
1-Methyl-4-(methylsulfonylmethyl)-pyridinium chloride 778

$C_8H_{12}ClN_3O$
6-Chloro-1,3-diethyl-4-imino-1,2,3,4-tetrahydro-2-pyrimidinone 138

$C_8H_{12}Cl_2$
1,4-Dichlorobicyclo(2.2.2)octane 6
3,4-Di

$C_8H_{12}Cl_2O_4$
O,*O*′-Ethylene bis(3-chloropropionate) 81

$C_8H_{12}Cl_3NO$
N-Cyclohexyl-2,2,2-trichloroacetamide 103

$C_8H_{12}Cl_3NO_4$
Isopropyl *N*-(1-acetoxy-2,2,2-trichloroethyl)carbamate 114
Methyl *N*-(1-butyryloxy-2,2,2-trichloroethyl)carbamate 114
Methyl *N*-(1-isobutyryloxy-2,2,2-trichloroethyl)carbamate 114
Propyl *N*-(1-acetoxy-2,2,2-trichloroethyl)carbamate 114

$C_8H_{12}F_3N_3S$
1-(3-Trifluoromethylcyclo-hexylidene)-3-thiosemicarbazide 1021

$C_8H_{12}HgN_2O_2$
1,1′-Mercuridi-2-pyrrolidinone 1138

$C_8H_{12}IN$
1-Ethyl-3-methylpyridinium iodide 958
1-Ethyl-4-methylpyridinium iodide 958
1,2,3-Trimethylpyridinium iodide 958
1,2,4-Trimethylpyridinium iodide 958
1,2,6-Trimethylpyridinium iodide 958

$C_8H_{12}INO$
4-Methoxy-1,3-dimethylpyridinium iodide 938

$C_8H_{12}NO_5PS$
(*p*-Toluenesulfonamidomethyl)-phosphonic acid 828

$C_8H_{12}N_2$
1-Allyl-3,5-dimethylpyrazole 864
2-(Dimethylaminomethyl)-pyridine 944
6-Methyl-2-picolylmethylamine 943
Phenethylhydrazine sulfate 235

$C_8H_{12}N_2O$
5-Cyano-4,4,5-trimethyl-2-pyrrolidinone 150
4-Ethoxy-1,3-phenylenediamine dihydrochloride 221
2-Ethoxy-1,3-phenylenediamine dihydrochloride 221
4-Ethoxy-1,3-phenylenediamine sulfate monohydrate 221
3,4,6,7,8,9-Hexahydro-2*H*-pyrido(1,2-a)pyrimidin-6-one 136

$C_8H_{12}N_2OS$
5,6-Dimethyl-2-ethylthio-4-pyrimidinol 970

$C_8H_{12}N_2O_2$
3,6-Dimethoxy-1,2-phenylenediamine dihydrochloride 254
Ethyl 2-cyano-3-dimethylaminoacrylate 146
Ethyl 1,5-dimethyl-1*H*-pyrazole-3-carboxylate 863
Ethyl 3,5-dimethyl-1*H*-pyrazole-4-carboxylate 863
3,3′-(Ethylenedioxy)dipropionitrile 144
4-(2-Hydroxyethoxy)-1,3-phenylenediamine dihydrochloride 221
4-(2-Hydroxyethoxy)-1,3-phenylenediamine sulfate 221
Pentamethylenehydantoin 135

$C_8H_{12}N_2O_2S$
3′-Amino-4′-methylmethanesulfonanilide 816
5-Butyl-2-thiobarbituric acid 976
N^1,N^1-Dimethylsulfanilamide 817

$C_8H_{12}N_2O_3$
3-Nitroso-1-oxa-3-azaspiro(4.5)-decan-2-one 134

$C_8H_{12}N_2O_3S$
2-(4-Aminophenylsulfonamido)-ethanol 811

$C_8H_{12}N_2O_4$
1,3-Bis(2-hydroxyethyl)uracil 131
2,6-Dioxo-*N*-methoxy-*N*-methyl-4-piperidine carboxamide 125
Ethyl 2-(ureidomethylene)-acetoacetate 127

$C_8H_{12}N_2O_4S$
2,5-Dimethoxysulfanilamide 812

$C_8H_{12}N_2O_5S$
2-Pyridinealdoxime dimethyl sulfate 1011

$C_8H_{12}N_2O_6S_2$
N,*N*′-Diformyl-L-cystine 118

$C_8H_{12}N_2S_2$
2,2′-Bis(5,6-dihydro-4*H*-1,3-thiazine) dihydrochloride 39

$C_8H_{12}N_4$
1-Amino-3-(3-methylphenyl)-guanidine 723
4-(Cyclohexylideneamino)-4*H*-1,2,4-triazole 1072

$C_8H_{12}N_4O$
4-Hydroxy-2-(isopropylidenehydrazino)-6-methylpyrimidine 971

$C_8H_{12}N_4O_2$
6-Amino-1,3-dimethyl-5-(methyliminomethyl)-uracil 1052
4-Hydroxy-2-propylamino-5-pyrimidinecarboxamide 971

$C_8H_{12}N_4O_3$
6-Amino-1,3-diethyl-5-nitrosouracil 132
2-(Furfurylureido)acetic acid hydrazide 851

$C_8H_{12}N_4O_4$
6-Amino-3-ethyl-1-(2-hydroxyethyl)-5-nitrosouracil 131
4-(Bis(carboxyethyl)amino)-4*H*-1,2,4-triazole 870

$C_8H_{12}N_6$
1-Phenylbiguanidine hydrochloride 726

$C_8H_{12}N_6O_2$
8-Hydrazinocaffeine 923

$C_8H_{12}O$
5,6-Epoxycyclooctene 21
1,3a,4,5,6,6a-Hexahydro-1-pentalenol 13
1-Methoxy-2-methyl-1,4-cyclohexadiene 17
3-Methyl-1-cyclohexene-1-carboxaldehyde 60
6-Methyl-3-cyclohexene-1-carboxaldehyde 60
Perhydro-2-pentalenone 55
3-Vinyl-7-oxabicyclo(4.1.0)heptane 20

$C_8H_{12}O_2$
Bicyclo(4.1.0)heptane-7-carboxylic acid 66
4-Cycloheptene-1-carboxylic acid 66
1-Cyclohexen-1-yl acetate 81
2-Cyclohexen-1-yl acetate 81
1,4-Cyclooctanedione 55
1-(1-Cyclopenten-1-yloxy)-2-propanone 52
5,6-Dihydro-4,6,6-trimethyl-α-pyrone 96
2,4-Dimethyl-1,3-pentadiene-1-carboxylic acid 61
2-Ethoxy-2-cyclohexen-1-one 52
Ethyl 1-cyclopentene-1-carboxylate 81
2,4-Hexadien-1-yl acetate 80
Methyl *trans*,*cis*-2,5-heptadienoate 80
Methyl 2-methyl-2-cyclopentene-1-carboxylate 81
4-Vinylcyclohexene dioxide 20

$C_8H_{12}O_3$
Methyl 3-oxo-1-cyclopentaneacetate 94

$C_8H_{12}O_4$
1,1-Cyclohexanedicarboxylic acid 66
cis-1,2-Cyclohexanedicarboxylic acid 66
cis-Dimethyl 1,2-cyclobutanedicarboxylate 77
5,10-Dioxa-1,4-cyclodecanedione 96
2-Ethoxycarbonyl-4-hydroxy-2-methylbutyric acid γ-lactone 96
trans-Norpinic acid 66

$C_8H_{12}O_5$
Diethyl 2-hydroxyfumarate, potassium salt 84
Diethyl 2-hydroxyfumarate, sodium salt 84

$C_8H_{12}O_6$
1,3,5-Pentanetricarboxylic acid 61

$C_8H_{12}O_6Si$
Triacetoxyvinylsilane 1109

$C_8H_{12}S$
2-*tert*-Butylthiophene 859

$C_8H_{13}Br$
5-Bromo-1,4-octadiene 5

$C_8H_{13}BrO_2$
6-Bromo-1,3-dioxaspiro(4.5)decane 23

$C_8H_{13}Br_2NS$
4-(Dimethylsulfonio)-1-methylpyridinium dibromide 941

$C_8H_{13}ClO$
1-Methyl-1-cyclohexanecarboxylic acid chloride 100

$C_8H_{13}ClO_4$
Diethyl 2-chloro-2-methylmalonate 83

$C_8H_{13}ClO_5$
1-(2-Acetoxyethoxy)-2-chloroethyl acetate 83

$C_8H_{13}F_3O_3Tl$
Ethyl 3-oxo-4,4,4-trifluorobutyrate, dimethylthallium salt 91

$C_8H_{13}NO$
3,5-Dimethyl-2-cyclohexen-1-one oxime 1013
5-Methyl-2,4-hexadiene-2-carboxamide 102
N,*N*,5-Trimethylfurfurylamine 850
Tropinone hydrobromide 59

$C_8H_{13}NOS$
3-(2-Tetrahydrofuryl)propyl thiocyanate 151

$C_8H_{13}NO_2$
2-Aminoethyl 2,4-hexadienoate hydrochloride 87
3-Amino-2-norbornanecarboxylic acid 72
N,*N*-Diallylglycine 72
Pentyl cyanoacetate 149

$C_8H_{13}NO_3$
1-Acetyl-4-piperidinecarboxylic acid 109
N-Butylmaleamic acid 122
Ethyl 2-oxo-1-pyrrolidineacetate 125
4-Hydroxyimino-2,2,5,5-tetramethyltetrahydrofuran-3-one 1009
γ-Oxo-1-pyrrolidinebutyric acid 109

$C_8H_{13}NO_6$
Diethyl 2-methyl-2-nitromalonate 75

$C_8H_{13}NS$
3-Amino-5,5-dimethyl-2-cyclohexene-1-thione 57
$C_8H_{13}N_2O_5P$
Pyridoxamine 5-phosphate monohydrochloride 932
$C_8H_{13}N_3$
2-(2-(Dimethylamino)ethyl)pyrazine 977
N^4,N^4-Dimethyl-1,2,4-benzenetriamine dihydrochloride 221
4,6-Dimethyl-2-(dimethylamino)-pyrimidine 975
1,2,6-Trimethyl-4(1*H*)-pyridone hydrazone dihydrochloride 1055
$C_8H_{13}N_3O$
1-Methyl-4,6,7,8-tetrahydro-1*H*-pyrimido(1,2-a)pyrimidin-2(3*H*)-one 136
8-Methyl-5,6,7,8-tetrahydro-4*H*-pyrimido(1,2-a)pyrimidin-2(3*H*)-one 133
3,4,5-Trimethyl-2-pyrrolecarboxylic hydrazide 841
$C_8H_{13}N_3O_2$
6-Amino-1,3-diethyluracil 132
6-Amino-1-isobutyluracil 133
1-Butyl-3-(cyanoacetyl)-urea 146
5-*tert*-Butyl-3-methyl-4-nitro-1*H*-pyrazole 862
$C_8H_{13}N_3O_2S$
Amino-N^2,N^4-dimethylsulfanilamide 816
$C_8H_{13}N_3S_2$
5-Cyclohexylamino-1,3,4-thiadiazole-2(3*H*)-thione 879
$C_8H_{13}N_5O$
2-(3,3-Dimethylguanidino)-6-hydroxy-4-methylpyrimidine 971
C_8H_{14}
Isopropylidenecyclopentane 2
1-Methyl-1-cycloheptene 3
3-Methyl-1,5-heptadiene 2
1,4-Octadiene (mixture of *cis* and *trans*) 2
2-Octyne 1015
3-Octyne 1015
$C_8H_{14}BrNO$
N-Allyl-2-bromoisovaleramide 106
$C_8H_{14}BrNO_3$
DL-2-Bromoisocaproylglycine 123
2-Bromopropionyl-DL-norvaline 123
$C_8H_{14}BrNO_4S$
4-Bromo-3-sulfolanyl *N*-propylcarbamate 153
$C_8H_{14}Br_2$
1,4-Bis(bromomethyl)cyclohexane 6
1,2-Dibromocyclooctane 5
$C_8H_{14}ClNO$
2-Chloro-*N*-cyclohexylacetamide 103
$C_8H_{14}ClNO_3$
Chloroacetyl-L-isoleucine 117
Chloroacetyl-DL-leucine 117
Chloroacetyl-L-leucine 117
Ethyl *N*-(2-chloroethyl)-*N*-propionylcarbamate 113
$C_8H_{14}ClNO_4S$
4-Chlorotetrahydrothiophene-3-ol-1,1-dioxide *N*-propylcarbamate 153
$C_8H_{14}ClN_5$
2-Chloro-4-ethylamino-6-isopropylamino-*s*-triazine 979
$C_8H_{14}Cl_2N_2O_2$
N,*N'*-Bis(3-chloropropionyl)-ethylenediamine 110
$C_8H_{14}Cl_2O_4S_2$
1,2-Bis(isopropylsulfonyl)-1,2-dichloroethylene 154
$C_8H_{14}Cl_2SSi_2$
2,5-Bis(dimethylsilyl)-3,4-dichlorothiophene 1110
$C_8H_{14}Cl_3NO_2S$
Ethyl *N*-(1-(propylthio)-2,2,2-trichloroethyl)carbamate 114
$C_8H_{14}Cl_3NO_3$
Butyl *N*-(1-methoxy-2,2,2-trichloroethyl)carbamate 113
Ethyl *N*-(1-isopropoxy-2,2,2-trichloroethyl)carbamate 113
Methyl *N*-(1-butoxy-2,2,2-trichloroethyl)carbamate 113
Methyl *N*-(1-*sec*-butoxy-2,2,2-trichloroethyl)carbamate 113
Methyl *N*-(1-isobutoxy-2,2,2-trichloroethyl)carbamate 113
$C_8H_{14}Cl_3NO_4$
Methyl *N*-(1-(1-ethoxyethoxy)-2,2,2-trichloroethyl)carbamate 113
$C_8H_{14}Cl_3O_5P$
Dimethyl 1-(butyryloxy)-2,2,2-trichloroethylphosphonate 159
$C_8H_{14}N_2$
4-(Aminomethyl)-cyclohexanecarbonitrile 145
(*N*-Hexamethyleneimino)-acetonitrile 145
$C_8H_{14}N_2O$
N-(1-Cyano-1-ethylpropyl)-acetamide 145
3,4-Dihydro-1,3,4,6-tetramethyl-2(1*H*)-pyrimidinone 133
cis-3a,4,5,6,7,7a-Hexahydro-2-(methylamino)-benzoxazole maleate 137
4-Morpholinebutyronitrile 147
$C_8H_{14}N_2O_2$
Ethyl 2-(3,4,5,6-tetrahydro-2-pyrimidinyl)acetate 91
2-Oxo-1-pyrrolidinebutyramide 124
$C_8H_{14}N_2O_3$
L-Prolyl-L-alanine monohydrate 124
$C_8H_{14}N_2O_4$
1-(1-Methyl-1-nitroethyl)-1-nitrocyclopentane 45
$C_8H_{14}N_2O_5$
L-2-Aminobutyryl-L-aspartic acid hydrate 119
$C_8H_{14}N_4$
5,6-Dihydrazonocyclooctene 769
$C_8H_{14}N_4OS$
5,6-Diamino-1,3-diethyl-2-thiouracil 139
$C_8H_{14}N_4O_2$
4-Cyclohexen-1,2-dicarboxylic dihydrazide 112
5,6-Diamino-1,3-diethyluracil 132
1,3-Diethyl-6-hydrazinouracil 134
$C_8H_{14}N_4O_3$
5,6-Diamino-3-ethyl-1-(2-hydroxyethyl)-uracil 131
$C_8H_{14}N_4S$
Bis(1-imidazolin-2-ylmethyl)sulfide dihydrochloride 38
$C_8H_{14}N_4S_2$
1,6-Diallyl-2,5-dithiobiurea 140
$C_8H_{14}O$
2-Allyl-1-cyclopentanol 13
2-Butenyl ether 17
2-Butenyl 1-methylallyl ether 17
1-Cyclohexen-1-yl ethyl ether 18
Cyclohexylidenemethyl methyl ether 18
Cyclohexyl methyl ketone 55
Cyclohexyl vinyl ether 18
3,5-Dimethylcyclohexanone (mixture of isomers) 55
2,5-Dimethyl-1,5-hexadien-3-ol 11
2,5-Dimethyl-3,5-hexadien-2-ol 11
2-Ethyl-2-hexenal 60
2-Methoxy-3-methyl-1-cyclohexene 18
1-Methyl-3-cyclohexene-1-methanol 14
2-Methyl-1,5-heptadien-4-ol 11
2-Methyl-1,6-heptadien-3-ol 11
4-Methyl-1,6-heptadien-4-ol 11
Methyl 2-methylcyclopentyl ketone 55
2-Methyl-2-norbornanol 15
2-Octyn-1-ol 1015
$C_8H_{14}O_2$
1-Acetylcyclohexanol 52
2-(Allyloxy)tetrahydropyran 22
Bicyclo(2.2.2)octane-1,4-diol 12
1,4-Bis-(vinyloxy)-butane 18
Cycloheptyl formate 77
cis-4-Cyclohexene-1,2-dimethanol 12
2,5-Dimethyl-2,3-dihydro-4*H*-pyran-2-methanol 21
3-Ethyl-2-hexenoic acid 61
Ethyl 4-methyl-3-pentenoate 80
4a,5,6,7,8,8a-Hexahydro-1,3-benzodioxane 23
2-(3-Hexyn-1-yloxy)ethanol 1016
9-Oxabicyclo(3.3.1)nonan-1-ol 22
2,2,3-Trimethyl-4-pentenoic acid 61
4-Vinyl-1,2-cyclohexanediol 13
$C_8H_{14}O_3$
4,4-Diethoxy-2-butyn-1-ol 1016
5,5-Diethyl-1,3-dioxan-2-one 97
2,4-Dimethyl-5-oxohexanoic acid 64
Ethyl 2,3-epoxy-3-methylvalerate 89
Ethyl 2-methyl-3-oxovalerate 94
1-Hydroxycycloheptane-carboxylic acid 64
cis-4-Hydroxycyclohexyl acetate 85
trans-4-Hydroxycyclohexyl acetate 85
Methyl hexahydrosalicylate 85
Methyl *cis*-2-hydroxycyclohexanecarboxylate 64
5-Methyl-5-propyl-1,3-dioxan-2-one 97
Tetrahydropyran-2-ylmethyl acetate 90
$C_8H_{14}O_3S$
3-(2-Butenyloxy)-tetrahydrothiophene 1,1-dioxide 152
3-(1-Methylallyloxy)-tetrahydrothiophene 1,1-dioxide 152
$C_8H_{14}O_4$
d-5-Acetoxy-4-methylpentanoic acid 95
1,3-Butanediol diacetate 75
2,5-Dimethyladipic acid 61
Ethylene glycol dipropionate 74
3-Ethyl-3-methylglutaric acid 61
$C_8H_{14}O_4S$
Diethyl thiodiglycolate 87
$C_8H_{14}O_4S_2$
1,4-Butanebis-(thioglycollic acid) 63
$C_8H_{14}O_5$
1-(2-Acetoxyethoxy)ethyl acetate 74
3-(Acetoxymethoxy)propyl acetate 74
3-Butylmalic acid monosodium salt hydrate 63
Diethylene glycol diacetate 92
Dimethyl 2-(methoxymethyl)-succinate 85
$C_8H_{14}O_5S_2$
2-(Carboxymethylthio)-ethyl ether 63
3-Sulfolanyl ether 153
$C_8H_{14}O_6$
1,3:4,6-Di-*O*-methylene-D-mannitol 16
$C_8H_{14}O_6S$
Dimethyl 3,3'-sulfonyldipropionate 153
$C_8H_{14}O_7S$
Diethyl sulfosuccinate, sodium salt 155
$C_8H_{14}S$
Methallyl sulfide 26
Octahydrocyclopenta(c)thiopyran 27
$C_8H_{15}BO_2$
4,4,6-Trimethyl-2-vinyl-1,3,2-dioxaborinane 1124
$C_8H_{15}BrO$
2-Bromo-2,4,4-trimethyl-3-pentanone 50
$C_8H_{15}BrO_2$
Butyl 2-bromo-2-methylpropionate 83
Ethyl 2-bromo-2-ethylbutyrate 83
$C_8H_{15}Cl$
Chlorocyclooctane 5
1-Chloro-1-ethylcyclohexane 6
$C_8H_{15}Cl_3N_2O_2$
Ethyl *N*-(1-(isopropylamino)-2,2,2-trichloroethyl)carbamate 114
Ethyl *N*-(1-(propylamino)-2,2,2-trichloroethyl)carbamate 114
Methyl *N*-(1-(butylamino)-2,2,2-trichloroethyl)carbamate 114
Methyl *N*-(1-(diethylamino)-2,2,2-trichloroethyl)carbamate 114
$C_8H_{15}IO$
3,4-Dimethyl-1-iodo-4-methoxy-1-pentene 17
$C_8H_{15}IO_2S$
3,3-Dimethyl-2-iodo-1-butenyl ethyl sulfone 154
$C_8H_{15}N$
2-Allylcyclopentylamine 32
Bis-(2-methylallyl)-amine 32
1-(3-Butenyl)pyrrolidine 40
N-Isopropyl-*N*-(3-methyl-2-buten-1-ylidene)-amine 1020
Triethylacetonitrile 150
$C_8H_{15}NO$
1-Butyl-2-pyrrolidinone 124
5,6-Dihydro-2,4,4,6-tetramethyl-4*H*-1,3-oxazine 977
2,6-Dimethyl-1-piperidinecarboxaldehyde 136
O-Ethylcyclohexanone oxime 1013
1-Isopropyl-4-piperidone 59
2-Methylcyclohexanecarboxamide 102
4-(2-Methyl-1-propenyl)morpholine 41
2-Methyl-3-quinuclidinol 37
$C_8H_{15}NO_2$
4-(Aminomethyl)-cyclohexanecarboxylic acid 67
N,*N*-Diethylacetoacetamide 107
Ethyl 3-(dimethylamino)-crotonate 88
Ethyl 3-(ethylamino)-crotonate 88
5-Ethyl-3-methyl-1-aza-4,6-dioxabicyclo(3.3.0)octane 41
Ethyl *N*-piperidinocarboxylate 110
6-Hydroxy-5,5,6-trimethyl-2-piperidone 131
Methyl 1-hydroxycyclohexane-carboximidate hydrochloride 1037
$C_8H_{15}NO_2S$
D-Penicillamine acetone adduct hydrochloride hydrate 68
$C_8H_{15}NO_3$
N-Acetyl-D-*allo*-isoleucine 123
N-Acetyl-L-isoleucine 123
N-Acetyl-D-leucine 123
N-Acetyl-DL-leucine 105
N-Acetyl-L-leucine 123
N-Acetyl-D-norleucine 123
N-Acetyl-DL-norleucine 123
N-*tert*-Butylsuccinamic acid 105
4-(Carbethoxymethyl)morpholine 90
6-Dimethylamino-4-ketohexanoic acid hydrochloride 72
Ethyl *N*,*N*-diethyloxamate 107
Ethyl 4-hydroxy-1-piperidinecarboxylate 89
1-(1-Nitrocyclohexyl)-1-ethanol 45
5-Nitro-2-octanone 48
2-Nitro-2,4,4-trimethyl-3-pentanone 48
$C_8H_{15}NO_3S$
N-Acetyl-D-ethionine 118
N-Acetyl-DL-ethionine 118
N-Acetyl-L-ethionine 118
DL-*N*-Acetyl-β-mercaptoisoleucine 118
$C_8H_{15}NO_3Si$
1-Vinyl-2,8,9-trioxa-5-aza-1-silabicyclo(3.3.3)undecane 1108
$C_8H_{15}NO_4$
tert-Butyl 2-methyl-2-nitropropionate 73
Ethyl 2,2-dimethyl-3-nitrobutyrate 73
Ethyl 2-nitrohexanoate 73
$C_8H_{15}NO_5S_2$
3-(Morpholinosulfonyl)-tetrahydrothiophene-1,1-dioxide 154
$C_8H_{15}N_3O_4$
L-2-Aminobutyrylglycylglycine 120
$C_8H_{15}N_3O_5$
L-Threonylglycylglycine 118
$C_8H_{15}N_3S_2$
Methyl 3-(3,4,5,6-tetrahydro-2*H*-azepin-7-yl)dithiocarbazate 140
$C_8H_{15}N_5O_4$
Glycylglycylglycylglycinamide hydrochloride 124
C_8H_{16}
cis-2,5-Dimethyl-3-hexene 1
trans-2,2-Dimethyl-3-hexene 1
3,4-Dimethyl-3-hexene 1
2-Methyl-2-heptene 1
trans-2-Methyl-3-heptene 1
trans-4-Methyl-2-heptene 1
4-Methyl-2-heptene (mixture of *cis* and *trans*) 1
cis-2-Octene 1
$C_8H_{16}BrNO_2$
1-Bromo-1-nitrooctane 46
$C_8H_{16}Br_4NO_4P$
Bis-(2,3-dibromopropyl) *N*-(2-hydroxyethyl)-phosphoramidate 160

$C_8H_{16}ClNO$
2-Chloro-*N,N*-diisopropylacetamide 106
2-Chloro-*N*-hexylacetamide 102
$C_8H_{16}ClNO_2$
N,N-Diethyl-2-ketomorpholinium chloride 96
$C_8H_{16}Cl_2O$
4-Chlorobutyl ether 17
$C_8H_{16}Cl_5N$
Tetrakis(2-chloroethyl)ammonium chloride 42
$C_8H_{16}N_2$
N-Butyl-*N*-ethylaminoacetonitrile 145
4-(Diethylamino)-butyronitrile 147
N,N'-Diethyl-2-butyne-1,4-diamine 1018
N,N-Diisopropylaminoacetonitrile 145
2-(Dipropylamino)acetonitrile 145
Perhydroquinoxaline 42
N,N,N',N'-Tetramethyl-2-butyne-1,4-diamine 1018
$C_8H_{16}N_2O$
1-Allyl-3-butylurea 128
Cycloheptylurea 127
1-(Cyclohexylmethyl)-urea 130
1-Cyclohexyl-3-methylurea 129
$C_8H_{16}N_2O_2$
N-Acetyl-DL-leucinamide 121
N-Acetyl-L-leucinamide 124
1,2-Dimethyl-1,2-dipropionylhydrazine 112
N,N'-Dipropyloxamide 110
*N*α-Acetyl-D-*allo*-isoleucinamide 124
2,*N,N,N',N'*-Pentamethylmalonamide 110
N,N,N',N'-Tetramethylsuccinamide 111
$C_8H_{16}N_2O_2S_2$
N,N'-Bis(2-hydroxypropyl)-dithiooxamide 139
4,4'-Dithiodimorpholine 42
$C_8H_{16}N_2O_3$
D-Alanyl-D-valine 123
β-Alanyl-L-valine 119
Glycyl-L-norleucine 119
DL-Leucylglycine 120
*N*α-Acetyl-D-lysine 120
*N*α-Acetyl-DL-lysine monohydrate 119
Nε-Acetyl-DL-lysine 119
Sarcosyl-L-valine 121
L-Valyl-L-alanine 119
$C_8H_{16}N_2O_4$
L-2-Aminobutyryl-L-threonine 118
Diethyl 1,2-dimethyl-1,2-hydrazinedicarboxylate 88
Diethyl ethylenedicarbamate 115
3,4-Dimethyl-3,4-dinitrohexane 45
1,3-Dinitro-2-isobutyl-2-methylpropane 46
1,1-Dinitrooctane 46
$C_8H_{16}N_2O_4S_2$
DL-Homocystine 71
$C_8H_{16}N_2O_6$
Bis-(2-hydroxyethyl) ethylenedicarbamate 114
$C_8H_{16}N_2O_7$
2-Acetamido-1,2-dideoxy-1-nitro-D-mannitol 103
1-Glucosyl-3-(hydroxymethyl)-urea 126
$C_8H_{16}N_2S$
5,5-Dimethyl-3-isopropylimidazolidine-2-thione 140
$C_8H_{16}N_4$
4-(Dipropylamino)-4*H*-1,2,4-triazole 870
$C_8H_{16}N_4OS$
1-(Cyclohexylcarbamoyl)-3-thiosemicarbazide 141
1-Cyclohexyl-5-thiobiurea 141
$C_8H_{16}N_4O_3$
N-α-Acetyl-L-arginine dihydrate 1060
*N*α-Acetyl-DL-arginine dihydrate 137
$C_8H_{16}N_6O_2$
Adipic dialdehyde disemicarbazone 112
$C_8H_{16}O$
2,4-Dimethylcyclohexanol (mixture of isomers) 13
2,4-Dimethyl-3-hexanone 48
2,2-Dimethyl-5-hexen-3-ol 11
3,5-Dimethyl-5-hexen-3-ol 11
4,4-Dimethyl-2-methylene-1-pentanol 11
1-Ethylcyclohexanol 13
cis-4-Ethylcyclohexanol 14
2-Ethyl-2-hexen-1-ol 11
3-Ethyl-5-hexen-3-ol 11
3-Isopropyl-2-pentanone 48
1-Methylcycloheptanol 12
2-Methyl-4-heptanone 48
3-Methyl-2-heptanone 48
6-Methyl-2-heptanone 48
2-Methyl-1-hepten-3-ol 11
4-Methyl-1-hepten-4-ol 11
5-Methyl-1-hepten-4-ol 11
4-Octanone 48
2-Octen-4-ol 11
1-Propyl-1-cyclopentanol 13
$C_8H_{16}OS$
6-(Methylthiomethoxy)-2-hexene 25
$C_8H_{16}OSi$
3-Hydroxy-3-methyl-1-trimethylsilyl-1-butyne 1015
$C_8H_{16}O_2$
Butyl isobutyrate 73
2-*tert*-Butyl-2-methyl-1,3-dioxolane 22
cis-1,4-Cyclohexanedimethanol 14
1,3-Cyclohexanedimethanol 12
2,2-Diethylbutyric acid 61
cis-5,5-Dimethyl-1,3-cyclohexanediol 12
2,5-Dimethyltetrahydropyran-2-methanol 21
Ethyl 3,3-dimethylbutyrate 73
2-(1-Hydroxyethyl)cyclohexanol 14
5-Hydroxy-4-octanone 50
Isobutyl butyrate 73
2-Methylheptanoic acid 61
Octanoic acid, octadecylamine salt 70
Pentyl propionate 73
Propyl isovalerate 73
$C_8H_{16}O_2S$
tert-Butyl 2-(methylthio)propionate 86
Hexyl mercaptoacetate 86
$C_8H_{16}O_2S_2$
O-Butyl *S*-(3-hydroxypropyl) dithiocarbonate 85
$C_8H_{16}O_3$
1-(2-Ethoxyethoxy)-2-(vinyloxy)-ethane 18
Ethyl 2,2-diethylglycolate 84
2-Hydroxyoctanoic acid 63
2-Methyl-4,6,9-trioxa-1-decene 18
$C_8H_{16}O_3S$
3-Butoxytetrahydrothiophene 1,1-dioxide 153
$C_8H_{16}O_3Tl$
Ethyl 3-oxobutyrate, dimethylthallium salt 93
$C_8H_{16}O_4S_2$
3-(Butylsulfonyl)-tetrahydrothiophene-1,1-dioxide 154
$C_8H_{16}O_4Si$
3,3'-(Dimethylsilylene)-bispropionic acid 1109
$C_8H_{16}O_8$
D-Gluco-L-talooctose 60
$C_8H_{16}S$
2-Butenyl butyl sulfide 26
$C_8H_{17}Br$
4-Bromooctane 4
$C_8H_{17}Br_2NO$
(3-(Bromomethyl)oxetan-3-ylmethyl)trimethylammonium bromide 42
$C_8H_{17}Cl$
3-(Chloromethyl)-heptane 4
2-Chlorooctane 4
$C_8H_{17}ClHg$
Chlorooctylmercury 1137
$C_8H_{17}ClN_2$
1-(3-Chloropropyl)-4-methylpiperazine dihydrochloride 36
$C_8H_{17}I$
2-Iodooctane 4
$C_8H_{17}N$
2,4-Dimethyl-3-ethylpyrrolidine 41
1-Ethyl-4-methylpiperidine 40
5-Ethyl-2-methylpiperidine 41
4-Propylpiperidine 40
$C_8H_{17}NO$
4-Butylmorpholine 42
2-Cyclohexylaminoethanol 33
N-Cyclohexyldimethylamine *N*-oxide 29
4-(Diethylamino)-2-butanone hydrochloride 57
1-Diethylamino-3-buten-2-ol 34
2-(Hexamethyleneimino)ethanol 36
4-Isobutylmorpholine 41
5-Methyl-3-heptanone oxime 1008
α-Methyl-1-piperidineethanol 37
2-Octanone oxime 1008
1-Piperidinepropanol 37
$C_8H_{17}NO_2$
4-Diethylaminobutyric acid hydrochloride 72
2,6-Dimethyl-4-morpholineethanol 36
L-Leucine ethyl ester 87
L-Leucine ethyl ester hydrochloride 89
1-(2-Methoxyethyl)-3-pyrrolidinemethanol 36
1-Nitrooctane 46
2-Nitrooctane 46
Octanohydroxamic acid, potassium salt 105
N,N,5,5-Tetramethyl-1,3-dioxan-2-amine 35
$C_8H_{17}NO_2Si$
(2-Cyanoethyl)-diethoxymethylsilane 1110
$C_8H_{17}NO_3$
3-Dimethylaminopropyl ethyl carbonate 88
Octyl nitrate 45
$C_8H_{17}NO_3S$
1-Piperidinepropanesulfonic acid 155
$C_8H_{17}NO_4$
Ethyl *N*-(1,1-bis(hydroxymethyl)-propyl)carbamate 114
$C_8H_{17}NS$
2-Cyclohexylaminoethanethiol 35
$C_8H_{17}N_3O$
Acetic acid 2-(1-methyl-4-piperidyl)-hydrazide 111
4-Carbamoyl-4-(dimethylamino)-piperidine dihydrochloride 108
4-(Ethylamino)-isonipecotamide 108
$C_8H_{17}N_3O_2$
N-Glycyl-L-leucinamide hydrochloride 124
L-Valyl-L-alaninamide acetate 121
$C_8H_{17}N_5$
N-Cyclohexylbiguanide hydrochloride 138
$C_8H_{17}O_6P$
Diethyl (methoxy(methoxycarbonyl)-methyl)phosphonate 159
C_8H_{18}
2,4-Dimethylhexane 1
3-Methylheptane 1
$C_8H_{18}BNO_3$
2-(2-Dimethylaminoethoxy)-4-methyl-1,3,2-dioxaborinane 1125
$C_8H_{18}Br_2N_2$
1,4-Dimethyl-1,4-diazoniabicyclo(2.2.2)octane dibromide 44
$C_8H_{18}ClO_2PS$
Dibutyl chlorothiophosphate 160
$C_8H_{18}F_2Sn$
Dibutyldifluorotin 1129
$C_8H_{18}INO_2$
Propionylcholine iodide 88
$C_8H_{18}N_2$
1,4-Bis-(aminomethyl)-cyclohexane (mixture of isomers) 31
1-Butylpiperazine dihydrobromide 39
1-Piperidinepropylamine 38
N-(2-Piperidylmethyl)-dimethylamine 39
2,3,5,6-Tetramethylpiperazine 39
$C_8H_{18}N_2O$
N-(3-Diethylaminopropyl)-formamide 104
Heptylurea 128
1-Hexyl-3-methylurea 128
$C_8H_{18}N_2O_2$
2-(2-Morpholinoethylamino)-ethanol 36
$C_8H_{18}N_4O_2$
Arginine ethyl ester hydrochloride 137
1,1'-Hexamethylenediurea 129
$C_8H_{18}N_4S_2$
S,S'-Hexamethylenebis-(isothiuronium bromide) 138
$C_8H_{18}O$
2,2-Dimethyl-3-hexanol 8
2,3-Dimethyl-2-hexanol 9
2,4-Dimethyl-3-hexanol 8
2,5-Dimethyl-2-hexanol 9
2,5-Dimethyl-3-hexanol 8
3,4-Dimethyl-2-hexanol (mixture of isomers) 8
3,4-Dimethyl-3-hexanol 9
3,5-Dimethyl-3-hexanol 9
4-Ethyl-3-hexanol 8
2-Ethyl-2-methyl-1-pentanol 8
2-Methyl-2-heptanol 9
2-Methyl-3-heptanol 8
3-Methyl-2-heptanol 8
3-Methyl-3-heptanol 9
3-Methyl-4-heptanol 8
4-Methyl-4-heptanol 9
5-Methyl-3-heptanol 8
4-Octanol 8
2,2,4-Trimethyl-1-pentanol 8
2,2,4-Trimethyl-3-pentanol 8
$C_8H_{18}OS$
Diisobutyl sulfoxide 154
Hexyl 2-hydroxyethyl sulfide 24
$C_8H_{18}O_2$
2-*sec*-Butyl-2-methyl-1,3-propanediol 10
2-Ethyl-2-methylhexanoic acid 61
3-Methyl-2,4-heptanediol 10
2,4,4-Trimethyl-2,3-pentanediol 11
$C_8H_{18}O_2Pb$
Methyl (triethylplumbyl)formate 1135
$C_8H_{18}O_3$
2,5-Dimethyl-1,2,6-hexanetriol 16
$C_8H_{18}O_4$
3-Ethoxy-1,1,3-trimethoxypropane 19
1,1,3,3-Tetramethoxybutane 19
Triethylene glycol ethyl ether 19
$C_8H_{18}O_8$
D-α,α-Glucooctitol 16
$C_8H_{18}S$
Butyl *sec*-butyl sulfide 26
Butyl *tert*-butyl sulfide 26
Butyl isobutyl sulfide 26
Heptyl methyl sulfide 26
2,4,4-Trimethyl-1-pentanethiol 24
$C_8H_{18}S_2$
2,9-Dithiadecane 26
3,8-Dithiadecane 26
4,7-Dithiadecane 26
$C_8H_{18}Si_2$
1,2-Divinyl-1,1,2,2-tetramethyldisilane 1096
$C_8H_{19}N$
2-Aminooctane 28
Dibutylamine hydrochloride 28
Diisobutylamine hydrochloride 28
N,N-Diisopropylethylamine *p*-toluenesulfonate 789
N,1-Dimethylhexylamine 28
$C_8H_{19}NO$
2-(Dipropylamino)-ethanol 34
$C_8H_{19}NO_2$
2,2'-(*tert*-Butylimino)diethanol 34
2-(2-Diethylaminoethoxy)-ethanol 32
3,3'-Iminodi-2-butanol 33
$C_8H_{19}NO_3$
N,N-Bis(2-hydroxypropyl)-ethanolamine 34
2-(Ethoxyhydroxymethyl)-butyranilide 410
$C_8H_{19}N_3$
1-(3-Aminopropyl)-4-methylpiperazine 38
N-(4-Piperidylmethyl)-ethylenediamine 38

$C_8H_{19}N_3O$
1-Ethyl-3-(3-dimethylaminopropyl)urea 126
$C_8H_{19}O_3P$
Di-*tert*-butyl phosphite 158
$C_8H_{20}AsF_6N$
Dibutylammonium hexafluoroarsenate 42
$C_8H_{20}BF_4N$
Dibutylammonium tetrafluoroborate 43
$C_8H_{20}BNaO_4$
Sodium tetraethoxyboron 1124
$C_8H_{20}BrN$
Diethylmethylpropylammonium bromide 43
$C_8H_{20}F_6NP$
Dibutylammonium hexafluorophosphate 42
$C_8H_{20}INO$
(2-Hydroxyethyl)-triethylammonium iodide 42
$C_8H_{20}NO_3P$
Diethyl *tert*-butylamidophosphate 160
$C_8H_{20}N_2$
N,N'-Diethyl-1,4-butanediamine 30
N,N'-Diethyl-*N,N'*-dimethylethylenediamine 30
N,N-Diethyl-*N',N'*-dimethylethylenediamine 30
N,N-Diethyl-*N'*-methyl-1,3-propanediamine 30
N,N-Diisopropylethylenediamine 29
N,N'-Dipropylethylenediamine 30
$C_8H_{20}N_2O$
2-(2-Diethylaminoethylamino)-ethanol 33
N'-(1,1-Dimethyl-2-hydroxyethyl)-2-methyl-1,2-propanediamine 33
$C_8H_{20}N_6NiO_2S_2$
Bis(1,3-diamino-2-propanol)-nickel(II) thiocyanate 1140
$C_8H_{20}O_2Si$
Bis(3-hydroxypropyl)dimethylsilane. 1104
Diethoxydiethylsilane 1096
$C_8H_{20}O_5P_2S_2$
Tetraethyl dithiopyrophosphate 160
$C_8H_{20}P_2S_2$
Tetraethyldiphosphine disulfide 160
$C_8H_{20}S_4Si$
Tetraethyl orthothiosilicate 1096
$C_8H_{21}NO_5$
Tetrakis(2-hydroxyethyl)-ammonium hydroxide 42
$C_8H_{22}N_4$
5-Methyl-1,4,8,11-tetraazaundecane tetrahydrochloride 29
1,4,9,12-Tetraazadodecane tetrahydrochloride 29
1,5,8,12-Tetraazadodecane tetrahydrochloride 29
$C_8H_{22}O_2Si_3$
1,1,1,2,3,3,3-Heptamethyl-2-trisilanecarboxylic acid 1109
$C_8H_{24}Si_3$
Octamethyltrisilane 1096
$C_8H_{26}Si_4$
2-(Dimethylsilyl)-2-ethyl-1,1,3,3-tetramethyltrisilane 1096
$C_8H_{28}N_4Si_4$
2,2,4,4,6,6,8,8-Octamethylcyclotetrasilazane 1096
$C_9H_2ClN_3$
2-Chloro-1,3,5-tricyanobenzene 760
$C_9H_2Cl_5NO$
2,3-Dichloro-3-(2,4,6-trichlorophenoxy)acrylonitrile 758
$C_9H_2F_{16}O_2$
2,2,3,3,4,4,5,5,6,6,7,7,8,8,9,9,-Hexadecafluorononanoic acid 62
$C_9H_3Cl_6NO_2$
1,4,5,6,7,7-Hexachlorobicyclo(2.2.1)-5-heptene-2,3-dicarboximide 126
$C_9H_3N_3$
1,3,5-Tricyanobenzene 768
$C_9H_4BrNO_4$
3-Bromo-6-nitrocoumarin 385
$C_9H_4Br_2F_5NO$
2',5'-Dibromo-2,2,3,3,3-pentafluoropropionanilide 396
$C_9H_4Br_2O_3$
5,7-Dibromo-2-benzofurancarboxylic acid 894
$C_9H_4Cl_3NO_2S$
N-(Trichloromethylthio)phthalimide. 715
$C_9H_4Cl_4F_3NO$
2,2,2,4'-Tetrachloro-2'-(trifluoromethyl)acetanilide 396
$C_9H_4Cl_4O_3$
2-Acetyl-3,4,5,6-tetrachlorobenzoic acid 337
$C_9H_4Cl_4O_4$
Methyl tetrachlorophthalate 366
$C_9H_4Cl_6S_2Si$
Bis(3,4,5-trichloro-2-thienyl)-methylsilane 1110
$C_9H_5BrClNO$
7-Bromo-5-chloro-8-quinolinol 983
$C_9H_5BrCl_4$
3-(4-Bromophenyl)-1,1,2,3-tetrachloro-1-propene 171
$C_9H_5BrCl_4O_3$
4-Bromophenyl 1,2,2,2-tetrachloroethyl carbonate 381
$C_9H_5BrN_2O_2$
4-Bromo-6-methyl-1,3-phenylene diisocyanate 768
$C_9H_5BrN_2O_3$
6-Bromo-4-hydroxycinnoline-3-carboxylic acid 998
$C_9H_5BrN_4S$
8-Bromo-2,3-dihydro-5*H*-*as*-triazino(5,6-b)indole-3-thione 919
$C_9H_5BrO_3$
5-Bromo-2-benzofurancarboxylic acid 893
$C_9H_5Br_2F_6NO_3S$
1,1,1,3,3,3-Hexafluoroisopropyl 4-amino-3,5-dibromobenzenesulfonate 796
$C_9H_5Br_3$
1,1,3-Tribromoindene 173
$C_9H_5Br_3N_2$
1-Phenyl-3,4,5-tribromo-1*H*-pyrazole. 864
2-Tribromomethylquinoxaline 998
$C_9H_5Br_5O$
Allyl pentabromophenyl ether 176
$C_9H_5Br_7O$
2,3-Dibromopropyl pentabromophenyl ether 176
$C_9H_5ClN_2O$
5-Chloro-2-(cyanomethyl)-benzoxazole 910
$C_9H_5ClN_2O_2$
2-Chloro-8-nitroquinoline 988
6-Chloro-5-nitroquinoline 993
6-Chloro-8-nitroquinoline 988
8-Chloro-6-nitroquinoline 989
$C_9H_5ClN_2O_3$
8-Chloro-4-hydroxycinnoline-3-carboxylic acid 998
C_9H_5ClOS
3-Chloro-2*H*-1-benzothiopyran-2-one 386
$C_9H_5ClO_2$
3-Chlorocoumarin 386
$C_9H_5Cl_2N$
4,8-Dichloroquinoline 995
$C_9H_5Cl_2NO$
2-(2,6-Dichlorobenzoyl)acetonitrile 757
$C_9H_5Cl_2NOS_2$
3-(3,4-Dichlorophenyl)rhodanine 732
$C_9H_5Cl_3$
1-Phenyl-1,3,3-trichloroallene 171
$C_9H_5Cl_3F_3NO$
2,2,2-Trichloro-3'-(trifluoromethyl)-acetanilide 397
$C_9H_5Cl_5$
1-Allyl-2,3,4,5,6-pentachlorobenzene 170
$C_9H_5Cl_5N_2O_2$
2,4-Dichlorobenzoic N^2-(2,2,2-trichloroacetyl)hydrazide 491
$C_9H_5Cl_5O_2$
1-(2,3,4,5,6-Pentachlorophenoxy)-2-propanone 282
$C_9H_5Cl_5O_2S$
S-(4-Chlorophenyl) 1,2,2,2-tetrachloroethyl thiocarbonate 381
$C_9H_5Cl_5O_3$
2-Chlorophenyl 1,2,2,2-tetrachloroethyl carbonate 381
$C_9H_5F_3N_2O$
N-(2-Trifluoromethylphenyl)-cyanoformamide 758
$C_9H_5F_3N_4O_4$
N-Cyano-2,6-dinitro-*N*-methyl-4-(trifluoromethyl)aniline 751
$C_9H_5F_5$
α-Methyl-2,3,4,5,6-pentafluorostyrene 170
$C_9H_5F_5O$
2,2,3,3,3-Pentafluoropropiophenone 288
$C_9H_5F_5O_2$
3-(Pentafluorophenyl)-propionic acid 325
$C_9H_5NO_4$
6-Nitrocoumarin 385
2-Nitro-1,3-indanedione 318
$C_9H_5NO_4S$
5-Nitrobenzo(b)thiophene-2-carboxylic acid 897
C_9H_5NS
2-Benzothiazolylacetylene copper(I) salt 1140
$C_9H_5N_3O_2$
2-Cyano-4-nitrophenylacetonitrile 755
$C_9H_5N_3O_5$
5,7-Dinitro-8-quinolinol 983
$C_9H_5N_5O_6S$
2-(Picrylamino)thiazole 876
$C_9H_6AsCl_2N$
Dichloro(8-quinolyl)arsine hydrochloride 1128
$C_9H_6BrNO_2$
5-Bromoindole-2-carboxylic acid 886
$C_9H_6BrNO_2S$
trans-3-(4-Bromophenylsulfonyl)-acrylonitrile 774
$C_9H_6Br_2N_2$
4,5-Dibromo-2-phenylimidazole 867
$C_9H_6Br_2O_2$
α, β-Dibromocinnamic acid 335
$C_9H_6ClF_3N_2O_2$
1-(3-Chlorophenyl)-3-(2,2,2-trifluoroacetyl)urea 505
$C_9H_6ClF_3N_4S$
3-(4-Chloro-2-(trifluoromethyl)-anilino)-5-mercapto-4*H*-1,2,4-triazole 869
$C_9H_6ClF_3O_2S$
2-(2-Chloro-1,1,2-trifluoroethylthio)-benzoic acid 336
$C_9H_6ClF_3O_3$
2-(2-Chloro-1,1,2-trifluoroethoxy)-benzoic acid 336
4-(2-Chloro-1,1,2-trifluoroethoxy)-benzoic acid 336
C_9H_6ClN
2-Chlorocinnamonitrile (mixture of *cis* and *trans*) 767
7-Chloroisoquinoline 998
6-Chloroquinoline 995
C_9H_6ClNO
3-(4-Chlorophenyl)-3-oxopropionitrile 752
C_9H_6ClNOS
4-Chlorobenzoylmethyl thiocyanate. 768
$C_9H_6Cl_2F_3NO$
2,2'-Dichloro-*α,α,α*-trifluoro-*m*-acetotoluidide 390
2,2-Dichloro-3'-(trifluoromethyl)-acetanilide 393
$C_9H_6Cl_2INO_3$
N-(2,2-Dichloroacetyl)-5-iodoanthranilic acid 394
$C_9H_6Cl_2N_2O$
2'-Cyano-2,4'-dichloroacetanilide 758
4,5-Dichloro-1-phenyl-4-pyrazolin-3-one 862
$C_9H_6Cl_2OS$
5,7-Dichloro-4-methyl-1-benzothiophen-3(2*H*)one 296
$C_9H_6Cl_3FN_2O_4$
2,2,2-Trichloroethyl *N*-(4-fluoro-2-nitrophenyl)carbamate 606
2,2,2-Trichloroethyl *N*-(4-fluoro-3-nitrophenyl)carbamate 606
$C_9H_6Cl_3NO_4$
4-(2,2,2-Trichloroacetamido)-salicylic acid 396
$C_9H_6Cl_4$
3-Phenyl-1,1,2,3-tetrachloro-1-propene 171
$C_9H_6Cl_4N_4O_3$
1-(2-Chloro-4-nitrophenyl)-3-(2,2,2-trichloroacetimidoyl)urea 723
$C_9H_6Cl_5NO_2$
2,2,2-Trichloroethyl *N*-(2,4-dichlorophenyl)carbamate 607
$C_9H_6Cl_8$
5,6-Bis-(chloromethyl)-1,2,3,4,7,7-hexachlorobicyclo-(2.2.1)-2-heptene 6
C_9H_6FNO
5-Fluoroindole-3-carboxaldehyde 888
$C_9H_6FNO_2S$
3-(4-Fluorophenyl)-2,4-thiazolidinedione 695
$C_9H_6F_3NO_4$
2-(2-Nitro-4-(trifluoromethyl)phenyl)-acetic acid 325
$C_9H_6F_3N_3O_6$
Methyl *N*-(2,6-dinitro-4-(trifluoromethyl)phenyl)carbamate.. 609
$C_9H_6F_3N_3S$
2-Amino-4-(2-thienyl)-6-(trifluoromethyl)pyrimidine 968
$C_9H_6F_5NO$
3-(Pentafluorophenyl)-propionamide 407
$C_9H_6F_6O_3S$
1,1,1,3,3,3-Hexafluoroisopropyl benzenesulfonate 796
C_9H_6IN
2-Iodocinnamonitrile 760
$C_9H_6I_3NO_3$
3-Acetamido-2,4,6-triiodobenzoic acid, bis(2-hydroxyethyl)-ammonium salt 405
3-Acetamido-2,4,6-triiodobenzoic acid, sodium salt dihydrate 405
$C_9H_6N_2O$
2-Benzoxazoleacetonitrile 909
$C_9H_6N_2OS_2$
5-(4-Pyridylmethylene)rhodanine 962
$C_9H_6N_2O_2$
4-Nitrocinnamonitrile 766
5-Nitroquinoline nitrate 989
$C_9H_6N_2O_3$
4-Nitroisocarbostyril 996
$C_9H_6N_2O_3S$
5-Furfurylidene-2-thiobarbituric acid 845
$C_9H_6N_2O_4$
1-Methyl-5-nitroisatin 703
5-Methyl-7-nitroisatin 702
$C_9H_6N_2O_5$
1-Methyl-6-nitroisatoic anhydride 703
$C_9H_6N_2O_5S$
N-(Hydroxymethyl)-4-nitrophthalimide 716
$C_9H_6N_2S$
2-(2-Benzothiazolyl)acetonitrile 917
$C_9H_6N_4$
Tetrazolo(1,5-a)quinoline 909
C_9H_6OS
4*H*-1-Benzothiopyran-4-one 304
$C_9H_6OS_2$
5-Phenyl-3*H*-1,2-dithiol-3-one 387
$C_9H_6O_2$
Phenylpropiolic acid, sodium salt 1018
$C_9H_6O_2S$
2-Furyl 2-thienyl ketone 855
$C_9H_6O_4$
3-Methoxyphthalic anhydride 388
$C_9H_6O_5$
2-Carboxy-*α*-oxobenzeneacetic acid 345

$C_9H_7BrCl_3NO$
4′-Bromo-3′-methyl-2-trichloroacetanilide 395
$C_9H_7BrN_2$
3-Amino-1-bromoisoquinoline 997
$C_9H_7BrN_2O_3$
2′-Bromo-2-diazo-3′-hydroxy-4′-methoxyacetophenone 1094
$C_9H_7BrN_4$
4-(3-Bromobenzylideneamino)-4*H*-1,2,4-triazole 1025
4-(4-Bromobenzylideneamino)-4*H*-1,2,4-triazole 1026
C_9H_7BrOSe
6-Bromo-2*H*-1-benzoselenin-4(3*H*)-one 213
$C_9H_7BrO_2$
α-Bromocinnamic acid 335
$C_9H_7BrO_3$
2-Bromo-5-norbornene-2,3-dicarboxylic anhydride 98
$C_9H_7BrO_5$
5-Bromo-2-methoxyisophthalic acid. 336
$C_9H_7Br_2ClO$
2,3-Dibromo-3-phenylpropionyl chloride 389
$C_9H_7Br_2N$
α,β-Dibromohydrocinnamonitrile 758
$C_9H_7Br_2NO$
2,6-Dibromo-4-ethylphenyl isocyanate 768
$C_9H_7Br_2NO_4$
α,β-Dibromo-2-nitrohydrocinnamic acid 330
α,β-Dibromo-4-nitrohydrocinnamic acid 330
$C_9H_7Br_3O$
Allyl 2,4,6-tribromophenyl ether 182
$C_9H_7ClFNO_5$
2-Chloro-4-fluoro-5-nitrophenyl ethyl carbonate 381
$C_9H_7ClF_3NO$
2-Chloro-α,α,α-trifluoro-*o*-acetotoluidide 390
2-Chloro-α,α,α-trifluoro-*m*-acetotoluidide 390
$C_9H_7ClF_3NO_2$
Methyl *N*-(2-chloro-5-trifluoromethylphenyl)carbamate 607
$C_9H_7ClF_3NO_3S_2$
2-Chloro-3-methylbenzothiazolium trifluoromethanesulfonate 916
$C_9H_7ClINO_3$
2′-Carboxy-2-chloro-4′-iodoacetanilide 390
$C_9H_7ClN_2$
7-Chloro-4-methylcinnoline 1001
2-Chloro-3-methylquinoxaline 1001
$C_9H_7ClN_2O$
2-Chloro-2′-cyanoacetanilide 758
2-Chloro-4-methoxyquinazoline 999
3-(*p*-Chlorophenyl)-5-methyl-1,2,4-oxadiazole 872
$C_9H_7ClN_2OS$
2-(2-Chloroacetamido)-benzothiazole 910
$C_9H_7ClN_2O_2$
7-Chloro-3*H*-1,4-benzodiazepin-2,5(1*H*,4*H*)-dione 701
7-Chloro-6-methyl-2,4(1*H*,3*H*)-quinazolinedione 1000
α-Chloro-4-nitrohydrocinnamonitrile 757
2-Chloro-5-nitrohydrocinnamonitrile 757
Methyl *N*-(3-chlorophenyl)-*N*-cyanocarbamate 752
$C_9H_7ClN_2O_3$
2-Chloro-5-methoxy-4-nitrophenylacetonitrile 753
$C_9H_7ClN_2O_4$
Ketomalonic acid 4-chlorophenylhydrazone 337
$C_9H_7ClN_2O_5$
2′-Carboxy-2-chloro-5′-nitroacetanilide 392
$C_9H_7ClN_2O_6$
Ethyl 2-chloro-3,5-dinitrobenzoate 366
Ethyl 2-chloro-4,5-dinitrobenzoate 372
$C_9H_7ClN_2S$
N-(2-Chloro-6-methylphenyl)-cyanothioformamide 759
N-(3-Chloro-2-methylphenyl)-cyanothioformamide 759
N-(5-Chloro-2-methylphenyl)-cyanothioformamide 759
$C_9H_7ClN_2S_2$
3-Benzylthio-5-chloro-1,2,4-thiadiazole 878
$C_9H_7ClN_4O$
2-(4-Chlorophenylhydrazono)-2-cyanoacetamide 1025
$C_9H_7ClN_4S$
3-Amino-3-mercapto-2-(4-chlorophenylazo)acrylonitrile 1080
C_9H_7ClO
4-Chlorocinnamaldehyde 321
C_9H_7ClOS
6-Chloro-4-methylbenzo(b)-thiophene-3-ol 897
5-Chloro-7-methylbenzo(b)thiophen-3(2*H*)-one 298
$C_9H_7ClO_2$
6-Chloro-4-chromanone 320
$C_9H_7ClO_2S$
3-Chloro-α-mercaptocinnamic acid.. 334
$C_9H_7ClO_3$
Methyl 4-chlorobenzoylformate 363
$C_9H_7Cl_2NO_3$
4-Acetamido-3,5-dichlorobenzoic acid 405
5-Chloro-*N*-(2-chloroacetyl)-anthranilic acid 390
$C_9H_7Cl_3N_2O_2$
1-(4-Chlorobenzoyl)-2-(dichloroacetyl)hydrazine 491
$C_9H_7Cl_3O_3$
2-(2,4,6-Trichlorophenoxy)-propionic acid 328
$C_9H_7Cl_4NO$
3′-Chloro-4′-methyl-2,2,2-trichloroacetanilide 395
$C_9H_7Cl_4NO_2$
2-Chloroethyl *N*-(2,4,5-trichlorophenyl)carbamate 606
2,2,2-Trichloroethyl *N*-(2-chlorophenyl)carbamate 607
$C_9H_7FN_2O_2S_2$
3-(2-Amino-4-thiazolyl)-benzenesulfonyl fluoride 876
$C_9H_7FN_4$
4-(4-Fluorobenzylideneamino)-4*H*-1,2,4-triazole 1026
$C_9H_7F_2NO_2S_3$
5-(Difluoromethylsulfonyl)-3-methyl-2(3*H*)-benzothiazolethione 727
5-(Difluoromethylsulfonyl)-2-(methylthio)benzothiazole 910
$C_9H_7F_3N_2O_3$
5-Acetamido-2-nitrobenzotrifluoride 397
$C_9H_7F_3N_2O_4$
4′-Methoxy-3′-nitro-2,2,2-trifluoroacetanilide 396
$C_9H_7F_3O_2$
4′-Methoxy-2,2,2-trifluoroacetophenone 288
$C_9H_7F_4NO$
4′-Methyl-2,2,2,3′-tetrafluoroacetanilide 395
$C_9H_7F_4NO_2$
2,2,2-Trifluoroethyl *N*-(2-fluorophenyl)carbamate 607
2,2,2-Trifluoroethyl *N*-(3-fluorophenyl)carbamate 607
$C_9H_7F_5O$
1-(Pentafluorophenyl)-2-propanol 197
$C_9H_7IS_2$
5′-Iodo-5-methyl-2,2′-bithiophene 859
C_9H_7N
Isoquinoline picrate 998
Quinoline picrate 995
C_9H_7NO
8-Hydroxyquinoline, zone refined 983
8-Hydroxyquinoline benzoate 983
8-Hydroxyquinoline picrate 983
8-Hydroxyquinoline, zinc salt 983
5-Phenylisoxazole 871
3-Phenylpropiolamide 1019
Quinoline *N*-oxide picrate 1007
$C_9H_7NOS_2$
3-Phenyl-2-thio-2,4-thiazolidinedione 571
$C_9H_7NO_2$
4-Aminocoumarin 384
6-Aminocoumarin hydrochloride 384
2-Benzofurancarboxamide 894
3-Cyanophenyl acetate 767
2,3-Dihydro-5-methylindole-2,3-dione 691
2-Furyl 2-pyrrolyl ketone 848
1,3-Isoquinolinediol 996
Methyl 2-cyanobenzoate 766
7-Methylisatin 706
N-Methylphthalimide 723
2-Phenyl-2-oxazolin-5-one 387
$C_9H_7NO_2S$
4-Nitrophenyl 1-propynyl sulfide 1017
3-Phenyl-2,4-thiazolidinedione 746
$C_9H_7NO_2S_2$
3-(3-Hydroxyphenyl)rhodanine 734
$C_9H_7NO_3$
3-Acetyl-2(3*H*)-benzoxazolone 709
N-Furfurylmaleimide 850
1-(Hydroxymethyl)isatin 697
$C_9H_7NO_3S$
8-Quinolinesulfonic acid 993
$C_9H_7NO_4$
3,4-Methylenedioxy-β-nitrostyrene 258
3-(2-Nitrovinyl)benzoic acid 343
4-(2-Nitrovinyl)benzoic acid 343
$C_9H_7NO_4S$
4-Nitrophenyl 1-propynyl sulfone 1018
$C_9H_7NO_5S$
N-(Methylsulfonyloxy)-phthalimide.. 801
$C_9H_7NO_6$
2-Formyl-4-nitrophenoxyacetic acid. 327
Methyl 2-carboxy-6-nitrobenzoate 373
Methyl 4-carboxy-2-nitrobenzoate 373
Methyl 4,5-methylenedioxy-2-nitrobenzoate 371
5-Methyl-4-nitroisophthalic acid 344
1-Methyl 3-nitrophthalate 373
Methyl 4-nitrophthalate 374
2-Methyl-3,4,6-pyridinetricarboxylic acid 940
C_9H_7NS
1-Isoquinolinethiol 997
4-Phenylthiazole 878
$C_9H_7N_3$
2-(2,2-Dicyanovinyl)-5-methylpyrrole. 839
$C_9H_7N_3O_2$
3-Amino-2-quinoxalinecarboxylic acid 1000
3-Nitro-3*H*-1,5-benzodiazepine 249
1-Nitro-3-phenyl-1*H*-pyrazole 862
1-Nitro-4-phenyl-1*H*-pyrazole 862
3-Nitro-5-phenyl-1*H*-pyrazole 862
$C_9H_7N_3O_2S$
2-Amino-4-(3-nitrophenyl)thiazole 876
$C_9H_7N_3O_3$
4-Hydroxy-3-methyl-8-nitrocinnoline. 999
$C_9H_7N_3O_3S$
2-(1,2,4-Benzotriazin-3-ylthio)acetic acid N^1-oxide 1005
2-(Methylthio)-5-(4-nitrophenyl)-1,3,4-oxadiazole 873
N-(6-Nitro-2-benzothiazolyl)-acetamide 915
$C_9H_7N_3O_5S$
2-(Methylsulfonyl)-5-(4-nitrophenyl)-1,3,4-oxadiazole 873
$C_9H_7N_3O_7$
2-Acetamido-4,5-dinitrobenzoic acid 423
$C_9H_7N_5O_2$
4-(3-Nitrobenzylideneamino)-4*H*-1,2,4-triazole 1058
4-(4-Nitrobenzylideneamino)-4*H*-1,2,4-triazole 1069
C_9H_8BrClO
Allyl 4-bromo-2-chlorophenyl ether.. 177
2-Bromo-4′-chloropropiophenone 289
4′-Bromo-3-chloropropiophenone 287
$C_9H_8BrCl_2NO_2$
2-Bromoethyl *N*-(2,3-dichlorophenyl)carbamate 605
2-Bromoethyl *N*-(2,4-dichlorophenyl)carbamate 605
2-Bromoethyl *N*-(2,6-dichlorophenyl)carbamate 605
2-Bromoethyl *N*-(3,4-dichlorophenyl)carbamate 604
C_9H_8BrFO
2-Bromomethyl-5-fluorocoumaran 183
$C_9H_8BrF_3$
2-(1-Bromoethyl)-α,α,α-trifluorotoluene 170
3-(1-Bromoethyl)-a,a,a-trifluorotoluene 169
C_9H_8BrN
5-Bromo-1-methylindole 887
$C_9H_8BrNO_3$
3′-Bromo-4′-methyl-2′-nitroacetophenone 289
$C_9H_8BrNO_4$
3-Bromo-3-(3-nitrophenyl)-propionic acid 330
Ethyl 2-bromo-3-nitrobenzoate 374
$C_9H_8BrNO_5$
3-Bromo-4,5-dimethoxy-2-nitrobenzaldehyde 322
$C_9H_8BrN_3O_3S$
2-(4′-Bromo-3′-nitrophenacyl)-2-thiopseudourea hydrobromide 731
$C_9H_8Br_2ClNO$
3-Chloro-2′,6′-dibromopropionanilide 391
$C_9H_8Br_2O_2$
5,6-Dibromo-2,2-dimethyl-1,3-benzodioxole 184
3,5-Dibromo-2-ethoxybenzaldehyde. 324
2,3-Dibromo-3-phenylpropionic acid. 332
$C_9H_8Br_2O_3$
2-(2,4-Dibromophenoxy)-propionic acid 328
$C_9H_8Br_3N$
Quinolinium bromide perbromide 990
$C_9H_8ClF_3N_2O$
1-(2-Chloro-5-trifluoromethylphenyl)-3-methylurea 506
$C_9H_8ClF_3O$
β-Chloro-2-methyl-α,α,β-trifluorophenetole 176
β-Chloro-3-methyl-α,α,β-trifluorophenetole 176
$C_9H_8ClF_5Si$
(Chloromethyl)-dimethyl(pentafluorophenyl)silane.. 1099
C_9H_8ClN
2-(2-Chloro-6-methylphenyl)-acetonitrile 758
4-Chloro-α-methylphenylacetonitrile 751
C_9H_8ClNO
5-Chloro-2,6-dimethylbenzoxazole 909
C_9H_8ClNOS
6-Chloro-3-ethyl-2-benzothiazolinone 746
$C_9H_8ClNO_2$
1-Chloro-4-(2-nitropropenyl)-benzene 257
$C_9H_8ClNO_3$
Benzyl 2-chloro-2-(hydroxyimino)-acetate 1009
2-(2-Chloroacetamido)benzoic acid.. 392
5-Chloro-2,4-dimethoxyphenyl isocyanate 768
2-Chlorohippuric acid 456
4-Chlorohippuric acid 475
2-Chloro-3-(3-nitrophenyl)-propionaldehyde 321
2-Chloro-3-(4-nitrophenyl)-propionaldehyde 321
5′-Chloro-2′-nitropropiophenone 289
$C_9H_8ClNO_4$
2-Chloroethyl 4-nitrobenzoate 365
C_9H_8ClNS
3-(4-Chlorophenylthio)propionitrile 754

$C_9H_8ClN_3O$
2-(2-Chloro-1-benzimidazolyl)-acetamide 904
$C_9H_8ClN_3O_3$
2-Chloro-4-(methoxymethyl)-6-methyl-5-nitronicotinonitrile 937
$C_9H_8ClN_3O_4S$
4-Chloro-*N*-(2-cyanoethyl)-3-nitrobenzenesulfonamide 818
$C_9H_8ClN_3S$
2-Amino-5-(4-chlorobenzyl)-1,3,4-thiadiazole 880
$C_9H_8ClN_3S_2$
5-(3-Chloro-*o*-toluidino)-1,3,4-thiadiazole-2(3*H*)-thione 879
$C_9H_8ClN_5$
2-Amino-4-(4-chloroanilino)-1,3,5-triazine hydrochloride 979
2-(2-Chlorophenyl)-4,6-diamino-1,3,5-triazine 979
$C_9H_8ClN_5O_5$
Oxamic N^2-(*N*-(2-chloro-4-nitrophenyl)carbamoyl)hydrazide 526
$C_9H_8Cl_2N_2$
5,6-Dichloro-1,2-dimethylbenzimidazole 906
$C_9H_8Cl_2N_2O_2$
1-Acetyl-3-(3,4-dichlorophenyl)urea 533
3-(3,4-Dichlorophenyl)-1-formyl-1-methylurea 533
Methyl *N*-(2,4-dichlorobenzylideneamino)-carbamate 492
$C_9H_8Cl_2N_2O_3$
2,2-Dichloro-4′-nitro-*o*-acetotoluidide 394
2′,4′-Dichloro-6′-nitropropionanilide 403
$C_9H_8Cl_2N_2O_4$
2-Chloroethyl *N*-(2-chloro-4-nitrophenyl)carbamate 604
2,2-Dichloro-4′-methoxy-2′-nitroacetanilide 394
2,2-Dichloro-4′-methoxy-3′-nitroacetanilide 394
$C_9H_8Cl_2N_2S$
2-(2,4-Dichloroanilino)-2-thiazoline 252
$C_9H_8Cl_2O$
Allyl 2,4-dichlorophenyl ether 182
2′,4′-Dichloropropiophenone 289
$C_9H_8Cl_2O_2$
2,6-Dichlorobenzyl acetate 362
Methyl 2,4-dichlorophenylacetate 362
$C_9H_8Cl_2O_2S$
1-Chloroethyl *S*-(4-chlorophenyl) thiocarbonate 381
2-(2,4-Dichloro-5-methylphenylthio)-acetic acid 325
$C_9H_8Cl_2O_3$
1-Chloroethyl 4-chlorophenyl carbonate 381
α,2-Dichlorobenzyl methyl carbonate 381
3,4-Dichloro-α-methoxyphenylacetic acid 325
2-(2,5-Dichlorophenoxy)-propionic acid 327
Methyl (2,4-dichlorophenoxy)-acetate 355
$C_9H_8Cl_3NO$
N-Benzyl-2,2,2-trichloroacetamide 397
2′-Methyl-2,2,2-trichloroacetanilide 398
2′-Methyl-2,2,5′-trichloroacetanilide 394
2′-Methyl-2,2,6′-trichloroacetanilide 394
3′-Methyl-2,2,2-trichloroacetanilide 397
4′-Methyl-2,2,2-trichloroacetanilide 398
4′-Methyl-2,2,3′-trichloroacetanilide 394
2′,4′,5′-Trichloropropionanilide 410
$C_9H_8Cl_3NO_2$
2-Chloroethyl *N*-(2,3-dichlorophenyl)carbamate 605
2-Chloroethyl *N*-(2,4-dichlorophenyl)carbamate 605
2-Chloroethyl *N*-(2,5-dichlorophenyl)carbamate 605
2-Chloroethyl *N*-(2,6-dichlorophenyl)carbamate 605
2-Chloroethyl *N*-(3,4-dichlorophenyl)carbamate 605
2-Chloroethyl *N*-(3,5-dichlorophenyl)carbamate 604
Ethyl *N*-(2,4,5-trichlorophenyl)-carbamate 687
3′-Methoxy-2,2,2-trichloroacetanilide 396
4′-Methoxy-2,2,2-trichloroacetanilide 396
$C_9H_8Cl_3N_3O_2S$
1-(2-(2,4,5-Trichlorophenoxy)acetyl)-3-thiosemicarbazide 729
$C_9H_8Cl_4N_2O$
4-Morpholino-2,3,5,6-tetrachloropyridine 946
$C_9H_8Cl_4OS$
1-(4-Chlorobenzylthio)-2,2,2-trichloroethanol 209
$C_9H_8Cl_5NO$
2,3,4,5,5-Pentachloro-1-pyrrolidino-2,4-pentadien-1-one 109
$C_9H_8Cl_5NO_2$
1-Morpholino-2,3,4,5,5-pentachloro-2,4-pentadien-1-one 109
$C_9H_8Cl_6O_2$
1,4,5,6,7,7-Hexachloro-5-norbornene-2,3-dimethanol 16
$C_9H_8Cl_6Si$
Chloromethyldimethyl(pentachlorophenyl)silane 1099
C_9H_8FN
4-Fluoro-α-methylphenylacetonitrile 765
$C_9H_8FNO_2$
4-Fluoro-β-nitropropenylbenzene 258
$C_9H_8F_3NO$
N-Methyl-3-trifluoromethylbenzamide 446
α,α,α-Trifluoro-*p*-acetotoluidide 398
α,α,α-Trifluoro-*o*-acetotoluidide 398
$C_9H_8F_3NO_2$
Methyl *N*-(2-trifluoromethylphenyl)-carbamate 611
Methyl *N*-(3-trifluoromethylphenyl)-carbamate 610
$C_9H_8F_3N_3O_2$
4-Nitrobenzaldehyde (2,2,2-trifluoroethyl)hydrazone 1021
$C_9H_8F_3N_3O_4$
2,6-Dinitro-*N*-ethyl-4-(trifluoromethyl)aniline 259
$C_9H_8F_3N_3O_6S$
2′,6′-Dinitro-*N*-methyl-4′-(trifluoromethyl)-methanesulfonanilide 805
$C_9H_8F_4O$
α,α-Bis(difluoromethyl)benzyl alcohol 197
$C_9H_8F_4O_3S$
2,2,3,3-Tetrafluoropropyl benzenesulfonate 796
$C_9H_8N_2$
3-Aminoisoquinoline 997
5-Aminoisoquinoline dihydrochloride 997
2-Aminoquinoline 988
5-Aminoquinoline hydrochloride 988
2-Phenylimidazole hydrochloride 869
3-Phenyl-1*H*-pyrazole 864
1-(2-Pyridyl)-pyrrole 954
$C_9H_8N_2O$
2-Acetylbenzimidazole 906
2-Amino-5-phenyloxazole 872
2-Cyanoacetanilide 757
2-Hydroxy-3-methylquinoxaline 999
Indole-3-carboxaldehyde oxime 1010
8-Methyl-4-hydroxycinnoline 999
7-Methyl-4(3*H*)-quinazolinone 1000
3-Phenyl-1*H*-pyrazol-5(4*H*)-one 862
N-(*o*-Tolyl)cyanoformamide 767
$C_9H_8N_2OS$
N-(2-Benzothiazolyl)acetamide 916
3-Phenyl-2-thioxo-4-imidazolidinone 743
$C_9H_8N_2O_2$
5-Benzyl-2,3-dihydro-1,3,4-oxadiazol-2-one 873
2,5-Dimethyl-4-nitrobenzonitrile 766
N-(2-Methoxyphenyl)-cyanoformamide 762
Methyl α-cyano-2-pyrroleacrylate 841
4-Methyl-1,8-naphthyridine-2,7-diol 999
2-Methyl-5-nitroindole 885
$C_9H_8N_2O_2S$
2-Amino-4-(3,4-dihydroxyphenyl)-thiazole hydrochloride 875
(2-Benzimidazolylthio)-acetic acid 903
$C_9H_8N_2O_2S_2$
2-(Methylsulfonyl)-5-phenyl-1,3,4-thiadiazole 879
N-(2-Pyridyl)-2-thiophenesulfonamide 965
$C_9H_8N_2O_3$
7,8-Dihydro-1,3-dioxolo(4,5-g)-quinazolin-6(5*H*)-one 711
2,3-Dihydro-2-oxo-5-benzimidazolyl acetate 907
1-Methyl-5-nitro-2-indolinone 885
6-Nitro-1-indolinecarboxaldehyde 702
N-(3-Pyridyl)-maleamic acid 944
$C_9H_8N_2O_4$
4-Acetamido-3-nitrobenzaldehyde 424
3-Hydroxy-3-nitromethyloxindole 689
Ketomalonic acid phenylhydrazone 343
$C_9H_8N_2O_5$
2-Acetamido-4-nitrobenzoic acid 426
Allyl 2,4-dinitrophenyl ether 267
β,2-Dinitro-4-methoxystyrene 267
3-Nitrohippuric acid 470
$C_9H_8N_2O_5S$
2-Cyanoethyl 3-nitrobenzenesulfonate 799
$C_9H_8N_2O_6$
3-Methyl-4-(2-nitrovinyl)-2,5-pyrroledicarboxylic acid 839
$C_9H_8N_2S$
2-Amino-4-phenylthiazole 876
N-(*p*-Tolyl)cyanothioformamide 767
$C_9H_8N_4$
4-(Benzylideneamino)-4*H*-1,2,4-triazole 1068
5-Methyl-3-(2-pyridyl)-1,2,4-triazine 979
$C_9H_8N_4O$
6-Amino-2-(4-pyridyl)-4-pyrimidinol 971
2-Cyano-2-(phenylhydrazono)-acetamide 765
4-(2-Hydroxybenzylideneamino)-4*H*-1,2,4-triazole 1037
$C_9H_8N_4O_2$
4-Amino-3-methyl-6-nitrocinnoline 999
4-((2,4-Dihydroxybenzylidene)-amino)-4*H*-1,2,4-triazole 1029
3-Phenyl-5-ureido-1,2,4-oxadiazole 873
$C_9H_8N_4O_5$
1,3-Dimethyl-2,4-dioxo-7-hydroxy-1,2,3,4-tetrahydro-6-pteridinecarboxylic acid 977
$C_9H_8N_4S_2$
1-(5-Phenyl-1,2,4-thiadiazol-3-yl)-2-thiourea 879
$C_9H_8N_6O_2$
2,4-Diamino-6-(3-nitrophenyl)-1,3,5-triazine 979
C_9H_8O
(3-Methoxyphenyl)acetylene 1017
1-Phenylpropargyl alcohol 1016
$C_9H_8OS_2$
2-(Thiobenzoyl)thiolacetic acid 331
$C_9H_8O_2$
Cinnamic acid, dodecylamine salt 351
Cinnamic acid, mercury(II) salt 336
Cinnamic acid, potassium salt 351
Cinnamic acid, sodium salt 351
2-Phenylacrylic acid 345
3-Vinylbenzoic acid 345
$C_9H_8O_2S$
2*H*-1-Benzothiopyran-1,1-dioxide 783
$C_9H_8O_2S_2$
Carboxymethyl dithiobenzoate 376
$C_9H_8O_3$
2-(4-Formylphenoxy)acetic acid 327
cis-4-Hydroxycinnamic acid 334
5-Hydroxy-6-methyl-2(3*H*)-benzofuranone 383
7-Hydroxy-3-methylphthalide monohydrate 382
6-Methoxyphthalide 383
3-Phenylglycidic acid, potassium salt 350
Vinyl salicylate 368
$C_9H_8O_3S_2$
Benzo(b)thiophene-4-yl methanesulfonate 898
$C_9H_8O_4$
3-Acetoxybenzoic acid 378
3-Formyl-4-hydroxyphenyl acetate 377
4-Formylphenoxyacetic acid 327
2-(4-Formylphenoxy)acetic acid, sodium salt 350
6-Methyl-1,3-benzodioxole-5-carboxylic acid 346
2-Methylisophthalic acid 345
Methyl piperonylate 375
2,5-Norbornadiene-2,3-dicarboxylic acid 65
$C_9H_8O_4S$
2-(2-Carboxyphenylthio)acetic acid 341
$C_9H_8O_5$
o-(Carbomethoxy)-salicylic acid 381
α-Carboxy-*o*-anisic acid 340
2-(3-Hydroxy-4-methoxyphenyl)-glyoxylic acid 326
4-Methoxyphthalic acid 340
Methyl 2,4-dihydroxy-3-formylbenzoate 367
5-Oxotricyclo(2.2.1.0(2,6))heptane-2,3-dicarboxylic acid 65
$C_9H_8O_5S$
6-Ethoxy-3*H*-2,1-benzothiol-3-one-1,1-dioxide 797
$C_9H_8O_6$
5-Carboxydehydroacetic acid 96
Dimethyl 4-oxo-4*H*-pyran-2,6-dicarboxylate 92
Ethyl 2,3,5-trioxo-1-cyclopentaneglyoxylate 93
C_9H_8S
Phenyl propargyl sulfide 1017
C_9H_9Br
β-Bromo-α-methylstyrene 169
$C_9H_9BrClNO$
4′-Bromo-2-chloro-*o*-acetotoluidide 391
2-Bromo-2′-chloropropionanilide 438
$C_9H_9BrClNO_2$
2-Bromo-4-chloro-6-methylphenyl *N*-methylcarbamate 640
2-Chloroethyl *N*-(4-bromophenyl)-carbamate 605
C_9H_9BrFNO
3-Bromo-3′-fluoropropionanilide 390
$C_9H_9BrN_2$
3-(4-Bromoanilino)propionitrile 751
$C_9H_9BrN_2O_4$
5-Bromo-3,6-dinitropseudocumene 261
1-Bromo-3,5-dinitro-2,4,6-trimethylbenzene 261
2-Bromo-*N*-(4-methoxy-2-nitrophenyl)acetamide 391
$C_9H_9BrO_2$
4-(2-Bromoethoxy)benzaldehyde 321
2-Bromomethyl-1,4-benzodioxane 183
2-(2-Bromophenyl)-1,3-dioxolane 184
4-Bromophenyl-2,3-epoxypropyl ether 189
$C_9H_9BrO_2S$
3-Bromo-1-propenyl phenyl sulfone 772
$C_9H_9BrO_3$
3-Bromo-2,5-dimethoxybenzaldehyde 322
3-Bromo-4,5-dimethoxybenzaldehyde 322
4-Bromo-2,5-dimethoxybenzaldehyde 336
2-Bromo-4,6-dimethyl-3-hydroxybenzoic acid 336
(2-Bromo-4-methylphenoxy)-acetic acid 325
$C_9H_9BrO_4$
3-Bromo-4,5-dimethoxybenzoic acid 336
C_9H_9BrS
Allyl 4-bromophenyl sulfide 210
$C_9H_9Br_2ClN_2S$
1-(2-Chloro-4,6-dibromophenyl)-3-ethyl-2-thiourea 733

$C_9H_9Br_2NO_2$
2-Amino-α,β-dibromohydrocinnamic acid ... 330
$C_9H_9Br_2NO_3$
3,5-Dibromo-L-tyrosine monohydrate ... 348
$C_9H_9Br_3$
1,2,3-Tribromo-4,5,6-trimethylbenzene ... 170
1,2,4-Tribromo-3,5,6-trimethylbenzene ... 170
1,3,5-Tribromo-2,4,6-trimethylbenzene ... 170
C_9H_9ClFNO
3-Chloro-4′-fluoropropionanilide ... 390
$C_9H_9ClFNO_2$
2-Chloroethyl *N*-(2-fluorophenyl)-carbamate ... 604
2-Chloroethyl *N*-(4-fluorophenyl)-carbamate ... 604
$C_9H_9ClFN_3S$
3-(4-Chloro-2-fluorobenzylideneamino)-2-methyl-2-thiopseudourea hydriodide ... 1026
$C_9H_9ClN_2$
3-(4-Chloroanilino)propionitrile ... 751
$C_9H_9ClN_2O$
1-Carbamoyl-5-chloroindoline ... 714
1-(3-Chlorophenyl)-2-imidazolidinone ... 711
$C_9H_9ClN_2OS$
4-(4-Chlorophenyl)-4-hydroxy-2-iminothiazolidine hydrobromide ... 728
$C_9H_9ClN_2O_2$
1-Acetyl-3-(4-chlorophenyl)urea ... 533
2′-Carbamoyl-2-chloroacetanilide ... 445
1-(2-Chloro-2-phenylacetyl)urea ... 579
3-(4-Chlorophenyl)-1-formyl-1-methylurea ... 533
Methyl (2-chlorobenzylidene)-carbazate ... 1025
$C_9H_9ClN_2O_3$
5′-Chloro-4′-methyl-2′-nitroacetanilide ... 403
2-Chloro-4′-nitro-*o*-acetotoluidide ... 392
2-Chloro-5′-nitro-*o*-acetotoluidide ... 392
4′-Chloro-6′-nitro-*m*-acetotoluidide ... 403
2′-Chloro-4′-nitropropionanilide ... 404
2′-Chloro-5′-nitropropionanilide ... 404
4′-Chloro-3′-nitropropionanilide ... 404
$C_9H_9ClN_2O_4$
2-Chloroethyl *N*-(3-nitrophenyl)-carbamate ... 604
2-Chloro-2′-methoxy-4′-nitroacetanilide ... 391
2-Chloro-4′-methoxy-2′-nitroacetanilide ... 391
2-Chloro-4′-methoxy-3′-nitroacetanilide ... 391
5′-Chloro-4′-methoxy-2′-nitroacetanilide ... 415
3-(4-Chloro-2-nitroanilino)-propionic acid ... 330
Ethyl *N*-(2-chloro-4-nitrophenyl)-carbamate ... 628
Ethyl *N*-(4-chloro-2-nitrophenyl)-carbamate ... 624
$C_9H_9ClN_2S$
2-(3-Chloroanilino)-2-thiazoline ... 252
$C_9H_9ClN_4O_2$
4-Amino-6-nitrocinnoline methochloride ... 999
C_9H_9ClO
Allyl 2-chlorophenyl ether ... 180
7-Chloro-4-indanol ... 192
p-Tolylacetyl chloride ... 389
C_9H_9ClOS
(4-Chlorophenylthio)acetone ... 284
2-(Phenylthio)propionyl chloride ... 389
$C_9H_9ClO_2$
Benzyl 2-chloroacetate ... 353
2-Chlorobenzaldehyde ethylene ketal ... 184
4-Chlorobenzyl acetate ... 353
1-(2-Chlorophenoxy)-2,3-epoxypropane ... 189
Ethyl 4-chlorobenzoate ... 367
Methyl 3-(chloromethyl)benzoate ... 365
(+/-)-Methyl α-chlorophenylacetate ... 361
Methyl 2-(4-chlorophenyl)acetate ... 353
2-Phenoxypropionyl chloride ... 389
$C_9H_9ClO_2S$
1-Chloroethyl *S*-phenyl thiocarbonate ... 382
Methyl 2-(4-chlorophenylthio)-acetate ... 357
$C_9H_9ClO_3$
2-Chloro-2′-hydroxy-5′-methoxyacetophenone ... 287
5-Chloromethyl-*o*-anisic acid ... 336
$C_9H_9ClO_3S$
4-Chlorophenylsulfonylacetone ... 781
2-(*p*-Tolylsulfonyl)-acetyl chloride ... 782
$C_9H_9ClO_4$
2-(4-Chloro-2-methoxyphenoxy)-acetic acid ... 325
C_9H_9ClS
2-Chloroallyl phenyl sulfide ... 211
3-Chloroallyl phenyl sulfide (*cis* and *trans*) ... 211
$C_9H_9Cl_2NO$
2-Chloro-*N*-(2-chlorobenzyl)-acetamide ... 390
2-Chloro-*N*-(4-chlorobenzyl)-acetamide ... 392
2,3′-Dichloro-*o*-acetotoluidide ... 393
2,3′-Dichloro-*p*-acetotoluidide ... 391
2,4′-Dichloro-*o*-acetotoluidide ... 393
2,5′-Dichloro-*o*-acetotoluidide ... 393
2,6′-Dichloro-*m*-acetotoluidide ... 393
N-(2,4-Dichlorobenzylidene)-ethanolamine ... 1022
N-(3,4-Dichlorobenzylidene)-ethanolamine ... 1023
3′,4′-Dichloropropioanilide ... 440
$C_9H_9Cl_2NOS$
2,6-Dichloro-*N*-(methoxymethyl)-thiobenzamide ... 738
$C_9H_9Cl_2NO_2$
2-Chloroethyl *N*-(2-chlorophenyl)-carbamate ... 605
2-Chloroethyl *N*-(3-chlorophenyl)-carbamate ... 604
2-Chloroethyl *N*-(4-chlorophenyl)-carbamate ... 605
2,2-Dichloro-2′-methoxyacetanilide ... 394
2,3′-Dichloro-2′-methoxyacetanilide ... 390
2,4′-Dichloro-2′-methoxyacetanilide ... 390
2,5′-Dichloro-2′-methoxyacetanilide ... 390
2,6-Dichlorophenyl *N*,*N*-dimethylcarbamate ... 683
Ethyl 2,3-dichlorophenylcarbamate ... 683
Ethyl *N*-(2,4-dichlorophenyl)-carbamate ... 637
Ethyl *N*-(2,5-dichlorophenyl)-carbamate ... 687
Ethyl *N*-(2,6-dichlorophenyl)-carbamate ... 635
Ethyl *N*-(3,4-dichlorophenyl)-carbamate ... 685
$C_9H_9Cl_2NO_3$
Ethyl *N*-(3,4-dichlorophenyl)-*N*-hydroxycarbamate ... 628
Methyl *N*-(3,4-dichlorophenyl)-*N*-methoxycarbamate ... 628
$C_9H_9Cl_2NO_4S$
Methyl *N*-(3,4-dichlorophenyl)-*N*-(methylsulfonyl)carbamate ... 808
$C_9H_9Cl_2N_3O$
1-(1-Amino-2-(3,4-dichlorophenyl)-ethylidene)urea hydrochloride ... 723
$C_9H_9Cl_2N_3O_2$
1-(2,5-Dichlorophenyl)-5-methylbiuret ... 533
Guanidino 2-(2,6-dichlorophenyl)-acetate ... 724
$C_9H_9Cl_3N_2O$
1,1-Dimethyl-3-(2,4,5-trichlorophenyl)urea ... 603
1,1-Dimethyl-3-(2,4,6-trichlorophenyl)urea ... 527
1-Ethyl-3-(2,4,5-trichlorophenyl)urea ... 546
$C_9H_9Cl_3O$
3-Chloropropyl-2,4-dichlorophenyl ether ... 176
3-Ethyl-5-methyl-2,4,6-trichlorophenol ... 190
$C_9H_9Cl_5Si$
(Pentachlorophenyl)trimethylsilane ... 1100
$C_9H_9Cl_5Sn$
(Pentachlorophenyl)trimethyltin ... 1129
$C_9H_9FN_2O$
Acetic (3-fluorobenzylidene)-hydrazide ... 1025
1-(*N*-(4-Fluorophenyl)carbamoyl)-aziridine ... 695
$C_9H_9FN_4O_2S_2$
2-(4-Fluorosulfonylphenyl)-5-hydrazono-4-methyl-1,3,4-thiadiazol-2-one dihydrochloride ... 803
$C_9H_9FN_4O_3$
Oxamic *N*2-(*N*-(4-fluorophenyl)-carbamoyl)hydrazide ... 532
C_9H_9FO
4′-Fluoro-2′-methylacetophenone ... 289
$C_9H_9FO_2$
4-Fluoro-α-methylphenylacetic acid ... 333
2-(4-Fluorophenyl)-1,3-dioxolane ... 183
$C_9H_9F_2NO$
N-(2,5-Difluorophenyl)propionamide ... 409
$C_9H_9F_2NOS$
4′-(Difluoromethylthio)-acetanilide ... 394
$C_9H_9F_3N_2O$
2′-Amino-4′-trifluoromethylacetanilide ... 396
1-Methyl-3-(2-trifluoromethylphenyl)urea ... 511
1-Methyl-3-(3-trifluoromethylphenyl)urea ... 510
$C_9H_9F_3N_2O_2$
N,*N*-Dimethyl-4-nitro-2-(trifluoromethyl)aniline ... 259
$C_9H_9F_3O_3S$
2,2,2-Trifluoroethyl α-toluenesulfonate ... 796
$C_9H_9F_5Sn$
(Pentafluorophenyl)trimethyltin ... 1129
$C_9H_9IO_2$
3,5-Dimethyl-2-iodobenzoic acid ... 346
$C_9H_9IO_2S$
2-Iodo-1-propenyl phenyl sulfone ... 783
$C_9H_9I_2NO_3$
3,5-Diiodo-DL-tyrosine ... 348
C_9H_9N
2,3-Dimethylbenzonitrile ... 767
2,4-Dimethylbenzonitrile ... 767
3,4-Dimethylbenzonitrile ... 767
C_9H_9NO
5-Amino-1-indanone ... 294
Cinnamaldehyde oxime ... 1013
3,4-Dihydro-2(1*H*)-quinolinone ... 692
2,3-Dimethylphenyl isocyanate ... 769
3-Ethylphenyl isocyanate ... 768
4-Ethylphenyl isocyanate ... 769
2-Indanone oxime ... 1014
1-Methyl-2-indolinone ... 693
2-Methylisoindolin-1-one ... 693
3-Phenyl-2-azetidinone ... 691
2,4-Xylyl isocyanate ... 769
C_9H_9NOS
2,5-Dihydro-1,4-benzothiazepin-3(4*H*)-one ... 700
2,3-Dimethyl-4-thiocyanophenol ... 768
2-Ethoxybenzothiazole ... 913
2,3,4,5-Tetrahydro-1,5-benzothiazepin-4-one ... 706
Thiochroman-4-one oxime ... 1013
$C_9H_9NO_2$
3′-Acetylformanilide ... 433
2-Aminocinnamic acid ... 335
4-Ethoxyphenyl isocyanate ... 769
4-(Formamido)acetophenone ... 392
N-Formyl-2-phenylacetamide ... 428
6-Methoxy-*m*-tolyl isocyanate ... 769
β-Methyl-β-nitrostyrene ... 258
4-Methyl-β-nitrostyrene ... 258
Methyl 4-pyridineacrylate ... 962
Pyruvanilide ... 434
$C_9H_9NO_2S$
3,4-Dimethoxyphenyl thiocyanate ... 768
$C_9H_9NO_3$
2-Acetamidobenzoic acid ... 433
N-Acetylsalicylamide ... 462
2-(Formylphenyl) *N*-methylcarbamate ... 674
3-Methoxy-β-nitrostyrene ... 257
4-Methoxy-β-nitrostyrene ... 257
2-(4-Methoxyphenyl)glyoxal 1-oxime ... 1009
3-(2-Nitrophenyl)propylene oxide ... 278
3′-Nitropropiophenone ... 295
$C_9H_9NO_3S$
2-(Carbamoylmethylthio)-benzoic acid ... 421
Cyanomethyl *p*-toluenesulfonate ... 798
2,3-Dihydro-*S*,*S*-dioxide-1,4-benzothiazepin-5(4*H*)-one ... 779
$C_9H_9NO_4$
4-Acetamidosalicylic acid ... 411
4-Acetamidosalicylic acid, calcium salt trihydrate ... 411
4-Acetamidosalicylic acid, sodium salt dihydrate ... 411
o-Benzamidoglycolic acid ... 475
2-Carbamoylphenoxyacetic acid ... 467
2-Carbamoylphenoxyacetic acid, sodium salt ... 464
2,5-Dimethyl-4-nitrobenzoic acid ... 344
3,5-Dimethyl-4-nitrobenzoic acid ... 344
Dimethyl 2,5-pyridinedicarboxylate ... 962
Dimethyl 3,4-pyridinedicarboxylate ... 940
Ethyl 2-nitrobenzoate ... 374
Ethyl 3-nitrobenzoate ... 373
N-Furfurylmaleamic acid ... 853
4-Hydroxy-3-methoxy-β-nitrostyrene ... 257
4′-Methoxy-3′-nitroacetophenone ... 292
Methyl 5-methyl-2-nitrobenzoate ... 374
Methyl 4-nitro-3-methylbenzoate ... 374
2-Methyl-3-nitrophenyl acetate ... 379
2-Methyl-6-nitrophenyl acetate ... 379
Methyl 4-nitrophenylacetate ... 360
4-Nitrobenzyl acetate ... 360
1-(3-Nitrophenoxy)acetone ... 283
2-(3-Nitrophenyl)-1,3-dioxolane ... 279
4-Nitrophenyl propionate ... 379
$C_9H_9NO_4S$
2-Acetamido-5-acetyl-3-thiophenecarboxylic acid, sodium salt ... 857
2-(2-Nitrobenzylthio)acetic acid ... 329
$C_9H_9NO_5$
2′,4′-Dihydroxy-3′-methyl-5′-nitroacetophenone ... 289
3,4-Dimethoxy-2-nitrobenzaldehyde ... 322
3,6-Dimethoxy-2-nitrobenzaldehyde ... 322
2-Methoxy-4-nitrophenyl acetate ... 377
Methyl 3-hydroxy-4-methyl-2-nitrobenzoate ... 368
Methyl 3-methoxy-5-nitrobenzoate ... 369
Methyl 4-methoxy-3-nitrobenzoate ... 369
C_9H_9NS
Benzylthioacetonitrile ... 754
2,3-Dihydro-3-methyl-2-methylenebenzothiazole ... 251
2,6-Dimethylbenzothiazole ... 917
α-Methylbenzyl isothiocyanate ... 768
3-Methylbenzyl thiocyanate ... 768
3-(Phenylthio)propionitrile ... 754
2,4-Xylyl isothiocyanate ... 768
3,4-Xylyl isothiocyanate ... 768
$C_9H_9N_3$
2-Amino-1,2-dihydro-1-isoquinolinimine ... 726
1-Amino-1,2-dihydro-2-quinolinimine ... 726
5-Amino-1-phenylpyrazole ... 860
(1-Methyl-2-benzimidazolyl)-acetonitrile ... 907
$C_9H_9N_3O$
1-Acetyl-2-aminobenzimidazole ... 904
2-Amino-5-benzyl-1,3,4-oxadiazole ... 873
5′-Amino-2′-cyanoacetanilide ... 763
3-Benzyl-4-hydroxy-4*H*-1,2,4-triazole ... 869
3-Hydrazinoisoquinoline *N*-oxide ... 1006
1-(*p*-Tolyl)-1,2,4-triazol-3-ol ... 869
$C_9H_9N_3OS$
5-Anilino-3-methoxy-1,2,4-thiadiazole ... 878
5,7-Dimethyl-4-hydroxy-2-mercapto-pyrido(2,3-d)-pyrimidine ... 1002

$C_9H_9N_3OS$ (Cont'd)
7-Methyl-3-(methylthio)-1,2,4-benzotriazine *N*¹-oxide ... 1005
$C_9H_9N_3O_2S$
1-Benzoyl-2-thiobiuret ... 745
2-(Methylthiomethyl)-5-nitrobenzimidazole ... 903
Piperonal thiosemicarbazone ... 1052
$C_9H_9N_3O_2S_2$
Benzenesulfonic (4-thiazolin-2-ylidene)hydrazide ... 821
$C_9H_9N_3O_3$
2-(2-Hydroxyethylamino)-5-nitrobenzonitrile ... 761
$C_9H_9N_3O_3S$
2-(3'-Nitrophenacyl)-2-thiopseudourea hydrobromide ... 743
2-(4'-Nitrophenacyl)-2-thiopseudourea hydrobromide ... 743
$C_9H_9N_3O_4$
1,2-Dihydro-4-(methoxymethyl)-6-methyl-5-nitro-2-oxonicotinonitrile ... 937
N,N'-(2-Nitrobenzylidene)bisformamide ... 426
$C_9H_9N_3O_5$
2',5'-Dinitro-*p*-acetotoluidide ... 423
4',5'-Dinitro-*o*-acetotoluidide ... 423
5',6'-Dinitro-*o*-acetotoluidide ... 424
$C_9H_9N_3O_5S$
2-Cyanoethyl 4-amino-3-nitrobenzenesulfonate ... 798
$C_9H_9N_3O_6$
4-Dimethylamino-3,5-dinitrobenzoic acid ... 342
2',4'-Dinitro-5'-methoxyacetanilide ... 414
N-(2,4-Dinitrophenyl)-L-alanine ... 349
$C_9H_9N_3O_7$
N-(2,4-Dinitrophenyl)-L-serine ... 348
$C_9H_9N_3O_7S$
2',6'-Dinitro-4'-(methylsulfonyl)acetanilide ... 777
$C_9H_9N_3O_8S$
Methyl *N*-(2,6-dinitro-4-(methylsulfonyl)phenyl)carbamate ... 777
$C_9H_9N_3S$
2-Amino-5-phenyl-6*H*-1,3,4-thiadiazine hydrobromide ... 248
2-Hydrazino-4-phenylthiazole ... 876
$C_9H_9N_3S_2$
5-(*o*-Toluidino)-1,3,4-thiadiazole-2(3*H*)-thione ... 879
5-(*p*-Toluidino)-1,3,4-thiadiazole-2(3*H*)-thione ... 879
$C_9H_9N_5$
2-Amino-4-phenylamino-1,3,5-triazine hydrochloride ... 979
2,4-Diamino-6-phenyl-1,3,5-triazine ... 979
$C_9H_9N_5O$
5-Amino-4-phenylazo-3-pyrazolol ... 1083
3-Phenyl-5-ureido-1,2,4-triazole ... 869
$C_9H_9N_5O_2$
5-(*o*-Vanillylideneamino)-1*H*-tetrazole ... 1031
$C_9H_9N_5O_3S$
N(4-Amino-*s*-triazin-2-yl)-sulfanilic acid ... 979
$C_9H_9N_5O_4$
1,3-Dimethyl-2,4-dioxo-7-hydroxy-1,2,3,4-tetrahydro-6-pteridinecarboxamide ... 977
$C_9H_9N_5S$
1-(Cyanoamidino)-3-phenyl-2-thiourea ... 747
C_9H_9P
1,2-Dihydrophosphinoline ... 830
$C_9H_{10}AsNO_8$
4-(Ethoxycarbonyloxy)-3-nitrobenzenearsonic acid ... 1126
$C_9H_{10}BF_4NO_2$
2-Ethyl-7-hydroxy-1,2-benzisoxazolium tetrafluoroborate ... 909
$C_9H_{10}BrClO_2S$
Bromomethyl 2-chloro-2-phenylethyl sulfone ... 772
$C_9H_{10}BrClS$
4-Bromophenyl 3-chloropropyl sulfide ... 209
$C_9H_{10}BrNO$
4'-Bromo-*o*-acetotoluidide ... 440
N-(2-Bromoethyl)benzamide ... 477
2'-Bromo-4'-methylacetanilide ... 440
4'-Bromo-*N*-methylacetanilide ... 410
N-(4-Bromophenyl)propionamide ... 409
3'-Bromopropionanilide ... 407
$C_9H_{10}BrNO_2$
2-Bromoethyl *N*-phenylcarbamate ... 605
1-Bromo-3-nitro-2,4,6-trimethylbenzene ... 262
Ethyl *N*-(4-bromophenyl)carbamate ... 634
$C_9H_{10}BrNO_4$
Ethyl 5-bromo-4-carboxy-3-methyl-2-pyrrolecarboxylate ... 840
$C_9H_{10}Br_2$
α-Bromo-2-(2-bromoethyl)toluene ... 169
$C_9H_{10}Br_2N_4$
1-((2,5-Dibromo-α-methylbenzylidene)amino)guanidine nitrate ... 1026
$C_9H_{10}Br_2O_2$
(2-Bromoethyl) (2-bromo-4-methoxyphenyl) ether ... 176
$C_9H_{10}Br_2O_2Te$
3-(Phenyltelluro)propionic acid, Te,Te-dibromide ... 332
$C_9H_{10}ClFO$
3-Chloropropyl 4-fluorophenyl ether. 176
$C_9H_{10}ClNO$
1-(2-Acetylvinyl)pyridinium chloride. 956
2-Chloro-*o*-acetotoluidide ... 393
2-Chloro-*m*-acetotoluidide ... 393
2-Chloro-*p*-acetotoluidide ... 393
3'-Chloro-*o*-acetotoluidide ... 439
3'-Chloro-*p*-acetotoluidide ... 409
4'-Chloro-*m*-acetotoluidide ... 409
N-(4-Chlorobenzylidene)ethanolamine ... 1025
2-Chloro-*N*-methylacetanilide ... 392
3'-Chloro-*N*-methylacetanilide ... 441
4'-Chloro-2'-methylacetanilide ... 409
5'-Chloro-2'-methylacetanilide ... 440
3-Chloropropionanilide ... 393
4'-Chloropropionanilide ... 409
$C_9H_{10}ClNOS$
S-(2-Chloroethyl) *N*-phenylthiocarbamate ... 606
6'-Chloro-2-mercapto-*o*-acetotoluidide ... 420
2-Chloro-3'-(methylthio)acetanilide ... 392
$C_9H_{10}ClNO_2$
β-Amino-4-chlorohydrocinnamic acid ... 348
2-Chloro-*m*-acetanisidide ... 391
2-Chloro-*p*-acetanisidide ... 392
4-Chlorobenzyl *N*-methylcarbamate. 634
2-Chloroethyl *N*-phenylcarbamate ... 605
5'-Chloro-2'-hydroxy-*p*-acetotoluidide ... 413
5'-Chloro-2'-hydroxypropionanilide ... 398
2-Chloro-2'-methoxyacetanilide ... 391
3'-Chloro-4'-methoxyacetanilide ... 420
4-Chloro-3-methylphenyl *N*-methylcarbamate ... 637
Ethyl 2-chlorophenylcarbamate ... 681
Ethyl 3-chlorophenylcarbamate ... 687
Ethyl 4-chlorophenylcarbamate ... 687
Methyl *N*-(2-chloro-6-methylphenyl)carbamate ... 635
Methyl *N*-(3-chloro-4-methylphenyl)carbamate ... 685
Methyl *N*-(4-chloro-2-methylphenyl)carbamate ... 685
Methyl *N*-(5-chloro-2-methylphenyl)carbamate ... 685
$C_9H_{10}ClNO_2S$
3-Chloropropyl 4-nitrophenyl sulfide ... 258
$C_9H_{10}ClNO_3$
5'-Chloro-2',4'-dimethoxyformanilide ... 417
3-Chloropropyl 4-nitrophenyl ether ... 258
Ethyl *N*-(3-chlorophenyl)-*N*-hydroxycarbamate ... 628
Methyl *N*-(3-chloro-2-methoxyphenyl)carbamate ... 619
Methyl *N*-(3-chlorophenyl)-*N*-methoxycarbamate ... 628
$C_9H_{10}ClNO_4S$
Methyl *N*-(3-chlorophenyl)-*N*-(methylsulfonyl)carbamate ... 808
Methyl *N*-(4-chlorophenyl)-*N*-(methylsulfonyl)carbamate ... 808
$C_9H_{10}ClNS$
4'-Chlorothiopropionanilide ... 746
$C_9H_{10}ClN_3$
N-(2-Chlorobenzylideneamino)acetamidine hydriodide ... 1026
$C_9H_{10}ClN_3O$
3-Amino-6-chloro-5-cyano-4-(methoxymethyl)-2-methylpyridine.. 938
1-(2-Carbamoylethyl)-3-cyanopyridinium chloride ... 948
$C_9H_{10}ClN_3O_2$
1-Acetyl-4-(3-chlorophenyl)semicarbazide ... 528
2-(4-Chlorophenylazo)-2-nitropropane ... 1026
$C_9H_{10}ClN_3O_2S$
1-(2-Chloro-4-nitrophenyl)-3-ethyl-2-thiourea ... 730
$C_9H_{10}ClN_3O_3$
3-(2-Chloro-4-nitrophenyl)-1,1-dimethylurea ... 568
$C_9H_{10}Cl_2$
4-Chloro-1-(1-chloropropyl)benzene.. 169
$C_9H_{10}Cl_2N_2O$
1-(4,5-Dichloro-2-methylphenyl)-3-methylurea ... 546
3-(2,3-Dichlorophenyl)-1,1-dimethylurea ... 593
3-(2,4-Dichlorophenyl)-1,1-dimethylurea ... 597
3-(2,5-Dichlorophenyl)-1,1-dimethylurea ... 597
3-(2,6-Dichlorophenyl)-1,1-dimethylurea ... 593
3-(3,4-Dichlorophenyl)-1,1-dimethylurea ... 543
3-(3,5-Dichlorophenyl)-1,1-dimethylurea ... 602
1-(2,3-Dichlorophenyl)-3-ethylurea ... 593
1-(2,4-Dichlorophenyl)-3-ethylurea ... 597
1-(2,5-Dichlorophenyl)-3-ethylurea ... 543
1-(2,6-Dichlorophenyl)-3-ethylurea ... 540
1-(3,4-Dichlorophenyl)-3-ethylurea ... 597
$C_9H_{10}Cl_2N_4O_2$
2-(3,4-Dichlorophenylureido)acetic acid hydrazide ... 534
$C_9H_{10}Cl_2O$
2,4-Dichloro-3-ethyl-5-methylphenol. 190
2,4-Dichloro-5-ethyl-3-methylphenol. 190
2,4-Dichloro-3,5,6-trimethylphenol ... 190
$C_9H_{10}Cl_2O_2S$
p-Chlorobenzyl 2-chloroethyl sulfone ... 772
$C_9H_{10}Cl_2O_2Te$
3-(Phenyltelluro)propionic acid Te,Te-dichloride ... 1142
$C_9H_{10}Cl_2S$
4-Chlorobenzyl 2-chloroethyl sulfide ... 209
4-Chlorophenyl 3-chloropropyl sulfide ... 209
$C_9H_{10}Cl_3NO$
1-(6-Methyl-2-pyridyl)-3,3,3-trichloro-2-propanol ... 925
$C_9H_{10}Cl_3N_3O$
1,1-Dimethyl-4-(2,4,5-trichlorophenyl)semicarbazide ... 579
$C_9H_{10}Cl_4Si$
(2,3,4,5-Tetrachlorophenyl)trimethylsilane ... 1100
(2,3,5,6-Tetrachlorophenyl)trimethylsilane ... 1100
$C_9H_{10}Cl_6O_4$
3,9-Bis(trichloromethyl)-2,4,8,10-tetraoxaspiro(5.5)undecane ... 183
$C_9H_{10}FNO$
3'-Fluoropropionanilide ... 407
4'-Fluoropropionanilide ... 438
$C_9H_{10}FNO_2$
4'-(Fluoromethoxy)acetanilide ... 391
p-Fluoro-D-phenylalanine ... 348
$C_9H_{10}FNO_4S$
N-Acetyl-4-methoxymetanilyl fluoride ... 802
$C_9H_{10}F_2N_2S$
1-(2,4-Difluorophenyl)-3-ethyl-2-thiourea ... 733
$C_9H_{10}F_3NO_2S$
6-Ethylsulfonyl-α,α,α-trifluoro-*m*-toluidine ... 772
$C_9H_{10}F_3N_3O_2$
3-(Methacrylamido)-1-(2,2,2-trifluoroethyl)-2-pyrazolin-5-one ... 693
$C_9H_{10}F_3N_5OS$
1-(*N*-(2-(Trifluoromethyl)phenyl)carbamoyl)-3-thiocarbohydrazide ... 728
$C_9H_{10}F_4Si$
(2,3,5,6-Tetrafluorophenyl)trimethylsilane ... 1100
$C_9H_{10}F_6O_4$
Ethyl perfluoroglutarate ... 83
$C_9H_{10}HgO_2$
Benzylmercuric acetate ... 1139
Phenylmercuric propionate ... 351
$C_9H_{10}HgO_3$
(4-Methoxyphenyl)mercuri acetate ... 1138
Phenylmercuric lactate ... 350
$C_9H_{10}INO$
4'-Iodopropionanilide ... 409
$C_9H_{10}INO_2$
4'-Iodo-2-methoxyacetanilide ... 401
$C_9H_{10}INO_4$
Ethyl 5-carboxy-2-iodo-4-methyl-3-pyrrolecarboxylate ... 843
$C_9H_{10}INS$
2,3-Dimethylbenzothiazolium iodide ... 916
3-Methyl-2-(methylthio)benzothiazolium iodide ... 914
$C_9H_{10}I_2O_2Te$
3-(Phenyltelluro)propionic acid Te,Te-diiodide ... 1142
$C_9H_{10}NO_3STl$
2-(Dimethylthallyl(III))saccharin ... 823
$C_9H_{10}N_2$
Cyclohexylidenemalononitrile ... 147
3-(Dimethylamino)benzonitrile ... 764
1,2-Dimethylbenzimidazole ... 906
2-Ethylbenzimidazole ... 905
3-Ethyl-1*H*-indazole ... 900
1-Phenyl-2-imidazoline hydriodide ... 726
$C_9H_{10}N_2O$
Acetic benzylidenehydrazide ... 502
2-Aminocinnamamide ... 422
2-Benzimidazoleethanol ... 902
Cinnamic acid hydrazide ... 502
5,6-Dimethyl-2-hydroxybenzimidazole ... 902
Pyruvic aldehyde 1-phenylhydrazone ... 1069
$C_9H_{10}N_2OS$
1-Acetyl-3-phenyl-2-thiourea ... 745
N-(Thiocarbamoylmethyl)benzamide ... 739
$C_9H_{10}N_2OS_2$
Methyl 3-benzoyldithiocarbazate ... 739
$C_9H_{10}N_2O_2$
Acetic (2-hydroxybenzylidene)hydrazide ... 496
2-Acetoacetamidopyridine ... 961
N-Acetylanthranilamide ... 469
1-Acetyl-2-benzoylhydrazine ... 500
1-Acetyl-3-phenylurea ... 579
(4-Acetylphenyl)urea ... 575
5-Cyano-1-ethyl-6-hydroxy-4-methyl-2(1*H*)-pyridone ... 144
4,7-Dimethoxybenzimidazole ... 903
α-Hydroxyimino-*o*-acetotoluidide ... 1010
α-Hydroxyimino-*p*-acetotoluidide ... 1012
1-(4-Nitrophenyl)azetidine ... 278
2,7-Nonadiynediamide ... 1019
$C_9H_{10}N_2O_2S$
*N*¹-(4-Methoxyphenyl)-1-thiooxamide ... 736
N-(*N*-Phenylthiocarbamoyl)glycine ... 745
$C_9H_{10}N_2O_3$
3'-Amino-4'-methyl-2'-nitroacetophenone ... 294

$C_9H_{10}N_2O_3$ (Cont'd)
5′-Amino-4′-methyl-2′-nitroacetophenone 294
2′-Amino-5′-nitropropiophenone 294
Ethyl 5-amino-4-cyano-2-methyl-4-furancarboxylate 849
Ethyl 2-pyridylglyoxylate oxime 1012
4′-Methyl-2′-nitroacetanilide 424
4′-Methyl-3′-nitroacetanilide 426
4-(3-Methylureido)benzoic acid 576
4′-Nitro-*o*-acetotoluidide 427
5′-Nitro-*o*-acetotoluidide 426
6′-Nitro-*o*-acetotoluidide 424
N-(3-Nitrobenzylidene)-ethanolamine 1037
o-(4-Nitrophenyl)acetone oxime 1011
N-(2-Pyridyl)succinamic acid 963
N-(3-Pyridyl)succinamic acid 963

$C_9H_{10}N_2O_3S$
2′-Methylthio-5′-nitroacetanilide 420

$C_9H_{10}N_2O_4$
4-Acetamido-3-nitroanisole 415
3-Amino-3-(2-nitrophenyl)-propionic acid 330
2,4-Dinitromesitylene 275
Ethyl 4-amino-3-nitrobenzoate 372
Ethyl *N*-(2-nitrophenyl)carbamate 668
Ethyl 3-nitrophenylcarbamate 670
Ethyl 4-nitrophenylcarbamate 671
4-Methoxy-2-methyl-3-nitrobenzamide 442
2′-Methoxy-4′-nitroacetanilide 415
Methyl *N*-(2-methyl-4-nitrophenyl)-carbamate 670
1-(1-Methyl-1-nitroethyl)-3-nitrobenzene 277
N-(3-Nitrophenyl)glycine methyl ester 358

$C_9H_{10}N_2O_4S$
Acetic acid (2-sulfobenzylidene)-hydrazide, sodium salt 794

$C_9H_{10}N_2O_5$
2′,4′-Dihydroxy-3′-methyl-5′-nitroacetophenone oxime 1008
4,5-Dimethoxy-2-nitrobenzaldehyde oxime 1009
2,4-Dinitro-6-isopropylphenol 264

$C_9H_{10}N_2O_6$
2-Ethoxymethyl-4,6-dinitrophenol 263
4-Methyl-3-(2-nitroethyl)-2,5-pyrroledicarboxylic acid 839

$C_9H_{10}N_2O_7$
3-(2,4-Dinitrophenoxy)-1,2-propanediol 262

$C_9H_{10}N_2S$
2-Anilino-2-thiazoline 250
2,3-Dihydro-4,6-dimethyl-2-iminobenzothiazole 749
4-(Dimethylamino)phenyl thiocyanate 768

$C_9H_{10}N_4$
3-Amino-1-benzyl-1*H*-1,2,4-triazole 869
6-Amino-2,3-dimethylpyrido(2,3-b)-pyrazine 1000
4-(Benzylamino)-4*H*-1,2,4-triazole 870
1-Cyano-3-(*p*-tolyl)-guanidine 758
3-Methylamino-5-phenyl-1,2,4-triazole 869

$C_9H_{10}N_4O$
1-Acetamido-2-aminobenzimidazole hydrobromide 904
1-Cyano-3-(4-methoxyphenyl)-guanidine 753

$C_9H_{10}N_4OS$
1-Amidino-3-benzoyl-2-thiourea 745
1-(Benzoylamidino)-2-thiourea 745

$C_9H_{10}N_4O_2$
1-(1-Azido-1-methylethyl)-4-nitrobenzene 769
1,6,7-Trimethyl-2,4(1*H*,3*H*)-pteridinedione 1003

$C_9H_{10}N_4O_3$
3-Allyl-8-hydroxy-1-methylxanthine 921

$C_9H_{10}N_4O_4$
2-Propanone N^2-(2,4-dinitrophenyl)-hydrazone 274
Theophylline-7-acetic acid 866

$C_9H_{10}N_6O_3S$
6-Amino-1,3-dimethyl-5-(2-mercapto-4-oxo-2-imidazolin-5-ylideneamino)uracil 137

$C_9H_{10}O$
2,3-Dihydro-2-methylbenzofuran 184
3-Methyl-1-indanone 297
α-Phenylallyl alcohol 204

$C_9H_{10}OS$
Allyl phenyl sulfoxide 771
2-(Methylthio)acetophenone 293
3-(Phenylthio)-1,2-epoxypropane 211
1-(Phenylthio)-2-propanone 284
2-Phenylthioxolane 214
S-(*p*-Tolyl) thioacetate 381

$C_9H_{10}OS_2$
o-Benzyl *S*-methyl dithiocarbonate 381
Spiro(bicyclo(3.2.0)hept-3-ene-6,2′-(1,3)dithiolan)-7-one 59

$C_9H_{10}O_2$
6,7-Dihydro-2-methylcyclopenta(b)-pyran-4(5*H*)-one 58
o-Ethylbenzoic acid 345
5,6-Indandiol 194
Phenethyl formate 361
3α,4,7,7α-tetrahydro-1,3-indandione 55
m-Tolyl acetate 380

$C_9H_{10}O_2S$
2-(Ethylthio)benzoic acid 342
Methyl styryl sulfone 783
2-(Phenylthio)-1,3-dioxolane 211
2-(Phenylthio)ethyl formate 357
2-(Phenylthio)propionic acid 330
2-(*p*-Tolylthio)acetic acid 330
p-Tolyl vinyl sulfone 785

$C_9H_{10}O_2Te$
3-(Phenyltelluro)propionic acid 1142

$C_9H_{10}O_3$
2′,4′-Dihydroxy-6′-methylacetophenone 289
3,5-Dimethyl-4-hydroxybenzoic acid 339
3,6-Dimethylsalicylic acid 338
Ethyl phenyl carbonate 381
4-Methoxybenzyl formate 355
4-(Methoxymethoxy)benzaldehyde 323
2-Methoxy-4-methylbenzoic acid 341
6-Methoxy-4-methylsalicylaldehyde 322
3-Methoxyphenyl acetate 378
α-Methoxyphenylacetic acid, sodium salt 350
3-Methoxy-*p*-toluic acid 340
Methyl 2-hydroxyphenylacetate 354
Methyl 3-methylsalicylate 368
Methyl 4-methylsalicylate 368
Methyl 5-methylsalicylate 368
3-Methylphenoxyacetic acid 327
5-Methyl-3a,4,7,7a-tetrahydrophthalic anhydride 99
Norbornane-2,3-dicarboxylic anhydride 98
L-3-Phenyllactic acid sodium salt monohydrate 350
o-Tolyloxyacetic acid 327

$C_9H_{10}O_3S$
2-(2-Hydroxyethylthio)benzoic acid 339
5-Indansulfonic acid, ammonium salt 796
2-(Methylsulfonyl)acetophenone 780

$C_9H_{10}O_4$
2′,4′-Dihydroxy-2-methoxyacetophenone 289
2,4′-Dihydroxy-3′-methoxyacetophenone 289
2,4-Dimethoxy-3-hydroxybenzaldehyde 322
3,6-Endoxo-3-methylhexahydrophthalic anhydride 98
4-Ethoxy-2-hydroxybenzoic acid 338
Ethyl 2,4-dihydroxybenzoate 367
Ethyl 3,5-dihydroxybenzoate 367
Ethyl 5-formyl-2-methyl-3-furancarboxylate 847
4-(2-Hydroxyethoxy)benzoic acid 338
5-Hydroxy-2,3-norbornanedicarboxylic acid γ-lactone 96
2-(2-Methoxyphenoxy)acetic acid 327
Methyl 2,4-dihydroxy-3-methylbenzoate 367
2′,4′,6′-Trihydroxypropiophenone monohydrate 289

$C_9H_{10}O_4S$
Acetyl *p*-toluenesulfonate 800
Methyl 4-(methylsulfonyl)benzoate 781
2-(*p*-Tolylsulfonyl)-acetic acid 781

$C_9H_{10}O_4S_2$
S-(*p*-Tosyl)mercaptoacetic acid 799

$C_9H_{10}O_5$
3,4-Dimethoxysalicylic acid 325
Methyl 2,5-dihydroxy-4-methoxybenzoate 367
Methyl 3-(methoxycarbonyl)-2-furanacetate 854

$C_9H_{10}O_5S$
Ethyl 2-sulfobenzoate, sodium salt 795

$C_9H_{10}O_7$
Dimethyl 3-hydroxy-4-methoxy-2,5-furandicarboxylate 845

$C_9H_{10}S$
Allyl phenyl sulfide 211

$C_9H_{10}S_2$
4-Phenyl-1,2-dithiolane sulfate 214

$C_9H_{11}AsO_2$
4-(Dimethylarsino)benzoic acid 1127

$C_9H_{11}Br$
β-Bromoisopropylbenzene 169

$C_9H_{11}BrN_2O$
3-(3-Bromophenyl)-1,1-dimethylurea 534
3-(4-Bromophenyl)-1,1-dimethylurea 589
1-(3-Bromophenyl)-3-ethylurea 534
1-(4-Bromophenyl)-3-ethylurea 589

$C_9H_{11}BrN_2O_2$
3-(4-Bromophenyl)-1-methoxy-1-methylurea 575

$C_9H_{11}BrN_4O_2$
8-Bromo-1,3-diethylxanthine 924

$C_9H_{11}BrO$
4-Bromo-α-ethylbenzyl alcohol 204
1-Bromo-4-isopropoxybenzene 177
2-Bromophenethyl methyl ether 186

$C_9H_{11}BrO_2$
β-Bromo-2-methoxyphenetole 176

$C_9H_{11}BrO_2S$
2-(Bromomethyl)benzyl methyl sulfone 772
3-Bromophenyl isopropyl sulfone 774

$C_9H_{11}BrO_3$
3-Bromo-4,5-dimethoxybenzyl alcohol 198
1-Bromo-2,4,5-trimethoxybenzene 177
1-Bromo-2,4,6-trimethoxybenzene 177

$C_9H_{11}BrO_3S$
2-Bromoethyl *p*-toluenesulfonate 796

$C_9H_{11}Br_2NO_2S$
2′,6′-Dibromo-4′-ethylmethanesulfonanilide 810

$C_9H_{11}Cl$
1-Chloro-2,4,5-trimethylbenzene 171

$C_9H_{11}ClHg$
Mesitylmercuric chloride 1139

$C_9H_{11}ClN_2O$
2-Amino-6′-chloro-*o*-acetotoluidide hydrochloride 423
1-(2-Chloro-6-methylphenyl)-3-methylurea 592
1-(3-Chloro-4-methylphenyl)-3-methylurea 597
1-(4-Chloro-2-methylphenyl)-3-methylurea 597
1-(5-Chloro-2-methylphenyl)-3-methylurea 597
3-(3-Chlorophenyl)-1,1-dimethylurea 580
3-(4-Chlorophenyl)-1,1-dimethylurea 571
1-(2-Chlorophenyl)-3-ethylurea 587
1-(3-Chlorophenyl)-3-ethylurea 580
1-(4-Chlorophenyl)-3-ethylurea 589

$C_9H_{11}ClN_2OS$
1-(3-Chlorophenyl)-3-(2-hydroxyethyl)-2-thiourea 734

$C_9H_{11}ClN_2O_2$
1-Acetonyl-3-carbamoylpyridinium chloride 946
3-Acetyl-1-(carbamoylmethyl)-pyridinium chloride 944
2-Amino-5-chloro-*N*-(2-hydroxyethyl)benzamide 447
1-(3-Chloro-2-methoxyphenyl)-3-methylurea 519
Ethyl *N*-(2-amino-4-chlorophenyl)-carbamate 624

$C_9H_{11}ClN_2O_2S$
Ethyl 4-chloro-2-(ethylthio)-5-pyrimidinecarboxylate 974

$C_9H_{11}ClN_2O_3$
1-(2-Carbamoylethyl)-3-carboxypyridinium chloride 948
3-Carbamoyl-1-(methoxycarbonylmethyl)-pyridinium chloride 946
5-Chloro-2,4-dimethoxyphenylurea 516
3-(4-Chloro-2-nitroanilino)-1-propanol 259

$C_9H_{11}ClN_2S$
3-(4-Chlorophenyl)-1,1-dimethyl-2-thiourea 748
1-(4-Chlorophenyl)-3-ethyl-2-thiourea 748

$C_9H_{11}ClN_4OS$
1-(*N*-(4-Chloro-2-methylphenyl)-carbamoyl)thiosemicarbazide 730

$C_9H_{11}ClN_4O_2$
((3-Chlorophenyl)ureido)acetic acid hydrazide 534

$C_9H_{11}ClN_4O_2S$
1-(5-Chloro-2-methoxyphenyl)-5-thiobiurea 729

$C_9H_{11}ClN_4O_3$
4-(2-Chloro-4-nitrophenyl)-1,1-dimethylsemicarbazide 568

$C_9H_{11}ClO$
4-Chloro-α-ethylbenzyl alcohol 204
4-Chloro-3-ethyl-5-methylphenol 192
1-Chloro-3-phenyl-2-propanol 197
3-Chloropropyl phenyl ether 176
4-Chloro-2,3,5-trimethylphenol 190
4-Chloro-2,3,6-trimethylphenol 190

$C_9H_{11}ClOS$
4-Chlorobenzyl 2-hydroxyethyl sulfide 210

$C_9H_{11}ClO_2$
1-Chloro-3-phenoxy-2-propanol 197
2-(4-Chloro-*m*-tolyloxy)-ethanol 203
2,5-Dimethoxybenzyl chloride 176

$C_9H_{11}ClO_2S$
Benzyl 2-chloroethyl sulfone 772

$C_9H_{11}ClO_3$
3-(2-Chlorophenoxy)-1,2-propanediol 198
3-(4-Chlorophenoxy)-1,2-propanediol 202
1-Chloro-2,4,6-trimethoxybenzene 182

$C_9H_{11}ClS$
2-Chloroethyl *p*-tolyl sulfide 209
3-Chloropropyl phenyl sulfide 209

$C_9H_{11}Cl_2N$
4-Chloro-*N*-(2-chloroethyl)-benzylamine hydrochloride 236

$C_9H_{11}Cl_2N_3O$
4-(2,3-Dichlorophenyl)-1,1-dimethylsemicarbazide 578
4-(2,4-Dichlorophenyl)-1,1-dimethylsemicarbazide 579
4-(2,5-Dichlorophenyl)-1,1-dimethylsemicarbazide 579
4-(2,6-Dichlorophenyl)-1,1-dimethylsemicarbazide 578
4-(3,4-Dichlorophenyl)-1,1-dimethylsemicarbazide 579
4-(3,5-Dichlorophenyl)-1,1-dimethylsemicarbazide 533

$C_9H_{11}Cl_2O_2P$
Mesityl dichlorophosphate 837

$C_9H_{11}Cl_3Si$
(2,3,6-Trichlorophenyl)-trimethylsilane 1104
(3,4,5-Trichlorophenyl)-trimethylsilane 1104

$C_9H_{11}F$
4-Fluorocumene 169

$C_9H_{11}FN_2O$
1,1-Dimethyl-3-(4-fluorophenyl)urea 589

$C_9H_{11}FN_2O_3S$
3-(3-Ethylureido)benzenesulfonyl fluoride ... 804
$C_9H_{11}FN_2S$
1-Ethyl-3-(2-fluorophenyl)-2-thiourea. 732
$C_9H_{11}FN_4O_2$
2-(4-Fluorophenylureido)-acethydrazide ... 534
$C_9H_{11}FO$
α-Ethyl-4-fluorobenzyl alcohol ... 207
4-Fluoro-2-propylphenol ... 192
$C_9H_{11}FO_2S$
2-Mesitylenesulfonyl fluoride ... 804
$C_9H_{11}FO_3$
3-(2-Fluorophenoxy)-1,2-propanediol ... 198
$C_9H_{11}F_3Si$
(2,4,6-Trifluorophenyl)-trimethylsilane ... 1104
$C_9H_{11}HgNS_2$
Phenylmercuric dimethyldithiocarbamate ... 352
$C_9H_{11}I$
1-Iodo-2,4,5-trimethylbenzene ... 171
$C_9H_{11}IN_2O$
1,1-Dimethyl-3-(4-iodophenyl)urea ... 538
$C_9H_{11}IN_4O$
1-((3-Iodo-4-methoxybenzylidene)-amino)guanidine nitrate ... 1024
$C_9H_{11}IN_4O_2$
1,3-Diethyl-8-iodoxanthine ... 924
$C_9H_{11}IO_2$
1,2-Dimethoxy-4-iodo-5-methylbenzene ... 177
1,4-Dimethoxy-2-iodo-5-methylbenzene ... 177
$C_9H_{11}IO_4$
4-Iodo-5-methyl-7-oxo-6-oxabicyclo(3.2.1)octane-2-carboxylic acid ... 97
$C_9H_{11}N$
N-Methylisoindoline ... 251
$C_9H_{11}NO$
o-Acetotoluidide ... 438
2-Allyl-4-aminophenol ... 217
N-Benzylacetamide ... 431
2,6-Dimethylbenzaldehyde oxime ... 1011
3′,5′-Dimethylformanilide ... 440
N-Ethylbenzamide ... 477
N-Ethylformanilide ... 441
2′-Ethylformanilide ... 438
4′-Ethylformanilide ... 438
2′,4′-Formoxylidide ... 440
2′,5′-Formoxylidide ... 440
2′,6′-Formoxylidide ... 439
3′,4′-Formoxylidide ... 440
1-Formyl-6-(dimethylamino)fulvene .. 60
N-(3-Methoxybenzylidene)-methylamine ... 1044
N-Methylacetanilide ... 441
4′-Methylacetophenone oxime ... 1014
N-(α-Methylbenzyl)-formamide ... 431
2-Methyl-2-phenylacetaldehyde oxime ... 1010
Methyl phenylacetimidate hydrochloride ... 358
3-Phenylpropionamide ... 436
Propionanilide ... 437
Propiophenone oxime *p*-tosylate ... 1011
2-(*p*-Tolyl)-acetamide ... 436
$C_9H_{11}NOS$
S-Benzoylcysteamine ... 372
S-Benzoylcysteamine hydrochloride ... 372
2-(Benzylthio)acetamide ... 421
o-Ethyl thiocarbanilate ... 750
N-Methyl-4-(methylthio)benzamide ... 468
3′-(Methylthio)acetanilide ... 421
4′-(Methylthio)acetanilide ... 421
2′-(Methylthio)acetophenone oxime. 1009
$C_9H_{11}NOS_2$
4-Ethoxydithiocarbanilic acid, ammonium salt ... 738
4-(2-Thienylthioformyl)-morpholine ... 856
$C_9H_{11}NO_2$
o-Acetanisidide ... 417
m-Acetanisidide ... 418
p-Acetanisidide ... 419
4-Amino-2,5-dimethylbenzoic acid ... 342
2-Aminoethyl benzoate hydrobromide ... 372
3′-Amino-4′-methoxyacetophenone .. 291
N-Benzylglycine hydrochloride ... 330
Benzyl *N*-methylcarbamate ... 665
2-Cyano-2-cyclohexylideneacetic acid ... 148
3-Cyano-5,6-dihydro-4,5,5,-trimethyl-2*H*-pyran-2-one ... 149
2-Dimethylaminobenzoic acid hydrochloride ... 343
4,5-Dimethylanthranilic acid ... 342
N-Ethylanthranilic acid ... 343
Ethyl *N*-phenylcarbamate ... 680
o-Formophenetidide ... 417
p-Formophenetidide ... 419
Glycine benzyl ester hydrochloride .. 361
Glycine benzyl ester *p*-toluenesulfonate ... 356
6′-Hydroxy-*m*-acetotoluidide ... 413
α-(2-Hydroxyethylimino)-*p*-cresol ... 1028
2-(Hydroxymethyl)-*N*-methylbenzamide ... 461
N-Hydroxymethyl-2-phenylacetamide ... 413
2-Hydroxy-2-phenylpropionamide ... 411
1-Isopropyl-3-nitrobenzene ... 278
2-Methoxyacetophenone, oxime ... 1009
4′-Methoxyacetophenone, oxime ... 1009
2′-Methoxy-5′-methylformanilide ... 417
4′-Methoxy-2′-methylformanilide ... 419
(*R*)-2-Methoxy-2-phenylacetamide ... 414
2-(4-Methoxyphenyl)acetamide ... 416
Methyl 3-amino-4-methylbenzoate ... 372
Methyl *N*-benzylcarbamate ... 673
Methyl 2-cyclopentylidene-2-cyanoacetate ... 148
3-(2-Methyl-1,3-dioxolan-2-yl)-pyridine ... 966
Methyl 4-(methylamino)benzoate ... 373
Methyl *N*-methylanthranilate ... 373
Methyl *N*-(*o*-tolyl)carbamate ... 681
Methyl *N*-(*m*-tolyl)carbamate ... 677
Methyl *N*-(*p*-tolyl)carbamate ... 682
2-Nitro-2-phenylpropane ... 258
1-(3-Nitropropyl)benzene ... 258
5-Nitropseudocumene ... 280
1-Nitro-2,3,5-trimethylbenzene ... 279
1-Nitro-3,4,5-trimethylbenzene ... 280
Phenyl *N*,*N*-dimethylcarbamate ... 680
N-Phenylglycine methyl ester ... 358
(*S*)-2-Phenylglycine methyl ester hydrochloride ... 358
Propyl nicotinate ... 962
$C_9H_{11}NO_2S$
2-(3-Methoxyphenylthio)acetamide .. 414
4-(Methylthio)phenyl *N*-methylcarbamate ... 667
$C_9H_{11}NO_3$
3-Acetyl-2,4-dimethylpyrrole-5-carboxylic acid ... 840
1,6-Dihydro-6-oxo-1,2,5-trimethyl-3-pyridinecarboxylic acid ... 134
2,3-Dimethoxy-*syn*-benzaldehyde oxime ... 1009
2,3-Dimethoxybenzamide ... 463
2′,5′-Dimethoxyformanilide ... 417
4-Dimethylaminosalicylic acid ... 338
4,7-Epoxy-2-methyl-2,3,3a,4,5,6,7,7a-octahydro-1*H*-isoindole-1,3-dione ... 133
Ethyl 5-formyl-4-methyl-3-pyrrolecarboxylate ... 839
4-Ethyl-5-formyl-3-methyl-2-pyrrolecarboxylic acid ... 842
Ethyl *N*-hydroxy-*N*-phenylcarbamate ... 674
Ethyl 2-(3-pyridyloxy)acetate ... 937
2-Formyl-4-methyl-3-pyrrolepropionic acid ... 840
5-Formyl-4-methyl-3-pyrrolepropionic acid ... 840
β-(Hydroxyamino)-hydrocinnamic acid ... 330
2-(2-Hydroxyethoxy)benzamide ... 461
2′-Hydroxy-4′-methoxyacetanilide ... 411
2-(2-Hydroxyphenoxy)-*N*-methylacetamide ... 411
4-Isopropyl-2-nitrophenol ... 264
Methyl 3-amino-*p*-anisate ... 369
Methyl *N*-(2-methoxyphenyl)-carbamate ... 653
Methyl *N*-(3-methoxyphenyl)-carbamate ... 654
Methyl *N*-(4-methoxyphenyl)-carbamate ... 654
2-Methyl-2-(4-nitrophenyl)-2-propanol ... 263
2-Nitro-1-phenyl-1-propanol ... 257
3-Phenyl-DL-serine ... 326
DL-*o*-Tyrosine ... 348
Veratraldehyde oxime ... 1009
$C_9H_{11}NO_3S$
4′-Acetylmethanesulfonanilide ... 823
4′-(Methylsulfonyl)acetanilide ... 782
2-(*p*-Tolylsulfonyl)-acetamide ... 779
N-(*p*-Tolylsulfonylmethyl)-formamide ... 782
N-(*p*-Tosyl)acetamide ... 828
$C_9H_{11}NO_4$
D-3-(3,4-Dihydroxyphenyl)alanine ... 348
Dimethyl 3-(cyanomethylene)-glutarate ... 149
Ethyl 2-carboxy-4-methyl-3-pyrrolecarboxylate ... 838
3-Nitrobenzaldehyde dimethyl acetal ... 269
Syringamide ... 461
$C_9H_{11}NO_4S$
3-Acetamido-4-methoxybenzenesulfinic acid, hydrate ... 786
3-Acetamido-4-methoxybenzenesulfinic acid, sodium salt ... 786
tert.-Butyl 5-nitro-2-thiophenecarboxylate ... 856
Dimethyl 5-amino-3-methyl-2,4-thiophenedicarboxylate ... 856
4-(*N*-Methyl(methylsulfonamido))-benzoic acid ... 824
Methyl *N*-(methylsulfonyl)-*N*-phenylcarbamate ... 824
N-Methyl-*N*-(phenylsulfonyl)glycine. 822
1-Nitropropyl phenyl sulfone ... 780
2-Nitro-2-propyl phenyl sulfone ... 780
$C_9H_{11}NO_5$
Diethyl 2-cyano-3-oxosuccinate ... 148
N-(2-Hydroxyethyl)-3,4,5-trihydroxybenzamide ... 461
5-Nitro-1,2,3-trimethoxybenzene ... 265
$C_9H_{11}NO_5S$
N-Ethyl-5-sulfoanthranilic acid ... 789
$C_9H_{11}NO_5S_2$
3-(2-Nitrophenylthio)-1-propanesulfonic acid, sodium salt .. 793
$C_9H_{11}NS$
2,2-Dimethylbenzothiazoline ... 250
Thiochroman-4-amine hydrochloride ... 241
$C_9H_{11}NS_2$
3,4-Dimethyldithiocarbanilic acid, ammonium salt ... 750
N-(2,4-Xylyl)-dithiocarbamic acid, ammonium salt ... 741
$C_9H_{11}N_2O_2P$
1,3-Dimethyl-1,2,3,4-tetrahydro-1,3,2-benzodiazaphosphorine-2,4-dione ... 837
$C_9H_{11}N_3$
2-Amino-1-ethylbenzimidazole ... 904
2-Amino-1-phenyl-2-imidazoline hydrobromide ... 866
2,5-Dimethyl-7-aminobenzimidazole. 904
4-(2-Imidazolin-2-yl)aniline ... 248
3-Imino-2-phenylpyrazolidine ... 725
2-(3-Picolyl)-2-imidazoline hydrochloride ... 953
2-(4-Pyridyl)-1,4,5,6-tetrahydropyrimidine ... 966
$C_9H_{11}N_3O$
2-(2-Benzimidazolylamino)-1-ethanol ... 902
Isonicotinic acid isopropylidenehydrazide ... 952
1-(α-Methylbenzylidene)-semicarbazide ... 574
$C_9H_{11}N_3OS$
1-Acetyl-4-phenyl-3-thiosemicarbazide ... 747
1-Benzoyl-4-methyl-3-thiosemicarbazide ... 745
4-Methoxy-3-methyl-2-benzothiazolinone hydrazone hydrochloride ... 736
$C_9H_{11}N_3O_2$
2-Acetoacetamido-4-methylpyrimidine ... 975
N-Acetylanthranilic acid hydrazide .. 499
2,6-Bis-(acetamido)-pyridine ... 959
1-(2-Carboxyphenyl)-3,3-dimethyltriazene ... 1095
3-Nitrobenzaldehyde dimethylhydrazone ... 277
$C_9H_{11}N_3O_2S$
2-(Carbamoylthio)-acetic acid 2-phenylhydrazide ... 666
1-Ethyl-3-(4-nitrophenyl)-2-thiourea ... 740
$C_9H_{11}N_3O_2S_2$
N,1-Dimethyl-2-mercapto-5-benzimidazolesulfonamide ... 903
Methanesulfonic (3-methyl-2-benzothiazolinylidene)hydrazide ... 820
$C_9H_{11}N_3O_3$
1,1-Dimethyl-3-(3-nitrophenyl)urea ... 564
4-Ethylamino-3-nitrobenzamide ... 469
1-Ethyl-3-(3-nitrophenyl)urea ... 564
1-Ethyl-3-(4-nitrophenyl)urea ... 567
1-Methyl-3-(2-methyl-4-nitrophenyl)urea ... 568
1-Methyl-3-(4-methyl-3-nitrophenyl)urea ... 564
2-Morpholino-5-nitropyridine ... 954
$C_9H_{11}N_3O_4S$
1-(*N*-Acetylsulfanilyl)urea ... 824
$C_9H_{11}N_3O_5$
α-((Ethoxycarbonylmethoxy)imino)-3-pyrazoleacetic acid ... 1074
$C_9H_{11}N_3O_6S$
2,6-Dinitro-*N*-ethyl-4-(methylsulfonyl)aniline ... 777
$C_9H_{11}N_3O_8S_2$
2′,6′-Dinitro-*N*-methyl-4′-(methylsulfonyl)-methanesulfonanilide ... 818
$C_9H_{11}N_3S$
1-Acetimido-3-phenyl-2-thiourea ... 741
1-Acetimido-3-phenyl-2-thiourea *p*-toluenesulfonate ... 747
Phenylacetaldehyde thiosemicarbazone ... 1060
$C_9H_{11}N_5$
2-Guanidino-5-methylbenzimidazole ... 905
$C_9H_{11}N_5O_2$
4′-Nitro-3,3′,5-trimethyl-1,5′-bi-1*H*-pyrazole ... 861
$C_9H_{11}N_5O_4$
2-(3-(3-Nitrophenyl)ureido)acetic acid hydrazide ... 575
$C_9H_{11}N_7O_2S$
N^1-(4,6-Diamino-1,3,5-triazin-2-yl)-sulfanilamide monohydrate ... 979
C_9H_{12}
Bicyclo(4.3.0)nona-3,7-diene ... 3
$C_9H_{12}AsNO_6$
Ethyl *N*-(4-arsono-3-hydroxyphenyl)-carbamate ... 1126
$C_9H_{12}AsNO_7$
4-(2-Hydroxypropoxy)-3-nitrobenzenearsonic acid ... 1126
$C_9H_{12}AsNO_8$
4-(2,3-Dihydroxypropoxy)-3-nitrobenzenearsonic acid, disodium salt ... 1126
$C_9H_{12}BF_4N_3$
4-Diazo-*N*-ethyl-*o*-toluidine fluoroborate ... 769
4,5-Dihydro-1-methyl-3-phenyl-1*H*-1,2,3-triazolium tetrafluoroborate ... 255

$C_9H_{12}BrN$
4-Bromo-*N*,*N*-dimethylbenzylamine .. 236
2-Bromo-*N*-ethyl-4-methylaniline ... 216
2-Bromo-3,4,6-trimethylaniline ... 219
2-Bromo-*N*,*N*,5-trimethylaniline ... 216
$C_9H_{12}BrNO$
1-(4-Bromophenoxymethyl)-ethylamine ... 240
$C_9H_{12}BrNO_2S$
N-(4-Bromo-2,6-dimethylphenyl)-methanesulfonamide ... 810
$C_9H_{12}BrNO_3$
Ethyl 3-(1-bromoethyl)-5-methyl-4-isoxazolecarboxylate ... 871
$C_9H_{12}BrNO_3S$
Dimethyl(2-hydroxy-5-nitrobenzyl)-sulfonium bromide ... 263
$C_9H_{12}BrN_3O$
4-(4-Bromophenyl)-1,1-dimethylsemicarbazide ... 533
$C_9H_{12}ClN$
2-Chloro-*N*,*N*-dimethyl-*p*-toluidine ... 220
5-Chloro-*N*,*N*-dimethyl-*o*-toluidine ... 220
N-(2-Chloroethyl)-benzylamine hydrochloride ... 236
3-Chloro-*N*-ethyl-4-methylaniline ... 216
$C_9H_{12}ClNO$
2-(4-Chlorobenzylamino)-ethanol hydrochloride ... 238
2-(2-Chlorophenoxy)-*N*-methylethylamine ... 240
2-(4-Chlorophenoxy)-*N*-methylethylamine ... 240
$C_9H_{12}ClNO_2$
1-(Carbethoxymethyl)-pyridinium chloride ... 956
$C_9H_{12}ClNO_4S$
4′-Chloro-2′,5′-dimethoxymethanesulfonanilide ... 806
5′-Chloro-2′,4′-dimethoxymethanesulfonanilide ... 806
$C_9H_{12}ClN_3$
4-(Propylamino)benzenediazonium chloride, zinc chloride double salt .. 1095
$C_9H_{12}ClN_3O$
4-(2-Chlorophenyl)-1,1-dimethylsemicarbazide ... 570
4-(3-Chlorophenyl)-1,1-dimethylsemicarbazide ... 577
4-(4-Chlorophenyl)-1,1-dimethylsemicarbazide ... 578
$C_9H_{12}ClN_5O_2S$
1-(*N*-(5-Chloro-2-methoxyphenyl)-carbamoyl)-3-thiocarbohydrazide ... 729
$C_9H_{12}Cl_2Si$
Dichloromethylphenethylsilane ... 1112
(2,6-Dichlorophenyl)trimethylsilane .. 1104
$C_9H_{12}Cl_4O_2$
7,7-Dimethoxy-1,4,5,6-tetrachloronorbornane ... 17
$C_9H_{12}FIN_2O_2$
(4-Fluoro-3-nitrophenyl)-trimethylammonium iodide ... 262
$C_9H_{12}F_2Si$
(2,6-Difluorophenyl)trimethylsilane .. 1104
$C_9H_{12}NO_4P$
Dimethyl *N*-benzoylphosphoramidate ... 835
$C_9H_{12}N_2$
N-Amino-1,2,3,4-tetrahydroisoquinoline ... 252
N,*N*-Dimethyl-*N*′-phenylformamidine ... 1062
3-Phenylpropionamidine hydrochloride ... 727
$C_9H_{12}N_2O$
Acetic acid (1-methyl-2-phenyl)-hydrazide ... 503
N-(2-Aminoethyl)benzamide ... 468
3′-Amino-4′-methylacetanilide ... 422
1-Cyano-4-hydroxy-1-cyclohexaneacetonitrile ... 144
4-Dimethylaminobenzaldehyde oxime ... 1010
1,3-Dimethyl-1-phenylurea ... 585
2,3-Dimethyl-4,5,6,7-tetrahydro-1*H*-pyrrolo(3,2-c)pyridin-4-one ... 841
1-(2-Ethylphenyl)urea ... 588
1-Ethyl-3-phenylurea ... 585
1-(4-Ethylphenyl)urea ... 589
Hydrocinnamic hydrazide ... 503
1-Methyl-3-(*o*-tolyl)urea ... 587
1-Methyl-3-(*m*-tolyl)urea ... 580
1-Methyl-3-(*p*-tolyl)urea ... 589
2-Phenylacetic *N*2-methylhydrazide. 503
Propionic acid 2-phenylhydrazide ... 501
2-Pyridinealdoxime *O*-propyl ether ... 1010
3-Pyridinealdoxime *O*-propyl ether ... 1010
1-(2,3-Xylyl)urea ... 579
1-(2,4-Xylyl)urea ... 597
1-(2,5-Xylyl)urea ... 597
1-(2,6-Xylyl)urea ... 592
1-(3,4-Xylyl)urea ... 597
$C_9H_{12}N_2OS$
1-(2-Hydroxyethyl)-3-phenyl-2-thiourea ... 734
2-(4-Methoxybenzyl)-2-thiopseudourea hydrochloride ... 738
3-(2-Methoxyphenyl)-2-methyl-2-thiopseudourea hydriodide ... 738
$C_9H_{12}N_2O_2$
2′-Amino-4′-methoxyacetanilide ... 414
1-(*o*-Anisyl)-3-methylurea ... 554
1-(*m*-Anisyl)-3-methylurea ... 557
1-(*p*-Anisyl)-3-methylurea ... 559
N,*N*-Dimethyl-4-nitrobenzylamine ... 277
1-(4-Ethoxyphenyl)urea ... 559
Ethyl 2,3-dicyano-3-methylbutyrate .. 147
N-Ethyl-5-nitro-*o*-toluidine ... 277
Ethyl 3-phenylcarbazate ... 503
4-Hydroxyhydrocinnamic acid hydrazide ... 497
1-Hydroxy-3-(2,4-xylyl)urea ... 575
4-Isopropyl-2-nitroaniline ... 271
4-Isopropyl-3-nitroaniline ... 271
1-Methoxy-1-methyl-3-phenylurea ... 575
4-Nitro-*N*-propylaniline ... 271
$C_9H_{12}N_2O_2S$
Benzenesulfonic isopropylidenehydrazide ... 821
1-(2,5-Dimethoxyphenyl)-2-thiourea .. 737
1-(3,4-Dimethoxyphenyl)-2-thiourea .. 736
$C_9H_{12}N_2O_3$
2-Butoxy-5-nitropyridine ... 937
3,5-Dimethoxybenzoic acid hydrazide ... 499
N,*N*-Dimethyl-4-nitrobenzylamine-*N*-oxide picrate ... 264
Ethyl *N*-(4-amino-2-hydroxyphenyl)-carbamate ... 644
3-Hydroxybutyric (furfurylidene)-hydrazide monohydrate ... 1036
N-(2-Hydroxyethyl)-*N*-methyl-4-nitroaniline ... 264
4-Methoxybenzyl hydrazinocarboxylate ... 654
$C_9H_{12}N_2O_3S$
Ethyl 3-ethyl-2-thiohydantoin-1-acetate ... 141
Ethyl 2-(ethylthio)-4-hydroxy-5-pyrimidinecarboxylate ... 970
3-(Morpholinosulfonyl)pyridine ... 966
2-(*p*-Tolylsulfonyl)-acetic acid hydrazide ... 779
$C_9H_{12}N_2O_4$
DL-*threo*-2-Amino-1-(*p*-nitrophenyl)-1,3-propanediol ... 263
3-(2-Nitroanilino)-1,2-propanediol ... 262
3-(4-Nitroanilino)-1,2-propanediol ... 262
$C_9H_{12}N_2O_4S$
N-(4-Sulfamoylphenyl)-β-alanine ... 778
$C_9H_{12}N_2S$
1,2-Dimethyl-3-phenyl-2-thiopseudourea ... 739
1,1-Dimethyl-3-phenyl-2-thiourea ... 747
2-Ethyl-1-phenyl-2-thiopseudourea hydriodide ... 726
1-Ethyl-3-phenyl-2-thiourea ... 747
2-(4-Methylbenzyl)-2-thiopseudourea hydrochloride ... 749
1-(2,6-Xylyl)-2-thiourea ... 749
$C_9H_{12}N_2S_2$
1-Methyl-3-(4-methylthiophenyl)-2-thiourea ... 739
$C_9H_{12}N_3P$
Tris-(2-cyanoethyl)-phosphine ... 158
$C_9H_{12}N_4$
3-Amino-2,4,5,6-tetrahydrocyclopentapyrazole-2-propionitrile ... 860
$C_9H_{12}N_4O$
N-(4,6-Dimethyl-5-(2-hydroxyethyl)-2-pyrimidinyl)-carbamonitrile ... 972
4-(3,3-Dimethyl-1-triazenyl)-benzamide ... 1095
1(2-Methoxybenzylideneamino)-guanidine nitrate ... 1049
$C_9H_{12}N_4OS_2$
1-(*N*-(4-(Methylthio)phenyl)-carbamoyl)thiosemicarbazide ... 739
$C_9H_{12}N_4O_2$
5-Carbamoyl-3-cyano-4,4-dimethyl-6-imino-2-piperidone ... 147
1,3-Dimethyl-7-ethylxanthine ... 923
3,7-Dimethyl-1-ethylxanthine ... 923
4-Hydroxy-6-methyl-2-(2-(3-oxo-*sec*-butylidene)hydrazino)pyrimidine ... 971
1,1′-(4-Methyl-1,3-phenylene)-bisurea ... 566
2-(Methylureido)benzoic acid hydrazide ... 577
$C_9H_{12}N_4O_2S$
1-(*N*-(3-Methoxyphenyl)carbamoyl)-3-thiosemicarbazide ... 736
$C_9H_{12}N_4O_3$
1-Allyl-6-amino-3-ethyl-5-nitrosouracil ... 132
6-Amino-1-cyclopropyl-3-ethyl-5-nitrosouracil ... 132
1,3-Diethyl-8-hydroxyxanthine ... 921
1,1-Dimethyl-4-(3-nitrophenyl)-semicarbazide ... 564
4-Ethyl-1-(4-nitrophenyl)-semicarbazide ... 567
$C_9H_{12}N_4O_4$
1-Ethyl-3-(2-hydroxyethyl)uric acid ... 921
$C_9H_{12}N_4O_4S$
4-(Semicarbazidosulfonyl)-acetanilide ... 824
$C_9H_{12}N_4S$
1-Amidino-3-(*p*-tolyl)-2-thiourea ... 741
6-(Butylthio)purine ... 923
4-(α-Methylbenzylideneamino)-3-thiosemicarbazide ... 1071
$C_9H_{12}N_6$
1,3,5-Tris(cyanomethyl)hexahydro-1,3,5-triazine ... 146
$C_9H_{12}O$
α,4-Dimethylbenzyl alcohol ... 207
Methyl phenethyl ether ... 187
m-Methylphenetole ... 178
1-Phenyl-1-propanol ... 204
2,4,5-Trimethylphenol ... 192
$C_9H_{12}OS$
2-(Benzylthio)-ethanol ... 210
2-Butyryl-3-methylthiophene ... 857
Ethyl *p*-tolyl sulfoxide ... 771
2-(Methylthio)ethyl phenyl ether ... 211
2-(Methylthiomethyl)anisole ... 211
4-(Methylthiomethyl)anisole ... 211
Phenyl propyl sulfoxide ... 771
3-(Phenylthio)propanol ... 210
2-(*p*-Tolylthio)-ethanol ... 210
$C_9H_{12}O_2$
6-Acetyl-3-methyl-2-cyclohexen-1-one ... 53
2-Allyl-4-hydroxy-3-methyl-2-cyclopenten-1-one ... 52
2,5-Dimethoxytoluene ... 179
2-Isopropylhydroquinone ... 193
4-Methoxybenzyl methyl ether ... 186
3-Methoxy-α-methylbenzyl alcohol ... 201
4-Methoxy-α-methylbenzyl alcohol ... 201
1-Methoxy-2-phenoxyethane ... 187
Methyl 5-norbornene-2-carboxylate .. 80
2-Methyl-5-norbornene-2-carboxylic acid ... 67
1-Phenyl-1,3-propanediol ... 199
4-Propylresorcinol ... 193
2-(*m*-Tolyloxy)-ethanol ... 200
2-(*p*-Tolyloxy)-ethanol ... 202
$C_9H_{12}O_2S$
Dimethoxymethyl phenyl sulfide ... 210
Ethyl *p*-tolyl sulfone ... 785
2-Hydroxy-3-methyl-5-(methylthio)-benzyl alcohol ... 210
Isopropyl phenyl sulfone ... 783
3-Phenylthio-1,2-propanediol ... 210
2,4,6-Trimethyl-benzenesulfinic acid ... 787
$C_9H_{12}O_3$
6-Acetoxy-3-oxatricyclo-[3,2,1,0^{2,4}]-octane ... 91
3-Butyl-1,2,4-cyclopentanetrione ... 54
5-*tert*-Butyl-2-furoic acid ... 848
2-(4-Methoxyphenoxy)ethanol ... 202
3-Phenoxy-1,2-propanediol ... 202
$C_9H_{12}O_3S$
2-(Methylsulfonyl)-1-phenyl-1-ethanol ... 774
2-(*p*-Tosyl)-1-ethanol ... 775
2,4,5-Trimethyl-benzenesulfonic acid ... 790
$C_9H_{12}O_4$
Allyl 2-acetylacetoacetate ... 93
Bicyclo(2.2.1)heptane-3,4-dicarboxylic acid ... 66
1-Carboxy-3-cyclohexene-1-acetic acid ... 66
Dimethyl 3-cyclopentene-1,1-dicarboxylate ... 78
Dimethyl 2-methylmuconate ... 78
Ethyl 1-cyclopentanone-2-glyoxylate ... 92
4-Methyl-4-cyclohexene-1,2-dicarboxylic acid ... 66
Methyl 6-methyl-2,4-cyclohexanedionecarboxylate ... 92
2,3-Norbornanedicarboxylic acid ... 65
2-Phenoxyethyl methanesulfonate ... 797
$C_9H_{12}O_5$
1-Carboxy-4-oxo-1-cyclohexaneacetic acid ... 65
Dimethyl 4-oxo-1,2-cyclopentanedicarboxylate ... 92
$C_9H_{12}S$
Benzyl ethyl sulfide ... 211
Ethyl *p*-tolyl sulfide ... 212
Isopropyl phenyl sulfide ... 211
Phenyl propyl sulfide ... 211
$C_9H_{12}S_2$
(Methylthio)(*p*-tolylthio)methane ... 212
$C_9H_{13}AsO_5$
4-(2-Hydroxypropoxy)-benzenearsonic acid ... 1126
$C_9H_{13}BF_4N_2O_3$
2-Ethoxy-1-ethyl-5-nitropyridinium tetrafluoroborate ... 937
$C_9H_{13}BrN_2O$
3-Carbamoyl-1-propylpyridinium bromide ... 947
$C_9H_{13}BrO_2$
2-Bromo-5,5-dimethyl-3-methoxy-2-cyclohexen-1-one ... 52
$C_9H_{13}BrSi$
(3-Bromophenyl)trimethylsilane ... 1101
(4-Bromophenyl)trimethylsilane ... 1104
$C_9H_{13}ClN_2O$
1-(2-Carbamoylethyl)-2-methylpyridinium chloride ... 948
1-(2-Carbamoylethyl)-3-methylpyridinium chloride ... 948
1-(2-Carbamoylethyl)-4-methylpyridinium chloride ... 948
$C_9H_{13}ClN_2O_2S$
N′-(3-Chloro-4-methylphenyl)-*N*,*N*-dimethylsulfamide ... 807
$C_9H_{13}ClN_2O_3$
N-(3-Chloro-2-methoxyphenyl)-*N*′-dimethylsulfamide ... 806
$C_9H_{13}ClOSi$
3-Chlorophenyl trimethylsilyl ether .. 1100
4-Chlorophenyl trimethylsilyl ether .. 1104
$C_9H_{13}ClO_2$
α-Chloro-2-norbornaneacetic acid ... 65
$C_9H_{13}ClO_5$
2,3,4,6-Tetramethylpyrylium perchlorate ... 935
$C_9H_{13}ClSi$
Benzylchlorodimethylsilane ... 1112
$C_9H_{13}FSi$
(2-Fluorophenyl)trimethylsilane ... 1104

$C_9H_{13}N$
2-*tert*-Butylpyridine ... 967
2,4-Dimethylbenzylamine ... 241
2,5-Dimethylbenzylamine ... 241
N,*N*-Dimethylbenzylamine hydrochloride ... 243
N-Ethyl-*o*-toluidine hydrochloride ... 220
N-Isopropylaniline ... 219
3-Isopropylaniline ... 219
4-Isopropylaniline hydrochloride ... 219
2,4,5-Trimethylaniline ... 219
N,3,5-Trimethylaniline ... 220

$C_9H_{13}NO$
2-Amino-4-isopropylphenol ... 217
3-Amino-4-isopropylphenol ... 217
2-Amino-1-phenoxypropane ... 239
4-*tert*-Butylpyridine-*N*-oxide ... 1006
2-(Dimethylamino)cresol ... 217
N,*N*-Dimethylbenzylamine-*N*-oxide (50% aqueous solution) ... 243
N,*N*-Dimethylbenzylamine-*N*-oxide picrate ... 277
4,5-Dimethyl-3-ethyl-2-pyrrolecarboxaldehyde ... 840
N,*N*-Dimethyl-4-methoxyaniline ... 218
2-(2-Ethoxyethyl)pyridine ... 939
4-(2-Ethoxyethyl)-pyridine ... 940
2-(2-Methoxyethyl)-6-methylpyridine. 940
N-Methyl-2-phenoxyethylamine ... 240

$C_9H_{13}NOS$
N-*tert*-Butyl-2-thiophenecarboxamide ... 856

$C_9H_{13}NOS_2$
3-Cyclohexylrhodanine ... 141

$C_9H_{13}NO_2$
3-Aminobicyclo(2.2.2)oct-5-ene-2-carboxylic acid ... 72
(1*R*,2*S*)-(+)-2-Amino-1-phenyl-1,3-propanediol ... 237
L-(+)-*threo*-2-Amino-1-phenyl-1,3-propanediol ... 238
1-Cyanocyclohexyl acetate ... 149
Cyclohexyl cyanoacetate ... 147
Ethyl 1-cyano-1-cyclopentanecarboxylate ... 149
Ethyl 2-cyano-2-hexenoate ... 148
Ethyl 2-cyano-3-methyl-2-pentenoate ... 149
Ethyl 3,4-dimethyl-2-pyrrolecarboxylate ... 840
3-(1-Hydroxybutyl)pyridine-*N*-oxide .. 1005
DL-4-Hydroxy-α-(methylaminomethyl)benzyl alcohol D-tartrate ... 237
3,3-Tetramethyleneglutarimide ... 125

$C_9H_{13}NO_2S$
Methanesulfono-2′,4′-xylidide ... 829

$C_9H_{13}NO_2Si$
(4-Nitrophenyl)trimethylsilane ... 1108

$C_9H_{13}NO_3$
(-)-Epinephrine bitartrate ... 236
Methyl 2,5-dimethyl-6-oxo-1,4,5,6-tetrahydro-3-pyridinecarboxylate ... 134

$C_9H_{13}NO_3S$
3-Amino-4-methoxyphenyl ethyl sulfone ... 775
3-Anilino-1-propanesulfonic acid ... 789
2′-Ethoxymethanesulfonanilide ... 814
N-Ethyl-4-methylmetanilic acid ... 789
2′-Methoxy-5′-methylmethanesulfonanilide ... 814
2-Methyl-1-(3-sulfopropyl)-pyridinium hydroxide, inner salt ... 957

$C_9H_{13}NO_3Si$
3-Nitrophenyl trimethylsilyl ether ... 1108

$C_9H_{13}NS$
N,*N*-Dimethyl-4-(methylthio)aniline .. 218

$C_9H_{13}NS_2O_2$
S,*S*-Dimethyl-*N*-(*p*-tolylsulfonyl)-sulfilimine ... 828

$C_9H_{13}N_2O_8P$
2′-Deoxyuridine 5′-monophosphate, disodium salt dihydrate ... 161

$C_9H_{13}N_3$
1-(4-Ethylphenyl)guanidine carbonate ... 727
3-Ethyl-1-*p*-tolyltriazene ... 1077
Phenethylguanidine hemisulfate ... 725
1-(2,4-Xylyl)guanidine carbonate ... 727

$C_9H_{13}N_3O$
Acetic 2-(2-amino-4-methylphenyl)-hydrazide ... 499
4-Dimethylaminobenzhydrazide ... 500
o-(Dimethylamino)-benzoic acid hydrazide ... 499
1,1-Dimethyl-4-phenylsemicarbazide ... 578
1-(4-Ethoxyphenyl)guanidine *p*-toluenesulfonate ... 724
4-(2,4-Xylyl)semicarbazide ... 579

$C_9H_{13}N_3OS$
4-(2-Methoxy-5-methylphenyl)-3-thiosemicarbazide ... 737
N-(1,3,4-Thiadiazol-2-yl)-1-cyclohexanecarboxamide ... 879

$C_9H_{13}N_3O_2$
1-Allyl-6-amino-3-ethyluracil ... 132
4-Methyl-1-(4-methylbenzoyl)-semicarbazide ... 574
1-(2-Tetrahydropyranyl)cytosine ... 133

$C_9H_{13}N_3O_2S_2$
1-Ethyl-3-(3-sulfamoylphenyl)-2-thiourea ... 826
1-Ethyl-3-(4-sulfamoylphenyl)-2-thiourea ... 826

$C_9H_{13}N_3O_3$
6-Amino-1,3-diethyl-5-formyluracil ... 132
2-(3-Methoxypropylamino)-5-nitropyridine ... 937

$C_9H_{13}N_3O_3S$
1-Ethyl-3-(3-sulfamoylphenyl)urea ... 825
1-Ethyl-3-(4-sulfamoylphenyl)urea ... 825

$C_9H_{13}N_3O_4$
2′-Deoxycytidine picrate ... 697
1,3-Diethyl-6-methyl-5-nitrouracil ... 133
N-(5-Nitro-2-pyridyl)-iminodiethanol. 932

$C_9H_{13}N_3S$
1,1-Dimethyl-4-phenyl-3-thiosemicarbazide ... 750
4-Ethyl-1-phenyl-3-thiosemicarbazide ... 741
4-(3,4-Xylyl)-3-thiosemicarbazide ... 750

$C_9H_{13}N_5$
6-(Dimethylamino)-9-ethylpurine ... 923
o-Tolylbiguanide hemihydrate ... 727

$C_9H_{13}N_5O$
2-Amino-9-butyl-6-hydroxypurine ... 921

$C_9H_{13}N_5O_2$
6-(Bis-(2-hydroxyethyl)amino)purine. 921

$C_9H_{13}N_5S$
2-Amino-9-butyl-6-mercaptopurine ... 922

$C_9H_{13}O_3P$
Dimethyl benzylphosphonate ... 835
(2-Isopropylphenyl)-phosphonic acid ... 836
(2,4,6-Trimethylphenyl)-phosphonic acid ... 837

$C_9H_{13}O_6$
4-(2,3-Dihydroxypropoxy)-benzenearsonic acid ... 1126

$C_9H_{14}AsF_6N$
Benzyldimethylammonium hexafluoroarsenate ... 255

$C_9H_{14}Br_2$
9,9-Dibromobicyclo(6.1.0)nonane ... 6

$C_9H_{14}Br_2O_4$
3,9-Bis(bromomethyl)-2,4,8,10-tetraoxa(5.5)spiroundecane ... 21

$C_9H_{14}Cl_3NO_4$
Butyl *N*-(1-acetoxy-2,2,2-trichloroethyl)carbamate ... 114

$C_9H_{14}F_3N_3O$
3-(Butylamino)-1-(2,2,2-trifluoroethyl)-2-pyrazolin-5-one ... 130

$C_9H_{14}F_6NP$
Benzyldimethylammonium hexafluorophosphate ... 244
Phenyltrimethylammonium hexafluorophosphate ... 255

$C_9H_{14}IN$
2,4-Dimethyl-1-ethylpyridinium iodide ... 958
3,4-Dimethyl-1-ethylpyridinium iodide ... 958
3,5-Dimethyl-1-ethylpyridinium iodide ... 958
4-Methyl-1-propylpyridinium iodide .. 958

$C_9H_{14}IPS$
Dimethylphenylphosphine sulfide methiodide ... 836

$C_9H_{14}NO_2P$
4-Dimethylamino-*m*-tolylphosphinic acid, monosodium salt ... 833

$C_9H_{14}NO_4P$
2-Aminoethyl benzyl phosphate ... 834

$C_9H_{14}N_2$
3-(Dimethylamino)benzylamine ... 241
2-(2-(Dimethylamino)-ethyl)-pyridine. 955
2-(2-Ethylaminoethyl)pyridine ... 946

$C_9H_{14}N_2O$
5-Amino-2-butoxypyridine hydrochloride ... 937
3-Phenyl-1,1,3-trimethylurea ... 585
3-(2-Pyridylmethylamino)-1-propanol ... 932
3-(3-Pyridylmethylamino)-1-propanol ... 933
3-(4-Pyridylmethylamino)-1-propanol ... 933

$C_9H_{14}N_2OS$
5-Butyl-4-hydroxy-2-mercapto-6-methylpyrimidine ... 970

$C_9H_{14}N_2O_2$
2,4-Diethoxy-5-methylpyrimidine ... 972
2,4-Diethoxy-6-methylpyrimidine ... 973
Ethyl 1-ethyl-3-methyl-1*H*-pyrazole-5-carboxylate ... 863
Ethyl 1-ethyl-5-methyl-1*H*-pyrazole-3-carboxylate ... 862
Ethyl 1,3,5-trimethyl-1*H*-pyrazole-4-carboxylate ... 863
Hexamethylenehydantoin ... 135
4-Hydroxy-2-isopropyl-6-methoxymethylpyrimidine ... 969
2-Nitroso-4,7,7-trimethyl-2-azabicyclo(2.2.1)heptan-3-one ... 134

$C_9H_{14}N_2O_2S$
3-Amino-*N*,*N*-dimethyl-4-methyl-benzenesulfonamide ... 817
N-(2-Aminoethyl)-*p*-toluenesulfonamide ... 816

$C_9H_{14}N_2O_4$
Butyl *N*-carbamoylmaleamate ... 127
2,2,6,6-Tetramethyltetrahydropyran-3,4,5-trione-3,5-dioxime ... 1011

$C_9H_{14}N_2O_4S$
N^4-(2,3-Dihydroxypropyl)-sulfanilamide ... 810

$C_9H_{14}N_2O_6S_2$
2-(*N*-(3-Aminophenyl)-*N*-methylsulfamoyl)ethyl sulfate, potassium salt ... 816

$C_9H_{14}N_4O$
2-(*sec*-Butylidenehydrazino)-4-hydroxy-6-methylpyrimidine ... 972
2-(Butylidenehydrazino)-4-hydroxy-6-methylpyrimidine ... 1037
4-Hydroxy-2-(isobutylidenehydrazino)-6-methylpyrimidine ... 1036

$C_9H_{14}N_4O_2$
1-Allyl-5,6-diamino-3-ethyluracil ... 132
1,5-Dimethyl-3-(2-(dimethylamino)-vinyl)-4-nitro-1*H*-pyrazole ... 862

$C_9H_{14}N_4O_3$
6-Amino-1-isobutyl-3-methyl-5-nitrosouracil ... 132

$C_9H_{14}N_4O_4$
L-Alanylglycylglycine anhydride ... 133
6-Amino-3-ethyl-1-(2-hydroxypropyl)-5-nitrosouracil ... 131
6-Amino-3-ethyl-1-(3-hydroxypropyl)-5-nitrosouracil ... 131

$C_9H_{14}O$
2-Butyl-5-methylfuran ... 854
2,5-Divinyltetrahydropyran ... 23
Epoxyethylbicyclo(2.2.1)heptane ... 20

$C_9H_{14}OS$
2-Norbornyl thiolacetate ... 78

$C_9H_{14}OSi$
(Dimethylphenylsilyl)methanol ... 1105
4-(Trimethylsilyl)phenol ... 1105

$C_9H_{14}O_2$
Bicyclo(2.2.2)octane-2-carboxylic acid ... 66
1-(1-Hydroxycyclohexyl)propyn-3-ol. 1016
2-(4-Methyl-1-cyclohexylidene)-acetic acid ... 66
Methyl 2-norbornanecarboxylate ... 78
α-Nonynoic acid ... 1019
5-Norbornene-2,2-dimethanol ... 15
2-Norbornyl acetate ... 78

$C_9H_{14}O_3$
Ethyl 2-acetyl-2-pentenoate ... 92
Ethyl 2-allylacetoacetate ... 93
Ethyl 3-oxo-6-heptenoate ... 93
Isopropyl 4,5-epoxy-2-hexenoate ... 90
Methyl 6-hydroxy-2,4-heptadiene-2-carboxylate ... 85
Methyl 3-oxo-1-cyclohexaneacetate. 94
2-Oxo-1-cyclohexanepropionic acid. 65
7-Oxo-8-nonenoic acid ... 64

$C_9H_{14}O_3Si$
Phenyltrimethoxysilane ... 1112

$C_9H_{14}O_4$
2-Ethyl-2-(2-hydroxyethyl)-glutaric acid γ-lactone ... 97
Hexahydro-4-methylphthalic acid ... 66
mono-Methyl hexahydrophthalate ... 95
Pinic acid ... 66

$C_9H_{14}O_5$
Diethyl 2-(hydroxymethylene)-succinate ... 85
Diethyl 2-oxoglutarate ... 93
2,5-Dihydro-2,5-dimethoxyfurfuryl acetate ... 89
Dimethyl 2-acetylglutarate ... 93
2,3-Tetrahydropyrandiol diacetate ... 89

$C_9H_{14}O_5S$
Methyl 1,4-dioxa-8-thiaspiro(4.5)-decane-6-carboxylate *S*-oxide ... 153

$C_9H_{14}O_6$
1,3/2,5/4,6-Tri-*O*-methylene-D-mannitol ... 16

$C_9H_{14}O_6S_3$
1,1,1-Propanetriyl tris-(mercaptoacetate) ... 86

$C_9H_{14}O_7$
2-Propylcitric acid, trisodium salt ... 69
Trimethyl citrate ... 84

$C_9H_{14}S$
2-Butyl-5-methylthiophene ... 859

$C_9H_{14}Si$
(3-Phenylpropyl)silane ... 1112

$C_9H_{15}AsN_2O_3$
4-(3-Aminopropylamino)-benzenearsonic acid ... 1127

$C_9H_{15}BO_2Si$
4-(Trimethylsilyl)-benzeneboronic acid ... 1125

$C_9H_{15}BO_3$
Triallyl borate ... 1124

$C_9H_{15}BrN_2O_4$
DL-*N*-(2-Bromo-3-methylbutyryl)-glycylglycine ... 124

$C_9H_{15}BrO$
2-Bromoallyl 5-hexen-1-yl ether ... 17

$C_9H_{15}BrO_2$
Ethyl 1-bromo-1-cyclohexanecarboxylate ... 84
Methyl α-bromo-1-cyclohexaneacetate ... 84

$C_9H_{15}Br_2NO$
3,5-Dibromo-2,2,6,6-tetramethyl-4-piperidone ... 58

$C_9H_{15}Br_3Cl_3O_4P$
Tris(1-bromomethyl-2-chloroethyl) phosphate ... 160

$C_9H_{15}ClN_2$
1-Benzyl-1,1-dimethylhydrazinium chloride ... 256

$C_9H_{15}ClN_2O_4$
Chloroacetylglycyl-DL-valine ... 117

$C_9H_{15}ClOS$
2-Chloromethyl-1-oxa-4-thiaspiro(4.5)decane ... 24

$C_9H_{15}ClO_2$
Chloromethyl 2-hydroxy-5-methylcyclohexyl ketone ... 51

$C_9H_{15}Cl_2NO$
N-Cyclohexyl-2,2-dichloro-*N*-methylacetamide 106
$C_9H_{15}Cl_2NO_3S$
2,2-Dichloro-4,4-dimethyl-3-morpholinothietane-1,1-dioxide 153
$C_9H_{15}Cl_3$
1,1,1-Trichloro-2-nonene 5
$C_9H_{15}N$
2,2,4,6-Tetramethyl-1,2-dihydropyridine 39
$C_9H_{15}NO$
3-Amino-2,5,5-trimethyl-2-cyclohexen-1-one 57
trans-Decahydroquinolin-2-one 125
5,6-Dihydro-4,4,6-trimethyl-2-vinyl-1,3-oxazine 42
5,5-Dimethyl-3-methoxy-2-cyclohexen-1-imine methosulfate 1050
5,5-Dimethyl-3-(methylamino)-2-cyclohexen-1-one hydrochloride 57
4,7-Epoxy-2-methyl-2,3,3a,4,5,6,7,7a-octahydro-1*H*-isoindole 40
4,7-Epoxy-7-methyl-2,3,3a,4,5,6,7,7a-octahydro-1*H*-isoindole hydrochloride 40
$C_9H_{15}NO_2$
1-Acetoacetylpiperidine 109
N-Acetyl-3-nortropanol 107
3-Aminobicyclo(2.2.2)octane-2-carboxylic acid 72
N-Cyclopentylacetoacetamide 105
Ethyl 3-(allylamino)-crotonate 88
Ethyl α-butylcyanoacetate 149
Ethyl α-*sec*.-butylcyanoacetate 149
Ethyl 6-cyanohexanoate 149
3-Quinuclidyl acetate hydrochloride 90
$C_9H_{15}NO_3$
Ethyl 2-acetyl-3-(dimethylamino)-acrylate 92
Ethyl 1,3-dimethyl-2-oxo-3-pyrrolidinecarboxylate 125
$C_9H_{15}NO_3S$
N-Formyl-DL-penicillamine acetone adduct 109
N-Formyl-L-penicillamine acetone adduct 109
$C_9H_{15}NO_4$
Ethyl 1-nitro-1-cyclohexanecarboxylate 77
$C_9H_{15}NO_5$
5-(Hydroxyimino)azelaic acid 1011
$C_9H_{15}N_3$
2-(Dimethylamino)-4,5,6-trimethylpyrimidine 975
$C_9H_{15}N_3O_3$
6-Amino-3-ethyl-1-(2-methoxyethyl)-uracil 131
$C_9H_{15}N_3O_6$
1,2,4-Tris(2-hydroxyethyl)-1,2,4-triazine-3,5,6(1*H*,2*H*,4*H*)-trione 131
$C_9H_{15}N_5$
2-Amino-4-cyclohexylamino-1,3,5-triazine 979
C_9H_{16}
Allylcyclohexane 3
Hexahydroindan 2
4-Methyl-1,3-octadiene 2
$C_9H_{16}BrNO$
N-Allyl-2-bromo-2-ethylbutyramide 106
$C_9H_{16}BrNO_2$
4-(2-Bromoisovaleryl)-morpholine 110
$C_9H_{16}ClNO$
2-Chloro-*N*-cyclohexyl-*N*-methylacetamide 103
N-(5-Chloro-2-methylcyclohexyl)-acetamide 106
$C_9H_{16}ClNO_2$
2-Chloroethyl *N*-cyclohexylcarbamate 113
$C_9H_{16}ClNO_3$
Ethyl *N*-(2-chloroethyl)-*N*-isobutyrylcarbamate 113
$C_9H_{16}Cl_3NO_3$
Ethyl *N*-(1-butoxy-2,2,2-trichloroethyl)carbamate 113
Ethyl *N*-(1-*sec*-butoxy-2,2,2-trichloroethyl)carbamate 113
Ethyl *N*-(1-isobutoxy-2,2,2-trichloroethyl)carbamate 113
Methyl *N*-(1-pentoxy-2,2,2-trichloroethyl)carbamate 113
$C_9H_{16}NO_4P$
Diethyl furfurylamidophosphate 854
$C_9H_{16}N_2$
3-(Cyclohexylamino)-propionitrile 146
1-(Dimethylamino)-cyclohexanecarbonitrile 145
1-(Ethylamino)-cyclohexanecarbonitrile 145
Hexahydro-1*H*-azepine-1-propionitrile 146
1-Piperidinebutyronitrile 147
1-Pyrrolidinevaleronitrile 147
2,3,6,7-Tetrahydro-5,7,7-trimethyl-5*H*-pyrrolo(1,2-a)imidazole 42
$C_9H_{16}N_2O$
2,6-Dimethyl-4-morpholinepropionitrile 147
4-Morpholinevaleronitrile 150
$C_9H_{16}N_2OS$
5-Ethyl-2-imino-5-isobutyl-4-thiazolidinone 138
$C_9H_{16}N_2O_2$
5-*endo*-Hydroxybicyclo(2.2.2)-octane-2-*endo*-carboxylic acid hydrazide 111
2-(1-Imidazolyl)acetaldehyde diethyl acetal 866
$C_9H_{16}N_2O_2S$
O-Methyl *N*-(*N*-cyclohexyl-carbamoyl)thiocarbamate 142
$C_9H_{16}N_2O_4$
1-(1-Methyl-1-nitroethyl)-1-nitrocyclohexane 46
1-(1-Methyl-1-nitropropyl)-1-nitrocyclopentane 45
$C_9H_{16}N_4O_2$
Malonic bis(isopropylidenehydrazide) 112
$C_9H_{16}N_4O_3$
L-Prolylglycylglycinamide 124
$C_9H_{16}N_4O_5$
L-Alanylglycylglycylglycine 120
$C_9H_{16}O$
1-Allyl-1-cyclohexanol 13
1-(2-Cyclohexen-1-yl)-1-propanol 14
2-(2-Cyclohexen-1-yl)-2-propanol 14
Cyclohexylacetone 55
2,6-Dimethyl-1,6-heptadien-4-ol 11
5-Ethyl-3-hepten-2-one 50
Hexahydro-5-indanol 13
3-Isopropyl-1-cyclohexanone 56
2-Norbornaneethanol 12
3,3,5-Trimethylcyclohexanone 56
$C_9H_{16}O_2$
3-Acetyl-2-heptanone 51
3-Acetyl-5-methylhexan-2-one 51
2,2-Dimethoxynorbornane 18
Ethyl cyclopentylacetate 77
6-Methyl-3-cyclohexene-1,1-dimethanol 14
2-Methylcyclohexyl acetate 78
4-Methyl-2,2-pentamethylene-1,3-dioxolane 23
trans-2-Nonenoic acid 61
4,4,6-Trimethyl-2-vinyl-1,3-dioxane 23
$C_9H_{16}O_3$
2-Ethoxy-3a,4,5,6,7,7a-hexahydro-1,3-benzodioxole 22
Ethyl 5-methyl-3-oxohexanoate 94
2-(1-Ethylpropyl)-5-methyl-1,3-dioxolan-4-one 97
2-Hydroxy-4-methoxycyclohexyl methyl ketone 51
2-Methoxycyclohexyl acetate 85
Methyl 3-*sec*-butyl-3-methylglycidate 89
Methyl 2,3-dimethyl-2-methoxy-4-pentenoate 85
2-Oxononanoic acid 64
$C_9H_{16}O_4$
Butyl ethyl malonate 75
Diethyl methylsuccinate 75
Dimethyl 2-methyladipate 76
Ethyl 2-methyl-1,3-dioxolane-2-propionate 90
2-Methyl-2,3-butanediol diacetate 75
Methyl 2-methyl-1,3-dioxolane-2-butyrate 91
$C_9H_{16}O_4S_2$
Diethyl 3,5-dithiapimelate 87
$C_9H_{16}O_5$
Diethyl 2-ethoxymalonate 85
2,5-Dihydro-2,5-dimethoxy-2-(dimethoxymethyl)furan 22
$C_9H_{16}O_6Si$
Propyltriacetoxysilane 1109
$C_9H_{17}Br$
1-(2-Bromoethyl)-1-ethylcyclopentane 5
3-Cyclohexylpropyl bromide 6
$C_9H_{17}BrO_2$
Isopropyl 2-bromohexanoate 83
$C_9H_{17}ClO_2$
Isopropyl 6-chlorohexanoate 82
$C_9H_{17}ClS$
3-Chloropropyl cyclohexyl sulfide 24
$C_9H_{17}Cl_2O_3P$
Bis(2-chloroisopropyl) isopropenylphosphonate 159
$C_9H_{17}Cl_3N_2O_2$
Ethyl *N*-(1-(butylamino)-2,2,2-trichloroethyl)carbamate 114
Ethyl *N*-(1-(diethylamino)-2,2,2-trichloroethyl)carbamate 114
Methyl *N*-(1-(butylmethylamino)-2,2,2-trichloroethyl)carbamate 114
Methyl *N*-(1-(pentylamino)-2,2,2-trichloroethyl)carbamate 114
$C_9H_{17}IN_4$
Hexamethylenetetramine, compound with allyl iodide 36
$C_9H_{17}N$
trans-Decahydroquinoline hydrobromide 42
3-Ethylheptanenitrile 150
1-Piperidino-1-isobutene 40
2,3,4,5-Tetrahydro-2,2,4,6-tetramethylpyridine 41
$C_9H_{17}NO$
N-Allylhexanamide 106
2-Azacyclodecanone 58
1-(2-Buten-1-yloxymethyl)-pyrrolidine 37
1-Butyl-4-piperidone 59
2-(Dimethylaminomethyl)-1-cyclohexanone 57
2-(Dimethylaminomethyl)-1-cyclohexanone hydrochloride 53
4-Piperidino-2-butanone hydrochloride 59
4-Propylcyclohexanone oxime 1013
$C_9H_{17}NO_2$
N-Cyclohexyl-β-alanine 72
1,1-Diallyloxytrimethylamine 35
trans-Ethyl 2-amino-1-cyclohexanecarboxylate 88
cis-2-(Ethylamino)-1-cyclohexanecarboxylic acid 65
Ethyl cyclohexanecarbamate 106
4-(*N*-(2-Hydroxyethyl)-*N*-isopropylamino)-2-butyn-1-ol 1016
$C_9H_{17}NO_2S$
2-(2-Butoxyethoxy)-ethyl thiocyanate 150
2-(1-Ethylpropyl)-4-thiazolidenecarboxylic acid 346
$C_9H_{17}NO_3$
N-Acetyl-L-leucine methyl ester 124
2-Carbamoyl-2-propylvaleric acid 122
2,2-Diisopropylmalonamic acid 122
Ethyl 4-morpholinepropionate 90
Methyl α-methyl-4-morpholinepropionate hydrochloride 90
$C_9H_{17}NO_4$
5-Aminoazelaic acid 71
Diethyl glutamate 87
Diethyl (methylimino)-diacetate 88
Ethyl 2,4-dimethyl-4-nitrovalerate 73
Ethyl 3-nitro-2,2,3-trimethylbutyrate 73
$C_9H_{17}NO_5$
D-Pantothenic acid 103
$C_9H_{17}N_3O$
Cyclooctanone semicarbazone 1070
$C_9H_{17}N_3O_2$
1-Acetyl-4-cyclohexylsemicarbazide 128
Methyl 2-amino-3,4,5,6-tetrahydro-4-pyrimidinebutyrate hydrochloride 137
2-(Methylamino)-3,4,5,6-tetrahydro-4-pyrimidinebutyric acid hydrochloride 137
$C_9H_{17}N_3O_4$
L-Alanyl-L-alanyl-L-alanine 120
Glycylglycyl-DL-valine 120
Glycylglycyl-L-valine 123
DL-Valylglycylglycine 110
$C_9H_{17}O_2Tl$
2,4-Heptanedione, dimethylthallium salt 51
$C_9H_{17}O_5P$
Diethyl (3-methoxycarbonyl-2-propenyl)phosphonate 159
C_9H_{18}
trans-2,2-Dimethyl-3-heptene 1
3,5-Dimethyl-3-heptene (mixture of *cis* and *trans*) 1
1,2,4-Trimethylcyclohexane 2
3,5,5-Trimethyl-1-hexene 1
$C_9H_{18}BrNOS$
3,4-Dihydro-3-(6-hydroxyhexyl)-2*H*-thiazol-1-ium bromide 36
$C_9H_{18}BrNO_2$
Methyl 1-methyl-3-piperidinecarboxylate methobromide 90
$C_9H_{18}ClNO$
2-Chloro-*N*-(1-methylhexyl)-acetamide 102
$C_9H_{18}ClNO_2$
1-Chloroethyl *N*,*N*-dipropylcarbamate 116
$C_9H_{18}Cl_3O_3P$
Bis(2-chloroisopropyl) 2-chloroisopropylphosphonate 159
$C_9H_{18}I_2$
1,9-Diiodononane 4
$C_9H_{18}N_2$
9-Aminononanenitrile 145
3-(Dipropylamino)propionitrile 146
1,1′-Methylenebis(pyrrolidine) 41
4-Pyrrolidinopiperidine dihydrochloride 39
$C_9H_{18}N_2O$
3-Cyclohexyl-1,1-dimethylurea 129
1-Cyclohexyl-3-ethylurea 129
Cyclooctylurea 127
2,2,6,6-Tetramethyl-4-piperidone oxime hydrochloride 1010
$C_9H_{18}N_2O_2$
N-Acetyl-L-leucine-*N*-methylamide 124
4-(Butylcarbamoyl)morpholine 127
4,4′-Methylenedimorpholine 40
$C_9H_{18}N_2O_3$
L-Alanyl-L-isoleucine 120
DL-Alanyl-DL-leucine 120
β-Alanyl-DL-leucine 119
L-Alanyl-L-leucine hydrate 120
L-Alanyl-L-norleucine 120
2-Aminobutyrylvaline 123
L-Leucyl-β-alanine 124
L-Norleucyl-L-alanine 120
Norvalylalanine methyl ester 120
Sarcosyl-L-isoleucine 121
$C_9H_{18}N_2O_3S$
DL-Methionylglycine ethyl ester 104
$C_9H_{18}N_2O_4$
2-Butyl-2-ethyl-1,3-dinitropropane 46
2,2-Dipropyl-1,3-dinitropropane 46
Propylene diurethane 115
$C_9H_{18}N_2O_8S_2$
N,*N*′-Methylenebis(2-carbamoylethyl) bis(methanesulfonate) 156
$C_9H_{18}N_2S$
3-Cyclohexyl-1,1-dimethyl-2-thiourea 139

$C_9H_{18}N_3O_4P$
Tris(2-carbamoylethyl)phosphine oxide 158
$C_9H_{18}N_4$
Perhydro-1,4,7,9b-tetraazaphenalene 39
$C_9H_{18}N_4O_3$
L-Valylglycylglycinamide hydrochloride hemihydrate 121
$C_9H_{18}O$
Amyl isopropyl ketone 48
1-Butyl-1-cyclopentanol 13
Decahydro-4-azulenol 13
2,6-Dimethylcyclohexanemethanol 13
2,4-Dimethyl-3-heptanone 48
2,6-Dimethyl-3-heptanone 48
3-Ethyl-4-heptanone 48
1-Methylcyclooctanol 12
3-Methyl-4-octanone 48
4-Nonanone 48
1-Nonen-3-ol 11
1-Nonen-4-ol 11
2-Propyl-1-cyclohexanol (mixture of *cis* and *trans*) 14
4-Propyl-1-cyclohexanol 14
3,3,5-Trimethylcyclohexanol 13
$C_9H_{18}OS$
S-Methyl thiooctanoate 87
$C_9H_{18}OSi$
4-(Trimethylsilyl)cyclohexanone 1109
$C_9H_{18}O_2$
2-(Butoxymethyl)tetrahydrofuran 22
2-*tert*-Butyl-2,4-dimethyl-1,3-dioxolane 22
2,2-Diisopropyl-1,3-dioxolane 22
Heptyl acetate 73
2-Isopropoxycyclohexanol 20
4-Methyl-1,2-cyclohexanedimethanol 12
3-Methyloctanoic acid 61
3,5,5-Trimethylhexanoic acid 61
$C_9H_{18}O_2S_2$
5,7-Bis(methylthio)-6-heptene-2,4-diol 24
$C_9H_{18}O_2Si$
Ethyl 2-(trimethylsilyl)-1-cyclopropane carboxylate 1109
$C_9H_{18}O_3$
2,6-Diethoxytetrahydropyran 22
2,6-Dihydroxy-2,6-dimethyl-4-heptanone 50
Diisobutyl carbonate 74
Ethyl 3-ethyl-3-hydroxyvalerate 85
$C_9H_{18}O_3Si_3$
2,4,6-Trimethyl-2,4,6-trivinylcyclotrisiloxane 1096
$C_9H_{18}O_5$
Bis-(2-ethoxyethyl) carbonate 85
2,5-Dimethoxytetrahydro-2-furaldehyde dimethyl acetal 22
$C_9H_{18}O_6$
4-Hydroxytetrahydropyran-3,3,5,5-tetramethanol, 70% solution 21
$C_9H_{19}Br$
2-Bromononane 4
$C_9H_{19}BrNP$
(2-Cyanoethyl)-triethylphosphonium bromide 158
$C_9H_{19}BrO_3$
2-Bromo-1,1,3-triethoxypropane 19
$C_9H_{19}N$
Allylhexylamine 32
1-(Aminomethyl)cyclooctane 31
N-Cyclooctylmethylamine 31
1-Propyl-4-methylpiperidine 40
3,3,5-Trimethylcyclohexylamine 31
$C_9H_{19}NO$
4-(Allylamino)-4-methyl-2-pentanol 33
1-Cyclohexylamino-2-propanol 33
1-Cyclohexylamino-2-propanol hydrochloride 33
2,2-Dimethyl-5-dimethylamino-3-pentanone hydrochloride 57
β,β-Dimethyl-1-pyrrolidinepropanol 37
1-Dipropylamino-2-propanone 57
4-Ethyl-2-imino-4-heptanol 1027
1-(Hexamethyleneimino)-2-propanol 36
2-Methyl-1-piperidinepropanol 37
Nonanoamide 102
$C_9H_{19}NO_2$
β,β-Dimethyl-4-morpholinepropanol 37
1-(2-Methoxyethyl)-3-methyl-3-pyrrolidinemethanol 36
α,2,6-Trimethyl-4-morpholineethanol 37
L-Valine *tert*-butyl ester hydrochloride 87
$C_9H_{19}NO_2S$
L-Methionine *tert*-butyl ester hydrochloride 87
$C_9H_{19}NO_3$
3,3-Dimethyl-3-(1,1-dimethyl-2-hydroxyethylamino)propionic acid 63
$C_9H_{19}NO_4$
D-2,4-Dihydroxy-3,3-dimethyl-*N*-(3-hydroxypropyl)butyramide 103
L-2,4-Dihydroxy-3,3-dimethyl-*N*-(3-hydroxypropyl)butyramide 103
$C_9H_{19}NS$
3-(Cyclohexylthio)propylamine 36
$C_9H_{19}N_3O$
4-Carbamoyl-4-(isopropylamino)-piperidine 108
4-Cyclohexyl-1,1-dimethylsemicarbazide 128
N-Isopropyl-4-methyl-1-piperazinecarboxamide 126
$C_9H_{19}N_3O_2$
L-Alanyl-L-isoleucinamide acetate 121
1-Ethyl-3-(2-morpholinoethyl)urea 126
Sarcosyl-L-isoleucinamide hydrochloride 121
$C_9H_{19}N_3O_3$
L-Isoleucyl-L-serinamide hydrochloride 118
L-Leucyl-L-serinamide hydrochloride 117
$C_9H_{19}N_5O_3$
3-Nitro-1-nitroso-1-octylguanidine 138
C_9H_{20}
2,2-Dimethylheptane 1
2-Methyloctane 1
2,2,5-Trimethylhexane 1
$C_9H_{20}BrN$
Allyltriethylammonium bromide 43
$C_9H_{20}BrNO_2$
Butyrylcholine bromide 88
$C_9H_{20}Br_2N_2$
1,5-Dimethyl-1,5-diazoniabicyclo(3.2.2)nonane dibromide 44
$C_9H_{20}ClNO_2$
(5-Carboxypentyl)-trimethylammonium chloride 67
$C_9H_{20}ClO_3P$
Dibutyl chloromethylphosphonate 159
$C_9H_{20}N_2$
3-Hexamethyleneimino-1-propylamine 38
$C_9H_{20}N_2O$
1,3-Dibutylurea 129
N-(3-Diethylaminopropyl)-acetamide 104
$C_9H_{20}N_2S$
1,3-Di-*tert*-butyl-2-thiourea 142
2-Octyl-2-thiopseudourea hydrobromide 137
$C_9H_{20}N_4$
2,2′-Methylenebis-(hexahydropyrimidine) 41
$C_9H_{20}N_4S$
1-(2-Diethylaminoethyl)-4-thioxohexahydro-1,3,5-triazine 139
$C_9H_{20}N_4S_2$
S,S′-Heptamethylene-diisothiuronium dibromide 138
$C_9H_{20}O$
2,2-Diethyl-1-pentanol 8
2,2-Dimethyl-3-heptanol 8
2,4-Dimethyl-3-heptanol 8
2,6-Dimethyl-2-heptanol 10
3,6-Dimethyl-3-heptanol 9
3-Ethyl-3-heptanol 9
5-Ethyl-2-heptanol 8
2-Methyl-2-octanol 9
2-Methyl-3-octanol 8
3-Methyl-3-octanol 10
3-Methyl-4-octanol 8
4-Methyl-4-octanol 9
4-Methyl-4-octanol 10
2-Methyl-2-propyl-1-pentanol 8
4-Nonanol 9
5-Nonanol 8
DL-2,2,5-Trimethyl-3-hexanol 8
$C_9H_{20}OS$
Heptyl 2-hydroxyethyl sulfide 24
$C_9H_{20}O_2$
3,5-Dimethyl-5-methoxy-3-hexanol 19
$C_9H_{20}O_3$
1,3-Diisopropoxy-2-propanol 19
2,4-Dimethyl-2-methoxymethyl-1,5-pentanediol 19
$C_9H_{20}O_3S$
Neopentyl 1-butanesulfonate 156
$C_9H_{20}O_4S$
2-Hydroxy-2-nonanesulfonic acid, sodium salt 155
Nonyl sulfate, sodium salt 157
$C_9H_{20}S$
Methyl *tert*-octyl sulfide 26
tert-Nonyl mercaptan 24
$C_9H_{20}Si$
3,3-Dimethyl-1-(trimethylsilyl)-1-butene 1096
$C_9H_{21}B_3O_6$
Triisopropoxyboroxine 1124
$C_9H_{21}NO$
3-Diethylamino-2,2-dimethyl-1-propanol 34
1-Dipropylamino-2-propanol 34
3-(2-Ethylbutoxy)-propylamine 35
3-Hexyloxypropylamine 35
$C_9H_{21}NO_4$
2,2′-Methyliminobis-(acetaldehyde dimethyl acetal) 35
$C_9H_{21}N_3$
N-(2-Dimethylaminoethyl) *N′*-methylpiperazine 39
$C_9H_{21}N_3O$
1-(3-Aminopropyl)-4-(2-hydroxyethyl)piperazine 36
$C_9H_{21}O_3P$
Tripropyl phosphite 158
$C_9H_{21}O_7P$
Tris-(2-methoxyethyl) phosphate 161
$C_9H_{22}AsF_6N$
Tripropylammonium hexafluoroarsenate 43
$C_9H_{22}F_6NP$
Tripropylammonium hexafluorophosphate 43
$C_9H_{22}NO_5P$
Diethyl bis-(2-hydroxyethyl)-aminomethylphosphonate 159
$C_9H_{22}N_2$
N,N,N′,N′-Tetraethylmethanediamine 30
$C_9H_{22}O_3Si$
Propyltriethoxysilane 1096
$C_9H_{24}NO_4P$
Tripropylamine phosphate 161
$C_9H_{24}N_2O_2Si$
N-(3-(Dimethoxymethylsilyl)-isobutyl)-ethylenediamine 1107
$C_9H_{24}O_2Si_2$
1,3-Bis(trimethylsilyloxy)propane 1107
$C_9H_{25}NOSi_2$
N-Trimethylsilyl-3-(trimethylsilyloxy)propylamine 1096
$C_9H_{26}N_2Si_2$
1,3-Bis(trimethylsilylamino)propane 1107
$C_9H_{26}Si_3$
Bis(trimethylsilylmethyl)-methylsilane 1096
$C_9H_{28}OSi_4$
Tris(trimethylsilyl)silanol 1109
$C_{10}Cl_8$
Perchlorofulvalene 173
$C_{10}Cl_8S$
2-(Pentachlorophenyl)-3,4,5-trichlorothiophene 855
$C_{10}Cl_{10}$
Bis(pentachlorocyclopentadienyl) 6
$C_{10}Cl_{12}$
Perchloropentacyclodecane 7
$C_{10}F_{19}I$
1-Iodononadecafluoro-1-decene 5
$C_{10}N_4S_4$
4′,4,5′,5-Tetracyanotetrathiafulvalene 880
$C_{10}HCl_6NO_2$
1,2,3,4,5,6-Hexachloro-7-nitronaphthalene 280
$C_{10}HCl_7S$
3,4-Dichloro-2-(pentachlorophenyl)-thiophene 855
2-(2,3,5,6-Tetrachlorophenyl)-3,4,5-trichlorothiophene 855
$C_{10}H_2Cl_8O_3$
8,8-Dihydroxyoctachloro-3a,4,7,7a-tetrahydro-4,7-methano-1*H*-inden-1-one 51
$C_{10}H_3Cl_3O_2$
1-(4-Methoxyphenyl)-2,3,3-trichloro-2-propen-1-one 301
$C_{10}H_4ClF_7N_2O_3$
2′-Chloro-2,2,3,3,4,4,4-heptafluoro-5′-nitrobutyranilide 395
$C_{10}H_4Cl_2N_2$
2,4-Dichlorobenzylidenemalonitrile 759
(2,4-Dichlorobenzylidene)-malononitrile 764
$C_{10}H_4Cl_4$
1,2,3,4-Tetrachloronaphthalene 174
$C_{10}H_4F_8N_2O_3$
3′-Nitro-2,2,3,3,4,4,4,4′-octafluorobutyranilide 395
$C_{10}H_5BrClNO_2$
N-(4-Bromo-2-chlorophenyl)-maleimide 722
$C_{10}H_5Br_3O_2$
1,4,6-Tribromo-2,3-naphthalenediol 195
$C_{10}H_5ClN_2$
4-Chlorobenzylidenemalononitrile 759
$C_{10}H_5ClN_2O_3S$
6-Diazo-5,6-dihydro-5-oxo-1-naphthalenesulfonyl chloride 803
$C_{10}H_5ClN_2O_4$
4-Chloro-1,5-dinitronaphthalene 277
$C_{10}H_5ClO_3$
2-Chloro-3-hydroxy-1,4-naphthoquinone potassium salt 316
$C_{10}H_5Cl_2NO_2$
N-(2,5-Dichlorophenyl)-maleimide 722
$C_{10}H_5Cl_2NO_3$
6,8-Dichloro-2-hydroxy-4-quinolinecarboxylic acid 983
$C_{10}H_5Cl_3S$
2-Phenyl-3,4,5-trichlorothiophene 859
$C_{10}H_5Cl_4N$
1-(2,3,5,6-Tetrachlorophenyl)pyrrole 841
$C_{10}H_5F_2NO_2$
N-(2,4-Difluorophenyl)maleimide 716
$C_{10}H_5F_7N_2O_3$
2,2,3,3,4,4,4-Heptafluoro-3′-nitrobutyranilide 397
$C_{10}H_5F_{15}$
3,4,5,5,6,6,7,7,8,8,9,9,10,10,10-Pentadecafluoro-3-decene 5
$C_{10}H_5F_{15}O_2$
Ethyl perfluorooctanoate 83
$C_{10}H_5NO_2$
2-Cyano-1,3-indanedione 757
$C_{10}H_5N_3O_2$
(3-Nitrobenzylidene)malononitrile 764
(4-Nitrobenzylidene)malononitrile 764
$C_{10}H_6BrF_3N_4OS$
1-(5-Bromo-1,3,4-thiadiazol-2-yl)-3-(3-(trifluoromethyl)phenyl)urea 880
$C_{10}H_6BrNO_2$
1-Bromo-5-nitronaphthalene 280
1-Bromo-8-nitronaphthalene 280
N-(4-Bromophenyl)maleimide 715
$C_{10}H_6BrNO_3$
2-(4-Bromophenyl)-3-hydroxymaleimide 716
2-(4-Bromophenyl)-*N*-hydroxymaleimide, sodium salt 721
$C_{10}H_6Br_2$
1,4-Dibromonaphthalene 174
1,5-Dibromonaphthalene 174

$C_{10}H_6Br_2$ (Cont'd)
2,3-Dibromonaphthalene ... 174
$C_{10}H_6Br_2FNO_2$
2,3-Dibromo-*N*-(2-fluorophenyl)-succinimide ... 715
$C_{10}H_6Br_2N_2O_2$
α-Cyano-3,5-dibromo-2-hydroxycinnamamide ... 761
2-(Dibromonitromethyl)quinoline ... 988
$C_{10}H_6Br_2O_2$
2,6-Dibromo-1,5-dihydroxynaphthalene ... 196
1,4-Dibromo-2,3-naphthalenediol ... 195
$C_{10}H_6ClNOS_2$
5-(4-Chlorobenzylidene)rhodanine ... 732
$C_{10}H_6ClNO_2$
N-(2-Chlorophenyl)-maleimide ... 720
N-(3-Chlorophenyl)-maleimide ... 721
N-(4-Chlorophenyl)-maleimide ... 720
$C_{10}H_6ClNO_3$
3-Carboxy-6-chloro-4-hydroxyquinoline ... 984
$C_{10}H_6ClN_3$
5-Chloro-*s*-triazolo(3,4-a)-isoquinoline ... 919
$C_{10}H_6ClN_3O_3$
2-Chloro-α-cyano-5-nitrocinnamamide ... 766
4-Chloro-α-cyano-3-nitrocinnamamide ... 759
$C_{10}H_6ClN_3S$
3-Chloro-10*H*-pyridazino(4,3-b)(1,4)-benzothiazine ... 968
$C_{10}H_6ClN_5O_2$
2-(5-Chloro-2-benzimidazolylimino)-4,5-imidazolidinedione ... 906
$C_{10}H_6Cl_2N_2O$
α-Cyano-2,4-dichlorocinnamamide ... 760
$C_{10}H_6Cl_2N_2OS$
2,4-Dichloro-*N*-(3-isothiazolyl)-benzamide ... 874
3,4-Dichloro-*N*-(2-thiazolyl)-benzamide ... 875
$C_{10}H_6Cl_2N_2O_2$
α-Cyano-2,4-dichloro-6-hydroxycinnamamide ... 761
$C_{10}H_6Cl_2N_2O_3$
2-Cyanoacetyl *N*-(2,3-dichlorophenyl)carbamate ... 752
$C_{10}H_6Cl_2N_2O_6S$
1-(2,5-Dichloro-4-sulfophenyl)-5-pyrazolone-3-carboxylic acid ... 790
$C_{10}H_6Cl_2N_4O_6$
2,3-Dichloro-4-(2,4-dinitrophenylhydrazono)-2-butenoic acid ... 342
$C_{10}H_6Cl_2O_2$
2,3-Dichloro-2,3-dihydro-1,4-naphthoquinone ... 319
$C_{10}H_6Cl_2O_4S_2$
2,6-Naphthalenedisulfonyl chloride. 804
$C_{10}H_6Cl_3NO_4$
4-Nitrobenzyl trichloroacrylate ... 374
$C_{10}H_6Cl_4N_2$
4,5-Dichloro-3-(dichloromethyl)-1-phenylpyrazole ... 859
$C_{10}H_6Cl_4N_2OS$
3-(4-Chlorophenyl)-5-imino-2-(trichloromethyl)-4-oxazolidinethione ... 727
$C_{10}H_6Cl_4O_4$
Dimethyl 2,3,5,6-tetrachloroterephthalate ... 367
$C_{10}H_6Cl_6S_2Si$
Bis(3,4,5-trichloro-2-thienyl)-dimethylsilane ... 1110
$C_{10}H_6FNO_2$
1-Fluoro-4-nitronaphthalene ... 280
N-(2-Fluorophenyl)maleimide ... 715
N-(4-Fluorophenyl)maleimide ... 723
$C_{10}H_6FNO_4S$
5-Nitro-1-naphthalenesulfonyl fluoride ... 803
$C_{10}H_6F_3NO_2$
2-(Trifluoromethyl)-4,7-quinolinediol. 981
$C_{10}H_6F_3NO_3$
2-(Trifluoromethyl)-4,5,8-quinolinetriol ... 981
$C_{10}H_6F_5O_3$
4-(1,1,2,3,3,3-Hexafluoro-1-propoxy)-benzoic acid ... 336
$C_{10}H_6F_6N_2O_2$
1-(2,2,2-Trifluoroacetyl)-3-(2-(trifluoromethyl)phenyl)urea ... 509
$C_{10}H_6INO_2$
N-(4-Iodophenyl)maleimide ... 720
$C_{10}H_6N_2$
1-Isoquinolinecarbonitrile ... 997
4-Quinolinecarbonitrile methiodide. 994
$C_{10}H_6N_2O$
2-Benzoylmalononitrile ... 757
3-Cyano-2-imino-2*H*-1-benzopyran ... 764
4-(Hydroxybenzylidene)-malononitrile ... 761
$C_{10}H_6N_2O_2$
N-(Cyanomethyl)-phthalimide ... 755
$C_{10}H_6N_2O_2S$
3-Nitro-1-azulenyl thiocyanate ... 768
$C_{10}H_6N_2O_3$
4-Hydroxybenzofuro(2,3-d)pyrimid-2(1*H*)-one, sodium salt ... 976
$C_{10}H_6N_2O_4$
2,4-Dinitroso-1,3-naphthalenediol tetrahydrate ... 257
N-(2-Nitrophenyl)maleimide ... 718
N-(4-Nitrophenyl)maleimide ... 719
5-Nitroquinaldic acid ... 989
5-Nitro-4-quinolinecarboxylic acid ... 988
$C_{10}H_6N_2O_5$
N-Acetyl-4-nitrophthalimide ... 719
$C_{10}H_6N_4O_5$
4,4′-Dinitro-2,2′-bipyridine-*N*-oxide ... 1006
$C_{10}H_6N_4O_6$
4,4′-Dinitro-2,2′-bipyridine-*N*,*N*′-dioxide ... 1006
$C_{10}H_6N_6O_4$
2-(5-Nitro-2-benzimidazolylimino)-4,5-imidazolidinedione ... 906
$C_{10}H_6O_2S_3$
4*H*-Dithieno(2,3-b:2′,3′-d)thiopyran-2-carboxylic acid ... 858
$C_{10}H_6O_4$
Phthalideneacetic acid ... 385
$C_{10}H_6O_5$
3a,8b-Dihydro-8b-hydroxy-4*H*-indeno(1,2-c)-1,2-dioxole-3,4-dione ... 307
2-Furoic anhydride ... 854
7-Hydroxy-3-coumarincarboxylic acid ... 382
$C_{10}H_7BrClNO_2$
5-Bromo-6-chloro-3-indolyl acetate ... 888
$C_{10}H_7BrClNO_3$
4′-Bromo-2′-chloromaleanilic acid ... 476
$C_{10}H_7BrN_2O_2$
2-Bromo-4-nitro-1-naphthylamine ... 271
$C_{10}H_7BrN_2O_3$
5-Bromo-6-methoxy-8-nitroquinoline ... 985
$C_{10}H_7BrN_4S$
8-Bromo-5-methyl-5*H*-1,2,4-triazino(5,6-b)indole-3(2*H*)-thione ... 919
$C_{10}H_7BrO_3S$
2-Bromo-1-naphthalenesulfonic acid ... 796
$C_{10}H_7BrS$
2-Bromo-5-phenylthiophene ... 859
$C_{10}H_7Br_2Cl_2NO_3$
2,3-Dibromo-3′,4′-dichlorosuccinanilic acid ... 405
$C_{10}H_7Br_2N$
2,4-Dibromo-1-naphthylamine ... 231
α,α-Dibromoquinaldine ... 981
$C_{10}H_7Br_3N_2$
1-Benzyl-3,4,5-tribromo-1*H*-pyrazole. 864
$C_{10}H_7Br_3Te$
(1-Naphthyl)tribromotellurium ... 1143
$C_{10}H_7ClFNO_3$
3′-Chloro-4′-fluoromaleanilic acid ... 455
$C_{10}H_7ClFN_3OS$
1-(3-Chloro-4-fluorophenyl)-3-(2-thiazolyl)urea ... 875
$C_{10}H_7ClN_2O$
2-Chloro-α-cyanocinnamamide ... 766
4-Chloro-α-cyanocinnamamide ... 759
2-Chloro-3-quinolinecarboxaldehyde oxime ... 989
2-Hydroxy-1-naphthalenediazonium chloride ... 1094
$C_{10}H_7ClN_2OS$
N-(4-Chlorobenzoyl)-*N*-methyl-cyanothioformamide ... 760
p-Chloro-*N*-(2-thiazolyl)-benzamide ... 877
$C_{10}H_7ClN_2O_2$
4-Chloro-3-hydroxy-1-phenyl-6-pyridazone ... 697
4-Chloro-2-methyl-6-nitroquinoline ... 989
$C_{10}H_7ClN_2O_4S$
1-Hydroxy-5-sulfo-2-naphthalenediazonium chloride ... 787
2-Hydroxy-4-sulfo-1-naphthalenediazonium chloride ... 1085
$C_{10}H_7ClN_4O_3S$
1-(2-Chloro-5-nitrophenyl)-3-(2-thiazolyl)urea ... 875
$C_{10}H_7ClO$
3-Chloro-2-naphthol ... 196
$C_{10}H_7ClO_3S$
Methyl 5-chloro-3(2*H*)oxobenzo(b)-thiophene-2-carboxylate ... 364
$C_{10}H_7Cl_2NO_2$
N-(2,3-Dichlorophenyl)succinimide ... 716
2-Propynyl *N*-(2,5-dichlorophenyl)-carbamate ... 1015
$C_{10}H_7Cl_2NO_3$
2′,3′-Dichloromaleanilic acid ... 454
2′,4′-Dichloromaleanilic acid ... 455
2′,5′-Dichloromaleanilic acid ... 450
2′,6′-Dichloromaleanilic acid ... 476
3′,4′-Dichloromaleanilic acid ... 476
$C_{10}H_7Cl_2N_3OS$
3,4-Dichloro-*N*-(5-methyl-1,3,4-thiadiazol-2-yl)benzamide ... 880
1-(3,4-Dichlorophenyl)-5-imino-3-methyl-4-thiohydantoin ... 732
1-(2,3-Dichlorophenyl)-3-(2-thiazolyl)urea ... 875
$C_{10}H_7Cl_3N_2$
1-Phenyl-3-(trichloromethyl)-pyrazole ... 859
$C_{10}H_7Cl_3O$
1-(*p*-Tolyl)-2,3,3-trichloro-2-propen-1-one ... 304
$C_{10}H_7Cl_3Te$
(1-Naphthyl)trichlorotellurium ... 1143
$C_{10}H_7FN_2OS_2$
3-(4-Fluorobenzylideneamino)-rhodanine ... 1070
$C_{10}H_7FN_2O_3$
2-Cyanoacetyl *N*-(4-fluorophenyl)-carbamate ... 752
$C_{10}H_7FN_4O_3$
1-(*N*-(4-Fluoro-3-nitrophenyl)-carbamoyl)imidazole ... 865
1-(*N*-(2-Fluoro-5-nitrophenyl)-carbamoyl)pyrazole ... 860
$C_{10}H_7F_2NO_3$
2′,4′-Difluoromaleanilic acid ... 450
$C_{10}H_7F_3O_2$
m-(Trifluoromethyl)-cinnamic acid ... 334
$C_{10}H_7F_6NO$
3′,5′-Bis(trifluoromethyl)acetanilide. 398
$C_{10}H_7F_{17}Si$
Dimethyl(perfluorooctyl)silane ... 1097
$C_{10}H_7I$
2-Iodonaphthalene ... 174
$C_{10}H_7IO$
1-Iodo-2-naphthol ... 196
$C_{10}H_7I_2NO$
5,7-Diiodo-8-hydroxy-2-methylquinoline ... 984
$C_{10}H_7NO$
1-Oxo-2-indancarbonitrile ... 757
2-Oxo-1-indancarbonitrile ... 757
$C_{10}H_7NOS_2$
5-Benzylidenerhodanine ... 749
$C_{10}H_7NO_2$
4-Isoquinolinecarboxylic acid ... 998
$C_{10}H_7NO_2S$
3-Nitro-2-phenylthiophene ... 857
$C_{10}H_7NO_2S_2$
5-(3-Hydroxybenzylidene)rhodanine. 734
$C_{10}H_7NO_3$
1-Acetylisatin ... 691
N-Acetylphthalimide ... 720
2,3-Dihydro-3-oxo-(1*H*)isoindol-1-ylideneacetic acid ... 691
N-Hydroxy-2-phenylmaleimide ... 721
2-Hydroxyquinoline-4-carboxylic acid ... 983
8-Hydroxyquinoline-4-carboxylic acid ... 983
3-Isoquinolinecarboxylic acid *N*-oxide ... 1006
Mordant dye (C.I. 10015) ... 1094
2-(2-Oxazolyl)benzoic acid, potassium salt ... 872
2-Phthalimidoacetaldehyde hydrate 721
4-Quinolinecarboxylic acid *N*-oxide. 1006
$C_{10}H_7NO_3S_2$
p-(4-Oxo-2-thioxothiazolidino)-benzoic acid ... 745
$C_{10}H_7NO_4$
2,3-Indoledicarboxylic acid ... 886
1-Isatinacetic acid ... 704
2-Phthalimidoacetic acid ... 721
$C_{10}H_7NO_5S$
4-Hydroxy-3-nitroso-1-naphthalenesulfonic acid, sodium salt ... 791
6-Nitro-1-naphthalenesulfonic acid, sodium salt ... 794
$C_{10}H_7NO_8S_2$
2-Nitroso-1-naphthol-3,6-disulfonic acid, disodium salt ... 791
$C_{10}H_7NO_9S_2$
1,8-Dihydroxy-2-nitroso-3,6-naphthalenedisulfonic acid, copper complex ... 791
$C_{10}H_7NS$
Benzo(b)thiophene-3-acetonitrile ... 898
$C_{10}H_7N_3$
(Anilinomethylene)-malononitrile ... 763
1*H*-Naphtho(1,8-d,e)triazine ... 1004
$C_{10}H_7N_3O_2S$
5-Picolinylidene-2-thiobarbituric acid ... 933
5-(3-Pyridylmethylene)-2-thiobarbituric acid ... 933
5-(4-Pyridylmethylene)-2-thiobarbituric acid ... 933
$C_{10}H_7N_3O_3$
α-Cyano-3-nitrocinnamamide ... 765
α-Cyano-4-nitrocinnamamide ... 766
4′-Nitro-2,2′-bipyridine-*N*-oxide ... 1006
$C_{10}H_7N_3O_3S$
4-Nitro-*N*-(2-thiazolyl)benzamide ... 877
N-(5-Nitrothiazol-2-yl)-benzamide ... 876
$C_{10}H_7N_3S_2$
4-Amino-5-cyano-2,3-dihydro-3-phenyl-2-thioxothiazole ... 739
$C_{10}H_7N_4S$
6-Mercapto-2-(4-pyridyl)purine ... 940
$C_{10}H_7N_5$
4-Amino-6-phenyl-*s*-triazine-2-carbonitrile ... 979
$C_{10}H_7N_5O_2$
2-(2-Benzimidazolylimino)-4,5-imidazolidinedione ... 906
$C_{10}H_7N_5O_5S_2$
N-(5-(2,4-Dinitrophenylthio)-1,3,4-thiadiazol-2-yl)acetamide ... 878
$C_{10}H_8AsCl_2N$
Dichloro(8-methyl-5-quinolyl)arsine hydrochloride ... 1128
$C_{10}H_8AsNO_5$
4-Nitro-1-naphthalenearsonic acid ... 1127
$C_{10}H_8AsN_5O_6$
4-(4-Arsono-2-nitrophenylazo)-3-methyl-5-pyrazolone ... 1127
$C_{10}H_8BrClF_3NO$
3-Bromo-2′-chloro-5′-(trifluoromethyl)propionanilide ... 390
$C_{10}H_8BrN$
8-(Bromomethyl)quinoline hydrobromide ... 995
1-Bromo-2-naphthylamine ... 231
$C_{10}H_8BrNO$
1-Acetyl-3-bromoindole ... 890
5-(Bromomethyl)-3-phenylisoxazole. 870
$C_{10}H_8BrNOS_2$
3-(4-Bromo-*m*-tolyl)-rhodanine ... 746
3-(5-Bromo-*o*-tolyl)-rhodanine ... 746

$C_{10}H_8BrNO_2$
6-Bromo-3,4-(methylenedioxy)-hydrocinnamonitrile 758
N-(4-Bromophenyl)succinimide 716
$C_{10}H_8BrNO_3$
N-(2-Bromoethoxy)phthalimide 714
4′-Bromomaleanilic acid 454
N-(2-Bromophenyl)maleamic acid 453
Ethyl 3-(5-bromo-2-furyl)-2-cyanoacrylate 853
$C_{10}H_8BrNS$
2-Amino-5-(4-bromophenyl)-thiophene hydrochloride 855
$C_{10}H_8Br_2ClNO_3$
3′-Chloro-2,3-dibromosuccinanilic acid 405
$C_{10}H_8Br_2FNO_3$
2,3-Dibromo-4′-fluorosuccinanilic acid 405
$C_{10}H_8Br_2N_2O$
4,4-Dibromo-3-methyl-1-phenyl-2-pyrazolin-5-one 705
$C_{10}H_8Br_3NO_3$
2,3,4′-Tribromosuccinanilic acid 405
$C_{10}H_8ClFN_4OS$
1-(3-Chloro-4-fluorophenyl)-3-(1,3,4-thiadiazol-2-yl)urea 878
$C_{10}H_8ClF_3N_4O_3$
Oxamic *N*2-(*N*-(2-chloro-5-(trifluoromethyl)phenyl)-carbamoyl)hydrazide 505
$C_{10}H_8ClF_3O_2$
2′-(2-Chloro-1,1,2-trifluoroethoxy)-acetophenone 288
α-Chloro-3-(trifluoromethyl)-hydrocinnamic acid 325
$C_{10}H_8ClF_3O_3$
1-Chloroethyl 3-(trifluoromethyl)-phenyl carbonate 381
Methyl 2-(2-chloro-1,1,2-trifluoroethoxy)benzoate 366
Methyl 4-(2-chloro-1,1,2-trifluoroethoxy)benzoate 366
$C_{10}H_8ClF_3O_4$
2-(2-Chloro-1,1,2-trifluoroethoxy)-*p*-anisic acid 336
$C_{10}H_8ClN$
6-Chlorolepidine 995
6-Chloro-8-methylisoquinoline 998
1-(4-Chlorophenyl)-1-cyclopropane-carbonitrile 756
α-Chloroquinaldine 981
$C_{10}H_8ClNO$
3-Chloro-1-methoxyisoquinoline 997
4-Chloro-6-methoxyquinoline 986
4-Chloro-8-methoxyquinoline 986
6-Chloro-2-methyl-4-quinolinol 984
$C_{10}H_8ClNO_2$
4-(2-Chloro-2-cyanoethyl)-benzoic acid 757
2-Chloro-α-cyanohydrocinnamic acid 757
N-(2-Chloroethyl)phthalimide 714
3-(2-Chlorophenyl)-2-cyanopropionic acid 752
2-(4-Chlorophenyl)succinimide 715
N-(3-Chlorophenyl)-succinimide 722
N-(4-Chlorophenyl)succinimide 715
5-Methyl-3-phenyl-4-isoxazolecarbonyl chloride 871
Propynyl *N*-(3-chlorophenyl)-carbamate 1019
$C_{10}H_8ClNO_3$
5-Chloro-*N*-(2-hydroxyethyl)-phthalimide 716
2′-Chloromaleanilic acid 476
3′-Chloromaleanilic acid 475
4′-Chloromaleanilic acid 476
$C_{10}H_8ClN_3OS$
4-Chloro-*N*-(5-methyl-1,3,4-thiadiazol-2-yl)benzamide 880
$C_{10}H_8ClN_3O_2$
1-(4-Chloro-2-nitrobenzyl)imidazole 865
$C_{10}H_8Cl_2F_3NO_2$
2-Chloroethyl *N*-(2-chloro-5-trifluoromethylphenyl)carbamate 604
$C_{10}H_8Cl_2FeHg_2$
1,1′-Bis(chloromercuri) ferrocene 1140
$C_{10}H_8Cl_2N_2O$
4,4-Dichloro-3-methyl-1-phenyl-3-pyrazolin-5-one 706
$C_{10}H_8Cl_2N_4O_2$
1-(Cyanoacetyl)-4-(2,3-dichlorophenyl)semicarbazide 752
1-(Cyanoacetyl)-4-(3,4-dichlorophenyl)semicarbazide 752
$C_{10}H_8Cl_2O_4$
4,6-Dichlororesorcinol diacetate 376
$C_{10}H_8Cl_3NO$
N-(Cyclopropylcarbonyl)-2,4,5-trichloroaniline 408
$C_{10}H_8Cl_3NO_2$
3′-Acetyl-2,2,2-trichloroacetanilide 397
Allyl *N*-(2,4,5-trichlorophenyl)-carbamate 687
$C_{10}H_8Cl_3NO_3$
2′,4′,5′-Trichlorosuccinanilic acid 407
$C_{10}H_8Cl_4N_2O_2$
N,*N*′-(4,5-Dichloro-1,2-phenylene)-bis(2-chloroacetamide) 390
$C_{10}H_8Cl_4N_2O_3S$
S-(2,3,5,6-Tetrachloro-4-pyridyl) *N*-(ethoxycarbonylmethyl)-thiocarbamate 964
$C_{10}H_8FNO_3$
2′-Fluoromaleanilic acid 461
3′-Fluoromaleanilic acid 453
4′-Fluoromaleanilic acid 454
$C_{10}H_8FN_3O$
2′-Fluoro-2-imidazolecarboxanilide 865
1-(*N*-(4-Fluorophenyl)carbamoyl)-pyrazole 860
$C_{10}H_8FN_3OS$
1-(3-Fluorophenyl)-3-(2-thiazolyl)urea 875
$C_{10}H_8F_3NO_3$
N-(3-(Trifluoromethyl)benzoyl)-glycine 446
$C_{10}H_8F_3N_3O_5$
2′,6′-Dinitro-*N*-methyl-4′-(trifluoromethyl)acetanilide 396
2′,6′-Dinitro-4′-(trifluoromethyl)-propionanilide 396
$C_{10}H_8F_3N_3O_6$
Ethyl *N*-(2,6-dinitro-4-(trifluoromethyl)phenyl)carbamate 609
Methyl *N*-(2,6-dinitro-4-(trifluoromethyl)phenyl)-*N*-methylcarbamate 609
$C_{10}H_8F_6MgO_4$
Magnesium trifluoroacetylacetonate monohydrate 70
$C_{10}H_8F_7NO_2$
3-(Heptafluoropropyl)-2-nitro-5-norbornene 47
$C_{10}H_8INO_2$
N-(4-Iodophenyl)succinimide 716
$C_{10}H_8INO_3$
4′-Iodomaleanilic acid 476
$C_{10}H_8N_2$
2,4-Dihydroindeno(1,2-c)pyrazole 864
$C_{10}H_8N_2O$
N-Benzoylimidazole 868
2,2′-Bipyridine-*N*-oxide 1006
α-Cyanocinnamamide 766
1-Cyano-1,2-dihydro-2-hydroxyquinoline 752
2,3-Dihydro-5*H*-imidazo(2,1-a)-isoindol-5-one 726
1-Isoquinolinecarboxamide 998
3-Isoquinolinecarboxamide 998
6-Phenyl-3(2*H*)-pyridazinone 968
$C_{10}H_8N_2OS$
5-Benzylidene-2-thiohydantoin 745
$C_{10}H_8N_2O_2$
N-(4-Acetylphenyl)cyanoformamide 763
2-Amino-1-nitronaphthalene 271
4-Amino-2-quinolinecarboxylic acid 988
N-Anilinomaleimide 721
2-Furaldehyde azine 1075
Indole-3-glyoxylamide 888
6-Methyl-5-nitroquinoline 989
7-Methyl-4-nitroquinoline 989
7-Methyl-6-nitroquinoline 989
5-Nitroquinaldine 989
3-(2-Nitrovinyl)indole 885
$C_{10}H_8N_2O_2S$
1-Methyl-5-oxo-3-phenyl-2-thiohydantoin 745
1-Methyl-3-phenyl-5-thioxohydantoin 745
$C_{10}H_8N_2O_3$
N-(2-Cyanoacetyl)anthranilic acid 755
2-Cyanoacetyl *N*-phenylcarbamate 757
1-(Hydroxyamino)-5-nitronaphthalene 1010
2-Methyl-5-(4-nitrophenyl)oxazole 872
1-Methyl-6-nitro-2(1*H*)-quinolone 691
2-Methyl-6-nitro-4(1*H*)-quinolone 320
3-Methyl-1-phenyl-2,4,5-imidazolidinetrione 708
2-Phthalimidoacetamide 718
$C_{10}H_8N_2O_4$
N-(3-Nitrophenyl)succinimide 718
$C_{10}H_8N_2O_5$
2′-Nitromaleanilic acid 469
3′-Nitromaleanilic acid 469
4′-Nitromaleanilic acid 470
$C_{10}H_8N_2O_6S$
5-Oxo-1-(4-sulfophenyl)-2-pyrazoline-3-carboxylic acid 790
5-Oxo-1-(4-sulfophenyl)-2-pyrazoline-3-carboxylic acid, lead salt 794
$C_{10}H_8N_2S_2$
3-(2-Benzothiazolylthio)propionitrile 914
2-Thiophenecarboxaldehyde azine 1076
p-Xylylene dithiocyanate 768
$C_{10}H_8N_4$
5-Amino-4-cyano-1-phenylpyrazole 860
4,4′-Azobis(pyridine) 1089
$C_{10}H_8N_4O_2$
4,4′-Azobis-(pyridine-1-oxide) 1089
$C_{10}H_8N_4O_3S$
N-(2-Thiazolyl)oxamic furfurylidenehydrazide 1069
$C_{10}H_8N_6O_3$
1,3-Dimethyl-9-hydroxypyrimido(4,5-g)-pteridine-2,4(1*H*,3*H*)-dione 1002
$C_{10}H_8NaO_5S$
Mordant green 4 (C.I. 10005) 1094
$C_{10}H_8O$
2-Naphthol, bismuth (III) salt 196
$C_{10}H_8OS$
2-Acetylbenzo(b)thiophene 898
$C_{10}H_8O_2$
1,7-Dihydroxynaphthalene 196
1-Indenecarboxylic acid 346
Indene-3-carboxylic acid 347
2-Methylchromone 320
$C_{10}H_8O_2S$
2-Acetyl-3-hydroxybenzo(b)-thiophene 897
2-Naphthylsulfinic acid 787
$C_{10}H_8O_2S_2$
2-(Phenylsulfonyl)thiophene 858
Thenoin 855
$C_{10}H_8O_3$
Bicyclo(2.2.2)octa-2,5-diene-2,3-dicarboxylic anhydride 98
3,6-Dimethylphthalic anhydride 388
4-Hydroxy-6-methylcoumarin 383
3-Oxo-1-indancarboxylic acid 331
$C_{10}H_8O_4$
5,7-Dihydroxy-4-methylcoumarin monohydrate 383
4-Methyldaphnetin 382
$C_{10}H_8O_4S$
5-Hydroxy-1-naphthalenesulfonic acid, sodium salt 792
7-Hydroxy-1-naphthalenesulfonic acid, sodium salt 792
7-Hydroxy-2-naphthalenesulfonic acid, sodium salt 792
8-Hydroxy-1-naphthalenesulfonic acid, sodium salt 792
$C_{10}H_8O_5S$
2,3-Dihydroxynaphthalene-6-sulfonic acid, sodium salt 791
$C_{10}H_8O_6S_2$
1,6-Naphthalenedisulfonic acid, disodium salt dihydrate 796
$C_{10}H_8O_7S_2$
7-Hydroxy-1,3-naphthalenedisulfonic acid 787
$C_{10}H_8O_8S_2$
4,5-Dihydroxy-2,7-naphthalenedisulfonic acid 787
$C_{10}H_8O_9S_3$
1,3,6-Naphthalenetrisulfonic acid, sodium salt 796
$C_{10}H_9AsN_4O_4$
4-(4-Arsonophenylazo)-3-methyl-5-pyrazolone 1088
$C_{10}H_9BO_2$
1-Naphthaleneboronic acid 1125
$C_{10}H_9BO_3$
6-Hydroxy-2-naphthaleneboronic acid 1125
$C_{10}H_9BrClNO_3$
4′-Bromo-2′-chlorosuccinanilic acid. 402
$C_{10}H_9BrCl_3NO$
N-(2-Bromo-4,5-dimethylphenyl)-2,2,2-trichloroacetamide 396
$C_{10}H_9BrF_3NO$
3-Bromo-2′-(trifluoromethyl)-propionanilide 390
$C_{10}H_9BrF_3NO_2$
2-Bromoethyl *N*-(3-trifluoromethylphenyl)carbamate 604
$C_{10}H_9BrHg_2O_7$
Methyl 3,4-bis((acetoxy)mercuri)-5-bromo-2-furoate 1138
$C_{10}H_9BrN_2$
3-Amino-1-bromo-4-methylisoquinoline 997
$C_{10}H_9BrN_2O$
5-Bromoindole-3-acetamide 887
$C_{10}H_9BrN_2O_2$
1-Bromo-5-methyl-5-phenylhydantoin 705
$C_{10}H_9BrN_2O_3$
2′-Bromo-4′-nitro-1-cyclopropanecarboxanilide 404
$C_{10}H_9BrO$
4-Bromophenyl cyclopropyl ketone 296
2-Bromotetralone 298
$C_{10}H_9BrO_2$
α-(Bromomethyl)cinnamic acid 334
2-Bromophenyl methacrylate 367
Methyl 4-bromocinnamate 364
$C_{10}H_9BrO_3$
3-(4-Bromobenzoyl)-propionic acid 331
3-Bromo-4-oxo-4-phenylbutyric acid. 330
$C_{10}H_9BrO_4$
6-Bromo-3,4-(methylenedioxy)-hydrocinnamic acid 332
$C_{10}H_9BrO_5$
2-Bromo-3-carboxy-6-methoxyphenyl acetate 376
$C_{10}H_9ClF_3NO$
2′-Chloro-5′-(trifluoromethyl)-propionanilide 396
$C_{10}H_9ClF_3NO_2$
2-Chloroethyl *N*-(2-trifluoromethylphenyl)carbamate 604
Ethyl *N*-(2-chloro-5-trifluoromethylphenyl)carbamate 607
$C_{10}H_9ClF_3N_3O_2$
1-Acetyl-4-(2-chloro-5-(trifluoromethyl)phenyl)-semicarbazide 505
1-Acetyl-4-(4-chloro-2-(trifluoromethyl)phenyl)-semicarbazide 504
$C_{10}H_9ClN_2$
1-(4-Pyridyl)pyridinium chloride 958
$C_{10}H_9ClN_2O$
2-Chloro-4-ethoxyquinazoline 999
4-Chloro-3-methyl-1-phenyl-2-pyrazolin-5-one 710
1-(3-Chlorophenyl)-3-methyl-2-pyrazolin-5-one 710
$C_{10}H_9ClN_2OS$
2-(2-Chlorophenyl)-5-imino-3-methyl-4-oxazolidinethione 729
2-(4-Chlorophenyl)-5-imino-3-methyl-4-oxazolidinethione 729

$C_{10}H_9ClN_2OS_2$
S-2-Thiazolin-2-yl *N*-(3-chlorophenyl)thiocarbamate 641
$C_{10}H_9ClN_2O_2$
2-Cyanoethyl *N*-(3-chlorophenyl)carbamate 757
Ethyl 2-chloro-3-cyano-6-methylisonicotinate 949
Ethyl *N*-(4-chlorophenyl)-*N*-cyanocarbamate 752
Methyl 2-(2-chloro-1-benzimidazolyl)acetate 907
$C_{10}H_9ClN_2O_2S$
N-(5-Chloro-2,4-dimethoxyphenyl)-cyanothioformamide 758
$C_{10}H_9ClN_2O_3$
2′-Chloro-4′-nitro-2-butenanilide 452
$C_{10}H_9ClN_2O_4$
Allyl *N*-(2-chloro-4-nitrophenyl)-carbamate 670
$C_{10}H_9ClN_2O_4S$
4-Chloro-3-(3-methyl-5-oxo-2-pyrazolin-1-yl)-benzenesulfonic acid 790
$C_{10}H_9ClN_4OS$
1-(3-Chlorophenyl)-3-(5-methyl-1,3,4-thiadiazol-2-yl)urea 878
$C_{10}H_9ClO$
4-(2-Chlorophenyl)-3-buten-2-one 286
4-(4-Chlorophenyl)-3-buten-2-one 286
$C_{10}H_9ClO_2$
2-Chloro-α-methylcinnamic acid 334
4-Chloro-α-methylcinnamic acid 334
Methyl 4-chlorocinnamate 364
$C_{10}H_9ClO_3$
3-Chloro-7-methoxy-3-methylphthalide 383
$C_{10}H_9Cl_2N$
N-(2,4-Dichlorobenzyl)-propargylamine 1015
$C_{10}H_9Cl_2NO$
2′,5′-Dichloro-1-cyclopropanecarboxanilide 408
N-(2,3-Dichlorophenyl)-cyclopropanecarboxamide 408
$C_{10}H_9Cl_2NO_2$
Acetoaceto-2,4-dichloroanilide 444
Allyl *N*-(2,3-dichlorophenyl)-carbamate 683
Allyl *N*-(2,4-dichlorophenyl)-carbamate 685
Allyl *N*-(2,5-dichlorophenyl)-carbamate 685
Allyl *N*-(2,6-dichlorophenyl)-carbamate 635
Allyl *N*-(3,4-dichlorophenyl)-carbamate 685
2′,4-Dichloroacetoacetanilide 442
2′,5′-Dichloroacetoacetanilide 444
3′,5′-Dichloro-2′-hydroxy-4′-methylacrylanilide 447
$C_{10}H_9Cl_2NO_3$
2′,3′-Dichlorosuccinanilic acid 407
2′,4′-Dichlorosuccinanilic acid 407
2′,5′-Dichlorosuccinanilic acid 407
3′,4′-Dichlorosuccinanilic acid 407
$C_{10}H_9Cl_2NO_4$
Methyl α,4-dichloro-3-nitrohydrocinnamate 360
$C_{10}H_9Cl_2NO_5$
N-(2,4-Dichlorophenyl)-2,3-dihydroxysuccinamic acid 398
$C_{10}H_9Cl_2N_3O$
1-Cyano-1-(3,4-dichlorophenyl)-3,3-dimethylurea 752
$C_{10}H_9Cl_2N_3OS$
1-(2,3-Dichlorophenyl)-3-(2-thiazolin-2-yl)urea 732
$C_{10}H_9Cl_3N_2O$
1-Allyl-3-(2,4,5-trichlorophenyl)urea 603
$C_{10}H_9Cl_3N_2O_2$
1-(2-Phenylacetyl)-2-(2,2,2-trichloroacetyl)hydrazine 491
O-(*N*-(2,4,5-Trichlorophenyl)-carbamoyl)acetone oxime 1012
$C_{10}H_9Cl_3O_3$
4-(2,4,5-Trichlorophenoxy)-butyric acid 325
$C_{10}H_9Cl_4NO_2$
2,2,2-Trichloroethyl *N*-(3-chloro-4-methylphenyl)carbamate 607
$C_{10}H_9Cl_4NO_3$
2′,5′-Dimethoxy-2,2,2,4′-tetrachloroacetanilide 394
Methyl *N*-(1-(4-chlorophenoxy)-2,2,2-trichloroethyl)carbamate 606
$C_{10}H_9CrNO_3$
m-Toluidine chromium tricarbonyl 1140
p-Toluidine chromium tricarbonyl 1140
$C_{10}H_9FN_2O$
1-(4-Fluorophenyl)-5-hydroxy-3-methylpyrazole 859
$C_{10}H_9FN_2OS$
2-(4-Fluorophenyl)-5-imino-3-methyl-4-oxazolidinethione 729
$C_{10}H_9FN_4OS$
1-(2-Fluorophenyl)-3-(5-methyl-1,3,4-thiadiazol-2-yl)urea 878
$C_{10}H_9F_3N_2$
5,6-Dimethyl-2-trifluoromethylbenzimidazole 904
$C_{10}H_9F_3N_2O_2$
N-Acetyl-*N*′-(2,2,2-trifluoroacetyl)-1,3-phenylenediamine 397
$C_{10}H_9F_3N_2O_3$
2′-Nitro-4′-(trifluoromethyl)-propionanilide 396
$C_{10}H_9F_3N_2O_5$
2,6-Dinitro-4-(trifluoromethyl)-phenyl isopropyl ether 259
2,6-Dinitro-4-(trifluoromethyl)-phenyl propyl ether 259
$C_{10}H_9F_3O$
o-(Trifluoromethyl)propiophenone 288
m-(Trifluoromethyl)propiophenone 288
$C_{10}H_9F_3O_2$
Ethyl α,α,α-trifluoro-*m*-toluate 366
(-)-1-Methyl-2,2,2-trifluoroethyl benzoate 366
3-(Trifluoromethyl)-hydrocinnamic acid 325
$C_{10}H_9F_3O_3$
2-(1-Phenyl-2,2,2-trifluoroethoxy)-acetic acid 325
$C_{10}H_9F_5$
1-Butyl-2,3,4,5,6-pentafluorobenzene 170
$C_{10}H_9F_6NO$
3-Amino-α,α-bis(trifluoromethyl)-4-methylbenzyl alcohol 215
$C_{10}H_9HgN_5O_2$
1-Benzylideneamino-1*H*-tetrazol-5-ylmercuri acetate 1138
$C_{10}H_9N$
2-Aminonaphthalene hydrobromide 231
trans-1-Cyano-2-phenylcyclopropane 758
4-Methylcinnamonitrile (mixture of *cis* and *trans*) 767
$C_{10}H_9NO$
N-Acetylindole 887
5-Amino-1-naphthol hydrochloride 226
2-Benzoylpropionitrile 757
3-Benzoylpropionitrile 757
Echinopsine hydrochloride 320
7-Hydroxy-8-methylisoquinoline 996
6-Methoxyquinoline picrate 986
5-Methylindole-3-carboxaldehyde 888
3-Methyl-5-phenylisoxazole 871
1-Methyl-2(1*H*)-quinolone 692
3-Oxo-3-(*p*-tolyl)propionitrile 757
5-Quinolinemethanol 984
8-Quinolinemethanol hydrochloride 984
5-(*p*-Tolyl)isoxazole 871
$C_{10}H_9NOS$
1-(2-Benzothiazolyl)acetone 916
1-(Methylthio)isoquinolin-3(4*H*)-one 701
$C_{10}H_9NOS_2$
3-Benzylrhodanine 745
2-Thioxo-3-(*o*-tolyl)-4-thiazolidinone 746
2-Thioxo-3-(*m*-tolyl)-4-thiazolidinone 746
$C_{10}H_9NO_2$
3-Amino-4-methoxy-1-indenone 291
3-(α-Cyanoethyl)benzoic acid 757
5,7-Dimethylisatin 704
6,7-Dimethylisatin 704
4-Hydroxy-6-methoxyquinoline 983
4-Hydroxy-1-methyl-2(1*H*)-quinolone 697
1-Indoleacetic acid 890
3-(4-Methoxyphenyl)-3-oxopropionitrile 753
5-Methylindole-2-carboxylic acid 886
N-Methyl-5-norbornene-2,3-dicarboximide 135
4-Methyl-5-phenyl-2-oxazolidinone 657
1-(4-Nitrophenyl)-1,3-butadiene 279
2-Phenylsuccinimide 719
Propynyl *N*-phenylcarbamate 1019
$C_{10}H_9NO_2S$
N-(2-Mercaptoethyl)phthalimide 717
6-(Methylsulfonyl)quinoline 994
$C_{10}H_9NO_2S_2$
3-(2-Benzothiazolylthio)-propionic acid 916
Ethyl (2-benzothiazolylthio)-formate 917
3-(2-Methoxyphenyl)-rhodanine 737
3-(3-Methoxyphenyl)-rhodanine 737
$C_{10}H_9NO_3$
2-Cyano-3-(2-hydroxyphenyl)-propionic acid 752
Ethyl α-cyano-β-(2-furyl)acrylate 853
N-Ethylisatoic anhydride 706
N-Furfuryl-2-furamide 853
7-Nitro-1-tetralone 295
1-Phenyl-1,2,3-butanetrione 2-oxime 1011
N-Phenylmaleamic acid 475
$C_{10}H_9NO_3S$
1-Amino-2-naphthalenesulfonic acid 788
2-Amino-7-naphthalenesulfonic acid 789
3-Amino-1-naphthalenesulfonic acid hydrate 788
5-Amino-1-naphthalenesulfonic acid 789
6-Amino-2-naphthalenesulfonic acid 789
5-Amino-1-naphthalenesulfonic acid, sodium salt 793
8-Amino-2-naphthalenesulfonic acid, sodium salt 793
2,3-Dihydro-3-oxo-4*H*-1,4-benzothiazine-2-acetic acid 708
$C_{10}H_9NO_4$
Ethyl 2,3-dihydro-2-oxo-3-benzoxazolecarboxylate 708
Methyl 2-nitrocinnamate 365
Methyl 3-nitrocinnamate 365
Methyl 4-nitrocinnamate 365
α-Methyl-4-nitrocinnamic acid 335
β-Methyl-3-nitrocinnamic acid 335
$C_{10}H_9NO_4S$
1-Amino-2-naphthol-4-sulfonic acid hemihydrate 787
Quinoliniummethyl sulfate inner salt 990
$C_{10}H_9NO_5$
2-Acetamidoisophthalic acid 433
2-Methoxy-4-nitrocinnamic acid 335
$C_{10}H_9NO_5S$
1,3-Dioxo-2-isoindolineethanesulfonic acid, potassium salt 794
$C_{10}H_9NO_6$
Dimethyl 5-nitroisophthalate 373
Dimethyl 2-nitroterephthalate 373
4-Formyl-2-methoxy-3-nitrophenyl acetate 377
2-(4-Nitrophenyl)succinic acid 330
$C_{10}H_9NO_6S_2$
2-Amino-1,5-naphthalenedisulfonic acid 788
4-Amino-1,7-naphthalenedisulfonic acid 789
4-Amino-2,7-naphthalenedisulfonic acid 789
6-Amino-1,3-naphthalenedisulfonic acid 788
4-Amino-1,5-naphthalenedisulfonic acid, monosodium salt 793
8-Amino-1,6-naphthalenedisulfonic acid, monosodium salt 793
$C_{10}H_9NO_7S_2$
2-Amino-8-naphthol-3,6-disulfonic acid, monosodium salt 791
$C_{10}H_9NS$
1-Methyl-2(1*H*)-quinolinethione 747
4-(2-Thienyl)aniline hydrochloride 856
$C_{10}H_9NS_2$
3-Methyl-4-phenyl-4-thiazolin-2-thione 743
$C_{10}H_9NSe$
2-(Methylseleno)cinnamonitrile 213
$C_{10}H_9N_2S$
Ethyl 2-cyano-3-(2-thienyl)acrylate 858
$C_{10}H_9N_3$
2-Amino-4-phenylpyrimidine 975
4-Amino-5-phenylpyrimidine 975
$C_{10}H_9N_3O$
Formic N^2-(3-quinolyl)hydrazide 988
2-Imidazolecarboxanilide 868
$C_{10}H_9N_3OS$
1-Phenyl-3-(2-thiazolin-2-yl)urea 581
3-Phenyl-1-thiocarbamoyl-2-pyrazolin-5-one 743
$C_{10}H_9N_3O_2$
1-Acetylisatin 3-hydrazone 1069
2-Cyanoacetic N^2-benzoylhydrazide 755
$C_{10}H_9N_3O_2S$
8-Amino-6-(methylthio)-5-nitroquinoline 987
4-Methoxy-2-(2-thiazolylazo)-phenol 1083
5-Methoxy-2-(2-thiazolylazo)-phenol 1083
6-Methyl-4-(2-thiazolylazo)-resorcinol 1084
$C_{10}H_9N_3O_3$
3-Methyl-1-(4-nitrophenyl)-2-pyrazolin-5-one 703
3-Methyl-2-(4-nitrophenyl)-3-pyrazolin-5-one 861
5-Phenyl-1,2,4-oxadiazol-3-yl *N*-methylcarbamate 873
$C_{10}H_9N_3O_3S$
5-Imino-3-methyl-2-(3-nitrophenyl)-4-oxazolidinethione 738
$C_{10}H_9N_3O_4S$
4-Nitro-1-(*p*-tosyl)imidazole 868
$C_{10}H_9N_3O_8$
N-(2,4-Dinitrophenyl)-L-aspartic acid 349
$C_{10}H_9N_5O$
4,6-Diamino-2-phenyl-5-nitrosopyrimidine 974
$C_{10}H_9N_5O_4$
1-(2-Cyanoacetyl)-4-(3-nitrophenyl)-semicarbazide 755
5-Nitro-2-furaldehyde (4-hydroxy-6-methylpyrimidin-2-yl)-hydrazone 1033
$C_{10}H_9O_3P$
(1-Naphthyl)phosphonic acid 837
$C_{10}H_{10}AsNO_3$
8-Methyl-5-quinolinearsonic acid 1128
$C_{10}H_{10}BrCl_2NO$
N-(2-Bromo-4,5-dimethylphenyl)-2,2-dichloroacetamide 394
$C_{10}H_{10}BrNO$
N-Allyl-2-bromobenzamide 457
1-(2-Bromoacetyl)indoline 693
4′-Bromo-2-butenanilide 461
3′-Bromo-cyclopropanecarboxanilide 408
4′-Bromo-cyclopropanecarboxanilide 408
$C_{10}H_{10}BrNO_2$
Allyl 4′-bromocarbanilate 634
4′-Bromoacetoacetanilide 444
$C_{10}H_{10}BrNO_2S$
3-(3-Bromoanilino)-2,3-dihydrothiophene-1,1-dioxide 773
$C_{10}H_{10}BrNO_2S_2$
2-(Carboxymethylthio)-3-methylbenzothiazolinium bromide 913
$C_{10}H_{10}BrNO_3$
6-Bromo-1,3-benzodioxole-5-propionamide 436
4′-Bromosuccinanilic acid 406
$C_{10}H_{10}BrNO_3S$
2-(3-Bromopropyl)-1,2-benzisothiazol-3(2*H*)-one-1,1-dioxide 804

$C_{10}H_{10}BrN_2S$
3-(4-Bromoanilino)-2,3-dihydrothiophene-1,1-dioxide 773
$C_{10}H_{10}BrO$
2-Bromo-1,2,3,4-tetrahydro-1-naphthol 196
$C_{10}H_{10}Br_2$
(2,2-Dibromo-1-methylcyclopropyl)-benzene 173
trans-1,2-Dibromo-1,2,3,4-tetrahydronaphthalene 174
$C_{10}H_{10}Br_2ClNO$
N-(4-Chloro-2,6-dibromophenyl)-butanamide 410
$C_{10}H_{10}Br_2O$
3,4-Dibromo-4-phenyl-2-butanone 285
$C_{10}H_{10}Br_2O_2$
Ethyl α,α-dibromo-*p*-toluate 366
$C_{10}H_{10}Br_2O_3$
2,2-Dibromo-2′,4′-dimethoxyacetophenone 288
$C_{10}H_{10}Br_3NO$
2-Bromo-*N*-(2,6-dibromo-4-ethylphenyl)acetamide 391
$C_{10}H_{10}Br_4$
1,3-Bis-(1,2-dibromoethyl)-benzene 169
1,4-Bis-(1,2-dibromoethyl)-benzene 169
1,2,3,4-Tetrakis(bromomethyl)-benzene 171
$C_{10}H_{10}ClFO$
3-Chloro-4-(4-fluorophenyl)-2-butanone 285
$C_{10}H_{10}ClF_3N_2O$
3-(2-Chloro-5-trifluoromethylphenyl)-1,1-dimethylurea 506
1-(2-Chloro-5-trifluoromethylphenyl)-3-ethylurea 506
$C_{10}H_{10}ClF_3N_2O_2$
1-(4-Chloro-2-(trifluoromethyl)-phenyl)-3-(2-hydroxyethyl)urea 504
$C_{10}H_{10}ClNO$
1-(2-Chloroacetyl)indoline 693
2′-Chloro-1-cyclopropanecarboxanilide 408
4′-Chloro-1-cyclopropanecarboxanilide 408
$C_{10}H_{10}ClNO_2$
2′-Acetyl-4′-chloroacetanilide 405
4′-Acetyl-2-chloroacetanilide 392
5′-Acetyl-2′-chloroacetanilide 405
Allyl 2′-chlorocarbanilate 681
Allyl 3′-chlorocarbanilate 677
Allyl *N*-(4-chlorophenyl)carbamate 682
2′-Chloroacetoacetanilide 444
3′-Chloroacetoacetanilide 444
$C_{10}H_{10}ClNO_2S_2$
2-(Carboxymethylthio)-3-methylbenzothiazolium chloride 913
$C_{10}H_{10}ClNO_3$
4′-Chlorosuccinanilic acid 406
Methyl *N*-(chloroacetyl)-anthranilate 393
$C_{10}H_{10}ClNO_4$
1-Carboxyethyl *N*-(3-chlorophenyl)-carbamate 665
Methyl α-chloro-3-nitrohydrocinnamate 359
Methyl α-chloro-4-nitrohydrocinnamate 359
$C_{10}H_{10}ClN_3$
5-(2-Amino-5-chlorophenyl)-3-methylpyrazole 861
$C_{10}H_{10}ClN_3O$
1-(2-Chloroethyl)-3-(2-cyanophenyl)urea 758
$C_{10}H_{10}ClN_3O_2$
5-Chloro-1-ethyl-2-methyl-6-nitrobenzimidazole 904
$C_{10}H_{10}ClN_3O_3$
1-Allyl-3-(2-chloro-4-nitrophenyl)urea 568
2-(2-Chloro-5-nitro-4-pyridyl)-3-(dimethylamino)acrolein 944
$C_{10}H_{10}ClN_3O_3S$
3-Amino-1-(2-chloro-4-methylsulfonylphenyl)-2-pyrazolin-5-one 776
$C_{10}H_{10}ClN_3O_4$
O-(*N*-(2-Chloro-4-nitrophenyl)-carbamoyl)acetone oxime 1010
$C_{10}H_{10}ClN_5$
2-Amino-4-anilino-6-(chloromethyl)-1,3,5-triazine 978
$C_{10}H_{10}ClN_5O_4S$
N-(4-Chloro-3-nitrophenyl)-*N*′-(3-methyl-2-thioureido)oxamide 730
$C_{10}H_{10}Cl_2N_2$
5,6-Dichloro-1-ethyl-2-methylbenzimidazole 906
$C_{10}H_{10}Cl_2N_2O$
1-Allyl-3-(2,3-dichlorophenyl)urea 593
1-Allyl-3-(2,4-dichlorophenyl)urea 597
1-Allyl-3-(2,5-dichlorophenyl)urea 597
1-Allyl-3-(2,6-dichlorophenyl)urea 593
1-Allyl-3-(3,4-dichlorophenyl)urea 597
1-Allyl-3-(3,5-dichlorophenyl)urea 545
4-Amino-*N*-cyclopropyl-3,5-dichlorobenzamide 450
1-Cyclopropyl-3-(2,6-dichlorophenyl)urea 534
5,6-Dichloro-1-(2-hydroxyethyl)-2-methylbenzimidazole 902
$C_{10}H_{10}Cl_2N_2O_2$
1-Acetyl-3-(3,4-dichlorophenyl)-1-methylurea 533
O-(*N*-(2,3-Dichlorophenyl)-carbamoyl)acetone oxime 1013
o-(*N*-(2,4-Dichlorophenyl)-carbamoyl)acetone oxime 1013
O-(*N*-(2,5-Dichlorophenyl)-carbamoyl)acetone oxime 1013
O-(*N*-(2,6-Dichlorophenyl)-carbamoyl)acetone oxime 1008
O-(*N*-(3,4-Dichlorophenyl)-carbamoyl)acetone oxime 1012
3-(3,4-Dichlorophenyl)-1-ethyl-1-formylurea 533
1-(3,4-Dichlorophenyl)-3-propionylurea 533
$C_{10}H_{10}Cl_2N_2O_3$
1-(3,5-Dichlorophenyl)-3-(ethoxycarbonyl)urea 532
Ethyl *N*-(*N*-(3,4-dichlorophenyl)-carbamoyl)carbamate 630
$C_{10}H_{10}Cl_2N_4$
2,6-Dichlorobenzaldehyde 2-imidazolin-2-ylhydrazone hydrochloride 1026
$C_{10}H_{10}Cl_2O$
3-Chloro-4-(2-chlorophenyl)-2-butanone 285
3-Chloro-4-(3-chlorophenyl)-2-butanone 285
3-Chloro-4-(4-chlorophenyl)-2-butanone 285
4,4′-Dichlorobutyrophenone 288
$C_{10}H_{10}Cl_2O_3$
4-(2,4-Dichlorophenoxy)butyric acid 329
Ethyl 2-(2,3-dichlorophenoxy)-acetate 355
$C_{10}H_{10}Cl_2O_4$
2,3′-Dichloro-4′,6′-dimethoxy-2′-hydroxyacetophenone 287
$C_{10}H_{10}Cl_3NO$
2′,3′-Dimethyl-2,2,2-trichloroacetanilide 398
2′,4′-Dimethyl-2,2,2-trichloroacetanilide 398
2′,5′-Dimethyl-2,2,2-trichloroacetanilide 398
3′,4′-Dimethyl-2,2,2-trichloroacetanilide 398
3′,5′-Dimethyl-2,2,2-trichloroacetanilide 398
2′-Ethyl-2,2,2-trichloroacetanilide 398
4′-Ethyl-2,2,2-trichloroacetanilide 398
N-Phenethyl-2,2,2-trichloroacetamide 397
2′,3′,4-Trichlorobutyranilide 391
2,2,2-Trichloro-*N*-(2,6-xylyl)-acetamide 398
$C_{10}H_{10}Cl_3NOS$
S-Isopropyl *N*-(2,4,5-trichlorophenyl)thiolcarbamate 642
S-Propyl *N*-(2,4,5-trichlorophenyl)-thiolcarbamate 642
$C_{10}H_{10}Cl_3NO_2$
2-Chloro-1-methylethyl *N*-(2,6-dichlorophenyl)carbamate 604
1,3-Dichloro-2-propyl *N*-(3-chlorophenyl)-carbamate 605
N-(2-Ethoxyphenyl)-2,2,2-trichloroacetamide 396
4′-Ethoxy-2,2,2-trichloroacetanilide 396
Isopropyl *N*-(2,4,5-trichlorophenyl)-carbamate 687
Propyl *N*-(2,4,5-trichlorophenyl)-carbamate 687
$C_{10}H_{10}Cl_3NO_2S$
Methyl *N*-(1-(phenylthio)-2,2,2-trichloroethyl)carbamate 609
$C_{10}H_{10}Cl_3NO_3$
2-Methoxyethyl *N*-(2,4,5-trichlorophenyl)carbamate 616
Methyl *N*-(1-phenoxy-2,2,2-trichloroethyl)carbamate 609
2,2,2-Trichloroethyl *N*-(2-methoxyphenyl)carbamate 608
$C_{10}H_{10}Co$
Cobaltocene (10% solution in benzene) 1142
Cobaltocene (7.5% solution in diethylbenzene) 1142
$C_{10}H_{10}FNO$
2′-Fluoro-2-butenanilide 457
4′-Fluorocyclopropane-carboxanilide 408
$C_{10}H_{10}FNO_2$
Allyl *N*-(4-fluorophenyl)carbamate 682
$C_{10}H_{10}FNO_3$
2-Fluorosuccinanilic acid 406
4′-Fluorosuccinanilic acid 406
$C_{10}H_{10}F_2N_4O_3S$
3-Amino-1-(2-amino-4-(difluoromethylsulfonyl)phenyl)-3-pyrazolin-5-one 772
$C_{10}H_{10}F_3NO$
4′-(Dimethylamino)-2,2,2-trifluoroacetophenone 288
$C_{10}H_{10}F_3NO_2$
Ethyl 3-(trifluoromethyl)carbanilate 610
$C_{10}H_{10}F_3N_3O_2$
1-Acetyl-4-(3-trifluoromethylphenyl)-semicarbazide 509
$C_{10}H_{10}F_3N_3O_4$
2,6-Dinitro-*N*-propyl-4-(trifluoromethyl)aniline 259
$C_{10}H_{10}F_3N_3O_6S$
2′,6′-Dinitro-*N*-ethyl-4′-(trifluoromethyl)-methanesulfonanilide 805
$C_{10}H_{10}F_4O_3S$
1-Methyl-2,2,3,3-tetrafluoropropyl benzenesulfonate 796
$C_{10}H_{10}HgO_4$
Methyl 4-(acetoxymercuri)benzoate 1138
$C_{10}H_{10}IN$
2-Methylisoquinolinium iodide 998
1-Methylquinolinium iodide 990
$C_{10}H_{10}INO$
1-Acetyl-5-iodoindoline 710
$C_{10}H_{10}INO_2$
2-Iodo-4,5-dimethoxyphenylacetonitrile 751
$C_{10}H_{10}INO_3$
4′-Iodosuccinanilic acid 406
$C_{10}H_{10}N_2$
3-Amino-4-methylisoquinoline 997
5-Amino-3-methylisoquinoline 997
5-Amino-6-methylquinoline 988
8-Aminoquinaldine 987
1,2-Diaminonaphthalene 230
β-Imino-*p*-methylhydrocinnamonitrile 755
3-Methyl-5-phenylpyrazole 862
2-(1-Propenyl)benzimidazole 907
$C_{10}H_{10}N_2O$
2-Acetyl-1-methylbenzimidazole 906
8-Amino-6-methoxyquinoline hydrochloride 985
8-Amino-6-methoxyquinoline picrate 985
4-Hydroxymethyl-1-phenyl-1*H*-pyrazole 860
5-(4-Methoxyphenyl)-1*H*-pyrazole 860
4-Methyl-1*H*-1,5-benzodiazepin-2-ol 1002
5-Methyl-1-phenyl-1*H*-pyrazol-3(2*H*)-one 709
5-Methyl-3-(*p*-tolyl)-1,2,4-oxadiazole 873
2-(2-Oxopropyl)benzimidazole 906
N-(2,3-Xylyl)cyanoformamide 763
N-(3,4-Xylyl)cyanoformamide 767
N-(3,5-Xylyl)cyanoformamide 763
$C_{10}H_{10}N_2OS$
2-Acetyl-3-hydroxybenzo(b)-thiophene hydrazone hydrate 1036
N-(2-Methoxy-5-methylphenyl)-cyanothioformamide 753
5-Methyl-3-phenyl-2-thioxo-4-imidazolidinone 743
α-(2-Thiazolylamino)-*o*-cresol 875
$C_{10}H_{10}N_2O_2$
5-Acetamido-2-methylbenzoxazole 909
N-(2-Aminophenyl)succinimide 718
Benzyl *N*-(cyanomethyl)carbamate 755
N-(2-Cyanoethyl)anthranilic acid 755
2-Cyanoethyl *N*-phenylcarbamate 757
2,3-Dimethyl-4-nitroindole 885
2,3-Dimethyl-5-nitroindole 886
2,3-Dimethyl-7-nitroindole 886
1,3-Dimethyl-2,4(1*H*,3*H*)-quinazolinedione 707
6,7-Dimethyl-2,3-quinoxalinediol 998
Ethyl *N*-cyano-*N*-phenylcarbamate 757
2-(2-Hydroxyethyl)-1(2*H*)-phthalazinone 697
1-Methyl-2-benzimidazoleacetic acid 907
4-(1-Methyl-1-nitroethyl)benzonitrile 765
$C_{10}H_{10}N_2O_2S$
4-Acetamido-3-methoxyphenyl thiocyanate 768
(2-Benzimidazolylmethylthio)-acetic acid 903
1-Mercapto-5-methyl-5-phenylhydantoin 705
$C_{10}H_{10}N_2O_2S_4$
3,3′-Diethyl-δ-5,5′-birhodanine 141
$C_{10}H_{10}N_2O_3$
1-Acetyl-6-nitroindoline 702
5-(4-Hydroxybenzyl)-hydantoin 697
3′-Nitro-1-cyclopropanecarboxanilide 426
$C_{10}H_{10}N_2O_3S$
4-(Acetamidophenylsulfonyl)-acetonitrile 782
$C_{10}H_{10}N_2O_4$
2′-Acetyl-4′-nitroacetanilide 427
2′-Acetyl-6′-nitroacetanilide 424
Allyl *N*-(3-nitrophenyl)carbamate 669
2,5-Dimethoxy-4-nitrophenylacetonitrile 752
4,5-Dimethoxy-2-nitrophenylacetonitrile 752
6,7-Dimethoxy-1,4(2*H*,3*H*)-phthalazinedione 697
Methyl *N*-(*N*-benzoylcarbamoyl)-carbamate 676
2,3,5a,6,7,8,10,10a-Octahydro-1*H*,5*H*-dipyrrolo[1,2-a:1′,2′-d]-pyrazine-3,5,8,10-tetraone 134
$C_{10}H_{10}N_2O_4S$
4′-Acetylthio-2′-nitroacetanilide 424
4-(3-Methyl-5-oxo-2-pyrazolin-1-yl)-benzenesulfonic acid 790
$C_{10}H_{10}N_2O_5$
Methyl *N*-acetyl-3-nitroanthranilate 424
Methyl *N*-acetyl-5-nitroanthranilate 427
5-Nitro-2-propionamidobenzoic acid 427
3′-Nitrosuccinanilic acid 426
4′-Nitrosuccinanilic acid 427
$C_{10}H_{10}N_2O_5S$
N^4-Maleylsulfanilamide 823
$C_{10}H_{10}N_2O_6$
Ethyl *N*-(2-carboxy-4-nitrophenyl)-carbamate 670
Ethyl *N*-(2-carboxy-5-nitrophenyl)-carbamate 669

$C_{10}H_{10}N_2O_7$
Methyl 3,5-dinitro-4-hydroxyhydrocinnamate 358
$C_{10}H_{10}N_2S$
1-Benzyl-2-mercaptoimidazole 866
N-(2,3-Xylyl)cyanothioformamide 763
N-(2,4-Xylyl)cyanothioformamide 767
N-(2,6-Xylyl)cyanothioformamide 763
$C_{10}H_{10}N_2S_2$
3-Amino-4-cyano-5,6,7,8-tetrahydro-1*H*-2-benzothiopyran-1-thione 763
2-(Methylthio)-5-(*p*-tolyl)-1,3,4-thiadiazole 878
$C_{10}H_{10}N_4$
(5,6-Dimethyl-2-benzimidazolyl)carbamonitrile 907
4-(4-Methylbenzylideneamino)-4*H*-1,2,4-triazole 1072
$C_{10}H_{10}N_4O$
2-Cyano-2-(*p*-tolylhydrazono)acetamide 765
4-Indolecarboxaldehyde semicarbazone 1059
4-(3-Methoxybenzylideneamino)-4*H*-1,2,4-triazole 1048
4-(4-Methoxybenzylideneamino)-4*H*-1,2,4-triazole 1049
4′-Methyl-1,2,4-triazole-5-carboxanilide 870
$C_{10}H_{10}N_4OS$
3-Amino-3-mercapto-2-(4-methoxyphenylazo)acrylonitrile 1086
1-Phenyl-3-thioureido-2-pyrazolin-5-one 710
$C_{10}H_{10}N_4O_2$
3,4-Bis((2-cyanoethyl)amino)-3-cyclobutene-1,2-dione 145
2-Cyano-2-(4-methoxyphenylhydrazono)acetamide 762
2-(Furfurylidenehydrazino)-4-hydroxy-6-methylpyrimidine 1036
3-Hydroxy-2-quinoxalineacetic acid hydrazide 999
5-Oxo-1-phenyl-2-pyrazoline-3-carboxylic acid hydrazide 710
$C_{10}H_{10}N_4O_3$
2-Nitrocinnamaldehyde semicarbazone 1069
$C_{10}H_{10}N_4O_3S$
1-(*N*-Ethylcarbamoyl)-2-mercapto-5-nitrobenzimidazole 903
$C_{10}H_{10}N_4O_4$
N,*N*′-Bis(5-methyl-3-isoxazolyl)oxamide 871
$C_{10}H_{10}N_4O_6$
N,*N*′-(2,3-Dinitro-1,4-phenylene)bisacetamide 423
1-(2,4,6-Trinitrophenyl)pyrrolidine 274
$C_{10}H_{10}N_4O_7$
N-(2,4-Dinitrophenyl)-L-asparagine 423
3-Hydroxy-2-oxobutyric acid 2,4-dinitrophenylhydrazone 339
4-(2,4,6-Trinitrophenyl)morpholine 273
$C_{10}H_{10}N_4S$
Indole-3-carboxaldehyde thiosemicarbazone 1071
$C_{10}H_{10}N_6O_3S$
N-(5-Azido-1,3,4-oxadiazol-2-ylmethyl)-*p*-toluenesulfonamide 873
$C_{10}H_{10}O$
5-Indancarboxaldehyde 324
Tricyclo[5.2.1.$0^{2,6}$]deca-4,8-dien-3-one 53
$C_{10}H_{10}OS$
2,3-Dihydro-1-benzothiepin-5(4*H*)-one 320
6-Methylbenzothiopyran-4(4*H*)-one 320
8-Methylbenzothiopyran-4(4*H*)-one 320
$C_{10}H_{10}OS_2$
5-Thenyl-2-thiophenemethanol 855
$C_{10}H_{10}O_2$
Allyl benzoate 375
Cinnamyl formate 361
1-(4-Hydroxybenzylidene)acetone 282
4-Hydroxy-7-methyl-1-indanone 290
7-Hydroxy-5-methyl-1-indanone 291
4-(2-Hydroxyphenyl)-3-buten-2-one 300
6-Hydroxy-1-tetralone 291
1-Indancarboxylic acid 334
6-Methyl-4-chromanone 320
β-Methylcinnamic acid 336
$C_{10}H_{10}O_2S$
2,5-Dihydro-3-phenylthiophene-1,1-dioxide 782
6-Ethoxy-3-hydroxythianaphthene 897
1-Propynyl *p*-tolyl sulfone 1019
$C_{10}H_{10}O_2Se$
2-(Methylseleno)cinnamic acid 213
$C_{10}H_{10}O_3$
2-Acetylphenyl acetate 379
3-Acetylphenyl acetate 379
4-Acetylphenyl acetate 379
2-Benzoylpropionic acid 330
2,3-Dihydro-8-methoxy-4*H*-1-benzopyran-4-one 291
3,3-Dimethyl-1,4-benzodioxan-2-one. 96
3-Ethoxyphthalide 383
3-Hydroxy-4′-methoxyacrylophenone, sodium salt 290
7-Methoxy-3-methylphthalide 383
Methyl 4-hydroxycinnamate 364
Phenacyl acetate 357
$C_{10}H_{10}O_3S$
3-Hydroxy-α-mercapto-β-methylcinnamic acid 334
α-Mercapto-2-methoxycinnamic acid 335
$C_{10}H_{10}O_4$
4-Acetoxyphenylacetic acid 380
4-Acetyl-3-hydroxyphenyl acetate 377
4-Allyloxy-2-hydroxybenzoic acid 338
5,6-Dimethoxyphthalide 383
2,5-Dimethoxyterephthaldehyde 323
Dimethyl terephthalate, zone refined 375
2,3-Epoxy-3-(2-methoxyphenyl)propionic acid, sodium salt 350
3-Hydroxy-7-methoxy-3-methylphthalide 382
Methyl 3-acetoxybenzoate 379
Methyl acetylsalicylate 380
3-Methylacetylsalicylic acid 380
$C_{10}H_{10}O_4S$
4-Hydroxy-α-mercapto-3-methoxycinnamic acid 334
$C_{10}H_{10}O_5$
4-Acetoxy-*m*-anisic acid 378
$C_{10}H_{10}O_6$
3,4-Dimethoxyphthalic acid 339
Dimethyl 2,5-dihydroxyterephthalate 367
Hydroquinone-*O*,*O*′-diacetic acid 327
Resorcinol-*O*,*O*′-diacetic acid 327
$C_{10}H_{10}O_8$
3-Hexyne-1,1,6,6-tetracarboxylic acid 1018
$C_{10}H_{10}S$
1,3-Butadien-2-yl phenyl sulfide 211
$C_{10}H_{11}AsO_4$
2,2′-(Phenylarsino)diacetic acid 1127
$C_{10}H_{11}BF_4N_2O_2$
1,4-Dihydro-2-methyl-5-nitroisoquinolinium tetrafluoroborate 277
$C_{10}H_{11}Br$
2-Bromo-3-phenyl-2-butene 169
$C_{10}H_{11}BrClNO$
4′-Bromo-2-chloro-2′,6′-dimethylacetanilide 391
$C_{10}H_{11}BrClNO_2$
2-Bromo-4-chloro-6-methylphenyl *N*-ethylcarbamate 640
$C_{10}H_{11}BrFNO$
4-Bromo-3′-fluorobutyranilide 390
$C_{10}H_{11}BrN_2O$
1-Allyl-3-(3-bromophenyl)urea 534
1-Allyl-3-(4-bromophenyl)urea 538
$C_{10}H_{11}BrN_2O_2$
1-(4-Bromo-2-nitrophenyl)pyrrolidine 274
O-(4-Bromophenylcarbamoyl)acetone oxime 1011
$C_{10}H_{11}BrN_2O_3$
2′-Bromo-2,2-dimethyl-4′-nitroacetanilide 404
5′-Bromo-4′-ethyl-2′-nitroacetanilide 403
$C_{10}H_{11}BrN_2O_4$
Ethyl 5-bromo-3-methyl-4-(2-nitrovinyl)-2-pyrrolecarboxylate 839
$C_{10}H_{11}BrN_4O_5$
8-Bromoinosine 920
$C_{10}H_{11}BrN_4O_6$
8-Bromoxanthosine 920
$C_{10}H_{11}BrO$
2-Bromobutyrophenone 296
4′-Bromobutyrophenone 297
2-Bromo-2′,4′-dimethylacetophenone 288
2-Bromomethyl-5-methylcoumaran.. 183
4-Bromophenyl cyclopropyl carbinol 203
$C_{10}H_{11}BrO_2$
Benzyl 2-bromopropionate 356
2-Bromomethyl-5-methoxycoumaran 183
2-(4-Bromophenyl)-2-methyl-1,3-dioxolane 183
5-(2-Bromopropyl)-1,3-benzodioxolane 185
Ethyl α-bromophenylacetate 361
Isopropyl 2-bromobenzoate 367
$C_{10}H_{11}BrO_2S$
4-Bromo-2-butenyl phenyl sulfone 772
$C_{10}H_{11}BrO_3$
4-(2-Bromoethoxy)-*m*-anisaldehyde.. 321
$C_{10}H_{11}BrO_4$
Ethyl 3-bromo-4,6-dimethyl-2*H*-pyran-5-carboxylate 96
$C_{10}H_{11}Br_2NO_2$
2,4-Dibromophenyl *N*-isopropylcarbamate 637
$C_{10}H_{11}Br_3O$
(3-Bromopropyl) (2,6-dibromo-4-methylphenyl) ether 176
$C_{10}H_{11}Cl$
4-Chloro-β,β-dimethylstyrene 170
$C_{10}H_{11}ClFNO_2$
2-Chloro-4-fluorophenyl *N*-isopropylcarbamate 637
$C_{10}H_{11}ClN_2$
5-Chloro-1-ethyl-2-methylbenzimidazole 906
$C_{10}H_{11}ClN_2O$
1-Allyl-3-(2-chlorophenyl)urea 587
1-Allyl-3-(3-chlorophenyl)urea 580
1-Allyl-3-(4-chlorophenyl)urea 589
4-Chloro-α-(2-imidazolin-2-yl)benzyl alcohol hydrochloride 244
1-(2-Chlorophenyl)-3-cyclopropylurea 581
$C_{10}H_{11}ClN_2O_2$
4′-Chloro-6′-formamido-*o*-acetotoluidide 425
O-(*N*-(2-Chlorophenyl)carbamoyl)acetone oxime 1011
O-(*N*-(3-Chlorophenyl)carbamoyl)acetone oxime 1012
O-(*N*-(4-Chlorophenyl)carbamoyl)acetone oxime 1012
3-(4-Chlorophenyl)-1-ethyl-1-formylurea 533
1-(4-Chlorophenyl)-3-propionylurea 533
$C_{10}H_{11}ClN_2O_3$
2-Chloro-4′,5′-dimethyl-2′-nitroacetanilide 392
2′-Chloro-4′-nitrobutyranilide 404
2′-Chloro-5′-nitrobutyranilide 404
$C_{10}H_{11}ClN_2O_4$
2-Chloroethyl *N*-(2-methyl-5-nitrophenyl)carbamate 604
Isopropyl *N*-(2-chloro-4-nitrophenyl)carbamate 670
$C_{10}H_{11}ClN_2O_5$
2-Methoxyethyl *N*-(2-chloro-5-nitrophenyl)carbamate 618
$C_{10}H_{11}ClN_2S$
1-Allyl-3-(4-chlorophenyl)-2-thiourea. 733
$C_{10}H_{11}ClN_4O_2$
3-Allyl-8-chloro-1-ethylxanthine 924
$C_{10}H_{11}ClO$
4′-Chlorobutyrophenone 297
3-Chloro-2-methylpropiophenone 287
4-Chlorophenyl cyclopropyl carbinol 201
4′-Methyl-3-chloropropiophenone 287
2-Phenylbutyryl chloride 389
$C_{10}H_{11}ClO_2$
4-Chloro-β-methylhydrocinnamic acid 333
2-(3-Chloropropoxy)benzaldehyde 321
4-(3-Chloropropoxy)-benzaldehyde 321
Ethyl 2-(chloromethyl)benzoate 366
Ethyl 4-chlorophenylacetate 353
Phenyl 4-chlorobutyrate 376
$C_{10}H_{11}ClO_2S$
2-(4-Chloro-2,5-dimethylphenylthio)acetic acid 325
3-Chloro-1-propenyl *p*-tolyl sulfone.. 772
Methyl 2-(4-chlorobenzylthio)acetate 358
Methyl 3-(4-chlorophenylthio)-2-methylpropionate 358
Methyl 3-(4-chlorophenylthio)propionate 358
$C_{10}H_{11}ClO_3$
4-(3-Chloropropoxy)-benzoic acid 336
$C_{10}H_{11}ClO_4S$
Ethyl 2-(4-chlorophenylsulfonyl)acetate 782
$C_{10}H_{11}Cl_2NO$
2′,3′-Dichlorobutyranilide 409
2,2-Dichloro-2′,3′-dimethylacetanilide 394
2,2-Dichloro-2′-ethylacetanilide 394
2,2-Dichloro-4′-ethylacetanilide 394
2,2-Dichloro-*N*-phenethylacetamide. 394
$C_{10}H_{11}Cl_2NOS$
S-Isopropyl *N*-(2,3-dichlorophenyl)thiolcarbamate 688
S-Isopropyl *N*-(2,4-dichlorophenyl)thiolcarbamate 688
S-Isopropyl *N*-(2,6-dichlorophenyl)thiolcarbamate 688
S-Isopropyl *N*-(3,4-dichlorophenyl)thiolcarbamate 688
S-Propyl *N*-(2,3-dichlorophenyl)thiolcarbamate 641
S-Propyl *N*-(2,6-dichlorophenyl)thiolcarbamate 641
$C_{10}H_{11}Cl_2NO_2$
2-Chloroethyl *N*-(2-chloro-6-methylphenyl)carbamate 604
2-Chloroethyl *N*-(3-chloro-4-methylphenyl)carbamate 605
2-Chloroethyl *N*-(4-chloro-2-methylphenyl)carbamate 605
2-Chloroethyl *N*-(5-chloro-2-methylphenyl)carbamate 606
1-Chloro-2-propyl *N*-(3-chlorophenyl)carbamate 605
1-Chloro-2-propyl *N*-(4-chlorophenyl)carbamate 605
2,2-Dichloro-2′-methoxy-5′-methylacetanilide 394
2,3-Dichlorophenyl *N*-isopropylcarbamate 635
3,4-Dichlorophenyl *N*-isopropylcarbamate 637
3,5-Dichlorophenyl *N*-isopropylcarbamate 640
3,4-Dichlorophenyl *N*-propylcarbamate 637
Isopropyl *N*-(2,3-dichlorophenyl)carbamate 683
Isopropyl *N*-(2,4-dichlorophenyl)carbamate 685
Isopropyl *N*-(2,5-dichlorophenyl)carbamate 685
Isopropyl *N*-(2,6-dichlorophenyl)carbamate 683
Isopropyl *N*-(3,4-dichlorophenyl)carbamate 685
Propyl *N*-(2,3-dichlorophenyl)carbamate 683
Propyl *N*-(2,4-dichlorophenyl)carbamate 685
Propyl *N*-(2,5-dichlorophenyl)carbamate 685

$C_{10}H_{11}Cl_2NO_2$ (Cont'd)
Propyl *N*-(2,6-dichlorophenyl)carbamate ... 683
Propyl *N*-(3,4-dichlorophenyl)carbamate ... 685
Propyl *N*-(3,5-dichlorophenyl)carbamate ... 640

$C_{10}H_{11}Cl_2NO_3$
2-Chloroethyl *N*-(3-chloro-2-methoxyphenyl)carbamate ... 604
2,5'-Dichloro-2',4'-dimethoxyacetanilide ... 390
Ethyl *N*-(3,4-dichlorophenyl)-*N*-methoxycarbamate ... 628
Isopropyl *N*-(3,4-dichlorophenyl)-*N*-hydroxycarbamate ... 628
2-Methoxyethyl *N*-(2,3-dichlorophenyl)carbamate ... 616
2-Methoxyethyl *N*-(2,4-dichlorophenyl)carbamate ... 620
2-Methoxyethyl *N*-(2,6-dichlorophenyl)carbamate ... 616
2-Methoxyethyl *N*-(3,4-dichlorophenyl)carbamate ... 616
2-Methoxyethyl *N*-(3,5-dichlorophenyl)carbamate ... 616

$C_{10}H_{11}Cl_2NO_4S$
Ethyl *N*-(3,4-dichlorophenyl)-*N*-(methylsulfonyl)carbamate ... 808

$C_{10}H_{11}Cl_2N_3$
4-(Diallylamino)-3,6-dichloropyridazine ... 968

$C_{10}H_{11}Cl_3N_2O$
3-(2-Chloroethyl)-1-(2,3-dichlorophenyl)-1-methylurea ... 504
1-Isopropyl-3-(2,4,5-trichlorophenyl)urea ... 603
1-Isopropyl-3-(2,4,6-trichlorophenyl)urea ... 546
1-Propyl-3-(2,4,5-trichlorophenyl)urea ... 603

$C_{10}H_{11}Cl_3N_2O_2$
Methyl *N*-(1-anilino-2,2,2-trichloroethyl)carbamate ... 609

$C_{10}H_{11}Cl_3O$
4-(Trichloromethyl)-2,4,5-trimethyl-2,5-cyclohexadien-1-one ... 51

$C_{10}H_{11}FN_2O$
1-Allyl-3-(4-fluorophenyl)urea ... 589
1-(Cyclopropyl)-3-(4-fluorophenyl)urea ... 534

$C_{10}H_{11}FN_2O_2$
O-(*N*-(4-Fluorophenyl)carbamoyl)acetone oxime ... 1013
N,*N'*-(4-Fluoro-1,2-phenylene)bisacetamide ... 404

$C_{10}H_{11}FO$
α-Cyclopropyl-4-fluorobenzyl alcohol ... 203
4'-Fluorobutyrophenone ... 298

$C_{10}H_{11}FO_3$
2-(4-Fluorophenoxy)-2-methylpropionic acid ... 325

$C_{10}H_{11}F_3N_2O$
1,1-Dimethyl-3-(2-trifluoromethylphenyl)urea ... 509
1,1-Dimethyl-3-(3-trifluoromethylphenyl)urea ... 510
1-Ethyl-3-(3-trifluoromethylphenyl)urea ... 510

$C_{10}H_{11}F_3N_4$
1-(α-Methyl-3-(trifluoromethyl)benzylideneamino)guanidine nitrate ... 1021

$C_{10}H_{11}F_3N_4OS$
1-Methyl-6-(3-(trifluoromethyl)phenyl)-2-thiobiurea ... 728

$C_{10}H_{11}F_3O$
2-(4-(Trifluoromethyl)phenyl)-2-propanol ... 197

$C_{10}H_{11}F_3S_2$
3-(Trifluoromethyl)benzaldehyde dimethyl mercaptal ... 209

$C_{10}H_{11}HgNO_3$
4-Acetamidophenylmercuric acetate ... 352

$C_{10}H_{11}IN_2O$
1-Cyclopropyl-3-(4-iodophenyl)urea ... 534

$C_{10}H_{11}IO$
4'-Iodobutyrophenone ... 289

$C_{10}H_{11}IO_2$
Methyl 3-ethyl-2-iodobenzoate ... 367

$C_{10}H_{11}IO_2S$
Ethyl α-iodostyryl sulfone ... 783
2-Iodo-1-propenyl *p*-tolyl sulfone ... 785

$C_{10}H_{11}IO_3$
2-(4-Iodophenoxy)-2-methylpropionic acid ... 325

$C_{10}H_{11}N$
2,7-Dimethylindole ... 891
3,4-Dimethylphenylacetonitrile ... 756
3,3-Dimethyl-2-phenyl-3*H*-azirine ... 252
1-Phenyl-3-pyrroline ... 251
N-(2-Propynyl)benzylamine hydrochloride ... 1018
2,4,6-Trimethylbenzonitrile ... 767

$C_{10}H_{11}NO$
5-Acetylindoline ... 298
N-Allylbenzamide ... 478
5,6-Dihydro-2-phenyl-4*H*-1,3-oxazine. 253
2,5-Dimethyl-*p*-anisonitrile ... 762
3,3-Dimethyl-2-indolinone ... 692
4-(1-Hydroxy-1-methylethyl)benzonitrile ... 761
4-Hydroxy-α-methylhydrocinnamonitrile ... 752
1-Indancarboxamide ... 429
Mesityl isocyanate ... 769
3-(Methylamino)acrylophenone ... 294
α-Methylcinnamaldoxime ... 1013
N-Phenylmethacrylamide ... 478
1,2,3,4-Tetrahydroquinoline-1-carboxaldehyde ... 710
α-Tetralone oxime ... 1014
2,4,6-Trimethylbenzonitrile *N*-oxide ... 1006
2,5,6-Trimethylbenzoxazole ... 910

$C_{10}H_{11}NOS$
3,4-Dihydro-1-benzothiepin-5(2*H*)-one oxime ... 1014
2,3-Dihydro-2-methyl-1,5-benzothiazepin-4(5*H*)-one ... 706
2,3-Dihydro-4-methyl-1,4-benzothiazepin-5(4*H*)-one ... 706
2,5-Dimethyl-6-methoxybenzothiazole ... 913
6-Ethoxy-2-methylbenzothiazole ... 913

$C_{10}H_{11}NOS_2$
5-Methoxy-2-methyl-6-(methylthio)benzothiazole ... 913

$C_{10}H_{11}NO_2$
3'-Acetylacetanilide ... 432
4'-Acetylacetanilide ... 432
O-Allylbenzohydroxamic acid ... 472
Allyl carbanilate ... 680
2-Allyloxybenzamide ... 467
N-Allylsalicylamide ... 462
4-Amino-2-methylcinnamic acid ... 335
3,5-Dimethoxybenzyl cyanide ... 753
(3,5-Dimethoxyphenyl)acetonitrile ... 752
3,5-Dimethoxyphenylacetonitrile ... 753
β,4-Dimethyl-β-nitrostyrene ... 258
Methyl 2-aminocinnamate ... 364

$C_{10}H_{11}NO_2S$
S-(4-Acetamidophenyl) thioacetate ... 436
O-Ethyl *N*-benzoylthiocarbamate ... 745

$C_{10}H_{11}NO_3$
2-Acetamidophenyl acetate ... 431
3-Acetamidophenylacetic acid ... 436
3-Acetamido-*p*-toluic acid ... 433
N-Acetonylanthranilic acid ... 343
4'-Acetyl-3'-hydroxyacetanilide ... 411
4-Acetylphenyl *N*-methylcarbamate. 675
4-Amino-2-methoxycinnamic acid ... 335
4-(3-Aminophenyl)-4-oxobutyric acid ... 329
3-(Anilinomethylene)-2,4-chromanedione ... 385
N-Benzoyl-D-alanine ... 349
N-Benzoyl-DL-alanine ... 349
N-Benzoyl-L-alanine ... 349
3-Benzoylalanine hydrochloride ... 349
Ethyl 4-formamidobenzoate ... 434
Ethyl β-oxo-3-pyridinepropionate ... 962
Ethyl 3-oxo-3-(2-pyridyl)propionate ... 961
Ethyl 3-oxo-3-(4-pyridyl)propionate ... 961
N-Formyl-2-phenylalanine ... 436
DL-*N*-Formylphenylalanine ... 429
p-Glyoxylophenetidide ... 418
1-(*p*-Methoxyphenyl)-2-nitro-1-propene ... 257
Methyl *N*-acetylanthranilate ... 436
3-Methyl-1,4-benzodioxan-2-carboxamide ... 420
Methyl hippurate ... 476
3'-Nitrobutyrophenone ... 295
N-(Phenylacetyl)glycine ... 429
Succinanilic acid ... 435

$C_{10}H_{11}NO_3S$
Methyl 2-amino-5-(cyclopropylcarbonyl)-3-thiophenecarboxylate ... 856

$C_{10}H_{11}NO_4$
2-(Acetoxymethyl)-3-acetoxypyridine ... 940
1-Carboxyethyl *N*-phenylcarbamate. 665
2,5-Dimethoxy-β-nitrostyrene ... 257
3,4-Dimethoxy-β-nitrostyrene ... 257
3,4-Dimethoxy-β-nitrostyrene ... 257
3,5-Dimethoxy-β-nitrostyrene ... 257
Dimethyl 4-aminophthalate ... 372
Dimethyl 6-methyl-2,4-pyridinedicarboxylate ... 960
Ethyl 4-hydroxyoxanilate ... 413
Ethyl 2-nitrophenylacetate ... 359
Ethyl 4-nitrophenylacetate ... 360
N-Formyl-L-tyrosine ... 412
β-Methyl-3-nitrohydrocinnamic acid. 330
β-Methyl-4-nitrohydrocinnamic acid. 330
2-Methyl-2-(3-nitrophenyl)-1,3-dioxolane ... 279
2-(4-Nitrophenyl)butyric acid ... 330
4-(2-Nitrophenyl)butyric acid ... 330
R-(+)-2-(Phenylcarbamoyloxy)propionic acid ... 665

$C_{10}H_{11}NO_4S$
4'-(2-Hydroxyethylthio)-3'-nitroacetophenone ... 290

$C_{10}H_{11}NO_5$
Ethyl 4-nitrophenoxyacetate ... 355

$C_{10}H_{11}NO_5S$
N-(4-Sulfophenyl)acetoacetamide, potassium salt ... 794

$C_{10}H_{11}NO_7$
2-(4-Hydroxy-2-methoxy-3-methyl-5-nitrophenoxy)acetic acid ... 325

$C_{10}H_{11}NS$
3-(Benzylthio)propionitrile ... 754
2-Isopropylphenyl isothiocyanate ... 768
4-Isopropylphenyl isothiocyanate ... 768
Mesityl thiocyanate ... 768
2-Methyl-2-(phenylthio)propionitrile. 754
2-Methyl-3-(phenylthio)propionitrile. 754
4-(Phenylthio)butyronitrile ... 754
2,4,6-Trimethylphenyl isothiocyanate ... 768

$C_{10}H_{11}N_3$
5-Amino-3-methyl-1-phenylpyrazole ... 861

$C_{10}H_{11}N_3O$
3-Amino-2,6-dimethyl-4(3*H*)-quinazolinone ... 711
3-Amino-2-methyl-1-phenyl-3-pyrazolin-5-one ... 701
2-(4-Aminophenyl)-3-methyl-3-pyrazolin-5-one ... 861
2-Azido-3-benzyloxypropionic acid, barium salt ... 350
1-Benzyl-5-imino-2-pyrazolin-3-ol ... 724

$C_{10}H_{11}N_3OS$
Acetic (3-methyl-2-benzothiazolinylidene)hydrazide ... 725
5-Anilino-3-ethoxy-1,2,4-thiadiazole ... 878

$C_{10}H_{11}N_3O_2$
2,4-Dicyano-3-ethyl-3-methylglutarimide ... 150
Isonicotinic acid (1-methyl-2-oxopropylidene)-hydrazide ... 952
3-Methyl-1-phenyl-4,5-pyrazolidinedione 4-oxime ... 1011

$C_{10}H_{11}N_3O_2S_2$
5-Carbethoxy-2-ethylthio-6-thiocyanopyrimidine ... 974

$C_{10}H_{11}N_3O_3$
1-Allyl-3-(3-nitrophenyl)urea ... 564
1-Allyl-3-(4-nitrophenyl)urea ... 568

$C_{10}H_{11}N_3O_3S$
3-(5-Imino-3-methyl-2-pyrazolin-1-yl)benzenesulfonic acid ... 861

$C_{10}H_{11}N_3O_4$
1-Acetyl-3-(4-methyl-3-nitrophenyl)urea ... 563
N,*N'*-Diacetyl-2-nitro-*p*-phenylenediamine ... 423
N,*N*-Dimethyl-2,4-dinitrostyrylamine ... 275
1-(2,4-Dinitrophenyl)-pyrrolidine ... 272
O-(*N*-(3-Nitrophenyl)carbamoyl)acetone oxime ... 1010

$C_{10}H_{11}N_3O_5$
1-(4-Pyridylcarbonyl)-2-(carboxymethoxyacetyl)hydrazine ... 939

$C_{10}H_{11}N_3O_6$
N-(2,4-Dinitrophenyl)-L-2-aminobutyric acid ... 349

$C_{10}H_{11}N_3O_7$
N-(2,4-Dinitrophenyl)-L-threonine ... 348

$C_{10}H_{11}N_3O_7S$
2',6'-Dinitro-*N*-methyl-4'-(methylsulfonyl)acetanilide ... 777

$C_{10}H_{11}N_3S$
4-Amino-2-(*N*-methylanilino)thiazole 876

$C_{10}H_{11}N_5$
2,4-Diamino-6-(*m*-tolyl)-1,3,5-triazine. 979
2-Guanidino-4-methylquinazoline ... 1001
2-Guanidino-4-methylquinazoline hydrochloride ... 1001

$C_{10}H_{11}N_5O$
5-Hydrazono-3-phenyl-2-pyrazoline-1-carboxamide ... 705

$C_{10}H_{11}N_5O_2S$
N^4-(6-Aminopyrimidin-4-yl)sulfanilamide monohydrochloride ... 974

$C_{10}H_{12}$
Cyclopropylphenylmethane ... 162
Isocrotylbenzene ... 164
Isotetralin ... 3
1-Phenyl-2-butene ... 164
1-Phenyl-3-butene ... 164

$C_{10}H_{12}BrCl$
4-Bromo-1-*tert*-butyl-2-chlorobenzene ... 169

$C_{10}H_{12}BrNO$
4'-Bromo-2',5'-acetoxylidide ... 410
3'-Bromo-2',4'-dimethylacetanilide ... 410
3'-Bromo-4'-ethylacetanilide ... 409
4'-Bromo-3'-ethylacetanilide ... 409
4-(4-Bromophenyl)-butyramide ... 437

$C_{10}H_{12}BrNOS$
S-Isopropyl *N*-(4-bromophenyl)thiolcarbamate ... 688
S-Propyl *N*-(4-bromophenyl)thiolcarbamate ... 688

$C_{10}H_{12}BrNO_2$
2-(4-Bromophenoxy)-2-methylpropionamide ... 400
4-Bromo-DL-phenylalanine methyl ester hydrochloride ... 358
Isopropyl *N*-(4-bromophenyl)carbamate ... 682

$C_{10}H_{12}BrNO_3$
2-Methoxyethyl *N*-(4-bromophenyl)carbamate ... 620

$C_{10}H_{12}Br_2$
1,2-Dibromo-3,4,5,6-tetramethylbenzene ... 169
1,3-Dibromo-2,4,5,6-tetramethylbenzene ... 169
1,4-Dibromo-2,3,5,6-tetramethylbenzene ... 170

$C_{10}H_{12}Br_2O$
(2-Bromo-4-methylphenyl) (3-bromopropyl) ether ... 176

$C_{10}H_{12}Br_2O_2$
1,2-Dibromo-3,6-dimethoxy-4,5-dimethylbenzene ... 177
α,α'-Dibromo-2,5-dimethoxy-*p*-xylene ... 176

$C_{10}H_{12}Br_2O_2Te$
3-(*p*-Tolyltelluro)propionic acid, te,te-dibromide ... 333

$C_{10}H_{12}Br_2O_4$
1,4-Dibromo-2,3,5,6-tetramethoxybenzene ... 177

$C_{10}H_{12}ClN$
1-(*p*-Chlorophenyl)-cyclopropanemethylamine hydrochloride 236

$C_{10}H_{12}ClNO$
N-Benzyl-3-chloropropionamide 392
2-Chloro-2′,3′-acetoxylidide 393
2-Chloro-2′,5′-acetoxylidide 393
2-Chloro-3′,4′-acetoxylidide 393
2-Chloro-4′-ethylacetanilide 393
3′-Chloro-4′-ethylacetanilide 409
5-Chloromethyl-2-phenyloxazolidine 244
3′-Chloro-*o*-propionotoluidide 439
3′-Chloro-*p*-propionotoluidide 409
4′-Chloro-*o*-propionotoluidide 409
5′-Chloro-*o*-propionotoluidide 409
6′-Chloro-*m*-propionotoluidide 409
Ethyl 2-chloro-2-phenylacetimidate hydrochloride 358
N-Ethyl-*N*-(*o*-tolyl)-carbamoyl chloride 430

$C_{10}H_{12}ClNOS$
S-(4-Chlorophenyl) *N*-(propyl)-thiocarbamate 641
S-Isopropyl *N*-(2-chlorophenyl)-thiolcarbamate 688
S-Isopropyl *N*-(3-chlorophenyl)-thiolcarbamate 688
S-Isopropyl *N*-(4-chlorophenyl)-thiolcarbamate 688
S-Propyl *N*-(3-chlorophenyl)-thiolcarbamate 688

$C_{10}H_{12}ClNO_2$
2-Chloro-2′-ethoxyacetanilide 391
2-Chloro-4′-ethoxyacetanilide 392
1-Chloroethyl *N*-methyl-*N*-phenylcarbamate 677
2-Chloroethyl *N*-(*o*-tolyl)carbamate 605
2-Chloroethyl *N*-(*m*-tolyl)carbamate 605
2-Chloroethyl *N*-(*p*-tolyl)carbamate 605
2-Chloro-2′-methoxy-5′-methylacetanilide 391
4′-Chloro-2′-methoxypropionanilide 401
5′-Chloro-2′-methoxypropionanilide 401
2-(4-Chlorophenoxy)-2-methylpropionamide 400
1-Chloro-2-propyl *N*-phenylcarbamate 605
Ethyl *N*-(2-chloro-6-methylphenyl)-carbamate 683
Ethyl *N*-(4-chloro-2-methylphenyl)-carbamate 685
Ethyl *N*-(5-chloro-2-methylphenyl)-carbamate 685
Isopropyl *N*-(2-chlorophenyl)-carbamate 681
Isopropyl *N*-(4-chlorophenyl)-carbamate 682
Propyl *N*-(2-chlorophenyl)-carbamate 681
Propyl *N*-(3-chlorophenyl)-carbamate 678
Propyl *N*-(4-chlorophenyl)-carbamate 682

$C_{10}H_{12}ClNO_2S$
S-(2-Hydroxyethyl) *N*-(3-chloro-2-methylphenyl)thiocarbamate 612

$C_{10}H_{12}ClNO_3$
2-Chloro-2′,5′-dimethoxyacetanilide 391
2-Chloroethyl *N*-(*o*-anisyl)-carbamate 604
2-Chloroethyl *N*-(*m*-anisyl)-carbamate 604
2-Chloroethyl *N*-(4-methoxyphenyl)-carbamate 604
Ethyl *N*-(2-chlorobenzyloxy)-carbamate 633
Ethyl *N*-(5-chloro-2-methoxyphenyl)-carbamate 619
Ethyl *N*-(3-chlorophenyl)-*N*-methoxycarbamate 628
Ethyl *N*-(4-chlorophenyl)-*N*-methoxycarbamate 628
Isopropyl *N*-(4-chlorophenyl)-*N*-hydroxycarbamate 628
2-Methoxyethyl *N*-(2-chlorophenyl)-carbamate 650
2-Methoxyethyl *N*-(3-chlorophenyl)-carbamate 649
2-Methoxyethyl *N*-(4-chlorophenyl)-carbamate 650

$C_{10}H_{12}ClNO_4$
4-Chloro-2,5-diethoxynitrobenzene 260
1-(2-(2-Chloroethoxy)-ethoxy)-4-nitrobenzene 258

$C_{10}H_{12}ClNO_4S$
2-Chloroethyl *N*-acetylsulfanilate 796
5-(Dimethylsulfamoyl)-2-methoxybenzoyl chloride 814
Ethyl *N*-(4-chlorophenyl)-*N*-(methylsulfonyl)carbamate 808

$C_{10}H_{12}ClN_3O_2$
1-Acetyl-4-(3-chloro-4-methylphenyl)semicarbazide 532
1-(4-Chlorobenzoyl)-4,4-dimethylsemicarbazide 528
6-Chloro-1,3-diethyl-7*H*-pyrrolo(2,3-d)pyrimidine-2,4(1*H*,3*H*)-dione 919
1-(2-Chloro-5-methylphenyl)-5-methylbiuret 533

$C_{10}H_{12}ClN_3O_3$
1-Acetyl-4-(5-chloro-2-methoxyphenyl)semicarbazide 518
1-(2-Chloro-4-nitrophenyl)-3-isopropylurea 568
Ethyl 2-(3-(5-chloro-2-pyridyl)ureido)-acetate 964

$C_{10}H_{12}Cl_2$
2,5-Bis-(chloromethyl)-*p*-xylene 170
4,6-Bis(chloromethyl)-*m*-xylene 170

$C_{10}H_{12}Cl_2N_2$
1-(3,4-Dichlorophenyl)-piperazine 248

$C_{10}H_{12}Cl_2N_2O$
3′,4′-Dichloro-2-(dimethylamino)-acetanilide hydrochloride 431
1-(2,3-Dichlorophenyl)-3-isopropylurea 593
1-(2,4-Dichlorophenyl)-3-isopropylurea 598
1-(2,5-Dichlorophenyl)-3-isopropylurea 597
1-(2,6-Dichlorophenyl)-3-isopropylurea 540
1-(3,4-Dichlorophenyl)-3-isopropylurea 597
1-(3,5-Dichlorophenyl)-3-isopropylurea 545
1-(2,3-Dichlorophenyl)-3-propylurea 593
1-(2,4-Dichlorophenyl)-3-propylurea 598
1-(2,5-Dichlorophenyl)-3-propylurea 597
1-(2,6-Dichlorophenyl)-3-propylurea 540
1-(3,4-Dichlorophenyl)-3-propylurea 597
1-(3,5-Dichlorophenyl)-3-propylurea 545
1-(3,4-Dichlorophenyl)-1,3,3-trimethylurea 543

$C_{10}H_{12}Cl_2N_2O_2$
1-(3,4-Dichlorophenyl)-3-(2-hydroxy-1-propyl)urea 548
1-(2,3-Dichlorophenyl)-3-(2-methoxyethyl)urea 556
1-(2,4-Dichlorophenyl)-3-(2-methoxyethyl)urea 556
1-(2,5-Dichlorophenyl)-3-(2-methoxyethyl)urea 556
1-(2,6-Dichlorophenyl)-3-(2-methoxyethyl)urea 520
1-(3,4-Dichlorophenyl)-3-(2-methoxyethyl)urea 556

$C_{10}H_{12}Cl_2N_2O_3S$
1-(3,4-Dichlorophenyl)-3,3-dimethyl-1-(methylsulfonyl)urea 809

$C_{10}H_{12}Cl_2N_4O_{10}$
4,4,7,7-Tetranitrosebacoyl chloride 100

$C_{10}H_{12}Cl_2O$
3-(4-Chlorophenyl)-3-methoxypropyl chloride 185

$C_{10}H_{12}Cl_2O_2$
2,5-Bis(chloromethyl)-1,4-dimethoxybenzene 176

$C_{10}H_{12}Cl_5N_2O_2P$
Pentachlorophenyl tetramethylphosphorodiamidate 833

$C_{10}H_{12}FNO$
N-(3-Fluorophenyl)butyramide 407

$C_{10}H_{12}FNO_2$
2-Fluorophenyl *N*-isopropylcarbamate 633
4-Fluorophenyl *N*-isopropylcarbamate 634
Isopropyl *N*-(4-fluorophenyl)-carbamate 682

$C_{10}H_{12}FNO_4S$
N-Acetyl-4-ethoxymetanilyl fluoride 802
4-(*tert*-Butyl)-3-nitrobenzenesulfonyl fluoride 803

$C_{10}H_{12}F_3N_3O$
1,1-Dimethyl-4-(3-trifluoromethylphenyl)-semicarbazide 509

$C_{10}H_{12}N_2$
5-Amino-2,3-dimethylindole 884
2-Benzyl-2-imidazoline thiosulfate 725
N-(2-Cyanoethyl)-*N*-methylaniline 755
3-(Dimethylaminomethyl)-benzonitrile 765
4-(Dimethylaminomethyl)-benzonitrile 765
2-Ethyl-1-methylbenzimidazole 907
2-Ethyl-4-methylbenzimidazole 907
N-Ethyl-*N*-phenylglycinonitrile 754
2-Propylbenzimidazole 905
3-(*o*-Toluidino)propionitrile 755
3-(*p*-Toluidino)propionitrile 755

$C_{10}H_{12}N_2O$
1-Acetyl-5-aminoindoline 701
1-Allyl-3-phenylurea 585
1-(4-Aminophenyl)-2-pyrrolidinone 690
1-Carbamoyl-1,2,3,4-tetrahydroquinoline 714
α,6-Dimethylbenzimidazole-methanol 902
5,6-Dimethyl-1-benzimidazolemethanol 902
5,6-Dimethyl-2-benzimidazolemethanol 902
1-Ethyl-5-hydroxy-2-methylbenzimidazole 902
2-(3-Hydroxypropyl)benzimidazole 902
4-Methyl-1-phenyl-3-pyrazolidinone 705
5-Methyl-1-phenyl-3-pyrazolidinone 710
2-(3-Methyl-2-pyrazolin-5-yl)phenol 860
4-(3-Methyl-2-pyrazolin-5-yl)phenol 246
4-Methyl-2,3,4,5-tetrahydro-1*H*-1,5-benzodiazepin-2-one 702
1-Nicotinoylpyrrolidine 949
1-(*m*-Tolyl)-2-imidazolidinone 711

$C_{10}H_{12}N_2OS$
1-(4-Acetylphenyl)-3-methyl-2-thiourea 744
1-Allyl-3-(3-hydroxyphenyl)-2-thiourea 735
1-Allyl-3-(4-hydroxyphenyl)-2-thiourea 735
2-(*p*-Anisidino)-2-thiazoline 247

$C_{10}H_{12}N_2OS_2$
Methyl 3-(*p*-toluoyl)-dithiocarbazate 739

$C_{10}H_{12}N_2O_2$
Acetic acid (2-methoxybenzylidene)-hydrazide 498
2-Acetoacetamido-3-methylpyridine 959
2-Acetoacetamido-4-methylpyridine 959
2-Acetoacetamido-6-methylpyridine 961
Acetone *O*-(phenylcarbamoyl)-oxime 1013
1-Acetyl-1-methyl-3-phenylurea 579
1-Acetyl-2-phenylacetylhydrazine 499
2,8-Decadiynediamide 1019
1,2-Diacetamidobenzene 425
N,*N*′-Diacetyl-1,4-phenylenediamine 428
4-Dimethylamino-β-nitrostyrene 258
N,*N*′-Dimethylphthalamide 471
N,*N*′-Dimethylterephthalamide 480
Ethyl 4-amidinobenzoate 724
Ethyl *N*-(benzylideneamino)-carbamate 502
2-Hydroximino-2′,5′-acetoxylidide 1012
2-(Hydroxyimino)-3′,4′-dimethylacetanilide 1012
Methyl (4-methylbenzylidene)-carbazate 1063
4-Nicotinoylmorpholine 949
1-(4-Nitrophenyl)pyrrolidine 276
1-Phenyl-3-propionylurea 579
N-(3-Pyridylmethyl)-acetoacetamide 949

$C_{10}H_{12}N_2O_2S$
1-Benzoyl-3-(2-hydroxyethyl)-2-thiourea 734
Tetrahydrofuroic (2-thenylidene)-hydrazide 1051

$C_{10}H_{12}N_2O_3$
Acetic (4-hydroxy-3-methoxybenzylidene)hydrazide 495
N-Anilinosuccinamic acid 503
1-(2-Carboxyphenyl)-3-ethylurea 576
2′,3′-Dimethyl-6′-nitroacetanilide 424
4′,5′-Dimethyl-2′-nitroacetanilide 425
4′,6′-Dimethyl-2′-nitroacetanilide 425
N-Ethyl-3′-nitroacetanilide 426
2-Isonitroso-2′-methoxy-3′-methylacetanilide 1009
Methyl *N*-methyl-*N*-(phenylcarbamoyl)carbamate 677
2′-Methyl-4′-nitropropionanilide 427
4′-Methyl-2′-nitropropionanilide 424
4′-Methyl-3′-nitropropionanilide 426
N-(3-Methyl-2-pyridyl)-succinamic acid 940
N-(4-Methyl-2-pyridyl)-succinamic acid 963
N-(6-Methyl-2-pyridyl)-succinamic acid 940
N-(4-Nitrophenethyl)acetamide 428
4-(4-Nitrophenyl)morpholine 276
N-(2-Pyridyl)glutaramic acid 963
N-(3-Pyridyl)glutaramic acid 945

$C_{10}H_{12}N_2O_3S$
S-Isopropyl *N*-(3-nitrophenyl)-thiolcarbamate 670

$C_{10}H_{12}N_2O_4$
4′-Amino-2′-hydroxysuccinanilic acid 411
1,4-Butanediol bis(cyanoacetate) 144
N-Carbamoyltyrosine 547
4-(Ethoxycarbonylamino)-anthranilic acid 668
Ethyl *N*-(4-methyl-2-nitrophenyl)-carbamate 668
2′-Hydroxy-5′-nitrobutyranilide 412
Isopropyl *N*-(3-nitrophenyl)-carbamate 669
2′-Methoxy-3′-nitropropionanilide 415
2′-Methoxy-5′-nitropropionanilide 415
Methyl (4-hydroxy-3-methoxybenzylidene)carbazate 495
1-(1-Methyl-1-nitropropyl)-4-nitrobenzene 278
2-Nitro-2-(4-nitrobenzyl)propane 277
Propyl *N*-(3-nitrophenyl)carbamate 669

$C_{10}H_{12}N_2O_4S$
2-(4-Acetamidophenylsulfonyl)-acetamide 779

$C_{10}H_{12}N_2O_5$
2-(*tert*-Butyl)-4,6-dinitrophenol 264
4-*sec*-Butyl-2,6-dinitrophenol 264
Ethyl *N*-(2-methoxy-5-nitrophenyl)-carbamate 652
Ethyl *N*-(4-methoxy-2-nitrophenyl)-carbamate 652
2-Methyl-2-nitro-1-(3-nitrophenyl)-1-propanol 263
3-Nitrotyrosine methyl ester hydrochloride 354

$C_{10}H_{12}N_2O_5S$
N^4-Succinoylsulfanilamide 823

$C_{10}H_{12}N_2O_9S_4$
3-Methyl-2-methylthio-6-nitro-5-sulfobenzothiazolium methyl sulfate 913

$C_{10}H_{12}N_2S$
1-Allyl-3-phenyl-2-thiourea 747
5,6-Dimethyl-2-(methylthio)-benzimidazole 903
2-(Ethylthiomethyl)benzimidazole 903
2-(Isopropylthio)benzimidazole 905

$C_{10}H_{12}N_2S_2$
3-Methyl-5-phenyl-3,4,5,6-tetrahydro-2*H*-1,3,5-thiadiazine-2-thione 742

$C_{10}H_{12}N_4$
1-Cyano-3-(3,4-xylyl)-guanidine 758
$C_{10}H_{12}N_4O$
9-Cyclopentylhypoxanthine 924
3-(Dimethylamino)-7-methyl-1,2,4-benzotriazine-1-oxide 1003
3-Methyl-4-phenylhydrazino-2-pyrazolin-5-one 861
$C_{10}H_{12}N_4OS$
3-Acetamidobenzaldehyde thiosemicarbazone 1070
4-Acetamidobenzaldehyde thiosemicarbazone 1070
3-(2-Ethoxyanilino)-5-mercapto-4*H*-1,2,4-triazole 869
$C_{10}H_{12}N_4O_2$
2,4-Bis(2-cyanoethylamino)-3-oxo-1-cyclobutenylium-1-olate 1094
$C_{10}H_{12}N_4O_2S$
6-(4-Carboxybutylthio)purine 922
Ethyl 5-methyl-2-(methylthio)(1,2,4)-triazolo(1,5-a)pyrimidine-7-carboxylate 922
$C_{10}H_{12}N_4O_3$
Adrenochrome monosemicarbazone 1034
3-Allyl-1-ethyl-8-hydroxyxanthine 921
6-Amino-1,3-diallyl-5-nitrosouracil 132
Diacetyl monoxime 4-nitrophenylhydrazone 1009
$C_{10}H_{12}N_4O_3S$
3-Amino-1-(2-amino-4-methylsulfonylphenyl)-2-pyrazolin-5-one 776
3-Amino-1-(2-amino-4-methylsulfonylphenyl)-2-pyrazolin-5-one hydrochloride 776
$C_{10}H_{12}N_4O_4$
Purine riboside 920
$C_{10}H_{12}N_6$
Ethylenediamine-*N,N,N',N'*-tetraacetonitrile 145
$C_{10}H_{12}N_6O_4$
2,4,6-Triacetamido-5-nitrosopyrimidine 975
$C_{10}H_{12}N_6S_2$
Terephthaldicarboxaldehyde dithiosemicarbazone 743
$C_{10}H_{12}O$
Allyl benzyl ether 186
Allyl *o*-tolyl ether 180
Allyl *p*-tolyl ether 181
2-(2-Butenyl)phenol 191
Ethyl styryl ether 187
8-Ketotricyclo[5.2.1.0^{2,6}]-4-decene 57
β-Methoxy-*β*-methylstyrene 187
2-(2-Methylallyl)phenol 191
α-Methylhydrocinnamaldehyde 321
3'-Methylpropiophenone 296
4-Phenoxy-2-butene 180
3-Phenyl-2-butanone 286
4-Phenyl-3-buten-2-ol 206
Tricyclo[5.2.1.0^{2,6}]deca-3,8-dien-10-ol 13
$C_{10}H_{12}OS$
Allyl 4-methoxyphenyl sulfide 211
1-(Benzylthio)acetone 211
2,3-Dihydro-4-hydroxy-8-methyl-4*H*-1-benzothiopyran 210
2-(Ethylthio)acetophenone 293
4-Phenylthio-2-butanone 284
3-(Phenylthio)butyraldehyde 321
2,3,4,5-Tetrahydro-1-benzothiepin-5-ol 208
$C_{10}H_{12}OS_2$
S-Methyl *O*-phenethyl dithiocarbonate 382
2-Phenyl-1,3-dithiane 1-oxide 771
$C_{10}H_{12}O_2$
2-Allyloxyanisole 178
Benzyl propionate 356
Dicyclopentadiene dioxide (mixture of *endo* and *exo*) 21
6,7-Dihydro-7-methyl-1,4(5*H*)-indandione 53
trans-2,3-Dihydroxy-1,2,3,4-tetrahydronaphthalene 205
2',5'-Dimethyl-4'-hydroxyacetophenone 291
2,3-Epoxy-3a-methyl-3a,4,7,7a-tetrahydroindan-1-one 58
Ethyl *p*-methylbenzoate 375
4-Ethylphenyl acetate 380
4'-Hydroxybutyrophenone 290
2'-Hydroxy-4',6'-dimethylacetophenone 290
2-Hydroxy-3,5,7-trimethyl-2,4,6-cycloheptatrien-1-one 310
2'-Methoxy-5'-methylacetophenone 291
4-Methylbenzyl acetate 362
1,2-Methylenedioxy-4-propylbenzene 185
Methyl 3-phenylpropionate 361
2-Methyl-5,6,7,8-tetrahydrochromone 59
5-Norbornene-2-acrylic acid 66
2-Phenylbutyric acid, magnesium salt monohydrate 351
2-Phenylbutyric acid, sodium salt 351
Phenyl isobutyrate 380
2-(*o*-Tolyl)-propionic acid 333
2,4,5-Trimethylbenzoic acid 346
2-(3,5-Xylyl)acetic acid 334
$C_{10}H_{12}O_2S$
Allyl *p*-tolyl sulfone 785
3-(Benzylthio)-propionic acid 330
Methyl 2-(benzylthio)acetate 357
5'-Methyl-2'-(methylsulfinyl)-acetophenone 771
Methyl 2-(phenylthio)propionate 357
Methyl 3-(phenylthio)propionate 357
2-(Phenethylthio)-acetic acid 330
2-(Phenylthio)ethyl acetate 357
3-(*o*-Tolylthio)propionic acid 330
3-(*p*-Tolylthio)propionic acid 330
$C_{10}H_{12}O_2Te$
3-(*p*-Tolyltelluro)propionic acid 333
$C_{10}H_{12}O_3$
Benzyl 2-hydroxypropionate 353
3,4'-Dihydroxybutyrophenone 289
2,4-Dimethoxy-5-methylbenzaldehyde 323
2,4-Dimethoxy-6-methylbenzaldehyde 323
4,5-Dimethoxy-2-methylbenzaldehyde 322
3,6-Dimethyl-4-cyclohexene-1,2-dicarboxylic anhydride 98
2,3-Dimethylphenoxyacetic acid 327
2,5-Dimethylphenoxyacetic acid 328
2,6-Dimethylphenoxyacetic acid 327
3,5-Dimethylphenoxyacetic acid 328
8,9-Epoxy-4-vinyl-3,5-dioxatricyclo-[5.2.1.0^{2,6}]decane 22
2-Ethoxy-3-methoxybenzaldehyde 323
2-Ethoxy-2-methyl-1,3-benzodioxole 183
2-Ethoxyphenylacetic acid 329
Ethyl 2-phenoxyacetate 355
Ethyl *m*-tolyl carbonate 381
4'-(2-Hydroxyethoxy)acetophenone 290
2-Isopropoxybenzoic acid 340
2-Isopropoxybenzoic acid, sodium salt 350
Isopropyl 4-hydroxybenzoate 368
3-Isopropylsalicylic acid 338
2-Methoxybenzyl acetate 355
3-Methoxybenzyl acetate 354
4'-(Methoxymethoxy)acetophenone 283
3-(3-Methoxyphenyl)propionic acid 328
3-(3-Methyl-2-buten-1-yl)-1,2,4-cyclopentanetrione 54
Methyl 3-phenoxypropionate 355
$C_{10}H_{12}O_3S$
3-(Phenoxy)tetrahydrothiophene-1,1-dioxide 775
1-(*p*-Tolylsulfonyl)-2-propanone 781
$C_{10}H_{12}O_4$
Diallyl fumarate 78
2,3-Dimethoxy-5,6-dimethyl-1,4-benzoquinone 52
4,5-Dimethoxy-2-methylbenzoic acid 339
2,6-Dimethoxyphenyl acetate 378
endo-3,6-Dimethyl-3,6-endoxohexahydrophthalic anhydride 98
4-Ethoxy-3-methoxybenzaldehyde 323
Ethyl 2,4-dihydroxy-3-methylbenzoate 367
Ethyl 2,5-dihydroxyphenylacetate 353
Ethyl 4,6-dimethyl-2-oxo-2*H*-pyran-5-carboxylate 386
Ethyl 4-hydroxy-DL-mandelate 353
Ethyl 4-hydroxy-1-methyl-6-oxo-2,4-cyclohexadiene-1-carboxylate 91
Ethyl vanillate 367
Glycerol 2-benzoate 367
3-(3-Methoxyphenoxy)-propionic acid 327
3-(4-Methoxyphenoxy)-propionic acid 329
Methyl 2,4-dihydroxy-3,6-dimethylbenzoate 367
Methyl 2,3-dimethoxybenzoate 369
Methyl 2,5-dimethoxybenzoate 369
mono-Methyl 2,4-dimethylmuconate 78
3-(3,4-Methylenedioxyphenyl)-1,2-propanediol 200
Methyl 3-methoxymandelate 353
Propylene glycol salicylate 368
2',4',5'-Trihydroxybutyrophenone 289
$C_{10}H_{12}O_4S$
Ethyl 2-(phenylsulfonyl)acetate 781
Methyl 3-(phenylsulfonyl)-propionate 781
3-Oxo-1-phenyl-1-butanesulfonic acid sodium salt 794
$C_{10}H_{12}O_5$
2,3-Dihydro-4-ethoxyphthalic acid 65
2,5-Dimethyl-4-(ethoxycarbonyl)-3-furoic acid 853
Methyl 4,6-dimethoxysalicylate 367
Methyl syringate 367
2,3,6-Trimethoxybenzoic acid 339
2,4,6-Trimethoxybenzoic acid 341
$C_{10}H_{12}O_5S$
α-(Ethoxycarbonyl)-*α*-toluenesulfonic acid, barium salt 795
$C_{10}H_{12}O_5S_2$
Phenyl 3-sulfolanesulfonate 799
$C_{10}H_{12}O_6S$
Diethyl 3,4-dihydroxy-2,5-thiophenedicarboxylate, disodium salt 855
$C_{10}H_{12}O_7$
Diethyl 3,4-dihydroxy-2,5-furandicarboxylate 845
Diethyl 3,4-dihydroxy-2,5-furandicarboxylate, disodium salt 845
$C_{10}H_{12}S$
Allyl *m*-tolyl sulfide 211
Allyl *p*-tolyl sulfide 212
Crotyl phenyl sulfide 211
2-Methylallyl phenyl sulfide 211
$C_{10}H_{13}AsO_3$
2-(2-Methylallyl)arsonic acid 1128
$C_{10}H_{13}AsO_6$
4-(Isopropoxycarbonyloxy)-benzenearsonic acid 1126
$C_{10}H_{13}Br$
2-Bromo-*p*-cymene 171
1-Bromo-4-phenylbutane 169
2,6-Diethylbromobenzene 171
$C_{10}H_{13}BrN_2O$
1-(4-Bromo-2,6-dimethylphenyl)-3-methylurea 546
1-(4-Bromophenyl)-3-isopropylurea 589
1-(4-Bromophenyl)-3-propylurea 538
$C_{10}H_{13}BrN_2O_2$
1-(4-Bromophenyl)-3-(2-methoxyethyl)urea 555
$C_{10}H_{13}BrN_2O_2S$
3,4-Dihydro-2-(methylsulfonamido)-isoquinolinium bromide 821
$C_{10}H_{13}BrN_2O_4$
Ethyl 5-bromo-3-methyl-4-(2-nitroethyl)-2-pyrrolecarboxylate 839
$C_{10}H_{13}BrO$
1-Bromo-3-(4-methoxyphenyl)-propane 176
DL-4-Bromo-*α*-propylbenzyl alcohol 205
$C_{10}H_{13}BrOS$
Dimethyl(phenacyl)sulfonium bromide 294
$C_{10}H_{13}BrO_2$
8-Bromo-2,3-bornanedione 51
$C_{10}H_{13}BrO_4$
Ethyl 3-bromo-5,6-dihydro-4,6-dimethyl-2*H*-pyran-5-carboxylate 96
$C_{10}H_{13}BrS$
4-Bromobutyl phenyl sulfide 209
$C_{10}H_{13}Br_2N$
N,N-Bis(2-bromoethyl)aniline 215
$C_{10}H_{13}Cl$
7-Chloro-*p*-cymene 169
1-Chloro-3-phenylbutane 169
α-Ethyl-4-methylbenzyl chloride 169
2,4,5-Trimethylbenzyl chloride 170
$C_{10}H_{13}ClHg$
2-Methyl-2-phenylpropylmercuric chloride 1139
$C_{10}H_{13}ClN_2O$
Butyric *N*2-(4-chlorophenyl)-hydrazide 494
1-(2-Chloroethyl)-3-(*p*-tolyl)urea 504
3-(2-Chloro-6-methylphenyl)-1,1-dimethylurea 593
3-(3-Chloro-2-methylphenyl)-1,1-dimethylurea 540
3-(3-Chloro-4-methylphenyl)-1,1-dimethylurea 597
3-(4-Chloro-2-methylphenyl)-1,1-dimethylurea 597
3-(5-Chloro-2-methylphenyl)-1,1-dimethylurea 597
1-(2-Chloro-6-methylphenyl)-3-ethylurea 540
1-(3-Chloro-2-methylphenyl)-3-ethylurea 540
1-(3-Chloro-4-methylphenyl)-3-ethylurea 597
1-(4-Chloro-2-methylphenyl)-3-ethylurea 597
1-(5-Chloro-2-methylphenyl)-3-ethylurea 597
1-(4-Chlorophenyl)-1-ethyl-3-methylurea 538
1-(2-Chlorophenyl)-3-isopropylurea 587
1-(3-Chlorophenyl)-3-isopropylurea 580
1-(4-Chlorophenyl)-3-isopropylurea 589
1-(2-Chlorophenyl)-3-propylurea 587
1-(3-Chlorophenyl)-3-propylurea 580
1-(4-Chlorophenyl)-3-propylurea 589
$C_{10}H_{13}ClN_2O_2$
2-Amino-5-chloro-*N*-(3-hydroxypropyl)benzamide 447
2-Chloro-*N,N*-diethyl-4-nitroaniline 276
3-(3-Chloro-2-methoxyphenyl)-1,1-dimethylurea 556
1-(3-Chloro-2-methoxyphenyl)-3-ethylurea 556
3-(3-Chloro-4-methylphenyl)-1-methoxy-1-methylurea 543
1-(2-Chlorophenyl)-3-(2-methoxyethyl)urea 555
1-(3-Chlorophenyl)-3-(2-methoxyethyl)urea 554
1-(4-Chlorophenyl)-3-(2-methoxyethyl)urea 519
$C_{10}H_{13}ClN_2O_2S$
1-(4-Chloro-2,5-dimethoxyphenyl)-3-methyl-2-thiourea 729
1-(5-Chloro-2,4-dimethoxyphenyl)-3-methyl-2-thiourea 729
$C_{10}H_{13}ClN_2O_3$
1-(4-Chloro-2,5-dimethoxyphenyl)-3-methylurea 516
1-(5-Chloro-2,4-dimethoxyphenyl)-3-methylurea 515
$C_{10}H_{13}ClN_4OS$
6-(3-Chloro-4-methyl)-1-methyl-2-thiobiurea 732
$C_{10}H_{13}ClN_4O_2S$
1-(*N*-(5-Chloro-2-methoxyphenyl)-carbamoyl)-4-methyl-3-thiosemicarbazide 729
$C_{10}H_{13}ClO$
4-*tert*-Butyl-2-chlorophenol 190
4-Chloro-5-isopropyl-*o*-cresol 192

$C_{10}H_{13}ClO$ (Cont'd)
1-(3-Chlorophenyl)-1-butanol 198
4-(2-Chlorophenyl)-1-butanol 198
DL-4-Chloro-α-propylbenzyl alcohol 205
1-Chloro-3-(p-tolyl)-2-propanol 197
3-Methoxy-3-phenylpropyl chloride 185

$C_{10}H_{13}ClO_2$
2-Chloro-1,4-diethoxybenzene 182
2-(4-Chloro-3,5-xylyloxy)-ethanol 203

$C_{10}H_{13}ClO_2S$
Benzyl 3-chloropropyl sulfone 772
3-Chloropropyl p-tolyl sulfone 772

$C_{10}H_{13}ClO_3$
3-(2-Chloro-6-methylphenoxy)-1,2-propanediol 198

$C_{10}H_{13}ClO_3S$
3-Chloropropyl p-toluenesulfonate 796

$C_{10}H_{13}ClS$
3-Chloro-1-methylpropyl phenyl sulfide 209
3-Chloropropyl m-tolyl sulfide 209

$C_{10}H_{13}Cl_2N_3O$
1-(2,5-Dichlorophenyl)-4-propylsemicarbazide 533

$C_{10}H_{13}F$
1,3-Diethyl-2-fluorobenzene 171

$C_{10}H_{13}FN_2O$
1-(2-Fluorophenyl)-3-isopropylurea 536
1-(4-Fluorophenyl)-3-isopropylurea 538

$C_{10}H_{13}FO$
DL-4-Fluoro-α-propylbenzyl alcohol 205

$C_{10}H_{13}F_3OSi$
3-(Trifluoromethyl)phenyl trimethylsilyl ether 1100

$C_{10}H_{13}HgNO_2$
4-(Dimethylamino)phenylmercuri acetate 1138

$C_{10}H_{13}IN_2$
3-Methyl-1-phenyl-2-imidazolinium iodide 255

$C_{10}H_{13}IN_2O_2$
1-Cyclohexyl-5-iodouracil 135

$C_{10}H_{13}IO_2$
1,2-Dimethoxy-3,4-dimethyl-5-iodobenzene 177
1,4-Dimethoxy-2,3-dimethyl-5-iodobenzene 176

$C_{10}H_{13}IO_3$
Ethylene glycol (4-iodophenyl)-methyl ether methyl ether 177

$C_{10}H_{13}N$
N-Allyl-N-methylaniline 220
N-Allyl-o-toluidine 219
11-Azatricyclo[4.4.1.0^{1,6}]undeca-3,8-diene 32
N-Benzylcyclopropylamine 242
N-Benzylideneisopropylamine 1063
N-Benzylidenepropylamine 1063
N-Crotylaniline 219
1,2-Dimethylindoline 252
2,3-Dimethylindoline (mixture of isomers) 250
1-Ethoxyisoindoline 247
N-Methallylaniline 219
6-Methyl-1,2,3,4-tetrahydroquinoline 250
1-Phenyl-1-cyclopropanemethylamine hydrochloride 241
1,2,3,4-Tetrahydroquinaldine 250

$C_{10}H_{13}NO$
3′,4′-Acetoxylidide 440
3′,5′-Acetoxylidide 440
N-Allyl-p-anisidine 218
N-Allyloxy-N-methylaniline 220
5′-Amino-2′,4′-dimethylacetophenone 294
2′,3′-Dimethylacetanilide 439
2′,4′-Dimethylacetanilide 440
N,N-Dimethyl-2-phenylacetamide 441
N-Ethylacetanilide 441
2′-Ethylacetanilide 438
4′-Ethylacetanilide 438
6-Methoxy-1,2,3,4-tetrahydroquinoline 247
N-Methylpropionanilide 441
Methyl 2,4,6-trimethyl-3-pyridyl ketone 961
N-Phenethylacetamide 430
2-Phenylbutyramide 429
4-Phenylbutyramide 437
N-Phenylisobutyramide 437
p-Propionotoluidide 438
N-Propylbenzamide 478
4-(Tetrahydropyran-4-yl)-pyridine 966
2-(p-Tolyl)oxazolidine 253
2,4,6-Trimethylbenzamide 477
4,N,N-Trimethylbenzamide 482
2′,4′,6′-Trimethylformanilide 440

$C_{10}H_{13}NOS$
S-Benzyl N,N-dimethylthiocarbamate 688
3-(Benzylthio)propionamide 421
S-Isopropyl N-phenylthiolcarbamate 688
2-Mercapto-p-propionotoluidide 420
3-Mercapto-p-propionotoluidide 420
2-Methyl-3-(phenylthio)-propionamide 421
N-Methyl-p-thioacetanisidide 736
S-Propyl N-phenylthiolcarbamate 688
O-(m-Tolyl) N,N-dimethyl-thiocarbamate 741

$C_{10}H_{13}NOSe$
N-(2-(Methylseleno)ethyl)-benzamide 1142

$C_{10}H_{13}NOSi$
3-(Trimethylsilyloxy)benzonitrile 1110

$C_{10}H_{13}NO_2$
Butyl nicotinate 964
1-(tert.-Butyl)-3-nitrobenzene 278
1-tert-Butyl-4-nitrobenzene 278
Butyl picolinate 962
(2,5-Dimethoxybenzylidene)-methylamine 1040
(3,4-Dimethoxybenzylidene)-methylamine 1040
3′-Ethoxyacetanilide 418
Ethyl N-benzylcarbamate 673
Ethyl 2-cyano-2-cyclopentylideneacetate 148
Ethyl N-methylcarbanilate 680
3-Ethylphenyl N-methylcarbamate 677
Ethyl N-(o-tolyl)carbamate 682
Ethyl N-(p-tolyl)carbamate 682
4′-Hydroxybutyranilide 413
α-(2-Hydroxyethylimino)-α-methyl-o-cresol 1028
3-Hydroxy-4′-methylpropionanilide 413
2-Isopropoxybenzamide 466
Isopropyl carbanilate 680
2-Isopropyl-5-methyl-4-nitrosophenol 257
3-(4-Methoxyphenyl)propionamide 418
Methyl 3-anilinopropionate 358
Methyl 2-cyano-2-(1-cyclohexen-1-yl)acetate 149
Methyl 2-(dimethylamino)benzoate 373
Methyl 4-(dimethylamino)benzoate 373
3-(2-Methyl-1,3-dioxolan-2-yl)aniline 219
Methyl N-(2-ethylphenyl)carbamate 681
α-Methyl-2-pyridinebutyric acid 963
Methyl N-(2,3-xylyl)carbamate 683
Methyl N-(2,4-xylyl)carbamate 685
Methyl N-(2,5-xylyl)carbamate 685
Methyl N-(2,6-xylyl)carbamate 683
DL-Phenylalanine methyl ester hydrochloride 358
2-Phenylglycine ethyl ester hydrochloride 358
Propyl N-phenylcarbamate 680
N-Propylsalicylamide 462
4-Tolyl N-ethylcarbamate 682

$C_{10}H_{13}NO_2S$
3-Anilinotetrahydrothiophene-1,1-dioxide 778
α,α-Dimethyl-4-nitrobenzyl methyl sulfide 270
3-Methyl-4-(methylthio)phenyl N-methylcarbamate 668
4-Phenylthiomorpholine-1,1-dioxide 778
1-(p-Tosyl)azetidine 828

$C_{10}H_{13}NO_3$
2-Amino-3′,4′-dimethoxyacetophenone hydrochloride 291
O-Benzyl-DL-serine 348
N-Benzyl-L-serine 348
1-Butoxy-2-nitrobenzene 267
4-sec-Butyl-2-nitrophenol 264
Butyl 4-nitrophenyl ether 269
3-Carbamoylbicyclo(2.2.2)oct-5-ene-2-carboxylic acid, ammonium salt 101
2′,4′-Dimethoxyacetanilide 417
3′,4′-Dimethoxyacetanilide 414
exo-2,7-Dimethyl-4,7-epoxy-2,3,3a,4,5,6,7,7a-octahydro-1H-isoindole-1,3-dione 133
Ethyl N-(o-anisyl)carbamate 653
Ethyl 2,4-dimethyl-5-formyl-3-pyrrolecarboxylate 840
Ethyl 3,5-dimethyl-4-formyl-2-pyrrolecarboxylate 839
Ethyl N-methoxy-N-phenylcarbamate 674
Ethyl N-(4-methoxyphenyl)-carbamate 654
Ethyl 6-methyl-3-pyridyloxyacetate 937
β-Hydroxy-p-acetophenetidide 411
4′-(2-Hydroxyethoxy)acetanilide 411
N-(2-Hydroxyethyl)-phenoxyacetamide 411
β-(2-Hydroxy-3-methylphenyl)-alanine 348
β-(2-Hydroxy-4-methylphenyl)-alanine 348
β-(2-Hydroxy-5-methylphenyl)-alanine 348
Isopropyl N-hydroxy-N-phenylcarbamate 674
2-Methoxyethyl N-phenylcarbamate 649
4-Methoxyhydrocinnamo-hydroxamic acid 417
Methyl 5-formyl-4-methyl-3-pyrrolepropionate 839
Methyl N-(2-methoxy-5-methylphenyl)carbamate 655
DL-α-Methyl-m-tyrosine monohydrate 348
DL-α-Methyltyrosine hydrochloride 348
Propyl N-hydroxy-N-phenylcarbamate 674
DL-Tyrosine methyl ester hydrochloride 354

$C_{10}H_{13}NO_3S$
3-Benzylsulfonylpropionamide 782
Ethyl 5-acetyl-2-amino-4-methyl-3-thiophenecarboxylate 856

$C_{10}H_{13}NO_4$
N-(tert-Butoxycarbonyl)-L-norvaline dicyclohexylamine salt 122
4,6-Diethoxy-2-pyridinecarboxylic acid 936
L-3,4-Dihydroxyphenylalanine methyl ester hydrochloride 353
Ethyl 2-carboxy-4,5-dimethyl-3-pyrrolecarboxylate 838
Ethyl 5-carboxy-2,4-dimethyl-3-pyrrolecarboxylate 840
Ethyl 5-carboxy-3,4-dimethyl-2-pyrrolecarboxylate 841
3-Ethyl 2-methyl 4-methyl-2,3-pyrroledicarboxylate 838
N-Furfurylglutaramic acid 854
N-(2-Hydroxyethyl)-2-(2-hydroxyphenoxy)acetamide 410
L-3-Methoxytyrosine hydrate 348
Methyl 4,5-dimethoxyanthranilate 369
Methyl N-(2,5-dimethoxyphenyl)-carbamate 653
Methyl 1,2-dimethyl-5-formyl-6-oxo-1,4,5,6-tetrahydro-3-pyridinecarboxylate 131
mono-Methyl 3-(1-pyrrolyl)glutarate 841
2,4,6-Trimethoxybenzaldehyde oxime 1009
3,4,5-Trimethoxybenzamide 464

$C_{10}H_{13}NO_4S$
tert-Butyl 3-nitrophenyl sulfone 779
Ethyl 4-(methylsulfonamido)-benzoate 824
Ethyl 4-(methylsulfonamido)-benzoate 823
Ethyl N-(methylsulfonyl)-N-phenylcarbamate 824
Ethyl N-(phenylsulfonylmethyl)-carbamate 781
1-Methyl-1-nitroethyl p-tolyl sulfone 780

$C_{10}H_{13}NO_4S_2$
2,3-Dimethylbenzothiazolium methyl sulfate 916

$C_{10}H_{13}NO_5$
3,4-Dimethoxy-α-(nitromethyl)-benzyl alcohol 263

$C_{10}H_{13}NO_5S$
4-(2-(Methylsulfonamido)ethoxy)-benzoic acid 814

$C_{10}H_{13}NS$
N-Phenylthiobutyramide 746

$C_{10}H_{13}NSe$
2-Methyl-3-phenylselenazolidine 213

$C_{10}H_{13}N_3$
5-Amino-1-ethyl-2-methylbenzimidazole dihydrochloride 903
2-Amino-1-isopropylbenzimidazole 904
2-Aminomethyl-5,6-dimethylbenzimidazole hydrochloride 904

$C_{10}H_{13}N_3O$
4′-Methylacetophenone semicarbazone 574
Propiophenone semicarbazone 578

$C_{10}H_{13}N_3OS$
1-(4-Acetamidophenyl)-3-methyl-2-thiourea 740
3-(4-Methoxybenzylideneamino)-2-methyl-2-thiopseudourea 1048

$C_{10}H_{13}N_3O_2$
2-Butanone 4-nitrophenylhydrazone 276
1-(Ethoxycarbonyl)-4-phenylsemicarbazide 577
4′-(3-Methylureido)acetanilide 579
5-Nitro-2-piperidinopyridine 943
1-Propyl-3-nicotinoylurea 949

$C_{10}H_{13}N_3O_2S$
3-Amino-1-(p-toluenesulfonyl)-3-pyrazoline 817
4-Ethyl-1-(4-hydroxybenzoyl)-3-thiosemicarbazide 734

$C_{10}H_{13}N_3O_2S_2$
N,1-Dimethyl-2-methylthio-5-benzimidazolesulfonamide 903

$C_{10}H_{13}N_3O_3$
N-(Benzyloxycarbonyl)glycine hydrazide 664
2,3-Dimethoxybenzaldehyde semicarbazone 1040
1,1-Dimethyl-3-(2-methyl-4-nitrophenyl)urea 568
1,1-Dimethyl-3-(4-methyl-3-nitrophenyl)urea 565
1-Ethyl-3-(2-methyl-4-nitrophenyl)urea 568
1-Isopropyl-3-(3-nitrophenyl)urea 564
1-Isopropyl-3-(4-nitrophenyl)urea 568
1-(3-Nitrophenyl)-3-propylurea 564
1-(4-Nitrophenyl)-3-propylurea 568

$C_{10}H_{13}N_3O_4$
N-Butyl-2,4-dinitroaniline 273
N,N-Diethyl-2,4-dinitroaniline 274
1-(2-Methoxyethyl)-3-(4-nitrophenyl)urea 552
1-Methoxy-1-methyl-3-(2-methyl-5-nitrophenyl)urea 564

$C_{10}H_{13}N_3O_6S$
2,6-Dinitro-4-(methylsulfonyl)-N-propylaniline 777

$C_{10}H_{13}N_3O_8S_2$
2′,6′-Dinitro-N-ethyl-4′-(methylsulfonyl)-methanesulfonanilide 818

$C_{10}H_{13}N_3S$
1-Isopropylidene-4-phenyl-3-thiosemicarbazide 771
p-Tolylacetaldehyde thiosemicarbazone 1072

$C_{10}H_{13}N_4O$
3-(Butyryl)pyridine semicarbazone 948

$C_{10}H_{13}N_4O_7P$
2′-Deoxyinosine 5′-monophosphate, disodium salt dihydrate ... 920
$C_{10}H_{13}N_5$
N^6-(Δ2-Isopentenyl)-adenine ... 923
6-Piperidinopurine ... 924
Triacanthine ... 923
$C_{10}H_{13}N_5O_2$
7-Amino-1,3-diethyl-2,4(1*H*,3*H*)-pteridinedione ... 977
$C_{10}H_{13}N_5O_3S$
1-Methyl-6-(4-methyl-2-nitrophenyl)-2-thiobiurea ... 739
$C_{10}H_{13}N_5O_5$
Adenosine-*N*1-oxide ... 1005
$C_{10}H_{13}N_5O_5S$
8-Mercaptoguanosine ... 920
$C_{10}H_{14}$
Tricyclo[5.2.1.0^{2,6}]dec-8-ene ... 2
$C_{10}H_{14}AsNO_2$
3-(Dimethylarsino)phenyl *N*-methylcarbamate methiodide ... 1127
$C_{10}H_{14}AsNO_6$
4-Butoxy-3-nitrobenzenearsonic acid ... 1126
$C_{10}H_{14}BeO_4$
Beryllium acetylacetonate ... 70
$C_{10}H_{14}BrNO$
2-(*N*-(2-Bromoethyl)anilino)-ethanol .. 215
$C_{10}H_{14}BrNO_2$
2-Bromo-4,5-dimethoxy-*N*-methylbenzylamine ... 236
$C_{10}H_{14}Br_2O$
3,8-Dibromocamphor ... 51
$C_{10}H_{14}ClN$
3-Chloro-*N*,*N*-diethylaniline ... 220
N-(2-Chloroethyl)-*N*-methylbenzylamine hydrochloride .. 236
4-Chloro-α-ethyl-*N*-methylbenzylamine hydrochloride .. 243
$C_{10}H_{14}ClNO$
2-(4-Chlorophenoxy)-*N*,1-dimethylethylamine ... 240
2-(4-Chloro-*m*-tolyloxy)-*N*-methylethylamine ... 240
$C_{10}H_{14}ClN_3O$
4-(3-Chloro-4-methylphenyl)-1,1-dimethylsemicarbazide ... 579
4-(4-Chloro-2-methylphenyl)-1,1-dimethylsemicarbazide ... 579
4-(5-Chloro-2-methylphenyl)-1,1-dimethylsemicarbazide ... 533
4-Diazo-*N*-ethyl-*N*-(2-hydroxyethyl)-aniline, zinc chloride double salt ... 1094
$C_{10}H_{14}ClN_3O_2$
4-(3-Chloro-2-methoxyphenyl)-1,1-dimethylsemicarbazide ... 519
$C_{10}H_{14}ClN_3O_4$
2,2′-(4-Chloro-6-nitro-1,3-phenylenediamino)-diethanol ... 259
4,5-Dihydro-1-methyl-3-(*p*-tolyl)-1*H*-1,2,3-triazolium perchlorate ... 256
$C_{10}H_{14}ClN_3O_5$
4,5-Dihydro-1-(4-methoxyphenyl)-3-methyl-1*H*-1,2,3-triazolium perchlorate ... 254
$C_{10}H_{14}Cl_4Si_2$
(2,3,5,6-Tetrachloro-1,4-phenylene)-bis(dimethylsilane) ... 1100
$C_{10}H_{14}CuO_6$
Methyl acetoacetate, copper(II) derivative ... 1142
$C_{10}H_{14}F_{12}Si_2$
Perfluorohexamethylenebis-(dimethylsilane) ... 1097
$C_{10}H_{14}N$
N-Benzylisopropylamine ... 243
$C_{10}H_{14}N_2$
Acetophenone dimethylhydrazone ... 235
Butyraldehyde phenylhydrazone ... 235
1,2-Cyclohexanediacetonitrile ... 150
1-Phenylpiperazine dihydrochloride. 250
4-Pyrrolidinoaniline ... 250
$C_{10}H_{14}N_2O$
N-*tert*-Butyl-*N*-nitrosoaniline ... 257
4′-(Dimethylamino)acetanilide ... 427
1,1-Dimethyl-3-(3-methylphenyl)urea ... 580
1,1-Dimethyl-3-(*o*-tolyl)urea ... 587
1,1-Dimethyl-3-(*p*-tolyl)urea ... 589
3′-(Ethylamino)acetanilide ... 425
1-(2-Ethylphenyl)-3-methylurea ... 587
1-Ethyl-1-(*o*-tolyl)urea ... 587
1-Ethyl-3-(*o*-tolyl)urea ... 587
1-Ethyl-3-(*m*-tolyl)urea ... 580
1-Ethyl-3-(*p*-tolyl)urea ... 589
3-Hydroxy-2-(pyrrolidinomethyl)-pyridine ... 932
1-Isopropyl-3-phenylurea ... 585
6-Methyl-2-pyridinealdoxime *O*-propyl ether ... 1054
1-Methyl-3-(2,3-xylyl)urea ... 593
1-Methyl-3-(2,4-xylyl)urea ... 597
1-Methyl-3-(2,5-xylyl)urea ... 597
1-Methyl-3-(2,6-xylyl)urea ... 593
1-Methyl-3-(3,4-xylyl)urea ... 597
1-Methyl-3-(3,5-xylyl)urea ... 602
4-Morpholinoaniline hydrochloride .. 248
2-(Morpholinomethyl)-pyridine ... 944
1-Phenyl-3-propylurea ... 585
(2,4,6-Trimethylphenyl)urea ... 571
$C_{10}H_{14}N_2OS$
1,1-Dimethyl-3-(2-methoxyphenyl)-2-thiourea ... 736
1,1-Dimethyl-3-(4-methoxyphenyl)-2-thiourea ... 736
1,1-Dimethyl-3-(4-(methylthio)-phenyl)urea ... 561
1-(4-Hydroxyphenyl)-3-propyl-2-thiourea ... 735
$C_{10}H_{14}N_2O_2$
3′-Amino-*p*-acetophenetidide ... 414
1-(*m*-Anisyl)-3-ethylurea ... 557
1-(*p*-Anisyl)-3-ethylurea ... 559
2-Butyl-5-nitroaniline ... 271
N-Butyl-4-nitroaniline ... 276
4-(*tert*-Butyl)-3-nitroaniline ... 271
N,*N*-Diethylnicotinamide 1-oxide ... 1006
N,*N*-Diethyl-3-nitroaniline ... 275
N,*N*-Diethyl-4-nitroaniline ... 276
1,1-Dimethyl-3-(2-methoxyphenyl)urea ... 554
1,1-Dimethyl-3-(3-methoxyphenyl)urea ... 557
1,1-Dimethyl-3-(4-methoxyphenyl)urea ... 559
1-Ethyl-3-(2-methoxyphenyl)urea ... 554
3-Hydroxy-2-(morpholinomethyl)-pyridine ... 932
1-(2-Hydroxy-1-propyl)-3-phenylurea. 548
1-(2-Methoxyethyl)-3-phenylurea ... 554
1-(2-Methoxy-5-methylphenyl)-3-methylurea ... 556
Methyl 2-cyano-3-piperidinoacrylate ... 147
4-Nitro-2,3,5,6-tetramethylaniline ... 271
5,5-Pentamethylene-3-vinylhydantoin ... 135
N-(Propoxymethyl)-nicotinamide ... 939
$C_{10}H_{14}N_2O_2O_2$
1-(*o*-Anisyl)-3-ethylurea ... 555
$C_{10}H_{14}N_2O_2S$
1-(2,5-Dimethoxyphenyl)-3-methyl-2-thiourea ... 735
Ethyl 6,7-dihydro-5*H*-thiazolo(3,2-a)-pyrimidine-3-acetate hydrochloride. 90
5-Ethyl-5-(2-methylallyl)-2-thiobarbituric acid ... 141
1-Methoxy-1-methyl-3-(4-(methylthio)phenyl)urea ... 561
$C_{10}H_{14}N_2O_2S_2$
1-Ethyl-3-(*p*-tolylsulfonyl)-2-thiourea. 826
$C_{10}H_{14}N_2O_3$
1-(2,5-Dimethoxyphenyl)-3-methylurea ... 556
N-(3-Methoxypropyl)-4-nitroaniline ... 267
$C_{10}H_{14}N_2O_3S$
1,1-Dimethyl-3-(*p*-tolylsulfonyl)urea .. 825
1-Ethyl-3-(*p*-tolylsulfonyl)-urea ... 825
3′-Sulfamoylbutyranilide ... 824
$C_{10}H_{14}N_2O_3S_2$
1-(2-Mercaptoethyl)-3-(*p*-tolylsulfonyl)-urea ... 815
$C_{10}H_{14}N_2O_4$
2,3-Dimethoxy-5-(2-nitroethyl)-aniline ... 257
3,4,5-Trimethoxybenzoic acid hydrazide ... 497
$C_{10}H_{14}N_2O_4S$
Propyl *N*-(4-sulfamoylphenyl)-carbamate ... 824
$C_{10}H_{14}N_2O_5S$
4-Amino-3-nitrophenyl 1-butanesulfonate ... 798
$C_{10}H_{14}N_2O_7$
5-(Hydroxymethyl)uridine ... 130
$C_{10}H_{14}N_2S$
3-Benzyl-1,1-dimethyl-2-thiourea ... 741
1,1-Dimethyl-3-(*o*-tolyl)-2-thiourea ... 748
1,1-Dimethyl-3-(*m*-tolyl)-2-thiourea ... 747
1,1-Dimethyl-3-(*p*-tolyl)-2-thiourea ... 741
1-(2-Ethylphenyl)-3-methyl-2-thiourea ... 748
1-Ethyl-3-(*m*-tolyl)-2-thiourea ... 747
1-Ethyl-3-(*p*-tolyl)-2-thiourea ... 748
1-Isopropyl-3-phenyl-2-thiourea ... 747
1-Methyl-3-(2,6-xylyl)-2-thiourea ... 749
1-Phenyl-3-propyl-2-thiourea ... 747
$C_{10}H_{14}N_3O$
4-(2-Chloro-6-methylphenyl)-1,1-dimethylsemicarbazide ... 578
$C_{10}H_{14}N_4$
1-(1-Benzylethylideneamino)-guanidine nitrate ... 1072
$C_{10}H_{14}N_4O$
2-(Cyclopentylidenehydrazino)-4-hydroxy-6-methylpyrimidine ... 971
$C_{10}H_{14}N_4OS$
1-(*N*-(2,6-Dimethylphenyl)-carbamoyl)-3-thiosemicarbazide ... 745
$C_{10}H_{14}N_4OSi$
4,5-Dihydro-1-phenyl-4-trimethylsilyl-1*H*-tetrazol-5-one ... 1110
$C_{10}H_{14}N_4O_2$
2,4-Dicyano-3-ethyl-3-methylglutaramide ... 105
1,3-Diethyl-7-methylxanthine ... 924
1,7-Diethyl-3-methylxanthine ... 923
3-Ethyl-1-propylxanthine ... 924
4-Isopropyl-1-(4-pyridylcarbonyl)-semicarbazide ... 952
$C_{10}H_{14}N_4O_2S$
1-(2-Ethoxyphenyl)-5-thiobiurea ... 737
1-(4-Ethoxyphenyl)-5-thiobiurea ... 737
3-Isobutyl-8-mercapto-1-methylxanthine ... 922
4-Isopropyl-1-(4-nitrophenyl)-3-thiosemicarbazide ... 740
1-(2-Methoxy-5-methylphenyl)-5-thiobiurea ... 736
$C_{10}H_{14}N_4O_3$
6-Amino-3-ethyl-1-(2-methylallyl)-5-nitrosouracil ... 132
1,1-Dimethyl-4-(4-methyl-3-nitrophenyl)semicarbazide ... 564
1-Ethyl-3-(2-methoxyethyl)xanthine .. 921
7-(2-Hydroxypropyl)theophylline ... 866
$C_{10}H_{14}N_4O_3S$
1-(2,5-Dimethoxyphenyl)-5-thiobiurea ... 735
$C_{10}H_{14}N_4O_4S$
1,1′-Thiobis(5,5-dimethylhydantoin).. 702
$C_{10}H_{14}N_4O_6$
Ethyl 6-amino-1-ethyl-5-nitro-3-uracilacetate ... 132
$C_{10}H_{14}N_4S$
6-(Pentylthio)purine ... 922
$C_{10}H_{14}N_5O_7P$
Adenosine 2′-monophosphoric acid. 920
Adenosine 2′(3′)-monophosphoric acid ... 920
Adenosine-5′-monophosphoric acid, disodium salt ... 920
$C_{10}H_{14}N_5O_8P$
Guanosine 2′(3′)-monophosphate, disodium salt monohydrate, mixture of isomers ... 922
$C_{10}H_{14}N_{10}$
trans-1,2-Cyclobutanedicarboguanamine ... 979
$C_{10}H_{14}O$
5-Allyl-4,5-dimethyl-2-cyclopenten-1-one ... 53
Benzyl isopropyl ether ... 186
α,α-Dimethylbenzyl methyl ether ... 178
3,4-Dimethyl-6-ethylphenol ... 192
α,α-Dimethylphenethyl alcohol ... 206
2,5-Dimethylphenethyl alcohol ... 208
α-Ethyl-α-methylbenzyl alcohol ... 204
α-Ethyl-4-methylbenzyl alcohol ... 205
Ethyl phenethyl ether ... 187
Ethyl 3,5-xylyl ether ... 182
8-Hydroxytricyclo[5.2.1.0^{2,6}]-4-decene ... 13
3-Isopropylanisole ... 178
α-Isopropylbenzyl alcohol ... 204
1-Phenyl-1-butanol ... 204
1-Phenyl-2-butanol ... 206
4-Propylanisole ... 181
6-Propyl-*o*-cresol ... 192
α,α,4-Trimethylbenzyl alcohol ... 205
α,2,5-Trimethylbenzyl alcohol ... 205
2,4,5-Trimethylbenzyl alcohol ... 205
$C_{10}H_{14}O_2$
1,4-Benzenedimethanol dimethyl ether ... 187
1-Benzyloxy-2-propanol ... 201
3-(Benzyloxy)propanol ... 201
1,5-Decalindione ... 55
DL-*trans*-2,3-Decalindione ... 54
1,2-Diethoxybenzene ... 178
1,3-Diethoxybenzene ... 178
1,4-Diethoxybenzene ... 179
1,4-Dimethoxy-2,3-dimethylbenzene. 179
3,4-Epoxytricyclo[5.2.1.0^{2,6}]decanol .. 20
4-Ethoxyphenethyl alcohol ... 202
α-Ethyl-4-methoxybenzyl alcohol ... 201
Ethyl 5-norbornene-2-carboxylate (mixture of *endo* and *exo*) ... 81
3-Isopropyl-6-methylcatechol ... 193
(+)-1-Methoxy-1-phenyl-2-propanol .. 201
Methyl 1-octen-3-yne-1-carboxylate .. 1018
1-Phenyl-1,4-butanediol ... 199
Spiro(4.5)decan-1,6-dione ... 55
Spiro(4.5)decan-1,7-dione ... 55
$C_{10}H_{14}O_2S$
Butyl phenyl sulfone ... 783
tert-Butyl phenyl sulfone ... 783
Isopropyl *p*-tolyl sulfone ... 785
4-(2-(Methylthio)ethoxy)anisole ... 210
$C_{10}H_{14}O_2Si$
4-(Trimethylsilyl)benzoic acid ... 1109
$C_{10}H_{14}O_3$
3-Benzyloxy-1,2-propanediol ... 202
3,5-Diethoxyphenol ... 191
2,6-Dimethoxyphenethyl alcohol ... 201
1-Formyl-3-cyclohexene-1-propionic acid ... 65
Furfuryl 2,2-dimethylpropionate ... 849
(-)-*cis*-Isoketopinic acid ... 65
4-(Methoxymethoxy)-α-methylbenzyl alcohol ... 200
7-Methyl-1,4-dioxaspiro(4.5)dec-7-ene-8-carboxaldehyde ... 60
2-Methyl-8-oxaspiro(4.5)decane-7,9-dione ... 98
8-Methyl-2-oxaspiro(4.5)decane-1,3-dione ... 98
$C_{10}H_{14}O_3S$
Butyl benzenesulfonate ... 800
4-Phenyl-1-butanesulfonic acid sodium salt ... 795
$C_{10}H_{14}O_4$
6-Carboxy-6-methyl-3-cyclohexene-1-acetic acid ... 66
2-(2-Cyclohexen-1-yl)-2-methylmalonic acid ... 65
Diethyl 2-(2-propynyl)malonate ... 1018
4,5-Dimethoxy-1,2-benzenedimethanol ... 199
Dimethyl 4-cyclohexene-1,2-dicarboxylate ... 81
3,5-Dimethyl-1-cyclohexene-1,2-dicarboxylic acid ... 65
Ethyl 2,4-dioxo-6-methyl-1-cyclohexanecarboxylate ... 92
3-Methoxy-5-methylhexahydrophthalic

$C_{10}H_{14}O_4$ (Cont'd)
anhydride 388
3-(2-Methoxyphenoxy)-1,2-propanediol 202
Methyl 2,2-dimethyl-4,6-dioxo-1-cyclohexanecarboxylate 93
Methyl 3,6-dimethyl-4-hydroxy-2-oxo-3-cyclohexene-1-carboxylate 91
4-Oxo-4-(2-oxocyclohexyl)-butyric acid 65
Resorcinol bis(2-hydroxyethyl) ether 201
1,1,2,2-Tetraacetylethane 51
1,2,3,5-Tetramethoxybenzene 178
$C_{10}H_{14}O_4S$
Benzylacetone sodium bisulfite compound 792
2-Methoxyethyl *p*-toluenesulfonate 797
$C_{10}H_{14}O_5$
2,2-Dimethyl-hexahydro-7-hydroxy-4,7-methano-1,3-dioxolo(4,5-c)-oxepin-6-one 95
3-(Methoxycarbonyl)propionic anhydride 75
$C_{10}H_{14}O_5S_2$
2,4-Dihydroxy-5-(isobutylthio)-benzenesulfonic acid 787
$C_{10}H_{14}O_6$
Ethyl 2,3-diacetoxycrotonate 78
Ethyleneglycol bis-(2,3-epoxybutyrate) 89
$C_{10}H_{14}O_8S_2$
1,2-Ethyleneglycol bis(dithioglycolate) 95
$C_{10}H_{14}S$
Butyl phenyl sulfide 211
tert-Butyl phenyl sulfide 211
4-*tert*-Butylthiophenol 209
Isobutyl phenyl sulfide 211
Mesityl methyl sulfide 213
Propyl *p*-tolyl sulfide 212
$C_{10}H_{14}S_2$
α,α'-Bis-(methylthio)-*p*-xylene 212
$C_{10}H_{14}S_3$
2,4,6-Tris(methylthio)toluene 211
$C_{10}H_{15}AsO_6$
4-(2-(2-Hydroxyethoxy)ethoxy)-benzenearsonic acid 1126
$C_{10}H_{15}Br$
2-(Bromomethyl)-6,6-dimethylbicyclo(3.1.1)hept-2-ene 6
$C_{10}H_{15}BrN_2O$
1-Butyl-3-carbamoylpyridinium bromide 964
1-Butyl-4-carbamoylpyridinium bromide 951
$C_{10}H_{15}BrO$
2-Bromoallyl 2-cyclohexen-1-ylmethyl ether 18
L-3-Bromocamphor 55
$C_{10}H_{15}BrO_3$
2-(Bromomethyl)-2,3-dimethyl-4-oxocyclohexanecarboxylic acid 64
$C_{10}H_{15}Cl$
Chlorocamphene 6
$C_{10}H_{15}ClN_2$
2-Chloro-*N*[1],*N*[1]-diethyl-1,4-phenylenediamine monohydrochloride 221
$C_{10}H_{15}ClN_2O_2$
(3-Nitrobenzyl)trimethylammonium chloride 279
$C_{10}H_{15}Cl_2NO$
3-(2,2-Dichloroacetyl)-3-azabicyclo(3.2.2)nonane 107
$C_{10}H_{15}Cl_2N_3O_6$
2,2'-(Iminobiscarboxamido)-diethanol bis(2-chloroacetate) 126
$C_{10}H_{15}Cl_3Si_2$
2,4,6-Trichloro-1,3-phenylenebis(dimethylsilane) 1100
$C_{10}H_{15}FNO_3P$
Diethyl 4-fluorophenylamidophosphate 833
$C_{10}H_{15}F_3O_2$
Cyclooctyl trifluoroacetate 82
$C_{10}H_{15}IN_2O_2$
(3-Hydroxycarbamoylphenyl)-trimethylammonium iodide 471
$C_{10}H_{15}N$
2-Amino-*p*-cymene 219
Benzylisopropylamine hydrobromide 243
Benzylisopropylamine salicyclate 243
N-*sec*-Butylaniline 219
2-(Cyclohexylidene)-butyronitrile 150
2-(1-Ethylpropyl)-pyridine 967
4-Isopropylbenzylamine 241
N-Isopropylbenzylamine hydrochloride 243
4-Isopropyl-3-methylaniline hydrochloride 219
N-Isopropyl-*m*-toluidine 219
1-Methyl-3-phenylpropylamine hydrochloride 241
$C_{10}H_{15}NO$
2-Amino-4-*sec*.-butylphenol 217
4-Amino-2-isopropylanisole 218
4-Aminothymol hydrochloride 217
4-Azatricyclo[4.3.1.1^{3,8}]undecan-5-one 135
(-)-*N*-Benzylalaninol 238
(+)-*N*-Benzylalaninol 238
DL-1-Benzylamino-2-propanol 238
DL-2-Benzylamino-1-propanol 238
3-Benzylamino-1-propanol 238
3-Benzylamino-1-propanol hydrochloride 238
2-Benzyloxypropylamine 239
4-(2-(Dimethylamino)ethyl)phenol sulfate dihydrate 238
2-(2-Ethoxyethyl)-6-methylpyridine 940
2-(2-Ethylanilino)ethanol 217
N-Ethyl-4-methoxybenzylamine 240
Eucarvone oxime 1008
5-Methoxy-1-pyrrolidino-1-pentene 838
(+)-2-Methylamino-3-phenyl-1-propanol 238
4-(2-(Methylamino)propyl)phenol hemisulfate 239
4α-Methyl-1,2,3,4,4a,5,6,7-octahydro-2-quinolinone 135
N-Methyl-2-(*o*-tolyloxy)-ethylamine 240
N-Methyl-2-(*m*-tolyloxy)-ethylamine 240
N-(2-Norbornyl)-acrylamide 105
3-Pentyloxypyridine 940
Perillartine 1013
2-Phenoxydiethylamine 243
α,2,4,6-Tetramethyl-3-pyridinemethanol 934
$C_{10}H_{15}NO_2$
2-(1-Aminoethylidene)-5,5-dimethyl-1,3-cyclohexanedione 57
2-(Benzylmethylamino)-ethanol *N*-oxide 238
Camphorimide (1,8,8-trimethyl-3-azabicyclo(3.2.1)octane-2,4-dione) 135
N-Cyclohexylsuccinimide 125
2,3-Dimethoxyphenethylamine picrate 265
Ethyl 2-cyano-3-ethyl-2-pentenoate 148
Ethyl 3-ethyl-5-methyl-2-pyrrolecarboxylate 841
DL-*N*-Ethylnorphenylephrine hydrochloride 237
Ethyl 3-nortricyclylcarbamate 115
Ethyl 3,4,5-trimethyl-2-pyrrolecarboxylate 842
3-(1-Hydroxypentyl)pyridine-*N*-oxide 1005
4-(2-(Methylamino)propoxy)phenol hydrochloride 237
3,3-Pentamethyleneglutarimide 126
5,8,8-Trimethyl-6-azabicyclo(3.2.1)-octane-4,7-dione 134
$C_{10}H_{15}NO_2S$
N-Butylbenzenesulfonamide 827
2,3,5,6-Tetramethylbenzene-sulfonamide 827
2',4',5'-Trimethylmethanesulfonanilide 829
$C_{10}H_{15}NO_3$
3-*tert*-Butyl-5-isoxazolepropionic acid 871
3-Carbamoylbicyclo(2.2.2)octane-2-carboxylic acid, ammonium salt 101
(-)-4-Cyano-5,5-dimethyl-6-oxoheptanoic acid benzyl isothiuronium salt 738
N-Cyclohexylmaleamic acid 122
3,5-Dimethoxy-4-hydroxyphenethylamine 237
(-)-3-(1,1-Dimethyl-2-oxopropyl)-2,6-piperidinedione 125
Ethyl 3-oxo-2-(2-pyrrolidinylidene)-butyrate 93
Ethyl 3-quinuclidinone-2-carboxylate hydrochloride 92
$C_{10}H_{15}NO_3O_3$
(+)-3-Nitrocamphor, sodium salt 53
$C_{10}H_{15}NO_3S$
S-Butyl 2,6-dioxopiperidine-4-thiocarboxylate 125
2-(*N*-Ethylanilino)-1-ethanesulfonic acid, potassium salt 793
3-(*N*-Methylanilino)-1-propanesulfonic acid 789
$C_{10}H_{15}NO_4$
Dimethyl 1-amino-1-cyclohexene-2,3-dicarboxylate 87
Dimethyl 2-amino-1-cyclohexene-1,5-dicarboxylate 87
5-(Hydroxyamino)-2-phenyl-1,3-dioxane-5-methanol 245
$C_{10}H_{15}NO_4S$
1,2-Dimethyl-5-vinylpyridinium methosulfate 958
$C_{10}H_{15}NO_4S_2$
S-(Ethoxycarbonylmethyl) 4-morpholine-1,1-dithioglyoxylate 357
$C_{10}H_{15}N_2O_8P$
Thymidine-5'-monophosphate, disodium salt dihydrate 131
$C_{10}H_{15}N_3$
2-Methyl-3-piperidinopyrazine sulfate 977
$C_{10}H_{15}N_3O$
1,1-Dimethyl-4-(*o*-tolyl)-semicarbazide 578
1,1-Dimethyl-4-(*m*-tolyl)-semicarbazide 577
1,1-Dimethyl-4-(*p*-tolyl)-semicarbazide 578
4-Hydroxy-6-methyl-2-piperidinopyrimidine 971
$C_{10}H_{15}N_3OS$
1-Cyclohexyl-3-(2-thiazolyl)urea 877
1,1-Dimethyl-4-(4-(methylthio)-phenyl)semicarbazide 561
$C_{10}H_{15}N_3O_2$
6-Amino-3-ethyl-1-(2-methylallyl)-uracil 133
4-(*o*-Anisyl)-1,1-dimethylsemicarbazide 553
4-(*m*-Anisyl)-1,1-dimethylsemicarbazide 557
4-(*p*-Anisyl)-1,1-dimethylsemicarbazide 557
5-Butyl-2-furaldehyde semicarbazone 1070
$C_{10}H_{15}N_3O_3S$
Ethyl 4-hydroxy-2-(2-(methylthio)-ethylamino)-5-pyrimidinecarboxylate 970
1-Propyl-3-(4-sulfamoylphenyl)urea 826
$C_{10}H_{15}N_3O_4$
Ethyl 4-hydroxy-2-(*N*-(2-hydroxyethyl)-*N*-methyl-amino)-5-pyrimidinecarboxylate 969
Ethyl 4-hydroxy-2-(2-methoxyethylamino)-5-pyrimidinecarboxylate 969
$C_{10}H_{15}N_3O_4S$
N-Ethyl-4-ethylamino-3-nitrobenzenesulfonamide 817
$C_{10}H_{15}N_3S$
1-Butyl-3-(2-pyridyl)-2-thiourea 965
$C_{10}H_{15}N_5$
*N*1-Phenethylbiguanide hydrochloride 725
$C_{10}H_{15}N_5O$
N[1]-(4-Ethoxyphenyl)biguanide hydrochloride 724
$C_{10}H_{15}N_5O_4$
Glycyl-L-histidylglycine 866
$C_{10}H_{16}$
Cyclopentylidenecyclopentane 2
1,5-Dimethyl-5-ethyl-1,3-cyclohexadiene 3
7,7-Dimethyl-2-methylenebicyclo(3.1.1)heptane 2
3,7-Dimethyl-1,3,6-octatriene 2
α-Pyronene 3
3-Pyronene 3
$C_{10}H_{16}Cl_3NO_3$
2,2,2-Trichloroethyl 4-hydroxy-1-piperidinepropionate 89
$C_{10}H_{16}Cl_6O_4$
1,1'-(Hexamethylenedioxy)bis(2,2,2-trichloroethanol) 19
$C_{10}H_{16}F_3O_2Tl$
6-Methyl-1,1,1-trifluoro-2,4-heptanedione, dimethylthallium salt 50
1,1,1-Trifluoro-2,4-octanedione, dimethylthallium salt 50
$C_{10}H_{16}F_6NP$
Benzyltrimethylammonium hexafluorophosphate 255
$C_{10}H_{16}F_6NSb$
Benzyltrimethylammonium hexafluoroantimonate 255
$C_{10}H_{16}HgO_4$
3-Acetoxy-4-(acetoxymercuri)-3-hexene 1138
$C_{10}H_{16}IN$
3,4-Dimethyl-1-propylpyridinium iodide 958
3,5-Dimethyl-1-propylpyridinium iodide 958
$C_{10}H_{16}NO_3P$
Diethyl phenylamidophosphate 836
$C_{10}H_{16}N_2$
4-Amino-*N*,*N*-diethylaniline monohydrochloride 221
4-Amino-*N*,*N*-diethylaniline sulfur dioxide complex 219
2-(Diethylaminomethyl)-pyridine 944
2,5-Dimethyl-1,4-xylylenediamine dihydrochloride 241
2-(2-(Isopropylamino)-ethyl)-pyridine 946
3-(4-Methylaminobutyl)pyridine 954
2-Methyl-*N*[1]-phenyl-1,2-propanediamine 241
(1-Methyl-1-phenylpropyl)hydrazine oxalate 235
2-Phenyl-1,4-butanediamine dihydrochloride 241
2-(2-(Propylamino)-ethyl)-pyridine 945
Spiro(diaziridine-3,2'-adamantane) 42
$C_{10}H_{16}N_2O$
Bis(4-cyanobutyl) ether 144
2-(Diethylaminomethyl)-3-hydroxypyridine 932
2-(4-(Ethylamino)anilino)ethanol dihydrochloride 221
N-(3-Methoxypropyl)-1,4-phenylenediamine dihydrochloride 221
$C_{10}H_{16}N_2O_2$
1-Butyl-3-furfurylurea 854
1-*tert*-Butyl-3-furfurylurea 854
1-Cyclohexyl-5,6-dihydro-2,4(1*H*,3*H*)-pyrimidinedione 135
5-Isobutyl-5-methyl-3-vinylhydantoin 133
5,8,8-Trimethyl-6-azabicyclo(3.2.1)-octane-4,7-dione 4-oxime 1012
$C_{10}H_{16}N_2O_3$
4,7,10-Trioxatridecanedinitrile 144
$C_{10}H_{16}N_2O_4$
1,2-Bis-(acetoacetamido)-ethane 110
1,1'-Dinitrodicyclopentyl 45
L-Pyroglutamyl-L-valine 125
$C_{10}H_{16}N_2O_4S$
4-Amino-*N*,*N*-bis(2-hydroxyethyl)-benzenesulfonamide 810
N[1],*N*[1]-Bis(2-hydroxyethyl)-metanilamide 810
$C_{10}H_{16}N_2O_6$
N,*N*'-Ethylenebis(succinamic acid) 121

$C_{10}H_{16}N_3O_7P$
5-Methyl-2′-deoxycytidine-5′-monophosphoric acid ... 161
$C_{10}H_{16}N_4$
1-Amino-3-(2,4,5-trimethylphenyl)-guanidine ... 726
1-(*N*-(3-Phenylpropyl)amino)-guanidine nitrate ... 726
$C_{10}H_{16}N_4O$
4-Hydroxy-6-methyl-2-(2-pentylidenehydrazino)-pyrimidine ... 972
$C_{10}H_{16}N_4O_2$
6-Amino-1,3-diethyl-5-(methyliminomethyl)uracil ... 1052
$C_{10}H_{16}N_4O_3$
5-Amino-3-butyl-1-ethyl-6-nitrosouracil ... 132
6-Amino-3-ethyl-1-isobutyl-5-nitrosouracil ... 132
L-Histidyl-L-2-aminobutyric acid ... 867
L-(+)-Homocarnosine sulfate ... 866
$C_{10}H_{16}N_4O_4$
6-Amino-3-ethyl-1-(3-methoxypropyl)-5-nitrosouracil ... 131
$C_{10}H_{16}N_5OP$
Bis(2-methyl-1-aziridinyl)-2-pyrimidinylaminophosphine oxide ... 976
$C_{10}H_{16}O$
2-Allyl-2,3-dimethyl-1-cyclopentanone ... 55
tert-Butyl 1,4-cyclohexadien-1-yl ether ... 17
4-Cyclodecen-1-one ... 56
2,2-Dimethyl-3,4-octadienal ... 60
2,7-Dimethyl-2,5-octadien-4-one ... 50
3,6-Divinyl-2-methyltetrahydropyran ... 22
4,5-Epoxy-1,9-decadiene ... 20
d-Fenchone ... 55
(+)-Limonen-10-ol ... 14
2,4-Nonadiene-2-carboxaldehyde ... 60
Tricyclo[5.2.1.0^{2,6}]decan-8-ol ... 15
2,4,6-Trimethyl-3-cyclohexene-1-carboxaldehyde ... 60
3,5,6-Trimethyl-3-cyclohexene-1-carboxaldehyde ... 60
$C_{10}H_{16}O_2$
1,3-Adamantanediol ... 12
Alloocimene dioxide (mixture of *cis* and *trans*) ... 20
2-Allyl-2,3-dimethyl-4-hydroxy-1-cyclopentanone ... 52
2-Buten-1-yl 3-hexenoate ... 80
2-(1-Cyclohexenyl)-butyric acid ... 67
2-Cyclohexylidenebutyric acid ... 66
4,5 9,10-Diepoxy-1-decene ... 20
5,5-Dimethyl-3-ethoxy-2-cyclohexen-1-one ... 52
7,12-Dioxaspiro(5,6)dodec-9-ene ... 22
Dipentene dioxide ... 21
Ethyl 1-cyclohexene-1-acetate ... 81
Ethyl 2-octynoate ... 1019
3-Isobutoxy-2-cyclohexen-1-one ... 52
2-(1-Methylallyl)-3-hexenoic acid ... 61
trans-3-Methylbicyclo(2.2.2)octane-2-carboxylic acid ... 66
2-Methyl-3-butyn-2-yl 2-tetrahydropyranyl ether ... 1017
6-Methyl-3-cyclohexenylmethyl acetate ... 81
Methyl 4-ethyl-2-heptenoate ... 79
2-Methyl-2-norbornyl acetate ... 77
1,2,3,4,6,7,8,8a-Octahydro-1,6-naphthalenediol ... 12
α-Propyl-1-cyclopentene-1-acetic acid ... 67
Tricyclo[5.2.1.0^{2,6}]decane-3,4-diol ... 12
$C_{10}H_{16}O_3$
Butyl 2-cyclopentanone-1-carboxylate ... 93
1,4-Dioxaspiro(4.5)decane-7-acetaldehyde ... 60
3,4-Epoxy-1-methyl-6-acetoxymethylcyclohexane ... 91
Ethyl 2-(3-buten-1-yl)acetoacetate ... 94
Ethyl 4-hydroxy-2-methyl-1-cyclohexene-1-carboxylate ... 85
4-Hydroxy-4-methyl-3-(3-oxobutyl)-valeric acid γ-lactone ... 96
Methyl 6-hydroxy-6-methyl-2,4-heptadiene-2-carboxylate ... 85
Methyl 7-methyl-3-oxo-6-octenoate ... 94
Methyl β-oxocyclohexane-propionate ... 94
Methyl 2-oxocyclohexane-propionate ... 94
1-Methyl-2-oxocyclohexanepropionic acid ... 65
1,2,4,5,9,10-Triepoxydecane ... 20
$C_{10}H_{16}O_3S$
5,6-Dihydro-4-hydroxy-2*H*-thiopyran-3-yl 2-ethoxyethyl ketone ... 58
$C_{10}H_{16}O_4$
Butyl 2-acetylacetoacetate ... 93
trans-1,2-Cyclohexanediacetic acid ... 66
trans-1,4-Cyclohexanediol diacetate ... 77
Diethyl 1,1-cyclobutanedicarboxylate ... 77
Dimethyl *cis*-1,2-cyclohexanedicarboxylate ... 77
Dipropyl maleate ... 78
2-Ethoxycarbonyl-2-(2-hydroxyethyl)-valeric acid γ lactone ... 96
Ethyl 2-acetyl-5-oxohexanoate ... 94
Ethyl 5-hydroxy-3-oxo-6-octenoate ... 91
2-(2-Hydroxyethyl)-2-propylglutaric acid γ-lactone ... 97
3-Methyl-1,1-cyclopentanediacetic acid ... 66
$C_{10}H_{16}O_4S_2$
Cyclohexane-1,1-bis-(mercaptoacetic acid) ... 65
$C_{10}H_{16}O_5$
1-Ethyl 5-methyl 2-acetylglutarate ... 93
3-Oxo-1,8-octanedicarboxylic acid ... 63
Propyl tetrahydrofurfuryl oxalate ... 77
$C_{10}H_{16}O_6$
3-(1-Carboxy-4-hydroxycyclohexyl)-2-hydroxypropionic acid, barium salt ... 70
Diethyl 2-(ethoxycarbonyl)malonate ... 75
$C_{10}H_{16}S$
2-Hexylthiophene ... 859
$C_{10}H_{16}Si$
tert-Butylphenylsilane ... 1112
(*o*-Tolyl)trimethylsilane ... 1122
$C_{10}H_{17}Br$
Bornyl bromide ... 6
$C_{10}H_{17}BrO_2$
4-(5-Bromo-5-hexen-1-yloxy)-2-buten-1-ol ... 20
2-Buten-1-yl 2-bromohexanoate ... 83
$C_{10}H_{17}ClN_2O_4$
Chloroacetyl-DL-leucylglycine ... 117
$C_{10}H_{17}ClO$
1-(2-Chloroethoxy)spiro(2.5)octane ... 21
4-Decenoyl chloride ... 100
$C_{10}H_{17}ClO_3$
Ethyl 2-chloro-2-(cyclohexyloxy)-acetate ... 85
$C_{10}H_{17}N$
1-Adamantanamine citrate ... 31
Cyclohexanebutyronitrile ... 150
2-Cyclohexyl-2-methylpropionitrile ... 150
$C_{10}H_{17}NO$
3-Aminocamphor hydrochloride ... 57
Bicyclo(3.3.0)octane-2-acetamide ... 102
Citral oxime ... 1008
2-(1-Cyclohexenyl)butyramide ... 102
cis-1-Decalone oxime ... 1014
7,7-Dimethyl-1-norbornanecarboxamide ... 102
4-Methylbicyclo(3.3.0)octane-2-carboxamide ... 102
$C_{10}H_{17}NO_2$
Acetoacetocyclohexylamide ... 105
1-Acetoacetyl-2-methylpiperidine ... 109
1-Acetoacetyl-3-methylpiperidine ... 109
1-Acetoacetyl-4-methylpiperidine ... 109
N-Butyl-2-oxo-1-cyclopentanecarboxamide ... 105
9-Cyanopelargonic acid ... 149
5,5-Dimethyl-3-(2-hydroxyethylamino)-2-cyclohexen-1-one hydrochloride ... 57
Ethyl 2-norbornylcarbamate ... 116
Ethyl 3-pyrrolidinocrotonate ... 90
N,N-Hexamethyleneacetoacetamide ... 109
Methyl *N*-isopropyl-*N*-(3-methyl-1,3-butadien-1-yl)carbamate ... 116
2-(Morpholinomethyl)-cyclopentanone hydrochloride ... 58
$C_{10}H_{17}NO_3$
3-Allyloxymethyl-5-methyl-1-aza-4,6-dioxabicyclo(3.3.0)octane ... 37
Ethyl β-morpholinocrotonate ... 91
Pinonic acid oxime ... 1011
$C_{10}H_{17}NO_4$
Methyl 1-nitrocyclohexane-propionate ... 89
$C_{10}H_{17}NO_5$
Methyl 4-hydroxy-2-(methoxycarbonyl)-1-pyrrolidinepropionate ... 89
$C_{10}H_{17}NS$
3,3,5-Trimethylcyclohexyl isothiocyanate ... 151
$C_{10}H_{17}N_3$
2-(2-(Diethylamino)ethyl)pyrazine ... 978
$C_{10}H_{17}N_3O$
1-(2-Cyanoethyl)-1-cyclohexylurea ... 146
6-(Diethylaminomethyl)-4-hydroxy-2-methylpyrimidine ... 971
2′-Methyl-1′,4′,5′,6′-tetrahydroacetophenone semicarbazone ... 1070
2′-Methyl-1′,4′,5′,6′-(and 3′,4′,5′,6′)-tetrahydroacetophenone semicarbazone ... 1070
$C_{10}H_{17}N_3OS$
1-Methyl-3-(methylthio)-1,2,4-triazaspiro(5,5)undec-2-en-5-one ... 124
$C_{10}H_{17}N_3O_2$
6-Amino-3-ethyl-1-isobutyluracil ... 133
$C_{10}H_{17}N_3O_3$
1,3-Diethyl-6-(2-hydroxyethylamino)-uracil ... 131
$C_{10}H_{17}N_3O_4$
L-Alanyl-L-prolylglycine monohydrate ... 119
$C_{10}H_{18}$
1-(*sec*-Butylidene)cyclohexane ... 3
cis-Cyclodecene ... 3
5-Decyne ... 1015
2,6-Dimethyl-2,*cis*-6-octadiene ... 2
(+)-*p*-Menth-1-ene ... 3
$C_{10}H_{18}BrNO$
1-Bromo-*N*-*tert*-butyl-1-cyclopentanecarboxamide ... 106
N-(4-Bromo-2,5-dimethylcyclohexyl)-acetamide ... 106
$C_{10}H_{18}BrNO_2$
N-(2-Bromo-2,2-diethylacetyl)-morpholine ... 110
$C_{10}H_{18}ClNO_5$
4-Cyclohexylidenemorpholinium perchlorate ... 44
$C_{10}H_{18}Cl_2SSi_2$
2,5-Bis(trimethylsilyl)-3,4-dichlorothiophene ... 1110
$C_{10}H_{18}Cl_3NO_3$
Ethyl *N*-(1-pentoxy-2,2,2-trichloroethyl)carbamate ... 113
$C_{10}H_{18}N_2$
4-(Hexamethyleneimino)-butyronitrile ... 147
1-Piperidinepentanenitrile ... 150
$C_{10}H_{18}N_2O$
1-Allyl-3-cyclohexylurea ... 129
5,5-Dimethyl-3-(2,2-dimethylhydrazino)-2-cyclohexen-1-one ... 58
$C_{10}H_{18}N_2O_2$
N-*tert*-Butyl-2-*aci*-nitrocyclohexaniminium hydroxide, inner salt ... 1054
O-(*N*-(Cyclohexyl)carbamoyl)-acetone oxime ... 1011
$C_{10}H_{18}N_2O_2S$
O-Ethyl *N*-(*N*-cyclohexylcarbamoyl)-thiocarbamate ... 142
$C_{10}H_{18}N_2O_4$
N-Acetylglycyl-L-leucine ... 123
Diethyl 1,4-piperazinedicarboxylate ... 117
1-(1-Methyl-1-nitropropyl)-1-nitrocyclohexane ... 46
$C_{10}H_{18}N_2O_5$
1,3-Bis(nitromethyl)-4,4-dimethylcyclohexanol ... 45
$C_{10}H_{18}N_2O_7$
N-(2-Hydroxyethyl)-ethylene-diaminetriacetic acid, trisodium salt ... 69
$C_{10}H_{18}N_2S$
3-Allyl-2,2,4,4-tetramethylimidazolidine-5-thione hydrochloride ... 139
$C_{10}H_{18}N_2S_4$
N,N′-(1,4-Cyclohexylenedimethyl)-bis-(dithiocarbamic acid, ammonium salt) ... 142
$C_{10}H_{18}N_4$
2,2′-Ethylenebis(1,4,5,6-tetrahydropyrimidine) ditosylate ... 39
1-Methyl-2-pyrrolidinone azine ... 151
$C_{10}H_{18}N_4O_3$
2,2′-Oxybis(acetic isopropylidenehydrazide) ... 111
$C_{10}H_{18}N_4S_2$
1,4-Bis(2-imidazolin-2-thio)butane dihydrobromide ... 38
$C_{10}H_{18}O$
2-Butylcyclohexanone ... 56
2-*sec*.-Butylcyclohexanone ... 56
2-*tert*.-Butylcyclohexanone ... 56
α-Citronellal ... 60
trans-5-Cyclodecenol ... 12
trans,cis-1-Decalol ... 15
trans-2-Decalol ... 15
3,7-Dimethyl-2,6-octadien-1-ol (mixture of *cis* and *trans*) ... 11
cis-1,2-Epoxycyclodecane ... 21
3-Ethyl-4-methyl-3-hepten-5-one ... 50
4-Ethyl-1-octyn-3-ol ... 1015
5-Isopropyl-2-methyl-2-cyclohexen-1-ol ... 14
2-(2-Methyl-6-methylenecyclohexyl)-1-ethanol ... 12
cis-Myrtanol ... 13
2,2,3,6-Tetramethylcyclohexanone ... 56
2,4,6-Trimethyl-1,6-heptadien-4-ol ... 12
$C_{10}H_{18}O_2$
3-Acetyl-2-octanone ... 51
1,1′-Bi-1-cyclopentanol ... 12
1-Cyclodecanol-2-one ... 52
2-Cyclohexylbutyric acid ... 67
Cyclooctyl acetate ... 77
2,3-Decalindiol ... 12
2,7-Dimethyl-6-hydroxy-2-octen-4-one ... 50
Ethyl 2-octenoate ... 79
5-Hexyl-2-tetrahydrofuranone ... 97
5-(1-Hydroxy-1-methylethyl)-2-methyl-2-vinyltetrahydrofuran ... 58
Linalool oxide ... 21
Lithium 4-cyclohexanebutyrate ... 70
Manganese(II) 4-cyclohexanebutyrate ... 1142
Methyl 2-cyclopentyl-2-methylpropionate ... 77
Methyl 2-nonenoate ... 79
Octahydro-2,2′-bi-2*H*-pyran ... 22
Sodium 4-cyclohexanebutyrate ... 70
2,2,5,5-Tetramethyl-3,4-hexanedione ... 51
$C_{10}H_{18}O_3$
2,5-Dimethyltetrahydropyran-2-ylmethyl acetate ... 90
Ethyl 2-butylacetoacetate ... 93
Ethyl 2-(1-hydroxycyclohexyl)-acetate ... 84
Ethyl 3-oxo-2,2,4-trimethylvalerate ... 93
Isovaleric anhydride ... 98
2-Methyl-7-oxononanoic acid ... 64
3-Methyl-7-oxononanoic acid ... 64
(+/-)-Pinolic acid ... 65
Tetrahydropyran-2-yl ether ... 22
$C_{10}H_{18}O_4$
1,3-Butanediol diglycidyl ether ... 20

$C_{10}H_{18}O_4$ (Cont'd)
3-(*tert*-Butyl)-adipic acid 61
Dimethyl 2,5-dimethyladipate 76
Ethyl 3,5-dihydroxy-6-octenoate 84
Ethyl 2-ethoxy-3-tetrahydrofuranacetate 89

$C_{10}H_{18}O_4S$
Diethyl 3,3′-thiodipropionate 87
3-Hexanoyloxy-tetrahydrothiophene 1,1-dioxide 153
3-Methyl-3-(2-methyl-1,3-dioxolan-2-yl)-1-(methylsulfinyl)-2-butanone 153

$C_{10}H_{18}O_5$
Diethyl (2-methoxyethyl)-malonate ... 85

$C_{10}H_{18}O_6$
Ethylenebis-(2-oxyethyl acetate) 74

$C_{10}H_{18}Si_2$
1-Phenyl-1,1,2,2-tetramethyldisilane 1112

$C_{10}H_{19}Br$
9-Bromo-2-methyl-2-nonene 5

$C_{10}H_{19}BrO_2$
10-Bromodecanoic acid 62
Ethyl 2-bromo-2-ethylhexanoate 83

$C_{10}H_{19}ClO_2$
Octyl chloroacetate 82

$C_{10}H_{19}Cl_3N_2O_2$
Methyl *N*-(1-(hexylamino)-2,2,2-trichloroethyl)carbamate 114

$C_{10}H_{19}N$
2-Aminosprio(4.5)decane hydrochloride 31
tert-Butylcyclohexylideneamine 1076
Dicyclopentylamine 31
N,*N*-Dimethyl-4-cyclooctene-1-amine 31
1-(2-Ethyl-1-butenyl)-pyrrolidine 40
1,8,8-Trimethyl-3-azabicyclo(3.2.1)-octane(camphidine) 41

$C_{10}H_{19}NO$
4-(*tert*-Butyl)caprolactam 125
7-Butyltetrahydro-2(3*H*)-azepinone ... 125
4-Cyclohexylmorpholine 40
3,7-Dimethyl-6-octenamide 102
4-(2-Ethyl-1-butenyl)-morpholine 40

$C_{10}H_{19}NO_2$
2-(*tert*-Butylamino)-ethyl methacrylate 88
Ditetrahydrofurfurylamine 35
Ethyl 3-(diethylamino)-crotonate 88
Isopropyl *N*-cyclohexylcarbamate 116
1-(4-Methoxy-2-buten-1-yloxymethyl)pyrrolidine 37
5-Methoxy-1-morpholino-1-pentene .. 38
Methyl α-methyl-1-piperidinepropionate hydrochloride 90
Propyl *N*-cyclohexylcarbamate 116

$C_{10}H_{19}NO_3$
Methyl 2,2-diisopropylmalamate 101
4-Methyl-4-(2-methyl-1,3-dioxolan-2-yl)valeramide 102
N-Octanoylglycine 123

$C_{10}H_{19}NO_3S$
3-(Cyclohexylamino)-4-hydroxytetramethylene sulfone 152
3-Hydroxy-4-(4-methylpiperidino)-tetramethylene sulfone 152

$C_{10}H_{19}NO_4$
N-(*tert*-Butoxycarbonyl)-ᴅ-valine dicyclohexylamine salt 115
N-(Ethoxycarbonylmethyl)-*N*-methyl-β-alanine ethyl ester 88

$C_{10}H_{19}NO_9$
N,*N*-Bis(carboxymethyl)glucamine, disodium salt 69

$C_{10}H_{19}NS_2$
N-(3,3,5-Trimethylcyclohexyl)-dithiocarbamic acid 139

$C_{10}H_{19}N_3O$
N-(Hexahydro-1-methyl-2-pyrimidinylidene)-2,2,2-trimethylacetamide hydrochloride .. 137

$C_{10}H_{19}N_3O_4$
Glycylglycyl-ʟ-isoleucine 120
ᴅ-Leucylglycylglycine 104
ᴅʟ-Leucylglycylglycine 120

$C_{10}H_{19}N_3O_7$
1-Methoxy-2,4,5-trimethyl-2,4,5-trinitrohexane 45

$C_{10}H_{19}O_2$
2-Methyl-2-(4-methyl-3-pentenyl)-1,3-dioxolane 184

$C_{10}H_{19}O_2Tl$
2,4-Octanedione, dimethylthallium salt 51

$C_{10}H_{19}O_4P$
1-Cyclohexen-1-yl diethyl phosphate 161

$C_{10}H_{20}$
sec-Butylcyclohexane 2
trans-2-Decene 1
Isobutylcyclohexane 2
2-Methyl-1-nonene 1

$C_{10}H_{20}Br_2$
1,2-Dibromodecane 4

$C_{10}H_{20}ClNO$
2-Chloro-*N*-(1,5-dimethylhexyl)-acetamide 102
2-Chloro-*N*-heptyl-*N*-methylacetamide 103
2-Chloro-*N*-octylacetamide 102

$C_{10}H_{20}ClNO_3$
(2-Hydroxy-3-(methacryloyloxy)-propyl)-trimethylammonium chloride 87

$C_{10}H_{20}N_2$
1,2-Bis(diethylamino)ethyne 1018
N,*N*-Dibutylglycinonitrile 145
Diisobutylaminoacetonitrile 145
4-(Diisopropylamino)-butyronitrile 147
4-(Dipropylamino)butyronitrile 147

$C_{10}H_{20}N_2O$
1-Cyclohexyl-3-isopropylurea 129
1-Cyclohexyl-3-propylurea 129
2-(2-Tetrahydro-furylaminomethyl)-piperidine 39

$C_{10}H_{20}N_2O_2$
4,4′-Ethylenedimorpholine 39
Ethyl 4-propyl-1-piperazinecarboxylate 115

$C_{10}H_{20}N_2O_3$
ʟ-2-Aminobutyryl-ʟ-isoleucine 120
ʟ-2-Aminobutyryl-ʟ-leucine 120
ʟ-Norvalyl-ʟ-norvaline 120
ʟ-Norvalyl-ʟ-valine 120
ʟ-Valyl-ʟ-norvaline 120
ʟ-Valyl-ʟ-valine 120
ᴅ-Valyl-ʟ-valine hydrochloride 123

$C_{10}H_{20}N_2O_4$
2,3-Dimethoxy-*N*,*N*,*N*′,*N*′-tetramethylsuccinamide 110
N,*N*′-Hexamethylenebis(methyl carbamate) 116

$C_{10}H_{20}N_2O_4S_2$
Cystine diethyl ester dihydrochloride 71

$C_{10}H_{20}N_2O_6$
1,4-Bis(1,3-dihydroxy-2-methyl-2-propyl)oxamide 110

$C_{10}H_{20}N_2O_8$
1,4-Bis(tris(hydroxymethyl)methyl)-oxamide 110

$C_{10}H_{20}N_4O_3$
ʟ-Isoleucylglycylglycinamide hydrochloride 121

$C_{10}H_{20}O$
1-Amylcyclopentanol 13
Amyl isobutyl ketone 48
1-Butylcyclohexanol 13
2-Butylcyclohexanol (mixture of isomers) 14
trans-4-Butylcyclohexanol 14
trans-4-*tert*-Butylcyclohexanol 14
3-*tert*-Butyl-2-methylcyclopentanol .. 13
Cyclodecanol 12
1-Cyclohexyl-1-butanol 13
(+)-1-Cyclohexyl-2-methylpropanol .. 13
1,2-Diethyl-1-cyclohexanol (*cis* and *trans*) 13
3,5-Dimethyl-4-octanone 48
2-Ethylcaprylaldehyde 60
Hexyl isopropyl ketone 48
Isooctyl vinyl ether 18
2-Isopentyl-3-methyl-2-buten-1-ol 12
3-Isopropyl-2-heptanone 48
3-Methyl-4-nonanone 48

$C_{10}H_{20}OS_2$
S-Methyl *O*-(1-methylheptyl) dithiocarbonate 87

$C_{10}H_{20}OSi$
2,3-Dimethyl-1-(trimethylsilyloxy)-1-cyclopentene 1110
2-Methyl-1-cyclohexen-1-yl trimethylsilyl ether 1111

$C_{10}H_{20}O_2$
1,1-Cyclohexanediethanol 12
Decanoic acid, octadecylamine salt 70
Ethyl 2,2-diethylbutyrate 73
2-Ethylhexyl acetate 73
Hexyl isobutyrate 73
2-(1-Hydroxybutyl)cyclohexanol 14
Hydroxycitronellal 60
2-Isopropyl-4,4,6-trimethyl-1,3-dioxane 23
2-Methylnonanoic acid 61
(*S*)-3-Methylnonanoic acid 61
3,3-Pentamethylene-1,5-pentanediol 13
2,2,5,5-Tetramethyl-4-hydroxy-3-hexanone 50

$C_{10}H_{20}O_2S$
2-Ethylhexyl mercaptoacetate 86
Isooctyl mercaptoacetate 86

$C_{10}H_{20}O_2Si$
Diallyldiethoxysilane 1096

$C_{10}H_{20}O_3$
Butyl butoxyacetate 85
9-Hydroxydecanoic acid 63
Methyl 3-hydroxy-2,3,4,4-tetramethylvalerate 85

$C_{10}H_{20}O_5$
2-Hydroxy-1,1,3,3-cyclohexanetetramethanol 13

$C_{10}H_{20}O_6Si$
Diacetoxydiisopropoxysilane 1109

$C_{10}H_{20}S_2$
1,1-Bis(methylthio)-1-octene 24
2,9-*p*-Menthanedithiol 24

$C_{10}H_{21}BO_3$
2-Butoxy-4,4,6-trimethyl-1,3,2-dioxaborinane 1124

$C_{10}H_{21}ClHg$
Chlorodecylmercury 1137

$C_{10}H_{21}F$
1-Fluorodecane 4

$C_{10}H_{21}N$
4-*tert*-Butylcyclohexylamine 31
1-Butyl-4-methylpiperidine 40
Cyclohexanebutylamine 31
N,*N*-Diisopropyl-2-methyl-1-propenylamine 32

$C_{10}H_{21}NO$
4-(2-Buten-1-ylamino)-4-methyl-2-pentanol 33
Decanamide 102
N,*N*-Dibutylacetamide 107
1-Diethylamino-5-methoxy-2-pentene 35
2,2-Diethyl-*N*,*N*-dimethylbutyramide 107
β,β-Dimethyl-1-piperidinepropanol 37

$C_{10}H_{21}NOS$
1-(2-(Diethylamino)ethylthio)-3-buten-2-ol 35

$C_{10}H_{21}NO_2$
N-Cyclohexyl-2,2′-iminodiethanol 33
ʟ-Leucine *tert*-butyl ester dibenzenesulfimide salt 817
1-Piperidinepropionaldehyde dimethyl acetal 38
1-Pyrrolidineacetaldehyde diethyl acetal 38

$C_{10}H_{21}NO_3$
2-Morpholinoacetaldehyde diethyl acetal 38

$C_{10}H_{21}NO_6$
2-(Dimethylamino)ethyl-β-ᴅ-glucoside hydrochloride 33

$C_{10}H_{21}NS$
1-(Butylthiomethyl)-piperidine 38

$C_{10}H_{21}N_3O$
N-*tert*-Butyl-4-methyl-1-piperazinecarboxamide 126
1-(*N*,*N*-Diethylcarbamoyl)-4-methylpiperazine citrate 126

$C_{10}H_{22}$
R-(-)-2,6-Dimethyloctane 1
2-Methylnonane 1
3-Methylnonane 1
4-Methylnonane 1

$C_{10}H_{22}AsF_6N$
Cyclohexyldiethylammonium hexafluoroarsenate 44

$C_{10}H_{22}ClN$
N-(2-Chloroethyl)-dibutylamine hydrochloride 32

$C_{10}H_{22}ClO_3P$
Dipentyl phosphorochloridate 160

$C_{10}H_{22}NO_3P$
Diethyl cyclohexylamidophosphate. 160

$C_{10}H_{22}N_2$
5-Amino-1,3,3-trimethylcyclohexanemethylamine. 30
1,2-Bis(diethylamino)ethylene 32
trans-*N*,*N*′-Diethyl-1,4-cyclohexylenediamine hydrochloride 31
N-(2-Piperidylmethyl)-diethylamine .. 39

$C_{10}H_{22}N_2O$
Decanoic hydrazide 112

$C_{10}H_{22}N_2O_2$
2-(Dimethylaminomethyl)-3-hydroxy-1-(2-hydroxyethyl)-piperidine 36

$C_{10}H_{22}N_4$
1,1′-Ethylenedipiperazine 39
1-(Nonylidene)guanidine nitrate 138

$C_{10}H_{22}N_4O_2$
1,1′-Hexamethylenebis(3-methylurea) 129

$C_{10}H_{22}O$
3-Decanol 9
3,4-Diethyl-3-hexanol 10
2,2-Dimethyl-3-octanol 9
3,5-Dimethyl-4-octanol 9
3,6-Dimethyl-3-octanol 10
6-Ethyl-3-octanol 9
2-Isopropyl-4-methylhexanol 8
2-Methyl-3-nonanol 9
2-Methyl-4-nonanol 9
3-Methyl-4-nonanol 9
4-Methyl-4-nonanol 10
5-Methyl-5-nonanol 10
2,3,4,5-Tetramethyl-3-hexanol 10
3,4,5-Trimethyl-4-heptanol 10

$C_{10}H_{22}OS$
Isopentyl sulfoxide 154
2-(Octylthio)-ethanol 25

$C_{10}H_{22}OSn$
Diisopentyloxotin 1129
Dipentyloxotin 1129

$C_{10}H_{22}O_2$
1,8-Dimethoxyoctane 17
2,7-Dimethyl-2,7-octanediol 10
3,6-Dimethyl-3,6-octanediol 10
2-(2-Ethylhexyloxy)-ethanol 19
5-Ethyl-3-methyl-2,4-heptanediol 10

$C_{10}H_{22}O_2S$
Pentyl sulfone 154

$C_{10}H_{22}O_2Sn$
Acetoxybutyldiethyltin 1131

$C_{10}H_{22}O_3$
1,1,3-Triethoxybutane 19

$C_{10}H_{22}O_4$
Triethylene glycol butyl ether 19

$C_{10}H_{22}Pb$
(3-Butenyl)triethyllead 1133

$C_{10}H_{22}S$
Ethyl octyl sulfide 26
Isopentyl sulfide 26
Pentyl sulfide 26

$C_{10}H_{22}S_2$
4,9-Dithiadodecane 26
Isopentyl disulfide 27

$C_{10}H_{22}S_2Si$
2-Isopropyl-2-(trimethylsilyl)-1,3-dithiane 1097

$C_{10}H_{23}Cl_6N_2OSb$
(*tert*-Butoxyimino)(2,2-dimethylpropyl)methylammonium hexachloroantimonate 1128
$C_{10}H_{23}I_2P$
(4-Iodobutyl)-triethylphosphonium iodide 158
$C_{10}H_{23}N$
Diisoamylamine 28
$C_{10}H_{23}NO$
4-Diisopropylamino-1-butanol 34
1-Dipropylamino-2-butanol 34
4-Dipropylamino-1-butanol 34
$C_{10}H_{23}NO_2$
5,5′-Iminodi-1-pentanol 33
$C_{10}H_{23}NO_3$
1,1′-Diethyl-2,2′,2″-nitrilotriethanol 33
$C_{10}H_{24}BF_4N$
Dipentylammonium tetrafluoroborate 42
$C_{10}H_{24}NO_3P$
Diethyl hexylamidophosphate 160
$C_{10}H_{24}N_2$
N,*N*′-Di-*tert*-butyl-ethylenediamine 30
N,*N*′-Diethyl-1,6-hexanediamine 30
$C_{10}H_{24}O_6P_2$
Tetraethyl ethylenediphosphonate 158
$C_{10}H_{24}O_6Si$
Methyltris(2-methoxyethoxy)silane 1106
$C_{10}H_{24}Si$
Dimethyloctylsilane 1096
$C_{10}H_{25}ClN_2O$
10-Undecenoic (4-chlorobenzylidene)hydrazide 1026
$C_{10}H_{26}OSi_2$
Trimethyl(4-(trimethylsilyl)butoxy)silane 1096
$C_{10}H_{26}Si_2$
Tetramethylenebis(trimethylsilane) 1096
$C_{10}H_{28}O_2Si_4$
1,1,1,3,3,3-Hexamethyl-2-(trimethylsilyl)-2-trisilanecarboxylic acid 1109
$C_{10}H_{30}Si_4$
Decamethyltetrasilane 1097
$C_{11}Cl_3F_5OS$
2-(Pentafluorobenzoyl)-3,4,5-trichlorothiophene 855
$C_{11}Cl_8N_2O$
Bis(2,3,5,6-tetrachloro-4-pyridyl) ketone 961
$C_{11}Cl_{10}OS$
S-Pentachlorophenyl pentachloro-1,3-butadiene-1-thiocarboxylate 377
$C_{11}H_4Br_4N_2O_3$
2-(1-Imidazolylcarbonyl)-3,4,5,6-tetrabromobenzoic acid 865
$C_{11}H_4Br_4N_2O_3S$
3,4,5,6-Tetrabromo-*N*-(2-thiazolyl)-phthalamic acid 875
$C_{11}H_4Cl_4N_2O_3$
2-(Imidazol-1-ylcarbonyl)-3,4,5,6-tetrachlorobenzoic acid 865
$C_{11}H_4Cl_5N$
2-(Pentachlorophenyl)pyridine 931
$C_{11}H_4Cl_6O_2$
1,2,3,4,9,9-Hexachloro-1,4,4a,8a-tetrahydro-1,4-methanonaphthalene-5,8-dione 57
$C_{11}H_4Cl_8N_2Si$
Bis(2,3,5,6-tetrachloro-4-pyridyl)-methylsilane 1123
$C_{11}H_5Cl_2F_3N_2O$
4,5-Dichloro-2-(3-trifluoromethylphenyl)pyridazin-3(2*H*)-one 693
$C_{11}H_5Cl_3O_2$
6-Phenyl-3,4,5-trichloro-2*H*-pyran-2-one 384
$C_{11}H_5Cl_4N_3O_4$
N-(2-Cyanoacetamido)-3,4,5,6-tetrachlorophthalamic acid 751
$C_{11}H_5Cl_5O$
4-Phenyl-1,2,3,4-tetrachloro-1,3-butadiene-1-carbonyl chloride 389
$C_{11}H_5FCl_2O_3S$
1-Chloro-4-fluorosulfonyl-2-naphthoyl chloride 804
$C_{11}H_6BrCl_4NO$
4-Bromo-1,2,3,4-tetrachloro-1,3-butadiene-1-carboxanilide 478
$C_{11}H_6BrNO_4$
6-Bromo-α-cyano-1,3-benzodioxole-5-acrylic acid 766
$C_{11}H_6Br_2ClNO_2$
2′-Chloro-4′,6′-dibromo-2-furanilide 844
4′-Chloro-2′,6′-dibromo-2-furanilide 844
$C_{11}H_6ClNO_4$
2-Chloro-5-maleimidobenzoic acid 720
4-Chloro-3-maleimidobenzoic acid 720
$C_{11}H_6ClNO_5$
2-Chloro-4-nitrophenyl 2-furoate 844
$C_{11}H_6ClN_5O_6$
N-(5-Chloro-2-pyridyl)-picrylamine 942
$C_{11}H_6Cl_2O_2$
3,5-Dichloro-2-phenyl-4*H*-pyran-4-one 940
$C_{11}H_6Cl_2O_3$
3-Acetyl-6,8-dichlorocoumarin 385
$C_{11}H_6Cl_3NS$
N-(2-Thienylmethylene)-2,4,5-trichloroaniline 1063
$C_{11}H_6Cl_4O$
2-Chloro-3-(trichlorovinyl)-1-indanone 296
$C_{11}H_6Cl_4O_2$
2,4,5,7-Tetrachlorotricyclo[6.2.1.$0^{2,7}$]undeca-4,9-diene-3,6-dione 53
$C_{11}H_6Cl_5NO$
1,2,3,4,4-Pentachloro-1,3-butadiene-1-carboxanilide 478
$C_{11}H_6F_3N_3$
3-(Trifluoromethyl)-1,2,4-triazolo(3,4-a)isoquinoline 908
$C_{11}H_6F_3N_3O_6$
N-(2,6-Dinitro-4-(trifluoromethyl)-phenyl)succinimide 714
$C_{11}H_6N_2O_2$
Piperonylidenemalononitrile 764
$C_{11}H_6N_6$
1,1,3,3,5,5-Hexacyanopentane 147
$C_{11}H_6O_{10}$
Benzenepentacarboxylic acid 345
$C_{11}H_7BrN_2$
Bromonorharman (Br position unspecified) 918
$C_{11}H_7BrN_2O_3$
2-Bromo-α-cyano-4,5-(methylenedioxy)cinnamamide 763
$C_{11}H_7BrO$
5-Bromo-1-naphthaldehyde 324
$C_{11}H_7BrO_2$
3-Bromo-2-naphthoic acid 346
4-Bromo-1-naphthoic acid 346
8-Bromo-1-naphthoic acid 347
$C_{11}H_7BrO_3$
3-Acetyl-6-bromocoumarin 385
4-Bromo-1-hydroxy-2-naphthoic acid 339
4-Bromo-3-hydroxy-2-naphthoic acid 338
$C_{11}H_7Br_2ClF_3NO_3$
2′-Chloro-2,3-dibromo-5′-(trifluoromethyl)succinanilic acid 395
$C_{11}H_7ClF_3NO$
4-Chloro-8-methoxy-2-(trifluoromethyl)quinoline 981
$C_{11}H_7ClF_3N_3OS$
1-(4-Chloro-2-(trifluoromethyl)-phenyl)-3-(2-thiazolyl)urea 874
$C_{11}H_7ClN_2O$
2-(4-Chlorobenzoyl)pyrazine 977
2-(4-Chlorobenzoyl)pyrimidine 968
$C_{11}H_7ClN_2O_2S$
4-Chloro-*N*-(2-thenylidene)-3-nitroaniline 1056
$C_{11}H_7ClN_2O_4$
5-Chloro-3-hydroxy-1-(3-nitrophenyl)-2(1*H*)-pyridone 696
4′-Chloro-3′-nitro-2-furanilide 844
$C_{11}H_7ClN_2O_5$
5′-Chloro-2′-hydroxy-4′-nitro-2-furanilide 845
$C_{11}H_7ClOS_2$
2-Chloro-2*H*-thiopyrano(2,3-b)(1)-benzothiophen-4(3*H*)-one 898
$C_{11}H_7Cl_2F_3N_2$
4,5-Dichloro-2,3-dihydro-2-(4-(trifluoromethyl)phenyl)pyridazine 244
$C_{11}H_7Cl_2NO$
2,3-Dichloro-*N*-(furfurylidene)aniline 1075
2,5-Dichloro-*N*-(furfurylidene)aniline 1073
$C_{11}H_7Cl_2NO_2$
3-(2-Chlorophenyl)-5-methyl-4-isoxazolecarboxylic acid chloride 871
2′,4′-Dichloro-2-furanilide 844
2′,5′-Dichloro-2-furanilide 844
2′,6′-Dichloro-2-furanilide 844
3′,4′-Dichloro-2-furanilide 844
3′,5′-Dichloro-2-furanilide 844
5,7-Dichloro-8-quinolyl acetate 994
$C_{11}H_7Cl_2NO_3$
4-(3,5-Dichlorosalicylidene)-2-methyl-2-oxazolin-5-one 382
$C_{11}H_7Cl_2NS$
2,4-Dichloro-*N*-(2-thienylmethylene)-aniline 1073
2,5-Dichloro-*N*-(2-thienylmethylene)-aniline 1073
3,4-Dichloro-*N*-(2-thienylmethylene)-aniline 1073
$C_{11}H_7Cl_2N_3OS_2$
1-(3,4-Dichlorobenzoyl)-3-(2-thiazolyl)-2-thiourea 875
$C_{11}H_7Cl_3N_2OS$
N-(4-(Chloromethyl)-2-thiazolyl)-3,4-dichlorobenzamide 874
$C_{11}H_7Cl_3N_4O$
1-(6-Chloro-4-pyrimidinyl)-3-(2,3-dichlorophenyl)urea 968
$C_{11}H_7Cl_6HgNO_2$
N-(Ethylmercuri)-1,4,5,6,7,7-hexachloro-bicyclo(2.2.1)hept-5-ene-2,3-dicarboximide 125
$C_{11}H_7FO_5S$
4-Fluorosulfonyl-1-hydroxy-2-naphthoic acid 801
$C_{11}H_7I_2NO_3$
2-(5,7-Diiodo-8-quinolyloxy)-acetic acid 985
$C_{11}H_7NO_4$
2-Benzoyl-5-nitrofuran 852
2-Benzoyl-5-nitrofuran 847
3-Maleimidobenzoic acid 721
5-Nitro-1-naphthalenecarboxylic acid 344
$C_{11}H_7NO_5$
4-Maleimidosalicylic acid 716
5-(4-Nitrophenyl)-2-furoic acid 847
$C_{11}H_7NO_7S$
8-Sulfo-2,4-quinolinedicarboxylic acid 991
$C_{11}H_7NS$
2-Naphthyl isothiocyanate 768
$C_{11}H_7N_3O_2$
2-Amino-3-cyanoquinoline-4-carboxylic acid, sodium salt 988
$C_{11}H_7N_3O_3$
3-(2,4-Dioxoimidazolidin-5-ylidene)-2-indolinone 700
$C_{11}H_7N_3O_5$
α-Cyano-4,5-(methylenedioxy)-2-nitrocinnamamide 764
3-(2,4-Dinitrophenoxy)-pyridine 938
$C_{11}H_7N_3O_5S$
N-(5-Nitro-2-thiazolyl)-phthalamic acid 877
$C_{11}H_7N_3O_6$
4-Methyl-1,3,8-trinitronaphthalene 276
$C_{11}H_7N_5O_6$
N-(2-Pyridyl)picrylamine 945
N-(3-Pyridyl)picrylamine 943
N-(4-Pyridyl)picrylamine 943
$C_{11}H_8BClF_4N_4$
3-(4-Chlorophenyl)-3*H*-tetrazolo(1,5-a)pyridin-4-ium bromide 1002
$C_{11}H_8BrNO_2$
N-(3-Bromoallyl)phthalimide 720
4′-Bromo-2-furanilide 844
$C_{11}H_8BrNS$
4-Bromo-*N*-(2-thienylmethylene)-aniline 1073
$C_{11}H_8BrN_2$
4-Bromo-3-methyl-1-naphthalenediazonium chloride, mercuric bromide complex 256
$C_{11}H_8Br_2O$
2,4-Dibromo-1-methoxynaphthalene 188
$C_{11}H_8ClF_3N_4O_2$
4-(2-Chloro-5-(trifluoromethyl)-phenyl)-1-(2-cyanoacetyl)-semicarbazide 751
$C_{11}H_8ClNO$
4-Chloro-*N*-(furfurylidene)aniline 1073
2-Chloro-6-phenoxypyridine 939
4-Chlorophenyl 2-pyrrolyl ketone 838
2-Chloro-*N*-(furfurylidene)aniline 1075
$C_{11}H_8ClNOS$
3-Chloro-4,5-dihydro(1)-benzothiepino(5,4-c)isoxazole 874
$C_{11}H_8ClNO_2$
4-(4-Chlorobenzylidene)-2-methyl-2-oxazolin-5-one 382
2-Chloro-6-methoxy-3-quinolinecarboxaldehyde 986
N-(3-Chloro-*p*-tolyl)-maleimide 722
N-(6-Chloro-*o*-tolyl)-maleimide 722
Methyl 4-chloro-α-cyanocinnamate 760
$C_{11}H_8ClNO_2S_2$
5-(4-Chloro-3-hydroxy-α-methylbenzylidene)rhodanine 728
$C_{11}H_8ClNO_3$
3-(2-Chlorophenyl)-5-methyl-4-isoxazolecarboxylic acid 871
3-Phthalimidopropionyl chloride 722
$C_{11}H_8ClNS$
4-Chloro-*N*-(2-thienylmethylene)-aniline 1073
$C_{11}H_8ClN_3O$
2-(Chloromethyl)(1,2,4)triazolo(4,3-a)quinolin-1(2*H*)-one 693
$C_{11}H_8ClN_3O_3$
8-Acetamido-6-chloro-5-nitroquinoline 988
4-Chloro-6-methyl-2-(4-nitrophenyl)-3(2*H*)-pyridazinone 703
$C_{11}H_8Cl_2N_2$
2-(3,4-Dichloroanilino)pyridine 931
2-(3,5-Dichloroanilino)pyridine 931
$C_{11}H_8Cl_2N_2O_2$
5-(2,4-Dichlorobenzylidene)-1-methylhydantoin 695
3,4-Dichloro-*N*-(5-methyl-3-isoxazolyl)benzamide 871
$C_{11}H_8Cl_2N_2O_2S$
3-(2-Chloroethyl)-1-(4-chlorophenyl)-5-thioxohydantoin 727
2-(2,4-Dichlorophenoxy)-*N*-(2-thiazolyl)acetamide 875
$C_{11}H_8Cl_2N_2O_3$
4-(2-Carbamoylphenylimino)-2,3-dichloro-2-butenoic acid 1074
$C_{11}H_8Cl_2N_4O$
1-(2,3-Dichlorophenyl)-3-(2-pyrimidinyl)urea 969
1-(2,5-Dichlorophenyl)-3-(2-pyrimidinyl)urea 969
$C_{11}H_8Cl_2O$
2,2-Dichloro-2,2a,7,7a-tetrahydro-1*H*-cyclobut(a)inden-1-one 286
$C_{11}H_8Cl_4O_4$
Isopropyl tetrachlorophthalate 366
$C_{11}H_8Cl_4O_5$
2-Methoxyethyl tetrachlorophthalate 366
$C_{11}H_8FNS$
4-Fluoro-*N*-(2-thienylmethylene)-aniline 1073
$C_{11}H_8FN_3O_3$
N-(2-Fluoro-5-nitrophenyl)-2-pyrrolecarboxamide 838
$C_{11}H_8F_3NO_2$
3,4-Dihydro-7-methoxy-4-(trifluoromethyl)-2(1*H*)-quinolinone 689
N-(3-Trifluoromethylphenyl)-succinimide 714
$C_{11}H_8F_3NO_3$
2′-(Trifluoromethyl)maleanilic acid 396
3′-(Trifluoromethyl)maleanilic acid 446

$C_{11}H_8F_3N_3$
- 2-Amino-4-phenyl-6-(trifluoromethyl)pyrimidine 968
- 5,6-Dihydro-3-(trifluoromethyl)-1,2,4-triazolo(3,4-a)isoquinoline 869

$C_{11}H_8F_3N_3OS$
- 1-(2-Thiazolyl)-3-(2-(trifluoromethyl)phenyl)urea 874
- 1-(2-Thiazolyl)-3-(3-(trifluoromethyl)phenyl)urea 874

$C_{11}H_8F_3N_3O_5$
- 1-(2,6-Dinitro-4-(trifluoromethyl)phenyl)-2-pyrrolidinone 689

$C_{11}H_8F_6N_2$
- 3-(3,5-Bis(trifluoromethyl)anilino)propionitrile 751

$C_{11}H_8F_7NO$
- 4′-(Heptafluoropropyl)acetanilide 438

$C_{11}H_8INS$
- 4-Iodo-*N*-(2-thienylmethylene)aniline 1073

$C_{11}H_8N_2$
- α-Methylbenzylidenemalononitrile 764
- 2-Methylbenzylidenemalononitrile 764
- 4-Methylbenzylidenemalononitrile 764

$C_{11}H_8N_2O$
- 2-Cyano-6-methoxyquinoline 985
- 4-Methoxybenzylidenemalononitrile 762

$C_{11}H_8N_2O_2$
- 2-(2-Cyanoethyl)phthalimide 755
- 2-(3-Nitrophenyl)pyridine 952
- 2-(4-Nitrophenyl)pyridine 960
- 3-(2-Nitrophenyl)pyridine 952
- 2-(2-Nitrophenyl)pyridine picrate 949
- 2-(3-Nitrophenyl)pyridine picrate 952
- 3-(2-Nitrophenyl)pyridine picrate 952
- 5-Phenyl-4-pyridazinecarboxylic acid 968
- Vanillylidenemalononitrile 760

$C_{11}H_8N_2O_2S$
- 5-Benzylidene-2-thiobarbituric acid 735
- 3-Nitro-*N*-(2-thienylmethylene)aniline 1056
- 4-Nitro-*N*-(2-thienylmethylene)aniline 1059

$C_{11}H_8N_2O_2Se$
- *N*-(2-Selenocyanatoethyl)phthalimide 213

$C_{11}H_8N_2O_3$
- α-Cyano-3,4-(methylenedioxy)cinnamamide 766
- 2-(1-Imidazolylcarbonyl)benzoic acid 868
- *N*-(5-Nitrofurfurylidene)aniline 1062

$C_{11}H_8N_2O_3S$
- 6-Norharmansulfonic acid 919

$C_{11}H_8N_2O_4$
- Methyl α-cyano-4-nitrocinnamate 765
- 3′-Nitro-2-furanilide 850
- 5-Nitro-2-furanilide 847
- *N*-(5-Nitro-*o*-tolyl)-maleimide 721
- 3-(Ureidomethylene)-2,4-chromanedione 570

$C_{11}H_8N_2O_5$
- *N*-(4-Methoxy-3-nitrophenyl)maleimide 716

$C_{11}H_8N_2S$
- 10*H*-Pyrido(3,2-b)(1,4)benzothiazine 966
- 4,5,6,7-Tetrahydrobenzo(b)thiophene-4-ylidenemalononitrile 856

$C_{11}H_8N_2SO_3$
- *N*-(2-Thiazolyl)phthalamic acid 877

$C_{11}H_8N_4O$
- 4-Hydroxy-1-phenylpyrazolo(3,4-d)pyrimidine 921
- 5-Phenyl-*s*-triazolo(1,5-a)pyrimidin-7-ol 918

$C_{11}H_8N_4O_3$
- 3-Amino-4-cyano-2-methyl-7-nitro-1(2*H*)-isoquinolinone 763
- 5-Hydroxy-2-(4-nitrophenylazo)pyridine 1084

$C_{11}H_8N_4O_4$
- 4-(2,4-Dinitroanilino)pyridine 943
- 2-(2,4-Dinitrophenyl)aminopyridine 942

$C_{11}H_8N_4O_7$
- *N*-Furfurylpicrylamine 850

$C_{11}H_8O_2$
- 2-Benzoylfuran 853
- 1-Hydroxy-2-naphthaldehyde 322
- 2,3-(Methylenedioxy)naphthalene 184
- 2-Naphthoic acid 347
- 1-Naphthoic acid, dodecylamine salt 351
- 1-Naphthoic acid, octadecylamine salt 351

$C_{11}H_8O_2S$
- Benzo(b)thiophene-3-acrylic acid 898
- 1-(2-Furyl)-3-(2-thienyl)-2-propen-1-one 858
- 3-(2-Furyl)-1-(2-thienyl)-2-propen-1-one 858

$C_{11}H_8O_3$
- 1,3-Di-(2-furyl)-2-propen-1-one 853
- 2-(1-Indenylidene)glycolic acid 334

$C_{11}H_8O_4$
- 6-Methyl-2*H*-1,3-dioxolo(4,5-*H*)(1)-benzopyran 386

$C_{11}H_8O_6$
- 7-Acetyl-4,9-dioxatricyclo(5.2.2.02,6)undec-10-ene-3,5,8-trione 98

$C_{11}H_8O_7$
- Methyl 2,3-dihydro-5,7-dihydroxy-2,4-dioxobenzopyran-8-carboxylate 382

$C_{11}H_9BrN_2O_2$
- 3-Bromo-α-cyano-4-methoxycinnamamide 762

$C_{11}H_9BrN_4O$
- 1-(3-Bromophenyl)-3-(2-pyrimidinyl)urea 969

$C_{11}H_9BrO$
- 1-Bromo-4-methoxynaphthalene 188
- 1-Bromo-5-methoxynaphthalene 182

$C_{11}H_9BrO_2$
- 3-(2-Bromoethyl)coumarin 382
- 2-Bromo-3-methyl-1,4-naphthalenediol 194

$C_{11}H_9Br_2Cl_2NO_3$
- 2,3-Dibromo-4′,5′-dichloro-2′-methylsuccinanilic acid 405

$C_{11}H_9ClF_3NO_3$
- 2′-Chloro-5′-(trifluoromethyl)succinanilic acid 395
- Ethyl 2-chloro-5-(trifluoromethyl)oxanilate 395

$C_{11}H_9ClF_3N_3O_3$
- 1-(2-Chloro-5-(trifluoromethyl)phenyl)-3-(2-oxo-3-oxazolidinyl)urea 606

$C_{11}H_9ClN_2$
- 3-Chloro-6-methyl-4-phenylpyridazine 968

$C_{11}H_9ClN_2O$
- 4-Chloro-6-methyl-2-phenyl-3(2*H*)-pyridazinone 708
- 4-Chlorophenyl 1-methyl-2-imidazolyl ketone 865
- 8-Chloro-1-tetrahydronorharmanone 884

$C_{11}H_9ClN_2O_3$
- 4′-Amino-5′-chloro-2′-hydroxy-2-furanilide 845

$C_{11}H_9ClN_2O_4$
- 3-Butyn-1-yl *N*-(2-chloro-4-nitrophenyl)carbamate 1015

$C_{11}H_9ClN_4$
- 4-Chloro-*N*-(2-pyridylazo)aniline 931

$C_{11}H_9ClN_4O_2$
- 4-(4-Chlorobenzylideneamino)-6-methyl-1,2,4-triazine-3,5(2*H*,4*H*)-dione 1070

$C_{11}H_9ClN_4O_3$
- 3′-Chloro-1-methyl-5-nitro-2-imidazolecarboxanilide 865

$C_{11}H_9ClN_4O_3S$
- 1-(2-Chloro-4-nitrophenyl)-3-(4-methyl-2-thiazolyl)urea 875

$C_{11}H_9ClO$
- 1-Chloro-3-methoxynaphthalene 188

$C_{11}H_9ClO_2$
- 4-Chloromethyl-6-methylcoumarin 382

$C_{11}H_9Cl_2NO_4$
- 3′,5′-Dichloro-2′-hydroxy-4′-methylmaleanilic acid 447

$C_{11}H_9Cl_3N_2O_2S$
- 5-Imino-3-(4-methoxyphenyl)-2-(trichloromethyl)-4-oxazolidinethione 728

$C_{11}H_9Cl_3N_6O_2$
- 2,4-Diamino-6-(2-(2,4,5-trichlorophenoxy)acetamido)-1,3,5-triazine 978

$C_{11}H_9Cl_4NO_4$
- Methyl *N*-(1-(4-chlorobenzoyloxy)-2,2,2-trichloroethyl)carbamate 606

$C_{11}H_9Cl_6NO_3$
- Ethyl *N*-(2,2,2-trichloro-1-(2,4,5-trichlorophenoxy)ethyl)carbamate 606

$C_{11}H_9FN_2$
- 2-(4-Fluoroanilino)pyridine 931
- 2-Fluoro-*N*-(2-pyridyl)aniline 928

$C_{11}H_9FN_2O$
- 1-(*N*-(4-Fluorophenyl)carbamoyl)pyrrole 838

$C_{11}H_9FN_4O$
- 1-(4-Fluorophenyl)-3-(2-pyrazinyl)urea 977
- 1-(2-Fluorophenyl)-3-(2-pyrimidinyl)urea 969
- 1-(4-Fluorophenyl)-3-(2-pyrimidinyl)urea 969

$C_{11}H_9F_2N_3O_2$
- 1-(2,4-Difluorophenyl)-3-(5-methyl-3-isoxazolyl)urea 871

$C_{11}H_9F_3N_2OS$
- 5,6-Dimethyl-2-(2,2,2-trifluoroacetamido)benzothiazole 910

$C_{11}H_9F_3N_4O_2$
- 1-(2-Cyanoacetyl)-4-(2-(trifluoromethyl)phenyl)semicarbazide 751
- 1-(2-Cyanoacetyl)-4-(3-(trifluoromethyl)phenyl)semicarbazide 751

$C_{11}H_9F_3N_4O_4$
- *N*-Cyano-2,6-dinitro-*N*-propyl-4-(trifluoromethyl)aniline 751

$C_{11}H_9F_3O_3$
- 4-Oxo-4-(2-trifluoromethylphenyl)butyric acid 325

$C_{11}H_9F_3O_4$
- (+)-*mono*-1-Methyl-2,2,2-trifluoroethyl phthalate 366

$C_{11}H_9F_6NO_2$
- 1,3-Bis(trifluoromethyl)-5-(1-methyl-1-nitroethyl)benzene 257

$C_{11}H_9F_6N_3S$
- *S*-Methyl-1-(2,5-bis(trifluoromethyl)benzylidene)isothiosemicarbazide hydriodide 728

$C_{11}H_9F_{17}Si$
- (Perfluorooctyl)trimethylsilane 1097

$C_{11}H_9IN_4O$
- 1-(4-Iodophenyl)-3-(2-pyrazinyl)urea 977

$C_{11}H_9NO$
- 3-Acetylquinoline 993
- 2-Benzoylpyrrole 842
- 2-(Cyanomethyl)-1-indanone 757
- 2-Cyano-1-tetralone 757
- *N*-(1-Naphthyl)-formamide 490
- 4-(2-Pyridyl)phenol hydrobromide 934

$C_{11}H_9NOS$
- *N*-Phenyl-2-thiophenecarboxamide 856
- 4-(2-Thienylmethyleneamino)phenol 1037

$C_{11}H_9NOS_3$
- 4-(4-Acetamidophenyl)-1,3-dithiole-2-thione 436

$C_{11}H_9NO_2$
- *N*-Allylphthalimide 720
- 2-Benzoylfuran oxime 1011
- 4-Benzylidene-2-methyl-2-oxazolin-5-one 385
- *N*-Benzylmaleimide 720
- 4-Ethylidene-2-phenyl-2-oxazolin-5-one 386
- 8-Hydroxy-5-quinolyl methyl ketone 983
- Methyl α-cyanocinnamate 766
- Methyl 4-cyanocinnamate 365
- 6-Methyl-1-nitronaphthalene 280
- Methyl 2-quinolinecarboxylate 993
- 2-Methyl-4-quinolinecarboxylic acid 993
- *N*-(*o*-Tolyl)-maleimide 720
- *N*-(*p*-Tolyl)-maleimide 722

$C_{11}H_9NO_2S_2$
- 5-(3-Hydroxy-α-methylbenzylidene)rhodanine 734

$C_{11}H_9NO_3$
- 6-Acetamidocoumarin 434
- 3-Acetyl-4-aminocoumarin 384
- 1,2-Dihydro-3-methyl-2-oxo-4-quinolinecarboxylic acid 993
- 4′-Hydroxy-2-furanilide 845
- *N*-(2-Methoxyphenyl)maleimide 717
- *N*-(3-Methoxyphenyl)maleimide 717
- *N*-(4-Methoxyphenyl)maleimide 717
- 6-Methoxy-4-quinolinecarboxylic acid 986
- Methyl β-cyano-α-hydroxycinnamate 761
- Methyl 3-isoquinolinecarboxylate *N*-oxide 1006
- 1-Propionylisatin 707
- 2-(8-Quinolyloxy)acetic acid hydrochloride 984

$C_{11}H_9NO_3S_2$
- 5-(4-Hydroxy-3-methoxybenzylidene)rhodanine 733

$C_{11}H_9NO_4$
- 3-Carboxy-4-hydroxy-6-methoxyquinoline 982
- 3-Carboxy-4-hydroxy-7-methoxyquinoline 982
- α-Cyano-4-hydroxy-3-methoxycinnamic acid 761
- 2,4-Diacetoxybenzonitrile 763
- Methyl 2-phthalimidoacetate 721
- 4-(3-Nitrophenyl)-1,3-butadiene-1-carboxylic acid 343
- 3-Phthalimidopropionic acid 720
- Phthaloyl-DL-alanine 721
- 2-Succinimidobenzoic acid 722
- 4-Succinimidobenzoic acid 722

$C_{11}H_9NO_5$
- 2′-Carboxymaleanilic acid 467
- *N*-(Ethoxycarbonyloxy)phthalimide 721

$C_{11}H_9NS_2$
- *N*-(1-Naphthyl)-dithiocarbamic acid 750

$C_{11}H_9N_3O$
- 1-(3-Cyanophenyl)-3-methyl-2-pyrazolin-5-one 765
- 4-Methylpyrimido(1,2-a)benzimidazol-2-ol 918

$C_{11}H_9N_3O_2$
- 2-Anilino-5-nitropyridine 943

$C_{11}H_9N_3O_3$
- 5-(Anilinomethylene)barbituric acid 700
- 1-(Cyclopropylcarbonyl)-5-nitrobenzimidazole 904
- 6-Methyl-2-(4-nitrophenyl)-3(2*H*)-pyridazinone 703
- 1-(α-Nitrofurfurylidene)-2-phenylhydrazine 1068

$C_{11}H_9N_3O_3S_2$
- 2-Acetamido-5-(4-nitrophenylthio)thiazole 876

$C_{11}H_9N_3O_5$
- Maleic acid (3-nitrobenzylidene)hydrazide 500

$C_{11}H_9N_3O_5S_2$
- 2-Acetamido-5-(4-nitrophenylsulfonyl)-thiazole 877

$C_{11}H_9N_5$
- 4-Amino-1-phenylpyrazolo(3,4-d)pyrimidine 923
- 4-(3-Indolylmethyleneamino)-4*H*-1,2,4-triazole 1054

$C_{11}H_{10}$
- 1,4-Dihydro-1,4-methanonaphthalene 162

$C_{11}H_{10}AsNO_5$
- 4-(2-Furamido)benzenearsonic acid 1126

$C_{11}H_{10}BrClF_3NO$
- 2-Bromo-4′-chloro-2′-(trifluoromethyl)butyranilide 396

$C_{11}H_{10}BrNO_2$
- 2-(2-Bromoethyl)-1,4(2*H*,3*H*)-isoquinolinedione 693
- *N*-(2-(Bromomethyl)phenyl)succinimide 714

$C_{11}H_{10}BrNO_3$
4′-Bromo-2′-methylmaleanilic acid 455
N-(3-Bromopropoxy)phthalimide 714

$C_{11}H_{10}Br_2ClNO_3$
3′-Chloro-2,3-dibromo-4′-methylsuccinanilic acid 407

$C_{11}H_{10}Br_2ClNO_4$
5′-Chloro-2,3-dibromo-2′-methoxysuccinanilic acid 401

$C_{11}H_{10}Br_2N_2O_5$
2,3-Dibromo-2′-methyl-4′-nitrosuccinanilic acid 427

$C_{11}H_{10}Br_2N_2O_6$
2,3-Dibromo-2′-methoxy-3′-nitrosuccinanilic acid 416
2,3-Dibromo-2′-methoxy-4′-nitrosuccinanilic acid 415
2,6-Dibromo-4-nitrophenyl *N*-(ethoxycarbonylmethyl)-carbamate 629

$C_{11}H_{10}ClF_3N_2O_2$
O-(*N*-(2-Chloro-5-trifluoromethylphenyl)carbamoyl)-acetone oxime 1008

$C_{11}H_{10}ClF_3O$
3-Chloro-4-(α,α,α-trifluoro-*m*-tolyl)-2-butanone 282

$C_{11}H_{10}ClF_3O_3$
4′-Methoxy-2′-(2-chloro-1,1,2-trifluoroethoxy)acetophenone 288

$C_{11}H_{10}ClN$
4-Chloro-6,8-dimethylquinoline 995
4-Chloro-8-ethylquinoline 995
1-(4-Chlorophenyl)-1-cyclobutane-carbonitrile 756
1-Phenylpyridinium chloride 957

$C_{11}H_{10}ClNO$
7-Chloro-4-ethoxyquinoline 987
5-Chloro-4-hydroxy-8-methylquinaldine 984
2-Chloro-6-methoxy-4-methylquinoline 987
4-Chloro-6-methoxy-2-methylquinoline 987
3-Chloro-2-nitroso-1,2,3,4-tetrahydro-1,4-methanonaphthalane dimer 257

$C_{11}H_{10}ClNO_2$
5-Chloro-2-methyl-3-indoleacetic acid 890
N-(3-Chloro-2-methylphenyl)-succinimide 716
N-(4-Chlorophenyl)glutarimide 715
N-(3-Chloropropyl)-phthalimide 714

$C_{11}H_{10}ClNO_3$
N-(3-Chloro-2-hydroxypropyl)-phthalimide 714
3′-Chloro-2′-methylmaleanilic acid 476
3′-Chloro-4′-methylmaleanilic acid 476
1-(4-Chlorophenyl)-5-oxo-3-pyrrolidinecarboxylic acid 521
N-(3-Chloropropoxy)-phthalimide 714
3-(*N*-(6-Chloro-*o*-tolyl)-carbamoyl)-acrylic acid, sodium salt 476

$C_{11}H_{10}ClNO_3S$
4-Chloro-*N*-furfurylbenzenesulfonamide 845

$C_{11}H_{10}ClNO_4$
3′-Chloro-6′-methoxymaleanilic acid 449
Ethyl 2-(chlorocarbonyl)oxanilate 434

$C_{11}H_{10}ClNO_5$
Ethyl 4-chloro-3-nitrobenzoylacetate 364

$C_{11}H_{10}ClN_3$
2-Anilino-4-chloro-6-methylpyrimidine 975
2-(4-Chlorobenzylamino)pyrazine 977
4-Chlorophenyl 2-pyrrolyl ketone hydrazone 1024

$C_{11}H_{10}ClN_3O$
3-Chloro-*N*-(α-methoxy-2-imidazolylmethylene)aniline tetrafluoroborate 865
3′-Chloro-1-methyl-2-imidazolecarboxanilide 865

$C_{11}H_{10}ClN_3O_2$
1-(*N*-(4-Chloro-2-methoxyphenyl)-carbamoyl)imidazole 865
1-(*N*-(5-Chloro-2-methoxyphenyl)-carbamoyl)imidazole 865

$C_{11}H_{10}ClN_3O_2S$
1-(2-Chloro-4-methoxyphenyl)-3-(2-thiazolyl)urea 875
1-(5-Chloro-2-methoxyphenyl)-3-(2-thiazolyl)urea 875

$C_{11}H_{10}ClN_3O_3$
2-Chloro-*N*-cyanomethyl-*N*-ethyl-4-nitrobenzamide 757
5-Phenyl-1,2,4-oxadiazol-3-yl *N*-(2-chloroethyl)carbamate 872

$C_{11}H_{10}Cl_2$
2,3-Dichloro-1,2,3,4-tetrahydro-1,4-methanonaphthalene 173

$C_{11}H_{10}Cl_2F_3NO$
2′,4-Dichloro-5′-(trifluoromethyl)-butyranilide 390

$C_{11}H_{10}Cl_2N_2O$
1-(*N*-(2,3-Dichlorophenyl)-carbamoyl)-2-pyrroline 695

$C_{11}H_{10}Cl_2N_2OS$
N-(3-Cyano-4,5,6,7-tetrahydrobenzo(b)thiophen-2-yl)-2,2-dichloroacetamide 855

$C_{11}H_{10}Cl_2N_2O_2$
Propyl *N*-cyano-*N*-(3,4-dichlorophenyl)carbamate 752

$C_{11}H_{10}Cl_2OS_2$
2′,5′-Dichloro-3,3-bis(methylthio)-acrylophenone 294

$C_{11}H_{10}Cl_2O_2$
Ethyl 2,6-dichlorocinnamate 365

$C_{11}H_{10}Cl_3NO_3$
2′,4′,5′-Trichloroglutaranilic acid 407

$C_{11}H_{10}Cl_3NO_4$
Methyl *N*-(1-(benzoyloxy)-2,2,2-trichloroethyl)carbamate 610

$C_{11}H_{10}CrO_3$
o-Xylene chromium tricarbonyl 1140
p-Xylene chromium tricarbonyl 1140

$C_{11}H_{10}FN_3OS$
5-Cyano-6-(2-fluorophenyl)-4-hydroxy-3,4,5,6-tetrahydro-2(1*H*)-pyrimidinethione potassium salt 751

$C_{11}H_{10}FN_3O_2$
1-(4-Fluorophenyl)-3-(5-methylisoxazol-3-yl)urea 871

$C_{11}H_{10}F_3NO$
3′-(Trifluoromethyl)-2-butenanilide 446

$C_{11}H_{10}F_3NO_2$
Allyl *N*-(3-trifluoromethylphenyl)-carbamate 610

$C_{11}H_{10}F_3NO_3$
N-Benzyl-*N*-(trifluoroacetyl)glycine 397
2′-(Trifluoromethyl)-succinanilic acid 396
3′-(Trifluoromethyl)-succinanilic acid 397

$C_{11}H_{10}F_3N_3O_2S$
1-(1-Thioacetamido)-4-(3-(trifluoromethyl)phenyl)oxamide 728

$C_{11}H_{10}F_3N_3O_5$
2′,6′-Dinitro-*N*-ethyl-4′-(trifluoromethyl)acetanilide 396
2′,6′-Dinitro-*N*-methyl-4′-(trifluoromethyl)propionanilide 396
2′,6′-Dinitro-4′-(trifluoromethyl)-butyranilide 396

$C_{11}H_{10}F_3N_3O_6$
Ethyl *N*-(2,6-dinitro-4-(trifluoromethyl)phenyl)-*N*-methylcarbamate 609
Isopropyl *N*-(2,6-dinitro-4-(trifluoromethyl)phenyl)carbamate 609
Methyl *N*-(2,6-dinitro-4-(trifluoromethyl)phenyl)-*N*-ethylcarbamate 609
Propyl *N*-(2,6-dinitro-4-(trifluoromethyl)phenyl)carbamate 609

$C_{11}H_{10}F_6O$
2,5-Bis(trifluoromethyl)-α-ethylbenzyl alcohol 197

$C_{11}H_{10}INO_2$
N-(4-Iodophenyl)glutarimide 715

$C_{11}H_{10}N_2$
1-Indolepropionitrile 885
3-Indolepropionitrile 891
3-Methyl-4-indoleacetonitrile 891
2-Styrylimidazole 869

$C_{11}H_{10}N_2O$
3-Acetamidoquinoline 988
4-Acetyl-1-phenyl-1*H*-pyrazole 863
α-Cyano-4-methylcinnamamide 764
4α,5-Dihydro-2*H*-indeno(1,2-c)-pyridazin-3(4*H*)-one 711
3,4-Dihydropyrimido(2,1-a)isoindol-6(2*H*)-one 726
2-Methylcinchoninamide 993
1-(1-Naphthyl)urea 603
1-(2-Naphthyl)urea 579
1-Tetrahydronorharmanone 884

$C_{11}H_{10}N_2OS$
2-Acetamido-4-phenylthiazole 877
3-Allyl-1,2-dihydro-2-thioxo-4(3*H*)-quinazolinone 746
4-Benzyl-6-hydroxy-2-mercaptopyrimidine 970
3-Methyl-*N*-(2-thiazolyl)benzamide 877

$C_{11}H_{10}N_2O_2$
1-Benzyluracil 705
α-Cyano-2-methoxycinnamamide 762
α-Cyano-4-methoxycinnamamide 762
1,2-Dihydro-1-methyl-2-phenyl-3,6-pyridazinedione 709
4-Hydroxy-6-methyl-2-phenyl-3(2*H*)-pyridazinone 697
1-Hydroxy-3-(1-naphthyl)urea 575
1-(2-Methoxybenzoyl)imidazole 866
Methyl 1-phenylimidazole-5-carboxylate 867
5-Methyl-1-phenyl-3-pyrazolecarboxylic acid monohydrate 863
3-(2-Nitro-1-propenyl)indole 885
8-Quinolinyl *N*-methylcarbamate 992

$C_{11}H_{10}N_2O_2S$
1-Acetyl-5-phenyl-2-thiohydantoin 709
2-Amino-4-phenyl-5-thiazolylacetic acid 876
N-(2-Benzothiazolyl)-acetoacetamide 916
5,6-Dihydro-3-(3,4-dihydroxyphenyl)-imidazo(2,1-b)thiazole hydrochloride 879
3-(3,4-Dihydroxyphenyl)-5,6-dihydroimidazo(2,1-b)thiazole hydrochloride 245
6-Methoxy-α-(2-thiazolylimino)-*o*-cresol 1030

$C_{11}H_{10}N_2O_2S_2$
5,6-Dihydro-3-(2-thienyl)imidazo(2,1-b)thiazole-2-acetic acid hydrobromide 858

$C_{11}H_{10}N_2O_3$
2-(Carboxymethylene)acetic acid N^2-benzylidenehydrazide 501
2-Cyanoacetyl *N*-(*p*-tolyl)carbamate 757
N-(4-Ethoxycarbonylphenyl)-cyanoformamide 763
6-Methoxy-3-indoleglyoxylamide 882
6-Methoxy-2-methyl-8-nitroquinoline 985
6-Methoxy-4-methyl-8-nitroquinoline 985

$C_{11}H_{10}N_2O_3S$
N-(3-Cyanopropyl)saccharin 819
1,3-Dimethyl-5-(2-thenylidene)-barbituric acid 858
1-(2-Naphthylsulfonyl)-urea 826
4-Oxo-3-phenyl-2-thioxo-1-imidazolidineacetic acid 745

$C_{11}H_{10}N_2O_4$
2-Cyanoacetyl *N*-(3-methoxyphenyl)carbamate 753
2-Cyanoacetyl *N*-(4-methoxyphenyl)carbamate 753
2-Cyano-1-methylethyl 3-nitrobenzoate 757
Maleic acid (3-hydroxybenzylidene)-hydrazide 496
N-(3-Nitrophenyl)glutarimide 718

$C_{11}H_{10}N_2O_4S$
3′-Sulfamoyl-2-furanilide 851

$C_{11}H_{10}N_2O_5$
2′-Methyl-5′-nitromaleanilic acid 473
4′-Methyl-2′-nitromaleanilic acid 469
4′-Methyl-3′-nitromaleanilic acid 473

$C_{11}H_{10}N_2O_6$
2′-Methoxy-4′-nitromaleanilic acid 465

$C_{11}H_{10}N_2S$
5,6-Dihydro-3-phenylimidazo(2,1-b)-thiazole 879
3-Methyl-6-phenyl-4(3*H*)-pyrimidinethione 740
2-Thiophenecarboxaldehyde phenylhydrazone 1068

$C_{11}H_{10}N_3O$
2-Anilino-*N*-nitrosopyridine 954

$C_{11}H_{10}N_4O$
1-Phenyl-3-(2-pyrimidinyl)urea 976

$C_{11}H_{10}N_4OS$
3-Methyl-8-phenyl-2-thiohypoxanthine 745

$C_{11}H_{10}N_4O_2$
4-Benzylideneamino-6-methyl-1,2,4-triazine-3,5(2*H*,4*H*)-dione 1070

$C_{11}H_{10}N_4O_3$
4-Amino-6-methyl-2-(4-nitrophenyl)-3(2*H*)-pyridazinone 701

$C_{11}H_{10}N_4O_3S$
1-(2-Methyl-5-nitrophenyl)-3-(2-thiazolyl)urea 876
1-(4-Methyl-2-nitrophenyl-3-(2-thiazolyl)urea 876

$C_{11}H_{10}N_4O_4$
3,5-Dimethyl-1-(2,4-dinitrophenyl)-pyrazole 861

$C_{11}H_{10}N_4O_4S$
1-(2-Methoxy-5-nitrophenyl)-3-(2-thiazolyl)urea 875
1-(2-Methyl-5-nitrophenyl)-3-(4-oxo-2-thiazolidinylidene)urea 740
N^1-(5-Nitro-2-pyridyl)-sulfanilamide 941

$C_{11}H_{10}N_6$
2,6-Diamino-8-phenylpurine 923

$C_{11}H_{10}O$
1-Methyl-2-naphthol 196

$C_{11}H_{10}OS$
Benzo(b)thiophene-3-yl ethyl ketone 898

$C_{11}H_{10}O_2$
α-Benzylidene-γ-butyrolactone 386
Cinnamylideneacetic acid 345
2,6-Dimethyl-2-chromen-4-one 320
4,6-Dimethylcoumarin 386
2-Hydroxy-3-methoxynaphthalene 195
1-Hydroxy-1-methylnaphthalen-2(1*H*)-one 282
1-Methylindene-2-carboxylic acid 345
Pentacyclo[6.2.1.0^{2,7}.0^{4,10}.0^{5,9}]-undecane-3,6-dione 57

$C_{11}H_{10}O_2S$
Methyl 2-naphthyl sulfone 786

$C_{11}H_{10}O_2S_2$
2-Methyl-5-(phenylsulfonyl)-thiophene 858

$C_{11}H_{10}O_3$
2-Benzyl-3,4-dihydroxycrotonic acid γ-lactone 383
2-Benzylsuccinic anhydride 388
4,8-Dimethyl-7-hydroxycoumarin 383
α-(2-Hydroxybenzylidene)-γ-butyrolactone 383
Methyl 2-formylcinnamate 365
Methyl 3-formylcinnamate 365
Methyl 4-formylcinnamate 365
2-(3-Oxo-1-indanyl)acetic acid 331
1,4,4a,8a-Tetrahydro-1,4-methano-naphthalene-5,8-dione-6,7-epoxide 20

$C_{11}H_{10}O_4$
3-Acetoxycinnamic acid 365
5,8-Dioxo-1,4,4a,5,8,8a-hexahydro-1-naphthalenecarboxylic acid 65
(3-Methoxystyryl)glyoxylic acid 335
2-Phenyl-1,1-cyclopropanedicarboxylic acid hydrate 346

$C_{11}H_{10}O_4S$
3-Benzoyloxy-2,3-dihydrothiophene-1,1-dioxide 781
4,5-Dimethoxybenzo(b)thiophene-2-carboxylic acid 897
2,2-Dimethyl-5-(2-thienylidene)-1,3-dioxane-4,6-dione 858
α-Hydroxy-1-naphthalenemethane-sulfonic acid, sodium salt 792
4-Methoxy-1-naphthalenesulfonic acid, sodium salt 792
Methyl 5-furfurylidene-4-oxo-3-tetrahydrothiophenecarboxylate 849

$C_{11}H_{10}O_8S_2$
Menadiol disulfate disodium salt trihydrate 796

$C_{11}H_{10}S$
2-(Mercaptomethyl)naphthalene 209

$C_{11}H_{11}Br$
5-Bromo-1-phenyl-1-pentyne 1015

$C_{11}H_{11}BrClNO_3$
4′-Bromo-2′-chloroglutaranilic acid 402

$C_{11}H_{11}BrN_2$
1-(2-Pyridylmethyl)pyridinium bromide 944

$C_{11}H_{11}BrN_2O$
4-Bromoantipyrine 710

$C_{11}H_{11}BrO_2$
2-Bromo-3,4-dihydro-7-methoxy-1(2*H*)-naphthalenone 298

$C_{11}H_{11}BrO_2S_2$
5-Bromo-6-(1,3-dithian-2-yl)-1,3-benzodioxole 213

$C_{11}H_{11}BrO_3$
4-Bromo-6,7-dimethoxy-1-indanone 291

$C_{11}H_{11}BrO_4$
2-Bromo-4,5-(dimethoxy)-cinnamic acid 334

$C_{11}H_{11}Br_2NO_4$
N-Acetyl-3,5-dibromo-L-tyrosine hemihydrate 398

$C_{11}H_{11}ClF_3NO$
4′-Chloro-2′-(trifluoromethyl)-butyranilide 396

$C_{11}H_{11}ClF_3NOS$
S-Propyl *N*-(2-chloro-5-(trifluoromethyl)phenyl)-thiocarbamate 611

$C_{11}H_{11}ClF_3NO_2$
Isopropyl *N*-(2-chloro-5-(trifluoromethyl)phenyl)carbamate 607
Isopropyl *N*-(4-chloro-3-(trifluoromethyl)phenyl)carbamate 607
Propyl *N*-(2-chloro-5-(trifluoromethyl)phenyl)carbamate 607

$C_{11}H_{11}ClF_3N_3O_3$
4-(4-Chloro-2-(trifluoromethyl)-phenyl)-1-(ethoxycarbonyl)-semicarbazide 606

$C_{11}H_{11}ClIN$
4-Chloro-1,2-dimethylquinolinium iodide 990

$C_{11}H_{11}ClINO_2S$
3-(2-Carboxyethyl)-5-chloro-2-methylbenzothiazolium iodide 915

$C_{11}H_{11}ClN_2$
7-Chloro-4-(dimethylamino)-quinoline dipicrate 990
2-(2-Chloroethylamino)quinoline hydrochloride 981
7-Chloro-1,2,3,4-tetrahydropyrido(1,2-a)-benzimidazole 251

$C_{11}H_{11}ClN_2O$
7-Chloro-4-(2-hydroxyethylamino)-quinoline 983

$C_{11}H_{11}ClN_2OS$
3-(2-(2-Chloroacetamido)-phenylthio)propionitrile 751
5-(4-Chlorophenyl)-3-methyl-2-methylimino-4-thiazolidinone 732

$C_{11}H_{11}ClN_2O_2$
2-Cyano-2-propyl *N*-(3-chlorophenyl)carbamate 757
Isopropyl *N*-(4-chlorophenyl)-*N*-cyanocarbamate 752
Propyl *N*-(4-chlorophenyl)-*N*-cyanocarbamate 752

$C_{11}H_{11}ClN_2O_5$
2′-Chloro-5′-nitroglutaranilic acid 404

$C_{11}H_{11}ClN_2S$
2-Amino-5-(2-chloroethyl)-4-phenylthiazole hydrobromide 874

$C_{11}H_{11}ClN_4O_2$
4-(3-Chloro-4-methylphenyl)-1-(2-cyanoacetyl)semicarbazide 752

$C_{11}H_{11}ClN_4O_3$
4-(5-Chloro-2-methoxyphenyl)-1-(2-cyanoacetyl)semicarbazide 751
1-(4-Chloro-3-nitrophenyl)-3-(2-cyanoethyl)-3-methylurea 751

$C_{11}H_{11}ClO$
1-(2-Chlorophenyl)-1-penten-3-one 286

$C_{11}H_{11}ClO_2S$
5-Chloro-3,4-dihydro-6-phenyl-2*H*-thiopyran-1,1-dioxide 783

$C_{11}H_{11}ClO_3$
4-(4-Chlorophenyl)-3-methyl-4-oxobutyric acid 325
Ethyl 4-chlorobenzoylacetate 364

$C_{11}H_{11}ClO_3S$
5-Chloromethyl-2-(3,4-methylenedioxyphenyl)-1,3-oxathiolane 209

$C_{11}H_{11}ClO_4$
5-(Chloromethyl)meconin 382
2-(4-Chlorophenyl)-3-methylsuccinic acid 325
Methyl 3-(5-chloro-2-methoxyphenyl)-2,3-epoxypropionate 354

$C_{11}H_{11}Cl_2NO$
N-(4,5-Dichloro-2-methylphenyl)-cyclopropanecarboxamide 408
4-(2,4-Dichlorophenylimino)-2-pentanone 1074

$C_{11}H_{11}Cl_2NO_2$
4-(2,4-Dichlorobenzoyl)-morpholine 707

$C_{11}H_{11}Cl_2NO_3$
2′,3′-Dichloroglutaranilic acid 407
2′,4′-Dichloroglutaranilic acid 407
2′,5′-Dichloroglutaranilic acid 407
3′,4′-Dichloroglutaranilic acid 407

$C_{11}H_{11}Cl_2NO_4$
3,5-Dichlorophenyl *N*-(ethoxycarbonylmethyl)-carbamate 640

$C_{11}H_{11}Cl_2NO_5$
2-(Carboxymethoxy)-3′,5′-dichloro-2′-hydroxy-4′-methylacetanilide 398

$C_{11}H_{11}Cl_3N_2O_2$
4-(*N*-(2,4,5-Trichlorophenyl)-carbamoyl)morpholine 714

$C_{11}H_{11}Cl_3N_2O_3$
Ethyl 2-(2,4,5-trichlorophenylureido)acetate 532

$C_{11}H_{11}Cl_3N_2O_5$
Ethyl *N*-(1-(4-nitrophenoxy)-2,2,2-trichloroethyl)carbamate 608

$C_{11}H_{11}Cl_4NO_2$
1,2,2,2-Tetrachloroethyl *N*-ethyl-*N*-phenylcarbamate 610

$C_{11}H_{11}Cl_4NO_3$
Ethyl *N*-(1-(4-chlorophenoxy)-2,2,2-trichloroethyl)carbamate 606

$C_{11}H_{11}CrNO_3$
N,*N*-Dimethylaniline chromium tricarbonyl 1140
2,6-Xylidine chromium tricarbonyl 1140

$C_{11}H_{11}FN_2O_5$
4′-Fluoro-3′-nitroglutaranilic acid 404

$C_{11}H_{11}FO$
Cyclobutyl 4-fluorophenyl ketone 289

$C_{11}H_{11}FO_4$
3-(3-Fluoro-*p*-anisoyl)-propionic acid 328

$C_{11}H_{11}FO_5S$
Ethyl 3-(fluorosulfonyl)-benzoylacetate 803

$C_{11}H_{11}F_2NO_3$
2′,4′-Difluoroglutaranilic acid 407

$C_{11}H_{11}F_3N_2O$
1-Allyl-3-(3-trifluoromethylphenyl)urea 510

$C_{11}H_{11}F_3N_2O_2$
O-(*N*-(3-(Trifluoromethyl)phenyl)-carbamoyl)acetone oxime 1008

$C_{11}H_{11}F_3N_2O_3$
1-(2-Methoxy-5-methylphenyl)-3-(2,2,2-trifluoroacetyl)urea 508

$C_{11}H_{11}F_3N_2O_5$
Butyl 2,6-dinitro-4-(trifluoromethyl)-phenyl ether 259

$C_{11}H_{11}F_3N_4O$
3-(Guanidinoimino)-4,4,4-trifluorobutyrophenone nitrate 1021

$C_{11}H_{11}F_3O$
3-Methyl-4,4,4-trifluorobutyrophenone 288
4-(4-(Trifluoromethyl)phenyl)-2-butanone 282

$C_{11}H_{11}F_3O_2$
(+/-)-1-(Trifluoromethyl)ethyl phenylacetate 353

$C_{11}H_{11}IO_3$
Ethyl *o*-iodobenzoylacetate 363

$C_{11}H_{11}I_2NO_4$
N-Acetyl-3,5-diiodo-L-tyrosine 413

$C_{11}H_{11}N$
1-Amino-2-methylnaphthalene hydrochloride 231
(2-Naphthylmethyl)amine hydrochloride 241
1-(*p*-Tolyl)-1-cyclopropanecarbonitrile 756

$C_{11}H_{11}NO$
4-Amino-2-methyl-1-naphthol hydrochloride 226
3-Amino-2-naphthalenemethanol 226
4,6-Dimethyl-2-hydroxyquinoline 984
4,8-Dimethyl-2-hydroxyquinoline 984
4,7-Dimethyl-2-quinolinol 984
6-Ethoxyquinoline 987
8-Ethoxyquinoline 986
6-Ethoxyquinoline hydrochloride 986
8-Ethyl-4-quinolinol 984
4-Methoxy-2-methylquinoline 987
8-Methoxy-2-methylquinoline 986
6-Methoxy-4-methylquinoline hydrochloride 986
1-(4-Methoxyphenyl)-1-cyclopropanecarbonitrile 753
4-Methyl-3-phenyl-2-pyrrolin-5-one 709
3-Oxo-3-(3,5-xylyl)propionitrile 758
1-Propionylindole 890
3-Propionylindole 888

$C_{11}H_{11}NO_2$
α-Acetyl-2-methoxyphenylacetonitrile 753
1-Benzoyl-2-pyrrolidinone 693
2,3-Dimethyl-7-indolecarboxylic acid 889
5-Ethyl-2-indolecarboxylic acid 889
1-Indolepropionic acid 887
N-Isopropylphthalimide 723
3-(4-Methoxybenzoyl)propionitrile 753
7-Methoxy-2-methyl-4-quinolinol 983
2-Methylindole-3-acetic acid 887
Methyl 3-indolylacetate 887
Methyl 1-methylindole-2-carboxylate 887
Methyl 3-methyl-1-indolecarboxylate 890
Methyl 3-methyl-2-indolecarboxylate 889
N-Phenylglutarimide 723
N-(*o*-Tolyl)succinimide 720
N-(*p*-Tolyl)succinimide 720

$C_{11}H_{11}NO_2S$
2-Benzothiazolebutyric acid 916
6-Methoxybenzo(b)thiophene-3-acetamide 897

$C_{11}H_{11}NO_2S_2$
4-(2-Benzothiazolylthio)-butyric acid 913
3-(4-Ethoxyphenyl)-rhodanine 737

$C_{11}H_{11}NO_3$
2-Acetamidocinnamic acid 433
4-Acetamidocinnamic acid 433
N-(*p*-Anisyl)-succinimide 717
2-Benzamidocrotonic acid 475
N-Benzylmaleamic acid 472
N-(Cinnamoyl)glycine 349
2-(5,6-Dihydro-4*H*-1,3-oxazin-2-yl)-benzoic acid 341
2-(5,6-Dihydro-4*H*-1,3-oxazin-2-yl)-benzoic acid hydrochloride 341
6,7-Dimethoxy-1(2*H*)-isoquinolone 690
3-(3,5-Dimethoxyphenyl)-3-oxopropionitrile 752
2,3-Dimethoxystyryl isocyanate 768
2-(2-Hydroxyethyl)-1,4-(2*H*,3*H*)-isoquinolinedione 697
3′-Methylmaleanilic acid 475
4′-Methylmaleanilic acid 476
2-Oxo-4-phenyl-3-pyrrolidinecarboxylic acid 692

$C_{11}H_{11}NO_3S$
Methyl 2,3-dihydro-3-oxo-4*H*-1,4-benzothiazine-2-acetate 709

$C_{11}H_{11}NO_4$
N-(Acetoacetyl)anthranilic acid 443
Ethyl 3-nitrocinnamate 365
cis-α-Ethyl-3-nitrocinnamic acid 335
2′-Hydroxy-5′-methylmaleanilic acid 461
4-(2-Methoxyanilino)-4-oxo-2-butenoic acid 465
3′-Methoxymaleanilic acid 466
4′-Methoxymaleanilic acid 466
3,4,5-Trimethoxybenzoyl cyanide 761

$C_{11}H_{11}NO_4S$
8-Ethoxy-5-quinolinesulfonic acid 986
2-Ethyl-5-phenylisoxazolium 4′-sulfonate monohydrate 871
α-Methyl-2-(methylthio)-5-nitrocinnamic acid 335

$C_{11}H_{11}NO_5$
N-Benzoyl-D-aspartic acid 473
4-(3-Carboxypropionamido)-benzoic acid 433
2′-Carboxysuccinanilic acid 420
Ethyl 3-(2-nitrophenyl)-2-oxiranecarboxylate 356

$C_{11}H_{11}NO_6$
2-Carboxyphenyliminodiacetic acid 343
N-(2-Hydroxybenzoyl)-L-aspartic acid 461
2-Nitrotoluene-α,α-diol diacetate 356
4-Nitrotoluene-α,α-diol diacetate 356

$C_{11}H_{11}NS_3$
4-(4-(Dimethylamino)phenyl-1,3-dithiole-2-thione 385

$C_{11}H_{11}N_3$
5-Cyano-1-ethyl-2-methylbenzimidazole 905
2-(2,2-Dicyanovinyl)-3,4,5-trimethylpyrrole 839
3-Pyrrolecarboxaldehyde phenylhydrazone 1056

$C_{11}H_{11}N_3O$
2-*N*-Anilino-4-hydroxy-6-methylpyrimidine 971
Formic (1,2-dihydro-2-methyl-1-isoquinolylidene)hydrazide 725
Formic N^1-methyl-N^2-(2-quinolyl)-hydrazide 993
5-Methyl-3-phenyl-4-pyrazolecarboxamide 861

$C_{11}H_{11}N_3OS$
6-Amino-1-benzyl-4-hydroxy-2-pyrimidinethione 734
S-Methyl 5-oxo-3-phenyl-2-pyrazoline-1-thioimidate hydriodide 739

$C_{11}H_{11}N_3O_2$
2-(Acetoacetamido)-benzimidazole 907
5-Anilino-3-methyl-4-pyrazolecarboxylic acid 861
1,2-Dihydro-1-(2-(2-pyridyl)-ethyl)-3,6-pyridazinedione 945
4-Ethyl-1-(4-nitrophenyl)pyrazole 861
Ethyl 2-(phenylazo)cyanoacetate 1088
4-Nitrosoantipyrine 706

$C_{11}H_{11}N_3O_3S_2$
1-(2-Thiazolyl)-3-(*p*-tolylsulfonyl)urea 877

$C_{11}H_{11}N_3O_4$
Ethyl *N*-(2-cyano-4-methyl-6-nitrophenyl)carbamate 762

$C_{11}H_{11}N_3O_7$
N-(2,4-Dinitrophenyl)-L-hydroxyproline 348
$C_{11}H_{11}N_3O_8$
N-Acetyl-3,5-dinitro-L-tyrosine 413
N-(2,4-Dinitrophenyl)-DL-glutamic acid hemihydrate 349
$C_{11}H_{11}N_3S$
6,7-Dihydro-3-imino-1-phenyl-3*H*,5*H*-imidazo(1,2-c)thiazole hydrochloride 749
$C_{11}H_{11}N_5$
2,6-Diamino-3-(phenylazo)pyridine 1087
2,6-Diamino-3-phenylazopyridine ethanedisulfonate 1087
2,6-Diamino-3-(phenylazo)-pyridine hydrochloride 1087
2,6-Diamino-3-(phenylazo)pyridine 2-hydroxyethanesulfonate 1088
$C_{11}H_{11}N_5O$
2,4-Diamino-6-phenylacetyl-1,3,5-triazine 979
4-(2,6-Diamino-3-pyridylazo)-phenol .. 1083
$C_{11}H_{11}N_5O_4$
1-(2-Cyanoacetyl)-4-(4-methyl-2-nitrophenyl)semicarbazide 754
$C_{11}H_{12}$
3-Methyl-1-phenyl-1,3-butadiene 164
1-Phenyl-1-cyclopentene 164
1-(*o*-Tolyl)-1,3-butadiene 164
$C_{11}H_{12}BF_4N$
1-Ethylquinolinium tetrafluoroborate 995
$C_{11}H_{12}BF_4NO$
2-Ethyl-5-phenylisoxazolium tetrafluoroborate 871
$C_{11}H_{12}BrClO_2$
4-Bromophenyl 3-chloro-2,2-dimethylpropionate 376
$C_{11}H_{12}BrCl_2NO$
2-Bromo-2′,3′-dichloro-*N*-methylbutyranilide 410
$C_{11}H_{12}BrCl_2NO_2$
N-(2-(4-Bromophenoxy)-1-methylethyl)-2,2-dichloroacetamide 394
$C_{11}H_{12}BrNO$
2-Ethyl-5-hydroxyisoquinolinium bromide 996
$C_{11}H_{12}BrNO_2$
Methyl 4-(2-bromoanilino)-2-butenoate 366
$C_{11}H_{12}BrNO_3$
3′-Bromoglutaranilic acid 405
4′-Bromo-2′-methylsuccinanilic acid 407
4′-Bromo-3′-methylsuccinanilic acid 402
N-(4-Bromophenyl)glutaranilic acid .. 402
dl-*N*-(2-Bromo-3-phenylpropionyl)-glycine 436
$C_{11}H_{12}BrN_3$
7-Bromo-2,3,6,8-tetramethylpyrido(2,3-b)pyrazine 998
$C_{11}H_{12}Br_2ClNO$
N-(4-Chloro-2,6-dibromophenyl)-2,2,2-trimethylacetamide 410
$C_{11}H_{12}Br_2O_2$
Ethyl 2,3-dibromo-3-phenylpropionate 361
$C_{11}H_{12}Br_2O_4$
2,3-Dibromo-3-(3,4-dimethoxyphenyl)propionic acid 327
$C_{11}H_{12}ClF_3N_2O$
1-(2-Chloro-5-trifluoromethylphenyl)-3-isopropylurea 506
1-(4-Chloro-2-trifluoromethylphenyl)-3-*N*-propylurea 506
$C_{11}H_{12}ClNO$
2-Chloroacetyl-1,2,3,4-tetrahydroisoquinoline 693
6′-Chloro-*o*-crotonotoluidide 481
2′-Chloro-6′-methyl-1-cyclopropanecarboxanilide 408
3′-Chloro-4′-methyl-1-cyclopropanecarboxanilide 408
5′-Chloro-2′-methyl-1-cyclopropanecarboxanilide 408
N-(3-Chloro-2-methylphenyl)-cyclopropanecarboxamide 408
4-(4-Chlorophenylimino)-2-pentanone 1074
4-(4-Chlorophenyl)-4-methoxybutyronitrile 752
$C_{11}H_{12}ClNO_2$
Allyl *N*-(2-chloro-6-methylphenyl)-carbamate 635
Allyl *N*-(3-chloro-4-methylphenyl)-carbamate 685
Allyl *N*-(4-chloro-2-methylphenyl)-carbamate 685
Allyl *N*-(5-chloro-2-methylphenyl)-carbamate 687
3′-Chloro-*o*-acetoacetotoluidide 444
4′-Chloro-*o*-acetoacetotoluidide 444
4-Chloro-*o*-acetoacetotoluidide 442
5′-Chloro-*o*-acetoacetotoluidide 444
4-(2-Chlorobenzoyl)-morpholine 710
4′-Chloro-2′-methoxycyclopropane-carboxanilide 401
Hydrastinine chloride monohydrate 255
$C_{11}H_{12}ClNO_3$
N-Acetyl-4-chloro-DL-phenylalanine . 404
N-Butyryl-5-chloroanthranilic acid 405
5′-Chloro-*o*-acetoacetanisidide 442
N-(Chloroacetyl)-DL-phenylalanine 392
N-(2-Chloroacetyl)-L-phenylalanine 392
5′-Chloro-2′-methoxyacetoacetanilide 442
2′-Chloro-6′-methylsuccinanilic acid 405
3′-Chloro-2′-methylsuccinanilic acid 402
3′-Chloro-4′-methylsuccinanilic acid 402
4-(*N*-(3-Chlorophenyl)carbamoyl)-butyric acid 405
N-(2-Chlorophenyl)glutaramic acid 406
N-(4-Chlorophenyl)glutaramic acid 407
3-(*N*-(6-Chloro-*o*-tolyl)-carbamoyl)-propionic acid, sodium salt 436
Ethyl *N*-(2-chloroacetyl)anthranilate 393
Ethyl 4-chlorohippurate 452
$C_{11}H_{12}ClNO_4$
4-Chlorobutyl 4-nitrobenzoate 365
1-Ethylquinolinium perchlorate 990
Methyl 4-acetamido-5-chloro-2-methoxy-benzoate 417
$C_{11}H_{12}ClN_3O$
1-(4-Chloro-2-cyanophenyl)-3-isopropylurea 760
$C_{11}H_{12}ClN_3OS$
4-Allyl-1-(2-chlorobenzoyl)-3-thiosemicarbazide 731
$C_{11}H_{12}ClN_3O_2$
Ethyl 2-(4-chloro-3-nitrophenylureido)acetate 524
$C_{11}H_{12}ClN_3O_3$
1-(3-Chloro-4-methylphenyl)-3-(2-oxo-3-oxazolidinyl)urea 630
$C_{11}H_{12}ClN_3O_4$
4-(*N*-(2-Chloro-4-nitrophenyl)-carbamoyl)morpholine 703
$C_{11}H_{12}Cl_2$
5,5-Dichlorotetracyclo[6.2.1.0^{2,7}.0^{4,6}]-undec-9-ene 6
9,9-Dichloro-1,4,5,8-tetrahydro-4a,8a-methanonaphthalene 6
$C_{11}H_{12}Cl_2N_2O$
Butyric (2,6-dichlorobenzylidene)-hydrazide 494
$C_{11}H_{12}Cl_2N_2O_2$
1-Butyryl-2-(2,4-dichlorobenzoyl)-hydrazine 493
1-Butyryl-3-(3,4-dichlorophenyl)urea . 534
4-(*N*-(2,3-Dichlorophenyl)-carbamoyl)morpholine 713
4-(*N*-(2,4-Dichlorophenyl)-carbamoyl)morpholine 705
4-(*N*-(2,5-Dichlorophenyl)-carbamoyl)morpholine 705
4-(*N*-(2,6-Dichlorophenyl)-carbamoyl)morpholine 705
4-(*N*-(3,4-Dichlorophenyl)-carbamoyl)morpholine 705
3-(3,4-Dichlorophenyl)-1-formyl-1-propylurea 534
3-(3,4-Dichlorophenyl)-1-methyl-1-propionylurea 534
2-Hydroxyisobutyric (2,4-dichlorobenzylidene)hydrazide 491
$C_{11}H_{12}Cl_2N_2O_3$
N-(4-Acetamido-3-methoxyphenyl)-2,2-dichloroacetamide 394
Ethyl *N*-(*N*-(3,4-dichlorophenyl)-carbamoyl)-*N*-methylcarbamate 631
$C_{11}H_{12}Cl_2N_4$
1-(2,4-Dichloro-α-methylcinnamylideneamino)-guanidine nitrate 1026
$C_{11}H_{12}Cl_2O$
3-Chloro-4-(4-chlorophenyl)-3-methyl-2-butanone 285
4α-(Dichloromethyl)-2,4a,5,6,7,8-hexahydro-2-naphthalenone 51
$C_{11}H_{12}Cl_2O_3$
Butyl 3,5-dichlorosalicylate 366
$C_{11}H_{12}Cl_2O_4$
2-(2,4-Dichlorophenoxy)ethyl lactate 353
$C_{11}H_{12}Cl_3NO$
2,2-Dimethyl-2′,4′,5′-trichloropropionanilide 410
N-Methyl-2′,3′,4-trichlorobutyranilide 391
2,2,2-Trichloro-2′,4′,6′-trimethylacetanilide 398
$C_{11}H_{12}Cl_3NOS$
S-Butyl *N*-(2,4,5-trichlorophenyl)-thiocarbamate 642
$C_{11}H_{12}Cl_3NO_2$
Butyl *N*-(2,4,5-trichlorophenyl)-carbamate 687
N-(2-(4-Chlorophenoxy)-1-methylethyl)-2,2-dichloroacetamide 394
2,2,2-Trichloroethyl *N*-(3-ethylphenyl)carbamate 610
$C_{11}H_{12}Cl_3NO_3$
2-Ethoxyethyl *N*-(2,4,5-trichlorophenyl)carbamate 616
Ethyl *N*-(1-phenoxy-2,2,2-trichloroethyl)carbamate 609
Methyl *N*-(1-(benzyloxy)-2,2,2-trichloroethyl)carbamate 608
$C_{11}H_{12}Cl_3NO_4$
(γ,4-Dichlorocinnamylidene)-dimethylammonium perchlorate 1062
Methyl *N*-(1-(4-methoxyphenoxy)-2,2,2-trichloroethyl)carbamate 608
$C_{11}H_{12}Cl_3N_3O_3$
1-(2,5-Dimethoxyphenyl)-3-(2,2,2-trichloroacetimidoyl)urea 723
$C_{11}H_{12}Cl_4OSi$
2′,3′,5′,6′-Tetrachloro-4′-(trimethylsilyl)acetophenone 1102
$C_{11}H_{12}FN$
4-(4-Fluorophenyl)-1,2,3,6-tetrahydropyridine hydrochloride 250
$C_{11}H_{12}FNO_2$
2-Butenyl *N*-(2-fluorophenyl)-carbamate 633
$C_{11}H_{12}FNO_3$
3′-Fluoroglutaranilic acid 405
4′-Fluoroglutaranilic acid 407
N-(2-Fluorophenyl)glutaramic acid 406
$C_{11}H_{12}FN_3O$
1-(2-Cyanoethyl)-3-(4-fluorophenyl)-1-methylurea 751
$C_{11}H_{12}F_2N_2OS$
4-(*N*-(2,4-Difluorophenyl)-thiocarbamoyl)morpholine 732
$C_{11}H_{12}F_2N_2O_3$
1-(2,4-Difluorophenyl)-3-(ethoxycarbonylmethyl)urea 532
$C_{11}H_{12}F_3NOS$
S-Isopropyl *N*-(3-(trifluoromethyl)-phenyl)thiocarbamate 611
S-Propyl *N*-(2-(trifluoromethyl)-phenyl)thiocarbamate 611
S-Propyl *N*-(3-(trifluoromethyl)-phenyl)thiocarbamate 611
$C_{11}H_{12}F_3NO_2$
Isopropyl *N*-(2-(trifluoromethyl)phenyl)carbamate .. 611
Isopropyl *N*-(3-(trifluoromethyl)phenyl)carbamate .. 610
Propyl *N*-(3-(trifluoromethyl)phenyl)-carbamate 610
2,2,2-Trifluoroethyl *N*-(2,5-xylyl)-carbamate 611
$C_{11}H_{12}F_3NO_3$
2-Methoxyethyl *N*-(3-(trifluoromethyl)phenyl)carbamate .. 608
$C_{11}H_{12}F_3NO_4$
2,2,2-Trifluoroethyl *N*-(2,5-dimethoxyphenyl)carbamate 608
$C_{11}H_{12}F_3N_3O_4$
N-Butyl-2,6-dinitro-4-(trifluoromethyl)aniline 259
$C_{11}H_{12}F_3N_3O_6$
Isopropyl *N*-(2,6-dinitro-4-(trifluoromethyl)phenyl)-*N*-methylcarbamate 609
$C_{11}H_{12}F_3N_3O_6S$
2′,6′-Dinitro-*N*-propyl-4′-(trifluoromethyl)-methanesulfonanilide 805
$C_{11}H_{12}IN$
2,3-Dimethylisoquinolinium iodide 998
2-Ethylisoquinolinium iodide 998
1-Ethylquinolinium iodide 990
1-Quinaldinium iodide 990
$C_{11}H_{12}INO$
6-Methoxy-1-methylquinolinium iodide 985
$C_{11}H_{12}INO_3$
N-Acetyl-4-iodo-L-phenylalanine 404
N-(4-Iodophenyl)glutaramic acid 407
$C_{11}H_{12}INO_4$
N-Acetyl-3-iodo-L-tyrosine 398
$C_{11}H_{12}I_2N_2$
1-Methyl-4-pyridiniopyridinium diiodide 958
$C_{11}H_{12}I_2O_3$
Butyl 3,5-diiodo-4-hydroxybenzoate . 368
$C_{11}H_{12}N_2$
1-Cyanomethyl-1,2,3,4-tetrahydro-quinoline 754
4,5-Dihydro-2-styrylimidazole 253
2-(Dimethylamino)quinoline 990
2,4-Dimethyl-1*H*-1,5-benzodiazepine hydrochloride 248
3,5-Dimethyl-1-phenylpyrazole 864
3,5-Dimethyl-4-phenyl-1*H*-pyrazole 864
1-Methyl-4-(*p*-tolyl)-1*H*-pyrazole 864
$C_{11}H_{12}N_2O$
2-Amino-5-ethyl-4-phenyloxazole 872
Antipyrine salicylate 696
4-(*N*-(2-Cyanoethyl)methylamino)-benzaldehyde 755
2-Cyano-2-phenylbutyramide 755
4,5-Dihydro-6-methyl-2-phenyl-3(2*H*)-pyridazinone 705
4,7-Dimethyl-1*H*-1,5-benzodiazepin-2(3*H*)-one 702
1,3-Dimethyl-4-phenyl-1,2-pyrazolin-5-one 863
2,3-Dimethyl-1-phenyl-3-pyrazolin-5-one 708
Formic acid (5-indanylmethylene)-hydrazide 503
N-ω-Formyltryptamine 885
1,2,3,4,6,10b-Hexahydropyrimido(2,1-a)isoindol-6-one 710
2-Hydroxy-3,6,7-trimethylquinoxaline 999
1-Indolepropionamide 885
2-(3-Indolylmethyleneamino)-ethanol 1033
5-Methylindole-3-acetamide 886
Salicylic (2-buten-1-ylidene)-hydrazide 1036
$C_{11}H_{12}N_2OS$
4-(4-(Amino)thiobenzoyl)morpholine. 739
1,2-Dihydro-3-isopropyl-2-thioxo-4(3*H*)-quinazolinone 746

$C_{11}H_{12}N_2OS$ (Cont'd)
1,2-Dihydro-3-propyl-2-thioxo-4(3*H*)-quinazolinone ... 746
$C_{11}H_{12}N_2OS_2$
2-(Morpholinothio)-benzothiazole ... 917
$C_{11}H_{12}N_2O_2$
2′-Carbamoyl-2-butenanilide ... 470
N-(2-Cyanoethyl)-*N*-phenylglycine ... 755
1-Cyano-1-methylethyl *N*-phenylcarbamate ... 757
N-(Dimethylaminomethyl)-phthalimide ... 720
5,6-Dimethylbenzimidazole-1-acetic acid hydrochloride ... 907
7,8-Dimethyl-1*H*-1,5-benzodiazepine-2,4(3*H*,5*H*)-dione ... 702
4-(1,1-Dimethyl-2-nitroethyl)-benzonitrile ... 765
3,5-Dimethyl-5-phenylhydantoin ... 705
3-Ethoxy-1-phenyl-2-pyrazolin-5-one ... 698
3-Ethyl-2-methyl-5-nitroindole ... 885
5-Methoxy-3-indoleacetamide ... 882
Methyl α-cyano-3,4-dimethyl-2-pyrroleacrylate ... 841
4-(1-Methyl-1-nitropropyl)-benzonitrile ... 756
4-(2-Methyl-2-nitropropyl)-benzonitrile ... 765
3-(Phenylazo)-2,4-pentanedione ... 1092
$C_{11}H_{12}N_2O_2S$
1,3-Difurfuryl-2-thiourea ... 852
(5,6-Dimethyl-2-benzimidazolylthio)-acetic acid ... 903
5-Imino-2-(4-methoxyphenyl)-3-methyl-4-oxazolidinethione ... 736
6-Methoxy-α-(2-thiazolylamino)-*o*-cresol ... 875
$C_{11}H_{12}N_2O_3$
1,3-Difurfurylurea ... 851
1,4-Dihydro-6,7-dimethoxy-1-methyl-4-cinnolinone picrate ... 58
2′,4′-Dimethoxy-2-cyanoacetanilide ... 753
5,8-Dimethoxy-3-methyl-2-quinoxalinol ... 998
3-Dimethylamino-2′-nitroacrylophenone ... 295
3-Dimethylamino-3′-nitroacrylophenone ... 295
3-Hydroxy-3-(2-(hydroxyimino)-propyl)-2-indolinone ... 1008
2′-Methyl-4′-nitro-2-butenanilide ... 470
4′-Methyl-3′-nitro-2-butenanilide ... 469
2′-Methyl-4′-nitro-1-cyclopropanecarboxanilide ... 427
$C_{11}H_{12}N_2O_3S$
N-(2-Thiazolyl)-1,2,3,6-tetrahydrophthalamic acid ... 877
$C_{11}H_{12}N_2O_4$
2-Acetamido-1-hydroxy-1-(4-nitrophenyl)-2-propene ... 412
3,5-Diacetamidobenzoic acid ... 426
5-(2,3-Dimethoxyphenyl)hydantoin ... 698
Dimethyl ketomalonate phenylhydrazone ... 373
2′-Methoxy-5′-nitro-2-butenanilide ... 465
4′-Methoxy-2′-nitro-1-cyclopropanecarboxanilide ... 415
*N*α-Benzoylasparagine ... 349
2′-Nitro-6′-propionylacetanilide ... 424
$C_{11}H_{12}N_2O_5$
4′-Methyl-3′-nitrosuccinanilic acid ... 420
4′-Methyl-3′-nitrosuccinanilic acid ... 426
N-(4-Nitrobenzoyl)glycine ethyl ester ... 473
3′-Nitroglutaranilic acid ... 426
4′-Nitroglutaranilic acid ... 427
$C_{11}H_{12}N_2O_6$
N-(2-Hydroxy-3-methoxy-5-nitrobenzylidene)-β-alanine ... 1032
$C_{11}H_{12}N_2O_7$
Ethyl 3,5-dinitro-4-hydroxyhydrocinnamate ... 358
$C_{11}H_{12}N_2S$
2-(Allylthiomethyl)benzimidazole ... 903
2-Amino-4-phenethylthiazole hydrochloride ... 876
2-(Dimethylamino)-4-phenylthiazole ... 876
4-Methyl-2-(*N*-methylanilino)-thiazole ... 876
$C_{11}H_{12}N_4O$
3-Indoleacetaldehyde semicarbazone ... 1059
$C_{11}H_{12}N_4OS$
1-Benzyl-5,6-diamino-2-thiouracil ... 739
$C_{11}H_{12}N_4O_2$
5-(Benzylamino)-1-methyl-4-nitroimidazole ... 866
4-(Benzylideneamino)-3,5-bis(hydroxymethyl)-4*H*-1,2,4-triazole ... 1029
2-(3-Indolyloxy)acetaldehyde semicarbazone ... 882
$C_{11}H_{12}N_4O_2S$
1-(2-Methoxy-5-methylphenyl)-3-(1,3,4-thiadiazol-2-yl)urea ... 878
Sulfamerazine ... 975
$C_{11}H_{12}N_4O_3S$
1-(2,5-Dimethoxyphenyl)-3-(1,3,4-thiadiazol-2-yl)urea ... 878
$C_{11}H_{12}N_4O_5$
1,3-Diethyl-2,4-dioxo-7-hydroxy-1,2,3,4-tetrahydro-6-pteridinecarboxylic acid ... 1002
$C_{11}H_{12}N_4O_6$
1-(2,4,6-Trinitrophenyl)piperidine ... 273
$C_{11}H_{12}N_4O_7$
Ethyl 2-(3-(3,5-dinitrophenyl)ureido)-acetate ... 563
$C_{11}H_{12}N_4S$
4-Benzylthio-5,6-diaminopyrimidine ... 973
$C_{11}H_{12}O$
5-Acetylindan ... 297
2-(2-Cyclopentenyl)phenol ... 191
Cyclopropyl *p*-tolyl ketone ... 296
4′-Methylcrotonophenone ... 306
2-Methyl-1-phenyl-1-buten-3-one ... 285
2-Methyl-4-phenyl-3-butyn-2-ol ... 1016
7-Methyl-1-tetralone ... 298
3-Phenyl-1-cyclopentanone ... 286
1-Phenyl-3-pentyn-2-ol ... 1016
4-(*p*-Tolyl)-3-buten-2-one ... 286
$C_{11}H_{12}OS_2$
3,3-Bis(methylthio)-1-phenyl-2-propen-1-one ... 293
$C_{11}H_{12}O_2$
2′-Allyloxyacetophenone ... 292
Ethyl atropate ... 376
α-Ethylcinnamic acid ... 336
1-Indanacetic acid ... 332
4-Methoxy-α-methylcinnamaldehyde ... 321
4-(3-Methoxyphenyl)-3-buten-2-one ... 283
Methyl α-methylcinnamate ... 365
Methyl β-methylcinnamate ... 365
DL-1-Phenylallyl acetate ... 361
1-Phenyl-1,4-pentanedione ... 308
1-(*p*-Tolyl)-1-cyclopropanecarboxylic acid ... 332
$C_{11}H_{12}O_2S$
Methyl 4-(methylthio)cinnamate ... 364
2-Phenylethyl 1-propynyl sulfone ... 1019
$C_{11}H_{12}O_3$
2-Acetyl-3,6-dimethylbenzoic acid ... 341
5′-Allyloxy-2′-hydroxyacetophenone ... 290
2-(2-Allylphenoxy)acetic acid ... 327
2-(1,3-Dioxolan-2-yl)acetophenone ... 293
4-Ethoxycinnamic acid ... 335
Ethyl benzoylacetate, thallium(I) salt ... 374
2a,5,5a,6,8a,8b-Hexahydro-6-hydroxy-2*H*-naphtho(1,8-bc)furan-2-one ... 95
1-(2-Hydroxy-4-methylphenyl)-1,3-butanedione ... 307
1-(2-Hydroxy-5-methylphenyl)-1,3-butanedione ... 307
3-Isopropoxyphthalide ... 383
2-Methoxy-α-methylcinnamic acid ... 335
4-Methoxy-α-methylcinnamic acid ... 335
4-Methoxy-β-methylcinnamic acid ... 335
1-(4-Methoxyphenyl)-1-cyclopropanecarboxylic acid ... 328
Methyl 2-oxo-3-phenylbutyrate ... 360
Methyl *m*-toluoylacetate ... 364
4-Oxo-4-(*o*-tolyl)butyric acid ... 329
4-Oxo-4-(*p*-tolyl)butyric acid ... 331
2-Phenyllevulinic acid ... 331
3-Propoxyphthalide ... 383
$C_{11}H_{12}O_3S$
α-Mercapto-2-methoxy-β-methylcinnamic acid ... 335
Methyl 3-hydroxy-α-mercapto-β-methylcinnamate ... 364
$C_{11}H_{12}O_4$
4-Acetoxyhydrocinnamic acid ... 380
2-Benzylsuccinic acid ... 332
2-(Carboxymethyl)-hydrocinnamic acid ... 332
2,6-Diacetyl-3-methoxyphenol ... 290
4,7-Dihydro-4,7a,7b-trimethyl-4,7-epoxyisobenzofuran-1,3(7a,7b)-dione ... 98
2,4-Dimethoxycinnamic acid ... 335
Dimethyl phenylmalonate ... 360
1,3-Dioxan-5-yl benzoate ... 375
Ethoxycarbonylmethyl benzoate ... 371
Ethyl acetylsalicylate ... 380
1,4,4a,5,8,8a-Hexahydro-5-hydroxy-8-oxo-1-naphthalene-carboxylic acid ... 64
3-(2-Hydroxy-5-methylbenzoyl)-propionic acid ... 326
2-Methyl-3-phenylsuccinic acid ... 331
2-Tetrahydrofuryl salicylate ... 371
$C_{11}H_{12}O_4S$
3-(Benzoyloxy)tetrahydrothiophene-1,1-dioxide ... 781
3,4-Dimethoxy-α-mercaptocinnamic acid ... 334
$C_{11}H_{12}O_5$
3-(2,3-Dimethoxyphenyl)glycidic acid, sodium salt ... 350
$C_{11}H_{12}O_5S$
Methyl 4-(phenylsulfonyl)-acetoacetate ... 781
$C_{11}H_{12}O_6$
Acetyl syringic acid ... 377
Ethyl 2,5-dioxo-3-ethyl-4-hydroxy-3-cyclopentene-1-glyoxylate ... 91
4-Methylcatechol-*O*,*O*-diacetic acid ... 327
$C_{11}H_{12}S$
1-Butyn-1-yl *p*-tolyl sulfide ... 1017
$C_{11}H_{13}BrClNO$
2-Bromo-6′-chloro-*o*-butyrotoluidide ... 439
2′-Bromo-4′-chloro-2,2-dimethylpropionanilide ... 410
2-Bromo-3′-chloro-2′-methylbutyranilide ... 409
2-Bromo-3′-chloro-4′-methylbutyranilide ... 410
$C_{11}H_{13}BrClNO_2$
2-Bromo-4-chlorophenyl *N*-butylcarbamate ... 637
$C_{11}H_{13}BrN_2$
5-Bromogramine ... 885
$C_{11}H_{13}BrN_2O_2$
1-(4-Bromo-2-nitrophenyl)-piperidine ... 274
4-(4-Bromophenylcarbamoyl)-morpholine ... 713
$C_{11}H_{13}BrN_2O_3$
N-(4-Acetamido-3-methoxyphenyl)-2-bromoacetamide ... 391
2′-Bromo-2,2-dimethyl-4′-nitropropionanilide ... 404
$C_{11}H_{13}BrO$
2-Bromo-4′-methylisobutyrophenone ... 296
1-(2-Bromo-4-methylphenoxy)-2-butene (mixture of isomers) ... 177
3′-Bromovalerophenone ... 289
$C_{11}H_{13}BrO_2$
2-(2-Bromophenoxy)-tetrahydropyran ... 183
2-(3-Bromophenoxy)-tetrahydropyran ... 183
2-(4-Bromophenoxy)-tetrahydropyran ... 183
tert-Butyl 2-bromobenzoate ... 367
tert-Butyl 4-bromobenzoate ... 367
$C_{11}H_{13}BrO_4$
2-Bromo-4,5-dimethoxyhydrocinnamic acid ... 328
$C_{11}H_{13}BrO_5$
Methyl 2-bromo-3,4,5-trimethoxybenzoate ... 366
$C_{11}H_{13}Br_3N_2O_2$
2-(Dimethylamino)ethyl *N*-(2,4,6-tribromophenyl)carbamate ... 627
$C_{11}H_{13}Br_3O_2$
1-Bromo-2-(2,3-dibromopropyl)-4,5-dimethoxybenzene ... 176
$C_{11}H_{13}ClFNO_2$
2-Chloro-4-fluorophenyl *N*-butylcarbamate ... 377
$C_{11}H_{13}ClN_2O$
1-Allyl-3-(2-chloro-6-methylphenyl)urea ... 540
1-Allyl-3-(3-chloro-4-methylphenyl)urea ... 598
1-Allyl-3-(4-chloro-2-methylphenyl)urea ... 598
1-Allyl-3-(5-chloro-2-methylphenyl)urea ... 598
1-(4-Chloro-2-methylphenyl)-3-cyclopropylurea ... 534
1-(2-Chlorophenylcarbamoyl)-pyrrolidine ... 695
4-Chloro-α-(1,4,5,6-tetrahydro-2-pyrimidinyl)benzyl alcohol hydrochloride ... 244
$C_{11}H_{13}ClN_2OS$
4-(4-Chlorophenyl)-4-hydroxy-3-methyl-2-(methylimino)-thiazolidine hydrobromide ... 728
$C_{11}H_{13}ClN_2O_2$
1-Acetyl-3-(4-chlorophenyl)-1-ethylurea ... 533
1-Allyl-3-(3-chloro-2-methoxyphenyl)urea ... 520
1-Butyryl-3-(4-chlorophenyl)urea ... 533
4-Chlorobenzoic *N*²-butyrylhydrazide ... 494
O-(*N*-(2-Chloro-6-methylphenyl)-carbamoyl)acetone oxime ... 1008
O-(*N*-(3-Chloro-4-methylphenyl)-carbamoyl)acetone oxime ... 1012
O-(*N*-(4-Chloro-2-methylphenyl)-carbamoyl)acetone oxime ... 1013
O-(*N*-(5-Chloro-2-methylphenyl)-carbamoyl)acetone oxime ... 1013
4-(2-Chlorophenylcarbamoyl)-morpholine ... 713
4-(3-Chlorophenylcarbamoyl)-morpholine ... 705
4-(4-Chlorophenylcarbamoyl)-morpholine ... 585
3-(4-Chlorophenyl)-1-formyl-1-propylurea ... 533
3-(4-Chlorophenyl)-1-methyl-1-propionylurea ... 533
$C_{11}H_{13}ClN_2O_3$
N-Acetyl-*N*′-(2-chloroacetyl)-2-methoxy-1,4-phenylenediamine ... 442
4′-Chloro-2,2-dimethyl-2′-nitropropionanilide ... 403
4′-Chloro-*N*-ethyl-2′-nitropropionanilide ... 403
O-(*N*-(3-Chloro-2-methoxyphenyl)-carbamoyl)acetone oxime ... 1009
Ethyl 2-(3-(3-chlorophenyl)ureido)-acetate ... 529
$C_{11}H_{13}ClN_2O_4$
5′-Chloro-4′-nitro-2′-propoxyacetanilide ... 416
$C_{11}H_{13}ClN_2S$
1-Allyl-3-(4-chlorobenzyl)-2-thiourea ... 730
1-Allyl-3-(3-chloro-2-methylphenyl)-2-thiourea ... 733
$C_{11}H_{13}ClN_4$
1-((4-Chloro-α-cyclopropylbenzylidene)amino)-guanidine nitrate ... 1026
$C_{11}H_{13}ClN_4O_3$
1-(3-Chlorophenyl)-4-(3,3-dimethylureido)oxamide ... 523
$C_{11}H_{13}ClN_4O_3S$
3-Chloro-1,6-diaminopyridazinium tosylate ... 968

$C_{11}H_{13}ClO$
DL-4-Chloro-α-cyclopropyl-α-methylbenzyl alcohol 203
4-Chloro-4′-methylbutyrophenone 287
3-Chloro-3-methyl-4-phenyl-2-butanone 285
2-Chloro-5-methylphenyl methallyl ether 177
3-(4-Chlorophenyl)-2,2-dimethylpropionaldehyde 321
4′-Chlorovalerophenone 289

$C_{11}H_{13}ClO_2$
4-Chloro-4′-methoxybutyrophenone 287
4-(4-Chlorophenyl)-4-methyl-1,3-dioxane 185
5-(2-Chlorophenyl)pentanoic acid 325

$C_{11}H_{13}ClO_2S$
Methyl 3-(4-chlorobenzylthio)-propionate 358

$C_{11}H_{13}ClO_3$
2-(2-(2-Chloroethoxy)ethoxy)-benzaldehyde 321
3-(2-(2-Chloroethoxy)ethoxy)-benzaldehyde 321
4-(2-(2-Chloroethoxy)ethoxy)-benzaldehyde 321
4-(3-Chloropropoxy)-*m*-anisaldehyde 321
4-(4-Chloro-*m*-tolyloxy)-butyric acid 329
Ethyl α-chloro-4-methoxyphenylacetate 356
Ethyl 3-(3-chlorophenoxy)-propionate 355
Ethyl 3-(4-chlorophenoxy)-propionate 355
Methyl 4-(3-chloropropoxy)-benzoate 365

$C_{11}H_{13}ClO_5$
2,6-Dicyclopropylpyrylium perchlorate 935

$C_{11}H_{13}Cl_2NO$
4-(*N*,*N*-Bis(2-chloroethyl)amino)-benzaldehyde 321
2′,5′-Dichloro-2,2-dimethylpropionanilide 409
3′,4′-Dichloro-2,2-dimethylpropionanilide 409
3′,5′-Dichloro-2,2-dimethylpropionanilide 410

$C_{11}H_{13}Cl_2NOS$
S-Butyl *N*-(2,3-dichlorophenyl)-thiocarbamate 749
S-Butyl *N*-(2,6-dichlorophenyl)-thiocarbamate 641

$C_{11}H_{13}Cl_2NO_2$
Butyl *N*-(2,3-dichlorophenyl)-carbamate 684
Butyl *N*-(2,5-dichlorophenyl)-carbamate 637
Butyl *N*-(3,4-dichlorophenyl)-carbamate 686
Butyl *N*-(3,5-dichlorophenyl)-carbamate 640
2,3-Dichlorophenyl *N*-butylcarbamate 636
2,4-Dichlorophenyl *N*-*tert*-butylcarbamate 637
3,5-Dichlorophenyl *N*-butylcarbamate 640

$C_{11}H_{13}Cl_2NO_3$
2-Ethoxyethyl *N*-(2,3-dichlorophenyl)carbamate 616
2-Ethoxyethyl *N*-(2,4-dichlorophenyl)carbamate 621
2-Ethoxyethyl *N*-(2,6-dichlorophenyl)carbamate 616
2-Ethoxyethyl *N*-(3,4-dichlorophenyl)carbamate 616

$C_{11}H_{13}Cl_2NO_4$
(β-Chlorocinnamylidene)-dimethylammonium perchlorate 1062

$C_{11}H_{13}Cl_2NO_4S$
Isopropyl *N*-(3,4-dichlorophenyl)-*N*-(methylsulfonyl)carbamate 808
Propyl *N*-(3,4-dichlorophenyl)-*N*-(methylsulfonyl)carbamate 808

$C_{11}H_{13}Cl_2N_3O_2$
1-(3,4-Dichlorobenzoyl)-4-propylsemicarbazide 528

$C_{11}H_{13}Cl_3$
1,3-Dimethyl-2,4,6-tris(chloromethyl)benzene 170

$C_{11}H_{13}Cl_3N_2O$
1-Butyl-3-(2,4,5-trichlorophenyl)urea 603
1,1-Diethyl-3-(2,4,5-trichlorophenyl)urea 603
1-Isobutyl-3-(2,4,5-trichlorophenyl)urea 546

$C_{11}H_{13}Cl_3N_2O_2$
2-Dimethylaminoethyl *N*-(2,4,5-trichlorophenyl)carbamate 673
2-Dimethylaminoethyl *N*-(2,4,6-trichlorophenyl)carbamate 627
Methyl *N*-(1-(benzylamino)-2,2,2-trichloroethyl)carbamate 610
Methyl *N*-(1-(*N*-methylanilino)-2,2,2-trichloroethyl)carbamate 610

$C_{11}H_{13}FN_2$
5-Fluorogramine 885
6-Fluoro-α-methyltryptamine 884
5-Fluoro-α-methyltryptamine hydrochloride 884

$C_{11}H_{13}FN_2O_3$
2,2-Dimethyl-2′-fluoro-5′-nitropropionanilide 405

$C_{11}H_{13}FO$
α-Cyclopropyl-4-fluoro-α-methyl-benzyl alcohol 203
4′-Fluorovalerophenone 297

$C_{11}H_{13}F_3N_2O$
1-Isopropyl-3-(2-trifluoromethylphenyl)urea 511
1-Isopropyl-3-(3-trifluoromethylphenyl)urea 510
1-Propyl-3-(2-(trifluoromethyl)-phenyl)urea 511
1-Propyl-3-(3-(trifluoromethyl)-phenyl)urea 510

$C_{11}H_{13}F_3N_2O_2$
1-(2-Methoxyethyl)-3-(3-(trifluoro-methyl)phenyl)urea 508

$C_{11}H_{13}F_3N_2S$
1-Isopropyl-3-(3-(trifluoromethyl)-phenyl)-2-thiourea 728

$C_{11}H_{13}I_2NO_3$
3,5-Diiodo-L-tyrosine ethyl ester hydrochloride 354

$C_{11}H_{13}N$
3-Methyl-2-phenylbutyronitrile 758
3-Methyl-2-phenylbutyronitrile 756
(5-Phenyl-4-pentyn-1-yl)amine 1018
4-Phenyl-1,2,3,6-tetrahydropyridine 253
2-Phenylvaleronitrile 758
1-(*p*-Tolyl)-3-pyrroline 251
2,4,6-Trimethylbenzyl cyanide 758

$C_{11}H_{13}NO$
5-Acetylindan oxime 1013
3-Amino-6,7,8,9-tetrahydro-5*H*-benzocyclohepten-5-one 294
N-Benzylcrotonamide 472
N-Benzylmethacrylamide 472
N-Crotonyl-*p*-toluidine 480
5-(Dimethylamino)-1-indanone 295
3-Dimethylamino-1-phenyl-2-propen-1-one 302
2,3-Dimethyl-6-methoxyindole 883
2,3-Dimethyl-7-methoxyindole 883
N-Ethylacrylanilide 482
2-Isopropyl-6-methylphenyl isocyanate 769
DL-2-(4-Methoxyphenyl)butyronitrile 753
4-Methoxy-4-phenylbutyronitrile 752
3′-Methyl-1-cyclopropanecarboxanilide 437
4′-Methylcyclopropane-carboxanilide 437
N-Methylmethacrylanilide 482
7-Methyl-1-tetralone oxime 1014
4-Phenylimino-2-pentanone 1074
3,4,5,6-Tetrahydro-1-benzazocin-2(1*H*)-one 704
1,2,3,4-Tetrahydro-1-naphthalenecarboxamide 437
N-(1,2,3,4-Tetrahydro-1-naphthyl)-formamide 441

$C_{11}H_{13}NOS$
3-(2-Aminoethyl)-4-methoxybenzo(b)thiophene hydrochloride 897
3-(2-Aminoethyl)-6-methoxybenzo(b)thiophene hydrochloride 897
2-Isopropyl-2*H*-1,3-benzothiazin-4(3*H*)-one 691
4-(Thiobenzoyl)-morpholine 742

$C_{11}H_{13}NO_2$
Acetoacetobenzylamide 443
m-Acetoacetotoluidide 444
5′-Acetyl-2′-methylacetanilide 433
4′-Acetylpropionanilide 432
3′-Allyl-2′-hydroxyacetanilide 413
2′-(Allyloxy)acetanilide 417
Allyl *N*-(*o*-tolyl)carbamate 681
Allyl *N*-(*m*-tolyl)carbamate 678
Allyl *N*-(*p*-tolyl)carbamate 682
2-Butenyl *N*-phenylcarbamate 680
2-Cyclopropyl-2-phenylglycine 349
Methyl 4-amino-β-methylcinnamate 364
Methyl 1,2,3,4-tetrahydro-3-isoquinolinecarboxylate hydrochloride 360
2′-Propionylacetanilide 432
3′-Propionylacetanilide 432
4′-Propionylacetanilide 432
1-Salicyloylpyrrolidine 697

$C_{11}H_{13}NO_2S$
2-(1-Aziridinyl)vinyl *p*-tolyl sulfone 782
2,3-Dihydro-3-(4-methylanilino)-thiophene-1,1-dioxide 778
4-(4-Hydroxythiobenzoyl)-morpholine 734

$C_{11}H_{13}NO_2Si$
N-(Trimethylsilyl)phthalimide 1123

$C_{11}H_{13}NO_3$
α′-Acetyl-*o*-acetanisidide 416
N-Acetyl-D-phenylalanine 429
N-Acetyl-DL-phenylalanine 429
Allyl *N*-(2-methoxyphenyl)-carbamate 653
Allyl *N*-(3-methoxyphenyl)-carbamate 654
Allyl *N*-(4-methoxyphenyl)-carbamate 655
N-Benzoyl-β-alanine methyl ester 477
N-Benzoyl-L-alanine methyl ester 476
2,3-Dihydro-7-nitro-3,3,4-trimethylbenzofuran 279
Ethyl 4-(acetamido)benzoate 434
Ethyl hippurate 476
Ethyl 4′-methyloxanilate 434
Glutaranilic acid 435
N-(4-Methoxybenzylidene)glycine methyl ester 1050
Methyl *N*-ethylphthalamate 467
2-Methyl-5-morpholino-1,4-benzoquinone 53
Methyl 3-(phenylcarbamoyl)-propionate 436
4-Nitrovalerophenone 295
N-(2-Phenylethyl)malonamic acid 443
2-Propionamidophenyl acetate 436
4-Propionylphenyl *N*-methylcarbamate 675
N-(*o*-Tolyl)succinamic acid 436
N-(*m*-Tolyl)succinamic acid 434

$C_{11}H_{13}NO_3S$
2,3-Dihydro-*S*,*S*-dioxide-5-ethoxy-1,4-benzothiazepine 775
2,3-Dihydro-3-(2-methoxyanilino)-thiophene-1,1-dioxide 775
2,3-Dihydro-3-(3-methoxyanilino)-thiophene-1,1-dioxide 775
4′-(Methylthio)succinanilic acid 420

$C_{11}H_{13}NO_4$
N-Acetyl-DL-tyrosine 412
N-Benzyl-DL-aspartic acid 349
Benzyl *N*-(carboxymethyl)-*N*-methylcarbamate 646
4-*tert*-Butyl-3-nitrobenzoic acid 344
Carbobenzyloxy-β-alanine 661
Diethyl 2,6-pyridinedicarboxylate 964
Diethyl 3,5-pyridinedicarboxylate 962
2,5-Dimethoxy-β-methyl-β-nitrostyrene 257
1-(3,4-Dimethoxyphenyl)-2-nitro-1-propene 257
Ethyl 3-acetamido-4-hydroxybenzoate 413
Ethyl 4-methoxyoxanilate 418
4′-Hydroxyglutaranilic acid 413
N-(2-(Methoxycarbonyl)phenyl)-glycine methyl ester 373
4′-Methoxysuccinanilic acid 418
Methyl 4-acetamido-2-methoxybenzoate 417
Methyl 4-(1-methyl-1-nitroethyl)-benzoate 373

$C_{11}H_{13}NO_4S$
(4-Acetamidophenylsulfonyl)-acetone 781
S-Carbobenzyloxy-L-cysteine 381
S-(4-Hydroxyphenyl) *N*-(ethoxycarbonylmethyl)-thiocarbamate 646
1-Methylquinolinium methyl sulfate 990

$C_{11}H_{13}NO_5$
3-Butoxy-4-nitrobenzoic acid 340
Diethyl 1,4-dihydro-4-oxo-2,6-pyridinedicarboxylate 92
Ethyl (2,6-dimethyl-4-oxo-4*H*-pyran-3-ylcarbonyl)carbamate 115
Ethyl 3-(4-hydroxy-3-nitrophenyl)-propionate 354
Ethyl 2-(2-nitrophenoxy)propionate 360
β-Nitro-3,4,5-trimethoxystyrene 257

$C_{11}H_{13}NO_6$
3-(Methoxycarbonylmethoxy)-1-(methoxy-carbonylmethyl)-2(1*H*)-pyridone 131
4′-Methoxy-L-tartranilic acid 442

$C_{11}H_{13}NS$
4-(*tert*-Butyl)-phenyl isothiocyanate 768
4-(1-Methyl-1-(methylthio)ethyl)-benzonitrile 763
4-Methylthio-2-phenylbutyronitrile 754
1,3,3-Trimethyl-2-indolinethione 742

$C_{11}H_{13}N_2O_2$
1-(*o*-Anisyl)-3-propylurea 558

$C_{11}H_{13}N_3$
5-Amino-2,3-dihydro-7-phenyl-1*H*-1,4-diazepine 248
5-Amino-3-methyl-1-(*p*-tolyl)pyrazole 861

$C_{11}H_{13}N_3O$
1-(2-Aminobenzoyl)-1,2,3,6-tetrahydropyridazine 701
1-(4-Aminophenyl)-3,4-dimethyl-2-pyrazolin-5-one 701
1-Ethyl-2-methyl-5-benzimidazolecarboxamide 906
4-Hydrazino-6-methoxy-2-methylquinoline 985
3-Indolepropionic acid hydrazide 884

$C_{11}H_{13}N_3OS$
N-(4,5-Dihydro-5-methyl-5-phenyl-1,3,4-thiadiazol-2-yl)acetamide 705

$C_{11}H_{13}N_3OS_2$
1-(4-(Methylthio)phenyl)-3-(2-thiazolin-2-yl)urea 739

$C_{11}H_{13}N_3O_2$
3-Amino-1-(4-methoxybenzyl)-2-pyrazolin-5-one 698
3-(Dimethylaminomethyl)-6-nitroindole 885
1,2,3,4,5,6-Hexahydro-2-imino-5-pyrimidinyl benzoate hydrochloride 726
Tetrahydro-2-furoic (2-pyridylmethylene)hydrazide 1053

$C_{11}H_{13}N_3O_2S$
2,3-Dihydro-2-(ethoxycarbonylhydrazinylylidene)-3-methylbenzothiazole 1062

$C_{11}H_{13}N_3O_3$
1-Allyl-3-(4-methyl-3-nitrophenyl)urea 564
6-Amino-3-ethyl-1-furfuryluracil 850
N-(Benzoylglycyl)glycine amide 477
Methyl *N*-(4-acetamidobenzylidene)carbazate 1048

$C_{11}H_{13}N_3O_4$

1-(2,4-Dinitrophenyl)piperidine 273
Ethyl pyruvate 3-nitrophenylhydrazone 284
O-(*N*-(2-Methyl-4-nitrophenyl)carbamoyl)acetone oxime 1010
O-(*N*-(4-Methyl-3-nitrophenyl)carbamoyl)acetone oxime 1010
4-(*N*-(3-Nitrophenyl)carbamoyl)morpholine 702
4-(*N*-(4-Nitrophenyl)carbamoyl)morpholine 703

$C_{11}H_{13}N_3O_5$

Ethyl 2-(3-(3-nitrophenyl)ureido)acetate 563

$C_{11}H_{13}N_3O_6$

N-(2,4-Dinitrophenyl)-L-valine 349

$C_{11}H_{13}N_3O_6S$

N-(2,4-Dinitrophenyl)-L-methionine dicyclohexylamine salt 348

$C_{11}H_{13}N_3O_7S$

2′,6′-Dinitro-*N*-methyl-4′-(methylsulfonyl)propionanilide 777
2′,6′-Dinitro-4′-(methylsulfonyl)butyranilide 777

$C_{11}H_{13}N_3O_8S$

Ethyl *N*-(2,6-dinitro-4-(methylsulfonyl)phenyl)-*N*-methylcarbamate 777
Isopropyl *N*-(2,6-dinitro-4-(methylsulfonyl)phenyl)carbamate 777
Propyl *N*-(2,6-dinitro-4-(methylsulfonyl)phenyl)carbamate 777

$C_{11}H_{13}N_5$

2,4-Diamino-6-(2,3-xylyl)-1,3,5-triazine 978

$C_{11}H_{13}N_5O_2$

Isonicotinic acid (propionylhydrazonoethylidene)hydrazide 1074

$C_{11}H_{13}N_5O_4$

6-Carbamoyl-1,3-diethyl-7-hydroxy-2,4(1*H*,3*H*)-pteridinedione 977
3-Nitroso-2,4,6-triacetamidopyridine 942

$C_{11}H_{14}$

Cyclopentylbenzene 162
2-Methyl-1-phenyl-1-butene 164
1-Phenyl-1-pentene 164

$C_{11}H_{14}AsNO_8$

4-(Butoxycarbonyloxy)-3-nitrobenzenearsonic acid 1127

$C_{11}H_{14}BrN$

N-Allyl-2-bromo-*N*-methylbenzylamine 236

$C_{11}H_{14}BrNO$

3′-Bromo-2,2-dimethylpropionanilide 407
4′-Bromo-2′,6′-dimethylpropionanilide 410
2′-Bromo-4′-isopropylacetanilide 410

$C_{11}H_{14}BrNOS$

S-Butyl *N*-(4-bromophenyl)thiocarbamate 688

$C_{11}H_{14}BrNO_2$

2-Bromo-4,5-dimethylphenyl *N*-ethylcarbamate 641
3-Bromo-2′-methoxy-5′-methylpropionanilide 391
Butyl *N*-(4-bromophenyl)carbamate 682

$C_{11}H_{14}BrNO_3$

2-Ethoxyethyl *N*-(4-bromophenyl)carbamate 615

$C_{11}H_{14}BrNO_4$

Diethyl 5-bromo-3-methyl-2,4-pyrroledicarboxylate 841

$C_{11}H_{14}Br_2O_3Te$

3-(4-Ethoxyphenyltelluro)propionic acid, Te,Te-dibromide 329

$C_{11}H_{14}ClNO$

N-Benzyl-4-chlorobutyramide 392
4-(3-Chlorobenzyl)morpholine 244
4-Chloro-*m*-butyrotoluidide 393
N-(*p*-Chloro-α,α-dimethylphenethyl)formamide 404
4′-Chloro-2,2-dimethylpropionanilide 409
2-Chloro-*N*-ethyl-*o*-acetotoluidide 393
6′-Chloro-*o*-isobutyrotoluidide 439
3-Chloro-*N*-phenethylpropionamide 392
3-Chloro-2′,6′-propionoxylidide 393
2-Chloro-2′,4′,6′-trimethylacetanilide 393

$C_{11}H_{14}ClNOS$

S-Butyl *N*-(3-chlorophenyl)thiocarbamate 688
S-(4-Chlorophenyl) *N*-*tert*-butylthiocarbamate 641
S-Isopropyl *N*-(2-chloro-6-methylphenyl)thiocarbamate 641
S-Isopropyl *N*-(3-chloro-4-methylphenyl)thiocarbamate 689
S-Isopropyl *N*-(4-chloro-2-methylphenyl)thiocarbamate 689
S-Propyl *N*-(2-chloro-6-methylphenyl)thiocarbamate 641
S-Propyl *N*-(4-chloro-2-methylphenyl)thiocarbamate 689

$C_{11}H_{14}ClNO_2$

4-*tert*-Butyl-1-chloromethyl-2-nitrobenzene 258
Butyl *N*-(3-chlorophenyl)carbamate 631
Butyl *N*-(4-chlorophenyl)carbamate 682
1-Chloroethyl *N*-ethyl-*N*-phenylcarbamate 677
2-Chloroethyl *N*-(2-ethylphenyl)carbamate 605
2-Chloroethyl *N*-(2,3-xylyl)carbamate 605
2-Chloroethyl *N*-(2,4-xylyl)carbamate 605
2-Chloroethyl *N*-(2,5-xylyl)carbamate 605
2-Chloroethyl *N*-(2,6-xylyl)carbamate 605
2-Chloroethyl *N*-(3,4-xylyl)carbamate 605
2-Chloroethyl *N*-(3,5-xylyl)carbamate 605
4′-Chloro-2′-methoxybutyranilide 401
3-Chloro-*p*-propionophenetidide 391
Isopropyl *N*-(2-chloro-6-methylphenyl)carbamate 636
Isopropyl *N*-(3-chloro-4-methylphenyl)carbamate 686
Isopropyl *N*-(4-chloro-2-methylphenyl)carbamate 685
Isopropyl *N*-(5-chloro-2-methylphenyl)carbamate 686
Propyl *N*-(2-chloro-6-methylphenyl)carbamate 636
Propyl *N*-(3-chloro-4-methylphenyl)carbamate 686
Propyl *N*-(4-chloro-2-methylphenyl)carbamate 686

$C_{11}H_{14}ClNO_3$

Butyl *N*-(4-chlorophenyl)-*N*-hydroxycarbamate 628
1-(3-Carbethoxy-2-oxopropyl)pyridinium chloride 956
N-(4-Chlorobenzoyl)-L-threonine 349
4′-Chloro-3′,5′-dimethoxypropionanilide 399
5′-Chloro-2′,4′-dimethoxypropionanilide 399
2-Chloroethyl *N*-(2-methoxy-5-methylphenyl)carbamate 604
2-Ethoxyethyl *N*-(2-chlorophenyl)carbamate 650
2-Ethoxyethyl *N*-(3-chlorophenyl)carbamate 649
2-Ethoxyethyl *N*-(4-chlorophenyl)carbamate 650
Isopropyl *N*-(4-chlorophenyl)-*N*-methoxycarbamate 628
2-Methoxyethyl *N*-(2-chloro-6-methylphenyl)carbamate 650
2-Methoxyethyl *N*-(4-chloro-2-methylphenyl)carbamate 616
2-Methoxyethyl *N*-(5-chloro-2-methylphenyl)carbamate 621

$C_{11}H_{14}ClNO_3S_2$

4-(4-Chlorophenylthiomethylsulfonyl)morpholine 806

$C_{11}H_{14}ClNO_4$

1-Chloro-5-ethoxy-4-nitro-2-propoxybenzene 260
2-Chloroethyl *N*-(2,5-dimethoxyphenyl)carbamate 604
2-(4-Chlorophenyl)-5-(hydroxyamino)-1,3-dioxane-5-methanol 236

$C_{11}H_{14}ClNO_4S$

Isopropyl *N*-(3-chlorophenyl)-*N*-(methylsulfonyl)carbamate 808
Isopropyl *N*-(4-chlorophenyl)-*N*-(methylsulfonyl)carbamate 808
Propyl *N*-(3-chlorophenyl)-*N*-(methylsulfonyl)carbamate 808
Propyl *N*-(4-chlorophenyl)-*N*-(methylsulfonyl)carbamate 808

$C_{11}H_{14}ClN_3O$

Acetic (2-chloro-4-dimethylaminobenzylidene)hydrazide 1025

$C_{11}H_{14}ClN_3O_2$

1-(4-Chlorobenzoyl)-4-isopropylsemicarbazide 532
1-(4-Chlorobenzoyl)-4-propylsemicarbazide 528
4-(4-Chloro-2-methoxyphenylazo)morpholine 1080
1-(2-Chloro-4-nitrophenyl)-4-methylpiperazine 261

$C_{11}H_{14}ClN_3O_3$

4-(4-Chloro-2-methylphenyl)-1-(ethoxycarbonyl)semicarbazide 627
3-(2-Chloro-4-nitrophenyl)-1,1-diethylurea 568

$C_{11}H_{14}ClN_3O_4$

4-(5-Chloro-2-methoxyphenyl)-1-(ethoxycarbonyl)semicarbazide 618
2-(Dimethylamino)ethyl *N*-(2-chloro-4-nitrophenyl)carbamate 670

$C_{11}H_{14}ClN_5$

1-(4-Chlorophenyl)-4,6-diamino-1,2-dihydro-2,2-dimethyl-1,3,5-triazine hydrochloride 979

$C_{11}H_{14}Cl_2$

Bis-(chloromethyl)-mesitylene 170

$C_{11}H_{14}Cl_2N_2O$

N,*N*-Bis(2-chloroethyl)-3-methyl-4-nitrosoaniline 257
1-*sec*-Butyl-3-(3,4-dichlorophenyl)urea 543
1-Butyl-3-(2,3-dichlorophenyl)urea 593
1-Butyl-3-(2,4-dichlorophenyl)urea 598
1-Butyl-3-(2,5-dichlorophenyl)urea 598
1-Butyl-3-(2,6-dichlorophenyl)urea 540
1-Butyl-3-(3,4-dichlorophenyl)urea 598
1-Butyl-3-(3,5-dichlorophenyl)urea 545
3,4-Dichlorobenzoic acid *tert*-butylhydrazide 494
3-(2,3-Dichlorophenyl)-1,1-diethylurea 593
3-(2,4-Dichlorophenyl)-1,1-diethylurea 598
3-(2,5-Dichlorophenyl)-1,1-diethylurea 598
3-(2,6-Dichlorophenyl)-1,1-diethylurea 593
3-(3,4-Dichlorophenyl)-1,1-diethylurea 598
1-(2,3-Dichlorophenyl)-3-isobutylurea 593
1-(2,4-Dichlorophenyl)-3-isobutylurea 598
1-(2,5-Dichlorophenyl)-3-isobutylurea 543
1-(2,6-Dichlorophenyl)-3-isobutylurea 593
1-(3,4-Dichlorophenyl)-3-isobutylurea 598

$C_{11}H_{14}Cl_2N_2O_2$

2-(Dimethylamino)ethyl *N*-(2,3-dichlorophenyl)carbamate 672
2-(Dimethylamino)ethyl *N*-(2,4-dichlorophenyl)carbamate 627
2-Dimethylaminoethyl *N*-(2,5-dichlorophenyl)carbamate 673
2-(Dimethylamino)ethyl *N*-(2,6-dichlorophenyl)carbamate 627
2-Dimethylaminoethyl *N*-(3,4-dichlorophenyl)carbamate 672
2-(Dimethylamino)ethyl *N*-(3,5-dichlorophenyl)carbamate 627

$C_{11}H_{14}Cl_2N_2O_3$

3-(3,4-Dichlorophenyl)-1,1-bis(2-hydroxyethyl)urea 548

$C_{11}H_{14}Cl_2O$

2,6-Dichloro-4-(*tert*-pentyl)-phenol 192

$C_{11}H_{14}Cl_2O_3Te$

3-(4-Ethoxyphenyltelluro)propionic acid, Te,Te-dichloride 329

$C_{11}H_{14}FNO$

2,2-Dimethyl-4′-fluoropropionanilide 409
4-(3-Fluorobenzyl)morpholine 244
4-(4-Fluorophenyl)-4-piperidinol 245

$C_{11}H_{14}FNO_2$

4-Fluorophenyl *N*-butylcarbamate 634

$C_{11}H_{14}FN_5O_3$

1-(2-(3-(4-Fluorophenyl)ureido)acetyl)-4-methylsemicarbazide 523

$C_{11}H_{14}F_2N_2O$

1-Butyl-3-(2,5-difluorophenyl)urea 543
1-(2,4-Difluorophenyl)-3-(2-methylpropyl)urea 543

$C_{11}H_{14}HgO_3$

2-Hydroxy-2-phenylpropylmercuri acetate 1137

$C_{11}H_{14}I_2O_3Te$

3-(4-Ethoxyphenyltelluro)propionic acid, Te,Te-diiodide 329

$C_{11}H_{14}NO_3STl$

2-(Diethylthallyl(III))saccharin 823

$C_{11}H_{14}N_2$

4-(Diethylamino)benzonitrile 765
5,6-Dimethyl-2-ethylbenzimidazole 907
3-(*N*-Ethylanilino)-propionitrile 754
N-ω-Methyltryptamine hydrochloride 885
2,3,5,10-Tetrahydro-1*H*-pyrazolo(1,2-b)phthalazine 253
3-(3,5-Xylidino)-propionitrile 755

$C_{11}H_{14}N_2O$

1-Allyl-3-(*o*-tolyl)urea 587
1-Allyl-3-(*m*-tolyl)urea 580
1-Allyl-3-(*p*-tolyl)urea 590
1-(2-Aminobenzoyl)pyrrolidine 701
4,5-Dihydro-7,8-dimethyl-1*H*-1,5-benzodiazepin-2(3*H*)-one 702
5,6-Dimethyl-2-benzimidazoleethanol 902
5,5-Dimethyl-2-phenyl-3-pyrazolidinone 710
Hexahydro-1-(2-pyridyl)-2*H*-azepin-2-one 959
α,5,6-Trimethyl-2-benzimidazolemethanol 902

$C_{11}H_{14}N_2OS$

1-(3-Acetylphenyl)-3-ethyl-2-thiourea 744
1-Cyclopropyl-3-(4-(methylthio)phenyl)urea 561
1-Isobutyryl-3-phenyl-2-thiourea 745

$C_{11}H_{14}N_2O_2$

N-Acetyl-D-phenylalaninamide 429
N-Acetyl-DL-phenylalaninamide 429
1-(4-Acetylphenyl)-3-ethylurea 575
N-Acetyl-*N*′-propionyl-1,4-phenylenediamine 428
1-Allyl-3-(*o*-anisyl)urea 555
1-Allyl-3-(*m*-anisyl)urea 557
1-Allyl-3-(*p*-anisyl)urea 559
4-(2-Aminobenzoyl)morpholine hydrochloride 701
1-Butyryl-3-phenylurea 579
1-Carbamoyl-5-methoxy-2-methylindoline 700
1-Cyclopropyl-3-(2-methoxyphenyl)urea 554
1-Cyclopropyl-3-(4-methoxyphenyl)urea 558
4-Dimethylamino-β-methyl-β-nitrostyrene 258
Ethyl 2-(2-pyrrolyl)-1-pyrroline-5-carboxylate 841
6′-Formamido-2′,4′-acetoxylidide 425

$C_{11}H_{14}N_2O_2$ (Cont'd)
N-(6-Methyl-2-pyridyl)-acetoacetamide 944
1-(2-Nitrophenyl)piperidine 279
1-(4-Nitrophenyl)piperidine 277
4-(*N*-Phenylcarbamoyl)morpholine 585
3-Phenylglutaramide 436
O-(*N*-(*o*-Tolyl)carbamoyl)acetone oxime 1012
O-(*N*-(*m*-Tolyl)carbamoyl)acetone oxime 1011
O-(*N*-(*p*-Tolyl)carbamoyl)acetone oxime 1013
$C_{11}H_{14}N_2O_2S$
1-(3-Carboxyphenyl)-3-propyl-1-thiourea 744
1-Ethyl-2-methyl-5-(methylsulfonyl)-benzimidazole 906
$C_{11}H_{14}N_2O_3$
N-Acetyl-D-tyrosinamide 412
N-Acetyl-DL-tyrosinamide 412
N-Acetyl-L-tyrosinamide hydrochloride 413
N-Anilinoglutaramic acid 503
N-Butyl-4-nitrobenzamide 473
2-*tert*-Butyl-3-(4-nitrophenyl)-oxaziridine 278
Carbobenzyloxy-DL-alaninamide 664
3-(Dimethylamino)-4′-nitropropiophenone hydrochloride 295
4′,5′-Dimethyl-2′-nitropropionanilide 424
4′-Ethyl-5′-methyl-2′-nitroacetanilide 424
Ethyl *N*-methyl-*N*-(*N*-phenylcarbamoyl)carbamate 677
Glycylglycine benzyl ester *p*-toluenesulfonate 420
Glycyl-D-phenylalanine 422
Glycyl-DL-phenylalanine 422
Glycyl-L-phenylalanine 422
4′-Isopropyl-2′-nitroacetanilide 424
O-(*N*-(2-Methoxyphenyl)carbamoyl)-acetone oxime 1009
O-(*N*-(3-Methoxyphenyl)carbamoyl)-acetone oxime 1009
O-(*N*-(4-Methoxyphenyl)carbamoyl)-acetone oxime 1009
N-(3-Methyl-2-pyridyl)-glutaramic acid 963
N-(4-Methyl-2-pyridyl)-glutaramic acid 963
N-(6-Methyl-2-pyridyl)-glutaramic acid 963
1-(4-Nitrophenyl)-3-piperidinol 264
L-Phenylalanylglycine monohydrate 422
4-(3-Phenylureido)butyric acid 577
N-(3-Pyridylmethyl)glutaramic acid 963
$C_{11}H_{14}N_2O_3S$
1-(Ethoxycarbonylmethyl)-3-(2-mercaptophenyl)urea 560
cis-Hexahydro-*N*-(2-thiazolyl)-phthalamic acid 877
trans-1-(4-Hydroxy-3-sulfolanyl)-3-phenyl-2-thiourea 775
1-Phenyl-3-sulfolanylurea 581
$C_{11}H_{14}N_2O_4$
Butyl *N*-(3-nitrophenyl)carbamate 670
Carbobenzyloxy-DL-serinamide 643
Dimethyl 3,3-dicyanopimelate 146
Dimethyl 4-methyl-1,3-phenylenedicarbamate 670
1-(2,2-Dimethyl-1-nitropropyl)-4-nitrobenzene 278
Ethyl *N*-(4,5-dimethyl-2-nitrophenyl)-carbamate 668
Ethyl 3,5-dimethyl-4-(2-nitrovinyl)-2-pyrrolecarboxylate 839
Isopropyl *N*-(2-methyl-4-nitrophenyl)carbamate 670
Isopropyl *N*-(4-methyl-3-nitrophenyl)carbamate 670
2′-Methoxy-4′-nitrobutyranilide 415
2′-Methoxy-5′-nitrobutyranilide 415
5-Morpholino-2-nitroanisole 267
3-Nitrophenyl *N*-butylcarbamate 674
$C_{11}H_{14}N_2O_5$
DL-2-Acetamido-1-(4-nitrophenyl)-1,3-propanediol 412
Ethyl *N*-(4-ethoxy-2-nitrophenyl)-carbamate 652
Ethyl 5-formyl-3-methyl-4-(2-nitroethyl)-2-pyrrolecarboxylate 839
N-(4-Nitrobenzyl)-L-threonine 349
$C_{11}H_{14}N_2O_5S$
4′-Sulfamoylglutaranilic acid 824
$C_{11}H_{14}N_2O_7$
5′-*O*-Acetyluridine 131
Methyl 3,5-bis(hydroxyimino)-1-(methoxycarbonyl)-4-oxo-1-cyclohexaneacetate 1011
$C_{11}H_{14}N_2S$
1-Benzyl-4,5-dihydro-2-(methylthio)-imidazole hydrochloride 248
2-(Isopropylthiomethyl)-benzimidazole 903
1-(*N*-Phenylthiocarbamoyl)-pyrrolidine 742
$C_{11}H_{14}N_2S_2$
N-*tert*-Butyl-2-benzothiazolesulfenamide 917
$C_{11}H_{14}N_4O$
6,7-Dihydro-3-indoleacetaldehyde semicarbazone 1059
5,6-Dimethylbenzimidazole-1-acethydrazide 907
$C_{11}H_{14}N_4O_2$
3-Acetamidoacetophenone semicarbazone 1063
$C_{11}H_{14}N_4O_3$
N-(3,5-Dimethylphenylureido)-oxamide 574
$C_{11}H_{14}N_4O_4$
Ethyl 7-theophyllineacetate 866
3-Guanidinopropyl 4-nitrobenzoate hyrochloride 726
1-(2-Methoxy-5-methylphenylureido)oxamide 551
$C_{11}H_{14}N_4O_5$
1-(Ethoxycarbonyl)-4-(2-methyl-5-nitrophenyl)semicarbazide 669
4-(Ethoxycarbonylmethyl)-1-(4-nitrophenyl)semicarbazide 567
$C_{11}H_{14}O$
Benzyl methallyl ether 186
3-(1-Butenyl)-anisole 178
α-Cyclopropyl-α-methylbenzyl alcohol 203
α-Cyclopropyl-4-methylbenzyl alcohol 203
β,β-Dimethylhydrocinnamaldehyde 321
2,4′-Dimethylpropiophenone 297
1,4,4a,7,8,8a-Hexahydro-1,4-methanonaphthalen-5(6*H*)-one 55
(+)-1,2,4a,5,8,8a-Hexahydro-4a-methyl-2-naphthalenone 54
(-)-1,2,4a,5,8,8a-Hexahydro-4a-methyl-2-naphthalenone 54
Isovalerophenone 296
2-Methylallyl *o*-tolyl ether 180
3′-Methylbutyrophenone 296
4′-Methylbutyrophenone 297
4-(3-Methylcrotyl)phenol 191
3-Methyl-1-phenyl-2-butanone 286
2-Methyl-4-phenyl-3-buten-2-ol 206
5-Methyl-5,6,7,8-tetrahydro-2-naphthol 191
1-Phenyl-1-cyclopentanol 203
1-Phenyl-2-pentanone 286
1-Phenyl-3-pentanone 286
2-Phenyl-3-pentanone 286
2-Phenyl-4-penten-2-ol 204
2,3,4,6-Tetramethylbenzaldehyde 324
4-(*o*-Tolyl)-2-butanone 287
4-(*p*-Tolyl)-2-butanone 287
4-(*o*-Tolyl)-3-buten-2-ol 207
4-(*p*-Tolyl)-3-buten-2-ol 207
2′,4′,5′-Trimethylacetophenone 297
$C_{11}H_{14}OS$
2,3-Dihydro-3,4-dimethyl-4-hydroxy-4*H*-1-benzothiopyran 210
2-(Phenylthio)tetrahydropyran 211
$C_{11}H_{14}OS_2$
2-(2-Methoxyphenyl)-1,3-dithiane 211
$C_{11}H_{14}O_2$
Benzyl butyrate 356
2-(Benzyloxymethyl)-2-methyloxirane 189
sec-Butyl benzoate 375
2-*tert*-Butylbenzoic acid, sodium salt 351
2-*tert*-Butyl-5-methyl-1,4-benzoquinone 54
α-Cyclopropyl-4-methoxybenzyl alcohol 201
2,2-Dimethyl-4-phenyl-1,3-dioxolane 189
2,3-Epoxypropyl 4-ethylphenyl ether 189
Ethyl 3,4-dimethylbenzoate 376
2-Ethyl-4-phenyl-1,3-dioxolane 184
Ethyl 3-phenylpropionate 361
Ethyl *m*-tolylacetate 360
2-Isopropyl-2-phenylacetic acid 332
2-(4-Isopropylphenyl)acetic acid 333
2-Mesityleneacetic acid 334
4′-Methoxybutyrophenone 293
4′-Methoxy-2-methylpropiophenone 292
4-(4-Methoxyphenyl)-2-butanone 284
5-Methoxy-1,2,3,4-tetrahydro-1-naphthol 195
8α-Methyl-1,2,3,4,4a,5,8,8a-octahydro-1,3-naphthalenedione 53
β-Methylphenethyl acetate 361
Methyl 2-phenylbutyrate 361
2-Methyl-3-phenylbutyric acid 332
(+)-3-Methyl-2-phenylbutyric acid 332
3-Methyl-2-phenylbutyric acid, sodium salt 351
4-Methyl-4-phenyl-1,3-dioxane 185
Phenethyl propionate 361
3-Phenylpropyl acetate 361
4-Phenylvaleric acid 332
2-(*o*-Tolyl)-butyric acid 333
4-(*p*-Tolyl)-butyric acid 329
p-Tolyl isobutyrate 380
$C_{11}H_{14}O_2Pb$
1,1-Diethyl-3-oxo-3*H*-2,1-benzoxaplumbole 1135
$C_{11}H_{14}O_2S$
Ethyl 2-(phenylthio)propionate 357
2-Methylallyl *p*-tolyl sulfone 785
Methyl 3-(benzylthio)propionate 357
1-Methylcyclopropyl *p*-tolyl sulfone 782
Methyl 2-methyl-3-(phenylthio)-propionate 357
1-Penten-1-yl phenyl sulfone 783
3-(Phenylthio)propyl acetate 357
$C_{11}H_{14}O_3$
Butyl salicylate 368
5-*tert*-Butylsalicylic acid 338
3,4-Diethoxybenzaldehyde 323
2′,4′-Dimethoxy-6′-methylacetophenone 291
2′,5′-Dimethoxy-4′-methylacetophenone 291
4′,5′-Dimethoxy-2′-methylacetophenone 291
2′,4′-Dimethoxypropiophenone 293
2′,5′-Dimethoxypropiophenone 291
Ethyl 2,6-dimethyl-4-hydroxybenzoate 368
Ethyl 4-ethoxybenzoate 370
Ethyl β-hydroxyhydrocinnamate 354
Ethyl 4-hydroxyhydrocinnamate 354
Ethyl 3-phenoxypropionate 355
Ethyl 3,5-xylyl carbonate 381
5-(3-Hydroxyphenyl)valeric acid 326
Isobutyl 4-hydroxybenzoate 368
Isobutyl salicylate 368
α-Isopropylmandelic acid 326
3-(4-Methoxy-3-methylphenyl)-propionic acid 329
DL-2-(4-Methoxyphenyl)butyric acid 329
(-)-2-(4-Methoxyphenyl)butyric acid 329
2-Methyl-2-phenyl-1,3-dioxolane-4-methanol 206
2-Phenoxyethyl propionate 361
$C_{11}H_{14}O_3S$
3-(3-Methylphenoxy)sulfolane 775
Methyl 3-(phenylsulfonyl)propyl ketone 781
$C_{11}H_{14}O_3Te$
3-(4-Ethoxyphenyltelluro)-propionic acid 329
$C_{11}H_{14}O_4$
2,6-Diethoxybenzoic acid 339
3-(2,3-Dimethoxyphenyl)-propionic acid 327
Dimethyl 1,3,5-heptatriene-1,6-dicarboxylate 79
5,6-Dimethyl-1,3,3a,4,5,7a-hexahydro-3-oxo-4-isobenzofurancarboxylic acid 96
Dimethyl 5-norbornene-2,3-dicarboxylate 78
4-Ethoxy-3-methoxyphenylacetic acid 327
Methyl-4-(dimethoxymethyl)-benzoate 365
Methyl 2-(3,5-dimethoxyphenyl)acetate 356
2′,4′,5′-Trimethoxyacetophenone 291
2′,4′,6′-Trimethoxyacetophenone 293
$C_{11}H_{14}O_4S$
4-(*tert*-Butylsulfonyl)benzoic acid hydrate 781
Ethyl 2-(*p*-tolylsulfonyl)-acetate 782
$C_{11}H_{14}O_4Si$
Diacetoxymethylphenylsilane 1112
$C_{11}H_{14}O_5$
Butyl gallate 367
2′-Hydroxy-3′,4′,6′-trimethoxyacetophenone 289
Methyl 2,4,6-trimethoxybenzoate 370
$C_{11}H_{14}O_7S$
Methyl 4,5-bis(methoxycarbonyl)-3-oxo-2,3,4,5-tetrahydro-2-thiopheneacetate 93
$C_{11}H_{14}S$
Crotyl *o*-tolyl sulfide 212
2-Methylallyl *o*-tolyl sulfide 212
4-Pentenyl phenyl sulfide 212
$C_{11}H_{14}S_2$
1,4,5,7-Tetrahydro-3*H*-2,6-benzodithionin 214
$C_{11}H_{15}Br$
2-Bromo-4-*tert*-butyl-1-methylbenzene 169
1-Bromopentamethylbenzene 170
$C_{11}H_{15}BrN_2O$
1-(4-Bromophenyl)-3-butylurea 538
3-(4-Bromophenyl)-1,1-diethylurea 538
1-(4-Bromophenyl)-3-isobutylurea 538
3-(3-Bromophenyl)-1-methyl-1-propylurea 534
$C_{11}H_{15}BrN_2O_2$
2-Dimethylaminoethyl *N*-(4-bromophenyl)carbamate 627
$C_{11}H_{15}BrO$
Benzyl 4-bromobutyl ether 185
2-*tert*-Butyl-4-bromo-6-methylphenol 190
$C_{11}H_{15}BrO_2$
10-Bromo-8-methoxytricyclo(5.2.1.02,6)dec-4-en-3-ol 20
$C_{11}H_{15}BrO_2S$
2-(5-Bromo-2-thienyl)-2,5,5-trimethyl-1,3-dioxane 859
$C_{11}H_{15}BrO_4$
3-(2-Bromo-4,5-dimethoxyphenyl)-1,2-propanediol 197
4-Bromo-5-methyl-7-oxo-6-oxabicyclo(3.2.1)octan-2-ylmethyl acetate 96
$C_{11}H_{15}Cl$
(+)-1-Chloro-3-methyl-2-phenylbutane 169
2,3,5,6-Tetramethylbenzyl chloride 170
$C_{11}H_{15}ClN_2$
1-(4-Chlorophenyl)-2-methylpiperazine 244
1-(*p*-Chlorophenyl)-4-methylpiperazine 253
1-(5-Chloro-*o*-tolyl)-piperazine 248
$C_{11}H_{15}ClN_2O$
1-*sec*-Butyl-3-(2-chlorophenyl)urea 536
1-Butyl-3-(2-chlorophenyl)urea 587
1-Butyl-3-(3-chlorophenyl)urea 580

$C_{11}H_{15}ClN_2O$ (Cont'd)
1-Butyl-3-(4-chlorophenyl)urea ... 590
1-(4-Chlorobenzyl)-3-ethyl-1-methylurea ... 528
6′-Chloro-2-dimethylamino-*o*-acetotoluidide ... 430
6′-Chloro-2-dimethylamino-*o*-acetotoluidide hydrochloride ... 424
1-(4-Chloro-α,α-dimethylphenethyl)-urea ... 590
1-(2-Chloro-6-methylphenyl)-3-isopropylurea ... 593
1-(3-Chloro-4-methylphenyl)-3-isopropylurea ... 598
1-(4-Chloro-2-methylphenyl)-3-isopropylurea ... 598
1-(5-Chloro-2-methylphenyl)-3-isopropylurea ... 598
1-(2-Chloro-6-methylphenyl)-3-propylurea ... 540
1-(3-Chloro-2-methylphenyl)-3-propylurea ... 540
1-(3-Chloro-4-methylphenyl)-3-propylurea ... 598
1-(4-Chloro-2-methylphenyl)-3-propylurea ... 598
1-(5-Chloro-2-methylphenyl)-3-propylurea ... 598
3-(2-Chlorophenyl)-1,1-diethylurea ... 587
3-(3-Chlorophenyl)-1,1-diethylurea ... 580
3-(4-Chlorophenyl)-1,1-diethylurea ... 590
1-(2-Chlorophenyl)-3-isobutylurea ... 587
1-(3-Chlorophenyl)-3-isobutylurea ... 580
1-(4-Chlorophenyl)-3-isobutylurea ... 590

$C_{11}H_{15}ClN_2O_2$
1-(3-Chloro-2-methoxyphenyl)-3-isopropylurea ... 556
1-(2-Chloro-4-methoxyphenyl)-3-propylurea ... 521
1-(3-Chloro-2-methoxyphenyl)-3-propylurea ... 556
1-(5-Chloro-2-methoxyphenyl)-3-propylurea ... 520
3-(5-Chloro-2-methylphenyl)-1-(2-hydroxyethyl)-1-methylurea ... 548
1-(2-Chloro-6-methylphenyl)-3-(2-methoxyethyl)urea ... 521
1-(5-Chloro-2-methylphenyl)-3-(2-methoxyethyl)urea ... 556
3-(4-Chlorophenyl)-1-ethyl-1-(2-hydroxyethyl)urea ... 548
1-(6-Chloro-*o*-tolyl)-3-(3-hydroxypropyl)-urea ... 549
2-Dimethylaminoethyl *N*-(2-chlorophenyl)carbamate ... 671
2-Dimethylaminoethyl *N*-(3-chlorophenyl)carbamate ... 671
2-(Dimethylamino)ethyl *N*-(4-chlorophenyl)carbamate ... 672

$C_{11}H_{15}ClN_2O_3$
3-Carbamoyl-1-(isopropoxycarbonylmethyl)-pyridinium chloride ... 946
1-(5-Chloro-2,4-dimethoxyphenyl)-3-ethylurea ... 515

$C_{11}H_{15}ClN_2O_3S$
*N*2-(Butylsulfonyl)-4-chlorobenzoic hydrazide ... 808

$C_{11}H_{15}ClN_2O_4$
1-(4-Imino-2,5-cyclohexadien-1-ylidene)piperidinium diperchlorate ... 1057

$C_{11}H_{15}ClO$
4-Chloro-α-*tert*-butylbenzyl alcohol ... 198
4-Chloro-2-isopropyl-5-methylanisole ... 177
2-Chloro-4-(*tert*-pentyl)-phenol ... 192
3-Methoxy-3-(*p*-tolyl)propyl chloride ... 185
4-(2-Methylphenoxy)butyl chloride ... 176
4-(3-Methylphenoxy)butyl chloride ... 176
4-(4-Methylphenoxy)butyl chloride ... 176

$C_{11}H_{15}ClO_2S$
Benzyl 4-chlorobutyl sulfone ... 772
Pentamethylbenzenesulfonyl chloride ... 804

$C_{11}H_{15}ClS$
5-Chloropentyl phenyl sulfide ... 209

$C_{11}H_{15}Cl_2NO_2S$
N,*N*-Bis(2-chloroethyl)-*p*-toluenesulfonamide ... 805

$C_{11}H_{15}Cl_5OSi_2$
1-(Pentachlorophenyl)-1,1,3,3,3-pentamethyldisiloxane ... 1100

$C_{11}H_{15}FN_2O$
1-Butyl-3-(2-fluorophenyl)urea ... 536
1-*tert*-Butyl-3-(4-fluorophenyl)urea ... 538
1-Butyl-3-(4-fluorophenyl)urea ... 590
1,1-Diethyl-3-(3-fluorophenyl)urea ... 534
1,1-Diethyl-3-(4-fluorophenyl)urea ... 590
1,3-Diethyl-1-(3-fluorophenyl)urea ... 534
1-(3-Fluoro-4-methylphenyl)-3-propylurea ... 543

$C_{11}H_{15}FN_2O_2$
2-Dimethylaminoethyl *N*-(4-fluorophenyl)carbamate ... 627

$C_{11}H_{15}FN_2O_5S$
1-(4-Fluorosulfonylphenyl)-3-(2-(2-hydroxyethoxy)ethyl)urea ... 801

$C_{11}H_{15}N$
N-Allyl-*N*-methylbenzylamine ... 243
N-Benzylidenebutylamine ... 1076
N-Crotyl-*p*-toluidine ... 220
N,*N*-Dimethyl-3-phenyl-2-propenamine hydrochloride ... 243
1-Propylindoline ... 252
2,3,7-Trimethylindoline ... 253

$C_{11}H_{15}NO$
4-Benzylmorpholine ... 252
4-(*tert*-Butyl)benzamide ... 477
1-Cyano-2,5,6,6-tetramethylbicyclo(3.1.0)hexan-4-one ... 148
4-Cyclopentyloxyaniline ... 218
N-Cyclopropyl-4-methoxybenzylamine hydrochloride ... 240
N,*N*-Diethylbenzamide ... 482
2′,3′-Dimethylpropionanilide ... 439
2′,4′-Dimethylpropionanilide ... 440
Isobutyl 6-methyl-2-pyridyl ketone ... 959
4′-Isopropylacetanilide ... 438
2-(Isopropylamino)-4-methylbenzaldehyde ... 324
5′-Isopropyl-2′-methylformanilide ... 440
3-Methyl-2-phenylbutyramide ... 429
α-Methyl-α-(propylimino)-*o*-cresol ... 1035
3-Methyl-5-(2-pyridyl)-2-pentanone ... 961
2′,6′-Propionoxylidide ... 439
2′-Propylacetanilide ... 438
2′,4′,5′-Trimethylacetanilide ... 441

$C_{11}H_{15}NOS$
S-*tert*-Butyl benzothiohydroxamate ... 1009
S-Butyl *N*-phenylthiocarbamate ... 688
S-Isopropyl *N*-(*o*-tolyl)-thiolcarbamate ... 688
S-Isopropyl *N*-(*m*-tolyl)-thiolcarbamate ... 688
S-Isopropyl *N*-(*p*-tolyl)-thiolcarbamate ... 688
S-Propyl *N*-(*m*-tolyl)thiolcarbamate ... 688
S-Propyl *N*-(*p*-tolyl)thiolcarbamate ... 688

$C_{11}H_{15}NO_2$
3-Amino-4-*tert*-butylbenzoic acid ... 342
D-2-Aminobutyric acid benzyl ester *p*-toluenesulfonate ... 356
L-2-Aminobutyric acid benzyl ester *p*-toluenesulfonate ... 792
N-Benzylglycine ethyl ester hydrobromide ... 358
2-Butoxybenzamide ... 466
2-(Butylamino)benzoic acid ... 343
Butyl carbanilate ... 680
sec-Butyl carbanilate ... 677
2-(2-Cyanoethyl)-5,5-dimethyl-1,3-cyclohexanedione ... 148
N-(1,1-Dimethyl-2-hydroxyethyl)-benzamide ... 462
2′-Ethoxypropionanilide ... 417
4′-Ethoxypropionanilide ... 419
Ethyl 2-cyano-2-cyclohexylideneacetate ... 148
Ethyl 2-cyano-2-cyclohexylideneacetate ... 148
Ethyl α-(cyclohexylidene)-cyanoacetate ... 149
Ethyl *N*-(2-ethylphenyl)carbamate ... 681
Ethyl *N*-phenethylcarbamate ... 671
Ethyl *N*-(2,3-xylyl)carbamate ... 683
Ethyl *N*-(2,4-xylyl)carbamate ... 685
Ethyl *N*-(2,5-xylyl)carbamate ... 685
Ethyl *N*-(2,6-xylyl)carbamate ... 683
Ethyl *N*-(3,4-xylyl)carbamate ... 685
Isobutyl 4-aminobenzoate ... 372
2-Isopropylphenyl *N*-methylcarbamate ... 681
3-Isopropylphenyl *N*-methylcarbamate ... 678
Isopropyl *N*-(*o*-tolyl)carbamate ... 681
Isopropyl *N*-(*m*-tolyl)carbamate ... 678
Isopropyl *N*-(*p*-tolyl)carbamate ... 682
2-Methoxy-4-(methyliminomethyl)-phenetole ... 1040
N-(α-Methylbenzyl)-glycine methyl ester ... 358
Methyl *N*-(2-ethyl-6-methylphenyl)-carbamate ... 683
2-Phenylsarcosine ethyl ester ... 358
Propyl *N*-(*o*-tolyl)carbamate ... 681
Propyl *N*-(*m*-tolyl)carbamate ... 678
Propyl *N*-(*p*-tolyl)carbamate ... 682
m-Tolyl *N*-propylcarbamate ... 678

$C_{11}H_{15}NO_2S$
S-Benzyl-L-cysteine methyl ester hydrochloride ... 357
S-Benzyl-DL-homocysteine ... 348
S-(4-Hydroxyphenyl) *N*-butylthiocarbamate ... 646
S-Isopropyl *N*-(*p*-anisyl)-thiolcarbamate ... 655
4-(Methylthio)phenyl *N*-propylcarbamate ... 667
3-Morpholino-1-(2-thienyl)-1-propanone hydrochloride ... 857

$C_{11}H_{15}NO_2S_2$
N-(3,4-Dimethoxyphenethyl)-dithiocarbamic acid, ammonium salt ... 735

$C_{11}H_{15}NO_3$
4-Amino-2-butoxybenzoic acid ... 340
5-Amino-2-phenyl-1,3-dioxane-5-methanol ... 245
N-Benzyl-L-threonine ... 348
4-*tert*-Butyl-2-nitroanisole ... 268
3′,5′-Dihydroxy-2-(isopropylamino)-acetophenone hemisulfate ... 289
3′,4′-Dihydroxy-2-(isopropylamino)-acetophenone hydrochloride ... 290
2-(2,5-Dimethoxybenzylidene-amino)ethanol ... 1029
2′,5′-Dimethoxy-4′-methylacetanilide ... 414
2′,5′-Dimethoxypropionanilide ... 414
exo-4,7-Epoxy-2-ethyl-7-methyl-2,3,3a,4,5,6,7,7a-octahydro-1*H*-isoindole-1,3-dione ... 135
2-Ethoxyethyl *N*-phenylcarbamate ... 649
Ethyl 3-acetyl-2,4-dimethyl-5-pyrrolecarboxylate ... 842
Ethyl *N*-(2-ethoxyphenyl)carbamate ... 653
Ethyl *N*-(4-ethoxyphenyl)carbamate ... 655
Ethyl 5-formyl-1,2,4-trimethyl-3-pyrrolecarboxylate ... 841
Ethyl *N*-(2-methoxy-5-methylphenyl)carbamate ... 653
Isopropyl *N*-methoxy-*N*-phenylcarbamate ... 674
Isopropyl *N*-(2-methoxyphenyl)-carbamate ... 654
Isopropyl *N*-(3-methoxyphenyl)-carbamate ... 654
Isopropyl *N*-(4-methoxyphenyl)-carbamate ... 655
N-Lactoyl-*p*-phenetidine ... 411
2-Methoxyethyl *N*-(*o*-tolyl)-carbamate ... 650
2-Methoxyethyl *N*-(*m*-tolyl)-carbamate ... 649
2-Methoxyethyl *N*-(*p*-tolyl)-carbamate ... 650
3-Methoxy-α-methylphenylalanine ... 348
DL-α-Methyl-*m*-tyrosine hydrochloride ... 354
3-Nitrophenyl pentyl ether ... 268
Propyl *N*-methoxy-*N*-phenylcarbamate ... 674
Propyl *N*-(2-methoxyphenyl)-carbamate ... 653
Propyl *N*-(4-methoxyphenyl)-carbamate ... 655
N-(3,4,5-Trimethoxybenzylidene)-methylamine ... 1041
Tyrosine ethyl ester ... 354
L-Tyrosine ethyl ester hydrochloride ... 354

$C_{11}H_{15}NO_3S$
cis-3-Benzylamino-4-hydroxy-2,3,4,5-tetrahydrothiophene-1,1-dioxide ... 774
trans-3-(Benzylamino)-4-hydroxytetrahydrothiophene-1,1-dioxide ... 774
4-(Benzylsulfonyl)morpholine ... 827

$C_{11}H_{15}NO_4$
2-Butoxy-4-nitrobenzyl alcohol ... 263
Diethyl 3-methyl-2,4-pyrroledicarboxylate ... 841
3,4-Dimethoxy-DL-phenylalanine ... 348
Dimethyl 3-(1-pyrrolyl)glutarate ... 841
Ethyl 5-carboxy-4-ethyl-3-methyl-2-pyrrolecarboxylate ... 840
N-Methyl-3,4,5-trimethoxybenzamide ... 463
N-Phenyl-D-ribosamine ... 225
6-(Pyrrolidinomethyl)-kojic acid hydrochloride ... 58
2-(3,4,5-Trimethoxyphenyl)-acetamide ... 414

$C_{11}H_{15}NO_4S$
N,*N*-Diethyl-4-sulfamoylbenzoic acid ... 823
Isopropyl *N*-(methylsulfonyl)-*N*-phenylcarbamate ... 824
1-Methyl-1-nitropropyl *p*-tolyl sulfone ... 780
Propyl *N*-(methylsulfonyl)-*N*-phenylcarbamate ... 824

$C_{11}H_{15}NO_5$
6-(Morpholinomethyl)-kojic acid ... 58

$C_{11}H_{15}NO_5S$
2-Methyl-2-nitropropyl *p*-toluenesulfonate ... 798

$C_{11}H_{15}NO_6S$
N,*N*-Bis(2-hydroxyethyl)-3-carboxybenzenesulfonamide ... 812
4-(3-Methyl-4-nitrophenoxy)-1-butanesulfonic acid, potassium salt ... 792
4-(3-Methyl-4-nitrophenoxy)-1-butanesulfonic acid, sodium salt ... 792
4-(2-(Methylsulfonamido)-ethoxy)-*m*-anisic acid ... 813

$C_{11}H_{15}N_3$
2-Amino-1-butylbenzimidazole ... 904
4-(1,5-Diazabicyclo(3.2.1)oct-8-yl)-pyridine ... 955

$C_{11}H_{15}N_3OS$
1-(4-Acetamidophenyl)-3-ethyl-2-thiourea ... 740

$C_{11}H_{15}N_3O_2$
1-Acetyl-4-(2,6-dimethylphenyl)-semicarbazide ... 574
1-Acetyl-4-(3,4-xylyl)semicarbazide ... 578
2-Cyclohexylamino-5-nitropyridine ... 943
N^3,N^3-Diethyl-3,5-pyridinedicarboxamide ... 964
1-(2-Ethoxyphenyl)-1,2-propanedione 1-hydrazone 2-oxime ... 1009
Glycyl-D-phenylalaninamide ... 429
Glycyl-L-phenylalaninamide acetate ... 422
1-Methyl-4-(2-nitrophenyl)piperazine ... 274
1-Methyl-4-(4-nitrophenyl)piperazine ... 276
1-Methyl-5-(2,4-xylyl)biuret ... 579

$C_{11}H_{15}N_3O_2S$
1-Isopropyl-3-(4-methyl-3-nitrophenyl)-2-thiourea ... 740

$C_{11}H_{15}N_3O_3$
N-Acetyl-L-tyrosine hydrazide ... 496
1-Butyl-3-(3-nitrophenyl)urea ... 564
1-Butyl-3-(4-nitrophenyl)urea ... 568

$C_{11}H_{15}N_3O_3$ (Cont'd)
1-(4-Carboxyphenyl)-4-propylsemicarbazide 576
1,1-Diethyl-3-(3-nitrophenyl)urea 564
1,1-Diethyl-3-(4-nitrophenyl)urea 568
Glycyl-L-tyrosinamide acetate 412
Glycyl-L-tyrosinamide hydrochloride 412
1-(2-Hydroxybenzoyl)-4-isopropylsemicarbazide 548
1-(4-Hydroxybenzoyl)-4-isopropylsemicarbazide 548
1-(2-Hydroxybenzoyl)-4-propylsemicarbazide 548
1-(4-Hydroxybenzoyl)-4-propylsemicarbazide 548
1-Isobutyl-3-(3-nitrophenyl)urea 564
1-Isopropyl-3-(4-methyl-3-nitrophenyl)urea 564
1-(4-Methyl-3-nitrophenyl)-3-propylurea 564
2-(3-Phenylureido)ethyl *N*-methylcarbamate 674
L-Tyrosyl-L-glycinamide hydrochloride 412

$C_{11}H_{15}N_3O_3S$
1-(2,3,4-Trimethoxybenzylidene)-3-thiosemicarbazide 1040

$C_{11}H_{15}N_3O_4$
2-Dimethylaminoethyl *N*-(3-nitrophenyl)carbamate 669
1-Hydroxyethyl-1-methyl-3-(2-methyl-4-nitrophenyl)urea 547
1-(2-Methoxyethyl)-3-(4-methyl-3-nitrophenyl)urea 552
1-(2,3,4-Trimethoxybenzylidene)-semicarbazide 549

$C_{11}H_{15}N_3O_4S$
1-(4-(*N*-(Acetyl)sulfamoyl)phenyl)-3-ethylurea 824

$C_{11}H_{15}N_3O_4S_2$
6-Methoxy-3-(3-sulfopropyl)-3*H*-benzothiazolin-2-one hydrazone 737

$C_{11}H_{15}N_3O_6S$
N-Butyl-2,6-dinitro-4-(methylsulfonyl)aniline 777
6-Nitro-1,2,3-trimethyl-benzimidazolium methyl sulfate 906

$C_{11}H_{15}N_3S$
4-Isopropylbenzaldehyde thiosemicarbazone 1063
1-(*N*-Methylthiocarbamoyl)-2-phenylpyrazolidine 747

$C_{11}H_{15}N_5$
4-Amino-2-cyclohexylamino-5-pyrimidinecarbonitrile 974

$C_{11}H_{15}N_5O_3$
4-Methyl-1-(4-(3-methylureido)-benzoyl)semicarbazide 574

$C_{11}H_{15}O_3P$
2,2-Dimethyltrimethylene phenyl phosphite 833

$C_{11}H_{16}$
3,4-Dimethylcumene 163
5-Isopropyl-*m*-xylene 163
(1-Methylbutyl)-benzene 162
(*S*)-(+)-(2-Methylbutyl)benzene 162
1,2,3,4,4a,5,8,8a-Octahydro-1,4-methanonaphthalene 3

$C_{11}H_{16}BF_4N_3$
4-Diazo-*N*,*N*-diethyl-*m*-toluidine fluoroborate 769

$C_{11}H_{16}BrN$
(4-Bromobenzyl)diethylamine hydrochloride 236
N-(2-Bromoethyl)-*N*-ethyl-*m*-toluidine hydrobromide 215

$C_{11}H_{16}ClN$
2-Chloro-*N*,*N*-diethyl-*p*-toluidine 220
5-Chloro-*N*,*N*-diethyl-*o*-toluidine 220
N-(2-Chloroethyl)-*N*-ethylbenzylamine hydrochloride 236
N-(2-Chloroethyl)-*N*-ethyl-*m*-toluidine 215
N-(3-Chloropropyl)-*N*-methylbenzylamine hydrochloride 236

$C_{11}H_{16}ClNO$
1-Chloro-3-(4-(dimethylamino)-phenyl)-2-propanol 215

$C_{11}H_{16}ClN_3$
4-Diethylamino-2-methylbenzenediazonium chloride, zinc chloride double salt 1095

$C_{11}H_{16}IN$
2,2-Dimethyl-1,2,3,4-tetrahydroisoquinolinium iodide 256

$C_{11}H_{16}INO$
1,1-Dimethyl-8-hydroxy-1,2,3,4-tetrahydroquinolinium iodide 254

$C_{11}H_{16}INO_2$
tert-Butyl 3,4-dimethyl-5-iodo-2-pyrrolecarboxylate 841

$C_{11}H_{16}NO_2$
5,5-Dimethyl-2-(dimethylaminomethylene)-1,3-cyclohexanedione 58

$C_{11}H_{16}N_2$
2-Amino-2-adamantanecarbonitrile hydrochloride 145
1-Benzylpiperazine dihydrochloride 250
2-Cyclohexylaminopyridine 953
3-Methyl-2-butanone phenylhydrazone 236
1-Methyl-4-phenylpiperazine 253
2-Methyl-1-phenylpiperazine 250
2-(Piperidinomethyl)-pyridine 944
2-(2-Pyrrolidinoethyl)pyridine 946
4-(2-Pyrrolidinoethyl)pyridine 954
1-(*p*-Tolyl)-piperazine dihydrochloride 250

$C_{11}H_{16}N_2O$
3-Amino-*N*-butylbenzamide 468
3-Amino-*N*,*N*-diethylbenzamide 468
3-Amino-*N*,*N*-diethylbenzamide hydrochloride 468
1-Benzyl-3-ethyl-1-methylurea 585
1-Benzyl-3-isopropylurea 571
1-Butyl-3-phenylurea 585
1,1-Diethyl-3-phenylurea 585
1-(2,6-Diethylphenyl)urea 593
1,1-Dimethyl-3-(2-ethylphenyl)urea 587
1,1-Dimethyl-3-(3-ethylphenyl)urea 580
1,1-Dimethyl-3-(2,3-xylyl)urea 593
1,1-Dimethyl-3-(2,4-xylyl)urea 598
1,1-Dimethyl-3-(2,5-xylyl)urea 598
1,1-Dimethyl-3-(2,6-xylyl)urea 593
1,1-Dimethyl-3-(3,4-xylyl)urea 598
1,1-Dimethyl-3-(3,5-xylyl)urea 602
3'-(Ethylamino)-4'-methylacetanilide 425
1-Ethyl-3-(2-ethylphenyl)urea 587
1-Ethyl-3-(4-methylbenzyl)urea 590
1-(2-Ethyl-6-methylphenyl)-3-methylurea 593
1-Ethyl-3-methyl-1-(*o*-tolyl)urea 587
1-Ethyl-3-(2,3-xylyl)urea 593
1-Ethyl-3-(2,4-xylyl)urea 598
1-Ethyl-3-(2,6-xylyl)urea 593
1-Ethyl-3-(3,4-xylyl)urea 598
1-Ethyl-3-(3,5-xylyl)urea 602
3-Hydroxy-2-(piperidinomethyl)-pyridine 932
1-Isobutyl-3-phenylurea 585
1-Isopropyl-3-(*o*-tolyl)urea 587
1-Isopropyl-3-(*m*-tolyl)urea 580
1-Isopropyl-3-(*p*-tolyl)urea 590
Mesityloxyacetamidine hydrochloride 724
1-(3-Methoxyphenyl)-piperazine dihydrochloride 247
2-(2-Morpholinoethyl)pyridine 946
1-Propyl-3-(*o*-tolyl)urea 587
1-Propyl-3-(*m*-tolyl)urea 580
1-Propyl-3-(*p*-tolyl)urea 590
2-Pyridinealdoxime *O*-isoamyl ether 1054
3-Pyridinealdoxime *O*-isoamyl ether 1057
4-Pyridinealdoxime *O*-isoamyl ether 1060
1-(2-Pyridylmethyl)-3-piperidinol 932

$C_{11}H_{16}N_2OS$
1,1-Dimethyl-3-(4-ethoxyphenyl)-2-thiourea 736
1-(4-Ethoxyphenyl)-3-ethyl-2-thiourea 738
1-Ethyl-3-(2-methoxy-5-methylphenyl)-2-thiourea 737
1-(2-Hydroxyethyl)-3-(2,4-xylyl)-2-thiourea 735

$C_{11}H_{16}N_2O_2$
1-(*m*-Anisyl)-3-propylurea 557
1-(*p*-Anisyl)-3-propylurea 559
tert-Butyl 2-cyano-2-(2-pyrrolidinylidene)acetate 149
1-*tert*-Butyl-3-(3-hydroxyphenyl)urea 548
1-*tert*-Butyl-3-(4-hydroxyphenyl)urea 549
1-Butyl-3-(4-hydroxyphenyl)-urea 549
Diethyl(4-nitrobenzyl)amine 277
2-(Dimethylamino)ethyl 4-aminobenzoate 372
2-Dimethylaminoethyl *N*-phenylcarbamate 671
1,1-Dimethyl-3-(2-methoxy-5-methylphenyl)urea 556
1-Ethyl-1-(2-hydroxyethyl)-3-phenylurea 548
1-Ethyl-3-(2-methoxy-5-methylphenyl)urea 556
Isobutyl 3-phenylcarbazate 677
1-Isopropyl-1-(3-methoxyphenyl)urea 557
1-Isopropyl-3-(2-methoxyphenyl)urea 555
1-Isopropyl-3-(3-methoxyphenyl)urea 557
1-Isopropyl-3-(4-methoxyphenyl)urea 559
1-(2-Methoxyethyl)-3-(*o*-tolyl)urea 555
2-Methoxy-4-morpholinoaniline hydrochloride 247
1-(2-Methoxyphenyl)-3-propylurea 555

$C_{11}H_{16}N_2O_2S$
Mesitylenesulfonic ethylidenehydrazide 1071

$C_{11}H_{16}N_2O_3$
4'-(3-Amino-2-hydroxy-1-propoxy)-acetanilide 410
3-(2,5-Dimethoxyphenyl)-1,1-dimethylurea 556
1-(2,5-Dimethoxyphenyl)-3-ethylurea 556
Ethyl 3-(3,5-diamino-4-hydroxyphenyl)propionate 354
1-(2-Hydroxyethyl)-3-(2-methoxy-5-methylphenyl)urea 547
1-(2-Methoxyethyl)-3-(2-methoxyphenyl)urea 555

$C_{11}H_{16}N_2O_3S$
1-Butyl-3-(phenylsulfonyl)-urea 825
2,2-Dimethyl-4'-sulfamoylpropionanilide 825

$C_{11}H_{16}N_2O_3S_2$
2-(4-Aminophenylsulfonyl)-*N*-(2-(methylthio)-ethyl)-acetamide 775
1-(2-(Methylthio)-ethyl)-3-(*p*-tolylsulfonyl)-urea 816

$C_{11}H_{16}N_2O_4S$
3-(3-Butylureido)benzenesulfonic acid, sodium salt 795

$C_{11}H_{16}N_2O_5$
N-(4-Hydroxy-3-methoxy-5-nitrobenzylidene)-4-ethoxyaniline 1029

$C_{11}H_{16}N_2O_6S$
Sulfanilamide-N^4-L-arabinoside 810

$C_{11}H_{16}N_2S$
1-Butyl-3-phenyl-2-thiourea 748
1-(4-*tert*-Butylphenyl)-2-thiourea 748
1,1-Diethyl-3-phenyl-2-thiourea 748
1,1-Dimethyl-3-(2,6-xylyl)-2-thiourea 749
1,1-Dimethyl-3-(3,4-xylyl)-2-thiourea 749
1-Ethyl-3-(3,4-xylyl)-2-thiourea 749
1-Isobutyl-3-phenyl-2-thiourea 748

$C_{11}H_{16}N_4O$
2-(Cyclohexylidenehydrazino)-4-hydroxy-6-methylpyrimidine 972
1-(1-(2-Methoxybenzyl)-ethylideneamino)guanidine nitrate 1049

$C_{11}H_{16}N_4OS$
1-Ethyl-6-(*o*-tolyl)-5-thiobiurea 743
1-Methyl-6-(3,4-xylyl)-2-thiobiurea 745
6-Phenyl-1-propyl-5-thiobiurea 745

$C_{11}H_{16}N_4O_2$
1-Ethyl-3-isobutylxanthine 923
1,1'-(4-Methyl-1,3-phenylene)bis(3-methylurea) 566

$C_{11}H_{16}N_4O_3$
((2-Ethoxyphenyl)ureido)acetic acid hydrazide 554

$C_{11}H_{16}N_4S$
6-Hexylthiopurine 922

$C_{11}H_{16}O$
4-Amylphenol 191
Benzyl *tert*-butyl ether 186
Benzyl isobutyl ether 186
4-*tert*-Butylanisole 181
α-(*tert*-Butyl)-benzyl alcohol 204
3-*tert*-Butyl-5-methylphenol 192
4-*tert*-Butyl-2-methylphenol 192
Butyl *o*-tolyl ether 181
4-Isopentylphenol 191
2-Mesityleneethanol 208
2-(1-Methylbutyl)phenol 191
4a-Methyl-1,2,3,4,4a,5,8,8a-octahydro-2-naphthalenone 56
(-)-3-Methyl-2-phenyl-1-butanol 206
(+)-3-Methyl-2-phenyl-1-butanol 207
DL-4-Methyl-α-(propyl)benzyl alcohol 205
Pentyl phenyl ether 180
1-Phenyl-1-pentanol 206
3-Phenyl-3-pentanol 206
α,α,4-Trimethylphenethyl alcohol 208

$C_{11}H_{16}OS_4$
2,4-Bis(1,3-dithiolan-2-yl)-1-cyclopentanone 59

$C_{11}H_{16}O_2$
2-Benzyl-2-methyl-1,3-propanediol 208
4-Benzyloxy-1-butanol 201
5-*tert*-Butyl-3-methylcatechol 193
6-*tert*-Butyl-3-methylcatechol 193
2-Cyclohexyl-5-methyl-3(2*H*)-furanone 58
2,6-Diethoxytoluene 181
4-(3-Hydroxyisoamyl)phenol 208
2-(2-Isopropylphenoxy)ethanol 202
Methyl 4-methyl-2,4,6-octatriene-2-carboxylate 79
Methyl 5-methyl-2,4,6-octatriene-2-carboxylate 79
Methyl 2-nonen-4-yne-2-carboxylate 1018
2-Norbornyl methacrylate 79
1-Phenyl-1,5-pentanediol 199
Tricyclo[5.2.1.$0^{2,6}$]dec-8-yl formate 77
(-)-1,2,3-Trimethyl-2-cyclopenteneacrylic acid 67

$C_{11}H_{16}O_2S$
Benzyl butyl sulfone 783
Benzyl *tert*-butyl sulfone 783
Butyl *o*-tolyl sulfone 785
tert-Butyl *p*-tolyl sulfone 785
Butyl *p*-tolyl sulfone 785
4-(2,2-Dimethylpropyl)-benzenesulfinic acid 787
4-(2,2-Dimethylpropyl)-benzenesulfinic acid, sodium salt 787
5-Isopropyl-2-thiophenebutyric acid 858

$C_{11}H_{16}O_3$
Allyl 3,4-epoxy-6-methylcyclohexanecarboxylate 91
2-(2-Benzyloxy)-ethoxy)-ethanol 200
(+/-)-3-Camphorcarboxylic acid 65
4-Carbethoxy-3,5-dimethyl-2-cyclohexene-1-one 92
3-(2-Ethylphenoxy)-1,2-propanediol 199
1-Ethyl-2,4,6-trimethoxybenzene 182
3-(2-Methylbenzyloxy)-1,2-propanediol 201

$C_{11}H_{16}O_3S$
4-*tert*-Butylphenyl methanesulfonate 800
Butyl *p*-toluenesulfonate 800
Isobutyl tosylate 800
4-Phenylbutyl methanesulfonate 800

$C_{11}H_{16}O_4$
6-(Acetoxymethyl)-3-methyl-3-cyclohexene-1-carboxylic acid 95
Dimethyl 2-(3-methyl-2-buten-1-ylidene)succinate 78
3-(2-Methoxybenzyloxy)-1,2-propanediol 202

$C_{11}H_{16}O_4S$
2-Ethoxyethyl *p*-toluenesulfonate 797

$C_{11}H_{16}O_4S_2$
9-(Acetylmethylthio)-1,5-dioxa-8-thiaspiro(5.5)undecan-10-one 58
4′α-Hydroxyoctahydrospiro(1,3-dioxane-2,3′-thiopyrano(2,3-b)-thiopyran)-6′-one 58
$C_{11}H_{16}O_5$
Diethyl 3-allyl-2-oxosuccinate 92
Dihydro-6′-(hydroxymethyl)spiro-cyclohexane-1,2′-furo-(3,4-d)(1,3)-dioxole)-4′(3a′H)-one 95
Methyl 1-(methoxycarbonyl)-4-oxocyclohexaneacetate 94
2-Methyl-2-(4-methyl-3-oxocyclohexyl)malonic acid 65
$C_{11}H_{16}O_6$
4-Carboxy-1,1-cyclohexanediacetic acid 66
$C_{11}H_{16}O_7$
2-(Diacetoxymethyl)-2,5-dihydro-2,5-dimethoxyfuran 89
$C_{11}H_{16}S$
4-*tert*-Butylphenyl methyl sulfide 213
Butyl *p*-tolyl sulfide 212
$C_{11}H_{16}S_2$
Benzaldehyde diethyl mercaptal 214
$C_{11}H_{16}Si$
1-Phenylsilacyclohexane 1112
$C_{11}H_{17}Br$
1-Bromo-3-methyladamantane 6
$C_{11}H_{17}BrN_2O$
4-Carbamoyl-1-isopentylpyridinium bromide 951
4-Carbamoyl-1-pentylpyridinium bromide 951
$C_{11}H_{17}BrO$
4-Bromo-4-penten-1-yl 2,4-hexadien-1-yl ether 18
$C_{11}H_{17}Cl$
2-Chloro-3-methylene-4,7,7-trimethylnorbornane 6
$C_{11}H_{17}Cl_2NPt$
(+)-*trans*-Dichloro(ethylene)(α-methylphenethylamine)platinum(II) 1142
$C_{11}H_{17}N$
N-Butylbenzylamine 243
α-(*sec*-Butyl)-benzylamine 241
2-*tert*-Butyl-6-methylaniline 219
N-Butyl-*o*-toluidine 220
N,*N*-Diethylbenzylamine 244
N,*N*-Diethyl-*m*-toluidine 220
(2,5-Dimethylbenzyl)ethylamine 243
2,6-Dimethyl-3-isopropylaniline hydrochloride 219
α-Ethyl-1-cycloheptene-1-acetonitrile 150
2-Ethyl-6-isopropylaniline 219
N-Isopentylaniline 219
N-Methylcarvacrylamine 219
N-Methylcarvacrylamine oxalate 219
$C_{11}H_{17}NO$
2-Amino-4-*tert*-amylphenol 217
2-Amino-4-*sec*-butylanisole 218
2-Amino-4-*tert*-butylanisole 217
N-*tert*-Butyl-3-methoxyaniline 218
N,*N*-Diethyl-*m*-anisidine 218
3-(Dimethylamino)-4-isopropylphenol 217
4-(Dimethylamino)-3-isopropylphenol 217
(+)-3-(Dimethylamino)-1-phenylpropanol 238
3,6-Dimethyl-2-dimethylaminomethylphenol 239
2-Methoxy-α,*N*-dimethylphenethyl-amine hydrochloride 240
3-Pentyloxyaniline 217
$C_{11}H_{17}NO_2$
1-Adamantane-carbohydroxamic acid 101
2,2′-(Benzylimino)-diethanol 239
6-(2-Cyanoethyl)-1,4-dioxaspiro(4.5)-decane 150
3,4-Dimethoxy-*N*-methylphenethylamine hydriodide 239
3-(2,5-Dimethoxyphenyl)-propylamine 239
(1*S*,2*S*)-2-(Dimethylamino)-1-phenyl-1,3-propanediol 237
Ethyl 3,5-dimethyl-4-ethyl-2-pyrrolecarboxylate 841
Ethyl 4,5-dimethyl-3-ethyl-2-pyrrolecarboxylate 841
N-2-Hydroxyethylnorephedrine hydrochloride 237
N-Methyl-1-phenyl-2,2′-iminodiethanol 237
N-(*m*-Tolyl)-diethanolamine 217
$C_{11}H_{17}NO_2S$
N-Butyl-*p*-toluenesulfonamide 828
N,*N*-Diethyl-*p*-toluenesulfonamide 828
$C_{11}H_{17}NO_3$
2,2′-(Benzylimino)-diethanol *N*-oxide (40% aqueous solution) 239
3,5-Dihydroxy-α-(isopropylaminomethyl)benzyl alcohol hemisulfate 237
2,6-Dimethoxy-4-(dimethylaminomethyl)phenol 237
Isoproterenol 237
(-)-Methyl 4-cyano-5,5-dimethyl-6-oxoheptanoate 148
$C_{11}H_{17}NO_3S$
4-(4-Amino-3-methylphenoxy)-1-butanesulfinic acid 786
$C_{11}H_{17}NO_4$
6-(Diethylaminomethyl)-kojic acid hydrochloride 58
Diethyl 2-cyano-2,3-dimethylsuccinate 149
Ethyl 1-(*tert*-butyl)-4,5-dioxo-3-pyrrolidinecarboxylate 133
N-(7-Oxabicyclo(4.1.0)hept-1-ylcarbonyl)glycine ethyl ester 109
$C_{11}H_{17}NS$
1-Adamantanethiocarboxamide 141
2-(*N*-Benzyl-*N*-ethylamino)-ethanethiol 241
m-(Pentylthio)-aniline 218
$C_{11}H_{17}N_3$
1-(4-Aminophenyl)-4-methylpiperazine hydrochloride 248
3-Butyl-1-*p*-tolyltriazene 1077
$C_{11}H_{17}N_3O$
1-Butyl-3-(4-methyl-2-pyridyl)urea 965
1-Butyl-3-(3-pyridylmethyl)urea 951
1-*tert*-Butyl-3-(3-pyridylmethyl)urea 951
1,1-Dimethyl-4-(2-ethylphenyl)-semicarbazide 578
1,1-Dimethyl-4-(2,3-xylyl)-semicarbazide 578
1,1-Dimethyl-4-(2,4-xylyl)-semicarbazide 579
1,1-Dimethyl-4-(2,5-xylyl)-semicarbazide 579
1,1-Dimethyl-4-(3,5-xylyl)-semicarbazide 579
1-Methyl-1-phenyl-4-propylsemicarbazide 578
1-(2,6-Xylyl)-4-ethylsemicarbazide 578
$C_{11}H_{17}N_3O_2$
1-(4-Amino-2-hydroxyphenyl)-3-butylurea 547
1,1-Dimethyl-4-(2-methoxy-5-methylphenyl)semicarbazide 554
$C_{11}H_{17}N_3O_3$
4-(2,5-Dimethoxyphenyl)-1,1-dimethylsemicarbazide 554
Ethyl 2-diethylamino-4-hydroxy-5-pyrimidinecarboxylate 971
Ethyl 4-hydroxy-2-isobutylamino-5-pyrimidinecarboxylate 972
$C_{11}H_{17}N_3O_3S$
1,1-Dimethyl-3-(4-(*N*,*N*-dimethylsulfamoyl)phenyl)urea 826
4-Isopropyl-1-(*p*-tosyl)-semicarbazide 824
$C_{11}H_{17}N_3O_4$
N-*tert*-Butoxycarbonylhistidine 867
$C_{11}H_{17}O_4P$
4-(*tert*-Pentyl)phenyl phosphate 836
$C_{11}H_{18}Br_2O_4$
Diethyl 2,6-dibromoheptanedioate 83
$C_{11}H_{18}ClN$
3-Chloro-1-adamantane-methylamine hydrochloride 32
(4-Methylbenzyl)-trimethylammonium chloride 256
$C_{11}H_{18}IN$
(2-(2,5-Cyclohexadien-1-ylidene)-ethyl)trimethylammonium iodide 43
$C_{11}H_{18}INO$
Candicine iodide 254
$C_{11}H_{18}N_2$
N-Benzyl-*N*,*N*′-dimethylethylenediamine 241
2-(Diethylamino)-3-cyclohexene-1-carbonitrile perchlorate 146
2-(2-Diethylaminoethyl)-pyridine 946
4-(2-Diethylaminoethyl)-pyridine 955
2,6-Dimethyl-2,6-heptanedicarbonitrile 150
1-Methylspiro(diaziridine-3,2′-adamantane) 42
$C_{11}H_{18}N_2O$
1-Adamantanecarboxylic acid hydrazide 112
4-Amino-α-diethylamino-*o*-cresol 238
2-(4-Amino-*N*-ethyl-*o*-toluidino)-ethanol hydrochloride 221
2-(4-Amino-*N*[1]-ethyl-*m*-toluidino)-ethanol sulfate hydrate 221
Ethyl 2,4-dimethyl-5-ethyl-3-pyrrolecarboxylate 843
4-Ethyl-*N*,*N*,3,5-tetramethyl-2-pyrrolecarboxamide 843
$C_{11}H_{18}N_2O_2$
1-Cyclohexyl-1,2-dihydrothymine 36
2-Methyl-2-(1-(nitromethyl)-cyclohexyl)propionitrile 148
$C_{11}H_{18}N_2O_2S$
4′-(Diethylamino)-methanesulfonanilide 819
4′-(Diethylamino)-methanesulfonanilide hydrochloride 819
$C_{11}H_{18}N_2O_3$
Ethyl 2-(3,3-diallylureido)acetate 127
$C_{11}H_{18}N_2O_3S$
3-Amino-*N*-butyl-4-methoxybenzenesulfonamide 813
5-(2-Hydroxyethyl)-5-isopentyl-2-thiobarbituric acid 139
$C_{11}H_{18}N_2O_3SSi$
N-Nitroso-*N*-(trimethylsilylmethyl)-*p*-toluenesulfonamide 1108
$C_{11}H_{18}N_2O_4$
Ethyl 4-acetoacetyl-1-piperazinecarboxylate 115
4,4′-Malonyldimorpholine 111
$C_{11}H_{18}N_2O_4S_2$
S,*S*′-(*N*,*N*′-Methylenebis-(2-carbamoylethyl)) bis-(thioacetate) 111
$C_{11}H_{18}N_4O$
2-(1,3-Dimethylbutylidene-hydrazino)-4-hydroxy-6-methyl-pyrimidine 972
$C_{11}H_{18}N_4O_3$
1-Butyl-4,5-dihydro-3-ethyl-8-hydroxyxanthine 921
$C_{11}H_{18}O$
2-Acetyl-3,3-dimethylbicyclo(2.2.1)-heptane 55
7,7-Dimethyl-2-(methoxymethyl)-bicyclo(3.1.1)hept-2-ene 18
1,3-Dimethyl-4,5,6,7-tetrahydro-2-indanol 15
2-Hexylidenecyclopentanone 57
Homolimonenol 14
4-Hydroxymethyltricyclo[5.2.1.0[2,6]]-decane 13
3-Methyl-1-adamantanol 15
4a-Methyldecahydronaphthalen-2-one 56
3-Methyl-2-decalone (mixture of isomers) 57
DL-Patchenol 13
Spiro(5.5)undecan-2-one 56
Spiro(5.5)undecan-3-one 56
5-Undecyn-2-one 1018
$C_{11}H_{18}OSi$
Benzyldiethylsilanol 1112
4-(Trimethylsilyl)phenethyl alcohol 1105
$C_{11}H_{18}O_2$
3,7-Dimethyl-2,6-octadienyl formate 80
8-Hydroxy-4-(hydroxymethyl)-tricyclo[5.2.1.0[2,6]]decane 13
2-(Isopropoxymethylene)-6-methylcyclohexanone 53
4-Methoxy-7a-methyl-5,6,7,7a-tetrahydro-1-indanol 21
5-Methyl-1,3-adamantanediol 15
Methyl 4-decynoate 1019
8a-Methyl-1,2,3,4,6,7,8,8a-octahydro-1,6-naphthalenediol 12
1,3,5-Octatriene-1-carboxaldehyde dimethyl acetal 19
(+)-3-Pinanecarboxylic acid 66
(-)-3-Pinanecarboxylic acid 66
Spiro(4.5)decane-8-carboxylic acid 66
2,3,3-Trimethyl-1-cyclohexen-1-yl acetate 81
3,5,5-Trimethyl-2-cyclohexen-1-yl acetate 81
$C_{11}H_{18}O_2S_2$
2-(2,4,5-Trimethyl-1,3-dithiolan-2-yl)-4-pentenoic acid 68
$C_{11}H_{18}O_3$
Butyl 7-oxabicyclo-(4.1.0)-heptane-1-carboxylate 90
Ethyl 3-(2-oxocyclohexyl)-propionate 94
Methyl 3-(3-methyl-2-oxocyclohexyl)propionate 94
$C_{11}H_{18}O_3Si$
Triallyloxyvinylsilane 1097
$C_{11}H_{18}O_4$
1,1-Cycloheptanediacetic acid 66
Diethyl 2-allyl-2-methylmalonate 78
Diethyl 2-(1-butylidene)malonate 78
cis-Dimethyl pinate 77
2-Ethoxycarbonyl-2-(2-hydroxyethyl)-hexanoic acid γ-lactone 96
3-Methyl-1,1-cyclohexanediacetic acid 66
4-Methyl-1,1-cyclohexanediacetic acid 66
Methyl 1,4-dioxaspiro(4.5)decane-7-acetate 91
Methyl 5-(2-methyl-1,3-dioxolan-2-yl)-2-methylenepentanoate 90
$C_{11}H_{18}O_4S$
Methyl α-D-camphorsulfonate 156
$C_{11}H_{18}O_5$
Diethyl 2-methyl-3-oxoadipate 93
Diethyl 2-oxo-3-propylsuccinate 92
4-Oxo-1,9-nonanedicarboxylic acid 64
$C_{11}H_{18}O_6$
2,2-Bis-(acetoxymethyl)-propyl acetate 76
1,2,5-Triacetoxypentane 74
$C_{11}H_{18}O_6S_3$
2-Hydroxymethyl-2-methyl-1,3-propanediol tris-(mercaptoacetate) 86
$C_{11}H_{18}Si$
Dimethyl(3-phenylpropyl)silane 1112
$C_{11}H_{19}Br_2NO$
3,3-Dibromoazacyclododecan-2-one 125
$C_{11}H_{19}ClO$
2-Chloro-7,7-dimethyl-1-norbornaneethanol 16
$C_{11}H_{19}N$
1-(2-Methyl-1-pyrrolidinyl)-cyclohexene 40
1-Piperidino-1-cyclohexene 41
$C_{11}H_{19}NO$
2-(*N*-Isopropyl-*N*-(4-methylpent-4-en-2-yn-1-yl)amino)-ethanol 1016
2-Piperidinocyclohexanone 59
$C_{11}H_{19}NO_2$
Acetoacet-2,6-dimethylpiperidide 109
N-Cycloheptylacetoacetamide 105
5,5-Dimethyl-3-((2-hydroxyethyl)-methylamino)-2-cyclohexen-1-one hydrochloride 57
Ethyl *sec*-butylethylcyanoacetate 149
2-Ethylhexyl cyanoacetate 149
Ethyl β-methyl-1-piperidineacrylate 90

$C_{11}H_{19}NO_2SSi$
N-(Trimethylsilylmethyl)-*p*-toluenesulfonamide 1110
$C_{11}H_{19}NO_3$
N-Cyclohexylglutaramic acid 122
$C_{11}H_{19}NO_4$
Ethyl 4-methoxycarbonyl-1-piperidineacetate 90
Methyl 4-carbethoxy-1-piperidineacetate 90
$C_{11}H_{19}NO_6$
N-*tert*-Butoxycarbonyl-D-glutamic acid 5-methyl ester 122
Diethyl 4-nitropimelate 75
$C_{11}H_{19}NSi$
N,*N*-Dimethyl-3-(trimethylsilyl)-aniline 1107
N,*N*-Dimethyl-4-(trimethylsilyl)-aniline 1107
$C_{11}H_{19}N_3O$
2,6-Bis-(dimethylaminomethyl)-3-pyridinol 932
3,5-Di(*tert*-butyl)-4-nitroso-1*H*-pyrazole 861
8a-Methyl-2-decalone semicarbazone 127
$C_{11}H_{19}N_3O_9S$
3-Methylcytidine methosulfate 130
$C_{11}H_{19}N_5O_6$
Glycylglycylglycylglycylalanine hemihydrate 123
$C_{11}H_{20}$
4-(*tert*-Pentyl)-1-cyclohexene 3
$C_{11}H_{20}ClNO_3$
Ethyl 6-(3-chloropropionamido)-hexanoate 102
$C_{11}H_{20}ClNO_4$
(2-(1,4-Cyclohexadien-1-yl)ethyl)-trimethylammonium perchlorate 43
(2-(2,5-Cyclohexadien-1-yl)ethyl)-trimethylammonium perchlorate 43
$C_{11}H_{20}ClNO_5$
2,5-Di-*tert*-butylisoxazolium perchlorate 871
$C_{11}H_{20}IN$
(2-(1,4-Cyclohexadien-1-yl)ethyl)-trimethylammonium iodide 43
$C_{11}H_{20}N_2$
3,5-Di-*tert*-butyl-1*H*-pyrazole 864
$C_{11}H_{20}N_2O$
N-Ethyl-2-(pyrrolidinomethylene)-butyramide 108
$C_{11}H_{20}N_2O_2$
O-(Butylcarbamoyl)cyclohexanone oxime 1011
N-*tert*-Butyl-2-aci-nitrocycloheptaniminium hydroxide, inner salt 1054
4-(*N*-(Cyclohexyl)carbamoyl)-morpholine 127
N-(2-Diethylaminoethyl)-*p*-hydroxybenzamide hydrochloride ... 462
$C_{11}H_{20}N_2O_3$
3-Cyclohexylureidomethyl propionate 127
$C_{11}H_{20}N_8$
1,1-Bis(1*H*-1,2,4-triazol-4-ylamino)-heptane 870
$C_{11}H_{20}O$
2-*tert*-Butyl-4-methylcyclohexanone. 56
1-Cyclohexyl-3-methyl-1-butanone.... 56
2-Methyl-3-isopropyl-5-hepten-4-one. 50
2-Pentyl-1-cyclohexanone 56
4-(*tert*-Pentyl)-cyclohexanone 56
$C_{11}H_{20}OSi$
Ethoxytriallylsilane 1097
$C_{11}H_{20}O_2$
Citronellyl formate 80
Cyclopentylcyclohexane-1,1′-diol 12
2-Cyclopentylhexanoic acid 66
Methyl 2-cyclohexylbutyrate 77
Methyl 2-cyclohexyl-2-methylpropionate 77
Methyl 2-decenoate 79
2-Methyl-2-octen-1-yl acetate 80
3,9-Undecanedione 51
cis-9-Undecenoic acid 61
trans-9-Undecenoic acid 61
Vinyl 3,5,5-trimethylhexanoate 78
$C_{11}H_{20}O_2Si$
Methyl 1-(trimethylsilyloxy)-cyclohex-1-en-4(5)-yl ketone 1109
$C_{11}H_{20}O_3$
5-*tert*-Butyl-2-hydroxycyclohexane-carboxylic acid 65
Ethyl 2-hydroxycyclohexane-propionate 85
2-(1-Ethylpentyl)-5-methyl-1,3-dioxolan-4-one 97
Heptyl acetoacetate 93
Methyl 3-oxodecanoate 93
10-Oxoundecanoic acid 64
$C_{11}H_{20}O_4$
1-(Butoxycarbonyl)-ethyl butyrate..... 75
7,7-Diethoxy-5-heptyne-2,4-diol 1016
Diethyl 2,4-dimethylglutarate (mixture of isomers) 76
Diethyl isobutylmalonate 75
Diethyl 2-methyl-2-propylmalonate... 75
1,5-Dioxaspiro(5.5)undecane-3,3-dimethanol 21
mono-Methyl sebacate 95
$C_{11}H_{20}O_4S$
Diethyl 2-(3-(methylthio)propyl)-malonate 86
$C_{11}H_{20}O_5$
Diethyl 2-ethyl-2-(methoxymethyl)-malonate 85
$C_{11}H_{20}O_6$
1,3-Bis(1,3-dioxan-2-yloxy)propane ... 22
$C_{11}H_{20}Si_2$
1,1,2,2,2-Pentamethyl-1-phenyldisilane 1112
$C_{11}H_{21}ClN_2O_2$
2-Chloro-1,3-dimorpholinopropane dihydrochloride 39
$C_{11}H_{21}N$
3-Aminospiro(5.5)undecane 31
3-Aminospiro(5.5)undecane hydrochloride 31
tert-Butylcycloheptylideneamine 1020
N-Cyclohexylpiperidine 41
1-(2-Ethyl-1-butenyl)-piperidine 40
N-Methyldicyclopentylamine 31
(+)-3-Pinanemethylamine hydrochloride 31
(-)-3-Pinanemethylamine hydrochloride 31
$C_{11}H_{21}NO$
2-(Diethylaminomethyl)-cyclohexanone picrate 300
1-(2-Hexen-1-yloxymethyl)-pyrrolidine 38
5-Methoxy-1-piperidino-1-pentene 38
5-Methoxy-1-piperidino-2-pentene 38
3-Octyloxypropionitrile 144
3-Piperidino-1-cyclohexanol hydrochloride 37
$C_{11}H_{21}NO_2$
Butyl *N*-cyclohexylcarbamate 116
N-*tert*-Butyl-4,4-dimethyl-3-oxopentanamide 105
Butyl 1-methyl-2-pyrrolidineacetate. 90
N-(1,4-Dimethylpentyl)-acetoacetamide 105
Methyl 3-(hexamethyleneimino)-isobutyrate 90
5-Methyl-6-morpholino-4-hexen-2-ol.. 37
$C_{11}H_{21}NO_3$
Ethyl 3-(2,6-dimethylmorpholino)-propionate 90
3-(2-Isopropyloxazolidino)-propyl acetate 90
Methyl α,2,6-trimethyl-4-morpholinepropionate 90
$C_{11}H_{21}NO_4$
N-(*tert*.-Butoxycarbonyl)-L-leucine monohydrate 121
tert-Butyl 3-nitro-2,2,3-trimethylbutyrate 73
Diethyl 2-(diethylamino)-malonate.... 88
Diethyl (2-dimethylaminoethyl)-malonate 88
Diethyl 3,3′-(methylimino)-dipropionate 88
$C_{11}H_{21}NS$
Decyl thiocyanate 151
$C_{11}H_{22}$
Amylcyclohexane 2
1-*tert*-Pentylcyclohexane 2
$C_{11}H_{22}ClNO$
2-Chloro-*N*-(1-methyloctyl)-acetamide 103
2-Chloro-*N*-nonylacetamide 103
$C_{11}H_{22}INO_2S_2$
(1,4-Dioxa-8-thiaspiro(4.5)dec-7-ylthiomethyl)trimethylammonium iodide 42
$C_{11}H_{22}I_2$
1,11-Diiodoundecane 4
$C_{11}H_{22}N_2$
3-(Dibutylamino)propionitrile 146
N,*N*′-Methylenebis(piperidine) 40
2-(Piperidinomethyl)-piperidine 39
2-(2-Pyrrolidinoethyl)-piperidine 39
$C_{11}H_{22}N_2O$
1-Butyl-3-cyclohexylurea 129
1-*tert*-Butyl-3-cyclohexylurea 129
3-Cyclohexyl-1,1-diethylurea 129
1-Cyclohexyl-3-isobutylurea 129
1-Isopentylisonipecotamide 108
1-Isopentylnipecotamide 108
1-Pentylisonipecotamide 108
1-Pentylnipecotamide 108
$C_{11}H_{22}N_2O_2$
2-(Dimethylamino)ethyl *N*-cyclohexylcarbamate 115
Menthyl carbazate 116
$C_{11}H_{22}N_2O_2Si_2$
O,*O*′-Bis(trimethylsilyl)-thymine 1110
$C_{11}H_{22}N_2O_3$
1,3-Dimorpholino-2-propanol 36
1,3-Dimorpholino-2-propanol dihydrochloride 37
L-Norleucyl-L-norvaline 120
L-Norvalyl-L-isoleucine 120
L-Norvalyl-L-leucine acetate 120
$C_{11}H_{22}N_2O_3S$
Norvalylmethionine methyl ester..... 118
$C_{11}H_{22}N_2O_8S_2$
N,*N*′-(1-Methylethylene)bis(2-carbamoylethyl) bis(methanesulfonate) 156
$C_{11}H_{22}N_4O$
1,3-Dipiperidinourea 128
$C_{11}H_{22}O$
4-*tert*-Amylcyclohexanol 14
3-Butyl-2-heptanone 48
2-*tert*-Butyl-5-methylcyclohexanol.... 14
(+)-1-Cyclohexyl-2,2-dimethyl-1-propanol 14
(-)-1-Cyclohexyl-2,2-dimethyl-1-propanol 14
(-)-1-Cyclohexyl-3-methyl-1-butanol... 14
(-)-1-Cyclohexyl-1-pentanol 14
2,8-Dimethyl-5-nonanone 48
n-Heptyl isopropyl ketone 48
Hexyl isobutyl ketone 48
Menthyl methyl ether 18
4-Methyl-5-decanone 48
$C_{11}H_{22}O_2$
Isopentyl hexanoate 73
7-Methyldecanoic acid 61
Nonyl acetate 73
Pentyl hexanoate 73
3,5,5-Trimethylhexyl acetate 73
$C_{11}H_{22}O_2S$
2-Ethylhexyl 3-mercaptopropionate. 86
Nonyl mercaptoacetate 86
$C_{11}H_{22}O_3$
Ethyl 2-ethyl-2-hydroxy-5-methylhexanoate 84
3-(Octyloxy)-propionic acid 63
Pyruvic aldehyde dibutyl acetal........ 50
$C_{11}H_{22}O_4$
Ethyl 2-(diethoxymethyl)butyrate...... 85
$C_{11}H_{22}O_5$
Tetraethylene glycol methyl vinyl ether 18
$C_{11}H_{23}Br$
2-Bromoundecane 4
$C_{11}H_{23}BrO_2$
4-Bromo-1,7-diethoxyheptane 17
$C_{11}H_{23}ClO$
Chloromethyl decyl ether 17
$C_{11}H_{23}N$
1,5-Dimethyl-*N*-isopropyl-4-hexenylamine 32
$C_{11}H_{23}NO_2$
N,*N*-Dibutyllactamide 106
Ethyl 3-(dipropylamino)propionate... 88
Ethyl *tert*-octylcarbamate 117
2-Hydroxy-*N*-octylpropionamide....... 103
$C_{11}H_{23}NO_3$
o-*tert*-Butyl-L-serine *tert*-butyl ester dibenzenesulfimide salt 813
$C_{11}H_{23}N_2O_2$
L-Valyl-L-leucinamide hydrochloride 124
$C_{11}H_{23}N_3O_3$
Lysylvaline dicyclohexylamine salt.. 119
$C_{11}H_{23}N_3S$
1-(*tert*-Octyl)-4-thioxohexahydro-1,3,5-triazine 140
$C_{11}H_{23}N_5O_6$
L-Arginine L-glutamate 137
$C_{11}H_{24}NO_4P$
Diisopropyl (*N*,*N*-diethylcarbamoyl)phosphonate 159
$C_{11}H_{24}N_2O$
1,3-Dibutyl-1-ethylurea 129
2-Dimethylamino-*N*-(1-methylhexyl)-acetamide hydrochloride 104
$C_{11}H_{24}N_2O_2$
1,4-Bis(2-hydroxypropyl)-2-methylpiperazine 37
$C_{11}H_{24}O$
2,4-Dimethyl-4-nonanol 10
5-Ethyl-2-nonanol 9
2-Ethyl-2-propyl-1-hexanol 8
(3*R*,3*S*) (5*S*)-2,2,3,5-Tetramethyl-3-heptanol 10
3,3,5,5-Tetramethyl-4-heptanol 9
6-Undecanol 9
$C_{11}H_{24}O_2Pb$
Trimethylplumbyl 2-ethylhexanoate. 1135
Tripropylplumbyl acetate 1135
$C_{11}H_{24}O_3Si$
Triisopropoxyvinylsilane 1097
$C_{11}H_{24}O_4Si$
Diethoxy(3-glycidyloxypropyl)-methylsilane 1106
$C_{11}H_{24}S$
Decyl methyl sulfide 26
Ethyl nonyl sulfide 26
$C_{11}H_{25}ClN_2$
2-Chloro-*N*,*N*,*N*′,*N*′-tetraethyl-1,3-propanediamine 32
$C_{11}H_{25}N$
Diisopropyl(1-ethylpropyl)amine....... 29
N-Isopropyl-1,5-dimethylhexylamine 28
1-Methyldecylamine 28
$C_{11}H_{25}NO$
1-Dibutylamino-2-propanol 34
3-Dibutylamino-1-propanol 34
5-Dipropylamino-1-pentanol 34
3-(2-Ethylhexyloxy)-propylamine....... 35
2-(1-Methyloctylamino)-ethanol 33
$C_{11}H_{26}IN$
sec-Octyltrimethylammonium iodide 43
$C_{11}H_{26}N_2$
N,*N*-Dibutyl-*N*′-methylethylenediamine 30
N-(2-Ethylhexyl)-1,3-propanediamine 30
$C_{11}H_{26}N_2O$
1,3-Bis(*tert*-butylamino)-2-propanol hydrochloride 33
1,3-Bis-(diethylamino)-2-propanol..... 34
$C_{11}H_{26}OSi$
Ethoxytripropylsilane 1097
$C_{11}H_{26}O_3Si$
Triethoxypentylsilane 1097
$C_{11}H_{28}Br_2N_2O_2$
N,*N*′-(2,2-Bis(hydroxymethyl)-trimethylene)-bis(trimethylammonium bromide)... 42

$C_{11}H_{28}I_2N_2$
N,N-Pentamethylenebis(trimethylammonium iodide) 44
$C_{11}H_{30}OSi_4$
2-Acetyl-1,1,1,3,3,3-hexamethyl-2-(trimethylsilyl)trisilane 1097
$C_{11}H_{32}Si_4$
2-(Dimethylsilyl)-1,1,1-triethyl-2,3,3-trimethyltrisilane 1097
$C_{12}BrF_9$
2-Bromo-2′,3,3′,4,4′,5,5′,6,6′-nonafluorobiphenyl 173
$C_{12}Cl_{12}Si$
Bis(pentachlorophenyl)dichlorosilane 1100
$C_{12}F_8N_2O_4$
3,3′-Dinitrooctafluorobiphenyl 262
$C_{12}HCl_9S_3Si$
Tris(3,4,5-trichloro-2-thienyl)silane 1110
$C_{12}H_2Cl_6N_2O_2S$
1,4,5,6,7,7-Hexachloro-*N*-(2-thiazolyl)-5-norbornene-2,3-dicarboximide 879
$C_{12}H_4Cl_5N_3O_3S$
S-(2,3,5,6-Tetrachloro-4-pyridyl) *N*-(2-chloro-4-nitrophenyl)thiocarbamate 928
$C_{12}H_5Br_2NO_5$
1,9-Dibromo-3-nitro-2,8-dibenzofurandiol 893
$C_{12}H_5ClN_4O_6$
N-Picryl-2,4,5-trichloroaniline 274
$C_{12}H_5Cl_3N_4O_6$
N-(2,4,6-Trichlorophenyl)picrylamine 261
$C_{12}H_5Cl_4NO$
5-Chloro-2-(2,4,5-trichlorophenoxy)aniline 223
$C_{12}H_5Cl_4N_3O_3$
N-(2-Pyrazinyl)-3,4,5,6-tetrachlorophthalamic acid 977
N-(2-Pyrimidinyl)-3,4,5,6-tetrachlorophthalamic acid 969
$C_{12}H_5Cl_5O_3S$
2,4,6-Trichlorophenyl 3,4-dichlorobenzenesulfonate 797
$C_{12}H_5FO_3$
7-Fluoro-1,2-naphthalic anhydride 388
$C_{12}H_5F_5$
2,3,4,5,6-Pentafluorobiphenyl 172
$C_{12}H_5N_3O_8$
2,3,7-Trinitrodibenzo-*p*-dioxin 274
$C_{12}H_6BrClN_4O_6$
N-(4-Bromo-2-chlorophenyl)picrylamine 261
$C_{12}H_6BrNO_3$
2-Bromo-7-nitrodibenzofuran 895
3-Bromo-7-nitrodibenzofuran 895
$C_{12}H_6Br_2N_4O_6$
N-(2,4-Dibromophenyl)-picrylamine 272
2′,5′-Dibromo-2,4,6-trinitrodiphenylamine 272
$C_{12}H_6Br_2O$
2,8-Dibromodibenzofuran 896
$C_{12}H_6Br_2O_2$
2,7-Dibromodibenzo-*p*-dioxin 185
$C_{12}H_6Br_2O_2S$
2,8-Dibromodibenzothiophene-5,5-dioxide 898
$C_{12}H_6Br_2S$
2,8-Dibromodibenzothiophene 899
$C_{12}H_6Br_4O_2$
4,4′,6,6′-Tetrabromo-2,2′-biphenol 193
$C_{12}H_6ClN_5O_8$
N-(2-Chloro-5-nitrophenyl)picrylamine 272
N-(4-Chloro-3-nitrophenyl)picrylamine 272
2-Chloro-4-nitro-*N*-picrylaniline 274
$C_{12}H_6Cl_2Hg_2O$
2,8-Bis(chloromercuri)dibenzofuran 1139
$C_{12}H_6Cl_2N_2O$
2-(3,4-Dichlorophenyl)oxazolo(4,5-b)pyridine 917
$C_{12}H_6Cl_2N_2O_4S_2$
2-Chloro-5-nitrophenyl disulfide 270
4-Chloro-2-nitrophenyl disulfide 270
4-Chloro-3-nitrophenyl disulfide 270
$C_{12}H_6Cl_2N_2O_6S$
4-Chloro-3-nitrophenyl sulfone 773
$C_{12}H_6Cl_2N_4O_6$
N-(2,3-Dichlorophenyl)-picrylamine 261
N-(2,4-Dichlorophenyl)-picrylamine 261
N-(2,5-Dichlorophenyl)-picrylamine 261
N-(3,4-Dichlorophenyl)-picrylamine 272
N-(3,5-Dichlorophenyl)picrylamine 272
$C_{12}H_6Cl_2O$
2,8-Dichlorodibenzofuran 896
$C_{12}H_6Cl_2O_3S$
4-Benzoyl-2,5-dichloro-3-thiophenecarboxylic acid 857
$C_{12}H_6Cl_2O_5S_2$
2,8-Dibenzofurandisulfonyl chloride 896
$C_{12}H_6Cl_4N_2O_4$
N-(5-Methyl-3-isoxazolyl)-3,4,5,6-tetrachlorophthalamic acid 870
$C_{12}H_6Cl_4O_2S$
2,4,4′,5-Tetrachlorodiphenyl sulfone 785
3,3′,4,4′-Tetrachlorodiphenyl sulfone 786
$C_{12}H_6Cl_4S$
2,4,4′,5-Tetrachlorodiphenyl sulfide 213
$C_{12}H_6Cl_5NO_2S$
2′,3,4,4′,5′-Pentachlorobenzenesulfonanilide 810
$C_{12}H_6F_2O_2S$
3,7-Difluorodibenzothiophene-5,5-dioxide 782
$C_{12}H_6F_3NO_4$
5-(2-Nitro-4-(trifluoromethyl)phenyl)-2-furaldehyde 843
$C_{12}H_6F_8N_4$
4,4′-Dihydrazinooctafluorobiphenyl 234
$C_{12}H_6F_{18}$
2-Methyl-octadecafluoro-2-undecene 5
$C_{12}H_6I_2O$
2,8-Diiododibenzofuran 896
$C_{12}H_6I_2O_2$
1,6-Diiododibenzo-*p*-dioxin 185
$C_{12}H_6N_2$
2,7-Dicyanonaphthalene 768
$C_{12}H_6N_2O_4$
2,3-Dihydroxy-1,4-phenazinedione 998
$C_{12}H_6N_2O_4S$
2,8-Dinitrodibenzothiophene 897
$C_{12}H_6N_2O_4S_2$
3,4-Dithenoyl-1,2,5-oxadiazole N^2-oxide 1006
$C_{12}H_6N_2O_5$
2,7-Dinitrodibenzofuran 895
$C_{12}H_6N_2O_8$
4,5-Dinitro-1,8-naphthalenedicarboxylic acid, disilver salt 351
$C_{12}H_6N_4O_5$
3,7-Dinitrobenzo(b)(1,8)-naphthyridin-5(10*H*)-one 943
$C_{12}H_6N_4O_8S$
2,4-Dinitrophenyl sulfide 270
$C_{12}H_6O_3$
2,3-Naphthalic anhydride 388
$C_{12}H_7BrN_2$
2-Bromophenazine 1001
$C_{12}H_7BrN_4O_6$
N-(2-Bromophenyl)picrylamine 272
N-(4-Bromophenyl)-picrylamine 261
$C_{12}H_7BrO$
1-Bromodibenzofuran 896
2-Bromodibenzofuran 896
$C_{12}H_7BrO_2$
2-Bromo-6-phenyl-1,4-benzoquinone 318
$C_{12}H_7BrO_4S$
8-Bromo-2-dibenzofuransulfonic acid, sodium salt 896
$C_{12}H_7BrS$
2-Bromodibenzothiophene 899
3-Bromodibenzothiophene 899
$C_{12}H_7BrS_2$
2-Bromothianthrene 214
$C_{12}H_7Br_2I$
3,5-Dibromo-2-iodobiphenyl 172
$C_{12}H_7Br_3N_2O$
2′,4′,6′-Tribromonicotinanilide 929
$C_{12}H_7Br_3O$
2-(4-Bromophenyl)-4,6-dibromophenol 193
$C_{12}H_7Br_4N$
2,2′,4,4′-Tetrabromodiphenylamine 224
$C_{12}H_7ClF_3NO_2$
2′-Chloro-5′-(trifluoromethyl)-2-furanilide 843
$C_{12}H_7ClF_3N_5O_4$
1-(4-Chloro-3-nitrophenyl)-3-(4-hydroxy-6-(trifluoromethyl)-2-pyrimidinyl)urea 968
$C_{12}H_7ClHgO$
3-(Chloromercuri)dibenzofuran 1139
$C_{12}H_7ClHgO_4S$
8-(Chloromercuri)-2-dibenzofuransulfonic acid, sodium salt 896
$C_{12}H_7ClN_2O$
2-(4-Chlorophenyl)oxazolo(4,5-b)pyridine 917
$C_{12}H_7ClN_2O_4$
4-Chloro-3,5-dinitrobiphenyl 261
$C_{12}H_7ClN_2O_4S$
4-Chlorophenyl 2,4-dinitrophenyl sulfide 260
$C_{12}H_7ClN_2O_5$
4′-Chloro-2,4-dinitrodiphenyl ether 267
$C_{12}H_7ClN_4O_6$
N-(2-Chlorophenyl)-picrylamine 260
N-(3-Chlorophenyl)-picrylamine 261
N-(4-Chlorophenyl)-picrylamine 261
$C_{12}H_7ClO$
2-Chlorodibenzofuran 896
$C_{12}H_7ClO_3S$
2-Dibenzofuransulfonyl chloride 896
$C_{12}H_7Cl_2NO$
2,5-Dichloro-4-pyridyl phenyl ketone 961
$C_{12}H_7Cl_2NO_3$
4-Chlorophenyl 4-chloro-2-nitrophenyl ether 268
$C_{12}H_7Cl_2N_3O_3$
4-(4-Nitrophenylazo)-2,6-dichlorophenol 1080
$C_{12}H_7Cl_2N_3O_4$
5-Chloro-2-pyridyl *N*-(2-chloro-4-nitrophenyl)carbamate 929
$C_{12}H_7Cl_3FNO_2S$
2′-Fluoro-2,4,5-trichlorobenzenesulfonanilide 809
$C_{12}H_7Cl_3N_2O_2$
5-Chloro-2-pyridyl *N*-(2,5-dichlorophenyl)carbamate 929
2-Pyridyl *N*-(2,4,5-trichlorophenyl)carbamate 953
3-Pyridyl *N*-(2,4,5-trichlorophenyl)carbamate 960
$C_{12}H_7Cl_3N_2O_4S$
4′-Nitro-2′,3,4-trichlorosulfonanilide 807
$C_{12}H_7Cl_4N$
4-Benzyl-2,3,5,6-tetrachloropyridine 966
Bis(2,4-dichlorophenyl)amine 224
$C_{12}H_7Cl_4NO$
α-(2,3,5,6-Tetrachloro-4-pyridyl)benzyl alcohol 932
$C_{12}H_7Cl_4NO_2S$
3,3′,4,4′-Tetrachlorobenzenesulfonanilide 810
3,3′,4,5′-Tetrachlorobenzenesulfonanilide 810
$C_{12}H_7Cl_4N_3O$
1-(5-Chloro-2-pyridyl)-3-(2,4,5-trichlorophenyl)urea 930
1-(2,3-Dichlorophenyl)-3-(3,5-dichloro-2-pyridyl)urea 930
$C_{12}H_7FN_2O_6S$
4-Fluoro-3-nitrophenyl 3-nitrophenyl sulfone 780
$C_{12}H_7FN_4O_6$
N-(2-Fluorophenyl)picrylamine 272
N-(3-Fluorophenyl)picrylamine 272
N-(4-Fluorophenyl)picrylamine 272
5-Nitro-2-pyridyl *N*-(2-fluoro-5-nitrophenyl)carbamate 929
5-Nitro-2-pyridyl *N*-(4-fluoro-3-nitrophenyl)carbamate 929
$C_{12}H_7FO$
3-Fluorodibenzofuran 896
$C_{12}H_7F_4N_3O$
4-(4-Methoxyphenylazo)-2,3,5,6-tetrafluoropyridine 1087
$C_{12}H_7IN_2O_8S$
3,7-Dinitrodibenziodolium hydrogen sulfate 281
$C_{12}H_7IN_3O_4$
4′-Iodo-2,4-dinitrodiphenylamine 261
$C_{12}H_7IN_4O_6$
4′-Iodo-2,4,6-trinitrodiphenylamine 272
$C_{12}H_7IO$
2-Iododibenzofuran 896
3-Iododibenzofuran 896
4-Iododibenzofuran 896
$C_{12}H_7IO_2$
2-Iododibenzo-*p*-dioxin 185
$C_{12}H_7IO_3S$
2-Iodophenoxathiin-10,10-dioxide 783
$C_{12}H_7IS$
4-Iododibenzothiophene 899
$C_{12}H_7I_2NO_2$
3-Nitrodibenziodolium iodide 280
$C_{12}H_7NO$
4-Aza-9-fluorenone 962
$C_{12}H_7NO_2S$
2-Nitrodibenzothiophene 897
$C_{12}H_7NO_2S_2$
2-Nitrothianthrene 280
$C_{12}H_7NO_3$
2-(2-Furyl)-3,1-benzoxazin-4(4*H*)-one 854
N-Hydroxynaphthalimide 722
3-Nitrodibenzofuran 895
$C_{12}H_7NO_4S$
2-(2-Nitrophenylthio)-1,4-benzoquinone 317
$C_{12}H_7NO_5$
2-Phthalimidosuccinic anhydride 721
$C_{12}H_7NO_5S$
2-Nitrophenoxathiin-10,10-dioxide 780
$C_{12}H_7NO_6S$
8-Nitro-2-dibenzofuransulfonic acid 895
2-(4-Nitrophenylsulfonyl)-1,4-benzoquinone 779
$C_{12}H_7N_3$
α-Cyano-3-indoleacrylonitrile 885
$C_{12}H_7N_3O_4$
2′-Cyano-4′-nitro-2-furanilide 850
$C_{12}H_7N_3O_5S$
3,7-Dinitro-5-oxophenothiazine 771
$C_{12}H_7N_3S_2$
1-Amino-2,4-naphthalenedithiocyanate 768
$C_{12}H_7N_5O$
2-Azido-5-(1-naphthyl)-1,3,4-oxadiazole 874
$C_{12}H_7N_5O_8$
2,3′,4,6-Tetranitrodiphenylamine 272
$C_{12}H_8As_2Cl_4N_2$
4,4′-Bis(dichloroarsino)azobenzene 1128
$C_{12}H_8As_2O_5$
2,2′-Oxybis(1,3,2-benzodioxarsole) 1128
$C_{12}H_8BrCl_2N_3O$
1-(5-Bromo-2-pyridyl)-3-(2,3-dichlorophenyl)urea 930
$C_{12}H_8BrF$
4-Bromo-4′-fluorobiphenyl 174
$C_{12}H_8BrFN_4O_3$
1-(5-Bromo-2-pyridyl)-3-(2-fluoro-5-nitrophenyl)urea 929
$C_{12}H_8BrFO$
4-Bromo-4′-fluorodiphenyl ether 182
$C_{12}H_8BrI$
5-Bromo-2-iodobiphenyl 172
$C_{12}H_8BrNO$
3-Amino-2-bromodibenzofuran 894
3-Amino-2-bromodibenzofuran hydrochloride 894
$C_{12}H_8BrNO_2$
4-Bromo-2′-nitrobiphenyl 279
$C_{12}H_8BrN_3O_3$
4-Bromo-*N*-(5-nitro-2-pyridyl)benzamide 929
$C_{12}H_8BrN_3O_4$
4′-Bromo-2,4-dinitrodiphenylamine 261
$C_{12}H_8Br_2$
trans-1,2-Dibromoacenaphthene 174

$C_{12}H_8Br_2$ (Cont'd)
2,2′-Dibromobiphenyl ... 172
$C_{12}H_8Br_2Cl_6$
2,3-Dibromo-5,6,7,8,9,9-hexachloro-1,2,3,4,4a,5,8,8a-octahydro-1,4:5,8-dimethanonaphthalene ... 6
$C_{12}H_8Br_2Hg$
Bis(4-bromophenyl)mercury ... 1137
$C_{12}H_8Br_2N_2$
4,4′-Dibromoazobenzene ... 1082
$C_{12}H_8Br_2N_2O$
4,4′-Dibromoazoxybenzene ... 1082
(3,5-Dibromo-2-ethoxybenzylidene)-malononitrile ... 762
$C_{12}H_8Br_2O$
2-Bromophenyl ether ... 182
$C_{12}H_8Br_2O_2S$
4-Bromophenyl sulfone ... 774
$C_{12}H_8Br_2S_2$
4-Bromophenyl disulfide ... 213
$C_{12}H_8ClFN_2O_4S$
4-Chloro-2′-fluoro-3-nitrobenzenesulfonanilide ... 807
$C_{12}H_8ClFN_4O_3$
1-(5-Chloro-2-pyridyl)-3-(2-fluoro-5-nitrophenyl)urea ... 929
1-(5-Chloro-2-pyridyl)-3-(4-fluoro-3-nitrophenyl)urea ... 929
$C_{12}H_8ClFO_2S$
4-Chlorophenyl 4-fluorophenyl sulfone ... 785
$C_{12}H_8ClF_2N_3O$
1-(2-Chloro-3-pyridyl)-3-(2,4-difluorophenyl)urea ... 930
$C_{12}H_8ClF_3N_4O$
1-(2-Chloro-5-(trifluoromethyl)-phenyl)-3-(2-pyrazinyl)urea ... 977
1-(2-Chloro-5-(trifluoromethyl)-phenyl)-3-(2-pyrimidinyl)urea ... 968
$C_{12}H_8ClHgN_5O$
6-Benzamido-7-(chloromercuri)-purine ... 1138
$C_{12}H_8ClNO$
3-Amino-2-chlorodibenzofuran ... 894
4-Chlorophenyl 3-pyridyl ketone ... 929
$C_{12}H_8ClNOS$
2-Chlorophenothiazine-5-oxide ... 771
$C_{12}H_8ClNO_2$
4-Chloro-3-nitrobiphenyl ... 281
$C_{12}H_8ClNO_2S$
2-Chlorophenyl 4-nitrophenyl sulfide ... 260
$C_{12}H_8ClNO_3$
2-Acetamido-3-chloro-1,4-naphthoquinone ... 431
4-Chloro-4′-nitrodiphenyl ether ... 260
4-Chlorophenyl 2-nitrophenyl ether ... 268
$C_{12}H_8ClNO_4S$
2-Chlorophenyl 4-nitrophenyl sulfone ... 773
$C_{12}H_8ClNO_5S$
Phenyl 4-chloro-3-nitrobenzenesulfonate ... 799
$C_{12}H_8ClN_3O_4$
3′-Chloro-2,4-dinitrodiphenylamine ... 261
4′-Chloro-2,4-dinitrodiphenylamine ... 260
3-Pyridyl *N*-(4-chloro-2-nitrophenyl)-carbamate ... 927
$C_{12}H_8ClN_3O_6S$
N-(4-Chloro-3-nitrophenyl)-2-nitrobenzenesulfonamide ... 807
$C_{12}H_8Cl_2Hg$
Bis(4-chlorophenyl)mercury ... 1137
$C_{12}H_8Cl_2INO_2S$
3,4-Dichloro-4′-iodobenzenesulfonanilide ... 810
$C_{12}H_8Cl_2N_2$
4,4′-Dichloroazobenzene ... 1082
$C_{12}H_8Cl_2N_2O$
3,3′-Dichloroazoxybenzene ... 1082
4,4′-Dichloroazoxybenzene ... 1093
3′,4′-Dichloronicotinanilide ... 949
2,4-Dichloro-*N*-(3-pyridyl)benzamide ... 965
α-(3,5-Dichloro-2-pyridylimino)-*o*-cresol copper (II) salt ... 1038
4-(Phenylazo)-2,6-dichlorophenol ... 1086
$C_{12}H_8Cl_2N_2O_2$
2,4-Dichlorobenzoic (furfurylidene)-hydrazide ... 1025
3,4-Dichlorobenzoic (furfurylidene)-hydrazide ... 1025
3,3′-Dichloro-4,4′-dihydroxyazobenzene ... 1080
4-(2,5-Dichlorophenylazo)-resorcinol ... 1083
2-Pyridyl *N*-(2,3-dichlorophenyl)-carbamate ... 929
2-Pyridyl *N*-(2,4-dichlorophenyl)-carbamate ... 953
2-Pyridyl *N*-(3,4-dichlorophenyl)-carbamate ... 964
3-Pyridyl *N*-(2,3-dichlorophenyl)-carbamate ... 960
3-Pyridyl *N*-(2,4-dichlorophenyl)-carbamate ... 960
3-Pyridyl *N*-(2,5-dichlorophenyl)-carbamate ... 960
3-Pyridyl *N*-(3,4-dichlorophenyl)-carbamate ... 960
4-Pyridyl *N*-(3,4-dichlorophenyl)-carbamate ... 964
$C_{12}H_8Cl_2N_2O_4S$
3,4-Dichloro-3′-nitrobenzenesulfonanilide ... 807
$C_{12}H_8Cl_2N_4O_3$
1-(2-Chloro-4-nitrophenyl)-3-(5-chloro-2-pyridyl)urea ... 928
1-(2,3-Dichlorophenyl)-3-(5-nitro-2-pyridyl)urea ... 927
$C_{12}H_8Cl_2O_2$
Cinnamylidenemalonyl chloride ... 389
$C_{12}H_8Cl_2O_2S$
2,2′-Thiobis(4-chlorophenol) ... 210
$C_{12}H_8Cl_2O_5S_2$
4,4′-Oxybis(benzenesulfonyl chloride) ... 802
$C_{12}H_8Cl_2S_2$
4-Chlorophenyl disulfide ... 213
$C_{12}H_8Cl_3NO$
N-(1-Naphthyl)-2,2,2-trichloroacetamide ... 398
2-(2,4,5-Trichlorophenoxy)aniline ... 223
$C_{12}H_8Cl_3NO_2S$
4-Chloro-*N*-(3,4-dichlorophenyl)-benzenesulfonamide ... 810
2′,3′,4-Trichlorobenzene-sulfonanilide ... 809
2′,4,5′-Trichlorobenzene-sulfonanilide ... 810
$C_{12}H_8Cl_3NO_3$
Furfuryl *N*-(2,4,5-trichlorophenyl)-carbamate ... 844
$C_{12}H_8Cl_3NO_4S$
2-Amino-4-chloro-5-(2,4-dichlorophenoxy)-benzenesulfonic acid, sodium salt ... 790
2-Amino-6-chloro-3-(2,4-dichlorophenoxy)-benzenesulfonic acid, sodium salt ... 790
2-(2-Amino-4-chlorophenoxy)-3,5-dichlorobenzenesulfonic acid, sodium salt ... 790
2-(2-Amino-4-chlorophenoxy)-4,5-dichlorobenzenesulfonic acid, sodium salt ... 790
2-(4-Amino-2-chlorophenoxy)-4,5-dichlorobenzenesulfonic acid, sodium salt ... 791
6-(2,4,5-Trichlorophenoxy)-metanilic acid ... 787
$C_{12}H_8Cl_3N_3O$
1-(2-Chloro-3-pyridyl)-3-(2,3-dichlorophenyl)urea ... 930
1-(5-Chloro-2-pyridyl)-3-(2,3-dichlorophenyl)urea ... 930
$C_{12}H_8Cl_4N_2$
2,2′,5,5′-Tetrachlorohydrazobenzene ... 235
$C_{12}H_8Cl_4Pb$
Bis(4-chlorophenyl)dichlorolead ... 1134
$C_{12}H_8Cl_6I_2$
2,3-Diiodo-5,6,7,8,9,9-hexachloro-octahydro-1,4:5,8-dimethanonaphthalene ... 6
$C_{12}H_8FNO_6S_2$
4-Benzenesulfonyl-3-nitrobenzenesulfonyl fluoride ... 803
$C_{12}H_8FN_3O_2$
4-Fluoro-3-nitroazobenzene ... 1081
$C_{12}H_8FN_3O_4$
2-Pyridyl *N*-(2-fluoro-5-nitrophenyl)-carbamate ... 929
2-Pyridyl *N*-(4-fluoro-2-nitrophenyl)-carbamate ... 927
2-Pyridyl *N*-(4-fluoro-3-nitrophenyl)-carbamate ... 928
3-Pyridyl *N*-(4-fluoro-3-nitrophenyl)-carbamate ... 928
4-Pyridyl *N*-(4-fluoro-3-nitrophenyl)-carbamate ... 928
$C_{12}H_8FN_5O_5$
1-(4-Fluoro-3-nitrophenyl)-3-(5-nitro-2-pyridyl)urea ... 929
$C_{12}H_8F_2$
2,2′-Difluorobiphenyl ... 174
$C_{12}H_8F_2Hg$
Bis(2-fluorophenyl)mercury ... 1137
$C_{12}H_8F_2N_2O_2$
2-Pyridyl *N*-(2,5-difluorophenyl)-carbamate ... 929
$C_{12}H_8F_2O$
4-Fluorophenyl ether ... 183
$C_{12}H_8F_3NO_2$
2′-(Trifluoromethyl)-2-furanilide ... 843
$C_{12}H_8F_3NS$
N-(2-Thienylmethylene)-2-(trifluoromethyl)aniline ... 1021
$C_{12}H_8F_3N_3O$
N-Nitroso-*N*-(2-pyridyl)-3-(trifluoromethyl)aniline ... 925
$C_{12}H_8Hg_2$
Dibenzo(b,e)(1,4)dimercurin ... 1139
$C_{12}H_8INO_2$
4-Iodo-4′-nitrobiphenyl ... 262
$C_{12}H_8I_2$
2,2′-Diiodobiphenyl ... 172
$C_{12}H_8N_2O$
Benzo(c)cinnoline *N*-oxide ... 1006
9-Nitrosocarbazole ... 891
$C_{12}H_8N_2OS_2$
5-(3-Indolylmethylene)rhodanine ... 890
$C_{12}H_8N_2O_2$
3-Nitrocarbazole ... 892
$C_{12}H_8N_2O_3$
3-Amino-2-nitrodibenzofuran ... 894
7-Amino-2-nitrodibenzofuran ... 894
2-(Aci-nitromethyl)naphth(1,2-d)-oxazole, lithium salt ... 909
$C_{12}H_8N_2O_4$
2,4′-Dinitrobiphenyl ... 278
3,3′-Dinitrobiphenyl ... 279
3,4′-Dinitrobiphenyl ... 279
$C_{12}H_8N_2O_4S$
2-(2,4-Dinitrostyryl)thiophene ... 856
5-Piperonylidene-2-thiobarbituric acid ... 734
$C_{12}H_8N_2O_5$
α-Furfurylidene-2,4-dinitrotoluene ... 850
p-Nitrophenyl ether ... 268
$C_{12}H_8N_2O_5Sn$
Bis(3-nitrophenyl)tin oxide ... 1130
Bis(4-nitrophenyl)tin oxide ... 1131
$C_{12}H_8N_2O_6$
2-Furamido-4-nitrobenzoic acid ... 847
$C_{12}H_8N_2O_6S$
3-Nitrophenyl sulfone ... 779
$C_{12}H_8N_2O_7S$
Phenyl 2,4-dinitrobenzenesulfonate ... 798
$C_{12}H_8N_4$
(Cyclohexane-1,4-diylidene)-dimalononitrile ... 764
$C_{12}H_8N_4O_2$
6-Nitro-2-(2-pyridyl)benzimidazole ... 942
$C_{12}H_8N_4O_6$
2,4,6-Trinitrodiphenylamine ... 274
$C_{12}H_8N_4O_7$
2-(Picrylamino)phenol ... 263
3-(Picrylamino)phenol ... 263
4-(Picrylamino)phenol ... 263
$C_{12}H_8O$
1-Acenaphthenone ... 287
$C_{12}H_8OS$
Dibenzothiophene-5-oxide ... 898
Phenoxathiin ... 214
$C_{12}H_8O_2$
1*H*,3*H*-Naphtho(1,8-cd)pyran-1-one ... 386
Naphtho(1,2-c)furan-1(3*H*)-one ... 386
$C_{12}H_8O_2S$
Phenoxathiin-10-oxide ... 771
$C_{12}H_8O_2S_2$
Thianthrene-5,10-dioxide ... 771
$C_{12}H_8O_3$
2-(4-Hydroxyphenyl)-1,4-benzoquinone ... 316
$C_{12}H_8O_3S$
3-Benzoyl-2-thiophenecarboxylic acid ... 855
2-Dibenzofuransulfinic acid ... 1110
2-Dibenzofuransulfinic acid, sodium salt ... 896
Phenoxathiin-10,10-dioxide ... 783
2-(2-Thienylcarbonyl)benzoic acid ... 857
$C_{12}H_8O_4$
1,6-Naphthalenedicarboxylic acid ... 345
1,8-Naphthalenedicarboxylic acid ... 345
Naphtho(2,3-d)-1,3-dioxolene-2-carboxylic acid ... 329
$C_{12}H_8O_4S$
2-Dibenzofuransulfonic acid ... 790
2-Dibenzofuransulfonic acid, sodium salt ... 896
4-Dibenzofuransulfonic acid, sodium salt monohydrate ... 896
$C_{12}H_8O_5$
3-Benzoxepin-2,4-dicarboxylic acid ... 338
$C_{12}H_8O_5S_2$
2,8-Dibenzofurandisulfinic acid, disodium salt ... 896
$C_{12}H_8O_7S_2$
4,6-Dibenzofurandisulfonic acid ... 896
2,8-Dibenzofurandisulfonic acid, disodium salt ... 896
$C_{12}H_8O_7S_3$
2-Sulfothianthrene-5,5,10,10-tetraoxide, sodium salt ... 796
$C_{12}H_8S_2$
Dibenzo(c,e)(1,2)dithiin ... 214
$C_{12}H_9BO_2S_2$
1-Thianthreneboronic acid ... 1125
$C_{12}H_9BO_3$
4-Dibenzofuranboronic acid ... 1125
$C_{12}H_9BrClN_3O_4$
2-(4-Bromophenyl)pyrido(1,2-b)-(1,2,4)triazin-5-ium perchlorate ... 978
$C_{12}H_9BrCl_6$
2-Bromo-5,6,7,8,9,9-hexachloro-octahydro-1,4:5,8-dimethanonaphthalene ... 6
$C_{12}H_9BrN_2$
4-Bromoazobenzene ... 1082
$C_{12}H_9BrN_2O$
4-(4-Bromophenylazo)phenol ... 1080
$C_{12}H_9BrN_2O_2$
3-Pyridyl *N*-(4-bromophenyl)-carbamate ... 960
$C_{12}H_9BrO$
2-Bromo-4-phenylphenol ... 193
$C_{12}H_9BrO_2S$
4-Bromophenyl phenyl sulfone ... 774
$C_{12}H_9BrO_3$
2-Bromo-3-ethoxy-1,4-naphthoquinone ... 316
$C_{12}H_9BrS$
4-Bromophenyl phenyl sulfide ... 210
$C_{12}H_9Br_2N$
Bis(2-bromophenyl)amine ... 224
4,4′-Dibromo-2-biphenylamine ... 224
$C_{12}H_9ClF_3NO_2$
4-Chloro-6,8-dimethoxy-2-(trifluoromethyl)quinoline ... 981
N-(2-Chloro-5-(trifluoromethyl)-phenyl)glutarimide ... 714
$C_{12}H_9ClHgO_3$
Methyl 4-(chloromercuri)-5-phenyl-2-furancarboxylate ... 1138
$C_{12}H_9ClN_2$
4-Chloroazobenzene ... 1082
$C_{12}H_9ClN_2O$
2′-Chloropicolinanilide ... 963

$C_{12}H_9ClN_2O$ (Cont'd)
2-Chloro-*N*-(2-pyridyl)benzamide 963
$C_{12}H_9ClN_2O_2$
4-Chloro-2-nitrodiphenylamine 261
2-Pyridyl *N*-(2-chlorophenyl)-carbamate 964
2-Pyridyl *N*-(3-chlorophenyl)-carbamate 964
2-Pyridyl *N*-(4-chlorophenyl)-carbamate 953
3-Pyridyl *N*-(2-chlorophenyl)-carbamate 960
3-Pyridyl *N*-(3-chlorophenyl)-carbamate 929
3-Pyridyl *N*-(4-chlorophenyl)-carbamate 960
$C_{12}H_9ClN_2O_2S$
4-(Phenylazo)benzenesulfonyl chloride 1092
$C_{12}H_9ClN_2O_4$
Ethyl 4-chloro-α-cyano-3-nitrocinnamate 759
$C_{12}H_9ClN_2O_4S$
2′-Chloro-3-nitrobenzenesulfonanilide 807
2′-Chloro-4′-nitrobenzenesulfonanilide 807
4-Chloro-3-nitrobenzenesulfonanilide 822
$C_{12}H_9ClN_2O_5$
Furfuryl *N*-(2-chloro-4-nitrophenyl)-carbamate 844
Furfuryl *N*-(2-chloro-5-nitrophenyl)-carbamate 844
$C_{12}H_9ClN_4O_2$
(6 or 7 or 8 or 9)-Chloro-1,3-dimethylbenzo(g)pteridine-2,4(1*H*,3*H*)-dione 1001
$C_{12}H_9ClN_4O_3$
4-Chlorobenzoic N^2-(3-nitro-2-pyridyl)hydrazide 930
$C_{12}H_9ClN_4O_5$
4-Carbamoyl-1-(2,4-dinitrophenyl)-pyridinium chloride 943
$C_{12}H_9ClO$
2-(2-Chlorophenyl)phenol 193
2-Chloro-4-phenylphenol 196
$C_{12}H_9ClOS$
2-Chloroacetyl-5-phenylthiophene 857
$C_{12}H_9Cl_2NO$
5-Chloro-2-(2-chlorophenoxy)-aniline 223
5-Chloro-2-(4-chlorophenoxy)aniline 223
2-(2,4-Dichlorophenoxy)-aniline 223
2,5-Dichloro-α-phenyl-4-pyridinemethanol 932
$C_{12}H_9Cl_2NO_2$
Ethyl α-cyano-3,4-dichlorocinnamate 766
$C_{12}H_9Cl_2NO_2S$
3′,4-Dichlorobenzenesulfonanilide 809
$C_{12}H_9Cl_2NO_3$
Ethyl 6,8-dichloro-2-hydroxy-4-quinolinecarboxylate 983
Furfuryl *N*-(2,3-dichlorophenyl)-carbamate 849
Furfuryl *N*-(2,4-dichlorophenyl)-carbamate 849
Furfuryl *N*-(3,4-dichlorophenyl)-carbamate 844
$C_{12}H_9Cl_2NO_4S$
2-(4-Amino-2-chlorophenoxy)-5-chlorobenzenesulfonic acid, sodium salt 791
4-(2-Amino-4-chlorophenoxy)-3-chlorobenzenesulfonic acid, sodium salt 790
2-(4-Aminophenoxy)-3,5-dichlorobenzenesulfonic acid, sodium salt 791
4-(2,4-Dichlorophenoxy)-metanilic acid, sodium salt 790
$C_{12}H_9Cl_2NO_4S_2$
4,4′-Dichlorodibenzene-sulfonamide, sodium derivative 828
4,4′-Dichlorodiphenyldisulfimide 809
$C_{12}H_9Cl_2NS$
5-Chloro-2-(4-chlorophenylthio)-aniline 224
$C_{12}H_9Cl_2N_3$
2,6-Dichlorobenzaldehyde 3-pyridylhydrazone nitrate 1026
4,7-Dichloro-1,3-dimethyl-1*H*-pyrazolo(3,4-b)quinoline 1003
$C_{12}H_9Cl_2N_3O$
1-(2-Chlorophenyl)-3-(5-chloro-2-pyridyl)urea 930
1-(2,3-Dichlorophenyl)-3-(2-pyridyl)urea 930
1-(2,3-Dichlorophenyl)-3-(3-pyridyl)urea 927
1-(2,6-Dichlorophenyl)-3-(2-pyridyl)urea 965
1-(2,6-Dichlorophenyl)-3-(3-pyridyl)urea 930
1-(2,6-Dichlorophenyl)-3-(4-pyridyl)urea 928
1-(3,4-Dichlorophenyl)-3-(2-pyridyl)urea 930
1-(3,4-Dichlorophenyl)-3-(3-pyridyl)urea 927
$C_{12}H_9Cl_2N_3O_2$
1-(2,3-Dichlorophenyl)-3-(3-hydroxy-2-pyridyl)urea 926
$C_{12}H_9Cl_3N_2O_2$
1-Furfuryl-3-(2,4,5-trichlorophenyl)urea 851
$C_{12}H_9Cl_4NO_3$
3-Hydroxy-3-pyrrolidino-4,5,6,7-tetrachlorophthalan-1-one 383
$C_{12}H_9Cl_4NO_4$
3-Hydroxy-3-morpholino-4,5,6,7-tetrachlorophthalide 383
$C_{12}H_9Cl_7$
2,5,6,7,8,9,9-Heptachloro-1,2,3,4,4a,5,8,8a-octahydro-1,4:5,8-dimethanonaphthalene 5
$C_{12}H_9FN_2O_2S$
1-(4-Fluorophenyl)-3-furoyl-2-thiourea 844
$C_{12}H_9FN_2O_4S$
2′-Fluoro-3-nitrobenzenesulfonanilide 807
$C_{12}H_9FN_4O_3$
1-(2-Fluoro-5-nitrophenyl)-3-(3-pyridyl)urea 927
1-(4-Fluoro-3-nitrophenyl)-3-(3-pyridyl)urea 927
1-(2-Fluorophenyl)-3-(5-nitro-2-pyridyl)urea 927
$C_{12}H_9FN_4O_4$
1-(4-Fluoro-3-nitrophenyl)-3-(3-hydroxy-2-pyridyl)urea 926
$C_{12}H_9FN_4O_5$
1-(2-Furoyl)-4-(4-fluoro-2-nitrophenyl)semicarbazide 844
$C_{12}H_9FO_2S$
4-Fluorophenyl phenyl sulfone 784
$C_{12}H_9F_2N_3O$
1-(2,5-Difluorophenyl)-3-(3-pyridyl)urea 930
$C_{12}H_9F_3N_2$
2-(3-(Trifluoromethyl)anilino)-pyridine 925
$C_{12}H_9F_7N_2O_2$
N-Acetyl-*N*′-(2,2,3,3,4,4,4-heptafluorobutyryl)-1,4-phenylenediamine 397
$C_{12}H_9IO$
2-Iodo-4-phenylphenol 193
$C_{12}H_9NO$
3-Aminodibenzofuran 894
2-Aminodibenzofuran hydrochloride 894
3-Aminodibenzofuran hydrochloride 894
2-Methylnaphth(1,2-d)oxazole 910
2-Methylnaphth(2,1-d)oxazole 910
Phenoxazine nitrate 253
$C_{12}H_9NOS$
3-Hydroxy-10*H*-phenothiazine 246
Phenothiazine-5-oxide 771
$C_{12}H_9NO_2$
2-Aminodibenzo-*p*-dioxin 248
5*H*-(1)-Benzopyrano(2,3-b)-pyridin-5-ol 932
2-Cyano-5-phenyl-2,4-pentadienoic acid 766
Phenyl nicotinate 964
1-Phthalamido-2-butyne 1019
4-Quinolineacrylic acid 993
$C_{12}H_9NO_2S$
2-Aminodibenzothiophene sulfone 897
3-Carboxy-*N*-(2-thenylidene)aniline 1074
3-Nitrophenyl phenyl sulfide 270
4-Nitrophenyl phenyl sulfide 270
4-(2-Thienylmethyleneamino)-benzoic acid 1074
N-(2-Thienylmethylene)-anthranilic acid 1063
$C_{12}H_9NO_3$
2-Nitro-4-phenylphenol 264
$C_{12}H_9NO_3S$
2-Aminophenoxathiin-10,10-dioxide 777
1-Dibenzofuransulfonamide 896
4-(2-Thienylmethyleneamino)-salicylic acid 1032
$C_{12}H_9NO_4$
Methyl α-cyano-3,4-(methylenedioxy)cinnamate 766
Methyl 3-nitro-1-naphthoate 374
$C_{12}H_9NO_4S$
2-Nitrophenyl phenyl sulfone 779
4-Nitrophenyl phenyl sulfone 779
$C_{12}H_9NO_4S_2$
5-Nitro-2-(phenylthio)-benzenesulinic acid 786
$C_{12}H_9NO_5$
2-(4-Methoxybenzoyl)-5-nitrofuran 846
$C_{12}H_9NO_5S$
4′-Nitro-4-biphenylsulfonic acid, sodium salt 794
$C_{12}H_9NO_6S$
3-Nitro-4-phenoxybenzenesulfonic acid 787
5-Nitro-2-phenoxybenzenesulfonic acid, sodium salt 792
2-(2-Nitrophenylsulfonyl)-hydroquinone 774
2-(3-Nitrophenylsulfonyl)-hydroquinone 774
$C_{12}H_9NS$
4-Aminodibenzothiophene 897
$C_{12}H_9NS_2$
1-Aminothianthrene 248
$C_{12}H_9N_3$
2-(3-Pyridyl)benzimidazole 949
$C_{12}H_9N_3O$
α-Cyano-3-indoleacrylamide 885
α-Dicyanomethylene-*p*-acetotoluidide 764
N-(2,2-Dicyanovinyl)-acetanilide 764
$C_{12}H_9N_3O_2$
2-(2*H*-Benzotriazol-2-yl)-hydroquinone 908
3,4-Dihydro-3-furfuryl-4-oxo-1,2,3-benzotriazine 853
4-Keto-6-methyl-1-phenylpyrazolo-(3,4-d)-5,7-oxazine 863
3-Nitroazobenzene 1089
2-(3-Nitrobenzylideneamino)-pyridine 1059
6-Nitroharman nitrate 919
$C_{12}H_9N_3O_3$
4-(4-Nitrophenylazo)phenol 1084
N-(2-Pyrazinyl)phthalamic acid 978
N-(2-Pyrimidinyl)phthalamic acid 976
$C_{12}H_9N_3O_4$
2-(2,4-Dinitrobenzyl)pyridine 949
2,4-Dinitro-6-phenylaniline 270
2′-Hydroxy-4′-nitronicotinanilide 933
4-(4-Nitrophenylazo)catechol 1082
2-Pyridyl *N*-(3-nitrophenyl)-carbamate 949
3-Pyridyl *N*-(3-nitrophenyl)-carbamate 949
$C_{12}H_9N_3O_5$
o-(2,4-Dinitroanilino)-phenol 264
3-(2,4-Dinitroanilino)phenol 263
Disperse yellow 1 264
$C_{12}H_9N_3O_8S$
2,4-Dinitro-4′-hydroxydiphenylamine-3′-sulfonic acid, sodium salt 791
$C_{12}H_9N_3S_2$
2-(2-Thienylmethylenehydrazino)-benzothiazole 916
$C_{12}H_9N_5$
5-Amino-4-cyano-1-phenyl-3-pyrazoleacetonitrile 860
3-(4-Cyanophenyl)-1-(2-pyridyl)-triazene 965
$C_{12}H_9N_5O_5S$
N-(2-Thiazolyl)oxamic N^2-(4-nitrobenzoyl)hydrazide 877
$C_{12}H_9N_5O_6$
N-(3-Methyl-2-pyridyl)-picrylamine 943
N-(4-Methyl-2-pyridyl)-picrylamine 942
N-(6-Methyl-2-pyridyl)-picrylamine 942
$C_{12}H_9N_7O$
4-(2-Benzimidazolylamino)-6-hydroxy-1,3,5-triazine-2-acetonitrile 978
$C_{12}H_{10}AsClO_2$
(Chloro(2-naphthyl)arsino)-acetic acid 1127
$C_{12}H_{10}AsNO_3$
3-Carbazolearsonic acid 1127
$C_{12}H_{10}AsNO_5S$
4-(4-Nitrophenylthio)-benzenearsonic acid 1126
$C_{12}H_{10}BClF_4N_4$
3-(4-Chlorophenyl)-7-methyl-3*H*-tetrazolo(1,5-a)pyridin-4-ium tetrafluoroborate 1002
$C_{12}H_{10}BF_4I$
Diphenyliodonium tetrafluoroborate 169
$C_{12}H_{10}BrClN_2$
3-(4-Bromo-2-chloroanilinomethyl)-pyridine 928
$C_{12}H_{10}BrClN_2O_2$
1-(2-Chloro-6-nitrobenzyl)-pyridinium bromide 946
$C_{12}H_{10}BrClN_4$
3-(4-Chlorophenyl)-6-methyl-3*H*-tetrazolo(1,5-a)pyridin-4-ium bromide 1002
$C_{12}H_{10}BrI$
Diphenyliodonium bromide 169
$C_{12}H_{10}BrNO$
N-(4-Bromo-1-naphthyl)-acetamide 490
$C_{12}H_{10}BrNO_2S$
4′-Bromobenzenesulfonanilide 828
4-(4-Bromophenylsulfonyl)-aniline 773
$C_{12}H_{10}BrNO_3$
Furfuryl *N*-(4-bromophenyl)-carbamate 849
$C_{12}H_{10}BrNS$
2-Bromo-*N*-(2-thienylmethylene)-*p*-toluidine 1073
4-Bromo-*N*-(2-thienylmethylene)-*o*-toluidine 1073
4-Bromo-*N*-(2-thienylmethylene)-*m*-toluidine 1076
$C_{12}H_{10}BrTl$
Bromodiphenylthallium 1142
$C_{12}H_{10}Br_2$
1,2-Bis(bromomethyl)naphthalene 172
2,3-Bis(bromomethyl)naphthalene 172
6,7-Dibromo-5,6-dihydro-5,9-methano-9*H*-benzocycloheptene 173
$C_{12}H_{10}Br_2Pb$
Diphenyllead dibromide 1135
$C_{12}H_{10}Br_4O_4$
Diethyl 3,4,5,6-tetrabromophthalate 367
$C_{12}H_{10}ClNO$
2-Amino-4-chlorophenyl phenyl ether 223
2-Chloro-*N*-(1-naphthyl)-acetamide 484
2-(4-Chlorophenoxy)-aniline 223
α-(4-Chlorophenyl)-2-pyridinemethanol 934
$C_{12}H_{10}ClNO_2$
3′-Chloro-4′-methyl-2-furanilide 844
Ethyl 2-chloro-α-cyanocinnamate 759
Ethyl 4-chloro-α-cyanocinnamate 766
$C_{12}H_{10}ClNO_2S$
4-Amino-4′-chlorodiphenyl sulfone 773
4-(3-Chlorophenylsulfonyl)aniline 773

$C_{12}H_{10}ClNO_3$
2-Chloro-3-(2-hydroxyethylamino)-1,4-naphthoquinone ... 316
4′-Chloro-2′-methoxy-2-furanilide ... 844
5′-Chloro-2′-methoxy-2-furanilide ... 844
3-(2-Chloro-5-nitrophenyl)-1-cyclopropyl-2-propen-1-one ... 299
Ethyl 6-chloro-2-hydroxy-4-quinolinecarboxylate ... 983
Ethyl 8-chloro-4-hydroxy-3-quinolinecarboxylate ... 984
Furfuryl *N*-(2-chlorophenyl)carbamate ... 849
Furfuryl *N*-(3-chlorophenyl)carbamate ... 849
Furfuryl *N*-(4-chlorophenyl)carbamate ... 849

$C_{12}H_{10}ClNO_3S$
4-Chloro-*N*-(3-hydroxyphenyl)benzenesulfonamide ... 806

$C_{12}H_{10}ClNO_4S$
2-Amino-5-(4-chlorophenoxy)benzenesulfonic acid, sodium salt ... 790
6-(2-Chlorophenoxy)-metanilic acid, sodium salt ... 790
6-(4-Chlorophenoxy)-metanilic acid, sodium salt ... 791

$C_{12}H_{10}ClNS$
4-(2-Chlorophenylthio)aniline ... 224
4-(4-Chlorophenylthio)aniline ... 224
5-Chloro-*N*-(2-thienylmethylene)-*o*-toluidine ... 1073

$C_{12}H_{10}ClN_3O_4$
1-(2,4-Dinitrobenzyl)pyridinium chloride ... 946

$C_{12}H_{10}ClN_3O_5$
2-Methoxy(1,2,4)triazino(6,1-a)isoquinolin-5-ium perchlorate ... 978

$C_{12}H_{10}ClN_3S$
1-(4-Chlorophenyl)-3-(2-pyridyl)-2-thiourea ... 959

$C_{12}H_{10}ClO_2P$
(4-Chlorophenyl)phenylphosphinic acid ... 833

$C_{12}H_{10}Cl_2CuN_2$
Dichloro(*N*-(2-pyridylmethylene)aniline-*N*,*N*′)copper(II) ... 1139

$C_{12}H_{10}Cl_2N_2O$
α-(3,5-Dichloro-2-pyridylamino)-*o*-cresol ... 933

$C_{12}H_{10}Cl_2N_2O_2$
1-(2,3-Dichlorophenyl)-3-furfurylurea ... 851
1-(2,4-Dichlorophenyl)-3-furfurylurea ... 852
1-(2,5-Dichlorophenyl)-3-furfurylurea ... 851
1-(2,6-Dichlorophenyl)-3-furfurylurea ... 851
1-(3,4-Dichlorophenyl)-3-furfurylurea ... 854

$C_{12}H_{10}Cl_2N_2O_3$
4-(4-Acetamidophenylimino)-2,3-dichloro-2-butenoic acid ... 1060

$C_{12}H_{10}Cl_2N_2S_2$
2-Amino-4-chlorophenyl disulfide ... 224

$C_{12}H_{10}Cl_2N_4$
2,6-Dichloro-4-(1,2,3,4-tetrahydro-1-quinolyl)-1,3,5-triazine ... 980

$C_{12}H_{10}Cl_2N_4O_2$
2-(4,6-Dichloro-1,3,5-triazin-2-yl)-*o*-acetanisidide ... 978

$C_{12}H_{10}Cl_2O$
3-(2,4-Dichlorophenyl)-1-cyclopropyl-2-propen-1-one ... 285

$C_{12}H_{10}Cl_2O_5S$
3-(2-(2,4-Dichlorophenoxy)acetoxy)-2,3-dihydrothiophene-1,1-dioxide ... 773

$C_{12}H_{10}Cl_2Pb$
Dichlorodiphenyllead ... 1135

$C_{12}H_{10}Cl_3NO$
2′,4′,5′-Trichloro-2,4-hexadienanilide ... 460
α-(Trichloromethyl)-2-quinolineethanol ... 981

$C_{12}H_{10}Cl_4O_5$
3-Hydroxybutyl tetrachlorophthalate ... 366

$C_{12}H_{10}Cl_6O$
5,6,7,8,9,9-Hexachloro-1,2,3,4,4a,5,8,8a-octahydro-1,4:5,8-dimethanonaphthalen-2-ol ... 16

$C_{12}H_{10}FN$
4-Amino-4′-fluorobiphenyl ... 224

$C_{12}H_{10}FNO_2S$
4′-Amino-4-biphenylsulfonyl fluoride ... 803
4′-Fluorobenzenesulfonanilide ... 809

$C_{12}H_{10}FNO_3$
Furfuryl *N*-(3-fluorophenyl)carbamate ... 844

$C_{12}H_{10}FN_3O$
1-(3-Fluorophenyl)-3-(3-pyridyl)urea ... 930

$C_{12}H_{10}F_2Si$
Difluorodiphenylsilane ... 1112

$C_{12}H_{10}F_3N_3O_4$
3-(2-Nitro-4-trifluoromethylphenylhydrazono)-2,4-pentanedione ... 282

$C_{12}H_{10}INO_3$
Diphenyliodonium nitrate ... 169

$C_{12}H_{10}IN_3O_2$
Ethyl 2-(2,2-dicyanovinyl)-5-iodo-4-methyl-3-pyrrolecarboxylate ... 839

$C_{12}H_{10}I_2Pb$
Diiododiphenyllead ... 1135

$C_{12}H_{10}I_2Sn$
Diiododiphenyltin ... 1132

$C_{12}H_{10}N_2$
5,6-Dihydrobenzo(*h*)quinazoline hydrochloride ... 977
α-Ethylbenzylidenemalononitrile ... 764
1-(2-Pyridyl)-2-(3-pyridyl)-ethylene ... 967
1-(3-Pyridyl)-2-(4-pyridyl)-ethylene ... 967

$C_{12}H_{10}N_2O$
2-Cyano-5-phenyl-2,4-pentadienoamide ... 764
2,7-Diaminodibenzofuran ... 894
3,4-Dihydronaphtho(1,8-ef)-1,4-diazepin-2(1*H*)-one ... 709
(4-Ethoxybenzylidene)malononitrile ... 762
Nicotinanilide ... 949
N-Nitrosodiphenylamine ... 257
2-(Phenylazo)phenol ... 1086
N-(4-Pyridyl)benzamide ... 949
α-(2-Pyridylimino)-*o*-cresol ... 1037
α-(2-Pyridylimino)-*o*-cresol copper(II) salt ... 1036
4-Pyridyl 4-pyridylmethyl ketone ... 961

$C_{12}H_{10}N_2OS$
2-Benzylidene-5,6-dihydroimidazo(2,1-b)thiazol-3(2*H*)-one ... 746
N-(3-Cyano-2-benzo(b)thiophen-2-yl)propionamide ... 898

$C_{12}H_{10}N_2O_2$
5-Amino-4-nitroacenaphthene ... 271
4,4′-Dihydroxyazobenzene ... 1083
2,5-Dimethoxybenzylidenemalononitrile ... 761
3,4-Dimethoxybenzylidenemalononitrile ... 761
Methyl 2,3-dicyano-3-phenylpropionate ... 755
Methyl 5-phenyl-4-pyridazinecarboxylate ... 968
2-(4-Nitrobenzyl)pyridine ... 960
4-(4-Nitrobenzyl)pyridine perchlorate ... 961
2′-Nitro-4-biphenylamine ... 271
5-Nitro-2-biphenylamine ... 271
2-Pyridyl *N*-phenylcarbamate ... 964
3-Pyridyl *N*-phenylcarbamate ... 960

$C_{12}H_{10}N_2O_2S$
2-(Carboxymethylthio)-5-phenylpyrimidine ... 974
3,7-Diaminodibenzothiophene-5,5-dioxide ... 777
3,7-Diaminodibenzothiophene-5,5-dioxide dihydrochloride ... 897
3-Nitro-*N*-(2-thienylmethylene)-*p*-toluidine ... 1056
4-Nitro-*N*-(2-thienylmethylene)-*o*-toluidine ... 1059
5-Nitro-*N*-(2-thienylmethylene)-*o*-toluidine ... 1056

$C_{12}H_{10}N_2O_3$
4,4′-Dihydroxyazoxybenzene ... 1083
Salicylic (furfurylidene)hydrazide ... 1032

$C_{12}H_{10}N_2O_3S$
5-Anisylidene-2-thiobarbituric acid ... 733
Harman-*N*-sulfonic acid ... 918
5-Nitro-*N*-(2-thienylmethylene)-*o*-anisidine ... 1043

$C_{12}H_{10}N_2O_4$
4-Amino-4′-nitrodiphenyl sulfone ... 776
2,4-Dihydroxybenzoic (furfurylidene)hydrazide ... 1028
Ethyl α-cyano-3-nitrocinnamate ... 766
Ethyl α-cyano-4-nitrocinnamate ... 766
N-(5-Methyl-3-isoxazolyl)phthalamic acid ... 871
Methyl 2-methyl-5-nitro-4-quinolinecarboxylate ... 989
2′-Methyl-4′-nitro-2-furanilide ... 850
4′-Methyl-2′-nitro-2-furanilide ... 847
4′-Methyl-3′-nitro-2-furanilide ... 847

$C_{12}H_{10}N_2O_4S$
4-Hydroxyazobenzene-4′-sulfonic acid, disodium salt dihydrate ... 1085
4-(4-Hydroxyphenylazo)benzenesulfonic acid, sodium salt ... 1085
4′-Nitrobenzenesulfonanilide ... 819
5-Vanillylidene-2-thiobarbituric acid ... 733
5-(*o*-Vanillylidene)-2-thiobarbituric acid ... 733

$C_{12}H_{10}N_2O_5$
Ethyl 4-hydroxy-7-nitro-3-quinolinecarboxylate ... 983
Furfuryl *N*-(3-nitrophenyl)carbamate ... 847
5-(4-Hydroxy-3-methoxybenzylidene)barbituric acid ... 699
2′-Methoxy-4′-nitro-2-furanilide ... 846
N-Phthaloylglycylglycine ... 718

$C_{12}H_{10}N_2O_5S$
2′-Hydroxy-4′-nitrobenzenesulfonanilide ... 811

$C_{12}H_{10}N_2O_6$
1-(2-Furoyl)-2-(3,4,5-trihydroxybenzoyl)hydrazine ... 845

$C_{12}H_{10}N_2S$
2,3-Dihydro-2-imino-3-methylnaphtho(1,2-d)thiazole ... 749

$C_{12}H_{10}N_3OP$
Diphenylphosphinyl azide ... 836

$C_{12}H_{10}N_4$
5-Amino-2-(2-pyridyl)benzimidazole ... 941
2,3-Diaminophenazine ... 999
2-Pyridinealdazine ... 1053
3-Pyridinealdazine ... 1057
4-Pyridinealdazine ... 1079

$C_{12}H_{10}N_4O$
Indole-3-carboxaldehyde cyanoacetylhydrazone ... 499

$C_{12}H_{10}N_4OS$
3-Methyl-8-phenyl-2-thiohypoxanthine ... 868

$C_{12}H_{10}N_4O_2$
4-Amino-3-nitroazobenzene ... 1087
3-Amino-2-phenyl-2*H*-pyrazolo(4,3-c)-pyridine-4,6-diol ... 921

$C_{12}H_{10}N_4O_3$
Benzoic N^2-(3-nitro-2-pyridyl)hydrazide ... 961
1-(5-Nitro-2-pyridyl)-3-phenylurea ... 943

$C_{12}H_{10}N_4O_4$
1,2-Bis(4-nitrophenyl)hydrazine ... 276
3,3′-Dinitrobenzidine ... 270
Oxalic bis(furfurylidene)hydrazide ... 1053

$C_{12}H_{10}N_4O_6S$
N,*N*′-Bis(4-nitrophenyl)sulfamide ... 819

$C_{12}H_{10}N_4S$
1-Benzyl-8-mercapto-1*H*-1,2,4-triazolo(4,3-b)pyridazin-4-ium hydroxide,inner salt ... 909

$C_{12}H_{10}N_6$
2,4-Diamino-6-(1-isoquinolyl)-1,3,5-triazine ... 997

$C_{12}H_{10}N_6O_5$
Mordant Brown 12 ... 1083

$C_{12}H_{10}O$
1′-Acetonaphthone picrate ... 298
3-Methyl-2-naphthaldehyde ... 324
6-Methyl-2-naphthaldehyde ... 324
7-Methyl-2-naphthaldehyde ... 344
Methyl α-naphthyl ketone ... 298
1-Phenyl-2,4-hexadiyn-1-ol ... 1016

$C_{12}H_{10}OS$
2-Acetyl-5-phenylthiophene ... 857
2-Phenoxythiophenol ... 209

$C_{12}H_{10}OSn$
Diphenyltin oxide ... 1132

$C_{12}H_{10}O_2$
3,3′-Biphenol ... 195
2-Methyl-1-naphthalenecarboxylic acid ... 346
Methyl 1-naphthoate ... 376
1-Methyl-2-naphthoic acid ... 346
1-Naphthaleneacetic acid, sodium salt ... 352
3-Phenylcatechol ... 193
4-Phenylresorcinol ... 194

$C_{12}H_{10}O_2S$
2-Methyl-3-(methylthio)-1,4-naphthoquinone ... 317

$C_{12}H_{10}O_3$
4,4′-Dihydroxydiphenyl ether ... 194
3-(Hydroxymethyl)-2-naphthoicacid ... 339
2-Hydroxy-1-naphthaleneacetic acid ... 326
2-Methoxy-3-methyl-1,4-naphthoquinone ... 316
1-Methoxy-2-naphthoic acid ... 340
2-Methoxy-1-naphthoic acid ... 341
3-Methoxy-2-naphthoic acid ... 341
6-Methoxy-1-naphthoic acid ... 341
Methyl 1-hydroxy-2-naphthoate ... 368
10-Oxatetracyclo[6.3.0.^{2,7}.0.^{3,6}]-trideca-4,12-diene-9,11-dione, isomer I ... 99
10-Oxatetracyclo[6.3.0^{2,7}.0^{3,6}]-trideca-4,12-diene-9,11-dione, isomer II ... 99

$C_{12}H_{10}O_3S$
Acenaphthene-5-sulfonic acid, potassium salt ... 796
4-Biphenylsulfonic acid ... 790

$C_{12}H_{10}O_4$
3-Acetyl-5-hydroxy-2-methylchromone ... 319
Cinnamylidenemalonic acid ... 335
Ethyl 3-hydroxy-1-oxo-2-indenecarboxylate, sodium salt ... 371
4-(3,4-(Methylenedioxy)phenyl)-1,3-butadiene-1-carboxylic acid ... 345

$C_{12}H_{10}O_4S$
2-(Phenylsulfonyl)hydroquinone ... 774

$C_{12}H_{10}O_4S_3$
Benzenesulfonothioic acid anhydrosulfide ... 798

$C_{12}H_{10}O_5$
2-(4-Acetoxyphenyl)succinic anhydride ... 388
Methyl 5-hydroxy-4-methylcoumarin-6-carboxylate ... 383

$C_{12}H_{10}O_5S$
1-Acetoxynaphthalene-6-sulfonic acid ... 790
5-Sulfo-1-naphthyl acetate ... 790

$C_{12}H_{10}O_6S_4$
4,4′-Dithiobis(benzenesulfonic acid, potassium salt) ... 793

$C_{12}H_{10}S_2$
2,2′-Biphenyldithiol ... 209

$C_{12}H_{11}BN_2$
2-Phenyl-1,3,2-benzodiazaborole ... 1125

$C_{12}H_{11}BO_2$
2-Biphenylylboronic acid hydrate ... 1125

$C_{12}H_{11}BO_3$
4-(Phenoxy)benzeneboronic acid ... 1125

$C_{12}H_{11}Br$
2-Bromomethyl-6-methylnaphthalene ... 172

$C_{12}H_{11}BrN_2$
4-Bromo-1-cinnamylpyrazole ... 864

$C_{12}H_{11}BrN_2O_2$
1-(4-Bromophenyl)-3-furfurylurea ... 852

$C_{12}H_{11}BrN_2O_2$ (Cont'd)
1-(4-Nitrobenzyl)pyridinium bromide 956
$C_{12}H_{11}BrN_2O_3$
2-Bromo-α-cyano-4,5-dimethoxycinnamamide 758
$C_{12}H_{11}BrO_2$
2-Bromo-1-phenyl-1-hexen-4-yne-3,6-diol 1016
$C_{12}H_{11}Br_2NO_5$
2,3-Dibromo-3-(2-(methoxycarbonyl)phenylcarbamoyl)propionic acid 433
$C_{12}H_{11}ClF_3NO_3$
2′-Chloro-5′-(trifluoromethyl)glutaranilic acid 395
$C_{12}H_{11}ClN_2O$
α-(5-Chloro-2-pyridylamino)-*o*-cresol 933
$C_{12}H_{11}ClN_2O_2$
2-Chloro-5-(3,4-dimethoxyphenyl)pyrimidine 973
1-(2-Chlorophenyl)-3-furfurylurea 851
1-(3-Chlorophenyl)-3-furfurylurea 854
1-(4-Chlorophenyl)-3-furfurylurea 854
1-(3-Nitrobenzyl)pyridinium chloride 951
1-(4-Nitrobenzyl)pyridinium chloride 956
$C_{12}H_{11}ClN_2O_3$
1-(2-Hydroxy-5-nitrobenzyl)pyridinium chloride 933
$C_{12}H_{11}ClN_2O_3S$
4′-Amino-5′-chloro-2′-hydroxybenzenesulfonanilide 811
$C_{12}H_{11}ClN_4$
2-Amino-3-(4-chlorophenylazo)-6-methylpyridine 1080
3-(4-Chlorophenyl)-1-(5-methyl-2-pyridyl)triazene 931
$C_{12}H_{11}ClN_4O$
1-(3-Chloro-4-methylphenyl)-3-(2-pyrimidinyl)urea 969
$C_{12}H_{11}ClN_4O_2$
1-(4-Chloro-2-methoxyphenyl)-3-(2-pyrazinyl)urea 977
1-(5-Chloro-2-methoxyphenyl)-3-(2-pyrazinyl)urea 977
1-(5-Chloro-2-methoxyphenyl)-3-(2-pyrimidinyl)urea 968
$C_{12}H_{11}ClO_4$
Dimethyl 4-chlorobenzylidenemalonate 364
$C_{12}H_{11}ClO_5S$
3-(2-(2-Chlorophenoxy)acetoxy)-2,3-dihydrothiopene-1,1-dioxide 773
$C_{12}H_{11}ClSi$
(4-Chlorophenyl)phenylsilane 1103
$C_{12}H_{11}Cl_2N_3O$
1-(*N*-(2,3-Dichlorophenyl)carbamoyl)-3,5-dimethylpyrazole 860
$C_{12}H_{11}Cl_2N_3OS$
3-(2-Chlorophenyl)-1-(3-chloropropyl)-5-imino-4-thiohydantoin 727
3-(4-Chlorophenyl)-1-(3-chloropropyl)-5-imino-4-thiohydantoin 727
$C_{12}H_{11}Cl_3N_2O_6$
Ethyl *N*-(1-(4-nitrobenzoyloxy)-2,2,2-trichloroethyl)carbamate 610
$C_{12}H_{11}Cl_3SSi$
2-(Dimethylphenylsilyl)-3,4,5-trichlorothiophene 1111
$C_{12}H_{11}Cl_4NO_3$
N-tert-Butyl-3,4,5,6-tetrachlorophthalamic acid 453
$C_{12}H_{11}Cl_4NO_4$
Ethyl *N*-(1-(4-chlorobenzoyloxy)-2,2,2-trichloroethyl)carbamate 606
$C_{12}H_{11}Cl_4NO_5$
N-(1,1-Bis(hydroxymethyl)ethyl)-3,4,5,6-tetrachlorophthalamic acid 446
$C_{12}H_{11}FN_2O$
4-(4-Fluorophenyl)-5-methyl-1*H*-1,2-diazepin-6(7*H*)-one 282
$C_{12}H_{11}F_3N_2OS$
2-Phenyl-2-(4-(trifluoromethyl)-2-thiazolylamino)-1-ethanol 874
$C_{12}H_{11}F_3N_4O_2$
3-(3-Aminobenzamido)-1-(2,2,2-trifluoroethyl)-2-pyrazolin-5-one 693
$C_{12}H_{11}IN_2$
Norharman methiodide 919
$C_{12}H_{11}IN_2O_2S$
4-Iodosulfanilanilide 807
$C_{12}H_{11}N$
1-Allylisoquinoline 998
5-Aminoacenaphthene 231
3-Aminoacenaphthene hydrochloride 231
2-Phenyl-4-picoline 966
2-(1-Propenyl)quinoline 995
2-(*p*-Tolyl)pyridine picrate 933
$C_{12}H_{11}NO$
2-Acetamidonaphthalene 490
7-Amino-1,4-dihydrodibenzofuran hydrochloride 894
N-Furfurylidene-*m*-toluidine 1075
N-Furfurylidene-*p*-toluidine 852
4-Hydroxydiphenylamine 227
2-(4-Methoxyphenyl)pyridine 939
4-(4-Methoxyphenyl)pyridine 939
2-(1-Oxo-1,2,3,4-tetrahydro-2-naphthyl)acetonitrile 757
$C_{12}H_{11}NOS$
2-Mercapto-*N*-(2-naphthyl)acetamide 420
$C_{12}H_{11}NO_2$
1-Acetamido-6-naphthol 486
4-(4-Aminophenoxy)-phenol 225
3-Benzyl-3-azabicyclo(3.1.0)hexane-2,4-dione 721
N-(2-Butenyl)phthalimide 723
Ethyl α-cyanocinnamate 766
N-(2-Ethylphenyl)maleimide 720
N-(4-Ethylphenyl)maleimide 720
Ethyl 7-quinolinecarboxylate 993
Ethyl 7-quinolinecarboxylate hydrochloride 993
N-Furfurylidene-*m*-anisidine 1049
N-(Furfurylidene)-*p*-anisidine 1049
3-(α-Hydroxybenzyl)pyridine-*N*-oxide 1005
N-(7-Hydroxy-1-naphthyl)acetamide 697
4,4′-Iminodiphenol 225
4-Isopropylidene-2-phenyl-5(4*H*)-oxazole 387
2-(4-Methoxyphenyl)pyridine-*N*-oxide 1005
N-(2-Methylallyl)-phthalimide 723
Methyl α-cyano-4-methylcinnamate 766
Methyl *N*-(1-naphthyl)carbamate 687
1-Naphthyl *N*-methylcarbamate 687
2-Naphthyl *N*-methylcarbamate 687
N-(2,3-Xylyl)maleimide 723
N-(2,4-Xylyl)maleimide 723
N-(2,6-Xylyl)maleimide 720
N-(3,4-Xylyl)maleimide 720
$C_{12}H_{11}NO_2S$
3-Aminodiphenyl sulfone 776
(2-Amino-1-naphthylthio)acetic acid, sodium salt 351
Benzenesulfonanilide 827
α-Cyano-β-methyl-2-(methylthio)cinnamic acid 763
S-Ethyl 3-indolethioglyoxylate 889
Phenyl 2-pyridylmethyl sulfone 960
Phenyl 4-pyridylmethyl sulfone 965
$C_{12}H_{11}NO_3$
1-Acetyl-3-indolyl acetate 890
1-Cyclopropyl-3-(3-nitrophenyl)-2-propen-1-one 285
7,8-Dimethyl-4-hydroxy-3-quinolinecarboxylic acid 984
N-(2-Ethoxyphenyl)maleimide 717
Ethyl 2-(4-cyanobenzoyl)acetate 766
Ethyl 3-cyano-3-phenylpyruvate 755
Ethyl 3-indolylglyoxylate 889
Ethyl 2-(2-oxo-3-indolinylidene)acetate 692
Furfuryl *N*-phenylcarbamate 848
4-Hydroxy-3-(1-(phenylimino)ethyl)-2(5*H*)-furanone 1032
2′-Methoxy-2-furanilide 846
3′-Methoxy-2-furanilide 846
4′-Methoxy-2-furanilide 846
1-(2-Methoxy-5-methylphenyl)maleimide 717
4-(Methylamino)phenyl 2-furoate 847
Methyl 6-methoxy-4-quinolinecarboxylate 986
7-Nitro-1,2,3,4-tetrahydrodibenzofuran 895
$C_{12}H_{11}NO_3S$
Diphenylamine-4-sulfonic acid 789
Diphenylamine-4-sulfonic acid, magnesium salt 793
3′-Hydroxybenzenesulfonanilide 812
4′-Hydroxybenzenesulfonanilide 812
Methyl 2-(3,4-dihydro-3-oxo-2*H*-1,4-benzothiazin-2-yl)acrylate 707
Phenyl sulfanilate 798
$C_{12}H_{11}NO_3S_2$
5-(2,3-Dimethoxybenzylidene)rhodanine 735
5-(3,4-Dimethoxybenzylidene)rhodanine 736
5-Ethoxymethylene-3-(4-hydroxyphenyl)-2-thioxo-4-thiazolidinone 734
5-(α-Methylvanillylidene)rhodanine 733
$C_{12}H_{11}NO_4$
N-(3-Acetylphenyl) maleanilic acid 464
N-(4-Acetylphenyl) maleanilic acid 464
2-Cyano-2-(2,3-dimethoxybenzylidene)acetic acid 761
1,2-Dihydro-6-methoxy-1-methyl-2-oxo-4-quinolinecarboxylic acid 690
Ethyl 1,3-dioxo-2-isoindolineacetate 722
Ethyl 2-oxo-3-indolineglyoxylate 690
2-Methyl-3-butyn-2-yl 4-nitrobenzoate 1018
4-Methyl-2-oxo-2*H*-1-benzopyran-7-yl *N*-methylcarbamate 676
3-(2-Nitrobenzylidene)-2,4-pentanedione 308
2-Phthalimidoethyl acetate 721
N-Phthaloylalanine methyl ester 721
$C_{12}H_{11}NO_4S$
1-Acetamidonaphthalene-4-sulfonic acid, sodium salt 792
5-Amino-2-phenoxybenzene-sulfonic acid sodium salt 792
$C_{12}H_{11}NO_4S_2$
Dibenzenesulfonamide 827
$C_{12}H_{11}NO_5$
2-(1,3-Benzodioxol-5-yl)-5-oxopyrrolidine-3-carboxylic acid 692
Carbobenzyloxy-L-aspartic anhydride 658
2-Nitro-3-(2,4-pentadien-1-yloxy)benzoic acid 340
2-Nitro-6-(2,4-pentadien-1-yloxy)benzoic acid 340
$C_{12}H_{11}NO_5S$
6-Acetamido-1-naphthol-3-sulfonic acid, sodium salt 791
$C_{12}H_{11}NO_8S_2$
4-Acetamido-5-hydroxynaphthalene-2,7-disulfonic acid disodium salt 791
$C_{12}H_{11}NS$
2-Anilinothiophenol 230
N-(1-Naphthyl)-thioacetamide 746
4-(Phenylthio)aniline 230
N-(2-Thienylmethylene)-*p*-toluidine 1073
$C_{12}H_{11}N_3$
α-(4-(Dimethylamino)benzylidene)malononitrile 763
2-Pyridinecarboxaldehyde phenylhydrazone 1053
$C_{12}H_{11}N_3O$
3,4-Dihydro-1(2*H*)-phenazinone oxime 1009
1,3-Dimethyl-1*H*-pyrazolo(3,4-b)quinol-4(9*H*)-one 861
Formic N^1-phenyl-N^2-(2-pyridyl)hydrazide 965
1-Phenyl-3-(2-pyridyl)urea 965
1-Phenyl-3-(3-pyridyl)urea 945
1-Phenyl-3-(4-pyridyl)urea 950
$C_{12}H_{11}N_3OS$
1-Furfurylidene-4-phenyl-3-thiosemicarbazide 1071
$C_{12}H_{11}N_3O_2$
2-Benzylamino-5-nitropyridine 943
Benzyl 3-aminopyrazine-2-carboxylate 977
8,9-Dimethoxy-1,2,4-triazolo(3,4-a)isoquinoline 908
1,2,3,5,6,7-Hexahydro-6-phenylimidazo(1,2-c)pyrimidine-5,7-dione 709
1-Methyl-5-nitro-2-styrylimidazole 867
2-Nitrobenzidine 271
$C_{12}H_{11}N_3O_2S$
1-(3-Acetylphenyl)-3-(2-thiazolyl)urea 877
2-(3-Aminobenzoyl)-*N*-(2-thiazolyl)acetamide 876
3-(4-Nitrophenyl)-5,6-dihydro-7*H*-thiazolo(3,2-a)pyrimidine 278
$C_{12}H_{11}N_3O_3$
2-Amino-4-nitro-4′-hydroxydiphenylamine 263
2-(1-Methyl-5-nitro-2-imidazolyl)acetophenone 867
$C_{12}H_{11}N_3O_3S$
4-(4-Aminophenylazo)benzenesulfonic acid, sodium salt 1088
$C_{12}H_{11}N_3O_4$
1-Furfuryl-3-(2-nitrophenyl)urea 850
1-Furfuryl-3-(3-nitrophenyl)urea 850
1-Furfuryl-3-(4-nitrophenyl)urea 850
$C_{12}H_{11}N_3O_5$
α-Cyano-4,5-dimethoxy-2-nitrocinnamamide 761
$C_{12}H_{11}N_3O_5S$
4-Amino-2′-nitro-4′-sulfodiphenylamine 788
$C_{12}H_{11}N_3S$
1-Phenyl-3-(2-pyridyl)-2-thiourea 965
1-Phenyl-3-(3-pyridyl)-2-thiourea 945
$C_{12}H_{11}N_5$
2-(4-Methyl-2-pyrimidinylamino)benzimidazole 975
$C_{12}H_{11}N_5O$
2-(2-Benzimidazolylamino)-4-hydroxy-6-methylpyrimidine 972
9-Benzylguanine 923
2,3-Dihydro-3-imino-5-methyl-2-phenyl-*s*-triazolo(4,3-a)pyrimidin-6-ol 918
$C_{12}H_{11}N_5OS$
3-Amino-1-(5-phenyl-1,3,4-thiadiazol-2-ylmethyl)-2-pyrazolin-5-one 878
$C_{12}H_{11}N_5O_3$
3-(2-Methoxy-5-nitrophenyl)-1-(2-pyridyl)triazene 939
1-(4-Methyl-3-nitrophenyl)-3-(2-pyrimidinyl)urea 975
$C_{12}H_{11}N_5O_4$
1-(2-Methoxy-5-nitrophenyl)-3-(2-pyrimidinyl)urea 973
$C_{12}H_{11}N_5S$
2-Amino-6-benzylthiopurine 922
$C_{12}H_{12}BrN$
6-Bromo-1,2,3,4-tetrahydrocarbazole 887
$C_{12}H_{12}BrNO$
N-(2-Bromophenyl)-1,3-pentadiene-1-carboxamide 457
$C_{12}H_{12}BrNO_2$
N-(2-(1-Bromoethyl)phenyl)succinimide 722
5-Bromo-3-indolebutyric acid 890
$C_{12}H_{12}Br_2O_6$
Dimethyl 2,5-dibromo-3,6-dimethoxyterephthalate 366
$C_{12}H_{12}ClF_3O$
3-Chloro-3-methyl-4-(α,α,α-trifluoro-*m*-tolyl)-2-butanone 282
$C_{12}H_{12}ClN$
1-(4-Chlorophenyl)-1-cyclopentanecarbonitrile 756
1-(4-Chlorophenyl)-2,5-dimethylpyrrole 838

$C_{12}H_{12}ClNO_2$
N-(3-Chloro-2-methylphenyl)-glutarimide 716
N-(3-Chloro-4-methylphenyl)-glutarimide 715

$C_{12}H_{12}ClNO_4$
4-Methyl-1-phenylpyridinium perchlorate 957

$C_{12}H_{12}ClNO_4S$
5-Oxo-1-(*p*-tosyl)-2-pyrrolidinecarboxylic acid chloride 824

$C_{12}H_{12}ClNO_5$
5′-Chloro-2′,4′-dimethoxymaleanilic acid 448
Dimethyl 5-(2-chloroacetamido)-isophthalate 392

$C_{12}H_{12}ClN_3$
Basic dye (C.I. 49400) 1052
3,3′-(4-Chlorophenylimino)-dipropionitrile 751

$C_{12}H_{12}ClN_3O$
3-Chloro-*N*-(α-methoxy-1-methyl-2-imidazolylmethylene)aniline 865

$C_{12}H_{12}ClN_3OS$
1-(3-Chloropropyl)-5-imino-3-phenyl-4-thiohydantoin 727

$C_{12}H_{12}ClN_3O_2$
1-(*N*-(4-Chloro-2-ethoxyphenyl)-carbamoyl)imidazole 865

$C_{12}H_{12}ClN_3O_3$
1-(*N*-(5-Chloro-2,4-dimethoxyphenyl)carbamoyl)-imidazole 864
1-(*N*-(5-Chloro-2,4-dimethoxyphenyl)carbamoyl)-pyrazole 859
Ethyl 2-(5-chloro-1-indazolylcarboxamido)acetate 900

$C_{12}H_{12}ClN_3S$
2-(4-Chlorophenyl)-3-imino-2,3,6,7-tetrahydro-5*H*-thiazolo(3,2-a)-pyrimidine hydrochloride 733

$C_{12}H_{12}Cl_2N_2O_2$
Butyl *N*-cyano-*N*-(3,4-dichlorophenyl)carbamate 752

$C_{12}H_{12}Cl_3NO_4$
Ethyl *N*-(1-(benzoyloxy)-2,2,2-trichloroethyl)carbamate 610

$C_{12}H_{12}Cl_3NO_5$
Methyl *N*-(1-(4-methoxybenzoyloxy)-2,2,2-trichloroethyl)carbamate 608

$C_{12}H_{12}Cl_5N_3O$
1-Methyl-4-(2,3,4,5,6-pentachlorophenylcarbamoyl)-piperazine 694

$C_{12}H_{12}Cl_6$
1,9,10,11,12,12-Hexachlorotricyclo[7.2.1.0[2,8]]-dodec-10-ene 7

$C_{12}H_{12}Cl_6O_5$
Bis(1-(ethoxycarbonyl)-2,2,2-trichloroethyl) ether 84

$C_{12}H_{12}F_3NO_2$
2-Butenyl *N*-(2-(trifluoromethyl)-phenyl)carbamate 611

$C_{12}H_{12}F_3N_3O$
2-Methyl-*N*-(3-(trifluoromethyl)-phenyl)-2-imidazoline-1-carboxamide 723

$C_{12}H_{12}F_3N_3O_5$
2′,6′-Dinitro-*N*-ethyl-4′-(trifluoromethyl)propionanilide 396
2′,6′-Dinitro-*N*-methyl-4′-(trifluoromethyl)butyranilide 396
2′,6′-Dinitro-*N*-propyl-4′-(trifluoromethyl)acetanilide 396
Ethyl 2-(3-(2-nitro-4-(trifluoromethyl)phenyl)ureido)-acetate 508

$C_{12}H_{12}F_3N_3O_6$
Ethyl *N*-(2,6-dinitro-4-(trifluoromethyl)phenyl)-*N*-ethylcarbamate 609
Propyl *N*-(2,6-dinitro-4-(trifluoromethyl)phenyl)-*N*-methylcarbamate 609

$C_{12}H_{12}FeN_2O_2$
3-(Ferrocenylmethylene)-carbazic acid 1142

$C_{12}H_{12}FeO_2$
Methyl ferrocenecarboxylate 1142

$C_{12}H_{12}INS$
2-(2-Thienylmethylene)picoline methiodide 957

$C_{12}H_{12}N_2$
2-Amino-4-benzylpyridine 942
4,5-Diaminoacenaphthene 230
2,2′-Diaminobiphenyl 230
3,4′-Diaminobiphenyl 219
6,6′-Dimethyl-2,2′-dipyridyl 967
2,2′-Ethylenedipyridine 953
1-Methyl-2-styrylimidazole 867
N-Phenyl-1,4-phenylenediamine sulfate 230
1-(3-Pyridyl)-2-(4-pyridyl)-ethane 967

$C_{12}H_{12}N_2O$
4-Acetyl-5-methyl-1-phenylpyrazole 862
2-(3-Amino-4-methoxyphenyl)-pyridine 937
Glycine-2-naphthylamide monohydrate 487
Harmalol hydrobromide 918
1-Methyl-3-(1-naphthyl)urea 603
1-Methyl-3-(2-naphthyl)urea 604
5-Methyl-4-phenyl-1*H*-1,2-diazepin-6(7*H*)-one 285
5-Methyl-6-phenyl-1*H*-1,2-diazepin-4(5*H*)-one 286
2-Methyl-3,4,4a,5-tetrahydroindeno(1,2-c)pyridazin-3(2*H*)-one 711
8-Methyl-1-tetrahydronorharmanone 884
1-Naphthylacetic hydrazide 503
2-Naphthylacetic hydrazide 504
2-(1-Naphthyl)acetic hydrazide 503
α-(3-Pyridylamino)-*o*-cresol 932

$C_{12}H_{12}N_2OS$
1-Furfuryl-3-phenyl-2-thiourea 852

$C_{12}H_{12}N_2O_2$
8-Acetamido-6-methoxyquinoline 985
meso-1,2-Bis(4-pyridyl)-1,2-ethanediol trihydrate 931
5-Cyclopropyl-5-phenylhydantoin 705
1,3,7,9-Decatetraen-5-yne-1,10-dicarboxaldehyde dioxime 1015
2,5-Dimethyl-1-(3-nitrophenyl)-pyrrole 839
α,α-Dimethyl-α,4-phenethyl diisocyanate 769
1,2-Di-2-pyridyl-1,2-ethanediol 932
2-Ethyl-4-methyl-8-nitroquinoline 988
1-Furfuryl-3-phenylurea 854
1-(5-Hydroxy-1-naphthyl)-3-methylurea 549
6-Methoxy-1-tetrahydronorharmanone 882
3,3′-(1,4-Phenylenedioxy)-dipropionitrile 753

$C_{12}H_{12}N_2O_3$
Antipyric acid 707
α-Cyano-2,3-dimethoxycinnamamide 761
α-Cyano-2,4-dimethoxycinnamamide 762
α-Cyano-2,5-dimethoxycinnamamide 762
α-Cyano-3,4-dimethoxycinnamamide 762
2′-Cyanoglutaranilic acid 764
Ethyl 3,4-dihydro-4-oxo-1-phthalazineacetate 1000
Ethyl 3-hydroxy-2-quinoxalineacetate 999
Ethyl 5-oxo-1-phenyl-2-pyrazoline-3-carboxylate 705
4-Glycidyloxy-2-indolecarboxamide, compound with 1,4-dioxane 882
5-(4-Methoxyphenyl)-3-methyl-4-pyrazolecarboxylic acid 860
Methyl α-cyano-3,4-dimethyl-5-formyl-2-pyrroleacrylate 842

$C_{12}H_{12}N_2O_3S$
4-Aminodiphenylamine-2-sulfonic acid 788
3-Amino-4-hydroxybenzenesulfonanilide 811
4′-Amino-2′-hydroxybenzenesulfonanilide 811
4-(4-Hydrazinophenyl)-benzenesulfonic acid, sodium salt 794
6-Methyl-2-(phenylsulfonylmethyl)-4(3*H*)-pyrimidinone 976

$C_{12}H_{12}N_2O_4$
N,*N*′-Difurfuryloxamide 854
1-Ethyl-3-(4-methoxyphenyl)-parabanic acid 699
N-(4-Methyl-3-nitrophenyl)-glutarimide 718

$C_{12}H_{12}N_2O_4S$
3-Amino-4-hydroxyphenyl sulfone 774

$C_{12}H_{12}N_2O_5$
Methyl α-benzamido-β-nitromethacrylate 469

$C_{12}H_{12}N_2O_5S_2$
N-Sulfanilylsulfanilic acid 817

$C_{12}H_{12}N_2O_6S$
N[1]-Acetyl-*N*[4]-maleylsulfanilamide 823

$C_{12}H_{12}N_2O_6S_2$
2,2′-Benzidinedisulfonic acid 788

$C_{12}H_{12}N_2O_8$
Ethyl methyl 2,4-dinitrophenylmalonate 357

$C_{12}H_{12}N_2S$
6,7-Dihydro-3-phenyl-5*H*-thiazolo(3,2-a)pyrimidine hydrobromide 879
4,4′-Thiodianiline 230

$C_{12}H_{12}N_2S_2$
N,*N*′-Bis(2-thenylidene)-ethylenediamine 1054

$C_{12}H_{12}N_4$
2,4-Diaminoazobenzene hydrochloride 1087
4-(5-Indanylmethyleneamino)-4*H*-1,2,4-triazole 1068

$C_{12}H_{12}N_4O$
2-(2,4-Diaminophenylazo)-phenol 1083

$C_{12}H_{12}N_4OS$
1-(4-(Methylthio)phenyl)-3-(2-pyrimidinyl)urea 974

$C_{12}H_{12}N_4O_2$
4-Hydroxy-2-(2-hydroxybenzylidenehydrazino)-6-methylpyrimidine 1037
4-Hydroxy-2-(4-hydroxybenzylidenehydrazino)-6-methylpyrimidine 1037
1,1′-(1,5-Naphthalenediyl)diurea 604

$C_{12}H_{12}N_4O_2S$
4-(Phenylazo)benzenesulfonic acid hydrazide 1092

$C_{12}H_{12}N_4O_3S$
2-(4-(Phenylazo)phenyl)-1-hydrazinesulfonic acid, 4-methylaniline salt 1091
4-(Phenylazo)-phenylhydrazinosulfonic acid 1091

$C_{12}H_{12}N_4O_4$
4,6-Dihydroxy-1,3,5-triazine-2-acetic acid *o*-anisidide 978
1-(5-Methyl-3-isoxazolyl)-3-(2-methyl-6-nitrophenyl)urea 871

$C_{12}H_{12}N_4O_4S$
N-(4,6-Dimethyl-2-pyrimidinyl)-4-nitrobenzenesulfonamide 976

$C_{12}H_{12}N_4O_4S_2$
N-(2-Thiazolyl)oxamic *N*[2]-toluenesulfonylhydrazide 876

$C_{12}H_{12}N_4O_5$
Ethyl 2-(5-nitro-1-benzimidazolylcarbonylamino)-acetate 905
Ethyl 2-(6-nitro-1*H*-indazol-1-ylcarbonylamino)acetate 900

$C_{12}H_{12}N_4O_5S_2$
N-(2-Thiazolyl)oxamic *N*[2]-(4-methoxyphenylsulfonyl)hydrazide 875

$C_{12}H_{12}N_4S$
1-(α-(2-Thienyl)benzylideneamino)-guanidine nitrate 1071

$C_{12}H_{12}N_6O_4$
1,3,7,9-Tetramethylpyrimido(5,4-g)-pteridine-2,4,6,8-(1*H*,3*H*,7*H*,9*H*)-tetraone 1003

$C_{12}H_{12}O$
2-Ethoxynaphthalene 188
4-Methoxy-1-methylnaphthalene 188
4-Methyl-1-naphthalenemethanol 208
6-Phenyl-3,5-hexadien-2-one 285

$C_{12}H_{12}OS$
Ethyl 2-naphthyl sulfoxide 771
Methyl 1-naphthylmethyl sulfoxide 771

$C_{12}H_{12}O_2$
Allyl cinnamate 365
3-Benzylidene-2,4-pentanedione 308
2-Benzyloxy-2*H*-pyran 183
1,4-Dimethoxynaphthalene 188
1,5-Dimethoxynaphthalene 188
1,7-Dimethoxynaphthalene 188
Ethyl 3-methyl-1,4-benzodioxane-2-carboxylate 363
5-Indanacrylic acid 336
6-Methoxy-2-methylene-1-tetralone 302
2-(2-Naphthyloxy)ethanol 202

$C_{12}H_{12}O_2S$
Methyl 1-naphthylmethyl sulfone 786

$C_{12}H_{12}O_2S_2$
2,5-Dimethyl-3-(phenylsulfonyl)-thiophene 858

$C_{12}H_{12}O_3$
Allyl benzoylacetate 364
3,4-Dihydro-6-methoxy-1-naphthoic acid 341
7-Ethoxy-4-methylcoumarin 384
3-Oxo-2-phenyl-1-cyclopentanecarboxylic acid 331
4-Oxo-6-phenyl-5-hexenoic acid 344

$C_{12}H_{12}O_4$
5,8-Dimethoxy-2-methylchromone 319
Dimethyl benzylidenemalonate 365
Methyl 5,8-dioxo-1,4,4a,5,8,8a-hexahydro-1-naphthalenecarboxylate 92
Methyl 2,4-dioxo-5-phenylpentanoate 374
Methyl 2-(methoxycarbonyl)-cinnamate 365
Tricyclo[4.2.2.0[2,5]]deca-3,9-diene-7,8-dicarboxylic acid 65

$C_{12}H_{12}O_5$
4-Acetoxy-3-methoxycinnamic acid 377
4-Carboxy-4-chromanacetic acid 331
5,7-Dimethoxy-1-oxo-2-indancarboxylic acid 327
2-(2,3-Dimethoxystyryl)-glyoxylic acid 334
2-(3,4-Dimethoxystyryl)-glyoxylic acid 334
5-Hydroxy-7-methyl-4,6-indandicarboxylic acid 338

$C_{12}H_{12}O_6$
4-(Diacetoxymethyl)benzoic acid 356
Pyrogallol triacetate 378

$C_{12}H_{12}O_6S_2$
Piperonal bis(carboxymethyl) mercaptal 330

$C_{12}H_{12}O_7$
Trimethyl 2-hydroxy-1,3,5-benzenetricarboxylate 367

$C_{12}H_{12}O_9$
Dimethyl 3,4-diacetoxy-2,5-furandicarboxylate 846

$C_{12}H_{12}S$
Methyl 2-naphthylmethyl sulfide 213

$C_{12}H_{12}S_2$
2,5-Dimethyl-3-(phenylthio)-thiophene 856
2-Methyl-5-(*o*-tolylthio)thiophene 856

$C_{12}H_{12}Sn$
Diphenyltin 1132

$C_{12}H_{13}Br$
1-Bromo-4-methyl-1-phenyl-1,4-pentadiene 169

$C_{12}H_{13}BrN_2$
4-Amino-1-benzylpyridinium bromide 942

$C_{12}H_{13}BrO$
2-Bromo-4,4-dimethyl-1-tetralone 298

$C_{12}H_{13}BrO_2$
4-(5-Bromo-2-ethoxyphenyl)-3-buten-2-one 284
$C_{12}H_{13}BrO_5$
Dimethyl 2-bromo-2-(3-methoxyphenyl)malonate 355
$C_{12}H_{13}Br_2NO_4$
2,3-Dibromo-2′-ethoxysuccinanilic acid 416
2,3-Dibromo-4′-ethoxysuccinanilic acid 417
2,3-Dibromo-2′-methoxy-5′-methylsuccinanilic acid 416
$C_{12}H_{13}ClFNO_2$
Cyclopentyl *N*-(3-chloro-4-fluorophenyl)carbamate 631
$C_{12}H_{13}ClN_2OS$
2-Chloro-3-phenyl-2,3,6,7-tetrahydro-5*H*-thiazolo(3,2-a)pyrimidin-3-ol hydrochloride 735
$C_{12}H_{13}ClN_2O_2$
Butyl *N*-(4-chlorophenyl)-*N*-cyanocarbamate 752
7-Chloro-5-ethoxy-1-methyl-1*H*-1,4-benzodiazepin-2(3*H*)-one 690
Pyridinium-1-acetate compound with pyridine hydrochloride 956
$C_{12}H_{13}ClN_2O_3S$
1-Amino-2-chloropyridinium *p*-toluenesulfonate 955
$C_{12}H_{13}ClN_2O_5$
Tetrahydrofurfuryl *N*-(4-chloro-3-nitrophenyl)carbamate 625
$C_{12}H_{13}ClN_6O_2$
6-(2-(4-Chloro-2-methylphenoxy)-acetamido)-2,4-diamino-1,3,5-triazine 978
$C_{12}H_{13}ClO$
5-(3-Chloropropionyl)indan 288
$C_{12}H_{13}ClO_3$
Ethyl 3-(4-chlorobenzoyl)propionate. 353
$C_{12}H_{13}ClO_3S$
5-Chloromethyl-2-methyl-2-(3,4-methylenedioxyphenyl)-1,3-oxathiolane 209
$C_{12}H_{13}Cl_2NO_2$
Cyclopentyl *N*-(2,3-dichlorophenyl)-carbamate 631
$C_{12}H_{13}Cl_2NO_3$
Ethyl 3-(2,6-dichlorophenylimino)-3-methoxypropionate 1050
$C_{12}H_{13}Cl_2N_3O_2S$
2-(2-Thioxo-1-imidazolidinyl)ethyl *N*-(2,3-dichlorophenyl)carbamate 731
$C_{12}H_{13}Cl_3N_2O$
1-(*N*-(2,4,5-Trichlorophenyl)-carbamoyl)piperidine 711
$C_{12}H_{13}FN_4OS_2$
2-(1,3-Dimethylureido)-5-(3-fluorobenzylthio)-1,3,4-thiadiazole 880
$C_{12}H_{13}F_3N_2O_2$
1-Butyryl-3-(3-(trifluoromethyl)-phenyl)urea 509
4-(*N*-(3-(Trifluoromethyl)phenyl)-carbamoyl)morpholine 693
$C_{12}H_{13}F_3N_2O_3$
Ethyl 2-(3-(3-(trifluoromethyl)-phenyl)ureido)acetate 509
$C_{12}H_{13}F_3O_2$
1-(Trifluoromethyl)ethyl 2-phenylpropionate 353
$C_{12}H_{13}FeNO$
N-Methylferrocenecarboxamide 1142
$C_{12}H_{13}IN_2$
3,4-Dihydro-2-methyl-9*H*-pyrido(3,4-b)indolium iodide 886
$C_{12}H_{13}N$
1,2-Dihydro-2-(1-propenyl)quinoline 253
2,5-Dimethyl-1-phenylpyrrole 841
N-Ethyl-2-naphthylamine 233
N-(1-Indanyl)propargylamine 1018
4-Isopropylcinnamonitrile (mixture of *cis* and *trans*) 767
2-Propylquinoline 995
2,4,6-Trimethylquinoline sulfate monohydrate 995
$C_{12}H_{13}NO$
1-Acetyl-2,3-dimethylindole 890
3-Acetyl-2-ethylindole 888
3-Butyrylindole 889
4-Cyano-2-methyl-2-phenylbutyraldehyde 757
2,2-Dimethyl-4-oxo-4-phenylbutyronitrile 757
α,α-Dimethyl-3-quinolinemethanol 984
α,α-Dimethyl-4-quinolinemethanol 984
α,α-Dimethyl-6-quinolinemethanol perchlorate 984
6-Ethoxyquinaldine 987
2-Methyl-3-propionylindole 888
3-Methyl-1-propionylindolizine 888
2-(1-Naphthylamino)ethanol 227
4-Phenyltetrahydropyran-4-carbonitrile 756
$C_{12}H_{13}NOS$
2-Oxo-1-cyclopentanethiocarboxanilide 744
$C_{12}H_{13}NO_2$
6-Acetamido-1-tetralone 432
1-(Acetoacetyl)indoline 706
6,7-Dimethoxy-1-methylisoquinoline 997
6,7-Dimethoxy-4-methylquinoline 984
Ethyl 2-(4-cyanophenyl)propionate 767
Ethyl 5-methylindole-2-carboxylate 890
2,3,4,5,6,7-Hexahydro-1*H*-1-benzazonine-2,7-dione 706
3-Indolyl butyrate 888
Methyl 1-indolepropionate 887
Methyl 3-indolepropionate 887
2-Oxo-1-cyclopentanecarboxanilide. 444
1-(Phenylethyl)succinimide 719
N-(*p*-Tolyl)glutarimide 720
N-(2,6-Xylyl)succinimide 720
$C_{12}H_{13}NO_2S$
4-(Benzoylthioformyl)-morpholine 743
$C_{12}H_{13}NO_3$
2-Benzamido-3-methylcrotonic acid. 475
1-Benzyl-5-oxo-3-pyrrolidinecarboxylic acid 692
2′,3′-Dimethylmaleanilic acid 474
2′,4′-Dimethylmaleanilic acid 474
2′,6′-Dimethylmaleanilic acid 476
3′,4′-Dimethylmaleanilic acid 476
3′,5′-Dimethylmaleanilic acid 474
2′-Ethylmaleanilic acid 474
4′-Ethylmaleanilic acid 483
Ethyl 5-methoxy-2-indolecarboxylate 882
N-(4-Ethylphenyl)maleamic acid 474
N-(2-Methoxy-5-methylphenyl)-succinimide 717
N-(4-Methoxyphenyl)glutarimide 717
Methyl 2-(cyclopropylcarboxamido)benzoate 420
N-Phenethylmaleamic acid 471
$C_{12}H_{13}NO_3S$
3-Acetyl-2-phenyl-4-thiazolidinecarboxylic acid 343
5,6-Dimethoxybenzo(b)thiophene-3-acetamide 897
N,*N*-Dimethyl-8-hydroxynaphthalenesulfonamide 812
Ethyl 2-(2,3-dihydro-3-oxo-4*H*-1,4-benzothiazin-2-yl)acetate 709
$C_{12}H_{13}NO_4$
N-Allyl-2-(carboxymethoxy)-benzamide 464
1,1-Dimethyl-2-propen-1-yl 4-nitrobenzoate 374
2′-Ethoxymaleanilic acid 465
4′-Ethoxymaleanilic acid 466
Ethyl 2,3-dihydro-α-hydroxy-2-oxo-3-indoleacetate 881
2′-Methoxy-5′-methylmaleanilic acid 465
3-Methyl-2-propen-1-yl 4-nitrobenzoate 374
Phthalimidoacetaldehyde dimethyl acetal 717
N-Piperonylacetoacetamide 444
5,6,7-Trimethoxyisocarbostyril 996
$C_{12}H_{13}NO_4S$
1-Amino-2-ethoxynaphthalene-6-sulfonic acid, sodium salt 792
$C_{12}H_{13}NO_5$
2′,4′-Dimethoxymaleanilic acid 463
2′,5′-Dimethoxymaleanilic acid 463
3′,5′-Dimethoxymaleanilic acid 463
trans-2-Hydroxycyclopentyl 4-nitrobenzoate 368
2-(4-Nitrophenoxy)ethyl methacrylate 369
$C_{12}H_{13}NO_6$
Carbobenzyloxy-DL-aspartic acid 659
N-(2-Hydroxybenzoyl)glutamic acid 461
2-Methoxyethyl 4-nitrobenzoylacetate 363
5-(5-Methyl-5-nitro-1,3-dioxan-2-yl)-1,3-benzodioxole 277
$C_{12}H_{13}NO_7$
3-(3,4-Dimethoxy-6-nitrobenzoyl)-propionic acid 327
$C_{12}H_{13}NO_9$
4-Nitrophenyl-β-D-glucuronide 326
$C_{12}H_{13}N_2O_3P$
Diphenyl aminophosphoramidate 836
$C_{12}H_{13}N_3$
2,4-Diaminodiphenylamine 230
2,2′-Iminodi-3-picoline 954
7-Imino-3,4,6,7-tetrahydro-2*H*-pyrimido(2,1-a)isoquinoline 725
3,3′-(Phenylimino)dipropionitrile 755
$C_{12}H_{13}N_3O$
α-Cyano-4-(dimethylamino)-cinnamamide 764
2,10-Dihydro-10-ethylimidazo(2,1-b)-quinazoline-5(3*H*)-one 726
4-(2-Keto-1-benzimidazolinyl)-1,2,3,6-tetrahydropyridine 904
2-(4-Methoxybenzylamino)-pyrimidine 973
$C_{12}H_{13}N_3OS$
8-Acetamido-5-amino-6-(methylthio)quinoline 987
1-(2-Thiazolyl)-3-(3,4-xylyl)urea 877
$C_{12}H_{13}N_3O_2$
2,4-Dicyano-3,3-pentamethyleneglutarimide 146
5-(4-(Dimethylamino)benzylidene)-hydantoin 867
N-(1,3-Dimethyl-5-pyrazolyl)-anthranilic acid 861
Ethyl 5-amino-1-phenylpyrazole-4-carboxylate 861
2-Ethyl-5-oxo-1-phenyl-3-pyrazoline-3-carboxamide 707
4-Formylaminoantipyrine 707
Oxamic acid (5-indanylmethylene)-hydrazide 499
$C_{12}H_{13}N_3O_3$
1-Acetyl-3-(hydroxyimino)-*N*-methyl-2-indolinecarboxamide 1012
1-(*N*-(2,4-Dimethoxyphenyl)-carbamoyl)pyrazole 860
1-(2,2-Dimethylpropionyl)-5-nitrobenzimidazole 904
Ethyl 2-(2-cyanophenylureido)-acetate 767
Ethyl *N*-(1-phenyl-5-oxo-2-pyrazolin-3-yl)carbamate 702
2-(1-Methyl-5-nitro-2-imidazolyl)-1-phenyl-1-ethanol 865
$C_{12}H_{13}N_3O_3S$
5-Amino-2-(4-aminoanilino)-benzenesulfonic acid 788
$C_{12}H_{13}N_3O_4S$
1-(Mesitylenesulfonyl)-4-nitroimidazole 868
$C_{12}H_{13}N_3O_5$
Ethyl 2,3-dioxobutyrate 2-(4-nitrophenylhydrazone) 373
Methyl methyl-octahydro-3,5,8-trioxo-2a,4,8a-triazaacenaphthylene-6-carboxylate 134
$C_{12}H_{13}N_3Pb$
Tris(1-pyrrolyl)lead 1135
$C_{12}H_{13}N_3S$
3-Imino-2-phenyl-2,3,6,7-tetrahydro-5*H*-thiazolo(3,2-a)pyrimidine hydrochloride 749
$C_{12}H_{14}$
4-Methyl-1-phenyl-1,3-pentadiene 164
(4-(2-Methylpropyl)phenyl)acetylene. 1019
3-Phenyl-1-cyclohexene 164
6-Phenyl-2-hexyne 1019
$C_{12}H_{14}AsN_3$
3,3′-(4-Aminophenylarsino)-dipropionitrile 1127
$C_{12}H_{14}BF_4N$
1-Ethyl-2-methylquinolinium tetrafluoroborate 990
$C_{12}H_{14}BrNO$
2-(4-Bromobenzyl)-4,4-dimethyl-2-oxazoline 244
$C_{12}H_{14}BrNO_2$
2-Bromo-*N*-(3-formyl-1-methylpropyl)benzamide 453
$C_{12}H_{14}BrNO_3$
4′-Bromo-2′,6′-dimethylsuccinanilic acid 407
4′-Bromo-2′-methylglutaranilic acid. 407
4′-Bromo-3′-methylglutaranilic acid. 402
2-Bromopropionyl-DL-phenylalanine 429
3-(2-Carboxyethyl)-2,5-dimethylbenzoxazolium bromide 910
$C_{12}H_{14}Br_2N_2$
1,1′-Ethylenebis(pyridinium bromide) 944
$C_{12}H_{14}Br_2O_2$
Isopropyl α,β-dibromohydrocinnamate 362
$C_{12}H_{14}ClFO_2$
2-(3-Chloropropyl)-2-(4-fluorophenyl)-1,3-dioxolane 183
$C_{12}H_{14}ClF_3N_2O_2$
2-Dimethylaminoethyl *N*-(2-chloro-5-trifluoromethylphenyl)carbamate. 606
$C_{12}H_{14}ClN$
N-Benzyl-*N*-(2-chloroethyl)-propargylamine hydrochloride 1015
6-Chloro-1,2-dihydro-2,2,4-trimethylquinoline 253
5-Chloro-2-(*o*-tolyl)-valeronitrile 751
$C_{12}H_{14}ClNO$
4-(4-Chlorophenyl)-4-ethoxybutyronitrile 752
$C_{12}H_{14}ClNO_2$
Ethyl 3-(3-chloroanilino)crotonate 366
Ethyl 3-(4-chloroanilino)crotonate 366
$C_{12}H_{14}ClNO_2S$
4-(2-Chloro-5-methoxythiobenzoyl)-morpholine 738
$C_{12}H_{14}ClNO_3$
4′-Chloro-2′,5′-dimethoxy-2-butenanilide 448
5′-Chloro-2′,4′-dimethoxy-2-butenanilide 448
N-(3-Chloro-2-methylphenyl)-glutaramic acid 407
N-(3-Chloro-4-methylphenyl)-glutaramic acid 407
Tetrahydrofurfuryl *N*-(3-chlorophenyl)carbamate 622
$C_{12}H_{14}ClNO_4$
4′-Chloro-2′,5′-dimethoxyacetoacetanilide 442
2-Chloroethyl lactate carbanilate 604
5-Chloropentyl 4-nitrobenzoate 365
1-(Methoxycarbonyl)-1-methylethyl *N*-(2-chlorophenyl)carbamate 630
$C_{12}H_{14}ClNO_4S$
4-Chloro-1,2-dimethylquinolinium methyl sulfate 991
$C_{12}H_{14}ClNS$
1-((2-Chlorophenyl)thiocarbamoyl)-piperidine 746
$C_{12}H_{14}ClN_3O_2$
2-Chlorocyclohexanone 4-nitrophenylhydrazone 276
$C_{12}H_{14}ClN_3O_4$
1-(3-Chlorobenzoyl)-4-(ethoxycarbonylmethyl)-semicarbazide 531
4,5-Dihydro-2-imidazolyl *N*-(5-chloro-2,4-dimethoxyphenyl)-carbamate 613

$C_{12}H_{14}Cl_2N_2O$
1-Cyclopentyl-3-(2,6-dichlorophenyl)urea 535
1-(*N*-(2,3-Dichlorophenyl)-carbamoyl)piperidine 711
1-(*N*-(2,4-Dichlorophenyl)-carbamoyl)piperidine 711
1-(*N*-(2,5-Dichlorophenyl)-carbamoyl)piperidine 711
1-(*N*-(2,6-Dichlorophenyl)-carbamoyl)piperidine 711
1-(*N*-(3,4-Dichlorophenyl)-carbamoyl)piperidine 711

$C_{12}H_{14}Cl_2N_2O_2$
1-Butyl-3-(3,4-dichlorophenyl)-1-formylurea 534
1-Butyryl-3-(3,4-dichlorophenyl)-1-methylurea 534
2,6-Dichloro-3-(morpholinomethyl)-benzamide 452
N,N'-(4,5-Dichloro-1,2-phenylene)-bispropionamide 404

$C_{12}H_{14}Cl_2N_2O_3$
Butyl *N*-(*N*-(3,4-dichlorophenyl)-carbamoyl)carbamate 630
Ethyl *N*-(3,4-dichlorophenyl)-*N*-(*N,N*-dimethylcarbamoyl)carbamate 630
Methyl *N*-(*N*-(3,4-dichlorophenyl)-carbamoyl)-*N*-propylcarbamate 630
Propyl *N*-(*N*-(3,4-dichlorophenyl)-carbamoyl)-*N*-methylcarbamate 630
2-(*p*-Tolylureido)ethyl 2,2-dichloroacetate 504

$C_{12}H_{14}Cl_3NO$
N-Benzyl-*N*-isopropyl-2,2,2-trichloroacetamide 397
4'-Butyl-2,2,2-trichloroacetanilide 398

$C_{12}H_{14}Cl_3NO_3$
Ethyl *N*-(1-benzyloxy-2,2,2-trichloroethyl)carbamate 608

$C_{12}H_{14}FNO_3S$
3-(Piperidinocarbonyl)-benzenesulfonyl fluoride 804

$C_{12}H_{14}F_2N_2O$
1-(*N*-(2,5-Difluorophenyl)-carbamoyl)piperidine 535

$C_{12}H_{14}F_3NO$
2,2-Dimethyl-3'-(trifluoromethyl)-propionanilide 397
4-(3-(Trifluoromethyl)phenyl)-3-piperidinol hydrochloride 244
4-(α,α,α-Trifluoro-*m*-tolyl)-4-piperidinol 244

$C_{12}H_{14}F_3NO_2$
Butyl *N*-(3-trifluoromethylphenyl)-carbamate 610

$C_{12}H_{14}F_3NO_3$
2-Ethoxyethyl *N*-(3-trifluoromethylphenyl)carbamate 608

$C_{12}H_{14}F_3N_3O_4$
2,6-Dinitro-*N*-pentyl-4-(trifluoromethyl)aniline 259

$C_{12}H_{14}INO$
1,2-Dimethyl-4-methoxyquinolinium iodide 985

$C_{12}H_{14}N_2$
6-Dimethylaminoquinaldine 989
2,3,6,7-Tetramethylquinoxaline 1001

$C_{12}H_{14}N_2O$
6-Amino-4,5-dimethyl-8-methoxyquinoline 985
1-(2-(Benzyloxy)ethyl)imidazole 866
4-*tert*-Butyl-1(2*H*)-phthalazinone 1001
2'-Cyano-2,2-dimethylpropionanilide 763
3-Cyano-*N*-phenethylpropionamide 755
8-(Ethylamino)-6-methoxyquinoline 985
1-Ethyl-2-methylbenzimidazol-5-yl methyl ketone 906
2-(4-Methoxystyryl)-2-imidazoline 724
2-(Morpholinomethyl)benzonitrile 764
α-Phenyl-4-morpholineacetonitrile 754
4,7,8-Trimethyl-1*H*-1,5-benzodiazepin-2-(3*H*)-one 702

$C_{12}H_{14}N_2OS$
1-(3-Acetylphenyl)-3-allyl-2-thiourea 744
5-Benzyl-5-ethyl-2-imino-4-thiazolidinone 744

$C_{12}H_{14}N_2OS_2$
3-Morpholinomethyl-2(3*H*)-benzothiazolethione 742

$C_{12}H_{14}N_2O_2$
3'-Acetamidocyclopropane-carboxanilide monohydrate 426
Butyl *N*-cyano-*N*-phenylcarbamate 757
3-Butyl-3,4-dihydro-2*H*-pyrido-(1,2-a)pyrimidine-2,4-dione 700
2-Butyl-1,4(2*H*,3*H*)-phthalazinedione 502
1-Isopropyl-7-methyl-2,4(1*H*,3*H*)-quinazolinedione 707
4-Methoxybenzoic (2-buten-1-ylidene)hydrazide 1048
DL-6-Methyltryptophan 884
Salicylic (cyclopentylidene)-hydrazide 1036
DL-Tryptophan methyl ester 884
DL-Tryptophan methyl ester hydrochloride 886

$C_{12}H_{14}N_2O_2S$
Methyl 3-amino-2-(*N*-phenylthiocarbamoyl)-2-butenoate 739

$C_{12}H_{14}N_2O_2S_2$
1-(4-Isothiocyanatophenylsulfonyl)-piperidine 827
2-(5-Methyl-1,3-dithian-2-ylidenehydrazino)benzoic acid 346

$C_{12}H_{14}N_2O_3$
5,8-Dimethoxy-2-ethyl-3-hydroxy-quinoxaline 998
Ethyl 2-(phenylazo)acetoacetate 1092
4-Hydroxy-2-(3-hydroxybutyl)-1(2*H*)-phthalazinone 696
O-(Morpholinocarbonyl)-benzaldehyde oxime 1011
2-Nitro-1-nitroso-1-phenylcyclohexane 258
2-(3-(2-Phenylcyclopropyl)ureido)-acetic acid 570

$C_{12}H_{14}N_2O_3S$
1-Aminopyridinium *p*-toluenesulfonate 955
1,1-Diethyl-3-(methylsulfonyl)-3-phenylurea 825
2-(3,4-Dimethoxyphenyl)-5-imino-3-methyl-4-oxazolidinethione 735
4'-Sulfamoyl-2,4-hexadienanilide 477

$C_{12}H_{14}N_2O_4$
4'-Acetamidosuccinanilic acid 428
(*N*-Benzoyl)glycylglycine methyl ester 434
2'-Carbamoylglutaranilic acid 474
6,7-Diethoxy-2,4(1*H*,3*H*)-quinazolinedione 999
Nα-(2-Phenylacetyl)-L-asparagine 434
3-Nitro-4-piperidinobenzoic acid 342

$C_{12}H_{14}N_2O_5$
N-(4-Aminobenzoyl)-glutamic acid, zinc salt 468
Carbobenzyloxy-DL-asparagine 657
Dimethyl 5-(3-methylureido)-isophthalate 576
N-(4-Methyl-3-nitrophenyl)-glutaramic acid 426

$C_{12}H_{14}N_2O_6$
Methyl 4-*tert*-butyl-3,5-dinitrobenzoate 372
2-Nitrobenzyl *N*-(ethoxycarbonylmethyl)-carbamate 659

$C_{12}H_{14}N_2O_6S$
N[1]-Acetyl-*N*[4]-succinoylsulfanilamide 824

$C_{12}H_{14}N_2S$
2-Piperidinobenzothiazole 917

$C_{12}H_{14}N_4$
4-(4-Isopropylbenzylideneamino)-4*H*-1,2,4-triazole 1072

$C_{12}H_{14}N_4O_2$
1-(2-Cyanoacetyl)-4-(3,4-xylyl)-semicarbazide 756

$C_{12}H_{14}N_4O_2S$
1-(Cyanoacetyl)-4-(2-methoxy-5-methylphenyl)-3-thiosemicarbazide 756

$C_{12}H_{14}N_4O_3$
Ethyl 2-(3-(1*H*-indazol-5-yl)ureido)-acetate 900
4-(3,4,5-Trimethoxybenzylidene-amino)-4*H*-1,2,4-triazole 1068

$C_{12}H_{14}N_4O_4$
7-Amino-1,3-diethyl-6-nitro-2,4-(1*H*,3*H*)-quinazolinedione 701

$C_{12}H_{14}N_4O_6$
N-Cyclohexylpicrylamine 273
N,N',N'',N'''-Tetraacetylglycoluril 135

$C_{12}H_{14}N_4O_6S$
N-Butyl-*N*-cyano-2,6-dinitro-4-(methylsulfonyl)aniline 777

$C_{12}H_{14}N_4O_6S_2$
4,4'-Dihydrazino-2,2'-biphenyldisulfonic acid 789

$C_{12}H_{14}N_4O_6S_3$
3,3'-Sulfonyldibenzenesulfonic acid dihydrazide 827

$C_{12}H_{14}O$
6-Acetyltetralin 298
Cyclopropyl 2,4-dimethylphenyl ketone 296
Cyclopropyl 2,5-dimethylphenyl ketone 296
Cyclopropyl 3,4-dimethylphenyl ketone 296
Cyclopropyl 4-ethylphenyl ketone 296
4,4-Dimethyl-1-tetralone 298
6,7-Dimethyl-1-tetralone 298
4-Methyl-1-phenyl-1-penten-3-one 285
6-Phenyl-5-hexyn-3-ol 1016
1-(*p*-Tolyl)-1-penten-3-one 286
2,2,4-Trimethyl-2*H*-1-benzopyran 185

$C_{12}H_{14}OS$
3-(Phenylthio)-1-cyclohexanone 284
2-(*p*-Tolylthio)cyclopentanone 284

$C_{12}H_{14}O_2$
2,3-Benzo-1,4-dioxaspiro(4.6)undec-2-ene 184
trans-1-Carbethoxy-2-phenylcyclopropane 361
Cyclopentyl benzoate 375
1,3-Diacetyl-5-ethylbenzene 308
3,4,6,7,8,9-Hexahydro-1(2*H*)-dibenzofuranone 853
6-Hydroxy-5-isopropenyl-3-methyl-2,3-dihydrobenzofuran 191
Isopropyl cinnamate 365
4-Isopropylcinnamic acid 336
5-Methoxy-1-methyl-2-tetralone 284
6-Methoxy-2-methyl-1-tetralone 293
6-Methoxy-5-methyl-1-tetralone 293
1-(4-Methoxyphenyl)-1-penten-3-one 283
trans-4-Methyl-2-(2-phenylvinyl)-1,3-dioxolane 184
2-Phenyl-1-cyclopentanecarboxylic acid 332

$C_{12}H_{14}O_2S$
β,5-Dimethyl-2-(methylthio)-cinnamic acid 335

$C_{12}H_{14}O_3$
4-(4-Acetylphenyl)butyric acid 331
5-Benzoylvaleric acid 331
Benzyl levulinate 356
Bicyclo(4.2.2)dec-7-ene-9,10-dicarboxylic anhydride 98
3-(*tert*-Butyl)-3-hydroxyphthalide 382
4-(3,4-Dimethoxyphenyl)-3-buten-2-one 283
3-(2,4-Dimethylbenzoyl)-propionic acid 331
3-(2,5-Dimethylbenzoyl)-propionic acid 331
Ethyl 4-methoxycinnamate 365
Ethyl 3-methyl-3-phenylglycidate 357
Ethyl 4-oxo-4-phenylbutyrate 360
Eugenyl acetate 378
1-(2-Hydroxy-5-methylphenyl)-1,3-pentanedione 307
Isoeugenyl acetate 378
α-Methylbenzyl 3-oxobutyrate 363
1-Methylpropyl benzoylformate 374
4-Phenyltetrahydropyran-4-carboxylic acid 332

$C_{12}H_{14}O_3S$
Ethyl 3-oxo-4-(phenylthio)butyrate 364

$C_{12}H_{14}O_4$
sec-Butyl phthalate 375
2,5-Dimethylhydroquinone diacetate 379
2,3-Dimethyl-2-phenylsuccinic acid 329
4-Ethoxy-3-methoxycinnamic acid 335
Ethyl 4-(allyloxy)salicylate 367
Ethyl *m*-anisoylacetate 363
1,3,3a,6,7,7a-Hexahydro-1-oxo-4,5,6-trimethyl-7-isobenzofuran-carboxylic acid 96
3-Methyl-3-phenylglutaric acid 332
1,3-Phenylenedipropionic acid 332

$C_{12}H_{14}O_5$
4'-Acetoxy-3',5'-dimethoxyacetophenone 377
Diethyl furfurylidenemalonate 853
3-(2,5-Dimethoxybenzoyl)-propionic acid 328
3-(3,4-Dimethoxybenzoyl)-propionic acid 327
Dimethyl 2-(3-methoxyphenyl)-malonate 355
3-(2-Hydroxy-*p*-tolyl)-glutaric acid 326
2,3,4-Trimethoxycinnamic acid 334

$C_{12}H_{14}O_5S$
Ethyl 2-(methylsulfonyl)-benzoylacetate 780
3-(3,4,5-Trimethoxyphenyl)-2-thiopyruvic acid 327

$C_{12}H_{14}O_6$
4'-Acetoxy-3',5'-dimethoxy-2-hydroxyacetophenone 377
2-Acetyl-3,4-bis-(acetoxymethyl)-furan 846
1-*O*-Benzoyl-D-ribose 367
3,5-Dimethoxy-4-(ethoxycarbonyl)-benzoic acid 369
2-Ethyl 3,4-dimethoxyphthalate 369
3-(3,4,5-Trimethoxyphenyl)-pyruvic acid 328

$C_{12}H_{14}O_6S$
(4-(Methylsulfonyl)phenyl)-methylene diacetate 781

$C_{12}H_{14}O_6Si$
Phenyltriacetoxysilane 1109

$C_{12}H_{14}S$
2-Cyclohexen-1-yl phenyl sulfide 211

$C_{12}H_{15}Br$
(4-Bromophenyl)-cyclohexane 169

$C_{12}H_{15}BrN_2O$
1-((4-Bromophenyl)carbamoyl)-piperidine 584
1-(3-Bromophenyl)-3-cyclopentylurea 534

$C_{12}H_{15}BrN_2O_2$
Piperidino *N*-(4-bromophenyl)-carbamate 641

$C_{12}H_{15}BrN_2O_3$
4'-Acetamido-3-bromo-3'-methoxypropionanilide 391

$C_{12}H_{15}BrN_2O_6$
5-Bromo-2',3'-isopropylideneuridine 970

$C_{12}H_{15}BrO$
4'-(Bromomethyl)-2,2-dimethylpropiophenone 287

$C_{12}H_{15}BrO_4$
4-(4-Bromo-2,5-dimethoxyphenyl)-butyric acid 325

$C_{12}H_{15}Cl$
1-Chloro-3-phenylcyclohexane 173

$C_{12}H_{15}ClIN_3O_2$
5-Chloro-1,3-diethyl-2-methyl-6-nitrobenzimidazolium iodide 904

$C_{12}H_{15}ClN_2O$
1-(*N*-(4-Chloro-2-methylphenyl)-carbamoyl)pyrrolidine 695
1-(2-Chlorophenylcarbamoyl)-piperidine 584
1-(3-Chlorophenylcarbamoyl)-piperidine 584
1-(4-Chlorophenylcarbamoyl)-piperidine 584
1-(2-Chlorophenyl)-3-cyclopentylurea 535
1-(3-Chlorophenyl)-3-cyclopentylurea 534

$C_{12}H_{15}ClN_2O$ (Cont'd)
4′-Chloro-1-pyrrolidineacetanilide hydrochloride 431
$C_{12}H_{15}ClN_2OS$
2-(2-Chlorophenyl)-*N*-ethyl-3-oxazolidinethiocarboxamide 729
$C_{12}H_{15}ClN_2O_2$
1-Acetyl-3-(4-chlorophenyl)-1-propylurea 533
1-Butyl-3-(4-chlorophenyl)-1-formylurea 533
1-Butyryl-3-(4-chlorophenyl)-1-methylurea 533
4-(*N*-(3-Chloro-4-methylphenyl)carbamoyl)morpholine 713
4-(*N*-(4-Chloro-2-methylphenyl)carbamoyl)morpholine 713
4-(*N*-(5-Chloro-2-methylphenyl)carbamoyl)morpholine 706
4-(*N*-(6-Chloro-2-methylphenyl)carbamoyl)morpholine 713
$C_{12}H_{15}ClN_2O_3$
4-(*N*-(3-Chloro-2-methoxyphenyl)carbamoyl)morpholine 699
$C_{12}H_{15}ClN_2S$
3-Allyl-1-(3-chlorobenzyl)-1-methyl-2-thiourea 732
$C_{12}H_{15}ClN_4O_3$
1-(2-Chloro-4-nitrophenyl)-3-piperidinourea 526
$C_{12}H_{15}ClO$
4-Chloro-2-cyclohexylphenol 191
4-Chloro-3′,4′-dimethylbutyrophenone 288
4-Chloro-4′-ethylbutyrophenone 287
4′-Chlorohexanophenone 289
1-(4-Chlorophenyl)-cyclohexanol 206
$C_{12}H_{15}ClO_2$
5-Chloropentyl benzoate 366
$C_{12}H_{15}ClO_2S$
trans-2-Chlorocyclopentyl *p*-tolyl sulfone 782
2-Chloro-1-hexen-1-yl phenyl sulfone 783
$C_{12}H_{15}ClO_3$
4-Chloro-2′,4′-dimethoxybutyrophenone 287
4-(2-(2-Chloromethyl-1,3-dioxolan-2-yl)ethyl)phenol 190
2-Chloro-5-(2,4,5-trimethyl-1,3-dioxolan-2-yl)phenol 190
$C_{12}H_{15}ClO_4$
4-(2-(2-Chloroethoxy)-ethoxy)-3-methoxybenzaldehyde 321
$C_{12}H_{15}Cl_2NO$
2,2-Dichloro-5′-isopropyl-2′-methylacetanilide 394
$C_{12}H_{15}Cl_2NO_2$
1-Dimethylamino-2-propyl 2,4-dichlorobenzoate hydrochloride 366
$C_{12}H_{15}Cl_2NO_3$
4,4′-Dichloro-2′,5′-dimethoxybutyranilide 390
4,5′-Dichloro-2′,4′-dimethoxybutyranilide 390
$C_{12}H_{15}Cl_2NO_4S$
Butyl *N*-(3,4-dichlorophenyl)-*N*-(methylsulfonyl)carbamate 808
$C_{12}H_{15}Cl_2NO_5$
(γ-Chloro-4-methoxycinnamylidene)-dimethylammonium perchlorate 1044
$C_{12}H_{15}Cl_2N_3$
3-(3,6-Dichloro-4-pyridazinyl)-3-azabicyclo(3.2.2)nonane 968
$C_{12}H_{15}Cl_2N_5O_3$
2-(3,4-Dichlorophenylureido)acetic acid *N*²-(ethylcarbamoyl)hydrazide 528
$C_{12}H_{15}Cl_3N_2O_2$
Ethyl *N*-(1-(benzylamino)-2,2,2-trichloroethyl)carbamate 610
$C_{12}H_{15}FN_2O$
1-Cyclopentyl-3-(2-fluorophenyl)urea 535
$C_{12}H_{15}FN_2O_2$
1-(4-Fluorophenyl)-3-valerylurea 533
$C_{12}H_{15}FN_4O_3$
1-(*N*-(2-Fluoro-5-nitrophenyl)carbamoyl)-4-methylpiperazine 694
$C_{12}H_{15}FO$
4-Fluoro-α-methyl-α-(1-methylcyclopropyl)-benzyl alcohol 205
1-(4-Fluorophenyl)cyclohexanol 206
3-(4-Fluorophenyl)-4-hexanone 287
$C_{12}H_{15}F_3N_2O$
1-Butyl-3-(3-(trifluoromethyl)phenyl)urea 510
1-*tert*-Butyl-3-(2-(trifluoromethyl)phenyl)urea 511
1-*tert*-Butyl-3-(3-(trifluoromethyl)phenyl)urea 510
1,1-Diethyl-3-(3-(trifluoromethyl)phenyl)urea 510
1-Isobutyl-3-(2-(trifluoromethyl)phenyl)urea 511
1-Isobutyl-3-(3-(trifluoromethyl)phenyl)urea 510
$C_{12}H_{15}F_3N_2O_2$
2-Dimethylaminoethyl *N*-(2-(trifluoromethyl)phenyl)carbamate 610
2-Dimethylaminoethyl *N*-(3-(trifluoromethyl)phenyl)carbamate 610
$C_{12}H_{15}F_5Si$
Butyldimethyl(pentafluorophenyl)silane 1100
$C_{12}H_{15}IN_2O_2S$
2-Amino-6-ethoxycarbonylmethyl-3-methylbenzothiazolium iodide 915
$C_{12}H_{15}IO_2S$
2-Iodo-1-hexen-1-yl phenyl sulfone 783
$C_{12}H_{15}N$
α-(*tert*-Butyl)phenylacetonitrile 756
N,N-Diallylaniline 220
2-Ethyl-2-phenylbutyronitrile 756
3-Methyl-2-phenylvaleronitrile 756
4-Methyl-2-phenylvaleronitrile 756
2-Methyl-3-propylindole 891
2-Phenylhexanenitrile 756
2,3,6,7-Tetrahydro-1*H*,5*H*-benzo(ij)quinolizine 253
4-(*p*-Tolyl)-1,2,3,6-tetrahydropyridine hydrochloride 250
$C_{12}H_{15}NO$
1-Acetamido-5,6,7,8-tetrahydronaphthalene 441
2-Acetamido-5,6,7,8-tetrahydronaphthalene 441
6-Acetyltetralin oxime 1013
2-Allyl-2-benzylacetamide 431
1-Benzoylpiperidine 707
1-Benzyl-3-piperidone 285
2′,6′-Crotonoxylidide 481
7,8-Dihydro-3,7,7-trimethyl-5(6*H*)-quinolinone hydrochloride 961
2-(Dimethylaminomethyl)-1-indanone hydrochloride 295
6-(Dimethylamino)-1-tetralone 295
2′,3′-Dimethyl-2-butenanilide 481
2′,4′-Dimethyl-1-cyclopropanecarboxanilide 437
2′,6′-Dimethylcyclopropane-carboxanilide 437
4-Ethoxy-4-phenylbutyronitrile 752
4′-Ethyl-2-butenanilide 480
N-Ethylcrotonanilide 482
2′-Ethyl-1-cyclopropanecarboxanilide 437
4′-Ethyl-1-cyclopropanecarboxanilide 437
3-Ethyl-6,7-dihydro-2-methyl-5-methylene-4(5*H*)indolone 58
N-Ethylmethacrylanilide 482
1-Ethyl-3-methyl-4-phenyl-2-azetidinone 692
3-Ethyl-1-methyl-4-phenyl-2-azetidinone 693
2′,6′-Methacryloxylidide 481
4-Methoxy-4-(*p*-tolyl)-butyronitrile 753
N-Methyl-*o*-methacrylotoluidide 482
N-Methyl-*p*-methacrylotoluidide 482
5-Methyl-5-phenyl-2-piperidone 692
2-Phenylcyclohexanone oxime 1010
1-Phenyl-1-cyclopentanecarboxamide 445
4-Phenylisonipecotaldehyde 324
4-(*p*-Tolylimino)-2-pentanone 1074
$C_{12}H_{15}NOS$
3-(2-Aminopropyl)-4-methoxybenzo(b)thiophene hydrochloride 897
N-*cis*-(2-Hydroxycyclopentyl)-thiobenzamide 734
$C_{12}H_{15}NOSe$
4,4-Dimethyl-2-(2-(methylseleno)-phenyl)-2-oxazoline 214
$C_{12}H_{15}NO_2$
2′-Acetamidobutyrophenone 432
2′,3′-Acetoacetoxylidide 444
2′,5′-Acetoacetoxylidide 444
2′,6′-Acetoacetoxylidide 444
3′,4′-Acetoacetoxylidide 444
4′-Acetylbutyranilide 432
Allyl *N*-(2,3-xylyl)carbamate 684
Allyl *N*-(2,4-xylyl)carbamate 686
Allyl *N*-(2,5-xylyl)carbamate 686
Allyl *N*-(2,6-xylyl)carbamate 684
Allyl *N*-(3,4-xylyl)carbamate 686
Allyl *N*-(3,5-xylyl)carbamate 686
β-Amino-5-indanpropionic acid 330
N-Benzyl-*N*-methylacetoacetamide 443
N-Benzylproline hydrochloride 343
2-(Butylcarbamoyl)benzoic acid 476
Cyclopentyl *N*-phenylcarbamate 678
3-(Dimethylamino)-2-(4-methoxyphenyl)acrolein 321
N-Ethylacetoacetanilide 444
Ethyl 3-anilinocrotonate 373
(4*S*,5*S*)-(-)-2-Ethyl-5-phenyl-2-oxazoline-4-methanol 245
2′-Methoxy-5′-methyl-2-butenanilide 465
Methyl 4-(dimethylamino)-cinnamate 364
1-Nitro-2-phenylcyclohexane 279
Phenyl 1-piperidinecarboxylate 709
N-(Tetrahydropyran-2-yl)benzamide 477
$C_{12}H_{15}NO_2S$
4-(4-Methoxythiobenzoyl)-morpholine 738
$C_{12}H_{15}NO_3$
o-Acetoacetophenetidide 442
N-Acetyl-L-phenylalanine methyl ester 429
N-Allyl-2′-hydroxy-4′-methoxyacetanilide 411
Allyl *N*-(2-methoxy-5-methylphenyl)-carbamate 653
5-Benzamidovaleric acid 475
N-Benzoyl-D-valine 349
N-Benzylglutaramic acid 431
N-*tert*-Butylphthalamic acid 474
4-Cyclohexyl-2-nitrophenol 265
N,N-Diethylphthalamic acid 471
3-(2,2-Dimethylpropionamido)-benzoic acid 433
2′,4′-Dimethylsuccinanilic acid 435
2′,5′-Dimethylsuccinanilic acid 435
3′,4′-Dimethylsuccinanilic acid 435
4′-Ethoxycarbonylpropionanilide 434
4′-Ethylsuccinanilic acid 435
2-Formylphenyl *N*-butylcarbamate 674
N-(4-Methoxybenzyl)-acetoacetamide 442
Methyl 2-butyramidobenzoate 434
2′-Methylglutaranilic acid 435
3′-Methylglutaranilic acid 434
4′-Methylglutaranilic acid 435
DL-*N*-(Phenylacetyl)-alanine methyl ester 429
N-(Phenylethyl)succinamic acid 429
3-(3,4,5-Trimethoxyphenyl)-propionitrile 752
3-(*N*-(2,6-Xylyl)carbamoyl)-propionic acid 436
$C_{12}H_{15}NO_3S$
N-Acetyl-*S*-benzyl-L-cysteine 421
N-Allyl-2-(*p*-tolylsulfonyl)-acetamide 779
N-Benzoyl-DL-methionine 468
1-Cyano-2,2-dimethylpropyl benzenesulfonate 799
4-(4-Hydroxy-3-methoxyphenyl-thiocarbonyl)morpholine 733
$C_{12}H_{15}NO_3S_2$
S-(Ethoxycarbonylmethyl)-1,1-dithio-4′-methyloxanilate 436
$C_{12}H_{15}NO_4$
3-Acetoxy-1-phenylpropyl carbamate 666
N-Acetyl-*O*-methyl-L-tyrosine 416
N-Acetyl-DL-tyrosine methyl ester 412
Allyl *N*-(2,5-dimethoxyphenyl)-carbamate 653
3-Butyn-1-yl *N*-(2,5-dimethoxyphenyl)carbamate 1017
Carbobenzyloxy-D-alanine methyl ester 663
Carbobenzyloxy-DL-alanine methyl ester 663
Carbobenzyloxy-D-2-aminobutyric acid 660
Carbobenzyloxy-L-2-aminobutyric acid 661
3′,5′-Dimethoxyacetoacetanilide 442
3-(4-Ethoxyphenylcarbamoyl)-propionic acid 418
2′-Ethoxysuccinanilic acid 416
Ethyl 4-nitro-3-phenylbutyrate 358
D-Glutamic acid γ-benzyl ester 356
2-Isobutyl-2-methyl-5-nitro-1,3-benzodioxole 279
2′-Methoxy-5′-methylsuccinanilic acid 416
N-(3-Methoxyphenyl)-glutaramic acid 418
N-(4-Methoxyphenyl)-glutaramic acid 418
4-(Propoxycarbonyl)phenyl *N*-methylcarbamate 676
$C_{12}H_{15}NO_4S$
1,4-Dimethylquinolinium methosulfate 991
1-Nitrocyclohexyl phenyl sulfone 780
$C_{12}H_{15}NO_5$
Carbobenzyloxy-DL-serine methyl ester 643
Carbobenzyloxy-L-serine methyl ester 643
Carbobenzyloxy-DL-threonine 643
Diethyl 5-formyl-3-methyl-2,4-pyrroledicarboxylate 838
$C_{12}H_{15}NO_5S$
Ethyl 2-(4-acetamidophenylsulfonyl)acetate 781
$C_{12}H_{15}NO_6$
N-(3,4,5-Trimethoxybenzoyl)glycine 475
$C_{12}H_{15}NO_6S$
2-Nitrophenyl-β-D-thiofucopyranoside 262
N-(*p*-Tosyl)-D-glutamic acid 824
N-(*p*-Tosyl)-L-glutamic acid 824
$C_{12}H_{15}NO_7$
4-Nitrophenyl-β-D-fucopyranoside 262
$C_{12}H_{15}NO_7S$
4-Nitrophenyl-β-D-thiogalactopyranoside 262
$C_{12}H_{15}NO_8$
4-Nitrophenyl-α-D-galactopyranoside hydrate 265
2-Nitrophenyl-β-D-glucopyranoside 263
$C_{12}H_{15}NS$
3-(2-(Dimethylamino)ethyl)benzo(b)-thiophene hydrochloride 897
2,2-Pentamethylenebenzothiazoline 250
$C_{12}H_{15}N_2O_4$
N-(2,4-Dinitro-6-methylphenyl)-piperidine 274
$C_{12}H_{15}N_3$
1-Anilinomethyl-3,5-dimethylpyrazole 861
4-Cyano-1-methyl-4-(2-pyridyl)-piperidine 945
$C_{12}H_{15}N_3O$
1-(4-Aminophenyl)-2-ethyl-3-methyl-3-pyrazolin-5-one 701
1-Benzyl-3-dimethylamino-2-pyrazolin-5-one 702
$C_{12}H_{15}N_3O_2$
1-(4-Aminophenyl)-4-(2-hydroxyethyl)-3-methyl-2-pyrazolin-5-one 696

$C_{12}H_{15}N_3O_2$ (Cont'd)

6-(Diethylamino)-1,4-(2*H*,3*H*)-phthalazinedione 500
6-Methoxy-α-tetralone semicarbazone 553

$C_{12}H_{15}N_3O_3$

1-(*N*-(3-Nitrophenyl)carbamoyl)-piperidine 702
1-(*N*-(4-Nitrophenyl)carbamoyl)-piperidine 703

$C_{12}H_{15}N_3O_3S$

1,2-Diaminopyridinium *p*-toluenesulfonate 942
1-(2,4-Dimethoxyphenyl)-3-(2-thiazolin-2-yl)urea 735

$C_{12}H_{15}N_3O_4$

1-(2,4-Dinitrophenyl)-2-methylpiperidine 274
1-(2,4-Dinitrophenyl)-3-methylpiperidine 273
1-(2,4-Dinitrophenyl)-4-methylpiperidine 273
4-(*N*-(2-Methyl-4-nitrophenyl)-carbamoyl)morpholine 703
4-(*N*-(4-Methyl-3-nitrophenyl)-carbamoyl)morpholine 702
1-(2-Phenyltriazolyl)-*meso*-erythritol 1028
Piperidino *N*-(3-nitrophenyl)-carbamate 669

$C_{12}H_{15}N_3O_5$

L-Glutamine 4-nitrobenzyl ester hydrobromide 420

$C_{12}H_{15}N_3O_6$

1-*tert*-Butyl-3,5-dimethyl-2,4,6-trinitrobenzene 275
N-(2,4-Dinitrophenyl)-L-isoleucine 349
N-(2,4-Dinitrophenyl)-L-leucine 349

$C_{12}H_{15}N_3O_7S$

2′,6′-Dinitro-*N*-ethyl-4′-(methylsulfonyl)propionanilide 778
2′,6′-Dinitro-*N*-methyl-4′-(methylsulfonyl)butyranilide 777
2′,6′-Dinitro-4′-(methylsulfonyl)-*N*-propylacetanilide 778

$C_{12}H_{15}N_3O_8S$

Ethyl *N*-(2,6-dinitro-4-(methylsulfonyl)phenyl)-*N*-ethylcarbamate 777
Isopropyl *N*-(2,6-dinitro-4-(methylsulfonyl)phenyl)-*N*-methylcarbamate 777
Methyl *N*-(2,6-dinitro-4-(methylsulfonyl)phenyl)-*N*-propylcarbamate 777

$C_{12}H_{15}N_3S$

5-Indancarboxaldehyde 4-methyl-3-thiosemicarbazone 1071

$C_{12}H_{15}O_4$

2′,5′-Dimethoxyacetoacetanilide 442

$C_{12}H_{16}$

5,6-Dimethyl-5-decen-1,9-diyne 1015
2,3-Dimethyl-1,4,5,8-tetrahydronaphthalene 3
1-Isopropenyl-4-isopropylbenzene 164
4-Methyl-2-phenyl-2-pentene 164

$C_{12}H_{16}BF_4NO$

1-Ethoxy-2-ethylisoindolium tetrafluoroborate 882

$C_{12}H_{16}BrNO$

4′-Bromo-2′,6′-dimethylbutyranilide 410

$C_{12}H_{16}BrNO_2$

4-Bromo-2,5-dimethylphenyl *N*-propylcarbamate 641
4-Bromo-2′-methoxy-5′-methylbutyranilide 391
2-(4-Bromophenoxy)-*N*,*N*,2-trimethylpropionamide 420

$C_{12}H_{16}BrNO_4$

Diethyl 2-(bromomethyl)-4-methyl-3,5-pyrroledicarboxylate 838

$C_{12}H_{16}ClN$

1-Benzyl-4-chloropiperidine hydrochloride 252
2-*tert*-Butyl-3-(4-chlorophenyl)-aziridine 244

$C_{12}H_{16}ClNO$

4′-Butyl-2-chloroacetanilide 393
5′-Chloro-2,2-dimethyl-2′-methoxypropionanilide 401
4-(*N*-(2-Chloroethyl)-*N*-ethylamino)-2-methylbenzaldehyde 321
(-)-3-Chloro-*N*-(α-methylphenethyl)-propionamide 392
3-Chlorophenyl 4-methylpiperidino ketone 695
3′-Chloro-2,2,2′-trimethylpropionanilide 409
3′-Chloro-2,2,4′-trimethylpropionanilide 410
4′-Chloro-2,2,2′-trimethylpropionanilide 410

$C_{12}H_{16}ClNOS$

S-Butyl *N*-(4-chloro-2-methylphenyl)thiocarbamate 689

$C_{12}H_{16}ClNO_2$

4′-Chloro-2,2-dimethyl-2′-methoxypropionanilide 401
4-Chloro-4′-ethoxybutyranilide 392
2-Chloroethyl *N*-(2-ethyl-6-methylphenyl)carbamate 605
1-Dimethylamino-2-propyl 4-chlorobenzoate hydrochloride 366
Pentyl 4-chlorocarbanilate 682

$C_{12}H_{16}ClNO_2S$

1-(4-Chlorophenylsulfonyl)-4-methylpiperidine 809

$C_{12}H_{16}ClNO_3$

5′-Chloro-2′,4′-dimethoxybutyranilide 399
4′-Chloro-3′,5′-dimethoxyisobutyranilide 399
2-Ethoxyethyl *N*-(2-chloro-6-methylphenyl)carbamate 650
2-Ethoxyethyl *N*-(4-chloro-2-methylphenyl)carbamate 616
2-Ethoxyethyl *N*-(5-chloro-2-methylphenyl)carbamate 616

$C_{12}H_{16}ClNO_4$

1-Butoxy-5-chloro-4-ethoxy-2-nitrobenzene 260
4-Carbethoxy-1-(carbethoxymethyl)-pyridinium chloride 956
(4-(2,5-Cyclohexadien-1-ylidene)-2-buten-1-ylidene)-dimethylammonium perchlorate 43
1,2,3,3-Tetramethyl-3*H*-indolium perchlorate 888

$C_{12}H_{16}ClNO_4S$

Butyl *N*-(3-chlorophenyl)-*N*-(methylsulfonyl)carbamate 808
Butyl *N*-(4-chlorophenyl)-*N*-(methylsulfonyl)carbamate 808

$C_{12}H_{16}ClNO_5S$

3-(3-Chloropropylsulfonyl)propyl 4-nitrophenyl ether 772

$C_{12}H_{16}ClN_3O_2$

4-Butyl-1-(4-chlorobenzoyl)-semicarbazide 532

$C_{12}H_{16}ClN_3O_2S$

4-(3-Methoxypropyl)-1-(4-chlorobenzoyl)-3-thiosemicarbazide 729

$C_{12}H_{16}Cl_2$

1,2-Bis(chloromethyl)-3,4,5,6-tetramethylbenzene 170
α1,α3-Dichlorohexamethylbenzene 170

$C_{12}H_{16}Cl_2F_4Si_2$

(2,3,5,6-Tetrafluoro-1,4-phenylene)-bis((chloromethyl)dimethylsilane) 1099

$C_{12}H_{16}Cl_2N_2O$

1-Butyl-3-(2,3-dichlorophenyl)-1-methylurea 594
1-Butyl-3-(2,4-dichlorophenyl)-1-methylurea 599
1-Butyl-3-(2,5-dichlorophenyl)-1-methylurea 599
1-Butyl-3-(2,6-dichlorophenyl)-1-methylurea 594
1-Butyl-3-(3,4-dichlorophenyl)-1-methylurea 599

$C_{12}H_{16}Cl_2N_2O_3S$

1-(3,4-Dichlorophenyl)-3,3-diethyl-1-(methylsulfonyl)urea 809

$C_{12}H_{16}Cl_2O_2$

1,4-Bis(chloromethyl)-2,5-diethoxybenzene 176

$C_{12}H_{16}Cl_2O_3$

2-(2,4-Dichlorophenoxy)-acetaldehyde diethylacetal 189

$C_{12}H_{16}FN_3O$

1-(*N*-(3-Fluorophenyl)carbamoyl)-4-methylpiperazine 694
1-(4-Fluorophenyl)-3-piperidinourea 533

$C_{12}H_{16}F_3NO$

N-(1-Adamantyl)trifluoroacetamide 103
N-(2-Adamantyl)-2,2,2-trifluoroacetamide 103

$C_{12}H_{16}F_3NO_4S_2$

o-(*m*-Tolyl) *N*,*N*-dimethylthiocarbamate methyl trifluoromethanesulfonate 742

$C_{12}H_{16}F_4Si$

Butyldimethyl(2,3,4,5-tetrafluorophenyl)silane 1100
Butyldimethyl(2,3,5,6-tetrafluorophenyl)silane 1100

$C_{12}H_{16}HgO_2$

2-Methyl-2-phenylpropylmercuric acetate 1139

$C_{12}H_{16}N_2$

Cyclohexanone phenylhydrazone 235
trans,*trans*-2,9-Dicyanodecalin 150
2,3-Dihydro-2,2,4-trimethyl-1*H*-1,5-benzodiazepine 250
α-(2-Dimethylaminoethyl)-benzyl cyanide 755
3-(*N*-Ethyl-*m*-toluidino)-propionitrile 755
2-Methylgramine 884
5-Methylgramine 885
Spiro(cyclohexane-1,2′-benzimidazoline) 253

$C_{12}H_{16}N_2O$

Acetic (4-isopropylbenzylidene)-hydrazide 502
1-Allyl-3-(2-ethylphenyl)urea 587
1-Allyl-3-(2,3-xylyl)urea 593
1-Allyl-3-(2,4-xylyl)urea 598
1-Allyl-3-(2,5-xylyl)urea 598
1-Allyl-3-(2,6-xylyl)urea 593
1-Allyl-3-(3,4-xylyl)urea 598
1-Allyl-3-(3,5-xylyl)urea 602
5-Cyano-3-(2-cyanoethyl)-3-isopropenyl-2-pentanone 148
1-Cyclopropyl-3-(2,6-dimethylphenyl)urea 581
1-Cyclopropyl-3-(2-ethylphenyl)urea 581
1-Cyclopropyl-3-(2,4-xylyl)urea 581
4,5-Dihydro-4,7,8-trimethyl-1*H*-1,5-benzodiazepin-2(3*H*)-one 702
5,6-Dimethyl-2-benzimidazolepropanol 902
4-Isopropylbenzaldehyde acetylhydrazone 500
2-Oxo-1,3-cyclohexanedipropionitrile 148
4-Phenylisonipecotamide hydrochloride 427
α-Phenyl-1-pyrrolidineacetamide 428
1,3,4,5-Tetrahydro-3,7,8-trimethyl-2*H*-1,5-benzodiazepin-2-one 702
α,α,5,6-Tetramethyl-2-benzimidazolemethanol 902

$C_{12}H_{16}N_2OS$

1-(3-Acetylphenyl)-3-propyl-2-thiourea 744
1-Allyl-3-(2-ethoxyphenyl)-2-thiourea 737
1-Allyl-3-(2-(1-hydroxyethyl)phenyl)-2-thiourea 734
1-Allyl-3-(3-(1-hydroxyethyl)phenyl)-2-thiourea 733
5-Benzyl-4-hydroxy-3-methyl-2-(methylimino)thiazolidine 734

$C_{12}H_{16}N_2O_2$

4′-Acetamidobutyranilide 428
1-Allyl-3-(2-methoxy-5-methylphenyl)urea 556
Benzaldehyde (*tert*-butoxycarbonyl)hydrazone 674
1-Cyclopropyl-3-(2-ethoxyphenyl)urea 554
1-Cyclopropyl-3-(2-methoxy-5-methylphenyl)urea 554
O-(*N*-(2-Ethylphenyl)carbamoyl)-acetone oxime 1012
Hexahydro-1-(4-nitrophenyl)-1*H*-azepine 276
2-Methyl-1-(4-nitrophenyl)piperidine 276
4-Methyl-1-(2-nitrophenyl)piperidine 274
4-Methyl-1-(4-nitrophenyl)piperidine 277
N,*N*′-(1,2-Phenylene)bis(*N*-ethylformamide) 425
α-Phenyl-4-morpholineacetamide 422
1-Phenyl-3-valerylurea 579
N,*N*,*N*′,*N*′-Tetramethylphthalic diamide 475
4-(*N*-(*o*-Tolyl)carbamoyl)morpholine 705
4-(*N*-(*m*-Tolyl)carbamoyl)morpholine 713
4-(*N*-(*p*-Tolyl)carbamoyl)morpholine 706
O-(*N*-(2,3-Xylyl)carbamoyl)acetone oxime 1012
O-(*N*-(2,4-Xylyl)carbamoyl)acetone oxime 1012
O-(*N*-(2,5-Xylyl)carbamoyl)acetone oxime 1012
O-(*N*-(2,6-Xylyl)carbamoyl)acetone oxime 1012
O-(*N*-(3,4-Xylyl)carbamoyl)acetone oxime 1012
O-(*N*-(3,5-Xylyl)carbamoyl)acetone oxime 1012

$C_{12}H_{16}N_2O_2S$

2-(1-Aziridinyl)ethyl *N*-(4-(methylthio)phenyl)carbamate 667
N-α-Benzoyl-L-methionine amide 468
1-(4-Ethoxycarbonyl)-3-ethyl-2-thiourea 744
4-(*N*-(4-Methoxyphenyl)-thiocarbamoyl)morpholine 738
*N*α-Benzoyl-DL-methioninamide 468

$C_{12}H_{16}N_2O_3$

4′-Acetamido-3′-methoxypropionanilide 416
4-Acetamidophenyl *N*-isopropylcarbamate 677
D-Alanyl-L-phenylalanine 422
L-Alanyl-D-phenylalanine 422
1-Allyl-3-(2,5-dimethoxyphenyl)urea 556
5-(4-Aminobenzamido)valeric acid 468
4-(*o*-Anisylcarbamoyl)morpholine 699
4-(*m*-Anisylcarbamoyl)morpholine 699
4-(*p*-Anisylcarbamoyl)morpholine 699
1-Butyl-3-(2-carboxyphenyl)urea 576
4′-*tert*-Butyl-2′-nitroacetanilide 425
1-Cyclopropyl-3-(2,5-dimethoxyphenyl)urea 550
2′,6′-Diethyl-3′-nitroacetanilide 432
2,6-Dimethyl-4-(4-nitrophenyl)-morpholine 276
Ethyl 2,2-bis(2-cyanoethyl)-acetoacetate 148
O-(*N*-(2-Methoxy-5-methylphenyl)-carbamoyl)acetone oxime 1009
1-(2-Methoxy-4-nitrophenyl)-piperidine 268
1-(4-Methoxy-2-nitrophenyl)-piperidine 267
N-(2-(2-Pyridyl)ethyl)glutaramic acid 949
Sarcosyl-L-phenylalanine 428

$C_{12}H_{16}N_2O_4$

Acetic acid (3,4,5-trimethoxybenzylidene)-hydrazide 502
N,*N*′-Bis(2-hydroxyethyl)-terephthalamide 462
O-(*N*-(2,5-Dimethoxyphenyl)-carbamoyl)acetone oxime 1009
2,2-Dimethyl-4′-methoxy-2′-nitropropionanilide 415
Ethyl 2-(3-(4-methoxyphenyl)ureido)-acetate 557
2-(*N*-(Hydroxymethyl)carbamoyl)-phenyl *N*-propylcarbamate 644
L(+)-Leucine 4-nitrophenyl ester hydrobromide 379
Methyl 3-amino-4-*tert*-butyl-5-nitrobenzoate 372
1-Nitro-4-(2-nitro-1,1,2-trimethylpropyl)benzene 277
L-Tyrosyl-β-alanine 411
L-Tyrosyl-L-alanine 411

$C_{12}H_{16}N_2O_4$ (Cont'd)
L-Valine 4-nitrobenzyl ester hydrobromide ... 348
$C_{12}H_{16}N_2O_4S$
4-(2-(4-Aminophenylsulfonyl)-acetyl)-morpholine ... 776
$C_{12}H_{16}N_2O_5$
2-*tert*-Butyl-4,6-dinitro-5-methylanisole ... 267
4-(3,4-Dimethoxy-6-nitroanilino)-2-butanone ... 283
$C_{12}H_{16}N_2O_6$
2′,3′-Isopropylideneuridine ... 131
$C_{12}H_{16}N_2O_{10}$
Bis(nitromethyl)methyldihydrodimethylspiro(dioxolanfuro(2,3-d)-dioxol)-2-one ... 96
$C_{12}H_{16}N_2S$
1-(Butylthiomethyl)-benzimidazole ... 903
1-Cyclopentyl-3-phenyl-2-thiourea ... 747
1,1-Pentamethylene-3-phenyl-2-thiourea ... 747
$C_{12}H_{16}N_4$
1-(α,4-Dimethylcinnamylideneamino)guanidine nitrate ... 1072
$C_{12}H_{16}N_4O_3$
2-(3-Furfurylureido)acetic (2-buten-1-ylidene)hydrazide ... 851
$C_{12}H_{16}N_4O_3S$
N-(4-Ethoxyphenyl)-*N*′-(3-methyl-2-thioureido)oxamide ... 737
$C_{12}H_{16}N_4O_6S$
2-Carbethoxyacetamide 4-methylsulfonyl-2-nitrophenylhydrazone ... 777
$C_{12}H_{16}N_6O$
6-Cyclohexylureidopurine ... 923
$C_{12}H_{16}N_6O_2$
1,2-Phenylenediacetaldehyde disemicarbazone ... 574
$C_{12}H_{16}N_6O_5$
4-Ethyl-1-(3-nitrophenylureidoacetyl)-semicarbazide ... 563
$C_{12}H_{16}O$
2-Cyclohexylphenol ... 191
DL-α-Cyclopropyl-α,4-dimethylbenzyl alcohol ... 203
Cyclopropyl 2,5-dimethylphenyl carbinol ... 203
6,8a-Dimethyl-1,2,4a,5,8,8a-hexahydro-2-naphthalenone ... 54
2,2-Dimethyl-4′-methylhydrocinnamaldehyde ... 321
DL-α-Methyl-α-(1-methylcyclopropyl)-benzyl alcohol ... 203
4-Methyl-4-phenyl-2-pentanone ... 286
4′-Methylvalerophenone ... 298
4-Methylvalerophenone ... 296
1,2,3,4,4a,5,8,8a-Octahydro-1,4:5,8-dimethano-2-naphthol ... 15
3,3a,4,6,7,8,8a,8b-Octahydro-1(2*H*)-*as*-indacenone ... 59
1-Phenyl-1-cyclohexanol ... 206
cis-2-Phenyl-1-cyclohexanol ... 206
2-Phenyl-1-cyclohexanol ... 206
1-Phenyl-1-cyclopentanemethanol ... 205
1-Phenyl-2-hexanone ... 286
1-Phenyl-3-hexanone ... 286
3-Phenyl-4-hexanone ... 286
Tricyclo[5.2.1.$0^{2,6}$]dec-3-en-8-yl acetate ... 81
2,2,2′-Trimethylpropiophenone ... 297
2,2,4′-Trimethylpropiophenone ... 297
$C_{12}H_{16}OS$
2,2-Dimethyl-4′-(methylthio)-propiophenone ... 294
$C_{12}H_{16}O_2$
4′-Butoxyacetophenone ... 292
4-*tert*-Butylphenyl acetate ... 380
4-Cyclohexylresorcinol ... 193
6,7-Dimethoxytetralin ... 178
4a,8-Dimethyl-1,2,3,4,4a,5,6,7-octahydro-1,7-naphthalenedione ... 53
α,α-Dimethylphenethyl acetate ... 361
4-(2,4-Dimethylphenyl)butyric acid ... 333
4-(2,5-Dimethylphenyl)butyric acid ... 334
4-(3,4-Dimethylphenyl)butyric acid ... 334
5,8a-Dimethyl-3,4,8,8a-tetrahydro-1,6(2*H*,7*H*)-naphthalenedione ... 57
Ethyl 2-phenylbutyrate ... 361
Ethyl 4-phenylbutyrate ... 361
Eugenyl ethyl ether ... 178
Glycidyl mesityl ether ... 189
1,4,4a,5,8,8a-Hexahydro-3-methoxy-8a-methyl-1-naphthalenone ... 52
Hydrocinnamyl propionate ... 362
4′-Hydroxy-5′-isopropyl-2′-methylacetophenone ... 300
Isobutyl phenylacetate ... 362
Isoeugenyl ethyl ether ... 179
Isopentyl benzoate ... 375
4-Isopropylbenzyl acetate ... 362
2-Isopropyl-4-phenyl-1,3-dioxolane ... 184
4′-Methoxyvalerophenone ... 293
DL-Methyl 3-methyl-2-phenylbutyrate ... 362
1-Methyl-1-phenylpropyl acetate ... 356
1-Methyl-3-phenylpropyl acetate ... 361
Methyl 5-phenylvalerate ... 361
3-Methyl-2-phenylvaleric acid, magnesium salt monohydrate ... 352
3-Methyl-2-phenylvaleric acid, piperazinium salt ... 351
3-Methyl-2-phenylvaleric acid, sodium salt ... 352
3-Methyl-2-(*o*-tolyl)-butyric acid ... 333
1,2,5a,6,7,8,9,9a-Octahydrocyclopenta(c)(1)benzopyran-4(3*H*)-one ... 96
4-(Pentyloxy)benzaldehyde ... 323
Phenethyl isobutyrate ... 362
4-(Phenoxymethyl)tetrahydropyran ... 183
DL-1-Phenylbutyl acetate ... 356
Phenyl hexanoate ... 380
2-Phenylhexanoic acid, sodium salt ... 352
4-Phenyl-2-propyl-1,3-dioxolane ... 184
β-Propylhydrocinnamic acid ... 333
2-(*o*-Tolyl)-valeric acid ... 333
4-(Trimethylacetyl)benzyl alcohol ... 290
$C_{12}H_{16}O_2S$
Cyclohexyl phenyl sulfone ... 782
Methyl 3-(benzylthio)-2-methylpropionate ... 357
$C_{12}H_{16}O_3$
4-Butoxyphenylacetic acid ... 329
5-(*tert*-Butyl)-4-hydroxy-*m*-toluic acid ... 339
3-*tert*-Butyl-5-methylsalicylic acid ... 338
3-*tert*-Butyl-6-methylsalicylic acid ... 338
2-(4-*tert*-Butylphenoxy)acetic acid ... 327
3′,4′-Dimethoxybutyrophenone ... 291
(*S*)-(-)-Ethyl 2-benzyloxypropionate ... 354
Ethyl 3-hydroxy-2-methyl-3-phenylpropionate ... 353
3-Hydroxy-1-(4-methoxyphenyl)-3-methyl-2-butanone ... 282
Isopentyl salicylate ... 368
4-Methoxybenzyl butyrate ... 354
5-Methoxy-8a-methyl-3,4,8,8a-tetrahydro-1,6(2*H*,7*H*)-naphthalenedione ... 52
2-(2-Methoxy-4-propenylphenoxy)-ethanol ... 202
2-Phenoxyethyl isobutyrate ... 355
Thymoxyacetic acid ... 328
3-(2,4,5-Trimethyl-1,3-dioxolan-2-yl)-phenol ... 191
$C_{12}H_{16}O_3S$
1-(Phenylsulfonyl)pinacolone ... 781
$C_{12}H_{16}O_4$
mono-Benzalpentaerythritol ... 200
Perhydrocyclobuta(1,2:3,4)-dicyclopentene-2,5-dicarboxylic acid ... 66
$C_{12}H_{16}O_4S$
Ethyl 3-(benzylsulfonyl)propionate ... 781
$C_{12}H_{16}O_5$
1,3-*O*-Benzylidene-D-arabitol ... 200
Diethyl 2,5-dimethyl-3,4-furandicarboxylate ... 853
$C_{12}H_{16}O_6$
Dimethyl 2,2-dimethyl-4-hydroxy-6-oxo-4-cyclohexene-1,3-dicarboxylate ... 91
Phenyl α-D-glucopyranoside ... 199
$C_{12}H_{16}O_7$
Arbutin ... 205
$C_{12}H_{16}O_7S$
2-*O*-(*p*-Tosyl)-D-arabinose ... 797
$C_{12}H_{16}O_8$
l-Glucosan triacetate ... 89
$C_{12}H_{16}O_8S_6$
2,2,5,5-Tetrakis(carboxymethylthio)-1,4-dithiane ... 68
$C_{12}H_{17}Br$
Triethylbromobenzene ... 170
$C_{12}H_{17}BrO$
2-Bromo-1-cyclododecanone ... 53
1-(2-Bromoethoxy)-4-butylbenzene ... 176
$C_{12}H_{17}BrO_2$
α-Bromo-1-adamantaneacetic acid ... 67
3-Bromo-1-adamantaneacetic acid ... 67
$C_{12}H_{17}BrO_3$
3-Bromo-α-hydroxy-1-adamantaneacetic acid ... 64
$C_{12}H_{17}Cl$
(+)-1-Chloro-3,3-dimethyl-2-phenylbutane ... 169
$C_{12}H_{17}ClN_2O$
1-*tert*-Butyl-3-(2-chlorobenzyl)urea ... 536
1-*tert*-Butyl-3-(3-chlorobenzyl)urea ... 527
1-Butyl-3-(2-chloro-6-methylphenyl)urea ... 594
1-*tert*-Butyl-3-(2-chloro-6-methylphenyl)urea ... 540
1-Butyl-3-(3-chloro-4-methylphenyl)urea ... 599
1-*tert*-Butyl-3-(5-chloro-2-methylphenyl)urea ... 543
1-Butyl-3-(4-chloro-2-methylphenyl)urea ... 599
1-Butyl-3-(5-chloro-2-methylphenyl)urea ... 599
1-Butyl-3-(2-chlorophenyl)-1-methylurea ... 587
1-Butyl-3-(3-chlorophenyl)-1-methylurea ... 580
1-Butyl-3-(4-chlorophenyl)-1-methylurea ... 590
1-(3-Chloro-4-methylphenyl)-1,3-diethylurea ... 543
3-(2-Chloro-6-methylphenyl)-1,1-diethylurea ... 593
3-(3-Chloro-4-methylphenyl)-1,1-diethylurea ... 599
3-(5-Chloro-2-methylphenyl)-1,1-diethylurea ... 599
1-(2-Chloro-6-methylphenyl)-3-isobutylurea ... 594
1-(3-Chloro-4-methylphenyl)-3-isobutylurea ... 543
1-(4-Chloro-2-methylphenyl)-3-isobutylurea ... 543
1-(5-Chloro-2-methylphenyl)-3-isobutylurea ... 543
6′-Chloro-2-propylamino-*o*-acetotoluidide hydrochloride ... 430
1,1-Diethyl-3-(4-chloro-2-methylphenyl)urea ... 599
$C_{12}H_{17}ClN_2OS$
2′-Chloro-2-(2-(dimethylamino)-ethylthio)-acetanilide hydrochloride ... 421
$C_{12}H_{17}ClN_2O_2$
1-Butyl-3-(2-chloro-4-methoxyphenyl)urea ... 521
1-Butyl-3-(3-chloro-2-methoxyphenyl)urea ... 556
1-(5-Chloro-2-methylphenyl)-3-(1,1-dimethyl-2-hydroxyethyl)urea ... 548
3-(5-Chloro-2-methylphenyl)-1-ethyl-1-(2-hydroxyethyl)urea ... 548
2-Dimethylaminoethyl *N*-(2-chloro-6-methylphenyl)carbamate ... 672
2-Dimethylaminoethyl *N*-(3-chloro-4-methylphenyl)carbamate ... 672
2-(Dimethylamino)ethyl *N*-(4-chloro-2-methylphenyl)carbamate ... 672
2-Dimethylaminoethyl *N*-(5-chloro-2-methylphenyl)carbamate ... 672
$C_{12}H_{17}ClN_2O_3$
2-Dimethylaminoethyl *N*-(3-chloro-2-methoxyphenyl)carbamate ... 652
$C_{12}H_{17}ClN_2O_3S$
1-Butyl-3-(α-chloro-*p*-tolylsulfonyl)-urea ... 805
1-(4-Chlorophenylsulfonyl)-3-isopentylurea ... 826
$C_{12}H_{17}ClO$
3-Isopropoxy-3-phenylpropyl chloride ... 185
$C_{12}H_{17}ClO_2$
1-Chloro-3-mesityloxy-2-propanol ... 197
$C_{12}H_{17}Cl_2N_3O_4$
N,*N*-Bis(2-hydroxyethyl)-*N*′-(4,5-dichloro-2-nitrophenyl)-ethylenediamine ... 259
$C_{12}H_{17}IN_2$
(3-Indolylmethyl)-trimethylammonium iodide ... 885
$C_{12}H_{17}N$
1-Butylindoline ... 252
N-Cyclohexylaniline ... 219
3-Methyl-3-phenylpiperidine ... 250
$C_{12}H_{17}NO$
4-Amino-2-cyclohexylphenol ... 217
3′-Amino-2,2,4′-trimethylpropiophenone ... 294
1-Benzyl-3-hydroxypiperidine ... 246
4-Benzyl-4-hydroxypiperidine ... 253
N-Benzyl-*N*-methylisobutyramide ... 441
4′-*tert*-Butylacetanilide ... 438
2-(1-Butyliminoethyl)phenol ... 1035
tert-Butyl(3-methoxybenzylidene)-amine ... 1039
α-(*sec*-Butyl)-phenylacetamide ... 429
3-Cyclohexyl-4-hydroxyaniline hydrochloride ... 217
3′,4′-Diethylacetanilide ... 440
3-Dimethylamino-2-methylpropiophenone ... 295
3-Dimethylamino-2-methylpropiophenone hydrochloride ... 295
3-(Dimethylamino)-4′-methylpropiophenone hydrochloride ... 295
2′,6′-Isobutyroxylidide ... 439
1-(3-Methoxyphenyl)piperidine ... 247
N-Methyl-*N*-(α-methylphenethyl)-acetamide ... 430
α-Phenyl-2-piperidinemethanol ... 205
4-(*p*-Tolyl)-4-piperidinol ... 245
N-(2,4,5-Trimethylbenzyl)-acetamide ... 431
2,2,4′-Trimethylpropionanilide ... 438
2′,4′,5′-Trimethylpropionanilide ... 441
$C_{12}H_{17}NOS$
S-Butyl *N*-(*m*-tolyl)thiocarbamate ... 688
2,2-Dimethyl-4′-(methylthio)-propionanilide ... 421
S-Isopropyl *N*-(2,4-xylyl)thiocarbamate ... 689
S-Propyl *N*-(2,4-xylyl)thiocarbamate ... 689
$C_{12}H_{17}NO_2$
N-Benzylalanine ethyl ester ... 358
Benzyl *N*-butylcarbamate ... 665
N-Benzyl-L-valine ... 349
4′-Butoxyacetanilide ... 419
2-Butylamino-4′-hydroxyacetophenone hydrochloride ... 290
N-*tert*-Butyl-3-methoxybenzamide ... 464
5-*tert*-Butyl-2-nitro-*m*-xylene ... 279
4-(*tert*-Butyl)-phenyl *N*-methylcarbamate ... 677
Butyl *N*-(*o*-tolyl)carbamate ... 681
Butyl *N*-(*p*-tolyl)carbamate ... 682
trans,*trans*-9-Cyano-2-decalincarboxylic acid ... 149
(-)-4-Cyano-2,3,3-trimethyl-1-cyclohexen-1-yl acetate ... 149
N-Cyclopropylveratrylamine hydrochloride ... 239
3,5-Diacetyl-2,4,6-trimethyl-1,4-dihydropyridine ... 58
N,*N*-Diethyl-*p*-anisamide ... 467
3-(Dimethylamino)-3′-methoxypropiophenone hydrochloride ... 292

$C_{12}H_{17}NO_2$ (Cont'd)

3-(Dimethylamino)-4′-methoxypropiophenone hydrochloride 292
1-Dimethylamino-2-propyl benzoate hydrochloride 373
2,2-Dimethyl-4′-methoxypropionanilide 419
2-(*N*-Ethylanilino)-ethyl acetate 359
Ethyl 3-benzylaminopropionate 358
Ethyl α-cyano-2-norbornaneacetate 149
Ethyl *N*-(2-ethyl-6-methylphenyl)-carbamate 684
Ethyl *N*-mesitylcarbamate 687
N-Furfuryl-1-cyclohexanecarboxamide 851
Isopropyl *N*-(2-ethylphenyl)-carbamate 681
Isopropyl *N*-(2,3-xylyl)carbamate 684
Isopropyl *N*-(2,4-xylyl)carbamate 686
Isopropyl *N*-(2,5-xylyl)carbamate 686
Isopropyl *N*-(2,6-xylyl)carbamate 684
Isopropyl *N*-(3,5-xylyl)carbamate 686
2′-Methoxy-5′-methylbutyranilide 417
Methyl *N*-(2,6-diethylphenyl)-carbamate 684
L-Norvaline benzyl ester *p*-toluenesulfonate 356
Pentyl carbanilate 680
α-Phenyl-4-morpholineethanol 246
Propyl *N*-(2-ethylphenyl)carbamate 681
Propyl *N*-(2,3-xylyl)carbamate 684
Propyl *N*-(2,4-xylyl)carbamate 686
Propyl *N*-(2,5-xylyl)carbamate 686
Propyl *N*-(2,6-xylyl)carbamate 684
Propyl *N*-(3,4-xylyl)carbamate 686
Propyl *N*-(3,5-xylyl)carbamate 686
4-(4-Pyridyl)-4-tetrahydropyranethanol 934
D-Valine benzyl ester *p*-toluenesulfonate 356
DL-Valine benzyl ester *p*-toluenesulfonate 356

$C_{12}H_{17}NO_2S$

S-Benzyl-L-cysteine ethyl ester hydrochloride 357
1-(Benzylsulfonyl)piperidine 827
N-Cyclohexylbenzenesulfonamide 827
S-Isopropyl *N*-(2-methoxy-5-methylphenyl)thiolcarbamate 653
N-*p*-Tosylpiperidine 827

$C_{12}H_{17}NO_3$

Butyl *N*-(*m*-anisyl)carbamate 654
3-*tert*-Butyl-4,5,6,7-tetrahydro-2,1-benzisoxazole-7-carboxylic acid 871
(-)-4-Cyano-1,2-epoxy-2,3,3-trimethylcyclohexyl acetate 149
N,*N*-Diethylvanillamide 461
4-(2,5-Dimethoxyphenyl)butyramide 414
DL-α,3-Dimethyltyrosine methyl ester hydrochloride 354
2-Ethoxyethyl *N*-(*o*-tolyl)carbamate 650
2-Ethoxyethyl *N*-(*m*-tolyl)carbamate 649
2-Ethoxyethyl *N*-(*p*-tolyl)carbamate 650
Ethyl 4-acetyl-3-ethyl-5-methyl-2-pyrrolecarboxylate 842
3-Hydroxy-*p*-butyrophenetidide 411
Isopropyl *N*-(2-methoxy-5-methylphenyl)carbamate 653
2-Methoxyethyl *N*-(2-ethylphenyl)-carbamate 650
2-Methoxyethyl *N*-(2,3-xylyl)-carbamate 650
2-Methoxyethyl *N*-(2,4-xylyl)-carbamate 650
2-Methoxyethyl *N*-(2,5-xylyl)-carbamate 650
2-Methoxyethyl *N*-(2,6-xylyl)-carbamate 650
2-Methoxyethyl *N*-(3,5-xylyl)-carbamate 651
3-(*N*-Propylcarbamoyl)-5-norbornene-2-carboxylic acid 122
Propyl *N*-(2-methoxy-5-methylphenyl)carbamate 653

$C_{12}H_{17}NO_3S$

3,3-Dimethyl-1-(phenylsulfonyl)-2-butanone oxime 1011
S-Isopropyl *N*-(2,5-dimethoxyphenyl)thiolcarbamate 648
N-Isopropyl-2-(*p*-tolylsulfonyl)-acetamide 779
S-Propyl *N*-(2,5-dimethoxyphenyl)-thiolcarbamate 648
N-Propyl-2-(*p*-tolylsulfonyl)-acetamide 779

$C_{12}H_{17}NO_3S_2$

N-(2-(Methylthio)-ethyl)-2-(*p*-tolylsulfonyl)-acetamide 775

$C_{12}H_{17}NO_4$

2,6-Dimethoxyphenyl *N*-isopropylcarbamate 647
2,4-Dimethyl-5-ethoxycarbonyl-3-pyrrolepropionic acid 840
Ethyl 5-(acetoxymethyl)-3,4-dimethyl-2-pyrrolecarboxylate 840
Isopropyl *N*-(2,5-dimethoxyphenyl)-carbamate 653
N-(4-Methoxybenzyl)-L-threonine 349
Methyl 2-amino-4,5-diethoxybenzoate 369
6-(Piperidinomethyl)-kojic acid 58

$C_{12}H_{17}NO_4S$

Butyl *N*-(methylsulfonyl)-*N*-phenylcarbamate 824

$C_{12}H_{17}NO_5S$

2-Aminophenyl-β-D-thiogalactopyranoside 216

$C_{12}H_{17}NO_6$

2-Aminophenyl β-D-galactopyranoside monohydrate 216
4-Aminophenyl α-D-mannopyranoside 216

$C_{12}H_{17}NS$

Ethyl *N*-phenylthiobutyrimidate 1051

$C_{12}H_{17}N_2$

4-(4-Methylpiperidino)aniline hydrochloride 248

$C_{12}H_{17}N_3O$

4-Anilino-4-carbamylpiperidine 425
1-Cyclohexyl-3-(2-pyridyl)urea 965
3,3-Dimethyl-2-hydrazonobutyrophenone oxime 1009
Pentanophenone semicarbazone 571

$C_{12}H_{17}N_3OS$

1-(4-Acetamidophenyl)-3-propyl-2-thiourea 741

$C_{12}H_{17}N_3O_2$

L-Alanyl-L-phenylalaninamide hydrochloride 422
1-Ethyl-3-(4-(*N*-methylacetamido)-phenyl)urea 569
4′-(3-Ethylureido)-*N*-methylacetanilide 569
Sarcosyl-L-phenylalaninamide acetate 428

$C_{12}H_{17}N_3O_2S$

N[1]-Diethyl-2-methyl-5-benzimidazolesulfonamide 906

$C_{12}H_{17}N_3O_2S_2$

1,*N*-Diethyl-2-methylthio-5-benzimidazolesulfonamide 903

$C_{12}H_{17}N_3O_3$

4-Butyl-1-(4-hydroxybenzoyl)-semicarbazide 496
1-Butyl-3-(4-methyl-3-nitrophenyl)urea 564
1,1-Diethyl-3-(2-methyl-4-nitrophenyl)urea 568
1,1-Diethyl-3-(4-methyl-3-nitrophenyl)urea 565
1-(Ethoxycarbonyl)-4-(2,6-xylyl)-semicarbazide 673
Ethyl 2-cyclopentylamino-4-hydroxy-5-pyrimidinecarboxylate 972
Ethyl 4-hydroxy-2-piperidino-5-pyrimidinecarboxylate 972
Ethyl 2-(3-(2-(2-pyridyl)ethyl)ureido)-acetate 949
Methyl 4-oxo-4,6,7,8-tetrahydro-1*H*-pyrimido(1,2-a)pyrimidine-2-butyrate hydrochloride 134
L-Phenylalanyl-L-serinamide hydrochloride 411
L-Seryl-L-phenylalaninamide hydrochloride 411

$C_{12}H_{17}N_3O_4$

1-Ethoxycarbonyl-4-(2-methoxy-5-methylphenyl)semicarbazide 553

$C_{12}H_{17}N_3O_4S$

1-(4-Acetamidophenylsulfonyl)-3-isopropylurea 824
1-(4-Acetamidophenylsulfonyl)-3-propylurea 824

$C_{12}H_{17}N_3O_5$

4-(2,5-Dimethoxyphenyl)-1-(ethoxycarbonyl)semicarbazide 648
2′,3′-Isopropylidenecytidine hydrochloride 131

$C_{12}H_{17}N_3O_6S_2$

N-Butyl-2′,6′-dinitro-4′-(methylsulfonyl)-methanesulfonanilide 818

$C_{12}H_{17}N_5O_2$

7-(Pyrrolidinomethyl)-theophylline 867

$C_{12}H_{17}N_5O_3$

7-(Morpholinomethyl)-theophylline 867

$C_{12}H_{17}N_5O_4$

6-Dimethylaminopurine-9-riboside 920

$C_{12}H_{17}N_5O_5S_3$

Methanesulfonic (4-aminophenyl-sulfopropyl-thiadiazol-2-in-5-ylidene)hydrazide 817

$C_{12}H_{17}N_7O_5$

L-Nitroarginine 4-nitroanilidehydrobromide 724

$C_{12}H_{17}O_3P$

5-Ethyl-5-methyl-2-phenyl-1,3,2-dioxaphosphorinane oxide 834

$C_{12}H_{18}$

1-(*tert*-Butyl)-4-ethylbenzene 163
2-Phenylhexane 162

$C_{12}H_{18}BrNO$

α-Bromo-1-adamantanecarboxamide 101
2-(4-Bromophenoxy)-*N*,*N*,2-trimethylpropylamine 236

$C_{12}H_{18}BrNO_6$

Diethyl 2-acetamido-2-(3-bromo-2-oxopropyl)malonate 117

$C_{12}H_{18}Br_2$

1-Bromo-3-(1-bromoethyl)-adamantane 6
1,3-Dibromo-5,7-dimethyladamantane 6

$C_{12}H_{18}ClNO$

2-Chloro-*N*,1′-dimethyl-2′-phenoxy-diethylamine hydrochloride 236
N-(2-Chloroethyl)-*N*-(2-methoxy-ethyl)-benzylamine hydrochloride 236

$C_{12}H_{18}Cl_4Si_2$

(2,3,5,6-Tetrachloro-1,4-phenylene)-bis(trimethylsilane) 1100
(3,4,5,6-Tetrachloro-1,2-phenylene)-bis(trimethylsilane) 1100

$C_{12}H_{18}Cl_6NO_2SSb$

N-Methyl-*N*-(*p*-toluenesulfonyl)-pyrrolidinium hexachloroantimonate 821

$C_{12}H_{18}Cl_6N_2O_5$

Diisopropyl *N*,*N*′-(1,1′-oxybis(2,2,2-trichloroethyl))biscarbamate 113
Dipropyl *N*,*N*′-(1,1′-oxybis(2,2,2-trichloroethyl))biscarbamate 113

$C_{12}H_{18}CuO_6$

Ethyl acetoacetate, copper derivative 70

$C_{12}H_{18}F_4Si_2$

1,4-Bis(trimethylsilyl)-2,3,5,6-tetrafluorobenzene 1100

$C_{12}H_{18}F_5NSi_2$

N,*N*-Bis(trimethylsilyl)-2,3,4,5,6-pentafluoroaniline 1102

$C_{12}H_{18}F_{12}Si_2$

Perfluorohexamethylenebis-(trimethylsilane) 1097

$C_{12}H_{18}INO$

(2-Benzoylethyl)-trimethylammonium iodide 295
α-Iodo-1-adamantaneacetamide 101

$C_{12}H_{18}I_3NO_4S$

(2-(2,4,6-Triiodophenoxy)-ethyl)-trimethylammonium methanesulfonate 794

$C_{12}H_{18}N_2$

4-Amino-1-benzylpiperidine dihydrochloride hydrate 249
2-Methyl-1-*m*-tolylpiperazine 253
2-Methyl-1-*p*-tolylpiperazine 248
4-(2-Piperidinoethyl)pyridine 955
1-(2,4-Xylyl)piperazine 248
1-(3,4-Xylyl)piperazine 248

$C_{12}H_{18}N_2O$

1-(Benzyl)-3-(*tert*-butyl)urea 585
1-Benzyl-3-isopropyl-1-methylurea 571
1-Butyl-1-methyl-3-phenylurea 585
3-Butyl-1-methyl-1-phenylurea 585
1-Butyl-3-(*o*-tolyl)urea 587
1-Butyl-3-(*m*-tolyl)urea 580
1-*tert*-Butyl-3-(*m*-tolyl)urea 580
1-Butyl-3-(*p*-tolyl)urea 590
3-((2-Cyanoethyl)methylamino)-5,5-dimethyl-2-cyclohexen-1-one 147
3′-Diethylaminoacetanilide 425
4′-Diethylaminoacetanilide 427
1,1-Diethyl-3-methyl-3-phenylurea 585
1-(2,6-Diethylphenyl)-3-methylurea 593
1,1-Diethyl-3-(*o*-tolyl)urea 587
1,1-Diethyl-3-(*m*-tolyl)urea 580
1,1-Diethyl-3-(*p*-tolyl)urea 590
N,*N*-Diisopropylisonicotinamide 952
N,*N*-Diisopropylnicotinamide 950
N-(2-(Dimethylaminoethyl))-2-methylbenzamide 469
1,1-Dimethyl-3-(2-ethyl-6-methylphenyl)urea 593
1,3-Dimethyl-2-(2-methoxyphenyl)-imidazolidine 247
N,*N*-Dipropylisonicotinamide 953
N,*N*-Dipropylnicotinamide 950
1-Ethyl-3-(2-ethyl-6-methylphenyl)urea 593
1-(2-Ethylphenyl)-3-isopropylurea 587
1-(2-Ethylphenyl)-3-propylurea 587
1-Isobutyl-3-(*o*-tolyl)urea 587
1-Isobutyl-3-(*m*-tolyl)urea 580
1-Isobutyl-3-(*p*-tolyl)urea 590
1-Isopropyl-3-(2,3-xylyl)urea 593
1-Isopropyl-3-(2,4-xylyl)urea 598
1-Isopropyl-3-(2,5-xylyl)urea 598
1-Isopropyl-3-(2,6-xylyl)urea 593
1-Isopropyl-3-(3,4-xylyl)urea 598
1-Isopropyl-3-(3,5-xylyl)urea 602
1-(4-Methoxyphenyl)-2-methylpiperazine 247
6-Methyl-2-pyridinealdoxime *o*-isoamyl ether 1054
N-Methyl-*N*-(2-(2-pyridyl)ethyl)-2-tetrahydrofurylamine 946
1-Propyl-3-(2,3-xylyl)urea 593
1-Propyl-3-(2,4-xylyl)urea 598
1-Propyl-3-(2,5-xylyl)urea 599
1-Propyl-3-(2,6-xylyl)urea 593
1-Propyl-3-(3,4-xylyl)urea 598
1-Propyl-3-(3,5-xylyl)urea 602

$C_{12}H_{18}N_2OS$

1-Methyl-3-(3-(methylthio)phenyl)-1-propylurea 561

$C_{12}H_{18}N_2O_2$

1-(*o*-Anisyl)-3-butylurea 558
1-(*m*-Anisyl)-3-butylurea 558
3-(*m*-Anisyl)-1,1-diethylurea 557
1-(*o*-Anisyl)-3-isobutylurea 555
1-(*m*-Anisyl)-3-isobutylurea 558
1-(*p*-Anisyl)-3-isobutylurea 559
1-Butyl-3-(4-methoxyphenyl)urea 559
1-Cyclohexyl-3-furfurylurea 854
3-(Diethylamino)phenyl *N*-methylcarbamate 673
1,1-Diethyl-3-(2-methoxyphenyl)urea 558
1,1-Diethyl-3-(4-methoxyphenyl)urea 559
2-Dimethylaminoethyl *N*-(*o*-tolyl)-carbamate 671
2-Dimethylaminoethyl *N*-(*m*-tolyl)-carbamate 671
2-Dimethylaminoethyl *N*-(*p*-tolyl)-carbamate 672
1,3-Divinyl-5-isobutyl-5-methylhydantoin 135

$C_{12}H_{18}N_2O_2$ (Cont'd)
1-(2-Ethoxyphenyl)-3-propylurea 555
N,*N*,2,3,5,6-Hexamethyl-4-nitroaniline 276
3'-(Isopropylamino)-4'-methoxyacetanilide 415
1-Isopropyl-3-(2-methoxy-5-methylphenyl)urea 560
1-(2-Methoxyethyl)-3-(2,4-xylyl)urea 556
1-(2-Methoxyethyl)-3-(2,6-xylyl)urea 556
1-(2-Methoxy-5-methylphenyl)-3-propylurea 556
Methyl 4-*tert*-butyl-3,5-diaminobenzoate dihydrochloride 372
4-(2-Nitro-1,1,2-trimethylpropyl)-aniline 258

$C_{12}H_{18}N_2O_2S$
3-Amino-*N*[1]-cyclohexylbenzenesulfonamide 817
1-(2,5-Dimethoxyphenyl)-3-propyl-2-thiourea 735
Mesitylenesulfonic isopropylidenehydrazide 822
1-Methyl-4-(*p*-tolylsulfonyl)-piperazine 818
p-Toluenesulfonic acid (2,2-dimethylpropylidene)hydrazide 1092

$C_{12}H_{18}N_2O_3$
3-Acetyl-4a,5,6,7,8,8a-hexahydro-2-methyl-4*H*-1-benzopyran-4-one dioxime 1010
1-(2,5-Dimethoxyphenyl)-1-isopropylurea 556
1-(2,5-Dimethoxyphenyl)-3-isopropylurea 556
1-(2,5-Dimethoxyphenyl)-3-propylurea 556
2-(Dimethylamino)ethyl *N*-(2-methoxyphenyl)carbamate 652
2-(Dimethylamino)ethyl *N*-(3-methoxyphenyl)carbamate 652
2-(Dimethylamino)ethyl *N*-(4-methoxyphenyl)carbamate 652
Ethyl 2,4-dimethyl-5-(dimethylcarbamoyl)-3-pyrrolecarboxylate 843

$C_{12}H_{18}N_2O_3S$
1-Benzylsulfonyl-3-butylurea 825
1-Butyl-3-(*p*-tolylsulfonyl)-urea 826
1,1-Diethyl-3-(*p*-tolylsulfonyl)urea 826
1-Isobutyl-3-(*p*-tolylsulfonyl)-urea 825
1-(1-Methylpropyl)-3-(*p*-tolylsulfonyl)-urea 826

$C_{12}H_{18}N_2O_3S_2$
1-(2-(Ethylthio)-ethyl)-3-(*p*-tolylsulfonyl)-urea 815

$C_{12}H_{18}N_2O_4$
1,4-Bis-(acetoacetyl)-piperazine 109
Diethyl 1-amino-2,5-dimethyl-3,4-pyrroledicarboxylate 842
1-(2,5-Dimethoxyphenyl)-3-(2-methoxyethyl)urea 556
6-(4-Methylpiperazinomethyl)-kojic acid 58

$C_{12}H_{18}N_2O_4S$
2-Amino-*N*-methyl-*N*-(2-methyl-1,3-dioxolan-2-ylmethyl)-benzenesulfonamide 816
1-(3-Methoxypropyl)-3-(*p*-tolylsulfonyl)-urea 815

$C_{12}H_{18}N_2O_4S_2$
1-(2-(Ethylsulfinyl)-ethyl)-3-(*p*-tolylsulfonyl)-urea 822

$C_{12}H_{18}N_2O_5$
1-Butyl-3-(ethoxycarbonylmethyl)-barbituric acid 134
Diethyl 2-acetamido-2-(2-cyanoethyl)-malonate 104

$C_{12}H_{18}N_2O_5S_2$
1-(3-(Methylsulfonyl)-propyl)-3-(*p*-tolylsulfonyl)-urea 825

$C_{12}H_{18}N_2O_6$
Gluconic phenylhydrazide 494

$C_{12}H_{18}N_2O_7S$
Sulfanilamide-*N*[4]-D-glucoside 811

$C_{12}H_{18}N_4O$
1-(α-Butyl-4-hydroxybenzylideneamino)-guanidine nitrate 1037
4-Hydroxy-6-methyl-2-(2-methylcyclohexylidenehydrazino)-pyrimidine 972

$C_{12}H_{18}N_4OS$
1-(4-(Dimethylamino)benzoyl)-4-ethyl-3-thiosemicarbazide 742
4-(2-Methylphenyl)-1-(*N*-propylcarbamoyl)-3-thiosemicarbazide 743

$C_{12}H_{18}N_4O_2$
4-(*tert*-Butylureido)benzoic acid hydrazide 577

$C_{12}H_{18}N_4O_2S$
3-(4-Amino-2-methyl-5-pyrimidyl)-4,5-dimethyl-5-(2-hydroxy-ethyl-)-2-thiazolidinone 971

$C_{12}H_{18}N_4O_4$
N,*N*-Diethyl-*N'*-(2,4-dinitrophenyl)-ethylenediamine 272

$C_{12}H_{18}N_4O_8$
Triethylamine *N*-oxide picrate 277

$C_{12}H_{18}N_4S$
6-Heptylmercaptopurine 923
6-(Heptylthio)purine 923

$C_{12}H_{18}N_4S_2$
2,2'-(2,5-Dimethyl-1,4-xylylene)bis(2-thiopseudourea) dihydrochloride 749
2,2'-(4,5-Dimethyl-1,2-xylylene)bis(2-thiopseudourea) dihydrochloride 739

$C_{12}H_{18}O$
3-Benzyl-3-methyl-2-butanol 207
3-Benzyl-3-pentanol 207
4-*tert*-Butylbenzyl methyl ether 186
4-(*tert*-Butyl)-α-methylbenzyl alcohol 208
α-*tert*-Butylphenethyl alcohol 207
β-*sec*-Butylphenethyl alcohol 207
4-*tert*-Butylphenetole 181
α,α-Diethylbenzyl methyl ether 186
2,5-Diisopropylphenol 192
1,4a-Dimethyl-4,4a,5,6,7,8-hexahydro-2(3*H*)-naphthalenone 54
α,α-Dimethyl-4-isopropylbenzyl alcohol 208
2,2-Dimethyl-3-(4-tolyl)-1-propanol 208
1,2-Epoxy-5,6-*trans*-9,10-*cis*-cyclododecadiene 20
Ethyl methyl β-phenylethyl carbinol. 207
α-Isopropyl-α-methylbenzyl methyl ether 186
β-Isopropyl-β-methylphenethyl alcohol 207
4-Isopropyl-2-methylphenethyl alcohol 208
2-Methyl-1-phenyl-2-pentanol 207
4-Methyl-4-phenyl-2-pentanol 207
3-Methyl-3-(4-tolyl)-2-butanol 208
4-*tert*-Pentylanisole 181
1-Phenyl-2-hexanol 207

$C_{12}H_{18}O_2$
1,4-Benzenedimethanol diethyl ether 187
2,5-Bis-(methoxymethyl)-*p*-xylene 187
2-(4-(*tert*-Butyl)-phenoxy)-ethanol 202
2,3-5,6-Diepoxy-1,2,3,4,5,6-hexamethylbicyclo(2.2.0)hexane 22
3,8-Dihydroxy-3,8-dimethyl-4,6-decadiyne 1015
3,3-Dimethyl-1-phenyl-1,2-butanediol 199
4-Tricyclo[4.3.1.$1^{3,8}$]-undecanecarboxylic acid 67

$C_{12}H_{18}O_2S$
tert-Butyl 2,4-xylyl sulfone 786
1,1-Diethoxy-1-(phenylthio)ethane 210
2,2-Dimethylpropyl *p*-tolyl sulfone 785
Isopropyl mesityl sulfone 786
Pentyl *p*-tolyl sulfone 785
2-(Phenylthio)acetaldehyde diethyl acetal 211

$C_{12}H_{18}O_2S_2$
α,α-Bis(2-hydroxyethylthio)-*p*-xylene 210

$C_{12}H_{18}O_3$
Butyl 3-oxatricyclo[3.2.1.$0^{2,4}$]-octane-7-carboxylate 91
2,4-Dimethylphenoxyacetaldehyde dimethyl acetal 189
(2,5-Dimethylphenoxy)-acetaldehyde dimethyl acetal 178
2,6-Dipropionyl-1-cyclohexanone 55
3,3-(Ethylenedioxy)camphor 53
2,2,4,4,6,6-Hexamethyl-1,3,5-cyclohexanetrione 55
3-Hydroxy-1-adamantaneacetic acid 65
3-Mesityloxy-1,2-propanediol 202
2-Methoxy-2-(2-phenoxyethoxy)-propane 187
2-(1-Octenyl)succinic anhydride 98
8-Oxo-2-decahydronaphthyl acetate 92
1,2,4-Triethoxybenzene 179

$C_{12}H_{18}O_3S$
2,2-Dimethylpropyl *p*-toluenesulfonate 800
Neopentyl *p*-toluenesulfonate 800
Pentyl *p*-toluenesulfonate 800
6-Phenyl-1-hexanesulfonic acid, sodium salt 795

$C_{12}H_{18}O_4$
(-)-4-Carboxy-2,3,3-trimethyl-1-cyclohexen-1-yl acetate 95
3-Carboxy-1,2,2-trimethylcyclopentaneacrylic acid 66
2,6-Diacetoxybicyclo(3.3.0)octane 77
Diethyl 2-cyclopentenylmalonate 79
Diethyl cyclopentylidenemalonate 78
Ethyl 2-(5,5-dimethyl-3-oxo-1-cyclohexenyloxy)acetate 91
Ethyl 4,6-dioxo-1,2,2-trimethyl-1-cyclohexanecarboxylate 93
1,4-Perhydropentalenediol diacetate 77

$C_{12}H_{18}O_4S$
2-Methoxybutyl *p*-toluenesulfonate 797

$C_{12}H_{18}O_5$
Diethyl 7-oxatricyclo(4.1.0)-heptane-3,4-dicarboxylate 90

$C_{12}H_{18}O_5Si$
3-Methoxy-1,2,3,6-tetrahydro-5-(trimethylsilyloxy)phthalic anhydride 1106

$C_{12}H_{18}O_6$
1,4-Butanediol diacetoacetate 93
Triethyl 1,2,3-propenetricarboxylate 79

$C_{12}H_{18}S$
Isoamyl *p*-tolyl sulfide 213

$C_{12}H_{19}Br$
1-(2-Bromoethyl)adamantane 6

$C_{12}H_{19}BrN_2O$
3-Carbamoyl-1-hexylpyridinium bromide 947
4-Carbamoyl-1-hexylpyridinium bromide 951

$C_{12}H_{19}BrO$
3-Bromo-1-adamantaneethanol 16
5-Bromo-5-hexen-1-yl 2,4-hexadien-1-yl ether 18

$C_{12}H_{19}BrO_2$
Methyl 1-bromo-2-decalincarboxylate 84

$C_{12}H_{19}BrPb$
(4-Bromophenyl)triethyllead 1134

$C_{12}H_{19}Br_2NO_5$
Diethyl 2-acetamido-2-(2,3-dibromopropyl)-malonate 117

$C_{12}H_{19}Br_4O_6P$
Bis-(2,3-dibromopropyl) 2-(methacryloyloxy)-ethyl phosphate 160

$C_{12}H_{19}ClN_2$
N'-(3-Chlorophenyl)-*N*,*N*-diethylethylenediamine 236
N'-(4-Chlorophenyl)-*N*,*N*-diethylethylenediamine 236

$C_{12}H_{19}Cl_3Si_2$
(2,3,5-Trichloro-1,4-phenylene)-bis(trimethylsilane) 1100

$C_{12}H_{19}F_3O_2$
2-Isopropyl-5-methylcyclohexyl trifluoroacetate 83

$C_{12}H_{19}N$
2-*sec*-Butyl-6-ethylaniline 219
α-(*sec*-Butyl)-*N*-methylbenzylamine 243
N,*N*-Diethyl-3,5-dimethylaniline 220
Diethyl (1-phenylethyl)amine picrate 277
N,*N*-Di-*N*-propylaniline 220

$C_{12}H_{19}NO$
2-Amino-4-*tert*-amylanisole 218
3-(Cyclohexylamino)-2-cyclohexen-1-one 57
2-(Diethylamino)-1-phenylethanol 238
5,5-Dimethyl-3-pyrrolidino-2-cyclohexen-1-one 58
N-Methyl-1-adamantanecarboxamide 106
2-Methyl-2-adamantanecarboxamide 102
β-(Phenoxy)triethylamine 240
2-(4,5,6,7-Tetrahydro-3*H*-azepin-2-yl)-1-cyclohexanone 59

$C_{12}H_{19}NOS$
β-(2-Dimethylaminoethylthio)-phenethyl alcohol 238

$C_{12}H_{19}NO_2$
2-Acetyl-8-methyldecahydroisoquinolin-7-one. 109
1-Adamantylacetohydroxamic acid 105
5,5-Dimethyl-3-morpholino-2-cyclohexen-1-one 59
Ethyl 2-cyano-3-ethyl-2-heptenoate 148
Ethyl 2-cyano-3-methyl-2-octenoate. 148
α-Hydroxy-1-adamantaneacetamide 103
3-(Morpholinomethyl)-2-norbornanone 59

$C_{12}H_{19}NO_3$
tert-Butyl 3,5-dimethyl-4-methoxy-2-pyrrolecarboxylate 838

$C_{12}H_{19}NO_3S$
2-(2,2-Dimethyl-1-hydroxy)propyl-*N*-methylbenzenesulfonamide 811

$C_{12}H_{19}NO_4$
2,2'-Iminobis(1-cyclopentanecarboxylic acid) 65
Methyl 4-hydroxy-2-(pyrrolidinocarbonyl)-cyclopentanecarboxylate 131

$C_{12}H_{19}NO_4S$
N-(2-Hydroxyethyl)-*N*-(2-hydroxypropyl)-*p*-toluenesulfonamide 812
N-(2-Hydroxyethyl)-*N*-(3-hydroxypropyl)-*p*-toluenesulfonamide 812

$C_{12}H_{19}NO_6$
Diethyl 2-acetamido-2-(2-formylethyl)malonate 104
Diethyl 2-acetamido-2-(2-oxopropyl)malonate 121

$C_{12}H_{19}N_3$
1-Benzyl-4-hydrazinopiperidine 249
1-Benzyl-4-hydrazinopiperidine dihydrochloride 249

$C_{12}H_{19}N_3O$
2-Cyanoacetic (3,3,5-trimethylcyclohexylidene)-hydrazide 147
1-(3-(Dimethylamino)-propyl)-3-phenylurea 562
1,1-Dimethyl-4-(2-ethyl-6-methylphenyl)semicarbazide 578

$C_{12}H_{19}N_3O_2$
1-(2-Cyanoacetyl)-3-cyclohexyl-1-ethylurea 149

$C_{12}H_{19}N_3O_3$
Ethyl 2-(1-ethylpropylamino)-4-hydroxy-5-pyrimidinecarboxylate 971
Ethyl 4-hydroxy-2-neopentylamino-5-pyrimidinecarboxylate 972

$C_{12}H_{19}N_3O_3S$
1-(*p*-Aminophenylsulfonyl)-3-isopentylurea 817

$C_{12}H_{19}N_3O_4S$
N-(2-(*N*-Ethyl-3-methyl-4-nitroanilino)ethyl)-methanesulfonamide 817

$C_{12}H_{19}O_2P$
Diisopropyl phenylphosphonite 833

$C_{12}H_{19}O_3P$
Diisopropyl phenylphosphonate 836

$C_{12}H_{19}P$
(Dipropyl)-phenylphosphine 830

$C_{12}H_{20}$
Cyclohexylidenecyclohexane 3
Dimethyltricyclo[5.2.1.0$^{2.6}$]decane, mixture of isomers 2
$C_{12}H_{20}Br_2O$
2,12-Dibromocyclododecanone 53
$C_{12}H_{20}INO$
(4-Hydroxy-2-isopropylphenyl)-trimethylammonium iodide 256
o-Methylcandicine iodide 254
$C_{12}H_{20}INO_5$
4,4-Bis-(acetoxymethyl)-3-ethyl-2-methyl-2-oxazolinium iodide 90
$C_{12}H_{20}IP$
Phenyltriethylphosphonium iodide 831
$C_{12}H_{20}NO_4P$
3-(Benzyloxy)propyl ethyl phosphoramidate 834
$C_{12}H_{20}N_2$
2-Adamantanone dimethylhydrazone 1072
N,N-Diethyl-*N',N'*-dimethyl-1,3-phenylenediamine 221
N,N-Diethyl-*N'*-phenylethylenediamine 241
α,α'-(Dimethylamino)-*p*-xylene dihydrobromide 243
2,7-Dimethyl-2,7-octanedicarbonitrile 150
N',N',2,3,5,6-Hexamethyl-1,4-phenylenediamine dihydrochloride 221
2,3,5,6-Tetramethyl-1,4-xylylenediamine dihydrochloride 241
$C_{12}H_{20}N_2O$
N-(1-Cyanocyclohexyl)-*N*-methylbutyramide 150
6,6'-Oxydihexanenitrile 144
$C_{12}H_{20}N_2O_2$
3,4-Bis(butylamino)-3-cyclobutene-1,2-dione 57
1,2-Bis(2-methyl-4-pentenoyl)hydrazine 112
1,6-Bis(1-pyrrolidinyl)-1,6-hexanedione 109
(2-Pyridylmethylamino)-acetaldehyde diethyl acetal 937
(3-Pyridylmethylamino)-acetaldehyde diethyl acetal 940
(4-Pyridylmethylamino)-acetaldehyde diethyl acetal 940
$C_{12}H_{20}N_2O_2S$
4-(2-Diethylaminoethylsulfonyl)-aniline 776
$C_{12}H_{20}N_2O_4$
1,1'-Dinitrodicyclohexyl 46
$C_{12}H_{20}N_2S$
1-(1-Adamantyl)-3-methyl-2-thiourea 142
$C_{12}H_{20}N_4$
1-(2-(2-Pyridylmethylamino)-ethyl)-piperazine 943
1-(2-(3-Pyridylmethylamino)-ethyl)-piperazine 943
$C_{12}H_{20}N_4O_3$
6-Amino-1,3-dibutyl-5-nitrosouracil 132
L-Histidyl-L-isoleucine hemihydrate 866
$C_{12}H_{20}N_6O_{14}$
3,3,5,8,10,10-Hexanitro-1,12-dodecanediol 45
$C_{12}H_{20}O$
4-Cyclohexylcyclohexanone 56
Diethyldimethyl-2-cyclohexen-1-one 54
1,4a-Dimethyldecahydro-naphthalen-2-one 57
4a,8a-Dimethyl-2-decalone 56
7,7-Dimethyl-3-(methoxymethyl)-2-methylenebicyclo(3.1.1)heptane 18
2-(2-Ethylbutylidene)-1-cyclohexanone 56
$C_{12}H_{20}OS$
2-(Butylthiomethylene)-6-methyl-1-cyclohexanone 53
$C_{12}H_{20}OSi$
4-Methyl-3-(1-(trimethylsilyl)vinyl)-2-cyclohexen-1-one 1109
1-Phenyl-3-(trimethylsilyl)-2-propanol 1105
$C_{12}H_{20}O_2$
1,3-Adamantanedimethanol 12
4-*cis*-8-*trans*-Cyclododecadiene-1,2-*trans*-diol 12
5-*cis*-9-*trans*-Cyclododecadiene-1,2-*cis*-diol 12
1-Cycloundecene-1-carboxylic acid 65
Decahydro-2-naphthyl acetate 78
5,7-Dimethyl-1,3-adamantanediol 15
5,9-Dimethyl-2,8-decadienoic acid 61
Ethyl 2-(1-cyclohexenyl)butyrate 81
Ethyl 2,5,5-trimethyl-2-cyclohexene-1-carboxylate 80
Geranyl acetate 80
DL-Isobornyl acetate 77
Methyl 2-methyl-2,4-decadienoate 79
Methyl 2-methylene-4-decenoate 79
1,1,5-Trimethyl-4,6-heptadienyl acetate 80
$C_{12}H_{20}O_3$
4-(4,4-Dimethyl-2-oxocyclohexyl)-butyric acid 65
7-Oxo-11-dodecenoic acid 64
3-(2-Oxo-1,3,3-trimethylcyclohexyl)-propionic acid 65
$C_{12}H_{20}O_4$
1,6-Bis-(1,3-dioxolan-2-yl)-3-hexene 23
1,4-Cyclohexanedione bis(1,3-propylene ketal) 58
Dibutyl fumarate 79
Diethyl 2-ethyl-2-propenylmalonate 79
Diethyl hexahydrophthalate 77
4a,9a-Dihydroxyperhydro-dibenzo(b,e)(1,4)dioxin 12
2,5-Dimethyl-2-hexene-1,6-diol diacetate 81
Dimethyl 1,2,2-trimethyl-1,3-cyclopentanedicarboxylate 77
2-Ethoxycarbonyl-2-(2-hydroxyethyl)-heptanoic acid γ-lactone 96
2-Ethoxycarbonyl-2-(2-hydroxyethyl)-5-methylhexanoic acid γ-lactone 96
2-(2-Hydroxyethyl)-2-isopentylglutaric acid γ-lactone 97
Methyl 4-(1-ethoxyethoxy)-2-cyclopenteneacetate 85
2-Methyl-3,5,5-triethoxy-2-cyclopenten-1-one 52
1,1,4,4-Tetraethoxy-2-butyne 1017
$C_{12}H_{20}O_4S_2$
Cyclohexane-1,1-bis-(3-mercaptopropionic acid) 65
$C_{12}H_{20}O_5$
3,3-Dimethyl-6-(2,2-dimethyl-1,3-dioxolan-4-yl)-2,4-dioxabicyclo-(3.1.1)heptan-1-ol 21
$C_{12}H_{20}O_6$
1,2-*O*-Cyclohexylideneinositol 21
Diethyl 2-(2-(carbomethoxy)-propyl)-malonate 75
2,3:4,5-Diisopropylidene-β-D-fructopyranoside 16
1,2:4,6-Di-*O*-isopropylidene-α-D-glucopyranoside 205
2,6-Dimethoxy-4-(2-hydroxyacetyl)-cyclohexyl acetate 91
Triethyl 1,1,2-propanetricarboxylate 75
$C_{12}H_{20}O_6S_3$
1,1,1-Propanetriyl tris-(3-mercaptopropionate) 86
$C_{12}H_{20}S$
7,7-Dimethyl-2-methylene-3-(methylthiomethyl)bicyclo(3.1.1)-heptane 27
$C_{12}H_{20}Si$
Tetraallylsilane 1097
$C_{12}H_{21}BO_3$
Trimethallyl borate 1124
$C_{12}H_{21}BrO$
2-Bromocyclododecanone 53
$C_{12}H_{21}ClO_2$
2-Menthoxyacetyl chloride 100
$C_{12}H_{21}FSi_2$
1,3-Bis(trimethylsilyl)-2-fluorobenzene 1104
$C_{12}H_{21}N$
1-(1-Adamantyl)ethylamine dibenzoyltartrate 371
1-(2,4-Dimethyl-1,4-pentadien-1-yl)-piperidine 40
Dodecahydrocarbazole 40
1-(5-Methyl-2,4-hexadien-1-yl)-piperidine 40
$C_{12}H_{21}NO$
(+/-)-β-Amino-1-adamantaneethanol hydrochloride 36
(2-(1,4-Cyclohexadien-1-yloxy)ethyl)-diethylamine 32
4-(2,5-Dimethyl-2,5-hexadien-1-yl)-morpholine 41
4-Isopropyl-7-methyl-5-oxooctanenitrile 148
4-(2,7-Octadien-1-yl)morpholine 41
$C_{12}H_{21}NO_2$
3-(Morpholinomethyl)-2-norbornanol, mixture of *endo* and *exo* 37
$C_{12}H_{21}NO_3$
cis-*N*-*tert*-Butylhexahydro-phthalamic acid 122
2,5-Dimethyl-1-hexylpyrrole 843
Methyl 9-cyano-2-methoxynonanoate 144
N-Octylmaleamic acid 122
$C_{12}H_{21}NO_4$
Ethyl 2-methyl-2-(1-nitrocyclohexyl)-propionate 77
$C_{12}H_{21}NO_5Si$
4-(2,3-Epoxy-1-hydroxy-4-(trimethylsilyloxy)butyl)glutarimide 1105
$C_{12}H_{21}NO_6$
Ethyl *N*-(ethoxycarbonyl)-*N*-(ethoxycarbonylmethyl)-3-aminopropionate 115
$C_{12}H_{21}N_3O_2$
5-Amino-1,3-dibutyluracil 133
$C_{12}H_{21}N_3O_3$
Hexahydro-1,3,5-tripropionyl-1,3,5-triazine 111
$C_{12}H_{21}N_3O_4$
N,N'-Bis-(acetoacetyl)-diethylenetriamine 110
$C_{12}H_{22}BrN$
1-Ethyl-1-(2,4-pentadien-1-yl)-piperidinium bromide 44
$C_{12}H_{22}BrNO_3$
DL-2-Bromoisocaproyl-DL-leucine 123
$C_{12}H_{22}Cl_2N_2O_4$
N,N'-Hexamethylenebis(2-chloroethyl carbamate) 113
$C_{12}H_{22}Hg_2O_6$
meso-1,4-Bis-(acetoxymercuri)-2,3-diethoxybutane 1138
$C_{12}H_{22}IN$
4-(4-Iodocyclohexyl)-cyclohexylamine 32
$C_{12}H_{22}N_2O$
1-(Cyclohexylcarbamoyl)piperidine 130
N-Nitrosodicyclohexylamine 45
$C_{12}H_{22}N_2O_2$
6-Amino-4-oxo-7,7,*N,N*-tetramethyl-5-octenamide 107
2,4-Bis(butylamino)-3-oxo-1-cyclobutenylium-1-olate 1095
1,2-Bis(1-methylpropyl)-4,5-dihydro-3,6(1*H*,2*H*)-pyridazinedione 136
1,2-Butyl-4,5-dihydro-3,6(1*H*,2*H*)-pyridazinedione 134
3-(1-Hydroxycyclohexyl)-1-methyl-4-piperidone oxime 1008
N-(3-Piperidinopropyl)-acetoacetamide 104
$C_{12}H_{22}N_2O_4$
N-ε-Cyclopentyloxycarbonyl-L-lysine 119
1,1-Dinitrocyclododecane 46
2-Methyl-2-propyl-1,3-propanediol carbamate cyclopropylcarbamate 115
$C_{12}H_{22}N_4O_5$
Glycyl-L-leucylglycylglycine 120
$C_{12}H_{22}N_6O_4$
1,1'-Azobis(2-(nitromethyl)-piperidine) 1088
$C_{12}H_{22}O$
1-Cyclohexylcyclohexanol 14
2-Cyclohexylcyclohexanol 14
4-Cyclohexylcyclohexanol 14
2-(2-Ethylbutyl)-1-cyclohexanone 56
2-Heptylcyclopentanone 55
2-(2-Methylpentyl)-cyclohexanone 56
2,2,6-Triethyl-1-cyclohexanone 56
$C_{12}H_{22}O_2$
(1,1'-Bicyclohexyl)-1,1'-diol 12
2-*tert*-Butylcyclohexyl acetate 78
Cycloundecanecarboxylic acid 66
3,4-Diisopropyl-2,5-hexanedione 51
4,7-Dimethyl-5-decyne-4,7-diol 1016
3,7-Dimethyl-6-octenyl acetate 80
trans-4-Ethoxy-1-decalol 20
Ethyl 2-methyl-2-nonenoate 79
Heptyl isovalerate 73
2-Hydroxycyclododecanone 52
Methyl 10-undecenoate 80
2,2,4,6,6-Pentamethyl-3,5-heptanedione 50
Vinyl decanoate 80
Vinyl neodecanoate (inhibited with 5 ppm hydroquinone) 78
$C_{12}H_{22}O_3$
Butyl 3-butoxymethacrylate 78
Ethyl 2-hexylacetoacetate 94
trans-4-(2-Hydroxyethoxy)-1-decalol 20
Methyl 4-*tert*-butyl-2-hydroxycyclohexanecarboxylate 84
Methyl 10,11-epoxyundecanoate 91
Methyl 5-hydroxy-2-methyl-3-decenoate 85
3-Pentyltetrahydropyran-4-yl acetate 91
$C_{12}H_{22}O_4$
Diethyl 2-butyl-2-methylmalonate 75
Diethyl 2,3-diethylsuccinate 75
Diethyl isopentylmalonate 75
Diethyl pentylmalonate 75
Diethyl *tert*-pentylmalonate 75
Diisopropyl adipate 76
2,5-Dimethyl-1,6-hexanediol diacetate 76
2,5-Dimethyl-2,5-hexanediol diacetate 75
Dimethyl 2,2,5,5-tetramethyladipate 76
mono-Ethyl sebacate 360
L-Menthyl 2,2-dihydroxyacetate 84
$C_{12}H_{22}O_4S_2$
Dibutyl dithiodiglycolate 87
$C_{12}H_{22}O_4Si$
1-Carboxy-4-(trimethylsilyl)-cyclohexaneacetic acid 1109
$C_{12}H_{22}O_6$
Diethyl 2,3-diethoxysuccinate 85
1,2:3,4-Di-*O*-isopropylidene-D-mannitol 16
Dimethyl 2-(dimethoxymethyl)-2-ethylglutarate 74
1,2-*O*-Isopropylidene-3,5,6-tri-*O*-methyl-α-D-glucofuranoside 16
$C_{12}H_{22}O_7S$
Dibutyl sulfosuccinate, sodium salt 155
$C_{12}H_{22}O_{11}$
D(+)-Treholose 16
$C_{12}H_{22}Si_2$
1,4-Bis(trimethylsilyl)benzene 1122
1,2-Phenylenebis(trimethylsilane) 1122
$C_{12}H_{23}BrO$
12-Bromo-2-dodecen-1-ol 15
$C_{12}H_{23}BrO_2$
2-Bromolauric acid 62
Ethyl 2-bromodecanoate 83
$C_{12}H_{23}ClO_2$
Decyl 2-chloroacetate 82
$C_{12}H_{23}N$
2-Bicyclohexylamine 31
tert-Butylcyclooctylideneamine 1020
Dicyclohexylamine phosphate 31
N-(1,5-Dimethyl-4-hexenylidene)-isobutylamine 31
$C_{12}H_{23}NO$
Cyclododecanone oxime 1008
2-Methyl-5-piperidino-3-hexen-1-ol 37
5-Methyl-3-piperidino-4-hexen-1-ol 37
4-Methyl-3-(piperidinomethyl)-2-pentanone 59

$C_{12}H_{23}NO_2$
Ethyl 1-(dimethylamino)-1-cyclohexaneacetate hydrochloride 88
N-(2-Ethylhexyl)-acetoacetamide 105
N-(Hydroxymethyl)-10-undecenoamide 104
N-Octylacetoacetamide 105
N-(*tert*-Octyl)acetoacetamide 105

$C_{12}H_{23}NO_3$
12-Amino-7-oxododecanoic acid 71
12-Amino-7-oxododecanoic acid hydrochloride 71
5-Morpholino-3-hexenal dimethyl acetal 38

$C_{12}H_{23}NO_4$
L-Aspartic acid di-*tert*-butyl ester hydrochloride 358
3,3′-Diallyloxy-2,2′-dihydroxydipropylamine 32
Diethyl (3-dimethylaminopropyl)-malonate 88
3,3′-Ethyliminobis(ethyl propionate) 88

$C_{12}H_{23}N_3O$
4-(Cyclohexylamino)-isonipecotamide 108

$C_{12}H_{23}PS_2$
Dicyclohexyldithiophosphinic acid, ammonium salt hydrate 160

$C_{12}H_{24}$
Cyclododecane 2
1-Cyclohexylhexane 2
1,3-Diisopropylcyclohexane 2
1,4-Diisopropylcyclohexane 2
1,1-Dineopentylethylene 2
2,2,4,6,6-Pentamethyl-3-heptene 2
Triisobutylene 2

$C_{12}H_{24}ClNO$
2-Chloro-*N*-methyl-*N*-(1-methyloctyl)-acetamide 103
N-Decyl-2-chloroacetamide 103

$C_{12}H_{24}ClNO_2S$
2-Chloro-3-(diisopropylamino)-2,4,4-trimethylthietane-1,1-dioxide 153

$C_{12}H_{24}N_2$
4-(Dibutylamino)-butyronitrile 147
1,2′-Ethylenedipiperidine 39
N,N,N′,N′-Tetraethyl-2-butyne-1,4-diamine 1018

$C_{12}H_{24}N_2O$
2,6-Bis-(dimethylaminomethyl)-cyclohexanone dihydrochloride 57
3-*tert*-Butyl-1-cyclohexyl-1-methylurea 129
1-Hexylisonipecotamide 108

$C_{12}H_{24}N_2OSi_2$
5,5,7,7-Tetramethyl-6-oxa-5,7-disilaundecanedinitrile 1110

$C_{12}H_{24}N_2O_2$
1,2-Bis(2-methylpentanoyl)-hydrazine 112
N,N,N′,N′-Tetraethylsuccinamide 111

$C_{12}H_{24}N_2O_2S$
11-(Amidinothio)-undecanoic acid 138

$C_{12}H_{24}N_2O_3$
D-Leucyl-L-leucine 123
DL-Leucyl-DL-leucine 120
L-Leucyl-D-leucine 123
N-L-Leucyl-L-leucine 123

$C_{12}H_{24}N_2O_4$
Diethyl hexamethylenedicarbamate 117

$C_{12}H_{24}N_2O_5$
*N′,N*²-Dicarbethoxy-*N′*-(2-hydroxy-ethyl)-2-methyl-1,2-propanediamine 87

$C_{12}H_{24}N_3O_3P$
Phosphoric trimorpholide 158

$C_{12}H_{24}O$
α,α-Dimethyl-4-isopropylcyclohexanemethanol 14
2-Dodecanone 49
3-Dodecanone 49
5-Dodecanone 49
2-(2-Ethylbutyl)-1-cyclohexanol 14
7-Ethyl-2-methyl-4-nonanone 48
Heptyl isobutyl ketone 48
1-Methoxy-4-(*tert*-pentyl)cyclohexane 17
2-Methyl-3-undecanone 49
2-Methyl-4-undecanone 49

$C_{12}H_{24}OSi$
tert-Butyldimethylsilyl 1-cyclohexen-1-yl ether 1111

$C_{12}H_{24}O_2$
tert-Butyl 2,2,3,3-tetramethylbutyrate 74
cis-1,2-Cyclododecanediol 12
Decyl acetate 74
3,7-Dimethyloctyl acetate 74
α-(1-Ethylpropyl)-2-hydroxycyclohexanemethanol 14
Lauric acid, octadecylamine salt 70
3-Methylundecanoic acid 61

$C_{12}H_{24}O_2S$
Decyl mercaptoacetate 86

$C_{12}H_{24}O_3$
1,4-Dibutoxy-2,3-epoxybutane 20
2-Hydroxydodecanoic acid 63

$C_{12}H_{24}O_3S$
4-(2,5-Dimethylcyclohexyl)-1-butanesulfonic acid, sodium salt 156

$C_{12}H_{24}O_4Pb$
Dibutylplumbylene diacetate 1135

$C_{12}H_{24}O_4Si$
Diethyl 3,3′-(dimethylsilylene)-dipropionate 1109

$C_{12}H_{24}O_5$
3,6,9,12,15-Pentaoxa-1-heptadecene 18

$C_{12}H_{24}S_2$
2-Octyl-1,3-dithiane 24

$C_{12}H_{24}Si_3$
1,1,1,3,3,3-Hexamethyl-2-phenyltrisilane 1112

$C_{12}H_{24}Sn$
Dibutyldivinyltin 1129

$C_{12}H_{25}Cl$
1-Chlorododecane 4

$C_{12}H_{25}ClHg$
Chlorododecylmercury 1137

$C_{12}H_{25}F$
1-Fluorododecane 4

$C_{12}H_{25}N$
N-Butyl-1,5-dimethyl-4-hexenylamine 32
Dodecamethyleneimine 39
1-Hexyl-4-methylpiperidine 40

$C_{12}H_{25}NO$
1-Dibutylamino-3-buten-2-ol 34
1-Diisopropylamino-5-methoxy-2-pentene 35
Dodecanamide 102

$C_{12}H_{25}NO_2$
1,1′-Cyclohexyliminodi-2-propanol 34

$C_{12}H_{25}NO_3$
O-*tert*-Butyl-L-threonine *tert*-butyl ester dibenzenesulfonamide salt 813

$C_{12}H_{25}N_3O_2$
2-(4-(2-Hydroxyethyl)-piperazinomethyl)-3-hydroxypiperidine 36
L-Isoleucyl-L-isoleucinamide hydrochloride 124
L-Leucyl-L-isoleucinamide hydrochloride 121
Leucylleucinamide hydrochloride 121
L-Norleucyl-L-norleucinamide hydrochloride 121

$C_{12}H_{26}$
2,2,4,6,6-Pentamethylheptane 1

$C_{12}H_{26}HgS_2$
Mercury(II) (*n*-hexylsulfide) 1137

$C_{12}H_{26}IN$
Cyclohexyltriethylammonium iodide 44

$C_{12}H_{26}N_2$
N,N,N′,N′-Tetraethyl-2-butene-1,4-diamine 32

$C_{12}H_{26}N_2OS$
β-(Morpholinoethylthio)-triethylamine dihydrochloride 38

$C_{12}H_{26}N_2O_2$
1,4-Bis-(2-hydroxypropyl)-2,5-dimethylpiperazine 36

$C_{12}H_{26}N_4O_2$
3,3′-Hexamethylenebis(1,1-dimethylurea) 129
1,1′-Hexamethylenebis(3-ethylurea) 129

$C_{12}H_{26}N_4O_6$
Neamine 32

$C_{12}H_{26}O$
6-Ethyl-3-decanol 9

$C_{12}H_{26}OS$
2-Hydroxyethyl decyl sulfide 25

$C_{12}H_{26}OSi$
Octyl trimethylsilyl ketone 1109

$C_{12}H_{26}OSn$
Dihexyloxotin 1129

$C_{12}H_{26}O_2$
2-Butyl-1,3-octanediol 10
5-Ethyl-3-methyl-2,4-nonanediol 10
7-Ethyl-2-methyl-4,6-nonanediol 10

$C_{12}H_{26}O_2PbS$
Triethylplumbyl cyclohexanesulfinate 1135

$C_{12}H_{26}O_2S$
2-(Decylsulfinyl)-ethanol 152

$C_{12}H_{26}O_2Sn$
Acetoxydiethylhexyltin 1131

$C_{12}H_{26}O_3$
Hydroxycitronellal dimethyl acetal 20

$C_{12}H_{26}O_4S_2$
1,4-Bis(*tert*-butylsulfonyl)butane 154

$C_{12}H_{26}O_6$
5-Hydroxymethyl-1,5,6,7,11-undecanepentol 16

$C_{12}H_{26}S$
Decyl ethyl sulfide 26

$C_{12}H_{26}S_2$
1,1-Dimethylbutyl disulfide 27
Hexyl disulfide 27

$C_{12}H_{27}Br_2Sb$
Dibromotributylantimony 1128

$C_{12}H_{27}ClGe$
Chlorotributylgermane 1097

$C_{12}H_{27}N$
Bis-(1,3-dimethylbutyl)-amine 28
tert-Butyl-*tert*-octylamine 28
Dihexylamine hydrochloride 29

$C_{12}H_{27}NO$
1-Dibutylamino-2-butanol 34
4-Dibutylamino-1-butanol 34
4-(Diisobutylamino)-1-butanol 34
6-Dipropylamino-1-hexanol 34

$C_{12}H_{27}NO_2$
Diisopropylaminoacetaldehyde diethyl acetal 35
1,1′-Dimethyl-1,1′-dipropyl-2,2′-iminodiethanol 33

$C_{12}H_{27}NO_3$
3,6,9-Trioxapentadecylamine 34

$C_{12}H_{27}NO_4$
2,2′-Iminobis(acetaldehyde diethyl acetal) 35

$C_{12}H_{27}O_4P$
Triisobutyl phosphate 160

$C_{12}H_{28}AsF_6N$
Tetrapropylammonium hexafluoroarsenate 43

$C_{12}H_{28}BF_4N$
Tetrapropylammonium tetrafluoroborate 43
Tributylammonium tetrafluoroborate 43

$C_{12}H_{28}ClN$
Tetrapropylammonium chloride 43

$C_{12}H_{28}ClN_2OP$
Tetraisopropylphosphorodiamidic chloride 160

$C_{12}H_{28}F_6NP$
Tributylammonium hexafluorophosphate 43

$C_{12}H_{28}Ge$
Tetrapropylgermane 1097
Tributylgermane 1097

$C_{12}H_{28}N_2$
N,N-Dibutyl-*N′*-methyl-1,3-propanediamine 30
N,N′-Dipropyl-1,6-hexanediamine 30

$C_{12}H_{28}N_2O_2S$
1-Dodecanesulfonic acid hydrazide 157

$C_{12}H_{28}N_2O_4$
Bi-(3-(2-methoxyethoxy)-propylamine) 34

$C_{12}H_{28}N_2S_2$
2-(Diethylamino)ethyl disulfide dihydrochloride 36

$C_{12}H_{28}N_6O_2$
4,4′-Hexamethylenebis(1,1-dimethylsemicarbazide) 577

$C_{12}H_{28}O_3Si$
Hexyltriethoxysilane 1097

$C_{12}H_{28}O_4Si$
Tetraisopropyl orthosilicate 1097

$C_{12}H_{28}O_8Si$
Tetrakis(2-methoxyethyl) orthosilicate 1106

$C_{12}H_{28}Sn$
Diethyldiisobutyltin 1129
Tetraisopropyltin 1129
Tetrapropyltin 1129

$C_{12}H_{29}N_3$
1,7-Di-(*sec*-butyl)-diethylene-triamine 30

$C_{12}H_{30}Cl_2N_2$
N,N′-Hexamethylenebis(trimethyl-ammonium chloride) 43

$C_{12}H_{30}Ge_2$
Hexaethyldigermane 1097

$C_{12}H_{30}Si_2$
Hexaethyldisilane 1097

$C_{12}H_{30}Si_4$
1,4-Divinyl-1,1,2,2,3,3,4,4-octamethyltetrasilane 1097

$C_{12}H_{36}Si_4Sn$
Tetrakis(trimethylsilyl)tin 1129

$C_{12}H_{36}Si_5$
2,2-Bis(dimethylsilyl)-1,1-dimethyl-3,3,3-triethyltrisilane 1097

$C_{12}H_{38}Si_6$
1,1,2,2,3,3,4,4,5,5,6,6-Dodecamethylhexasilane 1097

$C_{12}H_{48}Cr_2N_{12}O_{12}S_3$
Tris(ethylenediamine)chromium(III) sulfate 1141

$C_{12}H_{48}Fe_2N_{12}O_{12}S_3$
Tris(ethylenediamine)iron(II) sulfate 1141

$C_{13}Cl_4F_4N_2O_2S$
3,4,5,6-Tetrachloro-*N*-((2,3,5,6-tetrafluoro-4-pyridyl)thio)-phthalimide 967

$C_{13}Cl_6F_6O_2S_2$
1,5-Bis(3,4,5-trichloro-2-thienyl)-2,2,3,3,4,4-hexafluoro-1,5-pentanedione 857

$C_{13}HBrF_{10}$
Decafluorobenzhydryl bromide 173

$C_{13}H_3Cl_9S_3Si$
Methyltris(3,4,5-trichloro-2-thienyl)-silane 1110

$C_{13}H_4Cl_5F_2NOS$
S-(Pentachlorophenyl) *N*-(2,5-difluorophenyl)thiocarbamate 731

$C_{13}H_4Cl_{10}Si$
Bis(pentachlorophenyl)-methylsilane 1101

$C_{13}H_4F_{10}Si$
Bis(pentafluorophenyl)-methylsilane 1101

$C_{13}H_5BrFNO_3$
3-Bromo-7-fluoro-2-nitro-9-fluorenone 313

$C_{13}H_5Br_2NO_3$
2,7-Dibromo-4-nitro-9-fluorenone 313

$C_{13}H_5Cl_2IO$
2,7-Dichloro-4-iodo-9-fluorenone 315

$C_{13}H_5Cl_2NO_3$
2,3-Dichloro-7-nitro-9-fluorenone 313
2,7-Dichloro-4-nitro-9-fluorenone 313

$C_{13}H_5Cl_4NO$
2-Amino-1,3,4,7-tetrachloro-9-fluorenone 312

$C_{13}H_5Cl_4N_3O_5$
N-(5-Nitro-2-pyridyl)-3,4,5,6-tetrachlorophthalamic acid 929

$C_{13}H_5Cl_6NO_2$
2,4,5-Trichlorophenyl *N*-(2,4,5-trichlorophenyl)carbamate 641

$C_{13}H_5F_2NO_3$
1,3-Difluoro-7-nitro-9-fluorenone 313

$C_{13}H_5F_3O$
1,4,7-Trifluoro-9-fluorenone ... 315
$C_{13}H_6BrCl_2N_3O_6$
2-Bromo-6-chloro-4-nitrophenyl *N*-(4-chloro-3-nitrophenyl)-carbamate ... 626
$C_{13}H_6BrCl_3$
9-Bromo-2,4,7-trichlorofluorene ... 174
$C_{13}H_6BrNO_3$
2-Bromo-7-nitro-9-fluorenone ... 313
$C_{13}H_6Br_2ClFN_2O_4$
2,6-Dibromo-4-fluorophenyl *N*-(4-chloro-3-nitrophenyl)carbamate ... 626
$C_{13}H_6Br_2ClN_3O_6$
2,6-Dibromo-4-nitrophenyl *N*-(2-chloro-4-nitrophenyl)carbamate ... 627
$C_{13}H_6Br_2Cl_2$
2,7-Dibromo-9,9-dichlorofluorene ... 174
$C_{13}H_6Br_2F_2N_2O_4$
2,6-Dibromo-4-nitrophenyl *N*-(2,5-difluorophenyl)carbamate ... 628
$C_{13}H_6Br_3Cl_2NO_2$
2,4,6-Tribromophenyl *N*-(2,6-dichlorophenyl)carbamate ... 636
$C_{13}H_6Br_3NO$
2-Amino-1,3,7-tribromo-9-fluorenone ... 312
$C_{13}H_6Br_4Cl_2N_2O$
1,3-Bis(2-chloro-4,6-dibromophenyl)urea ... 546
1,3-Bis(4-chloro-2,6-dibromophenyl)urea ... 546
$C_{13}H_6ClF_3N_4O_6$
6-Chloro-*N*-picryl-α,α,α-trifluoro-*m*-toluidine ... 259
$C_{13}H_6ClNO_3$
4-Chloro-2-nitro-9-fluorenone ... 313
$C_{13}H_6Cl_2F_3NO_3$
2,4-Dichlorophenyl 2-nitro-α,α,α-trifluoro-*p*-tolyl ether ... 259
$C_{13}H_6Cl_2N_2O_2$
6,9-Dichloro-2-nitroacridine ... 996
$C_{13}H_6Cl_2O$
1,3-Dichloro-9-fluorenone ... 314
2,4-Dichloro-9-fluorenone ... 314
$C_{13}H_6Cl_3NO$
4-Amino-2,5,7-trichloro-9-fluorenone ... 312
2,4,7-Trichloro-9-fluorenone oxime ... 1014
$C_{13}H_6Cl_3NO_2$
5-Nitro-2,4,7-trichlorofluorene ... 280
$C_{13}H_6Cl_4N_2O_2$
N-(2-Chloro-5-nitrobenzylidene)-2,4,5-trichloroaniline ... 1058
$C_{13}H_6Cl_4O$
2,3,5,6-Tetrachlorobenzophenone ... 310
$C_{13}H_6Cl_4O_2$
2,2′,5′,6-Tetrachloro-4′-hydroxybenzophenone ... 310
2,3′,5′6-Tetrachloro-4′-hydroxybenzophenone ... 310
$C_{13}H_6Cl_5N$
N-(3,4-Dichlorobenzylidene)-2,4,5-trichloroaniline ... 1067
$C_{13}H_6Cl_5NO_2$
3,4-Dichlorophenyl *N*-(2,4,5-trichlorophenyl)carbamate ... 638
2,4,5-Trichlorophenyl *N*-(2,4-dichlorophenyl)carbamate ... 686
2,4,5-Trichlorophenyl *N*-(3,4-dichlorophenyl)carbamate ... 638
2,4,6-Trichlorophenyl *N*-(2,6-dichlorophenyl)carbamate ... 636
2,4,6-Trichlorophenyl *N*-(3,4-dichlorophenyl)carbamate ... 639
$C_{13}H_6Cl_6N_2O$
1,3-Bis(2,4,5-trichlorophenyl)urea ... 603
$C_{13}H_6FNO_3$
2-Fluoro-6-nitro-9-fluorenone ... 313
5-Fluoro-2-nitro-9-fluorenone ... 313
$C_{13}H_6F_2O$
1,4-Difluoro-9-fluorenone ... 314
2,4-Difluoro-9-fluorenone ... 314
2,5-Difluoro-9-fluorenone ... 314
$C_{13}H_6N_2O_5$
2,5-Dinitro-9-fluorenone ... 313
$C_{13}H_6N_4O_2$
4-Hydroxy-6*H*-indeno(2,1-g)pteridin-6-one ... 1002
$C_{13}H_7Br$
7-Bromo-7*H*-cyclopropa(a)-acenaphthylene ... 173
$C_{13}H_7BrClNO_3$
3-Bromo-4′-chloro-4-nitrobenzophenone ... 313
$C_{13}H_7BrCl_3NO$
2-Bromo-2′,4′,5′-trichlorobenzanilide ... 457
$C_{13}H_7BrCl_3NO_2$
2-Bromo-4-chlorophenyl *N*-(2,4-dichlorophenyl)carbamate ... 638
$C_{13}H_7BrFNO$
2-Amino-3-bromo-7-fluoro-9-fluorenone ... 312
$C_{13}H_7BrF_3NO_3$
2-Bromophenyl 2-nitro-4-(trifluoromethyl)phenyl ether ... 259
$C_{13}H_7BrN_2O_3$
2-Bromo-4-nitro-6(5*H*)-phenanthridinone ... 690
$C_{13}H_7BrOS$
2-Bromo-9-thioxanthenone ... 313
$C_{13}H_7BrO_2$
2-Bromo-9-xanthenone ... 320
$C_{13}H_7Br_2ClFNO_2$
2,4-Dibromophenyl *N*-(3-chloro-4-fluorophenyl)carbamate ... 638
$C_{13}H_7Br_2ClN_2O_4$
2,4-Dibromophenyl *N*-(2-chloro-4-nitrophenyl)carbamate ... 626
2,4-Dibromophenyl *N*-(4-chloro-2-nitrophenyl)carbamate ... 624
$C_{13}H_7Br_2Cl_2NO$
2′,6′-Dibromo-3,4-dichlorobenzanilide ... 459
$C_{13}H_7Br_2Cl_2NO_2$
2-Chloro-4,6-dibromophenyl *N*-(2-chlorophenyl)carbamate ... 633
2,4-Dibromophenyl *N*-(3,4-dichlorophenyl)carbamate ... 638
$C_{13}H_7Br_2NO$
1-Amino-2,4-dibromo-9-fluorenone ... 312
2-Amino-1,3-dibromo-9-fluorenone ... 312
4-Amino-1,3-dibromo-9-fluorenone ... 312
4-Amino-2,7-dibromo-9-fluorenone ... 312
$C_{13}H_7Br_2NO_2$
2,7-Dibromo-4-nitrofluorene ... 280
$C_{13}H_7Br_3$
2,7,9-Tribromofluorene ... 175
$C_{13}H_7Br_3ClNO$
3-Chloro-2′,4′,6′-tribromobenzanilide ... 460
4-Chloro-2′,4′,6′-tribromobenzanilide ... 480
$C_{13}H_7Br_3ClNOO$
2-Chloro-2′,4′,6′-tribromobenzanilide ... 479
$C_{13}H_7Br_3ClNO_2$
2,4,6-Tribromophenyl *N*-(2-chlorophenyl)carbamate ... 633
$C_{13}H_7Br_4NO_4$
N-Furfuryl-3,4,5,6-tetrabromophthalamic acid ... 844
$C_{13}H_7ClF_2N_2O_4$
2-Chloro-4-fluorophenyl *N*-(2-fluoro-5-nitrophenyl)carbamate ... 625
$C_{13}H_7ClN_2O_2$
9-Chloro-2-nitroacridine ... 996
$C_{13}H_7ClN_2O_2S_2$
2-((2-Chloro-4-nitrophenyl)thio)-benzothiazole ... 911
$C_{13}H_7ClN_2O_3$
10-Chloro-3-nitro-6(5*H*)-phenanthridinone ... 691
$C_{13}H_7ClN_2O_6$
2-Chloro-4-nitrophenyl 3-nitrobenzoate ... 377
$C_{13}H_7ClN_4O_8$
2,4-Dinitro-6-(2,4-dinitrobenzyl)-chlorobenzene ... 276
$C_{13}H_7ClO_3$
8-Chloro-4-dibenzofurancarboxylic acid ... 894
$C_{13}H_7Cl_2FN_2O_3$
3,4-Dichloro-4′-fluoro-3′-nitrobenzanilide ... 451
$C_{13}H_7Cl_2FN_2O_4$
2-Chloro-4-fluorophenyl *N*-(2-chloro-4-nitrophenyl)carbamate ... 626
2-Chloro-4-fluorophenyl *N*-(2-chloro-5-nitrophenyl)carbamate ... 625
2,5-Dichlorophenyl *N*-(4-fluoro-2-nitrophenyl)carbamate ... 624
$C_{13}H_7Cl_2FO$
2,4-Dichloro-4′-fluorobenzophenone 314
$C_{13}H_7Cl_2FO_2$
2,4-Dichlorophenyl 2-fluorobenzoate ... 377
$C_{13}H_7Cl_2F_2NO_2$
2-Chloro-4-fluorophenyl *N*-(3-chloro-4-fluorophenyl)carbamate ... 637
2,4-Dichlorophenyl *N*-(2,5-difluorophenyl)carbamate ... 637
3,4-Dichlorophenyl *N*-(2,5-difluorophenyl)carbamate ... 637
3,5-Dichlorophenyl *N*-(2,5-difluorophenyl)carbamate ... 639
$C_{13}H_7Cl_2N$
3,9-Dichloroacridine ... 996
$C_{13}H_7Cl_2NO$
4-Amino-2,7-dichloro-9-fluorenone ... 312
$C_{13}H_7Cl_2NO_2$
9,9-Dichloro-2-nitrofluorene ... 280
$C_{13}H_7Cl_2NO_3$
2,4′-Dichloro-4-nitrobenzophenone ... 313
$C_{13}H_7Cl_2NO_4$
2,4-Dichlorophenyl 3-nitrobenzoate .. 377
$C_{13}H_7Cl_3FNO_2$
2-Chloro-4-fluorophenyl *N*-(3,4-dichlorophenyl)carbamate ... 637
2,4-Dichlorophenyl *N*-(3-chloro-4-fluorophenyl)carbamate ... 637
$C_{13}H_7Cl_3F_2N_2O$
1-(2,5-Difluorophenyl)-3-(2,4,5-trichlorophenyl)urea ... 545
$C_{13}H_7Cl_3F_3NO_2S$
3,4,4′-Trichloro-2′-(trifluoromethyl)-benzenesulfonanilide ... 805
$C_{13}H_7Cl_3F_3N_2S$
2,4,5-Trichloro-3′-(trifluoromethyl)-benzenesulfonanilide ... 805
$C_{13}H_7Cl_3N_2O_2$
4-Chloro-*N*-(2,4-dichlorobenzylidene)-3-nitroaniline. 1024
N-(2-Chloro-5-nitrobenzylidene)-2,4-dichloroaniline ... 1058
N-(2-Chloro-5-nitrobenzylidene)-2,5-dichloroaniline ... 1051
N-(2-Chloro-5-nitrobenzylidene)-3,4-dichloroaniline ... 1058
N-(2-Nitrobenzylidene)-2,4,5-trichloroaniline ... 1055
$C_{13}H_7Cl_3N_2O_3$
4′-Nitro-2,2′,4-trichlorobenzanilide ... 451
$C_{13}H_7Cl_3N_2O_4$
2,3-Dichlorophenyl *N*-(4-chloro-3-nitrophenyl)carbamate ... 625
2,4-Dichlorophenyl *N*-(2-chloro-4-nitrophenyl)carbamate ... 626
2,5-Dichlorophenyl *N*-(4-chloro-3-nitrophenyl)carbamate ... 625
3,4-Dichlorophenyl *N*-(4-chloro-3-nitrophenyl)carbamate ... 625
$C_{13}H_7Cl_3O$
9-Hydroxy-2,4,7-trichlorofluorene ... 208
2,4,6-Trichlorobenzophenone ... 310
$C_{13}H_7Cl_3O_2$
2,4-Dichlorophenyl 2-chlorobenzoate ... 367
$C_{13}H_7Cl_4N$
2-Amino-1,3,4,7-tetrachlorofluorene. 231
2,3-Dichloro-*N*-(3,4-dichlorobenzylidene)aniline ... 1067
2,4-Dichloro-*N*-(2,4-dichlorobenzylidene)aniline ... 1067
2,4-Dichloro-*N*-(3,4-dichlorobenzylidene)aniline ... 1067
2,5-Dichloro-*N*-(2,4-dichlorobenzylidene)aniline ... 1067
2,5-Dichloro-*N*-(3,4-dichlorobenzylidene)aniline ... 1067
3,4-Dichloro-*N*-(2,4-dichlorobenzylidene)aniline ... 1067
$C_{13}H_7Cl_4NO$
2,2′,3′,4-Tetrachlorobenzanilide ... 481
2,2′,4,5′-Tetrachlorobenzanilide ... 483
2,2′,4,6′-Tetrachlorobenzanilide ... 481
2,2′,4′,6′-Tetrachlorobenzanilide ... 458
2,2′,5,6′-Tetrachlorobenzanilide ... 459
2,3′,4,4′-Tetrachlorobenzanilide ... 483
2′,3,4,6′-Tetrachlorobenzanilide ... 459
2′,3,4′,6′-Tetrachlorobenzanilide ... 460
3,3′,4,4′-Tetrachlorobenzanilide ... 481
3,3′,4,5′-Tetrachlorobenzanilide ... 460
$C_{13}H_7Cl_4NO_2$
4-Chlorophenyl *N*-(2,4,5-trichlorophenyl)carbamate ... 635
2,5-Dichlorophenyl *N*-(2,3-dichlorophenyl)carbamate ... 636
2,5-Dichlorophenyl *N*-(2,6-dichlorophenyl)carbamate ... 636
2,5-Dichlorophenyl *N*-(3,4-dichlorophenyl)carbamate ... 637
3,4-Dichlorophenyl *N*-(2,3-dichlorophenyl)carbamate ... 636
3,4-Dichlorophenyl *N*-(3,4-dichlorophenyl)carbamate ... 637
3,5-Dichlorophenyl *N*-(3,4-dichlorophenyl)carbamate ... 639
2,4,5-Trichlorophenyl *N*-(3-chlorophenyl)carbamate ... 631
2,4,5-Trichlorophenyl *N*-(4-chlorophenyl)carbamate ... 634
2,4,6-Trichlorophenyl *N*-(4-chlorophenyl)carbamate ... 635
$C_{13}H_7Cl_4NO_4$
N-Furfuryltetrachlorophthalamic acid ... 844
$C_{13}H_7Cl_4N_3O_3$
1-(2-Chloro-4-nitrophenyl)-3-(2,4,5-trichlorophenyl)urea ... 569
$C_{13}H_7Cl_5N_2O$
1-(2,3-Dichlorophenyl)-3-(2,4,5-trichlorophenyl)urea ... 541
1-(2,4-Dichlorophenyl)-3-(2,4,6-trichlorophenyl)urea ... 545
1-(2,5-Dichlorophenyl)-3-(2,4,5-trichlorophenyl)urea ... 545
1-(2,6-Dichlorophenyl)-3-(2,4,5-trichlorophenyl)urea ... 543
1-(2,6-Dichlorophenyl)-3-(2,4,6-trichlorophenyl)urea ... 543
1-(3,4-Dichlorophenyl)-3-(2,4,5-trichlorophenyl)urea ... 545
$C_{13}H_7Cl_5N_2OS$
S-(2,3,5,6-Tetrachloro-4-pyridyl) *N*-(3-chloro-2-methylphenyl)-thiocarbamate ... 930
$C_{13}H_7Cl_5O$
2,3,4,5,6-Pentachlorobenzhydrol ... 198
$C_{13}H_7FO$
1-Fluoro-9-fluorenone ... 314
3-Fluoro-9-fluorenone ... 314
4-Fluoro-9-fluorenone ... 314
$C_{13}H_7F_2NO$
2-Amino-4,7-difluoro-9-fluorenone ... 312
7-Amino-1,3-difluoro-9-fluorenone ... 312
$C_{13}H_7F_3N_2O_2S$
7-Nitro-2-(trifluoromethyl)-phenothiazine ... 259
$C_{13}H_7F_3N_4O_6$
N-Picryl-α,α,α-trifluoro-*o*-toluidine ... 259
N-Picryl-α,α,α-trifluoro-*m*-toluidine ... 259
$C_{13}H_7IO$
1-Iodo-9-fluorenone ... 315
2-Iodo-9-fluorenone ... 315
$C_{13}H_7I_3N_2O_3$
4′-Nitro-3,4,5-triiodobenzanilide ... 451
$C_{13}H_7NO$
2-Cyanodibenzofuran ... 896
$C_{13}H_7NO_3$
4-Nitro-9-fluorenone ... 313
$C_{13}H_7N_5O$
4-Amino-6*H*-indeno(2,1-g)pteridin-6-one ... 1003
$C_{13}H_7N_5O_6S$
2-(2,4,6-Trinitrophenylamino)-benzothiazole ... 914

$C_{13}H_8BrClFN$
2-Amino-1-bromo-3-chloro-7-fluorofluorene 231
$C_{13}H_8BrClN_2O_2$
3-Bromo-*N*-(2-chloro-5-nitrobenzylidene)aniline 1058
$C_{13}H_8BrClO$
4-Bromo-4′-chlorobenzophenone 314
$C_{13}H_8BrCl_2N$
4-Bromo-*N*-(2,4-dichlorobenzylidene)aniline 1066
4-Bromo-*N*-(3,4-dichlorobenzylidene)aniline 1066
$C_{13}H_8BrCl_2NO$
2-Bromo-2′,6′-dichlorobenzanilide 457
3-Bromo-2′,5′-dichlorobenzanilide 460
3-Bromo-3′,4′-dichlorobenzanilide 460
4′-Bromo-2,4-dichlorobenzanilide 480
$C_{13}H_8BrCl_2NO_2$
4-Bromophenyl *N*-(2,6-dichlorophenyl)carbamate 634
4-Bromophenyl *N*-(3,4-dichlorophenyl)carbamate 634
$C_{13}H_8BrCl_3N_2O$
1-(3-Bromophenyl)-3-(2,4,5-trichlorophenyl)urea 534
1-(4-Bromophenyl)-3-(2,4,5-trichlorophenyl)urea 540
$C_{13}H_8BrFN_2O_4$
4-Bromophenyl *N*-(4-fluoro-2-nitrophenyl)carbamate 624
$C_{13}H_8BrF_2NO_2$
4-Bromophenyl *N*-(2,5-difluorophenyl)carbamate 634
$C_{13}H_8BrF_3N_2O_2$
2-Pyridyl *N*-(4-bromo-2-(trifluoromethyl)phenyl)carbamate 925
$C_{13}H_8BrNO$
2-Bromo-6(5*H*)-phenanthridinone 692
$C_{13}H_8BrNO_2$
8-Bromo-2-dibenzofurancarboxamide 895
3-Bromo-2-nitrofluorene 280
9-Bromo-2-nitrofluorene 280
9-Bromo-3-nitrofluorene 280
$C_{13}H_8Br_2ClN$
2-Amino-7-chloro-1,3-dibromofluorene 231
$C_{13}H_8Br_2ClNO_2$
2,4-Dibromophenyl *N*-(3-chlorophenyl)carbamate 638
$C_{13}H_8Br_2FNO_2$
2,4-Dibromophenyl *N*-(2-fluorophenyl)carbamate 633
2,4-Dibromophenyl *N*-(3-fluorophenyl)carbamate 638
$C_{13}H_8Br_2F_2N_2O$
1-(3,4-Dibromophenyl)-3-(2,5-difluorophenyl)urea 544
$C_{13}H_8Br_2N_2$
Bis(4-bromophenyl)carbodiimide 769
$C_{13}H_8Br_2O_2$
1,3-Dibromo-4-methoxydibenzofuran 893
$C_{13}H_8Br_2O_3$
1,3-Dibromo-4-hydroxy-6-methoxydibenzofuran 893
$C_{13}H_8Br_3N$
2-Amino-1,3,7-tribromofluorene 231
$C_{13}H_8Br_3NO$
α-(4-Bromophenylimino)-4,6-dibromo-*o*-cresol 1036
2,2′,6′-Tribromobenzanilide 478
2′,4′,6′-Tribromobenzanilide 457
$C_{13}H_8Br_3NO_2$
2,4,6-Tribromophenyl *N*-phenylcarbamate 633
$C_{13}H_8ClFN_2O_2$
N-(4-Chloro-3-nitrobenzylidene)-4-fluoroaniline 1024
$C_{13}H_8ClFN_2O_3$
2′-Chloro-4-fluoro-5′-nitrobenzanilide 451
4-Chloro-4′-fluoro-3′-nitrobenzanilide 451
N-(4-Chloro-3-nitrophenyl)-2-fluorobenzamide 451
$C_{13}H_8ClFN_2O_4$
2-Chloro-4-nitrophenyl *N*-(2-fluorophenyl)carbamate 628
2-Fluorophenyl *N*-(4-chloro-2-nitrophenyl)carbamate 624
2-Fluorophenyl *N*-(4-chloro-3-nitrophenyl)carbamate 625
4-Fluorophenyl *N*-(2-chloro-4-nitrophenyl)carbamate 626
4-Fluorophenyl *N*-(4-chloro-3-nitrophenyl)carbamate 625
$C_{13}H_8ClFN_4O$
5-Chloro-1-(*N*-(2-fluorophenyl)carbamoyl)-1*H*-benzotriazole 908
$C_{13}H_8ClFN_4O_5$
1-(2-Chloro-4-nitrophenyl)-3-(4-fluoro-3-nitrophenyl)urea 523
1-(2-Chloro-5-nitrophenyl)-3-(4-fluoro-3-nitrophenyl)urea 525
$C_{13}H_8ClFO_2$
4-Fluorophenyl 2-chlorobenzoate 377
$C_{13}H_8ClF_2NOS$
S-(4-Chlorophenyl) *N*-(2,5-difluorophenyl)thiocarbamate 641
$C_{13}H_8ClF_2NO_2$
2-Chloro-4-fluorophenyl *N*-(3-fluorophenyl)carbamate 637
2-Chloro-4-fluorophenyl *N*-(4-fluorophenyl)carbamate 634
$C_{13}H_8ClF_2N_3O_3$
1-(2-Chloro-5-nitrophenyl)-3-(2,5-difluorophenyl)urea 525
$C_{13}H_8ClF_3N_2O_2$
4-Pyridyl *N*-(4-chloro-2-(trifluoromethyl)phenyl)carbamate 925
$C_{13}H_8ClF_3N_2O_4S$
2′-Chloro-2-nitro-5′-(trifluoromethyl)benzenesulfonanilide 805
$C_{13}H_8ClF_3N_4O_3$
1-(2-Chloro-5-(trifluoromethyl)phenyl)-3-(5-nitro-2-pyridyl)urea 925
1-(4-Chloro-2-(trifluoromethyl)phenyl)-3-(5-nitro-2-pyridyl)urea 925
1-(4-Chloro-3-(trifluoromethyl)phenyl)-3-(5-nitro-2-pyridyl)urea 925
$C_{13}H_8ClNO$
2-Chloro-6(5*H*)-phenanthridinone 691
3-Chloro-6(5*H*)-phenanthridinone 1000
5-Chloro-3-phenylanthranil 910
α-(4-Chlorophenyl)-β-(2-furyl)acrylonitrile 844
$C_{13}H_8ClNOS$
Phenothiazine-10-carbonyl chloride. 709
$C_{13}H_8ClNO_4$
3-(2-Chloro-5-nitrophenyl)-1-(2-furyl)-2-propen-1-one 844
3-(4-Chloro-3-nitrophenyl)-1-(2-furyl)-2-propen-1-one 844
2-Chlorophenyl 3-nitrobenzoate 376
$C_{13}H_8ClNS$
2-(4-Chlorophenyl)benzothiazole 912
α-(4-Chlorophenyl)-β-(2-thienyl)acrylonitrile 855
$C_{13}H_8ClN_3O_3$
4-(4-Nitrophenylazo)benzoyl chloride 1090
$C_{13}H_8ClN_5O_3$
5-Chloro-1-(*N*-(3-nitrophenyl)carbamoyl)-1*H*-benzotriazole 909
$C_{13}H_8Cl_2$
1,3-Dichlorofluorene 174
$C_{13}H_8Cl_2BrNO$
N-(2,3-Dichlorophenyl)-3-bromobenzamide 483
N-(2,4-Dichlorophenyl)-3-bromobenzamide 483
$C_{13}H_8Cl_2FN$
N-(2,4-Dichlorobenzylidene)-4-fluoroaniline 1027
2,3-Dichloro-*N*-(4-fluorobenzylidene)-aniline 1078
2,5-Dichloro-*N*-(4-fluorobenzylidene)-aniline 1078
$C_{13}H_8Cl_2FNO$
2′,3′-Dichloro-2-fluorobenzanilide 457
2,4-Dichloro-4′-fluorobenzanilide 459
2′,5′-Dichloro-3-fluorobenzanilide 460
3,4-Dichloro-2′-fluorobenzanilide 479
$C_{13}H_8Cl_2FNO_2$
2-Chloro-4-fluorophenyl *N*-(2-chlorophenyl)carbamate 633
2,5-Dichlorophenyl *N*-(2-fluorophenyl)carbamate 633
3,4-Dichlorophenyl *N*-(2-fluorophenyl)carbamate 633
$C_{13}H_8Cl_2F_2N_2O$
1-(2,5-Dichlorophenyl)-3-(2,4-difluorophenyl)urea 543
1-(2,5-Dichlorophenyl)-3-(2,5-difluorophenyl)urea 544
1-(3,4-Dichlorophenyl)-3-(2,5-difluorophenyl)urea 544
$C_{13}H_8Cl_2F_3NO_2S$
4-Chloro-*N*-(2-chloro-5-(trifluoromethyl)phenyl)-benzenesulfonamide 805
$C_{13}H_8Cl_2F_3NO_4S$
2-(2-Amino-α,α,α-trifluoro-*p*-tolyloxy)-3,5-dichlorobenzene-sulfonic acid, sodium salt 790
$C_{13}H_8Cl_2F_3NS$
2-Amino-3′,4′-dichloro-4-(trifluoro-methyl)phenyl) sulfide 222
$C_{13}H_8Cl_2F_3N_3O$
1-(5-Chloro-2-pyridyl)-3-(4-chloro-2-(trifluoromethyl)phenyl)urea 925
$C_{13}H_8Cl_2IN$
N-(2,4-Dichlorobenzylidene)-4-iodoaniline 1027
$C_{13}H_8Cl_2N_2$
Bis(4-chlorophenyl)carbodiimide 769
$C_{13}H_8Cl_2N_2O$
1-Amino-3,8-dichloro-6(5*H*)-phenanthridinone 999
7,10-Dichloro-2-methoxypyrido(3,2-b)quinoline 1002
$C_{13}H_8Cl_2N_2O_2$
3-Chloro-*N*-(2-chloro-5-nitrobenzylidene)aniline 1058
4-Chloro-*N*-(2-chloro-5-nitrobenzylidene)aniline 1058
N-(2,4-Dichlorobenzylidene)-3-nitroaniline 1024
N-(2,4-Dichlorobenzylidene)-4-nitroaniline 1024
2,3-Dichloro-*N*-(3-nitrobenzylidene)-aniline 1059
2,4-Dichloro-*N*-(3-nitrobenzylidene)-aniline 1059
2,5-Dichloro-*N*-(2-nitrobenzylidene)-aniline 1055
2,5-Dichloro-*N*-(3-nitrobenzylidene)-aniline 1059
2,6-Dichloro-*N*-(2-nitrobenzylidene)-aniline 1055
$C_{13}H_8Cl_2N_2O_3$
2,2′-Dichloro-4′-nitrobenzanilide 470
2′,5′-Dichloro-4-nitrobenzanilide 405
3,4-Dichloro-3′-nitrobenzanilide 473
4,4′-Dichloro-2′-nitrobenzanilide 469
$C_{13}H_8Cl_2N_2O_3S$
S-(4-Chlorophenyl) *N*-(4-chloro-3-nitrophenyl)thiocarbamate 626
$C_{13}H_8Cl_2N_2O_4$
3,4-Dichloro-2′-hydroxy-5′-nitrobenzanilide 447
$C_{13}H_8Cl_2N_4O_4$
N,*N*′-Bis(4-chloro-3-nitrophenyl)-formamidine 724
$C_{13}H_8Cl_2N_4O_5$
1,3-Bis-(2-chloro-4-nitrophenyl)urea.. 569
$C_{13}H_8Cl_2O$
2,4-Dichlorobenzophenone 310
2,6-Dichlorobenzophenone 310
$C_{13}H_8Cl_2OS_2$
S,*S*′-Bis-(4-chlorophenyl) dithiocarbonate 381
$C_{13}H_8Cl_2O_2$
2-Chlorophenyl 2-chlorobenzoate 377
3-(2,4-Dichlorophenyl)-1-(2-furyl)-2-propen-1-one 852
$C_{13}H_8Cl_3FN_2O$
1-(4-Fluorophenyl)-3-(2,4,5-trichlorophenyl)urea 540
$C_{13}H_8Cl_3N$
4-Amino-2,5,7-trichlorofluorene 231
N-(2-Chlorobenzylidene)-2,3-dichloroaniline 1064
N-(2-Chlorobenzylidene)-2,4-dichloroaniline 1064
N-(2-Chlorobenzylidene)-2,5-dichloroaniline 1064
N-(2-Chlorobenzylidene)-3,4-dichloroaniline 1064
N-(4-Chlorobenzylidene)-2,3-dichloroaniline 1065
N-(4-Chlorobenzylidene)-2,4-dichloroaniline 1066
N-(4-Chlorobenzylidene)-2,5-dichloroaniline 1066
N-(4-Chlorobenzylidene)-3,4-dichloroaniline 1065
2-Chloro-*N*-(2,4-dichlorobenzylidene)aniline 1064
2-Chloro-*N*-(3,4-dichlorobenzylidene)aniline 1064
3-Chloro-*N*-(3,4-dichlorobenzylidene)aniline 1067
4-Chloro-*N*-(2,4-dichlorobenzylidene)aniline 1066
4-Chloro-*N*-(3,4-dichlorobenzylidene)aniline 1066
$C_{13}H_8Cl_3NO$
4-Chloro-2-(2,4-dichlorobenzylideneamino)phenol.. 1022
4-Chloro-2-(2,6-dichlorobenzylideneamino)phenol.. 1034
2,2′,4-Trichlorobenzanilide 479
2,2′,4′-Trichlorobenzanilide 479
2,2′,5′-Trichlorobenzanilide 483
2,2′,6′-Trichlorobenzanilide 457
2,3′,4′-Trichlorobenzanilide 457
2′,3′,4-Trichlorobenzanilide 458
2′,3,5′-Trichlorobenzanilide 460
2,3′,5′-Trichlorobenzanilide 458
2′,3,6′-Trichlorobenzanilide 459
2,4,4′-Trichlorobenzanilide 480
2′,4,5′-Trichlorobenzanilide 480
2′,4′,6′-Trichlorobenzanilide 457
2′,4,6′-Trichlorobenzanilide 458
3′,4,4′-Trichlorobenzanilide 483
3′,4,5′-Trichlorobenzanilide 459
$C_{13}H_8Cl_3NOS$
S-(4-Chlorophenyl) *N*-(2,4-dichlorophenyl)thiocarbamate 641
S-(4-Chlorophenyl) *N*-(3,4-dichlorophenyl)thiocarbamate 641
$C_{13}H_8Cl_3NO_2$
2-Chlorophenyl *N*-(2,4-dichlorophenyl)carbamate 682
2-Chlorophenyl *N*-(2,5-dichlorophenyl)carbamate 682
2-Chlorophenyl *N*-(2,6-dichlorophenyl)carbamate 633
2-Chlorophenyl *N*-(3,4-dichlorophenyl)carbamate 633
3-Chlorophenyl *N*-(2,4-dichlorophenyl)carbamate 686
3-Chlorophenyl *N*-(2,6-dichlorophenyl)carbamate 636
3-Chlorophenyl *N*-(3,4-dichlorophenyl)carbamate 637
4-Chlorophenyl *N*-(2,3-dichlorophenyl)carbamate 683
4-Chlorophenyl *N*-(2,4-dichlorophenyl)carbamate 683
4-Chlorophenyl *N*-(2,6-dichlorophenyl)carbamate 634
4-Chlorophenyl *N*-(3,4-dichlorophenyl)carbamate 683
2,3-Dichlorophenyl *N*-(3-chlorophenyl)carbamate 636
2,4-Dichlorophenyl *N*-(3-chlorophenyl)carbamate 637
2,5-Dichlorophenyl *N*-(3-chlorophenyl)carbamate 637
3,4-Dichlorophenyl *N*-(3-chlorophenyl)carbamate 637
Phenyl *N*-(2,4,5-trichlorophenyl)-carbamate 633
2,3,4-Trichlorophenyl *N*-phenylcarbamate 681
2,4,5-Trichlorophenyl *N*-phenylcarbamate 632

$C_{13}H_8Cl_3NO_2$ (Cont'd)
2,4,6-Trichlorophenyl *N*-phenylcarbamate ... 633
2′,4′,5′-Trichlorosalicylanilide ... 447
$C_{13}H_8Cl_3NO_2S$
S-(4-Hydroxyphenyl) *N*-(2,4,5-trichlorophenyl)thiocarbamate ... 611
$C_{13}H_8Cl_3N_3O_3$
1-(2-Chloro-4-nitrophenyl)-3-(2,3-dichlorophenyl)urea ... 530
1-(3-Nitrophenyl)-3-(2,4,5-trichlorophenyl)urea ... 546
1-(4-Nitrophenyl)-3-(2,4,5-trichlorophenyl)urea ... 530
$C_{13}H_8Cl_4N_2$
N,*N*′-Bis(3,4-dichlorophenyl)formamidine ... 724
$C_{13}H_8Cl_4N_2O$
1,3-Bis(2,3-dichlorophenyl)urea ... 594
1,3-Bis(2,4-dichlorophenyl)urea ... 544
1,3-Bis(2,6-dichlorophenyl)urea ... 541
1,3-Bis(3,4-dichlorophenyl)urea ... 600
1-(2-Chlorophenyl)-3-(2,4,5-trichlorophenyl)urea ... 536
1-(2-Chlorophenyl)-3-(2,4,6-trichlorophenyl)urea ... 538
1-(3-Chlorophenyl)-3-(2,4,5-trichlorophenyl)urea ... 546
1-(3-Chlorophenyl)-3-(2,4,6-trichlorophenyl)urea ... 546
1-(4-Chlorophenyl)-3-(2,4,5-trichlorophenyl)urea ... 540
1-(4-Chlorophenyl)-3-(2,4,6-trichlorophenyl)urea ... 540
1-(2,3-Dichlorophenyl)-3-(2,4-dichlorophenyl)urea ... 541
1-(2,3-Dichlorophenyl)-3-(2,5-dichlorophenyl)urea ... 541
1-(2,3-Dichlorophenyl)-3-(2,6-dichlorophenyl)urea ... 541
1-(2,3-Dichlorophenyl)-3-(3,4-dichlorophenyl)urea ... 541
1-(2,3-Dichlorophenyl)-3-(3,5-dichlorophenyl)urea ... 542
1-(2,4-Dichlorophenyl)-3-(2,6-dichlorophenyl)urea ... 541
1-(2,4-Dichlorophenyl)-3-(3,4-dichlorophenyl)urea ... 544
1-(2,5-Dichlorophenyl)-3-(2,6-dichlorophenyl)urea ... 542
1-(2,5-Dichlorophenyl)-3-(3,4-dichlorophenyl)urea ... 544
1-(2,5-Dichlorophenyl)-3-(3,5-dichlorophenyl)urea ... 545
1-(3,4-Dichlorophenyl)-3-(2,6-dichlorophenyl)urea ... 542
$C_{13}H_8Cl_4N_2O_4$
N-(3,4-Dimethyl-5-isoxazolyl)-tetrachlorophthalamic acid ... 870
$C_{13}H_8Cl_4N_2S$
1,3-Bis(2,6-dichlorophenyl)-2-thiourea ... 733
1,3-Bis(3,4-dichlorophenyl)-2-thiourea ... 749
$C_{13}H_8Cl_6O_3$
Hexachlorohexahydro-1,4-methanonaphthalene-6,7-dicarboxylic anhydride ... 99
$C_{13}H_8FNO$
2-Amino-7-fluoro-9-fluorenone ... 312
3-Amino-7-fluoro-9-fluorenone ... 312
$C_{13}H_8FNO_2$
1-Fluoro-2-nitrofluorene ... 280
1-Fluoro-7-nitrofluorene ... 280
3-Fluoro-2-nitrofluorene ... 280
$C_{13}H_8FNO_3$
4-Fluoro-4′-nitrobenzophenone ... 313
$C_{13}H_8FNO_4$
4-Fluorophenyl 3-nitrobenzoate ... 376
$C_{13}H_8F_2INO_2$
2-Iodophenyl *N*-(2,5-difluorophenyl)-carbamate ... 633
$C_{13}H_8F_2N_2O_3$
2,2′-Difluoro-5′-nitrobenzanilide ... 451
$C_{13}H_8F_2N_2O_4$
2-Fluorophenyl *N*-(2-fluoro-5-nitrophenyl)carbamate ... 625
2-Fluorophenyl *N*-(4-fluoro-3-nitrophenyl)carbamate ... 625
$C_{13}H_8F_2N_4O_5$
1-(2,5-Difluorophenyl)-3-(3,5-dinitrophenyl)urea ... 525
1-(2-Fluoro-5-nitrophenyl)-3-(4-fluoro-3-nitrophenyl)urea ... 525
$C_{13}H_8F_2O$
2,3′-Difluorobenzophenone ... 314
$C_{13}H_8F_3N_3O_3$
1-(2,5-Difluorophenyl)-3-(4-fluoro-3-nitrophenyl)urea ... 524
$C_{13}H_8F_3N_3O_4$
2,4-Dinitro-3′-(trifluoromethyl)-diphenylamine ... 259
$C_{13}H_8INO$
2-Amino-3-iodo-9-fluorenone ... 312
$C_{13}H_8INO_2$
2-Iodo-7-nitrofluorene ... 280
$C_{13}H_8I_2N_2$
Bis(4-iodophenyl)carbodiimide ... 769
$C_{13}H_8N_2O_2S$
5-Nitro-α-(2-thenylidene)-*o*-tolunitrile ... 856
4-Nitro-α-(2-thienylmethylene)-phenylacetonitrile ... 856
$C_{13}H_8N_2O_2S_2$
2-(4-Nitrophenylimino)-1,3-benzodithiole ... 1059
$C_{13}H_8N_2O_3$
2-Amino-3-nitro-9-fluorenone ... 311
2-Amino-7-nitro-9-fluorenone ... 311
5-Amino-2-nitro-9-fluorenone ... 311
3-Nitro-6(5*H*)-phenanthridinone ... 691
$C_{13}H_8N_2O_4$
2,5-Dinitrofluorene ... 280
2,6-Dinitrofluorene ... 280
7-Nitro-4-dibenzofurancarboxamide ... 894
$C_{13}H_8N_2O_5$
4,4′-Dinitrobenzophenone ... 313
$C_{13}H_8N_2O_7$
1,5-Bis-(5-nitro-2-furyl)-1,4-pentadien-3-one ... 852
4,4′-Dihydroxy-3,3′-dinitrobenzophenone ... 310
4-Nitro-2-(3-nitrophenoxy)-benzoic acid ... 340
$C_{13}H_8N_4O_3$
5-Nitro-3-(2-pyridyl)-2-quinoxalinol ... 1000
$C_{13}H_8N_4O_4$
N,*N*′-Bis-(4-nitrophenyl)-carbodiimide ... 769
$C_{13}H_8N_4O_8$
Bis(2,4-dinitrophenyl)methane ... 275
N-Picrylanthranilic acid ... 342
3-(2,4,6-Trinitrophenylamino)-benzoic acid ... 342
4-(2,4,6-Trinitrophenylamino)-benzoic acid ... 342
$C_{13}H_8OSe$
9*H*-Selenoxanthen-9-one ... 213
$C_{13}H_8O_2$
Dibenzofuran-x-carboxaldehyde ... 895
3-Hydroxy-9-fluorenone ... 310
7-Hydroxy-1*H*-phenalen-1-one ... 310
$C_{13}H_8O_2S$
4-Dibenzothiophenecarboxylic acid ... 898
$C_{13}H_8O_2S_2$
1-Thianthrenecarboxylic acid ... 346
$C_{13}H_8O_3$
Dibenzofuran-2-carboxylic acid ... 895
6-Hydroxy-3,4-benzocoumarin ... 382
$C_{13}H_8O_3S$
1-Phenoxathiincarboxylic acid ... 345
Thioxanthen-9-one-10,10-dioxide ... 780
$C_{13}H_8O_4$
2-Hydroxy-3-dibenzofurancarboxylic acid ... 893
$C_{13}H_8O_4S$
4-Carboxydibenzothiophene-5,5-dioxide ... 898
$C_{13}H_9AsClNO_2$
10-Chloro-5,10-dihydro-1-phenarsazinecarboxylic acid ... 1127
$C_{13}H_9BBrF_4NS$
3-(4-Bromophenyl)thiazolo(3,2-a)-pyridinium tetrafluoroborate ... 445
$C_{13}H_9BrClFN_2O$
1-(3-Bromophenyl)-3-(3-chloro-4-fluorophenyl)urea ... 544
$C_{13}H_9BrClN$
2-Amino-1-bromo-3-chlorofluorene ... 231
4-Bromo-*N*-(2-chlorobenzylidene)-aniline ... 1064
4-Bromo-*N*-(4-chlorobenzylidene)-aniline ... 1065
$C_{13}H_9BrClNO$
2-Amino-5-bromo-2′-chlorobenzophenone ... 309
4′-Bromo-2-chlorobenzanilide ... 479
4′-Bromo-4-chlorobenzanilide ... 480
4-Bromo-2-chloro-*N*-(2-hydroxybenzylidene)aniline ... 1022
$C_{13}H_9BrClNO_2$
5-Bromo-4′-chlorosalicylanilide ... 463
4-Chlorophenyl *N*-(4-bromophenyl)-carbamate ... 683
$C_{13}H_9BrClN_3O_3$
1-(2-Bromo-4-nitrophenyl)-3-(4-chlorophenyl)urea ... 526
$C_{13}H_9BrCl_2N_2$
2,6-Dichlorobenzaldehyde (4-bromophenyl)hydrazone ... 234
$C_{13}H_9BrCl_2N_2O$
1-(3-Bromophenyl)-3-(3,5-dichlorophenyl)urea ... 545
1-(4-Bromophenyl)-3-(3,4-dichlorophenyl)urea ... 591
$C_{13}H_9BrFN$
4-Bromo-*N*-(4-fluorobenzylidene)-aniline ... 1077
$C_{13}H_9BrFN_3O_3$
1-(4-Bromophenyl)-3-(2-fluoro-5-nitrophenyl)urea ... 524
$C_{13}H_9BrF_2N_2O$
1-(3-Bromophenyl)-3-(2,5-difluorophenyl)urea ... 544
1-(4-Bromophenyl)-3-(2,5-difluorophenyl)urea ... 540
$C_{13}H_9BrN_2$
2-(3-Bromophenyl)benzimidazole ... 901
3-Bromo-2-phenyl-2*H*-indazole ... 900
$C_{13}H_9BrN_2O$
4-Amino-2-bromo-6(5*H*)-phenanthridinone ... 690
$C_{13}H_9BrN_2O_2S$
3′-Bromo-*N*-cyanobenzenesulfonanilide ... 809
$C_{13}H_9BrN_2O_4$
4-Nitrophenyl *N*-(4-bromophenyl)-carbamate ... 628
$C_{13}H_9BrN_4O_4$
4-Bromobenzaldehyde, 2,4-dinitrophenylhydrazone ... 1052
$C_{13}H_9BrN_4O_6$
4-Bromo-*N*-picryl-*m*-toluidine ... 272
N-(4-Bromo-*o*-tolyl)-picrylamine ... 272
$C_{13}H_9BrO$
2-Bromobenzophenone ... 310
$C_{13}H_9BrO_2$
4-Bromophenyl benzoate ... 377
3-(3-Bromophenyl)benzoic acid ... 338
$C_{13}H_9BrO_3$
4-Bromo-2-(phenoxy)benzoic acid ... 336
$C_{13}H_9Br_2NO$
3,4′-Dibromobenzanilide ... 458
$C_{13}H_9Br_2NO_2$
4-Bromophenyl *N*-(3-bromophenyl)-carbamate ... 634
$C_{13}H_9Br_3O$
Benzyl 2,4,6-tribromophenyl ether ... 185
$C_{13}H_9Cl$
9-Chlorofluorene ... 174
$C_{13}H_9ClFN$
3-Chloro-*N*-(4-fluorobenzylidene)-aniline ... 1077
4-Chloro-*N*-(4-fluorobenzylidene)-aniline ... 1077
$C_{13}H_9ClFNO$
3-Chloro-2′-fluorobenzanilide ... 457
3-Chloro-3′-fluorobenzanilide ... 456
3-Chloro-4′-fluorobenzanilide ... 458
4-Chloro-2′-fluorobenzanilide ... 457
$C_{13}H_9ClFNO_2$
4-Fluorophenyl *N*-(3-chlorophenyl)-carbamate ... 634
$C_{13}H_9ClFN_3O_3$
1-(2-Chloro-4-nitrophenyl)-3-(3-fluorophenyl)urea ... 526
1-(2-Chloro-5-nitrophenyl)-3-(3-fluorophenyl)urea ... 524
1-(4-Chloro-2-nitrophenyl)-3-(4-fluorophenyl)urea ... 523
$C_{13}H_9ClFN_3O_4$
1-(5-Chloro-2-hydroxyphenyl)-3-(2-fluoro-5-nitrophenyl)urea ... 513
1-(5-Chloro-2-hydroxyphenyl)-3-(4-fluoro-3-nitrophenyl)urea ... 513
$C_{13}H_9ClF_2N_2O$
1-(3-Chlorophenyl)-3-(2,5-difluorophenyl)urea ... 543
$C_{13}H_9ClF_3NO$
6-(4-Chlorophenoxy)-α,α,α-trifluoro-*m*-toluidine ... 222
$C_{13}H_9ClF_3NS$
6-(4-Chlorophenylthio)-α,α,α-trifluoro-*m*-toluidine ... 222
$C_{13}H_9ClF_3N_3O$
1-(2-Chloro-3-pyridyl)-3-(2-(trifluoromethyl)phenyl)urea ... 926
1-(5-Chloro-2-pyridyl)-3-(2-(trifluoromethyl)phenyl)urea ... 926
1-(5-Chloro-2-pyridyl)-3-(3-(trifluoromethyl)phenyl)urea ... 926
1-(2-Chloro-5-(trifluoromethyl)-phenyl)-3-(3-pyridyl)urea ... 925
$C_{13}H_9ClF_3N_3O_2$
1-(4-Chloro-3-(trifluoromethyl)-phenyl)-3-(3-hydroxy-2-pyridyl)urea ... 925
$C_{13}H_9ClF_4O_2$
1-Chloro-3-(5,6,7,8-tetrafluoro-1-naphthoxy)-2-propanol ... 197
$C_{13}H_9ClIN$
N-(4-Chlorobenzylidene)-4-iodoaniline ... 1077
$C_{13}H_9ClN_2$
α-(4-Chlorophenyl)-2-pyridineacetonitrile ... 945
2-Diazofluorene chloride ... 1095
$C_{13}H_9ClN_2O$
4-(Phenylazo)benzoyl chloride ... 1092
$C_{13}H_9ClN_2OS$
2-(2-Chlorobenzoyl)-4-methyl-2-thiazolylacetonitrile ... 875
$C_{13}H_9ClN_2O_2$
2-Chloro-*N*-(2-nitrobenzylidene)-aniline ... 1055
2-Chloro-*N*-(3-nitrobenzylidene)-aniline ... 1059
N-(2-Chloro-5-nitrobenzylidene)-aniline ... 1058
3-Chloro-*N*-(2-nitrobenzylidene)-aniline ... 1055
3-Chloro-*N*-(3-nitrobenzylidene)-aniline ... 1059
4-Chloro-*N*-(2-nitrobenzylidene)-aniline ... 1055
4-Chloro-*N*-(3-nitrobenzylidene)-aniline ... 1059
$C_{13}H_9ClN_2O_2S$
4′-Chloro-*N*-cyanobenzene-sulfonanilide ... 809
4-Chloro-4′-cyanobenzene-sulfonanilide ... 809
$C_{13}H_9ClN_2O_3$
2-Chloro-2′-nitrobenzanilide ... 450
2-Chloro-3′-nitrobenzanilide ... 469
2′-Chloro-3-nitrobenzanilide ... 452
2-Chloro-4′-nitrobenzanilide ... 470
2′-Chloro-4-nitrobenzanilide ... 452
3′-Chloro-3-nitrobenzanilide ... 451
4-Chloro-2′-nitrobenzanilide ... 469
4-Chloro-3-nitrobenzanilide ... 470
4-Chloro-4′-nitrobenzanilide ... 470
N-(4-Chloro-3-nitrobenzylidene)-2-hydroxyaniline ... 1022
4-Chloro-3-nitro-*N*-(4-hdyroxybenzylidene)aniline ... 1022
α-(2-Chloro-4-nitrophenylimino)-*o*-cresol ... 1033
α-(2-Chloro-5-nitrophenylimino)-*o*-cresol ... 1033

$C_{13}H_9ClN_2O_3$ (Cont'd)
α-(4-Chloro-3-nitrophenylimino)-*o*-cresol ... 1022
$C_{13}H_9ClN_2O_4$
2-Chloro-2′-hydroxy-5′-nitrobenzanilide ... 447
4-Chloro-2′-hydroxy-5′-nitrobenzanilide ... 447
5′-Chloro-2′-hydroxy-4-nitrobenzanilide ... 447
4-Chloro-*N*-(4-nitrophenyl)anthranilic acid ... 343
Phenyl *N*-(2-chloro-4-nitrophenyl)carbamate ... 671
$C_{13}H_9ClN_2O_4S$
S-(4-Hydroxyphenyl) *N*-(2-chloro-4-nitrophenyl)thiocarbamate ... 611
$C_{13}H_9ClN_2S_2$
7-Chloro-4-(4-methyl-2-thiazolylthio)quinoline hydrochloride ... 987
$C_{13}H_9ClN_4O_4$
2-Chlorobenzaldehyde 2,4-dinitrophenylhydrazone ... 1052
3-Pyridylmethyleneamino *N*-(2-chloro-4-nitrophenyl)carbamate ... 1026
$C_{13}H_9ClN_4O_5$
1-(2-Chloro-4-nitrophenyl)-3-(3-nitrophenyl)urea ... 562
$C_{13}H_9ClN_4O_6$
5-Chloro-*N*-picryl-*o*-toluidine ... 261
N-(3-Chloro-*o*-tolyl)-picrylamine ... 261
N-(3-Chloro-*p*-tolyl)picrylamine ... 272
N-(4-Chloro-*o*-tolyl)-picrylamine ... 261
$C_{13}H_9ClN_4O_7$
N-(3-Chloro-2-methoxyphenyl)picrylamine ... 266
$C_{13}H_9ClO_2$
2-(2-Chlorophenyl)benzoic acid ... 337
3-(2-Chlorophenyl)-1-(2-furyl)-2-propen-1-one ... 847
3-(4-Chlorophenyl)-1-(2-furyl)-2-propen-1-one ... 847
$C_{13}H_9ClO_3$
4-Chlorophenyl salicylate ... 376
$C_{13}H_9ClO_4$
1,4-Dihydro-1,4-dioxo-3-methyl-2-naphthylmethyl chloroformate ... 389
$C_{13}H_9ClO_4S$
2-Chloro-6-(phenylsulfonyl)benzoic acid ... 773
4-(4-Chlorophenylsulfonyl)benzoic acid ... 773
$C_{13}H_9Cl_2FN_2O$
1-(2,3-Dichlorophenyl)-3-(4-fluorophenyl)urea ... 539
1-(2,4-Dichlorophenyl)-3-(2-fluorophenyl)urea ... 537
1-(2,4-Dichlorophenyl)-3-(3-fluorophenyl)urea ... 543
1-(2,5-Dichlorophenyl)-3-(3-fluorophenyl)urea ... 543
1-(2,5-Dichlorophenyl)-3-(4-fluorophenyl)urea ... 540
1-(2,6-Dichlorophenyl)-3-(4-fluorophenyl)urea ... 539
1-(3,4-Dichlorophenyl)-3-(4-fluorophenyl)urea ... 539
$C_{13}H_9Cl_2FN_4O_3$
1-(2-Amino-4-fluoro-6-nitrophenyl)-3-(2,5-dichlorophenyl)urea ... 522
$C_{13}H_9Cl_2N$
N-Benzylidene-2,3-dichloroaniline ... 1067
N-Benzylidene-2,4-dichloroaniline ... 1067
N-(Benzylidene)-2,5-dichloroaniline ... 1025
N-Benzylidene-3,4-dichloroaniline ... 1067
2-Chloro-*N*-(2-chlorobenzylidene)aniline ... 1077
2-Chloro-*N*-(3-chlorobenzylidene)aniline ... 1064
2-Chloro-*N*-(4-chlorobenzylidene)aniline ... 1064
3-Chloro-*N*-(2-chlorobenzylidene)aniline ... 1064
3-Chloro-*N*-(4-chlorobenzylidene)aniline ... 1065
4-Chloro-*N*-(2-chlorobenzylidene)aniline ... 1025
4-Chloro-*N*-(4-chlorobenzylidene)aniline ... 1065
2,6-Dichlorobenzophenone imine ... 236
2,6-Dichlorobenzophenone imine hydrochloride ... 236
N-(2,4-Dichlorobenzylidene)aniline ... 1067
N-(3,4-Dichlorobenzylidene)aniline ... 1067
$C_{13}H_9Cl_2NO$
4-Amino-2,4′-dichlorobenzophenone ... 309
2-Amino-3,7-dichloro-9-hydroxyfluorene ... 226
4-Amino-2,7-dichloro-9-hydroxyfluorene ... 226
2,2′-Dichlorobenzanilide ... 478
2′,3′-Dichlorobenzanilide ... 481
2′,4-Dichlorobenzanilide ... 479
2,4′-Dichlorobenzanilide ... 479
2,4-Dichlorobenzanilide ... 481
2′,4′-Dichlorobenzanilide ... 481
2′,5′-Dichlorobenzanilide ... 481
2′,6′-Dichlorobenzanilide ... 481
3,3′-Dichlorobenzanilide ... 456
3′,4′-Dichlorobenzanilide ... 481
N-(2,4-Dichlorobenzylidene)-2-hydroxyaniline ... 1022
α-(2,3-Dichlorophenylimino)-*o*-cresol ... 1035
α-(2,4-Dichlorophenylimino)-*o*-cresol ... 1035
α-(2,5-Dichlorophenylimino)-*o*-cresol ... 1035
α-(3,4-Dichlorophenylimino)-*o*-cresol ... 1035
$C_{13}H_9Cl_2NOS$
S-(4-Chlorophenyl) *N*-(3-chlorophenyl)thiocarbamate ... 641
$C_{13}H_9Cl_2NO_2$
2-Chlorophenyl *N*-(3-chlorophenyl)carbamate ... 687
2-Chlorophenyl-*N*-(4-chlorophenyl)carbamate ... 682
3-Chlorophenyl *N*-(4-chlorophenyl)carbamate ... 634
4-Chlorophenyl *N*-(2-chlorophenyl)carbamate ... 682
4-Chlorophenyl *N*-(3-chlorophenyl)carbamate ... 682
4-Chlorophenyl *N*-(4-chlorophenyl)carbamate ... 683
2,4-Dichloro-2′-hydroxybenzanilide ... 462
2,4-Dichloro-3′-hydroxybenzanilide ... 447
2,4-Dichloro-4′-hydroxybenzanilide ... 463
3′,4′-Dichloro-2-hydroxybenzanilide ... 447
3,4-Dichloro-2′-hydroxybenzanilide ... 447
3,4-Dichloro-3′-hydroxybenzanilide ... 447
2,6-Dichloroindo-*m*-cresol, sodium salt ... 1037
N-(2,5-Dichlorophenyl)anthranilic acid ... 337
Phenyl *N*-(2,3-dichlorophenyl)carbamate ... 684
Phenyl *N*-(2,4-dichlorophenyl)carbamate ... 686
Phenyl *N*-(2,5-dichlorophenyl)carbamate ... 686
Phenyl *N*-(2,6-dichlorophenyl)carbamate ... 681
Phenyl *N*-(3,4-dichlorophenyl)carbamate ... 686
Phenyl *N*-(3,5-dichlorophenyl)carbamate ... 633
$C_{13}H_9Cl_2N_3O$
Isonicotinic 3,4-dichlorobenzylidenehydrazide ... 1025
4-Pyridinecarboxylic (2,4-dichlorobenzylidene)hydrazide ... 928
$C_{13}H_9Cl_2N_3O_2$
(2-Pyridyl)methyleneamino *N*-(2,5-dichlorophenyl)carbamate ... 1027
$C_{13}H_9Cl_2N_3O_3$
1-(2-Chlorophenyl)-3-(2-chloro-4-nitrophenyl)urea ... 568
1-(3-Chlorophenyl)-3-(2-chloro-4-nitrophenyl)urea ... 569
1-(4-Chlorophenyl)-3-(2-chloro-4-nitrophenyl)urea ... 526
3,4-Dichlorobenzoic *N*2-(4-nitrophenyl)hydrazide ... 492
N-(3,4-Dichlorophenyl)-*N*′-(furfurylideneamino)oxamide ... 1024
1-(2,6-Dichlorophenyl)-3-(3-nitrophenyl)urea ... 524
1-(2,6-Dichlorophenyl)-3-(4-nitrophenyl)urea ... 530
1-(3,5-Dichlorophenyl)-3-(3-nitrophenyl)urea ... 525
$C_{13}H_9Cl_2N_3O_4$
1-(2,3-Dichlorophenyl)-3-(2-hydroxy-5-nitrophenyl)urea ... 513
1-(2,5-Dichlorophenyl)-3-(2-hydroxy-4-nitrophenyl)urea ... 547
$C_{13}H_9Cl_3N_2O$
1-(2-Chlorophenyl)-3-(2,3-dichlorophenyl)urea ... 537
1-(2-Chlorophenyl)-3-(2,4-dichlorophenyl)urea ... 589
1-(2-Chlorophenyl)-3-(2,5-dichlorophenyl)urea ... 537
1-(2-Chlorophenyl)-3-(2,6-dichlorophenyl)urea ... 537
1-(2-Chlorophenyl)-3-(3,4-dichlorophenyl)urea ... 589
1-(2-Chlorophenyl)-3-(3,5-dichlorophenyl)urea ... 538
1-(3-Chlorophenyl)-3-(2,3-dichlorophenyl)urea ... 541
1-(3-Chlorophenyl)-3-(2,4-dichlorophenyl)urea ... 543
1-(3-Chlorophenyl)-3-(2,5-dichlorophenyl)urea ... 543
1-(3-Chlorophenyl)-3-(3,4-dichlorophenyl)urea ... 600
1-(3-Chlorophenyl)-3-(3,5-dichlorophenyl)urea ... 545
1-(4-Chlorophenyl)-3-(2,3-dichlorophenyl)urea ... 539
1-(4-Chlorophenyl)-3-(2,4-dichlorophenyl)urea ... 540
1-(4-Chlorophenyl)-3-(2,5-dichlorophenyl)urea ... 540
1-(4-Chlorophenyl)-3-(2,6-dichlorophenyl)urea ... 539
1-(4-Chlorophenyl)-3-(3,4-dichlorophenyl)urea ... 539
1-Phenyl-3-(2,4,5-trichlorophenyl)urea ... 586
1-Phenyl-3-(2,4,6-trichlorophenyl)urea ... 536
2,3,4′-Trichlorocarbanilide ... 591
2,4,4′-Trichlorocarbanilide ... 591
$C_{13}H_9Cl_3N_2O_2$
1-(5-Chloro-2-hydroxyphenyl)-3-(3,4-dichlorophenyl)urea ... 513
$C_{13}H_9Cl_3N_2O_5S$
2′-Methoxy-5′-nitro-2,4,5-trichlorobenzenesulfonanilide ... 806
$C_{13}H_9Cl_4NO_2S$
2′-Methyl-2,4,5,5′-tetrachlorobenzenesulfonanilide ... 810
$C_{13}H_9Cl_4NO_3S$
2′-Methoxy-2,4,5,5′-tetrachlorobenzenesulfonanilide ... 806
$C_{13}H_9F$
2-Fluorofluorene ... 174
$C_{13}H_9FIN$
N-(4-Fluorobenzylidene)-4-iodoaniline ... 1077
$C_{13}H_9FN_2O_2$
2-Amino-6-fluoro-7-nitrofluorene ... 271
N-(4-Fluorobenzylidene)-3-nitroaniline ... 1074
N-(4-Fluorobenzylidene)-4-nitroaniline ... 1074
4-Fluoro-*N*-(3-nitrobenzylidene)aniline ... 1074
$C_{13}H_9FN_2O_3$
2′-Fluoro-3-nitrobenzanilide ... 452
2-Fluoro-3′-nitrobenzanilide ... 451
3′-Fluoro-4-nitrobenzanilide ... 452
4′-Fluoro-3-nitrobenzanilide ... 452
4′-Fluoro-4-nitrobenzanilide ... 452
$C_{13}H_9FN_2O_4$
4-Nitrophenyl *N*-(4-fluorophenyl)carbamate ... 628
$C_{13}H_9FN_2O_5S$
4-(*N*-(4-Nitrophenyl)carbamoyl)benzenesulfonyl fluoride ... 803
$C_{13}H_9FN_4O_5$
1-(4-Fluoro-3-nitrophenyl)-3-(3-nitrophenyl)urea ... 525
$C_{13}H_9FN_4O_6$
1-(2-Fluoro-5-nitrophenyl)-3-(2-hydroxy-5-nitrophenyl)urea ... 513
$C_{13}H_9FO$
4-Fluoro-9-hydroxyfluorene ... 208
$C_{13}H_9FO_2$
4′-Fluoro-3-(2-furyl)acrylophenone ... 853
$C_{13}H_9F_2IN_2O$
1-(2,5-Difluorophenyl)-3-(4-iodophenyl)urea ... 540
$C_{13}H_9F_2N$
4-Fluoro-*N*-(4-fluorobenzylidene)aniline ... 1077
$C_{13}H_9F_2NO_2$
2-Fluorophenyl *N*-(4-fluorophenyl)carbamate ... 633
4-Fluorophenyl *N*-(3-fluorophenyl)carbamate ... 634
$C_{13}H_9F_2NO_2S$
S-(2-Hydroxyphenyl) *N*-(2,5-difluorophenyl)thiocarbamate ... 728
S-(4-Hydroxyphenyl) *N*-(2,5-difluorophenyl)thiocarbamate ... 612
$C_{13}H_9F_2N_3O_3$
1-(2,5-Difluorophenyl)-3-(3-nitrophenyl)urea ... 524
1-(4-Fluoro-3-nitrophenyl)-3-(2-fluorophenyl)urea ... 524
1-(3-Fluorophenyl)-3-(2-fluoro-5-nitrophenyl)urea ... 524
1-(3-Fluorophenyl)-3-(4-fluoro-3-nitrophenyl)urea ... 524
1-(4-Fluorophenyl)-3-(2-fluoro-5-nitrophenyl)urea ... 524
$C_{13}H_9F_3N_2O$
1-(2,5-Difluorophenyl)-3-(4-fluorophenyl)urea ... 539
$C_{13}H_9F_3N_2O_2$
3-Pyridyl *N*-(3-trifluoromethylphenyl)carbamate ... 925
$C_{13}H_9F_3N_2O_4S$
3-Nitro-3′-(trifluoromethyl)benzenesulfonanilide ... 805
$C_{13}H_9F_3N_2S$
1-(2,4-Difluorophenyl)-3-(2-fluorophenyl)-2-thiourea ... 733
$C_{13}H_9F_5Si$
Methyl(pentafluorophenyl)phenylsilane ... 1102
$C_{13}H_9I$
2-Iodofluorene ... 174
$C_{13}H_9IN_2O_2$
4-Iodo-*N*-(3-nitrobenzylidene)aniline ... 1074
$C_{13}H_9IO_2$
Diphenyliodonium-2-carboxylate monohydrate ... 350
$C_{13}H_9I_2NO_3$
1,4-Dihydro-3,5-diiodo-4-oxo-α-phenyl-1-pyridineacetic acid ... 331
$C_{13}H_9N$
Acridine hydrochloride ... 996
Acridine picrate ... 996
Benzo(f)quinoline ... 995
9-Iminofluorene ... 1079
$C_{13}H_9NO$
9-Fluorenone oxime ... 1014
Phenanthridine *N*-oxide ... 1007
3-Phenyl-1,2-benzisoxazole ... 910
3-Phenyl-2,1-benzisoxazole ... 909
α-Phenyl-2-furanacrylonitrile ... 850
$C_{13}H_9NOS$
4-Dibenzothiophenecarboxamide ... 898
2-(2-Hydroxyphenyl)benzothiazole ... 912
2-Naphthoylmethyl thiocyanate ... 768
3-Phenyl-2(3*H*)-benzothiazolone ... 746
5-Phenyl-2-benzoxazolethiol ... 909
$C_{13}H_9NO_2$
7-Amino-3,4-benzocoumarin ... 385
2-Amino-3-hydroxy-1*H*-phenalen-1-one ... 310
3-Nitrofluorene ... 280
4-Phenoxyphenyl isocyanate ... 768

$C_{13}H_9NO_2S$
2-(2,4-Dihydroxyphenyl)benzothiazole 912
4-(Phenylsulfonyl)benzonitrile 782
2-Phthalimidomethylthiophene 858

$C_{13}H_9NO_3$
α-Cyanofurfuryl benzoate 851
1-Hydroxy-2-nitrofluorene 265
2-Hydroxy-7-nitrofluorene 265
2-Nitrobenzophenone 313
2-(2-Pyridylcarbonyl)benzoic acid 959

$C_{13}H_9NO_3S$
4′-Nitro-3-(2-thienyl)acrylophenone 856
S-Phenyl 4-nitrothiobenzoate 379

$C_{13}H_9NO_4$
1-(2-Furyl)-3-(3-nitrophenyl)-2-propen-1-one 852
2-(2-Nitrophenoxy)benzaldehyde 323
2-Nitrophenyl benzoate 379

$C_{13}H_9NO_5$
N-Phthaloyl-L-glutamic anhydride 722

$C_{13}H_9NS$
2-Biphenylyl isothiocyanate 768
4-Biphenylyl isothiocyanate 768
α-Phenyl-β-(2-thienyl)acrylonitrile 856
2-(Phenylthio)benzonitrile 763
2-(2-Thienyl)quinoline 993

$C_{13}H_9NS_2$
3-Phenylbenzothiazoline-2-thione 749
2-(Phenylthio)benzothiazole 914

$C_{13}H_9N_3$
trans-2,3-Bis(4-pyridyl)acrylonitrile 965

$C_{13}H_9N_3OS$
1-(2-Thiazolylazo)-2-naphthol 1085

$C_{13}H_9N_3O_2$
6-Nitro-1-phenyl-1*H*-indazole 900

$C_{13}H_9N_3O_3S$
3-(2-Benzothiazolylamino)-5-nitrophenol 912

$C_{13}H_9N_3O_4$
Methyl 2,4-dihydroxypyrimido(4,5-b)quinoline-5-carboxylate 1002
3-Nitro-*N*-(3-nitrobenzylidene)aniline 1054

$C_{13}H_9N_3O_4S$
N-(2,4-Dinitrobenzylidene)-2-mercaptoaniline 1051

$C_{13}H_9N_3O_5$
2′,4-Dinitrobenzanilide 469
2′,6′-Dinitrobenzanilide 469
N-(5-Nitro-3-pyridyl)phthalamic acid 944

$C_{13}H_9N_3O_6$
4-Anilino-3,5-dinitrobenzoic acid 342
4-Carboxy-2′,4′-dinitrodiphenylamine 342
N-(2,4-Dinitrophenyl)anthranilic acid 342

$C_{13}H_9N_3O_8S_3$
4,5-Dihydroxy-3-(2-thiazolylazo)-2,7-naphthalenedisulfonic acid, sodium salt 1085

$C_{13}H_9N_5O_3$
1-(*N*-(3-Nitrophenyl)carbamoyl)-1*H*-benzotriazole 908

$C_{13}H_9N_5O_6$
4-Nitrobenzaldehyde 2,4-dinitrophenylhydrazone 1052

$C_{13}H_9N_5O_8$
3-Nitro-*N*-picryl-*p*-toluidine 272
5-Nitro-*N*-picryl-*o*-toluidine 272

$C_{13}H_9N_5O_9$
N-(2-Methoxy-5-nitrophenyl)picrylamine 266

$C_{13}H_{10}BF_3N_2O_3$
2-Hydroxy-5-(3-(trifluoromethyl)phenylazo)benzeneboronic acid 1124

$C_{13}H_{10}BrCl$
4-Bromo-α-chlorodiphenylmethane 174
(4-Bromophenyl)(4-chlorophenyl)methane 174

$C_{13}H_{10}BrClN_2$
1-(4-Bromophenyl)-3-(2-chlorophenyl)urea 537

$C_{13}H_{10}BrClN_2O$
1-(4-Bromophenyl)-3-(4-chlorophenyl)urea 538

$C_{13}H_{10}BrF_3N_2O_2S$
Benzenesulfonic acid 2-(2-bromo-5-(trifluoromethyl)phenyl)hydrazide 805

$C_{13}H_{10}BrN$
2-Amino-3-bromofluorene 231
N-Benzylidene-3-bromoaniline 1027
N-Benzylidene-4-bromoaniline 1065
N-(2-Bromobenzylidene)aniline 1027
N-(4-Bromobenzylidene)aniline 1027
N-(3-Bromobenzylidene)aniline picrate 1027

$C_{13}H_{10}BrNO$
2-Amino-5-bromobenzophenone 311
3-Bromobenzanilide 483
4′-Bromobenzanilide 480
α-(3-Bromophenylimino)-*o*-cresol 1035
α-(4-Bromophenylimino)-*o*-cresol 1035
α-(4-Bromophenylimino)-*p*-cresol 1023

$C_{13}H_{10}BrNO_2$
3-Amino-1-bromo-4-methoxydibenzofuran 893
N-(5-Bromo-2-hydroxybenzylidene)-2-hydroxyaniline 1028
Phenyl *N*-(4-bromophenyl)carbamate 683

$C_{13}H_{10}BrNO_4$
Ethyl 6-bromo-α-cyano-(3,4-methylenedioxy)cinnamate 766

$C_{13}H_{10}BrN_3O$
4-Bromophenylazo phenyl ketone oxime 1008

$C_{13}H_{10}BrN_3O_2$
1-(4-Bromo-α-nitrobenzylidene)-2-phenylhydrazine 1025

$C_{13}H_{10}BrN_3O_4$
4-Bromo-2′,4′-dinitro-3-methyldiphenylamine 261

$C_{13}H_{10}Br_2N_2$
N,*N*′-Bis(4-bromophenyl)formamidine 724

$C_{13}H_{10}Br_2N_2S$
1,3-Bis(3-bromophenyl)-2-thiourea 732
1,3-Bis(4-bromophenyl)-2-thiourea 733
1-(3,4-Dibromophenyl)-3-phenyl-2-thiourea 732

$C_{13}H_{10}ClFN_2O$
1-(2-Chlorophenyl)-3-(3-fluorophenyl)urea 588
1-(2-Chlorophenyl)-3-(4-fluorophenyl)-urea 537
1-(3-Chlorophenyl)-3-(2-fluorophenyl)urea 536
1-(3-Chlorophenyl)-3-(3-fluorophenyl)urea 534
1-(3-Chlorophenyl)-3-(4-fluorophenyl)urea 538
1-(4-Chlorophenyl)-3-(2-fluorophenyl)urea 537
1-(4-Chlorophenyl)-3-(3-fluorophenyl)urea 538
1-(4-Chlorophenyl)-3-(4-fluorophenyl)urea 590

$C_{13}H_{10}ClF_3N_2O_2$
1-(2-Chloro-5-trifluoromethylphenyl)-3-furfurylurea 843
1-(4-Chloro-2-(trifluoromethyl)phenyl)-3-furfurylurea 843
1-(4-Chloro-3-(trifluoromethyl)phenyl)-3-furfurylurea 843

$C_{13}H_{10}ClN$
2-Amino-3-chlorofluorene 231
N-Benzylidene-2-chloroaniline 1064
N-Benzylidene-3-chloroaniline 1062
N-Benzylidene-4-chloroaniline, tech. 1077
N-(2-Chlorobenzylidene)aniline 1064
N-(4-Chlorobenzylidene)aniline 1065
2-(2-Chlorostyryl)pyridine 931
4-(4-Chlorostyryl)pyridine 931

$C_{13}H_{10}ClNO$
2-Amino-4′-chlorobenzophenone 309
2′-Chlorobenzanilide 456
2-Chlorobenzanilide 478
3-Chlorobenzanilide 456
3′-Chlorobenzanilide 456
2-Chloro-*N*-(2-hydroxybenzylidene)aniline 1022
3-Chloro-*N*-(2-hydroxybenzylidene)aniline 1022
N-(5-Chloro-2-hydroxybenzylidene)-2-hydroxyaniline 1028
α-(2-Chlorophenylimino)-*o*-cresol 1035
α-(3-Chlorophenylimino)-*o*-cresol 1035
α-(3-Chlorophenylimino)-*p*-cresol 1023
α-(4-Chlorophenylimino)-*o*-cresol 1035
α-(4-Chlorophenylimino)-*p*-cresol 1023

$C_{13}H_{10}ClNOS$
2-Chloro-7-methoxy-10*H*-phenothiazine 247
4-Chloro-1-methoxy-10*H*-phenothiazine 247
4′-Chloro-2-(2-pyridylthio)-acetophenone 927

$C_{13}H_{10}ClNO_2$
2-Chloro-2′-hydroxybenzanilide 462
2-Chloro-3′-hydroxybenzanilide 462
2-Chloro-4′-hydroxybenzanilide 462
4-Chloro-2′-hydroxybenzanilide 462
4-Chloro-3′-hydroxybenzanilide 462
4-Chloro-4′-hydroxybenzanilide 463
5′-Chloro-2′-hydroxybenzanilide 447
α-(5-Chloro-2-hydroxyphenylimino)-*o*-cresol 1021
4-Chloro-*N*-phenylanthranilic acid 343
5-Chloro-*N*-phenylanthranilic acid 337
4-Chloro-*N*-phenylbenzohydroxamic acid 452
2-Chlorophenyl carbanilate 632
4-Chlorophenyl carbanilate 677
Phenyl *N*-(2-chlorophenyl)carbamate 681
Phenyl *N*-(3-chlorophenyl)carbamate 678
Phenyl *N*-(4-chlorophenyl)carbamate 682

$C_{13}H_{10}ClNO_3$
4-Chloro-2-nitrophenyl *p*-tolyl ether 268
4-Chloro-*m*-tolyl 4-nitrophenyl ether 269

$C_{13}H_{10}ClNO_4$
Benzo(b)quinolizinium perchlorate 1003

$C_{13}H_{10}ClNO_4S_2$
3-(4-Chlorophenyl)thiazolo(3,2-a)-pyridinium bisulfate 911

$C_{13}H_{10}ClNS$
2-Chloro-10-methyl-10*H*-phenothiazine 253
2-(4-Chlorophenyl)-3-(2-thienyl)-propionitrile 855

$C_{13}H_{10}ClN_3O$
4-Pyridinecarboxylic acid *N*-(4-chlorobenzylidene)hydrazide 493

$C_{13}H_{10}ClN_3O_2$
2-Chlorobenzaldehyde (4-nitrophenyl)hydrazone 261

$C_{13}H_{10}ClN_3O_3$
4-Chlorobenzoic *N*2-(4-nitrophenyl)-hydrazide 492
1-(2-Chloro-4-nitrophenyl)-3-phenylurea 568
1-(2-Chlorophenyl)-3-(3-nitrophenyl)urea 524
1-(2-Chlorophenyl)-3-(4-nitrophenyl)urea 568
1-(3-Chlorophenyl)-3-(3-nitrophenyl)urea 565
1-(4-Chlorophenyl)-3-(3-nitrophenyl)urea 524
1-(4-Chlorophenyl)-3-(4-nitrophenyl)urea 526

$C_{13}H_{10}ClN_3O_4$
3-Chloro-2′,4′-dinitro-4-methyldiphenylamine 261
4-Chloro-2′,4′-dinitro-2-methyldiphenylamine 261
1-(4-Chloro-2-nitrophenyl)-3-(3-hydroxyphenyl)urea 512
5-Nitro-2-pyridyl *N*-(5-chloro-2-methylphenyl)carbamate 927
2-Phenylpyrido(1,2-b)(1,2,4)triazin-5-ium perchlorate 1003
3-Pyridylmethyl *N*-(2-chloro-4-nitrophenyl)carbamate 929

$C_{13}H_{10}ClN_3O_5$
5-Nitro-2-pyridyl *N*-(4-chloro-2-methoxyphenyl)carbamate 927

$C_{13}H_{10}ClN_3O_6S$
4-Chloro-3,5′-dinitro-2′-methylbenzenesulfonanilide 807

$C_{13}H_{10}ClN_3O_7S$
4-Chloro-3,5′-dinitro-2′-methoxybenzenesulfonanilide 806

$C_{13}H_{10}ClN_5O_2$
4-(2-Chloro-4-nitrophenyl)-1-(4-pyridylcarbonyl)semicarbazide 929

$C_{13}H_{10}ClN_5O_5$
4-(3-Chlorophenyl)-1-(2,4-dinitrophenyl)semicarbazide 522
4-(4-Chlorophenyl)-1-(2,4-dinitrophenyl)semicarbazide 522

$C_{13}H_{10}Cl_2N_2$
2,4-Dichlorobenzaldehyde phenylhydrazone 234

$C_{13}H_{10}Cl_2N_2O$
1,3-Bis(2-chlorophenyl)urea 536
1,3-Bis(3-chlorophenyl)urea 534
1,3-Bis(4-chlorophenyl)urea 590
1-(2-Chlorophenyl)-3-(3-chlorophenyl)urea 588
1-(2-Chlorophenyl)-3-(4-chlorophenyl)urea 588
1-(3-Chlorophenyl)-3-(4-chlorophenyl)urea 590
1-(2,3-Dichlorophenyl)-3-phenylurea 594
1-(2,4-Dichlorophenyl)-3-phenylurea 599
1-(2,5-Dichlorophenyl)-3-phenylurea 599
1-(2,6-Dichlorophenyl)-3-phenylurea 594
1-(3,4-Dichlorophenyl)-3-phenylurea 599

$C_{13}H_{10}Cl_2N_2O_2$
4-Chloro-*N*-(4-chloro-3-nitrophenyl)-benzylamine 261
3,4-Dichlorobenzoic acid *N*2-(3-hydroxyphenyl)hydrazide 492
2-Pyridylmethyl *N*-(2,5-dichlorophenyl)carbamate 929
3-Pyridylmethyl *N*-(3,4-dichlorophenyl)carbamate 929
4-Pyridylmethyl *N*-(2,3-dichlorophenyl)carbamate 929

$C_{13}H_{10}Cl_2N_2O_3S$
3,4-Dichlorobenzoic N^2-(phenylsulfonyl)hydrazide 808

$C_{13}H_{10}Cl_2N_2O_4S$
4,5′-Dichloro-2′-methyl-3-nitrobenzenesulfonanilide 807

$C_{13}H_{10}Cl_2N_2O_5S$
4,5′-Dichloro-2′-methoxy-2-nitrobenzenesulfonanilide 806
4,5′-Dichloro-2′-methoxy-3-nitrobenzenesulfonanilide 806

$C_{13}H_{10}Cl_2N_2S$
1,3-Bis(2-chlorophenyl)-2-thiourea 733
1,3-Bis(4-chlorophenyl)-2-thiourea 733
1-(2,5-Dichlorophenyl)-3-phenyl-2-thiourea 749
1-(3,4-Dichlorophenyl)-3-phenyl-2-thiourea 732

$C_{13}H_{10}Cl_2N_4O_3$
1-(3,5-Dichloro-2-pyridyl)-3-(4-methyl-3-nitrophenyl)urea 949

$C_{13}H_{10}Cl_2N_6O_2$
8-(2,6-Dichlorobenzylidene-hydrazino)xanthine 1025

$C_{13}H_{10}Cl_2O_2$
Dichlorodiphenoxymethane 182

$C_{13}H_{10}Cl_2O_3S$
2,4-Dichlorophenyl *p*-toluenesulfonate 801

$C_{13}H_{10}Cl_3NO_2$
2,2,2-Trichloroethyl *N*-(1-naphthyl)-carbamate 611

$C_{13}H_{10}Cl_3N_3O$
1-(2-Pyridylmethyl)-3-(2,4,5-trichlorophenyl)urea 943
1-(3-Pyridylmethyl)-3-(2,4,5-trichlorophenyl)urea 930

$C_{13}H_{10}Cl_3N_3O_2$
1-(4-Chloro-2-methoxyphenyl)-3-(3,5-dichloro-2-pyridyl)urea 927

$C_{13}H_{10}Cl_6O_2$
5,6,7,8,9,9-Hexachloro-octahydro-1,4:5,8-dimethanonaphthalene-2-carboxylic acid 66

$C_{13}H_{10}Cl_7Sb$
Chlorodiphenylcarbenium hexachloroantimonate ... 1143
$C_{13}H_{10}FIN_2O$
1-(4-Fluorophenyl)-3-(4-iodophenyl)urea ... 538
$C_{13}H_{10}FN$
7-Amino-1-fluorofluorene ... 231
N-(4-Fluorobenzylidene)aniline ... 1077
$C_{13}H_{10}FNO$
2′-Fluorobenzanilide ... 456
2-(4-Fluorobenzylideneamino)-phenol ... 1035
3-Fluoro-*N*-(2-hydroxybenzylidene)-aniline ... 1022
α-(2-Fluorophenylimino)-*o*-cresol ... 1035
α-(4-Fluorophenylimino)-*p*-cresol ... 1023
N-(2-Hydroxybenzylidene)-4-fluoroaniline ... 1022
$C_{13}H_{10}FNO_2$
2-Fluoro-4′-hydroxybenzanilide ... 447
$C_{13}H_{10}FNO_2S$
S-(2-Hydroxyphenyl) *N*-(2-fluorophenyl)thiocarbamate ... 611
S-(2-Hydroxyphenyl) *N*-(4-fluorophenyl)thiocarbamate ... 611
S-(4-Hydroxyphenyl) *N*-(3-fluorophenyl)thiocarbamate ... 611
$C_{13}H_{10}FN_3O_3$
1-(2-Fluorophenyl)-3-(3-nitrophenyl)urea ... 524
1-(3-Fluorophenyl)-3-(3-nitrophenyl)urea ... 524
$C_{13}H_{10}FN_3O_5S$
3-(3-(3-Nitrophenyl)ureido)-benzenesulfonyl fluoride ... 803
$C_{13}H_{10}FN_5O_5$
1-(2,4-Dinitrophenyl)-4-(4-fluorophenyl)semicarbazide ... 522
$C_{13}H_{10}F_2N_2$
N,N′-Bis(4-fluorophenyl)-formamidine ... 724
$C_{13}H_{10}F_2N_2O$
1,3-Bis(3-fluorophenyl)urea ... 534
1,3-Bis(4-fluorophenyl)urea ... 590
$C_{13}H_{10}F_2N_2O_2$
1-(2,5-Difluorophenyl)-3-(3-hydroxyphenyl)urea ... 514
1-(2,5-Difluorophenyl)-3-(4-hydroxyphenyl)urea ... 514
3-Pyridylmethyl *N*-(2,5-difluorophenyl)carbamate ... 964
$C_{13}H_{10}F_2N_2O_5S_2$
1,3-Bis(3-fluorosulfonylphenyl)urea ... 804
$C_{13}H_{10}F_2N_2S$
1,3-Bis(2-fluorophenyl)-2-thiourea ... 733
$C_{13}H_{10}F_2N_4O_2$
4-(2,4-Difluorophenyl)-1-(4-pyridylcarbonyl)semicarbazide ... 928
$C_{13}H_{10}F_3N$
N-Phenyl-α,α,α-trifluoro-*m*-toluidine ... 215
$C_{13}H_{10}F_3NO_3$
Furfuryl *N*-(3-trifluoromethylphenyl)carbamate ... 843
$C_{13}H_{10}F_3N_3O$
1-(4-Pyridyl)-3-(2-(trifluoromethyl)-phenyl)urea ... 925
$C_{13}H_{10}HgO_3$
Phenylmercuric salicylate ... 1137
$C_{13}H_{10}IN$
N-Benzylidene-4-iodoaniline ... 1025
$C_{13}H_{10}INO$
N-(2-Hydroxybenzylidene)-2-iodoaniline ... 1022
$C_{13}H_{10}I_2N_2O$
1,3-Bis(3-iodophenyl)urea ... 534
1,3-Bis(4-iodophenyl)urea ... 538
$C_{13}H_{10}I_2N_2S$
1,3-Bis(3-iodophenyl)-2-thiourea ... 732
1,3-Bis(4-iodophenyl)-2-thiourea ... 733
$C_{13}H_{10}N_2$
9-Aminoacridine ... 995
α-Cyclopropylbenzylidene-malononitrile ... 764
1-(Dicyanomethylene)tetralin ... 764
(5-Indanylmethylene)-malononitrile ... 764
Perinaphthenone hydrazone ... 236
1-Phenyl-1*H*-indazole ... 900
2-Phenylindazole ... 900
α-Phenyl-2-pyridineacetonitrile ... 951
2-(2-Pyridyl)indole ... 945
$C_{13}H_{10}N_2O$
6-Amino-2-phenylbenzoxazole ... 909
2-(2-Benzimidazolyl)-phenol ... 902
N-(4-Cyano-1-naphthyl)-acetamide ... 767
2,4-Diamino-9-fluorenone ... 311
2,7-Diamino-9-fluorenone ... 311
$C_{13}H_{10}N_2O_2$
2-Amino-3-nitrofluorene ... 271
2-Amino-7-nitrofluorene ... 271
3-Amino-2-nitrofluorene ... 271
4-Amino-7-nitrofluorene ... 271
N-Benzylidene-3-nitroaniline ... 1056
N-Benzylidene-4-nitroaniline ... 1059
2-Dibenzofurancarboxylic hydrazide ... 895
Methyl α-cyano-3-indoleacrylate ... 889
N-(3-Nitrobenzylidene)aniline ... 1059
N-(4-Nitrobenzylidene)aniline ... 1073
5-Phenylazo-1,3-benzodioxole ... 1092
3-(4-Pyridylamino)phthalide ... 945
$C_{13}H_{10}N_2O_2S$
5-Cinnamylidene-2-thiobarbituric acid ... 734
$C_{13}H_{10}N_2O_3$
5-Cinnamylidenebarbituric acid ... 696
2′-Nitrobenzanilide ... 469
4′-Nitrobenzanilide ... 470
4-Nitrobenzanilide ... 472
2-(3-Nitrobenzylideneamino)phenol ... 1033
α-(3-Nitrophenylimino)-*o*-cresol ... 1033
α-(3-Nitrophenylimino)-*p*-cresol ... 1034
2-Nitro-α-phenylimino-*p*-cresol ... 1034
5-(Phenylazo)salicylic acid ... 1085
N-(2-Pyridyl)phthalamic acid ... 962
N-(3-Pyridyl)phthalamic acid ... 944
$C_{13}H_{10}N_2O_3S$
4-(2-Benzimidazolyl)-benzenesulfonic acid ... 905
2-(Methylsulfonyl)-5-(1-naphthyl)-1,3,4-oxadiazole ... 873
S-(4-Nitrophenyl) benzothiohydroxamate ... 1009
$C_{13}H_{10}N_2O_4$
Bis(3-nitrophenyl)methane ... 279
Bis(4-nitrophenyl)methane ... 278
2′-Hydroxy-4′-nitrobenzanilide ... 462
2′-Hydroxy-4-nitrobenzanilide ... 462
N-(2-Nitrophenyl)anthranilic acid ... 342
N-(3-Nitrophenyl)anthranilic acid ... 342
5-Nitro-*N*-phenylanthranilic acid ... 343
4-Nitro-*N*-phenyl-benzohydroxamic acid ... 472
Phenyl *N*-(3-nitrophenyl)carbamate ... 670
$C_{13}H_{10}N_2O_4S$
2,4-Dinitrophenyl *m*-tolyl sulfide ... 270
$C_{13}H_{10}N_2O_4S_2$
2,4-Dinitrophenyl *p*-tolyl disulfide ... 270
$C_{13}H_{10}N_2O_5$
2,4-Dinitro-2′-methyldiphenyl ether ... 267
$C_{13}H_{10}N_2O_6$
Bis(4-nitrophenoxy)methane ... 265
Dimethyl 5-nitro-2,4-quinolinedicarboxylate ... 988
$C_{13}H_{10}N_2O_6S$
2,4-Dinitrophenyl *m*-tolyl sulfone ... 778
$C_{13}H_{10}N_2S$
6-Amino-2-phenylbenzothiazole ... 914
$C_{13}H_{10}N_4$
1-(4-Dimethylaminophenyl)-1,2,2-tricyanoethylene ... 763
4-(2-Naphthylmethyleneamino)-4*H*-1,2,4-triazole ... 1072
$C_{13}H_{10}N_4O$
1-(Phenylazo)-1*H*-indazol-3(2*H*)-one ... 1092
$C_{13}H_{10}N_4O_3$
3-Nitrobenzoic (3-pyridylmethylene)hydrazide ... 1069
Picolinic acid (4-nitrobenzylidene)-hydrazide ... 1053
4-Pyridinecarboxylic (3-nitrobenzylidene)hydrazide ... 1058
$C_{13}H_{10}N_4O_4$
Benzaldehyde (2,4-dinitrophenyl)-hydrazone ... 273
1-Benzoyl-3-(3-nitro-4-pyridyl)urea ... 944
3-Nitrobenzaldehyde (4-nitrophenyl)hydrazone ... 277
*N*1-(2-Nitrobenzylidene)-5-nitro-1,2-phenylenediamine ... 1052
1-(5-Nitrofurfurylidene-2-imino)-2,4-dimethyl-5-cyano-6-pyridone ... 1069
$C_{13}H_{10}N_4O_4S$
1,3-Bis(3-nitrophenyl)-2-thiourea ... 740
$C_{13}H_{10}N_4O_5$
2-(2,4-Dinitrophenylazo)-*p*-cresol ... 1084
$C_{13}H_{10}N_4O_6$
N-Picrylbenzylamine ... 273
N-Picryl-*o*-toluidine ... 273
N-Picryl-*m*-toluidine ... 272
N-Picryl-*p*-toluidine ... 272
$C_{13}H_{10}N_4O_6S$
N-(3-Methylthiophenyl)-2,4,6-trinitroaniline ... 270
4-(Methylthio)-*N*-picrylaniline ... 270
$C_{13}H_{10}N_4O_7$
N-(2-Hydroxy-4-methylphenyl)-picrylamine ... 263
N-(2-Hydroxy-5-methylphenyl)-picrylamine ... 263
N-(2-Methoxyphenyl)picrylamine ... 273
N-(3-Methoxyphenyl)picrylamine ... 266
N-(4-Methoxyphenyl)picrylamine ... 267
$C_{13}H_{10}N_4S$
2-Pyridylcarboxaldehyde (2-benzothiazolyl)hydrazone ... 1072
$C_{13}H_{10}O$
4-Hydroxyfluorene ... 197
6-Methoxy-2-naphthylacetylene ... 1017
2-Methyldibenzofuran ... 897
4-Methyldibenzofuran ... 897
$C_{13}H_{10}OS$
4-Methoxydibenzothiophene ... 897
2*H*-Naphtho(1,2-b)thiopyran-4(3*H*)-one ... 296
3*H*-Naphtho(2,1-b)thiopyran-1(2*H*)-one ... 296
$C_{13}H_{10}OS_2$
1-Methoxythianthrene ... 211
$C_{13}H_{10}O_2$
2-Furfurylideneacetophenone ... 852
3-Methoxydibenzofuran ... 893
4-Methoxydibenzofuran picrate ... 893
3-(1-Naphthyl)acrylic acid ... 346
(*p*-Tolyl)-*p*-benzoquinone ... 318
$C_{13}H_{10}O_2S$
1-Phenyl-3-(2-thienyl)-1,3-propanedione, copper(I) derivative ... 855
$C_{13}H_{10}O_3$
3-Acetyl-2-naphthoic acid ... 344
Catechol monobenzoate ... 377
2′-Hydroxy-2-biphenylcarboxylic acid ... 339
3-Hydroxy-4-methoxydibenzofuran ... 893
3-Hydroxyphenyl benzoate ... 377
2-(4-Methoxyphenyl)-1,4-benzoquinone ... 317
2-Methoxy-5-phenyl-1,4-benzoquinone ... 316
3-Phenylsalicylic acid ... 339
$C_{13}H_{10}O_3S$
4-Methoxydibenzothiophene-5,5-dioxide ... 897
$C_{13}H_{10}O_4$
1-Acetoxy-2-naphthoic acid ... 380
$C_{13}H_{10}O_6$
Maclurin ... 310
$C_{13}H_{10}O_6S$
4-(4-Sulfophenoxy)benzoic acid, magnesium salt ... 792
$C_{13}H_{11}AsClN$
10-Chloro-5,10-dihydro-2-methylphenarsazine ... 1128
$C_{13}H_{11}AsCl_2N_2O_4$
4-(3-(2,3-Dichlorophenyl)ureido)-phenylarsonic acid ... 1126
$C_{13}H_{11}BN_2O_5$
4-(3-Borono-4-hydroxyphenylazo)-benzoic acid ... 1092
$C_{13}H_{11}Br$
4-Bromomethylbiphenyl ... 172
4-Bromo-3-methylbiphenyl ... 174
$C_{13}H_{11}BrClN_3O$
1-(5-Bromo-2-pyridyl)-3-(5-chloro-2-methylphenyl)urea ... 930
$C_{13}H_{11}BrN_2$
Benzaldehyde (4-bromophenyl)-hydrazone ... 234
4-Bromobenzaldehyde phenylhydrazone ... 234
$C_{13}H_{11}BrN_2O$
Bromoharmine hydrobromide ... 918
1-(2-Bromophenyl)-3-phenylurea ... 527
1-(4-Bromophenyl)-3-phenylurea ... 590
$C_{13}H_{11}BrN_2O_2$
3-Pyridyl *N*-(2-bromo-4-methylphenyl)carbamate ... 929
$C_{13}H_{11}BrN_2O_3$
4-(Benzyloxy)-5-bromo-2-nitroaniline ... 260
$C_{13}H_{11}BrO$
Benzyl 3-bromophenyl ether ... 176
Benzyl 4-bromophenyl ether ... 176
4-Bromobenzhydrol ... 208
2-Bromobenzyl phenyl ether ... 177
4-Bromobenzyl phenyl ether ... 178
$C_{13}H_{11}BrO_2$
2-Bromo-6-methoxy-4-phenylphenol. 193
$C_{13}H_{11}BrO_2S$
3-Bromo-4-methylphenyl phenyl sulfone ... 774
4-Bromophenyl *p*-tolyl sulfone ... 774
$C_{13}H_{11}BrO_3$
2-(6-Bromo-2-naphthyloxy)-propionic acid ... 328
$C_{13}H_{11}BrS$
Benzyl 4-bromophenyl sulfide ... 210
$C_{13}H_{11}Br_2NO$
1-(4′-Bromophenacyl)pyridinium bromide ... 956
$C_{13}H_{11}Cl$
(4-Chlorophenyl)phenylmethane ... 172
$C_{13}H_{11}ClN_2$
2-Chlorobenzaldehyde phenylhydrazone ... 234
4-Chlorobenzaldehyde phenylhydrazone ... 1072
α-(3-Chloropropyl)-benzylidenemalononitrile ... 758
$C_{13}H_{11}ClN_2O$
2-Amino-5-chlorobenzophenone oxime, mixture of *syn* and *anti* isomers ... 1008
*N*1-(4-Chlorobenzoyl)-*N*1-phenylhydrazine hydrochloride ... 492
1-(2-Chlorophenyl)-3-phenylurea ... 535
1-(3-Chlorophenyl)-3-phenylurea ... 535
1-(4-Chlorophenyl)-3-phenylurea ... 535
Isonicotinaldehyde *o*-(4-chlorobenzyl)oxime ... 1060
$C_{13}H_{11}ClN_2O_2$
4′-Amino-4-chloro-2′-hydroxybenzanilide ... 447
2-Pyridyl *N*-(4-chloro-2-methylphenyl)carbamate ... 960
2-Pyridyl *N*-(5-chloro-2-methylphenyl)carbamate ... 964
3-Pyridyl *N*-(3-chloro-4-methylphenyl)carbamate ... 960
3-Pyridyl *N*-(4-chloro-2-methylphenyl)carbamate ... 960
3-Pyridyl *N*-(5-chloro-2-methylphenyl)carbamate ... 960
3-Pyridylmethyl *N*-(3-chlorophenyl)-carbamate ... 929
4-Pyridylmethyl *N*-(3-chlorophenyl)-carbamate ... 929
$C_{13}H_{11}ClN_2O_2S$
1-(3-Chloro-4-methylphenyl)-3-(2-furoyl)-2-thiourea ... 844
3-(4-Chlorophenyl)-5,6-dihydroimidazo(2,1-b)thiazole-2-acetic acid hydrobromide ... 325
$C_{13}H_{11}ClN_2O_3$
4-Chloro-*N*-(6-methoxy-3-pyridyl)-anthranilic acid ... 937
α-(4-Chloro-3-nitroanilino)-*p*-cresol ... 223
5-Chloro-2-pyridyl *N*-(2-methoxyphenyl)carbamate ... 939
2-Pyridyl *N*-(4-chloro-2-methoxyphenyl)carbamate ... 927

$C_{13}H_{11}ClN_2O_3$ (Cont'd)
2-Pyridyl *N*-(5-chloro-2-methoxyphenyl)carbamate 927
$C_{13}H_{11}ClN_2O_3S$
N-(4-Chlorobenzylideneamino)sulfanilic acid, pyridinium salt 794
$C_{13}H_{11}ClN_2O_4S$
3′-Chloro-2′-methyl-3-nitrobenzenesulfonanilide 807
4′-Chloro-2′-methyl-3-nitrobenzenesulfonanilide 807
5′-Chloro-2′-methyl-3-nitrobenzenesulfonanilide 807
4-Chloro-*N*-(2-methyl-4-nitrophenyl)benzenesulfonamide 807
N-(5-Chloro-2-methyl-4-nitrophenyl)benzenesulfonamide 819
$C_{13}H_{11}ClN_2O_5S$
4-Chloro-2′-methoxy-5′-nitrobenzenesulfonanilide 806
5′-Chloro-2′-methoxy-3-nitrobenzenesulfonanilide 806
$C_{13}H_{11}ClN_2S$
1-(2-Chlorophenyl)-3-phenyl-2-thiourea 748
1-(3-Chlorophenyl)-3-phenyl-2-thiourea 747
1-(4-Chlorophenyl)-3-phenyl-2-thiourea 748
$C_{13}H_{11}ClN_4OS$
4-(3-Chlorophenyl)-1-picolinoylthiosemicarbazide 943
$C_{13}H_{11}ClN_4O_3$
1-(5-Chloro-2-methylphenyl)-3-(5-nitro-2-pyridyl)urea 927
1-(4-Chloro-2-nitrophenyl)-3-(6-methyl-2-pyridyl)urea 927
4-(2-Chloro-4-nitrophenyl)-1-phenylsemicarbazide 568
1-(2-Chloro-4-nitrophenyl)-3-(3-pyridylmethyl)urea 945
4-(3-Chlorophenyl)-1-(4-nitrophenyl)semicarbazide 567
1-(5-Chloro-2-pyridyl)-3-(4-methyl-3-nitrophenyl)urea 949
$C_{13}H_{11}ClN_4O_4$
1-(4-Chloro-2-methoxyphenyl)-3-(5-nitro-2-pyridyl)urea 927
$C_{13}H_{11}ClN_4O_4S$
4-(4-Chlorophenylsulfonyl)-1-nicotinoylsemicarbazide 928
4-(4-Chlorophenylsulfonyl)-1-(4-pyridylcarbonyl)semicarbazide 929
$C_{13}H_{11}ClN_4O_5$
1-(4-Chloro-2-nitrophenyl)-3-(4-sulfamoylphenyl)urea 807
$C_{13}H_{11}ClN_4O_5S$
1-(4-Chloro-3-nitrophenyl)-3-(3-sulfamoylphenyl)urea 524
$C_{13}H_{11}ClO$
2-Benzyl-4-chlorophenol 196
2-Chlorobenzhydrol 198
$C_{13}H_{11}ClO_2S$
Benzyl 4-chlorophenyl sulfone 774
4-Chlorophenyl *p*-tolyl sulfone 785
4-Chloro-2-(2-thienylethyl)benzoic acid 855
$C_{13}H_{11}ClS$
Benzyl 4-chlorophenyl sulfide 210
4-Chlorobenzyl phenyl sulfide 210
$C_{13}H_{11}ClSi$
9-Chloro-9-methyl-9*H*-9-silafluorene 1123
$C_{13}H_{11}Cl_2NO$
5-Chloro-2-(4-chloro-*m*-tolyloxy)aniline 223
N-(2-Hydroxybenzyl)-3,5-dichloroaniline 226
$C_{13}H_{11}Cl_2NO_2S$
3′,4-Dichloro-2′-methylbenzenesulfonanilide 809
4,5′-Dichloro-2′-methylbenzenesulfonanilide 810
$C_{13}H_{11}Cl_2NO_3S$
4-Chloro-*N*-(2-chloro-4-methoxyphenyl)benzenesulfonamide 806
2′,5′-Dichloro-4-methoxybenzenesulfonanilide 806
4,5′-Dichloro-2′-methoxybenzenesulfonanilide 806
$C_{13}H_{11}Cl_2N_2O_2$
1-(5-Chloro-2-methoxyphenyl)-3-(5-chloro-2-pyridyl)urea 927
$C_{13}H_{11}Cl_2N_3O$
1-(5-Chloro-2-pyridyl)-3-(5-chloro-2-methylphenyl)urea 930
1-(2,3-Dichlorophenyl)-3-(6-methyl-2-pyridyl)urea 930
1-(2,4-Dichlorophenyl)-3-(6-methyl-2-pyridyl)urea 930
1-(2,6-Dichlorophenyl)-3-(6-methyl-3-pyridyl)urea 930
4-(2,6-Dichlorophenyl)-1-phenylsemicarbazide 533
4-(3,4-Dichlorophenyl)-1-phenylsemicarbazide 533
1-(2,3-Dichlorophenyl)-3-(3-pyridylmethyl)urea 950
1-(2,4-Dichlorophenyl)-3-(3-pyridylmethyl)urea 950
1-(2,5-Dichlorophenyl)-3-(3-pyridylmethyl)urea 950
1-(2,6-Dichlorophenyl)-3-(3-pyridylmethyl)urea 950
1-(3,4-Dichlorophenyl)-3-(3-pyridylmethyl)urea 950
1-(3,4-Dichlorophenyl)-3-(4-pyridylmethyl)urea 930
$C_{13}H_{11}Cl_2N_3OS$
1-(3,5-Dichloro-2-pyridyl)-3-(4-(methylthio)phenyl)urea 941
$C_{13}H_{11}Cl_2N_3O_2$
1-(4-Chloro-2-methoxyphenyl)-3-(5-chloro-2-pyridyl)urea 927
$C_{13}H_{11}Cl_2N_3O_3S$
1-(2,3-Dichlorophenyl)-3-(4-sulfamoylphenyl)urea 809
$C_{13}H_{11}Cl_4NO_3$
3-Hydroxy-3-piperidino-4,5,6,7-tetrachlorophthalide 383
$C_{13}H_{11}Cl_4NSi$
4-(Dimethylphenylsilyl)-2,3,5,6-tetrachloropyridine 1121
$C_{13}H_{11}FN_2O$
1-(2-Fluorophenyl)-3-phenylurea 535
1-(4-Fluorophenyl)-3-phenylurea 536
$C_{13}H_{11}FN_2O_2$
4-Fluoro-*N*-(4-nitrobenzyl)aniline 261
1-(3-Fluorophenyl)-3-(3-hydroxyphenyl)urea 513
1-(4-Fluorophenyl)-3-(3-hydroxyphenyl)urea 514
3-Pyridylmethyl *N*-(3-fluorophenyl)carbamate 929
3-Pyridylmethyl *N*-(4-fluorophenyl)carbamate 929
$C_{13}H_{11}FN_2O_3S$
4′-Amino-4-(fluorosulfonyl)benzanilide 803
1-(3-Fluorosulfonylphenyl)-3-phenylurea 804
$C_{13}H_{11}FN_2O_4S$
4′-Amino-3-(fluorosulfonyl)-2′-hydroxybenzanilide 801
$C_{13}H_{11}FN_4OS$
4-(2-Fluorophenyl)-1-(4-pyridylcarbonyl)-3-thiosemicarbazide 928
$C_{13}H_{11}FN_4O_3$
1-(4-Fluoro-3-nitrophenyl)-3-(4-methyl-2-pyridyl)urea 928
1-(2-Fluoro-5-nitrophenyl)-3-(3-pyridylmethyl)urea 928
4-(4-Fluorophenyl)-1-(4-nitrophenyl)semicarbazide 526
$C_{13}H_{11}FN_4O_5S$
1-(2-Fluoro-5-nitrophenyl)-3-(4-sulfamoylphenyl)urea 807
$C_{13}H_{11}FO_2S$
4-Fluorophenyl *p*-tolyl sulfone 785
$C_{13}H_{11}FO_3S$
4-Fluorophenyl 4-methoxyphenyl sulfone 775
$C_{13}H_{11}F_2N_3O$
1-(2,5-Difluorophenyl)-3-(6-methyl-2-pyridyl)urea 930
$C_{13}H_{11}F_3N_2O_2$
1-Furfuryl-3-(2-trifluoromethylphenyl)urea 843
1-Furfuryl-3-(3-trifluoromethylphenyl)urea 843
$C_{13}H_{11}IO$
4-Iodo-3-phenylanisole 177
$C_{13}H_{11}N$
Benzophenone imine hydrochloride 1077
2-Stilbazole 966
4-Styrylpyridine 967
$C_{13}H_{11}NO$
1-Amino-9-hydroxyfluorene 226
2-Amino-3-hydroxyfluorene 226
3-Amino-2-hydroxyfluorene 226
4-(Benzylideneamino)phenol 1035
N-Benzylideneaniline *N*-oxide 257
Benzyl 2-pyridyl ketone 959
4-Biphenylcarboxaldehyde oxime 1014
9-Carbazolemethanol 891
N,α-Diphenylnitrone 1063
3-(2-Furyl)-2-phenylpropionitrile 850
2-Methyl-1*H*-benzene(e)isoindol-3-one 692
2-Methyl-1*H*-benzene(f)isoindol-3(2*H*)-one 692
2′-Phenylformanilide 489
α-(Phenylimino)-*o*-cresol 1035
α-(Phenylimino)-*p*-cresol 1035
α-(2-Pyridylmethylene)-*o*-cresol 934
$C_{13}H_{11}NOS$
S-Benzyl 3-pyridinethiocarboxylate 962
2-Methoxyphenothiazine 247
2-(Salicylideneamino)thiophenol 1032
$C_{13}H_{11}NO_2$
1-Ethylpyrano(3,4-b)indol-3(9*H*)-one 888
2-(2-Hydroxybenzylideneamino)phenol 1028
α-(3-Hydroxyphenylimino)-*o*-cresol 1029
α-(4-Hydroxyphenylimino)-*o*-cresol 1029
α-(4-Hydroxyphenylimino)-*p*-cresol 1029
4-Phenacylpyridine-*N*-oxide monohydrate 1006
4-(Phenylamino)benzoic acid 343
Phenyl carbanilate 680
α-Phenyl-2-nitrotoluene 278
α-Phenyl-4-nitrotoluene 279
$C_{13}H_{11}NO_2S$
Benzyl 2-nitrophenyl sulfide 270
Benzyl 4-nitrophenyl sulfide 270
2-Nitrophenyl *o*-tolyl sulfide 270
2-Nitrophenyl *m*-tolyl sulfide 277
4-Nitrophenyl *o*-tolyl sulfide 270
4-Nitrophenyl *m*-tolyl sulfide 270
$C_{13}H_{11}NO_2S_2$
3-Cyclopropyl-5-(2-hydroxybenzylidene)rhodanine 734
4-Nitrophenyl *p*-tolyl disulfide 270
$C_{13}H_{11}NO_3$
N-Acetyl-3-hydroxy-2-naphthalenecarboxamide 461
4-Amino-3-hydroxyphenyl benzoate 377
Benzyl 3-nitrophenyl ether 268
Benzyl 4-nitrophenyl ether 269
4′-Hydroxysalicylanilide 461
2-Methoxyphenyl nicotinate 937
6-Methoxy-4-quinolineacrylic acid 986
3-(4-Methyl-5-phenyl-2-oxazolyl)acrylic acid 872
3-(4-Methyl-5-phenyl-2-oxazolyl)acrylic acid, potassium salt 872
DL-4-Nitrobenzhydrol 263
4-Nitrobenzyl phenyl ether 265
2-Nitro-α-phenyl-*p*-cresol 264
2-Nitrophenyl *p*-tolyl ether 267
$C_{13}H_{11}NO_3S$
9-Amino-9-fluorenesulfonic acid, sodium salt 793
Benzyl 2-nitrophenyl sulfoxide 771
Methyl 2-amino-5-benzoyl-3-thiophenecarboxylate 856
2-(2-Thienyl)hippuric acid 858
$C_{13}H_{11}NO_4$
4-(3-Acetoxybenzylidene)-2-methyl-2-oxazolin-5-one 385
Dimethyl 2,4-quinolinedicarboxylate 991
Ethyl α-cyano-3,4-methylenedioxycinnamate 766
Ethyl 4-maleimidobenzoate 720
N-Furfurylphthalamic acid 851
2-(4-Nitrophenoxy)anisole 266
5-Phthalimido-2-pentenoic acid 721
$C_{13}H_{11}NO_4S$
4,5-Dihydro-4-phenyl-1,3,2,5-benzodioxathiazepine-2,2-dioxide 801
4-Hydroxy-α-phenylimino-*m*-toluenesulfonic acid 1034
α-Nitrobenzyl phenyl sulfone 778
2-Nitrophenyl *o*-tolyl sulfone 779
4-Nitrophenyl *m*-tolyl sulfone 779
$C_{13}H_{11}NO_5$
1-Acetoxy-3-phthalimido-2-propanone 717
2-Methyl-3-furylmethyl 4-nitrobenzoate 849
$C_{13}H_{11}NO_5S$
4-(Benzenesulfonamido)salicylic acid 812
Methyl 4-formyl-1-(phenylsulfonyl)-2-pyrrolecarboxylate 842
4-Nitrophenyl *p*-toluenesulfonate 799
$C_{13}H_{11}NO_6$
L-2-Phthalimidoglutaric acid 720
N-Phthaloyl-DL-glutamic acid 722
$C_{13}H_{11}NS$
10-Methylphenothiazine 251
Phenyl benzothioimidate 746
Thiobenzanilide 746
$C_{13}H_{11}N_2O_5P$
Bis-(3-nitrophenyl)-methylphosphine oxide 835
$C_{13}H_{11}N_3$
2-Amino-9-fluorenone hydrazone 1052
2-Anilinobenzimidazole 905
3,6-Diaminoacridine hemisulfate hemihydrate 995
$C_{13}H_{11}N_3O$
8-Amino-7-methyl-2-phenazinol 998
Benzoic (3-pyridylmethylene)hydrazide 1069
Benzoic (4-pyridylmethylene)hydrazide 1069
2-(2-Hydroxy-5-methylphenyl)benzotriazole 908
Isonicotinic acid benzylidenehydrazide 1059
1-Phenyl-3-(phenylimino)urea 586
$C_{13}H_{11}N_3OS$
5-Amino-4-cyano-3-methyl-2-thiophenecarboxanilide 856
1,2-Dihydro-4,6-dimethyl-2-oxo-1-(2-thienylmethyleneamino)nicotinonitrile 1069
$C_{13}H_{11}N_3OS_2$
1-Cinnamoyl-3-(2-thiazolyl)-2-thiourea 877
$C_{13}H_{11}N_3O_2$
Benzaldehyde 4-nitrophenylhydrazone 1057
1,2-Dihydro-4,6-dimethyl-1-furfurylideneamino-2-oxonicotinonitrile 1069
2,3-Dihydro-2-(3-nitrophenyl)benzimidazole 277
4-Hydroxybenzoic (3-pyridylmethylene)hydrazide 933
Isonicotinic acid salicylidenehydrazide 1033
2-Nitrobenzaldehyde phenylhydrazone 1055
3-Nitrobenzaldehyde phenylhydrazone 1058
2-Phenyl-2,3,5,8-tetrahydro-5,8-methano-1*H*-1,2,4-triazolo(1,2-a)pyridazine-1,3-dione 709
4-Pyridinecarboxylic (4-hydroxybenzylidene)hydrazide 1036
$C_{13}H_{11}N_3O_2S$
1-(2-Nitrophenyl)-3-phenyl-2-thiourea 739
1-(3-Nitrophenyl)-3-phenyl-2-thiourea 740
1-(4-Nitrophenyl)-3-phenyl-2-thiourea 740

$C_{13}H_{11}N_3O_3$
4-Methoxy-3′-nitroazobenzene 1089
4-Methoxy-4′-nitroazobenzene 1087
N-(4-Methyl-3-nitro-2-pyridyl)-benzamide 944
1-(3-Nitrophenyl)-3-phenylurea 564
1-(4-Nitrophenyl)-3-phenylurea 568

$C_{13}H_{11}N_3O_3S$
1-(3-Hydroxyphenyl)-3-(4-nitrophenyl)-2-thiourea 734

$C_{13}H_{11}N_3O_4$
2,4-Dinitro-2′-methyldiphenylamine 272
2,4-Dinitro-3′-methyldiphenylamine 273
2,4-Dinitro-4′-methyldiphenylamine 272
2-Hydroxy-4-nitrocarbanilide 547
4-(4-Nitrophenylazo)orcinol 1083

$C_{13}H_{11}N_3O_4S$
Benzenesulfonic N^2-benzylidene-N^1-nitrohydrazide 1069
5-Nitro-2-pyridyl *N*-(4-(methylthio)-phenyl)carbamate 941

$C_{13}H_{11}N_3O_5$
2,4-Dinitro-4′-hydroxy-2′-methyldiphenylamine 264
2,4-Dinitro-5′-hydroxy-2′-methyldiphenylamine 264
2,6-Dinitro-3′-hydroxy-4-methyldiphenylamine 264
2,4-Dinitro-2′-methoxydiphenylamine 266
2,4-Dinitro-3′-methoxydiphenylamine 266
2,4-Dinitro-4′-methoxydiphenylamine 266

$C_{13}H_{11}N_3O_6S$
3,5′-Dinitro-2′-methylbenzenesulfonanilide 818

$C_{13}H_{11}N_5O_2$
3-(4-Hydroxy-6-methyl-2-pyrimidinylhydrazono)-2-indolinone 1036
2-(3-(2-Thiazolyl)ureido)acetic acid N^2-(2,6-dichlorobenzylidene)-hydrazide 1026

$C_{13}H_{11}N_5O_4S$
1-(2,4-Dinitrophenyl)-4-phenyl-3-thiosemicarbazide 739

$C_{13}H_{11}N_5O_5$
1-(2,4-Dinitrophenyl)-4-phenylsemicarbazide 561

$C_{13}H_{12}$
1-Allylnaphthalene 166

$C_{13}H_{12}BrNO$
α-(2-Bromoanilino)-*o*-cresol 226
α-(3-Bromoanilino)-*o*-cresol 223
α-(4-Bromoanilino)-*o*-cresol 223
α-(4-Bromoanilino)-*p*-cresol 223
1-Phenacylpyridinium bromide 955

$C_{13}H_{12}BrNO_2$
Ethyl 7-bromo-2-methylquinoline-4-carboxylate 993

$C_{13}H_{12}BrNO_2S$
4′-Bromo-*N*-methylbenzenesulfonanilide 809

$C_{13}H_{12}BrNO_3$
Ethyl 3-bromo-α-cyano-4-methoxycinnamate 762
Ethyl 5-bromo-α-cyano-2-methoxycinnamate 758
(4*H*-Pyran-2-yl)methyl *N*-(4-bromophenyl)carbamate 622

$C_{13}H_{12}BrNO_4$
5-Bromo-2-carboxy-3-indolebutyric acid 889

$C_{13}H_{12}BrN_3$
4-(4-Bromophenylazo)-3-methylaniline tetrafluoroborate 1081

$C_{13}H_{12}BrN_3O$
4-(4-Bromophenyl)-1-phenylsemicarbazide 533
1-(3-Bromophenyl)-3-(3-pyridylmethyl)urea 928
1-(4-Bromophenyl)-3-(3-pyridylmethyl)urea 950

$C_{13}H_{12}BrN_3OS$
1-(5-Bromo-2-pyridyl)-3-(4-(methylthio)phenyl)urea 941

$C_{13}H_{12}BrN_3O_3$
3-Carbamoyl-1-(4-nitrobenzyl)-pyridinium bromide 946

$C_{13}H_{12}BrN_3O_4S$
3-Nitrobenzenesulfonic acid 2-(2-bromo-*p*-tolyl)hydrazide 822
3-Nitrobenzenesulfonic acid 2-(4-bromo-*o*-tolyl)hydrazide 822

$C_{13}H_{12}Br_2N_2O_2$
2,4-Dibromo-8-quinolyl *N*-isopropylcarbamate 992
2,4-Dibromo-8-quinolyl *N*-propylcarbamate 994

$C_{13}H_{12}Br_2O_2Te$
3-(1-Naphthyltelluro)propionic acid Te,Te-dibromide 1142

$C_{13}H_{12}Br_4N_2O_3$
2-(4-Methylpiperazinocarbonyl)-3,4,5,6-tetrabromobenzoic acid 694

$C_{13}H_{12}ClN$
9-Chloro-1,2,3,4-tetrahydroacridine 994

$C_{13}H_{12}ClNO$
α-(3-Chloroanilino)-*o*-cresol 223
α-(4-Chloroanilino)-*o*-cresol 223
α-(4-Chloroanilino)-*p*-cresol 223
α-(4-Chlorophenyl)-α-methyl-2-pyridinemethanol 926
2-(4-Chloro-*m*-tolyloxy)-aniline 223
4-(4-Chloro-*m*-tolyloxy)-aniline 223
5-Chloro-2-(*p*-tolyloxy)-aniline 223

$C_{13}H_{12}ClNO_2$
2-Chloroethyl *N*-(1-naphthyl)-carbamate 606

$C_{13}H_{12}ClNO_2S$
2-Amino-4-chloro-4′-methyldiphenyl sulfone 773
5-Chloro-2,4-dimethoxy-*N*-(2-thienylmethylene)aniline 1049
3′-Chloro-2′-methylbenzenesulfonanilide 809
4′-Chloro-2′-methylbenzenesulfonanilide 809
N-(5-Chloro-2-methylphenyl)-benzenesulfonamide 809

$C_{13}H_{12}ClNO_3$
Furfuryl *N*-(4-chloro-2-methylphenyl)carbamate 849
Furfuryl *N*-(5-chloro-2-methylphenyl)carbamate 849

$C_{13}H_{12}ClNO_3S$
4-Chloro-2′-methoxybenzenesulfonanilide 806
4′-Chloro-2′-(phenoxy)-methanesulfonanilide 806

$C_{13}H_{12}ClNO_4$
4′-Chloro-2′,5′-dimethoxy-2-furanilide 843
Furfuryl *N*-(4-chloro-2-methoxyphenyl)carbamate 843
Furfuryl *N*-(5-chloro-2-methoxyphenyl)carbamate 843

$C_{13}H_{12}ClNO_4S$
6-(4-Chloro-*m*-tolyloxy)-metanilic acid, sodium salt 791

$C_{13}H_{12}ClNS$
3-Chloro-2-(*p*-tolylthio)aniline 224

$C_{13}H_{12}ClN_3O$
2-Amino-5-chloro-*N*-(2-pyridylmethyl)benzamide 927
1-(2-Chloro-6-methylphenyl)-3-(3-pyridyl)urea 930
1-(3-Chloro-4-methylphenyl)-3-(3-pyridyl)urea 927
1-(4-Chloro-2-methylphenyl)-3-(3-pyridyl)urea 945
1-(2-Chlorophenyl)-3-(4-methyl-2-pyridyl)urea 930
1-(2-Chlorophenyl)-3-(6-methyl-2-pyridyl)urea 930
1-(3-Chlorophenyl)-3-(4-methyl-2-pyridyl)urea 930
4-(4-Chlorophenyl)-1-phenylsemicarbazide 578
1-(2-Chlorophenyl)-3-(3-pyridylmethyl)urea 950
1-(3-Chlorophenyl)-3-(3-pyridylmethyl)urea 951
1-(4-Chlorophenyl)-3-(3-pyridylmethyl)urea 950
1-(4-Chlorophenyl)-3-(4-pyridylmethyl)urea 953
4-Diazo-4′-methoxydiphenylamine chloride 1087

$C_{13}H_{12}ClN_3OS$
1-(5-Chloro-2-pyridyl)-3-(4-(methylthio)phenyl)urea 941

$C_{13}H_{12}ClN_3O_2$
1-(4-Chloro-2-methoxyphenyl)-3-(2-pyridyl)urea 927
1-(4-Chloro-2-methoxyphenyl)-3-(3-pyridyl)urea 927
1-(5-Chloro-2-methoxyphenyl)-3-(3-pyridyl)urea 926
1-(3-Chlorophenyl)-3-methacrylamido-2-pyrazolin-5-one 706

$C_{13}H_{12}ClN_3O_2S$
4-(4-Chlorophenyl)-1-(phenylsulfonyl)-3-thiosemicarbazide 810

$C_{13}H_{12}ClN_3O_3$
3-Carbamoyl-1-(4-nitrobenzyl)-pyridinium chloride 946
Ethyl 2-(4-chloroanilino)-4-hydroxy-5-pyrimidinecarboxylate 972

$C_{13}H_{12}ClN_3O_3S$
1-(3-Chlorophenyl)-3-(4-sulfamoylphenyl)urea 809

$C_{13}H_{12}ClN_3S$
4-(4-Chlorophenyl)-1-phenyl-3-thiosemicarbazide 750

$C_{13}H_{12}Cl_4N_2O_3$
2-(4-Methylpiperazinocarbonyl)-3,4,5,6-tetrachlorobenzoic acid 337

$C_{13}H_{12}FNO$
α-(4-Fluoroanilino)-*o*-cresol 223
α-(4-Fluoroanilino)-*p*-cresol 223

$C_{13}H_{12}FN_3O$
1-(2-Fluorophenyl)-3-(4-methyl-2-pyridyl)urea 930
4-(4-Fluorophenyl)-1-phenylsemicarbazide 578

$C_{13}H_{12}FN_3O_3S$
1-(3-Fluorophenyl)-3-(4-sulfamoylphenyl)urea 808
1-(4-Fluorophenyl)-3-(4-sulfamoylphenyl)urea 808

$C_{13}H_{12}F_3N_3$
2-(Dimethylamino)-4-phenyl-6-(trifluoromethyl)pyrimidine 968

$C_{13}H_{12}INO$
α-(4-Iodoanilino)-*o*-cresol 223

$C_{13}H_{12}INS$
2,3-Dimethylnaphtho(1,2-d)-thiazolium iodide 916

$C_{13}H_{12}N_2$
Benzaldehyde phenylhydrazone 1072
4-Methylazobenzene 1093
N-Phenylbenzamidine 727
α-Propylbenzylidenemalononitrile 764
2-(2-(2-Pyridyl)vinylene)aniline 942

$C_{13}H_{12}N_2O$
Acetic (1-naphthylmethylene)-hydrazide 502
2′-Aminobenzanilide 468
2-Aminobenzanilide 468
4′-Aminobenzanilide 468
2-Aminobenzophenone oxime 1009
Benzoic acid N^2-phenylhydrazide 503
N-Benzylnicotinamide 959
2′-Cyano-2,4-hexadienanilide 764
α-Cyano-5-indanacrylamide 764
3,4-Diaminobenzophenone monohydrochloride 311
1,1-Diphenylurea 585
Harmine methiodide 918
Harmine methosulfate 918
4-Methoxyazobenzene 1087
2-Methyl-4-(phenylazo)phenol 1086
3-Methyl-4-(phenylazo)phenol 1086
4-Methyl-2-(phenylazo)phenol 1086
3-Methyl-*N*-(2-pyridyl)benzamide hydrochloride 963
α-(4-Methyl-2-pyridylimino)-*o*-cresol 1038
α-(6-Methyl-2-pyridylimino)-*o*-cresol 1038
α-(4-Methyl-2-pyridylimino)-*o*-cresol copper(II) salt 1036
α-(6-Methyl-2-pyridylimino)-*o*-cresol copper(II) salt 1036
2-Phenyl-2-(2-pyridyl)acetamide 962
2-Pyridinealdoxime *O*-benzyl ether 1010
4-Pyridinealdoxime *O*-benzyl ether 1010
α-(3-Pyridylmethylimino)-*o*-cresol 1036
Salicylaldehyde phenylhydrazone 234

$C_{13}H_{12}N_2OS$
1-(2-Hydroxyphenyl)-3-phenyl-2-thiourea 734
1-(3-Hydroxyphenyl)-3-phenyl-2-thiourea 735
5-(5-Indanylmethylene)-2-thiohydantoin 744

$C_{13}H_{12}N_2O_2$
Acetic (2-hydroxy-1-naphthylmethylene)hydrazide 497
4-Amino-2-carboxydiphenylamine 342
4′-Amino-2′-hydroxybenzanilide 462
3,6-Dimethylene-1-methyl-4-phenyl-2,5-piperazinedione 710
Ethyl α,β-dicyanohydrocinnamate 755
6-Methoxy-α-(2-pyridylimino)-*o*-cresol 1030
1-Methyl-4-(4-nitrobenzylidene)-1,4-dihydropyridine 277
N-(4-Nitrophenyl)benzylamine 276
2-Pyridyl *N*-(*o*-tolyl)carbamate 964
2-Pyridyl *N*-(*m*-tolyl)carbamate 964
2-Pyridyl *N*-(*p*-tolyl)carbamate 964
3-Pyridyl *N*-(*o*-tolyl)carbamate 960
3-Pyridyl *N*-(*p*-tolyl)carbamate 960
8-Quinolyl *N*-allylcarbamate 992

$C_{13}H_{12}N_2O_2S$
Benzenesulfonic benzylidenehydrazide 821
4-Pyridyl *N*-(4-methylthio)phenyl)-carbamate 941

$C_{13}H_{12}N_2O_3$
4′-Acetamido-2-furanilide 847
4-Benzyloxy-2-nitroaniline 265
4-Hydroxy-3-(α-iminobenzyl)-1-methyl-2,6(1*H*,5*H*)-pyridinedione 1037
α-(3-Nitroanilino)-*p*-cresol 264
α-(4-Nitroanilino)-*o*-cresol 264
3-Pyridyl *N*-(*m*-anisyl)carbamate 939
2-Pyridyl *N*-(2-methoxyphenyl)-carbamate 939
3-Pyridyl *N*-(2-methoxyphenyl)-carbamate 939

$C_{13}H_{12}N_2O_3S$
2-Amino-5-nitro-4-(*p*-tolylthio)-phenol 263

$C_{13}H_{12}N_2O_3S_2$
3-Isopropyl-5-(3-nitrobenzylidene)-rhodanine 744

$C_{13}H_{12}N_2O_4$
N-(3,4-Dimethyl-5-isoxazolyl)-phthalamic acid 871
N-(4,5-Dimethyl-2-nitrophenyl)-2-furamide 850
3-(Ethoxycarbonyl)-5-phenyluracil 708
2-Hydroxy-*N*-(3-hydroxyphenyl)-5-nitrobenzylamine 262

$C_{13}H_{12}N_2O_4S$
4′-Hydroxy-2′-methyl-1,1′-azobenzene-4-sulfonic acid, sodium salt 1085
4-(4-Methoxyphenylazo)-benzenesulfonic acid, sodium salt 1087
4-(3-Methyl-4-hydroxyphenylazo)-benzenesulfonic acid 1085
4-Methyl-2′-nitrobenzenesulfonanilide 818
N-(2-Methyl-4-nitrophenyl)-benzenesulfonamide 819

$C_{13}H_{12}N_2O_5$
Ethyl α-cyano-4-methoxy-3-nitrocinnamate 762
2-(Furfuryliminomethyl)-6-methoxy-4-nitrophenol 1021
Furfuryl *N*-(2-methyl-5-nitrophenyl)-carbamate 847
Furfuryl *N*-(4-methyl-2-nitrophenyl)-carbamate 847

$C_{13}H_{12}N_2O_5S$
N-(3-Amino-4-hydroxyphenylsulfonyl)-anthranilic acid ... 811
2′-Hydroxy-4′-nitro-p-toluenesulfonanilide ... 811

$C_{13}H_{12}N_4$
1-(Anilinomethyl)-1H-benzotriazole ... 909

$C_{13}H_{12}N_4O$
2-Amino-6-(2H-benzotriazol-2-yl)-p-cresol ... 908
6-Methoxy-3-methyl-8-phenylpurine ... 921

$C_{13}H_{12}N_4OS$
2-Benzylthio-5-methyl-s-triazolo(1,5-a)pyrimidin-7-ol ... 918

$C_{13}H_{12}N_4OS_3$
3-Ethyl-2-thioxo-4,5-thiazolidinedione 5-azine w/3-methylbenzothiazol-2(3H)-one ... 742

$C_{13}H_{12}N_4O_2$
4-Amino-2-methyl-4′-nitroazobenzene ... 1088
6,7-Dihydro-5(H)-5-methyl-3-(3-nitrophenyl)cyclopenta(e)-1,2,4-triazine ... 979
3-(3-Nitrophenyl)-5,6,7,8-tetrahydro-1,2,4-benzotriazine ... 979

$C_{13}H_{12}N_4O_2S$
1-(4-Nitrophenyl)-3-(6-methyl-2-pyridyl)-2-thiourea ... 952

$C_{13}H_{12}N_4O_3$
3-Amino-4-cyano-2-isopropyl-7-nitro-1(2H)-isoquinolinone ... 763
1-(4-Methyl-3-nitrophenyl)-3-(3-pyridyl)urea ... 943
1-(3-Nitrophenyl)-3-(3-pyridylmethyl)urea ... 949
1-(4-Nitrophenyl)-3-(3-pyridylmethyl)urea ... 945

$C_{13}H_{12}N_4O_3S$
N-(2-Thiazolyl)oxamic (4-methoxybenzylidene)hydrazide ... 1048

$C_{13}H_{12}N_4O_4$
N,N′-Methylenebis(2-nitroaniline) ... 274
N,N′-Methylenebis(3-nitroaniline) ... 275
N,N′-Methylenebis(4-nitroaniline) ... 276

$C_{13}H_{12}N_4O_5$
1-(α-2-Nitrobenzyl)-2-(4-nitrophenyl)-hydrazine ... 263

$C_{13}H_{12}N_4S$
1-(3-Pyridylmethylene)-4-phenyl-3-thiosemicarbazide ... 1056

$C_{13}H_{12}N_6O_5$
1,5-Bis(4-nitrophenyl)carbazide ... 568

$C_{13}H_{12}O$
3-Methoxybiphenyl ... 180
Phenyl o-tolyl ether ... 182

$C_{13}H_{12}OS$
Benzyl phenyl sulfoxide ... 771
4-Methoxyphenyl phenyl sulfide ... 211

$C_{13}H_{12}O_2$
1-Acetyl-4-methoxynaphthalene ... 293
2-Ethoxy-1-naphthaldehyde ... 323
Ethyl 1-naphthoate ... 376
Ethyl 2-naphthoate ... 376
6-Methoxy-2-naphthyl methyl ketone ... 293
2-Methoxyphenyl phenyl ether ... 182
2-Methyl-1-naphthylacetic acid ... 334
Methyl 5-phenyl-2-penten-4-yne-2-carboxylate ... 1018
2-Naphthalenepropionic acid ... 334

$C_{13}H_{12}O_2S$
Benzyl phenyl sulfone ... 783
2-(Ethylthio)-3-methyl-1,4-naphthoquinone ... 317
3-(1-Naphthylthio)propionic acid ... 330
Phenyl o-tolyl sulfone ... 784
Phenyl p-tolyl sulfone ... 785
p-Tolyl benzenesulfinate ... 786

$C_{13}H_{12}O_2S_2$
Bis(phenylsulfinyl)methane ... 771
Methyl 4-(phenylsulfonyl)phenyl sulfide ... 776
S-Phenyl p-toluenethiosulfonate ... 801

$C_{13}H_{12}O_2Te$
3-(1-Naphthyltelluro)propionic acid ... 1142

$C_{13}H_{12}O_3$
7-Allyloxy-4-methylcoumarin ... 384
1-Cyclopropyl-3-(3,4-methylenedioxyphenyl)-2-propen-1-one ... 283
4-Ethoxyphenyl 2-furyl ketone ... 846
Ethyl 2-naphthyl carbonate ... 381
2-Methyl-3-furfuryl benzoate ... 849
(+)-2-(1-Naphthoxy)propionic acid ... 329
(-)-2-(1-Naphthoxy)propionic acid ... 328
3-(1-Naphthoxy)propionic acid ... 328
3-(2-Naphthoxy)propionic acid ... 328
2-(1-Naphthyloxy)propionic acid ... 328
2-Propoxy-1,4-naphthoquinone ... 316

$C_{13}H_{12}O_3S$
3-Methoxyphenyl phenyl sulfone ... 775
4-Methoxyphenyl phenyl sulfone ... 775
Phenyl p-toluenesulfonate ... 800

$C_{13}H_{12}O_4$
5,6-O-Benzylidene-2,3-dideoxy-L-ascorbic acid ... 384
1,2,3,4-Tetrahydro-1,4-methanonaphthalene-2,3-dicarboxylic acid ... 329

$C_{13}H_{12}O_5$
α-Acetyl-4-hydroxy-β-(hydroxymethyl)-3-methoxycinnamic acid γ-lactone ... 382
Dimethyl 1-oxo-2,7-indandicarboxylate ... 374
Ethyl 4-hydroxy-8-methylcoumarin-3-carboxylate ... 382

$C_{13}H_{12}O_6S$
Ethyl 4-sulfo-1-naphthyl carbonate ... 790

$C_{13}H_{12}S$
Phenyl p-tolyl sulfide ... 212

$C_{13}H_{13}AsO_3$
10-Methoxy-2,3,6,7-tetrahydro-1H,5H-arsenino(3,2,1-ij)arsinoline-1,7-dione ... 1126

$C_{13}H_{13}AsO_4$
4-(4-Tolyloxy)benzenearsonic acid ... 1126

$C_{13}H_{13}BrN_2O$
1-Benzyl-3-carbamoylpyridinium bromide ... 947

$C_{13}H_{13}BrN_4O$
1-(4-Bromo-2,6-dimethylphenyl)-3-(2-pyrazinyl)urea ... 977

$C_{13}H_{13}BrO_2$
3-(5-Bromo-2-methoxyphenyl)-1-cyclopropyl-2-propen-1-one ... 299

$C_{13}H_{13}BrO_2S$
3-Bromo-5-(phenylsulfonyl)tricyclo-[2.2.1.0^{2,6}]heptane ... 782

$C_{13}H_{13}BrSi$
(4-Bromophenyl)-methylphenylsilane ... 1103

$C_{13}H_{13}Br_2NO_5$
2,3-Dibromo-4′-(ethoxycarbonyl)-succinanilic acid ... 433

$C_{13}H_{13}ClF_3NO_4$
Methyl 2-(N-(2-chloro-5-(trifluoromethyl)phenyl)-carbamoyloxy)-2-methylpropionate ... 606

$C_{13}H_{13}ClN_2$
2-(2-(2-Chloroanilino)ethyl)pyridine ... 946
2-(2-(3-Chloroanilino)ethyl)pyridine ... 946
2-(2-(4-Chloroanilino)ethyl)pyridine ... 946

$C_{13}H_{13}ClN_2O$
6-Chloro-6,7,8,9,10,12-hexahydroazepino(2,1-b)-quinazolin-12-one ... 710
6′-Chloro-1-pyrroleaceto-o-toluidide ... 841

$C_{13}H_{13}ClN_2O_2$
N-(2-Aminoethyl)-4-chloro-1-hydroxy-2-naphthamide ... 485
4-(2-Chloroacetyl)-antipyrine ... 693
1-(2-Chloro-6-methylphenyl)-3-furfurylurea ... 852
1-(3-Chloro-4-methylphenyl)-3-furfurylurea ... 851
1-(4-Chloro-2-methylphenyl)-3-furfurylurea ... 854
1-(5-Chloro-2-methylphenyl)-3-furfurylurea ... 851
4-Chloro-8-quinolyl N-isopropylcarbamate ... 992

$C_{13}H_{13}ClN_2O_2S$
2′-Amino-4-chloro-4′-methylbenzenesulfonanilide ... 806

$C_{13}H_{13}ClN_2O_3$
Chloroacetyl-DL-tryptophan ... 881
1-(3-Chloro-2-methoxyphenyl)-3-(furfuryl)urea ... 846
Nα-Chloroacetyl-L-tryptophan ... 881

$C_{13}H_{13}ClN_2O_5$
1-(5-Chloro-2,4-dimethoxyphenyl)-3-ethyl-2,4,5-imidazolidinetrione ... 694

$C_{13}H_{13}ClN_4O_2$
4-(5-Chloro-2-methoxyphenyl)-1-(2-pyridyl)semicarbazide ... 927

$C_{13}H_{13}ClN_4O_3$
1-(5-Chloro-2,4-dimethoxyphenyl)-3-(2-pyrazinyl)urea ... 977

$C_{13}H_{13}ClN_4O_6$
(5,5-Dimethyl-2,4-dioxo-1-imidazolidinylmethyl) N-(2-chloro-4-nitrophenyl)carbamate ... 695

$C_{13}H_{13}ClO$
2-(2-Chlorobenzylidene)-cyclohexanone ... 282

$C_{13}H_{13}ClO_2$
9-Chloro-1,2,3,4-tetrahydro-1,4-methano-2-naphthyl acetate ... 362

$C_{13}H_{13}ClO_3$
3-Chloro-2-(2,5-dimethoxyphenyl)-2-cyclopenten-1-one ... 283

$C_{13}H_{13}ClO_5$
2,6-Dimethyl-4-phenylpyrylium perchlorate ... 935

$C_{13}H_{13}ClSi$
(2-Chlorobenzyl)phenylsilane ... 1119

$C_{13}H_{13}Cl_2NO$
2,6-Dichloro-N-furfuryl-N-methylbenzylamine ... 844

$C_{13}H_{13}Cl_3N_2O$
1,1-Diallyl-3-(2,4,5-trichlorophenyl)urea ... 603

$C_{13}H_{13}Cl_3N_2O_2$
2,3-Dimethyl-4-(1-hydroxy-2,2,2-trichloroethyl)-1-phenyl-3-pyrazolin-5-one ... 693
o-(N-(2,4,5-Trichlorophenyl)-carbamoyl)cyclohexanone oxime ... 1011

$C_{13}H_{13}Cl_4NO_4$
Isopropyl N-(1-(4-chlorobenzoyloxy)-2,2,2-trichloroethyl)carbamate ... 606
Propyl N-(1-(4-chlorobenzoyloxy)-2,2,2-trichloroethyl)carbamate ... 606

$C_{13}H_{13}FN_2$
2-(2-(4-Fluoroanilino)ethyl)pyridine ... 946

$C_{13}H_{13}FN_2O_2$
1-(2-Fluorophenyl)-3-furfuryl-3-methylurea ... 529

$C_{13}H_{13}IN_2$
1-Methyl-2-(2-(2-pyridyl)vinyl)-pyridinium iodide ... 952

$C_{13}H_{13}I_2NO_5$
N,O-Diacetyl-3,5-diiodo-L-tyrosine ... 429

$C_{13}H_{13}N$
Aminodiphenylmethane hydrobromide ... 241
N-Benzylaniline hydrochloride ... 242
4-(2-Phenylethyl)pyridine ... 966

$C_{13}H_{13}NO$
3-Amino-4-hydroxydiphenylmethane oxalate ... 226
2-(Anilinomethyl)phenol ... 227
4-Benzyloxyaniline ... 227
N,2-Dimethyl-1-naphthamide ... 488
2,6-Dimethyl-1-phenyl-4(1H)-pyridone hydrochloride ... 320
3,3-Dimethyl-1-(2-propyn-1-yl)-2-indolinone ... 1019
N-(Furfurylidene)-2,3-xylidine ... 1073
N-(Furfurylidene)-2,4-xylidine ... 1073
N-(Furfurylidene)-2,5-xylidine ... 1073
N-(Furfurylidene)-3,4-xylidine ... 1073
N-(Furfurylidene)-3,5-xylidine ... 1073
4-Hydroxy-4′-methyldiphenylamine ... 227
N-(1-Naphthyl)-propionamide ... 490
3-Phenyl-4,5,6,7-tetrahydro-1,2-benzisoxazole ... 871
3-(o-Toluidino)phenol ... 227
2-(o-Tolyloxy)-aniline ... 228

$C_{13}H_{13}NO_2$
2′,4′-Dimethyl-2-furanilide ... 847
2′,6′-Dimethyl-2-furanilide ... 853
Ethyl α-cyano-4-methylcinnamate ... 766
4′-Ethyl-2-furanilide ... 851
Ethyl N-(1-naphthyl)carbamate ... 687
N-Furfurylidene-p-phenetidine ... 1049
N-Furfuryl-3-methylbenzamide ... 851
N-Furfuryl-4-methylbenzamide ... 851
1-((2-Hydroxyethyl)iminomethyl)-2-naphthol ... 1029
4-Methoxyphenyl 2-pyridyl carbinol ... 931
N-(3-Methyl-2-buten-1-yl)-phthalimide ... 723
2-Methyl-3-(2-oxopropyl)-1(2H)-isoquinolinone ... 706
5-Phenyl-2-pyrrolepropionic acid ... 841

$C_{13}H_{13}NO_2S$
N-(Benzyl)benzenesulfonamide ... 821
α-Cyano-β,5-dimethyl-2-(methylthio)cinnamic acid ... 763
2,5-Dimethoxy-N-(2-thienylmethylene)aniline ... 1049
p-Toluenesulfonanilide ... 828
N-(o-Tolyl)benzenesulfonamide ... 827
2-(p-Tosyl)aniline ... 776

$C_{13}H_{13}NO_2S_2$
3-Benzyl-5-(ethoxymethylene)-2-thioxo-4-thiazolidinone ... 738
5-Ethoxymethylene-2-thioxo-3-(o-tolyl)-4-thiazolidinone ... 738
5-Ethoxymethylene-2-thioxo-3-(m-tolyl)-4-thiazolidinone ... 738

$C_{13}H_{13}NO_3$
4′-Ethoxy-2-furanilide ... 846
Ethyl β-cyano-α-hydroxy-2-methylcinnamate ... 761
Ethyl β-cyano-α-hydroxy-4-methylcinnamate ... 761
Ethyl α-cyano-2-methoxycinnamate ... 762
Ethyl α-cyano-4-methoxycinnamate ... 762
Ethyl 4-hydroxy-6-methyl-3-quinolinecarboxylate ... 984
Ethyl 2-methyl-3-indolylglyoxylate ... 888
Furfuryl N-(o-tolyl)carbamate ... 849
Furfuryl N-(m-tolyl)carbamate ... 849
Furfuryl N-(p-tolyl)carbamate ... 849
1,2,3,5,6,10b-Hexahydro-3-oxopyrrolo(2,1-a)isoquinoline-1-carboxylic acid ... 708
2-Hydroxy-3-naphthoic acid ethanolamide ... 486
Isopropyl β-cyano-α-hydroxycinnamate ... 761
2′-Methoxy-5′-methyl-2-furanilide ... 846
Methyl 1-allyl-3-hydroxy-2-indolecarboxylate ... 881
2-Propionyl-3-indoleacetic acid ... 888
Propyl β-cyano-α-hydroxycinnamate ... 761

$C_{13}H_{13}NO_3S_2$
5-(Ethoxymethylene)-3-(2-methoxyphenyl)-rhodanine ... 737
5-(Ethoxymethylene)-3-(3-methoxyphenyl)-rhodanine ... 737

$C_{13}H_{13}NO_4$
4-(3-Acetoxybenzylidene)-2-methyl-5-oxazolidinone ... 385
5-(Anilinomethylene)-2,2-dimethyl-1,3-dioxane-4,6-dione ... 384
2-Carboxy-3-indolebutyric acid ... 889
2′,5′-Dimethoxy-2-furanilide ... 845
Ethyl α-cyano-4-hydroxy-3-methoxycinnamate ... 761
Ethyl 3-(formylmethoxy)-2-indolecarboxylate ... 882
Ethyl 4-hydroxy-6-methoxy-3-quinolinecarboxylate ... 982
Ethyl 4-hydroxy-7-methoxy-3-quinolinecarboxylate ... 982
Ethyl 3-hydroxy-3-(2-oxo-3-indolinylidene)propionate ... 690
Ethyl 3-phthalimidopropionate ... 722
Furfuryl N-(2-methoxyphenyl)-carbamate ... 846
Methyl 1-benzyl-4,5-dioxo-3-pyrrolidinecarboxylate ... 700

$C_{13}H_{13}NO_4S$
N-Phthaloyl-DL-methionine 718
N-Phthaloyl-L-methionine 718

$C_{13}H_{13}NO_5$
2-(1,3-Benzodioxol-5-yl)-1-methyl-5-oxo-3-pyrrolidinecarboxylic acid 692
Carbobenzyloxy-L-glutamic anhydride 677
α,4-Diacetoxy-3-methoxyphenylacetonitrile 752
4′-(Ethoxycarbonyl)-maleanilic acid 475
Methyl 3-hydroxy-2-(methoxycarbonyl)-1-indoleacetate 881

$C_{13}H_{13}NO_6$
Dimethyl 2-(4-nitrobenzylidene)-succinate 371

$C_{13}H_{13}NO_8$
3-Methyl-5-nitro-1,2,4-benzenetriol triacetate 378

$C_{13}H_{13}NS$
2-(Benzylthio)aniline 230
N-(2-Thienylmethylene)-3,4-xylidine 1073
2-(*p*-Tolylthio)-aniline 230
4-(*m*-Tolylthio)aniline 230

$C_{13}H_{13}N_2O_3Tl$
(2,2′-Dipyridylthallyl(III)) lactate 932

$C_{13}H_{13}N_3$
2-Acetylpyridine phenylhydrazone 1072
1,3-Diphenylguanidine phthalate 727
Nitrin 235

$C_{13}H_{13}N_3O$
4-Amino-3-methoxyazobenzene 1086
Cyanoacetic acid 5-indanylidenehydrazide 1056
N-(2,4-Diamino-5-methylphenyl)-1,4-benzoquinoneimine trihydrate 1052
1,4-Diphenylsemicarbazide 578
1-(6-Methyl-2-pyridyl)-3-phenylurea 965
1-Phenyl-3-(3-pyridylmethyl)urea 965
1-Phenyl-3-(4-pyridylmethyl)urea 953

$C_{13}H_{13}N_3OS$
1-(4-Methoxyphenyl)-3-(2-pyridyl)-2-thiourea 939
1-(4-Methoxyphenyl)-3-(2-pyridyl)-2-thiourea 939
1-(4-(Methylthio)phenyl)-3-(2-pyridyl)urea 941

$C_{13}H_{13}N_3O_2$
4′-Amino-2′-hydroxycarbanilide 547
8,9-Dimethoxy-3-methyl-1,2,4-triazolo(3,4-a)isoquinoline 908
3,9-Dimethyl-6-methoxy-1*H*-pyrazolo(3,4-b)quinol-4(9*H*)-one 860
Ethyl 4-(2,2-dicyanovinyl)-3,5-dimethyl-2-pyrrolecarboxylate 840
N-Methylanthranilic furfurylidenehydrazide 1051
3-Nitro-4,4′-methylenedianiline 271

$C_{13}H_{13}N_3O_2S$
5-(4-(Dimethylamino)benzylidene)-2-thiobarbituric acid 734
1-(3-Hydroxy-2-pyridyl)-3-(4-(methylthio)phenyl)urea 932
Nicotinaldehyde (*p*-tolylsulfonyl)-hydrazone 1071

$C_{13}H_{13}N_3O_2S_2$
4-Phenyl-1-(phenylsulfonyl)-3-thiosemicarbazide 829

$C_{13}H_{13}N_3O_3$
5-(Anilinomethylene)-1,3-dimethylbarbituric acid 700
Ethyl 4-(1-imidazolylcarbonyl-amino)benzoate 868
6-Nitroharmaline nitrate 918

$C_{13}H_{13}N_3O_3S$
Ethyl 4-(2-thiazolylureido)benzoate 877
4-Furfuryl-1-(2-hydroxybenzoyl)-3-thiosemicarbazide 845
4-Phenyl-1-(phenylsulfonyl)-3-thiosemicarbazide 824
1-Phenyl-3-(4-sulfamoylphenyl)urea 826

$C_{13}H_{13}N_3O_4$
1-Furfuryl-1-methyl-3-(3-nitrophenyl)urea 850
1-Furfuryl-3-(2-methyl-4-nitrophenyl)urea 850
1-Furfuryl-3-(4-methyl-3-nitrophenyl)urea 850

$C_{13}H_{13}N_3O_5S$
3-Amino-4-methyl-2′-nitro-4′-sulfodiphenylamine 788
3-Nitrobenzenesulfonic acid 2-(2-methoxyphenyl)hydrazide 814
3-Nitrobenzenesulfonic acid 2-(3-methoxyphenyl)hydrazide 814

$C_{13}H_{13}N_3O_5S_2$
N[1]-(2-Thiazolyl)-*N*[4]-succinoylsulfanilamide 877

$C_{13}H_{13}N_3S$
1,4-Diphenyl-3-thiosemicarbazide 750
1-(3-Methylphenyl)-3-(2-pyridyl)-2-thiourea 965
1-(1-(2-Naphthyl)ethylidene)-3-thiosemicarbazide 1072
1-Phenyl-3-(3-pyridylmethyl)-2-thiourea 951
1-(2-Pyridyl)-3-(*o*-tolyl)-2-thiourea 959

$C_{13}H_{13}N_3Si$
Diphenylmethylsilyl azide 1112

$C_{13}H_{13}N_5$
2-(4,6-Dimethyl-2-pyrimidinylamino)-benzimidazole 977

$C_{13}H_{13}N_5O_2S$
Ethyl 2-(7-methyl(1,2,4)triazolo(3,4-c)(1,2,4)benzotriazin-1-ylthio)-acetate 919

$C_{13}H_{13}N_5O_4S$
2-(3-(2-Thiazolyl)ureido)acetic acid *N*[2]-(3,4-dihydroxybenzylidene)-hydrazide 875

$C_{13}H_{13}O_2P$
(Diphenylphosphinyl)-methanol 833

$C_{13}H_{13}O_3P$
(2-Methoxyphenyl)-phenylphosphinic acid 834
(3-Methoxyphenyl)-phenylphosphinic acid 834
(4-Methoxyphenyl)-phenylphosphinic acid 834

$C_{13}H_{14}$
Isopropylnaphthalene, mixture of isomers 166
2-Isopropylnaphthalene 166
2-Phenyl-2-norbornene 164
4,6,8-Trimethylazulene 168

$C_{13}H_{14}AsNO_4$
4-(Benzylamino)-2-hydroxybenzenearsonic acid 1126

$C_{13}H_{14}BrN$
1-(2-Methylbenzyl)pyridinium bromide 955

$C_{13}H_{14}BrNO_2$
N-(5-Bromopentyl)phthalimide 714

$C_{13}H_{14}BrNO_4$
2-(Methacryloyloxy)ethyl *N*-(3-bromophenyl)carbamate 622

$C_{13}H_{14}ClFN_2OS$
2-(2-Chloroethyl)-3-(4-fluorophenyl)-3-hydroxy-2,3,5,6-4*H*-imidazo(2,1-b)-thiazole 727
2-(2-Chloroethyl)-3-(4-fluorophenyl)-3-hydroxy-2,3,5,6-tetrahydro-imidazo(2,1-b)thiazole hydro-bromide 727

$C_{13}H_{14}ClF_3N_2O$
1-(4-Chloro-2-(trifluoromethyl)-phenyl)-3-cyclopentylurea 505

$C_{13}H_{14}ClF_3N_2O_2$
Piperidino *N*-(2-chloro-5-(trifluoromethyl)phenyl)carbamate 607

$C_{13}H_{14}ClN$
1-(4-Chlorophenyl)-1-cyclohexane-carbonitrile 756

$C_{13}H_{14}ClNO$
3′-Chloro-4′-methyl-2,4-hexadienanilide 460

$C_{13}H_{14}ClNO_3$
2-(4-Chlorophenyl)-2,5,5-trimethyltetrahydrofuran-3,4-dione 3-oxime 1008
Ethyl 3-(4-chlorobenzamido)-crotonate 472

$C_{13}H_{14}ClNS$
4-(*tert*-Butylthio)-7-chloroquinoline 987

$C_{13}H_{14}ClN_3O_2$
1-(α,α-Dimethyl-4-nitrobenzyl)-pyrazinium chloride 978

$C_{13}H_{14}ClN_3O_2S$
Metanilic acid 2-(5-chloro-*o*-tolyl)-hydrazide 807

$C_{13}H_{14}ClN_3O_3$
3-(2-Chloro-4-nitrophenyl)-1,1-diallylurea 569

$C_{13}H_{14}ClN_3O_4$
1-(5-Chloro-2,4-dimethoxyphenyl)-3-(5-methyl-3-isoxazolyl)urea 870
O-(*N*-(2-Chloro-4-nitrophenyl)-carbamoyl)cyclohexanone oxime 1010

$C_{13}H_{14}ClN_3O_5$
Dimethyl 2-(2-amino-5-chlorobenzoylhydrazono)succinate 492

$C_{13}H_{14}Cl_2N_2O$
1,1-Diallyl-3-(2,3-dichlorophenyl)urea 594
1,1-Diallyl-3-(2,4-dichlorophenyl)urea 599
1,1-Diallyl-3-(2,5-dichlorophenyl)urea 543
1,1-Diallyl-3-(2,6-dichlorophenyl)urea 594
1,1-Diallyl-3-(3,4-dichlorophenyl)urea 599

$C_{13}H_{14}Cl_2N_2O_2$
o-(*N*-(2,3-Dichlorophenyl)-carbamoyl)cyclohexanone oxime 1013
o-(*N*-(2,4-Dichlorophenyl)-carbamoyl)cyclohexanone oxime 1013
O-(*N*-(2,5-Dichlorophenyl)-carbamoyl)cyclohexanone oxime 1013
O-(*N*-(2,6-Dichlorophenyl)-carbamoyl)cyclohexanone oxime 1013
O-(*N*-(3,4-Dichlorophenyl)-carbamoyl)cyclohexanone oxime 1012
6-(3,4-Dichlorophenyl)-4,5-dihydro-2-(3-hydroxypropyl)-3(2*H*)-pyridazinone 694

$C_{13}H_{14}Cl_2N_2O_4$
Diethyl (2,6-dichloro-3-pyridylaminomethylene)malonate 954

$C_{13}H_{14}Cl_2O$
1-(2,4-Dichlorophenyl)-4,4-dimethyl-1-penten-3-one 282

$C_{13}H_{14}Cl_2O_2$
Methyl 1-(2,4-dichlorophenyl)-1-cyclopentanecarboxylate 361

$C_{13}H_{14}Cl_3NO_2$
Cyclohexyl *N*-(2,4,5-trichlorophenyl)carbamate 678

$C_{13}H_{14}Cl_3NO_4$
Isopropyl *N*-(1-benzoyloxy-2,2,2-trichloroethyl)carbamate 610
Propyl *N*-(1-(benzoyloxy)-2,2,2-trichloroethyl)carbamate 610

$C_{13}H_{14}Cl_3NO_5$
Ethyl *N*-(1-(4-methoxybenzoyloxy)-2,2,2-trichloroethyl)carbamate 608

$C_{13}H_{14}FNO_3$
2-(4-Fluorophenyl)-1-methyl-6-oxo-3-piperidinecarboxylic acid 689

$C_{13}H_{14}F_3NO_4$
Methyl 2-(*N*-(5-(trifluoromethyl)-phenyl)carbamoyloxy)-2-methylpropionate 610

$C_{13}H_{14}F_3N_3O_5$
N-Butyl-2′,6′-dinitro-4′-(trifluoromethyl)acetanilide 396
2′,6′-Dinitro-*N*-ethyl-4′-(trifluoromethyl)butyranilide 397
2′,6′-Dinitro-*N*-propyl-4′-(trifluoromethyl)propionanilide 396

$C_{13}H_{14}F_3N_3O_6$
Ethyl *N*-(2,6-dinitro-4-(trifluoromethyl)phenyl)-*N*-propylcarbamate 609
Isopropyl *N*-(2,6-dinitro-4-(trifluoromethyl)phenyl)-*N*-ethylcarbamate 609
Methyl *N*-butyl-*N*-(2,6-dinitro-4-(trifluoromethyl)phenyl)carbamate 609
Propyl *N*-(2,6-dinitro-4-(trifluoromethyl)phenyl)-*N*-ethylcarbamate 609

$C_{13}H_{14}INO_2S$
1-Methyl-2-(phenylsulfonylmethyl)-pyridinium iodide 956

$C_{13}H_{14}INS$
1-Ethyl-2-(2-thenylidenemethyl)-pyridinium iodide 957
1-Ethyl-4-(2-thenylidenemethyl)-pyridinium iodide 957

$C_{13}H_{14}N_2$
5-Amino-1,2,3,4-tetrahydroacridine 987
2-(2-Anilinoethyl)pyridine 945
N-Benzyl-1,4-phenylenediamine dihydrochloride 230
2-(4-(Dimethylamino)phenyl)-pyridine 955
3-(4-Methylbenzylamino)pyridine 943

$C_{13}H_{14}N_2O$
β-Alanine 2-naphthylamide 487
DL-Alanine-2-naphthylamide 487
L-Alanine-2-naphthylamide 487
4-(4-Amino-*m*-toluidino)-phenol 221
1,1-Dimethyl-3-(1-naphthyl)urea 603
2-(3,5-Dimethyl-1-pyrazolyl)-acetophenone 863
1-Ethyl-3-(1-naphthyl)urea 603
Harmaline methosulfate 882
α-(3-Methyl-2-pyridylamino)-*o*-cresol 933
α-(4-Methyl-2-pyridylamino)-*o*-cresol 1034
α-(6-Methyl-2-pyridylamino)-*o*-cresol 933
2-Morpholinoquinoline 994

$C_{13}H_{14}N_2OS$
1-Benzyl-3-furfuryl-2-thiourea 852

$C_{13}H_{14}N_2OS_2$
3-(*p*-Isopropylbenzylideneamino)-rhodanine 743

$C_{13}H_{14}N_2O_2$
N-(2-Aminoethyl)-1-hydroxy-2-naphthamide 485
2′-Carbamoyl-2,4-hexadienanilide 470
5-Cyclobutyl-5-phenylhydantoin 702
Ethyl 5-methyl-1-phenylpyrazole-4-carboxylate 862
1-Furfuryl-3-(*o*-tolyl)urea 854
1-Furfuryl-3-(*m*-tolyl)urea 854
1-Furfuryl-3-(*p*-tolyl)urea 851
6-Methoxy-α-(2-pyridylamino)-*o*-cresol 931
4-Methyl-4-(2-furoyl)pimelonitrile 847
N-Methyl-2-(4-methyl-2*H*-1,3-oxazin-6-yl)benzamide 477
Methyl 2-(5-methyl-1-pyrazolylmethyl)benzoate 862
8-Quinolinyl *N*-propylcarbamate 992
8-Quinolyl *N*-isopropylcarbamate 992
1,2,3,4-Tetrahydroharmane-3-carboxylic acid dihydrate 884

$C_{13}H_{14}N_2O_2S$
N-(2-Aminophenyl)-*p*-toluenesulfonamide 816
5-Amino-*o*-toluenesulfonanilide 817
1-Furfuryl-3-(3-(methylthio)-phenyl)urea 850
5-(4-Methoxybenzylidene)-3-methyl-2-(methylimino)-4-thiazolidinone 737
N′-Methyl-*N*′-phenylmetanilamide 817
4-(*m*-Tosyl)-1,3-phenylenediamine 776

$C_{13}H_{14}N_2O_3$
N-Acetyl-D-tryptophan 890
1-(*o*-Anisyl)-3-furfurylurea 846
1-(*p*-Anisyl)-3-furfurylurea 846
2,3-Dimethyl-5-oxo-1-phenyl-3-pyrazoline-4-acetic acid 704
Ethyl 6,7-dimethyl-3-hydroxy-2-quinoxalinecarboxylate 999
1-(Furfuryl)-3-(3-methoxyphenyl)urea 846
1-(Morpholinomethyl)isatin 691
N-(2-Pyridyl)-1,2,5,6-tetrahydrophthalamic acid 963
2-(2,4,5-Trimethoxyphenyl)-succinonitrile 752

$C_{13}H_{14}N_2O_3S$
2-Acetoacetamido-6-ethoxybenzothiazole 913
5-Amino-2-*p*-toluidinobenzenesulfonic acid 788
2-Ethoxycarbonyl-3,4,5,6,7,8-hexahydro(1)benzothieno(2,3-d)-pyrimidin-4-one 919

$C_{13}H_{14}N_2O_3S$ (Cont'd)
pyrimidin-4-one ... 919
4-Methoxymetanilanilide ... 813
Methyl 2-(3-(2-butenoyl)-2-thioureido)benzoate ... 744
$C_{13}H_{14}N_2O_4$
α-Cyano-3,4,5-trimethoxycinnamide. 764
$C_{13}H_{14}N_2O_4S$
1-Furfuryl-3-(p-tolylsulfonyl)-urea ... 854
$C_{13}H_{14}N_2O_7$
4'-(Diacetoxymethyl)-2'-nitroacetanilide ... 420
$C_{13}H_{14}N_2S$
4-(m-Tolylthio)-1,3-phenylenediamine ... 230
$C_{13}H_{14}N_4O$
1,3-Bis(3-pyridylmethyl)urea ... 951
3-(Diazoacetyl)-4,5-dihydro-3-ethyl-4-phenyl-3H-pyrazole ... 1095
1-(2,6-Dimethyl-4-pyrimidinyl)-3-phenylurea ... 975
1-(2-Pyrimidinyl)-3-(2,6-xylyl)urea ... 976
1-(2-Pyrimidinyl)-3-(3,4-xylyl)urea ... 976
$C_{13}H_{14}N_4O_3$
1-(2,5-Dimethoxyphenyl)-3-(2-pyrazinyl)urea ... 977
$C_{13}H_{14}N_4O_3S$
2-(3-Furfurylureido)acetic (2-thienylmethylene)hydrazide ... 1068
$C_{13}H_{14}N_4O_4$
1-(2-Cyanoacetyl)-4-(4-ethoxycarbonylphenyl)-semicarbazide ... 756
Diethyl 1-(benzotriazolylimino)-malonate ... 909
6-Methyl-4-veratrylideneamino-1,2,4-triazine-3,5(2H,4H)-dione ... 1042
$C_{13}H_{14}N_6O_2$
Isonicotinic acid 2,2'-methylenedihydrazide ... 952
$C_{13}H_{14}N_8O$
5-Acetamido-6-amino-2-(2-benzimidazolylamino)-4-pyrimidinol ... 970
$C_{13}H_{14}O$
α,α-Dimethyl-1-naphthalenemethanol ... 208
1-Ethyl-6-methoxynaphthalene ... 188
Ethyl 1-naphthylmethyl ether ... 188
4-Methyl-4-phenyl-2-cyclohexen-1-one ... 285
2-Naphthyl propyl ether ... 188
7-Phenyl-4,6-heptadien-3-one ... 286
$C_{13}H_{14}O_2$
1,4-Dihydro-6,7-dimethoxy-1,4-methanonaphthalene ... 187
6,7-Dimethoxy-1-methylnaphthalene ... 187
$C_{13}H_{14}O_3$
3-Acetyl-4-hydroxy-8a-methyl-1,5,8,8a-tetrahydro-1-naphthalenone ... 52
6,7-Dimethoxy-1-methyl-2-naphthol .. 195
Ethyl 3-oxo-5-phenyl-4-pentenoate ... 364
Methyl 2-methoxy-5-phenyl-2,4-pentadienoate ... 369
4-Oxo-1,2,3,4-tetrahydro-2-naphthalenepropionic acid ... 331
$C_{13}H_{14}O_4$
Cinnamylidene diacetate ... 356
Dimethyl 2-benzylidenesuccinate ... 371
Dimethyl bicyclo(3.2.2)nona-2,6,8-triene-6,7-dicarboxylate ... 79
4,7-Dioxo-7-phenylheptanoic acid ... 331
Ethyl 1-carboxy-2-phenyl-1-cyclopropanecarboxylate, potassium salt ... 364
$C_{13}H_{14}O_5$
6-Acetyl-2-methylresorcinol diacetate ... 378
Dimethyl 3-oxobicyclo(3.2.2)nona-6,8-diene-6,7-dicarboxylate ... 92
2,3-Epoxypropyl 4-(2,3-epoxypropoxy)benzoate ... 369
Ethyl 3-(3-methoxybenzoyl)pyruvate. 369
Ethyl 2-(methoxycarbonyl)-benzoylacetate ... 374
Methyl 2-(2,3-dimethoxystyryl)-glyoxylate ... 369
$C_{13}H_{14}O_5S$
3-(p-Tosyl)allyl pyruvate ... 781
$C_{13}H_{14}O_6$
2,3,5-Triacetoxytoluene ... 378
2,4,5-Triacetoxytoluene ... 378
$C_{13}H_{14}Si$
Benzylphenylsilane ... 1112
Phenyl(o-tolyl)silane ... 1119
$C_{13}H_{15}BrClNO_3S$
4-(α-(Carboxymethylthio)-2-chlorobenzylidene)morpholinium bromide ... 1051
$C_{13}H_{15}BrN_2O$
3-(4-Bromophenyl)-1,1-diallylurea ... 590
$C_{13}H_{15}BrN_2O_2$
O-(N-(3-Bromophenyl)carbamoyl)-cyclohexanone oxime ... 1008
O-(4-Bromophenylcarbamoyl)-cyclohexanone oxime ... 1012
3-(α-Carboxybenzyl)-1,2-dimethylimidazolium bromide ... 867
$C_{13}H_{15}BrN_2O_4$
(2-Bromo-3-phenylpropionyl)-glycylglycine ... 436
$C_{13}H_{15}BrO$
1-Benzoyl-1-bromocyclohexane ... 289
$C_{13}H_{15}BrO_2$
2-Bromo-4,4-dimethyl-6-methoxy-1-tetralone ... 287
$C_{13}H_{15}BrO_4$
4-(2-Bromomethyl)-2,5-dihydroxytoluene diacetate ... 376
$C_{13}H_{15}Br_2NO_5$
2,4-Dibromophenyl N-(butoxycarbonyloxymethyl)-carbamate ... 630
$C_{13}H_{15}ClF_3N_3O$
1-(2-Chloro-5-(trifluoromethyl)-phenyl)-3-piperidinourea ... 505
$C_{13}H_{15}ClN_2O$
3-(2-Chlorophenyl)-1,1-diallylurea ... 588
3-(3-Chlorophenyl)-1,1-diallylurea ... 580
3-(4-Chlorophenyl)-1,1-diallylurea ... 590
$C_{13}H_{15}ClN_2O_2$
O-(N-(2-Chlorophenyl)carbamoyl)-cyclohexanone oxime ... 1011
O-(N-(3-Chlorophenyl)carbamoyl)-cyclohexanone oxime ... 1012
O-(N-(4-Chlorophenyl)carbamoyl)-cyclohexanone oxime ... 1012
$C_{13}H_{15}ClN_2O_4$
Chloroacetylglycyl-D,L-phenylalanine ... 392
Cyclohexyl N-(2-chloro-4-nitrophenyl)carbamate ... 670
$C_{13}H_{15}ClO$
4-Chloro-α,α-dicyclopropylbenzyl alcohol ... 203
$C_{13}H_{15}ClO_2$
1-(4-Chlorophenyl)-1-cyclohexane-carboxylic acid ... 332
$C_{13}H_{15}Cl_2NO$
N-Cyclohexyl-2,4-dichlorobenzamide ... 456
2',3'-Dichloro-1-cyclohexanecarboxanilide ... 437
$C_{13}H_{15}Cl_2NO_2$
Cyclohexyl N-(2,3-dichlorophenyl)-carbamate ... 679
Cyclohexyl N-(2,4-dichlorophenyl)-carbamate ... 680
Cyclohexyl N-(2,5-dichlorophenyl)-carbamate ... 680
Cyclohexyl N-(3,4-dichlorophenyl)-carbamate ... 680
Cyclohexyl N-(3,5-dichlorophenyl)-carbamate ... 632
2,4-Dichlorophenyl N-cyclohexylcarbamate ... 680
$C_{13}H_{15}Cl_2NO_5$
2-(2,4-Dichlorophenoxyethyl) N-(ethoxycarbonylmethyl)-carbamate ... 614
$C_{13}H_{15}Cl_2NS$
3-Benzyl-5-(2-chloroethyl)-4-methylthiazolium chloride ... 874
$C_{13}H_{15}Cl_3N_2O$
1-Cyclohexyl-3-(2,4,5-trichlorophenyl)urea ... 582
$C_{13}H_{15}Cl_3N_2O_2$
2,6-Dimethyl-4-(N-(2,4,5-trichlorophenyl)carbamoyl)-morpholine ... 713
$C_{13}H_{15}FN_2O$
1,1-Diallyl-3-(4-fluorophenyl)urea ... 590
$C_{13}H_{15}FN_2O_2$
O-(N-(4-Fluorophenyl)carbamoyl)-cyclohexanone oxime ... 1008
$C_{13}H_{15}FN_2O_3$
4'-Fluoro-3'-nitrocyclohexanecarboxanilide ... 404
$C_{13}H_{15}FO$
α,α-Dicyclopropyl-4-fluorobenzyl alcohol ... 203
$C_{13}H_{15}F_2NO_2$
1-(4-(Difluoromethoxy)benzoyl)-piperidine ... 693
$C_{13}H_{15}F_3N_2O$
1-Cyclopentyl-3-(3-(trifluoromethyl)-phenyl)urea ... 510
$C_{13}H_{15}I_2NO_4$
N-Acetyl-3,5-diiodo-L-tyrosine ethyl ester ... 398
$C_{13}H_{15}N$
2-Butylquinoline ... 995
2-Butylquinoline picrate ... 995
1,2,3,4,4a,10b-Hexahydrophenanthridine ... 253
6-Methyl-1,2,3,4-tetrahydrocarbazole ... 887
1-Phenyl-1-cyclohexanecarbonitrile. 756
1-(p-Tolyl)-1-cyclopentanecarbonitrile ... 756
$C_{13}H_{15}NO$
3'-(Cyanomethyl)-2,2-dimethylpropiophenone ... 757
4'-(Cyanomethyl)-2,2-dimethylpropiophenone ... 757
1-Dimethylaminomethyl-2-naphthol. 239
2,4-Dimethyl-6-ethoxyquinoline ... 987
4,8-Dimethyl-2-ethoxyquinoline ... 987
1,2,3,4,4a,9b-Hexahydro-2-methyl-5H-indeno(1,2-c)pyridin-5-one ... 295
N-(3-Hydroxypropyl)-1-naphthylamine ... 227
1-(4-Methoxyphenyl)-1-cyclopentanecarbonitrile ... 753
4-Methoxy-1,2,3,4-tetrahydrocarbazole ... 883
3-Methyl-5-(3-phenylpropyl)-isoxazole ... 871
2,3,6,7-Tetrahydro-1H,5H-benzo(ij)-quinolizin-9-carboxaldehyde ... 324
$C_{13}H_{15}NO_2$
N-Cyclohexylideneanthranilic acid .. 345
N-(2,5-Dimethylphenyl)glutarimide ... 720
N-(3,5-Dimethylphenyl)glutarimide ... 720
2-(2-Ethoxyethyl)-1(2H)-isoquinolinone ... 698
N-1-Indanylacetoacetamide ... 444
Methyl 1-indolebutyrate ... 887
Methyl 3-indolebutyrate ... 887
2-Oxo-1-cyclohexanecarboxanilide .. 444
$C_{13}H_{15}NO_3$
tert-Butyl 3-nitrostyryl ketone ... 285
4-Cyano-4-(3-methoxyphenyl)-valeric acid ... 753
4,4-Dimethyl-1-(4-nitrophenyl)-1-penten-3-one ... 303
4'-Ethoxycarbonyl-2-butenanilide ... 477
Ethyl (2-oxo-4-phenyl)-3-pyrrolidinecarboxylate ... 692
4-(1-Methoxy-1-methylethyl)-2-phenyl-2-oxazolin-5-one ... 872
7-Methoxy-1-methyl-4-oxo-1,2,3,4-tetrahydro-1-naphthalenecarboxamide ... 417
2-Methyl-1-(2-nitrophenyl)-1-hexen-3-one ... 285
2-Oxo-2,3,4,5-tetrahydro-1H-benzazepine-4-propionic acid ... 692
2',4',6'-Trimethylmaleanilic acid ... 474
$C_{13}H_{15}NO_3S$
Ethyl 2-(3-hydroxyimino-1-indanylthio)acetate ... 1009
$C_{13}H_{15}NO_3S_2$
3-(3,4-Dimethoxyphenethyl)-rhodanine ... 735
$C_{13}H_{15}NO_4$
4'-Acetylglutaranilic acid ... 432
N-(3-Acetylphenyl)glutaramic acid ... 432
3-Butyn-1-yl N-(2,5-dimethoxyphenyl)carbamate ... 1017
N-Furfuryl-1,2,3,6-tetrahydrophthalamic acid ... 851
2-(α-Hydroxy-2-nitrobenzyl)-1-cyclohexanone ... 282
1-(2-Hydroxy-3-(phenoxy)propyl)-succinimide ... 716
$C_{13}H_{15}NO_4S$
6-Methoxy-5-(methoxymethoxy)-benzo(b)thiophene-3-acetamide ... 897
$C_{13}H_{15}NO_4S_2$
S-(Ethoxycarbonylmethyl)-1,1-dithio-4'-methoxyoxanilate ... 418
$C_{13}H_{15}NO_5$
N,O-Diacetyl-DL-tyrosine ... 429
N,O-Diacetyl-L-tyrosine ... 429
Dimethyl 5-(propionamido)-isophthalate ... 433
4'-(Ethoxycarbonyl)-succinanilic acid ... 420
Methyl 2'-(methoxycarbonyl)-succinanilate ... 433
$C_{13}H_{15}NO_6$
Carbobenzyloxy-DL-glutamic acid ... 659
$C_{13}H_{15}NO_6S$
1-Phenylsulfonyl-4,4-piperidinedicarboxylic acid ... 823
$C_{13}H_{15}NO_7$
Diethyl 2-(4-nitrophenoxy)malonate. 355
$C_{13}H_{15}NS$
2-(3,5-Dimethyl-4-ethyl-2H-pyrrol-2-ylidenemethyl)thiophene tetrafluoroborate ... 859
$C_{13}H_{15}NS_2$
4-tert-Butyl-3-phenyl-4-thiazolin-2-thione ... 743
$C_{13}H_{15}N_3$
3-Anilino-4,5,6,7-tetrahydro-1H-indazole ... 862
2-Dimethylamino-4-methyl-6-phenylpyrimidine ... 976
3,3'-(m-Tolylimino)dipropionitrile ... 755
$C_{13}H_{15}N_3O$
2-Cyanoacetic (α-propylbenzylidene)hydrazide ... 1071
1,1-Dimethyl-4-(1-naphthyl)-semicarbazide ... 570
2'-Ethyl-6'-methyl-1-imidazolecarboxanilide ... 868
$C_{13}H_{15}N_3OS$
Benzoic (3,4,5-trimethyl-4-thiazolin-2-ylidene)hydrazide ... 725
$C_{13}H_{15}N_3O_2$
N-Acetyl-D-tryptophanamide ... 886
Nα-Benzoyl-L-histidinol ... 865
N,N'-Diallyl-3,5-pyridinedicarboxamide ... 949
Ethyl 5-anilino-3-methyl-4-pyrazolecarboxylate ... 862
2-Nitroso-1,2-dihydroharmaline ... 918
$C_{13}H_{15}N_3O_2S$
1-(4-Acetamidophenyl)-3-(2-butenoyl)-2-thiourea ... 740
1,1-Diallyl-3-(4-nitrophenyl)-2-thiourea ... 740
1-(3,4-Dimethyl-5-isoxazolyl)-3-(4-(methylthio)phenyl)urea ... 871
1-(2-Methoxy-5-methylphenyl)-3-(4-methyl-2-thiazolyl)urea ... 875
3,3'-(p-Tosylimino)dipropionitrile ... 819
$C_{13}H_{15}N_3O_3$
1,1-Diallyl-3-(3-nitrophenyl)urea ... 564
1,1-Diallyl-3-(4-nitrophenyl)urea ... 568
Glycyl-DL-tryptophan ... 884
Glycyl-L-tryptophan ... 884
$C_{13}H_{15}N_3O_4$
2,6-Bis(acetoacetamido)pyridine ... 959
Cyanoacetic acid (3,4,5-trimethoxybenzylidene)-hydrazide ... 1043

$C_{13}H_{15}N_3O_4$ (Cont'd)
O-(N-(3-Nitrophenyl)carbamoyl)-cyclohexanone oxime ... 1010
$C_{13}H_{15}N_3O_4S$
N4-Acetyl-N1-(4,5-dimethyloxazol-2-yl)-sulfanilamide ... 872
1-Amino-2-carbamoylpyridinium p-toluenesulfonate ... 955
$C_{13}H_{15}N_3O_6$
Diethyl ketomalonate, 2-nitrophenylhydrazone ... 372
$C_{13}H_{15}N_3O_7$
Methyl 4,4-dimethoxy-2,3-dioxobutyrate 2-(4-nitrophenylhydrazone) ... 369
$C_{13}H_{15}N_3O_8$
N-Acetyl-3,5-dinitro-L-tyrosine ethyl ester ... 412
$C_{13}H_{15}N_3S$
3-Imino-2,3,5,6-tetrahydro-2-(2,6-xylyl)imidazo(2,1-b)thiazole hydrochloride ... 750
$C_{13}H_{15}N_5O_2$
4-(4-(Dimethylamino)-benzylideneamino)-6-methyl-1,2,4-triazine-3,5(2H,4H)-dione ... 1062
$C_{13}H_{15}N_5O_4$
2-(3-(3-Nitrophenyl)ureido)acetic (2-buten-1-ylidene)hydrazide ... 563
$C_{13}H_{16}$
α-Cyclohexylidenetoluene ... 164
1-Methylene-2-phenylcyclohexane ... 164
3-Phenyl-1-cycloheptene ... 164
1-(o-Tolyl)-1-cyclohexene ... 164
1-(m-Tolyl)-1-cyclohexene ... 164
1-(p-Tolyl)-1-cyclohexene ... 164
$C_{13}H_{16}BF_4NO$
2-(tert-Butyl)-5-phenylisoxazolium tetrafluoroborate ... 871
$C_{13}H_{16}BrClN_2O_2S$
3-(4-Chlorophenyl)-3-hydroxy-7-(2-hydroxyethyl)-2,3,5,6-tetrahydro-imidazo(2,1-b)thiazol-4-ium bromide ... 254
$C_{13}H_{16}BrNO_2$
Cyclohexyl N-(4-bromophenyl)-carbamate ... 679
$C_{13}H_{16}BrNO_3$
6-(Benzoylamino)-2-bromohexanoic acid ... 476
$C_{13}H_{16}Br_2N_2$
1,1'-Trimethylenebis(pyridinium bromide) ... 948
$C_{13}H_{16}Br_2N_2O$
1-Cyclohexyl-3-(2,5-dibromophenyl)urea ... 535
$C_{13}H_{16}Br_2O$
1,2-Dibromo-5-methyl-1-phenyl-3-hexanone ... 282
$C_{13}H_{16}ClF_3N_2O_2$
3-(Dimethylamino)propyl N-(2-chloro-5-(trifluoromethyl)phenyl)-carbamate ... 606
$C_{13}H_{16}ClN$
2-(2-Chloroethyl)-2-phenylpentanenitrile ... 751
$C_{13}H_{16}ClNO$
1-(4-Chlorobenzoyl)-3-methylpiperidine ... 695
4-Chloro-N-cyclohexylbenzamide ... 456
4-(4-Chlorophenyl)-4-isopropoxybutyronitrile ... 752
$C_{13}H_{16}ClNO_2$
Cyclohexyl 2'-chlorocarbanilate ... 679
Cyclohexyl 3'-chlorocarbanilate ... 678
Cyclohexyl 4'-chlorocarbanilate ... 679
Cyclopentyl N-(2-chloro-6-methylphenyl)carbamate ... 631
Cyclopentyl N-(3-chloro-2-methylphenyl)carbamate ... 631
$C_{13}H_{16}ClNO_3$
3-Chloro-N-(1-piperonylethyl)-propionamide ... 393
Cyclopentyl N-(2-chloro-4-methoxyphenyl)carbamate ... 620
$C_{13}H_{16}ClNO_3S$
Ethyl 2-(2-chloroacetamido)-4,5,6,7-tetrahydrobenzo(b)thiophene-3-carboxylate ... 855
$C_{13}H_{16}ClNO_4$
N-Chloroacetyl-L-tyrosine ethyl ester ... 391
$C_{13}H_{16}ClNO_5$
N-(5-Chloro-2,4-dimethoxyphenyl)-glutaramic acid ... 399
$C_{13}H_{16}ClN_3O_3$
N-(4-Chloro-2-nitrophenyl)-2-methyl-1-piperidinecarboxamide ... 694
N-(4-Chloro-2-nitrophenyl)-4-methyl-1-piperidinecarboxamide ... 694
$C_{13}H_{16}ClN_3O_4$
1-Methyl-3-piperidinyl N-(2-chloro-4-nitrophenyl)carbamate ... 626
1-Methyl-4-piperidinyl N-(2-chloro-4-nitrophenyl)carbamate ... 626
$C_{13}H_{16}ClN_5O_2$
4-(Benzylaminomethyleneamino)-2-(ethoxycarbonyl)-4H-1,2,4-triazol-2-ium chloride ... 1060
$C_{13}H_{16}Cl_2N_2O$
1-Cyclohexyl-3-(2,3-dichlorophenyl)urea ... 583
1-Cyclohexyl-3-(2,4-dichlorophenyl)urea ... 583
1-Cyclohexyl-3-(2,5-dichlorophenyl)urea ... 583
1-Cyclohexyl-3-(2,6-dichlorophenyl)urea ... 583
1-Cyclohexyl-3-(3,4-dichlorophenyl)urea ... 583
1-(N-(2,6-Dichlorophenyl)-carbamoyl)-2-methylpiperidine ... 696
1-(N-(2,6-Dichlorophenyl)-carbamoyl)-3-methylpiperidine ... 712
1-(N-(2,6-Dichlorophenyl)-carbamoyl)-4-methylpiperidine ... 696
1-(N-(3,4-Dichlorophenyl)-carbamoyl)-2-methylpiperidine ... 712
1-(N-(3,4-Dichlorophenyl)-carbamoyl)-3-methylpiperidine ... 712
1-(N-(3,4-Dichlorophenyl)-carbamoyl)-4-methylpiperidine ... 712
$C_{13}H_{16}Cl_2N_2O_2$
1-Acetyl-1-butyl-3-(3,4-dichlorophenyl)urea ... 534
4-(N-(2,3-Dichlorophenyl)-carbamoyl)-2,6-dimethylmorpholine ... 713
4-(N-(2,4-Dichlorophenyl)-carbamoyl)-2,6-dimethylmorpholine ... 714
4-(N-(2,6-Dichlorophenyl)-carbamoyl)-2,6-dimethylmorpholine ... 713
2,6-Dimethyl-4-(N-(2,5-dichlorophenyl)carbamoyl)-morpholine ... 714
2,6-Dimethyl-4-(N-(3,4-dichlorophenyl)carbamoyl)-morpholine ... 714
N,N'-(2,4,6-Trimethyl-1,3-phenylene)bis(2-chloroacetamide) ... 392
$C_{13}H_{16}Cl_2N_2O_3$
Butyl N-(N-(3,4-dichlorophenyl)-carbamoyl)-N-methylcarbamate ... 630
$C_{13}H_{16}Cl_2N_2O_4$
N,N'-(4-Methyl-1,3-phenylene)bis(2-chloroethyl carbamate) ... 604
$C_{13}H_{16}FNO_2$
Cyclohexyl N-(4-fluorophenyl)-carbamate ... 679
$C_{13}H_{16}F_2N_2O$
1-(N-(2,5-Difluorophenyl)-carbamoyl)-3-methylpiperidine ... 535
$C_{13}H_{16}F_2N_2O_2$
1-Methyl-4-piperidinyl N-(2,5-difluorophenyl)carbamate ... 627
1-Methyl-4-piperidyl N-(2,4-difluorophenyl)carbamate ... 627
$C_{13}H_{16}F_3IN_2$
1,3-Diethyl-2-methyl-5-(trifluoromethyl)-benzimidazolium iodide ... 901
$C_{13}H_{16}F_3NO$
N-Ethyl-2'-(trifluoromethyl)-butyranilide ... 398
$C_{13}H_{16}F_3NO_3$
2,2,2-Trifluoroethyl N-(4-butoxyphenyl)carbamate ... 609
$C_{13}H_{16}F_3N_3S$
4-Methyl-2'-(trifluoromethyl)-1-piperazinethiocarboxanilide ... 728
$C_{13}H_{16}IN$
1-Naphthyltrimethylammonium iodide ... 256
$C_{13}H_{16}INO$
4'-Iodo-1-cyclohexanecarboxanilide. ... 408
$C_{13}H_{16}N_2$
7-Amino-5-isopropyl-8-methylquinoline hydrochloride ... 988
1-(α-Cyanobenzyl)piperidine ... 754
4-(α-Cyanobenzyl)piperidine hydrochloride ... 1060
2,4,7,8-Tetramethyl-1,5-benzodiazepine sulfate ... 249
$C_{13}H_{16}N_2O$
Benzoic acid cyclohexylidenehydrazide ... 502
1-Benzyl-4-hydroxyisonipecotonitrile ... 752
1,1-Diallyl-3-phenylurea ... 585
3-(3-(Dimethylamino)propionyl)-indole ... 886
5,5-Dimethyl-3-(3-pyridylamino)-2-cyclohexen-1-one ... 944
α-(4-Methoxyphenyl)-1-pyrrolidineacetonitrile ... 753
$C_{13}H_{16}N_2OS$
2,3,5,6,7,8-Hexahydro-3-phenylthiazolo(3,2-a)(1,3)diazepin-3-ol hydrobromide ... 245
5-(4-Methoxybenzylidene)-3-methyl-2-(methylimino)thiazolidine ... 736
$C_{13}H_{16}N_2O_2$
4-Butyl-3,5-diketo-1-phenylpyrazolidine ... 708
1-(Diethylaminomethyl)-isatin ... 691
4,4-Diethyl-1-phenyl-3,5-pyrazolidinedione ... 708
Ethyl 5,6-dimethylbenzimidazole-1-acetate ... 907
Ethyl (5-indanylmethylene)-carbazate ... 502
4-Methoxybenzoic (cyclopentylidene)hydrazide ... 498
α-(4-Methoxyphenyl)-4-morpholineacetonitrile ... 753
Methyl N-(2-cyano-5-methylphenyl)-N-isopropylcarbamate ... 767
4-(2-Nitro-1,1,2-trimethylpropyl)-benzonitrile ... 765
O-(N-Phenylcarbamoyl)-cyclohexanone oxime ... 1011
Salicylic (cyclohexylidene)-hydrazide ... 1036
Tryptophan ethyl ester hydrochloride ... 358
$C_{13}H_{16}N_2O_2S_3$
2-Methylthio-N,N-pentamethylene-5-benzothiazolesulfonamide ... 914
$C_{13}H_{16}N_2O_3$
4'-Acetamido-3'-methoxy-2-butenanilide ... 465
Carbobenzyloxy-L-prolinamide ... 700
4'-(Cyclopropylcarboxamido)-2'-methoxyacetanilide ... 415
cis-Hexahydro-N-(2-pyridyl)-phthalamic acid ... 963
N-(3-Pyridyl)-hexahydrophthalamic acid ... 963
$C_{13}H_{16}N_2O_4$
N-Acetylglycyl-L-phenylalanine ... 436
Cyclohexyl N-(3-nitrophenyl)-carbamate ... 670
Cyclopentyl N-(4-methyl-2-nitrophenyl)carbamate ... 668
Diethyl N,N'-(4-methyl-1,3-phenylene)biscarbamate ... 670
Diethyl (2-pyridylaminomethylene)-malonate ... 954
N-(2-Hydroxycyclohexyl)-4-nitrobenzamide ... 462
$C_{13}H_{16}N_2O_5$
Carbobenzyloxy-β-alanylglycine ... 660
Carbobenzyloxy-DL-alanylglycine ... 676
Carbobenzyloxy-L-alanylglycine ... 659
Carbobenzyloxy-DL-glutamine ... 657
Carbobenzyloxyglycylglycine methyl ester ... 661
Carbobenzyloxysarcosylglycine ... 659
Ethyl 3-ethoxy-3-(3-nitrophenylimino)propionate ... 1049
Ethyl 3-ethoxy-3-(4-nitrophenylimino)propionate ... 1059
Methyl 4,4-dimethoxy-2,3-dioxobutyrate 2-phenylhydrazone ... 369
5-Methyl-5-(3,4,5-trimethoxyphenyl)-hydantoin ... 866
$C_{13}H_{16}N_2O_7$
2-(4-Nitrophenoxy)ethyl N-(ethoxycarbonylmethyl)-carbamate ... 648
$C_{13}H_{16}N_2O_8$
3',5'-Di-o-acetyluridine ... 131
$C_{13}H_{16}N_2S$
1,1-Diallyl-3-phenyl-2-thiourea ... 748
1,4-Dihydro-1-phenyl-4,4,6-trimethyl-2-pyrimidinethiol ... 248
$C_{13}H_{16}N_2S_2$
N-Cyclohexyl-2-benzothiazolesulfenamide ... 917
3-(Piperidinomethyl)benzothiazole-2(3H)-thione ... 749
$C_{13}H_{16}N_4$
4-(4-tert-Butylbenzylideneamino)-4H-1,2,4-triazole ... 1072
$C_{13}H_{16}N_4OS$
5-Cyano-6-(4-(dimethylamino)-phenyl)-4-hydroxy-3,4,5,6-4H-2(1H)-pyrimidinethione potassium salt ... 752
1-(1,5-Dimethyl-3-oxo-2-phenyl-4-pyrazolin-4-yl)-3-methyl-2-thiourea ... 747
$C_{13}H_{16}N_4O_2$
1-Methyl-3-piperidinone (4-nitrobenzylidene)hydrazone ... 1053
$C_{13}H_{16}N_4O_4$
2-Methyl-1-cyclohexanone (2,4-dinitrophenyl)hydrazone ... 273
3-Methylcyclohexanone (2,4-dinitrophenyl)hydrazone ... 273
$C_{13}H_{16}N_4O_4S$
2',3'-Isopropylidene-6-mercaptopurine riboside ... 920
$C_{13}H_{16}N_4O_5$
2',3'-Isopropylideneinosine ... 920
$C_{13}H_{16}O$
α-Butylcinnamaldehyde ... 321
Cyclopropyl 4-isopropylphenyl ketone ... 296
2,6-Diallyl-m-cresol ... 192
α,α-Dicyclopropylbenzyl alcohol ... 203
2,2-Dimethyl-5-phenyl-4-pentenal ... 321
4-Mesityl-3-buten-2-one ... 305
1-(4-Methoxyphenyl)-1-cyclohexene ... 180
3-(2-Methoxyphenyl)-1-cyclohexene ... 180
5-Methyl-1-phenyl-1-hexen-3-one ... 286
7-Phenyl-6-heptyn-3-ol ... 1016
2-Phenyl-2-norbornanol, mixture of endo and exo ... 204
1,2,3,4-Tetrahydro-3,4,4-trimethyl-1-naphthalenone ... 298
$C_{13}H_{16}O_2$
Cinnamyl butyrate ... 362
2,6-Diallyloxytoluene ... 181
Ethyl 2-benzyl-1-cyclopropanecarboxylate ... 361
1-(2-Methoxy-4-methylphenyl)-3-methyl-2-buten-1-one ... 301
2-(4-Methoxyphenyl)-1-cyclohexanone ... 284
4-(4-Methoxyphenyl)-2-cyclohexen-1-ol ... 202
1-(3-Methoxyphenyl)-4-methyl-1-penten-3-one ... 283
1-(4-Methoxyphenyl)-4-methyl-1-penten-3-one ... 283
4-Phenyl-2,2-tetramethylene-1,3-dioxolane ... 184
1,2,3,4-Tetrahydro-2-naphthalenepropionic acid ... 334
1-(m-Tolyl)-1-cyclopentanecarboxylic acid ... 332

$C_{13}H_{16}O_2S$
1-Cyclohexenyl *p*-tolyl sulfone 783
$C_{13}H_{16}O_3$
1-Acetoxy-1,2,4a,5,8,8a-hexahydro-4a-methyl-2-naphthalenone 92
3-*tert*-Butyl-3-hydroxy-5-methylphthalide 382
1-(3,4-Dimethoxyphenyl)-1-penten-3-one 283
6,7-Dimethoxy-1,2,3,4-tetrahydro-1,4-methano-2-naphthol 201
trans-4a,7-Dimethyl-1,4,4a,5,8,8a-hexahydro-2-methoxy-1,4-naphthalenedione 52
Ethyl 5-oxo-5-phenylvalerate 360
Ethyl 3-oxo-3-(3,5-xylyl)propionate 364
6-Methoxy-2-methyl-1,2,3,4-tetrahydro-1-naphthoic acid 329
3-Methyl-2-butyl benzoylformate 374
Phenethyl levulinate 360
$C_{13}H_{16}O_3S_3$
2-(2-(Phenylsulfonyl)ethylthio)-tetrahydrothiopyran-4-one 776
$C_{13}H_{16}O_4$
3-Benzyladipic acid 332
2,3-Dihydro-6,7-dimethoxy-2,2-dimethyl-4*H*-1-benzopyran-4-one 291
1-(4,5-Dimethoxy-2-hydroxyphenyl)-3-methyl-2-buten-1-one 300
Dimethyl 3-phenylglutarate 360
mono(1,2-Dimethylpropyl) phthalate 375
Ethyl 2,3-epoxy-3-(4-methoxyphenyl)butyrate 354
3-Ethyl-3-phenylglutaric acid 332
Methyl 2-(2-methyl-4-phenyl-1,3-dioxolan-2-yl)acetate 356
mono-Neopentyl phthalate 371
2-Phenyl-2-propylsuccinic acid 332
$C_{13}H_{16}O_4Se$
Ethyl 2-(ethoxycarbonylmethylseleno)benzoate 213
$C_{13}H_{16}O_5$
4-(3,4-Dimethoxybenzoyl)butyric acid 327
Methyl 4-ethoxy-3-methoxybenzoylacetate 363
$C_{13}H_{16}O_6$
4,6-*O*-Benzylidene-D-glucopyranoside 200
Ethyl 3-butyl-2,4,5-trioxo-1-cyclopentaneglyoxylate 93
$C_{13}H_{16}O_6Pb$
p-Tolyltriacetoxylead 1135
$C_{13}H_{16}O_7$
Triethyl 2,3,4-furantricarboxylate 853
$C_{13}H_{16}Si$
(1-Naphthyl)trimethylsilane 1123
$C_{13}H_{17}AsO_5$
3,3′-(3-Methoxyphenylarsino)dipropionic acid 1126
$C_{13}H_{17}BrN_2O$
1-(*N*-(3-Bromophenyl)carbamoyl)-4-methylpiperidine 695
1-(4-Bromophenyl)-3-cyclohexylurea 582
$C_{13}H_{17}BrN_2O_2$
4′-Bromo-3′-methyl-2-morpholinoacetanilide 408
4-(4-Bromophenylcarbamoyl)-2,6-dimethylmorpholine 713
1-Methyl-4-piperidinyl *N*-(3-bromophenyl)carbamate 627
$C_{13}H_{17}BrN_2O_3$
4′-Acetamido-4-bromo-2′-methoxybutyranilide 391
1-(4-Bromo-2,6-dimethylphenyl)-3-(ethoxycarbonylmethyl)urea 532
$C_{13}H_{17}BrO$
4-Bromo-2′,4′,6′-trimethylbutyrophenone 288
$C_{13}H_{17}BrO_2$
2-(4-Bromophenyl)-2-*tert*-butyl-1,3-dioxolane 184
Isoamyl α-bromophenylacetate 362
$C_{13}H_{17}BrO_3S$
cis-2-Bromocyclohexanol *p*-tosylate 799
$C_{13}H_{17}ClN_2O$
N-(4-Chlorobenzylidene)-2-morpholinoethylamine 1024
1-(*N*-(2-Chloro-6-methylphenyl)carbamoyl)piperidine 712
1-(*N*-(3-Chloro-4-methylphenyl)carbamoyl)piperidine 711
1-(*N*-(4-Chloro-3-methylphenyl)carbamoyl)piperidine 711
1-(*N*-(5-Chloro-2-methylphenyl)carbamoyl)piperidine 711
1-(2-Chlorophenylcarbamoyl)-2-methylpiperidine 711
1-(2-Chlorophenylcarbamoyl)-3-methylpiperidine 711
1-(2-Chlorophenylcarbamoyl)-4-methylpiperidine 711
1-(3-Chlorophenylcarbamoyl)-2-methylpiperidine 712
1-(3-Chlorophenylcarbamoyl)-3-methylpiperidine 712
1-(3-Chlorophenylcarbamoyl)-4-methylpiperidine 712
1-(4-Chlorophenylcarbamoyl)-2-methylpiperidine 711
1-(4-Chlorophenylcarbamoyl)-3-methylpiperidine 711
1-(4-Chlorophenylcarbamoyl)-4-methylpiperidine 711
1-(2-Chlorophenyl)-3-cyclohexylurea 582
1-(3-Chlorophenyl)-3-cyclohexylurea 582
1-(4-Chlorophenyl)-3-cyclohexylurea 582
6′-Chloro-1-pyrrolidineaceto-*o*-toluidide maleate 425
6′-Chloro-1-pyrrolidineaceto-*o*-toluidide methanesulfonate 423
6′-Chloro-1-pyrrolidineaceto-*o*-toluidide nitrate 423
6′-Chloro-1-pyrrolidineaceto-*o*-toluidide perchlorate 423
6′-Chloro-1-pyrrolidineaceto-*o*-toluidide succinate 403
$C_{13}H_{17}ClN_2O_2$
1-Acetyl-1-butyl-3-(4-chlorophenyl)urea 533
1-(*N*-(3-Chloro-2-methoxyphenyl)carbamoyl)piperidine 699
1-(*N*-(5-Chloro-2-methoxyphenyl)carbamoyl)piperidine 694
4-(*N*-(2-Chlorophenyl)carbamoyl)-2,6-dimethylmorpholine 696
4-(*N*-(3-Chlorophenyl)carbamoyl)-2,6-dimethylmorpholine 713
4-(*N*-(4-Chlorophenyl)carbamoyl)-2,6-dimethylmorpholine 713
1-Methyl-4-piperidyl *N*-(3-chlorophenyl)carbamate 627
$C_{13}H_{17}ClN_2O_3$
2-Morpholinoethyl *N*-(4-chlorophenyl)carbamate 627
Piperidino *N*-(4-chloro-2-methoxyphenyl)carbamate 619
Piperidino *N*-(5-chloro-2-methoxyphenyl)carbamate 620
$C_{13}H_{17}ClN_2O_3S$
4-(*N*-(5-Chloro-2,4-dimethoxyphenyl)thiocarbamoyl)morpholine 729
$C_{13}H_{17}ClN_2O_4$
Ethyl 2-(3-(4-chloro-2-methoxyphenyl)-3-methylureido)acetate 520
$C_{13}H_{17}ClN_2O_5$
Ethyl 2-(4-chloro-2,5-dimethoxyphenylureido)acetate 516
$C_{13}H_{17}ClO$
4-Chloro-4′-isopropylbutyrophenone 288
1-(4-Chlorophenyl)-1-cyclohexanemethanol 198
$C_{13}H_{17}ClOS$
5-Chloromethyl-2-methyl-2-(2,4-xylyl)-1,3-oxathiolane 209
$C_{13}H_{17}ClO_2$
β-Butyl-4-chlorohydrocinnamic acid 333
5′-Chloro-2′-hydroxy-4′-methylhexanophenone 291
$C_{13}H_{17}Cl_2NO$
3-Chloro-*N*-(4-chloro-α,α-dimethylphenethyl)-propionamide 393
3,4-Dichloro-*N*,*N*-diethylhydrocinnamamide 410
$C_{13}H_{17}Cl_2N_3O$
4-(3,4-Dichlorophenyl)-1,1-(hexamethylene)semicarbazide 533
$C_{13}H_{17}Cl_2N_3O_3$
4-(3,4-Dichlorophenyl)-1-(6-hydroxyhexanoyl)semicarbazide 513
$C_{13}H_{17}Cl_3N_2O$
1,1-Diisopropyl-3-(2,4,5-trichlorophenyl)urea 603
1,1-Dipropyl-3-(2,4,5-trichlorophenyl)urea 603
$C_{13}H_{17}Cl_3N_2O_2$
2-Diethylaminoethyl *N*-(2,4,5-trichlorophenyl)carbamate 673
$C_{13}H_{17}FN_2O$
1-Cyclohexyl-3-(4-fluorophenyl)urea 535
1-(*N*-(2-Fluorophenyl)carbamoyl)-3-methylpiperidine 696
1-(*N*-(3-Fluorophenyl)carbamoyl)-2-methylpiperidine 695
$C_{13}H_{17}FN_2O_2$
2,6-Dimethyl-4-(*N*-(4-fluorophenyl)carbamoyl)morpholine 696
1-Methyl-4-piperidinyl *N*-(4-fluorophenyl)carbamate 627
$C_{13}H_{17}FN_2O_4$
Hexyl *N*-(4-fluoro-3-nitrophenyl)carbamate 625
$C_{13}H_{17}FO$
1-(4-Fluorophenyl)cycloheptanol 205
$C_{13}H_{17}F_3N_2O$
1-Butyl-1-methyl-3-(3-trifluoromethylphenyl)urea 510
$C_{13}H_{17}N$
N-(2,2-Dimethylcyclopentylidene)aniline 1075
2-Mesityl-2-methylpropionitrile 756
3-Methyl-2-phenylhexanenitrile 756
4-(*tert*-Pentyl)phenylacetonitrile 756
$C_{13}H_{17}NO$
2-(Dimethylaminomethyl)-α-tetralone hydrochloride 295
4-Ethoxy-4-(*p*-tolyl)-butyronitrile 752
N-Ethyl-*o*-methacrylotoluidide 482
N-Ethyl-*m*-methacrylotoluidide 482
N-Ethyl-*p*-methacrylotoluidide 482
2-Ethyl-2-pentenanilide 478
4-Isopropoxy-4-phenylbutyronitrile 752
N-Methyl-1-phenylcyclopentanecarboxamide 437
N-(1-Phenylcyclohexyl)formamide 437
4-Phenyl-1,2,3,6-tetrahydro-1-pyridineethanol hydrochloride hydrate 246
3-Pyrrolidinopropiophenone hydrochloride 295
$C_{13}H_{17}NOS$
3-(2-(Dimethylamino)ethyl)-4-methoxybenzo(b)thiophene hydrochloride 897
1-(4-Methoxythiobenzoyl)piperidine 738
S-Piperidinomethyl thiobenzoate hydrochloride 373
$C_{13}H_{17}NO_2$
N-(1-Acetyl-2-methyl-1-propyl)benzamide 473
Allyl *N*-(2-ethyl-6-methylphenyl)carbamate 684
Cyclohexyl carbanilate 678
N-Ethyl-*o*-acetoacetotoluidide 444
5-Ethyl-3-phenyl-1-aza-4,6-dioxabicyclo(3.3.0)octane 248
N-(2-Formyl-2-methylpropyl)-*N*-methylbenzamide 473
4′-(4-Hydroxypiperidino)acetophenone 290
2-(4-Methoxyanilino)cyclohexanone 283
4′-Morpholinopropiophenone 295
3-Morpholinopropiophenone hydrochloride 295
Phenyl *N*-cyclohexylcarbamate 679
N-(3-Phenylpropyl)acetoacetamide 444
$C_{13}H_{17}NO_2S$
Cyclopentyl *N*-(4-(methylthio)phenyl)carbamate 667
Ethyl 3-(4-(methylthio)anilino)crotonate 372
Phenyl (3-pyrrolidino-1-propenyl) sulfone hydrochloride 782
$C_{13}H_{17}NO_3$
N-Acetyl-L-phenylalanine ethyl ester 471
N-Acetyl-D-tyrosine ethyl ester monohydrate 412
6-Benzamidohexanoic acid 476
N-Benzoyl-2,2-dimethylglycine ethyl ester 476
N-Benzoyl-L-valine methyl ester 476
4-(*tert*-Butyl)benzamidoacetic acid 475
2′,4′-Dimethylglutaranilic acid 435
3′,4′-Dimethylglutaranilic acid 435
Ethyl *N*-acetyl-*N*-phenethylcarbamate 671
Ethyl 4-butyramidobenzoate 434
Ethyl 4-(2,2-dimethylacetamido)benzoate 434
4′-Ethylglutaranilic acid 435
N-(2-Ethylphenyl)glutaramic acid 435
6-(2-Hydroxybenzylideneamino)hexanoic acid 326
N-(4-Methoxyphenethyl)acetoacetamide 442
Methyl 3-(pivaloylamino)benzoate 434
3-Nitro-2,2,3-trimethylbutyrophenone 295
N-(2-Phenethyl)glutaramic acid 420
2′,4′,6′-Trimethylsuccinanilic acid 436
N-(2,3-Xylyl)glutaramic acid 435
N-(2,5-Xylyl)glutaramic acid 435
N-(2,6-Xylyl)glutaramic acid 435
N-(3,5-Xylyl)glutaramic acid 435
$C_{13}H_{17}NO_3S$
Ethyl 2-acetamido-4,5,6,7-tetrahydrobenzo(b)thiophene-3-carboxylate 858
$C_{13}H_{17}NO_4$
N-Acetyl-*o*-methyl-DL-tyrosine methyl ester 416
N-Acetyl-D-tyrosine ethyl ester 412
N-Acetyl-DL-tyrosine ethyl ester 412
4-(Butoxycarbonyl)phenyl *N*-methylcarbamate 675
Carbobenzyloxy-D-norvaline 661
Carbobenzyloxy-DL-norvaline 661
Carbobenzyloxy-L-norvaline 661
Carbobenzyloxy-DL-valine 661
Dipropyl 3,5-pyridinedicarboxylate 962
4′-Ethoxyglutaranilic acid 418
2-(*N*-Furfurylcarbamoyl)cyclohexanecarboxylic acid 854
cis-*N*-Furfurylhexahydrophthalamic acid 854
4′-(2-Hydroxyethyl)glutaranilic acid 413
2′-Methoxy-5′-methylglutaranilic acid 416
$C_{13}H_{17}NO_4S$
Carbobenzyloxy-D-methionine 656
Carbobenzyloxy-DL-methionine 656
Carbobenzyloxy-L-methionine 656
1-Nitrocyclohexyl *p*-tolyl sulfone 780
$C_{13}H_{17}NO_4S_2$
3-(3-Sulfooxypropyl)-2,5,6-trimethylbenzothiazolium hydroxide, inner salt 916
$C_{13}H_{17}NO_5$
Carbobenzyloxy-L-threonine methyl ester 643
Diethyl 3-ethyl-5-formyl-2,4-pyrroledicarboxylate 842
2-Methoxyethyl *N*-(4-ethoxycarbonylphenyl)carbamate 649
3′,4′,5′-Trimethoxyacetoacet-anilide 442
$C_{13}H_{17}NO_5S$
Diethyl 2-acetamido-4-methyl-3,5-thiophenedicarboxylate 858
4-(4,4,4-Trimethylacetoacetamido)benzenesulfonic acid, barium salt 795
$C_{13}H_{17}NO_6$
4-(Carboxymethyl)-2-(ethoxycarbonyl)-5-methyl-3-

$C_{13}H_{17}NO_6$ (Cont'd)
pyrrolepropionic acid 842
Ethyl 4-carboxy-2-(ethoxycarbonyl)-5-methyl-3-pyrroleacetate 324
$C_{13}H_{17}NO_6S$
2-(*p*-Tosyloxy)cyclohexyl nitrate 798
$C_{13}H_{17}NS$
1-(Phenylthioacetyl)piperidine 741
$C_{13}H_{17}N_3$
2-(Piperidinomethyl)benzimidazole 904
$C_{13}H_{17}N_3O$
Formic (3,3-dimethyl-1-ethyl-2-indolinylidene)hydrazide 704
$C_{13}H_{17}N_3O_2$
*N*¹-Acetyl-*N*³-(2-cyanoethyl)-*N*³-(2-hydroxyethyl)-1,3-phenylenediamine 752
1-(2-Cyanoethyl)-3-(2-methoxy-5-methylphenyl)-1-methylurea 753
$C_{13}H_{17}N_3O_2S_2$
2-Butyl-5-*p*-toluenesulfonamido-1,3,4-thiadiazole 879
2-Isobutyl-5-*p*-toluenesulfonamido-1,3,4-thiadiazole 879
$C_{13}H_{17}N_3O_3$
1-(2-Cyanoethyl)-3-(2,5-dimethoxyphenyl)-1-methylurea 753
1-Cyclohexyl-3-(3-nitrophenyl)urea 565
1-Cyclohexyl-3-(4-nitrophenyl)urea 567
2′,5′-Dimethoxy-2-methyl-2-imidazoline-1-carboxanilide 698
1-(*N*-(2-Methyl-4-nitrophenyl)carbamoyl)piperidine 703
1-(*N*-(4-Methyl-3-nitrophenyl)carbamoyl)piperidine 702
2-Methyl-1-(*N*-(4-nitrophenyl)carbamoyl)piperidine 703
3-Methyl-1-(*N*-(4-nitrophenyl)carbamoyl)piperidine 703
$C_{13}H_{17}N_3O_4$
Carbobenzyloxy-*β*-alanylglycinamide 664
Carbobenzyloxyglycylsarcosinamide 658
N-Cyclohexyl-2,6-dinitro-4-methylaniline 274
Diethyl (6-amino-2-pyridylaminomethylene)-malonate 942
2,6-Dimethyl-4-(*N*-(4-nitrophenyl)carbamoyl)morpholine 703
4-(Ethoxycarbonylmethyl)-1-(phenylacetyl)semicarbazide 574
Glycylglycyl-DL-phenylalanine 422
$C_{13}H_{17}N_3O_4S$
N-Antipyrinyl-*N*-methylaminomethanesulfonic acid, sodium salt monohydrate 794
$C_{13}H_{17}N_3O_5$
Piperidino *N*-(2-methoxy-5-nitrophenyl)carbamate 652
$C_{13}H_{17}N_3O_7S$
N-Butyl-2′,6′-dinitro-4′-(methylsulfonyl)acetanilide 778
2′,6′-Dinitro-*N*-ethyl-4′-(methylsulfonyl)butyranilide 778
2′,6′-Dinitro-4′-(methylsulfonyl)-*N*-propylpropionanilide 778
$C_{13}H_{17}N_3O_8S$
Propyl *N*-(2,6-dinitro-4-(methylsulfonyl)phenyl)-*N*-ethylcarbamate 777
$C_{13}H_{17}N_3S$
4-*tert*-Butyl-3-phenyl-4-thiazolin-2-one hydrazone 725
$C_{13}H_{17}N_5$
2-Amino-1,4-dihydro-4,4,7,8-tetramethyl-*s*-triazino(1,2-a)-benzimidazole 907
$C_{13}H_{17}N_5O_7$
N-(3-Morpholinopropyl)-2,4,6-trinitroaniline 273
$C_{13}H_{18}$
4-*tert*-Butyl-*α*-methylstyrene 165
α-Cyclohexyltoluene 162
$C_{13}H_{18}BrN_3O$
1-(3-Bromophenyl)-3-(hexamethyleneimino)urea 534
$C_{13}H_{18}BrN_3O_2$
1-(4-Bromophenyl)-3-(2-morpholinoethyl)urea 523
$C_{13}H_{18}Br_2O_2$
3,5-Dibromo-7-methyl-1-adamantaneacetic acid 67
$C_{13}H_{18}ClN$
N-(Cyclopropylmethyl)-4-chloro-*α*-methylphenethylamine 242
$C_{13}H_{18}ClNO$
N-(Chloroacetyl)-*N*,*α*,*α*-trimethylphenethylamine 392
2-Chloro-*N*,*N*-diethylhydrocinnamamide 441
4-Chloro-*N*,*N*-diisopropylbenzamide 461
$C_{13}H_{18}ClNO_2$
2-Chloroethyl *N*-(2,6-diethylphenyl)-carbamate 605
$C_{13}H_{18}ClNO_3S$
N-*p*-Tosyl-L-valine chloromethyl ketone 804
$C_{13}H_{18}ClNO_4$
4′-Chloro-2′,5′-dimethoxy-2-propoxyacetanilide 399
4-Phenyl-perhydropyrrolizinium perchlorate 255
$C_{13}H_{18}ClN_3O$
1-(2-Chlorophenyl)-3-(2-pyrrolidinoethyl)urea 523
1-(4-Methyl-2-pentylidene)-4-(2-chlorophenyl)semicarbazide 1025
$C_{13}H_{18}ClN_3O_2$
1-(*N*-(5-Chloro-2-methoxyphenyl)-carbamoyl)-4-methylpiperazine 694
1-(2-Chlorophenyl)-3-(2-morpholinoethyl)urea 523
$C_{13}H_{18}ClN_3O_3$
3-(2-Chloro-4-nitrophenyl)-1,1-dipropylurea 569
1-(6-Hydroxyhexanoyl)-4-(2-chlorophenyl)semicarbazide 513
1-(6-Hydroxyhexanoyl)-4-(3-chlorophenyl)semicarbazide 513
$C_{13}H_{18}ClN_3O_4$
1-(*tert*-Butoxycarbonyl)-4-(5-chloro-2-methoxyphenyl)semicarbazide 519
2-Diethylaminoethyl *N*-(2-chloro-4-nitrophenyl)carbamate 665
$C_{13}H_{18}Cl_2N_2O$
1-Butyl-3-(2,3-dichlorophenyl)-1-ethylurea 541
1-Butyl-3-(2,6-dichlorophenyl)-1-ethylurea 594
1-Butyl-3-(3,4-dichlorophenyl)-1-ethylurea 600
3-(2,3-Dichlorophenyl)-1,1-diisopropylurea 594
3-(2,4-Dichlorophenyl)-1,1-diisopropylurea 599
3-(2,5-Dichlorophenyl)-1,1-diisopropylurea 599
3-(2,6-Dichlorophenyl)-1,1-diisopropylurea 541
3-(3,4-Dichlorophenyl)-1,1-diisopropylurea 577
3-(3,5-Dichlorophenyl)-1,1-diisopropylurea 527
1-(2,6-Dichlorophenyl)-3-(1,3-dimethylbutyl)urea 541
3-(2,3-Dichlorophenyl)-1,1-dipropylurea 594
3-(2,4-Dichlorophenyl)-1,1-dipropylurea 600
3-(2,5-Dichlorophenyl)-1,1-dipropylurea 600
3-(3,4-Dichlorophenyl)-1,1-dipropylurea 599
$C_{13}H_{18}Cl_2N_2O_2$
2-(Diethylamino)ethyl *N*-(2,3-dichlorophenyl)carbamate 672
2-(Diethylamino)ethyl *N*-(2,4-dichlorophenyl)carbamate 672
2-(Diethylamino)ethyl *N*-(2,6-dichlorophenyl)carbamate 627
2-(Diethylamino)ethyl *N*-(3,4-dichlorophenyl)carbamate 672
2-(Diethylamino)ethyl *N*-(3,5-dichlorophenyl)carbamate 627
$C_{13}H_{18}FN_3S$
1-(4-Fluorophenyl)-3-(2-pyrrolidinoethyl)-2-thiourea 730
$C_{13}H_{18}F_3NO$
N-(1-Adamantylmethyl)-trifluoroacetamide 103
$C_{13}H_{18}N_2$
3-(*N*-Butylanilino)-propionitrile 755
4-(1,5-Diazabicyclo(3.2.1)oct-8-yl)-toluene 250
3-(Diethylaminomethyl)-indole hydrochloride 884
1-(3-Dimethylaminopropyl)indole 885
N,*N*-Dimethylhomotryptamine 887
Leptocladine 39
Leptocladine hydrochloride 884
Mesityl oxide *p*-tosylhydrazone 236
5,*N*,*N*-Trimethyl tryptamine 887
$C_{13}H_{18}N_2O$
1-Allyl-3-(2-ethyl-6-methylphenyl)urea 594
1-Cyclohexyl-3-phenylurea 582
1-Cyclopentyl-3-(*o*-tolyl)urea 581
1-Cyclopentyl-3-(*m*-tolyl)urea 581
1-Cyclopentyl-3-(*p*-tolyl)urea 581
2,6-Dimethyl-1-isonicotinoylpiperidine 965
2,6-Dimethyl-1-nicotinoylpiperidine 954
2-(4-Methoxyanilino)azacyclohept-1-ene hydrochloride 698
8-(3-Methoxyphenyl)-1,5-diazabicyclo(3.2.1)octane 247
1-Methyl-4-(3-methylbenzoyl)-piperazine hydrochloride 704
1-(*N*-Phenylcarbamoyl)-2-methylpiperidine 711
1-(*N*-Phenylcarbamoyl)-3-methylpiperidine 711
1-(*N*-Phenylcarbamoyl)-4-methylpiperidine 711
α-Phenyl-1-piperidineacetamide 428
1-(*N*-(*o*-Tolyl)carbamoyl)piperidine 712
1-(*N*-(*m*-Tolyl)carbamoyl)piperidine 711
1-(*N*-(*p*-Tolyl)carbamoyl)piperidine 711
1-(*N*-(2,4-Xylyl)carbamoyl)-pyrrolidine 711
$C_{13}H_{18}N_2OS$
1-Cyclopentyl-3-(4-(methylthio)-phenyl)urea 561
1-(2,5-Dimethylmorpholino)-3-phenyl-2-thiourea 747
$C_{13}H_{18}N_2O_2$
1-(*N*-(*o*-Anisyl)carbamoyl)piperidine 699
1-(*N*-(*m*-Anisyl)carbamoyl)piperidine 699
N-Benzoyl-DL-leucinamide 477
N-Benzoyl-L-leucinamide 477
3-Benzyladipamide 436
Cyclohexyl (4-nitrobenzyl)amine 277
1-Cyclopentyl-3-(2-methoxyphenyl)urea 554
2-(2-(Diethylamino)ethoxy)-benzoxazole hydrochloride 909
3,5-Dimethyl-1-(4-nitrophenyl)-piperidine 276
1-(2,2-Dimethylpropionyl)-2-(4-methylbenzoyl)hydrazine 500
O-(*N*-(2-Ethyl-6-methylphenyl)-carbamoyl)acetone oxime 1012
4-(*N*-(2-Ethylphenyl)carbamoyl)-morpholine 713
1-(*N*-(4-Methoxyphenyl)carbamoyl)-piperidine 558
α-(4-Methoxyphenyl)-1-pyrrolidineacetamide 416
1-(4-Nitrophenyl)-octahydro-azocine 276
4-(*N*-Phenylcarbamoyl)-2,6-dimethylmorpholine 713
4-(*N*-(2,3-Xylyl)carbamoyl)-morpholine 713
4-(*N*-(2,5-Xylyl)carbamoyl)-morpholine 714
4-(*N*-(2,6-Xylyl)carbamoyl)-morpholine 713
4-(*N*-(3,4-Xylyl)carbamoyl)-morpholine 714
4-(*N*-(3,5-Xylyl)carbamoyl)-morpholine 714
$C_{13}H_{18}N_2O_2S$
Cyclohexanone (*p*-tosyl)hydrazone 1071
Piperidino *N*-(4-(methylthio)phenyl)-carbamate 667
$C_{13}H_{18}N_2O_3$
L-2-Aminobutyryl-L-phenylalanine dihydrate 422
tert-Butyl 3-(4-methoxybenzylidene)carbazate 652
Butyl 4-(3-methylureido)benzoate 577
2′,5′-Dimethoxy-1-pyrrolidinecarboxanilide 698
4′-(Dimethylamino)glutaranilic acid 427
Ethyl 2-(3-(3,4-xylyl)ureido)acetate 574
4-(*N*-(2-Methoxy-5-methylphenyl)-carbamoyl)morpholine 699
α-(4-Methoxyphenyl)-4-morpholineacetamide 416
Nα-Benzoyl-L-ornithine methyl ester hydrochloride 469
Nε-Benzoyl-DL-lysine 469
$C_{13}H_{18}N_2O_3S$
Carbobenzyloxy-L-methioninamide 657
1-Cyclopentyl-3-(*p*-tosyl)urea 825
S-(2-Diethylaminoethyl) 4-nitrothiobenzoate hydrochloride 372
N-(*p*-Tosyl)-1-piperidinecarboxamide 820
$C_{13}H_{18}N_2O_4$
L-2-Aminobutyryl-L-tyrosine 411
N,*N*′-Bis(ethoxycarbonyl)-4-methyl-1,3-phenylenediamine 669
5-*tert*-Butyl-4,6-dinitro-1,2,3-trimethylbenzene 275
Diethyl bis-(2-cyanoethyl)-malonate 148
4-(*N*-(2,5-Dimethoxyphenyl)-carbamoyl)morpholine 699
1-(1,2-Dimethyl-1-ethyl-2-nitropropyl)-4-nitrobenzene 277
2-(Isobutylamino)ethyl 3-nitrobenzoate hydrochloride 373
$C_{13}H_{18}N_2O_4S$
2,2-Dimethyl-3-(dimethylamino)-4-(4-nitrophenyl)thietane-1,1-dioxide 778
1-Ethyl-2-methyl-3-(3-sulfooxypropyl)-benzimidazolium hydroxide, inner salt 906
$C_{13}H_{18}N_2O_5$
Ethyl 2-(3-(2,5-dimethoxyphenyl)-ureido)acetate 550
Ethyl (3,4,5-trimethoxybenzylidene)-carbazate 498
Methyl 5-carboxy-2-(dimethylcarbamoyl)-4-methyl-3-pyrrolepropionate 842
$C_{13}H_{18}N_2O_5S_2$
4-Methylthio-2-(3-(*p*-tolylsulfonyl)-ureido)-butyric acid 815
$C_{13}H_{18}N_2S$
1-Cyclohexyl-3-phenyl-2-thiourea 747
$C_{13}H_{18}N_2S_2$
3-Benzyl-5-propyltetrahydro-2*H*-1,3,5-thiadiazine-2-thione 743
$C_{13}H_{18}N_4O_3$
Glycylglycyl-L-phenylalaninamide hydrochloride 422
Nα-Benzoyl-DL-arginine 725
L-Phenylalanylglycylglycinamide 422
$C_{13}H_{18}N_4O_4$
N-(3,3-Dimethylureido)-*N*′-(4-ethoxyphenyl)oxamide 557
Glycylglycyl-L-tyrosinamide hydrochloride 412
Heptaldehyde 2,4-dinitrophenylhydrazone 1052
$C_{13}H_{18}O$
Allyloxytricyclo[5.2.1.0^2,6]dec-3-ene 18
1-Benzylcyclohexanol 206
6-*tert*-Butyl-2,3-dihydro-5-hydroxyindene 192
4-Cyclohexylanisole 180
Cyclopropyl 4-isopropylphenyl carbinol 203
2,5-Diallyl-4,5-dimethyl-2-cyclopenten-1-one 53
DL-*α*,4-Dimethyl-*α*-(1-methylcyclopropyl)benzyl alcohol 203

$C_{13}H_{18}O$ (Cont'd)
- 1,2,4a,5,8,8a-Hexahydro-1,1,4a-trimethyl-2-naphthalenone 54
- 3-(4-Isopropylphenyl)-2-methylpropionaldehyde 321
- 5-Methyl-1-phenyl-3-hexanone 287
- 4-Methyl-4-(*p*-tolyl)-2-pentanone 298
- *trans*-2-Phenylcycloheptanol 207
- 1-Phenyl-1-cyclohexanemethanol 206
- 2-Phenyl-1-cyclohexanemethanol 206
- β-Phenylcyclopentaneethanol 205
- 1-(*o*-Tolyl)cyclohexanol 206
- 1-(*m*-Tolyl)cyclohexanol 206
- 4-(*p*-Tolyl)-3-hexanone 287

$C_{13}H_{18}O_2$
- Allyloxy-3,4-epoxytricyclo(5.2.1.0 2,6)decane 20
- β-Butylhydrocinnamic acid 333
- 2-Butyl-4-phenyl-1,3-dioxolane 184
- α,α-Dimethylhydrocinnamyl acetate 362
- 2-(1-Hydroxycyclohexyl)benzyl alcohol 199
- 1,1-Indandiethanol 208
- Isopentyl phenylacetate 362
- 2-(4-Methoxyphenyl)cyclohexanol (mixture of *cis* and *trans*) 202
- 4-(2-Methoxyphenyl)-3-hexanone 284
- 4-(3-Methoxyphenyl)-3-hexanone 283
- Methyl 5-cyclohexyl-2-penten-4-yne-2-carboxylate 1018
- 3-Methyl-2-phenylhexanoic acid 333
- 4-Methyl-2-(*o*-tolyl)-valeric acid 333
- 4′-(Pentyloxy)acetophenone 293
- Pentyl phenylacetate 362
- 4-(*tert*-Pentyl)phenyl acetate 381
- 2-Phenylheptanoic acid 333

$C_{13}H_{18}O_2S$
- Cyclohexyl *p*-tolyl sulfone 783

$C_{13}H_{18}O_3$
- (+)-7-Acetoxy-4a-methyl-1,2,4a,5,6,7,8,8a-octahydro-2-naphthalenone 92
- 2,4-Dihydroxy-5-hexylbenzaldehyde 322
- 3′,5′-Dimethoxyvalerophenone 291
- 4′-Hydroxy-2′-methoxy-5′-propylpropiophenone 290
- Isoamyl mandelate 353
- 1-Methoxy-4-(4-methoxybenzyloxy)-2-butene 186
- (+/-)-1-Methylpropyl α-methoxyphenylacetate 354
- 8α′-Methyl-3′,4′,8′,8a′-tetra-hydrospiro[1,3-dioxolane-2,1′-(2′*H*)-naphthalene]-6′-(7′*H*)one 59

$C_{13}H_{18}O_3S$
- Cyclohexyl *p*-toluenesulfonate 799
- 4-Methyl-4-(*p*-tosyl)-2-pentanone 781
- 1-(*p*-Tolylsulfonyl)pinacolone 781

$C_{13}H_{18}O_4$
- Diethylene glycol benzyl ether acetate 354
- 3,5-Diisopropoxybenzoic acid 341
- 5,5-Dimethyl-2-(4-hydroxy-3-methoxyphenyl)-1,3-dioxane 195
- 2,4,5-Triethoxybenzaldehyde 323
- 2,4,6-Triethoxybenzaldehyde 324

$C_{13}H_{18}O_4S$
- Butyl 2-(*p*-tolylsulfonyl)-acetate 782

$C_{13}H_{18}O_5$
- Dimethyl 9-oxo-2-cyclononene-1,2-dicarboxylate 92

$C_{13}H_{18}O_5S$
- 2,2-Dimethyl-1,3-dioxolan-4-ylmethyl *p*-toluenesulfonate 798

$C_{13}H_{18}S$
- Benzyl cyclohexyl sulfide 211
- Methyl 2-phenylcyclohexyl sulfide 211

$C_{13}H_{18}Si$
- 4-Phenyl-1-(trimethylsilyl)-1-butyne 1112

$C_{13}H_{19}BrN_2O$
- 3-(4-Bromophenyl)-1,1-diisopropylurea 590
- 3-(4-Bromophenyl)-1,1-dipropylurea 590

$C_{13}H_{19}BrN_2O_6$
- 5-((5-Bromo-4-ethyl-2-nitro)anilino)-1,2,3,4-pentanetetraol 259

$C_{13}H_{19}BrO$
- 4-(4-Bromophenyl)-4-heptanol 198

$C_{13}H_{19}BrO_2$
- Methyl α-bromo-1-adamantaneacetate 82

$C_{13}H_{19}BrO_2Pb$
- Triethylplumbyl 3-bromobenzoate 1133

$C_{13}H_{19}ClN_2$
- 2-(2-(2-Chloroanilino)ethyl)-piperidine 249
- 2-Chloro-*N*1-cyclohexyl-*N*1-methyl-1,4-phenylenediamine hydrochloride 221
- 6-Chloro-*N*-(2-pyrrolidinoethyl)-*o*-toluidine 248

$C_{13}H_{19}ClN_2O$
- 2-Butylamino-6′-chloro-*o*-acetotoluidide hydrochloride 430
- 1-*tert*-Butyl-3-(2-chlorobenzyl)-3-methylurea 528
- 1-Butyl-3-(2-chloro-6-methylphenyl)-1-methylurea 594
- 1-Butyl-3-(3-chloro-4-methylphenyl)-1-methylurea 599
- 1-Butyl-3-(5-chloro-2-methylphenyl)-1-methylurea 599
- 1-Butyl-3-(2-chlorophenyl)-1-ethylurea 588
- 1-Butyl-3-(3-chlorophenyl)-1-ethylurea 580
- 1-Butyl-3-(4-chlorophenyl)-1-ethylurea 590
- 6′-Chloro-2-diethylamino-*o*-acetotoluidide hydrochloride 423
- 6′-Chloro-2-diethylamino-*m*-acetotoluidide hydrochloride 424
- 2-Chloro-3′-(diethylamino)-propionanilide 425
- 3-(2-Chlorophenyl)-1,1-diisopropylurea 588
- 3-(3-Chlorophenyl)-1,1-diisopropylurea 580
- 3-(4-Chlorophenyl)-1,1-diisopropylurea 590
- 1-(4-Chlorophenyl)-3-(1,3-dimethylbutyl)urea 538
- 3-(2-Chlorophenyl)-1,1-dipropylurea 588
- 3-(3-Chlorophenyl)-1,1-dipropylurea 580
- 3-(4-Chlorophenyl)-1,1-dipropylurea 590
- 1-(2-Chlorophenyl)-3-hexylurea 536

$C_{13}H_{19}ClN_2O_2$
- 2-Chloroprocaine hydrochloride 366
- 2-(Diethylamino)ethyl *N*-(2-chlorophenyl)carbamate 672
- 2-(Diethylamino)ethyl *N*-(3-chlorophenyl)carbamate 671
- 2-(Diethylamino)ethyl *N*-(4-chlorophenyl)carbamate 673
- 1-Methyl-1-(4-nitrobenzyl)-piperidinium chloride 277

$C_{13}H_{19}ClN_2O_3$
- 2-(Bis-(2-hydroxyethyl)-amino)-6′-chloro-*o*-acetotoluidide hydrochloride 412

$C_{13}H_{19}ClN_2O_6$
- 5-((4-Chloro-5-ethyl-2-nitro)anilino)-1,2,3,4-pentanetetraol 259

$C_{13}H_{19}ClN_4$
- 1-(4-Chloro-α-pentylbenzylideneamino)-guanidine nitrate 1026

$C_{13}H_{19}ClO$
- 3-Isopropoxy-3-(*p*-tolyl)-propyl chloride 185

$C_{13}H_{19}ClO_2$
- 3-(*p*-Chlorophenyl)-2,2-diethyl-1,3-propanediol 198

$C_{13}H_{19}FN_2O$
- 1,1-Diisopropyl-3-(4-fluorophenyl)urea 590
- 1,1-Dipropyl-3-(4-fluorophenyl)urea 590

$C_{13}H_{19}FN_2O_3S$
- 1-(3-Fluorosulfonylphenyl)-3-hexylurea 804

$C_{13}H_{19}IO_2Pb$
- Triethylplumbyl 2-iodobenzoate 1133
- Triethylplumbyl 3-iodobenzoate 1133

$C_{13}H_{19}IO_8$
- 6-Iodo-α-methylglucoside triacetate 89

$C_{13}H_{19}N$
- *N*-Methyl-1-phenylcyclohexylamine 232
- 4-(2-Phenylethyl)piperidine hydrochloride 253
- 2-Phenyl-1,4,4-trimethylpyrrolidine 251

$C_{13}H_{19}NO$
- 3-(Benzylamino)-1-cyclohexanol 238
- 4-(Benzylamino)-4-methyl-2-pentanol 239
- *N*-(*sec*-Butyl)-*o*-acetotoluidide 441
- *N*-*tert*-Butyl-3-methoxy-*N*-methylbenzamide 464
- *N*,*N*-Diisopropylbenzamide 482
- 2-(Dimethylaminomethyl)-butyrophenone hydrochloride 295
- 2,2-Dimethyl-4′-ethylpropionanilide 438
- *N*-Ethyl-*o*-isobutyrotoluidide 441
- *N*-Isopropyl-2′,6′-acetoxylidide 441
- 5′-Isopropyl-2′-methylpropionanilide 440
- 2,2,3′,4′-Tetramethylpropionanilide 440

$C_{13}H_{19}NOS$
- *S*-Butyl *N*-(2,4-xylyl)thiocarbamate 689
- *S*-Propyl *N*-(2-ethyl-6-methylphenyl)-thiolcarbamate 749

$C_{13}H_{19}NO_2$
- 1-Benzyl-4-hydroxy-3-piperidinemethanol fumarate 245
- *N*-Benzyl-DL-leucine 349
- *N*-Benzyl-L-leucine 349
- Butyl *N*-(2-ethylphenyl)carbamate 681
- Butyl *N*-(2,3-xylyl)carbamate 684
- Butyl *N*-(2,4-xylyl)carbamate 686
- Butyl *N*-(2,5-xylyl)carbamate 686
- Butyl *N*-(2,6-xylyl)carbamate 684
- Butyl *N*-(3,5-xylyl)carbamate 687
- Carnegine hydrochloride 247
- 8α-Cyano-2-decalone ethylene ketal 150
- 4-(2-(Diethylamino)-ethoxy)-benzaldehyde 323
- 1-Dimethylamino-2-propyl *o*-toluate hydrochloride 373
- 1-Dimethylamino-2-propyl *p*-toluate hydrochloride 373
- 2,2-Dimethyl-4′-ethoxypropionanilide 419
- 2,2-Dimethyl-4′-(2-hydroxyethyl)-propionanilide 413
- Ethyl *N*-butyl-*N*-phenylcarbamate 680
- Ethyl α-cyano-α-1-cyclohexenylbutyrate 149
- Ethyl *N*-(2,6-diethylphenyl)-carbamate 684
- Ethyl *N*-(α,α-dimethylphenethyl)-carbamate 671
- L-Isoleucine benzyl ester *p*-toluenesulfonate 792
- Isoleucine benzyl ester *p*-toluenesulfonate 356
- Isopropyl *N*-(2-ethyl-6-methylphenyl)carbamate 684
- L-Leucine benzyl ester hydrochloride 358
- DL-Leucine benzyl ester, *p*-toluenesulfonate 788
- L-Phenylalanine *tert*-butyl ester dibenzenesulfimide salt 816
- Propyl *N*-(2-ethyl-6-methylphenyl)-carbamate 684

$C_{13}H_{19}NO_2S$
- *S*-Butyl *N*-(2-methoxy-5-methylphenyl)thiolcarbamate 653
- *N*-Cyclohexyl-*p*-toluenesulfonamide 827
- 3-Methyl-4-(methylthio)phenyl *N*-butylcarbamate 668

$C_{13}H_{19}NO_3$
- 3-Acetamido-1-adamantanecarboxylic acid 105
- Butyl *N*-(2-methoxy-5-methylphenyl)carbamate 653
- 3-(*N*,*N*-Diethylcarbamoyl)-5-norbornene-2-carboxylic acid 107
- 2-Ethoxyethyl *N*-(2-ethylphenyl)-carbamate 650
- 2-Ethoxyethyl *N*-(2,3-xylyl)-carbamate 650
- 2-Ethoxyethyl *N*-(2,4-xylyl)-carbamate 650
- 2-Ethoxyethyl *N*-(2,5-xylyl)-carbamate 650
- 2-Ethoxyethyl *N*-(2,6-xylyl)-carbamate 650
- Heptyl 4-nitrophenyl ether 269
- *N*-(2-Hydroxypropyl)-*N*-(2-phenoxyethyl)-acetamide 411
- 2-Methoxyethyl *N*-(2-ethyl-6-methylphenyl)carbamate 650
- 2-(4-Methoxyphenyl)-DL-glycine *tert*-butyl ester *p*-toluenesulfonate 355

$C_{13}H_{19}NO_3S$
- 1-Benzyl-4-piperidyl methanesulfonate hydrochloride 800
- *S*-Butyl *N*-(2,5-dimethoxyphenyl)-thiocarbamate 648
- *N*-Butyl-2-(*p*-tolylsulfonyl)-acetamide 779
- *N*,*N*-Diethyl-2-(*p*-tolylsulfonyl)-acetamide 779
- 4-(3-Phenylpropylsulfonyl)-morpholine 827
- *N*-(*p*-Tolylsulfonyl)hexanoamide 828

$C_{13}H_{19}NO_4$
- Diethyl 3-ethyl-5-methyl-2,4-pyrroledicarboxylate 840
- 2,6-Dimethoxyphenyl *N*-butylcarbamate 647
- Ethyl 2,4-dimethyl-5-(ethoxycarbonyl)-3-pyrroleacetate 842
- Methyl 2,4-dimethyl-5-(ethoxycarbonyl)-3-pyrrolepropionate 841
- 4-(3,4,5-Trimethoxyphenyl)-morpholine 178

$C_{13}H_{19}NO_4S$
- *N*-(*p*-Tosyl)-L-isoleucine 824
- *N*-(*p*-Tosyl)-L-leucine 824

$C_{13}H_{19}NO_4S_2$
- 2,5-Dimethyl-3-ethylbenzothiazolium ethyl sulfate 916

$C_{13}H_{19}NO_5$
- 1-Ethyl 5-methyl 2-acetyl-2-(2-cyanoethyl)-glutarate 148
- 3,4,5-Trimethoxybenzyl *N*-ethylcarbamate 646

$C_{13}H_{19}N_2O_3S$
- 1-Ethyl-2-methyl-3-(3-sulfopropyl)-benzimidazole 906

$C_{13}H_{19}N_3O$
- 1-Cyclohexyl-3-(4-methyl-2-pyridyl)urea 965
- 1-Cyclohexyl-3-(3-pyridylmethyl)urea 950

$C_{13}H_{19}N_3O_2$
- 1-(4-Acetamidophenyl)-3-*tert*-butylurea 569
- Ethyl 4-(2-pyridylmethyl)-1-piperazinecarboxylate 944
- 1-(5-Isopropyl-2-methylphenyl)-5-methylbiuret 579
- 1-(2-(4-Methoxyphenyl)-1-methylbutylidene)semicarbazide 553

$C_{13}H_{19}N_3O_2S$
- 1-Methyl-4-piperidinone (*p*-tosyl)-hydrazone 1055

$C_{13}H_{19}N_3O_3$
- 1-Butyl-1-ethyl-3-(4-nitrophenyl)urea 568
- 1,1-Diisopropyl-3-(3-nitrophenyl)urea 564
- 1,1-Diisopropyl-3-(4-nitrophenyl)urea 568
- 1,1-Dipropyl-3-(3-nitrophenyl)urea 565
- 1,1-Dipropyl-3-(4-nitrophenyl)urea 568

$C_{13}H_{19}N_3O_3S$
- 1-Cyclohexyl-3-(2-sulfamoylphenyl)urea 822
- 1-Cyclohexyl-3-(4-sulfamoylphenyl)urea 825

$C_{13}H_{19}N_3O_4$
- 4-Nitrophenyl *N*-(2-(diethylamino)-ethyl)carbamate 668

$C_{13}H_{19}N_3O_4S$
N-Acetyl-4-(3-butylureido)-benzenesulfonamide 826

$C_{13}H_{19}N_3O_8S_2$
2′,6′-Dinitro-*N*-pentyl-4′-(methylsulfonyl)-methanesulfonanilide 818

$C_{13}H_{19}N_3S$
4-Cyclohexyl-1-phenyl-3-thiosemicarbazide 750

$C_{13}H_{19}N_5O_2$
*N*α-Benzoyl-L-arginine amide hydrochloride 725
7-(Piperidinomethyl)-theophylline 867

$C_{13}H_{20}$
3-*tert*-Butyl-5-ethyltoluene 163
3,5-Diisopropyltoluene 163

$C_{13}H_{20}ClNO_4$
(4-(2,5-Cyclohexadien-1-ylidene)-2-buten-1-yl)trimethylammonium perchlorate 255

$C_{13}H_{20}ClN_3O_2$
1-(2-Carbamoylethyl)-3-(diethylcarbamoyl)pyridinium chloride 947

$C_{13}H_{20}Cl_2INO$
(2-(3,4-Dichlorophenyl)-2-hydroxyethyl)-dimethylisopropylammonium iodide 254

$C_{13}H_{20}F_2N_2S_2$
Triethylammonium *N*-(2,4-difluorophenyl)dithiocarbamate 730

$C_{13}H_{20}IN$
(4-(2,5-Cyclohexadien-1-ylidene)-2-buten-1-yl)trimethylammonium iodide 43

$C_{13}H_{20}N_2$
2-(2-Anilinoethyl)piperidine 249
3,6-Diethyl-4,5-dihydro-2,7-dimethyl-1*H*-azepine-4-carbonitrile 147
1-(3,7-Dimethyl-2,6-octadien-1-yl)-imidazole 867

$C_{13}H_{20}N_2O$
1-Benzyl-3-*tert*-butyl-1-methylurea 585
1-Benzyl-3-*tert*-butyl-1-methylurea 571
N-Benzyl-*N*-(2-dimethylaminoethyl)-acetamide 423
1-Butyl-1-ethyl-3-phenylurea 585
1-Butyl-3-(2-ethylphenyl)urea 587
1-*tert*-Butyl-3-(2-ethylphenyl)urea 588
1-*tert*-Butyl-3-(3-ethylphenyl)urea 580
1-*tert*-Butyl-3-(4-ethylphenyl)urea 590
1-*tert*-Butyl-3-(α-methylbenzyl)urea 571
1-Butyl-1-methyl-3-(*o*-tolyl)urea 587
1-Butyl-1-methyl-3-(*m*-tolyl)urea 580
1-Butyl-1-methyl-3-(*p*-tolyl)urea 590
3-(4-*tert*-Butylphenyl)-1,1-dimethylurea 590
1-Butyl-3-(2,3-xylyl)urea 594
1-Butyl-3-(2,4-xylyl)urea 599
1-Butyl-3-(2,5-xylyl)urea 599
1-Butyl-3-(2,6-xylyl)urea 594
1-*tert*-Butyl-3-(2,3-xylyl)urea 594
1-*tert*-Butyl-3-(2,4-xylyl)urea 599
1-(*tert*-Butyl)-3-(2,5-xylyl)urea 599
1-*tert*-Butyl-3-(2,6-xylyl)urea 594
1-Butyl-3-(3,4-xylyl)urea 599
1-Butyl-3-(3,5-xylyl)urea 602
2-(Cyclohexylideneamino)-1-cyclohexene-1-carboxamide 1056
N-Cyclohexylidene-2-carbamylcyclohex-1-enylamine 101
1,1-Diethyl-3-(2-ethylphenyl)urea 587
3-(2,6-Diethylphenyl)-1,1-dimethylurea 594
1-(2,6-Diethylphenyl)-3-ethylurea 594
1,1-Diethyl-3-(2,3-xylyl)urea 594
1,1-Diethyl-3-(2,4-xylyl)urea 599
1,1-Diethyl-3-(2,5-xylyl)urea 599
1,1-Diethyl-3-(2,6-xylyl)urea 594
1,1-Diethyl-3-(3,4-xylyl)urea 599
1,1-Diethyl-3-(3,5-xylyl)urea 602
1,1-Diisopropyl-3-phenylurea 585
1-(2,6-Diisopropylphenyl)urea 594
4′-(3-(Dimethylamino)propylamino)-acetophenone 294
2,2-Dimethyl-4′-(dimethylamino)-propionanilide 427
2,2-Dimethyl-4′-(dimethylamino)-propionanilide hydrochloride 427
1,1-Dipropyl-3-phenylurea 585
1-(2-Ethyl-6-methylphenyl)-3-isopropylurea 594
1-(2-Ethyl-6-methylphenyl)-3-propylurea 594
1-(2-Ethylphenyl)-3-isobutylurea 588
1-Isobutyl-3-(2,3-xylyl)urea 594
1-Isobutyl-3-(2,4-xylyl)urea 599
1-Isobutyl-3-(2,5-xylyl)urea 599
1-Isobutyl-3-(2,6-xylyl)urea 594
1-Isobutyl-3-(3,4-xylyl)urea 599
1-Isobutyl-3-(3,5-xylyl)urea 602
1-Isopropyl-3-mesitylurea 602

$C_{13}H_{20}N_2O_2$
4-Acetamido-2-diethylaminomethylphenol 412
1-Butyl-3-(4-ethoxyphenyl)urea 559
1-(*tert*-Butyl)-3-(3-(1-hydroxyethyl)-phenyl)urea 547
1-*tert*-Butyl-3-(3-(1-hydroxyethyl)-phenyl)urea 547
1-Butyl-3-(2-methoxy-5-methylphenyl)urea 556
1-*tert*-Butyl-3-(2-methoxy-5-methylphenyl)urea 556
3-Cyclohexyl-1-furfuryl-1-methylurea 852
2-(Diethylamino)ethyl *N*-phenylcarbamate 671
2-(Diethylamino)ethyl *N*-phenylcarbamate hydrochloride 671
1,1-Diethyl-3-(2-methoxy-5-methylphenyl)urea 560
1-(2,6-Diisopropylphenyl)-3-hydroxyurea 575
2-(Dimethylamino)ethyl *N*-(2-ethylphenyl)carbamate 672
2-Dimethylaminoethyl *N*-(2,3-xylyl)-carbamate 672
2-(Dimethylamino)ethyl *N*-(2,4-xylyl)carbamate 672
2-(Dimethylamino)ethyl *N*-(2,5-xylyl)carbamate 672
2-Dimethylaminoethyl *N*-(2,6-xylyl)-carbamate 672
2-(Dimethylamino)ethyl *N*-(3,4-xylyl)carbamate 672
2-(Dimethylamino)ethyl *N*-(3,5-xylyl)carbamate 673
3-(Dimethylamino)-4-isopropylphenyl *N*-methylcarbamate 673
4-(Dimethylamino)-3-isopropylphenyl *N*-methylcarbamate 673
4-Dimethylaminomethyl-3,5-xylyl *N*-methylcarbamate 673
Ethyl 4-(diethylamino)-carbanilate 670

$C_{13}H_{20}N_2O_2S$
3,3-Dimethyl-2-butanone (*p*-tosyl)-hydrazone 1092

$C_{13}H_{20}N_2O_3$
1-Butyl-3-(2,4-dimethoxyphenyl)urea 549
1-Butyl-3-(2,5-dimethoxyphenyl)urea 556
1-*tert*-Butyl-3-(2,5-dimethoxyphenyl)urea 550
1,1-Diethyl-3-(2,5-dimethoxyphenyl)urea 556
1-(2,5-Dimethoxyphenyl)-3-isobutylurea 556
2-Dimethylaminoethyl *N*-(2-methoxy-5-methylphenyl)-carbamate 652

$C_{13}H_{20}N_2O_3S$
2-(4-Aminophenylsulfonyl)-*N*-butylpropionamide 776
4-Hydroxy-2-hexanone (*p*-tosyl)-hydrazone 1036

$C_{13}H_{20}N_2O_4$
1,4-Bis(acetoacetyl)-2-methyl-piperazine 111
2-(Dimethylamino)ethyl *N*-(2,5-dimethoxyphenyl)carbamate 652
Ethyl *N*-(2,6-dioxo-1,8,8-trimethyl-7-azabicyclo(3.2.1)oct-7-yl)-carbamate 134

$C_{13}H_{20}N_2O_4S$
*N*6-(*p*-Tosyl)-L-lysine 816

$C_{13}H_{20}N_2O_5$
Dimethyl 2-(piperidinomethyleneacetamido)-malonate 108

$C_{13}H_{20}N_2S$
1,1-Diisopropyl-3-phenyl-2-thiourea 748
1,1-Dipropyl-3-phenyl-2-thiourea 748

$C_{13}H_{20}N_4O$
1-(3-(Isopentyloxy)-benzylideneamino)guanidine nitrate 1048

$C_{13}H_{20}N_4O_2$
3,3′-(4-Methyl-1,3-phenylene)bis(1,1-dimethylurea) 566
1,1′-(4-Methyl-1,3-phenylene)bis(3-ethylurea) 566

$C_{13}H_{20}N_4O_4S$
*N*α-(*p*-Tosyl)-D-arginine 819
*N*α-*p*-Tosyl-L-arginine trihydrate 819
*N*α-*p*-Tosyl-L-arginine hydrochloride 819

$C_{13}H_{20}N_4S$
6-(Octylthio)purine 922

$C_{13}H_{20}O$
Benzyl hexyl ether 186
4-*tert*-Butyl-α,α-dimethylbenzyl alcohol 205
2,6-Diisopropyl-4-methylphenol 192
2,6-Dimethyl-4-(1,1-dimethylpropyl)-phenol 192
4-(1-Ethyl-1-methylpropyl)-*o*-cresol 192
4,4a,5,6,7,8-Hexahydro-4a,8,8-trimethyl-2(3*H*)-naphthalenone 56
4-(4-Methyl-2-pentenyl)-3-cyclohexene-1-carboxaldehyde 60
Pseudoionone 50

$C_{13}H_{20}OS$
2-(4-(*tert*.-Butyl)-*o*-tolylthio)-ethanol 210

$C_{13}H_{20}O_2$
1-(4-(*tert*-Butyl)phenoxy)-2-propanol 202
2-(6-(*tert*.-Butyl)-*o*-tolyloxy)-ethanol 202
1,1-Dimethoxy-2,2-dimethyl-1-phenylpropane 186
2,6-Dipropoxytoluene 182
Methyl 2,4,4-undecatriene-2-carboxylate 79
2-(4-(*tert*-Pentyl)phenoxy)ethanol 202
Propylene glycol cuminyl ether 201

$C_{13}H_{20}O_2S$
tert-Butyl 2-isopropylphenyl sulfone 785
tert-Butyl 2,4,6-trimethylphenyl sulfone 786

$C_{13}H_{20}O_3$
9-Isopropyl-3-oxaspiro(5.5)-undecane-2,4-dione 99
4-Isopropylphenoxyacetaldehyde dimethyl acetal 189
Methyl 5-oxocyclopentene-heptanoate 92
3-Phenoxypropionaldehyde diethyl acetal 189

$C_{13}H_{20}O_4$
tert-Butyl 4-(2-tetrahydropyranyloxy)-2-butynoate 1017
Methyl 3-(10-methyl-1,4-dioxaspiro(4.5)dec-7-en-8-yl)-propionate 91
Methyl 2-oxo-1-(3-oxopentyl)-1-cyclohexanecarboxylate 93
Methyl 4-oxo-1,8,8-trimethyl-3-oxabicyclo(3.2.1)octane-2-acetate 96

$C_{13}H_{20}O_4S$
2-Butoxyethyl *p*-toluenesulfonate 797
6-Hydroxyhexyl *p*-toluenesulfonate 797

$C_{13}H_{20}O_5S$
2-(2-(2-Ethoxy)ethoxy)ethyl *p*-toluenesulfonate 797

$C_{13}H_{20}O_6$
Diethyl 3-acetyl-4-oxopimelate 93

$C_{13}H_{20}O_8$
2-Deoxy-1-*O*-methyl-3,5,6-tri-*O*-acetyl-α,β-D-galactofuranoside 74
β-Methylisorhamnoside 2,3,4-triacetate 89
Pentaerythritol tetraacetate 75

$C_{13}H_{21}BF_4S$
2,6-Di-*tert*-butylthiopyrylium tetrafluoroborate 966

$C_{13}H_{21}Br$
1-Bromo-3,5,7-trimethyladamantane 6

$C_{13}H_{21}BrN_2O$
4-Carbamoyl-1-heptylpyridinium bromide 953

$C_{13}H_{21}Cl$
6-Chloro-1,2,3,5,6,7,8,8a-octahydro-1,2,8a-trimethylnaphthalene 7

$C_{13}H_{21}ClN_2$
N′-(4-Chlorobenzyl)-*N*,*N*-diethylethylenediamine 241
N′-(6-Chloro-*o*-tolyl)-*N*,*N*-diethylethylenediamine 242

$C_{13}H_{21}ClN_2O_2$
(3-Nitrobenzyl)triethylammonium chloride 276

$C_{13}H_{21}ClOS$
5′-Chloromethyl-1,7,7-trimethylspiro(norbornane-2,2′-(1,3)oxathiolane) 24

$C_{13}H_{21}ClO_5$
2,6-Di-*tert*-butylpyrylium perchlorate 935

$C_{13}H_{21}NO$
2-(Diethylaminomethyl)-3,6-dimethylphenol 239
α-((Diethylamino)methyl)phenethyl alcohol 239
N,*N*-Dimethyl-1-adamantanecarboxamide 107
N-(1-Methyl-2-phenoxyethyl)-butylamine 240
2-(Octyloxy)pyridine 939
7-Phenoxyheptylamine 240
3-(Piperidinomethyl)-2-norbornanone 58

$C_{13}H_{21}NOS$
1-(Diethylamino)-3-(phenylthio)-2-propanol 238

$C_{13}H_{21}NO_2$
(+/-)-α-Amino-1-adamantanepropionic acid 72
tert-Butyl 3,5-dimethyl-4-ethyl-2-pyrrolecarboxylate 842
N-(Camphanylidene)urethane 116
Ethyl 2-cyano-3-methyl-2-nonenoate 148
Methyl α-amino-1-adamantaneacetate dibenzoyltartrate 356
Methyl α-amino-1-adamantaneacetate hydrochloride 88
2-Morpholinobicyclo(3.3.1)nonan-9-one 59

$C_{13}H_{21}NO_2S$
N,*N*-Diethyl-2,4,6-trimethylbenzenesulfonamide 829

$C_{13}H_{21}NO_3$
Ethyl 3-(5,5-dimethyl-3-oxo-1-cyclohexen-1-ylamino)propionate 88
N-(3,3,5-Trimethylcyclohexyl)-maleamic acid 122

$C_{13}H_{21}NO_3S_2$
S,*S*-Diisopropyl-*N*-(*p*-tolylsulfonyl)-sulfoximine 828

$C_{13}H_{21}NS$
β-(*p*-Tolylthio)triethylamine 241

$C_{13}H_{21}N_3O$
4-Amino-*N*-(2-(diethylamino)ethyl)-benzamide sulfate 468
1-*tert*-Butyl-3-(4-(dimethylamino)-phenyl)urea 567
1-Butyl-3-(4-(dimethylamino)-phenyl)urea 567
4-(2,6-Diethylphenyl)-1,1-dimethylsemicarbazide 579

$C_{13}H_{21}N_3OS$
1-(3-Methyl-1-adamantylcarbonyl)-3-thiosemicarbazide 140

$C_{13}H_{21}N_3O_2$
1-(3-Methyl-1-adamantylcarbonyl)-semicarbazide ... 127
N-(5-Nitro-2-pyridyl)octylamine ... 943
$C_{13}H_{21}N_3O_3S$
*N*α-(*p*-Tosyl)lysinamide acetate monohydrate ... 817
$C_{13}H_{21}N_3S$
1-Adamantyl methyl ketone thiosemicarbazone ... 142
$C_{13}H_{21}N_5O_3S$
*N*α-Tosyl-L-argininamide hydrochloride ... 819
$C_{13}H_{22}BrN$
Trimethyl(2,4,5-trimethylbenzyl)-ammonium bromide ... 256
$C_{13}H_{22}BrP$
Benzyltriethylphosphonium bromide ... 831
$C_{13}H_{22}ClNO_2$
O,O'-Dimethylcoryneine chloride ... 254
$C_{13}H_{22}ClP$
Benzyltriethylphosphonium chloride ... 831
$C_{13}H_{22}F_3O_2Tl$
1,1,1-Trifluoro-2,4-nonanedione, diethylthallium salt ... 50
$C_{13}H_{22}INO$
1-Methyl-1-(8-oxo-2-bicyclo(3.2.1)-octyl)-pyrrolidinium iodide ... 59
$C_{13}H_{22}INO_2$
O,O'-Dimethylcoryneine iodide ... 254
4-(5,5-Dimethyl-3-methoxy-2-cyclohexen-1-ylidene)-morpholinium iodide ... 42
$C_{13}H_{22}N_2$
N'-Benzyl-*N,N*-diethylethylenediamine ... 241
1-Cyano-*N*-(cyclohexyl)-cyclohexylamine ... 145
N,N-Diethyl-*N'*-methyl-*N'*-phenylethylenediamine ... 242
$C_{13}H_{22}N_2O$
1-Adamantanecarboxylic acid 2,2-dimethylhydrazide ... 112
1-(1-Adamantyl)-3-ethylurea ... 130
N-(1-Cyanocyclohexyl)-*N*-methylvaleramide ... 145
3-Cyclohexyl-1,1-diallylurea ... 130
3-(4-Methylpiperazinomethyl)-2-norbornanone ... 58
$C_{13}H_{22}N_2OS$
1-(4-Aminophenylthio)-3-(diethylamino)-2-propanol ... 238
$C_{13}H_{22}N_2O_2$
O-(*N*-(Cyclohexyl)carbamoyl)-cyclohexanone oxime ... 1012
2-(2,5-Dimethoxyphenyl)-2-methyl-1,4-butanediamine dihydrochloride. 239
$C_{13}H_{22}N_2O_3$
2-(2,4,5-Trimethoxyphenyl)-1,4-butanediamine dihydrochloride ... 239
$C_{13}H_{22}N_2O_3S$
N^1-Butyl-N^1-(3-sulfopropyl)-1,4-phenylenediamine ... 788
$C_{13}H_{22}N_2O_4$
1-Nitro-1-(1-nitrocyclohexyl)-cycloheptane ... 46
6-(D-Ribitylamino)-3,4-xylidine monohydrochloride ... 221
$C_{13}H_{22}N_2S$
7,14-Diazadispiro(5.1.5.2)-pentadeca-15-thione ... 141
$C_{13}H_{22}N_2S_4$
4,4'-Methylenebis(1-piperidinedithiocarboxylic acid, ammonium salt) ... 142
$C_{13}H_{22}N_4O$
4-(4-Cyclohexylcyclohexyl)-4,5-dihydro-1*H*-tetrazol-5-one ... 870
$C_{13}H_{22}N_6O_2$
4,4'-(4-Methyl-1,3-phenylene)bis(1,1-dimethylsemicarbazide) ... 565
$C_{13}H_{22}O$
2-Butyl-2-ethyl-5-methyl-3,4-hexadienal ... 60
2-Butyl-3,5,5-trimethyl-2-cyclohexen-1-one ... 54
Cyclohexyl dicyclopropyl carbinol ... 13
7-Isopropyl-2,3,4,4a,5,6,7,8-octahydro-2-naphtol ... 15
(-)-1,2,3,4,4a,5,8,8a-Octahydro-1,1,4a-trimethyl-2-naphthol ... 14
Perhydro-9b-phenalenol, mixture of isomers ... 15
3,5,7-Trimethyl-1-adamantanol ... 15
$C_{13}H_{22}OSi$
(4-Methoxyphenyl)triethylsilane ... 1107
$C_{13}H_{22}O_2$
trans-4-Allyloxydecahydro-1-naphthol ... 20
3,7-Dimethyl-2,6-octadienyl propionate ... 80
4-(4-Hydroxy-4-methylpentyl)-3-cyclohexene-1-carboxaldehyde ... 60
1-(1-Hydroxy-2,2,6-trimethylcyclohexyl)-1-butyn-3-ol ... 1016
2-Methoxy-5,9-cyclododecadien-1-ol ... 19
$C_{13}H_{22}O_2PbS_2$
S-(Triethylplumbyl) *p*-toluenethiosulfonate ... 1135
$C_{13}H_{22}O_3$
Methyl 2-oxocyclopentane-heptanoate ... 94
$C_{13}H_{22}O_3Si$
Benzyltriethoxysilane ... 1112
$C_{13}H_{22}O_4$
Dibutyl methylenesuccinate ... 79
Methyl 4-(1,3-dioxolan-2-yl)-4-ethyl-6-heptenoate ... 91
$C_{13}H_{22}O_5$
Diethyl 5-oxoazelate ... 93
6-Oxo-1,11-undecanedicarboxylic acid ... 64
$C_{13}H_{22}O_5S$
3-(1,3-Bis-(allyloxy)-2-propoxy)-tetrahydrothiophene 1,1-dioxide ... 152
$C_{13}H_{22}Si$
(*p*-Tolyl)triethylsilane ... 1122
$C_{13}H_{23}Br$
2-Bromo-1,12-tridecadiene ... 5
$C_{13}H_{23}ClOS$
2-Chloromethyl-6-isopropyl-9-methyl-1-oxa-4-thiaspiro(4.5)-decane ... 24
$C_{13}H_{23}N$
7,7-Dimethyl-2-(2-(dimethylamino)-ethyl)bicyclo(3.1.1)hept-2-ene ... 32
1-(2,5-Dimethyl-1,4-hexadien-1-yl)-piperidine ... 40
1-(2,5-Dimethyl-2,5-hexadien-1-yl)-piperidine ... 40
$C_{13}H_{23}NO$
4-(2,6-Dimethyl-2,5-heptadien-1-yl)-morpholine ... 41
3-(Piperidinomethyl)-2-norbornanol (mixture of *endo* and *exo*) ... 37
$C_{13}H_{23}NO_2$
Cyclohexyl *N*-cyclohexylcarbamate. 116
Decyl cyanoacetate ... 149
8,10-Diethylnorlobelidione hydrochloride ... 58
Ethyl 2-butyl-2-cyanohexanoate ... 149
$C_{13}H_{23}NO_2SSn$
N-Triethylstannyl-*p*-toluenesulfonamide ... 1131
$C_{13}H_{23}NO_4$
6-(*tert*-Butoxycarboxamido)-2,3-epoxy-2-methylcyclohexane-methanol ... 103
Ethyl 3-methyl-3-(2-nitrocyclohexyl)-butyrate ... 78
2-Methoxy-6-(2,2-dimethylpropionamido)-cyclohexanecarboxylic acid ... 104
$C_{13}H_{23}N_2O_2P$
1-(2-(*N*-Ethylanilino)-ethylamino)-isopropylphosphinic acid ... 834
$C_{13}H_{23}N_3O_2$
2,6-Bis(2-dimethylaminoethoxy)-pyridine ... 938
$C_{13}H_{23}N_5O$
2-Azido-5-undecyl-1,3,4-oxadiazole ... 873
2-Decanoyl-4,6-diamino-1,3,5-triazine ... 979
$C_{13}H_{24}ClNO$
2-Chloro-*N*-(10-undecenyl)-acetamide ... 103
$C_{13}H_{24}N_2O$
3-(4-Methylpiperazinomethyl)-2-norbornanol dihydrochloride ... 37
$C_{13}H_{24}N_2O_2$
4-(*N*-(Cyclohexyl)carbamoyl)-2,6-dimethylmorpholine ... 130
N-(3-(Hexamethyleneimino)-propyl)-acetoacetamide ... 104
$C_{13}H_{24}N_2O_4S$
1,3-Bis-(5-carboxypentyl)-2-thiourea. 141
$C_{13}H_{24}N_2O_5$
1,3-Bis-(5-carboxypentyl)-urea ... 127
$C_{13}H_{24}N_2S$
1,3-Dicyclohexyl-2-thiourea ... 142
$C_{13}H_{24}O$
Dicyclohexyl carbinol ... 15
1-Methoxy-1-cyclododecene ... 18
$C_{13}H_{24}O_2$
Citronellyl propionate ... 80
Ethyl 2,3-dimethyl-2-nonenoate ... 79
1-(2-Methoxy-2-cyclohexen-1-yl)-1-hexanol ... 20
Methyl 4-*tert*-butyl-2-methylcyclohexanecarboxylate ... 77
2-Methyl-2-nonen-1-yl propionate ... 74
2-(*tert*-Pentyl)-cyclohexyl acetate ... 78
$C_{13}H_{24}O_3$
Ethyl 2-heptylacetoacetate ... 94
Methyl 3-oxododecanoate ... 93
7-Oxotridecanoic acid ... 64
Vinyl 3-(2-ethylhexyloxy)-propionate 85
$C_{13}H_{24}O_3Si$
5-Norbornen-2-yltriethoxysilane ... 1110
$C_{13}H_{24}O_4$
Di-*tert*-butyl glutarate ... 76
Diethyl dipropylmalonate ... 75
Diethyl (2-ethylbutyl)-malonate ... 75
Diethyl (2-methylpentyl)-malonate ... 75
Dimethyl undecanedioate ... 76
Methyl 4-(dimethoxymethyl)-4-ethyl-6-heptenoate ... 86
$C_{13}H_{24}O_4S_2$
Dibutyl 2,2'-(methylenedithio)-diacetate ... 87
$C_{13}H_{24}O_6$
1,5-Bis-(5-methyl-1,3-dioxan-5-yl)-2,4-dioxapentane ... 22
$C_{13}H_{25}BrO_2$
Ethyl 11-bromoundecanoate ... 82
$C_{13}H_{25}ClN_2$
2-Chloro-1,3-dipiperidinopropane dihydrochloride ... 39
$C_{13}H_{25}N$
Tridecanenitrile ... 150
$C_{13}H_{25}NO$
N-Allyl-*N*-hexylisobutyramide ... 107
3-Azaspiro(5.5)undecanepropanol ... 37
N-Cyclohexyl-2,2-dimethyl-*N*-ethylpropionamide ... 107
2,5-Dimethyl-5-piperidino-3-hexen-1-ol ... 37
6-Methyl-4-piperidino-5-hepten-2-ol .. 37
$C_{13}H_{25}NO_2$
N-Allyl-*N*-(1,1-dimethyl-3-hydroxybutyl)isobutyramide ... 106
$C_{13}H_{25}NO_3$
13-Amino-7-oxotridecanoic acid hydrochloride ... 71
o-*tert*-Butyl-L-hydroxyproline *tert*-butyl ester dibenzenesulfonamide salt ... 814
Ethyl 2-butyl-2-carbamoylhexanoate ... 101
$C_{13}H_{25}NO_6$
3,3'-(*N*-(2,3-Dihydroxypropyl)imino)-bis(ethyl propionate) ... 87
$C_{13}H_{25}NS$
Dodecyl thiocyanate ... 151
$C_{13}H_{25}N_3O$
Cyclododecanone semicarbazone ... 127
1-Cyclohexyl-3-(2-pyrrolidinoethyl)urea ... 126
$C_{13}H_{25}N_3O_2$
Ethyl 4-(2-piperidylmethyl)-1-piperazinecarboxylate ... 90
$C_{13}H_{25}N_3S$
Cyclododecanone thiosemicarbazone ... 140
$C_{13}H_{26}$
1-Cyclohexylheptane ... 2
$C_{13}H_{26}ClNO$
2-Chloro-*N*-(1-methyldecyl)-acetamide ... 103
2-Chloro-*N*-undecylacetamide ... 103
$C_{13}H_{26}INO$
4-Methyl-4-(2-octen-1-yl)-morpholinium iodide ... 44
$C_{13}H_{26}N_2$
cis,trans-4,4'-Methylenebis(cyclohexylamine) ... 31
trans,trans-4,4'-Methylenebis(cyclohexylamine) ... 30
$C_{13}H_{26}N_2O$
3-*tert*-Butyl-1-cyclohexyl-1-ethylurea ... 130
3-Cyclohexyl-1,1-diisopropylurea ... 130
3-Cyclohexyl-1,1-dipropylurea ... 130
N,N-Diethyl-1-propylnipecotamide hydrochloride ... 108
1-Heptylnipecotamide ... 108
N-(1-Methylhexyl)-1-pyrrolidineacetamide hydrochloride ... 109
$C_{13}H_{26}N_2OS$
1-Dodecanoyl-2-thiourea ... 142
$C_{13}H_{26}N_2O_2$
1-Dodecanoylurea ... 128
$C_{13}H_{26}N_2O_4$
3,9,Bis(3-aminopropyl)-2,4,8,10-tetraoxaspiro(5,5)undecane ... 35
N',N^2-Dicarbethoxy-*N'*-isopropyl-2-methyl-1,2-propanediamine ... 115
Diethyl heptamethylenedicarbamate ... 116
$C_{13}H_{26}N_2O_5$
N',N^2-Dicarbethoxy-*N'*-(3-hydroxy-propyl)-2-methyl-1,2-propanediamine ... 115
$C_{13}H_{26}N_4$
Decanal (2-imidazolin-2-yl)-hydrazone hydriodide ... 1072
$C_{13}H_{26}N_4O_2$
1-Cyclododecyl-3-nitroguanidine ... 137
$C_{13}H_{26}O$
2,2-Dimethyl-3-undecanone ... 49
3-Tridecanone ... 49
$C_{13}H_{26}O_2$
2,6-Dimethyl-8-(1-methoxyethoxy)-2-octene ... 18
2-Methyldodecanoic acid ... 61
4-Methyldodecanoic acid ... 61
$C_{13}H_{26}Si_3$
1,1,1,2,3,3,3-Heptamethyl-2-phenyltrisilane ... 1112
$C_{13}H_{27}N$
N-Methylcyclododecylamine ... 31
$C_{13}H_{27}NO$
N,N-Dibutyl-2,2-dimethylpropionamide ... 107
1-(Octyloxymethyl)pyrrolidine ... 38
Tridecanamide ... 102
$C_{13}H_{27}NO_2$
Ethyl 3-(dibutylamino)propionate ... 88
$C_{13}H_{27}NSSn$
Tributyltin isothiocyanate ... 1131
$C_{13}H_{28}Cl_2Si$
Dichlorododecylmethylsilane ... 1097
$C_{13}H_{28}NO_4P$
Diisobutyl (*N,N*-diethylcarbamoyl)-phosphonate ... 159
$C_{13}H_{28}NO_5P$
Dibutyl (1-nitropentyl)phosphonate.. 158
$C_{13}H_{28}N_2O$
1,1,3-Tributylurea ... 129
$C_{13}H_{28}N_2S$
1,3-Dihexyl-2-thiourea ... 142
2-Dodecyl-2-thiopseudourea hydrochloride ... 138
1-Dodecyl-2-thiourea ... 142
$C_{13}H_{28}O$
2-Tridecanol ... 9
7-Tridecanol ... 9

$C_{13}H_{28}O_2$
1,9-Diethoxynonane ... 20

$C_{13}H_{28}O_3$
Tributyl orthoformate ... 19

$C_{13}H_{28}O_3S$
Dodecyl methanesulfonate ... 156

$C_{13}H_{28}O_4S$
Tridecyl sulfate, sodium salt ... 157

$C_{13}H_{28}S$
1-Tridecanethiol ... 24

$C_{13}H_{29}ISn$
(Iodomethyl)tributyltin ... 1129

$C_{13}H_{29}N$
N-Methyldihexylamine ... 29

$C_{13}H_{29}NO_2$
2-(2-(Nonyloxy)-ethoxy)-ethylamine ... 34

$C_{13}H_{29}NO_2S$
N-Dodecylmethanesulfonamide ... 157

$C_{13}H_{29}N_3$
1-Dodecylguanidine acetate ... 138

$C_{13}H_{30}IN$
Methyltributylammonium iodide ... 43

$C_{13}H_{30}IP$
Methyltributylphosphonium iodide ... 158

$C_{13}H_{30}N_2O_4$
N-(2-Hydroxyethyl)-*N*,*N'*,*N'*-tris(3-hydroxypropyl)ethylenediamine ... 32

$C_{13}H_{30}N_2S$
β-(3-Diethylaminopropylthio)-triethylamine dihydrochloride ... 36

$C_{13}H_{30}Si_5$
2,2-Bis(dimethylsilyl)-1-phenyl-1,3,3-trimethyltrisilane ... 1112

$C_{13}H_{31}N_3$
4-Methyl-1,1,7,7-tetraethyldiethylenetriamine ... 30

$C_{13}H_{32}N_3OP$
N-Methyl-*N'*,*N'*,*N''*,*N''*-tetraisopropylphosphoric triamide ... 160

$C_{14}Cl_5F_5O_2$
(2,3,4,5,6-Pentachloro)(2',3',4',5',6'-pentafluoro)benzil ... 306

$C_{14}Cl_{10}$
1,2-Bis(pentachlorophenyl)-acetylene ... 1015

$C_{14}Cl_{10}O_2$
Decachlorobenzil ... 307

$C_{14}F_{10}$
1,2-Bis(pentafluorophenyl)-acetylene ... 1015

$C_{14}F_{10}O_2$
Decafluorobenzil ... 306

$C_{14}H_4Br_4Cl_3NO_3$
3,4,5,6-Tetrabromo-2',4',5'-trichlorophthalanilic acid ... 455

$C_{14}H_4Br_4N_2O_2$
4,5-Dinitroso-1,3,6,8-tetrabromophenanthrene ... 257

$C_{14}H_4Cl_2F_{15}NO$
N-(2,3-Dichlorophenyl)-2,2,3,3,4,4,5,5,6,6,7,7,8,8,8-pentadecylfluorooctanilide ... 395

$C_{14}H_4Cl_6N_2O_5$
2',5'-Dichloro-4'-nitro-3,4,5,6-tetrachlorophthalanilic acid ... 452

$C_{14}H_4Cl_6O_4$
Bis(2,4,5-trichlorophenyl) oxalate ... 377

$C_{14}H_4Cl_7NO_3$
2',3,4,4',5,5',6-Heptachlorophthalanilic acid ... 455

$C_{14}H_4F_{10}O$
1,1-Bis(pentafluorophenyl)-1-ethanol ... 198

$C_{14}H_5BrCl_5NO_3$
4'-Bromo-2'-chloro-3,4,5,6-tetrachlorophthalanilic acid ... 455

$C_{14}H_5Br_4ClN_2O_5$
2'-Chloro-4'-nitro-3,4,5,6-tetrabromophthalanilic acid ... 451
2'-Chloro-5'-nitro-3,4,5,6-tetrabromophthalanilic acid ... 451
4'-Chloro-3'-nitro-3,4,5,6-tetrabromophthalanilic acid ... 451

$C_{14}H_5Br_4Cl_2NO_3$
2',3'-Dichloro-3,4,5,6-tetrabromophthalanilic acid ... 454
2',5'-Dichloro-3,4,5,6-tetrabromophthalanilic acid ... 455
3',4'-Dichloro-3,4,5,6-tetrabromophthalanilic acid ... 455

$C_{14}H_5Br_4FN_2O_5$
4'-Fluoro-3'-nitro-3,4,5,6-tetrabromophthalanilic acid ... 451

$C_{14}H_5Cl_4NO_3$
N-(2-Hydroxyphenyl)-3,4,5,6-tetrachlorophthalimide ... 716

$C_{14}H_5Cl_4N_3O_3$
2-(1-Benzotriazolylcarbonyl)-3,4,5,6-tetrachlorobenzoic acid ... 908

$C_{14}H_5Cl_5FNO_3$
4'-Fluoro-3,3',4,5,6-pentachlorophthalanilic acid ... 454

$C_{14}H_5Cl_5N_2O_5$
3'-Nitro-3,4,4',5,6-pentachlorophthalanilic acid ... 451
4'-Nitro-2',3,4,5,6-pentachlorophthalanilic acid ... 470

$C_{14}H_5Cl_6NO_3$
2',3,3',4,5,6-Hexachlorophthalanilic acid ... 454
3,3',4,4',5,6-Hexachlorophthalanilic acid ... 450
3,3',4,5,5',6-Hexachlorophthalanilic acid ... 455

$C_{14}H_6BrCl_3F_3NO_2$
2-Bromo-4,6-dichlorophenyl *N*-(4-chloro-2-(trifluoromethyl)phenyl)-carbamate ... 608

$C_{14}H_6BrCl_4NO_3$
3'-Bromo-3,4,5,6-tetrachlorophthalanilic acid ... 453
4'-Bromo-3,4,5,6-tetrachlorophthalanilic acid ... 456

$C_{14}H_6Br_2ClF_3N_2O_4$
2,6-Dibromo-4-nitrophenyl *N*-(4-chloro-2-(trifluoromethyl)phenyl)-carbamate ... 606

$C_{14}H_6Br_2Cl_2F_3NO_2$
(2-Chloro-4,6-dibromophenyl) *N*-(4-chloro-2-(trifluoromethyl)phenyl)-carbamate ... 608

$C_{14}H_6Br_4FNO_3$
2'-Fluoro-3,4,5,6-tetrabromophthalanilic acid ... 454
3'-Fluoro-3,4,5,6-tetrabromophthalanilic acid ... 453

$C_{14}H_6Br_4N_2O_5$
3'-Nitro-3,4,5,6-tetrabromophthalanilic acid ... 451

$C_{14}H_6Br_5NO_3$
3,3',4,5,6-Pentabromophthalanilic acid ... 453

$C_{14}H_6ClN_3O_6$
2-(2-Chloro-3,5-dinitrophenyl)-4-*H*-3,1-benzoxazin-4-one ... 385

$C_{14}H_6Cl_2N_2O_4$
N-(2,6-Dichloro-4-nitrophenyl)-phthalimide ... 719

$C_{14}H_6Cl_2O_2$
2,3-Dichloroanthraquinone ... 318

$C_{14}H_6Cl_3NO_2$
N-(2,4-Dimethylbenzoyl)-4-methylanthranilic acid ... 387

$C_{14}H_6Cl_4FNO_3$
2'-Fluoro-3,4,5,6-tetrachlorophthalanilic acid ... 453
3'-Fluoro-3,4,5,6-tetrachlorophthalanilic acid ... 453
4'-Fluoro-3,4,5,6-tetrachlorophthalanilic acid ... 454

$C_{14}H_6Cl_4F_3NO_2$
2,4,5-Trichlorophenyl *N*-(4-chloro-3-(trifluoromethyl)phenyl)carbamate ... 608

$C_{14}H_6Cl_4N_2O_5$
3'-Nitro-3,4,5,6-tetrachlorophthalanilic acid ... 450

$C_{14}H_6Cl_4O_3$
2-Benzoyl-3,4,5,6-tetrachlorobenzoic acid ... 337
Bis(2,6-dichlorobenzoic) anhydride ... 388

$C_{14}H_6Cl_5NO_3$
2'-Chloro-3,4,5,6-tetrachlorophthalanilic acid ... 454
3'-Chloro-3,4,5,6-tetrachlorophthalanilic acid ... 453
4'-Chloro-3,4,5,6-tetrachlorophthalanilic acid ... 454

$C_{14}H_6Cl_{10}Si$
Bis(pentachlorophenyl)-dimethylsilane ... 1101

$C_{14}H_6F_2O_3$
2,7-Difluoro-9-oxo-4-fluorenecarboxylic acid ... 344

$C_{14}H_6F_6N_2O_4S$
3-Nitro-4-(trifluoromethyl)phenyl sulfide ... 259

$C_{14}H_6F_6N_2O_4S_2$
2-Nitro-4-(trifluoromethyl)phenyl disulfide ... 259

$C_{14}H_6F_{10}Si$
Bis(pentafluorophenyl)-dimethylsilane ... 1101

$C_{14}H_6N_2O_4$
1,4,5,8-Naphthalenetetracarboxylic diimide ... 723

$C_{14}H_6O_4$
1,4,9,10-Anthracenetetraone ... 319

$C_{14}H_7BrClNO_2$
2-Amino-1-bromo-3-chloroanthraquinone ... 317

$C_{14}H_7BrCl_2F_3NO_2$
2-Bromo-4,6-dichlorophenyl *N*-(3-(trifluoromethyl)phenyl)carbamate ... 608

$C_{14}H_7BrCl_2N_2O$
α-(Bromo-chloro-4-hydroxyimino-cyclohexadienylidene)-4-chlorophenylacetonitrile ... 1013

$C_{14}H_7Br_2NO_2$
1-Amino-2,4-dibromoanthraquinone ... 317
6,7-Dibromo-2-phenyl-3,1-benzoxazin-4(4*H*)-one ... 387

$C_{14}H_7Br_3F_3NO_2$
2,4,6-Tribromophenyl *N*-(3-(trifluoromethyl)phenyl)carbamate ... 608

$C_{14}H_7Br_3N_2O_2$
4-Cyanophenyl *N*-(2,4,6-tribromophenyl)carbamate ... 760

$C_{14}H_7Br_4NO_3$
3,4,5,6-Tetrabromophthalanilic acid ... 453

$C_{14}H_7Br_4NO_4$
3'-Hydroxy-3,4,5,6-tetrabromophthalanilic acid ... 447

$C_{14}H_7ClN_2O_4$
4-Chloro-*N*-(4-nitrobenzoyl)-anthranilic acid ... 385
2-(2-Chloro-4-nitrophenyl)-4*H*-3,1-benzoxazin-4-one ... 384
2-(4-Chloro-2-nitrophenyl)-3,1-benzoxazin-4(4*H*)-one ... 384

$C_{14}H_7Cl_2FO_3$
2,4-Dichloro-3-(3-fluorobenzoyl)-benzoic acid ... 345

$C_{14}H_7Cl_2F_3O$
3,4-Dichloro-4'-(trifluoromethyl)-benzophenone ... 309

$C_{14}H_7Cl_2NO_2$
1-Amino-6,7-dichloroanthraquinone ... 317
4-(3,4-Dichlorophenyl)-1*H*-2,3-benzoxazin-1-one ... 382
N-(3,4-Dichlorophenyl)-phthalimide ... 723

$C_{14}H_7Cl_3F_3NO$
3,4,4'-Trichloro-2'-(trifluoromethyl)-benzanilide ... 446

$C_{14}H_7Cl_3F_3NO_2$
2,5-Dichlorophenyl *N*-(4-chloro-2-(trifluoromethyl)phenyl)carbamate ... 607
2,4,5-Trichlorophenyl *N*-(3-(trifluoromethyl)phenyl)carbamate ... 607
2,4,6-Trichlorophenyl *N*-(3-(trifluoromethyl)phenyl)carbamate ... 608

$C_{14}H_7Cl_3N_2O_2$
4-Cyanophenyl *N*-(2,4,5-trichlorophenyl)carbamate ... 760
4-Cyanophenyl *N*-(2,4,6-trichlorophenyl)carbamate ... 760

$C_{14}H_7Cl_4NO_3$
3,4,5,6-Tetrachlorophthalanilic acid ... 453

$C_{14}H_7Cl_5N_2O_3$
1-(6-Carboxy-2,4-dichlorophenyl)-3-(2,4,6-trichlorophenyl)urea ... 531

$C_{14}H_7NO_4$
1-Nitroanthraquinone ... 318
2-Nitrophenanthraquinone ... 308

$C_{14}H_7NO_5$
7-Nitro-9-oxo-1-fluorenecarboxylic acid ... 343
7-Nitro-9-oxo-2-fluorenecarboxylic acid ... 343

$C_{14}H_7NO_6$
Alizarin orange (C.I. 58015) ... 316

$C_{14}H_7NO_7S$
5-Nitroanthraquinone-1-sulfonic acid ... 789

$C_{14}H_8BrClN_2O$
α-(3-Bromo-4-hydroxyimino-2,5-cyclohexadien-1-ylidene)4-chlorophenylacetonitrile ... 1013

$C_{14}H_8BrF_3O_2$
4-Bromo-4'-(trifluoromethoxy)-benzophenone ... 309

$C_{14}H_8BrNO_2$
1-Amino-4-bromoanthraquinone ... 317
10-Bromo-3-nitrophenanthrene ... 280
9-Bromo-10-nitrophenanthrene ... 280
9-Bromo-3-nitrophenanthrene ... 280
N-(2-Bromophenyl)phthalimide ... 716

$C_{14}H_8BrNO_3$
1-Amino-2-bromo-4-hydroxyanthraquinone ... 316

$C_{14}H_8BrNO_5S$
1-Amino-4-bromoanthraquinone-2-sulfonic acid, sodium salt ... 793

$C_{14}H_8Br_2$
9,10-Dibromophenanthrene ... 175

$C_{14}H_8Br_2F_3NO_2$
4-Bromophenyl *N*-(4-bromo-2(trifluoromethyl)phenyl)-carbamate ... 607

$C_{14}H_8Br_2N_2O$
2,5-Bis-(4-bromophenyl)-1,3,4-oxadiazole ... 873

$C_{14}H_8Br_2O_3$
4-Bromobenzoic anhydride ... 367

$C_{14}H_8Br_4N_2O_2$
1,2-Bis(2,5-dibromobenzoyl)-hydrazine ... 501
3,5-Dibromosalicylaldehyde azine ... 1029

$C_{14}H_8ClFN_4O_3$
5-Chloro-1-(*N*-(2-fluoro-5-nitrophenyl)carbamoyl)-1*H*-indazole ... 899
5-Chloro-1-(*N*-(4-fluoro-3-nitrophenyl)carbamoyl)-1*H*-indazole ... 899
3-Chloro-1-(*N*-(2-fluorophenyl)-carbamoyl)-6-nitro-1*H*-indazole ... 899

$C_{14}H_8ClFN_4O_3S$
1-(4-Chloro-2-benzothiazolyl)-3-(4-fluoro-3-nitrophenyl)urea ... 911

$C_{14}H_8ClF_2N_3O$
1-(4-Chloro-2-cyanophenyl)-3-(2,5-difluorophenyl)urea ... 525

$C_{14}H_8ClF_2N_3OS$
1-(4-Chloro-2-benzothiazolyl)-3-(2,5-difluorophenyl)urea ... 911
1-(7-Chloro-2-benzothiazolyl)-3-(2,5-difluorophenyl)urea ... 911

$C_{14}H_8ClF_3N_2O_2$
N-(4-Chloro-3-nitrobenzylidene)-3-(trifluoromethyl)aniline ... 1020

$C_{14}H_8ClF_3N_2O_3$
2'-Chloro-3-nitro-5'-(trifluoromethyl)-benzanilide ... 445

$C_{14}H_8ClF_3N_4O$
5-Chloro-1-(*N*-(3-(trifluoromethyl)-phenyl)carbamoyl)-1*H*-benzotriazole ... 908

$C_{14}H_8ClF_3O$
4-Chloro-4'-(trifluoromethyl)-benzophenone ... 309

$C_{14}H_8ClF_4NO_2$
2-Chloro-4-fluorophenyl *N*-(3-(trifluoromethyl)phenyl)carbamate ... 607
2-Fluorophenyl *N*-(4-chloro-2-(trifluoromethyl)phenyl)carbamate ... 607

$C_{14}H_8ClF_4N_3O_3$
1-(2-Chloro-5-(trifluoromethyl)-phenyl)-3-(4-fluoro-3-nitrophenyl)urea ... 504

$C_{14}H_8ClF_5N_2O$
1-(2-Chloro-5-(trifluoromethyl)phenyl)-3-(2,5-difluorophenyl)urea ... 506

$C_{14}H_8ClNO_2$
1-Amino-5-chloroanthraquinone ... 317
2-Amino-1-chloroanthraquinone ... 317
N-(3-Chlorobenzoyl)anthranilic acid. 384
9-Chloro-10-nitrophenanthrene ... 280
N-(3-Chlorophenyl)-phthalimide ... 723
N-(4-Chlorophenyl)-phthalimide ... 723

$C_{14}H_8ClN_3O_4$
4-Cyanophenyl *N*-(2-chloro-4-nitrophenyl)carbamate ... 759

$C_{14}H_8ClN_3O_7$
3,5-Dinitrobenzoyl *N*-(2-chlorophenyl)carbamate ... 625

$C_{14}H_8ClN_5O_2$
5-Chloro-2-(cyanoamino)-1-(4-nitrophenyl)benzimidazole ... 905

$C_{14}H_8Cl_2$
2,3-Dichloroanthracene ... 175
2,6-Dichloroanthracene ... 175

$C_{14}H_8Cl_2F_2N_2OS$
1-(3,4-Dichlorobenzoyl)-3-(2,4-difluorophenyl)-2-thiourea ... 731

$C_{14}H_8Cl_2F_3N$
N-(2,6-Dichlorobenzylidene)-3-(trifluoromethyl)aniline ... 1021

$C_{14}H_8Cl_2F_3NO$
2,4-Dichloro-α,α,α-trifluoro-*o*-benzotoluidide ... 446
2,4-Dichloro-α,α,α-trifluoro-*m*-benzotoluidide ... 445
2,6′-Dichloro-α,α,α-trifluoro-*m*-benzotoluidide ... 445
3,4-Dichloro-a,a,a-trifluoro-*o*-benzotoluidide ... 446
4,6′-Dichloro-α,α,α-trifluoro-*m*-benzotoluidide ... 445
2′,3-Dichloro-5′-(trifluoromethyl)benzanilide ... 446

$C_{14}H_8Cl_2F_3NO_2$
2,4-Dichlorophenyl *N*-(2-(trifluoromethyl)phenyl)carbamate.. 607
2,4-Dichlorophenyl *N*-(3-(trifluoromethyl)phenyl)carbamate.. 607
3,4-Dichlorophenyl *N*-(3-(trifluoromethyl)phenyl)carbamate.. 607
3-(Trifluoromethyl)phenyl *N*-(2,3-dichlorophenyl)carbamate ... 607

$C_{14}H_8Cl_2F_3N_3O_3$
1-(4-Chloro-2-nitrophenyl)-3-(2-chloro-5-(trifluoromethyl)phenyl)urea ... 504

$C_{14}H_8Cl_2N_2O$
2,5-Bis-(2-chlorophenyl)-1,3,4-oxadiazole ... 873
4-Chloro-a-(3-chloro-4-hydroxyiminocyclohexadien-1-ylidene)-phenylacetonitrile ... 1013
2,3-Dichloro-4-hydroxyimino-a-phenyl-2,5-cyclohexadiene-δ1-a-acetonitrile ... 1013
2-(3,4-Dichlorophenyl)-5-phenyl-1,3,4-oxadiazole ... 873

$C_{14}H_8Cl_2N_2OS$
3′,4′-Dichloro-2-benzothiazolecarboxanilide ... 911

$C_{14}H_8Cl_2N_2O_2$
4-Cyanophenyl *N*-(2,3-dichlorophenyl)carbamate ... 767
4-Cyanophenyl *N*-(2,4-dichlorophenyl)carbamate ... 760
4-Cyanophenyl *N*-(2,5-dichlorophenyl)carbamate ... 760
4-Cyanophenyl *N*-(2,6-dichlorophenyl)carbamate ... 760
4-Cyanophenyl *N*-(3,4-dichlorophenyl)carbamate ... 760
4-Cyanophenyl *N*-(3,5-dichlorophenyl)carbamate ... 760

$C_{14}H_8Cl_2N_2O_4$
1,1-Bis(4-nitrophenyl)-2,2-dichloroethene ... 278
2,4-Dichloro-2′,4′-dinitrostilbene ... 275
2,6-Dichloro-2′,4′-dinitrostilbene ... 275

$C_{14}H_8Cl_2N_4O_3$
1-(*N*-(2,3-Dichlorophenyl)carbamoyl)-6-nitrobenzimidazole ... 901

$C_{14}H_8Cl_2N_4O_5S$
1-(3,4-Dichlorophenyl)-3-(3,5-dinitrobenzoyl)-2-thiourea ... 730

$C_{14}H_8Cl_2N_4O_6$
N,N′-Bis(4-chloro-3-nitrophenyl)oxamide ... 403

$C_{14}H_8Cl_2O_2$
3-Benzoyl-2,4-dichlorobenzaldehyde ... 322
2-(4-Chlorobenzoyl)benzoyl chloride 389
2,2′-Dichlorobenzil ... 309

$C_{14}H_8Cl_2O_3$
2-(3,4-Dichlorobenzoyl)benzoic acid. 337

$C_{14}H_8Cl_2O_4$
Bis(2-chlorophenyl) oxalate ... 377
Bis(4-chlorophenyl) oxalate ... 377

$C_{14}H_8Cl_3F_3N_2O$
1-(2-Chloro-5-(trifluoromethyl)phenyl)-3-(2,3-dichlorophenyl)urea... 506
1-(2-Chloro-5-(trifluoromethyl)phenyl)-3-(2,4-dichlorophenyl)urea... 506
1-(2,4,5-Trichlorophenyl)-3-(3-(trifluoromethyl)phenyl)urea ... 507
1-(2,4,6-Trichlorophenyl)-3-(3-(trifluoromethyl)phenyl)urea ... 507

$C_{14}H_8Cl_3NO$
3-Phenyl-4,5,6-trichloro-2-indolinone ... 692

$C_{14}H_8Cl_3N_3$
1-(4-Cyanophenyl)-3-(2,4,5-trichlorophenyl)urea ... 760

$C_{14}H_8Cl_3N_3OS$
1-(2-Benzothiazolyl)-3-(2,4,5-trichlorophenyl)urea ... 912
1-(4-Chloro-2-benzothiazolyl)-3-(2,3-dichlorophenyl)urea ... 911

$C_{14}H_8Cl_3N_3O_3S$
1-(2-Chloro-4-nitrophenyl)-3-(3,4-dichlorobenzoyl)-2-thiourea ... 730

$C_{14}H_8Cl_4N_2$
2,4-Dichlorobenzaldehyde azine ... 1079
2,6-Dichlorobenzaldehyde azine ... 1079
3,4-Dichlorobenzaldehyde azine ... 1079

$C_{14}H_8Cl_4N_2O$
3,4-Dichlorobenzoic (3,4-dichlorobenzylidene)hydrazide ... 493

$C_{14}H_8Cl_4N_2OS$
1-(3,4-Dichlorobenzoyl)-3-(3,4-dichlorophenyl)-2-thiourea ... 731

$C_{14}H_8Cl_4N_2O_2$
1,2-Bis-(3,4-dichlorobenzoyl)hydrazine ... 501
1-(2,4-Dichlorobenzoyl)-2-(3,4-dichlorobenzoyl)hydrazine ... 493

$C_{14}H_8Cl_4N_2O_3$
N-(6-Methyl-2-pyridyl)-3,4,5,6-tetrachlorophthalamic acid ... 940

$C_{14}H_8Cl_4N_2O_4$
2′-Chloro-4′-nitro-2-(2,4,5-trichlorophenoxy)acetanilide ... 401

$C_{14}H_8Cl_4O_2$
2,2′,6,6′-Tetrachlorobenzoin ... 282

$C_{14}H_8Cl_4O_3$
3,4-Dichlorophenyl 2-(2,4-dichlorophenoxy)acetate ... 376

$C_{14}H_8Cl_5NO_2$
2′,5′-Dichloro-2-(2,4,5-trichlorophenoxy)acetanilide ... 401

$C_{14}H_8Cl_6$
1,1-Bis(4-chlorophenyl)-1,2,2,2-tetrachloroethane ... 172

$C_{14}H_8Cl_6N_4$
Glyoxal bis(2,4,6-trichlorophenylhydrazone) ... 234

$C_{14}H_8Cl_8Si$
Bis(2,3,4,5-tetrachlorophenyl)dimethylsilane ... 1101
Bis(2,3,5,6-tetrachlorophenyl)dimethylsilane ... 1101

$C_{14}H_8FNO_2$
2-(*p*-Fluorophenyl)-3,1-benzoxazin-4(4*H*)-one ... 387

$C_{14}H_8FN_5O_5$
1-(*N*-(4-Fluoro-3-nitrophenyl)carbamoyl)-6-nitrobenzimidazole ... 901
1-(*N*-(4-Fluoro-3-nitrophenyl)carbamoyl)-6-nitro-1*H*-indazole ... 899

$C_{14}H_8F_3IO$
4-Iodo-4′-(trifluoromethyl)benzophenone ... 309

$C_{14}H_8F_3NO_4$
3-(Trifluoromethyl)phenyl 3-nitrobenzoate ... 376

$C_{14}H_8F_3NS$
2-(Phenylthio)-5-(trifluoromethyl)benzonitrile ... 758

$C_{14}H_8F_5N_3O_3$
1-(2,5-Difluorophenyl)-3-(4-nitro-2-(trifluoromethyl)phenyl)urea ... 504

$C_{14}H_8F_6$
3,3′-Bis(trifluoromethyl)biphenyl ... 172

$C_{14}H_8F_8Si$
Bis(2,3,4,5-tetrafluorophenyl)dimethylsilane ... 1101
Bis(2,3,5,6-tetrafluorophenyl)dimethylsilane ... 1101

$C_{14}H_8INO_2$
2-(2-Iodophenyl)-3,1-benzoxazin-4(4*H*)-one ... 387

$C_{14}H_8N_2O$
1,9-Pyrazoloanthrone ... 900

$C_{14}H_8N_2O_2S$
4,4′-Sulfonylbis(benzonitrile) ... 782

$C_{14}H_8N_2O_4$
2-(4-Nitrophenyl)-3,1-benzoxazin-4(4*H*)-one ... 385
N-(4-Nitrophenyl)phthalimide ... 719

$C_{14}H_8N_2O_6$
4,4′-Dinitrobenzil ... 308
3,7-Dinitro-2-methoxy-9-fluorenone.. 311

$C_{14}H_8N_2O_8$
5,5′-Dinitro-2,2′-biphenyldicarboxylic acid ... 344

$C_{14}H_8N_2S_2$
2,2′-Bibenzothiazole ... 914
2,2′-Dithiobis(benzonitrile) ... 763

$C_{14}H_8N_2S_3$
2-Benzothiazolyl sulfide ... 914

$C_{14}H_8N_4$
Δ2,2′-Bi-2*H*-benzimidazole ... 253

$C_{14}H_8N_4O_{11}$
2,8-Dimethoxy-1,3,7,9-tetranitrodibenzofuran ... 893

$C_{14}H_8O_2$
9-Oxo-1-fluorenecarboxaldehyde ... 324

$C_{14}H_8O_4$
2,7-Dihydroxyanthraquinone ... 315

$C_{14}H_8O_5$
6-Hydroxy-3-oxo-3*H*-xanthene-9-carboxylic acid ... 339
1,2,7-Trihydroxyanthraquinone ... 315

$C_{14}H_8O_5S$
Anthraquinone-2-sulfonic acid, thallium salt ... 794
4,6-Phenoxathiindicarboxylic acid... 341

$C_{14}H_8O_7S$
1,9-Dicarboxyphenoxathiin-10,10-dioxide ... 781
Quinizarin-2-sulfonic acid ... 787

$C_{14}H_8O_{10}$
Δ4,4′-Bi-4*H*-pyran-2,2′,6,6′-tetracarboxylic acid ... 67

$C_{14}H_9BrClFN_2O_4$
2-Bromo-4-chloro-6-methylphenyl *N*-(4-fluoro-3-nitrophenyl)carbamate ... 626

$C_{14}H_9BrClF_2NO_2$
2-Bromo-4-chloro-6-methylphenyl *N*-(2,4-difluorophenyl)carbamate ... 639
4-Bromo-2-chloro-5-methylphenyl *N*-(2,4-difluorophenyl)carbamate ... 638

$C_{14}H_9BrClNO_3$
4′-Bromo-2′-chlorophthalanilic acid. 455

$C_{14}H_9BrCl_2N_2O_4$
2-Bromo-4-chloro-5-methylphenyl *N*-(4-chloro-2-nitrophenyl)carbamate ... 628
2-Bromo-4-chloro-6-methylphenyl *N*-(4-chloro-2-nitrophenyl)carbamate ... 629
2-Bromo-4-chloro-6-methylphenyl *N*-(4-chloro-3-nitrophenyl)carbamate ... 626

$C_{14}H_9BrCl_2O$
4-(Bromomethyl)-3,5-dichlorobenzophenone ... 309

$C_{14}H_9BrCl_3NO_2$
2-Bromo-4-chloro-5-methyl *N*-(2,3-dichlorophenyl)carbamate ... 636
2-Bromo-4-chloro-5-methylphenyl *N*-(3,4-dichlorophenyl)carbamate ... 640
2-Bromo-4-chloro-5-methylphenyl *N*-(3,5-dichlorophenyl)carbamate ... 640
2-Bromo-4-chloro-6-methylphenyl *N*-(2,3-dichlorophenyl)carbamate ... 636
2-Bromo-4-chloro-6-methylphenyl *N*-(2,5-dichlorophenyl)carbamate ... 639
2-Bromo-4-chloro-6-methylphenyl *N*-(3,4-dichlorophenyl)carbamate ... 639

$C_{14}H_9BrF_3NO$
3-Bromo-3′-(trifluoromethyl)benzanilide ... 445

$C_{14}H_9BrN_2O$
2-(2-Bromophenyl)-5-phenyl-1,3,4-oxadiazole ... 874

$C_{14}H_9BrN_2O_2$
3-(4-Bromophenyl)-2,4-(1*H*,3*H*)-quinazolinedione ... 695
4-Cyanophenyl *N*-(4-bromophenyl)carbamate ... 760

$C_{14}H_9BrN_2O_3$
9-(Bromoacetyl)-3-nitrocarbazole ... 891

$C_{14}H_9BrN_2O_4$
2′-Bromo-2,4-dinitrostilbene ... 275
3′-Bromo-2,4-dinitrostilbene ... 275
4′-Bromo-2,4-dinitrostilbene ... 275

$C_{14}H_9BrN_2O_5$
2′-Bromo-4′-nitrophthalanilic acid ... 452

$C_{14}H_9BrO_2$
2-(Bromoacetyl)dibenzofuran ... 893

$C_{14}H_9BrO_3$
5-Bromo-1,2,3,4-tetrahydro-1,4-ethenonaphthalene-2,3-dicarboxylic anhydride ... 388
Methyl 8-bromo-4-dibenzofurancarboxylate ... 894

$C_{14}H_9Br_2Cl_3$
Di-(*p*-bromophenyl)trichloroethane... 172

$C_{14}H_9Br_2NO_3$
3′,4′-Dibromophthalanilic acid ... 454

$C_{14}H_9Br_2NO_5$
1,9-Dibromo-2,8-dimethoxy-3-nitrodibenzofuran ... 893

$C_{14}H_9ClFNO_3$
3′-Chloro-4′-fluorophthalanilic acid.. 454

$C_{14}H_9ClFN_3OS$
1-(4-Chloro-2-benzothiazolyl)-3-(2-fluorophenyl)urea ... 911
1-(4-Chloro-2-benzothiazolyl)-3-(3-fluorophenyl)urea ... 912
1-(4-Chloro-2-benzothiazolyl)-3-(4-fluorophenyl)urea ... 911

$C_{14}H_9ClF_2N_2OS$
1-(2-Chlorobenzoyl)-3-(2,4-difluorophenyl)-2-thiourea ... 731

$C_{14}H_9ClF_3IN_2O$
1-(2-Chloro-5-(trifluoromethyl)phenyl)-3-(4-iodophenyl)urea ... 506

$C_{14}H_9ClF_3N$
N-(4-Chlorobenzylidene)-3-(trifluoromethyl)aniline ... 1021

$C_{14}H_9ClF_3NO$
3-Chloro-α,α,α-trifluoro-*o*-benzotoluidide ... 445
4-Chloro-α,α,α-trifluoro-*o*-benzotoluidide ... 445
4-Chloro-α,α,α-trifluoro-*m*-benzotoluidide ... 445
6′-Chloro-α,α,α-trifluoro-*m*-benzotoluidide ... 445
3-Chloro-3′-(trifluoromethyl)benzanilide ... 445
α-(2-Chloro-5-(trifluoromethyl)phenylimino)-*o*-cresol ... 1020

$C_{14}H_9ClF_3NO_2$
4-Chlorophenyl *N*-(3-trifluoromethylphenyl)carbamate.... 611

$C_{14}H_9ClF_3NO_2S$
S-(2-Hydroxyphenyl) *N*-(4-chloro-2-(trifluoromethyl)phenyl)thiocarbamate ... 606

$C_{14}H_9ClF_3N_3O_2S$
1-(2-Chloro-5-(trifluoromethyl)phenyl)-3-(4-nitrophenyl)-2-thiourea 727
$C_{14}H_9ClF_3N_3O_3$
1-(2-Chloro-4-nitrophenyl)-3-(3-trifluoromethylphenyl)urea 506
$C_{14}H_9ClF_4N_2O$
1-(2-Chloro-5-(trifluoromethyl)phenyl)-3-(3-fluorophenyl)urea 506
1-(2-Chloro-5-(trifluoromethyl)phenyl)-3-(4-fluorophenyl)urea 506
$C_{14}H_9ClN_2$
N-(2-Chlorobenzylidene)-4-cyanoaniline 1063
$C_{14}H_9ClN_2O$
2-(2-Chlorobenzoyl)-2-pyridylacetonitrile 928
2-(4-Chlorophenyl)-5-phenyl-1,3,4-oxadiazole 873
$C_{14}H_9ClN_2O_2$
3-(4-Chlorophenyl)-2,4-(1*H*,3*H*)-quinazolinedione 695
4-Cyanophenyl *N*-(2-chlorophenyl)carbamate 760
4-Cyanophenyl *N*-(3-chlorophenyl)carbamate 760
4-Cyanophenyl *N*-(4-chlorophenyl)carbamate 767
$C_{14}H_9ClN_2O_3$
9-(Chloroacetyl)-3-nitrocarbazole 891
$C_{14}H_9ClN_2O_4$
2′-Chloro-2,4-dinitrostilbene 275
3′-Chloro-2,4-dinitrostilbene 275
2-(4-Chloro-3-nitrobenzoyl)benzamide 452
2′-Chloro-5′-nitrophthalanilic acid 473
$C_{14}H_9ClN_2O_5$
2-(4-Chloro-3-nitrobenzamido)benzoic acid 385
4′-Chloro-3′-nitrophthalanilic acid 451
$C_{14}H_9ClN_4O_3$
1-(2-Chloro-4-nitrophenyl)-3-(4-cyanophenyl)urea 759
$C_{14}H_9ClN_4O_3S$
1-(2-Benzothiazolyl)-3-(4-chloro-3-nitrophenyl)urea 911
$C_{14}H_9ClO$
6b,7a-Dihydro-7*H*-cyclopropa(a)acenaphthylene-7-carbonyl chloride 389
$C_{14}H_9ClO_2$
2-(Chloroacetyl)dibenzofuran 892
$C_{14}H_9ClO_3$
2-(2-Chlorobenzoyl)benzoic acid 337
Methyl 2-chloro-4-dibenzofurancarboxylate 894
$C_{14}H_9Cl_2FN_2O$
2,4-Dichlorobenzoic (3-fluorobenzylidene)hydrazide 493
$C_{14}H_9Cl_2FN_2OS$
1-(3,4-Dichlorobenzoyl)-3-(2-fluorophenyl)-2-thiourea 731
$C_{14}H_9Cl_2FN_2O_4$
2,4-Dichloro-6-methylphenyl *N*-(4-fluoro-2-nitrophenyl)carbamate 624
$C_{14}H_9Cl_2F_2N_3O_2$
1-(2,4-Dichlorobenzoyl)-4-(2,4-difluorophenyl)semicarbazide 528
1-(3,4-Dichlorobenzoyl)-4-(2,4-difluorophenyl)semicarbazide 528
$C_{14}H_9Cl_2F_3N_2O$
1-(2-Chlorophenyl)-3-(2-chloro-5-(trifluoromethyl)phenyl)urea 505
1-(3-Chlorophenyl)-3-(2-chloro-5-(trifluoromethyl)phenyl)urea 506
1-(4-Chlorophenyl)-3-(2-chloro-5-(trifluoromethyl)phenyl)urea 506
1-(2,3-Dichlorophenyl)-3-(3-trifluoromethylphenyl)urea 506
1-(2,4-Dichlorophenyl)-3-(3-trifluoromethylphenyl)urea 506
1-(2,5-Dichlorophenyl)-3-(3-trifluoromethylphenyl)urea 506
1-(2,6-Dichlorophenyl)-3-(3-trifluoromethylphenyl)urea 506
1-(3,4-Dichlorophenyl)-3-(3-trifluoromethylphenyl)urea 506
$C_{14}H_9Cl_2F_3O$
3,4-Dichloro-4′-(trifluoromethyl)benzhydrol 197
$C_{14}H_9Cl_2N$
6,9-Dichloro-2-methylacridine 996
$C_{14}H_9Cl_2NO$
5,6-Dichloro-3-phenyl-2-indolinone 692
$C_{14}H_9Cl_2NO_2$
3-(2,4-Dichlorobenzylideneamino)benzoic acid 1026
4-(2,4-Dichlorobenzylideneamino)benzoic acid 1026
4-(2,6-Dichlorobenzylideneamino)benzoic acid 1074
2,3-Dichloro-*N*-piperonylideneaniline 1076
2,4-Dichloro-*N*-piperonylideneaniline 1076
2,5-Dichloro-*N*-piperonylideneaniline 1076
3,4-Dichloro-*N*-piperonylideneaniline 1041
$C_{14}H_9Cl_2NO_3$
2′,3′-Dichlorophthalanilic acid 454
2′,4′-Dichlorophthalanilic acid 461
2′,5′-Dichlorophthalanilic acid 483
3′,4′-Dichlorophthalanilic acid 454
3′,5′-Dichlorophthalanilic acid 454
$C_{14}H_9Cl_2N_3O$
1-(2-Cyanophenyl)-3-(2,3-dichlorophenyl)urea 759
1-(2-Cyanophenyl)-3-(2,4-dichlorophenyl)urea 759
1-(2-Cyanophenyl)-3-(2,5-dichlorophenyl)urea 759
1-(2-Cyanophenyl)-3-(2,6-dichlorophenyl)urea 759
1-(3-Cyanophenyl)-3-(2,3-dichlorophenyl)urea 759
1-(3-Cyanophenyl)-3-(2,4-dichlorophenyl)urea 759
1-(3-Cyanophenyl)-3-(2,5-dichlorophenyl)urea 759
1-(3-Cyanophenyl)-3-(2,6-dichlorophenyl)urea 759
1-(4-Cyanophenyl)-3-(2,3-dichlorophenyl)urea 760
1-(4-Cyanophenyl)-3-(2,4-dichlorophenyl)urea 760
1-(4-Cyanophenyl)-3-(2,5-dichlorophenyl)urea 760
1-(*N*-(2,3-Dichlorophenyl)carbamoyl)benzimidazole 901
$C_{14}H_9Cl_2N_3OS$
1-(2-Benzothiazolyl)-3-(2,3-dichlorophenyl)urea 911
1-(2-Benzothiazolyl)-3-(2,5-dichlorophenyl)urea 912
1-(2-Benzothiazolyl)-3-(2,6-dichlorophenyl)urea 912
1-(2-Benzothiazolyl)-3-(3,4-dichlorophenyl)urea 917
$C_{14}H_9Cl_2N_3O_3S$
1-(3,4-Dichlorobenzoyl)-3-(4-nitrophenyl)-2-thiourea 730
$C_{14}H_9Cl_3$
1,1-Bis(4-chlorophenyl)-2-chloroethene 173
$C_{14}H_9Cl_3FNO_2$
2-Fluorobenzyl *N*-(2,4,5-trichlorophenyl)carbamate 621
4-Fluorobenzyl *N*-(2,4,5-trichlorophenyl)carbamate 622
2′-Fluoro-2-(2,4,5-trichlorophenoxy)acetanilide 400
$C_{14}H_9Cl_3N_2O$
2-Chlorobenzoic (3,4-dichlorobenzylidene)hydrazide 493
$C_{14}H_9Cl_3N_2OS$
1-(2-Chlorobenzoyl)-3-(3,4-dichlorophenyl)-2-thiourea 731
1-(2-Chlorophenyl)-3-(2,4-dichlorobenzoyl)-2-thiourea 731
1-(3-Chlorophenyl)-3-(3,4-dichlorobenzoyl)-2-thiourea 731
$C_{14}H_9Cl_3N_2O_2$
1-(4-Chlorobenzoyl)-2-(3,4-dichlorobenzoyl)hydrazide 493
1-(2-Chlorobenzoyl)-2-(3,4-dichlorobenzoyl)hydrazine 494
1-(4-Chlorobenzoyl)-2-(2,4-dichlorobenzoyl)hydrazine 494
$C_{14}H_9Cl_3N_2O_3$
3-(3-(2,4,5-Trichlorophenyl)ureido)benzoic acid 531
$C_{14}H_9Cl_3N_2O_4$
4′-Chloro-2-(2,4-dichlorophenoxy)-3′-nitroacetanilide 399
N-(3-Nitrophenyl)-2-(2,4,5-trichlorophenoxy)acetamide 401
2,4,5-Trichlorophenyl *N*-(4-methyl-3-nitrophenyl)carbamate 626
$C_{14}H_9Cl_3O_2S$
S-(4-Chlorophenyl) α,2-dichlorobenzyl thiocarbonate 382
$C_{14}H_9Cl_3O_3$
2-Chlorophenyl α,2-dichlorobenzyl carbonate 381
$C_{14}H_9Cl_4NO$
3,4-Dichloro-*N*-(2,4-dichlorobenzyl)benzamide 460
2′-Methyl-2,4,4′,5′-tetrachlorobenzanilide 460
$C_{14}H_9Cl_4NOS$
2-(4-Chlorophenylthio)-2′,4′,5′-trichloroacetanilide 403
$C_{14}H_9Cl_4NO_2$
N-(3-Chlorophenyl)-2,4,5-trichlorophenoxyacetamide 401
4′-Chloro-2-(2,4,5-trichlorophenoxy)acetanilide 400
2,4-Dichloro-6-methylphenyl *N*-(2,3-dichlorophenyl)carbamate 636
2,4-Dichlorophenoxy-*N*-(2,3-dichlorophenyl)acetamide 402
$C_{14}H_9Cl_4NO_3$
2,4,5-Trichlorophenyl *N*-(4-chloro-2-methoxyphenyl)carbamate 627
$C_{14}H_9Cl_4N_3O_2$
1-(2,4-Dichlorobenzoyl)-4-(3,4-dichlorophenyl)semicarbazide 528
$C_{14}H_9FN_2O$
2-(2-Fluorophenyl)-5-phenyl-1,3,4-oxadiazole 873
$C_{14}H_9FN_2O_2$
4-Cyanophenyl *N*-(4-fluorophenyl)carbamate 760
$C_{14}H_9FN_2O_4$
2,4-Dinitro-2′-fluorostilbene 275
2,4-Dinitro-3′-fluorostilbene 275
2,4-Dinitro-4′-fluorostilbene 275
4-Fluoro-*N*-(4,5-(methylenedioxy)-2-nitrobenzylidene)aniline 1024
$C_{14}H_9FN_2O_5$
4′-Fluoro-3′-nitrophthalanilic acid 452
$C_{14}H_9FN_4O_3$
1-(*N*-(4-Fluoro-3-nitrophenyl)carbamoyl)benzimidazole 901
$C_{14}H_9FN_4O_3S$
1-(2-Benzothiazolyl)-3-(2-fluoro-5-nitrophenyl)urea 911
$C_{14}H_9F_2NO_2$
2-Fluorobenzoic imide 456
$C_{14}H_9F_2NO_3$
2′,4′-Difluorophthalanilic acid 454
$C_{14}H_9F_2N_3O$
1-(*N*-(2,5-Difluorophenyl)carbamoyl)benzimidazole 901
$C_{14}H_9F_3N_2O_2$
N-(3-Nitrobenzylidene)-3-(trifluoromethyl)aniline 1021
N-(4-Nitrobenzylidene)-3-(trifluoromethyl)aniline 1021
$C_{14}H_9F_3N_2O_3$
3-Nitro-2′-(trifluoromethyl)benzanilide 446
3-Nitro-3′-(trifluoromethyl)benzanilide 446
$C_{14}H_9F_3N_2O_4$
2,6-Dihydroxy-3-(3-(trifluoromethyl)phenylazo)benzoic acid 1080
$C_{14}H_9F_3N_4O$
1-(*N*-(2-(Trifluoromethyl)phenyl)carbamoyl)-1*H*-benzotriazole 908
1-(*N*-(3-(Trifluoromethyl)phenyl)carbamoyl)-1*H*-benzotriazole 908
$C_{14}H_9F_4N$
N-(4-Fluorobenzylidene)-3-(trifluoromethyl)aniline 1021
$C_{14}H_9F_4NO$
2′-Fluoro-3-(trifluoromethyl)benzanilide 445
2-Fluoro-4′-(trifluoromethyl)benzanilide 445
4-Fluoro-2′-(trifluoromethyl)benzanilide 445
4-Fluoro-3′-(trifluoromethyl)benzanilide 445
4′-Fluoro-3-(trifluoromethyl)benzanilide 445
4-Fluoro-4′-(trifluoromethyl)benzanilide 445
$C_{14}H_9F_4NO_2$
3-(Trifluoromethyl)phenyl *N*-(2-fluorophenyl)carbamate 607
$C_{14}H_9F_4N_3O_2$
4-(2,4-Difluorophenyl)-1-(2,4-difluorobenzoyl)semicarbazide 528
$C_{14}H_9F_4N_3O_3$
1-(2-Fluoro-5-nitrophenyl)-3-(2-(trifluoromethyl)phenyl)urea 504
1-(4-Fluoro-3-nitrophenyl)-3-(3-(trifluoromethyl)phenyl)urea 504
$C_{14}H_9F_5$
2,2-Bis-(4-fluorophenyl)-1,1,1-trifluoroethane 172
$C_{14}H_9F_5O$
4,4′-Difluoro-α-(trifluoromethyl)benzhydrol 197
$C_{14}H_9F_6N$
3,5-Bis(trifluoromethyl)diphenylamine 222
$C_{14}H_9I_3O_3$
2-Phenyl-2-(2,4,6-triiodophenoxy)acetic acid 329
$C_{14}H_9NO$
4-Cyanobenzophenone 766
2-(Cyanomethyl)dibenzofuran 896
$C_{14}H_9NOS$
2-Benzoylbenzothiazole 916
$C_{14}H_9NO_2$
4-Cyanophenyl benzoate 766
2-Formamido-9-fluorenone 488
N-(1-Naphthyl)-maleimide 722
3-Nitrophenanthrene 280
9-Oxo-4-fluorenecarboxamide 488
Phenanthrenequinone 9-oxime 1011
2-Phenyl-4*H*-1,3-benzoxazin-4-one 700
4-Phenyl-1*H*-2,3-benzoxazin-1-one 386
N-Phenylphthalimide 720
$C_{14}H_9NO_3$
4-Furfurylidene-2-phenyl-2-oxazolin-5-one 854
2-(2-Furyl)-4-quinolinecarboxylic acid 991
N-(4-Hydroxyphenyl)phthalimide 716
$C_{14}H_9NO_4$
2-Acetyl-7-nitrodibenzofuran 895
7-Nitrofluorene-1-carboxylic acid 344
$C_{14}H_9NO_4S$
2-Methoxy-7-nitro-9-thioxanthenone 311
$C_{14}H_9NO_5$
1-Amino-4,5,8-trihydroxyanthraquinone 315
Mordant Dye (C.I. 60875) 1093
5-Nitro-1,2,3,4-tetrahydro-1,4-ethenonaphthalene-2,3-dicarboxylic anhydride 388
$C_{14}H_9NO_5S$
1-Aminoanthraquinone-2-sulfonic acid 788
$C_{14}H_9NO_6$
5-Nitro-10-oxo-1,2,3,4-tetrahydro-1,4-ethanonaphthalene-2,3-dicarboxylic anhydride 388
$C_{14}H_9N_3O$
6,12-Dihydrobenzimidazo(2,1-b)quinazolin-12-one 726
$C_{14}H_9N_3O_4$
4-Cyanophenyl *N*-(3-nitrophenyl)carbamate 764
1,4-Diamino-5-nitroanthraquinone 317
6-Nitro-1-phenyl-1*H*-indazole-3-carboxylic acid 900

$C_{14}H_9N_3O_6$
2,2′,4-Trinitrostilbene ... 275
2,3′,4-Trinitrostilbene ... 275
2,4,6-Trinitrostilbene ... 275
$C_{14}H_9N_3O_6S$
10-Acetyl-3,7-dinitrophenothiazine-5-oxide ... 771
$C_{14}H_{10}$
4-Ethynylbiphenyl ... 1019
$C_{14}H_{10}BCl_2F_4N_3O_2$
2-(2,5-Dichlorophenylsulfonylazo)-3-methylbenzothiazolium tetrafluoroborate ... 911
$C_{14}H_{10}BF_4N_3$
4-(2,2-Dicyanovinyl)-1-methylquinolinium tetrafluoroborate ... 990
$C_{14}H_{10}BrClFNO_2$
2-Bromo-4-chloro-6-methylphenyl *N*-(4-fluorophenyl)carbamate ... 635
$C_{14}H_{10}BrClN_2OS$
1-(3-Bromobenzoyl)-3-(3-chlorophenyl)-2-thiourea ... 731
$C_{14}H_{10}BrCl_2NOS$
4′-Bromo-2′-chloro-2-(4-chlorophenylthio)acetanilide ... 403
$C_{14}H_{10}BrCl_2NO_2$
2-Bromo-4-chloro-6-methylphenyl *N*-(2-chlorophenyl)carbamate ... 633
$C_{14}H_{10}BrCl_2NO_3$
2-Bromo-4-chlorophenyl *N*-(4-chloro-2-methoxyphenyl)-carbamate ... 619
$C_{14}H_{10}BrCl_3$
1-(4-Bromophenyl)-1-phenyl-2,2,2-trichloroethane ... 172
$C_{14}H_{10}BrF_3N_2O$
1-(4-Bromophenyl)-3-(3-trifluoromethylphenyl)urea ... 506
$C_{14}H_{10}BrN$
Aminobromophenanthrene (mixed isomers) ... 231
9-Amino-10-bromophenanthrene ... 231
2-Bromo-6-methylphenanthridine nitrate ... 1001
(4-Bromophenyl)phenylacetonitrile ... 751
$C_{14}H_{10}BrNO$
9-Bromo-2-formamidofluorene ... 490
$C_{14}H_{10}BrNO_2$
3-Acetamido-2-bromodibenzofuran ... 895
7-Acetamido-2-bromodibenzofuran ... 895
4-Bromo-*N*-piperonylideneaniline ... 1076
$C_{14}H_{10}BrNO_3$
3′-Bromophthalanilic acid ... 453
4′-Bromophthalanilic acid ... 456
$C_{14}H_{10}BrNO_3S$
4′-Bromo-2-(4-nitrophenylthio)-acetophenone ... 293
$C_{14}H_{10}BrNO_4$
2-Anilino-5-bromoterephthalic acid ... 337
$C_{14}H_{10}BrNO_4S$
Methyl 2-(4-bromo-2-nitrophenylthio)benzoate ... 372
$C_{14}H_{10}Br_2$
4,5-Dibromo-9,10-dihydrophenanthrene ... 175
1,1-Dibromo-2,2-diphenylethylene ... 174
α,β-Dibromostilbene ... 171
$C_{14}H_{10}Br_2ClNO_2$
2,4-Dibromophenyl *N*-(2-chloro-4-methylphenyl)carbamate ... 638
$C_{14}H_{10}Br_2ClNO_3$
2,4-Dibromophenyl *N*-(2-chloro-4-methoxyphenyl)carbamate ... 621
$C_{14}H_{10}Br_2Cl_2$
1,2-Bis(4-bromophenyl)-1,2-dichloroethane ... 173
$C_{14}H_{10}Br_2N_2O_2$
5-Bromosalicylaldehyde azine ... 1029
$C_{14}H_{10}Br_2N_2O_4$
2,2′-Dibromo-4,4′-dimethyl-6,6′-dinitrobiphenyl ... 261
$C_{14}H_{10}Br_2N_2O_4S$
2,6-Dibromo-4-nitrophenyl *N*-(4-(methylthio)phenyl)carbamate ... 623
$C_{14}H_{10}Br_2N_2O_5$
2,4-Dibromophenyl *N*-(2-methoxy-4-nitrophenyl)carbamate ... 618
$C_{14}H_{10}Br_2O_3$
1,9-Dibromo-2,8-dimethoxydibenzofuran ... 893
$C_{14}H_{10}Br_3NO$
2-Methyl-2′,4′,6′-tribromobenzanilide ... 458
3-Methyl-2′,4′,6′-tribromobenzanilide ... 460
4-Methyl-2′,4′,6′-tribromobenzanilide ... 480
$C_{14}H_{10}Br_3NO_2$
2-Methoxy-2′,4′,6′-tribromobenzanilide ... 448
4-Methoxy-2′,4′,6′-tribromobenzanilide ... 450
$C_{14}H_{10}Br_4$
2,2′-Bis(bromomethyl)-4,4′-dibromobiphenyl ... 171
1-(2-Bromophenyl)-2-(4-bromophenyl)-1,2-dibromoethane ... 172
$C_{14}H_{10}ClFN_2OS$
1-(2-Chlorobenzoyl)-3-(2-fluorophenyl)-2-thiourea ... 731
1-(3-Chlorophenyl)-3-(2-fluorobenzoyl)-2-thiourea ... 731
$C_{14}H_{10}ClFN_2O_2$
1-(4-Chlorobenzoyl)-2-(2-fluorobenzoyl)hydrazine ... 493
$C_{14}H_{10}ClFN_2O_3$
4-(3-(3-Chloro-4-fluorophenyl)-ureido)benzoic acid ... 531
$C_{14}H_{10}ClFN_2O_4$
4-Chloro-2-methylphenyl *N*-(4-fluoro-3-nitrophenyl)carbamate ... 625
2-Fluorobenzyl *N*-(2-chloro-4-nitrophenyl)carbamate ... 621
3-Fluorobenzyl *N*-(2-chloro-4-nitrophenyl)carbamate ... 621
4-Fluorobenzyl *N*-(2-chloro-4-nitrophenyl)carbamate ... 626
$C_{14}H_{10}ClFN_2O_5$
2-Chloro-4-fluorophenyl *N*-(2-methoxy-4-nitrophenyl)carbamate ... 618
2-Chloro-4-fluorophenyl *N*-(2-methoxy-5-nitrophenyl)carbamate ... 618
2-Chloro-4-methoxyphenyl *N*-(4-fluoro-3-nitrophenyl)carbamate ... 617
$C_{14}H_{10}ClF_2NO_3$
2-Chloro-4-methoxyphenyl *N*-(2,4-difluorophenyl)carbamate ... 617
$C_{14}H_{10}ClF_2N_3O_2$
1-(4-Chlorobenzoyl)-4-(2,4-difluorophenyl)semicarbazide ... 532
$C_{14}H_{10}ClF_3N_2O$
1-(3-Chlorophenyl)-3-(2-(trifluoromethyl)phenyl)urea ... 505
1-(3-Chlorophenyl)-3-(3-trifluoromethylphenyl)urea ... 510
1-(4-Chlorophenyl)-3-(2-(trifluoromethyl)phenyl)urea ... 505
1-(4-Chlorophenyl)-3-(3-trifluoromethylphenyl)urea ... 506
1-(2-Chlorophenyl)-3-(α,α,α-trifluoro-*m*-tolyl)urea ... 505
$C_{14}H_{10}ClF_3N_2O_2$
1-(2-Chloro-5-(trifluoromethyl)-phenyl)-3-(4-hydroxyphenyl)urea ... 504
$C_{14}H_{10}ClN$
Aminochlorophenanthrene(mixed isomers) ... 231
(4-Chlorophenyl)phenylacetonitrile ... 751
$C_{14}H_{10}ClNO$
9-(Chloroacetyl)carbazole ... 891
5-Chloro-6-methyl-3-phenyl-2,1-benzisoxazole ... 909
1-(4-Chlorophenyl)-3-oxoisoindoline ... 691
$C_{14}H_{10}ClNO_2$
6-Chloro-1,2-dihydro-2-phenyl-4*H*-3,1-benzoxazin-4-one ... 385
2-Chloro-4-hydroxy-4-phenyl-4*H*-1,3-benzoxazine ... 1002
1-(4-Chlorophenyl)-1-hydroxy-3-oxoisoindoline ... 690
2-Chloro-*N*-piperonylideneaniline ... 1076
3-Chloro-*N*-piperonylideneaniline ... 1041
$C_{14}H_{10}ClNO_3$
2-(2-Chlorobenzamido)benzoic acid ... 387
4-(4-Chlorobenzylideneamino)-salicylic acid ... 1022
N-(3-Chlorophenyl)phthalamic acid ... 453
2′-Chlorophthalanilic acid ... 453
4′-Chlorophthalanilic acid ... 454
$C_{14}H_{10}ClN_3O$
1-(2-Chlorophenyl)-3-(2-cyanophenyl)urea ... 759
1-(2-Chlorophenyl)-3-(3-cyanophenyl)urea ... 759
1-(3-Chlorophenyl)-3-(2-cyanophenyl)urea ... 760
$C_{14}H_{10}ClN_3OS$
1-(2-Benzothiazolyl)-3-(4-chlorophenyl)urea ... 912
$C_{14}H_{10}ClN_3O_2$
4-Quinazolinyl *N*-(3-chlorophenyl)-carbamate ... 998
$C_{14}H_{10}ClN_3O_3S$
1-Benzoyl-3-(4-chloro-2-nitrophenyl)-2-thiourea ... 729
1-(2-Chlorobenzoyl)-3-(4-nitrophenyl)-2-thiourea ... 730
$C_{14}H_{10}ClN_3O_5$
3′-Chloro-3,5-dinitro-4′-methylbenzanilide ... 451
4-Chloro-2′,6′-dinitro-4′-methylbenzanilide ... 450
$C_{14}H_{10}ClN_3O_6$
2-(3-Chloro-4-nitrophenyliminomethyl)-6-methoxy-4-nitrophenol ... 1021
2-Nitrobenzyl *N*-(2-chloro-4-nitrophenyl)carbamate ... 621
$C_{14}H_{10}ClN_5O_3$
5-Chloro-1-(*N*-(4-methyl-2-nitrophenyl)carbamoyl)-1*H*-benzotriazole ... 908
$C_{14}H_{10}Cl_2$
1,1-Bis(4-chlorophenyl)ethylene ... 173
2,3-Dichloro-9,10-dihydroanthracene ... 175
1,1-Dichloro-2,2-diphenylethylene ... 174
α,α′-Dichlorostilbene ... 171
$C_{14}H_{10}Cl_2FNO_2$
2-Chloro-4-fluorophenyl *N*-(4-chloro-2-methylphenyl)carbamate ... 638
2-Fluorobenzyl *N*-(3,4-dichlorophenyl)carbamate ... 622
3-Fluorobenzyl *N*-(3,4-dichlorophenyl)carbamate ... 622
4-Fluorobenzyl *N*-(2,3-dichlorophenyl)carbamate ... 622
4-Fluorobenzyl *N*-(2,5-dichlorophenyl)carbamate ... 622
4-Fluorobenzyl *N*-(3,4-dichlorophenyl)carbamate ... 622
$C_{14}H_{10}Cl_2FNO_3$
2-Chloro-4-fluorophenyl *N*-(2-chloro-4-methoxyphenyl)carbamate ... 621
$C_{14}H_{10}Cl_2F_2$
1,1-Bis-(4-fluorophenyl)-2,2-dichloroethane ... 172
$C_{14}H_{10}Cl_2N_2$
Benzoyl chloride azine ... 770
2-Chlorobenzaldehyde azine ... 1077
4-Chlorobenzaldehyde azine ... 1077
5,6-Dichloro-2-methyl-1-phenylbenzimidazole ... 906
$C_{14}H_{10}Cl_2N_2O$
2-(2,6-Dichlorophenyl)-1,2-dihydro-4(3*H*)-quinazolinone ... 694
2-(3,4-Dichlorophenyl)-1,2-dihydro-4(3*H*)-quinazolinone ... 695
$C_{14}H_{10}Cl_2N_2OS$
1-Benzoyl-3-(2,6-dichlorophenyl)-2-thiourea ... 731
1-(2-Chlorobenzoyl)-3-(3-chlorophenyl)-2-thiourea ... 731
$C_{14}H_{10}Cl_2N_2O_2$
1-Benzoyl-2-(2,4-dichlorobenzoyl)-hydrazine ... 500
1,2-Bis(4-chlorobenzoyl)-hydrazine ... 500
1-(2-Chlorobenzoyl)-2-(4-chlorobenzoyl)hydrazine ... 494
1-(4-Chlorobenzoyl)-3-(3-chlorophenyl)urea ... 532
3-Chloro-*N*-(2-chloro-5-nitrobenzylidene)-4-methylaniline ... 1058
5-Chloro-*N*-(2-chloro-5-nitrobenzylidene)-2-methylaniline ... 1058
N-(2,4-Dichlorobenzylidene)-2-methyl-4-nitroaniline ... 1025
N-(2,4-Dichlorobenzylidene)-3-nitro-*p*-toluidine ... 1024
3-Hydroxybenzoic acid (2,4-dichlorobenzylidene)hydrazide ... 496
$C_{14}H_{10}Cl_2N_2O_3$
1-(2-Carboxyphenyl)-3-(2,3-dichlorophenyl)urea ... 531
1-(2-Carboxyphenyl)-3-(2,4-dichlorophenyl)urea ... 531
1-(2-Carboxyphenyl)-3-(2,5-dichlorophenyl)urea ... 531
1-(2-Carboxyphenyl)-3-(3,4-dichlorophenyl)urea ... 531
1-(3-Carboxyphenyl)-3-(3,4-dichlorophenyl)urea ... 531
1-(4-Carboxyphenyl)-3-(2,3-dichlorophenyl)urea ... 531
1-(4-Carboxyphenyl)-3-(2,4-dichlorophenyl)urea ... 531
1-(4-Carboxyphenyl)-3-(2,5-dichlorophenyl)urea ... 531
1-(4-Carboxyphenyl)-3-(3,4-dichlorophenyl)urea ... 531
N-(2,4-Dichlorobenzylidene)-2-methoxy-5-nitroaniline ... 1023
2,5-Dichloro-2′-methyl-6′-nitrobenzanilide ... 450
3,4-Dichloro-2′-methyl-6′-nitrobenzanilide ... 450
4,4′-Dichloro-3-methyl-3′-nitrobenzanilide ... 451
4′,5′-Dichloro-2′-methyl-3-nitrobenzanilide ... 453
5-(2,5-Dichlorophenylazo)-4-methylsalicylic acid ... 1083
3-(3-(2,3-Dichlorophenyl)ureido)-benzoic acid ... 531
$C_{14}H_{10}Cl_2N_2O_3S$
2-(4-Chlorophenylthio)-4′-chloro-3′-nitroacetanilide ... 402
$C_{14}H_{10}Cl_2N_2O_4$
4-Chloro-2-methylphenyl *N*-(4-chloro-2-nitrophenyl)carbamate ... 624
2-Chloro-4-nitrophenyl *N*-(3-chloro-2-methylphenyl)carbamate ... 628
3,4-Dichloro-2′-methoxy-5′-nitrobenzanilide ... 448
2,4-Dichloro-3′-nitro-*o*-benzanisidide ... 448
2,4-Dichlorophenyl *N*-(2-methyl-4-nitrophenyl)carbamate ... 626
2,4-Dichlorophenyl *N*-(4-methyl-2-nitrophenyl)carbamate ... 624
3,4-Dichlorophenyl *N*-(2-methyl-6-nitrophenyl)carbamate ... 624
2-Nitrobenzyl *N*-(3,4-dichlorophenyl)carbamate ... 621
$C_{14}H_{10}Cl_2N_4O$
5-Chloro-1-(*N*-(3-chloro-4-methylphenyl)carbamoyl)-1*H*-benzotriazole ... 908
1-(2,3-Dichlorophenyl)-3-(1*H*-indazol-5-yl)urea ... 899
$C_{14}H_{10}Cl_2N_4O_2$
5-Chloro-1-(*N*-(5-chloro-2-methoxyphenyl)carbamoyl)-1*H*-benzotriazole ... 908
$C_{14}H_{10}Cl_2N_4O_4$
1-(2-Chlorobenzoyl)-4-(2-chloro-4-nitrophenyl)semicarbazide ... 526
$C_{14}H_{10}Cl_2O$
Bis-(2-chlorophenyl)-acetaldehyde ... 321
4′-Chloro-2-(4-chlorophenyl)-acetophenone ... 289
$C_{14}H_{10}Cl_2OS$
4′-Chloro-2-(4-chlorophenylthio)-acetophenone ... 288
$C_{14}H_{10}Cl_2O_3$
2,2′-Dichlorobenzilic acid ... 325
$C_{14}H_{10}Cl_2O_4S_2$
1,2-Bis(phenylsulfonyl)-1,2-dichloroethylene ... 784

$C_{14}H_{10}Cl_3N$
2-Chloro-*N*-(3,4-dichlorobenzylidene)-6-methylaniline ... 1067
3-Chloro-*N*-(2,4-dichlorobenzylidene)-4-methylaniline ... 1067
5-Chloro-*N*-(3,4-dichlorobenzylidene)-2-methylaniline ... 1067

$C_{14}H_{10}Cl_3NO$
4-Chloro-*N*-(2,4-dichlorobenzyl)-benzamide ... 452
2′-Methyl-2,4,4′-trichlorobenzanilide ... 460
2′-Methyl-2,4′,5′-trichlorobenzanilide ... 458
2-Methyl-2′,4′,6′-trichlorobenzanilide ... 458
3-Methyl-2′,4′,6′-trichlorobenzanilide ... 460
4′-Methyl-2,3′,4-trichlorobenzanilide ... 460
5′-Methyl-2,2′,4-trichlorobenzanilide ... 460
5′-Methyl-2′,3,4-trichlorobenzanilide ... 460
2,2′,6-Trichloro-*o*-benzotoluidide ... 459
2,3′,4-Trichloro-*o*-benzotoluidide ... 483
2,4,5′-Trichloro-*o*-benzotoluidide ... 460
2,4,6′-Trichloro-*o*-benzotoluidide ... 459
3,3′,4-Trichloro-*o*-benzotoluidide ... 481
3,4,6′-Trichloro-*o*-benzotoluidide ... 459
3,4,6′-Trichloro-*m*-benzotoluidide ... 460

$C_{14}H_{10}Cl_3NOS$
2-(4-Chlorophenylthio)-2′,5′-dichloroacetanilide ... 403
2-(4-Chlorophenylthio)-3′,4′-dichloroacetanilide ... 403

$C_{14}H_{10}Cl_3NO_2$
2-(2-Chlorophenoxy)-2′,5′-dichloroacetanilide ... 400
2-(4-Chlorophenoxy)-2′,5′-dichloroacetanilide ... 399
4′-(4-Chlorophenoxy)-2,3′-dichloroacetanilide ... 483
2,4-Dichlorophenyl *N*-(3-chloro-4-methylphenyl)carbamate ... 638
2,5-Dichlorophenyl *N*-(3-chloro-2-methylphenyl)carbamate ... 636
3,4-Dichlorophenyl *N*-(4-chloro-2-methylphenyl)carbamate ... 638
3,5-Dichlorophenyl *N*-(3-chloro-2-methylphenyl)carbamate ... 636
N-(2-Hydroxy-3-methoxybenzylidene)-2,4,5-trichloroaniline ... 1022
2′-Methoxy-2,4,5′-trichlorobenzanilide ... 449
4′-Methoxy-2,2′,4-trichlorobenzanilide ... 450
4-Methoxy-2′,4′,6′-trichlorobenzanilide ... 450

$C_{14}H_{10}Cl_3NO_2S$
2,4,5-Trichlorophenyl *N*-(4-(methylthio)phenyl)carbamate ... 623

$C_{14}H_{10}Cl_3NO_3$
2,4-Dichlorophenyl *N*-(2-chloro-4-methoxyphenyl)carbamate ... 621
2,5-Dichlorophenyl *N*-(5-chloro-2-methoxyphenyl)carbamate ... 619
3,4-Dichlorophenyl *N*-(5-chloro-2-methoxyphenyl)carbamate ... 619

$C_{14}H_{10}Cl_3N_3O_2$
4-(3-Chlorophenyl)-1-(2,4-dichlorobenzoyl)semicarbazide ... 532

$C_{14}H_{10}Cl_3N_3O_3$
1-(4-Hydroxybenzoyl)-4-(2,4,5-trichlorophenyl)semicarbazide ... 548

$C_{14}H_{10}Cl_3N_5O_4$
4-(2,6-Dichloro-4-nitrophenylazo)-3,5-dimethoxybenzenediazonium chloride, zinc chloride double salt ... 1094
Fast Dark Blue R salt (C.I. 37195) ... 1094

$C_{14}H_{10}Cl_4$
1,2-Bis(4-chlorophenyl)-1,2-dichloroethane ... 173
1,2-Diphenyl-1,1,2,2-tetrachloroethane ... 173

$C_{14}H_{10}Cl_4N_2O$
1-(2-Chloro-6-methylphenyl)-3-(2,4,6-trichlorophenyl)urea ... 543

$C_{14}H_{10}Cl_4N_4O_2$
1,6-Bis(2,5-dichlorophenyl)biurea ... 529
1,6-Bis(3,4-dichlorophenyl)biurea ... 529

$C_{14}H_{10}Cl_4O$
2,6-Dichlorobenzyl ether ... 188

$C_{14}H_{10}Cl_4O_2$
1,2-Bis(2,4-dichlorophenoxy)ethane ... 185

$C_{14}H_{10}FN$
2-(4-Fluorophenyl)-2-phenylacetonitrile ... 756

$C_{14}H_{10}FNO_3$
4-(3-Fluorobenzamido)benzoic acid ... 456
2′-Fluorophthalanilic acid ... 476
3′-Fluorophthalanilic acid ... 453
4′-Fluorophthalanilic acid ... 454

$C_{14}H_{10}FN_3O$
1-(*N*-(3-Fluorophenyl)carbamoyl)-benzimidazole ... 901

$C_{14}H_{10}FN_3OS$
1-(2-Benzothiazolyl)-3-(2-fluorophenyl)urea ... 911

$C_{14}H_{10}FN_3O_3S$
1-(3-Fluorophenyl)-3-(3-nitrobenzoyl)-2-thiourea ... 731
1-(4-Fluorophenyl)-3-(3-nitrobenzoyl)-2-thiourea ... 731

$C_{14}H_{10}FN_3O_5$
2-Carbamoylphenyl *N*-(2-fluoro-5-nitrophenyl)carbamate ... 625
2-Carbamoylphenyl *N*-(4-fluoro-3-nitrophenyl)carbamate ... 622
4-(3-(2-Fluoro-5-nitrophenyl)ureido)-benzoic acid ... 524

$C_{14}H_{10}FN_3O_6$
2-Nitrobenzyl *N*-(4-fluoro-2-nitrophenyl)carbamate ... 621

$C_{14}H_{10}FN_5O_3$
1-(4-Fluoro-3-nitrophenyl)-3-(1*H*-indazol-5-yl)urea ... 899
1-(4-Fluoro-3-nitrophenyl)-3-(1*H*-indazol-6-yl)urea ... 899

$C_{14}H_{10}F_2N_2OS$
1-(2-Fluorobenzoyl)-3-(4-fluorophenyl)-2-thiourea ... 731

$C_{14}H_{10}F_2N_2O_2$
1,2-Bis(2-fluorobenzoyl)hydrazine ... 501
N,N′-Bis(4-fluorophenyl)oxamide ... 403

$C_{14}H_{10}F_2N_2O_3$
3-(2,5-Difluorophenylureido)-benzoic acid ... 531

$C_{14}H_{10}F_2N_2O_4$
2-Nitrobenzyl *N*-(2,5-difluorophenyl)carbamate ... 621
4-Nitrobenzyl *N*-(2,4-difluorophenyl)carbamate ... 621
4-Nitrobenzyl *N*-(2,5-difluorophenyl)carbamate ... 621

$C_{14}H_{10}F_2N_4O$
1-(2-Benzimidazolyl)-3-(2,5-difluorophenyl)urea ... 901

$C_{14}H_{10}F_2N_4O_3$
4-Acetamido-2,6-difluoro-4′-nitroazobenzene ... 1081

$C_{14}H_{10}F_2N_4O_4$
4-(2,4-Difluorophenyl)-1-(3-nitrobenzoyl)semicarbazide ... 528

$C_{14}H_{10}F_2O_3$
4,4′-Difluorobenzilic acid hemihydrate ... 326

$C_{14}H_{10}F_3NO$
α,α,α-Trifluoro-*o*-benzotoluidide ... 446
2-(Trifluoromethyl)benzanilide ... 446
3′-(Trifluoromethyl)benzanilide ... 446
α-(2-(Trifluoromethyl)phenylimino)-*o*-cresol ... 1020
α-(3-(Trifluoromethyl)phenylimino)-*o*-cresol ... 1020

$C_{14}H_{10}F_3NO_2$
2-Fluorobenzyl *N*-(2,5-difluorophenyl)carbamate ... 622
3-Fluorobenzyl *N*-(2,5-difluorophenyl)carbamate ... 622
Phenyl *N*-(3-(trifluoromethyl)-phenyl)carbamate ... 610

$C_{14}H_{10}F_3NO_2S$
S-(2-Hydroxyphenyl) *N*-(2-(trifluoromethyl)phenyl)-thiocarbamate ... 728

$C_{14}H_{10}F_3N_3O_2S$
1-(4-Nitrophenyl)-3-(3-(trifluoromethyl)phenyl)-2-thiourea ... 728

$C_{14}H_{10}F_3N_3O_3$
1-(3-Nitrophenyl)-3-(2-(trifluoromethyl)phenyl)urea ... 508

$C_{14}H_{10}F_3N_5O_5$
1-(2,4-Dinitrophenyl)-4-(3-(trifluoromethyl)phenyl)-semicarbazide ... 508

$C_{14}H_{10}F_4N_2O$
1-(2-Fluorophenyl)-3-(3-(trifluoromethyl)phenyl)urea ... 505
1-(4-Fluorophenyl)-3-(2-(trifluoromethyl)phenyl)urea ... 505
1-(4-Fluorophenyl)-3-(3-(trifluoromethyl)phenyl)urea ... 506

$C_{14}H_{10}F_4O$
4-Fluoro-4′-(trifluoromethyl)-benzhydrol ... 197

$C_{14}H_{10}HgO_3$
(4-Dibenzofurylmercuri) acetate ... 1139

$C_{14}H_{10}INO$
9-(2-Iodoacetyl)carbazole ... 891

$C_{14}H_{10}INO_3$
N-(4-Iodophenyl)phthalamic acid ... 454

$C_{14}H_{10}N_2$
4-Phenylquinazoline ... 1001

$C_{14}H_{10}N_2O$
2-Benzoylbenzimidazole ... 905
2-Cyano-3-(1-naphthyl)acrylamide ... 764
α-(4-Cyanophenylimino)-*o*-cresol ... 1034
4-(Hydroxyimino)-α-phenyl-2,5-cyclohexadiene-δ1,α-acetonitrile ... 1010
2-Phenyl-4(3*H*)-quinazolinone ... 1000
3-Phenyl-4(3*H*)-quinazolinone ... 710

$C_{14}H_{10}N_2OS$
α-(2-Benzothiazolylimino)-*o*-cresol ... 1035
1,2-Dihydro-3-phenyl-2-thioxo-4(3*H*)-quinazolinone ... 746
2-(2-Hydroxy-1-naphthylmethyleneamino)thiazole ... 1038

$C_{14}H_{10}N_2O_2$
10-Amino-5*H*-dibenz(b,e)azepine-6,11-dione ... 701
9-Amino-10-nitroanthracene ... 271
N-Anilinophthalimide ... 721
α-Cyano-2-hydroxy-1-naphthaleneacrylamide ... 761
4-Cyanophenyl *N*-phenylcarbamate ... 767
3,4-Dihydro-3-phenyl-2*H*-pyrido(1,2-a)pyrimidine-2,4-dione ... 700
1,3-Diphenyluretidine-2,4-dione ... 709
Phenanthrenequinone dioxime ... 1010
2-Phenylimidazo(1,2-a)pyridine-6-carboxylic acid ... 919
3-Phenyl-2,4(1*H*,3*H*)-quinazolinedione ... 707

$C_{14}H_{10}N_2O_3$
2-Formamido-7-nitrofluorene ... 488
2-(3-Nitrocinnamoyl)pyridine ... 960
3-(Phenylcarbamoyl)-2(3*H*)-benzoxazolone ... 709

$C_{14}H_{10}N_2O_4$
3-Acetamido-2-nitrodibenzofuran ... 894
1,8-Diamino-4,5-dihydroxyanthraquinone ... 316
4,4′-Dinitrostilbene ... 279
Diphenyl azodicarboxylate ... 677
2-Nitro-*N*-piperonylideneaniline ... 1041
3-Nitro-*N*-piperonylideneaniline ... 1056
4-Nitro-*N*-piperonylideneaniline ... 1059

$C_{14}H_{10}N_2O_5$
4,4′-Dicarboxyazoxybenzene ... 1092
2-(2,4-Dinitrostyryl)phenol ... 264
4-Formylphenyl *N*-(4-nitrophenyl)-carbamate ... 670
2-(4,5-(Methylenedioxy)-2-nitrobenzylideneamino)phenol ... 264
N-(4-Nitrobenzoyl)anthranilic acid ... 473
2′-Nitrophthalanilic acid ... 469
3′-Nitrophthalanilic acid ... 469
4′-Nitrophthalanilic acid ... 470
N-(2-Phthalimidoacetyl)-succinimide ... 739

$C_{14}H_{10}N_2O_6$
4-Nitrobenzyl 4-nitrobenzoate ... 370

$C_{14}H_{10}N_2O_{10}S_2$
4,4′-Dinitrostilbene-2,2′-disulfonic acid, disodium salt ... 794

$C_{14}H_{10}N_4$
5,12-Dihydroquinoxalino(2,3-b)-quinoxaline ... 1004
5-Phenyl-3-(2-pyridyl)-1,2,4-triazine ... 980

$C_{14}H_{10}N_4O_2S_2$
N,N′-Bis(2-thiazolyl)isophthalamide ... 877

$C_{14}H_{10}N_4O_3$
Isatin 3-(4-nitrophenylhydrazone) ... 1069

$C_{14}H_{10}N_4O_4$
3-Nitrobenzaldehyde azine ... 1059
4-Nitrobenzaldehyde azine ... 1073

$C_{14}H_{10}N_4O_6$
N,N′-Bis(2-nitrophenyl)oxamide ... 424

$C_{14}H_{10}N_4O_7$
N-(4-Acetylphenyl)-2,4,6-trinitroaniline ... 294
3-Hydroxybenzoic acid (3,5-dinitro-4-hydroxybenzylidene)hydrazide ... 494
3′-(2,4,6-Trinitroanilino)-acetophenone ... 294

$C_{14}H_{10}N_4O_8$
Methyl *N*-(2,4,6-trinitrophenyl)-anthranilate ... 372

$C_{14}H_{10}N_4S_2$
N,N′-(1,4-Phenylenedimethylidyne)-bis(2-aminothiazole) ... 1075

$C_{14}H_{10}O$
7-Methyl-1*H*-phenalen-1-one ... 305

$C_{14}H_{10}OS$
2-Acetyldibenzothiophene ... 898
Methylthioxanthen-9-one, mixture of 2- and 4- isomers ... 314

$C_{14}H_{10}OSe$
3-Methyl-9*H*-selenoxanthen-9-one ... 213

$C_{14}H_{10}O_2$
2-Acetyldibenzofuran ... 895
Dibenz(c,e)oxepin-5(7*H*)-one ... 386
1,4-Dihydroanthraquinone ... 319
1,9-Dihydroxyphenanthrene ... 194
4,5-Dihydroxyphenanthrene ... 195
4-Methoxy-9-fluorenone ... 311
3-Methoxy-1*H*-phenalene-1-one ... 311
3-Phenylphthalide ... 387

$C_{14}H_{10}O_2S$
Benzoyl sulfide ... 388
Methyl 4-dibenzothiophene-carboxylate ... 898

$C_{14}H_{10}O_2S_2$
Benzoyl disulfide ... 372

$C_{14}H_{10}O_3$
4-Dibenzofuranacetic acid ... 894
4-Dibenzofuryl acetate ... 894
Disalicylaldehyde ... 185
2-Ethylnaphtho(1,2-b)furan-4,5-dione ... 320
6-Hydroxy-9-methyl-3*H*-xanthen-3-one ... 319
Methyl 1-dibenzofurancarboxylate ... 894
Methyl 2-dibenzofurancarboxylate ... 895
Methyl 4-dibenzofurancarboxylate ... 894

$C_{14}H_{10}O_3S$
1-Anthracenesulfonic acid, sodium salt ... 796
2-Anthracenesulfonic acid, sodium salt ... 796
3-(3,4-Methylenedioxyphenyl)-1-(2-thienyl)-2-propen-1-one ... 857
3-Phenanthrenesulfonic acid, potassium salt ... 796

$C_{14}H_{10}O_4$
4-Carboxyphenyl benzoate ... 380
2-(4-Hydroxybenzoyl)-benzoic acid ... 339
2-Methoxy-1-dibenzofurancarboxylic acid ... 893
4-Phenylphthalic acid ... 345

$C_{14}H_{10}O_4S$
3-Benzoyl-3*H*-1,2-benzoxathiole 2,2-dioxide ... 799
4-(2-Hydroxyethylthio)-1,8-naphthalic anhydride ... 388

$C_{14}H_{10}O_5$
2-Carboxyphenyl salicylate ... 377
9-Methyl-2,3,7-trihydroxy-6-fluorone ... 290
2,2′-Oxydibenzoic acid ... 340

$C_{14}H_{10}O_6$
4,4′-(1,4-Phenylene)bis(2-oxo-3-butenoic acid) ... 344

$C_{14}H_{10}O_6S$
5,5′-Thiodisalicylic acid ... 338

$C_{14}H_{10}S$
3-Mercaptophenanthrene ... 209

$C_{14}H_{10}S_2$
1,2-Bis(phenylthio)ethyne ... 1017

$C_{14}H_{11}Br$
2-Bromo-1,1-diphenylethylene ... 174
1-Bromomethylfluorene ... 172
4-Bromostilbene ... 172

$C_{14}H_{11}BrClN$
3-Bromo-6-chloro-9-ethylcarbazole ... 892

$C_{14}H_{11}BrClNO$
2-Bromo-2′-chloro-6′-methylbenzanilide ... 457
3-Bromo-2′-chloro-6′-methylbenzanilide ... 459
4-Bromo-2′-chloro-6′-methylbenzanilide ... 458

$C_{14}H_{11}BrClNOS$
4′-Bromo-2-(4-chlorophenylthio)-acetanilide ... 403

$C_{14}H_{11}BrClNO_2$
4-Bromo-2-chloro-*N*-(2-hydroxy-3-methoxybenzylidene)aniline ... 1022

$C_{14}H_{11}BrN_2$
2-(4-Bromophenyl)-1-methylbenzimidazole ... 901

$C_{14}H_{11}BrN_2O$
2-Bromo-4′-(phenylazo)-acetophenone ... 1079

$C_{14}H_{11}BrN_2O_2$
1-Benzoyl-2-(3-bromobenzoyl)-hydrazine ... 500
Ethyl 5-bromo-α-cyano-3-indoleacrylate ... 889

$C_{14}H_{11}BrN_2O_2S$
3′-Bromo-*N*-cyano-*p*-toluenesulfonanilide ... 809
4′-Bromo-*N*-cyano-*p*-toluenesulfonanilide ... 809

$C_{14}H_{11}BrN_2O_3$
3-Bromo-2′-methyl-6′-nitrobenzanilide ... 450

$C_{14}H_{11}BrN_2O_3S$
2′-(4-Bromo-2-nitrophenylthio)-acetanilide ... 421

$C_{14}H_{11}BrN_2O_4$
4-Bromophenyl *N*-(4-methyl-2-nitrophenyl)carbamate ... 624

$C_{14}H_{11}BrO$
2-(2-Bromoethyl)dibenzofuran ... 893
4-(Bromomethyl)benzophenone ... 309

$C_{14}H_{11}BrO_3$
1-Bromo-3,4-dimethoxydibenzofuran ... 893

$C_{14}H_{11}Br_2Cl$
1,1-Bis(4-bromophenyl)-2-chloroethane ... 171

$C_{14}H_{11}Br_2ClN_2O$
1-(3-Chloro-4-methylphenyl)-3-(2,5-dibromophenyl)urea ... 544

$C_{14}H_{11}Br_2ClN_2O_2$
1-(5-Chloro-2-methoxyphenyl)-3-(2,5-dibromophenyl)urea ... 520
1-(5-Chloro-2-methoxyphenyl)-3-(3,4-dibromophenyl)urea ... 520

$C_{14}H_{11}Br_2ClN_2O_3S$
4-Acetamido-*N*-(4-chloro-2,6-dibromophenyl)-benzenesulfonamide ... 808

$C_{14}H_{11}Br_2FN_2O_3S$
4-Acetamido-*N*-(2,6-dibromo-4-fluorophenyl)benzenesulfonamide ... 808

$C_{14}H_{11}Br_2N$
2,8-Dibromo-9-ethylcarbazole ... 892

$C_{14}H_{11}Br_2NO_3$
2,3-Dibromo-*N*-(1-naphthyl)-succinamic acid ... 433
2,4-Dibromophenyl *N*-(2-methoxyphenyl)carbamate ... 627

$C_{14}H_{11}Br_2N_3O_3$
1-(2,5-Dibromophenyl)-3-(4-methyl-2-nitrophenyl)urea ... 523

$C_{14}H_{11}ClFN$
3-Chloro-*N*-(4-fluorobenzylidene)-*o*-toluidine ... 1078
3-Chloro-*N*-(4-fluorobenzylidene)-*p*-toluidine ... 1078
4-Chloro-*N*-(4-fluorobenzylidene)-*o*-toluidine ... 1078
5-Chloro-*N*-(4-fluorobenzylidene)-*o*-toluidine ... 1078

$C_{14}H_{11}ClFNO$
3′-Chloro-3-fluoro-4′-methylbenzanilide ... 460

$C_{14}H_{11}ClFNO_2$
5′-Chloro-4-fluoro-2′-methoxybenzanilide ... 449
4-Chloro-3-methylphenyl *N*-(2-fluorophenyl)carbamate ... 633
2-(2-Chlorophenoxy)-2′-fluoroacetanilide ... 400
2-(2-Chlorophenoxy)-4′-fluoroacetanilide ... 399
2-(4-Chlorophenoxy)-4′-fluoroacetanilide ... 399
2-Fluorobenzyl *N*-(3-chlorophenyl)-carbamate ... 621
4-Fluorobenzyl *N*-(2-chlorophenyl)-carbamate ... 621
4-Fluorobenzyl *N*-(3-chlorophenyl)-carbamate ... 622
2-Fluorophenyl *N*-(2-chloro-4-methylphenyl)carbamate ... 633
4-Fluorophenyl *N*-(2-chloro-6-methylphenyl)carbamate ... 634
4-Fluorophenyl *N*-(4-chloro-2-methylphenyl)carbamate ... 635
4-Fluorophenyl *N*-(5-chloro-2-methylphenyl)carbamate ... 635

$C_{14}H_{11}ClFNO_3$
2-Fluorophenyl *N*-(2-chloro-4-methoxyphenyl)carbamate ... 620
2-Fluorophenyl *N*-(4-chloro-2-methoxyphenyl)carbamate ... 618
4-Fluorophenyl *N*-(2-chloro-4-methoxyphenyl)carbamate ... 620
4-Fluorophenyl *N*-(5-chloro-2-methoxyphenyl)carbamate ... 618

$C_{14}H_{11}ClFN_3O_3$
1-(3-Chloro-2-methylphenyl)-3-(4-fluoro-3-nitrophenyl)urea ... 524
1-(5-Chloro-2-methylphenyl)-3-(2-fluoro-5-nitrophenyl)urea ... 525
3-(2-Chloro-4-nitrophenyl)-1-(4-fluorophenyl)-1-methylurea ... 530

$C_{14}H_{11}ClFN_3O_4$
1-(2-Chloro-4-methoxyphenyl)-3-(4-fluoro-3-nitrophenyl)urea ... 518
1-(5-Chloro-2-methoxyphenyl)-3-(4-fluoro-3-nitrophenyl)urea ... 517

$C_{14}H_{11}ClF_2N_2O$
1-(3-Chloro-2-methylphenyl)-3-(2,5-difluorophenyl)urea ... 542
1-(3-Chloro-4-methylphenyl)-3-(2,5-difluorophenyl)urea ... 544
1-(5-Chloro-2-methylphenyl)-3-(2,4-difluorophenyl)urea ... 544

$C_{14}H_{11}ClF_2N_2O_2$
1-(4-Chloro-2-methoxyphenyl)-3-(2,5-difluorophenyl)urea ... 520
1-(5-Chloro-2-methoxyphenyl)-3-(2,5-difluorophenyl)urea ... 520

$C_{14}H_{11}ClF_2O$
α-(Chlorodifluoromethyl)benzhydrol. 197

$C_{14}H_{11}ClF_3N_3O$
1-(2-Chloro-5-trifluoromethylphenyl)-3-(3-pyridylmethyl)urea ... 925
1-(4-Chloro-2-trifluoromethylphenyl)-3-(3-pyridylmethyl)urea ... 925

$C_{14}H_{11}ClF_3N_3O_3S$
1-(2-Chloro-5-(trifluoromethyl)-phenyl)-3-(4-sulfamoylphenyl)urea ... 805

$C_{14}H_{11}ClIN$
9-Chloro-*N*-methylacridinium iodide 996

$C_{14}H_{11}ClN_2$
2-(4-Chlorophenyl)-1-methylbenzimidazole ... 901

$C_{14}H_{11}ClN_2O$
Benzoic (2-chlorobenzylidene)-hydrazide ... 493
2-(3-Chlorophenyl)-1,2-dihydro-4(3*H*)-quinazolinone ... 704
2-(4-Chlorophenyl)-1,2-dihydro-4(3*H*)-quinazolinone ... 694

$C_{14}H_{11}ClN_2OS$
1-Benzoyl-3-(2-chlorophenyl)-2-thiourea ... 741

$C_{14}H_{11}ClN_2O_2$
1-Benzoyl-2-(2-chlorobenzoyl)-hydrazine ... 500
1-Benzoyl-2-(3-chlorobenzoyl)-hydrazine ... 500
1-Benzoyl-2-(4-chlorobenzoyl)-hydrazine ... 500
4-Chlorobenzoic (2-hydroxybenzylidene)hydrazide ... 491
N-(2-Chloro-5-nitrobenzylidene)-*m*-toluidine ... 1058
N-(2-Chloro-5-nitrobenzylidene)-*p*-toluidine ... 1058
3-Chloro-*N*-(2-nitrobenzylidene)-*p*-toluidine ... 1055
5-Chloro-*N*-(3-nitrobenzylidene)-*o*-toluidine ... 1059
6-Chloro-*N*-(2-nitrobenzylidene)-*o*-toluidine ... 1055
Salicylic (2-chlorobenzylidene)-hydrazide ... 491

$C_{14}H_{11}ClN_2O_2S$
4′-Chloro-*N*-cyano-*p*-toluenesulfonanilide ... 809

$C_{14}H_{11}ClN_2O_3$
1-(2-Carboxyphenyl)-3-(2-chlorophenyl)urea ... 530
1-(2-Carboxyphenyl)-3-(3-chlorophenyl)urea ... 531
1-(2-Carboxyphenyl)-3-(4-chlorophenyl)urea ... 531
1-(4-Carboxyphenyl)-3-(2-chlorophenyl)urea ... 531
1-(4-Carboxyphenyl)-3-(3-chlorophenyl)urea ... 531
1-(4-Carboxyphenyl)-3-(4-chlorophenyl)urea ... 531
2′-Chloro-6′-methyl-3-nitrobenzanilide ... 452
2′-Chloro-6′-methyl-4-nitrobenzanilide ... 452
3′-Chloro-2′-methyl-3-nitrobenzanilide ... 451
3′-Chloro-2′-methyl-4-nitrobenzanilide ... 452
3-Chloro-2′-methyl-6′-nitrobenzanilide ... 450
3′-Chloro-4′-methyl-3-nitrobenzanilide ... 452
3′-Chloro-4′-methyl-4-nitrobenzanilide ... 452
4′-Chloro-2′-methyl-4-nitrobenzanilide ... 452
4-Chloro-2′-methyl-6′-nitrobenzanilide ... 450
5′-Chloro-2′-methyl-4-nitrobenzanilide ... 452
4′-Chloro-3-nitrobenzo-*o*-toluidide ... 453
5′-Chloro-3-nitrobenzo-*o*-toluidide ... 453
N-(2-Chloro-5-nitrobenzylidene)-*o*-anisidine ... 1043
N-(2-Chloro-5-nitrobenzylidene)-*p*-anisidine ... 1043
2′-Chloro-4′-nitro-2-phenylacetanilide ... 404

$C_{14}H_{11}ClN_2O_3S$
S-Benzyl *N*-(2-chloro-4-nitrophenyl)-thiocarbamate ... 627
S-Benzyl *N*-(2-chloro-5-nitrophenyl)-thiocarbamate ... 626
S-(4-Chlorophenyl) *N*-(4-methyl-2-nitrophenyl)thiocarbamate ... 625
2-(4-Chlorophenylthio)-2′-nitroacetanilide ... 402
2-(4-Chlorophenylthio)-3′-nitroacetanilide ... 402
S-(*o*-Tolyl) *N*-(2-chloro-4-nitrophenyl)thiocarbamate ... 627
S-(*o*-Tolyl) *N*-(2-chloro-5-nitrophenyl)thiocarbamate ... 626
S-(*m*-Tolyl) *N*-(2-chloro-4-nitrophenyl)thiocarbamate ... 627

$C_{14}H_{11}ClN_2O_4$
4-Chloro-2′-methoxy-5′-nitrobenzanilide ... 448
5′-Chloro-2′-methoxy-3-nitrobenzanilide ... 448
4-Chloro-2-methyl-6-nitrophenyl *N*-phenylcarbamate ... 627
2′-Chloro-4-nitro-*p*-benzanisidide ... 449
3-Chloro-5′-nitro-*o*-benzanisidide ... 448
α-(2-Chloro-5-nitrophenylimino)-6-methoxy-*o*-cresol ... 1021
α-(4-Chloro-3-nitrophenylimino)-6-methoxy-*o*-cresol ... 1030
2-(4-Chlorophenoxy)-3′-nitroacetanilide ... 399
4-Chlorophenyl *N*-(2-methyl-5-nitrophenyl)carbamate ... 625
4-Chlorophenyl *N*-(4-methyl-2-nitrophenyl)carbamate ... 624
3-Methyl-4-nitrophenyl *N*-(2-chlorophenyl)carbamate ... 628

$C_{14}H_{11}ClN_2O_4S$
2-Chloro-4-nitrophenyl *N*-(3-(methylthio)phenyl)carbamate ... 623
S-(4-Methoxyphenyl) *N*-(2-chloro-5-nitrophenyl)thiocarbamate ... 618

$C_{14}H_{11}ClN_2O_5$
3-Methoxyphenyl *N*-(4-chloro-3-nitrophenyl)carbamate ... 616

$C_{14}H_{11}ClN_2S$
2-Chloro-10-(2-cyanoethyl)-10*H*-phenothiazine ... 755

$C_{14}H_{11}ClN_2S_2$
3-(2-Chloroanilinomethyl)-2-benzothiazolinethione ... 741

$C_{14}H_{11}ClN_4$
5-(2-Amino-5-chlorophenyl)-3-phenyl-1*H*,1,2,4-triazole ... 869

$C_{14}H_{11}ClN_4O$
1-(*N*-(3-Chloro-4-methylphenyl)-carbamoyl)-1*H*-benzotriazole ... 908

$C_{14}H_{11}ClN_4O_4$
1-(2-Carbamoylphenyl)-3-(2-chloro-4-nitrophenyl)urea ... 526
2-Chlorobenzaldehyde (2,6-dinitro-4-methylphenyl)hydrazone ... 261

$C_{14}H_{11}ClN_4O_5$
4-(4-Chloro-3-nitrophenyl)-1-(4-hydroxybenzoyl)semicarbazide ... 513

$C_{14}H_{11}ClN_4O_6$
1-(5-Chloro-2-methoxyphenyl)-3-(3,5-dinitrophenyl)urea ... 517
1-(4-Chloro-2-nitrophenyl)-3-(2-methoxy-4-nitrophenyl)urea ... 518

$C_{14}H_{11}ClN_4O_8$
N-(5-Chloro-2,4-dimethoxyphenyl)-picrylamine ... 260

$C_{14}H_{11}ClO$
4-Chloro-2′-methylbenzophenone ... 310

$C_{14}H_{11}ClO_2$
4′-Chlorobenzoin ... 288
4-Chlorobenzyl benzoate ... 366
2-(4-Chlorobenzyl)benzoic acid ... 337

$C_{14}H_{11}ClO_2S$
4-Chlorophenyl *trans*-styryl sulfone. 774

$C_{14}H_{11}ClO_3$
4′-Chloro-2-hydroxy-4-methoxybenzophenone ... 310
4-Methoxyphenyl 2-chlorobenzoate. 376

$C_{14}H_{11}ClO_3S$
2-Phenyl-2-(phenylsulfonyl)acetyl chloride ... 782

$C_{14}H_{11}Cl_2FN_2O$
1-(4,5-Dichloro-2-methylphenyl)-3-(3-fluorophenyl)urea ... 546

$C_{14}H_{11}Cl_2FO$
4,4′-Dichloro-α-(fluoromethyl)-benzhydrol ... 197

$C_{14}H_{11}Cl_2N$
N-(2-Chlorobenzylidene)-6-chloro-2-methylaniline ... 1064
N-(4-Chlorobenzylidene)-3-chloro-4-methylaniline ... 1066

$C_{14}H_{11}Cl_2N$ (Cont'd)
N-(4-Chlorobenzylidene)-5-chloro-2-methylaniline 1066
3-Chloro-*N*-(2-chlorobenzylidene)-4-methylaniline 1064
5-Chloro-*N*-(2-chlorobenzylidene)-2-methylaniline 1064
N-(2,4-Dichlorobenzylidene)-*o*-toluidine 1064
N-(2,4-Dichlorobenzylidene)-*m*-toluidine 1067
N-(2,4-Dichlorobenzylidene)-*p*-toluidine 1066
N-(3,4-Dichlorobenzylidene)-*o*-toluidine 1064
N-(3,4-Dichlorobenzylidene)-*m*-toluidine 1067
N-(3,4-Dichlorobenzylidene)-*p*-toluidine 1066
2,3-Dichloro-*N*-(4-methylbenzylidene)aniline 1065
2,5-Dichloro-*N*-(4-methylbenzylidene)aniline 1066
3,4-Dichloro-*N*-(4-methylbenzylidene)aniline 1066
N-(4-Methylbenzylidene)-2,4-dichloroaniline 1066

$C_{14}H_{11}Cl_2NO$
N-Benzyl-2,4-dichlorobenzamide 483
2,3′-Dichloro-*o*-benzotoluidide 483
2,4-Dichlorobenzo-*o*-toluidide 479
2,4-Dichlorobenzo-*p*-toluidide 480
3′,4-Dichloro-*o*-benzotoluidide 483
2-(2,4-Dichlorobenzylideneamino)-*p*-cresol 1023
N-(2,4-Dichlorobenzylidene)-*o*-anisidine 1045
N-(2,4-Dichlorobenzylidene)-*m*-anisidine 1048
N-(3,4-Dichlorobenzylidene)-*o*-anisidine 1045
N-(3,4-Dichlorobenzylidene)-*m*-anisidine 1048
N-(3,4-Dichlorobenzylidene)-*p*-anisidine 1047
N-(2,4-Dichlorobenzylidene)-4-methoxyaniline 1024
2,3-Dichloro-*N*-(2-methoxybenzylidene)aniline 1046
2,3-Dichloro-*N*-(4-methoxybenzylidene)aniline 1047
2,2′-Dichloro-5′-methylbenzanilide 458
2,2′-Dichloro-6′-methylbenzanilide 457
2,3′-Dichloro-4′-methylbenzanilide 458
2′,3-Dichloro-6′-methylbenzanilide 459
2′,4′-Dichloro-2-methylbenzanilide 479
2′,4′-Dichloro-3-methylbenzanilide 481
2′,5′-Dichloro-2-methylbenzanilide 479
2,5′-Dichloro-2′-methylbenzanilide 458
2′,5′-Dichloro-3-methylbenzanilide 481
2′,6′-Dichloro-4-methylbenzanilide 458
3′,4′-Dichloro-3-methylbenzanilide 481
3′,4-Dichloro-4′-methylbenzanilide 458
4,4′-Dichloro-2′-methylbenzanilide 458
4,5′-Dichloro-2′-methylbenzanilide 458
2′,4′-Dichloro-2-phenylacetanilide 404
N-(4-Methoxybenzylidene)-2,4-dichloroaniline 1047
N-(4-Methoxybenzylidene)-2,5-dichloroaniline 1047
N-(4-Methoxybenzylidene)-3,4-dichloroaniline 1047

$C_{14}H_{11}Cl_2NOS$
4′-Chloro-2-(4-chlorophenylthio)-acetanilide 403
S-(4-Chlorophenyl) *N*-(4-chloro-2-methylphenyl)thiocarbamate 641
2,4-Dichloro-4′-(methylthio)-benzanilide 450
S-(*o*-Tolyl) *N*-(2,3-dichlorophenyl)-thiocarbamate 641
S-(*p*-Tolyl) *N*-(2,6-dichlorophenyl)-thiolcarbamate 688

$C_{14}H_{11}Cl_2NO_2$
Benzyl *N*-(2,6-dichlorophenyl)-carbamate 685
2′-Chloro-2-(2-chlorophenoxy)-acetanilide 400
4′-Chloro-2-(2-chlorophenoxy)-acetanilide 400
4′-Chloro-2-(4-chlorophenoxy)-acetanilide 402
4-Chloro-2-methylphenyl *N*-(4-chlorophenyl)carbamate 635
4-Chloro-3-methylphenyl *N*-(3-chlorophenyl)carbamate 638
4-Chloro-3-methylphenyl *N*-(4-chlorophenyl)carbamate 635
4-Chlorophenyl *N*-(2-chloro-6-methylphenyl)carbamate 683
4-Chlorophenyl *N*-(3-chloro-4-methylphenyl)carbamate 635
4-Chlorophenyl *N*-(4-chloro-2-methylphenyl)carbamate 635
4-Chlorophenyl *N*-(5-chloro-2-methylphenyl)carbamate 635
2,4-Dichlorobenzo-*o*-anisidide 465
2,4-Dichlorobenzo-*m*-anisidide 467
2,4-Dichlorobenzo-*p*-anisidide 467
2,6-Dichlorobenzo-*o*-anisidide 449
2,5′-Dichloro-2′-methoxybenzanilide 449
2′,6′-Dichloro-3-methoxybenzanilide 448
2′,6′-Dichloro-4-methoxybenzanilide 450
3′,4′-Dichloro-4-methoxybenzanilide 450
2,4′-Dichloro-2′-phenoxyacetanilide. 390
α-(2,4-Dichlorophenylimino)-6-methoxy-*o*-cresol 1021
α-(2,5-Dichlorophenylimino)-6-methoxy-*o*-cresol 1021
α-(3,4-Dichlorophenylimino)-6-methoxy-*o*-cresol 1021
p-Tolyl *N*-(3,4-dichlorophenyl)-carbamate 635

$C_{14}H_{11}Cl_2NO_2S$
S-(4-Chlorophenyl) *N*-(4-chloro-2-methoxyphenyl)thiocarbamate 620
3,7-Dichloro-10-ethylphenothiazine-5,5-dioxide 778
2,3-Dichlorophenyl *N*-(4-(methylthio)phenyl)carbamate 623
2,4-Dichlorophenyl *N*-(4-(methylthio)phenyl)carbamate 623
2,5-Dichlorophenyl *N*-(4-(methylthio)phenyl)carbamate 623
3,4-Dichlorophenyl *N*-(4-(methylthio)phenyl)carbamate 623
3,5-Dichlorophenyl *N*-(4-(methylthio)phenyl)carbamate 623
S-(4-Methoxyphenyl) *N*-(2,3-dichlorophenyl)thiocarbamate 621

$C_{14}H_{11}Cl_2NO_3$
2-(2,4-Dichlorophenoxy)-*N*-(4-hydroxyphenyl)acetamide 398
2,4-Dichlorophenyl *N*-(2-methoxyphenyl)carbamate 618
2,4-Dichlorophenyl *N*-(3-methoxyphenyl)carbamate 621
2,4-Dichlorophenyl *N*-(4-methoxyphenyl)carbamate 620
3-Methoxyphenyl *N*-(3,4-dichlorophenyl)carbamate 617
4-Methoxyphenyl *N*-(2,3-dichlorophenyl)carbamate 617
4-Methoxyphenyl *N*-(3,4-dichlorophenyl)carbamate 617

$C_{14}H_{11}Cl_2NO_4S$
2,4-Dichlorophenyl *N*-(*p*-tosyl)-carbamate 808

$C_{14}H_{11}Cl_2N_3O$
Anthranilic (2,6-dichlorobenzylidene)hydrazide 492

$C_{14}H_{11}Cl_2N_3O_2$
1-Benzoyl-4-(2,6-dichlorophenyl)-semicarbazide 532
1-(4-Chlorobenzoyl)-4-(2-chlorophenyl)semicarbazide 532

$C_{14}H_{11}Cl_2N_3O_3$
4-(2,5-Dichlorophenyl)-1-(2-hydroxybenzoyl)semicarbazide 512
4-(2,6-Dichlorophenyl)-1-(4-hydroxybenzoyl)semicarbazide 513
4-(3,4-Dichlorophenyl)-1-(2-hydroxybenzoyl)semicarbazide 512
4-(3,4-Dichlorophenyl)-1-(4-hydroxybenzoyl)semicarbazide 513
1-(2,3-Dichlorophenyl)-3-(2-methyl-5-nitrophenyl)urea 524
1-(2,5-Dichlorophenyl)-3-(2-methyl-5-nitrophenyl)urea 525
1-(2,6-Dichlorophenyl)-3-(2-methyl-4-nitrophenyl)urea 526
1-(3,4-Dichlorophenyl)-3-(2-methyl-6-nitrophenyl)urea 523

$C_{14}H_{11}Cl_2N_3O_4$
1-(5-Chloro-2-methoxyphenyl)-3-(4-chloro-2-nitrophenyl)urea 517

$C_{14}H_{11}Cl_3$
2,2-Diphenyl-1,1,1-trichloroethane.... 172

$C_{14}H_{11}Cl_3N_2O$
4′-Anilino-2,2,2-trichloroacetanilide.. 397
1-Benzyl-3-(2,4,5-trichlorophenyl)urea 528
1-(2-Chloro-6-methylphenyl)-3-(2,3-dichlorophenyl)urea 541
1-(2-Chloro-6-methylphenyl)-3-(2,4-dichlorophenyl)urea 542
1-(2-Chloro-6-methylphenyl)-3-(2,5-dichlorophenyl)urea 542
1-(2-Chloro-6-methylphenyl)-3-(2,6-dichlorophenyl)urea 541
1-(2-Chloro-6-methylphenyl)-3-(3,4-dichlorophenyl)urea 542
1-(3-Chloro-2-methylphenyl)-3-(3,4-dichlorophenyl)urea 542
1-(3-Chloro-4-methylphenyl)-3-(2,6-dichlorophenyl)urea 542
1-(3-Chloro-4-methylphenyl)-3-(3,4-dichlorophenyl)urea 544
1-(4-Chloro-2-methylphenyl)-3-(2,6-dichlorophenyl)urea 542
1-(4-Chloro-2-methylphenyl)-3-(3,4-dichlorophenyl)urea 544
1-(2-Chlorophenyl)-3-(3,4-dichlorobenzyl)urea 538
1-(2-Chlorophenyl)-3-(4,5-dichloro-2-methylphenyl)urea 538
1-(4-Chlorophenyl)-3-(4,5-dichloro-2-methylphenyl)urea 540

$C_{14}H_{11}Cl_3N_2O_2$
1-(2-Methoxyphenyl)-3-(2,4,5-trichlorophenyl)urea 519
1-(3-Methoxyphenyl)-3-(2,4,5-trichlorophenyl)urea 521
1-(4-Methoxyphenyl)-3-(2,4,5-trichlorophenyl)urea 521

$C_{14}H_{11}Cl_5N_2$
1,1-Bis(4-chloroanilino)-2,2,2-trichloroethane 222

$C_{14}H_{11}Cl_5Si$
Dimethyl(pentachlorophenyl)-phenylsilane 1102

$C_{14}H_{11}FCl_2N_2O$
1-(4,5-Dichloro-2-methylphenyl)-3-(2-fluorophenyl)urea 536

$C_{14}H_{11}FN_2O_2$
1-Benzoyl-2-(2-fluorobenzoyl)-hydrazine 500
1-Benzoyl-2-(3-fluorobenzoyl)-hydrazine 500
1-Benzoyl-2-(4-fluorobenzoyl)-hydrazine 500

$C_{14}H_{11}FN_2O_3$
2-Carbamoylphenyl *N*-(2-fluorophenyl)carbamate 622
2-Carbamoylphenyl *N*-(3-fluorophenyl)carbamate 622
2-Carbamoylphenyl *N*-(4-fluorophenyl)carbamate 622
1-(2-Carboxyphenyl)-3-(3-fluorophenyl)urea 530
1-(3-Carboxyphenyl)-3-(2-fluorophenyl)urea 530
1-(3-Carboxyphenyl)-3-(4-fluorophenyl)urea 531
1-(4-Carboxyphenyl)-3-(2-fluorophenyl)urea 531
4-(3-(3-Fluorophenyl)ureido)-benzoic acid 531

$C_{14}H_{11}FN_2O_4$
1-(4-Carboxy-3-hydroxyphenyl)-3-(3-fluorophenyl)urea 512
2-Fluorophenyl *N*-(4-methyl-2-nitrophenyl)carbamate 624
4-Fluorophenyl *N*-(4-methyl-2-nitrophenyl)carbamate 624

$C_{14}H_{11}FN_2O_4S$
4-(Methylthio)phenyl *N*-(4-fluoro-2-nitrophenyl)carbamate 623

$C_{14}H_{11}FN_2O_5$
2-Fluorophenyl *N*-(2-methoxy-4-nitrophenyl)carbamate 618
4-Fluorophenyl *N*-(2-methoxy-5-nitrophenyl)carbamate 617
3-Methoxyphenyl *N*-(4-fluoro-2-nitrophenyl)carbamate 616
4-Methoxyphenyl *N*-(2-fluoro-5-nitrophenyl)carbamate 617

$C_{14}H_{11}FN_2O_5S$
3-(2-Carboxyphenylureido)-benzenesulfonyl fluoride 804

$C_{14}H_{11}FN_2O_6S$
4-(3-(3-(Fluorosulfonyl)phenyl)-ureido)salicylic acid 801

$C_{14}H_{11}FN_4O_2S_2$
1-(3-Fluorophenyl)-5-(4-nitrophenyl)-2,4-dithiobiuret 730

$C_{14}H_{11}FN_4O_4$
1-(2-Carbamoylphenyl)-3-(4-fluoro-3-nitrophenyl)urea 524

$C_{14}H_{11}FN_4O_5$
1-(4-Fluoro-3-nitrophenyl)-3-(4-methyl-3-nitrophenyl)urea 525

$C_{14}H_{11}FN_4O_6$
1-(4-Fluoro-3-nitrophenyl)-3-(2-methoxy-3-nitrophenyl)urea 517
1-(4-Fluoro-3-nitrophenyl)-3-(2-methoxy-5-nitrophenyl)urea 517

$C_{14}H_{11}FO$
4′-Fluoro-α-phenylacetophenone 297

$C_{14}H_{11}FOS$
α-Cyclopropyl-α-(4-fluorophenyl)-2-thiophenemethanol 855

$C_{14}H_{11}F_2NO_2$
2-Fluorobenzyl *N*-(2-fluorophenyl)-carbamate 633
4-Fluorobenzyl *N*-(2-fluorophenyl)-carbamate 621

$C_{14}H_{11}F_2NO_2S$
N′-(4-Difluoromethylthiophenyl)-anthranilic acid 336
4-(Methylthio)phenyl *N*-(2,5-difluorophenyl)carbamate 623

$C_{14}H_{11}F_2NO_3$
3-Methoxyphenyl *N*-(2,4-difluorophenyl)carbamate 617

$C_{14}H_{11}F_2N_3O_3$
4-(2,4-Difluorophenyl)-1-(2-hydroxybenzoyl)semicarbazide 521
4-(2,4-Difluorophenyl)-1-(4-hydroxybenzoyl)semicarbazide 513
1-(2,5-Difluorophenyl)-3-(4-methyl-3-nitrophenyl)urea 525

$C_{14}H_{11}F_2N_3O_4$
1-(2,5-Difluorophenyl)-3-(2-methoxy-5-nitrophenyl)urea 517
1-(2,5-Difluorophenyl)-3-(4-methoxy-3-nitrophenyl)urea 517

$C_{14}H_{11}F_2N_5O_5$
4-(2,4-Difluorophenyl)-1-(2,4-dinitrophenyl)-1-methylsemicarbazide 522

$C_{14}H_{11}F_3N_2O$
1-Phenyl-3-(3-(trifluoromethyl)-phenyl)urea 510

$C_{14}H_{11}F_3N_2O_2$
1-(3-Hydroxyphenyl)-3-(2-(trifluoromethyl)phenyl)urea 507
1-(4-Hydroxyphenyl)-3-(2-(trifluoromethyl)phenyl)urea 507

$C_{14}H_{11}F_3N_4O_2$
1-(4-Pyridylcarbonyl)-4-(2-(trifluoromethyl)phenyl)-semicarbazide 925

$C_{14}H_{11}F_3N_4O_3$
1-(4-Nitrophenyl)-4-(2-(trifluoromethyl)phenyl)-semicarbazide 508

$C_{14}H_{11}F_3O$
α-(Trifluoromethyl)benzhydrol 197
3-(Trifluoromethyl)benzhydrol 197
4-(Trifluoromethyl)benzhydrol 197

$C_{14}H_{11}F_3O_3S$
α-(Trifluoromethyl)benzyl benzenesulfonate ... 796
$C_{14}H_{11}F_5Si$
Dimethyl(pentafluorophenyl)phenylsilane ... 1102
$C_{14}H_{11}IN_2O_3$
2-Iodo-*N*-methyl-4′-nitrobenzanilide ... 452
$C_{14}H_{11}IO_2$
o-Tolyl 2-iodobenzoate ... 377
$C_{14}H_{11}I_2N$
2,8-Diiodo-9-ethylcarbazole ... 892
3,6-Diiodo-9-ethylcarbazole ... 892
$C_{14}H_{11}N$
2-Aminophenanthrene ... 231
3-Aminophenanthrene hydrochloride ... 231
3-Phenylindole ... 891
2-Phenylindolizine ... 887
$C_{14}H_{11}NO$
2-Acetylcarbazole ... 892
3-Acetylcarbazole ... 892
9-Acetylcarbazole ... 892
9-Amino-3-hydroxyphenanthrene hydrochloride ... 226
2-Formamidofluorene ... 490
2-Methylamino-9-fluorenone ... 312
4-Methylbenzo(*H*)quinolin-2(1*H*)one ... 993
11-Methyldibenz(b,f)oxazepine ... 247
3-Phenyloxindole ... 692
2-(4-Pyridylmethylene)acetophenone ... 963
$C_{14}H_{11}NOS$
2-Acetamidodibenzothiophene ... 898
10-Acetylphenothiazine ... 710
3-Benzyl-2(3*H*)-benzothiazolone ... 743
2-(Benzyloxy)benzothiazole ... 912
2-(Benzylthio)benzoxazole ... 909
α-Phenyl-2-benzothiazolemethanol ... 912
$C_{14}H_{11}NO_2$
2-Acetamidodibenzofuran ... 895
3-Acetamidodibenzofuran ... 895
10-Acetylphenoxazine ... 710
2-Benzoylbenzamide ... 473
4-(Benzylideneamino)benzoic acid ... 1063
9,10-Dihydro-2-nitrophenanthrene ... 280
1,2-Dihydro-2-phenyl-4*H*-3,1-benzoxazin-4-one ... 385
1,4-Dimethyl-2,3-naphthalenedicarboximide ... 721
1,1-Diphenyl-2-nitroethylene ... 258
β-(2-Furyl)-α-(4-methoxyphenyl)acrylonitrile ... 846
3-Hydroxy-2-phenyl-1-isoindolinone ... 690
2-Methoxy-9(10*H*)-acridone ... 319
4-Nitrostilbene ... 278
N-Piperonylideneaniline ... 1076
$C_{14}H_{11}NO_2S$
2-(4-Hydroxy-3-methoxyphenyl)benzothiazole ... 912
$C_{14}H_{11}NO_3$
3-Benzamidobenzoic acid ... 476
2-Formylphenyl *N*-phenylcarbamate ... 666
4-Formylphenyl *N*-phenylcarbamate ... 674
N-(2-Hydroxyethyl)-1,8-naphthalimide ... 716
2-Methoxy-3-nitrofluorene ... 268
N-(1-Naphthyl)-maleamic acid ... 489
2-(4-Nitrophenyl)-3-phenyloxirane ... 269
2-(Piperonylideneamino)phenol ... 1037
$C_{14}H_{11}NO_3S$
2-(2-Nitrophenylthio)acetophenone ... 293
2-(4-Nitrophenylthio)acetophenone ... 293
S-(*p*-Tolyl) 4-nitrothiolbenzoate ... 379
$C_{14}H_{11}NO_4$
4-Benzamidosalicylic acid, calcium salt ... 462
4-(2-Hydroxybenzylideneamino)salicylic acid ... 1028
4-(4-Hydroxybenzylideneamino)salicylic acid ... 1028
2-Hydroxy-4′-phenylacetophenone nitrate ester ... 257
3′-Hydroxyphthalanilic acid ... 462
4-Nitrobenzyl benzoate ... 370
4-(4-Nitrobenzyloxy)benzaldehyde ... 322
$C_{14}H_{11}NO_5$
4-Methoxyphenyl 3-nitrobenzoate ... 378
$C_{14}H_{11}NO_6$
5-Nitro-1,2,3,4-tetrahydro-1,4-ethenonaphthalene-2,3-dicarboxylic acid ... 330
$C_{14}H_{11}NS$
2-Methyl-6-phenylbenzothiazole ... 917
4-Methyl-2-phenylbenzothiazole ... 917
6-Methyl-2-phenylbenzothiazole ... 917
2-(*o*-Tolyl)benzothiazole ... 917
2-(*p*-Tolyl)benzothiazole ... 917
$C_{14}H_{11}NS_2$
2-(*p*-Tolylthio)benzothiazole ... 914
$C_{14}H_{11}N_3$
4-Cyanobenzaldehyde phenylhydrazone ... 765
1-Cyano-*N*,*N*′-diphenylformamidine ... 1054
3,4-Diphenyl-4*H*-1,2,4-triazole ... 870
$C_{14}H_{11}N_3O$
4-Cyano-3,5-dimethylpyrido(1,2-a)benzimidazol-1(5*H*)-one ... 905
1-(2-Cyanophenyl)-3-phenylurea ... 764
1-(4-Cyanophenyl)-3-phenylurea ... 767
$C_{14}H_{11}N_3OS$
1-(2-Benzothiazolyl)-3-phenylurea ... 916
$C_{14}H_{11}N_3O_2$
2-(α-Cyano-4-nitrobenzylidene)-1,2-dihydro-1-methylpyridine ... 960
2,4-Dicyano-3-methyl-3-phenylglutarimide ... 757
3,4-Dihydro-4-(2-nitrophenyl)quinazoline ... 275
3-(2-Methoxyphenyl)-1,2,3-benzotriazin-4(3*H*)-one ... 698
N-(2-Pyridylaminomethyl)phthalimide ... 954
1,4,5-Triaminoanthraquinone ... 317
$C_{14}H_{11}N_3O_2S$
6,7-Dihydro-2′,6′-dioxospiro[thianaphthene-4(5*H*)-4′-piperidine]-3′,5′-dicarbonitrile ... 858
2-(*p*-Tosylazo)benzothiazole ... 917
$C_{14}H_{11}N_3O_2S_2$
3-(3-Nitroanilinomethyl)-2(3*H*)-benzothiazolethione ... 739
3-(4-Nitroanilinomethyl)-2(3*H*)-benzothiazolethione ... 741
$C_{14}H_{11}N_3O_3$
Benzoic (3-nitrobenzylidene)hydrazide ... 1069
1,2-Dihydro-2-(2-nitrophenyl)-4(3*H*)-quinazolinone ... 705
N-(3-(1-Imidazolyl)-2-oxopropyl)phthalimide ... 867
$C_{14}H_{11}N_3O_3S$
1-Benzoyl-3-(4-nitrophenyl)-2-thiourea ... 740
3-(3-Nitroanilinomethyl)-2(3*H*)-benzoxazolethione ... 739
3-(4-Nitroanilinomethyl)-2(3*H*)-benzoxazolethione ... 740
$C_{14}H_{11}N_3O_4$
2-Benzoylacetamido-5-nitropyridine ... 944
2′-Carbamoyl-4-nitrobenzanilide ... 473
4-Hydroxybenzoic (4-nitrobenzylidene)hydrazide ... 1034
N-Phthaloyl-DL-histidine ... 866
$C_{14}H_{11}N_3O_4S$
N-Cyano-4′-methyl-3-nitrobenzenesulfonanilide ... 822
$C_{14}H_{11}N_3O_5$
3-Acetyl-2′,4′-dinitrodiphenylamine ... 294
4-Acetyl-2′,4′-dinitrodiphenylamine ... 294
$C_{14}H_{11}N_3O_5S$
3,5-Dinitro-4′-(methylthio)benzanilide ... 467
$C_{14}H_{11}N_3O_6$
3′,4-Dinitro-2′-methoxybenzanilide ... 465
$C_{14}H_{11}N_3S$
5-Anilino-3-phenyl-1,2,4-thiadiazole ... 879
$C_{14}H_{11}N_5O_2S_2$
2,4-Dinitrobenzenesulfonic (3-methyl-2-benzothiazolinylidene)hydrazide ... 820
$C_{14}H_{11}N_5O_4S$
4-(5-Hydroxy-1-phenyl-1,2,3-triazol-4-ylazo)benzenesulfonic acid, sodium salt ... 1085
$C_{14}H_{11}N_5O_6$
3-Nitrobenzaldehyde (2,4-dinitrophenyl)methylhydrazone ... 1052
$C_{14}H_{12}As_2N_2O_2$
4,4′-Arsenobenzamide ... 1127
$C_{14}H_{12}BF_4N$
10-Methylacridine tetrafluoroborate ... 996
$C_{14}H_{12}BrN$
3-Bromo-9-ethylcarbazole ... 892
N-(4-Methylbenzylidene)-4-bromoaniline ... 1065
$C_{14}H_{12}BrNO$
4-Bromo-*N*-(3-methoxybenzylidene)aniline ... 1046
2′-Bromo-3′-methylbenzanilide ... 461
3-Bromo-2′-methylbenzanilide ... 457
4′-Bromo-2-methylbenzanilide ... 479
4′-Bromo-3-methylbenzanilide ... 480
4′-Bromo-4-methylbenzanilide ... 480
α-(4-Bromo-3-methylphenylimino)-*o*-cresol ... 1022
4′-Bromo-2-phenylacetanilide ... 430
4′-Bromo-2′-phenylacetanilide ... 484
α-(2-Bromo-*p*-tolylimino)-*o*-cresol ... 1035
N-(4-Methoxybenzylidene)-4-bromoaniline ... 1046
$C_{14}H_{12}BrNO_2$
3-Bromo-2′-methoxybenzanilide ... 449
α-(4-Bromophenylimino)-6-methoxy-*o*-cresol ... 1021
$C_{14}H_{12}BrN_3$
1-(4-(4-Bromophenylazo)phenyl)aziridine ... 1082
$C_{14}H_{12}BrN_3O_3$
4-Pyridinecarboxylic (3-bromo-4-hydroxy-5-methoxybenzylidene)hydrazide ... 926
$C_{14}H_{12}BrN_3O_4$
1-(2-Bromo-4-nitrophenyl)-3-(2-methoxyphenyl)urea ... 518
1-(2-Bromo-4-nitrophenyl)-3-(4-methoxyphenyl)urea ... 518
$C_{14}H_{12}BrN_5O_5$
4-(4-Bromophenyl)-1-(2,4-dinitrophenyl)-1-methylsemicarbazide ... 522
$C_{14}H_{12}Br_2ClN$
4-Chloro-2,2′-dibromo-*N*-ethyldiphenylamine ... 225
$C_{14}H_{12}Br_2N_2$
1,10-Ethano-1,10-phenanthrolin-1,10-diium dibromide ... 1002
$C_{14}H_{12}Br_2N_2O$
2,6-Dibromo-4-(4-(dimethylamino)phenylimino)-2,5-cyclohexadien-1-one ... 1057
$C_{14}H_{12}Br_2O_2$
5,5′-Dibromo-2,2′-dimethoxybiphenyl ... 177
$C_{14}H_{12}ClFN_2O$
1-(2-Chlorobenzyl)-3-(2-fluorophenyl)urea ... 536
1-(2-Chlorobenzyl)-3-(3-fluorophenyl)urea ... 528
1-(2-Chlorobenzyl)-3-(4-fluorophenyl)urea ... 528
1-(4-Chlorobenzyl)-3-(3-fluorophenyl)urea ... 528
1-(2-Chloro-5-methylphenyl)-3-(2-fluorophenyl)urea ... 537
1-(2-Chloro-5-methylphenyl)-3-(3-fluorophenyl)urea ... 544
1-(2-Chloro-6-methylphenyl)-3-(4-fluorophenyl)urea ... 539
1-(3-Chloro-2-methylphenyl)-3-(2-fluorophenyl)urea ... 537
1-(3-Chloro-4-methylphenyl)-3-(2-fluorophenyl)urea ... 537
1-(4-Chloro-2-methylphenyl)-3-(2-fluorophenyl)urea ... 537
1-(4-Chloro-2-methylphenyl)-3-(3-fluorophenyl)urea ... 544
1-(5-Chloro-2-methylphenyl)-3-(2-fluorophenyl)urea ... 537
1-(5-Chloro-2-methylphenyl)-3-(4-fluorophenyl)urea ... 540
$C_{14}H_{12}ClFN_2O_2$
1-(5-Chloro-2-methoxyphenyl)-3-(3-fluorophenyl)urea ... 520
$C_{14}H_{12}ClF_3N_2$
2-(2-(2-Chloro-4-trifluoromethylanilino)ethyl)pyridine ... 925
$C_{14}H_{12}ClF_3N_2O_2$
1,2-Dihydro-1-methyl-4-pyridyl *N*-(4-chloro-2-(trifluoromethyl)phenyl)carbamate ... 608
$C_{14}H_{12}ClF_3N_6O_3S$
2-(3-(2-Thiazolyl)ureido)acetic acid N^2-(*N*-(2-chloro-5-(trfluoromethyl)phenyl)carbamoyl)hydrazide ... 874
$C_{14}H_{12}ClIN_2O_2$
1-(5-Chloro-2-methoxyphenyl)-3-(2-iodophenyl)urea ... 529
$C_{14}H_{12}ClN$
N-Benzylidene-2-chloro-6-methylaniline ... 1067
N-Benzylidene-3-chloro-4-methylaniline ... 1067
N-(2-Chlorobenzylidene)-*o*-toluidine ... 1064
N-(2-Chlorobenzylidene)-*m*-toluidine ... 1064
N-(2-Chlorobenzylidene)-*p*-toluidine ... 1064
N-(3-Chlorobenzylidene)-*m*-toluidine ... 1063
N-(4-Chlorobenzylidene)-*o*-toluidine ... 1064
N-(4-Chlorobenzylidene)-*m*-toluidine ... 1065
N-(4-Chlorobenzylidene)-*p*-toluidine ... 1065
3-Chloro-9-ethylcarbazole ... 892
N-(4-Methylbenzylidene)-2-chloroaniline ... 1064
N-(4-Methylbenzylidene)-3-chloroaniline ... 1065
N-(4-Methylbenzylidene)-4-chloroaniline ... 1065
$C_{14}H_{12}ClNO$
N-Benzyl-4-chlorobenzamide ... 480
2-Chloro-*o*-benzotoluidide ... 478
2-Chloro-*m*-benzotoluidide ... 478
2-Chloro-*p*-benzotoluidide ... 479
4-Chloro-*o*-benzotoluidide ... 479
4-Chloro-*m*-benzotoluidide ... 480
4-Chloro-*p*-benzotoluidide ... 483
N-(2-Chlorobenzylidene)-*m*-anisidine ... 1046
N-(2-Chlorobenzylidene)-*p*-anisidine ... 1046
N-(3-Chlorobenzylidene)-*o*-anisidine ... 1045
N-(4-Chlorobenzylidene)-*o*-anisidine ... 1045
N-(4-Chlorobenzylidene)-*p*-anisidine ... 1050
3-Chloro-*N*-(2-methoxybenzylidene)aniline ... 1046
3-Chloro-*N*-(3-methoxybenzylidene)aniline ... 1044
3-Chloro-*N*-(4-methoxybenzylidene)aniline ... 1024
4-Chloro-*N*-(2-methoxybenzylidene)aniline ... 1046
4-Chloro-*N*-(3-methoxybenzylidene)aniline ... 1046
4-Chloro-*N*-(4-methoxybenzylidene)aniline ... 1046
2′-Chloro-6′-methylbenzanilide ... 481
3′-Chloro-2′-methylbenzanilide ... 456
3′-Chloro-4′-methylbenzanilide ... 481
4′-Chloro-2′-methylbenzanilide ... 456
5′-Chloro-2′-methylbenzanilide ... 481
α-(3-Chloro-4-methylphenylimino)-*p*-cresol ... 1023
α-(3-Chloro-6-methylphenylimino)-*p*-cresol ... 1023
α-(4-Chloro-2-methylphenylimino)-*p*-cresol ... 1023
α-(6-Chloro-2-methylphenylimino)-*o*-cresol ... 1034
3′-Chloro-2-phenylacetanilide ... 408
2′-Chloro-*o*-toluanilide ... 478
2′-Chloro-*m*-toluanilide ... 478
2′-Chloro-*p*-toluanilide ... 479
3′-Chloro-*m*-toluanilide ... 477
4′-Chloro-*o*-toluanilide ... 457
4′-Chloro-*m*-toluanilide ... 472
4′-Chloro-*p*-toluanilide ... 480
α-(3-Chloro-*o*-tolylimino)-*o*-cresol ... 1035

$C_{14}H_{12}ClNO$ (Cont'd)
α-(3-Chloro-*p*-tolylimino)-*o*-cresol 1035
α-(5-Chloro-*o*-tolylimino)-*o*-cresol 1035
2′,4-Dichloro-6′-methylbenzanilide ... 458
N-(3-Methoxybenzylidene)-2-chloroaniline 1046

$C_{14}H_{12}ClNOS$
4-Chloro-4′-(methylthio)benzanilide .. 450

$C_{14}H_{12}ClNOS_2$
S-(4-Chlorophenyl) *N*-(4-(methylthio)phenyl)thiocarbamate .. 624

$C_{14}H_{12}ClNO_2$
2-Chloro-*m*-benzanisidide 466
4-Chloro-*o*-benzanisidide 465
2-Chloro-*N*-(2-hydroxy-3-methoxybenzylidene)aniline 1022
2-Chloro-2′-methoxybenzanilide 465
2-Chloro-4′-methoxybenzanilide 466
3-Chloro-2′-methoxybenzanilide 449
3′-Chloro-4-methoxybenzanilide 450
α-(3-Chlorophenylimino)-6-methoxy-*o*-cresol 1021
α-(4-Chlorophenylimino)-2-methoxy-*p*-cresol 1030
α-(4-Chlorophenylimino)-6-methoxy-*o*-cresol 1021
4-Chlorophenyl *N*-(*o*-tolyl)-carbamate 682
4-Chlorophenyl *N*-(*m*-tolyl)-carbamate 683
4-Chlorophenyl *N*-(*p*-tolyl)-carbamate 683
4′-Chloro-*o*-salicylotoluidide 462
4-Chloro-*N*-(4-tolyl)anthranilic acid ... 343
Phenyl *N*-(2-chloro-6-methylphenyl)-carbamate 632
Phenyl *N*-(3-chloro-4-methylphenyl)-carbamate 683
Phenyl *N*-(4-chloro-2-methylphenyl)-carbamate 686
Phenyl *N*-(5-chloro-2-methylphenyl)-carbamate 686
m-Tolyl *N*-(3-chlorophenyl)-carbamate 631
m-Tolyl *N*-(4-chlorophenyl)-carbamate 634

$C_{14}H_{12}ClNO_2S$
S-(2-Hydroxyphenyl) *N*-(3-chloro-2-methylphenyl)thiocarbamate 611
S-(4-Hydroxyphenyl) *N*-(2-chloro-6-methylphenyl)thiocarbamate 611
S-(2-Hydroxyphenyl) *N*-(2-chloro-6-methyl)thiocarbamate 611
4-(Methylthio)phenyl *N*-(3-chlorophenyl)carbamate 623
4-(Methylthio)phenyl *N*-(4-chlorophenyl)carbamate 623

$C_{14}H_{12}ClNO_3$
3-Chloro-2-(4-methoxyanilino)-benzoic acid 336
2-Chloro-4-methoxyphenyl *N*-phenylcarbamate 617
2-Chlorophenyl *N*-(*o*-anisyl)-carbamate 654
4-Methoxyphenyl *N*-(2-chlorophenyl)carbamate 617

$C_{14}H_{12}ClNO_3S$
4-Acetamido-4′-chlorodiphenyl sulfone 773
N-(3-Acetylphenyl)-4-chlorobenzenesulfonamide 807
N-(2-Benzoyl-4-chlorophenyl)-methanesulfonamide 807
S-(2-Hydroxyphenyl) *N*-(4-chloro-2-methoxyphenyl)thiocarbamate 611

$C_{14}H_{12}ClNO_4S$
N-(4-(4-Chlorophenylsulfonyl)-phenyl)-glycine 773
Methyl 2-(4-chlorophenylsulfonamido)-benzoate 807

$C_{14}H_{12}ClNS$
N-(4-Chlorobenzylidene)-3-(methylthio)aniline 1051
10-(2-Chloroethyl)phenothiazine 244
3-Chloro-10-ethylphenothiazine 251
2′-Chloro-6′-methylthiobenzanilide ... 730
4-Chloro-*N*-methylthiobenzanilide 742

$C_{14}H_{12}ClN_3O$
4-Aminobenzoic (2-chlorobenzylidene)hydrazide 492
1-(4-Chlorobenzylidene)-4-phenylsemicarbazide 529

$C_{14}H_{12}ClN_3O_2$
4-Pyridylmethyleneamino *N*-(4-chloro-2-methylphenyl)carbamate ... 1027

$C_{14}H_{12}ClN_3O_2S$
4-(4-Chlorophenyl)-1-(4-hydroxybenzoyl)thiosemicarbazide. 728

$C_{14}H_{12}ClN_3O_3$
1-(2-Chloro-6-methylphenyl)-3-(3-nitrophenyl)urea 524
1-(2-Chloro-6-methylphenyl)-3-(4-nitrophenyl)urea 526
1-(2-Chlorophenyl)-3-(2-methyl-4-nitrophenyl)urea 529
1-(4-Chlorophenyl)-3-(2-methyl-4-nitrophenyl)urea 526
4-Pyridylmethyleneamino *N*-(5-chloro-2-methoxyphenyl)-carbamate 929

$C_{14}H_{12}ClN_3O_3S$
1-(4-Chloro-2-nitrophenyl)-3-(4-(methylthio)phenyl)urea 521

$C_{14}H_{12}ClN_3O_4$
1-(5-Chloro-2-hydroxyphenyl)-3-(2-methyl-5-nitrophenyl)urea 513
1-(3-Chloro-2-methylphenyl)-3-(2-hydroxy-5-nitrophenyl)urea 513
1-(2-Chlorophenyl)-3-(2-methoxy-3-nitrophenyl)urea 517
1-(2-Chlorophenyl)-3-(2-methoxy-5-nitrophenyl)urea 517
1-(3-Chlorophenyl)-3-(2-methoxy-3-nitrophenyl)urea 518
1-(3-Chlorophenyl)-3-(2-methoxy-5-nitrophenyl)urea 552
1-(3-Chlorophenyl)-3-(4-methoxy-3-nitrophenyl)urea 517

$C_{14}H_{12}ClN_3O_6$
5-Chloro-2,4-dimethoxy-2′,4′-dinitrodiphenylamine 260
5-Nitro-2-pyridyl *N*-(5-chloro-2,4-dimethoxyphenyl)carbamate 927

$C_{14}H_{12}ClN_5O_5$
4-(3-Chlorophenyl)-1-(2,4-dinitrophenyl)-1-methylsemicarbazide 522

$C_{14}H_{12}Cl_2$
4,4′-Dichlorobibenzyl 172
1,2-Dichloro-1,2-diphenylethane 173

$C_{14}H_{12}Cl_2IN$
2-(2,6-Dichlorostyryl)-1-methylpyridinium iodide 958
2-(3,4-Dichlorostyryl)-1-methylpyridinium iodide 958
4-(2,6-Dichlorostyryl)-1-methylpyridinium iodide 958
4-(3,4-Dichlorostyryl)-1-methylpyridinium iodide 958

$C_{14}H_{12}Cl_2N_2O$
1-Benzyl-3-(2,3-dichlorophenyl)urea .. 573
1-Benzyl-3-(2,4-dichlorophenyl)urea .. 573
1-Benzyl-3-(2,5-dichlorophenyl)urea .. 600
1-Benzyl-3-(2,6-dichlorophenyl)urea .. 573
1-Benzyl-3-(3,4-dichlorophenyl)urea .. 573
4-Chlorobenzoic N^2-(4-chlorobenzyl)hydrazide 492
1-(3-Chlorobenzyl)-3-(3-chlorophenyl)urea 527
1-(2-Chloro-5-methylphenyl)-3-(2-chlorophenyl)urea 538
1-(3-Chloro-2-methylphenyl)-3-(2-chlorophenyl)urea 537
1-(5-Chloro-2-methylphenyl)-3-(2-chlorophenyl)urea 538
1-(2-Chlorophenyl)-3-(2-chloro-6-methylphenyl)urea 537
1-(2-Chlorophenyl)-3-(3-chloro-4-methylphenyl)urea 537
1-(2-Chlorophenyl)-3-(4-chloro-2-methylphenyl)urea 537
1-(3-Chlorophenyl)-3-(2-chloro-6-methylphenyl)urea 541
1-(3-Chlorophenyl)-3-(3-chloro-4-methylphenyl)urea 600
1-(3-Chlorophenyl)-3-(4-chloro-2-methylphenyl)urea 600
1-(3-Chlorophenyl)-3-(5-chloro-2-methylphenyl)urea 600
1-(4-Chlorophenyl)-3-(2-chloro-6-methylphenyl)urea 539
1-(4-Chlorophenyl)-3-(4-chloro-2-methylphenyl)urea 591
1-(2,6-Dichlorophenyl)-3-(4-methylphenyl)urea 539
3-(2,3-Dichlorophenyl)-1-methyl-1-phenylurea 595
3-(2,4-Dichlorophenyl)-1-methyl-1-phenylurea 600
3-(2,5-Dichlorophenyl)-1-methyl-1-phenylurea 600
3-(2,6-Dichlorophenyl)-1-methyl-1-phenylurea 536
3-(3,4-Dichlorophenyl)-1-methyl-1-phenylurea 600
1-(2,6-Dichlorophenyl)-3-(*o*-tolyl)urea 537
1-(2,6-Dichlorophenyl)-3-(*m*-tolyl)urea 541

$C_{14}H_{12}Cl_2N_2OS$
1-(2,4-Dichlorophenyl)-3-(4-(methylthio)phenyl)urea 522

$C_{14}H_{12}Cl_2N_2O_2$
3-Amino-3′,5′-dichloro-2′-hydroxy-4′-methylbenzanilide 447
1-(2-Chloro-4-methoxyphenyl)-3-(2-chlorophenyl)urea 520
1-(2-Chloro-4-methoxyphenyl)-3-(3-chlorophenyl)urea 544
1-(4-Chloro-2-methoxyphenyl)-3-(2-chlorophenyl)urea 519
1-(4-Chloro-2-methoxyphenyl)-3-(3-chlorophenyl)urea 529
1-(2,3-Dichlorophenyl)-3-(2-methoxyphenyl)urea 519
1-(2,3-Dichlorophenyl)-3-(4-methoxyphenyl)urea 521
1-(2,6-Dichlorophenyl)-3-(2-methoxyphenyl)urea 519
1-(2,6-Dichlorophenyl)-3-(3-methoxyphenyl)urea 521
1-(2,6-Dichlorophenyl)-3-(4-methoxyphenyl)urea 521

$C_{14}H_{12}Cl_2N_2O_3S$
N^4-Acetyl-N^1-(2,5-dichlorophenyl)-sulfanilamide 808
1-(3,4-Dichlorophenyl)-3-(*p*-tosyl)urea 809

$C_{14}H_{12}Cl_2N_2S$
1-(3,5-Dichlorobenzyl)-3-phenyl-2-thiourea 730

$C_{14}H_{12}Cl_2N_4OS$
1-(2,3-Dichlorophenylcarbamoyl)-4-phenyl-3-thiosemicarbazide 730
1-(3,4-Dichlorophenyl)-6-phenyl-5-thiobiurea 732

$C_{14}H_{12}Cl_2N_4O_2$
1,6-Bis(2-chlorophenyl)diurea 532
1,6-Bis(4-chlorophenyl)diurea 532
4-(3-Chlorophenyl)-1-(*N*-(3-chlorophenyl)carbamoyl)-semicarbazide 528

$C_{14}H_{12}Cl_2O_2$
1,2-Bis(2-chlorophenoxy)ethane 179
1,2-Bis(4-chlorophenoxy)ethane 179
1,2-Bis(4-chlorophenyl)-1,2-ethanediol 198

$C_{14}H_{12}Cl_2O_2S$
2,2′-Thiobis(4-chloro-6-methylphenol) 210
2,2′-Thiobis(5-chloro-3-methylphenol) 210

$C_{14}H_{12}Cl_3NO_2$
6-Methoxy-α-(2,4,5-trichloroanilino)-*o*-cresol 222

$C_{14}H_{12}Cl_3NO_2S$
2′,5′-Dimethyl-2,4,5-trichlorobenzenesulfonanilide 810

$C_{14}H_{12}Cl_3NO_4S$
2′,5′-Dimethoxy-3,4,4′-trichlorobenzenesulfonanilide 806

$C_{14}H_{12}Cl_4O_4$
Cyclohexyl tetrachlorophthalate 375

$C_{14}H_{12}Cl_6O_2$
5,6,7,8,9,9-Hexachloro-1,2,3,4,4a,5,8,8a-octahydro-1,4:5,8-dimethanonaphth-2-yl acetate 74

$C_{14}H_{12}FN$
N-(4-Fluorobenzylidene)-*p*-toluidine. 1077

$C_{14}H_{12}FNO$
N-(4-Fluorobenzylidene)-*p*-anisidine. 1050
4-Fluoro-*N*-(4-methoxybenzylidene)-aniline 1024
2-Fluoro-2′-methylbenzanilide 457
4′-Fluoro-3-methylbenzanilide 458
4′-Fluoro-4-methylbenzanilide 458
4′-(4-Fluorophenyl)acetanilide 485

$C_{14}H_{12}FNOS$
S-Benzyl *N*-(2-fluorophenyl)-thiocarbamate 641
S-Benzyl *N*-(3-fluorophenyl)-thiocarbamate 641

$C_{14}H_{12}FNO_2$
2-Fluoro-2′-methoxybenzanilide 449
4′-Fluoro-4-methoxybenzanilide 449
α-(4-Fluorophenylimino)-2-methoxy-*o*-cresol 1021
N-Piperonyl-4-fluoroaniline 247
p-Tolyl *N*-(2-fluorophenyl)-carbamate 633

$C_{14}H_{12}FNO_2S$
4-Fluorophenyl *N*-(3-(methylthio)-phenyl)carbamate 623
4-Fluorophenyl *N*-(4-(methylthio)-phenyl)carbamate 623

$C_{14}H_{12}FNO_3$
3-Methoxyphenyl *N*-(3-fluorophenyl)carbamate 617
4-Methoxyphenyl *N*-(2-fluorophenyl)carbamate 617
4-Methoxyphenyl *N*-(4-fluorophenyl)carbamate 617

$C_{14}H_{12}FNO_4S$
N-(*p*-Anisoyl)-sulfanilyl fluoride 802

$C_{14}H_{12}FNS$
2-Amino-7-fluoro-3-(methylthio)-fluorene 230

$C_{14}H_{12}FN_3O_2$
1-(2-Carbamoylphenyl)-3-(2-fluorophenyl)urea 532
1-(2-Carbamoylphenyl)-3-(3-fluorophenyl)urea 532
1-(2-Carbamoylphenyl)-3-(4-fluorophenyl)urea 532

$C_{14}H_{12}FN_3O_2S$
1-Benzyl-3-(4-fluoro-3-nitrophenyl)-2-thiourea 730

$C_{14}H_{12}FN_3O_3$
4-(2-Fluorophenyl)-1-(4-hydroxybenzoyl)semicarbazide 513
4-(4-Fluorophenyl)-1-(4-hydroxybenzoyl)semicarbazide 513
1-(2-Fluorophenyl)-3-(2-methyl-5-nitrophenyl)urea 524
1-(3-Fluorophenyl)-3-(2-methyl-4-nitrophenyl)urea 526
1-(3-Fluorophenyl)-3-(2-methyl-5-nitrophenyl)urea 525

$C_{14}H_{12}FN_3O_4$
1-(2-Fluoro-5-nitrophenyl)-3-(2-methoxyphenyl)urea 517
1-(2-Fluoro-5-nitrophenyl)-3-(4-methoxyphenyl)urea 517
1-(2-Fluorophenyl)-3-(2-methoxy-4-nitrophenyl)urea 518
1-(3-Fluorophenyl)-3-(2-methoxy-4-nitrophenyl)urea 518
1-(3-Fluorophenyl)-3-(2-methoxy-5-nitrophenyl)urea 517

$C_{14}H_{12}F_2N_2OS$
1-(2,5-Difluorophenyl)-3-(4-(methylthio)phenyl)urea 522

$C_{14}H_{12}F_2N_2O_2$
1-(2,5-Difluorophenyl)-3-(4-methoxyphenyl)urea 560

$C_{14}H_{12}F_2N_4O_2$
1,6-Bis(4-fluorophenyl)biurea 529

$C_{14}H_{12}F_2O_2$
4-Benzyloxy-α,α-difluoroanisole 176

$C_{14}H_{12}F_3N$
N-Benzyl-α,α,α-trifluoro-*m*-toluidine. 222

$C_{14}H_{12}F_3NO$
6-(*p*-Tolyloxy)-α,α,α-trifluoro-*m*-toluidine 222
$C_{14}H_{12}F_3N_3O$
1-Phenyl-4-(3-(trifluoromethyl)phenyl)semicarbazide 509
1-(3-Pyridylmethyl)-3-(3-(trifluoromethyl)phenyl)urea 925
1-(4-Pyridylmethyl)-3-(3-(trifluoromethyl)phenyl)urea 925
$C_{14}H_{12}F_3N_3O_3S$
1-(4-Sulfamoylphenyl)-3-(2-(trifluoromethyl)phenyl)urea 805
$C_{14}H_{12}IN$
9-Ethyl-3-iodocarbazole 892
10-Methylacridinium iodide 996
$C_{14}H_{12}INO$
4-Iodo-*N*-(4-methoxybenzylidene)aniline 1024
4′-(4-Iodophenyl)acetanilide 409
$C_{14}H_{12}INO_2$
N-(2-Hydroxy-3-methoxybenzylidene)-4-iodoaniline 1022
α-(4-Iodophenylimino)-2-methoxy-*p*-cresol 1031
$C_{14}H_{12}IN_3O_4$
1-(4-Iodophenyl)-3-(2-methoxy-5-nitrophenyl)urea 517
$C_{14}H_{12}I_2O_2$
4,4′-Diiodo-3,3′-dimethoxybiphenyl 186
$C_{14}H_{12}N_2$
2-(4-Aminophenyl)-2-(phenyl)acetonitrile hydrochloride 754
2-Benzylbenzimidazole 905
N,2-Diphenylglycinonitrile 754
$C_{14}H_{12}N_2O$
2-(2-Benzimidazolylmethyl)phenol 902
Benzoic benzylidenehydrazide 501
2-(Phenoxymethyl)benzimidazole 903
3′-(Phenylazo)acetophenone 1092
4′-(Phenylazo)acetophenone 1092
α-Phenyl-2-benzimidazolemethanol 902
2-Phenyl-1,2,3,4-tetrahydro-4-quinazolinone 1000
Terephthaldicarboxaldehyde *mono*-(phenylhydrazone) 324
$C_{14}H_{12}N_2OS$
α-(2-Benzothiazolylamino)-*o*-cresol 912
1-Benzoyl-3-phenyl-2-thiourea 745
3-Cyano-1-methyl-4-(methylthio)-6-phenyl-2(1*H*)-pyridinone 763
$C_{14}H_{12}N_2OS_2$
3-(4-Hydroxyanilinomethyl)-benzothiazoline-2-thione hydrochloride 734
$C_{14}H_{12}N_2O_2$
Benzoic acid benzylidenehydrazide 500
N,*N*′-Bis(4-hydroxyphenyl)-1,4-diazabutadiene 1033
6,6′-Dimethyl-2,2′-pyridil 952
9-Ethyl-3-nitrocarbazole 891
4-Hydroxybenzaldehyde azine 1029
α-(3-Hydroxy-2-oxo-3-indolinyl)-2-methylpyridine 932
4′-(4-Hydroxyphenylazo)-acetophenone 1085
N-(4-Methylbenzylidene)-3-nitroaniline 1055
N-(4-Methylbenzylidene)-4-nitroaniline 1057
4-Methyl-*N*-(4-nitrobenzylidene)-aniline 1073
Methyl 4-(phenylazo)benzoate 1092
N-(2-Nitrobenzylidene)-*p*-toluidine 1055
N-(3-Nitrobenzylidene)-*o*-toluidine 1059
N-(3-Nitrobenzylidene)-*m*-toluidine 1059
N-(3-Nitrobenzylidene)-*p*-toluidine 1059
Oxanilide 434
Salicylaldehyde azine 769
$C_{14}H_{12}N_2O_2S$
1-(2-Carboxyphenyl)-3-phenyl-2-thiourea 744
N-Cyano-2′-methylbenzenesulfonanilide 826
N-Cyano-4′-methylbenzenesulfonanilide 826
N-Cyano-*p*-toluenesulfonanilide 826
$C_{14}H_{12}N_2O_2S_2$
1-(*p*-Tolylsulfonyl)-2(3*H*)-benzimidazolethione 823
$C_{14}H_{12}N_2O_3$
5-Acetamido-4-nitroacenaphthene 487
Benzoic (2,4-dihydroxybenzylidene)-hydrazide 494
2-Carbamoylphenyl *N*-phenylcarbamate 666
2,3,5,6,10,10a-Hexahydro-1*H*,4*H*-acenaphtho(1,8a-c)pyrrole-1,3,10-trione 10-oxime 1012
4-Hydroxybenzoic (2-hydroxybenzylidene)hydrazide 494
N-(4-Methoxybenzylidene)-3-nitroaniline 1043
N-(4-Methoxybenzylidene)-4-nitroaniline 1043
4-Methoxy-*N*-(3-nitrobenzylidene)-aniline 1043
3-Methyl-4′-nitrobenzanilide 470
N-Methyl-4′-nitrobenzanilide 470
4-Methyl-4′-nitrobenzanilide 470
N-(3-Methyl-2-pyridyl)-phthalamic acid 962
N-(4-Methyl-2-pyridyl)-phthalamic acid 962
N-(6-Methyl-2-pyridyl)-phthalamic acid 962
2-Nitrobenzo-*m*-toluidide 469
N-(2-Nitrobenzylidene)-*o*-anisidine 1043
N-(2-Nitrobenzylidene)-*m*-anisidine 1043
N-(3-Nitrobenzylidene)-*p*-anisidine 1044
2′-Nitro-4′-phenylacetanilide 487
4′-Nitro-2′-phenylacetanilide 427
α-(3-Nitro-*p*-tolylimino)-*o*-cresol 1033
α-(4-Nitro-*o*-tolylimino)-*o*-cresol 1033
α-(5-Nitro-*o*-tolylimino)-*o*-cresol 1033
2-(3-Phenylureido)benzoic acid 576
4-(3-Phenylureido)benzoic acid 576
N-(3-Pyridylmethyl)phthalamic acid 949
Salicylic (2-hydroxybenzylidene)-hydrazide 1028
$C_{14}H_{12}N_2O_3S$
2-(*p*-Aminophenyl)-6-methylbenzothiazole-7-sulfonic acid 914
10-Ethyl-3-nitrophenothiazine-5-oxide 771
4′-(Methylthio)-3-nitrobenzanilide 468
4′-(Methylthio)-4-nitrobenzanilide 468
$C_{14}H_{12}N_2O_4$
2,2′-Benzidinedicarboxylic acid 342
4,6′-Bianthranilic acid 342
2,2′-Dinitrobibenzyl 279
4-Hydroxybenzoic (2,4-dihydroxybenzylidene)hydrazide 1028
4-Methoxy-3-nitrobenzanilide 465
4′-Methoxy-3-nitrobenzanilide 465
4′-Methoxy-4-nitrobenzanilide 466
α-(2-Methoxy-4-nitrophenylimino)-*o*-cresol 1031
α-(2-Methoxy-5-nitrophenylimino)-*o*-cresol 1031
6-Methoxy-α-(3-nitrophenylimino)-*o*-cresol 1029
6-Methoxy-α-(4-nitrophenylimino)-*o*-cresol 1030
4-Nitrobenzyl carbanilate 670
$C_{14}H_{12}N_2O_4S$
S-(2-Hydroxyphenyl) *N*-(2-methyl-5-nitrophenyl)thiocarbamate 644
S-(4-Hydroxyphenyl) *N*-(2-methyl-5-nitrophenyl)thiocarbamate 644
S-(4-Hydroxyphenyl) *N*-(2-methyl-6-nitrophenyl)thiocarbamate 644
4′,4″-Sulfonyldiformanilide 782
$C_{14}H_{12}N_2O_5$
2-Hydroxy-*N*-(2-hydroxy-3-methoxy-5-nitrobenzylidene)aniline 1028
2′-Hydroxy-4-methoxy-4′-nitrobenzanilide 461
2-(4-Methoxyanilino)-5-nitrobenzoic acid 340
$C_{14}H_{12}N_2O_5S$
4-Acetamido-4′-nitrodiphenyl sulfone 780
4′-Sulfamoylphthalanilic acid 823
$C_{14}H_{12}N_2O_6$
1,2-Bis(2-nitrophenoxy)ethane 266
1,2-Bis(3-nitrophenoxy)ethane 266
1,2-Bis(4-nitrophenoxy)ethane 266
N,*N*′-(1,2-Phenylene)-dimaleamic acid 475
N,*N*′-(1,3-Phenylene)-dimaleamic acid 475
$C_{14}H_{12}N_2O_6S$
4-Amino-3-carboxy-*N*-(2-carboxyphenyl)-benzenesulfonamide 817
$C_{14}H_{12}N_2O_8S_4$
4-Methylsulfonyl-2-nitrophenyl disulfide 776
$C_{14}H_{12}N_2S$
2-(Benzylthio)benzimidazole 903
2-(Phenylthiomethyl)benzimidazole 903
$C_{14}H_{12}N_2S_2$
2-(4,5-Dimethyl-2-thiazolylthio)-quinoline 987
$C_{14}H_{12}N_3$
5-Anilino-4-phenyl-1*H*-1,2,3-triazole 869
$C_{14}H_{12}N_4$
5-Amino-1,4-diphenyl-1,2,3-triazole 869
5-Anilino-1-phenyl-1*H*-1,2,4-triazole 869
$C_{14}H_{12}N_4O$
1,2-Dihydro-4,6-dimethyl-2-oxo-1-(2-pyridylmethyleneamino)-nicotinonitrile 1053
3-Hydrazino-4-phenyl-1*H*-pyrido(1,2-c)pyrimidin-1-one 705
$C_{14}H_{12}N_4OS$
1-Cinnamoyl-3-(2-pyrimidinyl)-2-thiourea 976
$C_{14}H_{12}N_4O_2$
4-(1*H*-Benzotriazol-1-ylmethylamino)benzoic acid 909
$C_{14}H_{12}N_4O_2S$
1-(3-Nitroanilinomethyl)-2(3*H*)-benzimidazolethione 740
1-(4-Nitroanilinomethyl)-2(3*H*)-benzimidazolethione 905
1-(3-Nitrobenzylidene)-4-phenyl-3-thiosemicarbazide 1069
$C_{14}H_{12}N_4O_2S_2$
4-Nitrobenzenesulfonic (3-methyl-2-benzothiazolinylidene)hydrazide 820
$C_{14}H_{12}N_4O_3$
Anthranilic (3-nitrobenzylidene)-hydrazide 499
7,8-Dimethyl-10-(formylmethyl)-benzo(g)pteridine-2,4(3*H*,10*H*)-dione 1003
$C_{14}H_{12}N_4O_4$
Acetophenone (2,4-dinitrophenyl)-hydrazone 273
4′-Nitroacetophenone 4-nitrophenylhydrazone 277
Phenylacetaldehyde (2,4-dinitrophenyl)hydrazone 274
p-Tolualdehyde, 2,4-dinitrophenylhydrazone 1052
$C_{14}H_{12}N_4O_4S_2$
2-Nitrobenzenesulfonic (3-methyl-2-benzothiazolinylidene)hydrazide 820
3-Nitrobenzenesulfonic (3-methyl-2-benzothiazolinylidene)hydrazide 820
$C_{14}H_{12}N_4O_5$
4-Acetamido-2′,4′-dinitrodiphenylamine 423
Benzaldehyde(2,4-dinitrophenyl-α-hydroxymethyl)hydrazone 264
1-(4-Carboxyphenyl)-4-(3-nitrophenyl)semicarbazide 563
2,6-Dinitro-4′-ethoxyazobenzene 1086
Salicyaldehyde 2,6-dinitro-4-methylphenylhydrazone 1036
$C_{14}H_{12}N_4O_5S_2$
1-(3-Nitrobenzoyl)-3-(4-sulfamoylphenyl)-2-thiourea 822
$C_{14}H_{12}N_4O_6$
N-(2-Ethylphenyl)picrylamine 272
N-(4-Ethylphenyl)picrylamine 272
1-(2-Hydroxy-5-nitrophenyl)-3-(2-methyl-5-nitrophenyl)urea 547
N-Phenethylpicrylamine 273
N-Picryl-2,3-xylidine 272
N-Picryl-2,4-xylidine 273
N-Picryl-2,6-xylidine 272
N-Picryl-3,4-xylidine 273
N-(2,5-Xylyl)picrylamine 272
N-(3,5-Xylyl)picrylamine 272
$C_{14}H_{12}N_4O_7$
N-(4-Ethoxyphenyl)picrylamine 266
N-(4-Methoxy-*o*-tolyl)-picrylamine 267
N-(6-Methoxy-*m*-tolyl)picrylamine 266
N-(2,4,6-Trinitrophenyl)-*o*-phenetidine 266
$C_{14}H_{12}N_4O_8$
N-(2,4-Dimethoxyphenyl)-picrylamine 266
N-(2,5-Dimethoxyphenyl)-picrylamine 266
$C_{14}H_{12}N_6O_2$
Oxalic acid bis((2-pyridylmethylene)hydrazide) 1053
$C_{14}H_{12}N_6O_8$
N,*N*′-Bis(2,4-dinitrophenyl)-ethylenediamine 273
$C_{14}H_{12}O$
2-Ethyldibenzofuran 897
1-Fluorenemethanol 208
2-Fluorenemethanol 208
1-Methoxyfluorene 188
Phenyl 4-vinylphenyl ether 183
2-Stilbenol 191
4-Stilbenol 191
1,2,3,4-Tetrahydrophenanthren-4-one 298
$C_{14}H_{12}OS$
4-Acetyldiphenyl sulfide 294
4′-Methyl-3-(2-thienyl)-acrylophenone 857
$C_{14}H_{12}O_2$
4-Biphenylyl acetate 380
2-(2-Hydroxyethyl)dibenzofuran 893
4-(2-Hydroxyethyl)dibenzofuran 893
4-Hydroxy-3-methylbenzophenone 310
Methyl 4-biphenylcarboxylate 375
2,5-Norbornadien-7-yl benzoate 375
2-Phenoxyacetophenone 292
Phenyl 2-methylbenzoate 380
$C_{14}H_{12}O_2S$
Benzyl *S*-phenyl thiocarbonate 381
2,2-Diphenyl-2-mercaptoacetic acid 329
3-(4-Methoxyphenyl)-1-(2-thienyl)-2-propen-1-one 855
Phenyl *trans*-styryl sulfone 783
α-(Phenylthio)-*o*-toluic acid 342
$C_{14}H_{12}O_3$
2-(Benzyloxy)benzoic acid 339
4-Benzyloxybenzoic acid 339
Benzyl salicylate 368
2-(2-Buten-1-yloxy)-1,4-naphthoquinone 316
2,8-Dimethoxydibenzofuran 893
4,6-Dimethoxydibenzofuran 893
1-(2-Furyl)-3-(4-methoxyphenyl)-2-propen-1-one 846
4-Hydroxy-4′-methoxybenzophenone 310
2′-(Hydroxymethyl)-2-biphenylcarboxylic acid 338
DL-2-(4-Hydroxyphenyl)-2-phenylacetic acid 326
3-(4-Methoxy-1-naphthyl)-acrylic acid 335
2-Methoxyphenyl benzoate 378
4-Methoxyphenyl benzoate 378
γ-Oxo-1-naphthalenebutyric acid 331
γ-Oxo-2-naphthalenebutyric acid 331
2-(Phenoxymethoxy)benzaldehyde 323
2-(Phenoxymethyl)benzoic acid 340
$C_{14}H_{12}O_5$
Benzyl 3,5-dioxo-2-hydroxy-1-cyclopentene-1-acetate 367
9-Oxo-1,2,3,4-tetrahydro-1,4-ethanonaphthalene-2,3-dicarboxylic acid 331
$C_{14}H_{12}O_6S$
2-Carboxy-6-methoxyphenyl benzenesulfonate 797
2-Hydroxy-4-methoxybenzophenone-5-sulfonic acid 787

$C_{14}H_{12}O_6S$ (Cont'd)
2-Hydroxy-4-methoxybenzophenone-5-sulfonic acid, sodium salt 791
$C_{14}H_{12}S$
Phenyl styryl sulfide 212
$C_{14}H_{12}S_2$
1,2-Bis(phenylthio)ethene 212
$C_{14}H_{13}AsClNO_4$
5-Chloro-2,3-dihydro-2-hydroxy-2-phenyl-1*H*-isoarsindolium nitrate 1128
$C_{14}H_{13}Br$
Bromophenyl(*o*-tolyl)methane 173
$C_{14}H_{13}BrClNO_3$
4′-Bromo-2′-chloro-1,2,3,6-tetrahydrophthalanilic acid 407
$C_{14}H_{13}BrClNO_4S$
4-Bromo-*N*-(5-chloro-2,4-dimethoxyphenyl)-benzenesulfonamide 806
$C_{14}H_{13}BrN_2O$
1-Benzyl-3-(3-bromophenyl)urea 527
1-Benzyl-3-(4-bromophenyl)urea 572
3-Bromo-4-methoxybenzaldehyde phenylhydrazone 235
$C_{14}H_{13}BrN_2O_2$
1-(4-Bromophenyl)-3-(2-methoxyphenyl)urea 519
1-(4-Bromophenyl)-3-(4-methoxyphenyl)urea 521
$C_{14}H_{13}BrN_2O_3S$
1-(4-Bromophenyl)-3-(*p*-tosyl)urea 809
$C_{14}H_{13}BrO$
2-Bromo-1,1-diphenyl-1-ethanol 197
$C_{14}H_{13}BrO_2$
1-(Benzyloxy)-4-bromo-2-methoxybenzene 185
$C_{14}H_{13}ClHgO_2$
α-(Chloromercurimethyl)-4-hydroxybenzhydrol 1137
$C_{14}H_{13}ClN_2O$
4′-Anilino-2-chloroacetanilide 392
1-Benzyl-3-(2-chlorophenyl)urea 572
1-Benzyl-3-(3-chlorophenyl)urea 571
1-Benzyl-3-(4-chlorophenyl)urea 572
1-(3-Chlorobenzyl)-3-phenylurea 527
1-(2-Chloro-3-methylphenyl)-3-phenylurea 536
1-(3-Chloro-2-methylphenyl)-3-phenylurea 536
1-(3-Chloro-4-methylphenyl)-3-phenylurea 600
1-(4-Chloro-2-methylphenyl)-3-phenylurea 600
1-(5-Chloro-2-methylphenyl)-3-phenylurea 600
1-(2-Chlorophenyl)-3-(2-methylphenyl)urea 536
1-(3-Chlorophenyl)-3-(3-methylphenyl)urea 534
3-(2-Chlorophenyl)-1-methyl-1-phenylurea 588
3-(3-Chlorophenyl)-1-methyl-1-phenylurea 581
3-(4-Chlorophenyl)-1-methyl-1-phenylurea 591
1-(2-Chlorophenyl)-3-(*m*-tolyl)urea 588
1-(2-Chlorophenyl)-3-(*p*-tolyl)urea 588
1-(3-Chlorophenyl)-3-(*o*-tolyl)urea 588
1-(4-Chlorophenyl)-3-(*o*-tolyl)urea 588
1-(4-Chlorophenyl)-3-(*m*-tolyl)urea 591
1-(4-Chlorophenyl)-3-(*p*-tolyl)urea 591
6-Methyl-2-pyridinealdoxime *o*-4-chlorobenzyl ether 1054
$C_{14}H_{13}ClN_2OS$
1-(5-Chloro-2-methoxyphenyl)-3-phenyl-2-thiourea 729
1-(3-Chloro-2-methylphenyl)-3-(2-mercaptophenyl)urea 521
1-(4-Chloro-2-methylphenyl)-3-(2-mercaptophenyl)urea 521
1-(3-Chlorophenyl)-3-(3-(methylthio)-phenyl)urea 521
1-(4-Chlorophenyl)-3-(3-(methylthio)-phenyl)urea 522
$C_{14}H_{13}ClN_2O_2$
1-(*m*-Anisyl)-3-(3-chlorophenyl)urea 558
1-(5-Chloro-2-methoxyphenyl)-3-phenylurea 519
1-(2-Chloro-6-methylphenyl)-3-(2-hydroxyphenyl)urea 513
1-(2-Chloro-6-methylphenyl)-3-(3-hydroxyphenyl)urea 514
1-(3-Chloro-2-methylphenyl)-3-(3-hydroxyphenyl)urea 514
1-(5-Chloro-2-methylphenyl)-3-(2-hydroxyphenyl)urea 513
1-(2-Chlorophenyl)-3-(2-methoxyphenyl)urea 555
1-(2-Chlorophenyl)-3-(3-methoxyphenyl)urea 559
1-(*o*-Chlorophenyl)-3-(*p*-methoxyphenyl)urea 559
1-(3-Chlorophenyl)-3-(2-methoxyphenyl)urea 519
1-(4-Chlorophenyl)-3-(3-methoxyphenyl)urea 520
1-(4-Chlorophenyl)-3-(4-methoxyphenyl)urea 520
$C_{14}H_{13}ClN_2O_2S$
p-Toluenesulfonic acid (4-chlorobenzylidene)hydrazide 1071
$C_{14}H_{13}ClN_2O_3$
1-(2-Chloro-4-methoxyphenyl)-3-(3-hydroxyphenyl)urea 512
1-(4-Chloro-2-methoxyphenyl)-3-(2-hydroxyphenyl)urea 512
1-(5-Chloro-2-methoxyphenyl)-3-(3-hydroxyphenyl)urea 512
2-Pyridylmethyl *N*-(5-chloro-2-methoxyphenyl)carbamate 927
3-Pyridylmethyl *N*-(5-chloro-2-methoxyphenyl)carbamate 927
4-Pyridylmethyl *N*-(5-chloro-2-methoxyphenyl)carbamate 927
$C_{14}H_{13}ClN_2O_3S$
4-Acetamido-3′-chlorobenzenesulfonanilide 808
p-Toluenesulfonic acid (5-chlorosalicylidene)-hydrazide 1034
$C_{14}H_{13}ClN_2O_4$
4-Chloro-*N*-(2-hydroxy-3-methoxybenzyl)-3-nitroaniline 259
5-Chloro-2-pyridyl *N*-(2,5-dimethoxyphenyl)carbamate 936
2-Pyridyl *N*-(5-chloro-2,4-dimethoxyphenyl)carbamate 926
4-Pyridyl *N*-(5-chloro-2,4-(dimethoxy)phenyl)carbamate 926
$C_{14}H_{13}ClN_2O_5S$
5′-Chloro-4′-nitro-*p*-toluenesulfon-*o*-anisidide 813
$C_{14}H_{13}ClN_2S$
1-(3-Chlorobenzyl)-3-phenyl-2-thiourea 732
1-(2-Chloro-6-methylphenyl)-3-phenyl-2-thiourea 732
$C_{14}H_{13}ClN_4OS$
4-(3-Chlorophenyl)-1-(cyclohexylthiocarbamoyl)-semicarbazide 730
$C_{14}H_{13}ClN_4O_2$
4-(3-Chlorophenylureido)-benzhydrazide 532
$C_{14}H_{13}ClN_4O_3$
4-(3-Chloro-6-methoxyphenyl)-1-(4-pyridylcarbonyl)semicarbazide 926
1-(2-(3-Nitrobenzylidene)-hydrazinocarbonylmethyl)-pyridinium chloride 1057
$C_{14}H_{13}ClO$
Benzyl 4-chlorobenzyl ether 188
2-Chloro-1,1-diphenylethanol 197
Chloro(4-methoxyphenyl)-phenylmethane 180
$C_{14}H_{13}ClO_3S$
4-Chloro-4′-ethoxydiphenyl sulfone 773
$C_{14}H_{13}Cl_2NO$
2-Chloro-5-(4-chloro-3,5-xylyloxy)-aniline 224
$C_{14}H_{13}Cl_2NO_2$
α-(2,4-Dichloroanilino)-6-methoxy-*o*-cresol 222
α-(2,5-Dichloroanilino)-6-methoxy-*o*-cresol 223
α-(3,4-Dichloroanilino)-6-methoxy-*o*-cresol 222
$C_{14}H_{13}Cl_2NO_3$
N-(2,5-Dichlorophenyl)-1,2,3,6-tetrahydrophthalamic acid 402
2′,4′-Dichloro-1,2,3,6-tetrahydrophthalanilic acid 406
3′,4′-Dichloro-1,2,3,6-tetrahydrophthalanilic acid 406
$C_{14}H_{13}Cl_2NO_3S$
3,4-Dichloro-4′-ethoxybenzenesulfonanilide 806
3,4-Dichloro-2′-methoxy-5′-methylbenzenesulfonanilide 806
$C_{14}H_{13}Cl_2N_3O_2$
1-Carbamoylmethyl-3-(4-chlorophenylcarbamoyl)-pyridinium chloride 947
$C_{14}H_{13}Cl_2N_3O_3$
1-(5-Chloro-2,4-dimethoxyphenyl)-3-(5-chloro-2-pyridyl)urea 926
1-(3,5-Dichloro-2-pyridyl)-3-(2,4-dimethoxyphenyl)urea 936
$C_{14}H_{13}Cl_2N_3S$
1-(2,5-Dichlorophenyl-4-(*o*-tolyl)-3-thiosemicarbazide 750
$C_{14}H_{13}Cl_2N_5$
1,5-Bis(4-chlorophenyl)biguanide hydrochloride 724
$C_{14}H_{13}FN_2O$
1-Benzyl-3-(3-fluorophenyl)urea 527
1-(4-Fluorophenyl)-3-(*o*-tolyl)urea 537
1-(4-Fluorophenyl)-3-(*m*-tolyl)urea 538
$C_{14}H_{13}FN_2OS$
1-(2-Fluorophenyl)-3-(4-(methylthio)-phenyl)urea 522
1-(3-Fluorophenyl)-3-(4-(methylthio)-phenyl)urea 522
1-(4-Fluorophenyl)-3-(4-(methylthio)-phenyl)urea 522
$C_{14}H_{13}FN_2O_2$
1-(4-Fluorophenyl)-3-(2-methoxyphenyl)urea 519
1-(4-Fluorophenyl)-3-(4-methoxyphenyl)urea 559
$C_{14}H_{13}FN_2O_3S$
N^4-Acetyl-N^1-(2-fluorophenyl)-sulfanilamide 808
N^4-Acetyl-N^1-(4-fluorophenyl)-sulfanilamide 808
1-Benzyl-3-(3-fluorosulfonylphenyl)urea 803
$C_{14}H_{13}FN_2O_6S_2$
4′-Acetamido-4-fluorosulfonyl-2′-hydroxybenzenesulfonanilide 812
4-Fluorosulfonyl-3′-hydroxy-4′-methanesulfonamidobenzanilide 811
$C_{14}H_{13}FN_4O_2$
4-(3-(2-Fluorophenyl)ureido)benzoic hydrazide 532
$C_{14}H_{13}F_2N_3O$
1-(4,6-Dimethyl-2-pyridyl)-3-(2,5-difluorophenyl)urea 930
$C_{14}H_{13}F_3N_2$
2-(2-(3-(Trifluoromethyl)anilino)-ethyl)pyridine 925
$C_{14}H_{13}F_3N_2O_2$
1-Furfuryl-1-methyl-3-(2-(trifluoromethyl)phenyl)urea 843
$C_{14}H_{13}F_3N_6O_3S$
2-(3-(2-Thiazolyl)ureido)acetic N^2-(*N*-(2-trifluoromethylphenyl)-carbamoyl)hydrazide 874
$C_{14}H_{13}IN_2O$
2-Anilino-4′-iodoacetanilide 404
$C_{14}H_{13}IN_2O_2$
1-Methyl-4-(3-nitrostyryl)pyridinium iodide 956
$C_{14}H_{13}IO_2$
6-Hydroxy-5-iodo-3-phenyl-2-norbornanecarboxylic acid γ-lactone 386
$C_{14}H_{13}N$
1-Amino-9,10-dihydroanthracene 231
2-Aminostilbene 230
N-Benzylidene-*o*-toluidine 1064
2-Methylaminofluorene hydrochloride 233
2-Methyl-6-styrylpyridine hydrochloride 966
2-Phenylisoindoline 252
$C_{14}H_{13}NO$
5-Acetamidoacenaphthene 491
4-(2-Aminoethyl)dibenzofuran hydrochloride 894
3-Amino-2-methoxyfluorene 228
2-Amino-5-methylbenzophenone 311
2-Amino-6-methylbenzophenone 311
2-Amino-4′-phenylacetophenone hydrobromide 294
Benzophenone *O*-methyloxime 1013
o-Benzotoluidide 478
m-Benzotoluidide 477
p-Benzotoluidide 480
N-Benzylidene-*o*-anisidine 1045
N-Benzylidene-*m*-anisidine 1044
α-Benzylimino-*o*-cresol 1032
2,2-Diphenylacetamide 429
3-(Ethylamino)dibenzofuran hydrochloride 894
10-Ethylphenoxazine 251
N-(2-Methoxybenzylidene)aniline 1046
N-(4-Methoxybenzylidene)aniline 1046
N-Methylbenzanilide 482
2-(6-Methyl-2-pyridyl)acetophenone hydrochloride 961
4-(2-Naphthylimino)-2-butanone 1074
2′-Phenylacetanilide 489
2-Phenylacetanilide 430
o-Toluanilide 478
m-Toluanilide 477
p-Toluanilide 480
α-(*o*-Tolylimino)-*o*-cresol 1034
α-(*m*-Tolylimino)-*o*-cresol 1035
α-(*p*-Tolylimino)-*o*-cresol 1035
α-(4-Tolylimino)-*p*-cresol 1036
$C_{14}H_{13}NOS$
2′-(Methylthio)benzanilide 467
3′-(Methylthio)benzanilide 468
α-(3-(Methylthio)phenylimino)-*o*-cresol 1032
α-(4-(Methylthio)phenylimino)-*o*-cresol 1032
S-(*p*-Tolyl) *N*-phenylthiocarbamate 688
4′-(*m*-Tolylthio)formanilide 421
$C_{14}H_{13}NO_2$
Allyl *N*-(1-naphthyl)carbamate 687
m-Benzanisidide 466
p-Benzanisidide 466
N-Benzylanthranilic acid 349
2-Benzyloxybenzamide 463
4-Benzylphenyl carbamate 681
2-Biphenylyl *N*-methylcarbamate 682
2,6-Dimethyl-4-phenylnitrobenzene 278
Ethyl 2-cinnamylidene-2-cyanoacetate 766
α-(2-Hydroxy-5-methylphenylimino)-*o*-cresol 1029
Mandelanilide 411
2′-Methoxybenzanilide 465
4-Methoxy-*N*-phenylbenzamide 466
α-(2-Methoxyphenylimino)-*o*-cresol 1031
α-(3-Methoxyphenylimino)-*o*-cresol 1031
α-(4-Methoxyphenylimino)-*o*-cresol 1031
α-(4-Methoxyphenylimino)-*p*-cresol 1031
4-Methylbenzyl nicotinate 940
N-Methyl-*N*-phenylanthranilic acid 343
N-(1-Naphthyl)-acetoacetamide 489
Phenyl *N*-(*o*-tolyl)carbamate 681
Phenyl *N*-(*m*-tolyl)carbamate 678
Phenyl *N*-(*p*-tolyl)carbamate 683
1-Propylpyrano(3,4-b)indol-3(9*H*)-one 888
8-Quinolyl 3-methylcrotonate 993
N-(*o*-Tolyl)anthranilic acid 343
$C_{14}H_{13}NO_2S$
10-Ethylphenothiazine-5,5-dioxide 778
Ethyl 4-(2-thienylmethyleneamino)-benzoate 1073
4-(Methylthio)phenyl carbanilate 667
$C_{14}H_{13}NO_2S_2$
S-(4-Hydroxyphenyl) *N*-(3-(methylthio)phenyl)thiocarbamate 644
$C_{14}H_{13}NO_3$
Benzyl 4-nitro-*m*-tolyl ether 269

$C_{14}H_{13}NO_3$ (Cont'd)
4-(2,5-Dimethyl-3-formyl-1-pyrrolyl)-benzoic acid ... 841
Ethyl 2-cyano-3-oxo-5-phenyl-4-pentenoate ... 754
2-(2-Hydroxyphenoxy)acetanilide ... 411
α-(2-Hydroxyphenylimino)-6-methoxy-o-cresol ... 1028
4-Methoxy-N-phenylbenzohydroxamic acid ... 466
Methyl 6-methoxy-4-quinolineacrylate ... 986
N-(4-Oxocyclohexyl)phthalimide ... 720
Phenyl N-(o-anisyl)carbamate ... 653
Phenyl N-(m-anisyl)carbamate ... 654
Phenyl N-(p-anisyl)carbamate ... 655
p-Tolyl 4-aminosalicylate ... 377

$C_{14}H_{13}NO_3S$
2-(4-Nitrophenoxy)ethyl phenyl sulfide ... 266

$C_{14}H_{13}NO_4$
3,4-Dimethoxyphenyl 2-pyridinecarboxylate ... 936
2-Methoxyphenyl 4-nitrobenzyl ether ... 265
4-(4-Nitrobenzyloxy)benzyl alcohol ... 263
1-(4-Nitrophenoxy)-2-phenoxyethane ... 266
2-Propyl-3,4-quinolinedicarboxylic acid ... 987

$C_{14}H_{13}NO_4S$
4-(Benzylsulfonamido)-benzoic acid ... 823
4-Nitrobenzyl p-tolyl sulfone ... 779
N-(p-Tosyl)anthranilic acid ... 823

$C_{14}H_{13}NO_6$
Dimethyl 7-hydroxy-8-methoxy-2,4-quinolinedicarboxylate ... 982

$C_{14}H_{13}NS$
N-Ethylphenothiazine ... 251
N-Methylthiobenzanilide ... 746
N-Phenylthioacetanilide ... 746

$C_{14}H_{13}NS_2$
p-Tolyl N-phenyldithiocarbamate ... 749

$C_{14}H_{13}N_3$
5-Amino-2-methyl-1-phenylbenzimidazole ... 904
4-Amino-1-phenyl-3-cyclohexene-1,3-dicarbonitrile ... 763
5,6-Dimethyl-2-(4-pyridyl)-benzimidazole ... 951
3-Phenyl-1,3,5-pentanetricarbonitrile ... 756

$C_{14}H_{13}N_3O$
3-Amino-1-(1-naphthylmethyl)-2-pyrazolin-5-one ... 701
Benzil oxime hydrazone ... 1009
3,6-Dimethyl-1-phenyl-1H-pyrazolo(3,4-b)-pyridin-4-ol ... 918
4'-(Phenylazo)acetanilide ... 1092
1-Phenyl-3-(o-tolylimino)urea ... 586
p-Toluic (4-pyridylmethylene)-hydrazide ... 1069

$C_{14}H_{13}N_3OS$
1-Benzoyl-4-phenyl-3-thiosemicarbazide ... 745
1-Ethyl-3-(naphtho(1,2-d)thiazol-2-yl)urea ... 917
1-(p-Hydroxyanilinomethyl)-2-(3H)-benzimidazolethione hydrochloride ... 734
1-(2-Hydroxybenzylidene)-4-phenyl-3-thiosemicarbazide ... 1037
1-(3-Hydroxybenzylidene)-4-phenyl-3-thiosemicarbazide ... 734

$C_{14}H_{13}N_3O_2$
4-Aminobenzoic (2-hydroxybenzylidene)hydrazide ... 496
1-(2-Aminobenzoyl)-2-benzoylhydrazine ... 499
Anthranilic (2-hydroxybenzylidene)-hydrazide ... 496
1-(Benzoyl)-4-phenylsemicarbazide ... 577
1-(2-Carbamoylphenyl)-3-phenylurea ... 577
1-(4-Methylbenzoyl)-2-(4-pyridylcarbonyl)hydrazine hydrochloride ... 959
1-(4-Methyl-α-nitrobenzylidene)-2-phenylhydrazine ... 258
Nicotinic (4-methoxybenzylidene)-hydrazide ... 938
3-Nitrobenzaldehyde methylphenylhydrazone ... 277

$C_{14}H_{13}N_3O_2S$
1-(4-Carboxyphenyl)-4-phenyl-3-thiosemicarbazide ... 744

$C_{14}H_{13}N_3O_2S_2$
Benzenesulfonic (3-methyl-2-benzothiazolinylidene)hydrazide ... 820

$C_{14}H_{13}N_3O_3$
1-Benzyl-3-(3-nitrophenyl)urea ... 563
1-Benzyl-3-(4-nitrophenyl)urea ... 567
3,5-Dimethyl-4-(2-nitrophenylazo)-phenol ... 1084
Hippuric (furfurylidene)hydrazide ... 1068
1-(2-Methyl-4-nitrophenyl)-3-phenylurea ... 569
Nicotinic (4-hydroxy-3-methoxybenzylidene)hydrazide ... 1031
2-Pyridylmethyleneamino N-(2-methoxyphenyl)carbamate ... 1049
3-Pyridylmethyleneamino N-(2-methoxyphenyl)carbamate ... 1050

$C_{14}H_{13}N_3O_4$
2,4-Dimethyl-2',4'-dinitrodiphenylamine ... 272
2,5-Dimethyl-2',4'-dinitrodiphenylamine ... 272
3,4-Dimethyl-2',4'-dinitrodiphenylamine ... 272
3,5-Dimethyl-2',4'-dinitrodiphenylamine ... 273
1-(2-Hydroxyphenyl)-3-(2-methyl-5-nitrophenyl)urea ... 547
1-(3-Hydroxyphenyl)-3-(2-methyl-5-nitrophenyl)urea ... 547
1-(3-Hydroxyphenyl)-3-(4-methyl-2-nitrophenyl)urea ... 547
2-Methoxy-6-(3-methyl-2-pyridyliminomethyl)-4-nitrophenol ... 1032
2-Methoxy-4-nitro-6-(3-pyridylmethyliminomethyl)phenol ... 1031
1-(3-Methoxyphenyl)-3-(3-nitrophenyl)urea ... 552
1-(4-Methoxyphenyl)-3-(3-nitrophenyl)urea ... 552
4-Pyridylmethyl N-(2-methyl-5-nitrophenyl)carbamate ... 949

$C_{14}H_{13}N_3O_5$
2,4-Dinitro-2'-ethoxydiphenylamine ... 267
2,4-Dinitro-4'-ethoxydiphenylamine ... 266
1-(3-Hydroxyphenyl)-3-(2-methoxy-5-nitrophenyl)urea ... 547

$C_{14}H_{13}N_3O_5S$
4-Acetamido-N-(3-nitrophenyl)-benzenesulfonamide ... 822
N^4-Acetyl-N^1-(4-nitrophenyl)-sulfanilamide ... 819

$C_{14}H_{13}N_3S$
1-Anilinomethyl-2(3H)-benzimidazolethione ... 742
Benzophenone thiosemicarbazide ... 1072

$C_{14}H_{13}N_5O_4$
1-Isonicotinoyl-4-(2-methyl-6-nitrophenyl)semicarbazide ... 959

$C_{14}H_{13}N_5O_4S$
2-(3-(3-Nitrophenyl)ureido)acetic (2-thienylmethylene)hydrazide ... 1053

$C_{14}H_{14}$
7,7'-Bi(1,3,5-cycloheptatriene) ... 3
Phenyl(p-tolyl)methane ... 162

$C_{14}H_{14}AsN$
5,10-Dihydro-10-ethylphenarsazine ... 1128

$C_{14}H_{14}AsNO_3$
9-Ethyl-3-carbazolearsonic acid ... 1127

$C_{14}H_{14}AsNO_4$
2,3-Dihydro-2-hydroxy-2-phenyl-1H-isoarsindolium nitrate ... 1128

$C_{14}H_{14}BrIO_2$
Bis(4-methoxyphenyl)iodonium bromide ... 181

$C_{14}H_{14}BrN$
N-(4-Bromophenyl)-4-methylbenzylamine ... 232

$C_{14}H_{14}BrNO$
N-(4-Bromophenyl)-4-methoxybenzylamine ... 229
1-(4'-Methylphenacyl)pyridinium bromide ... 956

$C_{14}H_{14}BrNO_2$
4-Bromo N-(2-hydroxy-3-methoxybenzyl)aniline ... 222
6-Bromo-2-naphthyl N-isopropylcarbamate ... 687
6-Bromo-2-naphthyl N-propylcarbamate ... 687
1-(4-Carboxybenzyl)-2-methylpyridinium bromide ... 956

$C_{14}H_{14}BrNO_3$
4'-Bromo-1,2,3,6-tetrahydrophthalanilic acid ... 406

$C_{14}H_{14}BrNO_4$
2-(5-Bromo-1-methyl-3-indolylmethyl)succinic acid ... 883

$C_{14}H_{14}BrN_3$
4-(4-Bromophenylazo)-N,N-dimethylaniline ... 1081

$C_{14}H_{14}BrN_3O$
N-(4-(Bromophenylazo)phenyl)-ethanolamine ... 1080
1-(3-Bromophenyl)-3-(4,6-dimethyl-2-pyridyl)urea ... 930

$C_{14}H_{14}ClF_3N_2O_2$
O-(N-(2-Chloro-5-trifluoromethylphenyl)carbamoyl)-cyclohexanone oxime ... 1008

$C_{14}H_{14}ClN$
N-(4-Chlorophenyl)-4-methylbenzylamine ... 232
N-(α-Chloro-o-tolyl)-benzylamine hydrochloride ... 222

$C_{14}H_{14}ClNO$
N-(4-Chlorophenyl)-4-methoxybenzylamine ... 229
α-(3-Chloro-p-toluidino)-o-cresol ... 223
α-(5-Chloro-o-toluidino)-o-cresol ... 223

$C_{14}H_{14}ClNO_2$
α-(2-Chloroanilino)-6-methoxy-o-cresol ... 222
α-(3-Chloroanilino)-6-methoxy-o-cresol ... 222
α-(4-Chloroanilino)-6-methoxy-o-cresol ... 223
4-(1,2-Dihydro-2-methyl-1-oxo-3-isoquinolyl)butyryl chloride ... 707

$C_{14}H_{14}ClNO_2S$
4-Chloro-2'-ethylbenzenesulfonanilide ... 809
4-Chloro-4'-ethylbenzenesulfonanilide ... 809
4-Chloro-N-methylbenzenesulfono-m-toluidide ... 809
4'-Chloro-N-methyl-p-toluenesulfonanilide ... 828

$C_{14}H_{14}ClNO_3$
N-(3-Chlorophenyl)-1,2,5,6-tetrahydrophthalamic acid ... 406

$C_{14}H_{14}ClNO_3S$
5-(2-Chlorobenzylidene)-3-(3-methoxypropyl)-2,4-thiazolidinedione ... 694
4-Chloro-4'-ethoxybenzenesulfonanilide ... 806
5'-Chloro-4-methoxy-2'-methylbenzenesulfonanilide ... 806

$C_{14}H_{14}ClN_3$
4-(3-Chlorophenylazo)-N,N-dimethylaniline ... 1081
4-(4-Chlorophenylazo)-N,N-dimethylaniline ... 1089

$C_{14}H_{14}ClN_3O$
N-(4-Chloro-2-methylphenyl)-carbamic acid phenylhydrazide ... 579
1-(2-Chloro-6-methylphenyl)-3-(3-pyridylmethyl)urea ... 950
1-(3-Chloro-4-methylphenyl)-3-(3-pyridylmethyl)urea ... 950
1-(4-Chloro-2-methylphenyl)-3-(3-pyridylmethyl)urea ... 950
1-(5-Chloro-2-methylphenyl)-3-(3-pyridylmethyl)urea ... 950
1-(5-Chloro-2-pyridyl)-3-(3-ethylphenyl)urea ... 965

$C_{14}H_{14}ClN_3OS$
4-(4-Chlorophenyl)-dihydro-dimethyl-1H-pyrazolo(3,4-e)(1,4)-thiazepin-7(6H)-one ... 860
1-(5-Chloro-2-pyridyl)-3-(2-methoxy-5-methylphenyl)-2-thiourea ... 939

$C_{14}H_{14}ClN_3O_2$
1-Carbamoylmethyl-3-(phenylcarbamoyl)-pyridinium chloride ... 947
3-Carbamoyl-1-(phenylcarbamoylmethyl)-pyridinium chloride ... 942
1-(5-Chloro-2-methoxyphenyl)-3-(4-methyl-2-pyridyl)urea ... 927
1-(3-Chloro-2-methoxyphenyl)-3-(3-pyridylmethyl)urea ... 938
1-(4-Chloro-2-methoxyphenyl)-3-(3-pyridylmethyl)urea ... 927
1-(5-Chloro-2-methoxyphenyl)-3-(2-pyridylmethyl)urea ... 926

$C_{14}H_{14}ClN_3O_3$
1-(5-Chloro-2,4-dimethoxyphenyl)-3-(3-pyridyl)urea ... 926
1-(2-Chloro-3-pyridyl)-3-(2,5-dimethoxyphenyl)urea ... 936

$C_{14}H_{14}ClN_3O_3S$
1-(3-Chloro-2-methylphenyl)-3-(4-sulfamoylphenyl)urea ... 809

$C_{14}H_{14}ClN_3O_4$
1-(5-Chloro-2,4-dimethoxyphenyl)-3-(3-hydroxy-2-pyridyl)urea ... 926
4,5-Dihydro-1,3-diphenyl-1H-1,2,3-triazolium perchlorate ... 255

$C_{14}H_{14}ClN_3O_5S$
4-(4-Chlorophenylsulfonyl)-1-(p-tosyl)semicarbazide ... 808

$C_{14}H_{14}ClO_3P$
Di-o-tolyl chlorophosphate ... 836
Di-p-tolyl chlorophosphate ... 836

$C_{14}H_{14}Cl_2N_2O_2$
2,4-Dichloro-8-quinolyl N-butylcarbamate ... 992

$C_{14}H_{14}Cl_2N_2O_4S_2$
N,N'-Ethylenebis(4-chlorobenzenesulfonamide) ... 807

$C_{14}H_{14}Cl_2N_2O_5$
2-(Methacryloyloxy)ethyl N-(N-(3,5-dichlorophenyl)carbamoyl)-carbamate ... 521

$C_{14}H_{14}Cl_2Si$
Bis(2-chlorobenzyl)silane ... 1104

$C_{14}H_{14}Cl_4O_4$
2-Ethylbutyl tetrachlorophthalate ... 366
Hexyl tetrachlorophthalate ... 366
2-Methylpentyl tetrachlorophthalate ... 366

$C_{14}H_{14}FN$
N-(4-Fluorophenyl)-4-methylbenzylamine ... 232

$C_{14}H_{14}FNO$
N-(4-Fluorophenyl)-4-methoxybenzylamine ... 229

$C_{14}H_{14}FNO_2$
N-(4-Fluorophenyl)-2-hydroxy-3-methoxybenzylamine ... 222

$C_{14}H_{14}FNO_2S$
3-Butyl-5-(4-fluorobenzylidene)-2,4-thiazolidinedione ... 695

$C_{14}H_{14}FNO_3$
4'-Fluoro-1,2,3,6-tetrahyrophthalanilic acid ... 406

$C_{14}H_{14}FN_3$
4-(Dimethylamino)-4'-fluoroazobenzene ... 1081

$C_{14}H_{14}FN_3O_3S$
4-(2-Fluorophenyl)-1-(4-tolylsulfonyl)semicarbazide ... 808

$C_{14}H_{14}F_2N_2$
2,2'-Diamino-4,4'-difluorobibenzyl ... 224

$C_{14}H_{14}F_3NO_4$
2-(Isopropenyloxycarbonyl)ethyl N-(3-trifluoromethylphenyl)-carbamate ... 609

$C_{14}H_{14}FeN_2S$
4-(Ferrocenylmethylene)imidazolidine-2-thione 1142
$C_{14}H_{14}FeO_3$
γ-Oxoferrocenebutyric acid 1141
$C_{14}H_{14}Hg$
Dibenzylmercury 1139
Di(*p*-tolyl)mercury 1139
$C_{14}H_{14}IN$
N-(4-Iodophenyl)-4-methylbenzylamine 232
1-Methyl-2-styrylpyridinium iodide 957
1-Methyl-3-styrylpyridinium iodide 957
$C_{14}H_{14}INO$
2-(2-Hydroxystyryl)-1-methylpyridinium iodide 933
2-(3-Hydroxystyryl)-1-methylpyridinium iodide 933
$C_{14}H_{14}INOS$
1-(4′-(Methylthio)phenacyl)pyridinium iodide 941
$C_{14}H_{14}INO_2$
1-(2-Hydroxy-5-methylbenzoylmethyl)pyridinium iodide 933
α-(4-Iodoanilino)-6-methoxy-*o*-cresol 222
$C_{14}H_{14}INO_3$
4′-Iodo-1,2,3,6-tetrahydrophthalanilic acid 407
$C_{14}H_{14}NO_2P$
9-Ethyl-3-carbazolephosphonous acid 892
$C_{14}H_{14}N_2$
Benzaldehyde 2-methyl-2-phenylhydrazone 1072
1,2-Bis(6-methyl-2-pyridyl)ethylene 953
α-Butylbenzylidenemalononitrile 764
4,4′-Dimethylazobenzene 1093
N,*N*′-Diphenylacetamidine 1076
α-Isobutylbenzylidenemalononitrile 764
2-Methylbenzaldehyde phenylhydrazone 235
5,6,11,12-Tetrahydrodibenzo(b,f)(1,5)diazocine 254
$C_{14}H_{14}N_2O$
1-Allyl-3-(1-naphthyl)urea 603
3,3′-Azoxytoluene 1092
4,4′-Azoxytoluene 1093
1-Benzyl-3-phenylurea 571
1,3-Diphenyl-1-methylurea 586
2-(3-Ethyl-5-methyl-4-isoxazolyl)indole 885
4-Methoxybenzaldehyde phenylhydrazone 234
3′-Methylcarbanilide 580
4′-Methylcarbanilide 591
N-Methylharmine 918
3-Methyl-*N*-(4-methyl-2-pyridyl)benzamide hydrochloride 963
4-Methyl-*N*-(4-methyl-2-pyridyl)benzamide hydrochloride 963
6-Methyl-2-pyridinealdoxime *o*-benzyl ether 1051
α-Methyl-α-(3-pyridylmethylimino)-*o*-cresol 1033
N-Nitrosodibenzylamine 257
6-Phenylazo-3,4-xylenol 1085
1-Phenyl-3-(*o*-tolyl)urea 586
1-Phenyl-3-(*m*-tolyl)urea 586
1-Phenyl-3-(*p*-tolyl)urea 586
2-Pyridinealdoxime *O*-phenethyl ether 1010
3-Pyridinealdoxime *O*-phenethyl ether 1010
4-Pyridinealdoxime *O*-phenethyl ether 1010
3-(3-Pyridylamino)propiophenone 943
$C_{14}H_{14}N_2OS$
4-(Methylthio)-carbanilide 561
1-(4-(*m*-Tolylthio)phenyl)urea 561
$C_{14}H_{14}N_2OS_2$
1-(2-Mercaptophenyl)-3-(4-(methylthio)phenyl)urea 560
$C_{14}H_{14}N_2O_2$
4′-Amino-2′-hydroxy-*p*-toluanilide 462
3-Amino-4-methoxybenzanilide 465
2-(4-Aminophenoxy)acetanilide 414
4-(Anilino)phenyl *N*-methylcarbamate 671
1-(*m*-Anisyl)-3-phenylurea 558
Benzilic acid hydrazide 495
Benzyl 3-phenylcarbazate 664
6,6′-Dimethyl-2,2′-pyridoin 932
2-Methoxy-α-(3-methyl-2-pyridylimino)-*o*-cresol 1031
2-Methoxy-α-(6-methyl-2-pyridylimino)-*o*-cresol 1032
6-Methoxy-α-(4-methyl-2-pyridylimino)-*o*-cresol 1030
2-(4-Methoxyphenylazo)-4-methylphenol 1083
1-(2-Methoxyphenyl)-3-phenylurea 555
1-(4-Methoxyphenyl)-3-phenylurea 559
3-Methyl-*N*-(4-nitrobenzyl)aniline 277
N-(4-Methyl-3-nitrophenyl)benzylamine 275
4-Methyl-*N*-(4-nitrophenyl)benzylamine 277
o-(*N*-(1-Naphthyl)carbamoyl)acetone oxime 1012
4-Nitrodibenzylamine 277
α-Nitromethyl-*N*-phenylbenzylamine 258
N-(3-Nitrophenyl)-4-methylbenzylamine 275
N-(4-Nitrophenyl)phenethylamine 276
2-Pyridyl *N*-(3-ethylphenyl)carbamate 964
2-Pyridyl *N*-(2,4-xylyl)carbamate 964
3-Pyridyl *N*-(2,4-xylyl)carbamate 960
$C_{14}H_{14}N_2O_2S$
2,4,6-Cycloheptatrien-1-one *p*-tosylhydrazone hydrochloride 1075
1-(3-Hydroxyphenyl)-3-(4-(methylthio)phenyl)urea 547
4-Pyridylmethyl *N*-(4-(methylthio)phenyl)carbamate 941
$C_{14}H_{14}N_2O_2S_2$
1-Phenyl-3-(*p*-tolylsulfonyl)-2-thiourea 826
$C_{14}H_{14}N_2O_3$
4′-Amino-2′-hydroxy-*o*-anisanilide 461
2,2′-Azoxyanisole 1087
4,4′-Dimethoxyazobenzene *N*-oxide 1086
Ethyl 4-acetamido-α-cyanocinnamate 766
2-Hydroxy-*N*-(2-methyl-5-nitrophenyl)benzylamine 264
2-Hydroxy-*N*-(4-methyl-3-nitrophenyl)benzylamine 264
N-(2-Hydroxy-4-methyl-7-quinolyl)acetoacetamide 983
2-Methoxy-*N*-(4-nitrophenyl)benzylamine 267
4-Methoxy-*N*-(3-nitrophenyl)benzylamine 267
4-Methoxy-*N*-(4-nitrophenyl)benzylamine 267
3-(*N*-(2-Pyridyl)carbamoyl)-5-norbornene-2-carboxylic acid 963
3-(*N*-(3-Pyridyl)carbamoyl)-5-norbornene-2-carboxylic acid 963
4-Pyridylmethyl *N*-(2-methoxyphenyl)carbamate 939
$C_{14}H_{14}N_2O_3O$
2,2′-Dichloroazoxybenzene 1093
$C_{14}H_{14}N_2O_3S$
4-Acetamidobenzenesulfonanilide 825
1-(4-Ethoxyphenyl)-3-furoyl-2-thiourea 846
Ethyl 2-benzylthio-4-hydropyrimidine-5-carboxylate 972
Methyl 2-amino-4-methyl-5-(*N*-phenylcarbamoyl)-3-thiophenecarboxylate 856
1-Phenyl-3-(*p*-tosyl)urea 820
$C_{14}H_{14}N_2O_4$
4′-Acetamido-3′-methoxy-2-furanilide 846
4-Amino-3,3′-dimethoxy-4′-nitrobiphenyl 265
4-(Diacetoxymethyl)-1-phenyl-1*H*-pyrazole 863
2-Hydroxy-*N*-(2-methoxy-4-nitrophenyl)benzylamine 263
2-Hydroxy-*N*-(2-methoxy-5-nitrophenyl)benzylamine 263
6-Methoxy-α-(3-nitroanilino)-*o*-cresol 263
6-Methoxy-α-(4-nitroanilino)-*o*-cresol 263
N,*N*′-(1,5-Naphthalenediyl)bis(methyl carbamate) 687
2-Pyridyl *N*-(2,5-dimethoxyphenyl)carbamate 939
3-Pyridyl *N*-(2,5-dimethoxyphenyl)carbamate 939
$C_{14}H_{14}N_2O_4S$
N-(4-(4-Aminophenylsulfonyl)phenyl)glycine hydrochloride 776
N-(4-(4-Aminophenylsulfonyl)phenyl)glycine, morpholine salt 776
4-(2,6-Dimethyl-4-hydroxyphenylazo)benzenesulfonic acid 1085
N-Methyl-2′-nitro-*p*-toluenesulfonanilide 818
$C_{14}H_{14}N_2O_5$
N-(3-Nitrophenyl)-1,2,5,6-tetrahydrophthalamic acid 426
N-Phthaloylglycylglycine ethyl ester 718
$C_{14}H_{14}N_2O_6S$
2′,4-Dimethoxy-5′-nitrobenzenesulfonanilide 812
$C_{14}H_{14}N_2O_6S_2$
4,4′-Diaminostilbene-2,2′-disulfonic acid 788
$C_{14}H_{14}N_2O_6S_4$
Diethyl Δ5,5′-birhodanine-3,3′-diacetate 141
$C_{14}H_{14}N_2S$
1-Allyl-3-(1-naphthyl)-2-thiourea 749
2-Benzhydryl-2-thiopseudourea hydrobromide 749
1-Benzyl-3-phenyl-2-thiourea 748
1-Phenyl-3-(*p*-tolyl)-2-thiourea 748
$C_{14}H_{14}N_4$
4-Acetylpyridine azine 953
Benzil dihydrazone 235
$C_{14}H_{14}N_4O$
2,6-Bis-(2,2-dicyanoethyl)cyclohexanone 146
2-(Cinnamylidenehydrazino)-4-hydroxy-6-methylpyrimidine 1037
1-(4-(Phenoxy)benzylideneamino)guanidine nitrate 1049
$C_{14}H_{14}N_4OS$
4-(3-Phenyl-2-thioureido)benzoic hydrazide 745
$C_{14}H_{14}N_4O_2$
2-(*o*-Aminophenyl)-5,6-dimethoxy-2*H*-benzotriazole 908
N,*N*′-Bis(2-aminophenyl)oxamide 422
1,2-Bis(phenylazo)-1,2-ethanediol 234
4-(Dimethylamino)-3′-nitroazobenzene 1088
4-(Dimethylamino)-4′-nitroazobenzene 1089
$C_{14}H_{14}N_4O_2S$
4-Benzyl-1-(4-nitrophenyl)-3-thiosemicarbazide 740
$C_{14}H_{14}N_4O_3$
2-Acetyl-5-(diazoacetyl)-3-hydroxy-5-methyl-4-phenyl-3-pyrazoline 1094
4′-Amino-2-methoxy-3′-methyl-4-nitroazobenzene 1086
Disperse Red 31 (C.I. 11250) 1086
Ethyl 4-(3-(2-pyrazinyl)ureido)benzoate 977
1-(4-Methyl-2-nitrophenyl)-3-(6-methyl-2-pyridyl)urea 944
1-(2-Methyl-5-nitrophenyl)-3-(3-pyridylmethyl)urea 949
1-(4-Methyl-3-nitrophenyl)-3-(2-pyridylmethyl)urea 943
$C_{14}H_{14}N_4O_3S$
4-(3-(Methylthio)phenyl)-1-(4-nitrophenyl)semicarbazide 560
$C_{14}H_{14}N_4O_4$
N,*N*′-Bis-(2-nitrophenyl)ethylenediamine 274
4,4′-Diamino-3,3′-dimethyl-2,2′-dinitrobiphenyl 271
5,5′-Dimethyl-2,2′-dinitrobenzidine 271
N,*N*-Dimethyl-*N*′-(2,4-dinitrophenyl)-1,4-phenylenediamine 271
1-(2-Methoxy-4-nitrophenyl)-3-(4-methyl-2-pyridyl)urea 938
1-(2-Methoxy-4-nitrophenyl)-3-(4-pyridylmethyl)urea 938
$C_{14}H_{14}N_4O_4S_2$
1-(Benzylsulfonyl)-4-(3-nitrophenyl)-3-thiosemicarbazide 818
$C_{14}H_{14}N_4O_5S$
1-(2-Methyl-5-nitrophenyl)-3-(4-sulfamoylphenyl)urea 818
1-(4-Methyl-2-nitrophenyl)-3-(4-sulfamoylphenyl)urea 818
$C_{14}H_{14}N_4O_6$
N-(2-Hydroxy-3-methoxy-5-nitrobenzylidene)histidine 866
$C_{14}H_{14}N_4O_6S$
4-(2-Methoxy-4-nitrophenylureido)benzenesulfonamide 814
$C_{14}H_{14}O$
2-Benzyl-4-methylphenol 195
Benzyl phenyl carbinol 204
(4-Benzylphenyl) methyl ether 180
1-(4-Biphenylyl)-1-ethanol 205
p-Cresyl benzyl ether 178
Diphenylmethyl methyl ether 186
2-Hydroxy-5,6,7,8-tetrahydrophenanthrene 195
6-Methoxy-2-(1-methylvinyl)naphthalene 188
2-(4-Methylbenzyl)phenol 196
4-(α-Methylbenzyl)phenol 195
4-Methylbenzyl phenyl ether 178
α-Methyl-(4-biphenylyl)methanol 207
1,2,3,4-Tetrahydro-4-hydroxyphenanthrene 208
$C_{14}H_{14}OS$
4-(Methylthio)benzhydrol 210
2-(Phenylthio)ethyl phenyl ether 211
$C_{14}H_{14}OSi$
10,10-Dimethyl-10*H*-phenoxasilin 1123
$C_{14}H_{14}O_2$
6-Acetyl-2-methoxy-1-methylnaphthalene 293
2-Benzyl-5-methoxyphenol 195
4-Benzyloxyanisole 181
2-(4-Biphenylyloxy)-ethanol 202
2,2′-Dimethoxybiphenyl 179
1,2-Dimethylacenaphthene-1,2-diol 195
3,3′-Dimethyl-4,4′-biphenol 194
1,1-Diphenyl-1,2-ethanediol 199
Ethyl 1-naphthaleneacetate 362
Ethyl 2-naphthylacetate 362
2-(Hydroxymethyl)benzhydrol 199
DL-2-(2-Hydroxyphenyl)-2-phenyl-1-ethanol 200
DL-4-Methoxybenzhydrol 201
4-(2-Naphthyl)butyric acid 334
2-Phenoxy-2-phenyl-1-ethanol 201
$C_{14}H_{14}O_2S$
Benzyl *p*-tolyl sulfone 784
4-Ethylphenyl phenyl sulfone 784
Phenethyl phenyl sulfone 783
$C_{14}H_{14}O_2S_2$
S-Benzyl *p*-toluenethiosulfonate 801
p-Tolyl disulfoxide 771
S-(*p*-Tolyl) *p*-toluenethiosulfonate 801
$C_{14}H_{14}O_2Si$
Diphenylsilyl acetate 1112
$C_{14}H_{14}O_2Te_2$
Bis(4-methoxyphenyl) ditelluride 210
$C_{14}H_{14}O_3$
7-Allyloxy-4,8-dimethylcoumarin 383
2-Butoxy-1,4-naphthoquinone 316
4,4′-Oxydibenzyl alcohol 202
Phenoxymethyl ether 187
$C_{14}H_{14}O_3S$
5,6,7,8-Tetrahydro-2-phenanthrenesulfonic acid sodium salt 796
2,3-Xylyl benzenesulfonate 800
$C_{14}H_{14}O_4$
Ethyl 3-cinnamoylpyruvate 365

$C_{14}H_{14}O_4$ (Cont'd)
Methyl 2,4-dioxo-6-phenyl-1-cyclohexanecarboxylate 364
$C_{14}H_{14}O_4S$
1,2-Bis(phenylsulfonyl)ethane 783
4,4'-Sulfonyldibenzyl alcohol 775
$C_{14}H_{14}O_4S_2$
p-Tolyl disulfone 776
$C_{14}H_{14}O_4S_5$
Bis(*p*-tolylsulfonyl) trisulfide 798
Bis(*p*-tosyl) trisulfide 798
$C_{14}H_{14}O_5$
7-Acetoxy-3-acetyl-2-methyl-4-chromanone 379
5-(2-Methoxybenzylidene)-2,2-dimethyl-1,3-dioxane-4,6-dione 383
Purpurogallin trimethyl ether 290
$C_{14}H_{14}O_6S_2$
Diphenyl 1,2-ethanedisulfonate 800
Ethylene glycol bis(benzenesulfonate) 801
$C_{14}H_{14}O_8$
Tetramethyl 1,2,4,5-benzenetetracarboxylate 375
$C_{14}H_{14}S$
Benzyl *p*-tolyl sulfide 212
4-Methylbenzyl phenyl sulfide 212
Phenethyl phenyl sulfide 212
$C_{14}H_{14}S_2$
1,2-Bis(phenylthio)ethane 212
$C_{14}H_{14}Se$
Benzyl selenide 214
p-Tolyl selenide 1142
$C_{14}H_{14}Si$
9,9-Dimethyl-9*H*-9-silafluorene 1123
$C_{14}H_{14}Te_2$
p-Tolyl ditelluride 214
$C_{14}H_{15}BF_4S$
Diphenylethylsulfonium fluoborate 213
$C_{14}H_{15}BrClNO_3$
4'-Bromo-2'-chlorohexahydrophthalanilic acid 402
$C_{14}H_{15}ClFNO_2S$
1-(4-Fluorobenzyl)-4-(methylsulfonylmethyl)pyridinium chloride 928
$C_{14}H_{15}ClF_3NO$
2'-Chloro-5'-(trifluoromethyl)-1-cyclohexanecarboxanilide 395
4'-Chloro-2'-(trifluoromethyl)-1-cyclohexanecarboxanilide 395
$C_{14}H_{15}ClF_3NO_2$
Cyclohexyl *N*-(2-chloro-5-trifluoromethylphenyl)carbamate 611
$C_{14}H_{15}ClN_2O$
1-(4-Chlorophenyl)-3-(4-methyl-1-piperazinyl)-2-propyn-1-one 1015
$C_{14}H_{15}ClN_2O_2$
1-(3-Chloro-4-methylphenyl)-3-furfuryl-3-methylurea 845
Ethyl 2-chloro-α-cyano-4-(dimethylamino)-cinnamate 765
$C_{14}H_{15}ClN_2O_2S$
5-Amino-2-chloro-2',4'-dimethylbenzenesulfonanilide 807
$C_{14}H_{15}ClN_2O_3$
2-Chloroethyl antipyrate 693
3-(5-Chloro-2-methoxyphenyl)-1-furfuryl-1-methylurea 844
$C_{14}H_{15}ClN_2O_4$
1-(5-Chloro-2,4-dimethoxyphenyl)-3-furfurylurea 843
$C_{14}H_{15}ClN_2O_5$
cis-Hexahydro-*N*-(4-chloro-3-nitrophenyl)phthalamic acid 405
$C_{14}H_{15}ClO_4$
Diethyl (4-chlorobenzylidene)-malonate 364
$C_{14}H_{15}ClSi$
(4-Chlorobenzyl)-methylphenylsilane 1103
(Chloromethyl)-diphenylmethylsilane 1099
$C_{14}H_{15}Cl_2NO_3$
2-(3,4-Dichlorophenylcarbamoyl)-cyclohexanecarboxylic acid 406
cis-*N*-(2,3-Dichlorophenyl)-hexahydrophthalamic acid 406
cis-*N*-(2,5-Dichlorophenyl)-hexahydrophthalamic acid 406
cis-Hexahydro-*N*-(2,4-dichlorophenyl)phthalamic acid 406
$C_{14}H_{15}Cl_4NO_4$
Butyl *N*-(1-(4-chlorobenzoyloxy)-2,2,2-trichloroethyl)carbamate 606
$C_{14}H_{15}FN_2O_5$
4'-Fluoro-3'-nitrohexahydrophthalanilic acid 404
$C_{14}H_{15}F_3N_2O$
1,1-Diallyl-3-(3-trifluoromethylphenyl)urea 510
$C_{14}H_{15}F_3N_2O_2$
O-(*N*-(2-(Trifluoromethyl)phenyl)-carbamoyl)cyclohexanone oxime 1008
O-(*N*-(3-(Trifluoromethyl)phenyl)-carbamoyl)cyclohexanone oxime 1008
$C_{14}H_{15}N$
Dibenzylamine hydrobromide 243
Dibenzylamine hydrochloride 243
N-Ethyldiphenylamine 233
4-Methyl-*N*-phenylbenzylamine 232
4-(1-Penten-1-yl)quinoline 995
N-(1-Phenylethyl)aniline 232
5,6,7,8-Tetrahydro-2-phenanthreneamine hydrochloride 231
$C_{14}H_{15}NO$
N-(4-Acetylphenyl)-2,5-dimethylpyrrole 841
2-Amino-α-methyl-α-phenylbenzyl alcohol 226
2-(Benzylamino)-benzyl alcohol 226
5-Benzylidene-3-ethyl-4-methyl-3-pyrrolin-2-one 691
5-Benzylidene-4-ethyl-3-methyl-3-pyrrolin-2-one 691
2-Benzylidene-3-quinuclidinone 284
4-Methoxy-4'-methyldiphenylamine 229
4-Methoxy-*N*-phenylbenzylamine 229
2-Methyl-*N*-(1-naphthyl)-propionamide 490
4'-Methyl-3-pyrrolidinopropiophenone 842
N-(2-Naphthyl)butyramide 490
4-(1-Naphthylimino)-2-butanol 1039
α-(*m*-Toluidino)-*o*-cresol 226
α-(*p*-Toluidino)-*o*-cresol 226
α-(*p*-Toluidino)-*p*-cresol 226
$C_{14}H_{15}NOS$
2-(4-Aminophenylthio)-1-phenylethanol 226
S-Isopropyl *N*-(1-naphthyl)-thiolcarbamate 689
α-(4-(Methylthio)anilino)-*o*-cresol 226
$C_{14}H_{15}NO_2$
α-(*m*-Anisidino)-*o*-cresol 225
α-(*p*-Anisidino)-*o*-cresol 225
N-Cyclohexylphthalimide 720
N-(2,3-Dimethyl-2-buten-1-yl)-phthalimide 723
Ethyl α-cyano-β-ethylcinnamate 766
N-(5-Hydroxy-1-naphthyl)-butyramide 486
Isopropyl *N*-(1-naphthyl)carbamate 687
α-(4-Methoxyanilino)-*p*-cresol 225
3-(2-Phenoxyethoxy)aniline 228
4-(2-Phenoxyethoxy)aniline 227
Propyl *N*-(1-naphthyl)carbamate 687
2',4',5'-Trimethyl-2-furanilide 851
$C_{14}H_{15}NO_2S$
N-Benzyl-*p*-toluenesulfonamide 821
N-(Cyclohexylthio)phthalimide 750
2',3'-Dimethylbenzenesulfonanilide 828
2',4'-Dimethylbenzenesulfonanilide 828
2',6'-Dimethylbenzenesulfonanilide 828
N-Methyl-α-toluenesulfonanilide 827
N-Methyl-*p*-toluenesulfonanilide 828
p-Toluenesulfono-*o*-toluidide 828
$C_{14}H_{15}NO_3$
Butyl β-cyano-α-hydroxycinnamate 761
tert-Butyl 8-quinolinyl carbonate 994
1,2-Dihydro-2-methyl-1-oxo-3-isoquinolinebutyric acid 707
Ethyl β-cyano-2,3-dimethyl-α-hydroxycinnamate 761
Ethyl β-cyano-2,4-dimethyl-α-hydroxycinnamate 761
Ethyl β-cyano-2,5-dimethyl-α-hydroxycinnamate 761
Ethyl β-cyano-3,4-dimethyl-α-hydroxycinnamate 761
Ethyl β-cyano-2 and 4-ethyl-α-hydroxycinnamate 761
Ethyl β-cyano-4-ethyl-α-hydroxycinnamate 761
Furfuryl *N*-(2-ethylphenyl)-carbamate 849
Furfuryl *N*-(2,3-xylyl)carbamate 849
Furfuryl *N*-(2,4-xylyl)carbamate 849
Furfuryl *N*-(2,6-xylyl)carbamate 849
Furfuryl *N*-(3,4-xylyl)carbamate 849
Furfuryl *N*-(3,5-xylyl)carbamate 849
N-(3-Hydroxycyclohexyl)-phthalimide 716
$C_{14}H_{15}NO_3S$
2-(4-Aminophenoxy)ethyl phenyl sulfone 775
2'-Methoxy-5'-methylbenzenesulfonanilide 814
N-(3-Methoxy-4-methylphenyl)-benzenesulfonamide 815
$C_{14}H_{15}NO_4$
2-Carboxy-3-phenyl-3-pyrroline-2-propionic acid 341
N-(3,4-Dimethoxyphenethyl)-maleimide 717
Ethyl 3-acetoxy-2-cyano-2-phenylpropionate 754
Ethyl 1-benzyl-4,5-dioxo-3-pyrrolidinecarboxylate 692
Ethyl α-cyano-2,3-dimethoxycinnamate 762
Ethyl α-cyano-2,4-dimethoxycinnamate 762
Ethyl α-cyano-2,5-dimethoxycinnamate 762
Ethyl α-cyano-3,4-dimethoxycinnamate 762
Ethyl 1,2-dihydro-6-methoxy-1-methyl-2-oxo-4-quinolinecarboxylate 690
3-(*N*-(3-Furfuryl)carbamoyl)-5-norbornene-2-carboxylic acid 854
Furfuryl *N*-(2-methoxy-5-methylphenyl)carbamate 846
N-Phthaloylglycine *tert*.-butyl ester 721
N-Phthaloyl-DL-leucine 721
N-Phthaloyl-L-leucine 722
$C_{14}H_{15}NO_4S$
5-(3,4-Dimethoxybenzylidene)-3-ethyl-2,4-thiazolidinedione 698
2-Phenoxyethyl metanilate 797
$C_{14}H_{15}NO_4S_2$
1,2-Dimethylnaphtho(1,2-d)-thiazolium methyl sulfate 916
Di-toluene-4-sulfonamide 828
$C_{14}H_{15}NO_5$
2-(1,3-Benzodioxol-5-yl)-1-methyl-6-oxo-3-piperidinecarboxylic acid 692
$C_{14}H_{15}NO_6$
Methyl 2-methoxy-2-(methoxycarbonyl)-3-oxo-1-indolineacetate 363
$C_{14}H_{15}NO_7$
N-(2-Acetoxybenzoyl)glutamic acid 467
Diethyl 2-(4-nitrobenzoyl)malonate 359
$C_{14}H_{15}NS$
N-(2-Thienylmethylene)-2,4,6-trimethylaniline 1073
$C_{14}H_{15}N_2$
4-Amino-3,5-dimethylazobenzene 1088
$C_{14}H_{15}N_3$
4-Amino-2,6-dimethylazobenzene 1088
2-(*p*-(Dimethylamino)-benzylideneamino)-pyridine 1062
3-Methyl-4'-(methylamino)-azobenzene 1089
2-(*p*-Tolylazo)-*p*-toluidine 1088
4-(*m*-Tolylazo)-*m*-toluidine 1088
$C_{14}H_{15}N_3O$
1-(4-Anilinophenyl)-3-methylurea 567
2-(2-Benzimidazoyl)-2-valerylacetonitrile 905
4-(Dimethylamino)azobenzene *N*-oxide 1090
1-(2,6-Dimethylphenyl)-3-(3-pyridyl)urea 945
1-(3-Ethylphenyl)-3-(2-pyridyl)urea 965
1-Phenyl-4-(*o*-tolyl)semicarbazide 578
1-Phenyl-4-(*m*-tolyl)semicarbazide 578
4-Phenyl-1-(*m*-tolyl)semicarbazide 578
4-Phenyl-1-(*p*-tolyl)semicarbazide 578
4-Pyridinecarboxylic (5-norbornen-2-ylmethylene)hydrazide 1069
1-(3-Pyridylmethyl)-3-(*o*-tolyl)urea 950
1-(3-Pyridylmethyl)-3-(*m*-tolyl)urea 951
1-(3-Pyridylmethyl)-3-(*p*-tolyl)urea 950
1-(2-Pyridyl)-3-(3,4-xylyl)urea 965
$C_{14}H_{15}N_3OS$
1-(4-Ethoxyphenyl)-3-(2-pyridyl)-2-thiourea 939
1-(2-Methoxy-5-methylphenyl)-3-(3-pyridyl)-2-thiourea 937
1-(3-Methyl-2-pyridyl)-3-(4-(methylthio)phenyl)urea 941
1-(4-Methyl-2-pyridyl)-3-(3-methylthiophenyl)urea 941
1-(4-Methyl-2-pyridyl)-3-(4-(methylthio)phenyl)urea 941
1-(6-Methyl-2-pyridyl)-3-(4-(methylthio)phenyl)urea 941
$C_{14}H_{15}N_3O_2$
1-(*p*-Anisyl)-3-(3-pyridylmethyl)urea 938
4'-(Dimethylamino)azoxybenzene *N*-oxide 1090
1-(3-Hydroxy-2-pyridyl)-3-(3,4-xylyl)urea 933
1-(4-Methoxy-2-methylphenyl)-3-(2-pyridyl)urea 939
1-(2-Methoxyphenyl)-3-(3-pyridylmethyl)urea 938
1-(3-Methoxyphenyl)-3-(3-pyridylmethyl)urea 938
1-(4-Methoxyphenyl)-3-(3-pyridylmethyl)urea 938
N-Methylanthranilic (2-hydroxybenzylidene)hydrazide 1033
$C_{14}H_{15}N_3O_2S_2$
1-(Phenylsulfonyl)-4-(*p*-tolyl)-3-thiosemicarbazide 829
$C_{14}H_{15}N_3O_3$
3-(4-Amino-1,3-dimethyl-6-oxo-1,2,3,6-tetrahydropyrimidin-5-yl)-phthalide 701
1-(2,5-Dimethoxyphenyl)-3-(3-pyridyl)urea 937
N-(5,7-Dimethyl-1,8-naphthyridin-2-yl)-succinamic acid 1000
$C_{14}H_{15}N_3O_3S$
1-Anthraniloyl-2-(*p*-tolylsulfonyl)-hydrazine 819
1-(4-Sulfamoylphenyl)-3-(*o*-tolyl)urea 825
1-(4-Sulfamoylphenyl)-3-(*m*-tolyl)urea 826
$C_{14}H_{15}N_3O_3S_2$
1-(4-(Methythio)phenyl)-3-(4-sulfamoylphenyl)urea 816
$C_{14}H_{15}N_3O_4$
N-*sec*-Butyl-2,4-dinitro-1-naphthylamine 275
1-Furfuryl-1-methyl-3-(2-methyl-6-nitrophenyl)urea 850
$C_{14}H_{15}N_3O_4S$
1-(2-Methoxyphenyl)-3-(4-sulfamoylphenyl)urea 814
$C_{14}H_{15}N_3O_5$
4-(2,5-Dimethoxyphenyl)-1-(2-furoyl)-semicarbazide 845
5,5-Dimethyl-2-nitro-3-(4-nitroanilino)-2-cyclohexen-1-one 284
$C_{14}H_{15}N_3O_5S$
4-Acetyl-*N*-(2,6-dimethoxy-4-pyrimidinyl)-benzenesulfonamide 973
$C_{14}H_{15}N_3O_5S_2$
4'-(4-(Aminosulfonyl)-anilinosulfonyl)-acetanilide 825
$C_{14}H_{15}N_3S$
1-Phenyl-3-(2-(2-pyridyl)ethyl)-2-thiourea 951
1-Phenyl-4-(*o*-tolyl)-3-thiosemicarbazide 750
1-(2-Pyridyl)-3-(3,4-xylyl)-2-thiourea 954

$C_{14}H_{15}N_5$
2-(2-Benzimidazolylamino)-4,7-dimethyl-5*H*-1,3-diazepine ... 907
2-(4,6-Dimethyl-2-pyrimidinylamino)-6-methylbenzimidazole ... 975
$C_{14}H_{15}N_5O$
2-(5,6-Dimethyl-2-benzimidazolylamino)-6-methyl-4-pyrimidinol ... 972
2-(5,6-Dimethyl-2-benzimidazolyliminomethyl)-4-methyl-5-imidazolol ... 1037
$C_{14}H_{15}N_5O_3$
2-(3-Furfurylureido)acetic (3-pyridylmethylene)hydrazide ... 944
$C_{14}H_{15}N_5O_3S$
2-(3-(2-Thiazolyl)ureido)acetic acid *N*2-(2-methoxybenzylidene)-hydrazide ... 875
$C_{14}H_{15}N_5O_4$
Caffeine nicotinate ... 924
$C_{14}H_{15}N_5S_2$
1-Anilino-6-phenyl-2,5-dithiobiurea ... 743
$C_{14}H_{15}O_2P$
α-(Diphenylphosphinyl)-ethanol ... 834
Phenyl(3,5-xylyl)phosphinic acid ... 836
$C_{14}H_{15}O_3P$
(4-Ethoxyphenyl)-phenylphosphinic acid ... 834
$C_{14}H_{16}$
2-Isobutylnaphthalene ... 167
$C_{14}H_{16}B_2F_8N_2$
2,2′-Vinylenebis(1-methylpyridinium tetrafluoroborate) ... 952
4,4′-Vinylenebis(1-methylpyridinium tetrafluoroborate) ... 958
$C_{14}H_{16}BrN$
2,3-Dihydro-2,2-dimethyl-1*H*-benzene(f)isoindolium bromide ... 255
2,3-Dihydro-2,2-dimethyl-1*H*-benzene(e)isoindolium bromide ... 255
$C_{14}H_{16}BrNO_2$
2-(6-Bromohexyl)phthalimide ... 714
Ethyl 5-bromo-3-indolebutyrate ... 890
$C_{14}H_{16}BrNO_3$
4′-Bromohexahydrophthalanilic acid ... 406
$C_{14}H_{16}ClF_3N_2O$
1-(2-Chloro-5-trifluoromethylphenyl)-3-cyclohexylurea ... 510
N-(4-Chloro-3-(trifluoromethyl)-phenyl)-3-methyl-1-piperidinecarboxamide ... 693
$C_{14}H_{16}ClF_3N_2O_2$
4-(4-Chloro-2-(trifluoromethyl)-phenylcarbamoyl)-2,6-dimethylmorpholine ... 693
$C_{14}H_{16}ClNO$
3-(3-Chloroanilino)-5,5-dimethyl-2-cyclohexen-1-one ... 282
$C_{14}H_{16}ClNO_3$
5′-Chloro-2′,4′-dimethoxysorbanilide ... 447
cis-*N*-(2-Chlorophenyl)-hexahydrophthalamic acid ... 406
cis-*N*-(3-Chlorophenyl)-hexahydrophthalamic acid ... 406
cis-Hexahydro-*N*-(4-chlorophenyl)-phthalamic acid ... 406
$C_{14}H_{16}ClNO_5$
Ethyl 5-(chloroformyl)-3-(ethoxycarbonyl)-4-methyl-2-pyrroleacrylate ... 840
$C_{14}H_{16}ClNSi$
Chloro(dimethylamino)-diphenylsilane ... 1112
$C_{14}H_{16}Cl_2N_2O_2$
6-(4,5-Dichloro-2-methylphenyl)-dihydro-2-(3-hydroxypropyl)-3(2*H*)-pyridazinone ... 694
6-(3,4-Dichlorophenyl)-4,5-dihydro-2-(2-hydroxybutyl)-3(2*H*)-pyridazinone ... 697
$C_{14}H_{16}Cl_2OSi_2$
1,3-Dichloro-1,3-dimethyl-1,3-diphenyldisiloxane ... 1112
$C_{14}H_{16}Cl_3NO_2$
2-Methylcyclohexyl *N*-(2,4,5-trichlorophenyl)carbamate ... 678
$C_{14}H_{16}Cl_3NO_4$
Butyl *N*-(1-benzoyloxy)-2,2,2-trichloroethyl)carbamate ... 610
$C_{14}H_{16}Cl_3NO_5$
Isopropyl *N*-(1-(4-methoxybenzoyloxy)-2,2,2-trichloroethyl)carbamate ... 609
Propyl *N*-(1-(4-methoxybenzoyloxy)-2,2,2-trichloroethyl)carbamate ... 609
$C_{14}H_{16}FNO_3$
3′-Fluorohexahydrophthalanilic acid ... 406
cis-*N*-(2-Fluorophenyl)-hexahydrophthalamic acid ... 406
cis-*N*-(4-Fluorophenyl)-hexahydrophthalamic acid ... 406
$C_{14}H_{16}FNO_4$
Diethyl 2-(2-fluoroanilinomethylene)malonate ... 366
$C_{14}H_{16}F_3NO_2$
Cyclohexyl *N*-(3-trifluoromethylphenyl)carbamate ... 610
$C_{14}H_{16}F_3N_3O$
1-(2-Cyanoethyl)-1-propyl-3-(3-(trifluoromethyl)phenyl)urea ... 751
$C_{14}H_{16}F_3N_3O_5$
N-Butyl-2′,6′-dinitro-4′-(trifluoromethyl)propionanilide ... 397
2′,6′-Dinitro-*N*-propyl-4′-(trifluoromethyl)butyranilide ... 397
$C_{14}H_{16}F_3N_3O_6$
Propyl *N*-(2,6-dinitro-4-(trifluoromethyl)phenyl)-*N*-propylcarbamate ... 609
$C_{14}H_{16}F_6$
1,3-Bis(trifluoromethyl)-5-(1,1,2-trimethylpropyl)benzene ... 170
$C_{14}H_{16}INO_3$
cis-Hexahydro-*N*-(4-iodophenyl)-phthalamic acid ... 406
$C_{14}H_{16}NO_3P$
Dibenzyl phosphoramidate ... 834
$C_{14}H_{16}N_2$
5-Amino-9-methyl-1,2,3,4-tetrahydroacridine ... 987
2-(2-Benzylaminoethyl)pyridine ... 945
2,2′-Diamino-6,6′-dimethylbiphenyl ... 230
2,2′-Dimethylbenzidine hydrochloride ... 230
1,2-Di-*o*-tolylhydrazine ... 235
3-Ethyl-2-imino-3-phenyl-1-cyclopentanecarbonitrile ... 756
1,4,5,6,7,8-Hexahydro-2-phenylcycloheptimidazole ... 869
2-(2-(2-Methylanilino)ethyl)pyridine ... 946
2-(2-(3-Methylanilino)ethyl)pyridine ... 946
2-(2-(4-Methylanilino)ethyl)pyridine ... 946
4-Methyl-2-(4-methylbenzylamino)-pyridine ... 954
2-Piperidinoquinoline ... 994
2,3,5,6-Tetramethyl-1,4-benzenediacetonitrile ... 765
2,2′,6,6′-Tetramethyl-4,4′-dipyridyl ... 967
o-Tolidine ... 231
$C_{14}H_{16}N_2O$
N-(2-Aminoethyl)-1-naphthaleneacetamide hydrochloride ... 487
3-(2-(Benzylideneamino)ethyl)-4-methyl-3-pyrrolin-2-one ... 691
1,1-Bis(2-pyridylmethyl)ethanol ... 934
1-Ethyl-3-(1-naphthylmethyl)urea ... 603
1-Isopropyl-3-(1-naphthyl)urea ... 603
1-(1-Naphthyl)-3-propylurea ... 603
$C_{14}H_{16}N_2OS$
3-Benzoyl-1,1-diallyl-2-thiourea ... 745
3-Cyclohexyl-1,2-dihydro-2-thioxo-4(3*H*)-quinazolinone ... 741
$C_{14}H_{16}N_2O_2$
*N*ω-(Acetylacetyl)tryptamine ... 886
L-Alanine-4-methoxy-2-naphthylamide hydrochloride ... 486
1,2-Bis-(6-methyl-2-pyridyl)-1,2-ethanediol ... 932
4-Butyryl-3-methyl-1-phenyl-2-pyrazolin-5-one ... 710
4,4′-Diamino-3,3′-dimethoxybiphenyl ... 228
5,5-Dimethyl-3-hydroxy-2-(phenylazo)-2-cyclohexen-1-one ... 1083
5-Ethoxy-3-methyl-2-pyrrolecarboxanilide ... 838
Ethyl α-cyano-4-(dimethylamino)-cinnamate ... 765
2,2′-(Ethylenedioxy)dianiline ... 227
3,3′-(Ethylenedioxy)dianiline ... 227
1-(2-Ethylphenyl)-3-furfurylurea ... 852
1-Furfuryl-3-(2,3-xylyl)urea ... 852
1-Furfuryl-3-(2,4-xylyl)urea ... 851
1-Furfuryl-3-(2,5-xylyl)urea ... 851
1-Furfuryl-3-(2,6-xylyl)urea ... 852
1-Furfuryl-3-(3,4-xylyl)urea ... 852
1-Furfuryl-3-(3,5-xylyl)urea ... 852
2-Hydroxy-3-methoxy-*N*-(4-methyl-2-pyridyl)benzylamine ... 931
2-Isopropyl-6-phenoxymethyl-4-pyrimidinol ... 969
1-(2-Methoxyethyl)-3-(1-naphthyl)urea ... 557
N-Methyl-2-propionyl-3-indoleacetamide ... 888
1-(Piperidinomethyl)isatin ... 704
8-Quinolyl *N*-butylcarbamate ... 992
$C_{14}H_{16}N_2O_2S$
2-Amino-*N*-ethylbenzenesulfonanilide ... 816
$C_{14}H_{16}N_2O_3$
*N*α-Acetyl-6-methyl-DL-tryptophan ... 890
*N*α-Acetyl-D-tryptophan methyl ester ... 890
*N*α-Acetyl-L-tryptophan methyl ester ... 886
Antipyrine-4-propionic acid ... 707
5,5-Dimethyl-3-(3-nitroanilino)-2-cyclohexen-1-one ... 284
5,5-Dimethyl-3-(4-nitroanilino)-2-cyclohexen-1-one ... 284
Ethyl 1,4-dihydro-2-hydroxy-6-methyl-4-phenyl-5-pyrimidinecarboxylate ... 368
Ethyl 6,7-dimethyl-3-hydroxy-2-quinoxalineacetate ... 999
Ethyl 2-(2-imidazolidinylidene)-3-oxo-3-phenylpropionate ... 374
1-Furfuryl-3-(4-methoxy-2-methylphenyl)urea ... 846
α-Morpholino-*N*-phenylsuccinimide ... 722
*N*α-Acetyl-5-methyl-DL-tryptophan ... 886
*N*α-Acetyl-DL-tryptophan methyl ester ... 886
$C_{14}H_{16}N_2O_3S$
6′-Amino-*p*-toluenesulfon-*m*-anisidide ... 813
2-(4-Sulfanilylanilino)ethanol ... 774
$C_{14}H_{16}N_2O_4$
1-(2,5-Dimethoxyphenyl)-3-furfurylurea ... 846
N,*N*′-(1,4-Phenylene)-bis(acetoacetamide) ... 443
3,3′,4,4′-Tetramethyl-(2,2′-bipyrrole)-5,5′-dicarboxylic acid ... 840
$C_{14}H_{16}N_2O_4S$
2,5-Dimethoxysulfanilanilide ... 812
Harman methosulfate ... 918
$C_{14}H_{16}N_2O_4S_2$
1,2-Bis(*p*-tolylsulfonyl)hydrazine ... 829
N,*N*′-Ethylenebis(benzenesulfonamide) ... 818
$C_{14}H_{16}N_2O_4S_4$
4-(Methylsulfonamido)phenyl disulfide ... 816
$C_{14}H_{16}N_2O_5$
cis-Hexahydro-*N*-(3-nitrophenyl)-phthalamic acid ... 426
cis-Hexahydro-*N*-(4-nitrophenyl)-phthalamic acid ... 427
Methyl maleate veratrylidenehydrazide ... 497
$C_{14}H_{16}N_2O_5S_2$
4,4′-Oxybis(methanesulfonanilide) ... 815
$C_{14}H_{16}N_2O_6S_2$
4,4′-Diamino-5,5′-dimethyl-2,2′-biphenyldisulfonic acid ... 789
$C_{14}H_{16}N_2O_6S_4$
Diethyl 5,5′-birhodanine-3,3′-diacetate ... 141
$C_{14}H_{16}N_2O_7$
Methyl 4,4-dimethoxy-2,3-dioxobutyrate 2-(4-carboxyphenylhydrazone) ... 369
$C_{14}H_{16}N_2S$
1-Isopropyl-3-(1-naphthyl)-2-thiourea ... 749
$C_{14}H_{16}N_4O$
4,4′-Diamino-5-methoxy-2-methylazobenzene ... 1086
$C_{14}H_{16}N_4O_2$
1,1′-(1,5-Naphthalenediyl)bis(3-methylurea) ... 604
$C_{14}H_{16}N_4O_6$
1,2-Diacetyl-1-(2,4-dinitrophenyl)-2-(1-methylpropenyl)-hydrazine ... 502
$C_{14}H_{16}N_4O_6S_2$
4-Hydrazinosulfonyl-3′-hydroxy-4′-methylsulfonamidobenzanilide ... 811
$C_{14}H_{16}N_4O_7$
Methyl 4-oxo-3-tetrahydropyranacetate 2,4-dinitrophenylhydrazone ... 358
$C_{14}H_{16}N_6O$
1,3-Dihydrazino-1-hydroxy-4-phenyl-1*H*-pyrido(1,2-c)pyrimidine ... 246
$C_{14}H_{16}O$
2-Benzylidene-5-methyl-1-cyclohexanone ... 304
1-(2-Butoxy)naphthalene ... 188
2,3,3a,4,5,9b-Hexahydro-9b-methyl-1*H*-benzene(e)inden-5-one ... 287
2-Isopropyl-6-methoxynaphthalene ... 188
$C_{14}H_{16}O_2$
Benzophenone dimethyl ketal ... 189
α,α-Dimethyl-6-methoxy-2-naphthalenemethanol ... 203
Ethyl 1a,2,3,7b-tetrahydro-1*H*-cyclopropa(a)naphthalene-1-carboxylate ... 361
Ethyl 1a,2,7,7a-tetrahydro-1*H*-cyclopropa(b)naphthalene-1-carboxylate ... 361
4-Methylenecyclohexyl benzoate ... 375
2-(2-Phenyl-1-cyclohexenyl)-acetic acid ... 332
2,2′,6,6′-Tetramethyl-Δ4,4′-bi-4*H*-pyran ... 23
$C_{14}H_{16}O_2S$
2-(*p*-Tosyl)-5-norbornene ... 782
$C_{14}H_{16}O_2Si$
Bis(4-methoxyphenyl)silane ... 1106
Dimethyldiphenoxysilane ... 1106
$C_{14}H_{16}O_3$
1-Cyclopropyl-3-(3,4-dimethoxyphenyl)-2-propen-1-one ... 283
Ethyl 2-(4-methylbenzylidene)-acetoacetate ... 365
$C_{14}H_{16}O_4$
Diethyl 2-phenylmaleate ... 365
Ethyl 2-(4-methoxybenzylidene)-acetoacetate ... 364
$C_{14}H_{16}O_6$
Dimethyl 2-(dimethoxybenzylidene)-malonate ... 364
1,4-Dimethyl-2,3,5-triacetoxybenzene ... 378
3-(3-Methoxybenzoyl)adipic acid ... 328
3-Phenyl-1,3,5-pentanetricarboxylic acid ... 332
$C_{14}H_{16}P_2S_2$
1,2-Dimethyl-1,2-diphenyldiphosphine disulfide ... 836
$C_{14}H_{16}Si$
Benzylmethylphenylsilane ... 1112
Dibenzylsilane ... 1112
Dimethyldiphenylsilane ... 1112
Diphenylethylsilane ... 1112
Di(*p*-tolyl)silane ... 1122
Methylphenyl(*m*-tolyl)silane ... 1112
$C_{14}H_{17}BrClNO_2S$
1-(α-Carboxymethylthio)-3-chlorobenzylidene)piperidinium

$C_{14}H_{17}BrClNO_2S$ (Cont'd)
bromide 1051
$C_{14}H_{17}BrCl_2N_2O_2$
1-(2-Acetoxyethyl)-5,6-dichloro-3-ethyl-2-methylbenzimidazolium bromide 906
$C_{14}H_{17}BrN_2$
4-Bromo-3-(piperidinomethyl)indole. 885
$C_{14}H_{17}BrO_4$
Diethyl 4-bromobenzylmalonate 353
$C_{14}H_{17}ClN_2O$
3-(2-Chloro-6-methylphenyl)-1,1-diallylurea 595
3-(3-Chloro-4-methylphenyl)-1,1-diallylurea 600
3-(4-Chloro-2-methylphenyl)-1,1-diallylurea 600
3-(5-Chloro-2-methylphenyl)-1,1-diallylurea 600
Harmaline methochloride 918
$C_{14}H_{17}ClN_2OS$
2-(2-Chloroethyl)-3-hydroxy-3-phenyl-tetrahydro-5*H*-thiazolo(3,2-a)pyrimidine hydrobromide 727
3-(4-Chlorophenyl)-2-ethyl-3-hydroxytetrahydro-5*H*-thiazolo(3,2-a)pyrimidine hydrobromide 245
$C_{14}H_{17}ClN_2O_2$
3-(3-Chloro-2-methoxyphenyl)-1,1-diallylurea 521
O-(*N*-(2-Chloro-6-methylphenyl)-carbamoyl)cyclohexanone oxime 1012
O-(*N*-(3-Chloro-4-methylphenyl)-carbamoyl)cyclohexanone oxime 1013
O-(*N*-(4-Chloro-2-methylphenyl)-carbamoyl)cyclohexanone oxime 641
O-(*N*-(5-Chloro-2-methylphenyl)-carbamoyl)cyclohexanone oxime 1013
2-(4-Chlorophenyl)-*N*-methyl-6-oxo-3-piperidinecarboxamide 689
$C_{14}H_{17}ClN_2O_3$
O-(*N*-(3-Chloro-2-methoxyphenyl)-carbamoyl)cyclohexanone oxime 1009
O-(*N*-(5-Chloro-2-methoxyphenyl)-carbamoyl)cyclohexanone oxime 619
$C_{14}H_{17}ClN_2O_4$
2-Methylcyclohexyl *N*-(2-chloro-4-nitrophenyl)carbamate 670
$C_{14}H_{17}ClN_2O_6$
2-(*sec*-Butyl)-4,6-dinitrophenyl 2-chlorobutyrate 379
$C_{14}H_{17}ClO$
3′-Chloro-4′-cyclohexylacetophenone 289
$C_{14}H_{17}ClO_2S$
3-Chloro-2-norbornyl *p*-tolyl sulfone. 782
$C_{14}H_{17}ClO_4$
Ethyl 2-(4-chlorophenyl)-1,3-dioxolane-2-propionate 353
$C_{14}H_{17}ClO_6$
Diethyl (4-chloro-1,2-phenylenedioxy)diacetate 353
$C_{14}H_{17}Cl_2NO_2$
2-Methylcyclohexyl *N*-(2,3-dichlorophenyl)carbamate 678
2-Methylcyclohexyl *N*-(2,4-dichlorophenyl)carbamate 679
2-Methylcyclohexyl *N*-(2,5-dichlorophenyl)carbamate 679
2-Methylcyclohexyl *N*-(2,6-dichlorophenyl)carbamate 631
2-Methylcyclohexyl *N*-(3,4-dichlorophenyl)carbamate 679
$C_{14}H_{17}Cl_2NO_3$
2-(2,4-Dichlorophenoxy)-*N*-methyl-*N*-(tetrahydrofurfuryl)acetamide 402
$C_{14}H_{17}Cl_2N_3S$
4-Cyclohexyl-1-(2,4-dichlorobenzylidene)-3-thiosemicarbazide 1026
$C_{14}H_{17}Cl_3N_2O$
1-Cyclohexyl-1-methyl-3-(2,4,5-trichlorophenyl)urea 582
$C_{14}H_{17}F_3N_2O$
1-Cyclohexyl-3-(3-trifluoromethylphenyl)urea 510
2-Methyl-1-(*N*-(3-trifluoromethylphenyl)carbamoyl)-piperidine 693
3-Methyl-1-(*N*-(3-trifluoromethylphenyl)carbamoyl)-piperidine 693
4-Methyl-1-(*N*-(3-trifluoromethylphenyl)carbamoyl)-piperidine 693
α,α,α-Trifluoro-1-piperidineaceto-*m*-toluidide 398
$C_{14}H_{17}F_3N_2O_2$
4-(*N*-(3-(Trifluoromethyl)phenyl)-carbamoyl)-2,6-dimethylmorpholine 693
$C_{14}H_{17}IN_2OS$
3-Methyl-2-(2-(*N*-phenylacetamido)-vinyl)-2-thiazolinium iodide 431
$C_{14}H_{17}N$
Cyclohexylphenylacetonitrile 756
2,3-Cyclooctenoindole 891
Diethyl(2-naphthyl)amine 233
1-(2-(Dimethylamino)ethyl)-naphthalene hydrochloride 243
1-(*p*-Tolyl)-1-cyclohexanecarbonitrile 756
$C_{14}H_{17}NO$
1-Acetyl-2,2,4-trimethyl-1,2-dihydroquinoline 705
N-(α-Benzylfurfuryl)-ethylamine 850
N-(3-Cyclohexen-1-yl)acetanilide 441
5-Cyclohexyl-2-methylbenzoxazole .. 910
7-Diethylamino-1-naphthol 227
1-(4-Methoxyphenyl)-1-cyclohexanecarbonitrile 753
$C_{14}H_{17}NO_2$
N,*N*-Diacetyl-5,6,7,8-tetrahydro-1-naphthylamine 441
5,5-Dimethyl-3-(2-hydroxyanilino)-2-cyclohexen-1-one 282
5,5-Dimethyl-3-(3-hydroxyanilino)-2-cyclohexen-1-one 282
5,5-Dimethyl-3-(4-hydroxyanilino)-2-cyclohexen-1-one 282
4′-Ethoxy-2,4-hexadienanilide 466
Ethyl 4-cyano-4-phenylvalerate 755
Ethyl 3-indolebutyrate 887
2′-Methoxy-5′-methyl-2,4-hexadienanilide 465
N-(2,4,6-Trimethylphenyl)-glutarimide 720
$C_{14}H_{17}NO_2S$
4-(4-Methoxythiocinnamoyl)-morpholine 738
$C_{14}H_{17}NO_3$
4′-Butylmaleanilic acid 467
N-Cyclohexylphthalamic acid 474
Cyclopentyl *N*-(3-acetylphenyl)-carbamate 675
Fagaramide 477
cis-Hexahydro-*N*-phenylphthalamic acid 435
Methyl 1-methyl-6-oxo-2-phenyl-3-piperidinecarboxylate 692
$C_{14}H_{17}NO_3S$
1,4-Dimethylpyridinium *p*-toluenesulfonate 956
1-Ethylpyridinium *p*-toluenesulfonate 957
$C_{14}H_{17}NO_4$
2-Butenyl *N*-(4-(ethoxycarbonyl)-phenyl)carbamate 675
4′-Carboxy-2-pivaloylacetanilide 443
Ethyl 3-cyano-2-hydroxy-1,4,4a,5,6,7,8,8a-octa-*H*-4-oxo-1-naphthalenecarboxylate 144
Ethyl 3,4-dihydro-6,7-dimethoxy-1-isoquinolinecarboxylate 369
3′-Hydroxyhexahydro-phthalanilic acid 413
2-(2-Methoxyphenyl)-1-methyl-6-oxo-3-piperidinecarboxylic acid 690
$C_{14}H_{17}NO_4S$
2-Oxo-*N*-(*p*-tosyl)-1-cyclohexanecarboxamide 823
$C_{14}H_{17}NO_5$
N-(3,4-Diacetoxyphenethyl)-acetamide 420
Diethyl benzamidomalonate 473
Diethyl 2-(phenylcarbamoyl)-malonate 443
N-(3,4-Dimethoxyphenethyl)-maleamic acid 464
2-(2,4-Dimethoxyphenyl)-1-methyl-5-oxopyrrolidine-3-carboxylic acid 327
2-(3,4-Dimethoxyphenyl)-1-methyl-5-oxopyrrolidine-3-carboxylic acid 690
4′-(Ethoxycarbonyl)glutaranilic acid. 433
Ethyl 3-(2-carboxyphenylimino)-3-ethoxypropionate 1049
2-(2-Propen-2-yloxycarbonyl)ethyl *N*-(2-methoxyphenyl)carbamate 651
$C_{14}H_{17}NO_6$
N-Carbobenzyloxy-2-methyl-L-glutamic acid dicyclohexylamine salt 659
Diethyl 2-(4-nitrobenzyl)malonate 359
Ethyl 5-carboxy-3-(ethoxycarbonyl)-4-methyl-2-pyrroleacrylate 840
Hexyl 3-nitrophthalate 374
Methyl 5-methoxycarbonyl-4-methoxycarbonylmethyl-2-methyl-3-pyrroleacrylate 839
$C_{14}H_{17}N_3$
3-(1,5-Diazabicyclo(3.2.1)oct-8-yl-indole 885
$C_{14}H_{17}N_3O$
2-(Bis-(2-pyridylmethyl)-amino)-ethanol 932
1-Cyclohexylideneamino-1,2-dihydro-4,6-dimethyl-2-oxonicotinonitrile 1069
1-(*N*-(2,6-Diethylphenyl)carbamoyl)-imidazole 868
$C_{14}H_{17}N_3OS$
1-(2,6-Diethylphenyl)-3-(2-thiazolyl)urea 877
$C_{14}H_{17}N_3O_2$
3,5-Dimethyl-2-cyclohexen-1-one (4-nitrophenyl)hydrazone 276
Hippuric acid *N*-cyclopentylidenehydrazide 501
$C_{14}H_{17}N_3O_2S_2$
Benzenesulfonic (3-methyl-4,5,6,7-tetrahydro-2-benzothiazolinylidene)hydrazide 820
$C_{14}H_{17}N_3O_3$
1,1-Diallyl-3-(2-methyl-6-nitrophenyl)urea 562
1-(*N*-(2,5-Dimethoxyphenyl)-carbamoyl)-3,5-dimethylpyrazole 860
Ethyl *N*-(4-antipyryl)-carbamate 708
Ethyl 5-(4-methoxyanilino)-3-methyl-4-pyrazolecarboxylate 860
$C_{14}H_{17}N_3O_4$
O-(*N*-(2-Methyl-4-nitrophenyl)-carbamoyl)cyclohexanone oxime 1010
O-(*N*-(4-Methyl-3-nitrophenyl)-carbamoyl)cyclohexanone oxime 1010
$C_{14}H_{17}N_3O_4S$
5-Cyano-4-hydroxy-3,4,5,6-4*H*-6-(3,4,5-trimethoxyphenyl)-2(1*H*)-pyrimidinethione potassium salt 752
$C_{14}H_{17}N_3O_7$
4-(3,5-Bis(methoxycarbonyl)phenyl)-1-(ethoxycarbonyl)semicarbazide 673
$C_{14}H_{17}N_5O$
2-(4-(Dimethylamino)-benzylidenehydrazino)-4-hydroxy-6-methylpyrimidine 1034
$C_{14}H_{17}N_5O_3$
Carbobenzyloxyhistidine hydrazide. 866
$C_{14}H_{17}N_5O_6$
2′,3′-Diacetyladenosine 920
$C_{14}H_{17}N_5O_9$
4-(2,3-Dihydroxypropylamino)-6,7,8-trimethoxyquinazoline dinitrate ester maleate 999
$C_{14}H_{18}$
α-(4-Methylcyclohexylidene)toluene. 165
1,2,3,4,5,6,7,8-Octahydroanthracene 164
3-Phenyl-1-cyclooctene 164
1-(*o*-Tolyl)-cycloheptene 164
1-(*p*-Tolyl)-cycloheptene 165
$C_{14}H_{18}As_2N_2O_8$
4,4′-(Ethylenedioxy)bis(3-aminobenzenearsonic acid) 1126
$C_{14}H_{18}BrNO_2$
8,9-Dimethoxy-2,3,5,6-tetrahydro-1*H*-pyrrolo(2,1-a)isoquinolinium bromide 254
2-Methylcyclohexyl *N*-(3-bromophenyl)carbamate 631
2-Methylcyclohexyl *N*-(4-bromophenyl)carbamate 678
$C_{14}H_{18}BrNO_3$
DL-2-Bromoisovaleryl-DL-phenylalanine 429
$C_{14}H_{18}ClF_3N_2O_2$
2-(Diethylamino)ethyl *N*-(2-chloro-5-trifluoromethylphenyl)carbamate 606
$C_{14}H_{18}ClN$
5-Chloro-2-isopropyl-2-phenylvaleronitrile 751
$C_{14}H_{18}ClNO_2$
4-Chloro-2-methylphenyl *N*-cyclohexylcarbamate 632
Cyclohexyl *N*-(2-chloro-6-methylphenyl)carbamate 679
Cyclohexyl *N*-(3-chloro-4-methylphenyl)carbamate 632
Cyclohexyl *N*-(4-chloro-2-methylphenyl)carbamate 680
Cyclohexyl *N*-(5-chloro-2-methylphenyl)carbamate 680
2-Methylcyclohexyl *N*-(2-chlorophenyl)carbamate 678
2-Methylcyclohexyl *N*-(3-chlorophenyl)carbamate 678
2-Methylcyclohexyl *N*-(4-chlorophenyl)carbamate 631
$C_{14}H_{18}ClNO_4$
1-(Carbobutoxy)ethyl *N*-(3-chlorophenyl)carbamate 622
$C_{14}H_{18}ClNO_5$
2-(4-Chloro-3-methylphenoxyethyl) *N*-(ethoxycarbonylmethyl)-carbamate 614
4-Chloro-2-methylphenyl *N*-(butoxycarbonyloxymethyl)-carbamate 630
Tetrahydrofurfuryl *N*-(5-chloro-2,4-dimethoxyphenyl)carbamate 612
$C_{14}H_{18}ClN_3O_3$
3-(2-Chloro-4-nitrophenyl)-1-cyclohexyl-1-methylurea 568
3-(4-Chloro-2-nitrophenyl)-1-cyclohexyl-1-methylurea 523
$C_{14}H_{18}ClN_3O_4$
1-Chloro-2,4-dimorpholino-5-nitrobenzene 273
$C_{14}H_{18}ClN_3O_6$
2-(2-Oxo-1-imidazolidinyloxy)ethyl *N*-(5-chloro-2,4-dimethoxyphenyl)-carbamate 694
$C_{14}H_{18}ClN_5O_3$
4-(4-Chlorophenyl)-1-(pyrrolidinocarboxamidoacetyl)-semicarbazide 695
$C_{14}H_{18}Cl_2N_2$
1,1′-Tetramethylenebis(pyridinium chloride) 952
$C_{14}H_{18}Cl_2N_2O$
1-Cyclohexyl-3-(2,3-dichlorophenyl)-1-methylurea 583
1-Cyclohexyl-3-(2,4-dichlorophenyl)-1-methylurea 583
1-Cyclohexyl-3-(2,5-dichlorophenyl)-1-methylurea 583
1-Cyclohexyl-3-(2,6-dichlorophenyl)-1-methylurea 583
1-Cyclohexyl-3-(3,4-dichlorophenyl)-1-methylurea 583
1-Cyclohexyl-3-(3,5-dichlorophenyl)-1-methylurea 535
$C_{14}H_{18}FNO_2$
2-Methylcyclohexyl *N*-(4-fluorophenyl)carbamate 631
$C_{14}H_{18}F_2N_2O$
1-Cyclohexyl-3-(2,4-difluorophenyl)-1-methylurea 535

$C_{14}H_{18}F_3N_3O_3$
1-(6-Hydroxyhexanoyl)-4-(3-(trifluoromethyl)phenyl)-semicarbazide ... 507
$C_{14}H_{18}F_3N_3O_6S$
2′,6′-Dinitro-*N*-hexyl-4′-(trifluoromethyl)-methanesulfonanilide ... 805
$C_{14}H_{18}F_3N_3S$
1-(Hexamethyleneimino)-3-(3-(trifluoromethyl)phenyl)-2-thiourea ... 727
$C_{14}H_{18}F_4$
1,4-Di-*tert*-butyl-2,3,5,6-tetrafluorobenzene ... 170
$C_{14}H_{18}F_{14}O_2Si$
Bis(3,3,4,4,5,5,5-heptafluoropentyl)-diethoxysilane ... 1100
$C_{14}H_{18}IN$
(2-Naphthylmethyl)-trimethylammonium iodide ... 256
$C_{14}H_{18}N_2$
6-(Diethylamino)-5-methylquinoline ... 989
$C_{14}H_{18}N_2O$
3-(2-Aminoanilino)-5,5-dimethyl-2-cyclohexen-1-one ... 284
3-(4-Aminoanilino)-5,5-dimethyl-2-cyclohexen-1-one ... 284
Benzoic (cyclohexylmethylene)-hydrazide ... 501
1,1-Diallyl-3-(*o*-tolyl)urea ... 588
1,1-Diallyl-3-(*m*-tolyl)urea ... 580
1,1-Diallyl-3-(*p*-tolyl)urea ... 591
α-(4-Methoxyphenyl)-1-piperidineacetonitrile ... 753
2-Methyl-3-(morpholinomethyl)-indole ... 884
N-Methyltetrahydroharmine ... 918
Valeric acid cinnamylidenehydrazide ... 502
$C_{14}H_{18}N_2O_2$
3-(*m*-Anisyl)-1,1-diallylurea ... 558
1,1-Diallyl-3-(2-methoxyphenyl)urea ... 559
1,1-Diallyl-3-(4-methoxyphenyl)urea ... 559
4-(1,2-Dimethyl-1-ethyl-2-nitropropyl)benzonitrile ... 765
4-(2-Nitro-1,1,2-trimethylbutyl)-benzonitrile ... 765
O-(*N*-(*o*-Tolyl)carbamoyl)-cyclohexanone oxime ... 1011
O-(*N*-(*m*-Tolyl)carbamoyl)-cyclohexanone oxime ... 1011
O-(*N*-(*p*-Tolyl)carbamoyl)-cyclohexanone oxime ... 1013
$C_{14}H_{18}N_2O_2S$
Dicyclopropyl ketone *p*-tosylhydrazone ... 1075
2-Norbornanone (*p*-tosyl)hydrazone ... 1071
p-Toluenesulfonic (3-methyl-2-cyclohexen-1-ylidene)hydrazide ... 1071
$C_{14}H_{18}N_2O_3$
O-(*N*-(*o*-Anisyl)carbamoyl)-cyclohexanone oxime ... 1009
O-(*N*-(*m*-Anisyl)carbamoyl)-cyclohexanone oxime ... 1009
O-(*N*-(*p*-Anisyl)carbamoyl)-cyclohexanone oxime ... 1009
2-Carbamoylphenyl *N*-cyclohexylcarbamate ... 666
1-(2-Carboxyphenyl)-3-cyclohexylurea ... 576
1-(4-Carboxyphenyl)-3-cyclohexylurea ... 576
4,5-Dihydro-2-(3-hydroxypropyl)-6-(4-methoxyphenyl)-3(2*H*)-pyridazinone ... 696
Ethyl 2-amino-3-cyano-1,4,4a,5,6,7,8,8a-octahydro-4-oxo-1-naphthalenecarboxylate ... 145
cis-Hexahydro-*N*-(4-methyl-2-pyridyl)phthalamic acid ... 963
cis-Hexahydro-*N*-(3-pyridylmethyl)-phthalamic acid ... 949
L-Phenylalanyl-L-proline ... 701
$C_{14}H_{18}N_2O_4$
N-Acetylglycyl-L-phenylalanine methyl ester monohydrate ... 429
Cyclohexyl *N*-(4-methyl-3-nitrophenyl)carbamate ... 670
Diethyl (5-methyl-2-pyridylaminomethylene)-malonate ... 959
Diethyl (6-methyl-2-pyridylaminomethylene)-malonate ... 953
Diethyl (6-methyl-2-pyridylaminomethylene)malonate ... 954
2,5-Dimorpholino-1,4-benzoquinone ... 53
4,6-Dinitro-1,1,3,3,5-pentamethylindane ... 276
4′-(*N*-Methylacetamido)-glutaranilic acid ... 428
2-Methylcyclohexyl *N*-(3-nitrophenyl)carbamate ... 670
$C_{14}H_{18}N_2O_5$
4′-*tert*-Butyl-2′,6′-dimethyl-3′,5′-dinitroacetophenone ... 294
Carbobenzyloxy-β-alanyl-L-alanine ... 660
Carbobenzyloxy-β-alanylglycine methyl ester ... 664
Carbobenzyloxy-glycyl-glycine ethyl ester ... 677
Carbobenzyloxyglycylsarcosine methyl ester ... 677
Carbobenzyloxysarcosyl-L-alanine ... 660
Carbobenzyloxysarcosylglycine methyl ester ... 661
L-Phenylalanyl-L-glutamic acid ... 429
$C_{14}H_{18}N_2O_5S$
Hexahydro-4′-sulfamoylphthalanilic acid ... 824
$C_{14}H_{18}N_2O_6$
Carbobenzyloxyglycyl-L-threonine dicyclohexylamine salt hydrate ... 643
Carbobenzyloxyserylglycine methyl ester ... 642
$C_{14}H_{18}N_2O_7$
5′-Acetyl-2′,3′-isopropylideneuridine ... 131
$C_{14}H_{18}N_4$
1,2-Bis(4-pyridyl)-*N*,*N*′-dimethylethylenediamine ... 943
$C_{14}H_{18}N_4O_2$
β-Alanyl-L-tryptophan ... 883
$C_{14}H_{18}N_4O_2S$
4-Cyclohexyl-1-(3-nitrobenzylidene)-3-thiosemicarbazide ... 1063
$C_{14}H_{18}N_4O_4S_2$
N,*N*′-Bis(3-aminophenylsulfonyl)-ethylenediamine ... 816
$C_{14}H_{18}N_4S_2$
α,α′-Bis((4,5-dihydro-2-imidazolyl)-thio)-*o*-xylene dihydrobromide ... 248
$C_{14}H_{18}N_6O$
2′-Butoxy-2,6-diamino-3,5′-azopyridine ... 1086
$C_{14}H_{18}O$
1-Allyloxy-2,4-dimethyl-6-propenylbenzene ... 182
4-*tert*-Butylphenyl cyclopropyl ketone ... 296
2-(1-Cyclohepten-1-yl) benzyl alcohol ... 204
α,α-Dicyclopropyl-4-methylbenzyl alcohol ... 203
4-(4-(2-Methylpropyl)phenyl)-3-butyn-1-ol ... 1017
1,4,4a,4b,5,8,8a,9a-Octahydro-9(10*H*)-anthracenone ... 56
3,3,6,8-Tetramethyl-1-tetralone ... 298
$C_{14}H_{18}O_2$
1,3-Diallyloxy-4-ethylbenzene ... 178
Ethyl 6,7,8,9-tetrahydro-5*H*-benzocycloheptene-7-carboxylate ... 363
Isopentyl cinnamate ... 365
α-(2-Methoxy-2-cyclohexen-1-yl)-benzyl alcohol ... 201
1,2,3,4,4a,5,8,8a-Octahydro-1,4:5,8-dimethanonaphth-2-yl acetate ... 81
2,2-Pentamethylene-4-phenyl-1,3-dixolane ... 184
Pentyl cinnamate ... 365
1,2,3,4-Tetrahydro-2-naphthalenebutyric acid ... 334
1-(*p*-Tolyl)-1-cyclohexanecarboxylic acid ... 332
$C_{14}H_{18}O_3$
Ethyl 3-hydroxy-3-methyl-5-phenyl-4-pentenoate ... 354
Ethyl 6-oxo-6-phenylhexanoate ... 360
Methyl 4-(*tert*-butyl)benzoylacetate ... 364
Methyl α-phenylcyclopentane-glycolate ... 353
$C_{14}H_{18}O_4$
Diethyl 2,2′-(1,4-phenylene)-diacetate ... 361
Diethyl 2-phenylsuccinate ... 360
Dimethyl 2,4,6,8-decatetraene-2,9-dicarboxylate ... 79
Dimethyl 1,4,4a,5,8,8a-hexahydro-4a,8a-naphthalenedicarboxylate ... 77
Ethyl 5-(2-methoxyphenyl)-3-oxopentanoate ... 363
Ethyl 5-(4-methoxyphenyl)-5-oxovalerate ... 355
Ethyl 2-(2-methyl-4-phenyl-1,3-dioxolan-2-yl)acetate ... 356
3-Phenyl-3-propylglutaric acid ... 332
2,3,5,6-Tetramethylhydroquinone diacetate ... 380
(+)-*mono*-(1,2,2-Trimethylpropyl) phthalate ... 375
$C_{14}H_{18}O_4S$
Diethyl 2-methyl-2-(phenylthio)-malonate ... 357
1-Methyl-2-oxocyclopentylmethyl *p*-toluenesulfonate ... 799
$C_{14}H_{18}O_5$
2,2-Dimethyl-5,7,8-trimethoxy-4-chromanone ... 319
$C_{14}H_{18}O_6$
Diethyl 2-hydroxy-2-(4-methoxyphenyl)-malonate ... 353
Ethyl (3,4,5-trimethoxybenzoyl)-acetate ... 363
Methyl 4,6-*O*-benzylidenegalactoside ... 205
mono-Methyl 2-(2,5-dimethoxybenzyl)succinate ... 355
4-(3,4,5-Trimethoxybenzoyl)-butyric acid ... 327
$C_{14}H_{18}O_9$
Tetramethyl 4-oxo-1,1,3,5-cyclohexanetetracarboxylate ... 93
$C_{14}H_{18}S$
2-Norbornyl *p*-tolyl sulfide ... 213
$C_{14}H_{19}BrN_2O$
4′-Bromo-3′-methyl-2-piperidinoacetanilide ... 404
1-(*N*-(3-Bromophenyl)carbamoyl)-2,6-dimethylpiperidine ... 696
3-(4-Bromophenyl)-1-cyclohexyl-1-methylurea ... 582
$C_{14}H_{19}BrN_2O_2$
4′-Bromo-3′-methyl-2-(tetrahydrofurfurylamino)-acetanilide ... 404
$C_{14}H_{19}ClN_2O$
1-(3-Chlorobenzyl)-3-cyclohexylurea ... 527
1-(4-Chlorobenzyl)-3-cyclohexylurea ... 535
3′-Chloro-3,4′-dimethyl-1-piperidinecarboxanilide ... 696
3′-Chloro-4,4′-dimethyl-1-piperidinecarboxanilide ... 696
1-(*N*-(3-Chloro-2-methylphenyl)-carbamoyl)-4-methylpiperidine ... 696
1-(2-Chloro-6-methylphenyl)-3-cyclohexylurea ... 583
1-(3-Chloro-4-methylphenyl)-3-cyclohexylurea ... 583
1-(4-Chloro-2-methylphenyl)-3-cyclohexylurea ... 535
1-(5-Chloro-2-methylphenyl)-3-cyclohexylurea ... 583
N-(3-Chloro-2-methylphenyl)-3-methyl-1-piperidinecarboxamide ... 696
6′-Chloro-α-methyl-1-pyrrolidine-*o*-acetotoluidide hydrochloride ... 430
1-(*N*-(2-Chlorophenyl)carbamoyl)-2,6-dimethylpiperidine ... 696
1-(*N*-(4-Chlorophenyl)carbamoyl)-2,6-dimethylpiperidine ... 696
3-(2-Chlorophenyl)-1-cyclohexyl-1-methylurea ... 582
3-(3-Chlorophenyl)-1-cyclohexyl-1-methylurea ... 582
3-(4-Chlorophenyl)-1-cyclohexyl-1-methylurea ... 582
$C_{14}H_{19}ClN_2OS$
1-(2-Chloro-4-methoxyphenyl)-3-cyclohexyl-2-thiourea ... 729
$C_{14}H_{19}ClN_2O_2$
1-(*N*-(5-Chloro-2-methoxyphenyl)-carbamoyl)-2-methylpiperidine ... 694
1-(*N*-(5-Chloro-2-methoxyphenyl)-carbomoyl)-3-methylpiperidine ... 694
1-(2-Chloro-4-methoxyphenyl)-3-cyclohexylurea ... 520
1-(3-Chloro-2-methoxyphenyl)-3-cyclohexylurea ... 558
4-(*N*-(2-Chloro-6-methylphenyl)-carbamoyl)-2,6-dimethylmorpholine ... 713
4-(*N*-(3-Chloro-4-methylphenyl)-carbamoyl)-2,6-dimethylmorpholine ... 714
4-(*N*-(4-Chloro-2-methylphenyl)-carbamoyl)-2,6-dimethylmorpholine ... 714
4-(*N*-(5-Chloro-2-methylphenyl)-carbamoyl)-2,6-dimethylmorpholine ... 714
1-(3-Chloropropionyl)-4-(2-methoxyphenyl)piperazine ... 693
$C_{14}H_{19}ClN_2O_2S$
1-(*N*-(5-Chloro-2,4-dimethoxyphenyl)thiocarbamoyl)-piperidine ... 728
$C_{14}H_{19}ClN_2O_3$
4-(*N*-(3-Chloro-2-methoxyphenyl)-carbamoyl)-2,6-dimethylmorpholine ... 699
$C_{14}H_{19}ClN_2O_4$
Piperidino *N*-(5-chloro-2,4-dimethoxyphenyl)carbamate ... 613
$C_{14}H_{19}ClO_2$
3′-Chloro-2,2-dimethyl-4′-propoxypropiophenone ... 288
$C_{14}H_{19}Cl_2NO_2$
Chlorambucil ... 325
$C_{14}H_{19}Cl_2N_3$
1,1′-(Iminodiethylene)-bis(pyridinium chloride) ... 944
$C_{14}H_{19}Cl_2N_3O_2$
1-(2,3-Dichlorophenyl)-3-(3-morpholinopropyl)urea ... 566
1-(2,6-Dichlorophenyl)-3-(3-morpholinopropyl)urea ... 566
1-(3,4-Dichlorophenyl)-3-(3-morpholinopropyl)urea ... 566
$C_{14}H_{19}FN_2O$
2,6-Dimethyl-1-(*N*-(3-fluorophenyl)-carbamoyl)piperidine ... 695
$C_{14}H_{19}F_3N_2$
2-(2-(3-(Trifluoromethyl)anilino)-ethyl)piperidine ... 244
$C_{14}H_{19}F_3N_2O$
1,1-Diisopropyl-3-(3-trifluoromethylphenyl)urea ... 510
1,1-Dipropyl-3-(3-trifluoromethylphenyl)urea ... 510
$C_{14}H_{19}F_3N_4$
1-(α-Pentyl-3-(trifluoromethyl)-benzylideneamino)guanidine nitrate ... 1021
$C_{14}H_{19}F_3O_3$
Triethyl *o*-(3-(trifluoromethyl)-benzoate) ... 185
$C_{14}H_{19}FeN$
(*R*)-1-(Dimethylamino)-ethylferrocene tartrate ... 1141
(*S*)-1-(Dimethylamino)-ethylferrocene tartrate ... 1142
$C_{14}H_{19}IN_2O$
5-Acetyl-1,3-diethyl-2-methylbenzimidazolium iodide ... 906
$C_{14}H_{19}N$
2-(Benzylamino)norbornane ... 242
1-(2-Piperidinyl)-1-indanol hydrochloride ... 246
5-(2-Pyridyl)-2,7-nonadiene ... 967
$C_{14}H_{19}NO$
2-Benzyl-5,6-dihydro-4,4,6-trimethyl-1,3-oxazine ... 250
4-Butoxy-2-phenylbutyronitrile ... 753
2-(1-(Cyclohexylimino)ethyl)phenol ... 1038

$C_{14}H_{19}NO$ (Cont'd)
2′,6′-Diethyl-1-cyclopropanecarboxanilide ... 437
1,2-Dihydro-6-ethoxy-2,2,4-trimethylquinoline ... 247
4-Isopropoxy-4-(p-tolyl)-butyronitrile. 753
5′-Isopropyl-2′-methyl-1-cyclopropanecarboxanilide ... 437
2-Methyl-3-pyrrolidinopropiophenone hydrochloride ... 295
4-Phenyl-4-propionylpiperidine hydrochloride ... 286
3-Piperidinopropiophenone ... 295
3-Piperidinopropiophenone hydrochloride ... 295

$C_{14}H_{19}NO_2$
3-tert-Butyl-2-ethyl-3-hydroxy-1-isoindolinone ... 690
Cyclohexyl N-(o-tolyl)carbamate ... 679
Cyclohexyl N-(m-tolyl)carbamate ... 678
Cyclohexyl N-(p-tolyl)carbamate ... 679
Cyclopentyl N-(2,5-xylyl)carbamate ... 678
Ethyl N-(α-allylphenethyl)-carbamate ... 671
N-Ethyl-N-(2-phenoxyethyl)-crotonamide ... 417
2-Methylcyclohexyl N-phenylcarbamate ... 678
N-(1-Methyl-3-phenylpropyl)-acetoacetamide ... 444
p-Tolyl N-cyclohexylcarbamate ... 679

$C_{14}H_{19}NO_3$
N-Benzoyl-L-leucine methyl ester ... 476
Benzyl cis-N-(4-hydroxycyclohexyl)-carbamate ... 644
Benzyl trans-N-(4-hydroxycyclohexyl)carbamate ... 644
3-tert-Butyl-3-hydroxy-5-methoxy-2-methyl-1-isoindolinone ... 689
Cyclohexyl N-(o-anisyl)carbamate ... 653
Cyclohexyl N-(m-anisyl)carbamate ... 654
Cyclohexyl N-(4-methoxyphenyl)-carbamate ... 654
2,3-Dihydro-3,3-dimethyl-5-hydroxy-2-(4-morpholino)benzofuran ... 195
N-Ethyl-N-(2-phenoxyethyl)-acetoacetamide ... 442
N-(2,4,6-Trimethylphenyl)-glutaramic acid ... 436

$C_{14}H_{19}NO_3S$
3-Quinuclidyl p-toluenesulfonate hydrochloride ... 798

$C_{14}H_{19}NO_3Se$
2-Acetamido-3-(benzylseleno)-3-methylbutyric acid ... 213

$C_{14}H_{19}NO_4$
N-Benzyloxycarbonyl-L-isoleucine ... 661
N,N-Bis(2-acetoxyethyl)aniline ... 358
Carbobenzyloxy-DL-leucine ... 663
Carbobenzyloxy-L-leucine ... 661
Carbobenzyloxy-L-leucine dicyclohexylamine salt ... 676
Carbobenzyloxy-DL-norleucine ... 661
1-(Carbobutoxy)ethyl N-phenylcarbamate ... 665
Cyclopentyl N-(2,4-dimethoxyphenyl)carbamate ... 648
N-(2,2-Dicarbethoxyethyl)aniline ... 358
2′,5′-Diethoxyacetoacetanilide ... 442
Diethyl 2,4,6-trimethyl-3,5-pyridinedicarboxylate ... 962
Ethyl 3-acetamido-2-(hydroxymethyl)-2-phenylpropionate ... 411
Ethyl 2,4-dimethyl-5-(ethoxycarbonyl)-3-pyrroleacrylate .. 840
4-Nitrophenyl octanoate ... 379
Tetrahydrofurfuryl N-(2-methoxy-5-methylphenyl)carbamate ... 649

$C_{14}H_{19}NO_5$
N-(2,3-Dimethoxyphenylacetyl)-glycine ethyl ester ... 414
Ethyl 3-(2-amino-4,5-dimethoxybenzoyl)propionate ... 354
4-(3,4,5-Trimethoxybenzoyl)-morpholine ... 699

$C_{14}H_{19}NO_6$
Diethyl 5-(1,3-dioxolan-2-yl)-3-methyl-2,4-pyrroledicarboxylate ... 838
N-(4-Methoxybenzylidene)-α-glucosamine ... 1028
4-Methoxyphenyl N-(butoxycarbonyloxymethyl)-carbamate ... 651

$C_{14}H_{19}NO_8$
Tetraacetyl-L-rhamnonitrile ... 144

$C_{14}H_{19}NS$
N-Cyclohexyl-2-phenylthioacetamide ... 746

$C_{14}H_{19}N_3$
1-Benzyl-4-cyano-4-(methylamino)-piperidine ... 754

$C_{14}H_{19}N_3O$
N,N-Dimethyl-N′-furfuryl-N′-(2-pyridyl)ethylenediamine fumarate ... 943

$C_{14}H_{19}N_3O_2S_2$
Benzenesulfonic (4-tert-butyl-3-methyl-4-thiazolin-2-ylidene)-hydrazide ... 821
Benzenesulfonic (4-tert-butyl-3-methyl-4-thiazolin-2-ylidene)-hydrazide hydrobromide ... 820

$C_{14}H_{19}N_3O_3$
1-(4-Carboxyphenyl)-4-cyclohexylsemicarbazide ... 576
1-Cyclohexyl-1-methyl-3-(3-nitrophenyl)urea ... 565
1-Cyclohexylmethyl-3-(4-nitrophenyl)urea ... 567
1-Cyclohexyl-3-(4-methyl-3-nitrophenyl)urea ... 563
2-Methyl-1-(N-(4-methyl-3-nitrophenyl)carbamoyl)piperidine ... 702
3-Methyl-1-(N-(4-methyl-3-nitrophenyl)carbamoyl)piperidine ... 702
4-Methyl-1-(N-(4-methyl-3-nitrophenyl)carbamoyl)piperidine ... 702
L-Propyl-L-tyrosinamide hydrochloride ... 412

$C_{14}H_{19}N_3O_3S$
4-Allyl-1-(3,4,5-trimethoxybenzylidene)-3-thiosemicarbazide ... 1049

$C_{14}H_{19}N_3O_4$
Carbobenzyloxy-β-alanyl-β-alaninamide ... 664
Carbobenzyloxysarcosyl-L-alaninamide ... 664
2,6-Dimethyl-4-(N-(4-methyl-3-nitrophenyl)carbamoyl)morpholine. 694
1-(3,4-Dimethyl-2-tetrahydrofuryl)-3-(4-methyl-2-nitrophenyl)urea ... 562
1-(2,4-Dinitrophenyl)-4-propylpiperidine ... 274

$C_{14}H_{19}N_3O_5$
Carbobenzyloxy-L-alanyl-L-serinamide ... 643

$C_{14}H_{19}N_3O_6S$
1-Morpholino-2-(morpholinosulfonyl)-4-nitrobenzene ... 819

$C_{14}H_{19}N_3O_7S$
N-Butyl-2′,6′-dinitro-4′-(methylsulfonyl)propionanilide ... 778
2′,6′-Dinitro-4′-(methylsulfonyl)-N-propylbutyranilide ... 778

$C_{14}H_{19}N_3O_8S$
Ethyl N-butyl-N-(2,6-dinitro-4-(methylsulfonyl)phenyl)carbamate .. 777
Isopropyl N-(2,6-dinitro-4-(methylsulfonyl)phenyl)-N-propylcarbamate ... 777
Propyl N-(2,6-dinitro-4-(methylsulfonyl)phenyl)-N-propylcarbamate ... 777

$C_{14}H_{19}N_5O_5$
N-(2,4-Dinitrophenylhydrazonoethyl)-hexanamide ... 423

$C_{14}H_{19}N_7O_6$
Nα-Acetylnitroarginine 4-nitroanilide ... 725

$C_{14}H_{19}O_2$
2,3,3a,4,5,7,8,9,9a,9b-Decahydro-3-hydroxy-3a-methyl-1H-benzene(e)-inden-7-one ... 52

$C_{14}H_{20}$
1,2-Bis-(3-cyclohexen-1-yl)-ethylene (mixture of cis and trans) ... 3

$C_{14}H_{20}BrN_3O_2$
1-(4-Bromophenyl)-3-(3-morpholinopropyl)urea ... 566

$C_{14}H_{20}Br_2O_2$
2-Bromo-2-(3-bromo-1-adamantyl)-ethyl acetate ... 84

$C_{14}H_{20}ClFN_2O_2$
1-(3-Butoxypropyl)-3-(3-chloro-4-fluorophenyl)urea ... 521

$C_{14}H_{20}ClNO_2$
2-Chloro-6-isopropyl-3-methylphenyl N-propylcarbamate ... 640
Heptyl N-(4-chlorophenyl)-carbamate ... 683

$C_{14}H_{20}ClNO_3$
2-Chloromethyl-4-nitrophenyl heptyl ether ... 258

$C_{14}H_{20}ClNO_4$
1-Chloro-2,5-dibutoxy-4-nitrobenzene ... 265
Diethyl 4-chloromethyl-1,4-dihydro-2,6-dimethyl-3,5-pyridinedicarboxylate ... 89

$C_{14}H_{20}ClNO_5S$
1-(2-Chloro-α-methoxybenzylidene)-piperidinium methyl sulfate ... 1044

$C_{14}H_{20}ClN_3O$
1-(4-Methyl-2-pentylidene)-4-(3-chloro-2-methylphenyl)-semicarbazide ... 1025

$C_{14}H_{20}ClN_3O_2$
1-(2-Chlorophenyl)-3-(3-morpholinopropyl)urea ... 566
1-(3-Chlorophenyl)-3-(3-morpholinopropyl)urea ... 566
1-(4-Chlorophenyl)-3-(3-morpholinopropyl)urea ... 566

$C_{14}H_{20}ClN_3O_3$
N-(5-Chloro-2,4-dimethoxyphenyl)-4-methyl-1-piperazinecarboxamide ... 694
4-(4-Chloro-2-methylphenyl)-1-(6-hydroxyhexanoyl)semicarbazide ... 513

$C_{14}H_{20}ClN_3O_4$
4-(5-Chloro-2-methoxyphenyl)-1-(6-hydroxyhexanoyl)semicarbazide ... 512

$C_{14}H_{20}ClN_5O_3$
4-(3-Chlorophenyl)-1-(diethylureidoacetyl)-semicarbazide ... 529

$C_{14}H_{20}Cl_2N_2O$
2′,6′-Dichloro-2-(dipropylamino)-acetanilide hydrochloride ... 431

$C_{14}H_{20}Cl_2N_2OS$
2′,4′-Dichloro-2-(2-(diethylamino)-ethylthio)-acetanilide hydrochloride ... 421
2′,5′-Dichloro-2-(2-(diethylamino)-ethylthio)-acetanilide hydrochloride ... 421

$C_{14}H_{20}FNO$
2′-Fluorooctananilide ... 408

$C_{14}H_{20}INO_3$
8,9-Dimethoxy-4-methyl-2,3,4,5,6,10b-hexahydrooxazolo-(2,3-a)isoquinolinium iodide ... 254

$C_{14}H_{20}N_2$
3-Benzylaminoquinuclidine ... 248
Dicyclopropyl ketone azine ... 151
5-Methyl-5,6,6a,7,8,9,10,11-octahydropyrido(2,1-c)(1,4)-benzodiazepine di-hydrochloride ... 252

$C_{14}H_{20}N_2O$
4-Acetamidomethyl-4-phenylpiperidine ... 427
1-Allyl-3-(2,6-diethylphenyl)urea ... 594
1-Benzyl-3-cyclohexylurea ... 571
1-Cyclohexyl-1-methyl-3-phenylurea. 582
1-Cyclohexyl-3-(o-tolyl)urea ... 572
1-Cyclohexyl-3-(m-tolyl)urea ... 582
1-Cyclohexyl-3-(p-tolyl)urea ... 582
1-Cyclopentyl-3-(3,4-xylyl)urea ... 581
1-(N-(2-Ethylphenyl)carbamoyl)-piperidine ... 712
1-(N-(3-Ethylphenyl)carbamoyl)-piperidine ... 712
2′-(Hexahydro-1H-azepin-1-yl)-acetanilide ... 424
3-(4-Methylpiperazino)-propiophenone ... 295
2-Methyl-1-(N-(m-tolyl)carbamoyl)-piperidine ... 584
4-Methyl-1-(N-(m-tolyl)carbamoyl)-piperidine ... 585
4-(N-Propionylanilino)-piperidine ... 427
1-(N-(2,3-Xylyl)carbamoyl)piperidine. 712
1-(N-(2,4-Xylyl)carbamoyl)piperidine. 712
1-(N-(2,5-Xylyl)carbamoyl)piperidine. 712
1-(N-(2,6-Xylyl)carbamoyl)piperidine. 712
1-(N-(3,4-Xylyl)carbamoyl)piperidine. 712
1-(N-(3,5-Xylyl)carbamoyl)piperidine. 712

$C_{14}H_{20}N_2OS$
2,6-Dimethyl-4-(3-methylphenylthiocarbamoyl)-morpholine ... 747
4-Methyl-1-(N-(4-methylthio)phenyl)-carbamoyl)piperidine ... 701
2-Methyl-N-(3-(methylthio)phenyl)-1-piperidinecarboxamide ... 701

$C_{14}H_{20}N_2O_2$
1-(p-Anisyl)-3-cyclohexylurea ... 558
4-(N-Benzylcarbamoyl)-2,6-dimethylmorpholine ... 705
1-Cyclohexyl-3-(2-methoxyphenyl)urea ... 554
1-Cyclohexyl-3-(3-methoxyphenyl)urea ... 558
1-Cyclopentyl-3-(2-ethoxyphenyl)urea ... 554
O-(N-(2,6-Diethylphenyl)carbamoyl)-acetone oxime ... 1012
N,N′-(4,6-Dimethyl-1,3-xylylene)-bisacetamide ... 441
1,6-Dimorpholino-2,4-hexadiyne ... 1018
4-(N-(2-Ethyl-6-methylphenyl)-carbamoyl)morpholine ... 713
4′-(4-(2-Hydroxyethyl)piperazino)-acetophenone ... 290
1-(N-(2-Methoxy-5-methylphenyl)-carbamoyl)piperidine ... 698
2′-Methoxy-2-methyl-1-piperidinecarboxanilide ... 698
1-(N-(2-Methoxyphenyl)carbamoyl)-3-methylpiperidine ... 699
1-(N-(4-Methoxyphenyl)carbamoyl)-2-methylpiperidine ... 699
1-(N-(4-Methoxyphenyl)carbamoyl)-3-methylpiperidine ... 699
1-(N-(4-Methoxyphenyl)carbamoyl)-4-methylpiperidine ... 558
α-(4-Methoxyphenyl)-1-piperidineacetamide ... 416
4′-Methoxy-2-piperidinoacetanilide .. 416
3-Methyl-N-(2-morpholinoethyl)-benzamide hydrochloride ... 469
1-(1-Methyl-1-(4-nitrophenyl)ethyl)-piperidine ... 277
4,4′-(1,4-Phenylene)dimorpholine ... 250
1-Pyrrolidineaceto-o-phenetidide hydrochloride ... 415
4-(N-(o-Tolyl)carbamoyl)-2,6-dimethylmorpholine ... 713
4-(N-(m-Tolyl)carbamoyl)-2,6-dimethylmorpholine ... 713
4-(N-(p-Tolyl)carbamoyl)-2,6-dimethylmorpholine ... 713

$C_{14}H_{20}N_2O_2S$
2-Methylcyclohexanone (p-tosyl)-hydrazone ... 1071

$C_{14}H_{20}N_2O_3$
4′-Acetamido-2,2-dimethyl-3′-methoxypropionanilide ... 416
4-(N-(o-Anisyl)carbamoyl)-2,6-dimethylmorpholine ... 699
4-(N-(m-Anisyl)carbamoyl)-2,6-dimethylmorpholine ... 699
4-(N-(p-Anisyl)carbamoyl)-2,6-dimethylmorpholine ... 699
Carbobenzyloxy-DL-leucinamide ... 664
Carbobenzyloxy-L-leucinamide ... 664

$C_{14}H_{20}N_2O_3$ (Cont'd)
1-(*N*-(2,5-Dimethoxyphenyl)-carbamoyl)piperidine ... 699
Ethyl 4-(2-hydroxyanilino)-1-piperidinecarboxylate ... 696
L-Norvalylphenylalanine ... 429
L-Phenylalanyl-L-valine ... 422
Piperidino *N*-(2-methoxy-5-methylphenyl)carbamate ... 653
L-Valyl-L-phenylalanine ... 422

$C_{14}H_{20}N_2O_4$
N-(5-Acetamido-2-methoxyphenyl)-*N*-methylglycine ethyl ester ... 415
2-Morpholinoethyl *N*-(4-methoxyphenyl)carbamate ... 653
L-Valyl-L-tyrosine ... 412

$C_{14}H_{20}N_2O_4S$
1-Butyl-3-(2-(*p*-tolylsulfonyl)-acetyl)-urea ... 822

$C_{14}H_{20}N_2O_5$
4-(2,5-Diethoxy-4-nitrophenyl)-morpholine ... 267
4-(2,5-Dimethoxy-4-nitrophenyl)-2,6-dimethylmorpholine ... 265
Hexyl *N*-(2-methoxy-4-nitrophenyl)-carbamate ... 652

$C_{14}H_{20}N_2O_6$
N-(*tert*-Butoxycarbonyl)-L-proline *N*-hydroxysuccinimide ester ... 125

$C_{14}H_{20}N_2O_7$
Gluconic (4-methoxybenzylidene)-hydrazide ... 494

$C_{14}H_{20}N_2S$
1-Cyclohexyl-1-methyl-3-phenyl-2-thiourea ... 747

$C_{14}H_{20}N_4O_2$
4-(Cyclohexylureido)benzoic acid hydrazide ... 577

$C_{14}H_{20}N_4O_3$
Hexamethylenetetramine mandelate ... 42

$C_{14}H_{20}N_4O_4$
N^2-Carbobenzyloxy-L-arginine hydrobromide ... 657
Octyl aldehyde 2,4-dinitrophenylhydrazone ... 1052

$C_{14}H_{20}O$
2-Benzylidene-1-heptanol ... 207
4′-*tert*-Butyl-2′,6′-dimethylacetophenone ... 297
β-(4-*tert*.-Butylphenyl)-α-methylpropionaldehyde ... 321
α-Cyclohexyl-α-methylbenzyl alcohol ... 204
1,2,3,4a,4b,5,6,7,8,10,10a-Decahydrophenanthren-4(3*H*)-one ... 56
3-Mesityl-3-methyl-2-butanone ... 287
1-(4-Methyl-2-furyl)-2,6,6-trimethylcyclohexene ... 854
1-(*m*-Tolyl)-cycloheptanol ... 203

$C_{14}H_{20}OSi$
4-(3-Methoxyphenyl)-1-(trimethylsilyl)-1-butyne ... 1105

$C_{14}H_{20}O_2$
p-(*tert*.-Butyl)-β-methylhydrocinnamic acid ... 333
5,5-Diethyl-2-phenyl-1,3-dioxane ... 185
α,α-Dimethylphenethyl butyrate ... 362
4-(Heptyloxy)benzaldehyde ... 323
3-Methyl-2-(*o*-tolyl)-hexanoic acid ... 333
5-Methyl-2-(*o*-tolyl)-hexanoic acid ... 333
2,3,4,5,6-Pentamethylbenzyl acetate ... 360
Phenethyl hexanoate ... 362
2-Phenyl-1-cyclopropanecarboxaldehyde diethyl acetal ... 189
2-(*o*-Tolyl)-heptanoic acid ... 333

$C_{14}H_{20}O_2S$
Cyclohexyl 4-methylbenzyl sulfone ... 783

$C_{14}H_{20}O_3$
6,7-Dimethoxy-1,1-dimethyl-1,2,3,4-tetrahydro-2-naphthol ... 201
4-(2,4-Dimethoxyphenyl)-3-hexanone ... 284
Ethyl α-(*sec*.-butyl)-mandelate ... 353
Ethyl 4-(pentyloxy)benzoate ... 370
4-(Hexyloxy)-*m*-anisaldehyde ... 323

$C_{14}H_{20}O_4$
Decahydrocyclobuta(1,2:3,4)-dicyclopentene-1,4-diol diacetate ... 77
Hexahydrodispiro(1,3-dioxolane-2,1′(4′*H*)-naphthalene-4′,2″-(1,3)-dioxolane) ... 22
4,4a,5,6,8,8a-Hexahydro-3,6,6,8a-tetramethyl-3,5-ethano-1*H*,3*H*-pyrano(3,4-c)pyran-1,8-dione ... 97
Methyl 4-(2-butoxyethoxy)benzoate ... 369
2′,4′,5′-Triethoxyacetophenone ... 291

$C_{14}H_{20}O_5$
Dibutyl 3,4-furandicarboxylate ... 853
Isopropyl 4,6-dihydroxy-1,7-epoxy-1,4,4a,5,6,7,8,8a-octahydro-naphthalene-8-carboxylate ... 89

$C_{14}H_{20}O_9S$
1-Thio-D-galactopyranose tetraacetate ... 86

$C_{14}H_{20}O_{10}$
myo-Inositol 3,4,5,6-tetraacetate ... 84

$C_{14}H_{20}S_4$
Hexahydrodispiro(1,3-dithiolane-2,1′(4′*H*)-naphthalene-4′,2″-(1,3)-dithiolane) ... 27

$C_{14}H_{21}AsBrN$
(3-Cyanopropyl)dimethyl(2-phenylethyl)arsonium bromide ... 1127

$C_{14}H_{21}BrN_4O_2S$
1-(4-(Methylsulfonyl)benzyl)-hexamethylenetetraminium bromide ... 777

$C_{14}H_{21}ClN_2O$
3-(2-Chloro-6-methylphenyl)-1,1-diisopropylurea ... 541
3-(3-Chloro-4-methylphenyl)-1,1-diisopropylurea ... 600
3-(4-Chloro-2-methylphenyl)-1,1-diisopropylurea ... 600
3-(5-Chloro-2-methylphenyl)-1,1-diisopropylurea ... 600
3-(2-Chloro-6-methylphenyl)-1,1-dipropylurea ... 595
3-(3-Chloro-4-methylphenyl)-1,1-dipropylurea ... 600
3-(4-Chloro-2-methylphenyl)-1,1-dipropylurea ... 600
3-(5-Chloro-2-methylphenyl)-1,1-dipropylurea ... 600

$C_{14}H_{21}ClN_2O_2$
1-(3-Butoxypropyl)-3-(3-chlorophenyl)urea ... 520
2-Diethylaminoethyl *N*-(2-chloro-6-methylphenyl)carbamate ... 672
2-Diethylaminoethyl *N*-(3-chloro-4-methylphenyl)carbamate ... 672
2-Diethylaminoethyl *N*-(4-chloro-2-methylphenyl)carbamate ... 672

$C_{14}H_{21}ClN_2O_3$
3-Carbamoyl-1-(hexyloxycarbonylmethyl)-pyridinium chloride ... 947
4-Carbamoyl-1-(hexyloxycarbonylmethyl)-pyridinium chloride ... 951
2-Diethylaminoethyl *N*-(3-chloro-2-methoxyphenyl)carbamate ... 652

$C_{14}H_{21}ClO$
4-Chloro-2,6-di-(*tert*-butyl)phenol ... 192

$C_{14}H_{21}ClO_2$
α-Chloro-3,5-dimethyl-1-adamantaneacetic acid ... 65

$C_{14}H_{21}F_5Si_2$
(3-(Dimethyl(pentafluorophenyl)-silyl)propyl)trimethylsilane ... 1101

$C_{14}H_{21}N$
Cyclohexylphenethylamine hydrochloride ... 242
N,*N*-Dimethyl-1-phenylcyclohexylamine ... 233
4-(3-Phenylpropyl)-piperidine ... 250

$C_{14}H_{21}NO$
2-Cyclohexylamino-1-phenylethanol ... 238
N,*N*-Diisopropyl-*o*-toluamide ... 482
2,2-Dimethyl-4′-(dimethylaminomethyl)-propiophenone ... 295
4-Hydroxy-4-(3-phenylpropyl)-piperidine hydrochloride ... 246
3-(Phenethylamino)-1-cyclohexanol hydrobromide ... 238
4-Phenyl-4-propoxypiperidine hydrochloride ... 247
2-Phenyl-*N*,*N*,3-trimethylvaleramide ... 430
6-(2-Pyridyl)-2,4,4-trimethyl-3-hexanone ... 961
4′,5′,6′,7′-Tetrahydrospiro(cyclohexane-1,1′-indan)-3′-oxime ... 1013

$C_{14}H_{21}NOS$
S-Butyl *N*-(2-ethyl-6-methylphenyl)-thiocarbamate ... 688

$C_{14}H_{21}NO_2$
1-(2-Benzyloxyethyl)-3-pyrrolidinemethanol ... 245
Butyl *N*-(2-ethyl-6-methylphenyl)-carbamate ... 684
2,6-Di-*tert*.-butyl-4-nitrosophenol ... 257
2-(Diethylamino)-ethyl phenylacetate ... 359
Heptyl anthranilate ... 372
Heptyl carbanilate ... 680
Isopropyl *N*-(2,6-diethylphenyl)-carbamate ... 684
Methyl *N*-(2,6-diisopropylphenyl)-carbamate ... 684
Propyl *N*-(2,6-diethylphenyl)-carbamate ... 684

$C_{14}H_{21}NO_2S$
S-Benzyl-L-cysteine *tert*-butyl ester hydrochloride ... 357
Hexyl *N*-(4-(methylthio)phenyl)-carbamate ... 667

$C_{14}H_{21}NO_3$
2,4-Di-*tert*.-butyl-6-nitrophenol ... 264
2-Ethoxyethyl *N*-(2-ethyl-6-methylphenyl)carbamate ... 650
N-(2-Hydroxypropyl)-*N*-(2-(*o*-tolyloxy)-ethyl)-acetamide ... 411
N-(2-Hydroxypropyl)-*N*-(2-(*m*-tolyloxy)-ethyl)-acetamide ... 411
N-(2-Hydroxypropyl)-*N*-(2-(*p*-tolyloxy)ethyl)acetamide ... 411
Methyl 3-acetamido-1-adamantanecarboxylate ... 105
3-Nitrophenyl octyl ether ... 268

$C_{14}H_{21}NO_3S$
N-Butyl-2-(*p*-tolylsulfonyl)-propionamide ... 779

$C_{14}H_{21}NO_4$
Benzyl *N*-(2,2-diethoxyethyl)-carbamate ... 664
2,4-Dimethyl-5-(*tert*-butoxycarbonyl)-3-pyrrolepropionic acid ... 842

$C_{14}H_{21}NO_5$
2,3,4-Trimethoxybenzyl *N*-propylcarbamate ... 646
3,4,5-Trimethoxybenzyl *N*-propylcarbamate ... 646

$C_{14}H_{21}NO_5S$
tert-Butyl *N*-(mesitylenesulfonyloxy)-carbamate ... 799

$C_{14}H_{21}NO_6$
p-Phenetidine D-fructoside ... 216
p-Phenetidine D-galactoside ... 216
p-Phenetidino-L-sorboside ... 216

$C_{14}H_{21}N_3O$
1-Benzyl-4-methylamino-4-piperidinecarboxamide ... 425
1-(4-Methyl-4-phenylcyclohexyl)-semicarbazide ... 578
4-Methyl-1-(*N*-(2,5-xylyl)carbamoyl)-piperazine ... 705
4-Methyl-1-(*N*-(2,6-xylyl)carbamoyl)-piperazine ... 705
3′,4,4′-Trimethyl-1-piperazinecarboxanilide ... 705

$C_{14}H_{21}N_3OS$
1-Benzyl-3-(2-morpholinoethyl)-2-thiourea ... 739

$C_{14}H_{21}N_3O_2$
1-(3-Methoxyphenyl)-3-(2-pyrrolidinoethyl)urea ... 551
1-(3-Morpholinopropyl)-3-phenylurea ... 566
L-Norvalyl-L-phenylalaninamide hydrochloride ... 422
L-Valyl-L-phenylalaninamide acetate ... 422

$C_{14}H_{21}N_3O_3$
Carbobenzyloxy-L-leucine hydrazide ... 646
1,1-Diisopropyl-3-(4-methyl-3-nitrophenyl)urea ... 565
3′,5′-Dimethoxyvalerophenone semicarbazone ... 549
1,1-Dipropyl-3-(2-methyl-4-nitrophenyl)urea ... 569
1,1-Dipropyl-3-(2-methyl-6-nitrophenyl)urea ... 562
1,1-Dipropyl-3-(4-methyl-3-nitrophenyl)urea ... 564
1-(2-Methoxyphenyl)-3-(2-morpholinoethyl)urea ... 551

$C_{14}H_{21}N_3O_3S$
1-(6-Hydroxyhexanoyl)-4-(4-(methylthio)phenyl)semicarbazide ... 547

$C_{14}H_{21}N_3O_4$
Ethyl 3-cyano-2-hydroxy-1,4,4a,5,6,7,8,8a-octahydro-4-oxo-1-naphthalenecarboxylate hydrazine salt ... 144

$C_{14}H_{21}N_3O_4S$
1-(4-Acetamidophenylsulfonyl)-3-isopentylurea ... 824
4-Morpholino-3-(morpholinosulfonyl)-aniline ... 816

$C_{14}H_{21}N_3S$
1-(1-Ethyl-3-piperidinyl)-3-phenyl-2-thiourea ... 739

$C_{14}H_{21}N_5O_3$
2-(3,3-Diethylureido)acetic acid N^2-(*N*-phenylcarbamoyl)hydrazide ... 574

$C_{14}H_{22}$
1-*tert*-Butyl-2,3,4,5-tetramethylbenzene ... 162
4,6-Diisopropyl-1,3-dimethylbenzene ... 163
2,5-Diisopropyl-*p*-xylene ... 163
1-(2,2-Dimethylpropyl)-2,4,5-trimethylbenzene ... 163

$C_{14}H_{22}Br_2$
1-Bromo-3-(1-bromoethyl)-5,7-dimethyladamantane ... 6

$C_{14}H_{22}ClN_3O$
1-(6-Chloro-*o*-tolyl)-3-(2-(diethylamino)-ethyl)-urea ... 562

$C_{14}H_{22}ClN_3O_2$
4-Chloro-6-nitro-*N*,*N*,*N*′,*N*′-tetraethyl-1,3-phenylenediamine ... 273

$C_{14}H_{22}Cl_6N_2O_5$
Dibutyl *N*,*N*′-(1,1′-oxybis(2,2,2-trichloroethyl))biscarbamate ... 113

$C_{14}H_{22}NO_3P$
Ethyl (*N*,*N*-diethylcarbamoylmethyl)-phenylphosphinate ... 835

$C_{14}H_{22}N_2$
1-Benzyl-4-dimethylaminopiperidine dihydrochloride ... 249
2-Butanone (1-methyl-1-phenylpropyl)hydrazone ... 235
2-(2-(4-Methylanilino)ethyl)-piperidine ... 249

$C_{14}H_{22}N_2O$
N-(1-(Butylamino)-2-phenylethyl)-acetamide ... 428
1-*tert*-Butyl-3-(2,4-dimethylbenzyl)urea ... 600
1-(*tert*-Butyl)-3-(2,5-dimethylbenzyl)urea ... 573
1-*tert*-Butyl-3-(3,4-dimethylbenzyl)urea ... 599
1-Butyl-3-(2-ethyl-6-methylphenyl)urea ... 595
1-Butyl-3-(3-ethylphenyl)-1-methylurea ... 581
1-*tert*-Butyl-3-mesitylurea ... 602
1-Butyl-1-methyl-3-(2,4-xylyl)urea ... 600

$C_{14}H_{22}N_2O$ (Cont'd)
1-*tert*-Butyl-3-(2,4,5-trimethylphenyl)urea 603
2-Diethylamino-2′,6′-acetoxylidide 431
1,1-Diethyl-3-(2-ethyl-6-methylphenyl)urea 594
N,*N*-Diethyl-*N*′-(4-methoxybenzylidene)-ethylenediamine 1043
1-(2,6-Diethylphenyl)-3-isopropylurea 595
1-(2,6-Diethylphenyl)-3-propylurea 594
1-(2,6-Diisopropylphenyl)-3-methylurea 594
1,1-Diisopropyl-3-(*o*-tolyl)urea 588
1,1-Diisopropyl-3-(*m*-tolyl)urea 581
1,1-Diisopropyl-3-(*p*-tolyl)urea 591
3′-(Dipropylamino)acetanilide 425
1,1-Dipropyl-3-(*o*-tolyl)urea 588
1,1-Dipropyl-3-(*m*-tolyl)urea 580
1,1-Dipropyl-3-(*p*-tolyl)urea 591
1-(2-Ethyl-6-methylphenyl)-3-isobutylurea 594
$C_{14}H_{22}N_2OS$
1-(3-Butoxypropyl)-3-phenyl-2-thiourea 738
$C_{14}H_{22}N_2O_2$
3-(*o*-Anisyl)-1,1-diisopropylurea 559
3-(*m*-Anisyl)-1,1-diisopropylurea 558
3-(*p*-Anisyl)-1,1-diisopropylurea 559
3-(*m*-Anisyl)-1,1-dipropylurea 558
3-(*p*-Anisyl)-1,1-dipropylurea 559
2-Diethylamino-*p*-acetophenetidide hydrochloride 415
2-Diethylaminoethyl *N*-(*o*-tolyl)-carbamate 672
2-Diethylaminoethyl *N*-(*m*-tolyl)-carbamate 671
2-Diethylaminoethyl *N*-(*p*-tolyl)-carbamate 672
N,*N*′-Dimethyl-1,3-adamantanedicarboxamide 101
2-Dimethylaminoethyl *N*-(2-ethyl-6-methylphenyl)carbamate 672
4-(Dimethylamino)-3-isopropylphenyl *N*,*N*-dimethylcarbamate 673
3-(*N*-Ethyl-*N*-methylamino)-*N*-(2-phenoxyethyl)-propionamide 415
$C_{14}H_{22}N_2O_3$
2-(Diethylamino)ethyl *N*-(*o*-anisyl)-carbamate 652
2-(Diethylamino)ethyl *N*-(*m*-anisyl)-carbamate 652
2-(Diethylamino)ethyl *N*-(*p*-anisyl)-carbamate 652
$C_{14}H_{22}N_2O_3S$
1,1-Dipropyl-3-(*p*-tolylsulfonyl)urea 826
$C_{14}H_{22}N_2O_4$
α,α′-Diamino-1,3-adamantanediacetic acid 71
$C_{14}H_{22}N_2O_4S$
1-(3-Isopropoxypropyl)-3-(*p*-tolylsulfonyl)-urea 815
6-Methyl-*N*-ε-(*p*-tosyl)lysine hydrochloride 817
$C_{14}H_{22}N_2O_6$
tert-Butoxycarbonyl-L-valine *N*-hydroxysuccinimide ester 125
$C_{14}H_{22}N_2O_6S$
tert-Butoxycarbonyl-L-methionine *N*-hydroxysuccinimide ester 125
$C_{14}H_{22}N_2O_7$
D-Manno-L-mannooctose phenylhydrazone 234
$C_{14}H_{22}N_2S$
1-Butyl-3-(4-isopropylphenyl)-2-thiourea 749
$C_{14}H_{22}N_4$
1-(3-Methyl-α-pentylbenzylideneamino)-guanidine nitrate 1068
1-(4-Methyl-α-pentylbenzylideneamino)-guanidine nitrate 1072
$C_{14}H_{22}O$
2,4-Di-*sec*.-butylphenol 192
1-Methyl-α-ionone 54
1-Methyl-β-ionone 54
3-Methyl-α-ionone 54
3-Methyl-β-ionone 54
$C_{14}H_{22}O_2$
3,6-Bis-(methoxymethyl)-durene 186
1,4-Dibutoxybenzene 181
3,7-Dimethyl-1-adamantaneacetic acid 67
1,1′-Ethynylene-bis-cyclohexanol 1016
4-*tert*.-Octylcatechol 192
$C_{14}H_{22}O_3$
9-*tert*-Butyl-3-oxaspiro(5.5)-undecane-2,4-dione 99
2-Decen-1-ylsuccinic anhydride 98
Ethyl 2,3-dihydro-4-ethoxy-2,2,6-trimethylbenzoate 78
$C_{14}H_{22}O_3S$
2,2-Dimethylpropyl 2-(*p*-tolyl)-ethanesulfonate 800
$C_{14}H_{22}O_4$
Dicyclohexyl oxalate 77
$C_{14}H_{22}S$
2-(Allylthioethyl)-6,6-dimethylbicyclo(3.1.1)hept-2-ene 27
$C_{14}H_{23}BF_4O$
2,6-Di-*tert*-butyl-4-methylpyrylium tetrafluoroborate 935
$C_{14}H_{23}BrN_2O$
4-Carbamoyl-1-octylpyridinium bromide 953
$C_{14}H_{23}ClO_5$
2,6-Di-*tert*-butyl-4-methylpyrylium perchlorate 935
$C_{14}H_{23}N$
2-(1-Butylpentyl)-pyridine 967
2,6-Di-*tert*.-butyl-4-methylpyridine 967
(β-Methylphenethyl)pentylamine hydrochloride 242
N-Phenyloctylamine 219
1-(1-Piperidinocyclohexyl)-1-propyne 1018
$C_{14}H_{23}NO$
2-(Azacyclonon-1-en-2-yl)-1-cyclohexanone 59
6-Benzylamino-2-methyl-2-heptanol 239
o-(Octyloxy)aniline 217
m-(Octyloxy)aniline 217
4-(Octyloxy)aniline 218
3,5,7-Trimethyl-1-adamantanecarboxamide 102
$C_{14}H_{23}NOS$
α-(2-(Diethylamino)-ethylthiomethyl)-benzyl alcohol 238
β-(2-(Diethylamino)-ethylthio)-phenethyl alcohol 238
2-(2-(Diethylamino)ethylthio)-1-phenylethanol 237
1-(Diethylamino)-3-(*p*-tolylthio)-2-propanol 238
$C_{14}H_{23}NO_2$
1-Adamantylmethyl *N*-ethylcarbamate 116
α-Amino-3,5-dimethyl-1-adamantaneacetic acid 72
9-*tert*-Butyl-3-azaspiro(5.5)-undecane-2,4-dione, ammonium salt 135
2-(2-(Diethylamino)ethoxy)-1-phenylethanol 237
1-(Diethylamino)-3-(*o*-tolyloxy)-2-propanol 237
N,*N*-Diethyl-2-oxo-4,6,6-trimethyl-3-cyclohexene-1-carboxamide 107
N,*N*-Diethyl-5-pentyl-2-furamide 848
2,4-Dimethoxy-*N*-hexylaniline 217
$C_{14}H_{23}NO_3$
cis-*N*-Cyclohexylhexahydro-phthalamic acid 122
$C_{14}H_{23}NO_7S$
4-(3,5-Dimethoxyphenyl)-4-methylmorpholinium methosulfate 254
$C_{14}H_{23}NS$
2-(Octylthio)aniline hydrochloride 218
$C_{14}H_{23}NS_2$
N,*N*-Bis-(2-ethylthio)-ethylaniline 218
$C_{14}H_{23}N_3O$
1-(4-(Diethylamino)phenyl)-3-isopropylurea 567
1-(2-(*N*-Ethylanilino)ethyl)-3-propylurea 570
$C_{14}H_{23}N_3OS$
1-(3-Methyl-1-adamantylacetyl)-3-thiosemicarbazide 140
$C_{14}H_{23}N_3O_2$
1-(6-Aminohexyl)-3-(4-methoxyphenyl)urea 551
1-(2-(Diethylamino)ethyl)-3-(4-methoxyphenyl)urea 551
1-(3-Methyl-1-adamantylacetyl)-semicarbazide 127
$C_{14}H_{23}O_3P$
Di-*sec*-butyl phenylphosphonate 836
$C_{14}H_{24}$
1,2-Dicyclohexylethylene 3
Perhydrophenanthrene 2
$C_{14}H_{24}BF_4NO$
1-(5,5-Dimethyl-3-ethoxy-2-cyclohexen-1-ylidene)-pyrrolidinium tetrafluoroborate 42
$C_{14}H_{24}N_2$
N′-Benzyl-*N*,*N*-diethyl-1,3-propanediamine 242
1,6-Bis(pyrrolidino)-2-hexyne 1018
Tetradecanedinitrile 150
$C_{14}H_{24}N_2O$
1-(1-Adamantyl)-3-isopropylurea 130
1-(1-Adamantylmethyl)-3-ethylurea 130
1-(1-Adamantyl)-3-propylurea 130
$C_{14}H_{24}N_2O_2$
1,6-Bis(morpholino)-2-hexyne 1018
$C_{14}H_{24}N_2O_4$
N,*N*′-Hexamethylenebis(allyl carbamate) 116
$C_{14}H_{24}N_2S_2$
3,3′-Dithiobis(2,2,3,3-tetramethylpropionitrile) 145
$C_{14}H_{24}O$
2-(2-Ethylhexylidene)-1-cyclohexanone 54
$C_{14}H_{24}O_2$
1-Cyclododecen-1-yl acetate 78
2-Cyclohexyl cyclohexylacetate 77
2,4-Dibutyl-3-ethoxy-2-cyclobuten-1-one 52
7,7-Dimethyl-2-(2-methoxyethyl)-bicyclo(3.1.1)heptan-3-yl acetate 77
3,7-Dimethyl-2,6-octadienyl isobutyrate 80
$C_{14}H_{24}O_2S_2$
2-Acetoxy-1-methyl-1-propyl-6,10-dithiaspiro(4.5)decane 91
$C_{14}H_{24}O_3$
C14 acid derived from apoaromadendrone 65
Cyclohexyl cyclohexyloxyacetate 86
Decylsuccinic anhydride 98
Methyl 2-oxo-1-cyclododecanecarboxylate 92
$C_{14}H_{24}O_4$
Diethyl 1,1-cyclohexanediacetate 77
1,4-Dioxacyclohexadecane-5,16-dione 97
$C_{14}H_{24}O_5$
Diethyl 2-ethyl-2-(1-methyl-3-oxobutyl)malonate 93
6-Oxo-1,12-dodecanedicarboxylic acid 64
$C_{14}H_{24}O_6$
Triethyl 1,3,4-pentanetricarboxylate. 76
Triethyl 1,3,5-pentanetricarboxylate. 76
Triethyl 2,2,4-pentanetricarboxylate. 75
$C_{14}H_{24}O_6S_3$
2-Hydroxymethyl-2-methyl-1,3-propanediol tris-(3-mercaptopropionate) 86
$C_{14}H_{24}O_7$
Diethyl 2-(diethoxymethyl)-3-formylsuccinate 92
$C_{14}H_{24}Pb$
Phenethyltriethyllead 1135
$C_{14}H_{25}BrO$
2-Bromoallyl 10-undecen-1-yl ether 18
$C_{14}H_{25}BrO_4$
Diethyl 2-(3-bromopropyl)-2-butylmalonate 81
$C_{14}H_{25}IN_2$
1,3-Dicyclohexylcarbodiimide methiodide 151
$C_{14}H_{25}NO$
4-(2,9-Decadien-1-yl)morpholine 41
$C_{14}H_{25}NOSi_2$
2,2,6,6-Tetramethyl-4-(3-methylbenzyl)-1-oxa-4-aza-2,6-disilacyclohexane hydrochloride 1112
$C_{14}H_{25}NO_2$
8,10-Diethyllobelidione 59
2-Methylcyclohexyl *N*-cyclohexylcarbamate 116
3-Methylcyclohexyl *N*-cyclohexylcarbamate 116
$C_{14}H_{25}NO_2Pb$
N-(Triethylplumbyl)-hexahydrophthalimide 1133
$C_{14}H_{25}NO_3$
N-Decylmaleamic acid 121
$C_{14}H_{25}NO_6$
2-Isopropyl-2-nitro-1,3-trimethylene bis(butyrate) 75
$C_{14}H_{25}N_3O_9$
Kasugamycin hydrochloride hydrate 136
$C_{14}H_{26}$
1,2-Dicyclohexylethane 2
5-Ethyl-6-methyl-1,10-undecadiene 2
$C_{14}H_{26}IN$
N,*N*-Dimethyl-4-(4-iodocyclohexyl)-cyclohexylamine 32
$C_{14}H_{26}N_2$
6-Diethylamino-1-pyrrolidino-2-hexyne 1018
$C_{14}H_{26}N_2O$
Cyclohexanecarboxylic N^2-(cyclohexylmethyl)hydrazide 112
1,1-Dicyclohexyl-3-methylurea 130
1,3-Dicyclohexyl-1-methylurea 130
6-Diethylamino-1-morpholino-2-hexyne 1018
$C_{14}H_{26}N_2OS$
2-(Methylthio)-5-undecyl-1,3,4-oxadiazole 873
$C_{14}H_{26}N_2O_4$
2,5-Bis-(morpholinmethyl)-1,4-dioxane 40
$C_{14}H_{26}N_2O_5S$
N-(*tert*-Butoxycarbonyl)-L-methionyl-L-alanine methyl ester 115
$C_{14}H_{26}N_4O_2$
1,1′-Hexamethylenebis(3-allylurea) 129
N,*N*′-Hexamethylenebis(2-methyl-1-aziridinecarboxamide) 129
$C_{14}H_{26}N_4O_2S_2$
N,*N*′-Bis-(2-morpholinoethyl)-dithiooxamide 139
$C_{14}H_{26}N_4O_4$
N,*N*′-Hexamethylenebis(*O*-(carbamoyl)acetone oxime) 1011
$C_{14}H_{26}O$
2,6-Di-*sec*.-butylcyclohexanone 56
2-(2-Ethylhexyl)-cyclohexanone 56
2-Octyl-1-cyclohexanone 56
$C_{14}H_{26}O_2$
4-Butoxydecahydro-1-naphthol 20
3,7-Dimethyl-6-octenyl isobutyrate 80
2-(2-Hydroxyethyl)-2,6,6-triethyl-1-cyclohexanone 52
Methyl 2,4-dimethyl-3-hexyl-4-pentenoate 80
4α,5,5,8α-Tetramethyl-1,6-decalindiol 12
$C_{14}H_{26}O_3S_4$
2,2′-Oxydiethyl bis(*O*-butylxanthate) 86
$C_{14}H_{26}O_4$
Diethyl 2-ethyl-2-isoamylmalonate 76
Diethyl heptylmalonate 76
Dipentyl succinate 76
$C_{14}H_{26}O_4S$
11-(Methoxycarbonylmethylthio)-undecanoic acid 95
$C_{14}H_{26}O_5$
Tetraethylene glycol diallyl ether 18
$C_{14}H_{26}O_6$
Diisoamyl tartrate 84

$C_{14}H_{26}O_9Si_2$
1,3-Bis(2,2-dicarboxypropyl)-1,1,3,3-tetramethyldisiloxane 1109
$C_{14}H_{26}Si_4$
1,2,4,5-Benzenetetrayltetrakis(dimethylsilane) 1123
$C_{14}H_{27}BrO_5S$
3-Sulfopropyl 11-bromoundecanoate, sodium salt 155
$C_{14}H_{27}Cl_3O_2Sn$
Tributyltin trichloroacetate 1129
$C_{14}H_{27}N$
2-Amino-2-butyldecalin 31
5-Methyl-2-piperidino-3-octene 40
Tetradecanenitrile 150
$C_{14}H_{27}NO$
6-Methyl-7-piperidino-5-octen-3-ol 37
6-(Piperidinomethyl)-5-octen-3-ol 37
$C_{14}H_{27}NO_2$
8,10-Diethyllobelionol hydrochloride 58
5-Methyl-5-piperidino-3-hexenal dimethyl acetal 38
$C_{14}H_{27}NO_3$
14-Amino-7-oxotetradecanoic acid 71
$C_{14}H_{27}NO_4$
Diethyl (3-diethylaminopropyl)-malonate 88
Ethyl 2,2-dimethyl-3-nitrodecanoate. 73
$C_{14}H_{27}N_3O$
1-Cyclohexyl-3-(1-ethyl-3-piperidinyl)urea 126
$C_{14}H_{28}ClNO$
2-Chloro-*N*-dodecylacetamide 103
$C_{14}H_{28}N_2$
1,6-Bis-(diethylamino)-2-hexyne 1018
$C_{14}H_{28}N_2O$
Acetic dodecylidenehydrazide 112
$C_{14}H_{28}N_2O_2$
4,4′-Dihydroxy-1,1′-dimethyl-4,4′-ethylenedipiperidine 36
N,*N*,*N*′,*N*′-Tetraethyladipamide 111
$C_{14}H_{28}N_2O_4$
N,*N*′-Hexamethylenebis(isopropyl carbamate) 116
N,*N*′-Hexamethylenebis(propyl carbamate) 116
$C_{14}H_{28}N_2O_5$
Diethyl 1,1-dimethyl-*N*2-(1,1-dimethyl-2-hydroxyethyl)-ethylenedicarbamate 115
$C_{14}H_{28}N_2S$
3-Allyl-1,1-dipentylurea 142
$C_{14}H_{28}N_6S_2$
N,*N*′-Bis-(2-piperazinoethyl)-dithiooxamide 139
$C_{14}H_{28}O$
1-Ethyl-1-cyclododecanol 12
2-(2-Ethylhexyl)-1-cyclohexanol 14
7-Ethyl-2-methyl-4-undecanone 49
3-Tetradecanone 49
Tetradecyl aldehyde (50% sol in diethyl phthalate) 60
$C_{14}H_{28}O_2$
Ethyl laurate 74
1-Methylheptyl hexanoate 74
$C_{14}H_{28}O_2S$
Dodecyl mercaptoacetate 86
$C_{14}H_{28}O_3$
Ethyl 11-methoxyundecanoate 86
2-Hydroxymyristic acid 63
$C_{14}H_{28}O_4S$
α,α′-Di-*tert*-butyl-2,5-sulfolanedimethanol 152
$C_{14}H_{29}BrHg$
Bromotetradecylmercury 1137
$C_{14}H_{29}ClHg$
Chlorotetradecylmercury 1137
$C_{14}H_{29}ClO_2Sn$
Tributyltin chloroacetate 1129
$C_{14}H_{29}F$
1-Fluorotetradecane 4
$C_{14}H_{29}IOSi$
3-(*tert*-Butyldimethylsilyloxy)-1-iodo-1-octene 1097
$C_{14}H_{29}NO$
N,*N*-Dihexylacetamide 107
N,*N*-Dimethyldodecanamide 107
N-Dodecylacetamide 106
Tetradecanamide 102
$C_{14}H_{29}NO_2$
N-(2-Hydroxyethyl)-lauramide 104
2-Methyl-2-nitrotridecane 46
$C_{14}H_{30}HgS_2$
Mercury(II) (*N*-heptylsulfide) 1137
$C_{14}H_{30}I_2N_2O_4$
Succinyldicholine diiodide 88
$C_{14}H_{30}NO_4P$
Di-*sec*-butyl (*N*,*N*-diethylcarbamoylmethyl)-phosphonate 159
$C_{14}H_{30}N_2O_2$
Ethyl *N*-(3-(dibutylamino)propyl)-carbamate 115
$C_{14}H_{30}N_4O_2$
1,1′-Hexamethylenebis(3-isopropylurea) 129
1,1′-Hexamethylenebis(3-propylurea) 129
$C_{14}H_{30}O$
3-Ethyl-3-dodecanol 10
7-Ethyl-2-methylundecanol 8
7-Ethyl-2-methyl-4-undecanol 9
Heptyl ether 17
2-Pentyl-1-nonanol 8
$C_{14}H_{30}OS$
Dodecyl 2-hydroxyethyl sulfide 25
$C_{14}H_{30}O_2Pb$
Tributylplumbyl acetate 1135
Triethylplumbyl 2-ethylhexanoate 1135
$C_{14}H_{30}O_2S$
Hexyl octyl sulfone 154
2-Hydroxyethyl dodecyl sulfoxide 152
$C_{14}H_{30}O_2Sn$
Acetoxydiethyloctyltin 1131
$C_{14}H_{30}S$
Dodecyl ethyl sulfide 26
Heptyl sulfide 26
$C_{14}H_{30}S_2$
tert-Heptyl disulfide 27
$C_{14}H_{31}BO_2$
Tetradecaneboronic acid 1124
$C_{14}H_{31}N$
Diheptylamine 28
2-Methyl-2-tridecylamine hydrochloride 28
$C_{14}H_{31}NS$
2-(Dodecylthio)-ethylamine 36
$C_{14}H_{32}N_2$
N,*N*-Diisobutyl-1,6-hexanediamine 30
$C_{14}H_{32}N_2OS$
1-(Diethylamino)-3-(3-(diethylamino)propylthio)-2-propanol 35
$C_{14}H_{32}N_6NiS_2$
Bis(*N*,*N*-diethylethylenediamine)-nickel(II) thiocyanate 1140
$C_{14}H_{36}Br_3N_3O$
2-Hydroxyethylidynetris-(tetramethylammonium bromide) 42
$C_{14}H_{42}O_2Si_6$
2,2,3,6,7,7-Hexamethyl-3,6-bis(trimethylsilyl)-4,5-dioxa-2,3,6,7-tetrasilaoctane 1097
$C_{15}H_3Cl_7N_2O_4$
N-(2-Chloro-5-nitrophenyl)-1,4,5,6,7,7-hexachloro-5-norbornene-2,3-dicarboximide 715
$C_{15}H_3Cl_8NO_2$
N-(2,3-Dichlorophenyl)-1,4,5,6,7,7-hexachloro-5-norbornene-2,3-dicarboximide 715
$C_{15}H_4Cl_{12}O$
1,2,3,4,5,6,7,810,10,11,11-Dodecachloro-1,4,4a,4b,5,8,8a,8b-octahydro-1,4:5,8-dimethanofluoren-9-one 55
$C_{15}H_5Br_3F_3NO_2$
1,3,7-Tribromo-2-(2,2,2-trifluoroacetamido)-9-fluorenone 484
$C_{15}H_5Br_4ClF_3NO_3$
4′-Chloro-3,4,5,6-tetrabromo-2′-(trifluoromethyl)phthalanilic acid 445
$C_{15}H_5Cl_6FN_2O_4$
2-(4-Fluoro-3-nitrophenyl)-hexachloro-hexahydro-4,7-methanoisoindole-1,3-dione 721
$C_{15}H_5Cl_7N_2O_4$
2-(2-Chloro-5-nitrophenyl)-hexachloro-1,3,3a,4,7,7a-hexahydro-4,7-methanoisoindole-1,3-dione 715
$C_{15}H_5Cl_8NO_2$
N-(3,4-Dichlorophenyl)-1,4,5,6,7,7-hexachloro-5-norbornene-2,3-dicarboximide 715
$C_{15}H_6BrCl_6NO_2$
N-(4-Bromophenyl)-1,4,5,6,7,7-hexachloro-5-norbornene-2,3-dicarboximide 715
$C_{15}H_6Br_4F_3NO_3$
3,4,5,6-Tribromo-3′-(trifluoromethyl)-phthalanilic acid 445
2′-(Trifluoromethyl)-3,4,5,6-tetrabromophthalanilic acid 445
$C_{15}H_6Br_4N_2O_3$
2′-Cyano-3,4,5,6-tetrabromophthalanilic acid 759
$C_{15}H_6ClF_3N_2O_4$
3-Chloro-7-nitro-2-(2,2,2-trifluoroacetamido)-9-fluorenone 484
$C_{15}H_6Cl_4F_3NO_3$
3,4,5,6-Tetrachloro-3′-(trifluoromethyl)phthalanilic acid 445
$C_{15}H_6Cl_4N_2O_3$
N-(2-Cyanophenyl)-tetrachlorophthalamic acid 759
$C_{15}H_6Cl_5NO_2$
4-Cyanobenzyl pentachlorobenzoate 758
$C_{15}H_6Cl_6N_2O_2$
1,4,5,6,7,7-Hexachloro-*N*-(6-methyl-2-pyridyl)-5-norbornene-2,3-dicarboximide 964
$C_{15}H_6Cl_6N_2O_4S$
1,4,5,6,7,7-Hexachloro-*N*-(4-sulfamoylphenyl)-5-norbornene-2,3-dicarboximide 824
$C_{15}H_6F_3N_3O_6$
N-(2,6-Dinitro-4-(trifluoromethyl)-phenyl)phthalimide 714
$C_{15}H_6N_2O$
2-(Dicyanomethylene)-1-acenaphthenone 766
$C_{15}H_7BrF_3NO_2$
3-Bromo-2-(2,2,2-trifluoroacetamido)-9-fluorenone 484
$C_{15}H_7Br_2ClO_2$
2-Chloro-3-(dibromomethyl)-anthraquinone 315
$C_{15}H_7Cl_2NO_4$
4-(*N*-(2,4-Dichlorophenyl)-carbamoyl)phthalic anhydride 455
4-(*N*-(3,4-Dichlorophenyl)-carbamoyl)phthalic anhydride 455
$C_{15}H_7Cl_4NO_2$
2-Acetamido-1,3,4,7-tetrachloro-9-fluorenone 488
$C_{15}H_7Cl_4N_3O_3$
N-(1*H*-Indazol-6-yl)-3,4,5,6-tetrachlorophthalamic acid 899
$C_{15}H_7Cl_6HgNO_2$
1,4,5,6,7,7-Hexachloro-*N*-(phenylmercuri)bicyclo(2.1.1)hept-5-ene-2,3-dicarboximide 1138
$C_{15}H_7F_{13}O_2$
2-(Perfluoroheptanoyl)-acetophenone 306
$C_{15}H_7NO_6$
5 or 6 or 7 or 8-Nitro-2-anthraquinonecarboxylic acid 343
$C_{15}H_8BrCl_4NO_3$
4′-Bromo-3′-methyl-3,4,5,6-tetrachlorophthalanilic acid 455
$C_{15}H_8Br_2N_2O_3$
N-(3,5-Dibromosalicylideneamino)-phthalimide 1036
$C_{15}H_8Br_2O$
10,11-Dibromo-5*H*-dibenzo(a,d)-cyclohepten-5-one 314
$C_{15}H_8Br_4ClNO_3$
5′-Chloro-2′-methyl-3,4,5,6-tetrabromophthalanilic acid 455
$C_{15}H_8Br_4ClNO_4$
2′-Chloro-4′-methoxy-3,4,5,6-tetrabromophthalanilic acid 455
4′-Chloro-2′-methoxy-3,4,5,6-tetrabromophthalanilic acid 449
$C_{15}H_8Br_4N_2O_4$
2′-Carbamoyl-3,4,5,6-tetrabromophthalanilic acid 453
$C_{15}H_8Br_4N_2O_6$
2′-Methoxy-3′-nitro-3,4,5,6-tetrabromophthalanilic acid 448
2′-Methoxy-4′-nitro-3,4,5,6-tetrabromophthalanilic acid 449
$C_{15}H_8ClFN_2O_4S$
4-Chloro-6-fluorosulfonyl-2-(3-nitrophenyl)quinoline 989
4-Chloro-6-fluorosulfonyl-2-(4-nitrophenyl)quinoline 991
4-(4-Chloro-6-nitro-2-quinolyl)-benzenesulfonyl fluoride 989
$C_{15}H_8ClF_2NO_4S_2$
4-Chloro-2-phenylquinoline-4′,6-disulfonyl fluoride 994
$C_{15}H_8ClF_3N_2O_2$
4-Cyanophenyl *N*-(2-chloro-5-trifluoromethylphenyl)carbamate 758
$C_{15}H_8ClF_3N_4O_3$
1-(*N*-(4-Chloro-3-(trifluoromethyl)-phenyl)carbamoyl)-6-nitro-1*H*-indazole 899
$C_{15}H_8ClNO_3$
2-Chloro-6a,11-dihydro-5*H*-isoindolo(2,1-a)(3,1)benzoxazine-5,11-dione 709
$C_{15}H_8ClNO_4$
4-Chloro-2-phthalimidobenzoic acid. 715
$C_{15}H_8ClN_3O_4$
N-(2-Chloro-5-nitrobenzylideneamino)-phthalimide 1069
$C_{15}H_8ClN_3O_6$
3,3-Bis(4-nitrophenoxy)-2-chloroacrylonitrile 761
$C_{15}H_8Cl_2F_3N_3OS$
1-(4-Chloro-2-benzothiazolyl)-3-(2-chloro-5-(trifluoromethyl)-phenyl)urea 910
1-(4-Chloro-2-benzothiazolyl)-3-(4-chloro-2-(trifluoromethyl)-phenyl)urea 910
1-(4-Chloro-2-benzothiazolyl)-3-(4-chloro-3-(trifluoromethyl)-phenyl)urea 910
$C_{15}H_8Cl_2N_2O_2$
N-(2,4-Dichlorobenzylideneamino)-phthalimide 1070
N-(3,4-Dichlorobenzylideneamino)-phthalimide 1071
2-(2,4-Dichlorostyryl)-5-nitrobenzonitrile 759
$C_{15}H_8Cl_2N_2O_3$
N-(3,5-Dichlorosalicylideneamino)-phthalimide 1036
$C_{15}H_8Cl_2N_2O_4$
Phthalimido *N*-(2,4-dichlorophenyl)-carbamate 715
$C_{15}H_8Cl_3N$
α-(3-Chlorophenyl)-2,6-dichlorocinnamonitrile 767
$C_{15}H_8Cl_3N_3OS$
3-(4-Chlorophenyl)-1-(3,4-dichlorophenyl)-5-imino-4-thiohydantoin 732
$C_{15}H_8Cl_4F_3NO_2$
2′-Chloro-2-(2,4,5-trichlorophenoxy)-5′-(trifluoromethyl)acetanilide 394
$C_{15}H_8Cl_4N_2O_4$
2′-Carbamoyl-3,4,5,6-tetrachlorophthalanilic acid 451
$C_{15}H_8Cl_4N_2O_5$
2′-Methyl-3′-nitro-3,4,5,6-tetrachlorophthalanilic acid 451
4′-Methyl-2′-nitro-3,4,5,6-tetrachlorophthalanilic acid 450
$C_{15}H_8Cl_4O$
2,3′,4′,6-Tetrachlorochalcone 306

$C_{15}H_8Cl_5NO_3$
2′-Methyl-3,3′,4,5,6-pentachlorophthalanilic acid ... 454
2′-Methyl-3,4,5,6,6′-pentachlorophthalanilic acid ... 454
4′-Methyl-3,3′,4,5,6-pentachlorophthalanilic acid ... 455

$C_{15}H_8Cl_5NO_4$
2′-Methoxy-3,4,4′,5,6-pentachlorophthalanilic acid ... 449
2′-Methoxy-3,4,5,5′,6-pentachlorophthalanilic acid ... 449

$C_{15}H_8Cl_6N_2O_5$
1,4,5,6,7,7-Hexachloro-3-(*N*-(3-nitrophenyl)carbamoyl)-5-norbornene-2-carboxylic acid ... 432

$C_{15}H_8Cl_7NO_3$
3-(4-Chlorophenylcarbamoyl)-1,4,5,6,7,7-hexachloro-5-norbornene-2-carboxylic acid ... 434

$C_{15}H_8F_3NO_2$
2-(2,2,2-Trifluoroacetamido)-9-fluorenone ... 484

$C_{15}H_8F_3NO_3$
6-(2,2,2-Trifluoroacetamido)-3,4-benzocoumarin ... 397
7-(2,2,2-Trifluoroacetamido)-3,4-benzocoumarin ... 397

$C_{15}H_8F_3N_3O_5$
3,7-Dinitro-2-(2,2,2-trifluoroacetamido)fluorene ... 484

$C_{15}H_8F_6N_4O_5$
1,3-Bis(4-nitro-2-(trifluoromethyl)-phenyl)urea ... 508

$C_{15}H_8F_6O$
2,5-Bis(trifluoromethyl)-benzophenone ... 309

$C_{15}H_8F_{16}O_3S$
2,2,3,3,4,4,5,5,6,6,7,7,8,8,9,9-Hexadecylfluorononyl benzenesulfonate ... 796

$C_{15}H_8N_2O_6$
4-(*N*-(3-Nitrophenyl)carbamoyl)-phthalic anhydride ... 469
4-Nitro-2-phthalimidobenzoic acid ... 718

$C_{15}H_9BrClNO_2$
2-Acetamido-3-bromo-1-chloro-9-fluorenone ... 488

$C_{15}H_9BrCl_2F_3NO_2$
2-Bromo-4-chloro-6-methylphenyl *N*-(4-chloro-2-(trifluoromethyl)-phenyl)carbamate ... 608

$C_{15}H_9BrF_3NO$
9-Bromo-2-trifluoroacetamido-fluorene ... 484

$C_{15}H_9BrN_2O_3$
N-(5-Bromosalicylideneamino)-phthalimide ... 1036

$C_{15}H_9BrN_2S$
3-(4-Bromophenyl)imidazo(2,1-b)-benzothiazole ... 912
3-(4-Bromophenyl)thiazolo(3,2-a)-benzimidazole ... 901

$C_{15}H_9BrO_2$
10-Bromo-9-anthracenecarboxylic acid ... 338
2-(4-Bromophenyl)-1,3-indandione ... 319

$C_{15}H_9BrO_2S$
3-Bromobenzo(b)thiophene-4-yl benzoate ... 898

$C_{15}H_9ClFNO_2S$
4-Chloro-2-phenyl-6-quinolinesulfonyl fluoride ... 994

$C_{15}H_9ClFNO_3$
4-Chloro-4′-fluoro-3-nitrochalcone ... 303

$C_{15}H_9ClF_3N_3O$
1-(*N*-(4-Chloro-2-(trifluoromethyl)-phenyl)carbamoyl)benzimidazole ... 901
1-(4-Chloro-3-(trifluoromethyl)-phenyl)-3-(2-cyanophenyl)urea ... 758

$C_{15}H_9ClF_3N_3OS$
1-(2-Benzothiazolyl)-3-(4-chloro-2-(trifluoromethyl)phenyl)urea ... 910

$C_{15}H_9ClF_6N_2O$
1-(2-Chloro-5-(trifluoromethyl)-phenyl)-3-(3-(trifluoromethyl)-phenyl)urea ... 507

$C_{15}H_9ClF_6N_2S$
1-(2-Chlorophenyl)-3-(3,5-bis(trifluoromethyl)phenyl)-2-thiourea ... 728

$C_{15}H_9ClF_6O$
4-Chloro-(2′,3′ or 2′,4′ or 2′,5′ or 2′,6′)-bis-(trifluoromethyl)-benzhydrol ... 197

$C_{15}H_9ClN_2O_2$
α-(4-Chlorobenzylidene)-5-nitro-*o*-tolunitrile ... 759
2-Chloro-*α*-(4-nitrophenyl)-cinnamonitrile ... 765

$C_{15}H_9ClN_2O_4$
1,3-Dioxoisoindolin-2-yl *N*-(3-chlorophenyl)carbamate ... 715

$C_{15}H_9ClN_2O_5$
2-Chloro-5-nitro-3′-nitrochalcone ... 299

$C_{15}H_9ClO_2$
2′-Chloroflavone ... 320

$C_{15}H_9ClO_3$
Methyl 7-chloro-9-oxo-1-fluorenecarboxylate ... 374

$C_{15}H_9Cl_2FO$
3,4-Dichloro-4′-fluorochalcone ... 305

$C_{15}H_9Cl_2N$
2-Chloro-*α*-(3-chlorophenyl)-cinnamonitrile ... 767
4-Chloro-*α*-(4-chlorophenyl)-cinnamonitrile ... 759
4,7-Dichloro-2-phenylquinoline ... 994

$C_{15}H_9Cl_2NO$
2-(3,4-Dichlorophenyl)-5-phenyloxazole ... 872

$C_{15}H_9Cl_2NO_2$
N-(3,4-Dichlorobenzoyl)-5-methylanthranilic acid ... 385

$C_{15}H_9Cl_2NO_3$
2,4-Dichloro-3′-nitrochalcone ... 303
2,4-Dichloro-4′-nitrochalcone ... 303
2,4′-Dichloro-5-nitrochalcone ... 299
2,6-Dichloro-3′-nitrochalcone ... 303
2,6-Dichloro-4′-nitrochalcone ... 303
3,4-Dichloro-4′-nitrochalcone ... 303
4,4′-Dichloro-3-nitrochalcone ... 299

$C_{15}H_9Cl_2N_3OS$
1-(3,4-Dichlorophenyl)-5-imino-3-phenyl-4-thiohydantoin ... 732

$C_{15}H_9Cl_2N_3O_2$
Isatin, azine with 3,5-dichlorosalicylaldehyde ... 1039

$C_{15}H_9Cl_2N_3O_3$
5-Phenyl-1,2,4-oxadiazol-3-yl *N*-(3,4-dichlorophenyl)carbamate ... 872

$C_{15}H_9Cl_3FNO$
2-Acetamido-7-fluoro-1,3,4-trichlorofluorene ... 491

$C_{15}H_9Cl_3F_3NO_2$
2′-Chloro-2-(2,4-dichlorophenoxy)-5′-(trifluoromethyl)acetanilide ... 395
2-(2,4,5-Trichlorophenoxy)-2′-(trifluoromethyl)acetanilide ... 394
2-(2,4,5-Trichlorophenoxy)-3′-(trifluoromethyl)acetanilide ... 394

$C_{15}H_9Cl_3F_3N_3O_2$
4-(4-Chloro-2-(trifluoromethyl)-phenyl)-1-(2,4-dichlorobenzoyl)-semicarbazide ... 504

$C_{15}H_9Cl_3O$
2,4,4′-Trichlorochalcone ... 306
2,4′,6-Trichlorochalcone ... 306

$C_{15}H_9Cl_4NO$
2-Acetamido-1,3,4,7-tetrachlorofluorene ... 491

$C_{15}H_9Cl_4NO_2$
2′-Benzoyl-2,2,2,4′-tetrachloroacetanilide ... 395

$C_{15}H_9Cl_4NO_3$
N-(Benzyl)-3,4,5,6-tetrachlorophthalamic acid ... 450
N-(4-Methylphenyl)-tetrachlorophthalamic acid ... 483
2′-Methyl-3,4,5,6-tetrachlorophthalanilic acid ... 454
3′-Methyl-3,4,5,6-tetrachlorophthalanilic acid ... 456

$C_{15}H_9Cl_4NO_3S$
4′-(Methylthio)-3,4,5,6-tetrachlorophthalanilic acid ... 450

$C_{15}H_9Cl_4NO_4$
N-(3-Methoxyphenyl)-tetrachlorophthalamic acid ... 449
N-(4-Methoxyphenyl)-tetrachlorophthalamic acid ... 449

$C_{15}H_9Cl_5O$
2,2-Dichloro-2-phenyl-4′-(trichloromethyl)acetophenone ... 288

$C_{15}H_9Cl_5O_2$
4-Methylbenzyl pentachlorobenzoate ... 366

$C_{15}H_9Cl_6NO_3$
1,4,5,6,7,7-Hexachloro-3-(*N*-phenylcarbamoyl)-5-norbornene-2-carboxylic acid ... 434

$C_{15}H_9FN_2OS_2$
1-(3-Fluorophenyl)-5-oxo-3-phenyl-2,4-dithiohydantoin ... 732

$C_{15}H_9FN_2O_4$
2-Acetamido-7-fluoro-3-nitro-9-fluorenone ... 487

$C_{15}H_9FN_2O_5S$
2-(4-Fluorosulfonylphenyl)-4-hydroxy-6-nitroquinoline ... 983

$C_{15}H_9F_3N_2O_2$
4-Cyanophenyl *N*-(2-trifluoromethylphenyl)carbamate ... 758

$C_{15}H_9F_4NO$
7-Acetamido-1,2,3,4-tetrafluorofluorene ... 490

$C_{15}H_9F_6NO$
2-Amino-2′,5′-bis(trifluoromethyl)-benzophenone ... 309
3,4′-Bis(trifluoromethyl)benzanilide ... 446
4,4′-Bis(trifluoromethyl)benzanilide ... 446
α-(3,5-Bis(trifluoromethyl)-phenylimino)-*o*-cresol ... 1020

$C_{15}H_9NO_2$
4-Cyanobenzil ... 763
2-(4-Pyridylmethylene)-1,3-indandione ... 961

$C_{15}H_9NO_3$
N-Benzoylphthalimide ... 720
2-(Formamido)anthraquinone ... 488

$C_{15}H_9NO_4$
1-Amino-2-anthraquinonecarboxylic acid ... 342
3-Benzal-6-nitrophthalide ... 384
2-Methyl-1-nitroanthraquinone ... 318
2-Phthalimidobenzoic acid ... 720
4-Phthalimidobenzoic acid ... 722

$C_{15}H_9NO_4S$
Benzo(b)thiophen-4-yl 4-nitrobenzoate ... 897

$C_{15}H_9NO_5$
Methyl 7-nitro-9-oxo-1-fluorenecarboxylate ... 374

$C_{15}H_9N_3O_4$
N-(4-Nitrobenzylideneamino)-phthalimide ... 1069
5-Nitro-*α*-(3-nitrobenzylidene)-*o*-tolunitrile ... 764

$C_{15}H_9N_3O_6$
2-Acetamido-3,7-dinitro-9-fluorenone ... 487
2-(3,5-Dinitro-2-methylphenyl)-3,1-benzoxazin-4(4*H*)-one ... 385

$C_{15}H_{10}BrClF_3NO_2$
(2-Bromo-4-chloro-5-methyl)phenyl *N*-(2-(trifluoromethyl)phenyl)-carbamate ... 607
2-Bromo-4-chloro-6-methylphenyl *N*-(2-(trifluoromethyl)phenyl)-carbamate ... 607
2-Bromo-4-chloro-6-methylphenyl *N*-(3-(trifluoromethyl)phenyl)-carbamate ... 608

$C_{15}H_{10}BrClN_2O$
3-Bromo-6-chloro-4-hydroxyimino-*α*-(*p*-tolyl)-2,5-cyclohexadieneacetonitrile ... 1013

$C_{15}H_{10}BrClO$
3-Bromo-4′-chlorochalcone ... 306
4′-Bromo-2-chlorochalcone ... 305
4′-Bromo-4-chlorochalcone ... 305
4-Bromo-4′-chlorochalcone ... 306

$C_{15}H_{10}BrNO$
2-(2-Bromophenyl)-5-phenyloxazole ... 872

$C_{15}H_{10}BrNO_2$
2-Acetamido-3-bromo-9-fluorenone ... 488
2-(2-Bromobenzyl)phthalimide ... 715
2-Bromo-1-(methylamino)-anthraquinone ... 315

$C_{15}H_{10}BrNO_3$
3-Bromo-4′-nitrochalcone ... 303
4′-Bromo-3-nitrochalcone ... 303
4-Bromo-4′-nitrochalcone ... 303

$C_{15}H_{10}BrN_3OS$
1-(4-Bromophenyl)-5-imino-3-phenyl-4-thiohydantoin ... 695

$C_{15}H_{10}Br_4N_2O_3$
N-(2-(2-Pyridyl)ethyl)-3,4,5,6-tetrabromophthalamic acid ... 949

$C_{15}H_{10}ClFO$
2-Chloro-4′-fluorochalcone ... 305
4′-Chloro-4-fluorochalcone ... 306

$C_{15}H_{10}ClF_3N_2O_3$
1-(2-Carboxyphenyl)-3-(2-chloro-5-trifluoromethylphenyl)urea ... 505
1-(4-Carboxyphenyl)-3-(2-chloro-5-trifluoromethylphenyl)urea ... 505
3-(3-(4-Chloro-3-(trifluoromethyl)-phenyl)ureido)benzoic acid ... 505

$C_{15}H_{10}ClF_4NO_2$
3-Fluorobenzyl *N*-(4-chloro-2-(trifluoromethyl)phenyl)carbamate ... 607

$C_{15}H_{10}ClN$
α-(3-Chlorophenyl)cinnamonitrile ... 760
α-(4-Chlorophenyl)cinnamonitrile ... 759
4-Chloro-*α*-phenylcinnamonitrile ... 759
4-Chloro-2-phenylquinoline ... 994
7-Chloro-3-phenylquinoline ... 994
6-Chloro-4-phenylquinoline picrate ... 994

$C_{15}H_{10}ClNO$
2-(2-Chlorophenyl)-5-phenyloxazole ... 872

$C_{15}H_{10}ClNOS$
2-(4-Chlorobenzylidene)-3-oxo-4*H*-1,4-benzothiazine ... 706

$C_{15}H_{10}ClNO_2$
2-Acetamido-7-chloro-9-fluorenone ... 488
1-Amino-4-chloro-2-methylanthraquinone ... 317
N-(3-Chlorobenzoyl)-5-methylanthranilic acid ... 386
1-(4-Chlorobenzyl)isatin ... 689
2-Chloro-5-methyl-5*H*-dibenz(b,e)-azepine-6,11-dione ... 691

$C_{15}H_{10}ClNO_3$
2-Chloro-3-(furfurylamino)-1,4-naphthoquinone ... 850
4-Chloro-*N*-(3-methoxybenzoyl)-anthranilic acid ... 384
2-Chloro-3′-nitrochalcone ... 303
2-Chloro-4′-nitrochalcone ... 303
3-Chloro-4′-nitrochalcone ... 303
4-Chloro-3′-nitrochalcone ... 303
4′-Chloro-3-nitrochalcone ... 303
4-Chloro-4′-nitrochalcone ... 303
4′-Chloro-4-nitrochalcone ... 303

$C_{15}H_{10}ClNO_4$
2-Chloro-4-nitrophenyl cinnamate ... 377
α-(4-Chlorophenyl)-4-nitrocinnamic acid ... 335

$C_{15}H_{10}ClN_3O$
Isatin, azine with 2-chlorobenzaldehyde ... 1060
Isatin, azine with 4-chlorobenzaldehyde ... 1074

$C_{15}H_{10}ClN_3O_2$
2-Chloro-4,6-diphenoxy-1,3,5-triazine ... 978

$C_{15}H_{10}ClN_3O_3$
5-Phenyl-1,2,4-oxadiazol-3-yl *N*-(2-chlorophenyl)carbamate ... 872

$C_{15}H_{10}Cl_2F_3NO_2$
2′-Chloro-2-(2-chlorophenoxy)-5′-(trifluoromethyl)acetanilide ... 394
2′-Chloro-2-(4-chlorophenyl)-5′-(trifluoromethyl)acetanilide ... 395

$C_{15}H_{10}Cl_2F_3N_3O_2$
1-(4-Chlorobenzoyl)-4-(4-chloro-2-(trifluoromethyl)phenyl)-semicarbazide ... 504

$C_{15}H_{10}Cl_2F_3N_3O_2$ (Cont'd)
1-(2,4-Dichlorobenzoyl)-4-(2-trifluoromethylphenyl)-semicarbazide ... 504

$C_{15}H_{10}Cl_2N_2O_2$
α-(2,5-Dichloro-4-hydroxyimino-2,5-cyclohexadienylidene)-4-methoxyphenylacetonitrile ... 1009
N-(2,6-Dichlorophenyl-aminomethyl)phthalimide ... 719

$C_{15}H_{10}Cl_2N_2O_3$
2-Acetamido-1,3-dichloro-7-nitrofluorene ... 488
2-(2,2-Dichloroacetamido)-7-nitrofluorene ... 484

$C_{15}H_{10}Cl_2N_2O_4$
Phthalic *mono*-(3,4-dichlorobenzoyl)hydrazide ... 494
Piperonylic acid (3,5-dichlorosalicylidene)-hydrazide ... 496

$C_{15}H_{10}Cl_2N_4O_3$
3-Chloro-1-(*N*-(3-chloro-4-methylphenyl)carbamoyl)-6-nitro-1*H*-indazole ... 899

$C_{15}H_{10}Cl_2O$
2′,4-Dichlorochalcone ... 299
2,4′-Dichlorochalcone ... 306
2,6-Dichlorochalcone ... 306
3,4′-Dichlorochalcone ... 306
3,4-Dichlorochalcone ... 299
4,4′-Dichlorochalcone ... 305

$C_{15}H_{10}Cl_2O_2$
1,3-Bis(4-chlorophenyl)-1,3-propanedione ... 307
2,4-Dichloro-2′-hydroxychalcone ... 299
2,6-Dichloro-2′-hydroxychalcone ... 299
3,4-Dichloro-2′-hydroxychalcone ... 299
2,4-Dichlorophenyl cinnamate ... 377

$C_{15}H_{10}Cl_3N_3O_2$
N-(3-Chlorophenyl)-*N*′-(2,4-dichlorobenzylideneamino)-oxamide ... 492

$C_{15}H_{10}Cl_3N_5O_5$
N-(4-Chloro-3-nitrophenyl)-*N*′-(3,4-dichlorophenylureido)oxamide ... 527

$C_{15}H_{10}Cl_4N_2O_3$
N-(2-(2-Pyridyl)ethyl)-3,4,5,6-tetrachlorophthalamic acid ... 928

$C_{15}H_{10}Cl_4N_4O_3$
1-(3,4-Dichlorophenyl)-4-(3,4-dichlorophenylureido)oxamide ... 523

$C_{15}H_{10}FNO$
2-(3-Fluorophenyl)-5-phenyloxazole ... 872

$C_{15}H_{10}FNO_3$
3-Fluoro-4′-nitrochalcone ... 303
4-Fluoro-4′-nitrochalcone ... 303

$C_{15}H_{10}FNO_3S$
4-Hydroxy-2-phenyl-6-quinolinesulfonyl fluoride ... 984

$C_{15}H_{10}FN_3OS$
1-(4-Fluorophenyl)-5-imino-3-phenyl-4-thiohydantoin ... 695

$C_{15}H_{10}FN_3O_3$
1-(*N*-(4-Fluoro-3-nitrophenyl)-carbamoyl)indole ... 881
5-Phenyl-1,2,4-oxadiazol-3-yl *N*-(4-fluorophenyl)carbamate ... 872

$C_{15}H_{10}F_2O_2$
1,3-Bis(4-fluorophenyl)-1,3-propanedione ... 307

$C_{15}H_{10}F_3NO$
2-Acetamido-1,4,7-trifluorofluorene ... 490
2-(2,2,2-Trifluoroacetamido)-fluorene ... 484
9-(2,2,2-Trifluoroacetamido)-fluorene ... 484

$C_{15}H_{10}F_3NO_2$
N-Piperonylidene-3-(trifluoromethyl)aniline ... 1020

$C_{15}H_{10}F_3NO_3$
2′-(Trifluoromethyl)-phthalanilic acid ... 446
3′-(Trifluoromethyl)-phthalanilic acid ... 446

$C_{15}H_{10}F_3N_3O$
1-(*N*-(2-(Trifluoromethyl)phenyl)-carbamoyl)benzimidazole ... 901

$C_{15}H_{10}F_3N_3OS$
1-(2-Benzothiazolyl)-3-(2-(trifluoromethyl)phenyl)urea ... 911
1-(2-Benzothiazolyl)-3-(3-trifluoromethylphenyl)urea ... 911

$C_{15}H_{10}F_6N_2$
N,*N*′-Bis(3-(trifluoromethyl)phenyl)-formamidine ... 723

$C_{15}H_{10}F_6N_2O$
1,3-Bis(3-trifluoromethylphenyl)urea ... 510
1-(2-(Trifluoromethyl)phenyl)-3-(3-(trifluoromethyl)phenyl)urea ... 511

$C_{15}H_{10}INO_2$
N-(3-Iodobenzyl)phthalimide ... 721

$C_{15}H_{10}N_2$
11*H*-Indeno(1,2-b)quinoxaline ... 1000

$C_{15}H_{10}N_2OS_2$
1,3-Diphenyl-5-oxo-2,4-dithiohydantoin ... 745

$C_{15}H_{10}N_2O_2$
4-Cyano-*N*-(piperonylidene)aniline ... 1045
α-(4-Nitrophenyl)cinnamonitrile ... 765
2-Nitro-α-phenylcinnamonitrile ... 765
5-Nitro-2-styrylbenzonitrile ... 764
Spiro(fluoren-9,4′-imidazolidine)-2′,5′-dione ... 709

$C_{15}H_{10}N_2O_3$
4-(4-Carboxyphenyl)phthalazin-1(2*H*)-one ... 1000
1,3-Diphenylparabanic acid ... 708

$C_{15}H_{10}N_2O_4$
5-Methyl-*N*-(4-nitrobenzoyl)-anthranilic acid ... 385
N-(2-Methyl-4-nitrophenyl)-phthalimide ... 719

$C_{15}H_{10}N_2O_5$
3,3′-Dinitrochalcone ... 303
3,4′-Dinitrochalcone ... 303
4,4′-Dinitrochalcone ... 303

$C_{15}H_{10}N_2O_6$
2,4-Dinitro-3′,4′-(methylenedioxy)-stilbene ... 266

$C_{15}H_{10}N_4O$
1-(3-Cyanophenyl)-3-(4-cyanophenyl)urea ... 765

$C_{15}H_{10}N_4O_2$
2-(4-Nitrophenylazo)quinoline ... 1090

$C_{15}H_{10}N_4O_3$
Isatin, azine with 4-nitrobenzaldehyde ... 1073

$C_{15}H_{10}N_4O_3S$
5-Imino-1-(3-nitrophenyl)-3-phenyl-4-thiohydantoin ... 740

$C_{15}H_{10}N_4S$
1,3-Bis(3-cyanophenyl)-2-thiourea ... 765
1,3-Bis(4-cyanophenyl)-2-thiourea ... 767

$C_{15}H_{10}O$
2-Benzoyl-1-phenylacetylene ... 304

$C_{15}H_{10}O_2$
2-Anthracenecarboxylic acid ... 346
Biphenyleneacrylic acid ... 336
3-Phenanthrenecarboxylic acid ... 347
9-Phenanthrenecarboxylic acid ... 346
3-Phenyl-1*H*-2-benzopyran-1-one ... 384
3-Phenylcoumarin ... 386

$C_{15}H_{10}O_2S$
Benzo(b)thiophene-4-yl benzoate ... 898

$C_{15}H_{10}O_3$
3-Benzoylphthalide ... 384
4-Hydroxy-3-phenylcoumarin ... 382
7-Hydroxy-4-phenylcoumarin ... 383
Methyl 9-oxo-1-fluorenecarboxylate ... 374
9-Oxo-3-fluorenyl acetate ... 379

$C_{15}H_{10}O_4$
3-(2-Carboxyphenyl)phthalide ... 385

$C_{15}H_{10}O_5$
2,4′-Carbonyldibenzoic acid ... 344

$C_{15}H_{11}Br$
9-(Bromomethyl)anthracene ... 172
9-Bromo-10-methylanthracene ... 175

$C_{15}H_{11}BrCl_3NO_2$
2-Bromo-4,5-dimethylphenyl *N*-(2,4,5-trichlorophenyl)carbamate ... 641
4-Bromo-3,5-dimethylphenyl *N*-(2,4,5-trichlorophenyl)carbamate ... 640

$C_{15}H_{11}BrFNO$
2-Acetamido-3-bromo-7-fluorofluorene ... 491

$C_{15}H_{11}BrF_3N_3O_3$
4-(4-Bromo-2-(trifluoromethyl)-phenyl)-1-(4-hydroxybenzoyl)-semicarbazide ... 504

$C_{15}H_{11}BrN_2O_2$
1-(4-Bromoanilinomethyl)isatin ... 704
N-(4-Bromoanilinomethyl)-phthalimide ... 719

$C_{15}H_{11}BrN_2O_3$
N-(2-Bromophenyl)-2-nitrocinnamamide ... 452

$C_{15}H_{11}BrN_2O_4$
2′-Bromo-2-(3-nitrobenzoyl)-acetanilide ... 432
Piperonylic acid (5-bromosalicylidene)-hydrazide ... 496

$C_{15}H_{11}BrO$
4′-Bromochalcone ... 304
2-Bromo-3-phenylacrylophenone ... 304

$C_{15}H_{11}BrO_2$
2-Bromo-1,3-diphenyl-1,3-propanedione ... 308
2-Bromo-2′-hydroxychalcone ... 299
4-Bromo-2′-hydroxychalcone ... 299

$C_{15}H_{11}BrO_3$
3′-Benzoyloxy-2-bromoacetophenone ... 376
2-(Bromoacetyl)-phenyl benzoate ... 376
4-(Bromoacetyl)-phenyl benzoate ... 376

$C_{15}H_{11}Br_2Cl_2NO$
α,β-Dibromo-3′,4′-dichlorohydrocinnamanilide ... 405

$C_{15}H_{11}Br_2NO$
2-Acetamido-1,3-dibromofluorene ... 491

$C_{15}H_{11}ClFNO_3$
2-Chloro-4-fluorophenyl *N*-(3-acetylphenyl)carbamate ... 629

$C_{15}H_{11}ClFNO_4$
2-Chloro-4-fluorophenyl *N*-(2-(methoxycarbonyl)phenyl)-carbamate ... 630
4-Formyl-2-methoxyphenyl *N*-(3-chloro-4-fluorophenyl)carbamate ... 615

$C_{15}H_{11}ClF_3NO$
2′-Chloro-3-methyl-5′-(trifluoromethyl)benzanilide ... 446
2′-Chloro-4-methyl-5′-(trifluoromethyl)benzanilide ... 445
2′-Chloro-2-phenyl-5′-(trifluoromethyl)acetanilide ... 395

$C_{15}H_{11}ClF_3NOS$
2-(4-Chlorophenylthio)-3′-trifluoromethylacetanilide ... 395
S-(*o*-Tolyl) *N*-(2-chloro-5-(trifluoromethyl)phenyl)-thiocarbamate ... 608
S-(*m*-Tolyl) *N*-(2-chloro-6-(trifluoromethyl)phenyl)-thiocarbamate ... 608

$C_{15}H_{11}ClF_3NO_2$
4-Chloro-3-methylphenyl *N*-(3-(trifluoromethyl)phenyl)carbamate ... 607
2-(2-Chlorophenoxy)-2′-(trifluoromethyl)acetanilide ... 395
2-(2-Chlorophenoxy)-3′-(trifluoromethyl)acetanilide ... 394
2-(4-Chlorophenoxy)-2′-(trifluoromethyl)acetanilide ... 394
α-(2-Chloro-5-(trifluoromethyl)-phenylimino)-6-methoxy-*o*-cresol ... 1020

$C_{15}H_{11}ClF_3N_3O_2$
1-Benzoyl-4-(4-chloro-2-(trifluoromethyl)phenyl)-semicarbazide ... 505
1-(2-Carbamoylphenyl)-3-(4-chloro-3-(trifluoromethyl)phenyl)urea ... 504
1-(2-Chlorobenzoyl)-4-(3-(trifluoromethyl)phenyl)-semicarbazide ... 505

$C_{15}H_{11}ClF_3N_3O_3$
1-(4-Carboxyphenyl)-4-(2-chloro-5-(trifluoromethyl)phenyl)-semicarbazide ... 505
4-(4-Chloro-2-(trifluoromethyl)-phenyl)-1-(2-hydroxybenzoyl)-semicarbazide ... 504
4-(4-Chloro-2-(trifluoromethyl)-phenyl)-1-(4-hydroxybenzoyl)-semicarbazide ... 504

$C_{15}H_{11}ClF_3N_3O_4$
1-(5-Chloro-2-methoxyphenyl)-3-(4-nitro-3-(trifluoromethyl)phenyl)urea ... 504

$C_{15}H_{11}ClINO_2$
2′-Benzoyl-4′-chloro-2-iodoacetanilide ... 392

$C_{15}H_{11}ClN_2O$
7-Chloro-4-(4-hydroxyanilino)-quinoline ... 983

$C_{15}H_{11}ClN_2OS$
3-(4-Chlorophenyl)-2,3-dihydro-3-hydroxythiazolo(3,2-a)-benzimidazole hydrobromide ... 901

$C_{15}H_{11}ClN_2O_2$
N-(4-Chloroanilinomethyl)-phthalimide ... 719
4-Cyanophenyl *N*-(2-chloro-6-methylphenyl)carbamate ... 760
4-Cyanophenyl *N*-(3-chloro-4-methylphenyl)carbamate ... 760
4-Cyanophenyl *N*-(4-chloro-2-methylphenyl)carbamate ... 760
4-Cyanophenyl *N*-(5-chloro-2-methylphenyl)carbamate ... 760

$C_{15}H_{11}ClN_2O_3$
2-Carbamoyl-3-(4-chlorophenyl)-3-hydroxy-1-isoindolinone ... 694

$C_{15}H_{11}ClN_2O_4$
9-Chloro-6,7-dimethoxy-3-nitroacridine ... 995

$C_{15}H_{11}ClN_2O_6$
4-Formyl-2-methoxyphenyl *N*-(2-chloro-4-nitrophenyl)carbamate ... 615
4-(Methoxycarbonyl)phenyl *N*-(2-chloro-5-nitrophenyl)carbamate ... 622
Piperonyl *N*-(2-chloro-4-nitrophenyl)carbamate ... 657

$C_{15}H_{11}ClN_4O_3$
1-(*N*-(3-Chloro-4-methylphenyl)-carbamoyl)-6-nitro-1*H*-indazole ... 899
1-(*N*-(5-Chloro-2-methylphenyl)-carbamoyl)-6-nitro-1*H*-indazole ... 899

$C_{15}H_{11}ClN_4O_5$
N-(4-Chloro-3-nitrophenyl)-*N*′-(2-hydroxybenzylideneamino)-oxamide ... 491

$C_{15}H_{11}ClO$
2-Chlorochalcone ... 306
3′-Chlorochalcone ... 299
4-Chlorochalcone ... 306
4′-Chlorochalcone ... 305

$C_{15}H_{11}ClO_2$
2-Chloro-2′-hydroxychalcone ... 299
4-Chloro-4′-hydroxychalcone ... 299
1-(4-Chlorophenyl)-3-phenyl-1,3-propanedione ... 307

$C_{15}H_{11}Cl_2FN_4O_3$
N-(3,4-Dichlorophenylureido)-*N*′-(4-fluorophenyl)oxamide ... 532

$C_{15}H_{11}Cl_2F_3N_2O$
1-(2,4-Dichlorobenzyl)-3-(2-(trifluoromethyl)phenyl)urea ... 504
1-(2,4-Dichlorobenzyl)-3-(3-(trifluoromethyl)phenyl)urea ... 507
1-(4,5-Dichloro-2-methylphenyl)-3-(3-(trifluoromethyl)phenyl)urea ... 507

$C_{15}H_{11}Cl_2N$
4-Chloro-α-(4-chlorophenyl)-hydrocinnamonitrile ... 751
N-(Cinnamylidene)-2,3-dichloroaniline ... 1079
N-(Cinnamylidene)-2,4-dichloroaniline ... 1079
N-(Cinnamylidene)-3,4-dichloroaniline ... 1079

$C_{15}H_{11}Cl_2NO$
4-Acetamido-2,7-dichlorofluorene ... 490
N-(2,2-Dichloroacetyl)-2-fluorenamine ... 484
3′-(2,4-Dichlorobenzylideneamino)-acetophenone ... 1026
4′-(2,4-Dichlorobenzylideneamino)-acetophenone ... 1026
2′,5′-Dichlorocinnamanilide ... 457

$C_{15}H_{11}Cl_2NO$ (Cont'd)
2,11-Dichloro-5,11-dihydro-5-methyl-6*H*-dibenz(b,e)azepin-6-one 692
$C_{15}H_{11}Cl_2NO_2$
2′-Benzoyl-2,4′-dichloroacetanilide 392
2′-Benzoyl-2,5′-dichloroacetanilide 390
$C_{15}H_{11}Cl_2NO_3$
2-Acetylphenyl *N*-(2,3-dichlorophenyl)carbamate 629
3-Acetylphenyl *N*-(2,3-dichlorophenyl)carbamate 623
4-Acetylphenyl *N*-(2,3-dichlorophenyl)carbamate 629
Methyl *N*-(2,4-dichlorobenzoyl)-anthranilate 472
$C_{15}H_{11}Cl_2NO_4$
4-Chlorophenoxyacetyl *N*-(2-chlorophenyl)carbamate 621
3,4-Dichlorophenyl *N*-(2-methoxycarbonylphenyl)-carbamate 630
2-(Methoxycarbonyl)phenyl *N*-(2,3-dichlorophenyl)carbamate 622
Piperonyl *N*-(2,3-dichlorophenyl)-carbamate 646
Piperonyl *N*-(2,4-dichlorophenyl)-carbamate 655
Piperonyl *N*-(2,5-dichlorophenyl)-carbamate 665
Piperonyl *N*-(2,6-dichlorophenyl)-carbamate 631
Piperonyl *N*-(3,4-dichlorophenyl)-carbamate 665
$C_{15}H_{11}Cl_2N_3O$
5-Chloro-1-(*N*-(5-chloro-2-methylphenyl)carbamoyl)-1*H*-indazole 899
$C_{15}H_{11}Cl_2N_3OS$
1-(4-Chloro-2-benzothiazolyl)-3-(3-chloro-4-methylphenyl)urea 911
$C_{15}H_{11}Cl_2N_3O_2$
5-Chloro-1-(*N*-(5-chloro-2-methoxyphenyl)carbamoyl)-1*H*-indazole 899
$C_{15}H_{11}Cl_2N_3O_2S$
1-(4-Chloro-2-benzothiazolyl)-3-(5-chloro-2-methoxyphenyl)urea 911
2-(*N*-(3,4-Dichlorophenyl)-carbamoyl)amino-6-methoxybenzothiazole 913
1-(2,3-Dichlorophenyl)-3-(6-methoxy-2-benzothiazolyl)urea 911
$C_{15}H_{11}Cl_3N_2O_2S$
1-(5-Chloro-2-methoxyphenyl)-3-(3,4-dichlorobenzoyl)-2-thiourea 729
$C_{15}H_{11}Cl_3N_2O_4$
2′-Methyl-5′-nitro-2-(2,4,5-trichlorophenoxy)acetanilide 399
4′-Methyl-2′-nitro-2-(2,4,5-trichlorophenoxy)acetanilide 399
$C_{15}H_{11}Cl_3N_2O_5$
N-(2-Methoxy-5-nitrophenyl)-2-(2,4,5-trichlorophenoxy)acetamide 399
$C_{15}H_{11}Cl_4NO_2$
3′-Chloro-2′-methyl-2-(2,4,5-trichlorophenoxy)acetanilide 401
3′-Chloro-4′-methyl-2-(2,4,5-trichlorophenoxy)acetanilide 401
2,4-Dichloro-3,5-dimethylphenyl *N*-(2,6-dichlorophenyl)carbamate 636
$C_{15}H_{11}Cl_4NO_3$
N-(5-Chloro-2-methoxyphenyl)-2-(2,4,5-trichlorophenoxy)acetamide 399
2-(2,4-Dichlorophenoxy)ethyl *N*-(2,3-dichlorophenyl)carbamate 650
2-(2,4-Dichlorophenoxy)ethyl *N*-(2,6-dichlorophenyl)carbamate 622
2-(2,4-Dichlorophenoxy)ethyl *N*-(3,4-dichlorophenyl)carbamate 651
$C_{15}H_{11}Cl_4NO_4$
2,4,5-Trichlorophenyl *N*-(5-chloro-2,4-dimethoxyphenyl)carbamate 613
$C_{15}H_{11}FN_2O_2$
1-(4-Fluoroanilinomethyl)-isatin 704
3-(4-Fluoro-2-methylphenyl)-2,4(1*H*,3*H*)-quinazolinedione 695
$C_{15}H_{11}FN_2O_3$
2-Acetamido-7-fluoro-3-nitrofluorene 487
$C_{15}H_{11}FN_2O_3S$
6-Amino-2-(4-fluorosulfonylphenyl)-4-hydroxyquinoline 983
$C_{15}H_{11}FN_2O_4$
2-(2-Fluorobenzoyl)-4′-nitroacetanilide 443
$C_{15}H_{11}FN_2O_5$
4-Acetylphenyl *N*-(2-fluoro-5-nitrophenyl)carbamate 625
$C_{15}H_{11}FN_2O_6$
4-Formyl-2-methoxyphenyl *N*-(2-fluoro-5-nitrophenyl)carbamate 622
4-Formyl-2-methoxyphenyl *N*-(4-fluoro-3-nitrophenyl)carbamate 616
2-(Methoxycarbonyl)phenyl *N*-(2-fluoro-5-nitrophenyl)carbamate 625
4-(Methoxycarbonyl)phenyl *N*-(4-fluoro-3-nitrophenyl)carbamate 625
$C_{15}H_{11}FO$
4′-Fluorochalcone 305
$C_{15}H_{11}FO_2$
3-Fluoro-2′-hydroxychalcone 299
4-Fluoro-2′-hydroxychalcone 299
$C_{15}H_{11}F_2NO$
7-Acetamido-1,4-difluorofluorene 490
$C_{15}H_{11}F_2NO_3$
4-Acetylphenyl *N*-(2,4-difluorophenyl)carbamate 629
4-Acetylphenyl *N*-(2,5-difluorophenyl)carbamate 629
$C_{15}H_{11}F_2N_5O$
1-(2,5-Difluorophenyl)-3-(5-phenyl-1,2,4-triazol-3-yl)urea 869
$C_{15}H_{11}F_3N_2OS$
1-Benzoyl-3-(α,α,α-trifluoro-*o*-tolyl)-2-thiourea 728
$C_{15}H_{11}F_3N_2O_3$
3-(3-(3-(Trifluoromethyl)phenyl)-ureido)benzoic acid 509
4-(3-(3-(Trifluoromethyl)phenyl)-ureido)benzoic acid 509
$C_{15}H_{11}F_3N_2O_4$
2-Methoxy-4-nitro-6-(3-(trifluoromethyl)-phenyliminomethyl)phenol 1020
$C_{15}H_{11}F_3N_4O$
1-(2-Benzimidazolyl)-3-(2-(trifluoromethyl)phenyl)urea 901
1-(1*H*-Indazol-5-yl)-3-(2-(trifluoromethyl)phenyl)urea 899
$C_{15}H_{11}N$
1-Phenylisoquinoline 998
7-Phenylquinoline picrate 994
$C_{15}H_{11}NO$
1-Benzoylindole 889
3-Benzoylindole 888
3-Benzylidene-2-indolinone 707
5*H*-Dibenzo(a,d)cyclohepten-5-one oxime 1014
4-Hydroxy-2-phenylquinoline 983
4-Hydroxy-3-phenylquinoline 984
$C_{15}H_{11}NO_2$
1-Acetamido-9-fluorenone 488
2-Acetamido-9-fluorenone 488
3-Acetamido-9-fluorenone 488
N-Benzoyl-5-methylanthranilic acid 387
2-Hydroximino-3-phenyl-1-indanone 1010
N-(2-Methylbenzoyl)anthranilic acid 387
2-(*N*-Methylformamido)-9-fluorenone 488
N-(1-Naphthyl)-2-furamide 847
3-Phenyl-1,2-indandione 2-oxime 1011
2-(*m*-Tolyl)-3,1-benzoxazin-4(4*H*)-one 387
$C_{15}H_{11}NO_2S$
N-(Benzylthio)phthalimide 750
$C_{15}H_{11}NO_3$
7-Acetamido-3,4-benzocoumarin 434
Acridone-10-acetic acid 349
1-Amino-4-methoxyanthraquinone 317
4-Benzilcarboxamide 467
N-Benzyloxyphthalimide 717
6,11-Dihydro-1-methoxy-5*H*-dibenz(b,e)azepin-6,11-dione 690
N-(5-Hydroxy-1-naphthyl)-2-furamide 845
2-Nitrochalcone 303
3′-Nitrochalcone 303
3-Nitrochalcone 303
$C_{15}H_{11}NO_4$
9-Acetoxy-2-nitrofluorene 360
9-Acetoxy-3-nitrofluorene 360
Disperse Red 4 (C.I. 60755) 1093
2′-Hydroxy-3-nitrochalcone 300
Methyl 7-nitro-1-fluorenecarboxylate 374
3-Nitro-α-phenylcinnamic acid 335
$C_{15}H_{11}NO_4S$
4-Hydroxy-2-phenyl-6-quinolinesulfonic acid 983
$C_{15}H_{11}NO_4S_2$
4-(2,3-Dihydroxybenzylidene)-2-furfuryl-5-thioxo-3-isothiazolidinone 845
$C_{15}H_{11}NO_5$
4-(3,4-(Methylenedioxy)-benzylideneamino)salicylic acid 1032
$C_{15}H_{11}NS$
2-(Phenylthio)quinoline 987
$C_{15}H_{11}NS_2$
4,5-Diphenyl-2-mercaptothiazole 875
$C_{15}H_{11}N_3$
9-Anthracenylmethyl azide 770
5,6-Diphenyl-1,2,4-triazine 980
2-(Phenylazo)quinoline 1093
$C_{15}H_{11}N_3O$
Benzoic (4-cyanobenzylidene)-hydrazide 1069
5,6-Diphenyl-3-hydroxy-1,2,4-triazine 978
$C_{15}H_{11}N_3OS$
1,3-Diphenyl-5-imino-4-thiohydantoin 746
$C_{15}H_{11}N_3O_2$
1,3-Diphenyl-5-iminohydantoin 726
2-(4-Nitroanilino)quinoline 989
5-Phenyl-2-(2-pyridyl)-4,6-pyrimidinediol 969
$C_{15}H_{11}N_3O_3$
4-(3,4-Dihydro-4-oxo-1,2,3-benzotriazin-3-yl)phenylacetic acid 707
3-(3-Nitrophenyl)-1-phenyl-2-pyrazolin-5-one 703
$C_{15}H_{11}N_3O_3S$
4-(4-Quinolylazo)-benzenesulfonic acid 1089
$C_{15}H_{11}N_3O_4$
N-Cinnamylidene-2,4-dinitroaniline 1053
3-(4-Methyl-3-nitrophenyl)-2,4(1*H*,3*H*)-quinazolinedione 706
$C_{15}H_{11}N_3O_5$
2-Acetamido-3,7-dinitrofluorene 487
$C_{15}H_{11}N_3O_5S$
4-(3-(3-Nitrobenzoyl)-2-thioureido)-benzoic acid 744
$C_{15}H_{11}N_3O_7$
N,*O*-Bis(4-nitrobenzoyl)-methylhydroxylamine 472
$C_{15}H_{11}N_3O_8$
4′,6-Dinitro-2′-methoxyphthalanilic acid 466
$C_{15}H_{11}N_3O_9S$
3,5-Dinitrobenzoyl *N*-(*m*-tolylsulfonyl)carbamate 818
$C_{15}H_{11}N_3S$
5,6-Diphenyl-3-mercapto-1,2,4-triazine 978
$C_{15}H_{11}N_5O_2$
6-Phenyl-4-(2-pyridylmethyleneamino)-1,2,4-triazine-3,5(2*H*,4*H*)-dione 1053
$C_{15}H_{11}N_5O_5$
1-(*N*-(4-Methyl-3-nitrophenyl)-carbamoyl)-6-nitro-1*H*-indazole 900
$C_{15}H_{12}AsN_3$
Tris(2-pyridyl)arsine 1128
$C_{15}H_{12}BrClN_2O_4$
2-Bromo-4-chloro-5-methylphenyl *N*-(2-methyl-5-nitrophenyl)-carbamate 628
2-Bromo-4-chloro-5-methylphenyl *N*-(4-methyl-3-nitrophenyl)-carbamate 625
2-Bromo-4-chloro-6-methylphenyl *N*-(2-methyl-5-nitrophenyl)-carbamate 626
2-Bromo-4,5-dimethylphenyl *N*-(2-chloro-4-nitrophenyl)carbamate 628
2-Bromo-4,5-dimethylphenyl *N*-(4-chloro-2-nitrophenyl)carbamate 628
4-Bromo-2,5-dimethylphenyl *N*-(4-chloro-2-nitrophenyl)carbamate 624
$C_{15}H_{12}BrCl_2NO$
4′-Bromo-2,4-dichloro-2′,6′-dimethylbenzanilide 460
$C_{15}H_{12}BrCl_2NO_2$
2-Bromo-4-chloro-5-methylphenyl *N*-(3-chloro-4-methylphenyl)-carbamate 638
2-Bromo-4-chloro-5-methylphenyl *N*-(4-chloro-2-methylphenyl)-carbamate 638
2-Bromo-4-chloro-6-methylphenyl *N*-(3-chloro-4-methylphenyl)-carbamate 639
2-Bromo-4-chloro-6-methylphenyl *N*-(4-chloro-2-methylphenyl)-carbamate 639
2-Bromo-4,5-dimethylphenyl *N*-(2,3-dichlorophenyl)carbamate 636
2-Bromo-4,5-dimethylphenyl *N*-(2,5-dichlorophenyl)carbamate 638
2-Bromo-4,5-dimethylphenyl *N*-(3,4-dichlorophenyl)carbamate 638
4-Bromo-2,5-dimethylphenyl *N*-(2,3-dichlorophenyl)carbamate 636
4-Bromo-2,5-dimethylphenyl *N*-(2,5-dichlorophenyl)carbamate 638
4-Bromo-2,5-dimethylphenyl *N*-(3,4-dichlorophenyl)carbamate 638
4-Bromo-3,5-dimethylphenyl *N*-(2,3-dichlorophenyl)carbamate 637
4-Bromo-3,5-dimethylphenyl *N*-(3,4-dichlorophenyl)carbamate 639
$C_{15}H_{12}BrCl_2NO_3$
2-Bromo-4-chloro-5-methylphenyl *N*-(4-chloro-2-methoxyphenyl)-carbamate 621
2-Bromo-4-chloro-6-methylphenyl *N*-(4-chloro-2-methoxyphenyl)-carbamate 619
2-Bromo-4-chloro-6-methylphenyl *N*-(5-chloro-2-methoxyphenyl)-carbamate 619
2-(4-Bromophenoxy)ethyl *N*-(2,6-dichlorophenyl)carbamate 615
2-(2,4-Dichlorophenoxy)ethyl *N*-(4-bromophenyl)carbamate 650
$C_{15}H_{12}BrF_2NO_2$
2-Bromo-4,5-dimethylphenyl *N*-(2,4-difluorophenyl)carbamate 638
$C_{15}H_{12}BrF_3N_4$
1-(4-Bromo-4(trifluoromethyl)-diphenylmethyleneamino)-guanidine *p*-toluenesulfonate 1020
$C_{15}H_{12}BrN$
4-Bromo-*N*-(cinnamylidene)aniline 1077
$C_{15}H_{12}BrNO$
2-Acetamido-9-bromofluorene 491
N-(2-Bromophenyl)cinnamamide 456
$C_{15}H_{12}BrNO_2$
2-Benzoyl-2-bromoacetanilide 443
$C_{15}H_{12}BrNO_3$
4′-Bromo-2′-methylphthalanilic acid 454
4′-Bromo-3′-methylphthalanilic acid 454
$C_{15}H_{12}BrNO_4$
4-(Methoxycarbonyl)phenyl *N*-(4-bromophenyl)carbamate 629
Piperonyl *N*-(4-bromophenyl)-carbamate 646
$C_{15}H_{12}BrN_3O$
4′-Bromo-2′,6′-dimethyl-3,5-dinitrobenzanilide 451
$C_{15}H_{12}BrN_3O_3S$
4-Acetamido-4′-bromo-*N*-cyanobenzenesulfonanilide 808
$C_{15}H_{12}Br_2ClNO_4$
2,4-Dibromophenyl *N*-(5-chloro-2,4-dimethoxyphenyl)carbamate 613
$C_{15}H_{12}Br_2FNO$
α,β-Dibromo-4′-fluorohydrocinnamanilide 405

$C_{15}H_{12}Br_2FNO$ (Cont'd)
2,3-Dibromo-2′-fluoro-3-phenylpropionanilide ... 405
$C_{15}H_{12}Br_2N_2O_6$
2,6-Dibromo-4-nitrophenyl *N*-(3,5-dimethoxyphenyl)carbamate ... 616
$C_{15}H_{12}Br_2O$
4-Benzyloxy-β,β-dibromostyrene ... 178
1,3-Dibromo-1,3-diphenyl-2-propanone ... 285
2,3-Dibromo-3-phenylpropiophenone ... 296
$C_{15}H_{12}Br_2O_3$
3,8-Dibromo-3,4-dihydro-2,2-dimethylnaphtho(1,2-b)-2*H*-pyran-5,6-dione ... 320
$C_{15}H_{12}Br_3NO$
3-Phenyl-2,3,4′-tribromopropionanilide ... 405
$C_{15}H_{12}ClFN_2O_3$
4-Acetamidophenyl *N*-(3-chloro-4-fluorophenyl)carbamate ... 631
Methyl 2-(3-(3-chloro-4-fluorophenyl)ureido)benzoate ... 532
$C_{15}H_{12}ClFN_2O_4$
4-Chloro-3,5-dimethyl *N*-(2-fluoro-5-nitrophenyl)carbamate ... 626
4-Chloro-2,6-dimethylphenyl *N*-(4-fluoro-2-nitrophenyl)carbamate ... 624
4-Chloro-3,5-dimethylphenyl *N*-(4-fluoro-2-nitrophenyl)carbamate ... 624
4-Chloro-3,5-dimethylphenyl *N*-(4-fluoro-3-nitrophenyl)carbamate ... 626
$C_{15}H_{12}ClF_2NO_2$
4-Chloro-2,5-dimethylphenyl *N*-(2,5-difluorophenyl)carbamate ... 640
4-Chloro-2,6-dimethylphenyl *N*-(2,4-difluorophenyl)carbamate ... 639
4-Chloro-3,5-dimethylphenyl *N*-(2,5-difluorophenyl)carbamate ... 639
$C_{15}H_{12}ClF_2NO_3$
2-(2-Chlorophenoxy)ethyl *N*-(2,4-difluorophenyl)carbamate ... 615
2-(4-Chlorophenoxy)ethyl *N*-(2,4-difluorophenyl)carbamate ... 615
$C_{15}H_{12}ClF_3N_2O$
1-Benzyl-3-(2-chloro-5-(trifluoromethyl)phenyl)urea ... 504
1-(2-Chloro-5-methylphenyl)-3-(3-(trifluoromethyl)phenyl)urea ... 507
1-(2-Chloro-6-methylphenyl)-3-(3-trifluoromethylphenyl)urea ... 506
1-(3-Chloro-2-methylphenyl)-3-(2-(trifluoromethyl)phenyl)urea ... 505
1-(3-Chloro-2-methylphenyl)-3-(3-(trifluoromethyl)phenyl)urea ... 506
1-(3-Chloro-4-methylphenyl)-3-(2-(trifluoromethyl)phenyl)urea ... 506
1-(3-Chloro-4-methylphenyl)-3-(3-(trifluoromethyl)phenyl)urea ... 506
1-(4-Chloro-2-methylphenyl)-3-(3-trifluoromethylphenyl)urea ... 511
1-(5-Chloro-2-methylphenyl)-3-(2-trifluoromethylphenyl)urea ... 506
1-(5-Chloro-2-methylphenyl)-3-(3-(trifluoromethyl)phenyl)urea ... 506
1-(4-Chlorophenyl)-3-(2-methyl-3-(trifluoromethyl)phenyl)urea ... 506
$C_{15}H_{12}ClF_3N_2O_2$
2-(2-Pyridyl)ethyl *N*-(4-chloro-2-(trifluoromethyl)phenyl)carbamate ... 925
$C_{15}H_{12}ClF_3N_2S$
1-Benzyl-3-(4-chloro-2-(trifluoromethyl)phenyl)-2-thiourea ... 728
$C_{15}H_{12}ClF_3N_4OS$
6-(2-Chloro-5-(trifluoromethyl)-phenyl)-1-phenyl-2-thiobiurea ... 728
$C_{15}H_{12}ClIN_2O_4$
2,5-Dimethyl-4-iodophenyl *N*-(4-chloro-2-nitrophenyl)carbamate ... 628
$C_{15}H_{12}ClN$
2-Chloro-*N*-cinnamylideneaniline ... 1077
3-Chloro-*N*-cinnamylideneaniline ... 1027
4-Chloro-*N*-cinnamylideneaniline ... 1077
α-(4-Chlorophenyl)-hydrocinnamonitrile ... 751
4-Chloro-α-phenylhydrocinnamonitrile ... 751
$C_{15}H_{12}ClNO$
2-Acetamido-4-chlorofluorene ... 490
2-Acetamido-6-chlorofluorene ... 490
4′-(4-Chlorobenzylideneamino)-acetophenone ... 1026
2′-Chlorocinnamanilide ... 456
1-(4-Chlorophenyl)-4-phenyl-2-azetidinone ... 689
4-(2-Chlorophenyl)-1-phenyl-2-azetidinone ... 689
$C_{15}H_{12}ClNOS$
α-(4-Chlorophenyl)-α-methyl-2-benzothiazolemethanol ... 911
$C_{15}H_{12}ClNO_2$
2-Acetamido-3-chloro-9-hydroxyfluorene ... 486
3′-Acetyl-3-chlorobenzanilide ... 453
2-Chloro-2-benzoylacetanilide ... 443
1-(4-Chlorophenyl)-1-methoxy-3-oxoisoindoline ... 690
3-Chloro-*N*-piperonylidene-*o*-toluidine ... 1076
3-Chloro-*N*-piperonylidene-*p*-toluidine ... 1076
4-Chloro-*N*-piperonylidene-*o*-toluidine ... 1076
5-Chloro-*N*-piperonylidene-*o*-toluidine ... 1076
$C_{15}H_{12}ClNO_3$
3-Acetylphenyl *N*-(4-chlorophenyl)-carbamate ... 623
4-Acetylphenyl *N*-(4-chlorophenyl)-carbamate ... 629
3-Chloro-3′-(methoxycarbonyl)-benzanilide ... 456
N-(5-Chloro-2-methylphenyl)-phthalamic acid ... 454
2′-Chloro-6′-methylphthalanilic acid ... 454
3′-Chloro-2′-methylphthalanilic acid ... 476
3′-Chloro-4′-methylphthalanilic acid ... 454
4′-Chloro-2′-methylphthalanilic acid ... 454
Methyl *N*-(3-chlorobenzoyl)-anthranilate ... 456
Methyl *N*-(4-chlorobenzoyl)-anthranilate ... 476
$C_{15}H_{12}ClNO_4$
4′-Chloro-2′-methoxyphthalanilic acid ... 448
4-(Methoxycarbonyl)phenyl *N*-(3-chlorophenyl)carbamate ... 629
1-Phenylquinolinium perchlorate ... 990
Piperonyl *N*-(2-chlorophenyl)-carbamate ... 664
Piperonyl *N*-(3-chlorophenyl)-carbamate ... 665
Piperonyl *N*-(4-chlorophenyl)-carbamate ... 646
$C_{15}H_{12}ClNS$
10-Allyl-2-chlorophenothiazine ... 253
3-(4-Chlorophenylthio)-1-methylindole ... 881
$C_{15}H_{12}ClN_3O$
1-(*N*-(5-Chloro-2-methylphenyl)-carbamoyl)benzimidazole ... 901
1-(2-Chloro-6-methylphenyl)-3-(3-cyanophenyl)urea ... 759
1-(2-Chloro-6-methylphenyl)-3-(4-cyanophenyl)urea ... 760
1-(3-Chloro-4-methylphenyl)-3-(4-cyanophenyl)urea ... 760
1-(4-Chloro-2-methylphenyl)-3-(4-cyanophenyl)urea ... 760
1-(5-Chloro-2-methylphenyl)-3-(4-cyanophenyl)urea ... 760
$C_{15}H_{12}ClN_3OS$
1-(4-Chloro-2-cyanophenyl)-3-(4-(methylthio)phenyl)urea ... 751
$C_{15}H_{12}ClN_3O_2$
1-(*N*-(5-Chloro-2-methoxyphenyl)-carbamoyl)benzimidazole ... 901
$C_{15}H_{12}ClN_3O_2S$
1-(2-Benzothiazolyl)-3-(5-chloro-2-methoxyphenyl)urea ... 911
1-(4-Chlorophenyl)-3-(6-methoxy-2-benzothiazolyl)urea ... 911
$C_{15}H_{12}ClN_3O_4$
1-(4-Acetylphenyl)-3-(4-chloro-3-nitrophenyl)urea ... 523
$C_{15}H_{12}ClN_3O_5$
3-Acetamidophenyl *N*-(2-chloro-4-nitrophenyl)carbamate ... 626
4-Acetamidophenyl *N*-(2-chloro-4-nitrophenyl)carbamate ... 626
4-Acetamidophenyl *N*-(4-chloro-2-nitrophenyl)carbamate ... 624
4-Acetamidophenyl *N*-(4-chloro-3-nitrophenyl)carbamate ... 625
$C_{15}H_{12}ClN_5$
2-(2-Chloro-4-methyl-6-pyrimidyl)-2-(1-methyl-2-benzimidazolyl)-acetonitrile ... 975
$C_{15}H_{12}ClN_5O_5$
N-(2-Aminobenzylideneamino)-*N*′-(4-chloro-3-nitrophenyl)oxamide ... 492
$C_{15}H_{12}ClN_5S$
2-(4-Chloro-2-(methylthio)-6-pyrimidyl)-2-(1-methyl-2-benzimidazolyl)acetonitrile ... 974
$C_{15}H_{12}Cl_2FNO_3$
2-(2,4-Dichlorophenoxy)ethyl *N*-(4-fluorophenyl)carbamate ... 616
$C_{15}H_{12}Cl_2FNO_4$
2-Chloro-4-fluorophenyl *N*-(5-chloro-2,4-dimethoxyphenyl)carbamate ... 613
$C_{15}H_{12}Cl_2F_6O_5$
Ethyl 2,4-bis(2-chloro-1,1,2-trifluoroethoxy)benzoylacetate ... 363
$C_{15}H_{12}Cl_2INO$
6,9-Dichloro-2-methoxy-*N*-methylacridinium iodide ... 995
$C_{15}H_{12}Cl_2N_2O$
3,8-Dichloro-6-(2-hydroxyethylamino)phenanthridine ... 999
4-Methylbenzoic (2,4-dichlorobenzylidene)hydrazide ... 493
$C_{15}H_{12}Cl_2N_2O_2$
4′-Acetamido-2,4-dichlorobenzanilide ... 471
4′-Acetamido-3,4-dichlorobenzanilide ... 471
3,4-Dichlorobenzoic *N*2-(2-phenylacetyl)hydrazide ... 492
4-Methoxybenzoic (2,6-dichlorobenzylidene)hydrazide ... 1024
4′,4″-Methylenebis(2′-chloroformanilide) ... 440
$C_{15}H_{12}Cl_2N_2O_2S$
1-(2-Chlorobenzoyl)-3-(5-chloro-2-methoxyphenyl)-2-thiourea ... 731
1-(2,4-Dichlorobenzoyl)-3-(4-(methylthio)phenyl)urea ... 522
$C_{15}H_{12}Cl_2N_2O_3$
4-Acetamidophenyl *N*-(2,3-dichlorophenyl)carbamate ... 631
$C_{15}H_{12}Cl_2N_2O_4$
2′-Chloro-2-(4-chloro-2-methylphenoxy)-4′-nitroacetanilide monohydrate ... 401
5-Chloro-*N*-(2-chloro-5-nitrobenzylidene)-2,4-dimethoxy-aniline ... 1043
4-Chloro-2,5-dimethylphenyl *N*-(4-chloro-3-nitrophenyl)carbamate ... 625
2-(2,4-Dichlorophenoxy)-2′-methyl-5′-nitroacetanilide ... 399
2,6-Dimethyl-4-nitrophenyl *N*-(2,6-dichlorophenyl)carbamate ... 674
$C_{15}H_{12}Cl_2N_2O_6$
2-Chloro-4-nitrophenyl *N*-(5-chloro-2,4-dimethoxyphenyl)carbamate ... 614
$C_{15}H_{12}Cl_3N$
2-(*N*,*N*-Bis(chloromethyl)amino)-7-chlorofluorene ... 222
$C_{15}H_{12}Cl_3NO$
N-Benzyl-2,2,2-trichloroacetanilide ... 398
$C_{15}H_{12}Cl_3NOS$
2-(Benzylthio)-2′,4′,5′-trichloroacetanilide ... 403
$C_{15}H_{12}Cl_3NO_2$
5-Chloro-*N*-(2,4-dichlorobenzylidene)-2,4-dimethoxyaniline ... 1023
3′-Chloro-2-(2,4-dichlorophenoxy)-2′-methylacetanilide ... 400
3′-Chloro-2-(2,4-dichlorophenoxy)-4′-methylacetanilide ... 400
5′-Chloro-2-(2,4-dichlorophenoxy)-2′-methylacetanilide ... 400
2-Chloro-4,5-dimethylphenyl *N*-(2,3-dichlorophenyl)carbamate ... 637
4-Chloro-2,6-dimethylphenyl *N*-(2,3-dichlorophenyl)carbamate ... 636
4-Chloro-3,5-dimethylphenyl *N*-(2,3-dichlorophenyl)carbamate ... 636
2-(4-Chlorophenoxy)-4′,5′-dichloro-2′-methylacetanilide ... 400
2′,5′-Dichloro-2-(4-chloro-2-methylphenoxy)acetanilide ... 401
2′-Methyl-2-(2,4,5-trichlorophenoxy)-acetanilide ... 400
2,4,5-Trichlorophenyl *N*-(3,5-dimethylphenyl)carbamate ... 640
2,4,5-Trichlorophenyl *N*-(2,4-xylyl)-carbamate ... 640
$C_{15}H_{12}Cl_3NO_2S$
2-(4-Chlorophenylthio)ethyl *N*-(3,4-dichlorophenyl)carbamate ... 623
$C_{15}H_{12}Cl_3NO_3$
2-(2-Chlorophenoxy)ethyl *N*-(2,6-dichlorophenyl)carbamate ... 615
2-(2,4-Dichlorophenoxy)ethyl *N*-(2-chlorophenyl)carbamate ... 650
2-(2,4-Dichlorophenoxy)ethyl *N*-(3-chlorophenyl)carbamate ... 651
2-(2,4-Dichlorophenoxy)ethyl *N*-(4-chlorophenyl)carbamate ... 665
2′,5′-Dimethoxy-2,4,4′-trichlorobenzanilide ... 448
$C_{15}H_{12}Cl_3NO_4$
2,6-Dichlorophenyl *N*-(5-chloro-2,4-dimethoxyphenyl)carbamate ... 613
2,4,5-Trichlorophenyl *N*-(2,5-dimethoxyphenyl)carbamate ... 617
$C_{15}H_{12}Cl_3N_3O_2$
1-(4-Acetamidophenyl)-3-(2,4,5-trichlorophenyl)urea ... 526
$C_{15}H_{12}Cl_3N_3O_3$
4-(5-Chloro-2-methoxyphenyl)-1-(2,4-dichlorobenzoyl)semicarbazide ... 518
$C_{15}H_{12}Cl_4N_2O$
1-(3-Chloro-4-methylphenyl)-3-(4-(trichloromethyl)phenyl)urea ... 506
$C_{15}H_{12}FN$
N-Cinnamylidene-4-fluoroaniline ... 1027
$C_{15}H_{12}FNO$
2-Acetamido-5-fluorofluorene ... 490
3′-Fluorocinnamanilide ... 456
$C_{15}H_{12}FNO_2$
2′-Acetyl-2-fluorobenzanilide ... 453
3′-Acetyl-3-fluorobenzanilide ... 453
3′-Acetyl-4-fluorobenzanilide ... 453
2-Benzoyl-4-fluoro-*N*-methylbenzamide ... 478
$C_{15}H_{12}FNO_3$
4-Acetylphenyl *N*-(3-fluorophenyl)-carbamate ... 629
2-Fluorophenyl *N*-(3-acetylphenyl)-carbamate ... 629
Methyl 2-(3-fluorobenzamido)-benzoate ... 456
Methyl 2-(4-fluorophenylcarboxamido)-benzoate ... 456
$C_{15}H_{12}FNO_4$
2-(Methoxycarbonyl)phenyl *N*-(4-fluorophenyl)carbamate ... 622
4-(Methoxycarbonyl)phenyl *N*-(2-fluorophenyl)carbamate ... 629
$C_{15}H_{12}FNO_4S$
3-(2-Benzoylacetamido)-benzenesulfonyl fluoride ... 803
$C_{15}H_{12}FN_3O_2S$
1-(3-Fluorophenyl)-3-(6-methoxy-2-benzothiazolyl)urea ... 911
$C_{15}H_{12}FN_3O_4$
1-(3-Acetylphenyl)-3-(4-fluoro-3-nitrophenyl)urea ... 524
$C_{15}H_{12}FN_3O_5$
4-Acetamidophenyl *N*-(4-fluoro-3-nitrophenyl)carbamate ... 625
Methyl 2-(3-(4-fluoro-3-nitrophenyl)-ureido)benzoate ... 524

$C_{15}H_{12}FN_5O$
1-(3-Fluorophenyl)-3-(5-phenyl-1,2,4-triazol-3-yl)urea ... 869
$C_{15}H_{12}F_2N_2O_2$
1-(3-Acetylphenyl)-3-(2,5-difluorophenyl)urea ... 530
1-(4-Acetylphenyl)-3-(2,5-difluorophenyl)urea ... 530
$C_{15}H_{12}F_2N_2O_5$
2-(4-Nitrophenoxy)ethyl *N*-(2,4-difluorophenyl)carbamate ... 615
2-(4-Nitrophenoxy)ethyl *N*-(2,5-difluorophenyl)carbamate ... 615
$C_{15}H_{12}F_3N$
N-(4-Methylbenzylidene)-3-(trifluoromethyl)aniline ... 1021
$C_{15}H_{12}F_3NO$
N-(4-Methoxybenzylidene)-3-(trifluoromethyl)aniline ... 1020
2-Phenyl-3′-(trifluoromethyl)-acetanilide ... 397
$C_{15}H_{12}F_3NOS$
S-Benzyl *N*-(2-(trifluoromethyl)-phenyl)thiocarbamate ... 611
S-(*o*-Tolyl) *N*-(3-(trifluoromethyl)-phenyl)thiocarbamate ... 611
S-(*p*-Tolyl) *N*-(3-(trifluoromethyl)-phenyl)thiolcarbamate ... 611
$C_{15}H_{12}F_3NO_2$
Benzyl *N*-(3-trifluoromethylphenyl)-carbamate ... 609
4-Methoxy-3′-(trifluoromethyl)-benzanilide ... 446
2-Methoxy-α-(3-(trifluoromethyl)-phenylimino)-*o*-cresol ... 1020
2-Methoxy-α-(3-(trifluoromethyl)-phenylimino)-*p*-cresol ... 1020
3-(Trifluoromethyl)phenyl *N*-(*m*-tolyl)carbamate ... 610
$C_{15}H_{12}F_3NO_2S$
4-(Methylthio)phenyl *N*-(3-(trifluoromethyl)phenyl)carbamate ... 609
$C_{15}H_{12}F_3NO_3$
4-Methoxyphenyl *N*-(3-(trifluoromethyl)phenyl)carbamate ... 609
$C_{15}H_{12}F_3N_3O_3$
1-(4-Carboxyphenyl)-4-(2-(trifluoromethyl)phenyl)-semicarbazide ... 509
1-(2-Hydroxybenzoyl)-4-(2-(trifluoromethyl)phenyl)-semicarbazide ... 507
1-(4-Hydroxybenzoyl)-4-(2-(trifluoromethyl)phenyl)-semicarbazide ... 507
1-(2-Methyl-5-nitrophenyl)-3-(2-(trifluoromethyl)phenyl)urea ... 508
1-(2-Methyl-5-nitrophenyl)-3-(3-(trifluoromethyl)phenyl)urea ... 508
$C_{15}H_{12}F_3N_3O_3S$
1-(4-(Methylthio)phenyl)-3-(4-nitro-2-(trifluoromethyl)phenyl)urea ... 508
$C_{15}H_{12}F_3N_3O_4$
1-(2-Methoxy-4-nitrophenyl)-3-(3-trifluoromethylphenyl)urea ... 508
1-(2-Methoxy-5-nitrophenyl)-3-(2-(trifluoromethyl)phenyl)urea ... 507
1-(2-Methoxy-5-nitrophenyl)-3-(3-(trifluoromethyl)phenyl)urea ... 507
$C_{15}H_{12}F_6N_2$
2-(2-(3,5-Bis(trifluoromethyl)anilino)-ethyl)pyridine ... 925
$C_{15}H_{12}F_9FeO_6$
Ferric trifluoroacetylacetonate ... 70
$C_{15}H_{12}INO$
2-Acetamido-7-iodofluorene ... 491
$C_{15}H_{12}I_2O_3$
3-(3,5-Diiodo-4-hydroxyphenyl)-2-phenylpropionic acid ... 326
$C_{15}H_{12}I_2O_4$
3,5-Diiodo-4-(4-hydroxyphenoxy)-hydrocinnamic acid ... 325
$C_{15}H_{12}N_2$
8-Anilinoquinoline ... 988
9-Carbazolepropionitrile ... 891
3,5-Diphenyl-1*H*-pyrazole ... 864
3-Methyl-4-phenylcinnoline ... 1001
2-Styrylbenzimidazole ... 908
$C_{15}H_{12}N_2O$
3-(Benzylimino)-2-indolinone ... 1063
N-(4-Cyanobenzylidene)-*p*-anisidine ... 1050
4,5-Diphenyl-4-imidazolin-2-one ... 867
4,5-Diphenyl-2(3*H*)-imidazolone ... 582
1,3-Diphenyl-5-pyrazolone ... 706
2-(3-Methylphenyl)-5-phenyl-1,3,4-oxadiazole ... 873
2-Methyl-3-phenyl-4(3*H*)-quinazolinone ... 707
$C_{15}H_{12}N_2O_2$
4-Cyanophenyl *N*-(*o*-tolyl)-carbamate ... 767
4-Cyanophenyl *N*-(*m*-tolyl)-carbamate ... 767
4-Cyanophenyl *N*-(*p*-tolyl)-carbamate ... 767
2,7-Diformamidofluorene ... 490
3,4-Dihydro-8-methyl-3-phenyl-2*H*-pyrido(1,2-a)pyrimidine-2,4-dione ... 710
3,5-Diphenylhydantoin ... 704
Disperse Violet 4 (C.I. 61105) ... 1093
α-(4-Hydroxyimino-2,5-cyclohexadien-1-ylidene)-4-methoxyphenylacetonitrile ... 1009
4-(4-Hydroxyphenyl)-2-methyl-1(2*H*)-phthalazinone ... 697
Methyl 2-phenylimidazo(1,2-a)-pyridine-7-carboxylate ... 919
$C_{15}H_{12}N_2O_2S_2$
4-(2-Thioxobenzothiazolin-3-ylmethylamino)benzoic acid ... 741
N-(2-Thioxobenzothiazolin-3-ylmethyl)-anthranilic acid ... 741
$C_{15}H_{12}N_2O_3$
3-Acetamido-2-nitrofluorene ... 487
4-Cyanophenyl *N*-(2-methoxyphenyl)carbamate ... 762
4-Cyanophenyl *N*-(3-methoxyphenyl)carbamate ... 762
4-Cyanophenyl *N*-(4-methoxyphenyl)carbamate ... 762
1,2-Dihydro-2-(3,4-(methylenedioxy)-phenyl)-4(3*H*)-quinazolinone ... 700
Disperse Red 11 (C.I. 62015) ... 1093
Hydrofuramide ... 1075
2,4,5-Tri-(2-furyl)-2-imidazoline ... 852
$C_{15}H_{12}N_2O_4$
2-Acetamido-9-hydroxy-3-nitrofluorene ... 485
2-Benzoyl-4′-nitroacetanilide ... 443
9,9-Dimethyl-2,7-dinitrofluorene ... 280
2′-Hydroxy-5′-nitrocinnamanilide ... 462
2-Methyl-4-nitro-*N*-(piperonylidene)-aniline ... 1051
2-Methyl-5-nitro-*N*-(piperonylidene)-aniline ... 1051
3′-Nitro-2-benzoylacetanilide ... 443
$C_{15}H_{12}N_2O_4S$
4-(2,3-Dihydro-2-thioxo-3-benzoxazolylmethylamino)-salicylic acid ... 734
5-(2,3-Dihydro-2-thioxo-3-benzoxazolylmethylamino)-salicylic acid ... 734
$C_{15}H_{12}N_2O_5$
2,4-Dinitro-2′-methoxystilbene ... 267
2,4-Dinitro-3′-methoxystilbene ... 267
2,4-Dinitro-4′-methoxystilbene ... 267
2-Methoxy-5-nitro-*N*-piperonylideneaniline ... 1043
Methyl 4-(*N*-(3-nitrophenyl)-carbamoyl)benzoate ... 469
N-(2-Methyl-4-nitrophenyl)-phthalamic acid ... 470
4′-Methyl-3′-nitrophthalanilic acid ... 469
$C_{15}H_{12}N_2O_5S$
2,3-Dihydro-1,1-dioxo-*o*-(2-nitrophenyl)-4*H*-benzothiopyran-4-one oxime ... 1011
$C_{15}H_{12}N_2O_6$
Ethyl 4,6-dinitro-2-biphenylcarboxylate ... 372
4-(Methoxycarbonyl)phenyl *N*-(2-nitrophenyl)carbamate ... 668
4-(Methoxycarbonyl)phenyl *N*-(3-nitrophenyl)carbamate ... 669
$C_{15}H_{12}N_2S$
2-(4-Aminophenylthio)quinoline ... 987
2-(4-Aminophenylthio)quinoline hydrochloride ... 987
10-Phenothiazinepropionitrile ... 756
$C_{15}H_{12}N_3Sb$
Tris(2-pyridyl)antimony ... 1129
$C_{15}H_{12}N_4$
Di-2-benzimidazolylmethane ... 907
2-Pyridinecarboxaldehyde 2-quinolinyl-hydrazone ... 991
$C_{15}H_{12}N_4O$
1-(1-Naphthyl)-3-(2-pyrimidinyl)urea ... 976
Picolinic acid (3-indolylmethylene)-hydrazide ... 945
$C_{15}H_{12}N_4O_2S$
3-Methyl-2-benzothiazolinone (4-nitrobenzylidene)hydrazone ... 1061
$C_{15}H_{12}N_4O_2S_2$
*N*4-(2-Pyridylmethylene)-*N*1-(2-thiazolyl)-sulfanilamide ... 1071
$C_{15}H_{12}N_4O_3$
1,2-Dihydro-4,6-dimethyl-1-(2-nitrobenzylideneamino)-2-oxonicotinonitrile ... 1069
1,2-Dihydro-4,6-dimethyl-1-(3-nitrobenzylideneamino)-2-oxonicotinonitrile ... 1069
2,3-Dihydro-3-methyl-2-benzoxazolone (4-nitrobenzylidene)hydrazone ... 1061
$C_{15}H_{12}N_4O_3S$
1-(*N*-(4-(Methylthio)phenyl)-carbamoyl)-6-nitrobenzimidazole ... 903
$C_{15}H_{12}N_4O_8$
Ethyl 4-(picrylamino)benzoate ... 372
$C_{15}H_{12}N_4S_2$
1-Phenyl-3-(5-phenyl-1,2,4-thiadiazol-3-yl)-2-thiourea ... 879
$C_{15}H_{12}N_6S_2$
2,3-Dihydro-3-methyl-2-(1-phenyl-1*H*-tetrazol-5-ylthioimino)-benzothiazole ... 870
$C_{15}H_{12}O$
2-Methoxyphenanthrene ... 189
1-(4-Methoxyphenyl)-2-phenylacetylene ... 1017
4-Phenanthrenemethanol ... 208
3-Phenyl-1-indanone ... 297
2-Stilbenecarboxaldehyde ... 324
$C_{15}H_{12}OS$
2-Phenyl-2*H*-1-benzothiopyran-4(3*H*)-one ... 296
1-Phenyl-5-(2-thienyl)-1,4-pentadien-3-one ... 857
5-Phenyl-1-(2-thienyl)-2,4-pentadien-1-one ... 857
$C_{15}H_{12}OS_2$
Ethyl 2-dibenzofurancarbodithioate ... 895
$C_{15}H_{12}O_2$
3-Allyl-4-hydroxydibenzofuran ... 893
9,10-Dihydro-9-anthroic acid ... 334
1,2-Dihydro-4-phenanthroic acid ... 346
3,4-Dihydro-3-phenyl-1*H*-2-benzopyran-1-one ... 384
1,3-Diphenyl-2,3-epoxy-1-propanone ... 293
α,β-Epoxychalcone ... 293
1-Fluorenyl acetate ... 381
2-Fluorenyl acetate ... 381
3-Fluorenyl acetate ... 381
2′-Hydroxychalcone ... 300
2-Hydroxychalcone ... 300
4-Hydroxychalcone ... 282
4′-Hydroxychalcone ... 300
3-Methoxy-2-anthracenol ... 195
3-Methyl-3-phenylphthalide ... 384
Phenyl cinnamate ... 380
cis-α-Phenylcinnamic acid ... 336
β-Phenylcinnamic acid ... 336
2-Phenylcinnamic acid ... 336
3-(1,2-Propadien-1-yl)-2,2′-biphenol ... 194
4-Stilbenecarboxylic acid ... 345
3-(*p*-Tolyl)phthalide ... 386
$C_{15}H_{12}O_2S$
2-Dibenzothiophenepropionic acid ... 898
2-Phenylethynyl *p*-tolyl sulfone ... 1019
$C_{15}H_{12}O_3$
2-Acetylphenyl benzoate ... 379
3-Acetylphenyl benzoate ... 379
4-Acetylphenyl benzoate ... 379
3-Benzoyl-4-methylbenzoic acid ... 344
2,3-Dihydro-3-phenyl-2-benzofurancarboxylic acid ... 329
Ethyl 2-dibenzofurancarboxylate ... 895
4-Hydroxy-α-phenylcinnamic acid ... 334
4-Hydroxy-α-phenylcinnamic acid, sodium salt monohydrate ... 350
2-Isopropylnaphtho(1,2-b)furan-4,5-dione ... 320
3-Methyl-1,8,9-anthracenetriol ... 195
Methyl 2-benzoylbenzoate ... 371
Phenyl 3-oxo-3-phenylpropionate ... 379
$C_{15}H_{12}O_4$
mono-Benzyl phthalate ... 375
3-Cinnamoyl-4-hydroxy-6-methyl-2-pyrone ... 382
Methyl benzoylsalicylate ... 380
$C_{15}H_{12}O_4S$
3-Benzoyl-3,3a-dihydrocyclohept(d)-1,2-oxathiole-2,2-dioxide ... 799
$C_{15}H_{12}O_5$
2-(2-Benzoyloxyphenoxy)-acetic acid ... 378
2′-Carboxy-2-hydroxy-4-methoxybenzophenone ... 338
Methyl 5-(benzoyloxy)salicylate ... 377
$C_{15}H_{12}O_6$
2,4,6-Trihydroxybenzoylmethyl benzoate ... 367
$C_{15}H_{13}BrClFN_2O$
1-(2-Bromo-4,5-dimethylphenyl)-3-(3-chloro-4-fluorophenyl)urea ... 544
$C_{15}H_{13}BrClNOS$
2-(Benzylthio)-4′-bromo-2′-chloroacetanilide ... 402
4′-Bromo-2-(4-chlorophenylthio)-2′-methylacetanilide ... 403
4′-Bromo-2-(4-chlorophenylthio)-3′-methylacetanilide ... 403
$C_{15}H_{13}BrClNO_2$
2-Bromo-4-chloro-6-methylphenyl *N*-(3-methylphenyl)carbamate ... 640
2-Bromo-4,5-dimethylphenyl *N*-(3-chlorophenyl)carbamate ... 631
4-Bromo-2,5-dimethylphenyl *N*-(3-chlorophenyl)carbamate ... 631
4-Bromo-3,5-dimethylphenyl *N*-(2-chlorophenyl)carbamate ... 634
4-Chloro-3,5-dimethylphenyl *N*-(3-bromophenyl)carbamate ... 640
$C_{15}H_{13}BrClNO_3$
2-Bromo-4-chloro-6-methylphenyl *N*-(2-methoxyphenyl)carbamate ... 618
$C_{15}H_{13}BrFNO$
4′-Bromo-2′,6′-dimethyl-4-fluorobenzanilide ... 459
$C_{15}H_{13}BrFNO_2$
2-Bromo-4,5-dimethylphenyl *N*-(4-fluorophenyl)carbamate ... 634
4-Bromo-2,5-dimethylphenyl *N*-(2-fluorophenyl)carbamate ... 633
4-Bromo-2,5-dimethylphenyl *N*-(4-fluorophenyl)carbamate ... 635
$C_{15}H_{13}BrFNO_3$
2-(4-Bromophenoxy)ethyl *N*-(4-fluorophenyl)carbamate ... 615
$C_{15}H_{13}BrN_2$
N-(2-Bromo-3-(phenylimino)-propenyl)aniline ... 1054
N-(2-Bromo-3-(phenylimino)-propenyl)-aniline hydrobromide ... 1060
$C_{15}H_{13}BrN_2O$
2-(4-Bromophenoxymethyl)-1-methylbenzimidazole ... 901
$C_{15}H_{13}BrN_2O_3$
4′-Bromo-2′,6′-dimethyl-3-nitrobenzanilide ... 453
3-Hydroxybenzoic acid (5-bromo-2-methoxybenzylidene)hydrazide ... 495
2-Methoxybenzoic (5-bromo-2-hydroxybenzylidene)hydrazide ... 491
$C_{15}H_{13}BrN_2O_3S$
Cyssor(I) ... 1079
$C_{15}H_{13}BrOS$
4′-Bromo-2-(4-tolylthio)-acetophenone ... 288

$C_{15}H_{13}BrO_2$
2-Bromo-4′-methoxy-2-phenylacetophenone ... 292
2-(2-Bromo-4-methylphenoxy)-acetophenone ... 288
Methyl 2-bromo-2,2-diphenylacetate ... 362

$C_{15}H_{13}BrO_2S$
4-Bromophenyl 2-phenylcyclopropyl sulfone ... 774

$C_{15}H_{13}BrO_3$
4′-Bromo-2-(2-methoxyphenoxy)-acetophenone ... 288

$C_{15}H_{13}BrO_4$
8-Bromo-3,4-dihydro-2,2-dimethyl-3-hydroxynaphtho(1,2-b)-2*H*-pyran-5,6-dione ... 319

$C_{15}H_{13}BrO_4S$
4′-Bromophenacyl *p*-toluenesulfonate ... 796

$C_{15}H_{13}Br_2NO_2$
2,4-Dibromophenyl *N*-(2,5-dimethylphenyl)carbamate ... 638

$C_{15}H_{13}Br_2NO_4$
2,4-Dibromophenyl *N*-(2,4-dimethoxyphenyl)carbamate ... 613
2,4-Dibromophenyl *N*-(2,5-dimethoxyphenyl)carbamate ... 614

$C_{15}H_{13}ClFNO_2$
5-Chloro-2,4-dimethoxy-*N*-(4-fluorobenzylidene)-aniline ... 1050
4-Chloro-2,6-dimethylphenyl *N*-(4-fluorophenyl)carbamate ... 635
2-Chloro-4-fluorophenyl *N*-(2,3-dimethylphenyl)carbamate ... 636
2-Chloro-4-fluorophenyl *N*-(2,6-xylyl)carbamate ... 636
2-(4-Chloro-2-methylphenoxy)-2′-fluoroacetanilide ... 400
2-(4-Chloro-2-methylphenoxy)-4′-fluoroacetanilide ... 400

$C_{15}H_{13}ClFNO_3$
5′-Chloro-2′,4′-dimethoxy-4-fluorobenzanilide ... 447
2-Fluorobenzyl *N*-(5-chloro-2-methoxyphenyl)carbamate ... 612
4-Fluorobenzyl *N*-(5-chloro-2-methoxyphenyl)carbamate ... 612

$C_{15}H_{13}ClFNO_4$
2-Chloro-4-fluorophenyl *N*-(2,4-dimethoxyphenyl)carbamate ... 613
2-Chloro-4-fluorophenyl *N*-(2,5-dimethoxyphenyl)carbamate ... 614
2-Chloro-4-fluorophenyl *N*-(3,5-dimethoxyphenyl)carbamate ... 614
2-Fluorophenyl *N*-(5-chloro-2,4-dimethoxyphenyl)carbamate ... 613
4-Fluorophenyl *N*-(5-chloro-2,4-dimethoxyphenyl)carbamate ... 613

$C_{15}H_{13}ClFN_3O_2$
1-(4-Acetamidophenyl)-3-(3-chloro-4-fluorophenyl)urea ... 526

$C_{15}H_{13}ClFN_3O_3$
3-(2-Chloro-4-nitrophenyl)-1-ethyl-1-(3-fluorophenyl)urea ... 526

$C_{15}H_{13}ClFN_3O_5$
1-(5-Chloro-2,4-dimethoxyphenyl)-3-(4-fluoro-3-nitrophenyl)urea ... 514

$C_{15}H_{13}ClF_2N_2O_2S$
1-(5-Chloro-2,4-dimethoxyphenyl)-3-(2,4-difluorophenyl)-2-thiourea ... 729

$C_{15}H_{13}ClF_2N_2O_3$
1-(4-Chloro-2,5-dimethoxyphenyl)-3-(2,5-difluorophenyl)urea ... 516
1-(5-Chloro-2,4-dimethoxyphenyl)-3-(2,4-difluorophenyl)urea ... 515
1-(5-Chloro-2,4-dimethoxyphenyl)-3-(2,5-difluorophenyl)urea ... 515

$C_{15}H_{13}ClF_3NO_2$
α-(2-Chloro-5-(trifluoromethyl)-anilino)-6-methoxy-*o*-cresol ... 215

$C_{15}H_{13}ClF_3N_3O$
4-(2-Chloro-5-(trifluoromethyl)-phenyl)-1-methyl-1-phenylsemicarbazide ... 505

$C_{15}H_{13}ClINO_2$
4-Iodo-2,5-dimethylphenyl *N*-(3-chlorophenyl)carbamate ... 631

$C_{15}H_{13}ClN_2$
1-(4-Chlorobenzyl)-2-methylbenzimidazole hydrochloride ... 904
7-Chloro-2,3-dihydro-5-phenyl-1*H*-1,4-benzodiazepine ... 250
6-Chloro-8-(2,5-dimethyl-1-pyrrolyl)-quinoline ... 988
4-(4-Chlorophenyl)-1,2-dihydro-2-methylphthalazine ... 245

$C_{15}H_{13}ClN_2O$
3-(5-Chloro-2-(methylamino)phenyl)-phthalide ... 689
2-(4-Chlorophenoxymethyl)-1-methylbenzimidazole ... 901
4′-(3-Chlorophenyliminomethyl)-acetanilide ... 1027
p-Toluic (2-chlorobenzylidene)-hydrazide ... 1025
p-Toluic (4-chlorobenzylidene)-hydrazide ... 493

$C_{15}H_{13}ClN_2O_2$
4′-Acetamido-2-chlorobenzanilide ... 471
4′-Acetamido-4-chlorobenzanilide ... 471
1-(2-Benzoyl-4-chlorophenyl)-3-methylurea ... 530
1-Benzoyl-3-(3-chloro-*o*-tolyl)-urea ... 483
2-Chlorobenzoic (4-methoxybenzylidene)hydrazide ... 492
4-Chlorobenzoic N^2-(2-phenylacetyl)hydrazide ... 492
4-Chloro-2′-(*N*-methylcarbamoyl)-benzanilide ... 457
N-(4-Chloro-3-nitrobenzylidene)-4-ethylaniline ... 1025
N-(2-Chloro-5-nitrobenzylidene)-2,3-xylidine ... 1058
N-(2-Chloro-5-nitrobenzylidene)-3,4-xylidine ... 1058
N-(2-Chloro-5-nitrobenzylidene)-3,5-xylidine ... 1058
4-(3-Chlorophenyliminomethyl)-phenyl *N*-methylcarbamate ... 1026
2-Phenoxyacetic (4-chlorobenzylidene)hydrazide ... 1023

$C_{15}H_{13}ClN_2O_3$
4-Acetamidophenyl *N*-(2-chlorophenyl)carbamate ... 631
2-Carbamoylphenyl *N*-(2-chloro-6-methylphenyl)carbamate ... 622
1-(2-Carboxyphenyl)-3-(2-chloro-6-methylphenyl)urea ... 531
1-(2-Carboxyphenyl)-3-(4-chloro-2-methylphenyl)urea ... 531
1-(4-Carboxyphenyl)-3-(5-chloro-2-methylphenyl)urea ... 531
4-(4-Chlorobenzamido)phenyl *N*-methylcarbamate ... 630
4-Chlorobenzoic (4-hydroxy-3-methoxybenzylidene)hydrazide ... 491
3-(3-(2-Chloro-6-methylphenyl)-ureido)benzoic acid ... 531
N-(2-Chloro-5-nitrobenzylidene)-*o*-phenetidine ... 1043

$C_{15}H_{13}ClN_2O_3S$
2-(Benzylthio)-4′-chloro-3′-nitroacetanilide ... 402

$C_{15}H_{13}ClN_2O_4$
2-Carbamoylphenyl *N*-(2-chloro-4-methoxyphenyl)carbamate ... 617
2-Carbamoylphenyl *N*-(5-chloro-2-methoxyphenyl)carbamate ... 618
5-Chloro-2,4-dimethoxy-*N*-(3-nitrobenzylidene)-aniline ... 1044
4-Chloro-3,5-dimethyl *N*-(2-nitrophenyl)carbamate ... 624
4-(3-(5-Chloro-2-methoxyphenyl)-ureido)benzoic acid ... 519
4-Chloro-3-methylphenyl *N*-(4-methyl-2-nitrophenyl)carbamate ... 624
4-(3-(2-Chloro-6-methylphenyl)-ureido)salicylic acid ... 513
2-(4-Chlorophenoxy)-4′-methyl-2′-nitroacetanilide ... 399
2-(4-Chlorophenxoy)-2′-methyl-5′-nitroacetanilide ... 399
2,6-Dimethyl-4-nitrophenyl *N*-(4-chlorophenyl)carbamate ... 628
2,5-Dimethylphenyl *N*-(2-chloro-4-nitrophenyl)carbamate ... 626
2-(*N*-(Hydroxymethyl)carbamoyl)-phenyl *N*-(4-chlorophenyl)-carbamate ... 611
4-Methylbenzyl *N*-(4-chloro-3-nitrophenyl)carbamate ... 625
3-Methyl-4-nitrophenyl *N*-(3-chloro-2-methylphenyl)carbamate ... 628
3-Methyl-4-nitrophenyl *N*-(3-chloro-4-methylphenyl)carbamate ... 628

$C_{15}H_{13}ClN_2O_5$
2-Chloro-4-methoxyphenyl *N*-(2-methyl-5-nitrophenyl)carbamate ... 617
4-Chloro-2-methyl-6-nitrophenyl *N*-(4-methoxyphenyl)carbamate ... 620
4-Chloro-2-methylphenyl *N*-(2-methoxy-4-nitrophenyl)carbamate ... 618
3-(*N*-(4-Chloro-3-nitrophenyl)-carbamoyl)-5-norbornene-2-carboxylic acid ... 404
4-Methoxybenzyl *N*-(2-chloro-5-nitrophenyl)carbamate ... 612
2-Phenoxyethyl *N*-(2-chloro-5-nitrophenyl)carbamate ... 615
2-(Phenoxy)ethyl *N*-(4-chloro-3-nitrophenyl)carbamate ... 615

$C_{15}H_{13}ClN_2O_5S$
2-(4-Chlorophenyl)-1-methylethyl 2,4-dinitrobenzenesulfenate ... 771

$C_{15}H_{13}ClN_2O_6$
2-Chloro-4-nitrophenyl *N*-(2,5-dimethoxyphenyl)carbamate ... 614
2,6-Dimethoxyphenyl *N*-(4-chloro-3-nitrophenyl)carbamate ... 612

$C_{15}H_{13}ClN_2O_6S$
Ethyl 4-(4-chloro-3-nitrobenzenesulfonamido)-benzoate ... 807

$C_{15}H_{13}ClN_4O$
5-Chloro-1-(*N*-(3,4-xylyl)carbamoyl)-1*H*-benzotriazole ... 909

$C_{15}H_{13}ClN_4O_5$
1-(2-Chloro-4-nitrophenyl)-3-(2,5-dimethyl-4-nitrophenyl)urea ... 526

$C_{15}H_{13}ClN_6O_4$
2-(3-(2-Chloro-4-nitrophenyl)ureido)-acetic (2-pyridylmethylene)-hydrazide ... 928
2-(3-(2-Chloro-4-nitrophenyl)ureido)-acetic (4-pyridylmethylene)-hydrazide ... 928

$C_{15}H_{13}ClO$
2-Chloro-2′,5′-dimethylbenzophenone ... 310
4′-Chloro-4-methyldesoxybenzoin ... 297

$C_{15}H_{13}ClOS$
4′-Chloro-2-(4-tolylthio)-acetophenone ... 288

$C_{15}H_{13}ClO_2$
Benzyl α-chlorophenylacetate ... 356
Methyl 2-chloro-2,2-diphenylacetate ... 362

$C_{15}H_{13}Cl_2IN_2$
6-Chloro-3-(4-chlorophenyl)-3,4-dihydro-1-methylquinazolinium iodide ... 255

$C_{15}H_{13}Cl_2N$
N-(2,4-Dichlorobenzylidene)-2,3-dimethylaniline ... 1076
N-(2,4-Dichlorobenzylidene)-2,3-xylidine ... 1067
N-(2,4-Dichlorobenzylidene)-2,4-xylidine ... 1067
N-(2,4-Dichlorobenzylidene)-2,5-xylidine ... 1027
N-(2,4-Dichlorobenzylidene)-3,4-xylidine ... 1027
N-(2,4-Dichlorobenzylidene)-3,5-xylidine ... 1027
N-(3,4-Dichlorobenzylidene)-2,3-xylidine ... 1067
N-(3,4-Dichlorobenzylidene)-2,5-xylidine ... 1067
N-(3,4-Dichlorobenzylidene)-3,5-xylidine ... 1068

$C_{15}H_{13}Cl_2NO$
2,4-Dichlorobenzo-2′,3′-xylidide ... 481
2,4-Dichlorobenzo-2′,4′-xylidide ... 481
2,4-Dichlorobenzo-2′,5′-xylidide ... 482
2,4-Dichlorobenzo-3′,4′-xylidide ... 481
2,6-Dichlorobenzo-2′,6′-xylidide ... 459
N-(2,4-Dichlorobenzylidene)-*o*-phenetidine ... 1046
N-(2,4-Dichlorobenzylidene)-*p*-phenetidine ... 1047
2,4-Dichloro-2′,6′-dimethylbenzanilide ... 459
2,5-Dichloro-2′,6′-dimethylbenzanilide ... 459
3,4-Dichloro-2′,5′-dimethylbenzanilide ... 460
2,5-Dichloro-*N*-(2-ethoxybenzylidene)aniline ... 1024
2,4-Dichloro-4′-ethylbenzanilide ... 459
3,4-Dichloro-4′-ethylbenzanilide ... 459
2,4-Dichloro-*N*-phenethylbenzamide ... 471

$C_{15}H_{13}Cl_2NOS$
2-(Benzylthio)-2′,5′-dichloroacetanilide ... 402
2-(Benzylthio)-3′,4′-dichloroacetanilide ... 403
3′-Chloro-2-(4-chlorophenylthio)-4′-methylacetanilide ... 403
5′-Chloro-2-(4-chlorophenylthio)-2′-methylacetanilide ... 403
2-(4-Chlorophenylthio)-4′-chloro-2′-methylacetanilide ... 403

$C_{15}H_{13}Cl_2NO_2$
4′-Chloro-2-(4-chloro-2-methylphenoxy)acetanilide ... 400
2′-Chloro-2-(4-chlorophenoxy)-6′-methylacetanilide ... 400
3′-Chloro-2-(2-chlorophenoxy)-2′-methylacetanilide ... 400
3′-Chloro-2-(4-chlorophenoxy)-2′-methylacetanilide ... 399
3′-Chloro-2-(4-chlorophenoxy)-4′-methylacetanilide ... 400
4′-Chloro-2-(4-chlorophenoxy)-2′-methylacetanilide ... 399
5′-Chloro-2-(4-chlorophenoxy)-2′-methylacetanilide ... 399
4-Chloro-2-methylphenyl *N*-(3-chloro-4-methylphenyl)carbamate ... 638
4-Chloro-2-methylphenyl *N*-(5-chloro-2-methylphenyl)carbamate ... 638
4-Chloro-3-methylphenyl *N*-(3-chloro-4-methylphenyl)carbamate ... 638
2,4-Dichlorobenzo-*p*-phenetidide ... 467
N-(3,4-Dichlorobenzylidene)-2,5-dimethoxyaniline ... 1045
2,5-Dichloro-*N*-(2,5-dimethoxybenzylidene)aniline ... 1048
2,4-Dichloro-5′-methyl-*o*-benzanisidide ... 483
3,4-Dichloro-5′-methyl-*o*-benzanisidide ... 467
3,4-Dichloro-5′-methyl-*m*-benzanisidide ... 450
2,5-Dichlorophenyl *N*-(2,3-dimethylphenyl)carbamate ... 636
2,5-Dichlorophenyl *N*-(2,6-dimethylphenyl)carbamate ... 636
2,5-Dichlorophenyl *N*-(3,5-dimethylphenyl)carbamate ... 639
3,4-Dichlorophenyl *N*-(2,3-dimethylphenyl)carbamate ... 636
2,5-Dichlorophenyl *N*-(3-ethylphenyl)carbamate ... 638
2,4-Dichlorophenyl *N*-(2,5-xylyl)-carbamate ... 638
3,5-Dichlorophenyl *N*-(2,3-xylyl)-carbamate ... 636
2,3-Dichloro-*N*-(veratrylidene)-aniline ... 1042
2,5-Dichloro-*N*-(veratrylidene)-aniline ... 1042
3,4-Dichloro-*N*-(veratrylidene)-aniline ... 1042
3,4-Dimethylphenyl *N*-(2,5-dichlorophenyl)carbamate ... 638
Phenethyl *N*-(2,6-dichlorophenyl)-carbamate ... 632

$C_{15}H_{13}Cl_2NO_2S$
3-Methyl-4-(methylthio)phenyl *N*-(3,4-dichlorophenyl)carbamate..... 623
$C_{15}H_{13}Cl_2NO_3$
4′-Chloro-2-(4-chlorophenoxy)-2′-methoxyacetanilide.......... 399
5′-Chloro-2-(4-chlorophenoxy)-2′-methoxyacetanilide.......... 400
2-Chloro-4-methoxyphenyl *N*-(5-chloro-2-methylphenyl)carbamate... 617
2,4-Dichloro-2′,4′-dimethoxybenzanilide.......... 448
2,4-Dichloro-2′,5′-dimethoxybenzanilide.......... 448
2′,6′-Dichloro-2,6-dimethoxybenzanilide.......... 448
3,4-Dichloro-2′,5′-dimethoxybenzanilide.......... 448
2-(2,4-Dichlorophenoxy)ethyl *N*-phenylcarbamate.......... 650
3-(*N*-(3,4-Dichlorophenyl)-carbamoyl)-5-norbornene-2-carboxylic acid.......... 406
2,3-Dichlorophenyl *N*-(2-methoxy-5-methylphenyl)carbamate.......... 619
2,5-Dichlorophenyl *N*-(2-methoxy-5-methylphenyl)carbamate.......... 619
$C_{15}H_{13}Cl_2NO_4$
2-Chloro-4-methoxyphenyl *N*-(4-chloro-2-methoxyphenyl)-carbamate.......... 614
2-Chloro-4-methoxyphenyl *N*-(5-chloro-2-methoxyphenyl)-carbamate.......... 614
2,3-Dichlorophenyl *N*-(2,5-dimethoxyphenyl)carbamate.......... 623
3,4-Dichlorophenyl *N*-(2,5-dimethoxyphenyl)carbamate.......... 614
3,5-Dichlorophenyl *N*-(2,5-dimethoxyphenyl)carbamate.......... 614
2,6-Dimethoxyphenyl *N*-(2,3-dichlorophenyl)carbamate.......... 613
$C_{15}H_{13}Cl_2NO_4S$
Ethyl 4-((3,4-dichlorophenyl)-sulfonamido)benzoate.......... 808
$C_{15}H_{13}Cl_2N_3O$
N-Methylanthranilic (2,4-dichlorobenzylidene)hydrazide.......... 1025
$C_{15}H_{13}Cl_2N_3O_2$
1-(4-Acetamidophenyl)-3-(2,3-dichlorophenyl)urea.......... 569
1-(4-Acetamidophenyl)-3-(2,5-dichlorophenyl)urea.......... 569
1-(4-Acetamidophenyl)-3-(2,6-dichlorophenyl)urea.......... 526
1-(Benzylcarbonyl)-4-(3,4-dichlorophenyl)semicarbazide.......... 529
$C_{15}H_{13}Cl_2N_3O_3$
1-(4-Chloro-3-nitrophenyl)-3-(4-chlorophenyl)-3-ethylurea.......... 524
4,6-Dichloro-α-(2-(4-nitroanilino)-ethylimino)-*o*-cresol.......... 1033
4-(2,3-Dichlorophenyl)-1-(2-methoxybenzoyl)semicarbazide.......... 517
4-(2,6-Dichlorophenyl)-1-(4-methoxybenzoyl)semicarbazide.......... 520
4-(3,4-Dichlorophenyl)-1-(2-methoxybenzoyl)semicarbazide.......... 521
Phenylglycolic acid N^2-(*N*-(3,4-dichlorophenyl)carbamoyl)-hydrazide.......... 512
$C_{15}H_{13}Cl_2N_3O_4$
1-(4-Chloro-2-methoxyphenyl)-3-(2-chloro-4-nitrophenyl)-1-methylurea.. 518
1-(4,5-Dichloro-2-methylphenyl)-3-(2-methoxy-4-nitrophenyl)urea.......... 518
$C_{15}H_{13}Cl_2N_3O_4S$
N-Acetyl-4-(2,3-dichlorophenylureido)-benzenesulfonamide.......... 809
$C_{15}H_{13}Cl_2N_3O_5$
1-(4-Chloro-2,5-dimethoxyphenyl)-3-(4-chloro-2-nitrophenyl)urea.......... 516
1-(5-Chloro-2,4-dimethoxyphenyl)-3-(4-chloro-3-nitrophenyl)urea.......... 514
$C_{15}H_{13}Cl_3N_2O$
1-(4-Chlorobenzyl)-3-(2,3-dichlorophenyl)-1-methylurea.......... 539
1-(5-Chloro-2-methylphenyl)-3-(4,5-dichloro-2-methylphenyl)urea.......... 545
1-(2,6-Dimethylphenyl)-3-(2,4,5-trichlorophenyl)urea.......... 543
1-Phenethyl-3-(2,4,5-trichlorophenyl)urea.......... 527
$C_{15}H_{13}Cl_3N_2O_3$
1-(5-Chloro-2,4-dimethoxyphenyl)-3-(2,3-dichlorophenyl)urea.......... 515
$C_{15}H_{13}F$
4-Fluoro-α′-methylstilbene.......... 172
$C_{15}H_{13}FN_2O$
4′-(4-Fluorophenyliminomethyl)-acetanilide.......... 1027
$C_{15}H_{13}FN_2O_2$
N-Acetyl-*N*′-(3-fluorobenzoyl)-1,4-phenylenediamine.......... 452
N-Acetyl-*N*′-(4-fluorobenzoyl)-1,4-phenylenediamine.......... 452
1-(3-Acetylphenyl)-3-(2-fluorophenyl)urea.......... 530
1-(3-Acetylphenyl)-3-(3-fluorophenyl)urea.......... 530
1-(4-Acetylphenyl)-3-(3-fluorophenyl)urea.......... 530
4-(4-Fluorophenyliminomethyl)-phenyl *N*-methylcarbamate.......... 1026
$C_{15}H_{13}FN_2O_3$
4-Acetamidophenyl *N*-(2-fluorophenyl)carbamate.......... 631
4-Acetamidophenyl *N*-(3-fluorophenyl)carbamate.......... 631
4-Ethyl-2′-fluoro-5′-nitrobenzanilide.. 451
4-Fluorobenzoic (4-hydroxy-3-methoxybenzylidene)hydrazide.......... 491
Methyl 2-(2-fluorophenylureido)-benzoate.......... 531
$C_{15}H_{13}FN_2O_4$
3,4-Xylyl *N*-(4-fluoro-2-nitrophenyl)-carbamate.......... 624
$C_{15}H_{13}FN_2O_5$
4′-Acetamido-4-fluorosulfonyl-2′-hydroxybenzanilide.......... 801
2-(4-Nitrophenoxy)ethyl *N*-(4-fluorophenyl)carbamate.......... 615
$C_{15}H_{13}FN_4O$
1-(4-Fluorobenzyl)-2-(methylamino)-pyrido(2,3-d)pyrimidin-4(1*H*)-one.......... 928
$C_{15}H_{13}FN_4O_5$
N-(4-(4-Fluoro-3-nitrophenylureido)-phenyl)glycine.......... 525
$C_{15}H_{13}FO_3$
3,4-Dimethoxy-4′-fluorobenzophenone.......... 309
$C_{15}H_{13}FO_5S$
Methyl 2-fluoro-4-methoxy-6-(phenylsulfonyl)benzoate.......... 773
$C_{15}H_{13}F_2NOS$
2′,4′-Difluoro-2-(4-tolylthio)-acetanilide.......... 421
$C_{15}H_{13}F_2NO_3$
3-(*N*-(2,4-Difluorophenyl)-carbamoyl)-5-norbornene-2-carboxylic acid.......... 406
$C_{15}H_{13}F_2NO_4$
2,6-Dimethoxyphenyl *N*-(2,5-difluorophenyl)carbamate.......... 613
$C_{15}H_{13}F_2N_3O_3$
4-(2,4-Difluorophenyl)-1-(2-methoxybenzoyl)semicarbazide.......... 517
4-(2,5-Difluorophenyl)-1-(4-methoxybenzoyl)semicarbazide.......... 518
$C_{15}H_{13}F_3N_2O$
1-Benzyl-3-(3-trifluoromethylphenyl)urea.......... 504
1-Methyl-1-phenyl-3-(3-(trifluoromethyl)phenyl)urea.......... 510
1-(4-Tolyl)-3-(3-trifluoromethylphenyl)urea.......... 511
$C_{15}H_{13}F_3N_2OS$
1-(4-(Methylthio)phenyl)-3-(2-(trifluoromethyl)phenyl)urea.......... 508
1-(4-(Methylthio)phenyl)-3-(3-(trifluoromethyl)phenyl)urea.......... 508
$C_{15}H_{13}F_3N_2O_2$
1-(3-Methoxyphenyl)-3-(3-trifluoromethylphenyl)urea.......... 508
$C_{15}H_{13}F_3N_2O_3S$
N^4-Acetyl-N^1-(2-(trifluoromethyl)-phenyl)sulfanilamide.......... 805
1-(*p*-Tosyl)-3-(3-(trifluoromethyl)-phenyl)urea.......... 805
$C_{15}H_{13}F_3N_4O_2$
4-(Dimethylamino)-4′-nitro-2′-(trifluoromethyl)-1,1′-azobenzene..... 1080
4-(3-(3-(Trifluoromethyl)phenyl)-ureido)benzoic hydrazide.......... 509
$C_{15}H_{13}F_3N_4O_3$
2-(3-(Trifluoromethyl)phenylureido)-acetic furfurylidenehydrazide.......... 1021
$C_{15}H_{13}IN_2O_2$
1-(3-Acetylphenyl)-3-(4-iodophenyl)urea.......... 530
$C_{15}H_{13}IO_2S$
2-Iodo-2-phenylvinyl *p*-tolyl sulfone.. 784
α-Iodostyryl *p*-tolyl sulfone.......... 784
$C_{15}H_{13}I_2NO_3$
Ethyl 1,4-dihydro-3,5-diiodo-4-oxo-α-phenyl-1-pyridineacetate.......... 360
$C_{15}H_{13}I_2NO_4$
3,5-Diiodo-L-thyronine.......... 348
$C_{15}H_{13}N$
9-Allylcarbazole.......... 892
N-Cinnamylideneaniline.......... 1076
2,3-Diphenylpropionitrile.......... 756
2-(α-Methylbenzyl)benzonitrile.......... 767
3-Methyl-2-phenylindole.......... 891
5-Methyl-2-phenylindole.......... 891
2-(*p*-Tolyl)-indole.......... 887
$C_{15}H_{13}NO$
3-Acetamidofluorene.......... 441
4-Acetamidofluorene.......... 490
1-Benzoylindoline.......... 708
Cinnamanilide.......... 478
3-(Cinnamylideneamino)phenol.......... 1038
10,11-Dihydro-5*H*-dibenzo(a,d)-cyclohepten-5-one oxime.......... 1014
1,4-Dihydro-1-phenyl-3(2*H*)-isoquinolinone.......... 691
3,4-Dihydro-1-phenyl-2(1*H*)-quinolinone.......... 693
1,4-Diphenyl-2-azetidinone.......... 692
4-(1-Hydroxy-1-phenylethyl)-benzonitrile.......... 761
5-Methoxy-2-phenylindole.......... 883
N-(4-Methylbenzylidene)benzamide.. 478
2-(*N*-Methylformamido)fluorene.......... 491
3-Phenyl-1-indanone oxime.......... 1014
N-Styrylbenzamide.......... 478
$C_{15}H_{13}NOS$
2,3-Dihydro-2-phenyl-1,5-benzothiazepin-4(5*H*)-one.......... 709
2,3-Diphenyl-4-thiazolidinone.......... 710
$C_{15}H_{13}NO_2$
2-Acetamido-3-hydroxyfluorene.......... 486
2-Acetamido-7-hydroxyfluorene.......... 486
α-(3-Acetylphenylimino)-*o*-cresol.......... 1037
2-Benzoyl-*N*-methylbenzamide.......... 473
9-Carbazolepropionic acid.......... 892
7-(Dimethylamino)-3,4-benzocoumarin.......... 385
N,*N*-Dimethyl-4-dibenzofurancarboxamide.......... 894
4′-(2-Hydroxybenzylideneamino)-acetophenone.......... 1037
Methyl 7-amino-1-fluorenecarboxylate.......... 372
N-Phenacylbenzamide.......... 471
2-(Phenylacetyl)benzamide.......... 467
N-Piperonylidene-*o*-toluidine.......... 1076
N-Piperonylidene-*m*-toluidine.......... 1076
N-Piperonylidene-*p*-toluidine.......... 1076
$C_{15}H_{13}NO_2S$
10-Ethyl-4-phenothiazinecarboxylic acid.......... 343
4-Methylthio-*N*-piperonylideneaniline.......... 1051
10-Phenothiazinepropionic acid.......... 330
α-(2-Thenoylmethylene)-*p*-acetotoluidide.......... 857
$C_{15}H_{13}NO_3$
6-Acetyl-*N*-phenylanthranilic acid.......... 343
N-Benzylphthalamic acid.......... 471
3′-Carboxy-3-methylbenzanilide.......... 475
α-(2-Furoylmethylene)-*p*-acetotoluidide.......... 432
4-Glycylphenyl benzoate hydrochloride.......... 379
4-(4-Methoxybenzylideneamino)-benzoic acid.......... 1049
N-(4-Methoxybenzylidene)-anthranilic acid.......... 1049
4-Methoxy-β-nitrostilbene.......... 257
Methyl 4-benzamidobenzoate.......... 477
3-(4-Methylbenzamido)benzoic acid. 476
Methyl *N*-benzoylanthranilate.......... 472
Methyl *N*-(4-benzoylphenyl)-carbamate.......... 675
4-(4-Methylbenzylideneamino)-salicylic acid.......... 1032
Methyl 4-(*N*-phenylcarbamoyl)-benzoate.......... 474
2′-Methylphthalanilic acid.......... 474
3′-Methylphthalanilic acid.......... 474
4′-Methylphthalanilic acid.......... 474
4-Phenylhippuric acid.......... 488
N-Piperonylidene-*o*-anisidine.......... 1045
N-Piperonylidene-*m*-anisidine.......... 1045
N-Piperonylidene-*p*-anisidine.......... 1046
N-(*p*-Tolyl)phthalamic acid.......... 474
$C_{15}H_{13}NO_3S$
2,6-Dimethoxy-4-(2-benzothiazolyl)-phenol.......... 912
$C_{15}H_{13}NO_4$
4-Acetamidophenyl salicylate.......... 413
α-(3-Carboxyphenylimino)-6-methoxy-*o*-cresol.......... 1030
α-(4-Carboxyphenylimino)-6-methoxy-*o*-cresol.......... 1030
2-(4-Methoxyphenyl)-4′-nitroacetophenone.......... 292
2′-Methoxyphthalanilic acid.......... 465
3′-Methoxyphthalanilic acid.......... 466
4′-Methoxyphthalanilic acid.......... 466
4-Nitrobenzyl 4-methylbenzoate.......... 370
DL-4-Nitrodiphenylmethyl acetate.... 359
Piperonyl *N*-phenylcarbamate.......... 649
$C_{15}H_{13}NO_4S$
2-(5-Nitro-2-(*p*-tolylthio)phenyl)-acetic acid.......... 329
$C_{15}H_{13}NO_5$
α-(4-Carboxy-3-hydroxyphenylimino)-6-methoxy-*o*-cresol.......... 1028
Methyl 4-benzyloxy-3-nitrobenzoate. 369
$C_{15}H_{13}NO_5S_2$
4′-Methylsulfonyl-2-(4-nitrophenylthio)acetophenone.......... 776
$C_{15}H_{13}NO_6$
4′,5′-Dimethoxy-2-furfurylidene-2′-nitroacetophenone.......... 845
Dimethyl 5-(furamido)isophthalate... 851
$C_{15}H_{13}NS$
10-Allylphenothiazine.......... 251
$C_{15}H_{13}N_3$
3-Anilino-5-phenylpyrazole.......... 862
4-Cyanobenzaldehyde methylphenylhydrazone.......... 765
$C_{15}H_{13}N_3O$
1-(Benzylideneimino)-2,4-dimethyl-5-cyano-6-pyridone.......... 1056
N-(2-Cyanoacetyl)hydrazobenzene... 755
1-(4-Cyanophenyl)-3-(*p*-tolyl)urea.......... 767
3-(2-(4-Pyridyl)-ethyl)-4(3*H*)-quinazolinone.......... 953
3-(3,5-Xylyl)-1,2,3-benzotriazin-4(3*H*)-one.......... 711
$C_{15}H_{13}N_3OS$
1-(*N*-(4-(Methylthio)phenyl)-carbamoyl)benzimidazole.......... 903
1-(*N*-(4-(Methylthio)phenyl)-carbamoyl)-1*H*-indazole.......... 900
$C_{15}H_{13}N_3OS_2$
1-(2-Benzothiazolyl)-3-(4-(methylthio)phenyl)urea.......... 914
$C_{15}H_{13}N_3O_2$
1-(4-Cyanophenyl)-3-(2-methoxyphenyl)urea.......... 762
1-(4-Cyanophenyl)-3-(3-methoxyphenyl)urea.......... 762
1-(4-Cyanophenyl)-3-(4-methoxyphenyl)urea.......... 762

$C_{15}H_{13}N_3O_2$ (Cont'd)
1,2-Dihydro-4,6-dimethyl-2-oxo-1-(salicylideneamino)-nicotinonitrile.. 1036
2,3-Dihydro-2-(2-(2-pyridyl)-ethyl)-1,4-phthalazinedione 950
2,3-Dihydro-2-(2-(4-pyridyl)-ethyl)-1,4-phthalazinedione 963
1-(5-Methyl-3-isoxazolyl)-3-(1-naphthyl)urea 871

$C_{15}H_{13}N_3O_2S$
4-(2,3-Dihydro-2-thioxo-1-benzimidazolylmethylamino)-benzoic acid 742
1-(6-Methoxy-2-benzothiazolyl)-3-phenylurea 913

$C_{15}H_{13}N_3O_3$
1,3-Dimethyl-5-(3-indolylmethylene)-barbituric acid 889
4′-(3-Nitrophenyliminomethyl)-acetanilide 1056

$C_{15}H_{13}N_3O_3S$
4-Acetamido-*N*-cyanobenzenesulfonanilide 824
4-(2,3-Dihydro-2-thioxo-1-benzimidazolylmethylamino)-salicylic acid 734
5-(2,3-Dihydro-2-thioxo-1-benzimidazolylmethylamino)-salicylic acid 734

$C_{15}H_{13}N_3O_4$
8-(1,3-Dihydro-3-oxo-1-isobenzofuranyl)-1,2,3,5,6,7-hexahydro-6-methylimidazo(1,2-c)-pyrimidine-5,7-dione 709
Mandelic 3-nitrobenzylidenehydrazide 496
2-Methoxybenzoic (3-nitrobenzylidene)hydrazide 497
4-Methoxybenzoic (3-nitrobenzylidene)hydrazide 1045
4-(3-Nitrophenyliminomethyl)-phenyl *N*-methylcarbamate 1056
2-(Phenoxy)acetic (3-nitrobenzylidene)hydrazide 498

$C_{15}H_{13}N_3O_4S$
1-Benzoyl-3-(4-methoxy-2-nitrophenyl)-2-thiourea 737

$C_{15}H_{13}N_3O_5$
4-Hydroxy-3-methoxybenzoic 3-nitrobenzylidenehydrazide 495
3-(4-Methyl-3-nitrophenylureido)-benzoic acid 563
4-(3-(2-Methyl-5-nitrophenyl)ureido)-benzoic acid 563

$C_{15}H_{13}N_3O_6$
2-Carbamoylphenyl *N*-(2-methoxy-4-nitrophenyl)carbamate 652
N-(2,4-Dinitrophenyl)-L-phenylalanine 349
3-(3-(2-Methoxy-5-nitrophenyl)-ureido)benzoic acid 552
2-Methyl-4-nitrophenyl *N*-(4-methyl-3-nitrophenyl)carbamate 669

$C_{15}H_{13}N_3O_7$
O-(2,4-Dinitrophenyl)-L-tyrosine monohydrate 348

$C_{15}H_{13}N_3S$
3,4-Diphenyl-4-thiazolin-2-one hydrazone 726

$C_{15}H_{13}N_3S_2$
2-Methylthio-4-phenyl-5-phenylimino-1,3,4-thiadiazol-2-ine ... 739

$C_{15}H_{13}N_5S_2$
1-(5-Anilino-1,2,4-thiadiazol-3-yl)-3-phenyl-2-thiourea 879

$C_{15}H_{14}BrClN_4O$
1-(5-Bromo-2-(4-chlorobenzyloxy)-benzylideneamino)guanidine hydrochloride 1023

$C_{15}H_{14}BrN$
9-Bromo-2-dimethylaminofluorene hydrobromide 233

$C_{15}H_{14}BrNO$
2-Bromo-2′,6′-dimethylbenzanilide ... 457
3-Bromo-2′,6′-dimethylbenzanilide ... 459
4′-Bromo-2′,6′-dimethylbenzanilide .. 457

$C_{15}H_{14}BrNOS$
2-(Benzylthio)-4′-bromoacetanilide ... 402

$C_{15}H_{14}BrNO_2$
4-Bromo-*N*-(2-hydroxy-3-methoxybenzylidene)-2-methylaniline 1022
2-(4-Bromo-3-methylphenyliminomethyl)-6-methoxyphenol 1021
N-(3,4-Dimethoxybenzylidene)-4-bromoaniline 1041

$C_{15}H_{14}BrNO_3$
2′-Bromo-2,6-dimethoxybenzanilide. 448
4-Methoxybenzyl *N*-(4-bromophenyl)carbamate 612
2-Phenoxyethyl *N*-(4-bromophenyl)-carbamate 615

$C_{15}H_{14}BrNS$
3-Benzyl-2-methylbenzothiazolium bromide 915

$C_{15}H_{14}BrN_3O_2$
1-(4-Acetamidophenyl)-3-(4-bromophenyl)urea 569
2-(2-Bromoethyl)benzaldehyde (4-nitrophenyl)hydrazone 258

$C_{15}H_{14}BrN_3O_3$
Ethyl 4-(3-(5-bromo-2-pyridyl)-ureido)benzoate 962

$C_{15}H_{14}ClFN_2O$
1-Benzyl-3-(3-chloro-4-fluorophenyl)-1-methylurea 536
1-(4-Chlorophenyl)-1-ethyl-3-(4-fluorophenyl)urea 539

$C_{15}H_{14}ClFN_2O_2$
1-(3-Chloro-4-fluorophenyl)-3-(4-(2-hydroxyethyl)phenyl)urea 514

$C_{15}H_{14}ClFN_2O_2S$
1-(5-Chloro-2,4-dimethoxyphenyl)-3-(3-fluorophenyl)-2-thiourea 728

$C_{15}H_{14}ClFN_2O_3$
1-(3-Chloro-4-fluorophenyl)-3-(2,4-dimethoxyphenyl)urea 515

$C_{15}H_{14}ClN$
N-(2-Chlorobenzylidene)-2,3-xylidine 1064
N-(2-Chlorobenzylidene)-2,4-xylidine 1064
N-(2-Chlorobenzylidene)-2,5-xylidine 1064
N-(2-Chlorobenzylidene)-3,5-xylidine 1065
N-(4-Chlorobenzylidene)-2,3-xylidine 1065
N-(4-Chlorobenzylidene)-2,4-xylidine 1078
N-(4-Chlorobenzylidene)-2,5-xylidine 1078
N-(4-Chlorobenzylidene)-3,4-xylidine 1066
N-(4-Chlorobenzylidene)-3,5-xylidine 1066
2-Chloro-2′,5′-dimethoxybenzanilide 448
3-Chloro-4-methyl-*N*-(4-methylbenzylidene)aniline 1066
5-Chloro-2-methyl-*N*-(4-methylbenzylidene)aniline 1066

$C_{15}H_{14}ClNO$
N-Benzyl-4-chloro-*N*-methylbenzamide 483
2-Chloro-2′,3′-benzoxylidide 479
2-Chloro-2′,4′-benzoxylidide 479
2-Chloro-2′,5′-benzoxylidide 479
2-Chloro-3′,4′-benzoxylidide 479
4-Chloro-2′,3′-benzoxylidide 480
4-Chloro-2′,4′-benzoxylidide 480
4-Chloro-2′,5′-benzoxylidide 480
N-(2-Chlorobenzylidene)-*p*-phenetidine 1046
N-(4-Chlorobenzylidene)-*o*-phenetidine 1045
N-(4-Chlorobenzylidene)-*p*-phenetidine 1047
N-(4-Chlorobenzyl)-2-phenylacetamide 408
2-Chloro-2′,6′-dimethylbenzanilide ... 457
2′-Chloro-2,6′-dimethylbenzanilide ... 479
2′-Chloro-3,6′-dimethylbenzanilide ... 459
2′-Chloro-4,4′-dimethylbenzanilide ... 458
2′-Chloro-4,6′-dimethylbenzanilide ... 458
3-Chloro-2′,3′-dimethylbenzanilide ... 459
3-Chloro-2′,4′-dimethylbenzanilide ... 460
3′-Chloro-2,4′-dimethylbenzanilide ... 479
3-Chloro-2′,5′-dimethylbenzanilide ... 460
3-Chloro-2′,6′-dimethylbenzanilide ... 459
3′-Chloro-3,4′-dimethylbenzanilide ... 481
4-Chloro-2′,6′-dimethylbenzanilide ... 458
2-Chloro-*N*-(2-ethoxybenzylidene)-aniline 1024
2-Chloro-2′-ethylbenzanilide 457
2-Chloro-4′-ethylbenzanilide 457
3-Chloro-2′-ethylbenzanilide 457
3-Chloro-4′-ethylbenzanilide 458
3-Chloro-*N*-(4-methoxybenzylidene)-6-methylaniline 1047
4-Chloro-*N*-phenethylbenzamide 480

$C_{15}H_{14}ClNOS$
S-Benzyl *N*-(2-chloro-6-methylphenyl)thiocarbamate 641
2-(Benzylthio)-4′-chloroacetanilide ... 402
2-(4-Chlorophenylthio)-4′-methylacetanilide 403
S-(*o*-Tolyl) *N*-(2-chloro-6-methylphenyl)thiocarbamate 641
S-(*m*-Tolyl) *N*-(2-chloro-6-methylphenyl)thiocarbamate 642
S-(*m*-Tolyl) *N*-(3-chloro-2-methylphenyl)thiocarbamate 641

$C_{15}H_{14}ClNO_2$
N-(2-Chlorobenzylidene)-2,5-dimethoxyaniline 1045
N-(4-Chlorobenzylidene)-2,5-dimethoxyaniline 1045
2-Chloro-*N*-(2,5-dimethoxybenzylidene)aniline 1046
2-Chloro-*N*-(3,4-dimethoxybenzylidene)-aniline 1041
3-Chloro-*N*-(3,4-dimethoxybenzylidene)aniline 1042
4-Chloro-*N*-(2,5-dimethoxybenzylidene)aniline 1047
4-Chloro-*N*-(3,4-dimethoxybenzylidene)aniline 1041
2-Chloro-2′-ethoxybenzanilide 449
2-Chloro-4′-ethoxybenzanilide 449
3-Chloro-4′-ethoxybenzanilide 450
4-Chloro-4′-ethoxybenzanilide 466
2-Chloro-*N*-(3-ethoxy-4-hydroxybenzylidene)aniline 1021
2-Chloro-2′-methoxy-5′-methylbenzanilide 449
2′-Chloro-3-methoxy-6-methylbenzanilide 448
3′-Chloro-4-methoxy-4′-methylbenzanilide 450
4-Chloro-2′-methoxy-5′-methylbenzanilide 452
2-Chloro-5′-methyl-*o*-benzanisidide .. 466
4-Chloro-5′-methyl-*o*-benzanisidide .. 467
3-Chloro-2-methyl-*N*-(2-hydroxy-3-methoxybenzylidene)aniline 1022
3-Chloro-4-methyl-*N*-(2-hydroxy-3-methoxybenzylidene)aniline 1022
4-Chloro-2-methyl-*N*-(2-hydroxy-3-methoxybenzylidene)aniline 1022
5-Chloro-2-methyl-*N*-(2-hydroxy-3-methoxybenzylidene)aniline 1022
4-Chloro-2-methylphenyl *N*-(*o*-tolyl)-carbamate 633
4-Chloro-3-methylphenyl *N*-(*o*-tolyl)-carbamate 633
2-(2-Chlorophenoxy)-2′-methylacetanilide 400
4-Chlorophenyl *N*-(2-ethylphenyl)-carbamate 633
4-Chlorophenyl *N*-(2,3-xylyl)-carbamate 683
4-Chlorophenyl *N*-(2,4-xylyl)-carbamate 683
4-Chlorophenyl *N*-(2,5-xylyl)-carbamate 635
4-Chlorophenyl *N*-(2,6-xylyl)-carbamate 634
4-Chlorophenyl *N*-(3,4-xylyl)-carbamate 635
4-Chlorophenyl *N*-(3,5-xylyl)-carbamate 635
2-Chloro-3-piperidino-1,4-naphthoquinone 318
2,6-Xylyl *N*-(3-chlorophenyl)-carbamate 636
2,6-Xylyl *N*-(4-chlorophenyl)-carbamate 634
3,5-Xylyl *N*-(2-chlorophenyl)-carbamate 633

$C_{15}H_{14}ClNO_2S$
S-Benzyl *N*-(4-chloro-2-methoxyphenyl)thiocarbamate 620
2-(4-Chlorophenoxy)-4′-(methylthio)-acetanilide 399
2-(4-Chlorophenylthio)-4′-methoxyacetanilide 401
S-(4-Methoxyphenyl) *N*-(2-chloro-6-methylphenyl)thiocarbamate 621
3-Methyl-4-(methylthio)phenyl *N*-(4-chlorophenyl)carbamate 623
4-(Methylthio)phenyl *N*-(3-chloro-4-methylphenyl)carbamate 623

$C_{15}H_{14}ClNO_3$
2-Chloro-2′,4′-dimethoxybenzanilide 447
5′-Chloro-2′,4-dimethoxybenzanilide 448
α-(5-Chloro-2,4-dimethoxyphenylimino)-*o*-cresol 1031
α-(5-Chloro-2-methoxyphenylimino)-6-methoxy-*o*-cresol 1021
4-Chloro-3-methylphenyl *N*-(2-methoxyphenyl)carbamate 618
4-Chloro-3-methylphenyl *N*-(4-methoxyphenyl)carbamate 620
3-(*N*-(4-Chlorophenyl)carbamoyl)-5-norbornene-2-carboxylic acid 406
4-Chlorophenyl *N*-(2-methoxy-5-methylphenyl)carbamate 619
4-Methoxybenzyl *N*-(2-chlorophenyl)carbamate 620
4-Methoxyphenyl *N*-(2-chloro-4-methylphenyl)carbamate 617
4-Methoxyphenyl *N*-(3-chloro-4-methylphenyl)carbamate 617
4-Methoxyphenyl *N*-(5-chloro-2-methylphenyl)carbamate 617
2-(Phenoxy)ethyl *N*-(2-chlorophenyl)carbamate 615

$C_{15}H_{14}ClNO_3S$
S-(4-Chlorophenyl) *N*-(2,5-dimethoxyphenyl)thiocarbamate 729
4-(Methylthio)phenyl *N*-(5-chloro-2-methoxyphenyl)carbamate 617

$C_{15}H_{14}ClNO_4$
2-Chloro-4-methoxyphenyl *N*-(3-methoxyphenyl)carbamate 614
4-Methoxyphenyl *N*-(5-chloro-2-methoxyphenyl)carbamate 617

$C_{15}H_{14}ClNO_4S$
S-(4-Hydroxyphenyl) *N*-(5-chloro-2,4-dimethoxyphenyl)thiocarbamate 611

$C_{15}H_{14}ClNS$
4-Chloro-*N*-methylthiobenzo-*o*-toluidide 742

$C_{15}H_{14}ClN_2O_2S$
S-(*o*-Tolyl) *N*-(4-chloro-2-methoxyphenyl)thiocarbamate 620

$C_{15}H_{14}ClN_3O$
N-Methylanthranilic (2-chlorobenzylidene)hydrazide 492

$C_{15}H_{14}ClN_3O_2$
1-(3-Acetamidophenyl)-3-(2-chlorophenyl)urea 525
1-(4-Acetamidophenyl)-3-(4-chlorophenyl)urea 526
4-(3-Chlorophenyl)-1-(2-phenylacetyl)semicarbazide 528

$C_{15}H_{14}ClN_3O_3$
1-Benzyl-3-(4-chloro-2-nitrophenyl)-1-methylurea 523
1-(4-Carboxyphenyl)-4-(3-chloro-4-methyl)phenylsemicarbazide 530
1-(4-Chlorobenzyl)-3-(2-methyl-5-nitrophenyl)urea 523
4-(3-Chloro-2-methylphenyl)-1-(2-hydroxybenzoyl)semicarbazide 491
4-(3-Chloro-2-methylphenyl)-1-(4-hydroxybenzoyl)semicarbazide 513

$C_{15}H_{14}ClN_3O_3$ (Cont'd)
4-(4-Chloro-2-methylphenyl)-1-(2-hydroxybenzoyl)semicarbazide 491
1-(3-Chloro-2-methylphenyl)-3-(2-methyl-4-nitrophenyl)urea 526
1-(3-Chloro-2-methylphenyl)-3-(4-methyl-3-nitrophenyl)urea 524
α-(2-Chloro-6-nitrophenyl)-*N*-(4-dimethylaminophenyl)nitrone 1054
4-(2-Chlorophenyl)-1-(2-methoxybenzoyl)semicarbazide 517
4-(3-Chlorophenyl)-1-(2-methoxybenzoyl)semicarbazide 517
4-(3-Chlorophenyl)-1-(4-methoxybenzoyl)semicarbazide 520
Ethyl 4-(3-(5-chloro-2-pyridyl)ureido)-benzoate 962

$C_{15}H_{14}ClN_3O_4$
4-(4-Chloro-2-methoxyphenyl)-1-(2-hydroxybenzoyl)semicarbazide 512
4-(4-Chloro-2-methoxyphenyl)-1-(4-hydroxybenzoyl)semicarbazide 512
4-(5-Chloro-2-methoxyphenyl)-1-(4-hydroxybenzoyl)semicarbazide 512
1-(3-Chloro-2-methylphenyl)-3-(2-methoxy-5-nitrophenyl)urea 517
1-(5-Chloro-2-methylphenyl)-3-(2-methoxy-5-nitrophenyl)urea 517
1-(4-Chloro-2-nitrophenyl)-3-(4-(2-hydroxyethyl)phenyl)urea 512

$C_{15}H_{14}ClN_3O_4S$
1-(4-Chloro-2,5-dimethoxyphenyl)-3-(4-nitrophenyl)-2-thiourea 729

$C_{15}H_{14}ClN_3O_5$
1-(2-Chloro-4-methoxyphenyl)-3-(2-methoxy-4-nitrophenyl)urea 516
1-(2-Chloro-4-nitrophenyl)-3-(2,5-dimethoxyphenyl)urea 552

$C_{15}H_{14}ClN_3O_5S$
3-(2-Benzoylhydrazino)-2-methyl-1,2-benzisothiazole perchlorate 915

$C_{15}H_{14}Cl_2N_2$
N,*N*′-Bis(3-chloro-4-methylphenyl)-formamidine 724
N,*N*′-Bis(4-chloro-2-methylphenyl)-formamidine 724
α-(2,3-Dichlorophenylimino)-*N*,*N*-dimethyl-*p*-toluidine 1061
α-(2,4-Dichlorophenylimino)-*N*,*N*-dimethyl-*p*-toluidine 1061
α-(2,5-Dichlorophenylimino)-*N*,*N*-dimethyl-*p*-toluidine 1061
α-(3,4-Dichlorophenylimino)-*N*,*N*-dimethyl-*p*-toluidine 1061

$C_{15}H_{14}Cl_2N_2O$
1-Benzyl-3-(2,6-dichlorophenyl)-1-methylurea 527
1-Benzyl-3-(3,4-dichlorophenyl)-1-methylurea 573
1,3-Bis(3-chloro-4-methylphenyl)urea 544
1,3-Bis-(4-chloro-2-methylphenyl)urea 601
1,3-Bis-(5-chloro-2-methylphenyl)urea 601
1-(4-Chlorobenzyl)-3-(2-chlorophenyl)-1-methylurea 537
1-(2-Chloro-6-methylphenyl)-3-(4-chloro-2-methylphenyl)urea 542
1-(3-Chloro-4-methylphenyl)-3-(4-chloro-2-methylphenyl)urea 544
1-(3-Chloro-4-methylphenyl)-3-(5-chloro-2-methylphenyl)urea 601
1-(5-Chloro-2-methylphenyl)-3-(2-chloro-6-methylphenyl)urea 542
1-(3-Chloro-2-methylphenyl)-3-(3-chloro-4-methyl)urea 542
3-(3-Chlorophenyl)-1-(4-chlorophenyl)-1-ethylurea 539
1-(2,4-Dichlorophenyl)-3-(2,6-dimethylphenyl)urea 542
1-(2,6-Dichlorophenyl)-3-(2,4-dimethylphenyl)urea 542
1-(2,6-Dichlorophenyl)-3-(2,6-dimethylphenyl)urea 541
1-(3,4-Dichlorophenyl)-3-(4-ethylphenyl)urea 540
3-(2,6-Dichlorophenyl)-1-ethyl-1-phenylurea 595
3-(3,4-Dichlorophenyl)-1-ethyl-1-phenylurea 601
1-(2,3-Dichlorophenyl)-3-phenethylurea 570
1-(2,4-Dichlorophenyl)-3-phenethylurea 570
1-(2,5-Dichlorophenyl)-3-phenethylurea 570
1-(2,6-Dichlorophenyl)-3-phenethylurea 536
1-(3,4-Dichlorophenyl)-3-phenethylurea 570
1-(2,4-Dichlorophenyl)-3-(2,4-xylyl)urea 601
1-(2,5-Dichlorophenyl)-3-(2,6-xylyl)urea 542
1-(3,4-Dichlorophenyl)-3-(2,4-xylyl)urea 544
1-(3,4-Dichlorophenyl)-3-(2,5-xylyl)urea 544
1-(3,4-Dichlorophenyl)-3-(2,6-xylyl)urea 542

$C_{15}H_{14}Cl_2N_2OS$
1-(4,5-Dichloro-2-methylphenyl)-3-(4-(methylthio)phenyl)urea 522

$C_{15}H_{14}Cl_2N_2O_2$
1-(2-Chloro-4-methoxyphenyl)-3-(5-chloro-2-methylphenyl)urea 521
1-(5-Chloro-2-methoxyphenyl)-3-(2-chloro-5-methylphenyl)urea 520
1-(5-Chloro-2-methoxyphenyl)-3-(3-chloro-2-methylphenyl)urea 520
1-(5-Chloro-2-methoxyphenyl)-3-(5-chloro-2-methylphenyl)urea 520
1-(3,4-Dichlorobenzyl)-3-(2-methoxyphenyl)urea 519
1-(4,5-Dichloro-2-methylphenyl)-3-(4-methoxyphenyl)urea 521
1-(2,3-Dichlorophenyl)-3-(4-(2-hydroxyethyl)phenyl)urea 514
1-(3,4-Dichlorophenyl)-3-(1-hydroxyethyl)phenylurea 512
3-(Dimethylamino)phenyl *N*-(2,6-dichlorophenyl)carbamate 627
3-(4-Pyridyl)propyl *N*-(2,3-dichlorophenyl)carbamate 929
3-(4-Pyridyl)propyl *N*-(3,4-dichlorophenyl)carbamate 929

$C_{15}H_{14}Cl_2N_2O_2S$
1-(5-Chloro-2,4-dimethoxyphenyl)-3-(4-chlorophenyl)-2-thiourea 729

$C_{15}H_{14}Cl_2N_2O_3$
1-(4-Chloro-2,5-dimethoxyphenyl)-3-(3-chlorophenyl)urea 516
1-(2-Chloro-4-methoxyphenyl)-3-(5-chloro-2-methoxyphenyl)urea 521
1-(2-Chlorophenyl)-3-(4-chloro-2,5-dimethoxyphenyl)urea 516
1-(2,3-Dichlorophenyl)-3-(2,5-dimethoxyphenyl)urea 515
1-(2,6-Dichlorophenyl)-3-(2,5-dimethoxyphenyl)urea 550
1-(3,4-Dichlorophenyl)-3-(2,4-dimethoxyphenyl)urea 515

$C_{15}H_{14}Cl_2N_4O$
1-(4-(3,4-Dichlorobenzyloxy)-benzylideneamino)guanidine acetate 1023
1-(3-(3,4-Dichlorobenzyloxy)-benzylideneamino)guanidine hydrochloride 1023
1-(2-(2,4-Dichlorobenzyloxy)-benzylideneamino)guanidine nitrate 1023
1-(2-(3,4-Dichlorobenzyloxy)-benzylideneamino)guanidine nitrate 1023

$C_{15}H_{14}Cl_2N_4OS$
1-(*N*-Benzylthiocarbamoyl)-4-(3,4-dichlorophenyl)semicarbazide 730

$C_{15}H_{14}Cl_2N_4O_2$
4-(3,4-Dichlorophenyl)-1-(2-(methylamino)benzoyl)-semicarbazide 527

$C_{15}H_{14}Cl_2N_4O_3$
Disperse Orange 5 (C.I. 11100) 1084

$C_{15}H_{14}Cl_2N_6O_2$
8-(2,4-Dichlorophenylazo)-1,3-diethylxanthine 924

$C_{15}H_{14}Cl_2O_2$
2,2′-Methylenebis(4-chloro-6-methylphenol) 193

$C_{15}H_{14}FN$
N-(4-Fluorobenzylidene)-2,4-xylidine 1078
N-(4-Fluorobenzylidene)-2,5-xylidine 1078
N-(4-Fluorobenzylidene)-3,4-xylidine 1078

$C_{15}H_{14}FNO$
2′,3′-Dimethyl-3-fluorobenzanilide 459
4′-Fluoro-3-phenylpropionanilide 408

$C_{15}H_{14}FNOS$
2-(Benzylthio)-4′-fluoroacetanilide 402

$C_{15}H_{14}FNO_2$
2′-Ethoxy-3-fluorobenzanilide 449
3-Fluoro-4′-(2-hydroxyethyl)-benzanilide 447
4-Fluoro-4′-(2-hydroxyethyl)-benzanilide 447
3-Fluoro-2′-methoxy-5′-methylbenzanilide 449
4-Fluoro-2′-methoxy-5′-methylbenzanilide 449
4-Fluorophenyl *N*-(2,3-dimethylphenyl)carbamate 634
2-Fluorophenyl *N*-(2,5-xylyl)-carbamate 633
4-Fluorophenyl *N*-(2,5-xylyl)-carbamate 635
N-Veratrylidene-4-fluoroaniline 1042
2,5-Xylyl *N*-(2-fluorophenyl)-carbamate 633
2,5-Xylyl *N*-(4-fluorophenyl)-carbamate 635

$C_{15}H_{14}FNO_3$
2′,5′-Dimethoxy-3-fluorobenzanilide 448
2′,5′-Dimethoxy-4-fluorobenzanilide 448
N-(2,5-Dimethoxyphenyl)-2-fluorobenzamide 448
2-Ethoxyphenyl *N*-(2-fluorophenyl)-carbamate 615
3-(*N*-(4-Fluorophenyl)carbamoyl)-5-norbornene-2-carboxylic acid 406
4-Methoxybenzyl *N*-(4-fluorophenyl)carbamate 612
2-Phenoxyethyl *N*-(2-fluorophenyl)-carbamate 615
2-Phenoxyethyl *N*-(3-fluorophenyl)-carbamate 615

$C_{15}H_{14}FNO_4$
2,6-Dimethoxyphenyl *N*-(4-fluorophenyl)carbamate 613
2-Fluorophenyl *N*-(2,4-dimethoxyphenyl)carbamate 614
2-Fluorophenyl *N*-(2,5-dimethoxyphenyl)carbamate 614
4-Fluorophenyl *N*-(2,5-dimethoxyphenyl)carbamate 614

$C_{15}H_{14}FN_3O_2$
1-(4-Acetamidophenyl)-3-(2-fluorophenyl)urea 526
1-(4-Acetamidophenyl)-3-(3-fluorophenyl)urea 526
1-(4-Acetamidophenyl)-3-(4-fluorophenyl)urea 526
4-(4-Fluorophenyl)-1-(2-phenylacetyl)semicarbazide 528

$C_{15}H_{14}FN_3O_3$
1-Benzyl-3-(4-fluoro-3-nitrophenyl)-1-methylurea 523
1-(2-Fluoro-5-nitrophenyl)-3-(phenethyl)urea 523
1-(2-Fluoro-5-nitrophenyl)-3-(3,4-xylyl)urea 525
4-(4-Fluorophenyl)-1-(2-methoxybenzoyl)semicarbazide 517
4-(4-Fluorophenyl)-1-(4-methoxybenzoyl)semicarbazide 520

$C_{15}H_{14}FN_3O_4$
1-(2-Fluoro-5-nitrophenyl)-3-(2-methoxy-5-methylphenyl)urea 517

$C_{15}H_{14}FN_3O_5$
1-(2,5-Dimethoxyphenyl)-3-(2-fluoro-5-nitrophenyl)urea 516

$C_{15}H_{14}F_2N_2O$
1-(2,4-Difluorophenyl)-3-(3-ethylphenyl)urea 544
1-(2,5-Difluorophenyl)-3-(3-ethylphenyl)urea 544

$C_{15}H_{14}F_2N_2O_2$
1-(2,5-Difluorophenyl)-3-(2-ethoxyphenyl)urea 519
1-(2,5-Difluorophenyl)-3-(4-(2-hydroxyethyl)phenyl)urea 514
1-(2,5-Difluorophenyl)-3-(4-(2-hydroxyethyl)phenyl)urea 514
1-(2,5-Difluorophenyl)-3-(2-methoxy-5-methylphenyl)urea 520

$C_{15}H_{14}F_2N_2O_3$
1-(2,5-Difluorophenyl)-3-(2,4-dimethoxyphenyl)urea 515

$C_{15}H_{14}F_2N_4O_2$
4-(2,4-Difluorophenyl)-1-(2-methylaminobenzoyl)-semicarbazide 527

$C_{15}H_{14}F_3NO_2$
N-(3-(Trifluoromethyl)phenyl)-2-hydroxy-3-methoxybenzylamine 222

$C_{15}H_{14}F_3NO_3$
3′-(Trifluoromethyl)-1,2,3,6-tetrahydrophthalanilic acid 397

$C_{15}H_{14}F_3N_3$
N,*N*-Dimethyl-4-(4-(trifluoromethyl)-phenylazo)aniline 1080

$C_{15}H_{14}F_3N_3O$
1-(4,6-Dimethyl-2-pyridyl)-3-(3-(trifluoromethyl)phenyl)urea 926

$C_{15}H_{14}INOS$
2-(Benzylthio)-4′-iodoacetanilide 402

$C_{15}H_{14}INO_2$
4-Iodo-*N*-veratrylideneaniline 1042
1-Methyl-2-(3,4-(methylenedioxy)-styryl)-pyridinium iodide 957
1-Methyl-4-(3,4-(methylenedioxy)-styryl)-pyridinium iodide 957

$C_{15}H_{14}N_2$
(4-Aminophenyl)-(*m*-tolyl)-acetonitrile hydrochloride 754
(4-Aminophenyl)-(*p*-tolyl)-acetonitrile hydrochloride 754
N-(Cinnamylidene)phenylhydrazone 1068
Dibenzylcyanamide 756
4-(2,5-Dimethyl-1-pyrrolyl)-isoquinoline 997
1-Methyl-2-(*p*-tolyl)benzimidazole 907
N-(3-(Phenylimino)propenyl)aniline 1054
α-Phenyl-4-pyridinebutyronitrile 953
3-(2-(2-Pyridyl)ethyl)indole 967
3-(2-(4-Pyridyl)ethyl)indole 967

$C_{15}H_{14}N_2O$
2-Acetamido-4-aminofluorene 487
2-Acetamido-7-aminofluorene 487
Acetic acid diphenylmethylenehydrazide 502
Acetic N^2-benzylidene-N^1-phenylhydrazide 502
(4-Aminophenyl)-(*p*-anisyl)-acetonitrile hydrochloride 753
Benzil *mono*(methylhydrazone) 302
3,4′-Diaminochalcone 302
1,2-Dihydro-2-(*m*-tolyl)-4(3*H*)-quinazolinone 704
1,2-Dihydro-2-(*p*-tolyl)-4(3*H*)-quinazolinone 709
2-(5,6-Dimethyl-2-benzimidazolyl)-phenol 902
5-Hydroxymethyl-2-methyl-1-phenylbenzimidazole 902
2-(2-Methoxyphenyl)-1-methylbenzimidazole 903
2-(4-Methoxyphenyl)-1-methylbenzimidazole 903
1-Phenyl-1,2-propanedione 1-(phenylhydrazone) 285

$C_{15}H_{14}N_2OS$
1-Benzoyl-3-methyl-3-phenyl-2-thiourea 745
1-(2-Benzoylphenyl)-3-methyl-2-thiourea 744
2-Thiophenecarboxylic acid (5-indanylmethylene)-hydrazide 1073

$C_{15}H_{14}N_2OS_2$
5-(3-Indolylmethylene)-3-isopropylrhodanine 890

$C_{15}H_{14}N_2O_2$
α-(4-Acetamidophenylimino)-*o*-cresol 1033
Acetophenone 4-carboxyphenylhydrazone 343
Benzoic (2-methoxybenzylidene)-hydrazide 498
Benzoic (4-methoxybenzylidene)-hydrazide 498
4-(2-Benzoylacetamido)aniline 443
1-Benzoyl-2-(2-methylbenzoyl)-hydrazine 500
1-(2-Benzoylphenyl)-3-methylurea 575
Bis(3-methoxyphenyl)carbodiimide 769
1,2-Dibenzoyl-1-methylhydrazine 500
1,2-Dihydro-2-(3-methoxyphenyl)-4(3*H*)-quinazolinone 698
1,2-Dihydro-2-(4-methoxyphenyl)-4(3*H*)-quinazolinone 698
2-Dimethylamino-1-nitrofluorene 274
2-(Dimethylamino)-7-nitrofluorene 277
N-(3,5-Dimethyl-4-nitrobenzylidene)-aniline 278
N,*N*′-Diphenylmalonamide 443
3-Ethyl-5-methyl-6-phenyl-4(5*H*)-isoxazolo(4,5-c)pyridinone 874
3-Hydroxy-3-(6-methyl-2-pyridylmethyl)-2-indolinone 932
2-Methoxybenzoic benzylidenehydrazide 497
4-Methoxybenzoic benzylidenehydrazide 1045
4′,4″-Methylenebis(formanilide) 439
N-(2-Nitrobenzylidene)-2,5-xylidine 1055
N-(2-Nitrobenzylidene)-2,6-xylidine 1055
N-(2-Nitrobenzylidene)-3,4-xylidine 1055
N-(2-Nitrobenzylidene)-3,5-xylidine 1055
N-(3-Nitrobenzylidene)-2,4-xylidine 1059
N-(3-Nitrobenzylidene)-2,5-xylidine 1074
N-(3-Nitrobenzylidene)-3,4-xylidine 1059
Salicylic (α-methylbenzylidene)-hydrazide 496
Salicylic (2-phenylethylidene)-hydrazide 496
p-Toluic (2-hydroxybenzylidene)-hydrazide 1034

$C_{15}H_{14}N_2O_2S$
N-Cyano-2′,6′-dimethylbenzenesulfonanilide 826
N-Cyano-2′-methyl-*p*-toluenesulfonanilide 826
N-Cyano-3′-methyl-*o*-toluenesulfonanilide 826
N-Cyano-3′-methyl-*p*-toluenesulfonanilide 827
N-Cyano-4′-methyl-*p*-toluenesulfonanilide 827

$C_{15}H_{14}N_2O_2S_2$
5-Methyl-1-(*p*-toluenesulfonyl)-2-thioxobenzimidazoline 823

$C_{15}H_{14}N_2O_2S_3$
N-Benzyl-2-methylthio-5-benzothiazolesulfonamide 913
N-Benzyl-2-methylthio-6-benzothiazolesulfonamide 913

$C_{15}H_{14}N_2O_3$
4-(2-Aminobenzamido)-phenylacetic acid 468
Benzoic (4-hydroxy-3-methoxybenzylidene)hydrazide 1048
1-Benzoyl-2-(3-methoxybenzoyl)-hydrazine 498
N-Benzyl-*N*-methyl-3-nitrobenzamide 470
2-Carbamoylphenyl *N*-(*m*-tolyl)-carbamate 666
1-(2-Carboxyphenyl)-3-(*o*-tolyl)urea 576
1-(2-Carboxyphenyl)-3-(*m*-tolyl)urea 576
1-(2-Carboxyphenyl)-3-(*p*-tolyl)urea 576
1-(4-Carboxyphenyl)-3-(*o*-tolyl)urea 576
1-(4-Carboxyphenyl)-3-(*m*-tolyl)urea 576
1-(4-Carboxyphenyl)-3-(*p*-tolyl)urea 576
2′,3′-Dimethyl-3-nitrobenzanilide 469
2′,3-Dimethyl-4′-nitrobenzanilide 470
2′,4-Dimethyl-4′-nitrobenzanilide 470
4-Ethoxy-*N*-(3-nitrobenzylidene)-aniline 1044
N-(3-Ethyl-5-methyl-4-isoxazolylmethyl)phthalimide 872
2′-Ethyl-4-nitrobenzanilide 472
4′-Ethyl-4-nitrobenzanilide 473
Ethyl 4-(phenylazo)phenyl carbonate 1092
Ethyl 5-phenylazosalicylate 1083
4-Hydroxybenzoic (4-methoxybenzylidene)hydrazide 1031
4-(6-Hydroxy-3,4-xylylazo)-benzoic acid 1084
Mandelic acid *N*-(2-hydroxybenzylidene)hydrazide 494
2-Methoxybenzoic (2-hydroxybenzylidene)hydrazide 1031
N-(4-Methoxybenzylidene)-2-methyl-5-nitroaniline 1043
Methyl 4-(2-hydroxy-5-methylphenylazo)benzoate 1085
2′-Methyl-5′-nitro-2-phenylacetanilide 426
N-Methyl-*N*-(2-pyridylmethyl)-phthalamic acid 944
3-Nitrobenzo-2′,4′-xylidide 473
3-Nitrobenzo-2′,6′-xylidide 473
3-Nitrobenzo-3′,4′-xylidide 473
4-Nitrobenzo-2′,3′-xylidide 472
4-Nitrobenzo-2′,5′-xylidide 472
4-Nitrobenzo-2′,6′-xylidide 472
4-(3-Nitrobenzylideneamino)-phenethyl alcohol 1033
N-(2-Nitrobenzylidene)-*p*-phenetidine 1043
2-(*N*-Phenylcarbamoyl)phenyl *N*-methylcarbamate 666
Salicylic (2-methoxybenzylidene)-hydrazide 495
Salicylic (4-methoxybenzylidene)-hydrazide 495
p-Toluic (2,4-dihdyroxybenzylidene)hydrazide 494

$C_{15}H_{14}N_2O_3S$
S-Benzyl *N*-(4-methyl-2-nitrophenyl)-thiocarbamate 668
2-(Benzylthio)-3′-nitroacetanilide 420
2-Carbamoylphenyl *N*-(4-(methylthio)phenyl)carbamate 666
N-Cyano-4′-ethoxybenzenesulfonanilide 815
N-(2-Methyl-6-benzoxazolyl)-*p*-toluenesulfonamide 910
S-(*o*-Tolyl) *N*-(4-methyl-2-nitrophenyl)thiocarbamate 669
S-(*m*-Tolyl) *N*-(2-methyl-5-nitrophenyl)thiocarbamate 669

$C_{15}H_{14}N_2O_4$
1-(*o*-Anisyl)-3-(2-carboxyphenyl)urea 557
1-(*o*-Anisyl)-3-(4-carboxyphenyl)urea 557
1-(*m*-Anisyl)-3-(2-carboxyphenyl)urea 557
1-(*m*-Anisyl)-3-(4-carboxyphenyl)urea 557
1-(*p*-Anisyl)-3-(2-carboxyphenyl)urea 557
1-(*p*-Anisyl)-3-(4-carboxyphenyl)urea 557
2,5-Dimethoxy-*N*-(2-nitrobenzylidene)aniline 1043
2′-Hydroxy-4′-methacrylamido-2-furanilide 845
2-Methoxybenzoic (2,4-dihydroxybenzylidene)hydrazide 494
2′-Methoxy-3-methyl-4′-nitrobenzanilide 465
2-Methoxy-*N*-methyl-4′-nitrobenzanilide 465
2′-Methoxy-4-methyl-4′-nitrobenzanilide 465
2′-Methoxy-4-methyl-5′-nitrobenzanilide 465
6-Methoxy-α-(2-methyl-5-nitrophenylimino)-*o*-cresol 1030
6-Methoxy-α-(4-methyl-3-nitrophenylimino)-*o*-cresol 1030
N-(2-Methoxy-4-nitrophenyl)-2-phenylacetamide 416
N-(2-Methoxy-5-nitrophenyl)-2-phenylacetamide 416
3-Methyl-2′-methyl-6′-nitrobenzanilide 464
2-Pyridyl *N*-(4-(ethoxycarbonyl)-phenyl)carbamate 962

$C_{15}H_{14}N_2O_4S$
S-(4-Methoxyphenyl) *N*-(2-methyl-5-nitrophenyl)thiocarbamate 652
3-Methyl-4-nitrophenyl *N*-(4-(methylthio)phenyl)carbamate 667
4-(Methylthio)phenyl *N*-(4-methyl-2-nitrophenyl)carbamate 666
2-Nitrobenzyl *N*-(4-(methylthio)-phenyl)carbamate 667

$C_{15}H_{14}N_2O_5$
N-(4-Ethoxyphenyl)-4-nitroanthranilic acid 340
6-Methoxy-α-(2-methoxy-4-nitrophenylimino)-*o*-cresol 1030
6-Methoxy-α-(2-methoxy-5-nitrophenylimino)-*o*-cresol 1030
4-Methoxyphenyl *N*-(4-methyl-2-nitrophenyl)carbamate 651
3-(*N*-(3-Nitrophenyl)carbamoyl)-bicyclo(2.2.1)hept-5-ene-2-carboxylic acid 426
2-Phenoxyethyl *N*-(2-nitrophenyl)-carbamate 649

$C_{15}H_{14}N_2O_5S$
4-(Methylthio)phenyl *N*-(2-methoxy-5-nitrophenyl)carbamate 651

$C_{15}H_{14}N_2O_6$
1,3-Bis(2-nitrophenoxy)propane 267
1,3-Bis(3-nitrophenoxy)propane 268
1,3-Bis(4-nitrophenoxy)propane 269
N-(2,5-Dimethoxy-4-nitrophenyl)-anthranilic acid 340
N-(3,4-Dimethoxyphenyl)-4-nitroanthranilic acid, potassium salt 350
4′-(2-Furoylamino)-3′-hydroxymalonanilic acid 845
4-Methoxyphenyl *N*-(2-methoxy-4-nitrophenyl)carbamate 648
4-Methoxyphenyl *N*-(2-methoxy-5-nitrophenyl)carbamate 648

$C_{15}H_{14}N_2S_2$
3,5-Diphenyl-3,4,5,6-tetrahydro-2*H*-1,3,5-thiadiazine-2-thione 742
4,6-Diphenyl-3,4,5,6-tetrahydro-2*H*-1,3,5-thiadiazine-2-thione 741

$C_{15}H_{14}N_4$
3,5-Dianilino-1*H*-pyrazole 861
2,3-Dihydro-5,6-di-(2-pyridyl)-2-methylpyrazine 966
4-(4-(Dimethylamino)phenylazo)-benzonitrile 1089

$C_{15}H_{14}N_4O$
3-(3-Aminophenylamino)-1-phenyl-2-pyrazolin-5-one 701
1-(*N*-(2,3-Xylyl)carbamoyl)-1*H*-benzotriazole 909
1-(*N*-(3,4-Xylyl)carbamoyl)-1*H*-benzotriazole 909

$C_{15}H_{14}N_4OS$
1-(1*H*-Indazol-5-yl)-3-(4-(methylthio)-phenyl)urea 899

$C_{15}H_{14}N_4O_2$
5-(2-Carboxyphenyl)-3-methyl-1-phenylformazan 1091
Hippuric (4-pyridylmethylene)-hydrazide 1059

$C_{15}H_{14}N_4O_3$
N-Methylanthranilic (3-nitrobenzylidene)hydrazide 500

$C_{15}H_{14}N_4O_4$
1-(4-Acetamidophenyl)-3-(3-nitrophenyl)urea 562
N,*N*′-Bis(2-methyl-5-nitrophenyl)-formamidine 725
1-(2-Carbamoylphenyl)-3-(2-methyl-5-nitrophenyl)urea 563
1-(2-Carbamoylphenyl)-3-(4-methyl-2-nitrophenyl)urea 562
1-(2-Carbamoylphenyl)-3-(4-methyl-3-nitrophenyl)urea 563
N-(4-Nitrobenzylidene)-*N*′-(4-nitrophenyl)ethylenediamine 1057

$C_{15}H_{14}N_4O_5$
1-(2-Carbamoylphenyl)-3-(2-methoxy-5-nitrophenyl)urea 552
1-(4-Carboxyphenyl)-4-(4-methyl-2-nitrophenyl)semicarbazide 562
Ethyl 4-(3-(3-nitro-2-pyridyl)ureido)-benzoate 961
1-(2-Methyl-4-nitrophenyl)-3-(4-methyl-3-nitrophenyl)urea 562

$C_{15}H_{14}N_4O_6$
3-Hydroxy-4-methoxybenzaldehyde (2,4-dinitrophenyl)-methylhydrazone 1029
1-(4-Methoxy-3-nitrophenyl)-3-(2-methyl-5-nitrophenyl)urea 552
N-(2,4,5-Trimethylphenyl)-picrylamine 272
N-(2,4,6-Trimethylphenyl)-picrylamine 272

$C_{15}H_{14}N_4S$
5-Amino-3-(benzylphenylamino)-1,2,4-thiadiazole 878

$C_{15}H_{14}N_6$
1,1′-Trimethylenebis(benzotriazole) 909

$C_{15}H_{14}N_6O_4$
2-(3-(3-Nitrophenyl)ureido)acetic (3-pyridylmethylene)hydrazide 948

$C_{15}H_{14}O$
2,2′-Dimethylbenzophenone 314
4,4′-Dimethylbenzophenone 311
4-Ethylbenzophenone 313
4-(1-Indanyl)phenol 196
1-(4-Methoxyphenyl)-1-phenylethylene 180
4-Methoxy-α-phenylstyrene 180
cis-4-Methoxystilbene 180
2-Methoxystilbene 180
4-Methoxystilbene 180
α-Methyl-2-fluorenemethanol 208
4′-Methyl-2-phenylacetophenone 297
8-Methyl-1,2,3,4-tetrahydro-1-phenanthrenone 296
3-Phenylpropiophenone 296
4′-Phenylpropiophenone 297
1,2,3,4-Tetrahydro-9-phenanthrenecarboxaldehyde 324

$C_{15}H_{14}OS$
2-(Benzylthio)acetophenone 293

$C_{15}H_{14}OS_2$
S-Diphenylmethyl *S*′-methyl dithiocarbonate 381

$C_{15}H_{14}O_2$
3-Allyl-2,2′-biphenol 194
2-(2-Allyloxyphenyl)phenol 195
p-Anisyl benzyl ketone 293
3′-Benzyloxyacetophenone 292
4′-Benzyloxyacetophenone 293
Benzyl phenylacetate 356
(DL)-2-(4-Biphenylyl)propionic acid 333
(*S*)-(+)-2-(4-Biphenylyl)-propionic acid 333
2,3-Dihydro-4-phenyl-4*H*-1-benzopyran-4-ol 203
Diphenylmethyl acetate 356
4-Methoxy-4′-methylbenzophenone 311
2-(4-Methoxyphenyl)acetophenone 292
4′-Methoxy-2-phenylacetophenone 292
α-Methylbenzyl benzoate 370
Methyl 3-phenyl-2,5-norbornadiene-2-carboxylate 375
Phenethyl benzoate 375
2-Phenoxypropiophenone 291
1,2,3,4-Tetrahydro-9-methoxyphenanthren-1-one 291
m-Tolyl phenylacetate 380
p-Tolyl phenylacetate 381

$C_{15}H_{14}O_2S$
4,5-Dihydro-2-(ethoxycarbonyl)-naphtho(1,2-b)thiophene 858
trans-Phenyl 2-phenylcyclopropyl sulfone 782
Styryl *p*-tolyl sulfone 784

$C_{15}H_{14}O_3$
Benzyl mandelate 353
4′-Benzyloxy-2′-hydroxyacetophenone 289
2-(4-Biphenylyloxy)propionic acid 327
3-(4-Biphenylyloxy)propionic acid 328

$C_{15}H_{14}O_3$ (Cont'd)
1,5-Bis(5-methyl-2-furyl)-1,4-pentadien-3-one ... 853
Di-*p*-tolyl carbonate ... 381
2-Ethoxy-2-phenyl-1,3-benzodioxole. 183
Guaiacol phenylacetate ... 378
2-Hydroxy-4-methoxy-4′-methylbenzophenone ... 310
2-Hydroxy-4′-methoxy-2-phenylacetophenone ... 289
4-Hydroxy-α-phenylhydrocinnamic acid, sodium salt trihydrate ... 350
3-(4-Hydroxyphenyl)-2-phenylpropionic acid ... 326
2-(2-Methoxyphenoxy)-acetophenone ... 291
4′-Methoxy-2-phenoxyacetophenone ... 291
2-(4-Methoxyphenyl)-2-phenylacetic acid ... 328
Methyl 4-benzyloxybenzoate ... 369
α-Methylbenzyl salicylate ... 368
2-Methyl-3-(1-naphthoyl)-propionic acid ... 331
Phenethyl salicylate ... 368

$C_{15}H_{14}O_3S$
2-(Benzylsulfonyl)acetophenone ... 780
2-(*p*-Tolylsulfonyl)acetophenone ... 780

$C_{15}H_{14}O_4$
2-(1-Acetyl-2-naphthyloxy)-propionic acid ... 327
5-Cinnamylidene-2,2-dimethyl-1,3-dioxane-4,6-dione ... 386
2′,4′-Dihydroxy-2-(3-methoxyphenyl)acetophenone ... 289
4-Methoxybenzyl phenyl carbonate.. 381
3-(4-Methoxy-1-naphthoyl)-propionic acid ... 328
2-Methyl-1,4-naphthalenediol diacetate ... 379
4-(2-Phenoxyethoxy)benzoic acid ... 340

$C_{15}H_{14}O_4S$
3,6-Dimethyl-2-(4-tolylsulfonyl)-1,4-benzoquinone ... 781
1,3-Diphenyl-3-oxo-1-propanesulfonic acid, sodium salt.. 794
4′-Methoxy-2-(phenylsulfonyl)-acetophenone ... 775

$C_{15}H_{14}O_5$
4-Methoxybenzyl 2,4,6-trihydroxyphenyl ketone ... 289
2-(2-Methoxyphenyl)-2′,4,6′-trihydroxyacetophenone ... 289
Methyl 3-benzyloxy-4,5-dihydroxybenzoate ... 367

$C_{15}H_{14}S$
Styryl *o*-tolyl sulfide ... 212
Styryl *m*-tolyl sulfide ... 212

$C_{15}H_{14}S_2$
1-(Phenylthio)-2-(*p*-tolylthio)-ethylene ... 212

$C_{15}H_{15}BrClNO_3S$
1-(4-Chlorobenzoylmethyl)-4-(methylsulfonylmethyl)pyridinium bromide ... 928

$C_{15}H_{15}BrN_2$
α-(4-Bromophenylimino)-*N*,*N*-dimethyl-*p*-toluidine ... 1062

$C_{15}H_{15}BrN_2O$
α-(4-Bromoanilino)-*p*-acetotoluidide. 430
1-(3-Bromophenyl)-3-(3,5-dimethylphenyl)urea ... 545
1-(4-Bromophenyl)-3-(2-phenethyl)urea ... 535

$C_{15}H_{15}BrN_2O_3$
1-(3-Bromophenyl)-3-(3,5-dimethoxyphenyl)urea ... 515

$C_{15}H_{15}BrN_2O_3S$
p-Toluenesulfonic acid (5-bromo-2-methoxybenzylidene)-hydrazide ... 1048

$C_{15}H_{15}BrO_2$
1-(2-Bromo-4-methylphenoxy)-2-phenoxyethane ... 177
2-(6-Bromo-2-naphthoxy)-tetrahydropyran ... 187

$C_{15}H_{15}BrO_4$
Bromo-2-hydroxy-3-(3-hydroxy-3-methylbutyl)-1,4-naphthoquinone.... 315

$C_{15}H_{15}Br_2N$
2,2′-Dibromo-4,4′,*N*-trimethyldiphenylamine ... 225

$C_{15}H_{15}ClF_3NO_3$
2′-Chloro-hexahydro-5′-(trifluoromethyl)phthalanilic acid.... 395

$C_{15}H_{15}ClIN$
4-(4-Chlorostyryl)-1-ethylpyridinium iodide ... 957

$C_{15}H_{15}ClN_2$
α-(2-Chlorophenylimino)-*N*,*N*-dimethyl-*p*-toluidine ... 1061
α-(3-Chlorophenylimino)-*N*,*N*-dimethyl-*p*-toluidine ... 1061
α-(4-Chlorophenylimino)-*N*,*N*-dimethyl-*p*-toluidine ... 1061

$C_{15}H_{15}ClN_2O$
Basic dye (C.I. 45000) ... 1060
1-Benzyl-3-(2-chloro-6-methylphenyl)urea ... 573
1-Benzyl-3-(3-chloro-4-methylphenyl)urea ... 573
1-Benzyl-3-(4-chloro-2-methylphenyl)urea ... 573
1-Benzyl-3-(5-chloro-2-methylphenyl)urea ... 573
1-Benzyl-3-(3-chlorophenyl)-1-methylurea ... 571
α-(4-Chloroanilino)-*p*-acetotoluidide. 430
1-(2-Chlorobenzyl)-1-methyl-3-phenylurea ... 535
1-(3-Chlorobenzyl)-1-methyl-3-phenylurea ... 535
3-(2-Chloro-6-methylphenyl)-1-methyl-1-phenylurea ... 536
1-(3-Chloro-2-methylphenyl)-3-(*m*-tolyl)urea ... 541
1-(3-Chloro-4-methylphenyl)-3-(*m*-tolyl)urea ... 544
1-(3-Chloro-4-methylphenyl)-3-*p*-tolylurea ... 591
1-(2-Chlorophenyl)-3-(2,3-dimethylphenyl)urea ... 537
1-(3-Chlorophenyl)-3-(2-ethylphenyl)urea ... 536
1-(3-Chlorophenyl)-3-(4-ethylphenyl)urea ... 539
1-(4-Chlorophenyl)-3-(2-ethylphenyl)urea ... 537
1-(4-Chlorophenyl)-3-ethyl-3-phenylurea ... 535
1-(4-Chlorophenyl)-3-(4-ethylphenyl)urea ... 539
3-(3-Chlorophenyl)-1-ethyl-1-phenylurea ... 581
1-(3-Chlorophenyl)-3-(4-methylbenzyl)urea ... 539
1-(2-Chlorophenyl)-3-phenethylurea.. 570
1-(3-Chlorophenyl)-3-phenethylurea.. 570
1-(4-Chlorophenyl)-3-phenethylurea.. 570
1-(2-Chlorophenyl)-3-(1-phenylethyl)urea ... 535
1-(2-Chlorophenyl)-3-(2,4-xylyl)urea ... 589
1-(2-Chlorophenyl)-3-(2,6-xylyl)urea ... 537
1-(3-Chlorophenyl)-3-(2,3-xylyl)urea ... 541
1-(3-Chlorophenyl)-3-(2,4-xylyl)urea ... 601
1-(3-Chlorophenyl)-3-(2,5-xylyl)urea ... 544
1-(3-Chlorophenyl)-3-(2,6-xylyl)urea ... 541
1-(3-Chlorophenyl)-3-(3,4-xylyl)urea ... 544
1-(4-Chlorophenyl)-3-(2,5-xylyl)urea ... 540
1-(4-Chlorophenyl)-3-(2,6-xylyl)urea ... 539

$C_{15}H_{15}ClN_2OS$
1-(2-Chloro-5-methylphenyl)-3-(4-(methylthio)phenyl)urea ... 522
1-(3-Chloro-2-methylphenyl)-3-(4-(methylthio)phenyl)urea ... 522
1-(5-Chloro-2-methylphenyl)-3-(4-(methylthio)phenyl)urea ... 522

$C_{15}H_{15}ClN_2O_2$
1-(3-Chlorobenzyl)-3-(4-methoxyphenyl)urea ... 521
1-(2-Chloro-4-methoxyphenyl)-3-(*o*-tolyl)urea ... 520
1-(4-Chloro-2-methoxyphenyl)-3-(*o*-tolyl)urea ... 529
1-(2-Chloro-6-methylphenyl)-3-(2-methoxyphenyl)urea ... 519
1-(2-Chloro-6-methylphenyl)-3-(4-methoxyphenyl)urea ... 521
1-(3-Chloro-2-methylphenyl)-3-(4-methoxyphenyl)urea ... 521
1-(3-Chloro-4-methylphenyl)-3-(4-methoxyphenyl)urea ... 521
1-(4-Chloro-2-methylphenyl)-3-(4-methoxyphenyl)urea ... 521
1-(2-Chlorophenyl)-3-(2-ethoxyphenyl)urea ... 519
1-(2-Chlorophenyl)-3-(4-ethoxyphenyl)urea ... 520
3-(4-Pyridyl)propyl *N*-(2-chlorophenyl)carbamate ... 929
4-Pyridylpropyl *N*-(3-chlorophenyl)-carbamate ... 930

$C_{15}H_{15}ClN_2O_2S$
1-(2-Chloro-4-methoxyphenyl)-3-(4-(methylthio)phenyl)urea ... 517
1-(5-Chloro-2-methoxyphenyl)-3-(4-(methylthio)phenyl)urea ... 517

$C_{15}H_{15}ClN_2O_3$
2-Chloro-3-(2,6-dimethyl-3,4,5,6-tetrahydro-2*H*-1,3,4-oxadiazin-4-yl)-1,4-naphthoquinone ... 318
1-(2-Chloro-4-methoxyphenyl)-3-(2-methoxyphenyl)urea ... 516
1-(2-Chloro-4-methoxyphenyl)-3-(4-methoxyphenyl)urea ... 516
1-(4-Chloro-2-methoxyphenyl)-3-(2-methoxyphenyl)urea ... 516
1-(4-Chloro-2-methoxyphenyl)-3-(4-methoxyphenyl)urea ... 516
1-(5-Chloro-2-methoxyphenyl)-3-(4-methoxyphenyl)urea ... 516
1-(2-Chlorophenyl)-3-(2,4-dimethoxyphenyl)urea ... 515
1-(3-Chlorophenyl)-3-(2,4-dimethoxyphenyl)urea ... 515

$C_{15}H_{15}ClN_2O_4$
5-Norbornen-2-ylmethyl *N*-(4-chloro-2-nitrophenyl)carbamate ... 628
4-Pyridylmethyl *N*-(5-chloro-2,4-dimethoxyphenyl)carbamate ... 926

$C_{15}H_{15}ClN_2O_4S$
4′-(*N*-(5-Chloro-2-methoxyphenyl)-sulfamoyl)acetanilide ... 806

$C_{15}H_{15}ClN_2S$
1-(4-Chlorobenzyl)-1-methyl-3-phenyl-2-thiourea ... 730

$C_{15}H_{15}ClN_4O$
1-(2-(4-Chlorobenzyloxy)-benzylideneamino)guanidine nitrate ... 1023

$C_{15}H_{15}ClN_4OS$
1-(*N*-(3-Chlorophenyl)carbamoyl)-4-(2-methylphenyl)-3-thiosemicarbazide ... 732

$C_{15}H_{15}ClN_4O_2$
4-(3-Chlorophenyl)-1-(2-(methylamino)benzoyl)-semicarbazide ... 527

$C_{15}H_{15}ClN_4O_3$
2-(Furfurylureido)acetic acid(2-chlorobenzylidene)hydrazide ... 1026

$C_{15}H_{15}ClN_4O_4$
4-(5-Chloro-2,4-dimethoxyphenyl)-1-(2-pyridylcarbonyl)semicarbazide.... 927
4-(5-Chloro-2,4-dimethoxyphenyl)-1-(3-pyridylcarbonyl)semicarbazide.... 926

$C_{15}H_{15}ClO_2S$
4-(1-Chloro-1-methylethyl)phenyl phenyl sulfone ... 784
4-Chlorophenyl mesityl sulfone ... 774

$C_{15}H_{15}ClS$
4-Chloro-α-ethylbenzyl phenyl sulfide ... 210
4-Chloro-α-methylphenethyl phenyl sulfide ... 210

$C_{15}H_{15}Cl_2NO_2$
N-(2,3-Dichlorophenyl)-2,3-dimethoxybenzylamine ... 228

$C_{15}H_{15}Cl_2NO_3$
N-(2,4-Dichlorobenzyl)-1,2,3,6-tetrahydrophthalamic acid ... 406

$C_{15}H_{15}Cl_2N_3O_2$
1-(4-(Chlorophenyl-carbamoyl)methyl)-3-(methyl-carbamoyl)pyridinium chloride ... 948

$C_{15}H_{15}Cl_3N_2O$
1-(3,4-Dichlorobenzyl)-3-(dimethylcarbamoyl)-pyridinium chloride ... 946

$C_{15}H_{15}FN_2$
N-(4-(Dimethylamino)benzylidene)-4-fluoroaniline ... 1025

$C_{15}H_{15}FN_2O$
1-Benzyl-3-(2-fluorophenyl)-1-methylurea ... 527
1-(4-Ethylphenyl)-3-(2-fluorophenyl)urea ... 537
α-(4-Fluoroanilino)-*p*-acetotoluidide. 430
1-(2-Fluorophenyl)-3-(2-ethylphenyl)urea ... 536
1-(2-Fluorophenyl)-3-(2,3-xylyl)urea ... 537
1-(2-Fluorophenyl)-3-(2,5-xylyl)urea ... 538
1-(2-Fluorophenyl)-3-(2,6-xylyl)urea ... 537
1-(3-Fluorophenyl)-3-(3,4-xylyl)urea ... 544

$C_{15}H_{15}FN_2O_2$
1-(2-Ethoxyphenyl)-3-(3-fluorophenyl)urea ... 519
1-(4-Ethoxyphenyl)-3-(3-fluorophenyl)urea ... 521
1-(2-Fluorophenyl)-3-(4-(2-hydroxyethyl)phenyl)urea ... 514
1-(3-Fluorophenyl)-3-(4-(2-hydroxyethyl)phenyl)urea ... 514
1-(4-Fluorophenyl)-3-(4-(2-hydroxyethyl)phenyl)urea ... 514
1-(2-Fluorophenyl)-3-(2-methoxy-5-methylphenyl)urea ... 519
1-(3-Fluorophenyl)-3-(2-methoxy-5-methylphenyl)urea ... 520

$C_{15}H_{15}FN_2O_3$
1-(2,4-Dimethoxyphenyl)-3-(2-fluorophenyl)urea ... 515
1-(2,4-Dimethoxyphenyl)-3-(4-fluorophenyl)urea ... 515
1-(2,5-Dimethoxyphenyl)-3-(3-fluorophenyl)urea ... 515
1-(2,5-Dimethoxyphenyl)-3-(4-fluorophenyl)urea ... 515

$C_{15}H_{15}FN_4O_2$
4-(4-Fluorophenyl)-1-(2-(methylamino)benzoyl)-semicarbazide ... 527

$C_{15}H_{15}FN_4O_2S$
3-Ethyl-5-(4-(fluorosulfonyl)phenyl)-1-phenylformazan ... 1091

$C_{15}H_{15}FN_4O_3$
1-(4-(Dimethylamino)phenyl)-3-(2-fluoro-5-nitrophenyl)urea ... 523
2-(3-Furfurylureido)acetic (2-fluorobenzylidene)hydrazide ... 1024
2-(3-Furfurylureido)acetic (4-fluorobenzylidene)hydrazide ... 1024

$C_{15}H_{15}FO_2$
1-(4-Fluorophenoxy)-3-phenoxypropane ... 177

$C_{15}H_{15}FO_2S$
4-Fluorophenyl mesityl sulfone ... 774

$C_{15}H_{15}F_3N_2O_3$
Methyl 3-hydroxy-1-methyl-5-(3-(trifluoromethyl)benzyl)-4-pyrazoleacetate ... 859

$C_{15}H_{15}IN_2$
N,*N*-Dimethyl-α-(4-iodophenylimino)-*p*-toluidine ... 1025
2-Phenyl-1,3-dimethylbenzimidazolium iodide..... 906

$C_{15}H_{15}IN_2O$
α-(4-Iodoanilino)-*p*-acetotoluidide..... 430

$C_{15}H_{15}IN_2O_2$
1-Ethyl-2-(3-nitrostyryl)pyridinium iodide ... 956
1-Ethyl-4-(3-nitrostyryl)pyridinium iodide ... 956

$C_{15}H_{15}IN_2O_3$
1-(2,5-Dimethoxyphenyl)-3-(4-iodophenyl)urea ... 515

$C_{15}H_{15}N$
N-Benzylidene-4-ethylaniline ... 1077
N-Benzylidene-2,3-xylidine ... 1067
N-Benzylidene-2,4-xylidine ... 1067
N-Benzylidene-2,5-xylidine ... 1067
9,10-Dihydro-9,9-dimethylacridine.... 250
9-(Dimethylamino)fluorene ... 233

$C_{15}H_{15}N$ (Cont'd)
N-(4-Methylbenzylidene)-*o*-toluidine. 1064
N-(4-Methylbenzylidene)-*p*-toluidine. 1065
2-Phenyl-5,6,7,8-tetrahydroquinoline hydrochloride.. 966

$C_{15}H_{15}NO$
10-Aminomethyl-10,11-dihydro-5*H*-dibenzo(a,d)cyclohepten-5-one 312
2′,3′-Benzoxylidide 481
2′,4′-Benzoxylidide 481
2′,5′-Benzoxylidide 481
2′,6′-Benzoxylidide 481
3′,4′-Benzoxylidide 481
N-Benzylidene-*p*-phenetidine 1047
α-Benzylimino-α-methyl-*o*-cresol 1035
N-Benzyl-*N*-methylbenzamide 472
2-Benzyl-*N*-methylbenzamide 479
N-Benzyl-4-methylbenzamide 472
2-(Dimethylamino)-9-hydroxyfluorene 227
α-(2,3-Dimethylphenylimino)-*o*-cresol 1035
1,3-Diphenylacetone oxime 1010
2,2-Diphenyl-*N*-methylacetamide 430
N-(2-Ethoxybenzylidene)aniline 1046
2-(Ethylamino)benzophenone 312
α-(4-Ethylphenylimino)-*o*-cresol 1035
α-(4-Ethylphenylimino)-*p*-cresol 1038
1,2,3,3a,4,5-Hexahydropyrido(3,2,1-jk)carbazol-6(6*H*)-one 890
N-(2-Methoxybenzylidene)-*p*-toluidine 1046
N-(3-Methoxybenzylidene)-*o*-toluidine 1046
N-(3-Methoxybenzylidene)-*m*-toluidine 1044
N-(4-Methoxybenzylidene)-*o*-toluidine 1046
N-(4-Methoxybenzylidene)-*p*-toluidine 1047
N-(4-Methylbenzylidene)-*o*-anisidine 1045
N-(4-Methylbenzylidene)-*m*-anisidine 1047
N-(4-Methylbenzylidene)-*p*-anisidine 1047
N-Methyl-*o*-toluanilide 483
2′-Methyl-*m*-toluanilide 479
4′-Methyl-*o*-toluanilide 479
α-(Phenethylimino)-*o*-cresol 1033
3-(Propylamino)dibenzofuran 894
p-Tolu-*m*-toluidide 480
p-Tolu-*p*-toluidide 480
α-(2,4-Xylylimino)-*o*-cresol 1035
α-(2,5-Xylylimino)-*o*-cresol 1035
α-(3,4-Xylylimino)-*o*-cresol 1035
α-(3,4-Xylylimino)-*p*-cresol 1038
α-(3,5-Xylylimino)-*o*-cresol 1034

$C_{15}H_{15}NOS$
3-(4-(Dimethylamino)phenyl)-1-(2-thienyl)-2-propen-1-one 856
N-(4-Methoxybenzylidene)-3-(methylthio)aniline 1042
4-Methoxy-*N*-methylthiobenzanilide. 738
2′-(Methylthio)-4′-phenylacetanilide. 487

$C_{15}H_{15}NO_2$
p-Benzophenetidide 466
N-Benzylidene-2,5-dimethoxyaniline 1045
α-Benzylimino-6-methoxy-*o*-cresol 1030
3′-Benzyloxyacetanilide 418
4′-Benzyloxyacetanilide 419
N-(4-Biphenylyl)-2-methoxyacetamide 419
4-Cyclohexylidene-2-phenyl-5(4*H*)-oxazolone 387
N-(2,5-Dimethoxybenzylidene)-aniline 1048
N-(3,4-Dimethoxybenzylidene)-aniline 1042
3-(4-(Dimethylamino)phenyl)-1-(2-furyl)-2-propen-1-one 850
1-(α,α-Dimethylbenzyl)-4-nitrobenzene 278
1,3-Diphenyl-2-nitropropane 277
2′-Ethoxybenzanilide 465
4′-Ethoxybenzanilide 466
α-(4-Ethoxyphenylimino)-*o*-cresol 1031
α-(4-Ethoxyphenylimino)-*p*-cresol 1032
Ethyl α-cyano-5-indanacrylate 766
3′-(2-Hydroxybenzylamino)-acetophenone 290
N-(2-Methoxybenzylidene)-*p*-anisidine 1046
N-(3-Methoxybenzylidene)-*o*-anisidine 1045
N-(3-Methoxybenzylidene)-*m*-anisidine 1044
N-(4-Methoxybenzylidene)-*o*-anisidine 1045
N-(4-Methoxybenzylidene)-*m*-anisidine 1047
N-(4-Methoxybenzylidene)-*p*-anisidine 1047
α-(2-Methoxy-5-methylphenylimino)-*p*-cresol 1032
α-(4-Methoxy-2-methylphenylimino)-*o*-cresol 1031
6-Methoxy-α-(4-methylphenylimino)-*o*-cresol 1030
2′-Methoxy-2-phenylacetanilide 416
4′-Methoxy-2-phenylacetanilide 419
2′-Methoxy-*o*-toluanilide 465
4′-Methoxy-*o*-toluanilide 466
N-Methyl-*p*-anisanilide 467
Phenethyl anthranilate 372
Phenyl *N*-(2-ethylphenyl)carbamate. 680
Phenyl *N*-(2,3-xylyl)carbamate 684
Phenyl *N*-(2,4-xylyl)carbamate 686
Phenyl *N*-(2,5-xylyl)carbamate 686
Phenyl *N*-(2,6-xylyl)carbamate 684
Phenyl *N*-(3,4-xylyl)carbamate 686
Phenyl *N*-(3,5-xylyl)carbamate 682
2′,6′-Salicyloxylidide 462
p-Tolu-*m*-anisidide 466
m-Tolyl *N*-(*m*-tolyl)carbamate 678
2,6-Xylyl *N*-phenylcarbamate 681

$C_{15}H_{15}NO_2S$
α,α-Dimethyl-4-nitrobenzyl phenyl sulfide 270
S-(2-Hydroxyphenyl) *N*-(2-ethylphenyl)thiocarbamate 735
S-(2-Hydroxyphenyl) *N*-(3,4-xylyl)-thiocarbamate 646
4-Isopropylphenyl 4-nitrophenyl sulfide 270
2-(2-Mercapto-4,5,6-trimethyl-3-pyridyl)benzoic acid 940
1-Methyl-2-(4-nitrophenyl)ethyl phenyl sulfide 270
4′-(Methylthio)-2-phenoxyacetanilide 414
2-(*p*-Tolylsulfonyl)-isoindoline 827

$C_{15}H_{15}NO_3$
Benzyl 3,4-dimethyl-5-formyl-2-pyrrolecarboxylate 839
2′,4-Dimethoxybenzanilide 463
α-(2,5-Dimethoxyphenylimino)-*o*-cresol 1031
α-(2-Hydroxy-5-methylphenylimino)-6-methoxy-*o*-cresol 1028
4-(2-Hydroxy-1-naphthoyl)-morpholine 697
4-(4-Methoxybenzylamino)-benzoic acid 340
2-Methoxy-α-(4-methoxyphenylimino)-*p*-cresol 1030
N-(1-Naphthyl)glutaramic acid 489
3-(*N*-Phenylcarbamoyl)-5-norbornene-2-carboxylic acid 434
Phenyl *N*-(2-methoxy-5-methylphenyl)carbamate 653
m-Tolyl *N*-(2-methoxyphenyl)-carbamate 653
m-Tolyl *N*-(4-methoxyphenyl)-carbamate 655

$C_{15}H_{15}NO_3S$
N-Acetyl-*p*-toluenesulfonanilide 820
S-(2-Hydroxyphenyl) *N*-(2-methoxy-5-methylphenyl)thiocarbamate 642
S-(4-Hydroxyphenyl) *N*-(2-methoxy-5-methylphenyl)thiocarbamate 642
3-Methoxyphenyl *N*-(4-(methylthio)-phenyl)carbamate 651
4-Methoxyphenyl *N*-(4-(methylthio)-phenyl)carbamate 651
5-Methyl-2-methoxy-4-nitrophenyl *p*-tolyl sulfide 266
4-(Methylthio)phenyl *N*-(4-methoxyphenyl)carbamate 651

$C_{15}H_{15}NO_3S_2$
3-Methylbenzothiazolium *p*-toluenesulfonate 916

$C_{15}H_{15}NO_4$
4-Methoxyphenyl *N*-(2-methoxphenyl)carbamate 648
4-Methoxyphenyl *N*-(4-methoxyphenyl)carbamate 648

$C_{15}H_{15}NO_4S$
2′,5′-Dimethoxy-2-(2-thenoyl)-acetanilide 855
α,α-Dimethyl-4-nitrobenzyl phenyl sulfone 779
S-(4-Hydroxyphenyl) *N*-(2,5-dimethoxyphenyl)thiocarbamate 642
4-Isopropylphenyl 4-nitrophenyl sulfone 779
10-Methylacridinium methyl sulfate. 996
4-(1-Methyl-1-nitroethyl)phenyl phenyl sulfone 780

$C_{15}H_{15}NO_5$
Furfuryl *N*-(4-(ethoxycarbonyl)-phenyl)carbamate 848
2-Methoxy-5-nitrobenzyl 4-methoxyphenyl ether 265

$C_{15}H_{15}NO_5S$
2-(2-Nitrophenyl)ethyl *p*-toluenesulfonate 799

$C_{15}H_{15}NO_6S$
2-(4-Nitrophenoxy)ethyl *p*-toluenesulfonate 797

$C_{15}H_{15}NS$
10-Isopropylphenothiazine 251
10-Propylphenothiazine 251

$C_{15}H_{15}NS_2$
Dibenzyldithiocarbamic acid, zinc salt 743

$C_{15}H_{15}N_3$
5-Indancarboxaldehyde 2-pyridylhydrazone 1068

$C_{15}H_{15}N_3O$
4-Acetamidobenzaldehyde phenylhydrazone 431
4-Aminobenzoic (4-methylbenzylidene)hydrazide 499
6,9-Diamino-2-ethoxyacridine hydrochloride 995
4′-(Phenylazo)propionanilide 1092

$C_{15}H_{15}N_3OS$
1-(4-Methoxybenzylidene)-4-phenyl-3-thiosemicarbazide 737

$C_{15}H_{15}N_3O_2$
1-(4-Acetamidophenyl)-3-phenylurea 569
1-(3-Acetylphenyl)-3-(3-methyl-2-pyridyl)urea 961
4-Aminobenzoic (4-methoxybenzylidene)hydrazide 498
Anthranilic (4-methoxybenzylidene)hydrazide 1043
4-(4-(Dimethylamino)phenylazo)-benzoic acid, sodium salt 1089
N,*N*-Dimethyl-α-(3-nitrophenylimino)-*p*-toluidine 1054
N,*N*-Dimethyl-α-(4-nitrophenylimino)-*p*-toluidine 1057
2-(4-Methylphenylcarboxamido)-benzoic hydrazide 502
Propiophenone 4-nitrophenylhydrazone 276
4-(4-Tolylazo)phenyl *N*-methylcarbamate 1092

$C_{15}H_{15}N_3O_2S$
Benzenesulfonic (1-methyl-2-indolinylidene)hydrazide 820
1-(2-Carbamoylphenyl)-3-(3-(methylthio)phenyl)urea 561
1-(2-Carbamoylphenyl)-3-(4-(methylthio)phenyl)urea 560

$C_{15}H_{15}N_3O_2S_2$
p-Toluenesulfonic (3-methyl-2-benzothiazolinylidene)hydrazide 820

$C_{15}H_{15}N_3O_3$
Anthranilic (4-hydroxy-3-methoxybenzylidene)hydrazide 495
1-Benzyl-1-methyl-3-(4-nitrophenyl)urea 567
1-Benzyl-3-(2-methyl-4-nitrophenyl)urea 569
1-(2,4-Dimethylphenyl)-3-(4-nitrophenyl)urea 568
1-(2,6-Dimethylphenyl)-3-(3-nitrophenyl)urea 564
Ethyl 4-(3-(3-pyridyl)ureido)benzoate 945
1-(4-Methoxybenzoyl)-4-phenylsemicarbazide 557
α-(3-Nitroanilino)-*p*-acetotoluidide 425
1-(3-Nitrophenyl)-3-(2-phenethyl)urea 562
1-(4-Nitrophenyl)-3-(2-phenethyl)urea 566
1-(4-Nitrophenyl)-3-(2,6-xylyl)urea 568

$C_{15}H_{15}N_3O_3S$
1-(2-Hydroxybenzoyl)-4-(4-(methylthio)phenyl)semicarbazide.. 547
1-(2-Methyl-4-nitrophenyl)-3-(4-(methylthio)phenyl)urea 560

$C_{15}H_{15}N_3O_3S_2$
4-Methoxybenzenesulfonic (3-methyl-2-benzothiazolinylidene)-hydrazide 813

$C_{15}H_{15}N_3O_4$
Ethyl 4-(3-(3-hydroxy-2-pyridyl)-ureido)benzoate 933
1-(2-Methoxy-3-nitrophenyl)-3-(*o*-tolyl)urea 552

$C_{15}H_{15}N_3O_4S$
4-(4-(Ethoxycarbonyl)phenyl)-1-(2-thenoyl)semicarbazide 856
1-(2-Methoxy-3-nitrophenyl)-3-(4-(methylthio)phenyl)urea 551
1-(2-Methoxy-5-nitrophenyl)-3-(4-(methylthio)phenyl)urea 551

$C_{15}H_{15}N_3O_5$
1-(2-Methoxy-3-nitrophenyl)-3-(4-methoxyphenyl)urea 550

$C_{15}H_{15}N_3O_6$
1-(2,5-Dimethoxyphenyl)-3-(2-hydroxy-5-nitrophenyl)urea 546

$C_{15}H_{15}N_3O_6S$
N^4-Acetyl-N^1-(2-methoxy-5-nitrophenyl)sulfanilamide 813

$C_{15}H_{15}N_3S$
S-Methyl-1-(4-phenylbenzylidene)-isothiosemicarbazide 1071
4-Phenyl-1-(2-phenylethylidene)-3-thiosemicarbazide 1060

$C_{15}H_{15}N_5O_5$
1-(2,4-Dinitrophenyl)-4-(2-ethylphenyl)semicarbazide 561
1-(2,4-Dinitrophenyl)-1-methyl-4-(*o*-tolyl)semicarbazide 561
1-(2,4-Dinitrophenyl)-4-(2,4-xylyl)-semicarbazide 561

$C_{15}H_{15}N_7$
2,4-Bis(5,6-dimethyl-1,2,4-triazin-3-yl)pyridine 980
2,6-Bis(5,6-dimethyl-1,2,4-triazin-3-yl)pyridine 979

$C_{15}H_{16}$
1,3-Diphenylpropane 162
Ditolylmethane, mixture of isomers. 163
4-Propylbiphenyl 162

$C_{15}H_{16}AsNO_4$
2-Hydroxy-2-phenyl-1,2,3,4-tetrahydroisoarsinolinium nitrate 1127

$C_{15}H_{16}BrN$
2-Methyl-2-phenylisoindolinium bromide 255

$C_{15}H_{16}BrNO$
1-(Methoxymethyl)-4-styrylpyridinium bromide 938

$C_{15}H_{16}BrNO_2$
6-Bromo-2-naphthyl *N*-butylcarbamate 687
N-(4-Bromophenyl)-veratrylamine 228

$C_{15}H_{16}BrNO_3$
4′-Bromo-2′-methyl-1,2,3,6-tetrahydrophthalanilic acid 435
4′-Bromo-3′-methyl-1,2,3,6-tetrahydrophthalanilic acid 442

$C_{15}H_{16}BrNO_4$
5-Bromo-2-(ethoxycarbonyl)-3-indolebutyric acid 889

$C_{15}H_{16}BrN_3$
1-Benzyl-4-(2-imidazolin-2-yl)-pyridinium bromide 955
4-(4-Bromophenylazo)-*N*-ethyl-*N*-methylaniline 1081

$C_{15}H_{16}BrN_3O$
1-(2-Bromo-4,5-dimethylphenyl)-3-(4-pyridylmethyl)urea 928

$C_{15}H_{16}ClN$
4-Chloro-*N*-(3,4-dimethylphenyl)-benzylamine 225
N-(3-Chloro-*p*-tolyl)-4-methylbenzylamine 232

$C_{15}H_{16}ClNO$
4-Chloro-*N*-(4-ethoxyphenyl)-benzylamine 224
3-Chloro-6,7,8,9,10,11-hexahydrocycloocta(b)quinolin-12(5*H*)-one 320
4-Chloro-*N*-(2-(*p*-tolyloxy)-ethyl)-aniline 224

$C_{15}H_{16}ClNO_2$
α-(3-Chloro-2-methylanilino)-6-methoxy-*o*-cresol 222
α-(3-Chloro-4-methylanilino)-6-methoxy-*o*-cresol 223
α-(4-Chloro-2-methylanilino)-6-methoxy-*o*-cresol 222
α-(5-Chloro-2-methylanilino)-6-methoxy-*o*-cresol 223
N-(4-Chlorophenyl)-2,3-dimethoxybenzylamine 229
N-(4-Chlorophenyl)-3,4-dimethoxybenzylamine 228
N-(3-Chlorophenyl)veratrylamine 228

$C_{15}H_{16}ClNO_2S$
4-Chloro-2′,4′,5′-trimethylbenzenesulfonanilide 810
4-Chloro-2′,4′,6′-trimethylbenzenesulfonanilide 810

$C_{15}H_{16}ClNO_3$
α-(5-Chloro-2,4-dimethoxyanilino)-*o*-cresol 223
α-(5-Chloro-2-methoxyanilino)-6-methoxy-*o*-cresol 225
3′-Chloro-2′-methyl-1,2,3,6-tetrahydrophthalanilic acid 407

$C_{15}H_{16}ClN_3$
N-Benzyl-4-diazo-*N*-ethylaniline, tin chloride double salt 232

$C_{15}H_{16}ClN_3O$
4-(3-Chloro-2-methylphenyl)-1-methyl-1-phenylsemicarbazide 533
1-(2-Chloro-6-methylphenyl)-3-(2-(2-pyridyl)ethyl)urea 928
1-(3-Chloro-4-methylphenyl)-3-(2-(2-pyridyl)ethyl)urea 930

$C_{15}H_{16}ClN_3O_2$
1-(Benzylcarbamoylmethyl)-3-carbamoylpyridinium chloride 955
4-(5-Chloro-2-methoxyphenyl)-1-methyl-1-phenylsemicarbazide 519

$C_{15}H_{16}ClN_3O_4$
Diethyl 5-(3-chloroanilino)-3,4-pyrazoledicarboxylate 862

$C_{15}H_{16}ClN_3O_4S$
4-(5-Chloro-2-methoxyphenyl)-1-(*p*-tosyl)semicarbazide 806

$C_{15}H_{16}ClN_3O_5S$
1-(5-Chloro-2,4-dimethoxyphenyl)-3-(4-sulfamoylphenyl)urea 806

$C_{15}H_{16}ClN_5O_4$
4-(3-Chlorophenyl)-1-(3-furfurylureidoacetyl)semicarbazide 844

$C_{15}H_{16}Cl_2N_2$
*N*α-(2,3-Dichlorophenyl-*N*[4],*N*[4]-dimethyltoluene-α,4-diamine 224
*N*α-(3,4-Dichlorophenyl)-*N*[4],*N*[4]-dimethyltoluene-α,4-diamine 224

$C_{15}H_{16}Cl_3NO_2$
2,2,2-Trichloroethyl 3-(*N*-(2-propynyl)-benzylamino)-propionate. 1015

$C_{15}H_{16}FNO_2$
2,3-Dimethoxy-*N*-(4-fluorophenyl)-benzylamine 229
N-(4-Fluorophenyl)-3,4-dimethoxybenzylamine 228
5-Norbornen-2-ylmethyl *N*-(4-fluorophenyl)carbamate 631

$C_{15}H_{16}FN_3O$
1-(4-(Dimethylamino)phenyl)-3-(4-fluorophenyl)urea 526

$C_{15}H_{16}FN_5O_4$
4-(4-Fluorophenyl)-1-(3-(furfuryl)-ureidoacetyl)semicarbazide 844

$C_{15}H_{16}F_3NO_3$
cis-Hexahydro-*N*-(2-(trifluoromethyl)phenyl)-phthalamic acid 397
cis-Hexahydro-*N*-(3-(trifluoromethyl)phenyl)-phthalamic acid 397

$C_{15}H_{16}F_3NO_4$
Diethyl ((α,α,α-trifluoro-*m*-toluidino)-methylene)-malonate 366

$C_{15}H_{16}FeO_3$
Ethyl β-oxoferrocenepropionate 1142

$C_{15}H_{16}IN$
1-Ethyl-4-styrylpyridinium iodide 957
1-Methyl-2-(4-methylstyryl)-pyridinium iodide 957
1-Methyl-4-(4-methylstyryl)-pyridinium iodide 957
2-Methyl-2-phenylisoindolinium iodide 255

$C_{15}H_{16}INO$
1-Ethyl-4-(2-hydroxystyryl)-pyridinium iodide 933
4-(4-Methoxystyryl)-1-methylpyridinium iodide 938

$C_{15}H_{16}INO_2$
2,3-Dimethoxy-*N*-(4-iodophenyl)-benzylamine 229
N-(4-Iodophenyl)veratrylamine 228
1-Methyl-2-(vanillylidenemethyl)-pyridinium iodide 931

$C_{15}H_{16}INO_3$
1-(2′,5′-Dimethoxyphenacyl)-pyridinium iodide 938

$C_{15}H_{16}NO_4$
3′-Acetyl-1,2,3,6-tetrahydrophthalanilic acid 432
N-Benzoyl-2-(4-oxocyclohexyl)-glycine 473

$C_{15}H_{16}N_2$
3-Amino-2-dimethylaminofluorene 230
N′-Benzylidene-*N*,*N*-dimethyl-*p*-phenylenediamine 1057
9-(Dimethylaminomethyl)-carbazole. 892
N,*N*-Dimethyl-α-phenylimino-*p*-toluidine 1061
1,3-Diphenylimidazolidine 252
1,2-Diphenylpyrazolidine 253
N,*N*′-Di(*p*-tolyl)formamidine 727

$C_{15}H_{16}N_2O$
4′-Anilinopropionanilide 427
1-Benzyl-1-methyl-3-phenylurea 571
1-Benzyl-3-(*o*-tolyl)urea 572
1-Benzyl-3-(*m*-tolyl)urea 571
1-Benzyl-3-(*p*-tolyl)urea 572
1,3-Bis(*o*-tolyl)urea 588
1,2-Dihydro-2-(5-norbornen-2-yl)-4(3*H*)-quinazolinone 704
2-(4-(Dimethylamino)-benzylideneamino)phenol 1034
4-(4-(Dimethylamino)-benzylideneamino)phenol 1034
4-(4-(Dimethylamino)phenylimino)-2-methyl-2,5-cyclohexadien-1-one 1057
1,3-Di-*p*-tolylurea 591
1-Ethyl-1,3-diphenylurea 586
1-(2-Ethylphenyl)-3-phenylurea 586
1-(4-Ethylphenyl)-3-phenylurea 586
1-Methyl-1-phenyl-3-(*o*-tolyl)urea 588
6-Methyl-2-pyridinealdoxime *o*-phenethyl ether 1054
N-(1-Naphthyl)-1-pyrrolidinecarboxamide 711
1-Phenethyl-3-phenylurea 570
1-Phenyl-3-(2,3-xylyl)urea 586
1-Phenyl-3-(2,4-xylyl)urea 600
1-Phenyl-3-(2,5-xylyl)urea 586
1-Phenyl-3-(2,6-xylyl)urea 586
ʟ-Proline 2-naphthylamide hydrobromide 489

$C_{15}H_{16}N_2OS$
1-(4-(Methylthio)phenyl)-3-(*o*-tolyl)urea 561

$C_{15}H_{16}N_2O_2$
N-Acetyldihydroharmaline 882
N-Acetylharmaline 918
3′-Anilino-4′-methoxyacetanilide 415
1-(*o*-Anisyl)-3-benzylurea 553
1-(*m*-Anisyl)-3-benzylurea 553
1-(*p*-Anisyl)-3-benzylurea 553
3-(*o*-Anisyl)-1-methyl-1-phenylurea 555
N,*N*′-Bis(4-methoxyphenyl)-formamidine 724
3,4-Dimethoxybenzaldehyde phenylhydrazone 234
1-(2-Ethoxyphenyl)-3-phenylurea 554
1-(4-Ethoxyphenyl)-3-phenylurea 558
2-Hydroxyisobutyric (1-naphthylmethylene)hydrazide 496
1-(3-Hydroxyphenyl)-3-(2,6-xylyl)urea 549
1-(2-Methoxyphenyl)-3-(*o*-tolyl)urea ... 555
1-(3-Methoxyphenyl)-3-(*o*-tolyl)urea ... 559
4-(*N*-(1-Naphthyl)carbamoyl)-morpholine 579
2-Piperidinocinchoninic acid 993
4-Pyridylmethyl *N*-(2,4-dimethylphenyl)carbamate 964
4-Pyridylmethyl *N*-(2,6-dimethylphenyl)carbamate 964
3-Pyridylmethyl *N*-(3-ethylphenyl)-carbamate 964
4-Pyridylmethyl *N*-(3-ethylphenyl)-carbamate 964

$C_{15}H_{16}N_2O_2S$
1,3-Bis-(2-methoxyphenyl)-2-thiourea 738
1,3-Bis(3-methoxyphenyl)-2-thiourea 738
1-(2,5-Dimethoxyphenyl)-3-phenyl-2-thiourea 735
4,4′-Dimethoxythiocarbanilide 738
1-(4-Methoxyphenyl)-3-(3-(methylthio)phenyl)urea 551
Phenylacetaldehyde *p*-tosylhydrazone 821
p-Tolualdehyde *p*-tosylhydrazone 1071

$C_{15}H_{16}N_2O_3$
4′-Amino-2,2′-dimethoxybenzanilide. 464
1,3-Bis-(*m*-anisyl)urea 558
1,3-Bis(2-methoxyphenyl)urea 551
1,3-Bis-(4-methoxyphenyl)urea 559
2,2′-Dimethoxycarbanilide 555
1-(2,5-Dimethoxyphenyl)-3-phenylurea 557
α-(4-Ethylanilino)-3-nitrobenzyl alcohol 269
Ethyl 2-(3-(1-naphthyl)ureido)-acetate 574
1-(2-Methoxyphenyl)-3-(4-methoxyphenyl)urea 555
N-γ-(1-Naphthyl)-ᴅʟ-glutamine 487

$C_{15}H_{16}N_2O_3S$
4-Ethyl-3′-sulfamoylbenzanilide 823

$C_{15}H_{16}N_2O_4$
ʟ-Aspartic acid 4-methoxy-2-naphthylamide 486
2,3-Dimethoxy-*N*-(3-nitrophenyl)-benzylamine 266
2,3-Dimethoxy-*N*-(4-nitrophenyl)-benzylamine 266
1-(2,4-Dimethoxyphenyl)-3-(3-hydroxyphenyl)urea 546
1-(2,5-Dimethoxyphenyl)-3-(3-hydroxyphenyl)urea 546
1-(2,5-Dimethoxyphenyl)-3-(4-hydroxyphenyl)urea 546
Ethyl 3-(furfurylureido)benzoate 851
Ethyl 2-(3-(5-hydroxy-1-naphthyl)-ureido)acetate 548
Ethyl 2-(2-imidazolidinylidene)-2-phthalidylacetate 384
6-Methoxy-α-(4-methyl-3-nitroanilino)-*o*-cresol 263
3-Pyridylmethyl *N*-(2,5-dimethoxyphenyl)carbamate 936

$C_{15}H_{16}N_2O_4S$
N[4]-Acetyl-*N*[1]-(2-methoxyphenyl)-sulfanilamide 814
N[4]-Acetyl-*N*[1]-(4-methoxyphenyl)-sulfanilamide 815
2-Thiophenecarboxylic acid (3,4,5-trimethoxybenzylidene)-hydrazide... 1048

$C_{15}H_{16}N_2O_5$
N-(2-Hydroxy-3-methoxybenzyl)-2-methoxy-4-nitroaniline 263
6-Methoxy-α-(2-methoxy-5-nitroanilino)-*o*-cresol 263

$C_{15}H_{16}N_2O_5S$
1-(3,5-Dimethoxybenzoyl)-2-(phenylsulfonyl)hydrazine 815

$C_{15}H_{16}N_2O_6$
1-Methyl-2-norbornyl 3,5-dinitrobenzoate 374

$C_{15}H_{16}N_2S$
1-Allyl-3-(1-naphthylmethyl)-2-thiourea 747
1,3-Dibenzyl-2-thiourea 743
2,3-Dimethyl-1,3-diphenyl-2-thiopseudourea 738
1,3-Di-*o*-tolyl-2-thiourea 748
1-(4-Ethylphenyl)-3-phenyl-2-thiourea 748
1-(α-Methylbenzyl)-3-phenylthiourea. 748
1-(4-Methylbenzyl)-3-phenyl-2-thiourea 743
1-Phenethyl-3-phenyl-2-thiourea 741
1-Phenyl-3-(2,6-xylyl)-2-thiourea 748

$C_{15}H_{16}N_4$
1,3-Dimethyl-4,5-di-(4-pyridyl)-imidazoline 955

$C_{15}H_{16}N_4O$
5-Indancarboxaldehyde (4-hydroxy-6-methyl-2-pyrimidinyl)-hydrazone... 1037
4-Pyridinecarboxylic 4-(dimethylamino)-benzylidenehydrazide 1061

$C_{15}H_{16}N_4O_2$
4′-(Dimethylamino)-2-methyl-4-nitroazobenzene 1089
4-(Dimethylamino)-2-methyl-4′-nitroazobenzene 1089
4-(Dimethylamino)-3-methyl-4′-nitroazobenzene 1089
4′-(Dimethylamino)-3-methyl-4-nitroazobenzene 1089
N-Ethyl-*N*-methyl-4-(4-nitrophenylazo)aniline 1089
4-(3-Ethylphenyl)-1-(isonicotinoyl)-semicarbazide 959
4,4′-Methylenebis(phenylurea) 592

$C_{15}H_{16}N_4O_2S$
1-(4-(Dimethylamino)phenyl)-3-(4-nitrophenyl)-2-thiourea 740

$C_{15}H_{16}N_4O_3$
2-(3-Benzylureido)acetic (furfurylidene)hydrazide 1053
Disperse Dye (C.I. 11180) 1084
2-Methoxy-4′-(dimethylamino)-4-nitroazobenzene 1086
4-(2-Methoxy-5-methylphenyl)-1-nicotinoylsemicarbazide 939
4-(2-Methoxy-5-methylphenyl)-1-(4-pyridylcarbonyl)semicarbazide 939

$C_{15}H_{16}N_4O_4$
5,7-Dihydro-3-(2-nitro-phenyl-)-5,5,7,7-tetra-methyl-furo(3,4-e)-1,2,4-triazine-4-oxide 1005
5,7-Dihydro-3-(3-nitro-phenyl-)-5,5,7,7-tetra-methyl-furo(3,4-e)-1,2,4-triazine-4-oxide 1006
2-(Furfurylureido)acetic acid (2-hydroxybenzylidene)hydrazide 1036
2-(3-Furfurylureido)acetic (3-hydroxybenzylidene)hydrazide 1033
4-(2-Methoxy-5-methylphenyl)-1-(4-nitrophenyl)semicarbazide 552

$C_{15}H_{16}N_4O_6$
N,*N*′-Methylenebis(2-methoxy-4-nitroaniline) 265

$C_{15}H_{16}N_4S_3$
3-Benzyl-2,6,8-tris(methylthio)-purine 923

$C_{15}H_{16}N_6S_3$
1,9-Diphenyl-2,5,8-trithiotriurea 750
$C_{15}H_{16}O$
Benzyl 4-methylbenzyl ether 188
2-(4-Biphenylyl)-2-propanol 207
1,1-Diphenyl-1-propanol 204
1,3-Diphenyl-2-propanol 207
α-Ethyldiphenylmethanol 204
1,2,3,9,10,10a-Hexahydro-6-methylphenanthren-3-one 287
1,4,4a,4b,5,8,8a,9a-Octahydro-1,4:5,8-dimethanofluoren-9-one, isomer 1 55
1,4,4a,4b,5,8,8a,9a-Octahydro-1,4:5,8-dimethanofluoren-9-one, isomer 2 55
1,2,3,4-Tetrahydro-4-hydroxy-4-methylphenanthrene 208
$C_{15}H_{16}OSi$
Acetyldiphenylmethylsilane 1113
$C_{15}H_{16}O_2$
7-*syn*-Benzyloxymethyl-5-norbornen-2-one 283
Bis(*p*-tolyloxy)methane 179
1,3-Diphenoxypropane 178
1,1-Diphenyl-1,2-propanediol 203
8-Hydroxy-4a-methyl-4,4a,9,10-tetrahydro-2(3*H*)-phenanthrenone 282
3-(4-Hydroxyphenyl)-2-phenyl-1-propanol 200
1-(4-Methoxyphenyl)-2-phenylethanol 177
8-Methoxy-3,4,9,10-tetrahydro-1(2*H*)-phenanthrenone 283
8-Methoxy-4,4a,9,10-tetrahydro-2(3*H*)-phenanthrenone 283
2,2′-Methylenebis(4-methylphenol) 194
Methyl 4-(2-naphthyl)butyrate 362
4,4′-Propylidenediphenol 190
$C_{15}H_{16}O_2S$
α,α-Dimethoxybenzyl phenyl sulfide 210
4-Isopropylphenyl phenyl sulfone 784
4-Methylbenzyl *p*-tolyl sulfone 785
Phenethyl *p*-tolyl sulfone 784
Phenyl 3-phenylpropyl sulfone 784
$C_{15}H_{16}O_2Se$
5-Methyl-9-(phenylselenenyl)-2-oxatricyclo[4.2.1.0^{4,8}]nonan-3-one 1142
$C_{15}H_{16}O_2Si$
Methyl (methyldiphenylsilyl)carboxylate 1109
$C_{15}H_{16}O_3$
2,6-Dimethoxybenzhydrol 200
$C_{15}H_{16}O_3S$
2-Phenylethyl *p*-toluenesulfonate 800
Phenyl mesitylenesulfonate 800
p-Tolyl 2,4-xylenesulfonate 801
$C_{15}H_{16}O_4$
Ethyl 2,4-dioxo-6-phenyl-1-cyclohexanecarboxylate 364
2-Hydroxy-3-(3-hydroxy-3-methylbutyl)-1,4-naphthoquinone 315
endo-2-Norbornyl phthalate 370
$C_{15}H_{16}O_4S$
Ethyl 2,2-dimethylthieno(2,3-f)-1,3-benzodioxole-7-acetate 898
2-Phenoxyethyl *p*-toluenesulfonate 797
p-Tolyl 2-phenoxyethanesulfonate 797
$C_{15}H_{16}O_4S_2$
Bis(*p*-tosyl)methane 785
$C_{15}H_{16}O_5$
Methyl 5-(3,4-dimethoxyphenyl)-2-methyl-3-furancarboxylate 845
$C_{15}H_{16}S$
α,α-Dimethylbenzyl phenyl sulfide 212
Phenyl 1-phenylpropyl sulfide 212
Phenyl 3-phenylpropyl sulfide 212
$C_{15}H_{16}Si$
Allyldiphenylsilane 1112
$C_{15}H_{17}AsN_2O_6$
4-(3-(2,5-Dimethoxyphenyl)ureido)-phenylarsonic acid 1126
$C_{15}H_{17}BrNO_2P$
Dimethyl(4-nitrobenzyl)-phenylphosphonium bromide 830
$C_{15}H_{17}BrN_2O_2$
N-(4-Bromophenyl)-2-piperidinosuccinimide 715
$C_{15}H_{17}BrSi$
3-Bromo-3′-(trimethylsilyl)biphenyl 1123
$C_{15}H_{17}ClN_2$
*N*α-(4-Chlorophenyl)-*N*^{4},*N*^{4}-dimethyltoluene-α,4-diamine 224
$C_{15}H_{17}ClN_2O_5$
1-Butyl-3-(5-chloro-2,4-dimethoxyphenyl)-2,4,5-imidazolidinetrione 694
$C_{15}H_{17}FN_2$
N^4,*N*^{4}-Dimethyl-*N*α-(4-fluorophenyl)-toluene-α,4-diamine 224
$C_{15}H_{17}F_6NO_2$
1,3-Bis(trifluoromethyl)-5-(2-nitro-1,1,2-trimethylbutyl)benzene 257
$C_{15}H_{17}IN_2$
N^4,*N*^{4}-Dimethyl-*N*α-(4-iodophenyl)-toluene-α,4-diamine 224
$C_{15}H_{17}N$
N,*N*-Dibenzylmethylamine 243
Dimethyl(diphenylmethyl)amine hydrochloride 243
1,2-Diphenylethyl-*N*-methylamine 242
1,2-Diphenyl-*N*-methylethylamine hydrochloride 242
4-Methyl-α-phenylphenethylamine 241
4-Methyl-α-phenylphenethylamine hydrochloride 241
N-(*p*-Tolyl)-4-methylbenzylamine 232
$C_{15}H_{17}NO$
2-Cyclohexyl-5-phenyloxazole 872
4-Cyclohexyl-2(1*H*)-quinolone 993
α-(3,4-Dimethylanilino)-*p*-cresol 227
α-(4-Ethylanilino)-*o*-cresol 226
α-(4-Ethylanilino)-*p*-cresol 227
4-Isopropoxydiphenylamine 229
N-(2-Methoxyethyl)-4-biphenylylamine 218
N-(4-Methoxyphenyl)-4-methylbenzylamine 229
4-Methoxy-*N*-(*p*-tolyl)benzylamine 229
α-(*N*-Methylanilinomethyl)benzyl alcohol hydrochloride 237
N-(2-Naphthyl)-2,2,2-trimethylacetamide 490
1,4,4a,4b,5,8,8a,9a-Octahydro-1,4:5,8-dimethanofluoren-9-one oxime 1013
α-(Phenethylamino)-*o*-cresol 238
$C_{15}H_{17}NO_2$
α-(Benzylamino)-6-methoxy-*o*-cresol. 237
Benzyl 3,4,5-trimethylpyrrole-2-carboxylate 838
Butyl *N*-(1-naphthyl)carbamate 687
4-(Dimethylamino)-4′-hydroxybenzhydrol 225
2,5-Dimethyl-1-(4-ethoxycarbonylphenyl)pyrrole 841
α-(3,4-Dimethyl-5-ethoxy-2*H*-pyrrolylidene)-β-benzyl alcohol 838
α-(4-Ethoxyanilino)-*p*-cresol 226
3-Hydroxy-2-(1,2,3,4-tetrahydro-1-isoquinolyl)-2-cyclohexen-1-one 283
4-Methoxy-*N*-(4-methoxyphenyl)-benzylamine 229
α-(2-Methoxy-5-methylanilino)-*p*-cresol 226
6-Methoxy-α-(4-methylanilino)-*o*-cresol 225
Methyl 3-(2-methyl-1-naphthylamino)-propionate hydrochloride 358
1-Morpholinomethyl-2-naphthol 246
2-Naphthyl *N*-*tert*-butylcarbamate 687
α-(*p*-Phenetidino)-*o*-cresol 225
$C_{15}H_{17}NO_2S$
N-Ethyl-2′-methylbenzenesulfonanilide 828
4-(4-Isopropylphenylsulfonyl)aniline 776
N-Methyl-2-(phenylsulfonylmethyl)-benzylamine hydrobromide 778
N-Methyl-2,4-xylenesulfonanilide 828
3-(3-Morpholinopropionyl)benzo(b)-thiophene hydrochloride 898
$C_{15}H_{17}NO_3$
3-Benzamido-2-norbornanecarboxylic acid 475
Ethyl β-cyano-α-hydroxy-4-isopropylcinnamate 761
Ethyl 4-(2,4-hexadienamido)-benzoate 477
α-(2-Hydroxy-5-methylanilino)-6-methoxy-*o*-cresol 225
1-Methylbutyl β-cyano-α-hydroxycinnamate 761
$C_{15}H_{17}NO_3S_2$
5-(3,4-Dimethoxybenzylidene)-3-isopropylrhodanine 736
$C_{15}H_{17}NO_4$
Diethyl (α-cyanobenzyl)malonate 755
Ethyl 1-benzoyl-4-oxo-3-piperidinecarboxylate 706
N-(4-Methoxyphenyl)-1,2,5,6-tetrahydrophthalamic acid 418
2-Methyl-*exo*-norbornyl 4-nitrobenzoate 374
2-Methyl-*endo*-norbornyl 4-nitrobenzoate 374
N-(*o*-Tolyl)-diacetoacetamide 444
$C_{15}H_{17}NO_4S$
2′,4′-Dimethoxy-*p*-toluenesulfonanilide 813
$C_{15}H_{17}NO_4S_2$
5-(3,5-Dimethoxy-4-hydroxybenzylidene)-3-propylrhodanine 733
5-(4-Hydroxy-3-methoxybenzylidene)-3-(3-methoxypropyl)rhodanine 733
$C_{15}H_{17}NO_5$
Methyl 1,4-dihydro-6,7-dimethoxy-1-methyl-4-oxo-3-quinolineacetate 354
$C_{15}H_{17}NO_5S$
3-Methoxycarbonyl-1-methylpyridinium *p*-toluenesulfonate 956
3-(2-(3,4,5-Trimethoxyphenyl)ethyl)-4-isothiazolecarboxylic acid 874
$C_{15}H_{17}N_3$
4-Dimethylamino-2′-methylazobenzene 1090
4-Dimethylamino-3′-methylazobenzene 1090
4-(Dimethylamino)-4′-methylazobenzene 1091
4-(Dimethylamino)-*N*-phenylbenzamidine 726
4-(*N*-Ethyl-*N*-methylamino)-azobenzene 1090
$C_{15}H_{17}N_3O$
4-(Dimethylamino)-2-methoxyazobenzene 1087
4-(Dimethylamino)-3′-methoxyazobenzene 1086
4-(Dimethylamino)-3′-methylazobenzene *N*-oxide 1090
1-(3,5-Dimethylphenyl)-3-(4-methyl-2-pyridyl)urea 965
1-(2-Ethylphenyl)-3-(4-methyl-2-pyridyl)urea 965
1-(2-Ethylphenyl)-3-(3-pyridylmethyl)urea 950
2-(1-Methyl-2-benzimidazolyl)-2-valerylacetonitrile 905
1-(2-Pyridylmethyl)-3-(2,3-xylyl)urea 945
1-(2-Pyridylmethyl)-3-(2,4-xylyl)urea 965
1-(2-Pyridylmethyl)-3-(2,6-xylyl)urea 945
1-(2-Pyridylmethyl)-3-(3,4-xylyl)urea 945
1-(3-Pyridylmethyl)-3-(2,3-xylyl)urea 965
1-(3-Pyridylmethyl)-3-(2,4-xylyl)urea 950
1-(3-Pyridylmethyl)-3-(2,5-xylyl)urea 950
1-(3-Pyridylmethyl)-3-(2,6-xylyl)urea 950
1-(3-Pyridylmethyl)-3-(3,4-xylyl)urea 950
1-(3-Pyridylmethyl)-3-(3,5-xylyl)urea 951
$C_{15}H_{17}N_3OS$
1-(4,6-Dimethyl-2-pyridyl)-3-(3-(methylthio)phenyl)urea 941
1-(4,6-Dimethyl-2-pyridyl)-3-(4-(methylthio)phenyl)urea 941
$C_{15}H_{17}N_3O_2$
2,2′-Diamino-4′-methoxy-4-methylbenzanilide 464
1-(2,6-Diethylphenyl)-3-(2-pyrazinyl)urea 977
5,7-Dihydro-3-phenyl-5,5,7,7-tetramethylfuro(3,4-e)-1,2,4-triazine-4-oxide 1006
N^4,*N*^{4}-Dimethyl-*N*α-(3-nitrophenyl)-toluene-α,4-diamine 275
N-(2,3-Dimethyl-5-oxo-1-phenyl-3-pyrazolin-4-yl)-2-butenamide 707
1-(4,6-Dimethyl-2-pyridyl)-3-(2-methoxyphenyl)urea 939
1-(2-Methoxy-5-methylphenyl)-3-(3-pyridylmethyl)urea 938
1-(4-Methoxy-2-methylphenyl)-3-(3-pyridylmethyl)urea 939
$C_{15}H_{17}N_3O_2S$
N-(1-Ethyl-3-methyl-4-oxo-2-thioxoimidazolidin-5-ylidenemethyl)-acetanilide 742
$C_{15}H_{17}N_3O_3$
1-(2,4-Dimethoxyphenyl)-3-(5-methyl-2-pyridyl)urea 936
1-(2,5-Dimethoxyphenyl)-3-(5-methyl-2-pyridyl)urea 936
1-(2,4-Dimethoxyphenyl)-3-(3-pyridylmethyl)urea 936
1-(2,5-Dimethoxyphenyl)-3-(2-pyridylmethyl)urea 936
1-(2,5-Dimethoxyphenyl)-3-(3-pyridylmethyl)urea 938
$C_{15}H_{17}N_3O_3S$
1-(2-Ethylphenyl)-3-(4-sulfamoylphenyl)urea 825
1-(4-Sulfamoylphenyl)-3-(2,5-xylyl)urea 826
1-(4-Sulfamoylphenyl)-3-(2,6-xylyl)urea 826
$C_{15}H_{17}N_3O_3S_2$
1-(Benzylsulfonyl)-4-(4-methoxyphenyl)-3-thiosemicarbazide 815
$C_{15}H_{17}N_3O_4$
N-Butyl-2,4-dinitro-*N*-methyl-1-naphthylamine 274
$C_{15}H_{17}N_3O_4S$
4′-Amino-2′-hydroxy-*N*4-propionylsulfanilanilide 811
4-(3-Ethylphenyl)-1-(4-sulfophenyl)-semicarbazide 790
$C_{15}H_{17}N_3O_5S$
1-(2,4-Dimethoxyphenyl)-3-(4-sulfamoylphenyl)urea 812
$C_{15}H_{17}N_3S$
1-(4-(Dimethylamino)phenyl)-3-(phenyl)-2-thiourea 740
1-(α-Methylbenzyl)-3-(4-pyridylmethyl)-2-thiourea 952
1-(4-Pyridylmethyl)-3-(2,6-xylyl)-2-thiourea 953
$C_{15}H_{17}N_5OS_2$
1-Anilino-6-(4-methoxyphenyl)-2,5-dithiobiurea 738
$C_{15}H_{18}$
2,3,4,5,6,7,8,9-Octahydro-1*H*-trindene 162
$C_{15}H_{18}BrF_3N_4OS$
4-(4-Bromo-2-(trifluoromethyl)-phenyl)-1-(*N*-cyclohexylthiocarbamoyl)-semicarbazide 728
$C_{15}H_{18}BrNO_3$
4′-Bromo-3′-methylhexahydrophthalanilic acid 407
$C_{15}H_{18}Br_2N_2$
5-Bromo-2-(5-bromo-3-ethyl-4-methyl-pyrrol-2-ylmethylene)-3-ethyl-4-methyl-2*H*-pyrrole 839
$C_{15}H_{18}Br_2N_4O_2$
1,1′-Trimethylenebis(4-(hydroxyiminomethyl)pyridinium bromide) 1010
$C_{15}H_{18}ClN$
Benzyldimethylphenylammonium chloride 255
$C_{15}H_{18}ClNO_3$
cis-*N*-(3-Chloro-2-methylphenyl)-hexahydrophthalamic acid 406

$C_{15}H_{18}ClNO_3$ (Cont'd)
cis-Hexahydro-N-(3-chloro-4-methylphenyl)phthalamic acid 406
$C_{15}H_{18}ClNO_4$
cis-N-(5-Chloro-2-methoxyphenyl)-hexahydrophthalamic acid 401
$C_{15}H_{18}Cl_2N_4O_2$
1,1′-Trimethylenebis(4-(hydroxyiminomethyl)pyridinium chloride) 1010
$C_{15}H_{18}F_3NO_2$
2-Methylcyclohexyl N-(3-trifluoromethylphenyl)carbamate 610
$C_{15}H_{18}F_3N_3O$
1-Butyl-1-(2-cyanoethyl)-3-(3-(trifluoromethyl)phenyl)urea 751
$C_{15}H_{18}F_3N_3O_5$
N-Butyl-2′,6′-dinitro-4′-(trifluoromethyl)butyranilide 397
$C_{15}H_{18}F_3N_3O_6$
Propyl N-butyl-N-(2,6-dinitro-4-(trifluoromethyl)phenyl)carbamate 609
$C_{15}H_{18}F_6N_2$
2-(2-(3,5-Bis(trifluoromethyl)anilino)-ethyl)piperidine 244
$C_{15}H_{18}N_2$
3-(N-Cyclohexylformimidoyl)-indole 1057
3-(Diallylaminomethyl)-indole 884
N,N-Diethyl-4-(2-pyridyl)aniline 954
2-(2-(2,4-Dimethylanilino)ethyl)-pyridine 946
N^4,N^4-Dimethyl-Nα-phenyltoluene-α,4-diamine 232
2-(α-Ethylbenzylaminomethyl)-pyridine 943
4-(α-Ethylbenzylaminomethyl)-pyridine 951
3-Ethyl-2-imino-3-phenyl-1-cyclohexanecarbonitrile 756
2-(2-(N-Ethyl-N-phenylamino)ethyl)-pyridine hydrochloride 946
N-Isopropyl-N′-phenyl-1,4-phenylenediamine 232
2-(2-(Phenethylamino)ethyl)pyridine. 945
$C_{15}H_{18}N_2O$
1-Butyl-3-(1-naphthyl)urea 603
1-tert-Butyl-3-(2-naphthyl)urea 604
5-Cyano-2-ethoxy-6-methyl-4-phenyl-1,2,3,4-tetrahydropyridine 762
1,1-Diethyl-3-(1-naphthyl)urea 603
2-(4-(Dimethylamino)benzylamino)-phenol 226
$C_{15}H_{18}N_2O_2$
2-(N-Benzyl-N-(2-hydroxyethyl)-aminomethyl)-3-hydroxypyridine 932
1-tert-Butyl-3-(5-hydroxy-1-naphthyl)urea 549
2-Dimethylaminoethyl N-(1-naphthyl)carbamate 673
1-(2-Ethyl-6-methylphenyl)-3-furfurylurea 852
1-Furfuryl-1-methyl-3-(3,4-xylyl)urea.. 852
3-Methyl-1-phenyl-4-valeroyl-2-pyrazolin-5-one 711
N-Phenyl-2-piperidinosuccinimide 722
1-Propyl-1,2,3,4-tetrahydronorharman-3-carboxylic acid 919
2,2′-(Trimethylenedioxy)dianiline 227
4,4′-(Trimethylenedioxy)dianiline 227
$C_{15}H_{18}N_2O_2S$
5-Amino-N-ethyl-o-toluenesulfonanilide 817
5,6-Dihydro-8,9-dimethoxy-3-(ethylimino)-3H-thiazolo(4,3-a)-isoquinoline hydrochloride 735
N,N-Dimethyl-2-(p-tosyl)-1,4-phenylenediamine 776
$C_{15}H_{18}N_2O_3$
Nα-Acetyl-N-methyltryptophan methyl ester 885
$C_{15}H_{18}N_2O_4$
N,N′-Bis(allyloxycarbonyl)-4-methyl-1,3-phenylenediamine 669
1-Butyl-3-(2-methoxy-5-methylphenyl)-2,4,5-imidazolidinetrione 698
2,2′-Methylenebis(3,4-dimethyl-5-pyrrolecarboxylic acid) 840
$C_{15}H_{18}N_2O_4S$
4′-Amino-2′-hydroxy-6-methoxy-3,4-xylenesulfonanilide 810
$C_{15}H_{18}N_2O_5$
Carbobenzyloxy-L-norvalylglycine 659
cis-Hexahydro-N-(4-methyl-3-nitrophenyl)phthalamic acid 426
$C_{15}H_{18}N_2O_9$
2′,3′,5′-Triacetyluridine 131
$C_{15}H_{18}N_2S$
4-(4-Isopropylphenylthio)-1,3-phenylenediamine 230
$C_{15}H_{18}N_4O$
1-(4-Butylphenyl)-3-(2-pyrimidinyl)urea 976
$C_{15}H_{18}N_4O_2S_2$
4-(4-(Dimethylamino)phenyl)-1-(phenylsulfonyl)-3-thiosemicarbazide 819
$C_{15}H_{18}N_6O_3$
tert-Butyl N-(5-azido-α-benzyl-1,3,4-oxadiazol-2-ylmethyl)carbamate 873
$C_{15}H_{18}N_6O_3S$
2-(3-(2-Thiazolyl)ureido)acetic N^2-(N-(2,4-xylyl)carbamoyl)hydrazide 876
$C_{15}H_{18}O$
8a-Methyl-2,3,4,4a,4b,5,8,8a-octahydro-2-phenanthrenone 54
$C_{15}H_{18}OSi$
Diphenylethoxymethylsilane 1113
$C_{15}H_{18}O_2$
1-Phenyl-2-norbornyl acetate 361
$C_{15}H_{18}O_2S_2$
2,5-Dimethyl-3-(mesitylsulfonyl)-thiophene 859
3,5-Dimethyl-2-(mesitylsulfonyl)-thiophene 859
$C_{15}H_{18}O_3$
Ethyl 4,4-dimethyl-1-oxo-1,2,3,4-tetrahydro-2-naphthalenecarboxylate 363
(2-Hydroxynorbornan-2-yl)-phenylacetic acid 326
2-(2-Methoxyphenethyl)-1,3-cyclohexanedione 307
$C_{15}H_{18}O_4$
Diethyl 2-phenylcyclopropane-1,1-dicarboxylate 360
3-(4-(2,2-Dimethylpropionyl)-benzoyl)propionic acid 331
Ethyl 2-hydroxy-4-oxo-2-phenylcyclohexanecarboxylate 354
$C_{15}H_{18}O_5$
trans-2,5-Dihydro-2,5-dimethoxy-2-methyl-3-furylmethyl benzoate 369
Ethyl 2-(4-hydroxy-3-ethoxybenzylidene)acetoacetate 367
Ethyl 2-(1-(3-methoxyphenyl)-ethylidene)succinate 364
$C_{15}H_{18}Si$
(3-Biphenylyl)trimethylsilane 1121
Methyl(2-methylbenzyl)-phenylsilane 1119
Methyl(3-methylbenzyl)-phenylsilane 1113
Methyl(4-methylbenzyl)-phenylsilane 1120
Methyl(phenethyl)phenylsilane 1113
$C_{15}H_{19}BrO_2$
4-Bromo-3,5-dimethylcyclohexyl benzoate 375
3-Phenylallyl 2-bromohexanoate 362
$C_{15}H_{19}BrO_4$
Diethyl (2-bromo-2-phenylethyl)-malonate 360
$C_{15}H_{19}ClF_3NO$
6′-Chloro-α,α,α-trifluoro-m-octanotoluidide 395
$C_{15}H_{19}ClN_2O_2$
N-Acetyl-6′-chloro-1-pyrrolidineaceto-o-toluidide hydrochloride 430
6-(4-Chloro-3-methylphenyl)-2-(2-hydroxybutyl)-2,3,4,5-tetrahydro-3-pyridazinone 693
$C_{15}H_{19}ClN_4$
4-(4,6-Diamino-m-tolyl)imino-2,5-cyclohexadien-1-ylidene)di-methylammonium chloride 1052
$C_{15}H_{19}Cl_2NO_2$
N-Cyclohexyl-2-(2,4-dichlorophenoxy)-N-methylacetamide 402
$C_{15}H_{19}F_3N_2O$
1-Cyclohexyl-1-methyl-3-(3-trifluoromethylphenyl)urea 510
$C_{15}H_{19}N$
Diethyl(1-naphthylmethyl)amine 243
2-(1-Methylcyclohexyl)indole 891
$C_{15}H_{19}NO$
9-Benzyl-9-azabicyclo(3.3.1)nonan-3-one 285
N,N-Dimethallylbenzamide 482
5,5-Dimethyl-3-(3-methylanilino)-2-cyclohexen-1-one 284
3-(Methylphenethylamino)-2-cyclohexen-1-one hydrobromide 284
$C_{15}H_{19}NOS$
2-(Cyclohexylcarbonyl)-thioacetanilide 744
$C_{15}H_{19}NO_2$
Allyl 4-phenylisonipecotate hydrochloride 361
5,5-Dimethyl-3-(2-methoxyanilino)-2-cyclohexen-1-one 283
5,5-Dimethyl-3-(3-methoxyanilino)-2-cyclohexen-1-one 283
5,5-Dimethyl-3-(4-methoxyanilino)-2-cyclohexen-1-one 283
Isopropyl 1-indolebutyrate 887
$C_{15}H_{19}NO_3$
cis-N-Benzylhexahydro-phthalamic acid 431
cis-Hexahydro-N-(2-methylphenyl)-phthalamic acid 435
cis-Hexahydro-N-(3-methylphenyl)-phthalamic acid 435
cis-Hexahydro-N-(4-methylphenyl)-phthalamic acid 435
o-(4-Hydroxybenzoyl)tropine 368
o-(2-Hydroxybenzoyl)tropine hydrochloride 368
o-(3-Hydroxybenzoyl)tropine hydrochloride 368
$C_{15}H_{19}NO_3S$
Hexahydro-N-(4-(methylthio)-phenyl)phthalamic acid 421
$C_{15}H_{19}NO_3S_4$
1-(α,β-Bis(1,3-dithiolan-2-yl)butyl)-4-methoxy-3-nitrobenzene 268
$C_{15}H_{19}NO_4$
Anhydrocotarnine-acetone 284
Cyclopentyl N-(4-(ethoxycarbonyl)-phenyl)carbamate 675
Ethyl 2-(6,7-dimethoxy-1,2,3,4-tetrahydro-1-isoquinolylidene)-acetate 354
cis-Hexahydro-N-(4-methoxyphenyl)phthalamic acid 418
Methyl 3-(3-hydroxy-3-methylbutoxy)-2-indolecarboxylate. 881
$C_{15}H_{19}NO_5$
Diethyl 2-(2-methoxyphenylimino)-succinate 1049
$C_{15}H_{19}NO_6$
Carbobenzyloxy-L-glutamic acid α-ethyl ester dicyclohexylamine salt 659
$C_{15}H_{19}NO_7$
4-(Methoxycarbonyl)phenyl N-(butoxycarbonyloxymethyl)-carbamate 675
$C_{15}H_{19}N_3$
1,2-Dihydro-2-(4-(dimethylamino)-phenylimino)-1-ethylpyridine 725
1,4-Dihydro-4-(4-(dimethylamino)-phenylimino)-1-ethylpyridine hydrochloride 249
$C_{15}H_{19}N_3O_2$
1-(2,6-Diethylphenyl)-3-(5-methyl-3-isoxazolyl)urea 871
2-(2-(2-Formylcyclohexyl)ethyl)-2H-benzotriazole 1-oxide 1006
$C_{15}H_{19}N_3O_2S$
1-(1,3-Benzodioxol-6-ylmethylene)-4-cyclohexyl-3-thiosemicarbazide 1071
6-Hexyl-4-(2-thiazolylazo)resorcinol.. 1083
$C_{15}H_{19}N_3O_3$
Ethyl 3,4-benzo-2-oxo-1,5,9-triazaspiro(5.5)undec-3-ene-9-carboxylate 709
$C_{15}H_{19}N_3O_4$
5-(N-(8-Methoxy-4-quinolyl)amino)-pentyl nitrate 985
$C_{15}H_{19}N_3O_6$
Carbobenzyloxy-β-alanylglycylglycine 659
Carbobenzyloxy-L-alanylglycylglycine 676
Carbobenzyloxy-glycylglycylglycine methyl ester 661
Carbobenzyloxyglycylglycyl-sarcosine 660
$C_{15}H_{19}N_5O_6$
5′-O-Acetyl-2′,3′-O-isopropylideneguanosine 922
$C_{15}H_{20}BF_4N_3$
Dimethyl(2-(dimethylamino-methylene)-2-(3-indolyl)ethylidene)-ammonium tetrafluoroborate 1054
$C_{15}H_{20}BrN$
(4-Methyl-1-naphthylmethyl)-trimethylammonium bromide 256
$C_{15}H_{20}Br_2N_2$
1,1′-Trimethylenebis(3-methylpyridinium bromide) 948
1,1′-Trimethylenebis(4-methylpyridinium bromide) 948
$C_{15}H_{20}ClFN_2O_2$
1-(3-Chloro-4-fluorophenyl)-3-cyclohexyl-3-(2-hydroxyethyl)urea ... 513
$C_{15}H_{20}ClF_3N_2O_2$
1-(3-Butoxypropyl)-3-(4-chloro-2-(trifluoromethyl)phenyl)urea 504
$C_{15}H_{20}ClNO_2$
α-Acetyl-2-chloro-N,N-diethylhydrocinnamamide 444
2-Methylcyclohexyl N-(2-chloro-6-methylphenyl)carbamate 678
2-Methylcyclohexyl N-(4-chloro-2-methylphenyl)carbamate 679
2-Methylcyclohexyl N-(5-chloro-2-methylphenyl)carbamate 679
3-Methylcyclohexyl N-(2-chloro-6-methylphenyl)carbamate 632
$C_{15}H_{20}ClNO_3$
5′-Chloro-2′,4′-dimethoxy-1-cyclohexanecarboxanilide 399
4-Methylcyclohexyl N-(4-chloro-2-methoxyphenyl)carbamate 618
$C_{15}H_{20}ClN_5O_3$
4-(4-Chloro-2-methylphenyl)-1-(pyrrolidinocarboxamidoacetyl)-semicarbazide 695
$C_{15}H_{20}Cl_2N_2O_2$
1-Cyclohexyl-3-(2,3-dichlorophenyl)-1-(2-hydroxyethyl)urea 513
$C_{15}H_{20}FN_3O_4$
1-Cyclohexyl-3-(2-fluoro-5-nitrophenyl)-1-(2-hydroxyethyl)urea. 513
$C_{15}H_{20}F_3N_3O_2$
1-(3-Morpholinopropyl)-3-(3-trifluoromethylphenyl)urea 508
$C_{15}H_{20}F_3N_5O_3$
1-(2-(3,3-Diethylureido)acetyl)-4-(3-(trifluoromethyl)phenyl)-semicarbazide 509
$C_{15}H_{20}INO$
5,6-Benzo-2,3-dimethyl-8-ethoxy-2-azabicyclo(2.2.2)-oct-2-enium iodide 254
$C_{15}H_{20}N_2$
3-(Cyclohexylaminomethyl)-indole ... 884
1-(5-Indanylmethyleneamino)-piperidine 1072
3-(3-Methylpiperidinomethyl)indole.. 884
3-(4-Methylpiperidinomethyl)indole.. 884
2-Propyl-1,2,3,4-tetrahydroharmane hydrochloride 884
$C_{15}H_{20}N_2O$
1,1-Diallyl-3-(2-ethylphenyl)urea 588

$C_{15}H_{20}N_2O$ (Cont'd)
1,1-Diallyl-3-(2,3-xylyl)urea 595
1,1-Diallyl-3-(2,4-xylyl)urea 600
1,1-Diallyl-3-(2,5-xylyl)urea 600
1,1-Diallyl-3-(2,6-xylyl)urea 595
1,1-Diallyl-3-(3,4-xylyl)urea 600
1,1-Diallyl-3-(3,5-xylyl)urea 602
Thermopsine 136

$C_{15}H_{20}N_2O_2$
N-(3-Acetylphenyl)-2-methyl-1-piperidinecarboxamide 706
N-Benzyl-*N*-(3-cyanopropyl)-glycine ethyl ester 756
tert-Butyl (5-indanylmethylene)-carbazate 503
1,1-Diallyl-3-(2-methoxy-5-methylphenyl)urea 556
O-(*N*-(2,4-Dimethylphenyl)-carbamoyl)cyclohexanone oxime 1012
N-(2-(*N*-Ethyl-3-methylanilino)ethyl)-succinimide 718
O-(*N*-(2-Ethylphenyl)carbamoyl)-cyclohexanone oxime 1011
4-(*N*-(5-Hydroxypentyl)amino)-8-methoxyquinoline 982
Tetrahydro-2-furoic (α-propylbenzylidene)hydrazide 502
O-(*N*-(2,3-Xylyl)carbamoyl)-cyclohexanone oxime 1012
O-(*N*-(2,5-Xylyl)carbamoyl)-cyclohexanone oxime 1012
O-(*N*-(2,6-Xylyl)carbamoyl)-cyclohexanone oxime 1012
O-(*N*-(3,4-Xylyl)carbamoyl)-cyclohexanone oxime 1012
O-(*N*-(3,5-Xylyl)carbamoyl)-cyclohexanone oxime 1012

$C_{15}H_{20}N_2O_3$
5-(3-(Morpholino)-propyliminomethyl)-1,3-benzodioxole 1056

$C_{15}H_{20}N_2O_4$
O-(*N*-(2,5-Dimethoxyphenyl)-carbamoyl)cyclohexanone oxime 1009
3-Methylcyclohexyl *N*-(2-methyl-5-nitrophenyl)carbamate 669
Methyl 2-(*N*-(2-nitrobenzyl)amino)-cyclohexanecarboxylate 358
Piperidino *N*-(4-(ethoxycarbonyl)-phenyl)carbamate 676

$C_{15}H_{20}N_2O_5$
3-(3,5-Bis(methoxycarbonyl)phenyl)-1,1-diethylurea 576
Carbobenzyloxy-β-alanyl-L-alanine methyl ester 663
Carbobenzyloxy-L-alanyl-β-alanine methyl ester 664
Carbobenzyloxy-L-alanyl-2-aminobutyric acid 659
Carbobenzyloxyglycyl-L-norvaline 659
N-Carbobenzyloxylglycyl-L-valine 660
N-Carbobenzyloxy-DL-norvalylglycine 659
Carbobenzyloxy-L-norvalylglycine 659
Carbobenzyloxysarcosylalanine methyl ester 661
Carbobenzyloxyvalylglycine 659
Nα-Carbobenzyloxy-Nε-formyl-L-lysine 659

$C_{15}H_{20}N_2O_5S$
N-Benzyloxycarbonyl-glycyl-L-methionine 655
N-Benzyloxycarbonyl-DL-methionyl-glycine 656
N-Benzyloxycarbonyl-L-methionyl-glycine 656

$C_{15}H_{20}N_2O_6$
Carbobenzyloxy-L-alanyl-L-serine methyl ester 643
Carbobenzyloxyglycylthreonine methyl ester 642
Carbobenzyloxysarcosylserine methyl ester 642
Carbobenzyloxy-L-seryl-L-alanine methyl ester 642

$C_{15}H_{20}N_2O_7$
Carbobenzyloxy-L-seryl-L-serine methyl ester 642

$C_{15}H_{20}N_4$
N,*N*′-Bis(2-aminophenyl)-1,3-propanediamine 241
5,5′-Methylenebis(2-(dimethylamino)pyridine) 955

$C_{15}H_{20}N_4O_2$
4-Butyl-1-(3-indolylacetyl)-semicarbazide 888
1,1′-(4-Methyl-1,3-phenylene)bis(3-allylurea) 566

$C_{15}H_{20}N_4O_4$
N,*N*′-(4-Methyl-1,3-phenylene)bis(*o*-carbamoylacetone oxime) 1010
1,1′-(4-Methyl-1,3-phenylene)bis(3-propionylurea) 565

$C_{15}H_{20}N_4O_5$
Carbobenzyloxyglycylglycyl-L-alaninamide 664

$C_{15}H_{20}N_4S$
5-(4-Aminobutylimino)-7-phenyl-1,2,3,5-tetrahydroimidazo(1,2-c)-thiazole di-hydrochloride 248

$C_{15}H_{20}O$
2,3,4,4a,4b,5,8,8a,9,10-Decahydro-8a-methyl-2-phenanthrenone 54
2-(4-Isopropylphenyl)-cyclohexanone 286
2-(4-Methoxyphenyl)-6-methyl-2,5-heptadiene 181
1-Methyl-1-(4-methylbenzoyl)-cyclohexane 296

$C_{15}H_{20}O_2$
2,4-Dimethyl-6-hydroxycyclohexyl phenyl ketone 290
Ethyl 3-phenyl-2-heptenoate 365
Methyl 5,6,7,8-tetrahydro-2-naphthalenebutyrate 362
α-(*o*-Tolyl)-cyclohexaneacetic acid 332

$C_{15}H_{20}O_2S$
2-Cyclohexylvinyl *p*-tolyl sulfone 783

$C_{15}H_{20}O_3$
2,3,4,4a,4b,5,6,7,8,8a-Decahydro-6,7-dihydroxy-1-methyl-2-phenanthrenone 51
2,3,4,4a,4b,5,6,7,8,8a-Decahydro-6,7-dihydroxy-8a-methyl-2-phenanthrenone 51
6,7-Dimethoxy-1,2-dimethyl-1,2,3,4-tetrahydro-1,4-methano-2-naphthol 201
1-(3,4-Dimethoxyphenyl)-4,4-dimethyl-1-penten-3-one 283

$C_{15}H_{20}O_3S$
Bicyclo(3.2.1)oct-8-yl *p*-toluenesulfonate 758

$C_{15}H_{20}O_4$
Diethyl 2-ethyl-2-phenylmalonate 360
Diethyl 2-phenethylmalonate 360
(-)-*mono*-(2,2-Dimethyl-1-ethylpropyl) phthalate 375
Ethyl 5-(2-methoxyphenyl)-2-methyl-3-oxopentanoate 363
(+)-*mono*-1-Methylhexyl phthalate 371
DL-*mono*-1-Methylhexyl phthalate 371
Methyl 2-(4-pentyloxybenzoyl)-acetate 363

$C_{15}H_{20}O_4S_3$
2-(1,4-Dioxa-8-thiaspiro(4.5)dec-7-ylthio)ethyl phenyl sulfone 776

$C_{15}H_{20}O_5$
Diethyl 2-(4-methoxybenzyl)-malonate 355
Diethyl 2-(2-phenoxyethyl)-malonate 355
Monoethyl 2-(1-(3-methoxyphenyl)-ethyl)succinate, racemic mixture 355

$C_{15}H_{20}O_6$
Ethyl 4-acetoxy-1-(ethoxycarbonyl)-2,4-cyclopentadiene-1-acetate 79

$C_{15}H_{20}Si$
Phenyltriallylsilane 1113

$C_{15}H_{20}Si_2$
2,2-Diphenyl-1,1,1-trimethyldisilane 1113

$C_{15}H_{21}ClN_2O$
3-(2-Chloro-6-methylphenyl)-1-cyclohexyl-1-methylurea 583
3-(3-Chloro-4-methylphenyl)-1-cyclohexyl-1-methylurea 584
3-(4-Chloro-2-methylphenyl)-1-cyclohexyl-1-methylurea 584
3-(5-Chloro-2-methylphenyl)-1-cyclohexyl-1-methylurea 584
6′-Chloro-3-pyrrolidino-*o*-butyrotoluidide hydrochloride 426

$C_{15}H_{21}ClN_2O_2$
1-(3-Chloro-2-methoxyphenyl)-3-(cyclohexylmethyl)urea 558
1-(5-Chloro-2-methoxyphenyl)-3-(2-methylcyclohexyl)urea 519

$C_{15}H_{21}ClN_2S$
1-(2-Chlorobenzyl)-3-cyclohexyl-1-methyl-2-thiourea 730

$C_{15}H_{21}ClO_2$
3-Chloropropyl 3-methyl-2-phenylvalerate 353

$C_{15}H_{21}ClO_4$
α-Chloro-5-methyl-1,3-adamantanediacetic acid 65

$C_{15}H_{21}Cl_3N_2O$
1,1-Dibutyl-3-(2,4,5-trichlorophenyl)urea 603
1,1-Diisobutyl-3-(2,4,5-trichlorophenyl)urea 546

$C_{15}H_{21}FN_2O_2$
1-Cyclohexyl-3-(2-fluorophenyl)-1-(2-hydroxyethyl)urea 548
1-Cyclohexyl-3-(3-fluorophenyl)-1-(2-hydroxyethyl)urea 513

$C_{15}H_{21}F_3N_2O_2$
1-(3-Butoxypropyl)-3-(3-(trifluoromethyl)phenyl)urea 508

$C_{15}H_{21}IN_2O_2$
(3-(2-(Methoxycarbonyl)ethyl)-2-indolyl)trimethylammonium iodide 887

$C_{15}H_{21}N$
1,2,3,5,6,10b-Hexahydro-10b-propylpyrrolo(2,1-a)isoquinoline maleate 253
2-(Phenethylamino)-norbornane 242

$C_{15}H_{21}NO$
N-Allyl-*N*-phenylhexanamide 441
4-Butyryl-4-phenylpiperidine hydrochloride 286
N-Cyclohexylpropionanilide 441

$C_{15}H_{21}NOS$
3,5-Di-*tert*-butyl-4-hydroxyphenyl isothiocyanate 768

$C_{15}H_{21}NO_2$
Cyclohexyl *N*-(2,3-xylyl)carbamate 679
Cyclohexyl *N*-(2,4-xylyl)carbamate 680
Cyclohexyl *N*-(2,5-xylyl)carbamate 680
Cyclohexyl *N*-(2,6-xylyl)carbamate 679
Ethyl α-cyano-α-ethylbicyclo(3.2.1)-oct-2-ene-2-acetate 149
N-Ethyl-3-methyl-*N*-(2-phenoxyethyl)-crotonamide 417
Ethyl 1-methyl-3-phenyl-4-piperidinecarboxylate 359
2-Methylcyclohexyl *N*-(*o*-tolyl)-carbamate 678
2-Methylcyclohexyl *N*-(*m*-tolyl)-carbamate 678
2-Methylcyclohexyl *N*-(*p*-tolyl)-carbamate 678
6-(2-(2-Pyridyl)ethyl)-1,4-dioxaspiro(4.5)decane 966
2,3-Xylyl *N*-cyclohexylcarbamate 679
2,6-Xylyl *N*-cyclohexylcarbamate 679

$C_{15}H_{21}NO_2S$
3-Methyl-4-(methylthio)phenyl *N*-cyclohexylcarbamate 667

$C_{15}H_{21}NO_3$
4′-Butylglutaranilic acid 435
3-(*N*-Cyclohexylcarbamoyl)-5-norbornene-2-carboxylic acid 122
Cyclohexyl *N*-(2-methoxy-5-methylphenyl)carbamate 653
Ethyl 6-benzamidohexanoate 476
2-Methylcyclohexyl *N*-(*m*-anisyl)-carbamate 654
2-Methylcyclohexyl *N*-(*p*-anisyl)-carbamate 654
2-Methylcyclohexyl *N*-(2-methoxyphenyl)carbamate 653
2-Phenoxyethyl *N*-cyclohexylcarbamate 649

$C_{15}H_{21}NO_4$
4-(Butoxycarbonyl)phenyl *N*-propylcarbamate 676
tert-Butyl *N*-mesitoyloxycarbamate 654
Cyclohexyl *N*-(2,5-dimethoxyphenyl)carbamate 648
Dibutyl 3,5-pyridinedicarboxylate 962
Diethyl 2-(α-cyanocyclohexylmethylene)-malonate 148
Dimethyl 3,3′-(benzylimino)-dipropionate 358
Ethyl *N*-(2,4-dimethyl-α-(ethoxycarbonyl)benzyl)carbamate 677

$C_{15}H_{21}NO_6$
Methyl 4-(*tert*-butoxycarbonyl)-2-(methoxycarbonyl)-5-methyl-3-pyrroleacetate 842

$C_{15}H_{21}N_2O_2$
2-Methylcyclohexyl *N*-(*p*-tolyl)-carbamate 678

$C_{15}H_{21}N_3$
1-Benzyl-4-cyano-4-(ethylamino)-piperidine 754
5-*tert*-Butylamino-2,3-dihydro-7-phenyl-1*H*,-1,4-diazepine methanesulfonate 248
8-Dimethylamino-2-(4-pyridyl)-2-octenenitrile 954
5,6-Dimethyl-1-(piperidinomethyl)-benzimidazole 906

$C_{15}H_{21}N_3O_2$
4,4′-(Phenyliminomethylene)-bismorpholine 726

$C_{15}H_{21}N_3O_3$
1-Cyclohexyl-1-methyl-3-(4-methyl-3-nitrophenyl)urea 525
1-(Hexahydro-1*H*-azepin-1-yl)-3-(2-methoxycarbonyl)phenylurea 577

$C_{15}H_{21}N_3O_4$
Carbobenzyloxyglycyl-L-norvalinamide 664
Carbobenzyloxyglycyl-L-valinamide 664
1-Cyclohexyl-3-(2-methoxy-4-nitrophenyl)-1-methylurea 552

$C_{15}H_{21}N_3O_7S$
N-Butyl-2′,6′-dinitro-4′-(methylsulfonyl)butyranilide 778

$C_{15}H_{21}N_3S$
4,5-Dimethyl-2-(4-(dimethylamino)-phenylimino)-3-ethyl-4-thiazoline 725

$C_{15}H_{21}N_5O_3S$
4-(3-Morpholinopropyl)-1-(4-nitrobenzylidene)-thiosemicarbazide 1060

$C_{15}H_{21}N_5O_4$
Hippuryl-L-arginine acetate 725

$C_{15}H_{22}ClNO_2$
2-Chloroethyl *N*-(2,6-diisopropylphenyl)carbamate 605
Octyl *N*-(4-chlorophenyl)carbamate 683

$C_{15}H_{22}ClN_3O_2$
1-(2-Chloro-6-methylphenyl)-3-(3-morpholinopropyl)urea 525

$C_{15}H_{22}ClN_3O_3$
3-(2-Chloro-4-nitrophenyl)-1,1-dibutylurea 569

$C_{15}H_{22}Cl_2N_2O$
1,1-Dibutyl-3-(2,3-dichlorophenyl)urea 595
1,1-Dibutyl-3-(2,4-dichlorophenyl)urea 601
1,1-Dibutyl-3-(2,5-dichlorophenyl)urea 601
1,1-Dibutyl-3-(2,6-dichlorophenyl)urea 595
1,1-Dibutyl-3-(3,4-dichlorophenyl)urea 601
1,1-Dibutyl-3-(3,5-dichlorophenyl)urea 527
3-(2,3-Dichlorophenyl)-1,1-diisobutylurea 595
3-(2,4-Dichlorophenyl)-1,1-diisobutylurea 601
3-(2,6-Dichlorophenyl)-1,1-diisobutylurea 595
3-(3,4-Dichlorophenyl)-1,1-diisobutylurea 544

$C_{15}H_{22}Cl_2N_2OS$
3′,4′-Dichloro-2-(2-(diethylamino)-ethylthio)-propionanilide 421

$C_{15}H_{22}Cl_4N_4NiZn$
[2,12-Dimethyl-3,7,11,17-tetraazabicyclo[11.3.1]heptadeca-1¹⁷,2,11,13,15-pentane]nickel(II)-tetrachlorozincate 1140

$C_{15}H_{22}INO_2$
(6-Methoxy-1-oxo-1,2,3,4-tetrahydro-2-naphthylmethyl)-trimethylammonium iodide 292

$C_{15}H_{22}IN_3O$
(2,3-Dimethyl-5-oxo-1-phenyl-3-pyrazolin-4-ylmethyl)-trimethylammonium iodide 702

$C_{15}H_{22}N_2O$
1-Benzyl-3-cyclohexyl-1-methylurea 582
1-Cyclohexyl-3-(2-ethylphenyl)urea 582
1-Cyclohexyl-1-methyl-3-(*o*-tolyl)urea 582
1-Cyclohexyl-1-methyl-3-(*m*-tolyl)urea 582
1-Cyclohexyl-1-methyl-3-(*p*-tolyl)urea 583
1-Cyclohexyl-3-(2,3-xylyl)urea 583
1-Cyclohexyl-3-(2,5-xylyl)urea 584
1-Cyclohexyl-3-(2,6-xylyl)urea 583
1-Cyclohexyl-3-(3,4-xylyl)urea 584
1-Cyclohexyl-3-(3,5-xylyl)urea 584
1-Cyclopentyl-3-(4-isopropylphenyl)urea 581
1-Cyclopentyl-3-(2,4,5-trimethylphenyl)urea 581
1-(*N*-(2-Ethyl-6-methylphenyl)-carbamoyl)piperidine 712
1-(*N*-(2-Ethylphenyl)carbamoyl)-2-methylpiperidine 712
N-Methyl-*N*-phenethyl-1-pyrrolidineacetamide 430
2-Methyl-1-(*N*-(2,5-xylyl)carbamoyl)-piperidine 712
2-Methyl-1-(*N*-(2,6-xylyl)carbamoyl)-piperidine 712

$C_{15}H_{22}N_2OS$
1-Cyclohexyl-3-(2-ethoxyphenyl)-2-thiourea 737
1-Cyclohexyl-3-(4-ethoxyphenyl)-2-thiourea 738

$C_{15}H_{22}N_2O_2$
3-(*o*-Anisyl)-1-cyclohexyl-1-methylurea 554
3-(*m*-Anisyl)-1-cyclohexyl-1-methylurea 558
3-(*p*-Anisyl)-1-cyclohexyl-1-methylurea 558
4-*tert*-Butyl-5-(4-methoxyphenyl)-2-(methylamino)-2-oxazoline 247
3′-(Cyclohexylamino)-4′-methoxyacetanilide 415
1-Cyclohexyl-3-(2-methoxy-5-methylphenyl)urea 554
4-(*N*-(2,6-Diethylphenyl)carbamoyl)-morpholine 713
2,6-Dimethyl-4-(*N*-(2-ethylphenyl)-carbamoyl)morpholine 713
2,6-Dimethyl-2′-methoxy-1-piperidinecarboxanilide 699
2,6-Dimethyl-4′-methoxy-1-piperidinecarboxanilide 699
2,6-Dimethyl-4-(*N*-(2,3-xylyl)-carbamoyl)morpholine 713
2,6-Dimethyl-4-(*N*-(2,4-xylyl)-carbamoyl)morpholine 714
2,6-Dimethyl-4-(*N*-(2,5-xylyl)-carbamoyl)morpholine 714
2,6-Dimethyl-4-(*N*-(2,6-xylyl)-carbamoyl)morpholine 713
2,6-Dimethyl-4-(*N*-(3,4-xylyl)-carbamoyl)morpholine 714
2,6-Dimethyl-4-(*N*-(3,5-xylyl)-carbamoyl)morpholine 714
8-(4-Ethoxy-3-methoxyphenyl)-1,5-diazabicyclo(3.2.1)octane 247
N-(4-Ethoxyphenyl)-2-piperidinoacetamide 419
N-(1-Methyl-2-phenoxyethyl)-1-pyrrolidineacetamide 415
2-(1-(3-Morpholinopropylimino)-ethyl)phenol 1033
1-Piperidineaceto-*p*-phenetidide hydrochloride 415

$C_{15}H_{22}N_2O_2S$
p-Toluenesulfonic acid cyclooctylidenehydrazide 1071

$C_{15}H_{22}N_2O_2S_2$
2-Ethylimino-3-isopropyl-5-(phenylsulfonylmethyl)-thiazolidine hydrochloride 782

$C_{15}H_{22}N_2O_3$
Butyl 4-(3-propylureido)benzoate 577
1-Cyclohexyl-3-(2,5-dimethoxyphenyl)urea 554
2,6-Dimethyl-4-(*N*-(2-methoxy-5-methylphenyl)carbamoyl)-morpholine 699
L-Norleucyl-L-phenylalanine hydrate 429
1-(3,4,5-

$C_{15}H_{22}N_2O_4$
N,*N*′-Bis(isopropoxycarbonyl)-4-methyl-1,3-phenylenediamine 669
N-6-Carbobenzyloxy-L-lysine methyl ester hydrochloride 657
4-(*N*-(2,5-Dimethoxyphenyl)-carbamoyl)-2,6-dimethylmorpholine 699
Ethyl 3-(5-(acetamido)-2-methoxyanilino)butyrate 415
D-Leucyl-L-tyrosine 412
N,*N*′-(4-Methyl-1,3-phenylene)-bis(propyl carbamate) 669
1-Methyl-4-piperidinyl *N*-(2,5-dimethoxyphenyl)carbamate 652
L-Valyl-L-tyrosine methyl ester hydrobromide 412

$C_{15}H_{22}N_2O_5$
tert-Butyl (2,3,4-trimethoxybenzylidene)-carbazate 499
tert-Butyl (3,4,5-trimethoxybenzylidene)-carbazate 498

$C_{15}H_{22}N_2O_5S$
N-(*p*-Tosyl)-L-glutamine isopropyl ester 824

$C_{15}H_{22}N_2O_6$
N,*N*′-Bis(2-methoxyethoxycarbonyl)-4-methyl-1,3-phenylenediamine 647

$C_{15}H_{22}N_2O_8$
Gluconic (3-ethoxy-4-hydroxybenzylidene)hydrazide 1028

$C_{15}H_{22}N_2S$
2-(2-(Dimethylamino)ethyl)-2-(2-(methylthio)ethyl)-2-phenylacetonitrile 754
4-Dimethylamino-2-(2-(methylthio)-ethyl)-2-phenylbutyronitrile hydrochloride 754
2-(Octylthio)benzimidazole 903

$C_{15}H_{22}N_4$
1-(β-Pentylcinnamylideneamino)-guanidine 1072

$C_{15}H_{22}O$
2-(*p*-Cumenyl)-cyclohexanol 206
2′,4′-Diisopropylpropiophenone 297
6-Methoxy-1,2,3,4-tetrahydro-1,1,4,4-tetramethylnaphthalene 182
4,5,5,8,8-Pentamethyl-5,6,7,8-tetrahydro-2-naphthol 196

$C_{15}H_{22}O_2$
Ethyl 3-phenylheptanoate 362
2-Hexyl-4-phenyl-1,3-dioxolane 184
4′-Hydroxynonanophenone 290
p-Tolyl octanoate 381

$C_{15}H_{22}O_3$
4-(Heptyloxy)-*m*-anisaldehyde 323
5-*tert*-Octylsalicylic acid 339
2-(4-(*tert*-Pentyl)phenoxy)ethyl acetate 355
1,2,2-Trimethylpropyl α-methoxyphenylacetate 354
3-Veratryl-4-hexanone 283

$C_{15}H_{22}O_5$
tert-Butyl 2-(3,4,5-trimethoxyphenyl)acetate 354

$C_{15}H_{22}O_6S$
Diethyl 2,2′-(2-thenylidene)-bis-acetoacetate 855

$C_{15}H_{22}O_{10}$
1-*O*-Methyl-2,3,4,6-tetra-*O*-acetyl-α-D-glucoside 74

$C_{15}H_{22}O_{11}S_3$
1,3-*O*-Benzylidene-2,4,5-tris-*O*-(methylsulfonyl)-D-arabitol 798

$C_{15}H_{23}BrN_2O$
3-(4-Bromophenyl)-1,1-dibutylurea 539
3-(4-Bromophenyl)-1,1-diisobutylurea 539

$C_{15}H_{23}ClHg$
2,4,6-Triisopropylphenylmercuric chloride 1139

$C_{15}H_{23}ClN_2O$
6′-Chloro-3-diethylamino-*o*-butyrotoluidide 426
3-(2-Chlorophenyl)-1,1-dibutylurea 588
3-(3-Chlorophenyl)-1,1-dibutylurea 581
3-(4-Chlorophenyl)-1,1-dibutylurea 591
3-(2-Chlorophenyl)-1,1-diisobutylurea 588
3-(3-Chlorophenyl)-1,1-diisobutylurea 581
3-(4-Chlorophenyl)-1,1-diisobutylurea 591

$C_{15}H_{23}ClN_2O_2$
6′-Chloro-2-(2-diethylaminoethoxy)-*o*-acetotoluidide hydrochloride 414

$C_{15}H_{23}ClO_2$
α-Chloro-3,5,7-trimethyl-1-adamantaneacetic acid 67

$C_{15}H_{23}FN_2O$
1,1-Dibutyl-3-(4-fluorophenyl)urea 591

$C_{15}H_{23}IN_2$
2-Hexyl-1,3-dimethylbenzimidazolium iodide 906

$C_{15}H_{23}N$
α-Cyclohexyl-*N*-methylphenethylamine 242

$C_{15}H_{23}NO$
1-(1-Adamantanecarbonyl)-pyrrolidine 109
N-Benzyl-2,2-dimethyl-*N*-propylpropionamide 431
N,*N*-Diisobutylbenzamide 482
2-(1,2,3,4-Tetrahydro-2,2,4,7-tetramethyl-1-quinolyl)-1-ethanol 205

$C_{15}H_{23}NOS$
3-Methyl-*N*-(2-(methylthio)-ethyl)-2-phenylvaleramide 420

$C_{15}H_{23}NO_2$
1-(1-Adamantanecarbonyl)-morpholine 110
Butyl *N*-(2,6-diethylphenyl)-carbamate 684
Ethyl *N*-(2,6-diisopropylphenyl)-carbamate 684
4′-Hydroxynonananilide 413
1-Nitro-2,4,6-triisopropylbenzene 280
Octyl anthranilate 372
Octyl carbanilate 680
2-Octyl *N*-phenylcarbamate 680

$C_{15}H_{23}NO_4$
Methyl 2,4-dimethyl-5-(*tert*-butoxycarbonyl)-3-pyrrolepropionate 842

$C_{15}H_{23}NO_5$
3,4,5-Trimethoxybenzyl *N*-butylcarbamate 647

$C_{15}H_{23}NO_6$
Triethyl 1-cyano-3,3,4-pentanetricarboxylate 148
Triethyl 3-cyano-2,3,4-pentanetricarboxylate 144

$C_{15}H_{23}N_3O$
1-Benzyl-4-(dimethylamino)-isonipecotamide 423
1-Benzyl-4-ethylamino-4-piperidinecarboxamide 425
1-(*N*-4-(Dimethylamino)phenyl)-carbamoyl)-2-methylpiperidine 567

$C_{15}H_{23}N_3O_2$
L-Isoleucyl-L-phenylalaninamide hydrochloride 422
1-(4-Methyl-2-pentylidene)-4-(4-ethoxyphenyl)semicarbazide 1048
L-Phenylalanyl-L-isoleucinamide 422

$C_{15}H_{23}N_3O_3$
1-(*m*-Anisyl)-3-(3-morpholinopropyl)urea 552
1-(*p*-Anisyl)-3-(3-morpholinopropyl)urea 552
1,1-Dibutyl-3-(3-nitrophenyl)urea 565
1,1-Dibutyl-3-(4-nitrophenyl)urea 568
1,1-Diisobutyl-3-(4-nitrophenyl)urea 568
1-Heptyl-3-(2-methyl-5-nitrophenyl)urea 564
Isoleucyltyrosinamide 413
L-Norleucyl-L-tyrosinamide 412

$C_{15}H_{23}N_5$
1-(α-(2-Piperidinoethyl)-benzylideneamino)guanidine dihydrochloride 1072

$C_{15}H_{23}N_5O_3$
4-Propyl-1-(2-(3-propylureido)-benzoyl)semicarbazide 574

$C_{15}H_{24}$
β-Caryophyllene 2
Guaiene 3
1,5,9-Trimethyl-1,5,9-cyclododecatriene mixture of isomers 3
2,5,9-Trimethyl-1,5,9-cyclododecatriene 2

$C_{15}H_{24}Br_2O_2$
2-Acetoxy-6,7-dibromo-1,1,4a-trimethyldecahydronaphthalene 84

$C_{15}H_{24}ClN_3O$
2′-Chloro-2-(*N*-(2-diethylaminoethyl)-*N*-methylamino)-acetanilide dihydrochloride 424

$C_{15}H_{24}IN$
(1-Phenylcyclohexyl)-trimethylammonium iodide 255

$C_{15}H_{24}N_2$
2-(2-(2-Phenethylamino)ethyl)-piperidine 249

$C_{15}H_{24}N_2O$
N-Benzyl-3-(*N*-butyl-*N*-methylamino)-propionamide 425
1-Butyl-3-(2,6-diethylphenyl)urea 595
1-*tert*-Butyl-3-(2,6-diethylphenyl)urea 595
1,1-Dibutyl-3-phenylurea 586
2-Diethylamino-*N*-(α-methylphenethyl)-acetamide 428
1,1-Diethyl-3-(2,6-diethylphenyl)urea 595
1-(2,6-Diethylphenyl)-3-isobutylurea 595
1,1-Diisobutyl-3-phenylurea 586
1,1-Diisopropyl-3-(2-ethylphenyl)urea 588
3-(2,6-Diisopropylphenyl)-1,1-dimethylurea 571
1-(2,6-Diisopropylphenyl)-3-ethylurea 595
1,1-Diisopropyl-3-(2,3-xylyl)urea 595
1,1-Diisopropyl-3-(2,4-xylyl)urea 600
1,1-Diisopropyl-3-(2,5-xylyl)urea 601
1,1-Diisopropyl-3-(2,6-xylyl)urea 595
1,1-Diisopropyl-3-(3,4-xylyl)urea 601
1,1-Diisopropyl-3-(3,5-xylyl)urea 602
1,1-Dipropyl-3-(2-ethylphenyl)urea 588
1,1-Dipropyl-3-(3-ethylphenyl)urea 581
1,1-Dipropyl-3-(2,3-xylyl)urea 595
1,1-Dipropyl-3-(2,4-xylyl)urea 600
1,1-Dipropyl-3-(2,5-xylyl)urea 601
1,1-Dipropyl-3-(2,6-xylyl)urea 595
1,1-Dipropyl-3-(3,4-xylyl)urea 601

$C_{15}H_{24}N_2O_2$
4′-Amino-2′-hydroxyisononananilide 411
1-(3-Butoxypropyl)-3-(*m*-tolyl)urea 558
2-Diethylaminoethyl *N*-(3,5-dimethylphenyl)carbamate 673
2-Diethylaminoethyl *N*-(2-ethylphenyl)carbamate 672
2-Diethylaminoethyl *N*-(2,3-xylyl)-carbamate 672
2-Diethylaminoethyl *N*-(2,4-xylyl)-carbamate 672

$C_{15}H_{24}N_2O_2$ (Cont'd)
2-Diethylaminoethyl *N*-(2,6-xylyl)-carbamate 672
1,1-Diisopropyl-3-(2-methoxy-5-methylphenyl)urea 556
2-Dimethylaminoethyl *N*-(2,6-diethylphenyl)carbamate 672
1,1-Dipropyl-3-(2-methoxy-5-methylphenyl)urea 560
4′-Methoxy-3′-(1-methylpentylamino)acetanilide 415
1-(2-Methoxyphenyl)-4-(3-methoxypropyl)piperazine dihydrochloride 247
2,6-Xylyl *N*-(2-(diethylamino)-ethyl)-carbamate hydrochloride 668
$C_{15}H_{24}N_2O_2S$
3,5-Dimethyl-2-(dimorpholinomethyl)thiophene 856
$C_{15}H_{24}N_2O_3$
1-Butyl-3-(3,4-dimethoxyphenethyl)urea 549
2-Diethylaminoethyl *N*-(2-methoxy-5-methylphenyl)carbamate 652
1,1-Diisopropyl-3-(2,5-dimethoxyphenyl)urea 557
3-(2,5-Dimethoxyphenyl)-1,1-dipropylurea 557
2-Ethoxyethyl *N*-(3-(diethylamino)-phenyl)carbamate 649
$C_{15}H_{24}N_2O_4$
2-Diethylaminoethyl *N*-(2,5-dimethoxyphenyl)carbamate 652
$C_{15}H_{24}N_2O_5$
4-(Ethoxycarbonylhydrazino-carbonyl)-2,3,3-trimethyl-1-cyclo-hexen-1-yl acetate 116
$C_{15}H_{24}N_2O_6$
N-tert-Boc-L-isoleucine *N*-succinimide ester 125
$C_{15}H_{24}N_2O_{17}P_2$
Uridine-5′-diphosphoglucose, disodium salt 161
$C_{15}H_{24}N_2S$
1,1-Dibutyl-3-phenyl-2-thiourea 748
1,1-Diisobutyl-3-phenyl-2-thiourea 748
$C_{15}H_{24}N_4$
1-(α-Hexyl-2-methylbenzylideneamino)-guanidine nitrate 1072
1-(α-Hexyl-3-methylbenzylideneamino)-guanidine nitrate 1068
$C_{15}H_{24}N_4O_2$
1,1′-(4-Methyl-1,3-phenylene)bis(3-isopropylurea) 566
1,1′-(4-Methyl-1,3-phenylene)bis(3-propylurea) 566
$C_{15}H_{24}N_4O_2S_2$
Thiamine propyl disulfide 970
$C_{15}H_{24}N_4O_4$
1,1′-(4-Methyl-1,3-phenylene)bis(3-(2-methoxyethyl)urea) 552
$C_{15}H_{24}O$
Butyl 4-(*tert*-pentyl)-phenyl ether 181
Cedrene epoxide 20
4,6-Di-*tert*-butyl-2-methylphenol 192
6,6-Dimethyl-2-(2-(methallyloxy)-ethyl)bicyclo(3.1.1)hept-2-ene 18
2,2-Dimethyl-5-phenyl-3-heptanol 207
2-Methyl-4-*tert*-octylphenol 192
4-(*tert*-Octyl)anisole 181
$C_{15}H_{24}O_2$
4-Hydroxy-3,5-di-*tert*-butylbenzyl alcohol 200
Isoacorone 55
3-Methyl-5-*tert*-octylcatechol 193
5-Methyl-3-*tert*-octylcatechol 193
3,5,7-Trimethyl-1-adamantaneacetic acid 67
$C_{15}H_{24}O_2Pb$
Triethylplumbyl 3-phenylpropionate 1135
$C_{15}H_{24}O_2S$
Benzyl octyl sulfone 784
Dibutoxy(phenylthio)methane 210
$C_{15}H_{24}O_3$
4-(1-Hydroxyethyl)-4-phenyl-1,7-heptanediol 200
α-Hydroxy-3,5,7-trimethyl-1-adamantaneacetic acid 65
$C_{15}H_{24}O_3S$
2,2-Dimethylpropyl 2-methyl-1-phenyl-1-propane sulfonate 800
$C_{15}H_{24}O_4$
Spiro(5.5)undecane-2,2-diacetic acid 66
1,1,3,3-Tetraallyloxypropane 18
$C_{15}H_{24}O_4S$
2-(2-Ethylbutoxy)ethyl *p*-toluenesulfonate 797
$C_{15}H_{24}O_6$
7,7-Diethoxy-5-heptyne-2,4-diol diacetate 1017
$C_{15}H_{24}O_8$
Tetraethyl 1,1,2,3-propanetetracarboxylate 75
Tetraethyl propane-1,2,2,3-tetracarboxylate 75
$C_{15}H_{24}S$
6,6-Dimethyl-2-(methallylthioethyl)-bicyclo(3.1.1)hept-2-ene 27
$C_{15}H_{25}BrN_2O$
3-Carbamoyl-1-nonylpyridinium bromide 947
4-Carbamoyl-1-nonylpyridinium bromide 951
$C_{15}H_{25}NO$
N,N-Dibutyl-3-methoxyaniline 218
N,N-Diethyl-1-adamantanecarboxamide 107
$C_{15}H_{25}NOS$
2-(3-(Diethylamino)propylthio)-1-phenylethanol 237
$C_{15}H_{25}NO_2$
2-(1-Adamantyl)ethyl *N*-ethylcarbamate 116
α-Amino-3,5,7-trimethyl-1-adamantaneacetic acid 72
2-Pyrrolidinobicyclo(3.3.1)nonan-9-one ethylene ketal 41
$C_{15}H_{25}NO_7$
Diethyl 2-acetamido-2-(2-(ethoxycarbonyl)-1-methylethyl)-malonate 122
$C_{15}H_{25}N_3O$
1-Butyl-3-(4-(diethylamino)-phenyl)urea 567
4-(2,6-Diisopropylphenyl)-1,1-dimethylsemicarbazide 578
$C_{15}H_{25}N_3O_2$
N-(6-Butoxy-3-pyridyl)-2-(diethylamino)-acetamide 937
1-(3,5-Dimethyl-1-adamantylacetyl)-semicarbazide 127
$C_{15}H_{25}O_3$
Ethyl 2-oxocyclododecane-carboxylate 92
$C_{15}H_{26}$
Dehydrated dihydrocarotol 3
$C_{15}H_{26}ClN_4O_2P$
N-(*N*-(4-Chlorophenyl)carbamoyl)-*N′,N′,N″,N″*-tetraethylphosphoramide 833
$C_{15}H_{26}NO_5P$
Diethyl 4-(bis-(2-hydroxyethyl)-amino)-benzylphosphonate 834
$C_{15}H_{26}N_2$
2-(2-(Dibutylamino)ethyl)pyridine 946
$C_{15}H_{26}N_2O$
1-(1-Adamantyl)-3-butylurea 130
1-(1-Adamantylmethyl)-3-isopropylurea 130
$C_{15}H_{26}N_2O_2$
DL-*N*-(3-(Dimethylamino)propyl)-camphorimide hydrochloride 133
$C_{15}H_{26}N_2S$
Cyclohexyl-2-(pyrrolidinomethylene)-thiobutyramide 139
$C_{15}H_{26}N_2S_2$
3,5-Dicyclohexyl-3,4,5,6-tetrahydro-2*H*-1,3,5-thiadiazine-2-thione 140
$C_{15}H_{26}N_2S_4$
4,4′-Trimethylenebis(1-piperidinedithiocarboxylic acid, ammonium salt) 142
$C_{15}H_{26}N_4O_6$
N-Acetyl-L-alanyl-L-alanyl-L-alanyl-L-alanine methyl ester 124
$C_{15}H_{26}O$
Cedrol 13
9-Cyclohexylbicyclo(3.3.1)-nonan-9-ol 14
3,5,5,8,8-Pentamethyl-2-decalone 57
$C_{15}H_{26}O_4$
Diethyl allyl-1-(methylbutyl)-malonate 79
$C_{15}H_{26}O_5$
Diethyl 3-heptyl-2-oxosuccinate 92
$C_{15}H_{26}O_8Si$
Phenyltris(2-methoxyethoxy)silane 1106
$C_{15}H_{27}BrO_4$
Diethyl 2-(3-bromopropyl)-2-isopentylmalonate 81
$C_{15}H_{27}F_3Si_3$
1,3,5-Trifluoro-2,4,6-tris(trimethylsilyl)benzene 1101
$C_{15}H_{27}N$
N,N-3,5-Tetramethyl-1-adamantanemethylamine hydrochloride 29
$C_{15}H_{27}N_5O_6$
D-Alanyl-D-alanyl-D-alanyl-D-alanyl-D-alanine 120
$C_{15}H_{27}O_7P$
Tris-(tetrahydrofurfuryl)- phosphate 161
$C_{15}H_{28}F_2Si_3$
1,3-Difluoro-2,4,6-tris(trimethylsilyl)-benzene 1101
$C_{15}H_{28}N_2$
6-Diethylamino-1-piperidino-2-hexyne 1018
$C_{15}H_{28}N_2O$
1,3-Dicyclohexyl-1-ethylurea 130
$C_{15}H_{28}N_2O_5$
1,3-Bis(5-(carbomethoxy)-pentyl)urea 127
$C_{15}H_{28}O_2$
2-Hydroxycyclopentadecanone 52
4,4′-Isopropylidenedicyclohexanol 14
$C_{15}H_{28}O_2NP$
Dicyclopentyl(*N,N*-diethylcarbamoyl)phosphine oxide 158
$C_{15}H_{28}O_3$
2-Methoxy-2-nonen-1-yl pentanoate 78
13-Oxopentadecanoic acid 64
$C_{15}H_{28}O_4$
Dibutyl diethylmalonate 76
Diethyl dibutylmalonate 76
$C_{15}H_{29}NO$
3-(Dodecyloxy)-propionitrile 144
$C_{15}H_{29}NO_2$
5-(*N*-Butyl-*N*-(2-hydroxyethyl)-amino)-2,6-dimethyl-3-heptyn-2-ol 1016
6-Methyl-6-piperidino-4-heptenal dimethyl acetal 38
$C_{15}H_{29}NO_3$
N-Lauroylsarcosine 124
$C_{15}H_{29}N_3O_2$
4-Propyl-1-(10-undecenoyl)-semicarbazide 126
$C_{15}H_{29}N_3O_4$
D-Valyl-D-valyl-D-valine 120
$C_{15}H_{30}ClNO$
2-Chloro-*N*-tridecylacetamide 103
$C_{15}H_{30}IN$
1-Methyl-1-(2-methyl-2-octen-1-yl)-piperidinium iodide 44
$C_{15}H_{30}N_2$
D-*N*-(3-(Dimethylamino)propyl)-camphidine dihydrochloride 39
$C_{15}H_{30}N_2O$
3-Cyclohexyl-1,1-dibutylurea 130
3-Cyclohexyl-1,1-diisobutylurea 130
N-Ethyl-*N*-heptyl-1-pyrrolidineacetamide hydrochloride 108
N-Methyl-*N*-(1-methylhexyl)-1-piperidineacetamide hydrochloride 108
1-Nonylisonipecotamide 108
4-(3-(4-Piperidyl)propyl)-1-piperidineethanol 36
$C_{15}H_{30}N_2O_2$
1-Tetradecanoylurea 128
$C_{15}H_{30}O$
4-Pentadecanone 49
5-Pentadecanone 49
6-Pentadecanone 49
7-Pentadecanone 49
8-Pentadecanone 49
$C_{15}H_{30}O_2$
Glycidyl lauryl ether 20
$C_{15}H_{30}O_2S$
11-(Butylthio)-undecanoic acid 63
Dodecyl 3-mercaptopropionate 86
3-(Dodecylthio)-propionic acid 63
$C_{15}H_{30}O_3$
2-Hydroxypropyl dodecanoate 84
$C_{15}H_{30}Si_3$
1,3,5-Benzenetriyltris(trimethylsilane) 1123
$C_{15}H_{31}BrO_2$
1-Bromo-11-(1-ethoxyethoxy)-undecane 17
$C_{15}H_{31}ClS$
3-Chloropropyl dodecyl sulfide 24
$C_{15}H_{31}N_3OS$
1-(2-Morpholinoethyl)-3-*tert*-octyl-2-thiourea 139
$C_{15}H_{32}$
2,6,10-Trimethyldodecane 1
$C_{15}H_{32}NO_4P$
Dibutyl (1-(*N,N*-diethylcarbamoyl)-ethyl)phosphonate 159
Diisopropyl (*N,N*-diisobutylcarbamoyl)phosphonate 159
$C_{15}H_{32}N_2O$
1-Tetradecylurea 129
$C_{15}H_{32}N_2S$
1,3-Diheptyl-2-thiourea 142
$C_{15}H_{32}O$
3-Methyl-3-tetradecanol 10
2-Pentadecanol 9
8-Pentadecanol 9
3,7,11-Trimethyl-3-dodecanol 10
$C_{15}H_{32}OS$
3-Dodecylthio-1-propanol 25
$C_{15}H_{32}O_2Sn$
Tributyltin propionate 1131
$C_{15}H_{32}O_3S$
3-(Dodecylsulfonyl)propanol 152
Tetradecyl methanesulfonate 156
$C_{15}H_{32}S$
Dodecyl propyl sulfide 26
$C_{15}H_{32}Si_4$
1,1,1,3,3,3-Hexamethyl-2-phenyl-2-(trimethylsilyl)trisilane 1113
$C_{15}H_{33}BO_3$
Tripentyl borate 1124
$C_{15}H_{33}N$
1-Heptyloctylamine 28
$C_{15}H_{33}NO$
2-Dodecyloxy-*N*-methylethylamine 35
3-Dodecyloxypropylamine 35
$C_{15}H_{33}NS$
3-(Dodecylthio)-propylamine 35
$C_{15}H_{33}N_3O_3$
1-(2-(*N,N*-Bis-(2-hydroxypropyl)-amino)-ethyl)-4-(2-hydroxypropyl)-piperazine 36
$C_{15}H_{33}O_3P$
Diamyl amylphosphonate 158
$C_{15}H_{34}IP$
Isopropyltributylphosphonium iodide 158
$C_{15}H_{34}N_2$
3-(*tert*-Dodecylamino)-propylamine 30
$C_{15}H_{34}O_3Si$
Tripentyloxysilane 1097
$C_{15}H_{36}Si_4$
1,2-Propadiene-1,3-diylidenetetrakis(trimethylsilane) 1097
$C_{16}Cl_{12}S_4Si$
Tetrakis(3,4,5-trichloro-2-thienyl)-silane 1110
$C_{16}H_2Cl_{10}O_2$
1,4-Bis(pentachlorophenyl)-2-butene-1,4-dione 307
$C_{16}H_2F_{10}O$
2,5-Bis(pentafluorophenyl)furan 845

$C_{16}H_3F_{10}N$
2,5-Bis(pentafluorophenyl)pyrrole ... 838
$C_{16}H_4Cl_6N_2O_2$
N-(2-Cyanophenyl)-1,4,5,6,7,7-hexachloro-5-norbornene-2,3-dicarboximide ... 767
$C_{16}H_4Cl_{10}O_2$
1,4-Bis(pentachlorophenyl)-1,4-butanedione ... 307
$C_{16}H_4F_{10}O_2$
1,4-Bis(pentafluorophenyl)-1,4-butanedione ... 307
$C_{16}H_5Cl_7F_3NO_3$
3-[*N*-[2-Chloro-5-(trifluoromethyl)phenyl]carbamoyl]-1,4,5,6,7,7-hexachloro-5-norbornene-2-carboxylic acid ... 446
$C_{16}H_6Cl_4$
1,3,6,8-Tetrachloropyrene ... 175
$C_{16}H_6Cl_7NO_2$
N-(3-Chloro-4-methylphenyl)-1,4,5,6,7,7-hexachloro-5-norbornene-2,3dicarboximide ... 715
$C_{16}H_6F_{12}N_2O$
3,3′,5,5′-Tetrakis(trifluoromethyl)azoxybenzene ... 1080
$C_{16}H_7Cl_4NO_3$
2-(1-Indolylcarbonyl)-3,4,5,6-tetrachlorobenzoic acid ... 881
$C_{16}H_7Cl_5N_2O_2$
5,7-Dichloro-8-quinolyl *N*-(2,4,5-trichlorophenyl)carbamate ... 992
$C_{16}H_7Cl_6NO_2S$
1,4,5,6,7,7-Hexachloro-*N*-(3-(methylthio)phenyl)-5-norbornene-2,3-dicarboximide ... 718
$C_{16}H_7Cl_6NO_3$
1,4,5,6,7,7-Hexachloro-*N*-(3-methoxyphenyl)-5-norbornene-2,3-dicarboximide ... 717
$C_{16}H_7FN_2OS$
5-Cyano-9-fluoro(1)benzothieno(2,3-c)quinolizin-6-one ... 1003
$C_{16}H_8BrCl_2NO_3$
6-Bromo-3-(2,3-dichlorophenylcarbamoyl)coumarin ... 483
6-Bromo-3-(2,4-dichlorophenylcarbamoyl)coumarin ... 483
6-Bromo-3-(2,5-dichlorophenylcarbamoyl)coumarin ... 483
6-Bromo-3-(3,4-dichlorophenylcarbamoyl)coumarin ... 483
$C_{16}H_8Cl_2F_3NO$
3-(2,4-Dichlorobenzylidene)-6-(trifluoromethyl)-2-indolinone ... 689
$C_{16}H_8Cl_2F_6N_2$
N,N′-Bis(2-chloro-5-(trifluoromethyl)phenyl)-1,4-diazabutadiene ... 1020
$C_{16}H_8Cl_2N_4O_3$
3-Amino-4-cyano-2-(2,4-dichlorophenyl)-7-nitro-1(2*H*)-isoquinolinone ... 758
$C_{16}H_8Cl_3NOS_2$
3-(4-Chlorophenyl)-5-(3,4-dichlorobenzylidene)rhodanine ... 731
$C_{16}H_8Cl_3NO_2$
2-Chloro-3-(2,4-dichloroanilino)-1,4-naphthoquinone ... 315
3-Chloro-2-(3,4-dichloroanilino)-1,4-naphthoquinone ... 315
$C_{16}H_8Cl_4N_2O_2$
5-Chloro-8-quinolyl *N*-(2,4,5-trichlorophenyl)carbamate ... 992
5,7-Dichloro-8-quinolyl *N*-(2,4-dichlorophenyl)carbamate ... 982
5,7-Dichloro-8-quinolyl *N*-(3,4-dichlorophenyl)carbamate ... 982
$C_{16}H_8Cl_7NO_3$
N-(5-Chloro-2-methoxyphenyl)-1,4,5,6,7,7-hexachloro-5-norbornene-2,3-dicarboximide ... 715
$C_{16}H_8F_{12}N_2$
1,2-Bis(3,5-ditrifluoromethylphenyl)hydrazine ... 234
$C_{16}H_8N_2OS$
6-Cyanobenzothiazolo(3,2-a)quinolin-5-one ... 766
$C_{16}H_8N_2O_2$
2,2′-Bis(benzoylcyanide) ... 765
$C_{16}H_8N_2O_4$
Phthalic anhydride azine ... 723
$C_{16}H_8N_2O_4S$
N,N′-Thiodiphthalimide ... 723
$C_{16}H_8O_2$
1,6-Dihydro-1,6-pyrenedione ... 309
$C_{16}H_8O_6$
9,10-Dihydro-9,10-dioxo-2,3-anthracenedicarboxylic acid ... 344
$C_{16}H_9BF_4N_2$
1-Pyrenyldiazonium tetrafluoroborate ... 1095
$C_{16}H_9Br$
4-Bromopyrene ... 174
$C_{16}H_9BrN_2O_4$
2-(2-(2-Bromo-4,5-methylenedioxyphenyl)vinyl)-5-nitrobenzonitrile ... 764
$C_{16}H_9BrO_2$
4b-Bromo-4b,5,9b,10-tetrahydroindeno(2,1-a)indene-5,10-dione ... 309
$C_{16}H_9BrO_3$
2-(4-Bromophenyl)-3-hydroxy-1,4-naphthoquinone ... 315
$C_{16}H_9Br_2NO_3$
4-(3,5-Dibromosalicylidene)-2-phenyl-2-oxazolin-5-one ... 383
$C_{16}H_9Br_4NO_4$
3′-Acetyl-3,4,5,6-tetrabromophthalanilic acid ... 453
$C_{16}H_9ClFNO_2$
2-Chloro-3-(4-fluoroanilino)-1,4-naphthoquinone ... 315
$C_{16}H_9ClINO_2$
2-Chloro-3-(4-iodoanilino)-1,4-naphthoquinone ... 315
$C_{16}H_9ClN_2$
(8 or 9 or 10 or 11)-Chlorobenzo(a)phenazine ... 1001
$C_{16}H_9ClN_2O_4$
4-(4-Chlorobenzylidene)-2-(3-nitrophenyl)-2-oxazolin-5-one ... 382
2-Chloro-3-(3-nitroanilino)-1,4-naphthoquinone ... 317
3-Chloro-2-(4-nitroanilino)-1,4-naphthoquinone ... 317
$C_{16}H_9ClN_4O_4$
(1 or 2 or 3 or 4)-Chloro-9-nitro-6*H*-indolo(2,3-b)quinoxaline-6-acetic acid ... 1000
$C_{16}H_9Cl_2NO_2$
3-Chloro-2-(4-chloroanilino)-1,4-naphthoquinone ... 315
6-Chloro-2-(4-chlorophenyl)-4-quinolinecarboxylic acid ... 982
6,8-Dichlorocinchophen ... 993
$C_{16}H_9Cl_2NO_4$
2-Chloro-2-(α-chloro-4-nitrobenzyl)-1,3-indandione ... 308
$C_{16}H_9Cl_3F_3NO_4$
2-(2,4-Dichlorophenoxy)acetyl *N*-(2-chloro-4-(trifluoromethyl)phenyl)carbamate ... 606
$C_{16}H_9Cl_3N_2$
4,6-Bis(4-chlorophenyl)-2-chloropyrimidine ... 968
$C_{16}H_9Cl_3N_2O_2$
5-Chloro-8-quinolyl *N*-(2,3-dichlorophenyl)carbamate ... 991
5-Chloro-8-quinolyl *N*-(2,4-dichlorophenyl)carbamate ... 992
5-Chloro-8-quinolyl *N*-(2,5-dichlorophenyl)carbamate ... 992
5-Chloro-8-quinolyl *N*-(3,4-dichlorophenyl)carbamate ... 992
5,7-Dichloro-8-quinolyl *N*-(3-chlorophenyl)carbamate ... 992
5,7-Dichloro-8-quinolyl *N*-(4-chlorophenyl)carbamate ... 982
8-Quinolyl *N*-(2,4,5-trichlorophenyl)carbamate ... 992
$C_{16}H_9Cl_6NO_2S$
1,4,5,6,7,7-Hexachloro-*N*-(4-methylthiophenyl)-5-norbornene-2,3-dicarboximide ... 715
$C_{16}H_9Cl_7O$
2,2-Dichloro-3′-dichloromethyl-4′-trichloromethyl-2-phenylacetophenone ... 288
$C_{16}H_9F$
3-Fluorofluoranthene ... 173
$C_{16}H_9FN_2O_6$
4-Coumarinyl *N*-(4-fluoro-3-nitrophenyl)carbamate ... 628
$C_{16}H_9F_2NO_2$
4-(2,4-Difluorobenzylidene)-2-phenyl-2-oxazolin-5-one ... 386
4-(3,4-Difluorobenzylidene)-2-phenyl-2-oxazolin-5-one ... 386
$C_{16}H_9F_3N_2O_2$
4-Nitro-α-(3-(trifluoromethyl)phenyl)cinnamonitrile ... 758
$C_{16}H_9F_3N_2O_4$
Phthalimido *N*-(3-(trifluoromethyl)phenyl)carbamate ... 715
$C_{16}H_9F_9N_2O$
1-(3,5-Bis(trifluoromethyl)phenyl)-3-(2-(trifluoromethyl)phenyl)urea ... 511
$C_{16}H_9I_2NO_3$
4-(3,5-Diiodosalicylidene)-2-phenyl-2-oxazolin-5-one ... 383
$C_{16}H_9NO_6$
2-(5-Nitro-3-phthalidylidene)-2-phenylacetic acid ... 365
$C_{16}H_9NSO_2$
10,11-Dihydro-12*H*-benzo(c)phenothiazine-10,11-dione ... 308
$C_{16}H_9N_3O$
6-Cyanobenzimidazo(1,2-a)quinolin-5(7*H*)-one ... 989
$C_{16}H_{10}$
Pyrene, zone refined ... 167
$C_{16}H_{10}BrClN_2O_2$
5-Chloro-8-quinolyl *N*-(4-bromophenyl)carbamate ... 982
$C_{16}H_{10}BrCl_2F_3N_4O_3$
N-(4-Bromo-2-(trifluoromethyl)phenylureido)-*N′*-(3,4-dichlorophenyl)oxamide ... 505
$C_{16}H_{10}BrCl_4NO_3$
4′-Bromo-2′,6′-dimethyl-3,4,5,6-tetrachlorophthalanilic acid ... 455
$C_{16}H_{10}BrN$
11-Bromo-11*H*-indeno(1,2-b)quinoline ... 995
$C_{16}H_{10}BrNO_2$
2-(4-Bromophenyl)-4-quinolinecarboxylic acid ... 981
$C_{16}H_{10}BrNO_3$
N-(4′-Bromophenacyl)-phthalimide ... 721
4-(5-Bromosalicylidene)-2-phenyl-2-oxazolin-5-one ... 383
$C_{16}H_{10}BrN_3O_3$
N-(1-Bromo-4-nitro-3-isoquinolyl)benzamide ... 996
$C_{16}H_{10}Br_2O_2$
2-Bromo-2-(α-bromobenzyl)-1,3-indanedione ... 309
$C_{16}H_{10}Br_4ClNO_5$
4′-Chloro-2′,5′-dimethoxy-3,4,5,6-tetrabromophthalanilic acid ... 448
$C_{16}H_{10}Br_5NO_3$
2′,6′-Dimethyl-3,4,4′,5,6-pentabromophthalanilic acid ... 450
$C_{16}H_{10}ClNO$
α-Benzoyl-4-chlorocinnamonitrile ... 759
$C_{16}H_{10}ClNO_2$
4-(4-Chlorobenzylidene)-2-phenyl-2-oxazolin-5-one ... 386
2-(*p*-Chlorophenyl)-quinoline-4-carboxylic acid ... 982
$C_{16}H_{10}ClNO_3$
7-Chloro-4-hydroxy-2-phenylquinoline-3-carboxylic acid ... 984
$C_{16}H_{10}ClNO_5$
4-(*N*-(5-Chloro-2-methoxyphenyl)carbamoyl)phthalic anhydride ... 449
$C_{16}H_{10}ClN_3O_4$
5-Chloro-8-quinolyl *N*-(3-nitrophenyl)carbamate ... 988
5-Nitro-8-quinolyl *N*-(3-chlorophenyl)carbamate ... 982
8-Quinolyl *N*-(2-chloro-4-nitrophenyl)carbamate ... 989
8-Quinolyl *N*-(4-chloro-3-nitrophenyl)carbamate ... 981
$C_{16}H_{10}Cl_2$
11,12-Dichloro-9,10-dihydro-9,10-ethenoanthracene ... 174
2,3-Dichloro-4b,8b,8c,8d-tetrahydrodibenzo(a,f)cyclopropa(cd)pentalene ... 173
$C_{16}H_{10}Cl_2F_6N_4O_2$
1,6-Bis(4-chloro-2-(trifluoromethyl)phenyl)biurea ... 505
$C_{16}H_{10}Cl_2N_2O_2$
5-Chloro-8-quinolyl *N*-(2-chlorophenyl)carbamate ... 991
5-Chloro-8-quinolyl *N*-(3-chlorophenyl)carbamate ... 992
5-Chloro-8-quinolyl *N*-(4-chlorophenyl)carbamate ... 991
5,7-Dichloro-8-quinolyl *N*-phenylcarbamate ... 991
8-Quinolyl *N*-(2,3-dichlorophenyl)carbamate ... 991
8-Quinolyl *N*-(2,4-dichlorophenyl)carbamate ... 992
8-Quinolyl *N*-(2,5-dichlorophenyl)carbamate ... 992
8-Quinolyl *N*-(2,6-dichlorophenyl)carbamate ... 982
8-Quinolyl *N*-(3,4-dichlorophenyl)carbamate ... 992
$C_{16}H_{10}Cl_2O_2$
trans-1,2-Bis(4-chlorobenzoyl)ethylene ... 306
4,4′-Stilbenedicarbonyl chloride ... 389
$C_{16}H_{10}Cl_2O_3$
3′,4′-Dichloro-3,4-(methylenedioxy)chalcone ... 304
$C_{16}H_{10}Cl_4$
9,10-Dihydro-2,3,11,12-tetrachloro-9,10-ethanoanthracene ... 173
$C_{16}H_{10}Cl_4F_3NO_2S$
1-(4-Chlorophenylthio)-2,2,2-trichloroethyl *N*-(3-trifluoromethylphenyl)carbamate ... 606
$C_{16}H_{10}Cl_4N_4O_2$
3a,6a-Diphenyl-1,3,4,6-tetrachloroglycoluril ... 711
Oxalic bis((2,4-dichlorobenzylidene)hydrazide) ... 493
$C_{16}H_{10}Cl_6FNO_3$
3-[*N*-(4-Fluorophenyl)-*N*-methylcarbamoyl]-1,4,5,6,7,7-hexachloro-5-norbornene-2-carboxylic acid ... 407
$C_{16}H_{10}Cl_7NO_3$
3-[*N*-(3-Chloro-4-methylphenyl)carbamoyl]-1,4,5,6,7,7-hexachloro-5-norbornene-2-carboxylic acid ... 406
$C_{16}H_{10}Cr_2HgO_6$
Bis-(cyclopentadienylchromium tricarbonyl) mercury ... 1140
$C_{16}H_{10}FNO_2$
4-(2-Fluorobenzylidene)-2-phenyl-2-oxazolin-5-one ... 386
4-(4-Fluorobenzylidene)-2-phenyl-2-oxazolin-5-one ... 385
$C_{16}H_{10}FN_3O_4$
5-Nitro-8-quinolyl *N*-(4-fluorophenyl)carbamate ... 981
$C_{16}H_{10}F_2N_2O_2$
1-Isoquinolyl *N*-(2,5-difluorophenyl)carbamate ... 996
8-Quinolyl *N*-(2,5-difluorophenyl)carbamate ... 982
$C_{16}H_{10}F_3N$
α-(3-(Trifluoromethyl)phenyl)cinnamonitrile ... 758
$C_{16}H_{10}F_{16}O_3S$
2,2,3,3,4,4,5,5,6,6,7,7,8,8,9,9-Hexadecafluorononyl *p*-toluenesulfonate ... 796

$C_{16}H_{10}N_2$
α,β-Diphenylfumaronitrile 764
Diphenylmethylenemalononitrile 764
2-Phenyl-4-quinolinecarbonitrile 989

$C_{16}H_{10}N_2O_2$
5,10-Dihydroindeno(2,1-a)indene-5,10-dione dioxime 1013
Isoindigo 487

$C_{16}H_{10}N_2O_3$
α-Benzoyl-4-nitrocinnamonitrile 765

$C_{16}H_{10}N_2O_4$
4-(3-Nitrobenzylidene)-2-phenyl-2-oxazolin-5-one 385
4-(4-Nitrobenzylidene)-2-phenyl-2-oxazolin-5-one 385
5-Nitro-α-piperonylidene-*o*-tolunitrile 764

$C_{16}H_{10}N_2O_4S$
1,3-Dinitro-4-(phenylthio)-naphthalene 270
3-(3-Methyl-2-benzothiazolinylidene)-6-nitrophthalide 385

$C_{16}H_{10}N_2O_7$
2-Furamido-4-nitrophenyl 2-furoate 847
4-(*N*-(2-Methoxy-4-nitrophenyl)-carbamoyl)phthalic anhydride 466
4-(*N*-(2-Methoxy-5-nitrophenyl)-carbamoyl)phthalic anhydride 466

$C_{16}H_{10}N_2S_2$
δ1,1′-Biisoindoline-3,3′-dithione 746
2-(2-Quinolylthio)benzothiazole 987

$C_{16}H_{10}N_4O_3$
3-Amino-4-cyano-7-nitro-2-phenyl-1(2*H*)-isoquinolinone 763

$C_{16}H_{10}N_4O_5$
4-(2,4-Dinitrophenylazo)-1-naphthol 1084

$C_{16}H_{10}N_4O_6$
N-Picryl-1-naphthylamine 272

$C_{16}H_{10}O$
2,3-Benzodiphenylene oxide 896
2-Hydroxypyrene 196
4-Hydroxypyrene 196

$C_{16}H_{10}O_2$
2-Benzylidene-1,3-indandione 308
Dibenzoylacetylene 1018
1,6-Dihydroxypyrene 195
Indeno(2,1a)indene-5,10(4bh,9bh)-dione 308

$C_{16}H_{10}O_3$
4-Carboxy-5-phenanthrenecarboxaldehyde 344

$C_{16}H_{10}O_4$
2-(2-Benzofuranylcarbonyl)-benzoic acid 895

$C_{16}H_{10}O_5$
2-Acetoxy-1-hydroxyanthraquinone 353

$C_{16}H_{10}O_7$
2′,3,4-Benzophenone-tricarboxylic acid 344

$C_{16}H_{11}AsClN$
7-Chloro-7,12-dihydrobenzo(c)-phenarsazine 1128

$C_{16}H_{11}BClF_4N_3$
2-(4-Chlorophenyl)-1,2,4-triazolo(3,4-a)isoquinolinium tetrafluoroborate 1002

$C_{16}H_{11}BF_4N_4O_2$
2-(4-Nitrophenyl)-1,2,4-triazolo(3,4-a)isoquinolinium tetrafluoroborate 1003

$C_{16}H_{11}Br$
2-Bromo-1-phenylnaphthalene 169

$C_{16}H_{11}BrF_4N_4O_3$
N-(4-Bromo-2-(trifluoromethyl)-phenylureido)-*N*′-(4-fluorophenyl)-oxamide 505

$C_{16}H_{11}BrN_2O_2$
8-Quinolinyl *N*-(3-bromophenyl)-carbamate 982
8-Quinolyl *N*-(4-bromophenyl)-carbamate 991

$C_{16}H_{11}BrO$
3-Acetyl-9-bromophenanthrene 298
9-(2-Bromoacetyl)phenanthrene 171

$C_{16}H_{11}BrO_2$
Methyl 10-bromo-9-anthracenecarboxylate 376

$C_{16}H_{11}BrO_2S$
3-(Bromomethyl)benzo(b)thiophene-5-yl benzoate 897

$C_{16}H_{11}BrO_3$
4′-Bromo-3,4-(methylenedioxy)-chalcone 304

$C_{16}H_{11}Br_2Cl$
10-Chloro-11,12-dibromo-10,11-dihydro-5,10-methano-5*H*-dibenzo(a,d)cycloheptene 173

$C_{16}H_{11}Br_2ClF_3NO_2$
2,4-Dibromo-5,6-dimethylphenyl *N*-(4-chloro-2-(trifluoromethyl)-phenyl)carbamate 608

$C_{16}H_{11}Br_4NO_3$
2′,3′-Dimethyl-3,4,5,6-tetrabromophthalanilic acid 455
4′-Ethyl-3,4,5,6-tetrabromophthalanilic acid 454

$C_{16}H_{11}Br_4NO_4$
2′-Ethoxy-3,4,5,6-tetrabromophthalanilic acid 448
4′-(2-Hydroxyethyl)-3,4,5,6-tetrabromophthalanilic acid 462

$C_{16}H_{11}Br_4NO_5$
2′,4′-Dimethoxy-3,4,5,6-tetrabromophthalanilic acid 447

$C_{16}H_{11}Cl$
11-Chloro-9,10-dihydro-9,10-ethenoanthracene 174
2-Chloro-9,10-dihydro-9,10-ethenoanthracene 174

$C_{16}H_{11}ClFNO_2S$
4-Chloro-2-(*m*-tolyl)-6-quinolinesulfonyl fluoride 994
4-Chloro-2-(*p*-tolyl)-6-quinolinesulfonyl fluoride 994

$C_{16}H_{11}ClF_3NO_3$
Methyl *N*-(2-chloro-5-(trifluoromethyl)phenyl)-phthalamate 445

$C_{16}H_{11}ClF_3NO_4$
4-Formyl-2-methoxyphenyl *N*-(4-chloro-2-(trifluoromethyl)phenyl)-carbamate 606

$C_{16}H_{11}ClHgN_2O$
1-(4-(Chloromercuri)phenylazo)-2-naphthol 1137

$C_{16}H_{11}ClN_2$
2-(4-Chlorophenylazo)naphthalene 1082

$C_{16}H_{11}ClN_2O$
5-(4-Chlorophenyl)-5-hydroxy-5*H*-imidazo(2,1-a)isoindole *p*-toluenesulfonate 864

$C_{16}H_{11}ClN_2O_2$
2-Chloro-3-(4-methyl-2-pyridylamino)-1,4-naphthoquinone 954
2-Chloro-3-(6-methyl-2-pyridylamino)-1,4-naphthoquinone 954
3-Chloro-2-(3-methyl-2-pyridylamino)-1,4-naphthoquinone 954
5-Chloro-8-quinolyl *N*-phenylcarbamate 991
8-Quinolyl *N*-(2-chlorophenyl)-carbamate 991
8-Quinolyl *N*-(3-chlorophenyl)-carbamate 992
8-Quinolyl *N*-(4-chlorophenyl)-carbamate 991

$C_{16}H_{11}ClN_2O_3$
4-Chloro-2-(4-methoxyphenyl)-6-nitroquinoline 985

$C_{16}H_{11}ClN_2O_4$
Phthalimido *N*-(3-chloro-4-methylphenyl)carbamate 716

$C_{16}H_{11}ClN_2O_5$
Phthalimido *N*-(5-chloro-2-methoxyphenyl)carbamate 715

$C_{16}H_{11}ClN_2O_9S_2$
Calcium fast purple 1080

$C_{16}H_{11}ClN_4O_{10}S_2$
Gallion tetrahydrate 1083

$C_{16}H_{11}Cl_2NO$
2,4-Dichloro-α-(4-methoxyphenyl)-cinnamonitrile 763

$C_{16}H_{11}Cl_2NO_2S$
3,4-Dichloro-*N*-(1-naphthyl)-benzenesulfonamide 810

$C_{16}H_{11}Cl_2NO_4$
2,6-Dichloro-4′-methoxy-3′-nitrochalcone 301

$C_{16}H_{11}Cl_2N_3OS$
1-(2,3-Dichlorophenyl)-3-(4-phenyl-2-thiazolyl)urea 875

$C_{16}H_{11}Cl_2N_3O_2$
1-(3,4-Dichlorophenyl)-5-imino-3-(*p*-tolyl)hydantoin 724
1-(3,5-Dichlorophenyl)-5-imino-3-(*p*-tolyl)hydantoin 724

$C_{16}H_{11}Cl_3N_4O_6$
N,*N*′-(2,2,2-Trichloroethylidene)-bis(4-nitrobenzamide) 446

$C_{16}H_{11}Cl_4NO_3$
2′,4′-Dimethyl-3,4,5,6-tetrachlorophthalanilic acid 474
2′,5′-Dimethyl-3,4,5,6-tetrachlorophthalanilic acid 456
2′,6′-Dimethyl-3,4,5,6-tetrachlorophthalanilic acid 454
3′,4′-Dimethyl-3,4,5,6-tetrachlorophthalanilic acid 455
4′-Ethyl-3,4,5,6-tetrachlorophthalanilic acid 454
N-(2-Phenylethyl)-3,4,5,6-tetrachlorophthalamic acid 452

$C_{16}H_{11}Cl_4NO_4$
4′-Ethoxy-3,4,5,6-tetrachlorophthalanilic acid 449
2′-Methoxy-5′-methyl-3,4,5,6-tetrachlorophthalanilic acid 449

$C_{16}H_{11}Cl_6NO_3$
1,4,5,6,7,7-Hexachloro-3-(*N*-(*m*-tolyl)carbamoyl)-5-norbornene-2-carboxylic acid 434

$C_{16}H_{11}Cl_6NO_4$
1,4,5,6,7,7-Hexachloro-3-(*N*-(4-methoxyphenyl)carbamoyl)-5-norbornen-2-carboxylic acid 418

$C_{16}H_{11}FN_2O_2$
1-Isoquinolyl *N*-(4-fluorophenyl)-carbamate 996
8-Quinolyl *N*-(4-fluorophenyl)-carbamate 991

$C_{16}H_{11}FN_4O_3S$
1-(4-Fluoro-3-nitrophenyl)-3-(4-phenyl-2-thiazolyl)urea 875

$C_{16}H_{11}FO_3$
4′-Fluoro-3,4-(methylenedioxy)-chalcone 301

$C_{16}H_{11}F_3N_4OS_2$
1-(5-Mercapto-1,3,4-thiadiazol-2-yl)-3-(3-(trifluoromethyl)phenyl)urea 880

$C_{16}H_{11}F_3O$
3′-(Trifluoromethyl)chalcone 299

$C_{16}H_{11}N$
2-Aminopyrene 231
4-Aminopyrene 231
11*H*-Indeno(1,2-b)quinoline 995
6*H*-Indeno(2,1-b)quinoline 995

$C_{16}H_{11}NO$
2-Benzoylquinoline 989
11*H*-Indeno(1,2-b)quinoline *N*-oxide 1007
5*H*-Indeno(1,2-b)quinolin-10(11*H*)-one 320

$C_{16}H_{11}NO_2$
4-Benzylidene-2-phenyl-2-oxazolin-5-one 385
4-Benzylidene-2-phenyl-2-oxazolin-5-one 387
4-Methylbenzofuro(3,2-g)quinol-2(1*H*)-one 919

$C_{16}H_{11}NO_3$
N-(Benzoylmethyl)-phthalimide 719
3-Hydroxy-2-phenylcinchoninic acid 983
2-Hydroxy-3-phenyl-4-quinolinecarboxylic acid 983
4-Hydroxy-2-phenyl-3-quinolinecarboxylic acid 984
2-(3-Indolylcarbonyl)benzoic acid 888
N-(4-Phenoxyphenyl)maleimide 717

$C_{16}H_{11}NO_4$
4-(*N*-Benzylcarbamoyl)phthalic anhydride 472
N-(2-Carboxybenzyl)phthalimide 721
Methyl 10-nitro-9-anthracenecarboxylate 374
2-(4-Nitrobenzyl)-1,3-indanedione 308

$C_{16}H_{11}NO_5$
N-(Benzyloxycarbonyloxy)-phthalimide 717
3,4-Methylenedioxy-4′-nitrochalcone 303

$C_{16}H_{11}NO_5S$
4-(4-Sulfoanilino)-1,2-naphthoquinone, sodium salt 793

$C_{16}H_{11}NO_7S_2$
5,6-Dihydroxy-8-(2-nitrophenylthio)-2-naphthalenesulfonic acid, sodium salt 791

$C_{16}H_{11}NS$
7*H*-Benzo(c)phenothiazine 250
2-(3-Indenyl)benzothiazole 917

$C_{16}H_{11}N_3$
3-Phenyl-1*H*-1,2,4-triazolo(4,3-b)-isoquinoline perchlorate 909
3-(4-Pyridyl)-2*H*-benzene(g)indazole 953
3-(4-Pyridyl)-2*H*-benzene(g)indazole methanesulfonate 953

$C_{16}H_{11}N_3O$
2-Phenylpyrimido(1,2-a)-benzimidazol-4(3*H*)-one 1003

$C_{16}H_{11}N_3OS_2$
3-Methyl-2-benzothiazolinone (2,3-dihydro-2oxobenzo(b)-thiophen3ylidene)hydrazone 1061

$C_{16}H_{11}N_3O_2S$
2-(α-Cyano-4-nitrobenzylidene)-3-methylbenzothiazoline 765

$C_{16}H_{11}N_3O_4$
8-Quinolyl *N*-(3-nitrophenyl)-carbamate 988

$C_{16}H_{11}N_3O_5$
2-Methyl-4,5-bis(4-nitrophenyl)-oxazole 872
2′-Nitro-4′-phthalimidoacetanilide 718

$C_{16}H_{11}N_3O_6$
Phthalimido *N*-(2-methyl-5-nitrophenyl)carbamate 718

$C_{16}H_{11}N_3O_9S_2$
3-Hydroxy-4-(4-nitrophenylazo)-2,7-naphthalenedisulfonic acid, disodium salt 1084

$C_{16}H_{12}$
1,2-Dihydroaceanthrylene 167

$C_{16}H_{12}BF_4NO_2$
4-(1,3-Dioxo-2-indanylidenemethyl)-1-methylpyridinium tetrafluoroborate 956

$C_{16}H_{12}BrClF_3NO_2$
2-Bromo-4,5-dimethylphenyl *N*-(2-chloro-5-(trifluoromethyl)phenyl)-carbamate 608
2-Bromo-4,5-dimethylphenyl *N*-(4-chloro-2-(trifluoromethyl)phenyl)-carbamate 607
4-Bromo-3,5-dimethylphenyl *N*-(4-chloro-2-(trifluoromethyl)phenyl)-carbamate 608

$C_{16}H_{12}BrClO_2$
5-Bromo-4′-chloro-2-methoxychalcone 299

$C_{16}H_{12}BrNO$
2-(5-Phenyl-3-isoxazolyl)benzyl bromide 870

$C_{16}H_{12}BrN_3O_2$
Isatin, azine with 5-bromo-*o*-anisaldehyde 1042

$C_{16}H_{12}BrO$
Acetyl-9-bromophenanthrene, mixture of isomers 298

$C_{16}H_{12}Br_2$
9,10-Bis(bromomethyl)anthracene 172
4,5-Bis(bromomethyl)-phenanthrene 172

$C_{16}H_{12}Br_2Cl_2N_2O_3$
3,5-Dibromo-4-hydroxyhydrocinnamic acid (3,5-dichlorosalicylidene)-hydrazide 495

$C_{16}H_{12}Br_2F_3NO$
2,3-Dibromo-3-phenyl-3′-(trifluoromethyl)propionanilide 397

$C_{16}H_{12}Br_2F_3NO_2$
2,4-Dibromo-5,6-dimethylphenyl *N*-(3-(trifluoromethyl)phenyl)-carbamate 607

$C_{16}H_{12}Br_2O_2$
4,4′-Bis(bromoacetyl)biphenyl 306

$C_{16}H_{12}Br_2O_2S$
4′-Bromophenacyl sulfide 288

$C_{16}H_{12}Br_4N_2O_3$
4′-(Dimethylamino)-3,4,5,6-tetrabromophthalanilic acid 451

$C_{16}H_{12}ClF_3N_2O_3$
4-Acetamidophenyl *N*-(2-chloro-5-(trifluoromethyl)phenyl)carbamate .. 607
4-Acetamidophenyl *N*-(4-chloro-2-(trifluoromethyl)phenyl)carbamate .. 606

$C_{16}H_{12}ClF_3N_4O_3$
N-(3-Chlorophenylureido)-*N*′-(3-(trifluoromethyl)phenyl)oxamide 505

$C_{16}H_{12}ClN$
N-(3-Chlorophenyl)-2-naphthylamine 233
4-Chloro-α-(*p*-tolyl)-cinnamonitrile 767

$C_{16}H_{12}ClNO$
3-(4-Chlorobenzoyl)-2-methylindole .. 881
4-Chloro-α-(4-methoxyphenyl)-cinnamonitrile 758
α-(3-Chlorophenyl)-4-methoxycinnamonitrile 763
α-(4-Chlorophenyl)-4-methoxycinnamonitrile 758
2-(3-Chlorophenyl)-4-methyl-5-phenyloxazole 872

$C_{16}H_{12}ClNO_2$
N-(2-(4-Chlorophenyl)ethyl)-phthalimide 716

$C_{16}H_{12}ClNO_2S$
4-Chloro-*N*-(1-naphthyl)-benzenesulfonamide 810

$C_{16}H_{12}ClNO_3$
3-(4-Chlorobenzyl)-4-phenyl-2,5-oxazolidinedione 695
2-Chloro-4′-methyl-5-nitrochalcone .. 299

$C_{16}H_{12}ClNO_4$
4-(2-(2-Chlorobenzoyl)acetamido)-benzoic acid 443
4-(2-(4-Chlorobenzoyl)acetamido)-benzoic acid 443
2-Chloro-4′-methoxy-3′-nitrochalcone 301
4-Chloro-4′-methoxy-3-nitrochalcone 299

$C_{16}H_{12}ClN_3O_2$
1-(3-Chloro-2-methylphenyl)-5-imino-3-phenylhydantoin 724
1-(4-Chloro-2-methylphenyl)-5-imino-3-phenylhydantoin 724
1-(3-Chlorophenyl)-5-imino-3-(*p*-tolyl)hydantoin 724

$C_{16}H_{12}ClN_3O_4$
N-(3-Chlorophenyl)-*N*′-(piperonylideneamino)oxamide 492

$C_{16}H_{12}Cl_2$
11,12-Dichloro-9,10-dihydro-9,10-ethanoanthracene 173
2,12-Dichloro-9,10-dihydro-9,10-ethanoanthracene 173
11,12-Dichloro-10,11-dihydro-5,10-methano-5*H*-dibenzo(a,d)-cycloheptene 173

$C_{16}H_{12}Cl_2FN_3O_2$
2,4-Dichlorohippuric (3-fluorobenzylidene)hydrazide 493
2,4-Dichlorohippuric (4-fluorobenzylidene)hydrazide 493
3,4-Dichlorohippuric (2-fluorobenzylidene)hydrazide 493
3,4-Dichlorohippuric (3-fluorobenzylidene)hydrazide 493
3,4-Dichlorohippuric (4-fluorobenzylidene)hydrazide 493

$C_{16}H_{12}Cl_2F_3NO_2$
2′-Chloro-2-(4-chloro-2-methylphenoxy)-5′-(trifluoromethyl)acetanilide 395
4-Chloro-3,5-dimethylphenyl *N*-(2-chloro-5-(trifluoromethyl)phenyl)-carbamate 608

$C_{16}H_{12}Cl_2F_3NO_3$
2-(4-Chlorophenoxy)ethyl *N*-(4-chloro-3-(trifluoromethyl)phenyl)-carbamate 606
2-(2,4-Dichlorophenoxy)ethyl *N*-(3-trifluoromethylphenyl)carbamate 609

$C_{16}H_{12}Cl_2N_2O_2S$
2-(2-Aminophenylthio)-*N*-(3,4-dichlorophenyl)-succinimide 715

$C_{16}H_{12}Cl_2N_4O$
1-(2,3-Dichlorophenyl)-3-(1-phenyl-5-pyrazolyl)urea 860

$C_{16}H_{12}Cl_2N_4O_2$
Oxalic bis(4-chlorobenzylidene)-hydrazide 1024

$C_{16}H_{12}Cl_2N_4O_4$
3,4-Dichlorohippuric (3-nitrobenzylidene)hydrazide 493

$C_{16}H_{12}Cl_2N_4O_6S_2$
S,S′-Ethylene bis(*N*-(2-chloro-4-nitrophenyl)thiocarbamate) 627

$C_{16}H_{12}Cl_2O$
2,4-Dichloro-4′-methylchalcone 305
2,6-Dichloro-4′-methylchalcone 306
3,4-Dichloro-4′-methylchalcone 287
3-(2,4-Dichlorophenyl)-2-methylacrylophenone 305

$C_{16}H_{12}Cl_2O_2$
1,4-Bis-(4-chlorophenoxy)-2-butyne ... 1017

$C_{16}H_{12}Cl_3NO_3$
N-(3-Acetylphenyl)-2-(2,4,5-trichlorophenoxy)acetamide 401

$C_{16}H_{12}Cl_3NO_4$
N-(2-(Methoxycarbonyl)phenyl)-2,4,5-trichlorophenoxyacetamide 401

$C_{16}H_{12}Cl_3N_3O_2$
4-Chlorohippuric (3,4-dichlorobenzylidene)hydrazide 492

$C_{16}H_{12}Cl_4N_2$
N,N′-Bis-(3,4-dichlorobenzylidene)-ethylenediamine 1054

$C_{16}H_{12}Cl_4N_2O_2S_2$
S,S′-Ethylene bis(*N*-(2,3-dichlorophenyl)thiocarbamate) 642

$C_{16}H_{12}Cl_4N_2O_3$
4′-(Dimethylamino)-3,4,5,6-tetrachlorophthalanilic acid 451

$C_{16}H_{12}Cl_4N_2O_4S$
4-(2,2-Dichloroacetamido)phenyl sulfone 772

$C_{16}H_{12}Cl_4N_6$
2,5-Diimino-3a,6a-diphenyl-1,3,4,6-tetrachloroimidazo(4,5-d)imidazole. 727

$C_{16}H_{12}Cl_4O_2$
1,4-Bis-(2,4-dichlorophenoxy)-2-butene 182

$C_{16}H_{12}Cl_{10}OSi_2$
1,3-Bis(pentachlorophenyl)-1,1,3,3-tetramethyldisiloxane 1101

$C_{16}H_{12}Cl_{10}Si_2$
1,2-Bis(pentachlorophenyl)-1,1,2,2-tetramethyldisilane 1101

$C_{16}H_{12}FN_3OS$
1-(2-Fluorophenyl)-3-(4-phenyl-2-thiazolyl)urea 875
1-(4-Fluorophenyl)-3-(4-phenyl-2-thiazolyl)urea 875

$C_{16}H_{12}F_2O_8S_2$
Bis-(4-(fluorosulfonyl)-phenyl) succinate 804

$C_{16}H_{12}F_3NO$
2′-(Trifluoromethyl)cinnamanilide 446

$C_{16}H_{12}F_3NO_3$
3-Acetylphenyl *N*-(2-(trifluoromethyl)phenyl)carbamate .. 609
Methyl *N*-phenyl-*N*-(trifluoroacetyl)-anthranilate 366

$C_{16}H_{12}F_3NO_4$
Piperonyl *N*-(3-trifluoromethylphenyl)carbamate 610

$C_{16}H_{12}F_3N_5O$
1-(5-Phenyl-1*H*-1,2,4-triazol-3-yl)-3-(2-(trifluoromethyl)phenyl)urea 869

$C_{16}H_{12}F_5NO$
N-(1,1-Bis-(4-fluorophenyl)-2,2,2-trifluoroethyl)-acetamide 398

$C_{16}H_{12}F_{10}OSi_2$
1,3-Bis(pentafluorophenyl)-1,1,3,3-tetramethyldisiloxane 1101

$C_{16}H_{12}N_2$
meso-1,2-Dicyano-1,2-diphenylethane 755
2,3-Diphenylpyrazine 978
3,6-Diphenylpyridazine 968
4-Styrylcinnoline 1001

$C_{16}H_{12}N_2Cl_4$
N,N′-Bis(2,4-dichlorobenzylidene)-ethylenediamine 1024

$C_{16}H_{12}N_2O$
4-(Phenylazo)-1-naphthol 1086
4-(Phenylazo)-1-naphthol, potassium salt 1086
2-Phenyl-4-quinolinecarboxamide 988
1-(2-Pyridyliminomethyl)-2-naphthol. 1039
3-Pyridyl 2-quinolylmethyl ketone 993

$C_{16}H_{12}N_2OS$
N-Benzoyl-*N*-(*p*-tolyl)-cyanothioformamide 766
O-(8-Quinolyl) *N*-phenylthiocarbamate 994

$C_{16}H_{12}N_2O_2$
2-(2-Aminophenyl)-4-quinolinecarboxylic acid 988
(1,1′-Biisoindoline)-3,3′-dione 691
9,10-Bis-(formamido)-anthracene 487
1-Diazo-3-phenoxy-4-phenyl-3-buten-2-one 1094
3,3′-Dimethyl-4,4′-biphenylene diisocyanate 769
1-(2-Hydroxyphenylazo)-2-naphthol .. 1085
N-(4-Methylbenzylideneamino)-phthalimide 1070
α-(4-Methylbenzylidene)-5-nitro-*o*-tolunitrile 764
4-Methyl-α-(4-nitrophenyl)-cinnamonitrile 766
3-Nitro-α-(*p*-tolyl)cinnamonitrile 766
2-Phenyl-1-phthalimidoaziridine 721
8-Quinolyl *N*-phenylcarbamate 991

$C_{16}H_{12}N_2O_2S$
2-(4-Aminophenylthio)-4-quinolinecarboxylic acid 987
N-Benzoyl-*N*-(4-methoxyphenyl)-cyanothioformamide 753
1-Cyano-1,2-dihydro-2-(phenylsulfonyl)isoquinoline 818

$C_{16}H_{12}N_2O_3$
α-(4-Methoxybenzylidene)-5-nitro-*o*-tolunitrile 762
α-(4-Methoxyphenyl)-3-nitrocinnamonitrile 762
N-(4-Phthalimidophenyl)acetamide .. 723

$C_{16}H_{12}N_2O_4$
1,4-Bis(4-nitrophenyl)-1,3-butadiene. 278
N,N′-Bis(piperonylidene)hydrazine ... 1068
α-Cyano-3-nitrophenethyl benzoate. 757
3,3′-Dimethoxy-4,4′-biphenylene diisocyanate 768
1-(2,4-Dinitrophenyl)-4-phenyl-1,3-butadiene 275
N-(3,4-(Methylenedioxy)-anilinomethyl)-phthalimide 719
N-(Vanillylideneamino)-phthalimide. 1032

$C_{16}H_{12}N_2O_4S$
4-(4-Hydroxy-1-naphthylazo)-benzenesulfonic acid 1085
4-Hydroxy-3-(phenylazo)-1-naphthalenesulfonic acid, sodium salt 1084
Orange I (C.I. 14600) 1085

$C_{16}H_{12}N_2O_5$
3-Acetamido-8-acetyl-2-nitrodibenzofuran 894
1-Acetyl-1,2-dihydro-2-(3-nitrophenyl)-4*H*-3,1-benzoxazin-4-one 703
3-(Diacetylamino)-2-nitrodibenzofuran 894
Ethyl *N*-(3-nitro-9-oxo-2-fluorenyl)-carbamate 669
Piperonylic acid piperonylidenehydrazide 502

$C_{16}H_{12}N_2O_6$
3,3′-Dinitro-4′-methoxychalcone 301

$C_{16}H_{12}N_2O_6S$
4-(2,4-Dihydroxyphenylazo)-3-hydroxy-1-naphthalene-sulfonic acid 1085

$C_{16}H_{12}N_2O_7S_2$
8-Hydroxy-5-(4-sulfophenylazo)-2-naphthalenesulfonic acid 1085
Orange G 1084

$C_{16}H_{12}N_2S_2$
N,N′-Bis(2-thenylidene)-1,4-phenylenediamine 1059

$C_{16}H_{12}N_4$
2-(2-Aminophenyl)-2*H*-naphtho(1,2-d)triazole 908

$C_{16}H_{12}N_4OS_2$
1-Benzoyl-3-(5-phenyl-1,2,4-thiadiazol-3-yl)-2-thiourea 879

$C_{16}H_{12}N_4O_3$
N-(2-(2-Pyridyl)-5-benzimidazolyl)-maleamic acid 962

$C_{16}H_{12}N_4O_3S$
5-Imino-1-(4-methyl-3-nitrophenyl)-3-phenyl-4-thiohydantoin 740

$C_{16}H_{12}N_4O_4$
5-Imino-1-(2-methyl-5-nitrophenyl)-3-phenylhydantoin 725
5-Imino-1-(4-methyl-2-nitrophenyl)-3-phenylhydantoin 702

$C_{16}H_{12}O$
2-Acetylanthracene 298
2-Benzylidene-1-indanone 306
9,10-Dihydro-9,10-ethanoantracen-11-one 286
3,4-Diphenylfuran 854

$C_{16}H_{12}O_2$
5*H*-Dibenzo(a,d)cycloheptene-5-carboxylic acid 334
1,3-Dimethylanthraquinone 319
2,7-Dimethyl-9,10-phenanthrenequinone 318
4,5-Dimethyl-9,10-phenanthrenequinone 318
2,3-Diphenyl-2-cyclopropene-1-carboxylic acid 346
Methyl 9-anthracenecarboxylate 376
10-Methyl-9-anthracenecarboxylic acid 346
Methyl 4-phenanthrenecarboxylate. 376
4-Phenanthreneacetic acid 334

$C_{16}H_{12}O_2S$
1-Naphthyl phenyl sulfone 785
2-Naphthyl phenyl sulfone 785

$C_{16}H_{12}O_3$
2,3-Diphenylsuccinic anhydride 388
5-Hydroxy-7-methylflavone 319
2-Methoxy-3-methylanthraquinone ... 317
3,4-(Methylenedioxy)-chalcone 304

$C_{16}H_{12}O_3S$
2,8-Diacetylphenoxathiin 309

$C_{16}H_{12}O_4$
2-(Benzoylacetyl)benzoic acid disodium derivative 351
1,4-Dihydro-9-hydroxy-8-methoxy-10-methyl-1,4-anthracenedione 315
2′-Hydroxy-2-piperonylideneacetophenone 300
4,4′-Stilbenedicarboxylic acid 346
Techtochrysin 319

$C_{16}H_{12}O_5$
6-Hydroxy-3-oxo-3*H*-xanthene-9-propionic acid 326
5-(Methoxycarbonyl)-1,2,3,4-tetra-hydro-1,4-ethenonaphthalene-2,3-dicarboxylic anhydride 388

$C_{16}H_{12}O_5S$
4-(Ethoxycarbonylmethylthio)-1,8-naphthalic anhydride 388

$C_{16}H_{12}O_6$
5-(2-(1,3-Benzodioxol-5-yl)-2-(hydroxy)acetyl)-1,3-benzodioxole 290
3,4-Dicarboxy-1,2,3,4-tetrahydro-1-naphthalenesuccinic acid dianhydride 99

$C_{16}H_{12}O_7$
2,2′-(Dibenzofuran-2,8-diyldioxy)-diacetic acid 893
3-Oxo-4,5,6-trihydroxy-3*H*-xanthene-9-propionic acid 326

$C_{16}H_{12}O_8$
5,6,8-Triacetoxy-1,4-naphthoquinone 378
$C_{16}H_{12}PbS_4$
Tetrakis(2-thienyl)lead 1135
$C_{16}H_{13}Br$
9-Bromo-9,10-dihydro-9,10-ethanoanthracene 173
9-Bromomethyl-10-methylphenanthrene 172
$C_{16}H_{13}BrCl_2N_4O_3$
2-(3-(3,4-Dichlorophenyl)ureido)-acetic (5-bromo-2-hydroxybenzylidene)hydrazide 513
$C_{16}H_{13}BrF_3NO_2$
2-Bromo-4,5-dimethylphenyl *N*-(2-(trifluoromethyl)phenyl)carbamate 607
$C_{16}H_{13}BrF_3NO_3$
2-(4-Bromophenoxy)ethyl *N*-(2-(trifluoromethyl)phenyl)carbamate 606
$C_{16}H_{13}BrN_2O_2$
N-(4-Bromo-*o*-toluidinomethyl)-phthalimide 719
$C_{16}H_{13}BrN_2O_2S$
2-(2-Aminophenylthio)-*N*-(3-bromophenyl)-succinimide 717
$C_{16}H_{13}BrO$
3-Bromo-4′-methylchalcone 306
4-Bromo-4′-methylchalcone 306
$C_{16}H_{13}BrO_2$
3-Bromo-4-methoxychalcone 302
4′-Bromo-4-methoxychalcone 301
$C_{16}H_{13}Cl$
9-Chloro-9,10-dihydro-9,10-ethanoanthracene 173
12-Chloro-10,11-dihydro-5,10-methano-5*H*-dibenzo(a,d)-cycloheptene 173
9-(Chloromethyl)-10-methylanthracene 172
$C_{16}H_{13}ClFNO_3$
2-(4-Chloro-3-methylphenoxy)ethyl *N*-(4-fluorophenyl)carbamate 616
$C_{16}H_{13}ClFNO_4$
2-Chloro-4-fluorophenyl *N*-(4-(ethoxycarbonyl)phenyl)carbamate. 630
4-(Ethoxycarbonyl)phenyl *N*-(3-chloro-4-fluorophenyl)carbamate 630
$C_{16}H_{13}ClFN_3O_2$
N-(4-Chlorobenzoyl)glycine (3-fluorobenzylidene)hydrazide 493
2-Chlorohippuric (4-fluorobenzylidene)hydrazide 493
$C_{16}H_{13}ClFN_5O_4$
2-(3-(2-Chloro-4-nitrophenyl)ureido)-acetic (2-fluorobenzylidene)-hydrazide 522
2-(3-(2-Chloro-4-nitrophenyl)ureido)-acetic (3-fluorobenzylidene)-hydrazide 522
2-(3-(2-Chloro-4-nitrophenyl)ureido)-acetic (4-fluorobenzylidene)-hydrazide 522
$C_{16}H_{13}ClF_3NO$
4-Chloro-*N*-ethyl-2′-(trifluoromethyl)benzanilide 446
$C_{16}H_{13}ClF_3NO_2$
4′-Chloro-2′-methoxy-*N*-methyl-4-(trifluoromethyl)benzanilide 445
2-(4-Chloro-2-methylphenoxy)-2′-(trifluoromethyl)acetanilide 395
2-(4-Chloro-2-methylphenoxy)-3′-(trifluoromethyl)acetanilide 395
$C_{16}H_{13}ClF_3N_3O_4S$
N-Acetyl-4-(4-chloro-2-(trifluoromethyl)phenylureido)-benzenesulfonamide 805
$C_{16}H_{13}ClN_2O$
9β-(3-Chlorophenyl)-2,3,5,9b-tetrahydro-1*H*-imidazo(2,1-a)-isoindol-5-one 695
3-(3-Chloro-*o*-tolyl)-2-methyl-4(3*H*)-quinazolone 710
Cinnamic (2-chlorobenzylidene)-hydrazide 493
$C_{16}H_{13}ClN_2O_2$
N-(3-Chloro-*o*-toluidinomethyl)-phthalimide 719
N-(3-Chloro-*p*-toluidinomethyl)-phthalimide 719
N-(4-Chloro-*o*-toluidinomethyl)-phthalimide 719
N-(5-Chloro-*o*-toluidinomethyl)-phthalimide 719
$C_{16}H_{13}ClN_2O_2S$
2-(2-Aminophenylthio)-*N*-(2-chlorophenyl)-succinimide 717
2-(*o*-Aminophenylthio)-*N*-(*p*-chlorophenyl)-succinimide 717
$C_{16}H_{13}ClN_2O_4$
Methyl 6-chloro-5-cyano-4-(dimethoxyphenyl)-2-pyridinecarboxylate 936
$C_{16}H_{13}ClN_2O_6$
4-(Ethoxycarbonyl)phenyl *N*-(4-chloro-3-nitrophenyl)carbamate 625
$C_{16}H_{13}ClN_4O_3$
3-Chloro-6-nitro-1-(*N*-(3,4-xylyl)-carbamoyl)-1*H*-indazole 900
$C_{16}H_{13}ClN_4O_3S$
1-(2-Chloro-4-nitrophenyl)-3-(3-cyano-4,5,6,7-tetrahydrobenzo(b)-thiophen-2-yl)urea 855
$C_{16}H_{13}ClN_4O_5$
1-((5-Chloro-2,4-dimethoxyphenyl)-carbamoyl)-6-nitrobenzimidazole 901
1-(*N*-(5-Chloro-2,4-dimethoxyphenyl)carbamoyl)-6-nitro-1*H*-indazole 899
$C_{16}H_{13}ClO$
12-Chloro-10,11-dihydro-5,10-methano-5*H*-dibenzo(a,d)-cyclohexen-11-ol 205
2-Chloro-4′-methylchalcone 306
3-Chloro-4′-methylchalcone 306
4-Chloro-4′-methylchalcone 306
4′-Chloro-4-methylchalcone 306
$C_{16}H_{13}ClO_2$
4′-Chloro-2-methoxychalcone 301
4′-Chloro-3-methoxychalcone 301
4′-Chloro-4-methoxychalcone 302
4-Chloro-4′-methoxychalcone 299
1,2-Dibenzoyl-1-chloroethane 308
$C_{16}H_{13}ClO_4$
2-(Acetonyloxy)phenyl 4-chlorobenzoate 377
$C_{16}H_{13}ClO_5$
2,6-Diphenylpyrylium perchlorate 935
$C_{16}H_{13}Cl_2N$
Bis(2-chlorobenzyl)acetonitrile 752
2-(*N*,*N*-Bis(chloromethyl)amino)-anthracene 222
5-Chloro-3-(4-chlorophenyl)-1-ethylindole 881
$C_{16}H_{13}Cl_2NO$
4′,5′-Dichloro-2′-methylcinnamanilide 461
$C_{16}H_{13}Cl_2NO_2$
Ethyl 4-(2,4-dichlorobenzylideneamino)-benzoate 1026
$C_{16}H_{13}Cl_2NO_3$
N-(4-Acetylphenyl)-2-(2,4-dichlorophenoxy)acetamide 402
Ethyl 4-(2,4-dichlorobenzamido)-benzoate 477
Ethyl 4-(3,4-dichlorobenzamido)-benzoate 477
$C_{16}H_{13}Cl_2NO_4$
2,4-Dichlorophenyl *N*-(4-(ethoxycarbonyl)phenyl)carbamate. 630
$C_{16}H_{13}Cl_2N_3O_2$
N,*N*-Diacetyl-2,6-dichloro-4-(phenylazo)aniline 1082
Hippuric (2,4-dichlorobenzylidene)-hydrazide 493
$C_{16}H_{13}Cl_2N_3O_3$
3,4-Dichlorohippuric (2-hydroxybenzylidene)hydrazide 491
3,4-Dichlorohippuric (4-hydroxybenzylidene)hydrazide 491
1-(3,4-Dichlorophenyl)-4-(2-methoxybenzylideneamino)-oxamide 492
$C_{16}H_{13}Cl_2N_3O_4$
1-(2,4-Dichlorobenzoyl)-4-(2-(methoxycarbonyl)phenyl)-semicarbazide 528
$C_{16}H_{13}FN_2O$
9β-(3-Fluorophenyl)-1,2,3,9b-tetrahydro-5*H*-imidazo(2,1-a)-isoindol-5-one 695
$C_{16}H_{13}FN_2O_4$
5-Indanyl *N*-(4-fluoro-3-nitrophenyl)-carbamate 626
3-Phenylallyl *N*-(4-fluoro-3-nitrophenyl)carbamate 625
$C_{16}H_{13}FN_2O_7S$
5′-(Fluorosulfonyl)-2-(3-nitrobenzoyl)-*o*-acetanisidide 802
$C_{16}H_{13}FN_4O$
1-(3-Fluorophenyl)-3-(1-phenyl-5-pyrazolyl)urea 860
$C_{16}H_{13}FN_4O_3S_2$
Benzoic (2-(4-fluorosulfonylph)-4-methyl-1,3,4-thiadiazol-2-in-5-ylidene)hydrazide 803
$C_{16}H_{13}FO$
4′-Fluoro-4-methylchalcone 305
4-Fluoro-4′-methylchalcone 306
$C_{16}H_{13}FO_2$
4′-Fluoro-4-methoxychalcone 301
$C_{16}H_{13}F_2N_3OS$
1-(2,5-Difluorophenyl)-3-(5,6-dimethyl-2-benzothiazolyl)urea 912
$C_{16}H_{13}F_3N_2O$
α-(3-(Trifluoromethyl)phenylimino)-*p*-acetotoluidide 1021
$C_{16}H_{13}F_3N_2O_2$
1-(4-Acetylphenyl)-3-(2-(trifluoromethyl)phenyl)urea 509
$C_{16}H_{13}F_3N_2O_3$
Methyl 2-(3-(trifluoromethyl)-phenylureido)benzoate 509
$C_{16}H_{13}F_3N_2O_4$
2,6-Dimethyl-4-nitrophenyl *N*-(3-trifluoromethylphenyl)carbamate 610
$C_{16}H_{13}F_5Si$
Dimethyl(pentafluorophenyl)-styrylsilane 1102
$C_{16}H_{13}IN_2O_2S$
2-(2-Aminophenylthio)-*N*-(4-iodophenyl)-succinimide 715
$C_{16}H_{13}N$
5,6-Dihydroindolo(2,1-a)-isoquinoline 891
4-Methyl-2-phenylquinoline (in hexane) 991
1-(*p*-Tolyl)isoquinoline 998
2-(*p*-Tolyl)quinoline 994
$C_{16}H_{13}NO$
1-Acetamidoanthracene 491
2-Acetylphenanthrene oxime 1014
N-(2-Anthracenyl)acetamide 491
1-Benzoyl-3-methylindole 887
2-Benzoyl-3-methylindole 886
3-Benzoyl-2-phenylpropionitrile 757
2-Benzyl-8-hydroxyquinoline 983
1-Benzylindole-3-carboxaldehyde 887
2-(Benzyloxy)quinoline 987
2,4-Diphenyl-5-methyloxazole 872
2,4-Diphenyl-3-oxobutyronitrile 755
α-(4-Methoxyphenyl)cinnamonitrile 763
2-Methoxy-α-phenylcinnamonitrile 762
β-Methoxy-α-phenylcinnamonitrile 762
6-Methoxy-2-phenylquinoline 986
2-(2-Methoxyphenyl)quinoline picrate 986
Methyl 3-phenanthryl ketone oxime 1014
2-(2-Methylphenyl)-5-phenyloxazole 872
2-(3-Methylphenyl)-5-phenyloxazole 872
4-Methyl-1-phenyl-2(1*H*)-quinolone 692
N-(3-Phenanthryl)acetamide 441
N-(9-Phenanthryl)acetamide 441
5-Phenyl-2-(*p*-tolyl)-oxazole 872
$C_{16}H_{13}NO_2$
2-Benzamidocinnamaldehyde 473
4-Benzylidene-2-phenyl-5-oxazolidinone 385
1-Benzylindole-3-carboxylic acid 888
5-Benzyloxy-3-indolecarboxaldehyde 882
4-Benzyl-2-phenyl-2-oxazolin-5-one 385
2,9-Diacetylcarbazole 892
3,4-Dihydro-3′-(4-pyridyl)-1-oxospiro(naphthalene-2(1*H*),2′-oxirane) 967
2,2-Diphenylsuccinimide 718
Ethyl α-cyano-1-naphthaleneacrylate 766
N-Furfuryl-1-naphthamide 851
2-(3-Methoxyphenyl)-5-phenyloxazole 872
2-(*N*-Methylacetamido)-9-fluorenone 488
N-(4-Methylbenzyl)phthalimide 722
5-Methyl-*N*-(3-methylbenzoyl)-anthranilic acid 387
1-(4-Nitrophenyl)-4-phenyl-1,3-butadiene 278
3-Phenylacetyl-2-oxoindoline 691
$C_{16}H_{13}NO_2S$
2-(Benzylsulfonyl)quinoline 994
2-(Diacetylamino)dibenzothiophene. 898
$C_{16}H_{13}NO_3$
α-(Benzamido)cinnamic acid 476
2-(4-Ethoxyphenyl)-3,1-benzoxazin-4(4*H*)-one 384
Furfuryl *N*-(1-naphthyl)carbamate 849
N-(3-Methoxybenzoyl)-5-methylanthranilic acid 384
4′-Methyl-3-nitrochalcone 303
4′-Methyl-4-nitrochalcone 303
4-Methyl-4′-nitrochalcone 303
2-(4-Nitrobenzyl)-1-indanone 295
4-(Piperonylideneamino)-acetophenone 1074
$C_{16}H_{13}NO_4$
3′-Acetylphthalanilic acid 473
4′-Acetylphthalanilic acid 473
4-(2-Benzoylacetamido)-benzoic acid 443
Cinnamyl 4-nitrobenzoate 374
2-Methoxy-4′-nitrochalcone 301
3′-Methoxy-3-nitrochalcone 301
3-Methoxy-4′-nitrochalcone 301
4-Methoxy-3′-nitrochalcone 301
4-Methoxy-4′-nitrochalcone 301
4-Nitro-α-(*o*-tolyl)cinnamic acid 335
4′-Phenoxymaleanilic acid 487
$C_{16}H_{13}NO_4S$
7-Anilino-1-naphthol-3-sulfonic acid. 787
$C_{16}H_{13}NO_5S$
2-Phthalimidoethyl benzenesulfonate 799
$C_{16}H_{13}NO_6$
4-(*N*-(4-Methoxyphenyl)carbamoyl)-isophthalic acid, 4-methoxyaniline salt 466
$C_{16}H_{13}NS$
2-(Benzylthio)quinoline 987
4-Methyl-1-phenyl-2(1*H*)-quinolinethione 744
$C_{16}H_{13}N_3$
1-(Phenylazo)-2-naphthaleneamine 1088
1-Phenyl-2-(phenylazo)pyrrole 1091
$C_{16}H_{13}N_3O$
6-Cyano-7,8,9,10-tetrahydrobenzimidazolo(1,2-b)-isoquinolin-11(5*H*)-one 905
N-(5-Phenyl-3-pyrazolyl)benzamide 863
2-Phenyl-4-quinolinecarboxylic hydrazide 989
$C_{16}H_{13}N_3OS$
2-Benzothiazolinone (α-acetylbenzylidene)hydrazone 916
$C_{16}H_{13}N_3O_2$
1-Phenyl-3-(*N*-phenylcarbamoyl)-2-pyrazolin-4-one 423
$C_{16}H_{13}N_3O_2S$
5-Imino-1-(2-methoxyphenyl)-3-phenyl-4-thiohydantoin 737
5-Imino-1-(4-methoxyphenyl)-3-phenyl-4-thiohydantoin 737
2-(*p*-Tosylazo)quinoline 994
$C_{16}H_{13}N_3O_3$
Cinnamic (3-nitrobenzylidene)-hydrazide 501
Disperse Blue 5 (C.I. 62035) 1094

$C_{16}H_{13}N_3O_3$ (Cont'd)
5-Phenyl-1,2,4-oxadiazol-3-yl *N*-(*p*-tolyl)carbamate ... 873
$C_{16}H_{13}N_3O_3S$
4-(4-Amino-1-naphthylazo)-benzenesulfonic acid, thallium salt ... 1088
$C_{16}H_{13}N_3O_3S_2$
Ethyl 2-(2-benzamido-4-thiazolyl)-4-thiazolecarboxylate ... 877
$C_{16}H_{13}N_3O_4$
N-(3-Nitro-*p*-toluidinomethyl)-phthalimide ... 719
N-(5-Nitro-*o*-toluidinomethyl)-phthalimide ... 719
$C_{16}H_{13}N_3O_5$
3-Amino-2-methyl-7-nitro-4-phenyl-1(2*H*)-isoquinolinone ... 690
4′-(2,4-Dinitrostyryl)-acetanilide ... 426
$C_{16}H_{13}N_3O_5S$
Methyl 2-(3-(3-nitrobenzoyl)-2-thioureido)benzoate ... 744
1-Phenyl-3-(3-sulfobenzamido)-2-pyrazolin-5-one, barium salt ... 793
$C_{16}H_{13}N_3O_6S$
1-(Hydroxymethyl)-3-(4-phthalimidophenylsulfonyl)urea ... 812
$C_{16}H_{13}N_5O_3$
3-Methyl-4-(4-nitrophenylazo)-1-phenyl-1*H*-pyrazol-5(4*H*)-one ... 1092
$C_{16}H_{14}$
1-Benzylideneindan ... 162
1-Benzylindene ... 166
9,10-Dihydro-9,10-ethanoanthracene ... 162
2,3-Dimethylanthracene ... 168
$C_{16}H_{14}BF_4N_5S_2$
3-Methyl-2-((3-methyl-2-benzothiazolinylidene)aminazo)benzothiazolium tetrafluoroborate ... 915
$C_{16}H_{14}BrCl_2NO_4$
2-Bromo-4-chloro-5-methylphenyl *N*-(5-chloro-2,4-dimethoxyphenyl)-carbamate ... 613
2-Bromo-4-chloro-6-methylphenyl *N*-(5-chloro-2,4-dimethoxyphenyl)-carbamate ... 614
$C_{16}H_{14}BrF_3N_2O$
1-(4-Bromo-2,6-dimethylphenyl)-3-(2-(trifluoromethyl)phenyl)urea ... 506
$C_{16}H_{14}BrF_3N_4OS$
1-(4-Bromo-2-(trifluoromethyl)-phenyl)-6-(*o*-tolyl)-5-thiobiurea ... 727
$C_{16}H_{14}BrN$
1-Benzylquinolinium bromide ... 990
9-Bromo-3-(dimethylamino)-phenanthrene ... 234
1-(1-Naphthylmethyl)pyridinium bromide ... 959
$C_{16}H_{14}BrNO_2$
3-Phenylallyl *N*-(4-bromophenyl)-carbamate ... 632
$C_{16}H_{14}BrNO_3$
4′-Bromo-2′,6′-dimethylphthalanilic acid ... 454
$C_{16}H_{14}BrNO_4$
4-Formyl-2-methoxyphenyl *N*-(2-bromo-4-methylphenyl)carbamate ... 614
4-(Methoxycarbonyl)phenyl *N*-(2-bromo-4-methylphenyl)carbamate ... 630
$C_{16}H_{14}BrN_5$
5-Amino-4-(4-bromophenylazo)-3-methyl-1*H*-pyrazole ... 1080
$C_{16}H_{14}Br_2ClNO_2$
2′-Chloro-2,3-dibromo-4′-methoxy-3-phenylpropionanilide ... 402
$C_{16}H_{14}Br_2N_2O_2$
5-Bromo-*o*-anisaldehyde azine ... 1050
$C_{16}H_{14}Br_2N_2O_4$
α,β-Dibromo-2′-methoxy-3′-nitrohydrocinnamanilide ... 416
$C_{16}H_{14}Br_2O_4Pb$
Diphenylplumbylene bis(2-bromoacetate) ... 1133
$C_{16}H_{14}ClFN_2$
7-Chloro-2,3-dihydro-5-(4-fluorophenyl)-5-methyl-5*H*-imidazo(2,1-a)-isoindole ... 244
$C_{16}H_{14}ClFN_4O_2$
2-(4-Fluorophenylureido)acetic acid (2-chlorobenzylidene)-hydrazide ... 494
$C_{16}H_{14}ClFN_4O_4$
N-(3-(5-Chloro-2-methoxyphenyl)-ureido)-*N*′-(4-fluorophenyl)oxamide . 517
$C_{16}H_{14}ClF_3N_2$
N-(2-Chloro-4-(dimethylamino)-benzylidene)-α,α,α-trifluoromethyl)-aniline ... 1020
$C_{16}H_{14}ClF_3N_2O$
1-(4-Chlorophenyl)-1-ethyl-3-(3-(trifluoromethyl)phenyl)urea ... 506
1-(2-Chloro-5-(trifluoromethyl)-phenyl)-3-(3,4-xylyl)urea ... 507
$C_{16}H_{14}ClF_3N_2O_2S$
1-(5-Chloro-2,4-dimethoxyphenyl)-3-(3-(trifluoromethyl)phenyl)-2-thiourea ... 727
$C_{16}H_{14}ClF_3N_2O_3$
1-(5-Chloro-2,4-dimethoxyphenyl)-3-(2-(trifluoromethyl)phenyl)urea ... 504
$C_{16}H_{14}ClN$
1-Benzylquinolinium chloride ... 990
3-Chloro-*N*-(cinnamylidene)-4-methylaniline ... 1079
4-Chloro-β-methyl-α-phenylhydrocinnamonitrile ... 751
2-(4-Chlorophenyl)-1-ethylindole ... 881
$C_{16}H_{14}ClNO$
4-Chloro-α-(4-methoxyphenyl)-hydrocinnamonitrile ... 751
2′-Chloro-6′-methylcinnamanilide ... 478
3′-Chloro-2′-methylcinnamanilide ... 461
3′-Chloro-4′-methylcinnamanilide ... 457
5′-Chloro-2′-methylcinnamanilide ... 457
$C_{16}H_{14}ClNO_2$
5′-Chloro-2′-methoxycinnamanilide .. 449
4′-Chloro-*N*-methylbenzoylacetanilide ... 444
1-(4-Chlorophenyl)-4-(4-methoxyphenyl)-2-azetidinone ... 689
3-Phenylallyl *N*-(2-chlorophenyl)-carbamate ... 632
$C_{16}H_{14}ClNO_3$
Ethyl 4-(2-chlorobenzamido)-benzoate ... 483
Ethyl 4-(3-chlorobenzamido)-benzoate ... 456
Ethyl 4-(4-chlorobenzamido)-benzoate ... 456
$C_{16}H_{14}ClNO_4$
3-Acetoxyphenyl *N*-(3-chloro-2-methylphenyl)carbamate ... 622
2-Acetylphenyl *N*-(4-chloro-2-methoxyphenyl)carbamate ... 618
4-Acetylphenyl *N*-(5-chloro-2-methoxyphenyl)carbamate ... 620
2-Benzylisoquinolinium perchlorate. 997
4-Ethoxycarbonylphenyl *N*-(4-chlorophenyl)carbamate ... 630
4-Formyl-2-methoxyphenyl *N*-(2-chloro-6-methylphenyl)carbamate ... 615
4-Formyl-2-methoxyphenyl *N*-(3-chloro-2-methylphenyl)carbamate ... 615
4-Formyl-2-methoxyphenyl *N*-(5-chloro-2-methylphenyl)carbamate ... 614
4-(Methoxycarbonyl)phenyl *N*-(5-chloro-2-methylphenyl)carbamate ... 630
Piperonyl *N*-(2-chloro-6-methylphenyl)carbamate ... 665
Piperonyl *N*-(3-chloro-4-methylphenyl)carbamate ... 646
Piperonyl *N*-(4-chloro-2-methylphenyl)carbamate ... 646
Piperonyl *N*-(5-chloro-2-methylphenyl)carbamate ... 665
$C_{16}H_{14}ClNO_5$
5′-Chloro-2′,4′-dimethoxyphthalanilic acid ... 448
4-Formyl-2-methoxyphenyl *N*-(4-chloro-2-methoxyphenyl)-carbamate ... 612
4-Formyl-2-methoxyphenyl *N*-(5-chloro-2-methoxyphenyl)-carbamate ... 615
2-(Methoxycarbonyl)phenyl *N*-(2-chloro-4-methoxyphenyl)-carbamate ... 617
2-(Methoxycarbonyl)phenyl *N*-(5-chloro-2-methoxyphenyl)-carbamate ... 616
$C_{16}H_{14}ClNO_5S$
N-(3-(4-Chlorophenoxy)-2-hydroxypropyl)saccharin ... 806
$C_{16}H_{14}ClN_3O$
1-(3-Chlorophenyl)-3-(3-(2-cyanoethyl)phenyl)urea ... 752
$C_{16}H_{14}ClN_3OS$
1-(3-Chlorophenyl)-3-(3-cyano-4,5,6,7-tetrahydrobenzo(b)-thiophen-2-yl)urea ... 855
$C_{16}H_{14}ClN_3O_2$
2-(2-Chloro-1-benzimidazolyl)-4′-methoxyacetanilide ... 903
Hippuric (4-chlorobenzylidene)-hydrazide ... 493
$C_{16}H_{14}ClN_3O_3$
2-Chlorohippuric (2-hydroxybenzylidene)hydrazide ... 491
4-Chlorohippuric (2-hydroxybenzylidene)hydrazide ... 491
$C_{16}H_{14}ClN_3O_3S$
1-(2-Benzothiazolyl)-3-(5-chloro-2,4-dimethoxyphenyl)urea ... 911
1-(4-Chloro-2-benzothiazolyl)-3-(2,5-dimethoxyphenyl)urea ... 912
$C_{16}H_{14}ClN_3O_5$
Ethyl 4-(3-(4-chloro-2-nitrophenyl)-ureido)benzoate ... 523
$C_{16}H_{14}ClN_3O_6$
2-(2,6-Dinitro-4-methylanilino)ethyl 2-chlorobenzoate ... 366
$C_{16}H_{14}ClN_5O_4$
2-(3-(3-Nitrophenyl)ureido)acetic (2-chlorobenzylidene)hydrazide ... 523
$C_{16}H_{14}ClN_5O_4S_2$
3-Methyl-2-((3-methyl-2-benzothiazolinylidene)aminoazo)benzothiazolium perchlorate ... 915
$C_{16}H_{14}ClN_5O_5$
N-(4-Chloro-3-nitrophenyl)-*N*′-(3-methylphenylureido)oxamide ... 529
2-(3-(2-Chloro-4-nitrophenyl)ureido)-acetic (2-hydroxybenzylidene)-hydrazide ... 1022
$C_{16}H_{14}Cl_2$
1,1-Dichloro-2,2-di-(*p*-tolyl)-ethylene . 174
$C_{16}H_{14}Cl_2N_2$
N,*N*′-Bis(4-chlorobenzylidene)-ethylenediamine ... 1056
4′-Chloroacetophenone azine ... 1078
$C_{16}H_{14}Cl_2N_2O$
2-Butoxy-7,10-dichloropyrido(3,2-b)-quinoline ... 1002
$C_{16}H_{14}Cl_2N_2O_2$
2,4-Dichloro-4′-(*N*-methylacetamido)-benzanilide ... 477
$C_{16}H_{14}Cl_2N_2O_2S$
1-(3,4-Dichlorobenzoyl)-3-(4-ethoxyphenyl)-2-thiourea ... 520
1-(2,5-Dichlorophenyl)-3-(2-phenoxypropionyl)-2-thiourea ... 729
$C_{16}H_{14}Cl_2N_2O_3$
1-(2,4-Dichlorobenzoyl)-3-(2-methoxy-5-methylphenyl)urea ... 519
1-(2,4-Dichlorophenoxyacetyl)-2-(4-methylbenzoyl)hydrazine ... 492
2-(2,4-Dichlorophenoxy)ethyl *N*-(4-methyl-2-nitrophenyl)carbamate ... 615
Ethyl 4-(3-(2,4-dichlorophenyl)-ureido)benzoate ... 532
4-Hydroxyhydrocinnamic acid (3,5-dichlorosalicylidene)hydrazide ... 495
$C_{16}H_{14}Cl_2N_2O_4$
2,4-Dichloro-3,5-dimethylphenyl *N*-(4-methyl-2-nitrophenyl)-carbamate ... 624
N-(4,5-Dimethyl-2-nitrophenyl)-2-(2,4-dichlorophenoxy)acetamide ... 401
$C_{16}H_{14}Cl_2N_2O_4S$
4′,4″-Sulfonylbis(2-chloroacetanilide) ... 772
$C_{16}H_{14}Cl_2N_2O_5$
2-(2,4-Dichlorophenoxy)ethyl *N*-(4-methyl-3-nitrophenyl)carbamate ... 622
$C_{16}H_{14}Cl_2N_4$
N-(4-(2,5-Dichlorophenylazo)-phenyl)-*N*-ethylglycinonitrile ... 1088
$C_{16}H_{14}Cl_2N_4O_3$
2-(3-(2,3-Dichlorophenyl)ureido)-acetic (3-hydroxybenzylidene)-hydrazide ... 512
$C_{16}H_{14}Cl_2O_2$
1,4-Bis-(2-chlorophenoxy)-2-butene ... 181
1,4-Bis-(4-chlorophenoxy)-2-butene ... 181
$C_{16}H_{14}Cl_2O_4S_2$
1,2-Dichloro-1,2-di-*p*-tosylethylene ... 785
$C_{16}H_{14}Cl_3NO_2$
2,4-Dichloro-3,5-dimethylphenyl *N*-(4-chloro-2-methylphenyl)-carbamate ... 638
2′,3′-Dimethyl-2-(2,4,5-trichlorophenoxy)acetanilide ... 400
2′,4′-Dimethyl-2-(2,4,5-trichlorophenoxy)acetanilide ... 400
2′-Ethyl-2-(2,4,5-trichlorophenoxy)-acetanilide ... 400
$C_{16}H_{14}Cl_3NO_3$
2-(2,4-Dichlorophenoxy)ethyl *N*-(2-chloro-6-methylphenyl)carbamate ... 666
N-(2-Ethoxyphenyl)-2,4,5-trichlorophenoxyacetamide ... 399
$C_{16}H_{14}Cl_3NO_4$
2,3-Dimethoxybenzyl *N*-(2,4,5-trichlorophenyl)carbamate ... 612
2′,4′,6′-Trichloro-3,4,5-trimethoxybenzanilide ... 447
$C_{16}H_{14}Cl_3N_3O$
5′-Methyl-2′-(4-tolylazo)-2,2,2-trichloroacetanilide ... 1080
$C_{16}H_{14}Cl_3N_5O_3$
4-(3-Chlorophenyl)-1-(3,4-dichlorophenylureidoacetyl)-semicarbazide ... 528
$C_{16}H_{14}Cl_4O_3$
2-(2,4-Dichlorophenoxy)ethyl ether ... 185
$C_{16}H_{14}Cl_4O_4$
1,4-Bis(2,4-dichlorophenoxy)-2,3-butanediol ... 202
$C_{16}H_{14}FNO_3$
4′-(Ethoxycarbonyl)-4-fluorobenzanilide ... 456
Ethyl 4-(3-fluorobenzamido)-benzoate ... 456
$C_{16}H_{14}FNO_4$
4-(Ethoxycarbonyl)phenyl *N*-(4-fluorophenyl)carbamate ... 630
$C_{16}H_{14}FN_3O$
1,4-Dihydro-1-(*p*-fluorobenzyl)-2-methylamino-4(4*H*)-quinazolinone ... 704
$C_{16}H_{14}FN_3OS$
1-(3-Cyano-4,5,6,7-tetrahydrobenzo(b)thiophen-2-yl)-3-(4-fluorophenyl)urea ... 855
1-(5,6-Dimethyl-2-benzothiazolyl)-3-(3-fluorophenyl)urea ... 912
1-(4-Fluorophenyl)-3-(4-phenyl-4-thiazolin-2-yl)urea ... 535
$C_{16}H_{14}FN_3O_5$
Ethyl 2-(3-(2-fluoro-5-nitrophenyl)-ureido)benzoate ... 524
Ethyl 4-(3-(4-fluoro-3-nitrophenyl)-ureido)benzoate ... 524
$C_{16}H_{14}FN_5O_4$
2-(3-(4-Fluorophenyl)ureido)acetic (3-nitrobenzylidene)hydrazide ... 522
$C_{16}H_{14}F_2N_2$
N,*N*′-Bis-(4-fluorobenzylidene)-ethylenediamine ... 1054
$C_{16}H_{14}F_2N_2O_3$
1-(4-Acetyl-3-methoxyphenyl)-3-(2,5-difluorophenyl)urea ... 520
Ethyl 4-(2,5-difluorophenylureido)-benzoate ... 532
$C_{16}H_{14}F_3NO$
N-(1,1-Diphenyl-2,2,2-trifluoroethyl)-acetamide ... 397
$C_{16}H_{14}F_3NO_2$
N-(3-Ethoxy-4-hydroxybenzylidene)-3-trifluoromethylaniline ... 1020

$C_{16}H_{14}F_3NO_2$ (Cont'd)
3-(Trifluoromethyl)phenyl N-(2-ethylphenyl)carbamate ... 611
3-(Trifluoromethyl)phenyl N-(2,5-xylyl)carbamate ... 611
Veratrylidene-3-(trifluoromethyl)-aniline ... 1020

$C_{16}H_{14}F_3NO_2S$
3-Methyl-4-(methylthio)phenyl N-(3-(trifluoromethyl)phenyl)carbamate ... 609

$C_{16}H_{14}F_3N_3O_2$
1-(4-Acetamidophenyl)-3-(2-(trifluoromethyl)phenyl)urea ... 508
1-(4-Acetamidophenyl)-3-(3-trifluoromethylphenyl)urea ... 509
1-(2-Phenylacetyl)-4-(2-(trifluoromethyl)phenyl)-semicarbazide ... 509
1-(2-Phenylacetyl)-4-(3-(trifluoromethyl)phenyl)-semicarbazide ... 509

$C_{16}H_{14}F_3N_3O_3$
1-(2-Methoxybenzoyl)-4-(3-(trifluoromethyl)phenyl)-semicarbazide ... 507
1-(4-Methoxybenzoyl)-4-(2-(trifluoromethyl)phenyl)-semicarbazide ... 507

$C_{16}H_{14}F_3N_5O_6$
1-(2,6-Dinitro-4-(trifluoromethyl)-phenyl)-4-(2-methoxy-5-methyl)-semicarbazide ... 507

$C_{16}H_{14}F_4N_2O$
1-Ethyl-3-(4-fluorophenyl)-1-(2-(trifluoromethyl)phenyl)urea ... 505

$C_{16}H_{14}I_2N_2O_3$
Benzoyl-3,5-diiodo-DL-tyrosinamide ... 462

$C_{16}H_{14}N_2$
1-Allyl-2-phenyl-1H-benzimidazole ... 907
2,3-Dihydro-5,6-diphenylpyrazine ... 253
3,4-Dihydro-2-methylindolo(1,2-c)-quinazoline ... 1001
4,5-Diphenyl-2-methylimidazole ... 867

$C_{16}H_{14}N_2O$
4-(4,5-Dihydro-1-imidazolyl)-benzophenone ... 252
1-Ethyl-3-nitroso-2-phenylindole ... 884
4-(4-Methoxyphenyl)-3-methylcinnoline ... 999
9β-Phenyl-2,3,5,9b-tetrahydro-1H-imidazo(2,1-a)isoindol-5-one ... 710
1-(N-(p-Tolyl)carbamoyl)indole ... 890

$C_{16}H_{14}N_2OS$
1-Furfuryl-3-(1-naphthyl)-2-thiourea ... 852
4'-(6-Methyl-2-benzothiazolyl)-acetanilide ... 916

$C_{16}H_{14}N_2O_2$
1-(4-Aminophenyl)-4-(4-nitrophenyl)-1,3-butadiene ... 271
N-(2-Anilinoethyl)-phthalimide ... 718
3-Anilino-1-phenylsuccinimide ... 720
4-Cyanophenyl N-(2,4-dimethylphenyl)carbamate ... 767
4-Cyanophenyl N-(2,3-xylyl)-carbamate ... 767
4-Cyanophenyl N-(2,6-xylyl)-carbamate ... 767
4-Cyanophenyl N-(3,5-xylyl)-carbamate ... 767
2,3-Dihydro-9b-(4-hydroxyphenyl)-1H-imidazo(2,1-a)isoindol-5(9bh)-one ... 697
(3S,6S)-3,6-Diphenyl-2,5-piperazinedione ... 709
1-Furfuryl-3-(1-naphthyl)urea ... 852
4-Hydroxy-4-phenyl-3-(phenylazo)-3-buten-2-one ... 282
9β-(4-Hydroxyphenyl)-1,2,5,9b-tetrahydro-3H-imidazo(2,1a)-isoindol-5-one ... 697
4-(3-Indolylmethylamino)-benzoic acid ... 886
5-Methyl-1-(methylamino)-5H-dibenz(b,e)azepine-6,11-dione ... 691
3-Phenylacetyl-2-oximinoindoline ... 1011

$C_{16}H_{14}N_2O_2S$
2-(2-Aminophenylthio)-N-phenylsuccinimide ... 715
1-Cinnamoyl-3-(4-hydroxyphenyl)-2-thiourea ... 734
2,7-Dihydro-3,6-diphenyl-1,4,5-thiadiazepine-1,1-dioxide ... 784
5-(4-Hydroxybenzyl)-3-phenyl-2-thiohydantoin ... 734

$C_{16}H_{14}N_2O_3$
N-(4-Acetamidobenzylidene)-anthranilic acid ... 1074
2-(4-Acetamidophenyl)-2,3-dihydro-4H-1,3-benzoxazin-4-one ... 703
4'-Anilinomaleanilic acid ... 470
N-(o-Anisidinomethyl)-phthalimide ... 717
4-Cyanophenyl N-(2-methoxy-5-methylphenyl)carbamate ... 762
5-(4-Methoxyphenyl)-5-phenylhydantoin ... 709

$C_{16}H_{14}N_2O_3S$
4'-Amino-2'-hydroxy-2-naphthalenesulfonanilide ... 811

$C_{16}H_{14}N_2O_4$
2-(2-(Carboxymethoxy)phenyl)-1,2-dihydro-4(3H)-quinazolinone ... 698
2-Cyanophenyl N-(2,5-dimethoxyphenyl)carbamate ... 761
4-Cyanophenyl N-(2,5-dimethoxyphenyl)carbamate ... 761
Disperse Blue 26 (C.I. 63305) ... 1093
2-(4-Nitrobenzoyl)propionanilide ... 443
3-Phenylallyl N-(2-nitrophenyl)-carbamate ... 668

$C_{16}H_{14}N_2O_4S$
9-Cyano-10-methylacridinium methyl sulfate ... 996
4-(3-(2-Phenoxyacetyl)-2-thioureido)-benzoic acid ... 735

$C_{16}H_{14}N_2O_5$
2-Acetylphenyl N-(4-methyl-3-nitrophenyl)carbamate ... 669
2,4-Dinitro-2'-ethoxystilbene ... 267
3'-(2-Hydroxy-3-methoxy-5-nitrobenzylideneamino)-acetophenone ... 1032
N-(2-Methoxy-5-nitrophenyl)-benzoylacetamide ... 442
2-(4-Nitrobenzoyl)acet-o-anisidide ... 442
Piperonylic acid vanillylidenehydrazide ... 495

$C_{16}H_{14}N_2O_6$
3-Acetoxyphenyl N-(4-methyl-2-nitrophenyl)carbamate ... 666
3-Acetoxyphenyl N-(4-methyl-3-nitrophenyl)carbamate ... 666
Carbobenzyloxyglycine 4-nitrophenyl ester ... 658
2,3-Dimethoxy-2',4'-dinitrostilbene ... 266
2,4-Dimethoxy-2',4'-dinitrostilbene ... 269
2,5-Dimethoxy-2',4'-dinitrostilbene ... 267
3,4-Dimethoxy-2',4'-dinitrostilbene ... 266

$C_{16}H_{14}N_2O_6S$
4'-(Acetylsulfamoyl)-phthalanilic acid ... 823
4'-(Acetylsulfamoyl)phthalanilic acid, sodium salt ... 821

$C_{16}H_{14}N_2O_7$
(4-Formyl-3-methoxyphenyl) N-(2-methoxy-4-nitrophenyl)carbamate ... 651
4-(Methoxycarbonyl)phenyl N-(2-methoxy-4-nitrophenyl)carbamate ... 652

$C_{16}H_{14}N_4$
(4,6-Diphenyl-2-pyrimidinyl)-hydrazine ... 976
2,2'-Ethylenebibenzimidazole ... 907

$C_{16}H_{14}N_4O$
3-Methyl-2-phenyl-4-phenylazo-3-pyrazolin-5-one ... 1088

$C_{16}H_{14}N_4O_2$
3α,6α-Diphenylglycoluril ... 704
1,5-Diphenyl-2,3,6,7-tetrahydro-1H,5H-s-triazolo(1,2-a)-s-triazole-3,7-dione ... 705
1-(5-Oxo-1-phenyl-3-pyrazolin-3-yl)-3-phenylurea ... 702

$C_{16}H_{14}N_4O_3$
6-Nitro-1-(N-(3,4-xylyl)carbamoyl)-1H-indazole ... 900

$C_{16}H_{14}N_4O_4$
N,N'-Bis-(3-nitrobenzylidene)-ethylenediamine ... 1053
Dihydroxytartaric acid osazone, disodium salt ... 351
Hippuric (3-nitrobenzylidene)-hydrazide ... 499

$C_{16}H_{14}N_4O_6$
Ethyl benzoylformate 2,4-dinitrophenylhydrazone ... 372

$C_{16}H_{14}N_4O_8$
1,4-Bis(2-methoxy-4-nitrophenyl)-oxamide ... 414
2-(2,6-Dinitro-4-methylanilino)ethyl 4-nitrobenzoate ... 372

$C_{16}H_{14}N_6O_2$
5-Amino-3-methyl-4-(4-nitrophenylazo)-1H-pyrazole ... 1087

$C_{16}H_{14}N_6O_7$
4,4'-Diacetamido-3,3'-dinitroazoxybenzene ... 1088

$C_{16}H_{14}O$
9,10-Dihydro-11-hydroxy-9,10-ethanoanthracene ... 206
10,11-Dihydro-5,10-methano-5H-dibenzo(a,d)cyclohepten-11-ol ... 205
2,3-Dimethyl-9(10H)-anthracenone ... 314
10,10-Dimethyl-9(10H)-phenanthrenone ... 298
1-(2-Fluorenyl)allyl alcohol ... 208
2'-Methylchalcone ... 304
trans-3-Methylchalcone ... 304
3'-Methylchalcone ... 304
4'-Methylchalcone ... 304
4-Methylchalcone ... 304

$C_{16}H_{14}OS_2$
2,6-Bis(2-thienylidene)-cyclohexanone ... 857

$C_{16}H_{14}O_2$
1,2-Bis(4-methoxyphenyl)acetylene ... 1017
1,2-Dibenzoylethane ... 308
10,11-Dihydro-11,12-dihydroxy-5,10-methano-5H-dibenzo(a,d)-cycloheptene ... 200
2,3-Dimethoxyanthracene ... 178
2,7-Dimethoxyphenanthrene ... 187
2,4-Diphenyl-3-oxobutyraldehyde ... 324
4-Methoxychalcone ... 301
Methyl 1,2-dihydro-4-phenanthrenecarboxylate ... 376
Methyl β-phenylcinnamate ... 365
Methyl 4-stilbenecarboxylate ... 375
7-Phenoxy-1-tetralone ... 291
4-Styrylphenyl acetate ... 380

$C_{16}H_{14}O_2S$
2-Dibenzothiophenebutyric acid ... 898

$C_{16}H_{14}O_2S_2$
Benzoylmethyl disulfide ... 308

$C_{16}H_{14}O_2Se$
2-(Methylseleno)-α-phenylcinnamic acid ... 213

$C_{16}H_{14}O_3$
2-Benzoyl-3,6-dimethylbenzoic acid ... 345
1-Benzoylethyl benzoate ... 374
3-Benzoyl-2-phenylpropionic acid ... 331
4-Benzyloxycinnamic acid ... 334
4-(4-Biphenylyl)-4-oxobutyric acid ... 331
2-(2,4-Dimethylbenzoyl)-benzoic acid ... 344
5,8-Dimethyl-1,2,3,4-tetrahydro-1,4-ethenonaphthalene-2,3-dicarboxylic anhydride ... 388
Ethyl 2-benzoylbenzoate ... 374
2-(4-Ethylbenzoyl)-benzoic acid ... 344
2'-Hydroxy-2-methoxychalcone ... 300
2'-Hydroxy-3-methoxychalcone ... 300
2'-Hydroxy-4-methoxychalcone ... 300
4'-Hydroxy-4-methoxychalcone ... 300
4-Methoxyphenyl cinnamate ... 364
4-Methoxy-α-phenylcinnamic acid ... 335
α-Phenylphenacyl acetate ... 360

$C_{16}H_{14}O_4$
2-Acetylphenyl 4-methoxybenzoate ... 378
4-Benzoyloxyphenoxyacetone ... 378
2,2'-Biphenyldiacetic acid ... 332
9,10-Dihydro-1,4-dihydroxy-5-methoxy-9-methyl-10-anthracenone ... 310
Dimethyl 2,2'-biphenyldicarboxylate ... 375
Dimethyl 4,4'-biphenyldicarboxylate ... 376
Diphenyl succinate ... 380
2,3-Diphenylsuccinic acid ... 332
2,2'-(Ethylenedioxy)dibenzaldehyde ... 323
1-(2-Hydroxyphenyl)-3-(4-methoxyphenyl)-1,3-propanedione ... 307
Methoxy-3-(4-methoxyphenyl)-phthalide ... 383
α-Methylbenzyl phthalate ... 370
Methyl 4-phenoxybenzoylacetate ... 363
3-(4-Phenoxybenzoyl)propionic acid ... 328

$C_{16}H_{14}O_5$
2-(3,4-Dimethoxybenzoyl)-benzoic acid ... 339

$C_{16}H_{14}O_6$
2,2'-Dihydroxy-3,3'-dimethoxy-5,5'-biphenyldicarboxaldehyde ... 322
4,4'-(Ethylenedioxy)dibenzoic acid ... 339
1,2,4-Naphthalenetriol triacetate ... 378

$C_{16}H_{14}S$
9-Anthracenylmethyl methyl sulfide ... 213
8,9,10,11-Tetrahydrobenzo(b)-naphtho(1,2-d)-thiophene picrate ... 899

$C_{16}H_{14}S_2$
1,2-Bis(p-tolylthio)ethyne ... 1017

$C_{16}H_{15}BCl_2F_4N_2$
Bis(4-chlorobenzylidene)-ethylhydrazinium tetrafluoroborate ... 254

$C_{16}H_{15}BrClNO_2$
2-Bromo-4-chloro-6-methylphenyl N-(2,4-dimethylphenyl)carbamate ... 639
2-Bromo-4-chloro-6-methylphenyl N-(3,5-dimethylphenyl)carbamate ... 640
4-Bromo-2-chloro-5-methylphenyl N-(2,4-xylyl)carbamate ... 639
2-Bromo-4,5-dimethylphenyl N-(3-chloro-4-methylphenyl)carbamate ... 638
2-Bromo-4,5-dimethylphenyl N-(4-chloro-2-methylphenyl)carbamate ... 639
4-Bromo-2,5-dimethylphenyl N-(3-chloro-4-methylphenyl)carbamate ... 638
4-Bromo-2,5-dimethylphenyl N-(4-chloro-2-methylphenyl)carbamate ... 639
4-Bromo-3,5-dimethylphenyl N-(4-chloro-2-methyl)phenylcarbamate ... 639

$C_{16}H_{15}BrClNO_3$
2-Bromo-4-chloro-5-methylphenyl N-(2-methoxy-5-methylphenyl)-carbamate ... 619
2-Bromo-4-chloro-6-methylphenyl N-(2-methoxy-5-methylphenyl)-carbamate ... 619
2-Bromo-4,5-dimethylphenyl N-(4-chloro-2-methoxyphenyl)-carbamate ... 619
2-Bromo-4,5-dimethylphenyl N-(5-chloro-2-methoxyphenyl)-carbamate ... 621
4-Bromo-2,5-dimethylphenyl N-(4-chloro-2-methoxyphenyl)-carbamate ... 619
4-Bromo-2,5-dimethylphenyl N-(5-chloro-2-methoxyphenyl)-carbamate ... 619
4-Bromo-3,5-dimethylphenyl N-(4-chloro-2-methoxyphenyl)-carbamate ... 619

$C_{16}H_{15}BrN_2O_3$
4-Hydroxyhydrocinnamic acid (5-bromosalicylidene)hydrazide ... 495

$C_{16}H_{15}BrN_2O_4$
2-Bromo-4,5-dimethylphenyl N-(2-methyl-5-nitrophenyl)carbamate ... 625
2-Bromo-4,5-dimethylphenyl N-(4-methyl-3-nitrophenyl)carbamate ... 625
4-Bromo-2,5-dimethylphenyl N-(2-methyl-5-nitrophenyl)carbamate ... 625
4-Bromo-2,5-dimethylphenyl N-(4-methyl-3-nitrophenyl)carbamate ... 625
4-Bromo-3,5-dimethylphenyl N-(2-methyl-5-nitrophenyl)carbamate ... 629

$C_{16}H_{15}BrN_2O_5$
2-(4-Bromophenoxy)ethyl N-(4-methyl-3-nitrophenyl)carbamate ... 622

$C_{16}H_{15}BrO$
1-Bromo-2-(4-methoxyphenyl)indan 181
$C_{16}H_{15}BrO_2$
2-Bromo-3-cyclohexyl-1,4-naphthoquinone 319
Ethyl 2-bromo-2,2-diphenylacetate 362
$C_{16}H_{15}BrO_3S$
1,2,3,4-Tetrahydro-1-naphthyl 4-bromobenzenesulfonate 797
$C_{16}H_{15}Br_2NOS$
2,3-Dibromo-4′-(methylthio)-3-phenylpropionanilide 421
$C_{16}H_{15}ClFN_5O_3$
4-(3-Chlorophenyl)-1-(4-fluorophenylureidoacetyl)semicarbazide 528
$C_{16}H_{15}ClF_3N_5O_4$
2-(Furfurylureido)acetic N^2-(N-(2-chloro-5-trifluoromethyl-phenyl)carbamoyl)hydrazide 843
$C_{16}H_{15}ClINO_2$
4-Iodo-2,5-dimethylphenyl N-(3-chloro-4-methylphenyl)carbamate 639
$C_{16}H_{15}ClN_2OS$
2-(4-Chlorophenyl)-3-oxazolidinethiocarboxanilide 729
$C_{16}H_{15}ClN_2O_2$
2-(2-Aminoethyl)-1-(4-chlorophenyl)-1-hydroxy-3-oxoisoindoline 689
2-Chloro-4′-(N-methylacetamido)benzanilide 471
$C_{16}H_{15}ClN_2O_2S$
1-(2-Chlorobenzoyl)-3-(4-ethoxyphenyl)-2-thiourea 729
1-(2-Chlorobenzoyl)-3-(2-methoxy-5-methylphenyl)-2-thiourea 729
$C_{16}H_{15}ClN_2O_3$
4-Acetamidophenyl N-(3-chloro-2-methylphenyl)carbamate 631
1-(3-Acetylphenyl)-3-(5-chloro-2-methoxyphenyl)urea 518
N-(N-(Acrylamido)ethyl)-4-chloro-1-hydroxy-2-naphthamide 485
N-(3-(2-Chloroacetamido)-4-methoxyphenyl)benzamide 445
4-Chlorobenzoic (2,5-dimethoxybenzylidene)hydrazide 492
Ethyl 4-(3-(2-chlorophenyl)ureido)benzoate 531
Ethyl 4-(3-(3-chlorophenyl)ureido)benzoate 532
Ethyl 4-(3-(4-chlorophenyl)ureido)benzoate 532
Methyl 2-(3-(3-chloro-2-methylphenyl)ureido)benzoate 531
$C_{16}H_{15}ClN_2O_3S$
1-(2-Chlorobenzoyl)-3-(2,4-dimethoxyphenyl)-2-thiourea 728
$C_{16}H_{15}ClN_2O_4$
3-Acetamidophenyl N-(4-chloro-2-methoxyphenyl)carbamate 618
4-Acetamidophenyl N-(2-chloro-4-methoxyphenyl)carbamate 620
4-Acetamidophenyl N-(4-chloro-2-methoxyphenyl)carbamate 618
4-Acetamidophenyl N-(5-chloro-2-methoxyphenyl)carbamate 618
2-Chloro-4,5-dimethylphenyl N-(4-methyl-2-nitrophenyl)carbamate 624
4-Chloro-2,5-dimethylphenyl N-(4-methyl-2-nitrophenyl)carbamate 624
4-Chloro-2,6-dimethylphenyl N-(4-methyl-2-nitrophenyl)carbamate 624
Methyl 2-(3-(5-chloro-2-methoxyphenyl)ureido)benzoate 519
$C_{16}H_{15}ClN_2O_5$
3-(3-(5-Chloro-2,4-dimethoxyphenyl)ureido)benzoic acid 514
4-(3-(5-Chloro-2,4-dimethoxyphenyl)ureido)benzoic acid 514
4-Chloro-3,5-dimethylphenyl N-(2-methoxy-4-nitrophenyl)carbamate 618
2-(2-Chlorophenoxy)ethyl N-(4-methyl-3-nitrophenyl)carbamate 615
2-(4-Chlorophenoxy)ethyl N-(4-methyl-3-nitrophenyl)carbamate 615
4-(Methoxycarbonylamino)phenyl N-(5-chloro-2-methoxyphenyl)carbamate 619
$C_{16}H_{15}ClN_2O_6$
N-(4-Chloro-3-nitrophenyl)-3,4,5-trimethoxybenzamide 448
1-(3-Phthalimidopropyl)pyridinium perchlorate 948
$C_{16}H_{15}ClN_4O_2$
1-(N-(5-Chloro-2-methoxyphenyl)carbamoyl)-5,6-dimethyl-1H-benzotriazole 908
$C_{16}H_{15}ClN_4O_3$
1-(5-Chloro-2,4-dimethoxyphenyl)-3-(1H-indazol-5-yl)urea 899
2-(3-(3-Chlorophenyl)ureido)acetic (2-hydroxybenzylidene)hydrazide 513
2-(3-(3-Chlorophenyl)ureido)acetic (3-hydroxybenzylidene)hydrazide 512
$C_{16}H_{15}ClN_4O_4$
1-(2-Chloro-5-nitrophenyl)-3-(N-methylacetamido)urea 523
$C_{16}H_{15}ClO_2$
4-Chloro-4′-phenoxybutyrophenone 287
$C_{16}H_{15}Cl_2N$
N-(2,4-Dichlorobenzylidene)-2,4,5-trimethylaniline 1027
$C_{16}H_{15}Cl_2NO$
2,4-Dichloro-N-ethyl-2′-methylbenzanilide 461
2,4-Dichloro-2′,4′,5′-trimethylbenzanilide 483
2,4-Dichloro-2′,4′,6′-trimethylbenzanilide 460
$C_{16}H_{15}Cl_2NO_2$
3′-Chloro-2-(4-chloro-2-methylphenoxy)-2′-methylacetanilide 401
3′-Chloro-2-(4-chloro-2-methylphenoxy)-4′-methylacetanilide 401
4-Chloro-2,5-dimethylphenyl N-(3-chloro-2-methylphenyl)carbamate 636
4-Chloro-2,6-dimethylphenyl N-(3-chloro-2-methylphenyl)carbamate 637
4-Chloro-2,6-dimethylphenyl N-(5-chloro-2-methylphenyl)carbamate 639
4-Chloro-3,5-dimethylphenyl N-(5-chloro-2-methylphenyl)carbamate 639
3,4-Dichloro-N,5′-dimethyl-2′-methoxybenzanilide 449
2-(2,4-Dichlorophenoxy)-2′,4′-dimethylacetanilide 400
3-Ethyl-5-methylphenyl N-(3,4-dichlorophenyl)carbamate 639
4-Isopropylphenyl N-(2,6-dichlorophenyl)carbamate 634
$C_{16}H_{15}Cl_2NO_2S$
2-(4-Chlorophenylthio)ethyl N-(3-chloro-2-methylphenyl)carbamate 623
2,4-Dichloro-3,5-dimethylphenyl N-(3-(methylthio)phenyl)carbamate 623
$C_{16}H_{15}Cl_2NO_3$
4-Chloro-3,5-dimethylphenyl N-(5-chloro-2-methoxyphenyl)carbamate 619
2-(2,4-Dichlorophenoxy)ethyl N-(o-tolyl)carbamate 665
2-(2,4-Dichlorophenoxy)ethyl N-(m-tolyl)carbamate 666
2-(2,4-Dichlorophenoxy)ethyl N-(p-tolyl)carbamate 665
$C_{16}H_{15}Cl_2NO_3S$
2-(4-Chlorophenylthio)ethyl N-(4-chloro-2-methoxyphenyl)carbamate 617
$C_{16}H_{15}Cl_2NO_4$
2,4-Dichloro-6-methylphenyl N-(2,5-dimethoxyphenyl)carbamate 614
2′,6′-Dichloro-3,4,5-trimethoxybenzanilide 447
3,4-Dimethoxybenzyl N-(3,4-dichlorophenyl)carbamate 612
$C_{16}H_{15}Cl_2NO_5$
2-Chloro-4-methoxyphenyl N-(5-chloro-2,4-dimethoxyphenyl)carbamate 613
$C_{16}H_{15}Cl_2N_3OS$
1-(2,4-Dichlorobenzylidene)-4-(2-methoxy-5-methylphenyl)-3-thiosemicarbazide 1024
$C_{16}H_{15}Cl_2N_3O_2$
1-(2,4-Dichlorobenzoyl)-4-(2,6-dimethylphenyl)semicarbazide 528
$C_{16}H_{15}Cl_2N_3O_3$
1-(3-Chloro-4-methylphenyl)-3-(2-chloro-4-nitrophenyl)-1-ethylurea 526
1-(2,4-Dichlorobenzoyl)-4-(2-methoxy-5-methylphenyl)semicarbazide 492
2-(3-Phenylureido)ethyl N-(3,4-dichlorophenyl)carbamate 674
$C_{16}H_{15}Cl_2N_3O_4$
1-(3,4-Dichlorobenzoyl)-4-(2,5-dimethoxyphenyl)semicarbazide 515
$C_{16}H_{15}Cl_3$
1-(o-Tolyl)-1-(p-tolyl)-2,2,2-trichloroethane 172
$C_{16}H_{15}Cl_3N_2O$
1-(4-Trichloromethyl)phenyl)-3-(2,5-xylyl)urea 511
$C_{16}H_{15}Cl_5N_2$
1,1-Bis(3-chloro-4-methylanilino)-2,2,2-trichloroethane 222
$C_{16}H_{15}Cl_5OSi$
((Pentachlorophenyl)phenylmethoxy)trimethylsilane 1102
$C_{16}H_{15}FN_2O_2S$
1-(2-Fluorobenzoyl)-3-(4-ethoxyphenyl)-2-thiourea 729
$C_{16}H_{15}FN_2O_3$
4′-Acetamido-4-fluoro-3′-methoxybenzanilide 448
1-(4-(Ethoxycarbonyl)phenyl)-3-(2-fluorophenyl)urea 531
Ethyl 4-(4-fluorophenylureido)benzoate 532
$C_{16}H_{15}FN_2O_3S$
1-(2,5-Dimethoxyphenyl)-3-(2-fluorobenzoyl)-2-thiourea 729
$C_{16}H_{15}FN_2O_5S$
Ethyl 4-(3-(fluorosulfonyl)phenylureido)benzoate 804
$C_{16}H_{15}FN_2O_6$
N-(2-Fluoro-5-nitrophenyl)-3,4,5-trimethoxybenzamide 450
$C_{16}H_{15}FN_4O$
2-Dimethylamino-1-(4-fluorobenzyl)-4(1H)-pyrido(2,3-d)pyrimidinone 928
10-(4-Fluorobenzyl)-1,2,3,5,10,10a-hexa-H-imidazo(1,2-a)pyrido(2,3-d)-pyrimidin-5-one 930
$C_{16}H_{15}FN_4O_3$
2-(3-(4-Fluorophenyl)ureido)acetic (2-hydroxybenzylidene)hydrazide 512
2-(3-(4-Fluorophenyl)ureido)acetic (3-hydroxybenzylidene)hydrazide 513
$C_{16}H_{15}FN_4O_3S$
N-(4-Fluorophenyl)-N′-(3-(4-(methylthio)phenyl)ureido)oxamide 521
$C_{16}H_{15}FN_4O_4$
1-(2-Fluoro-5-nitrophenyl)-3-(4-(N-methylacetamido)phenyl)urea 523
$C_{16}H_{15}FO$
DL-2-(4-Fluorophenyl)-2-phenyltetrahydrofuran 184
$C_{16}H_{15}F_2NO_3$
2-(3-Methylphenoxy)ethyl N-(2,4-difluorophenyl)carbamate 616
$C_{16}H_{15}F_3N_2$
N-(4-(Dimethylamino)benzylidene)-3-(trifluoromethyl)aniline 1020
$C_{16}H_{15}F_3N_2O$
1-Benzyl-1-methyl-3-(2-(trifluoromethyl)phenyl)urea 509
1-(2-Ethylphenyl)-3-(2-(trifluoromethyl)phenyl)urea 511
1-(2-Ethylphenyl)-3-(3-(trifluoromethyl)phenyl)urea 511
1-(3-Ethylphenyl)-3-(2-(trifluoromethyl)phenyl)urea 511
1-(4-Ethylphenyl)-3-(2-(trifluoromethyl)phenyl)urea 511
1-(4-Ethylphenyl)-3-(3-(trifluoromethyl)phenyl)urea 511
1-Phenethyl-3-(3-trifluoromethylphenyl)urea 510
1-(2-(Trifluoromethyl)phenyl)-3-(2,3-xylyl)urea 511
1-(2-(Trifluoromethyl)phenyl)-3-(2,4-xylyl)urea 511
1-(2-(Trifluoromethyl)phenyl)-3-(2,6-xylyl)urea 511
1-(3-(Trifluoromethyl)phenyl)-3-(2,3-xylyl)urea 511
1-(3-(Trifluoromethyl)phenyl)-3-(2,4-xylyl)urea 512
1-(3-(Trifluoromethyl)phenyl)-3-(2,6-xylyl)urea 511
1-(3-(Trifluoromethyl)phenyl)-3-(3,4-xylyl)urea 512
$C_{16}H_{15}F_3N_2O_2$
1-(2-Ethoxyphenyl)-3-(2-(trifluoromethyl)phenyl)urea 508
1-(2-Ethoxyphenyl)-3-(3-(trifluoromethyl)phenyl)urea 508
1-(3-(1-Hydroxyethyl)phenyl)-3-(3-(trifluoromethyl)phenyl)urea 507
1-(4-(2-Hydroxyethyl)phenyl)-3-(2-(trifluoromethyl)phenyl)urea 507
1-(4-(2-Hydroxyethyl)phenyl)-3-(3-(trifluoromethyl)phenyl)urea 507
1-(2-Methoxy-5-methylphenyl)-3-(2-(trifluoromethyl)phenyl)urea 508
1-(2-Methoxy-5-methylphenyl)-3-(3-(trifluoromethyl)phenyl)urea 508
$C_{16}H_{15}F_3N_2O_3$
1-(2,4-Dimethoxyphenyl)-3-(3-(trifluoromethyl)phenyl)urea 507
1-(2,5-Dimethoxyphenyl)-3-(2-(trifluoromethyl)phenyl)urea 507
1-(2,5-Dimethoxyphenyl)-3-(3-(trifluoromethyl)phenyl)urea 507
$C_{16}H_{15}F_5OSi$
Dimethyl(pentafluorophenyl)(1-phenylethoxy)silane 1102
((Pentafluorophenyl)phenylmethoxy)trimethylsilane 1104
$C_{16}H_{15}HgNO_2$
(9-Ethyl-3-carbazolyl)mercuri acetate 1138
$C_{16}H_{15}IN_2OS$
1-(4-Iodophenyl)-3-(3-phenylpropionyl)-2-thiourea 732
$C_{16}H_{15}IN_2O_4$
2,5-Dimethyl-4-iodophenyl N-(2-methyl-5-nitrophenyl)carbamate 626
$C_{16}H_{15}I_2NO_3$
4-Amino-3,5-diiodo-α-(4-methoxyphenyl)hydrocinnamic acid 325
$C_{16}H_{15}N$
11-Amino-9,10-dihydro-9,10-ethanoanthracene 241
1-Benzyl-3,4-dihydroisoquinoline hydrochloride 250
N-(Cinnamylidene)-m-toluidine 1020
N-(Cinnamylidene)-p-toluidine 1078
1,2-Dihydro-1-methyl-2-phenylquinoline 251
2,2-Diphenylbutyronitrile 756
1-Ethyl-2-phenylindole 887
$C_{16}H_{15}NO$
3-Acetylphenyl N-(p-tolyl)carbamate 666
N-(Cinnamylidene)-o-anisidine 1049
N-(Cinnamylidene)-m-anisidine 1039
N-(Cinnamylidene)-p-anisidine 1050
2-(9,10-Dihydro-9-anthracenylamino)-1-ethanol 239
1,4-Diphenyl-3-methyl-2-azetidinone 692
2-(N-Methylacetamido)fluorene 491
4′-(4-Methylbenzylideneamino)acetophenone 1074
N-Methyl-5-benzyloxyindole 882
N-(1-Naphthyl)sorbamide 490
2-(4-Pyridylmethyl)-1-tetralone 961
$C_{16}H_{15}NOS_2$
3-Allyl-5-(5-indanylmethylene)rhodanine 742
$C_{16}H_{15}NO_2$
N-(Cinnamyl)-p-anisidine 466
2,3-Dihydro-5-methoxy-3-phenyl-2-indolecarboxaldehyde 324
9-(Ethoxycarbonylmethyl)carbazole 892

$C_{16}H_{15}NO_2$ (Cont'd)
4-Ethyl-N-(piperonylidene)aniline 1063
3-Hydroxy-2-(2-phenylethyl)-1-isoindolinone 690
1-(4-Methoxyphenyl)-4-phenyl-2-azetidinone 690
β-Methyl-N-phenylcinnamamide 478
3-Phenylallyl N-phenylcarbamate 680
N-Piperonylidene-2,3-xylidine 1051
N-Piperonylidene-2,4-xylidine 1045
N-Piperonylidene-2,5-xylidine 1051
N-Piperonylidene-3,4-xylidine 1076
N-Piperonylidene-3,5-xylidine 1041
2-(p-Toluoyl)acetanilide 443

$C_{16}H_{15}NO_2S$
2-(4-Ethoxy-3-methoxyphenyl)-benzothiazole 912
Methyl 10-phenothiazinepropionate 359

$C_{16}H_{15}NO_3$
N-Benzoylmonoethanolamine benzoate 373
N-Benzoyl-D-phenylalanine 471
N-Benzoyl-DL-phenylalanine 471
N-Benzoyl-L-phenylalanine 349
2′,4′-Dimethylphthalanilic acid 474
2′,5′-Dimethylphthalanilic acid 474
Ethyl 4-benzamidobenzoate 483
Ethyl 4-(2-hydroxybenzylideneamino)-benzoate 1034
Ethyl 4-(4-hydroxybenzylideneamino)-benzoate 1037
Ethyl 4′-phenyloxanilate 489
N-(2-Ethylphenyl)phthalamic acid 474
4′-Ethylphthalanilic acid 474
3′-(2-Hydroxy-3-methoxybenzylideneamino)-acetophenone 1031
4′-(2-Hydroxy-3-methoxybenzylideneamino)-acetophenone 1031
2-(4-Methylbenzylcarbamoyl)-benzoic acid 431
4-(1-Methyl-1-nitroethyl)-benzophenone 313
Methyl 2-(2-phenylacetamido)-benzoate 434
N-(2-Phenethyl)phthlalamic acid 471
N-Piperonylidene-p-phenetidine 1041
N-(2,3-Xylyl)phthalamic acid 474
N-(2,6-Xylyl)phthalamic acid 474
N-(3,4-Xylyl)phthalamic acid 474

$C_{16}H_{15}NO_3S$
4-Acetylphenyl N-(3-(methylthio)-phenyl)carbamate 667
4-Acetylphenyl N-(4-(methylthio)-phenyl)carbamate 667
Benzoylmethyl N-(4-(methylthio)-phenyl)carbamate 667
2-(Benzoylmethylsulfinyl)-acetanilide 771
2-(4-(3-Methylbenzamido)-phenylthio)acetic acid 468
2-(2,3,4-Trimethoxyphenyl)-benzothiazole 912

$C_{16}H_{15}NO_4$
4-Acetylphenyl N-(4-methoxyphenyl)carbamate 654
Carbobenzyloxy-D-(-)-phenylglycine 658
N-Carbobenzyloxy-DL-2-phenylglycine 661
2,5-Dimethoxy-N-(piperonylidene)-aniline 1041
α,α-Dimethyl-4-nitrobenzyl benzoate 370
2′-Ethoxyphthalanilic acid 465
4′-Ethoxyphthalanilic acid 467
(3-Hydroxybenzoyl)-o-acetanisidide 442
2′-Methoxy-5′-methylphthalanilic acid 467
4-Methoxyphenyl N-(3-acetylphenyl)carbamate 651
Piperonyl N-(o-tolyl)carbamate 664
Piperonyl N-(m-tolyl)carbamate 665
Piperonyl N-(p-tolyl)carbamate 665
4-(Veratrylideneamino)benzoic acid 1042
N-Veratrylideneanthranilic acid 1042

$C_{16}H_{15}NO_4S$
3-(Acetoxy)phenyl N-(4-(methylthio)-phenyl)carbamate 666
4-Formyl-2-methoxyphenyl N-(4-(methylthio)phenyl)carbamate 649
2-(Methoxycarbonyl)phenyl N-(4-(methylthio)phenyl)carbamate 667
4-(Methoxycarbonyl)phenyl N-(4-(methylthio)phenyl)carbamate 667
3-Phenyl-4-(phenylsulfonylmethyl)-2-oxazolidinone 781

$C_{16}H_{15}NO_5$
Piperonyl N-(o-anisyl)carbamate 646
Piperonyl N-(3-methoxyphenyl)-carbamate 646
Piperonyl N-(4-methoxyphenyl)-carbamate 646

$C_{16}H_{15}NO_6S$
4-(2-(p-Anisoyl)acetamido)-benzenesulfonic acid, barium salt 792
4-(Methoxycarbonyl)phenyl N-(p-tosyl)carbamate 823

$C_{16}H_{15}NO_7$
Dimethyl 5-nitro-10-oxo-1,2,3,4-tetrahydro-1,4-ethanonaphthalene-2,3-dicarboxylate 359

$C_{16}H_{15}NS$
N-Allylthiobenzanilide 746
N-(Cinnamylidene)-3-(methylthio)-aniline 1051
N-Cinnamylidene-4-(methylthio)-aniline 1051
4,5-Diphenyl-2-methyl-2-thiazoline 252

$C_{16}H_{15}N_3O$
4-Aminobenzoic cinnamylidenehydrazide 1052
4′-(4-(1-Aziridinyl)phenylazo)-acetophenone 1091
4-Cyano-3,5-dimethyl-2-ethylpyrido(1,2-a)benzimidazol-1(5H)-one 905
1-(3-Cyanophenyl)-3-(2,6-dimethylphenyl)urea 765
1-(4-Cyanophenyl)-3-(2,6-dimethylphenyl)urea 581
1-(4-Cyanophenyl)-3-(2,4-xylyl)urea 581
5,6-Dimethylbenzimidazole-1-carboxanilide 907
Isonicotinic acid 5-indanylidenehydrazide 1069
Nicotinic acid (5-indanylmethylene)-hydrazide 1056
Picolinic acid (5-indanylmethylene)-hydrazide 1071

$C_{16}H_{15}N_3O_2$
4-Acetoacetamidoazobenzene 1092
4-Cyanoanilino N-(3,5-xylyl)-carbamate 765
4′-(4-Oxo-1,2,3,4-tetrahydroquinazolin-2-yl)-acetanilide 704

$C_{16}H_{15}N_3O_2S$
1-(6-Methoxy-2-benzothiazolyl)-3-(4-(methylthio)phenyl)urea 913
p-Toluenesulfonic N^2-(2-quinolyl)-hydrazide 994

$C_{16}H_{15}N_3O_2S_2$
Benzenesulfonic (3-methyl-4-phenyl-4-thiazolin-2-ylidene)-hydrazide 820

$C_{16}H_{15}N_3O_3$
1-(4-Cyanophenyl)-3-(2,5-dimethoxyphenyl)urea 761
1,2-Dihydro-4,6-dimethyl-1-(3-methoxysalicylideneamino)-2-oxonicotinonitrile 1031
1,2-Dihydro-4,6-dimethyl-2-oxo-1-(vanillylideneamino)nicotinonitrile 1032
1-(N-(2,4-Dimethoxyphenyl)-carbamoyl)benzimidazole 903
Hippuric (2-hydroxybenzylidene)-hydrazide 496
6-(3-Hydroxypropylamino)-2-nitrophenanthridine 996

$C_{16}H_{15}N_3O_3S$
1-(2-Benzothiazolyl)-3-(2,4-dimethoxyphenyl)urea 912
1-(2-Benzothiazolyl)-3-(2,5-dimethoxyphenyl)urea 912
4-(3-Phenyl-2-thioureido)-hippuric acid 744

$C_{16}H_{15}N_3O_4$
N,N-Dimethyl-2′,4′-dinitro-4-stilbenamine 275

$C_{16}H_{15}N_3O_4S$
1-(4-Ethoxyphenyl)-3-(3-nitrobenzoyl)-2-thiourea 737

$C_{16}H_{15}N_3O_5$
4-Acetamidophenyl N-(4-methyl-2-nitrophenyl)carbamate 668
1-(4-Hydroxybenzoyl)-4-(2-(methoxycarbonyl)phenyl)-semicarbazide 548

$C_{16}H_{15}N_3O_5S$
1-(2,5-Dimethoxyphenyl)-3-(3-nitrobenzoyl)-2-thiourea 735

$C_{16}H_{15}N_3O_6$
4-Acetamidophenyl N-(2-methoxy-4-nitrophenyl)carbamate 652
Methyl 2-(3-(2-methoxy-5-nitrophenyl)ureido)benzoate 552

$C_{16}H_{15}N_3O_7$
2-(4-Nitrophenoxy)ethyl N-(4-methyl-2-nitrophenyl)carbamate 665

$C_{16}H_{15}N_3S_2$
1-Phenyl-4-(2,6-xylyl)-1H-1,2,4-triazole-3,5(2H,4H)-dithione 748

$C_{16}H_{15}N_5$
5-Amino-3-methyl-4-(phenylazo)-1H-pyrazole 1087

$C_{16}H_{15}N_5OS$
1-(4-(Methylthio)phenyl)-3-(5-phenyl-1,2,4-triazol-3-yl)urea 869

$C_{16}H_{15}N_5O_5$
4-Acetamidobenzaldehyde (2,6-dinitro-4-methylphenyl)hydrazone 424

$C_{16}H_{15}N_5S_2$
6-Amino-3-methyl-2-benzothiazolinone azine with 3-methyl-2-benzothiazolinone 739

$C_{16}H_{15}O_4PS$
2,3-Dihydro-3-(diphenylphosphinyloxy)-thiophene-1,1-dioxide 835

$C_{16}H_{16}$
4,4′-Dimethylstilbene 165
1,2-Diphenyl-2-butene 164
2,3-Diphenyl-1-butene 164

$C_{16}H_{16}BF_3N_2$
p-Tolualdehyde azine boron trifluoride complex 1078

$C_{16}H_{16}BrClFN_3$
4-(2-Bromo-6-chloro-4-fluorophenylazo)-N,N-diethylaniline 1081

$C_{16}H_{16}BrN$
N-(4-Isopropylbenzylidene)-4-bromoaniline 1027
1,1′,3,3′-Tetrahydro-2,2′-spirobi(2H-isoindolium)bromide 255

$C_{16}H_{16}BrNO$
3-Bromo-2′-ethyl-6′-methylbenzanilide 459
2-Bromo-2′,4′,6′-trimethylbenzanilide 458
3-Bromo-2′,4′,6′-trimethylbenzanilide 460

$C_{16}H_{16}BrNOS$
2-(Benzylthio)-4′bromo-3′-methylacetanilide 403
4′-Bromo-2-(benzylthio)-2′-methylacetanilide 421

$C_{16}H_{16}BrNO_2$
4-Bromo-2,5-dimethylphenyl N-(m-tolyl)carbamate 641

$C_{16}H_{16}BrNO_2S$
3-Methyl-4-(methylthio)phenyl N-(2-bromo-4-methylphenyl)carbamate 623

$C_{16}H_{16}BrNO_3$
2-Bromo-4,5-dimethylphenyl N-(2-methoxyphenyl)carbamate 618
2-Bromo-4,5-dimethylphenyl N-(3-methoxyphenyl)carbamate 620
4-Bromo-2,5-dimethylphenyl N-(2-methoxyphenyl)carbamate 618

$C_{16}H_{16}BrNO_3S$
2-(4-Bromophenoxy)ethyl N-(3-(methylthio)phenyl)carbamate 615

$C_{16}H_{16}BrNO_4$
2,4-Dimethoxybenzyl N-(4-bromophenyl)carbamate 612
Ethyl 5-bromo-α-(ethoxycarbonyl)-3-indoleacrylate 889

$C_{16}H_{16}BrN_3O_4$
1-(2-(4-Bromophenoxy)ethyl)-3-(4-methyl-2-nitrophenyl)urea 517

$C_{16}H_{16}Br_2N_2O_3S$
p-Toluenesulfonic acid (3,5-dibromo-2-ethoxybenzylidene)-hydrazide 1049

$C_{16}H_{16}Br_2O_4$
5,5′-Dibromo-4,4′-biveratrole 179
2,2′-Dibromo-4,4′,5,5′-tetramethoxybiphenyl 185

$C_{16}H_{16}Br_4Si$
Bis(α,α-dibromobenzyl)-dimethylsilane 1113

$C_{16}H_{16}ClFN_2O_2S$
1-(5-Chloro-2,4-dimethoxyphenyl)-3-(2-fluorophenyl)-3-methyl-2-thiourea 729

$C_{16}H_{16}ClN$
N-(4-Isopropylbenzylidene)-4-chloroaniline 1027

$C_{16}H_{16}ClNO$
2-Chloro-N,N-dibenzylacetamide 392
4-Chloro-2′-(dimethylaminomethyl)-benzophenone 309
2-Chloro-N-(1,2-diphenylethyl)-acetamide 392
3-Chloro-2′-ethyl-6′-methylbenzanilide 459
4-Chloro-2′-ethyl-6′-methylbenzanilide 458
2-Chloro-N-methyl-2-phenyl-o-acetotoluidide 441
5′-Chloro-2′-methyl-3-phenylpropionanilide 408
2-Chloro-2-phenyl-2′,6′-acetoxylidide 430
3-(4-Chlorophenyl)-2-isoindolineethanol 244
2-Chloro-2′,4′,6′-trimethylbenzanilide 458
3-Chloro-2′,4′,6′-trimethylbenzanilide 460
4-Chloro-2′,4′,6′-trimethylbenzanilide 459

$C_{16}H_{16}ClNOS$
2-(Benzylthio)-3′-chloro-4′-methylacetanilide 402
2-(Benzylthio)-4′-chloro-2′-methylacetanilide 421
2-(4-Chlorophenylthio)-2′,3′-dimethylacetanilide 403
2-(4-Chlorophenylthio)-3′,4′-dimethylacetanilide 403

$C_{16}H_{16}ClNO_2$
Benzyl N-(2-chloro-1-phenylethyl)-carbamate 604
2-Chloro-3-(cyclohexylamino)-1,4-naphthoquinone 318
4-Chloro-N,5′-dimethyl-2′-methoxybenzanilide 449
4-Chloro-3,5-dimethylphenyl N-(m-tolyl)carbamate 640
2-(4-Chloro-2-methylphenoxy)-2′-methylacetanilide 399
5′-Chloro-2′-methyl-2-(phenoxy)-propionanilide 399
4-Chloro-2-methylphenyl N-(2,5-dimethylphenyl)carbamate 638
4-Chloro-2-methylphenyl N-(2,6-xylyl)carbamate 636
4-Chloro-3-methylphenyl N-(2,5-xylyl)carbamate 638
3-Chloro-4-methyl-N-(veratrylidene)-aniline 1042
2-(4-Chlorophenoxy)-2′,6′-acetoxylidide 400
2-(2-Chlorophenoxy)-2′,4′-dimethylacetanilide 402

$C_{16}H_{16}ClNO_2$ (Cont'd)
2-(4-Chlorophenoxy)-2′,3′-dimethylacetanilide ... 402
2-(4-Chlorophenoxy)-2′,4′-dimethylacetanilide ... 400
2-(2-Chlorophenoxy)-2′-ethylacetanilide ... 400
2-(4-Chlorophenoxy)-2′-ethylacetanilide ... 400
4-Chlorophenyl *N*-(2-ethyl-6-methylphenyl)carbamate ... 634
3-Ethyl-5-methylphenyl *N*-(3-chlorophenyl)carbamate ... 640
3-Ethyl-5-methylphenyl *N*-(4-chlorophenyl)carbamate ... 635

$C_{16}H_{16}ClNO_2S$
2-Chloro-4,5-dimethylphenyl *N*-(4-(methylthio)phenyl)carbamate ... 668
4-Chloro-2,6-dimethylphenyl *N*-(3-methylthiophenyl)carbamate ... 624
4-Chloro-2,6-dimethylphenyl *N*-(4-(methylthio)phenyl)carbamate ... 623
4-Chloro-3,5-dimethylphenyl *N*-(3-methylthiophenyl)carbamate ... 624
2-(4-Chlorophenylthio)-4′-ethoxyacetanilide ... 401

$C_{16}H_{16}ClNO_3$
2′-Chloro-2,6-dimethoxy-6′-methylbenzanilide ... 448
2-Chloro-4-methoxyphenyl *N*-(2,4-dimethylphenyl)carbamate ... 621
2-Chloro-4-methoxyphenyl *N*-(2-ethylphenyl)carbamate ... 620
2-(4-Chlorophenoxy)-4′-ethoxyacetanilide ... 399
2-Ethoxyphenyl *N*-(3-chloro-4-methylphenyl)carbamate ... 615
4-Methoxybenzyl *N*-(3-chloro-2-methylphenyl)carbamate ... 612
4-Methoxybenzyl *N*-(5-chloro-2-methylphenyl)carbamate ... 612
2-Phenoxyethyl *N*-(5-chloro-2-methylphenyl)carbamate ... 615
2,6-Xylyl *N*-(5-chloro-2-methoxyphenyl)carbamate ... 619

$C_{16}H_{16}ClNO_3S$
S-Benzyl *N*-(5-chloro-2,4-dimethoxyphenyl)thiocarbamate ... 613
2-(2-Chlorophenoxy)ethyl *N*-(3-(methylthio)phenyl)carbamate ... 614
3-Methyl-4-(methylthio)phenyl *N*-(2-chloro-4-methoxyphenyl)-carbamate ... 617
3-Methyl-4-(methylthio)phenyl *N*-(5-chloro-2-methoxyphenyl)-carbamate ... 617

$C_{16}H_{16}ClNO_4$
α-(5-Chloro-2,4-dimethoxyphenylimino)-6-methoxy-*o*-cresol ... 1021
4-Chloro-2-methylphenyl *N*-(2,5-dimethoxyphenyl)carbamate ... 614

$C_{16}H_{16}ClNO_4S$
N-(3-Chlorophenyl)-*N*-tosyl-β-alanine ... 808
N-(4-Chlorophenyl)-*N*-tosyl-β-alanine ... 808
4-(Methylthio)phenyl *N*-(5-chloro-2,4-dimethoxyphenyl)carbamate ... 613

$C_{16}H_{16}ClNO_5$
2-Chloro-4-methoxyphenyl *N*-(2,5-dimethoxyphenyl)carbamate ... 614
2,6-Dimethoxyphenyl *N*-(5-chloro-2-methoxyphenyl)carbamate ... 612

$C_{16}H_{16}ClN_3O$
4-Chlorobenzoic (4-(dimethylamino)benzylidene)-hydrazide ... 492
2-Chloro-5′-methyl-2′-(*p*-tolylazo)-acetanilide ... 1080
4-(Dimethylamino)benzoic (2-chlorobenzylidene)hydrazide ... 492

$C_{16}H_{16}ClN_3O_2$
1-(4-Acetamidophenyl)-3-(3-chloro-2-methylphenyl)urea ... 526
1-(4-Acetamidophenyl)-3-(5-chloro-2-methylphenyl)urea ... 526
1-(4-Chlorobenzoyl)-4-(3-ethylphenyl)semicarbazide ... 528
1-(4-Chlorobenzoyl)-4-(2,6-xylyl)-semicarbazide ... 528
3-Hydroxybenzoic acid (2-chloro-4-(dimethylamino)benzylidene)-hydrazide ... 496

$C_{16}H_{16}ClN_3O_3$
1-(2-Chlorobenzoyl)-4-(2-methoxy-5-methylphenyl)semicarbazide ... 518
1-(4-Chlorobenzoyl)-4-(2-methoxy-5-methylphenyl)semicarbazide ... 519
1-(4-Chlorobenzyl)-1-methyl-3-(2-methyl-5-nitrophenyl)urea ... 523
4-(4-Chloro-2-methylphenyl)-1-(2-methoxybenzoyl)semicarbazide ... 517
1-(2-Chloro-4-nitrophenyl)-3-(2-ethyl-6-methylphenyl)urea ... 530
2,5-Dimethoxy-4-(phenylacetamido)benzene-diazonium chloride zinc chloride double salt ... 1094

$C_{16}H_{16}ClN_3O_4$
1-(2-Carbamoylphenyl)-3-(5-chloro-2,4-dimethoxyphenyl)urea ... 515
1-(4-Chlorobenzoyl)-4-(2,5-dimethoxyphenyl)semicarbazide ... 516
4-(3-(5-Chloro-2-methoxyphenyl)-ureido)phenyl *N*-methylcarbamate ... 618

$C_{16}H_{16}ClN_3O_5$
1-(4-Chloro-2,5-dimethoxyphenyl)-3-(2-methyl-5-nitrophenyl)urea ... 516
1-(5-Chloro-2,4-dimethoxyphenyl)-3-(2-methyl-5-nitrophenyl)urea ... 549
1-(5-Chloro-2,4-dimethoxyphenyl)-3-(4-methyl-3-nitrophenyl)urea ... 516

$C_{16}H_{16}ClN_3O_6$
1-(5-Chloro-2,4-dimethoxyphenyl)-3-(2-methoxy-4-nitrophenyl)urea ... 514

$C_{16}H_{16}Cl_2$
1,2-Bis(*p*-tolyl)-1,2-dichloroethane ... 174

$C_{16}H_{16}Cl_2N_2O$
1-Benzyl-3-(3,4-dichlorophenyl)-1-ethylurea ... 573
1-(2,3-Dichlorophenyl)-3-(2-ethyl-6-methylphenyl)urea ... 541
1-(2,4-Dichlorophenyl)-3-(2-ethyl-6-methylphenyl)urea ... 542
1-(2,5-Dichlorophenyl)-3-(2-ethyl-6-methylphenyl)urea ... 542
1-(2,6-Dichlorophenyl)-3-(2-ethyl-6-methylphenyl)urea ... 541
1-(2,5-Dichlorophenyl)-3-ethyl-3-(*p*-tolyl)urea ... 540
1-(3,4-Dichlorophenyl)-3-(2,4,5-trimethylphenyl)urea ... 545

$C_{16}H_{16}Cl_2N_2O_2$
5′-Chloro-2′-(4-chlorophenoxy)-2-(dimethylamino)acetanilide hydrochloride ... 486
1-(2,4-Dichlorobenzyl)-3-(2-methoxy-5-methylphenyl)urea ... 520

$C_{16}H_{16}Cl_2N_2O_3$
5-Chloro-2-(4-chlorophenoxy)-*N*,*N*-diethyl-4-nitroaniline ... 260
1-(5-Chloro-2,4-dimethoxyphenyl)-3-(3-chloro-2-methylphenyl)urea ... 515

$C_{16}H_{16}Cl_2N_4O_2$
1-(4-(Dimethylamino)benzoyl)-4-(2,6-dichlorophenyl)semicarbazide ... 527

$C_{16}H_{16}Cl_2N_4O_3$
Disperse Red 32 (C.I. 11190) ... 1084

$C_{16}H_{16}Cl_2N_4O_4$
1,6-Bis(4-chloro-2-methoxyphenyl)-biurea ... 518
4′-(Bis(2-hydroxyethyl)amino)-2,6-dichloro-4-nitroazobenzene ... 1080

$C_{16}H_{16}Cl_2O_3$
Bis-(2-chloro-2-phenoxyethyl) ether ... 187

$C_{16}H_{16}Cl_2O_4$
1,4-Bis-(2-chlorophenoxy)-2,3-butanediol ... 202
1,4-Bis(4-chlorophenoxy)-2,3-butanediol ... 202

$C_{16}H_{16}FN$
N-(4-Isopropylbenzylidene)-4-fluoroaniline ... 1027

$C_{16}H_{16}FNO$
3-Fluoro-2′,4′,5′-trimethylbenzanilide ... 460

$C_{16}H_{16}FNO_2$
2′,4′-Dimethyl-2-(4-fluorophenyl)-acetanilide ... 409
3-Fluorobenzyl *N*-(2,4-dimethylphenyl)carbamate ... 622

$C_{16}H_{16}FNO_2S$
2,6-Dimethyl-4-(methylthio)phenyl *N*-(3-fluorophenyl)carbamate ... 623

$C_{16}H_{16}FNO_3S$
(2-(Hydroxymethyl)-6-methyl-4-(methylthio))phenyl *N*-(2-fluorophenyl)carbamate ... 611

$C_{16}H_{16}FNO_4$
2′-Fluoro-3,4,5-trimethoxybenzanilide ... 449
4′-Fluoro-3,4,5-trimethoxybenzanilide ... 447
(2-(4-Methoxyphenoxy)ethyl) *N*-(2-fluorophenyl)carbamate ... 612

$C_{16}H_{16}FN_3O_2$
1-(2-Fluorophenyl)-3-(4-(*N*-methylacetamido)phenyl)urea ... 527
1-(4-Fluorophenyl)-3-(4-(*N*-methylacetamido)phenyl)urea ... 527

$C_{16}H_{16}FN_3O_3$
1-(4-Acetamido-3-methoxyphenyl)-3-(3-fluorophenyl)urea ... 518
1-(2-Fluoro-5-nitrophenyl)-3-isopropyl-3-phenylurea ... 524

$C_{16}H_{16}FN_5O_4$
1-(4-(Dimethylamino)phenyl)-5-(4-fluoro-3-nitrophenyl)biuret ... 523

$C_{16}H_{16}F_2N_4O_6S_2$
1,2-Bis(2-fluorosulfonylphenylureido)ethane ... 803

$C_{16}H_{16}F_3NO_2$
N-(α,α,α-Trifluoro-*m*-tolyl)-veratrylamine ... 222

$C_{16}H_{16}F_3NO_2S$
3′-(Trifluoromethyl)-2,4,6-trimethylbenzenesulfonanilide ... 805

$C_{16}H_{16}F_3N_3$
N-Ethyl-*N*-methyl-4-(4-(trifluoromethyl)phenylazo)aniline ... 1080

$C_{16}H_{16}F_3N_3O$
1-(4-(Dimethylamino)phenyl)-3-(2-(trifluoromethyl)phenyl)urea ... 508
1-(4-(Dimethylamino)phenyl)-3-(3-(trifluoromethyl)phenyl)urea ... 508

$C_{16}H_{16}F_3N_5O_4$
2-(Furfurylureido)acetic N^2-(*N*-(2-(trifluoromethyl)phenyl)-carbamoyl)hydrazide ... 843

$C_{16}H_{16}FeO_4$
Bis(di

$C_{16}H_{16}HgO_3$
(2,2-Diphenyl-2-hydroxyethyl)-mercuri acetate ... 1137

$C_{16}H_{16}INO_2$
1-Ethyl-2-(3,4-(methylenedioxy)-styryl)pyridinium iodide ... 957
1-Ethyl-4-(3,4-(methylenedioxy)-styryl)-pyridinium iodide ... 957

$C_{16}H_{16}I_2O_2$
3,3′-Diiodo-2,2′,6,6′-tetramethyl-4,4′-biphenol ... 193

$C_{16}H_{16}I_2O_4$
2,2′-Diiodo-4,4′,5,5′-tetramethoxybiphenyl ... 177

$C_{16}H_{16}N_2$
Acetophenone azine ... 1063
2-Benzyl-5,6-dimethylbenzimidazole ... 905
Cinnamaldehyde methylphenylhydrazone ... 235
N,*N*′-Dibenzylideneethylenediamine ... 1054
3-(4-(Methylamino)benzyl)indole ... 887
2-Methylbenzaldehyde azine ... 770
6α-Methyl-5,6,6a,11-tetrahydroisoindolo-(2,1-a)-quinazoline hydrochloride ... 250
6-Methyl-3-(*p*-tolyl)-3,4-dihydroquinazoline ... 251
p-Tolualdehyde azine ... 1078

$C_{16}H_{16}N_2O$
3-Acetamido-9-ethylcarbazole ... 891
4-(3-Aminophenyl)-3-butenanilide ... 468
1-(2-Carboxyphenyl)-3-(2,6-xylyl)urea ... 576
1,2-Dihydro-2-(2,4-xylyl)-4(3*H*)-quinazolinone ... 704
1,2-Dihydro-2-(2,5-xylyl)-4(3*H*)-quinazolinone ... 704
2-(α-Hydroxybenzyl)-5,6-dimethylbenzimidazole ... 902
1-Methyl-2-(*p*-tolyloxymethyl)-benzimidazole ... 903

$C_{16}H_{16}N_2OO_5$
2-(4-Bromo-3-methylphenyliminomethyl)-6-methoxy-4-nitrophenol ... 1022

$C_{16}H_{16}N_2OS$
α-(4-(Dimethylamino)phenyl)-2-benzothiazolemethanol ... 912
1,2,3,3a,4,10a-Hexahydro-3-phenyl-9a(*H*)pyrido(2,1-b)pyrrolo(3,2-e)(1,3)-thiazin-4-one ... 710

$C_{16}H_{16}N_2OS_2$
3-Butyl-5-(3-indolylmethylene)-rhodanine ... 889

$C_{16}H_{16}N_2O_2$
3-Acetyl-2-(3-ethyl-5-methyl-4-isoxazolyl)indole ... 885
2-(4-Aminobenzoyl)-propionanilide ... 443
p-Anisaldehyde azine ... 1050
N-Benzoyl-DL-phenylalaninamide ... 471
N,*N*′-Bis(2-hydroxy-5-methylphenyl)-1,4-diazabutadiene ... 1028
tert-Butyl 2-cyano-2-(1,2-dihydro-2-quinolylidene)acetate ... 766
1,2-Dibenzoyl-1,2-dimethylhydrazine ... 503
N,*N*′-Dibenzyloxamide ... 424
1,2-Dihydro-2-(2-ethoxyphenyl)-4(3*H*)-quinazolinone ... 698
1,2-Dihydro-2-(4-ethoxyphenyl)-4(3*H*)-quinazolinone ... 698
4-Dimethylamino-4′-nitrostilbene ... 277
N,*N*′-Di(*o*-tolyl)oxamide ... 434
4-Hydroxybenzoic (α,4-dimethylbenzylidene)hydrazide ... 547
3-(2-Hydroxyethylamino)-3-phenyl-2-indolinone ... 690
2-Methoxybenzaldehyde azine ... 234
2-Methoxybenzoic (4-methylbenzylidene)hydrazide ... 498
4-Methylbenzoic (4-methoxybenzylidene)hydrazide ... 498
N-(4-Nitrobenzylidene)-2,4,6-trimethylaniline ... 278
Phenylacetic acid (2-methoxybenzylidene)hydrazide ... 498
2-Phenylacetic (4-methoxybenzylidene)hydrazide ... 498

$C_{16}H_{16}N_2O_2S$
1-(2-Acetylphenyl)-3-(4-(methylthio)-phenyl)urea ... 560
1-(3-Acetylphenyl)-3-(4-(methylthio)-phenyl)urea ... 560
1-(4-Acetylphenyl)-3-(4-(methylthio)-phenyl)urea ... 560
Benzenesulfonic *N*′-allyl-*N*2-benzylidenehydrazide ... 821
N-Cyano-2′,6′-dimethyl-*p*-toluenesulfonanilide ... 827
N^4-(5-Indanylmethylene)-sulfanilamide ... 1075
3-Phenyl-1,5,6,10b-tetrahydro-3*H*-(1,2,3)thiadiazolo(4,3-a)-isoquinoline2,2dioxide ... 829

$C_{16}H_{16}N_2O_2S_2$
2-Acetamidophenyl disulfide ... 468
4-Acetamidophenyl disulfide ... 422

$C_{16}H_{16}N_2O_3$
1-(3-Acetylphenyl)-3-(3-methoxyphenyl)urea ... 557
4′-Anilinosuccinanilic acid ... 420
L-α-Benzamido-β-(4-hydroxyphenyl)-propionamide ... 412
N-Benzoyl-L-tyrosinamide ... 462
Benzyl 3-(4-methoxybenzylidene)-carbazate ... 646

$C_{16}H_{16}N_2O_3$ (Cont'd)
1,2-Dihydro-2-(2,3-dimethoxyphenyl)-4(3H)-quinazolinone 698
1,2-Dihydro-2-(2,5-dimethoxyphenyl)-4(3H)-quinazolinone 698
1,2-Dihydro-2-(3,4-dimethoxyphenyl)-4(3H)-quinazolinone 698
4-(4-(Dimethylamino)-benzylideneamino)salicylic acid 1032
2',4'-Dimethyl-3-nitro-4-toluanilide 470
Mandelic acid N-(4-methoxybenzylidene)hydrazide 495
2-Methoxybenzoic (2-methoxybenzylidene)hydrazide 497
4-Methoxybenzoic (4-methoxybenzylidene)hydrazide 1041
4-Methylbenzoic acid (4-hydroxy-3-methoxybenzylidene)hydrazide 1030
4-Nitro-2',4',6'-trimethylbenzanilide 473
Pyroglutamic acid 4-methoxy-2-naphthylamide 690
2-(3-(2,4-Xylyl)ureido)benzoic acid 576
4-(3-(2,3-Xylyl)ureido)benzoic acid 576
4-(3-(2,4-Xylyl)ureido)benzoic acid 576
4-(3-(2,5-Xylyl)ureido)benzoic acid 576
4-(3-(2,6-Xylyl)ureido)benzoic acid 576

$C_{16}H_{16}N_2O_3S$
4-Acetamidophenyl N-(4-(methylthio)phenyl)carbamate 667
2-(3-Anilinopropyl)-1,2-benzisothiazol-3(2H)-one-1,1-dioxide 818
2-(Benzylthio)-2'-methyl-5'-nitroacetanilide 420
Methyl 2-(3-(4-(methylthio)phenyl)-ureido)benzoate 560

$C_{16}H_{16}N_2O_4$
1-Benzoyl-3-(2,4-dimethoxyphenyl)urea 549
1-Benzoyl-3-(2,5-dimethoxyphenyl)urea 550
1,2-Bis(2-hydroxy-3-methoxybenzylidene)hydrazine 769
1-(2-Carboxyphenyl)-3-(2-methoxy-5-methylphenyl)urea 554
1-(4-Carboxyphenyl)-3-(2-methoxy-5-methylphenyl)urea 554
2,2'-Dimethyl-4,4'-dinitrobibenzyl 279
Ethyl 3-carboxymethyl-1,3-dihydro-2H-pyrrolo(3,4-b)quinoline-2-carboxylate 988
Ethyl 4-(3-hydroxyphenylureido)-benzoate 548
4-(3-(2-Ethylphenyl)ureido)-salicylic acid 547
2-Methoxybenzoic (4-hydroxy-3-methoxybenzylidene)hydrazide 495
2-Methoxy-4-nitro-6-(phenethyliminomethyl)phenol 1030
Nα-(2-(1-Naphthyl)acetyl)-asparagine 349
2-(Phenoxy)acetic (4-hydroxy-3-methoxybenzylidene)hydrazide 495
2,6-Xylyl N-(4-methyl-2-nitrophenyl)-carbamate 668
4-(3-(2,5-Xylyl)ureido)salicylic acid 548
4-(3-(2,6-Xylyl)ureido)salicylic acid 547

$C_{16}H_{16}N_2O_4S$
3'-Hydroxy-4'-(phenylsulfonylamino)-methacrylanilide 812

$C_{16}H_{16}N_2O_4S_2$
Dimethyl 4,4'-dithiobis(N-phenylcarbamate) 655

$C_{16}H_{16}N_2O_4Se_2$
6-Nitro-3,4-xylyl diselenide 213

$C_{16}H_{16}N_2O_5$
2-Carbamoylphenyl N-(2,4-dimethoxyphenyl)carbamate 647
1-(2-Carboxyphenyl)-3-(2,5-dimethoxyphenyl)urea 554
1-(4-Carboxyphenyl)-3-(2,5-dimethoxyphenyl)urea 554
3-(3-(2,4-Dimethoxyphenyl)ureido)-benzoic acid 549
2-(4-Ethoxyphenyliminomethyl)-6-methoxy-4-nitrophenol 1029
Methyl N-(α-methoxy-3-nitrobenzyl)-anthranilate 369
2-Phenoxyethyl N-(2-methyl-5-nitrophenyl)carbamate 649
2-(Phenoxy)ethyl N-(4-methyl-2-nitrophenyl)carbamate 665

$C_{16}H_{16}N_2O_5S$
Methyl-8-oxo-7-(phenoxy-acetamido)-5-thia-1-azabicyclo-(4.2.0)oct-2-ene-2-carboxylic acid, lithium salt 699
1-Methyl-2-phenylpropyl 2,4-dinitrobenzenesulfenate 771
2-(4-Nitrophenoxy)ethyl N-(3-(methylthio)phenyl)carbamate 648

$C_{16}H_{16}N_2O_5S_2$
2,3-Dimethyl-6-nitrobenzothiazolium p-toluenesulfonate 915

$C_{16}H_{16}N_2O_6$
1,4-Bis(2-nitrophenoxy)butane 267
1,4-Bis(4-nitrophenoxy)butane 269
4-Methoxybenzyl N-(2-methoxy-4-nitrophenyl)carbamate 646
4-Nitrobenzyl N-(2,5-dimethoxyphenyl)carbamate 648
2-Phenoxyethyl N-(2-methoxy-4-nitrophenyl)carbamate 649

$C_{16}H_{16}N_2O_6S$
Dimethyl 5-sulfanilamidoisophthalate 817

$C_{16}H_{16}N_2S_2$
N,N'-Dibenzyldithiooxamide 739

$C_{16}H_{16}N_4$
4-(4-(N-Ethyl-N-methylamino)-phenylazo)benzonitrile 1089

$C_{16}H_{16}N_4OS$
1-(4-Acetamidobenzylidene)-4-phenyl-3-thiosemicarbazide 1070

$C_{16}H_{16}N_4O_2$
1-(4-Acetamidobenzylidene)-4-phenylsemicarbazide 574
1-(3-(Benzoyloxy)-α-methylbenzylideneamino)-guanidine hydrochloride 1070
1-Methyl-6-(4-nitrophenylazo)-1,2,3,4-tetrahydroquinoline 1090

$C_{16}H_{16}N_4O_3$
2-(2H-Benzotriazol-2-yl)-4-(tert-butyl)-6-nitrophenol 908
1,3-Diethyl-5-nitroso-6-phenyl-7H-pyrrolo-(2,3-d)pyrimidine-2,4(1H,3H)-dione 919
4-(Dimethylamino)benzoic (3-nitrobenzylidene)hydrazide 499

$C_{16}H_{16}N_4O_3S$
4-(4-Acetamidophenyl)-1-(4-hydroxybenzoyl)-3-thiosemicarbazide 734
4-(2-Methoxy-5-methylphenyl)-1-(3-nitrobenzylidene)-3-thiosemicarbazide 1048

$C_{16}H_{16}N_4O_3S_2$
N4-Acetylsulfanilic acid (3-methyl-2-benzothiazolinylidene)-hydrazide 820

$C_{16}H_{16}N_4O_4$
1-(4-Acetamidophenyl)-3-(2-methyl-5-nitrophenyl)urea 562
1-(4-Acetamidophenyl)-3-(4-methyl-2-nitrophenyl)urea 561
1-(4-Acetamidophenyl)-3-(4-methyl-3-nitrophenyl)urea 562
1,4-Bis(4-nitrophenyl)piperazine 276
Tricyclo[5.2.1.$0^{2,6}$]dec-8-en-10-one 2,4-dinitrophenylhydrazone 1069

$C_{16}H_{16}N_4O_6$
N-(4-Butylphenyl)picrylamine 272
2,5-Dimethoxybenzaldehyde (2,4-dinitrophenyl)methylhydrazone 1040

$C_{16}H_{16}N_4S$
3'-Aminoacetophenone azine with 3-methyl-2-benzothiazolinone 1052

$C_{16}H_{16}N_6O_6$
1,6-Bis(2-methyl-5-nitrophenyl)-biurea 563
1,6-Bis(4-methyl-2-nitrophenyl)-biurea 562
1,6-Bis(4-methyl-3-nitrophenyl)-biurea 563

$C_{16}H_{16}O$
Acetyl-1,2,3,4-tetrahydroanthracene, mixture of isomers 296
9-Acetyl-1,2,3,4-tetrahydrophenanthrene 298
2,7-Dimethyl-9-hydroxymethylfluorene 205
3',4'-Dimethyl-2-phenylacetophenone 297
2,2-Diphenyltetrahydrofuran 184
3,4-Diphenyltetrahydrofuran 184
4-Isopropylbenzophenone 313
3-Phenylbutyrophenone 296
4-Phenylbutyrophenone 296
5-Phenyl-5,6,7,8-tetrahydro-2-naphthol 196
2,3',5'-Trimethylbenzophenone 314

$C_{16}H_{16}OSe$
3,4-Dihydro-6-methyl-4-phenyl-2H-1-benzoselenin-4-ol 1141

$C_{16}H_{16}O_2$
1,1-Bis(2-methoxyphenyl)ethylene 187
1,1-Bis(4-methoxyphenyl)ethylene 181
1,4,4a,4b,5,8,8a,8b,9,10-Decahydro-1,4:5,8-dimethanoanthracene-9,10-dione 55
9,10-Dihydro-9,10-dimethyl-9,10-anthracenediol 200
2,3-Dimethyl-1,4,4a,9a-tetrahydroanthraquinone 318
2,2-Diphenylbutyric acid 346
2,4-Diphenylbutyric acid 333
3,4-Diphenylbutyric acid 333
2,3-Diphenyl-1,4-dioxane 185
Di-p-tolylacetic acid 333
Ethyl diphenylacetate 362
Hydro-4,5-phenanthrenedimethanol. 200
4-(2-Hydroxyoctahydro-1,3,4-metheno-2H-cyclobuta(cd)-pentalen-2-yl)phenol 194
4'-Methoxy-3-phenylpropiophenone. 293
2-Methyl-4'-phenoxypropiophenone. 292
Phenethyl phenylacetate 362
(+/-)-1-Phenylethyl phenylacetate 362
2-Phenyl-2-(o-tolyl)-1,3-dioxolane 184
6,7,8,9-Tetrahydrodibenzo(b,d)(1,6)-dioxecin 185
3,3',5,5'-Tetramethyldiphenoquinone 56

$C_{16}H_{16}O_2S$
Ethyl 4-(benzylthio)benzoate 372
1-Methyl-2-phenylvinyl p-tolyl sulfone 784
1,5,7-Trimethylthioxanthene-10,10-dioxide 783
2,4,7-Trimethylthioxanthene-10,10-dioxide 783

$C_{16}H_{16}O_3$
Benzoin ethylene ketal 203
(4-(Benzyloxy)-phenoxy)-2-propanone 283
4-Benzyloxyphenyl propionate 377
3,4-Diphenyl-3-hydroxybutyric acid 326
Ethyl benzilate 356
2-(2-(4-Hydroxyphenyl)propyl)-benzoic acid 339
4-Hydroxy-α-(m-tolyl)-hydrocinnamic acid 326
4-(Methoxymethoxy)-4'-hydroxystilbene 191
Methyl 2-(4-benzyloxyphenyl)-acetate 353
Phenethyl mandelate 353
6,7,9,10-Tetrahydrodibenzo(h,j)-(1,4,7)trioxacycloundecin 189

$C_{16}H_{16}O_3S$
2-Methyl-2-(phenylsulfonyl)-propiophenone 780
3-(Methylsulfonyl)-3-phenylpropiophenone 780
4'-Methyl-2-(p-tolylsulfonyl)-acetophenone 780
3-Phenyl-3-(p-tosyl)-propionaldehyde 780

$C_{16}H_{16}O_4$
7-Acetoxy-6-allyl-4,8-dimethylcoumarin 384
4-Benzyloxy-3,5-dimethoxybenzaldehyde 322
2,3'-Dihydroxy-2',3,5,6'-tetramethyldiphenoquinone, sodium salt 51
2',5'-Dimethoxy-2-biphenylacetic acid 327
5-(α,α-Dimethyl-4-hydroxybenzyl)-salicylic acid 338
2,3-Dimethyl-1,4-naphthalenediol diacetate 379
2-Hydroxy-3-phenoxypropyl benzoate 367
4'-Methoxy-2-(2-methoxyphenoxy)-acetophenone 291
Methyl 4-benzyloxy-3-methoxybenzoate 369
2,4,5-Trimethoxybenzophenone 310
2,4,6-Trimethoxybenzophenone 311
3,4,4'-Trimethoxybenzophenone 311

$C_{16}H_{16}O_4Pb$
Diphenylplumbylene diacetate (contains acetic acid of crystallization) 1135

$C_{16}H_{16}O_4S_2$
1,2-Bis(benzylsulfonyl)ethene 784

$C_{16}H_{16}O_5$
3-(3,6-Dimethoxy-2-naphthoyl)-propionic acid 328
Dimethyl 9-oxo-1,2,3,4-tetrahydro-1,4-ethanonaphthalene-2,3-dicarboxylate 360

$C_{16}H_{16}S_2$
1,2-Bis(p-tolylthio)ethene 213
2,2-Diphenyl-1,3-dithiane 209
6,7,8,9-Tetrahydrodibenzo(b,d)(1,6)-dithiecin 214

$C_{16}H_{17}BrN_2$
2-Benzylamino-3,4-dihydroisoquinolinium bromide 255
4-Bromo-N-(4-(dimethylamino)-benzylidene)-3-methylaniline 1061

$C_{16}H_{17}BrN_2O$
2-(Benzylamino)-4'-bromo-3'-methylacetanilide 404

$C_{16}H_{17}BrN_2OS$
1-(4-Bromo-2,6-dimethylphenyl)-3-(4-(methylthio)phenyl)urea 522

$C_{16}H_{17}BrN_2O_2S$
2-(2-Bromoethyl)benzaldehyde (p-tosyl)hydrazone 1079

$C_{16}H_{17}BrN_2O_3S$
p-Toluenesulfonic acid (5-bromo-2-ethoxybenzylidene)-hydrazide 1049

$C_{16}H_{17}BrN_6O_3$
2-(5-Methyl-2-pyridylureido)acetic acid N^2-(N-(3-bromophenyl)-carbamoyl)hydrazide 928

$C_{16}H_{17}Br_2N_3$
4-(2,6-Dibromophenylazo)-N,N-diethylaniline 1081

$C_{16}H_{17}ClN_2$
1-(α-Amino-4-chlorobenzyl)-1,2,3,4-tetrahydroisoquinoline dihydrochloride 244
α-(3-Chloro-o-tolylimino)-N,N-dimethyl-p-toluidine 1061
α-(3-Chloro-p-tolylimino)-N,N-dimethyl-p-toluidine 1061
α-(4-Chloro-o-tolylimino)-N,N-dimethyl-p-toluidine 1061
α-(5-Chloro-o-tolylimino)-N,N-dimethyl-p-toluidine 1061

$C_{16}H_{17}ClN_2O$
1-Benzyl-3-(2-chlorophenyl)-1-ethylurea 572
1-Benzyl-3-(3-chlorophenyl)-1-ethylurea 571
1-(2-Chlorobenzyl)-3-(2,5-xylyl)urea 528
1-(3-Chloro-2-methylphenyl)-3-(2-ethylphenyl)urea 537
1-(3-Chloro-2-methylphenyl)-3-(4-ethylphenyl)urea 539

$C_{16}H_{17}ClN_2O$ (Cont'd)

1-(3-Chloro-4-methylphenyl)-3-(2-ethylphenyl)urea ... 538
1-(3-Chloro-4-methylphenyl)-3-(4-ethylphenyl)urea ... 540
1-(2-Chloro-6-methylphenyl)-3-phenethylurea ... 570
1-(3-Chloro-2-methylphenyl)-3-phenethylurea ... 527
1-(5-Chloro-2-methylphenyl)-3-phenethylurea ... 570
1-(3-Chloro-2-methylphenyl)-3-(3,4-xylyl)urea ... 542
1-(3-Chloro-4-methylphenyl)-3-(2,3-xylyl)urea ... 542
1-(3-Chloro-4-methylphenyl)-3-(2,4-xylyl)urea ... 601
1-(3-Chloro-4-methylphenyl)-3-(2,6-xylyl)urea ... 542
1-(4-Chloro-2-methylphenyl)-3-(2,4-xylyl)urea ... 601
1-(4-Chloro-2-methylphenyl)-3-(2,6-xylyl)urea ... 542
1-(5-Chloro-2-methylphenyl)-3-(2,4-xylyl)urea ... 601
1-(2-Chlorophenyl)-3-(2,4-dimethylbenzyl)urea ... 528
1-(2-Chlorophenyl)-3-(2,5-dimethylbenzyl)urea ... 538
1-(3-Chlorophenyl)-3-(3,4-dimethylbenzyl)urea ... 544
1-(2-Chlorophenyl)-3-(2-ethyl-6-methylphenyl)urea ... 537
1-(3-Chlorophenyl)-3-(2-ethyl-6-methylphenyl)urea ... 541
1-(4-Chlorophenyl)-3-(2-ethyl-6-methylphenyl)urea ... 539
3-(4-Chlorophenyl)-1-ethyl-1-(*o*-tolyl)urea ... 537
1-(2-Chlorophenyl)-3-isopropyl-3-phenylurea ... 535
1-(2-Chlorophenyl)-3-(2,4,5-trimethylphenyl)urea ... 538
1-(2-Chlorophenyl)-3-(2,4,6-trimethylphenyl)urea ... 538
1-(3-Chlorophenyl)-3-(2,4,6-trimethylphenyl)urea ... 546
1-(4-Chlorophenyl)-3-(2,4,6-trimethylphenyl)urea ... 540
α-(3-Chloro-*p*-toluidino)-*p*-acetotoluidide ... 430

$C_{16}H_{17}ClN_2O_2$

1-Benzyl-3-(4-chloro-2-methoxyphenyl)-1-methylurea ... 518
1-(2-Chloro-6-methylphenyl)-3-(4-(2-hydroxyethyl)phenyl)urea ... 514

$C_{16}H_{17}ClN_2O_3$

1-Benzyl-3-(5-chloro-2,4-dimethoxyphenyl)urea ... 514
1-(4-Chloro-2,5-dimethoxyphenyl)-3-(*o*-tolyl)urea ... 516
1-(5-Chloro-2-methoxyphenyl)-3-(4-(2-hydroxyethyl)phenyl)urea ... 512
1-(3-Chloro-2-methylphenyl)-3-(2,4-dimethoxyphenyl)urea ... 515
1-(3-Chloro-2-methylphenyl)-3-(2,5-dimethoxyphenyl)urea ... 515
1-(5-Chloro-2-methylphenyl)-3-(2,5-dimethoxyphenyl)urea ... 515
1-(6-Chloro-2-methylphenyl)-3-(2,5-dimethoxyphenyl)urea ... 515

$C_{16}H_{17}ClN_2O_3S$

1-(5-Chloro-2,4-dimethoxyphenyl)-3-(3-(methylthio)phenyl)urea ... 514

$C_{16}H_{17}ClN_2O_4$

1-(4-Chloro-2,5-dimethoxyphenyl)-3-(4-methoxyphenyl)urea ... 516
1-(5-Chloro-2,4-dimethoxyphenyl)-3-(4-methoxyphenyl)urea ... 515
1-(5-Chloro-2-methoxyphenyl)-3-(2,4-dimethoxyphenyl)urea ... 515

$C_{16}H_{17}ClN_2O_5S$

4-(*sec*-Butyl)-5′-chloro-2′-hydroxy-4′-nitrobenzenesulfonanilide ... 811

$C_{16}H_{17}ClN_2S$

2-Chloro-10-(2-dimethylaminoethyl)-10*H*-phenothiazine hydrochloride ... 249

$C_{16}H_{17}ClN_4O_2$

4-(4-Chloro-2-methylphenyl)-1-(2-(methylamino)benzoyl)-semicarbazide ... 527
1-(*N*-(4-Chlorophenyl)carbamoyl)-4-ethyl-1-phenylsemicarbazide ... 529

$C_{16}H_{17}ClN_4O_2S$

1-(3-Chlorophenyl)-6-(2-methoxy-5-methylphenyl)-5-thiobiurea ... 729

$C_{16}H_{17}ClN_4O_3$

2-(4-(2-Chloro-4-nitrophenylazo)-*N*-ethylanilino)-1-ethanol ... 1080

$C_{16}H_{17}ClN_4O_4$

4′-(Bis(2-hydroxyethyl)amino)-2′-chloro-4-nitroazobenzene ... 1080

$C_{16}H_{17}ClN_6O_3$

4-(3-Chlorophenyl)-1-(6-methyl-2-pyridylureidoacetyl)semicarbazide ... 928
4-(4-Chlorophenyl)-1-(5-methyl-2-pyridylureidoacetyl)semicarbazide ... 928
2-(5-Methyl-2-pyridylureido)acetic N^2-(*N*-(3-chlorophenyl)carbamoyl)-hydrazide ... 928

$C_{16}H_{17}ClO_2$

7-Chloro-4a,5,6,7,8,8a-hexahydro-4-methyl-3-phenylcoumarin ... 386

$C_{16}H_{17}Cl_2N_3$

4-(2,3-Dichlorophenylazo)-*N*,*N*-diethylaniline ... 1081
4-(2,6-Dichlorophenylazo)-*N*,*N*-diethylaniline ... 1081
4-(3,5-Dichlorophenylazo)-*N*,*N*-diethylaniline ... 1082

$C_{16}H_{17}Cl_2N_3O_2$

N,*N*-Bis-(2-hydroxyethyl)-4-(2,5-dichlorophenylazo)aniline ... 1084
1-(4-Chlorophenylcarbamoyl-methyl)-3-(dimethylcarbamoyl)-pyridinium chloride ... 947

$C_{16}H_{17}FN_2O$

3-(3-Ethylphenyl)-1-(4-fluorophenyl)-1-methylurea ... 539
3-(3-Fluorophenyl)-1-isopropyl-1-phenylurea ... 535
3-(4-Fluorophenyl)-1-isopropyl-1-phenylurea ... 536
1-(2-Fluorophenyl)-3-(2,4,5-trimethylphenyl)urea ... 536
1-(3-Fluorophenyl)-3-(2,4,6-trimethylphenyl)urea ... 546

$C_{16}H_{17}FN_2S$

1-(4-Fluorophenyl)-3-(2,4,6-trimethylphenyl)-2-thiourea ... 733

$C_{16}H_{17}FN_4O_2$

1-(4-(Dimethylamino)benzoyl)-4-(4-fluorophenyl)semicarbazide ... 527

$C_{16}H_{17}F_2N_3$

N,*N*-Diethyl-4-(2,4-difluorophenylazo)aniline ... 1081

$C_{16}H_{17}INO$

1-Ethyl-2-(4-methoxystyryl)-pyridinium iodide ... 938

$C_{16}H_{17}IN_2O$

2-(4-Acetamidostyryl)-1-methylpyridinium iodide ... 938

$C_{16}H_{17}IO$

4-Butoxy-4′-iodobiphenyl ... 173

$C_{16}H_{17}IO_4$

2-Iodo-3′,4,4′,5-tetramethoxybiphenyl ... 185

$C_{16}H_{17}N$

N-(α-Allylbenzyl)aniline ... 232
1-Amino-2-benzylindan ... 241
1-Benzyl-1,2,3,4-tetrahydroisoquinoline ... 250
N-(4-Methylbenzylidene)-3,4-dimethylaniline ... 1066
N-(4-Methylbenzylidene)-2,3-xylidine ... 1065
N-(4-Methylbenzylidene)-2,4-xylidine ... 1066
N-(4-Methylbenzylidene)-2,5-xylidine ... 1066

$C_{16}H_{17}NO$

1-Benzhydryl-3-hydroxyazetidine ... 245
N-(1-Benzyl-2-phenylethyl)-formamide ... 430
3-(Diethylamino)dibenzofuran ... 895
4-(Dimethylamino)-2′-methylbenzophenone ... 312
2′,3′-Dimethyl-2-phenylacetanilide ... 437
2′,5′-Dimethyl-2-phenylacetanilide ... 437
2,2-Diphenyl-*N*-ethylacetamide ... 430
N-Diphenylmethylpropionamide ... 438
N-(2-Ethoxybenzylidene)-*m*-toluidine ... 1046
Ethyl 2,2-diphenylacetimidate hydrochloride ... 363
2′-Ethyl-6′-methylbenzanilide ... 478
4-(4-Isopropylbenzylideneamino)-phenol ... 1038
N-(2-Methoxybenzylidene)-3,5-xylidine ... 1046
N-(3-Methoxybenzylidene)-2,3-xylidine ... 1048
N-(3-Methoxybenzylidene)-2,4-xylidine ... 1048
N-(3-Methoxybenzylidene)-2,5-xylidine ... 1048
N-(4-Methoxybenzylidene)-2,3-xylidine ... 1047
N-(4-Methoxybenzylidene)-2,5-xylidine ... 1047
N-(4-Methoxybenzylidene)-3,4-xylidine ... 1047
N-(4-Methoxybenzylidene)-3,5-xylidine ... 1047
(4-Methoxybenzyl)(4-methylbenzylidene)amine ... 1050
2-Methylbenzo-2′,3′-xylidide ... 479
2-Methylbenzo-2′,4′-xylidide ... 479
2-Methylbenzo-2′,5′-xylidide ... 479
3-Methylbenzo-2′,4′-xylidide ... 482
3-Methylbenzo-2′,5′-xylidide ... 482
4-Methylbenzo-2′,4′-xylidide ... 480
N-(4-Methylbenzylidene)-*o*-phenetidine ... 1045
N-(4-Methylbenzylidene)-*p*-phenetidine ... 1047
4′-(Phenethylamino)acetophenone ... 294
N-Phenethyl-2-phenylacetamide ... 430
2-Phenylaceto-2′,3′-xylidide ... 437
2-Phenylaceto-2′,5′-xylidide ... 437
2,2′,6′-Trimethylbenzanilide ... 479
2′,4,5′-Trimethylbenzanilide ... 480
2′,4′,6′-Trimethylbenzanilide ... 478
2′,4,6′-Trimethylbenzanilide ... 480
α-(2,4,6-Trimethylphenylimino)-*o*-cresol ... 1035
α-(2,4,6-Trimethylphenylimino)-*p*-cresol ... 1038

$C_{16}H_{17}NOS$

2-(Benzylthio)-4′-methylacetanilide ... 421
4′-(Methylthio)-3-phenylpropionanilide ... 421
4-(2-Naphthylthioacetyl)morpholine ... 746
S-(*m*-Tolyl) *N*-(2-ethylphenyl)-thiocarbamate ... 688

$C_{16}H_{17}NO_2$

(3-(Benzyloxy)-4-methoxybenzylidene)methylamine ... 1039
N-Benzyl-L-phenylalanine ... 349
2-Cyclohexyl-4-quinolinecarboxylic acid ... 993
N-(2,5-Dimethoxybenzylidene)-*m*-toluidine ... 1048
N-(2,5-Dimethoxybenzylidene)-*p*-toluidine ... 1047
N-(3,4-Dimethoxybenzylidene)-*o*-toluidine ... 1041
N-(3,4-Dimethoxybenzylidene)-*p*-toluidine ... 1041
2,5-Dimethoxy-*N*-(4-methylbenzylidene)aniline ... 1045
4-(Dimethylamino)-4′-methoxybenzophenone ... 311
2′,3′-Dimethyl-4-methoxybenzanilide ... 466
2′,6′-Dimethyl-2-methoxybenzanilide ... 464
2′,6′-Dimethyl-3-methoxybenzanilide ... 464
2′,6′-Dimethyl-4-methoxybenzanilide ... 467
α-(3,4-Dimethylphenylimino)-6-methoxy-*o*-cresol ... 1030
3-Ethyl-5-methylphenyl *N*-phenylcarbamate ... 681
α-(4-Ethylphenylimino)-2-methoxy-*o*-cresol ... 1030
2-Ethylphenyl *N*-(*o*-tolyl)carbamate ... 681
N-(2-Methoxybenzylidene)-*p*-phenetidine ... 1046
N-(4-Methoxybenzylidene)-*p*-phenetidine ... 1047
4-Methoxy-2′,6′-dimethylbenzanilide ... 467
3-((4-Methoxyphenyl)amino)-propiophenone ... 292
Methyl 3-(2-biphenylylamino)-propionate hydrochloride ... 358
4-Methyl-2′-(*N*-(methyl)carbamoyl)-benzhydryl alcohol ... 461
DL-Phenylalanine benzyl ester *p*-toluenesulfonate ... 792
L-Phenylalanine benzyl ester *p*-toluenesulfonate ... 792
Phenyl *N*-(2-ethyl-6-methylphenyl)-carbamate ... 684
o-Tolu-*p*-phenetidide ... 466
p-Tolu-*p*-phenetidide ... 466
N-Vanillylidene-2,4-xylidine ... 1030
N-Vanillylidene-2,5-xylidine ... 1030
N-Veratrylidene-*m*-toluidine ... 1042
2,6-Xylyl 2-methylcarbanilate ... 677
2,3-Xylyl *N*-(*o*-tolyl)carbamate ... 682
2,3-Xylyl *N*-(*m*-tolyl)carbamate ... 684
3,5-Xylyl *N*-(*o*-tolyl)carbamate ... 682

$C_{16}H_{17}NO_2S$

2-(Benzylthio)-4′-methoxyacetanilide ... 414
3,4-Dimethylphenyl *N*-(4-(methylthio)phenyl)carbamate ... 668
S-Diphenylmethyl-L-cysteine ... 348
3-Methyl-4-(methylthio)phenyl *N*-(2-methylphenyl)carbamate ... 667
4-(Methylthio)phenyl *N*-(2,3-dimethylphenyl)carbamate ... 668
4-(Methylthio)phenyl *N*-(3-ethylphenyl)carbamate ... 667
4-(Methylthio)phenyl *N*-(2,5-xylyl)-carbamate ... 668
1-Phenylsulfonyl-2,3-dimethylindoline (mixture of isomers) ... 786

$C_{16}H_{17}NO_3$

Benzyl 4-acetyl-3,5-dimethyl-2-pyrrolecarboxylate ... 838
3-(*N*-Benzylcarbamoyl)-5-norbornene-2-carboxylic acid ... 431
Benzyl 4-ethyl-5-formyl-3-methyl-2-pyrrolecarboxylate ... 838
Benzyl *N*-(2-hydroxy-1-phenylethyl)-carbamate ... 643
N-Benzyl-L-tyrosine ... 348
2,3-Dihydro-3-(1-(2,5-dimethylanilino)ethylidene)-6-methyl-4*H*-pyran-2,4-dione ... 385
2,3-Dihydro-3-(1-(4-ethylanilino)-ethylidene)-6-methyl-4*H*-pyran-2,4-dione ... 385
N-(2,5-Dimethoxybenzylidene)-*o*-anisidine ... 1045
2,5-Dimethoxy-*N*-(4-methoxybenzylidene)aniline ... 1045
2,5-Dimethyl-1-(4-ethoxycarbonylphenyl)-3-formylpyrrole ... 841
α-(4-Ethoxycarbonylanilino)-*p*-cresol ... 368
α-(2-Ethoxyphenylimino)-6-methoxy-*o*-cresol ... 1029
α-(4-Ethoxyphenylimino)-6-methoxy-*o*-cresol ... 1029
Ethyl 4-(2-hydroxybenzylamino)-benzoate ... 368
1,3,3a,4,7,7a-Hexahydro-2-methyl-3-phenyl-1-oxo-3a-isoindolecarboxylic acid ... 692
3′-(2-Hydroxy-3-methoxybenzylamino)-acetophenone ... 290
4′-(2-Hydroxy-3-methoxybenzylamino)-

$C_{16}H_{17}NO_3$ (Cont'd)
acetophenone ... 290
4-Methoxyphenyl N-(2,5-xylyl)carbamate ... 651
4-Methoxyphenyl N-(2,6-xylyl)carbamate ... 651
3-(N-(4-Methylphenyl)carbamoyl)-5-norbornene-2-carboxylic acid ... 434
2-Phenoxyethyl N-(p-tolyl)carbamate ... 650
3-(N-(m-Tolyl)carbamoyl)-5-norbornene-2-carboxylic acid ... 436
D-Tyrosine benzyl ester p-toluenesulfonate ... 353
L-Tyrosine benzyl ester p-toluenesulfonate ... 787
N-Veratrylidene-m-anisidine ... 1042
N-Veratrylidene-p-anisidine ... 1042
2,6-Xylyl N-(4-methoxyphenyl)carbamate ... 655
3,5-Xylyl N-(4-methoxyphenyl)carbamate ... 655

$C_{16}H_{17}NO_3S$
4-Methoxybenzyl N-(4-(methylthio)phenyl)carbamate ... 646
3-Methyl-4-(methylthio)phenyl N-(2-methoxyphenyl)carbamate ... 651
3-Methyl-4-(methylthio)phenyl N-(4-methoxyphenyl)carbamate ... 651
2-(Phenoxy)ethyl N-(4-(methylthio)phenyl)carbamate ... 649

$C_{16}H_{17}NO_3S_3$
3-Methyl-2-(methylthio)benzothiazolium p-toluenesulfonate ... 915

$C_{16}H_{17}NO_4$
Benzyl 3,5-dimethyl-4-(methoxycarbonyl)-2-pyrrolecarboxylate ... 842
2-Ethoxyphenyl N-(2-methoxyphenyl)carbamate ... 647
2-Ethoxyphenyl N-(4-methoxyphenyl)carbamate ... 647
Ethyl α-(ethoxycarbonyl)-3-indoleacrylate ... 889
3-(N-(4-Methoxyphenyl)carbamoyl)-5-norbornene-2-carboxylic acid ... 418
3,4,5-Trimethoxybenzanilide ... 463

$C_{16}H_{17}NO_4S$
2,6-Dimethoxyphenyl N-(3-(methylthio)phenyl)carbamate ... 647
2,6-Dimethoxyphenyl N-(4-(methylthio)phenyl)carbamate ... 647
4-(1,1-Dimethyl-2-nitroethyl)phenyl phenyl sulfone ... 778
N-Ethyl-N-(p-toluenesulfonyl)anthranilic acid ... 823
S-(4-Methoxyphenyl) N-(2,5-dimethoxyphenyl)thiocarbamate ... 648
4-(Methylthio)phenyl N-(2,5-dimethoxyphenyl)carbamate ... 648
3-(Phenyl(p-tosyl)amino)propionic acid ... 330

$C_{16}H_{17}NO_5$
3α-(2,3-Dimethoxyphenyl)-5,6-epoxyhexahydroisoindoline-1,3-dione ... 716
Dimethyl 2-ethoxy-3H-1-benzazepine-3,4-dicarboxylate ... 369
Dimethyl 2-(2-ethoxy-3-indolyl)maleate ... 882
3'-Hydroxy-3,4,5-trimethoxybenzanilide ... 461

$C_{16}H_{17}NO_5S$
2,3-Dimetyl-5-phenylbenzoxazolium methyl sulfate ... 910
Ethyl 4-(4-methoxyphenylsulfonamido)benzoate ... 814
4-Methoxybenzyl N-(p-tosyl)carbamate ... 812

$C_{16}H_{17}N_2O_5P$
Bis-(3-nitrophenyl)-butylphosphine oxide ... 835

$C_{16}H_{17}N_3O$
4-Aminobenzoic (α,4-dimethylbenzylidene)hydrazide ... 499
1,2-Dihydro-2-(4-dimethylaminophenyl)-4(3H)-quinazolinone ... 704
4'-(4-(Dimethylamino)phenylazo)acetophenone ... 1089
N-Methylanthranilic (2-phenylethylidene)hydrazide ... 1060

$C_{16}H_{17}N_3O_2$
2-Acetamido-4-(4-(dimethylamino)phenylimino)-2,5-cyclohexadien-1-one ... 1054
1-(4-Acetamidophenyl)-3-(p-tolyl)urea ... 569
1-(2-Carbamoylphenyl)-3-(2-ethylphenyl)urea ... 577
1-(2-Carbamoylphenyl)-3-(3-ethylphenyl)urea ... 565
1-(2-Carbamoylphenyl)-3-(2,5-xylyl)urea ... 577
1-(2-Carbamoylphenyl)-3-(2,6-xylyl)urea ... 577
4-(2-Cyanoethyl)-4-(5-methyl-2-furoyl)pimelonitrile ... 847
4-(Dimethylamino)benzoic (2-hydroxybenzylidene)hydrazide ... 496
N-(4-(Dimethylamino)benzylidene)-2-methyl-4-nitroaniline ... 1057
N-(4-(Dimethylamino)benzylidene)-2-methyl-5-nitroaniline ... 1055
N-(4-(Dimethylamino)benzylidene)-4-methyl-3-nitroaniline ... 1055
4-Hydroxybenzoic (4-dimethylaminobenzylidene)hydrazide ... 1034
N-Methylanthranilic (4-methoxybenzylidene)hydrazide ... 1044
N-Methyl-4'-(3-phenylureido)acetanilide ... 569
1-(4-Nitrophenyl)-4-phenylpiperazine ... 276
1-Phenyl-5-(2,4-xylyl)biuret ... 579
Salicylic (4-dimethylaminobenzylidene)hydrazide ... 1032

$C_{16}H_{17}N_3O_2S$
1-(4-Acetamidophenyl)-3-(4-(methylthio)phenyl)urea ... 560
1-(3,4-Dimethoxybenzylidene)-4-phenyl-3-thiosemicarbazide ... 735
1-(2-Hydroxybenzylidene)-4-(2-methoxy-5-methylphenyl)-3-thiosemicarbazide ... 733

$C_{16}H_{17}N_3O_3$
1-(3-Acetamidophenyl)-3-(4-methoxyphenyl)urea ... 552
1-(4-Acetamidophenyl)-3-(2-methoxyphenyl)urea ... 552
1-(4-Acetamidophenyl)-3-(4-methoxyphenyl)urea ... 553
4-Aminobenzoic (2,3-dimethoxybenzylidene)hydrazide ... 497
4-Aminobenzoic (2,4-dimethoxybenzylidene)hydrazide ... 497
N-Benzoyl-L-tyrosine hydrazide ... 497
1-Benzyl-1-ethyl-3-(4-nitrophenyl)urea ... 567
1-Benzyl-1-methyl-3-(4-methyl-2-nitrophenyl)urea ... 562
1-Benzyl-1-methyl-3-(4-methyl-3-nitrophenyl)urea ... 562
1-(4-Carboxyphenyl)-4-(3,4-xylyl)semicarbazide ... 576
1-(3,4-Dimethoxybenzylidene)-4-phenylsemicarbazide ... 549
N-(4-(Dimethylamino)benzylidene)-2-methoxy-4-nitroaniline ... 1043
N-(4-(Dimethylamino)benzylidene)-2-methoxy-5-nitroaniline ... 1043
4-(2,6-Dimethylphenyl)-1-(2-hydroxybenzoyl)semicarbazide ... 547
1-(2,4-Dimethylphenyl)-3-(4-methyl-3-nitrophenyl)urea ... 564
N-(4-Ethoxyphenyl)-N'-(4-nitrophenyl)acetamidine ... 1044
4-(2-Ethylphenyl)-1-(2-hydroxybenzoyl)semicarbazide ... 548
1-(2-Ethylphenyl)-3-(4-methyl-3-nitrophenyl)urea ... 563
1-(4-Ethylphenyl)-3-(4-methyl-3-nitrophenyl)urea ... 564
2-(Methylamino)benzoic vanillylidenehydrazide ... 495
N-Methylanthranilic (4-hydroxy-3-methoxybenzylidene)hydrazide ... 1031
1-(2-Methyl-4-nitrophenyl)-3-phenethylurea ... 566
1-(2-Methyl-5-nitrophenyl)-3-phenethylurea ... 562
1-(4-Methyl-3-nitrophenyl)-3-phenethylurea ... 562
1-(4-Methyl-3-nitrophenyl)-3-(2,3-xylyl)urea ... 564
1-(4-Methyl-3-nitrophenyl)-3-(2,6-xylyl)urea ... 564
2-(3-Phenylureido)ethyl N-phenylcarbamate ... 674

$C_{16}H_{17}N_3O_3S$
1,2-Diaminoisoquinolinium p-toluenesulfonate ... 997
1-(2-Methoxybenzoyl)-4-(4-(methylthio)phenyl)semicarbazide ... 551
1-(4-Methoxybenzoyl)-4-(4-(methylthio)phenyl)semicarbazide ... 551

$C_{16}H_{17}N_3O_4$
4-Butyl-2',4'-dinitrodiphenylamine ... 275
1-(2-Carbamoylphenyl)-3-(2,5-dimethoxyphenyl)urea ... 550
1,3-Diethyl-5-nitro-6-styryluracil ... 706
4-Ethoxy-3-methoxybenzaldehyde (4-nitrophenyl)hydrazone ... 1040
1-(4-(2-Hydroxyethyl)phenyl)-3-(2-methyl-5-nitrophenyl)urea ... 547
Isonicotinic acid (3,4,5-trimethoxybenzylidene)-hydrazide ... 1059
1-(2-Methoxy-4-nitrophenyl)-3-(phenethyl)urea ... 552
1-(2-Methoxy-5-nitrophenyl)-3-(phenethyl)urea ... 552
Nicotinic acid (3,4,5-trimethoxybenzylidene)-hydrazide ... 1042
Picolinic acid (3,4,5-trimethoxybenzylidene)-hydrazide ... 1042

$C_{16}H_{17}N_3O_5$
4-(2,5-Dimethoxyphenyl)-1-(4-hydroxybenzoyl)semicarbazide ... 546
1-(2,4-Dimethoxyphenyl)-3-(2-methyl-5-nitrophenyl)urea ... 549
1-(2,4-Dimethoxyphenyl)-3-(4-methyl-3-nitrophenyl)urea ... 549
1-(2,5-Dimethoxyphenyl)-3-(2-methyl-5-nitrophenyl)urea ... 550
1-(2,5-Dimethoxyphenyl)-3-(4-methyl-3-nitrophenyl)urea ... 550
1-(4-(2-Hydroxyethyl)phenyl)-3-(2-methoxy-5-nitrophenyl)urea ... 547
3,4,5-Trimethoxybenzaldehyde 2-nitrophenylhydrazone ... 1043
3,4,5-Trimethoxybenzaldehyde 4-nitrophenylhydrazone ... 1048

$C_{16}H_{17}N_3O_6$
1-(2,5-Dimethoxyphenyl)-3-(2-methoxy-4-nitrophenyl)urea ... 550
1-(2,5-Dimethoxyphenyl)-3-(2-methoxy-5-nitrophenyl)urea ... 550
1-(2,5-Dimethoxyphenyl)-3-(4-methoxy-3-nitrophenyl)urea ... 550

$C_{16}H_{17}N_5O_2$
2-(3-(5-Methyl-2-pyridyl)ureido)acetic benzylidenehydrazide ... 1070

$C_{16}H_{17}N_5O_4$
4'-(Diethylamino)-2,4-dinitroazobenzene ... 1088

$C_{16}H_{17}N_5O_5$
2-(4-(2,4-Dinitrophenylazo)-N-ethylanilino)-1-ethanol ... 1084

$C_{16}H_{17}N_5O_6$
4'-(Bis(2-hydroxyethyl)amino)-2,4-dinitroazobenzene ... 1083

$C_{16}H_{17}O_4P$
3-((4-Methoxyphenyl)phenylphosphinyl)propionic acid ... 834

$C_{16}H_{18}$
meso-2,3-Diphenylbutane ... 162
1,1-Diphenyl-2-methylpropane ... 165
1,2-Di-p-tolylethane ... 163
(+)-1-Phenyl-1-xylylethane ... 163

$C_{16}H_{18}AsN$
1,4-Diphenyl-1,2,3,4,5,6-hexahydro-1,4-azarsenine ... 1126

$C_{16}H_{18}AsNO_7$
Benzyl N-(4-arsono-3-(2-hydroxyethoxy)phenyl)carbamate ... 1126

$C_{16}H_{18}BrN_3$
4-(2-Bromophenylazo)-N,N-diethylaniline ... 1081
4-(3-Bromophenylazo)-N,N-diethylaniline ... 1081
4-(4-Bromophenylazo)-N,N-diethylaniline ... 1081
4-(4-Bromophenylazo)-N-isopropyl-N-methylaniline ... 1082

$C_{16}H_{18}BrN_5O_7$
8-Bromoadenosine 2',3',5'-triacetate ... 922

$C_{16}H_{18}BrN_5O_8$
8-Bromoguanosine 2',3',5'-triacetate ... 922

$C_{16}H_{18}ClNO_2$
N-(5-Chloro-2,4-dimethoxyphenyl)-4-methylbenzylamine ... 229
N-(3-Chloro-p-tolyl)-2,3-dimethoxybenzylamine ... 228

$C_{16}H_{18}ClNO_4$
α-(5-Chloro-2,4-dimethoxyanilino)-6-methoxy-o-cresol ... 225

$C_{16}H_{18}ClN_3$
Basic Orange 4-11(C.I. 46035) ... 1094
Basic Yellow 9 (C.I. 46040) ... 1094
4-(2-Chlorophenylazo)-N,N-diethylaniline ... 1081
4-(3-Chlorophenylazo)-N,N-diethylaniline ... 1081
4-(4-Chlorophenylazo)-N,N-diethylaniline ... 1081

$C_{16}H_{18}ClN_3O$
1-(3-Chloro-2-methylphenyl)-3-(4-(dimethylamino)phenyl)urea ... 526

$C_{16}H_{18}ClN_3O_2$
1-(5-Chloro-2-methoxyphenyl)-3-(4-(dimethylamino)phenyl)urea ... 518
3-Dimethylcarbamoyl-1-(phenylcarbamoylmethyl)pyridinium chloride ... 947

$C_{16}H_{18}ClN_3O_3$
1-(2-(3,4-Dimethoxybenzylidene)hydrazinocarbonylmethyl)pyridinium chloride ... 936

$C_{16}H_{18}ClN_3O_5$
4,5-Dihydro-1-(4-methoxyphenyl)-3-(p-tolyl)-1H-1,2,3-triazolium perchlorate ... 255

$C_{16}H_{18}ClO_3P$
Bis(3,5-dimethylphenyl) chlorophosphate ... 837

$C_{16}H_{18}Cl_2N_2$
N,N'-Bis(4-chlorobenzyl)ethylenediamine ... 242

$C_{16}H_{18}Cl_4O_4$
2-Octyl tetrachlorophthalate ... 366

$C_{16}H_{18}FN_3$
4-(Diethylamino)-3'-fluoroazobenzene ... 1082
4-(Diethylamino)-4'-fluoroazobenzene ... 1081
N,N-Diethyl-4-(2-fluorophenylazo)aniline ... 1081

$C_{16}H_{18}IN$
1-Ethyl-2-(4-methylstyryl)pyridinium iodide ... 957
1-Ethyl-4-(4-methylstyryl)pyridinium iodide ... 957

$C_{16}H_{18}INO$
2-(2-Ethoxystyryl)-1-methylpyridinium iodide ... 938
2-(4-Ethoxystyryl)-1-methylpyridinium iodide ... 938
1-Ethyl-4-(2-methoxystyryl)pyridinium iodide ... 938

$C_{16}H_{18}INO_2$
2-(2,3-Dimethoxystyryl)-1-methylpyridinium iodide ... 937
2-(2,5-Dimethoxystyryl)-1-methylpyridinium iodide ... 938

$C_{16}H_{18}INO_2$ (Cont'd)
2-(3,4-Dimethoxystyryl)-1-methylpyridinium iodide ... 937
4-(3,4-Dimethoxystyryl)-1-methylpyridinium iodide ... 937

$C_{16}H_{18}N_2$
N,N-Dimethyl-α-(o-tolylimino)-p-toluidine ... 1061
N,N-Dimethyl-α-(m-tolylimino)-p-toluidine ... 1061
N,N-Dimethyl-α-(p-tolylimino)-p-toluidine ... 1061
1,4-Diphenylpiperazine ... 253
4-Isopropylbenzaldehyde phenylhydrazone ... 235
2,2′,6,6′-Tetramethylazobenzene ... 1093

$C_{16}H_{18}N_2O$
3-Amino-2′,4′-dimethyl-p-toluanilide ... 468
2-Anilino-1-methoxy-1,2,3,4-tetrahydroisoquinoline ... 247
1-Benzyl-1-ethyl-3-phenylurea ... 571
1-Benzyl-3-(2-ethylphenyl)urea ... 572
1-Benzyl-1-methyl-3-(o-tolyl)urea ... 572
4-Benzyloxy-1-cyano-1-cyclohexaneacetonitrile ... 752
1-Benzyl-3-(2,3-xylyl)urea ... 573
1-Benzyl-3-(2,4-xylyl)urea ... 573
1-Benzyl-3-(2,5-xylyl)urea ... 573
1-Benzyl-3-(2,6-xylyl)urea ... 573
1-Benzyl-3-(3,4-xylyl)urea ... 573
1-Benzyl-3-(3,5-xylyl)urea ... 574
4-(4-tert-Butylphenylazo)phenol ... 1086
4,4′-Diethylazoxybenzene ... 1093
N-(4-(Dimethylamino)benzylidene)-p-anisidine ... 1044
1-(2,4-Dimethylbenzyl)-3-phenylurea ... 586
2,5-Dimethyl-4-(4-(dimethylamino)-phenylimino)-2,5-cyclohexadien-1-one ... 1057
3,4′-Dimethyl-4-ethoxyazobenzene ... 1087
1-(2,6-Dimethylphenyl)-3-(3-methylphenyl)urea ... 595
1-(2,6-Dimethylphenyl)-3-(4-methylphenyl)urea ... 591
3-(2,6-Dimethylphenyl)-1-methyl-1-phenylurea ... 586
1-(2,5-Dimethylphenyl)-3-(p-tolyl)urea ... 592
1-(2-Ethyl-6-methylphenyl)-3-phenylurea ... 595
1-Ethyl-1-phenyl-3-(o-tolyl)urea ... 588
1-(2-Ethylphenyl)-3-(m-tolyl)urea ... 588
1-Ethyl-3-phenyl-1-(o-tolyl)urea ... 586
1-(4-Ethylphenyl)-3-(o-tolyl)urea ... 588
1-(4-Ethylphenyl)-3-(m-tolyl)urea ... 591
1-Methyl-1-phenyl-3-(3,4-xylyl)urea ... 586
1-(N-(1-Naphthyl)carbamoyl)piperidine ... 712
3-(1-Naphthyl)-1,1-pentamethyleneurea ... 585
1-Phenethyl-3-(o-tolyl)urea ... 570
4-(Phenylazo)-2,3,5,6-tetramethylphenol ... 1093
1-Phenyl-3-(2,4,5-trimethylphenyl)urea ... 586
1-Phenyl-3-(2,4,6-trimethylphenyl)urea ... 586
α-(p-Toluidino)-p-acetotoluidide ... 430
1-(o-Tolyl)-3-(2,3-xylyl)urea ... 589
1-(o-Tolyl)-3-(2,4-xylyl)urea ... 589
1-(o-Tolyl)-3-(2,5-xylyl)urea ... 589
1-(m-Tolyl)-3-(2,4-xylyl)urea ... 601
1-(m-Tolyl)-3-(2,6-xylyl)urea ... 595
1-(p-Tolyl)-3-(2,4-xylyl)urea ... 591

$C_{16}H_{18}N_2OS$
α-(4-(Methylthio)anilino)-p-acetotoluidide ... 420
1-(4-(Methylthio)phenyl)-3-(2,3-xylyl)urea ... 561

$C_{16}H_{18}N_2O_2$
α-(p-Anisidino)-p-acetotoluidide ... 416
1-(o-Anisyl)-3-phenethylurea ... 553
1-(p-Anisyl)-3-phenethylurea ... 553
1-Benzyl-3-(2-methoxy-5-methylphenyl)urea ... 553
(2-tert-Butylphenyl)(2-nitrophenyl)amine ... 274
1-(2,6-Dimethylphenyl)-3-(4-methoxyphenyl)urea ... 559
1-(2-Ethylphenyl)-3-(2-methoxyphenyl)urea ... 555
1-(2-Ethylphenyl)-3-(4-methoxyphenyl)urea ... 559
1-(4-Ethylphenyl)-3-(2-methoxyphenyl)urea ... 555
1-(4-Methoxyphenyl)-3-(4-methylbenzyl)urea ... 559
1-(2-Methoxyphenyl)-3-(2,3-xylyl)urea ... 555
1-(2-Methoxyphenyl)-3-(2,5-xylyl)urea ... 555
1-(2-Methoxyphenyl)-3-(2,6-xylyl)urea ... 555
1-(2-Methoxyphenyl)-3-(3,4-xylyl)urea ... 555
1-(3-Methoxyphenyl)-3-(2,6-xylyl)urea ... 560
1-(4-Methoxyphenyl)-3-(2,3-xylyl)urea ... 559
1-(4-Methoxyphenyl)-3-(2,5-xylyl)urea ... 592
1-(4-Methoxyphenyl)-3-(3,4-xylyl)urea ... 560
4-Phenoxybutyric phenylhydrazide ... 498
4-Pyridyl N-(2,6-diethylphenyl)carbamate ... 964
8-Quinolyl N-cyclohexylcarbamate ... 991

$C_{16}H_{18}N_2O_2S$
Mesitylenesulfonic benzylidenehydrazide ... 1071
1-(2-Methoxy-5-methylphenyl)-3-(3-(methylthio)phenyl)urea ... 551
1-(2-Methoxy-5-methylphenyl)-3-(4-(methylthio)phenyl)urea ... 551
Propiophenone (p-tosyl)hydrazone ... 821
3-(4-Pyridyl)propyl N-(4-(methylthio)phenyl)carbamate ... 941

$C_{16}H_{18}N_2O_3$
4,4′-Azoxydiphenetole ... 1087
1-Benzyl-3-(2,5-dimethoxyphenyl)urea ... 553
2,6-Dimethoxy-4-(4-(dimethylamino)phenylimino)-2,5-cyclohexadien-1-one ... 1041
1-(2,4-Dimethoxyphenyl)-3-(o-tolyl)urea ... 549
1-(2-Ethoxyphenyl)-3-(2-methoxyphenyl)urea ... 555
1-(4-Ethoxyphenyl)-3-(2-methoxyphenyl)urea ... 551
2′,4′,5′-Trihydroxybutyrophenone phenylhydrazone ... 234

$C_{16}H_{18}N_2O_3S$
N^4-Acetyl-N^1-benzyl-N^1-methylsulfanilamide ... 825
N^4-Acetyl-N^1-(4-ethylphenyl)sulfanilamide ... 825
N^4-Acetyl-N^1-(2,3-xylyl)sulfanilamide ... 825
N^4-Acetyl-N^1-(2,5-xylyl)sulfanilamide ... 825
N^4-Acetyl-N^1-(2,6-xylyl)sulfanilamide ... 825
1-(2,5-Dimethoxyphenyl)-3-(4-(methylthio)phenyl)urea ... 550
p-Toluenesulfonic acid (4-ethoxybenzylidene)hydrazide ... 1048

$C_{16}H_{18}N_2O_4$
1-(m-Anisyl)-3-(2,5-dimethoxyphenyl)urea ... 557
1,3-Diacetyl-5-ethyl-5-phenylhexahydropyrimidine-4,6-dione ... 708
1-(2,4-Dimethoxyphenyl)-3-(2-methoxyphenyl)urea ... 549
1-(2,4-Dimethoxyphenyl)-3-(4-methoxyphenyl)urea ... 549
2′-Hydroxy-4′-pivalamido-2-furanilide ... 845
N,N′-(1,5-Naphthalenediyl)bis(ethyl carbamate) ... 671
N-(5-Nitro-o-tolyl)veratrylamine ... 266
2-(3-Pyridyl)ethyl N-(2,4-dimethoxyphenyl)carbamate ... 936

$C_{16}H_{18}N_2O_4S$
3-Cyano-1-(2-methoxyethyl)pyridinium p-toluenesulfonate ... 938
5′-Nitro-2,2′,4,6-tetramethylbenzenesulfonanilide ... 818

$C_{16}H_{18}N_2O_5$
5-Norbornen-2-ylmethyl N-(2-methoxy-4-nitrophenyl)carbamate ... 652

$C_{16}H_{18}N_2O_5S$
4-(sec-Butyl)-2′-hydroxy-4′-nitrobenzenesulfonanilide ... 811
3-(4-Nitroanilino)propyl p-toluenesulfonate ... 798

$C_{16}H_{18}N_2O_6S$
1-(Phenylsulfonyl)-2-(3,4,5-trimethoxybenzoyl)hydrazine ... 814

$C_{16}H_{18}N_2S$
1-Benzyl-1-ethyl-3-phenyl-2-thiourea ... 748
N,N-Dimethyl-α-(4-(methylthio)-phenylimino)-p-toluidine ... 1051

$C_{16}H_{18}N_4$
2,3-Butanedione bis(phenylhydrazone) ... 1068
4-(Dimethylamino)-2′,5′-dimethyl-4′-nitroazobenzene ... 1090
1-(α-Phenethylbenzylideneamino)-guanidine nitrate ... 1072

$C_{16}H_{18}N_4O$
Anthranilic (4-(dimethylamino)-benzylidene)hydrazide ... 499
1-(4-(Dimethylamino)benzylidene)-4-phenylsemicarbazide ... 570

$C_{16}H_{18}N_4O_2$
N,N-Diethyl-4-(4-nitrophenylazo)-aniline ... 1089
4,4′-Dimethoxybenzil dihydrazone ... 234
1-(4-(Dimethylamino)benzoyl)-4-phenylsemicarbazide ... 570
4′-(Dimethylamino)-2′,3-dimethyl-4-nitroazobenzene ... 1089
4-(Dimethylamino)-2,6-dimethyl-4′-nitroazobenzene ... 1089
1-(Diphenylcarbamoyl)-4-ethylsemicarbazide ... 578
Ethyl (diphenylamidino)carbazate ... 726
4-(2-Methylphenyl)-1-(2-methylphenylcarbamoyl)-semicarbazide ... 578

$C_{16}H_{18}N_4O_2S$
4-(4-Methoxyphenyl)-1-(2-(methylamino)benzoyl)-3-thiosemicarbazide ... 736

$C_{16}H_{18}N_4O_3$
3-Amino-4-cyano-2-hexyl-7-nitro-1(2H)-isoquinolinone ... 763
1-(4-(Dimethylamino)phenyl)-3-(2-methyl-5-nitrophenyl)urea ... 562
1-(4-(Dimethylamino)phenyl)-3-(4-methyl-2-nitrophenyl)urea ... 561
2-(3-Furfurylureido)acetic (2-phenylethylidene)hydrazide ... 1053

$C_{16}H_{18}N_4O_4$
2-Adamantanone (2,4-dinitrophenyl)hydrazone ... 274
Adipic bis(furfurylidenehydrazide) ... 1068
4′-(Bis(2-hydroxyethyl)amino)-4-nitroazobenzene ... 1083
1,6-Bis(2-methoxyphenyl)biurea ... 551
N,N′-Dimethyl-N,N′-bis(2-nitrophenyl)ethylenediamine ... 274
N,N′-Dimethyl-N,N′-bis(4-nitrophenyl)ethylenediamine ... 274

$C_{16}H_{18}N_4O_5$
Carbobenzyloxyglycyl-L-histidine hydrate ... 868
Carbobenzyloxy-L-histidylglycine ... 866
2-(Furfurylureido)acetic acid N^2-(4-hydroxy-3-methoxybenzylidene)-hydrazide ... 1030
2-(3-Furfurylureido)acetic (3-hydroxy-4-methoxybenzylidene)-hydrazide ... 1031
N′-(2-Hydroxyethyl)-N,N′-bis(4-nitrophenyl)ethylenediamine ... 264

$C_{16}H_{18}N_4O_8$
2′,3′,5′-Triacetylinosine ... 920

$C_{16}H_{18}N_6$
1,4-Bis-(2-pyridylmethyleneamino)-piperazine ... 1053

$C_{16}H_{18}N_6O_2S$
1-(5-Methyl-2-pyridylureidoacetyl)-4-phenyl-3-thiosemicarbazide ... 959

$C_{16}H_{18}N_6O_3$
4,4-Diphenyl-1-(N-(hydrazinocarbonylmethyl)-carbamoyl)semicarbazide ... 574

$C_{16}H_{18}N_6O_7$
1-(Furfurylureidoacetyl)-4-(2-methoxy-5-nitrophenyl)-semicarbazide ... 846

$C_{16}H_{18}N_{10}$
1,4-Bis(azidomethyl)-1,4-bis(p-tolyl)-2-tetrazene ... 769

$C_{16}H_{18}O$
4-(tert-Butyl)-phenyl phenyl ether ... 183
1,1-Dibenzylethanol ... 207
2,4-Di-(2-cyclopentenyl)-phenol ... 191
1,4a-Dimethyl-4,4a,9,10-tetrahydro-2(3H)-phenanthrenone ... 285
1,3-Diphenyl-1-butanol ... 204
Diphenylmethyl propyl ether ... 186
α-Methylbenzyl ether ... 186
6-(α-Methylbenzyl)-2,4-xylenol ... 196

$C_{16}H_{18}OS$
2-(4-Methoxyphenyl)-1-methylethyl phenyl sulfide ... 211
4-(Phenylthio)butyl phenyl ether ... 211

$C_{16}H_{18}OSi$
9-(Trimethylsilyl)xanthene ... 1111

$C_{16}H_{18}O_2$
1-(2-Benzylphenoxy)-2-propanol ... 202
2,2-Bis-(4-hyxroxyphenyl)-butane ... 196
1,2-Bis(o-tolyloxy)ethane ... 179
1,2-Bis(m-tolyloxy)ethane ... 179
2,2′-Diethoxybiphenyl ... 187
4,4′-Diethoxybiphenyl ... 181
5,5-Dimethyl-3-(4-hydroxystyryl)-2-cyclohexen-1-one ... 282
1,4-Diphenoxybutane ... 179
1,4-Diphenyl-2,3-butanediol ... 200
2,3-Diphenyl-2,3-butanediol ... 203
1,1-Diphenyl-2-methyl-1,2-propanediol ... 199
1,2-Di-p-tolyloxyethane ... 179
3,3′-Ethylenebis(anisole) ... 178
Ethylene glycol dibenzyl ether ... 186
4-Hydroxy-2-phenyl-2,3,5,6-tetramethyl-3,5-cyclohexadien-1-one ... 316
5-Phenyl-2,3,5,6-tetramethyl-2-cyclohexene-1,4-dione ... 308
3,3′,5,5′-Tetramethyl-4,4′-bisphenol ... 194

$C_{16}H_{18}O_2S$
2,4-Diethylphenyl phenyl sulfone ... 784
2,4-Dimethyl-6-(3-methylbenzyl)-benzenesulfinic acid, sodium salt ... 787
Mesityl o-tolyl sulfone ... 785
Mesityl m-tolyl sulfone ... 786
Mesityl p-tolyl sulfone ... 786
1-Methyl-1-phenylethyl o-tolyl sulfone ... 784

$C_{16}H_{18}O_2S_2$
α,α-Dimethyl-4-(phenylsulfonyl)-benzyl methyl sulfide ... 776

$C_{16}H_{18}O_2Te_2$
Bis(4-ethoxyphenyl) ditelluride ... 211

$C_{16}H_{18}O_3$
1-(4-(Benzyloxy)-phenoxy)-2-propanol ... 201
2-(4,4-Dimethyl-2-hydroxy-6-oxo-1-cyclohexen-1-yl)acetophenone ... 307
2,4,6-Trimethoxydiphenylmethane ... 182

$C_{16}H_{18}O_3S$
3-Phenylpropyl p-toluenesulfonate ... 800
p-Tolyl mesitylsulfonate ... 801

$C_{16}H_{18}O_4$
Butyl 3-cinnamoylpyruvate ... 365
Diethyl 2-cinnamylidenemalonate ... 365
Diethyl 2-phenyl-2-(2-propynyl)-malonate ... 1018
6,6′-Dimethoxy-4,4′-dimethyl-2,2′-biphenol ... 193
Ethyl 4,4-dimethyl-1-oxo-1,2,3,4-tetrahydro-2-naphthaleneglyoxylate ... 363

$C_{16}H_{18}O_4$ (Cont'd)
2,2′-(Ethylenedioxy)dianisole 179
4,4′-(Ethylenedioxy)dianisole 179
mono-1-Methyl-2-norbornyl phthalate 375
3,3′,4,4′-Tetramethoxybiphenyl 178

$C_{16}H_{18}O_4S$
2-(Benzyloxyethyl) *p*-toluenesulfonate 797

$C_{16}H_{18}O_4S_2$
2,3-Dimethoxyphenyl disulfide 210

$C_{16}H_{18}O_6$
Diethyl 4-acetoxybenzylidenemalonate 379
Hydrocoerulignone 194
2-Naphthyl-β-D-glucopyranoside 200
Terephthaloylbis-(ethyl acetate) 364

$C_{16}H_{18}O_{10}$
Tetramethyl 3,7-dioxobicyclo(3.3.0)-octane-2,4,6,8-tetracarboxylate 94

$C_{16}H_{18}S$
1-Phenylethyl sulfide 212

$C_{16}H_{18}S_2$
4,4′-Bis(ethylthio)biphenyl 213

$C_{16}H_{18}Si$
(9-Fluorenyl)trimethylsilane 1123

$C_{16}H_{19}BrN_2$
N-(4-Bromo-3-methylphenyl)-4-(dimethylamino)benzylamine 224

$C_{16}H_{19}BrN_2O_2$
1-(4-Bromophenyl)-3-(cyclohexylamino)succinimide 715

$C_{16}H_{19}BrN_2O_2S$
1-Benzyl-3-(pyrrolidinylsulfonyl)-pyridinium bromide 955

$C_{16}H_{19}ClIN$
12-Chloro-6,7,8,9,10,11-hexahydro-5-methylcycloocta(b)quinolinium iodide 991

$C_{16}H_{19}ClN_2$
*N*α-(3-Chloro-*o*-tolyl)-*N*[4],*N*[4]-dimethyl-α,4-toluenediamine 224
*N*α-(3-Chloro-*p*-tolyl)-*N*[4],*N*[4]-dimethyl-α,4-toluenediamine 224
*N*α-(4-Chloro-*o*-tolyl)-*N*[4],*N*[4]-dimethyl-α,4-toluenediamine 224

$C_{16}H_{19}ClPb$
tert-Butylchlorodiphenyllead 1135

$C_{16}H_{19}IN_2$
(4-(Benzylideneamino)phenyl)-trimethylammonium iodide 1057

$C_{16}H_{19}IN_2O$
2-Propylharmine iodide 918

$C_{16}H_{19}N$
N-Benzyl-*N*-ethyl-*m*-toluidine 233
Bis(α-methylbenzyl)amine 243
2-(α-Butylbenzyl)pyridine 953
N-(2,4,6-Trimethylbenzyl)aniline 242

$C_{16}H_{19}NO$
N,*N*-Dibenzylethanolamine 239
2-Dimethylamino-1-(2-naphthyl)-1-butanone 295
α-(4-(Dimethylamino)phenyl)-phenethyl alcohol 226
1,1-Diphenyl-3-(methylamino)-1-propanol hydrochloride dihydrate 237
N-Ethyl-*N*-(2-phenoxyethyl)aniline hydrochloride 229
3-Isobutyl-2-oxo-3-phenyl-1-cyclopentanecarbonitrile 757
3-Isopropyl-2-oxo-3-phenyl-1-cyclohexanecarbonitrile 757
N-(4-Methoxybenzyl)-phenethylamine hydrochloride 240
α-(4-Methoxyphenyl)-*N*-methylphenethylamine 240
N-Methyl-*N*-(2-phenoxyethyl)-benzylamine 240
1-Piperidinomethyl-2-naphthol 195
N-(2-(*p*-Tolyloxy)-ethyl)-*p*-toluidine 229
α-(2,4,6-Trimethylanilino)-*o*-cresol 226
N-(2,4-Xylyl)-4-methoxybenzylamine 229

$C_{16}H_{19}NOS$
Benzo(b)thiophene-3-yl 2-piperidinoethyl ketone 898

$C_{16}H_{19}NO_2$
Benzyl 3,5-dimethyl-4-ethyl-2-pyrrolecarboxylate 838
Bis-(2-phenoxyethyl)-amine 240
2,3-Dimethoxy-*N*-(*p*-tolyl)-benzylamine 229
3,4-Dimethoxy-*N*-(*p*-tolyl)-benzylamine 228
α-(3,4-Dimethylanilino)-6-methoxy-*o*-cresol 225
α-(4-Ethylanilino)-6-methoxy-*o*-cresol 225
Ethyl 4-benzyl-3,5-dimethyl-2-pyrrolecarboxylate 842
N-(*p*-Methoxybenzyl)-*p*-phenetidine 229
5-Norbornen-2-ylmethyl *N*-(*m*-tolyl)-carbamate 678
5-Norbornen-2-ylmethyl *N*-(*p*-tolyl)-carbamate 678
Pentyl *N*-(1-naphthyl)carbamate 687
N-(*m*-Tolyl)veratrylamine 228

$C_{16}H_{19}NO_2S$
2,3-Dimethoxy-*N*-(4-(methylthio)-phenyl)-benzylamine 228
N,*N*-Dimethyl-2-(phenylsulfonylmethyl)-benzylamine 778
5-Norbornen-2-ylmethyl *N*-(4-(methylthio)phenyl)carbamate 668
2,4,6,*N*-Tetramethylbenzene-sulfonanilide 828

$C_{16}H_{19}NO_3$
Benzyl 3,5-dimethyl-4-(2-hydroxyethyl)-2-pyrrolecarboxylate 838
2,3-Dimethoxy-*N*-(4-methoxyphenyl)benzylamine 228
3,4-Dimethoxy-*N*-(4-methoxyphenyl)benzylamine 228
N-(1,1-Dimethyl-2-hydroxyethyl)-2-methoxy-1-
N-(2,3-Dimethylphenyl)-1,2,3,6-tetrahydrophthalamic acid 420
N-(2,4-Dimethylphenyl)-1,2,3,6-tetrahydrophthalamic acid 420
N-(2,6-Dimethylphenyl)-1,2,3,6-tetrahydrophthalamic acid 420
N-(3,5-Dimethylphenyl)-1,2,3,6-tetrahydrophthalamic acid 436
3′,4′-Dimethyl-1,2,3,6-tetrahydrophthalanilic acid 435
α-(2-Ethoxyanilino)-6-methoxy-*o*-cresol 225
α-(4-Ethoxyanilino)-6-methoxy-*o*-cresol 225
Ethyl 4-(*tert*-butyl)-β-cyano-α-hydroxycinnamate 761
Hexyl β-cyano-α-hydroxycinnamate 761

$C_{16}H_{19}NO_3S_2$
3-Butyl-5-(3,4-dimethoxybenzylidene)rhodanine 735

$C_{16}H_{19}NO_4$
cis-*N*-(4-Acetylphenyl)-hexahydrophthalamic acid 432
N-(4-Ethoxyphenyl)-1,2,3,6-tetrahydrophthalamic acid 418

$C_{16}H_{19}NO_4S$
Diethyl 2-(2-benzothiazolyl)-2-ethylmalonate 916

$C_{16}H_{19}NO_5$
Diethyl 2-(2-benzoxazolyl)-2-ethylmalonate 910
Methyl 2-((2-carboxycyclohexyl)-carboxamido)benzoate 433

$C_{16}H_{19}NO_5S$
Methyl 3-(2-(3,4,5-trimethoxyphenyl)ethyl)-4-isothiazolecarboxylate 874

$C_{16}H_{19}N_3$
4-Dimethylamino-4′-ethylazobenzene 1090
2,3′-Dimethyl-4-(dimethylamino)-azobenzene 1090
3′,4′-Dimethyl-4-(dimethylamino)-azobenzene 1090
N-Ethyl-*N*-methyl-4-(*p*-tolylazo)-aniline 1090

$C_{16}H_{19}N_3O$
1-Cyclohexyl-3-(4-quinolyl)urea 989
4-(Diethylamino)-2-hydroxyazobenzene 1084
4-(Dimethylamino)-3′-ethoxyazobenzene 1086
4-(Dimethylamino)-4′-ethoxyazobenzene 1086
4-(Dimethylamino)-4′-methoxyazobenzene 1087
4-(2,4-Dimethylbenzyl)-1-phenylsemicarbazide 578
1-(2-Ethyl-6-methylphenyl)-3-(3-pyridylmethyl)urea 950
1,2,5,6,7,8-Hexahydro-6-isopropyl-4-phenylpyrido(4,3-d)pyrimidin-2-one 705
1-(2-(*N*-Methylanilino)ethyl)-3-phenylurea 562
1-(2-(2-Pyridyl)ethyl)-3-(3,4-xylyl)urea 950

$C_{16}H_{19}N_3OS$
1-(4-(Dimethylamino)phenyl)-3-(3-(methylthio)phenyl)urea 560
1-(4-(Dimethylamino)phenyl)-3-(4-(methylthio)phenyl)urea 560

$C_{16}H_{19}N_3O_2$
N-(2-(*N*-(2-Cyanoethyl)-3-methylanilino)ethyl)succinimide 754
N-(2-(*N*-(2-Cyanoethyl)-3-methylanilino)ethyl)succinimide 754
4,5-Dihydro-2-(2-hydroxybutyl)-6-(3-indolyl)-3(2*H*)-pyridazinone 881
4-(Dimethylamino)-*N*-(4-methyl-3-nitrophenyl)benzylamine 275
Ethyl 2-cyano-3-(4-(*N*-(2-cyanoethyl)-*N*-methylamino)phenyl)propionate 758

$C_{16}H_{19}N_3O_3S$
4-(4-(Diethylamino)-phenylazo)-benzenesulfonic acid, sodium salt 1090

$C_{16}H_{19}N_3O_4S$
4′-Amino-*N*[4]-butyryl-2′-hydroxy-sulfanilanilide 811
4′-Amino-*N*-isobutyryl-2′-hydroxysulfanilanilide 811

$C_{16}H_{19}N_3O_5$
Ethyl 1-(2-(methoxycarbonyl)-phenylcarbamoyl)-2-imidazoline-2-acetate 726

$C_{16}H_{19}N_3O_5S$
3-(4-Diethylamino-2-hydroxyphenylazo)-4-hydroxybenzenesulfonic acid 1083

$C_{16}H_{19}N_5$
1,5-Di-*p*-tolyl-3-ethylpentaazadiene 1078

$C_{16}H_{19}N_5O_4S$
2-(Furfurylureido)acetic acid *N*[2]-(*N*-(4-(methylthio)phenyl)carbamoyl)-hydrazide 850

$C_{16}H_{19}N_5O_6$
7-(2,3-Dihydroxypropyl)-theophylline nicotinate 921

$C_{16}H_{19}N_5S_2$
1-Anilino-6-(3,4-xylyl)-2,5-dithiobiurea 750

$C_{16}H_{19}OP$
Butyl diphenylphosphinite 833

$C_{16}H_{19}O_2P$
Ethyl phenyl(3,5-xylyl)phosphinate 836

$C_{16}H_{20}$
2,6-Diisopropylnaphthalene 166
1-Phenyladamantane 163

$C_{16}H_{20}BrNO$
1-Benzyloxy-4-(*tert*-butyl)-pyridinium bromide 940

$C_{16}H_{20}BrN_2Tl$
Bis(2-(dimethylamino)phenyl)-bromothallium(III) 1142

$C_{16}H_{20}Br_4$
6,7-Bis(dibromomethyl)-1,1,4,4-tetramethyl-1,2,3,4-tetrahydronaphthalene 172

$C_{16}H_{20}ClNO_3$
2-(*N*-(3-Chlorobenzyl)-*N*-methylcarbamoyl)-cyclohexanecarboxylic acid 407

$C_{16}H_{20}ClNO_4$
Cyclohexyl lactate 3′-chlorocarbanilate 665

$C_{16}H_{20}ClNO_5$
4′-Chloro-2′,5′-dimethoxyhexa-hydrophthalanilic acid 399

$C_{16}H_{20}ClN_3$
Bindschedlers Green, indicator grade 1057

$C_{16}H_{20}Cl_4O_2$
1,2:7,8-Bis(dichloromethano)-1,8-dimethoxydodecahydro-biphenylene 17

$C_{16}H_{20}FeN_2O_2$
tert-Butyl 3-(ferrocenylmethylene)-carbazate 1142

$C_{16}H_{20}INO_2S$
N-(2-(Benzo(b)thiophene-3-ylcarbonyl)ethyl)-*N*-methylmorpholinium iodide 897

$C_{16}H_{20}N_2$
1,2-Dibenzyl-1,2-dimethylhydrazine 235
N,*N*′-Dibenzylethylenediamine dihydrochloride 242
1,2-Diethyl-1,2-diphenylhydrazine 235
N[4],*N*[4]-Dimethyl-*N*α-(*o*-tolyl)-α,4-toluenediamine 232
2-(α-Ethylbenzylaminomethyl)-6-methylpyridine 943
2-(2-(4-Isopropylanilino)ethyl)-pyridine 946
4-Phenyl-1,2,3,6-tetrahydro-1-pyridinevaleronitrile hydrochloride 758
N-(*p*-Tolyl)-4-dimethylaminobenzylamine 232

$C_{16}H_{20}N_2O$
2,2′-Bis(6-methyl-2-pyridyl)-*tert*-butanol 934
1-Butyl-1-methyl-3-(1-naphthyl)urea 603
1-*tert*-Butyl-3-(1-naphthylmethyl)urea 603
N[4],*N*[4]-Dimethyl-*N*α-(*p*-methoxyphenyl)-α,4-toluenediamine 228
N,*N*-Dipropyl-3-quinolinecarboxamide hydrochloride 993
N-(2-Naphthyl)-L-leucinamide hydrochloride 487
4,4′-Oxybis(*N*,*N*-dimethylaniline) 229

$C_{16}H_{20}N_2OS_2$
3-Butyl-5-(4-(dimethylamino)-benzylidene)rhodanine 742

$C_{16}H_{20}N_2O_2$
N,*N*′-Bis(4-methoxyphenyl)-ethylenediamine 228
1-(2,6-Diethylphenyl)-3-furfurylurea 852
Ethyl α-cyano-4-(diethylamino)-cinnamate 765
4-Hexanoyl-3-methyl-1-phenyl-2-pyrazolin-2-one 710
1-Isobutyl-1,2,3,4-tetrahydronorharman-3-carboxylic acid 919
3,3′-(Tetramethylenedioxy)dianiline 227
4,4′-(Tetramethylenedioxy)dianiline 228
L-Valine-4-methoxy-2-naphthylamide 486

$C_{16}H_{20}N_2O_2S$
L-Methionine-4-methoxy-2-naphthylamide 486

$C_{16}H_{20}N_2O_2S_2$
5-(4-(Dimethylamino)benzylidene)-3-(3-methoxypropyl)rhodanine 736

$C_{16}H_{20}N_2O_3$
1-(4-Butoxyphenyl)-3-furfurylurea 846

$C_{16}H_{20}N_2O_3S$
4′-Amino-4-(*tert*-butyl)-2′-hydroxybenzenesulfonanilide 811

$C_{16}H_{20}N_2O_4$
N-(*tert*-Butoxycarbonyl)-L-tryptophan 886
N,*N*′-(2,5-Dimethyl-1,4-phenylene)-bis(acetoacetamide) 443

$C_{16}H_{20}N_2O_5$
Carbobenzyloxyprolylglycine methyl ester 700

$C_{16}H_{20}N_2O_5S$
d-(-)-Benzylpenicilloic acid hydrate 424

$C_{16}H_{20}N_2O_6S$
(+)-(Phenoxymethyl)penicilloic acid hydrate 415

$C_{16}H_{20}N_2S$
N[4],*N*[4]-Dimethyl-*N*α-(4-(methylthio)-phenyl)-α,4-toluenediamine 230

$C_{16}H_{20}N_2S_2$
4-(Dimethylamino)phenyl disulfide ... 230
$C_{16}H_{20}N_4O_2$
4-Amino-4′-(bis(2-hydroxyethyl)amino)azobenzene ... 1083
3,3′-(1,5-Naphthalenediyl)bis(1,1-dimethylurea) ... 569
$C_{16}H_{20}N_4O_4$
Phenylalanyl-L-alanylglycine anhydride ... 701
L-Phenylalanyl-D-alanylglycine anhydride ... 701
$C_{16}H_{20}N_4O_5$
L-Phenylalanyl-L-serylglycine anhydride hemihydrate ... 696
$C_{16}H_{20}O$
4-(4-Cyclohexylphenyl)-3-butyn-1-ol .. 1016
DL-1,8a-Dimethyl-2,3,4,4a,4b,5,8,8a-octahydro-2-phenanthrenone ... 54
8-Phenyltricyclo[5.2.1.0^{2,6}]-decan-8-ol ... 203
$C_{16}H_{20}OSi$
(4-(Benzyloxy)phenyl)-trimethylsilane ... 1105
(Diphenylmethoxy)trimethylsilane ... 1113
2-(Diphenylmethylsilyl)-2-propanol ... 1105
Di(*o*-tolyl)ethoxysilane ... 1122
4-(Trimethylsilyl)benzhydrol ... 1105
$C_{16}H_{20}O_2$
5-Norbornen-2-ylmethyl 5-norbornene-2-carboxylate ... 81
$C_{16}H_{20}O_2S$
2-(Butylthiomethylene)-6-methoxy-1-tetralone ... 291
$C_{16}H_{20}O_2Si$
Diethoxydiphenylsilane ... 1113
$C_{16}H_{20}O_3$
6-(4-*tert*-Butylphenyl)-4-oxo-5-hexenoic acid ... 344
2,7-Dihydroxy-1,2,4a,8a-tetrahydro-2,4,6,8-tetra-methyl-1,4-ethenonaphthalen-3(4*H*)-one ... 52
2,4-Dimethoxy-4b,5,6,8,8a-hexahydro-9(10*H*)-phenanthrone ... 284
Ethyl 1-oxo-1,2,3,4-tetrahydro-2,4,4-trimethyl-2-naphthalenecarboxylate ... 364
$C_{16}H_{20}O_4$
Ethyl 4,4-dimethyl-6-methoxy-1-oxo-1,2,3,4-tetrahydro-2-naphthalenecarboxylate ... 363
Ethyl 4,4-dimethyl-7-methoxy-1-oxo-1,2,3,4-tetrahydro-2-naphthalenecarboxylate ... 363
Methyl 3-(5-methoxy-1-methyl-2-oxo-1,2,3,4-tetrahydro-1-naphthyl)-propionate ... 355
$C_{16}H_{20}O_5$
Diethyl 3-oxo-2-phenyladipate ... 364
$C_{16}H_{20}O_6$
1-*tert*-Butyl-2,4,5-triacetoxybenzene. 378
Diethyl 2-(3,4-dimethoxy-benzylidene)malonate ... 364
Dimethyl 3-(3-methoxybenzoyl)-adipate ... 355
$C_{16}H_{20}O_8$
Ethyl (2,4-bis(ethoxycarbonyl)-3,5-dihydroxy)phenylacetate ... 367
$C_{16}H_{20}Pb$
Diethyldiphenyllead ... 1135
$C_{16}H_{20}S_4$
Δ2,2′-Bi(hexahydro-4,7-methano-1,3-benzodithiole) ... 27
4,4′,5,5′-Bis(pentamethylene)-tetrathiafulvalene ... 880
$C_{16}H_{20}Si$
Benzyl(3-phenylpropyl)silane ... 1113
Butyldiphenylsilane ... 1113
Dibenzyldimethylsilane ... 1113
$C_{16}H_{21}ClN_2OS$
2-Butyl-3-(4-chlorophenyl)-3-hydroxy-2,3,6,7-tetrahydro-5*H*-thiazolo(3,2-a)pyrimidine ... 245
2-(2-Chloroethyl)-3-hydroxy-2,3,6,7-tetrahydro-3-(3,4-xylyl)-5*H*-thiazolo(3,2-a)pyrimidine hydro-bromide ... 244
$C_{16}H_{21}ClN_2O_3$
Ethyl 4-(4-chlorophenyl)-2-(1,2,3,4,5,6-hexahydro-2-pyrimidinyl)-4-oxobutyrate ... 353
$C_{16}H_{21}ClO_2$
3-(3-Chloro-4-propoxyphenyl)-4,4-dimethyl-1-pentyn-3-ol ... 1015
$C_{16}H_{21}F_3N_2O_2$
1-Cyclohexyl-1-(2-hydroxyethyl)-3-(2-(trifluoromethyl)phenyl)urea ... 507
$C_{16}H_{21}N$
4-(1-Adamantyl)aniline hydrochloride ... 219
6-Cyano-1,1,4,4,7-pentamethyl-1,2,3,4-tetrahydronaphthalene ... 768
$C_{16}H_{21}NO$
3-(*N*-Benzyl-*N*-methylaminomethyl)-2-norbornanone ... 285
N-Cyclohexyl-*N*-methylcinnamamide ... 482
5,5-Dimethyl-3-(2,4-dimethylanilino)-2-cyclohexen-1-one ... 284
5,5-Dimethyl-3-(phenethylamino)-2-cyclohexen-1-one ... 284
$C_{16}H_{21}NOS_2$
3a,8b-Dimethyl-2,2-ethylenedithio-hexahydro-6-methoxycyclopent(b)-indole ... 247
$C_{16}H_{21}NO_2$
Hexyl 3-indoleacetate ... 890
4-(4-Hydroxy-4-phenylpiperidino)-3-penten-2-one ... 282
1-(Isopropylamino)-3-(1-naphthyloxy)-2-propanol hydrochloride ... 237
$C_{16}H_{21}NO_3$
α-(1-Benzyl-4-hydroxy-4-piperidyl)-γ-butyrolactone ... 382
cis-2′,4′-Dimethylhexahydro-phthalanilic acid ... 435
cis-2′,5′-Dimethylhexahydro-phthalanilic acid ... 420
cis-2′,6′-Dimethylhexahydro-phthalanilic acid ... 435
cis-3′,4′-Dimethylhexahydro-phthalanilic acid ... 435
cis-3′,5′-Dimethylhexahydro-phthalanilic acid ... 435
cis-2′-Ethylhexahydro-phthalanilic acid ... 436
cis-4′-Ethylhexahydro-phthalanilic acid ... 435
cis-Hexahydro-*N*-(2-phenethyl)-phthalamic acid ... 429
Hexahydro-*N*-(2,3-xylyl)-phthalamic acid ... 435
3-Methylcyclohexyl *N*-(3-acetylphenyl)carbamate ... 675
$C_{16}H_{21}NO_3S$
Phenyltrimethylammonium *p*-toluenesulfonate ... 798
$C_{16}H_{21}NO_4$
Convolvine ... 369
Cyclohexyl lactate carbanilate ... 665
Diethyl (2,4-xylidinomethylene)-malonate ... 373
2-(2-Ethoxyphenylcarbamoyl)-cyclohexanecarboxylic acid ... 416
cis-Hexahydro-*N*-(4-ethoxyphenyl)-phthalamic acid ... 418
cis-Hexahydro-*N*-(2-methoxy-5-methylphenyl)phthalamic acid ... 416
Homatropine-*N*-oxide hydrobromide ... 1005
2-(Methacryloyloxy)ethyl *N*-(4-isopropylphenyl)carbamate ... 665
$C_{16}H_{21}NO_4S$
3-(1-Hydroxyethylidene)-4-(phenylsulfonylmethyl)-1-isopropyl-2-pyrrolidinone ... 774
$C_{16}H_{21}NO_5$
Diethyl 2-acetamido-2-benzylmalonate ... 429
2′,4′-Dimethoxyhexahydro-phthalanilic acid ... 414
$C_{16}H_{21}NO_6$
Carbobenzyloxy-L-aspartic acid 1-*tert*-butyl ester dicyclohexylamine salt ... 659
$C_{16}H_{21}N_2O$
3-(2-Hydroxybenzylideneamino)-quinoline ... 983
$C_{16}H_{21}N_3$
2-Amino-4′-(dimethylamino)-5-ethyldiphenylamine ... 232
$C_{16}H_{21}N_3O_2$
Ethyl 2-(2-cyanoethyl)-2,3-dicyano-3,6-dimethyl-6-heptenoate ... 149
$C_{16}H_{21}N_3O_3$
N-(2-(3-Acetamido-*N*-ethylanilino)-ethyl)succinimide ... 718
$C_{16}H_{21}N_3O_6$
Carbobenzyloxy-β-alanylglycylglycine methyl ester ... 661
Carbobenzyloxy-L-alanylglycylglycine methyl ester ... 661
Carbobenzyloxyglycylglycyl-L-alanine methyl ester ... 661
$C_{16}H_{21}N_3O_7$
Carbobenzyloxyglycylglycyl-L-threonine ... 643
$C_{16}H_{21}N_5$
4-(3-(Dimethylamino)propylamino)-1-methyl-1*H*-pyrazolo(3,4-b)-quinoline triphosphate ... 1003
$C_{16}H_{22}Br_2O_4$
Dimethyl α,α′-dibromo-1,3-adamantanediacetate ... 82
$C_{16}H_{22}ClN_3$
7-Chloro-4-(3-(diethylamino)-propylamino)quinoline ... 988
$C_{16}H_{22}ClN_3O_2$
1-Chloro-2,4-dipiperidino-5-nitrobenzene ... 273
1-Chloro-3,5-dipiperidino-2-nitrobenzene ... 273
$C_{16}H_{22}Cl_3NO$
2′,4′,5′-Trichlorodecananilide ... 410
$C_{16}H_{22}IN_3O_4$
5-(*N*-(8-Methoxy-4-quinolyl)amino)-pentyl nitrate methiodide ... 985
$C_{16}H_{22}N_2$
5-Indancarboxaldehyde hexamethylenehydrazone ... 1072
$C_{16}H_{22}N_2O$
1,1-Diallyl-3-(2-ethyl-6-methylphenyl)urea ... 595
4-Phenyl-4-(pyrrolidinocarbonyl)-piperidine hydrochloride ... 703
2-Propyl-1,2,3,4-tetrahydroharmine hydrochloride ... 882
$C_{16}H_{22}N_2O_2$
3-(3-Acetylphenyl)-1-cyclohexyl-1-methylurea ... 575
O-(*N*-(2-Ethyl-6-methylphenyl)-carbamoyl)cyclohexanone oxime ... 1012
4-Morpholinocarbonyl-4-phenylpiperidine hydrochloride ... 703
$C_{16}H_{22}N_2O_3$
4′-(Dimethylamino)-hexahydrophthalanilic acid ... 427
Ethyl 4-(2-methylpiperidino-carboxamido)benzoate ... 707
$C_{16}H_{22}N_2O_3S$
O-(3,4-Dimethoxybenzylidene-amino) *N*-cyclohexylthiocarbamate 1040
$C_{16}H_{22}N_2O_5$
N-Benzyloxycarbonyl-glycyl-L-isoleucine ... 660
N-Benzyloxycarbonyl-L-isoleucyl-glycine ... 660
N-Benzyloxycarbonyl-L-leucyl-glycine ... 660
N-Benzyloxycarbonyl-DL-norvalylglycine methyl ester ... 661
Carbobenzyloxy-L-alanyl-L-alanine ethyl ester ... 661
Carbobenzyloxyalanyl-2-aminobutyric acid methyl ester ... 661
Carbobenzyloxy-L-alanyl-L-norvaline ... 660
Carbobenzyloxy-L-alanyl-L-valine ... 660
Carbobenzyloxyglycylnorvaline methyl ester ... 661
Carbobenzyloxyglycyl-L-valine methyl ester ... 661
Carbobenzyloxy-L-norvalyl-L-alanine ... 660
Carbobenzyloxy-sarcosyl-L-valine ... 660
Carbobenzyloxy-L-valyl-L-alanine ... 676
Nε-Carbobenzyloxy-*N*α-acetyl-DL-lysine ... 660
Nε-Carbobenzyloxy-*N*α-acetyl-L-lysine ... 660
$C_{16}H_{22}N_2O_5S$
N-Benzyloxycarbonyl-L-methionyl-L-alanine ... 656
trans-4-*tert*-Butylcyclohexyl 2,4-dinitrobenzenesulfenate ... 771
Carbobenzyloxyalanylmethionine ... 655
Carbobenzyloxy-L-methionylglycine methyl ester ... 667
Carbobenzyloxysarcosyl-L-methionine ... 655
$C_{16}H_{22}N_2O_6$
(+)-1-*tert*-Butylpentyl 3,5-dinitrobenzoate ... 372
Carbobenzyloxyalanylthreonine methyl ester ... 642
Carbobenzyloxy-L-norvalyl-L-serine .. 642
Carbobenzyloxy-L-valyl-L-serine ... 643
N,*N*′-(1,4-Cyclohexylenedimethylene)-dimaleamic acid ... 105
$C_{16}H_{22}N_2O_7$
Carbobenzyloxy-L-threonyl-L-serine methyl ester ... 642
$C_{16}H_{22}N_4O_3$
4-Nitrobenzoic *N*^{2}-(azacyclodec-1-en-2-yl)hydrazide ... 473
$C_{16}H_{22}N_4O_3S$
4-(3-Morpholinopropyl)-1-(piperonylidene)-3-thiosemicarbazide ... 1072
$C_{16}H_{22}N_4O_6$
5-((6,7,8-Trimethoxy-4-quinazolinyl)-amino)-1-pentanol nitrate ester ... 999
$C_{16}H_{22}O$
Dipentenemonophenol ... 191
9-(4-Hydroxyphenyl)-*p*-menth-1-ene .. 191
3,5,5,8,8-Pentamethyl-5,6,7,8-tetrahydro-2-naphthalenecarboxaldehyde ... 324
$C_{16}H_{22}O_2$
2-Benzylideneheptyl acetate ... 362
1,1′-Butadiynylenedicyclohexanol ... 1016
Methyl 2,5-diisopropylcinnamate ... 365
Methyl 4-(6,7,8,9-tetrahydro-5*H*-benzocyclohepten-5-yl)butyrate ... 363
5,5,8,8-Tetramethyl-5,6,7,8-tetrahydro-2-naphthoxyacetaldehyde ... 321
$C_{16}H_{22}O_3$
trans-2,3,4,4a,4b,5,6,7,8,8a-Decahydro-6,7-dihydroxy-1,8a-dimethyl-2-phenanthrenone ... 51
cis-2,3,4,4a,4b,5,6,7,8,8a-Decahydro-6,7-dihydroxy-1,8a-dimethyl-2-phenanthrenone ... 51
3,5-Di-*tert*-butyl-3-hydroxyphthalide. 383
10-Oxo-10-phenyldecanoic acid ... 330
3,3,5-Trimethylcyclohexyl salicylate ... 368
$C_{16}H_{22}O_3S$
Ethyl 2-acetyl-3-(benzylthio)-3,3-dimethylpropionate ... 363
$C_{16}H_{22}O_3Si$
(1-Naphthyl)triethoxysilane ... 1123
$C_{16}H_{22}O_4$
Diisobutyl phthalate ... 371
mono-2,2-Dimethyl-1-isopropylpropyl phthalate ... 371
(-)-*mono*-2,2-Dimethyl-1-isopropylpropyl phthalate ... 371
Ethyl 4-(pentyloxy)benzoylacetate ... 363
1-Methylheptyl 2-carboxybenzoate ... 375
(+)-*mono*(2-Octyl) phthalate ... 375
$C_{16}H_{22}O_6$
2,5-Dimethyl-3-hexyne-2,5-diyl bis(3-oxobutyrate) ... 1018
$C_{16}H_{22}O_7$
1-*O*-Mesitoyl-β-D-glucose ... 367

$C_{16}H_{22}O_{10}$
- Tetraethyl 1,4-dihydroxy-1,3-butadiene-1,2,3,4-tetracarboxylate, disodium salt ... 86

$C_{16}H_{22}Si_2$
- ((Diphenylsilyl)methyl)trimethylsilane ... 1113
- 1,2-Diphenyl-1,1,2,2-tetramethyldisilane ... 1113

$C_{16}H_{23}BrN_2O_3S$
- 1-(4-(Acetylsulfamoyl)butyl)-3,3-dimethyl-3*H*-indolium bromide ... 821

$C_{16}H_{23}ClN_2O$
- 6′-Chloro-2-(*N*-cyclohexyl-*N*-methylamino)-*o*-acetotoluidide hydrochloride ... 424
- 6′-Chloro-2-(2,6-dimethylpiperidino)-*o*-acetotoluidide hydrochloride ... 424
- 3-(2-Chlorophenyl)-1-cyclohexyl-1-isopropylurea ... 535
- 3-(3-Chlorophenyl)-1-cyclohexyl-1-isopropylurea ... 582
- 3-(4-Chlorophenyl)-1-cyclohexyl-1-isopropylurea ... 583

$C_{16}H_{23}ClN_2O_3$
- 1-(4-Chloro-2-methoxyphenyl)-3-cyclohexyl-3-(2-hydroxyethyl)urea ... 512
- 2′-Chloro-4′-nitrodecananilide ... 404

$C_{16}H_{23}ClO_4$
- α3-Chloro-5,7-dimethyl-1,3-adamantanediacetic acid ... 65
- Methyl 3-(carboxymethyl)-α-chloro-5-methyl-1-adamantaneacetate ... 95

$C_{16}H_{23}Cl_2NO$
- 3′,4′-Dichlorodecananilide ... 410
- 3′,5′-Dichlorodecananilide ... 410

$C_{16}H_{23}Cl_2N_3O$
- 1-(3-(2,3-Dichlorophenylureido)propyl)-2-methylpiperidine ... 565
- 1-(3-(2,6-Dichlorophenylureido)propyl)-2-methylpiperidine ... 565
- 1-(3-(3,4-Dichlorophenylureido)propyl)-2-methylpiperidine ... 565

$C_{16}H_{23}FN_2O$
- 1-Cyclohexyl-3-(3-fluorophenyl)-1-isopropylurea ... 535

$C_{16}H_{23}F_3N_2O$
- 1,1-Dibutyl-3-(3-trifluoromethylphenyl)urea ... 510
- 1,1-Dibutyl-3-(4-trifluoromethylphenyl)urea ... 511
- 1,1-Diisobutyl-3-(3-trifluoromethylphenyl)urea ... 581

$C_{16}H_{23}I_3N_2O_8$
- 3-Acetamido-2,4,6-triiodobenzoic acid, compound with 1-deoxy-1-(methylamino)glucitol ... 405

$C_{16}H_{23}N$
- *N*-(2-Norbornyl)-(+)-amphetamine hydrochloride ... 242

$C_{16}H_{23}NO$
- *N*-Allyl-*N*-hexylbenzamide ... 482
- 4-(2,3-Dimethyl-4-phenyl-2-buten-1-yl)morpholine ... 251

$C_{16}H_{23}NOS$
- 4-(4,4-Dimethyl-4-phenylthiobutyryl)morpholine ... 746

$C_{16}H_{23}NO_2$
- Allyl *N*-(2,6-diisopropylphenyl)carbamate ... 684
- Cyclohexyl *N*-(2-ethyl-6-methylphenyl)carbamate ... 679
- Ethyl 1-benzyl-4-piperidineacetate ... 359
- 2-Isopropylphenyl *N*-cyclohexylcarbamate ... 679
- 2-Methylcyclohexyl *N*-(2,3-xylyl)carbamate ... 678
- 2-Methylcyclohexyl *N*-(2,4-xylyl)carbamate ... 679
- 2-Methylcyclohexyl *N*-(2,6-xylyl)carbamate ... 679
- 3-(2-Methylpiperidino)-propyl benzoate hydrochloride ... 373

$C_{16}H_{23}NO_2S_2$
- 5,6-Dibutoxy-2-(methylthio)benzothiazole ... 913

$C_{16}H_{23}NO_3$
- 5-(*N*-(1-Benzylethyl)-*N*-methylcarbamoyl)-valeric acid, sodium salt ... 429
- 2-Methylcyclohexyl *N*-(2-methoxy-5-methylphenyl)carbamate ... 653
- 3-(Nonylamido)benzoic acid, branched isomers ... 433

$C_{16}H_{23}NO_6$
- Ethyl 4-(*tert*-butoxycarbonyl)-2-(ethoxycarbonyl)-5-methyl-3-pyrroleacetate ... 842

$C_{16}H_{23}NO_{10}$
- 2-Acetamido-2-deoxy-1,3,4,6-tetra-*O*-acetyl-α-D-glucopyranose ... 104

$C_{16}H_{23}N_2O$
- 7-Acetamido-1-ethyl-1,2,3,4-tetrahydro-2,2,4-trimethylquinoline ... 427

$C_{16}H_{23}N_3$
- 1-Benzyl-4-cyano-4-(isopropylamino)-piperidine hydrate ... 754
- 1-Benzyl-4-(1-cyanopropylamino)-piperidine ... 754
- 1-Benzyl-4-cyano-4-(propylamino)-piperidine hydrate ... 754

$C_{16}H_{23}N_3O_2$
- 1-Nitro-2,4-dipiperidinobenzene ... 274

$C_{16}H_{23}N_3O_2S$
- 4-Cyclohexyl-1-(3,4-dimethoxybenzylidene)-3-thiosemicarbazide ... 1040
- 4-(Cyclohexyl)-1-(3-ethoxy-2-hydroxybenzylidene)-3-thiosemicarbazide ... 1030

$C_{16}H_{23}N_3O_3$
- 1-*tert*-Butyl-3-(2-methoxy-5-(*N*-phenylcarbamoyl)phenyl)urea ... 554
- 1-Cyclohexyl-1-isopropyl-3-(4-nitrophenyl)urea ... 567
- *N*-(2-(Morpholinomethyl)phenyl)-4-morpholinecarboxamide ... 702

$C_{16}H_{23}N_3O_4$
- D-Alanyl-D-alanyl-D-alanine benzyl ester ... 420
- Carbobenzyloxy-L-alanyl-L-norvalinamide ... 664
- Carbobenzyloxy-β-alanylvalinamide ... 658
- Carbobenzyloxy-L-valyl-L-alaninamide ... 664
- 1-Cyclohexyl-1-(2-hydroxyethyl)-3-(2-methyl-5-nitrophenyl)urea ... 547
- 1-Cyclohexyl-1-(2-hydroxyethyl)-3-(4-methyl-2-nitrophenyl)urea ... 547

$C_{16}H_{23}N_3O_4S$
- Carbobenzyloxy-L-alanyl-L-methioninamide ... 656
- Carbobenzyloxymethionyl-L-alaninamide ... 656
- Carbobenzyloxysarcosyl-L-methioninamide ... 656

$C_{16}H_{23}N_5S$
- 7-Amino-2-(2-(diethylamino)ethylthio)-4-methyl-1,8-naphthyridine ... 999

$C_{16}H_{24}$
- 1,1,4,4,5,8-Hexamethyl-1,2,3,4-tetrahydronaphthalene ... 164

$C_{16}H_{24}BrNO$
- 1-(2-(1-Adamantyl)-2-bromoacetyl)pyrrolidine ... 109

$C_{16}H_{24}BrNO_3$
- α-Bromo-5-nitro-*o*-tolyl nonyl ether ... 258

$C_{16}H_{24}Br_2N_4O_8$
- 1,1′-(4,6-Dibromo-1,3-phenylene)bis(3-(1,1-bis(hydroxymethyl)-2-hydroxy-ethyl)urea) ... 512

$C_{16}H_{24}ClNO_4$
- 1-Chloro-2,5-dipentyloxy-4-nitrobenzene ... 268

$C_{16}H_{24}ClN_3O$
- 2′-Chloro-2-(*N*-ethyl-*N*-(2-pyrrolidinoethyl)-amino)acetanilide dihydrochloride ... 423
- 1-(2-Chlorophenyl)-3-(3-(2-methylpiperidino)propyl)urea ... 585
- 1-(3-Chlorophenyl)-3-(3-(2-methylpiperidino)propyl)urea ... 565
- 1-(4-Chlorophenyl)-3-(3-(2-methylpiperidino)propyl)urea ... 565

$C_{16}H_{24}Cl_2N_2OS$
- 2′,5′-Dichloro-2-(2-(dipropylamino)ethylthio)-acetanilide hydrochloride ... 421

$C_{16}H_{24}N_2$
- 1-Benzyl-4-pyrrolidinopiperidine dihydrochloride ... 249
- 5,5′-Diethyl-3,3′,4,4′-tetramethyl-(2,2′-bipyrrole) ... 839
- 2-Nonylbenzimidazole ... 907

$C_{16}H_{24}N_2O$
- 1-Allyl-3-(2,6-diisopropylphenyl)urea ... 595
- Benzoic (1-methyloctylidene)hydrazide ... 501
- 1-Cyclohexyl-3-(3,4-dimethylbenzyl)urea ... 584
- 1-Cyclohexyl-3-(2-ethyl-6-methylphenyl)urea ... 583
- 1-Cyclohexyl-3-(2-ethylphenyl)-1-methylurea ... 582
- 1-Cyclohexyl-1-isopropyl-3-phenylurea ... 582
- 1-Cyclohexyl-3-(4-isopropylphenyl)urea ... 583
- 1-Cyclohexyl-1-methyl-3-(2,3-xylyl)urea ... 583
- 1-Cyclohexyl-1-methyl-3-(2,4-xylyl)urea ... 584
- 1-Cyclohexyl-1-methyl-3-(2,5-xylyl)urea ... 584
- 1-Cyclohexyl-1-methyl-3-(2,6-xylyl)urea ... 583
- 1-Cyclohexyl-1-methyl-3-(3,4-xylyl)urea ... 584
- 1-Cyclohexyl-1-methyl-3-(3,5-xylyl)urea ... 584
- 1-Cyclohexyl-3-(2,4,5-trimethylphenyl)urea ... 582
- 1-Cyclohexyl-3-(2,4,6-trimethylphenyl)urea ... 584
- 1-Cyclopentyl-3-(2-isopropyl-6-methylphenyl)urea ... 581
- 1-(*N*-(2,6-Diethylphenyl)carbamoyl)piperidine ... 712
- *N*-(1-Ethyl-1,2,3,4-tetrahydro-2,2,4-trimethyl-7-quinolyl)acetamide ... 426
- *N*-Methyl-*N*-(α-methylphenethyl)-1-pyrrolidineacetamide ... 428
- 3-(2-Methylpiperidino)-butyranilide hydrochloride ... 426

$C_{16}H_{24}N_2OS$
- 2,6-Dimethyl-4-(*N*-(2-isopropylphenyl)thiocarbamoyl)morpholine ... 747
- 10-Undecenoic (thenylidene)hydrazide ... 1071

$C_{16}H_{24}N_2O_2$
- 2,3-Bis(4-isopropenyl-2-oxazolin-2-yl)butane ... 38
- 1-Cyclohexyl-3-(2-methoxy-5-methylphenyl)-1-methylurea ... 554
- *O*-(*N*-(2,6-Diisopropylphenyl)carbamoyl)acetone oxime ... 1012
- 2,6-Dimethyl-4-(*N*-(2-ethyl-6-methylphenyl)carbamoyl)morpholine ... 713
- Piperidino *N*-(4-butylphenyl)carbamate ... 687
- 2′-Propoxy-1-piperidineacetanilide hydrochloride ... 415
- Salicylic (1-methyloctylidene)hydrazide ... 1032
- *N*,*N*′-(2,3,5,6-Tetramethyl-1,4-xylylene)bisacetamide ... 431
- 10-Undecenoic (furfurylidene)hydrazide ... 1068

$C_{16}H_{24}N_2O_3$
- 3-Acetamidophenyl *N*-(*tert*-butyl)carbamate ... 677
- Butyl 4-(3-butylureido)benzoate ... 577
- 1-Cyclohexyl-3-(2,5-dimethoxyphenyl)-1-methylurea ... 554
- 3′-(Dipropylamino)succinanilic acid ... 425
- 4′-Nitrodecananilide ... 427
- 3,4,5-Trimethoxybenzaldehyde-*N*,*N*-hexamethylene hydrazone ... 1042

$C_{16}H_{24}N_2O_3S$
- 2-(Dimethylhydroxymethyl)cyclohexanone (*p*-tosyl)hydrazone ... 812

$C_{16}H_{24}N_2O_4$
- 2,5-Bis-(morpholinomethyl)hydroquinone ... 246
- *N*′,N^2-Dicarbethoxy-2-methyl-*N*′-phenyl-1,2-propanediamine ... 669

$C_{16}H_{24}N_2S$
- 1-Benzyl-3-cyclohexyl-1-ethyl-2-thiourea ... 747

$C_{16}H_{24}N_2S_2$
- *N*,*N*,*N*′,*N*′-Tetraethyldithioterephthalamide ... 742

$C_{16}H_{24}N_4O_2$
- Isopropyl 6-(guanidinoimino)-6-phenylhexanoate nitrate ... 1070

$C_{16}H_{24}N_4O_6$
- *N*-Decyl-2,4,6-trinitroaniline ... 273

$C_{16}H_{24}N_4S$
- 4-Cyclohexyl-1-(4-dimethylaminobenzylidene)-3-thiosemicarbazide ... 1062

$C_{16}H_{24}N_6O_4$
- 1,1′-Hexamethylenebis(3-(5-methyl-3-isoxazolyl)urea) ... 871

$C_{16}H_{24}O$
- 5-Methoxy-1,1,4,4,7-pentamethyl-1,2,3,4-tetrahydronaphthalene ... 188

$C_{16}H_{24}O_2$
- 3′,5′-Di-*tert*-butyl-4′-hydroxyacetophenone ... 291
- 5,7-Dimethoxy-1,2,3,4-tetrahydro-1,1,4,4-tetramethylnaphthalene ... 182
- 2-Hexyl-2-methyl-4-phenyl-1,3-dioxolane ... 184
- α,α,2,2-Tetramethyl-6-chromanpropanol ... 199

$C_{16}H_{24}O_2S$
- Phenyl 3,4,5-trimethylcyclohexylmethyl sulfone ... 783

$C_{16}H_{24}O_3$
- Methyl 3,5-di-*tert.*-butyl-4-hydroxybenzoate ... 368
- 4-(Octyloxy)-*m*-anisaldehyde ... 323
- 8-Oxo-2-perhydrophenanthrenol acetate ... 94

$C_{16}H_{24}O_4$
- 2,2-Bis(3-oxobutyl)-5,5-dimethyl-1,3-cyclohexanedione ... 55
- Dimethyl 1,3-adamantanediacetate ... 77
- Nonyl 2,3,4-trihydroxyphenyl ketone ... 289

$C_{16}H_{24}O_8$
- Dimethyl 2-(4,6-diacetoxyheptylidene)malonate ... 78

$C_{16}H_{24}O_{10}$
- 1,2-*O*-Isopropylidene-*myo*-inositol 3,4,5,6-tetraacetate ... 86

$C_{16}H_{25}ClN_2O$
- 3-(2-Chloro-6-methylphenyl)-1,1-dibutylurea ... 541
- 3-(3-Chloro-4-methylphenyl)-1,1-dibutylurea ... 601
- 3-(4-Chloro-2-methylphenyl)-1,1-dibutylurea ... 601
- 3-(5-Chloro-2-methylphenyl)-1,1-dibutylurea ... 601
- 3-(2-Chloro-6-methylphenyl)-1,1-diisobutylurea ... 595
- 3-(3-Chloro-4-methylphenyl)-1,1-diisobutylurea ... 544
- 3-(4-Chloro-2-methylphenyl)-1,1-diisobutylurea ... 544
- 3-(5-Chloro-2-methylphenyl)-1,1-diisobutylurea ... 544
- 4-Decyloxyphenyldiazonium chloride, zinc chloride double salt ... 1086

$C_{16}H_{25}ClO_2$
- 1-Chloro-2,5-dipentyloxybenzene ... 182

$C_{16}H_{25}N$
- 4-Benzyl-1-butylpiperidine ... 251

$C_{16}H_{25}NO$
- 1-(1-Adamantanecarbonyl)piperidine ... 109
- 1-(1-Adamantylacetyl)pyrrolidine ... 109
- *N*,*N*-Diethyl-3-methyl-2-phenylvaleramide ... 430

$C_{16}H_{25}NO$ (Cont'd)
2,2-Dimethyl-N-ethyl-N-(1-methyl-2-phenylethyl)propionamide 430
2′,5′-Dimethyloctananilide 440
2′,6′-Dimethyloctananilide 439
(3-Ethoxycyclohexyl)-phenethylamine hydrochloride 240
2′-Ethyloctananilide 438
2′,3′-Octanoxylidide 439
2′,4′-Octanoxylidide 440
3′,4′-Octanoxylidide 440
$C_{16}H_{25}NO_2$
3-Allyl-2-(2-diethylaminoethoxy)-anisole 239
2-(Diethylamino)-ethyl 2-phenylbutyrate citrate 359
5-(Diethylamino)-pentyl benzoate 373
2-(Dimethylamino)-ethyl 3-methyl-2-phenylvalerate 359
4′-Ethoxyoctananilide 419
3′-Hydroxydecananilide 413
Isopropyl N-(2,6-diisopropylphenyl)-carbamate 684
2′-(Octyloxy)acetanilide 417
3′-(Octyloxy)acetanilide 418
4′-(Octyloxy)acetanilide 419
Propyl N-(2,6-diisopropylphenyl)-carbamate 684
$C_{16}H_{25}NO_2S$
Ethyl α-(2-(diethylamino)-ethylthio)-phenylacetate citrate 357
$C_{16}H_{25}NO_3$
Decyl 4-nitrophenyl ether 269
2,4-Di-tert.-amyl-6-nitrophenol 265
$C_{16}H_{25}NO_5$
2-Diethylaminoethyl 3,4,5-trimethoxybenzoate 369
2-Diethylaminoethyl 3,4,5-trimethoxybenzoate hydrochloride 369
$C_{16}H_{25}N_3O$
1-Benzyl-4-isopropylamino-4-piperidinecarboxamide 425
1-Benzyl-4-propylamino-4-piperidinecarboxamide 425
1-(N-(4-Butylphenyl)carbamoyl)-4-methylpiperazine 705
Decanoic (3-pyridylmethylene)-hydrazide 1056
1-(1-Ethyl-3-piperidinyl)-3-(3,4-xylyl)urea 561
1-(3-(2-Methylpiperidino)propyl)-3-phenylurea 565
$C_{16}H_{25}N_3O_3$
1,1-Dibutyl-3-(4-methyl-3-nitrophenyl)urea 564
$C_{16}H_{25}N_5O_3$
1-(Diethylureidoacetyl)-4-(2,4-dimethylphenyl)semicarbazide 574
1-(Diethylureidoacetyl)-4-(2-ethylphenyl)semicarbazide 574
$C_{16}H_{26}ClN_3O$
6′-Chloro-2-(N-(2-diethylamino-ethyl)-N-methylamino)-o-acetotoluidide dihydrochloride 424
2′-Chloro-2-(N-(2-diethylamino-ethyl)-N-methylamino)-propionanilide dihydrochloride 424
N-(2-(Diethylamino)ethyl)-N-(ethyl)-glycyl-2-chloroanilide 423
$C_{16}H_{26}N_2$
2-(2-(4-Isopropylanilino)ethyl)-piperidine 249
5-(1-Methylheptylamino)indoline dihydrochloride 885
$C_{16}H_{26}N_2O$
1-Benzyl-1-butyl-3-tert-butylurea 586
1,1-Dibutyl-3-(o-tolyl)urea 588
1,1-Dibutyl-3-(m-tolyl)urea 581
1,1-Dibutyl-3-(p-tolyl)urea 591
2-Diethylamino-N-methyl-N-(α-methylphenethyl)-acetamide 428
1,1-Diisobutyl-3-(o-tolyl)urea 588
1,1-Diisobutyl-3-(m-tolyl)urea 581
1,1-Diisobutyl-3-(p-tolyl)urea 591
1,1-Diisopropyl-3-(2-ethyl-6-methylphenyl)urea 595
1-(2,6-Diisopropylphenyl)-3-isopropylurea 595
1-(2,6-Diisopropylphenyl)-3-propylurea 595
3-(2-Ethyl-6-methylphenyl)-1,1-dipropylurea 595
$C_{16}H_{26}N_2OS$
1-Methyl-1-(4-octyloxyphenyl)-2-thiourea 736
$C_{16}H_{26}N_2O_2$
3-(o-Anisyl)-1,1-diisobutylurea 559
3-(m-Anisyl)-1,1-diisobutylurea 558
3-(p-Anisyl)-1,1-diisobutylurea 559
4′-Butoxy-2-(butylamino)-acetanilide hydrochloride 414
4′-Butoxy-2-(diethylamino)-acetanilide hydrochloride 415
1,1-Dibutyl-3-(2-methoxyphenyl)urea 555
1,1-Dibutyl-3-(3-methoxyphenyl)urea 558
1,1-Dibutyl-3-(4-methoxyphenyl)urea 559
2-(Diethylamino)ethyl N-(2-ethyl-6-methylphenyl)carbamate 672
4′-Methoxy-3′-(1-methylhexylamino)acetanilide 415
N,N,N′,N′-Tetramethyl-1,3-adamantanedicarboxamide 101
$C_{16}H_{26}N_2O_3$
2-(3-(4-(2-Methoxyphenyl)-piperazino)propoxy)ethanol 245
$C_{16}H_{26}N_2O_3S$
4′-(Sulfamoyl)decananilide 825
$C_{16}H_{26}N_2O_4$
1,4-Bis-(acetoacetylaminomethyl)-cyclohexane 111
$C_{16}H_{26}N_2O_4S$
N(δ)-Tosyl-L-ornithine tert-butyl ester 816
$C_{16}H_{26}N_3O_3P$
Dimorpholinophenethylamino-phosphine oxide 834
$C_{16}H_{26}N_4O_2$
Oxalic bis((4-methylcyclohexylidene)hydrazide) 112
$C_{16}H_{26}O$
2,6-Di-tert-butyl-4-methylanisole 182
2,4-Dimethyl-6-isooctylphenol 192
$C_{16}H_{26}O_2$
Cedryl formate 78
2,5-Di-tert.-amylhydroquinone 193
2,5-Di-tert.-butyl-1,4-dimethoxybenzene 182
2,6-Di-tert.-butyl-α-methoxy-p-cresol 191
2,2′-Dimethyl-1-hydroxymethyl-1,1′,2,2′,3,3′,6,6′-octahydrobenzhydrol 14
6-Methyl-3-cyclohexenylmethyl 2-ethyl-2-hexenoate 79
$C_{16}H_{26}O_3$
2-Ethyl-3-propylacrylic anhydride 98
$C_{16}H_{26}O_4$
1,10-Bis-(1,3-dioxolan-2-yl)-3,7-decadiene 23
2-(3-Methyl-2-butenyl)-3,5,5-triethoxy-2-cyclopenten-1-one 52
$C_{16}H_{26}O_5$
Diethyl 2-oxo-1,3-cyclohexanedipropionate 94
$C_{16}H_{26}O_7$
Zinc borneol glucuronic acid (from urine) 69
$C_{16}H_{27}BrN_2O$
3-Carbamoyl-1-decylpyridinium bromide 947
4-Carbamoyl-1-decylpyridinium bromide 952
$C_{16}H_{27}Cl_4FeO$
2,6-Di-tert-butyl-4-isopropylpyrylium tetrachloroferrate 935
$C_{16}H_{27}IN_2O_2$
4-(5,5-Dimethyl-3-morpholino-2-cyclohexen-1-ylidene)-morpholinium iodide 42
$C_{16}H_{27}N$
4-Decylaniline hydrochloride 219
$C_{16}H_{27}NO$
2-(Decyloxy)-aniline 217
3-(Decyloxy)aniline 217
4-(Decyloxy)-aniline 218
$C_{16}H_{27}NO_2$
2-(1-Adamantyl)ethyl N-propylcarbamate 116
1-(1-Adamantyl)propyl N-ethylcarbamate 116
$C_{16}H_{27}NO_3$
N-Cyclododecylmaleamic acid 105
$C_{16}H_{27}NO_3S$
Cyclohexyltrimethylammonium p-toluenesulfonate 798
$C_{16}H_{27}O_3P$
Bis(1-methylbutyl) phenylphosphonate 836
$C_{16}H_{28}INO_3$
N-(9,9-Ethylenedioxybicyclo(3.3.1)-nonan-2-yl)-N-methylmorpholinium iodide 44
$C_{16}H_{28}N_2$
Cyclooctanone azine 151
$C_{16}H_{28}N_2O$
1-(1-Adamantylmethyl)-3-butylurea 130
$C_{16}H_{28}N_2O_6$
N,N′-Hexamethylenebis(methyl succinamate) 111
$C_{16}H_{28}O_2$
Ambrettolide 97
$C_{16}H_{28}O_3$
Dodecylsuccinic anhydride 98
L-Menthyl 3,3,3-trimethylpyruvate 94
4-Methoxycyclohexyl 4-methoxycyclohexylmethyl ketone 52
6-Oxo-15-hexadecenoic acid 64
$C_{16}H_{28}O_5$
6-Oxo-1,14-tetradecane-dicarboxylic acid 64
$C_{16}H_{28}Si$
Decylphenylsilane 1113
$C_{16}H_{29}IN_2$
(-)-Sparteine methiodide 42
$C_{16}H_{29}N$
1-Decyl-2,5-dimethylpyrrole 842
$C_{16}H_{29}NO_2$
2-Methyl-1-piperidino-5-(2-tetrahydropyranyloxy)-2-pentene 38
$C_{16}H_{29}NO_3$
N-Dodecylmaleamic acid 121
$C_{16}H_{30}Cl_4O_2Si_4$
1,1′-(2,3,5,6-Tetrachloro-1,4-phenylene)bis(1,1,3,3,3-pentamethyldisiloxane) 1101
$C_{16}H_{30}N_2O$
Dodecanoic (2-buten-1-ylidene)-hydrazide 1070
$C_{16}H_{30}N_2O_6$
N(α)-N(ε)-Di-t-boc-L-lysine dicyclohexylamine salt 115
$C_{16}H_{30}N_4O_4$
N,N′-Hexamethylenebis(4-carbamoylmorpholine) 713
$C_{16}H_{30}N_4O_5$
Glycyl-L-leucylglycyl-L-leucine 120
$C_{16}H_{30}O$
2,4-Di(tert-pentyl)cyclohexanone 56
$C_{16}H_{30}O_2$
2-Hydroxycyclohexadecanone 52
$C_{16}H_{30}O_3$
Hexyl 3-hexyloxycrotonate 85
10-Oxopalmitic acid 64
$C_{16}H_{30}O_4$
Dipentyl adipate 76
mono-Dodecyl succinate 95
$C_{16}H_{30}O_4PbS$
Triethylplumbyl (+)-10-camphorsulfonate 1135
$C_{16}H_{31}BrO_2$
Ethyl 2-bromomyristate 83
$C_{16}H_{31}NO_2$
N-(2-Buten-1-yl)-N-(1,1-dimethyl-3-hydroxybutyl)hexanamide 106
2-Diethylaminoethyl 2-cyclohexylbutyrate citrate 88
5-Piperidino-3-nonenal dimethyl acetal 38
$C_{16}H_{31}NO_3S$
N-(1,1-Dioxotetrahydrothiophen-3-yl)dodecanamide 153
$C_{16}H_{31}N_3O_4S$
N(α)-tert-BOC-L-isoleucyl-L-methionamide 118
tert-Butoxycarbonyl-L-leucyl-L-methioninamide 118
$C_{16}H_{32}ClNO$
2-Chloro-N-tetradecylacetamide 103
$C_{16}H_{32}ClNOS$
2-Chloro-N-(2-(dodecylthio)-ethyl)-acetamide 102
$C_{16}H_{32}Cl_2N_4NiO_8$
(5,7,7,12,14,14-Hexamethyl-1,4,8,11-tetraazacyclotetradeca-4,11-diene)nickel(II) perchlorate 1139
$C_{16}H_{32}N_2$
N,N′-Bis-(1-isopropyl-2-methylpropylidene)-ethylenediamine 1052
Octyl aldehyde azine 1020
$C_{16}H_{32}N_2O$
N-(1,3-Dimethylbutyl)-N-propyl-1-piperidineacetamide hydrochloride 109
N-(1,5-Dimethylhexyl)-N-methyl-1-piperidineacetamide hydrochloride 109
N-(1-Isobutyl-3-methylbutyl)-N-methyl-1-pyrrolidineacetamide hydrochloride 108
$C_{16}H_{32}N_2O_2$
N,N′-Ethylenebis(2,2-dimethyl-N-ethylpropionamide) 107
$C_{16}H_{32}N_2O_4$
N,N′-Hexamethylenebis(butyl carbamate) 116
$C_{16}H_{32}O$
3-Hexadecanone 49
7-Hexadecanone 49
$C_{16}H_{32}OSi_4$
2-Benzoyl-1,1,1,3,3,3-hexamethyl-2-(trimethylsilyl)trisilane 1109
$C_{16}H_{32}O_2$
2-Hydroxy-6-propyl-1,3,5-triethylcyclohexanemethanol 14
Palmitic acid, dodecylamine salt 70
Palmitic acid, octadecylamine salt 70
$C_{16}H_{32}O_2S$
4-(Dodecylthio)butyric acid 63
Ethyl 2-(dodecylthio)acetate 86
$C_{16}H_{32}O_3$
2-Hydroxypalmitic acid 63
$C_{16}H_{32}O_3S$
5-Tridecyl-1,2-oxathiolane-2,2-dioxide 156
$C_{16}H_{32}O_5$
Aleuritic acid 63
$C_{16}H_{33}Br$
Hexadecyl bromide (mixture of isomers) 4
$C_{16}H_{33}BrHg$
Bromohexadecylmercury 1137
$C_{16}H_{33}NO$
N,N-Dimethyltetradecanamide 107
Hexadecanamide 102
$C_{16}H_{33}N_3S$
N-Decyl-4-methyl-1-piperazinethiocarboxamide 140
$C_{16}H_{34}HgS_2$
Mercury(II) (N-octylsulfide) 1137
$C_{16}H_{34}N_2O$
N-Butyl-2-diethylamino-N-(1,3-dimethylbutyl)-acetamide hydrochloride 107
2-Diethylamino-N-(1,5-dimethylhexyl)-N-ethylacetamide hydrochloride 107
2-Diethylamino-N-ethyl-N-octylacetamide hydrochloride 107
2-Diethylamino-N-(1-isobutyl-3-methylbutyl)-N-methylacetamide hydrochloride 107
N-(2-(Dimethylamino)-ethyl)-lauramide 104
Hexdecanoic acid hydrazide 112
$C_{16}H_{34}N_4O_2$
1,1′-Hexamethylenebis(3-butylurea). 129

$C_{16}H_{34}N_4O_2$ (Cont'd)
3,3′-Hexamethylenebis(1,1-diethylurea) 129
1,1′-Hexamethylenebis(3-isobutylurea) 129
$C_{16}H_{34}N_4O_4$
Bis(2-(dimethylamino)ethyl) *N,N′*-hexamethylenebiscarbamate 115
$C_{16}H_{34}O$
5-Hexadecanol 9
$C_{16}H_{34}OS$
2-(Dodecylthiomethyl)-2-propanol 25
2-(Tetradecylthio)-ethanol 25
$C_{16}H_{34}OSn$
Dioctyloxotin 1129
$C_{16}H_{34}O_2$
1,2-Alkanediols, c15 to c18, mixed 11
$C_{16}H_{34}O_3$
Tripentyl orthoformate 19
$C_{16}H_{34}O_3S$
1-Hexadecanesulfonic acid, barium salt 156
2-Hexyl-1-decanesulfonic acid, sodium salt 156
$C_{16}H_{34}O_4S$
1-Hydroxyhexadecane-1-sulfonic acid sodium salt 155
$C_{16}H_{34}O_5Si_2$
1,3-Bis(2,3-epoxy-3-propoxypropyl)-1,1,3,3-tetramethyldisiloxane 1106
$C_{16}H_{34}S$
Butyl dodecyl sulfide 26
2-Ethylhexyl sulfide 26
$C_{16}H_{35}BO_2$
Hexadecaneboronic acid 1124
$C_{16}H_{35}NO_2$
2-(2-(Dodecyloxy)-ethoxy)-ethylamine 34
$C_{16}H_{35}OP$
Bis(2-ethylhexyl)phosphine oxide 158
$C_{16}H_{35}O_3P$
Bis(1,3-dimethylbutyl) *sec*-butylphosphonate 158
$C_{16}H_{36}BF_4N$
Bis-(2-ethylhexyl)ammonium tetrafluoroborate 43
$C_{16}H_{36}Cl_2N_4NiO_8$
(5,7,7,12,14,14-Hexamethyl-1,4,8,11-tetraazacyclotetradecane-N^1,N^4,N^8,N^{11})nickel(II) perchlorate 1139
$C_{16}H_{36}F_6NP$
Bis(2-ethylhexyl)ammonium hexafluorophosphate 43
$C_{16}H_{36}N_2$
N,N′-Bis-(2,4-dimethyl-3-pentyl)-ethylenediamine 30
N-Tetradecylethylenediamine 29
$C_{16}H_{36}N_2O_2$
3,3′-(Decamethylenedioxy)-bis-(propylamine) 35
$C_{16}H_{36}Si$
Tetrabutylsilane 1097
$C_{16}H_{36}Sn$
Tetra(*sec*-butyl)tin 1129
Tetraisobutyltin 1129
$C_{16}H_{40}F_6N_2Si$
Dibutylammonium hexafluorosilicate 43
$C_{16}H_{44}Si_5$
Tetrakis((trimethylsilyl)methyl)-silane 1097
$C_{17}H_6Cl_{10}O_2$
1,3-Bis(pentachlorophenyl)-2,2-dimethyl-1,3-propanedione 307
1,5-Bis(pentachlorophenyl)-1,5-pentanedione 307
$C_{17}H_8Br_2O$
3,6-Dibromo-7*H*-benzene(de)-anthracen-7-one 314
$C_{17}H_8Cl_{10}O_2$
2-(Benzyloxy)-decachloro-octahydro-1,3,4-metheno-2*H*-cyclobuta(cd)pentalen-2-ol 205
$C_{17}H_8N_2O$
10-(Dicyanomethylene)-9(10*H*)-phenanthrenone 766
$C_{17}H_9BrO$
3-Bromobenzanthrone 314
$C_{17}H_9ClF_3NO_2$
3-Chloro-2-(3-(trifluoromethyl)-anilino)-1,4-naphthoquinone 315
$C_{17}H_9Cl_2F_3N_2O_2$
5,7-Dichloro-8-quinolyl *N*-(3-trifluoromethylphenyl)carbamate 981
$C_{17}H_9Cl_3N_2O_4$
2,4-Dichloro-1-naphthyl *N*-(2-chloro-4-nitrophenyl)carbamate 629
$C_{17}H_9Cl_6NO_2$
1,4,5,6,7,7-Hexachloro-*N*-(2,3-xylyl)-5-norbornene-2,3-dicarboximide 722
$C_{17}H_9IO$
3-Iodobenzanthrone 310
$C_{17}H_9NO_3$
2-Nitrobenzanthrone 313
3-Nitrobenzanthrone 313
$C_{17}H_9N_5O_3$
5-Amino-7-cyano-10-nitro-12*H*-isoquino(2,3-a)quinazolin-12-one 763
$C_{17}H_{10}BrCl_2NO_2$
6-Bromo-2-naphthyl *N*-(3,4-dichlorophenyl)carbamate 640
$C_{17}H_{10}BrFN_2O_4$
6-Bromo-2-naphthyl *N*-(4-fluoro-2-nitrophenyl)carbamate 625
$C_{17}H_{10}BrF_2NO_2$
6-Bromo-2-naphthyl *N*-(2,4-difluorophenyl)carbamate 640
6-Bromo-2-naphthyl *N*-(2,5-difluorophenyl)carbamate 640
$C_{17}H_{10}ClF_3N_2O_2$
5-Chloro-8-naphthyl *N*-(3-trifluoromethylphenyl)carbamate 981
$C_{17}H_{10}ClN$
7-Chlorobenz(c)acridine 996
$C_{17}H_{10}ClNO$
9-Amino-3-chlorobenzanthrone 312
$C_{17}H_{10}ClNO_4$
2-(3-Carboxyanilino)-3-chloro-1,4-naphthoquinone 343
$C_{17}H_{10}Cl_2O_2$
3,4-Dichloro-5-(diphenylmethylene)-2(5*H*)-furanone 386
$C_{17}H_{10}Cl_3N$
2,4-Bis(4-chlorophenyl)-6-chloropyridine 928
$C_{17}H_{10}Cl_3NO$
7-Chloro-2-(4-chlorophenyl)-6-methyl-4-quinolinecarbonyl chloride 982
$C_{17}H_{10}Cl_3NO_2$
2,4-Dichloro-1-naphthyl *N*-(2-chlorophenyl)carbamate 634
$C_{17}H_{10}F_2O_7S_2$
Phenyl 4,7-bis-(fluorosulfonyl)-1-hydroxy-2-naphthoate 802
$C_{17}H_{10}F_6N_2O$
3-Amino-4-(2,5-bis(trifluoromethyl)-phenyl)-2(1*H*)-quinolinone 981
$C_{17}H_{10}N_2O_4$
N,N′-Methylenediisatin 706
$C_{17}H_{10}N_4O_3$
6-Cyano-5-methyl-9-nitrobenzimidazo(1,2-b)isoquinolin-11(5*H*)-one 765
$C_{17}H_{10}O$
4-Pyrenecarboxaldehyde 324
$C_{17}H_{10}O_9$
3,3′,4,4′-Benzophenone-tetracarboxylic acid 344
$C_{17}H_{11}BrClNO_2$
2-(4-Bromo-3-methylanilino)-3-chloro-1,4-naphthoquinone 315
$C_{17}H_{11}BrFNO_2$
6-Bromo-2-naphthyl *N*-(3-fluorophenyl)carbamate 641
$C_{17}H_{11}BrO_2$
3-Bromo-5-(diphenylmethylene)-2(5*H*)-furanone 386
$C_{17}H_{11}BrO_3$
Phenyl 4-bromo-1-hydroxy-2-naphthoate 377
$C_{17}H_{11}Br_2ClN_2O_2$
2,4-Dibromo-8-quinolyl *N*-(3-chloro-2-methylphenyl)carbamate 637
$C_{17}H_{11}Br_2NO_2$
2,4-Dibromophenyl *N*-(1-naphthyl)-carbamate 640
$C_{17}H_{11}Br_4NO_5$
N-((4-Ethoxycarbonyl)phenyl)-3,4,5,6-tetrabromophthalamic acid 452
$C_{17}H_{11}ClFNO$
3′-Chloro-4′-fluoro-1-naphthanilide 484
$C_{17}H_{11}ClFNO_2$
2-Naphthyl *N*-(3-chloro-4-fluorophenyl)carbamate 640
$C_{17}H_{11}ClFNO_4S$
2′-Chloro-5′-fluorosulfonyl-1-hydroxy-2-naphthanilide 486
3-(4-Chlorophenylcarbamoyl)-4-hydroxy-1-naphthalenesulfonyl fluoride 802
Methyl 4-(4-chloro-6-fluorosulfonyl-2-quinolyl)benzoate 994
$C_{17}H_{11}ClN_2O$
1-(4-Chlorobenzoyl)-2-cyano-1,2-dihydroquinoline 751
6-(Chloromethyl)benzo(a)phenazin-5(7*H*)-one 320
$C_{17}H_{11}ClN_2O_3$
4-(8-Chloro-5,6-dihydro-3-oxo-3*H*-pyrrolo(3,2,1-ij)quinazolin-1-yl)-benzoic acid 707
1-(2-Chloro-5-nitrophenyliminomethyl)-2-naphthol 1023
1-(4-Chloro-3-nitrophenyliminomethyl)-2-naphthol 1023
$C_{17}H_{11}ClN_2O_4$
2′-Chloro-1-hydroxy-4′-nitro-2-naphthanilide 486
4′-Chloro-1-hydroxy-2′-nitro-2-naphthanilide 486
4-Chloro-1-hydroxy-3′-nitro-2-naphthanilide 485
4-Chloro-1-hydroxy-4′-nitro-2-naphthanilide 485
1-Naphthyl *N*-(2-chloro-4-nitrophenyl)carbamate 629
2-Naphthyl *N*-(2-chloro-4-nitrophenyl)carbamate 629
2-Naphthyl *N*-(2-chloro-5-nitrophenyl)carbamate 629
2-Naphthyl *N*-(4-chloro-2-nitrophenyl)carbamate 629
2-Naphthyl *N*-(4-chloro-3-nitrophenyl)carbamate 629
$C_{17}H_{11}ClN_2O_6$
4-Methyl-7-coumarinyl *N*-(4-chloro-2-nitrophenyl)carbamate 628
$C_{17}H_{11}ClN_4O_3$
2-(2-Chloro-5-nitrobenzoyl)-2-(1-methyl-2-benzimidazolyl)-acetonitrile 905
$C_{17}H_{11}ClO_2$
3-Chloro-5-(diphenylmethylene)-2(5*H*)-furanone 386
$C_{17}H_{11}ClO_3$
Phenyl 4-chloro-1-hydroxy-2-naphthoate 377
$C_{17}H_{11}Cl_2N$
(2,4-Dichlorobenzylidene)(1-naphthyl)amine 1027
$C_{17}H_{11}Cl_2NO$
2,4-Dichloro-*N*-(1-naphthyl)-benzamide 489
2,4-Dichloro-*N*-(2-naphthyl)-benzamide 484
3,4-Dichloro-*N*-(1-naphthyl)-benzamide 489
3,4-Dichloro-*N*-(2-naphthyl)-benzamide 484
1-(2,3-Dichlorophenyliminomethyl)-2-naphthol 1038
1-(2,4-Dichlorophenyliminomethyl)-2-naphthol 1038
1-(2,5-Dichlorophenyliminomethyl)-2-naphthol 1037
1-(3,4-Dichlorophenyliminomethyl)-2-naphthol 1023
$C_{17}H_{11}Cl_2NO_2$
7-Chloro-2-(4-chlorophenyl)-6-methyl-4-quinolinecarboxylic acid 982
3,4-Dichloro-*N*-(5-hydroxy-1-naphthyl)benzamide 447
2,4-Dichloro-1-naphthyl *N*-phenylcarbamate 681
2,4-Dichlorophenyl *N*-(1-naphthyl)-carbamate 640
$C_{17}H_{11}Cl_3N_2O$
1-(1-Naphthyl)-3-(2,4,5-trichlorophenyl)urea 546
$C_{17}H_{11}Cl_3N_2O_2$
2,4-Dichloro-8-quinolyl *N*-(3-chloro-2-methylphenyl)carbamate 982
2-Methyl-8-quinolyl *N*-(2,4,5-trichlorophenyl)carbamate 982
$C_{17}H_{11}Cl_4NO_5$
N-(4-Ethoxycarbonylphenyl)-3,4,5,6-tetrachlorophthalamic acid 453
$C_{17}H_{11}FN_2O_4$
2-Naphthyl *N*-(2-fluoro-5-nitrophenyl)carbamate 629
2-Naphthyl *N*-(4-fluoro-3-nitrophenyl)carbamate 629
$C_{17}H_{11}FN_2O_6$
4-Methyl-2-oxo-2*H*-1-benzopyran-7-yl *N*-(4-fluoro-3-nitrophenyl)-carbamate 622
$C_{17}H_{11}FN_2O_6S$
4′-Fluorosulfonyl-1-hydroxy-2′-nitro-2-naphthanilide 801
$C_{17}H_{11}FO_5S$
Phenyl 4-fluorosulfonyl-1-hydroxy-2-naphthoate 801
$C_{17}H_{11}F_2NO_2$
1-Naphthyl *N*-(2,5-difluorophenyl)-carbamate 640
2-Naphthyl *N*-(2,4-difluorophenyl)-carbamate 640
$C_{17}H_{11}F_3N_2O_2$
8-Quinolyl *N*-(3-(trifluoromethyl)-phenyl)carbamate 981
$C_{17}H_{11}F_6NOS$
5-(2,5-Bis(trifluoromethyl)phenyl)-3,5-dihydro-4,1-benzothiazepin-2(1*H*)-one 693
$C_{17}H_{11}F_6NO_2$
α(2-Amino-4-(trifluoromethyl)-phenyl)-4-(trifluoromethyl)-cinnamic acid 334
$C_{17}H_{11}NO$
α-(2-Furyl)-β-(1-naphthyl)-acrylonitrile 854
2-Methyl-7*H*-dibenz(f,ij)isoquinolin-7-one 1003
$C_{17}H_{11}NO_2$
2-Maleimidofluorene 723
Methyl 10-cyano-9-anthracenecarboxylate 766
$C_{17}H_{11}NO_4$
2-Naphthyl 4-nitrobenzoate 379
2-Phenyl-4-piperonylidene-2-oxazolin-5-one 384
$C_{17}H_{11}NO_5$
2-(Carboxy-4-hydroxyphenyl)-quinoline-4-carboxylic acid 984
2-Methoxy-3-(4-nitrophenyl)-1,4-naphthoquinone 316
$C_{17}H_{11}N_3O_3$
2-(2-Nitroso-1-naphthylazo)benzoic acid, sodium salt 1092
$C_{17}H_{11}N_3S$
6-Cyano-7-methylbenzimidazo(1,2-a)quinoline-5(7*H*)-thione 766
$C_{17}H_{12}$
1,2-Benzofluorene 167
$C_{17}H_{12}BrCl$
13-Bromo-13-chloro-9,10,12,13-tetrahydro-11*H*-9,10-*endo*-cyclopropanthracene 173
$C_{17}H_{12}BrN$
Naphtho(2,1-b)quinolizinium bromide 1003
$C_{17}H_{12}BrNO$
1-(4-Bromophenyliminomethyl)-2-naphthol 1038
$C_{17}H_{12}BrNO_3$
4-(3-Bromo-4-methoxybenzylidene)-2-phenyl-2-oxazolin-5-one 384

$C_{17}H_{12}Br_4N_2O_4$
4′-(*N*-Methylacetamido)-3,4,5,6-tetrabromophthalanilic acid ... 452

$C_{17}H_{12}ClNO$
2-Chloro-*N*-(1-naphthyl)-benzamide ... 489
2-Chloro-*N*-(2-naphthyl)benzamide ... 484
3-Chloro-*N*-(1-naphthyl)benzamide ... 485
4-Chloro-*N*-(2-naphthyl)-benzamide ... 490
1-(2-Chlorophenyliminomethyl)-2-naphthol ... 1038
1-(3-Chlorophenyliminomethyl)-2-naphthol ... 1039
1-(4-Chlorophenyliminomethyl)-2-naphthol ... 1038

$C_{17}H_{12}ClNO_2$
4′-Chloro-3-hydroxy-2-naphthanilide ... 486
4-Chloro-*N*-(5-hydroxy-1-naphthyl)-benzamide ... 484
3-Chloro-2-(2-methylanilino)-1,4-naphthoquinone ... 318
3-Chloro-2-(3-methylanilino)-1,4-naphthoquinone ... 318
3-Chloro-2-(4-methylanilino)-1,4-naphthoquinone ... 318
2-(4-Chlorophenyl)-7-methylcinchoninic acid ... 993
1-Naphthyl *N*-(4-chlorophenyl)-carbamate ... 683
2-Naphthyl *N*-(2-chlorophenyl)-carbamate ... 634
2-Naphthyl *N*-(3-chlorophenyl)-carbamate ... 641
2-Naphthyl *N*-(4-chlorophenyl)-carbamate ... 635

$C_{17}H_{12}ClNO_3$
2-Chloro-3-(2-hydroxy-5-methylanilino)-1,4-naphthoquinone ... 316
3-Chloro-2-(2-methoxyanilino)-1,4-naphthoquinone ... 317
3-Chloro-2-(4-methoxyanilino)-1,4-naphthoquinone ... 317
2-(2-Chloro-5-nitrobenzylidene)-1-tetralone ... 299
2-(4-Chloro-3-nitrobenzylidene)-1-tetralone ... 299
4′-Chloro-2′-(phenoxy)-2-furanilide ... 844

$C_{17}H_{12}ClNO_4$
2′-Benzoyl-4′-chloromaleanilic acid ... 467

$C_{17}H_{12}ClNO_5$
4-Coumarinyl *N*-(2-chloro-4-methoxyphenyl)carbamate ... 617

$C_{17}H_{12}ClN_3O_4$
2-Methyl-8-quinolyl *N*-(2-chloro-4-nitrophenyl)carbamate ... 981

$C_{17}H_{12}Cl_2$
1-(2,4-Dichlorobenzylidene)-1,4-dihydronaphthalene ... 173
12,13-Dichloro-9,10-dihydro-9,10-propenoanthracene ... 174
13,13-Dichloro-9,10,12,13-tetrahydro-11*H*-9,10-*endo*-cyclopropanthracene ... 173

$C_{17}H_{12}Cl_2N_2O$
1-(2,3-Dichlorophenyl)-3-(1-naphthyl)urea ... 596
1-(2,4-Dichlorophenyl)-3-(1-naphthyl)urea ... 602
1-(2,6-Dichlorophenyl)-3-(1-naphthyl)urea ... 543
1-(3,4-Dichlorophenyl)-3-(1-naphthyl)urea ... 602

$C_{17}H_{12}Cl_2N_2O_2$
4-Chloro-8-quinolyl *N*-(4-chloro-2-methylphenyl)carbamate ... 982
4-Chloro-8-quinolyl *N*-(5-chloro-2-methylphenyl)carbamate ... 982
5-Chloro-8-quinolyl *N*-(3-chloro-4-methylphenyl)carbamate ... 992
1,3-Dichloro-5-(2-fluorenyl)-5-methylhydantoin ... 709
2-Methyl-8-quinolyl *N*-(2,3-dichlorophenyl)carbamate ... 992
2-Methyl-8-quinolyl *N*-(2,4-dichlorophenyl)carbamate ... 992
2-Methyl-8-quinolyl *N*-(2,5-dichlorophenyl)carbamate ... 992
2-Methyl-8-quinolyl *N*-(2,6-dichlorophenyl)carbamate ... 991
2-Methyl-8-quinolyl *N*-(3,4-dichlorophenyl)carbamate ... 992
2-Methyl-8-quinolyl *N*-(3,5-dichlorophenyl)carbamate ... 982

$C_{17}H_{12}Cl_2O$
1,5-Bis(4-chlorophenyl)-1,4-pentadien-3-one ... 282
2-(3,4-Dichlorobenzylidene)-1-tetralone ... 297
1-(2,4-Dichlorophenyl)-5-phenyl-1,4-pentadien-3-one ... 282

$C_{17}H_{12}Cl_4N_2O_2$
2,5-Bis(2,2-dichloroacetamido)-fluorene ... 484

$C_{17}H_{12}Cl_5F_5Si$
Dimethyl(3-(pentachlorophenyl)-propyl)(pentafluorophenyl)silane ... 1101

$C_{17}H_{12}FN$
N-(4-Fluorobenzylidene)-1-naphthylamine ... 1079

$C_{17}H_{12}FNO$
2-Fluoro-*N*-(1-naphthyl)benzamide ... 485
3-Fluoro-*N*-(2-naphthyl)benzamide ... 484
N-(4-Fluoro-1-naphthyl)benzamide ... 485
4-Fluoro-*N*-(2-naphthyl)benzamide ... 484
1-(2-Fluorophenyliminomethyl)-2-naphthol ... 1023

$C_{17}H_{12}FNO_2$
1-Naphthyl *N*-(2-fluorophenyl)-carbamate ... 634
2-Naphthyl *N*-(4-fluorophenyl)-carbamate ... 683

$C_{17}H_{12}FNO_4S$
N-(1-Hydroxy-2-naphthoyl)-sulfanilyl fluoride ... 802

$C_{17}H_{12}FN_3O$
2-(2-Fluorobenzoyl)-2-(1-methyl-2-benzimidazolyl)acetonitrile ... 901

$C_{17}H_{12}FN_3O_4$
1-(4-Fluoro-3-nitrophenyl)-3-(5-hydroxy-1-naphthyl)urea ... 513

$C_{17}H_{12}F_2N_2O$
1-(2,5-Difluorophenyl)-3-(1-naphthyl)urea ... 545

$C_{17}H_{12}F_2N_2O_2$
2-Methyl-8-quinolyl *N*-(2,5-difluorophenyl)carbamate ... 982

$C_{17}H_{12}F_3NO_2S$
3′-Trifluoromethyl-2-naphthalenesulfonanilide ... 805

$C_{17}H_{12}F_3N_3O_2$
5-Imino-3-(*p*-tolyl)-1-(3-trifluoromethylphenyl)hydantoin ... 723

$C_{17}H_{12}F_6O$
α-(2,6-Bis(trifluoromethyl)phenyl)-cinnamyl alcohol ... 197

$C_{17}H_{12}F_{10}Si$
Dimethyl(pentafluorophenyl)(3-(pentafluorophenyl)propyl)silane ... 1101

$C_{17}H_{12}INO$
1-(4-Iodophenyliminomethyl)-2-naphthol ... 1038

$C_{17}H_{12}IN_3O$
2-(2-Iodobenzoyl)-2-(1-methyl-2-benzimidazolyl)acetonitrile ... 901

$C_{17}H_{12}N_2$
2-(2-Phenyl-3-indolyl)acrylonitrile ... 886

$C_{17}H_{12}N_2O$
1-Benzoyl-2-cyano-1,2-dihydroquinoline ... 754
2-Benzoyl-1,2-dihydro-1-isoquinolinecarbonitrile ... 752
2-Indolyl ketone ... 889

$C_{17}H_{12}N_2O_2$
N-(3-Nitrobenzylidene)-1-naphthylamine ... 1059
2-(3-Nitrostyryl)quinoline ... 993

$C_{17}H_{12}N_2O_3$
1-(3-Nitrophenyliminomethyl)-2-naphthol ... 1033
1-(4-Nitrophenyliminomethyl)-2-naphthol ... 1034

$C_{17}H_{12}N_2O_4$
1-Hydroxy-4′-nitro-2-naphthanilide ... 486
3-Hydroxy-2′-nitro-2-naphthanilide ... 485
3-Hydroxy-3′-nitro-2-naphthanilide ... 485
3-Hydroxy-4′-nitro-2-naphthanilide ... 486
N-(1-Naphthyl)-5-nitroanthranilic acid ... 341

$C_{17}H_{12}N_2O_5$
4-(*N*-(4-Acetamidophenyl)-carbamoyl)phthalic anhydride ... 470
6-Methoxy-2-(3-nitrophenyl)-4-quinolinecarboxylic acid ... 985
Phthalimido *N*-(4-acetylphenyl)-carbamate ... 721

$C_{17}H_{12}N_2O_6S$
5-(6-Sulfo-2-naphthylazo)salicylic acid, disodium salt ... 1085

$C_{17}H_{12}N_4$
4-(9-Anthracenylmethyleneamino)-4*H*-1,2,4-triazole ... 1072

$C_{17}H_{12}N_4OS$
5,8-Quinolinedione 5-((3-methyl-2-benzothiazolinylidene)hydrazone) ... 954

$C_{17}H_{12}N_4O_3$
3-Amino-2-benzyl-4-cyano-7-nitro-1(2*H*)-isoquinolinone ... 763

$C_{17}H_{12}N_4O_5$
1-((2,4-Dinitrophenyl)azomethyl)-2-naphthol ... 1085

$C_{17}H_{12}N_4O_7$
2,7-Diacetamido-3,6-dinitro-9-fluorenone ... 487

$C_{17}H_{12}N_6O_3$
2-(5-Nitro-2-benzimidazolylamino)-6-phenyl-4-pyrimidinol ... 971

$C_{17}H_{12}O$
9,10-Dihydro-11*H*-9,10-*endo*-cyclopropanthracen-13(12*H*)-one ... 286
1,5-Diphenyl-1,4-pentadiyn-3-ol ... 1016
4-Hydroxymethylpyrene ... 208

$C_{17}H_{12}OS$
2,6-Diphenyl-4*H*-pyran-4-thione ... 293
1-(2-Naphthyl)-3-(2-thienyl)-2-propen-1-one ... 857

$C_{17}H_{12}O_2$
3-Benzylidene-2,3-dihydro-5-phenyl-2-furanone ... 386
2,2-Dimethyl-1*H*-cyclopenta(a)-biphenylene-1,3(2*H*,3*H*)-dione ... 308
5-(Diphenylmethylene)-2(5*H*)-furanone ... 386
2,6-Diphenyl-4*H*-pyran-4-one ... 940
4,6-Diphenyl-2*H*-pyran-2-one ... 384
3-Furyl-1-(2-naphthyl)-2-propen-1-one ... 854
2-Naphthyl benzoate ... 380

$C_{17}H_{12}O_2S$
2-(3-Methylphenylthio)-1,4-naphthoquinone ... 317

$C_{17}H_{12}O_3$
2,4-Dihydroxyphenyl 1-naphthyl ketone ... 310

$C_{17}H_{12}S_2$
2,6-Diphenyl-4*H*-thiopyran-4-thione ... 213

$C_{17}H_{13}AsClN$
12-Chloro-7,12-dihydro-10-methylbenzo(a)phenarsazine ... 1128

$C_{17}H_{13}BF_4O$
2,6-Diphenylpyrylium tetrafluoroborate ... 935

$C_{17}H_{13}Br$
11-Bromo-9,10-dihydro-12-methylene-9,10-ethanoanthracene ... 173

$C_{17}H_{13}BrN_2O$
4′-(4-Bromo-β-cyanostyryl)-acetanilide ... 767
1-(4-Bromophenyl)-3-(1-naphthyl)urea ... 540

$C_{17}H_{13}BrN_2O_2$
2-Methyl-8-quinolyl *N*-(4-bromophenyl)carbamate ... 991

$C_{17}H_{13}BrN_2O_3$
5-Nitro-2-phenacylisoquinolinium bromide ... 997

$C_{17}H_{13}BrO$
4′-Bromo-5-phenyl-2,4-pentadienophenone ... 305

$C_{17}H_{13}ClFNO_6$
2-Chloro-4-fluorophenyl *N*-(3,5-bis(methoxycarbonyl)phenyl)-carbamate ... 629

$C_{17}H_{13}ClN_2O$
1-(2-(Chlorophenyl)azomethyl)-2-naphthol ... 1080
1-(2-Chlorophenyl)-3-(1-naphthyl)urea ... 589
1-(3-Chlorophenyl)-3-(1-naphthyl)urea ... 603
1-(4-Chlorophenyl)-3-(1-naphthyl)urea ... 592

$C_{17}H_{13}ClN_2O_2$
3′-Amino-4-chloro-1-hydroxy-2-naphthanilide ... 485
4′-Amino-2′-chloro-1-hydroxy-2-naphthanilide ... 485
4′-Amino-4-chloro-1-hydroxy-2-naphthanilide ... 485
5-Chloro-8-quinolyl *N*-(*o*-tolyl)-carbamate ... 993
5-Chloro-8-quinolyl *N*-(*m*-tolyl)-carbamate ... 992
5-Chloro-8-quinolyl *N*-(*p*-tolyl)-carbamate ... 991
2-Methyl-8-quinolyl *N*-(2-chlorophenyl)carbamate ... 991
2-Methyl-8-quinolyl *N*-(3-chlorophenyl)carbamate ... 992
2-Methyl-8-quinolyl *N*-(4-chlorophenyl)carbamate ... 991
8-Quinolyl *N*-(2-chloro-6-methylphenyl)carbamate ... 982
8-Quinolyl *N*-(3-chloro-4-methylphenyl)carbamate ... 992
8-Quinolyl *N*-(4-chloro-2-methylphenyl)carbamate ... 992
8-Quinolyl *N*-(5-chloro-2-methylphenyl)carbamate ... 992

$C_{17}H_{13}ClN_2O_3$
1-Isoquinolyl *N*-(4-chloro-2-methoxyphenyl)carbamate ... 996
5-Isoquinolyl *N*-(4-chloro-2-methoxyphenyl)carbamate ... 996
8-Quinolyl *N*-(2-chloro-4-methoxyphenyl)carbamate ... 981
8-Quinolyl *N*-(3-chloro-2-methoxyphenyl)carbamate ... 981
8-Quinolyl *N*-(4-chloro-2-methoxyphenyl)carbamate ... 981

$C_{17}H_{13}ClN_2O_4S$
4-(4-Chlorobenzylidene)-3-methyl-1-(4-sulfophenyl)-2-pyrazolin-5-one ... 789

$C_{17}H_{13}ClO$
2-(2-Chlorobenzylidene)-1-tetralone ... 305
2-(3-Chlorobenzylidene)-1-tetralone ... 305
2-(4-Chlorobenzylidene)-1-tetralone ... 305
12-Chloro-10,11-dihydro-5,10-etheno-5*H*-dibenzo(a,d)-cyclohepten-11-ol ... 208
4′-Chloro-5-phenyl-2,4-pentadienophenone ... 306
1-(4-Chlorophenyl)-5-phenyl-1,4-pentadien-3-one ... 282

$C_{17}H_{13}ClO_2$
2-(4-Chlorobenzylidene)-1-phenyl-1,3-butanedione ... 307

$C_{17}H_{13}ClO_5$
2,6-Diphenylpyrylium perchlorate ... 935

$C_{17}H_{13}Cl_2NO$
1-(2,3-Dichloro-*N*-methylanilino)-2-naphthol ... 227
1-(2,5-Dichloro-*N*-methylanilino)-2-naphthol ... 227

$C_{17}H_{13}Cl_2NO_6$
3,4-Dichlorophenyl *N*-(3,5-bis(methoxycarbonyl)phenyl)-carbamate ... 629

$C_{17}H_{13}Cl_4NO_3$
3,4,5,6-Tetrachloro-2′,4′,5′-trimethylphthalanilic acid ... 455
3,4,5,6-Tetrachloro-2′,4′,6′-trimethylphthalanilic acid ... 455

$C_{17}H_{13}Cl_6NO_3$
1,4,5,6,7,7-Hexachloro-3-(*N*-(2,4-xylyl)carbamoyl)-5-norbornene-2-carboxylic acid ... 434

$C_{17}H_{13}Cl_6NO_4$
3-(*N*-(4-Ethoxyphenyl)carbamoyl)-1,4,5,6,7,7-hexachloro-5-norbornene-2-carboxylic acid ... 418

$C_{17}H_{13}FN_2O$
1-(2-Fluorophenyl)-3-(1-naphthyl)urea 538
1-(2-Fluorophenyl)-3-(2-naphthyl)urea 538
1-(3-Fluorophenyl)-3-(2-naphthyl)urea 546

$C_{17}H_{13}FN_2O_2$
1-(2-Fluorophenyl)-3-(7-hydroxy-1-naphthyl)urea 514
2-Methyl-8-quinolyl *N*-(4-fluorophenyl)carbamate 982

$C_{17}H_{13}FN_2O_3S$
1-(2-Fluorosulfonylphenyl)-3-(1-naphthyl)urea 803

$C_{17}H_{13}FO$
1-(4-Fluorophenyl)-5-phenyl-2,4-pentadien-1-one 305

$C_{17}H_{13}F_3N_2O_2$
2-Acetamido-7-(2,2,2-trifluoroacetamido)fluorene 484

$C_{17}H_{13}N$
N-Benzylidene-2-naphthylamine 1063
11-Methyl-11*H*-benzo(a)carbazole 892
4-Styrylquinoline nitrate 994

$C_{17}H_{13}NO$
α-Benzoyl-4-methylcinnamonitrile 766
3-Cinnamylidene-2(3*H*)-indolone 691
2,6-Diphenylpyridine *N*-oxide 1006
N-(1-Naphthyl)-benzamide 489
1-Phenyliminomethyl-2-naphthol 1037

$C_{17}H_{13}NOS$
4-Phenoxy-*N*-(2-thienylmethylene)-aniline 1050

$C_{17}H_{13}NO_2$
α-Benzoyl-4-methoxycinnamonitrile 762
N-Cinnamoyl-5-methylanthranilic acid 386
1-(2-Hydroxyphenyliminomethyl)-2-naphthol 1037
1-(4-Hydroxyphenyliminomethyl)-2-naphthol 1029
6-Methoxy-2-phenyl-4-quinolinecarboxaldehyde 986
4-(4-Methylbenzylidene)-2-phenyl-2-oxazolin-5-one 386
3,4-Methylenedioxy-α-(*p*-tolyl)-cinnamonitrile 765
N-(1-Naphthyl)anthranilic acid 343
2-Naphthyl *N*-phenylcarbamate 681
N-(3-Phenylallyl)phthalimide 720

$C_{17}H_{13}NO_3$
4-(3-Methoxybenzylidene)-2-phenyl-2-oxazolin-5-one 383
4-(4-Methoxybenzylidene)-2-phenyl-2-oxazolin-5-one 386
2-(4-Methoxyphenyl)-4-quinolinecarboxylic acid 985
6-Methoxy-2-phenyl-4-quinolinecarboxylic acid 985
4-(5-Methyl-4-phenyl-2-oxazolyl)-benzoic acid 872
2-(2-Nitrobenzylidene)-1-tetralone 303
2-(3-Nitrobenzylidene)-1-tetralone 303
4′-Nitro-5-phenyl-2,4-pentadienophenone 301
1-(3-Nitrophenyl)-5-phenyl-1,4-pentadien-3-one 285
2-Propionamidoanthraquinone 488

$C_{17}H_{13}NO_4$
3-Benzamido-8-methoxycoumarin 464
2-(4-Hydroxyphenyl)-6-methoxy-4-quinolinecarboxylic acid 982
2-(2-Nitrobenzylidene)-1-phenyl-1,3-butanedione 308
3-Phenyl-2-phthalimidopropionic acid 719
N-Phthaloyl-DL-phenylalanine 719

$C_{17}H_{13}NO_5$
4-(*N*-(2-Ethoxyphenyl)carbamoyl)-phthalic anhydride 465

$C_{17}H_{13}NO_7S_2$
4-Benzylideneamino-5-hydroxy-2,7-naphthalenedisulfonic acid monosodiumsalt 1034

$C_{17}H_{13}N_3O_2$
Isonicotinic (2-hydroxy-1-naphthylmethylene)hydrazide 933
α-Nitro-2-naphthaldehyde phenyl hydrazone 258

$C_{17}H_{13}N_3O_3$
2-Methoxy-1-(4-nitrophenylazo)-naphthalene 1087
1-(1-Naphthyl)-3-(3-nitrophenyl)urea 564
1-((4-Nitrophenyl)azomethyl)-2-naphthol 1085

$C_{17}H_{13}N_3O_4$
2-Methyl-8-quinolyl *N*-(3-nitrophenyl)carbamate 988
4′-Nitro-2-(3-(*p*-tolyl)-1,2,4-oxadiazol-5-yl)acetophenone 873
8-Quinolyl *N*-(4-methyl-2-nitrophenyl)carbamate 988

$C_{17}H_{13}N_3O_5$
8-Quinolyl *N*-(2-methoxy-4-nitrophenyl)carbamate 985
8-Quinolyl *N*-(2-methoxy-5-nitrophenyl)carbamate 985

$C_{17}H_{13}N_3O_6S$
N-(2,4-Dinitro-1-naphthyl)-*p*-toluenesulfonamide 817

$C_{17}H_{14}$
1-Benzylnaphthalene 166
9,10-Dihydro-11-methylene-9,10-ethanoanthracene 164
9,10,12,13-Tetrahydro-11*H*-9,10-*endo*-cyclopropanthracene 162

$C_{17}H_{14}BF_4N_3$
2-(*p*-Tolyl)-1,2,4-triazolo(3,4-a)-isoquinolinium tetrafluoroborate 1003

$C_{17}H_{14}BrF_4N_5O_3$
4-(4-Bromo-2-(trifluoromethyl)-phenyl)-1-(4-fluorophenyl-ureidoacetyl)semicarbazide 504

$C_{17}H_{14}BrN$
1-(4-Bromophenyl)-2-methyl-5-phenylpyrrole 838

$C_{17}H_{14}BrNO$
2-Phenacylisoquinolinium bromide 997

$C_{17}H_{14}BrNO_2S$
1-(4-Bromophenyl)-3-(*p*-tolylthio)-succinimide 715

$C_{17}H_{14}BrNO_6$
α-(4-Bromophenyl)-3,4-dimethoxy-2-nitrocinnamic acid 335

$C_{17}H_{14}BrN_3$
1-(4-Bromophenyl)-2-(1-(3-isoquinolinyl)ethylidene)hydrazine 996

$C_{17}H_{14}Br_2$
11-Bromo-10-bromomethyl-10,11-dihydro-5,10-methano-5*H*-dibenzo(a,d)cycloheptene 171

$C_{17}H_{14}Br_2O$
2-Bromo-2-(α-bromobenzyl)-1-tetralone 297

$C_{17}H_{14}Br_4O$
1,5-Diphenyl-1,2,4,5-tetrabromo-3-pentanone 285

$C_{17}H_{14}Br_4O_4$
4,4-Bis(3,5-dibromo-4-hydroxyphenyl)valeric acid 326

$C_{17}H_{14}ClFN_2O_5$
Dimethyl 5-(3-(3-chloro-4-fluorophenyl)ureido)isophthalate 530

$C_{17}H_{14}ClF_3N_2O_3$
Ethyl 4-(2-chloro-5-(trifluoromethyl)-phenylureido)benzoate 505

$C_{17}H_{14}ClNO$
1-(4-Chloroanilinomethyl)-2-naphthol 227
10-Chloro-9,10-dihydro-9,10-ethanoanthracene-11-carboxamide 436
Chloropropionylphenanthrene oxime, (mixed isomers) 1014
4-Chloro-α-(2-quinolinylmethyl)-benzyl alcohol 244

$C_{17}H_{14}ClNO_2$
4′-(2-(4-Chlorobenzoyl)-vinyl)-acetanilide 434
α-(4-Chlorophenyl)-3,4-dimethoxycinnamonitrile 762

$C_{17}H_{14}ClNO_2S$
5′-Chloro-2′-methyl-2-naphthalenesulfonanilide 810

$C_{17}H_{14}ClNO_3S$
5′-Chloro-2′-methoxy-2-naphthalenesulfonanilide 806

$C_{17}H_{14}ClNO_5S$
S-(4-Chlorophenyl) *N*-(3,5-bis(methoxycarbonyl)phenyl)-thiocarbamate 630

$C_{17}H_{14}ClN_3OS$
1-(4-Chloro-7-methoxy-2-quinolyl)-3-phenyl-2-thiourea 986

$C_{17}H_{14}ClN_3O_2$
1-(4-Chloro-7-methoxy-2-quinolyl)-3-phenylurea 986

$C_{17}H_{14}ClN_3O_2S$
1-(5-Chloro-2-methoxyphenyl)-3-(4-phenyl-2-thiazolyl)urea 875

$C_{17}H_{14}ClN_3O_3$
2-Chloro-*N*-(3,5-dioxo-1,2-diphenyl-4-pyrazolidinyl)-acetamide 693
Ethyl 4-(5-chloro-1*H*-indazol-1-ylcarbonylamino)benzoate 900

$C_{17}H_{14}ClN_3O_3S$
Ethyl 4-(3-(4-chloro-2-benzothiazolyl)ureido)benzoate 916

$C_{17}H_{14}ClN_3O_7$
Dimethyl 5-(3-(4-chloro-2-nitrophenyl)ureido)isophthalate 529

$C_{17}H_{14}Cl_2$
11,13-Dichloro-10,11-dihydro-5,10-ethano-5*H*-dibenzo(a,d)-cycloheptene 173

$C_{17}H_{14}Cl_2F_3NO_3$
2-(4-Chloro-3-methylphenoxy)ethyl *N*-(4-chloro-2(trifluoromethyl)-phenyl)carbamate 606

$C_{17}H_{14}Cl_2N_2O_2$
2,7-Diacetamido-1,3-dichlorofluorene 491

$C_{17}H_{14}Cl_2N_2O_3S$
Ethyl 4-(2,4-dichlorobenzoylthioureido)-benzoate 731
Ethyl 4-(3,4-dichlorobenzoylthioureido)-benzoate 731

$C_{17}H_{14}Cl_2N_4O_4$
2-(3,4-Dichlorophenylureido)acetic acid piperonylidene hydrazide 494

$C_{17}H_{14}Cl_2O$
11,12-Dichloro-9,10-dihydro-2-methoxy-9,10-ethanoanthracene 180
11,12-Dichloro-9,10-dihydro-9-methoxy-9,10-ethanoanthracene 187
2,6-Dichloro-2′,4′-dimethylchalcone 306
3,4-Dichloro-2′,4′-dimethylchalcone 305

$C_{17}H_{14}Cl_2O_3$
3′,4′-Dichloro-2,3-dimethoxychalcone 301
3′,4′-Dichloro-3,4-dimethoxychalcone 300

$C_{17}H_{14}Cl_3NO_4$
N-(4-(Ethoxycarbonyl)phenyl)-2-(2,4,5-trichlorophenoxy)acetamide 399

$C_{17}H_{14}Cl_3N_3O_5$
N-(5-Chloro-2,4-dimethoxybenzamido)-*N*′-(3,4-dichlorophenyl)oxamide 492

$C_{17}H_{14}Cl_4O_4$
4,4-Bis-(3,5-dichloro-4-hydroxyphenyl)-valeric acid 325

$C_{17}H_{14}FNO_2$
4-Acetamido-4′-fluorochalcone 432

$C_{17}H_{14}FNO_5$
Dimethyl 5-(2-fluorophenylcarboxamido)-isophthalate 455

$C_{17}H_{14}FN_3O_4S$
N-(1-Benzyl-5-oxo-2-pyrazolin-3-yl)-4-(fluorosulfonyl)benzamide 803

$C_{17}H_{14}FN_3O_7$
Dimethyl 5-(3-(4-fluoro-3-nitrophenyl)ureido)isophthalate 529

$C_{17}H_{14}F_3N_3OS$
1-(3-Cyano-4,5,6,7-tetrahydrobenzo(b)thiophen2yl)-3(3(trifluoromethyl)phenyl)urea 855
1-(5,6-Dimethyl-2-benzothiazolyl)-3-(2-(trifluoromethyl)phenyl)urea 911
1-(5,6-Dimethyl-2-benzothiazolyl)-3-(3-(trifluoromethyl)phenyl)urea 911

$C_{17}H_{14}F_6N_2O$
1,3-Bis(2-(trifluoromethyl)phenyl)-1-ethylurea 511
1-Ethyl-1-(2-(trifluoromethyl)phenyl)-3-(3-(trifluoromethyl)phenyl)urea 511

$C_{17}H_{14}N_2$
2,3-Diphenylglutaronitrile 756
1,1-Diphenyl-1,3-propanedicarbonitrile 756
1-Naphthaldehyde phenylhydrazone 235
2-Naphthaldehyde phenylhydrazone 235
2-Phenyl-3-(*p*-tolyl)succinonitrile 755

$C_{17}H_{14}N_2O$
2-(*p*-Anisyl)-3-phenylsuccinonitrile 753
4-Benzylidene-3-methyl-1-phenyl-2-pyrazolin-5-one 711
5-Benzyloxyindole-3-acetonitrile 882
3-(Cyanomethyl)-2-(4-methoxyphenyl)indole 882
4′-(β-Cyanostyryl)-acetanilide 765
5,6-Diphenyl-1-methyl-2(1*H*)-pyrimidinone 712
2-(2-Hydroxy-1-naphthylmethyleneamino)-3-methylpyridine 1039
2-(2-Hydroxy-1-naphthylmethyleneamino)-6-methylpyridine 1039
1-Methoxy-4-(phenylazo)-naphthalene 1087
4-(2-Methylimidazol-1-yl)-benzophenone 868
1-(4-Methyl-2-pyridyliminomethyl)-2-naphthol 1039
1-(1-Naphthyl)-3-phenylurea 586
1-(3-Pyridylmethyliminomethyl)-2-naphthol 1039

$C_{17}H_{14}N_2OS$
1-(5-Hydroxy-1-naphthyl)-3-phenyl-2-thiourea 735

$C_{17}H_{14}N_2O_2$
3′-Amino-1-hydroxy-2-naphthanilide 485
4′-Amino-1-hydroxy-2-naphthanilide 485
4′-Amino-1-hydroxy-2-naphthanilide sulfurous acid salt 485
3-(Cyanomethyl)-5-methoxy-3-phenyl-2(3*H*)-indolone 753
3,3′-Dimethyldiphenylmethane-4,4′-diisocyanate 769
N-Methyl-*N*-phenyl-3-indoleglyoxylamide 889
2-Methyl-8-quinolyl *N*-phenylcarbamate 991
N-(2-Phenylpropylideneamino)-phthalimide 1070
8-Quinolyl *N*-(*o*-tolyl)carbamate 991
8-Quinolyl *N*-(*m*-tolyl)carbamate 992
8-Quinolyl *N*-(*p*-tolyl)carbamate 991

$C_{17}H_{14}N_2O_2S$
2,4-Dicyano-1-(1-methyl-1-(phenylsulfonyl)ethyl)benzene 782
8-Quinolinyl *N*-(4-(methylthio)-phenyl)carbamate 987

$C_{17}H_{14}N_2O_3$
N-(4-Acetylanilinomethyl)-phthalimide 719
N-(Benzylcarbamoylmethyl)-phthalimide 721
3,6-Diacetamido-9-fluorenone 433
1,3-Dimethyl-5-(1-naphthylmethylene)barbituric acid 708
4-Ethoxy-α-(4-nitrophenyl)-cinnamonitrile 762
N-(4-Nitrophenyl)-2-hydroxy-1-naphthylmethylamine 264
8-Quinolyl *N*-(*p*-anisyl)carbamate 985
8-Quinolyl *N*-(2-methoxyphenyl)-carbamate 985
8-Quinolyl *N*-(3-methoxyphenyl)-carbamate 986

$C_{17}H_{14}N_2O_3S$
4-(3-Cinnamoyl-2-thioureido)-benzoic acid 744

$C_{17}H_{14}N_2O_4$
4′-Acetamido-3-nitrochalcone 432
N-(3-Ethoxy-4-hydroxybenzylideneamino)-phthalimide 1030
3-Ethoxy-4-hydroxy-α-(4-nitrophenyl)cinnamonitrile 760
N-(Veratrylideneamino)-phthalimide 1042

$C_{17}H_{14}N_2O_4S$
1-Nitro-2-p-toluenesulfonamidonaphthalene 818

$C_{17}H_{14}N_2O_5S$
2′-Methoxy-5′-nitro-2-naphthalenesulfonanilide 814

$C_{17}H_{14}N_2O_6$
Phthalimido N-(2,5-dimethoxyphenyl)carbamate 716

$C_{17}H_{14}N_2S$
1-(2-Naphthyl)-3-phenyl-2-thiourea 748
N-Phenyl-N′-(2-thienylmethylene)-1,4-phenylenediamine 1057

$C_{17}H_{14}N_4$
3-Amino-5,8-diphenyl-1,2,4-triazocine 248

$C_{17}H_{14}N_4O$
1,2-Dihydro-4,6-dimethyl-1-(3-indolylmethyleneamino)-2-oxonicotinonitrile 1071
6-Methyl-4-phenyl-5-(phenylazo)-2-pyrimidinol 1084

$C_{17}H_{14}N_4O_2$
1-(3-Phthalamidopropyl)-1H-benzotriazole 909

$C_{17}H_{14}N_4O_6$
N-(2,4-Dinitrophenyl)-L-tryptophan hydrate 884

$C_{17}H_{14}N_6O_7$
2,3-Dimethyl-1-phenyl-4-(2,4,6-trinitrophenylamino)-3-pyrazolin-5-one 701

$C_{17}H_{14}O$
2-Benzylidene-1-tetralone 304
Benzyl 1-naphthyl ether 186
Benzyl 2-naphthyl ether 186
9,10-Dihydro-9,10-ethanoanthracene-11-carboxaldehyde 324
9,10-Dihydro-9-methoxy-9,10-ethenoanthracene 188
9,10-Dihydro-12-methyl-9,10-ethanoanthracen-11-one 286
1,5-Diphenyl-1,4-pentadien-3-one 286
1-(α-Hydroxybenzyl)naphthalene 204
5-Phenyl-2,4-pentadienophenone 304

$C_{17}H_{14}O_2$
2-(Cinnamylidene)-2-phenylacetic acid 345
9,10-Dihydro-9,10-ethanoanthracene-11-carboxylic acid 332
3,4-Diphenyl-4-hydroxy-2-cyclopenten-1-one 282
3,4-Diphenyl-5-methyl-2(5H)-furanone 386
2-(Hydroxymethylene)-4-phenyl-1-tetralone 290
Methyl 10-methyl-9-anthracenecarboxylate 376
Methyl 3-methyl-2-anthracenecarboxylate 376

$C_{17}H_{14}O_2S$
2-Methyl-4-phenyl-1-buten-3-yn-1-yl phenyl sulfone 1019
4-Phenyl-1-buten-3-yn-1-yl p-tolyl sulfone 1019

$C_{17}H_{14}O_3$
3-Benzoyl-2-benzylidenepropionic acid 341
3-Benzoyl-3-benzylidenepropionic acid 341
4′-(Carboxymethyl)chalcone 331
1,1-Dibenzoylacetone 307
10,11-Dihydro-11-hydroxy-5,10-methano(5H)-dibenzo(a,d)-cycloheptene-10-carboxylic acid 326
4′-Methyl-3,4-(methylenedioxy)-chalcone 304
3-Oxo-1-phenyl-1-indanacetic acid 331

$C_{17}H_{14}O_4$
7-Benzyloxy-6-hydroxy-4-methylcoumarin 382
2-(4-Cinnamoylphenoxy)acetic acid 327
2′,5′-Dimethoxyflavone 319
2,2-Dimethyl-5-(1-naphthylmethylene)-1,3-dioxane-4,6-dione 386
4-Hydroxy-6-methyl-3-(5-phenyl-2,4-pentadienoyl)-2-pyrone 382
2-(4-(3-Oxo-3-phenyl-1-propen-1-yl)-phenoxy)acetic acid 328

$C_{17}H_{14}O_5$
1,3-Dihydroxyacetone dibenzoate 374

$C_{17}H_{14}O_6$
3,5-Dihydroxy-6-[5-(2,4,6-trihydroxy-phenyl)-2,4-pentadienylidene]-2,4-cyclohexadienone 282

$C_{17}H_{14}O_7$
Bis(2-(methoxycarbonyl)phenyl) carbonate 381

$C_{17}H_{15}Br$
11-(Bromomethyl)-9,10-dihydro-9,10-ethanoanthracene 171

$C_{17}H_{15}BrO_3$
5-Bromo-2,4′-dimethoxychalcone 299

$C_{17}H_{15}BrO_4$
5-Bromo-3,4′-dimethoxy-4-hydroxychalcone 300

$C_{17}H_{15}Br_2NO_3$
Methyl 2-(2,3-dibromo-3-phenylpropionamido)benzoate 433

$C_{17}H_{15}Cl$
11-(Chloromethyl)-9,10-dihydro-9,10-ethanoanthracene 171

$C_{17}H_{15}ClF_3NO_2$
2,3,5-Trimethylphenyl N-(4-chloro-2-(trifluoromethyl)phenyl)carbamate 608

$C_{17}H_{15}ClF_3NO_4$
2′-Chloro-5′-(trifluoromethyl)-3,4,5-trimethoxybenzanilide 445

$C_{17}H_{15}ClF_3N_3O_3$
1-(4-Acetamido-3-methoxyphenyl)-3-(4-chloro-3-(trifluoromethyl)-phenyl)urea 504

$C_{17}H_{15}ClHgO$
11-(Chloromercuri)-9,10-dihydro-12-methoxy-9,10-ethanoanthracene 1138

$C_{17}H_{15}ClN_2$
α-(4-Chlorophenyl)-4-(dimethylamino)cinnamonitrile 759

$C_{17}H_{15}ClN_2O_2S$
2-(2-Aminophenylthio)-N-(3-chloro-o-tolyl)-succinimide 715
2-(2-Aminophenylthio)-N-(3-chloro-p-tolyl)-succinimide 718

$C_{17}H_{15}ClN_2O_3$
2-Chloro-4-dimethylamino-4′-nitrochalcone 302

$C_{17}H_{15}ClN_2O_3S$
1-(2-Chlorobenzoyl)-3-(4-ethoxycarbonylphenyl)-2-thiourea 731

$C_{17}H_{15}ClN_2O_5$
4-Allyl-2-methoxyphenyl N-(2-chloro-5-nitrophenyl)carbamate 615

$C_{17}H_{15}ClN_2O_6$
Carbobenzyloxyglycine 2-chloromethyl-4-nitrophenyl ester 604

$C_{17}H_{15}ClN_4$
5-Chloro-3-methyl-4-(4-tolylazo)-1H-pyrazole 1088

$C_{17}H_{15}ClN_4O$
1-Benzyl-4-(4-chlorophenylhydrazono)-3-methyl-2-pyrazolin-5-one 1024

$C_{17}H_{15}ClN_6O_4S_2$
3-Methyl-2-[(3-methyl-2-benzo-thiazolinylidene)hydrazono-methylazo]benzothiazolium perchlorate 1091

$C_{17}H_{15}ClO$
2-Chloro-2′,4′-dimethylchalcone 306

$C_{17}H_{15}ClO_2$
4′-Chloro-2-ethoxychalcone 301

$C_{17}H_{15}ClO_3$
4′-Chloro-2,3-dimethoxychalcone 300
4′-Chloro-2,4-dimethoxychalcone 302
4′-Chloro-2,5-dimethoxychalcone 302

$C_{17}H_{15}Cl_2N$
α,β-Bis(4-chlorophenyl)valeronitrile 752

$C_{17}H_{15}Cl_2NO$
2,4′-Dichloro-4-(dimethylamino)-chalcone 302

$C_{17}H_{15}Cl_2NO_4$
2-(2,4-Dichlorophenoxy)acetyl N-(2,4-xylyl)carbamate 620

$C_{17}H_{15}Cl_2N_3O_2S$
3-(3-Allyl-2-thioureido)phenyl N-(3,4-dichlorophenyl)carbamate 731
4-(3-Allyl-2-thioureido)phenyl N-(3,4-dichlorophenyl)carbamate 731

$C_{17}H_{15}Cl_2N_5O_7$
N-(5-Chloro-2,4-dimethoxyphenylureido)-N′-(4-chloro-3-nitrophenyl)oxamide 516

$C_{17}H_{15}Cl_3$
3,3-Bis(p-tolyl)-1,1,2-trichloro-1-propene 174

$C_{17}H_{15}Cl_3O_2$
3,3-Bis(4-methoxyphenyl)-1,1,2-trichloro-1-propene 179

$C_{17}H_{15}FN_2O$
10b-(4-Fluorophenyl)-1,2,3,4,6,10b-hexahydropyrimido(2,1-a)isoindol-6-one 695

$C_{17}H_{15}FN_2O_5$
Dimethyl 5-(2-fluorophenylureido)-isophthalate 530

$C_{17}H_{15}FN_2O_6S$
4′-(3-Fluorosulfonylbenzamido)-succinanilic acid 803

$C_{17}H_{15}F_3N_2O_2S$
1-(4-Ethoxycarbonyl)-3-(3-(trifluoromethyl)phenyl)-2-thiourea 728

$C_{17}H_{15}F_3N_2O_3$
Ethyl 4-(3-(2-(trifluoromethyl)-phenyl)ureido)benzoate 509
Ethyl 4-(3-(3-(trifluoromethyl)-phenyl)ureido)benzoate 509

$C_{17}H_{15}FeN_3O$
Isonicotinic acid (ferrocenylmethylene)-hydrazide 1141
Picolinic acid (ferrocenylmethylene)-hydrazide 1141

$C_{17}H_{15}N$
2-(Phenethyl)quinoline 994
4-(Phenethyl)quinoline 994

$C_{17}H_{15}NO$
2-Benzyl-3-oxo-2-phenylbutyronitrile 755
2-(10,11-Dihydro-5H-dibenzo(a,d)-cyclohepten-5-ylidene)acetamide 477
2-(4-Methoxyphenyl)-7-methylquinoline 986
2-(4-Methoxyphenyl)-6-methylquinoline picrate 986

$C_{17}H_{15}NO_2$
2,3-Bis(4-methoxyphenyl)-acrylontirile 763
2-Diacetylaminofluorene 489
3,4-Dimethoxy-α-phenylcinnamonitrile 762
2-(4-(Dimethylamino)phenyl)-1,3-indandione 318
2-(N-Ethylacetamido)-9-fluorenone 488
4-(5-Indanylmethyleneamino)-benzoic acid 1074
N-(5-Indanylmethylene)-anthranilic acid 1074
α-(4-Methoxyphenyl)-1-isoquinolinemethanol 996

$C_{17}H_{15}NO_2S$
N-(2-Methyl-1-naphthyl)-benzenesulfonamide 827

$C_{17}H_{15}NO_3$
2-Acetamido-9-fluorenyl acetate 489
4-(Anisylideneamino)-cinnamic acid 1049
5-Benzyloxyindole-3-acetic acid 882
2′,4′-Dimethyl-3-nitrochalcone 303
5-(5-Indanylmethyleneamino)-salicylic acid 1037
N-(2-(4-Methoxyphenyl)ethyl)-phthalimide 717
N-(4-Phenoxyphenyl)glutarimide 717

$C_{17}H_{15}NO_4$
2-(Ethoxycarbonyl)benzil monoxime 1011
4-(Ethoxycarbonyl)-N-(piperonylidene)aniline 1051
2-Ethoxy-4′-nitrochalcone 301
α-Methylcinnamyl 4-nitrobenzoate 374
α-(1-Propen-1-yl)benzyl 4-nitrobenzoate 374

$C_{17}H_{15}NO_5$
4′-Carboxy-2-(o-anisoyl)acetanilide 442
4′-Carboxy-2-(p-anisoyl)acetanilide 442
2,3-Dimethoxy-4′-nitrochalcone 301
3,4-Dimethoxy-3′-nitrochalcone 300
3,4-Dimethoxy-4′-nitrochalcone 300
N-(4-Ethoxycarbonylphenyl)-phthalamic acid 475
4-Formyl-2-methoxyphenyl N-(3-acetylphenyl)carbamate 649
Melicopine 311
4-(Methoxycarbonyl)phenyl N-(3-acetylphenyl)carbamate 674
4′-(2-Methoxyphenoxy)-maleanilic acid 486
N-(3,4,5-Trimethoxybenzoyl)-anthranilic acid 387
4-(N-(2,3-Xylyl)carbamoyl)-isophthalic acid, 2,3-dimethylaniline salt 474

$C_{17}H_{15}NO_5S$
Ethyl 2-benzoyl-2-(4-nitrophenylthio)acetate 364
2-Phthalimidoethyl p-toluenesulfonate 799

$C_{17}H_{15}NO_6$
Ethyl 2-(3-nitrophenyl)-2-benzoyloxyacetate 371

$C_{17}H_{15}N_3O$
10-Benzyl-2,3-dihydroimidazo(2,1-b)-quinazolin-5(10H)-one 726
10-Benzyl-2,10-dihydroimidazo(2,1-b)quinazolin-5(3H)-one methanesulfonate 726
1-Cinnamylideneamino-1,2-dihydro-4,6-dimethyl-2-oxonicotinonitrile 1069
4-(1-Naphthyl)-1-phenylsemicarbazide 578
1-(1-Naphthyl)-3-(3-pyridylmethyl)urea 951
3-Phenyl-2,3,3a,10-tetrahydropyrrolo(2,3-b)(1,5)-benzodiazepin-4(5H)-one 726

$C_{17}H_{15}N_3OS_2$
1-(4-(Methylthio)phenyl)-3-(4-phenyl-2-thiazolyl)urea 876

$C_{17}H_{15}N_3O_2$
α-(4-(Dimethylamino)benzylidene)-4-nitrophenylacetonitrile 765
1,3-Di(p-tolyl)-5-iminohydantoin 726
2-Methoxybenzoic (3-indolylmethylene)hydrazide 1042
2-Phenyl-4-phenylacetamido-3-pyrazolin-5-one 707

$C_{17}H_{15}N_3O_2S$
2-(N-Ethylanilino)-4-(4-nitrophenyl)-thiazole 876

$C_{17}H_{15}N_3O_3$
3′,5′-Dicyano-3,4-dihydrospiro(naphthalene-2(1H),4′-piperidine)-2′,6′-dione 753
Ethyl 4-(1-benzimidazolylcarbonylamino)-benzoate 906
Ethyl 4-(1H-indazol-1-ylcarbonylamino)benzoate 900
2-Nitrobenzoic acid (5-indanylmethylene)hydrazide 499

$C_{17}H_{15}N_3O_3S$
2-Amino-1-cyanoisoquinolinium p-toluenesulfonate 998
1-(1-Naphthyl)-3-(4-sulfamoylphenyl)urea 826

$C_{17}H_{15}N_3O_4$
Hippuric (3,4-(methylenedioxy)-benzylidene)hydrazide 501

$C_{17}H_{15}N_3O_4S$
3′-(Hydrazinosulfonyl)-1-hydroxy-2-naphthanilide 812

$C_{17}H_{15}N_3O_5S$
4-(1-Benzyl-5-oxo-2-pyrazolin-3-ylcarbamoyl)benzenesulfonic acid, barium salt 793
$C_{17}H_{15}N_3S$
1-(6-Methyl-2-pyridyl)-3-(2-naphthyl)-2-thiourea 965
$C_{17}H_{15}N_5O_5$
5-Indancarboxaldehyde 4-(2,4-dinitrophenyl)-semicarbazone 1053
$C_{17}H_{16}$
9,10-Dihydro-11-methyl-9,10-ethanoanthracene 162
1,4-Diphenyl-2-methyl-1,3-butadiene 165
$C_{17}H_{16}BF_4N_3O$
2-(2-Benzoylhydrazino)-1-methylquinolinium tetrafluoroborate 990
$C_{17}H_{16}BF_4N_3O_2S$
1-Ethyl-2-(phenylsulfonylazo)-quinolinium tetrafluoroborate 990
1-Ethyl-4-(phenylsulfonylazo)-quinolinium tetrafluoroborate 988
$C_{17}H_{16}BF_4N_5OS_2$
6-Methoxy-3-methyl-2[(3-methyl-2-benzothiazolinylidene)aminoazo]-benzothiazolium tetrafluoroborate 913
$C_{17}H_{16}BrClN_2O_4$
2-Bromo-4-*tert*-butylphenyl *N*-(4-chloro-2-nitrophenyl)carbamate 624
$C_{17}H_{16}BrN$
2-Phenethylisoquinolinium bromide 997
$C_{17}H_{16}BrNO$
4′-Bromo-4-(dimethylamino)-chalcone 302
$C_{17}H_{16}Br_2ClNO$
2-Chloro-4′,6′-dibromo-*N*-ethyl-*o*-acetotoluidide 410
$C_{17}H_{16}ClF_3N_6O_3$
2-(5-Methyl-2-pyridylureido)acetic N^2-[*N*-(4-chloro-3-trifluoromethyl-phenyl)carbamoyl]hydrazide 925
$C_{17}H_{16}ClN$
4-Chloro-2,2-diphenyl-3-methylbutyronitrile 751
4-Chloro-*β*-ethyl-*α*-phenylhydrocinnamonitrile 751
3-(4-Chlorophenyl)-1-ethyl-5-methylindole 881
$C_{17}H_{16}ClNO$
4′-Chloro-4-(dimethylamino)-chalcone 302
$C_{17}H_{16}ClNO_2$
2′-Benzoyl-4′-chlorobutyranilide 405
2-Chloro-4-(dimethylamino)-2′-hydroxychalcone 300
Cinnamyl *N*-(2-chloro-4-methylphenyl)carbamate 632
Cinnamyl *N*-(3-chloro-2-methylphenyl)carbamate 632
$C_{17}H_{16}ClNO_3$
4-Allyl-2-methoxyphenyl *N*-(3-chlorophenyl)carbamate 616
4-Allyl-2-methoxyphenyl *N*-(4-chlorophenyl)carbamate 616
5′-Chloro-2′,4′-dimethoxycinnamanilide 448
5-Indanyl *N*-(5-chloro-2-methoxyphenyl)carbamate 619
2-Methoxy-4-(1-propenyl)phenyl *N*-(4-chlorophenyl)carbamate 616
2-Methoxy-4-propenylphenyl *N*-(3-chlorophenyl)carbamate 616
3-Phenylallyl *N*-(5-chloro-2-methoxyphenyl)carbamate 618
$C_{17}H_{16}ClNO_4$
4-Chloro-2-methylphenyl *N*-(4-(ethoxycarbonyl)phenyl)carbamate. 630
4-(Ethoxycarbonyl)phenyl *N*-(2-chloro-6-methylphenyl)carbamate 630
Methyl 2-(2-(4-chloro-2-methylphenxoy)acetamido)-benzoate 402
$C_{17}H_{16}ClNO_5$
2-Acetylphenyl *N*-(5-chloro-2,4-dimethoxyphenyl)carbamate 614
$C_{17}H_{16}ClNO_6$
4-Formyl-2-methoxyphenyl *N*-(5-chloro-2,4-dimethoxyphenyl)-carbamate 613
$C_{17}H_{16}ClN_3O_2S$
1-(5-Chloro-2-methoxyphenyl)-3-(5,6-dimethyl-2-benzothiazolyl)urea 911
$C_{17}H_{16}ClN_5O_4$
2-(3-(2-Chloro-4-nitrophenyl)ureido)-acetic (4-methylbenzylidene)-hydrazide 522
2-(3-(2-Chloro-4-nitrophenyl)ureido)-acetic (2-phenylethylidene)-hydrazide 522
$C_{17}H_{16}ClN_5O_5$
1-(2-(4-Chlorobenzamido)acetyl)-4-(2-methyl-5-nitrophenyl)-semicarbazide 528
2-(3-(2-Chloro-4-nitrophenyl)ureido)-acetic (4-methoxybenzylidene)-hydrazide 517
$C_{17}H_{16}ClN_5O_5S_2$
2-[(6-Methoxy-3-methyl-2-benzo-thiazolinylidene)aminoazo]-3-methylbenzothiazolium perchlorate 913
$C_{17}H_{16}ClN_5O_6$
2-(3-(2-Chloro-4nitrophenyl)ureido)-acetic (4
$C_{17}H_{16}Cl_2N_2O_2$
3-(2,2-Dichloroacetamido)-4,4′-dimethylbenzanilide 445
4′,4″-Methylenebis(2-chloroacetanilide) 393
$C_{17}H_{16}Cl_2N_2O_3$
Ethyl 4-(4,5-dichloro-2-methylphenylureido)benzoate 532
$C_{17}H_{16}Cl_2N_4O_3$
2-(*p*-Tolylureido)acetic acid 2,4-dichlorobenzylidenehydrazide 529
$C_{17}H_{16}Cl_3NO_2$
2-(2,4,5-Trichlorophenoxy)-2′,4′,6′-trimethylacetanilide 402
$C_{17}H_{16}Cl_3NO_3$
(2-(4-Chloro-3,5-dimethylphenoxy)-ethyl) *N*-(2,6-dichlorophenylcarbamate 616
$C_{17}H_{16}Cl_3NO_4$
2,4-Dichloro-3,5-dimethylphenyl *N*-(5-chloro-2,4-dimethoxyphenyl)-carbamate 613
$C_{17}H_{16}FNO$
4-Dimethylamino-4′-fluorochalcone. 302
$C_{17}H_{16}FNO_3$
3,4-Dihydro-2*H*-1-benzopyran-3-ylmethyl *N*-(4-fluorophenyl)-carbamate 622
$C_{17}H_{16}INO_2$
1-Allyl-2-(3,4-methylenedioxystyryl)-pyridinium iodide 957
$C_{17}H_{16}INS$
1-Methyl-2-methylthio-4-phenylquinolinium iodide 987
$C_{17}H_{16}N_2$
2-(2-(2-Aminophenyl)ethyl)quinoline. 988
3-Cyano-1-(diphenylmethyl)-azetidine 756
2-(4-(Dimethylamino)phenyl)-quinoline dipicrate 990
4,5-Diphenyl-2-ethylimidazole 868
$C_{17}H_{16}N_2O$
Benzoic acid (5-indanylmethylene)-hydrazide 502
9,10-Dihydro-9,10-ethanoanthracene-11-carboxylic hydrazide 503
1,2-Dihydro-2-(5-indanyl)-4(3*H*)-quinazolinone 704
7,8-Dimethyl-4-phenyl-1*H*-1,5-benzodiazepin-2(3*H*)-one 702
6,7a-Diphenyl-2,3,5,7a-tetrahydro-pyrazolo(5,1-b)oxazole 253
1,2,3,4,6,10b-Hexahydro-10b-phenylpyrimido(2,1-a)isoindol-6-one 710
4-Methylbenzoic cinnamylidenehydrazide 501
$C_{17}H_{16}N_2OS$
1-Benzyl-3-cinnamoyl-2-thiourea 743
$C_{17}H_{16}N_2O_2$
N-(4-Anilinophenyl)glutarimide 719
N-(3-Anilinopropyl)phthalimide 718
3-(2-Benzamidoethyl)-2-indolinone 691
5-Benzyloxyindole-3-acetamide 882
Cinnamic (4-methoxybenzylidene)-hydrazide 1048
4-Cyanophenyl *N*-(2-ethyl-6-methylphenyl)carbamate 767
2,5-Diacetamidofluorene 491
2,7-Diacetamidofluorene 491
4-Hydroxybenzoic acid (5-indanylmethylene)hydrazide 496
10b-(4-Hydroxyphenyl)-1,2,3,4,6,10b-hexahydropyrimido(2,1-a)isoindol-6-one 697
2-Methoxybenzoic cinnamylidenehydrazide 1048
2-Methyl-1-phenyl-5-benzimidazolyl propionate 907
N-(3,4-Xylidinomethyl)-phthalimide 719
$C_{17}H_{16}N_2O_2S$
2-(2-Aminophenylthio)-*N*-(*p*-tolyl)-succinimide 715
1,3-Bis(3-acetylphenyl)-2-thiourea 744
1,3-Bis(4-acetylphenyl)-2-thiourea 744
$C_{17}H_{16}N_2O_3$
1-(*N*-(2,5-Dimethoxyphenyl)-carbamoyl)indole 882
N-(*o*-Phenetidinomethyl)-phthalimide 717
N-(*p*-Phenetidinomethyl)-phthalimide 716
$C_{17}H_{16}N_2O_3S$
2-(2-Aminophenylthio)-*N*-(2-methoxyphenyl)-succinimide 716
2-(2-Aminophenylthio)-*N*-(3-methoxyphenyl)-succinimide 717
2-(2-Aminophenylthio)-*N*-(4-methoxyphenyl)-succinimide 718
Ethyl 4-(3-benzoyl-2-thioureido)-benzoate 744
$C_{17}H_{16}N_2O_4$
4-Acetamidophenyl *N*-(3-acetylphenyl)carbamate 675
N-(Benzylideneamino)-*N*-(benzyloxycarbonyl)glycine 499
4-Hydroxyhydrocinnamic acid piperonylidenehydrazide 496
$C_{17}H_{16}N_2O_4S$
4-(3-(2-Phenoxypropionyl)-2-thioureido)benzoic acid 736
4-Sulfobenzoic acid (5-indanylmethylene)hydrazide 1074
$C_{17}H_{16}N_2O_5$
N-(4-Acetamido-3-methoxyphenyl)-phthalamic acid 465
4′-Benzamido-3′-hydroxysuccinanilate 462
1,3-Bis(2-(methoxycarbonyl)-phenyl)urea 576
3-Phenylallyl *N*-(2-methoxy-4-nitrophenyl)carbamate 652
$C_{17}H_{16}N_2O_8$
4,4-Bis-(4-hydroxy-3-nitrophenyl)-valeric acid 325
$C_{17}H_{16}N_2S$
3-Ethyl-2-(2-(phenylimino)-ethylidene)benzothiazoline 1062
$C_{17}H_{16}N_2Se$
2,3-Dihydro-3-ethyl-2-(2-(phenylimino)ethylidene)-benzoselenazole 915
$C_{17}H_{16}N_4$
1-Phenyl-*N*,*N*′-di-2-pyridinylmethanediamine 953
$C_{17}H_{16}N_4O$
3-Methyl-1-phenyl-4-(4-tolylhydrazono)-2-pyrazolin-5-one 1053
$C_{17}H_{16}N_4OS$
1-(4-(Methylthio)phenyl)-3-(1-phenyl-5-pyrazolyl)urea 860
5,8-Quinolinedione 5-[(3-methyl-4,5,6,7-tetrahydro-2-benzothiazolinylidene)hydrazone] 954
$C_{17}H_{16}N_4O_3$
Ethyl 4-(3-(5-1*H*-indazolyl)ureido)-benzoate 900
$C_{17}H_{16}N_4O_6$
Ethyl 2-(*N*,*N*′-bis-(3-nitrophenyl)-amidino)acetate 1056
$C_{17}H_{16}O$
9,10-Dihydro-11-hydroxy-12-methyl-9,10-ethanoanthracene 206
9,10-Dihydro-11-(hydroxymethyl)-9,10-ethanoanthracene 206
2′,4′-Dimethylchalcone 304
2,6-Dimethylchalcone 304
4,4′-Dimethylchalcone 306
1,5-Diphenyl-4-penten-1-one 304
trans-1-(4-Methylbenzoyl)-2-phenylcyclopropane 296
2-Methyl-3-(*p*-tolyl)acrylophenone 305
$C_{17}H_{16}O_2$
6-Hydroxy-1-phenyl-1,2,3,4-tetrahydronaphthalene 195
3-Methoxy-4′-methylchalcone 301
4′-Methoxy-4-methylchalcone 301
4-Methoxy-4′-methylchalcone 302
Phenethyl cinnamate 365
$C_{17}H_{16}O_2S$
Ethyl *α*-(phenylthio)cinnamate 364
$C_{17}H_{16}O_3$
10,11-Dihydro-2,3-dimethoxy-5*H*-dibenzo(a,D)cyclohepten-5-one 311
3,4-Dimethoxychalcone 300
4,4′-Dimethoxychalcone 301
2-(2,4-Dimethylbenzoyl)-3-methylbenzoic acid 344
2-(4-Isopropylbenzoyl)benzoic acid 345
2-(2,4,6-Trimethylbenzoyl)-benzoic acid 345
$C_{17}H_{16}O_3S$
9,10-Dihydro-9,10-ethanoanthracen-11-yl methanesulfonate 799
$C_{17}H_{16}O_4$
4-Acetoxy-*α*-phenylhydrocinnamic acid 332
4-Benzyloxy-3-methoxycinnamic acid 334
Bis(4-methoxybenzoyl)methane 307
Di(*p*-tolyl) malonate 380
7-Methoxy-1,3,3a,3b,4,5,11,11a-octahydrophenanthro(1,2-c)furan-1,3-dione 388
2,2′-(Trimethylenedioxy)-dibenzaldehyde 323
$C_{17}H_{16}O_5$
2-(*p*-Anisyl)-3-phenylsuccinic acid 328
2-(*p*-Anisyl)-3-phenylsuccinic acid, diammonium salt 350
3-(3-Benzyloxy-4-methoxyphenyl)-glycidic acid, sodium salt 69
$C_{17}H_{16}O_6$
1,3-Bis(4-carboxyphenoxy)propane 339
$C_{17}H_{16}O_8$
1,5-Dimethyl-6-(ethoxycarbonyl)-bicyclo(2.2.2)oct-5-ene-2,3,7,8-tetra-carboxylic dianhydride 98
$C_{17}H_{16}Si$
Methyl(1-naphthyl)phenylsilane 1121
$C_{17}H_{17}BrClNO_2$
2-Bromo-4-chloro-6-methylphenyl *N*-(4-isopropylphenyl)carbamate 635
$C_{17}H_{17}BrClNO_4$
2-Bromo-4,5-dimethylphenyl *N*-(5-chloro-2,4-dimethoxyphenyl)-carbamate 613
4-Bromo-2,5-dimethylphenyl *N*-(5-chloro-2,4-dimethoxyphenyl)-carbamate 613
$C_{17}H_{17}BrClNO_5$
2-(4-Bromophenoxy)ethyl *N*-(5-chloro-2,4-dimethoxyphenyl)-carbamate 612
$C_{17}H_{17}BrN_2O$
1-Bromo-3-(2-hydroxy-3,4,5,6-tetra-hydrobenzylideneamino)-4-methyl-isoquinolinium perchlorate 1022
$C_{17}H_{17}BrO_2$
4′-Bromo-3′,5′-dimethyl-2-phenoxypropiophenone 288
$C_{17}H_{17}Br_2N$
2-(*N*,*N*-Bis(2-bromoethyl)amino)-fluorene 222

$C_{17}H_{17}Br_2NO_2$
α,β-Dibromo-2′-ethoxyhydrocinnamanilide 416

$C_{17}H_{17}ClFNO_2$
4-Chloro-2-isopropyl-5-methylphenyl *N*-(4-fluorophenyl)-carbamate 635

$C_{17}H_{17}ClN_2O_2$
1-(4-Chloro-α-hydroxybenzyl)-1,2,3,4-tetrahydro-2-isoquinolinecarboxamide 693

$C_{17}H_{17}ClN_2O_2S$
4-Chlorobenzaldehyde *N*-phenyl-*N*-(3-sulfolanyl)hydrazone 773

$C_{17}H_{17}ClN_2O_3$
2′-Benzoyl-4′-chloro-2-(2-hydroxyethylamino)acetanilide 398
4-*tert*-Butyl-4′-chloro-3′-nitrobenzanilide 451
4-Chloro-1-hydroxy-*N*-(*N*-(methacrylamido)ethyl)-2-naphthamide 485
Ethyl 4-(5-chloro-2-methylphenylureido)benzoate 532
Methyl 2-(3-(2-chloro-6-methylphenyl)-1-methylureido)-benzoate 531

$C_{17}H_{17}ClN_2O_3S$
S-(4-*tert*-Butylphenyl) *N*-(2-chloro-4-nitrophenyl)thiocarbamate 627

$C_{17}H_{17}ClN_2O_4$
1-(3-Acetylphenyl)-3-(4-chloro-2,5-dimethoxyphenyl)urea 516
Ethyl 4-(3-(2-chloro-4-methoxyphenyl)ureido)benzoate 520
Ethyl 4-(5-chloro-2-methoxyphenylureido)benzoate 519

$C_{17}H_{17}ClN_2O_5$
4-Acetamidophenyl *N*-(5-chloro-2,4-dimethoxyphenyl)carbamate 613
1-(4-Chlorobenzoyl)-2-(3,4,5-trimethoxybenzoyl)hydrazine 492

$C_{17}H_{17}ClN_2O_5S$
2-(4-(3-(5-Chloro-2,4-dimethoxyphenyl)ureido)-phenylthio)acetic acid 514

$C_{17}H_{17}ClN_2O_7$
2-(4-Nitrophenoxy)ethyl *N*-(5-chloro-2,4-dimethoxyphenyl)carbamate 612

$C_{17}H_{17}ClN_4O_2$
2-(3-Benzylureido)acetic (4-chlorobenzylidene)hydrazide 523
2-(3-(3-Chlorophenyl)ureido)acetic (4-methylbenzylidene)hydrazide 523
2-(3-(3-Chlorophenyl)ureido)acetic (2-phenylethylidene)hydrazide 522

$C_{17}H_{17}ClN_4O_3$
4′-Chloro-2′-nitro-4-phenyl-1-piperazinecarboxanilide 694

$C_{17}H_{17}ClN_4O_4$
N-(3-Chlorophenylureido)-*N*′-(4-ethoxyphenyl)oxamide 520

$C_{17}H_{17}ClN_6O_4S$
2-((1,3-Dimethyl-2-benzimidazolinylidene)aminoazo)-3-methylbenzothiazolium perchlorate 915

$C_{17}H_{17}Cl_2N$
2-(*N*,*N*-Bis(2-chloroethyl)amino)-fluorene 222

$C_{17}H_{17}Cl_2NO$
2,4-Dichloro-2′,6′-diethylbenzanilide 459
2,5-Dichloro-2′,6′-diethylbenzanilide 460
2,6-Dichloro-2′,6′-diethylbenzanilide 459

$C_{17}H_{17}Cl_2NOS$
S-(4-*tert*-Butylphenyl) *N*-(3,4-dichlorophenyl)thiocarbamate 641

$C_{17}H_{17}Cl_2NO_2$
2,4-Dichloro-3,5-dimethylphenyl *N*-(3,4-xylyl)carbamate 639

$C_{17}H_{17}Cl_2NO_3$
4-Butoxyphenyl *N*-(2,3-dichlorophenyl)carbamate 617
3,4-Dichloro-*N*-(3,4-dimethoxyphenethyl)benzamide 464
2-(2,4-Dichlorophenoxy)ethyl *N*-(3-ethylphenyl)carbamate 616
2-(2,4-Dichlorophenoxy)ethyl *N*-(2,3-xylyl)carbamate 650
2-(2,4-Dichlorophenoxy)ethyl *N*-(2,4-xylyl)carbamate 651

$C_{17}H_{17}Cl_2NO_4$
3-(4-Methoxyphenoxy)propyl *N*-(2,6-dichlorophenyl)carbamate 613

$C_{17}H_{17}Cl_2NO_4S$
2-(4-Chlorophenylthio)ethyl *N*-(5-chloro-2,4-dimethoxyphenyl)-carbamate 613

$C_{17}H_{17}Cl_2N_3O_3$
1-(3-Phenylureido)-2-propyl *N*-(3,4-dichlorophenyl)carbamate 529

$C_{17}H_{17}Cl_3N_2O$
1-(2,6-Diethylphenyl)-3-(2,4,5-trichlorophenyl)urea 543

$C_{17}H_{17}FN_2O_3$
4-*tert*-Butyl-2′-fluoro-5′-nitrobenzanilide 451
4-*tert*-Butyl-4′-fluoro-3′-nitrobenzanilide 451

$C_{17}H_{17}FN_2O_4$
2-Hydroxypropyl 4-(3-(4-fluorophenyl)ureido)benzoate 513

$C_{17}H_{17}FN_4O_2$
2-(3-(4-Fluorophenyl)ureido)acetic (4-methylbenzylidene)hydrazide 523

$C_{17}H_{17}FN_4O_3$
N-(4-Fluoro-3-nitrophenyl)-4-phenyl-1-piperazinecarboxamide 694
2-(3-(4-Fluorophenyl)ureido)acetic (4-methoxybenzylidene)hydrazide 517

$C_{17}H_{17}FN_4O_4$
2-(3-(4-Fluorophenyl)ureido)acetic (3-hydroxy-4-methoxybenzylidene)-hydrazide 512
2-(3-(4-Fluorophenyl)ureido)acetic (4-hydroxy-3-methoxybenzylidene)-hydrazide 512

$C_{17}H_{17}F_2NO_2$
2-Isopropyl-5-methylphenyl *N*-(2,4-difluorophenyl)carbamate 639

$C_{17}H_{17}F_3N_2O$
1-(3,4-Dimethylbenzyl)-3-(2-(trifluoromethyl)phenyl)urea 509
1-(3,4-Dimethylbenzyl)-3-(3-(trifluoromethyl)phenyl)urea 509
1-(2-Ethyl-6-methylphenyl)-3-(3-trifluoromethylphenyl)urea 511
1-Mesityl-3-(3-(trifluoromethyl)-phenyl)urea 512
1-(2-(Trifluoromethyl)phenyl)-3-(2,4,6-trimethylphenyl)urea 511
1-(3-(Trifluoromethyl)phenyl)-3-(2,4,5-trimethylphenyl)urea 512

$C_{17}H_{17}F_3N_6O_3$
1-(5-Methyl-2-pyridylureidoacetyl)-4-(3-(trifluoromethyl)phenyl)-semicarbazide 925

$C_{17}H_{17}N$
11-(Aminomethyl)-9,10-dihydro-9,10-ethanoanthracene hydrochloride hydrate 241
N-(Cinnamylidene)-2,3-xylidine 1079
N-(Cinnamylidene)-2,4-xylidine 1079
N-(Cinnamylidene)-3,4-xylidine 1079
N-(Cinnamylidene)-3,5-xylidine 1079
2,2-Diphenylisovaleronitrile 756
2,2-Diphenylvaleronitrile 756
2,3-Diphenylvaleronitrile 756

$C_{17}H_{17}NO$
N-Benzhydrylcyclopropane-carboxamide 437
N-Benzyl-*N*-methylcinnamamide 482
N-Cinnamylidene-*p*-phenetidine 1050
2-Diethylamino-9-fluorenone 312
4,5-Dihydro-3,3-diphenyl-5-methyl-2(3*H*)furanimine hydrochloride 1075
1,4-Dihydro-3-ethoxy-1-phenylisoquinoline 247
3,3-Dimethyl-2-(2-hydroxy-5-methylphenyl)-3*H*-indole 246
1,4-Diphenyl-3-ethyl-2-azetidinone 692
2′-Ethylcinnamanilide 478
4′-Ethylcinnamanilide 483
O-(2-Methylallyl)benzophenone oxime 1011
N-(5,6,7,8-Tetrahydro-1-naphthyl)-benzamide 478

$C_{17}H_{17}NO_2$
4-Carboxy-*N*-(4-isopropylbenzylidene)aniline 1063
Cinnamyl *N*-(*p*-tolyl)carbamate 681
9,10-Dihydro-9-(1-methyl-1-nitroethyl)anthracene 258
2-(3-(Dimethylamino)propionyl)-dibenzofuran hydrochloride 895
N-(Diphenylmethyl)-acetoacetamide 443
4-Methoxy-α-(4-methoxyphenyl)-hydrocinnamonitrile 753
4-Methoxy-α-(4-methoxyphenyl)-hydrocinnamonitrile 753
Methyl 1-phenyl-1,2,3,4-tetrahydro-3-isoquinolinecarboxylate hydrochloride 362
2-Morpholinobenzophenone 312
3-Phenylallyl *N*-(*m*-tolyl)carbamate 680

$C_{17}H_{17}NO_2S$
S-(3-Benzamidopropyl) thiobenzoate 477
5-Indanyl *N*-(4-(methylthio)phenyl)-carbamate 668

$C_{17}H_{17}NO_3$
2′,5′-Dimethoxycinnamanilide 463
2-(Methoxycarbonyl)-2′,6′-dimethylbenzanilide 474
4-(1-Methyl-1-nitropropyl)-benzophenone 313
4-Propionylphenyl *N*-(*p*-tolyl)-carbamate 675
2′,4′,6′-Trimethylphthalanilic acid 474
2,6-Xylyl *N*-(3-acetylphenyl)-carbamate 675

$C_{17}H_{17}NO_3S$
S-(2-Tolyl) *N*-(4-(ethoxycarbonyl)-phenyl)thiocarbamate 676

$C_{17}H_{17}NO_4$
2-Benzamido-3-methoxy-3-phenylpropionic acid 463
Carbobenzyloxy-DL-phenylalanine 657
α-(4-(Ethoxycarbonyl)phenylimino)-6-methoxy-*o*-cresol 1030
4-Formyl-2-methoxyphenyl *N*-(2,5-xylyl)carbamate 648
3-(4-Indolyl)-5-methyl-4-cyclohexene-1,2-dicarboxylic acid 889
4-Methoxybenzyl *N*-(3-acetylphenyl)carbamate 646
4-(Methoxycarbonyl)phenyl *N*-(3-ethylphenyl)carbamate 676
2-(4-Methoxy-1-naphthyl)-1-methyl-5-oxo-3-pyrrolidinecarboxylic acid 690
4-Nitrobenzyl 2,4,6-trimethylbenzoate 370
N-(4-Phenoxyphenyl)-glutaramic acid 418
Piperonyl *N*-(2,3-xylyl)carbamate 665
Piperonyl *N*-(2,4-xylyl)carbamate 646
Piperonyl *N*-(2,6-xylyl)carbamate 665
3-(Propionyloxy)phenyl *N*-(*o*-tolyl)-carbamate 666
4-Propionylphenyl *N*-(4-methoxyphenyl)carbamate 654

$C_{17}H_{17}NO_4S$
S-(4-Methoxyphenyl) *N*-(4-(ethoxycarbonyl)phenyl)-thiocarbamate 654
4-(Methylthio)phenyl *N*-(4-(ethoxycarbonyl)phenyl)carbamate 667

$C_{17}H_{17}NO_5$
3-Acetylphenyl *N*-(2,4-dimethoxyphenyl)carbamate 647
O-(Carbobenzyloxy)-L-tyrosine hydrochloride 381
4-(Ethoxycarbonyl)phenyl *N*-(2-methoxyphenyl)carbamate 653
2′-(2-Ethoxy-5-nitrobenzyloxy)-acetophenone 292
4-(Methoxycarbonyl)phenyl *N*-(2-ethoxyphenyl)carbamate 653
4-(Methoxycarbonyl)phenyl *N*-(4-ethoxyphenyl)carbamate 654
4-Methoxyphenyl *N*-(4-(ethoxycarbonyl)phenyl)carbamate 651

$C_{17}H_{17}NO_6$
4-Formyl-2-methoxyphenyl *N*-(2,5-dimethoxyphenyl)carbamate 647
2-(Methoxycarbonyl)phenyl *N*-(2,4-dimethoxyphenyl)carbamate 647
2-(Methoxycarbonyl)phenyl *N*-(2,5-dimethoxyphenyl)carbamate 647

$C_{17}H_{17}NO_8$
Tetramethyl 9ah-quinolizine-1,2,3,4-tetracarboxylate 90

$C_{17}H_{17}N_3$
N-Ethyl-*N*-(4-(phenyliminomethyl)-phenyl)aminoacetonitrile 1061
2,3,3a,4,5,6-Hexahydro-3-(4-pyridyl)-benzo(6,7)cyclohepta(1,2-c)-pyrazole 966
2-Methyl-3-(4-pyridyl)-3,3a,4,5-tetrahydro-2*H*-benzene(g)indazole 966

$C_{17}H_{17}N_3O$
Acetic acid, (9-ethyl-2-carbazolylmethylene)hydrazide 1062
3-(3-Aminophenyl)-2-ethyl-1-phenyl-3-pyrazolin-5-one 701
10β-(2-Aminophenyl)-1,2,3,4,6,10b-hexahydropyrimido(2,1-a)isoindol-6-one 701
5-Indancarboxaldehyde 4-phenylsemicarbazone 1071
N-Methylanthranilic cinnamylidenehydrazide 1060
2,2,2-Tris(2-cyanoethyl)-acetophenone 757

$C_{17}H_{17}N_3O_2$
Hippuric *N*-(4-methylbenzylidene)-hydrazide 1069
Hippuric (2-phenylethylidene)-hydrazide 499
p-Toluic (4-acetamidobenzylidene)-hydrazide 1063

$C_{17}H_{17}N_3O_2S$
2-(2-Hydroxyethoxy)benzaldehyde, 3-methyl-2-benzothiazolinone azine 1031

$C_{17}H_{17}N_3O_3$
1-(4-Acetamidophenyl)-3-(3-acetylphenyl)urea 569
2-(4-Acetamidophenyl)-7-nitro-1,2,3,4-tetrahydroisoquinoline 428
1,2-Dihydro-1-(2,3-dimethoxybenzylideneamino)-4,6-dimethyl-2-oxonicotinonitrile 1048
1,2-Dihydro-4,6-dimethyl-2-oxo-1-(veratrylideneamino)nicotinonitrile 1042
Hippuric (4-methoxybenzylidene)-hydrazide 498
2-Methoxybenzoic (4-acetamidobenzylidene)-hydrazide 497
2-(2-Nitro-6-phenanthridinylamino)-1-butanol 999

$C_{17}H_{17}N_3O_3S$
1-(4-Acetamidophenyl)-3-(2-phenoxyacetyl)-2-thiourea 736

$C_{17}H_{17}N_3O_4$
1-Acetyl-3-(2-methoxy-5-(*N*-phenylcarbamoyl)phenyl)urea 554
Hippuric (4-hydroxy-3-methoxybenzylidene)hydrazide 495
2-Methoxybenzoic (4-(*N*-methylcarbamoyloxy)benzylidene)-hydrazide 652

$C_{17}H_{17}N_3O_4S$
4-Acetamido-*N*-cyano-4′-ethoxybenzenesulfonanilide 814
2-Amino-1-carbamoylisoquinolinium *p*-toluenesulfonate 998
1-(2,5-Dimethoxyphenyl)-3-(6-methoxy-2-benzothiazolyl)urea 912
3-Nitrobenzaldehyde *N*-phenyl-*N*-(3-sulfolanyl)hydrazone 1058

$C_{17}H_{17}N_3O_5$
2′,6′-Diethyl-3,5-dinitrobenzanilide 469
Methyl 2-(1-methyl-3-(2-methyl-5-nitrophenyl)ureido)benzoate 563

$C_{17}H_{17}N_3O_5S_2$
Methanesulfonic acid (1-methyl-2-phenyl-6-sulfo-4(1*H*)-

$C_{17}H_{17}N_3O_5S_2$ (Cont'd)
quinolylidene) hydrazide ... 1056
$C_{17}H_{17}N_3O_6$
Ethyl 4-(3-(4-methoxy-3-nitrophenyl)-ureido)benzoate ... 552
$C_{17}H_{17}N_3S$
5-Indancarboxaldehyde 4-phenyl-3-thiosemicarbazone ... 1077
$C_{17}H_{17}N_5O_4$
2′,3′-*o*-Benzylideneadenosine ... 920
2-(3-(5-Methyl-2-pyridyl)ureido)-acetic (piperonylmethylene)-hydrazide ... 1070
2-(3-(3-Nitrophenyl)ureido)acetic (2-phenylethylidene)hydrazide ... 562
$C_{17}H_{17}N_5O_5$
2-(4-Methoxyphenylureido)acetic acid (3-nitrobenzylidene)hydrazide ... 553
$C_{17}H_{17}N_5O_6$
2-(3-(3-Nitrophenyl)ureido)acetic (4-hydroxy-3-methoxybenzylidene)-hydrazide ... 547
$C_{17}H_{18}BF_4N_3O_2S$
1-Ethyl-2-(2-(phenylsulfonyl)-hydrazino)quinolinium tetrafluoroborate ... 990
$C_{17}H_{18}BrNO$
2-Bromo-2′,6′-diethylbenzanilide ... 457
3-Bromo-2′,6′-diethylbenzanilide ... 459
4-Bromo-2′,6′-diethylbenzanilide ... 458
6′-Bromo-*N*-methyl-2-phenyl-*o*-propionotoluidide ... 410
$C_{17}H_{18}BrNO_2$
4-Bromo-2,5-dimethylphenyl *N*-(2,4-dimethylphenyl)carbamate ... 639
4-Bromo-3,5-dimethylphenyl *N*-(2,6-dimethylphenyl)carbamate ... 637
4-Bromo-2,5-dimethylphenyl *N*-(2-ethylphenyl)carbamate ... 633
$C_{17}H_{18}BrNO_3$
2-Bromo-4,5-dimethylphenyl *N*-(4-ethoxyphenyl)carbamate ... 620
2-Bromo-4,5-dimethylphenyl *N*-(2-methoxy-5-methylphenyl)-carbamate ... 619
4-Bromo-2,5-dimethylphenyl *N*-(2-methoxy-5-methylphenyl)-carbamate ... 619
4-Bromo-3,5-dimethylphenyl *N*-(2-methoxy-5-methylphenyl)-carbamate ... 619
1a,1b,3,4,5,5a,6,6a-Octahydro-2,5-methano-2*H*-indeno(1,2-b)oxiren-3-yl *N*-(3-bromophenyl)carbamate ... 631
$C_{17}H_{18}BrNO_4$
4-Bromo-2,5-dimethylphenyl *N*-(2,5-dimethoxyphenyl)carbamate ... 614
4-Bromo-3,5-dimethylphenyl *N*-(2,5-dimethoxyphenyl)carbamate ... 614
$C_{17}H_{18}ClFN_2O_2$
3-(3-Chloro-4-fluorophenyl)-1-ethyl-1-(2-methoxy-5-methylphenyl)urea ... 521
$C_{17}H_{18}ClN$
N-(2-Chlorobenzylidene)-2,6-diethylaniline ... 1064
$C_{17}H_{18}ClNO$
N-Benzyl-2′-chloro-2-methylpropionanilide ... 404
2-Chloro-2′,6′-diethylbenzanilide ... 457
3-Chloro-2′,6′-diethylbenzanilide ... 459
4-Chloro-2′,6′-diethylbenzanilide ... 458
3-Chloro-*N*-(1,2-diphenylethyl)-propionamide ... 392
2-Chloro-5′-isopropyl-2′-methylbenzanilide ... 405
2-Chloro-*N*-methyl-2-phenyl-2′,6′-acetoxylidide ... 441
$C_{17}H_{18}ClNOS$
2-(4-Chlorophenylthio)-2′,4′,6′-trimethylacetanilide ... 403
$C_{17}H_{18}ClNO_2$
N-(4-Chlorobenzylidene)-3,4-dimethoxyphenethylamine ... 1041
3-Chloro-4-biphenylyl *N*-*tert*-butylcarbamate ... 632
4-Chloro-2,5-dimethylphenyl *N*-(2,4-xylyl)carbamate ... 640
4-Chloro-3,5-dimethylphenyl *N*-(2,5-xylyl)carbamate ... 639
4-Chloro-3,5-dimethylphenyl *N*-(2,6-xylyl)carbamate ... 637
2-(4-Chloro-2-methylphenoxy)-2′,4′-dimethylacetanilide ... 400
2-(4-Chloro-2-methylphenoxy)-2′-ethylacetanilide ... 400
2-(2-(4-Chlorophenyl)-2-hydroxypropyl)-*N*-methylbenzamide ... 446
3-Ethyl-5-methylphenyl *N*-(5-chloro-2-methylphenyl)carbamate ... 639
4-Isobutylphenyl *N*-(4-chlorophenyl)carbamate ... 634
2-Isopropyl-5-methylphenyl *N*-(4-chlorophenyl)carbamate ... 635
3-Phenylpropyl *N*-(3-chloro-2-methylphenyl)carbamate ... 632
2,3,5-Trimethylphenyl *N*-(5-chloro-2-methylphenyl)carbamate ... 639
$C_{17}H_{18}ClNO_3$
cis-4-(3-Chloroacryloyl)-3a,8b-dimethyl-1,2,3,3a,4,8b-hexahydro-6-methoxycyclopent(b)indol-2-one ... 699
3-Chloro-*N*-(3,4-dimethoxyphenethyl)benzamide ... 464
4-Chloro-*N*-(3,4-dimethoxyphenethyl)benzamide ... 464
4-Chloro-3,5-dimethylphenyl *N*-(2-ethoxyphenyl)carbamate ... 618
2-Chloro-4,5-dimethylphenyl *N*-(2-methoxy-5-methylphenyl)-carbamate ... 619
4-Chloro-2,6-dimethylphenyl *N*-(2-methoxy-5-methylphenyl)-carbamate ... 619
1,2-Epoxy-octahydro-4,7-methano-1*H*-inden-5-yl *N*-(4-chlorophenyl)-carbamate ... 631
3-Ethyl-5-methylphenyl *N*-(5-chloro-2-methoxyphenyl)carbamate ... 619
2-Isopropylphenyl *N*-(5-chloro-2-methoxyphenyl)carbamate ... 618
3-Phenylpropyl *N*-(5-chloro-2-methoxyphenyl)carbamate ... 618
$C_{17}H_{18}ClNO_4$
2-Benzyl 4-ethyl 5-chloromethyl-3-methyl-2,4-pyrroledicarboxylate ... 838
2-(2-Chlorophenoxy)ethyl *N*-(2-methoxy-5-methylphenyl)-carbamate ... 613
$C_{17}H_{18}ClNO_6$
2,6-Dimethoxyphenyl *N*-(5-chloro-2,4-dimethoxyphenyl)carbamate ... 612
(4-Methoxyphenoxy)methyl *N*-(5-chloro-2,4-dimethoxyphenyl)-carbamate ... 613
$C_{17}H_{18}ClN_3O$
p-Toluic (2-chloro-4-(dimethylamino)benzylidene)-hydrazide ... 1025
$C_{17}H_{18}ClN_3O_2$
1-Allylcarbamoylmethyl-3-(phenylcarbamoyl)-pyridinium chloride ... 947
$C_{17}H_{18}ClN_3O_3S$
1-(4-Acetamidophenyl)-3-(5-chloro-2,4-dimethoxyphenyl)-2-thiourea ... 728
$C_{17}H_{18}ClN_3O_4$
1-(4-Acetamidophenyl)-3-(5-chloro-2,4-dimethoxyphenyl)urea ... 514
$C_{17}H_{18}ClN_3O_5$
1-(4-Chlorobenzoyl)-4-(3,4,5-trimethoxyphenyl)semicarbazide ... 516
$C_{17}H_{18}ClN_5O_6$
N-(4-(2-Chloro-4,6-dinitrophenylazo)-3-methylphenyl)diethanolamine ... 1081
$C_{17}H_{18}Cl_2N_2O$
1-Benzyl-3-(2,6-dichlorophenyl)-1-isopropylurea ... 528
1-Butyl-3-(2,3-dichlorophenyl)-1-phenylurea ... 596
1-Butyl-3-(2,4-dichlorophenyl)-1-phenylurea ... 536
1-Butyl-3-(2,5-dichlorophenyl)-1-phenylurea ... 601
1-Butyl-3-(2,6-dichlorophenyl)-1-phenylurea ... 536
1-(2,3-Dichlorophenyl)-3-(2,6-diethylphenyl)urea ... 541
1-(2,4-Dichlorophenyl)-3-(2,6-diethylphenyl)urea ... 542
1-(2,5-Dichlorophenyl)-3-(2,6-diethylphenyl)urea ... 542
1-(2,6-Dichlorophenyl)-3-(2,6-diethylphenyl)urea ... 541
1-(3,4-Dichlorophenyl)-3-(2,6-diethylphenyl)urea ... 542
α-(2,4-Dichlorophenylimino)-5-(diethylamino)-*o*-cresol ... 1022
$C_{17}H_{18}Cl_2N_2O_2$
3-Diethylaminophenyl *N*-(2,3-dichlorophenyl)carbamate ... 627
3-Diethylaminophenyl *N*-(2,6-dichlorophenyl)carbamate ... 673
$C_{17}H_{18}Cl_2N_4O_4$
4′-(Bis(2-hydroxyethyl)amino)-2,6-dichloro-2′-methyl-4-nitroazobenzene ... 1080
$C_{17}H_{18}Cl_2O_2$
2,2-Bis(4-methoxyphenyl)-1,3-dichloropropane ... 176
$C_{17}H_{18}FNOS$
S-(4-*tert*-Butylphenyl) *N*-(4-fluorophenyl)thiocarbamate ... 641
$C_{17}H_{18}FNO_2$
2-*tert*-Butylphenyl *N*-(4-fluorophenyl)carbamate ... 633
2-Isopropyl-5-methylphenyl *N*-(3-fluorophenyl)carbamate ... 639
$C_{17}H_{18}FNO_5$
3-Fluorobenzyl *N*-(3,4,5-trimethoxyphenyl)carbamate ... 612
3,4,5-Trimethoxybenzyl *N*-(2-fluorophenyl)carbamate ... 612
$C_{17}H_{18}FN_3O_5$
1-(2,5-Diethoxyphenyl)-3-(4-fluoro-3-nitrophenyl)urea ... 515
$C_{17}H_{18}HgO_4$
2-Hydroxy-2-(4-methoxyphenyl)-2-phenylethylmercuri acetate ... 1137
$C_{17}H_{18}INO$
1-Allyl-2-(4-methoxystyryl)-pyridinium iodide ... 938
2-Methyl-5-phenyl-3-propylbenzoxazolium iodide ... 910
$C_{17}H_{18}INO_4$
2,5-Dimethyl-4-iodophenyl *N*-(2,5-dimethoxyphenyl)carbamate ... 614
$C_{17}H_{18}N_2$
N-β-Benzyltryptamine hydrochloride ... 885
N-Cinnamylidene-*N*′,*N*′-dimethyl-1,4-phenylenediamine ... 1057
1,2,3,3a,8,8a-Hexahydro-8-methyl-3a-phenylpyrrolo(2,3-b)indole hydrochloride ... 251
3-(*N*-Methylbenzylaminomethyl)-indole ... 884
$C_{17}H_{18}N_2O$
2-Acetamido-9-(dimethylamino)-fluorene ... 488
Benzoic (α-propylbenzylidene)-hydrazide ... 1063
1,1-Diallyl-3-(1-naphthyl)urea ... 603
1,2-Dihydro-2-(4-isopropylphenyl)-4(3*H*)-quinazolinone ... 704
2-(4-Methoxyphenyl)tryptamine ... 882
2-(4-Methoxyphenyl)tryptamine hydrochloride ... 882
9-(Morpholinomethyl)carbazole ... 892
2-(4-Phenyl-1,2,3,6-tetrahydro-1-pyridinylmethyl)-3-pyridinol hydrochloride ... 932
$C_{17}H_{18}N_2OS$
3,6-Bis(dimethylamino)-9-xanthenethione ... 251
$C_{17}H_{18}N_2O_2$
1-Acetyl-3-(1,2-diphenylethyl)-urea ... 570
1-Acetyl-1,3-di(*o*-tolyl)urea ... 579
N-Acetyl-*N*-methyl-*N*′-phenylacetyl-1,4-phenylenediamine ... 428
3-(3-Acetylphenyl)-1-benzyl-1-methylurea ... 571
1-(3-Acetylphenyl)-3-phenethylurea ... 570
2,6-Diethyl-*N*-(2-nitrobenzylidene)-aniline ... 1055
2-(*N*-(3,4-Dimethoxybenzyl)amino)-2-phenylacetonitrile ... 753
4′,4″-Methylenebis(acetanilide) ... 439
α,α′-(1-Methylethylenediimino)di-*o*-cresol ... 1033
O-(*N*-(1-Naphthyl)carbamoyl)-cyclohexanone oxime ... 1012
2-Phenyl-3-(*m*-tolyl)succinamide ... 427
Salicylic (α-propylbenzylidene)-hydrazide ... 1036
$C_{17}H_{18}N_2O_2S$
1-Benzyl-3-(2-phenoxypropionyl)-2-thiourea ... 735
N-(2-(3-Indolyl)ethyl)-*p*-toluenesulfonamide ... 886
$C_{17}H_{18}N_2O_2S_2$
4-(2-Thienyl)-1-(*p*-tolylsulfonyl)-isonipecotonitrile ... 856
$C_{17}H_{18}N_2O_3$
4-Acetamidophenyl *N*-(3-ethylphenyl)carbamate ... 677
4-Acetamidophenyl *N*-(2,5-xylyl)-carbamate ... 677
4-Acetamidophenyl *N*-(2,6-xylyl)-carbamate ... 677
1-(4-Acetylphenyl)-3-(2-methoxy-5-methylphenyl)urea ... 553
4′-Anilinoglutaranilic acid ... 427
Benzoic acid (4-ethoxy-3-methoxybenzylidene)hydrazide ... 497
1-(2-Carboxyphenyl)-3-(2-ethyl-6-methylphenyl)urea ... 576
1-(4-Carboxyphenyl)-3-(2-ethyl-6-methylphenyl)urea ... 576
2′,6′-Diethyl-3-nitrobenzanilide ... 473
2′,6′-Diethyl-4-nitrobenzanilide ... 472
1,2-Dihydro-2-(2-ethoxy-3-methoxyphenyl)-4-(3*H*)-quinazolinone ... 698
1,2-Dihydro-2-(4-ethoxy-3-methoxyphenyl)-4(3*H*)-quinazolinone ... 698
5′-Isopropyl-2′-methyl-3-nitrobenzanilide ... 432
5′-Isopropyl-2′-methyl-4-nitrobenzanilide ... 473
$C_{17}H_{18}N_2O_3S$
Ethyl 4-(3-(4-(methylthio)phenyl)-ureido)benzoate ... 560
Salicylaldehyde *N*-phenyl-*N*-(3-sulfolanyl)hydrazone ... 774
$C_{17}H_{18}N_2O_4$
1-(3-Acetylphenyl)-3-(2,5-dimethoxyphenyl)urea ... 550
1-(4-Acetylphenyl)-3-(2,5-dimethoxyphenyl)urea ... 551
Benzoic acid (3,4,5-trimethoxybenzylidene)-hydrazide ... 498
4-(*tert*-Butyl)-2′-hydroxy-4′-nitrobenzanilide ... 462
Carbobenzyloxy-L-tyrosinamide ... 642
Dibenzyl 1-methyl-1,2-hydrazinedicarboxylate ... 664
1,2-Dihydro-2-(2,3,4-trimethoxyphenyl)-4(3*H*)-quinazolinone ... 704
1,2-Dihydro-2-(3,4,5-trimethoxyphenyl)-4(3*H*)-quinazolinone ... 700
Dimethyl 4′,4″-methylenedicarbanilate ... 648
4-Hydroxyhydrocinnamic acid vanillylidenehydrazide ... 495
4-Methoxybenzoic (2,3-dimethoxybenzylidene)hydrazide ... 1041
$C_{17}H_{18}N_2O_5$
3-Acetamidophenyl *N*-(2,4-dimethoxyphenyl)carbamate ... 648
3-Acetamidophenyl *N*-(2,5-dimethoxyphenyl)carbamate ... 648
4-Acetamidophenyl *N*-(2,4-dimethoxyphenyl)carbamate ... 648
4-Acetamidophenyl *N*-(2,5-dimethoxyphenyl)carbamate ... 648

$C_{17}H_{18}N_2O_5$ (Cont'd)

N-(3,4-Dimethoxyphenethyl)-4-nitrobenzamide 464
4-Hydroxybenzoic acid (3,4,5-trimethoxybenzylidene)hydrazide 495
4-Hydroxy-3-methoxybenzoic (3-ethoxy-4-hydroxybenzylidene)-hydrazide 495

$C_{17}H_{18}N_2O_5S$

4-Acetamido-*N*-(4-(ethoxycarbonyl)-phenyl)benzenesulfonamide 823

$C_{17}H_{18}N_2O_6$

1,5-Bis(2-nitrophenoxy)pentane 268
1,5-Bis(3-nitrophenoxy)pentane 268
2-(4-Methoxyphenoxy)ethyl *N*-(4-methyl-2-nitrophenyl)carbamate 648
2-(4-Methoxyphenoxy)ethyl *N*-(4-methyl-3-nitrophenyl)carbamate 647
N-(4-Methyl-3-nitrophenyl)-3,4,5-trimethoxybenzamide 463
4,4′-(Pentamethylenedioxy)-bis-(nitrobenzene) 269

$C_{17}H_{18}N_2O_7$

N-(2-Methoxy-5-nitrophenyl)-3,4,5-trimethoxybenzamide 464
N-(4-Methoxy-2-nitrophenyl)-3,4,5-trimethoxybenzamide 463

$C_{17}H_{18}N_2O_8$

1-(3,4,5-Trihydroxybenzoyl)-2-(3,4,5-trimethoxybenzoyl)hydrazine 494

$C_{17}H_{18}N_2O_8S$

2-Phenoxyethyl 5-acetamido-2-methoxy-3-nitrobenzenesulfonate 797

$C_{17}H_{18}N_2O_{12}$

1-*O*-(3,4-Dinitrophenyl)-2,3,4-tri-*O*-acetyl-β-D-xylopyranoside 355

$C_{17}H_{18}N_2S$

1-Allyl-3-(diphenylmethyl)-2-thiourea 748

$C_{17}H_{18}N_2S_2$

3,5-Dibenzyl-3,4,5,6-tetrahydro-2*H*-1,3,5-thiadiazine-2-thione 742
3,5-Dimethyl-4,6-diphenyl-3,4,5,6-tetrahydro-2*H*-1,3,5-thiadiazine-2-thione 742
3,5-Di-*p*-tolyl-3,4,5,6-tetrahydro-2*H*-1,3,5-thiadiazine-2-thione 742

$C_{17}H_{18}N_4$

4-(4-(*N*-Methyl-*N*-isopropylamino)-phenylazo)benzonitrile 1089

$C_{17}H_{18}N_4O$

1,2-Dihydro-4,6-dimethyl-1-(4-(dimethylamino)-benzylideneamino)-2-oxonicotino-nitrile 1061

$C_{17}H_{18}N_4O_3$

2-(3-Benzylureido)acetic (3-hydroxybenzylidene)hydrazide 547

$C_{17}H_{18}N_4O_4$

1-(4-(*N*-Methylacetamido)phenyl)-3-(2-methyl-5-nitrophenyl)urea 562
1-(4-(*N*-Methylacetamido)phenyl)-3-(4-methyl-2-nitrophenyl)urea 561
1-(4-(*N*-Methylacetamido)phenyl)-3-(4-methyl-3-nitrophenyl)urea 562

$C_{17}H_{18}N_4O_5$

1-(2-Methoxy-5-nitrophenyl)-3-(4-(*N*-methylacetamido)phenyl)urea 562

$C_{17}H_{18}N_4S$

3-Methyl-2-benzothiazolinone (4-(dimethylamino)benzylidene)-hydrazone 1060

$C_{17}H_{18}N_6O_2$

Glutaric acid bis-((2-pyridylmethylene)-hydrazide) 1053

$C_{17}H_{18}O$

4,4-Dimethyl-3-(2-naphthyl)-1-pentyn-3-ol 1017
1,1-Diphenyl-3-pentanone 287
3-Methyl-2-phenylbutyrophenone 296
5-Phenylvalerophenone 296
2,3,5,6-Tetramethylbenzophenone 313

$C_{17}H_{18}O_2$

Benzoin isopropyl ether 292
Benzyl 2-methoxy-4-propenylphenyl ether 178
Benzyl 2,4,6-trimethylbenzoate 370
4-(*tert*-Butyl)-phenyl benzoate 380
4,4′-Cyclopentylidenediphenol 194
α-Cyclopropyl-4-methoxydiphenylmethanol 201
2′,4′-Dimethyl-2-phenoxypropiophenone 292
Ethyl 2-(1,2,3,4-tetrahydro-1-fluorenylidene)acetate 375
2-(2-Methyl-1-phenylpropyl)-benzoic acid 345

$C_{17}H_{18}O_2S$

2-(2-(Butylthio)phenyl)benzoic acid 342
Mesityl styryl sulfone 784

$C_{17}H_{18}O_3$

2,6-Diethoxybenzophenone 311
Ethyl 4-benzyloxyphenylacetate 354
Ethyl 3,3-diphenyl-3-hydroxypropionate 354

$C_{17}H_{18}O_3S$

α,α-Diphenyl-2-sulfolanemethanol 774
Tricyclo[5.2.1.$0^{2,6}$]deca-4,8-dien-10-yl *p*-toluenesulfonate 799

$C_{17}H_{18}O_4$

4-(Benzyloxy)-3,5-dimethoxyacetophenone 291

$C_{17}H_{18}O_4S_2$

2,3-Bis(*p*-tosyl)-1-propene 785

$C_{17}H_{18}O_5$

6,7,8,8A,9,10-hexahydro-2-methoxyphenanthrene-7,8-dicarboxylic acid 328
2,3,3′,4′-Tetramethoxybenzophenone 311

$C_{17}H_{18}O_5S$

Ethyl α-(*p*-tolylsulfonyloxy)-phenylacetate 797

$C_{17}H_{18}O_6$

4,4′,5,6-Tetramethoxy-2-biphenylcarboxylic acid 340

$C_{17}H_{19}BrN_2O$

4′-Bromo-2,6-diethylcarbanilide 591

$C_{17}H_{19}BrN_2O_3$

1-(4-Bromo-2,6-dimethylphenyl)-3-(2,5-dimethoxyphenyl)urea 516

$C_{17}H_{19}Cl$

4-*tert*-Butyl-4′-chlorodiphenylmethane 173

$C_{17}H_{19}ClN_2O$

1-Benzyl-3-(3-chlorophenyl)-1-isopropylurea 571
1-Butyl-3-(2-chlorophenyl)-1-phenylurea 588
1-Butyl-3-(3-chlorophenyl)-1-phenylurea 581
1-Butyl-3-(4-chlorophenyl)-1-phenylurea 591
1-(4-Butylphenyl)-3-(4-chlorophenyl)urea 539
1-(4-Chlorobenzyl)-1-methyl-3-(2,5-xylyl)urea 528
1-(3-Chloro-4-methylphenyl)-3-(2-ethyl-6-methylphenyl)urea 542
1-(4-Chloro-2-methylphenyl)-3-(2-ethyl-6-methylphenyl)urea 542
1-(3-Chloro-4-methylphenyl)-3-(2,4,5-trimethylphenyl)urea 545
1-(3-Chloro-4-methylphenyl)-3-(2,4,6-trimethylphenyl)urea 545
1-(5-Chloro-2-methylphenyl)-3-(2,4,5-trimethylphenyl)urea 545
1-(2-Chlorophenyl)-3-(2,6-diethylphenyl)urea 537
1-(3-Chlorophenyl)-3-(2,6-diethylphenyl)urea 541
1-(4-Chlorophenyl)-3-(2,6-diethylphenyl)urea 539
1-(3-Chlorophenyl)-1-ethyl-3-(3-ethylphenyl)urea 534

$C_{17}H_{19}ClN_2OS$

2-Chloro-10-(3-dimethylaminopropyl)-10*H*-phenothiazine-*N*-oxide maleate 249
2-Chloro-10-(3-dimethylaminopropyl)-10*H*-phenothiazine-5-oxide picrate 275

$C_{17}H_{19}ClN_2O_2$

α-(5-Chloro-2,4-dimethoxyphenylimino)-*N*,*N*-dimethyl-*p*-toluidine 1044
1-(4-Chlorophenyl)-1-ethyl-3-(2-methoxy-5-methylphenyl)urea 519
1-(4-Chlorophenyl)-1-ethyl-3-(2-methoxy-5-methylphenyl)urea 519
3-Diethylaminophenyl *N*-(2-chlorophenyl)carbamate 673
3-Diethylaminophenyl *N*-(3-chlorophenyl)carbamate 673

$C_{17}H_{19}ClN_2O_2S$

2-Chloro-7,8-dihydroxy-10-(3dimethylaminopropyl)-10*H*-phenothiazine hydrochloride 245
1-(5-Chloro-2,4-dimethoxyphenyl)-3-(2,4-dimethylphenyl)-2-thiourea 729
2,5-Diethoxy-4-(*p*-tolylthio)-diazobenzene, zinc chloride double salt 1094

$C_{17}H_{19}ClN_2O_3$

1-Benzyl-3-(5-chloro-2,4-dimethoxyphenyl)-1-methylurea 514
α-(5-Chloro-2,4-dimethoxyanilino)-*p*-acetotoluidide 419
1-(2-Chlorophenyl)-3-(2,5-diethoxyphenyl)urea 515

$C_{17}H_{19}ClN_2O_4$

1-(5-Chloro-2,4-dimethoxyphenyl)-3-(4-(2-hydroxyethyl)phenyl)urea 512
1-(5-Chloro-2,4-dimethoxyphenyl)-3-(2-methoxy-5-methylphenyl)urea 515

$C_{17}H_{19}ClN_2O_4S$

1-(5-Chloro-2,4-dimethoxyphenyl)-3-(2,5-dimethoxyphenyl)-2-thiourea 728

$C_{17}H_{19}ClN_2O_5$

1-(4-Chloro-2,5-dimethoxyphenyl)-3-(2,5-dimethoxyphenyl)urea 516
1-(5-Chloro-2,4-dimethoxyphenyl)-3-(2,5-dimethoxyphenyl)urea 514

$C_{17}H_{19}ClN_4O_3$

4-(5-Chloro-2-methoxyphenyl)-1-(4-(dimethylamino)benzoyl)-semicarbazide 518

$C_{17}H_{19}ClN_4O_4$

4′-(Bis(2-hydroxyethyl)amino)-2-chloro-2′-methyl-4-nitroazobenzene 1083

$C_{17}H_{19}ClPb$

Chlorodiphenyl(4-pentenyl)lead 1135

$C_{17}H_{19}Cl_3N_2O$

1-(3,4-Dichlorobenzyl)-3-(diethylcarbamoyl)-pyridinium chloride 948
1-(3,4-Dichlorobenzyl)-4-(diethylcarbamoyl)-pyridinium chloride 952

$C_{17}H_{19}Cl_3N_2O_2$

2,2,2-Trichloroethyl 4-cyano-4-phenyl-1-piperidinepropionate hydrochloride 751

$C_{17}H_{19}FNO_3P$

Diethyl *N*-(7-fluoro-2-fluorenyl)-phosphoramidate 837

$C_{17}H_{19}FN_2O$

1-(4-Butylphenyl)-3-(2-fluorophenyl)urea 537
1-(4-Butylphenyl)-3-(4-fluorophenyl)urea 539
1-(2-Fluorophenyl)-3-(5-isopropyl-2-methylphenyl)urea 538

$C_{17}H_{19}FN_2O_2$

2-(*N*-Ethylanilino)ethyl *N*-(2-fluorophenyl)carbamate 627
2-(*N*-Ethylanilino)ethyl *N*-(4-fluorophenyl)carbamate 627

$C_{17}H_{19}F_2N_3O$

1-(4-(Diethylamino)phenyl)-3-(2,5-difluorophenyl)urea 526

$C_{17}H_{19}INO$

2-(2-Ethoxystyryl)-1-ethylpyridinium iodide 938

$C_{17}H_{19}IN_2O$

4-(4-Acetamidostyryl)-1-ethylpyridinium iodide 956

$C_{17}H_{19}N$

N-Benzylidene-2,6-diethylaniline 1067
2-Diethylaminofluorene 233
2,3-Diphenylpiperidine 250
N-(4-Isopropylbenzylidene)-*p*-toluidine 1065

$C_{17}H_{19}NO$

α-(4-Butylphenylimino)-*o*-cresol 1034
2-Diethylamino-9-fluorenol 227
2′,6′-Diethylbenzanilide 481
2,2′-Dimethyl-6′-ethylbenzanilide 479
2′,3-Dimethyl-6′-ethylbenzanilide 481
2′,4-Dimethyl-6′-ethylbenzanilide 480
(-)-*N*,2-Diphenyl-3-methylbutyramide 437
1-(Diphenylmethyl)-3-methoxyazetidine hydrochloride 247
N-(2-Ethoxybenzylidene)-2,3-xylidine 1046
N-(2-Ethoxybenzylidene)-2,5-xylidine 1046
N-(2-Ethoxybenzylidene)-2,6-xylidine 1046
2,3,4,6,7,11B-hexahydro-1*H*-dibenzo(a,f)quinolizin-13(12*H*)-one 287
N-(4-Isopropylbenzylidene)-*p*-anisidine 1050
N-Methyl-*N*-(1-methyl-2-phenylethyl)benzamide 471
4′-(α-Methylphenethylamino)-acetophenone hydrochloride 294
2-Phenyl-2′,4′,5′-trimethylacetanilide 438
2-Phenyl-2′,4′,6′-trimethylacetanilide 438
2,2′,4′,6′-Tetramethylbenzanilide 480
2′,3,4′,6′-Tetramethylbenzanilide 482
2′,4,4′,6′-Tetramethylbenzanilide 481

$C_{17}H_{19}NOS$

2-(Benzylthio)-2′,3′-dimethylacetanilide 421
2-(Benzylthio)-3′,4′dimethylacetanilide 421
N-Ethyl-2-methoxythiobenzo-*m*-toluidide 736
N-Ethyl-4-methoxythiobenzo-*m*-toluidide 736

$C_{17}H_{19}NO_2$

2-(*N*,*N*-Bis(2-hydroxyethyl)amino)-fluorene 225
N-Cyclohexyl-1-hydroxy-2-naphthamide 486
Cyclohexyl *N*-(1-naphthyl)-carbamate 680
N-(2,5-Dimethoxybenzylidene)-3,4-xylidine 1048
3-(Dimethylamino)-4′-phenoxypropiophenone hydrochloride 292
N-Ethyl-2-methoxybenzo-*m*-toluidide 467
2′-Ethyl-4-methoxy-6′-methylbenzanilide 467
Ethyl α-phenyl-2-pyridinebutyrate 964
α-(4-Hydroxyphenyl)-1-methyl-1,2,3,4-tetrahydro-6-quinolinemethanol 245
2-Methoxy-2′,4′,6′-trimethylbenzanilide 464
4-Methoxy-2′,4′,6′-trimethylbenzanilide 467
6-Methoxy-α-(2,4,6-trimethylphenylimino)-*o*-cresol 1031
2-Naphthyl *N*-cyclohexylcarbamate. 680
Phenyl *N*-(2,6-diethylphenyl)-carbamate 684
2-Phenyl-*N*-(*p*-tolyl)-glycine ethyl ester 358
o-Tolyl *N*-(2-ethyl-6-methylphenyl)-carbamate 682
N-Veratrylidene-2,3-xylidine 1042
N-Veratrylidene-2,4-xylidine 1042
N-Veratrylidene-2,5-xylidine 1042
N-Veratrylidene-3,4-xylidine 1042
2,6-Xylyl *N*-(2,6-xylyl)carbamate 684

$C_{17}H_{19}NO_2S$

2-(Benzylthio)-4′-ethoxyacetanilide 414
4-(2-(6-Methoxy-2-naphthyl)-thioacetyl)morpholine 738
N-(4-(Methylthio)phenyl)-2-phenylglycine ethyl ester 357
2,3,5-Trimethylphenyl *N*-(4-(methylthio)phenyl)carbamate 668

$C_{17}H_{19}NO_3$
o-Benzyl-DL-serine benzyl ester hydrochloride ... 354
4-*tert*-Butylphenyl 4-nitrobenzyl ether ... 265
2′,5′-Diethoxybenzanilide ... 465
N-(2,5-Dimethoxybenzylidene)-*p*-phenetidine ... 1047
3,5-Dimethoxy-2′,6′-dimethylbenzanilide ... 463
N-(3,4-Dimethoxyphenethyl)-benzamide ... 464
α-(2-(3,4-Dimethoxyphenyl)-ethylimino)-*p*-cresol ... 1029
3-(*N*-(3,4-Dimethylphenyl)-carbamoyl)-5-norbornene-2-carboxylic acid ... 434
2,2-Diphenyl-2-(2-(methylamino)-ethoxy)acetic acid hydrobromide ... 327
2-Isopropylphenyl *N*-(4-methoxyphenyl)carbamate ... 654
5-Norbornen-2-ylmethyl *N*-(3-acetylphenyl)carbamate ... 675
2-Phenoxyethyl *N*-(2,5-dimethylphenyl)carbamate ... 650
2-Phenoxyethyl *N*-(3-ethylphenyl)-carbamate ... 649
3-(*N*-(2-Phenylethyl)carbamoyl)-5-norbornene-2-carboxylic acid ... 429
3-(*N*-(2,5-Xylyl)carbamoyl)-5-norbornene-2-carboxylic acid ... 434
$C_{17}H_{19}NO_3S$
1-(Diphenylmethyl)-3-azetidinyl methanesulfonate ... 798
3-Methyl-4-(methylthio)phenyl *N*-(2-ethoxyphenyl)carbamate ... 651
3-Methyl-4-(methylthio)phenyl *N*-(4-ethoxyphenyl)carbamate ... 651
$C_{17}H_{19}NO_3S_3$
2-(Methylthio)benzothiazole ethyl *p*-toluenesulfonate ... 916
$C_{17}H_{19}NO_4$
Benzyl 5-acetoxymethyl-3,4-dimethyl-2-pyrrolecarboxylate ... 838
5-(Benzyloxycarbonyl)-2,4-dimethyl-3-pyrrolepropionic acid ... 842
N-(4-Benzyloxyphenoxy)glycine ethyl ester ... 354
2,5-Dimethoxy-*N*-(2,5-dimethoxybenzylidene)aniline ... 1045
2,5-Dimethoxy-*N*-(veratrylidene)-aniline ... 1041
3-(*N*-(4-Ethoxyphenyl)carbamoyl)-bicyclo(2.2.1)hept-5-ene-2-carboxylic acid ... 418
Ethyl 4-(2-hydroxy-3-methoxybenzylamino)benzoate ... 367
2-Phenoxyethyl *N*-(2-ethoxyphenyl)-carbamate ... 647
2-Phenoxyethyl *N*-(4-ethoxyphenyl)-carbamate ... 647
$C_{17}H_{19}NO_4S$
2-(4-Methoxyphenoxy)ethyl *N*-(3-(methylthio)phenyl)carbamate ... 647
$C_{17}H_{19}NO_4S_2$
S-Benzyl-*N*-(4-toluenesulfonyl)-L-cysteine ... 815
$C_{17}H_{19}NO_5$
2-(2-Naphthoxy)ethyl *N*-(ethoxycarbonylmethyl)-carbamate ... 649
2-(Phenoxy)ethyl *N*-(2,4-dimethoxyphenyl)carbamate ... 646
2-Phenoxyethyl *N*-(2,5-dimethoxyphenyl)carbamate ... 647
$C_{17}H_{19}NO_5S$
Diethyl 2-(benzo(b)thiophene-3-ylmethyl)-2-formamido)malonate ... 897
$C_{17}H_{19}NO_6$
Diethyl 2-phthalimidoglutarate ... 499
2,6-Dimethoxyphenyl *N*-(2,5-dimethoxyphenyl)carbamate ... 647
$C_{17}H_{19}N_2I$
Methyl(3-(*N*-methylanilino)-allylylidene)phenylammonium iodide ... 1054
$C_{17}H_{19}N_3$
1-(4-(Dimethylamino)phenyl)-6-ethylbenzimidazole ... 905
$C_{17}H_{19}N_3O$
4′-(4-(*N*-Ethyl-*N*-methylamino)-phenylazo)acetophenone ... 1089
p-Toluic (4-(dimethylamino)-benzylidene)hydrazide ... 1079
$C_{17}H_{19}N_3O_2$
1-((4-Acetylamino)phenyl)-3-(2,5-xylyl)urea ... 569
1-(3-Acetylphenyl)-3-(4-(dimethylamino)phenyl)urea ... 567
2-(Phenoxy)acetic (4-(dimethylamino)benzylidene)-hydrazide ... 498
$C_{17}H_{19}N_3O_2S_2$
1,4-Diphenyl-1-(3-sulfolanyl)-3-thiosemicarbazide ... 779
$C_{17}H_{19}N_3O_3$
Anthranilic (4-ethoxy-3-methoxybenzylidene)hydrazide ... 1040
1-Benzyl-1-isopropyl-3-(4-nitrophenyl)urea ... 571
4-(Dimethylamino)benzoic (4-hydroxy-3-methoxybenzylidene)-hydrazide ... 495
1-(3,4-Dimethylbenzyl)-3-(4-methyl-2-nitrophenyl)urea ... 562
4-(2,6-Dimethylphenyl)-1-(2-methoxybenzoyl)semicarbazide ... 551
4-(2-Methoxy-5-methylphenyl)-1-(2-phenylacetyl)semicarbazide ... 553
4-(2-Methoxy-5-methylphenyl)-1-(2-phenylacetyl)semicarbazide ... 553
1-(4-Methyl-3-nitrophenyl)-3-mesitylurea ... 564
1-(3-Phenylureido)-2-propyl *N*-phenylcarbamate ... 674
$C_{17}H_{19}N_3O_4$
1-(4-Acetamidophenyl)-3-(2,5-dimethoxyphenyl)urea ... 550
4-(Bis(2-hydroxyethyl)amino)-*N*-(4-nitrobenzylidene)aniline ... 1033
1-Isopropyl-3-(2-methoxy-4-nitrophenyl)-1-phenylurea ... 552
1-Isopropyl-3-(2-methoxy-5-nitrophenyl)-1-phenylurea ... 552
1-(4-Methoxybenzoyl)-4-(2-methoxy-5-methylphenyl)semicarbazide ... 550
$C_{17}H_{19}N_3O_4S_2$
Methyl 8-benzamido-2,2-dimethyl-2,3,6,7,8,8a-hexahydro-7-oxo-5-thioxo-5*H*-thiazolo(3,2-c)-pyrimidine-3-carboxylate ... 742
$C_{17}H_{19}N_3O_5$
4-(2,5-Dimethoxyphenyl)-1-(4-methoxybenzoyl)semicarbazide ... 550
$C_{17}H_{19}N_3O_5S$
N-(4-Ethoxyphenyl)-*N*′-(4-tolylsulfonamido)oxamide ... 815
$C_{17}H_{19}N_3O_6$
1-(2-Hydroxybenzoyl)-4-(3,4,5-trimethoxyphenyl)semicarbazide ... 546
$C_{17}H_{19}N_3S$
2-(4-(Dimethylamino)phenylimino)-3-ethylbenzothiazoline ... 725
$C_{17}H_{19}N_5O_3$
2-(3-(2-Ethoxyphenyl)ureido)acetic (2-pyridylmethylene)hydrazide ... 937
$C_{17}H_{19}N_5O_4$
4,5-Dihydro-2,4,5-tripropionyl(1,2,4)-triazolo(3,4-c)(1,2,4)benzotriazin-1(2*H*)one ... 980
$C_{17}H_{19}O_2P$
3-(Phenyl(3,5-xylyl)phosphino)-propionic acid ... 830
$C_{17}H_{20}$
1,1-Diphenylpentane ... 163
4,4′-Isopropylideneditoluene ... 163
$C_{17}H_{20}BrN$
N-(1-(3-Bromophenyl)pentyl)aniline hydrochloride ... 236
N-(1-(4-Bromophenyl)pentyl)aniline hydrochloride ... 224
1,3-Dihydrospiro(2*H*-benzene(f)-isoindole-2,1′-piperidinium) bromide ... 255
$C_{17}H_{20}BrNO$
(Benzoylmethyl)-benzyldimethylammonium bromide ... 295
$C_{17}H_{20}ClN$
α-Butyl-2-chloro-*N*-phenylbenzylamine ... 225
$C_{17}H_{20}ClNO$
5-Chloro-2-(4-pentylphenoxy)-aniline ... 223
$C_{17}H_{20}ClNO_4$
N-(5-Chloro-2,4-dimethoxyphenyl)-2,3-dimethoxybenzylamine ... 229
N-(5-Chloro-2,4-dimethoxyphenyl)-3,4-dimethoxybenzylamine ... 228
$C_{17}H_{20}ClN_3O_3$
1-(5-Chloro-2,4-dimethoxyphenyl)-3-(4-(dimethylamino)phenyl)urea ... 514
$C_{17}H_{20}ClN_3O_4$
1,3-Bis(*p*-tolyl)tetrahydro-1,2,3-triazinium perchlorate ... 255
$C_{17}H_{20}ClN_5O_3$
2-(3-Furfurylureido)acetic (2-chloro-4-(dimethylamino)benzylidene)-hydrazide ... 1025
$C_{17}H_{20}ClN_5O_4$
2-(3-Methyl-2*H*-pyran-6-ylureido)-acetic *N*2-(*N*-(3-chloro-2-methyl-phenyl)carbamoyl)hydrazide ... 532
$C_{17}H_{20}F_3N_3O$
1-(2-Cyanoethyl)-1-cyclohexyl-3-(3-(trifluoromethyl)phenyl)urea ... 508
$C_{17}H_{20}IN$
2-(4-Isopropylstyryl)pyridine methiodide ... 957
$C_{17}H_{20}INO_2$
2-(3,4-Dimethoxystyryl)-1-ethylpyridinium iodide ... 937
4-(3,4-Dimethoxystyryl)-1-ethylpyridinium iodide ... 937
$C_{17}H_{20}N_2$
1,3-Dibenzylimidazolidine ... 252
N-(4-(Diethylamino)benzylidene)-aniline ... 1061
9-(Diethylaminomethyl)-carbazole ... 892
N-(4-(Dimethylamino)benzylidene)-4-ethylaniline ... 1062
N-(4-(Dimethylamino)benzylidene)-2,5-xylidine ... 1061
N,*N*-Dimethyl-α-(2-ethylphenylimino)-*p*-toluidine ... 1062
N,*N*-Dimethyl-*N*′-(fluoren-9-yl)-ethylenediamine dihydrochloride ... 242
N,*N*-Dimethyl-α-(2,4-xylylimino)-*p*-toluidine ... 1061
N,*N*-Dimethyl-α-(3,4-xylylimino)-*p*-toluidine ... 1062
$C_{17}H_{20}N_2O$
3′-(*N*-Benzyl-*N*-ethylamino)-acetanilide ... 425
1-Benzyl-3-(2-ethyl-6-methylphenyl)urea ... 573
1-Benzyl-1-ethyl-3-(*o*-tolyl)urea ... 572
1-Benzyl-1-isopropyl-3-phenylurea ... 571
1,3-Bis(α-methylbenzyl)urea ... 571
1-Butyl-1,3-diphenylurea ... 586
1-Cyclohexyl-3-(1-naphthyl)urea ... 584
1,1-Dibenzyl-3-ethylurea ... 571
1-(2,6-Diethylphenyl)-3-phenylurea ... 586
N-(4-(Dimethylamino)benzylidene)-2-methyl-*p*-anisidine ... 1044
2,6-Dimethyl-4′-(dimethylamino)-4-hydroxybenzophenone imine ... 1034
1-(2,4-Dimethylphenyl)-3-(2,6-dimethylphenyl)urea ... 596
1,3-Di(2,4-xylyl)urea ... 601
1-Ethyl-1,2,3,4,5,6,7,12b-octahydroindolo(2,3-a)quinolizin-4-one ... 710
1-(2-Ethylphenyl)-3-(3,4-xylyl)urea ... 589
1-Mesityl-3-(*m*-tolyl)urea ... 602
1-(2-Methoxy-1-naphthylmethyleneamino)-piperidine ... 1049
2-Methyl-*N*-(1-naphthyl)-1-piperidinecarboxamide ... 712
3-Methyl-*N*-(1-naphthyl)-1-piperidinecarboxamide ... 712
4-Methyl-*N*-(1-naphthyl)-1-piperidinecarboxamide ... 712
1-Phenethyl-3-(2,4-xylyl)urea ... 570
1-Phenethyl-3-(2,6-xylyl)urea ... 586
α-(2,4-Xylidino)-*p*-acetotoluidide ... 430
$C_{17}H_{20}N_2OS$
1-(2,6-Diethylphenyl)-3-(4-hydroxyphenyl)-2-thiourea ... 735
10-(3-Dimethylaminopropyl)-10*H*-phenothiazine-5-oxide hydrochloride ... 771
1-(4-Ethoxyphenyl)-3-phenethyl-2-thiourea ... 736
2-(*N*-Ethylanilino)ethyl *N*-phenylthiocarbamate ... 742
1-(4-(Methylthio)phenyl)-3-(2,4,5-trimethylphenyl)urea ... 561
1-(4-(Methylthio)phenyl)-3-(2,4,6-trimethylphenyl)urea ... 561
$C_{17}H_{20}N_2O_2$
2-(Anilino)ethyl *N*-ethyl-*N*-phenylcarbamate ... 671
1-Benzyl-1-ethyl-3-(4-methoxyphenyl)urea ... 558
N-(4-(Dimethylamino)benzylidene)-*p*-phenetidine ... 1044
1-(2,4-Dimethylbenzyl)-3-(2-methoxyphenyl)urea ... 555
1-(2,5-Dimethylbenzyl)-3-(2-methoxyphenyl)urea ... 553
1-(3,4-Dimethylbenzyl)-3-(4-methoxyphenyl)urea ... 736
2-(*N*-Ethylanilino)ethyl *N*-phenylcarbamate ... 671
1-(2-Ethyl-6-methylphenyl)-3-(2-methoxyphenyl)urea ... 555
1-(2-Ethylphenyl)-3-(4-(2-hydroxyethyl)phenyl)urea ... 549
1-(4-(2-Hydroxyethyl)phenyl)-3-(2,5-xylyl)urea ... 549
1-(4-(2-Hydroxyethyl)phenyl)-3-(2,6-xylyl)urea ... 549
1-Mesityl-3-(2-methoxyphenyl)urea ... 552
1-(4-Methoxyphenyl)-3-(2,4,6-trimethylphenyl)urea ... 560
$C_{17}H_{20}N_2O_2S$
N-Benzoyl-2-morpholino-1-cyclopentene-1-thiocarboxamide ... 740
10-(3-Dimethylaminopropyl)-10*H*-phenothiazine-*N*,5-dioxide picrate ... 1005
2-Methylpropiophenone (*p*-tosyl)-hydrazone ... 822
$C_{17}H_{20}N_2O_3$
1-Benzyl-3-(2,5-dimethoxyphenyl)-1-methylurea ... 550
1,3-Bis-(2-ethoxyphenyl)-urea ... 559
2,2′-Dimethoxy-5,5′-dimethylcarbanilide ... 557
1-(2,5-Dimethoxyphenyl)-3-(2,6-dimethylphenyl)urea ... 550
1-(2,5-Dimethoxyphenyl)-3-(2-phenethyl)urea ... 553
$C_{17}H_{20}N_2O_3S$
4-Acetamido-2′,4′,5′-trimethylbenzenesulfonanilide ... 825
1-(3,5-Dimethoxyphenyl)-3-(2-ethoxyphenyl)-2-thiourea ... 735
$C_{17}H_{20}N_2O_4$
4,4-Bis-(3-amino-4-hydroxyphenyl)-valeric acid ... 325
1-(2,4-Dimethoxyphenyl)-3-(4-(2-hydroxyethyl)phenyl)urea ... 546
$C_{17}H_{20}N_2O_4S$
2′-Hydroxy-4′-trimethylacetamido-benzenesulfonanilide ... 812
$C_{17}H_{20}N_2O_5$
1,3-Bis(2,4-dimethoxyphenyl)urea ... 549
1,3-Bis(2,5-dimethoxyphenyl)urea ... 550
$C_{17}H_{20}N_2O_6$
Carbobenzyloxy-L-valine *N*-hydroxysuccinimide ester ... 717
$C_{17}H_{20}N_4$
Bis(4-(dimethylamino)phenyl)-carbodiimide ... 769
$C_{17}H_{20}N_4O$
1-((4-Benzyloxy-α-ethyl)-benzylideneamino)guanidine acetate ... 1040

$C_{17}H_{20}N_4O$ (Cont'd)
1-((4-Benzyloxy-α-ethyl)-benzylideneamino)guanidine nitrate 1040
1-(α-Methyl-3-(2-phenylethoxy)-benzylideneamino)guanidine nitrate 1049

$C_{17}H_{20}N_4O_2$
1-((4-Benzyloxy-3-ethoxybenzylidene)amino)-guanidine nitrate 1039
4-(Dimethylamino)-4′-nitro-2,2′,6-trimethylazobenzene 1089
4-(Dimethylamino)-4′-nitro-2,3′,6-trimethylazobenzene 1089
4-(2,6-Dimethylphenyl)-1-(2-(methylamino)benzoyl)-semicarbazide 570
4-(3,5-Dimethylphenyl)-1-(2-(methylamino)benzoyl)-semicarbazide 570
1-(2,6-Dimethylphenyl)-6-(2-methylphenyl)biurea 574
N-Ethyl-*N*-isopropyl-4-(4-nitrophenylazo)aniline 1089
6-Ethyl-3-phenyl-1-(*o*-tolyl)biurea 578
4,4′-Methylenebis(1-methyl-3-phenylurea) 592

$C_{17}H_{20}N_4O_3$
4′-Diethylamino-2-methoxy-4-nitroazobenzene 1091

$C_{17}H_{20}N_4O_3S$
Ethyl 2-(2-imidazolidinylidene)-3-(5-oxo-2-thioxo-4-imidazolidinyl)-3-phenylpropionate 741

$C_{17}H_{20}N_4O_4$
4′-(Bis(2-hydroxyethyl)amino)-2′-methyl-4-nitroazobenzene 1083
4,4′-Methylenebis(*N*,*N*-dimethyl-3-nitroaniline) 275

$C_{17}H_{20}N_4O_4S$
2-Morpholino(1,2,4)triazino(6,1-a)-isoquinolin-5-ium ethanesulfonate 980

$C_{17}H_{20}N_4O_5$
3-(2,4-Dinitrophenyl-hydrazonomethyl)camphor 1053
Disperse Red 16 (C.I. 11225) 1083
2-(3-Furfurylureido)acetic (3,4-dimethoxybenzylidene)hydrazide 1040

$C_{17}H_{20}O$
tert-Butyl diphenylmethyl ether 186
1,1-Dibenzyl-1-propanol 207
(+)-1,3-Diphenyl-1-pentanol 204
5,5-Diphenyl-1-pentanol 207
2-(2-(*o*-Tolyl)ethyl)phenethyl alcohol 207

$C_{17}H_{20}OS$
Benzyl 2-*tert*-butylphenyl sulfoxide 771

$C_{17}H_{20}O_2$
1-(2-(Benzyloxy)phenyl)-1-butanol 201
1,4a-Dimethyl-7-methoxy-4,4a,9,10-tetrahydro-2(3*H*)-phenanthrenone 283
4,4′-Methylenebis(2,6-dimethylphenol) 190

$C_{17}H_{20}O_2S$
2-Benzylphenyl *tert*-butyl sulfone 784

$C_{17}H_{20}O_3$
Ethyl 2,2-diallylbenzoylacetate 364

$C_{17}H_{20}O_3S$
4-*tert*-Butylphenyl *p*-toluenesulfonate 800
4-Phenylbutyl *p*-toluenesulfonate 800

$C_{17}H_{20}O_4$
3-(3-Benzyloxy-4-methoxyphenyl)-1,2-propanediol 199
Ethyl 2-acetyl-2-(5,6,7,8-tetrahydro-2-naphthoyl)-acetate 364
5-Methyl-7-oxo-4-phenyl-6-oxabicyclo(3.2.1)octan-2-ylmethyl acetate 384

$C_{17}H_{20}O_6S_2$
2,2-Dimethyl-1,3-propanediol bis(benzenesulfonate) 798
1,2-Propanediol ditosylate 798

$C_{17}H_{20}O_7S$
1-*S*-Phenyl-2,3,4-tri-*O*-acetyl-β-D-thioxyloside 356

$C_{17}H_{20}O_{10}$
Tetramethyl bicyclo(3.3.1)nonane-2,6-dione-1,3,5,7-tetracarboxylate 93

$C_{17}H_{20}Si$
9-Methyl-9-(trimethylsilyl)fluorene 1123

$C_{17}H_{21}ClN_2$
4,4′-(Chloromethylene)bis(*N*,*N*-dimethylaniline)perchlorate 233

$C_{17}H_{21}ClN_2O_2$
*N*α-(5-Chloro-2,4-dimethoxyphenyl)-*N*[4],*N*[4]-dimethyltoluene-α,4-diamine 224

$C_{17}H_{21}IN_2$
4-(4-(Dimethylamino)styryl)-1-ethylpyridinium iodide 954

$C_{17}H_{21}IN_2O$
9-Propylharmine methiodide 918

$C_{17}H_{21}N$
2-Benzylamino-4-phenylbutane 243

$C_{17}H_{21}NO$
2-Amino-4-*tert*.-amylphenyl phenyl ether 228
2-Benzylamino-1-(4-methoxyphenyl)propane 240
N-(2-Benzyloxy-1-methylethyl)-benzylamine 239
N-Ethyl-α-(4-methoxyphenyl)-phenethylamine 240
3,4,5,6,7,8-Hexahydro-1-phenethyl-2(1*H*)-quinolinone 704
N-(1-Naphthylmethyl)hexanoamide 490
N-(2-(*p*-Tolyloxy)-ethyl)-2,4-xylidine 229
N-(2-(*p*-Tolyloxy)-ethyl)-3,4-xylidine 229

$C_{17}H_{21}NO_2$
(Bis(4-methoxyphenyl)methyl)-dimethylamine 229
Cinnamoyltropine hydrochloride 365
2-(Diethylamino)ethyl 2-naphthalenecarboxylate 373
N-(2,3-Dimethoxybenzyl)-2,4-xylidine 228
α-(3,4-Dimethoxyphenyl)-*N*-methylphenethylamine 239
α-(3,4-Dimethoxyphenyl)-*N*-methylphenethylamine hydrochloride 239
N,*N*-Dimethylformamide dibenzyl acetal 240
N-(4-Ethylphenyl)veratrylamine 228
4-(5-Phenoxypentyloxy)aniline 228
N-Veratryl-3,4-xylidine 228

$C_{17}H_{21}NO_2S$
N-(3-(Cyclohexylthio)propyl)-phthalimide 718
2′,6′-Diethyl-*p*-toluenesulfonanilide 829
4′-Ethyl-2,4,6-trimethylbenzenesulfonanilide 829

$C_{17}H_{21}NO_3$
3-Benzyloxy-4,5-dimethoxyphenethylamine oxalate 239
N-(4-Ethoxyphenyl)veratrylamine 228

$C_{17}H_{21}NO_4$
Isobornyl 4-nitrobenzoate 374
9-(Phthalimido)nonanoic acid 721

$C_{17}H_{21}NO_4S_2$
3-Butyl-5-(2,3,4-trimethoxybenzylidene)rhodanine 735

$C_{17}H_{21}NO_5$
cis-*N*-(4-(Ethoxycarbonyl)phenyl)-hexahydrophthalamic acid 433

$C_{17}H_{21}NO_6$
Diethyl 2-acetamido-2-(benzoylmethyl)-malonate 432

$C_{17}H_{21}NO_8$
Ethyl 2-(2,2-bis(ethoxycarbonyl)-vinyl)-5-carboxy-4-methyl-3-pyrrolecarboxylate 838

$C_{17}H_{21}N_2O_2S$
2,3-Dihydroxypromazine hydrochloride 245

$C_{17}H_{21}N_3$
Basic Yellow 3 (C.I. 41005) 1060
N,*N*-Diethyl-4-(*p*-tolylazo)-aniline 1091
4-(Dimethylamino)-4′-ethyl-2-methylazobenzene 1090

$C_{17}H_{21}N_3O$
1-(4-Anilinophenyl)-3-butylurea 569
4-(2,6-Diethylphenyl)-1-phenylsemicarbazide 578
1-(2,6-Diethylphenyl)-3-(2-pyridylmethyl)urea 945
1-(2,6-Diethylphenyl)-3-(3-pyridylmethyl)urea 950
1-(4-(Dimethylamino)phenyl)-3-(2-ethylphenyl)urea 567
1-(4-(Dimethylamino)phenyl)-3-(2,5-xylyl)urea 567
1-(4-(Dimethylamino)phenyl)-3-(2,6-xylyl)urea 567
1-(4-(Dimethylamino)phenyl)-3-(3,4-xylyl)urea 567

$C_{17}H_{21}N_3O_2$
1-(2-Methoxyphenyl)-3-(2-(*N*-methylanilino)ethyl)urea 551
1-(3-Methoxyphenyl)-3-(2-(*N*-methylanilino)ethyl)urea 551
1-(4-Methoxyphenyl)-3-(2-(*N*-methylanilino)ethyl)urea 551

$C_{17}H_{21}N_3O_4S$
4′-Amino-2′-hydroxy-*N*[4]-isovalerylsulfanilanilide 811
4′-Amino-2′-hydroxy-*N*4-pivaloylsulfanilanilide 811
4′-Amino-2′-hydroxy-*N*4-valerylsulfanilanilide 811

$C_{17}H_{21}N_4O_9P$
Riboflavin-5′-phosphate 696

$C_{17}H_{21}N_5O_3$
2-(3-Furfurylureido)acetic (4-(dimethylamino)benzylidene)-hydrazide 1053

$C_{17}H_{21}N_5O_4$
2-(Furfurylureido)acetic acid *N*2-(*N*-(2,4-xylyl)carbamoyl)hydrazide 851

$C_{17}H_{21}N_5O_5$
2-(Furfurylureido)acetic acid *N*2-(*N*-(4-ethoxyphenyl)carbamoyl)-hydrazide 846

$C_{17}H_{21}N_5O_6$
4-(2,5-Dimethoxyphenyl)-1-(3-furfurylureidoacetyl)semicarbazide 845

$C_{17}H_{21}N_7O_4$
tert-Butyl *N*(*N*-(5-azido-α-benzyl1,3,4oxadiazol2-yl-methyl)-carbamoyl-methyl)carbamate 873

$C_{17}H_{22}NO_3P$
Diethyl diphenylmethyl-amidophosphate 835

$C_{17}H_{22}N_2$
N-Benzyl-*N*′,*N*′-dimethyl-*N*-phenylethylenediamine hydrochloride 242
Bis(dimethylamino)-diphenylmethane 220
4-(Dimethylamino)-*N*-(4-ethylphenyl)benzylamine 232
*N*4,*N*4-Dimethyl-*N*-α-(2,4-xylyl)-α,4-toluenediamine 232
*N*4,*N*4-Dimethyl-*N*-α-(3,4-xylyl)-α,4-toluenediamine 232

$C_{17}H_{22}N_2O$
N,*N*-Diethyl-β-(2-pyridyloxy)-phenethylamine 936
1,1-Diisopropyl-3-(1-naphthyl)urea 603
*N*4,*N*4-Dimethyl-*N*(α)-(4-ethoxyphenyl)toluene-α,4-diamine 228
1,1-Dipropyl-3-(1-naphthyl)urea 603

$C_{17}H_{22}N_2O_2$
N-(6-Aminohexyl)-1-hydroxy-2-naphthanilide 485
1-(Diethylmethyl)-1,2,3,4-tetrahydronorharman-3-carboxlic acid 919
L-Leucine-4-methoxy-2-naphthylamide hydrochloride 486
2,2′-Methylenebis(5-(dimethylamino)phenol) 227
N,*N*′-Methylenebis(4-ethoxyaniline) 228
2,2′-(Pentamethylenedioxy)dianiline 227
3,3′-(Pentamethylenedioxy)dianiline 227

$C_{17}H_{22}N_2O_3$
3-(4-Butoxyphenyl)-1-furfuryl-1-methylurea 846
N-(2-(*N*-Methylanilino)ethyl)-1,2,3,6-tetrahydrophthalamic acid 423

$C_{17}H_{22}N_2O_4$
5,5′-Methylenebis(4-ethyl-3-methyl-2-pyrrolecarboxylic acid) 840

$C_{17}H_{22}N_2O_5$
Carbobenzyloxyprolylalanine methyl ester 708

$C_{17}H_{22}N_2O_6$
4-(3,5-Bis(methoxycarbonyl)-phenylcarbamoyl)-2,6-dimethylmorpholine 707
Carbobenzyloxyserylproline methyl ester 696

$C_{17}H_{22}N_2O_7$
Dipropyl 2-(α-carbamoyl-2-nitrobenzyl)malonate 432

$C_{17}H_{22}N_2O_8$
T-boc-L-glutamic acid γ-(4-nitrobenzyl) ester dicyclohexylamine salt 674

$C_{17}H_{22}N_4$
4,4′-Bis(dimethylamino)-benzophenone hydrazone 235

$C_{17}H_{22}N_4S$
1,3-Bis(4-(dimethylamino)phenyl)-2-thiourea 740

$C_{17}H_{22}OSi$
Diphenyl(4-hydroxybutyl)-methylsilane 1105

$C_{17}H_{22}O_2$
Ethyl 2-(3,4-dihydro-5,8-dimethyl-1-naphthyl)propionate 362

$C_{17}H_{22}O_3$
4β-Methyl-4b,5,6,7,8,8a,9,10-octahydro-1,7-phenanthrenediol 7-acetate 354

$C_{17}H_{22}O_3S$
Tricyclo[5.2.1.0[2,6]]dec-10-yl *p*-toluenesulfonate 799

$C_{17}H_{22}O_4$
Ethyl 6-methoxy-1-oxo-1,2,3,4-tetrahydro-2,4,4-trimethyl-2-naphthalenecarboxylate 363
2-Phenyl-3,5,5-triethoxy-2-cyclopenten-1-one 283

$C_{17}H_{22}O_5$
Diethyl 2-(benzoylmethyl)-2-ethylmalonate 363

$C_{17}H_{22}O_6$
7,8-Dimethoxy-1-ethoxy-1-hydroxy-4-methyl-3a,4,5,9b-tetrahydro-naphtho(1,2-c)furan-3(1*H*)-one 383

$C_{17}H_{22}Si$
Butyldiphenylmethylsilane 1113
(4-*tert*-Butylphenyl)-methylphenylsilane 1113

$C_{17}H_{23}BrN_2O_4$
DL-2-Bromoisocaproylglycyl-DL-phenylalanine 429

$C_{17}H_{23}BrN_4O_3S$
N-Acetyl-4-(5-cyano-1-ethyl-2-methyl-3-benzimidazolium)-butanesulfonamide bromide 905

$C_{17}H_{23}ClO_2$
4-Chlorophenyl 10-undecenoate 362

$C_{17}H_{23}F_3N_2O$
1-Cyclohexyl-1-isopropyl-3-(3-(trifluoromethyl)phenyl)urea 510

$C_{17}H_{23}N$
6-Cyano-7-ethyl-1,2,3,4-tetrahydro-1,1,4,4-tetramethylnaphthalene 768
3,5,5,8,8-Pentamethyl-5,6,7,8-tetrahydro-2-naphthaleneacetonitrile 758

$C_{17}H_{23}NO$
5,5-Dimethyl-3-(methylphenethylamino)-2-cyclohexen-1-one 284

$C_{17}H_{23}NO_2Pb$
Triethylplumbyl 4-methyl-2-quinolinecarboxylate 1136

$C_{17}H_{23}NO_3$
Atropine 354
d-Camphoric *o*-toluidide 436
d-Camphoric *m*-toluidide 436
d-Camphoric *p*-toluidide 436
2-(*N*-(2,5-Dimethylbenzyl)-carbamoyl)-cyclohexanecarboxylic acid 503

$C_{17}H_{23}NO_3$ (Cont'd)
4-Propionylphenyl *N*-(4-methylcyclohexyl)carbamate 675
2-(*N*-(2,4,6-Trimethylphenyl)-carbamoyl)-cyclohexanecarboxylic acid 431
$C_{17}H_{23}NO_4$
Atropine-*N*-oxide hydrochloride 1005
2-*tert*-Butylcyclohexyl 4-nitrobenzoate 374
(-)-Hyoscyamine-*N*-oxide hydrochloride 1005
3-Methylcyclohexyl *N*-(4-(ethoxycarbonyl)phenyl)carbamate 675
$C_{17}H_{23}NO_4S$
2,4-Dimethyl-1-(2-methoxyethyl)-pyridinium *p*-toluenesulfonate 955
$C_{17}H_{23}NO_5$
1-Carbobenzyloxy-4-*tert*-butoxyproline dicyclohexylamine salt 697
Diethyl β-oxobenzyliminodipropionate 443
Ethyl 3-(2,2-diethoxyethoxy)-2-indolecarboxylate 882
$C_{17}H_{23}NO_6$
Carbobenzyloxy-L-glutamic acid 1-*tert*-butyl ester 659
Carbobenzyloxy-L-glutamic acid 5-*tert*-butyl ester dicyclohexylamine salt 659
$C_{17}H_{23}NO_6S$
Diethyl 1-phenylsulfonyl-4,4-piperidinedicarboxylate 823
$C_{17}H_{23}NO_7$
Ethyl 2-(ethoxycarbonyl)-4-(ethoxycarbonylmethyl)-5-formyl-3-pyrrolepropionate 842
$C_{17}H_{23}NO_9S$
1-*S*-(2-Cyanoethyl)-2,3,4,6-tetra-*O*-acetyl-1-thio-β-D-galactopyranoside 144
$C_{17}H_{23}NSi$
Phenyl(1-phenyl-2-(trimethylsilyl)-ethyl)amine 1107
$C_{17}H_{23}N_3O$
N,*N*-Dimethyl-*N′*-(4-methoxybenzyl)-*N′*-(2-pyridyl)ethylenediamine 937
1,8a-Dimethyl-2,3,4,4a,4b,5,8,8a-octahydrophenanthren-2-one semicarbazone 1070
Pyrilamine maleate 937
$C_{17}H_{23}N_3O_2$
2-Cyclohexylacetic (4-acetamidobenzylidene)-hydrazide 502
Methyl (2-aminoethyl)-methylpyrrol-2-ylmethyl-methyl-2*H*-pyrrole-propionate hydrobromide hydrate 842
$C_{17}H_{23}N_3O_3$
L-Norleucyl-L-tryptophan 884
$C_{17}H_{23}N_3O_6$
Carbobenzyloxyglycylglycylvaline 676
Carbobenzyloxy-L-valylglycylglycine 676
$C_{17}H_{23}N_3O_6S$
N-Benzyloxycarbonyl-L-methionyl-glycyl-glycine 656
Carbobenzyloxyglycylglycyl-L-methionine 655
$C_{17}H_{23}N_5$
1,3-Dimethyl-4-(3-(dimethylamino)-propylamino)-1*H*-pyrazolo(3,4-b)-quinoline 1003
$C_{17}H_{24}Br_2N_2$
1,1′-Trimethylenebis(2,4-dimethylpyridinium bromide) 948
1,1′-Trimethylenebis(3,4-dimethylpyridinium bromide) 948
1,1′-Trimethylenebis(3,5-dimethylpyridinium bromide) 948
$C_{17}H_{24}ClNO_2$
4-Chloro-α-(4-(4-tetrahydropyranyl)-1-piperidino)-*o*-cresol 244
$C_{17}H_{24}Cl_2O_4$
Dimethyl α,α′-dichloro-5-methyl-1,3-adamantanediacetate 83
$C_{17}H_{24}INO$
4-Methyl-4-(4-phenyl-2-cyclohexen-1-yl)morpholinium iodide 255
$C_{17}H_{24}N_2$
3,5-Dimethyl-2-(3,5-dimethyl-4-ethylpyrrol-2-ylmethylene)-4-ethyl-2*H*-pyrrole hydrobromide 839
2-(4,5-Dimethyl-3-ethylpyrrol-2-ylmethylene)-3,5-dimethyl-4-ethyl-2*H*-pyrrole hydrobromide 839
$C_{17}H_{24}N_2O$
Bis(3,5-dimethyl-4-ethylpyrrol-2-yl) ketone 840
1,1-Diallyl-3-(2,6-diethylphenyl)urea 595
4-Phenyl-1,4-(piperidinocarbonyl)-piperidine hydrochloride 703
$C_{17}H_{24}N_2O_2$
1-Cyano-*N*-(3,4-dimethoxyphenethyl)-cyclohexylamine 753
O-(*N*-(2,6-Diethylphenyl)carbamoyl)-cyclohexanone oxime 1012
$C_{17}H_{24}N_2O_3$
5-(3,5-Di-*tert*.-butylphenyl)-3-nitroso-2-oxazolidinone 657
Ethyl 4-(3-cyclohexyl-3-methylureido)benzoate 577
2-(*N*-(2-(*N*-Methylanilino)ethyl)-carbamoyl)-cyclohexanecarboxylic acid 423
$C_{17}H_{24}N_2O_5$
N-Benzyloxycarbonyl-L-isoleucyl-L-alanine 660
Carbobenzyloxy-β-alanyl-L-leucine 660
Carbobenzyloxy-L-alanyl-L-leucine 660
Carbobenzyloxy-L-alanyl-L-norleucine 660
Carbobenzyloxy-L-alanyl-L-norvaline methyl ester 661
Carbobenzyloxyalanylvaline methyl ester 661
Carbobenzyloxy-β-alanyl-L-valine methyl ester 677
Carbobenzyloxyglycyl-L-leucine methyl ester 661
Carbobenzyloxyglycyl-L-norleucine methyl ester 663
Carbobenzyloxy-L-leucylglycine methyl ester 661
Carbobenzyloxysarcosyl-L-valine methyl ester 661
Carbobenzyloxy-L-valyl-L-alanine methyl ester 663
Diethyl *N*-(4-methylaminobenzoyl)-L-glutamate 471
$C_{17}H_{24}N_2O_5S$
Carbobenzyloxy-L-alanyl-L-methionine methyl ester 656
Carbobenzyloxy-L-methionyl-L-alanine methyl ester 656
$C_{17}H_{24}N_2O_6$
Carbobenzyloxynorvalylserine methyl ester 643
$C_{17}H_{24}N_2O_6S$
Carbobenzyloxymethionyl-L-threonine 642
Carbobenzyloxyserylmethionine methyl ester 642
$C_{17}H_{24}N_4O_2$
*N*α-L-Leucyl-L-tryptophanamide acetate 884
*N*α-L-Isoleucyl-L-tryptophanamide hydrochloride 884
L-Tryptophyl-L-leucinamide acetate 884
$C_{17}H_{24}N_4O_4$
2,4-Bis(morpholinocarboxamido)-toluene 703
$C_{17}H_{24}N_4O_5$
Carbobenzyloxy-L-valylglycylglycinamide 664
$C_{17}H_{24}N_4O_5S$
Carbobenzyloxymethionyl-glycylglycinamide 656
$C_{17}H_{24}N_4O_6$
6-(*N*-(6,7,8-Trimethoxy-4-quinazolinyl)amino) hexyl nitrate 999
$C_{17}H_{24}N_6O$
2,6-Diamino-3-(4-(2-diethylaminoethoxy)-phenylazo)-pyridine 1086
2,6-Diamino-3-(4-(2-diethylaminoethoxy)-phenylazo)-pyridine methanesulfonate 1086
$C_{17}H_{24}O$
6-Acetyl-1,1,4,4,7-pentamethyl-1,2,3,4-tetrahydronaphthalene 298
3-Ethyl-5,6,7,8-tetrahydro-5,5,8,8-tetramethyl-2-naphthaldehyde 324
1,2,3,4,4a,9,10,10a-Octahydro-1,1,4a-trimethyl-6-phenanthrenol 192
$C_{17}H_{24}O_2$
Ethyl 2-(5,8-dimethyl-1,2,3,4-tetrahydro-1-naphthyl)propionate 362
Phenyl 10-undecenoate 362
$C_{17}H_{24}O_4$
(+)-*mono*-1-*tert*-Butyl-3-methylbutyl phthalate 371
(-)-*mono*-1-*tert*-Butyl-3-methylbutyl phthalate, brucine salt 371
DL-*mono*-1-*tert*-Butylpentyl phthalate 371
1-Ethyl 5-propyl 3-methyl-2-phenylglutarate 360
2,2′-Methylenebis(5,5-dimethyl-3-hydroxy-2-cyclohexen-1-one) 51
$C_{17}H_{24}O_4S$
2-Oxo-1,3,3-trimethylcyclohexylmethyl *p*-toluenesulfonate 798
$C_{17}H_{24}O_6$
Dimethyl shelloate 84
Ethyl 3-octyl-2,4,5-trioxo-1-cyclopentaneglyoxylate 93
$C_{17}H_{24}Si_2$
Methylphenyl(4-(trimethylsilylmethyl)phenyl)silane 1120
$C_{17}H_{25}Cl$
6-Chloromethyl-7-ethyl-1,2,3,4-tetrahydro-1,1,4,4-tetramethylnaphthalene 174
$C_{17}H_{25}ClN_2O$
1-(2-Chloro-6-methylphenyl)-3-cyclohexyl-3-isopropylurea 535
$C_{17}H_{25}NO$
α-Butylimino-2-methylallyloxymesitylene 1048
1-Phenyl-3-piperidino-1-cyclohexanol 245
$C_{17}H_{25}NO_2$
3-(Allylamino)-1,3-dimethylbutyl 2-phenylacetate 358
5-Isopropyl-2-methylphenyl *N*-cyclohexylcarbamate 680
2-Methylcyclohexyl *N*-(2-ethyl-6-methylphenyl)carbamate 679
α-(4-(4-Tetrahydropyranyl)-1-piperidino)-*o*-cresol 246
$C_{17}H_{25}NO_3$
2′,6′-Diisopropylglutaranilic acid 435
$C_{17}H_{25}NO_4$
N,*N*-Bis(2-carbethoxyethyl)-benzylamine 358
N-(2-Carbethoxyethyl)-*N*-(2-carbomethoxypropyl)benzylamine 359
$C_{17}H_{25}NO_5$
4-Butoxyphenyl *N*-(butoxycarbonylmethyl)-carbamate 651
$C_{17}H_{25}NO_6$
Ethyl 2-(ethoxycarbonyl)-4-(ethoxycarbonylmethyl)-5-methyl-3-pyrrolepropionate 842
$C_{17}H_{25}N_3$
1-Benzyl-4-butylamino-4-cyanopiperidine hydrate 754
$C_{17}H_{25}N_3O$
1-Benzyl-4-carbamyl-4-pyrrolidinopiperidine 425
Plasmocid dihydroiodide 985
$C_{17}H_{25}N_3O_3$
1-Cyclohexyl-1-isopropyl-3-(4-methyl-3-nitrophenyl)urea 563
Decanoic (3-nitrobenzylidene)-hydrazide 1063
$C_{17}H_{25}N_3O_4$
Carbobenzyloxy-β-alanyl-L-leucinamide 664
Carbobenzyloxy-L-alanyl-L-leucinamide 664
Carbobenzyloxy-L-alanyl-L-norleucinamide 664
Carbobenzyloxy-L-isoleucyl-L-alaninamide 658
DL-Leucylglycyl-DL-phenylalanine 349
$C_{17}H_{25}N_3O_5$
Carbobenzyloxy-L-leucyl-L-serinamide 643
Carbobenzyloxyserylleucinamide 643
$C_{17}H_{25}N_3O_5S$
Carbobenzyloxy-L-methionyl-L-threoninamide 642
$C_{17}H_{25}N_3O_6$
Valyltyrosylserine 410
$C_{17}H_{26}BrNO_3$
α-Bromo-5-nitro-*o*-tolyl decyl ether 258
$C_{17}H_{26}ClN_3O$
1-(2-Chloro-6-methylphenyl)-3-(3-(2-methylpiperidino)propyl)urea 525
$C_{17}H_{26}HgO_2$
2,4,6-Triisopropylphenylmercuri acetate 1139
$C_{17}H_{26}N_2O$
1-Benzyl-3-cyclohexyl-3-isopropylurea 571
1-Cyclohexyl-3-(2,6-diethylphenyl)urea 583
1-Cyclohexyl-3-(2-ethyl-6-methylphenyl)-1-methylurea 583
1-(*N*-(2,6-Diethylphenyl)carbamoyl)-2-methylpiperidine 585
N-(1-Ethyl-1,2,3,4-tetrahydro-2,2,4-trimethyl-7-quinolyl)propionamide 426
N-(1-Ethyl-1,2,3,4-tetrahydro-2,2,4-trimethyl-7-quinolyl)propionamide 433
$C_{17}H_{26}N_2OS$
1-Cyclohexyl-1-isopropyl-3-(4-(methylthio)phenyl)urea 561
$C_{17}H_{26}N_2O_2$
4′-Butoxy-1-piperidineacetanilide hydrochloride 415
4-(*N*-(2,6-Diethylphenyl)carbamoyl)-2,6-dimethylmorpholine 714
4-(*N*-(2,6-Diisopropylphenyl)-carbamoyl)morpholine 714
2,2′-Methylenebis(3-amino-5,5-dimethyl-2-cyclohexen-1-one) 57
$C_{17}H_{26}N_2O_2S$
2-*sec*-Butylcyclohexanone (*p*-tosyl)-hydrazone 1071
4-*tert*-Butylcyclohexanone (*p*-tosyl)-hydrazone 1091
$C_{17}H_{26}N_2O_4$
N,*N′*-Bis(butoxycarbonyl)-4-methyl-1,3-phenylenediamine 669
1-Cyclohexyl-3-(2,4-dimethoxyphenyl)-1-(2-hydroxyethyl)urea 546
$C_{17}H_{26}N_2O_5$
4-(2-(2-Butoxy-5-nitrobenzyloxy)-ethyl)-morpholine oxalate 268
$C_{17}H_{26}N_2O_6$
N,*N′*-Bis(2-ethoxyethoxycarbonyl)-4-methyl-1,3-phenylenediamine 647
$C_{17}H_{26}N_4O_3S$
1-(3,4-Dimethoxybenzylidene)-4-(3-morpholinopropyl)-3-thiosemicarbazide 1040
$C_{17}H_{26}O$
Allyl 4-*tert*-octylphenyl ether 181
2′,4′,6′-Triisopropylacetophenone 297
$C_{17}H_{26}O_2$
2-Octyl-4-phenyl-1,3-dioxolane 184
$C_{17}H_{26}O_2S$
11-(Phenylthio)undecanoic acid 330
$C_{17}H_{26}O_3$
3-(Decyloxy)benzoic acid 340
$C_{17}H_{27}NO$
2′,4′,6′-Trimethyloctananilide 440
$C_{17}H_{27}NO_2$
Butyl *N*-(2,6-diisopropylphenyl)-carbamate 684

$C_{17}H_{27}NO_2$ (Cont'd)

1-(3,5-Dimethyl-1-adamantanecarbonyl)morpholine ... 110

$C_{17}H_{27}NO_3$

o-(*tert*-Butyl)-L-tyrosine *tert*-butyl ester hydrochloride ... 355

N-Vanillylnonanoamide ... 411

$C_{17}H_{27}N_3O_2$

1-(*p*-Anisyl)-3-(3-(2-methylpiperidino)propyl)urea ... 552

4-(Decanoylamino)benzoic hydrazide ... 502

1-(2,6-Diethylphenyl)-3-(2-morpholinoethyl)urea ... 585

1-(2-Ethyl-6-methylphenyl)-3-(3-morpholinopropyl)urea ... 585

$C_{17}H_{28}N_2$

5-(1-Methyloctylamino)indoline dihydrochloride ... 249

$C_{17}H_{28}N_2O$

1-Butyl-3-(2,6-diisopropylphenyl)urea ... 596

1,1-Dibutyl-3-(3,4-dimethylphenyl)urea ... 601

1,1-Dibutyl-3-(2,3-xylyl)urea ... 596

1,1-Dibutyl-3-(2,4-xylyl)urea ... 601

1,1-Dibutyl-3-(2,5-xylyl)urea ... 601

1,1-Dibutyl-3-(2,6-xylyl)urea ... 596

1,1-Dibutyl-3-(3,5-xylyl)urea ... 602

1,1-Diethyl-3-(2,6-diisopropylphenyl)urea ... 596

3-(2,6-Diethylphenyl)-1,1-diisopropylurea ... 596

3-(2,6-Diethylphenyl)-1,1-dipropylurea ... 596

1,1-Diisobutyl-3-(2,3-xylyl)urea ... 596

1,1-Diisobutyl-3-(2,4-xylyl)urea ... 601

1,1-Diisobutyl-3-(2,5-xylyl)urea ... 601

1,1-Diisobutyl-3-(2,6-xylyl)urea ... 596

1,1-Diisobutyl-3-(3,4-xylyl)urea ... 601

1-(2,6-Diisopropylphenyl)-3-isobutylurea ... 596

$C_{17}H_{28}N_2O_2$

1,1-Dibutyl-3-(2-methoxy-5-methylphenyl)urea ... 557

3-(3,5-Di-*tert*.-butylphenyl)-3-hydroxypropionic acid hydrazide ... 495

2-(Diethylamino)ethyl *N*-(2,6-diethylphenyl)carbamate ... 673

1,1-Diisobutyl-3-(2-methoxy-5-methylphenyl)urea ... 557

2-Dimethylaminoethyl *N*-(2,6-diisopropylphenyl)carbamate ... 673

$C_{17}H_{28}N_2O_3$

1,1-Dibutyl-3-(2,5-dimethoxyphenyl)urea ... 557

1,1-Diisobutyl-3-(2,5-dimethoxyphenyl)urea ... 557

$C_{17}H_{28}N_4OS$

1-Dodecanoyl-3-(2-pyrimidinyl)-2-thiourea ... 976

$C_{17}H_{28}N_4O_2$

1,1′-(4-Methyl-1,3-phenylene)bis(3-butylurea) ... 566

3,3′-(4-Methyl-1,3-phenylene)bis(1,1-diethylurea) ... 566

1,1′-(4-Methyl-1,3-phenylene)bis(3-isobutylurea) ... 566

$C_{17}H_{28}N_4O_4$

N,*N*′-Bis(2-dimethylaminoethoxycarbonyl)-4-methyl-1,3-phenylenediamine ... 669

$C_{17}H_{28}O$

2,4-Dimethyl-6-isononylphenol ... 192

$C_{17}H_{28}O_2$

Guaiyl acetate ... 81

$C_{17}H_{28}O_4$

Methyl 2-(1,3-dioxo-2-cyclotetradecyl)acetate ... 92

$C_{17}H_{28}O_4S$

2-(4-(Iso-nonyl)phenoxy)-1-ethanesulfonic acid, sodium salt, mixture of isomers ... 792

$C_{17}H_{28}O_4S_2$

S-(Dimethoxy-thiopyranyl)trimethyl-2-oxabicyclo(2.2.1)heptane-1-thiocarboxylate ... 89

$C_{17}H_{28}O_8S_4$

Pentaerythritol tetrakis-(3-mercaptopropionate) ... 86

$C_{17}H_{29}BF_4O$

2,4,6-Tri-*tert*-butylpyrylium tetrafluoroborate ... 935

$C_{17}H_{29}BrN_2O$

3-Carbamoyl-1-undecylpyridinium bromide ... 947

4-Carbamoyl-1-undecylpyridinium bromide ... 951

$C_{17}H_{29}IN_2O$

4,7-Endomethylene2-methyl-2(3morpholinopropyl)-4*H*-isoindolinium iodide methiodide ... 44

$C_{17}H_{29}IO$

2,4,6-Tri-*tert*-butylpyrylium iodide ... 935

$C_{17}H_{29}N$

(3,5-Dimethyl-1-adamantylmethyl)-pyrrolidine hydrochloride ... 109

$C_{17}H_{29}NO$

2-(Azacyclododec-1-en-2-yl)-1-cyclohexanone ... 59

2,6-Di-*tert*-butyl-4-(dimethylaminomethyl)phenol ... 239

2-Dodecyloxypyridine ... 940

$C_{17}H_{29}NO_2$

1-(1-Adamantyl)propyl *N*-propylcarbamate ... 116

N-Dodecyl-2-furamide ... 848

3,5,7-Trimethyl-1-adamantyl *N*-propylcarbamate ... 116

$C_{17}H_{29}NO_3S$

N-Octyl-*N*-(3-sulfopropyl)aniline ... 789

$C_{17}H_{29}N_3O$

2-(*N*-(2-Diethylaminoethyl)-*N*-methylamino)-2′,6′-acetoxylidide dihydrochloride ... 424

$C_{17}H_{30}BrN$

1-Dodecylpyridinium bromide ... 959

$C_{17}H_{30}ClO_4$

Ethyl 13-acetoxy-2-chlorotridecanoate ... 83

$C_{17}H_{30}N_2OS$

1-((2-(Diethylaminoethylthio))-methyl)-1-(4-(dimethylamino)-phenyl)ethanol ... 237

$C_{17}H_{30}N_2O_4$

Dodecyl *N*-carbamoylmaleamate ... 127

$C_{17}H_{30}O_3$

7-Oxo-16-heptadecenoic acid ... 64

$C_{17}H_{31}N$

N,*N*-Diethyl-3,5-dimethyl-1-adamantylmethylamine hydrochloride ... 29

$C_{17}H_{31}NO_4$

Cyclododecyl *N*-(ethoxycarbonylmethyl)-carbamate ... 115

$C_{17}H_{32}N_2$

1,5-Dipiperidino-2-methyl-3-hexene ... 40

$C_{17}H_{32}N_2O$

3-Butyl-1,1-dicyclohexylurea ... 130

$C_{17}H_{32}N_2O_5S$

tert-Butoxycarbonyl-L-isoleucyl-L-methionine methyl ester ... 118

N-(*tert*-Butoxycarbonyl)-L-leucyl-L-methionine methyl ester ... 118

$C_{17}H_{32}OSi$

2,5,6,9-Tetramethyl-1-(trimethylsilyloxy)-1,5,9-decatriene ... 1098

$C_{17}H_{32}O_2Si$

tert-Butyldimethylsilyl 2-(1,5,5-trimethyl-2-cyclohexen-1-yl)acetate ... 1111

$C_{17}H_{32}O_3$

Methyl 3-oxohexadecanoate ... 94

$C_{17}H_{32}O_4$

Diethyl diisopentylmalonate ... 76

Diethyl tridecanedioate ... 76

Diheptyl malonate ... 76

$C_{17}H_{33}NO_2$

N-Allyl-*N*-(1,1-dimethyl-3-hydroxybutyl)octanamide ... 107

$C_{17}H_{33}NO_3$

17-Amino-7-oxo-heptadecanoic acid ... 71

$C_{17}H_{33}NO_3S$

S-Dodecyl *N*-(ethoxycarbonylmethyl)-thiocarbamate ... 116

$C_{17}H_{33}NO_4$

tert-Decyl 2,4-dimethyl-4-nitrovalerate ... 89

$C_{17}H_{34}ClNO_2$

2-Chloro-*N*-(2-dodecyloxyethyl)-*N*-methylacetamide ... 102

$C_{17}H_{34}N_2O$

N-Butyl-*N*-(1,3-dimethylbutyl)-1-piperidineacetamide hydrochloride ... 108

$C_{17}H_{34}N_2OS$

1-Hexadecanoyl-2-thiourea ... 141

$C_{17}H_{34}N_2O_2$

1-Hexadecanoylurea ... 128

$C_{17}H_{34}O$

2-Heptadecanone ... 49

4-Heptadecanone ... 49

6-Heptadecanone ... 49

7-Heptadecanone ... 49

8-Heptadecanone ... 49

$C_{17}H_{34}OSn$

5,6-Dihydro-2-(tributylstannyl)-4*H*-pyran ... 1132

$C_{17}H_{35}Br$

1-Bromoheptadecane ... 4

$C_{17}H_{35}ClO_2$

1-Chloro-3-(4-ethyl-1-isobutyloctyloxy)-2-propanol ... 19

$C_{17}H_{35}NO_2$

Ethyl 3-(dodecylamino)-propionate ... 88

$C_{17}H_{36}N_2O$

2-Diethylamino-*N*-(1-methyldecyl)-acetamide hydrochloride ... 104

N-(3-Dimethylaminopropyl)-dodecanamide ... 104

1-Hexadecylurea ... 581

1,1,3,3-Tetrabutylurea ... 128

$C_{17}H_{36}N_2S$

1,3-Dioctyl-2-thiourea ... 142

2-(*tert*.-Hexadecyl)-2-thiopseudourea hydrobromide ... 138

$C_{17}H_{36}O$

3,9-Diethyl-6-tridecanol ... 9

$C_{17}H_{36}O_2Pb$

Tripentylplumbyl acetate ... 1135

$C_{17}H_{36}Si$

Dodecamethyl-1-silacyclohexane ... 1097

$C_{17}H_{37}N$

N-Methylhexadecylamine ... 28

$C_{17}H_{37}NHCl$

N-Methylhexadecylamine hydrochloride ... 28

$C_{17}H_{38}BrN$

Pentyltributylammonium bromide ... 43

$C_{17}H_{38}N_2$

N,*N*,*N*′,*N*′-Tetrabutylmethanediamine ... 30

$C_{17}H_{38}O_2Si$

Diethoxydodecylmethylsilane ... 1098

$C_{17}H_{38}O_2Sn$

(Diethoxymethyl)tributyltin ... 1130

$C_{17}H_{38}O_6P_2$

Tetraisobutyl methylenediphosphonate ... 158

$C_{17}H_{44}Br_4N_4$

Methanetetrayltetrakis(tetramethyl-ammonium bromide) ... 42

$C_{17}H_{44}Si_8$

1-Phenyl-2,2,3,3-tetrakis(dimethylsilyl)-1,4,4-trimethyltetrasilane ... 1113

$C_{18}HCl_{15}Si$

Tris(pentachlorophenyl)silane ... 1101

$C_{18}HF_{15}Si$

Tris(pentafluorophenyl)silane ... 1101

$C_{18}H_4Cl_5NO_4$

2-Chloro-3-(3,4,5,6-tetrachlorophthalimido)-1,4-naphthoquinone ... 721

$C_{18}H_8Cl_{10}O_2$

1,6-Bis(pentachlorophenyl)-1,6-hexanedione ... 307

$C_{18}H_8F_{10}O_2$

1,6-Bis(pentafluorophenyl)-1,6-hexanedione ... 307

$C_{18}H_8HgN_2O_2$

Bis(2-(2-benzoxazolyl)ethynyl)-mercury ... 1139

$C_{18}H_8I_4N_2O_4$

4,4″-Dinitro-2,2′,5′,6″-tetraiodo-*p*-terphenyl ... 262

$C_{18}H_9Br_4NO_3$

N-(1-Naphthyl)-tetrabromophthalamic acid ... 484

$C_{18}H_9Br_4NO_4$

N-(5-Hydroxy-1-naphthyl)-3,4,5,6-tetrabromophthalamic acid ... 484

$C_{18}H_9ClO_2$

1-Chlorobenzo(5,6)-cyclohepta(1,2,3-de)naphthalene-7,12-dione ... 318

2-Chloro-7,12-dihydrobenz(a)-anthracene-7,12-dione ... 319

6-Chloronaphthacenequinone ... 318

$C_{18}H_9Cl_4NO_3$

N-(1-Naphthyl)-3,4,5,6-tetrachlorophthalamic acid ... 455

N-(2-Naphthyl)-3,4,5,6-tetrachlorophthalamic acid ... 484

$C_{18}H_9Cl_6O_4P$

Tris(2,6-dichlorophenyl) phosphate ... 837

$C_{18}H_9NO$

3-Cyanobenzanthrone ... 766

$C_{18}H_9N_3O_2$

9-Nitroacenaphtho(1,2-b)-quinoxaline ... 1000

$C_{18}H_9N_3O_9$

2,5-Bis(3-nitrobenzoyl)-3-nitrofuran ... 847

$C_{18}H_{10}BrClN_2O$

α-(Bromo-2-chloro-4-hydroxyimino-cyclohexadienylidene)-1-naphthaleneacetonitrile ... 1013

$C_{18}H_{10}Br_2$

6,12-Dibromochrysene ... 175

$C_{18}H_{10}Br_2S_4$

4,4′-Bis(4-bromophenyl)-tetrathiafulvalene ... 879

$C_{18}H_{10}ClF_3N_2O_3$

4-(2-Chloro-5-(trifluoromethyl)-phenylazo)-3-hydroxy-2-naphthoic acid ... 1080

$C_{18}H_{10}Cl_2N_2O$

α(2,5-Dichloro-4-hydroxyiminocyclohexadien-1-ylidene)-1-naphthaleneacetonitrile ... 1013

2-(2,4-Dichlorophenyl)-5-(2-naphthyl)-1,3,4-oxadiazole ... 874

2-(3,4-Dichlorophenyl)-5-(2-naphthyl)-1,3,4-oxadiazole ... 874

$C_{18}H_{10}Cl_2N_4O_6$

2,5-Bis(3-nitroanilino)-3,6-dichloro-1,4-benzoquinone ... 315

$C_{18}H_{10}Cl_2S_4$

4,4′-Bis(4-chlorophenyl)-tetrathiafulvalene ... 879

$C_{18}H_{10}Cl_3N_3O$

2-(4-Chlorophenyl)-4-(2-diazoacetyl)-6,8-dichloro-3-methylquinoline ... 1094

$C_{18}H_{10}Cl_4$

1,4-Diphenyl-2,3,5,6-tetrachlorobenzene ... 172

$C_{18}H_{10}Cl_4N_2O_2$

2,5-Bis(4-chloroanilino)-3,6-dichloro-1,4-benzoquinone ... 315

$C_{18}H_{10}Cl_5OP$

Diphenyl(pentachlorophenyl)-phosphine oxide ... 833

$C_{18}H_{10}F_2Cl_2N_2O_2$

2,5-Bis(4-fluoroanilino)-3,6-dichloro-1,4-benzoquinone ... 315

$C_{18}H_{10}F_2S_4$

4,4′-Bis(4-fluorophenyl)-tetrathiafulvalene ... 879

$C_{18}H_{10}F_3NO$

3-(2,2,2-Trifluoroacetamido)-fluoranthene ... 484

$C_{18}H_{10}I_2S_4$

4,4′-Bis(4-iodomethyl)-tetrathiafulvalene ... 879

$C_{18}H_{10}N_2$

Acenaphtho(1,2-b)quinoxaline ... 1001

$C_{18}H_{10}N_2O_4$

Dinitrochrysene ... 280

$C_{18}H_{10}N_2O_4$ (Cont'd)
6,12-Dinitrochrysene 280
$C_{18}H_{10}O_2$
11,12-Dihydrochrysene-11,12-dione 319
$C_{18}H_{10}O_3$
3,4-Benzo-9-oxofluorene-1-carboxylic acid 345
1,3-Dihydro-4-phenylnaphtho(2,3-c)-furan-1,3-dione 388
2-Hydroxybenz(a)anthracene-7,12-dione 316
1-Hydroxybenzo(5,6)cyclohepta(1,2,3-de)naphthalene-7,12-dione 316
$C_{18}H_{10}O_3S$
4-Phenylthio-1,8-naphthalic anhydride 388
$C_{18}H_{11}Br$
7-Bromobenz(a)anthracene 175
$C_{18}H_{11}Br_4NO_7$
N-(3,5-Bis-(methoxycarbonyl)-phenyl)-3,4,5,6-tetrabromophthalamic acid 452
$C_{18}H_{11}ClF_3NO$
2′-Chloro-5′-trifluoromethyl-1-naphthalenecarboxanilide 484
$C_{18}H_{11}ClN_2O$
2-(3-Chlorophenyl)-5-(1-naphthyl)-1,3,4-oxadiazole 873
2-(4-Chlorophenyl)-5-(2-naphthyl)-1,3,4-oxadiazole 873
$C_{18}H_{11}ClN_2O_2$
4-Chloro-4′-cyano-1-hydroxy-2-naphthanilide 761
$C_{18}H_{11}Cl_5Si$
Diphenyl(pentachlorophenyl)silane 1102
$C_{18}H_{11}F$
1-Fluorobenz(a)anthracene 175
6-Fluorochrysene 175
$C_{18}H_{11}FN_2O$
2-(2-Fluorobenzoyl)-2-quinolylacetonitrile 981
2-(2-Fluorophenyl)-5-(1-naphthyl)-1,3,4-oxadiazole 873
2-(4-Fluorophenyl)-5-(2-naphthyl)-1,3,4-oxadiazole 873
$C_{18}H_{11}FO_3$
2-(5-Fluoro-1-naphthoyl)-benzoic acid 344
$C_{18}H_{11}I_2NO_4$
2,6-Diiodo-4-(5-oxo-2-phenyl-2-oxazolin-4-ylidenemethyl)phenyl acetate 382
$C_{18}H_{11}NO_2$
6-Nitrochrysene 280
$C_{18}H_{11}N_3O_4$
2,7-Dinitro-9-phenylcarbazole 891
$C_{18}H_{11}N_5O_9$
2,4-Dinitro-4′-(2,4-dinitrophenoxy)-diphenylamine 267
$C_{18}H_{12}$
Benzo(c)phenanthrene 166
1-Ethenylpyrene 167
$C_{18}H_{12}BiBr_3$
Tris(4-bromophenyl)bismuthine 1141
$C_{18}H_{12}BiCl_3$
Tris(4-chlorophenyl)bismuthine 1141
$C_{18}H_{12}BiCl_5$
Dichlorotris(4-chlorophenyl)-bismuth 1141
$C_{18}H_{12}BiF_3$
Tris(4-fluorophenyl)bismuthine 1141
$C_{18}H_{12}Br_2N_2O_2$
2,2′-Dibromo-2″-nitrotriphenylamine 261
$C_{18}H_{12}Br_3Cl_2Sb$
Dichlorotris(4-bromophenyl)-antimony 1128
$C_{18}H_{12}Br_3Sb$
Tris(4-bromophenyl)antimony 1128
$C_{18}H_{12}ClNO$
2,6-Diphenylisonicotinoyl chloride 962
$C_{18}H_{12}ClNO_3$
2-(3-Acetylanilino)-3-chloro-1,4-naphthoquinone 318
$C_{18}H_{12}Cl_2N_2OS$
1-(2,4-Dichlorobenzoyl)-3-(1-naphthyl)-2-thiourea 731
$C_{18}H_{12}Cl_2N_2O_2$
5-Acetyl-3,4-bis(4-chlorophenyl)-6(1*H*)-pyridazinone 967
2,5-Dianilino-3,6-dichloro-1,4-benzoquinone 315
1-(2,4-Dichlorobenzoyl)-2-(1-naphthoyl)hydrazine 503
$C_{18}H_{12}Cl_2N_2O_2S$
1-(2,4-Dichlorobenzoyl)-3-(7-hydroxy-1-naphthyl)-2-thiourea 728
$C_{18}H_{12}Cl_2N_4$
1,2-Bis(7-chloro-4-quinolyl)-hydrazine 988
$C_{18}H_{12}Cl_3NO_2$
2,4-Dichloro-1-naphthyl *N*-(3-chloro-2-methylphenyl)carbamate 637
N-(1-Naphthyl)-2,4,5-trichlorophenoxyacetamide 484
$C_{18}H_{12}Cl_3O_4P$
Tris(2-chlorophenyl) phosphate 833
$C_{18}H_{12}Cl_3P$
Tris(3-chlorophenyl)phosphine 830
$C_{18}H_{12}Cl_4N_2O_4$
2-Butyne-1,4-diol bis-(3,4-dichlorocarbanilate) 1015
$C_{18}H_{12}Cl_4Pb$
Chlorotris(4-chlorophenyl)lead 1134
$C_{18}H_{12}FN$
12-Amino-6-fluorochrysene 231
$C_{18}H_{12}F_3NO$
1-(α,α,α-Trifluoro-*m*-tolyliminomethyl)-2-naphthol 1020
$C_{18}H_{12}F_3NO_4$
4-Methyl-2-oxo-2*H*-1-benzopyran-7-yl *N*-(3-(trifluoromethyl)phenyl)-carbamate 609
$C_{18}H_{12}N_2$
12*H*-benzene(5,6)Isoindolo(2,1-a)-benzimidazole 908
Benzo(b)naphtho(2,3-f)(1,4)-diazocine 1003
6,7′-Biquinoline 988
$C_{18}H_{12}N_2OS$
1-(2-Benzothiazolyliminomethyl)-2-naphthol 1035
$C_{18}H_{12}N_2O_2$
4-Cyanophenyl *N*-(1-naphthyl)-carbamate 767
3,6-Diphenylisoxazolo(4,5-c)pyridin-4(5*H*)-one 919
9-(2-Nitrophenyl)carbazole 891
9-(4-Nitrophenyl)carbazole 891
$C_{18}H_{12}N_2O_2S$
10-(2-Nitrophenyl)phenothiazine 274
$C_{18}H_{12}N_2O_4$
1-(2,4-Dinitrostyryl)naphthalene 276
N,N′-Ethylenediphthalimide 718
$C_{18}H_{12}N_2O_5$
N-(1-Naphthyl)-6-nitrophthalamic acid 472
$C_{18}H_{12}N_2O_6$
1,3-Bis-(4-nitrophenoxy)-benzene 269
N,N′-(Ethylenedioxy)diphthalimide 717
$C_{18}H_{12}N_2S_2$
2-Quinolyl disulfide 987
$C_{18}H_{12}N_3O_7P$
Tris(3-nitrophenyl)phosphine oxide 835
$C_{18}H_{12}N_4O_6$
N-(2-Biphenylyl)picrylamine 273
$C_{18}H_{12}N_4O_7$
4′-Phenoxy-2,4,6-trinitrodiphenylamine 266
$C_{18}H_{12}N_4O_9$
2-(*N*-(2,4,6-Trinitrophenoxy)anilino)-hydroquinone 262
$C_{18}H_{12}N_6O_8$
N,N′-Bis(2,4-dinitrophenyl)-1,2-phenylenediamine 273
N,N′-Bis(2,4-dinitrophenyl)-1,3-phenylenediamine 274
$C_{18}H_{12}O$
9-(2-Furfurylidene)fluorene 855
1-Hydroxybenzo(c)phenanthrene 196
2-Hydroxybenzo(c)phenanthrene 196
3-Hydroxybenzo(c)phenanthrene 196
$C_{18}H_{12}O_2$
2-Cinnamylidene-1,3-indandione 319
3-(1-Naphthyl)phthalide 385
$C_{18}H_{12}O_3$
Dibenzo(*H,K*)-4-oxatricyclo[5.2.2.0^{2,6}]undec-8,10-diene-3,5-dione 388
2-(1-Naphthoyl)benzoic acid 345
2-(2-Naphthoyl)benzoic acid 344
$C_{18}H_{12}O_4$
9,10-Dihydro-9,10-ethenoanthracene-11,12-dicarboxylic acid 341
9,10-Dihydro-9-hydroxy-9,10-ethanoanthracene-11,12-dicarboxylic anhydride 183
2-(9,10-Dihydro-10-oxo-9-anthracenyl)succinic anhydride 388
$C_{18}H_{12}O_6$
1,2-Diacetoxyanthraquinone 378
$C_{18}H_{12}O_7$
Di-*o*-benzoyltartaric anhydride 388
$C_{18}H_{12}S$
2-(2-Naphthyl)thianaphthene 899
9-(2-Thenylidene)fluorene 859
$C_{18}H_{13}Br$
2-Bromo-*p*-terphenyl 172
$C_{18}H_{13}BrClNO_2$
4-(2-Bromoacetyl)-2-(4-chlorophenyl)-6-methoxyquinoline hydrobromide 981
$C_{18}H_{13}BrClNO_3S$
4-Bromo-*N*-(4-chloro-2-phenoxy)-benzenesulfonamide 806
$C_{18}H_{13}BrN_2OS$
2-(4-Bromoanilino)-3-cyano-5,6-dihydro-6-phenyl-4*H*-thiopyran-4-one 759
$C_{18}H_{13}BrN_2O_2$
1-(3-Bromobenzoyl)-2-(1-naphthoyl)-hydrazine 501
1-(3-Bromobenzoyl)-2-(2-naphthoyl)-hydrazine 501
1-(4-Bromobenzoyl)-2-(1-naphthoyl)-hydrazine 500
$C_{18}H_{13}BrN_2O_4$
2-(4-(6-Bromo-1-hydroxy-2-naphthylazo)phenoxy)acetic acid 326
$C_{18}H_{13}Br_2N$
3-(*N,N*-Bis(bromomethyl)amino)-fluoranthene 222
2,2′-Dibromo-4-phenyldiphenylamine 224
$C_{18}H_{13}Br_3Si$
Tris(4-bromophenyl)silane 1104
$C_{18}H_{13}ClN_2$
4-Chloro-3-methyl-1-phenyl-1*H*-pyrrolo(2,3-b)quinoline 1003
$C_{18}H_{13}ClN_2O_2$
1-(2-Chlorobenzoyl)-2-(1-naphthoyl)-hydrazine 503
$C_{18}H_{13}ClN_2O_5$
4-Chloro-1-hydroxy-4′-nitro-2-naphth-*o*-anisidide 485
$C_{18}H_{13}Cl_2NO$
2,5-Dichloro-α,α-diphenyl-4-pyridinemethanol 932
$C_{18}H_{13}Cl_2NO_2$
2,4-Dichloro-1-naphthyl *N*-(*o*-tolyl)-carbamate 682
Ethyl 6,8-dichloro-2-phenyl-4-quinolinecarboxylate 993
$C_{18}H_{13}Cl_2NO_3$
2,4-Dichloro-1-naphthyl *N*-(4-methoxyphenyl)carbamate 655
$C_{18}H_{13}Cl_2N_3O$
3,5-Dichloro-α-(5,6-dimethyl-2-benzimidazolyl)-2-hydroxycinnamonitrile 902
$C_{18}H_{13}Cl_2O_4P$
2,2′-Dichlorotriphenyl phosphate 836
$C_{18}H_{13}Cl_4NSi$
4-((Diphenyl)methylsilyl)-2,3,5,6-tetrachloropyridine 1121
$C_{18}H_{13}FN_2O_2$
1-(2-Fluorobenzoyl)-2-(2-naphthoyl)-hydrazine 500
$C_{18}H_{13}F_2P$
2,6-Bis(4-fluorophenyl)-4-methylphosphorin 931
$C_{18}H_{13}F_3N_2O_2$
2-Methyl-8-quinolyl *N*-(3-trifluoromethylphenyl)carbamate 981
$C_{18}H_{13}I$
4-Iodo-*p*-terphenyl 172
$C_{18}H_{13}N$
1-Aminobenzo(c)phenanthrene 231
2-Aminobenzo(c)phenanthrene 231
3-Aminobenzo(c)phenanthrene 231
2-Aminobenzo(c)phenanthrene hydrochloride 231
3-Aminobenzo(c)phenanthrene hydrochloride 231
6-Aminobenzo(c)phenanthrene hydrochloride 231
$C_{18}H_{13}NO$
4-Biphenylyl 2-pyridyl ketone 961
2,6-Diphenyl-4-pyridinecarboxaldehyde 961
10-Phenylphenoxazine 253
$C_{18}H_{13}NOS$
10-Phenylphenothiazine-5-oxide 771
$C_{18}H_{13}NO_2$
5,6-Dihydrobenz(c)acridine-7-carboxylic acid 993
9,10-Dihydro-9,10-ethanoanthracene-11,12-dicarboximide 722
4′-Nitro-*m*-terphenyl 278
4-Nitro-*p*-terphenyl 278
N-Piperonylidene-1-naphthylamine 1076
$C_{18}H_{13}NO_2S$
10-Phenylphenothiazine-5,5-dioxide 778
$C_{18}H_{13}NO_3$
2′-Benzoyl-2-furanilide 852
2-(2-Hydroxy-1-naphthylmethyleneamino)-benzoic acid 1037
3-(2-Hydroxy-1-naphthylmethyleneamino)-benzoic acid 1034
4-(2-Hydroxy-1-naphthylmethyleneamino)-benzoic acid 1034
$C_{18}H_{13}NO_4$
8-Acetoxy-2-phenyl-4-quinolinecarboxylic acid 993
4-(2-Hydroxy-1-naphthylmethyleneamino)-salicylic acid 1037
$C_{18}H_{13}NO_5$
7-Acetoxy-3-(benzamido)coumarin 467
4-Coumarinyl *N*-(3-acetylphenyl)-carbamate 666
$C_{18}H_{13}NS$
N-Phenylphenothiazine 248
$C_{18}H_{13}N_3$
2,2′-Iminobis(quinoline) 989
$C_{18}H_{13}N_3OS$
3-Methyl-2-benzothiazolinone (1,2-dihydro-2-oxo-1-naphthylidene)-hydrazone 1061
$C_{18}H_{13}N_3O_5$
2,4-Dinitro-4′-phenoxydiphenylamine 266
$C_{18}H_{13}N_5O$
2-Amino-6,7-diphenyl-4-hydroxypyrimido-(4,5-b)pyrazine 1002
$C_{18}H_{14}$
5,12-Dihydrotetracene 167
trans-1-(1-Naphthyl)-2-phenylethylene 166
$C_{18}H_{14}AsBr$
(2-Bromophenyl)diphenylarsine 1126
$C_{18}H_{14}BiClS_2$
Bis(phenylthio)(4-chlorophenyl)-bismuthine 1141
$C_{18}H_{14}BrNO$
4′-Bromo-2-(1-naphthyl)acetanilide 485
$C_{18}H_{14}BrNO_2$
3-(2-Bromoacetyl)-6-methoxy-2-phenylquinoline hydrobromide 981
4-(2-Bromoacetyl)-6-methoxy-2-phenylquinoline hydrobromide 981
$C_{18}H_{14}BrNO_2S$
6-Bromo-2-naphthyl *N*-(4-(methylthio)phenyl)carbamate 668

$C_{18}H_{14}BrNO_3$
4-(5-Bromo-2-ethoxybenzylidene)-2-phenyl-2-oxazolin-5-one ... 384
$C_{18}H_{14}BrNO_3S$
4-Bromo-*N*-(4-phenoxyphenyl)-benzenesulfonamide ... 806
$C_{18}H_{14}Br_2N_2$
N,*N*-Bis(2-bromophenyl)-1,2-phenylenediamine ... 224
$C_{18}H_{14}ClF_3N_2O_5$
Dimethyl 5-(3-(2-chloro-5-(trifluoromethyl)phenyl)ureido)-isophthalate ... 505
$C_{18}H_{14}ClNO$
11-Chloro-6,7-dihydro-5*H*-benzo(6,7)cyclohepta(1,2-b)-quinolin-8(13*H*)-one ... 320
1-(3-Chloro-*o*-tolyliminomethyl)-2-naphthol ... 1038
1-(3-Chloro-*p*-tolyliminomethyl)-2-naphthol ... 1038
1-(4-Chloro-*o*-tolyliminomethyl)-2-naphthol ... 1038
1-(5-Chloro-*o*-tolyliminomethyl)-2-naphthol ... 1038
$C_{18}H_{14}ClNO_2$
3-Chloro-2-(2,4-dimethylanilino)-1,4-naphthoquinone ... 318
3-Chloro-2-(3,4-dimethylanilino)-1,4-naphthoquinone ... 317
2-Chloro-3-(4-ethylanilino)-1,4-naphthoquinone ... 317
4′-Chloro-3-hydroxy-2′-methyl-2-naphthanilide ... 486
5′-Chloro-3-hydroxy-2′-methyl-2-naphthanilide ... 486
1-(5-Chloro-2-methoxyphenyliminomethyl)-2-naphthol ... 1022
4-Chloro-2-methylphenyl *N*-(1-naphthyl)carbamate ... 640
4-Chloro-3-methylphenyl *N*-(1-naphthyl)carbamate ... 640
2-Chloro-3-(phenethylamino)-1,4-naphthoquinone ... 317
2-(4-Chlorophenyl)-4-(1,2-epoxyethyl)-6-methoxyquinoline ... 981
$C_{18}H_{14}ClNO_2S_2$
3-(4-Chlorophenyl)-5-(2-methoxy-3-methylbenzylidene)rhodanine ... 729
$C_{18}H_{14}ClNO_3$
3-Chloro-2-(4-ethoxyanilino)-1,4-naphthoquinone ... 317
1-Naphthyl *N*-(2-chloro-4-methoxyphenyl)carbamate ... 621
1-Naphthyl *N*-(4-chloro-2-methoxyphenyl)carbamate ... 619
2-Naphthyl *N*-(4-chloro-2-methoxyphenyl)carbamate ... 619
2-Naphthyl *N*-(5-chloro-2-methoxyphenyl)carbamate ... 619
$C_{18}H_{14}ClNO_4$
2-(4-Chlorophenyl)-4-(3-ethoxy-2-hydroxybenzylidene)-2-oxazolin-5-one ... 382
$C_{18}H_{14}ClNO_5$
4-Methyl-7-coumarinyl *N*-(4-chloro-2-methoxyphenyl)carbamate ... 618
$C_{18}H_{14}ClN_3O$
β-(4-Chlorophenyl)-*β*-hydroxy-2-(2-imidazolyl)hydrocinnamonitrile ... 864
$C_{18}H_{14}ClN_3O_4S$
3-(Methylthio)-2-phenyl-1,2,4-triazino(2,3-a)quinolin-11-ium perchlorate ... 1002
$C_{18}H_{14}Cl_2IN$
2-(2,6-Dichlorostyryl)-1-methylquinolinium iodide ... 990
2-(3,4-Dichlorostyryl)-1-methylquinolinium iodide ... 990
4-(2,6-Dichlorostyryl)-1-methylquinolinium iodide ... 990
4-(3,4-Dichlorostyryl)-1-methylquinolinium iodide ... 990
$C_{18}H_{14}Cl_2N_2O$
1-(4,5-Dichloro-2-methylphenyl)-3-(1-naphthyl)urea ... 546
$C_{18}H_{14}Cl_2N_2O_4$
2-Butyne-1,4-diol bis-(2-chlorocarbanilate) ... 1015
2-Butyne-1,4-diol bis-(3-chlorocarbanilate) ... 1015
2,4-Dichloro-8-quinolyl *N*-(2,5-dimethoxyphenyl)carbamate ... 984
$C_{18}H_{14}Cl_2O_2$
11,12-Dichloro-9,10-dihydro-2,3-dimethoxy-9,10-ethenoanthracene .. 178
$C_{18}H_{14}Cl_4N_4O_3$
2,2′-Oxybis(acetic acid (2,4-dichlorobenzylidene)hydrazide) ... 1023
2,2′-Oxybis(acetic (3,4-dichlorobenzylidene)hydrazide) ... 493
$C_{18}H_{14}FNO$
4-Cyclopropylformyl-4′-fluorodiphenylacetonitrile ... 755
$C_{18}H_{14}FNO_5S$
1-Hydroxy-5′-fluorosulfonyl-2-naphth-*o*-anisidide ... 801
$C_{18}H_{14}N_2$
N-Benzylidene-4-(2-pyridyl)aniline ... 1077
5,6-Dimethylbenzo(a)phenazine ... 1001
1-(1*H*-Indol-3-yl)-1-(3*H*-indol-3-yl)-ethene ... 891
$C_{18}H_{14}N_2O$
Benzoic (1-naphthylmethylene)-hydrazide ... 501
1,2-Dihydro-2-(2-naphthyl)-4(3*H*)-quinazolinone ... 704
$C_{18}H_{14}N_2O_2$
Benzoic (7-hydroxy-1-naphthylmethylene)hydrazide ... 496
(4-Benzyloxy-3-methoxybenzylidene)-malononitrile ... 762
3,6-Diphenyl-5-methylisoxazolo(4,5-c)pyridin-4(5*H*)-one ... 874
4-Hydroxybenzoic (1-naphthylmethylene)hydrazide ... 496
$C_{18}H_{14}N_2O_2S$
1-(3-Hydroxyphenyl)-3-(1-naphthylcarbonyl)-2-thiourea ... 734
$C_{18}H_{14}N_2O_3$
2-Carbamoylphenyl *N*-(1-naphthyl)-carbamate ... 666
3-Hydroxy-2-naphthoic (2-hydroxybenzylidene)hydrazide ... 495
1-(2-Methyl-5-nitrophenyliminomethyl)-2-naphthol ... 1037
1-(4-Methyl-3-nitrophenyliminomethyl)-2-naphthol ... 1037
$C_{18}H_{14}N_2O_4$
3-Benzamido-1-benzoyl-2,4-pyrrolidinedione ... 705
3,6-Bis(3-hydroxybenzylidene)-2,5-piperazinedione ... 696
1,6-Bis(2-nitrophenyl)-1,3,5-hexatriene ... 278
1,6-Bis(4-nitrophenyl)-1,3,5-hexatriene ... 278
1,4-Diacetamidoanthraquinone ... 488
2-(4-(2-Hydroxy-1-naphthylazo)-phenoxy)acetic acid ... 1083
1-(2-Methoxy-5-nitrophenyliminomethyl)-2-naphthol ... 1032
3-Methyl-4-nitrophenyl *N*-(1-naphthyl)carbamate ... 674
2-Naphthyl *N*-(4-methyl-2-nitrophenyl)carbamate ... 668
$C_{18}H_{14}N_2O_5$
1-Hydroxy-4′-nitro-2-naphtho-*o*-anisidide ... 485
2-Naphthyl *N*-(2-methoxy-4-nitrophenyl)carbamate ... 654
$C_{18}H_{14}N_2O_6$
4-Methyl-7-coumarinyl *N*-(2-methyl-5-nitrophenyl)carbamate ... 666
$C_{18}H_{14}N_4$
Indole-3-carboxaldehyde azine ... 1079
N,*N*′-(1,4-Phenylenedimethylidyne)-bis(2-aminopyridine) ... 1066
$C_{18}H_{14}N_4O$
1-(1-Methyl-2-benzimidazolylazo)-2-napthol ... 1085
$C_{18}H_{14}N_4OS$
5-Methyl-7,8-quinolinedione 7-((3-methyl-2-benzothiazolinylidene)-hydrazone) ... 954
7-Methyl-5,8-quinolinedione 5-((3-methyl-2-benzothiazolinylidene)-hydrazone) ... 954
$C_{18}H_{14}N_4O_7S$
Acid Orange 3 (C.I. 10385) ... 1094
$C_{18}H_{14}N_4S$
3-Methyl-2-benzothiazolinone (1,2-dihydro-2-imino-1-naphthylidene)-hydrazone hydriodide ... 1053
$C_{18}H_{14}O$
2-Biphenylyl phenyl ether ... 183
4,4-Diphenyl-2,5-cyclohexadien-1-one ... 306
1-(1-Hydroxyethyl)pyrene ... 208
1-(2-Indanylidene)-2-indanone ... 287
2-Indanylidene-1-indanone ... 298
4-(Methoxymethyl)pyrene ... 188
4-Phenoxybiphenyl ... 183
$C_{18}H_{14}OS$
6,7-Dihydrobenzo(b)naphtho(1,2-d)-thiepin-8(9*H*)-one ... 286
$C_{18}H_{14}OSi$
10-Phenyl-10*H*-phenoxasilin ... 1122
$C_{18}H_{14}O_2$
9-(Acetoacetyl)anthracene ... 307
5,6,6a,7,8,12b-Hexahydrobenzo(c)-phenanthrene-5,8-dione ... 309
Methyl 9,10-dihydro-9,10-ethenoanthracene-9-carboxylate ... 376
2-(1-Naphthyl)-2-phenylacetic acid ... 333
$C_{18}H_{14}O_3$
2-Acetonyl-2-phenyl-1,3-indandione . 318
3-Acetoxy-6-acetylphenanthrene ... 379
3-Acetyl-4,5-diphenyl-2(5*H*)-furanone ... 385
6-Acetyl-3-phenanthryl acetate ... 379
1-Benzylidene-3-piperonylideneacetone ... 286
6,7-Diphenyl-3-oxabicyclo(3.1.1)-heptane-2,4-dione ... 387
2-Methoxy-2-phenylnaphtho(2,3-d)-1,3-dioxole ... 183
γ-Oxo-2-phenanthrenebutyric acid ... 331
2-Piperonylidene-1-tetralone ... 296
2,4′,4″-Trihydroxy-*p*-terphenyl ... 194
$C_{18}H_{14}O_3S_2$
Phenyl 3-(phenylsulfonyl)phenyl sulfoxide ... 782
$C_{18}H_{14}O_4$
3-Acetoxyacetyl-4,5-phenanthrylene oxide, mixture of isomers ... 378
7-Acetoxy-2-methyl-3-phenyl-4*H*-1-benzopyran-4-one ... 378
9,10-Dihydro-9,10-dimethyl-9,10-ethanoanthracene-2,3,6,7-tetrol ... 194
9,10-Dihydro-9,10-ethanoanthracene-11,12-dicarboxylic acid ... 331
9,10-Dihydro-9,10-ethanoanthracene-11,12-dicarboxylic acid, disodium salt ... 351
9,10-Dihydro-9,10-ethanoanthracene-11,12-dicarboxylic acid, mercury(II) salt ... 1139
Dimethyl 4,5-phenanthrene-dicarboxylate ... 375
$C_{18}H_{14}O_4S_2$
1,4-Bis(phenylsulfonyl)benzene ... 784
$C_{18}H_{14}O_5$
Ethyl 5-methoxyanthraquinone-2-carboxylate ... 369
$C_{18}H_{14}O_6$
4,4′-Diacetoxybenzil ... 379
2-Formyl-1,3,8-trimethoxyanthraquinone ... 323
$C_{18}H_{14}O_8$
(+)-Dibenzoyl-L-tartaric acid disilver salt ... 351
$C_{18}H_{14}S$
4-Biphenylyl phenyl sulfide ... 212
$C_{18}H_{15}AsN_2O_5$
4-(Diphenylamino)-2-nitrobenzenearsonic acid ... 1127
$C_{18}H_{15}AsS_3$
Tris(phenylthio)arsine ... 1127
$C_{18}H_{15}BF_4O$
2,4-Diphenyl-6-methylpyrylium tetrafluoroborate ... 935
$C_{18}H_{15}BF_4OS$
2-(4-Methoxyphenyl)-6-phenylthiopyrylium tetrafluoroborate ... 937
$C_{18}H_{15}BO_3$
Triphenyl borate ... 1125
$C_{18}H_{15}B_3O_3$
2,4,6-Triphenylboroxin ... 1125
$C_{18}H_{15}BiBr_2$
Dibromotriphenylbismuth ... 1143
$C_{18}H_{15}BiF_2$
Difluorotriphenylbismuth ... 1142
$C_{18}H_{15}BiN_2O_6$
Triphenylbismuth dinitrate ... 1142
$C_{18}H_{15}BiS_2$
Bis(phenylthio)phenylbismuthine ... 1142
$C_{18}H_{15}BrO_2$
2-(5-Bromo-2-methoxybenzylidene)-1-tetralone ... 299
$C_{18}H_{15}BrPb$
Bromotriphenyllead ... 1133
$C_{18}H_{15}BrSn$
Bromotriphenyltin ... 1132
$C_{18}H_{15}Br_2Sb$
Triphenylantimony dibromide ... 1129
$C_{18}H_{15}ClFNO_4S$
4-Chloro-6-fluorosulfonyl-2-(4-ethoxy-3-methoxyphenyl)quinoline . 985
$C_{18}H_{15}ClHgO_2$
12-(Chloromercuri)-9,10-dihydro-9,10-ethanoanthracen-11-yl acetate ... 1138
$C_{18}H_{15}ClNO_5P$
4′-Chloro-3-hydroxy-2′-methyl-2-naphthalilide monophosphate ester ... 833
$C_{18}H_{15}ClN_2O$
1-(2-Chloro-6-methylphenyl)-3-(1-naphthyl)urea ... 543
1-(3-Chloro-2-methylphenyl)-3-(1-naphthyl)urea ... 543
1-(3-Chloro-4-methylphenyl)-3-(1-naphthyl)urea ... 545
1-(4-Chloro-2-methylphenyl)-3-(1-naphthyl)urea ... 545
1-(5-Chloro-2-methylphenyl)-3-(1-naphthyl)urea ... 545
$C_{18}H_{15}ClN_2O_2$
1-(5-Chloro-2-methoxyphenyl)-3-(1-naphthyl)urea ... 520
1-(3-Chloro-2-methylphenyl)-3-(5-hydroxy-1-naphthyl)urea ... 514
1-(5-Chloro-2-methylphenyl)-3-(5-hydroxy-1-naphthyl)urea ... 514
1-(5-Chloro-2-methylphenyl)-3-(7-hydroxy-1-naphthyl)urea ... 514
4-(4-Chlorophenyl)-6-(4-methoxyphenyl)-1-methyl-2(1*H*)-pyrimidinone ... 694
5-Chloro-8-quinolyl *N*-(2,4-xylyl)-carbamate ... 992
2-Methyl-8-quinolyl *N*-(4-chloro-2-methylphenyl)carbamate ... 992
$C_{18}H_{15}ClN_2O_3$
1-(5-Chloro-2-methoxyphenyl)-3-(5-hydroxy-1-naphthyl)urea ... 512
1-(5-Chloro-2-nitrobenzoyl)-4-phenyl-1,2,3,6-tetrahydropyridine ... 694
5-Chloro-8-quinolyl *N*-(2-methoxy-5-methylphenyl)carbamate ... 986
2-Methyl-4-quinolyl *N*-(4-chloro-2-methoxyphenyl)carbamate ... 981
2-Methyl-8-quinolyl *N*-(4-chloro-2-methoxyphenyl)carbamate ... 981
4-Methyl-2-quinolyl *N*-(4-chloro-2-methoxyphenyl)carbamate ... 981
$C_{18}H_{15}ClN_2O_4$
4-Chloro-8-quinolyl *N*-(2,5-dimethoxyphenyl)carbamate ... 984

$C_{18}H_{15}ClO_4$
Methyl 2-(2-(4-chlorobenzoylvinyl)-phenoxy)acetate ... 353
$C_{18}H_{15}ClO_5$
2,4-Diphenyl-6-methylpyrylium perchlorate ... 934
$C_{18}H_{15}ClPb$
Chlorotriphenyllead ... 1136
$C_{18}H_{15}ClSi$
(2-Chlorophenyl)diphenylsilane ... 1103
(4-Chlorophenyl)diphenylsilane ... 1103
$C_{18}H_{15}Cl_2NO_2$
Methyl 2,2-bis(2-chlorobenzyl)-2-cyanoacetate ... 752
$C_{18}H_{15}Cl_2N_3O$
4-(2,4-Dichlorobenzylideneamino)-antipyrine ... 1027
$C_{18}H_{15}Cl_2N_3O_2$
2-(Cyclopropylcarboxamido)-benzoic (2,4-dichlorobenzylidene)-hydrazide ... 493
2,4-Dichlorohippuric (3-phenylallylidene)hydrazide ... 493
3,4-Dichlorohippuric (3-phenyl-2-propenylidene)hydrazide ... 493
$C_{18}H_{15}Cl_2N_3O_6$
4-(3,5-Bis(methoxycarbonyl)phenyl)-1-(2,4-dichlorobenzoyl)-semicarbazide ... 528
$C_{18}H_{15}Cl_4NO_3$
5′-Isopropyl-2′-methyl-3,4,5,6-tetrachlorophthalanilic acid ... 455
$C_{18}H_{15}FIN$
2-(4-Fluorostyryl)-1-methylquinolinium iodide ... 990
$C_{18}H_{15}FN_2O_2$
1,2-Dihydro-2,*N*-dimethyl-3-(2-fluorophenyl)-1-oxo-4-isoquinolinecarboxamide ... 695
$C_{18}H_{15}FPb$
Fluorotriphenyllead ... 1136
$C_{18}H_{15}F_3N_2O_5$
Dimethyl 5-(3-(3-(trifluoromethyl)-phenyl)ureido)isophthalate ... 509
$C_{18}H_{15}IN_2O_2$
1-Methyl-2-(3-nitrostyryl)-quinolinium iodide ... 990
$C_{18}H_{15}IO$
2,4-Diphenyl-6-methylpyrylium iodide ... 935
$C_{18}H_{15}IPb$
Iodotriphenyllead ... 1136
$C_{18}H_{15}ISn$
Iodotriphenyltin ... 1132
$C_{18}H_{15}N$
11-Cyanomethyl-9,10-dihydro-9,10-ethanoanthracene ... 758
2,6-Diphenyl-4-methylpyridine ... 966
Diphenyl-3-pyridylmethane ... 966
$C_{18}H_{15}NO$
12-Cyano-9,10-dihydro-9-methoxy-9,10-ethanoanthracene ... 753
3,3-Diphenyl-2-oxo-1-cyclopentanecarbonitrile ... 757
α,α-Diphenyl-2-pyridinemethanol ... 933
4-Methyl-α-(4-methylbenzoyl)-cinnamonitrile ... 766
N-(1-Naphthyl)-2-phenylacetamide ... 489
1-(*o*-Tolyliminomethyl)-2-naphthol ... 1038
1-(*m*-Tolyliminomethyl)-2-naphthol ... 1039
1-(*p*-Tolyliminomethyl)-2-naphthol ... 1038
$C_{18}H_{15}NOS$
1-(4-(Methylthio)-phenyliminomethyl)-2-naphthol ... 1032
$C_{18}H_{15}NO_2$
4-Acetyl-6-methoxy-2-phenylquinoline ... 986
Ethyl α-cyano-β-phenylcinnamate ... 766
Ethyl α-cyano-4-phenylcinnamate ... 766
Ethyl 2-phenyl-4-quinolinecarboxylate ... 993
3-Hydroxy-2′-methyl-2-naphthanilide ... 486
1-(2-Hydroxy-5-methylphenyliminomethyl)-2-naphthol ... 1029
2-Hydroxy-1-naphtho-*o*-toluidide ... 486
4-Methoxy-*N*-(1-naphthyl)-benzamide ... 487
1-(2-Methoxyphenyliminomethyl)-2-naphthol ... 1032
1-(4-Methoxyphenyliminomethyl)-2-naphthol ... 1032
2-Naphthyl 2′-methylcarbanilate ... 677
N-(1-Naphthyl)-2-phenoxyacetamide ... 486
α-4-Pyridylbenzhydrol-*N*-oxide ... 1005
4,4′-(2-Pyridylmethylene)diphenol ... 934
m-Tolyl *N*-(1-naphthyl)carbamate ... 687
$C_{18}H_{15}NO_2S$
4-(Methylthio)phenyl *N*-(1-naphthyl)-carbamate ... 668
2-Naphthyl *N*-(4-(methylthio)-phenyl)carbamate ... 668
$C_{18}H_{15}NO_3$
4-(3-Ethoxybenzylidene)-2-phenyl-2-oxazolin-5-one ... 384
4-(4-Ethoxybenzylidene)-2-phenyl-2-oxazolin-5-one ... 384
Furfuryl *N*,*N*-diphenylcarbamate ... 849
3-Hydroxy-2-naphthoyl-*p*-anisidide ... 485
5-Methoxy-*N*-(1-naphthyl)-anthranilic acid ... 340
1-Methyl-2-(3-nitrobenzyloxy)-naphthalene ... 265
1-Methyl-2-(4-nitrobenzyloxy)-naphthalene ... 265
2-Naphthyl *N*-(2-methoxyphenyl)-carbamate ... 653
2-Naphthyl *N*-(4-methoxyphenyl)-carbamate ... 655
4a-Phenyl-2,3,4,4a-tetrahydro-1*H*,6*H*-pyrido(1,2-a)(3,1)-benzoxazine-1,6-dione ... 709
$C_{18}H_{15}NO_4$
4-(2,3-Dimethoxybenzylidene)-2-phenyl-2-oxazolin-5-one ... 383
4-(2,4-Dimethoxybenzylidene)-2-phenyl-2-oxazolin-5-one ... 383
4-(2,5-Dimethoxybenzylidene)-2-phenyl-2-oxazolin-5-one ... 384
4-(2,6-Dimethoxybenzylidene)-2-phenyl-2-oxazolin-5-one ... 383
4-(3,5-Dimethoxybenzylidene)-2-phenyl-2-oxazolin-5-one ... 383
Ethyl 1-hydroxy-2-oxo-1,2,3,10-tetrahydroindeno(2,1-e)indole-1-carboxylate ... 689
Methyl 3-phenyl-2-phthalimidopropionate ... 719
2-Phenyl-4-veratrylidene-2-oxazolin-5-one ... 383
4-(2-Phthalimidophenyl)butyric acid. 722
$C_{18}H_{15}NO_4S$
1-Hydroxy-4′-(methylsulfonyl)-2-naphthanilide ... 774
$C_{18}H_{15}NO_5$
2-(2-Acetoxyphenyl)-1-acetyl-1,2-dihydro-4*H*-3,1-benzoxazin-4-one ... 700
Ethyl α-benzoyl-4-nitrocinnamate ... 365
$C_{18}H_{15}NO_8$
4-Nitro-2′,4′,5′-triacetoxybiphenyl ... 378
$C_{18}H_{15}NS$
2,6-Diphenyl-1-methyl-4(1*H*)-pyridinethione ... 295
$C_{18}H_{15}NS_2$
N-(1-Naphthyl)-2-(thiophenoxy)-thioacetamide ... 739
$C_{18}H_{15}N_3$
1,2,3-Tris(4-pyridyl)cyclopropane ... 966
$C_{18}H_{15}N_3O$
4-Aminobenzoic (1-naphthylmethylene)hydrazide ... 1052
Anthranilic (1-naphthylmethylene)-hydrazide ... 499
$C_{18}H_{15}N_3OS$
1-(2-Hydroxy-1-naphthylmethylene)-4-phenyl-3-thiosemicarbazide ... 734
$C_{18}H_{15}N_3O_3$
1-(2-Methyl-4-nitrophenyl)-3-(1-naphthyl)urea ... 568
$C_{18}H_{15}N_3O_4$
1-(4-(Ethoxycarbonyl)phenyl)-5-imino-3-phenylhydantoin ... 726
1-(5-Hydroxy-1-naphthyl)-3-(2-methyl-5-nitrophenyl)urea ... 564
1-(7-Hydroxy-1-naphthyl)-3-(2-methyl-5-nitrophenyl)urea ... 564
$C_{18}H_{15}N_3O_4S$
Disperse Yellow 42 (C.I. 10338) ... 817
$C_{18}H_{15}N_3O_8S_2$
Lignin pink ... 1085
$C_{18}H_{15}N_3Si$
Triphenylsilyl azide ... 1113
$C_{18}H_{15}N_4O_3S$
Disperse Red 58 (C.I. 11135) ... 1082
$C_{18}H_{15}N_5O$
5,8-Quinolinedione 5-((1,3-dimethyl-2-benzimidazolinylidene)-hydrazone) ... 954
$C_{18}H_{15}N_5O_6$
2-(4-(Dimethylamino)-3,5-dinitrophenyl)-8-methyl-5-nitroquinoline ... 988
$C_{18}H_{15}P$
2,6-Diphenyl-4-methylphosphorin ... 966
$C_{18}H_{15}SSb$
Triphenylantimony sulfide ... 1129
$C_{18}H_{16}$
9,10-Dihydro-9,10-dimethyl-9,10-ethenoanthracene ... 164
2,3-Dimethyl-4b,8b,8c,8d-tetrahydrodibenzo(a,f)-cyclopropa(c,d)pentalene ... 162
9,10,11,12,13,14-Hexahydro-9,10-(1′,2′)-*endo*-cyclobutanthracene ... 162
$C_{18}H_{16}AsNO_3$
2-(Diphenylamino)-benzenearsonic acid ... 1127
$C_{18}H_{16}BF_4N$
2,6-Diphenyl-1-methylpyridinium tetrafluoroborate ... 957
$C_{18}H_{16}BrNO$
1-(Diphenylmethoxy)pyridinium bromide ... 957
$C_{18}H_{16}Br_2$
2-*tert*-Butyl-9,10-dibromoanthracene ... 175
$C_{18}H_{16}ClF_3N_2O_4$
2-(4-Acetamidophenoxy)ethyl *N*-(4-chloro-2-(trifluoromethyl)phenyl)-carbamate ... 606
$C_{18}H_{16}ClIN_2S$
7-Chloro-5,10-dihydro-3,10-dimethyl-5-phenylthiazolo(2,3-b)-quinazolin-10-ium iodide ... 255
$C_{18}H_{16}ClN$
2,6-Diphenyl-1-methylpyridinium chloride ... 957
$C_{18}H_{16}ClNO_2$
11b-(4-Chlorophenyl)-2,3,4,5,7,11b-hexahydro(1,3)oxazepino(2,3-a)-isoindol-7-one ... 695
$C_{18}H_{16}ClNO_4$
Ethyl 2-(3-(4-chlorophenyl)-3-hydroxy-1-oxo-2-isoindolinyl)-acetate ... 689
$C_{18}H_{16}ClN_3O$
4-(4-Chlorobenzylideneamino)-antipyrine ... 1027
1-(3-Chloro-4-methylphenyl)-3-(2-methyl-4-quinolyl)urea ... 981
N-(4-Chlorophenyl)-2-(2-methylimidazol-1-ylmethyl)-benzamide ... 865
$C_{18}H_{16}ClN_3O_2$
1-(5-Chloro-2-methoxyphenyl)-3-(2-methyl-4-quinolyl)urea ... 981
1-(4-Chloro-7-methoxy-2-quinolyl)-3-(*p*-tolyl)urea ... 986
1-(*N*-(4-Chloro-2-(phenoxy)phenyl)-carbamoyl)-3,5-dimethylpyrazole ... 859
N-(3-Chlorophenyl)-*N*′-(α-methylcinnamylideneamino)-oxamide ... 1025
$C_{18}H_{16}ClN_3O_3$
1-(*p*-Anisyl)-3-(4-chloro-7-methoxy-2-quinolyl)urea ... 986
$C_{18}H_{16}ClN_5O_4$
2-(3-(2-Chloro-4-nitrophenyl)ureido)-acetic (3-phenylallylidene)-hydrazide ... 529
$C_{18}H_{16}ClN_5O_4S$
2-((1,2-Dihydro-1-methyl-2-quinolylidene)aminoazo)-3-methyl-benzothiazolium perchlorate ... 915
2-((1,4-Dihydro-1-methyl-4-quinolylidene)aminoazo)-3-methyl-benzothiazolium perchlorate ... 915
$C_{18}H_{16}ClOSb$
Chlorohydroxytriphenylantimony ... 1129
$C_{18}H_{16}Cl_2$
11,12-Bis(chloromethyl)-9,10-dihydro-9,10-ethanoanthracene ... 172
11,12-Dichloro-9,10-dihydro-9,10-dimethyl-9,10-ethanoanthracene ... 173
$C_{18}H_{16}Cl_2N_2O_2$
3-Chloro-2-(chloromethyl)-6-methoxy-4-(4-methoxyanilino)-quinoline picrate ... 981
$C_{18}H_{16}Cl_2N_4O_2$
4-(3-Allylureido)benzoic (3,4-dichlorobenzylidene)hydrazide ... 529
$C_{18}H_{16}Cl_2N_4O_3$
2,2′-Oxydiacetic acid bis((4-chlorobenzylidene)hydrazide) ... 1023
$C_{18}H_{16}Cl_2O_2$
1,4-Bis(4-chloro-3-methylphenoxy)-2-butyne ... 1017
11,12-Dichloro-9,10-dihydro-2,3-dimethoxy-9,10-ethanoanthracene.. 178
$C_{18}H_{16}Cl_3NO_3$
4′-(2,5-Dimethoxystyryl)-2,2,2-trichloroacetanilide ... 396
$C_{18}H_{16}Cl_3NO_5$
2-(2,4-Dichlorophenoxy)ethyl *o*-(3-chlorophenyl)carbamoyllactate ... 615
$C_{18}H_{16}FN_3O$
11-(4-Fluorobenzyl)-2,3,4,11-tetrahydro-6*H*-pyrimido(2,1-b)-quinazolin-6-one ... 726
$C_{18}H_{16}F_3N_3O$
1-Benzyl-1-(2-cyanoethyl)-3-(3-(trifluoromethyl)phenyl)urea ... 751
$C_{18}H_{16}F_5N_3O_2$
4′-(4-(3-Methylureido)benzyl)-2,2,3,3,3-pentafluoropropionanilide. 510
$C_{18}H_{16}FeN_2O$
1,2-Dihydro-2-ferrocenyl-4(3*H*)-quinazolinone ... 1141
$C_{18}H_{16}Ge$
Triphenylgermane ... 1113
$C_{18}H_{16}GeO$
Triphenylgermanol ... 1113
$C_{18}H_{16}IN$
1-Methyl-2-(2-(1-naphthyl)vinyl)-pyridinium iodide ... 958
$C_{18}H_{16}NO_3P$
Diphenyl *N*-phenylphosphoramidate ... 836
$C_{18}H_{16}N_2$
2′-Acetonaphthone phenylhydrazone ... 1072
Cinnamaldehyde azine ... 1076
4,5-Dihydro-3-(*m*-tolyl)-2*H*-benzene(g)indazole ... 864
1,1,2-Triphenylhydrazine ... 235
$C_{18}H_{16}N_2O$
1-Benzyl-3-(1-naphthyl)urea ... 572
4′-(β-Cyano-4-methylstyryl)-acetanilide ... 767
4-(4-(Dimethylamino)phenylimino)-1(4*H*)-naphthalenone ... 1057
4,6-Diphenyl-1-ethyl-2(1*H*)-pyrimidinone ... 706
5,6-Diphenyl-1-ethyl-2(1*H*)-pyrimidinone ... 713
1-Ethoxy-4-(phenylazo)naphthalene.. 1087
1-Methyl-3-(1-naphthyl)-1-phenylurea ... 586
1-(1-Naphthyl)-3-(*o*-tolyl)urea ... 589
1-(1-Naphthyl)-3-(*m*-tolyl)urea ... 603
1-(1-Naphthyl)-3-(*p*-tolyl)urea ... 592
$C_{18}H_{16}N_2O_2$
4-Allyl-1,2-diphenyl-3,5-pyrazolidinedione ... 708
4,6-Bis(4-methoxyphenyl)pyrimidine 972
4′-(β-Cyano-4-methoxystyryl)-acetanilide ... 762

$C_{18}H_{16}N_2O_2$ (Cont'd)
3-(Cyanomethyl)-2-(3,4-dimethoxyphenyl)indole 882
1′,2′-Dihydro-2,2′-dimethylspiro(isoindoline-1,3′(4′*H*)-isoquinoline)-1′,3-dione 710
4-(4-(Dimethylamino)benzylidene)-2-phenyl-5(4*H*)-oxazolone 385
Ethyl 4,5-diphenyl-2-imidazolecarboxylate 867
2-Methyl-8-quinolyl *N*-(*m*-tolyl)-carbamate 994
N,*N*′-(*p*-Phenylenedimethylidyne)-bis(furfurylamine) 1076
1-Phenyl-1,2,3,4-tetrahydronorharman-3-carboxylic acid 919
8-Quinolyl *N*-(2,4-xylyl)carbamate 992
8-Quinolyl *N*-(2,6-xylyl)carbamate 994

$C_{18}H_{16}N_2O_2S$
1-(5-Hydroxy-1-naphthyl)-3-(4-(methylthio)phenyl)urea 547
1-(7-Hydroxy-1-naphthyl)-3-(3-(methylthio)phenyl)urea 547
2-Methyl-8-quinolinyl *N*-(4-(methylthio)phenyl)carbamate 987

$C_{18}H_{16}N_2O_3$
1-Isopropyl-6,7-methylenedioxy-4-phenyl-2(1*H*)-quinazolinone 705
2-Methyl-8-quinolyl *N*-(3-methoxyphenyl)carbamate 992
*N*α-Benzoyl-DL-tryptophan 886
8-Quinolinyl *N*-(4-ethoxyphenyl)-carbamate 986
8-Quinolyl *N*-(2-methoxy-5-methylphenyl)carbamate 986

$C_{18}H_{16}N_2O_3S$
N[4]-Acetyl-*N*[1]-(1-naphthyl)-sulfanilamide 825
2-(1-Naphthyl)-*N*-(3-sulfamoylphenyl)acetamide 825
p-Toluenesulfonic acid (2-hydroxy-1-naphthylmethylene)-hydrazide 1033

$C_{18}H_{16}N_2O_4$
2-Butyne-1,4-diol dicarbanilate 1015
Diethyl 3,3′-(1,4-phenylene)bis(2-cyanoacrylate) 766
N,*N*′-Dipiperonylidene-ethylenediamine 1053
Ethyl 4-(phthalimidomethylamino)-benzoate 719
8-Quinolinyl *N*-(2,4-dimethoxyphenyl)carbamate 984
8-Quinolyl *N*-(2,5-dimethoxyphenyl)-carbamate 986

$C_{18}H_{16}N_2O_4S$
2-(α-Ethoxycarbonyl-4-nitrobenzylidene)-3-methylbenzothiazoline 374
Scarlet *R* 1084

$C_{18}H_{16}N_2O_7S_2$
Scarlet R (C.I. 16150) 1084

$C_{18}H_{16}N_2O_8$
Diethyl 6,6′-dinitro-2,2′-biphenyldicarboxylate 373

$C_{18}H_{16}N_4$
3,5-Cyclohexadiene-1,2-dione bis(phenylhydrazone) 235

$C_{18}H_{16}N_4O_2$
N,*N*-Dimethyl-4-(4-nitro-1-naphthylazo)aniline 1089
N,*N*-Dimethyl-4-(4-nitrophenylazo)-1-naphthylamine 1090

$C_{18}H_{16}N_4O_3$
2,3-Dimethyl-4-(3-nitrobenzylideneamino)-1-phenyl-3-pyrazolin-5-one 1059
1-(4-Methyl-2-nitrophenyl)-3-(2-methyl-4-quinolyl)urea 993

$C_{18}H_{16}N_4O_6S$
4′-(3-(Hydrazinosulfonyl)-benzamido)-2′-hydroxy-2-furanilide 845

$C_{18}H_{16}N_6O_2$
3-(*N*-(4-(2-Cyano-4-nitrophenylazo)-phenyl)-*N*-ethylamino)propionitrile 1089

$C_{18}H_{16}N_6O_7$
2,2′-Oxydiacetic acid bis((3-nitrobenzylidene)hydrazide) 1041

$C_{18}H_{16}N_6O_{12}S_2$
N,*N*′-Bis(2,4-dinitrophenyl)-L-cystine 348

$C_{18}H_{16}O$
2-Acetyldibenzo(e,h)bicyclo(2.2.2)-octa-5,7-diene 287
11-Acetyl-9,10-dihydro-9,10-ethanoanthracene 286
Benzyl 1-methyl-2-naphthyl ether 186
5,6,6a,7,12,12a-Hexahydrobenz(a)-anthracen-5-one 298
2-(4-Methylbenzylidene)-1-tetralone .. 305
4′-Methyl-5-phenyl-2,4-pentadienophenone 304
1-(4-Methylphenyl)-5-phenyl-1,4-pentadien-3-one 286

$C_{18}H_{16}OSi$
(4-Phenoxyphenyl)phenylsilane 1106

$C_{18}H_{16}O_2$
9,10-Dihydro-9,10-ethanoanthracen-11-yl acetate 361
9,10-Dihydro-9,10-ethanoanthracen-9-yl acetate 361
10,11-Dihydro-5,10-methano-5*H*-benzo(a,d)cyclohepten-12-yl acetate 361
10,11-Dihydro-5,10-methano-5*H*-dibenzo(a,d)cyclohepten-11-yl acetate 361
4′-Isopropylisoflavone 320
2-(4-Methoxybenzylidene)-1-tetralone 293
Methyl 9,10-dihydro-9,10-ethanoanthracene-11-carboxylate 361
Methyl 4-(4-phenyl-1,3-butadien-1-yl)benzoate 375
Methyl 4-styrylcinnamate 365

$C_{18}H_{16}O_3$
2,8-Diallyloxydibenzofuran 893
Ethyl 3-hydroxy-1-phenylindene-2-carboxylate 368
2-(5,6,7,8-Tetrahydro-2-naphthoyl)-benzoic acid 344

$C_{18}H_{16}O_4$
2-Benzylidene-1-(2-hydroxy-4-methoxyphenyl)-1,3-butanedione 307
Bis(phenacyl)acetic acid 331
3′-(3,4-Dimethoxyphenyl)-1-oxospiro(indan-2,2′-oxirane) 291
mono-(1-Styrylethyl) phthalate 380
α-Truxillic acid 346

$C_{18}H_{16}O_5$
3′,4′5′-Trimethoxyflavone 320
2-(3,4,5-Trimethoxyphenyl)-1,3-indandione 307

$C_{18}H_{16}O_6$
7-Hydroxy-3′,4′,5′-trimethoxyflavone 319
2,4,5-Triacetoxybiphenyl 378

$C_{18}H_{16}O_7$
(-)-Usnic acid 319

$C_{18}H_{16}O_7S$
1,9-Dicarbethoxyphenoxathiin-10,10-dioxide 775

$C_{18}H_{16}Pb$
Triphenyllead 1136

$C_{18}H_{17}BiO_2$
Triphenylbismuth dihydroxide 1142

$C_{18}H_{17}Br$
11-(Bromomethyl)-9,10-dihydro-12-methyl-9,10-ethanoanthracene 172

$C_{18}H_{17}BrNP$
Aminotriphenylphosphonium bromide 831

$C_{18}H_{17}BrN_2O_2$
1-(4-Bromophenyl)-3-(phenethylamino)succinimide 715

$C_{18}H_{17}BrO$
4′-Bromo-4-isopropylchalcone 305

$C_{18}H_{17}Cl$
3-Chloro-7-isopropyl-1-methylphenanthrene 175
11-(Chloromethyl)-9,10-dihydro-12-methyl-9,10-ethanoanthracene 172

$C_{18}H_{17}ClFNO_4$
4-(Butoxycarbonyl)phenyl *N*-(3-chloro-4-fluorophenyl)carbamate 630

$C_{18}H_{17}ClF_3NO_2$
4-Chloro-2-isopropyl-5-methylphenyl *N*-(3-(trifluoromethyl)phenyl)carbamate .. 608

$C_{18}H_{17}ClNP$
Aminotriphenylphosphonium chloride 831

$C_{18}H_{17}ClN_2O$
2-Chloro-5,9,10,14b-tetrahydro-5-methylisoquino(2,1-d)(1,4)-benzodiazepin-6(7*H*)-one 707

$C_{18}H_{17}ClN_2O_2$
4-(4-Chlorophenyl)-2-(2-hydroxybutyl)phthalazin-1(2*H*)-one .. 319

$C_{18}H_{17}ClN_2O_3$
N-(6-Chloro-2-methoxy-9-acridinyl)-glycine ethyl ester 995
N-(1-(4-Chlorophenyl)-1-methoxy-1*H*-isoindol-3-yl)glycine methyl ester hydrochloride 723

$C_{18}H_{17}ClO$
4′-Chloro-4-isopropylchalcone 305

$C_{18}H_{17}Cl_2NO_4$
4-(Butoxycarbonyl)phenyl *N*-(2,3-dichlorophenyl)carbamate 630
2-(2,4-Dichlorophenoxy)acetyl *N*-(2,4,6-trimethylphenyl)carbamate 620

$C_{18}H_{17}FN_2O_2$
N[2]-Dimethyl-3-(2-fluorophenyl)-1-oxo-1,2,3,4-tetrahydro-4-isoquinolinecarboxamide 689

$C_{18}H_{17}FN_4O_2$
2-(3-(4-Fluorophenyl)ureido)acetic (3-phenylallylidene)hydrazide 523

$C_{18}H_{17}F_3N_2O_4$
2-(4-Acetamidophenoxy)ethyl *N*-(3-(trifluoromethyl)phenyl)carbamate .. 608

$C_{18}H_{17}F_3N_4O_2$
2-(3-(3-(Trifluoromethyl)phenyl)-ureido)acetic (4-methylbenzylidene)hydrazide 508

$C_{18}H_{17}F_3N_4O_4$
2-(3-(3-(Trifluoromethyl)phenyl)-ureido)acetic (3-hydroxy-4-methoxybenzylidene)hydrazide 507

$C_{18}H_{17}FeN$
N-Ferrocenylmethylene-*p*-toluidine .. 1142

$C_{18}H_{17}IN_2$
1-Allyl-2-(2-(3-indolyl)-vinyl)-pyridinium iodide 967

$C_{18}H_{17}I_2NO_4$
N-Benzoyl-3,5-diiodo-L-tyrosine ethyl ester 447

$C_{18}H_{17}N$
2-Mesitylquinoline 991

$C_{18}H_{17}NO$
9,10-Dihydro-9-morpholinoanthracene 251
10-(Dimethylaminomethyl)-5*H*-dibenzo(a,d)cyclohepten-5-one 313
N,*N*-Diphenylsorbamide 482
1-(*N*-Methyl-*p*-toluidino)-2-naphthol .. 233
1-(*m*-Toluidinomethyl)-2-naphthol 227

$C_{18}H_{17}NOS$
2-(4-Isopropylbenzylidene)-3-oxo-4*H*-1,4-benzothiazine 706
1-(4-(Methylthio)anilinomethyl)-2-naphthol 226

$C_{18}H_{17}NO_2$
4-(*m*-Anisoyl)-4-phenylbutyronitrile 753
4-Cyanobenzyl 2,4,6-trimethylbenzoate 763
3,6-Diacetyl-9-ethylcarbazole 892
3,4-Dihydro-4,4-dimethyl-α,2-epoxy-2-(4-pyridylmethyl)-1(2*H*)-naphthalenone 961
3,4-Dimethoxy-α-(*p*-tolyl)-cinnamonitrile 762
6-Methoxy-α-methyl-2-phenyl-4-quinolinemethanol 983
8-Methoxy-2-(4-pyridylmethylene)-1-benzosuberone 937
1-(*N*-Methyl-*p*-anisidino)-2-naphthol .. 225
4′-(2-(*p*-Toluoyl)vinyl)acetanilide 434

$C_{18}H_{17}NO_2S$
2,5-Dimethyl-1-(4-(phenylsulfonyl)-phenyl)pyrrole 841

$C_{18}H_{17}NO_3$
2-(3-Benzyloxyphenyl)-*N*-methylsuccinimide 716
2,6-Dimethoxy-α-(4-methoxyphenyl)-cinnamonitrile 763
3,4-Dimethoxy-α-(4-methoxyphenyl)-cinnamonitrile 762
2,3-Dimethoxy-5,6,8,12b-tetrahydroisoindolo(1,2-a)-isoquinolin-8-one 697
Ethyl 5-benzyloxyindole-2-carboxylate 882
4-Isopropyl-3′-nitrochalcone 303
4-Isopropyl-4′-nitrochalcone 303
N-(1-Naphthyl)-1,2,3,6-tetrahydrophthalamic acid 489

$C_{18}H_{17}NO_3S_2$
5-(2-Hydroxy-1-naphthylmethylene)-3-(3-methoxypropyl)rhodanine 733

$C_{18}H_{17}NO_4$
N-(2-(3,4-Dimethoxyphenyl)ethyl)-phthalimide 716
10b-(2-Hydroxy-4-methoxyphenyl)-3,4,6,10b-tetrahydro-2*H*-(1,3)-oxazino(2,3-a)isoindol-6-one 696

$C_{18}H_{17}NO_4S$
2,3-Dihydro-3-oxo-2-(3,4,5-trimethoxybenzylidene)-4*H*-1,4-benzothiazine 706

$C_{18}H_{17}NO_5$
3-Acetylphenyl *N*-(4-(ethoxycarbonyl)phenyl)carbamate. 666
4-Acetylphenyl *N*-(4-(ethoxycarbonyl)phenyl)carbamate. 674
α-(2,4-Dimethoxybenzylidene)-hippuric acid 464
α-(2,6-Dimethoxybenzylidene)-hippuric acid 463
4-(*N*-Mesitylcarbamoyl)isophthalic acid, 2,4,6-trimethylaniline salt 474

$C_{18}H_{17}NO_5S$
N-Carbobenzyloxy-*S*-benzoyl-L-cysteine 643

$C_{18}H_{17}NO_6$
4-Formyl-2-methoxyphenyl *N*-(4-ethoxycarbonylphenyl)carbamate 649
2-(Methoxycarbonyl)phenyl *N*-(4-(ethoxycarbonyl)phenyl)carbamate. 675
4-(Methoxycarbonyl)phenyl *N*-(4-(ethoxycarbonyl)phenyl)carbamate. 676
3′-Nitro-3,4,4′-trimethoxychalcone 300

$C_{18}H_{17}NS$
4-(2,5-Dimethyl-1-pyrrolyl)phenyl sulfide 839

$C_{18}H_{17}N_3$
N,*N*-Dimethyl-4-(2-naphthylazo)-aniline 1090

$C_{18}H_{17}N_3O$
Acetic acid (1-methyl-2-phenyl-3-indolylmethylene)hydrazide 1054
1,2-Dihydro-4,6-dimethyl-1-(5-indanylmethyleneamino)-2-oxonicotinonitrile 1069
N-(3,5-Dimethyl-1-phenyl-2-pyrazolin-4-yl)benzamide 477

$C_{18}H_{17}N_3O_2$
Cinnamic (4-acetamido-benzylidene)hydrazide 501
2,3-Dimethyl-4-(2-hydroxybenzylideneamino)-1-phenyl-3-pyrazolin-5-one 1037
2,3-Dimethyl-4-(4-hydroxybenzylideneamino)-1-phenyl-3-pyrazolin-5-one 1038

$C_{18}H_{17}N_3O_3$
α,3-Bis-(phenylcarbamoyl)-2-methyleneoxazolidine 707
*N*α-Furylacryloyl-L-tryptophanamide 889

$C_{18}H_{17}N_3O_4S$
Ethyl 4-(3-(6-methoxy-2-benzothiazolyl)ureido)benzoate 913

$C_{18}H_{17}N_3O_7$
Carbobenzyloxy-L-asparagine 4-nitrophenyl ester 674
Dimethyl 5-(3-(2-methyl-5-nitrophenyl)ureido)isophthalate 563

$C_{18}H_{17}N_3S$
3-Methyl-2-benzothiazolinone, azine with 5-indancarboxaldehyde.. 1062
$C_{18}H_{18}$
1,4-Bis(*p*-tolyl)-1,3-butadiene 165
9,10-Dihydro-11,12-dimethyl-9,10-ethanoanthracene 162
$C_{18}H_{18}BF_4N_3O$
2-(2-Benzoylhydrazino)-1-ethylquinolinium tetrafluoroborate. 990
$C_{18}H_{18}BrNO_3$
Bromomethoxymethyltetrahydroiminoethanophenanthro(4,5-bcd)-furan-5-one hydrobromide 319
$C_{18}H_{18}BrN_3O$
4-(4-(4-Bromophenylazo)phenyl)-2,6-dimethylmorpholine 1082
$C_{18}H_{18}Br_2O_4Pb$
Diethylplumbyl bis(3-bromobenzoate) 1133
$C_{18}H_{18}ClF_3N_4O_2$
1-(2-Chloro-5-(trifluoromethyl)-phenyl)-6-(2-ethylphenyl)-3-methylbiurea 505
$C_{18}H_{18}ClF_3N_4O_3$
1-(2-Chloro-5-(trifluoromethyl)-phenyl)-6-(2-methoxy-5-methyl-phenyl-)-3-methylbiurea 504
$C_{18}H_{18}ClN$
1-*tert*-Butyl-2-(4-chlorophenyl)indole 881
1-*tert*-Butyl-3-(4-chlorophenyl)indole 881
$C_{18}H_{18}ClNO$
2-Chloro-4-dimethylamino-4′-methylchalcone 302
$C_{18}H_{18}ClNO_3$
4-Allyl-2-methoxyphenyl *N*-(2-chloro-4-methylphenyl)carbamate... 616
4-Allyl-2-methoxyphenyl *N*-(3-chloro-2-methylphenyl)carbamate... 616
$C_{18}H_{18}ClNO_4$
4-Allyl-2-methoxyphenyl *N*-(5-chloro-2-methoxyphenyl)-carbamate 612
4-(Butoxycarbonyl)phenyl *N*-(3-chlorophenyl)carbamate 630
2-Chloro-4,5-dimethyl *N*-(4-(ethoxycarbonyl)phenyl)carbamate. 630
4-Chloro-2,6-dimethylphenyl *N*-(4-(ethoxycarbonyl)phenyl)carbamate. 630
4-Chloro-3,5-dimethylphenyl *N*-(4-(ethoxycarbonyl)phenyl)carbamate. 630
5-Indanyl *N*-(5-chloro-2,4-dimethoxyphenyl)carbamate 613
$C_{18}H_{18}ClNO_6$
4-(Ethoxycarbonyl)phenyl *N*-(5-chloro-2,4-dimethoxyphenyl)-carbamate 614
$C_{18}H_{18}Cl_2N_2O_2$
7-(*N*,*N*-Bis(chloromethyl)amino)-2-(ethoxycarboxamido)fluorene 604
N-(α-Cyano-3,4-dichlorobenzyl)-3,4-dimethoxyphenethylamine hydrochloride 753
$C_{18}H_{18}Cl_2N_2O_4S_2$
S,*S*′-Ethylene bis(*N*-(4-chloro-2-methoxyphenyl)thiocarbamate 614
$C_{18}H_{18}Cl_2N_4O_2$
4-(3-Propylureido)benzoic (2,4-dichlorobenzylidene)hydrazide 529
$C_{18}H_{18}Cl_2N_4O_4$
2-(3-(2,3-Dichlorophenyl)ureido)-acetic (3,4-dimethoxybenzylidene)-hydrazide 516
$C_{18}H_{18}Cl_2N_4O_5$
2-(1-Methyl-3-(2-methyl-4-nitrophenyl)ureido)ethyl *N*-(3,4-dichlorophenyl)carbamate 526
$C_{18}H_{18}Cl_2O_2$
1,4-Bis-(4-chloro-3-methylphenoxy)-*trans*-2-butene 182
2-(2,4-Dichlorophenyl)-5,5-dimethyl-2-phenyl-1,3-dioxane 183
$C_{18}H_{18}Cl_3NO_2$
N-(2,6-Diethylphenyl)-2-(2,4,5-trichlorophenoxy)acetamide 401
5′-Isopropyl-2′-methyl-2-(2,4,5-trichlorophenoxy)acetanilide 401
$C_{18}H_{18}Cl_{10}O_2Si_3$
1,5-Bis(pentachlorophenyl)-1,1,3,3,5,5-hexamethyltrisiloxane 1101
$C_{18}H_{18}Cl_{10}Si_3$
1,3-Bis(pentachlorophenyl)-1,1,2,2,3,3-hexamethyltrisilane 1101
$C_{18}H_{18}FNO_5S$
N-(4-Isopropoxybenzoylacetyl)-sulfanilyl fluoride 802
$C_{18}H_{18}FNO_6S$
2-(4-Ethoxy-3-methoxybenzoyl)-4′-(fluorosulfonyl)acetanilide 802
$C_{18}H_{18}FN_3O_5$
Butyl 4-(3-(2-fluoro-5-nitrophenyl)-ureido)benzoate 524
Butyl 4-(3-(4-fluoro-3-nitrophenyl)-ureido)benzoate 524
$C_{18}H_{18}F_3NO_2$
4-*tert*-Butylphenyl *N*-(3-(trifluoromethyl)phenyl)carbamate.. 611
$C_{18}H_{18}F_3NO_3$
4-Butoxyphenyl *N*-(2-(trifluoromethyl)phenyl)carbamate.. 608
Octahydro-2,5-methano-2*H*-indeno(1,2-b)oxiren-3-yl *N*(3-(trifluoromethyl)phenyl)carbamate.. 610
$C_{18}H_{18}F_{10}Si_3$
1,3-Bis(perfluorophenyl)-1,1,2,2,3,3-hexamethyltrisilane 1101
$C_{18}H_{18}INO_4$
N-(4-Iodophenyl)-2-(3,4,5-trimethoxybenzylidene)acetamide.. 448
$C_{18}H_{18}NSb$
(2,5-Dimethyl-3-pyrrolyl)-diphenylantimony 1129
$C_{18}H_{18}N_2$
2-(3-(Dimethylamino)phenyl)-8-methylquinoline picrate 989
4-Dimethylamino-α-(*p*-tolyl)-cinnamonitrile 765
4,5-Diphenyl-2-propylimidazole 869
$C_{18}H_{18}N_2O$
9,10-Dihydro-11-methyl-9,10-ethanoanthracene-11-carboxylic hydrazide 503
4-(Dimethylamino)-α-(4-methoxyphenyl)cinnamonitrile 762
5-Indancarboxaldehyde (phenylacetyl)-hydrazone 498
p-Toluic acid (5-indanylmethylene)-hydrazide 501
$C_{18}H_{18}N_2O_2$
5-Benzyl-2,3,5,6-tetrahydroimidazo(2,1-a)-isoquinoline-α,5-diol 245
4-Cyanophenyl *N*-(2,6-diethylphenyl)carbamate 767
1,4-Dibenzoylpiperazine 482
1,4-Dibenzyl-2,5-piperazinedione 709
N-(4-Isopropylanilinomethyl)-phthalimide 719
Mandelic acid (5-indanylmethylene)-hydrazide 496
4′-(*N*-Methylacetamido)-cinnamanilide 470
4-Methyl-3-phenyl-*N*-(*o*-tolyl)-2-isoxazoline-4-carboxamide 432
$C_{18}H_{18}N_2O_2S$
1-Cinnamoyl-3-(4-ethoxyphenyl)-2-thiourea 737
Ethyl β-amino-α-(*N*-phenylthiocarbamoyl)cinnamate 739
$C_{18}H_{18}N_2O_3$
Cinnamic (2,4-dimethoxybenzylidene)hydrazide 497
3′,4′-Dihydro-6′,7′-dimethoxyspiro(3*H*-indole-3,1′(2*H*)-isoquinolin)-2(1*H*)-one 690
4,4-Dimethyl-2-(diphenylnitromethyl)-2-oxazoline 258
10b-(2-Hydroxy-4-methoxyphenyl)-1,2,3,4-tetrahydropyrimido(2,1-a)isoindol-6(10-bh)-one 696
Octahydro-9-oxo-9*H*-pyrido(1,2,3-lm)pyrrolo(2,3-d)carbazole-12-carboxylic acid 708
$C_{18}H_{18}N_2O_3S$
2-(2-Aminophenylthio)-*N*-(4-ethoxyphenyl)-succinimide 716
Methyl 2-(3-(3-phenylpropionyl)-2-thioureido)benzoate 744
$C_{18}H_{18}N_2O_4$
Diethyl 4,4′-azodibenzoate 1092
Dimethyl 2-(1,2-diphenylhydrazino)-maleate 364
Disperse Blue 23 (C.I. 61545) 1093
Ethyl 4-(3-(3-acetylphenyl)ureido)-benzoate 575
Ethyl 2-(2-benzoylphenylureido)-acetate 575
$C_{18}H_{18}N_2O_4S$
Methyl 2-(3-(2-phenoxypropionyl)-2-thioureido)benzoate 737
$C_{18}H_{18}N_2O_5$
3-Acetamidophenyl *N*-(4-(ethoxycarbonyl)phenyl)carbamate. 675
4-Acetamidophenyl *N*-(4-(ethoxycarbonyl)phenyl)carbamate. 675
4-Allyl-2-methoxyphenyl *N*-(4-methyl-2-nitrophenyl)carbamate 649
4-Allyl-2-methoxyphenyl *N*-(4-methyl-3-nitrophenyl)carbamate 649
1-(4-(Ethoxycarbonyl)phenyl)-3-(2-(methoxycarbonyl)phenyl)urea 576
$C_{18}H_{18}N_2O_5S$
4-Cyano-7-diethylamino-3-(methoxycarbonylacetyl)coumarin.. 765
$C_{18}H_{18}N_2O_6$
4-Allyl-2-methoxyphenyl *N*-(2-methoxy-4-nitrophenyl)carbamate... 649
Carbobenzyloxy-L-alanine 4-nitrobenzyl ester 674
$C_{18}H_{18}N_2S_2$
1,4-Bis-(thiobenzoyl)-piperazine 746
$C_{18}H_{18}N_3OP$
Tris(3-aminophenyl)phosphine oxide 834
$C_{18}H_{18}N_4$
5,5′,6,6′-Tetramethyl-2,2′-bibenzimidazole 904
$C_{18}H_{18}N_4O_2$
4-Acetamidobenzaldehyde azine 1074
9,10-Dihydro-9,10-ethanoanthracene-11,12-dicarboxylic dihydrazide 499
Oxalic bis(2-phenylethylidene)-hydrazide 1060
$C_{18}H_{18}N_4O_3$
4-(3-Allylureido)benzoic (2-hydroxybenzylidene)hydrazide 548
2′,6′-Diethyl-6-nitro-1*H*-indazole-1-carboxanilide 900
Hippuric (4-acetamidobenzylidene)-hydrazide 501
$C_{18}H_{18}N_4O_4Cl_2$
2-(3,4-Dichlorophenylureido)acetic acid (2,3-dimethoxybenzylidene)-hydrazide 492
$C_{18}H_{18}N_4O_6S_2$
S,*S*′-Ethylene bis(*N*-(2-methyl-5-nitrophenyl)thiocarbamate) 669
$C_{18}H_{18}N_8$
Bismarck Brown B 1087
$C_{18}H_{18}O$
tert-Butyl 9-fluorenyl ketone 287
9,10-Dihydro-11-(hydroxymethyl)-11-methyl-9,10-ethanoanthracene 206
9,10-Dihydro-9,10-dimethyl-11-hydroxy-9,10-ethanoanthracene 206
10,11-Dihydro-5,11-dimethyl-11-hydroxy-5,10-methano-5*H*-dibenzo(a,d)cycloheptene 205
9,10-Dihydro-11-(1-hydroxyethyl)-9,10-ethanoanthracene 206
9,10-Dihydro-11-(2-hydroxyethyl)-9,10-ethanoanthracene 206
9,10-Dihydro-11-(hydroxymethyl)-12-methyl-9,10-ethanoanthracene 206
1-(4-Methoxyphenyl)-4-(*p*-tolyl)-1,3-butadiene 181
2′,4,4′-Trimethylchalcone 305
2′,4′,6′-Trimethylchalcone 304
$C_{18}H_{18}O_2$
1-(4-Biphenylyl)-1,4-hexanedione 309
1,4-Bis-(4-methoxyphenyl)-1,3-butadiene 181
9-*tert*-Butyl-9-hydroxy-10(9*H*)-anthrone 310
tert-Butyl 9-xanthenyl ketone 284
2-Ethoxy-4′-methylchalcone 301
4-Ethoxy-4′-methylchalcone 301
Hydrocinnamoin 205
2′-Hydroxy-4-isopropylchalcone 300
2-(1,2,3,4-Tetrahydro-6-naphthylmethyl)benzoic acid 345
$C_{18}H_{18}O_2S$
3,3′-Thiodipropiophenone 308
$C_{18}H_{18}O_3$
4-Allyl-2-methoxyphenyl phenylacetate 356
1,2-Bis(4-methoxyphenyl)-3-butyn-2-ol 1016
9,10-Dihydro-2,3-dimethoxy-11-hydroxy-9,10-ethanoanthracene 201
2,3-Dimethoxy-4′-methylchalcone 301
2,4-Dimethoxy-4′-methylchalcone 302
2,5-Dimethoxy-4′-methylchalcone 302
3,4-Dimethoxy-4′-methylchalcone 300
3,6-Dimethyl-2-(2,4-dimethylbenzoyl)benzoic acid 344
(+/-)-2-Methyl-1-phenylpropyl benzoylformate 370
$C_{18}H_{18}O_3S$
9,10-Dihydro-9,10-ethanoanthracene-11-methanol methanesulfonate 799
$C_{18}H_{18}O_4$
4,4′-Bis(2,3-epoxypropoxy)biphenyl.. 189
1,4-Butylene glycol dibenzoate 371
Dibenzyl succinate 356
Diethyl 4,4′-biphenyldicarboxylate 375
mono(α-Ethyl-α-methylbenzyl) phthalate 370
2,2,2′,2′-Tetramethyl-(5,5′-bi(1,3-benzodioxole)) 184
2,2′-(Tetramethylenedioxy)-dibenzaldehyde 323
4,4′-(Tetramethylenedioxy)-dibenzaldehyde 324
2,3,4′-Trimethoxychalcone 300
2,4,4′-Trimethoxychalcone 300
3,4,5-Trimethoxychalcone 300
$C_{18}H_{18}O_4S_2$
1,4-Bis(*p*-tosyl)-2-butyne 1019
$C_{18}H_{18}O_5$
2′-Hydroxy-3,4,5-trimethoxychalcone 300
Methyl 4-(2-phenoxyethoxy)-benzoylacetate 363
$C_{18}H_{18}O_6$
2-Acetyl-5-methoxyphenyl 3,4-dimethoxybenzoate 377
3,3′-(Ethylenedioxy)di-*p*-anisaldehyde 323
$C_{18}H_{18}O_8$
Dimethyl *cis*-3b,6b-diethyl-2,5-dihydroxy-1,4-dioxo-1,3a,3b,4,6a,6b-hexahydrocyclobuta(1,2:3,4)-dicyclopentene-3,6-dicarboxylate 91
$C_{18}H_{19}BrClNO_2$
2-Bromo-4-(*tert*-butyl)phenyl *N*-(3-chloro-2-methylphenyl)carbamate... 636
2-Bromo-4-*tert*-butylphenyl *N*-(5-chloro-2-methylphenyl)carbamate... 639
$C_{18}H_{19}BrN_2O$
4-Bromo-α-(4-piperidinophenylimino)-*o*-cresol 1038
$C_{18}H_{19}BrN_2O_4$
2-Bromo-4-*tert*-butylphenyl *N*-(4-methyl-2-nitrophenyl)carbamate 624
$C_{18}H_{19}ClN_2$
2-Chloro-5,6,7,9,10,14b-hexahydro-5-methylisoquino(2,1-d)(1,4)-benzodiazepine 252
6-Chloro-2,3,4,4a,9,9a-hexahydro-1-methyl-4a-phenyl-1*H*-pyrido(2,3-b)-indole 251
$C_{18}H_{19}ClN_2O$
11-(4-Chlorophenyl)1,2,3,4,6,11-hexahydro-11-hydroxypyridazino(1,2-b)-

$C_{18}H_{19}ClN_2O$ (Cont'd)
phthalazine 244

$C_{18}H_{19}ClN_2O_2S_2$
1-(2-Chlorothiobenzoyl)-4-(*p*-tolylsulfonyl)-piperazine 820

$C_{18}H_{19}ClN_2O_3$
Piperidino *N*-(5-chloro-2-phenoxyphenyl)carbamate 619

$C_{18}H_{19}ClN_2O_3S$
S-(4-*tert*-Butyl-2-methylphenyl) *N*-(2-chloro-5-nitrophenyl)-thiocarbamate 626
Ethyl 2-(3-(3-chlorophenyl)ureido)-4,5,6,7-tetrahydrobenzo(b)-thiophene-3-carboxylate 855

$C_{18}H_{19}ClN_2O_5$
1-(5-Chloro-2,4-dimethoxyphenyl)-3-(4-(ethoxycarbonyl)phenyl)urea 515
Ethyl 4-(3-(4-chloro-2,5-dimethoxyphenyl)ureido)benzoate 516

$C_{18}H_{19}ClN_4O_2$
2-Chlorohippuric (4-(dimethylamino)benzylidene)-hydrazide 492

$C_{18}H_{19}ClN_4O_3$
2-(3-(2-Ethoxyphenyl)ureido)acetic (4-chlorobenzylidene)hydrazide 518

$C_{18}H_{19}ClN_4S$
2-(4-Amino-1-naphthylazo)-3-methyl-4,5,6,7-tetrahydrobenzothiazolium chloride 1088

$C_{18}H_{19}Cl_2NO_3$
4-Chloro-2-isopropyl-5-methylphenyl *N*-(2-chloro-4-methoxyphenyl)carbamate 621
1,2-Epoxyoctahydro-4,7-ethano-1*H*-inden-5-yl *N*-(3,5-dichlorophenyl)-carbamate 631

$C_{18}H_{19}Cl_2NO_5$
2-(2-Chloro-4-methylphenoxy)ethyl *N*-(5-chloro-2,4-dimethoxyphenyl)-carbamate 612

$C_{18}H_{19}Cl_2N_3O_3$
2-(3-(5-Chloro-2-methylphenyl)-1-methyl-ureido)ethyl *N*-(3-chlorophenyl)carbamate 674
2-(3-(4-Chlorophenyl)-1-ethylureido)-ethyl *N*-(4-chlorophenyl)carbamate 674

$C_{18}H_{19}FN_2O_3$
Butyl 4-(2-fluorophenylureido)-benzoate 532
Butyl 4-(3-fluorophenylureido)-benzoate 532

$C_{18}H_{19}FN_2O_3S$
Ethyl 2-(4-fluorophenylureido)-4,5,6,7-tetrahydro-3-benzo(b)-thienylcarboxylate 855

$C_{18}H_{19}FN_3$
9-(3-Dimethylaminopropylamino)-2-fluoroacridine hydrochloride 996

$C_{18}H_{19}F_3N_2$
N,*N*-Diethyl-α-(3-(trifluoromethyl)-phenylimino)-*p*-toluidine 1020

$C_{18}H_{19}F_3N_2O$
1-Butyl-1-phenyl-3-(3-trifluoromethylphenyl)urea 510
1-(4-Butylphenyl)-3-(2-(trifluoromethyl)phenyl)urea 511
1-(4-Butylphenyl)-3-(3-(trifluoromethyl)phenyl)urea 511
1-(2,6-Diethylphenyl)-3-(3-trifluoromethylphenyl)urea 511

$C_{18}H_{19}F_3N_2O_2$
3-(Diethylamino)phenyl *N*-(2-(trifluoromethyl)phenyl)carbamate 610
3-Diethylaminophenyl *N*-(3-trifluoromethylphenyl)carbamate 610

$C_{18}H_{19}IN_2O$
1-Allyl-2-(4-acetamidostyryl)-pyridinium iodide 956

$C_{18}H_{19}I_3O_6$
3,3′,4,4′,5,5′-Hexamethoxy-2,2′,6-triiodobiphenyl 185

$C_{18}H_{19}N$
10-(2-Aminoethyl)-10,11-dihydro-5-methylene-5*H*-dibenzo(a,d)-cycloheptene 241
9,10-Dihydro-11-(dimethylamino)-9,10-ethanoanthracene 243
1,2-Dihydro-6-phenyl-1,2,2-trimethylquinoline 252
3,3-Diphenyl-2-ethyl-1-pyrroline 252
2,2-Diphenylhexanenitrile 756

$C_{18}H_{19}NO$
1,4-Dihydro-1-phenyl-2,4,4-trimethyl-3(2*H*)-isoquinolinone 691
4-Dimethylamino-4′-methylchalcone 302
N^3-Diphenylcyclo-pentanecarboxamide 437
2,6-Diphenyl-1-methyl-4-piperidinone 285
1,4-Diphenyl-3-propyl-2-azetidinone 691
β-Ethyl-4-methoxy-α-phenylhydrocinnamonitrile 753
2′-Ethyl-6′-methylcinnamanilide 478

$C_{18}H_{19}NO_2$
3-(3-Aminopropyl)-3-phenyl-1-isochromanone 384
4-Cyclohexyl-2-nitrobiphenyl 278
2-(Dimethylamino)-4′-methoxychalcone 301
4-(Dimethylamino)-4′-methoxychalcone 301
4′-Morpholinodeoxybenzoin 295
3-Phenylallyl *N*-(2,5-xylyl)carbamate 681

$C_{18}H_{19}NO_2S$
4-Phenyl-1,2,3,6-tetrahydro-1-(*p*-tolylsulfonyl)-pyridine 827

$C_{18}H_{19}NO_3$
4-Allyl-2-methoxyphenyl *N*-(*o*-tolyl)-carbamate 650
4-Allyl-2-methoxyphenyl *N*-(*m*-tolyl)-carbamate 651
N-(Diphenylmethyl)glutaramic acid 435
Ethyl 2-(4-(dimethylamino)benzoyl)-benzoate 373
2-(*N*-(1-Naphthyl)carbamoyl)-cyclohexanecarboxylic acid 489
3-Phenylallyl *N*-(4-ethoxyphenyl)-carbamate 654
3-Phenylpropyl *N*-(3-acetylphenyl)-carbamate 675

$C_{18}H_{19}NO_3S$
4-Allyl-2-methoxyphenyl *N*-(4-(methylthio)phenyl)carbamate 649

$C_{18}H_{19}NO_4$
N-Acetyl-*o*-benzyl-L-tyrosine 413
4-Allyl-2-methoxyphenyl *N*-(3-methoxyphenyl)carbamate 647
1-(4-Aminophenyl)-2-(2,5-diacetoxyphenyl)ethane hydrochloride 378
D-Aspartic acid dibenzyl ester *p*-toluenesulfonate 356
L-Aspartic acid dibenzyl ester *p*-toluenesulfonate 356
4-(Butoxycarbonyl)phenyl *N*-phenylcarbamate 675
4-(Butoxycarbonyl)phenyl *N*-phenylcarbamate 676
4-(Ethoxycarbonyl)phenyl *N*-(2-ethylphenyl)carbamate 675
4-(Ethoxycarbonyl)phenyl *N*-(2,5-xylyl)carbamate 676
Ethyl 4-(3-phenoxypropionamido)-benzoate 414
3-Phenylallyl *N*-(2,4-dimethoxyphenyl)carbamate 648
2,5-Xylyl *N*-(4-(ethoxycarbonyl)-phenyl)carbamate 676

$C_{18}H_{19}NO_4S$
3,4-Dihydro-6,7-dimethoxy-1-(phenylsulfonylmethyl)-isoquinoline 775
1-(2-Hydroxyethyl)quinolinium *p*-toluenesulfonate 983

$C_{18}H_{19}NO_5$
1,4-Benzodioxan-2-ylmethyl *N*-(2-methoxy-5-methylphenyl)-carbamate 416
N-Benzoyl-3-(2,4-dimethoxyphenyl)-alanine 463
N-Benzoyl-3-(2,6-dimethoxyphenyl)-alanine 464
Carbobenzyloxy-*O*-benzyl-L-serine 658
2-Ethoxyphenyl *N*-(4-(ethoxycarbonyl)phenyl)carbamate 649

$C_{18}H_{19}NO_5S$
4-Allyl-2-methoxyphenyl *N*-(*p*-tosyl)-carbamate 813

$C_{18}H_{19}NO_6$
Diethyl ((5-methyl-3-phenyl-4-isoxazolyl)formyl)malonate 871
Diethyl 5-nitro-1,2,3,4-tetrahydro-1,4-ethenonaphthalene-2,3-dicarboxylate 359
2,6-Dimethoxyphenyl *N*-(4-(ethoxycarbonyl)phenyl)carbamate 647
4-(Ethoxycarbonylphenyl) *N*-(2,4-dimethoxyphenyl)carbamate 648
4-(Ethoxycarbonyl)phenyl *N*-(2,5-dimethoxyphenyl)carbamate 648
Methyl 5-(benzyloxycarbonyl)-2-carboxy-4-methyl-3-pyrrolepropionate 843

$C_{18}H_{19}NO_7S$
4-(2-(4-Ethoxy-3-methoxybenzoyl)-acetamido)benzenesulfonic acid, barium salt 792

$C_{18}H_{19}N_2O_2$
1-*tert*-Butyl-1-hydroxy-2-phenyl-3-oxoisoindoline 689

$C_{18}H_{19}N_3O$
6-Amino-1-isopropyl-7-methyl-4-phenylquinazolin-2(1*H*)-one 701
1-Benzyl-2-hydroxy-2-phenyl-2,3,5,6-tetrahydro-1*H*-imidazo(1,2-a)-imidazole 245
5′-Methyl-2′-(4-tolylazo)-2-butenanilide 1088

$C_{18}H_{19}N_3O_2$
4-Diacetamido-2′,3-dimethylazobenzene 1091

$C_{18}H_{19}N_3O_2S$
4,6-Dihydro-8,9-dimethoxy-4-(2-aminophenylimino)-2*H*-(1,3)-thiazino(4,3-a)isoindole dihydrochloride 735
p-Toluenesulfonic acid (1,2-dihydro-1,4-dimethyl-2-quinolylidene)-hydrazide 820

$C_{18}H_{19}N_3O_3$
2-(3-Ethyl-5-methyl-4-isoxazolyl)-*N*,*N*-dimethyl-3-indoleglyoxylamide 886

$C_{18}H_{19}N_3O_3S$
1-(4-(*N*-Methylacetamido)phenyl)-3-(2-phenoxyacetyl)-2-thiourea 736
3-Methyl-2-benzothiazolinone, azine of 4-(2-hydroxyethoxy)-3-methoxybenzaldehyde 733

$C_{18}H_{19}N_3O_4$
N,*N*-Diethyl-4-(2,4-dinitrostyryl)-aniline 275
1,2-Dihydro-4,6-dimethyl-2-oxo-1-(3,4,5-trimethoxy-benzylideneamino)-nicotinonitrile 763
Ethyl 4-(3-acetamidophenylureido)-benzoate 565
Ethyl 4-(3-(4-acetamidophenyl)-ureido)benzoate 569

$C_{18}H_{19}N_3O_6$
2-(4-Acetamidophenoxy)ethyl *N*-(4-methyl-2-nitrophenyl)carbamate 665

$C_{18}H_{19}N_5O_3$
Disperse dye (C.I. 11230) 1084

$C_{18}H_{19}N_5O_4$
1-(2-Methoxy-4-nitrophenyl)-3-(1-methyl-2-phenyl-4-pyrazolin-4-yl)urea 551

$C_{18}H_{19}N_5O_6$
N^6-(4-Methoxybenzoyl)adenosine 921
2-(3-(3-Nitrophenyl)ureido)acetic (3,4-dimethoxybenzylidene)-hydrazide 550

$C_{18}H_{19}O_4P$
1-Cyclohexen-1-yl diphenyl phosphate 835

$C_{18}H_{20}$
1-Phenyl-1,3,3-trimethylindan 163

$C_{18}H_{20}BF_4O$
2,6-Diphenyl-4-methylpyrylium tetrafluoroborate 935

$C_{18}H_{20}BrNO$
2-(Benzoylmethyl)-2-methyl-1,2,3,4-tetrahydroisoquinolinium bromide 294

$C_{18}H_{20}BrNO_2$
2-Bromo-4,5-dimethylphenyl *N*-(4-isopropylphenyl)carbamate 635
4-Bromo-2,5-dimethylphenyl *N*-(4-isopropylphenyl)carbamate 634

$C_{18}H_{20}BrNO_2S$
2-Bromo-4-*tert*-butylphenyl *N*-(4-(methylthio)phenyl)carbamate 623

$C_{18}H_{20}BrNO_4S$
1-(Ethoxycarbonylallyl)-2-(phenylsulfonylmethyl)pyridinium bromide 956

$C_{18}H_{20}BrN_3$
1-(4-(4-(Bromophenylazo)phenyl)-hexahydro-1*H*-azepine 1082
1-(4-(4-Bromophenylazo)phenyl)-3-methylpiperidine 1082
1-(4-(4-Bromophenylazo)phenyl)-4-methylpiperidine 1082

$C_{18}H_{20}ClN$
(9-Anthracenylmethyl)-trimethylammonium chloride 256

$C_{18}H_{20}ClNO$
2-(Benzoylmethyl)-2-methyl-1,2,3,4-tetrahydroisoquinolinium chloride 294

$C_{18}H_{20}ClNOS$
4′-Butyl-2-(4-chlorophenylthio)-acetanilide 403
S-(4-*tert*-Butylphenyl) *N*-(3-chloro-2-methylphenyl)thiocarbamate 641

$C_{18}H_{20}ClNO_2$
4-*tert*-Butyl-*N*-(5-chloro-2-methoxyphenyl)benzamide 449
4-Chloro-2-isopropyl-5-methylphenyl *N*-(*m*-tolyl)-carbamate 631
4-Chloro-2-isopropyl-5-methylphenyl *N*-(*p*-tolyl)carbamate 635
2-(2-Chlorophenoxy)-5′-isopropyl-2′-methylacetanilide 400
2-Isopropyl-5-methylphenyl *N*-(5-chloro-2-methylphenyl)carbamate 639

$C_{18}H_{20}ClNO_2S$
S-(4-*tert*-Butylphenyl) *N*-(4-chloro-2-methoxyphenyl)thiocarbamate 620
4-Chloro-2-isopropyl-5-methylphenyl *N*-(4-(methylthio)-phenyl)carbamate 623

$C_{18}H_{20}ClNO_3$
4-Butoxyphenyl *N*-(5-chloro-2-methylphenyl)carbamate 617
4-*tert*-Butylphenyl *N*-(5-chloro-2-methoxyphenyl)carbamate 627
2-(4-Chloro-3-methylphenoxy)ethyl *N*-(2,6-xylyl)carbamate 616

$C_{18}H_{20}ClNO_3S$
2-(4-Chloro-3,5-dimethylphenoxy)-ethyl *N*-(3-(methylthio)phenyl)-carbamate 615

$C_{18}H_{20}ClNO_4$
4-Butoxyphenyl *N*-(5-chloro-2-methoxyphenyl)carbamate 617

$C_{18}H_{20}ClN_3O_2$
1-(*N*-(5-Chloro-2-methoxyphenyl)-carbamoyl)-4-phenylpiperazine 694

$C_{18}H_{20}ClN_3O_3$
2,5-Diethoxy-4-(2-phenylacetamido)-benzenediazonium chloride 416

$C_{18}H_{20}ClN_3O_4$
1-(5-Chloro-2,4-dimethoxyphenyl)-3-(4-(*N*-methylacetamido)phenyl)urea. 514

$C_{18}H_{20}ClN_5O_2$
2-(3-(3-Chlorophenyl)ureido)acetic (4-(dimethylamino)benzylidene)-hydrazide 522

$C_{18}H_{20}ClN_5O_4$
4-(4-Chloro-2-methylphenyl)-1-(4-methoxyphenylureidoacetyl)-semicarbazide 520

$C_{18}H_{20}Cl_2N_4$
2-Chloro-4-(dimethylamino)-benzaldehyde azine 1062

$C_{18}H_{20}FN_5O_5$
4-(2,5-Dimethoxyphenyl)-1-(3-(4-fluorophenyl)ureidoacetyl)-

$C_{18}H_{20}FN_5O_5$ (Cont'd)
semicarbazide 516
$C_{18}H_{20}F_3NO$
6-(4-(1-Ethylpropyl)-phenoxy)-α,α,α-trifluoro-*m*-toluidine 222
$C_{18}H_{20}FeO_4$
5-(Ferrocenylmethyl)-2,2,5-trimethyl-1,3-dioxane-4,6-dione 1139
$C_{18}H_{20}INO_2$
1-Allyl-2-(3,4-dimethoxystyryl)-pyridinium iodide 937
$C_{18}H_{20}I_2O_2$
2,2′-Diiodo-4,4′-dimethoxy-3,3′,5,5′-tetramethylbiphenyl 177
$C_{18}H_{20}I_2O_4$
2,2′-Diiodo-3,3′-dimethyl-4,4′,5,5′-tetramethoxybiphenyl 177
$C_{18}H_{20}I_2O_6$
2,2′-Diiodo-3,3′,4,4′,5,5′-hexamethoxybiphenyl 177
$C_{18}H_{20}N_2$
N,N′-Bis(2,5-dimethylphenyl)-1,4-diazabutadiene 1054
N,N′-Bis(4-methylbenzylidene)-ethylenediamine 1054
2,3,4,4a,9,9a-Hexahydro-1-methyl-4a-phenyl-1*H*-pyrido(2,3-b)indole hydrochloride 251
4′-Methylacetophenone azine 1078
1,1′-(1,3-Phenylene)bis(2,5-dimethylpyrrole) 839
$C_{18}H_{20}N_2O$
3-Allyl-1,1-dibenzylurea 571
3,4-Dihydro-1-isopropyl-7-methyl-4-phenyl-2(1*H*)-quinazolinone 713
2-Hydroxy-5-methyl-5,6,7,8,-tetrahydro-1-naphthalene-carboxaldehyde phenylhydrazone 234
6-Methoxy-1-tetralone (*p*-tolyl)-hydrazone 235
$C_{18}H_{20}N_2OS$
1-(3-Phenylpropionyl)-3-(2,3-xylyl)-2-thiourea 745
$C_{18}H_{20}N_2O_2$
3-(2-Aminoethyl)-2-(3,4-dimethoxyphenyl)indole 882
N,N′-Bis(4-ethoxyphenyl)-1,4-diazabutadiene 1040
N,N′-Bis(2-methoxybenzylidene)-ethylenediamine 1043
N,N′-Bis-(4-methoxybenzylidene)-ethylenediamine 1043
N,N′-Bis(2-phenylethyl)oxamide 423
N,N′-Dibenzoyl-1,4-butanediamine 471
N,N′-Dibenzyl-*N,N*′-dimethyloxamide 431
2,3-Dimethoxy-5,6,8,13a-tetrahydro-13*H*-isoquino(1,2-b)quinazoline 246
N-(4-(Dimethylamino)benzylidene)-4-(ethoxycarbonyl)aniline 1061
N,N′-Dimethyl-*N,N*′-di(*o*-tolyl)-oxamide 425
3-Ethoxybenzaldehyde azine 1039
4-Ethoxybenzaldehyde azine 1094
2,2′-Ethylenebis(*N*-methylbenzamide) 479
4-Methoxybenzoic (α-propylbenzylidene)hydrazide 1045
1-Methyl-1-(α-(4′-methylphenacyl)-benzyl)urea 571
4-(2,4,5-Trimethylphenyl-iminomethyl)phenyl *N*-methylcarbamate 1074
$C_{18}H_{20}N_2O_2Se$
2-Benzamidoethyl selenide 1142
$C_{18}H_{20}N_2O_3$
1-(2-Carboxyphenyl)-3-(2,6-diethylphenyl)urea 576
1-(4-Carboxyphenyl)-3-(2,6-diethylphenyl)urea 576
Cyanoethylmethyl-(*N*-methylanilinomethylene)-oxo-cyclopentanecarboxylic acid 754
4-(4-(Diethylamino)-benzylideneamino)salicylic acid 1032
4,4-Dimethyl-2-(1-(1-naphthyl)-1-nitropropyl)-2-oxazoline 258
4′,4″-Oxybis(propionanilide) 419
DL-Phenylalanyl-DL-phenylalanine 422
L-Phenylalanyl-L-phenylalanine 422
2-(Phenylcarbamoyl)phenyl *N*-*tert*-butylcarbamate 666
$C_{18}H_{20}N_2O_3S$
1-Benzoyl-3-(3,4-dimethoxyphenethyl)-2-thiourea 735
S-(4-*tert*-Butylphenyl) *N*-(2-methyl-5-nitrophenyl)thiocarbamate 669
o-(2′-(Dimethylamino)carbonyloxy)-2-biphenylyl *N,N*-dimethylthiocarbamate 738
2-Methoxybenzaldehyde *N*-phenyl-*N*-(3-sulfolanyl)hydrazone 1048
4-Methoxybenzaldehyde N^2-phenyl-N^2-(3-sulfolanyl)hydrazone 775
$C_{18}H_{20}N_2O_4$
1,6-Bis(4-nitrophenyl)hexane 278
2-(4-Butylphenyliminomethyl)-6-methoxy-4-nitrophenol 1030
4,4′-Diiminobis(ethyl benzoate) 373
2,4-Dimethoxybenzaldehyde azine 234
2′,2″-Dimethoxy-4′,4″-biacetanilide 414
2,3-Dimethyl-2,3-dinitro-1,4-diphenylbutane 258
N,N′-Divanillylidene-ethylenediamine 1029
1-(4-(Ethoxycarbonyl)phenyl)-3-(4-(2-hydroxyethyl)phenyl)urea 548
3-Ethoxy-4-hydroxybenzaldehyde azine 1094
3′,3″-(Ethylenedioxy)diacetanilide 414
4′,4″-(Ethylenedioxy)diacetanilide 414
Veratraldehyde azine 1042
$C_{18}H_{20}N_2O_4S_2$
2-(Ethoxycarboxamido)phenyl disulfide 668
$C_{18}H_{20}N_2O_4Se_2$
2-Acetamido-4-methoxyphenyl diselenide 210
$C_{18}H_{20}N_2O_5$
Ethyl 4-(2,4-dimethoxyphenyl-ureido)benzoate 549
Ethyl 4-(3-(2,5-dimethoxyphenyl)-ureido)benzoate 550
2-Isopropyl-5-methylphenyl *N*-(2-methoxy-4-nitrophenyl)carbamate 652
$C_{18}H_{20}N_2O_5S_2$
4,4′-Oxybis(*N*-allylbenzene-sulfonamide) 815
$C_{18}H_{20}N_2O_6$
1,6-Bis(2-nitrophenoxy)hexane 268
1,6-Bis(3-nitrophenoxy)hexane 268
1,6-Bis(4-nitrophenoxy)hexane 269
4-Butoxyphenyl *N*-(2-methoxy-4-nitrophenyl)carbamate 651
2′,5′-Diethoxy-4′-nitro-2-phenoxyacetanilide 414
$C_{18}H_{20}N_2O_{10}$
3,3′-Dinitro-2,2′,5,5′,6,6′-hexamethoxybiphenyl 265
$C_{18}H_{20}N_2S$
3-Allyl-1,1-dibenzyl-2-thiourea 748
1-Allyl-3-(1,2-diphenylethyl)-2-thiourea 741
10-(1-Methyl-2-pyrrolidinylmethyl)-phenothiazine hydrochloride 249
10-(2-Pyrrolidinoethyl)-phenothiazine 249
$C_{18}H_{20}N_3O_3$
2,5-Diethoxy-4-(2-phenylacetamido)-azobenzene, zinc chloride salt 1094
$C_{18}H_{20}N_3O_7P$
1,1-Diethyl-1*H*-phosphindolium picrate 832
$C_{18}H_{20}N_4$
4-Benzyl-α-(3-pyridyl)-1-piperazineacetonitrile 945
5,6,8,8a,13,14,16,16a-Octahydro-1,2,4,5-tetrazino(6,1-a:3,4-a′)-diisoquinoline 252
$C_{18}H_{20}N_4O_2$
Hippuric *N*-(4-dimethylaminobenzylidene)-hydrazide 1060
3-Methyl-1-(4-(4-nitrophenylazo)-phenyl)piperidine 1092
4-Methyl-1-(4-(4-nitrophenylazo)-phenyl)piperidine 1091
$C_{18}H_{20}N_4O_3$
4-(3-Propylureido)benzoic (3-hydroxybenzylidene)hydrazide 548
$C_{18}H_{20}N_4O_4$
2-(3-(2-Ethoxyphenyl)ureido)acetic (2-hydroxybenzylidene)hydrazide 546
N,N′-(1,5-Naphthalenediyl)bis(*O*-(carbamoyl)acetone oxime) 1012
$C_{18}H_{20}N_4O_5$
2-(1-Methyl-3-(2-methyl-4-nitrophenyl)ureido)ethyl *N*-phenylcarbamate 670
$C_{18}H_{20}N_6O_2$
Adipic acid bis-((2-pyridylmethylene)-hydrazide) 1071
$C_{18}H_{20}N_6O_4$
2-(3-(3-Nitrophenyl)ureido)acetic (4-(dimethylamino)benzylidene)-hydrazide 561
$C_{18}H_{20}N_6O_7$
4-(2,5-Dimethoxyphenyl)-1-(3-(3-nitrophenyl)ureidoacetyl)-semicarbazide 550
$C_{18}H_{20}O$
1,1-Bis(*p*-tolyl)-3-buten-1-ol 205
3,3-Dimethyl-1,1-diphenyl-2-butanone 287
4,4-Diphenylcyclohexanol 206
$C_{18}H_{20}O_2$
1,4-Bis(benzyloxy)-2-butene 186
2,4-Bis(4-hydroxyphenyl)-4-methyl-1-pentene 194
1,4-Bis-(*o*-tolyloxy)-*trans*-2-butene 181
1,4-Bis-(*p*-tolyloxy)-2-butene 181
9,10-Diethyl-9,10-dihydro-9,10-anthracenediol 200
trans-α,α′-Diethyl-*p,p*′-stilbenediol 196
9,10-Dihydro-9,10-dimethoxy-9,10-dimethylanthracene 187
3,4-Dihydro-4-(4-hydroxyphenyl)-2,2,4-trimethyl-2*H*-1-benzopyran 183
α,α′-Dimethyl-4,4′-dimethoxystilbene 179
1,4,5,8-Tetrahydro-2,3,6,7-tetramethyl-9,10(9*H*,10*H*)-anthracenedione 55
$C_{18}H_{20}O_3$
9,10-Diethyl-1,2,3,4-tetrahydro-1,4-ethanonaphthalene-2,3-dicarboxylic anhydride 388
Ethyl 3,4-diphenyl-3-hydroxybutyrate 353
4-Oxo-3-phenyl-2,3,5,6-tetramethyl-1,5-cyclohexadien-1-yl acetate 357
Phenethyl-2-(2-phenylethoxy)-acetate 355
$C_{18}H_{20}O_4$
4,4′-Bis(methoxymethoxy)stilbene 187
3-(3,4-Dihydroxyphenyl)-1-ethyl-2-methyl-5,6-indandiol 194
Methyl 4,4-bis(4-hydroxyphenyl)-valerate 354
$C_{18}H_{20}O_5$
2-(3,4-Dimethoxy-α-methylbenzyl)-6-methoxybenzoic acid 339
$C_{18}H_{20}O_6$
3,3′,4,4′,5-Pentamethoxy-benzophenone 311
$C_{18}H_{20}Si$
(1-Cyclohexen-3-yl)diphenylsilane 1111
Diallyldiphenylsilane 1113
$C_{18}H_{21}BF_4N_2$
Bis(4-methylbenzylidene)-ethylhydrazinium tetrafluoroborate 256
$C_{18}H_{21}BF_4N_2O_2$
Bis(4-methoxybenzylidene)-ethylhydrazinium tetrafluoroborate 254
$C_{18}H_{21}BrN_2O_3S$
3-(Morpholinosulfonyl)-1-(3-phenylallyl)pyridinium bromide 957
$C_{18}H_{21}ClN_2$
1-(4-Chlorodiphenylmethyl)-4-methylpiperazine hydrochloride 251
α-(4-Chlorophenyl)-2-(pyrrolidinomethyl)benzylamine dihydrochloride 244
$C_{18}H_{21}ClN_2O$
1-Benzyl-1-butyl-3-(3-chlorophenyl)urea 535
1-Butyl-3-(5-chloro-2-methylphenyl)-1-phenylurea 601
1-(4-Butylphenyl)-3-(3-chloro-2-methylphenyl)urea 539
1-(4-Butylphenyl)-3-(3-chloro-4-methylphenyl)urea 540
1-(4-Butylphenyl)-3-(4-chloro-2-methylphenyl)urea 540
1-(4-Butylphenyl)-3-(5-chloro-2-methylphenyl)urea 545
1-(4-Chloro-2-methylphenyl)-3-(2,6-diethylphenyl)urea 542
1-(5-Chloro-2-methylphenyl)-3-(2,6-diethylphenyl)urea 542
1-(3-Chloro-4-methylphenyl)-1-ethyl-3-(3-ethylphenyl)urea 545
$C_{18}H_{21}ClN_2O_2$
1-(3-Chloro-4-methylphenyl)-1-ethyl-3-(2-methoxy-5-methylphenyl)urea 520
$C_{18}H_{21}ClN_2O_3$
1-(3-Chloro-4-methylphenyl)-3-(2,5-dimethoxyphenyl)-1-ethylurea 516
$C_{18}H_{21}ClN_2O_4$
2-(5-Carboxy-2-ethyl-3-oxopentyl)-6-(4-chlorophenyl)-4,5-dihydro-3(2*H*)-pyridazinone 695
1-(5-Chloro-2-methoxyphenyl)-3-(2,5-diethoxyphenyl)urea 515
$C_{18}H_{21}ClN_6O_2$
2-(3-(6-Methyl-2-pyridyl)ureido)-acetic (2-chloro-4-(dimethylamino)-benzylidene hydrazide 928
$C_{18}H_{21}ClO$
α-*tert*-Butyl-4-chloro-4′-methylbenzhydrol 198
$C_{18}H_{21}Cl_2N_3O_2$
1-(4-Chlorophenyl-carbamoylmethyl)-3-(diethyl-carbamoyl)pyridinium chloride 947
$C_{18}H_{21}IO_2$
1-(2,6-Dimethoxy-3-iodophenyl)-2-methyl-1-phenylpropane 177
$C_{18}H_{21}NO$
N-Benzyl-*N*-*tert*-butylbenzamide 472
1-Benzyl-4-phenyl-4-piperidinol hydrochloride 245
2-(2-(Diethylamino)ethyl)-dibenzofuran 895
2-(2-(Diethylamino)ethyl)-dibenzofuran hydrochloride 895
4-(2-(Diethylamino)ethyl)-dibenzofuran hydrochloride 895
2-Diethylamino-4-methylbenzophenone 312
2′,6′-Diethyl-2-methylbenzanilide 479
2′,6′-Diethyl-4-methylbenzanilide 480
2′,6′-Diethyl-2-phenylacetanilide 437
4-(Dimethylamino)-1,1-diphenyl-2-butanone hydrochloride 285
2,2-Dimethyl-*N*-(diphenylmethyl)-propionamide 438
N-(4-Isopropylbenzylidene)-*p*-phenetidine 1050
N-(4-Methoxybenzylidene)-2,6-diethylaniline 1047
$C_{18}H_{21}NOS$
2-(Benzylthio)-2′,4′,6′-trimethylacetanilide 421
N-Isobutyl-4-methoxythiobenzanilide 736
$C_{18}H_{21}NO_2$
2-(4-Benzyloxyphenyl)-3-ethyloxazolidine 246
4-(*tert*-Butyl)-phenyl 2-methylcarbanilate 677
2′,6′-Diethyl-3-methoxybenzanilide 464
2′,6′-Diethyl-4-methoxybenzanilide 467
2-(Dimethylamino)-ethyl 2,2-diphenylacetate hydrochloride 359
2-Phenyl-*p*-butyrophenetidide 416
$C_{18}H_{21}NO_2S$
2,6-Dimethyl-4-(methylthio)phenyl *N*-(2,5-xylyl)carbamate 668
2-Isopropyl-5-methylphenyl *N*-(4-(methylthio)phenyl)carbamate 668

$C_{18}H_{21}NO_3$
N-Acetyl-L-leucine 1-naphthyl ester. 489
4-(Benzyloxy)phenyl *N*-butylcarbamate 646
2-*sec*-Butylphenyl *N*-(4-methoxyphenyl)carbamate 654
4-*tert*-Butylphenyl *N*-(2-methoxyphenyl)carbamate 653
4-*tert*-Butylphenyl *N*-(4-methoxyphenyl)carbamate 655
2,3-Dimethoxy-5,6,8,8a,9,12,12a,12b-octahydroisoindolo(1,2-b)isoquinolin-8-one 698
α-(3,4-Dimethoxyphenethylimino)-α-methyl-*o*-cresol 1030
5-Isopropyl-2-methylphenyl *N*-(4-methoxyphenyl)carbamate 655
2-(3-Methylphenoxy)ethyl *N*-(3-ethylphenyl)carbamate 649
1a,1b,3,4,5,5a,6,6a-Octahydro-2,5-methano-2*H*-indeno(1,2-b)oxiren-3-yl *N*-(*p*-tolyl)carbamate 678
Octahydro-6-oxo-6*H*-pyrido(3,2,1-jk)carbazole-11b-propionic acid 708
2-(*m*-Tolyloxy)ethyl *N*-(2,6-xylyl)carbamate 650
3-(*N*-(2,4,6-Trimethylphenyl)carbamoyl)-5-norbornene-2-carboxylic acid 434
$C_{18}H_{21}NO_3S$
4-Butoxyphenyl *N*-(4-(methylthio)phenyl)carbamate 651
2-Hydroxymethyl-6-methyl-4-(methylthio)phenyl *N*-(2,6-xylyl)carbamate 642
4-Hydroxy-4-phenyl-1-*p*-tolylsulfonylpiperidine 812
$C_{18}H_{21}NO_4$
Benzyl 5-acetoxymethyl-4-ethyl-3-methyl-2-pyrrolecarboxylate 838
4-Benzyl 2-ethyl 3-ethyl-2-methyl-2,4-pyrroledicarboxylate 843
2′,5′-Diethoxy-2-phenoxyacetanilide 414
2-(7,7-Dimethylbicyclo(3.1.1)hept-2-en-2-yl)ethyl 4-nitrobenzoate 373
Ethyl 5-(benzyloxycarbonyl)-2,4-dimethyl-3-pyrroleacetate 838
Ethyl 4-(3,4-dimethoxybenzylamino)benzoate 369
Ethyl 5-hydroxy-3-(2-oxocyclohexylmethyl)-2-indolecarboxylate 881
3-Ethyl-5-methylphenyl *N*-(2,5-dimethoxyphenyl)carbamate 648
4′-Ethyl-3,4,5-trimethoxybenzanilide 463
2-(4-Methoxyphenoxy)ethyl *N*-(3,4-xylyl)carbamate 647
Octahydro-2,5-methano-2*H*-indeno(1,2-b)oxiren-3-yl *N*-(4-methoxyphenyl)carbamate 654
$C_{18}H_{21}NO_4S$
4-Amino-4′-cyclohexyl-3-sulfodiphenyl ether 787
$C_{18}H_{21}NO_5$
4′-Ethoxy-3,4,5-trimethoxybenzanilide 463
$C_{18}H_{21}NO_5S$
Diethyl 2-acetamido-2-(benzo(b)thiophene-3-ylmethyl)malonate 898
3-Ethyl-2-methyl-5-phenylbenzoxazolium methyl sulfate 910
$C_{18}H_{21}N_3$
Hexahydro-1-(4-(phenylazo)phenyl)-1*H*-azepine 1091
4-Methyl-1-(4-(phenylazo)phenyl)-piperidine 1091
$C_{18}H_{21}N_3O$
4′-(4-(Diethylamino)phenylazo)-acetophenone 1089
5′-Methyl-2′-(*p*-tolylazo)butyranilide 1088
$C_{18}H_{21}N_3OS$
1-(4-Butoxybenzylidene)-4-phenyl-3-thiosemicarbazide 1049
$C_{18}H_{21}N_3O_2$
1-Benzyl-4-(2-nitroanilino)piperidine. 273
1-(4-Butoxybenzylidene)-4-phenylsemicarbazide 553
3-(2-(Dimethylamino)acetyl)-2-(3-ethyl-5-methyl-4-isoxazolyl)indole 885
1-(4-(*N*-Methylacetamido)phenyl)-3-(2,6-xylyl)urea 569
$C_{18}H_{21}N_3O_3$
1-Benzyl-1-isopropyl-3-(4-methyl-3-nitrophenyl)urea 562
1-Butyl-3-(2-methyl-4-nitrophenyl)-1-phenylurea 569
4-Dimethylaminobenzoic (3,4-dimethoxybenzylidene)hydrazide 1041
2-(1-Ethyl-3-phenylureido)ethyl *N*-phenylcarbamate 674
$C_{18}H_{21}N_3O_4$
1-(2,5-Dimethoxyphenyl)-3-(4-(*N*-methylacetamido)phenyl)urea 550
2-(*N*-Ethylanilino)ethyl *N*-(2-methyl-5-nitrophenyl)carbamate 669
cis-Hexahydro-*N*-(2-methyl-5-oxo-1-phenyl-3-pyrazolin-4-yl)-phthalamic acid 708
L-Tyrosyl-L-tyrosinamide hydrochloride 410
$C_{18}H_{21}N_3O_5$
1-(2-(Methylamino)benzoyl)-2-(3,4,5-trimethoxybenzoyl)hydrazine 497
$C_{18}H_{21}N_3O_6$
1-(3,4-Dimethoxyphenethyl)-3-(2-methoxy-4-nitrophenyl)urea 549
$C_{18}H_{21}N_3S$
4-Cyclohexyl-1-(1-naphthylmethylene)-3-thiosemicarbazide 1072
$C_{18}H_{21}N_5O_3$
2-(Pyrrolidinocarboxamido)acetic acid *N*²-(*N*-(1-naphthyl)carbamoyl)-hydrazide 709
$C_{18}H_{21}N_5O_4$
2-(3-(6-Methyl-2-pyridyl)ureido)-acetic (3,4-dimethoxybenzylidene)-hydrazide 936
$C_{18}H_{21}N_5O_5$
Disperse Violet 12 (C.I. 11120) 1084
$C_{18}H_{21}N_5O_6$
Carbobenzyloxy-L-histidylglycylglycine 868
$C_{18}H_{21}O_4P$
Cyclohexyl diphenyl phosphate 835
$C_{18}H_{22}$
1,4-Diphenylhexane 162
3,3′,5,5′-Tetramethylbibenzyl 163
$C_{18}H_{22}ClN_3O_2$
1-(5-Chloro-2-methoxyphenyl)-3-(2-(*N*-ethylanilino)ethyl)urea 522
3-Diethylcarbamoyl-1-(phenylcarbamoylmethyl)-pyridinium chloride 947
$C_{18}H_{22}ClN_3S$
1-(4-Chlorophenyl)-3-(4-diethylamino-*o*-tolyl)-2-thiourea 743
$C_{18}H_{22}Cl_2N_2O_4$
Ethylene glycol bis(3-pyridiniumpropionate) dichloride 940
$C_{18}H_{22}Hg$
Dimesitylmercury 1139
$C_{18}H_{22}INO_2$
(3-Oxo-3-(4-phenoxyphenyl)propyl)-trimethylammonium iodide 292
$C_{18}H_{22}N_2$
4-Anilino-1-benzylpiperidine 249
1,4-Dibenzylpiperazine dihydrochloride 251
N,*N*-Dimethyl-α-(2,4,6-trimethylphenylimino)-*p*-toluidine 1062
1-(Diphenylmethyl)-4-methylpiperazine 251
$C_{18}H_{22}N_2O$
1-Benzyl-3-(2,6-diethylphenyl)urea 573
1-(4-Butylphenyl)-3-(*m*-tolyl)urea 591
1-(4-Butylphenyl)-3-(*p*-tolyl)urea 591
1-Cyclohexyl-1-methyl-3-(1-naphthyl)urea 584
4′-(Diethylamino)-3-methylbenzanilide hydrochloride 470
4′-(Diethylamino)-4-methylbenzanilide hydrochloride 470
N-(2-(Dimethylamino)-ethyl)-2,2-diphenylacetamide hydrochloride 423
1-(3,4-Dimethylbenzyl)-3-(2-ethylphenyl)urea 572
1-(3,4-Dimethylbenzyl)-3-(2,6-xylyl)urea 573
2,6-Dimethyl-*N*-(1-naphthyl)-1-piperidinecarboxamide 712
1-(2,4-Dimethylphenyl)-3-(2-ethyl-6-methylphenyl)urea 596
1-Methyl-1-(1-methyl-2-phenylethyl)-3-(*o*-tolyl)urea 570
α-(Piperidinomethyl)-α-(2-pyridyl)-benzyl alcohol 931
$C_{18}H_{22}N_2OS$
1-(5-Isopropyl-2-methylphenyl)-3-(4-(methylthio)phenyl)urea 561
$C_{18}H_{22}N_2O_2$
3-(*m*-Anisyl)-1-butyl-1-phenylurea 558
1-(2,6-Diethylphenyl)-3-(2-methoxyphenyl)urea 555
1-(2,6-Diethylphenyl)-3-(4-methoxyphenyl)urea 560
2-(*N*-Ethylanilino)ethyl *N*-(*p*-tolyl)-carbamate 672
Ethyl 4-(4-(dimethylamino)-benzylamino)benzoate 373
3′-(*N*-Ethyl-*N*-(2-phenoxyethyl)-amino)acetanilide 415
2-(*N*-Ethyl-*N*-(*m*-tolyl)amino)ethyl *N*-phenylcarbamate 671
1-(5-Isopropyl-2-methylphenyl)-3-(2-methoxyphenyl)urea 555
1-(5-Isopropyl-2-methylphenyl)-3-(4-methoxyphenyl)urea 560
$C_{18}H_{22}N_2O_2S$
trans-3,4-Bis(benzylamino)-tetrahydrothiophene-1,1-dioxide 777
2-(*N*-Ethylanilino)ethyl *N*-(4-(methylthio)phenyl)carbamate 667
Valerophenone (*p*-tosyl)hydrazone 822
$C_{18}H_{22}N_2O_3$
2-Butoxy-5-nitro-*N*-(*m*-tolyl)-benzylamine 267
1-(2,5-Diethoxyphenyl)-3-(*o*-tolyl)urea 550
N,*N*-Diethyl-β-(4-nitrophenoxy)-phenethylamine 265
2-(*N*-Ethyl-*N*-phenylamino)ethyl *N*-(*p*-anisyl)carbamate 652
$C_{18}H_{22}N_2O_3S$
1-(2,6-Diethylphenyl)-3-tosylurea 826
3-Hydroxyisovalerophenone (*p*-tosyl)hydrazone 812
3-Hydroxyvalerophenone (*p*-tosyl)-hydrazone 1036
$C_{18}H_{22}N_2O_4$
1-(2,5-Diethoxyphenyl)-3-(4-methoxyphenyl)urea 550
N,*N*′-(1,5-Naphthalenediyl)-bis(isopropyl carbamate) 671
$C_{18}H_{22}N_2O_5$
2-(1-(3-Carboxypropionyloxy)butyl)-4,5-dihydro-6-phenylpyridazin-3(2*H*)-one 708
$C_{18}H_{22}N_2O_6$
Carbobenzyloxy-L-isoleucine *N*-hydroxysuccinimide ester 717
$C_{18}H_{22}N_2S$
1-Benzyl-1-butyl-3-phenyl-2-thiourea. 748
$C_{18}H_{22}N_4$
N,*N*′-Bis(4-(dimethylamino)-benzylidene)hydrazine 1061
$C_{18}H_{22}N_4O_2$
4,4′-Bis(dimethylamino)oxanilide 423
1,6-Bis(2,5-dimethylphenyl)biurea 579
4′-(Dimethylamino)-4-nitro-2,2′,5,6′-tetramethylazobenzene 1089
$C_{18}H_{22}N_4O_4$
1-Adamantyl methyl ketone (2,4-dinitrophenyl)hydrazone 274
1-(2-(*N*-Ethylanilino)ethyl)-3-(2-methoxy-5-nitrophenyl)urea 551
$C_{18}H_{22}N_4O_4S_2$
p-Toluenesulfonic acid 2,3-butanediylidenedihydrazide 1075
$C_{18}H_{22}N_4O_5$
Carbobenzyloxy-L-histidyl-L-alanine methyl ester 868
$C_{18}H_{22}N_4O_6$
N,*N*′-Bis(2-hydroxyethyl)-*N*,*N*′-bis(4-nitrophenyl)ethylenediamine 263
$C_{18}H_{22}N_4O_{10}S_2$
4,4′-Dithiobis(1-β-D-ribofuranosyl-2(1*H*)-pyrimidinone) 131
$C_{18}H_{22}N_6O_3$
4-(2,4-Dimethylphenyl)-1-(5-methyl-2-pyridylureidoacetyl)-semicarbazide 959
$C_{18}H_{22}N_6O_5$
2-(5-Methyl-2-pyridylureido)acetic *N*²-(*N*-(2,5-dimethoxyphenyl)-carbamoyl)hydrazide 937
$C_{18}H_{22}O$
2,2-Dimethyl-3-phenyl-1-(*p*-tolyl)-propanol 204
$C_{18}H_{22}OS$
2-(Benzylthiomethylene)-1-decalone. 284
$C_{18}H_{22}O_2$
2-Cyclohexylmethylene-6-furfurylidene-1-cyclohexanone 853
2,6-Di-*tert*-butyl-1,4-naphthoquinone 318
2,7-Di-*tert*-butyl-1,4-naphthoquinone 318
4,4′-Dimethoxy-3,3′,5,5′-tetramethylbiphenyl 187
4,4′-(1,3-Dimethylbutylidene)-diphenol 194
1,6-Diphenoxyhexane 179
meso-3,4-Diphenyl-3,4-hexanediol 200
4,4′-(1-Ethyl-2-methyltrimethylene)-bisphenol 190
$C_{18}H_{22}O_2S$
Benzyl 4-(neopentyl)phenyl sulfone 784
Benzyl 4-*tert*-pentylphenyl sulfone 784
Dimesityl sulfone 786
$C_{18}H_{22}O_3$
1-(2,6-Dimethoxyphenyl)-2-methyl-1-phenyl-1-propanol 200
2-(2-Hydroxy-2-adamantylmethyl)-benzoic acid 339
8a-Methyl-2,3,4,4a,4b,5,8,8a-octahydro-2-oxo-1-phenanthrenepropionic acid 65
$C_{18}H_{22}O_4$
1,2-Bis(3,5-dimethoxyphenyl)ethane. 178
3,4-Bis(4-hydroxyphenyl)-3,4-hexanediol 199
mono(Decahydronaphth-2-yl) phthalate 371
3,3′-Dimethyl-4,4′,5,5′-tetramethoxybiphenyl 178
4b,5,6,7,8,8a,9,10-Octahydro-1,7-phenanthrenediol diacetate 380
$C_{18}H_{22}O_5$
Diveratryl ether 186
$C_{18}H_{22}O_5S_2$
2-Mesitylenesulfonic anhydride 801
$C_{18}H_{22}O_6$
1,4-Bis-(4-methoxyphenoxy)-2,3-butanediol 202
3,3′,4,4′,5,5′-Hexamethoxybiphenyl 187
$C_{18}H_{22}O_6S_2$
2-Benzhydryl-1,3-propanediol bis(methanesulfonate) 800
$C_{18}H_{22}S_2$
Mesityl disulfide 213
$C_{18}H_{22}Si$
Cyclohexyldiphenylsilane 1111
$C_{18}H_{23}IN_2$
4-(4-(Diethylamino)-styryl)-1-methylpyridinium iodide 954
$C_{18}H_{23}IN_2O_3S$
4-Methyl-4-(2-phenylsulfonyl-2-(2-pyridyl)ethyl)morpholinium iodide 946
$C_{18}H_{23}N$
N-Butyl-α-phenylphenethylamine 242
$C_{18}H_{23}NO$
1,2-Diphenyl-3-methyl-4-(methylamino)-2-butanol hydrochloride 237
N-(1-Naphthylmethyl)heptanoamide 490
$C_{18}H_{23}NO_2$
Bis-(2-(benzyloxy)-ethyl)-amine 240

$C_{18}H_{23}NO_2Si$
1-Phenyl-2-(trimethylsilyl)ethyl *N*-phenylcarbamate 1107

$C_{18}H_{23}NO_3$
4-(2-(4-Aminophenoxy)ethoxy)-2-*tert*-butylphenol 225
4′-Butyl-1,2,3,6-tetrahydrophthalanilic acid 435
8,9-Dimethoxy-1,2,3,3a,5,6,10b,12a-octahydrobenzo(a)cyclopenta(f)-quinolizin-12(11*H*)-one 283

$C_{18}H_{23}NO_3S$
2-(α-(1,1-Dimethylethyl)-α-hydroxybenzyl)-*N*-methylbenzenesulfonamide 811

$C_{18}H_{23}NO_5S_2$
1-Methyl-3-(*p*-tosylamino)propyl *p*-toluenesulfonate 827

$C_{18}H_{23}NO_6$
Diethyl 2-acetamido-2-(3-oxo-3-phenylpropyl)malonate 432

$C_{18}H_{23}N_3$
4-(Diethylamino)-4′-ethylazobenzene 1090

$C_{18}H_{23}N_3O$
1-(2,6-Diethylphenyl)-3-(2-(2-pyridyl)-ethyl)urea 950
1-(2-(*N*-Methylanilino)ethyl)-3-(2,5-xylyl)urea 562

$C_{18}H_{23}N_3O_2$
1-(4-(Diethylamino)phenyl)-3-(2-methoxyphenyl)urea 552
N,*N*-Dimethyl-2-(3-ethyl-5-methyl-4-isoxazolyl)-β-hydroxytryptamine 881
1-(2-(*N*-Ethylanilino)ethyl)-3-(4-methoxyphenyl)urea 551

$C_{18}H_{23}N_3O_6$
Carbobenzyloxyprolylglycylglycine methyl ester 700

$C_{18}H_{23}N_3O_6S$
Dimethyl 3-(benzamidomethylcarbamoyl)-5,5-dimethyl-2,4-thiazolidinedicarboxylate 576

$C_{18}H_{23}N_5O_3$
1-(Diethylureidoacetyl)-4-(1-naphthyl)semicarbazide 574

$C_{18}H_{24}$
2,7-Di-*tert*-butylnaphthalene 167
1-(3-Methylbenzyl)adamantane 164

$C_{18}H_{24}As_2Br_2$
5,6,7,8-Tetrahydro-5,5,8,8-tetramethyldibenzo(e,g)(1,4)-diarsenium dibromide 1128

$C_{18}H_{24}Cl_2N_2O$
10-Undecenoic (2,6-dichlorobenzylidene)hydrazide 494

$C_{18}H_{24}Ge$
(2-Biphenylyl)triethylgermane 1120

$C_{18}H_{24}INO$
O-(2-Benzylphenyl)-choline iodide 255

$C_{18}H_{24}N_2$
N,*N*′-Bis(1-phenylethyl)-ethylenediamine 242
N,*N*′-Dibenzyl-*N*,*N*′-dimethylethylenediamine 242
N,*N*′-Diethyl-*N*,*N*′-diphenylethylenediamine 232
2,7-Diethyl-1,2,3,3a,4,6,7,8-octahydrobenzo(lmn)(3,8)-phenanthroline 251
4,4′-Hexamethylenebis(aniline) 230
N,*N*′-Tetramethylenebis(*N*-methylaniline) picrate 276

$C_{18}H_{24}N_2O$
Bis(4-dimethylaminophenyl)methyl methyl ether 229

$C_{18}H_{24}N_2O_2$
N,*N*′-Bis-(4-methoxybenzyl)-ethylenediamine dihydrochloride 240
1-(2,6-Diisopropylphenyl)-3-furfurylurea 852
2,2′-Ethylidenebis(5-(dimethylamino)phenol) 227
2,2′-(Hexamethylenedioxy)dianiline 227

$C_{18}H_{24}N_2O_4$
Diethyl 1,2,3,4,4a,5,10,10a-octahydrobenzo(b)(1,7)-naphthyridine-1,1dicarboxylate 358
Diethyl 3,3′,4,4′-tetramethyl-(2,2′-bipyrrole)-5,5′-dicarboxylate 841

$C_{18}H_{24}N_2O_4S_2$
N,*N*′-Ethylenebis(*N*-methyl-*p*-toluenesulfonamide) 818

$C_{18}H_{24}N_2O_5$
Carbobenzyloxy-L-prolyl-L-valine 708

$C_{18}H_{24}N_2O_6$
Carbobenzyloxythreonylproline methyl ester 696

$C_{18}H_{24}N_2O_{15}$
1-*O*-(3,4-Dinitrophenyl)-β-D-cellobioside 262

$C_{18}H_{24}N_4O_2$
1,1′-(1,5-Naphthalenediyl)bis(3-isopropylurea) 604
1,1′-(1,5-Naphthalenediyl)bis(3-propylurea) 604

$C_{18}H_{24}N_4O_3$
Disperse Black 4 (C.I. 11220) 1082

$C_{18}H_{24}O_2$
1,4,4a,9,10,10a-Hexahydro-7-methoxy-1,1,4a-trimethyl-2(3*H*)-phenanthrenone 284

$C_{18}H_{24}O_3$
2,3,4,4a,4b,5,8,8a,9,10-Decahydro-8a-methyl-2-oxo-1-phenanthrene-propionic acid 65

$C_{18}H_{24}O_3S$
7,7-Dimethyl-2-methylenebicyclo(3.1.1)heptan-3-ylmethyl *p*-tosylate 799

$C_{18}H_{24}O_4$
trans-*tert*-Butylcyclohexyl hydrogen phthalate 371

$C_{18}H_{24}Si$
(2-Biphenylyl)triethylsilane 1120
Bis(3-phenylpropyl)silane 1113

$C_{18}H_{25}NO$
3-(*N*-Methyl-*N*-(α-methylphenethyl)-aminomethyl)-2-norbornanone 242
2-(α-Piperidinobenzyl)-cyclohexanone 286

$C_{18}H_{25}NO_2$
1-(*tert*-Butyl)-1-hydroxy-2-phenylhexahydro-3-oxoisoindoline 690

$C_{18}H_{25}NO_3$
4′-Butylhexahydrophthalanilic acid 420
1′,4′,5′,6′,7′,8′-Hexahydro-4,4,7′,7′-tetramethylspiro(cyclohexane-1,3′(2′*H*)-quinoline)-2,5′6-trione 59

$C_{18}H_{25}NO_4$
1-Benzyl-4-(propionyloxy)-3-pyrrolidinylmethyl propionate 357
d-Camphoric *p*-phenetidide 418

$C_{18}H_{25}NO_6$
tert-Butyl 3,5-bis(ethoxycarbonyl)-4-methyl-2-pyrroleacrylate 840

$C_{18}H_{25}N_3O$
1-(1,2,3,4,5,6,7,8-Octahydro-3-phenanthryl)-1-propanone semicarbazone 1070

$C_{18}H_{25}N_3O_2$
tert-Butyl 2-(2-cyanoethyl)-2,3-dicyano-3,6-dimethyl-6-heptenoate 149

$C_{18}H_{25}N_3O_3$
1-(2-Cyanoethyl)-1-cyclohexyl-3-(2,5-dimethoxyphenyl)urea 752

$C_{18}H_{25}N_3O_4$
Carbobenzyloxy-L-prolyl-L-valinamide hydrate 708

$C_{18}H_{25}N_3O_6$
Carbobenzyloxyglycylglycyl-L-norleucine 660
Carbobenzyloxyglycylglycyl-L-norvaline methyl ester 662
Carbobenzyloxyglycylglycyl-L-valine methyl ester 662
Carbobenzyloxy-L-leucylglycylglycine 676
Carbobenzyloxy-L-norvalylglycylglycine methylester 662

$C_{18}H_{25}N_3O_6S$
Carbobenzyloxyglycylglycyl-L-methionine methyl ester 656

$C_{18}H_{25}N_3O_6S$
Carbobenzyloxymethionylglycylglycine methyl ester 667

$C_{18}H_{25}N_6O_{12}P$
Cytidyl (3′-5′)cytidine ammonium salt 161

$C_{18}H_{25}O_3S$
2-Oxo-1-cyclododecyl phenyl sulfone 781

$C_{18}H_{26}BBrS$
(2-(2-Benzylthioethyl)-6,6-dimethylbicyclo(3.1.1)hept-3-yl)-bromoborane 1125

$C_{18}H_{26}Br_2N_2$
1,1′-Tetramethylenebis(3,4-dimethylpyridinium bromide) 952
1,1′-Tetramethylenebis(3,5-dimethylpyridinium bromide) 952

$C_{18}H_{26}Br_2N_2O_4$
2,5-Dibromo-*N*,*N*′-di-*tert*-butyl-3,6-dimethoxyterephthalamide 447

$C_{18}H_{26}ClNO$
5′-Chloro-2′-methyl-10-undecenanilide 440

$C_{18}H_{26}ClNO_2$
4-(*tert*-Butyl)cyclohexyl *N*-(5-chloro-2-methylphenyl)carbamate 631

$C_{18}H_{26}ClN_3$
Chloroquine 989

$C_{18}H_{26}Cl_2N_2O_8$
1,1′-Tetramethylenebis(2,5-dimethylpyridinium perchlorate) 952
1,1′-Tetramethylenebis(3,4-dimethylpyridinium perchlorate) 952

$C_{18}H_{26}Cl_2O_4$
Dimethyl α,α′-dichloro-5,7-dimethyl-1,3-adamantanediacetate 83

$C_{18}H_{26}I_2N_2$
1,1′-Hexamethylenebis(3-methylpyridinium iodide) 958
1,1′-Hexamethylenebis(4-methylpyridinium iodide) 958

$C_{18}H_{26}N_2O$
10-Undecenoic acid *N*-benzylidenehydrazide 500

$C_{18}H_{26}N_2O_2$
10-Undecenoic (2-hydroxybenzylidene)hydrazide 1034

$C_{18}H_{26}N_2O_3$
10-Undecenoic (2,4-dihydroxybenzylidene)hydrazide 494

$C_{18}H_{26}N_2O_4$
4-*tert*-Butylcyclohexyl *N*-(4-methyl-2-nitrophenyl)carbamate 668

$C_{18}H_{26}N_2O_5$
Carbobenzyloxy-L-alanyl-L-isoleucine methyl ester 664
Carbobenzyloxy-L-alanyl-L-norleucine methyl ester 662
Carbobenzyloxy-L-isoleucyl-L-alanine methyl ester 662
Carbobenzyloxy-L-leucyl-L-alanine methyl ester 662
Carbobenzyloxy-L-leucylglycine ethyl ester 661
Carbobenzyloxy-DL-norleucylglycine ethyl ester 663
Carbobenzyloxy-L-norvalyl-L-norvaline 660
Carbobenzyloxysarcosylisoleucine methyl ester 662

$C_{18}H_{26}N_2O_5S$
Carbobenzyloxy-L-norvalyl-L-methionine 655
Carbobenzyloxy-L-valyl-L-methionine 667

$C_{18}H_{26}N_2O_6$
Carbobenzyloxyisoleucylserine methyl ester 642
Carbobenzyloxy-L-leucyl-L-serine methyl ester 643
Carbobenzyloxy-L-norleucyl-L-serine methyl ester 643
Carbobenzyloxythreonylvaline methyl ester 645
Ethyl 9-(2-hydroxy-4-nitrophenylcarbamoyl)-nonanoate 427
*N*Δ-(*tert*-Butoxycarbonyl)-*N*α-carbobenzyloxy-L-ornithine 657

$C_{18}H_{26}N_2O_6S$
Carbobenzyloxy-L-methionyl-L-threonine methyl ester 642

$C_{18}H_{26}N_4O_4$
1,1′-Hexamethylenebis(3-furfurylurea) 852

$C_{18}H_{26}O$
7-Methoxy-1,2,3,4,4a,9,10,10a-octahydro-1,1,4a-trimethylphenanthrene 180
8-Methoxy-1,2,3,4,4a,9,10,10a-octahydro-1,1,4a-trimethylphenanthrene 180

$C_{18}H_{26}OSi_2$
1,3-Dibenzyl-1,1,3,3-tetramethyldisiloxane 1113
(Oxydi-2,1-phenylene)-bis(trimethylsilane) 1106
4-(Trimethylsilyl)phenyl ether 1107

$C_{18}H_{26}O_2$
5,5-Dimethyl-2-(1,1-dimethyl-3-butenyl)-2-phenyl-1,3-dioxane 185

$C_{18}H_{26}O_2S$
Methyl 2-methyl-2-(phenylthio)-4-decenoate 357

$C_{18}H_{26}O_3$
L-Menthyl DL-mandelate 353

$C_{18}H_{26}O_4$
Bis-(6-methyl-3-cyclohexenylmethyl) oxalate 79
Diethyl 2-ethyl-2-(β-methylphenethyl)-malonate 360

$C_{18}H_{26}O_5$
3,4-*O*-Isopropylidene-2-*O*-mesityl-1-*O*-methyl-β-D-arabinoside 184

$C_{18}H_{26}O_9$
2-Oxo-1,1,3,3-cyclohexanetetrapropionic acid 65

$C_{18}H_{26}Si_2$
4,4′-Bis(trimethylsilyl)biphenyl 1123
1,2-Dibenzyl-1,1,2,2-tetramethyldisilane 1113

$C_{18}H_{27}BrO_2$
Benzyl 11-bromoundecanoate 353
o-Tolyl 10-bromodecanoate 376

$C_{18}H_{27}Br_3O$
Dodecyl 2,4,6-tribromophenyl ether 177

$C_{18}H_{27}Cl$
1-Chloro-1-phenyl-1-dodecene 171

$C_{18}H_{27}ClN_2O_2$
3-Diethylaminopropyl *N*-(4-chlorobenzyl)-*N*-cyclopropylcarbamate 671

$C_{18}H_{27}NOS$
4′-(Methylthio)-10-undecenanilide 421

$C_{18}H_{27}NO_2$
4′-Butoxy-3-piperidinopropiophenone hydrochloride 292
N,3-Dimethyl-*N*-(1-methyl-2-(2,4,6-trimethylphenoxy)-ethyl)-crotonamide 418
2-Methylcyclohexyl *N*-(2,6-diethylphenyl)carbamate 679

$C_{18}H_{27}NO_2S$
Isopropyl α-(2-piperidinoethylthio)-phenylacetate citrate 357

$C_{18}H_{27}NO_3$
Methyl *N*-(3,7-dimethyl-7-hydroxyoctylidene)-anthranilate 1037

$C_{18}H_{27}NO_4$
Diisobutyl 2,2′-(phenylimino)-diacetate 358

$C_{18}H_{27}NO_4S$
1-Nitrocyclododecyl phenyl sulfone 780

$C_{18}H_{27}NO_5$
7-Nitro-2-nonyloxymethyl-1,4-benzodioxan 266

$C_{18}H_{27}NO_6$
Ethyl 4-(*tert*-butoxycarbonyl)-2-(ethoxycarbonyl)-5-methyl-3-pyrrolepropionate 842

$C_{18}H_{27}N_3O$
Pentaquine thiodisalicylate 985

$C_{18}H_{27}N_3O_4S$
Carbobenzyloxy-L-methionyl-L-valinamide 656
Carbobenzyloxy-L-norvalyl-L-methioninamide 656

$C_{18}H_{27}N_3O_5$
Carbobenzyloxy-L-threonyl-L-leucinamide 643
$C_{18}H_{28}$
1,4-Dicyclohexyl-1,3-cyclohexadiene 3
$C_{18}H_{28}BrNO_3$
α-Bromo-5-nitro-*o*-tolyl undecyl ether 258
$C_{18}H_{28}Cl_2N_2OS$
2-(2-(Dibutylamino)-ethylthio)-2′,5′-dichloroacetanilide hydrochloride 421
$C_{18}H_{28}IN_3O_2S$
(3-Dansylamido)-propyltrimethylammonium iodide 818
$C_{18}H_{28}I_2N_2$
1,3-Bis(trimethylammoniomethyl)-azulene diiodide 256
$C_{18}H_{28}N_2$
2-(2-(Diethylamino)-ethyl)-3-methyl-2-phenylvaleronitrile 755
$C_{18}H_{28}N_2O$
1-(*N*-(4-Butylphenyl)carbamoyl)-2,6-dimethylpiperidine 585
1-Cyclohexyl-3-(2,6-diethylphenyl)-1-methylurea 583
$C_{18}H_{28}N_2O_4$
N,*N*′-Di-*tert*-butyl-2,5-dimethoxyterephthalamide 463
$C_{18}H_{28}N_2O_5$
4-(2,5-Dibutoxy-4-nitrophenyl)-morpholine 267
$C_{18}H_{28}N_2Si_2$
1,2-Bis(trimethylsilyl)-1,2-diphenylhydrazine 1108
$C_{18}H_{28}O$
6-Ethyl-7-methoxymethyl-1,2,3,4-tetrahydro-1,1,4,4-tetramethylnaphthalene 188
$C_{18}H_{28}O_2$
3-Butyl-5-(2,6,6-trimethyl-1-cyclohexen-1-yl)-2,4-pentadienoic acid 67
2′-Hydroxydodecanophenone 290
4′-Hydroxydodecanophenone 290
α-Pentylcinnamaldehyde diethyl acetal 182
$C_{18}H_{28}O_3$
4-(Decyloxy)-*m*-anisaldehyde 323
Methyl 3-(decyloxy)benzoate 370
Methyl 4-(decyloxy)benzoate 370
$C_{18}H_{28}O_4S$
Methyl 11-(phenylsulfonyl)-undecanoate 781
$C_{18}H_{28}O_6$
(-)-1,2:5,6-Di-*O*-cyclohexylidene-L-inositol 22
$C_{18}H_{28}O_7$
Ethyl 1,2-bis(methoxycarbonyl)-9-oxocyclononanepropionate 93
$C_{18}H_{28}Si_3$
2,2-Diphenyl-1,1,1,3,3,3-hexamethyltrisilane 1113
$C_{18}H_{29}ClN_2OS$
2′-Chloro-2-(2-(dibutylamino)-ethylthio)-acetanilide hydrochloride 421
$C_{18}H_{29}ClN_2O_3$
4-Carbamoyl-1-(decyloxycarbonylmethyl)-pyridinium chloride 951
$C_{18}H_{29}ClO$
2-Chlorophenyl dodecyl ether 177
$C_{18}H_{29}Cl_3OSn$
Tributyltin 2,4,5-trichlorophenoxide 1133
$C_{18}H_{29}N$
4-Benzyl-1-hexylpiperidine 251
$C_{18}H_{29}NO$
2′,3′-Decanoxylidide 439
2′,4′-Decanoxylidide 440
2-(1-(Decylimino)ethyl)phenol 1038
2′,6′-Dimethyldecananilide 439
2′-Ethyldecananilide 438
4′-Ethyldecananilide 438
5′-Isopropyl-*o*-octanotoluidide 440
$C_{18}H_{29}NO_2$
2′-(Decyloxy)acetanilide 417
3′-(Decyloxy)acetanilide 418
4′-(Decyloxy)acetanilide 419
2-(Diethylamino)-ethyl 3-methyl-2-phenylvalerate 359
1-Dodecyl-4-nitrobenzene 279
4′-Ethoxydecananilide 419
4′-Hydroxydodecananilide 413
2′-Methoxy-5′-methyldecananilide 417
$C_{18}H_{29}NO_3$
Dodecyl 2-nitrophenyl ether 268
Dodecyl 3-nitrophenyl ether 268
Dodecyl 4-nitrophenyl ether 269
$C_{18}H_{29}NO_4S$
Dodecyl 4-nitrophenyl sulfone 779
$C_{18}H_{29}NO_5$
N-Dodecylglucamine 32
$C_{18}H_{29}N_3S$
1-(3-(Dibutylamino)propyl)-3-phenyl-2-thiourea 740
$C_{18}H_{30}$
Hexaethylbenzene 163
3-Phenyldodecane 162
4-Phenyldodecane 162
5-Phenyldodecane 162
6-Phenyldodecane 162
1,2,4,5-Tetraisopropylbenzene 163
$C_{18}H_{30}Br_6O_2$
9,10,12,13,15,16-Hexabromooctadecanoic acid 62
$C_{18}H_{30}ClN_3O_2$
3-Carbamoyl-1-(decylcarbamoylmethyl)-pyridinium chloride 947
$C_{18}H_{30}Cl_2OSn$
Tributyltin 2,4-dichlorophenoxide 1133
$C_{18}H_{30}FNO_2S$
N-Dodecylsulfanilyl fluoride 803
$C_{18}H_{30}N_2O$
1-(1-Azacyclohept-2-en-1-yl)-2-morpholino-1-cyclooctene 39
1,1-Dibutyl-3-(2-ethyl-6-methylphenyl)urea 596
2-Diethylamino-*N*-methyl-*N*-(2-methyl-1-phenylbutyl)-acetamide hydrochloride 423
1,1-Diisobutyl-3-(2-ethyl-6-methylphenyl)urea 596
N-Dodecylnicotinamide 963
1-Nonyl-3-(2,5-xylyl)urea 601
$C_{18}H_{30}N_2O_5S$
N-Dodecyl-3-nitrosulfanilic acid, sodium salt 793
$C_{18}H_{30}N_4O$
1-((4-Decyloxybenzylidene)amino)-guanidine nitrate 1049
$C_{18}H_{30}O$
2-Dodecylphenol 191
Dodecyl phenyl ether 180
Ethyl 2,4,6-triisopropylbenzyl ether 186
$C_{18}H_{30}O_2$
1,1′-Bis-(3,5,5-trimethyl-2-cyclohexen-1-ol) 14
2-Dodecylhydroquinone 193
$C_{18}H_{30}O_3$
2-(1-Tetradecenyl)succinic anhydride 98
$C_{18}H_{31}BF_4O$
2,4,6-Tri-*tert*-butyl-3-methylpyrylium tetrafluoroborate 935
$C_{18}H_{31}BrN_2O$
3-Carbamoyl-1-dodecylpyridinium bromide 947
4-Carbamoyl-1-dodecylpyridinium bromide 956
$C_{18}H_{31}N$
N-Phenyldodecylamine 220
$C_{18}H_{31}NO$
2-(Azacyclotridec-1-en-2-yl)-1-cyclohexanone 59
4-(2-Butyloctyloxy)aniline 218
2-(Dodecyloxy)aniline 217
3-(Dodecyloxy)aniline 217
4-(Dodecyloxy)aniline 218
$C_{18}H_{31}NO_2$
3,5,7-Trimethyl-1-adamantyl *N*-butylcarbamate 116
$C_{18}H_{31}NO_2S$
4-(Dodecylsulfonyl)aniline 776
$C_{18}H_{31}NO_3$
1-(2-Aminophenoxy)-3-(*tert*-nonyloxy)-2-propanol 216
$C_{18}H_{31}NS$
2-(Dodecylthio)-aniline 218
$C_{18}H_{31}N_3O_2$
2,β-Bis-(diethylamino)-*p*-acetophenetidide 416
$C_{18}H_{32}Br_4O_2$
9,10,12,13-Tetrabromooctadecanoic acid 62
$C_{18}H_{32}N_2$
N,*N*-Dihexyl-1,4-phenylenediamine hydrochloride 221
$C_{18}H_{32}N_2O_3S$
3-Amino-4-(dodecylamino)-benzenesulfonic acid 788
$C_{18}H_{32}O_2$
2-Cyclopentene-1-tridecanoic acid 67
8-Hydroxyoleic acid lactone 97
Linoleic acid, urea adduct 61
$C_{18}H_{32}O_3$
Tetradecylsuccinic anhydride 98
$C_{18}H_{32}O_4$
trans-1,2-Cyclohexanediol *mono*((-)-menthoxyacetate) 84
$C_{18}H_{32}O_5$
Diethyl 2-butyl-2-heptanoylmalonate 93
$C_{18}H_{32}O_7$
Tributyl citrate 74
$C_{18}H_{32}Si$
Dodecylphenylsilane 1113
Phenyltributylsilane 1113
$C_{18}H_{33}Br$
18-Bromo-6,9-octadecadiene 5
$C_{18}H_{33}ClPb$
Chlorotricyclohexyllead 1135
$C_{18}H_{33}IPb$
Iodotricyclohexyllead 1135
$C_{18}H_{33}N$
2,5-Dimethyl-1-dodecylpyrrole 842
$C_{18}H_{33}NO_6$
2-(*N*-Dodecylcarbamoylmethyl)-2-hydroxysuccinic acid 117
$C_{18}H_{34}ClF_3S$
2-Chloro-1,1,2-trifluoroethyl hexadecyl sulfide 24
$C_{18}H_{34}N_2O_3$
Decyl 3-carbamoyl-1-piperidineacetate 108
Decyl 4-carbamoyl-1-piperidineacetate 108
$C_{18}H_{34}N_4O_2$
N,*N*′-Hexamethylenebis(1-piperidinecarboxamide) 130
Sebacic bis(isobutylidene)-hydrazide 1070
$C_{18}H_{34}O_2$
Oleic acid, calcium salt 70
Oleic acid, copper(II) salt 70
Oleic acid urea adduct 127
$C_{18}H_{34}O_2S$
9,10-Epithiooctadecanoic acid 68
$C_{18}H_{34}O_3$
9,10-Epoxyoctadecanoic acid 68
13-Oxooctadecanoic acid 64
$C_{18}H_{34}O_4$
Carboxymethyl palmitate 95
Dibutyl sebacate 76
$C_{18}H_{34}Pb$
Tricyclohexyllead 1135
$C_{18}H_{35}BrO_2$
2-Bromostearic acid 62
$C_{18}H_{35}ClO_2$
2-Chlorostearic acid 62
$C_{18}H_{35}NO_2$
2-(Diethylamino)ethyl 2-cyclohexyl-2-isobutylacetate 88
12-Oxooctadecanamide 101
$C_{18}H_{35}NO_3$
18-Amino-7-oxooctadecanoic acid 71
$C_{18}H_{35}N_3O_4$
L-Isoleucyl-L-isoleucyl-L-isoleucine hydrate 120
$C_{18}H_{35}N_3S_2$
2-Amino-5-hexadecylthio-1,3,4-thiadiazole 878
$C_{18}H_{36}B_2O_6$
2,2′-(2-Methyl-2,4-pentanediyldioxy)-bis(4,4,6-trimethyl-1,3,2-dioxaborinane) 1124
$C_{18}H_{36}N_2O$
1-Dodecylisonipecotamide 108
1-Dodecylnipecotamide 108
N-Ethyl-*N*-(1-isobutyl-3-methylbutyl)-1-piperidineacetamide hydrochloride 109
N-Methyl-*N*-(1-methyldecyl)-1-pyrrolidineacetamide hydrochloride 108
$C_{18}H_{36}N_2O_2$
N,*N*,*N*′,*N*′-Tetraisobutyloxamide 110
$C_{18}H_{36}N_2S_4$
Bis-(dibutylthiocarbamoyl) disulfide 139
$C_{18}H_{36}O$
2,6-Dimethyl-10-hexadecanone 49
9,10-Epoxyoctadecane 20
Hexadecyl vinyl ether 18
3-Octadecanone 49
5-Octadecanone 49
9-Octadecanone 49
$C_{18}H_{36}OS$
S-Hexadecyl thioacetate 87
$C_{18}H_{36}O_2$
Ricinoleyl alcohol 12
Stearic acid, calcium salt 70
Stearic acid, dihydroxyaluminum salt 70
Stearic acid, dodecylamine salt 70
Stearic acid, octadecylamine salt 70
Stearic acid, sodium salt 70
$C_{18}H_{36}O_3$
(D)-9-Hydroxystearic acid 63
$C_{18}H_{36}Sn$
(2-Cyclohexen-1-yl)tributyltin 1132
$C_{18}H_{37}Br$
tert-Octadecyl bromide 4
$C_{18}H_{37}BrHg$
Bromooctadecylmercury 1137
$C_{18}H_{37}ClHg$
Chlorooctadecylmercury 1137
$C_{18}H_{37}ClO_2S$
5,7,7-Trimethyl-2-(1,3,3-trimethylbutyl)-1-octanesulfonyl chloride 157
$C_{18}H_{37}HgNO_3$
Octadecylmercury(II) nitrate 1137
$C_{18}H_{37}NO_2$
18-Aminooctadecanoic acid 67
N-(2-Hydroxyethyl)-palmitamide 103
12-Hydroxyoctadecanamide 101
Octadecanohydroxamic acid 1008
$C_{18}H_{37}NO_4$
2-(2-(2-(Dodecyloxy)-ethoxy)-ethoxy)-acetamide 101
$C_{18}H_{37}NS$
Thiooctadecanamide 141
$C_{18}H_{38}NO_4P$
Dibutyl (*N*,*N*-dibutylcarbamoylmethyl)-phosphonate 159
$C_{18}H_{38}O$
2-Methyl-2-heptadecanol 10
$C_{18}H_{38}OS$
2-(Hexadecylthio)-ethanol 25
$C_{18}H_{38}O_2S$
2-(Hexadecylsulfinyl)-ethanol 152
$C_{18}H_{38}O_3$
9,10-Dihydroxyoctadecanol 16
$C_{18}H_{38}O_3S$
2-(Hexadecylsulfonyl)-ethanol 152
$C_{18}H_{38}O_4$
Glyoxal bis-(dibutyl acetal) 19
$C_{18}H_{38}O_4S$
1-Hydroxy-1-octadecanesulfonic acid, sodium salt 155
$C_{18}H_{38}S$
Nonyl sulfide 26
5,7,7-Trimethyl-2-(1,3,3-trimethylbutyl)-1-octanethiol 24
$C_{18}H_{38}Si_4$
1,2,4,5-Hexatetraene-1,6-diylidene-tetrakis(trimethylsilane) 1098

$C_{18}H_{39}BO_3$
Trihexyl borate ... 1124

$C_{18}H_{39}N$
Dinonylamine ... 29

$C_{18}H_{39}NO$
2-(Dioctylamino)-ethanol ... 34

$C_{18}H_{39}N_2O_9P$
Fructose-1-phosphate, dicyclohexylammonium salt ... 161

$C_{18}H_{40}Ge$
Trihexylgermane ... 1098

$C_{18}H_{40}N_2$
N,N,N′,N′-Tetrabutylethylenediamine ... 274

$C_{18}H_{40}Si_4$
2,5-Cyclohexadiene-1,2,4,5-tetrayltetrakis(trimethylsilane) ... 1098

$C_{18}H_{42}B_8O_{15}$
Isopropyl polyborate ... 1124

$C_{18}H_{42}Cl_4P_2Sn$
Tetrachlorobis(tripropylphosphine)tin(IV) ... 1129

$C_{18}H_{42}N_3O_4P$
Tris-(2-(diethylamino)ethyl)phosphate trihydrochloride ... 161

$C_{18}H_{42}N_3P$
Tris(dipropylamino)phosphine ... 158

$C_{18}H_{42}N_4O_{14}$
Ethylenediaminetetraacetic acid, dimorpholinium salt tetrahydrate ... 67

$C_{18}H_{42}OSn_2$
Bis(tripropyltin) oxide ... 1129

$C_{18}H_{54}Si_8$
1,1,1,4,4,4-Hexamethyl-2,2,3,3-tetrakis(trimethylsilyl)tetrasilane ... 1098

$C_{19}H_3Cl_{15}Si$
Methyltris(pentachlorophenyl)silane ... 1101

$C_{19}H_3F_{15}Si$
Methyltris(pentafluorophenyl)silane ... 1101

$C_{19}H_7Cl_6NO_2$
1,4,5,6,7,7-Hexachloro-*N*-(1-naphthyl)-5-norbornene-2,3-dicarboximide ... 722

$C_{19}H_8Cl_{10}Si$
Bis(pentachlorophenyl)methylphenylsilane ... 1102

$C_{19}H_{10}Cl_6N_2O_2$
1-(5-Chloro-2-(pentachlorophenoxy)phenyl)-3-phenylurea ... 554

$C_{19}H_{10}Cl_8Si$
Bis(2,3,4,5-tetrachlorophenyl)methylphenylsilane ... 1102

$C_{19}H_{10}O_{13}$
5,5′-Carbonylbis-(trimellitic acid) ... 344

$C_{19}H_{11}Br_2N_3O_2$
2,2′-(Pyrido(2,3-b)pyrazine-2,3-diyl)bis(4-bromophenol) ... 1002

$C_{19}H_{11}Cl_5$
2,3-Diphenyl-1,5,6,7,7-pentachlorobicyclo(2.2.1)hepta-2,5-diene ... 174

$C_{19}H_{11}Cl_5O$
2,3,4,5,6-Pentachlorotriphenylmethanol ... 198

$C_{19}H_{11}F_5O$
2,3,4,5,6-Pentafluorotriphenylmethanol ... 198

$C_{19}H_{11}NO_4$
4-(*N*-(2-Naphthyl)carbamoyl)phthalic anhydride ... 474

$C_{19}H_{11}N_3$
10-Azadibenzo(a,c)phenazine ... 1003

$C_{19}H_{12}Br_2O_5S$
Bromophenol Red ... 787

$C_{19}H_{12}ClN$
3-Chloro-α-(1-naphthylmethylene)phenylacetonitrile ... 765
α-(4-Chlorophenyl)-β-(1-naphthyl)acrylonitrile ... 759

$C_{19}H_{12}Cl_2O$
3-(2,4-Dichlorophenyl)-2′-acrylonaphthone ... 299
3-(3,4-Dichlorophenyl)-1′-acrylonaphthone ... 305

$C_{19}H_{12}Cl_3NO_2$
3-Chloro-1,1′-biphenyl-4-yl *N*-(2,4-dichlorophenyl)carbamate ... 632

$C_{19}H_{12}Cl_4N_2O_5S$
5-Chloro-2-(4-chloro-2-(3,4-dichlorophenylureido)-phenoxy)benzenesulfonic acid, sodium salt ... 792

$C_{19}H_{12}Cl_4O_5S$
α,α-Bis-(3,5-dichloro-2-hydroxyphenyl)-*o*-toluenesulfonic acid, sodium salt ... 792

$C_{19}H_{12}F_2N_4O_5S$
1-(2,4-Difluorophenyl)-3-(2-(2,4-dinitrophenylthio)phenyl)urea ... 523

$C_{19}H_{12}F_6N_2O$
1-(3,5-Bis(trifluoromethyl)phenyl)-3-(1-naphthyl)urea ... 512

$C_{19}H_{12}N_2$
9-Methylacenaphtho(1,2-b)quinoxaline ... 1001

$C_{19}H_{12}N_2O_2$
α-(1-Naphthylmethylene)-5-nitro-*o*-tolunitrile ... 766

$C_{19}H_{12}N_2O_3$
5-(9-Anthracenylmethylene)barbituric acid ... 709

$C_{19}H_{12}N_2O_4$
2-Methyl-3-phthalimido-4-quinolinecarboxylic acid ... 993
Phthalimido *N*-(1-naphthyl)carbamate ... 722

$C_{19}H_{12}OS$
4-Benzoyldibenzothiophene ... 898

$C_{19}H_{12}O_2$
Benzo(c)phenanthrene-1-carboxylic acid ... 346
Benzo(c)phenanthrene-3-carboxylic acid ... 346
Benzo(c)phenanthrene-5-carboxylic acid ... 346
Benzo(c)phenanthrene-6-carboxylic acid ... 346
2-Benzoyldibenzofuran ... 895
7,12-Dihydro-10-methylbenz(a)anthracene-7,12-dione ... 319

$C_{19}H_{12}O_3$
2-Benzoyldibenzo-*p*-dioxin ... 311
7,12-Dihydro-10-methoxybenz(a)anthracene-7,12-dione ... 317
7,12-Dihydro-9-methoxybenz(a)anthracene-7,12-dione ... 317
Methyl 7-oxo-7*H*-benzo(c)fluorene-6-carboxylate ... 374

$C_{19}H_{12}O_3S$
2-(7-Oxo-7*H*-benzene(de)anthracen-3-ylthio)acetic acid ... 329

$C_{19}H_{12}O_4$
9,10-Dihydro-10-formyl-9,10-ethanoanthracene-11,12-dicarboxylic anhydride ... 388
6-Hydroxy-9-(2-hydroxyphenyl)-3*H*-xanthen-3-one ... 319

$C_{19}H_{13}Br$
4-Bromo-7-methylbenz(a)anthracene ... 175
5-Bromo-7-methylbenz(a)anthracene ... 175

$C_{19}H_{13}Cl$
4-Chloro-7-methylbenz(a)anthracene ... 175
7-(Chloromethyl)benzene(a)anthracene ... 172
9-Chloro-7-methylbenz(a)anthracene ... 175

$C_{19}H_{13}ClF_2N_2O_2$
1-(4-Chloro-2-phenoxyphenyl)-3-(2,5-difluorophenyl)urea ... 519

$C_{19}H_{13}ClN_2O$
9-(*N*-(4-Chlorophenyl)carbamoyl)carbazole ... 891

$C_{19}H_{13}ClN_2O_2$
4′-Chloro-2-hydroxy-3-carbazolecarboxanilide ... 891

$C_{19}H_{13}ClO$
4′-Chloro-3-(1-naphthyl)acrylophenone ... 306
2-Chloro-4′-phenylbenzophenone ... 310

$C_{19}H_{13}Cl_2NO$
N-(2,4-Dichlorobenzylidene)-4-phenoxyaniline ... 1024

$C_{19}H_{13}Cl_2NO_2$
4-Biphenylyl *N*-(3,4-dichlorophenyl)carbamate ... 632
4,6-Dichloro-α-(4-phenoxyphenylimino)-*o*-cresol ... 1031

$C_{19}H_{13}Cl_2N_3O$
3,4-Dichloro-4′-(phenylazo)benzanilide ... 1082

$C_{19}H_{13}Cl_5Si$
Diphenylmethyl-(pentachlorophenyl)silane ... 1102

$C_{19}H_{13}F$
10-Fluoro-7-methylbenz(a)anthracene ... 175
3-Fluoro-7-methylbenz(a)anthracene ... 175
1-Fluoro-12-methylbenzo(c)phenanthrene ... 174

$C_{19}H_{13}F_5Si$
Diphenylmethyl-(pentafluorophenyl)silane ... 1102

$C_{19}H_{13}N$
β-(1-Naphthyl)-α-phenylacrylonitrile ... 767
9-Phenylacridine ... 996
2-Phenylbenzo(*H*)quinoline ... 1001
9-(Phenylimino)fluorene ... 1077
6-Phenylphenanthridine ... 998
9-Phenylphenanthridine ... 1004
6-Phenylphenanthridine picrate ... 998
9-Phenylphenanthridine picrate ... 1004

$C_{19}H_{13}NO$
Benzo(c)phenanthrene-1-carboxamide ... 489
2-(1-Naphthyl)-5-phenyloxazole ... 872
9-Phenoxyacridine ... 995

$C_{19}H_{13}NOS$
10-Benzoyl-10*H*-phenothiazine ... 709
N-(3-Dibenzothienyl)benzamide ... 898

$C_{19}H_{13}NO_3$
3-(1-Naphthyl)-4′-nitroacrylophenone ... 301
3-(3-Nitrophenyl)-2′-acrylonaphthone ... 302

$C_{19}H_{13}NO_4$
2-Maleimido-9-fluorenyl acetate ... 722
4′-Nitro-4-biphenylyl benzoate ... 379

$C_{19}H_{13}NO_5$
4-(*N*-(1-Naphthyl)carbamoyl)isophthalic acid ... 474

$C_{19}H_{13}N_3O_5$
1,3-Diphthalimidoacetone oxime ... 1010

$C_{19}H_{13}N_3O_6$
Tris(4-nitrophenyl)methane ... 279

$C_{19}H_{13}N_5O_2$
2-(4,5-Di-(2-furyl)-2-imidazolylmethyleneamino)benzimidazole ... 1073

$C_{19}H_{13}N_5O_{13}$
Dimethyl 2-(*N*-(2,4,5,7-tetranitrofluoren-9-ylidene)aminoxy)succinate ... 1009

$C_{19}H_{14}$
1-Methylbenz(a)anthracene ... 167
10-Methylbenz(a)anthracene ... 167
2-Methylbenz(a)anthracene ... 167
3-Methylbenz(a)anthracene ... 167
4-Methylbenz(a)anthracene ... 167
5-Methylbenz(a)anthracene ... 167
6-Methylbenz(a)anthracene ... 167
7-Methylbenz(a)anthracene ... 167
9-Methylbenz(a)anthracene ... 167
1-Methylbenzo(c)phenanthrene ... 166
2-Methylbenzo(c)phenanthrene ... 166
3-Methylbenzo(c)phenanthrene ... 166
4-Methylbenzo(c)phenanthrene ... 166
5-Methylbenzo(c)phenanthrene ... 166

$C_{19}H_{14}BrNO_2$
4-Biphenylyl *N*-(4-bromophenyl)carbamate ... 632

$C_{19}H_{14}Br_2$
4,4′-Dibromotriphenylmethane ... 172

$C_{19}H_{14}Br_2O$
4,4′-Dibromotriphenylmethanol ... 198

$C_{19}H_{14}ClN$
2-(4-Chlorophenyl)-3-(1-naphthyl)propionitrile ... 751

$C_{19}H_{14}ClNO$
N-(2-Chlorobenzylidene)-4-phenoxyaniline ... 1046
N-(4-Chlorobenzylidene)-4-phenoxyaniline ... 1047
N-(α-(4-Chlorophenoxy)benzylidene)aniline ... 1024

$C_{19}H_{14}ClNO_2$
2-(4-Chlorobenzoyl)-*N*-methyl-1-naphthamide ... 484

$C_{19}H_{14}ClN_3O$
4-Chloro-4′-(phenylazo)benzanilide ... 1082

$C_{19}H_{14}ClN_3O_4$
1-(4-Chloro-2-phenoxyphenyl)-3-(4-nitrophenyl)urea ... 518

$C_{19}H_{14}Cl_2N_2$
N,N′-Bis-(4-chlorophenyl)benzamidine ... 724
N-(2,4-Dichlorobenzylidene)-*N′*-phenyl-1,4-phenylenediamine ... 1024

$C_{19}H_{14}Cl_2N_2O$
2-(1-Naphthyl)acetic (2,4-dichlorobenzylidene)hydrazide ... 494

$C_{19}H_{14}Cl_2N_2OS$
1-(3,4-Dichlorophenyl)-3-(4-phenoxyphenyl)-2-thiourea ... 729

$C_{19}H_{14}Cl_2O$
2,5-Bis(4-chlorobenzylidene)-1-cyclopentanone ... 299

$C_{19}H_{14}FNO$
N-(4-Fluorobenzylidene)-4-phenoxyaniline ... 1024

$C_{19}H_{14}FN_5O_3$
1-(2-Fluoro-5-nitrophenyl)-3-(4-(phenylazo)phenyl)urea ... 1081

$C_{19}H_{14}F_2N_2OS$
1-(2,4-Difluorophenyl)-3-(4-phenoxyphenyl)-2-thiourea ... 729

$C_{19}H_{14}F_3Sb$
Diphenyl(4-(trifluoromethyl)phenyl)antimony ... 1128

$C_{19}H_{14}N_2$
N-2-Fluorenyl-3-pyridinealdimine ... 1079
N-2-Fluorenyl-4-pyridinealdimine ... 1079

$C_{19}H_{14}N_2O$
3-Hydroxy-2-(phenylazo)fluorene ... 1085
2-(2-Methylphenyl)-5-(1-naphthyl)-1,3,4-oxadiazole ... 873
2-(3-Methylphenyl)-5-(1-naphthyl)-1,3,4-oxadiazole ... 874
2-(3-Methylphenyl)-5-(2-naphthyl)-1,3,4-oxadiazole ... 874
2-(4-Methylphenyl)-5-(1-naphthyl)-1,3,4-oxadiazole ... 874

$C_{19}H_{14}N_2O_2$
2-(3-Methoxyphenyl)-5-(2-naphthyl)-1,3,4-oxadiazole ... 873
2-(4-Methoxyphenyl)-5-(1-naphthyl)-1,3,4-oxadiazole ... 873

$C_{19}H_{14}N_2O_2S$
N-Cyano-4′-phenylbenzenesulfonanilide ... 826

$C_{19}H_{14}N_2O_4$
4-Biphenylyl *N*-(2-nitrophenyl)carbamate ... 668
4-Biphenylyl *N*-(4-nitrophenyl)carbamate ... 670
N,N′-Trimethylenediphthalimide ... 719

$C_{19}H_{14}N_3O_7$
5-Methylbenz(a)anthracene picrate ... 167

$C_{19}H_{14}N_4O$
1,3-Di-6-quinolylurea ... 989
5,8-Quinolinedione 5-((1,2-dihydro-1-methyl-2-quinolylidene)hydrazone) ... 954

$C_{19}H_{14}N_4O_4$
Benzophenone (2,4-dinitrophenyl)hydrazone ... 273
N,N′-Bis-(3-nitrophenyl)benzamidine ... 725
N,N′-Bis-(4-nitrophenyl)benzamidine ... 1057

$C_{19}H_{14}O$
5-Methoxybenz(a)anthracene ... 188
1-Methoxybenzo(c)phenanthrene ... 188
2-Methoxybenzo(c)phenanthrene ... 188
3-Methoxybenzo(c)phenanthrene ... 188
6-Methoxychrysene ... 188

$C_{19}H_{14}O_2$
6,7-Dihydrocyclopenta(1,2-b:1,5-b′)-bis(1)benzopyran 184
3,5-Diphenylbenzoic acid 345
2′-Hydroxy-2-(1-naphthylmethylene)-acetophenone. 300
$C_{19}H_{14}O_3$
3-(4-Methoxy-1-naphthyl)phthalide 384
$C_{19}H_{14}O_3S$
2-(Phenylsulfonyl)benzophenone 780
$C_{19}H_{14}O_4$
2-(4-Methoxybenzoyl)-1-naphthoic acid 340
$C_{19}H_{14}S$
9-(Phenylthio)fluorene 212
$C_{19}H_{15}BiCl_2$
Bis(4-chlorophenyl)(*o*-tolyl)-bismuthine 1141
$C_{19}H_{15}BrN_2O$
3-(4-Bromophenyl)-1,1-diphenylurea. 536
$C_{19}H_{15}BrN_2O_2$
4-Anilinophenyl *N*-(4-bromophenyl)-carbamate 627
$C_{19}H_{15}BrN_4$
2,3,5-Triphenyl-2*H*-tetrazolium bromide 870
$C_{19}H_{15}Br_4NO_5$
4′-(Butoxycarbonyl)-3,4,5,6-tetrabromophthalanilic acid 453
$C_{19}H_{15}ClN_2$
N-(4-Chlorobenzylidene)-*N*′-phenyl-1,4-phenylenediamine 1024
$C_{19}H_{15}ClN_2O$
2-(1-Naphthyl)acetic 2-chlorobenzylidenehydrazide 494
2-(1-Naphthyl)acetic 4-chlorobenzylidenehydrazide 494
$C_{19}H_{15}ClN_2O_2$
1-(2-Chlorobenzoyl)-2-(1-naphthylacetyl)hydrazine 493
$C_{19}H_{15}ClO$
2-Chlorotriphenylmethanol 198
4-Chlorotriphenylmethanol 198
$C_{19}H_{15}Cl_2N_3O$
1-(3,4-Dichlorophenyl)-3-(phenylaminophenyl)urea 525
$C_{19}H_{15}Cl_2N_3O_2$
N,*N*′-(4-(α-Cyano-2,4-dichlorostyryl)-1,2-phenylene)-bisacetamide 767
$C_{19}H_{15}Cl_3IP$
Methyltris(4-chlorophenyl)-phosphonium iodide 832
$C_{19}H_{15}Cl_4FeO_2$
6,7-Dihydrocyclopenta(1,2-b:1,5-b′)-bis(1)benzopyrylium tetrachloroferrate 183
$C_{19}H_{15}Cl_6NO_2$
1,4,5,6,7,7-Hexachloro-*N*-(5-isopropyl-2-methylphenyl)-5-norbornene-2,3-dicarboximide 722
$C_{19}H_{15}FN_2O$
4′-Anilino-3-fluorobenzanilide 451
N-(4-Anilinophenyl)-4-fluorobenzamide 451
$C_{19}H_{15}FN_2O_5$
2-(2-(Naphthoxy)ethyl *N*-(2-fluoro-5-nitrophenyl)carbamate 620
$C_{19}H_{15}FO$
2-Fluorotriphenylmethanol 198
$C_{19}H_{15}F_2N_3O$
1-(4-Anilinophenyl)-3-(2,5-difluorophenyl)urea 527
$C_{19}H_{15}F_2N_3S$
1-(4-Anilinophenyl)-3-(2,4-difluorophenyl)-2-thiourea 730
$C_{19}H_{15}F_3IP$
Methyltris(4-fluorophenyl)-phosphonium iodide 832
$C_{19}H_{15}N$
N-Benzylidene-4-biphenylamine 1079
5,6-Dihydro-6-phenylphenanthridine 252
N-(Diphenylmethylene)aniline 1077
3-(2-Naphthyl)-2-phenylpropionitrile. 756
2-(4-Phenyl-1,3-butadien-1-yl)-quinoline 994
$C_{19}H_{15}NO$
2-Acetyl-2-cyanodibenzo(e,h)-bicyclo(2.2.2)octa-5,7-diene 757
2-Anilinobenzophenone 312
N-Benzylidene-4-phenoxyaniline 1050
10-Benzylphenoxazine 252
N,*N*-Diphenylbenzamide 482
N-(1-Naphthyl)cinnamamide 478
N-(2-Naphthyl)cinnamamide 489
$C_{19}H_{15}NOSi$
Triphenylsilyl isocyanate 1114
$C_{19}H_{15}NO_2$
1-Benzoyl-1,2,2a,3-tetrahydrobenz(cd)indole-5-carboxaldehyde 706
1-(Diphenylmethyl)-4-nitrobenzene 279
3′-(2-Hydroxy-1-naphthylmethyleneamino)-acetophenone 1074
4′-(2-Hydroxy-1-naphthylmethyleneamino)-acetophenone 1037
4-(β-Methylcinnamylidene)-2-phenyl-2-oxazolin-5-one 387
α-(4-Phenoxyphenylimino)-*o*-cresol 1031
α-(4-Phenoxyphenylimino)-*p*-cresol 1032
$C_{19}H_{15}NO_2S$
N-(2-Mercaptophenyl)-*N*-phenylanthranilic acid 341
$C_{19}H_{15}NO_2S_2$
4-Nitrobenzaldehyde diphenyl mercaptal 278
$C_{19}H_{15}NO_3$
Diphenyl(4-nitrophenyl)carbinol 263
2-Naphthyl *N*-(3-acetylphenyl)-carbamate 675
$C_{19}H_{15}NO_5$
4-(2-Acetoxy-3-methoxybenzylidene)-2-phenyl-2-oxazolin-5-one 383
4-Methyl-7-coumarinyl *N*-(3-acetylphenyl)carbamate 666
$C_{19}H_{15}NS$
10-Benzylphenothiazine 252
$C_{19}H_{15}NSSi$
Triphenylsilyl isothiocyanate 1110
$C_{19}H_{15}NSSn$
Triphenyltin isothiocyanate 1131
$C_{19}H_{15}N_3O_2$
N-(3-Nitrobenzylidene)-*N*′-phenyl-1,4-phenylenediamine 1057
α-(2-Nitrobenzylidene)-*N*-phenyl-2-pyridinemethylamine 943
α-(4-Nitrobenzylidene)-*N*-phenyl-2-pyridinemethylamine 953
$C_{19}H_{15}N_3O_3$
2-(1-Naphthyl)acetic 3-nitrobenzylidenehydrazide 501
$C_{19}H_{15}N_3O_5S$
1-(4-Biphenylyl)-3-(2-nitrophenylsulfonyl)urea 819
2-(2-Methoxybenzoyl)-*N*-(3-(3-nitrophenyl)-2-thiazolyl)acetamide 875
$C_{19}H_{16}$
2-Methyl-*p*-terphenyl 165
$C_{19}H_{16}BrNO$
1-Phenacyl-2-phenylpyridinium bromide 953
$C_{19}H_{16}BrNO_4$
6-Bromo-2-naphthyl *N*-(2,5-dimethoxyphenyl)carbamate 648
$C_{19}H_{16}ClF_2NO$
1-(4-Fluorobenzyl)-3-(4-fluorobenzyloxy)pyridinium chloride 926
$C_{19}H_{16}ClNO_2$
2-Chloro-3-(2,4,6-trimethylanilino)-1,4-naphthoquinone 318
$C_{19}H_{16}ClNO_3$
4′-Chloro-2′,5′-dimethoxy-1-naphthanilide 484
1-(5-Chloro-2,4-dimethoxyphenyliminomethyl)-2-naphthol 1037
$C_{19}H_{16}ClNO_4$
5′-Chloro-2′,4′-dimethoxy-2-hydroxy-3-naphthoylanilide 485
2-Naphthyl *N*-(5-chloro-2,4-dimethoxyphenyl)carbamate 614
$C_{19}H_{16}ClNO_5$
4-(2-Chlorobenzylidene)-2-(3,4,5-trimethoxyphenyl)-2-oxazolin-5-one. 382
$C_{19}H_{16}ClN_3O$
1-(4-Anilinophenyl)-3-(3-chlorophenyl)urea 526
$C_{19}H_{16}ClO_3P$
Diphenyl α-chlorobenzylphosphonate 836
$C_{19}H_{16}Cl_2F_3NO_4$
2-(4-(2-Chlorovinyl)-2-methoxyphenoxy)ethyl *N*-(4-chloro-2-(trifluromethyl)phenyl)carbamate 606
$C_{19}H_{16}Cl_2Si$
(Dichloromethyl)triphenylsilane 1100
$C_{19}H_{16}FN_3O$
1-(4-Anilinophenyl)-3-(2-fluorophenyl)urea 527
1-(4-Anilinophenyl)-3-(3-fluorophenyl)urea 525
1-(4-Anilinophenyl)-3-(4-fluorophenyl)urea 525
$C_{19}H_{16}GeO_2$
Triphenylgermanecarboxylic acid 1109
$C_{19}H_{16}INO_2$
1-Methyl-2-(3,4-(methylenedioxy)-styryl)-quinolinium iodide 987
1-(4′-Phenoxyphenacyl)pyridinium iodide 938
$C_{19}H_{16}N_2$
Benzophenone phenylhydrazone 235
N-Benzylidene-*N*′-phenyl-1,4-phenylenediamine 1057
9,9-Bis(2-cyanoethyl)fluorene 758
$C_{19}H_{16}N_2O$
5-(α-Cyanobenzylidene)-3-methyl-4-phenyl-2-pyrrolidinone 763
N^1,N^2-Diphenylbenzoic hydrazide 503
N-(2-Hydroxybenzylidene)-*N*′-phenyl-1,4-phenylenediamine 1033
4-Methoxy-4′-phenylazobenzene 1087
4-Methylbenzoic (1-naphthylmethylene)hydrazide 501
1,1,3-Triphenylurea 586
$C_{19}H_{16}N_2O_2$
1-(3-Acetylphenyl)-3-(1-naphthyl)urea 575
3-Hydroxy-2-naphthoic (2-phenylethylidene)hydrazide 496
Mandelic (1-naphthylmethylene)-hydrazide 495
4-Methylbenzoic (2-hydroxy-1-naphthylmethylene)hydrazide 496
1-(3-Methylbenzoyl)-2-(2-naphthoyl)-hydrazine 504
1-(4-Methylbenzoyl)-2-(1-naphthoyl)-hydrazine 500
1-(4-Methylbenzoyl)-2-(2-naphthoyl)-hydrazine 500
$C_{19}H_{16}N_2O_3$
4-Acetamidophenyl *N*-(1-naphthyl)-carbamate 677
1-(3-Acetylphenyl)-3-(5-hydroxy-1-naphthyl)urea 548
2-Methoxybenzoic (2-hydroxy-1-naphthylmethylene)hydrazide 495
1-(3-Methoxybenzoyl)-2-(1-naphthoyl)hydrazine 498
1-(4-Methoxybenzoyl)-2-(1-naphthoyl)hydrazine 498
3-(Methylcarbamoyloxy)-2-naphthanilide 489
$C_{19}H_{16}N_2O_4$
8-Quinolyl *N*-(4-(ethoxycarbonyl)-phenyl)carbamate 991
$C_{19}H_{16}N_2O_5$
N-Phthaloylglycyl-DL-phenylalanine. 691
N-Phthaloyl-DL-phenylalanylglycine. 719
$C_{19}H_{16}N_2O_7$
4-(3-Nitrobenzylidene)-2-(3,4,5-trimethoxyphenyl)-2-oxazolin-5-one. 383
$C_{19}H_{16}N_2O_8$
Dimethyl (2-(4-nitrobenzoyl)-acetamido)terephthalate 443
$C_{19}H_{16}N_2S$
1,1,3-Triphenyl-2-thiourea 741
$C_{19}H_{16}N_4$
Triphenylformazan 1092
$C_{19}H_{16}N_4O_2$
3-Methyl-4′-(4-nitrophenylazo)-diphenylamine 1089
$C_{19}H_{16}N_4O_3$
3-Methoxy-4′-(4-nitrophenylazo)-diphenylamine 1086
$C_{19}H_{16}N_4O_4$
N,*N*′-(4-Methyl-1,3-phenylene)bis(2-pyridyl carbamate) 948
N,*N*′-(4-Methyl-1,3-phenylene)bis(3-pyridyl carbamate) 949
$C_{19}H_{16}O$
2-(Cinnamylidene)-1-tetralone 304
2,5-Dibenzylidenecyclopentanone 287
9,10,14,15-Tetrahydro-13*H*-9,10-cyclopentanthracen-13-one 298
$C_{19}H_{16}O_2$
Ethyl 9,10-dihydro-9,10-ethenoanthracene-11-carboxylate 376
5,6,6a,7,8,12b-Hexahydro-1-methylbenzo(c)phenanthrene-5,8-dione 309
2-Hydroxytriphenylcarbinol 199
$C_{19}H_{16}O_2S$
Diphenylmethyl phenyl sulfone 785
2-Naphthyl 2-phenylcyclopropyl sulfone 782
$C_{19}H_{16}O_2Si$
Triphenylsilylcarboxylic acid 1109
$C_{19}H_{16}O_3$
4-(2-Anthranoyl)butyric acid 331
2,5-Bis(2-hydroxybenzylidene)-1-cyclopentanone 282
2,5-Bis(2-hydroxybenzylidene)-1-cyclopentanone, disodium salt 282
α-(4-Methoxy-1-naphthyl)-*o*-toluic acid 341
$C_{19}H_{16}O_4$
cis-3,5-Cyclopentenediol dibenzoate 371
4-Hydroxy-3-(3-oxo-1-phenylbutyl)-coumarin 382
$C_{19}H_{16}O_4S$
3-(2-Furyl)-3-(phenylsulfonyl)-propiophenone 852
$C_{19}H_{16}O_5$
10,11-Diacetoxy-10,11-dihydro-5*H*-dibenzo(a,d)cyclohepten-5-one 357
$C_{19}H_{16}S_2$
α,α-Bis(phenylthio)toluene 212
$C_{19}H_{17}BF_4O$
4-Ethyl-2,6-diphenylpyrylium tetrafluoroborate 935
$C_{19}H_{17}BiCl_2$
Dichlorodiphenyl(*p*-tolyl)bismuth 1143
$C_{19}H_{17}ClGe$
(Chloromethyl)triphenylgermane 1100
$C_{19}H_{17}ClHgO_2$
12-(Chloromercurimethyl)-10,11-dihydro-5,10-methano-5*H*-dibenzo(a,d)cyclohepten-11-yl acetate 1138
$C_{19}H_{17}ClIN$
2-(4-Chlorostyryl)-1-ethylquinolinium iodide 981
$C_{19}H_{17}ClN_2O_2$
2-(4′-Acetamidophenacyl)-isoquinolinium chloride 997
$C_{19}H_{17}ClN_2O_4$
2-(8-Quinolyloxy)ethyl *N*-(4-chloro-2-methoxyphenyl)carbamate 981
$C_{19}H_{17}ClO$
2-(4-Chlorobenzylidene)-4,4-dimethyl-α-tetralone 299
$C_{19}H_{17}ClO_3$
2-(4-Chlorobenzyl)-1-oxo-2-indanpropionic acid 325
$C_{19}H_{17}ClO_5$
2,4-Diphenyl-6-ethylpyrylium perchlorate 935
$C_{19}H_{17}ClPb$
Benzylchlorodiphenyllead 1136
$C_{19}H_{17}Cl_2N$
N-(2,6-Dichlorobenzyl)-*N*-methyl-1-naphthalenemethylamine 236
$C_{19}H_{17}FN_2O$
10-(3-Fluorobenzyl)-2,3,4,4a-tetrahydrobenzo(b)naphthyridin-

$C_{19}H_{17}FN_2O$ (Cont'd)
5(10*H*)-one ... 724
1-(4-Fluorophenyl)-4-isopropyl-2-phenylimidazole-5-carboxaldehyde. 865

$C_{19}H_{17}IN_2O_2$
1-Ethyl-2-(3-nitrostyryl)quinolinium iodide ... 990

$C_{19}H_{17}IN_2O_5S$
N-(3-(4-Iodophenyl)-4-phenyl-2-thiazolidinylidene)crotonamide ... 743

$C_{19}H_{17}N$
N-Benzhydrylaniline ... 232
2,4-Dibenzylpyridine ... 953
N,*N*-Diphenylbenzylamine ... 233
2,4-Diphenyl-6-methylaniline ... 230
2-Ethyl-*N*-(1-naphthylmethylene)aniline ... 1077

$C_{19}H_{17}NO$
3-Benzyloxydiphenylamine ... 227
4-Benzyloxydiphenylamine ... 227
α-(4-Biphenylyl)-α-methyl-3-pyridinemethanol ... 931
5,6-Dihydro-5,5-dimethylbenz(c)acridin-7-(12*H*)-one ... 320
3,3-Diphenyl-2-oxo-1-cyclohexanecarbonitrile ... 757
α,α-Diphenyl-2-pyridineethanol ... 932
α,α-Diphenyl-3-pyridineethanol hydrochloride ... 932
1,2-Diphenyl-2-(2-pyridyl)ethanol ... 931
4-Ethyl-*N*-(2-hydroxy-1-naphthylmethylene)aniline ... 1036
N-Ethyl-*N*-(1-naphthyl)benzamide ... 489
N-Methyl-*N*-(2-naphthyl)-*o*-toluamide ... 489
3-(2-Naphthylamino)propiophenone. 294
N-(1-Naphthyl)-3-phenylpropionamide ... 489
1-(2,4-Xylyliminomethyl)-2-naphthol.. 1039
1-(2,5-Xylyliminomethyl)-2-naphthol.. 1039
1-(3,4-Xylyliminomethyl)-2-naphthol.. 1039

$C_{19}H_{17}NO_2$
2-(4-Acetamidobenzylidene)-1-tetralone ... 433
4-Cyclopropylformyl-4′-methoxydiphenylacetonitrile ... 753
2′,4′-Dimethyl-2-hydroxy-3-naphthoylanilide ... 486
6-(2,2-Diphenyl-2-hydroxyethyl)-2(1*H*)-pyridone ... 931
1-(2-Ethoxyphenyliminomethyl)-2-naphthol ... 1032
1-(4-Ethoxyphenyliminomethyl)-2-naphthol ... 1032
Ethyl 6-methyl-2-phenylquinolinecarboxylate ... 993
3-Hydroxy-2-naphtho-2′,3′-xylidide.... 486
2′-Methoxy-5′-methyl-1-naphthanilide ... 465
1-(2-Methoxy-5-methylphenyliminomethyl)-2-naphthol ... 1032
6-Methoxy-2-phenyl-4-propionylquinoline ... 986
1-Naphthyl *N*-(2,5-dimethylphenyl)carbamate ... 686
1-Naphthyl *N*-(2,6-xylyl)carbamate.... 685
2-Naphthyl *N*-(2,5-xylyl)carbamate.... 686
2-Naphthyl *N*-(2,6-xylyl)carbamate.... 685
2,6-Xylyl *N*-(1-naphthyl)carbamate.... 685
3,5-Xylyl *N*-(1-naphthyl)carbamate.... 687

$C_{19}H_{17}NO_2S$
3-Methyl-4-(methylthio)phenyl *N*-(1-naphthyl)carbamate ... 668
N-Phenyl-*p*-toluenesulfonanilide ... 828

$C_{19}H_{17}NO_3$
1-Acetyl-1,2-dihydro-2-(5-indanyl)-3,1,4-benzoxazin-4-one ... 707
2′,5′-Dimethoxy-1-naphthalenecarboxanilide ... 486
4,4-Dimethyl-2-(2-nitrobenzylidene)-1-tetralone ... 303
Ethyl 3-benzyloxy-α-cyanocinnamate ... 762
Ethyl 4-benzyloxy-α-cyanocinnamate ... 761
Ethyl 6-methoxy-2-phenylquinoline-4-carboxylate ... 985
3-Hydroxy-2-naphthoyl-*o*-phenetidide ... 485
2-(2,2,2-Trimethylacetamido)anthraquinone ... 488

$C_{19}H_{17}NO_4$
N-Benzyl-2-(2,3-dimethoxyphenyl)maleimide ... 716
3,4-Dihydro-4,4-dimethyl-3-(3-nitrophenyl)spiro(naphthalene-2(1*H*)-2′-oxirane)-1-one ... 295
2-(3,4-Dimethoxyphenyl)-7-methylcinchoninic acid ... 985
2-Naphthyl *N*-(2,5-dimethoxyphenyl)carbamate ... 648
4-Phthalimidobutyl benzoate ... 721

$C_{19}H_{17}NO_5$
4-(2,4,5-Trimethoxybenzylidene)-2-phenyl-2-oxazolin-5-one ... 383

$C_{19}H_{17}NO_6$
4-(2-Hydroxybenzylidene)-2-(3,4,5-trimethoxyphenyl)-2-oxazolin-5-one. 382
4-Methyl-7-coumarinyl *N*-(2,5-dimethoxyphenyl)carbamate ... 647

$C_{19}H_{17}N_2$
Sempervirine nitrate ... 1003

$C_{19}H_{17}N_2O_2P$
Diphenylphosphinic (2-hydroxybenzylidene)hydrazide.... 496

$C_{19}H_{17}N_2O_3PS$
Diphenyl *N*-(*N*-(phenyl)thiocarbamoyl)phosphoramidate.... 835

$C_{19}H_{17}N_3$
4-(*p*-Tolylazo)diphenylamine ... 1089
1,2,3-Triphenylguanidine ... 727

$C_{19}H_{17}N_3O$
4-Aminobenzoic (1-(2-naphthyl)ethylidene)hydrazide ... 1052
3-(5-Indanylmethyleneamino)-2-methyl-4(3*H*)-quinazolone ... 1068

$C_{19}H_{17}N_3O_2$
1-(3-Acetamidophenyl)-3-(1-naphthyl)urea ... 565
5-(α-Aminostyryl)-*N*-methyl-3-phenylisoxazole-4-carboxamide ... 871
N-Methylanthranilic (2-hydroxy-1-naphthylmethylene)hydrazide ... 496

$C_{19}H_{17}N_3O_3$
4-(Piperonylideneamino)antipyrine ... 1054

$C_{19}H_{17}N_3O_3S$
4-(4-(Benzylamino)-phenylazo)benzenesulfonic acid, potassium salt ... 1089

$C_{19}H_{17}N_3O_5$
Ethyl α-cyano-3-methoxy-β-(4-nitrophenylimino)hydrocinnamate.. 1043

$C_{19}H_{17}N_3O_5S$
1-Methyl-5-(α-(phenylsulfonyl)-2-nitrohydrocinnamoyl)imidazole ... 868

$C_{19}H_{18}AsI$
(Methyl)-triphenylarsonium iodide.... 1127

$C_{19}H_{18}BrN$
1-Benzhydryl-4-methylpyridinium bromide ... 957

$C_{19}H_{18}BrNO_2S$
1-Benzyl-2-(phenylsulfonylmethyl)pyridinium bromide ... 955

$C_{19}H_{18}ClFN_2$
5-Chloromethyl-1-(4-fluorophenyl)-4-isopropyl-2-phenylimidazole ... 864
5-Chloromethyl-1-(4-fluorophenyl)-4-isopropyl-2-phenylimidazole hydrochloride ... 864

$C_{19}H_{18}ClFN_2O_4S_2$
4-Chloro-2-(4-(*N*,*N*-diethylaminosulfonyl)-phenyl)-6-quinolinesulfonyl fluoride ... 994

$C_{19}H_{18}ClF_3N_2O_3S$
Ethyl 2-(3-(2-chloro-5-(trifluoromethyl)phenyl)ureido)-4,5,6,7-tetrahydrobenzo(b)thiophene-3-carboxylate ... 855

$C_{19}H_{18}ClNO_2$
4-(4-Chlorobutyryl)-4′-methoxydiphenylacetonitrile ... 751

$C_{19}H_{18}ClNO_4$
3-(4-Chlorophenyl)-3-hydroxy-2-(tetrahydropyran-2-yloxy)-1-isoindolinone ... 689

$C_{19}H_{18}ClN_3O$
Cresyl Fast Violet ... 1095

$C_{19}H_{18}ClN_3O_2$
1-(4-Chloro-7-methoxy-2-quinolyl)-3-(2,6-xylyl)urea ... 986
4′-Chloro-4-oxo-4-(4-phenyl-2-pyrazolin-5-yl)butyranilide ... 404

$C_{19}H_{18}ClOP$
(Hydroxymethyl)-triphenylphosphonium chloride ... 830

$C_{19}H_{18}Cl_2N_2O_3$
1,3-Bis-(2-chloroethyl)-1,3-dibenzoylurea ... 504

$C_{19}H_{18}Cl_2N_2O_3S$
Butyl 4-(3-(2,4-dichlorobenzoyl)-2-thioureido)benzoate ... 731

$C_{19}H_{18}Cl_2O_4$
Dimethyl 2,2-bis(4-chlorobenzyl)-malonate ... 353

$C_{19}H_{18}F_3NO_4$
4-(Butoxycarbonyl)phenyl *N*-(2-(trifluoromethyl)phenyl)carbamate.. 610

$C_{19}H_{18}F_3NS$
9-(3-(Diethylamino)propylidene)-2-(trifluoromethyl)thioxanthene ... 244

$C_{19}H_{18}IN$
1-Ethyl-2-styrylquinolinium iodide.... 990
1-Methyl-2-(4-methylstyryl)-quinolinium iodide ... 990

$C_{19}H_{18}INO$
1-Ethyl-2-(2-hydroxystyryl)-quinolinium iodide ... 983
2-(4-Methoxystyryl)-1-methylquinolinium iodide ... 985

$C_{19}H_{18}NO_3P$
Diphenyl *N*-benzylphosphoramidate ... 835

$C_{19}H_{18}N_2$
2-(4-(Dimethylamino)styryl)-quinoline ... 990
N,*N*-Dimethyl-α-(1-naphthylimino)-*p*-toluidine ... 1062
2-(1,2-Diphenylethylamino)-pyridine. 953
α-Phenyl-*N*-(3-pyridyl)-phenethylamine ... 943

$C_{19}H_{18}N_2O$
4,6-Di(*p*-tolyl)-1-methyl-2(1*H*)-pyrimidinone ... 713
1-(2-Ethylphenyl)-3-(1-naphthyl)urea.. 589
1-(4-Ethylphenyl)-3-(1-naphthyl)urea.. 592
1-(1-Naphthyl)-3-(2-phenylethyl)urea. 570
1-(1-Naphthyl)-3-(2,3-xylyl)urea ... 596
1-(1-Naphthyl)-3-(2,4-xylyl)urea ... 602
1-(1-Naphthyl)-3-(2,5-xylyl)urea ... 602
1-(1-Naphthyl)-3-(2,6-xylyl)urea ... 596

$C_{19}H_{18}N_2O_2$
1-(2-Ethoxyphenyl)-3-(1-naphthyl)urea ... 555
1-(4-Ethoxyphenyl)-3-(1-naphthyl)urea ... 560

$C_{19}H_{18}N_2O_2S$
4′-Anilino-*p*-toluenesulfonanilide ... 819
5,6-Dihydro-8,9-dimethoxy-3-(phenylimino)-3*H*-thiazolo(4,3-a)-isoquinoline hydrochloride ... 735

$C_{19}H_{18}N_2O_3$
N-(2-Acetylphenyl)-5-methyl-3-phenyl-4-isoxazolin-4-carboxamide. 432
5-Benzyloxy-*N*,*N*-dimethyl-3-indoleglyoxylamide ... 881
1,3-Diphenyl-1-(2-oxo-1-cyclopentylcarbonyl)urea ... 575
1-(4-Methoxyphenyl)-1,2,3,4-tetrahydronorharman-3-carboxylic acid ... 919
2-Methyl-8-quinolyl *N*-(2-methoxy-5-methylphenyl)carbamate ... 986

$C_{19}H_{18}N_2O_4$
Carbobenzyloxy-DL-tryptophan ... 883

$C_{19}H_{18}N_2O_4S$
Ethyl 4-(2-(2-aminophenylthio)-succinimido)benzoate ... 718

$C_{19}H_{18}N_2O_4S_2$
N,*N*′-(2-Methyl-1,4-phenylene)-bis(benzenesulfonamide) ... 819
N,*N*′-(4-Methyl-1,2-phenylene)-bis(benzenesulfonamide) ... 818

$C_{19}H_{18}N_2O_5$
4′-(2-Benzoylacetamido)-succinanilic acid ... 443

$C_{19}H_{18}N_2O_5S$
1-(3-Nitrobenzyl)pyridinium *p*-toluenesulfonate ... 951

$C_{19}H_{18}N_2O_6$
N,*N*′-Bis(2-furylmethoxycarbonyl)-4-methyl-1,3-phenylenediamine ... 847
Carbobenzyloxy-L-proline 4-nitrophenyl ester ... 706
Dimethyl 5-(2-(3-aminobenzoyl)-acetamido)isophthalate ... 443

$C_{19}H_{18}N_2O_8$
3,9-Bis(3-nitrophenyl)-2,4,8,10-tetraoxaspiro(5.5)undecane ... 269

$C_{19}H_{18}N_4$
N,*N*-Bis(2-cyanoethyl)-α-phenylimino-*p*-toluidine ... 1055

$C_{19}H_{18}N_4O$
4-(4-(2,6-Dimethylmorpholino)-phenylazo)benzonitrile ... 1091

$C_{19}H_{18}N_4O_5$
6-Phenyl-4-(3,4,5-trimethoxybenzylideneamino)-1,2,4-triazine-3,5-(2*H*,4*H*)-dione ... 1042

$C_{19}H_{18}N_6O_2$
1,1′-(4-Methyl-1,3-phenylene)bis(3-(3-pyridyl)urea) ... 943

$C_{19}H_{18}O$
2-Benzyl-5-benzylidene-1-cyclopentanone ... 304
2-Benzylidene-4,4-dimethyl-1-tetralone ... 305
Bis-(4-methylstyryl) ketone ... 286

$C_{19}H_{18}OS_2$
O-(9,10-Dihydro-9,10-ethanoanthracen-11-ylmethyl) *S*-methyl dithiocarbonate ... 382

$C_{19}H_{18}OSi$
Diphenyl(4-methoxyphenyl)silane.... 1106
2-(Diphenylmethylsilyl)phenol ... 1119
Methoxytriphenylsilane ... 1098
Methyl(4-phenoxyphenyl)-phenylsilane ... 1106

$C_{19}H_{18}O_2$
9,10-Dihydro-9,10-ethanoanthracene-11-methanol acetate ... 361
2,5-Dimethyl-3,4-diphenyl-4-hydroxy-2-cyclopentenone ... 282
2-(2-Ethoxybenzylidene)-1-tetralone.. 301
2-(4-Ethoxybenzylidene)-1-tetralone.. 301
Ethyl 9,10-dihydro-9,10-ethanoanthracene-11-carboxylate... 361
3-Mesitoylacrylophenone ... 308
3-(4-Methoxyphenyl)-5-phenyl-2-cyclohexen-1-one ... 283
Methyl 9,10-dihydro-11-methyl-9,10-ethanoanthracene-11-carboxylate... 361

$C_{19}H_{18}O_2S$
Mesityl 2-naphthyl sulfone ... 786

$C_{19}H_{18}O_2SSi$
(Methylsulfonyl)triphenylsilane ... 1114

$C_{19}H_{18}O_3$
4-Allyl-2-methoxyphenyl cinnamate. 378
Bis(4-methoxybenzylidene)acetone.. 302
9,10-Dihydro-12-ethoxy-9,10-ethanoanthracene-11-carboxylic acid ... 328
2-(2,3-Dimethoxybenzylidene)-1-tetralone ... 300
3-(1-(4-Ethylphenyl)ethyl)-4-hydroxycoumarin ... 382
Methyl 3-(9,10-dihydro-2-phenanthroyl)propionate ... 360

$C_{19}H_{18}O_3S$
1-Naphthyl mesitylenesulfonate ... 801

$C_{19}H_{18}O_4$
2-(4-Methoxybenzylidene)-4-(4-methoxyphenyl)-3-butenoic acid ... 335

$C_{19}H_{18}Pb$
Methyltriphenyllead ... 1136

$C_{19}H_{18}PbS$
(Methylthio)triphenyllead ... 1136

$C_{19}H_{18}SSi$
(Methylthio)triphenylsilane ... 1114

$C_{19}H_{18}Si$
Benzyldiphenylsilane 1113
(2-Biphenylyl)methylphenylsilane 1120
(3-Biphenylyl)methylphenylsilane 1121
(4-Biphenylyl)methylphenylsilane 1121
Diphenyl(*m*-tolyl)silane 1113
Diphenyl(*p*-tolyl)silane 1120
Methyltriphenylsilane 1113

$C_{19}H_{19}BrF_3NO_2$
4-Bromo-2-*tert*-butyl-6-methylphenyl *N*-(3-(trifluoromethyl)phenyl)carbamate 608

$C_{19}H_{19}BrO_3$
1,2-Diphenyl-2-oxoethyl 5-bromovalerate 362

$C_{19}H_{19}ClHgO$
12-(Chloromercuri)-10,11-dihydro-5,11-dimethyl-11-methoxy-5,10-methano-5*H*-dibenzo(a,d)-cycloheptene 1138

$C_{19}H_{19}ClN_2$
6-Chloro-2-(4-(diethylamino)phenyl)-quinoline 990

$C_{19}H_{19}ClN_2OS$
4′-Chloro-3-morpholino-*N*-phenylthiocinnamamide 732

$C_{19}H_{19}ClN_2O_2$
1-(4-Chlorophenyl)-1,5,6,7,8,8a-hexahydro-1-hydroxy-2-phenylimidazo(1,5-a)pyridin-3(2*H*)one 694

$C_{19}H_{19}ClN_2O_4$
2-(2-Anilinovinyl)-1-ethylquinolinium perchlorate 1054
4-Cyclohexylphenyl *N*-(4-chloro-2-nitrophenyl)carbamate 624

$C_{19}H_{19}ClN_4O_2$
1-(2,3-Dihydro-1,5-dimethyl-3-oxo-2-phenyl-4-pyrazolyl)-3-(3-chloro-2-methylphenyl)urea 695

$C_{19}H_{19}Cl_2NO_2$
6-Chloro-3-(4-chlorophenyl)-3-(2-(dimethylamino)ethyl)-1*H*-2-benzpyran-1-one hydrochloride 382
4-Cyclohexylphenyl *N*-(3,4-dichlorophenyl)carbamate 631

$C_{19}H_{19}Cl_2NO_4$
2-(Isopentyloxycarbonyl)phenyl *N*-(2,3-dichlorophenyl)carbamate 622
2-(2-Methoxy-4-(1-propenyl)-phenoxy)ethyl *N*-(2,6-dichlorophenyl)carbamate 616

$C_{19}H_{19}Cl_3F_3N_3O_2$
(3,4-Dichlorophenylcarbamoyl-methyl)-dimethyl-(3-(trifluoro-methyl)phenylcarbamoylmethyl ammonium chloride 397

$C_{19}H_{19}Cl_3O_4$
3,3-Bis(3,4-dimethoxyphenyl)-1,1,2-trichloro-1-propene 178

$C_{19}H_{19}FN_2$
1-Benzyl-4-(4-fluorophenyl)-isonipecotonitrile 755
1-(4-Fluorophenyl)-4-isopropyl-5-methyl-2-phenylimidazole 865

$C_{19}H_{19}FN_2O$
1-(4-Fluorophenyl)-5-(hydroxymethyl)-4-isopropyl-2-phenylimidazole 864

$C_{19}H_{19}FN_2O_4$
4-Cyclohexylphenyl *N*-(4-fluoro-3-nitrophenyl)carbamate 625

$C_{19}H_{19}IN_2OS$
2-(2-Acetanilinovinyl)-3-ethylbenzothiazolium iodide 915

$C_{19}H_{19}IN_2O_2$
2-(2-*N*-Acetanilinovinyl)-benzoxazole ethiodide 909

$C_{19}H_{19}IN_2O_2S$
5-Isopropyl-6-(methylthio)-8-phenyl-1,3-dioxolo(4,5-g)quinazolin-5-ium iodide 999

$C_{19}H_{19}IN_2O_3$
1-Isopropyl-3-methyl-6,7-(methylenedioxy)-2-oxo-4-phenyl-3-quinazolinium iodide 705

$C_{19}H_{19}NO$
2-(2-Aminobenzylidene)-4,4-dimethyl-1-tetralone 302
2′-Methylspiro(benzosuberan-5,1′-isoindolin)-3′-one 692
1-(2,4-Xylidinomethyl)-2-naphthol 239
1-(2,5-Xylidinomethyl)-2-naphthol 227
1-(3,4-Xylidinomethyl)-2-naphthol 227

$C_{19}H_{19}NO_2$
1-(2-(Benzylcarbonyl)benzoyl)-pyrrolidine 706
4-(Dimethylamino)-α-(4-hydroxyphenyl)-1-naphthalenemethanol 225
Ethyl *N*-(9,10-dihydro-9,10-ethanoanthracen-11-yl)carbamate 671
N-(1-Indanyl)-4-oxo-4-phenylbutyramide 432
2-(4-Isopropyl-3-methylphenyl)-1,3-(2*H*,4*H*)-isoquinolinedione 723
2-(2-Piperidinoacetyl)dibenzofuran 894

$C_{19}H_{19}NO_2S$
4-(2,5-Dimethyl-1-pyrrolyl)phenyl *m*-tolyl sulfone 841
4-(2,5-Dimethyl-1-pyrrolyl)phenyl *p*-tolyl sulfone 841
1,5-Diphenyl-3-(2-(methylthio)-ethyl)-2,4-pyrrolidinedione 690

$C_{19}H_{19}NO_3$
4′-(*N*,*N*-Dimethyl-carbamoylmethoxy)chalcone 414
2,6-Diphenyl-4-oxo-3-piperidineacetic acid 331
4-(2-(Phenylacetyl)benzoyl)-morpholine 706

$C_{19}H_{19}NO_3S$
1-Benzylpyridinium *p*-toluenesulfonate 955

$C_{19}H_{19}NO_4$
2-(2-(3,4-Dimethoxyphenyl)ethyl)-isoquinoline-1,3(2*H*,4*H*)-dione 716
5-Indanyl *N*-(3-(ethoxycarbonyl)-phenyl)carbamate 676
Methyl 4,4-dibenzyl-4-nitrocrotonate 373
Protopapaverine hydrochloride 996

$C_{19}H_{19}NO_5$
α-(2,4-Dimethoxy-3-methylbenzylidene)hippuric acid 464
α-(2,4-Dimethoxy-6-methylbenzylidene)hippuric acid 464

$C_{19}H_{19}NS$
2-(2,5-Dimethyl-1-pyrrolyl)phenyl *o*-tolyl sulfide 838
4-(2,5-Dimethyl-1-pyrrolyl)phenyl *m*-tolyl sulfide 839

$C_{19}H_{19}N_2O_3P$
Diphenyl *N*-(*N*-methylanilino)-phosphoramidate 836

$C_{19}H_{19}N_3O$
2,3-Dimethyl-4-(4-methylbenzylideneamino)-1-phenyl-3-pyrazolin-5-one 1075
1-(2-Methyl-4-quinolyl)-3-(3,4-xylyl)urea 989

$C_{19}H_{19}N_3O_2$
2,3-Dimethyl-4-(2-methoxybenzylideneamino)-1-phenyl-3-pyrazolin-5-one 1050
2,3-Dimethyl-4-(4-methoxybenzylideneamino)-1-phenyl-3-pyrazolin-5-one 1050

$C_{19}H_{19}N_3O_3$
Benzyl *N*-(3-(3-phenyl-1,2,4-oxadiazol-5-yl)propyl)carbamate 873
Carbobenzyloxy-L-tryptophanamide 883
2-(Cyclopropylcarboxamido)-benzoic (4-methoxybenzylidene)-hydrazide 498
4-(2-Hydroxy-3-methoxybenzylideneamino)-antipyrine 1031
N-(4-Morpholinoanilinomethyl)-phthalimide 719

$C_{19}H_{19}N_3O_3S$
Ethyl 4-(3-(5,6-dimethyl-2-benzothiazolyl)ureido)benzoate 916
Ethyl 2-(3-(4-(6-methyl-2-benzothiazolyl)phenyl)ureido)-acetate 916

$C_{19}H_{19}N_3O_3Si$
1-(4-Nitrophenylazo)-6-(trimethylsilyl)-2-naphtol 1084

$C_{19}H_{19}N_3O_5S_2$
N^1-(2-Thiazolyl)-N^4-(3,4,5-trimethoxy-benzylidene)sulfanilamide 1051

$C_{19}H_{19}N_5O_4$
2-(Furfurylureido)acetic acid N^2-(*N*-(1-naphthyl)carbamoyl)hydrazide 851

$C_{19}H_{19}N_7O_6$
Folic acid 1002

$C_{19}H_{20}BrNO_3$
2-Bromo-4-*tert*-butylphenyl *N*-(3-acetylphenyl)carbamate 629

$C_{19}H_{20}Br_2N_2O_3$
2-Acetamido-7-(*N*,*N*-bis(2-hydroxyethyl)amino)-1,3-dibromofluorene 485

$C_{19}H_{20}ClN$
β-*tert*-Butyl-α-(4-chlorophenyl)-hydrocinnamonitrile 751

$C_{19}H_{20}ClNO$
4′-Chloro-4-diethylaminochalcone 302

$C_{19}H_{20}ClNO_2$
4-Cyclohexylphenyl *N*-(4-chlorophenyl)carbamate 631

$C_{19}H_{20}ClNO_4$
4-(Butoxycarbonyl)phenyl *N*-(2-chloro-6-methylphenyl)carbamate 630

$C_{19}H_{20}ClNO_5$
4-(Butoxycarbonyl)phenyl *N*-(2-chloro-4-methoxyphenyl)-carbamate 620

$C_{19}H_{20}ClN_3$
1-Benzyl-4-(3-chloroanilino)-4-cyanopiperidine 754
1-Benzyl-4-(4-chloroanilino)-4-cyanopiperidine 754

$C_{19}H_{20}ClN_5O_4S_2$
3-Isobutyl-2-((3-methyl-2-benzothiazolinylidene)aminoazo)-benzothiazolium perchlorate 915

$C_{19}H_{20}Cl_2FN$
2-(*N*,*N*-Bis(2-chloropropyl)amino)-7-fluorofluorene 233

$C_{19}H_{20}Cl_2N_2O_4$
Bis(2-chloroethyl) *N*,*N*′-(methylenedi-4,1-phenylene)-biscarbamate 604

$C_{19}H_{20}FNO_2$
4-Cyclohexylphenyl *N*-(2-fluorophenyl)carbamate 631

$C_{19}H_{20}F_3NOS$
S-(4-*tert*-Butyl-2-methylphenyl) *N*-(2-(trifluoromethyl)phenyl)-thiocarbamate 611

$C_{19}H_{20}F_3N_3$
3-Methyl-1-(4-(4-(trifluoromethyl)-phenylazo)phenyl)piperidine 1080
4-Methyl-1-(4-(4-(trifluoromethyl)-phenylazo)phenyl)piperidine 1080
1-(4-(4-(Trifluoromethyl)phenylazo)-phenyl)hexahydro-1*H*-azepine 1080

$C_{19}H_{20}F_3N_5O_2$
2-(3-(3(Trifluoromethyl)phenyl)-ureido)acetic (4-(dimethylamino)-benzylidene)hydrazide 508

$C_{19}H_{20}INO$
2-(4-Hydroxystyryl)-1,3,3-trimethyl-3*H*-indolium iodide 881

$C_{19}H_{20}N_2$
1-Benzyl-4-cyano-4-phenylpiperidine hydrochloride 755
2-(4-(Diethylamino)phenyl)quinoline 989

$C_{19}H_{20}N_2O$
2-(Cyclohexylideneamino)-*N*-phenylbenzamide 477
1,2,3,3a-Tetrahydro-3a-(2,4-xylyl)-pyrrolo(1,2-a)quinazolin-5(4*H*)-one 707

$C_{19}H_{20}N_2OS$
6-Ethoxy-3-ethyl-2-(2-(phenylimino)-ethylidene)benzothiazoline 1044

$C_{19}H_{20}N_2O_2$
4,7-Dihydro-4-ethyl-4-phenyl-1*H*-2,6-benzodiazonine-3,5(2*H*,6*H*)-dione 708
4-Ethoxybenzoic acid (5-indanylmethylene)-hydrazide 498
N-(2-(*N*-Ethyl-3-methylanilino)ethyl)-phthalimide 718
4-Methoxy-*N*-(5-(4-methoxyanilino)-2,4-pentadien-1-ylidene)aniline perchlorate 1041
β-Morpholino-*N*-phenylcinnamamide 471

$C_{19}H_{20}N_2O_3$
4-Diethylamino-4′-nitrochalcone 302

$C_{19}H_{20}N_2O_5$
Carbobenzyloxyglycyl-D-phenylalanine 657
Carbobenzyloxy-L-phenylalanylglycine 660

$C_{19}H_{20}N_2O_5S$
2-Amino-1-(ethoxycarbonyl)-isoquinolinium *p*-toluenesulfonate 998

$C_{19}H_{20}N_2O_6$
4-(Butoxycarbonyl)phenyl *N*-(4-methyl-2-nitrophenyl)carbamate 668
Carbobenzyloxyglycyl-L-tyrosine 645
Carbobenzyloxy-L-tyrosylglycine 645
Carbobenzyloxy-L-valine 4-nitrophenyl ester 674
1-(4-Hydroxybenzoyl)-2-(3,4,5-trimethoxycinnamoyl)hydrazine 495

$C_{19}H_{20}N_2O_6S$
Carbobenzyloxy-L-methionine 4-nitrophenyl ester 667

$C_{19}H_{20}N_2O_7$
4-(Butoxycarbonyl)phenyl *N*-(2-methoxy-4-nitrophenyl)carbamate 652

$C_{19}H_{20}N_4$
4-(4-(Hexahydro-1*H*-azepin-1-yl)-phenylazo)benzonitrile 1091
2,2′-Methylenebis(5,6-dimethylbenzimidazole) 907
4-(4-(4-Methylpiperidino)phenylazo)-benzonitrile 1091

$C_{19}H_{20}N_4O$
5-(Dimethylaminomethyleneamino)-1,3-diphenyl-4-pyrazolemethanol 860

$C_{19}H_{20}N_4O_2S$
1-(2,3-Dimethyl-5-oxo-1-phenyl-3-pyrazolin-4-yl)-3-((3-methylthio)-phenyl)urea 701

$C_{19}H_{20}N_4O_3$
4-(2-(Dimethylamino)-acetamido)-1,2-diphenyl-3,5-pyrazolidinedione 700

$C_{19}H_{20}N_4O_4$
Malonic bis(4-methoxybenzylidenehydrazide) 497
1,1′-(4-Methyl-1,3-phenylene)bis(3-furfurylurea) 850

$C_{19}H_{20}O$
4-Isopropyl-4′-methylchalcone 305

$C_{19}H_{20}O_2$
1,4-Bis(4-methoxyphenyl)-2-methyl-1,3-butadiene 179
2-(2,2-Dimethyl-1-indanylmethyl)-benzoic acid 345
4,4-Dimethyl-3-(4-phenoxyphenyl)-1-pentyn-3-ol 1016

$C_{19}H_{20}O_3$
2,2-Dimethyl-1-phenylpropyl benzoylformate 374

$C_{19}H_{20}O_4$
1,5-Bis(4-methoxyphenyl)-1,5-pentanedione 307
(+)-*mono*-(2,2-Dimethyl-1-phenylpropyl) phthalate 375
(-)-*mono*-(2,2-Dimethyl-1-phenylpropyl) phthalate 375
2,2-Dimethyl-1,3-propanediol dibenzoate 371
4,4′-Isopropylidenediphenol diacetate 380

$C_{19}H_{20}O_5$
2,3,4,5-Dibenzylidene-DL-xylitol 203

$C_{19}H_{20}O_6$
Phenyl-4,6-*o*-benzylidene-β-D-glucopyranoside 177

$C_{19}H_{20}S_2$
1,2-Bis(*p*-tolylthio)-1-cyclopentene 211

$C_{19}H_{21}BrClNO_4$
2-Bromo-4-*tert*-butylphenyl *N*-(5-chloro-2,4-dimethoxyphenyl)-carbamate 613

$C_{19}H_{21}Br_2N$
2-(*N*,*N*-Bis(2-bromopropyl)amino)-fluorene ... 233
$C_{19}H_{21}ClN_2O$
3-(4-Chlorophenyl)-1-cyclohexyl-1-phenylurea ... 535
$C_{19}H_{21}ClN_2OS$
5-Chloro-2-(4-(2-diethylaminoethoxy)-phenyl)-benzothiazole hydrochloride ... 913
$C_{19}H_{21}ClN_2O_2$
10-Chloro-2,3-dimethoxy-1,3-methyl-5,6,13,13a-tetrahydro-8*H*-isoquino(1,2-b)quinazoline ... 246
$C_{19}H_{21}ClN_2O_4$
Butyl 4-(3-(5-chloro-2-methoxyphenyl)ureido)benzoate ... 519
$C_{19}H_{21}Cl_2N$
2-(*N*,*N*-Bis(2-chloropropyl)amino)-fluorene ... 233
$C_{19}H_{21}Cl_2NO$
2,4-Dichloro-2′,6′-diisopropylbenzanilide ... 460
2,5-Dichloro-2′,6′-diisopropylbenzanilide ... 460
3,4-Dichloro-2′,6′-diisopropylbenzanilide ... 460
$C_{19}H_{21}Cl_2N_3O_3$
2-(Chloro-methylphenyl-1-methylureido)ethyl *N*-(5-chloro-2-methylphenyl)carbamate ... 674
$C_{19}H_{21}Cl_3N_2$
3,3-Bis(4-(dimethylamino)phenyl)-1,1,2-trichloro-1-propene ... 233
$C_{19}H_{21}I_2NO_2$
N,*N*-Diethyl-3,5-diiodo-4-hydroxy-α-phenylhydrocinnamamide ... 413
$C_{19}H_{21}N$
Benzylmethyl(5-phenyl-4-pentyn-1-yl)amine ... 1018
1-Isopropyl-3-(3,5-xylyl)indole ... 887
$C_{19}H_{21}NO$
6-(Benzylamino)-3,3-dimethyl-1-tetralone ... 294
1-Benzyl-4-phenyl-3-propyl-2-azetidinone ... 691
4-(*tert*-Butyl)-1,4-dihydro-1-phenyl-3(2*H*)-isoquinolinone ... 691
N-Cyclohexylbenzanilide ... 482
2′,4′-Dimethyl-4-(dimethylamino)-chalcone ... 302
5′-Isopropyl-2′-methylcinnamanilide ... 478
N-(2-Phenylcyclohexyl)benzamide ... 477
$C_{19}H_{21}NO_2$
3-Benzoyl-4-hydroxy-1-methyl-4-phenylpiperidine ... 290
N-(2-Benzoylphenyl)-3,3-dimethylbutyramide ... 432
4-Biphenylyl *N*-cyclohexylcarbamate ... 679
N,*N*-Bis-(2-benzoylethyl)-methylamine ... 312
1-*tert*-Butyl-1-hydroxy-2-methyl-6-phenyl-3-isoindolinone ... 690
4-Cyclohexylphenyl *N*-phenylcarbamate ... 679
4-(Diethylamino)-2′-hydroxychalcone ... 300
2,2-Diphenyl-4a,5,6,7,8,8a-hexahydro-4*H*-1,3-dioxino(5,4-b)-pyridine hydrochloride ... 248
4′-(4-Propoxybenzylideneamino)-propiophenone ... 1045
$C_{19}H_{21}NO_3$
4-(2-Nitro-1,1,2-trimethylpropyl)-benzophenone ... 312
$C_{19}H_{21}NO_4$
4-Butoxyphenyl *N*-(3-acetylphenyl)-carbamate ... 651
N-Carbobenzyloxy-*S*-benzyl-DL-homocysteine ... 656
2-(2-(3,4-Dimethoxyphenyl)ethyl)-1,3,3a,4,7,7a-hexahydro-4,7-methanoisoindole-1,3-dione ... 716
10a,10b-Dimethyl-9,9-ethylenedioxy-6,8a,9,10,10a,10b-hexahydro-3-methoxybenz(b)-cyclopent(hi)indolizin-6-one ... 699
L-Glutamic acid dibenzyl ester *p*-toluenesulfonate ... 356
2-Isopropylphenyl *N*-(4-(ethoxycarbonyl)phenyl)carbamate. 675
$C_{19}H_{21}NO_4S$
Carbobenzyloxy-*S*-benzyl-L-cysteine methyl ester ... 656
4-Carboxy-4-phenyl-1-*p*-tosylpiperidine ... 824
1-(2-Methoxyethyl)quinolinium *p*-toluenesulfonate ... 985
$C_{19}H_{21}NO_5$
4-Allyl-2-methoxyphenyl *N*-(2,4-dimethoxyphenyl)carbamate ... 646
$C_{19}H_{21}NS$
N-Cyclohexylthiobenzanilide ... 742
$C_{19}H_{21}N_3$
4-Anilino-1-benzyl-4-cyanopiperidine ... 754
1,2-Dihydro-2-(4-(dimethylamino)-phenylimino)-1-ethylquinoline hydrochloride ... 725
1,4-Dihydro-4-(4-(dimethylamino)-phenylimino)-1-ethylquinoline hydrochloride ... 1057
9-(4-Tolylazo)-2,3,5,6-tetrahydro-1*H*,7*H*-benzo(ij)quinolizine ... 1091
$C_{19}H_{21}N_3O_2$
N-(3-Acetylphenyl)-4-phenyl-1-piperazinecarboxamide ... 704
Hippuric (1-phenyl-1-butylidene)-hydrazide ... 501
$C_{19}H_{21}N_3O_4$
1,2-Dihydro-4,6-dimethyl-1-(α-methyl-3,4,5-trimethoxy-benzylideneamino)-2-oxonicotino-nitrile ... 1051
Ethyl 4-(3-(4-(*N*-methylacetamido)-phenyl)ureido)benzoate ... 569
Hippuric acid *N*-(4-ethoxy-3-methoxybenzylidene)hydrazide ... 497
$C_{19}H_{21}N_3O_5$
Benzoic (2,6-bis(ethoxycarbonyl)-1,4-dihydro-1-methyl-4-pyridylidene)hydrazide ... 499
Butyl 4-(3-(2-methyl-5-nitrophenyl)-ureido)benzoate ... 563
Diethyl 2-(4-(cyanomethyl)-3-indolylmethyl)-2-formamidomalonate ... 885
$C_{19}H_{21}N_3O_6$
Butyl 4-(3-(2-methoxy-4-nitrophenyl)ureido)benzoate ... 553
Isobutyl 4-(3-(2-methoxy-5-nitrophenyl)ureido)benzoate ... 553
$C_{19}H_{22}BrNO$
2-Bromo-2′,6′-diisopropylbenzanilide ... 457
3-Bromo-2′,6′-diisopropylbenzanilide ... 459
4-Bromo-2′,6′diisopropylbenzanilide ... 458
N-(4-Bromo-2,6-dimethylphenyl)-4-*tert*-butylbenzamide ... 459
$C_{19}H_{22}BrNO_2$
2-Bromo-4-*tert*-butylphenyl *N*-(3,5-dimethylphenyl)carbamate ... 639
$C_{19}H_{22}BrN_5O_5$
Disperse Violet 24 (C.I. 11200) ... 1088
$C_{19}H_{22}ClNO$
2-Chloro-2′,6′-diisopropylbenzanilide ... 457
3-Chloro-2′,6′-diisopropylbenzanilide ... 459
2-Chloro-*N*-isopropyl-2-phenyl-2′,5′-acetoxylidide ... 441
$C_{19}H_{22}ClNO_2$
4-Chloro-2-isopropyl-5-methylphenyl *N*-(3-ethylphenyl)-carbamate ... 631
4-Chloro-2-isopropyl-5-methylphenyl *N*-(2,5-xylyl)-carbamate ... 640
2-(4-Chloro-2-methylphenoxy)-2′,6′-diethylacetanilide ... 401
2,6-Diisopropylphenyl *N*-(3-chlorophenyl)carbamate ... 636
$C_{19}H_{22}ClNO_3$
N-(5-Chloro-2,4-dimethoxyphenyl)-4-*tert*-butylbenzamide ... 447
4-Chloro-2-isopropyl-5-methylphenyl *N*-(4-ethoxyphenyl)-carbamate ... 620
$C_{19}H_{22}ClNO_4$
Diethyl 4-(*p*-chlorophenyl)-1,4-dihydro-2,6-dimethyl-3,5-pyridine dicarboxylate ... 376
$C_{19}H_{22}ClNO_5$
4-Butoxyphenyl *N*-(5-chloro-2,4-dimethoxyphenyl)carbamate ... 613
$C_{19}H_{22}ClN_3O_2$
2-Methoxybenzoic (2-chloro-4-(diethylamino)benzylidene)-hydrazide ... 492
$C_{19}H_{22}ClN_5O_2$
2-(3-Benzylureido)acetic (2-chloro-4-(dimethylamino)benzylidene)-hydrazide ... 522
$C_{19}H_{22}Cl_2N_2O$
1-(2,3-Dichlorophenyl)-3-(2,6-diisopropylphenyl)urea ... 541
1-(2,4-Dichlorophenyl)-3-(2,6-diisopropylphenyl)urea ... 542
1-(2,6-Dichlorophenyl)-3-(2,6-diisopropylphenyl)urea ... 541
$C_{19}H_{22}Cl_2O$
α-*tert*-Butyl-3,3′-dichloro-4,4′-dimethylbenzhydrol ... 198
$C_{19}H_{22}Cl_2O_2$
2,2-Bis(4-methoxy-3-methylphenyl)-1,3-dichloropropane ... 176
$C_{19}H_{22}Cl_4O_4$
10-Undecenyl tetrachlorophthalate.. 366
$C_{19}H_{22}FNO_3$
2-(4-*tert*-Butylphenoxy)ethyl *N*-(4-fluorophenyl)carbamate ... 615
$C_{19}H_{22}INOS$
1-(α-Methylthio-4-phenoxybenzylidene)piperidinium iodide ... 254
$C_{19}H_{22}NO_4P$
Diethyl (2-acetamido-9-fluorenyl)-phosphonate ... 835
$C_{19}H_{22}N_2$
N-Cinnamylidene-*N*′,*N*′-diethyl-1,4-phenylenediamine ... 1057
α-(2-Dimethylaminoethyl)-α-phenylhydrocinnamonitrile ... 755
$C_{19}H_{22}N_2O$
5-Benzyloxy-*N*,*N*-dimethyltryptamine oxalate ... 881
Cinchonine sulfate hexahydrate ... 984
Cinchonine sulfate ... 983
Cinchotoxine ... 989
$C_{19}H_{22}N_2O_2$
12-Acetyldecahydro-9*H*-pyrido(1,2,3-lm)pyrrolo(2,3-d)-carbazol-9-one ... 706
4-Anilino-1-benzylisonipecotic acid hydrochloride ... 349
2-Butylmalonanilide ... 443
N,*N*′-Diphenylpimelamide ... 436
$C_{19}H_{22}N_2O_2S_2$
3,5-Bis-(4-ethoxyphenyl)-3,4,5,6-tetrahydro-2*H*-1,3,5-thiadiazine-2-thione ... 736
$C_{19}H_{22}N_2O_3$
1-(4-(Ethoxycarbonyl)phenyl)-3-ethyl-3-(*m*-tolyl)urea ... 577
2-(*N*-Ethylanilino)ethyl *N*-(3-acetylphenyl)carbamate ... 671
Ethyl 4-(3-isopropyl-3-phenylureido)benzoate ... 577
3-(Methoxycarbonyl)propionic *N*[1],*N*[2]-bis(*p*-tolyl)hydrazide ... 503
$C_{19}H_{22}N_2O_3S$
S-(4-*tert*-Butyl-2-methylphenyl) *N*-(2-methyl-5-nitrophenyl)-thiocarbamate ... 669
Butyl 4-(3-(4-(methylthio)phenyl)-ureido)benzoate ... 561
$C_{19}H_{22}N_2O_4$
Diethyl 4,4′-methylenebis(*N*-phenylcarbamate) ... 673
6,7-Dimethoxy-2-methyl-1-(2-nitrobenzyl)-1,2,3,4-tetrahydroisoquinoline ... 265
2′,2‴-(Trimethylenedioxy)-diacetanilide ... 417
3′,3‴-(Trimethylenedioxy)-diacetanilide ... 418
4′,4‴-(Trimethylenedioxy)-diacetanilide ... 419
$C_{19}H_{22}N_2O_4S$
3-Ethoxy-4-hydroxybenzaldehyde *N*-phenyl-*N*-(3-sulfolanyl)hydrazone. 774
$C_{19}H_{22}N_2O_6$
Diethyl 1,4-dihydro-2,6-dimethyl-4-(3-nitrophenyl)-3,5-pyridinedicarboxylate ... 359
$C_{19}H_{22}N_2S$
10-(1-Ethyl-3-pyrrolidinylmethyl)-phenothiazine hydrochloride ... 249
$C_{19}H_{22}N_2S_2$
3,5-Bis(2-phenylethyl)-3,4,5,6-tetrahydro-2*H*-1,3,5-thiadiazine-2-thione ... 741
$C_{19}H_{22}N_4O_2$
1-(4-(4-Nitrophenylazo)phenyl)-octahydroazocine ... 1090
6-(4-Nitrophenylazo)-1,2,3,4-tetrahydro-2,2,4,7-tetramethylquinoline ... 1091
$C_{19}H_{22}N_4O_3$
4-(3,3-Diethylureido)benzoic (4-hydroxybenzylidene)hydrazide ... 548
$C_{19}H_{22}N_4O_4$
2-(3-Benzylureido)acetic (3,4-dimethoxybenzylidene)hydrazide ... 549
2-(3-(2-Ethoxyphenyl)ureido)acetic (4-methoxybenzylidene)hydrazide ... 497
$C_{19}H_{22}N_6O_4$
5,5′-Benzylidenebis(6-amino-1,3-dimethyluracil) ... 701
$C_{19}H_{22}N_6O_6$
1-(3-(2-Ethoxyphenyl)ureidoacetyl)-4-(2-methyl-5-nitrophenyl)-semicarbazide ... 552
$C_{19}H_{22}O_2$
2,2-Dimethyl-3-(6-methoxy-2-naphthyl)-4-hexyn-3-ol ... 1016
2,2-Diphenylheptanoic acid ... 333
1-(4-Hydroxyphenyl)-2-phenyl-3-heptanone ... 283
$C_{19}H_{22}O_3$
Methyl 2-(1,4a-dimethyl-1,2,3,4,4a,9-hexahydro-2-oxo-1-phenanthryl)-acetate ... 360
2-Methyl-1-phenylpropyl α-methoxyphenylacetate ... 354
$C_{19}H_{22}O_4$
Ethyl 4,4-bis(4-hydroxyphenyl)-valerate ... 354
2,4,5-Triethoxybenzophenone ... 311
$C_{19}H_{22}S$
1-Phenylcyclohexyl *p*-tolyl sulfide ... 211
$C_{19}H_{23}BrN_2O$
1-(2-Benzamidoethyl)-1-phenylpyrrolidinium bromide ... 469
$C_{19}H_{23}ClN_2$
1-(4-Chloro-α-phenylbenzyl)-hexahydro-4-methyl-1*H*-1,4-diazepine ... 248
1-(4-Chloro-α-phenylbenzyl)-hexahydro-4-methyl-1*H*-1,4-diazepine octadecyl sulfate ... 248
$C_{19}H_{23}ClN_2O$
1-(3-Chlorophenyl)-3-(2,6-diisopropylphenyl)urea ... 541
1-(4-Chlorophenyl)-3-(2,6-diisopropylphenyl)urea ... 539
$C_{19}H_{23}ClN_2O_4$
2-(4-(4-(Dimethylamino)phenyl)-1,3-butadien-1-yl)-1-ethylpyridinium perchlorate ... 954
$C_{19}H_{23}Cl_2N_3O_2$
1-(4-Chlorobenzyl-carbamoylmethyl)-3-(diethyl-carbamoyl)-pyridinium chloride ... 948
$C_{19}H_{23}IN_2O$
(2-(4-Methoxyphenyl)indol-3-ylmethyl)trimethylammonium

$C_{19}H_{23}IN_2O$ (Cont'd)
iodide ... 887
(4-Methoxyphenyl)-methyl-tetrahydro-1*H*-pyrazolo(1,2-b)-phthalazin-4-ium iodide ... 254

$C_{19}H_{23}N$
10,11-Dihydro-*N*,*N*-dimethyl-5*H*-dibenzo(a,d)cycloheptene-10-ethanamine ... 243
2-Dipropylaminofluorene ... 233

$C_{19}H_{23}NO$
N-Benzyl-*N*-*tert*-butyl-*o*-toluamide ... 472
1-Benzyl-4-hydroxy-4-(*p*-tolyl)-piperidine hydrochloride ... 246
N-(*tert*-Butyl)-2-(4-methylbenzyl)-benzamide ... 479
2,2-Diphenyl-5-(dimethylamino)-3-pentanone hydrochloride ... 285
N-Ethyl-2-phenyl-*o*-butyrotoluidide ... 441

$C_{19}H_{23}NOS$
2-(Benzylthio)-4′-butylacetanilide ... 421

$C_{19}H_{23}NO_2$
2-(α-(Benzylamino)benzyl)-butyric acid ... 330
(3-Benzyloxy-4-methoxybenzylidene)*tert*-butylamine ... 1039
2,6-Diisopropylphenyl *N*-phenylcarbamate ... 681
2-Dimethylaminomethyl-4′-methoxy-2-phenylpropiophenone hydrochloride ... 292
2-Isopropyl-5-methylphenyl *N*-(3-ethylphenyl)carbamate ... 686
2-Isopropyl-5-methylphenyl *N*-(2,6-xylyl)carbamate ... 685
Phenyl *N*-(2,6-diisopropylphenyl)-carbamate ... 685

$C_{19}H_{23}NO_3$
4-Butoxyphenyl *N*-(2,5-xylyl)-carbamate ... 651

$C_{19}H_{23}NO_3S$
N-Butyl-1,3-diphenyl-3-oxopropanesulfonamide ... 823
N,*N*-Diethyl-1,3-diphenyl-3-oxopropanesulfonamide ... 822

$C_{19}H_{23}NO_4$
2-Benzyl 4-*tert*-butyl 3,5-dimethyl-2,4-pyrroledicarboxylate ... 843
Diethyl 4-phenyl-1,4-dihydro-2,6-lutidine-3,5-dicarboxylate ... 962
2,6-Dimethoxyphenyl *N*-(2,6-diethylphenyl)carbamate ... 647
Ethyl 5-(benzyloxycarbonyl)-2,4-dimethyl-3-pyrrolepropionate ... 842
2-Isopropyl-5-methylphenyl *N*-(2,5-dimethoxyphenyl)carbamate ... 648
Octahydro-2,5-methano-2*H*-indeno(1,2-b)oxiren-3-yl *N*-(3-ethoxyphenyl)carbamate ... 654

$C_{19}H_{23}NO_5$
4-Butoxyphenyl *N*-(2,4-dimethoxyphenyl)carbamate ... 647

$C_{19}H_{23}N_3$
Hexahydro-1-(4-(*p*-tolylazo)phenyl)-1*H*-azepine ... 1091

$C_{19}H_{23}N_3O$
4-(Benzoylhydrazino)-1-benzylpiperidine ... 499

$C_{19}H_{23}N_3O_2$
2-(Diethylamino)malondianilide ... 443

$C_{19}H_{23}N_5O_2$
2-(3-Benzylureido)acetic (4-(dimethylamino)benzylidene)-hydrazide ... 561

$C_{19}H_{23}N_5O_4$
2-(4-Methoxyphenylureido)acetic acid N^2-(*N*-(2-ethylphenyl)-carbamoyl)hydrazide ... 553
2-(4-Methoxyphenylureido)acetic acid N^2-(*N*-(2,4-xylyl)carbamoyl)-hydrazide ... 553

$C_{19}H_{23}N_5O_5$
Disperse Violet 13 (C.I. 11195) ... 1084

$C_{19}H_{23}N_5O_6$
2-(4-Methoxyphenylureido)acetic N^2-(*N*-(2,5-dimethoxyphenyl)-carbamoyl)hydrazide ... 550

$C_{19}H_{24}BrN$
1,1-Dibenzyl-3,3-dimethylazetidinium bromide ... 255

$C_{19}H_{24}BrNO_2$
(2-(Diphenylacetoxy)-ethyl)-trimethylammonium bromide ... 359

$C_{19}H_{24}INO$
(4-(1,3-Dihydro-1-methyl-1-isobenzofuranyl)benzyl)-trimethylammonium iodide ... 255

$C_{19}H_{24}N_2$
2,6-Diethyl-*N*-(4-(dimethylamino)-benzylidene)aniline ... 1061
1,3-Dimethyl-3-(diphenylaminomethyl)-pyrrolidine hydrochloride ... 249

$C_{19}H_{24}N_2O$
1-Benzyl-3-ethyl-1-methyl-3-(α-methylbenzyl)urea ... 572
4,4′-Bis(dimethylamino)-2,2′-dimethylbenzophenone ... 312
N,*N*′-Bis(2-ethyl-6-methylphenyl)urea ... 596
3-Butyl-1,1-dibenzylurea ... 572
1-(4-Butylphenyl)-3-(2-ethylphenyl)urea ... 589
1-(4-Butylphenyl)-3-(4-ethylphenyl)urea ... 591
1-Butyl-1-phenyl-3-(2,4-xylyl)urea ... 601
1-(4-Butylphenyl)-3-(2,3-xylyl)urea ... 591
1-(4-Butylphenyl)-3-(2,4-xylyl)urea ... 592
1-(4-Butylphenyl)-3-(2,6-xylyl)urea ... 591
1-Cyclohexyl-1-ethyl-3-(1-naphthyl)urea ... 584
2-(Diethylamino)-*N*-(diphenylmethyl)-acetamide hydrochloride ... 430
1-(2,6-Diethylphenyl)-3-(2,4-dimethylphenyl)urea ... 596
N-(2-(Dimethylamino)-ethyl)-2,2-diphenyl-*N*-methylacetamide hydrochloride ... 423
3,3-Dimethyl-1,7-di-2-pyridyl-4-heptanone ... 961
N-(1,2-Diphenylethyl)-2-(*N*-ethyl-*N*-methylamino)-acetamide ... 428
3′-(Dipropylamino)benzanilide ... 469

$C_{19}H_{24}N_2O_2$
N-((3-Amino-4-ethoxy-2-methyl-5-propyl)phenyl)benzamide ... 465
1-Benzyl-1-butyl-3-(4-methoxyphenyl)urea ... 558
6-(Dimethylamino)-1-(3-hydroxy-4-methoxybenzyl)-1,2,3,4-tetrahydroisoquinoline ... 245
Niquine dioxalate ... 982

$C_{19}H_{24}N_2O_2S$
N,*N*-Dimethyl-4-(phenylsulfonyl)-3-piperidinoaniline ... 778

$C_{19}H_{24}N_2O_3$
2-(Benzylamino)-*N*-(2-(3,4-dimethoxyphenyl)ethyl)acetamide hydrochloride ... 413
1-(2,6-Diethylphenyl)-3-(2,5-dimethoxyphenyl)urea ... 550
3′-(*N*-Ethyl-*N*-(2-(4-methoxyphenoxy)ethyl)amino)-acetanilide ... 414

$C_{19}H_{24}N_2O_4$
3-(2,5-Dimethoxyphenyl)-1-ethyl-1-(2-methoxy-5-methylphenyl)urea ... 550

$C_{19}H_{24}N_2O_5$
Diethyl 2-acetamido-2-(1-methylindol-3-ylmethyl)malonate ... 883
Diethyl (*N*-(2-(3-indolyl)ethyl)-carbamoylmethyl)malonate ... 886

$C_{19}H_{24}N_2O_6S$
N-(*tert*-Butoxycarbonyl)-*S*-benzylcysteine *N*-hydroxysuccinimide ester ... 718

$C_{19}H_{24}N_2O_7$
o-Benzyl-*N*-(*tert*-butoxycarbonyl)-L-serine *N*-hydroxysuccinimide ester ... 717

$C_{19}H_{24}N_2S$
10-(3-(Diethylamino)propyl)-phenothiazine ... 249
2-Pentamethylbenzyl-3-phenyl-2-thiopseudourea hydrochloride ... 749

$C_{19}H_{24}N_2S_2$
10-(3-(Dimethylamino)-2-methylpropyl)-2-(methylthio)-phenothiazine ... 248

$C_{19}H_{24}N_4$
1,3-Diethyl-2,3-dihydro-2-(4-(dimethylamino)phenylimino)-benzimidazole ... 725

$C_{19}H_{24}N_4O_2$
4′-(Dimethylamino)-4-nitro-2,2′,3,5,6-pentamethylazobenzene ... 1091
4,4′-Methylenebis(1,1-dimethyl-3-phenylurea) ... 592
4,4′-Methylenebis(1-ethyl-3-phenylurea) ... 592

$C_{19}H_{24}N_4O_4$
1-Adamantyl ethyl ketone (2,4-dinitrophenyl)hydrazone ... 274

$C_{19}H_{24}N_4O_4S_2$
2,4-Pentanedione bis((*p*-tosyl)-hydrazone) ... 1056

$C_{19}H_{24}O$
1,3-Di(4-tolyl)-2,2-dimethylpropanol ... 205

$C_{19}H_{24}OS$
3-(Benzylthiomethylene)-8a-methyl-2-decalone ... 284

$C_{19}H_{24}O_2$
3-(Benzyloxymethyl)-α-(*tert*-butyl)-benzyl alcohol ... 201
2,2-Bis-(3,5-dimethyl-4-hydroxyphenyl)-propane ... 194

$C_{19}H_{24}O_3$
1-(2-Ethoxy-6-methoxyphenyl)-2-methyl-1-phenyl-1-propanol ... 201
2-Hydroxy-3-(7-methyloctyl)-1,4-naphthoquinone ... 316
2-Hydroxy-3-(2-methyloctyl)-1,4-naphthoquinone sodium salt ... 316

$C_{19}H_{24}O_4$
Ethyl 2-acetyl-2-(4-cyclohexylbenzoyl)-acetate ... 364

$C_{19}H_{24}O_6S_2$
2,2-Diethyl-1,3-propanediol bis-(benzenesulfonate) ... 800

$C_{19}H_{25}BF_4O$
2,6-Di-*tert*-butyl-4-phenylpyrylium tetrafluoroborate ... 935

$C_{19}H_{25}BrN_2O$
(2-(Diphenylacetamido)-ethyl)-trimethylammonium bromide ... 423

$C_{19}H_{25}Cl_3N_2O$
1,1-Dicyclohexyl-3-(2,4,5-trichlorophenyl)carbamate ... 582

$C_{19}H_{25}IN_2$
2-(4-(Diethylamino)-styryl)-1-ethylpyridinium iodide ... 955

$C_{19}H_{25}NO$
6,7,8,9,10,11,12,13,14,15-Decahydro-16-hydroxycyclododeca(b)-quinoline ... 984
DL-4-Dimethylamino-1,2-diphenyl-3-methyl-2-butanol hydrochloride ... 238
L-4-Dimethylamino-1,2-diphenyl-3-methyl-2-butanol hydrochloride ... 131
2,2-Dimethyl-3-(1,2-diphenylethylamino)-1-propanol ... 238
7-Isopropyl-1,2,3,4,4a,9,10,10a-octahydro-1-phenanthrylmethyl isocyanate ... 769

$C_{19}H_{25}NO_2$
3,3′-Dimethoxy-α-methyldiphenethylamine hydrochloride ... 239
4-Hydroxy-α-(1-(1-phenethylethylamino)ethyl)benzyl alcohol hydrochloride ... 237

$C_{19}H_{25}NO_3$
1-(3,4-Dimethoxyphenethyl)-3,4,5,6,7,8-hexahydro-2(1*H*)-quinolinone ... 697

$C_{19}H_{25}NO_6$
Diethyl 2-acetamido-2-(4-acetylphenethyl)malonate ... 432
3,4,5-Trimethoxy-*N*-(3,4,5-trimethoxybenzyl)aniline ... 228

$C_{19}H_{25}N_3$
5-Amino-1,3-dibenzyl-5-methylhexahydropyrimidine ... 248

$C_{19}H_{25}N_3O$
1-(2,6-Diisopropylphenyl)-3-(3-pyridylmethyl)urea ... 950

$C_{19}H_{25}N_3OS$
1-(4-(Diethylamino)phenyl)-3-(2-methoxy-5-methylphenyl)-2-thiourea ... 740

$C_{19}H_{25}N_3O_3$
Methyl 5-(4-(2-acetamidoethyl)-3-methyl-2*H*-pyrrol-2-ylidenemethyl)-4-methyl-3-pyrrolepropionate hydrobromide ... 839

$C_{19}H_{25}N_3S$
1-(2-(Diethylamino)ethyl)-1,3-diphenyl-2-thiourea ... 739
1-(4-(Diethylamino)-*o*-tolyl)-2-thio-3-(*p*-tolyl)-urea ... 740

$C_{19}H_{26}As_2I_2$
6,7,8,9-Tetrahydro-5,5,9,9-tetramethyl-5*H*-dibenzo(f,h)(1,5)-diarsoninium diiodide ... 1128

$C_{19}H_{26}Br_2O_2$
Menthyl α,β-dibromo-hydrocinnamate ... 361

$C_{19}H_{26}ClNPb$
N-(3-(Chlorodiphenylplumbyl)-propyl)diethylamine hydrochloride ... 1134

$C_{19}H_{26}Cl_2N_2O$
3-(2,3-Dichlorophenyl)-1,1-dicyclohexylurea ... 583
3-(2,4-Dichlorophenyl)-1,1-dicyclohexylurea ... 584
3-(2,5-Dichlorophenyl)-1,1-dicyclohexylurea ... 584
3-(2,6-Dichlorophenyl)-1,1-dicyclohexylurea ... 582
3-(3,4-Dichlorophenyl)-1,1-dicyclohexylurea ... 584

$C_{19}H_{26}FN_3O_3$
1,1-Dicyclohexyl-3-(2-fluoro-5-nitrophenyl)urea ... 524

$C_{19}H_{26}INO_2$
2,3-Dimethoxy-7-methyl-octahydro-isoindolo(1,2-a)isoquinolinium iodide ... 254

$C_{19}H_{26}N_2O$
1,1-Dibutyl-3-(1-naphthyl)urea ... 603

$C_{19}H_{26}N_2O_4$
Diethyl 2,2′-methylenebis(3,4-dimethyl-5-pyrrolecarboxylate) ... 840
Ethyl 5-(4-(ethoxycarbonyl)-3-methylpyrrol-2-yl-methyl)-4-ethyl-3-methyl-2-pyrrolecarboxylate ... 840

$C_{19}H_{26}N_2O_5$
Carbobenzyloxy-L-prolyl-L-isoleucine ... 708
Carbobenzyloxy-L-prolyl-L-leucine ... 708
Carbobenzyloxyprolylvaline methyl ester ... 700

$C_{19}H_{26}N_2O_5S$
Carbobenzyloxy-L-prolyl-L-methionine methyl ester ... 700

$C_{19}H_{26}N_6O$
4-(Diethylamino)-4′-(dimethylamino)-2-(ureido)-azobenzene ... 1088

$C_{19}H_{26}O$
3-(4-Cyclohexylphenyl)-4,4-dimethyl-1-pentyn-3-ol ... 1016

$C_{19}H_{26}OSi_2$
4,4′-Bis(trimethylsilyl)-benzophenone ... 1109

$C_{19}H_{26}O_2$
2-Benzoyl-1-cyclododecanone ... 308
Menthyl cinnamate ... 365

$C_{19}H_{26}O_3$
Decahydro-4,6a,9,9-tetramethylphenanthro(2,3-d)(1,3)-dioxol-3-one ... 59

$C_{19}H_{26}O_4$
DL-*mono*-1-Cyclohexyl-3-methylbutyl phthalate ... 371

$C_{19}H_{26}Si_2$
9,9-Bis(trimethylsilyl)fluorene ... 1123

$C_{19}H_{27}ClN_2O$
1-(*p*-Chlorophenyl)-3,3-dicyclohexylurea ... 583

$C_{19}H_{27}ClN_2O$ (Cont'd)
3-(3-Chlorophenyl)-1,1-dicyclohexylurea ... 582
$C_{19}H_{27}Cl_2NO_2$
Cyclododecyl *N*-(2,6-dichlorophenyl)carbamate ... 636
$C_{19}H_{27}FN_2O$
1,1-Dicyclohexyl-3-(2-fluorophenyl)urea ... 535
1,1-Dicyclohexyl-3-(3-fluorophenyl)urea ... 535
1,1-Dicyclohexyl-3-(4-fluorophenyl)urea ... 535
$C_{19}H_{27}F_3N_2O_3$
11-(3-(3-(Trifluoromethyl)phenyl)ureido)undecanoic acid ... 509
$C_{19}H_{27}NO$
N-(4-Cyclohexylcyclohexyl)benzamide ... 477
$C_{19}H_{27}NOS$
S-Phenyl *N*-(4-cyclohexylcyclohexyl)thiocarbamate ... 688
$C_{19}H_{27}NO_2$
1,2,3,5,6,11,12,13,14,14a-Decahydro-8,9-dimethoxydibenzo(a,i)quinolizine hydrochloride ... 246
$C_{19}H_{27}NO_3$
4-*tert*-Butylcyclohexyl *N*-(3-acetylphenyl)carbamate ... 675
Hexahydro-1′,4,4,7′,7′-pentamethylspiro-(cyclohexane-1,3′(2′*H*)quinoline)-2,5′,6-trione ... 58
$C_{19}H_{27}NO_5S$
N-(3,4-Dimethoxybenzyl)-10-camphorsulfonamide ... 812
$C_{19}H_{27}N_3$
1-Benzyl-4-cyano-4-cyclohexylaminopiperidine ... 754
$C_{19}H_{27}N_3O$
Decanoic (3-indolylmethylene)hydrazide ... 1071
$C_{19}H_{27}N_3O_2S$
3-(3-Allylureido)-α-methylbenzyl *N*-cyclohexylcarbamate ... 738
$C_{19}H_{27}N_3O_3$
1,1-Dicyclohexyl-3-(3-nitrophenyl)urea ... 563
1,1-Dicyclohexyl-3-(4-nitrophenyl)urea ... 567
$C_{19}H_{27}N_3O_6$
Carbobenzyloxyglycylglycyl-L-isoleucine methyl ester ... 663
Carbobenzyloxyglycylglycyl-L-leucine methyl ester ... 663
Carbobenzyloxyglycylglycylnorleucine methyl ester ... 662
$C_{19}H_{27}N_5O_2S$
1-Antipyrinyl-3-(3-morpholinopropyl)-2-thiourea ... 740
$C_{19}H_{28}Br_2N_2$
1,1′-Pentamethylenebis(3,4-dimethylpyridinium bromide) ... 958
1,1′-Pentamethylenebis(3,5-dimethylpyridinium bromide) ... 958
$C_{19}H_{28}ClN_3O_3S$
1-(2-Chloro-5-nitrophenyl)-3-dodecanoyl-2-thiourea ... 730
$C_{19}H_{28}Cl_2N_2OS$
1-(2,5-Dichlorophenyl)-3-dodecanoyl-2-thiourea ... 732
$C_{19}H_{28}F_3NO_2S$
Dodecyl 2-nitro-4-trifluoromethylphenyl sulfide ... 259
$C_{19}H_{28}N_2O$
1,1-Diallyl-3-(2,6-diisopropylphenyl)urea ... 596
1,1-Dicyclohexyl-3-phenylurea ... 582
$C_{19}H_{28}N_2O_2$
O-(*N*-(2,6-Diisopropylphenyl)carbamoyl)cyclohexanone oxime ... 1012
10-Undecenoic acid *N*-(2-methoxybenzylidene)hydrazide ... 498
$C_{19}H_{28}N_2O_3$
10-Undecenoic (4-hydroxy-3-methoxybenzylidene)hydrazide ... 1032
$C_{19}H_{28}N_2O_5$
N-Benzyloxycarbonyl-L-isoleucyl-L-valine ... 660
Carbobenzyloxy-L-isoleucyl-L-norvaline ... 660
Carbobenzyloxynorleucyl-L-valine ... 660
Carbobenzyloxy-L-norvalyl-L-isoleucine ... 660
Carbobenzyloxy-L-norvalyl-L-leucine ... 660
Carbobenzyloxy-L-norvalyl-L-norleucine ... 660
Carbobenzyloxynorvalylvaline methyl ester ... 662
Carbobenzyloxy-L-valyl-L-isoleucine ... 676
$C_{19}H_{28}N_2O_5S$
Carbobenzyloxyisoleucylmethionine ... 655
Carbobenzyloxy-L-leucyl-L-methionine ... 656
Carbobenzyloxy-L-methionyl-L-isoleucine ... 655
Carbobenzyloxy-L-methionyl-L-leucine ... 655
Carbobenzyloxy-L-valyl-L-methionine methyl ester ... 656
$C_{19}H_{28}N_2O_6$
Nε-*tert*-Butoxycarbonyl-Nα-carbobenzyloxy-L-lysine ... 659
$C_{19}H_{28}N_4O_2$
N,*N*′-(4-Methyl-1,3-phenylene)bis(1-piperidinecarboxamide) ... 565
$C_{19}H_{28}O_2$
4-Methyl-2-(3,5,5,8,8-pentamethyl-5,6,7,8-tetrahydro-2-naphthyl)-1,3-dioxolane ... 184
$C_{19}H_{28}O_3$
Dodecahydro-4,6a,9,9-tetramethylphenanthro(2,3-d)(1,3)-dioxol-3-one ... 54
Menthyl α-methoxyphenylacetate ... 354
$C_{19}H_{28}O_4$
11-(4-Acetylphenoxy)undecanoic acid ... 328
3,9-Di-(3-cyclohexenyl)-2,4,8,10-tetraoxaspiro(5,5)undecane ... 23
Diethyl 4-ethyl-4-phenylpimelate ... 360
$C_{19}H_{28}O_5$
11-(4-(Methoxycarbonyl)phenoxy)undecanoic acid ... 370
$C_{19}H_{28}O_{12}$
Tetraethyl α,α′-dihydroxy-4-oxo-3,5-tetrahydropyrandimalonate ... 91
$C_{19}H_{29}ClN_2OS$
1-(2-Chlorophenyl)-3-dodecanoyl-2-thiourea ... 732
1-(3-Chlorophenyl)-3-dodecanoyl-2-thiourea ... 732
$C_{19}H_{29}ClN_2O_3S$
S-Dodecyl *N*-(2-chloro-4-nitrophenyl)thiocarbamate ... 730
$C_{19}H_{29}N$
N-Benzylidenecyclododecylamine ... 1063
$C_{19}H_{29}NO_2$
Cyclohexyl *N*-(2,6-diisopropylphenyl)carbamate ... 679
$C_{19}H_{29}NO_3$
Ethyl 4-(decanoylamino)benzoate ... 434
N-Vanillyl-10-undecenoamide ... 411
$C_{19}H_{29}N_3O$
1-Benzyl-4-cyclohexylamino-4-piperidinecarboxamide ... 425
Pamaquine ... 985
$C_{19}H_{29}N_3O_4$
Carbobenzyloxy-L-norvalyl-L-leucinamide ... 664
$C_{19}H_{29}N_3O_4S$
Carbobenzyloxyisoleucylmethioninamide ... 656
Carbobenzyloxy-L-methionyl-L-isoleucinamide ... 656
Carbobenzyloxy-L-methionyl-L-leucinamide ... 656
Carbobenzyloxy-L-norleucyl-L-methioninamide ... 656
$C_{19}H_{30}BrNO_2$
1,1-Dimethyl-4-(3-methyl-2-phenylvaleryloxy)-piperidinium bromide ... 359
$C_{19}H_{30}N_2O$
1-Cyclohexyl-3-(2,6-diisopropylphenyl)urea ... 583
$C_{19}H_{30}N_2O_2$
4-(*N*-(2,6-Diisopropylphenyl)carbamoyl)-2,6-dimethylmorpholine ... 714
2,2′-Methylenebis(5,5-dimethyl-3-(methylamino)-2-cyclohexen-1-one ... 57
$C_{19}H_{30}N_2O_3$
4′-Acetamido-3′-methoxydecananilide ... 416
$C_{19}H_{30}O_2$
Benzyl laurate ... 356
3-(Dodecyloxy)benzaldehyde ... 323
4′-Methoxydodecanophenone ... 293
$C_{19}H_{30}O_3$
2-(Dodecyloxy)benzoic acid ... 341
$C_{19}H_{30}O_4S$
11-(Phenylsulfonyl)undecyl acetate ... 781
$C_{19}H_{30}O_7$
Ethyl 1,2-bis(methoxycarbonyl)-10-oxocyclodecanepropionate ... 94
$C_{19}H_{31}ClN_2OS$
6′-Chloro-2-(2-dibutylamino)-ethylthio)-*o*-acetotoluidide hydrochloride ... 420
$C_{19}H_{31}NO$
N-Benzyldodecanamide ... 431
N-Dodecylbenzamide ... 483
$C_{19}H_{31}NO_2$
2,6-Di-*tert*-butyl-4-morpholinomethylphenol ... 246
3-(Diethylamino)-propyl 3-methyl-2-phenylvalerate hydrochloride ... 359
$C_{19}H_{32}BrNO_2$
Diethylmethyl(2-(3-methyl-2-phenylvaleryloxy)-ethyl)-ammonium bromide ... 359
$C_{19}H_{32}N_2$
2-(2-Piperidino-1-cyclohexen-1-yl)-azacyclonon-1-ene ... 39
$C_{19}H_{32}N_2O$
2-Diethylamino-*N*-methyl-*N*-(3-methyl-2-phenylpentyl)-acetamide hydrochloride ... 428
3-(2,6-Diethylphenyl)-1,1-diisobutylurea ... 596
1,1-Diisopropyl-3-(2,6-diisopropylphenyl)urea ... 596
3-(2,6-Diisopropylphenyl)-1,1-dipropylurea ... 596
Tridecanoic N^2-phenylhydrazide ... 503
$C_{19}H_{32}N_2O_2$
2-(Diethylamino)ethyl *N*-(2,6-diisopropylphenyl)carbamate ... 673
N-Dodecyl-*N*-methyl-4-nitroaniline ... 276
$C_{19}H_{32}N_2S$
1-Dodecyl-3-phenyl-2-thiourea ... 748
$C_{19}H_{32}N_4O_2$
3,3′-(4-Methyl-1,3-phenylene)bis(1-butyl-1-methylurea) ... 566
$C_{19}H_{32}O$
Dodecyl *p*-tolyl ether ... 181
$C_{19}H_{32}O_3Sn$
Tributyltin salicylate ... 1130
$C_{19}H_{32}O_6$
6,12-Dioxo-1,17-heptadecanedicarboxylic acid ... 64
$C_{19}H_{33}BrNO_2P$
(4-Nitrobenzyl)-tributylphosphonium bromide ... 830
$C_{19}H_{33}BrN_2O$
3-Dodecylcarbamoyl-1-methylpyridinium bromide ... 948
4-Dodecylcarbamoyl-1-methylpyridinium bromide ... 952
$C_{19}H_{33}ClN_2O$
3-Dodecylcarbamoyl-1-methylpyridinium chloride ... 948
4-Dodecylcarbamoyl-1-methylpyridinium chloride ... 952
$C_{19}H_{33}ClOSn$
Tributyltin 4-chloro-3-methylphenoxide ... 1133
$C_{19}H_{33}IN_2O$
3-Dodecylcarbamoyl-1-methylpyridinium iodide ... 948
4-Dodecylcarbamoyl-1-methylpyridinium iodide ... 952
$C_{19}H_{33}N$
Benzyldodecylamine ... 243
N-Dodecyl-*m*-toluidine ... 219
$C_{19}H_{33}NO_2$
N-Butyl-3,5,7-trimethyl-1-adamantylglycine ... 72
$C_{19}H_{33}NO_6S_2$
Diethylmethyl(2-(1-phenyl-2-propionyloxy)ethylthio)ethyl)-ammonium methyl sulfate ... 357
$C_{19}H_{34}N_2$
N-Dodecyl-*N*-methyl-1,4-phenylenediamine dihydrochloride ... 221
$C_{19}H_{34}N_2O$
1,1,3-Tricyclohexylurea ... 130
$C_{19}H_{34}N_2O_3S$
3,5-Di-*tert*-butyl-4-hydroxybenzaldehyde *N*-phenyl-*N*-(3-sulfolanyl)hydrazone ... 775
$C_{19}H_{34}O_6$
Dibutyl 3-(butoxycarbonyl)adipate ... 76
$C_{19}H_{34}Si_3$
1,4-Dihydro-1,1,4-tris(trimethylsilyl)-naphthalene ... 1123
$C_{19}H_{34}Si_5$
2,2-Bis(dimethylsilyl)-1,1-diphenyl-1,3,3-trimethyltrisilane ... 1114
$C_{19}H_{36}O_2$
Isohexadecyl acrylate, mixture of branched isomers ... 79
$C_{19}H_{37}HgN$
Octadecylmercury(II)cyanide ... 1138
$C_{19}H_{37}N$
Nonadecanenitrile ... 150
$C_{19}H_{37}NO_2$
Oleyl carbamate ... 117
$C_{19}H_{37}NO_4$
Dodecyl 4-methyl-4-nitrohexanoate ... 89
$C_{19}H_{38}N_2O$
N-Ethyl-*N*-(1-methyldecyl)-1-pyrrolidineacetamide hydrochloride ... 108
N-Methyl-*N*-(1-methyldecyl)-1-piperidineacetamide hydrochloride ... 109
$C_{19}H_{38}N_2O_2$
1-Octadecanoylurea ... 128
N,*N*,*N*′,*N*′-Tetrabutylmalonamide ... 111
$C_{19}H_{38}O$
2-Nonadecanone ... 49
9-Nonadecanone ... 49
$C_{19}H_{38}OSn$
(2-Methoxy-2-cyclohexen-1-yl)-tributyltin ... 1130
$C_{19}H_{38}O_2$
Ethyl heptadecanoate ... 74
2-Methyloctadecanoic acid ... 61
$C_{19}H_{38}O_2S$
2-Methylthioethyl palmitate ... 86
$C_{19}H_{38}O_5S$
Methyl 2-sulfooctadecanoate, sodium salt ... 156
2-Methyl-2-sulfooctadecanoic acid ... 156
2-Methyl-2-sulfooctadecanoic acid pyridine salt ... 963
$C_{19}H_{39}ClO_2$
1-Chloro-3-isohexadecyloxy-2-propanol ... 19
$C_{19}H_{39}NO$
N-Octadecylformamide ... 106
$C_{19}H_{39}NO_4S$
3-(Hexadecylsulfonamido)-propionic acid ... 157
$C_{19}H_{40}Cl_2Si$
Dichloromethyloctadecylsilane ... 1098
$C_{19}H_{40}N_2S$
2-(5,7,7-Trimethyl-2-(1,3,3-trimethylbutyl)octyl)-2-thiopseudourea hydrobromide ... 137
$C_{19}H_{40}N_4$
1-(Octadecylideneamino)guanidine nitrate ... 138
$C_{19}H_{40}O$
2-Nonadecanol ... 9

$C_{19}H_{41}N$
N-Isopropylhexadecylamine 28
$C_{19}H_{42}O_6P_2$
Tetrabutyl 1,3-trimethylenediphosphonate 158
$C_{19}H_{46}BN$
Cetyltrimethylammonium borohydride 43
$C_{20}H_4Cl_4I_4O_5$
Bengal Red B 383
$C_{20}H_4Cl_{10}O_2$
1,4-Bis(pentachlorobenzoyl)benzene 307
$C_{20}H_4F_{10}O_2$
1,4-Bis(pentafluorobenzoyl)benzene 307
$C_{20}H_6Br_2Cl_{12}$
2,3-Dibromonaphthalene-bis(hexachlorocyclopentadiene) adduct 173
$C_{20}H_6Cl_{12}INO_2$
2-Iodo-3-nitronaphthalene-bis(hexachlorocyclopentadiene) adduct 279
$C_{20}H_6Cl_{13}NO_2$
2-Chloro-3-nitronaphthalene-bis(hexachlorocyclopentadiene) adduct 279
$C_{20}H_7Cl_{12}I$
2-Iodonaphthalene-bis(hexachlorocyclopentadiene) adduct 173
$C_{20}H_7Cl_{12}NO_5S$
3-Nitro-2-naphthalenesulfonic acid-bis(hexachlorocyclopentadiene) adduct 789
$C_{20}H_8Br_4O_5$
3,4,5,6-Tetrabromofluorescein 339
$C_{20}H_8Br_4O_7$
4′,5′-Dihydroxy-4,5,6,7-tetrabromofluorescein 338
$C_{20}H_8Cl_4O_7$
4′,5′-Dihydroxy-4,5,6,7-tetrachlorofluorescein 338
$C_{20}H_8Cl_{12}O_3S$
2-Naphthalenesulfonic acid-bis(hexachlorocyclopentadiene) adduct 790
$C_{20}H_{10}Br_4O_5$
2-(3,6-Dihydroxy-9*H*-xanthen-9-yl)-3,4,5,6-tetrabromobenzoic acid 339
$C_{20}H_{10}Br_6N_2O_4$
Bis(2,4-dibromophenyl) *N*,*N*′-(4,6-dibromo-1,3-phenylene)-biscarbamate 639
$C_{20}H_{10}Cl_2F_6N_2O_2$
2,5-Bis(3-(trifluoromethyl)anilino)-3,6-dichloro-1,4-benzoquinone 315
$C_{20}H_{10}Cl_4O_5$
2-(3,6-Dihydroxy-9*H*-xanthen-9-yl)-3,4,5,6-tetrachlorobenzoic acid 339
$C_{20}H_{10}Cl_6N_2$
N,*N*′-Terephthalylidenebis(2,4,5-trichloroaniline) 1065
$C_{20}H_{10}Cl_6S_2Si$
Bis(3,4,5-trichloro-2-thienyl)-diphenylsilane 1111
$C_{20}H_{10}N_4O_8S$
2,4,4′,7-Tetranitrodi-1-naphthyl sulfide 270
$C_{20}H_{10}O_3$
3,9-Coeroxonone 309
Dinaphtho(1,2-b:2′,3-d)furan-7,12-dione 895
$C_{20}H_{10}O_4$
2,2′-Bis(1,4-naphthoquinone) 318
$C_{20}H_{11}Cl_3NO_3$
2-Benzoyl-4-chlorophenyl *N*-(2,4-dichlorophenyl)carbamate 675
$C_{20}H_{11}Cl_4F_3N_2O_2$
1-(5-Chloro-2-(3,4-dichlorophenoxy)-phenyl)-3-(6-chlorotrifluoro-*m*-tolyl)urea 508
$C_{20}H_{11}Cl_4NO_3$
2-Benzoyl-4-chlorophenyl *N*-(2,4,5-trichlorophenyl)carbamate 675
$C_{20}H_{11}NO_6S$
1,4-Dihydroxy-2-(2-nitrophenylthio)-anthraquinone 316
$C_{20}H_{11}N_3O_2$
11-Nitrodibenzo(a,c)phenazine 1000
$C_{20}H_{12}Br_2$
4,4′-Dibromo-1,1′-binaphthyl 174
$C_{20}H_{12}Br_2Cl_2N_2$
N,*N*′-Terephthalylidenebis(4-bromo-2-chloroaniline) 1027
$C_{20}H_{12}Br_2Cl_2N_2O_2S_2$
Bis(*S*-(4-chlorophenyl)) *N*,*N*′-(4,6-dibromo-1,3-phenylene)-bis(thiocarbamate) 626
$C_{20}H_{12}Br_4N_2O_3$
4′-Anilino-3,4,5,6-tetrabromophthalanilic acid 451
$C_{20}H_{12}ClF_3N_4O_5S$
1-(4-Chloro-2-(trifluoromethyl)-phenyl)-3-(2-(2,4-dinitrophenylthio)-phenyl)urea 504
$C_{20}H_{12}ClNO_2$
3-(2-Chlorobenzylideneamino)-6*H*-dibenzo(b,d)pyran-6-one 1025
9-(4-Chlorobenzylidene)-2-nitrofluorene 279
$C_{20}H_{12}Cl_2$
9-(3,4-Dichlorobenzylidene)fluorene. 174
$C_{20}H_{12}Cl_2N_4O_4$
N,*N*′-(1,4-Phenylenedimethylidyne)-bis(2-chloro-4-nitroaniline) 1058
N,*N*′-Terephthalylidenebis(4-chloro-3-nitroaniline) 1055
$C_{20}H_{12}Cl_2O_4$
9-(2,4-Dichlorophenoxymethyl)-6-hydroxy-3*H*-xanthen-3-one 319
$C_{20}H_{12}Cl_3F_3N_2O_4S_2$
2-(4-Chloro-2-sulfophenylthio)-4′,5-dichloro-3′(trifluoromethyl)-carbanilide, sodium salt 790
$C_{20}H_{12}Cl_3F_3N_2O_5S$
5-Chloro-2-(4-chloro-2-(4-chloro-trifluoro-*m*-tolylureido)-phenoxy)-benzenesulfonic acid 790
$C_{20}H_{12}Cl_3NO_2$
2′-Benzoyl-2,4,4′-trichlorobenzanilide 473
$C_{20}H_{12}Cl_3NO_3$
2-Benzoyl-4-chlorophenyl *N*-(2,3-dichlorophenyl)carbamate 665
2-Benzoyl-4-chlorophenyl *N*-(2,5-dichlorophenyl)carbamate 665
$C_{20}H_{12}Cl_4N_2$
N,*N*′-Bis-(2,4-dichlorobenzylidene)-1,4-phenylenediamine 1058
N,*N*′-(1,4-Phenylenedimethylidyne)-bis(2,3-dichloroaniline) 1078
N,*N*′-(1,4-Phenylenedimethylidyne)-bis(2,4-dichloroaniline) 1078
N,*N*′-(1,4-Phenylenedimethylidyne)-bis(2,5-dichloroaniline) 1066
N,*N*′-(1,4-Phenylenedimethylidyne)-bis(3,4-dichloroaniline) 1078
N,*N*′-(1,4-Phenylenedimethylidyne)-bis(3,5-dichloroaniline) 1066
$C_{20}H_{12}Cl_4N_2O_2$
1-(2-Benzoyl-4-chlorophenyl)-3-(2,4,5-trichlorophenyl)urea 575
N,*N*′-Bis(2,5-dichlorophenyl)-terephthalamide 455
$C_{20}H_{12}Cl_4N_2O_3$
4′-Anilino-3,4,5,6-tetrachlorophthalanilic acid 451
$C_{20}H_{12}FN$
12-Cyanomethyl-9-fluorobenz(a)-anthracene 758
$C_{20}H_{12}F_3N$
9-(4-(Trifluoromethyl)phenylimino)-fluorene 1079
$C_{20}H_{12}N_2$
Dibenzo(a,c)phenazine 1001
$C_{20}H_{12}N_2O_2$
5,8-Dihydro-2,3-diphenyl-5,8-quinoxalinedione 999
$C_{20}H_{12}N_2O_4$
N,*N*′-(1,1′-Biphenyl-4,4′-diyl)-bismaleimide 722
1,4-Diphthalimido-2-butyne 1018
$C_{20}H_{12}O$
6-Hydroxybenzo(a)pyrene 196
$C_{20}H_{12}O_2$
2-Phenyl-9,10-phenanthrenequinone 318
$C_{20}H_{12}O_3S$
2-(Dibenzothiophene-2-ylcarbonyl)-benzoic acid 898
$C_{20}H_{12}O_4$
5,6,11,12-Tetrahydroxydibenzo(b,h)-biphenylene 194
$C_{20}H_{12}O_5$
3-(6-Hydroxy-3-oxo-3*H*-xanthen-9-yl)benzoic acid 339
4-(6-Hydroxy-3-oxo-3*H*-xanthen-9-yl)benzoic acid 339
$C_{20}H_{13}BN_2O_5$
1-(3-Borono-4-hydroxyphenylazo)-anthraquinone 1125
$C_{20}H_{13}Br$
2-Bromotriptycene 174
$C_{20}H_{13}BrF_3NO_2$
(2-Bromo-4-phenyl)phenyl *N*-(2-(trifluoromethyl)phenyl)carbamate.. 607
$C_{20}H_{13}Br_2ClN_2$
Bis-(4-bromobenzylidene)-2-chloro-1,4-phenylenediamine 1058
$C_{20}H_{13}Cl$
9-(4-Chlorobenzylidene)fluorene 174
$C_{20}H_{13}ClN_2O$
1-(1-Chloro-2-naphthylazo)-2-naphthol 1085
$C_{20}H_{13}ClN_2O_4$
2′-Benzoyl-4′-chloro-4-nitrobenzanilide 452
$C_{20}H_{13}ClN_2O_5$
2-Benzoyl-4-chlorophenyl *N*-(3-nitrophenyl)carbamate 665
$C_{20}H_{13}ClO$
10-Chloro-10-phenyl-9(10*H*)-anthracenone 320
$C_{20}H_{13}ClO_2S$
4-(4-Chlorophenylthio)benzil 306
$C_{20}H_{13}ClO_3$
4-(4-Chlorophenoxy)benzil 306
$C_{20}H_{13}Cl_2NO$
6-(2,2-Dichloroacetamido)chrysene.. 484
$C_{20}H_{13}Cl_2NO_2$
2′-Benzoyl-2,4-dichlorobenzanilide... 473
2′-Benzoyl-4,4′-dichlorobenzanilide.. 453
$C_{20}H_{13}Cl_2NO_3$
2-Benzoyl-4-chlorophenyl *N*-(2-chlorophenyl)carbamate 675
2-Benzoyl-4-chlorophenyl *N*-(4-chlorophenyl)carbamate 622
2-Benzoyl-4-chorophenyl *N*-(3-chlorophenyl)carbamate 675
4-Benzoylphenyl *N*-(3,4-dichlorophenyl)carbamate 675
$C_{20}H_{13}Cl_2NO_4$
2-(Phenoxycarbonyl)phenyl *N*-(3,4-dichlorophenyl)carbamate 629
$C_{20}H_{13}Cl_2N_3$
6-Amino-2,3-bis(2-chlorophenyl)-quinoxaline 998
$C_{20}H_{13}Cl_3N_2O_2$
1-(2-Benzoyl)-4-chlorophenyl)-3-(2,3-dichlorophenyl)urea 530
1-(2-Benzoyl-4-chlorophenyl)-3-(2,4-dichlorophenyl)urea 575
1-(2-Benzoyl-4-chlorophenyl)-3-(2,5-dichlorophenyl)urea 530
1-(2-Benzoyl-4-chlorophenyl)-3-(2,6-dichlorophenyl)urea 575
$C_{20}H_{13}Cl_4N_3O_3$
4-(5-Chloro-4-(4-chlorophenoxy)-1-(3,4-dichlorobenzoyl)-semicarbazide 554
$C_{20}H_{13}N$
5-Cyano-7-methylbenz(a)-anthracene 768
13*H*-Dibenzo(a,i)carbazole 892
$C_{20}H_{13}NOS$
4b,11-Dihydro-4b-phenylisoindolo(1,2-b)-benzothiazol-11-one 710
$C_{20}H_{13}NO_2$
9-(3-Nitrobenzylidene)fluorene 280
9-(4-Nitrobenzylidene)fluorene 279
$C_{20}H_{13}NO_4$
1-Amino-4-hydroxy-2-phenoxyanthraquinone 316
$C_{20}H_{13}N_3O_2$
2,3-Diphenyl-7-nitroquinoxaline 1000
$C_{20}H_{13}N_3O_5$
2,5-Dinitro-9-(4-methoxyphenylimino)fluorene 1049
$C_{20}H_{13}N_7$
2-(4,5-Di-(2-pyridyl)-2*H*-imidazol-2-ylideneamino)-benzimidazole 966
$C_{20}H_{14}$
9-Benzylidenefluorene 166
2,2′-Binaphthyl 166
$C_{20}H_{14}BrClFNO_5$
Ethyl 1-(6-bromopiperonyl)-7-chloro-1,4-dihydro-6-fluoro-4-oxo-3-quinolinecarboxylate 364
$C_{20}H_{14}BrClN_2O_2$
1-(2-Benzoyl-4-chlorophenyl)-3-(4-bromophenyl)urea 575
$C_{20}H_{14}BrNO$
α-(5-Bromo-2-methoxybenzylidene)-1-naphthaleneacetonitrile 763
$C_{20}H_{14}BrNO_4$
Ethyl 6-bromodihydrodioxo-4-methyl-7*H*-dibenzo(f,i,j)-isoquinoline-1-carboxylate 993
$C_{20}H_{14}Br_2I_2N_4O_2$
1,1′-(4,6-Dibromo-1,3-phenylene)-bis(3-(4-iodophenyl)urea) 525
$C_{20}H_{14}Br_2N_2$
N,*N*′-Bis(4-bromobenzylidene)-1,4-phenylenediamine 1065
N,*N*′-(1,4-Phenylenedimethylidyne)-bis(4-bromoaniline) 1078
N,*N*′-Terephthalylidenebis(2-bromoaniline) 1064
N,*N*′-Terephthalylidenebis(3-bromoaniline) 1065
$C_{20}H_{14}Br_2N_6O_6$
1,1′-(4,6-Dibromo-1,3-phenylene)-bis(3-(3-nitrophenyl)urea) 525
$C_{20}H_{14}Br_2O_4$
2,5-Bis(4-methoxyphenyl)-3,6-dibromo-1,4-benzoquinone 317
$C_{20}H_{14}Br_2Sn$
Dibromodi-1-naphthyltin 1133
$C_{20}H_{14}Br_4N_4O_2$
1,1′-(4,6-Dibromo-1,3-phenylene)-bis(3-(3-bromophenyl)urea) 525
$C_{20}H_{14}ClN$
2-(2-Chlorobenzylideneamino)-fluorene 1065
2-(4-Chlorobenzylideneamino)-fluorene 1066
$C_{20}H_{14}ClNO_2$
5-Chloro-2-(2-hydroxybenzylideneamino)-benzophenone 1023
$C_{20}H_{14}ClNO_3$
2-Benzoyl-4-chlorophenyl *N*-phenylcarbamate 665
4-Benzoylphenyl *N*-(4-chlorophenyl)carbamate 629
$C_{20}H_{14}ClN_2O_3$
4-Benzoylphenyl *N*-(2-chlorophenyl)carbamate 629
4-Benzoylphenyl *N*-(3-chlorophenyl)carbamate 629
$C_{20}H_{14}ClN_3O$
6-(4-Chlorophenyl)-3,5-dicyano-4-methyl-4-phenyl-4*H*-pyran-2-amine.. 758
4-(3-Chlorophenyl)-1-(fluoren-9-ylidene)semicarbazide 1026
$C_{20}H_{14}ClN_3O_4$
1-(2-Benzoylphenyl)-3-(4-chloro-3-nitrophenyl)urea 529
$C_{20}H_{14}ClN_3O_5$
2-(*N*-Phenylcarbamoyl)phenyl *N*-(4-chloro-2-nitrophenyl)carbamate 628
$C_{20}H_{14}Cl_2N_2$
N,*N*′-Bis(2-chlorobenzylidene)-1,4-phenylenediamine 1058
N,*N*′-Bis(4-chlorobenzylidene)-1,4-phenylenediamine 1058

$C_{20}H_{14}Cl_2N_2$ (Cont'd)
N,N′-(1,4-Phenylenedimethylidyne)-bis(2-chloroaniline) ... 1077
N,N′-(1,4-Phenylenedimethylidyne)-bis(3-chloroaniline) ... 1078
N,N′-(1,4-Phenylenedimethylidyne)-bis(4-chloroaniline) ... 1078

$C_{20}H_{14}Cl_2N_2O_2$
1-(2-Benzoyl-4-chlorophenyl)-3-(2-chlorophenyl)urea ... 575
1-(2-Benzoyl-4-chlorophenyl)-3-(3-chlorophenyl)urea ... 575
1-(4-Benzoylphenyl)-3-(2,4-dichlorophenyl)urea ... 530
1-(4-Benzoylphenyl)-3-(2,5-dichlorophenyl)urea ... 530
1-(4-Benzoylphenyl)-3-(2,6-dichlorophenyl)urea ... 530
1-(4-Benzoylphenyl)-3-(3,5-dichlorophenyl)urea ... 575

$C_{20}H_{14}Cl_2N_2O_4$
2-Benzyl-4-chlorophenyl *N*-(2-chloro-4-nitrophenyl)carbamate ... 626

$C_{20}H_{14}Cl_2N_2O_5$
2,6-Bis(2-chloro-5-nitrobenzylidene)-1-cyclohexanone ... 299
2,6-Bis(4-chloro-3-nitrobenzylidene)-1-cyclohexanone ... 299

$C_{20}H_{14}Cl_2S_4$
4,4′-Bis(4-chlorophenyl)-5,5′-dimethyltetrathiafulvalene ... 880

$C_{20}H_{14}Cl_4O$
2,6-Bis(2,4-dichlorobenzylidene)-1-cyclohexanone ... 299

$C_{20}H_{14}FNO_4S_2$
N-(1-Anthracenyl)-4-fluorosulfonylbenzenesulfonamide. 827

$C_{20}H_{14}F_2N_2$
N,N′-Bis(4-fluorobenzylidene)-1,4-phenylenediamine ... 1065
N,N′-(1,4-Phenylenedimethylidyne)-bis(4-fluoroaniline) ... 1065
N,N′-Terephthalylidenebis(3-fluoroaniline) ... 1065

$C_{20}H_{14}F_2N_2O_2$
1-(2-Benzoylphenyl)-3-(2,4-difluorophenyl)urea ... 530
1-(2-Benzoylphenyl)-3-(2,5-difluorophenyl)urea ... 530
N,N′-Bis(2-fluorophenyl)-isophthalamide ... 455
N,N′-Bis(4-fluorophenyl)-isophthalamide ... 455

$C_{20}H_{14}Hg_2O_4$
Bis-(phenylmercuri) phthalate ... 1138

$C_{20}H_{14}I_2N_2$
N,N′-(1,4-Phenylenedimethylidyne)-bis(4-iodoaniline) ... 1078

$C_{20}H_{14}I_6N_2O_6$
N,N′-Adipoylbis-(3-amino-2,4,6-triiodobenzoic acid, sodium salt) ... 405

$C_{20}H_{14}N_2$
2,2′-Diazonaphthalene ... 1093
9,10-Dimethylacenaphtho(1,2-b)-quinoxaline ... 1001

$C_{20}H_{14}N_2O$
4b,5-Dihydro-4b-phenyl-11*H*-isoindolo(2,1-a)benzimidazol-11-one ... 707
2-(2-Naphthyl)-4-quinolinecarboxamide ... 991

$C_{20}H_{14}N_2O_2$
2-(Benzylideneamino)-7-nitrofluorene ... 1073
2,3-Diphenyl-5,8-quinoxalinediol ... 998
Disperse Blue 19 (C.I. 61110) ... 1093
4-(4-Hydroxyphenyl)-2-phenyl-1(2*H*)-phthalazinone ... 697
8-Quinolyl *N*-(1-naphthyl)carbamate. 994

$C_{20}H_{14}N_2O_3$
Rhodamine sulfate ... 1052

$C_{20}H_{14}N_2O_4S$
8-Hydroxy-5-(1-naphthylazo)-2-naphthalenesulfonic acid, sodium salt ... 1085

$C_{20}H_{14}N_2O_{10}S_3$
6-Hydroxy-5-(4-sulfo-1-naphthylazo)-1,3-naphthalenedisulfonic acid, trisodium salt ... 1084

$C_{20}H_{14}N_2O_{11}S_3$
4,5-Dihydroxy-3-(4-sulfo-1-naphthylazo)-2,7-naphthalenedisulfonic acid trisodium salt ... 1083

$C_{20}H_{14}N_4$
5,6-Diphenyl-3-(4-pyridyl)-1,2,4-triazine ... 980

$C_{20}H_{14}N_4O_2$
2-Amino-1-(4-nitrophenylazo)-anthracene ... 1088

$C_{20}H_{14}N_4O_3$
3,5-Dicyano-4-methyl-4-(4-nitrophenyl)-6-phenyl-4*H*-pyran-2-amine ... 763

$C_{20}H_{14}N_4O_4$
N,N′-Bis(3-nitrobenzylidene)-1,4-phenylenediamine ... 1057

$C_{20}H_{14}N_4O_6$
N,N′-Bis(3-nitrophenyl)-isophthalamide ... 469
N,N′-Bis(3-nitrophenyl)-terephthalamide ... 469
N,N′-Terephthalylidenebis(2-hydroxy-5-nitroaniline) ... 1033

$C_{20}H_{14}O$
2-(9-Fluorenylidenemethyl)phenol ... 196
4-(9-Fluoroenylidenemethyl)phenol.. 196

$C_{20}H_{14}OS$
S-(2-Fluorenyl) thiobenzoate ... 380

$C_{20}H_{14}O_2$
5-Acetoxybenz(a)anthracene ... 381
7-Benz(a)anthryl acetate ... 381
Dibenzo(h,l)tricyclo[4.3.3.0^{1,6}]dodec-3-ene-7,10-dione ... 309
1,3-Dibenzoylbenzene ... 313
3,3-Diphenylphthalide ... 386
7-Methylbenz(a)anthracene-5-carboxylic acid ... 346
7-Methylbenz(a)anthracene-9-carboxylic acid ... 346
Methyl benzo(c)phenanthrene-3-carboxylate ... 376
Methyl benzo(c)phenanthrene-4-carboxylate ... 376

$C_{20}H_{14}O_2S$
1-Naphthyl sulfone ... 786

$C_{20}H_{14}O_3$
2′-Benzoyl-2-biphenylcarboxylic acid ... 344
2,2-Diphenyl-4*H*-1,3-benzodioxin-4-one ... 384
3-((3,4-Methylenedioxy)phenyl)-2′-acrylonaphthone ... 304

$C_{20}H_{14}O_4$
2,4-Dibenzoylresorcinol ... 310
Diphenyl isophthalate ... 379
2-(2-(2-Hydroxy-1-naphthoyl)vinyl)-benzoic acid ... 339
3,3′-(1,4-Phenylene)bis(1-(2-furyl)-2-propen-1-one) ... 853

$C_{20}H_{14}S_2$
2-Naphthyl disulfide ... 213

$C_{20}H_{14}Te_2$
1-Naphthyl ditelluride ... 1143

$C_{20}H_{15}Br$
5-Bromo-7,12-dimethylbenz(a)-anthracene ... 175

$C_{20}H_{15}Br_2NO_2$
3-Bromo-4-biphenylyl *N*-(2-bromo-4-methylphenyl)carbamate ... 632

$C_{20}H_{15}Cl$
7-Chloromethyl-12-methylbenz(a)-anthracene ... 172
1-Chlorotriphenylethylene ... 174

$C_{20}H_{15}ClFNO_2$
2-Benzyl-4-chlorophenyl *N*-(2-fluorophenyl)carbamate ... 632
2-Benzyl-4-chlorophenyl *N*-(3-fluorophenyl)carbamate ... 632
2-Benzyl-4-chlorophenyl *N*-(4-fluorophenyl)carbamate ... 632

$C_{20}H_{15}ClN_2O_2$
1-(2-Benzoyl-4-chlorophenyl)-3-phenylurea ... 575
1-(4-Benzoylphenyl)-3-(2-chlorophenyl)urea ... 530
1-(4-Benzoylphenyl)-3-(3-chlorophenyl)urea ... 530
1-(4-Benzoylphenyl)-3-(4-chlorophenyl)urea ... 575

$C_{20}H_{15}ClN_2O_3$
N-(4-Benzyloxybenzylidene)-4-chloro-3-nitroaniline ... 1024

$C_{20}H_{15}ClN_2O_4$
2-Benzylphenyl *N*-(4-chloro-2-nitrophenyl)carbamate ... 624

$C_{20}H_{15}ClN_2O_5$
4-(Benzyloxy)phenyl *N*-(2-chloro-4-nitrophenyl)carbamate ... 612
3-Chloro-4-biphenylyl *N*-(2-methoxy-4-nitrophenyl)carbamate ... 618

$C_{20}H_{15}ClN_4O_2$
2-(2-Chloro-4-(4-nitrophenylazo)-phenyl)isoindoline ... 1082

$C_{20}H_{15}ClN_4O_5S$
1-(4-Chloro-2-methylphenyl)-3-(2-(2,4-dinitrophenylthio)phenyl)urea... 521

$C_{20}H_{15}ClO$
2,2,2-Triphenylacetyl chloride ... 389

$C_{20}H_{15}ClO_7$
2,4-Bis(4-methoxyphenyl)-6-methylpyrylium perchlorate ... 934

$C_{20}H_{15}Cl_2NO$
N-(4-Benzyloxybenzylidene)-2,4-dichloroaniline ... 1040
N-(4-Benzyloxybenzylidene)-3,4-dichloroaniline ... 1040
2′,3′-Dichloro-2,2-diphenylacetanilide ... 408
3′,4′-Dichloro-2,2-diphenylacetanilide ... 408

$C_{20}H_{15}Cl_2NO_2$
2-Benzyl-4-chlorophenyl *N*-(2-chlorophenyl)carbamate ... 632
2-Benzyl-4-chlorophenyl *N*-(3-chlorophenyl)carbamate ... 632
2-Benzyl-4-chlorophenyl *N*-(4-chlorophenyl)carbamate ... 632
3-Chloro-4-biphenylyl *N*-(2-chloro-6-methylphenyl)carbamate ... 637

$C_{20}H_{15}Cl_2NO_3$
3-Chloro-4-biphenylyl *N*-(5-chloro-2-methoxyphenyl)carbamate ... 618

$C_{20}H_{15}Cl_2N_3O$
3,4-Dichloro-2′-(4-methylphenylazo)-benzanilide ... 1081

$C_{20}H_{15}F$
6,8-Dimethyl-5-fluorobenz(a)-anthracene ... 175
7,12-Dimethyl-10-fluorobenz(a)-anthracene ... 175
7,12-Dimethyl-5-fluorobenz(a)-anthracene ... 175
7,12-Dimethyl-9-fluorobenz(a)-anthracene ... 175

$C_{20}H_{15}FN_2O_5$
4-(Benzyloxy)phenyl *N*-(2-fluoro-5-nitrophenyl)carbamate ... 612

$C_{20}H_{15}F_3N_2OS$
1-(4-Phenoxyphenyl)-2-thio-3-(3-(trifluoromethyl)phenyl)urea ... 728

$C_{20}H_{15}N$
2-(Benzylideneamino)fluorene ... 1064
2,3-Diphenylindole ... 887
6-(*p*-Tolyl)phenanthridine picrate ... 998

$C_{20}H_{15}NO$
1-Acetamidobenzo(c)phenanthrene. 490
2-Acetamidobenzo(c)phenanthrene. 490
3-Acetamidobenzo(c)phenanthrene. 490
4-Acetamidobenzo(c)phenanthrene. 490
5-Acetamidobenzo(c)phenanthrene. 490
6-Acetamidobenzo(c)phenanthrene. 490
4-((Diphenyl)hydroxymethyl)-benzonitrile ... 761
Triphenylmethyl isocyanate ... 769

$C_{20}H_{15}NOS$
α,α-Diphenyl-2-benzothiazolemethanol ... 912

$C_{20}H_{15}NO_2$
α-(3-Benzoylphenylimino)-*o*-cresol.... 1034
N-Phenyldibenzamide ... 477

$C_{20}H_{15}NO_3$
N-Benzoyl-*N*-phenylanthranilic acid. 471
4-Benzoylphenyl *N*-phenylcarbamate ... 675
3,3-Bis(4-hydroxyphenyl)-oxindole ... 689
Ethyl 1-pyreneoxamate ... 489
N-(3,4-Methylenedioxybenzylidene)-4-phenoxyaniline ... 1050

$C_{20}H_{15}NO_3S$
2-(*p*-Toluenesulfonamido)-9-fluorenone ... 823

$C_{20}H_{15}NS$
Triphenylmethyl thiocyanate ... 768

$C_{20}H_{15}N_3O$
3,5-Dicyano-4,6-diphenyl-4-methyl-4*H*-pyran-2-amine ... 763

$C_{20}H_{15}N_3O_3$
α-(4-Nitrobenzylidene)harmine ... 918

$C_{20}H_{15}N_3O_4S_2$
2′-(2-Benzothiazolyl)-4′-nitro-*p*-toluenesulfonanilide ... 915

$C_{20}H_{15}N_5O$
N-(2,3-Bis(2-pyridyl)quinoxalin-5-yl)-acetamide ... 1000

$C_{20}H_{16}$
3,9-Dimethylbenz(a)anthracene ... 167
6,8-Dimethylbenz(a)anthracene ... 167
1,12-Dimethylbenzo(c)-phenanthrene ... 167
5,8-Dimethylbenzo(c)phenanthrene. 167
5,6-Dimethylchrysene ... 168
9-Methyl-9-phenylfluorene ... 166

$C_{20}H_{16}As_2N_2$
5,5′-Arsenobis(8-methylquinoline) dihydrochloride ... 1128

$C_{20}H_{16}BrNO$
N-(4-Benzyloxybenzylidene)-3-bromoaniline ... 1047
N-(4-Benzyloxybenzylidene)-4-bromoaniline ... 1040
4′-Bromo-2,2-diphenylacetanilide ... 408

$C_{20}H_{16}BrNO_3$
(2-Bromo-4-phenyl)phenyl *N*-(2-methoxyphenyl)carbamate ... 618
(2-Bromo-4-phenyl)phenyl *N*-(4-methoxyphenyl)carbamate ... 620

$C_{20}H_{16}BrN_3$
2-(4-(4-Bromophenylazo)phenyl)-isoindoline ... 1082

$C_{20}H_{16}ClN$
10-Methyl-9-phenylacridinium chloride ... 996

$C_{20}H_{16}ClNO$
N-(4-Benzyloxybenzylidene)-3-chloroaniline ... 1023
N-(4-Benzyloxybenzylidene)-4-chloroaniline ... 1040
4′-Chloro-2,2-diphenylacetanilide ... 408

$C_{20}H_{16}ClNOS$
2-(2-Aminophenylthiomethyl)-4′-chlorobenzophenone hydrochloride ... 309

$C_{20}H_{16}ClNO_2$
2-Benzyl-4-chlorophenyl *N*-phenylcarbamate ... 632
2-Benzylphenyl *N*-(2-chlorophenyl)-carbamate ... 632
3-Chloro-4-biphenylyl *N*-(*o*-tolyl)-carbamate ... 632
5-Chloro-2-(2-hydroxybenzylamino)-benzophenone ... 309
N-(5-Chloro-2-methoxyphenyl)-benzanilide ... 466

$C_{20}H_{16}ClNO_2S$
3-(4-Chlorophenyl)-3,4-dihydro-2-phenyl-2*H*-1,4-benzothiazine-1,1-dioxide ... 774

$C_{20}H_{16}ClNO_3$
4-(Benzyloxy)phenyl *N*-(3-chlorophenyl)carbamate ... 612
2-(10-Chloro-5*H*-benzo(6,7)-cyclohepta(1,2-b)quinolin-8-yloxy)-acetic acid ... 985
3-Chloro-4-biphenylyl *N*-(4-methoxyphenyl)carbamate ... 620

$C_{20}H_{16}ClN_3O$
6-(4-Hydroxy-1-naphthylazo)-1-methylquinolinium chloride ... 1084

$C_{20}H_{16}ClN_3O_3$
1-(2-Chloro-4-nitrophenyl)-3-(diphenylmethyl)urea 526
$C_{20}H_{16}ClN_3O_4S$
2′-Chloro-5′-(N-(2-cyanoethyl)-sulfamoyl)-1-hydroxy-2-naphthanilide 812
$C_{20}H_{16}Cl_2N_2O_2$
2,5-Bis(2-methylanilino)-3,6-dichloro-1,4-benzoquinone 318
2,5-Bis(3-methylanilino)-3,6-dichloro-1,4-benzoquinone 318
2,5-Bis(4-methylanilino)-3,6-dichloro-1,4-benzoquinone 317
$C_{20}H_{16}Cl_2N_2O_3$
5-Acetyl-3,4-bis(4-chlorophenyl)-1-(2-hydroxyethyl)-6(1H)-pyridazinone. 693
$C_{20}H_{16}Cl_2N_2O_4$
2,5-Bis(2-methoxyanilino)-3,6-dichloro-1,4-benzoquinone 316
2,5-Bis(3-methoxyanilino)-3,6-dichloro-1,4-benzoquinone 315
2,5-Bis(4-methoxyanilino)-3,6-dichloro-1,4-benzoquinone 316
$C_{20}H_{16}Cl_2N_2S$
1-(3,4-Dichlorophenyl)-3-(diphenylmethyl)-2-thiourea 730
$C_{20}H_{16}Cl_2O$
2,6-Bis(2-chlorobenzylidene)-1-cyclohexanone 299
2,6-Bis(4-chlorobenzylidene)-1-cyclohexanone 282
α-(2,6-Dichloro-3-methylphenyl)-benzhydrol 198
$C_{20}H_{16}FNO$
N-(4-Benzyloxybenzylidene)-2-fluoroaniline 1046
N-(4-Benzyloxybenzylidene)-4-fluoroaniline 1040
2,2-Diphenyl-2′-fluoroacetanilide 408
2,2-Diphenyl-4′-fluoroacetanilide 408
$C_{20}H_{16}FNO_3$
4-(Benzyloxy)phenyl N-(2-fluorophenyl)carbamate 612
4-(Benzyloxy)phenyl N-(4-fluorophenyl)carbamate 612
$C_{20}H_{16}FN_3O_3$
1-Benzyl-3-(2-fluoro-5-nitrophenyl)-1-phenylurea 524
1-Benzyl-3-(4-fluoro-3-nitrophenyl)-1-phenylurea 523
$C_{20}H_{16}FN_5O_2$
4-(3-(2-Fluorophenyl)ureido)benzoic (2-pyridylmethylene)hydrazide 928
4-(3-(2-Fluorophenyl)ureido)benzoic (3-pyridylmethylene)hydrazide 929
$C_{20}H_{16}F_2N_2O$
1-(2,4-Difluorophenyl)-3-(diphenylmethyl)urea 536
$C_{20}H_{16}F_3N_3O$
1-(4-Anilinophenyl)-3-(2-(trifluoromethyl)phenyl)urea 508
1-(4-Anilinophenyl)-3-(3-(trifluoromethyl)phenyl)urea 508
$C_{20}H_{16}F_3N_3S$
1-(4-Anilinophenyl)-3-(2-(trifluoromethyl)phenyl)-2-thiourea 728
$C_{20}H_{16}F_4N_2$
N,N′-Dimethyl-N,N′-diphenyl-2,3,5,6-tetrafluoro-p-phenylenediamine 224
$C_{20}H_{16}IN$
1-(9-Anthracenylmethyl)pyridinium iodide dihydrate 959
$C_{20}H_{16}INO$
N-(4-Benzyloxybenzylidene)-4-iodoaniline 1047
$C_{20}H_{16}N_2$
N,N′-Dibenzylidene-1,4-phenylenediamine 1058
N,N′-Terephthalylidenedianiline 1078
$C_{20}H_{16}N_2O$
Benzil monophenylhydrazone 302
2-(3-Benzyloxyphenyl)-benzimidazole 902
$C_{20}H_{16}N_2O_2$
1-(4-Benzoylphenyl)-3-phenylurea 575
N,N′-Bis(2-hydroxybenzylidene)-1,4-phenylenediamine 1028
N,N′-Bis(4-hydroxybenzylidene)-1,4-phenylenediamine 1028
4,4′-Dimethyl-2,2′-dihydroxy-6,6′-biquinoline 984
1-Hydroxy-1-phenyl-2-(4-pyridylmethyl)-1H-isoindol-3(2H)-one 931
N,N′-(1,4-Phenylenedimethylidyne)-bis(3-aminophenol) 1029
N-Phenyl-N′-piperonylidene-1,4-phenylenediamine 1057
N,N′-Terephthalylidenebis(2-aminophenol) 1029
N,N′-Terephthalylidenebis(4-aminophenol) 1029
$C_{20}H_{16}N_2O_2S$
2-(2-Aminophenylthio)-N-(1-naphthyl)-succinimide 718
N-Cyano-2′-phenyl-p-toluenesulfonanilide 826
N-Cyano-4′-phenyl-p-toluenesulfonanilide 826
$C_{20}H_{16}N_2O_2S_2$
2′-(2-Benzothiazolyl)-p-toluenesulfonanilide 915
3,4-Bis(3-methyl-2(3H)-benzothiazolylidenemethyl)-1,3-cyclobutadienediylium-1,2-diolate 1095
$C_{20}H_{16}N_2O_3$
N-(4-Benzyloxybenzylidene)-3-nitroaniline 1039
2-Hydroxy-5-(indanylmethylene)-1-phenyl-4,6-(1H,5H)pyrimidinedione 697
$C_{20}H_{16}N_2O_4$
N,N′-(1,4-Phenylene)bis(anthranilic acid) 342
$C_{20}H_{16}N_2O_4S$
N-(7-Nitro-2-fluorenyl)-p-toluenesulfonamide 822
$C_{20}H_{16}N_2O_5$
2,6-Bis(3-nitrobenzylidene)-cyclohexanone 285
$C_{20}H_{16}N_2O_8S$
4′,4″-Sulfonyldimaleanilic acid 781
$C_{20}H_{16}N_2S_2$
4-Methyl-2-quinolyl disulfide 987
$C_{20}H_{16}N_4$
5,12-Dihydro-5,12-dimethylquinoxalino(2,3-b)-phenazine diperchlorate 249
$C_{20}H_{16}N_4O_2$
1-Cinnamoyl-3-(diazoacetyl)-4-phenyl-2-pyrazoline 1073
1,3-Dimethyl-6,7-diphenyllumazine 978
2-(4-(4-Nitrophenylazo)phenyl)-isoindoline 1090
$C_{20}H_{16}N_4O_3$
β-(4-Antipyryl)-α-cyano-4-nitrostyrene 765
$C_{20}H_{16}N_4O_5S$
Eriochrome Red B (C.I. 18760) 1085
$C_{20}H_{16}N_4O_5S_2$
1-(2-(2,4-Dinitrophenylthio)phenyl)-3-(4-(methylthio)phenyl)urea 560
$C_{20}H_{16}N_4O_7S_2$
1-(2-(2,4-Dinitrophenylthio)phenyl)-3-(p-tosyl)urea 815
$C_{20}H_{16}N_6O_2$
N,N′-Terephthalylidenebis-(nicotinic hydrazide) 1069
$C_{20}H_{16}O$
2,6-Bis(benzylidene)-1-cyclohexanone 304
7,12-Dihydro-6,8-dimethylbenz(a)-anthracen-7-one 315
4,7-Dihydro-1,3-diphenylisobenzofuran 846
7,12-Dimethyl-9-hydroxybenz(a)-anthracene 197
2,2-Diphenylacetophenone 296
5-Methoxy-7-methylbenz(a)-anthracene 188
6-Methoxy-12-methylbenz(a)-anthracene 189
8-Methoxy-7-methylbenz(a)-anthracene 189
4′-Methyl-3-(1-naphthyl)-acrylophenone 305
1,2,3,4-Tetrahydro-2-phenylphenanthren-1-one 298
3-(p-Tolyl)-1′-acrylonaphthone 305
$C_{20}H_{16}O_2$
1,6-Bis(4-formylphenyl)-1,3,5-hexatriene 324
7,8-Dihydro-6H(1)benzopyrano(3,2-d)xanthene compound with hydrochloric acid 184
Diphenylmethyl benzoate 370
9-(2-(Hydroxymethyl)phenyl)-9-fluorenol 199
2′-Methoxy-3-phenyl-1′-acrylonaphthone 301
β-Methyl-1-pyrenepropionic acid 334
2-Phenoxy-4′-phenylacetophenone 292
2-Phenyl-4′-phenoxyacetophenone 292
$C_{20}H_{16}O_2Pb$
Bis(2-furyl)diphenyllead 1135
$C_{20}H_{16}O_2S$
Phenyl α-phenylstyryl sulfone 784
$C_{20}H_{16}O_2S_4$
4,4′-Bis(4-methoxyphenyl)-tetrathiafulvalene 880
$C_{20}H_{16}O_3$
3-(4-Benzyloxyphenyl)-1-(2-furyl)-2-propen-1-one 845
3,6-Diphenyl-1,2,3,6-tetrahydrophthalic anhydride 388
$C_{20}H_{16}O_4$
Dimethyl 9,10-dihydro-9,10-ethenoanthracene-11,12-dicarboxylate 371
$C_{20}H_{16}O_4S_2$
2,5-Bis(4-methoxyphenylthio)-1,4-benzoquinone 316
$C_{20}H_{16}O_5$
6-(6-Hydroxy-3-oxo-3H-xanthen-9-yl)-3-cyclohexene-1-carboxylic acid. 326
$C_{20}H_{16}O_6$
5,7-Diacetoxy-2-methyl-3-phenyl-4H-1-benzopyran-4-one 378
$C_{20}H_{16}O_7$
3,4-Dihydroxy-2,5-dioxotetrahydrofuran di-p-toluate 371
$C_{20}H_{16}PbS_2$
Bis(2-thienyl)diphenyllead 1135
$C_{20}H_{16}S_2$
5,16-Dihydrotribenzo(b,d,h)(1,6)-dithiecin 214
$C_{20}H_{16}S_4$
4,4′-Bis(p-tolyl)tetrathiafulvalene 880
$C_{20}H_{16}S_6$
4,4′-Bis(4-(methylthio)phenyl)-tetrathiafulvalene 880
$C_{20}H_{17}As$
5,7-Dihydro-6-phenyldibenz(c,e)-arsepin 1128
$C_{20}H_{17}BrFNO$
4-Bromo-2-(4-(cyclopropylformyl)-phenyl)-2-(4-fluorophenyl)-butyronitrile 751
$C_{20}H_{17}BrO_2$
5-Bromo-2-methoxytriphenylmethanol 198
$C_{20}H_{17}ClF_3N$
3-tert-Butyl-1-chloro-4-(3-(trifluoromethyl)phenyl)-isoquinoline 996
$C_{20}H_{17}ClNP$
(Cyanomethyl)-triphenylphosphonium chloride 831
$C_{20}H_{17}ClN_2O$
1-Benzyl-3-(2-chlorophenyl)-1-phenylurea 527
$C_{20}H_{17}ClN_4O_3$
2-(3-(3-Chlorophenyl)ureido)acetic (2-hydroxy-1-naphthylmethylene)-hydrazide 513
$C_{20}H_{17}ClO_2Sn$
Methyl 2-(chlorodiphenylstannyl)-benzoate 1131
$C_{20}H_{17}ClS$
2-Chloro-1,2-diphenylethyl phenyl sulfide 212
$C_{20}H_{17}Cl_2NO_2$
2-(4-tert-Butylphenyl)-4-(2,4-dichlorobenzylidene)-2-oxazolin-5-one 382
$C_{20}H_{17}Cl_2N_3O_2$
3-(4-Chlorophenylcarbamoyl)-1-(phenylcarbamoylmethyl)-pyridinium chloride 956
$C_{20}H_{17}Cl_3IP$
Ethyltris(3-chlorophenyl)-phosphonium iodide 830
$C_{20}H_{17}Cl_4NOSi$
Dimethylphenyl(phenyl(2,3,5,6-tetrachloro-4-pyridyl)methoxy)-silane 1107
$C_{20}H_{17}FN_2O$
1-Benzyl-3-(3-fluorophenyl)-1-phenylurea 527
1-Benzyl-3-(4-fluorophenyl)-1-phenylurea 527
1-(Diphenylmethyl)-3-(2-fluorophenyl)urea 535
$C_{20}H_{17}FN_4O$
1-(2-Fluorophenyl)-3-(5-methyl-2-(phenylazo)phenyl)urea 1081
$C_{20}H_{17}NO$
2-(Benzylamino)benzophenone 312
N-Benzylbenzanilide 472
N-(4-Benzyloxybenzylidene)aniline 1039
2,2-Diphenylacetanilide 430
2,3-Diphenyl-1-(4-pyridyl)-1-cyclopropanol 934
2-(4-Methoxyphenyl)-3-(1-naphthyl)-propionitrile 753
N-(4-Methylbenzylidene)-4-phenoxyaniline 1050
3-Methyl-1-(2-naphthyl)-4-phenyl-2-azetidinone 691
N-Tritylformamide 438
$C_{20}H_{17}NO_2$
N-(4-Benzyloxybenzylidene)-4-hydroxyaniline 1029
2-Biphenylyl N-(o-tolyl)carbamate 680
4-Biphenylyl N-(m-tolyl)carbamate 681
Cinnamyl N-(1-naphthyl)carbamate 681
4-Cyclohexylphenyl N-(o-tolyl)-carbamate 679
N-(4-Methoxybenzylidene)-4-phenoxyaniline 1050
Methyl 3-(diphenylamino)benzoate 373
Methyl 4-(diphenylamino)benzoate 373
$C_{20}H_{17}NO_3$
2-Biphenylyl N-(2-methoxyphenyl)-carbamate 653
4-(Ethoxycarbonyl)-N-(2-hydroxy-1-naphthylmethylene)aniline 1034
6-Methoxy-α-(4-phenoxyphenylimino)-o-cresol 1029
DL-2-(4-Phenoxyphenyl)-2-phenylglycine 328
$C_{20}H_{17}NO_3S$
2′-Benzoyl-p-toluenesulfonanilide 823
N-(p-Tolylsulfonyl)-benzanilide 823
$C_{20}H_{17}NO_4$
4-Biphenylyl 2-methoxy-5-nitrobenzyl ether 269
1,4-Dihydro-1,4-dioxo-3-methyl-2-naphthylmethyl N-benzylcarbamate 654
2-Naphthyl N-(4-(ethoxycarbonyl)-phenyl)carbamate 676
$C_{20}H_{17}NS$
N-Benzylthiobenzanilide 743
$C_{20}H_{17}N_3O$
1-(Diphenylmethyl)-3-(3-nitrophenyl)urea 563
$C_{20}H_{17}N_3O_2$
4,4-Diphenyl-1-(2-hydroxybenzylidene)-semicarbazide 1036
Hippuric (1-naphthylmethylene)-hydrazide 501
N′-(4-Nitrophenyl)-N-(4-tolyl)-benzamidine 1055
$C_{20}H_{17}N_3O_2S$
Benzenesulfonic (9,10-dihydro-10-methyl-9-acridinylidene)hydrazide 1055

$C_{20}H_{17}N_3O_2S_2$
4′-Amino-2′-(2-benzothiazolyl)-*p*-toluenesulfonanilide 914

$C_{20}H_{17}N_3O_3$
Hippuric (2-hydroxy-1-naphthylmethylene)hydrazide 496

$C_{20}H_{17}N_5O_5$
4-(3-Furfurylureido)benzoic (3-nitrobenzylidene)hydrazide 1069

$C_{20}H_{18}$
1,4-Dibenzylbenzene 162
12,12a-Dihydro-3,9-dimethylbenz(a)-anthracene 167
1′,2′,3,4-Tetrahydro-1,1′-binaphthyl 165
1,1,1-Triphenylethane 163
1,1,2-Triphenylethane 163

$C_{20}H_{18}BrNO$
4′-Bromo-2′,6′-dimethyl-2-(1-naphthyl)acetanilide 485

$C_{20}H_{18}Br_2N_2$
α,α'-Bis(4-bromoanilino)-*p*-xylene 232

$C_{20}H_{18}Br_2N_6O_6S_2$
1,1′-(4,6-Dibromo-1,3-phenylene)-bis(3-(4-sulfamoylphenyl)urea) 807

$C_{20}H_{18}ClN_3O$
1-(4-Anilinophenyl)-3-(3-chloro-2-methylphenyl)urea 525

$C_{20}H_{18}ClN_3O_2$
1-(4-Anilinophenyl)-3-(5-chloro-2-methoxyphenyl)urea 518
4-Chloro-*N*,*N*′-dibenzyl-6-nitro-1,3-phenylenediamine 261
3-(*N*-Phenylcarbamoyl)-1-(*N*-phenylcarbamoylmethyl)-pyridinium chloride 947

$C_{20}H_{18}Cl_2N_2$
α,α'-Bis(4-chloroanilino)-*p*-xylene 232

$C_{20}H_{18}Cl_4N_2O_4$
O,*O*′-(1,4-Cyclohexylene) bis(*N*-(2,3-dichlorophenyl)carbamate) 623

$C_{20}H_{18}F_2N_2$
α,α'-Bis(4-fluoroanilino)-*p*-xylene 232

$C_{20}H_{18}F_3O_6PS$
Methyltriphenoxyphosphonium trifluoromethanesulfonate 830

$C_{20}H_{18}Fe_2$
1,1″-Biferrocene 1140

$C_{20}H_{18}INO_2$
1-Ethyl-2-(3,4-(methylenedioxy)-styryl)quinolinium iodide 990

$C_{20}H_{18}INO_4$
Berberinium iodide 997

$C_{20}H_{18}I_2N_2$
α,α'-Bis(4-iodoanilino)-*p*-xylene 232

$C_{20}H_{18}NO_4$
Berberine bisulfate 997

$C_{20}H_{18}N_2$
N-(4-Methylbenzylidene)-*N*′-phenyl-1,4-phenylenediamine 1057
N-(2-(2-Pyridyl)ethyl)-2-fluorenamine. 946

$C_{20}H_{18}N_2O$
1-Benzyl-1,3-diphenylurea 579
N-(1,2-Diphenylethyl)-isonicotinamide 951
N-(1,2-Diphenylethyl)-picolinamide 953
N-(4-Methoxybenzylidene)-*N*′-phenyl-1,4-phenylenediamine 1043
4-Phenyl-2-(4-pyridyl)-1,4,7,8-tetrahydroquinolin-5(6*H*)-one 952

$C_{20}H_{18}N_2O_2$
α-(4-Anilinophenylimino)-6-methoxy-*o*-cresol 1030
4-Anilinophenyl *N*-(*p*-tolyl)-carbamate 671
6-Benzyl-4-hydroxy-3-(α-iminobenzyl)pyridin-2(1*H*)-one 1033
N-(Diphenylmethoxymethyl)-nicotinamide 938
2-Methoxybenzoic (1-(1-naphthyl)-ethylidene)hydrazide 498
N-Vanillylidenebenzidine 1029

$C_{20}H_{18}N_2O_2S$
Deoxybenzoin (phenylsulfonyl)-hydrazone 1060
1-(4-Ethoxyphenyl)-3-(1-naphthylcarbonyl)-2-thiourea 737
1-(1-Naphthyl)-3-(2-phenoxypropionyl)-2-thiourea 736
N-(4-Nitrophenyl)-α-(*p*-tolylthio)-benzylamine 270

$C_{20}H_{18}N_2O_3$
4-Hydroxy-3-(α-iminobenzyl)-6-(4-methoxyphenyl)-1-methylpyridin-2(1*H*)-one 931
2-(1-Naphthyl)acetic 4-hydroxy-3-methoxybenzylidenehydrazide 495

$C_{20}H_{18}N_2O_4$
1-(4-(Ethoxycarbonyl)phenyl)-3-(5-hydroxy-1-naphthyl)urea 548
3-(4-Methoxyphenylcarbamoyl)-2-naphthyl *N*-methylcarbamate 654

$C_{20}H_{18}N_2O_5$
2-(1-Naphthoxy)ethyl *N*-(4-methyl-2-nitrophenyl)carbamate 665

$C_{20}H_{18}N_2O_6$
1,4-Bis(2-carboxybenzoyl)-piperazine 707

$C_{20}H_{18}N_2O_6S_2$
N-(4-Nitrophenyl)di-α-toluenesulfonamide 819

$C_{20}H_{18}N_2O_7S_2$
5-Nitro-2-(*p*-toluenesulfonamido)-phenyl *p*-toluenesulfonate 819

$C_{20}H_{18}N_2O_{10}$
Methyl 2-deoxy-3,5-bis-*O*-(4-nitro-benzoyl)-α,β-D-riboside 370

$C_{20}H_{18}N_2Pb$
Diphenyldi(1-pyrrolyl)lead 1135

$C_{20}H_{18}N_4OS$
1-(4-(Methylthio)phenyl)-3-(4-(phenylazo)phenyl)urea 1087

$C_{20}H_{18}N_4O_2$
3-Hydroxy-2-quinoxalineacetic acid (5-indanylmethylene)-hydrazide 999
1-Methyl-6-(4-nitrophenylazo)-1,2,3,4-tetrahydrobenzo(*H*)-quinoline 1090
Oxalic bis((3-phenylallylidene)-hydrazide) 501

$C_{20}H_{18}N_4O_3$
1-(4-Anilinophenyl)-3-(2-methyl-5-nitrophenyl)urea 562
1-(4-Anilinophenyl)-3-(4-methyl-2-nitrophenyl)urea 561
Basic Orange 15 (C.I. 46045) 1094

$C_{20}H_{18}N_4O_4$
1-(4-Anilinophenyl)-3-(2-methoxy-5-nitrophenyl)urea 552
1,3-Bis((2-nitrophenyl)-aminomethyl)benzene 274
4-(3-Furfurylureido)benzoic (2-hydroxybenzylidene)hydrazide 1036

$C_{20}H_{18}O$
α-Benzylbenzhydrol 204
2,6-Dibenzylidenecyclohexanone 285
4-Methoxytriphenylmethane 180
Methyl trityl ether 186
(1-Naphthyl) (2,4,6-trimethylphenyl) ketone 314
α-(*p*-Tolyl)benzhydrol 245
2-(2,4,6-Trimethylbenzoyl)-naphthalene 314
1,1,2-Triphenylethanol 204
2,2,2-Triphenyl-1-ethanol 207

$C_{20}H_{18}OGe$
Acetyltriphenylgermane 1114

$C_{20}H_{18}OPbS$
S-Triphenylplumbyl thioacetate 1135

$C_{20}H_{18}OSi$
Acetyltriphenylsilane 1114
Benzoyldiphenylmethylsilane 1109

$C_{20}H_{18}O_2$
p-Anisyldiphenylcarbinol 201
1,4-Bis(4-acetylphenyl)-1,3-butadiene 308
7,12-Dihydro-7,12-dihydroxy-7,12-dimethylbenz(a)anthracene 200
Diphenyl(2-methoxyphenyl)-methanol 201
α,α'-Ethynylenebis(cinnamyl alcohol) 1016
5,6,6a,7,12,12a-Hexahydro-1,12-dimethylbenz(a)anthracene-5,7-dione 309
5,6,6a,7,12,12a-Hexahydro-3,9-dimethylbenz(a)-anthracene-5,7-dione 309
α-(3-Methoxy-2-naphthyl)cinnamyl alcohol 202
1-(4-Methoxyphenyl)-7-phenyl-1,4,6-heptatrien-3-one 284
2-(α-(2-Naphthyl)benzyl)-propionic acid 333
2-Phenyl-5-(2,4-xylyl)-1,4-hydroquinone 193

$C_{20}H_{18}O_2S$
3-(2-Furyl)-3-(*p*-tolylthio)-propiophenone 850

$C_{20}H_{18}O_3$
2-Hydroxy-3-(3-(4-tolyl)propyl)-1,4-naphthoquinone 316
Phenolphthalol 199

$C_{20}H_{18}O_3Si$
Triphenoxyvinylsilane 1114

$C_{20}H_{18}O_4$
1,4-Bis(4-acetoxyphenyl)-1,3-butadiene 380
10,11-Dihydro-5,11-methano-5*H*-dibenzo(a,d)cyclohepten-10,12-diyl diacetate 357
Ethyl 12-carboxy-9,10-dihydro-9,10-ethanoanthracene-11-carboxylate 360
2-(5-Oxo-6,7,8,9-tetrahydro-5*H*-benzocyclohepten-6-ylidene-methyl-)phenoxyacetic acid 327

$C_{20}H_{18}O_5$
2-(6-Hydroxy-3-oxo-3*H*-xanthen-9-yl)-cyclohexanecarboxylic acid 326

$C_{20}H_{18}S_2$
α,α'-Bis(phenylthio)-*m*-xylene 212

$C_{20}H_{18}Si$
Diphenylstyrylsilane 1114

$C_{20}H_{19}BF_4O$
2,6-Diphenyl-4-isopropylpyrylium tetrafluoroborate 935

$C_{20}H_{19}BrNO_2P$
(Diphenyl)methyl(4-nitrobenzyl)-phosphonium bromide 831

$C_{20}H_{19}BrSi$
(2-Bromoethyl)triphenylsilane 1099

$C_{20}H_{19}ClHgO_2$
12-(Chloromercuri)-9,10-dihydro-9,10-dimethyl-9,10-ethanoanthracen-11-yl acetate 1138

$C_{20}H_{19}ClNOP$
Carbamoylmethyl-triphenylphosphonium chloride 831

$C_{20}H_{19}ClN_2$
4β-(4-Chlorophenyl)-4b,5-dihydro-11*H*-isoindolo(2,1-a)benzimidazole 244

$C_{20}H_{19}ClN_4O$
4-Chloro-2′-(2-imidazolyl)-α-(1-methyl-2-imidazolin-2-yl)-benzhydrol 864

$C_{20}H_{19}ClO_5$
2,4-Diphenyl-6-isopropylpyrylium perchlorate 935

$C_{20}H_{19}ClSi$
(1-Chloroethyl)triphenylsilane 1114

$C_{20}H_{19}Cl_4NO_3$
2′,6′-Diisopropyl-3,4,5,6-tetrachlorophthalanilic acid 455

$C_{20}H_{19}IN_2O$
4-(4-Acetamidostyryl)-1-methylquinolinium iodide 990

$C_{20}H_{19}N$
N,*N*-Dibenzylaniline 233

$C_{20}H_{19}NO$
3-Benzyl-2-oxo-3-phenyl-1-cyclohexanecarbonitrile 757
3,3-Diphenyl-1-methyl-2-oxo-1-cyclohexanecarbonitrile 757
N-Methyl-*N*-(2-phenylethyl)-2-naphthalenecarboxamide 488
1-(2,4,6-Trimethylphenyl-iminomethyl)-2-naphthol 1039

$C_{20}H_{19}NO_2$
Ethyl 4,5-diphenyl-2-methyl-3-pyrrolecarboxylate 843

$C_{20}H_{19}NO_2S$
N,*N*-Diphenyl-2,4-xylenesulfonamide 828

$C_{20}H_{19}NO_3$
α,α-Dimethyl-4-nitrobenzyl 1-methyl-2-naphthyl ether 268
1-(4-(2,2-Dimethylpropionyl)benzyl)-isatin 691

$C_{20}H_{19}NO_4$
9-Desoxyberberine 247
4,4-Dimethyl-2-(5-methoxy-2-nitrobenzylidene)-1-tetralone 301
Ethyl 4-benzyloxy-α-cyano-3-methoxycinnamate 762

$C_{20}H_{19}NO_5$
Berberine acetone adduct 997
Berberine sulfate 997

$C_{20}H_{19}NS$
2-(2-(Diphenylmethylthio)ethyl)-pyridine picrate 941

$C_{20}H_{19}N_3O$
1-(4-Anilinophenyl)-3-(*m*-tolyl)urea 567
1-(4-Anilinophenyl)-3-(*p*-tolyl)urea 570
4-Cinnamylideneamino-2,3-dimethyl-1-phenyl-3-pyrazolin-5-one 1075

$C_{20}H_{19}N_3OS$
1-(4-Anilinophenyl)-3-(4-(methylthio)-phenyl)urea 560
4-(2-Methoxy-5-methylphenyl)-1-(1-naphthylmethylene)-3-thiosemicarbazide 1048

$C_{20}H_{19}N_3O_2$
1-(4-Anilinophenyl)-3-(2-methoxyphenyl)urea 553
N′-Dibenzyl-L-histidine 867

$C_{20}H_{19}N_3O_3S$
5,6-Diaminophenanthridinium *p*-toluenesulfonate 793

$C_{20}H_{19}N_5$
4-(Dimethylamino)-4′(phenylazo)-azobenzene 1090

$C_{20}H_{19}P$
Di-*m*-tolylphenylphosphine 830

$C_{20}H_{20}BF_4P$
Ethyltriphenylphosphonium fluoroborate 831

$C_{20}H_{20}Br_2O_4$
Dimethyl 6,6′-bis(1-bromoethyl)-2,2′-biphenyldicarboxylate 375

$C_{20}H_{20}Br_2O_6S_2$
Bicyclo(2.2.2)octane-2,3-diol bis(4-bromobenzenesulfonate) 796

$C_{20}H_{20}ClF_6N_3O_2$
Bis-(a,a,a-trifluoro-*m*-tolylcarbamoylmethyl)-dimethylammonium chloride 397

$C_{20}H_{20}ClNO_2$
2-(4-Chlorobenzyl)-1-oxo-1,2,3,4-tetrahydro-2-naphthalenepropionamide 405
3-(4-Chlorophenyl)-2-cyclohexyl-3-hydroxy-1-isoindolinone 689

$C_{20}H_{20}ClNO_3$
Methyl *N*-(2-(2-(4-chlorobenzyl)-1-oxo-2-indanyl)ethyl)carbamate 629

$C_{20}H_{20}ClN_7O_7$
2′-(2-Chloro-4,6-dinitrophenylazo)-5′-((2-cyanoethyl)(2-hydroxyethyl)-amino)-4′methoxyacetanilide 1080

$C_{20}H_{20}ClPS$
(Methylthiomethyl)-triphenylphosphonium chloride 830

$C_{20}H_{20}Cl_2N_4O_2$
Adipic bis(2-chlorobenzylidene)-hydrazide 493

$C_{20}H_{20}F_3NO_2$
4-Cyclohexylphenyl *N*-(2-(trifluoromethyl)phenyl)carbamate 610

$C_{20}H_{20}GeO$
2-(Triphenylgermyl)-1-ethanol 1105

$C_{20}H_{20}HgO_3$
10,11-Dihydro-5,11-dimethyl-11-hydroxy-5,10-methano-5*H*-dibenzo(a,d)cyclohepten-12-ylmercuric acetate 1137

$C_{20}H_{20}IN$
1-Ethyl-2-(4-methylstyryl)-quinolinium iodide 990

$C_{20}H_{20}INO$
1-Ethyl-2-(4-methoxystyryl)-quinolinium iodide ... 985
$C_{20}H_{20}INO_2$
2-(3,4-Dimethoxystyryl)-1-methylquinolinium iodide ... 984
$C_{20}H_{20}I_2N_2O_5$
N-Acetyl-D-phenylalanyl-3,5-diiodo-L-tyrosine ... 398
$C_{20}H_{20}N_2$
1-Benzyl-4-(α-cyanobenzylidene)-piperidine hydrochloride ... 765
5-Indancarboxaldehyde azine ... 1079
$C_{20}H_{20}N_2O$
2,4-Di(3-indolyl)-1-butanol ... 881
1-Ethyl-3-(1-naphthyl)-1-(*o*-tolyl)urea ... 589
1-Isopropyl-3-(1-naphthyl)-1-phenylurea ... 587
1-(4-Isopropylphenyl)-3-(1-naphthyl)urea ... 592
1-(1-Naphthyl)-3-(2,4,6-trimethylphenyl)urea ... 602
$C_{20}H_{20}N_2OS$
S-Benzyl-L-cysteine-β-naphthylamide ... 487
1-(2-Methoxy-5-methylphenyl)-1-methyl-3-(2-naphthyl)-2-thiourea ... 737
$C_{20}H_{20}N_2O_2$
N,*N'*-Bis-(isocoumaran-5-ylidene)-ethylenediamine ... 1054
9-Fluorenylideneamino *N*-cyclohexylcarbamate ... 1013
3-(9-Hydroxyimino-6,7,8,9-tetrahydro-5*H*-benzocyclohepten-8-yl)-2-methyl-1-isoindolinone ... 1010
$C_{20}H_{20}N_2O_2S$
5-Methyl-7-phenyl-8-(phenylsulfonyl)-1,2,3,7-tetrahydroimidazo(1,2-a)pyridine ... 779
4-(4-(*p*-Tolylsulfonyl)benzylamino)-aniline ... 776
$C_{20}H_{20}N_2O_3$
1,3-Diphenyl-1-(2-oxo-1-cyclohexylcarbonyl)urea ... 575
$C_{20}H_{20}N_2O_3S$
4'-Amino-*N*-benzyl-2'-hydroxy-*p*-toluenesulfanilide ... 811
3-Amino-*N*-ethyl-4-phenoxybenzenesulfonanilide ... 813
$C_{20}H_{20}N_2O_3Si$
4-(2-Hydroxy-6-(trimethylsilyl)-1-naphthylazo)benzoic acid ... 1084
$C_{20}H_{20}N_2O_4$
2-Butyne-1,4-diol bis-(2-methylcarbanilate) ... 1015
Diethyl 9,10-dihydroanthracene-9,10-biimine-11,12-dicarboxylate ... 673
2-(4-(Methacrylamido)benzoyl)acet-*o*-aniside ... 442
$C_{20}H_{20}N_2O_4S_2$
N,*N'*-(1,2-Phenylene)bis(*p*-toluenesulfonamide) ... 818
N,*N'*-(1,4-Phenylene)bis(*p*-toluenesulfonamide) ... 819
$C_{20}H_{20}N_2O_5$
2-(2,3-Dimethoxyphenyl)-*N*-(2-hydroxy-2-(2-oxindolin-3-ylidene)-ethyl)acetamide ... 689
$C_{20}H_{20}N_2O_5S_2$
5-Amino-2-(*p*-toluenesulfonamido)-phenyl *p*-toluenesulfonate ... 815
$C_{20}H_{20}N_2O_6$
2-Butyne-1,4-diol bis-(2-methoxycarbanilate) ... 1017
4'-(2-(4-Methoxybenzoyl)-acetamido)succinanilic acid ... 442
$C_{20}H_{20}N_4O_2$
4-(4-Acetamidobenzylideneamino)-2,3-dimethyl-1-phenyl-3-pyrazolin-5-one ... 1074
$C_{20}H_{20}N_4O_4$
Ethyl 4-(3-(1-methyl-3-oxo-2-phenyl-4-pyrazolin-4-yl)ureido)benzoate ... 707
$C_{20}H_{20}N_6O_3$
1-(5-Methyl-2-pyridylureidoacetyl)-4-(1-naphthyl)semicarbazide ... 964
$C_{20}H_{20}N_6O_6$
Adipic bis((3-nitrobenzylidene)-hydrazide) ... 499
$C_{20}H_{20}O$
2-(4-Isopropylbenzylidene)-1-tetralone ... 305
1-(9-Phenanthryl)-1-cyclohexanol ... 195
5-Phenyl-1-(2,4,6-trimethylphenyl)-2,4-pentadien-1-one ... 305
$C_{20}H_{20}OSi$
Ethoxytriphenylsilane ... 1098
1-(Triphenylsilyl)ethanol ... 1105
$C_{20}H_{20}OSn$
2-(Triphenylstannyl)ethanol ... 1130
$C_{20}H_{20}O_2$
Ethyl 9,10-dihydro-12-methyl-9,10-ethanoanthracene-11-carboxylate ... 361
2-Hydroxymethyl-3-(1-naphthyl)-3-phenyl-1-propanol ... 207
3-(4-Methoxyphenyl)-6-methyl-6-phenyl-2-cyclohexen-1-one ... 283
$C_{20}H_{20}O_2Si$
(2-Hydroxyethoxy)triphenylsilane ... 1105
$C_{20}H_{20}O_4$
cis-1,3-Cyclohexanediol dibenzoate ... 371
Cyclohexane-1,4-diol dibenzoate ... 371
trans-1,3-Cyclohexanediol dibenzoate ... 371
1,3-Dimethoxy-7-(3,4-dimethoxyphenyl)naphthalene ... 187
5,5-Dimethyl-4-hydroxy-4-(4-phenoxyphenyl)-2-hexynoic acid ... 1017
3-Phenyl-1,1,3-trimethylindan-4',5-dicarboxylic acid ... 345
2-(3,4,5-Trimethoxybenzylidene)-1-tetralone ... 302
$C_{20}H_{20}O_6$
Cubebin ... 206
1,2:4,6-Di-*O*-benzylidene-α-D-glucopyranoside ... 205
$C_{20}H_{20}Pb$
Ethyltriphenyllead ... 1136
$C_{20}H_{20}PbS$
(Ethylthio)triphenyllead ... 1136
$C_{20}H_{20}Si$
Dibenzylphenylsilane ... 1114
Diphenylmethyl(*o*-tolyl)silane ... 1119
Ethyltriphenylsilane ... 1114
Phenyl(*o*-tolyl)(*m*-tolyl)silane ... 1119
$C_{20}H_{20}Sn$
Ethyltriphenyltin ... 1132
Tetracyclopentadienyltin ... 1131
$C_{20}H_{21}BF_4N_4S$
2-(4-Anilinophenylazo)-3-methyl-4,5,6,7-tetrahydrobenzothiazolium tetrafluoroborate ... 1091
$C_{20}H_{21}Br_3O_{10}$
2,4,6-Tribromophenyl tetracetyl-β-D-glucopyranoside ... 353
$C_{20}H_{21}ClN_2OS$
2-(4-Chlorobenzoyl)-1-(α-(methylthio)phenyliminomethyl)-piperidine hydriodide ... 1024
4'-Chloro-3-morpholino-*N*-(4-tolyl)-thiocinnamamide ... 732
$C_{20}H_{21}Cl_4FeO$
2-*tert*-Butyl-4,6-diphenylpyrylium tetrachloroferrate ... 935
$C_{20}H_{21}F_2NO$
2-(Bis(4-fluorophenyl)methyl)-3-quinuclidinol ... 244
2-(Bis(4-fluorophenyl)methyl)-3-quinuclidinol fumarate ... 244
$C_{20}H_{21}IN_2$
4-(4-(Dimethylamino)styryl)-1-methylquinolinium iodide ... 989
$C_{20}H_{21}NO_2$
9-(2,2-Dimethylpropionyl)-*N*-methyl-1-fluorenecarboxamide ... 488
1,2,3,4,12,12a-Hexahydro-12-hydroxy-12-phenyl-pyrido(1,2-b)(2)-benzazepin-6(11*H*)-one ... 696
4-(2-(Phenylacetyl)benzoyl)-piperidine ... 706
$C_{20}H_{21}NO_3$
1-(2-Benzoylbenzoyl)-2-piperidinemethanol ... 697
8,9-Dimethoxy-10b-phenyl-1,5,6,10b-tetrahydropyrrolo(2,1-a)-isoquinolin-3(2*H*)-one ... 698
Ethyl 4-cyano-4-(4-methoxyphenyl)-3-phenylbutyrate ... 753
$C_{20}H_{21}NO_4$
1-Benzoyl-3-hydroxy-2-piperidinylmethyl benzoate ... 696
N-Methylnorpapaverine, betaine salt ... 997
Papaverine ... 997
$C_{20}H_{21}NO_4S$
Dihydro-1'-methyl-4-phenylsulfonylspiro(3*H*-2-benzopyran-3,4'-piperidin)-1-one ... 778
$C_{20}H_{21}NO_5$
4-Allyl-2-methoxyphenyl *N*-(4-(ethoxycarbonyl)phenyl)carbamate ... 649
2-(Isobutoxycarbonyl)phenyl *N*-(3-acetylphenyl)carbamate ... 666
2-Methoxy-4-(1-propenyl)phenyl *N*-(4-(ethoxycarbonyl)phenyl)-carbamate ... 665
$C_{20}H_{21}NO_7$
(L)-2,3-Diacetoxy-*N*-(1-(1-naphthyl)-ethyl)succinamic acid ... 487
$C_{20}H_{21}NO_7S$
Papaverine sulfonic acid (sulfonic acid unspecified) ... 997
$C_{20}H_{21}NSi$
N,*N*-Dimethyl-3-(diphenylsilyl)-aniline ... 1107
1-(Triphenylsilyl)ethylamine ... 1107
$C_{20}H_{21}N_3O_2$
5,5-Diethyl-1,3-diphenyl-2-iminobarbituric acid ... 726
4'-(4-(2,6-Dimethylmorpholino)-phenylazo)acetophenone ... 1090
$C_{20}H_{21}N_3O_2S$
p-Toluenesulfonic N^2-(1,2-dihydro-1-ethylpyridylidene)-N^1-phenylhydrazide ... 821
$C_{20}H_{21}N_3O_3$
4-(3,4-2,3-Dimethyl-4-(3-ethoxy-4-hydroxybenzylideneamino)-1-phenyl-3-pyrazolin-5-one ... 1031
N-(4-Ethoxyphenyl)-*N'*-(α-methylcinnamylideneamino)-oxamide ... 1043
$C_{20}H_{22}$
9-Cyclohexyl-9-methylfluorene ... 166
α-Cyclohexylstilbene ... 165
$C_{20}H_{22}Br_2N_2O_2$
N,*N'*-(3,3'-Dimethyl-1,1'-biphenyl-4,4'-diyl)bis(3-bromopropionamide) ... 393
$C_{20}H_{22}ClN$
N-(4-Chlorobenzhydryl)-norbornanamine hydrochloride ... 243
α-(3-Chloropropyl)-β-ethyl-α-phenylhydrocinnamonitrile ... 751
$C_{20}H_{22}ClNO$
4-Chlorophenyl 4-phenethylpiperidino ketone ... 695
$C_{20}H_{22}ClNO_2$
2-(4-Chlorophenoxy)-*N*-cyclohexylacetanilide ... 401
$C_{20}H_{22}ClNO_4$
2-(2-Methoxy-4-(1-propenyl)-phenoxy)ethyl *N*-(4-chloro-2-methylphenyl)carbamate ... 616
Norpapaverine methochloride ... 996
$C_{20}H_{22}ClNO_5$
2-(Isopentyloxycarbonyl)phenyl *N*-(5-chloro-2-methoxyphenyl)-carbamate ... 617
$C_{20}H_{22}ClN_5O_5S_2$
3-Isobutyl-2-((6-methoxy-3-methyl-2-benzothiazolinylidene)aminoazo) benzothiazolium perchlorate ... 913
$C_{20}H_{22}Cl_2N_4O_3$
3-(2-(2,5-Dichlorophenylazo)-5-(diethylamino)-phenylcarbamoyl)-propionic acid ... 1088
$C_{20}H_{22}Cl_4N_4O_2$
N,*N'*-Bis(*N*-(3,4-dichlorophenyl)-carbamoyl)hexamethylenediamine ... 602
$C_{20}H_{22}FN$
10,11-Dihydro-10-(2-dimethylaminoethyl)-3-fluoro-5-methylene-5*H*-dibenzo(a,d)-cycloheptene maleate ... 243
$C_{20}H_{22}FN_5O_2$
4-(4-Methyl-1-piperazinamido)-benzoic (4-fluorobenzylidene)-hydrazide ... 694
$C_{20}H_{22}N_2$
1-Benzyl-4-(α-cyanobenzyl)-piperidine ... 755
$C_{20}H_{22}N_2O$
Hexahydro-spiro(3-cyclohexene-1,1'-(2'*H*-indolo(2,3-a)quinolizin))-4'-one ... 710
$C_{20}H_{22}N_2OS$
4-Diphenylmethylene-1-(*N*-methylthiocarbamoyl)-3-piperidinol ... 734
$C_{20}H_{22}N_2O_2$
4-Cyanophenyl *N*-(2,6-diisopropylphenyl)carbamate ... 767
Ethyl 5-phenyl-1,2,3,9b-tetrahydro-5*H*-imidazo(2,1-a)isoindole-5-acetate hydrochloride ... 359
3-(4-Pentyloxyphenyl)-1-phenyl-2-pyrazolin-5-one ... 698
4',4'''-(1,4-Piperazinediyl)-diacetophenone ... 308
Quininone ... 986
$C_{20}H_{22}N_2O_3$
4'-Anilinohexahydro-phthalanilic acid ... 427
$C_{20}H_{22}N_2O_4S$
Butyl 4-(3-(2-phenoxyacetyl)-2-thioureido)benzoate ... 736
$C_{20}H_{22}N_2O_5$
N-Acetyl-L-tyrosyl-L-phenylalanine ... 412
N-Benzyloxycarbonyl-L-alanyl-L-phenylalanine ... 657
Carbobenzyloxy-β-alanylglycine benzyl ester ... 663
Carbobenzyloxy-β-alanyl-L-phenylalanine ... 657
Carbobenzyloxyglycylsarcosine benzyl ester ... 663
Carbobenzyloxyphenyl-alanylglycine methyl ester ... 658
$C_{20}H_{22}N_2O_6$
N-(Benzyloxycarbonyl)-*N*-(benzyloxycarboxamido)glycine ethyl ester ... 654
Carbobenzyloxyglycyl-L-serine benzyl ester ... 643
Carbobenzyloxy-L-leucine 4-nitrophenyl ester ... 659
Carbobenzyloxy-L-phenylalanyl-L-serine ... 644
Carbobenzyloxyserylglycine benzyl ester ... 643
Carbobenzyloxy-L-seryl-L-phenylalanine ... 644
Carbobenzyloxy-L-tyrosyl-L-alanine ... 644
2-(2-Methoxy-4-(1-propenyl)-phenoxy)ethyl *N*-(4-methyl-3-nitrophenyl)carbamate ... 647
$C_{20}H_{22}N_2O_7$
Carbobenzyloxy-L-seryl-L-tyrosine ... 642
$C_{20}H_{22}N_4O$
1,2-Bis-(5,6-dimethyl-2-benzimidazolyl)-ethanol hydrochloride ... 902
2,3-Dimethyl-4-(4-(dimethylamino)-benzylideneamino)-1-phenyl-3-pyrazolin-5-one ... 1062
$C_{20}H_{22}N_4O_2$
Adipic bis(benzylidenehydrazide) ... 1070
4-(3-Allylureido)benzoic (3-phenylpropylidene)hydrazide ... 574
α,α'-(Ethylenedinitrilo)-bis-(*p*-acetotoluidide) ... 1054
$C_{20}H_{22}N_4O_4$
Adipic bis(2-hydroxybenzylidenehydrazide) ... 1028
4-(3-Allylureido)benzoic (3-ethoxy-4-hydroxybenzylidene)hydrazide ... 547
$C_{20}H_{22}N_4O_5$
2,2'-Oxydiacetic bis((2-methoxybenzylidene)hydrazide) ... 1041
2,2'-Oxydiacetic bis((4-methoxybenzylidene)hydrazide) ... 1041

$C_{20}H_{22}N_4O_7$
2,2′-Oxydiacetic acid vanillylidenehydrazide ... 1027

$C_{20}H_{22}N_6O_4$
4-(4-Methyl-1-piperazinylcarboxamido)benzoic (3-nitrobenzylidene)hydrazide ... 1061

$C_{20}H_{22}O$
α-Cyclopentyl-4-methoxystilbene ... 179
4-(4-Methoxyphenyl)-4-methyl-1-phenyl-1-cyclohexene ... 180

$C_{20}H_{22}O_2$
(1,9 or 2,9 or 3,9 or 4,9)-Diisopropyl-9-hydroxy-10(9*H*)-anthrone ... 310

$C_{20}H_{22}O_2S_2$
3,3′-(Ethylenedithio)-dipropiophenone ... 308

$C_{20}H_{22}O_3$
1,2-Bis(4-methoxyphenyl)-2-ethoxy-3-butyne ... 1017

$C_{20}H_{22}O_4$
2-Allyl-4-(2-(1,3-benzodioxol-5-yl)-1-methylethyl)-5-methoxyphenol ... 191
Diethyl 2,3-diphenylsuccinate ... 360
9,10-Dimethyl-2,3,6,7-tetramethoxyphenanthrene ... 178
2,2′-(Hexamethylenedioxy)-dibenzaldehyde ... 323
2,2′,6,6′-Tetramethyl-4,4′-biphenol diacetate ... 380

$C_{20}H_{22}O_4S_2$
cis-4,5-Di-*p*-tosyl-1-cyclohexene ... 783

$C_{20}H_{22}O_6$
1,3:4,6-Di-*O*-benzylidenedulcitol ... 199
Triethylene glycol dibenzoate ... 369

$C_{20}H_{22}O_6S_2$
cis-1-Cyclohexene-1,2-diol di-*p*-tosylate ... 798
trans-1-Cyclohexene-1,2-diol di-*p*-tosylate ... 798

$C_{20}H_{22}O_8$
Populin dihydrate ... 368

$C_{20}H_{22}Si_2$
1,1-Dimethyl-1,2,2-triphenyldisilane ... 1114

$C_{20}H_{23}BrN_2O_4S$
3-Acetamidosulfonylbutyl-2-methyl-5-phenylbenzoxazolium bromide ... 910

$C_{20}H_{23}ClN_2O$
2-Chloroacetyl-1-(2-(dimethylaminomethyl)phenyl)-1,2,3,4-tetrahydroisoquinoline hydrochloride ... 703

$C_{20}H_{23}ClN_2O_3$
Basic Red 11 (C.I. 45050) ... 1062

$C_{20}H_{23}ClN_4O_3$
2-(3-(3-Chlorophenyl)ureido)acetic (4-butoxybenzylidene)hydrazide ... 517

$C_{20}H_{23}ClN_6O_4S$
2-((1,3-Dimethyl-2-benzimidazolinylidene)aminoazo)-3-isobutylbenzothiazolium perchlorate ... 915

$C_{20}H_{23}Cl_2N_3O_3$
2-(3-(5-Chloro-2-methylphenyl)-1-ethyl-ureido)ethyl *N*-(5-chloro-2-methyl-phenyl)carbamate ... 674
2-(3-(5-Chloro-2-methylphenyl)-ureido)isobutyl *N*-(5-chloro-2-methylphenyl)carbamate ... 674

$C_{20}H_{23}F_3N_2O$
1-(2,6-Diisopropylphenyl)-3-(3-trifluoromethylphenyl)urea ... 511

$C_{20}H_{23}N$
N-(2-Norbornyl)-benzhydrylamine hydrochloride ... 243

$C_{20}H_{23}NO$
N-Cyclohexyl-2,2-diphenylacetamide ... 430
4-Diethylamino-4′-methylchalcone ... 302
1-(2,2-Diphenylacetyl)-4-methylpiperidine ... 710
9-Octanoylcarbazole ... 892

$C_{20}H_{23}NO_2$
4′-(4-Butoxybenzylideneamino)-propiophenone ... 1049
3-(1,1-Dimethyl-2-(dimethylamino)-ethyl)-3-phenylphthalide ... 384
2-(2,2-Diphenyl-1,3-dioxolan-4-yl)-piperidine hydrochloride ... 248

$C_{20}H_{23}NO_3$
4-Cyclohexylphenyl *N*-(2-methoxyphenyl)carbamate ... 653
4-Cyclohexylphenyl *N*-(4-methoxyphenyl)carbamate ... 654
2-(Diethylamino)-ethyl 9-xanthenecarboxylate ... 359
4-(1,2-Dimethyl-1-ethyl-2-nitropropyl)benzophenone ... 312
1-Methyl-4-piperidinyl benzilate hydrochloride ... 353
4-(2-Nitro-1,1,2-trimethylbutyl)-benzophenone ... 312

$C_{20}H_{23}NO_3S$
S-(4-*tert*-Butylphenyl) *N*-(4-(ethoxycarbonyl)phenyl)-thiocarbamate ... 676
3-Phenyl-3-(piperidinosulfonyl)-propiophenone ... 822

$C_{20}H_{23}NO_4$
4-(Butoxycarbonyl)phenyl *N*-(2,5-xylyl)carbamate ... 676
4-(Butoxycarbonyl)phenyl *N*-(2,6-xylyl)carbamate ... 676
2-*tert*-Butylphenyl *N*-(4-(ethoxycarbonyl)phenyl)carbamate ... 675
3,4-Dihydropapaverine ... 246
2-Isopropyl-5-methylphenyl *N*-(4-(ethoxycarbonyl)phenyl)carbamate ... 676
5-Isopropyl-2-methylphenyl *N*-(4-(ethoxycarbonyl)phenyl)carbamate ... 676
DL-Pavine hydrochloride ... 246

$C_{20}H_{23}NO_4S$
(2-(4-Allyl-2-methoxyphenoxy)ethyl) *N*-(3-(methylthio)phenyl)carbamate ... 647

$C_{20}H_{23}NO_5$
4-Butoxyphenyl *N*-(4-(ethoxycarbonyl)phenyl)carbamate ... 651

$C_{20}H_{23}NO_5S$
1-Carbobenzyloxy-4-piperidinyl *p*-toluenesulfonate ... 797

$C_{20}H_{23}NO_6$
4-(Butoxycarbonylphenyl) *N*-(2,4-dimethoxyphenyl)carbamate ... 647
4-(Butoxycarbonyl)phenyl *N*-(2,5-dimethoxyphenyl)carbamate ... 648

$C_{20}H_{23}NO_{12}$
4-Nitrophenyl-α-D-galactopyranoside tetraacetate ... 356
4-Nitrophenyl-β-D-galactopyranoside tetraacetate ... 269

$C_{20}H_{23}N_3$
1-Benzyl-4-cyano-4-(*o*-toluidino)-piperidine ... 754
1-Benzyl-4-cyano-4-(*m*-toluidino)-piperidine ... 754
1-Benzyl-4-cyano-4-(*p*-toluidino)-piperidine ... 754

$C_{20}H_{23}N_3O_3$
1-Furfuryl-1-methyl-3-(2-(*N*-methylfurfurylaminomethyl)-phenyl)urea ... 850

$C_{20}H_{23}N_3O_4$
N-Acetyl-L-phenylalanyl-L-tyrosinamide ... 412
N-Benzylglycylglycylphenylalanine ... 428
Carbobenzyloxy-β-alanyl-L-phenylalaninamide ... 658
Carbobenzyloxysarcosyl-L-phenylalaninamide ... 658

$C_{20}H_{23}N_3O_5$
1-(2-(4-Allyl-2-methoxyphenoxy)-ethyl)-3-(4-methyl-2-nitrophenyl)urea ... 551
Carbobenzyloxy-sarcosyltyrosinamide ... 645
Carbobenzyloxy-L-tyrosyl-L-alaninamide ... 645

$C_{20}H_{23}N_3O_6$
*N*α-(*tert*-Butoxycarbonyl)-L-tryptophan *N*-hydroxysuccinimide ester ... 890
Carbobenzyloxy-L-seryl-L-tyrosinamide ... 645

$C_{20}H_{23}N_5O_2$
4-(3-Allylureido)benzoic (4-(dimethylamino)benzylidene)-hydrazide ... 500

$C_{20}H_{23}N_5O_4$
2-(3-(2-Ethoxyphenyl)ureido)acetic (4-acetamidobenzylidene)-hydrazide ... 551

$C_{20}H_{24}$
2,3,6,7-Tetraethylbiphenylene ... 168

$C_{20}H_{24}BrNO_9$
1-*O*-(4-Bromophenyl)-2,3,4,6-tetraacetyl-β-D-glucosamine ... 401

$C_{20}H_{24}ClNO_5$
2-(3,4,5-Trimethylphenoxy)ethyl *N*-(5-chloro-2,4-dimethoxyphenyl)-carbamate ... 612

$C_{20}H_{24}ClN_5O_3$
2-(3-(2-Ethoxyphenyl)ureido)acetic (2-chloro-4-(dimethylamino)-benzylidene)hydrazide ... 517

$C_{20}H_{24}Cl_2$
1,2-Dichloro-1,2-dimesitylethane ... 174

$C_{20}H_{24}Cl_{10}Si_4$
1,4-Bis(pentachlorophenyl)-1,1,2,2,3,3,4,4-octamethyltetrasilane ... 1101

$C_{20}H_{24}F_2N_4O_2$
1,1′-Hexamethylenebis(3-(2-fluorophenyl)urea) ... 536

$C_{20}H_{24}I_2O_2$
Thymol iodide ... 197

$C_{20}H_{24}N_2O$
4-(*N*-Acetylanilino)-1-benzylpiperidine ... 426
5-(1-Adamantyl)-2,3-dihydro-5-hydroxyimidazo(2,1-a)isoindole ... 246
1-Benzyl-4-formamidomethyl-4-phenylpiperidine ... 427
5-Benzyloxy-*N*-ethyl-*N*-methyltryptamine oxalate ... 882
N-(2-Cyclohexyl-1-phenylethyl)-nicotinamide ... 959
N-(1,2-Diphenylethyl)-1-pyrrolidineacetamide ... 428
3-(4-Methylpiperazinyl)-3-phenylpropiophenone ... 295

$C_{20}H_{24}N_2O_2$
N,*N*′-Diethyl-*N*,*N*′-di(*o*-tolyl)oxamide ... 425
Epiquinidine ... 982
Ethyl 1,2,3,5,6,7,8,9-octahydro-5-phenylimidazo(1,2-a)quinoline-4-carboxylate ... 375
N,*N*′-Hexamethylenebis(benzamide) ... 478
Isoquinidine ... 37
Quinine bisulfate ... 982
Quinine dihydrochloride monohydrate ... 982
Quinine salicylate ... 982
2′-(*o*-Tolyloxy)-1-piperidineacetanilide hydrochloride ... 415

$C_{20}H_{24}N_2O_2S$
Butyl 4-(3-phenethyl-2-thioureido)-benzoate ... 741

$C_{20}H_{24}N_2O_2S_2$
S,*S*′-Ethylene bis(*N*-(2-ethylphenyl)-thiocarbamate) ... 688

$C_{20}H_{24}N_2O_3$
1-(4-Butoxycarbonylphenyl)-3-(2,6-dimethylphenyl)urea ... 577
1-(4-(Butoxycarbonyl)phenyl)-3-(2-ethylphenyl)urea ... 577
1-(4-Butoxycarbonylphenyl)-3-(2,5-xylyl)urea ... 577
Quinine-*N*-oxide ... 1005

$C_{20}H_{24}N_2O_4$
3-(Diethylamino)phenyl *N*-(4-(ethoxycarbonyl)phenyl)carbamate ... 673
N,*N*′-Diveratrylideneethylenediamine ... 1041
4-Ethoxy-3-methoxybenzaldehyde azine ... 234
2-(*N*-Ethylanilino)ethyl *N*-(4-(ethoxycarbonyl)phenyl)carbamate ... 671
2′,2‴-(Tetramethylenedioxy)-diacetanilide ... 417
3′,3‴-(Tetramethylenedioxy)-diacetanilide ... 418
4′,4‴-(Tetramethylenedioxy)-diacetanilide ... 419

$C_{20}H_{24}N_2O_4S$
Ethyl 2-(3-(2-methoxy-5-methylphenyl)ureido)-tetrahydrobenzo(b)thiophene-3-carboxylate ... 855
4-Morpholinophenyl sulfone ... 778

$C_{20}H_{24}N_2O_4S_2$
L-Cystine dibenzyl ester ditosylate ... 356
S,*S*′-Ethylene bis(*N*-(2-methoxy-5-methylphenyl)thiocarbamate) ... 654

$C_{20}H_{24}N_2O_6$
2,4,6-Trimethoxybenzaldehyde azine ... 234

$C_{20}H_{24}N_2O_6S$
*N*δ-Carbobenzyloxy-*N*α-(*p*-tosyl)-L-ornithine ... 815

$C_{20}H_{24}N_2O_6S_2$
S,*S*′-Ethylene bis(*N*-(2,5-dimethoxyphenyl)thiocarbamate) ... 648

$C_{20}H_{24}N_2O_7$
Diethyl 2-acetamido-2-(3-phthalimidopropyl)malonate ... 719

$C_{20}H_{24}N_2S$
10-(1,2,5-Trimethyl-3-pyrrolidinylmethyl)-phenothiazine hydrochloride ... 249

$C_{20}H_{24}N_4O_4$
4-(3,3-Diethylureido)benzoic (4-hydroxy-3-methoxybenzylidene)-hydrazide ... 546

$C_{20}H_{24}N_4O_5$
2-(3-(2-Ethoxyphenyl)ureido)acetic (3,4-dimethoxybenzylidene)-hydrazide ... 549

$C_{20}H_{24}N_4O_6$
2-(3-(4-Methoxyphenyl)ureido)-acetic (2,3,4-trimethoxybenzylidene)hydrazide ... 549

$C_{20}H_{24}N_6O_2$
Suberic acid bis(2-pyridylmethylenehydrazide) ... 1075

$C_{20}H_{24}O$
Benzyl triethylphenyl ketone ... 296

$C_{20}H_{24}O_2$
1,4-Bis-(2,5-dimethylphenoxy)-*trans*-2-butene ... 182
1,1-Bis-(4-hydroxy-3-methylphenyl)-cyclohexane ... 195
1,1-Bis(4-methoxyphenyl)-2-ethyl-1-butene ... 181
Diethylstilbestrol dimethyl ether ... 181
9,10-Dihydro-9,10-dipropyl-9,10-anthracenediol ... 200
2,3-Dihydro-2-(2-hydroxy-4-methylphenyl)-2,4,4,7-tetramethyl-4*H*-1-benzopyran ... 195

$C_{20}H_{24}O_2Pb$
Triethylplumbyl 9-fluorenecarboxylate ... 1135

$C_{20}H_{24}O_2S$
Benzyl 4-(cyclohexylmethyl)phenyl sulfone ... 783

$C_{20}H_{24}O_2S_2$
cis-*p*-Tolyl 2-(*p*-tosyl)cyclohexyl sulfide ... 776

$C_{20}H_{24}O_3$
2,2-Dimethyl-1-phenylpropyl α-methoxyphenylacetate ... 354

$C_{20}H_{24}O_5$
Dimethyl 1,2,3,4,9,10-hexahydro-7-methoxy-1-methyl-1,2-phenanthrenedicarboxylate ... 355

$C_{20}H_{24}O_6$
9,10-Dihydro-1,2,3,5,6,7-hexamethoxyanthracene ... 187

$C_{20}H_{25}BF_4N_2O_4$
Bis(2,4-dimethoxybenzylidene)-ethylhydrazinium tetrafluoroborate ... 254

$C_{20}H_{25}BrO$
4-Bromophenyl 2,4-di-*tert*-butylphenyl ether ... 182

$C_{20}H_{25}ClN_2O$
1-(4-Chloro-2-methylphenyl)-3-(2,6-diisopropylphenyl)urea ... 542

$C_{20}H_{25}Cl_2N_3O_3S$
4-(*N*-Butyl-4-(2,5-dichlorophenylazo)-anilino)-1-butanesulfonic acid, sodium salt 1090
$C_{20}H_{25}IN_2$
2-Ethyl-11-methyl-2-phenyl-2,3,5,10-tetrahydro-1*H*-pyrazolo(1,2-b)-phthalazin-11-ium iodide 256
$C_{20}H_{25}N$
Dihydro-10-(2-(dimethylamino)-ethyl)-5-methyl-5*H*-dibenzo(a,d)-cycloheptene maleate 243
$C_{20}H_{25}NO$
1-Benzyl-2-(dimethylaminomethyl)-1,2,3,4-tetrahydro-1-naphthol 239
N-Benzyl-2,2-dimethyl-*N*-(2-phenylethyl)propionamide 428
3-*tert*-Butyl-6-methoxy-4-phenyl-1,2,3,4-tetrahydroisoquinoline 247
N,*N*-Diethyl-2,2-diphenylbutyramide 441
2′,6′-Diisopropyl-2-methylbenzanilide 479
3-(Phenethylamino)-1-phenyl-1-cyclohexanol hydrochloride 237
1-Phenyl-3-pyrrolidino-1-(*p*-tolyl)-1-propanol 245
$C_{20}H_{25}NO_2$
2′,6′-Diisopropyl-2-methoxybenzanilide 464
2′,6′-Diisopropyl-3-methoxybenzanilide 464
Ethyl 2-(3-(dimethylamino)-1-phenylpropyl)benzoate 373
$C_{20}H_{25}NO_3$
1-(4-(Benzyloxy)phenyl)-4,4-dimethyl-1-hydroxy-3-pentanone oxime 1008
2,6-Diisopropylphenyl *N*-(2-methoxyphenyl)carbamate 653
Octahydro-2,5-methano-2*H*-indeno(1,2-b)oxiren-3-yl *N*-(2,4,6-trimethylphenyl)carbamate 678
$C_{20}H_{25}NO_3S$
2-(4-*tert*-Butylphenoxy)ethyl *N*-(3-(methylthio)phenyl)carbamate 649
$C_{20}H_{25}NO_4$
9,9-(Ethylenedioxy)-3-methoxy-6,7,8,8a,9,10,10a,10b-octahydro-8,10a,10b-trimethylbenz(b)-cyclopent(hi)indolizin-6-one 699
DL-Norlaudanosine hydrochloride 997
DL-Pseudolaudanine 245
$C_{20}H_{25}NO_6S$
5-Nitro-2-(4-(1,1,3,3-tetramethylbutyl)phenoxy)-benzenesulfonic acid 788
$C_{20}H_{25}N_3$
9-(3-Diethylaminopropylamino)-acridine dihydrochloride 995
3,3-Dimethyl-2-(4-(dimethylamino)-phenylimino)-1-ethylindoline hydrochloride 725
$C_{20}H_{25}N_3O$
N-(1-(4-Aminophenyl)-4-piperidinyl)-propionanilide 422
1-Benzyl-4-(*p*-toluidino)-isonipecotamide 423
$C_{20}H_{25}N_3S$
10-(3-(4-Methyl-1-piperazinyl)propyl)-10*H*-phenothiazine 249
$C_{20}H_{25}N_5O_2$
4-(3-Propylureido)benzoic (4-(dimethylamino)benzylidene)-hydrazide 570
$C_{20}H_{25}N_5O_6$
4-(2,5-Dimethoxyphenyl)-1-(3-(2-ethoxyphenyl)ureidoacetyl)-semicarbazide 550
$C_{20}H_{25}O_5P$
trans-1,2-Cyclohexanediol *mono*(dibenzyl phosphate) 833
$C_{20}H_{26}BF_4N_4P$
Bis(1-ethyl-3-methyl-2-benzimidazolinyl)phosphonium tetrafluoroborate 831
$C_{20}H_{26}ClF_3N_2O$
3-(2-Chloro-5-(trifluoromethyl)-phenyl)-1,1-dicyclohexylurea 505
$C_{20}H_{26}N_2$
2,7-Bis(dimethylamino)-9,10-dihydro-9,9-dimethylanthracene 233
$C_{20}H_{26}N_2O$
1-Benzyl-3-(2,6-diisopropylphenyl)urea 572
1-(4-Butylphenyl)-3-(2,4,5-trimethylphenyl)urea 592
1-(4-Butylphenyl)-3-(2,4,6-trimethylphenyl)urea 592
1-Cyclohexyl-1-isopropyl-3-(1-naphthyl)urea 584
1-(2,6-Diisopropylphenyl)-3-(2-methylphenyl)urea 589
3-(2,6-Diisopropylphenyl)-1-methyl-1-phenylurea 596
N-(2-(Dimethylamino)-ethyl)-2,2-diphenyl-*N*-ethylacetamide hydrochloride 423
2-((4-Dimethylamino-1-phenyl)-butyl)-*N*-methylbenzamide 471
N-(1,2-Diphenylethyl)-2-(*N*-methyl-*N*-propylamino)-acetamide 428
$C_{20}H_{26}N_2O_2$
1-(2,6-Diisopropylphenyl)-3-(2-methoxyphenyl)urea 555
$C_{20}H_{26}N_2O_6$
Diethyl 5-(3-ethoxycarbonyl-4-methyl-pyrrol-2-ylmethyl)3-methyl-2,4-pyrroledicarboxylate 839
$C_{20}H_{26}N_4$
N,*N*′-Bis-(4-dimethylaminobenzylidene)-ethylenediamine 1053
$C_{20}H_{26}N_4O_3$
N-Acetylglycyl-L-lysine-β-naphthylamide 487
$C_{20}H_{26}N_4O_4$
4,4′-Bis(*N*,*N*-dimethylureido)-3,3′-dimethoxybiphenyl 551
1,1′-Hexamethylenebis(3-(3-hydroxyphenyl)urea) 546
$C_{20}H_{26}N_4O_6S_2$
1,1′-Hexamethylenebis(3-phenylsulfonylurea) 825
$C_{20}H_{26}N_4O_6S_4$
N,*N*′-Bis-(*p*-tolylsulfonylcarbamoyl)-cystamine 816
1,1′-Dithiodiethylbis(3-(*p*-tolylsulfonyl)urea, sodium derivative) 816
$C_{20}H_{26}N_{10}O_4$
7,7′-(1,4-Piperazinediyldimethyl)-ditheophylline 867
$C_{20}H_{26}O_2$
2,2-Bis-(4-hydroxy-3-methylphenyl)-4-methylpentane 196
1,6-Bis(4-methoxyphenyl)hexane 179
$C_{20}H_{26}O_3$
4-(α-Hydroxybenzyl)-4-phenyl-1,7-heptanediol 199
$C_{20}H_{26}O_3Pb$
Triethylplumbyl 4-(2-naphthyl)-4-oxobutyrate 1134
$C_{20}H_{26}O_4$
1,4-Bis(2,5-dimethylphenoxy)-2,3-butanediol 199
1,4-Bis(3,4-dimethylphenoxy)-2,3-butanediol 202
2,2′,5,5′-Tetramethoxy-3,3′,4,4′-tetramethylbiphenyl 179
$C_{20}H_{26}O_4S$
2-(4-*tert*-Amylphenoxy)ethyl *p*-toluenesulfonate 797
$C_{20}H_{26}O_5S_2$
D-Glucose dibenzyl mercaptal 211
$C_{20}H_{26}O_6$
Decahydro-6,7-diacetoxy-1,8a-dimethyl-9,10-epoxyphenanthren-2-one 92
D-Manno-L-manno-octose phenylosazone 234
$C_{20}H_{26}O_7$
Diethyl 2,2′-(4-methoxybenzylidene)-bis-acetoacetate 363
$C_{20}H_{26}S_2$
4-*tert*-Butylphenyl disulfide 213
$C_{20}H_{27}BF_4O$
4-Benzyl-2,6-di-*tert*-butylpyrylium tetrafluoroborate 935
$C_{20}H_{27}BF_4O_2$
2,4-Di-*tert*-butyl-6-(2-methoxyphenyl)pyrylium tetrafluoroborate 934
$C_{20}H_{27}BrN_2O_3S$
1-Cinnamyl-3-(*N*-(3-(1-methylethoxy)propyl)sulfamoyl)-pyridinium bromide 938
$C_{20}H_{27}F_3N_2O$
1,1-Dicyclohexyl-3-(3-trifluoromethylphenyl)urea 510
$C_{20}H_{27}N$
N-Ethylbis-(3-phenylpropyl)-amine citrate 243
$C_{20}H_{27}NO$
2-(Benzyloxy)-*N*,*N*-diethyl-3-phenylpropylamine 239
N-(1-Naphthyl)decanamide 490
4-Octyloxydiphenylamine 229
$C_{20}H_{27}NO_2$
(-)-3-(Dimethylamino)-1-(2-ethoxyphenyl)-2-methyl-1-phenyl-1-propanol 237
$C_{20}H_{27}NO_3$
1-(4-Benzyloxy-3-hydroxymethyl-phenyl)-2-(*tert*. butylamino)ethanol .. 237
$C_{20}H_{27}N_3O_2$
1-(2-(Diethylamino)ethyl)-3-(4-methoxyphenyl)-1-phenylurea 551
$C_{20}H_{27}N_3O_4S$
4′-Amino-N^4-(2-ethylhexanoyl)-2′-hydroxysulfanilanilide 811
$C_{20}H_{27}N_5O_4$
Carbobenzyloxy-L-histidyl-L-leucinamide 866
$C_{20}H_{28}$
9-(*p*-Tolyl)perhydrofluorene 163
$C_{20}H_{28}N_2$
Tricyclo[5.2.1.0^{2,6}]decan-8-one azine. 1073
$C_{20}H_{28}N_2O$
2-(2,6-Diethyl-α-(2-(dimethylamino)-ethoxy)benzyl)pyridine hydrochloride 938
$C_{20}H_{28}N_2O_2$
N,*N*′-Dicyclohexylterephthalamide .. 474
1-Octyl-1,2,3,4-tetrahydronorharman-3-carboxylic acid 919
$C_{20}H_{28}N_2O_4$
Ethyl 1-(3-aminopropyl)-3-(ethoxycarbonylmethyl)-2-indolepropionate hydrochloride 883
$C_{20}H_{28}N_4O_2$
1,1′-(1,5-Naphthalenediyl)bis(3-butylurea) 604
3,3′-(1,5-Naphthalenediyl)bis(1,1-diethylurea) 569
$C_{20}H_{28}N_6O_2$
4,4′-Hexamethylenebis(1-phenylsemicarbazide) 578
1,1′-Hexamethylenebis(3-(3-pyridylmethyl)urea) 951
$C_{20}H_{28}O$
9-Hydroxy-9-(*o*-tolyl)-perhydrofluorene 204
6,16-Methanobenzocyclo-pentadecan-18-one 298
$C_{20}H_{28}O_3$
9,10-Dihydroxy-4,6a-dimethyl-dodecahydrobenz(a)anthracen-3(2*H*)-one 51
$C_{20}H_{28}O_{13}$
2′,4′,6′-Trihydroxyacetophenone 4′-β-neohesperidoside trihydrate 290
$C_{20}H_{28}Sn$
Dibutyldiphenyltin 1132
Diisobutyldiphenyltin 1132
$C_{20}H_{29}ClN_2O$
3-(3-Chloro-4-methylphenyl)-1,1-dicyclohexylurea 584
3-(5-Chloro-2-methylphenyl)-1,1-dicyclohexylurea 584
$C_{20}H_{29}ClN_2O_2$
1-(5-Chloro-2-methoxyphenyl)-3,3-dicyclohexylurea 519
$C_{20}H_{29}F_3N_2OS$
1-Dodecanoyl-3-(2-(trifluoromethyl)-phenyl)-2-thiourea 728
1-Dodecanoyl-3-(3-(trifluoromethyl)-phenyl)-2-thiourea 728
$C_{20}H_{29}N_3O_2$
10-Undecenoic (4-acetamidobenzylidene)-hydrazide 1070
$C_{20}H_{29}N_3O_3$
1,1-Dicyclohexyl-3-(2-methyl-4-nitrophenyl)urea 568
1,1-Dicyclohexyl-3-(2-methyl-5-nitrophenyl)urea 563
1,1-Dicyclohexyl-3-(4-methyl-3-nitrophenyl)urea 563
$C_{20}H_{29}N_3O_4$
1,1-Dicyclohexyl-3-(2-methoxy-4-nitrophenyl)urea 552
1,1-Dicyclohexyl-3-(2-methoxy-5-nitrophenyl)urea 552
$C_{20}H_{30}$
1,1′-(1,4-Phenylene)bis(1-methylcyclohexane) 162
$C_{20}H_{30}F_3NOS$
S-Dodecyl *N*-(2-(trifluoromethyl)-phenyl)thiocarbamate 611
$C_{20}H_{30}I_2N_2$
1,1′-Hexamethylenebis(2,5-dimethylpyridinium iodide) 958
1,1′-Hexamethylenebis(3,4-dimethylpyridinium iodide) 958
1,1′-Hexamethylenebis(3,5-dimethylpyridinium iodide) 958
$C_{20}H_{30}N_2$
N,*N*′-(Terephthalylidene)-dodecamethylenediamine 1079
$C_{20}H_{30}N_2O$
3-Benzyl-1,1-dicyclohexylurea 571
1,1-Dicyclohexyl-3-(*o*-tolyl)urea 582
1,1-Dicyclohexyl-3-(*m*-tolyl)urea 582
1,1-Dicyclohexyl-3-(*p*-tolyl)urea 583
$C_{20}H_{30}N_2O_2$
3-(*o*-Anisyl)-1,1-dicyclohexylurea 558
3-(*p*-Anisyl)-1,1-dicyclohexylurea 558
$C_{20}H_{30}N_2O_3$
10-Undecenoic (2,3-dimethoxybenzylidene)hydrazide 497
$C_{20}H_{30}N_2O_3S$
4-(3-Dodecanoyl-2-thioureido)-benzoic acid 744
$C_{20}H_{30}N_2O_4$
1,1′-(1,3-Adamantanedicarbonyl)-bismorpholine 110
$C_{20}H_{30}N_2O_5$
N-Benzyloxycarbonyl-L-isoleucyl-L-isoleucine 658
N-Benzyloxycarbonyl-L-leucyl-L-isoleucine 660
N-Benzyloxycarbonyl-L-leucyl-L-leucine 660
Carbobenzyloxyisoleucylnorvaline methyl ester 662
Carbobenzyloxynorleucylvaline methyl ester 661
Carbobenzyloxy-L-norvalyl-L-leucine methyl ester 662
Carbobenzyloxynorvalylnorleucine methyl ester 662
$C_{20}H_{30}N_2O_5S$
Carbobenzyloxy-L-isoleucyl-L-methionine methyl ester 656
Carbobenzyloxy-L-methionyl-L-isoleucine methyl ester 656
$C_{20}H_{30}O_2$
D-α,α'-Bicamphor 55
$C_{20}H_{30}O_4$
1,8-Bis(2-oxocyclohexyl)-1,8-octanedione 55
5-Hydroxy-2-(1-octynyl)-3-oxo-1-cyclopentene-1-heptanoic acid 1016
$C_{20}H_{30}Si_2$
1,2-Bis(4-(trimethylsilyl)phenyl)-ethane 1122
1,2-Diphenyl-1,1,2,2-tetraethyldisilane 1114

$C_{20}H_{31}NO_2$
2-(Diethylamino)-ethyl α-phenylcyclohexaneacetate hydrochloride 359
2-Methylcyclohexyl *N*-(2,6-diisopropylphenyl)carbamate 679

$C_{20}H_{31}NO_2S$
Cyclododecyl *N*-(3-(methylthio)-phenyl)carbamate 667

$C_{20}H_{31}NO_3$
Menthyl *N*-(2-hydroxy-1-methyl-2-phenylethyl)carbamate 643

$C_{20}H_{31}NO_4S$
Diethyl (2-(*N*-benzyl-*N*-methylamino)-ethyl)(2-(methylthio)-ethyl)-malonate hydrochloride 357

$C_{20}H_{31}N_3O$
10-Undecenoic 4-(dimethylamino)-benzylidenehydrazide 500

$C_{20}H_{31}N_3O_3S$
1-Dodecanoyl-3-(2-methyl-4-nitrophenyl)-2-thiourea 740
1-Dodecanoyl-3-(2-methyl-5-nitrophenyl)-2-thiourea 740

$C_{20}H_{31}N_3O_4$
Carbobenzyloxy-L-isoleucyl-L-leucinamide 664
Carbobenzyloxy-L-leucyl-L-isoleucinamide 664
Carbobenzyloxyleucylleucinamide 664

$C_{20}H_{31}N_3O_4S$
1-Dodecanoyl-3-(2-methoxy-4-nitrophenyl)-2-thiourea 736

$C_{20}H_{32}BrNO$
4′-Bromo-2′,6′-dimethyldodecananilide 410

$C_{20}H_{32}Br_2Si_4$
1,4-Bis(4-bromophenyl)-1,1,2,2,3,3,4,4-octamethyltetrasilane 1104

$C_{20}H_{32}ClNO$
N-(2-Chlorobenzylidene)-tetradecanamide 438

$C_{20}H_{32}ClNO_3$
4′-Chloro-2′,5′-dimethoxydodecananilide 399

$C_{20}H_{32}N_2$
Camphor azine 1079

$C_{20}H_{32}N_2O$
1-Cyclohexyl-3-(2,6-diisopropylphenyl)-1-methylurea 583
N-(2-(Dodecyloxy)-phenyl)-glycinonitrile 753

$C_{20}H_{32}N_2O_2$
4′-Acetamidododecananilide 428
N,*N*′-Dihexylterephthalamide 475
Dodecanoic (2-methoxybenzylidene)hydrazide 1045

$C_{20}H_{32}N_2O_2S$
1-Dodecanoyl-3-(4-methoxyphenyl)-2-thiourea 737

$C_{20}H_{32}N_2O_3S$
S-Dodecyl *N*-(2-methyl-5-nitrophenyl)thiocarbamate 669
1-Ethyl-2-undecyl-5-benzimidazolesulfonic acid 906

$C_{20}H_{32}N_2O_5$
2,5-Dibutoxy-*N*-methyl-4-nitro-*N*-(tetrahydrofurfuryl)aniline 265

$C_{20}H_{32}N_2O_6$
N,*N*′-Dodecamethylenebis-(maleamic acid) 121

$C_{20}H_{32}N_2Si_2$
1,2-Bis(4-(dimethylamino)phenyl)-1,1,2,2-tetramethyldisilane 1108

$C_{20}H_{32}O_3$
Methyl 3-(dodecyloxy)benzoate 370
Methyl 4-(dodecyloxy)benzoate 370

$C_{20}H_{32}Si_3$
Bis(trimethylsilylmethyl)-diphenylsilane 1114

$C_{20}H_{33}ClN_2O_3$
3-Carbamoyl-1-(dodecyloxycarbonylmethyl)-pyridinium chloride 947
4-Carbamoyl-1-(dodecyloxycarbonylmethyl)-pyridinium chloride 951

$C_{20}H_{33}NO_2$
3-(Diethylamino)-propyl 3-methyl-2-(*o*-tolyl)-valerate hydrochloride 359
2′-(Dodecyloxy)acetanilide 417
3′-(Dodecyloxy)acetanilide 418
4′-(Dodecyloxy)acetanilide 419
N-(2-Methoxy-5-methylphenyl)-dodecanamide 417

$C_{20}H_{33}NO_3$
2-(2-Diethylaminoethoxy)-ethyl 2-ethyl-2-phenylbutyrate citrate 355
2-Nitrophenyl tetradecyl ether 268
3-Nitrophenyl tetradecyl ether 268
4-Nitrophenyl tetradecyl ether 269

$C_{20}H_{33}NO_6S$
1-Methyl-4-piperidinyl 3-methyl-2-phenylvalerate, dimethyl sulfate salt 359

$C_{20}H_{34}$
9-(2-Methylcyclohexyl)-perhydrofluorene 2
Phenyl-*N*-tetradecane (mixture of isomers) 162

$C_{20}H_{34}ClN_3O_2$
4-Carbamoyl-1-(dodecylcarbamoylmethyl)-pyridinium chloride 951
1-Decylcarbamoylmethyl-3-(dimethylcarbamoyl)-pyridinium chloride 947

$C_{20}H_{34}N_2O$
1-Dodecyl-3-(*p*-tolyl)urea 591

$C_{20}H_{34}N_4O_2$
Adipic bis(benzylidenehydrazide) 1070
3,3′-Hexamethylenebis(1,1-diallylurea) 129
Sebacic bis(cyclopentylidenehydrazide) 1070

$C_{20}H_{34}OS$
2-(Dodecylthio)-1-phenylethanol 210

$C_{20}H_{34}O_2$
7α-Hydroxymanool 15

$C_{20}H_{34}O_3$
1-Hexadecenylsuccinic anhydride 98

$C_{20}H_{34}O_4$
((2,3,5,6-Tetramethyl-1,4-phenylene)dimethylene)bis(*tert*-butyl peroxide) 307

$C_{20}H_{34}S$
Phenyl tetradecyl sulfide 212

$C_{20}H_{35}BrN_2O$
4-Carbamoyl-1-tetradecylpyridinium bromide 956

$C_{20}H_{35}N$
N-Ethyl-*N*-phenyldodecylamine 220

$C_{20}H_{35}NO$
4-Hexyloxy-*N*-(1-methylheptyl)-aniline 218
2-(Tetradecyloxy)aniline 217
3-(Tetradecyloxy)aniline 217
4-(Tetradecyloxy)aniline 218

$C_{20}H_{35}NO_2S$
N-Tetradecylbenzenesulfonamide 827

$C_{20}H_{35}NO_6S_2$
(2-(2-Butyryloxy-1-phenylethylthio)-ethyl)diethylmethylammonium methyl sulfate 357

$C_{20}H_{36}ClP$
(4-Methylbenzyl)-tributylphosphonium chloride 832

$C_{20}H_{36}N_2O_4$
N,*N*′-Bis(cyclohexyloxycarbonyl)-1,6-hexanediamine 116

$C_{20}H_{36}N_2O_4S$
3-Dodecylcarbamoyl-1-methylpyridinium methanesulfonate 948
4-Dodecylcarbamoyl-1-methylpyridinium methanesulfonate 952

$C_{20}H_{36}OSn$
Benzyl (tributylstannyl)methyl ether 1130

$C_{20}H_{36}O_2$
Ethyl 2-cyclopentene-1-tridecanoate 81

$C_{20}H_{36}O_3$
Hexadecylsuccinic anhydride 98

$C_{20}H_{36}O_4$
Bis-(2-ethylhexyl) fumarate 79
Bis(2-ethylhexyl) maleate 79
1-Hexadecenylsuccinic acid 61
Hexadecyl maleate 95

$C_{20}H_{36}S$
2-Hexadecylthiophene 859

$C_{20}H_{37}N$
N,*N*-Dibutyl-2-(1-adamantyl)-ethylamine hydrochloride 29

$C_{20}H_{37}NO_3$
cis-*N*-(Dodecyl)-hexahydrophthalamic acid 122
N-Hexadecylmaleamic acid 121

$C_{20}H_{37}NO_3S$
Methyltributylammonium *p*-toluenesulfonate 794

$C_{20}H_{38}N_2O_3$
Dodecyl 4-carbamoyl-1-piperidineacetate 108

$C_{20}H_{38}N_4O_2$
1,1′-Hexamethylenebis(3-cyclohexylurea) 582
N,*N*′-Hexamethylenebis(4-methyl-1-piperidinecarboxamide) 136

$C_{20}H_{38}O_2$
Ethyl elaidate 80
Vinyl stearate 80

$C_{20}H_{38}O_4$
Hexadecyl succinate 95

$C_{20}H_{39}N$
Bis-(4-cyclohexylbutyl)-amine 31

$C_{20}H_{39}NO_2$
14-Oxoeicosanamide 101

$C_{20}H_{39}N_3O$
4-Amino-1,2-dimethyl-3-pentadecyl-3-pyrazolin-5-one hydrochloride 133

$C_{20}H_{39}N_3O_2$
3-Carbamoyl-*N*-dodecyl-1-piperidineacetamide 108

$C_{20}H_{40}ClNO$
2-Chloro-*N*-octadecylacetamide 103

$C_{20}H_{40}N_2O$
1-Tetradecylnipecotamide 109

$C_{20}H_{40}N_6O_4$
1,1′-Hexamethylenebis(3-(2-morpholinoethyl)urea) 126

$C_{20}H_{40}O$
3-Eicosanone 49
Octadecyl vinyl ether 18

$C_{20}H_{40}O_2S$
Octadecyl mercaptoacetate 86

$C_{20}H_{40}O_4Sn$
Diacetoxydiisooctyltin 1131

$C_{20}H_{41}NO$
N,*N*-Dibutyllauramide 107

$C_{20}H_{41}N_3O$
4-Amino-1,2-dimethyl-5-(pentadecyl)pyrazolidinone 132

$C_{20}H_{42}NO_4P$
Dioctyl (*N*,*N*-dimethylcarbamoylmethyl)-phosphonate 159

$C_{20}H_{42}N_2O$
2-Diethylamino-*N*-tetradecylacetamide citrate 104

$C_{20}H_{42}N_2O_2$
N^2-(2-Hydroxyethyl)-stearic hydrazide 111

$C_{20}H_{42}N_4O_2$
3,3′-Hexamethylenebis(1,1-diisopropylurea) 129
3,3′-Hexamethylenebis(1,1-dipropylurea) 581

$C_{20}H_{42}N_4O_4$
N,*N*′-Bis(2-diethylaminoethoxycarbonyl)-1,6-hexamethylenediamine 115

$C_{20}H_{42}O$
Decyl ether 17
3-Ethyl-3-octadecanol 10
2-Methyl-2-nonadecanol 10
3,7,11,15-Tetramethyl-3-hexadecanol 10

$C_{20}H_{42}OS$
Decyl sulfoxide 154
2-Hydroxyethyl octadecyl sulfide 25

$C_{20}H_{42}O_2S$
2-(Octadecylsulfinyl)-ethanol 152

$C_{20}H_{42}O_2Sn$
Tributyltin 2-ethylhexanoate 1131

$C_{20}H_{42}S$
Decyl sulfide 26

$C_{20}H_{44}Ge$
Tetrapentylgermane 1098

$C_{20}H_{44}O_4Si$
Tetraamyl orthosilicate 1098

$C_{20}H_{44}Si$
Butylhexadecylsilane 1098
Didecylsilane 1098

$C_{21}H_3EuF_{30}O_6$
Tris(1,1,1,5,5,6,6,7,7,7-decafluoro-2,4-heptanedionato)europium dihydrate 1141

$C_{21}H_3F_{30}O_6Pr$
Tris(1,1,1,5,5,6,6,7,7,7-decafluoro-2,4-heptanedionato)-praseodymium dihydrate 1141

$C_{21}H_3F_{30}O_6Yb$
Tris(1,1,1,5,5,6,6,7,7,7-decafluoro-2,4-heptanedionato)ytterbium dihydrate 1141

$C_{21}H_6Cl_5NO_3$
Vat Red 39 (C.I. 67900) 1094

$C_{21}H_6Cl_{12}O_4S$
3-Sulfo-2-naphthoic acid, cyclic anhydride bis(hexachloro-cyclopentadiene) adduct 799

$C_{21}H_8Cl_{12}O_5S$
3-Sulfo-2-naphthoic acid-bis(hexachlorocyclopentadiene) adduct, disodium salt 794

$C_{21}H_{10}Cl_{12}O_3S$
3-Methyl-2-naphthalenesulfonic acid-bis(hexachlorocyclo-pentadiene) adduct, calcium salt 795
3-Methyl-2-naphthalenesulfonic acid-bis(hexachlorocyclo-pentadiene) adduct 790

$C_{21}H_{10}F_{10}Si$
Bis(pentafluorophenyl)-methylstyrylsilane 1102
Bis(pentafluorophenyl)methyl(α-styryl)silane 1102

$C_{21}H_{12}ClNO_3$
1-Benzamido-4-chloroanthraquinone 473
N-(5-Chloro-1-anthraquinonyl)-benzamide 484
1-Chloro-4-benzamidoanthraquinone 488
2-(4-Chlorophenyl)spiro(isoindoline-1,1′-phthalan)-3,3′-dione 695

$C_{21}H_{12}Cl_2O$
10-(2,4-Dichlorobenzylidene)-anthrone 314
10-(3,4-Dichlorobenzylidene)-anthrone 314

$C_{21}H_{12}Cl_6N_2O_4$
N,*N*′-Bis(2,4,5-trichlorophenoxycarbonyl)-2-methyl-1,3-phenylenediamine 630

$C_{21}H_{12}Cl_6O_5$
3-(3,6-Dihydroxy-9*H*-xanthen-9-yl)-hexachloro-5-norbornene-2-carboxylic acid 326

$C_{21}H_{12}F_9Sb$
Tris(3-(trifluoromethyl)phenyl)-antimony 1128

$C_{21}H_{12}O$
1-Benzo(a)pyrenecarboxaldehyde 324
6-Benzo(a)pyrenecarboxaldehyde 324
7*H*-Dibenz(a,k)anthracen-7-one 315

$C_{21}H_{13}ClF_3NO_3$
2-Benzoyl-4-chlorophenyl *N*-(3-trifluoromethylphenyl)carbamate 610

$C_{21}H_{13}ClN_2O_2$
3-Chloro-6a-phenylisoindolo(2,1-a)-quinazoline-5,11(6*H*,6a*H*)-dione 705

$C_{21}H_{13}ClO$
10-(2-Chlorobenzylidene)-9-anthrone 314
10-(4-Chlorobenzylidene)-9-anthrone 314

$C_{21}H_{13}Cl_2N$
N-(9-Anthracenylmethylene)-2,3-dichloroaniline ... 1051
N-(9-Anthracenylmethylene)-2,4-dichloroaniline ... 1068
N-(9-Anthracenylmethylene)-2,5-dichloroaniline ... 1068
N-(9-Anthracenylmethylene)-3,4-dichloroaniline ... 1051
N-(9-Anthracenylmethylene)-3,5-dichloroaniline ... 1051

$C_{21}H_{13}Cl_3N_2O_2S$
1-(2-Benzoyl-4-chlorophenyl)-3-(2,4-dichlorobenzoyl)-2-thiourea ... 731

$C_{21}H_{13}Cl_4NO_3$
N-(Diphenylmethyl)-3,4,5,6-tetrachlorophthalamic acid ... 453

$C_{21}H_{13}NO_2$
N-(2-Fluorenyl)phthalimide ... 723

$C_{21}H_{13}NO_3$
N-(2-Anthraquinonyl)benzamide ... 488
10-(3-Nitrobenzylidene)-9-anthrone ... 313
10-(4-Nitrobenzylidene)-9-anthrone ... 313
Vat Dye (C.I. 60515) ... 1094

$C_{21}H_{13}NO_4$
Pigment Red 89 (C.I. 60745) ... 1093

$C_{21}H_{14}$
10-Methylbenzo(a)pyrene ... 166
8-Methylbenzo(a)pyrene ... 166
9-Methylbenzo(a)pyrene ... 166

$C_{21}H_{14}BrN$
N-(9-Anthracenylmethylene)-3-bromoaniline ... 1068
N-(9-Anthracenylmethylene)-4-bromoaniline ... 1067

$C_{21}H_{14}BrNO_3$
N-(7-Bromo-2-fluorenyl)-phthalamic acid ... 489

$C_{21}H_{14}BrN_3$
3-(4-Bromophenyl)-5,6-diphenyl-1,2,4-triazine ... 978

$C_{21}H_{14}Br_4N_2O_4$
Bis(2,4-dibromophenyl) *N,N'*-(4-methyl-1,3-phenylene)-biscarbamate ... 626

$C_{21}H_{14}ClF_3N_2O_2$
1-(2-Benzoyl-4-chlorophenyl)-3-(3-trifluoromethylphenyl)urea ... 509

$C_{21}H_{14}ClN$
N-(9-Anthracenylmethylene)-2-chloroaniline ... 1065
N-(9-Anthracenylmethylene)-3-chloroaniline ... 1068
N-(9-Anthracenylmethylene)-4-chloroaniline ... 1067

$C_{21}H_{14}ClNO_4S$
2-(4-Chlorophenyl)-4-phenylsulfonyl-1,3(2*H*,4*H*)-isoquinolinedione ... 773

$C_{21}H_{14}Cl_2F_2N_2O_4$
Bis(2-chloro-4-fluorophenyl) *N,N'*-(4-methyl-1,3-phenylene)-biscarbamate ... 626

$C_{21}H_{14}Cl_2F_3NO_2$
2-Benzyl-4-chlorophenyl *N*-(2-chloro-5-(trifluoromethyl)phenyl)-carbamate ... 607

$C_{21}H_{14}Cl_2N_2O_2$
Benzil, azine with 3,5-dichlorosalicylaldehyde ... 1094

$C_{21}H_{14}Cl_2O$
3,4-Dichloro-4'-phenylchalcone ... 305

$C_{21}H_{14}Cl_2O_3$
DL-Benzoin 2,4-dichlorobenzoate ... 370

$C_{21}H_{14}FN$
N-(9-Anthracenylmethylene)-4-fluoroaniline ... 1066

$C_{21}H_{14}N_2$
7-Methyl-1,2,3,4-dibenzophenazine ... 1001

$C_{21}H_{14}N_2O_2$
N-(9-Anthracenylmethylene)-3-nitroaniline ... 1056
N-(9-Anthracenylmethylene)-4-nitroaniline ... 1058

$C_{21}H_{14}N_2O_3$
1-Amino-5-benzamidoanthraquinone ... 487
2-(4-Nitrophenyl)-4,5-diphenyloxazole ... 872

$C_{21}H_{14}N_2S$
10-(2-Quinolyl)phenothiazine ... 994

$C_{21}H_{14}O$
9-Benzylideneanthrone ... 314
1-(Hydroxymethyl)benzo(a)pyrene ... 208
10-Phenyl-5*H*-dibenzo(a,d)-cyclohepten-5-one ... 314

$C_{21}H_{14}O_2$
1,2-Dihydrobenz(j)aceanthrylene-12-carboxylic acid ... 346

$C_{21}H_{14}O_2S$
2-Methyl-3-(2-naphthylthio)-1,4-naphthoquinone ... 317

$C_{21}H_{14}O_3$
2,3-Diphenyl-7-hydroxy-4*H*-1-benzopyran-4-one ... 319
Phenyl 1-hydroxy-2-anthracenoate ... 377

$C_{21}H_{14}O_4$
1-(3-Methoxyphenoxy)-anthraquinone ... 317

$C_{21}H_{15}BF_4O$
1,3-Diphenyl-2-benzopyrylium tetrafluoroborate ... 1002

$C_{21}H_{15}ClF_3NO_2$
2-Benzyl-4-chlorophenyl *N*-(2-(trifluoromethyl)phenyl)carbamate ... 607

$C_{21}H_{15}ClF_3NO_3$
4-Benzyloxyphenyl *N*-(2-chloro-5-(trifluoromethyl)phenyl)carbamate ... 606

$C_{21}H_{15}ClN_2$
2-(2-Chlorophenyl)-4,5-diphenylimidazole ... 865
2-(4-Chlorophenyl)-4,5-diphenylimidazole ... 865

$C_{21}H_{15}ClN_2O$
Benzil, azine with 4-chlorobenzaldehyde ... 1074

$C_{21}H_{15}ClN_2O_3$
N-(4-(4-Chlorophenoxy)-anilinomethyl)-phthalimide ... 717

$C_{21}H_{15}ClN_2O_6$
2-(Benzyloxycarbonyl)phenyl *N*-(4-chloro-2-nitrophenyl)carbamate ... 628
4-(Benzyloxycarbonyl)phenyl *N*-(4-chloro-2-nitrophenyl)carbamate ... 621

$C_{21}H_{15}ClO_4$
mono-(3-Chlorophenyl)-phenylmethyl phthalate ... 366

$C_{21}H_{15}Cl_2FN_4O_2$
4-(3-(2-Fluorophenyl)ureido)benzoic (3,4-dichlorobenzylidene)hydrazide ... 529

$C_{21}H_{15}Cl_2NO_3$
2'-Benzoyl-4'-chloro-2-(4-chlorophenoxy)acetanilide ... 402
2-Benzoyl-4-chlorophenyl *N*-(4-chloro-2-methylphenyl)carbamate ... 622

$C_{21}H_{15}Cl_2NO_4$
4-(Benzyloxycarbonyl)phenyl *N*-(2,3-dichlorophenyl)carbamate ... 621

$C_{21}H_{15}Cl_3N_2O_3$
4-(3,4-Dichlorophenyliminomethyl)-2-methoxyphenyl *N*-(3-chlorophenyl)carbamate ... 616

$C_{21}H_{15}Cl_3N_4O_2$
4-(3-(3-Chlorophenyl)ureido)-benzoic (3,4-dichlorobenzylidene)-hydrazide ... 529

$C_{21}H_{15}CrO_6$
Salicylaldehyde chromium complex ... 1142

$C_{21}H_{15}FN_2O_6$
2-Benzoyl-5-methoxyphenyl *N*-(2-fluoro-5-nitrophenyl)carbamate ... 620
2-(Benzyloxycarbonyl)phenyl *N*-(4-fluoro-3-nitrophenyl)carbamate ... 621

$C_{21}H_{15}N$
1-(4-Biphenylyl)isoquinoline ... 998
2-(4-Biphenylyl)quinoline ... 994
5-Cyano-7,12-dimethylbenz(a)-anthracene ... 768
9-(Phenyliminomethyl)anthracene ... 1064
Triphenylacrylonitrile ... 765

$C_{21}H_{15}NO$
N-(9-Anthracenylmethylene)-2-hydroxyaniline ... 1034
N-(9-Anthracenylmethylene)-3-hydroxyaniline ... 1036
N-(9-Anthracenylmethylene)-4-hydroxyaniline ... 1036
2-(4-Dibenzofuryl)-1,2-dihydroquinoline ... 896
1,2-Diphenyl-4(1*H*)-quinolinone ... 320
3,4-Diphenyl-2(1*H*)-quinolinone ... 993
1-(1-Naphthyliminomethyl)-2-naphthol ... 1039
N-(2-Naphthyl)-1-naphthalenecarboxamide ... 482
Triphenylacrylonitrile epoxide ... 758

$C_{21}H_{15}NO_2$
Cyanodiphenylmethyl benzoate ... 753
2-Hydroxy-*N*-(1-naphthyl)-1-naphthamide ... 486
2-Naphthyl *N*-(1-naphthyl)-carbamate ... 687
2,4,4-Triphenyl-2-oxazolin-5-one ... 384

$C_{21}H_{15}NO_3$
N-(3-Benzyloxyphenyl)phthalimide ... 717
Boronal (1-hydroxy-4-(*p*-toluidino)-anthraquinone) ... 316

$C_{21}H_{15}NO_4$
5-Benzamido-6-benzoyl-1,3-benzodioxole ... 473

$C_{21}H_{15}NO_4S$
N-Benzyl-9,10-dihydro-9,10-dioxo-2-anthracenesulfonamide ... 815
2-Phenyl-4-(phenylsulfonyl)-isoquinoline-1,3(2*H*,4*H*)-dione ... 781

$C_{21}H_{15}NO_6$
2-(2-Hydroxyphenoxy)-4'-nitroacetophenone, benzoate ester 377

$C_{21}H_{15}NO_6S$
Acid Violet 43 (C.I. 60730) ... 1093

$C_{21}H_{15}NS$
2,4,5-Triphenylthiazole ... 878

$C_{21}H_{15}N_3$
3,5,6-Triphenyl-1,2,4-triazine ... 980
3,5,6-Triphenyl-1,2,4-triazine methiodide ... 980

$C_{21}H_{15}N_3O$
5,6-Diphenyl-3-(4-hydroxyphenyl)-1,2,4-triazine ... 978

$C_{21}H_{15}N_3O_3$
3-Amino-2,4-diphenyl-7-nitro-1(2*H*)-isoquinolone ... 271
Benzil, azine with 4-nitrobenzaldehyde ... 1073

$C_{21}H_{15}N_3O_4$
2,4-Bis(3-nitrostyryl)pyridine ... 961
2,6-Bis(3-nitrostyryl)pyridine ... 961

$C_{21}H_{15}N_3S$
3-Methyl-2-benzothiazolinone (1*H*-phenalen-1-ylidene)hydrazone ... 1062

$C_{21}H_{15}N_4O_2$
5,7-Di-*tert*-butyl-3-(3-nitrophenyl)-1,2,4-benzotriazine ... 1003

$C_{21}H_{16}$
1,2-Diphenylindene ... 166
1,3-Diphenylindene ... 165
1,2,3-Triphenyl-1-cyclopropene ... 165

$C_{21}H_{16}BrNO_7$
Ethyl 5-(6-bromppiperonyl)-5,8-dihydro-8-oxo-1,3-dioxolo(4,5g)-quinoline-7-carboxylate ... 364

$C_{21}H_{16}Br_2O$
2-(4,4'-Dibromobiphenyl-2-yl)-propiophenone ... 289

$C_{21}H_{16}ClFN_4O_2$
4-(3-(2-Fluorophenyl)ureido)benzoic (2-chlorobenzylidene)hydrazide ... 529

$C_{21}H_{16}ClF_3N_2O$
1-(2-Chloro-5-(trifluoromethyl)-phenyl)-3-(diphenylmethyl)urea ... 505

$C_{21}H_{16}ClN$
6-(3-Chlorophenethyl)-phenanthridine ... 998

$C_{21}H_{16}ClNO_2$
2'-Benzoyl-4'-chloro-2-phenylacetanilide ... 405

$C_{21}H_{16}ClNO_3$
2-Benzoyl-4-chlorophenyl *N*-(*m*-tolyl)carbamate ... 629
5-Chloro-2-(*o*-vanillylideneamino)-benzophenone ... 1022
Methyl *N*-(4-chlorobenzoyl)-*N*-phenylanthranilate ... 476

$C_{21}H_{16}ClNO_4$
2-Benzoyl-4-chlorophenyl *N*-(*o*-anisyl)carbamate ... 652
2-Benzoyl-4-chlorophenyl *N*-(4-methoxyphenyl)carbamate ... 649
2-Benzoyl-5-methoxyphenyl *N*-(4-chlorophenyl)carbamate ... 654
2-(Benzyloxycarbonyl)phenyl *N*-(3-chlorophenyl)carbamate ... 621

$C_{21}H_{16}ClN_3O_2$
4-(5-Chloro-2-methoxyphenyl)-1-(9-fluorenylidene)semicarbazide ... 518
6-(α-(4-Chlorophenyl)-α-hydroxy-2-(2-imidazolyl)benzyl)pyridin-2(1*H*)-one. 865

$C_{21}H_{16}ClN_3O_3$
N-(2-Chlorobenzoyl)anthranilic (2-hydroxybenzylidene)hydrazide ... 491

$C_{21}H_{16}Cl_2N_2O_2$
1-(2-Benzoyl-4-chlorophenyl)-3-(2-chloro-6-methylphenyl)urea ... 530

$C_{21}H_{16}Cl_2N_2O_3$
4-(3-Chlorophenyliminomethyl)-2-methoxyphenyl *N*-(3-chlorophenyl)-carbamate ... 1023

$C_{21}H_{16}Cl_2N_2O_4$
N,N'-Bis(4-chlorophenoxycarbonyl)-4-methyl-1,3-phenylenediamine ... 669

$C_{21}H_{16}Cl_4O_5S$
α,α-Bis-(3,5-dichloro-2-methoxyphenyl)-*o*-toluenesulfonic acid, sodium salt ... 792

$C_{21}H_{16}F_2N_2O_4$
Bis(4-fluorophenyl) *N,N'*-(4-methyl-1,3-phenylene)biscarbamate ... 626

$C_{21}H_{16}F_3NO$
N-(4-Benzyloxybenzylidene)-2-trifluoromethylaniline ... 1020
N-(4-Benzyloxybenzylidene)-3-trifluoromethylaniline ... 1020
2,2-Diphenyl-2'-(trifluoromethyl)-acetanilide ... 397

$C_{21}H_{16}F_3N_5O_2$
4-(3-(3-(Trifluoromethyl)phenyl)-ureido)benzoic (4-pyridylmethylene)hydrazide ... 925

$C_{21}H_{16}F_4N_2O$
4-Fluoro-4'-(trifluoromethyl)-benzophenone 2-methoxyphenylhydrazone hydrochloride ... 234

$C_{21}H_{16}N_2$
N,N'-Di-1-naphthylformamidine ... 1060
2,3-Diphenyl-6-methylquinoxaline ... 1000
1,3,5-Triphenylpyrazole ... 862

$C_{21}H_{16}N_2O$
4'-(9-Fluorenylideneamino)-acetanilide ... 1060

$C_{21}H_{16}N_2OS$
α-(4-(6-Methyl-2-benzothiazolyl)-phenylimino)-*o*-cresol ... 1038

$C_{21}H_{16}N_2O_2$
2-Benzoyl-1,2-dihydro-2-phenyl-4(3*H*)-quinazolinone ... 703
2-Benzoyl-1,2-dihydro-2-phenyl-4(3*H*)-quinazolinone zinc chloride salt ... 703
3-(4,5-Diphenyl-2-imidazolyl)-catechol ... 865
4-Ethyl-*N*-(2-nitro-9-fluorenylidene)-aniline ... 1059

$C_{21}H_{16}N_2O_3$
Disperse Dye (C.I. 61115) ... 1093

$C_{21}H_{16}N_2O_5$
4'-Benzyloxy-2,4-dinitrostilbene ... 265
Methyl *N*-(4-nitrobenzoyl)-*N*-phenylanthranilate ... 472

$C_{21}H_{16}N_2O_7S_2$
1-Amino-4-(*p*-toluenesulfonamido)-anthraquinone-2-sulfonic acid sodium salt ... 793

$C_{21}H_{16}N_4$
4-(4-(2-Isoindolinyl)phenylazo)-benzonitrile ... 1090

$C_{21}H_{16}N_6O_2S$
3-Methyl-2-benzothiazolinone (α-(4-nitrophenylazo)benzylidene)-

$C_{21}H_{16}N_6O_2S$ (Cont'd)
hydrazone 742
$C_{21}H_{16}O$
9,10-Dihdyro-9,10(1',2')-benzenoanthracene-9-methanol 208
9-(4-Methoxybenzylidene)fluorene 188
4-Methoxy-9-phenylphenanthrene 180
2-(1-Naphthylmethylene)-1-tetralone 305
2-(2-Naphthylmethylene)-α-tetralone 286
1-(2-Naphthyl)-5-phenyl-2,4-pentadien-1-one 305
α-Phenylchalcone 304
7,8,9,10-Tetrahydro-10-methylbenzo(a)pyren-7-one 298
1,3,3-Triphenyl-2-propen-1-one 304
$C_{21}H_{16}O_2$
12-Acetoxy-7-methylbenz(a)anthracene 381
3,4-Dihydro-3,3-diphenyl-1*H*-2-benzopyran-1-one 387
7,12-Dimethylbenz(a)anthracene-5-carboxylic acid 346
2,6-Distyryl-4*H*-pyran-4-one 284
9-Methoxy-9-phenylanthrone 310
12-Methyl-7-benzene(a)anthracenyl acetate 381
7-Methyl-5-benzene(a)anthracenyl acetate 381
1,1'-Methylenedi-2-naphthol 197
Methyl β-methyl-1-pyreneacrylate 365
$C_{21}H_{16}O_3$
12-Acetoxy-10-methoxybenz(a)anthracene 378
DL-Benzoin benzoate 370
$C_{21}H_{16}O_3S$
α-(Phenylsulfonyl)chalcone 780
$C_{21}H_{16}O_4$
Benzyl 4-(benzoyloxy)benzoate 378
$C_{21}H_{16}O_7$
3-(3-Oxo-4,5,6-trihydroxy-3*H*-xanthen-9-yl)-5-norbornene-2-carboxylic acid 326
$C_{21}H_{16}S$
9-Anthracenylmethyl phenyl sulfide. 212
$C_{21}H_{17}BrI_2N_4O_2$
1,1'-(2-Bromo-6-methylphenylene)-bis(3-(4-iodophenyl)urea)mixture of. 539
$C_{21}H_{17}BrN_2O_4S_2$
N,N'-(2-Bromo-6-methylphenylene)-bis(*S*-(4-hydroxyphenyl) thiocarbamate) 611
$C_{21}H_{17}BrN_6O_6$
1,1'-(2-Bromo-6-methylphenylene)-bis(3-(3-nitrophenyl)urea)mixture of isomers 563
$C_{21}H_{17}BrO$
3-Bromo-1,1,1-triphenylacetone 282
$C_{21}H_{17}BrO_2$
2-Bromo-1,2-diphenylethyl benzoate 370
$C_{21}H_{17}Br_3N_4O_2$
1,1'-(2-Bromo-6-methyl-1,4-phenylene)bis(3-(3-bromophenyl)-urea) 546
1,1'-(2-Bromo-6-methylphenylene)-bis(3-(4-bromophenyl)urea)mixture of isomers 539
$C_{21}H_{17}ClN_2O_2$
Benzilic (4-chlorobenzylidene)-hydrazide 491
1-(4-Benzoylphenyl)-3-(3-chloro-4-methylphenyl)urea 530
1-(4-Benzoylphenyl)-3-(4-chloro-2-methylphenyl)urea 530
$C_{21}H_{17}ClN_2O_2S$
1-(2-Benzoyl-4-chlorophenyl)-3-(4-(methylthio)phenyl)urea 521
$C_{21}H_{17}ClN_2O_3$
1-(*o*-Anisyl)-3-(2-benzoyl-4-chlorophenyl)urea 518
1-(*m*-Anisyl)-3-(2-benzoyl-4-chlorophenyl)urea 518
1-(*p*-Anisyl)-3-(2-benzoyl-4-chlorophenyl)urea 518
$C_{21}H_{17}ClO$
3-Chloro-2,3-diphenylpropiophenone 296
$C_{21}H_{17}Cl_2NO_2$
2-(2,4-Dichlorophenoxy)-*N*-(diphenylmethyl)acetamide 401
$C_{21}H_{17}Cl_2N_3O$
2,4-Dichloro-5'-methyl-2'-(4-tolylazo)benzanilide 1081
$C_{21}H_{17}Cl_3N_2O$
1,1-Dibenzyl-3-(2,4,5-trichlorophenyl)urea 528
$C_{21}H_{17}F_3N_2O$
1-Benzyl-1-phenyl-3-(3-(trifluoromethyl)phenyl)urea 509
1-(Diphenylmethyl)-3-(2-(trifluoromethyl)phenyl)urea 510
1-(Diphenylmethyl)-3-(3-(trifluoromethyl)phenyl)urea 510
$C_{21}H_{17}F_5OSi$
Dimethyl(diphenylmethoxy)-(pentafluorophenyl)silane 1102
$C_{21}H_{17}FeN$
N-Ferrocenylmethylene-2-naphthylamine 1142
$C_{21}H_{17}IN_2$
1,3-Diphenyl-5-(4-iodophenyl)-2-pyrazoline 252
$C_{21}H_{17}N$
1,2-Dihydro-2,2-diphenylquinoline 252
α,β-Diphenylhydrocinnamonitrile 756
2,4-Distyrylpyridine 966
2,6-Distyrylpyridine 966
3,3,3 Triphenylpropionitrile 758
$C_{21}H_{17}NO$
1-(2-Amino-1-naphthylmethyl)-2-naphthol 226
2-(*N*-*p*-Anisylideneamino)-fluorene 1050
N-(Cinnamylidene)-4-phenoxyaniline 1050
N-(2-Fluorenyl)-*p*-toluamide 489
N-Phenylcinnamanilide 483
2'-Phenylcinnamanilide 478
1,3,4-Triphenyl-2-azetidinone 692
$C_{21}H_{17}NO_2$
2-Acetylphenyl *N*-phenylbenzimidate 1051
2,6-Dibenzyloxybenzonitrile 761
N-(2-Fluorenyl)mandelamide 485
α-(1-Naphthoylmethylene)-*p*-acetotoluidide 433
2-(*N*-Vanillylideneamino)-fluorene 1031
$C_{21}H_{17}NO_2S$
2-Benzoyl-2-(phenylthio)-acetanilide 443
$C_{21}H_{17}NO_3$
4-(4-Benzyloxybenzylideneamino)-benzoic acid 1039
N-(Diphenylmethyl)phthalamic acid. 476
$C_{21}H_{17}NO_3S$
2-(*N*-Methyl-*p*-toluenesulfonamido)-9-fluorenone 823
$C_{21}H_{17}NO_4$
3-(Benzoyloxy)phenyl *N*-(*o*-tolyl)-carbamate 666
$C_{21}H_{17}NO_4S$
2-Benzoyl-2-(phenylsulfonyl)-acetanilide 780
$C_{21}H_{17}NO_5$
3-(Benzoyloxy)phenyl *N*-(2-methoxyphenyl)carbamate 651
3-(Benzoyloxy)phenyl *N*-(4-methoxyphenyl)carbamate 651
9-(4-(Dimethylamino)phenyl)-2,6,7-trihydroxy-3*H*-xanthen-3-one 290
9-(4-(Dimethylamino)phenyl)-2,6,7-trihydroxy-3*H*-xanthen-3-one sulfate 290
Dimethyl 5-((1-naphthyl)-carboxamido)isophthalate 488
$C_{21}H_{17}N_3O$
3-((3-((3-Aminobenzylidene)amino)-benzylidene)amino)benzaldehyde 1052
4'-(Phenylazo)cinnamanilide 1091
$C_{21}H_{17}N_3O_2$
3,5-Dicyano-4-(4-methoxyphenyl)-4-methyl-6-phenyl-4*H*-pyran-2-amine 761
$C_{21}H_{17}N_3O_3$
N-Benzoylanthranilic (3-hydroxybenzylidene)hydrazide 496
9-Ethyl-3-(3-nitrophenylcarboxamido)carbazole. 892
$C_{21}H_{17}N_3O_6$
1,2,3-Propanetriyl trinicotinate 940
$C_{21}H_{17}N_5O_4$
2-(2-(2,4-Dinitroanilino)vinyl)-benzaldehyde phenylhydrazone 274
$C_{21}H_{18}$
5-Phenyldibenzosuberane 163
8,9,11-Trimethylbenz(a)anthracene 168
$C_{21}H_{18}AsBr$
(2-Propynyl)-triphenylarsonium bromide 1127
$C_{21}H_{18}BrN$
2-(2-Biphenylyl)-3,4-dihydroisoquinolinium bromide 256
$C_{21}H_{18}BrNO$
N-(4-Benzyloxybenzylidene)-2-bromo-4-methylaniline 1047
$C_{21}H_{18}BrN_3$
2-(4-(4-Bromophenylazo)-3-methylphenyl)isoindoline 1082
$C_{21}H_{18}BrN_3O$
2-(4-(4-Bromophenylazo)-2-methoxyphenyl)isoindoline 1080
2-(4-(4-Bromophenylazo-3-methoxyphenyl)isoindoline 1080
$C_{21}H_{18}ClNO$
N-(4-Benzyloxybenzylidene)-4-chloro-2-methylaniline 1023
3'-Chloro-2,2-diphenyl-2'-methylacetanilide 408
3'-Chloro-2,2-diphenyl-4'-methylacetanilide 404
4'-Chloro-2,2-diphenyl-2'-methylacetanilide 408
5'-Chloro-2,2-diphenyl-2'-methylacetanilide 404
$C_{21}H_{18}ClNO_2$
2-Benzylphenyl *N*-(3-chloro-2-methylphenyl)carbamate 632
3-Chloro-4-biphenylyl *N*-(2,5-dimethylphenyl)carbamate 632
$C_{21}H_{18}ClNO_3$
2-Chloro-4-biphenylyl *N*-(2-methoxy-5-methylphenyl)carbamate 618
$C_{21}H_{18}ClNO_4$
4-(Benzyloxy)phenyl *N*-(2-chloro-4-methoxyphenyl)carbamate 612
4-Benzyloxyphenyl *N*-(5-chloro-2-methoxyphenyl)carbamate 612
3-Chloro-4-biphenylyl *N*-(2,5-dimethoxyphenyl)carbamate 614
$C_{21}H_{18}ClNO_4S$
2-(*N*-(2-Phenoxyethyl)-anilinosulfonyl)benzoyl chloride 815
$C_{21}H_{18}ClN_3O$
2-Chloro-5'-methyl-2'-(4-methylphenylazo)benzanilide 1081
4-Chloro-5'-methyl-2'-(4-methylphenylazo)benzanilide 1082
$C_{21}H_{18}ClN_3O_3$
3-(2-Chloro-4-nitrophenyl)-1,1-dibenzylurea 567
$C_{21}H_{18}ClN_3O_5S$
1-(5-Chloro-2,4-dimethoxyphenyl)-3-(2-(4-nitrophenylthio)phenyl)urea 514
$C_{21}H_{18}Cl_2N_2O$
1,1-Dibenzyl-3-(2,3-dichlorophenyl)urea 573
1,1-Dibenzyl-3-(2,4-dichlorophenyl)urea 573
1,1-Dibenzyl-3-(2,5-dichlorophenyl)urea 573
1,1-Dibenzyl-3-(2,6-dichlorophenyl)urea 573
1,1-Dibenzyl-3-(3,4-dichlorophenyl)urea 573
$C_{21}H_{18}FN_3$
3-(1-(4-Fluorophenyl)-4-isopropyl-2-phenylimidazol-5-yl)acrylonitrile 865
$C_{21}H_{18}FN_3O_3$
1-(1,2-Diphenylethyl)-3-(4-fluoro-3-nitrophenyl)urea 523
$C_{21}H_{18}FN_5O_3$
1-(2-Fluoro-5-nitrophenyl)-3-(5-methyl-2-(*p*-tolylazo)phenyl)urea 1081
$C_{21}H_{18}F_2N_2$
2,4-Bis(4-fluorophenyl)-7-methyl-5,6,7,8-tetrahydroquinazoline 969
$C_{21}H_{18}F_2N_4O_2$
1,1'-(4-Methyl-1,3-phenylene)bis(3-(4-fluorophenyl)urea) 525
$C_{21}H_{18}N_2$
α,α-Bis(benzylidenimino)toluene 1063
N-Cinnamylidene-*N'*-phenyl-1,4-phenylenediamine 1057
4,5-Dihydro-2,4,5-triphenylimidazole 726
3-(4-(Dimethylamino)phenyl)-benzo(f)quinoline 989
DL-Isoamarine 250
$C_{21}H_{18}N_2O$
3-Benzamido-9-ethylcarbazole 891
10,11-Dihydro-5*H*-dibenz(b,f)azepin-5-yl 2-pyridylmethyl ketone 965
1,2-Dihydro-2-diphenylmethyl-4(3*H*)-quinazolinone 704
7,12-Dimethylbenz(a)anthracene-5-carboxylic acid hydrazide 503
$C_{21}H_{18}N_2O_2$
N-(4-Acetamidobenzylidene)-4-phenoxyaniline 1050
2-(Benzoylphenylmethylamino)-benzamide 471
1-(2-Benzoylphenyl)-3-(2-methylphenyl)urea 575
1-(4-Benzoylphenyl)-3-(3-methylphenyl)urea 575
1-(4-Benzoylphenyl)-3-(*o*-tolyl)urea 575
2-(3-Benzyloxyphenyl)-1,2-dihydro-4(3*H*)-quinazolinone 698
3-Hydroxy-2-naphthoic acid (5-indanylmethylene)-hydrazide 497
$C_{21}H_{18}N_2O_2S$
Anthrone (*p*-tosyl)hydrazone 1071
4-Ethylthio-*N*-(2-(2-pyridyl)ethyl)-1,8-naphthalimide 941
$C_{21}H_{18}N_2O_3$
Benzilic acid *N*-(2-hydroxybenzylidene)hydrazide 1028
N-(4-Benzyloxybenzylidene)-2-methyl-5-nitroaniline 1039
N-(4-Benzyloxybenzylidene)-4-methyl-3-nitroaniline 1039
$C_{21}H_{18}N_2O_4$
N,N'-Bis(phenoxycarbonyl)-4-methyl-1,3-phenylenediamine 669
$C_{21}H_{18}N_2O_5$
4-(Benzyloxy)phenyl *N*-(4-methyl-2-nitrophenyl)carbamate 646
$C_{21}H_{18}N_2O_6$
4,4'-Methylenebis(maleanilic acid) 475
$C_{21}H_{18}N_2Sn$
1-(Triphenyltin)imidazole 1132
$C_{21}H_{18}N_3O_{10}P$
Tris(3-methyl-4-nitrophenyl) phosphate 835
Tris(5-methyl-2-nitrophenyl) phosphate 835
$C_{21}H_{18}N_4O_2$
2-(2-Methyl-4-(4-nitrophenylazo)-phenyl)isoindoline 1090
$C_{21}H_{18}N_4O_3$
2-(3-Methoxy-4-(4-nitrophenylazo)-phenyl)isoindoline 1087
$C_{21}H_{18}N_4O_5S$
1-(2-(2,4-Dinitrophenylthio)phenyl)-3-(2-ethylphenyl)urea 560
$C_{21}H_{18}N_4O_6S$
1-(2-(2,4-Dinitrophenylthio)phenyl)-3-(2-methoxy-5-methylphenyl)urea 551
$C_{21}H_{18}N_6$
2,6,4-Trianilino-1,3,5-triazine 979
$C_{21}H_{18}N_6O_2S_2$
4,4'-Methylenebis(1-phenyl-3-(2-thiazolyl)urea) 878
$C_{21}H_{18}O$
7,12-Dihydro-6,8,12-trimethylbenz(a)anthracen-7-one 315
7,12-Dimethyl-5-hydroxymethylbenz(a)anthracene 208
7,12-Dimethyl-9-methoxybenz(a)anthracene 189
1,9-Diphenyl-1,3,6,8-nonatetraen-5-one 286

$C_{21}H_{18}O$ (Cont'd)
2,2-Diphenylpropiophenone 297
3,3-Diphenylpropiophenone 297
1,1,1-Triphenylacetone 287
1,1,3-Triphenylallyl alcohol 204
3,3,3-Triphenylpropionaldehyde 321
$C_{21}H_{18}OS$
3-Phenyl-3-(phenylthio)-propiophenone 293
$C_{21}H_{18}O_2$
2,2-Diphenyl-4′-methoxyacetophenone 292
Methyl 2-benzhydrylbenzoate 376
Methyl 2,2,2-triphenylacetate 362
2,3,3-Triphenylpropionic acid 333
3,3,3-Triphenylpropionic acid 333
$C_{21}H_{18}O_3$
1,5-Diphenyl-3-(2-furyl)-1,5-pentanedione 853
Methyl 4-(diphenylhydroxymethyl)-benzoate 368
Methyl 6-(9-fluorenyl)-3-oxo-1-cyclohexene-1-carboxylate 374
$C_{21}H_{18}O_3S$
4,5-Dihydro-5,5-diphenyl-2*H*-3,1-benzoxathiepin-1,1-dioxide 785
3-Phenyl-3-(phenylsulfonyl)-propiophenone 780
$C_{21}H_{18}O_4$
2,5-Di(benzyloxy)benzoic acid 339
$C_{21}H_{18}O_6$
3-Benzyl-5,7-diacetoxy-2-methyl-4*H*-1-benzopyran-4-one 378
3-Benzyl-7,8-diacetoxy-2-methyl-4*H*-1-benzopyran-4-one 378
$C_{21}H_{18}Si$
1-Propynyltriphenylsilane 1114
$C_{21}H_{19}BF_4O$
2,4-Diphenyl-5,6,7,8-tetrahydrobenzopyrylium tetrafluoroborate 936
$C_{21}H_{19}BrN_2O$
3-(4-Bromophenyl)-1,1-dibenzylurea 572
$C_{21}H_{19}BrO_3$
5-Bromo-2,4-dimethoxytriphenylmethanol 198
$C_{21}H_{19}Br_2P$
(3-Bromoallyl)-triphenylphosphonium bromide 831
$C_{21}H_{19}ClN_2O$
1-(3-Chloro-4-methylphenyl)-3-(diphenylmethyl)urea 536
3-(2-Chlorophenyl)-1,1-dibenzylurea 572
3-(3-Chlorophenyl)-1,1-dibenzylurea 571
3-(4-Chlorophenyl)-1,1-dibenzylurea 572
$C_{21}H_{19}ClN_2O_2$
1-(5-Chloro-2-methoxyphenyl)-3-(diphenylmethyl)urea 519
$C_{21}H_{19}ClN_2O_3$
5-Chloro-*N*,*N*-dibenzyl-2-methoxy-4-nitroaniline 260
$C_{21}H_{19}ClN_2O_4$
1-(4-Chloro-2-phenoxyphenyl)-3-(2,5-dimethoxyphenyl)urea 515
$C_{21}H_{19}ClO_2S$
2-Chloro-1,2-diphenylethyl *p*-tolyl sulfone 784
$C_{21}H_{19}Cl_2N_3O_2$
3-Benzylcarbamoyl-1-(4-chlorophenylcarbamoylmethyl)-pyridinium chloride 955
1-Benzylcarbamoylmethyl-3-(4-chlorophenylcarbamoyl)-pyridinium chloride 955
$C_{21}H_{19}Cl_3Si$
Tris(2-chlorobenzyl)silane 1104
$C_{21}H_{19}FN_2O$
1,1-Dibenzyl-3-(2-fluorophenyl)urea 527
1,1-Dibenzyl-3-(3-fluorophenyl)urea 527
$C_{21}H_{19}FN_4O$
1-(2-Fluorophenyl)-3-(5-methyl-2-(4-tolylazo)phenyl)urea 1082
1-(3-Fluorophenyl)-3-(5-methyl-2-(4-tolylazo)phenyl)urea 1081
$C_{21}H_{19}N$
5-Amino-10,11-dihydro-10-phenyldibenzo(a,d)cycloheptene 230
N-Benzylidene-α-phenylphenethylamine 1060
1-Tritylaziridine 252
$C_{21}H_{19}NO$
3-Anilino-3-phenylpropiophenone 294
4-Benzoyl-4′-(dimethylamino)-biphenyl 312
N-Benzyl-2,2-diphenylacetamide 430
N-(4-Benzyloxybenzylidene)-*p*-toluidine 1040
N,*N*-Dibenzylbenzamide 472
3-(4-(Dimethylamino)phenyl)-1′-acrylonaphthone 302
4-(4-(Dimethylamino)-α-phenylbenzylidene)-2,5-cyclohexadien-1-one 285
$C_{21}H_{19}NOS$
2,2-Diphenyl-4′-(methylthio)-acetanilide 421
$C_{21}H_{19}NO_2$
7-Acetoxy-5,6-dihydro-5,5-dimethylbenz(c)acridine 994
N-(4-Benzyloxybenzylidene)-*p*-anisidine 1040
4-Biphenylyl *N*-(2,5-xylyl)carbamate 681
4-(2-Methoxyphenoxy)-*N*-(4-methylbenzylidene)aniline 1042
$C_{21}H_{19}NO_2S$
N-(2-Fluorenyl)-*N*-methyl-*p*-toluenesulfonamide 829
$C_{21}H_{19}NO_3$
4-Allyl-2-methoxyphenyl *N*-(1-naphthyl)carbamate 655
2-Benzylphenyl *N*-(4-methoxyphenyl)carbamate 654
4-Biphenylyl *N*-(2-ethoxyphenyl)-carbamate 653
N-(3,4-Dimethoxybenzylidene)-4-phenoxyaniline 1040
$C_{21}H_{19}NO_3S_3$
2-Methylthio-3-phenylbenzothiazolium *p*-toluenesulfonate 916
$C_{21}H_{19}NO_4$
4-(Benzyloxy)phenyl *N*-(2-methoxyphenyl)carbamate 646
Ethyl 3-(6-methoxy-2-phenyl-4-quinolyl)-3-oxopropionate 986
$C_{21}H_{19}NO_4S$
2′-(4-Methoxybenzoyl)-*p*-toluenesulfonanilide 814
$C_{21}H_{19}NS$
N,*N*-Dibenzylthiobenzamide 743
$C_{21}H_{19}N_3$
2-(4-(*p*-Tolylazo)phenyl)isoindoline 1091
$C_{21}H_{19}N_3O$
2-(3-Methoxy-4-(phenylazo)phenyl)-isoindoline 1087
$C_{21}H_{19}N_3O_2$
1-(3-Acetylphenyl)-3-(4-anilinophenyl)urea 567
Hippuric (1-(2-naphthyl)ethylidene)-hydrazide 501
$C_{21}H_{19}N_3O_3$
1,1-Dibenzyl-3-(3-nitrophenyl)urea 565
1,1-Dibenzyl-3-(4-nitrophenyl)urea 567
$C_{21}H_{19}N_3O_4$
1-Benzyl-3-(2-methoxy-4-nitrophenyl)-1-phenylurea 553
1-Benzyl-3-(2-methoxy-5-nitrophenyl)-1-phenylurea 552
$C_{21}H_{20}$
1,1,2-Triphenylpropane 163
$C_{21}H_{20}BrOP$
(Acetonyl)triphenylphosphonium bromide 831
$C_{21}H_{20}Br_2N_2$
2,2′-Trimethylenebis-(isoquinolinium bromide) 997
$C_{21}H_{20}Br_4O_2$
2,2-Bis(4-allyloxy-3,5-dibromophenyl)-propane 182
$C_{21}H_{20}Br_8O_2$
4,4′-Isopropylidenebis(2,6-dibromophenyl 2,3-dibromopropyl ether) 176
$C_{21}H_{20}ClNO$
6-(4-Chlorobenzyl)-5-oxo-6,7,8,9-tetrahydro-5*H*-benzocycloheptene-6-propionitrile 752
5-Chloro-*N*,*N*-dibenzyl-2-methoxyaniline 224
$C_{21}H_{20}ClNO_5$
2-(2-Naphthoxy)ethyl *N*-(5-chloro-2,4-dimethoxyphenyl)carbamate 614
$C_{21}H_{20}ClN_3O_2$
1-Benzylcarbamoylmethyl-3-(phenylcarbamoyl)-pyridinium chloride 955
3-Benzylcarbamoyl-1-(phenylcarbamoylmethyl)-pyridinium chloride 955
$C_{21}H_{20}ClN_3O_3$
1-(4-Anilinophenyl)-3-(5-chloro-2,4-dimethoxyphenyl)urea 516
$C_{21}H_{20}Cl_2N_2O_4$
2,4-Diacetyl-2′,2″-dichloroglutaranilide 444
$C_{21}H_{20}Cl_2N_2O_8$
2,2′-Trimethylenebis-(isoquinolinium perchlorate) 997
$C_{21}H_{20}Cl_2O_3$
Permethrin 355
$C_{21}H_{20}Cl_4O_4$
Diethyl bis-(3,4-dichlorobenzyl)-malonate 353
$C_{21}H_{20}Ge$
Allyltriphenylgermane 1114
$C_{21}H_{20}INO$
2-(4-Benzyloxystyryl)-1-methylpyridinium iodide 936
$C_{21}H_{20}N_2$
9,10-Dihydro-9-(4-(dimethylamino)-phenyl)acridine 250
3,4-Dihydro-3-(4-(dimethylamino)-phenyl)benzo(f)quinoline 989
N-(4-Dimethylaminobenzylidene)-4-aminobiphenyl 1061
4,4′-Dimethyl-2-(4-tolyl)azobenzene 1093
2,4-Diphenyl-7-methyl-5,6,7,8-tetrahydroquinazoline 253
1,2,3-Triphenylimidazolidine 251
$C_{21}H_{20}N_2O$
1,1-Dibenzyl-3-phenylurea 574
N-(4-Dimethylaminobenzylidene)-4-phenoxyaniline 1044
$C_{21}H_{20}N_2OS$
1-Benzyl-3-(4-(methylthio)phenyl)-1-phenylurea 560
$C_{21}H_{20}N_2O_2$
N-(3,4-Dimethoxybenzylidene)-*N*′-phenyl-1,4-phenylenediamine 1040
1-Ethyl-7-(methacrylamido)-3-phenyl-2(1*H*)-quinolinone 691
Methyl (5,6-dihydro-5,5-dimethylbenz(c)acridine-7-yl)-carbamate 989
$C_{21}H_{20}N_2O_2S$
Deoxybenzoin (*p*-tosyl)hydrazone 819
2-Phenylpropiophenone (phenylsulfonyl)hydrazone 821
$C_{21}H_{20}N_2O_2S_3$
5,6-Dihydro-7-(4-methoxyphenylazo)-3-(4-methoxyphenylthio)-4*H*-1,2-benzodithiole 1086
$C_{21}H_{20}N_2O_3$
Alstonine perchlorate 940
Carbobenzyloxy-D-alanine β-naphthylamide 664
4′,4‴-(3-Oxo-1,4-pentadien-1,5-diyl)-bisacetanilide 434
$C_{21}H_{20}N_2O_3S$
2-Hydroxy-1-naphthaldehyde *N*-phenyl-*N*-(3-sulfolanyl)hydrazone 775
$C_{21}H_{20}N_2O_4$
Dibenzyl 2,3-diazabicyclo(2.2.1)-hept-5-ene-2,3-dicarboxylate 700
Ethyl 4-benzoyl-2-hydroxy-2-phenyl-1,7-diazabicyclo(4.1.0)hept-5-ene-5-carboxylate 368
$C_{21}H_{20}N_2S$
1,1-Dibenzyl-3-phenyl-2-thiourea 741
1-(1,2-Diphenylethyl)-3-phenyl-2-thiourea 741
$C_{21}H_{20}N_4$
1-(4-(3-Biphenylyl)-α-methylbenzylideneamino)-guanidine nitrate 1072
$C_{21}H_{20}N_4O$
4,4-Dibenzyl-1-(3-pyridylmethylene)-semicarbazide 1063
$C_{21}H_{20}N_4OS$
1-(4-(Methylthio)phenyl)-3-(2-(*p*-tolylazo)phenyl)urea 1087
$C_{21}H_{20}N_4S$
1,3-Bis-(anilinomethyl)-2(3*H*)-benzimidazolethione 742
$C_{21}H_{20}O$
2,7-Dibenzylidene-1-cycloheptanone 304
2,6-Dibenzyl-4-methylphenol 196
2,4′-Dimethyltriphenylmethanol 204
4-(1,1-Diphenylpropyl) phenol 191
Ethyl triphenylmethyl ether 187
$C_{21}H_{20}OSi$
(Allyloxy)triphenylsilane 1114
$C_{21}H_{20}O_2$
4,4′-Benzylidenedianisole 180
1,1-Diphenyl-2-(3-methoxyphenyl)-ethanol 201
$C_{21}H_{20}O_2S$
Dibenzylmethyl phenyl sulfone 784
2,2-Diphenylethyl *p*-tolyl sulfone 784
$C_{21}H_{20}O_2Si$
3-(Triphenylsilyl)propionic acid 1109
$C_{21}H_{20}O_3$
2,5-Bis(4-methoxybenzylidene)-cyclopentanone 302
$C_{21}H_{20}O_3S$
4-Biphenylyl mesitylenesulfonate 801
$C_{21}H_{20}O_4S_2$
4-Benzylsulfonyl-3-(phenylsulfonylmethyl)toluene 784
$C_{21}H_{20}O_{11}$
Quercitrin 319
$C_{21}H_{20}S$
S-(1,1-(Diphenyl)ethyl)-*p*-thiocresol 212
$C_{21}H_{20}Si$
(1-Propen-1-yl)triphenylsilane 1114
$C_{21}H_{20}Sn$
Phenyltricyclopentadienyltin 1132
$C_{21}H_{21}BO_3$
Tribenzyl borate 1125
Tri-*o*-tolyl borate 1125
Tritolyl borate (mixture of *m*- and *p*-) 1124
$C_{21}H_{21}BO_6$
Tris-(hydroxybenzyl) borate 1125
$C_{21}H_{21}BiCl_2$
Dichloritri(*o*-tolyl)bismuth 1143
$C_{21}H_{21}BiO_3$
Tris(2-methoxyphenyl)bismuthine 1141
Tris(4-methoxyphenyl)bismuthine 1141
$C_{21}H_{21}BrSi$
(3-Bromopropyl)triphenylsilane 1100
$C_{21}H_{21}ClN_2O$
5-Chloro-*N*¹,*N*¹-dibenzyl-2-methoxy-1,4-phenylenediamine 220
$C_{21}H_{21}ClO_3Pb$
Chlorotris(4-methoxyphenyl)lead 1134
$C_{21}H_{21}ClO_3Si$
Chlorotris(2-methoxyphenyl)silane 1106
$C_{21}H_{21}ClO_5$
2,6-Diphenyl-4-*tert*-butylpyrylium perchlorate 936
$C_{21}H_{21}ClSi$
(3-(2-Chlorophenyl)propyl)-diphenylsilane 1103
Chlorotri(*o*-tolyl)silane 1122
$C_{21}H_{21}ClSn$
Chlorotribenzyltin 1132
$C_{21}H_{21}Cl_2Sb$
Dichlorotribenzylantimony 1129
$C_{21}H_{21}FN_2O_2$
(1-(4-Fluorophenyl)-4-isopropyl-2-phenylimidazol-5-yl)methyl acetate 865
$C_{21}H_{21}N$
N-Benzyl-α-phenylphenethylamine 242
$C_{21}H_{21}NO$
1-(4-Butylphenyliminomethyl)-2-naphthol 1038

$C_{21}H_{21}NO$ (Cont'd)
2-Oxo-3-phenethyl-3-phenyl-1-cyclohexanecarbonitrile 757
Tribenzylamine *N*-oxide 234
$C_{21}H_{21}NO_2$
2-*sec*-Butylphenyl *N*-(1-naphthyl)carbamate 682
$C_{21}H_{21}NO_4S_2$
4-(2-(4-(*N*-Methyl-*N*-phenylsulfamoyl)phenyl)ethyl)benzenesulfinic acid, sodium salt 828
$C_{21}H_{21}NS_2Sn$
Triphenyltin *N*,*N*-dimethyldithiocarbamate 1131
$C_{21}H_{21}N_2$
3-(Cyclohexylamino)-2-phenylquinoline picrate 989
$C_{21}H_{21}N_3O$
1-(4-Anilinophenyl)-3-(2,5-xylyl)urea 570
2-(1-Naphthyl)acetic (4-(dimethylamino)benzylidene)hydrazide 500
$C_{21}H_{21}N_3O_2$
1,3-Bis(4-methoxyphenyl)-2-phenylguanidine 724
Glycyl-L-phenylalanine-2-naphthylamide, hemihydrate 487
$C_{21}H_{21}N_3O_3$
1-(4-Anilinophenyl)-3-(2,5-dimethoxyphenyl)urea 550
$C_{21}H_{21}N_3O_4$
2-Nitrostrychnine 698
$C_{21}H_{21}N_3O_5$
Carbobenzyloxyglycyl-D-tryptophan 886
Carbobenzyloxy-L-tryptophylglycine 889
$C_{21}H_{21}N_3O_6S$
4′-Butylsulfamoyl-1-hydroxy-2′-nitro-2-naphthanilide 812
$C_{21}H_{21}N_3O_6S_3$
1,3,5-Tris(phenylsulfonyl)hexahydro-1,3,5-triazine 827
$C_{21}H_{21}N_5O_4$
2-(4-Methoxyphenylureido)acetic acid N^2-(*N*-(1-naphthyl)carbamoyl)hydrazide 553
$C_{21}H_{21}OP$
Tribenzylphosphine oxide 836
$C_{21}H_{21}O_3P$
Tris(2-methoxyphenyl)phosphine 830
$C_{21}H_{21}PS$
Tri-*p*-tolylphosphine sulfide 836
$C_{21}H_{21}Sb$
Tris(*o*-tolyl)antimony 1129
Tris(*p*-tolyl)antimony 1129
$C_{21}H_{22}BrNP$
Propylaminotriphenylphosphonium bromide 831
$C_{21}H_{22}BrOP$
(3-Hydroxypropyl)triphenylphosphonium bromide 830
$C_{21}H_{22}ClNO$
1-(4-Methylbenzyl)-3-(4-methylbenzyloxy)pyridinium chloride 936
$C_{21}H_{22}ClN_3S$
Basic Yellow 12 (C.I. 48065) 1076
$C_{21}H_{22}ClN_5O_4S$
2-((1,4-Dihydro-1-methyl-4-quinolylidene)aminoazo)-3-isobutylbenzothiazolium perchlorate 915
$C_{21}H_{22}Cl_2O_4$
Diethyl bis-(4-chlorobenzyl)malonate 353
$C_{21}H_{22}Ge$
Propyltriphenylgermane 1114
$C_{21}H_{22}INO_2$
2-(3,4-Dimethoxystyryl)-1-ethylquinolinium iodide 984
$C_{21}H_{22}IOP$
(3-Hydroxypropyl)triphenylphosphonium iodide 830
$C_{21}H_{22}N_2$
2,3-Diphenyl-4,5,6,7,8,9-hexahydro-2*H*-cyclooctapyrazole 864
3,3-Diphenyl-4,5,6,7,8,9-hexahydro-3*H*-cyclooctapyrazole 864
2,2′-(Phenylmethylenebis)bis(4-methylaniline) 230
$C_{21}H_{22}N_2O$
1-Butyl-3-(1-naphthyl)-1-phenylurea 587
1-(2,6-Diethylphenyl)-3-(1-naphthyl)urea 597
1-(5-Isopropyl-2-methylphenyl)-3-(1-naphthyl)urea 602
$C_{21}H_{22}N_2O_2S$
Ethyl 2-(4-(1-methyl-6-phenyl-1,2,3,6-tetrahydro-2-thioxo-4-pyrimidinyl)phenyl)acetate 745
$C_{21}H_{22}N_2O_3S$
N,5-Dimethyl-2-(2-hydroxy-2-phenyl-2-(2-pyridyl)ethyl)benzenesulfonamide 931
$C_{21}H_{22}N_2O_4$
2,4-Diacetylglutaranilide 433
4′,4‴-Methylenebis(acetoacetanilide) 444
4,4′-Methylenebis(allyl *N*-phenylcarbamate) 673
$C_{21}H_{22}N_2O_8$
Diethyl bis(4-nitrobenzyl)malonate 359
$C_{21}H_{22}N_4O_4$
Carbobenzyloxyglycyl-L-tryptophanamide 883
Ethyl 4-(3-(1,5-dimethyl-3-oxo-2-phenyl-4-pyrazolin-4-yl)ureido)benzoate 702
$C_{21}H_{22}N_6O_2$
4,4′-(4-Methyl-1,3-phenylene)bis(1-phenylsemicarbazide) 565
1,1′-(4-Methyl-1,3-phenylene)bis(3-(3-pyridylmethyl)urea) 950
1,1′-(4-Methyl-1,3-phenylene)bis(3-(4-pyridylmethyl)urea) 945
$C_{21}H_{22}OSi$
Dimethyl(diphenylmethoxy)phenylsilane 1107
Diphenyl(2-hydroxy-2-phenylethyl)methylsilane 1105
1-(Triphenylsilyl)-2-propanol 1105
2-(Triphenylsilyl)-2-propanol 1105
3-(Triphenylsilyl)-1-propanol 1105
Tri(*o*-tolyl)silanol 1109
$C_{21}H_{22}OSn$
Tribenzyltin hydroxide 1132
$C_{21}H_{22}O_2$
Ethyl 10,11-dihydro-5,10-ethano-13-methyl-5*H*-dibenzo(a,d)cycloheptene-12-carboxylate 362
Methyl 9-cyclohexyl-9-fluorenecarboxylate 361
$C_{21}H_{22}O_3$
3-(8-Benzyl-9-oxo-6,7,8,9-tetrahydro-5*H*-benzocyclohepten-8-yl)propionic acid 331
$C_{21}H_{22}O_3Si$
3-(Triphenylsilyloxy)-1,2-propanediol 1104
$C_{21}H_{22}O_4Si$
Tris(2-methoxyphenyl)silanol 1109
$C_{21}H_{22}O_5$
1,5-Bis(3,4-dimethoxyphenyl)-1,4-pentadien-3-one 283
$C_{21}H_{22}O_7S$
Diethyl (α-(*p*-tosyloxy)benzylidene)malonate 798
$C_{21}H_{22}O_9$
Aloin 310
$C_{21}H_{22}O_{10}S$
2-(2-(2-Carboxyethyl)-5-methoxy-6-tosyloxyphenyl)succinic acid 797
$C_{21}H_{22}Pb$
Propyltriphenyllead 1136
Tri(*o*-tolyl)lead 1136
Tri(*p*-tolyl)lead 1136
$C_{21}H_{22}PbS$
(Propylthio)-triphenyllead 1136
$C_{21}H_{22}Si$
Diphenyl(3-phenylpropyl)silane 1114
Isopropyltriphenylsilane 1114
Tri(*o*-tolyl)silane 1122
$C_{21}H_{23}F_3N_4O_3$
2-(3-(3-(Trifluoromethyl)phenyl)ureido)acetic (4-butoxybenzylidene)hydrazide 507
$C_{21}H_{23}IN_2O_2$
2-(2-Acetanilinovinyl)-5,6-dimethyl-3-ethylbenzoxazolium iodide 910
$C_{21}H_{23}NO$
3-*tert*-Butyl-2-methyl-4-(*o*-tolyl)-1(2*H*)-isoquinolinone 693
2-(4-(Diethylamino)benzylidene)-1-tetralone 302
$C_{21}H_{23}NO_2$
trans-Norlobelanine 308
trans-Norlobelanine hydrochloride 308
$C_{21}H_{23}NO_3$
Ethyl 2,6-diphenyl-4-oxo-3-piperidineacetate 358
$C_{21}H_{23}NO_4$
1,7-Diphenyl-4-ethyl-4-nitro-1,7-heptanedione 308
$C_{21}H_{23}NO_5$
Cryptopine 319
$C_{21}H_{23}NO_6$
2-(Isobutoxycarbonyl)phenyl *N*-(4-(ethoxycarbonyl)phenyl)carbamate 666
$C_{21}H_{23}N_3$
2,6-Bis(4-(dimethylamino)phenyl)pyridine 954
$C_{21}H_{23}N_3O$
2,3-Dimethyl-4-(4-isopropylbenzylideneamino)-1-phenyl-3-pyrazolin-5-one 1075
$C_{21}H_{23}N_3O_3$
3-Amino-2-hexyl-7-nitro-4-phenyl-1(2*H*)-isoquinolinone 690
$C_{21}H_{23}N_3O_6$
Carbobenzyloxyglycylglycylglycine benzyl ester 662
Carbobenzyloxyglycylglycyl-L-phenylalanine 657
$C_{21}H_{24}BrNO_5$
Ethyl 4-bromo-3-(1-(ethoxycarbonyl)-2-oxocyclohexylmethyl)-2-indolecarboxylate 888
Ethyl 6-bromo-3-(1-(ethoxycarbonyl)-2-oxocyclohexylmethyl)-2-indolecarboxylate 888
$C_{21}H_{24}ClNO$
N-(Cyclohexyl)-α-chloro-3,5,7-trimethyl-1-adamantane-acetamide 103
$C_{21}H_{24}ClNO_6$
2-(2-Methoxy-4-(1-propenyl)phenoxy)ethyl *N*-(5-chloro2,4dimethoxyphenyl)carbamate 613
$C_{21}H_{24}ClN_3O_4$
1-(2-Chloro-4-nitrophenyl)-3-cyclohexyl-3-(2-hydroxy-2-phenylethyl)urea 512
$C_{21}H_{24}FN_2O_3P$
Dimethyl (1-(4-fluorophenyl)-4-isopropyl-2-phenylimidazol-5-yl)methylphosphonate 867
$C_{21}H_{24}N_2$
1-Cyano-*N*-(1-(1,2-diphenyl)ethyl)cyclohexylamine 754
α,α-Diphenyl-1-piperidinebutyronitrile hydrochloride 755
$C_{21}H_{24}N_2O$
N-Benzyl-3-(4-phenyl-1,2,3,6-tetrahydropyridino)-propionamide 427
Bis-(4-(dimethylamino)benzylidene)acetone 302
β-(Cyclohexylamino)cinnamanilide 471
1-(2-(Diethylamino)-1-phenylethoxy)isoquinoline picrate 997
N,*N*-Diethyl-β-(1-isoquinolyloxy)phenethylamine 997
N,*N*,*N*′,*N*′-Tetramethyl-4,4′-furfurylidenedianiline 850
$C_{21}H_{24}N_2OS_2$
10-(2-(2-Piperidinoethylthio)-acetyl)phenothiazine hydrochloride 700
$C_{21}H_{24}N_2O_2$
1,4-Bis(benzoylmethyl)-hexahydro-1*H*-1,4-diazepine dihydrochloride 294
4-Butoxybenzoic acid (5-indanylmethylene)-hydrazide 498
3-(3-(Dimethylamino)-propyl)-1,5-diphenyl-2,4-pyrrolidinedione 691
3,3-Diphenyl-*N*,*N*,*N*′,*N*′-tetramethyl-1,2-cyclopropanedicarboxamide 441
5-(4-Hydroxyimino-1-methyl-3-piperidyl)dibenzosuberol 1008
Vindolinine 359
$C_{21}H_{24}N_2O_3$
Isostrychnic acid monohydrate 328
$C_{21}H_{24}N_2O_5$
N-Acetyl-L-phenylalanyl-L-tyrosine methyl ester 412
Carbobenzyloxy-β-alanyl-β-alanine benzyl ester 664
Carbobenzyloxy-β-alanyl-L-alanine benzyl ester 663
Carbobenzyloxy-L-alanyl-L-alanine benzyl ester 662
Carbobenzyloxy-D-alanyl-L-phenylalanine methyl ester 658
Carbobenzyloxy-β-alanyl-L-phenylalanine methyl ester 658
Carbobenzyloxysarcosylphenylalanine methyl ester 657
*N*ε-Carbobenzyloxy-*N*α-benzoyl-L-lysine 659
$C_{21}H_{24}N_2O_6$
Carbobenzyloxy-L-alanyl-L-tyrosine methyl ester 645
N-Carbobenzyloxyglycyl-*o*-benzyl-L-threonine 646
Carbobenzyloxyglycyl-L-threonine benzyl ester 643
Carbobenzyloxy-L-phenylalanyl-L-serine methyl ester 644
Carbobenzyloxysarcosyl-L-tyrosine methyl ester 643
Carbobenzyloxy-L-seryl-L-alanine benzyl ester 645
Carbobenzyloxy-L-threonylglycine benzyl ester 643
Carbobenzyloxy-L-threonylphenylalanine 642
N,*N*′-Dicarbobenzyloxy-L-ornithine 657
$C_{21}H_{24}N_2O_6S$
N-(*tert*-Butoxycarbonyl)-*S*-benzyl-L-cysteine 4-nitrophenyl ester 667
$C_{21}H_{24}N_2O_7$
Carbobenzyloxy-L-seryl-L-tyrosine methyl ester 642
Carbobenzyloxy-L-tyrosyl-L-threonine 644
$C_{21}H_{24}N_4O_2$
Ethyl 2-((2,2-dicyanovinyl)-4-ethyl--3-methylpyrrolylmethyl)-4,5-dimethyl-3-pyrrolecarboxylate 840
Ethyl 5-((2,2-dicyanovinyl)-3-ethyl-4-methylpyrrolylmethyl)-2,4-dimethyl)-3-pyrrolecarboxylate 840
4,4′-Methylenebis(1-allyl-3-phenylurea) 592
$C_{21}H_{24}N_4O_4$
N,*N*′-(Methylenedi-4,1-phenylene)bis(*O*-(carbamoyl)acetone oxime) 1013
$C_{21}H_{24}N_4O_5$
Carbobenzyloxy-L-phenylalanylglycylglycinamide 658
$C_{21}H_{24}N_4O_6$
Carbobenzyloxy-L-tyrosylglycylglycinamide 645
$C_{21}H_{24}O$
5-(4-Biphenylyl)-2-propylcyclohexanone 306
Diphenylmethyl 1-methylcyclohexyl ketone 286
$C_{21}H_{24}OSi_2$
1,1,1-Trimethyl-3,3,3-triphenyldisiloxane 1114
$C_{21}H_{24}O_2$
1,4-Bis(4-methoxyphenyl)-4-methyl-1-cyclohexene 179
α-Cyclopentyl-4,4′-dimethoxystilbene 179
2,5-Dimethyl-2-(4-methoxyphenyl)-5-phenyl-1-cyclohexanone 284
$C_{21}H_{24}O_2S$
9,10-Dihydro-9,10-ethanoanthracen-11-ylmethyl

$C_{21}H_{24}O_2S$ (Cont'd)
1,1-dimethylethanesulfinate 786

$C_{21}H_{24}O_3$
2-Phenyl-2-(4-(tetrahydropyran-4-ylmethyl)phenyl)-1,3-dioxolane hydrochloride 184

$C_{21}H_{24}O_3S$
9,10-Dihydro-9,10-ethanoanthracene-11-methanol 1-butanesulfonate 800

$C_{21}H_{24}O_5$
8a-(Benzoylmethyl)-4,4a,5,6,8,8a-hexahydro-3,6,6-trimethyl-3,5-ethano-1*H*,3*H*-pyrano(3,4-c)pyran-1,8-dione 385

$C_{21}H_{24}O_8S$
Methyl 4,6-*O*-benzylideneglucoside, 2-*p*-toluenesulfonate 797

$C_{21}H_{24}S$
9-Anthracenylmethyl hexyl sulfide 213

$C_{21}H_{24}Si_2$
(4-(Diphenylsilyl)phenyl)-trimethylsilane 1120
1,1,1-Trimethyl-2,2,2-triphenyldisilane 1114
1,1,2-Trimethyl-1,2,2-triphenyldisilane 1114

$C_{21}H_{25}ClN_2OS$
2-Chloro-10-(3-(2-methoxyethyl)-1-methyl-3-pyrrolidinyl-methyl)-phenothiazine hydrochloride 247

$C_{21}H_{25}FN_4O$
1-(3-Fluorophenyl)-3-(5-methyl-2-(4-methylcyclohexylazo)phenyl)urea 1081

$C_{21}H_{25}NO_2$
4-Cyclohexylphenyl *N*-(2,6-xylyl)-carbamate 679
4-(2-Diethylaminoethoxy)chalcone 301
Ethyl 2,2-diphenyl-2-piperidinoacetate 359
4′-((5-Hydroxy-1,2,3,4-tetrahydro-2-isoquinolinyl)methyl)-2,2-dimethylpropiophenone 290

$C_{21}H_{25}NO_2S$
O-[4-(2,3-Dihydro-2,2,4-trimethyl-4*H*-1-benzopyran-4-yl)phenyl] *N*,*N*-dimethylthiocarbamate 750

$C_{21}H_{25}NO_3$
5,5-Dimethyl-2-(3-morpholino-1-phenyl-2-propenylidene)-1,3-cyclohexanedione 309
2-((3,3-Dimethyl-2-oxo-1-phenyl)-butyl)-4-methoxy-*N*-methylbenzamide 466
N-(2-Methyl-5-(2,5,5-trimethyl-1,3-dioxan-2-yl)phenyl)benzamide 477

$C_{21}H_{25}NO_3S$
S-(4-*tert*-Butyl-2-methylphenyl) *N*-(4-(ethoxycarbonyl)phenyl)-thiocarbamate 676

$C_{21}H_{25}NO_4$
2-(2-Methoxy-4-propenylphenoxy)-ethyl *N*-(3-ethylphenyl)carbamate 647
2-(2-Methoxy-4-(1-propenyl)-phenoxy)ethyl *N*-(3,4-xylyl)-carbamate 647
D-Tetrahydropalmatine hydrochloride 246
L-Tetrahydropalmatine hydrochloride 246

$C_{21}H_{25}NO_5$
Carbobenzyloxy-*O*-(*tert*-butyl)-L-tyrosine dicyclohexylamine salt 645
2-(2-Methoxy-4-(1-propenyl)-phenoxy)ethyl *N*-(2-methoxy-5-methylphenyl)carbamate 647

$C_{21}H_{25}NO_5S$
Benzyl *N*-(4-(*p*-tosyloxy)cyclohexyl)-carbamate 799

$C_{21}H_{25}N_3O$
4′-(4-(Octahydro-1-azocinyl)-phenylazo)acetophenone 1091

$C_{21}H_{25}N_3O_5$
Carbobenzyloxy-L-threonyl-L-phenylalaninamide 644
N-(3-(3,4-Dimethoxyphenyl)-1-methyl-4-piperidyl)-4-nitrobenzamide 463

$C_{21}H_{25}N_3O_6$
Carbobenzyloxy-L-threonyl-L-tyrosinamide 645
Carbobenzyloxy-L-tyrosyl-L-threoninamide 642
Ethyl 7-acetoxymethyl-6-benzamido-1,2,3,5,6,7-hexahydro-7-methyl-5-oxoimidazo(1,2-a)-pyridine-8-carboxylate 707

$C_{21}H_{25}N_5O_8$
*N*ε-(2,4-dinitrophenyl)-L-lysyl-L-tyrosine hydrobromide monohydrate 412

$C_{21}H_{26}BrNO_3$
Diethylmethyl(2-(9*H*-xanthen-9-ylcarbonyloxy)-ethyl)-ammonium bromide 355

$C_{21}H_{26}BrNO_4$
DL-6′-Bromolaudanosine 246

$C_{21}H_{26}ClNO_2$
2′-Chloro-3,5-di-*tert*-butyl-4-hydroxybenzanilide 447

$C_{21}H_{26}ClNO_4$
4-Chloro-*o*-tolyl 2-heptyloxy-5-nitrobenzyl ether 269

$C_{21}H_{26}ClN_3O_2$
6-Chloro-9-(3-diethylamino-2-hydroxypropylamino)-2-methoxyacridine di-hydrochloride 995

$C_{21}H_{26}Cl_2N_2O$
(4-Chlorophenylcarbamoylmethyl)-dimethyl(tetrahydro-2-naphthyl-methyl)-ammonium chloride 431

$C_{21}H_{26}N_2$
N,*N*′-Bis-(2,6-diethylphenyl)-carbodiimide 769

$C_{21}H_{26}N_2O$
4-Acetamidomethyl-1-benzyl-4-phenylpiperidine 427
1-Benzyl-4-(*N*-propionylanilino)-piperidine 426
3-Cyclohexyl-1,1-dibenzylurea 571

$C_{21}H_{26}N_2O_2$
N,*N*′-Diphenyl-*N*,*N*′,3,3-tetramethylglutaramide 436

$C_{21}H_{26}N_2O_2S_2$
Bis(*S*-propyl) *N*,*N*′-(methylenedi-4,1-phenylene)bisthiocarbamate 688
4,4′-Methylenebis(*S*-isopropyl *N*-phenylthiolcarbamate) 688

$C_{21}H_{26}N_2O_3$
2-Acetyl-6-(dimethylamino)-1-(3-hydroxy-4-methoxybenzyl)-1,2,3,4-tetrahydro-isoquinoline 696

$C_{21}H_{26}N_2O_4$
N,*N*′-Bis(3,4-dimethoxybenzylidene)-1,3-propanediamine 1041
*N*ε-Carbobenzyloxy-L-lysine benzyl ester hydrochloride 657
4,4′-Methylenebis(isopropyl *N*-phenylcarbamate) 673
4,4′-Methylenebis(propyl carbanilate) 673
2′,2″-(Pentamethylenedioxy)-diacetanilide 417
3′,3″-(Pentamethylenedioxy)-diacetanilide 418
4′,4″-(Pentamethylenedioxy)-diacetanilide 419

$C_{21}H_{26}N_2O_5$
4-Butoxybenzoic acid (3,4,5-trimethoxybenzylidene)-hydrazide 497
2-(2-*tert*-Butyl-6-methylphenoxy)-ethyl *N*-(4-methyl-2-nitrophenyl)-carbamate 649
N-(α-Cyano-3,4,5-trimethoxybenzyl)-3,4-dimethoxyphenethylamine hydrochloride 753
2′,5′-Dibutoxy-4′-nitrobenzanilide 463

$C_{21}H_{26}N_2O_6S$
Carbobenzyloxy-*N*ε-tosyl-L-lysine 815

$C_{21}H_{26}N_2O_8$
Dimethyl 2,2′-methylenebis(5-carboxy-4-methyl-3-pyrrolepropionate) 839

$C_{21}H_{26}N_2S$
10-(1,3-Diethyl-3-pyrrolidinylmethyl)-phenothiazine hydrochloride 249

$C_{21}H_{26}N_2S_2$
3,5-Bis(4-isopropylphenyl)-3,4,5,6-tetrahydro-2*H*-1,3,5-thiadiazine-2-thione 743

$C_{21}H_{26}N_4O_4$
L-Alanyl-L-tyrosyl-L-phenylalaninamide hydrochloride 412
4-(3,3-Diethylureido)benzoic (3,4-dimethoxybenzylidene)hydrazide 549
4-(3-Propylureido)benzoic (4-ethoxy-3-methoxybenzylidene)hydrazide 549

$C_{21}H_{26}O_2$
3,5-Di-*tert*-butyl-4-hydroxybenzophenone 310
α,4-Dimethylbenzyl 3-methyl-2-phenylvalerate 362
4′-Octyl-4-biphenylcarboxylic acid 345
4-(1,1,3,3-Tetramethylbutyl)phenyl benzoate 380

$C_{21}H_{26}O_2S$
Benzyl 4-(1-methylcyclohexylmethyl)phenyl sulfone 783

$C_{21}H_{26}O_2S_2$
p-Tolyl 2-(3,4-xylylsulfonyl)-cyclohexyl sulfide 776

$C_{21}H_{26}O_3$
2-Hydroxy-4-octyloxybenzophenone 310

$C_{21}H_{26}O_4$
2,2′-Dihydroxy-4-octyloxybenzophenone 310

$C_{21}H_{26}O_8$
Tetraallyl *cis*,*cis*,*cis*,*cis*-1,2,3,4-cyclopentanetetracarboxylate 81

$C_{21}H_{27}Br$
4-Bromo-4′-(*tert*.-octyl)-diphenylmethane 174

$C_{21}H_{27}BrN_4O$
1-(3-Bromo-α-heptyl-4-hydroxy-2-phenylbenzylideneamino)-guanidine hydrochloride 1023

$C_{21}H_{27}NO$
N-Benzyl-2,2-dimethyl-*N*-(1-methyl-2-phenylethyl)propionamide 428
1,2-Diphenyl-3-(pyrrolidinomethyl)-2-butanol hydrochloride 203

$C_{21}H_{27}NO_2$
2′-(2-Diethylaminoethoxy)-dihydrochalcone hydrochloride 292

$C_{21}H_{27}NO_2Pb$
Triethylplumbyl 9-ethyl-3-carbazolecarboxylate 1134

$C_{21}H_{27}NO_3$
4-Butoxyphenyl *N*-(2,6-diethylphenyl)carbamate 651
2-(4-*tert*-Butylphenoxy)ethyl *N*-(3-ethylphenyl)carbamate 650
4-Nitrobenzyl 4-(*tert*-octyl)phenyl ether 265

$C_{21}H_{27}NO_{11}S$
N,1,3,4-Tetraacetyl-6-*O*-(*p*-tosyl)-β-D-galactosamine 798

$C_{21}H_{27}N_3O_3$
2-(1-Methyl-3-(2,4-xylyl)ureido)ethyl *N*-(2,4-xylyl)carbamate 574

$C_{21}H_{27}N_3O_3S$
Ethyl 2-(diethylamino)-4-(3-(4-(methylthio)phenyl)ureido)-benzoate 560

$C_{21}H_{27}N_5O_2$
4-(3,3-Diethylureido)benzoic (4-(dimethylamino)benzylidene)-hydrazide 570

$C_{21}H_{27}N_7O_{14}P_2$
β-Nicotinamide adenine dinucleotide, oxidized, free acid (nad,dpn) 922

$C_{21}H_{28}B_2F_8N_2$
N-Cinnamylidene-*N*,*N*′-diethyl-*N*′,*N*′-dimethyl-1,4-phenylene-diammonium bis(tetrafluoroborate) 255

$C_{21}H_{28}BrNO_2$
Diethyl(2-(diphenylacetoxy)-ethyl)-methylammonium bromide 359

$C_{21}H_{28}N_2$
N,*N*′-Bis(4-butylphenyl)-formamidine 727

$C_{21}H_{28}N_2O$
2-(*N*-Butyl-*N*-methylamino)-*N*-(1,2-diphenylethyl)-acetamide 428
N-(2-(Diethylamino)ethyl)-2,2-diphenyl-*N*-methylacetamide hydrochloride 423
1-(2,6-Diisopropylphenyl)-3-(2,4-xylyl)urea 596

$C_{21}H_{28}N_2O_2$
7,7-Dimethyl-3,4,5,6,7,8-hexahydro-2-phenyl-4-piperazino-2*H*-1-benzopyran-5-one 284

$C_{21}H_{28}N_4O_2$
4,4′-Methylenebis(1-isopropyl-3-phenylurea) 592
4,4′-Methylenebis(3-phenyl-1-propylurea) 592
3,3′-(4-Methyl-1,3-phenylene)bis(1,1-diallylurea) 566

$C_{21}H_{28}N_4O_4$
1,1′-(Methylenedi-4,1-phenylene)-bis(3-(2-methoxyethyl)urea) 555
N,*N*′-(4-Methyl-1,3-phenylene)bis(*O*-carbamoylcyclohexanone oxime) 1010

$C_{21}H_{28}O$
Benzyl 4-*tert*-octylphenyl ether 178

$C_{21}H_{28}O_3S$
4-(1,1,3,3-Tetramethylbutyl)phenyl *p*-toluenesulfonate 800

$C_{21}H_{28}O_6S_2$
2-Butyl-2-ethyl-1,3-propanediol bis-(benzenesulfonate) 800
2,2-Diethyl-1,3-propanediol di-*p*-toluenesulfonate 801

$C_{21}H_{29}BrN_2O$
Diethyl-(2-(diphenylacetamido)-ethyl)-methylammonium bromide 423
(3-(2,2-Diphenyl-*N*-methylacetamido)-propyl)-trimethylammonium bromide 425

$C_{21}H_{29}NO$
3-(Dimethylaminomethyl)-2-phenyl-1-(3,5-xylyl)-2-butanol hydrochloride 238

$C_{21}H_{29}NO_3$
2-Decylaminomethyl-3-hydroxy-1,4-naphthoquinone 316
1-(3,5-Dimethoxyphenyl)-3-(dimethylaminomethyl)-2-phenyl-2-butanol hydrochloride 237

$C_{21}H_{29}NO_6$
Diethyl 2-acetamido-2-(4-(trimethylacetyl)benzyl)malonate 429
Methyl 6-(3-(3,4-dimethoxycinnamoyl)-propionamido)hexanoate 414

$C_{21}H_{29}N_3O_4S$
4′-Amino-2′-hydroxy-*N*4-isononanoylsulfanilanilide 811
4′-Amino-2′-hydroxy-*N*4-nonanoylsulfanilanilide 811

$C_{21}H_{29}N_3O_5S_2$
1-(4-Methylphenylsulfonyl)-3-(6-(4-methylphenylsulfonamido)-hexyl)urea 826

$C_{21}H_{30}N_2$
N,*N*-Diethyl-*N*′-(1,2-diphenylethyl)-1,3-propanediamine 242
4,4′-Methylenebis(*N*,*N*,2,6-tetramethylaniline) 233
4,4′-Methylenebis(*N*,*N*,3,5-tetramethylaniline) 233

$C_{21}H_{30}N_2O$
N,*N*-Diethyl-*N*′-(α-(3-methoxyphenyl)-phenethyl)-ethylenediamine 240

$C_{21}H_{30}N_2O_2$
4,4′-Bis-(dimethylamino)-α-(1-ethoxyethyl)-diphenylmethanol 225

$C_{21}H_{30}N_2O_2S$
6-(Dodecylthio)-5-nitroquinoline 987

$C_{21}H_{30}N_2O_4$
N,*N*′-Bis(cyclohexyloxycarbonyl)-4-methyl-1,3-phenylenediamine 669

$C_{21}H_{30}O$
9-Hydroxy-9-(2,6-xylyl)-perhydrofluorene 205

$C_{21}H_{30}OSi_2$
1,3-Diphenyl-1-propenyloxy(dimethyl)-(trimethylsilylmethyl)silane 1107

$C_{21}H_{31}NO$
N-(2-Adamantyl)-1-adamantanecarboxamide 106

$C_{21}H_{31}NO_3$
3-(*N*-Methyl-*N*-(1-(4-methylbenzyl)-ethyl)carbamoyl)-1,2,2-trimethyl-cyclopentane carboxylic acid, sodium salt 429

$C_{21}H_{31}NO_5$
Diisopentyl (4-formylphenyl)-iminodiacetate 358

$C_{21}H_{32}N_2O$
1,1-Dicyclohexyl-3-(2,4-xylyl)urea 584
1,1-Dicyclohexyl-3-(2,5-xylyl)urea 584

$C_{21}H_{32}N_2O_3$
1,1-Dicyclohexyl-3-(2,5-dimethoxyphenyl)urea 554

$C_{21}H_{32}N_2O_3S$
Methyl 2-(3-dodecanoyl-2-thioureido)benzoate 744

$C_{21}H_{32}N_2O_5$
Carbobenzyloxy-L-isoleucyl-L-leucine methyl ester 663
Carbobenzyloxy-L-leucyl-L-leucine methyl ester 662
Carbobenzyloxy-L-norleucyl-L-leucine methyl ester 662
Carbobenzyloxy-L-norleucyl-L-norleucine methyl ester 662

$C_{21}H_{32}N_2S$
5-Amino-6-(dodecylthio)quinoline 987

$C_{21}H_{32}N_4O_2$
N,N'-(4-Methyl-1,3-phenylene)bis(1-carbamoyl-2-methylpiperidine) 703
1,1'-(4-Methyl-1,3-phenylene)bis(3-cyclohexylurea) 565

$C_{21}H_{32}N_4O_4$
2,4-Bis(2,6-dimethylmorpholinocarboxamido)-toluene 703

$C_{21}H_{32}O_2$
Cinnamyl dodecanoate 287
Methyl abietate 80

$C_{21}H_{32}O_4$
Ethyl 11-(4-acetylphenoxy)-undecanoate 355

$C_{21}H_{33}BrO_4S$
4-(Phenylsulfonyl)butyl 11-bromoundecanoate 772

$C_{21}H_{33}NOSn$
Tributyltin 8-hydroxyquinolate 1130

$C_{21}H_{33}NO_2$
3-(Diethylamino)-propyl α-phenylcyclohexaneacetate hydrochloride 359

$C_{21}H_{33}NO_3$
4-(Tetradecanamido)benzoic acid 433

$C_{21}H_{34}ClN_3O_2$
3-Allylcarbamoyl-1-(decylcarbamoylmethyl)-pyridinium chloride 948

$C_{21}H_{34}N_2OS$
1-Cyclododecyl-3-(4-ethoxyphenyl)-2-thiourea 738

$C_{21}H_{34}N_2O_2$
4'-(*N*-Methylacetamido)-dodecananilide 428

$C_{21}H_{34}N_2O_2S$
1-(Dodecanoyl)-3-(2-ethoxyphenyl)-2-thiourea 737

$C_{21}H_{34}N_2O_3S$
1-(2,5-Dimethoxyphenyl)-3-dodecanoyl-2-thiourea 735

$C_{21}H_{34}N_4O_2$
2,7-Bis(2-(diethylamino)-ethoxy)-4-methyl-1,8-naphthyridine hydrochloride 999

$C_{21}H_{34}N_4S_2$
2,7-Bis-(2-(diethylamino)-ethylthio)-4-methyl-1,8-naphthyridine 999

$C_{21}H_{34}O_3$
3-(Tetradecyloxy)benzoic acid 340
4-(Tetradecyloxy)benzoic acid 341

$C_{21}H_{35}BrO$
3-(3-Bromophenyl)-3-pentadecanol 203

$C_{21}H_{35}ClN_2O_3$
1-Dodecyloxycarbonylmethyl-3-(methylcarbamoyl)-pyridinium chloride 956

$C_{21}H_{35}NO_3$
2-Nitro-5-pentadecylphenol 264
4-Nitro-3-pentadecylphenol 265

$C_{21}H_{35}NO_3Sn$
Tributyltin 4-acetamidobenzoate 1131

$C_{21}H_{35}N_3OS$
1-(4-(Dimethylamino)phenyl)-3-dodecanoyl-2-thiourea 740

$C_{21}H_{36}$
1-Phenylpentadecane 163

$C_{21}H_{36}BrNO_2$
Diethylmethyl(2-(3-methyl-2-(3,4-xylyl)-valeryloxy)-ethyl)-ammonium bromide 359
Diethylmethyl(3-(2-phenylheptanoyloxy)-propyl)-ammonium bromide 359

$C_{21}H_{36}ClN_3O_2$
1-Dodecylcarbamoylmethyl-3-(methylcarbamoyl)-pyridinium chloride 948

$C_{21}H_{36}IN$
(1,1,2,2,3,3,4,4-Octamethyl-1,2,3,4-tetrahydro-5-naphthyl)trimethyl-ammonium iodide 256

$C_{21}H_{36}N_2O$
1,1-Dibutyl-3-(2,6-diisopropylphenyl)urea 596
1,1-Diisobutyl-3-(2,6-diisopropylphenyl)urea 596

$C_{21}H_{36}N_2O_2$
Acetic acid N^2-(3-(dodecyloxy)-benzyl)hydrazide 498
(+/-)-2-Isopropyl-5-methylcyclohexyl *N*-decahydronaphth-2-ylidenecarbazate 1048

$C_{21}H_{36}N_2O_5S$
Hexocyclium 254

$C_{21}H_{36}N_2S$
1-Phenyl-3-tetradecyl-2-thiourea 748

$C_{21}H_{36}N_4O_2$
3,3'-(4-Methyl-1,3-phenylene)bis(1-butyl-1-ethylurea) 566
3,3'-(4-Methyl-1,3-phenylene)bis(1,1-diisopropylurea) 566
1,1'-(4-Methyl-1,3-phenylene)bis(3-(1,3-dimethylbutyl)urea) 566
3,3'-(4-Methyl-1,3-phenylene)bis(1,1-dipropylurea) 566

$C_{21}H_{36}N_4O_4$
N,N'-Bis(2-diethylaminoethoxycarbonyl)-4-methyl-1,3-phenylenediamine 669

$C_{21}H_{36}O_2$
5-Pentadecylresorcinol 193

$C_{21}H_{36}O_3S$
Tetradecyl *p*-toluenesulfonate 800

$C_{21}H_{37}BF_4N_2$
1,3-Dipyrrolidinododecahydro-1*H*-phenalenium tetrafluoroborate 40

$C_{21}H_{37}ClN_2O_2S$
6-Chloro-*N*-hexadecyl-3-pyridinesulfonamide 960

$C_{21}H_{37}NO$
2-(Hexadecyloxy)pyridine 940

$C_{21}H_{38}ClN_3$
4-Chloro-2-hexadecylamino-6-methylpyrimidine 975

$C_{21}H_{38}N_2S_2$
3,5-Bis-(3,3,5-trimethylcyclohexyl)-tetrahydro-1,3,5-thiadiazine-2)2*H*)-thione 140

$C_{21}H_{38}O_3$
Di-L-menthyl carbonate 78

$C_{21}H_{38}O_4$
Methyl acetylricinoleate 80

$C_{21}H_{39}N_3$
1,3,5-Tricyclohexylhexahydro-*s*-triazine 40

$C_{21}H_{39}N_3O$
2-Hexadecylamino-4-hydroxy-6-methylpyrimidine 971

$C_{21}H_{40}O$
3-Pentadecylcyclohexanone 56

$C_{21}H_{40}O_3$
Methyl 14-oxoeicosanoate 94

$C_{21}H_{40}O_3S$
Methyl 9-(acetylthio)-octadecanoate 75

$C_{21}H_{40}O_4$
Diethyl 2-tetradecylmalonate 76

$C_{21}H_{40}O_5$
2,3-Dihydroxypropyl 4-oxostearate 91

$C_{21}H_{41}NO$
3-(Octadecyloxy)propionitrile 144
3-(*tert*-Octadecyloxy)propionitrile 144

$C_{21}H_{41}N_3S$
2-Heptadecyl-4-(hydrazinomethyl)-thiazole hydrochloride 876

$C_{21}H_{42}O$
10-Heneicosanone 49
11-Heneicosanone 49

$C_{21}H_{42}O_2$
Ethyl nonadecanoate 74
Isopropyl stearate 74

$C_{21}H_{42}O_3$
2-Hydroxypropyl stearate 84

$C_{21}H_{42}O_5S$
3-Sulfopropyl isooctadecanoate, sodium salt 156

$C_{21}H_{43}ClO_2$
1-Chloro-3-isooctadecyloxy-2-propanol 19

$C_{21}H_{43}IO_2Sn$
4,5-Diethyl-6-(iodomethyl)-2-(tributylstannyl)-1,3-dioxane 1129

$C_{21}H_{43}NO$
N-Octadecylpropionamide 106

$C_{21}H_{43}NO_2$
N-Hexadecyl-*N*-methylglycine ethyl ester 88

$C_{21}H_{43}NO_4S$
N-Methyl-*N*-stearoyltaurine 155

$C_{21}H_{44}NO_4P$
Dihexyl (1-(*N,N*-dipropylcarbamoyl)-ethyl)phosphonate 159
Dioctyl (*N,N*-diethylcarbamoyl)-phosphonate 159

$C_{21}H_{44}N_2S$
1,3-Didecyl-2-thiourea 142

$C_{21}H_{44}O$
11-Heneicosanol 9

$C_{21}H_{44}O_2$
Bis(decyloxy)methane 17

$C_{21}H_{45}BO_3$
Tris(1-isopropyl-2-methylpropyl) borate 1124

$C_{21}H_{46}N_2$
N-Octadecyl-1,3-propanediamine 30

$C_{21}H_{46}O_6P_2$
1,1-Dibutyl 3,3-dihexyl methylenediphosphonate 159

$C_{22}H_8BrIO_2$
Vat Dye (C.I. 59310) 1095

$C_{22}H_8Br_2O_2$
Vat Orange 3 (C.I. 59300) 1095

$C_{22}H_{11}Cl_5O_2$
9-Anthracenylmethyl pentachlorobenzoate 367

$C_{22}H_{12}BF_4N_9S_3$
Tris(6-nitro-2-benzothiazolinylideneamino)-carbenium tetrafluoroborate 1073

$C_{22}H_{12}Cl_2F_6N_2$
N,N'-(1,4-Phenylenedimethylidyne) bis(6-chloro-3-trifluoromethyl)-aniline 1021

$C_{22}H_{12}Cl_2F_6N_2O_2$
N,N'-Bis(2-chloro-5-(trifluoromethyl)phenyl)-isophthalamide 445

$C_{22}H_{12}Cl_{12}O_2$
10,11-Dimethoxy-dodecachloro-octahydro-1,4:5,8-dimethanotriphenylene 186

$C_{22}H_{12}N_2$
8,9-Benzoacenaphtho(1,2-b)-quinoxaline 1002

$C_{22}H_{12}O_3$
1*H*-Dibenz(a,kl)-7-anthracenone-13-carboxylic acid 345

$C_{22}H_{14}$
1,2-Bis(phenylethynyl)benzene 1019
1,4-Bis(phenylethynyl)benzene 1019
1-Vinylbenzo(a)pyrene 167

$C_{22}H_{14}Br_2N_4O_{14}S_4$
3,6-Bis(4-bromo-2-sulfophenylazo)-4,5-dihydroxy-2,7-naphthalenedisulfonic acid 1084

$C_{22}H_{14}F_3N$
N-(9-Anthracenylmethylene)-2-(trifluoromethyl)aniline 1020
N-(9-Anthracenylmethylene)-3-(trifluoromethyl)aniline 1020

$C_{22}H_{14}F_6N_2$
N,N'-(1,4-Phenylenedimethylidyne) bis((3-trifluoromethyl)aniline) 1021
N,N'-(1,4-Phenylenedimethylidyne) bis((3-trifluoromethyl)aniline) 1021

$C_{22}H_{14}F_6N_2O_2$
N,N'-Bis(2-(trifluoromethyl)phenyl)-isophthalamide 446

$C_{22}H_{14}N_2O_2$
5,7,12,14-Tetrahydro-5,14(1',4')-benzenophthalazino(2,3-b)-phthalazine-7,12-dione 708

$C_{22}H_{14}N_2O_4$
β-(9-Anthracenyl)-2,4-dinitrostyrene 276

$C_{22}H_{14}N_6O_{14}S_2$
3,6-Bis(2-hydroxy-5-nitrophenylazo)-4,5-dihydroxy-2,7-naphthalenedisulfonic acid 1084

$C_{22}H_{14}O$
1-Acetylbenzo(a)pyrene 298

$C_{22}H_{14}O_2$
6-Benzo(a)pyrenyl acetate 381
7-Benzo(a)pyrenyl acetate 381
2,3-Diphenyl-1,4-naphtoquinone 318

$C_{22}H_{14}O_3$
10-Piperonylidene-9-anthrone 314

$C_{22}H_{14}O_4$
1,1'-Binaphthyl-8,8'-dicarboxylic acid 345

$C_{22}H_{15}BiCl_2$
Bis(4-chlorophenyl)-1-naphthylbismuthine 1141

$C_{22}H_{15}BrO$
3-Bromo-2,4,5-triphenylfuran 854

$C_{22}H_{15}Br_4N_3O_3$
N-(3-Methyl-5-(4-methylphenylazo)-phenyl)-3,4,5,6-tetrabromophthalamic acid 1082
5'-Methyl-2'-(*p*-tolylazo)-3,4,5,6-tetrabromophthalanilic acid 1081

$C_{22}H_{15}ClN_2O_2$
2-(4-Chlorophenyl)-6-methoxy-4-(2-pyridylcarbonyl)quinoline 981
4b-(4-Chlorophenyl)-5-methylisoindolo(1,2-b)quinazoline-10,12(4b*H*,5*H*)-dione 694

$C_{22}H_{15}Cl_3SSi$
(3,4,5-Trichloro-2-thienyl)-triphenylsilane 1111

$C_{22}H_{15}NO_2$
9-(2-Nitrocinnamylidene)fluorene 278
2-Phenyl-4-(4-phenylbenzylidene)-2-oxazolin-5-one 386

$C_{22}H_{15}NO_4$
1-(2-Methylanilino)-2-anthraquinonecarboxylic acid 343
Vat Dye (C.I. 60750) 1093

$C_{22}H_{15}N_3O_6$
1,3-Distyryl-2,4,6-trinitrobenzene 275

$C_{22}H_{15}N_5$
2-(4,5-Diphenyl-2-imidazolylideneamino)-benzimidazole 908

$C_{22}H_{16}$
9-Cinnamylidenefluorene 166
7,10-Dimethylbenzo(a)pyrene 166
trans-1,2-Di(2-naphthyl)ethylene 167
2,3-Diphenylnaphthalene 166

$C_{22}H_{16}BrN$
- N-(9-Anthracenylmethylene)-4-bromo-2-methylaniline 1068
- N-(9-Anthracenylmethylene)-4-bromo-3-methylaniline 1068

$C_{22}H_{16}ClN$
- N-(9-Anthracenylmethylene)-3-chloro-4-methylaniline 1068
- N-(9-Anthracenylmethylene)-4-chloro-2-methylaniline 1068
- N-(9-Anthracenylmethylene)-5-chloro-2-methylaniline 1068

$C_{22}H_{16}ClNO$
- 2-Chloro-4-(4-methoxyphenyl)-3-phenylquinoline 986

$C_{22}H_{16}ClNO_2$
- 2′-Benzoyl-4′-chlorocinnamanilide 453

$C_{22}H_{16}ClNO_5$
- 2,4,6-Triphenyl-1,3-oxazin-1-ium perchlorate 972

$C_{22}H_{16}Cl_4N_2O_2$
- *N,N′*-Bis(4,5-dichloro-2-methylphenyl)isophthalamide 456

$C_{22}H_{16}FNO_4S$
- 4-Hydroxy-2-(4-(*p*-tolyloxy)phenyl)-6-quinolinesulfonyl fluoride 982

$C_{22}H_{16}FN_3O_3$
- 3-Amino-2-benzyl-4-(4-fluorophenyl)-7-nitro-1(2*H*)-isoquinolinone 694

$C_{22}H_{16}F_3NO_4$
- 2-Benzoyl-5-methoxyphenyl *N*-(2-(trifluoromethyl)phenyl)carbamate 608
- 4-(Benzyloxycarbonyl)phenyl *N*-(2-(trifluoromethyl)phenyl)carbamate 609

$C_{22}H_{16}F_3N_5O_4$
- 4-(3-(3-(Trifluoromethyl)phenyl)ureido)benzoic (3-nitrobenzylidene)hydrazide 509

$C_{22}H_{16}F_4N_4O_2$
- 4-(3-(3-(Trifluoromethyl)phenyl)ureido)benzoic (2-fluorobenzylidene)hydrazide 505

$C_{22}H_{16}N_2$
- 10,11-Dimethyldibenzo(a,c)phenazine 1001
- 11,12-Dimethyldibenzo(a,c)phenazine 1001
- 2,6-Diphenyl-4,4′-dipyridyl 966

$C_{22}H_{16}N_2OS_2$
- 4-(2-Thienylmethyleneamino)phenyl ether 1049

$C_{22}H_{16}N_2O_2$
- *N*-(9-Anthracenylmethylene)-2-methyl-4-nitroaniline 1033
- *N*-(9-Anthracenylmethylene)-2-methyl-5-nitroaniline 1056
- *N*-(9-Anthracenylmethylene)-4-methyl-3-nitroaniline 1055
- 2,2′-Dihydroxy-1,1′-naphthalazine 1029
- 2,7′-Dihydroxy-1,1′-naphthalazine 1029

$C_{22}H_{16}N_2O_2S_2$
- 2,4-Bis(3-methyl-2(3*H*)-benzothiazolylidenemethyl)-13cyclobutadienediylium13diolate 1095

$C_{22}H_{16}N_2O_2S_3$
- *p*-(*N*-(2-Thenylidene)amino)phenyl sulfone 1075

$C_{22}H_{16}N_2O_3$
- *N*-(9-Anthracenylmethylene)-2-methoxy-5-nitroaniline 1031

$C_{22}H_{16}N_2O_4$
- *N,N′*-Bis(piperonylidene)-1,4-phenylenediamine 1058
- 1,1′-(3,3′-Dimethyl-1,1′-biphenyl-4,4′-diyl)bismaleimide 716
- *N,N′*-(1,4-Phenylenedimethylidyne)-bis(4-carboxyaniline) 1062
- *N,N′*-Terephthalylidenebis(3-carboxyaniline) 1062
- *N,N′*-Terephthalylidene-dianthranilic acid 1062

$C_{22}H_{16}N_2O_5$
- *N,N′*-(Oxydi-4,1-phenylene)bis(2-furamide) 846

$C_{22}H_{16}N_2O_6$
- *N,N′*-Terephthalylidenebis(4-aminosalicylic acid) 1028

$C_{22}H_{16}N_2O_6S$
- 4′,4″-Sulfonylbis(2-furanilide) 851

$C_{22}H_{16}N_2O_8S_2$
- *N,N′*-Diphthalyl-L-cystine 718

$C_{22}H_{16}N_2S_2$
- 4-Phenyl-2-(α-(phenylazo)benzylidene)-1,3-dithiole 1093

$C_{22}H_{16}O$
- 2-Benzylidene-3-phenyl-1-indanone 304
- 2,4-Diphenyl-1-naphthol 196
- 1-(1-Hydroxyethyl)benzo(a)pyrene 208
- 10-(4-Methylbenzylidene)anthrone 314
- 5,9,10,14b-Tetrahydrobenzo(4,5)cyclohept(1,2,3-de)anthracen-5-one 315

$C_{22}H_{16}O_2$
- 3-(4-Biphenylcarbonyl)acrylophenone 307
- 10,11-Dihydrospiro[5*H*-dibenzo(a,d)cycloheptene-5,1′-(3′*H*)isobenzofuranphthalan]-3′-one 386
- 10-(2-Methoxybenzylidene)-9-anthrone 311
- 10-(3-Methoxybenzylidene)-9-anthrone 311
- 10-(4-Methoxybenzylidene)-9-anthrone 311
- 3,4,5-Triphenyl-2(3*H*)-furanone 846

$C_{22}H_{16}O_3$
- 2-(4-Benzyloxyphenyl)chromone 319
- 3,4-Methylenedioxy-4′-phenylchalcone 304

$C_{22}H_{16}O_5$
- 2-Acetylresorcinol dibenzoate 378

$C_{22}H_{17}ClN_2$
- 5-(4-Chlorophenyl)-2,3-dihydro-5-phenyl-5*H*-imidazo(2,1-a)isoindole 244

$C_{22}H_{17}ClO_2$
- 4-Benzyloxy-4′-chlorochalcone 300

$C_{22}H_{17}ClS$
- 11-Chloro-9,10-dihydro-12-(phenylthio)-9,10-ethanoanthracene 211

$C_{22}H_{17}Cl_2NO$
- α,β-Bis(4-chlorophenyl)-4-methoxyhydrocinnamonitrile 751

$C_{22}H_{17}FO_2$
- *o*-12-Methyl-7-benzene(a)-anthracenylmethyl acetate 363

$C_{22}H_{17}F_6NO$
- 3,6-Bis(trifluoromethyl)-9-phenanthryl 2-piperidinyl ketone 288

$C_{22}H_{17}N$
- *N*-(9-Anthracenylmethylene)-*o*-toluidine 1077
- *N*-(9-Anthracenylmethylene)-*m*-toluidine 1068
- *N*-(9-Anthracenylmethylene)-*p*-toluidine 1063
- 2-(Cinnamylideneamino)fluorene 1077

$C_{22}H_{17}NO$
- *N*-(9-Anthracenylmethylene)-*o*-anisidine 1049
- *N*-(9-Anthracenylmethylene)-*m*-anisidine 1048
- *N*-(9-Anthracenylmethylene)-*p*-anisidine 1048
- α,α-Diphenyl-2-quinolinemethanol 983
- α,α-Diphenyl-5-quinolinemethanol 983

$C_{22}H_{17}NO_3$
- Ethyl β-cyano-α-hydroxy-2-(1-naphthyl)cinnamate 761

$C_{22}H_{17}N_2O$
- 4-Benzyl-1,3-diphenyl-1*H*-pyrazol-5(4*H*)-one 862

$C_{22}H_{17}N_3O$
- 2-Picolinylidenemethyl-3-*o*-tolyl-4(3*H*)-quinazolinone 945

$C_{22}H_{17}N_3O_3$
- 3-Amino-2-benzyl-7-nitro-4-phenyl-1(2*H*)-isoquinolinone 701

$C_{22}H_{17}O_3P$
- (1-(Carboxymethyl)-2,2-dihydroxyvinyl)-triphenylphosphonium hydroxide 831

$C_{22}H_{18}$
- 1,2-Dihydro-3,5-dimethylbenz(j)-aceanthrylene 165
- 1,2-Dihydro-2,3-diphenylnaphthalene 163
- *p*-Distyrylbenzene 165

$C_{22}H_{18}Br_2N_2$
- *N,N′*-(1,4-Phenylenedimethylidyne)-bis(4-bromo-*o*-toluidine) 1066
- *N,N′*-Terephthalylidenebis(2-bromo-4-methylaniline) 1066
- *N,N′*-Terephthalylidenebis(4-bromo-3-methylaniline) 1066

$C_{22}H_{18}Br_2N_6O_4$
- 1,1′-(4,6-Dibromo-1,3-phenylene)-bis(3-(2-carbamoylphenyl)urea) 525

$C_{22}H_{18}Br_4$
- 1,2-Bis(α,β-dibromophenethyl)-benzene 173

$C_{22}H_{18}ClF_3N_4O_2$
- 6-(4-Chloro-3-methylphenyl)-3-phenyl-1-(2-(trifluoromethyl)phenyl)biurea 505

$C_{22}H_{18}ClNO_2$
- 2′-Benzoyl-4′-chloro-3-phenylpropionanilide 405

$C_{22}H_{18}ClNO_4$
- 2-(Benzyloxycarbonyl)phenyl *N*-(5-chloro-2-methylphenyl)carbamate 621
- 4-(Benzyloxycarbonyl)phenyl *N*-(5-chloro-2-methylphenyl)carbamate 621

$C_{22}H_{18}ClN_3O_2$
- α-(4-Chlorophenyl)-α-(1,2-dihydro-2-oxo-6-pyridylmethyl)-2-(2-imidazolyl)benzyl alcohol dihydrochloride 864

$C_{22}H_{18}ClN_3O_3$
- *N*-(2-Chlorobenzoyl)anthranilic (4-methoxybenzylidene)hydrazide 492

$C_{22}H_{18}Cl_2N_2$
- *N,N′*-Bis-(4-chlorobenzylidene)-4,5-dimethyl-1,2-phenylenediamine 1054
- *N,N′*-(1,4-Phenylenedimethylidyne)-bis(3-chloro-*o*-toluidine) 1078
- *N,N′*-(1,4-Phenylenedimethylidyne)-bis(3-chloro-*p*-toluidine) 1078
- *N,N′*-(1,4-Phenylenedimethylidyne)-bis(4-chloro-*o*-toluidine) 1078
- *N,N′*-(1,4-Phenylenedimethylidyne)-bis(5-chloro-*o*-toluidine) 1066
- *N,N′*-(1,4-Phenylenedimethylidyne)-bis(6-chloro-*o*-toluidine) 1065

$C_{22}H_{18}Cl_2N_2O_2$
- *N,N′*-Bis(2-chloro-5-methylphenyl)-isophthalamide 455
- *N,N′*-Bis(3-chloro-2-methylphenyl)-isophthalamide 455
- *N,N′*-Bis(5-chloro-2-methylphenyl)-isophthalamide 456
- *N,N′*-Bis(2-chloro-5-methylphenyl)-terephthalamide 455
- *N,N′*-Bis(2-chloro-6-methylphenyl)-terephthalamide 458
- *N,N′*-Bis(3-chloro-4-methylphenyl)-terephthalamide 455
- *N,N′*-(1,4-Phenylenedimethylidyne)-bis(3-chloro-*o*-anisidine) 1024

$C_{22}H_{18}Cl_2N_2O_4$
- *N,N′*-Bis(5-chloro-2-methoxyphenyl)terephthalamide 449

$C_{22}H_{18}F_3NO_3$
- 2-(4-Biphenylyloxy)ethyl *N*-(3-(trifluoromethyl)phenyl)carbamate 608

$C_{22}H_{18}F_3N_3$
- 2-(3-Methyl-4-(4-(trifluoromethyl)-phenylazo)phenyl)isoindoline 1080

$C_{22}H_{18}F_3N_3O_3$
- 1-(2-Methoxy-5-(*N*-phenylcarbamoyl)phenyl)-3-(2-(trifluoromethyl)phenyl)urea 508

$C_{22}H_{18}F_6N_2$
- *N,N′*-(1,4-Phenylenedimethylene)-bis(3-(trifluoromethyl)aniline) 222

$C_{22}H_{18}INO$
- 2-(2-Hydroxy-1-naphthylvinyl)-1-methylquinolinium iodide 983

$C_{22}H_{18}N_2$
- 6,7-Dimethyl-2,3-diphenylquinoxaline 1001
- 2-Phenyl-3-(3,4-xylyl)quinoxaline 1000

$C_{22}H_{18}N_2O$
- 4,5-Diphenyl-2-(4-methoxyphenyl)-imidazole 866

$C_{22}H_{18}N_2O_2S$
- 9-Anthraldehyde (*p*-tosyl)hydrazone 1071

$C_{22}H_{18}N_2O_2S_2$
- 4,4′-Bis(4-acetamidophenyl)-tetrathiafulvalene 880

$C_{22}H_{18}N_2O_3$
- 4-(3-Acetylphenyliminomethyl)-phenyl *N*-phenylcarbamate 1074
- 2-(Benzocyclobuten-1-ylmethylamino)-5-nitrobenzophenone 312

$C_{22}H_{18}N_2O_5S$
- *N*-(4-Amino-3-methoxyanthraquinon-1-yl)-*p*-toluenesulfonamide 813

$C_{22}H_{18}N_4$
- 4-(4-(2-Isoindolinyl)-2-methylphenylazo)benzonitrile 1092
- 4-(4-(2-Isoindolinyl)-3-methylphenylazo)benzonitrile 1092

$C_{22}H_{18}N_4O$
- 4-(4-(2-Isoindolinyl)-3-methoxyphenylazo)benzonitrile 1086

$C_{22}H_{18}N_4O_4$
- *N,N′*-(1,4-Phenylenedimethylidyne)-bis(3-nitro-*p*-toluidine) 1056
- *N,N′*-(1,4-Phenylenedimethylidyne)-bis(5-nitro-*o*-toluidine) 1056
- *N,N′*-Terephthalylidenebis(4-hydroxybenzoic hydrazide) 496
- *N,N′*-Terephthalylidenebis(2-methyl-4-nitroaniline) 1058

$C_{22}H_{18}N_4O_6$
- *N,N′*-(1,4-Phenylenedimethylidyne)-bis(5-nitro-*o*-anisidine) 1043
- *N,N′*-Terephthalylidenebis(2-methoxy-4-nitroaniline) 1043

$C_{22}H_{18}N_4O_8$
- *N,N′*-Bis(2-methoxy-3-nitrophenyl)-isophthalamide 463
- *N,N′*-Bis(2-methoxy-3-nitrophenyl)-terephthalamide 463

$C_{22}H_{18}O$
- 1,2-Dihydro-1,2-diphenyl-1-naphthol 204
- 9,10-Dihydro-11-hydroxy-11-phenyl-9,10-ethanoanthracene 204
- 2,3-Diphenyl-α-tetralone 297
- 9-Hydroxy-10-phenyl-4b,9,9a,10-tetrahydroindeno(1,2-a)indene 207

$C_{22}H_{18}OPb$
- (2-Furyl)triphenyllead 1135

$C_{22}H_{18}OSn$
- (2-Furyl)triphenyltin 1132

$C_{22}H_{18}O_2$
- 1,2-Dihydro-1,2-diphenyl-1,4-naphthalenediol 199
- 2,2′-Dimethoxy-1,1′-binaphthyl 187
- 6,6′-Dimethoxy-1,2′-binaphthyl 187
- 6,6′-Dimethoxy-2,2′-binaphthyl 188
- 1,10-Diphenyl-1,3,7,9-decatetraene-5,6-dione 309
- 4-Methoxy-4′-phenylchalcone 301
- Methyl 7,12-dimethylbenz(a)-anthracene-10-carboxylate 376
- 3-Phenyl-3-(2,4-xylyl)phthalide 386
- 1,2,4-Triphenylbutan-1,4-dione 308

$C_{22}H_{18}O_3$
- 4-Benzyloxy-2′-hydroxychalcone 299
- 2′-(2,3-Dimethylbenzoyl)-2-biphenylcarboxylic acid 344
- *O*-(2-Methylbenzoyl)-DL-benzoin 357
- 4-Oxo-2,3,4-triphenylbutyric acid 331

$C_{22}H_{18}O_5$
- 2,6-Bis(3,4-methylenedioxybenzylidene)-1-cyclohexanone 285
- 3′,4′,5′-Trimethoxy-α-naphthoflavone 319
- 3′,4′,5′-Trimethoxy-β-naphthoflavone 319

$C_{22}H_{18}O_6$
- 7-Hydroxy-3′,4′,5′-trimethoxy-α-naphthoflavone 319

$C_{22}H_{18}O_7$
7-Methyl-3-(3-oxo-4,5,6-trihydroxy-3H-xanthen-9-yl)-5-norbornene-2-carboxylic acid ... 326
$C_{22}H_{18}S$
9-Anthracenylmethyl p-tolyl sulfide ... 213
$C_{22}H_{18}SSi$
(2-Thienyl)triphenylsilane ... 1111
$C_{22}H_{18}Si$
Diphenyl(1-naphthyl)silane ... 1122
$C_{22}H_{19}BiCl_2O_4$
Triphenylbismuth bis(2-chloroacetate) ... 1141
$C_{22}H_{19}BiO_4S_2$
Bis(2-(methoxycarbonyl)phenylthio)phenylbismuth ... 1142
$C_{22}H_{19}ClN_2O_2$
Bis-(4-methoxybenzylidene)-2-chloro-1,4-phenylenediamine ... 1044
$C_{22}H_{19}ClN_2O_4$
1-(2-Benzoyl-4-chlorophenyl)-3-(2,5-dimethoxyphenyl)urea ... 554
Ethyl 4-(5-chloro-2-phenoxyphenylureido)benzoate ... 519
2-Methoxy-4-(4-methoxyphenyliminomethyl)phenyl N-(3-chlorophenyl)carbamate ... 613
$C_{22}H_{19}ClO_2$
3-Chloro-3-(4-methoxyphenyl)-2-phenylpropiophenone ... 292
$C_{22}H_{19}ClO_3$
2-(4-Biphenylyl)-4-(4-chlorophenoxymethyl)-1,3-dioxolane ... 183
$C_{22}H_{19}FN_4O_2$
4-(3-(2-Fluorophenyl)ureido)benzoic (4-methylbenzylidene)hydrazide ... 529
$C_{22}H_{19}FN_4O_3$
4-(3-(2-Fluorophenyl)ureido)benzoic (4-methoxybenzylidene)hydrazide ... 518
$C_{22}H_{19}FN_4O_4$
4-(3-(2-Fluorophenyl)ureido)benzoic (4-hydroxy-3-methoxybenzylidene)hydrazide ... 518
$C_{22}H_{19}FN_4O_4S_2$
3-Aminobenzenesulfonic (6-(fluoro-sulfonyl-)-1-methyl-2-phenyl-4(1H)-quinolylidene)hydrazide ... 1052
$C_{22}H_{19}F_3N_2O$
1,1-Dibenzyl-3-(3-trifluoromethylphenyl)urea ... 509
$C_{22}H_{19}F_3N_2S$
1,1-Dibenzyl-3-(2-(trifluoromethyl)phenyl)-2-thiourea ... 728
$C_{22}H_{19}F_3N_4O$
1-(5-Methyl-2-(p-tolylazo)-phenyl)-3-(2-(trifluoromethyl)-phenyl)urea ... 1080
$C_{22}H_{19}N$
α-Benzyl-α-phenylhydrocinnamonitrile ... 756
1,2-Dihydro-2-phenyl-2-(p-tolyl)quinoline ... 252
9-(4-(Dimethylamino)benzylidene)fluorene ... 233
α,β-Diphenyl-α-methylhydrocinnamonitrile ... 756
2,3,4-Triphenylbutyronitrile ... 758
$C_{22}H_{19}NO$
2-Benzoyl-1-benzyl-3-phenylaziridine ... 298
N-Benzylcinnamanilide ... 482
1,2-Diphenyl-3-(4-methylbenzoyl)aziridine ... 296
$C_{22}H_{19}NO_2$
N-Benzyl-2-(phenylacetyl)benzamide ... 471
9-(α,α-Dimethyl-4-nitrobenzyl)fluorene ... 278
5,7-Diphenyl-8-nitro-1,2,3,4-tetrahydronaphthalene ... 279
$C_{22}H_{19}NO_2Pb$
1-(Triphenylplumbyl)succinimide ... 1135
$C_{22}H_{19}NO_3$
9-(2-(N-Methylcarbamoyl)benzyl)xanthen-9-ol ... 462
Methyl N-phenyl-N-toluoylanthranilate ... 471
$C_{22}H_{19}NO_3S$
2-(N-Ethyl-p-toluenesulfonamido)-9-fluorenone ... 820
$C_{22}H_{19}NO_4$
2-(Benzyloxycarbonyl)phenyl N-(o-tolyl)carbamate ... 655
4-Biphenylyl N-(4-(ethoxycarbonyl)phenyl)carbamate ... 675
Methyl 4-(4-methoxy-N-phenylbenzamido)benzoate ... 466
$C_{22}H_{19}NO_4S$
2-(Benzyloxycarbonyl)phenyl N-(4-(methylthio)phenyl)carbamate ... 655
4-(Benzyloxycarbonyl)phenyl N-(4-(methylthio)phenyl)carbamate ... 655
2-Imino-3-phenyl-2H-1-benzopyran p-toluenesulfonic acid salt ... 385
$C_{22}H_{19}NO_5$
2-(Benzyloxycarbonyl)phenyl N-(3-methoxyphenyl)carbamate ... 651
$C_{22}H_{19}N_3O$
2,3-Dimethyl-4-(1-naphthylmethyleneamino)-1-phenyl-3-pyrazolin-5-one ... 1075
4′-(4-(2-Isoindolinyl)phenylazo)acetophenone ... 1090
$C_{22}H_{19}N_3O_2$
6-Amino-2,3-bis(4-methoxyphenyl)quinoxaline ... 999
4-(2-Hydroxy-1-naphthylmethyleneamino)antipyrine ... 1038
$C_{22}H_{19}N_3O_3$
N,N-Dimethyl-2-(5-methyl-3-phenyl-4-isoxazolyl)-α-oxo-3-indoleacetamide ... 885
$C_{22}H_{19}N_3O_4$
N-Benzoylanthranilic (4-hydroxy-3-methoxybenzylidene)hydrazide ... 495
$C_{22}H_{19}N_3O_5$
N,N-Bis(2-phthalimidoethyl)acetamide ... 739
$C_{22}H_{19}N_5O_3S$
1-(4-(4-Methyl-2-pyrimidinylsulfamoyl)phenyl)-3-(1-naphthyl)urea ... 976
$C_{22}H_{19}N_5O_6$
Ethyl 6-amino-1-benzamido-5-cyano-4-(4-nitrophenyl)-2-oxo-1,2,3,4-tetrahydro-3-pyridinecarboxylate ... 763
$C_{22}H_{19}O_2P$
2-(Triphenylphosphoranylidene)-γ-butyrolactone ... 832
$C_{22}H_{20}$
1,1-Bis(o-tolyl)-2-phenylethylene ... 165
1,3-Diphenyl-2-(o-tolyl)-1-propene ... 165
$C_{22}H_{20}BrClN_4O_2$
1-(4-Bromo-2,6-dimethylphenyl)-6-(3-chlorophenyl)-3-phenylbiurea ... 529
$C_{22}H_{20}BrN_3$
2-(4-(4-Bromophenylazo)-2,5-dimethylphenyl)isoindoline ... 1082
2-(4-(4-Bromophenylazo)-3,5-dimethylphenyl)isoindoline ... 1082
$C_{22}H_{20}BrOP$
(4-Oxo-2-buten-1-yl)triphenylphosphonium bromide ... 831
$C_{22}H_{20}Br_2O_3$
2,6-Bis(5-bromo-2-methoxybenzylidene)-1-cyclohexanone ... 299
$C_{22}H_{20}ClNO$
N-Benzyl-2-chloro-2-phenyl-o-acetotoluidide ... 431
$C_{22}H_{20}ClNO_2$
2-Benzyl-4-chlorophenyl N-(2,5-dimethylphenyl)carbamate ... 632
2-Benzyl-4-chlorophenyl N-(2,6-xylyl)carbamate ... 632
$C_{22}H_{20}ClNO_3$
N-(4-Benzyloxybenzylidene)-5-chloro-2,4-dimethoxyaniline ... 1039
$C_{22}H_{20}Cl_2CuO_6$
Ethyl 2-chlorobenzoylacetate, copper(II) derivative ... 351
$C_{22}H_{20}Cl_2F_3NO_2$
N-tert-Butyl-3-hydroxy-3-(1,3-dichloro-6-trifluoromethyl-9-phenanthryl)propionamide hydrate ... 396
$C_{22}H_{20}Cl_2N_2O_2$
2,5-Bis(3,4-dimethylanilino)-3,6-dichloro-1,4-benzoquinone ... 318
2,5-Bis(2-ethylanilino)-3,6-dichloro-1,4-benzoquinone ... 317
2,5-Bis(4-ethylanilino)-3,6-dichloro-1,4-benzoquinone ... 318
$C_{22}H_{20}Cl_2N_2O_4$
2,5-Bis(4-ethoxyanilino)-3,6-dichloro-1,4-benzoquinone ... 316
$C_{22}H_{20}Cl_2N_4O_3$
1-(5-Chloro-2-ethoxyphenyl)-6-(3-chlorophenyl)-3-phenylbiurea ... 518
$C_{22}H_{20}CuF_2O_6$
Ethyl 2-fluorobenzoylacetate, copper(II) derivative ... 351
$C_{22}H_{20}CuF_2O_{10}S_2$
Ethyl 3-(fluorosulfonyl)benzoylacetate, copper(II) derivative ... 351
$C_{22}H_{20}I_6N_4O_6$
3,3′-Hexamethylenediureidobis-(2,4,6-triiodobenzoic acid) ... 530
$C_{22}H_{20}N_2$
N,N′-Bis(4-methylbenzylidene)-1,4-phenylenediamine ... 1058
2-(4-(Dimethylamino)benzylideneamino)fluorene ... 1062
N,N′-(1,4-Phenylenedimethylidyne)di-o-toluidine ... 1077
N,N′-(1,4-Phenylenedimethylidyne)di-m-toluidine ... 1065
N,N′-Terephthalylidenedi-p-toluidine ... 1065
$C_{22}H_{20}N_2O$
1,2-Dihydro-3,3-dimethyl-methyl-3H-benzo(a)pyrano(2,3-c)phenazine ... 1001
9-Ethyl-3-(4-methylbenzamido)carbazole ... 891
1-Naphthaleneacetic acid (5-indanylmethylene)-hydrazide ... 502
$C_{22}H_{20}N_2O_2$
1-(4-Benzoylphenyl)-3-(2,4-xylyl)urea ... 575
6,6′-Bi(4,8-dimethyl-2-quinolinol) ... 984
N,N′-Bis(2-hydroxybenzylidene)-m-xylylenediamine ... 1038
N,N′-Bis(4-methoxybenzylidene)-1,4-phenylenediamine ... 1044
N,N′-(1,4-Phenylenedimethylidyne)bis(2-hydroxy-4-methylaniline) ... 1034
N,N′-(1,4-Phenylenedimethylidyne)bis(2-hydroxy-5-methylaniline) ... 1028
N,N′-(1,4-Phenylenedimethylidyne)di-o-anisidine ... 1049
N,N′-Terephthalylidenedi-m-anisidine ... 1050
N,N′-Terephthalylidenedi-p-anisidine ... 1050
$C_{22}H_{20}N_2O_2S_2$
N,N′-Bis(4-(methylthio)phenyl)isophthalamide ... 468
N,N′-Bis(4-(methylthio)phenyl)terephthalamide ... 468
$C_{22}H_{20}N_2O_3$
Benzilic (4-methoxybenzylidene)hydrazide ... 495
$C_{22}H_{20}N_2O_4$
Benzilic (4-hydroxy-3-methoxybenzylidene)hydrazide ... 494
1-(4-Benzoylphenyl)-3-(2,5-dimethoxyphenyl)urea ... 553
N,N′-Bis(3-methoxyphenyl)isophthalamide ... 464
N,N′-Bis(4-methoxyphenyl)isophthalamide ... 463
N,N′-Bis(3-methoxyphenyl)terephthalamide ... 463
2,5-Bis(p-toluidino)terephthalic acid ... 343
3,6-Dimethylpyrazine-2,5-diyl bis(phenylacetate) ... 978
$C_{22}H_{20}N_2O_4S$
N-Acetyl-4′-(p-tolylsulfonamido)benzanilide ... 819
$C_{22}H_{20}N_2O_4Se_2$
1,1′-Diselenobis(3-phthalimidopropane) ... 1142
$C_{22}H_{20}N_2O_5$
O-Benzyl-N-benzyloxycarbonyl-L-tyrosine 4-nitrophenyl ester ... 646
2-(4-Biphenylyloxy)ethyl N-(4-methyl-2-nitrophenyl)carbamate ... 651
2-(4-Phenylphenoxy)ethyl N-(4-methyl-3-nitrophenyl)carbamate ... 652
$C_{22}H_{20}N_4O_2$
2-(3,5-Dimethyl-4-(4-nitrophenylazo)phenyl)isoindoline ... 1090
$C_{22}H_{20}N_4O_4$
2-(3,5-Dimethoxy-4-(4-nitrophenylazo)phenyl)isoindoline ... 1086
1,1′-(1,5-Naphthalenediyl)bis(3-furfurylurea) ... 850
$C_{22}H_{20}N_6O_4S_2$
4,4′-Bis(2-thiazolylureido)-3,3′-dimethoxybiphenyl ... 875
$C_{22}H_{20}O$
7,12-Dimethyl-5-ethoxybenz(a)anthracene ... 189
2,3-Diphenyl-1,2,3,4-tetrahydro-1-naphthol ... 207
3-Phenyl-3-(p-tolyl)propiophenone ... 297
$C_{22}H_{20}O_2$
1,6-Bis(4-acetylphenyl)-1,3,5-hexatriene ... 308
Ethyl 2,2,2-triphenylacetate ... 362
2,3,4-Triphenylbutyric acid ... 333
$C_{22}H_{20}O_2S_4$
4,4′-Bis(4-methoxyphenyl)-5,5′-dimethyltetrathiafulvalene ... 880
$C_{22}H_{20}O_3S$
3-Phenyl-3-(p-tosyl)propiophenone ... 780
$C_{22}H_{20}O_4$
Methyl 3,5-dibenzyloxybenzoate ... 368
$C_{22}H_{20}O_4Pb$
Di(2-furyl)bis(4-methoxyphenyl)lead ... 1134
$C_{22}H_{20}S_4$
4,4′-Bis(p-tolyl)-5,5′-dimethyltetrathiafulvalene ... 880
$C_{22}H_{20}Si$
(1-Butyn-1-yl)triphenylsilane ... 1114
$C_{22}H_{20}Sn$
Dicyclopentadienyldiphenyltin ... 1132
$C_{22}H_{21}BiO_4$
Triphenylbismuth diacetate ... 1143
$C_{22}H_{21}Cl$
Chlorotribenzylmethane ... 173
$C_{22}H_{21}ClN_2O$
3-(2-Chloro-6-methylphenyl)-1,1-dibenzylurea ... 573
3-(3-Chloro-4-methylphenyl)-1,1-dibenzylurea ... 574
3-(4-Chloro-2-methylphenyl)-1,1-dibenzylurea ... 573
3-(5-Chloro-2-methylphenyl)-1,1-dibenzylurea ... 602
$C_{22}H_{21}ClN_2O_2$
1-(5-Chloro-2-methoxyphenyl)-3-(1,2-diphenylethyl)urea ... 518
$C_{22}H_{21}ClN_4O$
1-(5-Chloro-2-methylphenyl)-3-(5-methyl-2-(4-methylphenylazo)phenyl)urea ... 1081
$C_{22}H_{21}ClN_4O_2$
1-(5-Chloro-2-methoxyphenyl)-3-(5-methyl-2-(4-tolylazo)phenyl)urea ... 1080
$C_{22}H_{21}ClN_4O_3$
6-(4-Chloro-3-methoxyphenyl)-3-phenyl-1-(o-tolyl)biurea ... 518
$C_{22}H_{21}ClN_4O_4$
6-(5-Chloro-2,4-dimethoxyphenyl)-1,1-diphenylbiurea ... 517
$C_{22}H_{21}IN_2S$
2-(4-(Dimethylamino)styryl)naphtho(1,2-d)thiazole methiodide ... 915
$C_{22}H_{21}N$
N-Styryldibenzylamine ... 242
$C_{22}H_{21}NO$
N-(4-Benzyloxybenzylidene)-4-ethylaniline ... 1039
N-(4-Benzyloxybenzylidene)-3,4-xylidine ... 1040

$C_{22}H_{21}NO$ (Cont'd)
2′,3′-Dimethyl-2,2-diphenylacetanilide ... 430
2,2-Diphenyl-2′,4′-acetoxylidide ... 437
2,2-Diphenyl-2′,5′-acetoxylidide ... 437
2,2-Diphenyl-2′,6′-acetoxylidide ... 437
2,2-Diphenyl-2′-ethylacetanilide ... 437
2,3,4-Triphenylbutyramide ... 436

$C_{22}H_{21}NO_2$
Benzyliminobis(4-methoxyphenyl)-methane ... 1046
N-(4-Benzyloxybenzylidene)-*p*-phenetidine ... 1040
2-(4-*tert*-Butylphenyl)-4-cinnamylidene-2-oxazolin-5-one ... 386
(3,4-Dimethoxybenzylidene)-(diphenylmethyl)amine ... 1040
2,2-Diphenyl-2′-ethoxyacetanilide ... 416
2,2-Diphenyl-4′-ethoxyacetanilide ... 419
2-(α-Hydroxy-2-methyl-α-phenylbenzyl)-*N*-methylbenzamide ... 462

$C_{22}H_{21}NO_2S$
N-Ethyl-*N*-(2-fluorenyl)-*p*-toluenesulfonamide ... 829

$C_{22}H_{21}NO_3$
N-(3,4-Dibenzyloxyphenyl)-acetamide ... 413

$C_{22}H_{21}NO_3S$
2-(4-Biphenylyloxy)ethyl *N*-(3-(methylthio)phenyl)carbamate ... 665
3,5-Diphenyl-2-methyl-4-(phenylsulfonyl)isoxazolidine ... 784

$C_{22}H_{21}NO_5$
4-(Benzyloxy)phenyl *N*-(2,4-dimethoxyphenyl)carbamate ... 646
4-(Benzyloxy)phenyl *N*-(2,5-dimethoxyphenyl)carbamate ... 646

$C_{22}H_{21}N_3$
2-(3,5-Dimethyl-4-(phenylazo)-phenyl)isoindoline ... 1091
2-(3-Methyl-4-(*p*-tolylazo)phenyl)-isoindoline ... 1091

$C_{22}H_{21}N_3O$
2-(4-(4-Methoxyphenylazo)-3-methylphenyl)isoindoline ... 1087
2-(3-Methoxy-4-(*p*-tolylazo)phenyl)-isoindoline ... 1087

$C_{22}H_{21}N_3O_2$
(Dimethylaminomethyl)phenyl-5-methyl-3-phenylisoxazolo(4,5-c)-pyridin-4(5*H*)-one ... 874
2-(3-Methoxy-4-(4-methoxyphenylazo)phenyl)-isoindoline ... 1086

$C_{22}H_{21}N_3O_3$
1-(4-Anilinophenyl)-3-(4-ethoxycarbonylphenyl)urea ... 567
1,1-Dibenzyl-3-(2-methyl-5-nitrophenyl)urea ... 563
1,1-Dibenzyl-3-(4-methyl-2-nitrophenyl)urea ... 562
1,1-Dibenzyl-3-(4-methyl-3-nitrophenyl)urea ... 563
4,4-Diphenyl-1-(3-ethoxy-4-hydroxybenzylidene)-semicarbazide ... 1030

$C_{22}H_{21}N_3O_3S$
1-(2,2-Diphenyl-2-hydroxyacetyl)-4-(4-methoxyphenyl)-3-thiosemicarbazide ... 733

$C_{22}H_{21}N_5O_3$
1-(2-Methyl-5-nitrophenyl)-3-(5-methyl-2-(4-tolylazo)phenyl)urea ... 1088

$C_{22}H_{21}N_5O_4$
4-(3-Furfurylureido)benzoic (4-acetamidobenzylidene)-hydrazide ... 1068

$C_{22}H_{21}O_4Sb$
Triphenylantimony diacetate ... 1128

$C_{22}H_{22}$
1-*tert*-Butyl-3,5-diphenylbenzene ... 165
Tribenzylmethane ... 162
Tri(*p*-tolyl)methane ... 163

$C_{22}H_{22}B_2F_8N_4$
4,4′-Azobis(1-ethylquinolinium tetrafluoroborate) ... 988

$C_{22}H_{22}BrP$
(3-Buten-1-yl)-triphenylphosphonium bromide ... 831

$C_{22}H_{22}Br_2N_2$
2,2′-Tetramethylenebis-(isoquinolinium bromide) ... 997

$C_{22}H_{22}ClN_3O_2$
3-Benzylcarbamoyl-1-(benzylcarbamoylmethyl)-pyridinium chloride ... 955

$C_{22}H_{22}ClP$
(2-Methylallyl)-triphenylphosphonium chloride ... 831

$C_{22}H_{22}Cl_4N_2O_2$
α,α′-(1,4-Cyclohexylenebis-(methylimino)bis-(4,6-dichloro-*o*-cresol) ... 1038

$C_{22}H_{22}Cl_4Si_2$
2,3,5,6-Tetrachloro-1,4-phenylene

$C_{22}H_{22}CuO_6$
Ethyl benzoylacetate, copper(II) derivative ... 351

$C_{22}H_{22}CuO_8$
Methyl 3-methoxybenzoylacetate, copper(II) complex ... 350

$C_{22}H_{22}F_5NSi_2$
N,*N*-Bis(dimethylphenylsilyl)-2,3,4,5,6-pentafluoroaniline ... 1102

$C_{22}H_{22}Fe_2N_2$
Ferrocenecarboxaldehyde (ferrocenylmethyl)hydrazone ... 1139

$C_{22}H_{22}INO$
2-(4-Benzyloxystyryl)-1-ethylpyridinium iodide ... 936

$C_{22}H_{22}N_2$
N-(4-Isopropylbenzylidene)-*N*′-phenyl-1,4-phenylenediamine ... 1057

$C_{22}H_{22}N_2O$
1,1-Dibenzyl-3-(*o*-tolyl)urea ... 572
1,1-Dibenzyl-3-(*m*-tolyl)urea ... 571
1,1-Dibenzyl-3-(*p*-tolyl)urea ... 572
1-(Diphenylmethyl)-3-(3,4-xylyl)urea ... 586
1,1,3-Tribenzylurea ... 572

$C_{22}H_{22}N_2O_2$
3-(*m*-Anisyl)-1,1-dibenzylurea ... 558
3-(*p*-Anisyl)-1,1-dibenzylurea ... 553
7a-Benzamido-2,3,3a,4,5,7a-hexahydro-2-methyl-5-phenyl-1*H*-isoindol-1-one ... 707
1,1-Dibenzyl-3-(2-methoxyphenyl)urea ... 574
3,4-Dimethoxybenzaldehyde benzylphenylhydrazone ... 234
6-(3-Hydroxy-3-methylbutyl)-methyl-benzo(a)phenazin-5(7*H*)-one ... 319

$C_{22}H_{22}N_2O_3$
N,*N*-Dibenzyl-2-ethoxy-4-nitroaniline ... 267
4-(4,5-Dihydro-4,4-dimethyl-2-oxazolyl)-5-(2-hydroxy-2-phenylethyl)-3-pheny lisoxazole ... 872

$C_{22}H_{22}N_2O_6S$
Carbobenzyloxy-*S*-benzyl-L-cysteine *N*-hydroxy succinimide ... 718
Phenyl 2-(3-cyclohexylhexahydro-2,4,6-trioxopyrimidin-1-yl)-benzenesulfonate ... 799

$C_{22}H_{22}N_2S$
1-(1,2-Diphenylethyl)-3-(*o*-tolyl)-2-thiourea ... 741
1-Diphenylmethyl-3-(2-phenethyl)-2-thiourea ... 741

$C_{22}H_{22}N_4OS$
1-(4-(Methylthio)phenyl)-3-(5-methyl-2-(*p*-tolylazo)phenyl)urea ... 1087

$C_{22}H_{22}N_4O_2$
1-(3,4-Bis(benzyloxy)-benzylideneamino)guanidine acetate ... 1039
Oxalic acid (5-indanylmethylene)-hydrazide ... 503
3-Phenyl-6-(*m*-tolyl)-1-(*o*-tolyl)biurea ... 578

$C_{22}H_{22}N_4O_2S_2$
2-(4-(Dimethylamino)-α-(4-nitrophenylazo)benzylidene)-4,5,6-tetrahydro-1,3-benzodithiole ... 277

$C_{22}H_{22}N_4O_3$
2,2′-Oxydiacetic bis(cinnamylidenehydrazide) ... 1042

$C_{22}H_{22}N_4O_5$
4-(3-Furfurylureido)benzoic (3,4-dimethoxybenzylidene)hydrazide ... 1041

$C_{22}H_{22}N_4S_2$
1,1′-(4,5-Dimethyl-1,2-phenylene)-bis(3-phenyl-2-thiourea) ... 739

$C_{22}H_{22}O$
2,6-Bis(4-methylbenzylidene)-1-cyclohexanone ... 285
4,6-Diphenyl-2,3,5,6-tetramethyl-2,4-cyclohexadien-1-one ... 285
Tribenzylmethanol ... 207
2,3′,4″-Trimethyltriphenylmethanol ... 204
Tris(*o*-tolyl)methanol ... 204
Tris(*p*-tolyl)methanol ... 195

$C_{22}H_{22}OSi$
(2-Buten-1-yloxy)triphenylsilane ... 1114

$C_{22}H_{22}O_2S$
1,1-Dibenzylethyl phenyl sulfone ... 784

$C_{22}H_{22}O_2Sn$
Ethyl 2-(triphenylstannyl)acetate ... 1131

$C_{22}H_{22}O_3$
9-Acetyl-7-isopropyl-1-methyl-3-phenanthryl acetate ... 379
2,6-Bis(2-methoxybenzylidene)-1-cyclohexanone ... 283
2,6-Bis(4-methoxybenzylidene)-cyclohexanone ... 283
3-Trityloxy-1,2-propanediol ... 199

$C_{22}H_{22}O_4$
3,4-Bis-(4-acetoxyphenyl)-2,4-hexadiene ... 380
Diethyl 9,10-dihydro-9,10-ethanoanthracene-11,12-dicarboxylate ... 357

$C_{22}H_{22}O_5$
2,6-Divanillylidenecyclohexanone ... 282

$C_{22}H_{22}O_6S_2$
3,5-Octadiyne-1,8-diyl di-*p*-toluenesulfonate ... 1019

$C_{22}H_{22}O_{14}$
Tetramethyl 3-(bis(methoxycarbonyl)-cyclopropenyl)-4-methoxy-7-oxabicyclo(2.2.1)hepta-2,5-diene-1,2,5,6-tetracarboxylate ... 89

$C_{22}H_{22}S$
Tribenzylmethanethiol ... 209

$C_{22}H_{22}S_2$
α,α′-Bis(benzylthio)-*o*-xylene ... 212
α,α′-Bis(benzylthio)-*p*-xylene ... 212

$C_{22}H_{22}Si$
(2-Buten-1-yl)triphenylsilane ... 1114
1-Phenyl-1,2,3,4-tetrahydro-1-(*o*-tolyl)-1-silanaphthalene ... 1120

$C_{22}H_{23}AsBrN$
10-Benzyl-5,10-dihydro-10-propyl-1--phenarsazinium bromide ... 1128

$C_{22}H_{23}ClN_2O_3$
α-(4-Chlorophenyl)-1-methyl-α-(2-(2-(tetrahydropyranyloxy)phenyl)-2-imidazolemethanol ... 864

$C_{22}H_{23}ClSi$
(2-Chlorophenethyl)dibenzylsilane ... 1103

$C_{22}H_{23}Cl_3Si_2$
(2,3,6-Trichloro-1,4-phenylene)-bis(dimethylphenylsilane) ... 1102

$C_{22}H_{23}N$
1-(α-(1-Naphthyl)-benzyl)-piperidine ... 250

$C_{22}H_{23}NO$
N-Benzyl-α-(3-methoxyphenyl)-phenethylamine ... 240
2-(4-Phenyl-1,2,5,6-tetrahydro-1-pyridylmethyl)-1-tetralone ... 295

$C_{22}H_{23}NOSi$
Morpholinotriphenylsilane ... 1112

$C_{22}H_{23}NO_2$
1-Cyclohexyl-2,3-dibenzoylaziridine ... 293

$C_{22}H_{23}NO_5$
Diethyl 2-acetamido-2-(9-fluorenyl)-malonate ... 488
Pseudocoralyne ... 997

$C_{22}H_{23}NO_7$
(-)-β-Narcotine ... 292

$C_{22}H_{23}NO_8$
Narcotine-*N*-oxide hydrochloride ... 1005

$C_{22}H_{23}NO_8S_3$
2,2′-(Phenylsulfonylimino)diethyl bis(benzenesulfonate) ... 822

$C_{22}H_{23}N_3O$
3-Methyl-1-(2-(*N*-methylanilinomethyl)phenyl)-3-phenylurea ... 562

$C_{22}H_{23}N_3O_2$
Alanylphenylalanine β-naphthylamide ... 487
1-(3,4-Dibenzyloxyphenyl)-3,3-dimethyltriazene ... 1094

$C_{22}H_{23}N_3O_3S$
3-Methyl-2-benzothiazolinone, azine with 4′-(2-(methylacryloyloxy)ethoxy)acetophenone ... 1044

$C_{22}H_{23}N_3O_4S$
3-Methyl-2-benzothiazolinone, azine with 4-(2-(methylacryloyloxy)-ethoxy)-*m*-anisaldehyde ... 1045

$C_{22}H_{23}N_3O_5$
Carbobenzyloxy-L-alanyl-L-tryptophan ... 889
Carbobenzyloxyglycyl-L-tryptophan methyl ester ... 883
Carbobenzyloxy-sarcosyl-L-tryptophan ... 889
Carbobenzyloxy-L-tryptophylglycine methyl ester ... 883

$C_{22}H_{23}N_3O_6$
*N*α-*tert*-BOC-L-tryptophan 4-nitrophenyl ester ... 888

$C_{22}H_{23}O_5$
Diethyl (1,3-diphenyl-3-oxopropyl)-malonate ... 360

$C_{22}H_{24}BrNP$
Butylaminotriphenylphosphonium bromide ... 831

$C_{22}H_{24}ClNO_5$
2-(4-Chlorophenyl)-1-(2-(3,4-dimethoxyphenyl)ethyl)-6-oxo-3-piperidinecarboxylic acid ... 689

$C_{22}H_{24}ClN_3$
(4-(Bis(*p*-tolyl)hydrazono)-2,5-cyclohexadien-1-ylidene)-dimethylammonium chloride ... 255

$C_{22}H_{24}F_6N_4O_2$
1,1′-Hexamethylenebis(3-(3-(trifluoromethyl)phenyl)urea) ... 510

$C_{22}H_{24}Ge$
Butyltriphenylgermane ... 1115

$C_{22}H_{24}GeO$
(4-Hydroxybutyl)triphenylgermane ... 1105

$C_{22}H_{24}N_2$
N,*N*′-(1,4-Phenylenedimethylidyne)-di-*p*-toluidine ... 232

$C_{22}H_{24}N_2O$
1,2-Dihydro-1-(1-(4-methoxyphenyl)-4-piperidyl)-2-methylenequinoline ... 247

$C_{22}H_{24}N_2O_2$
1,2-Bis(2-phenyl-4-pentenoyl)-hydrazine ... 501
N,*N*′-Diallyl-*N*,*N*′-di(*o*-tolyl)oxamide ... 425
N-(2-(3-Indolyl)ethyl)-6-oxo-6-phenylhexanamide ... 888
N,*N*′-(1,4-Phenylenedimethylidyne)-di-*p*-anisidine ... 229

$C_{22}H_{24}N_2O_4$
4′,4″-Bis(*o*-acetoacetotoluidide) ... 444
Ethyl (2-(2-(3,4-dimethoxyphenyl)-indol-3-yl)ethylimino)acetate ... 1040
Vomicine ... 697

$C_{22}H_{24}N_2O_4S_2$
N,*N*′-(1,4-Phenylene)bis(*N*-methyl-*p*-toluenesulfonamide) ... 819

$C_{22}H_{24}N_2O_5$
N-Benzyloxycarbonyl-L-phenylalanyl-L-proline ... 708
Carbobenzyloxyglycyl-L-proline benzyl ester ... 708
Carbobenzyloxyprolylglycine benzyl ester ... 700

$C_{22}H_{24}N_2O_6$
4,4′-Bis(acetoacetamido)-3,3′-dimethoxybiphenyl ... 486
Carbobenzyloxy-L-prolyl-L-tyrosine ... 697

$C_{22}H_{24}N_2O_6S_2$
S,*S*′-Ethylene bis(*N*-(4-(ethoxycarbonyl)phenyl)-

$C_{22}H_{24}N_2O_6S_2$ (Cont'd)
thiocarbamate) 676
$C_{22}H_{24}N_2O_7$
Carbobenzyloxy-α-L-glutamyl-L-phenylalanine 659
$C_{22}H_{24}N_2O_8S_2$
N,N'-Dicarbobenzyloxy-L-cystine 657
$C_{22}H_{24}N_2O_{10}$
3,3'-Dimethoxybenzidine-*N,N,N',N'*-tetraacetic acid, tetrasodium salt 350
$C_{22}H_{24}N_4O$
8-(3-Anilinopropylamino)-6-phenyl-2,6-diazabicyclo(5.2.0)nona-1,7-dien-9-one perchlorate 294
$C_{22}H_{24}N_4O_4$
Carbobenzyloxysarcosyl-L-tryptophanamide 883
$C_{22}H_{24}N_4O_5$
L-Phenylalanyl-D-tyrosylglycine anhydride 696
$C_{22}H_{24}N_4O_6S$
*N*4-(3,4,5-Trimethoxybenzoyl)-sulfamethazine 972
$C_{22}H_{24}N_6$
Neutral Violet 999
$C_{22}H_{24}N_6O_5$
2,2'-Oxydiacetic acid bis((4-acetamidobenzylidene)hydrazide) 1041
$C_{22}H_{24}OSi$
α(4-(Trimethylsilyl)phenyl)-benzhydrol 1105
4-(Triphenylsilyl)-1-butanol 1105
$C_{22}H_{24}O_6$
3,3',5,5'-Tetramethyl-2,4,4'-triacetoxybiphenyl 380
$C_{22}H_{24}O_{10}$
Tetraethyl δ4,4'-bi-4*H*-pyran-2,2',6,6'-tetracarboxylate 89
3,3',5,5'-Tetramethoxy-2,4,4'-triacetoxybiphenyl 377
$C_{22}H_{24}Pb$
Butyltriphenyllead 1136
tert-Butyltriphenyllead 1136
Isobutyltriphenyllead 1136
$C_{22}H_{24}Si$
(4-*tert*-Butylphenyl)diphenylsilane 1120
Butyltriphenylsilane 1114
Isobutyltriphenylsilane 1114
$C_{22}H_{24}Sn$
Butyltriphenyltin 1132
tert-Butyltriphenyltin 1132
Isobutyltriphenyltin 1132
$C_{22}H_{25}NO_2$
Lobelanine hydrochloride 308
$C_{22}H_{25}NO_3$
N,N-Dibenzylhexahydro-phthalamic acid 431
1,2,3,4,5,6,7,8-Octahydro-9-(4-pyridyl)-3,3,6,6-tetramethyl-1,8-xanthenedione 961
$C_{22}H_{25}NO_4$
2-(1-Benzoyl-2-piperidyl)-2',5'-dimethoxyacetophenone 698
$C_{22}H_{25}NO_9$
N-Oxynornarceine 340
N-Oxynornarceine hydrochloride 340
$C_{22}H_{25}NSi$
N,N-Dimethyl-1-(triphenylsilyl)-ethylamine 1108
$C_{22}H_{25}N_3O_2$
N,N'-Diphenyl-1-methyl-2,6-piperidinediacetamide 431
$C_{22}H_{25}N_3O_4$
Carbobenzyloxy-L-phenylalanyl-L-prolinamide 700
Carbobenzyloxy-L-prolyl-L-phenylalaninamide 704
$C_{22}H_{25}N_3O_5$
Carbobenzyloxy-L-prolyl-L-tyrosinamide 645
$C_{22}H_{25}N_3O_6$
Carbobenzyloxyalanyl-glycylglycine benzyl ester 662
Carbobenzyloxy-β-alanylglycylglycine benzyl ester 662
Carbobenzyloxy-L-asparaginyl-L-phenylalanine methyl ester 671
Carbobenzyloxyglycylglycyl-β-alanine benzyl ester 664
Carbobenzyloxyglycylglycyl-L-alanine benzyl ester 662
Carbobenzyloxyglycylglycyl-L-phenylalanine methyl ester 671
Carbobenzyloxyglycylglycyl-sarcosine benzyl ester 661
Carbobenzyloxyphenylalanyl-glycylglycine methyl ester 657
$C_{22}H_{25}N_3O_7$
Carbobenzyloxyglycylglycyl tyrosine methyl ester 644
Carbobenzyloxytyrosyl-glycylglycine methyl ester 645
$C_{22}H_{26}Cl_2N_2$
2-(4-*N*(2-Chloroethyl)methylamino)-styryl-1,3,3-trimethyl-3*H*-pseudoindolium chloride 881
$C_{22}H_{26}Cl_2N_2O_5$
4,4'-Oxybis(2-chloro-6-(morpholmethyl)phenol) 244
$C_{22}H_{26}Cl_2N_2O_6$
N,N'-Ethylenebis(*N*-(chloromethyl)-2,6-dimethoxybenzamide) 445
$C_{22}H_{26}F_2N_2O_2$
N,N'-Bis(4-fluorophenyl)-decanediamide 236
$C_{22}H_{26}N_2O$
10,11-Dihydro-10-(2-(4-methyl-1-piperazinyl)ethyl)dibenzosuberone dihydrochloride 285
$C_{22}H_{26}N_2O_2$
α,α-(1,4-Cyclohexylenebis-(methylimino))-di-*o*-cresol 1038
$C_{22}H_{26}N_2O_4$
Tilorone analog R 11,043DA 316
$C_{22}H_{26}N_2O_5$
Carbobenzyloxy-L-alanyl-L-2-aminobutyric acid benzyl ester 662
Carbobenzyloxy-L-phenylalanyl-L-norvaline 657
Carbobenzyloxy-L-valylglycine benzyl ester 663
Carbobenzyloxy-L-valyl-L-phenylalanine 671
Methylaminocolchicine 414
$C_{22}H_{26}N_2O_5S$
Carbobenzyloxy-L-alanyl-*S*-benzyl-L-cysteine methyl ester 656
Carbobenzyloxy-L-methionyl-L-phenylalanine 666
Carbobenzyloxy-L-phenylalanyl-L-methionine 667
$C_{22}H_{26}N_2O_6$
Carbobenzyloxy-L-threonyl-L-alanine benzyl ester 643
Carbobenzyloxy-L-threonyl-L-phenylalanine methyl ester 657
Carbobenzyloxy-L-tyrosyl-L-valine 645
Carbobenzyloxy-L-valyl-L-tyrosine 645
N,N'-Dicarbobenzyloxy-DL-lysine 659
N,N'-Dicarbobenzyloxy-L-lysine 659
$C_{22}H_{26}N_2O_7$
Carbobenzyloxythreonyltyrosine methyl ester 642
$C_{22}H_{26}N_4O$
4-(4-(Diethylamino)-benzylideneamino)antipyrine 1053
$C_{22}H_{26}N_4O_2$
Adipic bis((2-phenylethylidene)-hydrazide) 1060
Oxalic bis((α-propylbenzylidene)-hydrazide) 499
$C_{22}H_{26}N_4O_3$
Diglycolic bis((3-phenylpropylidene)hydrazide) 498
$C_{22}H_{26}N_4O_4$
4,4'-Bis(cyclopropylureido)-3,3'-dimethoxybiphenyl 551
$C_{22}H_{26}N_4O_6$
Adipic bis((4-hydroxy-3-methoxybenzylidene)hydrazide) 1030
$C_{22}H_{26}N_4O_7$
2,2'-Oxydiacetic acid bis((3,4-dimethoxybenzylidene)hydrazide) 497
2,2'-Oxydiacetic acid bis((3-ethoxy-4-hydroxybenzylidene)hydrazide) 1027
$C_{22}H_{26}N_4O_8$
Oxalic acid (3,4,5-trimethoxybenzylidene)hydrazide 503
$C_{22}H_{26}O$
α-Cyclohexyl-4-ethoxystilbene 180
$C_{22}H_{26}O_3$
2-(2-(Decahydronaphth-2-yl)ethyl)-3-hydroxy-1,4-naphthoquinone 316
$C_{22}H_{26}O_4$
6,6-Bis-(4-hydroxyphenyl)-3-(1-hydroxy-1-methyl-ethyl)heptanoic acid γ-lactone 383
$C_{22}H_{26}O_4Si$
Ethyl 2-(2,2-diphenyl-6-propenyl-1,3-dioxa-2-silacyclohex-4-yl)acetate 1109
$C_{22}H_{26}O_6$
1,6-Di-*O*-benzyl-2,4:3,5-di-*O*-methylene-L-iditol 183
Dimethyl 3,4-bis(4-methoxyphenyl)-adipate 355
$C_{22}H_{26}O_{12}S_2$
2,4:3,5-Di-*O*-methylene-1,6-bis-*O*-(*p*-tosyl)-L-iditol 798
$C_{22}H_{26}Si_2$
(Trimethylsilyl)(triphenylsilyl)-methane 1114
$C_{22}H_{27}ClN_4O_2$
6-Chloro-9-(4-(diethylamino)-1-methylbutylamino)-2-nitroacridine 996
$C_{22}H_{27}NO$
α-Benzyl-β-methyl-α-phenyltetrahydro-1-pyridinepropanol-2-naphthalenesulfonate 246
1,2-Diphenyl-3-methyl-4-(1,2,5,6-tetrahydro-1-pyridyl)-2-butanol hydrochloride 245
$C_{22}H_{27}NO_3$
Ethyl α-(*N*-hexanoylanilino)-phenylacetate 360
$C_{22}H_{27}NO_4$
α-Coralydine 246
α-Coralydine hydrochloride 246
Corydaline 247
2,2'-(4-Pyridylmethylene)bis(5,5-dimethyl-3-hydroxy-2-cyclohexen-1-one) 931
$C_{22}H_{27}N_3O$
6-(4-Acetylphenylazo)-1-ethyl-1,2,3,4-tetrahydro-2,2,4-trimethylquinoline 1091
$C_{22}H_{27}N_3O_4$
Carbobenzyloxy-L-norvalyl-L-phenylalaninamide 658
1-Cyclohexyl-1-(2-hydroxy-2-phenylethyl)-3-(2-methyl-5-nitrophenyl)urea 547
$C_{22}H_{27}N_3O_4S$
Carbobenzyloxy-L-methionyl-L-phenylalaninamide 666
$C_{22}H_{27}N_3O_5$
N,N'-Bis(carbobenzyloxy)-L-lysinamide 664
Carbobenzyloxy-L-valyl-L-tyrosinamide 643
$C_{22}H_{27}N_3O_5S$
Carbobenzyloxy-L-methionyl-L-tyrosinamide 644
$C_{22}H_{28}ClN_3$
3-Chloro-9-(4-(diethylamino-1-methylbutylamino)acridine dihydrochloride 988
$C_{22}H_{28}Cl_2N_2$
1,4-Bis(2-chlorobenzylaminomethyl)-cyclohexane dihydrochloride 236
$C_{22}H_{28}Cl_2N_4O$
1-(4-(3,4-Dichlorobenzyloxy)-α-heptylbenzylideneamino)-guanidine nitrate 1023
$C_{22}H_{28}Cl_2N_4O_2$
1,1'-Hexamethylenebis(3-(3-chloro-2-methylphenyl)urea) 541
1,1'-Hexamethylenebis(3-(3-chloro-4-methylphenyl)urea) 545
$C_{22}H_{28}INO_4$
DL-Coreximine dimethyl ether methiodide 254
$C_{22}H_{28}N_2OS$
1-(*N*-(4-(Methylthio)phenyl)-carbamoyl)-4-(3-phenylpropyl)-piperidine 701
$C_{22}H_{28}N_2O_2$
N,N'-Diisopropyl-*N,N'*-di(*o*-tolyl)-oxamide 425
$C_{22}H_{28}N_2O_3$
2-Acetyl-1-(3,4-dimethoxybenzyl)-6-(dimethylamino)1,2,3,4-tetrahydroisoquinoline 697
Voacangine 355
$C_{22}H_{28}N_2O_4$
2',2''-(Hexamethylenedioxy)-diacetanilide 417
4',4''-(Hexamethylenedioxy)-diacetanilide 419
$C_{22}H_{28}N_2O_5$
Ethyl 3-(4,4-dimethyl)-2,6-dioxodicyclohexyl)-3-(2-hydroxyphenyl)-(2-imidazolidinylidene)-cyclohexanepropionate 368
$C_{22}H_{28}N_2O_6$
1,10-Bis(2-nitrophenoxy)decane 268
1,10-Bis(3-nitrophenoxy)decane 268
1,10-Bis(4-nitrophenoxy)decane 269
N,N'-Bis-(3,4,5-trimethoxybenzylidene)-ethylenediamine 1043
1-(2,4-Dinitrophenoxy)-2-(4-*tert*-octylphenoxy)ethane 266
$C_{22}H_{28}N_2S$
10-(1-Ethyl-3-propyl-3-pyrrolidinylmethyl)-phenothiazine hydrochloride 249
$C_{22}H_{28}N_4O_2$
N,N'-(1,5-Naphthalenediyl)bis(1-piperidinecarboxamide) 569
$C_{22}H_{28}N_4O_5$
N-Benzyloxycarbonyl-L-valyl-L-tyrosine hydrazide 645
$C_{22}H_{28}N_6$
N,N'-Bis(α-cyano-4-dimethylaminobenzyl)-ethylenediamine 754
$C_{22}H_{28}N_6O_3$
Diglycolic bis((4-(dimethylamino)-benzylidene)hydrazide) 498
$C_{22}H_{28}O$
4-*tert*-Butyl-2,6-diphenyl-1-cyclohexanol 207
$C_{22}H_{28}O_2$
2,9-Bis-(4-hydroxyphenyl)-*p*-menthane 195
3,3'-Di-*tert*-butyl-5,5'-dimethyldiphenoquinone 54
2-(2,3-Dihydro-7-ethyl-2,4,4-trimethyl-4*H*-1-benzopyran-2-yl)-5-ethylphenol 191
4,5-Dimethyl-2-(3,4-dihydro-2,4,4,6,7-pentamethyl-2*H*-1-benzopyran-2-yl)phenol 191
Dipentenebisphenol 195
$C_{22}H_{28}O_4$
2,6-Di-*tert*-butyl-1,4-naphthalenediol diacetate 379
Ethyl 4-(2-(4-*tert*-amylphenoxy)-ethoxy)benzoate 369
$C_{22}H_{28}O_4S$
3-Cyclohexyl-3-hydroxy-3-phenylpropyl *p*-toluenesulfonate 797
$C_{22}H_{29}Cl_2N_3$
N,N-Dibutyl-4-(2,5-dichlorophenylazo)-3,5-xylidine 1090
$C_{22}H_{29}NO$
2-(1-(Dimethylamino)cyclohexyl)-1,1-diphenyl-1-ethanol 238
2-(1-(Dimethylamino)cyclohexyl)-1,1-diphenyl-1-ethanol hydrochloride 238
$C_{22}H_{29}NOS$
10-Decylphenothiazine-5-oxide 771
$C_{22}H_{29}NO_2$
(+)-1-Benzyl-3-(dimethylamino)-2-methyl-1-phenylpropionate hydrochloride 359

$C_{22}H_{29}NO_2$ (Cont'd)
1-*tert*-Butyl-1,3-dihydro-1(4-dimethylaminomethyl)phenyl)-3-methyl-3-isobenzofuranol 238
Lobelanidine hydrochloride 246
$C_{22}H_{29}NO_2S$
10-Decylphenothiazine-5,5-dioxide 778
$C_{22}H_{29}NO_4$
2-Heptyloxy-5-nitrobenzyl 2,6-xylyl ether 269
$C_{22}H_{29}NO_6S$
2-(4-*tert*-Octylphenoxy)ethyl 3-nitrobenzenesulfonate 797
$C_{22}H_{29}NS$
10-Decylphenothiazine 251
$C_{22}H_{29}N_3O_3$
2-(1-Ethyl-3-(2,4-xylyl)ureido)ethyl *N*-(2,4-xylyl)carbamate 674
$C_{22}H_{30}$
1-(4-Cyclohexylphenyl)adamantane 162
Di-*tert*-butylbibenzyl 163
$C_{22}H_{30}ClN_3O$
2-(*N*-Benzyl-*N*-(2-diethylaminoethyl)-amino)-6'-chloro-*o*-acetotoluidide 423
$C_{22}H_{30}INO_4$
Laudanosine methiodide 254
$C_{22}H_{30}NO_4P$
Diisopropyl (*N*,*N*-dibenzylcarbamoylmethyl)-phosphonate 835
$C_{22}H_{30}N_2$
4,4'-Cyclohexylidenebis-(*N*,*N*-dimethylaniline) 233
$C_{22}H_{30}N_2O$
4-(Decyloxy)azobenzene 1087
$C_{22}H_{30}N_2O_2$
4-(4-Decyloxyphenylazo)phenol 1083
$C_{22}H_{30}N_2O_4$
1-(2-Amino-4-nitrophenoxy)-2-(4-*tert*-octylphenoxy)ethane 265
$C_{22}H_{30}N_4O_2$
4,4'-Bis(3-propylureido)-3,3'-dimethylbiphenyl 602
1,1'-Hexamethylenebis(3-benzylurea) 572
3,3'-Hexamethylenebis(1-methyl-1-phenylurea) 586
$C_{22}H_{30}N_4O_2S_2$
1,1'-Hexamethylenebis(3-(4-(methylthio)phenyl)urea) 561
$C_{22}H_{30}N_4O_4$
1,1'-Hexamethylenebis(3-(2-methoxyphenyl)urea) 551
$C_{22}H_{30}N_4O_6S_2$
1,1'-Hexamethylenebis(3-(*p*-toluenesulfonamido)urea) 826
$C_{22}H_{30}N_4O_6S_4$
1,1'-(Dithiobis(trimethylene))bis(3-(*p*-tolylsulfonyl)-urea) 816
$C_{22}H_{30}O$
4-Methylbenzyl 4-(*tert*-octyl)phenyl ether 178
1-Naphthyl undecyl ketone 298
$C_{22}H_{30}OSi_2$
1,1'-Oxybis(1-phenylsilacyclohexane) 1115
$C_{22}H_{30}O_2$
6,6'-Di-*tert*.butyl-4,4'-bi-*o*-cresol 194
1,2-Dibutyl-1,2-diphenyl-1,2-ethanediol 200
Ethylene glycol 1,2-bis-(dihydrodicyclopentadienyl ether) 18
2-Naphthyl laurate 381
$C_{22}H_{30}O_2S$
3-*tert*-Butyl-2-hydroxy-5-methylphenyl sulfide 210
$C_{22}H_{30}O_3S$
p-Tolyl 2,4,6-triisopropylbenzenesulfonate 801
$C_{22}H_{30}O_4S$
4,4'-Sulfonylbis(6-(*tert*-butyl)-*m*-cresol) 774
$C_{22}H_{30}O_6$
3,4,4a,4b,5,6,7,8,8a,9-Decahydro-1,8a-dimethyl-2,6,7-triacetoxyphenanthrene 78
$C_{22}H_{30}O_6S$
1,4,4a,5,8,8a-Hexahydro-2-methoxy-4a-methyl-4-oxo-1-naphthyl 10-camphorsulfonate 156
$C_{22}H_{31}NO$
N-(2-Naphthyl)dodecanamide 490
$C_{22}H_{31}NO_2$
2-(2-(4-*tert*-Octylphenoxy)ethoxy)-aniline 228
$C_{22}H_{31}NO_5$
7-(Butoxycarbonylmethoxy)-3-(2-diethylaminoethyl)-4-methylcoumarin 383
$C_{22}H_{31}N_3O_4$
Ethyl 9-(4-(3-methyl-5-oxo-3-pyrazolin-2-yl)-phenylcarbamoyl)-nonanoate 861
$C_{22}H_{31}N_3O_4S$
4'-Amino-N^4-decanoyl-2'-hydroxysulfanilanilide 811
$C_{22}H_{32}Cl_2N_4O_2$
1,1'-(*N*,*N*'-Hexamethylenebis(2-carbamoylethyl))bis(pyridinium chloride) 948
$C_{22}H_{32}N_2O_2$
3,3'-(Decamethylenedioxy)dianiline 228
N,*N*-Diethyl-*N*'-(α-(3,4-dimethoxyphenyl)-phenethyl)-ethylenediamine 239
8-Quinolyl *N*-dodecylcarbamate 993
$C_{22}H_{32}N_2O_3$
Ethyl 4-(3,3-dicyclohexylureido)-benzoate 577
$C_{22}H_{32}O$
9-Hydroxy-9-(2,4,6-trimethylphenyl)-perhydrofluorene 205
$C_{22}H_{32}O_2$
1-Adamantyl 1-adamantaneacetate 78
Vitamin A acetate 81
$C_{22}H_{32}O_3S_2$
Δ^4-2,2-(Propylenedithio)-4,6a,9,9-(tetramethyl-perhydro)phenanthro(2,3-d)(1,3)-dioxol-3-one 53
$C_{22}H_{32}O_4$
1,10-Bis(2-oxocyclohexyl)-1,10-decanedione 55
$C_{22}H_{32}O_5$
3-Oxotetramethyl-perhydrophenanthro(2,3-d)(1,3)-dioxole-4-propionic acid 67
$C_{22}H_{32}O_6$
4a,5-Epoxy-3-oxotetrahydro-4,6a,9,9-tetramethyl-phenanthro(2,3-d)(1,3)dioxole-4-propionic acid 67
$C_{22}H_{32}Si$
Decyldiphenylsilane 1114
$C_{22}H_{33}N$
1'-(4-Methylbenzyl)-3,4,4a,5,6,7,8,8a-octa-*H*-spiro[naphthalene-2(1*H*),4'-piperidine] 250
$C_{22}H_{33}NO$
N-Acetyldehydroabietylamine 441
$C_{22}H_{33}N_3O_8$
*N*α,*N*ε-Bis(*tert*-butoxycarbonyl)-L-lysine 4-nitrophenyl ester 674
$C_{22}H_{34}ClNO_2$
Cyclotridecyl *N*-(5-chloro-2,4-dimethoxyphenyl)carbamate 613
$C_{22}H_{34}Cl_2N_2O_8$
1,1'-Decamethylenebis(4-methylpyridinium perchlorate) 959
$C_{22}H_{34}N_2O_2$
1,1'-(1,3-Adamantanedicarbonyl)-bispiperidine 110
$C_{22}H_{34}N_2O_3S$
Ethyl 4-(3-dodecanoyl-2-thioureido)-benzoate 744
$C_{22}H_{34}O$
Tri-2-norbornylmethanol 13
$C_{22}H_{34}O_2$
Ethyl abietate 81
$C_{22}H_{34}O_3$
Methyl 4-dodecylbenzoylacetate 364
$C_{22}H_{34}O_3SSn$
Tributyltin 2-naphthalenesulfonate 1131
$C_{22}H_{34}O_4$
Methyl 4-(dodecyloxy)-benzoylacetate 363
Pinan-10-ylmethyl 2,2-dimethyl-3-(vinyloxycarbonyl)-1-cyclobutaneacetate 77
$C_{22}H_{35}BrO$
2-Bromohexadecanophenone 296
$C_{22}H_{35}ClO$
4'-Chlorohexadecanophenone 297
$C_{22}H_{35}Cl_2NO$
N-(3,4-Dichlorophenyl)-hexadecanamide 410
$C_{22}H_{35}Cl_3N_2O$
Hexadecanoic acid (2,4,6-trichlorophenyl)-hydrazide 501
$C_{22}H_{35}NO_3S$
S-Dodecyl *N*-(4-(ethoxycarbonyl)-phenyl)thiocarbamate 676
$C_{22}H_{35}N_3O_2S$
1-Dodecanoyl-3-(4-(*N*-methylacetamido)phenyl)-2-thiourea 740
$C_{22}H_{36}ClNO$
N-(4-Chlorophenyl)-hexadecanamide 409
$C_{22}H_{36}ClNO_2S$
4-(Hexadecylthio)-3-nitrochlorobenzene 260
$C_{22}H_{36}ClNO_4S$
4-(Hexadecylsulfonyl)-3-nitrochlorobenzene 773
$C_{22}H_{36}Cl_2N_8O_2$
1,1'-(2,5-Dimethoxy-1,4-phenylene-dimethylene bis(hexamethylene-tetramminium) dichloride 254
$C_{22}H_{36}Cl_2O_2S$
3,4-Dichlorophenyl hexadecyl sulfone 774
$C_{22}H_{36}FNO_5S$
4-(Hexadecyloxy)-3-nitrobenzenesulfonyl fluoride 802
$C_{22}H_{36}FNO_6S_2$
4-Hexadecylsulfonyl-3-nitrobenzenesulfonyl fluoride 803
$C_{22}H_{36}N_2$
2-Pentadecylbenzimidazole 907
$C_{22}H_{36}N_2O$
Dodecanoic (α-propylbenzylidene)-hydrazide 502
$C_{22}H_{36}N_2O_3$
4'-Nitrohexadecananilide 427
3'-Nitropalmitanilide 426
$C_{22}H_{36}N_4O_5$
Hexadecanoic N^2-(2,4-dinitrophenyl)hydrazide 499
$C_{22}H_{36}N_4O_{10}$
Tetraethyl 2,2'-(1,4-piperazinediyldimethyl)bis(2-formamidomalonate) 108
$C_{22}H_{36}O_2$
4'-Hydroxyhexadecanophenone 291
$C_{22}H_{36}O_3$
Methyl 3-(tetradecyloxy)benzoate 370
Methyl 4-(tetradecyloxy)benzoate 370
$C_{22}H_{36}O_5$
Kojyl palmitate 91
$C_{22}H_{37}ClO$
4-Chlorophenyl hexadecyl ether 181
$C_{22}H_{37}ClO_2S$
4-Chlorophenyl hexadecyl sulfone 785
$C_{22}H_{37}ClO_3S$
4-(Hexadecyloxy)benzenesulfonyl chloride 802
$C_{22}H_{37}NO$
4-Aminophenyl pentadecyl ketone 294
N-(2,3-Dimethylphenyl)-tetradecanamide 439
N-(3,5-Dimethylphenyl)-tetradecanamide 440
5'-Isopropyl-*o*-laurotoluidide 440
$C_{22}H_{37}NO_2$
4'-Hydroxypalmitanilide 413
2'-(Tetradecyloxy)acetanilide 417
3'-(Tetradecyloxy)acetanilide 418
4'-(Tetradecyloxy)acetanilide 419
$C_{22}H_{37}NO_2S$
N-(2-(Dodecylthio)-phenyl)-glycine ethyl ester 357
Hexadecyl 2-nitrophenyl sulfide 270
$C_{22}H_{37}NO_3$
N-(2-(Dodecyloxy)-phenyl)-glycine ethyl ester 355
Hexadecyl 2-nitrophenyl ether 267
Hexadecyl 3-nitrophenyl ether 268
Hexadecyl 4-nitrophenyl ether 269
$C_{22}H_{37}NO_4S$
Hexadecyl 2-nitrophenyl sulfone 779
Hexadecyl 4-nitrophenyl sulfone 780
1-Nitrohexadecyl phenyl sulfone 780
$C_{22}H_{37}NO_4S_2$
N-Hexadecyl-1,2-benzenedisulfonimide 829
$C_{22}H_{37}NO_5S_2$
4-Hexadecylthio-3-nitrobenzenesulfonic acid 788
$C_{22}H_{37}NO_6S$
4-(Hexadecyloxy)-3-nitrobenzenesulfonic acid 788
$C_{22}H_{38}$
Hexadecylbenzene 162
$C_{22}H_{38}BrNO_2S$
4-Bromo-*N*-hexadecylbenzenesulfonamide 828
$C_{22}H_{38}ClNO$
5-Chloro-2-(hexadecyloxy)-aniline 215
$C_{22}H_{38}ClNO_2S$
4-Chloro-*N*-hexadecylbenzenesulfonamide 828
5-Chloro-2-(hexadecylsulfonyl)-aniline 773
$C_{22}H_{38}ClNS$
5-Chloro-2-hexadecylthioaniline hydrochloride 215
$C_{22}H_{38}ClN_3O_2$
1-Decylcarbamoylmethyl-3-(diethylcarbamoyl)-pyridinium chloride 947
3-Dimethylcarbamoyl-1-(dodecylcarbamoylmethyl)-pyridinium chloride 947
$C_{22}H_{38}FNO_2S$
4-Fluoro-*N*-hexadecylbenzenesulfonamide 828
$C_{22}H_{38}FNO_2S_2$
4-(Hexadecylthio)metanilyl fluoride 802
$C_{22}H_{38}FNO_3S$
4-(Hexadecyloxy)metanilyl fluoride 802
$C_{22}H_{38}FNO_4S_2$
3-(Hexadecylsulfamoyl)-benzenesulfonyl fluoride 827
4-(Hexadecylsulfonyl)-metanilyl fluoride 803
$C_{22}H_{38}N_2O$
3'-Aminohexadecananilide 422
4'-Aminohexadecananilide 422
1-(1-Azacyclohept-2-en-1-yl)-2-morpholino-1-cyclododecene 39
Hexadecanoic N^2-phenylhydrazide 503
$C_{22}H_{38}N_2O_2$
3-Amino-*N*-(3-(dodecyloxy)propyl)-benzamide hydrochloride 464
4-Amino-*N*-(3-(dodecyloxy)propyl)-benzamide hydrochloride 465
$C_{22}H_{38}N_2O_4S$
N-Hexadecyl-2-nitrobenzenesulfonamide 822
N-Hexadecyl-3-nitrobenzenesulfonamide 822
$C_{22}H_{38}N_2O_4S_2$
4-(Hexadecylthio)-3-nitrobenzenesulfonamide 815
$C_{22}H_{38}N_2O_5S$
3-(*N*-Dodecyl-3-methyl-4-nitroanilino)-1-propanesulfonic acid, ammonium salt 793
4-Hexadecyloxy-3-nitrobenzenesulfonamide 814
$C_{22}H_{38}N_2O_6S_2$
4-Hexadecylsulfonyl-3-nitrobenzenesulfonamide 822
$C_{22}H_{38}O$
4-*tert*-Butylphenyl dodecyl ether 181
Hexadecyl phenyl ether 180
1-Phenyl-1-hexadecanol 204
$C_{22}H_{38}OS$
2-Octadecanoylthiophene 857

$C_{22}H_{38}O_2$
1,4-Bis(octyloxy)benzene ... 179
3-(Dodecyloxy)-α-ethyl-α-methylbenzyl alcohol ... 200
1-(2-Furyl)-1-octadecanone ... 847
4-(Hexadecyloxy)phenol ... 191

$C_{22}H_{38}O_3$
2-(1-Octadecenyl)succinic anhydride ... 98

$C_{22}H_{38}O_3S$
Hexadecyl benzenesulfonate ... 800
Iso-hexadecyl benzenesulfonate (mixture of isomers) ... 800

$C_{22}H_{38}O_3S_2$
4-(Hexadecylthio)-benzenesulfonic acid ... 788

$C_{22}H_{38}O_5S_2$
4-(Hexadecylsulfonyl)-benzenesulfonic acid, sodium salt ... 795

$C_{22}H_{38}O_6$
6,15-Dioxo-1,20-eicosanedicarboxylic acid ... 64

$C_{22}H_{38}S$
Hexadecyl phenyl sulfide ... 212

$C_{22}H_{38}Si_4$
1,2-Bis(4-(trimethylsilyl)phenyl)-1,1,2,2-tetramethyldisilane ... 1122

$C_{22}H_{39}BO_3$
2,6-Di-*tert*-butylphenyl dibutyl borate ... 1125

$C_{22}H_{39}BrN_2O$
4-Carbamoyl-1-hexadecylpyridinium bromide ... 952

$C_{22}H_{39}ClN_2O_2S$
6-Chloro-*N*-hexadecyl-*N*-methyl-3-pyridinesulfonamide ... 966
2-Chloro-4-hexadecylsulfonylphenylhydrazine. 778

$C_{22}H_{39}N$
N,N-Dioctylaniline ... 220
N-Phenylhexadecylamine ... 220

$C_{22}H_{39}NO$
2-(Hexadecyloxy)aniline ... 218
3-(Hexadecyloxy)aniline ... 217
4-(Hexadecyloxy)aniline ... 218
4-(Hexadecyloxy)aniline hydrochloride ... 218

$C_{22}H_{39}NO_2S$
2-Aminophenyl hexadecyl sulfone ... 776

$C_{22}H_{39}NO_3S$
3-(*N*-Dodecyl-*m*-toluidino)-1-propanesulfonic acid, sodium salt ... 794

$C_{22}H_{39}NO_3S_2$
4-(Hexadecylthio)-metanilic acid, potassium salt ... 793

$C_{22}H_{39}NS$
3-Hexadecylthioaniline ... 218
N-Octyl-4-(octylthio)aniline hydrochloride ... 218

$C_{22}H_{40}ClN_3O_2S$
3-Chloro-*N*¹-hexadecyl-4-hydrazinobenzenesulfonamide ... 821

$C_{22}H_{40}NO_2S$
3-Amino-*N,N*-bis(2-ethylhexyl)-benzenesulfonamide ... 817

$C_{22}H_{40}N_2$
N,N'-Bis-(1-methylheptyl)-1,4-phenylenediamine ... 221
N,N'-Bis(1-methylheptyl)-1,4-phenylenediamine dihydrochloride. 222
N,N'-Dioctyl-1,4-phenylenediamine dihydrochloride ... 222

$C_{22}H_{40}N_2O_2S$
2-Amino-*N*-hexadecylbenzenesulfonamide ... 817
*N*1-Hexadecylsulfanilamide ... 817
4-(Hexadecylsulfonyl)-phenylhydrazine ... 778

$C_{22}H_{40}N_2O_4$
N,N'-Hexamethylenebis(2-methylcyclohexyl carbamate) ... 677

$C_{22}H_{40}N_2O_6$
N,N'-Dodecamethylenebis(glutaramic acid) ... 122

$C_{22}H_{40}N_2S$
3-(Hexadecylthio)-phenylhydrazine hydrochloride ... 235

$C_{22}H_{40}O_3$
2-Octadecylsuccinic anhydride ... 98

$C_{22}H_{40}O_4$
2-(1-Octadecenyl)succinic acid ... 61

$C_{22}H_{40}Si$
Hexadecylphenylsilane ... 1114

$C_{22}H_{40}Si_5$
2,2-Bis(trimethylsilyl)-1,1-diphenyl-1,3,3,3-tetramethyltrisilane ... 1115
1,1,2,2,3,3,4,4,5,5-Decamethyl-1,5-diphenylpentasilane ... 1115

$C_{22}H_{41}NO_3$
N-Octadecylmaleamic acid ... 121

$C_{22}H_{41}N_3O_2S$
N-Hexadecyl-4-hydrazinobenzenesulfonamide ... 821

$C_{22}H_{41}N_3O_4S_2$
4-(*N*-Hexadecylaminosulfonyl)-benzenesulfonic acid hydrazide ... 827

$C_{22}H_{42}N_4O_2S$
1-Hexadecanesulfonic acid (3,6-dimethyl-4(3*H*)-pyrimidinylidene)-hydrazide ... 821
N-Hexadecyl-6-hydrazino-*N*-methyl-3-pyridinesulfonamide ... 959

$C_{22}H_{42}O_2$
7,16-Docosanedione ... 51
21-Docosenoic acid ... 61
13-Docosenoic acid urea adduct ... 127
Stearyl methacrylate ... 79

$C_{22}H_{42}O_3$
Ethyl stearoylacetate ... 94

$C_{22}H_{42}O_4$
Diethyl 1,18-octadecanedioate ... 76
Octadecyl succinate ... 95

$C_{22}H_{44}$
1-Docosene ... 1

$C_{22}H_{44}N_2O$
1-Hexadecylisonipecotamide ... 108
Octadecanoic butylidenehydrazide.. 112

$C_{22}H_{44}N_2O_2$
N,N'-Didecyloxamide ... 110
N-(2-(Methylamino)ethyl)-2-oxononadecanamide ... 104

$C_{22}H_{44}N_2S$
1-Allyl-3-octadecyl-2-thiourea ... 142

$C_{22}H_{44}O_2$
Butyl stearate ... 74

$C_{22}H_{44}O_4Pb_2$
Bis(triethylplumbyl) (+)-camphorate ... 1135

$C_{22}H_{44}O_5S0$
2-Sulfodocosanoic acid, pyridinium salt ... 963

$C_{22}H_{45}N$
N-Cyclohexylhexadecylamine ... 31

$C_{22}H_{45}O_5P$
Ethyl 2-(diethylphosphono)-hexadecanoate ... 159

$C_{22}H_{46}NO_4P$
Bis(1,3-dimethylbutyl) (*N,N*-diisobutylcarbamoylmethyl)-phosphonate ... 159
Bis(2-ethylhexyl) (*N,N*-diethylcarbamoylmethyl)-phosphonate ... 159
Bis(2,2,4-trimethylpentyl) (*N,N*-diethylcarbamoylmethyl)-phosphonate ... 159

$C_{22}H_{46}O$
3-Ethyl-3-eicosanol ... 10

$C_{22}H_{46}S$
Undecyl sulfide ... 26

$C_{22}H_{46}S_2$
Undecyl disulfide ... 27

$C_{22}H_{47}IS$
Methyloctadecylpropylsulfonium iodide ... 27

$C_{22}H_{47}N$
Dibutyltetradecylamine ... 330
Diundecylamine ... 28

$C_{22}H_{48}BrN$
Hexadecyltriethylammonium bromide ... 44

$C_{22}H_{48}IN$
Dibutylmethyltridecylammonium iodide ... 44

$C_{22}H_{54}Si_6$
1,1,1,4,4,4-Hexakis(trimethylsilyl)-2-butyne ... 1098

$C_{22}H_{66}Si_{10}$
2,2,5,5-Tetrakis(trimethylsilyl)-3,3,4,4-tetramethylhexasilane ... 1098

$C_{23}H_{13}BrO$
3-Bromo-6-phenylbenzanthrone ... 310

$C_{23}H_{14}BF_7O$
2,4,6-Tris(4-fluorophenyl)pyrylium tetrafluoroborate ... 926

$C_{23}H_{14}ClF_2P$
2,6-Bis(4-fluorophenyl)-4-(4-chlorophenyl)phosphorin ... 931

$C_{23}H_{14}ClNS_2$
3,3-Bis(2-naphthylthio)-2-chloroacrylonitrile ... 763

$C_{23}H_{14}Cl_2FN_3O_4S$
4-(2,5-Dichlorophenylazo)-4'-fluorosulfonyl-1-hydroxy-2-naphthanilide ... 1085

$C_{23}H_{14}F_3P$
2,4,6-Tris(4-fluorophenyl)-phosphorin ... 931

$C_{23}H_{14}O$
6-Phenylbenzanthrone ... 314

$C_{23}H_{15}ClO$
3-(9-Anthryl)-4'-chloroacrylophenone ... 305

$C_{23}H_{15}ClO_3S$
Phenyl 4-(4-chlorophenylthio)-1-hydroxy-2-naphthoate ... 377

$C_{23}H_{15}ClO_5S$
Phenyl 4-(4-chlorophenylsulfonyl)-1-hydroxy-2-naphthoate ... 775

$C_{23}H_{15}NO_3$
3-(9-Anthracenyl)-1-(4-nitrophenyl)-2-propen-1-one ... 303

$C_{23}H_{16}BF_5O$
2,6-Diphenyl-4-(4-fluorophenyl)-pyrylium tetrafluoroborate ... 926

$C_{23}H_{16}BrNO_2$
3-Bromo-4-biphenylyl *N*-(1-naphthyl)carbamate ... 632

$C_{23}H_{16}ClN_3$
N-Benzylidene-*N*-(4-(4-chlorophenylazo)-1-naphthyl)-amine tetrafluoroborate ... 1082

$C_{23}H_{16}Cl_2N_2$
5-(4-Chlorobenzyl)-5-(4-chlorophenyl)-5*H*-imidazo(2,1-a)-isoindole ... 865

$C_{23}H_{16}N_2$
N-2-Fluorenyl-4-quinolinealdimine ... 1079

$C_{23}H_{16}N_2O_3$
Disperse Green 1 (C.I. 56060) ... 1093
Phenyl 1-hydroxy-4-phenylazo-2-naphthoate ... 1085

$C_{23}H_{16}N_4O_5S$
1-(2-(2,4-Dinitrophenylthio)phenyl)-3-(1-naphthyl)urea ... 560

$C_{23}H_{16}O_2$
3-(9-Anthryl)-2'-hydroxyacrylophenone ... 300

$C_{23}H_{16}O_3$
2-(Benzoylmethyl)-2-phenyl-1,3-indandione ... 318

$C_{23}H_{16}O_4$
3-Benzoyl-2-(2-methoxyphenyl)-4*H*-1-benzopyran-4-one ... 319
3-(4-Methoxybenzoyl)-2-phenyl-4*H*-1-benzopyran-4-one ... 319

$C_{23}H_{16}O_5S$
Phenyl 4-(phenylsulfonyl)-1-hydroxy-2-naphthoate ... 775

$C_{23}H_{16}O_7$
Dimethyl 4-(6-hydroxy-3-oxo-3*H*-xanthen-9-yl)isophthalate ... 339

$C_{23}H_{17}BrF_6N_4O_2$
1,1'-(2-Bromo-6-methylphenylene)-bis(3-(3-(trifluoromethyl)phenyl)-urea) ... 505

$C_{23}H_{17}ClO_4S$
2,4,6-Triphenylthiopyrylium perchlorate ... 980

$C_{23}H_{17}Cl_4FeO$
2,4,6-Triphenylpyrylium tetrachloroferrate ... 935

$C_{23}H_{17}F_3N_2$
1-(3-(Trifluoromethyl)trityl)-imidazole hydrochloride ... 864

$C_{23}H_{17}IS$
2,4,6-Triphenylthiopyrylium iodide ... 940

$C_{23}H_{17}N$
N-(2-Naphthyl)benzophenone imine. 244
2,4,6-Triphenylpyridine ... 959

$C_{23}H_{17}NO$
N-(2-Biphenylyl)-1-naphthalenecarboxamide ... 489
2,6-Diphenyl-4-(4-hydroxyphenyl)-pyridine ... 934
N-(α-Phenoxybenzylidene)-2-naphthylamine ... 724

$C_{23}H_{17}NO_2$
10-(4-Acetamidobenzylidene)-9-anthrone ... 433
1-(4-Phenoxyphenyliminomethyl)-2-naphthol ... 1032

$C_{23}H_{17}NO_3$
4-(4-Benzyloxybenzylidene)-2-phenyl-2-oxazolin-5-one ... 383

$C_{23}H_{17}NO_5$
1-(4-Ethoxyanilino)-2-anthraquinonecarboxylic acid ... 340

$C_{23}H_{17}N_3O_3$
2-(4-(2-Benzoylacetamido)phenyl-4(3*H*)quinazolinone ... 1000

$C_{23}H_{17}N_3O_3S$
1-(1-Naphthyl)-3-(2-(4-nitrophenylthio)phenyl)urea ... 560

$C_{23}H_{17}N_5O_4$
2,6-Bis-(phthalimidomethylamino)-pyridine ... 953

$C_{23}H_{17}P$
2,4,6-Triphenylphosphorin ... 966

$C_{23}H_{18}$
1,2,3-Triphenyl-1,3-cyclopentadiene. 165

$C_{23}H_{18}ClNO$
10-(2-Chloro-4-dimethylaminobenzylidene)-9-anthrone ... 312

$C_{23}H_{18}ClNO_2$
N-(9-Anthracenylmethylene)-5-chloro-2,4-dimethoxyaniline ... 1020

$C_{23}H_{18}F_6N_4O_2$
1,1'-(4-Methyl-1,3-phenylene)bis(3-(3-(trifluoromethyl)phenyl)urea) ... 508

$C_{23}H_{18}NO_3$
Ethyl *N*-phenyl-*N*-*p*-toluoylanthranilate ... 476

$C_{23}H_{18}N_2O$
1-(4-Anilinophenyliminomethyl)-2-naphthol ... 1034

$C_{23}H_{18}N_2O_2$
1-(4-(Benzyloxy)phenylazo)-2-naphthol ... 1083

$C_{23}H_{18}N_2S_2$
4,4'-Methylenebis(*N*-(2-thenylidene)-aniline) ... 1076

$C_{23}H_{18}N_3O_2$
1,1'-(4-Methyl-1,3-phenylene)bis(3-(2-cyanophenyl)urea) ... 764

$C_{23}H_{18}N_4O$
3-(4-Acetamidophenyl)-5,6-diphenyl-1,2,4-triazine ... 980

$C_{23}H_{18}N_4OS_2$
1,3-Diphenyl-5-(*N*-(*m*-tolyl)-carbamoylimino)dithiohydantoin ... 746

$C_{23}H_{18}N_6O_2$
1,1'-(4-Methyl-1,3-phenylene)bis(3-(3-cyanophenyl)urea) ... 763
1,1'-(4-Methyl-1,3-phenylene)bis(3-(4-cyanophenyl)urea) ... 764

$C_{23}H_{18}N_6O_2S_2$
1,1'-(4-Methyl-1,3-phenylene)bis(3-(2-benzothiazolyl)urea) ... 915

$C_{23}H_{18}O$
α,α-Diphenyl-1-naphthalenemethanol ... 204
4',5-Diphenyl-2,4-pentadienophenone ... 305
2-(4-Methylbenzylidene)-3-phenyl-1-indanone ... 304

$C_{23}H_{18}O_2$
9,10-Dihydro-12-phenyl-9,10-ethanoanthracene-11-

$C_{23}H_{18}O_2$ (Cont'd)
carboxylic acid ... 332
10,10-Dimethylspiro(anthracene-9(10*H*),1′(3′*H*)-isobenzofuran)-3′-one ... 386
10-(2-Ethoxybenzylidene)-9-anthrone ... 311
10-(4-Ethoxybenzylidene)-9-anthrone ... 311
$C_{23}H_{18}O_2S$
2,4-Diphenyl-1-buten-3-yn-1-yl *p*-tolyl sulfone ... 1019
$C_{23}H_{18}O_3$
4′-Acetoxy-3,3-diphenylacrylophenone ... 379
10-(2,5-Dimethoxybenzylidene)-9-anthrone ... 311
10-(3,4-Dimethoxybenzylidene)-9-anthrone ... 311
$C_{23}H_{18}O_4$
Esculetin dibenzyl ether ... 384
$C_{23}H_{18}O_9S$
Mordant Blue (C.I. 43820) ... 1093
$C_{23}H_{19}BrO_3S$
9,10-Dihydro-9,10-propanoanthracen-11-yl 4-bromobenzenesulfonate ... 796
$C_{23}H_{19}Br_2NaO_5S$
Bromoxylenol Blue, water soluble ... 1093
$C_{23}H_{19}ClN_2OS$
1-Benzyl-4-(4,5-dihydro-(1)-benzothiepino(5,4-c)isoxazol-3-yl)-pyridinium chloride ... 955
$C_{23}H_{19}Cl_5O_5S$
α,α-Bis-(3,5-dichloro-2-ethoxyphenyl)-5-chloro-*o*-toluenesulfonic acid,sodium salt ... 791
$C_{23}H_{19}FN_2O_7S$
2-(2-Fluorobenzoyl)-4′-(2-sulfobenzamido)acet-*o*-anisidide, barium salt ... 792
$C_{23}H_{19}FN_4O_2$
4-(3-(2-Fluorophenyl)ureido)benzoic (3-phenylallylidene)hydrazide ... 529
$C_{23}H_{19}F_3N_4O_2$
4-(3-(3-(Trifluoromethyl)phenyl)-ureido)benzoic (4-methylbenzylidene)hydrazide ... 509
$C_{23}H_{19}F_3N_4O_4$
4-(3-(3-(Trifluoromethyl)phenyl)-ureido)benzoic (4-hydroxy-3-methoxybenzylidene)hydrazide ... 508
$C_{23}H_{19}F_5OSi$
1,3-Diphenyl-1-propenyloxy(dimethyl)-(pentafluorophenyl)silane ... 1102
$C_{23}H_{19}FeNO$
N,*N*-Diphenyl-1-ferrocenecarboxamide ... 1140
$C_{23}H_{19}FeN_3$
N-Ferrocenylmethylene-4-(phenylazo)aniline ... 1141
$C_{23}H_{19}N$
N-(9-Anthracenylmethylene)-2-ethylaniline ... 1036
N-(9-Anthracenylmethylene)-4-ethylaniline ... 1067
N-(9-Anthracenylmethylene)-2,3-xylidine ... 1067
N-(9-Anthracenylmethylene)-2,4-xylidine ... 1079
N-(9-Anthracenylmethylene)-2,5-xylidine ... 1068
N-(9-Anthracenylmethylene)-2,6-xylidine ... 1079
N-(9-Anthracenylmethylene)-3,4-xylidine ... 1068
N-(9-Anthracenylmethylene)-3,5-xylidine ... 1068
$C_{23}H_{19}NO$
N-(9-Anthracenylmethylene)-2-ethoxyaniline ... 1046
N-(9-Anthracenylmethylene)-4-ethoxyaniline ... 1048
N-(9-Anthracenylmethylene)-2-methoxy-5-methylaniline ... 1045
N-(9-Anthracenylmethylene)-4-methoxy-2-methylaniline ... 1031
9,10-Dihydro-9,10-ethanoanthracene-11-carboxanilide ... 437
10-(4-(Dimethylamino)benzylidene)-9-anthrone ... 313
$C_{23}H_{19}NO_2$
2-(5-Hydroxy-5*H*-dibenzo(a,d)-cyclohepten-5-yl)-*N*-methylbenzamide ... 462
Hydroxymethyltetrahydrospiro-5*H*-dibenzo(a,d)cycloheptene-(1*H*)-isoindol-3′-one ... 697
3-Methoxy-α-phenyl-α-(4-pyridyl)-2-naphthalenemethanol ... 931
α-(4′-Phenylphenacylidene)-*p*-acetotoluidide ... 433
$C_{23}H_{19}NSn$
(3-Pyridyl)triphenyltin ... 1133
$C_{23}H_{19}N_3$
2-Benzyl-4,5-dihydro-3-(4-pyridyl)-2*H*-benzene(g)indazole hydrobromide ... 959
$C_{23}H_{19}N_3O$
1-(4-Anilinophenyl)-3-(1-naphthyl)urea ... 567
$C_{23}H_{19}N_5O_6S$
4-(4-(Bis(cyanoethyl)-amino)benzylidene)-5-oxo-1-(4-sulfophenyl)-3-pyrazoline-carboxylic acid, sodium salt ... 794
$C_{23}H_{20}ClN_3O_2$
N-(2-Chlorobenzoyl)anthranilic (3-phenylpropylidene)hydrazide ... 493
$C_{23}H_{20}Cl_2N_2O_4$
Bis(4-chloro-2-methylphenyl) *N*,*N*′-(4-methyl-1,3-phenylene)-biscarbamate ... 626
$C_{23}H_{20}Cl_2N_4O_2$
2-(2-Chlorobenzamido)benzoic (2-chloro-4-(dimethylamino)-benzylidene)hydrazide ... 492
$C_{23}H_{20}Cl_4O_5S$
α,α-Bis-(3,5-dichloro-2-ethoxyphenyl)-*o*-toluenesulfonic acid, sodium salt ... 791
$C_{23}H_{20}FN_3O_4S$
2-Benzoyl-2-(4-dimethylaminophenylimino)-4′-(fluorosulfonyl)acetanilide ... 1057
$C_{23}H_{20}FN_5O_3$
4-(3-(2-Fluorophenyl)ureido)benzoic (4-acetamidobenzylidene)-hydrazide ... 529
$C_{23}H_{20}N_2$
Benzophenone azine with 5-indancarboxaldehyde ... 1079
$C_{23}H_{20}N_2O$
1-(2,2-Diphenylvinyl)-2-phenyl-3-pyrazolidinone ... 704
$C_{23}H_{20}N_2O_2$
2-(2,5-Dimethoxyphenyl)-4,5-diphenylimidazole ... 866
N-(1,2-Diphenylethylaminomethyl)-phthalimide ... 719
$C_{23}H_{20}N_2O_4$
4-(4-(Ethoxycarbonyl)-phenyliminomethyl)phenyl *N*-phenylcarbamate ... 1074
$C_{23}H_{20}N_2O_7$
Carbobenzyloxy-L-tyrosine 4-nitrophenyl ester ... 644
$C_{23}H_{20}N_2O_8$
2′,3′-Dibenzoyluridine ... 696
$C_{23}H_{20}N_4$
4-(2,6-Dimethyl-4-(2-isoindolinyl)-phenylazo)benzonitrile ... 1091
$C_{23}H_{20}N_4O_5S$
3-Carboxy-4-hydroxybenzene-sulfonic (aminomethylphenyl-quinolylidene)hydrazide ... 1033
$C_{23}H_{20}O$
2,5-Dicinnamylidene-1-cyclopentanone ... 306
4-Hydroxy-2,3,4-triphenyl-1-cyclopentanone ... 282
$C_{23}H_{20}O_2$
1-Benzoyl-2-(2,4,6-trimethylbenzoyl)-benzene ... 313
4-Benzyloxy-4′-methylchalcone ... 300
1,2-Dihydro-1,2-diphenyl-3-methyl-1,4-naphthalenediol ... 199
1,1-Diphenyl-4-(4-methoxyphenyl)-3-buten-2-one ... 283
$C_{23}H_{20}O_3$
3-(4-(Benzyloxy)phenyl)-3-methyl-1*H*-2-benzopyran-1(4*H*)-one ... 383
Methyl 4-oxo-2,3,4-triphenylbutyrate ... 360
3-(2,4,6-Trimethylbenzoyl)-2-biphenylcarboxylic acid ... 344
$C_{23}H_{20}O_3S$
9,10-Dihydro-9,10-ethanoanthracen-11-ol *p*-tosylate ... 799
$C_{23}H_{20}O_5S$
3-((3,4-Methylenedioxy)phenyl)-3-(*p*-tosyl)propiophenone ... 780
$C_{23}H_{21}ClN_2O_2S$
N-Methyl-4′-chloro-2′-(3,4-dihydro-1-isoquinolinyl)-*p*-toluenesulfonanilide ... 818
$C_{23}H_{21}ClN_2O_3$
1-(2-Chlorophenyl)-3-(4-(2,5-dimethoxystyryl)phenyl)urea ... 516
$C_{23}H_{21}ClN_4O_4$
4-(3-(3-Chlorophenyl)ureido)-benzoic (3-ethoxy-4-hydroxybenzylidene)hydrazide ... 512
$C_{23}H_{21}ClO_4S$
4′-Chloro-3-(4-methoxyphenyl)-3-(*p*-tosyl)propiophenone ... 773
3-(2-Chlorophenyl)-4′-methoxy-3-(*p*-tosyl)propiophenone ... 773
$C_{23}H_{21}Cl_2N_5O_2$
4-(3-(3-Chlorophenyl)ureido)-benzoic-(2-chloro-4-(dimethylamino)benzylidene)-hydrazide ... 527
$C_{23}H_{21}FN_4O_4$
4-(3-(2-Fluorophenyl)ureido)benzoic (3,4-dimethoxybenzylidene)-hydrazide ... 516
$C_{23}H_{21}NO$
10,11-Dihydro-5-(2-(*N*-methylcarbamoyl)phenyl)-5*H*-dibenzo(a,d)cycloheptene ... 480
$C_{23}H_{21}NO_2$
o-(10,11-Dihydro-5-hydroxydibenzo(a,d)cyclohepten-5-yl)-*N*-methylbenzamide ... 462
3-(2-(Dimethylaminomethyl)phenyl)-3-phenylphthalide ... 385
(4-(Dimethylamino)-1-naphthyl)(4-hydroxy-1-naphthyl)methanol ... 225
3-(1,1-Diphenylethyl)-3-hydroxy-2-methyl-1-isoindolinone ... 690
2-(Phenylacetyl)-*N*-(2-phenylethyl)-benzamide ... 471
$C_{23}H_{21}NO_3$
N-(4-Benzyloxybenzylidene)-4-(ethoxycarbonyl)aniline ... 1039
Phenylalanine α-phenylphenacyl ester ... 356
$C_{23}H_{21}NO_4$
4-(Carbobenzyloxy)phenyl *N*-(2,5-xylyl)carbamate ... 659
Diethyl (3-fluoranthenylamino)-malonate ... 358
$C_{23}H_{21}NO_4S$
4′-(1,3-Diphenyl-3-oxopropylsulfonyl)acetanilide ... 780
$C_{23}H_{21}NO_6$
4-(Benzyloxycarbonyl)phenyl *N*-(2,5-dimethoxyphenyl)carbamate ... 646
$C_{23}H_{21}NO_7$
Diethyl (2-(phthalimidomethyl)-benzoyl)malonate ... 721
$C_{23}H_{21}N_3O$
4′-(4-(2-Isoindolinyl)-2-methylphenylazo)acetophenone ... 1092
N-(5-Methyl-2-(4-methylphenylazo)-phenyl)cinnamamide ... 1092
$C_{23}H_{21}N_3O_2$
2-Benzamidobenzoic (3-phenylpropylidene)hydrazide ... 501
4′-(4-(2-Isoindolinyl)-3-methoxyphenylazo)acetophenone ... 1087
$C_{23}H_{21}N_3O_3$
Disperse Dye (C.I. 61540) ... 1093
$C_{23}H_{21}N_3O_3S$
1-(2,4-Dimethoxyphenyl)-3-(4-(6-methyl-2-benzothiazolyl)-phenyl)urea ... 912
$C_{23}H_{21}N_5O$
4-(4-(*N*,*N*-Bis(2-cyanoethyl)amino)-benzylidene)-3-methyl-1-phenyl-2-pyrazolin-5-one ... 755
$C_{23}H_{22}$
9-Butyl-9-phenylfluorene ... 166
$C_{23}H_{22}BrO_2P$
(3-(Methoxycarbonyl)-2-propen-1-yl)-triphenylphosphonium bromide ... 831
$C_{23}H_{22}ClNO_2$
4-*tert*-Butyl-*N*-(4-chloro-2-phenoxyphenyl)benzamide ... 449
$C_{23}H_{22}ClNO_5$
2-(4-Cyanophenyl)-5-(4-pentylphenyl)pyrylium perchlorate ... 935
$C_{23}H_{22}Cl_2N_2O$
Basic Dye (C.I. 45105) ... 1094
$C_{23}H_{22}Cl_2N_4O_2$
1,1′-(4-Methyl-1,3-phenylene)bis(3-(3-chloro-2-methylphenyl)urea) ... 525
1,1′-(4-Methyl-1,3-phenylene)bis(3-(3-chloro-4-methylphenyl)urea) ... 525
1,1′-(4-Methyl-1,3-phenylene)bis(3-(5-chloro-2-methylphenyl)urea) ... 525
$C_{23}H_{22}Cl_2O_3S$
4,6-Dichloro-α-(2,4-dimethyl-6-(3-methylbenzyl)phenylsulfonyl)-*o*-cresol ... 773
$C_{23}H_{22}IN_3$
2-(4-Dimethylaminophenylimino-methyl)-1-methylbenzo(f)-quinolinium iodide ... 1053
$C_{23}H_{22}N_2O_3$
Carbobenzyloxyproline-2-naphthylamide ... 709
$C_{23}H_{22}N_2O_4S_2$
Bis(4-(methylthio)phenyl) *N*,*N*′-(4-methyl-1,3-phenylene)-biscarbamate ... 666
$C_{23}H_{22}N_2O_6$
3-Hydroxy-2-naphthoic acid N^2-(3,4,5-trimethoxycinnamoyl)-hydrazide ... 495
$C_{23}H_{22}N_2O_8$
Carbobenzyloxy-L-aspartic acid 4-benzyl 1-(*N*-hydroxysuccinimide) esters ... 722
$C_{23}H_{22}N_4O_2$
N-Benzoylanthranilic (4-(dimethylamino)benzylidene)-hydrazide ... 500
$C_{23}H_{22}O$
2,2-Diphenyl-3-methylbutyrophenone ... 297
14-(2-Methylallyl)-9,10,14,15-tetrahydro-13*H*-9,10-cyclopentanthracen-13-one ... 298
$C_{23}H_{22}O_2S$
1,2-Diphenylethenyl mesityl sulfone ... 784
$C_{23}H_{22}O_4$
2′,4′-Bis(benzyloxy)-2-methoxyacetophenone ... 291
$C_{23}H_{23}BrN_4O_2S_2$
1,1′-(4-Bromo-2-methyl-1,3-phenylene)bis(3-(3-(methylthio)-phenyl)urea) ... 522
$C_{23}H_{23}ClN_4O_2$
4-(4-Chloro-3-methylphenyl)-1-(2,4-dimethylphenylcarbamoyl)-1-phenylsemicarbazide ... 529
$C_{23}H_{23}ClN_4O_3$
6-(4-Chloro-3-methylphenyl)-1-(2-methoxy-5-methylphenyl)-3-phenylbiurea ... 518
$C_{23}H_{23}ClN_4O_4$
1-(*N*-(5-Chloro-2,4-dimethoxyphenyl)carbamoyl)-4-(3-methyl-phenyl)-1-phenyl-semicarbazide ... 514

$C_{23}H_{23}FN_2O_2$
Ethyl 3-(1-(4-fluorophenyl)-4-isopropyl-2-phenylimidazol-5-yl)-acrylate 865
$C_{23}H_{23}IN_2S_2$
Dithiazanine iodide 915
$C_{23}H_{23}N$
2,2-Diphenyl-3-methylbutyrophenone imine hydrochloride 1077
$C_{23}H_{23}NO$
N,*N*-Bis(α-methylbenzyl)benzamide 482
1-(4-(Dimethylamino)phenyl)-1,3-diphenylallyl alcohol 226
3-(4-(Dimethylamino)phenyl)-1,1-diphenylallyl alcohol 226
3-(4-(Dimethylamino)phenyl)-3-phenylpropiophenone 295
2-(1-(3,5-Dimethylphenyl)-2-naphthyl)-4,5-dihydro-5,5-dimethyloxazole 253
$C_{23}H_{23}NO_2$
4-Cyclohexylphenyl *N*-(1-naphthyl)-carbamate 679
N-Tritylglycine ethyl ester 358
$C_{23}H_{23}NO_3$
Benzyl 1-(3-indolylmethyl)-2-oxo-1-cyclohexanecarboxylate 883
Ethyl α-(4-benzyloxyanilino)-phenylacetate 354
$C_{23}H_{23}NO_4$
2-(4-Biphenylyloxy)ethyl *N*-(2-methoxy-5-methylphenyl)-carbamate 648
$C_{23}H_{23}NO_5$
threo-3-(3,4-Dibenzyloxyphenyl)-serine hydrate 348
erythro-3-(3,4-Dibenzyloxyphenyl)-serine 348
$C_{23}H_{23}NO_5S$
Benzyl *N*-(1-phenyl-2-(*p*-tosyloxy)-ethyl)carbamate 797
$C_{23}H_{23}NO_6$
8-(2-Oxopropyl)neooxyberberine 292
$C_{23}H_{23}N_3O_3S$
Benzilic acid *N*[2]-(*N*-(2-methoxy-5-methylphenyl)thiocarbamoyl)-hydrazide 733
$C_{23}H_{23}N_3O_6$
N',*N*α-Bis(carbobenzyloxy)-L-histidine methyl ester 866
$C_{23}H_{24}$
1,1,1-Triphenylpentane 163
$C_{23}H_{24}As_2I_2$
5,10-Dihydro-5,7,10-trimethyl-5,10-(*o*-xylylene)arsanthrene-5,10-diium diiodode 1128
$C_{23}H_{24}BrNO_5$
2-Benzyl-5-bromo-3a-(2,3-dimethoxyphenyl)-hexahydro-hydroxy-1,3-isoindolinedione 716
$C_{23}H_{24}BrN_3O_2$
4,4'-(4-Bromo-3-nitrophenylmethylene)bis(*N*,*N*-dimethylaniline) 261
$C_{23}H_{24}BrN_3O_3$
2-Bromo-6-(bis(4-(dimethylamino)-phenyl)methyl)-4-nitrophenol 259
$C_{23}H_{24}BrO_2P$
(1-(Ethoxycarbonyl)ethyl)-triphenylphosphonium bromide 831
$C_{23}H_{24}BrP$
(4-Penten-1-yl)-triphenylphosphonium bromide 831
$C_{23}H_{24}Br_2N_2O$
2,4-Dibromo-6-bis(4-(dimethylamino)phenyl)-methylphenol 223
$C_{23}H_{24}ClO_2P$
(2-(Ethoxycarbonyl)ethyl)-triphenylphosphonium chloride 831
$C_{23}H_{24}ClP$
(3-Methyl-2-buten-1-yl)-triphenylphosphonium chloride 831
$C_{23}H_{24}Cl_2N_2$
4,4'-(2,3-Dichlorophenylmethylene)-bis(*N*,*N*-dimethylaniline) 225
4,4'-((2,4-Dichlorophenyl)-methylene)bis(*N*,*N*-dimethylaniline) 225
4,4'-((3,4-Dichlorophenyl)-methylene)bis(*N*,*N*-dimethylaniline) 225
$C_{23}H_{24}Cl_2N_2O$
4-(Bis(4-(dimethylamino)phenyl)-methyl)-2,6-dichlorophenol 223
$C_{23}H_{24}FN_2O_3P$
5-(Diacetylphosphinylmethyl)-1-(4-fluorophenyl)-4-isopropyl-2-phenylimidazole 865
$C_{23}H_{24}FN_3O_2$
1-(2-(4-Fluorophenyl)-3-(3-phenyl-5-isoxazolyl)propionyl)-4-methylpiperazine hydrochloride 870
$C_{23}H_{24}Ge$
(4-Penten-1-yl)triphenylgermane 1115
$C_{23}H_{24}IOP$
(2-Ethoxy-1-propenyl)-triphenylphosphonium iodide 830
$C_{23}H_{24}N_2O$
2,2-Bis(2-(2-pyridyl)ethyl)-propiophenone 961
1-Cyclohexyl-3-(1-naphthyl)-1-phenylurea 582
1,1-Dibenzyl-3-(2-ethylphenyl)urea 572
1,1-Dibenzyl-3-(2,3-xylyl)urea 573
1,1-Dibenzyl-3-(2,4-xylyl)urea 574
1,1-Dibenzyl-3-(2,5-xylyl)urea 574
1,1-Dibenzyl-3-(2,6-xylyl)urea 573
1,1-Dibenzyl-3-(3,4-xylyl)urea 574
1,1-Dibenzyl-3-(3,5-xylyl)urea 574
N-(4-Diethylaminobenzylidene)-4-phenoxyaniline 1044
$C_{23}H_{24}N_2OS$
1-(1,2-Diphenylethyl)-3-(4-ethoxyphenyl)-2-thiourea 736
$C_{23}H_{24}N_2O_2$
N-Benzoyl-DL-leucine-2-naphthylamide 489
N-Benzoyl-L-leucine-2-naphthylamide 489
α-(4-Biphenylyl)-α-(morpholinomethyl)-2-pyridinemethanol 934
1,1-Dibenzyl-3-(2-methoxy-5-methylphenyl)urea 553
α-(α-Morpholinobenzyl)-α-phenyl-2-pyridinemethanol 940
4-(Morpholinomethyl)-α-(2-pyridyl)-benzhydrol 931
$C_{23}H_{24}N_2O_3$
1,1-Dibenzyl-3-(2,5-dimethoxyphenyl)urea 560
$C_{23}H_{24}N_2O_3S$
2-(*N*-Methyl-*p*-toluenesulfonamido)-*N*-phenethylbenzamide 819
$C_{23}H_{24}N_2S$
1-(1,2-Diphenylethyl)-3-(2,4-xylyl)-2-thiourea 741
$C_{23}H_{24}N_4O$
1-(5-Methyl-2-(*p*-tolylazo)phenyl)-3-(2,5-xylyl)urea 1088
$C_{23}H_{24}N_4O_2$
Malonic acid (5-indanylmethylene)-hydrazide 503
1,1'-(4-Methyl-1,3-phenylene)bis(3-benzylurea) 563
$C_{23}H_{24}N_4O_6$
Carbobenzyloxy-L-histidyl-L-tyrosine monohydrate 866
Carbobenzyloxy-L-tryptophylglycylglycine 889
$C_{23}H_{24}O$
2,4-Bis(1-phenylethyl)anisole 180
$C_{23}H_{24}O_2Sn$
Ethyl 2-(triphenylstannyl)-propionate 1131
Tribenzyltin acetate 1131
$C_{23}H_{24}O_3$
2,6-Bis(2-hydroxy-5-methylbenzyl)-4-methylphenol 194
$C_{23}H_{24}O_5S_2$
p-Tolyl 2-(2-(5-methyl-2-sulfinophenyl)ethyl)-*p*-toluenesulfonate, sodium salt 801
$C_{23}H_{24}Si$
Cyclopentyltriphenylsilane 1110
$C_{23}H_{24}Sn$
(4-Penten-1-yl)triphenyltin 1132
$C_{23}H_{25}BrNOP$
(1-(Dimethylcarbamoyl)ethyl)-triphenylphosphonium bromide 831
$C_{23}H_{25}BrN_2$
4,4'-((2-Bromophenyl)methylene)-bis(*N*,*N*-dimethylaniline) 224
4,4'-((4-Bromophenyl)methylene)-bis(*N*,*N*-dimethylaniline) 225
$C_{23}H_{25}ClN_2$
4,4'-((2-Chlorophenyl)methylene)-bis(*N*,*N*-dimethylaniline) 224
$C_{23}H_{25}ClN_2O_5$
1-(2-Hydroxy-5-(1,2,3,4-tetrahydroquinol-1-yl-pentadien-1-ylidene)-1,2,3,4-tetra-hydroquinolinium perchlorate 1034
$C_{23}H_{25}ClN_4$
Basic Red 10 (C.I. 50250) 1095
$C_{23}H_{25}ClO_2Si$
1-(2-Chloroethoxy)-3-(triphenylsilyl)-2-propanol 1099
$C_{23}H_{25}FN_2$
4,4'-((3-Fluorophenyl)methylene)-bis(*N*,*N*-dimethylaniline) 225
$C_{23}H_{25}N$
N-Benzyl-*N*-isopropylbenzhydrylamine 243
$C_{23}H_{25}NO$
Dibenzyl(3-phenoxypropyl)amine hydrochloride 240
$C_{23}H_{25}NO_2$
N,*N*-Bis(2-hydroxyethyl)tritylamine 237
6-Methoxy-2-(4-phenyl-1,2,5,6-tetrahydro-1-pyridylmethyl)-1-tetralone 292
$C_{23}H_{25}NSi$
1-(Triphenylsilyl)piperidine 1111
$C_{23}H_{25}N_3$
N-(4-(Diethylamino)benzylidene)-*N'*-phenyl-1,4-phenylenediamine 1060
$C_{23}H_{25}N_3O$
7-(3-Dimethylaminopropylamino)-9-methoxybenz(c)acridine dihydrochloride 985
$C_{23}H_{25}N_3O_2$
Bis(4-dimethylaminophenyl)(4-nitrophenyl)methane 277
4,4'-((3-Nitrophenyl)methylene)-bis(*N*,*N*-dimethylaniline) 277
$C_{23}H_{25}N_3O_3$
1,3-Diphenyl-1-(2-morpholino-1-cyclopenten-1-ylcarbonyl)urea 565
Ethyl α-(2-imidazolidinylidene)-3-oxo-2-(2-phenylethyl)-1-isoindolineacetate 691
$C_{23}H_{25}N_3O_5$
Carbobenzyloxysarcosyl-L-tryptophan methyl ester 883
Carbobenzyloxytryptophylalanine methyl ester 883
$C_{23}H_{25}N_3O_6$
Carbobenzyloxytryptophylserine methyl ester 881
$C_{23}H_{25}N_5O_5$
Carbobenzyloxy-L-histidyl-L-tyrosinamide 865
$C_{23}H_{26}BrNO_4$
5-Bromo-6-(3,5-dimethoxy-4-methyl-β-piperidinostyryl)-1,3-benzodioxole 246
$C_{23}H_{26}ClNO_2$
3-(4-Chlorophenyl)-3-(2-(dimethylaminomethyl)cyclohexyl)-phthalide 382
$C_{23}H_{26}IP$
(Isoamyl)triphenylphosphonium iodide 831
$C_{23}H_{26}N_2O$
3-(Bis(4-(dimethylamino)phenyl)-methyl)phenol 227
$C_{23}H_{26}N_2O_4$
2,4-Diacetyl-4',4''-dimethylglutaranilide 444
$C_{23}H_{26}N_2O_5$
Carbobenzyloxyphenyl-alanylproline methyl ester 704
Carbobenzyloxyprolyl-L-alanine benzyl ester 708
6,8,9,10,10a,10b,11,12,12a,12b-Decahydro-11,11-(ethylenedioxy)-3-methoxy-8,12a,12b-trimethylbenzo-[b]cyclopenta-[hi]pyrido-[3,4-f]indolizine-6,9-dione 699
$C_{23}H_{26}N_2O_6$
Carbobenzyloxy-L-prolyl-L-tyrosine methyl ester 697
Carbobenzyloxy-L-seryl-L-proline benzyl ester 697
$C_{23}H_{26}N_2O_8$
N(*tert*-Butoxycarbonyl)-L-glutamic acid γ-benzyl ester α-(4-nitrophenyl) ester 659
$C_{23}H_{26}O$
1,5-Bis-(4-isopropylphenyl)-1,4-pentadien-3-one 287
$C_{23}H_{26}OSi$
Ethoxytri(*o*-tolyl)silane 1122
$C_{23}H_{26}O_4Si$
Ethoxytris(2-methoxyphenyl)silane 1106
$C_{23}H_{26}Si$
Pentyltriphenylsilane 1115
$C_{23}H_{27}BrN_2O_2$
2-Bromo-4-(bis(4-(dimethylamino)-phenyl)methyl)-6-methoxyphenol 223
$C_{23}H_{27}NO_2$
6-Methoxy-2-(4-phenyl-1,2,5,6-tetrahydro-1-pyridylmethyl)-1,2,3,4-tetrahdyro-1-naphthol 245
$C_{23}H_{27}NO_3$
5,5-Diphenyl-2-(2-piperidinylethyl)-1,3-dioxan-4-one hydrochloride 387
$C_{23}H_{27}NO_5$
Octaverine hydrochloride 997
$C_{23}H_{27}N_2O_2$
6-Benzamido-*N*-(2-(3-indolyl)ethyl)-hexanamide 889
$C_{23}H_{27}N_2P$
4-*tert*-Butyl-1-dimethoxy-2,6-diphenylphosphorin 966
$C_{23}H_{27}N_3O_3$
5-(3-(Dimethylamino)propyl)-*N*-(2-methoxybenzyl)-3-phenyl-4-isoxazolecarboxamide 871
$C_{23}H_{27}N_3O_5$
3-Amino-4-(3,4-dimethoxyphenyl)-2-hexyl-7-nitro-1(2*H*)-isoquinolinone 690
$C_{23}H_{27}N_3O_7$
Carbobenzyloxy-L-threonylglycylglycine benzylester 643
$C_{23}H_{28}N_2O_2$
3-Morpholino-4-phenyl-4-(1,2,3,4-tetrahydro-1-quinolinyl)-2-butanone 284
$C_{23}H_{28}N_2O_3$
2,3-Dimorpholino-3-phenylpropiophenone 294
$C_{23}H_{28}N_2O_5$
Carbobenzyloxyalanylnorvaline benzyl ester 662
Carbobenzyloxy-β-alanyl-L-valine benzyl ester 663
Carbobenzyloxy-L-alanyl-L-valine benzyl ester 677
Carbobenzyloxyglycylleucine benzyl ester 663
Carbobenzyloxy-L-isoleucylglycine benzyl ester 662
Carbobenzyloxy-L-isoleucyl-L-phenylalanine 660
Carbobenzyloxynorvalyl-phenylalanine methyl ester 657
Carbobenzyloxy-L-phenylalanyl-L-isoleucine 660
Carbobenzyloxy-L-phenylalanyl-L-norvaline methyl ester 657
Carbobenzyloxy-L-phenylalanyl-L-valine methyl ester 662
Carbobenzyloxysarcosyl-L-valine benzyl ester 662
Carbobenzyloxyvalylalanine benzyl ester 662
Carbobenzyloxy-L-valyl-L-phenylalanine methyl ester 658

$C_{23}H_{28}N_2O_5S$
Carbobenzyloxy-L-phenylalanyl-L-methionine methyl ester ... 655
$C_{23}H_{28}N_2O_6$
Carbobenzyloxynorleucyltyrosine ... 642
Carbobenzyloxynorvalyltyrosine methyl ester ... 644
Carbobenzyloxy-L-seryl-L-valine benzyl ester ... 643
Carbobenzyloxy-L-valyl-L-tyrosine methyl ester ... 645
$C_{23}H_{28}N_2O_6S$
Carbobenzyloxymethionyltyrosine methyl ester ... 644
$C_{23}H_{28}N_4O_4$
N,*N*′-(Methylenedi-4,1-phenylene)-bis(4-morpholinecarboxamide) ... 585
$C_{23}H_{28}N_4O_8$
Malonic acid (3,4,5-trimethoxybenzylidene)hydrazide ... 1074
$C_{23}H_{28}O$
9-Hydroxy-9-(1-naphthyl)-perhydrofluorene ... 208
$C_{23}H_{28}O_2$
6,6′-Dihydroxy-5,5′-dimethyl-2,2′,3,3′-tetrahydro-3,3,3′3′-tetramethyl1,1′-spirobi[1*H*-indene] ... 195
$C_{23}H_{28}O_3$
2-(3-(Decahydronaphth-1-yl)propyl)-3-hydroxy-1,4-naphthoquinone ... 316
Ethyl 3,3-bis(4-methylphenethyl)-2,3-epoxypropionate ... 357
$C_{23}H_{28}O_4$
2,2′-(Phenylmethylene)bis(5,5-dimethyl-3-hydroxy-2-cyclohexen-1-one) ... 307
$C_{23}H_{29}NO_2$
Ethyl 1-*tert*-butyl-4-phenyl-1,2,3,4-tetrahydro-2-isoquinolineacetate hydrochloride ... 359
$C_{23}H_{29}N_3$
2-Piperidino-*N*-(2-piperidinobenzylidene)aniline ... 1053
$C_{23}H_{29}N_3O_2$
6-(4-Acetylphenylazo)-1,2,3,4-tetrahydro-2,2,4,7-tetramethyl-1-quinolineethanol ... 1084
$C_{23}H_{29}N_3O_4$
Carbobenzyloxy-L-isoleucyl-L-phenylalaninamide ... 658
Carbobenzyloxyphenyl-alanylisoleucinamide ... 658
Carbobenzyloxy-L-phenylalanyl-L-leucinamide ... 658
Carbobenzyloxyphenyl-alanylnorleucinamide ... 658
$C_{23}H_{29}N_3O_4S$
Carbobenzyloxy-*S*-benzyl-L-cysteinyl-L-valinamide ... 656
1-Methyl-3-(4-nitrobenzoyl)-1-(4-octyloxyphenyl)-2-thiourea ... 737
$C_{23}H_{29}N_3O_5$
Carbobenzyloxy-L-isoleucyl-L-tyrosinamide ... 645
Carbobenzyloxy-L-leucyl-L-tyrosinamide ... 645
Carbobenzyloxynorleucyl-tyrosinamide ... 645
$C_{23}H_{29}N_5O_3$
N-(1-Ethyl-6-(4-nitrophenylazo)-1,2,3,4-tetrahydro-2,2,4-trimethyl-7-quinolyl)propionamide ... 1091
$C_{23}H_{30}ClNO_3$
2,6-Di-*tert*-butyl-4-(methoxymethyl)-phenyl *N*-(4-chlorophenyl)-carbamate ... 620
$C_{23}H_{30}ClNO_4$
3-Chlorophenyl 2-decyloxy-5-nitrobenzyl ether ... 269
4-Chlorophenyl 2-decyloxy-5-nitrobenzyl ether ... 269
$C_{23}H_{30}ClN_3$
6-Chloro-9-(4-diethylamino)-1-methylbutylamino)-2-methylacridine dihydrochloride ... 995
$C_{23}H_{30}N_2O$
1,1-Dicyclohexyl-3-(1-naphthyl)urea ... 584
$C_{23}H_{30}N_2O_2$
1-Cyclohexyl-1-(2-hydroxy-2-phenylethyl)-3-(2,5-xylyl)urea ... 547
N,*N*′-Dibenzyl-*N*,*N*′-diisopropylmalonamide ... 431
$C_{23}H_{30}N_2O_2S_2$
4,4′-Methylenebis(*S*-butyl *N*-phenylthiolcarbamate) ... 688
$C_{23}H_{30}N_2O_4$
4,4′-Methylenebis(butyl *N*-phenylcarbamate) ... 673
$C_{23}H_{30}N_2O_4S_2$
3,5-Bis(3,4-dimethoxyphenethyl)-3,4,5,6-tetrahydro-2*H*-1,3,5-thiadiazine-2-thione ... 735
$C_{23}H_{30}N_2O_5$
3-Nitro-4-(2-(4-(1,1,3,3-tetramethylbutyl)phenoxy)ethoxy)-benzamide ... 467
$C_{23}H_{30}N_2O_8$
2,2′-Methylenebis(diethyl 4-methyl-3,5-pyrroledicarboxylate) ... 842
$C_{23}H_{30}O_4$
4-(2-(4-*tert*-Octylphenoxy)ethoxy)-benzoic acid ... 340
$C_{23}H_{31}NO$
N-(Diphenylmethyl)decanamide ... 438
$C_{23}H_{31}NO_4S$
Veratryl 2-(2-(diethylamino)-ethylthio)-2-phenylacetate, citrate salt ... 355
$C_{23}H_{31}NO_5$
3,4,5-Trimethoxybenzyl *N*-(2,6-diisopropylphenyl)carbamate ... 646
$C_{23}H_{32}N_2O$
2,6-Di-*tert*-butyl-4-(4-(dimethylamino)-phenyliminomethyl)phenol ... 1033
$C_{23}H_{32}N_2OS$
1-Dodecanoyl-3-(2-naphthyl)-2-thiourea ... 745
$C_{23}H_{32}N_2O_3$
4-(2-(Diethylamino)ethoxy)-α-methyl-α-(2-(methylcarbamoyl)-benzyl)benzyl alcohol ... 461
$C_{23}H_{32}N_2SO$
1-Dodecanoyl-3-(1-naphthyl)-2-thiourea ... 745
$C_{23}H_{32}N_4O_2$
4,4′-Methylenebis(1-butyl-3-phenylurea) ... 592
4,4′-Methylenebis(1,1-diethyl-3-phenylurea) ... 592
4,4′-Methylenebis(1-isobutyl-3-phenylurea) ... 592
$C_{23}H_{32}N_4O_4$
Bis(2-dimethylaminoethyl) 4,4′-methylenebis(*N*-phenylcarbamate) ... 672
$C_{23}H_{32}O_4S$
2-(4-*tert*-Octylphenoxy)ethyl *p*-toluenesulfonate ... 797
$C_{23}H_{33}BrN_2O$
Diethyl(3-(2,2-diphenyl-*N*-methylacetamido)-propyl)-methylammonium bromide ... 425
$C_{23}H_{33}NO$
1-(4-*tert*-Butylphenyl)-3-(dimethylaminomethyl)-2-phenyl-2-butanol hydrochloride ... 238
$C_{23}H_{34}N_2O_4$
N,*N*′-Bis(2-methylcyclohexyloxycarbonyl)-4-methyl-1,3-phenylenediamine ... 669
Di-*tert*-butyl 2,2′-methylenebis(3,4-dimethyl-5-pyrrolecarboxylate) ... 840
$C_{23}H_{35}NS$
4-Phenyl-5-tetradecylthiazole ... 878
$C_{23}H_{35}N_3OS$
8-Acetamido-5-amino-6-(dodecylthio)quinoline ... 987
$C_{23}H_{35}N_3O_6$
Carbobenzyloxyisoleucyl-leucylglycine methyl ester ... 662
$C_{23}H_{36}ClFO_4S$
5-Fluorosulfonyl-2-(hexadecyloxy)-benzoyl chloride ... 802
$C_{23}H_{36}Cl_2O_5S_2$
5-Chlorosulfonyl-2-(hexadecylsulfonyl)benzoyl chloride ... 804
$C_{23}H_{36}F_3NO_3$
Hexadecyl 4-nitro-3-(trifluoromethyl)phenyl ether ... 259
$C_{23}H_{36}N_2O_4S$
N-Dodecyl-*N*-methyl-2-phthalimidoethanesulfonamide ... 822
$C_{23}H_{36}N_4O_2$
3,3′-(4-Methyl-1,3-phenylene)bis(1-cyclohexyl-1-methylurea) ... 565
$C_{23}H_{36}O_2$
1,2-Propanediyldioxybis-(tricyclo[5.2.1.0^{2,6}]decane) ... 17
$C_{23}H_{36}O_4$
Ethyl 2-(4-(dodecyloxy)benzoyl)-acetate ... 363
$C_{23}H_{37}ClO_5S$
3-Chlorosulfonyl-4-hexadecyloxybenzoic acid ... 802
5-(Chlorosulfonyl)-2-(hexadecyloxy)-benzoic acid ... 802
$C_{23}H_{37}ClO_6S_2$
5-(Chlorosulfonyl)-2-(hexadecylsulfonyl)benzoic acid ... 804
$C_{23}H_{37}FO_4S_2$
5-Fluorosulfonyl-2-(hexadecylthio)-benzoic acid ... 802
$C_{23}H_{37}FO_5S$
5-(Fluorosulfonyl)-2-(hexadecyloxy)-benzoic acid ... 802
$C_{23}H_{37}FO_6S_2$
5-Fluorosulfonyl-2-(hexadecylsulfonyl)-benzoic acid ... 804
$C_{23}H_{37}NOS_2$
6-Hexadecyloxy-2-mercaptobenzothiazole ... 913
$C_{23}H_{37}NO_3$
3-(Hexadecanamido)benzoic acid ... 433
4-Palmitamidobenzoic acid ... 436
$C_{23}H_{37}NO_4$
4-Hexadecyloxy-3-nitrobenzaldehyde ... 323
$C_{23}H_{37}NO_5$
3-Hexadecyloxy-4-nitrobenzoic acid ... 340
$C_{23}H_{37}NO_6S$
2-Hexadecylsulfonyl-5-nitrobenzoic acid ... 779
4-Hexadecylsulfonyl-3-nitrobenzoic acid ... 779
$C_{23}H_{38}ClNO$
N-(2-Chloro-6-methylphenyl)-hexadecanamide ... 409
$C_{23}H_{38}ClN_3O_3S$
1-Dodecanesulfonic acid (2-chloro-α-morpholinobenzylidene)-hydrazide ... 821
$C_{23}H_{38}FNO_3S$
4-Fluorosulfonyl-*N*-hexadecylbenzamide ... 804
$C_{23}H_{38}HgO_2$
p-Tolylmercuric hexadecanoate ... 1139
$C_{23}H_{38}N_2O_3$
N,*N*-Bis(2-ethylhexyl)-3-nitrobenzamide ... 470
$C_{23}H_{38}N_2S$
3-Hexadecyl-2-iminobenzothiazoline ... 743
$C_{23}H_{38}N_6O_4$
1,1′-(4-Methyl-1,3-phenylene)bis(3-(3-morpholinopropyl)urea) ... 566
$C_{23}H_{38}O$
4′-Methylhexadecanophenone ... 297
4′-Pentadecylacetophenone ... 298
$C_{23}H_{38}O_2$
2-(Hexadecyloxy)benzaldehyde ... 323
4-(Hexadecyloxy)benzaldehyde ... 324
$C_{23}H_{38}O_2S$
2-(Hexadecylthio)benzoic acid ... 342
3-(Hexadecylthio)benzoic acid ... 341
4-(Hexadecylthio)benzoic acid ... 342
$C_{23}H_{38}O_3$
2-(Hexadecyloxy)benzoic acid ... 341
3-(Hexadecyloxy)benzoic acid ... 340
$C_{23}H_{38}O_4S$
2-(Hexadecylsulfonyl)benzoic acid ... 781
4-(Hexadecylsulfonyl)benzoic acid ... 781
$C_{23}H_{38}O_5S_2$
2-Hexadecylthio-5-sulfobenzoic acid ... 788
2-Hexadecylthio-5-sulfobenzoic acid, pyridine salt ... 793
$C_{23}H_{38}O_6S$
2-Hexadecyloxy-5-sulfobenzoic acid ... 788
$C_{23}H_{38}O_7S_2$
2-(Hexadecylsulfonyl)-5-sulfobenzoic acid ... 789
2-Hexadecylsulfonyl-5-sulfobenzoic acid, lead salt ... 794
$C_{23}H_{39}ClN_2O_4S$
4-Chloro-*N*-hexadecyl-*N*-methyl-3-nitrobenzenesulfonamide ... 822
$C_{23}H_{39}Cl_2NO_2S$
3,4-Dichloro-*N*-hexadecyl-*N*-methylbenzenesulfonamide ... 829
$C_{23}H_{39}F_3N_2O_2S$
4-Hexadecylsulfonyl-2-trifluoromethylphenylhydrazine ... 772
$C_{23}H_{39}NO_2$
Hexadeca-*p*-anisidide ... 419
4-(Hexadecyloxy)benzaldehyde oxime ... 1009
$C_{23}H_{39}NO_3$
4-Amino-3-hexadecyloxybenzoic acid ... 340
$C_{23}H_{39}NO_5S$
4-(Hexadecyloxy)-3-nitrophenyl methyl sulfone ... 775
$C_{23}H_{40}BrN$
3-Bromo-*N*-hexadecyl-5-methylaniline ... 220
$C_{23}H_{40}BrNO_2S$
4-Bromo-*N*-hexadecyl-*N*-methylbenzenesulfonamide ... 828
$C_{23}H_{40}ClN$
N-(5-Chloro-*m*-tolyl)-hexadecylamine ... 216
$C_{23}H_{40}FNO_4S_2$
3-(*N*-Hexadecyl-*N*-methylaminosulfonyl)-benzenesulfonyl fluoride ... 827
$C_{23}H_{40}N_2O$
3-Amino-*N*,*N*-bis(2-ethylhexyl)-benzamide ... 468
p-Amino-*N*,*N*-bis(2-ethylhexyl)-benzamide ... 468
4-Amino-*N*-hexadecylbenzamide ... 468
3-Amino-*N*-hexadecylbenzamide hydrochloride ... 468
N^1-Hexadecylanthranilamide ... 469
$C_{23}H_{40}N_2O_2$
N-Hexadecyl-*N*-methyl-4-nitroaniline ... 276
$C_{23}H_{40}N_2O_2S_2$
1-(4-(Hexadecylsulfonyl)-phenyl)-2-thiourea ... 782
$C_{23}H_{40}N_2O_4S$
N-Hexadecyl-*N*-methyl-4-nitrobenzenesulfonamide ... 822
$C_{23}H_{40}N_2O_5S$
4-Hexadecyloxy-3-(hydrazinosulfonyl)-benzoic acid ... 815
$C_{23}H_{40}N_2S$
1-Hexadecyl-3-phenyl-2-thiourea ... 748
$C_{23}H_{40}N_2S_2$
1-(2-(Hexadecylthio)phenyl)-2-thiourea ... 739
$C_{23}H_{40}O$
Hexadecyl *p*-tolyl ether ... 181
$C_{23}H_{40}O_2$
1-Hexadecyloxy-4-methoxybenzene ... 179
$C_{23}H_{40}O_2S$
Hexadecyl *p*-tolyl sulfone ... 786
$C_{23}H_{40}O_3$
Furfuryl octadecanoate ... 849
$C_{23}H_{40}O_3S$
Hexadecyl *p*-toluenesulfonate ... 801
2-Hexyldecyl *p*-toluenesulfonate ... 801
$C_{23}H_{40}O_{12}S_4$
2,6-Bis(2,2(dihydroxymethyl)propyl dimethanesulfonate)-1,3,5-trimethylbenzene ... 798
$C_{23}H_{41}NO$
2-(Octadecyloxy)pyridine ... 940

$C_{23}H_{41}NO_2$
N-Octadecyl-2-furamide 848
$C_{23}H_{41}NO_2S$
N-Hexadecyl-*p*-toluenesulfonamide. 828
$C_{23}H_{41}NO_3S$
4-(*N*-Dodecyl-3-methylanilino)-1-butanesulfonic acid, sodium salt.... 794
2-(Hexadecyloxy)-5-(methylsulfonyl)aniline 775
$C_{23}H_{41}N_3O_2S_2$
1-Hexadecanesulfonic acid (α-anilino-α-mercaptomethylene)hydrazide 815
$C_{23}H_{42}ClNO$
Benzyldimethyl(2-dodecyloxyethyl)ammonium chloride 254
$C_{23}H_{42}N_2$
4-(Hexadecylamino)benzylamine tartrate (1 to 2) 216
$C_{23}H_{42}N_2O$
N-(7-Ethoxyheptyl)-*N*′-(1-methylheptyl)-1,4-phenylenediamine 221
N-(3-Isopropoxypropyl)-*N*′-(1-methyldecyl)-1,4-phenylenediamine 221
$C_{23}H_{42}N_2O_2S$
Ethyl 2-(2-amino-4-thiazolyl)octadecanoate 877
*N*¹-Hexadecyl-*N*¹-methylmetanilamide 817
$C_{23}H_{42}N_2O_3S$
*N*¹-Hexadecyl-4-methoxymetanilamide 813
$C_{23}H_{42}Si$
Benzylhexadecylsilane 1115
$C_{23}H_{43}N_3$
N-(4-(Diethylamino)-1-methylbutyl)-*N*′-(1-methylheptyl)-1,4-phenylenediamine trihydrochloride 221
$C_{23}H_{43}N_3O_2S$
4-Hydrazino-*N*¹-hexadecyl-*N*¹-methylbenzenesulfonamide hydrochloride 821
$C_{23}H_{43}N_3O_4S_2$
4-(*N*-Hexadecyl-*N*-methylaminosulfonyl)benzenesulfonic acid hydrazide 820
$C_{23}H_{44}N_4O$
Octadecanoic cyclopentylidenehydrazide 112
$C_{23}H_{44}O_2$
Methyl 13-docosenoate 80
$C_{23}H_{44}O_3$
6-Oxotricosanoic acid 64
$C_{23}H_{44}O_4$
Diethyl ethyl(tetradecyl)malonate 94
Diethyl 2-hexadecylmalonate 76
$C_{23}H_{45}NO_3$
N-Octadecylglutaramic acid 105
$C_{23}H_{46}O$
10-Tricosanone 49
12-Tricosanone 49
6-Tricosanone 50
$C_{23}H_{46}O_2Sn$
Tributyltin 10-undecylenate 1131
$C_{23}H_{47}NO$
N-(1,1-Diethyloctadecyl)-formamide. 106
12-Tricosanone oxime 1008
$C_{23}H_{48}N_2O$
1-(*tert*-Butyl)-3-octadecylurea 129
N-(3-Dimethylaminopropyl)octadecanamide 104
$C_{23}H_{48}Si$
Allyldidecylsilane 1098
$C_{23}H_{49}N$
Dibutylpentadecylamine 29
$C_{23}H_{50}O_2Si$
Diethoxymethyloctadecylsilane 1098
$C_{24}F_{20}Si$
Tetrakis(pentafluorophenyl)silane.... 1101
$C_{24}F_{20}Sn$
Tetrakis(pentafluorophenyl)tin 1130
$C_{24}H_5F_{15}Sn$
Phenyltris(2,3,4,5,6-pentafluorophenyl)tin 1130
$C_{24}H_{10}Br_2O_2$
Vat Orange 1 (C.I. 59105) 1095
$C_{24}H_{10}Cl_{10}Ge$
Bis(pentachlorophenyl)diphenylgermane 1103
$C_{24}H_{10}Cl_{10}Sn$
Bis(pentachlorophenyl)diphenyltin.. 1130
$C_{24}H_{10}F_{10}Ge$
Bis(pentafluorophenyl)diphenylgermane 1103
$C_{24}H_{10}F_{10}Si$
Bis(2,3,4,5,6-pentafluorophenyl)diphenylsilane 1103
$C_{24}H_{10}N_2O_4$
3,4,9,10-Perylenetetracarboxylic diimide 723
$C_{24}H_{10}O_3$
Benzo(ghi)perylene-1,2-dicarboxylic anhydride 388
$C_{24}H_{12}Cl_8Si$
Bis(2,3,4,5-tetrachlorophenyl)diphenylsilane 1103
$C_{24}H_{12}O_2$
Vat Yellow 4 (C.I. 59100) 1095
$C_{24}H_{12}S$
Diacenaphtho(1,2-b)(1′,2′-d)thiophene 859
$C_{24}H_{14}N_2$
Benzo(a)naphtho(2,1-c)phenazine 1001
1,2,3,4,6,7-Tribenzophenazine 249
$C_{24}H_{14}N_2O_2$
3,3′-Azobis(dibenzofuran) 1092
$C_{24}H_{14}N_2O_4$
2-Phenyl-3-phthalimidoquinoline-4-carboxylic acid 988
$C_{24}H_{14}N_4O_4$
α,α′-Terephthalylidenebis(5-nitro-*o*-tolunitrile) 764
$C_{24}H_{14}N_8O_{13}$
4-(Picrylamino)phenyl ether 266
$C_{24}H_{14}O_2$
3-Benzoyl benzanthrone 314
$C_{24}H_{14}O_7$
8-(3-Oxo-4,5,6-trihydroxy-3*H*-xanthen-9-yl)-1-naphthoic acid 338
$C_{24}H_{14}S_2$
4,4′-Bi(dibenzothiophene) 899
$C_{24}H_{15}Cl_5Ge$
(Pentachlorophenyl)triphenylgermane 1103
$C_{24}H_{15}Cl_5Pb$
(Pentachlorophenyl)triphenyllead 1134
$C_{24}H_{15}Cl_5Si$
(Pentachlorophenyl)triphenylsilane. 1103
$C_{24}H_{15}F_5Ge$
(Pentafluorophenyl)triphenylgermane 1103
$C_{24}H_{15}F_5Pb$
(Pentafluorophenyl)triphenyllead 1134
$C_{24}H_{15}NO_9$
4-Nitrofluorescein 379
$C_{24}H_{15}N_3O_3$
4,5-Diphenyl-2-(4-nitrobenzylideneamino)-3-furancarbonitrile 1056
$C_{24}H_{15}O_8S_2$
2-Dibenzofuransulfonic acid, barium salt 896
$C_{24}H_{16}$
1,1′-Bi-(acenaphthenylidene) 167
$C_{24}H_{16}Br_2O_6$
Bis(4′-bromophenacyl) phthalate 366
$C_{24}H_{16}ClNO_2$
5-Chloro-2-(2-hydroxy-1-naphthylmethyleneamino)benzophenone 1023
$C_{24}H_{16}ClN_3O_5$
3-(*N*-Phenylcarbamoyl)-2-naphthyl *N*-(4-chloro-2-nitrophenyl)carbamate 629
$C_{24}H_{16}Cl_2$
2,4-Dichloro-1,3,5-triphenylbenzene. 172
$C_{24}H_{16}Cl_2N_2O_2$
4-(1-Naphthyliminomethyl)phenyl *N*-(3,4-dichlorophenyl)carbamate 635
$C_{24}H_{16}Cl_2O_2$
3,3″-(1,4-Phenylene)bis(4′-chloroacrylophenone) 309
$C_{24}H_{16}Cl_4Fe$
2,4-Diphenyl-6-(*p*-tolyl)pyrylium tetrachloroferrate 936
$C_{24}H_{16}Cl_4Ge$
(2,3,4,5-Tetrachlorophenyl)triphenylgermane 1103
$C_{24}H_{16}Cl_4Pb$
Tetrakis(4-chlorophenyl)lead 1134
$C_{24}H_{16}Cl_4Si$
(2,3,4,5-Tetrachlorophenyl)triphenylsilane 1103
Tetrakis(3-chlorophenyl)silane 1101
$C_{24}H_{16}Cl_4Sn$
(2,3,4,5-Tetrachlorophenyl)triphenyltin 1130
Tetrakis(4-chlorophenyl)tin 1130
$C_{24}H_{16}F_4Si$
(2,3,4,5-Tetrafluorophenyl)triphenylsilane 1103
Tetrakis(3-fluorophenyl)silane 1101
$C_{24}H_{16}N_2$
2,4-Diphenylbenzo(h)quinazoline 1001
2,3-Diphenyl-5,6-benzoquinoxaline... 1001
1,4-Phenylenebis(2-quinoline) 995
$C_{24}H_{16}N_2O_2$
9-Fluorenylideneamino *N*-(1-naphthyl)carbamate 687
$C_{24}H_{16}N_2O_6$
3,3″-(1,4-Phenylene)bis(3′-nitroacrylophenone) 308
3,3″-(1,4-Phenylene)bis(4′-nitroacrylophenone) 308
$C_{24}H_{16}O_2$
1,8-Dibenzoylnaphthalene 314
3,3-Diphenyl-1*H*,3*H*-naphtho(1,8-cd)pyran-1-one 386
$C_{24}H_{16}O_2Sn$
10,10′-Spirobi(10*H*-phenoxastannin). 1133
$C_{24}H_{16}Si$
9,9′-Spirobi(9*H*-9-silafluorene) 1123
$C_{24}H_{17}Br$
1-Bromo-2,4,6-triphenylbenzene 172
$C_{24}H_{17}BrO$
4-(4-Bromophenyl)-2,6-diphenylphenol 193
$C_{24}H_{17}ClN_2OS$
N-(3-(4-Chlorophenyl)-4-phenyl-4-thiazolin-2-ylidene)cinnamamide..... 1026
$C_{24}H_{17}ClO$
3-Chloro-2,4,6-triphenylphenol 193
$C_{24}H_{17}ClO_2$
2-(2-(4-Chlorobenzoyl)benzylidene)-α-tetralone 310
$C_{24}H_{17}FN_2OS$
N-(3-(2-Fluorophenyl)-4-phenyl-4-thiazolin-2-ylidene)cinnamamide..... 731
N-(3-(3-Fluorophenyl)-4-phenyl-4-thiazolin-2-ylidene)cinnamamide..... 731
$C_{24}H_{17}F_3Si$
Phenyltris(3-fluorophenyl)silane 1102
$C_{24}H_{17}N$
2,5-Di(2-naphthyl)pyrrole 839
4-Methyl-2-(2-naphthyl)benz(f)isoquinoline 996
4-Methyl-2-(2-naphthyl)benzo(h)quinoline 995
$C_{24}H_{17}NS$
10-(4-Biphenylyl)phenothiazine 251
$C_{24}H_{17}N_3O_7S_3$
Thiazine Red R 1085
$C_{24}H_{18}AsClO$
(4-Biphenylyl)(4-chlorophenyl)phenylarsine oxide 1128
$C_{24}H_{18}ClNO_2$
2-Benzyl-4-chlorophenyl *N*-(1-naphthyl)carbamate 633
$C_{24}H_{18}Cl_2Ge$
(2,6-Dichlorophenyl)triphenylgermane 1104
$C_{24}H_{18}Cl_2N_4O_2$
α-(5-Chloro-2-pyridylimino)-*o*-cresol copper(II) salt 1035
$C_{24}H_{18}Cl_2Pb$
Bis(4-chlorophenyl)diphenyllead 1134
$C_{24}H_{18}Cl_2Si$
Bis(3-chlorophenyl)diphenylsilane ... 1103
Bis(4-chlorophenyl)diphenylsilane ... 1103
$C_{24}H_{18}Cl_2Sn$
(2,6-Dichlorophenyl)triphenyltin 1130
$C_{24}H_{18}F_2Si$
Bis(3-fluorophenyl)diphenylsilane.... 1102
$C_{24}H_{18}FeO_2$
1,1′-Dibenzoylferrocene 1142
$C_{24}H_{18}N_2$
2-Cyano-3,4-diphenyl-1-(*p*-tolyl)pyrrole 839
$C_{24}H_{18}N_2O_3S$
2′-(Naphth(2,3-d)oxazol-2-yl)-*p*-toluenesulfonanilide 909
$C_{24}H_{18}N_2O_{12}$
2,5-Bis(3-nitro-4-methoxyphenyl)-3,6-diacetoxy-1,4-benzoquinone 355
$C_{24}H_{18}N_2S_2$
3-(*N*-(1-Naphthyl)-anilinomethyl)-2(3*H*)-benzothiazolethione 742
3-(*N*-(2-Naphthyl)-*N*-phenylaminomethyl)-2(3*H*)benzothiazolethione ... 742
$C_{24}H_{18}N_4O_2$
Oxalic bis(1-naphthylmethylene)hydrazide 1053
$C_{24}H_{18}N_4O_4$
Oxalic bis((2-hydroxy-1-naphthylmethylene)hydrazide) 496
N,*N*′-(1,3-Phenylene)bis(*N*-(aminomethyl)phthalimide) 718
$C_{24}H_{18}N_4O_{12}$
2,5-Bis(*N*-(2-methoxy-5-nitrophenyl)carbamoyl)terephthalic acid 463
$C_{24}H_{18}O$
4-Benzylidene-2,6-diphenyl-4*H*-pyran 966
2-Biphenylyl 4-biphenylyl ether 183
2-Methyl-3,4,5-triphenyl-2,4-cyclopentadienone 285
$C_{24}H_{18}OSi$
10,10-Diphenyl-10*H*-phenoxasilin 1122
$C_{24}H_{18}OSn$
10,10-Diphenyl-10*H*-phenoxastannin 1133
$C_{24}H_{18}O_2$
3,3″-(1,4-Phenylene)diacrylophenone 309
$C_{24}H_{18}O_2SSi$
10,10-Diphenyl-10*H*-phenothiasilin-5,5-dioxide 1110
$C_{24}H_{18}O_4$
3,3″-(1,4-Phenylene)bis(4′-hydroxyacrylophenone) 307
α,α′-(1,4-Phenylenedimethylidyne)bis(2′-hydroxyacetophenone) 307
$C_{24}H_{18}SSi$
10,10-Diphenyl-10*H*-phenothiasilin... 1122
$C_{24}H_{19}BF_4O$
4-Benzyl-2,6-diphenylpyrylium tetrafluoroborate 935
2,6-Diphenyl-4-(*p*-tolyl)pyrylium tetrafluoroborate 936
$C_{24}H_{19}BF_4O_2$
2,4-Diphenyl-6-(4-methoxyphenyl)pyrylium tetrafluoroborate 934
$C_{24}H_{19}BrGe$
(4-Bromophenyl)triphenylgermane ... 1103
$C_{24}H_{19}BrPb$
(3-Bromophenyl)triphenyllead 1133
(4-Bromophenyl)triphenyllead 1134
$C_{24}H_{19}BrSi$
(4-Bromophenyl)triphenylsilane 1103
$C_{24}H_{19}BrSn$
(4-Bromophenyl)triphenyltin 1130
$C_{24}H_{19}ClNP$
N-(4-Chlorophenyl)-*p*,*p*,*p*-triphenylphosphine imide 836
$C_{24}H_{19}ClO_6$
2,4-Diphenyl-6-(4-methoxyphenyl)pyrylium perchlorate 934
$C_{24}H_{19}ClPb$
(4-Chlorophenyl)triphenyllead 1134
$C_{24}H_{19}ClSi$
(3-Chlorophenyl)triphenylsilane 1102
$C_{24}H_{19}ClSn$
(4-Chlorophenyl)triphenyltin 1130
$C_{24}H_{19}Cl_4FeO_2$
2,6-Diphenyl-4-(4-methoxyphenyl)pyrylium tetrachloroferrate 934
2,6-Diphenyl-6-(4-methoxyphenyl)pyrylium tetrachloroferrate 934

$C_{24}H_{19}NO$
4-(10,11-Dihydro-5-hydroxy-5*H*-dibenzo(a,d)cyclohepten-5-yl)-isoquinoline ... 996

$C_{24}H_{19}NO_2$
2-Benzylphenyl *N*-(1-naphthyl)-carbamate ... 680
Ethyl 4-(9-anthracenylmethylene-amino)benzoate ... 1063

$C_{24}H_{19}NO_4$
Benzyl 1,2-diphenyl-4-hydroxy-5-oxo-3-pyrroline-3-carboxylate ... 689

$C_{24}H_{19}NO_5$
3,3-Bis(4-acetoxyphenyl)oxindole ... 692

$C_{24}H_{19}N_2O_2P$
N-(2-Nitrophenyl)-*p*,*p*,*p*-triphenylphosphine imide ... 835
N-(3-Nitrophenyl)-*p*,*p*,*p*-triphenylphosphine imide ... 835
N-(4-Nitrophenyl)-*p*,*p*,*p*-triphenylphosphine imide ... 835

$C_{24}H_{19}N_3O$
N-Benzylidene-*N*-(1-(4-methoxyphenylazo)-2-naphthyl)-amine tetrafluoroborate ... 1088

$C_{24}H_{19}N_3O_3$
3-Hydroxy-4-(2-methoxyphenylazo)-2-naphthanilide ... 1083

$C_{24}H_{20}$
trans-1,2-Di(7-methyl-2-naphthyl)-ethylene ... 167
trans-1-(6-Methyl-2-naphthyl)-2-(7-methyl-2-naphthyl)ethylene ... 167

$C_{24}H_{20}AsBr$
Tetraphenylarsonium bromide ... 1127

$C_{24}H_{20}BF_4N_3O_2$
2,3-Bis(4-methoxyphenyl)-3*H*-naphtho(1,2-d)triazolium tetrafluoroborate ... 908

$C_{24}H_{20}BrN_3O$
2-(4-Methoxyphenyl)-3-(*p*-tolyl)-3*H*-naphtho(1,2-d)triazolium bromide ... 908

$C_{24}H_{20}Br_2Cl_2N_2O_4S_2$
N,*N*′-(4,6-Dibromo-1,3-phenylene)-bis(2-(4-chlorophenylthio)ethyl carbamate) ... 623

$C_{24}H_{20}F_3N_5O_3$
4-(3-(3-(Trifluoromethyl)phenyl)-ureido)benzoic (4-acetamidobenzylidene)-hydrazide ... 509

$C_{24}H_{20}N_2$
1′-Acetonaphthone azine ... 1079
N,*N*′-Bis(cinnamylidene)-1,4-phenylenediamine ... 1058
4,5-Diphenyl-2-(5-indanyl)-imidazole ... 867

$C_{24}H_{20}N_2O$
5-Indancarboxaldehyde azine with benzil ... 1072
1,4-Naphthoquinone bis(*p*-tolyl)-hydrazone ... 302

$C_{24}H_{20}N_2O_2$
α,α′-(Ethylenedinitrilo)-bis-(1-methyl-2-naphthol) ... 1033

$C_{24}H_{20}N_2O_3$
4-(Diphenylhydroxymethyl)-*N*-methyl-3-phenyl-5-isoxazolecarboxamide ... 871
2-(3,4,5-Trimethoxyphenyl)-dibenzo(e,g)benzimidazole ... 903

$C_{24}H_{20}N_2O_4$
2-(4-(3,4-Dihydro-4-oxo-3-phenylphthalazin-1-yl)phenoxy)-propionic acid ... 699

$C_{24}H_{20}N_2O_6$
N,*N*′-Bis(2-(methoxycarbonyl)-phenyl)terephthalamide ... 474

$C_{24}H_{20}N_4O_2S$
3-Phenyl-1-(*m*-tolyl)-5-(*m*-tolylcarbamoylimino)-4-thiohydantoin ... 1075

$C_{24}H_{20}N_6O_2$
N,*N*′-(*p*-Phenylene-dimethylidyne)bis(1-amino-1,2-dihydro-4,6-dimethyl-2-oxonicotinonitrile ... 1069

$C_{24}H_{20}O$
10-(4-Isopropylbenzylidene)-9-anthrone ... 314

$C_{24}H_{20}OPb$
(2-Hydroxyphenyl)triphenyllead ... 1134

$C_{24}H_{20}OSi$
Diphenyl(4-phenoxyphenyl)silane ... 1106
(Phenoxy)triphenylsilane ... 1115
2-(Triphenylsilyl)phenol ... 1119
4-(Triphenylsilyl)phenol ... 1105

$C_{24}H_{20}OSn$
2-(Triphenylstannyl)phenol ... 1130
3-(Triphenylstannyl)phenol ... 1130

$C_{24}H_{20}O_2$
2-(4-Benzyloxybenzylidene)-1-tetralone ... 300
1,4-Bis-(2-naphthoxy)-*trans*-2-butene ... 188
4-Hydroxy-5-methyl-2,3,4-triphenyl-2-cyclopentenone ... 282

$C_{24}H_{20}O_2Sn$
2,2′-(Diphenylstannylene)diphenol ... 1130
3,3′-(Diphenylstannylene)diphenol ... 1130

$C_{24}H_{20}O_3Sn$
3,3′,3″-(Phenylstannylidyne)-triphenol ... 1130

$C_{24}H_{20}O_4Si$
Tetraphenyl orthosilicate ... 1115

$C_{24}H_{20}O_4Sn$
Tetrakis(3-hydroxyphenyl)tin ... 1130

$C_{24}H_{20}O_9$
Diethyl 3,4-bis(benzoyloxy)-2,5-furandicarboxylate ... 846

$C_{24}H_{20}PbS$
(Thiophenoxy)triphenyllead ... 1136

$C_{24}H_{20}Si$
(4-Biphenylyl)diphenylsilane ... 1121

$C_{24}H_{21}ClN_2O_5$
4-(4-(Ethoxycarbonyl)-phenyliminomethyl)-2-methoxyphenyl *N*-(3-chlorophenyl)-carbamate ... 620

$C_{24}H_{21}F_3N_4O_4$
4-(3-(3-(Trifluoromethyl)phenyl)-ureido)benzoic (3,4-dimethoxybenzylidene)hydrazide ... 507

$C_{24}H_{21}N$
N-(9-Anthracenylmethylene)-2,4,6-trimethylaniline ... 1068

$C_{24}H_{21}NO_2$
3-Ethoxy-α-phenyl-α-(2-pyridyl)-2-naphthalenemethanol ... 931
2′-(*N*-Methylanilino)-1-oxo-1,2′,3′4-tetrahydrospiro[3*H*-2-benzopyran-3,1′-(1*H*)indene] ... 385

$C_{24}H_{21}NO_3$
5,6-Dihydro-2,3-dimethoxy-12b-phenylisoindolo(1,2-a)isoquinolin-8(12bh)-one ... 698
α-(2,3-Dimethoxy-1-naphthyl)-α-(3-pyridyl)benzyl alcohol ... 931

$C_{24}H_{21}NO_4$
Methyl 9-(α,α-dimethyl-4-nitrobenzyl)-9-fluorenecarboxylate ... 359

$C_{24}H_{21}NO_6$
2-(Benzyloxycarbonyl)phenyl *N*-(4-(ethoxycarbonyl)phenyl)carbamate ... 655

$C_{24}H_{21}NO_6S$
3-(4-Acetamidophenylsulfonyl)-3-((3,4-methylenedioxy)phenyl)-propiophenone ... 780

$C_{24}H_{21}NSn$
4-(Triphenylstannyl)aniline ... 1130

$C_{24}H_{21}N_2O_3P$
Diphenyl (α-anilino-2-pyridylmethyl)phosphonate ... 834
Diphenyl (α-anilino-3-pyridylmethyl)phosphonate ... 946

$C_{24}H_{22}ClNO_2$
3-(α-(4-Chlorophenyl)-2-(dimethylamino)benzyl)-3-methylphthalide ... 382

$C_{24}H_{22}ClNO_5S$
3-(4-(Acetamido)phenylsulfonyl)-3-(2-chlorophenyl)-4′-methoxypropiophenone ... 773
4′-(3-(4-Chlorophenyl)-1-(4-methoxyphenyl)-3-oxopropylsulfonyl)acetanilide ... 773

$C_{24}H_{22}ClN_3O_4$
2-Methoxy-4-(4-(*N*-methylacetamido)-phenyliminomethyl)phenyl *N*-(3-chlorophenyl)carbamate ... 1023

$C_{24}H_{22}Cl_2N_2O_4$
N,*N*′-(1,4-Phenylenedimethylidyne)-bis(5-chloro-2,4-dimethoxyaniline) ... 1049

$C_{24}H_{22}F_3N_5O_2$
4-(3-(3-(Trifluoromethyl)phenyl)-ureido)benzoic (4-(dimethylamino)-benzylidene)hydrazide ... 509

$C_{24}H_{22}N_2O$
3-(2-(Benzylideneamino)ethyl)-2-(4-methoxyphenyl)indole ... 1046

$C_{24}H_{22}N_2O_3$
1-(4-Methoxybenzylideneamino)-3-phenoxy-4-(*p*-tolyl)-2-azetidinone ... 1040

$C_{24}H_{22}N_2O_3S$
5,6-Dimethyl-11*H*-indolo(3,2-c)-quinolinium *p*-tosylate ... 989

$C_{24}H_{22}N_2O_4$
Benzil, azine with 3,4,5-trimethoxybenzaldehyde ... 1042
2-Benzyl-6-cyano-3a-(2,3-dimethoxy-phenyl)-3a,4,7,7a-tetrahydro-1,3-isoindolinedione ... 761

$C_{24}H_{22}N_2O_5$
1-(4-Methoxybenzylideneamino)-4-(4-methoxyphenyl)-3-phenoxy-2-azetidinone *N*-oxide ... 697

$C_{24}H_{22}N_2O_6$
2-(3-Nitrobenzoyl)-4′-(2-(*p*-tolyloxy)-ethoxy)acetanilide ... 442

$C_{24}H_{22}N_2O_6S$
S-Benzyl-*N*-carbobenzyloxy-ʟ-cysteine 4-nitrophenyl ester ... 667
3′,5′-Dibenzoyl-4-thiothymidine ... 738

$C_{24}H_{22}N_4O_6$
2,5-Bis(*N*-(4,6-dimethyl-2-pyridyl)-carbomyl)terephthalic acid monohydrate ... 962

$C_{24}H_{22}N_6O_2$
1,1′-(1,5-Naphthalenediyl)bis(3-(3-pyridylmethyl)urea) ... 953

$C_{24}H_{22}O$
2,6-Bis(cinnamylidene)-1-cyclohexanone ... 304

$C_{24}H_{22}O_2$
Bis(3,4-dihydro-6-methoxy-1-naphthyl)acetylene ... 1017
1,2-Dihydro-2,3-dimethyl-1,2-diphenyl-1,4-naphthalenediol ... 199

$C_{24}H_{22}O_3$
2,4-Dimethyl-6-methoxy-α,β-diphenylcinnamic acid ... 335
1,5-Diphenyl-3-(4-methoxyphenyl)-1,5-pentanedione ... 307

$C_{24}H_{22}O_3S$
9,10-Dihydro-9,10-ethanoanthracene-11-methanol *p*-tosylate ... 799

$C_{24}H_{22}O_3Si_2$
Tetraphenyl-1,3-disiloxanediol ... 1115

$C_{24}H_{22}O_4$
1,4-Bis-(2-naphthoxy)-2,3-butanediol 202

$C_{24}H_{22}O_6$
2,6-Dibenzyloxyhydroquinone diacetate ... 377

$C_{24}H_{22}Si_2$
1,1,2,2-Tetraphenyldisilane ... 1115

$C_{24}H_{23}ClN_2O_3$
Basic Red 8 (C.I. 45150) ... 1060

$C_{24}H_{23}ClN_2S$
10-(1-Benzyl-3-piperidyl)-2-chlorophenothiazine fumarate ... 249

$C_{24}H_{23}ClN_4O_4$
4-(3-(3-Chlorophenyl)ureido)-benzoic (4-ethoxy-3-methoxybenzylidene)hydrazide ... 516

$C_{24}H_{23}N$
1-(10,11-Dihydro-5*H*-dibenzo(a,d)-cyclohepten-5-yl)-2-methylisoindoline ... 251

$C_{24}H_{23}NO_2$
Benzyl 2-(diphenylmethyl-aminomethyl)acrylate hydrobromide ... 370
Benzyl 1-(diphenylmethyl)-2-azetidinecarboxylate ... 373
2-(10,11-Dihydro-5-hydroxy-5*H*-dibenzo(a,d)cyclohepten-5-yl)-4,*N*-dimethylbenzamide ... 461
2-(10,11-Dihydro-5-hydroxy-5*H*-dibenzo(a,d)cyclohepten-5-yl)-*N*-ethylbenzamide ... 461

$C_{24}H_{23}NO_3$
N-4-*tert*-Butylbenzoyl-*N*-phenylanthranilic acid ... 471
5-(5-Methoxy-2-(*N*-methylcarbamoyl)phenyl)-dibenzosuberol ... 461

$C_{24}H_{23}NO_4$
2-Benzoyl-*N*-(2-(3,4-dimethoxyphenyl)ethyl)benzamide ... 463
Dimethyl 5-pyrrolidinodibenzo(a,c)-cyclooctene-6,7-dicarboxylate ... 371

$C_{24}H_{23}NO_5$
Carbobenzyloxy-ʟ-tyrosine benzyl ester ... 645

$C_{24}H_{23}N_3O$
4′-(2,6-Dimethyl-4-(2-isoindolinyl)-phenylazo)acetophenone ... 1091

$C_{24}H_{23}N_3O_4$
N-Benzoylanthranilic (3-ethoxy-4-methoxybenzylidene)hydrazide ... 497

$C_{24}H_{23}N_3O_5$
Ethyl 4-(2-methoxy-5-(*N*-phenylcarbamoyl)phenylureido)-benzoate ... 554

$C_{24}H_{24}BrO_2P$
(2-(Ethoxycarbonyl)allyl)-triphenylphosphonium bromide ... 831

$C_{24}H_{24}ClN_3O_6$
Diethyl 6-amino-1-benzamido-4-(4-chlorophenyl)-2-oxo-1,2,3,4-tetrahydro-3,5-pyridinedicarboxylate ... 694

$C_{24}H_{24}INS$
3-Ethyl-2-(2-(4,6,8-trimethyl-1-azulenyl)vinyl)benzothiazolium iodide ... 915

$C_{24}H_{24}N_2$
1-*tert*-Butyl-3-(triphenylmethyl)-carbodiimide ... 769
N,*N*′-(1,4-Phenylenedimethylidyne)-bis(4-ethylaniline) ... 1065
N,*N*′-(1,4-Phenylenedimethylidyne)-di-2,3-xylidine ... 1078
N,*N*′-(1,4-Phenylenedimethylidyne)-di-2,4-xylidine ... 1078
N,*N*′-(1,4-Phenylenedimethylidyne)-di-2,6-xylidine ... 1078
N,*N*′-(1,4-Phenylenedimethylidyne)-di-3,4-xylidine ... 1078
N,*N*′-Terephthalylidenebis(3,5-dimethylaniline) ... 1066
N,*N*′-Terephthalylidenebis(2-ethylaniline) ... 1064

$C_{24}H_{24}N_2O$
2-Benzoyl-1-(2-(methylaminomethyl)phenyl)-1,2,3,4-tetrahydroisoquinoline hydrochloride ... 703

$C_{24}H_{24}N_2O_2$
N,*N*′-Bis(2-ethylphenyl)-isophthalamide ... 475
N,*N*′-Bis(2,3-xylyl)isophthalamide ... 475
N,*N*′-Bis(3,4-xylyl)isophthalamide ... 475
N,*N*′-Dibenzyl-*N*,*N*′-dimethylisophthalamide ... 471
N,*N*′-Dibenzyl-*N*,*N*′-dimethylphthalamide ... 472
N,*N*′-Dibenzyl-*N*,*N*′-dimethylterephthalamide ... 475
N,*N*′-Diethyl-*N*,*N*′-diphenylterephthalamide ... 475
N,*N*′-Di-(2,4-xylyl)isophthalamide ... 475
N,*N*′-(1,4-Phenylenedimethylidyne)-bis(5-methyl-*o*-anisidine) ... 1045
N,*N*′-(1,4-Phenylenedimethylidyne)-di-*o*-phenetidine ... 1045
N,*N*′-(1,4-Phenylenedimethylidyne)-di-*p*-phenetidine ... 1050
N,*N*′-Terephthalylidenebis(2-methyl-*p*-anisidine) ... 1047

$C_{24}H_{24}N_2O_3$
2,6-Bis(4-acetamidobenzylidene)-1-cyclohexanone 432
1-(4-Ethoxycarbonylphenyl)-3-(α-phenylphenethyl)urea 577
$C_{24}H_{24}N_2O_3S$
N-(2-(2-(4-Methoxyphenyl)indol-3-yl)-ethyl)-*p*-toluenesulfonamide 882
$C_{24}H_{24}N_2O_4$
N,N'-Bis(3,4-dimethoxybenzylidene)-1,4-phenylenediamine 1041
N,N'-Bis(4-ethoxyphenyl)-isophthalamide 463
N,N'-(1,4-Phenylenedimethylidyne)-bis(2,4-dimethoxyaniline) 1045
$C_{24}H_{24}N_4O_8$
Diethy 6-amino-1-benzamido-4-(4-nitrophenyl)-2-oxo-1,2,3,4-tetrahydro-3,5-pyridinedicarboxylate 701
$C_{24}H_{24}O$
4-(3,5-Diphenylcyclohexyl)phenol 195
$C_{24}H_{24}O_2Si$
Bis(1-naphthyl)diethoxysilane 1123
$C_{24}H_{24}Si$
(1-Cyclohexenyl)triphenylsilane 1111
(1-Cyclohexen-3-yl)triphenylsilane 1111
$C_{24}H_{25}BF_4O_2$
2-*tert*-Butyl-5,6-dihydro-4-(4-methoxyphenyl)naphtho(1,2-b)-pyrylium tetrafluoroborate 1002
$C_{24}H_{25}N$
5,6-Diphenyl-1,2,3,4,5,6-hexahydro-2-methyl-2-benzazocine 252
$C_{24}H_{25}NO$
N-Benzyl-2,2-dimethyl-2'-phenylpropionanilide 488
$C_{24}H_{25}NO_2$
1,3-Dihydro-3-(4-(dimethylaminomethyl)phenyl)-1-methyl-3-phenyl-1isobenzofuranol 245
$C_{24}H_{25}NO_3$
2-(2-(4-Benzyloxyphenyl)-2-hydroxypropyl)-*N*-methylbenzamide 461
$C_{24}H_{25}NO_4S$
3-(4-(Dimethylamino)phenyl)-4'-methoxy-3-(phenylsulfonyl)-propiophenone 775
$C_{24}H_{25}NO_9S$
Coralyne sulfoacetate 997
$C_{24}H_{25}N_3O_2$
N,N-Bis-(2-salicylideneaminoethyl)-aniline 1033
$C_{24}H_{25}N_3O_3S_2$
Benzenesulfonic *N*[1]-(3,5-diethyl-4-hydroxyphenyl)-*N*[2]-(3-methyl-2-benzothiazolinylidene)hydrazide 812
$C_{24}H_{26}$
4,4'-Bis(1-cyclohexen-1-yl)biphenyl 165
$C_{24}H_{26}BrO_2P$
(1-(Ethoxycarbonyl)-propyl)-triphenylphosphonium bromide 831
$C_{24}H_{26}BrP$
Cyclohexyltriphenylphosphonium bromide 831
$C_{24}H_{26}Br_2N_6O_2$
1,1'-(4,6-Dibromo-1,3-phenylene)-bis(3-(4-(dimethylamino)phenyl)-urea) 523
$C_{24}H_{26}ClO_3P$
(Ethoxy(ethoxycarbonyl)methyl)-triphenylphosphonium chloride 830
$C_{24}H_{26}Cl_4N_4O_2$
Sebacic bis((3,4-dichlorobenzylidene)hydrazide) 493
$C_{24}H_{26}F_6N_2O_2$
N,N'-Bis(3-(trifluoromethyl)phenyl)-1,10-decanediamide 397
$C_{24}H_{26}INO$
2-(1-(3,5-Dimethylphenyl)-2-naphthyl)-4,5-dihydro-3,5,5-trimethyloxazolium iodide 252
$C_{24}H_{26}I_2N_2$
2,2'-Hexamethylene-bis(isoquinolinium iodide) 998
$C_{24}H_{26}N_2O$
4-Butoxybenzaldehyde benzylphenylhydrazone 234
1,1-Dibenzyl-3-(2-ethyl-6-methylphenyl)urea 573
$C_{24}H_{26}N_2O_3$
N-(4-(Dimethylamino)benzylidene)-3-methoxy-4-(2-phenoxyethoxy)-aniline 1041
1,3-Diphenyl-2-(3,4,5-trimethoxyphenyl)-imidazolidine 246
$C_{24}H_{26}N_4$
N,N'-Bis-(4-dimethylaminobenzylidene)-1,4-phenylenediamine 1057
$C_{24}H_{26}N_4O_2$
Adipic bis-(cinnamylidene-hydrazide) 1070
$C_{24}H_{26}N_4O_5$
Carbobenzyloxyhistidyl-phenylalanine methyl ester 866
$C_{24}H_{26}N_4O_6$
Carbobenzyloxyglycylglycyl-L-tryptophan methyl ester 883
Carbobenzyloxy-L-histidyl-L-tyrosine methyl ester 866
Carbobenzyloxy-L-tryptophylglycylglycine methyl ester 890
$C_{24}H_{26}OSi$
1-(Triphenylsilyl)cyclohexanol 1105
$C_{24}H_{26}O_2$
3',6'-Dimethoxy-2,4,4',6''-tetramethyl-*p*-terphenyl 187
2,2-Dimethyl-1-(4-phenoxyphenyl)-3-(*p*-tolyl)-1-propanol 201
$C_{24}H_{26}O_2Si$
1-Allyloxy-3-(triphenylsilyl)-2-propanol 1104
$C_{24}H_{26}O_4$
4'-Methoxy-4-(1-(4-methoxyphenyl)-3-oxo-1-cyclohexen-4-yl)-butyrophenone 291
$C_{24}H_{26}O_5$
2,6-Bis(2,4-dimethoxybenzylidene)-1-cyclohexanone 300
2,6-Bis(3,4-dimethoxybenzylidene)-1-cyclohexanone 283
$C_{24}H_{26}Si$
Cyclohexyltriphenylsilane 1111
$C_{24}H_{27}BrN_6O_{10}$
5'-(Bis(2-acetoxyethyl)amino)-2'-(2-bromo-4,6-dinitrophenylazo)-4'-ethoxyacetanilide 1081
$C_{24}H_{27}INP$
(2-Piperidinylmethyl)-triphenylphosphonium iodide 831
$C_{24}H_{27}N_3O_3$
1,3-Diphenyl-1-(2-morpholino-1-cyclohexen-1-ylcarbonyl)urea 565
$C_{24}H_{27}N_3O_5$
Carbobenzyloxy-L-norvalyl-L-tryptophan 889
$C_{24}H_{27}N_3O_5S$
Carbobenzyloxy-L-methionyl-L-tryptophan 883
$C_{24}H_{27}N_3O_6$
Carbobenzyloxyglycylglycyl-L-proline benzyl ester 700
$C_{24}H_{27}N_3O_6S_3$
1,3,5-Tris(*p*-tolylsulfonyl)hexahydro-1,3,5-triazine 820
$C_{24}H_{27}N_7O_5$
2'-(2,6-Dicyano-4-nitrophenylazo)-5'-(*N*-(2-(2-ethoxyethoxy)ethyl)-*N*-ethylamino)acetanilide 1086
$C_{24}H_{27}O_4P$
4-Ethyl-3',3'',5',5''-tetramethyltriphenyl phosphate 836
Tris(2,3-dimethylphenyl) phosphate 837
Tris(2,4-dimethylphenyl) phosphate 837
Tris(2,5-dimethylphenyl) phosphate 837
Tris-(2,6-dimethylphenyl)phosphate 837
Tris(3,4-dimethylphenyl) phosphate 837
$C_{24}H_{28}As_2Br_2$
5,6,15,16-Tetrahydro-6,6,15,15-tetramethyltribenzo(b,d,h)(1,6)-diarsenicinium dibromide 1128
$C_{24}H_{28}ClNO_2$
2'-Benzoyl-4'-chloro-10-undecenanilide 453
$C_{24}H_{28}Cl_2N_4O_2$
Sebacic acid bis((2-chlorobenzylidene)hydrazide) 1025
Sebacic bis((4-chlorobenzylidene)-hydrazide) 493
$C_{24}H_{28}N_2$
α,α'-(4-Ethylanilino)-*p*-xylene 232
4,4'-(*o*-Tolylmethylene)bis(*N,N*-dimethylaniline) 233
$C_{24}H_{28}N_2O$
2,6-Bis(4-(dimethylamino)-benzylidene)-1-cyclohexanone 285
$C_{24}H_{28}N_2O_2$
N,N'-(1,4-Phenylenedimethylidyne)-di-*p*-phenetidine 229
$C_{24}H_{28}N_2O_3S$
2-(Bis(4-(dimethylamino)phenyl)-methyl)phenyl methanesulfonate 798
$C_{24}H_{28}N_2O_4$
Dimethyl-(hydroxy-methyl-imidazolylmethyl-phenoxy)-propiophenone ethylene ketal 865
$C_{24}H_{28}N_2O_8$
N,N'-Hexamethylenebis(4-(methoxycarbonyl)-phenylcarbamate) 676
$C_{24}H_{28}N_4O_2$
3,3'-(1,5-Naphthalenediyl)bis(1,1-diallylurea) 569
$C_{24}H_{28}N_4O_4$
2,4-Bis(3-benzamidopropylamino)-3-oxo-1-cyclobutenylium-1-olate 1095
Carbobenzyloxynorvalyl-tryptophanamide 883
N,N'-(1,5-Naphthalenediyl)bis(*O*-(carbamoyl)cyclohexanone oxime) 1012
$C_{24}H_{28}N_4O_4S$
Carbobenzyloxy-L-methionyl-L-tryptophanamide 883
$C_{24}H_{28}N_6O_6$
Sebacic acid bis((3-nitro-benzylidene)hydrazide) 1063
$C_{24}H_{28}O_2$
2,6-Di-*tert*-butyl-4-hydroxy-4-phenyl-1(4*H*)-naphthalenone 300
$C_{24}H_{28}O_3$
2-(4,4-Dimethyl-1-(4-methoxyphenyl)-3-oxopentyl)-1-tetralone 284
$C_{24}H_{28}O_3Pb$
Tris(2-ethoxyphenyl)lead 1134
Tris(4-ethoxyphenyl)lead 1134
$C_{24}H_{28}O_4$
α,α'-Diethyl-4,4'-stilbenediol dipropionate 380
$C_{24}H_{28}Si$
Hexyltriphenylsilane 1115
Tris(phenethyl)silane 1115
$C_{24}H_{29}ClN_2O$
10-Undecenoic (α-(3-chlorophenyl)-benzylidene)hydrazide 494
10-Undecenoic (4-chloro-α-phenylbenzylidene)hydrazide 1026
$C_{24}H_{29}NO_4$
p,p'-Dimethoxylobelanine hydrochloride 307
$C_{24}H_{29}N_3O_6$
Carbobenzyloxyglycylglycyl-L-valine benzyl ester 662
Carbobenzyloxyvalylglycylglycine benzyl ester 661
$C_{24}H_{30}GeSi$
Triethyl(triphenylgermyl)silane 1115
$C_{24}H_{30}N_2O_3$
Ethyl 4-(4-(3-phenylpropyl)-piperidinocarboxamido)benzoate 707
$C_{24}H_{30}N_2O_3S$
N,N-Tetramethylene-4-(*m*-tolyl)-1-(*p*-tolylsulfonyl)-isonipecotamide 819
$C_{24}H_{30}N_2O_5$
Carbobenzyloxy-L-alanyl-L-isoleucine benzyl ester 662
Carbobenzyloxy-L-alanyl-L-leucine benzyl ester 662
Carbobenzyloxy-L-isoleucyl-L-alanine benzyl ester 662
Carbobenzyloxy-L-isoleucyl-L-phenylalanine methyl ester 657
Carbobenzyloxyleucylalanine benzyl ester 662
Carbobenzyloxy-L-norleucyl-L-phenylalanine methyl ester 657
Carbobenzyloxy-L-phenylalanyl-L-isoleucine methyl ester 657
Carbobenzyloxy-L-phenylalanyl-L-norleucine methyl ester 657
$C_{24}H_{30}N_2O_5S$
Carbobenzyloxy-*S*-benzyl-L-cysteinyl-L-leucine dicyclohexylamine salt 656
Carbobenzyloxy-*S*-benzyl-L-cysteinyl-L-valine methyl ester 667
$C_{24}H_{30}N_2O_6$
Carbobenzyloxy-L-isoleucyl-L-tyrosine methyl ester 644
Carbobenzyloxy-L-norleucyl-L-tyrosine methyl ester 645
Carbobenzyloxy-L-threonyl-L-valine benzyl ester 643
Carbobenzyloxy-L-tyrosyl-L-isoleucine methyl ester 645
Carbobenzyloxy-L-tyrosyl-L-leucine methyl ester 645
$C_{24}H_{30}N_3OP$
Tris(4-(dimethylamino)phenyl)-phosphine oxide 835
$C_{24}H_{30}N_3Sb$
Tris(4-(dimethylamino)phenyl)-antimony 1128
$C_{24}H_{30}N_4O_2$
Adipic bis(α,4-dimethylbenzylidene)hydrazide 1070
$C_{24}H_{30}N_4O_4$
Sebacic bis(4-hydroxybenzylidene)-hydrazide 1028
$C_{24}H_{30}N_4O_6$
N,N'-Bis(carbobenzyloxy)-L-lysylglycinamide 663
4,4'-Bis(ethoxycarbonylmethyl)-ureido)-3,3'-dimethylbiphenyl 577
$C_{24}H_{30}N_4O_7$
Diglycolic bis((4-ethoxy-3-methoxybenzylidene)hydrazide) 497
$C_{24}H_{30}N_4O_8$
N,N'-Bis(2-methoxy-5-nitrophenyl)-1,10-decanediamide 486
Succinic acid bis((3,4,5-trimethoxy-benzylidene)hydrazide) 1060
$C_{24}H_{30}N_6O_7$
Carbobenzyloxynitroarginylalanine benzyl ester 724
$C_{24}H_{30}N_8O_{12}$
N,N'-Dodecamethylenebis(2,4,6-trintroaniline) 273
$C_{24}H_{30}O$
2',4',6'-Triisopropylchalcone 304
$C_{24}H_{30}O_2$
1,6-Dimesityl-1,6-hexanedione 309
$C_{24}H_{30}O_5$
2-(4-*tert*-Octylphenoxy)ethyl phthalate 371
$C_{24}H_{30}O_{11}$
1-*O*-Mesitoyl-2,3,4,6-tetra-*O*-acetyl-α-D-glucoside 370
$C_{24}H_{30}Si_2$
4,4''-Bis(trimethylsilyl)-*p*-terphenyl 1122
1,1,1-Triethyl-2,2,2-triphenyldisilane 1115
$C_{24}H_{31}NO_6$
Methyl 3-nitro-4-(2-(4-*tert*-octylphenoxy)ethoxy)benzoate 369
$C_{24}H_{31}N_3OSi$
Tris(4-(dimethylamino)phenyl)-silanol 1108
$C_{24}H_{31}N_3O_4S$
Carbobenzyloxy-*S*-benzyl-L-cysteinyl-L-leucinamide 667
$C_{24}H_{31}N_3O_5$
N-(6-Aminohexanoyl)-L-phenylalanyl-L-tyrosine 411
$C_{24}H_{31}N_5O_5$
2-(3-(3-Nitrophenyl)ureido)acetic (3,5-di-*tert*-butyl-4-hydroxy-

$C_{24}H_{31}N_5O_5$ (Cont'd)
benzylidene)hydrazide 548
$C_{24}H_{32}$
1,1-Diphenyl-1-dodecene 164
$C_{24}H_{32}ClN_3O_7S$
1-Phenyl-1-(3-(2-(pyrrolidinocarbonyl)phenylsulfonamido)propyl)pyrrolidinium perchlorate 820
$C_{24}H_{32}Cl_2N_2O_3$
3-(4-Chlorophenylcarbamoyl)-1-(decyloxycarbonylmethyl)pyridinium chloride 947
$C_{24}H_{32}Cl_2N_4O_2$
1,1'-Hexamethylenebis(3-(4-chlorobenzyl)-3-methylurea) 539
$C_{24}H_{32}N_2O_2$
N,N'-Dibenzyl-*N,N'*-di-*tert*-butyloxamide 424
$C_{24}H_{32}N_4O_2$
1,1'-(1,5-Naphthalenediyl)bis(3-cyclohexylurea) 584
$C_{24}H_{32}N_4O_4$
2,4-Bis(3-(4-methoxyanilino)propylamino)-3-oxo-1-cyclobutenylium-1-olate 1043
$C_{24}H_{32}N_4O_6$
1,4-Bis-(3,4,5-trimethoxybenzylideneamino-piperazine 1051
$C_{24}H_{32}N_4O_7P_2$
Tetraanilinium pyrophosphate 219
$C_{24}H_{32}N_6O_2$
Adipic bis(4-(dimethylamino)benzylidenehydrazide) 500
$C_{24}H_{32}N_6O_4$
1,1'-Hexamethylenebis(3-(4-acetamidophenyl)urea) 569
$C_{24}H_{32}N_6O_6$
1,2,3,4,5,6-Hexakis-(2-cyanoethoxy)hexane 144
$C_{24}H_{32}O_2$
2,9-Bis(4-hydroxy-3-methylphenyl)hexahydro-*p*-cymene 195
4'-Phenoxydodecanophenone 292
2,3,7,8-Tetraisopropyldibenzo-*p*-dioxin 185
$C_{24}H_{33}BO_3$
Tris(1-ethynylcyclohexyl) borate 1125
$C_{24}H_{33}ClN_2O_3$
N-(3-Chloro-*o*-tolyl)-2-decyloxy-5-nitrobenzylamine 267
1-Decyloxycarbonylmethyl-3-(phenylcarbamoyl)-pyridinium chloride 947
$C_{24}H_{33}Cl_2N_3O_2$
3-(4-Chlorophenylcarbamoyl)-1-(decylcarbamoylmethyl)pyridinium chloride 948
$C_{24}H_{33}NO_4$
Methyl 3-amino-4-(2-(4-*tert*-octylphenoxy)ethoxy)benzoate 369
$C_{24}H_{34}ClN_3O_2$
1-Decylcarbamoylmethyl-3-(phenylcarbamoyl)-pyridinium chloride 947
$C_{24}H_{34}N_2O_2$
3-(2-(Diethylamino)ethylthio)-3(4(dimethylamino)phenyl)-4'methoxypropiophenone hydrochloride 291
$C_{24}H_{34}N_2O_5S$
2'-Dodecyloxy-3-nitrobenzenesulfonanilide 814
$C_{24}H_{34}N_4O_2$
1,1'-Hexamethylenebis(3-(4-methylbenzyl)urea) 572
$C_{24}H_{34}N_4O_5S$
4-Dodecanamido-3-nitrobenzenesulfonic 2-phenylhydrazide 818
$C_{24}H_{34}N_4O_6$
1,1'-Hexamethylenebis(3-(2,5-dimethoxyphenyl)urea) 551
$C_{24}H_{34}N_4S_2$
1,1'-(3,3'-Dimethylbiphenyl-4,4'-diyl)-bis(3-(2-methylpropyl)-2-thiourea) 749
$C_{24}H_{34}O_2$
1-Naphthyl myristate 381
2-Naphthyl myristate 381
$C_{24}H_{34}O_{10}$
Tetraethyl 2,2'-(2,5-dimethoxy-1,4-phenylenedimethylene)dimalonate 354
$C_{24}H_{35}BrSi$
3-Bromo-3'-(tributylsilyl)biphenyl 1104
$C_{24}H_{35}N$
N-Dodecyldiphenylamine 233
$C_{24}H_{35}NO_6S_2$
Diethylmethyl(2-(1-phenyl-2-(phenylacetoxy)ethylthio)ethyl)ammonium methyl sulfate 357
$C_{24}H_{36}ClN$
2-Chloro-4-pentadecylquinoline 995
$C_{24}H_{36}N_2O_3S$
2'-(Dodecyloxy)-metanilanilide 813
$C_{24}H_{36}N_4O_2$
3,3'-(1,5-Naphthalenediyl)bis(1,1-diisopropylurea) 569
$C_{24}H_{36}N_4O_2O_3$
3,3'-(1,5-Naphthalenediyl)bis(1,1-dipropylurea) 569
$C_{24}H_{36}OSn$
Tributyltin 2-phenylphenoxide 1133
$C_{24}H_{37}F_4NO_5S_2$
4-Hexadecylsulfonyl-*N*-(trifluoroacetyl)-metanilyl fluoride 772
$C_{24}H_{38}BrClO_2$
2-Bromo-5'-chloro-2'-(hexadecyloxy)-acetophenone 287
$C_{24}H_{38}N_2OS$
2-Amino-4-(4-methoxyphenyl)-5-tetradecylthiazole 875
$C_{24}H_{38}N_2O_2$
N,N'-Dicyclohexyl-1,3-adamantanedicarboxamide 105
$C_{24}H_{38}N_2O_4S$
4-(5-Oxo-3-pentadecyl-2-pyrazolin-1-yl)benzenesulfonic acid, sodium salt 795
$C_{24}H_{38}N_2O_5$
2-Hexadecylcarbamoyl-4-nitrobenzoic acid 470
$C_{24}H_{38}N_2S_2$
2-Methylthio-5-(4-pentadecylphenyl)-1,3,4-thiadiazole 878
$C_{24}H_{38}O_2$
Bis-(6-methyl-3-cyclohexenylmethoxy)-6-methyl-3-cyclohexenylmethane 18
$C_{24}H_{39}ClO$
4'-Chlorooctadecanophenone 289
$C_{24}H_{39}ClO_5S$
Methyl 3-chlorosulfonyl-4-(hexadecyloxy)benzoate 802
$C_{24}H_{39}Cl_2NS$
3,4-Dichloro-*N*-hexadecyl-*N*-methylthiobenzamide 742
$C_{24}H_{39}NO_2$
4'-Acetylpalmitanilide 432
$C_{24}H_{39}NO_4S$
Methyl 4-hexadecylthio-3-nitrobenzoate 372
$C_{24}H_{39}NO_6S$
Methyl 4-hexadecylsulfonyl-3-nitrobenzoate 779
$C_{24}H_{40}ClNOS$
4'-Chloro-2-mercaptostearanilide 402
$C_{24}H_{40}ClNO_3S$
2-Chloro-4'-hexadecylsulfonylacetanilide 772
$C_{24}H_{40}Cl_2S$
3,4-Dichlorophenyl 5,7,7-trimethyl-2-(1,3,3-trimethylbutyl)-octylsulfide 210
$C_{24}H_{40}FNO_5S$
3-Nitro-4-(octadecyloxy)-benzenesulfonyl fluoride 802
$C_{24}H_{40}FNO_5S_2$
3-Acetamido-4-hexadecylsulfonyl-benzenesulfonyl fluoride 803
$C_{24}H_{40}HgO_2$
Phenylmercuric stearate 1138
$C_{24}H_{40}N_2OS_2$
Methyl 3-(4-pentadecylbenzoyl)-dithiocarbazate 745
$C_{24}H_{40}N_2O_2$
N,N,N',N'-Tetraisobutylisophthalamide 475
$C_{24}H_{40}N_2O_2S$
4-(3-Hexadecyl-2-thioureido)-benzoic acid 744
$C_{24}H_{40}N_2O_2S_3$
N-Hexadecyl-2-methylthio-6-benzothiazolesulfonamide 914
$C_{24}H_{40}N_2O_3$
N-(3-Nitrophenyl)octadecanamide 426
$C_{24}H_{40}N_2O_4$
4-(*N*-Hexadecyl-*N*-methylamino)-3-nitrobenzoic acid 342
$C_{24}H_{40}N_2Si_2$
1,2-Bis(triethylsilyl)-1,2-diphenylhydrazine 1108
$C_{24}H_{40}N_4O_5$
Stearic *N*2-(2,4-dinitrophenyl)-hydrazide 499
$C_{24}H_{40}O$
Octadecanophenone 296
$C_{24}H_{40}O_2$
2'-(Hexadecyloxy)acetophenone 292
4'-(Hexadecyloxy)acetophenone 293
Phenyl stearate 380
$C_{24}H_{40}O_3$
4-Hexadecyloxy-*m*-anisaldehyde 323
2-(Hexadecyloxy)phenyl acetate 378
Methyl 3-(hexadecyloxy)benzoate 370
Methyl 4-(hexadecyloxy)benzoate 370
$C_{24}H_{40}O_3S$
4'-(Hexadecylsulfonyl)-acetophenone 780
$C_{24}H_{40}O_4$
2-(Hexadecyloxy)-*p*-anisic acid 341
4-Hexadecyloxy-*m*-anisic acid 340
$C_{24}H_{40}O_5S$
5-Acetyl-2-(hexadecyloxy)-benzenesulfonic acid, potassium salt 792
$C_{24}H_{41}ClN_2O$
4-Chloro-N^1-hexadecyl-N^1-methyl-3-aminobenzamide hydrochloride 450
$C_{24}H_{41}ClN_2O_3$
3-Carbamoyl-1-(hexadecyloxycarbonylmethyl)-pyridinium chloride 947
4-Carbamoyl-1-(hexadecyloxycarbonylmethyl)-pyridinium chloride 951
$C_{24}H_{41}ClN_2O_4S$
3-Nitro-4-(octadecylamino)-benzenesulfonyl chloride 803
$C_{24}H_{41}FO_3S$
4-(Octadecyloxy)benzenesulfonyl fluoride 802
$C_{24}H_{41}NO$
2'-Ethylhexadecananilide 438
4'-Ethylhexadecananilide 439
2',3'-Hexadecanoxylidide 439
2',4'-Hexadecanoxylidide 440
2',6'-Hexadecanoxylidide 439
Octadecananilide 437
$C_{24}H_{41}NOS$
4'-(Hexadecylthio)-acetanilide 421
$C_{24}H_{41}NO_2$
Ethyl 4-((*N*-dodecyl)anilino)butyrate. 359
2'-(Hexadecyloxy)acetanilide 417
3'-(Hexadecyloxy)acetanilide 418
4'-(Hexadecyloxy)acetanilide 419
2'-(Hexadecyloxy)acetophenone oxime 1009
$C_{24}H_{41}NO_3$
Methyl 3-amino-4-hexadecyloxybenzoate 369
2-Nitrophenyl octadecyl ether 268
3-Nitrophenyl octadecyl ether 268
4-Nitrophenyl octadecyl ether 269
2-Nitrophenyl 2-(1,3,3-trimethylbutyl)-5,7,7-trimethyloctyl ether 268
$C_{24}H_{41}NO_3S$
4'-(Hexadecylsulfonyl)acetanilide 782
$C_{24}H_{41}NO_4S$
Methyl 3-amino-4-(hexadecylsulfonyl)-benzoate 776
$C_{24}H_{41}NO_6S$
3-Nitro-4-octadecyloxybenzenesulfonic acid. 788
$C_{24}H_{41}NS$
N-Hexadecyl-*N*-methylthiobenzamide 742
$C_{24}H_{41}N_3O$
Stearic acid (4-pyridylmethylene)-hydrazide 502
$C_{24}H_{41}N_3O_2S_2$
1-Hexadecanesulfonic (3-methyl-2-benzothiazolinylidene)hydrazide 821
$C_{24}H_{41}N_3O_5S_3$
1-Hexadecanesulfonic (3-methyl-5-sulfo-2-benzothiazolinylidene)-hydrazide, pyridine salt 820
$C_{24}H_{42}ClN_3O_2$
3-Diethylcarbamoyl-1-(dodecylcarbamoylmethyl)-pyridinium chloride 947
$C_{24}H_{42}N_2$
1-(1-Methylheptyl)-5-(1-methylheptylamino)indoline dihydrochloride 249
$C_{24}H_{42}N_2O$
3-Amino-*N*-hexadecyl-*N*-methylbenzamide hydrochloride 468
1-(1-Azacyclohept-2-en-1-yl)-2-morpholino-1-cyclotetradecene 39
Palmitic N^2-(2,6-xylyl)hydrazide 503
Stearic N^2-phenylhydrazide 503
$C_{24}H_{42}N_2O_3S$
N^4-Acetyl-N^1-hexadecyl-sulfanilamide 825
$C_{24}H_{42}N_2O_4S$
N-Hexadecyl-*N*-methyl-4-methylsulfonyl-2-nitroaniline 777
$C_{24}H_{42}O$
Octadecyl phenyl ether 180
3-Phenyl-3-octadecanol 207
$C_{24}H_{42}OS$
2-(Hexadecylthio)-1-phenylethanol 210
$C_{24}H_{42}O_4S$
2,6-Dimethyl-4-hexadecyloxy-benzenesulfonic acid 788
$C_{24}H_{43}BrN_2O$
4-Carbamoyl-1-octadecylpyridinium bromide 952
$C_{24}H_{43}N$
4-Hexadecyl-α-methylbenzylamine 241
N-Hexadecyl-3,5-xylidine hydrochloride 220
$C_{24}H_{43}NO$
2-Hexadecyloxy-α-methylbenzylamine 240
4-Hexadecyloxy-α-methylbenzylamine 240
3-(Octadecyloxy)aniline 217
4-(Octadecyloxy)aniline 218
2-(2-(1,3,3-Trimethylbutyl)-5,7,7-trimethyloctyloxy)aniline 217
$C_{24}H_{43}NO_2$
N-(3,5-Dimethoxyphenyl)-hexadecylamine 217
$C_{24}H_{43}NS$
2-(Octadecylthio)aniline hydrochloride 218
$C_{24}H_{43}N_3O_5S_2$
N^3-Acetyl-4-(hexadecylsulfonyl)-metanilic hydrazide 824
$C_{24}H_{44}N_2$
N,N'-Bis(1-methyloctyl)-1,4-phenylenediamine dihydrochloride. 222
$C_{24}H_{44}N_2O_2S$
N^1-Hexadecyl-N^1-methyl-4-methylsulfonyl-1,2-phenylenediamine 776
N^1-Octadecylsulfanilamide hydrochloride 817
$C_{24}H_{44}N_2O_2S_2$
N^1,N^1-Dimethyl-4-(hexadecylthio)-metanilamide 815
$C_{24}H_{44}N_2O_3S$
N^1,N^1-Dimethyl-4-hexadecyloxy-metanilamide 813
$C_{24}H_{44}N_2O_4S_2$
N^1-Hexadecyl-N^1-methyl-4-methylsulfonylmetanilamide 816

$C_{24}H_{44}N_2S$
(2-(Octadecylthio)phenyl)-hydrazine monohydrochloride ... 235
$C_{24}H_{44}O_2$
2-Decyl-1,3-cyclotetradecanedione ... 54
$C_{24}H_{44}O_8$
Dodecyloxyethoxyethoxy-ethoxyethyl maleate ... 94
$C_{24}H_{44}Si$
Tetracyclohexylsilane ... 1111
$C_{24}H_{44}Sn$
Tetracyclohexyltin ... 1132
$C_{24}H_{45}N$
2,5-Dimethyl-1-octadecylpyrrole ... 842
$C_{24}H_{45}NO_3$
cis-*N*-(Hexadecyl)-hexahydrophthalamic acid ... 122
$C_{24}H_{46}O_2$
Ethyl 13-docosenoate ... 80
$C_{24}H_{46}O_3$
13-Oxotetracosanoic acid ... 64
7-Oxotetracosanoic acid ... 64
3-Oxotetracosanoic acid, sodium salt ... 69
$C_{24}H_{46}O_4$
Dimethyl docosanedioate ... 76
$C_{24}H_{47}N_3O$
1-(2-(Diethylamino)-ethyl)-3-pentadecyl-2-pyrazolin-5-one ... 133
$C_{24}H_{48}N_2O$
N-(2-Diethylaminoethyl)-oleamide ... 104
1-Octadecylisonipecotamide ... 108
1-Octadecylnipecotamide ... 108
$C_{24}H_{48}O_5S$
Ethyl 2-sulfodocosanoate, sodium salt ... 156
$C_{24}H_{49}NO$
N-Dodecyldodecanamide ... 106
$C_{24}H_{50}Cl_2SnN$
Dichlorodidodecyltin ... 1129
$C_{24}H_{50}Hg$
Didodecylmercury ... 1137
$C_{24}H_{50}NO_4P$
Bis(2-ethylhexyl) (*N*,*N*-diisopropylcarbamoylmethyl)-phosphonate ... 159
$C_{24}H_{50}N_2O$
N-(2-(Diethylamino)ethyl)-stearamide ... 104
$C_{24}H_{50}N_4O_2$
3,3′-Hexamethylenebis(1,1-dibutylurea) ... 129
$C_{24}H_{50}O$
Dodecyl ether ... 17
$C_{24}H_{50}OS$
Dodecyl sulfoxide ... 154
$C_{24}H_{50}O_2Sn$
Lauroyloxytributyltin ... 1131
$C_{24}H_{50}O_4S$
Didodecyl sulfate ... 157
$C_{24}H_{50}S$
tert.-Dodecyl sulfide ... 26
$C_{24}H_{51}BO_3$
2-Octyl borate ... 1124
Trioctyl borate ... 1124
Tris(2-ethylhexyl) borate ... 1124
$C_{24}H_{51}N$
Dibutylhexadecylamine ... 29
Didodecylamine ... 28
$C_{24}H_{51}O_4P$
Didodecyl phosphate ... 160
$C_{24}H_{51}P$
Trioctylphosphine ... 158
$C_{24}H_{52}Ge$
Octadecyltriethylgermane ... 1098
Tetrahexylgermane ... 1098
$C_{24}H_{52}NO_3P$
Dilauryl amidophosphate ... 160
$C_{24}H_{52}O_4Si$
Tetrahexyl orthosilicate ... 1098
Tetrakis(2-ethyl-1-butyl) orthosilicate ... 1098
$C_{24}H_{52}Si$
Trioctylsilane ... 1098
$C_{24}H_{54}Ge_2O$
Tributylgermyl oxide ... 1098
$C_{24}H_{54}O_4SSn_2$
Bis-(tributylstannyl) sulfate ... 1131
$C_{24}H_{54}Si_6$
1,2,4,5-Hexatetraene-3,4-diyl-1,6-diylidenehexakis(trimethylsilane) ... 1098
$C_{24}H_{60}Ag_4I_4P_4$
Triethylphosphine-silver iodide complex ... 1141
$C_{24}H_{64}Si_8$
1,1,1,8,8,8-Hexakis(trimethylsilyl)-1,8-disilaoctane ... 1098
$C_{25}H_{14}O_3$
2-Methoxydibenzo(b,def)chrysene-7,14-dione ... 311
3-Methoxydibenzo(b,def)chrysene-7,14-dione ... 311
$C_{25}H_{16}Cl_6O_5$
Hexachloro-5-(6-hydroxy-3-oxo-3*H*-xanthen-9-yl)-tricycloundec-9-ene-4-carboxylic acid ... 326
$C_{25}H_{16}Cl_6O_7$
Hexachloro-5-(3-oxo-4,5,6-trihydroxy-3*H*-xanthen-9-yl)-tricycloundec-9-ene-4-carboxylic acid ... 326
$C_{25}H_{16}N_8O_{12}$
4,4′-Methylenebis(*N*-(2,4,6-trinitrophenyl)aniline) ... 273
$C_{25}H_{16}O_3$
3-(4-Methoxybenzoyl)benzanthrone ... 311
$C_{25}H_{16}O_6$
2,2′-(1,3-Indenediformyl)-dibenzoic acid ... 344
$C_{25}H_{16}S_2$
2,2′-Methylenebis(dibenzo-thiophene) ... 899
$C_{25}H_{17}BF_4O$
1,3-Diphenylnaphtho(2,1-b)-pyrylium tetrafluoroborate ... 1002
2,4-Diphenylnaphtho(1,2-b)-pyrylium tetrafluoroborate ... 1002
$C_{25}H_{17}NO$
6-(4-Dibenzofuryl)-5,6-dihydrophenanthridine ... 896
$C_{25}H_{18}$
9,9-Diphenylfluorene ... 166
$C_{25}H_{18}BF_4Cl_3N_6S_3$
Tris(6-chloro-3-methyl-2-benzothiazolinylideneamino)-carbenium tetrafluoroborate ... 1027
$C_{25}H_{18}BrClNO_2S$
4-Bromo-*N*-(3-(3-chloro-2-methylphenyl)-4-phenyl-4-thiazolin-2ylidene)cinnamamide ... 731
$C_{25}H_{18}F_4OPb$
(4-Methoxy-2,3,5,6-tetrafluorophenyl)triphenyllead ... 1133
$C_{25}H_{18}F_4OSn$
(4-Methoxy-2,3,5,6-tetrafluorophenyl)triphenyltin ... 1130
$C_{25}H_{18}N_2O_2$
Benzil, azine with 2-hydroxy-1-naphthaldehyde ... 1037
$C_{25}H_{18}N_4O_2S$
3-Methyl-2-benzothiazolinone (1,2-dihydro-2-oxo-3-(phenylcarbamoyl)-1-naphthylidenehydrazone ... 1061
$C_{25}H_{18}O$
4-(4-Biphenylyl)benzophenone ... 313
Bis(biphenylyl) ketone ... 314
9,9-Diphenylxanthene ... 185
$C_{25}H_{18}O_2$
2,4,6-Triphenylbenzoic acid ... 345
$C_{25}H_{18}O_6$
7,8-Diacetoxy-2,3-diphenyl-4*H*-1-benzopyran-4-one ... 378
$C_{25}H_{19}ClN_2O_3S$
3-(4-Chlorophenylcarbamoyl)-2-naphthyl *N*-(4-(methylthio)phenyl)-carbamate ... 623
$C_{25}H_{19}ClN_2O_4$
4-(5-Hydroxy-1-naphthyliminomethyl)-2-methoxyphenyl *N*-(3-chlorophenyl)-carbamate ... 1022
3-(*N*-Phenylcarbamoyl)-2-naphthyl *N*-(5-chloro-2-methoxyphenyl)-carbamate ... 620
$C_{25}H_{19}ClO_5$
2,4-Diphenyl-6-styrylpyrylium perchlorate ... 936
$C_{25}H_{19}Cl_2OP$
(4-Chlorophenoxymethyl)-triphenylphosphonium chloride ... 830
$C_{25}H_{19}F_3Si$
(3-(Trifluoromethyl)phenyl)-triphenylsilane ... 1100
$C_{25}H_{19}NO_2$
3,5-Diphenyl-4′-methyl-2-nitrobiphenyl ... 278
$C_{25}H_{19}NSSi$
2-(Triphenylsilyl)benzothiazole ... 1111
$C_{25}H_{19}NSSn$
2-(Triphenylstannyl)benzothiazole ... 1132
$C_{25}H_{20}$
Tetraphenylmethane ... 163
$C_{25}H_{20}ClN$
2-Chloro-4-tritylaniline ... 224
$C_{25}H_{20}Cl_3P$
2,4-Dichlorobenzyltriphenyl-phosphonium chloride ... 832
$C_{25}H_{20}N_2$
9-(4-(2,5-Dimethyl-1-pyrrolyl)phenyl)-acridine ... 996
$C_{25}H_{20}N_2O$
1,1,3,3-Tetraphenylurea ... 586
$C_{25}H_{20}N_2O_3S$
3-(*N*-Phenylcarbamoyl)-2-naphthyl *N*-(4-(methylthio)phenyl)carbamate ... 667
$C_{25}H_{20}N_2O_5S_2$
Mordant Dye - Alizarin Blue R (C.I. 52060) ... 1094
$C_{25}H_{20}N_4O_4$
4,4′-Methylenebis(*N*-(2-pyridyloxycarbonyl)aniline) ... 959
4,4′-Methylenebis(*N*-(3-pyridyloxycarbonyl)aniline) ... 960
$C_{25}H_{20}O$
α,α-Diphenyl-4-biphenylmethanol ... 204
Methyl triphenylphenyl ether ... 180
Phenyl trityl ether ... 186
$C_{25}H_{20}OSSi$
S-(Triphenylsilyl)thiobenzoate ... 1109
$C_{25}H_{20}OSi$
Benzoyltriphenylsilane ... 1109
$C_{25}H_{20}O_2$
3,4-Diphenyl-2-methyl-1-naphthyl acetate ... 380
2-Phenoxytriphenylmethanol ... 201
$C_{25}H_{20}O_2Si$
4-(Triphenylsilyl)benzoic acid ... 1109
$C_{25}H_{20}O_2Sn$
Triphenylstannyl benzoate ... 1135
4-(Triphenylstannyl)benzoic acid ... 1131
$C_{25}H_{20}O_6$
3-Benzoyl-2-(3,4,5-trimethoxyphenyl)-4*H*-1-benzopyran-4-one ... 319
$C_{25}H_{20}Si$
9-(2-Biphenylyl)-9-methyl-9*H*-9-silafluorene ... 1120
Diphenyl(9-fluorenyl)silane ... 1122
$C_{25}H_{21}BF_4O_3$
2,4-Bis(4-methoxyphenyl)-6-phenylpyrylium tetrafluoroborate ... 934
$C_{25}H_{21}BrFP$
(4-Fluorobenzyl)-triphenylphosphonium bromide ... 830
$C_{25}H_{21}BrPb$
(4-Bromobenzyl)triphenyllead ... 1134
$C_{25}H_{21}BrSi$
(α-Bromobenzyl)triphenylsilane ... 1115
(4-(Bromomethyl)phenyl)-triphenylsilane ... 1100
$C_{25}H_{21}ClFP$
(4-Fluorobenzyl)-triphenylphosphonium chloride ... 830
$C_{25}H_{21}ClSi$
(2-Chlorobenzyl)triphenylsilane ... 1103
(4-Chlorobenzyl)triphenylsilane ... 1103
Chlorodiphenyl(diphenylmethyl)-silane ... 1121
$C_{25}H_{21}Cl_2P$
(3-Chlorobenzyl)-triphenylphosphonium chloride ... 831
$C_{25}H_{21}Cl_4FeO$
2,6-Bis(*p*-tolyl)-4-phenylpyrylium tetrachloroferrate ... 936
$C_{25}H_{21}FSi$
(4-Fluorobenzyl)triphenylsilane ... 1103
$C_{25}H_{21}N$
3,5-Diphenyl-4′-methyl-2-biphenylamine ... 230
N-(2-Phenylbenzhydryl)aniline ... 242
N-Tritylaniline ... 232
$C_{25}H_{21}NO$
2-Anilinotriphenylmethanol ... 226
10-(4-(Dimethylamino)-cinnamylidene)-9-anthrone ... 312
4-(4-(Dimethylamino)-α-phenylbenzylidene)-1(4*H*)-naphthalenone ... 302
10-(1-Methyl-1,2,3,4-tetrahydro-6-quinolylmethylene)-9-anthrone ... 312
$C_{25}H_{21}NO_2$
12β-Benzyl-2,3-diformyl-5,6,8,12b-tetrahydroisoindolo(1,2-a)-isoquinoline ... 324
$C_{25}H_{21}N_3O$
5′-Methyl-2′-((*p*-tolyl)azo)-1-naphthalene carboxanilide ... 1088
$C_{25}H_{21}N_3O_2$
1-(4-Anilinophenyl)-3-(4-phenoxyphenyl)urea ... 553
N-(4-Methoxybenzylidene)-*N*-(4-(4-methoxyphenylazo)-1-naphthyl)-amine ... 1086
N-(4-Methoxybenzylidene)-*N*-(1-(4-methoxyphenylazo)-2-naphthyl)-amine tetrafluoroborate ... 1086
$C_{25}H_{21}N_3O_6$
*N*α-Carbobenzyloxy-DL-tryptophan 4-nitrophenyl ester ... 888
*N*α-Carbobenzyloxy-L-tryptophan 2-nitrophenyl ester ... 888
$C_{25}H_{21}O_2P$
2,6-Bsi(4-methoxyphenyl)-4-phenylphosphorin ... 937
$C_{25}H_{22}AsBr$
Benzyltriphenylarsonium bromide ... 1127
$C_{25}H_{22}As_2$
Methylenebis-(diphenylarsine) ... 1127
$C_{25}H_{22}BF_4NOS$
2-(3-Ethoxy-1*H*-phenalen-1-ylidenemethyl)-3-ethylbenzo-thiazolium tetrafluoroborate ... 913
$C_{25}H_{22}BrP$
Benzyltriphenylphosphonium bromide ... 832
$C_{25}H_{22}ClPS$
(Phenylthiomethyl)-triphenylphosphonium chloride ... 830
$C_{25}H_{22}Ge$
Benzyltriphenylgermane ... 1115
$C_{25}H_{22}NO_3P$
Diphenyl(α-anilinobenzyl)-phosphonate ... 834
$C_{25}H_{22}NP$
N-(*o*-Tolyl)-*p*,*p*,*p*-triphenyl-phosphine imide ... 836
N-(*m*-Tolyl)-*p*,*p*,*p*-triphenylphosphine imide ... 836
N-(*p*-Tolyl)-*p*,*p*,*p*-triphenylphosphine imide ... 836
$C_{25}H_{22}N_2O$
1,1-Dibenzyl-3-(1-naphthyl)urea ... 572
$C_{25}H_{22}N_2O_4$
Dibenzyl 2,2′-methylenebis(5-pyrrolecarboxylate) ... 838
$C_{25}H_{22}N_2O_4S$
5-(3,4-Dimethoxybenzylidene)-2-(2-methoxyphenylimino)-3-phenyl-4-thiazolidinone ... 737
$C_{25}H_{22}N_2O_6$
Bis(2-acetylphenyl) *N*,*N*′-(4-methyl-1,3-phenylene)biscarbamate ... 666
4,4′-Methylenebis(furfuryl *N*-phenylcarbamate) ... 847
$C_{25}H_{22}N_2O_8$
Bis(3-acetoxyphenyl) *N*,*N*′-(4-methyl-1,3-phenylene)-biscarbamate ... 666
Bis(4-(methoxycarbonyl)phenyl) *N*,*N*′-(4-methyl-1,3-phenylene)-

$C_{25}H_{22}N_2O_8$ (Cont'd)
biscarbamate 670
2,2-Bis-(phthalimidomethyl)-1,3-propanediol diacetate 717
$C_{25}H_{22}N_4O$
1,1,5,5-Tetraphenylcarbazide 574
$C_{25}H_{22}N_6O_2$
4,4'-Bis(3-(3-pyridyl)ureido)-diphenylmethane 950
$C_{25}H_{22}N_6O_4S_2$
1,1'-(4-Methyl-1,3-phenylene)bis(3-(6-methoxy-2-benzothiazolyl)urea) 913
$C_{25}H_{22}O$
1,2-Dimethyl-3,4,5-triphenyl-2,4-cyclopentadienol 205
$C_{25}H_{22}OPb$
(2-(Hydroxymethyl)phenyl)-triphenyllead 1134
(2-(Hydroxymethyl)phenyl)-triphenyllead 1134
(3-(Hydroxymethyl)phenyl)-triphenyllead 1134
(4-(Hydroxymethyl)phenyl)-triphenyllead 1134
(2-Methoxyphenyl)triphenyllead 1134
(4-Methoxyphenyl)triphenyllead 1134
$C_{25}H_{22}OSi$
(Benzyloxy)triphenylsilane 1107
(4-Methoxyphenyl)triphenylsilane 1106
(4-Phenoxyphenyl)(phenyl)(*o*-tolyl)-silane 1106
4-(Triphenylsilyl)benzyl alcohol 1105
$C_{25}H_{22}OSn$
(4-Methoxyphenyl)triphenyltin 1130
2-(Triphenylstannyl)benzyl alcohol 1130
4-(Triphenylstannyl)benzyl alcohol 1130
$C_{25}H_{22}O_2$
9-Anthracenylmethyl 2,4,6-trimethylbenzoate 376
Ethyl 9,10-dihydro-12-phenyl-9,10-ethanoanthracene-11-carboxylate 361
$C_{25}H_{22}O_2Si$
(4-Methoxyphenyl)(4-phenoxyphenyl)phenylsilane 1106
$C_{25}H_{22}Pb$
Benzyltriphenyllead 1136
(*p*-Tolyl)triphenyllead 1136
$C_{25}H_{22}Si$
Benzyltriphenylsilane 1115
Diphenyl(diphenylmethyl)silane 1121
(*o*-Tolyl)triphenylsilane 1119
(*m*-Tolyl)triphenylsilane 1115
(*p*-Tolyl)triphenylsilane 1120
$C_{25}H_{22}Sn$
(*o*-Tolyl)triphenyltin 1132
(*p*-Tolyl)triphenyltin 1132
$C_{25}H_{23}BrCl_2N_2O_4S_2$
N,N'(4-Bromo-2-methyl-*m*-phenylene)bis(2-(4-chlorophenylthio)ethyl carbamate) 623
$C_{25}H_{23}ClN_2O_3$
Benzyl *N*-(3-(5-chloro-2-oxo-3-phenyl-3-indolinyl)propyl)-carbamate 690
$C_{25}H_{23}NO$
2-(1-Benzyl-3,3-dimethyl-2-indolinylidene)acetophenone 302
$C_{25}H_{23}NO_7$
N,O-Bis(benzyloxycarbonyl)-L-tyrosine 659
N,O-Bis(carbobenzyloxy)-DL-tyrosine 671
$C_{25}H_{23}N_3O_2$
4-(4-Benzyloxybenzylideneamino)-2,3-dimethyl-1-phenyl-3-pyrazolin-5-one 1050
$C_{25}H_{24}ClNO_2$
12β-(4-Chlorobenzyl)2,3-dimethoxy-5,6,8,12b-tetrahydro-isoindolo(1,2-a)isoquinoline 244
$C_{25}H_{24}Cl_2N_2O_4$
Bis(4-chloro-3,5-dimethylphenyl) *N,N'*-(4-methyl-1,3-phenylene)-biscarbamate 626
$C_{25}H_{24}N_2Si$
1,2-Diphenyl-1-((diphenyl)-methylsilyl)hydrazine 1108
$C_{25}H_{24}N_4O_4$
4,4'-Methylenebis(1-furfuryl-3-phenylurea) 854
$C_{25}H_{24}N_4O_6$
N,N'-Bis(4-acetamidophenoxycarbonyl)-2,4-toluenediamine 669
$C_{25}H_{24}O_2$
2,2'-Diethoxy-1,1'-methylenedinaphthalene 188
1-(2,4-Dimethylbenzoyl)-2-(2,4,6-trimethylbenzoyl)benzene 313
$C_{25}H_{24}O_3S$
9,10-Dihydro-9,10-ethanoanthracene-11-ethanol *p*-tosylate 799
9,10-Dihydro-11-methyl-9,10-ethanoanthracene-11-methanol *p*-toluenesulfonate 799
9,10-Dihydro-12-methyl-9,10-ethanoanthracene-11-methanol *p*-tosylate 799
9,10-Dihydro-α-methyl-9,10-ethanoanthracene-11-methanol *p*-tosylate 799
$C_{25}H_{25}ClN_2O_2S$
4'-Chloro-2'-(3,4-dihydro-4,4-dimethyl-1-isoquinolinyl) *N*,4-dimethylbenzenesulfonanilide hydrochloride 807
$C_{25}H_{25}ClN_2O_3$
Basic Red 3 (C.I. 45210) 1043
$C_{25}H_{25}NO_2$
12b-Benzyl-2,3-dimethoxy-5,6,8,12b-tetrahydroisoindolo(1,2-a)-isoquinoline 246
$C_{25}H_{25}NO_3$
Methyl *N*-(4-*tert*-butylbenzoyl)-*N*-phenylanthranilate 476
$C_{25}H_{25}NO_4$
2,6-Bis(2,3-dimethoxystyryl)pyridine. 936
$C_{25}H_{25}NO_5$
Diethyl 5-benzamido-1,2,3,4-tetrahydro-1,4-ethenonaphthalene-2,3-dicarboxylate 474
$C_{25}H_{25}N_3O_3$
(8-(4-Hydroxy-1-naphthylazo)-2-naphthyl)trimethylammonium acetate 1084
$C_{25}H_{25}O_4P$
Diethyl 2-(triphenylphosphoranylidene)-malonate 360
$C_{25}H_{26}ClNO_3$
12b-(4-Chlorobenzyl)-2,3-dimethoxy-5,6,9,12,12a,12b-hexahydro-isoindolo(1,2-a)isoquinolin-8(8ah)-one 694
$C_{25}H_{26}N_2O_2$
2-Benzyl-*N,N'*-dibenzoyl-1,4-butanediamine 471
N-(2-*tert*-Butyl-1,2-diphenyl-2-nitrosoethyl)benzamide, mixture of isomers 471
$C_{25}H_{26}N_2O_4$
Bis(4-methylbenzyl) *N,N'*-(4-methyl-1,3-phenylene)biscarbamate 657
Bis(2,3-xylyl) *N,N'*-(4-methyl-1,3-phenylene)biscarbamate 670
$C_{25}H_{26}N_2O_6$
Bis(2-phenoxyethyl) *N,N'*-(4-methyl-1,3-phenylene)biscarbamate 649
$C_{25}H_{26}N_4O_3$
1,3-Bis(2-(1,2-dihydro-2-methyl-1-oxo-3-isoquinolinyl)ethyl)urea 703
4,8-Diphenyl-1,2,4,5,6,7-hexahydro-6-morpholino-3*H*,8*H*-2a,4,8a-triazaacenaphthylene-3,5-dione 709
$C_{25}H_{26}O$
1,2,2-Triphenyl-1-heptanone 297
$C_{25}H_{27}FO_4$
Ro-hexahydro-1-methylbenzo(c)-phenanthrene-5,8-diol dipropionate 361
$C_{25}H_{27}NO_2$
N-Trityl-L-leucine 349
$C_{25}H_{27}NO_3$
N-Trityl-DL-serine isopropyl ester 354
$C_{25}H_{27}N_3O_7$
Diethyl 6-amino-1-benzamido-4-(4-methoxyphenyl)-2-oxo-1,4,5,6-tetrahydro-3,5-pyridine-dicarboxylate 698
$C_{25}H_{28}FNO_4$
Methyl 3,5-dihydroxy-7-(3-(4-fluorophenyl)-1-isopropyl-2-indolyl)-6-heptenoate 881
Methyl 7-(3-(4-fluorophenyl)-1-isopropyl-2-indolinyl)-5-hydroxy-3-oxo-6-heptenoate 363
$C_{25}H_{28}N_2O$
1,1-Dibenzyl-3-(2,6-diethylphenyl)urea 573
$C_{25}H_{28}N_4O_2$
Glutaric acid (5-indanylmethylene)-hydrazide 1075
3,3'-(4-Methyl-1,3-phenylene)bis(1-benzyl-1-methylurea) 563
3,3'-(4-Methyl-1,3-phenylene)bis(1-ethyl-1-phenylurea) 566
1,1'-(4-Methyl-1,3-phenylene)bis(3-phenethylurea) 562
1,1'-(4-Methyl-1,3-phenylene)bis(3-(2,4-xylyl)urea) 566
*N*α-Trityl-L-arginine 725
*N*α-Trityl-L-arginine monohydrate 725
$C_{25}H_{28}N_4O_5$
Methyl 5-(5-(dicyanovinyl)-3-ethyl-4-methylpyrrol-2-yl-methyl)-3-carbethoxy-4-methyl-γ-oxo-2-pyrrolebutyrate 840
$C_{25}H_{28}O_3$
Tri-*p*-anisylmethane 179
$C_{25}H_{29}NO$
1-(α-(Benzylamino)benzyl)-2,2-dimethyl-1-phenyl-1-propanol hydrochloride 237
1-(4-Biphenylyl)-3-(dimethylaminomethyl)-2-phenyl-2-butanol hydrochloride 238
DL-1,2-Diphenyl-3-methyl-4-(*N*-methylbenzylamino)-2-butanol hydrochloride 237
$C_{25}H_{29}NO_5S_2$
2'-(*N*-Methyl-*N*-phenylsulfamoyl)-3,3',5,5'tetramethyl-2-bibenzyl-sulfonic acid sodium salt 795
$C_{25}H_{29}N_3O_5$
Carbobenzyloxynorvalyltryptophan methyl ester 883
Carbobenzyloxytryptophylvaline methyl ester 883
$C_{25}H_{29}N_3O_5S$
Carbobenzyloxymethionyl-tryptophan methyl ester 883
$C_{25}H_{29}N_3O_6$
Carbobenzyloxy-L-alanyl-L-prolylglycine benzyl ester 700
Carbobenzyloxyglycylprolylalanine benzyl ester 700
$C_{25}H_{30}N_2$
4,4'-(Phenylmethylene)bis(*N,N*,3-trimethylaniline) 233
$C_{25}H_{30}N_2O_2$
4,4'-Bis(dimethylamino)-4''-methoxy-2''-methyltritylalcohol 225
6,6-Dimethyl-2-oxo-*N*-phenethyl-4-(phenethylamino)-3-cyclohexene-1-carboxamide 443
$C_{25}H_{30}N_2O_4$
Bis(5-norbornen-2-ylmethyl) *N,N'*-(4-methyl-1,3-phenylene)-biscarbamate 670
$C_{25}H_{30}N_2O_5$
Carbobenzyloxyprolylvaline benzyl ester 708
$C_{25}H_{30}N_4O_4$
Carbobenzyloxy-L-isoleucyl-L-tryptophan 889
Carbobenzyloxy-L-isoleucyl-L-tryptophanamide 886
Carbobenzyloxy-L-tryptophyl-L-leucinamide 883
$C_{25}H_{30}N_4O_7$
Ethyl 1-(2-acetamidoethyl)-5-(1,3-diacetoxy-2-imidazolidinylidene)-2,6-dioxo-4-phenyl-3-piperidine carboxylate 722
$C_{25}H_{30}N_6O_2$
N,N'-Methylenebis(4-(methylamino)antipyrine) 708
$C_{25}H_{30}N_6O_8$
4,4'-Trimethylenebis(1-(2,4-dinitrophenyl)piperidine) 273
$C_{25}H_{30}O_2$
5,6,6a,6b,7,8,12b,13,14,14a-Decahydro-3-ethoxy-10-methoxypicene 187
$C_{25}H_{30}Pb$
Triethyl(triphenylmethyl)lead 1136
$C_{25}H_{31}NPb$
N-(3-(Triphenylplumbyl)propyl)-diethylamine 1134
N-(3-(Triphenylplumbyl)propyl)-diethylamine methiodide 1134
$C_{25}H_{31}N_3O_6$
Carbobenzyloxyglycylglycyl-L-isoleucine benzyl ester 663
Carbobenzyloxyglycylglycyl-leucine benzyl ester 663
Carbobenzyloxyglycylleucyl-glycine benzyl ester 663
$C_{25}H_{31}N_3O_7$
N,N'-Bis(carbobenzyloxy)-L-lysylglycine methyl ester 663
Carbobenzyloxy-DL-norvalylglycyl-DL-tyrosine methyl ester 645
$C_{25}H_{31}N_3O_8$
N,N'-Bis(carbobenzyloxy)-L-lysyl-L-serine 645
*N*α-*tert*-Butoxycarbonyl-*N*ε-carbobenzyloxy-L-lysine 4-nitrophenyl ester 674
$C_{25}H_{32}N_2O_5S$
Carbobenzyloxy-*S*-benzyl-L-cysteinyl-L-leucine methyl ester 656
$C_{25}H_{32}N_2O_6$
Carbobenzyloxy-L-threonyl-L-leucine benzyl ester 643
$C_{25}H_{32}N_4O_2$
4,4'-Methylenebis(1,1-pentamethylene-3-phenylurea) 712
$C_{25}H_{32}N_4O_4$
4,4'-Trimethylenebis(1-(2-nitrophenyl)piperidine) 274
$C_{25}H_{33}ClN_2O$
Lauric (4-chloro-α-phenylbenzylidene)hydrazide 494
$C_{25}H_{33}NO$
2',4',6'-Trimethylhexadecananilide 440
$C_{25}H_{33}NO_4$
3-(2-(2,4-Di-*tert*-pentylphenoxy)-acetamido)benzoic acid 414
$C_{25}H_{34}Cl_2N_4O_4$
1,1'-Trimethylenebis(1-(4-nitrobenzyl)pyrrolidinium chloride). 275
$C_{25}H_{34}N_2O_3$
Tilorone 311
$C_{25}H_{34}O_3$
4-Dodecyloxy-2-hydroxybenzophenone 310
$C_{25}H_{35}ClN_2O_3$
3-(*N*-Phenylcarbamoyl)-1-(undecyloxycarbonylmethyl)-pyridinium chloride 947
$C_{25}H_{35}ClO_2S$
4-Chlorophenyl 4-(*tert*-dodecyl)-benzyl sulfone 786
$C_{25}H_{35}Cl_2NO$
2,4-Dichloro-*N*-(2-dodecyloxyphenyl)-benzylamine 229
$C_{25}H_{35}Cl_3N_2O$
1-(3,4-Dichlorobenzyl)-3-(dodecylcarbamoyl)-pyridinium chloride 946
1-(3,4-Dichlorobenzyl)-4-(dodecylcarbamoyl)-pyridinium chloride 951
$C_{25}H_{35}NO_4$
2-Decyloxy-5-nitrobenzyl 2,5-xylyl ether 269
$C_{25}H_{35}NO_5S$
2,4-Di(*tert*-pentyl)phenyl 3-acetamido-4-methoxybenzene

$C_{25}H_{35}NO_5S$ (Cont'd)
sulfonate ... 797
$C_{25}H_{35}N_3O_3S$
2-(*m*-Aminophenyl)-1-dodecylbenzimidazole-5-sulfonic acid,sodium salt ... 904
$C_{25}H_{36}ClN_3O_2$
3-Benzylcarbamoyl-1-(decylcarbamoylmethyl)-pyridinium chloride ... 946
$C_{25}H_{36}INO$
1-(4′-Dodecylphenacyl)-pyridinium iodide ... 955
$C_{25}H_{36}N_2O_2$
3-Amino-2′-(dodecyloxy)-benzanilide hydrochloride ... 465
$C_{25}H_{36}N_2O_3$
2-Decyloxy-5-nitro-*N*-(2,5-xylyl)-benzylamine ... 267
$C_{25}H_{36}N_4O_2$
4,4′-Methylenebis(1-butyl-1-methyl-3-phenylurea) ... 592
$C_{25}H_{36}O_2$
1,1-Bis(2-hydroxy-3,5-dimethylphenyl)isononane ... 194
$C_{25}H_{36}Si_4$
1,1,1,2,3,3,3-Heptamethyl-2-(triphenylsilyl)trisilane ... 1115
$C_{25}H_{37}ClN_2O$
1-Benzyl-4-(dodecylcarbamoyl)-pyridinium chloride ... 951
$C_{25}H_{37}NO$
α-Diamylamino-1,2,3,4-tetrahydro-9-phenanthrenemethanol hydrochloride ... 239
$C_{25}H_{37}NO_8S$
Troxonium tosylate ... 369
$C_{25}H_{38}BrNOS$
2-Bromo-4-(4-(hexadecyloxy)-phenyl)-thiazole ... 875
$C_{25}H_{38}ClNO$
4-Chloro-6-methoxy-2-pentadecylquinoline ... 987
$C_{25}H_{38}N_3O_3PS_2$
Bis(4-*tert*-butylphenyl) *N*-(3-(3-methoxypropyl)-2-thioureido)-thiophosphoramidate ... 834
$C_{25}H_{38}O_3$
ent-15α-Angeloyloxykaur-16-en-3β-ol, 90% & 16β,17-epoxykaurane-3β-ol ... 85
$C_{25}H_{38}O_7$
Laserpitin ... 91
$C_{25}H_{39}FN_2O_3S_2$
N-(3-Hexadecyl-4-oxo-2-thiazolidinylidene)sulfanilyl fluoride ... 1075
$C_{25}H_{39}NO$
1-Methyl-4-pentadecyl-2(1*H*)-quinolone ... 693
$C_{25}H_{39}NO_2$
4-Hydroxy-6-methoxy-2-pentadecylquinoline ... 983
$C_{25}H_{39}NO_2S$
4-(Hexadecyloxy)benzoylmethyl thiocyanate ... 768
$C_{25}H_{39}NO_5$
4-(Hexadecyloxy)-3-nitrocinnamic acid ... 335
$C_{25}H_{39}N_3O_2$
N-(5-Oxo-3-pentadecyl-1-phenyl-2-pyrazolin-4-yl)-formamide ... 702
$C_{25}H_{39}N_3O_3$
2-Methyl-1-(3-nitrophenyl)-3-pentadecyl-3-pyrazolin-5-one ... 702
$C_{25}H_{40}Cl_2N_2O$
Stearic acid (3,4-dichlorobenzylidene)hydrazide ... 492
Stearic (2,4-dichlorobenzylidene)-hydrazide ... 492
$C_{25}H_{40}N_2OS$
3-Hexadecyl-2-phenylimino-4-thiazolidinone ... 1074
$C_{25}H_{40}N_2O_2S_3$
2-(4-Hexadecylsulfonylphenyl)-5-methylthio-1,3,4-thiadiazole ... 878
$C_{25}H_{40}O_3$
2-(Hexadecyloxy)cinnamic acid ... 335
$C_{25}H_{41}ClN_2O$
Octadecanoic (4-chlorobenzylidene)hydrazide ... 1026
Stearic (2-chlorobenzylidene)-hydrazide ... 494
$C_{25}H_{41}FN_4O_5$
1-(2,6-Dinitro-4-fluorophenyl)-3-octadecylurea ... 523
$C_{25}H_{41}NO_2$
N-Methyl-3-oxooctadecananilide ... 445
$C_{25}H_{41}NO_3$
p-Hexadecanoacetanisidide ... 442
2-Hexadecanoyl-*o*-acetanisidide ... 442
4-(Octadecanamido)benzoic acid ... 433
$C_{25}H_{41}NO_4$
3-Acetamido-4-hexadecyloxybenzoic acid ... 416
$C_{25}H_{41}NS_2$
2-(Octadecylthio)benzothiazole ... 914
$C_{25}H_{41}N_3O_3S$
3-Amino-1-(4-hexadecyl-sulfonyl)-phenyl-2-pyrazolin-5-one, mixture of branched isomers ... 776
$C_{25}H_{42}ClNO_3S_2$
5-(Dimethylsulfamoyl)-2-(hexadecylthio)benzoyl chloride ... 815
$C_{25}H_{42}Cl_2N_2O$
1-(2,3-Dichlorophenyl)-3-octadecylurea ... 541
1-(3,4-Dichlorophenyl)-3-octadecylurea ... 545
1-(3,5-Dichlorophenyl)-3-octadecylurea ... 545
$C_{25}H_{42}N_2$
4,4′-Trimethylenebis(1-(1-cyclohexen-1-yl)piperidine) ... 41
$C_{25}H_{42}N_2O$
Octadecanoic benzylidenehydrazide ... 502
$C_{25}H_{42}N_2O_2$
Octadecanoic (4-hydroxybenzylidene)hydrazide ... 496
Stearic 2-hydroxybenzylidenehydrazide ... 496
$C_{25}H_{42}N_2O_2S_2$
Methyl 3-(4-(hexadecyloxy)benzoyl)-dithiocarbazate ... 737
$C_{25}H_{42}N_2O_4$
Methyl 4-(*N*-hexadecyl-*N*-methylamino)-3-nitrobenzoate ... 372
$C_{25}H_{42}O$
Nonadecanophenone ... 297
$C_{25}H_{42}O_2$
2′-Hexadecyloxy-5′-methylacetophenone ... 293
$C_{25}H_{42}O_2S$
Ethyl 2-(hexadecylthio)-benzoate ... 372
$C_{25}H_{42}O_3$
Ethyl (3-pentadecylphenoxy)-acetate ... 357
2-(Hexadecyloxy)-hydrocinnamic acid ... 329
2′-Hexadecyloxy-4′-methoxyacetophenone ... 293
3-(Octadecyloxy)benzoic acid ... 340
4-(Octadecyloxy)benzoic acid ... 341
2-(4-(*tert*-Pentyl)phenoxy)ethyl laurate ... 355
$C_{25}H_{42}O_5S$
Iso-octadecyl 2-sulfobenzoate, sodium salt, mixture of branched isomers ... 795
$C_{25}H_{42}O_6S$
3-(2-Hexadecyloxy-5-sulfophenyl)-propionic acid ... 788
$C_{25}H_{43}BrN_2O$
1-(3-Bromophenyl)-3-octadecylurea ... 534
$C_{25}H_{43}ClN_2O$
1-(3-Chlorophenyl)-3-octadecylurea ... 534
1-(4-Chlorophenyl)-3-octadecylurea ... 539
$C_{25}H_{43}Cl_2NO_2S$
3,4-Dichloro-*N*-methyl-*N*-octadecylbenzenesulfonamide ... 829
$C_{25}H_{43}FN_2O$
1-(2-Fluorophenyl)-3-octadecylurea ... 536
1-(4-Fluorophenyl)-3-octadecylurea ... 539
$C_{25}H_{43}N$
N,*N*-Dimethyl-4-(1-heptadecen-1-yl)-aniline ... 220
$C_{25}H_{43}NO$
N-Octadecylbenzamide ... 478
2′,4′,6′-Trimethylhexadecananilide ... 440
$C_{25}H_{43}NOS$
N-Hexadecyl-4-methoxy-*N*-methylthiobenzamide ... 736
$C_{25}H_{43}NO_2$
4′-Ethoxyheptadecananilide ... 419
2′-Hexadecyloxy-5′-methylacetophenone oxime ... 1009
2′-(Hexadecyloxy)-propiophenone oxime ... 1009
$C_{25}H_{43}NO_3$
3-(*N*-Hexadecylcarbamoyl)-5-norbornene-2-carboxylic acid ... 122
$C_{25}H_{43}NO_3S$
N-(*p*-Tosyl)octadecanamide ... 828
$C_{25}H_{44}$
1-Phenylnonadecane ... 163
$C_{25}H_{44}ClN$
Dimethyldodecyl(5,6,7,8-tetrahydro-2-naphthylmethyl)ammonium chloride ... 256
$C_{25}H_{44}N_2O$
1-Octadecyl-3-phenylurea ... 586
$C_{25}H_{44}N_2O_2$
1-(3-Hydroxyphenyl)-3-octadecylurea ... 548
$C_{25}H_{44}N_2O_2S$
N-(4-Hexadecylsulfonylphenyl)-propionaldehyde hydrazone ... 779
$C_{25}H_{44}N_2O_4S$
N^3-Acetyl-N^1,N^1-bis-(2-ethylhexyl)-4-methoxymetanilamide ... 814
$C_{25}H_{44}N_2S$
1-Octadecyl-3-phenyl-2-thiourea ... 732
$C_{25}H_{44}N_4O_2$
3,3′-(4-Methyl-1,3-phenylene)bis(1,1-dibutylurea) ... 566
3,3′-(4-Methyl-1,3-phenylene)bis(1,1-diisobutylurea) ... 566
$C_{25}H_{44}O_2$
α,α-Dimethyl-4-hexadecyloxybenzyl alcohol ... 202
$C_{25}H_{44}O_3S$
2-(Octyl)decyl *p*-toluenesulfonate, branched isomers ... 801
$C_{25}H_{45}ClN_2O$
Benzyldimethyl(tetradecyl-carbamoylmethyl)ammonium chloride ... 431
$C_{25}H_{45}N$
N-Cyclohexyltricyclohexyl-methylamine ... 31
$C_{25}H_{45}NO_2S$
N-Octadecyl-*p*-toluenesulfonamide ... 828
$C_{25}H_{45}N_3O_2S_2$
1-Hexadecanesulfonic acid ((methylthio)(*p*-toluidino)-methylene)-hydrazide ... 816
$C_{25}H_{46}BrN$
Benzyldimethyl(hexadecyl)-ammonium bromide ... 255
$C_{25}H_{46}BrNO_3$
Benzylbis(2-hydroxyethyl)(2-dodecyloxyethyl)ammonium bromide ... 254
$C_{25}H_{46}ClNO_3$
Benzylbis(2-hydroxyethyl)(2-dodecyloxyethyl)ammonium chloride ... 254
$C_{25}H_{46}N_2O_2S$
N-(2-(*N*-Hexadecylanilino)-ethyl)-methanesulfonamide ... 818
$C_{25}H_{48}O_4$
Diethyl ethyl(hexadecyl)malonate ... 94
Diethyl 2-octadecylmalonate ... 76
Dioctyl azelate ... 76
$C_{25}H_{49}NO_4$
2-(Hydroxymethyl)-tetrahydrofurfuryl *N*-octadecylcarbamate ... 114
5-(Hydroxymethyl)-tetrahydrofurfuryl *N*-octadecylcarbamate ... 115
$C_{25}H_{50}N_2O$
4-Methyl-*N*-octadecyl-1-piperidinecarboxamide ... 136
$C_{25}H_{50}O_2$
Pentacosanoic acid ... 61
$C_{25}H_{50}O_3$
Didodecyl carbonate ... 74
$C_{25}H_{52}N_2S$
1,3-Didodecyl-2-thiourea ... 142
$C_{25}H_{52}O$
9-Octyl-9-heptadecanol ... 10
$C_{25}H_{52}O_3$
Trioctyl orthoformate ... 19
$C_{25}H_{53}N$
Dibutylheptadecylamine ... 29
$C_{25}H_{54}O_6P_2$
Tetrahexyl methylenediphosphonate ... 159
$C_{25}H_{58}BN$
Methyltrioctylammonium borohydride ... 44
$C_{25}H_{58}Si_7$
1,1,1,2,3,3,3-Heptamethyl-2-(2,3,4,5-tetrakis(trimethylsilyl)phenyl)-trisilane ... 1098
$C_{26}H_7F_{15}Si$
trans-Styryltris(pentafluorophenyl)-silane ... 1103
$C_{26}H_{10}O_3$
3,4-Coronenedicarboxylic anhydride ... 388
$C_{26}H_{14}$
Rubicene ... 166
$C_{26}H_{14}Cl_4N_2O_4$
α,α′-Piperonylidenebis(3,5-dichloro-2-hydroxycinnamonitrile) ... 761
$C_{26}H_{15}Br$
3-Bromophenanthro(3,4-c)-phenanthrene ... 175
$C_{26}H_{16}$
9,9′-Bifluorenylidene ... 165
Phenanthro(3,4-c)phenanthrene ... 167
$C_{26}H_{16}Cl_4F_2N_4O_2S_2$
4-Chloro-2-(3-chloro-4-fluorophenylureido)phenyl disulfide ... 522
$C_{26}H_{16}Cl_4N_2O_2S$
4-(*N*-(2,4-Dichlorobenzylidene)-amino)phenyl sulfone ... 1075
$C_{26}H_{16}N_2$
Perinaphthenone (9-fluorenylidene)hydrazone ... 236
1*H*-Phenalen-1-one azine ... 769
$C_{26}H_{16}N_2O_6S_2$
4-Benzoyl-2-nitrophenyl disulfide ... 308
$C_{26}H_{16}O_3$
2,8-Dibenzoyldibenzofuran ... 895
$C_{26}H_{17}ClO_5S$
4-(4-((4-Chlorophenyl)sulfonyl)-phenoxy)benzil ... 773
$C_{26}H_{18}Br_2Cl_2N_4O_2S_2$
2-(4-Bromophenylureido)-4-chlorophenyl disulfide ... 522
$C_{26}H_{18}Cl_2N_2O_2$
6-(4-Chlorophenoxy)-3-(4-(4-chlorophenoxy)-phenyl)-3,4-dihydroquinazoline ... 247
$C_{26}H_{18}Cl_2N_2O_2S$
4,4′-Bis(4-chlorobenzylideneamino)-diphenyl sulfone ... 1063
4-(2-Chlorobenzylideneamino)-phenyl sulfone ... 1075
$C_{26}H_{18}Cl_2N_2S_2$
4,4′-Dithiobis(*N*-(4-chlorobenzylidene)aniline) ... 1024
$C_{26}H_{18}F_6Si$
Bis(3-(trifluoromethyl)phenyl)-diphenylsilane ... 1100
$C_{26}H_{18}N_2$
6,7-Diphenyldibenzo(e,g)(1,4)-diazocine ... 1003
Tetrabenzo(b,d,h,j)(1,6)-diazacyclododecine ... 254
$C_{26}H_{18}N_4O_6S$
4-(*N*-(3-Nitrobenzylidene)amino)-phenyl sulfone ... 1059
$C_{26}H_{18}O_4$
4,4′-Bi(6-methoxydibenzofuran) ... 893

$C_{26}H_{18}O_4S_2$
2,2′-(Dithiodi-2,1-phenylene)-bis(benzoic acid) ... 342
$C_{26}H_{18}O_6S$
4-(4-((4-Hydroxyphenyl)sulfonyl)-phenoxy)benzil ... 775
$C_{26}H_{19}BF_4O_2$
5,6-Dihydro-2-(4-methoxyphenyl)-4-phenylnaphtho(1,2-b)pyrylium tetrafluoroborate ... 934
$C_{26}H_{19}NO$
4-((10-Methyl-9(10*H*)-acridinylidene)-ethylidene)-1(4*H*)-naphthalenone ... 302
$C_{26}H_{20}$
6β,7,7a,13b,14,14a-Hexahydroacenaphtho(1,2-potassium)fluoranthene ... 167
$C_{26}H_{20}BF_4N$
5,6-Dihydro-2,4-diphenyl-1-methylbenzo(h)quinolinium tetrafluoroborate ... 957
$C_{26}H_{20}Br_2N_4O_4S$
1,1′-(Sulfonyldi-4,1-phenylene)bis(3-(4-bromophenyl)urea) ... 773
$C_{26}H_{20}ClNO_2$
N-(4-Chloro-2-phenoxyphenyl)-2,2-diphenylacetamide ... 401
$C_{26}H_{20}N_2$
Benzophenone azine ... 1077
9,9′-Biacridan ... 253
2,2′-Bis(benzylideneamino)biphenyl. 1060
2,2′-Bis(phenyliminomethyl)-biphenyl ... 1077
$C_{26}H_{20}N_2O_2$
N,*N*′-Bis(4-phenoxyphenyl)-1,4-diazabutadiene ... 1040
1,2-Dibenzoyl-1,2-diphenylhydrazine ... 500
$C_{26}H_{20}N_2O_2S_2$
N,*N*′-Bis(2-hydroxybenzylidene)-4-aminophenyl disulfide ... 1032
$C_{26}H_{20}N_2O_3$
6-(4-(Benzyloxy)phenyl)-5-methyl-3-phenylisoxazolo(4,5-c)pyridin-4(5*H*)-one ... 874
4,4′-Bis(2-hydroxybenzylideneamino)-diphenyl ether ... 1028
$C_{26}H_{20}N_2O_4S$
4,4′-Bis(4-hydroxybenzylideneamino)-diphenyl sulfone ... 1028
4-(2-Hydroxybenzylideneamino)-phenyl sulfone ... 1028
$C_{26}H_{20}N_2S_2$
4-(Benzylideneamino)phenyl disulfide ... 1052
$C_{26}H_{20}N_4$
2,5-Dibenzyl(cyclohexane-1,4-diylidene)dimalononitrile ... 764
$C_{26}H_{20}OS_2$
2,2-Bis(phenylthio)-2-phenylacetophenone ... 293
$C_{26}H_{20}O_2$
Triphenylmethyl benzoate ... 375
$C_{26}H_{20}O_4$
Ethyl 2-(9-anthracenylmethyl)-2,3-dihydro-3-oxo-2-benzofurancarboxylate ... 363
$C_{26}H_{20}O_6$
3-Benzyl-7,8-diacetoxy-2-phenyl-4*H*-1-benzopyran-4-one ... 378
$C_{26}H_{20}Pb$
1-Phenyl-2-(triphenylplumbyl)-acetylene ... 1136
$C_{26}H_{20}Si$
1-Phenyl-2-(triphenylsilyl)acetylene.. 1115
$C_{26}H_{21}BF_4O$
2,6-Bis(*p*-tolyl)-4-phenylpyrylium tetrafluoroborate ... 936
$C_{26}H_{21}BrNO_3P$
(4′-Nitrophenacyl)-triphenylphosphonium bromide ... 830
$C_{26}H_{21}ClN_2O_5$
3-(*N*-(2-Methoxyphenyl)carbamoyl)-2-naphthyl *N*-(5-chloro-2methoxyphenyl)carbamate ... 620
$C_{26}H_{21}FN_4O_6S$
4-(4-Dimethylaminophenylimino)-5-(3-fluorosulfonylbenzamido)-2-(2-furamido)-2,5-cyclohexadien-1-one.. 1051
$C_{26}H_{21}NO_2$
3,5-Bis(*p*-tolyl)-4-nitrobiphenyl ... 278
Ethyl 2-cyano-3,3-di(1-naphthyl)-propionate ... 757
$C_{26}H_{21}NO_3$
N-Benzoyl-DL-phenylalanine β-naphthyl ester ... 488
$C_{26}H_{21}NSi$
4-(Triphenylsilyl)benzyl cyanide ... 1107
$C_{26}H_{21}N_3O_4S$
N-(3-(2-Methoxy-5-methylphenyl)-4-(4-nitrophenyl)-4thiazolin2ylidene)-cinnamamide ... 737
$C_{26}H_{21}OP$
(Phenacylidene)-triphenylphosphorane ... 832
$C_{26}H_{22}$
1,4-Bis(4-phenyl-1,3-butadienyl)-benzene ... 165
1,1,1,2-Tetraphenylethane ... 163
1,1,2,2-Tetraphenylethane ... 164
$C_{26}H_{22}BF_4O_2$
5,6-Dihydro-4-(4-methoxyphenyl)-2-phenylnaphtho(1,2-b)pyrylium tetrafluoroborate ... 935
$C_{26}H_{22}ClO_2P$
(1,3-Benzodioxol-5-ylmethyl)-triphenylphosphonium chloride ... 831
$C_{26}H_{22}ClO_4P$
(2-(2,5-Cyclohexadien-1-ylidene)-vinyl)triphenylphosphonium perchlorate ... 832
$C_{26}H_{22}N_2O$
N-(4-Benzyloxybenzylidene)-*N*′-phenyl-1,4-phenylenediamine ... 1043
$C_{26}H_{22}N_2O_2$
N-Benzoyl-DL-phenylalanine-2-naphthylamide ... 488
$C_{26}H_{22}N_2O_4$
2,5-Bis(*N*-(2-methoxycarbonylphenyl)-carbamoyl)terephthalic acid ... 473
$C_{26}H_{22}N_4$
N,*N*′-Bis(4-anilinophenyl)-1,4-diazabutadiene ... 1053
$C_{26}H_{22}N_4O_2$
Oxalic bis((1-(1-naphthyl)-ethylidene)hydrazide) ... 501
$C_{26}H_{22}N_4O_4S$
1,1′-(Sulfonyldi-4,1-phenylene)bis(3-phenylurea) ... 782
$C_{26}H_{22}N_4O_5$
2,2′-Oxybis(acetic acid (2-hydroxy-1-naphthylmethylene)hydrazide) ... 1030
$C_{26}H_{22}O$
Dibenzhydryl ether ... 186
$C_{26}H_{22}OSi$
10,10-Dibenzyl-10*H*-phenoxasilin ... 1122
$C_{26}H_{22}OSn$
2-(Triphenylstannyl)acetophenone ... 1131
$C_{26}H_{22}O_2$
1,1-Diphenyl-3-ethoxy-1*H*,3*H*-naphtho(1,8-cd)pyran ... 183
1,1,2,2-Tetraphenyl-1,2-ethyleneglycol ... 199
$C_{26}H_{22}O_2S$
p-Tolyl triphenylmethyl sulfone ... 784
$C_{26}H_{22}O_2Si$
(4-Methoxybenzoyl)triphenylsilane ... 1106
$C_{26}H_{22}O_8$
2,3,4-Tribenzoyl-D-xylose ... 368
$C_{26}H_{22}S$
p-Tolyl triphenylmethyl sulfide ... 212
$C_{26}H_{22}S_2$
2,2′-Bis(benzylthio)biphenyl ... 212
$C_{26}H_{22}Si$
Styryltriphenylsilane ... 1115
$C_{26}H_{22}Si_2$
9,9′-Dimethyl-9,9′-bi(9*H*-9-silafluorene) ... 1123
$C_{26}H_{23}BF_4O_3S$
2,4,6-Tris(4-methoxyphenyl)-thiopyrylium tetrafluoroborate ... 937
$C_{26}H_{23}BF_4O_4$
2,4,6-Tris(4-methoxyphenyl)-pyrylium tetrafluoroborate ... 934
$C_{26}H_{23}BrNP$
(2-(Phenylimino)ethyl)-triphenylphosphonium bromide ... 1076
$C_{26}H_{23}Br_2P$
2-(Bromomethyl)-benzyltriphenylphosphonium bromide ... 830
3-(Bromomethyl)-benzyltriphenylphosphonium bromide ... 830
4-(Bromomethyl)-benzyltriphenylphosphonium bromide ... 830
$C_{26}H_{23}FeCl_4O$
2,4-Diphenyl-6-(2,4,6-trimethylphenyl)pyrylium tetrachloroferrate ... 936
$C_{26}H_{23}GeN$
5,10-Dihydro-10,10-diphenyl-5-ethylphenazagermine ... 1108
$C_{26}H_{23}NO$
2-Anilino-1,1,2-triphenyl-1-ethanol ... 203
$C_{26}H_{23}N_3O_2$
α-(1,4-Diacetyl-1,4-dihydropyridin-3-yl)-3,3′-methylenediindole ... 888
1-(4-(Diphenylamino)phenyl)-3-(2-methoxyphenyl)urea ... 552
$C_{26}H_{23}O_3P$
2,4,6-(4-Methoxyphenyl)phosphorin.. 937
$C_{26}H_{24}BF_4P$
(2-(2,5-Cyclohexadien-1-ylidene)-ethyl)triphenylphosphonium tetrafluoroborate ... 832
$C_{26}H_{24}BrP$
(4-Methylbenzyl)-triphenylphosphonium bromide ... 832
$C_{26}H_{24}ClOP$
(4-Methoxybenzyl)-triphenylphosphonium chloride ... 830
$C_{26}H_{24}ClO_4P$
(2-(2,5-Cyclohexadien-1-ylidene)-ethyl)triphenylphosphonium perchlorate ... 832
$C_{26}H_{24}ClP$
(2-Methylbenzyl)-triphenylphosphonium chloride ... 832
(3-Methylbenzyl)-triphenylphosphonium chloride ... 832
(4-Methylbenzyl)-triphenylphosphonium chloride ... 832
$C_{26}H_{24}CuF_2O_{12}S_2$
Ethyl α-acetyl-3-(fluorosulfonyl)-benzoylacetate, copper(II) derivative ... 351
$C_{26}H_{24}Ge$
(Phenethyl)triphenylgermane ... 1115
$C_{26}H_{24}GeO$
1-Phenyl-2-(triphenylgermyl)ethanol. 1105
$C_{26}H_{24}INO$
2-(4-Benzyloxystyryl)-1-ethylquinolinium iodide ... 984
$C_{26}H_{24}N_2O_4$
N,*N*′-(1,4-Phenylenedimethylidyne)-bis(ethyl 4-aminobenzoate) ... 1062
$C_{26}H_{24}N_2O_6$
2,5-Bis(*N*-(2,5-xylyl)carbamoyl)-terephthalic acid ... 474
$C_{26}H_{24}N_2O_8$
2,5-Bis(*N*-(4-ethoxyphenyl)-carbamoyl)terephthalic acid hydrate ... 463
$C_{26}H_{24}N_4O_2$
α-(3-Methyl-2-pyridylimino)-*o*-cresol copper(II) salt ... 1036
1,1′-(1,5-Naphthalenediyl)bis(3-benzylurea) ... 572
$C_{26}H_{24}OPb$
(4-(1-Hydroxyethyl)phenyl)-triphenyllead ... 1134
(4-(2-Hydroxyethyl)phenyl)-triphenyllead ... 1134
$C_{26}H_{24}OSi$
(2-Phenylethoxy)triphenylsilane ... 1115
1-Phenyl-2-(triphenylsilyl)-1-ethanol.. 1105
$C_{26}H_{24}OSn$
(3-Methoxybenzyl)triphenyltin ... 1130
(2-(Methoxymethyl)phenyl)-triphenyltin ... 1130
$C_{26}H_{24}O_2SSi$
p-Tolyl triphenylsilylmethyl sulfone. 1110
$C_{26}H_{24}O_6$
2-Hydroxymethyl-2-methyl-1,3-propanediol tribenzoate ... 371
$C_{26}H_{24}O_8$
2,3,5-Tri-*O*-benzoyl-D-xylitol ... 367
$C_{26}H_{24}O_9$
4-Acetylresorcinol bis(3,4-dimethoxybenzoate) ... 377
$C_{26}H_{24}Pb$
Diphenyldi(*p*-tolyl)lead ... 1136
(Phenethyl)triphenyllead ... 1136
$C_{26}H_{24}Si$
Diphenyldi(*o*-tolyl)silane ... 1119
Diphenyldi(*m*-tolyl)silane ... 1115
Phenethyltriphenylsilane ... 1115
$C_{26}H_{24}Sn$
Dibenzyldiphenyltin ... 1132
(3-Methylbenzyl)triphenyltin ... 1132
Triphenyl(2,4-xylyl)tin ... 1133
Triphenyl(2,5-xylyl)tin ... 1133
Triphenyl(2,6-xylyl)tin ... 1132
$C_{26}H_{25}F_3N_4O_3$
4-(3-(3-(Trifluoromethyl)phenyl)-ureido)benzoic (4-butoxybenzylidene)hydrazide ... 507
$C_{26}H_{25}NO$
2′-*tert*-Butyl-10,11-dihydrospiro-[5*H*-dibenzo(a,d)cycloheptene-5,1′-(1*H*)-isoindol]-3′(2′*H*)one ... 707
$C_{26}H_{25}NO_2$
1′-Benzyl-1,4-dihydro-4-phenylspiro(3*H*-2-benzopyran-3,4′-piperidine)-1-one hydrochloride ... 385
$C_{26}H_{25}NO_3$
2,3-Diethoxy-α-phenyl-α-(2-pyridyl)-1-naphthalenemethanol ... 931
$C_{26}H_{25}NPb$
N,*N*-Dimethyl-3-(triphenylplumbyl)-aniline ... 1134
N,*N*-Dimethyl-4-(triphenylplumbyl)-aniline ... 1134
N,*N*-Dimethyl-4-(triphenylplumbyl)-aniline ... 1134
$C_{26}H_{25}NSi$
N,*N*-Dimethyl-2-(triphenylsilyl)-aniline ... 1107
N,*N*-Dimethyl-3-(triphenylsilyl)-aniline ... 1107
$C_{26}H_{25}NSn$
N,*N*-Dimethyl-2-(triphenylstannyl)-aniline ... 1130
N,*N*-Dimethyl-3-(triphenylstannyl)-aniline ... 1130
N,*N*-Dimethyl-4-(triphenylstannyl)-aniline ... 1130
$C_{26}H_{26}N_2O_3$
1-(3-Dibenzofuryl)-3-(4-(2-methoxyphenyl)piperazino)-1-propanone ... 893
$C_{26}H_{26}N_2O_3S$
4-(2-Anilinovinyl)-1-ethylquinolinium *p*-toluenesulfonate ... 1060
$C_{26}H_{26}N_2O_5$
Carbobenzyloxyglycyl-L-phenylalanine benzyl ester ... 657
Carbobenzyloxyphenyl-alanylglycine benzyl ester ... 657
$C_{26}H_{26}N_2O_6$
Carbobenzyloxyglycyl-L-tyrosine benzyl ester ... 644
Carbobenzyloxy-L-tyrosyl-L-phenylalanine ... 644
$C_{26}H_{26}N_2O_7$
Carbobenzyloxy-L-tyrosyl-L-tyrosine 671
$C_{26}H_{26}N_4O_2S_2$
L-Cystine bis(*N*-(2-naphthyl)amide)... 487
L-Cystine bis(*N*-(2-naphthyl)amide) dihydrochloride ... 487
$C_{26}H_{26}N_4O_4$
Bis(8-quinolyl) *N*,*N*′-hexamethylenebiscarbamate ... 984

$C_{26}H_{26}N_4O_6$
4,4′-Bis(furfurylureido)-3,3′-dimethoxybiphenyl ... 846
$C_{26}H_{26}O_2$
1,2-Bis(2,4,6-trimethylbenzoyl)-benzene ... 313
$C_{26}H_{26}O_4$
2,5-Bis(2,4-xylyl)hydroquinone diacetate ... 380
$C_{26}H_{26}Si_2$
1,2-Dimethyl-1,1,2,2-tetraphenyldisilane ... 1115
$C_{26}H_{28}ClN_3$
Basic Dye (C.I. 44520) ... 1060
$C_{26}H_{28}N_2$
N,N′-Bis(4-isopropylbenzylidene)-1,4-phenylenediamine ... 1058
N,N′-(1,4-Phenylenedimethylidyne)-bis(2,4,5-trimethylaniline) ... 1065
N,N′-Terephthalylidenebis(2,4,6-trimethylaniline) ... 1066
$C_{26}H_{28}N_2O_2$
N,N′-Diethyl-*N,N′*-di(*o*-tolyl)-isophthalamide ... 475
N,N′-Diethyl-*N,N′*-di(*m*-tolyl)-isophthalamide ... 483
N,N′-Diethyl-*N,N′*-di(*m*-tolyl)-terephthalamide ... 475
N,N′-Diisopropyl-*N,N′*-diphenylterephthalamide ... 482
$C_{26}H_{28}N_2O_4$
α,α′-Bis(4-carbethoxyanilino)-*p*-xylene ... 373
$C_{26}H_{28}N_2S_2O_6$
1,1′-Ethylenedipyridinium di-*p*-toluenesulfonate ... 944
$C_{26}H_{28}N_4O_4$
1,4-Bis(3-phthalimidopropyl)-piperazine ... 718
$C_{26}H_{29}ClNP$
(2-(Cyclohexylimino)ethyl)-triphenylphosphonium chloride ... 1076
$C_{26}H_{29}ClO_3S$
2-(10-(4-Chlorophenylthio)decyl)-3-hydroxy-1,4-naphthoquinone ... 315
$C_{26}H_{29}NO$
N-Benzyl-*N*-*tert*-butyl-2,2-diphenylpropionamide ... 432
$C_{26}H_{29}NO_2$
13-Benzyl-10,11-dimethoxy-6-methyl-5,7,8,13-tetrahydro-6*H*-dibenz(c,f)azonine ... 246
$C_{26}H_{30}IN_3$
2,4-Bis-(4-dimethylaminostyryl)-1-methylpyridinium iodide ... 954
$C_{26}H_{30}I_2N_2$
2,2′-(1,3-Butadiene-1,4-diyl)bis(1,3,3-trimethyl-3*H*-indolium iodide) ... 888
$C_{26}H_{30}N_2O$
4-Methyl-α,α,γ-triphenyl-1-piperazinepropanol ... 245
$C_{26}H_{30}N_4O_2$
Adipic acid (5-indanylmethylene)-hydrazide ... 1075
$C_{26}H_{30}N_4O_5$
Carbobenzyloxy-histidyl-L-valine benzyl ester ... 868
$C_{26}H_{30}O$
2,6-Bis(4-iso-propylbenzylidene)-1-cyclohexanone ... 304
$C_{26}H_{30}O_2$
3,10-Dimethoxy-6a,12b-dimethyl-5,6,6a,6b,7,8,12b,13-octahydropicene ... 187
$C_{26}H_{30}O_3$
9-(5-Indanyl)-2,3,6,7-tetrahydro-2,2,7,7-tetramethyl-9*H*-xanthene-4,5(1*H*,8*H*)-dione ... 320
$C_{26}H_{30}O_{11}$
4,6:4′,6′-Di-*O*-benzylidenetrehalose .. 199
$C_{26}H_{30}Si$
(2-Cyclohexylethyl)triphenylsilane ... 1111
$C_{26}H_{31}NO$
1,2-Diphenyl-3-methyl-4-(methylphenethylamino)-2-butanol hydrochloride ... 237
$C_{26}H_{31}N_3O$
1,1-Bis(2-(2-pyridyl)ethyl)-3,3-dimethyl-5-(2-pyridyl)-2-pentanone ... 961
N-(2-(Diethylamino)-ethyl)-*N*-(1,2-diphenylethyl)-nicotinamide hydrochloride ... 942
$C_{26}H_{31}N_3O_5$
Carbobenzyloxy-L-isoleucyl-L-tryptophan methyl ester ... 890
Carbobenzyloxy-L-norleucyl-L-tryptophan methyl ester ... 883
Carbobenzyloxytryptophylleucine methyl ester ... 883
$C_{26}H_{32}Ge$
Octyltriphenylgermane ... 1115
$C_{26}H_{32}N_2O$
4,4′-Bis(dimethylamino)-2,2′,4″-trimethyltrityl alcohol ... 226
4,4′-((4-Methoxyphenyl)methylene)-bis(*N,N*,3-trimethylaniline) ... 229
$C_{26}H_{32}N_2O_2$
1,4-Bis-(4-*sec*.-butylaminophenoxy)-benzene ... 229
$C_{26}H_{32}N_2O_5$
Carbobenzyloxyprolylleucine benzyl ester ... 700
Carbobenzyloxyprolylnorleucine benzyl ester ... 700
$C_{26}H_{32}O_3$
1,4a-Dimethyl-hexahydro-7-methoxy-1-(3-methoxyphenethyl)-2(3*H*)-phenanthrenone ... 283
$C_{26}H_{32}O_4$
2,2′-(5-Indanylmethylene)-bis-(5,5-dimethyl-1,3-cyclohexanedione) ... 309
$C_{26}H_{32}O_6$
Methyl 3,3-bis(4,4-dimethyl-2-hydroxy-6-oxo-1-cyclohexenyl)-2-phenylpropionate ... 353
$C_{26}H_{32}Si$
(2-Ethylhexyl)triphenylsilane ... 1115
(1-Methylheptyl)triphenylsilane ... 1115
Octyltriphenylsilane ... 1115
$C_{26}H_{32}Sn$
Octyltriphenyltin ... 1132
$C_{26}H_{33}Cl_2N_3$
Methyl Green (C.I. 42585) ... 1060
$C_{26}H_{33}NO_7$
Ethyl 1-(β-phenethyl)-4-(3,4,5-trimethoxybenzoyloxy)-3-piperidinecarboxylate ... 369
Methyl 2-(3-nitro-4-(2-(4-*tert*-octylphenoxy)ethoxy)benzoyl)-acetate ... 363
$C_{26}H_{33}N_3O_8$
N,N′-Bis(carbobenzyloxy)-lysylserine methyl ester ... 643
$C_{26}H_{34}N_2O_3$
2-(4-(Decanoylamino)benzoyl)-propionanilide ... 444
N-(1-(1-Hydroxy-6-methoxy-1,2,3,4-tetrahydro-2-naphthylmethyl)-4-piperidinyl)propionanilide ... 411
$C_{26}H_{34}N_2O_5$
Carbobenzyloxyisoleucylvaline benzyl ester ... 663
$C_{26}H_{34}N_4O_4$
Sebacic bis((2-methoxybenzylidene)hydrazide) ... 497
Sebacic bis((4-methoxybenzylidene)hydrazide) ... 497
$C_{26}H_{34}N_4O_6$
1,1′-Hexamethylenebis(3-(4-(ethoxycarbonyl)phenyl)urea) ... 577
Sebacic acid bis(3-hydroxy-4-methoxybenzylidenehydrazide) ... 1027
$C_{26}H_{34}N_4O_8$
Adipic acid (3,4,5-trimethoxybenzylidene)-hydrazide ... 1051
$C_{26}H_{34}O_3$
4′-Phenylphenacyl dodecanoate ... 357
$C_{26}H_{34}O_6$
3,3′-(Decamethylenedioxy)di-*p*-anisaldehyde ... 323
$C_{26}H_{35}BrO_{17}$
Acetobromocellobiose ... 89
$C_{26}H_{35}NO_2$
2,2-Dimethyl-4-(1,2,3,4-tetrahydroisoquinolinylmethyl)-propiophenone-2,2-dimethyltrimethyleneketal hydrochloride ... 252
$C_{26}H_{35}NO_6$
Ethyl 4-(2-decyloxy-5-nitrobenzyloxy)benzoate ... 369
$C_{26}H_{36}Cl_2N_2O_3$
3-(4-Chlorophenylcarbamoyl)-1-(dodecyloxycarbonylmethyl)-pyridinium chloride ... 955
$C_{26}H_{36}N_2O$
9-(Dodecylamino)-2-methoxyacridine ... 995
$C_{26}H_{36}N_2O_4$
N,N′-Bis(2-ethoxyphenyl)-decanediamide ... 414
N,N′-Bis(4-ethoxyphenyl)-1,10-decanediamide ... 414
2′,2″-(Decamethylenedioxy)-diacetanilide ... 417
3′,3″-(Decamethylenedioxy)-diacetanilide ... 419
4′,4″-(Decamethylenedioxy)-diacetanilide ... 419
$C_{26}H_{36}N_4O_2$
1,1′-(1,5-Naphthalenediyl)bis(3-cyclohexyl-3-methylurea) ... 584
3,3′-(1,5-Naphthalenediyl)bis(1-cyclohexyl-1-methylurea) ... 567
$C_{26}H_{37}ClN_2O_3$
1-Dodecyloxycarbonylmethyl-3-(phenylcarbamoyl)-pyridinium chloride ... 947
$C_{26}H_{37}Cl_2N_3O_2$
3-(4-Chlorophenylcarbamoyl)-1-(dodecylcarbamoylmethyl)-pyridinium chloride ... 948
$C_{26}H_{38}ClN_3O_2$
1-Dodecylcarbamoylmethyl-3-(phenylcarbamoyl)-pyridinium chloride ... 948
$C_{26}H_{38}N_2$
N,N-Dibenzyl-*N′,N′*-di-*tert*-butyl-1,4-diamino-2-butene ... 1057
$C_{26}H_{38}N_2O_4Pb$
Diethylplumbyl bis(2-(butylamino)-benzoate) ... 1134
$C_{26}H_{38}O_2$
2,2′-Butylidenebis(6-*tert*-butyl-*p*-cresol) ... 194
4,4′-Butylidenebis(6-*tert*-butyl-*m*-cresol) ... 196
$C_{26}H_{38}O_4$
Bis-(3,3,5-trimethylcyclohexyl) phthalate ... 375
$C_{26}H_{39}Cl_2N_3O_2S$
S-(5-Heptadecyl-1,3,4-oxadiazol-2-yl) *N*-(3,4-dichlorophenyl)-thiocarbamate ... 872
$C_{26}H_{39}N$
Dibenzyldodecylamine ... 243
$C_{26}H_{39}NO$
9-(2-(Diamylamino)-1-hydroxyethyl)-1,2,3,4-tetrahydrophenanthrene hydrochloride ... 239
N-(1-Naphthyl)hexadecanamide ... 490
$C_{26}H_{40}ClNO_2$
2-Chloro-5,8-dimethoxy-4-(pentadecyl)quinoline ... 986
$C_{26}H_{40}N_2O_2$
N,N′-Dodecamethylenebis(2-hydroxybenzylamine) ... 237
4-Hexadecanoyl-3-methyl-1-phenyl-2-pyrazolin-5-one ... 706
$C_{26}H_{41}FO_6S$
Methyl 2-(3-fluorosulfonyl-4-(hexadecyloxy)benzoyl)acetate ... 802
Methyl 2-(5-fluorosulfonyl-2-(hexadecyloxy)benzoyl)acetate ... 802
$C_{26}H_{41}N$
2-Hexadecyl-6-methylquinoline ... 995
N-(1-Naphthyl)hexadecylamine ... 233
α-Pentadecyl-2-naphthalenemethylamine ... 241
$C_{26}H_{41}NO_3$
5,8-Dimethoxy-4-pentadecyl-2-hydroxyquinoline ... 982
$C_{26}H_{41}NS$
1,6-Dimethyl-4-pentadecyl-2(1*H*)-quinolinethione ... 747
1,7-Dimethyl-4-pentadecyl-2(1*H*)-quinolinethione ... 742
$C_{26}H_{42}ClF_3N_2O$
1-(2-Chloro-5-trifluoromethylphenyl)-3-octadecylurea ... 507
$C_{26}H_{42}N_2$
N,N′-Bis(1-methylheptyl)-1,5-naphthalenediamine ... 232
$C_{26}H_{42}N_2O$
4-Heptadecyl-1*H*-1,5-benzodiazepin-2(3*H*)-one ... 702
$C_{26}H_{42}N_4O$
1-Methyl-4-pentadecyl-2(1*H*)-quinolone semicarbazone ... 571
$C_{26}H_{42}O_4$
Bis-(3,5,5-trimethylhexyl) phthalate .. 375
Ethyl 4-(pentadecyloxy)-benzoylacetate ... 363
3-(4-(Hexadecyloxy)-benzoyl)-propionic acid ... 328
4-Hexadecyloxy-3-methoxycinnamic acid ... 335
Methyl 2-(hexadecyloxy)-benzoylacetate ... 363
Methyl 4-(hexadecyloxy)-benzoylacetate ... 363
Methyl 2-(isohexadecyloxy)-benzoylacetate ... 363
$C_{26}H_{43}BrO_3$
4-(3-Bromo-4-(hexadecyloxy)-phenyl)butyric acid ... 329
$C_{26}H_{43}F_3N_2O$
1-Octadecyl-3-(2-(trifluoromethyl)-phenyl)urea ... 511
1-Octadecyl-3-(3-(trifluoromethyl)-phenyl)urea ... 510
$C_{26}H_{43}NO_3$
2′-(Hexadecyloxy)acetoacetanilide .. 442
$C_{26}H_{43}N_3O_2S$
1-Hexadecanesulfonic acid (1-methyl-2(1*H*)-quinolylidene)-hydrazide ... 821
$C_{26}H_{44}BrNO$
4′-Bromo-2′,6′-dimethylstearanilide .. 410
$C_{26}H_{44}N_2O$
Stearic (4-methylbenzylidene)-hydrazide ... 502
$C_{26}H_{44}N_2O_2$
1-(*m*-Anisyl)-3-oleylurea ... 558
4-(4-Hexadecylphenyl)-4-hydrazonobutyric acid, sodium salt ... 351
Stearic (2-methoxybenzylidene)-hydrazide ... 498
Stearic (4-methoxybenzylidene)-hydrazide ... 498
$C_{26}H_{44}N_2O_3$
Stearic (4-hydroxy-3-methoxybenzylidene)hydrazide ... 495
4-Stearoylaminophenyl *N*-methylcarbamate ... 677
$C_{26}H_{44}N_2O_3S_2$
Methyl 3-(2-hexadecyloxy-5-methoxybenzoyl)dithiocarbazate ... 737
$C_{26}H_{44}N_2O_6S$
4-(4-Hexadecyloxy-3-nitrophenylsulfonyl)-morpholine ... 814
$C_{26}H_{44}O_2$
4-(4-Hexadecylphenyl)butyric acid ... 333
$C_{26}H_{44}O_3$
2′,5′-Dimethoxyoctadecanophenone ... 293
4-(4-(Hexadecyloxy)phenyl)-butyric acid ... 329
Methyl 3-(octadecyloxy)benzoate ... 370
Methyl 4-(octadecyloxy)benzoate ... 370
$C_{26}H_{44}O_6S$
4-(4-Hexadecyloxy-3-sulfophenyl)-butyric acid ... 788
$C_{26}H_{45}ClN_2O$
1-(3-Chloro-4-methylphenyl)-3-octadecylurea ... 545
1-(4-Chloro-2-methylphenyl)-3-octadecylurea ... 545

$C_{26}H_{45}NO$
1-(1-Adamantylcarbonyl)-decamethylpiperidine ... 136
$C_{26}H_{45}NO_2$
4′-Hydroxyeicosananilide ... 413
2′-(Octadecyloxy)acetanilide ... 417
3′-(Octadecyloxy)acetanilide ... 419
4′-(Octadecyloxy)acetanilide ... 419
$C_{26}H_{45}NO_2S$
3,4-Dimethoxy-*N*-hexadecyl-*N*-methylthiobenzamide ... 736
$C_{26}H_{45}NO_3$
2,5-Dimethoxy-*N*-hexadecyl-*N*-methylbenzamide ... 467
Methyl 3-amino-4-(hexadecyloxy)-hydrocinnamate ... 355
$C_{26}H_{45}N_3O_5S$
N,*N*-Bis(2-ethylhexyl)-4-morpholino-3-nitrobenzenesulfonamide ... 818
$C_{26}H_{46}INOS$
Hexadecyl(2-methoxy-α-(methylthio)-benzylidene)-methylammonium iodide ... 1042
$C_{26}H_{46}N_2O$
1-Octadecyl-3-(*o*-tolyl)urea ... 588
1-Octadecyl-3-(*m*-tolyl)urea ... 581
1-Octadecyl-3-(*p*-tolyl)urea ... 591
$C_{26}H_{46}N_2OS$
1-(4-(Methylthio)phenyl)-3-octadecylurea ... 561
$C_{26}H_{46}N_2O_2$
1-(*o*-Anisyl)-3-octadecylurea ... 555
1-(*p*-Anisyl)-3-octadecylurea ... 559
1-(3-Methoxyphenyl)-3-octadecylurea ... 558
$C_{26}H_{46}N_2O_3$
Acetic acid *N*²-(4-hexadecyloxy-3-methoxybenzyl)hydrazide ... 497
$C_{26}H_{46}N_2O_4S$
*N*³-Acetyl-*N*¹-hexadecyl-4-methoxy-*N*¹-methylmetanilamide ... 814
2-Hexadecyloxy-5-(morpholinosulfonyl)aniline ... 813
$C_{26}H_{46}N_2O_5S$
N-(*tert*-Butyl)-4-hexadecyloxy-3-nitrobenzenesulfonamide ... 814
$C_{26}H_{46}O_3$
2-(1-Docosenyl)succinic anhydride ... 98
$C_{26}H_{47}N_3O_3S$
*N*¹-Hexadecyl-4-morpholino-metanilamide ... 816
$C_{26}H_{48}N_2O_3S$
*N*¹-(*tert*-Butyl)-4-(hexadecyloxy)-metanilamide ... 813
$C_{26}H_{49}NO_3$
cis-Hexahydro-*N*-(octadecyl)-phthalamic acid ... 122
$C_{26}H_{50}O_4$
Didodecyl oxalate ... 75
$C_{26}H_{50}O_6$
Triethylene glycol bis(decanoate) ... 91
$C_{26}H_{52}N_2O$
1-Cyclohexyl-1-methyl-3-octadecylurea ... 130
$C_{26}H_{53}NO$
N-Octadecyloctanamide ... 106
$C_{26}H_{54}O$
5-Butyl-5-docosanol ... 10
3-Ethyl-3-tetracosanol ... 10
$C_{26}H_{54}O_2$
1,2-Didodecyloxyethane ... 17
$C_{26}H_{55}N$
Dibutyloctadecylamine ... 29
Ditridecylamine ... 28
$C_{27}H_{18}$
3-Methylphenanthro(3,4-c)-phenanthrene ... 167
Truxene ... 167
$C_{27}H_{18}Cl_2N_4O_4$
N,*N*′-(4-Methyl-1,3-phenylene)bis(5-chloro-8-quinolyl carbamate) ... 989
$C_{27}H_{18}Cl_4N_2$
4,4′-Methylenebis(*N*-(2,4-dichlorobenzylidene)aniline) ... 1027
$C_{27}H_{18}O$
10-Benzhydrylidene-9-anthrone ... 313
$C_{27}H_{19}ClO_2$
1,1′-(4-Chlorobenzylidene)di(2-naphthol) ... 193
$C_{27}H_{19}Cl_3O_5$
7-(2,4-Dichlorophenyl)-5,6,8,9-tetrahydrodibenzo(c,h)xanthylium perchlorate ... 926
$C_{27}H_{19}NO_4$
1,1′-(3-Nitrobenzylidene)di(2-naphthol) ... 263
$C_{27}H_{20}$
1,1,3,3-Tetraphenylallene ... 165
1,1,3-Triphenylindene ... 166
$C_{27}H_{20}BClF_4O$
7-(4-Chlorophenyl)-5,6,8,9-tetrahydrodibenzo(c,h)xanthylium tetrafluoroborate ... 926
$C_{27}H_{20}ClNO_2$
2′-Benzoyl-4′-chloro-2,2-diphenylacetanilide ... 405
$C_{27}H_{20}Cl_2N_2$
4,4′-Methylenebis(*N*-(2-chlorobenzylidene)aniline) ... 1025
4,4′-Methylenebis(*N*-(4-chlorobenzylidene)aniline) ... 1027
$C_{27}H_{20}Cl_2O_5$
7-(4-Chlorophenyl)-5,6,8,9-tetrahydrodibenzo(c,h)xanthylium perchlorate ... 931
$C_{27}H_{20}F_2N_2$
4,4′-Methylenebis(*N*-(4-fluorobenzylidene)aniline) ... 1027
$C_{27}H_{20}N_4$
5-(2-Anthracenyl)-1,3-diphenylformazan ... 726
$C_{27}H_{20}N_4O_4$
N,*N*′-Bis(8-quinolyloxycarbonyl)-4-methyl-1,3-phenylenediamine ... 988
$C_{27}H_{20}O$
10-Benzhydryl-9-anthrone ... 315
1,2,3-Triphenyl-1-indenol ... 196
$C_{27}H_{20}O_4$
1,1′-Methylenebis(4-methoxydibenzofuran) ... 893
$C_{27}H_{21}BF_4O$
5,6-Dihydro-2,4-(4-methoxyphenyl)-naphtho(1,2-b)pyrylium tetrafluoroborate ... 934
$C_{27}H_{21}Br$
2-Bromo-1,1,3,3-tetraphenyl-1-propene ... 174
$C_{27}H_{21}ClO_5$
7-Phenyl-5,6,8,9-tetrahydrodibenzo(c,h)xanthylium perchlorate ... 936
$C_{27}H_{21}N$
7-Phenyl-5,6,8,9-tetrahydrodibenz(c,h)acridine ... 966
$C_{27}H_{21}NO_5S_2$
N-(9-Oxo-2-fluorenyl) di-*p*-toluenesulfimide ... 823
$C_{27}H_{21}NO_7$
2,3,5-Tri-*O*-benzoyl-β-D-ribofuranosyl cyanide ... 753
$C_{27}H_{22}$
4a,5,6,12c-Tetrahydro-3-methylphenanthro(3,4-c)-phenanthrene ... 167
1,1,2,2-Tetraphenylcyclopropane ... 165
1,1,3,3-Tetraphenyl-1-propene ... 164
$C_{27}H_{22}BF_4O$
8-Benzylidene-2,4-diphenyl-5,6,7,8-tetrahydro-benzopyrylium tetrafluoroborate ... 936
$C_{27}H_{22}ClNO_2$
N-(4-Chlorophenyl)-2-((2,2-diphenyl-2-hydroxy)ethyl)benzamide ... 446
$C_{27}H_{22}N_2$
4,4′-Methylenebis(*N*-benzylideneaniline) ... 1063
$C_{27}H_{22}N_2O_2$
4,4′-Methylenebis(*N*-(2-hydroxybenzylidene)aniline) ... 1029
4,4′-Methylenebis(*N*-(4-hydroxybenzylidene)aniline) ... 1036
$C_{27}H_{22}N_2O_4$
Diphenyl 4,4′-methylenedicarbanilate ... 673
4,5-Diphenyl-2-(3,4,5-trimethoxybenzylideneamino)-3-furancarbonitrile ... 1043
$C_{27}H_{22}O$
2,3,3-Triphenylpropiophenone ... 297
$C_{27}H_{22}O_2$
Ethyl 2,4,6-triphenylbenzoate ... 370
4,4,5,5-Tetraphenyl-1,3-dioxolane ... 184
$C_{27}H_{22}Si$
1-(Triphenylsilyl)indene ... 1121
$C_{27}H_{23}ClN_2O_6$
3-(*N*-(2,5-Dimethoxyphenyl)-carbamoyl)-2-naphthyl *N*-(5-chloro-2-methoxyphenyl)carbamate ... 620
$C_{27}H_{23}F_3Si$
Phenyltris(4-fluorobenzyl)silane ... 1103
$C_{27}H_{23}NO$
2-(Benzylideneamino)-1,1,2-triphenyl-1-ethanol ... 203
$C_{27}H_{23}NO_3Sn$
Triphenyltin 4-acetamidobenzoate ... 1131
$C_{27}H_{24}F_4Si_2$
2,3,5,6-Tetrafluoro-1-(trimethylsilyl)-4-(triphenylsilyl)benzene ... 1103
$C_{27}H_{24}N_2O_7S_2$
2-(*N*-Benzyl-*p*-toluenesulfonamido)-5-nitrophenyl *p*-toluenesulfonate ... 821
$C_{27}H_{24}N_2O_9S_3$
2-(Di-*p*-tosylamino)-5-nitrophenyl *p*-toluenesulfonate ... 822
$C_{27}H_{24}O_3$
1,3,5-Tris(4-methoxyphenyl)benzene 179
$C_{27}H_{24}O_7$
3-*O*-Benzoyl-1,2:4,6-di-*O*-benzylidene-α-D-glucopyranoside ... 370
$C_{27}H_{24}O_8$
1-*O*-Methyl-2,3,5-tri-*O*-benzoyl-α-D-arabinofuranoside ... 371
$C_{27}H_{24}O_{10}S$
2-*O*-(Methylsulfonyl)-1,3,4-tri-*O*-benzoyl-D-arabinose ... 798
$C_{27}H_{25}BrN_2O_8$
N,*N*′-(2-Bromo-6-methylphenylene)-bis(1,4-benzodioxan-2-ylmethyl carbamate) ... 680
$C_{27}H_{25}BrSi$
(4-Bromophenyl)tribenzylsilane ... 1104
$C_{27}H_{25}ClSn$
(3-(2-Chlorophenyl)propyl)-triphenyltin ... 1130
$C_{27}H_{25}NO_3S$
2-(Diphenylhydroxymethyl)-*N*-methyl-*N*-phenyl-*p*-toluenesulfonamide ... 811
$C_{27}H_{25}NO_4$
Ethyl 2′-methyl-3′-oxo-′.3′,10,11-tetrahydrospiro[5*H*-dibenzo(a,d)-cycloheptene-5,1′-(1*H*)isoindole]-5′-acetate ... 699
$C_{27}H_{25}NSi$
5,10-Dihydro-10,10-diphenyl-2,5,8-trimethylphenazasiline ... 1108
$C_{27}H_{25}N_3O$
4-(5-Indanylmethyleneamino)-benzoic acid (5-indanylmethylene)-hydrazide ... 501
N-(5-Indanylmethylene)-anthranilic acid (5-indanylmethylene)-hydrazide ... 1056
$C_{27}H_{25}N_3O_2$
3-(1,1-Bis(3-indolyl)ethyl)-1,4-diacetyl-1,4-dihydropyridine ... 888
$C_{27}H_{26}BrP$
(3-Phenylpropyl)-triphenylphosphonium bromide ... 832
$C_{27}H_{26}ClOP$
(4-Ethoxybenzyl)-triphenylphosphonium chloride ... 830
$C_{27}H_{26}NO_3P$
Dibenzyl (α-anilinobenzyl)-phosphonate ... 834
$C_{27}H_{26}N_2O$
4-(Bis(4-(dimethylamino)phenyl)-methylene)-1(4*H*)-naphthalenone ... 302
$C_{27}H_{26}N_2O_6$
Bis(4-propionylphenyl) *N*,*N*′-(4-methyl-1,3-phenylene)-biscarbamate ... 670
$C_{27}H_{26}N_6O_2$
4,4′-Methylenebis(1-phenyl-3-(3-pyridylmethyl)urea) ... 950
4,4′-Methylenebis(1-phenyl-3-(4-pyridylmethyl)urea) ... 953
$C_{27}H_{26}OSi$
1-Phenyl-3-(triphenylsilyl)-2-propanol ... 1105
$C_{27}H_{26}Si$
Diphenylphenethyl(*o*-tolyl)silane ... 1119
Phenyl(*o*-tolyl)(*m*-tolyl)(*p*-tolyl)-silane ... 1119
Phenyltribenzylsilane ... 1115
Phenyltri(*o*-tolyl)silane ... 1119
Phenyltri(*m*-tolyl)silane ... 1116
Phenyltri(*p*-tolyl)silane ... 1120
$C_{27}H_{26}Sn$
Mesityltriphenyltin ... 1133
$C_{27}H_{27}NSi$
N,*N*-Dimethyl-α-(triphenylsilyl)-benzylamine ... 1108
$C_{27}H_{28}ClO_4P$
(2-(1-Methyl-2,5-cyclohexadien-1-yl)-ethyl)triphenylphosphonium perchlorate ... 832
$C_{27}H_{28}Cl_2N_2O_6$
1,1′-Trimethylenebis(2-(4-chlorophenyl)-6-oxo-3-piperidinecarboxylic acid) ... 689
$C_{27}H_{28}N_2O_5$
Carbobenzyloxy-β-alanyl-L-phenylalanine benzyl ester ... 658
Carbobenzyloxy-L-alanyl-L-phenylalanine benzyl ester ... 658
Carbobenzyloxyphenyl-alanylalanine benzyl ester ... 658
Carbobenzyloxy-L-phenylalanyl-L-phenylalanine methyl ester ... 658
Carbobenzyloxysarcosyl-phenylalanine benzyl ester ... 658
$C_{27}H_{28}N_2O_6$
Carbobenzyloxy-L-alanyl-L-tyrosine benzyl ester ... 644
Carbobenzyloxy-L-phenylalanyl-L-tyrosine methyl ester ... 644
Carbobenzyloxysarcosyltyrosine benzyl ester ... 642
Carbobenzyloxy-L-seryl-L-phenylalanine benzyl ester ... 645
Carbobenzyloxy-L-tyrosyl-L-phenylalanine methyl ester ... 644
$C_{27}H_{28}N_2O_6S$
S-Benzyl-*N*-carbobenzyloxy-L-cysteinyl-L-tyrosine ... 644
$C_{27}H_{28}N_2O_7$
O-Benzyl-*N*-*tert*-BOC-L-tyrosine 4-nitrophenyl ester ... 652
Carbobenzyloxyseryltyrosine benzyl ester ... 644
Carbobenzyloxytyrosyltyrosine methyl ester ... 642
$C_{27}H_{28}N_2O_9S$
Glycinecresol Red ... 350
$C_{27}H_{28}N_4O$
2,5-Bis(4-(*N*-2-cyanoethyl-*N*-methylamino)benzylidene)-cyclopentanone ... 755
$C_{27}H_{28}Si_2$
(4-(Trimethylsilyl)phenyl)-triphenylsilane ... 1120
$C_{27}H_{29}BrN_2O_6S_2$
N,*N*′-(2-Bromo-6-methylphenylene)-bis(2-(4-(methylthio)phenoxy)ethyl carbamate) ... 617
$C_{27}H_{29}N_3O_5S$
Carbobenzyloxy-*S*-benzyl-L-cysteinyltyrosinamide ... 642
$C_{27}H_{29}N_3O_7$
N,*N*′-Bis-(3,4,5-trimethoxybenzylidene)-anthranilic acid hydrazide ... 499
$C_{27}H_{30}N_2O$
N-Benzyl-*N*-(2-cyclohexyl-1-phenylethyl)-nicotinamide ... 948
$C_{27}H_{30}N_2O_4$
Bis(3-phenylpropyl) *N*,*N*′-(4-methyl-1,3-phenylene)biscarbamate ... 670

$C_{27}H_{31}FN_2O_4$
Ethyl (4-fluorophenyl-isopropyl-2-phenylimidazol-5-yl)dihydroxy-6-heptenoate 865
$C_{27}H_{31}NO_2$
N-Trityl-L-leucine ethyl ester 359
$C_{27}H_{31}NO_{11}S$
N-(3-Methoxybenzylidene)-6-O-(p-tosyl)-1,3,4-tri-O-acetyl-β-D-galactosamine 797
$C_{27}H_{32}ClNO_2$
4-Chloro-α-(2-(4-diethylamino-phenyl)ethoxy)-α-(p-tolyl)-phenethyl alcohol 226
$C_{27}H_{32}Cl_2N_3O_3PS$
O,O'-Bis(4-tert-butylphenyl) (3,4-dichlorophenylureido)-thiophosphoramidate 833
$C_{27}H_{32}N_2O$
1,1-Dibenzyl-3-(2,6-diisopropylphenyl)urea 573
$C_{27}H_{32}N_4O_2$
4,4'-Methylenebis(1,1-diallyl-3-phenylurea) 592
3,3'-(4-Methyl-1,3-phenylene)bis(1-benzyl-1-ethylurea) 563
$C_{27}H_{32}N_4O_4$
4,4'-Methylenebis(O-(carbamoyl)-cyclohexanone oxime) 1012
$C_{27}H_{32}N_4S_2$
4,4'-Trimethylenebis(1-(2-benzothiazolyl)piperidine) 917
$C_{27}H_{32}O_3S$
2-Hydroxy-3-(10-(4-methylphenylthio)decyl)-1,4-naphthoquinone 316
$C_{27}H_{32}Si$
Cyclohexyltribenzylsilane 1111
$C_{27}H_{33}Bi$
Tri(mesityl)bismuthine 1143
$C_{27}H_{33}IPb$
Iodotris(mesityl)lead 1136
$C_{27}H_{33}NO_3$
2,6-Di-tert-butyl-4-methoxymethylphenyl N-(2-naphthyl)carbamate 655
$C_{27}H_{33}N_2O_3PS$
Bis(4-tert-butylphenyl) N-(2-hydroxybenzylideneamino)-thiophosphoramidate 834
$C_{27}H_{33}OP$
Tris(3-phenylpropyl)phosphine oxide 836
$C_{27}H_{34}N_2$
4,4'-((4-tert-Butylphenyl)methylene)-bis(N,N-dimethylaniline 233
$C_{27}H_{34}N_2O$
N'-(α-(4-Benzyloxyphenyl)-phenethyl)-N,N-diethylethylenediamine di-hydro-chloride 240
$C_{27}H_{34}N_2O_4$
4,4'-Methylenebis(cyclohexyl N-phenylcarbamate) 673
$C_{27}H_{34}N_4O_8$
N,N'-Bis(carbobenzyloxy)-L-lysylglycylglycine methyl ester 663
$C_{27}H_{34}OSi$
Tris(2,4,6-trimethylphenyl)silanol 1109
$C_{27}H_{34}Pb$
Trimesityllead 1136
$C_{27}H_{35}N_3$
2,6-Dimethyl-4,4',4''-methylidynetris(N,N-dimethylaniline) 233
$C_{27}H_{36}N_2O_5$
Carbobenzyloxyisoleucyliso-leucine benzyl ester 663
Carbobenzyloxy-L-isoleucyl-L-leucine benzyl ester 663
Carbobenzyloxy-L-leucyl-L-leucine benzyl ester 663
$C_{27}H_{36}N_4O_2$
4,4'-Methylenebis(1-cyclohexyl-3-phenylurea) 584
N,N'-(Methylenedi-4,1-phenylene)-bis(2-methyl-1-piperidinecarboxamide) 712
N,N'-(Methylenedi-4,1-phenylene)-bis(3-methyl-1-piperidinecarboxamide) 712
N,N'-(Methylenedi-4,1-phenylene)-bis(4-methyl-1-piperidinecarboxamide) 712
$C_{27}H_{36}N_4O_4$
4,4'-Methylenebis(N-(2,6-dimethylmorpholinocarboxamido)-aniline) 713
$C_{27}H_{36}N_4O_6$
N,N'-Bis(carbobenzyloxy)-L-lysyl-L-valinamide 663
$C_{27}H_{36}N_4O_8$
Pimelic acid (3,4,5-trimethoxybenzylidene)-hydrazide 1051
$C_{27}H_{36}N_4S_2$
4,4'-Trimethylenebis(1-(N-phenylthiocarbamoyl)piperidine) 747
$C_{27}H_{37}NO_4PbS$
Diethylmethyl(3-(triphenylplumbly)-propyl)ammonium methosulfate 1134
$C_{27}H_{38}F_2N_4O_6S_4$
4-(tert-Dodecylsulfamoyl)benzene-sulfonic (5-(difluoromethylsulfonyl-3-methyl-2-benzene-thiazolinylidene)hydrazide 805
$C_{27}H_{39}ClN_2O_3$
3-Benzylcarbamoyl-1-(dodecyloxycarbonylmethyl)-pyridinium chloride 955
$C_{27}H_{40}N_2O_3$
2-Decyloxy-N-(2,6-diethylphenyl)-5-nitrobenzylamine 267
$C_{27}H_{40}N_4O_2$
4,4'-Methylenebis(1,1-diisopropyl-3-phenylurea) 592
4,4'-Methylenebis(3-(1,3-dimethylbutyl)-1-phenylurea) 592
4,4'-Methylenebis(1,1-dipropyl-3-phenylurea 592
$C_{27}H_{40}O_8$
2,6-Bis(2,2-(diethoxycarbonyl)-propyl)mesitylene 360
$C_{27}H_{40}Si_4$
Tris(4-(trimethylsilyl)phenyl)silane 1122
$C_{27}H_{42}ClNO$
Benzyldiethyl(2-(4-(1,1,3,3-tetramethylbutyl)phenoxy)ethyl)-ammonium chloride 254
$C_{27}H_{42}N_2O_6$
4-Hexadecyloxy-3-(2-(hydroxyimino)-acetoacetamido)-benzoic acid 1009
$C_{27}H_{42}N_4O_5S$
3-Carboxy-4-hydroxybenzenesulfonic (methyl-pentadecyl-pyrimidinylidene)-hydrazide 774
$C_{27}H_{42}Si_5$
2,2-Bis(trimethylsilyl)-1,1,1-trimethyl-3,3,3-triphenyltrisilane 1116
$C_{27}H_{43}NO$
1-Hexadecyloxy-2-naphthalenemethylamine 240
$C_{27}H_{43}NO_3S$
4-(Morpholinothiocarbonyl)phenyl palmitate 742
$C_{27}H_{43}NO_5$
3-Acetoacetamido-4-(hexadecyloxy)-benzoic acid 442
$C_{27}H_{43}NS$
2-(Octadecylthio)quinoline 987
$C_{27}H_{44}ClNO_2S$
4-(3-Chloro-4-(hexadecyloxy)-thiobenzoyl)-morpholine 738
$C_{27}H_{44}N_2O$
Octadecanoic cinnamylidenehydrazide 502
$C_{27}H_{44}N_2O_2$
1,2-Dimethyl-3-(4-(hexadecyloxy)-phenyl)-3-pyrazolin-5-one 699
$C_{27}H_{44}N_4O_2$
3,3'-(4-Methyl-1,3-phenylene)bis(1-cyclohexyl-1-isopropylurea) 565
$C_{27}H_{44}O_3$
Ethyl 2-benzoyloctadecanoate 364
$C_{27}H_{44}O_5$
Methyl 2-(4-hexadecyloxy-3-methoxybenzoyl)acetate 363
$C_{27}H_{45}FN_4O_5S_3$
1-Hexadecanesulfonic (fluorophenyl)(sulfopropyl)-thiadiazol-2-ylidene)hydrazide 821
$C_{27}H_{45}IN_2O_4$
(4'-Acetamidophenacyl)diethyl(10-ethoxycarbonyldecyl)ammonium iodide 431
$C_{27}H_{45}N_3O$
1-(4-Aminophenyl)-3-heptadecyl-2-methyl-3-pyrazolin-5-one 701
$C_{27}H_{45}N_3O_2$
Stearic (4-acetamidobenzylidene)-hydrazide 502
$C_{27}H_{46}ClN_3O_3S$
1-Hexadecanesulfonic acid (2-chloro-α-morpholinobenzylidene)hydrazide 821
$C_{27}H_{46}N_2O$
Stearic (3-phenylpropylidene)-hydrazide 502
$C_{27}H_{46}N_2O_2$
Acetic acid (3-isooctadecyloxybenzylidene)-hydrazide 498
$C_{27}H_{46}N_2O_3$
Stearic (2,5-dimethoxybenzylidene)-hydrazide 497
Stearic (3,4-dimethoxybenzylidene)-hydrazide 497
Stearic (3-ethoxy-4-hydroxybenzylidene)hydrazide 495
$C_{27}H_{46}N_6O_2$
1,1'-(4-Methyl-1,3-phenylene)bis(3-(3-(2-methylpiperidino)propyl)urea) 565
$C_{27}H_{46}O_2S$
Ethyl 2-(octadecylthio)benzoate 372
$C_{27}H_{46}O_4$
2,4-Di(decyloxy)benzoic acid 341
$C_{27}H_{47}ClN_2O$
Dimethyl(dodecylcarbamoyl-methyl)(5,6,7,8-tetrahydro-2-naphthylmethyl)ammonium chloride 431
$C_{27}H_{47}N_3O_2$
1-(3-Acetamidophenyl)-3-octadecylurea 565
$C_{27}H_{47}N_3O_4S$
1-Hexadecanesulfonic acid (m-hydroxy-α-morpholinobenzylidene)-hydrazide 812
$C_{27}H_{47}N_3O_5S_3$
1-Hexadecanesulfonic acid (3-(4-sulfobutyl)-2-benzothiazolinylidene)hydrazide 820
$C_{27}H_{48}ClNO_3$
Bis-(2-hydroxyethyl)-cinnamyl(2-dodecyloxyethyl)ammonium chloride 254
$C_{27}H_{48}N_2O$
1-Benzyl-1-methyl-3-octadecylurea 572
1-Ethyl-3-octadecyl-1-phenylurea 579
1-(4-Ethylphenyl)-3-octadecylurea 591
1-Octadecyl-3-(2-phenylethyl)urea 570
1-Octadecyl-3-(2,3-xylyl)urea 596
1-Octadecyl-3-(2,4-xylyl)urea 602
1-Octadecyl-3-(2,5-xylyl)urea 602
1-Octadecyl-3-(2,6-xylyl)urea 596
1-Octadecyl-3-(3,4-xylyl)urea 602
$C_{27}H_{48}N_2O_2$
1-(2-Methoxy-5-methylphenyl)-3-octadecylurea 557
$C_{27}H_{48}N_2O_3$
1-(2,4-Dimethoxyphenyl)-3-octadecylurea 550
$C_{27}H_{48}N_4O_3S$
1-Hexadecanesulfonic acid 2-(3-amino-α-morpholinobenzylidene)-hydrazide 816
$C_{27}H_{49}ClN_2O$
Benzyldimethyl(hexadecyl-carbamoylmethyl)ammonium chloride 424
$C_{27}H_{49}NO_3S$
N-(Hexadecyl)-N-(4-sulfobutyl)-m-toluidine 789
$C_{27}H_{49}N_3O_3S$
3-Amino-N-hexadecyl-4-(4-morpholino)-N-methylbenzenesulfonamide 816
$C_{27}H_{50}N_2O_4S$
N-(2-(2,5-Dimethoxy-N-hexadecylanilino)-ethyl)-methanesulfonamide 813
$C_{27}H_{50}N_3O_2S$
2,6-Dimethyl-N^4-hexadecyl-N^4-(2-(methylsulfonylamino)ethyl)-1,4-phenylenediamine 816
$C_{27}H_{50}O_4$
Diethyl ethyl(oleyl)malonate 94
$C_{27}H_{51}N_3O_2S$
3-Amino-N^4-octyl-N^1,N^1,N^4-trimethylsulfanilamide 816
$C_{27}H_{52}O_4$
Diethyl ethyl(octadecyl)malonate 94
$C_{27}H_{54}BrNO_4$
3-(4-Bromobutoxy)-2-methoxypropyl N-octadecylcarbamate 112
$C_{27}H_{54}O$
14-Heptacosanone 50
$C_{27}H_{55}NO_5$
3-(4-Hydroxybutoxy)-2-methoxypropyl N-octadecylcarbamate 114
$C_{27}H_{57}BO_3$
Tris(1-isobutyl-3-methylbutyl) borate 1124
Tris(3,3,5-trimethylhexyl) borate 1124
$C_{27}H_{64}Si_8$
1,2,3,4-Tetrakis(trimethylsilyl)-5-(tris(trimethylsilyl)silyl)benzene 1098
$C_{27}H_{68}Si_8$
3,4,5,6-Tetrakis(trimethylsilyl)-1-(tris(trimethylsilyl)silyl)-1-cyclohexene 1111
$C_{28}H_{15}NO_4$
1,1'-Dianthrimide 318
$C_{28}H_{16}F_{12}Si$
Tetrakis(3-(trifluoromethyl)phenyl)-silane 1100
$C_{28}H_{17}NO_2$
10-Nitro-9,9'-biphenanthrene 280
$C_{28}H_{17}N_3O_4$
1,1'-Iminobis(4-aminoanthraquinone) 317
$C_{28}H_{18}$
9,9'-Biphenanthrene 168
$C_{28}H_{18}Cl_2N_2O_8S_2$
Solubilised Vat Green 21 (C.I. 59051) 1094
$C_{28}H_{18}N_2O_4$
1,8-Bis(benzamido)anthraquinone 488
$C_{28}H_{18}N_2O_5$
Algol Red (1,5-dibenzamido-4-hydroxyanthraquinone) 486
$C_{28}H_{18}O_2$
Bianthronyl 314
$C_{28}H_{18}O_6$
1,1'-Oxalylbis(4-methoxydibenzofuran) 893
$C_{28}H_{18}O_7$
4,4'-Oxydibenzoic dianhydride 388
$C_{28}H_{19}N_3O_6S_2$
Azocarmine G (C.I. 50085) 1000
$C_{28}H_{20}$
1,2,3,-Triphenylazulene 166
1,2,3-Triphenylnaphthalene 166
$C_{28}H_{20}Cl_2N_2O_2$
1,4-Bis(p-tolylamino)-6,7-dichloroanthraquinone 317
$C_{28}H_{20}F_4N_2O$
N-(1,2-Bis(2-fluorophenyl)-2-(2-fluorobenzylideneamino)ethyl)-2-fluorobenzamide 450
$C_{28}H_{20}F_6N_4O_4S$
1,1'-(Sulfonyldi-4,1-phenylene)bis(3-(2-trifluoromethylphenyl)urea) 772
$C_{28}H_{20}Hg_4O_{13}$
2',4',5',7'-Tetrakis-(acetoxymercuri)-fluorescein 350

$C_{28}H_{20}N_2$
2,3,5,6-Tetraphenylpyrazine (amaron) 978
2,2,3,3-Tetraphenylsuccinonitrile 755
$C_{28}H_{20}N_2O_2$
N,N'-Bis(2-hydroxy-1-naphthylmethylene)-1,4-phenylenediamine 1033
$C_{28}H_{20}N_2O_4S_2$
4-(Piperonylideneamino)phenyl disulfide 1052
$C_{28}H_{20}N_2O_6S$
4-(Piperonylideneamino)phenyl sulfone 1063
$C_{28}H_{20}N_4$
3,4,7,8-Tetraphenyl-1,2,5,6-tetrazocine 249
$C_{28}H_{20}O$
Tetraphenylfuran 846
$C_{28}H_{20}O_2$
1,2,3,4-Tetraphenyl-2-buten-1,4-dione 313
2,3,5,6-Tetraphenyl-1,4-dioxin 977
$C_{28}H_{20}O_4$
1,8-Dibenzyloxyanthraquinone 316
$C_{28}H_{20}O_6$
Anthrarufin bis-(3-methoxyphenyl)-ether 317
3,3'-(5-Indanylmethylene)bis(4-hydroxycoumarin) 382
$C_{28}H_{20}S$
Tetraphenylthiophene 859
$C_{28}H_{20}S_4$
4,4'-Bis(2-naphthyl)-5,5'-dimethyltetrathiafulvalene 880
$C_{28}H_{20}Si$
Bis(phenylethynyl)diphenylsilane 1116
$C_{28}H_{21}ClO_2$
9,10-Bis(4-methoxyphenyl)-1-chloroanthracene 183
$C_{28}H_{22}$
1,2,3,4-Tetraphenyl-1,3-butadiene 165
$C_{28}H_{22}Cl_4N_4O_2$
4,4'-Bis(3,5-dichlorophenylureido)-3,3'-dimethylbiphenyl 545
$C_{28}H_{22}Cl_7NO_8$
N,N-Bis(2-(2-(2,4,5-trichlorophenoxy)acetoxy)ethyl)-2-(4-chlorophenoxy)acetamide 398
$C_{28}H_{22}NO_2P$
2-Hydroxy-1-phenyl-3-(triphenylphosphonium)-2-pyrrolin-5-one, inner salt 830
$C_{28}H_{22}N_2O_2$
4',4'''-Vinylenedibenzanilide 478
$C_{28}H_{22}N_2O_3$
Sudan Green (C.I. 61565) 317
$C_{28}H_{22}N_2O_4$
1,4-Dihydroxy-5,8-di-*p*-toluidinoanthraquinone 316
$C_{28}H_{22}N_4O_2$
2,3,6,7-Tetrahydro-1,2,5,6-tetraphenyl-1*H*,5*H*-*s*-triazolo(1,2-a)-*s*-triazole-3,7-dione 705
$C_{28}H_{22}O$
9-Anthracenylmethyl 4-benzylphenyl ether 187
$C_{28}H_{22}OSi$
6-(Triphenylsilyl)-2-naphthol 1105
$C_{28}H_{22}OSn$
6-(Triphenylstannyl)-2-naphthol 1130
$C_{28}H_{22}O_2$
1,1'-(4-Methylbenzylidene)di(2-naphthol) 194
1,1'-Terephthalylidenebis(4-phenyl-3-buten-2-one) 287
1,1,4,4-Tetraphenyl-2-butyne-1,4-diol 1016
$C_{28}H_{22}O_2S$
1-Phenylphenacyl sulfide 308
$C_{28}H_{22}O_3$
1,1'-(2-Methoxybenzylidene)di-2-naphthol 194
1,1'-(4-Methoxybenzylidene)di-2-naphthol 194
$C_{28}H_{22}O_6$
3,3'-(4-Isopropylbenzylidene)bis(4-hydroxycoumarin) 382
$C_{28}H_{22}O_9$
3,3'-(3,4,5-Trimethoxybenzylidene)-bis(4-hydroxycoumarin) 382
$C_{28}H_{22}Pb$
(1-Naphthyl)triphenyllead 1136
$C_{28}H_{22}Si$
(2-Naphthyl)triphenylsilane 1122
$C_{28}H_{22}Sn$
(1-Naphthyl)triphenyltin 1133
$C_{28}H_{23}BF_4O$
7-(4-Methylphenyl)-5,6,8,9-tetrahydrodibenzo(c,h)xanthylium tetrafluoroborate 936
$C_{28}H_{23}BF_4O_2$
7-(4-Methoxyphenyl)-5,6,8,9-tetrahydrodibenzo(c,h)xanthylium tetrafluoroborate 934
$C_{28}H_{23}BrN_4O_4S_2$
2-(4-Bromobenzyl)isothiouronium 8-(4-hydroxy-1-naphthylazo)-2-naphthalenesulfonate 1080
$C_{28}H_{23}Br_2ClN_4O_2$
1,1'-((2-Chlorobenzyl)-2-methylphenylene)bis(3-(4-bromophenyl)urea) 537
$C_{28}H_{23}ClI_2N_4O_2$
1,1'-((2-Chlorobenzyl)-2-methylphenylene)bis(3-(4-iodophenyl)urea) 537
$C_{28}H_{23}ClO_5$
7-(4-Methylphenyl)-5,6,8,9-tetrahydrodibenzo(c,h)xanthylium perchlorate 980
$C_{28}H_{23}ClO_6$
7-(2-Methoxyphenyl)-5,6,8,9-tetrahydrodibenzo(c,h)xanthylium perchlorate 934
7-(4-Methoxyphenyl)-5,6,8,9-tetrahydrodibenzo(c,h)xanthylium perchlorate 934
$C_{28}H_{23}Cl_3N_4O_2$
1,1'-((2-Chlorobenzyl)-2-methylphenylene)bis(3-(3-chlorophenyl)urea) 536
$C_{28}H_{23}NO$
7-(4-Methoxyphenyl)-5,6,8,9-tetrahydrodibenzo(c,h)acridine 939
$C_{28}H_{23}P$
8-Benzylidene-2,4-diphenyl-5,6,7,8-tetrahydrophosphinoline 966
$C_{28}H_{24}Cl_2N_4O_2$
4,4'-Bis(3-chlorophenylureido)-3,3'-dimethylbiphenyl 545
$C_{28}H_{24}Cl_2N_4O_4S$
1,1'-(Sulfonyldi-4,1-phenylene)bis(3-(3-chloro-4-methylphenyl)urea) 774
1,1'-(Sulfonyldi-4,1-phenylene)bis(3-(4-chloro-2-methylphenyl)urea) 773
$C_{28}H_{24}Fe_2N_2$
N,N'-Bis-(ferrocenylmethylene)-1,4-phenylenediamine 1141
$C_{28}H_{24}N_2$
N,N'-Bis(diphenylmethylene)-ethylenediamine 242
$C_{28}H_{24}N_2O$
N-(2-Benzylideneamino-1,2-diphenylethyl)-benzamide 1053
$C_{28}H_{24}N_2O_2$
3-Benzyloxybenzaldehyde azine 1050
3,4-Bis(1-ethyl-2(1*H*)-quinolinylidenemethyl)-1,3-cyclobutadienediylium-1,2-diolate 1095
2,4-Bis(1-ethyl-2(1*H*)-quinolylidenemethyl)-1,3-cyclobutadienediylium-1,3-diolate 982
1,2-Diphenyl-4-(3-(1-naphthyl)-propyl)-3,5-pyrazolidinedione 708
$C_{28}H_{24}N_2O_2S$
4-(4-Methylbenzylideneamino)-phenyl sulfone 1063
$C_{28}H_{24}N_2O_2Te_2$
Bis(2-(4-methoxybenzylideneamino)-phenyl) ditelluride 1041
$C_{28}H_{24}N_2O_3$
2-(3,4-Dibenzyloxyphenyl)-1,2-dihydro-4(3*H*)-quinazolinone 698
$C_{28}H_{24}N_2O_4S$
4-(*N*-(4-Methoxybenzylidene)amino)-phenyl sulfone 1050
$C_{28}H_{24}N_2O_6$
4,4'-Bis-(2-carboxyphenylamino)-3,3'-dimethoxybiphenyl 340
$C_{28}H_{24}N_2O_6S$
4,4'-Sulfonylbis(*N*-(2-hydroxy-3-methoxybenzylidene)aniline) 1030
$C_{28}H_{24}N_2S_2$
4,4'-Dithiobis(*N*-(4-methylbenzylidene)aniline) 1052
$C_{28}H_{24}O$
1,2,3,4-Tetrahydro-1,2,3-triphenyl-1-naphthol 204
$C_{28}H_{24}O_2$
9,10,19,20-Tetrahydrotetrabenzo(a,c,g,l)-cyclododecene-9,19-diol 200
$C_{28}H_{24}O_2Si$
Bis(9-xanthenyl)dimethylsilane 1111
$C_{28}H_{24}O_9$
D-Lyxofuranose 1-acetate 2,3,5-tribenzoate 370
$C_{28}H_{25}NO$
N,3,4,4-Tetraphenylbutyramide 438
$C_{28}H_{25}NO_2$
N-Tritylphenylalanine diethylamine salt 349
$C_{28}H_{26}$
1,1,4,4-Tetraphenylbutane 163
$C_{28}H_{26}BF_4P$
(4-(2,5-Cyclohexadien-1-ylidene)-2-buten-1-yl)triphenylphosphonium tetrafluoroborate 832
$C_{28}H_{26}CuN_4O_4$
Bis(2-methoxy-6-(4-methyl-2-pyridinyliminomethyl)phenolato)-copper 1049
$C_{28}H_{26}IP$
(4-(2,5-Cyclohexadien-1-ylidene)-2-buten-1-yl)triphenylphosphonium iodide 832
$C_{28}H_{26}N_2O_6$
5'-Trityluridine 700
$C_{28}H_{26}N_2O_7$
Carbobenzyloxy-*O*-benzyl-L-tyrosine *N*-hydroxysuccinimide ester 717
$C_{28}H_{26}N_4O_2$
Adipic bis((1-naphthylmethylene)-hydrazide) 501
$C_{28}H_{26}O_2$
3-(3-Benzylidene-2-oxocyclohexyl)-3-phenylpropiophenone 304
1,1-Diphenyl-2,2-di(*p*-tolyl)-1,2-ethanediol 200
1,2-Diphenyl-1,2-di-*p*-tolyl-1,2-ethanediol 203
$C_{28}H_{26}O_3$
4-Methoxy-α-phenylbenzyl ether 186
$C_{28}H_{26}O_3Sn$
Ethyl 2-(3-(triphenylstannyl)-phenoxy)acetate 1130
$C_{28}H_{27}ClN_6O_6S_2$
1,1'-((2-Chlorobenzyl)-2-methylphenylene)bis(3-(4-sulfamoylphenyl)urea) 581
$C_{28}H_{27}N_3O_5$
Carbobenzyloxy-L-tryptophyl-L-phenylalanine 889
$C_{28}H_{28}BrOP$
(2,5-Dimethyl-4-methoxybenzyl)-triphenylphosphonium bromide 830
(3-(3-Methoxyphenyl)propyl)-triphenylphosphonium bromide 830
$C_{28}H_{28}Br_8N_2O_6$
N,N'-Dodecamethylenebis(3,4,5,6-tetrabromophthalamic acid) 484
$C_{28}H_{28}Ge$
Tetrabenzylgermane 1116
$C_{28}H_{28}N_2$
N,N'-Dimethyl-1,2,*N,N'*-tetraphenyl-1,2-ethylenediamine 232
$C_{28}H_{28}N_2O_2$
Bis(1,3,3-trimethyl-2-indolinylidenemethyl)-1,3-cyclobutadienediylium-1,3-diolate 245
$C_{28}H_{28}N_2O_6$
2,5-Bis(*N*-mesitylcarbamoyl)-terephthalic acid 467
$C_{28}H_{28}N_4O_4$
Carbobenzyloxy-L-phenylalanyl-L-tryptophanamide 886
Carbobenzyloxy-L-tryptophyl-L-phenylalaninamide 886
$C_{28}H_{28}N_4O_5$
Carbobenzyloxy-L-tyrosyl-L-tryptophanamide 881
$C_{28}H_{28}OSi$
(4-Methoxyphenyl)tribenzylsilane 1106
(4-Methoxyphenyl)tri(*o*-tolyl)silane 1107
(*o*-Tolyloxy)tri(*o*-tolyl)silane 1122
$C_{28}H_{28}O_2P_2$
Ethylenebis((2-methoxyphenyl)-phenylphosphine) (dipamp) 830
$C_{28}H_{28}O_4Pb$
Tetrakis(2-methoxyphenyl)lead 1134
Tetrakis(4-methoxyphenyl)lead 1134
$C_{28}H_{28}O_4Si$
Tetrakis(2-methoxyphenyl)silane 1107
Tetratolyl orthosilicate, mixture of isomers 1098
$C_{28}H_{28}Pb$
Benzyltri(*p*-tolyl)lead 1136
Tetrakis(*m*-tolyl)lead 1133
$C_{28}H_{28}S_4Si$
Tetrakis(phenylthiomethyl)silane 1107
$C_{28}H_{28}Si$
Dibenzyldi(*m*-tolyl)silane 1116
Diphenethyldiphenylsilane 1116
Tetrakis(*m*-tolyl)silane 1098
Tetrakis(*p*-tolyl)silane 1122
$C_{28}H_{28}Sn$
Diphenethyldiphenyltin 1132
Tetrakis(*o*-tolyl)tin 1133
Tetrakis(*m*-tolyl)tin 1129
Tetrakis(*p*-tolyl)tin 1133
$C_{28}H_{29}Bi$
Dimesityl(1-naphthyl)bismuthine 1143
$C_{28}H_{29}NPb$
N,N-Diethyl-4-(triphenylplumbyl)-aniline 1134
$C_{28}H_{29}NSi$
N,*N*,α-Trimethyl-α-(triphenylsilyl)-benzylamine 1108
N,*N*,4-Trimethyl-α-(triphenylsilyl)-benzylamine 1108
$C_{28}H_{29}N_3O_6$
Carbobenzyloxyglycylglycyl-L-phenylalanine benzyl ester 658
$C_{28}H_{29}N_3O_7$
Carbobenzyloxyglycylglycyl-L-tyrosine benzyl ester 645
$C_{28}H_{29}N_3O_8$
N,N'-Bis(benzyloxycarbonyl)-L-lysine 4-nitrophenyl ester 659
$C_{28}H_{30}Cl_2N_6$
4,4'-(1,2-Ethanediyldi-4,1-piperidinediyl)bis(7-chloroquinazoline) 1000
$C_{28}H_{30}I_6N_2O_6$
N,N'-Adipoylbis-(3-amino-α-ethyl-2,4,6-triiodohydrocinnamic acid) 405
$C_{28}H_{30}N_2O_5S$
Carbobenzyloxyalanyl-*S*-benzylcysteine benzyl ester 655
Carbobenzyloxy-*S*-benzyl-L-cysteinyl-L-alanine benzyl ester 656
$C_{28}H_{30}N_2O_6$
Carbobenzyloxy-L-threonyl-L-phenylalanine benzyl ester 644
$C_{28}H_{30}N_2O_6S$
Carbobenzyloxy-*S*-benzyl-L-cysteinyl-L-tyrosine methyl ester 644
$C_{28}H_{30}N_2O_7$
Carbobenzyloxythreonyltyrosine benzyl ester 644
$C_{28}H_{30}N_2Pb$
Bis(4-(dimethylamino)phenyl)-diphenyllead 1134
$C_{28}H_{30}O_2Si_2$
1,1,2,2-Tetra(*o*-tolyl)-1,2-disilanediol 1122
$C_{28}H_{30}Si_2$
4,4'-Bis(dimethylphenylsilyl)-biphenyl 1121

$C_{28}H_{30}Si_2$ (Cont'd)
1,1-Diethyl-1,2,2,2-tetraphenyldisilane ... 1116
Diphenyl(*o*-tolyl)(4-trimethylsilyl)phenyl)silane ... 1119

$C_{28}H_{31}BN_2O$
Dimethyl(ethoxyimino)ammonium tetraphenylborate ... 44

$C_{28}H_{31}NO_4SSn$
Trimethyl(4-(triphenylstannyl)phenyl)ammonium methosulfate ... 1130

$C_{28}H_{31}N_3O_4S$
Carbobenzyloxy-L-prolyl-L-methionine-2-naphthylamide ... 700

$C_{28}H_{32}N_2$
N,*N*′-(1,4-Phenylenedimethylidyne)bis(4-butylaniline) ... 1065

$C_{28}H_{32}N_2O_2$
N,*N*′ Bis(carvacryl)terephthalamide ... 480

$C_{28}H_{32}N_2O_3$
2-(3,6-Bis-(diethylamino)-9*H*-xanthen-9-yl)benzoic acid sulfate ... 343

$C_{28}H_{32}N_2Si_2$
1,2-Bis(dimethylphenylsilyl)-1,2-diphenylhydrazine ... 1108

$C_{28}H_{32}N_4Si_2$
3,3,6,6-Tetramethyl-1,2,4,5-tetraphenyl-1,2,4,5-tetraaza-3,6-disilacyclohexane ... 1116

$C_{28}H_{32}O_2$
2-Hydroxy-3-phenyl-3-(2,3,5,6-tetramethylphenyl)-2′,4′,6′-trimethylpropiophenone ... 290

$C_{28}H_{32}O_5$
2-Deoxy-1-*O*-methyl-3,5,6-tri-*O*-benzyl-α,β-D-galactofuranoside ... 178

$C_{28}H_{33}F_3N_3O_2PS_2$
O,*O*′-Bis(4-*tert*-butylphenyl) (3-(trifluoromethyl)-phenylthioureido)-thiophosphoramidate ... 833

$C_{28}H_{33}N_2O_4PS$
O,*O*′-Bis(4-*tert*-butylphenyl) (piperonylideneamino)-thiophosphoramidate ... 835

$C_{28}H_{34}N_2O_6S$
4-Nitro-4′-(2-(4-(1,1,3,3-tetramethylbutyl)phenoxy)ethoxy)-benzenesulfonanilide ... 813

$C_{28}H_{34}N_4O_2$
Sebacic bis(cinnamylidenehydrazide) ... 501
Suberic acid bis(5-indanylmethylenehydrazide) ... 1075

$C_{28}H_{35}ClN_3O_4PS$
O,*O*′-Bis(4-*tert*-butylphenyl) (3-(2-chloro-6-methoxyphenyl)ureido)-thiophosphoramidate ... 833

$C_{28}H_{35}N_4O_6PS$
O,*O*′-Bis(4-*tert*-butylphenyl) (2-methoxy-4-nitrophenylureido)-thiophosphoramidate ... 834

$C_{28}H_{36}N_2O_4S$
4-Amino-4′-(2-(4-*tert*-octylphenoxy)-ethoxy)benzenesulfonanilide ... 812

$C_{28}H_{36}N_2O_6$
N,*N*′-Bis(4-(ethoxycarbonyl)phenyl)-1,10-decanediamide ... 433

$C_{28}H_{36}N_2O_7$
N-(*tert*-Butoxycarbonyl)-L-valyl-L-aspartic acid dibenzyl ester ... 677

$C_{28}H_{36}N_3O_3PS_2$
Bis(4-*tert*-butylphenyl) *N*-(4-(methylthio)phenylureido)-thiophosphoramidate ... 834

$C_{28}H_{36}O_{15}$
4-Methoxy-2′,3,4′,6′-tetrahydroxy-hydrocinnamophenone 4′-β-neohesperidoside ... 290

$C_{28}H_{36}Si$
Decyltriphenylsilane ... 1116

$C_{28}H_{37}ClN_3O_3PS$
Bis(4-t-butylphenyl) (chloro-methylphenyl-methylthio-ureido)-thiophosphoramidate ... 833

$C_{28}H_{37}NO_8$
3,3′,4,4′,5,5′-Hexamethoxylobelanine ... 291

$C_{28}H_{37}N_3O_7$
N,*N*′-Bis(carbobenzyloxy)-L-lysyl-L-isoleucine ... 659
N,*N*′-Bis(carbobenzyloxy)-lysylvaline methyl ester ... 663

$C_{28}H_{38}F_2N_4O_5S_3$
3-(Alkylcarbamoyl)benzenesulfonic acid (5-difluoromethylsulfonyl-3-methyl-2-benzo-thiazolinylidene)hydrazide ... 805
3(*tert*-Dodecylcarbamoyl)benzene-sulfonic acid (5-difluoromethyl-sulfonyl-3-methyl-2-benzo-thiazolinylidene)hydrazide ... 805

$C_{28}H_{38}N_2O_5S$
N-(2-(2-Dodecyloxyphenylsulfamoyl)-ethyl)-phthalimide ... 815

$C_{28}H_{38}N_4O_4$
1,2-Bis(4-nitro-1,1,3,3,5-pentamethyl-6-indanyl)hydrazine ... 275

$C_{28}H_{38}N_4O_6$
N,*N*′-Bis(carbobenzyloxy)-L-lysyl-L-isoleucinamide ... 663
N,*N*′-Bis(carbobenzyloxy)-L-lysyl-L-leucinamide ... 663
Sebacic acid bis(2,5-dimethoxybenzylidenehydrazide) ... 1040
Sebacic bis((3,4-dimethoxybenzylidene)hydrazide) ... 497

$C_{28}H_{38}N_4O_8$
Pactamycin ... 546

$C_{28}H_{38}O_3$
4′-Phenylphenacyl tetradecanoate ... 357

$C_{28}H_{40}N_2O_4$
N,*N*′-Hexamethylenebis(4-*tert*-butylphenyl carbamate) ... 677
N,*N*′-Hexamethylenebis(2-isopropyl-5-methylphenylcarbamate) ... 686

$C_{28}H_{40}N_2O_6$
N,*N*′-Hexamethylenebis(2-(4,5-dimethoxy-2-methylphenyl)-acetamide) ... 413

$C_{28}H_{40}O$
4′-Phenylhexadecanophenone ... 297

$C_{28}H_{40}O_2$
4′-Phenoxyhexadecanophenone ... 292
3,3′,5,5′-Tetra-*tert*-butyldiphenoquinone ... 54

$C_{28}H_{40}O_4Pb$
Bis-(2-ethylhexanoyloxy)-diphenyllead ... 1135

$C_{28}H_{40}Si$
Bis(2-cyclohexylethyl)-diphenylsilane ... 1111

$C_{28}H_{41}ClN_2O$
1-(3-Chloro-*p*-tolyl)-4-(3-(4-*tert*-octyl)-phenoxy)propyl)piperazine-hydrochloride ... 247
1-(3-Chloro-*p*-tolyl)-4-(6-(4-*tert*-pentylphenoxy)-hexyl)-piperazine hydrochloride ... 247

$C_{28}H_{41}N$
1,3,6,8-Tetra-*tert*.-butylcarbazole ... 892

$C_{28}H_{41}NO$
4′-Phenylhexadecananilide ... 489

$C_{28}H_{41}NO_3$
4-Hexadecyloxy-3-nitrobiphenyl ... 269

$C_{28}H_{41}N_3O_6S$
2′-Hydroxy-4′-(*p*-nitrobenzenesulfonamido)-hexadecananilide ... 812

$C_{28}H_{42}N_2$
N,*N*-Diamyl-*N*′,*N*′-dibenzyl-1,4-diamino-2-butene dihydrochloride ... 242

$C_{28}H_{42}N_2O_4S$
N-Hexadecyl-3-nitrobenzenesulfonanilide ... 822
N-Hexadecyl-4-nitrobenzenesulfonanilide ... 822

$C_{28}H_{42}O$
Heptadecyl 2-naphthyl ketone ... 298

$C_{28}H_{42}O_2$
4,4′-Dihydroxy-3,3′,5,5′-tetra-*tert*-butylbiphenyl ... 194
2-(Octadecanoyl)-1-naphthol ... 294

$C_{28}H_{42}O_2S_2$
2,2′-Dithiobis-(4-(1,1,3,3-tetramethylbutyl)phenol) ... 210

$C_{28}H_{42}O_4S$
2,2′-Sulfonylbis(4-(1,1,3,3-tetramethylbutyl)phenol) ... 774

$C_{28}H_{43}FN_2O_5S_2$
3-(3-Amino-4-(hexadecyloxy)-benzenesulfonamido)-benzenesulfonyl fluoride ... 813

$C_{28}H_{43}NO$
N-(2-Naphthyl)octadecanamide ... 490

$C_{28}H_{43}NO_2$
2-Octadecylamino-1,4-naphthoquinone ... 318

$C_{28}H_{43}NO_3S$
3-Amino-4-(hexadecyloxy)phenyl phenyl sulfone ... 775

$C_{28}H_{44}N_2O_2$
8-Quinolyl *N*-octadecylcarbamate ... 993

$C_{28}H_{44}N_2O_2S$
N[1]-Hexadecylmetanilanilide ... 817
N[1]-Hexadecylsulfanilanilide ... 817

$C_{28}H_{44}N_2O_3S$
4-Amino-4′-hexadecyloxybenzene-sulfonanilide ... 813

$C_{28}H_{44}N_2O_4$
N,*N*′-Dodecamethylenebis(2-hydroxy-3-methoxybenzylamine) ... 236

$C_{28}H_{44}N_2O_4S$
4′-Amino-4-(hexadecyloxy)-2′-hydroxybenzenesulfonanilide ... 810

$C_{28}H_{44}N_2O_5$
N-(2-Hydroxy-4-nitrophenyl)-2-octadecylsuccinimide ... 716

$C_{28}H_{44}Si$
Bis(2-ethylhexyl)diphenylsilane ... 1116
Bis(1-methylheptyl)diphenylsilane ... 1116
Diphenylhexadecylsilane ... 1116

$C_{28}H_{45}NO_3S$
1-Hexadecyl-4-methylpyridinium benzenesulfonate ... 957

$C_{28}H_{45}NO_5$
Methyl 3-acetoacetamido-4-(hexadecyloxy)benzoate ... 444

$C_{28}H_{46}N_2O$
7,8-Dimethyl-4-heptadecyl-1*H*-1,5-benzodiazepin-2(3*H*)-one ... 702
Stearic acid (5-indanylmethylene)-hydrazide ... 503

$C_{28}H_{46}N_2O_6$
2′-Hydroxy-4′-nitro-2-octadecylsuccinanilic acid ... 412

$C_{28}H_{46}O_5$
2-Methoxyethyl 4-(hexadecyloxy)-benzoylacetate ... 363

$C_{28}H_{46}Si_6$
1,6-Bis(2-phenyl-1-ethynyl)-1,1,2,2,3,3,4,4,5,5,6,6-dodecamethylhexasilane ... 1116

$C_{28}H_{47}NO$
N-Dodecylhexadecanamide ... 106

$C_{28}H_{48}ClN_3O_2S$
1-Hexadecanesulfonic acid (2-chloro-α-piperidinobenzylidene)-hydrazide ... 821
1-Hexadecanesulfonic acid (2-chloro-α-piperidinobenzylidene)-hydrazide hydrochloride ... 820
1-Hexadecanesulfonic acid (4-chloro-α-piperidinobenzylidene)-hydrazide ... 821

$C_{28}H_{48}N_2O_6$
N,*N*′-Dodecamethylenebis-(hexahydrophthalamic acid) ... 122

$C_{28}H_{48}O_3$
2′,4′-Di(decyloxy)acetophenone ... 293

$C_{28}H_{49}ClN_4O_2S$
Hexadecanesulfonic acid (4-amino-2-chloro-α-piperidinobenzylidene)-hydrazide ... 816
1-Hexadecanesulfonic (5-amino-2-chloro-α-piperidino-benzylidene)hydrazide ... 816

$C_{28}H_{50}N_2$
1-(1-Methylnonyl)-5-(1-methylnonylamino)indoline dihydrochloride ... 249

$C_{28}H_{50}N_2O$
1-(2-Ethyl-6-methylphenyl)-3-octadecylurea ... 596
1-Ethyl-3-octadecyl-1-(*o*-tolyl)urea ... 579
1-(4-Isopropylphenyl)-3-octadecylurea ... 591
1-Mesityl-3-octadecylurea ... 602

$C_{28}H_{50}N_2O_3S$
N[1]-Cyclohexyl-4-(hexadecyloxy)-metanilamide ... 813

$C_{28}H_{50}N_4O_2S$
1-Hexadecanesulfonic 2-(3-amino-α-piperidinobenzylidene)hydrazide ... 816

$C_{28}H_{50}Si_6$
1,6-Distyryl-1,1,2,2,3,3,4,4,5,5,6,6-dodecamethylhexasilane ... 1116

$C_{28}H_{51}N_3O$
1-(2-(*N*-Methylanilino)ethyl)-3-octadecylurea ... 562

$C_{28}H_{52}O_{12}$
Sucrose monopalmitate ... 84

$C_{28}H_{53}N_3O_2S$
3-Amino-*N*[4],*N*[4]-diethyl-*N*[1]-octadecyl-sulfanilamide ... 816

$C_{28}H_{54}O$
1,1-Diallyl-1-docosanol ... 12

$C_{28}H_{56}N_2Si_6$
1,6-Bis(4-(dimethylamino)phenyl)-dodecamethylhexasilane ... 1108

$C_{28}H_{56}O_2$
Decyl stearate ... 74

$C_{28}H_{56}O_4Sn$
Diethyldilauroyloxytin ... 1131

$C_{28}H_{57}NS$
S-Tetradecyl tetradecanothioimidate hydrochloride ... 141

$C_{28}H_{58}O$
Tetradecyl ether ... 17

$C_{28}H_{58}S$
Tetradecyl sulfide ... 26

$C_{28}H_{59}N$
Ditetradecylamine ... 28

$C_{28}H_{59}O_4P$
Dimyristyl phosphate ... 160

$C_{28}H_{60}Sn$
Tetraheptyltin ... 1129

$C_{29}H_{14}O_2$
Vat Orange 9 (C.I. 59700) ... 1095

$C_{29}H_{20}N_4O_{12}$
4,4′-Isopropylidenediphenol bis-3,5-dinitrobenzoate ... 379

$C_{29}H_{20}O_2$
Tetraphenyl-2-pyrone ... 384

$C_{29}H_{21}BF_4O$
2,3,4,6-Tetraphenylpyrylium tetrafluoroborate ... 935
2,3,5,6-Tetraphenylpyrylium tetrafluoroborate ... 936

$C_{29}H_{21}BrN_2O_8$
N,*N*′-(2-Bromo-6-methyl-phenylene)-bis(4-methylcoumarin-7-yl carbamate), isomers ... 630

$C_{29}H_{22}BF_4N$
1,2,4,6-Tetraphenylpyridinium tetrafluoroborate ... 957

$C_{29}H_{22}N_2O_4$
Bis(2-naphthyl) *N*,*N*′-(4-methyl-1,3-phenylene)biscarbamate ... 670
4,4′-Methylenebis(*N*-(3,4-(methylenedioxy)benzylidene)-aniline) ... 1076

$C_{29}H_{22}N_2O_8$
4,4′-Methylenebis(3,4-(methylenedioxy)phenyl carbanilate) ... 678

$C_{29}H_{22}N_6O_2$
1,1′-(Methylenedi-4,1-phenylene)-bis(3-(2-cyanophenyl)urea) ... 764
1,1′-(Methylenedi-4,1-phenylene)-bis(3-(3-cyanophenyl)urea) ... 765
1,1′-(Methylenedi-4,1-phenylene)-bis(3-(4-cyanophenyl)urea) ... 581

$C_{29}H_{22}N_6O_2S_2$
4,4′-Methylenebis(*N*-(*N*-(2-benzothiazolyl)carbamoyl)aniline) ... 916

$C_{29}H_{22}O$
2,3,4,5-Tetraphenyl-2,4-cyclopentadienol 206
$C_{29}H_{23}N$
N-(2-Naphthyl)-2-phenylbenzhydrylamine 232
N-(2-Naphthyl)tritylamine 232
$C_{29}H_{23}NO$
10-(4-(Dimethylamino)-α-phenylbenzylidene)-9-anthrone 312
$C_{29}H_{24}BrP$
(2-Naphthylmethyl)-triphenylphosphonium bromide 832
$C_{29}H_{24}ClP$
1-(Naphthylmethyl)triphenyl-phosphonium chloride 832
2-(Naphthylmethyl)triphenyl-phosphonium chloride 832
$C_{29}H_{24}N_4O_4$
N,*N'*-(4-Methyl-1,3-phenylene)bis(2-methyl-8-quinolyl carbamate) 988
$C_{29}H_{24}OSi$
2-Methoxy-1-(triphenylsilyl)-naphthalene 1106
$C_{29}H_{24}O_2S$
S-Phenyl 2'-(2,4,6-trimethylbenzoyl)-2-biphenylthiocarboxylate 379
$C_{29}H_{24}O_4$
1,1'-(3,4-Dimethoxybenzylidene)di-2-naphthol 195
$C_{29}H_{24}O_4Si$
2-Methoxy-3-(triphenylsilyl)-naphthalene 1106
$C_{29}H_{24}Si$
Diphenyl(1-naphthyl)(*p*-tolyl)silane ... 1121
$C_{29}H_{25}BrO_3$
Diphenylmethyl 2-(4-bromomethylphenyl)-2-(*p*-tolyl)-glycolate 353
$C_{29}H_{25}N_3O_3$
Carbobenzyloxytryptophan β-naphthylamide 890
$C_{29}H_{26}FeIP$
(Ferrocenylmethyl)-triphenylphosphonium iodide 1140
$C_{29}H_{26}N_2$
4,4'-Methylenebis(*N*-(4-methylbenzylidene)aniline) 1066
$C_{29}H_{26}N_2O_2$
4,4'-Methylenebis(*N*-(4-methoxybenzylidene)aniline) 1041
$C_{29}H_{26}N_2O_4$
4,4'-Methylenebis(*N*-(4-hydroxy-3-methoxybenzylidene)aniline) 1027
$C_{29}H_{26}N_2O_4S_2$
3,4-Dihydro-1-(2-(di-*p*-tosylamino)-phenyl)isoquinoline 819
$C_{29}H_{26}N_6O_6$
4',4''-Methylenebis(1-(4-hydroxybenzoyl)-4-phenylsemicarbazide) 548
4,4'-(Methylenedi-4,1-phenylene)-bis(1-(2-hydroxybenzoyl)-semicarbazide) 548
$C_{29}H_{26}O$
1,1,5,5-Tetraphenyl-3-pentanone 287
$C_{29}H_{26}O_2$
4,4-Diphenyl-3-(4-methoxyphenyl)-butyrophenone 292
$C_{29}H_{26}O_3$
Diphenylmethyl bis(*p*-tolyl)-glycolate 353
$C_{29}H_{27}NO$
N-Benzyl-2,2-diphenyl-*o*-propionotoluidide 431
$C_{29}H_{27}NO_3$
2-(2-(4-Benzyloxy)phenyl)-2-hydroxy-1-phenylethyl)-*N*-methylbenzamide. 461
$C_{29}H_{28}N_2O_5$
5'-*O*-Tritylthymidine 698
5'-*O*-Tritylthymidinebenzenate 696
$C_{29}H_{28}N_4O_2$
4,4'-Bis(3-methyl-3-phenylureido)-diphenylmethane 592
4,4'-Methylenebis(1-benzyl-3-phenylurea) 572
$C_{29}H_{28}O_{12}S_4$
Pentaerythritol tetrakis(benzenesulfonate) 800
$C_{29}H_{29}NSi$
10,10-Dibenzyl-5,10-dihydro-2,5,8-trimethylphenazasiline 1108
$C_{29}H_{29}N_3O_4S$
4-(4-Diethylamino-*o*-tolylimino)-1,4-dihydro-4'-methyl-sulfonyl-1-oxonaphthanilide 1062
$C_{29}H_{29}N_3O_5$
Carbobenzyloxyphenylalanyl-tryptophan methyl ester 886
Carbobenzyloxytryptophylalanine benzyl ester 890
Carbobenzyloxytryptophyl-phenylalanine methyl ester 886
$C_{29}H_{29}N_3O_6$
Carbobenzyloxytryptophyltyrosine methyl ester 881
$C_{29}H_{30}N_2O_2$
2-Oxo-*N*-phenethyl-4-(phenethylamino)-6-phenyl-3-cyclohexene-1-carboxamide 443
$C_{29}H_{30}N_2O_4$
Dibenzyl 5,5'-methylenebis(3,4-dimethyl-2-pyrrolecarboxylate) 838
$C_{29}H_{30}N_2O_5$
Carbobenzyloxy-L-prolyl-L-phenylalanine benzyl ester 704
$C_{29}H_{30}N_2O_6$
Carbobenzyloxy-L-prolyl-L-tyrosine benzyl ester 697
$C_{29}H_{30}N_2O_7$
Carbobenzyloxyalanylaspartic acid dibenzyl ester 659
$C_{29}H_{30}O_7$
2,3-Di-*O*-acetyl-1-*O*-methyl-5-*O*-trityl-α-D-xylofuranoside 356
2,3-Di-*O*-acetyl-1-*O*-methyl-5-*O*-trityl-β-D-xylofuranoside 355
$C_{29}H_{30}Si$
(4-*tert*-Butylphenyl)diphenyl(*o*-tolyl)-silane 1120
$C_{29}H_{32}N_2O_5$
Carbobenzyloxyphenylalanyl-norvaline benzyl ester 658
Carbobenzyloxy-L-phenylalanyl-L-valine benzyl ester 658
$C_{29}H_{32}N_2O_9S$
Sarcosine Cresol Red 1093
$C_{29}H_{32}N_8O_4$
4,4'-(1,3-Propanediyldi-4,1-piperidindiyl)-bis(7-nitro-quinazoline) bis(methyl sulfate) 1000
$C_{29}H_{32}Si_2$
Pentamethylenebis(diphenylsilane). 1116
$C_{29}H_{33}ClN_4O_3$
4-(3-(3-Chlorophenyl)ureido)-benzoic (3,5-di-*tert*-butyl-4-hydroxybenzylidene)hydrazide 548
$C_{29}H_{33}N_3$
5,5-Dimethyl-3,7,11-tris(dimethylamino)-5*H*-naphth(3,2,1-de)anthracene 233
$C_{29}H_{35}NO_2$
1-Phenethylnorlobelanidine 245
$C_{29}H_{36}BrO_2P$
(4-(4,4,5,5-Tetramethyl-1,3-dioxolan-2-yl)butyl)triphenylphosphonium bromide 831
$C_{29}H_{36}N_2O_3$
4'-Amino-4-(2-(4-*tert*-octylphenoxy)-ethoxy)benzanilide 464
$C_{29}H_{36}N_4O_2$
3,3'-(4-Methyl-1,3-phenylene)bis(1-benzyl-1-isopropylurea) 563
$C_{29}H_{37}ClN_3O_4PS_2$
Bis(4-t-butyl-phenyl) *N*-(5-chloro-2,4-dimethoxy-phenyl-)-2-thioureidothiophosphoramidate 833
$C_{29}H_{37}N_2O_4PS$
Bis(4-*tert*-butylphenyl) *N*-(2,3-dimethoxybenzylideneamino)-thiophosphoramidate 834
$C_{29}H_{38}N_2O_4$
4,4'-Methylenebis(2-methylcyclohexyl *N*-phenylcarbamate) 673
$C_{29}H_{38}N_3O_4PS$
O,*O'*-Bis(4-*tert*-butylphenyl) (2-methoxy-5-methylphenylureido)-thiophosphoramidate 834
$C_{29}H_{39}N_3O_7$
N,*N'*-Bis(carbobenzyloxy)-L-lysyl-L-isoleucine methyl ester 663
$C_{29}H_{40}N_4O_2S_2$
4,4'-Trimethylenebis(1-(*N*-(2-methoxyphenyl)thiocarbamoyl)-piperidine) 738
$C_{29}H_{40}N_4O_8$
Azelaic acid (3,4,5-trimethoxybenzylidene)-hydrazide ... 1051
$C_{29}H_{42}ClN_3O_3$
4-Nitrobenzaldehyde 2-(5-chloro-2-(hexadecyloxy)phenyl)hydrazone 234
$C_{29}H_{42}N_2O_3$
3-(5-Isoquinolylcarbamoyl)-5-nonadecenoic acid 997
$C_{29}H_{42}N_2O_6$
2-Diethylaminoethyl 3-nitro-4-(2-(4-*tert*-octylphenoxy)ethoxy)benzoate. 369
$C_{29}H_{42}N_6O_4$
4,4'-Methylenebis(3-(3-morpholinopropyl)-1-phenylurea) 566
$C_{29}H_{43}Cl_2NO$
N-(2,5-Dichlorophenyl)-2-(hexadecyloxy)-benzylamine 229
N-(3,4-Dichlorophenyl)-4-(hexadecyloxy)-benzylamine 229
$C_{29}H_{43}Cl_2N_3$
N-(4-(2,5-Dichlorophenylazo)-phenyl)-*N*-methylhexadecylamine ... 1090
$C_{29}H_{43}NS$
N-Hexadecylthiobenzanilide 746
$C_{29}H_{44}ClNO_2S$
4-Chloro-*N*-(4-(hexadecylsulfonyl)-phenyl)-benzylamine 778
$C_{29}H_{44}FNO_3S$
3-(2-(Hexadecyloxy)-benzylamino)-benzenesulfonyl fluoride 802
$C_{29}H_{44}N_2O$
Stearic (1-naphthylmethylene)-hydrazide 502
$C_{29}H_{44}N_2O_2$
Stearic (2-hydroxy-1-naphthylmethylene)hydrazide 497
$C_{29}H_{44}N_2O_3$
4'-Amino-3-hexadecyloxy-2'-hydroxybenzanilide 461
$C_{29}H_{46}N_2O$
1-(1-Naphthyl)-3-octadecylurea 603
$C_{29}H_{46}N_2O_2S$
N'-Benzyl-*N*'-hexadecylmetanilamide 816
$C_{29}H_{46}N_2O_4$
Bis(4-*tert*-butylcyclohexyl) *N*,*N'*-(4-methyl-1,3-phenylene)-biscarbamate 670
$C_{29}H_{47}NO$
3-(2-Hexadecyloxy-1-naphthyl)-propylamine 240
$C_{29}H_{52}N_2O$
1-(2,6-Diethylphenyl)-3-octadecylurea 596
1-(5-Isopropyl-2-methylphenyl)-3-octadecylurea 602
$C_{29}H_{58}O_4Sn$
Bis(tributylstannyl) methylenesuccinate 1131
$C_{29}H_{60}N_2S$
1,3-Ditetradecyl-2-thiourea 142
$C_{30}H_{14}O_2$
8,16-Pyranthrenedione 309
$C_{30}H_{18}N_2O_4$
N,*N'*-Bis(2-anthraquinonyl)-acetamidine 725
$C_{30}H_{20}$
1,2,3-Triphenylacenaphthylene 166
$C_{30}H_{20}N_2$
4,4'-Bis(2-quinolyl)biphenyl 994
N,*N'*-(1,4-Phenylene)dicarbazole 891
$C_{30}H_{20}N_2O_2$
6,6'-Bi(4-hydroxy-2-phenylquinoline). 983
$C_{30}H_{20}N_4O_{11}S_3$
Fast Sulphon Black F (C.I. 26990) 1083
$C_{30}H_{20}N_4O_{14}S_4$
3,6-Bis-(4-sulfo-1-naphthylazo)-4,5-di-hydroxy-2,7-naphthalenedisulfonic acid,sodium salt 1085
$C_{30}H_{20}O_2$
2,2'-Dimethylbianthrone 315
4,4'-Dimethylbianthrone 315
Hexahydro-4'b-phenyl-spiro(indene-1,10'-indeno(1,2-a)indene)-3,9'-dione 308
$C_{30}H_{20}S_4$
4,4'-Bis(biphenylyl)-tetrathiafulvalene 880
4,4',5,5'-Tetraphenyltetrathiafulvalene 880
$C_{30}H_{21}As$
Tris(1-naphthyl)arsine 1128
$C_{30}H_{21}ClSi$
Chlorotris(1-naphthyl)silane 1123
$C_{30}H_{21}Sb$
Tris(1-naphthyl)antimony 1129
$C_{30}H_{22}$
5-Benzylidene-1,2,3-triphenyl-1,3-cyclopentadiene 165
$C_{30}H_{22}Cl_2F_6N_4O_2$
4,4'-Bis((4-chloro-3-trifluoromethyl)-phenylureido)-3,3'-dimethylbiphenyl 507
$C_{30}H_{22}F_4Si_2$
1,4-Bis(diphenylsilyl)-2,3,5,6-tetrafluorobenzene 1103
$C_{30}H_{22}O$
4-(4-Biphenylyl)-2,6-diphenylphenol.. 196
$C_{30}H_{22}OSi$
2-(Triphenylsilyl)dibenzofuran 1110
4-(Triphenylsilyl)dibenzofuran 1110
Tris(1-naphthyl)silanol 1110
$C_{30}H_{22}O_2$
10,10'-Bis(2-dimethylanthrone) 313
3-(4-Methoxyphenyl)-2,4,5-triphenylcyclopentadienone 284
$C_{30}H_{22}Pb$
Tris(1-naphthyl)lead 1136
$C_{30}H_{22}SSi$
2-(Triphenylsilyl)dibenzothiophene... 1111
4-(Triphenylsilyl)dibenzothiophene... 1111
$C_{30}H_{22}Si$
Tris(1-naphthyl)silane 1123
$C_{30}H_{23}BrSi$
2-Bromo-2'-(triphenylsilyl)biphenyl ... 1103
$C_{30}H_{23}ClOSi$
(5-Chloro-2-(phenoxy)phenyl)-triphenylsilane 1102
$C_{30}H_{23}Cl_4FeO$
4-(*p*-Tolyl)-2,3,6-triphenylpyrylium tetrachloroferrate 935
$C_{30}H_{23}N$
2,3,4,6-Tetraphenylaniline 230
$C_{30}H_{24}$
Δ5,5'-Bidibenzosuberane 165
$C_{30}H_{24}F_6N_4O_2$
4,4'-Bis(2-(trifluoromethyl)-phenylureido)-3,3'-dimethylbiphenyl 511
$C_{30}H_{24}N_2O_2S$
4-(Cinnamylideneamino)phenyl sulfone 1075
$C_{30}H_{24}N_2S_2$
4,4'-Dithiobis(*N*-cinnamylideneaniline) 1052
$C_{30}H_{24}OSi$
(2-Phenoxyphenyl)triphenylsilane..... 1106
$C_{30}H_{24}O_4$
3,3-Diphenyl-6,6-bis(*p*-tolyl)-1,4-dioxane-2,5-dione 384
3,6-Diphenyl-3,6-bis(*o*-tolyl)-1,4-dioxane-2,5-dione 384
3,6-Diphenyl-3,6-bis(*m*-tolyl)-1,4-dioxane-2,5-dione 386
$C_{30}H_{24}Pb$
(4-Biphenylyl)triphenyllead 1136
$C_{30}H_{24}Si$
(3-Biphenylyl)triphenylsilane 1121
4-Biphenylyltriphenylsilane 1121
$C_{30}H_{24}Sn$
Bis(1-indenyl)diphenyltin 1133

$C_{30}H_{25}ClSi$
1-Chloro-1,1,2,2,2-pentaphenyldisilane 1116
$C_{30}H_{25}N_3O_7$
2′,3′-*O*-Benzylidene-*N*⁴,5′-*O*-dibenzoylcytidine 700
$C_{30}H_{26}Br_2Sn$
Bromo((4-bromo-1,2,3,4-tetraphenyl)-1,3-butadienyl)-dimethyltin 1132
$C_{30}H_{26}Fe_2$
1,4-Bis-(2-ferrocenylvinyl)-benzene 1142
$C_{30}H_{26}GeN_2$
1,2-Diphenyl-1-(triphenylgermyl)-hydrazine 1108
$C_{30}H_{26}N_2Ge$
1,2-Diphenyl-1-(triphenylgermyl)-hydrazine 1108
$C_{30}H_{26}N_2O_6$
1,4-Diamino-2,3-bis(2-phenoxyethoxy)anthraquinone 316
$C_{30}H_{26}N_2Si$
1,2-Diphenyl-1-(triphenylsilyl)-hydrazine 1108
$C_{30}H_{26}N_4O_2S_2$
4,4′-Dithiobis(*N*-(4-acetamidobenzylidene)aniline) 1052
$C_{30}H_{26}N_4O_4S$
4,4′-Sulfonylbis(*N*-(4-acetamidobenzylidene)aniline) 1063
$C_{30}H_{26}N_6O_4$
8,16-Bis(4-nitrophenyl)-5,6,8,8a,13,14,16,16a-octahydro[1,2,4,5]-tetrazino[6,1a:3,4a′]diisoquinoline 278
$C_{30}H_{26}O_2$
1,1′-(4-Isopropylbenzylidene)di-2-naphthol 194
$C_{30}H_{26}Si_2$
1,1,1,2,2-Pentaphenyldisilane 1116
$C_{30}H_{27}NO_2$
3,4-Dihydro-3-(α-(*N*-methylbenzylamino)benzyl)-3-phenyl-1*H*-2-benzopyran-1-one 384
$C_{30}H_{28}N_2O_2$
N,*N*′-Dibenzyl-*N*,*N*′-dimethyl-2,2′-biphenyldicarboxamide 472
$C_{30}H_{28}N_2O_4$
4-Benzyloxy-3-methoxybenzaldehyde azine 1039
$C_{30}H_{28}N_2O_4S$
4-(*N*-(2-Ethoxybenzylidene)amino)-phenyl sulfone 1050
$C_{30}H_{28}N_2O_5$
Carbobenzyloxyglutamic acid γ-benzyl ester 2-naphthylamide 676
$C_{30}H_{28}N_4$
8,16-Diphenyl-5,6,8,8a,13,14,16,16a-octahydro-[1,2,4,5]-tetrazino[6,1a:,34a′]diisoquinoline 253
1,1′,3,3′-Tetraphenyl-Δ2,2′-biimidazolidine 253
$C_{30}H_{28}N_4O_5$
Carbobenzyloxy-L-tryptophyl-L-tryptophan 890
$C_{30}H_{28}N_4O_8S_2$
4,4′-Bis(4-ethoxyphenylazo)-3,3′-disulfostilbene disodium salt 1087
$C_{30}H_{28}O_4S$
2,5-Bis(diphenylhydroxymethyl)-tetrahydrothiophene-1,1-dioxide 774
$C_{30}H_{28}O_9$
1,2-*O*-Isopropylidene-3,5,6-tri-*O*-benzoyl-α-D-glucofuranoside 370
$C_{30}H_{28}S_4$
Tetrakis(*p*-tolylthio)ethylene 213
$C_{30}H_{29}ClN_4O_4S$
4-(1-Hydroxy-2-naphthamido)-benzenesulfonic *N*-(α-cyclohexylamino-2-chloro-benzylidene)hydrazide 812
$C_{30}H_{30}CuO_{12}$
Ethyl α-acetyl-3-(methoxycarbonyl)-benzoylacetate, copper(II) derivative 350
Ethyl α-acetyl-4-(methoxycarbonyl)-benzoylacetate, copper(II) derivative 350
$C_{30}H_{30}N_2O_2$
2-(Bis(4-(dimethylamino)phenyl)-methyl)phenyl benzoate 379
$C_{30}H_{30}N_4O_2$
4,4′-Bis(*p*-tolylureido)-3,3′-dimethylbiphenyl 592
1,1′-(3,3′-Dimethyl-1,1′-biphenyl-4,4′-diyl)bis(3-(*o*-tolyl)urea) 589
$C_{30}H_{30}N_4O_2S$
4-(4-(Dimethylamino)-benzylideneamino)phenyl sulfone 1062
$C_{30}H_{30}N_4O_3$
Bis(4-(2,5-dimethylphenylureido)-phenyl) ether 560
$C_{30}H_{30}N_4O_4$
4,4′-Bis(3-(4-methoxyphenyl)ureido)-3,3′-dimethylbiphenyl 551
$C_{30}H_{30}N_4O_4S$
1,1′-(Sulfonyldi-4,1-phenylene)bis(3-(2,4-xylyl)urea) 782
$C_{30}H_{30}N_4O_5$
Carbobenzyloxyhistidyl-phenylalanine benzyl ester 867
$C_{30}H_{30}N_4O_6$
Carbobenzyloxytryptophyl-glycylglycine benzyl ester 883
$C_{30}H_{30}N_4O_6S_2$
4,4′-Bis(3-(*p*-toluenesulfonyl)-ureido)-3,3′-dimethylbiphenyl 826
$C_{30}H_{30}N_4S_2$
4,4′-Dithiobis(*N*-(4-dimethylaminobenzylidene)aniline). 1052
$C_{30}H_{30}O_2$
1,1,6,6-Tetraphenyl-1,6-hexanediol 194
$C_{30}H_{30}O_8P_2$
(+/-)-1,2-Cyclohexanediol bis(diphenyl phosphate) 834
trans-1,4-Cyclohexanediol bis(diphenyl phosphate) 834
$C_{30}H_{32}ClP$
(Pentamethylbenzyl)-triphenylphosphonium chloride 832
$C_{30}H_{32}N_2O_7$
Carbobenzyloxy-L-alanyl-L-glutamic acid dibenzyl ester 659
$C_{30}H_{32}Si$
Bis(3-phenylpropyl)diphenylsilane 1116
Phenyltriphenethylsilane 1116
$C_{30}H_{34}N_2O_5$
Carbobenzyloxyisoleucyl-phenylalanine benzyl ester 658
Carbobenzyloxy-L-leucyl-L-phenylalanine benzyl ester 671
Carbobenzyloxy-L-phenylalanyl-L-isoleucine benzyl ester 658
$C_{30}H_{34}N_2O_6$
Carbobenzyloxyisoleucyltyrosine benzyl ester 643
$C_{30}H_{34}N_4O_2$
1,1′-Hexamethylenebis(3-(1-naphthylmethyl)urea 603
$C_{30}H_{34}O_2$
2,6-Di-*tert*-butyl-3,4-diphenyl-4-hydroxy-α-tetralone 290
$C_{30}H_{34}O_2Si_2$
1,2-Dipropoxy-1,1,2,2-tetraphenyldisilane 1116
$C_{30}H_{35}ClN_2O_2$
N,*N*′-Bis-(4-pentyloxybenzylidene)-2-chloro-1,4-phenylenediamine 1044
$C_{30}H_{36}N_2O_2$
N,*N*′-Bis(4-pentyloxybenzylidene)-1,4-phenylenediamine 1047
$C_{30}H_{36}Si_3$
Bis(4-(trimethylsilyl)phenyl)-diphenylsilane 1120
$C_{30}H_{37}Cl_3Si$
Dodecyltris(3-chlorophenyl)silane 1101
$C_{30}H_{37}F_3Si$
Dodecyltris(3-fluorophenyl)silane 1101
$C_{30}H_{38}N_4O_2$
Sebacic acid bis(5-indanylmethylenehydrazide) 1075
$C_{30}H_{38}O_2Pb$
Triphenylplumbyl laurate 1136
$C_{30}H_{38}Si$
9-Dodecyl-9-phenyl-9*H*-9-silafluorene 1122
$C_{30}H_{39}O_4P$
Tris(4-*tert*-butylphenyl) phosphate 836
$C_{30}H_{40}Si$
Dodecyltriphenylsilane 1116
$C_{30}H_{42}BrNO$
3-Bromo-10-(2-diheptylamino-1-hydroxyethyl)phenanthrene 239
$C_{30}H_{42}Cl_2NO_3$
Ethyl 11-(4-(4-(bis-(2-chloroethyl)-amino)benzylideneamino)-phenoxy)undecanoate 1044
$C_{30}H_{42}N_3Sb$
Tris(4-(diethylamino)phenyl)-antimony 1128
$C_{30}H_{42}N_4O$
3-Pentadecyl-1-phenyl-4-phenylazo-2-pyrazolin-5-one 1092
$C_{30}H_{42}N_4O_6$
Sebacic bis((4-ethoxy-3-methoxybenzylidene)hydrazide) 497
$C_{30}H_{42}O_{14}$
Dimethyl 2,5-dimethoxy-β,β,β′,β′-tetrakis(ethoxycarbonyl)-1,4-benzenedibutyrate 363
$C_{30}H_{43}ClF_3NO$
6-Chloro-*N*-(2-(hexadecyloxy)-benzyl)-α,α,α-trifluoro-*m*-toluidine 222
$C_{30}H_{43}ClO$
4′-(4-Chlorophenyl)-octadecanophenone 289
$C_{30}H_{43}NO$
9-Octadecanoylcarbazole 892
$C_{30}H_{43}NOS$
10-Octadecanoylphenothiazine 710
$C_{30}H_{43}NO_4$
4′-(4-Nitrophenoxy)-octadecanophenone 292
$C_{30}H_{43}N_3O_3S_2$
3-Hexadecyl-2-benzothiazolinone, azine with 2-formylbenzene-sulfonic acid, sodium salt 739
$C_{30}H_{43}N_3O_4S_2$
3-Carboxybenzenesulfonic acid (3-hexadecyl-2-benzothiazolinylidene)hydrazide 821
$C_{30}H_{44}ClNS$
3-Chloro-10-octadecylphenothiazine 252
$C_{30}H_{44}O$
4′-Phenyloctadecanophenone 297
$C_{30}H_{44}O_2$
4-Biphenylyl octadecanoate 380
2′-Hydroxy-4′-phenyloctadecanophenone 290
4′-Phenoxyoctadecanophenone 292
$C_{30}H_{45}NO$
4′-Phenyloctadecananilide 489
$C_{30}H_{45}NOS$
10-Octadecylphenothiazine-5-oxide 771
$C_{30}H_{45}NO_2S$
10-Octadecylphenothiazine-5,5-dioxide 778
$C_{30}H_{45}NS$
10-Octadecylphenothiazine 252
$C_{30}H_{45}N_3O_3S_2$
4-(Hexadecyloxy)benzenesulfonic (3-methyl-2-benzothiazolinylidene)-hydrazide 814
$C_{30}H_{46}$
4-Octadecylbiphenyl 165
$C_{30}H_{46}N_2$
N,*N*-Dibenzyl-*N*′,*N*′-dihexyl-1,4-diamino-2-butene dihydrochloride 242
$C_{30}H_{46}N_2O$
4′-Anilinostearanilide 430
$C_{30}H_{46}N_2O_2$
4-Hexadecyloxy-3-methoxybenzaldehyde phenylhydrazone 234
$C_{30}H_{46}N_2O_3$
4′-Amino-2-hexadecyloxy-2′-hydroxy-*m*-toluanilide 461
$C_{30}H_{46}N_2O_4$
4′-Amino-4-hexadecyloxy-2′-hydroxy-3-methoxybenzanilide 461
$C_{30}H_{46}N_2O_4S$
4-Acetamido-4′-hexadecyloxy-benzenesulfonanilide 815
$C_{30}H_{46}N_4O_2$
N,*N*′-Terephthalylidenebis(10-undecenoic hydrazide) 502
$C_{30}H_{46}N_4O_4S_3$
4-(Hexadecylsulfamoyl)-benzenesulfonic (3-methyl-2-benzothiazolinylidene)hydrazide 820
$C_{30}H_{46}N_4O_7S_4$
4-(Hexadecylsulfamoyl)-benzenesulfonic (3-methyl-5-sulfo-2-benzothiazolinylidene)hydrazide 820
$C_{30}H_{47}ClN_2O_2S$
4-Chloro-*N*¹-hexadecyl-*N*¹-phenethylmetanilamide 807
$C_{30}H_{47}NOSi$
3-(Decyldimethylsilyl)-*N*-(1-phenyl-2-(*o*-tolyl)ethyl)propionamide 1107
3-(Decyldimethylsilyl)-*N*-(1-phenyl-2-(*m*-tolyl)ethyl)propionamide 1107
3-(Decyldimethylsilyl)-*N*-(1-phenyl-2-(*p*-tolyl)ethyl)propionamide 1107
$C_{30}H_{48}N_4O_4$
Methyl4′-(4-amino-2-methyl-5-oxo-3-pentadecyl-3-pyrazolin-1-yl)-succinanilate 701
$C_{30}H_{49}NO_3S$
1-Octadecylpyridinium *p*-toluenesulfonate 957
$C_{30}H_{50}N_2O_4S_2Sn$
Dioctyldi-*p*-toluenesulfonamidotin 1131
$C_{30}H_{50}Si$
Phenyltris(2-cyclohexylethyl)silane 1111
$C_{30}H_{52}INO_2S$
1-(4-Hexadecyloxy-3-methoxy-α-(methylthio)-benzylidene)-piperidinium iodide 254
$C_{30}H_{52}N_2O_3S$
*N*⁴-Acetyl-*N*¹-cyclohexyl-*N*¹-hexadecylsulfanilamide 825
$C_{30}H_{53}BF_4O_2S$
(2-(Ethoxycarbonyl)phenyl)-octadecylpropylsulfonium tetrafluoroborate 372
$C_{30}H_{53}ClN_2O_4S$
4-Chloro-*N*,*N*-didodecyl-3-nitrobenzenesulfonamide 822
$C_{30}H_{53}N_3O_4S$
1-Hexadecanesulfonic acid (α-piperidinoveratrylidene)-hydrazide 813
$C_{30}H_{54}BrNO_2S$
4-Bromo-*N*,*N*-didodecylbenzenesulfonamide 829
$C_{30}H_{54}N_2O_4S$
N,*N*-Didodecyl-3-nitrobenzenesulfonamide 822
$C_{30}H_{55}N$
N,*N*-Didodecylaniline 220
$C_{30}H_{56}N_2O_2S$
*N*¹,*N*¹-Didodecylmetanilamide 817
$C_{30}H_{56}O_2$
2-Hexadecyl-1,3-cyclotetradecanedione 54
$C_{30}H_{56}Si$
Phenyltrioctylsilane 1116
Phenyltris(2-ethylhexyl)silane 1116
$C_{30}H_{57}BO_3$
(-)-Trimenthyl borate 1125
(+/-)-Trimenthyl borate 1125
$C_{30}H_{58}O_3$
13-Oxotriacontanoic acid 64
$C_{30}H_{58}O_4S$
Dilauryl 3,3′-thiodipropionate 87
$C_{30}H_{60}N_2O_2$
N,*N*′-Ditetradecyloxamide 110
N,*N*′-Ethylenebis(tetradecanamide). 110
$C_{30}H_{60}O_2Sn$
Tributyl(oleoyloxy)stannane 1131
$C_{30}H_{61}NO$
N-Decyloctadecanamide 106
N-Octadecyldodecanamide 106
$C_{30}H_{62}O$
12-Triacontanol 9
$C_{30}H_{62}OS_2$
Bis(3-(dodecylthio)propyl)ether 25
$C_{30}H_{62}O_2Sn$
Tributyltin stearate 1131

$C_{30}H_{62}O_4S_2$
1,6-Bis(dodecylsulfonyl)hexane ... 154
$C_{30}H_{62}S_2$
1,6-Bis(dodecylthio)hexane ... 26
$C_{30}H_{63}BO_3$
Tris(decyl) borate ... 1124
$C_{30}H_{63}O_4P$
Dipentadecyl phosphate ... 160
$C_{30}H_{64}Si$
Tris(decyl)silane ... 1098
$C_{31}H_{20}Cl_2N_2$
6,6′-Methylenebis(4-chloro-2-phenylquinoline) ... 995
$C_{31}H_{20}Cl_3P$
(2,4,7-Trichloro-9-fluorenylidene)-triphenylphosphorane ... 833
$C_{31}H_{20}NO_2$
4-Nitro-1,3,9-triphenylfluorene ... 279
$C_{31}H_{21}NO$
4-Hydroxy-2,3,5,6-tetraphenylbenzonitrile ... 761
$C_{31}H_{22}NO_2P$
(2-Nitro-9-fluorenylidene)-triphenylphosphorane ... 832
$C_{31}H_{22}O$
2,4,6-Triphenylbenzophenone ... 314
$C_{31}H_{23}NO_3$
4′-Methoxy-4-nitro-2,3,5-triphenylbiphenyl ... 268
$C_{31}H_{23}P$
(9-Fluorenylidene)-triphenylphosphorane ... 837
$C_{31}H_{24}$
9,9′,10,10′-Tetrahydro-11,11′-spirobi(9,10-ethanoanthracene) ... 162
$C_{31}H_{24}BF_4N_3S_6$
Tris(4-(*p*-tolyl)-1,3-dithiolan-2-ylideneamino)carbenium tetrafluoroborate ... 1076
$C_{31}H_{24}BrP$
9-Fluorenyltriphenylphosphonium bromide ... 832
$C_{31}H_{24}N_4O_8$
2′,3′,5′-Tribenzoylinosine ... 922
$C_{31}H_{24}O$
5,12-Diphenyl-5,5a,6,11,11a,12-hexahydro-5,12-epoxy-6,11-methanonaphthacene ... 183
12-Hydroxy-9,9′,10,10′-tetrahydro-11,11′-spirobi(9,10-ethanoanthracene) ... 206
$C_{31}H_{24}OSi$
9-(Triphenylsilyl)xanthene ... 1111
$C_{31}H_{25}N$
4-(*p*-Tolyl)-2,3,6-triphenylaniline ... 230
$C_{31}H_{25}NO$
4′-Methoxy-2,3,5-triphenyl-4-biphenylamine ... 228
$C_{31}H_{25}N_2P$
((Diphenylmethylene)hydrazono)-triphenylphosphorane ... 1072
$C_{31}H_{25}N_5O_7S$
2′,3′,5′-Tribenzoylthioguanosine ... 920
$C_{31}H_{26}Cl_4N_2O_6$
Bis(2-(2,4-dichlorophenoxy)ethyl) *N*,*N*′-(methylenedi-4,1-phenylene)-biscarbamate ... 614
$C_{31}H_{26}GeO$
α-(Triphenylgermyl)benzhydrol ... 1105
$C_{31}H_{26}N_2$
4,4′-Methylenebis(*N*-cinnamylideneaniline) ... 1077
$C_{31}H_{26}N_6O_4S_2$
4,4′-Methylenebis(3-(6-methoxy-2-benzothiazolyl)-1-phenylurea) ... 913
$C_{31}H_{26}OSi$
(Diphenylmethoxy)triphenylsilane ... 1107
$C_{31}H_{26}O_3P_2$
Diphenyl (triphenylphosphoranylidene-methyl)phosphonate ... 836
$C_{31}H_{26}Si$
(Diphenylmethyl)triphenylsilane ... 1116
$C_{31}H_{27}NSi$
9,10-Dihydro-10-methyl-9-(triphenylsilyl)acridine ... 1108
$C_{31}H_{28}N_2O$
10-(Bis(4-(dimethylamino)phenyl)-methylene)-9-anthrone ... 312
$C_{31}H_{28}OSi_2$
1-Methoxy-1,1,2,2,2-pentaphenyldisilane ... 1116
$C_{31}H_{28}O_3$
α,α-Bis(3-ethoxy-2-naphthyl)benzyl alcohol ... 195
$C_{31}H_{28}Si_2$
1-Methyl-1,1,2,2,2-pentaphenyldisilane ... 1116
$C_{31}H_{30}Cl_2N_4$
4,4′-Methylenebis(*N*-(2-chloro-4-(dimethylamino)benzylidene)-aniline) ... 1025
$C_{31}H_{30}N_2O$
5-Benzyloxy-*N*,*N*-dibenzyltryptamine hydrochloride ... 882
$C_{31}H_{30}N_2O_4$
4,4′-Methylenebis(*N*-(3-ethoxy-4-hydroxybenzylidene)aniline) ... 1029
$C_{31}H_{30}N_2O_6$
2′,3′-Isopropylidene-5′-trityluridine ... 700
$C_{31}H_{30}N_4O_5$
Carbobenzyloxytryptophyl-tryptophan methyl ester ... 890
$C_{31}H_{31}BrN_8O_4$
1,1′-(2-Bromo-6-methylphenylene)-bis(3-antipyrinylurea) mixture of isomers ... 710
$C_{31}H_{32}N_4$
4,4′-Methylenebis(*N*-(4-(dimethylamino)benzylidene)-aniline) ... 1062
$C_{31}H_{32}N_4O_2$
4,4′-Methylenebis(1-benzyl-1-methyl-3-phenylurea) ... 572
4,4′-Methylenebis(1,3-diphenyl-1-ethylurea) ... 592
4,4′-Methylenebis(3-(2-phenethyl)-1-phenylurea) ... 570
$C_{31}H_{33}N_3O_5$
Carbobenzyloxytryptophylvaline benzyl ester ... 890
$C_{31}H_{33}N_3O_8$
Carbobenzyloxyglycylglycyl-L-glutamic acid dibenzyl ester ... 663
$C_{31}H_{34}N_2O_4$
Dibenzyl 2,2′-methylenebis(3-ethyl-4-methyl-5-pyrrolecarboxylate) ... 838
$C_{31}H_{35}NO_4$
N-(3,4-Dimethoxyphenethyl)-norlobelanine ... 307
$C_{31}H_{35}N_3O_7$
N,*N*′-Bis(carbobenzyloxy)-L-lysyl-L-phenylalanine ... 659
$C_{31}H_{36}N_2O_5S$
Carbobenzyloxy-*S*-benzyl-L-cysteinyl-L-leucine benzyl ester ... 667
$C_{31}H_{36}N_2O_6$
2-(3-Nitrobenzoyl)-2′-(2-(4-*tert*-octylphenoxy)ethoxy)acetanilide ... 442
$C_{31}H_{36}N_4O_4$
4,4′-Trimethylenebis(*N*-(piperidinomethyl)phthalimide) ... 721
$C_{31}H_{36}N_4O_7$
N,*N*′-Bis(carbobenzyloxy)-L-lysyl-L-tyrosinamide ... 645
$C_{31}H_{37}N_3OSi$
(4-Methoxyphenyl)tris(4-(dimethylamino)phenyl)silane ... 1106
$C_{31}H_{38}ClN_3O_4$
5,9-Dihydro-5,5,9,9-tetramethyl-3,7,11-tris(dimethylamino)-5*H*-naphth(3,2,1-de)anthracen-13b-ylium perchlorate ... 234
$C_{31}H_{38}O_4$
2,7-Bis(benzoyloxy)-1,4a,4b-trimethylperhydrophenanthrene ... 369
$C_{31}H_{40}O_3$
2,6-Bis(3-*tert*-butyl-2-hydroxy-5-methylbenzyl)-4-methylphenol ... 196
$C_{31}H_{41}NO_3$
2-Hydroxy-3-(α-(tetradecylamino)-benzyl)-1,4-naphthoquinone ... 316
$C_{31}H_{43}ClN_2O$
2-(4-Chlorophenyl)-α-(dihexylaminomethyl)-6,8-dimethyl-4-quinoline methanol hydrochloride ... 981
$C_{31}H_{43}Cl_2N_3O_5S_2$
4-(Hexadecyloxy)-3-sulfobenzoic acid 3-((4,6-dichloro-3-methyl-2-benzothiazolinylidene)hydrazide) ... 814
$C_{31}H_{44}BrNO_3$
2′-Bromo-2-(4-(hexadecyloxy)-benzoyl)-acetanilide ... 443
$C_{31}H_{44}N_2O_5S$
4-(4-(2-Hexadecyloxyphenyl)-5-oxo-2-pyrazolin-1-yl)benzenesulfonic acid sodium salt ... 792
$C_{31}H_{44}N_4O_2S_2$
4,4′-Trimethylenebis(1-(*N*-(4-ethoxyphenyl)thiocarbamoyl)-piperidine) ... 736
$C_{31}H_{44}N_4S_2$
4,4′-Trimethylenebis(1-(*N*-(2,4-xylyl)-thiocarbamoyl)piperidine) ... 741
4,4′-Trimethylenebis(1-(*N*-(2,6-xylyl)-thiocarbamoyl)piperidine) ... 747
4,4′-Trimethylenebis(1-(*N*-(3,4-xylyl)-thiocarbamoyl)piperidine) ... 747
$C_{31}H_{44}O_3$
4′-Phenylphenacyl heptadecanoate ... 357
$C_{31}H_{45}NO_7S$
Methyl 3-(4-carboxyphenylsulfonamido)-4-hexadecyloxybenzoate ... 814
$C_{31}H_{45}N_3O_5S_2$
Carboxy-hexadecyloxybenzenesulfonic (3-methyl-2-benzothiazolinylidene)-hydrazide ... 814
$C_{31}H_{45}N_3O_6S_3$
2-(Hexadecylsulfonyl)-5-sulfobenzoic acid 5-(3-methyl-2-benzothiazolinylidene)hydrazide ... 820
$C_{31}H_{45}N_3O_8S_3$
5-Carboxy-2-hexadecyloxybenzenesulfonic (3-methyl-5-sulfobenzo-thiazolinylidene)hydrazide ... 814
$C_{31}H_{45}N_3O_9S_4$
3-Carboxy-4-(hexadecylsulfonyl)-benzenesulfonic (3-methyl-5-sulfo-benzothiazolinylidene)hydrazide ... 823
$C_{31}H_{45}N_7O_8$
Inosine 2′,3′,5′-tris(*N*-cyclohexylcarbamate) ... 922
$C_{31}H_{46}F_2N_4O_6S_4$
3-(Hexadecylsulfamoyl)benzene-sulfonic acid (5-(difluoro-methylsulfonyl)-3-methyl-2-benzo-thiazolinylidene)hydrazide ... 805
4-(Hexadecylsulfamoyl)benzene-sulfonic acid (5-(difluoro-methylsulfonyl)-3-methyl-2-benzo-thiazolinylidene)hydrazide ... 805
$C_{31}H_{46}N_2$
9-(Octadecylamino)acridine ... 995
$C_{31}H_{46}N_2O_3$
2-(3-Aminobenzoyl)-4′-hexadecyloxyacetanilide ... 442
$C_{31}H_{46}O$
4′-(*p*-Tolyl)octadecanophenone ... 297
$C_{31}H_{46}O_2$
2′-Methoxy-4′-phenyloctadecanophenone ... 291
4′-(4-Methylphenoxy)-octadecanophenone ... 293
$C_{31}H_{46}O_3$
4-Benzyloxyphenyl stearate ... 378
$C_{31}H_{47}NO_3S_3$
3-Hexadecyl-2-methylthiobenzothiazolium *p*-toluenesulfonate ... 913
$C_{31}H_{47}N_3O_2S$
4-Nitrobenzaldehyde 2-(octadecylthio)phenylhydrazone ... 1051
$C_{31}H_{48}N_2O$
N-(4-(Dimethylamino)benzylidene)-4-hexadecyloxyaniline ... 1044
$C_{31}H_{48}N_2O_3$
*N*α-Oleoyl-DL-tryptophan ethyl ester ... 887
$C_{31}H_{48}N_2O_3S$
4-Acetamido-*N*-benzyl-*N*-hexadecylbenzenesulfonamide ... 821
$C_{31}H_{48}N_4O_2$
4,4′-Methylenebis(1,1-dibutyl-3-phenylurea) ... 592
$C_{31}H_{48}N_4O_4S_3$
4-(*N*-Hexadecyl-*N*-methylsulfamoyl)benzenesulfonic (3-methyl-2-benzothiazolinylidene)-hydrazide ... 820
$C_{31}H_{48}N_4O_7S_4$
3-(*N*-Hexadecyl-*N*-methylsulfamoyl)benzenesulfonic (3-methyl-5-sulfo-2-benzenethiazolinylidene)hydrazide ... 820
$C_{31}H_{50}N_2O_3S$
N^1-Hexadecyl-4-methoxy-N^2-(α-methylbenzyl)-metanilamide ... 813
$C_{31}H_{54}N_2O$
1-Cyclohexyl-3-octadecyl-1-phenylurea ... 582
$C_{31}H_{54}N_2O_3$
N,*N*-Didodecyl-3-nitrobenzamide ... 473
$C_{31}H_{55}N_3O_3S$
2-(3-Amino-3-methylpentyl)-1-octadecyl-5-benzimidazolesulfonic acid, sodium salt ... 904
$C_{31}H_{56}N_2O$
3-Amino-*N*,*N*-didodecylbenzamide ... 468
4-Amino-*N*,*N*-didodecylbenzamide ... 468
$C_{31}H_{56}N_4O_3S$
1-Hexadecanesulfonic (4-(diethylamino)-α-morpholinobenzylidene)hydrazide ... 821
$C_{31}H_{58}N_2O$
Stearic (2-butyl-2-ethyl-5-methyl-3,4-hexadien-1-ylidene)hydrazide ... 112
$C_{31}H_{60}O_4$
Diethyl didodecylmalonate ... 76
$C_{32}H_{20}O_3$
Tetraphenylphthalic anhydride ... 388
$C_{32}H_{22}O_3$
Tetraphenyl-1,2-dihydrophthalic anhydride ... 388
$C_{32}H_{22}O_6$
Dimethyl 10,10′-bis(2-anthronecarboxylate) ... 374
$C_{32}H_{24}N_2O_2$
N-(4-Nitrobenzylidene)-4-tritylaniline ... 1063
N,*N*′-Terephthalylidenebis(4-phenoxyaniline) ... 1041
$C_{32}H_{24}O$
6,6-Diphenyl-2-diphenylmethylenebicyclo(3.2.0)-hept-3-en-7-one ... 286
$C_{32}H_{24}OS$
4-(1-Benzothiophen-3-ylmethyl)-2,4,6-triphenyl-4*H*-pyran ... 940
$C_{32}H_{24}Pb$
Di(1-naphthyl)diphenyllead ... 1136
$C_{32}H_{24}S_4$
5,5′-Dimethyl-4,4′-(4-biphenylyl)-tetrathiafulvalene ... 880
$C_{32}H_{25}BiO_6$
Triphenylbismuth disalicylate ... 1141
$C_{32}H_{25}N_3O$
1-(Diphenylmethylene)-2,4,4-triphenylsemicarbazide ... 574
$C_{32}H_{26}ClNSi$
3-Chloro-9-ethyl-6-(triphenylsilyl)-carbazole ... 1108
$C_{32}H_{26}Cl_4N_4O_2$
Sebacic acid bis(3,4-dichlorobenzylidenehydrazide) ... 1025
$C_{32}H_{26}Cl_4Si_2$
2,3,5,6-Tetrachloro-1,4-phenylene-bis(diphenylmethylsilane) ... 1103
$C_{32}H_{26}F_5NSi_2$
N,*N*-Bis((diphenyl)methylsilyl)-2,3,4,5,6-pentafluoroaniline ... 1102
$C_{32}H_{26}N_4$
N,*N*′-Terephthalylidenebis(4-anilinoaniline) ... 1060

$C_{32}H_{26}Si$
(9,10-Dihydro-9-anthracenyl)triphenylsilane ... 1122
$C_{32}H_{27}BiN_2O_4$
Triphenylbismuth bis(4-aminobenzoate) ... 1141
$C_{32}H_{27}Cl_3Si_2$
(2,3,6-Trichloro-1,4-phenylene)bis(diphenylmethylsilane) ... 1103
$C_{32}H_{27}NSi$
5,10-Dihydro-5-ethyl-2,10,10-triphenylphenazasiline ... 1108
9-Ethyl-1-(triphenylsilyl)carbazole ... 1108
9-Ethyl-3-(triphenylsilyl)carbazole ... 1108
$C_{32}H_{28}BrN_3Sn$
4-Bromo-4′-(dimethylamino)-2′-(triphenylstannyl)azobenzene ... 1130
$C_{32}H_{28}Cl_4N_6O_4S_2Zn$
1-Methyl-2-(phenylsulfonylazo)quinolinium tetrachlorozincate ... 990
$C_{32}H_{28}Ge$
(2,2-Diphenylethyl)triphenylgermane ... 1116
$C_{32}H_{28}N_2O_2$
N,N′-Diethyl-*N,N′*-di(2-naphthyl)phthalamide ... 488
N,N′-(3,3′-Dimethylbiphenyl-4,4′-diyl)biscinnamamide ... 483
$C_{32}H_{28}N_2O_2S$
4,4′-Sulfonylbis(*N*-(5-indanylmethylene)aniline) ... 1075
$C_{32}H_{28}N_4O_2Sn$
4-(Dimethylamino)-4′-nitro-2-(triphenylstannyl)azobenzene ... 1130
N,N-Dimethyl-2-(4-nitrophenylazo)-4-(triphenylstannyl)aniline ... 1130
$C_{32}H_{28}OSi$
(4-Phenoxyphenyl)phenyl(*o*-tolyl)(*p*-tolyl)silane ... 1106
$C_{32}H_{28}O_2Si$
Bis(diphenylmethoxy)phenylsilane ... 1121
$C_{32}H_{28}Si$
(1,1-Diphenylethyl)triphenylsilane ... 1116
(1,2-Diphenylethyl)triphenylsilane ... 1116
(2,2-Diphenylethyl)triphenylsilane ... 1116
$C_{32}H_{29}N_3$
2-(2-Pyridyl)-*N,N,N′,N′*-tetraphenyl-1,3-propanediamine ... 946
$C_{32}H_{29}N_3O_4S$
(α-(4-Amino-*m*-tolyl)-α-(*p*-anilinophenyl)-α-hydroxy-*p*-tolyl)sulfanilic acid, sodium salt ... 791
$C_{32}H_{30}N_4O_5$
2′,3′-Isopropylidene-5′-tritylinosine ... 921
$C_{32}H_{31}N_5O_4$
2′,3′-Isopropylidene-5′-trityladenosine ... 921
$C_{32}H_{32}N_2O_2$
N,N′-Dibenzyl-*N,N′*-di(2,6-xylyl)oxamide ... 424
$C_{32}H_{32}N_2O_2S$
4-(*N*-(4-Isopropylbenzylidene)amino)phenyl sulfone ... 1075
$C_{32}H_{32}N_2S_2$
4,4′-Dithiobis(*N*-(4-isopropylbenzylidene)aniline) ... 1052
$C_{32}H_{34}Cl_2N_2O_4$
N,N′-Dodecamethylenebis(2-amino-3-chloro-1,4-naphthoquinone) ... 318
$C_{32}H_{34}N_2O_9$
α,α′-Dibenzyl carbobenzyloxy-L-glutamyl-L-glutamate ... 657
$C_{32}H_{34}N_4O_2$
4,4′-Bis(2,6-dimethylphenylureido)-3,3′-dimethylbiphenyl ... 596
1,1′-(3,3′-Dimethyl-1,1′-biphenyl-4,4′-diyl)bis(3-(2,5-xylyl)urea) ... 602
$C_{32}H_{34}N_4O_4$
Sebacic acid bis(2-hydroxy-1-naphthylmethylenehydrazide) ... 1029
$C_{32}H_{34}O_2$
α,α′-Hexamethylenebis(benzhydrol) ... 200
$C_{32}H_{35}N_3O_5$
Carbobenzyloxy-L-tryptophyl-L-leucine benzyl ester ... 890
$C_{32}H_{36}N_2O_2$
N,N′-Dicyclohexyl-*N,N′*-diphenylisophthalamide ... 475
N,N′-Dicyclohexyl-*N,N′*-diphenylterephthalamide ... 475
$C_{32}H_{36}N_2O_8$
N,O-Bis(carbobenzyloxy)tyrosylleucine methyl ester ... 661
$C_{32}H_{36}O_4Pb$
Tetrakis(4-ethoxyphenyl)lead ... 1134
$C_{32}H_{36}S_4Si$
Tetrakis((*p*-tolyl)thiomethyl)silane ... 1107
$C_{32}H_{36}Si$
Tetra(phenethyl)silane ... 1116
$C_{32}H_{37}N_3O_7$
N,N′-Bis(carbobenzyloxy)lysylphenylalanine methyl ester ... 658
$C_{32}H_{37}N_3O_8$
N,N′-Bis(carbobenzyloxy)-L-lysyl-L-tyrosine methyl ester ... 645
$C_{32}H_{39}ClN_2O_2$
N,N′-Bis-(4-hexyloxybenzylidene)-2-chloro-1,4-phenylenediamine ... 1044
$C_{32}H_{40}N_2O_4$
Tilorone Analog R 9536-DA ... 373
$C_{32}H_{40}N_4Pb$
Tetrakis(4-(dimethylamino)phenyl)lead ... 1134
$C_{32}H_{40}N_4Si$
Tetrakis(4-(dimethylamino)phenyl)silane ... 1108
$C_{32}H_{40}N_4Si_2$
3,3,6,6-Tetraethyl-1,2,4,5-tetraphenyl-1,2,4,5-tetraaza-3,6-disilacyclohexane ... 1116
$C_{32}H_{42}CuO_8$
Ethyl 2-(4-(pentyloxy)benzoyl)acetate, copper(II) derivative ... 1139
$C_{32}H_{42}N_4O_2$
Dodecanedioic acid (5-indanylmethylene)-hydrazide ... 1075
$C_{32}H_{44}N_2OS$
5-Benzylidene-3-hexadecyl-2-phenylimino-4-thiazolidinone ... 744
$C_{32}H_{44}Si_5$
Tetrakis(dimethylphenylsilyl)silane ... 1117
$C_{32}H_{46}FN_3O_5S_2$
1-Hexadecanesulfonic acid (2-(4-fluorophenyl)-1-methyl-6-sulfo-4(1*H*)-quinolylidene)hydrazide ... 1055
1-Hexadecanesulfonic acid (1-methyl-2-(*o*-fluorophenyl)-6-sulfo-4(1*H*)-quinolylidene hydrazide ... 1056
$C_{32}H_{46}N_2O_3$
4′-(2-(*N*-Phenylcarbamoyl)propionyl)hexadecananilide ... 444
$C_{32}H_{46}N_4O_7S_2$
1-Hexadecanesulfonic (1-methyl-2-(4-nitrophenyl)-6-sulfo-4(1*H*)-quinolylidene)hydrazide ... 1054
$C_{32}H_{46}N_4O_8$
Dodecanedioic acid (3,4,5-trimethoxybenzylidene)-hydrazide ... 1051
$C_{32}H_{46}O_3$
4′-Phenylphenacyl stearate ... 357
$C_{32}H_{46}O_4$
4,4′-Diacetoxy-3,3′,5,5′-tetra-*tert*-butylbiphenyl ... 380
$C_{32}H_{47}N_3O_2S$
1-Hexadecanesulfonic acid (4-methyl-1-phenyl-2(1*H*)-quinolylidene)hydrazide ... 821
1-Hexadecanesulfonic (1-methyl-2-phenyl-4(1*H*)-quinolylidene)hydrazide hydrochloride ... 1055
$C_{32}H_{47}N_3O_3S$
3′-(2-Methyl-5-oxo-1-phenyl-3-pyrazolin-3-yl)-1-hexadecanesulfonanilide ... 819
$C_{32}H_{47}N_3O_4S_2$
3-Methyl-2-benzothiazolinone, azine with 4′-hexadecyloxy-3′-sulfoacetophenone, sodium salt ... 1044
$C_{32}H_{47}N_3O_5S_2$
1-Hexadecanesulfonic (1-methyl-2-phenyl-5-sulfo-4(1*H*)-quinolylidene)hydrazide ... 1056
$C_{32}H_{48}N_2O_7S$
Dimethyl 5-(4-(hexadecyloxy)metanilamido)-isophthalate ... 813
$C_{32}H_{48}N_4O_2S$
1-Hexadecanesulfonic acid 2-(4-aminophenyl)-1-methyl-4(1*H*)-quinolylidene)hydrazide ... 1052
$C_{32}H_{48}N_4O_8S_4$
3-Acetamido-4-(hexadecylsulfonyl)benzenesulfonic (3-methyl-5-sulfo-2-benzothiazolinylidene)hydrazide ... 820
$C_{32}H_{48}N_4O_9S_4$
6(*N*-Hexadecyl-*N*-methylaminosulfonyl)-3-methyl-2-(methylthio)benzothiazolium salt ... 915
$C_{32}H_{49}N_3O_4$
3-(4-(2-Ethyl-3-methyl-5-oxo-3-pyrazolin-1-yl)-phenyl-carbamoyl)-5-nonadecenoic acid ... 703
$C_{32}H_{50}NNiO_2S$
2,2′-Thiobis(4-*tert*-octylphenolato)-*N*-butylamine nickel(II) ... 1141
$C_{32}H_{50}N_2O_4S$
5′-(*N*-Benzyl-*N*-hexadecylsulfamoyl)-*o*-acetanisidide ... 814
$C_{32}H_{50}N_4O_6S$
3-Nitrobenzenesulfonic acid (α-(*N*-methylhexadecylamino)veratrylidene)hydrazide ... 812
$C_{32}H_{51}N_3O_4$
Methyl 4′-(3-heptadecyl-2-methyl-5-oxo-3-pyrazolin-1-yl)succinanilate ... 709
$C_{32}H_{52}O_4$
Tetramethylene-2,2′-bis(1,3-cyclotetradecanedione) ... 54
$C_{32}H_{56}Br_2N_4O_2$
1,1′-(4,6-Dibromo-1,3-phenylene)bis(3-dodecylurea) ... 525
$C_{32}H_{58}O_6$
13,20-Dioxodotriacontanedioic acid ... 64
$C_{32}H_{59}N_3O_4S$
4-(Dimethylamino)-*N,N*-didodecyl-3-nitrobenzenesulfonamide ... 818
$C_{32}H_{60}O_6$
Tris-(2-ethylhexyl) 1,3,5-pentanetricarboxylate ... 76
$C_{32}H_{60}Si$
Tetrakis(2-cyclohexylethyl)silane ... 1111
$C_{32}H_{64}O_4Sn$
Bis-(2-ethylhexanoyloxy)bis(2-ethylhexyl)tin ... 1131
Bis-(2-ethylhexanoyloxy)dioctyltin ... 1131
$C_{32}H_{66}O$
Hexadecyl ether ... 17
$C_{32}H_{66}PbS_2$
Bis-(hexadecylthio)-lead (II) ... 1133
$C_{32}H_{66}S$
Hexadecyl sulfide ... 26
$C_{32}H_{68}O_4Si$
Tetraoctyl orthosilicate ... 1099
$C_{32}H_{68}Si$
Tetrakis(2-ethylhexyl)silane ... 1098
Tetraoctylsilane ... 1098
$C_{32}H_{68}Sn$
Tetraoctyltin ... 1129
$C_{33}H_{18}N_4O_9S_2$
6-Diazo-5,6-dihydro-5-oxo-1-naphthalenesulfonic 4-benzoyl-1,3-phenylene ester ... 798
$C_{33}H_{21}NS_3$
2,3,3-Tris(2-naphthylthio)acrylonitrile ... 763
$C_{33}H_{22}Cl_2N_4O_4$
4,4′-Methylenebis(*N*-(5-chloro-8-quinolyloxycarbonyl)aniline) ... 991
$C_{33}H_{24}N_4O_4$
4,4′-Methylenebis(*N*-(8-quinolyloxycarbonyl)aniline) ... 991
$C_{33}H_{24}N_4O_{17}S_4$
Direct Blue 10 (C.I. 24340) ... 1082
$C_{33}H_{26}N_2$
4,5-Diphenyl-1-tritylimidazole ... 869
$C_{33}H_{26}O_9$
β-L-Arabinopyranose tetrabenzoate ... 372
β-D-Arabinose tetrabenzoate ... 371
β-D-Ribopyranose tetrabenzoate ... 371
α-D-Xylofuranose tetrabenzoate ... 370
$C_{33}H_{27}N_3O$
2-Benzyl-4,4-diphenyl-1-(diphenylmethylene)semicarbazide ... 571
$C_{33}H_{28}NP$
(2-(Dimethylamino)-9-fluorenylidene)triphenylphosphorane ... 832
$C_{33}H_{28}OSi$
(1,3-Diphenyl-1-propenyloxy)triphenylsilane ... 1117
$C_{33}H_{29}BrSi$
(4-Bromo-4′-biphenylyl)tribenzylsilane ... 1104
$C_{33}H_{30}OSi$
(Di(*p*-tolyl)methoxy)triphenylsilane ... 1121
(3-Phenoxyphenyl)tribenzylsilane ... 1106
$C_{33}H_{30}Si$
4-Biphenylyltribenzylsilane ... 1121
$C_{33}H_{32}N_2O_5$
Carbobenzyloxy-L-phenylalanyl-L-phenylalanine benzyl ester ... 658
$C_{33}H_{32}N_2O_6$
Carbobenzyloxyphenylalanyltyrosine benzyl ester ... 645
$C_{33}H_{32}N_2O_7$
Carbobenzyloxy-L-tyrosyl-L-tyrosine benzyl ester ... 645
$C_{33}H_{32}N_4O_4$
Dibenzyl 5,5′-methylenebis(3-methyl-4-(2-cyanoethyl)-2-pyrrolecarboxylate ... 838
$C_{33}H_{32}O_6$
1-*O*-Benzoyl-2,3,4-tri-*O*-benzyl-β-L-arabinose ... 371
$C_{33}H_{34}BNO$
1-Ethyl-2-ethoxypyridinium tetraphenylborate ... 938
$C_{33}H_{34}N_2O_8$
Dimethyl 2,2′-methylenebis(5-(benzyloxycarbonyl)-4-methyl-3-pyrroleacetate) ... 839
$C_{33}H_{34}Si_3$
1,1,1,2,2-Pentaphenyl-3,3,3-trimethyltrisilane ... 1117
$C_{33}H_{36}N_2O_6$
1-Hydroxy-5′-nitro-2′-(2-(4-*tert*-octylphenoxy)ethoxy)-2-naphthanilide ... 485
$C_{33}H_{36}N_2O_{13}S$
Methyl Xylenol Blue ... 1093
$C_{33}H_{36}N_4O_2$
4,4′-Methylenebis(1-benzyl-1-ethyl-3-phenylurea) ... 572
$C_{33}H_{36}O_{12}S_4$
Pentaerythritol tetrakis(*p*-toluenesulfonate) ... 798
$C_{33}H_{37}NO_8$
1-*O*-(4-Nitrobenzoyl)-2,3,5-tri-*O*-benzyl-L-arabinofuranose ... 370
$C_{33}H_{38}BrOP$
(3,5-Di-*tert*-butyl-4-hydroxybenzyl)triphenylphosphonium bromide ... 830
$C_{33}H_{38}Si$
Phenyltris(3-phenylpropyl)silane ... 1117
$C_{33}H_{40}N_2O_7$
Dimethyl 5-(3-amino-4-(2-(4-*tert*-octylphenoxy)ethoxy)benzamido)isophthalate ... 464
$C_{33}H_{40}N_2O_9S$
Glycinethymol Blue ... 1093
$C_{33}H_{40}N_4O_8S_4$
Tetrakis(*p*-tosylaminomethyl)methane ... 829
$C_{33}H_{44}NO_7P$
Thymolphthalein monophosphate 2-amino-2-ethyl-1,3-propanediol salt ... 1093
$C_{33}H_{44}O_2Sn$
Tribenzyltin laurate ... 1131
$C_{33}H_{44}Si_4$
Phenyltris(3-(trimethylsilyl)phenyl)silane ... 1117
Phenyltris(4-(trimethylsilyl)phenyl)silane ... 1120
$C_{33}H_{45}NO_2S$
N-(2-Fluorenyl)-2-(hexadecylthio)succinimide ... 718
$C_{33}H_{46}Si$
Dodecyltribenzylsilane ... 1117
Dodecyltris(*m*-tolyl)silane ... 1099

$C_{33}H_{47}N_3O_5S$
4-(Carboxymethoxy)benzenesulfonic (1-methyl-4-pentadecyl-2-quinolylidene)hydrazide 812

$C_{33}H_{48}ClN_3O_5S_2$
1-Hexadecanesulfonic (1-(2-chloroethyl)-2-phenyl-6-sulfo-4(1*H*)-quinolylidene)hydrazidee 1079

$C_{33}H_{48}N_4O_2$
4,4′-Methylenebis(1-cyclohexyl-1-isopropyl-3-phenylurea) 583

$C_{33}H_{48}N_4O_6S_2$
5-Carboxy-2-hexadecyloxybenzenesulfonic 5-(4-methoxyphenyl-3-methyl-2-thiadiazolinylidene)hydrazide 814

$C_{33}H_{49}N_3O_5S_2$
1-Hexadecanesulfonic (1-methyl-6-sulfo-2-(*o*-tolyl)-4(1*H*)-quinolylidene)hydrazide 1055

$C_{33}H_{49}N_3O_6S_2$
1-Hexadecanesulfonic (2-(4-methoxy-phenyl)-1-methyl-6-sulfo-4(1*H*)quinolylidene)hydrazide 813

$C_{33}H_{49}N_3O_7S_3$
1-Hexadecanesulfonic (1-methyl-2-(4-methylsulfonylphenyl)-6-sulfo-4(1*H*)quinolylidene)hydrazide 1055

$C_{33}H_{50}N_6O_2$
4,4′-Bis(3-(3-(2-methylpiperidino)-propyl)ureido)diphenylmethane 565

$C_{33}H_{51}N_3O_6S$
4-Sulfobenzoic acid 4-(α-(*N*-hexadecyl-*N*-methylamino)-veratrylidene)hydrazide 812

$C_{33}H_{52}N_4O_2$
3,3′-(4-Methyl-1,3-phenylene)bis(1,1-dicyclohexylurea) 565

$C_{33}H_{56}N_2O$
Stearic (β-hexylcinnamylidene)-hydrazide 502

$C_{33}H_{62}O_6$
Tricaprin 74

$C_{33}H_{66}O$
17-Tritriacontanone 50

$C_{33}H_{68}N_2S$
1,3-Dihexadecyl-2-thiourea 142

$C_{33}H_{68}O_2$
Dihexadecyloxymethane 17

$C_{33}H_{72}Br_2N_2$
N,N′-Methylenebis(dimethyltetradecylammonium bromide) 44

$C_{34}H_{16}O_2$
Isoviolanthrone 314

$C_{34}H_{18}O_2$
13,13′-Dibenzanthronyl 309

$C_{34}H_{18}O_2S$
3,3′-Thiobis(7*H*-benz(de)anthracen-7-one) 308

$C_{34}H_{20}Cl_4F_6N_4O_4$
1,1′-(1,3-Phenylene)bis(3-(6-(3,4-dichlorophenoxy)-α,α,α-trifluoro-*m*-tolyl)urea) 507
1,1′-(1,4-Phenylene)bis(3-(6-(3,4-dichlorophenoxy)-α,α,α-trifluoro-*m*-tolyl)urea) 507

$C_{34}H_{24}B_2F_8O_2$
4,4′-Bis(2,6-diphenylpyrylium tetrafluoroborate) 935

$C_{34}H_{24}B_2F_8S_2$
4,4′-Bis(2,6-diphenylthiopyrylium tetrafluoroborate) 213

$C_{34}H_{24}N_2O_2S_2$
4,4′-Dithiobis(*N*-(2-hydroxy-1-naphthylmethylene)aniline) 1028

$C_{34}H_{24}N_2O_4S$
4,4′-Bis(*N*-(2-hydroxy-1-naphthylmethylene)amino)-diphenyl sulfone 1034

$C_{34}H_{24}N_2S_2$
2,6-Diphenyl-4*H*-thiopyran-4-one azine 769

$C_{34}H_{24}N_4$
3,3′-Bis(2-(2-quinolyl)vinyl)-azobenzene 1093

$C_{34}H_{24}O$
2,2,3,4-Tetraphenyl-1(2*H*)-naphthalenone 297

$C_{34}H_{24}O_2$
2,2′,6,6′-Tetraphenyl-Δ4,4′-bi-4*H*-pyran 185

$C_{34}H_{24}S_2$
2,2′,6,6′-Tetraphenyl-Δ4,4′-bi-4*H*-thiopyran 209

$C_{34}H_{25}N_7O_8S_2$
Direct Green 1 (C.I. 30280) 1084

$C_{34}H_{26}N_2O_6$
Acid Violet 9 (C.I. 45190) 1003

$C_{34}H_{26}N_4O_4S$
1,1′-(Sulfonyldi-4,1-phenylene)bis(3-(1-naphthyl)urea) 786

$C_{34}H_{26}N_4O_6$
Bis(8-quinolyl) *N,N*′-(3,3′-dimethoxybiphenyl-4,4′-diyl)-biscarbamate 984

$C_{34}H_{26}N_4O_{10}S_2$
Chlorazol azurine 1083

$C_{34}H_{26}O_3$
α,α-Diphenylfurfuryl ether 845

$C_{34}H_{28}N_2O_2$
N,N′-Bis(4-benzyloxybenzylidene)-1,4-phenylenediamine 1039

$C_{34}H_{28}N_6O_6S_2$
Benzopurpurin B 1088

$C_{34}H_{28}N_6O_{16}S_4$
Direct Blue 1 (C.I. 24410) 1083

$C_{34}H_{30}N_2O_2Sn$
4-((4-(Dimethylamino)-2-(triphenylstannyl)phenyl)-iminomethyl)benzoic acid 1130

$C_{34}H_{30}N_4$
Benzil, azine with 5-indancarboxaldehyde 1054

$C_{34}H_{32}O_4$
1,2,5,6-Tetrakis(4-methoxyphenyl)-3-hexyne 1017

$C_{34}H_{34}N_2O_5S$
Carbobenzyloxy-*S*-benzylcysteinylphenylalanine benzyl ester 666

$C_{34}H_{34}N_2O_6S$
Carbobenzyloxy-*S*-benzyl-L-cysteinyl-L-tyrosine benzyl ester 644

$C_{34}H_{34}N_4O_6$
4,4′-Bis(4-(carbethoxy)-phenylureido)-3,3′-dimethylbiphenyl 577

$C_{34}H_{34}Si_2$
1-Butyl-1,1,2,2,2-pentaphenyldisilane 1117

$C_{34}H_{36}O_6$
2,3,4,6-Tetra-*O*-benzyl-α-D-glucopyranose 200

$C_{34}H_{40}Si$
Benzyltris(3-phenylpropyl)silane 1117

$C_{34}H_{42}N_2O_4$
Belladonnine bisulfate 360

$C_{34}H_{43}N_5O_6S$
Dibucaine, *N*-(4-sulfanilylphenyl)-glycine salt 985

$C_{34}H_{44}$
4,4′-Bis(4-pentyl-1-cyclohexen-1-yl)-biphenyl 165

$C_{34}H_{44}N_2O_2$
N,N′-Bis(4-heptyloxybenzylidene)-1,4-phenylenediamine 1047

$C_{34}H_{44}O_{18}$
4-Ethylphenylhepta-*O*-acetyl-β-maltoside 355

$C_{34}H_{45}Cl_3Si$
Hexadecyltris(3-chlorophenyl)-silane 1101
Hexadecyltris(4-chlorophenyl)-silane 1104

$C_{34}H_{46}O_2$
α2,α2′-(2,3,5,6-Tetramethyl-1,4-phenylene)bis(6-*tert*-butyl-2,4-xylenol) 194

$C_{34}H_{48}ClN_3$
Tris(4-(diethylamino)-2-methylphenyl)carbenium chloride 233

$C_{34}H_{48}Si$
Hexadecyltriphenylsilane 1117

$C_{34}H_{49}N_3O_2$
N-Hexadecyl-3′-(2-methyl-5-oxo-1-phenyl-3-pyrazolin-3-yl)acetanilide 702

$C_{34}H_{51}N_3O_7S_2$
1-Hexadecanesulfonic (2(3,4-dimethoxy-phenyl)-1-methyl-6-sulfo-4-quinolylidene)hydrazide 1041

$C_{34}H_{61}N_3O_5S$
N,N-Didodecyl-4-morpholino-3-nitrobenzenesulfonamide 818

$C_{34}H_{68}O_4Sn$
Bis-(2-ethylhexanoyloxy)-bis-(3,5,5-trimethylhexyl)-tin 1131

$C_{34}H_{69}NO$
N-Octadecylpalmitamide 106

$C_{34}H_{70}O_4$
Glyoxal bis-(bis-(2-ethylhexyl)-acetal) 19

$C_{34}H_{70}O_4Sn_2$
Bis(tributyltin) sebacate 1131

$C_{34}H_{71}O_4P$
Diheptadecyl phosphate 160

$C_{35}H_{20}N_2O_4$
N,N′-Bis(2-anthraquinonyl)-benzamidine 725

$C_{35}H_{23}N_7$
2,6-Bis(5,6-diphenyl-1,2,4-triazin-3-yl)pyridine 980

$C_{35}H_{25}BF_4S$
2,3,4,5,6-Pentaphenylthiopyrylium tetrafluoroborate 941

$C_{35}H_{26}Cl_2N_4O_4$
1,1′-(4-Methyl-1,3-phenylene)bis(3-(2-benzoyl-4-chlorophenyl)urea) 525

$C_{35}H_{26}N_2O_2$
4,4′-Methylenebis(*N*-(2-hydroxy-1-naphthylmethylene)aniline) 1028

$C_{35}H_{26}O$
6-Methoxy-1,2,3,4-tetraphenylnaphthalene 180

$C_{35}H_{28}N_4O_4$
Bis(2-methyl-8-quinolyl) 4,4′-methylenebis(*N*-phenylcarbamate) 991
1,1′-(4-Methyl-1,3-phenylene)bis(3-(4-benzoylphenyl)urea) 565

$C_{35}H_{28}N_4O_6$
Bis(2-(*N*-phenylcarbamoyl)phenyl) *N,N*′-(4-methyl-1,3-phenylene)-biscarbamate 670

$C_{35}H_{30}$
6,7,8,9-Tetrahydro-1,2,3,4-tetraphenyl-5*H*-benzocycloheptene 165

$C_{35}H_{30}N_4O_3$
1,3-Bis(3,4-diphenyl-5-isoxazolylethyl)urea 871

$C_{35}H_{30}O_{10}$
1-*O*-Methyl-2,3,4,6-tetra-*O*-benzoyl-β-D-glucopyranoside 371
Methyl 2,3,4,6-tetrabenzoyl-α-D-mannopyranoside 184

$C_{35}H_{30}S_3$
1,1,1-Tris(2-naphthylthiomethyl)-ethane 209

$C_{35}H_{32}O_{11}S$
α-Methylmannoside 6-(*p*-tosylate) 2,3,4-tribenzoate 156

$C_{35}H_{33}N_3O_5$
Carbobenzyloxy-L-tryptophyl-L-phenylalanine benzyl ester 890

$C_{35}H_{33}N_3O_6$
Carbobenzyloxytryptophyltyrosine benzyl ester 881

$C_{35}H_{38}O_6$
2,5-*O*-Methylene-1,3,4,6-tetra-*O*-benzyl-D-mannitol 183

$C_{35}H_{40}N_4$
4,4′-Methylenebis(*N*-(4-(diethylamino)benzylidene)aniline) 1062

$C_{35}H_{40}N_4O_2$
4,4′-Methylenebis(1-benzyl-1-isopropyl-3-phenylurea) 573
4,4′-Methylenebis(*N*-(*N*-butyl-*N*-phenylcarbamoyl)aniline) 592

$C_{35}H_{48}O_6$
Dibutyl bis(4-(2,2-dimethylpropionyl)benzyl)-malonate 360

$C_{35}H_{50}Si$
Heptadecyltriphenylsilane 1117

$C_{35}H_{53}N_3O_5S$
3-Carboxybenzenesulfonic (4-hexadecyloxy-α-piperidinobenzylidene)hydrazide 814

$C_{35}H_{56}N_2O_2$
N-(2-Diethylaminoethyl)-4-(hexadecyloxy)-benzanilide 465

$C_{35}H_{70}$
17-Pentatriacontene 1

$C_{35}H_{70}O$
18-Pentatriacontanone 50

$C_{35}H_{72}O$
18-Pentatriacontanol 9

$C_{36}H_{20}O_4$
Vat Green 1 (C.I. 59825) 1094

$C_{36}H_{21}O_3Sb$
Tris(4-dibenzofuryl)antimony 1128

$C_{36}H_{24}B_3Br_3N_6O_6$
2,4,6-Tris(5-(4-bromophenylazo)-2-hydroxyphenyl)boroxin 1125

$C_{36}H_{24}B_3N_9O_{12}$
2,4,6-Tris(5-(2-nitrophenylazo)-2-hydroxyphenyl)boroxin 1125

$C_{36}H_{24}N_2$
N,N′-(4,4′-Biphenylene)dicarbazole 892
N,N′-Bis(9-anthracenylmethylene)-1,4-phenylenediamine 1058

$C_{36}H_{24}N_2O_2$
3,4-Bis[1-phenyl-2(1*H*)-quinolinylidenemethyl]-1,3-cyclobutadienediylium-1,2-diolate 1095

$C_{36}H_{24}N_2O_3$
2,6-Diphenyl-4-pyridinecarboxylic anhydride 962

$C_{36}H_{24}N_2S_2$
10,10′-(1,1′-Biphenyl-4,4′-diyl)-bis(phenothiazine 253

$C_{36}H_{25}N_6O_{12}S_3$
Direct Red 26 (C.I. 29190) 1085

$C_{36}H_{26}$
Pentaphenylbenzene 165

$C_{36}H_{26}O_2$
3,3′′-(1,4-Phenylene)bis(4′-phenylacrylophenone) 309

$C_{36}H_{26}O_2Si_2$
10,10′-Bis(10-phenyl-10*H*-phenoxasilin) 1122

$C_{36}H_{27}ClSi$
Chlorotri(2-biphenylyl)silane 1120

$C_{36}H_{28}N_6O_{12}S_3$
Acid Brown 13 (C.I. 10410) 1094

$C_{36}H_{28}O$
5-Methoxy-1,2,3,4,5-pentaphenyl-1,3-cyclopentadiene 178

$C_{36}H_{28}OSi$
Tris(4-biphenylyl)silanol 1110

$C_{36}H_{28}O_4S_2Sn$
Bis(2-(phenylsulfonyl)phenyl)-diphenyltin 1131

$C_{36}H_{28}Si$
Bis(4-biphenylyl)diphenylsilane 1121
Tris(2-biphenyl)silane 1120
Tris(3-biphenylyl)silane 1121

$C_{36}H_{28}Sn$
4-(Triphenylstannyl)-*p*-terphenyl 1132

$C_{36}H_{30}B_2F_8N_2$
4,4′-Bis(2,6-diphenyl-1-methylpyridinium tetrafluoroborate) 957

$C_{36}H_{30}B_2F_8N_4$
4,4′-Azobis(2,6-diphenyl-1-methylpyridinium tetrafluoroborate) 1088

$C_{36}H_{30}Cl_2N_4O_8$
4,4′-Azobis(2,6-diphenyl-1-methylpyridinium perchlorate) 1088

$C_{36}H_{30}GeSn$
Triphenyl(triphenylstannyl)germane 1132

$C_{36}H_{30}Ge_2$
Hexaphenyldigermane 1099

$C_{36}H_{30}Ge_2O$
Hexaphenyldigermoxane 1117

$C_{36}H_{30}N_4$
2,6-Diphenyl-1-methyl-4(1*H*)-pyridone azine 769

$C_{36}H_{30}OPbSi$
Triphenylplumbyl triphenylsilyl oxide 1136

$C_{36}H_{30}OSi_2$
1,1,1,3,3,3-Hexaphenyldisiloxane 1117
$C_{36}H_{30}O_3Si_3$
Hexaphenylcyclotrisiloxane 1099
$C_{36}H_{30}Pb_2$
Hexaphenyldilead 1136
$C_{36}H_{30}SSn_2$
Triphenyltin sulfide 1132
$C_{36}H_{30}Si_2$
Hexaphenyldisilane 1117
$C_{36}H_{30}Sn_2$
Hexaphenylditin 1132
$C_{36}H_{32}$
5,6,7,8,9,10-Hexahydro-1,2,3,4-tetraphenylbenzocyclooctene 165
$C_{36}H_{32}N_2O_2$
N,*N*′-Dibenzyl-*N*,*N*′-di(*o*-tolyl)phthalamide 472
$C_{36}H_{32}N_6$
N,*N*,*N*′,*N*′-Tetrakis(4-aminophenyl)benzidine 230
$C_{36}H_{32}O$
4-Benzyl-3,5,5-triphenylvalerophenone 297
$C_{36}H_{32}Si_3$
1,1,2,2,3,3-Hexaphenyltrisilane 1117
$C_{36}H_{39}N_3O_3$
N,*N*′,*N*″-Triethyl-*N*,*N*′,*N*″-tri(*o*-tolyl)-1,3,5-benzenetricarboxamide 475
$C_{36}H_{42}As_2N_6O_{14}$
Ethylenebis(butylmethylphenylarsonium) dipicrate 1128
$C_{36}H_{42}N_4O_2$
3,3′-Hexamethylenebis(1,1-dibenzylurea) 572
$C_{36}H_{43}N_3O_5$
N-(3-Amino-4-(2-(4-*tert*-octylphenoxy)ethoxy)benzoyl-*N*′-(1-hydroxy-2-naphthoyl)ethylenediamine 485
$C_{36}H_{44}Si$
Tetrakis(3-phenylpropyl)silane 1117
$C_{36}H_{47}ClN_2O_2$
N,*N*′-Bis-(4-octyloxybenzylidene)-2-chloro-1,4-phenylenediamine 1044
$C_{36}H_{48}N_2O_2$
N,*N*′-Bis(4-octyloxybenzylidene)-1,4-phenylenediamine 1047
$C_{36}H_{49}Cl_3Si$
Octadecyltris(3-chlorophenyl)silane 1102
Octadecyltris(4-chlorophenyl)silane 1104
$C_{36}H_{49}F_3Si$
Octadecyltris(3-fluorophenyl)silane 1102
$C_{36}H_{49}N_3O_4$
3-(3-(2-Methyl-5-oxo-1-phenyl-3-pyrazolin-3-yl)phenylcarbamoyl)-5-nonadecenoic acid 703
$C_{36}H_{50}$
1,2,5b,6,7,10b,10c,10f-Octaethyl-5a,5b,10a,10b,10c,10d,10e,10f-octahydropentaleno(2,1,6-cde:5,4,3-c′d′e′)diazulene 3
$C_{36}H_{50}N_2O_5S$
N,*N*-Dibenzyl-4-hexadecyloxy-3-nitrobenzenesulfonamide 814
$C_{36}H_{52}Ge$
Octadecyltriphenylgermane 1117
$C_{36}H_{52}GeSi_4$
Tetrakis(4-(trimethylsilyl)phenyl)germane 1123
$C_{36}H_{52}N_2O_3S$
N^1,*N*^{1}-Dibenzyl-4-(hexadecyloxy)metanilamide 813
$C_{36}H_{52}OSi$
3-(Triphenylsilyl)-3-octadecanol 1105
$C_{36}H_{52}O_6$
Bis(4-pivaloylbenzyl)acetic acid 2,2-dimethyltrimethylene acetal 332
$C_{36}H_{52}Si$
Octadecyltriphenylsilane 1117
$C_{36}H_{52}Si_5$
Tetrakis(4-(trimethylsilyl)phenyl)silane 1122
$C_{36}H_{52}Sn$
Octadecyltriphenyltin 1132
$C_{36}H_{58}Cl_2Si$
Bis(3-chlorophenyl)didodecylsilane 1102
Bis(4-chlorophenyl)didodecylsilane 1104
$C_{36}H_{58}F_2Si$
Bis(3-fluorophenyl)didodecylsilane 1102
$C_{36}H_{58}OSi$
10,10-Didodecyl-10*H*-phenoxasilin 1123
$C_{36}H_{58}Si$
9,9-Didodecyl-9*H*-9-silafluorene 1123
$C_{36}H_{60}O_2$
Vitamin A palmitate 81
$C_{36}H_{60}Si$
Didodecyldiphenylsilane 1117
$C_{36}H_{62}Si_2$
3,3′-Bis(tributylsilyl)biphenyl 1123
$C_{36}H_{63}BO_3$
Tris(2-cyclohexylcyclohexyl) borate 1125
$C_{36}H_{70}Si$
Dodecyltris(2-cyclohexylethyl)silane 1111
$C_{36}H_{72}O_2$
Stearyl stearate 74
$C_{36}H_{72}O_4Sn$
Dibutyltin dimyristate 1131
Dihexyldilauroyloxytin 1131
$C_{36}H_{73}NO$
N-Octadecyloctadecanamide 106
$C_{36}H_{73}N_3S$
18-Pentatriacontanone thiosemicarbazone 140
$C_{36}H_{74}Cl_2Si$
Dichlorodioctadecylsilane 1099
$C_{36}H_{74}Hg$
Dioctadecylmercury 1137
$C_{36}H_{74}O$
n-Octadecyl ether 17
$C_{36}H_{74}S$
Octadecyl sulfide 27
$C_{36}H_{75}ClPb$
Chlorotridodecyllead 1133
$C_{36}H_{75}N$
Dioctadecylamine hydrochloride 29
Tridodecylamine hydrochloride 29
$C_{36}H_{75}O_4P$
Dioctadecyl phosphate 160
$C_{36}H_{75}P$
Tri(dodecyl)phosphine 158
$C_{36}H_{75}PS_3$
Trilauryl trithiophosphite 158
$C_{36}H_{76}Si$
Dioctadecylsilane 1099
Tridodecylsilane 1099
$C_{37}H_{22}O_4$
Tris(4-dibenzofuryl)methanol 893
$C_{37}H_{28}O$
5-(4-Methoxybenzylidene)-1,2,3,4-tetraphenyl-1,3-cyclopentadiene 180
$C_{37}H_{28}Si$
(*p*-Tolyl)tris(1-naphthyl)silane 1123
$C_{37}H_{30}N_2O_8$
N,*N*′-(4-Methyl-1,3-phenylene)bis(4-(benzyloxycarbonyl)phenyl carbamate) 657
$C_{37}H_{30}O_3Pb_2$
Bis(triphenylplumbyl) carbonate 1135
$C_{37}H_{30}Pb$
Triphenyl(triphenylmethyl)lead 1136
$C_{37}H_{30}Si$
Triphenyl(triphenylmethyl)silane 1117
$C_{37}H_{30}Sn$
Triphenyl(triphenylmethyl)tin 1132
$C_{37}H_{31}NSi$
N-(Diphenylmethyl)-*N*-(triphenylsilyl)aniline 1108
N,*N*-Diphenyl-α-(triphenylsilyl)benzylamine 1108
N-(Diphenyl(triphenylsilyl)methyl)aniline 1107
$C_{37}H_{32}Si_2$
Methylenebis(triphenylsilane) 1117
1,1,1,2,2-Pentaphenyl-2-(*p*-tolyl)disilane 1120
$C_{37}H_{36}N_4O_2$
3,3′-(4-Methyl-1,3-phenylene)bis(1,1-dibenzylurea) 563
$C_{37}H_{37}N_3O_5$
N,2′,3′,5′-Tetrabenzylcytidine 697
$C_{37}H_{40}N_2O_6$
Oxyacanthine hydrochloride tetrahydrate 245
$C_{37}H_{42}N_2O_9S$
2-Phenoxyethyl 4-(4-(2-(4-isooctylphenoxy)ethoxy)-3-nitrobenzamido)benzenesulfonate 797
$C_{37}H_{51}F_3Si$
Hexadecyltris(3-fluorobenzyl)silane 1102
$C_{37}H_{54}Si$
Hexadecyltribenzylsilane 1117
$C_{37}H_{55}N_3O_8S$
3,5-Dicarboxybenzenesulfonic [4-(hexadecyloxy-3-methoxy-α-piperidinobenzylidene)hydrazide] 812
$C_{37}H_{57}N_5O_7S_4$
3-Acetamido-4-(hexadecylsulfonyl)benzenesulfonic 3-(3-methyl-5-piperidinosulfonyl)-2-benzothiazolinylidene)hydrazide 820
$C_{38}H_{22}O_6$
3,7-Dibenzoyl-2,8-diphenyl-4*H*,6*H*-benzo(1,2-b:5,4-b′)dipyran-4,6-dione 320
$C_{38}H_{27}N_9O_{13}S_3$
Owens Blue 1083
$C_{38}H_{28}O_2Si$
Bis(9-xanthenyl)diphenylsilane 1111
$C_{38}H_{28}Si$
Bis(9-fluorenyl)diphenylsilane 1122
$C_{38}H_{30}O_2$
α,α,α′,α′-Tetraphenyl-(1,1′-biphenyl)-2,2′-dimethanol 200
$C_{38}H_{30}Si_2$
1,2-Bis(triphenylsilyl)acetylene 1117
$C_{38}H_{32}N_2$
1,2-Bis(trityl)hydrazine 235
$C_{38}H_{32}OSi$
Diphenyl(diphenylmethoxy)(diphenylmethyl)silane 1107
$C_{38}H_{32}O_6$
2′-Acetoxy-2,4,4′-tris(benzyloxy)chalcone 377
$C_{38}H_{32}Si$
Bis(2-biphenylyl)dibenzylsilane 1122
Bis(diphenylmethyl)diphenylsilane 1117
Bis(*p*-terphenyl-4-yl)dimethylsilane 1121
Triphenyl(1,2,2-triphenylethyl)silane 1117
$C_{38}H_{34}Br_2P_2$
Ethylenebis(triphenylphosphonium bromide) 832
$C_{38}H_{34}N_2O_3$
2′-(Dibenzylamino)-6′-(diethylamino)spiro[isobenzofuran-1(3*H*),9′-(9*H*)xanthene]-3-one 385
$C_{38}H_{34}OSi_2$
1,1-Bis(triphenylsilyl)ethanol 1105
$C_{38}H_{34}Si_2$
1,2-Bis(triphenylsilyl)ethane 1117
1,1-Di(*p*-tolyl)-1,2,2,2-tetraphenyldisilane 1120
1,2-Di(*p*-tolyl)-1,1,2,2-tetraphenyldisilane 1120
$C_{38}H_{36}N_2$
1,1′-Bis(diphenylmethyl)-4,4′-dimethyl-1,1′,4,4′-tetrahydro-4,4′-dipyridyl 251
$C_{38}H_{36}N_2Si_2$
1,2-Bis(diphenylmethylsilyl)-1,2-diphenylhydrazine 1108
$C_{38}H_{36}N_4Si_2$
3,6-Dimethyl-1,2,3,4,5,6-hexaphenyl-1,2,4,5-tetraaza-3,6-disilacyclohexane 1117
$C_{38}H_{36}N_8$
Tetrakis-(5,6-dimethyl-2-benzimidazolyl)-ethylene 905
$C_{38}H_{37}ClN_8O_4$
1,1′-((2-Chlorobenzyl)-2-methylphenylene)bis(3-antipyrinylurea), mixture of isomers 695
$C_{38}H_{38}N_4O_2Si_2$
1,1′-(1,1′-Biphenyl-4,4′-diylbisazo)-bis(6-(trimethylsilyl)-2-naphthol) 1086
$C_{38}H_{51}N_3O_5S_2$
1-Hexadecanesulfonic (2-(4-biphenyl)-1-methyl-6-sulfo-4(1*H*)-quinolylidene)hydrazide 1055
$C_{38}H_{52}N_2O_2$
N,*N*′-Bis-(4-nonyloxybenzylidene)-1,4-phenylenediamine 1044
$C_{38}H_{52}O_4S_4$
4,4′-Bis(2,5-dibutoxy-4-ethylphenyl)tetrathiafulvalene 880
$C_{38}H_{57}N_3O_5S_2$
1-Hexadecanesulfonic (2-(4-cyclohexyl-phenyl)-1-methyl-6-sulfo-4-quinolylidene)hydrazide 1055
$C_{38}H_{58}N_4O_8$
N,*N*″-Hexamethylenebis(*N*′-(2-(3,4-dimethoxyphenyl)ethyl)hexanediamide) 413
$C_{38}H_{62}F_2Si$
Bis(2-fluorobenzyl)didodecylsilane 1104
Bis(3-fluorobenzyl)didodecylsilane 1102
$C_{38}H_{64}Si$
Bis(*m*-tolyl)didodecylsilane 1099
$C_{38}H_{66}ClN_3O_6$
N,*N*-Bis(2-(*N*-dodecylcarbamoyloxy)ethyl)-2-(4-chlorophenoxy)acetamide 615
$C_{38}H_{71}N_3O_4S$
3-Nitro-*N*^{1},*N*^{1},*N*^{4},*N*^{4}-tetrakis-(2-ethylhexyl)-sulfanilamide 817
$C_{38}H_{76}N_2O_2$
N,*N*′-Ethylenebisoctadecanamide 110
$C_{38}H_{76}O_2$
Ethyl 2-(octadecyl)octadecanoate 74
$C_{38}H_{78}O_2Pb$
Tridodecylplumbyl acetate 1133
$C_{38}H_{82}HgSi_2$
Bis(dimethylhexadecylsilylmethyl)mercury 1099
$C_{39}H_{26}N_7O_{10}S_2$
Direct Dye (C.I. 31793) 1084
$C_{39}H_{30}N_2$
Bis-(triphenylmethyl)-carbodiimide 769
$C_{39}H_{30}O_3$
2,2,4,4,6,6-Hexaphenyl-1,3,5-trioxane 185
$C_{39}H_{34}Si_2$
1,2-Bis(triphenylsilyl)-1-propene 1117
1,3-Bis(triphenylsilyl)-1-propene 1117
$C_{39}H_{36}GeSi$
Triphenyl(3-(triphenylgermyl)propyl)silane 1118
$C_{39}H_{36}Ge_2$
Trimethylenebis(triphenylgermane) 1118
$C_{39}H_{36}OSi_2$
1,3-Bis(triphenylsilyl)-2-propanol 1105
1,1,1-Triphenyl-3,3,3-tris(*m*-tolyl)disiloxane 1117
$C_{39}H_{36}Pb_2$
1,3-Bis(triphenylplumbyl)propane 1136
$C_{39}H_{36}Si_2$
1,2-Propanediylbis(triphenylsilane) 1117
1,3-Trimethylenebis(triphenylsilane) 1117
$C_{39}H_{42}Si$
(3-Biphenylyl)tris(3-phenylpropyl)silane 1121
$C_{39}H_{49}FN_2O_6S_2$
2-Hexadecyloxy-5-(4-(1-naphthylsulfamoyl)-phenylcarbamoyl)benzenesulfonyl fluoride 815
4-Hexadecyloxy-3-(4-(1-naphthylsulfamoyl)-phenylcarbamoyl)benzenesulfonyl fluoride 815
$C_{39}H_{50}N_2O_7S_2$
4-Hexadecyloxy-3-(4-(1-naphthylsulfamoyl)phenylcarbamoyl)benzenesulfonic acid, potassium salt 815
2-Hexadecyloxy-5-(4-(1-naphthylsulfamoyl)phenylcarbamoyl)benzenesulfonic acid, sodium salt 815
$C_{39}H_{51}N_3O_9S_2$
3-Carboxy-4-hydroxybenzenesulfonic (2-(4-(hexadecyloxy)phenyl)-1-methyl-6-sulfoquinolylidene)hydrazide 1031
$C_{39}H_{54}O_6$
Glycerol trilaurate 74
$C_{39}H_{56}N_4O_2$
4,4′-Methylenebis(1,1-dicyclohexyl-3-phenylurea) 583

$C_{39}H_{56}O_2Si$
Tri(*o*-tolyl)silyl stearate ... 1122
$C_{39}H_{58}Si$
Octadecyltribenzylsilane ... 1117
$C_{39}H_{76}O_5$
Glycerol 1,2-distearate ... 74
$C_{40}H_{26}$
1,3,6,8-Tetraphenylpyrene ... 166
$C_{40}H_{26}O_3$
6,7-Dibenzoyl-1,3,5-triphenylcyclopenta(c)pyran ... 307
$C_{40}H_{26}O_8S$
1,1′-(Sulfonylbis)4,1-phenyleneoxy-4,1-phenylene)bis(2-phenylethanedione) ... 775
$C_{40}H_{28}B_2F_8O_2$
1,4-Phenylene-4,4′-bis(2,6-diphenyl-4-pyrylium tetrafluoroborate) ... 935
$C_{40}H_{28}Si$
Tetrakis(2-naphthyl)silane ... 1123
$C_{40}H_{30}Si_2$
1,4-Bis(triphenylsilyl)-1,3-butadiyne.. 1118
$C_{40}H_{32}N_2O_4S$
4,4′-Sulfonylbis(*N*-(4-benzoyloxybenzylidene)aniline) ... 1039
$C_{40}H_{32}O_{10}$
Xylitol pentabenzoate ... 370
$C_{40}H_{34}N_2Si$
Bis(9-ethyl-3-carbazolyl)-diphenylsilane ... 1108
$C_{40}H_{34}O_4Sn_2$
Bis-(triphenylstannyl) succinate ... 1131
$C_{40}H_{36}Cl_2P_2$
trans-2-Butylene-1,4-bis(triphenylphosphonium chloride) ... 832
$C_{40}H_{36}N_2S_4Sn_2$
Bis(triphenyltin) ethylenebis(dithiocarbamate) ... 1130
$C_{40}H_{36}N_4O_2$
3,3′-(1,5-Naphthalenediyl)bis(1,1-dibenzylurea) ... 567
$C_{40}H_{36}O_4Si$
Bis(2-(2-phenoxyethoxy)phenyl)-diphenylsilane ... 1106
$C_{40}H_{38}Br_2P_2$
Tetramethylenebis(triphenyl-phosphonium bromide) ... 830
$C_{40}H_{38}N_2Si_2$
1,4-Piperazinediylbis(triphenyl-silane) ... 1112
$C_{40}H_{38}O_2Si_2$
1,4-Bis(triphenylsilyl)-2,3-butanediol ... 1104
$C_{40}H_{38}Pb_2$
1,4-Bis(triphenylplumbyl)butane ... 1136
$C_{40}H_{38}Si_2$
1,2-Butanediylbis(triphenylsilane) ... 1118
1,1-Diphenyl-1,2,2,2-tetra(*p*-tolyl)-disilane ... 1120
1,2-Diphenyl-1,1,2,2-tetra(*p*-tolyl)-disilane ... 1120
Tetramethylenebis(triphenylsilane).. 1118
$C_{40}H_{42}Si_5$
1,1,2,2,3,3-Hexaphenyl-4,4,5,5-tetramethylcyclopentasilane ... 1118
$C_{40}H_{43}N_3O_9$
N,*N*′-Bis(carbobenzyloxy)-L-lysyl-L-aspartic acid dibenzyl ester ... 676
$C_{40}H_{46}O_{11}$
2,2′,3,3′-Tetra-*O*-benzyltrehalose ... 199
$C_{40}H_{52}Si_2$
9-((Dibenzylmethylsilyl)methyl)-9-dodecyl-9*H*-9-silafluorene ... 1122
$C_{40}H_{53}N_3O_8S_2$
5-Carboxy-2-hexadecyloxybenzenesulfonic (1-methyl-6-sulfo-2-(*m*-tolyl)-4(1*H*)quinolylidene)hydrazide ... 1055
$C_{40}H_{55}ClN_2O_2$
Bis-(4-*N*-decyloxybenzylidene)-2-chloro-1,4-phenylenediamine ... 1044
$C_{40}H_{56}N_2O_2$
Bis-(4-*N*-decyloxybenzylidene)-1,4-phenylenediamine ... 1044
$C_{40}H_{56}N_4O_2$
Sebacic bis((α-hexylcinnamylidene)hydrazide) ... 502
$C_{40}H_{62}N_4O_4$
Sebacic bis((3,5-di-*tert*-butyl-4-hydroxybenzylidene)hydrazide) ... 496
$C_{40}H_{63}N_3O_5S_2$
1-Hexadecanesulfonic (2-(3,5-di-*tert*-butylphenyl)-1-methyl-6-sulfo-4(1*H*)quinolylidene)hydrazide ... 818
$C_{40}H_{68}Si$
Diphenylditetradecylsilane ... 1118
$C_{40}H_{75}N$
N,*N*-Dihexadecyl-3,5-xylidine ... 220
$C_{40}H_{77}NO_5$
N,*N*-Bis(2-(dodecanoyl)ethyl)-dodecanamide ... 107
$C_{40}H_{80}O_4Sn$
Bis(lauroyloxy)dioctyltin ... 1129
$C_{41}H_{30}Cl_2N_4O_4$
1,1′-(Methylenedi-4,1-phenylene)-bis(3-(2-benzoyl-4-chlorophenyl)urea ... 530
$C_{41}H_{32}N_4O_4$
4,4′-Methylenebis(*N*-(*N*-(4-benzoylphenyl)carbamoyl)aniline) ... 575
$C_{41}H_{32}O_{11}$
β-D-Glucose pentabenzoate ... 371
$C_{41}H_{36}O_3$
1,3-Bis(trityloxy)-2-propanol ... 200
$C_{41}H_{39}NO_9$
1-*O*-(4-Nitrobenzoyl)-2,3,4,6-tetra-*O*-benzyl-α-D-glucopyranose ... 372
1-*O*-(4-Nitrobenzoyl)-2,3,4,6-tetra-*O*-benzyl-β-D-glucose ... 371
$C_{41}H_{40}N_2Si_2$
1,4-Bis(triphenylsilyl)-2,3,4,5,6,7-hexahydro-1*H*-1,4-diazepine ... 1099
$C_{41}H_{40}O_2Si_2$
1-(Triphenylsilyl)-3-(2-(triphenylsilyl)ethoxy)-2-propanol ... 1104
$C_{41}H_{40}Si_2$
1-(Methylphenyl(*p*-tolyl)silyl)-4-(tribenzylsilyl)benzene ... 1120
Pentamethylenebis(triphenylsilane). 1118
$C_{41}H_{44}O_4Si$
4-(*tert*-Butyldiphenylsilyloxy)-2-methoxy-6-(trityloxy)-tetrahydropyran ... 1105
$C_{41}H_{45}N_3O_9$
4′-Carbethoxy-3-(2-(3-nitrobenzoyl)-acetamido)-4-(2-(4-*tert*-octylphenoxy)ethoxy)benzanilide ... 442
$C_{41}H_{54}O_3Si_3$
3,5-Bis(*tert*-butyldiphenylsilyloxy)-1-(trimethylsilyloxy)-1-cyclohexene ... 1107
$C_{42}H_{22}N_2O_6$
1,4-Bis-(1-anthraquinonylamino)-anthraquinone ... 317
$C_{42}H_{25}N_3O_6$
4,4′-Dibenzamido-1,1′-dianthrimide.. 487
4,5′-Dibenzamido-1,1′-dianthrimide.. 487
$C_{42}H_{28}N_2O_4$
3,4′-Dinitro-2,3′,4,5′,6-pentaphenylbiphenyl ... 278
$C_{42}H_{29}NO_3$
2,6-Diphenyl-4-(4-nitro-α,3,5-triphenylbenzylidene)-4*H*-pyran ... 960
$C_{42}H_{30}F_4Si_2$
(2,3,5,6-Tetrafluoro-1,4-phenylene)-bis(triphenylsilane) ... 1103
$C_{42}H_{30}O_2$
4,4″-Dihydroxy-3,3″,5,5″-tetraphenyl-*p*-terphenyl ... 194
$C_{42}H_{30}O_6$
Hexakis(phenoxy)benzene ... 182
$C_{42}H_{31}N$
2,4,6-Tris(4-biphenylyl)aniline ... 230
$C_{42}H_{32}Cl_2N_{12}O_6S_2$
Thiocarbamyl Nitro Blue tetrazolium chloride ... 1094
$C_{42}H_{32}N_2$
2,3′,4,5′,6-Pentaphenyl-3,4′-biphenyldiamine ... 230
$C_{42}H_{34}Pb_2$
1,4-Bis(triphenylplumbyl)benzene ... 1136
$C_{42}H_{34}Si_2$
1,3-Bis(triphenylsilyl)benzene ... 1118
1,4-Phenylenebis(triphenylsilane) ... 1120
$C_{42}H_{36}F_6OSi_2$
1,1,1,3,3,3-Hexakis(3-fluorobenzyl)-disiloxane ... 1102
$C_{42}H_{42}Si_2$
Hexabenzyldisilane ... 1118
Hexakis(*p*-tolyl)disilane ... 1123
$C_{42}H_{42}Sn_2$
Hexa(*o*-tolyl)ditin ... 1133
$C_{42}H_{44}N_2O_3S$
2-Methyl-1-octadecyl-5-benzimidazolesulfonic acid ... 906
$C_{42}H_{62}O_{16}$
Glycyrrhizic acid, monosodium salt. 67
$C_{42}H_{65}NO_{16}$
Ammonium glycyrrhizinate ... 67
$C_{42}H_{72}Si$
(2-Biphenylyl)tris(decyl)silane ... 1120
$C_{42}H_{76}N_2O_2$
N,*N*′-(1,4-Phenylene)-bisoctadecanamide ... 428
$C_{42}H_{76}Si_3$
Bis(4-(trimethylsilyl)phenyl)-didodecylsilane ... 1123
$C_{42}H_{78}CuO_6$
Methyl 3-oxoeicosanoate, copper(II) derivative ... 70
$C_{42}H_{79}N$
N,*N*-Dioctadecylaniline ... 220
$C_{42}H_{80}Si$
Phenyltridodecylsilane ... 1118
$C_{42}H_{82}O_4S$
Dioctadecyl 3,3′-thiodipropionate ... 87
$C_{42}H_{82}Si$
Octadecyltris(2-cyclohexylethyl)-silane ... 1111
$C_{42}H_{86}Si$
Cyclohexyltridodecylsilane ... 1111
$C_{42}H_{87}ClSn$
Chlorotritetradecyltin ... 1129
$C_{43}H_{32}OSi$
9,9-Diphenyl-4-(triphenylsilyl)-xanthene ... 1111
$C_{43}H_{38}$
1,1,1,6,6,6-Hexaphenyl-3-methyl-3-hexene ... 164
$C_{43}H_{40}N_4O_2$
4,4′-Methylenebis(1,1-dibenzyl-3-phenylurea) ... 572
$C_{43}H_{56}N_2O_8$
2,5-Dioxo-3-tetrahydrofurandocosa-β,δ-dienoic acid, α-substituted ... 444
$C_{43}H_{66}Si$
Hexadecyltris(3-phenylpropyl)-silane ... 1118
$C_{44}H_{26}O_{12}$
3,3′,3″,3‴-(1,4-Phenylenedimethylidyne)-tetrakis(4-hydroxycoumarin) ... 382
$C_{44}H_{30}N_4$
5,10,15,20-Tetraphenyl-21*H*,23*H*-porphine hydrochloride ... 839
$C_{44}H_{38}B_2F_8N_2$
4,4′-Bis(2,6-distyryl-1-methylpyridinium tetrafluoroborate) ... 957
$C_{44}H_{38}Cl_2P_2$
p-Xylylenebis(triphenyl-phosphonium chloride) ... 832
$C_{44}H_{46}Si_2$
Ethylenebis(tribenzylsilane) ... 1118
$C_{44}H_{63}NO_4$
5-(Trityloxymethyl)-tetrahydrofurfuryl *N*-octadecylcarbamate ... 649
$C_{44}H_{64}N_2O_2$
N,*N*′-Bis(4-dodecyloxybenzylidene)-1,4-phenylenediamine ... 1044
$C_{44}H_{76}Si$
Dihexadecyldiphenylsilane ... 1118
$C_{44}H_{80}Cr_2N_2O_7$
1-Hexadecyl-4-methylpyridinium dichromate ... 959
$C_{44}H_{84}O_4Sn$
Dibutyltin dioleate ... 1131
$C_{44}H_{86}N_2O_5$
(Tetrahydrofuran-2,5-diyldimethyl) bis(*N*-octadecylcarbamate) ... 115
$C_{44}H_{88}O_4Sn$
Dibutyltin distearate ... 1131
$C_{45}H_{32}Si$
Phenyltris(9-fluorenyl)silane ... 1122
$C_{45}H_{36}Si_2$
1,1-Bis(triphenylsilyl)indene ... 1121
$C_{46}H_{30}BF_4N_3S_6$
Tris(4,5-diphenyl-1,3-dithiolan-2-ylideneamino)carbenium tetrafluoroborate ... 1076
$C_{46}H_{42}Cl_2P_2$
(2,5-Dimethyl-1,4-xylylene)-bis(triphenylphosphonium chloride) ... 832
(4,5-Dimethyl-1,2-xylylene)-bis(triphenylphosphonium chloride) ... 832
$C_{46}H_{82}O_4$
Octadecamethylene-2,2′-bis(1,3-cyclotetradecanedione) ... 54
$C_{46}H_{86}O_6$
13,34-Dioxohexatetra-contanedioic acid ... 64
$C_{46}H_{88}Si$
(3-*tert*-Butylphenyl)tridodecylsilane. 1099
$C_{48}H_{36}N_2$
2,3-Bis(triphenylmethylazo)-naphthalene ... 1054
$C_{48}H_{36}Si$
Tetrakis(4-biphenylyl)silane ... 1121
$C_{48}H_{38}OSi_2$
2-(Triphenylsilyl)phenyl ether ... 1106
4-(Triphenylsilyl)phenyl ether ... 1106
$C_{48}H_{38}Si_2$
3,3′-Bis(triphenylsilyl)biphenyl ... 1121
4,4′-Bis(triphenylsilyl)biphenyl ... 1121
$C_{48}H_{40}Br_2Si_4$
1,4-Dibromo-1,1,2,2,3,3,4,4-octaphenyltetrasilane ... 1118
$C_{48}H_{40}Cl_2Si_4$
1,4-Dichloro-1,1,2,2,3,3,4,4-octaphenyltetrasilane ... 1118
$C_{48}H_{40}N_4$
2,3-Bis(2-(triphenylmethyl)-hydrazino)naphthalene ... 235
$C_{48}H_{40}O_2Si_4$
2,2,3,3,5,5,6,6-Octaphenyl-1,4-dioxa-2,3,5,6-tetrasilacyclohexane ... 1118
2,2,4,4,5,5,6,6-Octaphenyl-1,3-dioxa-2,4,5,6-tetrasilacyclohexane ... 1118
$C_{48}H_{40}O_3Si_4$
2,2,4,4,6,6,7,7-Octaphenyl-1,3,5-trioxa-2,4,6,7-tetrasilacycloheptane ... 1118
$C_{48}H_{40}SiSn_2$
Bis(triphenylstannyl)diphenylsilane. 1132
$C_{48}H_{40}Si_3$
Octaphenyltrisilane ... 1118
$C_{48}H_{40}Si_4$
Octaphenylcyclotetrasilane ... 1118
$C_{48}H_{41}BrSi_4$
1-Bromo-1,1,2,2,3,3,4,4-octaphenyltetrasilane ... 1118
$C_{48}H_{41}ClSi_4$
1-Chloro-1,1,2,2,3,3,4,4-octaphenyltetrasilane ... 1118
$C_{48}H_{41}N_3Si$
Phenyltris(9-ethyl-3-carbazolyl)-silane ... 1108
$C_{48}H_{42}O_2Si_4$
1,1,2,2,3,3,4,4-Octaphenyl-1,4-tetrasilanediol ... 1104
$C_{48}H_{42}O_{10}$
1,6-Di-*O*-benzyl-2,3,4,5-tetra-*O*-benzoyldulcitol ... 371
$C_{48}H_{42}S_6$
Hexakis(phenylthiomethyl)benzene. 212
$C_{48}H_{42}Si_4$
1,1,2,2,3,3,4,4-Octaphenyltetrasilane ... 1118
$C_{48}H_{46}Si_2$
1,4-Bis(tribenzylsilyl)benzene ... 1120
$C_{48}H_{48}N_6$
Hexakis(anilinomethyl)benzene ... 242
$C_{48}H_{48}N_6S_6$
Hexakis(4-aminophenylthiomethyl)-benzene ... 230
$C_{48}H_{48}O_2$
1-(5-*tert*-Butyl-1,3-dihydro-1,3,3-triphenyl-1-isobenzofuryl)-3,3-

$C_{48}H_{48}O_2$ (Cont'd)
dimethyl-1,2-diphenyl-2-butanol....... 203
$C_{48}H_{52}Si$
Dodecyltris(2-biphenylyl)silane......... 1120
$C_{48}H_{54}O_6$
Hexakis(4-methoxyphenoxymethyl)-benzene....... 179
$C_{48}H_{60}N_6OSi$
1,1,1,3,3,3-Hexakis(4-(dimethylamino)phenyl)disiloxane.. 1107
$C_{48}H_{68}O_2Si$
Bis(2-phenoxyphenyl)-didodecylsilane....... 1106
Bis(4-phenoxyphenyl)-didodecylsilane....... 1106
$C_{48}H_{68}Si$
Bis(2-biphenylyl)didodecylsilane...... 1120
Bis(3-biphenylyl)didodecylsilane...... 1121
$C_{48}H_{80}N_2O_2$
N,N'-(1,1'-Biphenyl-4,4'-diyl)-bisoctadecamide....... 489
$C_{48}H_{84}Ge$
Dioctadecyldiphenylgermane........... 1118
$C_{48}H_{84}Si$
(2-Biphenylyl)tridodecylsilane........... 1120
Dioctadecyldiphenylsilane................ 1118
$C_{48}H_{92}Si$
Phenyltritetradecylsilane................ 1118
$C_{48}H_{94}O_2$
1,1,4,4-Tetraundecyl-2-butyne-1,4-diol....... 1016
$C_{48}H_{94}O_4$
Didocosyl succinate....... 76
$C_{48}H_{99}BO_3$
Trihexadecyl borate....... 1124
$C_{48}H_{100}Ge$
Tetradodecylgermane....... 1099
$C_{48}H_{100}Si$
Tetradodecylsilane....... 1099
Tris(hexadecyl)silane....... 1099
$C_{48}H_{100}Sn$
Tetralauryltin....... 1129
$C_{49}H_{43}ClSi_4$
1-Chloro-4-methyl-1,1,2,2,3,3,4,4-octaphenyltetrasilane....... 1118
$C_{50}H_{40}Si$
Bis(*p*-terphenylyl)dibenzylsilane....... 1121
$C_{50}H_{42}Cl_2P_2$
1,1'-Biphenyl-4,4'-diylbis(methyl)-bis(triphenylphosphonium chloride)....... 832
$C_{50}H_{46}Si_4$
1,4-Dimethyl-1,1,2,2,3,3,4,4-octaphenyltetrasilane....... 1118
$C_{50}H_{80}O_4$
Diethylstilbestrol dipalmitate....... 380
$C_{50}H_{84}O_2$
2,4-Bis-(4-(hexadecyloxy)-phenyl)-4-methyl-1-pentene....... 181
$C_{50}H_{88}Si$
Bis(*p*-tolyl)dioctadecylsilane....... 1123
Dibenzyldioctadecylsilane....... 1118
$C_{50}H_{92}Cr_2N_2O_7$
Benzyldimethylhexadecyl-ammonium dichromate....... 255
$C_{51}H_{48}OSi_4$
1,1,2,2,3,3,4,4-Octaphenyl-1-propoxytetrasilane....... 1118
$C_{52}H_{40}$
Octaphenylcyclobutane....... 165
$C_{52}H_{40}O_4$
2,2,4,4,6,6,8,8-Octaphenyl-1,3,5,7-tetroxocane....... 185
$C_{52}H_{48}N_4Si_2$
3,3,6,6-Tetrabenzyl-1,2,4,5-tetraphenyl-1,2,4,5-tetraaza-3,6-disilacyclohexane....... 1099
$C_{52}H_{50}O_{10}$
1,6-Di-*O*-trityl-2,3,4,5-tetra-*O*-acetyl-D-mannitol....... 356
$C_{52}H_{108}Si$
Butyltrihexadecylsilane....... 1099
$C_{53}H_{106}$
18-Heptadecyl-17-hexatriacontene... 2
$C_{53}H_{108}$
18-Heptadecylhexatriacontane....... 1
$C_{53}H_{108}O$
18-Heptadecyl-18-hexatriacontanol.. 10
$C_{54}H_{42}N_4$
1,1'-(1,4-Phenylene)bis(2-(5-indanyl)-4,5-diphenylimidazole)....... 867
$C_{54}H_{50}Si_2$
3,3'-Bis(tribenzylsilyl)biphenyl....... 1121
4,4'-Bis(tribenzylsilyl)biphenyl....... 1121
$C_{54}H_{54}O_6$
Hexakis(benzoxymethyl)benzene..... 186
$C_{54}H_{54}Se_6$
Hexakis(benzylselenomethyl)-benzene....... 1143
$C_{54}H_{58}Si_6$
1,1,1,6,6,6-Hexamethyl-2,2,3,3,4,4,5,5-octaphenylhexasilane....... 1119
$C_{54}H_{60}N_6$
Hexakis(*N*-methylanilinomethyl)-benzene....... 242
$C_{54}H_{64}Si$
Octadecyltris(2-biphenylyl)silane..... 1119
$C_{54}H_{78}O_3$
1,3,5-Trimethyl-2,4,6-tris(3,5-di-*tert*.-butyl-4-hydroxybenzyl)-benzene....... 196
$C_{54}H_{96}Si$
Bis(3-phenylpropyl)-dioctadecylsilane....... 1118
$C_{54}H_{103}ClSi$
(4-Chlorophenyl)trihexadecylsilane.. 1104
$C_{54}H_{104}Si$
Phenyltrihexadecylsilane....... 1119
$C_{54}H_{111}BO_3$
Trioctadecyl borate....... 1124
$C_{54}H_{111}ClPb$
Chlorotrioctadecyllead....... 1133
$C_{54}H_{111}N$
Trioctadecylamine....... 29
Trioctadecylamine hydrochloride..... 29
$C_{54}H_{111}O_3P$
Trioctadecyl phosphite....... 158
$C_{55}H_{42}Si$
Methyltris(*p*-terphenyl-4-yl)silane..... 1120
$C_{55}H_{112}O_3$
Trioctadecyl orthoformate....... 18
$C_{56}H_{40}Cl_4Pd_2$
Chloro(1,2,3,4-tetraphenylcyclobutadiene)-μ-chloropalladium(II) dimer....... 1143
$C_{57}H_{46}Si_3$
1,3,3-Tris(triphenylsilyl)allene....... 1119
$C_{57}H_{48}Si_3$
1,3,3-Tris(triphenylsilyl)-1-propene.... 1119
$C_{57}H_{112}Si_2$
Trihexadecyl(4-(trimethylsilyl)-phenyl)silane....... 1123
$C_{60}H_{44}Si$
Phenyltris(*p*-terphenylyl)silane....... 1121
$C_{60}H_{50}Si_4$
Decaphenyltetrasilane....... 1119
$C_{60}H_{52}O_2Si_5$
Decaphenyl-1,5-pentasilanediol....... 1119
$C_{60}H_{52}Si_5$
1,1,2,2,3,3,4,4,5,5-Decaphenylpentasilane....... 1119
$C_{60}H_{108}OSi$
(4-Phenoxyphenyl)-trihexadecylsilane....... 1106
$C_{60}H_{108}O_5$
2',5'-Dihydroxyocta-decanophenone dioctadecanoate... 378
$C_{60}H_{108}Si$
(2-Biphenylyl)trihexadecylsilane....... 1120
(4-Biphenylyl)trihexadecylsilane....... 1121
$C_{60}H_{110}O_4$
Octadecylhydroquinone dioctadecanoate....... 379
$C_{61}H_{46}Si$
Benzyltris(*p*-terphenylyl)silane....... 1121
$C_{61}H_{118}Si$
Benzyltrioctadecylsilane....... 1119
$C_{62}H_{56}Si_6$
1,1,2,2,3,3,4,4,5,5-Decaphenyl-6,6-dimethylcyclohexasilane....... 1099
$C_{62}H_{130}Si_2$
Ethylenebis(tris(decyl)silane)....... 1099
$C_{63}H_{66}N_4O_{15}$
γ-Pentabenzyl *N*-carbobenzyloxy-L-glutamyl-L-glutamyl-L-glutamyl-L-glutamate....... 420
$C_{64}H_{52}O_8$
1,5-Di-*O*-trityl-2,3,4-tri-*O*-benzoylxylitol....... 370
$C_{64}H_{132}Ge$
Tetra(hexadecyl)germane....... 1099
$C_{64}H_{132}Pb$
Tetrahexadecyllead....... 1133
$C_{66}H_{64}O_2Si_5$
1,1,2,2,3,3,4,4,5,5-Decaphenyl-1,5-dipropoxypentasilane....... 1119
$C_{66}H_{78}O_6$
Hexakis(4-isopropylphenoxymethyl)benzene... 179
$C_{66}H_{114}Si_2$
1,1,1-Trihexadecyl-2,2,2-triphenyldisilane....... 1119
$C_{66}H_{130}Si_2$
1,3-Phenylenebis(trisdecylsilane)..... 1099
$C_{67}H_{58}Si_5$
1-Methylundecaphenylpentasilane.. 1119
$C_{68}H_{54}O_{19}$
Sucrose benzoate....... 372
$C_{68}H_{68}Si_5$
1,1,2,2,3,3,4,4,5,5-Decaphenyl-1,5-dibutylpentasilane....... 1119
$C_{72}H_{54}S_6$
Hexakis(2-naphthylthiomethyl)-benzene....... 214
$C_{72}H_{60}SiSn_4$
Tetrakis(triphenylstannyl)silane....... 1132
$C_{72}H_{60}Si_4Sn$
Tetrakis(triphenylsilyl)tin....... 1132
$C_{72}H_{60}Sn_5$
Tetrakis(triphenylstannyl)tin....... 1132
$C_{72}H_{62}Si_6$
1,1,2,2,3,3,4,4,5,5,6,6-Dodecaphenylhexasilane....... 1119
$C_{72}H_{90}O_6$
Hexakis(4-*tert*-butylphenoxymethyl)benzene....... 179
$C_{72}H_{90}S_6$
Hexakis(4-*tert*-butylphenylthiomethyl)benzene....... 213
$C_{72}H_{148}Si$
Tetraoctadecylsilane....... 1099
$C_{72}H_{148}Sn$
Tetraoctadecyltin....... 1129
$C_{74}H_{66}Si_6$
1,6-Dimethyl-1,1,2,2,3,3,4,4,5,5,6,6-dodecaphenylhexasilane....... 1119
$C_{76}H_{72}Si_7$
1,2,2,3,3,4,4,5,5,6,6,7-Dodecaphenyl-1,1,7,7-tetramethylheptasilane....... 1119
$C_{84}H_{70}Ge_2Si_4$
1,4-Bis(triphenylgermyl)-1,1,2,2,3,3,4,4-octaphenyltetrasilane....... 1119
$C_{84}H_{72}Si_7$
1,1,2,2,3,3,4,4,5,5,6,6,7,7-Tetradecaphenylheptasilane....... 1119
$C_{96}H_{80}Si_7$
Perphenylheptasilane....... 1119

Product Catalog Number

To

Page Number

Cross Reference

Product Page

S34014-6 491
S34015-4 773
S34016-2 487
S34019-7 210
S34020-0 427
S34021-9 791
S34024-3 412
S34027-8 441
S34028-6 909
S34029-4 105
S34030-8 444
S34032-4 442
S34033-2 384
S34041-3 842
S34044-8 52
S34046-4 412
S34047-2 82
S34049-9 51
S34051-0 819
S34052-9 51
S34056-1 579
S34058-8 824
S34066-9 119
S34069-3 119
S34070-7 218
S34071-5 106
S34072-3 464
S34073-1 239
S34076-6 848
S34078-2 178
S34079-0 383
S34081-2 939
S34083-9 228
S34084-7 1088
S34086-3 487
S34087-1 249
S34089-8 922
S34090-1 412
S34094-4 813
S34095-2 317
S34097-9 860
S34098-7 860
S34101-0 219
S34102-9 230
S34105-3 817
S34106-1 468
S34109-6 817
S34111-8 788
S34112-6 792
S34113-4 224
S34115-0 291
S34118-5 904
S34120-7 311
S34127-4 337
S34129-0 1138
S34131-2 975
S34132-0 971
S34133-9 787
S34137-1 1050
S34139-8 765
S34140-1 937
S34144-4 200
S34147-9 786
S34148-7 622
S34153-3 499
S34156-8 376
S34157-6 707
S34158-4 239
S34159-2 349
S34160-6 754
S34162-2 425
S34164-9 425
S34165-7 425
S34166-5 425
S34170-3 425
S34171-1 425
S34175-4 667
S34176-2 644
S34177-0 249
S34179-7 232
S34180-0 905
S34181-9 242
S34185-1 970
S34189-4 1064
S34191-6 1046
S34192-4 882
S34193-2 377
S34194-0 378
S34196-7 424
S34197-5 681
S34199-1 426
S34200-9 249
S34201-7 823
S34202-5 250
S34203-3 815
S34204-1 255
S34206-8 255
S34208-4 997
S34209-2 109
S34210-6 380
S34212-2 196
S34213-0 358
S34214-9 127
S34215-7 836
S34216-5 758
S34218-1 326
S34220-3 1018
S34221-1 34
S34223-8 57
S34225-4 812
S34226-2 196
S34229-7 34
S34230-0 37
S34232-7 176
S34233-5 763
S34234-3 176
S34235-1 1018
S34237-8 6
S34239-4 180
S34240-8 943
S34241-6 1018
S34242-4 1043
S34244-0 769
S34245-9 1130
S34246-7 1129
S34247-5 710
S34248-3 314
S34250-5 971
S34252-1 233
S34256-4 123
S34258-0 429
S34259-9 123
S34260-2 124
S34261-0 110
S34262-9 429
S34265-3 280
S34269-6 436
S34271-8 436
S34272-6 123
S34273-4 121
S34274-2 123
S34275-0 123
S34276-9 123
S34277-7 429
S34278-5 118
S34279-3 484
S34283-1 20
S34285-8 40
S34286-6 1086
S34287-4 340
S34292-0 43
S34294-7 93
S34295-5 297
S34296-3 294
S34297-1 275
S34299-8 267
S34300-5 145
S34301-3 60
S34302-1 970
S34306-4 1124
S34308-0 10
S34309-9 964
S34310-2 296
S34311-0 181
S34312-9 962
S34314-5 19
S34315-3 286
S34318-8 418
S34319-6 436
S34321-8 436
S34322-6 436
S34323-4 147
S34325-0 108
S34326-9 948
S34327-7 947
S34328-5 948
S34329-3 948
S34330-7 948
S34331-5 948
S34332-3 108
S34333-1 122
S34337-4 975
S34339-0 92
S34340-4 359
S34342-0 974
S34347-1 956
S34352-8 661
S34354-4 661
S34355-2 661
S34358-7 657
S34359-5 658
S34362-5 665
S34363-3 96
S34364-1 823
S34366-8 665
S34367-6 665
S34369-2 984
S34372-2 974
S34376-5 824
S34378-1 144
S34384-6 43
S34388-9 117
S34389-7 693
S34390-0 1126
S34391-9 117
S34392-7 392
S34393-5 117
S34394-3 117
S34395-1 117
S34397-8 117
S34399-4 102
S34403-6 1065
S34404-4 1066
S34405-2 174
S34406-0 260
S34407-9 968
S34409-5 287
S34410-9 485
S34411-7 995
S34414-1 288
S34416-8 469
S34417-6 277
S34419-2 751
S34420-6 287
S34421-4 606
S34422-2 974
S34423-0 881
S34424-9 192
S34427-3 697
S34428-1 288
S34430-3 287
S34431-1 440
S34433-8 287
S34434-6 123
S34435-4 974
S34436-2 819
S34437-0 809
S34438-9 244
S34439-7 82
S34440-0 17
S34441-9 470
S34442-7 260
S34443-5 1127
S34444-3 752
S34445-1 752
S34447-8 752
S34448-6 185
S34450-8 873
S34452-4 873
S34454-0 873
S34455-9 872
S34459-1 982
S34462-1 190
S34465-6 289
S34470-2 1076
S34471-0 1077
S34478-8 219
S34481-8 220
S34482-6 206
S34483-4 754
S34485-0 757
S34486-9 885
S34487-7 885
S34489-3 767
S34490-7 127
S34491-5 102
S34493-1 884
S34495-8 827
S34496-6 1057
S34497-4 150
S34501-6 6
S34502-4 193
S34503-2 923
S34504-0 18
S34505-9 151
S34506-7 77
S34507-5 82
S34508-3 127
S34509-1 142
S34510-5 191
S34511-3 105
S34512-1 129
S34514-8 129
S34515-6 203
S34516-4 296
S34517-2 296
S34518-0 296
S34519-9 296
S34520-2 203
S34521-0 296
S34523-7 17
S34524-5 151
S34528-8 859
S34529-6 488
S34531-8 1091
S34533-4 486
S34537-7 429
S34538-5 441
S34539-3 58
S34540-7 220
S34541-5 147
S34542-3 78
S34543-1 32
S34545-8 181
S34548-2 6
S34550-4 353
S34551-2 650
S34552-0 327
S34553-9 198
S34555-5 874
S34557-1 874
S34559-8 873
S34564-4 873
S34566-0 872
S34567-9 605
S34568-7 823
S34569-5 817
S34570-9 790
S34571-7 190
S34573-3 150
S34576-8 757
S34578-4 146
S34580-6 15
S34581-4 142
S34582-2 191
S34583-0 938
S34584-9 74
S34585-7 465
S34587-3 1094
S34588-1 181
S34590-3 1094
S34591-1 765
S34593-8 369
S34594-6 443
S34595-4 932
S34596-2 944
S34597-0 1018
S34598-9 1018
S34600-4 473
S34601-2 13
S34602-0 79
S34603-9 19
S34604-7 76
S34605-5 360
S34607-1 117
S34610-1 96
S34612-8 1049
S34613-6 962
S34614-4 117
S34615-2 181
S34616-0 380
S34618-7 9
S34622-5 490
S34623-3 118
S34626-8 100
S34627-6 702
S34628-4 702
S34629-2 96
S34630-6 1015
S34632-2 290
S34634-9 918
S34635-7 310
S34640-3 194
S34641-1 875
S34643-8 326
S34644-6 35
S34645-4 145
S34646-2 147
S34648-9 331
S34651-9 1094
S34652-7 284
S34653-5 257
S34659-4 1062
S34660-8 233
S34661-6 1034
S34662-4 147
S34665-9 88
S34666-7 295
S34667-5 39
S34670-5 481
S34671-3 907
S34672-1 907
S34674-8 907
S34675-6 902
S34676-4 902
S34678-0 902
S34679-9 903
S34680-2 702
S34681-0 331
S34682-9 331
S34683-7 243
S34684-5 972
S34685-3 1016
S34686-1 1001
S34688-8 239
S34691-8 23
S34692-6 769
S34693-4 1001
S34696-9 955
S34697-7 139
S34698-5 107
S34699-3 907
S34700-0 39
S34701-9 902
S34702-7 33
S34703-5 1002
S34705-1 486
S34707-8 918
S34709-4 973
S34710-8 937
S34711-6 903
S34712-4 37
S34713-2 147
S34715-9 425
S34717-5 470
S34720-5 333
S34721-3 334
S34724-8 208
S34729-9 841
S34730-2 747
S34731-0 44
S34734-5 998
S34735-3 480
S34736-1 367
S34738-8 107
S34739-6 27
S34741-8 190
S34742-6 441
S34743-4 913
S34744-2 1018
S34746-9 499
S34747-7 271
S34748-5 568
S34750-7 276
S34751-5 561
S34752-3 794
S34755-8 195
S34756-6 195
S34757-4 191
S34758-2 789
S34759-0 1001
S34760-4 756
S34761-2 868
S34762-0 242
S34763-9 252
S34764-7 430
S34765-5 582
S34767-1 1000
S34769-8 708
S34770-1 297
S34772-8 869
S34774-4 145
S34775-2 147
S34777-9 34
S34778-7 146
S34779-5 159
S34780-9 182
S34781-7 989
S34782-5 26
S34783-3 26
S34784-1 26
S34786-8 26
S34787-6 25
S34788-4 141
S34790-6 126
S34791-4 135
S34792-2 98
S34793-0 17
S34794-9 180
S34795-7 98
S34796-5 181
S34797-3 139
S34798-1 19
S34800-7 752
S34801-5 340
S34803-1 752
S34805-8 93
S34807-4 973
S34808-2 861
S34809-0 126
S34810-4 708
S34811-2 673
S34812-0 361
S34813-9 354
S34814-7 889
S34815-5 889
S34816-3 149
S34817-1 889
S34822-8 149
S34823-6 147
S34824-4 755
S34825-2 81
S34826-0 92
S34827-9 88
S34828-7 115
S34829-5 683
S34830-9 81
S34833-3 907
S34834-1 999
S34836-8 843
S34837-6 88
S34840-6 74
S34841-4 19
S34844-9 995

Product	Page
S34845-7	342
S34846-5	974
S34847-3	974
S34854-6	769
S34855-4	1094
S34856-2	769
S34858-9	487
S34859-7	487
S34860-0	503
S34861-9	657
S34867-8	348
S34868-6	348
S34877-5	196
S34879-1	328
S34885-6	34
S34886-4	146
S34887-2	42
S34888-0	43
S34889-9	87
S34890-2	1125
S34891-0	1131
S34892-9	115
S34893-7	87
S34894-5	115
S34895-3	115
S34896-1	669
S34899-6	657
S34901-1	1092
S34903-8	804
S34906-2	500
S34908-9	759
S34910-0	809
S34911-9	190
S34912-7	190
S34913-5	147
S34917-8	14
S34918-6	54
S34919-4	969
S34922-4	889
S34925-9	93
S34926-7	1002
S34927-5	973
S34928-3	922
S34930-5	141
S34931-3	61
S34932-1	10
S34934-8	888
S34935-6	207
S34936-4	357
S34938-0	862
S34939-9	190
S34940-2	9
S34941-0	49
S34942-9	110
S34943-7	90
S34944-5	188
S34945-3	670
S34946-1	671
S34949-6	117
S34951-8	1015
S34952-6	1092
S34953-4	1088
S34954-2	361
S34955-0	332
S34956-9	705
S34961-5	41
S34962-3	141
S34963-1	317
S34966-6	187
S34967-4	127
S34968-2	1016
S34969-0	42
S34970-4	1079
S34971-2	1079
S34972-0	1079
S34976-3	287
S34977-1	198
S34983-6	873
S34984-4	873
S34987-9	756
S34989-5	872
S34991-7	297
S34992-5	490
S34993-3	488
S34994-1	707
S34996-8	327
S34997-6	104
S34999-2	438
S35002-8	32
S35003-6	1018
S35005-2	437
S35006-0	438
S35007-9	412
S35008-7	853
S35010-9	1036
S35017-6	1008
S35019-2	104
S35020-6	422
S35021-4	120
S35026-5	118
S35028-1	884
S35033-8	104
S35034-6	23
S35037-0	5
S35038-9	11
S35039-7	48
S35040-0	48
S35041-9	923
S35043-5	102
S35044-3	98
S35045-1	17
S35047-8	940
S35048-6	98
S35050-8	13
S35051-6	109
S35052-4	36
S35055-9	85
S35056-7	48
S35057-5	48
S35058-3	922
S35060-5	205
S35062-1	327
S35064-8	1085
S35065-6	282
S35066-4	282
S35067-2	155
S35068-0	14
S35069-9	60
S35070-2	20
S35072-9	893
S35074-5	921
S35075-3	33
S35076-1	860
S35079-6	411
S35085-0	1037
S35086-9	1010
S35087-7	136
S35088-5	208
S35089-3	1036
S35090-7	191
S35091-5	920
S35093-1	791
S35095-8	195
S35099-0	972
S35100-8	486
S35101-6	972
S35104-0	933
S35105-9	932
S35106-7	326
S35107-5	486
S35109-1	485
S35110-5	485
S35112-1	310
S35116-4	983
S35117-2	334
S35119-9	326
S35121-0	246
S35122-9	921
S35124-5	812
S35125-3	932
S35126-1	227
S35128-8	866
S35129-6	787
S35131-8	1085
S35132-6	13
S35134-2	999
S35137-7	196
S35139-3	1079
S35140-7	499
S35141-5	1071
S35142-3	922
S35147-4	4
S35148-2	807
S35150-4	362
S35151-2	353
S35152-0	25
S35154-7	77
S35155-5	145
S35156-3	368
S35157-1	164
S35159-8	12
S35163-6	378
S35165-2	362
S35167-9	752
S35168-7	185
S35170-9	753
S35171-7	345
S35172-5	48
S35173-3	135
S35174-1	921
S35177-6	700
S35178-4	193
S35180-6	1051
S35181-4	137
S35183-0	863
S35187-3	91
S35188-1	411
S35190-3	358
S35191-1	120
S35194-6	349
S35195-4	120
S35196-2	127
S35197-0	823
S35198-9	823
S35201-2	140
S35203-9	208
S35204-7	219
S35205-5	115
S35207-1	973
S35209-8	355
S35210-1	188
S35213-6	311
S35214-4	311
S35216-0	939
S35218-7	814
S35219-5	815
S35220-9	247
S35221-7	292
S35222-5	465
S35227-6	344
S35228-4	149
S35230-6	752
S35231-4	202
S35232-2	180
S35233-0	284
S35234-9	283
S35235-7	247
S35236-5	240
S35237-3	416
S35238-1	753
S35240-3	257
S35243-8	872
S35245-4	416
S35246-2	753
S35248-9	185
S35249-7	931
S35250-0	416
S35251-9	753
S35253-5	554
S35255-1	753
S35258-6	1001
S35260-8	80
S35263-2	84
S35266-7	186
S35267-5	354
S35268-3	106
S35269-1	89
S35270-5	297
S35271-3	219
S35274-8	372
S35275-6	373
S35276-4	766
S35277-2	889
S35278-0	148
S35279-9	12
S35280-2	77
S35281-0	12
S35285-3	755
S35286-1	90
S35289-6	63
S35291-8	258
S35294-2	885
S35295-0	383
S35297-7	366
S35299-3	127
S35303-5	973
S35304-3	90
S35305-1	317
S35307-8	379
S35309-4	334
S35310-8	89
S35311-6	374
S35312-4	819
S35313-2	872
S35315-9	193
S35316-7	193
S35317-5	192
S35319-1	327
S35320-5	176
S35321-3	285
S35322-1	332
S35325-6	873
S35326-4	874
S35327-2	874
S35328-0	874
S35331-0	873
S35332-9	872
S35333-7	872
S35335-3	692
S35338-8	710
S35339-6	680
S35341-8	941
S35342-6	37
S35343-4	37
S35345-0	888
S35346-9	863
S35347-7	1051
S35348-5	1054
S35349-3	1054
S35350-7	1054
S35353-1	71
S35355-8	184
S35356-6	251
S35361-2	703
S35362-0	142
S35364-7	145
S35366-3	147
S35367-1	946
S35369-8	41
S35370-1	267
S35372-8	487
S35375-2	742
S35377-9	872
S35378-7	603
S35380-9	1000
S35382-5	567
S35383-3	280
S35384-1	279
S35387-6	280
S35390-6	344
S35391-4	45
S35394-9	989
S35398-1	356
S35401-5	61
S35402-3	1019
S35406-6	98
S35407-4	17
S35408-2	940
S35411-2	997
S35412-0	98
S35413-9	922
S35414-7	939
S35417-1	117
S35419-8	160
S35420-1	1011
S35422-8	420
S35423-6	62
S35424-4	800
S35425-2	250
S35427-9	126
S35428-7	135
S35429-5	135
S35430-9	922
S35433-3	372
S35434-1	361
S35436-8	362
S35437-6	361
S35438-4	415
S35439-2	291
S35440-6	429
S35444-9	1092
S35446-5	1091
S35453-8	376
S35454-6	799
S35455-4	212
S35456-2	350
S35457-0	286
S35458-9	350
S35459-7	1138
S35462-7	212
S35463-5	212
S35464-3	428
S35468-6	286
S35469-4	247
S35470-8	361
S35471-6	444
S35472-4	332
S35473-2	428
S35474-0	710
S35475-9	162
S35476-7	163
S35477-5	873
S35478-3	869
S35479-1	470
S35481-3	823
S35482-1	721
S35484-8	142
S35485-6	147
S35487-2	955
S35488-0	40
S35489-9	749
S35490-2	39
S35491-0	944
S35493-7	295
S35494-5	104
S35496-1	39
S35500-3	181
S35503-8	922
S35504-6	290
S35505-4	978
S35506-2	1010
S35508-9	1010
S35509-7	1011
S35510-0	1054
S35511-9	1057
S35512-7	1060
S35513-5	1010
S35514-3	1010
S35515-1	1010
S35517-8	1010
S35518-6	1010
S35520-8	1056
S35521-6	1060
S35522-4	956
S35524-0	943
S35525-9	943
S35527-5	932
S35528-3	933
S35529-1	933
S35531-3	967
S35532-1	967
S35535-6	38
S35536-4	946
S35537-2	954
S35538-0	39
S35539-9	146
S35540-2	295
S35542-9	327
S35546-1	824
S35548-8	823
S35552-6	790
S35554-2	461
S35556-9	1050
S35557-7	1050
S35558-5	1033
S35559-3	364
S35562-3	144
S35565-8	53
S35566-6	24
S35567-4	102
S35569-0	17
S35570-4	26
S35571-2	1124
S35576-3	702
S35582-8	859
S35583-6	1129
S35585-2	902
S35586-0	249
S35588-7	39
S35589-5	125
S35590-9	135
S35591-7	1001
S35592-5	431
S35593-3	241
S35594-1	1129
S35596-8	388
S35597-6	43
S35599-2	768
S35606-9	727
S35607-7	206
S35609-3	164
S35610-7	287
S35612-3	380
S35613-1	381
S35615-8	164
S35616-6	164
S35617-4	217
S35618-2	816
S35620-4	74
S35624-7	1131
S35626-3	1132
S35627-1	1131
S35629-8	288
S35633-6	836
S35634-4	1131
S35635-2	1133
S35636-0	1133
S35637-9	1131
S35638-7	1130
S35639-5	1129
S35640-9	1133
S35641-7	1131
S35642-5	1131
S35643-3	1131
S35644-1	1133
S35646-8	1131
S35649-2	150
S35650-6	148
S35652-2	19
S35653-0	75
S35655-7	484
S35657-3	814
S35658-1	228
S35661-1	477
S35663-8	902
S35664-6	702
S35665-4	431
S35666-2	368
S35667-0	253
S35671-9	995
S35672-7	270
S35673-5	1131
S35675-1	43
S35679-4	757
S35685-9	127
S35687-5	110
S35700-6	1001
S35701-4	487
S35702-2	431
S35704-9	487
S35705-7	791
S35706-5	427
S35707-3	487
S35708-1	415

Product	Page	Product	Page	Product	Page	Product	Page	Product	Page	Product	Page	Product	Page
S35711-1	876	S35816-9	500	S35920-3	210	S36039-2	1141	S36148-8	373	S36265-4	383	S36379-0	353
S35713-8	422	S35817-7	447	S35923-8	1037	S36040-6	1013	S36149-6	217	S36266-2	195	S36380-4	22
S35714-6	959	S35818-5	500	S35924-6	193	S36041-4	501	S36151-8	882	S36269-7	548	S36381-2	891
S35715-4	913	S35819-3	500	S35925-4	14	S36042-2	88	S36153-4	14	S36272-7	37	S36382-0	8
S35717-0	443	S35820-7	500	S35928-9	321	S36044-9	359	S36154-2	149	S36273-5	972	S36384-7	1054
S35718-9	940	S35821-5	498	S35931-9	108	S36045-7	955	S36155-0	680	S36274-3	932	S36385-5	964
S35719-7	365	S35822-3	500	S35932-7	956	S36046-5	159	S36156-9	687	S36277-8	1084	S36386-3	692
S35722-7	426	S35825-8	905	S35934-3	63	S36048-1	126	S36157-7	687	S36279-4	885	S36393-6	823
S35724-3	413	S35827-4	754	S35935-1	974	S36050-3	75	S36158-5	127	S36280-8	1033	S36394-4	44
S35725-1	51	S35828-2	392	S35937-8	741	S36051-1	115	S36159-3	106	S36281-6	269	S36395-2	246
S35727-8	95	S35829-0	765	S35938-6	1014	S36053-8	8	S36161-5	71	S36282-4	269	S36396-0	422
S35728-6	53	S35830-4	755	S35941-6	169	S36054-6	366	S36162-3	83	S36283-2	720	S36397-9	754
S35729-4	862	S35831-2	754	S35942-4	182	S36055-4	461	S36163-1	82	S36284-0	117	S36398-7	295
S35730-8	105	S35832-0	754	S35943-2	913	S36058-9	247	S36166-6	687	S36285-9	25	S36399-5	917
S35732-4	1034	S35833-9	754	S35944-0	972	S36059-7	167	S36167-4	90	S36286-7	952	S36400-2	796
S35735-9	431	S35834-7	754	S35945-9	442	S36060-0	461	S36168-2	110	S36287-5	185	S36403-7	504
S35736-7	106	S35835-5	754	S35946-7	182	S36063-5	316	S36170-4	50	S36288-3	241	S36404-5	487
S35737-5	79	S35836-3	754	S35947-5	268	S36064-3	1002	S36171-2	61	S36289-1	82	S36405-3	899
S35738-3	182	S35837-1	754	S35948-3	268	S36065-1	1002	S36172-0	79	S36294-8	286	S36406-1	746
S35739-1	25	S35839-8	427	S35950-5	392	S36070-8	205	S36173-9	115	S36295-6	232	S36408-8	426
S35740-5	139	S35840-1	348	S35954-8	486	S36071-6	282	S36174-7	115	S36298-0	314	S36409-6	739
S35741-3	1140	S35843-6	246	S35955-6	17	S36073-2	354	S36177-1	125	S36304-9	86	S36411-8	803
S35742-1	97	S35846-0	356	S35957-2	787	S36074-0	163	S36179-8	890	S36305-7	420	S36412-6	249
S35745-6	978	S35847-9	178	S35959-9	822	S36076-7	442	S36180-1	144	S36306-5	1002	S36413-4	379
S35746-4	788	S35848-7	273	S35960-2	803	S36077-5	768	S36184-4	41	S36307-3	334	S36414-2	1000
S35747-2	793	S35849-5	269	S35961-0	201	S36078-3	268	S36185-2	335	S36308-1	815	S36415-0	277
S35748-0	317	S35850-9	181	S35962-9	910	S36081-3	39	S36186-0	8	S36310-3	498	S36416-9	791
S35750-2	937	S35851-7	882	S35963-7	206	S36082-1	944	S36187-9	361	S36311-1	498	S36417-7	342
S35751-0	861	S35852-5	882	S35965-3	789	S36084-8	104	S36188-7	756	S36314-6	939	S36419-3	356
S35753-7	787	S35853-3	717	S35971-8	1037	S36085-6	104	S36189-5	110	S36317-0	328	S36420-7	989
S35754-5	970	S35856-8	327	S35973-4	972	S36086-4	104	S36190-9	117	S36318-9	327	S36421-5	706
S35755-3	979	S35858-4	110	S35974-2	1060	S36087-2	915	S36191-7	8	S36319-7	310	S36422-3	791
S35758-8	971	S35859-2	959	S35976-9	752	S36090-2	769	S36192-5	284	S36322-7	873	S36423-1	876
S35759-6	342	S35860-6	325	S35978-5	1054	S36091-0	159	S36193-3	141	S36323-5	873	S36425-8	73
S35762-6	816	S35861-4	312	S35979-3	358	S36092-9	149	S36194-1	820	S36325-1	182	S36426-6	289
S35763-4	139	S35862-2	141	S35981-5	52	S36094-5	21	S36197-6	378	S36328-6	104	S36428-2	890
S35765-0	813	S35863-0	805	S35982-3	77	S36096-1	192	S36199-2	500	S36329-4	185	S36429-0	11
S35766-9	817	S35864-9	159	S35983-1	65	S36098-8	49	S36202-6	385	S36330-8	288	S36430-4	105
S35769-3	971	S35865-7	159	S35985-8	238	S36099-6	81	S36203-4	169	S36331-6	71	S36431-2	329
S35771-5	971	S35866-5	111	S35987-4	972	S36100-3	887	S36204-2	173	S36333-2	366	S36432-0	124
S35772-3	787	S35867-3	501	S35988-2	140	S36101-1	383	S36206-9	888	S36334-0	954	S36435-5	311
S35773-1	813	S35868-1	902	S35989-0	971	S36104-6	902	S36208-5	180	S36335-9	312	S36436-3	57
S35775-8	775	S35870-3	837	S35991-2	44	S36105-4	131	S36209-3	181	S36337-5	743	S36440-1	429
S35778-2	970	S35871-1	487	S35992-0	833	S36106-2	766	S36211-5	206	S36340-5	500	S36443-6	680
S35779-0	271	S35875-4	36	S35993-9	78	S36108-9	1095	S36215-8	95	S36341-3	500	S36444-4	443
S35781-2	250	S35876-2	221	S35994-7	74	S36109-7	105	S36216-6	119	S36342-1	882	S36445-2	475
S35782-0	787	S35877-0	165	S35996-3	98	S36110-0	327	S36221-2	55	S36345-6	832	S36446-0	183
S35783-9	979	S35878-9	139	S35997-1	26	S36111-9	1062	S36222-0	24	S36346-4	90	S36447-9	248
S35784-7	923	S35879-7	1033	S36002-3	426	S36112-7	906	S36224-7	1138	S36347-2	327	S36448-7	707
S35785-5	817	S35880-0	125	S36003-1	422	S36113-5	863	S36226-3	61	S36348-0	993	S36449-5	835
S35786-3	788	S35882-7	477	S36005-8	885	S36115-1	748	S36229-8	959	S36350-2	77	S36450-9	250
S35789-8	869	S35884-3	501	S36006-6	955	S36117-8	267	S36230-1	144	S36351-0	1001	S36451-7	703
S35790-1	788	S35885-1	501	S36007-4	250	S36118-6	276	S36232-8	36	S36352-9	1037	S36454-1	799
S35792-8	24	S35890-8	860	S36009-0	769	S36119-4	771	S36233-6	104	S36354-5	193	S36456-8	953
S35793-6	48	S35891-6	276	S36011-2	500	S36120-8	273	S36234-4	24	S36356-1	111	S36457-6	112
S35794-4	48	S35892-4	985	S36013-9	739	S36122-4	87	S36239-5	902	S36359-6	90	S36458-4	195
S35795-2	25	S35894-0	830	S36014-7	243	S36123-2	142	S36240-9	92	S36361-8	91	S36459-2	58
S35799-5	249	S35897-5	199	S36015-5	196	S36124-0	22	S36241-7	14	S36362-6	973	S36462-2	39
S35800-2	425	S35901-7	872	S36016-3	314	S36125-9	793	S36243-3	812	S36363-4	887	S36463-0	38
S35801-0	791	S35902-5	172	S36017-1	4	S36126-7	251	S36244-1	36	S36364-2	90	S36464-9	1095
S35802-9	1049	S35903-3	22	S36019-8	182	S36128-3	980	S36247-6	413	S36365-0	318	S36465-7	100
S35803-7	444	S35904-1	868	S36020-1	79	S36129-1	1018	S36248-4	1037	S36367-7	79	S36466-5	22
S35804-5	16	S35906-8	63	S36023-6	1131	S36130-5	160	S36249-2	13	S36368-5	23	S36467-3	962
S35805-3	137	S35907-6	150	S36024-4	439	S36131-3	42	S36250-6	1036	S36369-3	187	S36470-3	964
S35806-1	412	S35908-4	267	S36025-2	170	S36134-8	40	S36251-4	970	S36370-7	176	S36473-8	940
S35808-8	902	S35910-6	181	S36026-0	170	S36135-6	25	S36253-0	971	S36371-5	176	S36474-6	940
S35809-6	972	S35911-4	827	S36027-9	6	S36137-2	33	S36254-9	908	S36372-3	355	S36475-4	948
S35811-8	916	S35912-2	14	S36031-7	50	S36139-9	940	S36255-7	124	S36373-1	431	S36476-2	976
S35812-6	500	S35913-0	56	S36034-1	425	S36140-2	939	S36258-1	918	S36375-8	250	S36477-0	145
S35813-4	972	S35914-9	708	S36036-8	787	S36141-0	724	S36260-3	13	S36376-6	241	S36478-9	1069
S35814-2	500	S35916-5	275	S36037-6	22	S36144-5	117	S36261-1	16	S36377-4	943	S36482-7	161
S35815-0	500	S35917-3	155	S36038-4	1141	S36145-3	946	S36264-6	265	S36378-2	90	S36484-3	877

Product	Page
S36485-1	1098
S36488-6	288
S36489-4	170
S36491-6	98
S36492-4	15
S36493-2	1098
S36495-9	90
S36497-5	904
S36498-3	30
S36499-1	1003
S36500-9	1099
S36501-7	855
S36503-3	110
S36504-1	741
S36506-8	161
S36507-6	328
S36508-4	823
S36509-2	755
S36511-4	827
S36512-2	228
S36513-0	250
S36514-9	589
S36515-7	819
S36516-5	819
S36518-1	1130
S36521-1	328
S36523-8	76
S36524-6	197
S36525-4	852
S36526-2	1097
S36527-0	16
S36528-9	41
S36529-7	32
S36530-0	196
S36532-7	196
S36533-5	758
S36534-3	837
S36535-1	837
S36537-8	837
S36538-6	1141
S36539-4	1141
S36540-8	1141
S36541-6	1141
S36542-4	157
S36544-0	161
S36547-5	1031
S36548-3	283
S36549-1	720
S36551-3	362
S36554-8	1077
S36555-6	1059
S36557-2	197
S36558-0	1029
S36559-9	1029
S36560-2	696
S36562-9	366
S36563-7	70
S36565-3	166
S36566-1	70
S36568-8	86
S36573-4	212
S36574-2	756
S36576-9	14
S36578-5	124
S36579-3	84
S36580-7	444
S36581-5	707
S36583-1	295
S36585-8	500
S36586-6	500
S36587-4	503
S36588-2	932
S36589-0	462
S36590-4	754
S36591-2	962
S36592-0	381
S36593-9	444
S36597-1	917
S36601-3	373
S36604-8	30
S36608-0	886
S36609-9	790
S36611-0	356
S36612-9	671
S36613-7	1139
S36614-5	352
S36616-1	1110
S36618-8	172
S36619-6	1142
S36621-8	780
S36623-4	58
S36625-0	382
S36626-9	887
S36630-7	25
S36631-5	115
S36635-8	978
S36637-4	974
S36638-2	248
S36640-4	1126
S36641-2	1126
S36642-0	226
S36644-7	971
S36652-8	226
S36653-6	876
S36661-7	230
S36666-8	154
S36669-2	707
S36670-6	285
S36671-4	975
S36674-9	916
S36675-7	888
S36678-1	386
S36685-4	159
S36686-2	325
S36687-0	932
S36689-7	836
S36690-0	921
S36691-9	194
S36692-7	194
S36694-3	397
S36695-1	932
S36704-4	152
S36706-0	830
S36716-8	193
S36719-2	25
S36720-6	368
S36722-2	948
S36724-9	361
S36726-5	604
S36727-3	665
S36728-1	665
S36731-1	2
S36737-0	793
S36738-9	81
S36743-5	986
S36745-1	897
S36747-8	189
S36752-4	804
S36754-0	362
S36756-7	975
S36757-5	755
S36758-3	985
S36759-1	149
S36761-3	65
S36765-6	872
S36766-4	142
S36770-2	140
S36771-0	995
S36774-5	317
S36775-3	979
S36784-2	873
S36786-9	43
S36788-5	191
S36790-7	192
S36793-1	1131
S36809-1	329
S36822-9	204
S36823-7	251
S36829-6	971
S36831-8	791
S36836-9	51
S36837-7	192
S36839-3	163
S36840-7	160
S36841-5	1009
S36846-6	885
S36847-4	10
S36848-2	1001
S36851-2	972
S36856-3	952
S36857-1	951
S36860-1	768
S36863-6	241
S36866-0	261
S36867-9	136
S36873-3	940
S36874-1	940
S36876-8	897
S36883-0	81
S36890-3	10
S36891-1	967
S36895-4	9
S36897-0	1006
S36899-7	171
S36902-0	21
S36903-9	67
S36904-7	53
S36907-1	145
S36909-8	146
S36910-1	1075
S36912-8	1126
S36915-2	25
S36924-1	902
S36926-8	885
S36928-4	154
S36929-2	959
S36930-6	133
S36932-2	14
S36935-7	237
S36937-3	163
S36941-1	853
S36948-9	38
S36949-7	209
S36950-0	1138
S36954-3	940
S36958-6	247
S36959-4	328
S36960-8	293
S36963-2	191
S36964-0	149
S36967-5	31
S36971-3	1
S36972-1	135
S36974-8	368
S36975-6	368
S36976-4	887
S36980-2	885
S36985-3	241
S36986-1	862
S36988-8	960
S36994-2	866
S36997-7	246
S37000-2	1006
S37004-5	269
S37013-4	295
S37014-2	6
S37028-2	725
S37035-5	964
S37036-3	255
S37039-8	823
S37045-2	888
S37046-0	1019
S37047-9	1019
S37048-7	790
S37050-9	1057
S37052-5	965
S37056-8	967
S37057-6	36
S37060-6	201
S37067-3	98
S37068-1	160
S37070-3	8
S37082-7	206
S37083-5	250
S37085-1	1124
S37087-8	1129
S37093-2	289
S37100-9	333
S37103-3	412
S37104-1	80
S37106-8	1056
S37107-6	53
S37110-6	1069
S37111-4	442
S37112-2	142
S37114-9	982
S37118-1	1131
S37128-9	273
S37131-9	822
S37134-3	191
S37137-8	1131
S37141-6	977
S37146-7	127
S37147-5	932
S37152-1	983
S37158-0	62
S37159-9	30
S37162-9	67
S37176-9	967
S37180-7	882
S37186-6	250
S37187-4	101
S37189-0	789
S37204-8	344
S37205-6	54
S37206-4	54
S37208-0	12
S37210-2	139
S37213-7	944
S37222-6	71
S37223-4	110
S37229-3	219
S37230-7	1005
S37236-6	1082
S37237-4	192
S37238-2	206
S37241-2	1125
S37242-0	166
S37244-7	296
S37250-1	891
S37252-8	289
S37257-9	128
S37259-5	946
S37260-9	237
S37261-7	940
S37262-5	105
S37273-0	5
S37274-9	159
S37278-1	998
S37285-4	983
S37286-2	703
S37295-1	434
S37298-6	249
S37300-1	877
S37304-4	422
S37305-2	403
S37306-0	1002
S37307-9	465
S37309-5	194
S37311-7	249
S37312-5	754
S37313-3	497
S37316-8	315
S37317-6	923
S37318-4	979
S37319-2	55
S37321-4	163
S37322-2	196
S37324-9	486
S37325-7	871
S37326-5	829
S37328-1	788
S37330-3	4
S37331-1	5
S37334-6	853
S37336-2	886
S37338-9	747
S37340-0	33
S37341-9	972
S37342-7	366
S37343-5	88
S37344-3	48
S37345-1	755
S37347-8	386
S37348-6	126
S37350-8	147
S37353-2	972
S37354-0	63
S37355-9	48
S37356-7	917
S37357-5	289
S37359-1	488
S37361-3	891
S37362-1	782
S37364-8	250
S37366-4	741
S37368-0	902
S37370-2	13
S37372-9	934
S37373-7	741
S37375-3	101
S37376-1	837
S37379-6	420
S37383-4	72
S37386-9	320
S37387-7	32
S37388-5	256
S37389-3	15
S37391-5	986
S37398-2	249
S37401-6	158
S37402-4	208
S37406-7	907
S37409-1	972
S37411-3	13
S37412-1	380
S37415-6	907
S37417-2	9
S37419-9	8
S37425-3	21
S37426-1	149
S37428-8	150
S37436-9	84
S37438-5	91
S37439-3	20
S37441-5	89
S37446-6	91
S37447-4	22
S37450-4	90
S37452-0	1124
S37453-9	80
S37454-7	85
S37455-5	482
S37456-3	79
S37457-1	79
S37465-2	374
S37472-5	126
S37473-3	386
S37477-6	215
S37481-4	1063
S37485-7	111
S37488-1	57
S37498-9	263
S37499-7	984
S37500-4	1014
S37501-2	1078
S37504-7	126
S37506-3	968
S37507-1	231
S37516-0	89
S37517-9	11
S37520-9	14
S37523-3	18
S37524-1	20
S37526-8	80
S37527-6	91
S37528-4	81
S37530-6	368
S37531-4	76
S37532-2	14
S37536-5	21
S37538-1	1011
S37540-3	489
S37541-1	279
S37543-8	484
S37546-2	891
S37548-9	442
S37551-9	852
S37555-1	879
S37557-8	879
S37562-4	486
S37563-2	115
S37565-9	29
S37569-1	126
S37571-3	91
S37572-1	86
S37574-8	20
S37576-4	86
S37577-2	185
S37578-0	1124
S37579-9	1124
S37580-2	14
S37582-9	1124
S37583-7	98
S37584-5	90
S37585-3	40
S37586-1	91
S37588-8	18
S37589-6	26
S37591-8	111
S37592-6	111
S37593-4	706
S37596-9	702
S37597-7	702
S37598-5	1092
S37599-3	109
S37601-9	444
S37603-5	110
S37604-3	702
S37607-8	702
S37609-4	21
S37612-4	1056
S37613-2	1069
S37618-3	984
S37619-1	984
S37626-4	32
S37631-0	238
S37633-7	393
S37634-5	420
S37639-6	393
S37649-3	83
S37651-5	103
S37652-3	970
S37655-8	969
S37656-6	36
S37660-4	393
S37661-2	329
S37662-0	201
S37663-9	1063
S37664-7	380
S37666-3	34
S37673-6	378
S37675-2	243
S37677-9	333
S37678-7	949
S37682-5	1063
S37683-3	102
S37684-1	236
S37686-8	103
S37687-6	366
S37688-4	75
S37689-2	75
S37695-7	960
S37697-3	63
S37698-1	934
S37700-7	967
S37706-6	125
S37707-4	103
S37708-2	240
S37710-4	240
S37711-2	240
S37717-1	103
S37723-6	34
S37724-4	949
S37726-0	952
S37727-9	365
S37729-5	439
S37730-9	333
S37733-3	202
S37734-1	103
S37736-8	766
S37740-6	83
S37742-2	776
S37744-9	756

Product	Page
S37747-3	934
S37750-3	241
S37751-1	34
S37753-8	756
S37754-6	197
S37756-2	1053
S37757-0	949
S37758-9	34
S37759-7	238
S37760-0	971
S37761-9	950
S37762-7	429
S37763-5	972
S37764-3	207
S37769-4	972
S37774-0	438
S37776-7	910
S37782-1	355
S37784-8	971
S37785-6	368
S37786-4	37
S37787-2	236
S37790-2	236
S37791-0	84
S37792-9	34
S37800-3	393
S37801-1	240
S37803-8	393
S37807-0	393
S37808-9	431
S37809-7	393
S37810-0	746
S37813-5	937
S37814-3	368
S37817-8	390
S37819-4	84
S37820-8	393
S37821-6	283
S37823-2	411
S37824-0	197
S37828-3	467
S37829-1	238
S37831-3	959
S37838-0	105
S37842-9	33
S37848-8	101
S37849-6	1010
S37852-6	11
S37854-2	101
S37856-9	216
S37858-5	972
S37867-4	119
S37872-0	104
S37877-1	155
S37890-9	89
S37897-6	130
S37899-2	105
S37901-8	972
S37908-5	102
S37918-2	973
S37922-0	19
S37923-9	84
S37925-5	30
S37926-3	942
S37930-1	34
S37940-9	137
S37942-5	40
S37949-2	34
S37951-4	37
S37954-9	67
S37955-7	130
S37963-8	202
S37965-4	106
S37966-2	25
S37968-9	336
S37969-7	822
S37972-7	40
S37977-8	954
S37985-9	34
S37989-1	112
S37991-3	104
S37994-8	102
S37997-2	28
S38002-4	328
S38005-9	279
S38007-5	766
S38011-3	392
S38012-1	392
S38014-8	781
S38015-6	423
S38016-4	390
S38019-9	104
S38020-2	340
S38024-5	420
S38025-3	102
S38029-6	961
S38032-6	467
S38033-4	37
S38034-2	970
S38035-0	102
S38040-7	239
S38045-8	34
S38046-6	353
S38047-4	240
S38052-0	240
S38053-9	431
S38055-5	466
S38062-8	825
S38064-4	257
S38068-7	103
S38071-7	430
S38072-5	383
S38073-3	85
S38079-2	355
S38082-2	714
S38084-9	241
S38085-7	476
S38086-5	430
S38087-3	482
S38091-1	439
S38094-6	758
S38096-2	365
S38097-0	437
S38099-7	84
S38100-4	758
S38101-2	466
S38103-9	473
S38104-7	356
S38105-5	200
S38106-3	32
S38107-1	34
S38109-8	90
S38110-1	416
S38112-8	241
S38113-6	355
S38116-0	754
S38117-9	393
S38118-7	220
S38123-3	468
S38124-1	393
S38126-8	354
S38127-6	941
S38128-4	263
S38130-6	34
S38132-2	1076
S38134-9	692
S38136-5	90
S38137-3	103
S38138-1	1083
S38140-3	340
S38141-1	826
S38143-8	436
S38144-6	439
S38147-0	972
S38148-9	962
S38149-7	333
S38150-0	433
S38152-7	243
S38153-5	972
S38154-3	332
S38157-8	430
S38158-6	766
S38159-4	82
S38160-8	236
S38161-6	720
S38163-2	756
S38164-0	31
S38165-9	333
S38166-7	189
S38167-5	950
S38168-3	75
S38169-1	481
S38170-5	436
S38171-3	549
S38172-1	328
S38175-6	358
S38177-2	481
S38178-0	942
S38180-2	174
S38181-0	333
S38183-7	103
S38184-5	222
S38186-1	241
S38188-8	945
S38189-6	372
S38191-8	242
S38192-6	704
S38193-4	138
S38195-0	756
S38197-7	333
S38198-5	103
S38199-3	288
S38200-0	392
S38203-5	884
S38204-3	34
S38205-1	841
S38207-8	949
S38209-4	241
S38210-8	972
S38211-6	423
S38212-4	382
S38214-0	103
S38216-7	248
S38219-1	240
S38220-5	668
S38221-3	782
S38222-1	473
S38224-8	562
S38225-6	238
S38226-4	245
S38228-0	463
S38229-9	353
S38230-2	364
S38232-9	707
S38236-1	238
S38238-8	137
S38239-6	333
S38241-8	333
S38242-6	86
S38243-4	226
S38245-0	693
S38246-9	754
S38247-7	333
S38250-7	398
S38251-5	825
S38252-3	774
S38255-8	431
S38258-2	757
S38261-2	342
S38266-3	1010
S38267-1	753
S38270-1	443
S38272-8	259
S38273-6	757
S38274-4	873
S38275-2	218
S38276-0	718
S38277-9	710
S38279-5	368
S38280-9	774
S38282-5	755
S38284-1	269
S38286-8	813
S38288-4	891
S38289-2	804
S38290-6	799
S38291-4	426
S38294-9	1133
S38295-7	802
S38296-5	917
S38297-3	817
S38300-7	778
S38302-3	788
S38303-1	292
S38307-4	776
S38308-2	336
S38311-2	351
S38312-0	803
S38313-9	1021
S38314-7	804
S38315-5	259
S38316-3	377
S38317-1	877
S38320-1	801
S38322-8	802
S38324-4	336
S38326-0	803
S38327-9	443
S38328-7	103
S38329-5	333
S38330-9	145
S38331-7	693
S38333-3	769
S38334-1	753
S38337-6	295
S38339-2	242
S38342-2	751
S38345-7	103
S38346-5	754
S38347-3	144
S38348-1	200
S38350-3	76
S38351-1	332
S38355-4	937
S38356-2	353
S38357-0	710
S38358-9	242
S38360-0	420
S38362-7	327
S38364-3	242
S38368-6	430
S38369-4	362
S38370-8	103
S38371-6	240
S38372-4	332
S38373-2	357
S38374-0	358
S38375-9	239
S38376-7	240
S38377-5	239
S38380-5	693
S38381-3	102
S38382-1	756
S38385-6	250
S38386-4	392
S38388-0	359
S38390-2	240
S38391-0	34
S38392-9	963
S38393-7	251
S38396-1	102
S38398-8	373
S38399-6	242
S38400-3	240
S38401-1	201
S38403-8	392
S38404-6	326
S38405-4	240
S38406-2	325
S38410-0	755
S38411-9	753
S38412-7	101
S38413-5	242
S38415-1	227
S38417-8	443
S38418-6	428
S38419-4	755
S38420-8	943
S38422-4	428
S38423-2	359
S38424-0	428
S38425-9	103
S38427-5	283
S38429-1	690
S38431-3	249
S38432-1	242
S38435-6	357
S38437-2	691
S38439-9	240
S38440-2	360
S38441-0	359
S38442-9	353
S38443-7	353
S38444-5	242
S38445-3	428
S38446-1	801
S38451-8	239
S38452-6	240
S38453-4	240
S38458-5	956
S38459-3	444
S38461-5	444
S38463-1	442
S38465-8	444
S38468-2	444
S38475-5	444
S38476-3	444
S38477-1	444
S38485-2	442
S38487-9	444
S38488-7	442
S38493-3	443
S38499-2	984
S38501-8	984
S38502-6	1035
S38504-2	1035
S38505-0	463
S38510-7	271
S38518-2	103
S38519-0	138
S38523-9	137
S38524-7	36
S38525-5	149
S38529-8	359
S38530-1	393
S38536-0	956
S38542-5	137
S38547-6	32
S38548-4	84
S38549-2	984
S38553-0	37
S38558-1	204
S38561-1	708
S38563-8	905
S38566-2	137
S38569-7	156
S38571-9	773
S38572-7	76
S38573-5	240
S38575-1	426
S38581-6	747
S38584-0	324
S38585-9	240
S38586-7	351
S38588-3	804
S38590-5	323
S38591-3	803
S38594-8	351
S38595-6	259
S38597-2	336
S38598-0	785
S38600-6	813
S38601-4	804
S38604-9	772
S38606-5	813
S38607-3	469
S38610-3	829
S38611-1	813
S38613-8	828
S38615-4	828
S38616-2	878
S38617-0	1032
S38618-9	101
S38619-7	973
S38634-0	240
S38636-7	818
S38647-2	916
S38656-1	291
S38659-6	297
S38666-9	970
S38669-3	802
S38672-3	328
S38673-1	815
S38678-2	876
S38679-0	738
S38680-4	274
S38683-9	1009
S38685-5	796
S38686-3	180
S38687-1	803
S38690-1	790
S38693-6	1009
S38694-4	803
S38697-9	341
S38700-2	817
S38701-0	817
S38702-9	801
S38704-5	781
S38705-3	268
S38709-6	227
S38711-8	240
S38714-2	822
S38715-0	793
S38717-7	176
S38719-3	802
S38720-7	87
S38721-5	296
S38722-3	323
S38725-8	287
S38726-6	215
S38728-2	959
S38730-4	241
S38732-0	790
S38734-7	240
S38735-5	782
S38736-3	241
S38740-1	217
S38742-8	169
S38744-4	787
S38745-2	425
S38756-8	875
S38757-6	768
S38758-4	697
S38760-6	267
S38761-4	204
S38763-0	381
S38765-7	793
S38766-5	803
S38767-3	693
S38773-8	89
S38782-7	1097
S38784-3	1098
S38785-1	1097
S38791-6	1076
S38792-4	39
S38794-0	39
S38801-7	933
S38802-5	933
S38803-3	735
S38810-6	910
S38811-4	267
S38814-9	668
S38815-7	812
S38816-5	296
S38818-1	251
S38823-8	247
S38824-6	825
S38825-4	439
S38828-9	877
S38829-7	268
S38832-7	1008
S38833-5	1008
S38834-3	30
S38837-8	1008
S38839-4	241
S38840-8	242
S38842-4	1018
S38844-0	30
S38846-7	30
S38847-5	30
S38848-3	30
S38850-5	29
S38852-1	30
S38855-6	30

Product	Page
S38857-2	1018
S38858-0	32
S38860-2	30
S38861-0	1054
S38862-9	33
S38863-7	1033
S38864-5	29
S38868-8	30
S38872-6	29
S38873-4	1142
S38876-9	1141
S38877-7	1142
S38878-5	1141
S38879-3	475
S38881-5	300
S38882-3	1043
S38883-1	1035
S38885-8	36
S38886-6	1098
S38887-4	772
S38888-2	785
S38889-0	892
S38892-0	499
S38894-7	176
S38895-5	176
S38896-3	791
S38904-8	379
S38905-6	292
S38906-4	413
S38907-2	413
S38908-0	413
S38912-9	260
S38916-1	774
S38919-6	966
S38922-6	308
S38924-2	944
S38926-9	779
S38927-7	739
S38928-5	1006
S38929-3	959
S38930-7	394
S38931-5	413
S38935-8	851
S38936-6	111
S38937-4	109
S38938-2	110
S38940-4	853
S38942-0	97
S38946-3	960
S38952-8	88
S38953-6	250
S38954-4	251
S38956-0	149
S38958-7	851
S38960-9	851
S38968-4	418
S38969-2	264
S38970-6	990
S38978-1	916
S38980-3	915
S38983-8	442
S38984-6	180
S38986-2	760
S38990-0	910
S38991-9	252
S39005-4	915
S39006-2	235
S39007-0	107
S39011-9	439
S39015-1	876
S39021-6	740
S39022-4	746
S39023-2	710
S39024-0	324
S39026-7	1037
S39027-5	393
S39028-3	421
S39030-5	393
S39031-3	1000
S39032-1	94
S39040-2	752
S39042-9	333
S39043-7	351
S39044-5	794
S39046-1	342
S39053-4	341
S39054-2	775
S39055-0	234
S39057-7	350
S39058-5	1062
S39060-7	710
S39061-5	350
S39062-3	351
S39067-4	775
S39068-2	804
S39076-3	959
S39077-1	949
S39082-8	944
S39083-6	109
S39084-4	109
S39085-2	109
S39090-9	1142
S39091-7	136
S39092-5	969
S39093-3	393
S39094-1	953
S39101-8	999
S39103-4	907
S39104-2	977
S39107-7	907
S39110-7	1002
S39122-0	139
S39129-8	209
S39133-6	767
S39134-4	490
S39135-2	327
S39139-5	171
S39140-9	337
S39142-5	683
S39144-1	768
S39148-4	75
S39149-2	797
S39150-6	268
S39154-9	24
S39155-7	210
S39156-5	32
S39157-3	796
S39158-1	859
S39160-3	24
S39163-8	363
S39166-2	796
S39167-0	380
S39168-9	212
S39169-7	374
S39171-9	957
S39172-7	393
S39173-5	393
S39181-6	985
S39184-0	987
S39185-9	304
S39187-5	304
S39188-3	484
S39189-1	34
S39190-5	240
S39191-3	426
S39194-8	755
S39195-6	803
S39197-2	139
S39198-0	115
S39199-9	851
S39202-2	35
S39205-7	4
S39206-5	9
S39207-3	35
S39208-1	35
S39210-3	35
S39214-6	441
S39215-4	192
S39218-9	36
S39227-8	671
S39228-6	404
S39230-8	215
S39232-4	1140
S39234-0	1087
S39236-7	873
S39241-3	1006
S39242-1	76
S39244-8	932
S39245-6	380
S39246-4	979
S39247-2	867
S39250-2	979
S39251-0	867
S39252-9	867
S39254-5	237
S39259-6	190
S39261-8	320
S39263-4	45
S39265-0	115
S39270-7	89
S39271-5	89
S39272-3	89
S39275-8	720
S39276-6	720
S39277-4	719
S39279-0	718
S39280-4	722
S39282-0	721
S39284-7	722
S39285-5	722
S39292-8	982
S39294-4	1000
S39296-0	632
S39297-9	804
S39298-7	781
S39299-5	814
S39300-2	813
S39305-3	774
S39308-8	778
S39314-2	213
S39315-0	181
S39316-9	323
S39318-5	774
S39321-5	236
S39326-6	386
S39330-4	953
S39332-0	166
S39333-9	166
S39336-3	1046
S39337-1	1077
S39339-8	561
S39340-1	302
S39343-6	10
S39345-2	14
S39346-0	10
S39349-5	795
S39351-7	823
S39354-1	443
S39356-8	963
S39358-4	1034
S39362-2	214
S39363-0	768
S39364-9	768
S39367-3	768
S39373-8	214
S39379-7	469
S39380-0	476
S39381-9	475
S39383-5	472
S39384-3	476
S39385-1	476
S39387-8	489
S39388-6	122
S39389-4	122
S39392-4	122
S39395-9	30
S39396-7	33
S39397-5	202
S39398-3	33
S39399-1	33
S39400-9	29
S39401-7	797
S39408-4	153
S39409-2	978
S39412-2	74
S39415-7	807
S39423-8	1138
S39424-6	440
S39426-2	440
S39429-7	591
S39446-7	90
S39447-5	377
S39449-1	1133
S39456-4	904
S39458-0	906
S39460-2	903
S39462-9	910
S39464-5	903
S39469-6	915
S39472-6	906
S39480-7	906
S39485-8	909
S39487-4	103
S39488-2	913
S39489-0	906
S39490-4	90
S39494-7	916
S39501-3	906
S39504-8	914
S39510-2	904
S39511-0	906
S39515-3	991
S39516-1	910
S39519-6	917
S39523-4	901
S39524-2	906
S39528-5	990
S39531-5	1097
S39532-3	1098
S39537-4	1136
S39542-0	1135
S39543-9	1136
S39545-5	1099
S39547-1	1135
S39552-8	1136
S39553-6	1136
S39554-4	1135
S39557-9	1135
S39558-7	1135
S39559-5	1131
S39560-9	1131
S39561-7	1129
S39562-5	1129
S39564-1	1131
S39567-6	1132
S39568-4	1131
S39569-2	1131
S39570-6	1131
S39571-4	1133
S39572-2	1131
S39573-0	1129
S39574-9	1129
S39575-7	1131
S39577-3	1131
S39581-1	1131
S39583-8	1132
S39584-6	1131
S39585-4	1129
S39586-2	1131
S39588-9	1131
S39589-7	1131
S39590-0	58
S39604-4	1030
S39606-0	931
S39609-5	1057
S39611-7	372
S39614-1	250
S39617-6	1035
S39618-4	1025
S39627-3	181
S39635-4	317
S39636-2	257
S39638-9	140
S39642-7	264
S39662-1	1125
S39664-8	1125
S39665-6	439
S39671-0	369
S39672-9	372
S39674-5	371
S39675-3	371
S39678-8	75
S39686-9	917
S39692-3	915
S39701-6	870
S39702-4	914
S39715-6	446
S39719-9	536
S39720-2	459
S39721-0	505
S39722-9	445
S39725-3	537
S39734-2	446
S39740-7	445
S39741-5	456
S39742-3	540
S39743-1	457
S39746-6	535
S39747-4	390
S39751-2	438
S39752-0	445
S39760-1	536
S39763-6	408
S39765-2	242
S39766-0	242
S39767-9	243
S39769-5	773
S39770-9	288
S39774-1	339
S39779-2	97
S39780-6	76
S39782-2	1016
S39783-0	1016
S39785-7	1016
S39786-5	97
S39787-3	97
S39788-1	983
S39793-8	570
S39798-9	237
S39799-7	964
S39800-4	40
S39801-2	335
S39804-7	335
S39805-5	219
S39812-8	805
S39818-7	430
S39819-5	431
S39820-9	430
S39821-7	67
S39822-5	815
S39828-4	239
S39832-2	1074
S39834-9	826
S39835-7	816
S39836-5	940
S39837-3	816
S39841-1	486
S39848-9	355
S39849-7	956
S39851-9	88
S39852-7	956
S39853-5	462
S39861-6	815
S39863-2	792
S39866-7	397
S39867-5	790
S39868-3	397
S39872-1	792
S39875-6	791
S39876-4	792
S39877-2	791
S39878-0	1015
S39879-9	1017
S39880-2	1015
S39881-0	1076
S39882-9	334
S39888-8	949
S39893-4	949
S39894-2	369
S39896-9	790
S39897-7	431
S39901-9	834
S39902-7	272
S39903-5	303
S39904-3	1059
S39905-1	1076
S39907-8	1076
S39908-6	925
S39911-6	1076
S39912-4	1051
S39913-2	1076
S39914-0	1035
S39916-7	1076
S39917-5	1076
S39918-3	1037
S39919-1	440
S39921-3	983
S39924-8	1056
S39925-6	261
S39926-4	1076
S39927-2	390
S39928-0	1076
S39929-9	1076
S39930-2	1015
S39931-0	1015
S39935-3	93
S39936-1	235
S39938-8	35
S39941-8	361
S39944-2	93
S39945-0	910
S39949-3	146
S39950-7	144
S39956-6	987
S39962-0	339
S39964-7	339
S39965-5	339
S39966-3	260
S39969-8	122
S39970-1	18
S39972-8	18
S39973-6	18
S39975-2	272
S39976-0	990
S39977-9	272
S39978-7	957
S39986-8	81
S39991-4	265
S40001-7	335
S40008-4	69
S40010-6	779
S40011-4	413
S40012-2	411
S40013-0	411
S40014-9	136
S40015-7	146
S40016-5	947
S40018-1	777
S40023-8	42
S40028-9	429
S40030-0	351
S40033-5	951
S40034-3	964
S40043-2	943
S40046-7	951
S40048-3	951
S40050-5	775
S40051-3	825
S40056-4	947
S40057-2	951
S40059-9	825
S40060-2	822
S40063-7	815
S40064-5	826
S40065-3	825
S40068-8	824
S40069-6	824
S40071-8	775
S40072-6	779
S40073-4	779
S40074-2	826
S40076-9	776
S40080-7	352
S40081-5	352
S40082-3	779
S40083-1	815
S40086-6	962
S40087-4	794
S40090-4	947
S40091-2	942
S40093-9	947
S40096-3	776
S40101-3	779
S40104-8	815

Product	Page
S40105-6	824
S40106-4	779
S40107-2	951
S40108-0	947
S40109-9	822
S40114-5	430
S40115-3	962
S40123-4	955
S40125-0	946
S40126-9	948
S40127-7	368
S40128-5	350
S40129-3	947
S40132-3	951
S40133-1	947
S40135-8	428
S40138-2	428
S40140-4	947
S40142-0	947
S40144-7	952
S40145-5	428
S40146-3	951
S40147-1	947
S40149-8	362
S40150-1	362
S40155-2	952
S40156-0	948
S40159-5	776
S40160-9	423
S40161-7	947
S40162-5	359
S40163-3	947
S40166-8	249
S40170-6	947
S40171-4	951
S40173-0	420
S40175-7	423
S40178-1	423
S40180-3	430
S40181-1	359
S40183-8	948
S40185-4	249
S40190-0	351
S40193-5	938
S40194-3	947
S40195-1	956
S40199-4	952
S40200-1	948
S40202-8	952
S40203-6	948
S40204-4	952
S40205-2	948
S40207-9	359
S40208-7	825
S40209-5	423
S40210-9	249
S40217-6	951
S40218-4	947
S40219-2	948
S40220-6	952
S40221-4	955
S40222-2	955
S40223-0	955
S40225-7	755
S40226-5	700
S40227-3	247
S40233-8	249
S40235-4	948
S40236-2	359
S40237-0	359
S40238-9	956
S40240-0	948
S40242-7	359
S40243-5	425
S40244-3	423
S40247-8	423
S40249-4	359
S40253-2	947
S40254-0	947
S40257-5	816
S40260-5	249
S40266-4	951
S40267-2	947
S40268-0	947
S40269-9	948
S40270-2	947
S40271-0	947
S40272-9	352
S40275-3	254
S40278-8	425
S40282-6	947
S40283-4	952
S40285-0	947
S40286-9	946
S40287-7	951
S40288-5	946
S40290-7	951
S40291-5	256
S40292-3	254
S40293-1	431
S40296-6	955
S40297-4	947
S40298-2	948
S40299-0	948
S40301-6	955
S40303-2	254
S40304-0	431
S40305-9	254
S40306-7	424
S40307-5	351
S40309-1	985
S40312-1	825
S40314-8	817
S40315-6	828
S40317-2	955
S40319-9	243
S40323-7	441
S40325-3	373
S40331-8	147
S40333-4	462
S40334-2	370
S40341-5	106
S40342-3	112
S40346-6	462
S40348-2	202
S40350-4	36
S40352-0	292
S40353-9	1070
S40354-7	142
S40356-3	35
S40370-9	107
S40374-1	979
S40376-8	779
S40379-2	781
S40381-4	963
S40383-0	947
S40385-7	202
S40386-5	949
S40387-3	405
S40394-6	461
S40398-9	405
S40401-2	951
S40402-0	947
S40403-9	955
S40404-7	952
S40406-3	947
S40438-1	369
S40440-3	320
S40444-6	58
S40447-0	319
S40449-7	21
S40450-0	245
S40451-9	924
S40458-6	55
S40461-6	91
S40475-6	3
S40476-4	65
S40480-2	359
S40501-9	1029
S40503-5	389
S40505-1	10
S40512-4	213
S40518-3	192
S40521-3	103
S40527-2	273
S40529-9	914
S40531-0	56
S40534-5	279
S40536-1	321
S40538-8	1096
S40547-7	66
S40550-7	192
S40557-4	386
S40575-2	837
S40579-5	126
S40580-9	160
S40581-7	2
S40583-3	833
S40591-4	31
S40593-0	1142
S40610-4	874
S40612-0	141
S40620-1	796
S40623-6	426
S40625-2	141
S40628-7	840
S40630-9	118
S40633-3	1052
S40636-8	979
S40639-2	903
S40640-6	487
S40641-4	970
S40642-2	970
S40643-0	970
S40647-3	998
S40648-1	263
S40650-3	921
S40651-1	349
S40655-4	1003
S40656-2	1002
S40659-7	917
S40660-0	372
S40662-7	754
S40663-5	196
S40664-3	755
S40665-1	850
S40667-8	314
S40668-6	285
S40669-4	195
S40670-8	959
S40671-6	881
S40672-4	882
S40673-2	265
S40674-0	70
S40675-9	167
S40677-5	127
S40678-3	340
S40679-1	398
S40680-5	462
S40681-3	195
S40682-1	158
S40686-4	171
S40687-2	171
S40688-0	887
S40690-2	440
S40692-9	174
S40693-7	203
S40695-3	13
S40696-1	368
S40698-8	116
S40699-6	109
S40700-3	889
S40701-1	706
S40703-8	982
S40704-6	95
S40705-4	954
S40706-2	1128
S40707-0	243
S40712-7	261
S40713-5	201
S40714-3	197
S40717-8	197
S40723-2	933
S40724-0	67
S40725-9	101
S40726-7	41
S40732-1	224
S40734-8	1003
S40735-6	230
S40736-4	193
S40737-2	318
S40738-0	1087
S40739-9	309
S40741-0	743
S40746-1	239
S40748-8	393
S40749-6	62
S40751-8	5
S40752-6	197
S40753-4	170
S40754-2	170
S40755-0	170
S40756-9	223
S40758-5	105
S40759-3	23
S40761-5	151
S40762-3	107
S40765-8	76
S40766-6	87
S40771-2	709
S40775-5	289
S40777-1	920
S40780-1	1085
S40786-0	87
S40787-9	269
S40788-7	307
S40789-5	812
S40790-9	755
S40792-5	87
S40793-3	36
S40794-1	239
S40797-6	33
S40798-4	315
S40799-2	315
S40804-2	61
S40805-0	8
S40806-9	1002
S40809-3	887
S40810-7	344
S40811-5	886
S40812-3	886
S40813-1	11
S40815-8	28
S40816-6	328
S40817-4	328
S40820-4	487
S40821-2	487
S40822-0	273
S40824-7	346
S40825-5	978
S40827-1	184
S40829-8	397
S40830-1	28
S40832-8	836
S40834-4	102
S40840-9	33
S40841-7	987
S40842-5	83
S40844-1	149
S40846-8	13
S40851-4	917
S40852-2	363
S40853-0	362
S40855-7	83
S40856-5	1051
S40857-3	22
S40858-1	360
S40867-0	137
S40868-9	161
S40869-7	161
S40870-0	161
S40872-7	854
S40877-8	81
S40879-4	57
S40881-6	254
S40883-2	734
S40884-0	812
S40885-9	24
S40888-3	310
S40891-3	934
S40892-1	461
S40897-2	315
S40898-0	1069
S40899-9	115
S40900-6	904
S40903-0	884
S40905-7	310
S40908-1	411
S40911-1	796
S40913-8	24
S40914-6	970
S40915-4	118
S40918-9	504
S40920-0	376
S40922-7	148
S40924-3	185
S40927-8	8
S40928-6	8
S40929-4	54
S40930-8	54
S40932-4	78
S40936-7	213
S40937-5	919
S40938-3	255
S40940-5	977
S40941-3	962
S40942-1	934
S40944-8	367
S40945-6	920
S40947-2	1106
S40948-0	298
S40949-9	59
S40950-2	1125
S40952-9	489
S40953-7	166
S40958-8	279
S40959-6	1033
S40961-8	873
S40962-6	900
S40963-4	944
S40965-0	265
S40967-7	98
S40968-5	110
S40969-3	61
S40970-7	7
S40971-5	723
S40972-3	242
S40973-1	422
S40976-6	387
S40978-2	919
S40979-0	887
S40981-2	745
S40983-9	918
S40985-5	866
S40987-1	145
S40993-6	39
S40994-4	40
S40996-0	884
S40997-9	882
S40999-5	900
S41000-4	1079
S41005-5	141
S41010-1	250
S41012-8	93
S41013-6	139
S41015-2	866
S41016-0	856
S41017-9	1141
S41018-7	739
S41019-5	63
S41023-3	580
S41024-1	16
S41027-6	249
S41031-4	1010
S41032-2	1
S41038-1	943
S41040-3	227
S41043-8	2
S41044-6	861
S41045-4	733
S41046-2	710
S41048-9	226
S41050-0	211
S41052-7	172
S41055-1	1062
S41057-8	828
S41058-6	933
S41059-4	1078
S41060-8	734
S41061-6	733
S41062-4	733
S41063-2	845
S41064-0	734
S41065-9	734
S41069-1	261
S41072-1	173
S41075-6	170
S41076-4	887
S41091-8	205
S41100-0	287
S41101-9	286
S41103-5	264
S41104-3	932
S41107-8	872
S41110-8	399
S41115-9	197
S41117-5	407
S41120-5	255
S41125-6	1014
S41126-4	377
S41127-2	203
S41128-0	49
S41131-0	1107
S41132-9	282
S41133-7	250
S41134-5	369
S41138-8	22
S41141-8	111
S41148-5	768
S41150-7	145
S41151-5	762
S41152-3	215
S41153-1	945
S41155-8	239
S41156-6	590
S41157-4	801
S41158-2	516
S41159-0	597
S41160-4	592
S41164-7	146
S41166-3	1036
S41167-1	274
S41169-8	443
S41170-1	115
S41172-8	111
S41173-6	442
S41175-2	76
S41177-9	22
S41178-7	1083
S41179-5	48
S41182-5	48
S41186-8	128
S41187-6	13
S41189-2	9
S41190-6	9
S41192-2	1
S41194-9	12
S41195-7	11
S41197-3	8
S41198-1	1
S41200-7	76
S41201-5	9
S41202-3	1
S41203-1	9
S41205-8	1
S41206-6	1
S41208-2	204
S41209-0	158
S41211-2	943
S41212-0	1060
S41214-7	188
S41215-5	945
S41216-3	950
S41219-8	99
S41222-8	238
S41223-6	165
S41224-4	372
S41227-9	836
S41228-7	160
S41229-5	133
S41231-7	329
S41232-5	325
S41234-1	379
S41236-8	379

Product	Page
S41238-4	749
S41239-2	426
S41240-6	1035
S41242-2	2
S41243-0	8
S41245-7	13
S41253-8	968
S41254-6	227
S41256-2	194
S41258-9	919
S41259-7	39
S41261-9	317
S41262-7	207
S41263-5	247
S41264-3	190
S41265-1	834
S41267-8	501
S41268-6	163
S41269-4	250
S41270-8	280
S41271-6	750
S41272-4	754
S41274-0	56
S41276-7	328
S41277-5	971
S41279-1	796
S41282-1	15
S41285-6	308
S41288-0	141
S41289-9	141
S41290-2	141
S41291-0	488
S41292-9	26
S41296-1	158
S41298-8	293
S41301-1	169
S41305-4	714
S41312-7	139
S41313-5	26
S41314-3	235
S41322-4	589
S41327-5	138
S41329-1	138
S41332-1	988
S41340-2	166
S41344-5	1008
S41352-6	833
S41353-4	344
S41360-7	297
S41365-8	20
S41366-6	138
S41369-0	169
S41371-2	39
S41374-7	973
S41375-5	79
S41379-8	12
S41396-8	877
S41397-6	35
S41401-8	95
S41402-6	412
S41403-4	1054
S41409-3	14
S41410-7	1019
S41411-5	41
S41416-6	328
S41419-0	215
S41425-5	956
S41427-1	20
S41430-1	80
S41435-2	342
S41436-0	310
S41444-1	172
S41447-6	1001
S41455-7	8
S41464-6	38
S41468-9	271
S41469-7	294
S41475-1	884
S41478-6	138
S41482-4	192
S41484-0	21
S41485-9	937
S41491-3	193
S41497-2	21
S41498-0	137
S41500-6	61
S41515-4	790
S41533-2	760
S41534-0	1097
S41537-5	63
S41542-1	2
S41545-6	298
S41549-9	285
S41550-2	783
S41551-0	795
S41553-7	975
S41563-4	1085
S41567-7	339
S41569-3	40
S41571-5	970
S41575-8	264
S41576-6	2
S41577-4	72
S41581-2	977
S41586-3	62
S41599-5	703
S41604-5	334
S41605-3	347
S41606-1	295
S41612-6	887
S41614-2	319
S41616-9	677
S41617-7	92
S41621-5	1112
S41622-3	3
S41626-6	850
S41627-4	60
S41629-0	43
S41631-2	383
S41632-0	235
S41635-5	35
S41654-1	324
S41656-8	12
S41663-0	4
S41664-9	80
S41665-7	89
S41667-3	1083
S41668-1	437
S41670-3	233
S41671-1	867
S41676-2	487
S41679-7	884
S41680-0	1001
S41681-9	1001
S41682-7	248
S41684-3	904
S41692-4	444
S41693-2	823
S41697-5	8
S41698-3	8
S41699-1	13
S41700-9	76
S41702-5	793
S41704-1	349
S41706-8	702
S41707-6	169
S41708-4	1053
S41709-2	1002
S41710-6	881
S41713-0	13
S41714-9	203
S41715-7	203
S41716-5	754
S41717-3	754
S41718-1	203
S41720-3	203
S41725-4	1135
S41726-2	1136
S41727-0	1136
S41728-9	1135
S41730-0	273
S41731-9	393
S41732-7	392
S41733-5	392
S41735-1	13
S41737-8	11
S41738-6	11
S41739-4	11
S41740-8	11
S41741-6	11
S41742-4	9
S41743-2	9
S41744-0	9
S41745-9	10
S41746-7	8
S41747-5	11
S41748-3	326
S41749-1	993
S41750-5	940
S41751-3	21
S41752-1	164
S41754-8	1136
S41755-6	247
S41757-2	918
S41762-9	352
S41764-5	221
S41765-3	754
S41766-1	908
S41771-8	804
S41772-6	882
S41773-4	503
S41774-2	973
S41776-9	791
S41777-7	375
S41779-3	11
S41780-7	1
S41781-5	4
S41782-3	471
S41783-1	909
S41785-8	111
S41786-6	15
S41789-0	323
S41793-9	891
S41794-7	1012
S41797-1	250
S41800-5	246
S41801-3	919
S41805-6	183
S41806-4	183
S41807-2	183
S41808-0	205
S41809-9	203
S41810-2	753
S41813-7	429
S41814-5	115
S41816-1	867
S41823-4	31
S41826-9	17
S41830-7	28
S41832-3	905
S41833-1	11
S41837-4	33
S41838-2	145
S41849-8	134
S41850-1	891
S41852-8	1127
S41856-0	4
S41859-5	218
S41860-9	250
S41863-3	20
S41864-1	219
S41866-8	69
S41867-6	190
S41868-4	373
S41869-2	80
S41872-2	323
S41877-3	2
S41880-3	764
S41883-8	1087
S41885-4	37
S41892-7	257
S41898-6	2
S41900-1	13
S41905-2	55
S41910-9	793
S41911-7	1
S41912-5	866
S41913-3	1028
S41922-2	13
S41928-1	207
S41930-3	98
S41934-6	787
S41935-4	755
S41941-9	57
S41942-7	891
S41945-1	46
S41949-4	8
S41950-8	93
S41960-5	1057
S41962-1	62
S41964-8	336
S41965-6	192
S41966-4	61
S41967-2	21
S41968-0	95
S41970-2	495
S41972-9	169
S41974-5	208
S41976-1	697
S41979-6	138
S41981-8	160
S41983-4	166
S41985-0	38
S41987-7	69
S41988-5	164
S41989-3	326
S41990-7	36
S41996-6	94
S41998-2	21
S42001-8	952
S42002-6	444
S42003-4	132
S42004-2	833
S42005-0	464
S42011-5	126
S42012-3	971
S42013-1	828
S42015-8	146
S42021-2	415
S42022-0	112
S42024-7	69
S42026-3	147
S42035-2	159
S42038-7	424
S42041-7	394
S42042-5	394
S42047-6	202
S42048-4	952
S42049-2	108
S42050-6	108
S42051-4	108
S42052-2	108
S42055-7	724
S42056-5	974
S42057-3	201
S42058-1	202
S42061-1	431
S42063-8	137
S42064-6	794
S42065-4	421
S42067-0	444
S42068-9	482
S42069-7	482
S42070-0	424
S42071-9	430
S42072-7	965
S42073-5	954
S42075-1	444
S42077-8	482
S42078-6	482
S42079-4	405
S42080-8	410
S42081-6	423
S42082-4	708
S42084-0	987
S42085-9	812
S42088-3	331
S42089-1	447
S42090-5	108
S42091-3	104
S42092-1	150
S42096-4	953
S42098-0	417
S42099-9	441
S42102-2	993
S42105-7	462
S42108-1	359
S42110-3	109
S42111-1	108
S42113-8	953
S42117-0	107
S42118-9	953
S42120-0	430
S42121-9	415
S42122-7	421
S42123-5	421
S42124-3	442
S42125-1	502
S42128-6	421
S42130-8	417
S42131-6	426
S42133-2	430
S42135-9	360
S42136-7	350
S42137-5	335
S42142-1	335
S42146-4	483
S42147-2	483
S42148-0	483
S42149-9	483
S42150-2	35
S42154-5	108
S42156-1	108
S42159-6	108
S42162-6	668
S42163-4	424
S42165-0	999
S42166-9	754
S42168-5	415
S42169-3	415
S42170-7	415
S42171-5	426
S42172-3	421
S42176-6	423
S42177-4	424
S42178-2	424
S42179-0	429
S42180-4	921
S42183-9	104
S42184-7	107
S42185-5	107
S42186-3	107
S42187-1	107
S42192-8	108
S42194-4	109
S42195-2	109
S42199-5	415
S42201-0	424
S42202-9	424
S42204-5	415
S42206-1	1086
S42208-8	403
S42209-6	425
S42212-6	104
S42215-0	108
S42218-5	424
S42221-5	103
S42222-3	108
S42225-8	109
S42226-6	108
S42227-4	108
S42228-2	67
S42229-0	108
S42230-4	108
S42231-2	416
S42234-7	423
S42237-1	421
S42239-8	1086
S42242-8	421
S42244-4	418
S42245-2	359
S42246-0	413
S42247-9	700
S42251-7	956
S42253-3	953
S42254-1	951
S42256-8	405
S42258-4	108
S42259-2	109
S42262-2	357
S42263-0	428
S42264-9	359
S42265-7	359
S42268-1	415
S42270-3	959
S42273-8	109
S42274-6	357
S42275-4	108
S42276-2	108
S42280-0	355
S42283-5	530
S42287-8	429
S42290-8	108
S42291-6	999
S42293-2	357
S42295-9	423
S42297-5	357
S42298-3	357
S42300-9	108
S42301-7	108
S42303-3	104
S42304-1	948
S42307-6	405
S42312-2	412
S42313-0	425
S42314-9	816
S42315-7	362
S42317-3	430
S42321-1	921
S42324-6	423
S42325-4	942
S42326-2	104
S42327-0	444
S42338-6	444
S42340-8	463
S42344-0	108
S42347-5	971
S42356-4	1033
S42359-9	423
S42361-0	414
S42366-1	423
S42368-8	994
S42371-8	132
S42372-6	87
S42374-2	788
S42379-3	482
S42380-7	482
S42381-5	787
S42383-1	939
S42385-8	482
S42388-2	482
S42393-9	949
S42394-7	924
S42401-3	415
S42402-1	415
S42405-6	483
S42407-2	351
S42410-2	882
S42411-0	918
S42412-9	107
S42413-7	384
S42416-1	284
S42419-6	317
S42423-4	1005
S42425-0	360
S42428-5	382
S42432-3	246
S42433-1	918
S42439-0	254
S42440-4	247
S42443-9	989
S42444-7	365
S42445-5	309
S42449-8	246
S42450-1	246
S42453-6	997
S42455-2	254
S42462-5	892
S42464-1	59
S42466-8	919
S42467-6	884
S42468-4	58
S42471-4	246
S42473-0	307

Product	Page
S42476-5	892
S42477-3	254
S42478-1	254
S42484-6	317
S42485-4	477
S42488-9	882
S42491-9	918
S42492-7	918
S42493-5	918
S42495-1	918
S42497-8	918
S42500-1	291
S42508-7	319
S42509-5	368
S42510-9	368
S42511-7	368
S42517-6	932
S42520-6	319
S42522-2	1005
S42526-5	704
S42530-3	919
S42531-1	37
S42542-7	39
S42555-9	311
S42558-3	919
S42559-1	414
S42560-5	884
S42562-1	254
S42565-6	918
S42566-4	918
S42567-2	254
S42569-9	996
S42570-2	997
S42571-0	58
S42572-9	884
S42573-7	884
S42574-5	918
S42575-3	887
S42579-6	918
S42582-6	892
S42583-4	58
S42584-2	884
S42586-9	1005
S42589-3	292
S42590-7	982
S42591-5	918
S42592-3	918
S42593-1	919
S42595-8	918
S42597-4	919
S42598-2	919
S42601-6	997
S42603-2	308
S42604-0	308
S42606-7	919
S42608-3	245
S42610-5	340
S42611-3	340
S42615-6	997
S42616-4	246
S42620-2	216
S42621-0	919
S42622-9	704
S42623-7	58
S42624-5	368
S42625-3	918
S42626-1	919
S42628-8	996
S42632-6	245
S42637-7	58
S42638-5	867
S42641-5	1003
S42642-3	42
S42657-1	319
S42659-8	319
S42668-7	355
S42677-6	73
S42679-2	886
S42681-4	239
S42682-2	60
S42683-0	269
S42686-5	9
S42691-1	853
S42699-7	5
S42706-3	344
S42707-1	233
S42710-1	12
S42717-9	236
S42721-7	905
S42722-5	385
S42724-1	38
S42726-8	789
S42728-4	1003
S42730-6	933
S42731-4	332
S42733-0	343
S42734-9	918
S42735-7	236
S42736-5	922
S42746-2	973
S42747-0	872
S42748-9	486
S42749-7	873
S42754-3	690
S42759-4	70
S42764-0	202
S42765-9	1000
S42772-1	354
S42775-6	768
S42776-4	323
S42777-2	22
S42780-2	1
S42783-7	76
S42785-3	8
S42786-1	209
S42788-8	347
S42792-6	1135
S42793-4	1093
S42795-0	922
S42796-9	787
S42797-7	835
S42802-7	61
S42803-5	768
S42807-8	243
S42808-6	170
S42810-8	767
S42811-6	163
S42813-2	220
S42820-5	1112
S42822-1	36
S42825-6	157
S42827-2	441
S42828-0	243
S42832-9	788
S42833-7	271
S42834-5	785
S42838-8	276
S42839-6	703
S42847-7	53
S42848-5	31
S42850-7	348
S42852-3	3
S42859-0	884
S42860-4	389
S42861-2	220
S42862-0	321
S42865-5	53
S42869-8	33
S42877-9	82
S42880-9	141
S42881-7	98
S42886-8	958
S42890-6	868
S42893-0	768
S42895-7	158
S42898-1	705
S42902-3	383
S42903-1	885
S42907-4	486
S42909-0	1014
S42910-4	13
S42911-2	867
S42912-0	356
S42913-9	1008
S42920-1	1053
S42931-7	155
S42932-5	74
S42933-3	45
S42934-1	107
S42937-6	4
S42938-4	161
S42939-2	158
S42940-6	769
S42942-2	189
S42947-3	787
S42948-1	914
S42953-8	322
S42954-6	1
S42955-4	67
S42959-7	9
S42970-8	31
S42972-4	11
S42978-3	257
S42980-5	1019
S42982-1	12
S42984-8	257
S42985-6	916
S42988-0	146
S42989-9	16
S42996-1	59
S43000-5	170
S43001-3	2
S43002-1	748
S43004-8	217
S43006-4	91
S43009-9	13
S43011-0	915
S43013-7	1015
S43014-5	9
S43015-3	60
S43016-1	145
S43021-8	178
S43028-5	194
S43031-5	31
S43032-3	917
S43033-1	747
S43036-6	791
S43041-2	55
S43045-5	109
S43050-1	52
S43054-4	159
S43057-9	48
S43059-5	81
S43061-7	176
S43063-3	152
S43071-4	18
S43072-2	5
S43073-0	201
S43080-3	957
S43085-4	8
S43087-0	982
S43089-7	342
S43091-9	90
S43093-5	1
S43095-1	12
S43098-6	943
S43100-1	379
S43102-8	1014
S43103-6	149
S43104-4	149
S43105-2	377
S43106-0	190
S43107-9	53
S43109-5	702
S43111-7	48
S43114-1	148
S43116-8	232
S43117-6	954
S43118-4	746
S43121-4	325
S43123-0	783
S43124-9	78
S43126-5	205
S43127-3	147
S43128-1	150
S43131-1	75
S43134-6	361
S43135-4	597
S43136-2	847
S43137-0	83
S43138-9	877
S43140-0	1084
S43141-9	804
S43142-7	1141
S43143-5	149
S43144-3	81
S43145-1	82
S43147-8	48
S43148-6	48
S43149-4	82
S43151-6	4
S43152-4	3
S43153-2	8
S43154-0	1
S43155-9	3
S43156-7	9
S43159-1	1
S43162-1	9
S43164-8	1
S43165-6	8
S43166-4	2
S43167-2	203
S43170-2	356
S43171-0	73
S43176-1	937
S43178-8	943
S43179-6	768
S43182-6	963
S43183-4	146
S43184-2	300
S43185-0	239
S43186-9	1141
S43187-7	205
S43188-5	238
S43190-7	60
S43193-1	174
S43195-8	160
S43196-6	187
S43197-4	966
S43198-2	835
S43199-0	203
S43200-8	489
S43201-6	1135
S43203-2	787
S43204-0	476
S43205-9	768
S43207-5	260
S43209-1	266
S43210-5	905
S43211-3	30
S43212-1	116
S43215-6	1
S43216-4	9
S43217-2	1
S43218-0	2
S43219-9	183
S43220-2	238
S43221-0	131
S43223-7	1140
S43226-1	997
S43229-6	124
S43231-8	462
S43232-6	755
S43233-4	283
S43234-2	208
S43235-0	176
S43236-9	481
S43237-7	236
S43238-5	773
S43240-7	88
S43241-5	88
S43242-3	88
S43243-1	125
S43245-8	953
S43247-4	282
S43248-2	354
S43249-0	762
S43250-4	933
S43251-2	329
S43252-0	934
S43253-9	40
S43254-7	40
S43255-5	93
S43256-3	891
S43257-1	1037
S43260-1	163
S43262-8	204
S43263-6	150
S43266-0	203
S43268-7	26
S43271-7	257
S43272-5	257
S43276-8	219
S43277-6	932
S43278-4	356
S43279-2	375
S43280-6	325
S43281-4	297
S43282-2	109
S43283-0	1085
S43284-9	264
S43286-5	884
S43287-3	361
S43290-3	861
S43291-1	180
S43293-8	279
S43294-6	145
S43296-2	16
S43297-0	789
S43298-9	28
S43299-7	704
S43300-4	32
S43302-0	30
S43303-9	148
S43304-7	181
S43305-5	380
S43306-3	577
S43307-1	294
S43309-8	134
S43310-1	77
S43312-8	974
S43313-6	290
S43317-9	48
S43318-7	9
S43319-5	11
S43321-7	9
S43323-3	237
S43326-8	884
S43328-4	267
S43329-2	19
S43330-6	111
S43331-4	37
S43333-0	259
S43335-7	187
S43336-5	11
S43337-3	8
S43338-1	871
S43343-8	1140
S43344-6	414
S43345-4	70
S43346-2	163
S43347-0	272
S43348-9	296
S43349-7	944
S43350-0	66
S43352-7	785
S43354-3	775
S43355-1	35
S43358-6	74
S43359-4	279
S43360-8	280
S43361-6	785
S43362-4	27
S43363-2	272
S43364-0	308
S43366-7	370
S43367-5	182
S43368-3	182
S43369-1	179
S43370-5	291
S43371-3	311
S43372-1	427
S43374-8	146
S43375-6	872
S43376-4	43
S43379-9	191
S43380-2	145
S43382-9	160
S43384-5	939
S43385-3	792
S43386-1	202
S43388-8	35
S43389-6	332
S43391-8	843
S43392-6	1036
S43393-4	994
S43394-2	938
S43395-0	342
S43396-9	1
S43397-7	305
S43398-5	305
S43399-3	301
S43400-0	146
S43401-9	213
S43402-7	275
S43404-3	336
S43405-1	12
S43407-8	20
S43408-6	20
S43409-4	20
S43410-8	11
S43411-6	4
S43412-4	303
S43413-2	767
S43414-0	294
S43415-9	767
S43420-5	3
S43421-3	860
S43422-1	124
S43424-8	376
S43425-6	145
S43426-4	35
S43427-2	933
S43431-0	59
S43432-9	1124
S43433-7	1125
S43434-5	1124
S43435-3	762
S43436-1	764
S43438-8	764
S43439-6	189
S43442-6	65
S43444-2	1085
S43446-9	77
S43449-3	979
S43452-3	338
S43456-6	698
S43459-0	328
S43460-4	710
S43461-2	368
S43462-0	38
S43463-9	29
S43464-7	2
S43466-3	238
S43467-1	1011
S43472-8	755
S43473-6	59
S43475-2	974
S43476-0	345
S43477-9	65
S43479-5	350
S43481-7	37
S43483-3	268
S43486-8	241
S43487-6	207
S43488-4	204
S43489-2	204
S43490-6	201
S43492-2	381
S43493-0	833
S43500-7	374
S43501-5	835
S43503-1	450
S43506-6	235
S43507-4	248
S43508-2	284
S43510-4	174
S43511-2	830
S43512-0	296
S43513-9	147
S43514-7	74
S43516-3	198
S43517-1	145
S43519-8	148

Product	Page
S43522-8	67
S43523-6	66
S43524-4	66
S43526-0	1001
S43527-9	993
S43528-7	13
S43529-5	174
S43530-9	815
S43531-7	188
S43532-5	213
S43533-3	250
S43537-6	966
S43541-4	370
S43542-2	949
S43544-9	38
S43545-7	336
S43547-3	298
S43550-3	937
S43554-6	1127
S43555-4	8
S43559-7	331
S43560-0	361
S43561-9	170
S43562-7	952
S43563-5	859
S43564-3	587
S43565-1	76
S43567-8	56
S43571-6	132
S43572-4	136
S43573-2	270
S43575-9	11
S43576-7	164
S43578-3	879
S43581-3	900
S43589-9	503
S43603-8	158
S43611-9	317
S43613-5	786
S43615-1	768
S43617-8	77
S43623-2	175
S43624-0	127
S43629-1	236
S43639-9	1090
S43640-2	2
S43645-3	375
S43648-8	2
S43649-6	67
S43656-9	979
S43674-7	1
S43682-8	13
S43687-9	793
S43703-4	983
S43710-7	345
S43712-3	887
S43717-4	337
S43720-4	921
S43722-0	389
S43723-9	4
S43724-7	869
S43732-8	135
S43734-4	48
S43742-5	8
S43744-1	1
S43746-8	1
S43750-6	11
S43753-0	368
S43755-7	4
S43757-3	832
S43760-3	124
S43763-8	34
S43764-6	984
S43766-2	151
S43772-7	999
S43773-5	971
S43774-3	765
S43779-4	1085
S43780-8	999
S43781-6	235
S43782-4	31
S43783-2	241
S43784-0	297
S43786-7	375
S43787-5	337
S43788-3	317
S43798-0	270
S43799-9	360
S43801-4	90
S43807-3	860
S43808-1	338
S43810-3	799
S43816-2	888
S43818-9	984
S43819-7	983
S43821-9	344
S43825-1	920
S43830-8	255
S43831-6	61
S43832-4	33
S43833-2	310
S43835-9	87
S43836-7	339
S43838-3	15
S43847-2	18
S43855-3	115
S43859-6	264
S43865-0	877
S43867-7	220
S43870-7	132
S43875-8	4
S43877-4	787
S43881-2	907
S43892-8	147
S43893-6	146
S43895-2	351
S43898-7	917
S43902-9	973
S43905-3	160
S43909-6	1097
S43911-8	915
S43916-9	2
S43917-7	11
S43918-5	42
S43919-3	63
S43927-4	290
S43942-8	910
S43943-6	59
S43945-2	319
S43947-9	29
S43948-7	239
S43949-5	61
S43950-9	220
S43952-5	156
S43959-2	170
S43961-4	977
S43964-9	1002
S43965-7	353
S43966-5	164
S43970-3	231
S43971-1	15
S43975-4	1109
S43976-2	916
S43977-0	836
S43978-9	290
S43980-0	719
S43981-9	61
S43983-5	787
S43984-3	966
S43985-1	70
S43990-8	890
S43992-4	312
S43995-9	150
S43996-7	219
S44000-0	262
S44004-3	970
S44005-1	1143
S44009-4	761
S44011-6	276
S44015-9	252
S44019-1	5
S44021-3	984
S44022-1	194
S44025-6	768
S44032-9	157
S44057-4	119
S44065-5	158
S44066-3	1135
S44067-1	983
S44075-2	429
S44076-0	230
S44082-5	217
S44087-6	827
S44088-4	787
S44089-2	983
S44091-4	1
S44094-9	743
S44097-3	154
S44100-7	1083
S44106-6	34
S44107-4	774
S44108-2	954
S44110-4	817
S44111-2	955
S44112-0	956
S44115-5	201
S44117-1	1094
S44119-8	51
S44120-1	96
S44122-8	579
S44123-6	65
S44125-2	756
S44126-0	169
S44131-7	14
S44134-1	199
S44137-6	677
S44139-2	122
S44141-4	767
S44144-9	236
S44145-7	2
S44148-1	101
S44150-3	902
S44151-1	1090
S44155-4	368
S44162-7	197
S44168-6	238
S44172-4	5
S44174-0	1092
S44177-5	164
S44180-5	104
S44181-3	3
S44190-2	312
S44194-5	979
S44195-3	317
S44196-1	489
S44198-8	174
S44199-6	235
S44200-3	174
S44201-1	314
S44203-8	213
S44204-6	223
S44206-2	11
S44212-7	827
S44213-5	247
S44214-3	160
S44217-8	350
S44218-6	319
S44219-4	857
S44224-0	63
S44226-7	939
S44227-5	221
S44228-3	260
S44230-5	140
S44236-4	280
S44237-2	979
S44239-9	219
S44243-7	241
S44252-6	8
S44253-4	375
S44258-5	788
S44259-3	1
S44263-1	88
S44270-4	167
S44272-0	159
S44275-5	333
S44282-8	975
S44283-6	29
S44287-9	983
S44292-5	74
S44303-4	367
S44306-9	343
S44313-1	835
S44317-4	158
S44318-2	97
S44327-1	297
S44336-0	317
S44338-7	158
S44339-5	100
S44342-5	1142
S44343-3	955
S44344-1	62
S44348-4	345
S44350-6	962
S44354-9	167
S44356-5	979
S44361-1	995
S44365-4	40
S44366-2	374
S44371-9	973
S44373-5	21
S44378-6	192
S44382-4	66
S44384-0	321
S44387-5	827
S44397-2	321
S44404-9	8
S44405-7	253
S44407-3	836
S44414-6	50
S44423-5	798
S44429-4	338
S44431-6	257
S44432-4	160
S44439-1	320
S44441-3	66
S44444-8	188
S44445-6	324
S44447-2	830
S44451-0	45
S44452-9	339
S44455-3	354
S44458-8	215
S44463-4	144
S44464-2	225
S44465-0	156
S44468-5	1077
S44469-3	350
S44475-8	40
S44477-4	326
S44481-2	220
S44497-9	293
S44504-5	38
S44506-1	372
S44512-6	187
S44517-7	348
S44519-3	241
S44520-7	219
S44521-5	976
S44523-1	878
S44526-6	260
S44530-4	767
S44534-7	870
S44539-8	766
S44540-1	800
S44542-8	215
S44544-4	369
S44546-0	933
S44547-9	888
S44548-7	81
S44550-9	38
S44552-5	951
S44557-6	338
S44559-2	433
S44560-6	702
S44566-5	354
S44567-3	169
S44570-3	353
S44574-6	998
S44578-9	294
S44579-7	287
S44581-9	150
S44587-8	1018
S44592-4	767
S44593-2	279
S44596-7	855
S44597-5	488
S44598-3	1143
S44603-3	57
S44610-6	348
S44612-2	37
S44614-9	221
S44616-5	970
S44618-1	727
S44624-6	341
S44627-0	756
S44628-9	924
S44632-7	253
S44633-5	940
S44634-3	335
S44637-8	356
S44639-4	6
S44651-3	790
S44655-6	932
S44658-0	71
S44661-0	294
S44663-7	289
S44671-8	1093
S44674-2	1089
S44676-9	837
S44677-7	2
S44682-3	142
S44683-1	33
S44686-6	758
S44687-4	297
S44688-2	325
S44690-4	383
S44691-2	160
S44692-0	1052
S44693-9	705
S44694-7	280
S44695-5	205
S44696-3	233
S44697-1	1128
S44700-5	13
S44701-3	170
S44704-8	905
S44705-6	885
S44706-4	933
S44707-2	488
S44708-0	798
S44709-9	258
S44710-2	905
S44711-0	657
S44712-9	203
S44714-5	932
S44715-3	115
S44716-1	197
S44718-8	659
S44719-6	203
S44721-8	755
S44722-6	205
S44723-4	233
S44725-0	976
S44726-9	237
S44727-7	705
S44728-5	348
S44729-3	203
S44731-5	937
S44732-3	994
S44733-1	784
S44735-8	1083
S44736-6	724
S44737-4	1083
S44739-0	933
S44740-4	1142
S44742-0	25
S44744-7	125
S44745-5	8
S44746-3	943
S44747-1	802
S44749-8	1093
S44750-1	1127
S44752-8	325
S44753-6	884
S44754-4	1065
S44755-2	50
S44756-0	196
S44760-9	311
S44761-7	739
S44762-5	276
S44763-3	313
S44764-1	426
S44766-8	1010
S44767-6	295
S44768-4	22
S44770-6	1016
S44771-4	700
S44772-2	360
S44773-0	889
S44774-9	1012
S44775-7	780
S44776-5	734
S44777-3	1128
S44778-1	993
S44780-3	937
S44781-1	292
S44783-8	131
S44784-6	984
S44785-4	335
S44786-2	302
S44788-9	66
S44789-7	787
S44790-0	762
S44792-7	149
S44793-5	373
S44794-3	756
S44799-4	417
S44802-8	811
S44803-6	417
S44806-0	765
S44807-9	148
S44808-7	241
S44809-5	298
S44810-9	975
S44811-7	910
S44813-3	57
S44814-1	756
S44816-8	765
S44818-4	332
S44819-2	753
S44820-6	332
S44821-4	150
S44822-2	158
S44823-0	756
S44824-9	489
S44825-7	150
S44828-1	773
S44831-1	358
S44833-8	11
S44834-6	62
S44835-4	906
S44837-0	434
S44838-9	374
S44841-9	141
S44842-7	254
S44843-5	1082
S44845-1	360
S44848-6	326
S44849-4	886
S44852-4	920
S44856-7	341
S44857-5	464
S44860-5	382
S44861-3	13
S44862-1	1125
S44864-8	756
S44865-6	227
S44866-4	307
S44870-2	1
S44872-9	916
S44874-5	63
S44878-8	316
S44879-6	923
S44882-6	125
S44886-9	323
S44887-7	192
S44891-5	924
S44892-3	348
S44895-8	370
S44897-4	206
S44899-0	263
S44903-2	888
S44905-9	1005
S44906-7	246

Product	Page	Product	Page	Product	Page	Product	Page	Product	Page	Product	Page	Product	Page
S44907-5	1084	S45019-7	846	S45133-9	661	S45296-3	244	S45475-3	876	S45637-3	170	S45864-3	337
S44908-3	136	S45020-0	786	S45134-7	661	S45297-1	924	S45479-6	313	S45644-6	1093	S45865-1	1113
S44910-5	1002	S45021-9	342	S45135-5	119	S45299-8	65	S45481-8	12	S45645-4	356	S45867-8	931
S44912-1	247	S45022-7	67	S45136-3	468	S45300-5	20	S45482-6	205	S45648-9	48	S45869-4	706
S44914-8	133	S45023-5	1124	S45137-1	471	S45302-1	887	S45483-4	198	S45649-7	910	S45870-8	327
S44916-4	978	S45024-3	870	S45140-1	487	S45304-8	133	S45485-0	207	S45653-5	9	S45871-6	35
S44917-2	246	S45025-1	994	S45142-8	422	S45308-0	472	S45489-3	433	S45654-3	10	S45873-2	127
S44918-0	884	S45027-8	717	S45145-2	977	S45311-0	944	S45490-7	66	S45658-6	1	S45875-9	9
S44919-9	319	S45030-8	654	S45146-0	339	S45312-9	1083	S45491-5	205	S45660-8	13	S45876-7	193
S44921-0	84	S45032-4	230	S45151-7	885	S45313-7	860	S45493-1	221	S45661-6	311	S45877-5	159
S44922-9	71	S45033-2	1085	S45152-5	884	S45314-5	817	S45495-8	58	S45663-2	389	S45878-3	192
S44923-7	78	S45035-9	333	S45153-3	884	S45315-3	985	S45499-0	131	S45665-9	982	S45881-3	788
S44924-5	131	S45036-7	161	S45154-1	118	S45321-8	112	S45500-8	920	S45666-7	896	S45885-6	23
S44925-3	5	S45037-5	850	S45156-8	117	S45322-6	1083	S45501-6	920	S45671-3	18	S45886-4	185
S44926-1	257	S45038-3	828	S45157-6	422	S45324-2	283	S45502-4	120	S45674-8	301	S45887-2	206
S44929-6	314	S45039-1	786	S45158-4	120	S45326-9	753	S45503-2	120	S45676-4	206	S45891-0	3
S44931-8	172	S45041-3	756	S45159-2	921	S45327-7	1085	S45504-0	119	S45678-0	32	S45892-9	245
S44932-6	164	S45042-1	753	S45162-2	923	S45328-5	902	S45506-7	349	S45680-2	138	S45894-5	4
S44933-4	1142	S45044-8	756	S45163-0	422	S45329-3	206	S45507-5	119	S45682-9	119	S45896-1	210
S44934-2	343	S45045-6	753	S45164-9	161	S45330-7	698	S45509-1	234	S45683-7	207	S45904-6	62
S44936-9	696	S45046-4	328	S45165-7	920	S45339-0	996	S45510-5	170	S45685-3	833	S45906-2	1029
S44937-7	293	S45047-2	332	S45166-5	161	S45342-0	1093	S45511-3	1126	S45693-4	310	S45907-0	922
S44938-5	338	S45048-0	999	S45167-3	161	S45345-5	32	S45512-1	8	S45694-2	155	S45908-9	335
S44940-7	284	S45049-9	48	S45168-1	738	S45346-3	901	S45515-6	60	S45696-9	224	S45909-7	1009
S44941-5	192	S45050-2	785	S45170-3	131	S45347-1	1	S45517-2	37	S45698-5	164	S45910-0	308
S44942-3	43	S45051-0	280	S45171-1	920	S45352-8	271	S45520-2	295	S45700-0	345	S45917-8	64
S44945-8	117	S45053-7	290	S45173-8	131	S45353-6	391	S45521-0	286	S45701-9	1009	S45918-6	1013
S44950-4	385	S45054-5	577	S45174-6	462	S45355-2	119	S45522-9	773	S45702-7	961	S45924-0	1005
S44953-9	35	S45055-3	331	S45177-0	117	S45362-5	209	S45523-7	743	S45704-3	318	S45925-9	1127
S44957-1	1129	S45056-1	172	S45179-7	349	S45372-2	388	S45524-5	773	S45710-8	31	S45926-7	367
S44959-8	1128	S45058-8	657	S45181-9	381	S45374-9	1082	S45526-1	38	S45715-9	163	S45927-5	1085
S44960-1	100	S45059-6	338	S45182-7	922	S45375-7	71	S45528-8	250	S45717-5	866	S45928-3	996
S44964-4	905	S45063-4	1050	S45183-5	659	S45376-5	775	S45529-6	198	S45727-2	2	S45931-3	1140
S44971-7	49	S45064-2	943	S45185-1	119	S45377-3	993	S45531-8	221	S45733-7	375	S45935-6	1004
S44972-5	300	S45065-0	973	S45187-8	866	S45380-3	190	S45532-6	354	S45734-5	87	S45936-4	104
S44973-3	924	S45067-7	250	S45188-6	422	S45384-6	174	S45533-4	975	S45739-6	98	S45937-2	1009
S44974-1	922	S45069-3	11	S45189-4	487	S45385-4	1010	S45535-0	93	S45743-4	1020	S45939-9	756
S44976-8	169	S45070-7	24	S45190-8	422	S45390-0	751	S45536-9	234	S45748-5	48	S45940-2	87
S44977-6	150	S45071-5	920	S45192-4	119	S45400-1	767	S45539-3	304	S45750-7	96	S45941-0	262
S44978-4	126	S45072-3	57	S45193-2	725	S45409-5	247	S45540-7	1110	S45752-3	876	S45955-0	41
S44979-2	101	S45075-8	58	S45203-3	850	S45410-9	188	S45541-5	153	S45757-4	262	S45958-5	345
S44980-6	253	S45076-6	37	S45207-6	155	S45423-0	126	S45546-6	31	S45758-2	39	S45960-7	316
S44981-4	151	S45078-2	1125	S45208-4	71	S45428-1	145	S45550-4	13	S45760-4	115	S45970-4	154
S44982-2	943	S45079-0	841	S45211-4	191	S45430-3	279	S45551-2	3	S45764-7	220	S45971-2	371
S44983-0	170	S45080-4	881	S45212-2	45	S45431-1	1079	S45555-5	174	S45769-8	789	S45974-7	22
S44984-9	314	S45081-2	411	S45214-9	864	S45433-8	978	S45556-3	12	S45772-8	75	S45975-5	369
S44985-7	186	S45082-0	921	S45223-8	246	S45434-6	959	S45560-1	991	S45774-4	292	S45977-1	206
S44986-5	974	S45083-9	117	S45225-4	238	S45436-2	196	S45564-4	162	S45776-0	709	S45983-6	79
S44987-3	854	S45084-7	475	S45227-0	932	S45439-7	32	S45565-2	920	S45777-9	812	S45985-2	1077
S44988-1	754	S45086-3	718	S45230-0	349	S45441-9	427	S45571-7	710	S45778-7	354	S45991-7	793
S44990-3	1002	S45090-1	718	S45231-9	891	S45444-3	201	S45576-8	62	S45781-7	221	S45992-5	263
S44991-1	203	S45093-6	1126	S45232-7	43	S45447-8	98	S45577-6	236	S45782-5	208	S45994-1	288
S44994-6	937	S45095-2	970	S45233-5	761	S45448-6	717	S45582-2	336	S45783-3	14	S46003-6	206
S44996-2	871	S45101-0	923	S45235-1	170	S45449-4	896	S45583-0	184	S45788-4	31	S46006-0	1083
S44997-0	290	S45102-9	1060	S45240-8	706	S45450-8	970	S45584-9	334	S45792-2	60	S46011-7	136
S44998-9	257	S45111-8	118	S45248-3	152	S45451-6	238	S45586-5	51	S45796-5	215	S46012-5	901
S44999-7	121	S45112-6	487	S45255-6	1083	S45452-4	882	S45591-1	791	S45805-8	1010	S46013-3	60
S45000-6	769	S45113-4	118	S45257-2	4	S45453-2	260	S45596-2	235	S45810-4	982	S46014-1	920
S45001-4	196	S45115-0	119	S45261-0	20	S45458-3	1010	S45598-9	102	S45812-0	920	S46016-8	920
S45002-2	785	S45116-9	118	S45264-5	757	S45459-1	35	S45600-4	1	S45815-5	936	S46017-6	923
S45003-0	579	S45119-3	120	S45266-1	3	S45460-5	974	S45603-9	1080	S45819-8	769	S46018-4	105
S45004-9	286	S45120-7	120	S45269-6	339	S45461-3	974	S45604-7	1095	S45827-9	221	S46020-6	663
S45005-7	5	S45122-3	118	S45273-4	412	S45464-8	110	S45607-1	987	S45830-9	337	S46021-4	15
S45006-5	150	S45123-1	883	S45275-0	412	S45465-6	36	S45614-4	920	S45832-5	1124	S46022-2	237
S45007-3	720	S45125-8	119	S45276-9	486	S45466-4	787	S45622-5	491	S45836-8	124	S46023-0	1140
S45008-1	310	S45126-6	489	S45277-7	422	S45467-2	912	S45627-6	955	S45842-2	146	S46024-9	910
S45011-1	920	S45127-4	489	S45278-5	701	S45468-0	71	S45630-6	3	S45844-9	490	S46025-7	231
S45013-8	911	S45128-2	488	S45279-3	119	S45469-9	9	S45631-4	725	S45850-3	931	S46030-3	12
S45014-6	286	S45129-0	496	S45280-7	120	S45470-2	691	S45632-2	900	S45853-8	73	S46031-1	262
S45015-4	867	S45130-4	886	S45291-2	906	S45472-9	199	S45633-0	973	S45854-6	192	S46033-8	920
S45017-0	235	S45131-2	105	S45292-0	696	S45473-7	775	S45634-9	45	S45856-2	215	S46037-0	66
S45018-9	218	S45132-0	656	S45294-7	383	S45474-5	781	S45636-5	1014	S45862-7	191	S46039-7	10

Product	Page
S46045-1	69
S46048-6	1087
S46049-4	136
S46051-6	136
S46052-4	65
S46053-2	725
S46055-9	1140
S46056-7	1001
S46057-5	361
S46065-6	757
S46066-4	1095
S46069-9	245
S46073-7	202
S46074-5	1140
S46076-1	201
S46078-8	133
S46081-8	309
S46082-6	476
S46085-0	334
S46087-7	10
S46089-3	1016
S46095-8	14
S46100-8	270
S46106-7	879
S46107-5	1140
S46108-3	1140
S46114-8	1137
S46116-4	53
S46117-2	25
S46118-0	369
S46120-2	66
S46121-0	169
S46123-7	942
S46124-5	37
S46128-8	135
S46131-8	903
S46132-6	934
S46134-2	804
S46135-0	50
S46136-9	260
S46137-7	389
S46140-7	1002
S46145-8	209
S46146-6	801
S46147-4	365
S46151-2	920
S46153-9	313
S46156-3	779
S46163-6	289
S46166-0	794
S46170-9	111
S46174-1	817
S46176-8	1140
S46177-6	910
S46186-5	758
S46187-3	1094
S46191-1	314
S46194-6	248
S46201-2	3
S46210-1	131
S46213-6	314
S46217-9	977
S46218-7	320
S46219-5	338
S46220-9	70
S46222-5	1094
S46223-3	24
S46224-1	337
S46228-4	290
S46232-2	866
S46234-9	89
S46235-7	169
S46236-5	1015
S46240-3	929
S46244-6	242
S46247-0	218
S46249-7	423
S46250-0	1093
S46252-7	119
S46253-5	96
S46255-1	924
S46257-8	922
S46259-4	487
S46261-6	123
S46262-4	125
S46263-2	351
S46264-0	97
S46265-9	50
S46266-7	70
S46267-5	92
S46268-3	40
S46269-1	1140
S46275-6	38
S46276-4	24
S46279-9	22
S46280-2	13
S46281-0	845
S46283-7	1142
S46285-3	15
S46286-1	196
S46289-6	162
S46292-6	265
S46293-4	264
S46294-2	966
S46295-0	799
S46296-9	348
S46297-7	3
S46298-5	743
S46300-0	176
S46301-9	1002
S46302-7	311
S46303-5	134
S46304-3	1100
S46310-8	40
S46311-6	14
S46313-2	869
S50001-1	296
S50003-8	144
S50004-6	369
S50005-4	144
S50008-9	734
S50009-7	377
S50011-9	811
S50013-5	799
S50014-3	114
S50015-1	914
S50018-6	1079
S50019-4	718
S50020-8	814
S50021-6	819
S50022-4	798
S50024-0	442
S50025-9	470
S50026-7	809
S50027-5	431
S50028-3	1069
S50030-5	739
S50032-1	276
S50034-8	21
S50035-6	215
S50037-2	766
S50039-9	797
S50040-2	1056
S50041-0	372
S50043-7	485
S50046-1	757
S50049-6	140
S50051-8	330
S50052-6	424
S50053-4	137
S50055-0	13
S50056-9	467
S50057-7	749
S50058-5	749
S50059-3	273
S50060-7	984
S50062-3	749
S50065-8	274
S50067-4	813
S50068-2	439
S50069-0	339
S50070-4	326
S50071-2	326
S50072-0	326
S50073-9	338
S50074-7	338
S50075-5	326
S50076-3	338
S50077-1	319
S50079-8	326
S50082-8	326
S50083-6	319
S50084-4	319
S50085-2	339
S50087-9	701
S50088-7	702
S50090-9	702
S50091-7	432
S50092-5	834
S50093-3	709
S50094-1	861
S50099-2	938
S50102-6	725
S50104-2	724
S50106-9	391
S50109-3	914
S50110-7	486
S50111-5	788
S50112-3	279
S50115-8	913
S50116-6	336
S50118-2	341
S50122-0	959
S50123-9	827
S50124-7	739
S50126-3	739
S50130-1	796
S50133-6	322
S50134-4	749
S50135-2	1034
S50141-7	792
S50142-5	801
S50144-1	956
S50146-8	817
S50149-2	1092
S50151-4	989
S50153-0	985
S50156-5	1014
S50157-3	160
S50158-1	160
S50160-3	160
S50161-1	836
S50163-8	835
S50164-6	160
S50168-9	101
S50170-0	346
S50174-3	1054
S50177-8	1053
S50179-4	1075
S50180-8	764
S50182-4	742
S50183-2	502
S50184-0	503
S50185-9	503
S50186-7	764
S50188-3	462
S50189-1	87
S50190-5	872
S50191-3	812
S50194-8	813
S50195-6	329
S50196-4	25
S50198-0	741
S50200-6	443
S50201-4	798
S50202-2	1088
S50203-0	181
S50204-9	373
S50205-7	814
S50207-3	746
S50210-3	701
S50211-1	436
S50213-8	861
S50214-6	742
S50215-4	703
S50216-2	701
S50218-9	703
S50219-7	915
S50221-9	827
S50223-5	436
S50224-3	710
S50225-1	736
S50228-6	746
S50229-4	1092
S50232-4	254
S50233-2	861
S50235-9	462
S50236-7	355
S50237-5	461
S50238-3	427
S50239-1	910
S50240-5	997
S50241-3	997
S50242-1	997
S50245-6	547
S50246-4	432
S50248-0	22
S50249-9	82
S50251-0	333
S50252-9	208
S50255-3	128
S50256-1	21
S50258-8	22
S50261-8	19
S50262-6	113
S50266-9	14
S50268-5	20
S50269-3	18
S50270-7	20
S50271-5	41
S50272-3	41
S50276-6	248
S50277-4	82
S50278-2	113
S50279-0	20
S50281-2	20
S50282-0	220
S50283-9	220
S50284-7	194
S50286-3	571
S50287-1	946
S50290-1	34
S50292-8	17
S50293-6	17
S50296-0	37
S50297-9	504
S50298-7	20
S50299-5	17
S50300-2	20
S50301-0	707
S50302-9	333
S50305-3	83
S50306-1	83
S50309-6	333
S50313-4	333
S50315-0	333
S50317-7	35
S50318-5	273
S50319-3	273
S50320-7	1015
S50321-5	272
S50322-3	438
S50323-1	261
S50325-8	272
S50326-6	438
S50327-4	379
S50328-2	270
S50329-0	261
S50330-4	1051
S50331-2	952
S50332-0	266
S50333-9	229
S50334-7	261
S50335-5	272
S50336-3	744
S50337-1	499
S50339-8	502
S50340-1	999
S50342-8	497
S50344-4	1054
S50345-2	1079
S50346-0	503
S50349-5	958
S50350-9	69
S50352-5	255
S50353-3	958
S50357-6	44
S50359-2	1129
S50360-6	44
S50362-2	43
S50363-0	43
S50364-9	43
S50367-3	326
S50368-1	326
S50371-1	339
S50373-8	326
S50375-4	1059
S50377-0	42
S50380-0	43
S50381-9	43
S50382-7	43
S50383-5	43
S50389-4	42
S50391-6	244
S50392-4	747
S50393-2	993
S50397-5	933
S50399-1	224
S50401-7	486
S50402-5	229
S50403-3	990
S50404-1	486
S50406-8	879
S50408-4	490
S50411-4	264
S50412-2	485
S50414-9	220
S50415-7	437
S50416-5	350
S50420-3	329
S50421-1	416
S50424-6	220
S50426-2	957
S50427-0	486
S50428-9	34
S50429-7	88
S50431-9	774
S50435-1	210
S50438-6	132
S50439-4	873
S50440-8	743
S50441-6	749
S50442-4	248
S50443-2	710
S50445-9	74
S50446-7	938
S50447-5	261
S50448-3	229
S50449-1	957
S50450-5	301
S50451-3	984
S50454-8	157
S50455-6	261
S50456-4	937
S50458-0	301
S50459-9	290
S50462-9	290
S50463-7	547
S50466-1	203
S50469-6	48
S50471-8	1
S50474-2	1
S50475-0	48
S50476-9	48
S50478-5	10
S50479-3	11
S50482-3	11
S50487-4	48
S50488-2	3
S50491-2	11
S50493-9	14
S50494-7	11
S50495-5	9
S50496-3	12
S50499-8	9
S50500-5	9
S50501-3	11
S50502-1	10
S50504-8	11
S50505-6	788
S50507-2	780
S50508-0	800
S50510-2	802
S50511-0	380
S50512-9	816
S50513-7	786
S50514-5	963
S50515-3	790
S50518-8	772
S50519-6	1085
S50521-8	1031
S50522-6	1035
S50523-4	1035
S50524-2	953
S50525-0	292
S50526-9	275
S50528-5	944
S50529-3	277
S50533-1	956
S50535-8	720
S50536-6	218
S50537-4	254
S50539-0	363
S50540-4	1009
S50542-0	428
S50543-9	807
S50544-7	233
S50545-5	363
S50546-3	913
S50547-1	816
S50549-8	984
S50550-1	829
S50552-8	745
S50553-6	956
S50554-4	913
S50555-2	1088
S50557-9	232
S50558-7	955
S50559-5	793
S50560-9	670
S50562-5	958
S50564-1	933
S50566-8	957
S50567-6	957
S50568-4	958
S50569-2	937
S50570-6	937
S50571-4	836
S50573-0	738
S50576-5	740
S50578-1	666
S50581-1	1035
S50583-8	742
S50585-4	816
S50586-2	802
S50587-0	815
S50589-7	28
S50593-5	121
S50595-1	716
S50597-8	720
S50598-6	716
S50602-8	475
S50603-6	475
S50605-2	720
S50607-9	720
S50608-7	720
S50610-9	1068
S50611-7	1073
S50613-3	1072
S50614-1	1075
S50616-8	1075
S50617-6	1075
S50618-4	1075
S50620-6	503
S50621-4	503
S50622-2	1074
S50623-0	503
S50624-9	1060
S50625-7	1071
S50626-5	1077
S50627-3	502
S50628-1	1062
S50630-3	1051
S50631-1	1051

Product	Page
S50633-8	501
S50634-6	440
S50635-4	820
S50636-2	913
S50637-0	914
S50640-0	1062
S50641-9	279
S50642-7	220
S50645-1	146
S50647-8	295
S50648-6	295
S50649-4	943
S50650-8	38
S50652-4	1074
S50653-2	867
S50654-0	1074
S50656-7	501
S50657-5	697
S50658-3	330
S50659-1	496
S50660-5	1072
S50661-3	496
S50664-8	1068
S50665-6	1056
S50666-4	1051
S50667-2	1051
S50668-0	1037
S50670-2	739
S50671-0	330
S50672-9	734
S50673-7	740
S50674-5	734
S50675-3	742
S50676-1	742
S50678-8	734
S50679-6	740
S50681-8	742
S50682-6	734
S50684-2	742
S50685-0	734
S50687-7	742
S50688-5	861
S50689-3	905
S50690-7	339
S50691-5	341
S50692-3	774
S50696-6	305
S50697-4	1038
S50698-2	1037
S50699-0	1037
S50700-8	1033
S50701-6	1038
S50702-4	1038
S50703-2	1039
S50704-0	1038
S50705-9	485
S50707-5	485
S50708-3	932
S50711-3	1038
S50712-1	1039
S50714-8	1038
S50715-6	1038
S50716-4	1032
S50717-2	1038
S50718-0	1035
S50719-9	1038
S50720-2	1039
S50721-0	486
S50723-7	917
S50724-5	903
S50725-3	369
S50726-1	75
S50728-8	834
S50731-8	740
S50732-6	816
S50733-4	432
S50734-2	372
S50735-0	702
S50736-9	974
S50737-7	25
S50741-5	440
S50742-3	994
S50745-8	1090
S50762-8	1062
S50763-6	1034
S50764-4	1061
S50765-2	1051
S50766-0	1061
S50767-9	1057
S50768-7	1061
S50769-5	1061
S50770-9	271
S50771-7	937
S50772-5	954
S50774-1	14
S50777-6	907
S50780-6	795
S50782-2	818
S50783-0	217
S50784-9	995
S50785-7	133
S50790-3	223
S50793-8	790
S50794-6	791
S50795-4	222
S50797-0	223
S50798-9	215
S50799-7	269
S50800-4	223
S50803-9	790
S50805-5	790
S50806-3	790
S50809-8	223
S50814-4	483
S50815-2	268
S50816-0	791
S50818-7	508
S50820-9	791
S50822-5	223
S50823-3	224
S50824-1	790
S50826-8	791
S50827-6	223
S50830-6	224
S50836-5	268
S50837-3	507
S50843-8	507
S50847-0	787
S50851-9	554
S50857-8	223
S50859-4	791
S50860-8	790
S50861-6	222
S50862-4	790
S50863-2	176
S50864-0	222
S50865-9	176
S50866-7	790
S50867-5	222
S50868-3	223
S50869-1	481
S50870-5	790
S50871-3	223
S50872-1	554
S50875-6	259
S50878-0	4
S50883-7	789
S50891-8	602
S50892-6	265
S50894-2	33
S50899-3	850
S50900-0	5
S50904-3	381
S50907-8	33
S50908-6	88
S50909-4	33
S50910-8	33
S50914-0	279
S50916-7	278
S50919-1	1015
S50921-3	73
S50927-2	236
S50928-0	236
S50930-2	16
S50931-0	111
S50933-7	800
S50934-5	360
S50935-3	362
S50936-1	30
S50938-8	31
S50939-6	813
S50941-8	362
S50942-6	74
S50943-4	204
S50945-0	76
S50947-7	364
S50948-5	993
S50949-3	204
S50950-7	76
S50952-3	1033
S50953-1	301
S50955-8	303
S50956-6	303
S50957-4	176
S50958-2	28
S50959-0	302
S50963-9	303
S50964-7	1076
S50965-5	692
S50967-1	976
S50969-8	736
S50975-2	772
S50976-0	499
S50977-9	1071
S50979-5	1074
S50981-7	763
S50982-5	903
S50983-3	707
S50986-8	497
S50987-6	1051
S50989-2	290
S50994-9	748
S50995-7	819
S50996-5	442
S50997-3	1090
S51000-9	736
S51001-7	816
S51004-1	822
S51008-4	738
S51011-4	815
S51012-2	746
S51014-9	699
S51015-7	234
S51016-5	470
S51017-3	140
S51018-1	110
S51020-3	814
S51024-6	155
S51025-4	812
S51027-0	918
S51028-9	744
S51029-7	413
S51030-0	155
S51033-5	792
S51034-3	80
S51037-8	791
S51038-6	814
S51039-4	764
S51041-6	742
S51042-4	331
S51044-0	124
S51047-5	742
S51051-3	814
S51054-8	822
S51055-6	138
S51056-4	834
S51061-0	813
S51062-9	736
S51063-7	742
S51064-5	742
S51072-6	236
S51074-2	363
S51075-0	94
S51076-9	828
S51077-7	38
S51078-5	944
S51079-3	1088
S51083-1	6
S51085-8	906
S51086-6	1053
S51087-4	1069
S51088-2	1071
S51089-0	1069
S51090-4	1071
S51091-2	1069
S51092-0	1059
S51093-9	1059
S51094-7	1059
S51095-5	306
S51096-3	1059
S51097-1	1059
S51099-8	1059
S51100-5	1039
S51101-3	1059
S51102-1	302
S51104-8	1059
S51105-6	1059
S51106-4	1053
S51107-2	305
S51108-0	264
S51109-9	434
S51110-2	225
S51112-9	918
S51113-7	56
S51114-5	86
S51116-1	56
S51119-6	104
S51123-4	108
S51124-2	819
S51126-9	248
S51130-7	192
S51131-5	324
S51132-3	1060
S51133-1	385
S51135-8	216
S51136-6	444
S51142-0	150
S51146-3	1090
S51147-1	380
S51149-8	251
S51150-1	41
S51155-2	905
S51158-7	768
S51161-7	371
S51166-8	33
S51170-6	86
S51171-4	86
S51173-0	33
S51174-9	162
S51176-5	221
S51178-1	17
S51180-3	191
S51181-1	155
S51183-8	742
S51186-2	745
S51187-0	30
S51189-7	33
S51191-9	34
S51192-7	30
S51193-5	33
S51194-3	155
S51198-6	1061
S51204-4	1053
S51206-0	1061
S51207-9	1069
S51208-7	908
S51209-5	1056
S51210-9	499
S51211-7	382
S51214-1	140
S51216-8	117
S51217-6	142
S51218-4	791
S51219-2	982
S51220-6	335
S51221-4	991
S51222-2	820
S51223-0	988
S51224-9	983
S51225-7	993
S51226-5	1124
S51227-3	359
S51230-3	94
S51231-1	794
S51233-8	467
S51234-6	736
S51235-4	482
S51236-2	742
S51237-0	179
S51238-9	987
S51239-7	746
S51240-0	743
S51241-9	953
S51243-5	958
S51244-3	223
S51245-1	990
S51248-6	856
S51249-4	169
S51251-6	742
S51252-4	738
S51253-2	990
S51254-0	990
S51255-9	956
S51256-7	989
S51257-5	1038
S51258-3	1076
S51259-1	1054
S51261-3	1129
S51262-1	782
S51264-8	743
S51266-4	419
S51267-2	1042
S51268-0	503
S51269-9	736
S51270-2	707
S51272-9	746
S51274-5	693
S51275-3	816
S51276-1	733
S51278-8	364
S51279-6	742
S51282-6	364
S51285-0	1124
S51286-9	31
S51287-7	31
S51288-5	31
S51290-7	1124
S51291-5	1124
S51292-3	1124
S51293-1	232
S51295-8	1061
S51296-6	1125
S51297-4	1124
S51298-2	1124
S51299-0	1124
S51300-8	1124
S51301-6	1125
S51302-4	1125
S51303-2	1125
S51304-0	1124
S51305-9	1035
S51306-7	1035
S51307-5	1033
S51308-3	1076
S51309-1	1061
S51310-5	1061
S51311-3	1054
S51312-1	224
S51314-8	275
S51315-6	1061
S51316-4	813
S51317-2	86
S51319-9	995
S51322-9	224
S51324-5	235
S51325-3	997
S51328-8	329
S51329-6	737
S51331-8	779
S51332-6	789
S51335-0	1125
S51336-9	744
S51340-7	737
S51342-3	182
S51343-1	822
S51345-8	268
S51346-6	555
S51348-2	444
S51350-4	956
S51351-2	221
S51353-9	819
S51354-7	814
S51355-5	444
S51356-3	813
S51359-8	819
S51362-8	216
S51363-6	742
S51364-4	816
S51365-2	1090
S51367-9	221
S51368-7	472
S51371-7	743
S51372-5	746
S51373-3	816
S51374-1	482
S51376-8	229
S51377-6	738
S51381-4	743
S51382-2	915
S51383-0	707
S51384-9	870
S51386-5	335
S51388-1	140
S51390-3	70
S51391-1	802
S51395-4	796
S51396-2	960
S51399-7	276
S51400-4	829
S51401-2	425
S51402-0	742
S51404-7	425
S51406-3	829
S51407-1	223
S51409-8	223
S51410-1	223
S51412-8	272
S51413-6	225
S51414-4	1061
S51415-2	224
S51416-0	224
S51417-9	232
S51418-7	226
S51419-5	287
S51420-9	1061
S51421-7	1061
S51422-5	1061
S51423-3	224
S51424-1	232
S51428-4	182
S51429-2	279
S51431-4	94
S51432-2	364
S51433-0	106
S51434-9	366
S51436-5	363
S51438-1	23
S51441-1	309
S51445-4	498
S51446-2	1075
S51447-0	1071
S51448-9	966
S51449-7	498
S51450-0	1075
S51451-9	1051
S51452-7	1037
S51453-5	1072
S51454-3	1042
S51455-1	1032
S51458-6	1068
S51460-8	1051
S51461-6	766
S51462-4	266
S51463-2	267
S51464-0	266
S51465-9	303
S51466-7	303
S51467-5	303
S51468-3	1044
S51470-5	224
S51471-3	1037
S51475-6	772
S51476-4	444
S51477-2	467

Product	Page
S51478-0	465
S51479-9	191
S51480-2	358
S51481-0	290
S51483-7	467
S51484-5	962
S51485-3	472
S51493-4	330
S51494-2	359
S51495-0	359
S51496-9	360
S51497-7	330
S51500-0	411
S51501-9	411
S51502-7	411
S51503-5	411
S51507-8	58
S51508-6	58
S51509-4	6
S51511-6	125
S51512-4	823
S51514-0	143
S51515-9	485
S51516-7	815
S51517-5	746
S51518-3	814
S51520-5	443
S51521-3	815
S51525-6	471
S51526-4	995
S51528-0	993
S51531-0	743
S51533-7	796
S51538-8	739
S51539-6	985
S51542-6	744
S51543-4	478
S51544-2	111
S51545-0	469
S51547-7	802
S51548-5	955
S51550-7	1077
S51551-5	485
S51555-8	938
S51557-4	472
S51558-2	739
S51559-0	320
S51560-4	271
S51561-2	828
S51562-0	803
S51563-9	990
S51565-5	439
S51567-1	259
S51575-2	948
S51576-0	329
S51579-5	918
S51580-9	792
S51581-7	1062
S51582-5	802
S51583-3	941
S51584-1	972
S51586-8	804
S51588-4	828
S51589-2	1044
S51590-6	1077
S51593-0	254
S51595-7	956
S51596-5	226
S51600-7	223
S51601-5	223
S51602-3	226
S51603-1	223
S51606-6	957
S51607-4	226
S51608-2	1059
S51609-0	1039
S51610-4	1062
S51611-2	36
S51612-0	146
S51614-7	807
S51615-5	360
S51616-3	797
S51619-8	61
S51620-1	293
S51625-2	370
S51628-7	363
S51629-5	467
S51631-7	163
S51634-1	1109
S51636-8	45
S51638-4	1109
S51641-4	29
S51643-0	30
S51644-9	789
S51645-7	156
S51646-5	1061
S51647-3	1071
S51651-1	28
S51654-6	35
S51656-2	1070
S51659-7	145
S51660-0	418
S51661-9	1059
S51662-7	419
S51664-3	128
S51665-1	417
S51667-8	419
S51668-6	417
S51669-4	1037
S51670-8	1035
S51672-4	1059
S51673-2	1038
S51674-0	261
S51676-7	188
S51677-5	793
S51678-3	1109
S51680-5	364
S51684-8	801
S51685-6	817
S51686-4	268
S51687-2	268
S51688-0	755
S51689-9	364
S51691-0	370
S51692-9	444
S51694-5	141
S51696-1	769
S51698-8	870
S51700-3	904
S51701-1	273
S51703-8	904
S51704-6	130
S51706-2	1124
S51707-0	1059
S51708-9	990
S51709-7	305
S51711-9	485
S51712-7	1076
S51713-5	1035
S51714-3	1076
S51720-8	907
S51721-6	995
S51724-0	691
S51725-9	304
S51726-7	932
S51729-1	226
S51730-5	227
S51731-3	232
S51732-1	224
S51734-8	223
S51735-6	232
S51736-4	224
S51737-2	230
S51741-0	707
S51742-9	978
S51743-7	701
S51749-6	151
S51751-8	379
S51752-6	239
S51753-4	363
S51754-2	887
S51755-0	691
S51758-5	338
S51759-3	761
S51761-5	708
S51762-3	852
S51763-1	934
S51765-8	847
S51766-6	188
S51767-4	307
S51768-2	867
S51769-0	320
S51770-4	700
S51772-0	1053
S51773-9	1031
S51776-3	447
S51777-1	571
S51782-8	183
S51783-6	1037
S51784-4	1069
S51785-2	1069
S51787-9	1073
S51788-7	382
S51805-0	1048
S51806-9	227
S51807-7	225
S51808-5	71
S51809-3	227
S51810-7	75
S51811-5	267
S51817-4	351
S51820-4	955
S51821-2	1062
S51822-0	227
S51823-9	261
S51824-7	261
S51825-5	227
S51826-3	228
S51827-1	687
S51830-1	261
S51832-8	260
S51833-6	259
S51834-4	261
S51835-2	273
S51836-0	31
S51838-7	267
S51839-5	90
S51840-9	41
S51843-3	788
S51844-1	187
S51847-6	181
S51850-6	788
S51852-2	50
S51853-0	180
S51855-7	60
S51856-5	351
S51857-3	16
S51860-3	744
S51864-6	96
S51865-4	67
S51866-2	800
S51868-9	823
S51870-0	97
S51871-9	761
S51872-7	360
S51877-8	822
S51878-6	96
S51880-8	761
S51881-6	761
S51882-4	761
S51883-2	761
S51884-0	138
S51885-9	76
S51886-7	761
S51888-3	761
S51889-1	726
S51891-3	761
S51892-1	761
S51894-8	761
S51897-2	761
S51899-9	721
S51900-6	721
S51902-2	761
S51904-9	800
S51907-3	823
S51911-1	332
S51918-9	350
S51919-7	69
S51920-0	77
S51923-5	696
S51924-3	195
S51925-1	96
S51928-6	75
S51930-8	794
S51932-4	10
S51933-2	378
S51936-7	85
S51937-5	268
S51939-1	75
S51940-5	1070
S51942-1	5
S51946-4	430
S51947-2	75
S51948-0	75
S51951-0	716
S51954-5	827
S51955-3	84
S51959-6	801
S51962-6	959
S51964-2	182
S51965-0	193
S51967-7	182
S51969-3	6
S51976-6	5
S51977-4	5
S51978-2	76
S51981-2	53
S51982-0	17
S51984-7	747
S51985-5	94
S51987-1	671
S51997-9	31
S52002-0	279
S52003-9	230
S52004-7	284
S52006-3	489
S52007-1	74
S52009-8	65
S52010-1	69
S52012-8	1043
S52013-6	978
S52015-2	1053
S52016-0	1042
S52017-9	1056
S52019-5	1070
S52021-7	1042
S52022-5	1070
S52023-3	1062
S52024-1	907
S52026-8	258
S52029-2	441
S52030-6	246
S52032-2	391
S52034-9	431
S52044-6	89
S52048-9	261
S52049-7	967
S52053-5	81
S52054-3	476
S52057-8	42
S52058-6	245
S52059-4	1014
S52060-8	372
S52061-6	941
S52062-4	100
S52063-2	221
S52066-7	92
S52071-3	761
S52075-6	385
S52077-2	661
S52080-2	231
S52083-7	343
S52084-5	160
S52085-3	85
S52096-9	748
S52097-7	5
S52100-0	354
S52101-9	86
S52102-7	86
S52104-3	131
S52105-1	468
S52115-9	139
S52117-5	96
S52118-3	46
S52119-1	141
S52122-1	430
S52124-8	46
S52125-6	265
S52126-4	142
S52127-2	97
S52129-9	96
S52141-8	151
S52144-2	76
S52145-0	756
S52146-9	139
S52152-3	85
S52158-2	79
S52161-2	430
S52164-7	360
S52166-3	151
S52167-1	97
S52174-4	151
S52175-2	233
S52177-9	46
S52183-3	151
S52197-3	1070
S52202-3	1071
S52203-1	792
S52205-8	379
S52210-4	324
S52211-2	31
S52222-8	265
S52227-9	347
S52240-6	345
S52244-9	33
S52246-5	1138
S52247-3	351
S52256-2	1071
S52258-9	1074
S52259-7	303
S52261-9	1038
S52262-7	303
S52263-5	303
S52265-1	1074
S52267-8	1030
S52268-6	1061
S52269-4	1059
S52270-8	1044
S52274-0	29
S52306-2	46
S52321-6	939
S52326-7	341
S52327-5	483
S52328-3	358
S52331-3	747
S52334-8	750
S52335-6	1038
S52336-4	765
S52337-2	750
S52338-0	722
S52339-9	1032
S52340-2	1078
S52341-0	1074
S52342-9	306
S52344-5	1050
S52345-3	1077
S52346-1	1078
S52348-8	1054
S52349-6	1077
S52351-8	495
S52352-6	719
S52353-4	745
S52355-0	767
S52358-5	496
S52359-3	717
S52361-5	229
S52362-3	1078
S52363-1	229
S52365-8	1039
S52366-6	229
S52367-4	766
S52368-2	719
S52369-0	761
S52371-2	141
S52372-0	142
S52373-9	740
S52374-7	766
S52376-3	1004
S52377-1	1077
S52379-8	1079
S52380-1	1074
S52382-8	1078
S52383-6	229
S52384-4	1063
S52387-9	722
S52388-7	497
S52390-9	745
S52391-7	746
S52393-3	1071
S52394-1	1078
S52396-8	1050
S52397-6	1039
S52398-4	1074
S52399-2	1077
S52403-4	1013
S52405-0	53
S52409-3	336
S52411-5	1035
S52412-3	1078
S52413-1	1078
S52415-8	1077
S52416-6	302
S52417-4	734
S52419-0	1033
S52420-4	240
S52421-2	142
S52424-7	762
S52425-5	830
S52427-1	303
S52429-8	302
S52430-1	301
S52433-6	1061
S52434-4	385
S52435-2	385
S52436-0	1063
S52437-9	761
S52438-7	750
S52442-5	721
S52443-3	495
S52444-1	1077
S52446-8	302
S52447-6	300
S52448-4	306
S52449-2	300
S52450-6	303
S52451-4	495
S52452-2	496
S52454-9	722
S52457-3	1070
S52458-1	1036
S52461-1	1036
S52463-8	164
S52467-0	1032
S52468-9	1070
S52469-7	1042
S52470-0	1036
S52472-7	1069
S52473-5	1070
S52478-6	855
S52481-6	696
S52482-4	709
S52483-2	201
S52484-0	265
S52487-5	386
S52489-1	762
S52494-8	385
S52496-4	269
S52497-2	292
S52498-0	269
S52499-9	269
S52500-6	258
S52501-4	268
S52502-2	269
S52503-0	269
S52504-9	52
S52505-7	1075
S52506-5	761
S52507-3	764
S52508-1	879
S52511-1	723
S52513-8	766
S52514-6	137
S52515-4	746
S52525-1	382

Product Page

S52527-8........266
S52530-8........819
S52531-6........267
S52532-4........267
S52533-2........797
S52534-0........184
S52535-9........279
S52536-7........711
S52537-5........827
S52538-3........945
S52540-5........714
S52541-3........758
S52542-1........800
S52544-8........253
S52545-6........856
S52546-4........827
S52548-0........411
S52549-9........185
S52550-2........126
S52552-9........755
S52553-7........751
S52554-5........751
S52555-3........751
S52556-1........240
S52558-8........753
S52559-6........239
S52561-8........243
S52562-6........359
S52564-2........208
S52565-0........41
S52566-9........904
S52569-3........393
S52571-5........306
S52572-3........306
S52573-1........1031
S52575-8........306
S52576-6........306
S52577-4........301
S52578-2........430
S52579-0........306
S52580-4........1078
S52584-7........698
S52586-3........698
S52589-8........1044
S52594-4........1069
S52595-2........698
S52596-0........1141
S52597-9........704
S52598-7........495
S52599-5........762
S52604-5........502
S52605-3........762
S52606-1........1038
S52609-6........752
S52614-2........496
S52615-0........495
S52616-9........358
S52617-7........872
S52618-5........1013
S52623-1........183
S52626-6........690
S52627-4........267
S52629-0........755
S52630-4........858
S52631-2........272
S52632-0........267
S52633-9........320
S52634-7........818
S52635-5........361
S52636-3........756
S52637-1........258
S52639-8........269
S52640-1........258
S52642-8........269
S52644-4........258
S52645-2........369
S52654-1........413
S52661-4........430
S52664-9........761
S52665-7........46
S52666-5........46
S52671-1........148
S52673-8........355
S52678-9........761
S52685-1........430
S52687-8........83
S52691-6........792
S52700-9........81
S52704-1........498
S52706-8........748
S52707-6........746
S52708-4........384
S52709-2........721
S52710-6........743
S52711-4........383
S52712-2........384
S52714-9........717
S52715-7........386
S52716-5........762
S52717-3........1069
S52718-1........719
S52720-3........750
S52723-8........386
S52724-6........386
S52725-4........762
S52726-2........766
S52727-0........383
S52729-7........1071
S52730-0........138
S52731-9........719
S52732-7........142
S52733-5........965
S52734-3........385
S52735-1........385
S52737-8........183
S52738-6........181
S52740-8........24
S52743-2........326
S52744-0........15
S52745-9........1090
S52748-3........36
S52750-5........206
S52754-8........824
S52757-2........380
S52758-0........796
S52761-0........35
S52762-9........61
S52764-5........202
S52765-3........344
S52766-1........871
S52768-8........138
S52769-6........1091
S52775-0........579
S52776-9........757
S52777-7........156
S52778-5........156
S52779-3........247
S52781-5........247
S52782-3........216
S52783-1........216
S52785-8........792
S52786-6........718
S52789-0........721
S52790-4........718
S52792-0........719
S52794-7........472
S52796-3........932
S52799-8........89
S52802-1........361
S52804-8........375
S52805-6........80
S52806-4........80
S52807-2........74
S52808-0........362
S52809-9........18
S52810-2........60
S52811-0........362
S52812-9........207
S52814-5........356
S52815-3........80
S52816-1........362
S52818-8........20
S52819-6........365
S52821-8........182
S52822-6........80
S52825-0........78
S52826-9........362
S52827-7........94
S52828-5........80
S52829-3........80
S52830-7........80
S52832-3........361
S52836-6........60
S52837-4........91
S52838-2........60
S52839-0........78
S52840-4........80
S52843-9........2
S52844-7........56
S52846-3........361
S52847-1........381
S52849-8........362
S52850-1........380
S52852-8........362
S52853-6........73
S52854-4........362
S52855-2........73
S52858-7........208
S52860-9........436
S52862-5........56
S52863-3........671
S52864-1........23
S52867-6........150
S52868-4........105
S52878-1........393
S52898-6........427
S52909-5........757
S52911-7........383
S52912-5........385
S52916-8........383
S52917-6........383
S52919-2........295
S52922-2........246
S52925-7........237
S52927-3........287
S52928-1........355
S52933-8........393
S52943-5........892
S52944-3........279
S52947-8........102
S52948-6........393
S52949-4........392
S52951-6........392
S52954-0........390
S52955-9........392
S52956-7........391
S52959-1........472
S52960-5........393
S52964-8........751
S52965-6........1015
S52969-9........694
S52972-9........11
S52975-3........138
S52976-1........763
S52981-8........268
S52984-2........738
S52986-9........1042
S52987-7........139
S52989-3........765
S52990-7........765
S52992-3........953
S52993-1........765
S52996-6........743
S52997-4........385
S52998-2........743
S53002-6........762
S53004-2........1050
S53005-0........763
S53006-9........762
S53007-7........762
S53008-5........766
S53010-7........1006
S53012-3........1042
S53015-8........933
S53016-6........1042
S53017-4........1030
S53018-2........1078
S53019-0........1030
S53020-4........1042
S53021-2........1050
S53022-0........1078
S53024-7........199
S53025-5........195
S53029-8........191
S53030-1........144
S53032-8........81
S53034-4........220
S53038-7........833
S53041-7........428
S53042-5........367
S53044-1........1143
S53058-1........122
S53061-1........813
S53064-6........906
S53065-4........485
S53066-2........443
S53067-0........780
S53071-9........63
S53072-7........82
S53073-5........763
S53077-8........323
S53081-6........185
S53083-2........738
S53084-0........745
S53085-9........739
S53087-5........940
S53091-3........673
S53094-8........379
S53099-9........270
S53100-6........270
S53102-2........265
S53104-9........979
S53105-7........280
S53106-5........777
S53107-3........705
S53110-3........779
S53111-1........782
S53113-8........782
S53114-6........340
S53115-4........385
S53116-2........766
S53118-9........854
S53119-7........57
S53120-0........738
S53121-9........820
S53125-1........15
S53127-8........330
S53128-6........41
S53129-4........1000
S53130-8........1075
S53132-4........269
S53133-2........763
S53134-0........735
S53135-9........734
S53138-3........870
S53139-1........1014
S53140-5........1042
S53142-1........420
S53144-8........228
S53146-4........430
S53147-2........430
S53148-0........954
S53151-0........387
S53152-9........416
S53153-7........419
S53155-3........667
S53156-1........1089
S53159-6........228
S53168-5........360
S53169-3........764
S53171-5........764
S53172-3........762
S53173-1........782
S53175-8........719
S53176-6........1057
S53177-4........440
S53178-2........1079
S53180-4........762
S53181-2........764
S53182-0........440
S53183-9........766
S53184-7........762
S53186-3........764
S53187-1........766
S53189-8........762
S53192-8........1033
S53194-4........295
S53196-0........346
S53204-5........857
S53209-6........909
S53214-2........1074
S53215-0........767
S53217-7........765
S53218-5........1042
S53219-3........1039
S53222-3........1074
S53223-1........1073
S53227-4........1094
S53228-2........1042
S53229-0........1074
S53230-4........1074
S53231-2........1073
S53232-0........766
S53233-9........1037
S53234-7........765
S53235-5........767
S53236-3........762
S53237-1........476
S53240-1........473
S53243-6........719
S53244-4........719
S53245-2........721
S53246-0........473
S53247-9........432
S53251-7........1074
S53252-5........1074
S53253-3........216
S53257-6........466
S53258-4........762
S53260-6........762
S53261-4........476
S53262-2........372
S53263-0........298
S53264-9........329
S53266-5........1074
S53267-3........331
S53270-3........111
S53277-0........219
S53279-7........218
S53281-9........306
S53282-7........1020
S53283-5........1021
S53284-3........1020
S53285-1........430
S53287-8........1020
S53288-6........1077
S53289-4........229
S53290-8........88
S53295-9........797
S53300-9........861
S53302-5........192
S53303-3........360
S53306-8........1020
S53307-6........1020
S53309-2........1021
S53310-6........1021
S53311-4........1020
S53312-2........385
S53313-0........1021
S53314-9........430
S53315-7........266
S53316-5........1021
S53317-3........379
S53318-1........1020
S53320-3........943
S53323-8........943
S53324-6........951
S53325-4........90
S53326-2........944
S53327-0........266
S53329-7........1013
S53330-0........487
S53331-9........1060
S53332-7........1009
S53333-5........486
S53334-3........954
S53341-6........749
S53342-4........993
S53344-0........985
S53346-7........1013
S53349-1........945
S53352-1........1092
S53356-4........133
S53357-2........1093
S53358-0........1140
S53359-9........416
S53366-1........996
S53368-8........980
S53371-8........970
S53373-4........1042
S53374-2........1042
S53375-0........495
S53377-7........210
S53379-3........210
S53386-6........376
S53389-0........749
S53390-4........304
S53391-2........274
S53392-0........306
S53393-9........306
S53394-7........306
S53395-5........258
S53396-3........1074
S53397-1........307
S53399-8........1009
S53400-5........1013
S53401-3........1013
S53402-1........1013
S53404-8........372
S53408-0........155
S53409-9........443
S53410-2........793
S53414-5........789
S53416-1........757
S53418-8........789
S53422-6........761
S53425-0........755
S53426-9........794
S53427-7........800
S53429-3........158
S53430-7........273
S53433-1........1140
S53435-8........1140
S53438-2........1087
S53439-0........1093
S53441-2........1093
S53442-0........179
S53443-9........174
S53444-7........577
S53447-1........990
S53449-8........1140
S53450-1........1140
S53461-7........722
S53468-4........743
S53469-2........301
S53470-6........761
S53474-9........126
S53476-5........258
S53480-3........971
S53481-1........191
S53483-8........996
S53484-6........125
S53485-4........79
S53486-2........13
S53487-0........22
S53488-9........69
S53489-7........897
S53492-7........379
S53493-5........467
S53497-8........56
S53498-6........267
S53502-8........1045
S53525-7........1065
S53527-3........1065
S53528-1........1047
S53531-1........1047
S53533-8........1047
S53534-6........1044
S53535-4........1044
S53536-2........1044
S53538-9........1058
S53539-7........1044
S53540-0........1044
S53541-9........1044

Product	Page
S53542-7	1044
S53544-3	1044
S53549-4	423
S53550-8	423
S53551-6	752
S53552-4	183
S53553-2	39
S53554-0	427
S53556-7	85
S53557-5	145
S53558-3	288
S53559-1	248
S53560-5	88
S53561-3	358
S53562-1	39
S53566-4	248
S53568-0	248
S53571-0	176
S53572-9	245
S53573-7	245
S53574-5	244
S53575-3	293
S53576-1	248
S53579-6	104
S53581-8	184
S53585-0	108
S53586-9	108
S53589-3	116
S53592-3	94
S53596-6	1008
S53598-2	50
S53599-0	82
S53603-2	301
S53606-7	956
S53607-5	304
S53609-1	704
S53610-5	957
S53611-3	267
S53612-1	385
S53614-8	579
S53620-2	274
S53621-0	272
S53622-9	494
S53623-7	496
S53624-5	495
S53625-3	496
S53628-8	719
S53631-8	105
S53632-6	741
S53633-4	139
S53635-0	954
S53636-9	735
S53639-3	741
S53640-7	836
S53641-5	831
S53643-1	1127
S53645-8	767
S53648-2	1031
S53649-0	1070
S53650-4	1049
S53651-2	300
S53653-9	340
S53654-7	1026
S53655-5	1052
S53656-3	257
S53659-8	763
S53660-1	301
S53663-6	1026
S53664-4	1060
S53665-2	716
S53667-9	1022
S53668-7	1063
S53669-5	767
S53671-7	717
S53673-3	279
S53676-8	436
S53677-6	741
S53678-4	769
S53679-2	736
S53680-6	1062
S53682-2	832
S53683-0	830
S53685-7	738
S53686-5	741
S53687-3	736
S53688-1	734
S53697-0	104
S53701-2	313
S53703-9	741
S53704-7	741
S53705-5	1049
S53706-3	750
S53709-8	763
S53710-1	819
S53713-6	979
S53721-7	915
S53723-3	95
S53724-1	827
S53726-8	364
S53728-4	827
S53730-6	63
S53732-2	709
S53733-0	749
S53734-9	916
S53736-5	824
S53744-6	191
S53747-0	820
S53749-7	804
S53750-0	157
S53751-9	340
S53753-5	824
S53754-3	447
S53761-6	973
S53765-9	63
S53767-5	364
S53774-8	434
S53775-6	757
S53778-0	743
S53780-2	762
S53782-9	1050
S53783-7	719
S53788-8	738
S53791-8	1029
S53792-6	384
S53794-2	717
S53797-7	1013
S53798-5	741
S53799-3	1079
S53800-0	1039
S53801-9	1059
S53802-7	138
S53804-3	1039
S53805-1	287
S53807-8	1075
S53809-4	945
S53811-6	1042
S53812-4	975
S53814-0	1053
S53817-5	1053
S53818-3	1053
S53820-5	832
S53821-3	719
S53822-1	466
S53824-8	383
S53825-6	750
S53826-4	1072
S53827-2	469
S53828-0	762
S53829-9	383
S53830-2	750
S53831-0	476
S53832-9	879
S53834-5	769
S53839-6	769
S53844-2	769
S53846-9	768
S53850-7	769
S53852-3	202
S53853-1	202
S53855-8	200
S53859-0	202
S53860-4	210
S53861-2	203
S53862-0	202
S53863-9	202
S53864-7	202
S53865-5	203
S53866-3	411
S53872-8	303
S53873-6	306
S53874-4	425
S53875-2	227
S53876-0	387
S53878-7	1042
S53879-5	387
S53880-9	1078
S53881-7	239
S53882-5	957
S53883-3	430
S53884-1	430
S53886-8	384
S53893-0	677
S53895-7	306
S53896-5	303
S53897-3	306
S53898-1	303
S53900-7	704
S53902-3	719
S53906-6	1056
S53909-0	1070
S53910-4	743
S53911-2	939
S53912-0	750
S53913-9	476
S53914-7	274
S53916-3	303
S53917-1	384
S53919-8	277
S53922-8	718
S53924-4	250
S53925-2	386
S53927-9	368
S53929-5	358
S53931-7	126
S53933-3	743
S53936-8	953
S53937-6	358
S53939-2	84
S53940-6	136
S53946-5	1010
S53948-1	499
S53956-2	767
S53957-0	1008
S53958-9	274
S53959-7	1042
S53960-0	1074
S53961-9	753
S53962-7	720
S53963-5	719
S53964-3	301
S53967-8	288
S53968-6	766
S53969-4	475
S53970-8	731
S53971-6	743
S53972-4	387
S53974-0	1094
S53975-9	959
S53976-7	242
S53980-5	1030
S53982-1	359
S53984-8	835
S53987-2	760
S53989-9	1042
S53990-2	959
S54011-0	49
S54012-9	152
S54013-7	25
S54014-5	25
S54015-3	25
S54016-1	26
S54018-8	49
S54021-8	49
S54022-6	483
S54023-4	28
S54024-2	101
S54025-0	49
S54026-9	152
S54028-5	33
S54029-3	28
S54030-7	101
S54031-5	142
S54032-3	25
S54033-1	152
S54035-8	154
S54036-6	152
S54038-2	827
S54039-0	887
S54041-2	887
S54042-0	887
S54043-9	887
S54045-5	182
S54049-8	802
S54050-1	30
S54052-8	701
S54053-6	94
S54054-4	152
S54055-2	427
S54056-0	413
S54058-7	146
S54059-5	704
S54060-9	758
S54062-5	423
S54064-1	776
S54066-8	1005
S54067-6	85
S54069-2	86
S54070-6	19
S54071-4	19
S54075-7	125
S54076-5	1052
S54081-1	22
S54083-8	73
S54084-6	877
S54086-2	968
S54087-0	862
S54089-7	970
S54090-0	150
S54093-5	724
S54095-1	264
S54098-6	794
S54099-4	264
S54100-1	88
S54102-8	756
S54103-6	50
S54104-4	664
S54106-0	178
S54113-3	253
S54114-1	93
S54119-2	749
S54120-6	748
S54121-4	88
S54122-2	332
S54123-0	360
S54126-5	291
S54130-3	228
S54131-1	266
S54133-8	228
S54134-6	1020
S54135-4	750
S54137-0	156
S54139-7	146
S54140-0	82
S54141-9	418
S54142-7	815
S54144-3	750
S54148-6	387
S54150-8	177
S54151-6	367
S54152-4	879
S54155-9	1029
S54156-7	330
S54157-5	330
S54158-3	252
S54159-1	215
S54160-5	741
S54161-3	853
S54168-0	870
S54170-2	821
S54172-9	746
S54183-4	772
S54188-5	389
S54193-1	272
S54195-8	273
S54196-6	1038
S54197-4	273
S54198-2	273
S54199-0	277
S54200-8	717
S54201-6	274
S54202-4	247
S54207-5	704
S54208-3	704
S54209-1	246
S54215-6	704
S54216-4	499
S54222-9	1140
S54224-5	704
S54225-3	698
S54226-1	698
S54228-8	704
S54229-6	694
S54232-6	1059
S54234-2	382
S54236-9	906
S54237-7	906
S54238-5	382
S54239-3	971
S54241-5	1071
S54263-6	1078
S54264-4	717
S54265-2	749
S54266-0	715
S54267-9	738
S54268-7	247
S54270-9	939
S54271-7	741
S54273-3	747
S54274-1	738
S54277-6	736
S54278-4	736
S54279-2	464
S54280-6	476
S54282-2	981
S54283-0	462
S54284-9	292
S54285-7	502
S54288-1	887
S54291-1	305
S54293-8	329
S54294-6	180
S54297-0	981
S54301-2	1021
S54302-0	228
S54303-9	1021
S54304-7	1020
S54305-5	1077
S54306-3	854
S54310-1	235
S54316-0	830
S54321-7	832
S54324-1	831
S54327-6	830
S54329-2	836
S54330-6	160
S54331-4	160
S54333-0	836
S54334-9	158
S54335-7	831
S54336-5	831
S54343-8	832
S54344-6	1127
S54345-4	831
S54346-2	830
S54350-0	832
S54352-7	835
S54353-5	835
S54354-3	835
S54355-1	836
S54357-8	836
S54358-6	836
S54359-4	836
S54360-8	326
S54361-6	830
S54366-7	958
S54367-5	753
S54368-3	387
S54370-5	1020
S54371-3	677
S54375-6	562
S54376-4	962
S54377-2	228
S54378-0	49
S54385-3	49
S54389-6	9
S54392-6	49
S54393-4	49
S54398-5	49
S54404-3	49
S54407-8	49
S54408-6	49
S54409-4	49
S54410-8	49
S54411-6	49
S54412-4	49
S54418-3	50
S54420-5	9
S54421-3	9
S54424-8	755
S54433-7	802
S54435-3	909
S54439-6	1060
S54443-4	222
S54444-2	1079
S54446-9	94
S54448-5	994
S54449-3	908
S54450-7	994
S54458-2	858
S54459-0	261
S54461-2	81
S54466-3	887
S54467-1	887
S54472-8	158
S54474-4	180
S54475-2	192
S54478-7	800
S54479-5	800
S54480-9	331
S54486-8	334
S54488-4	353
S54494-9	853
S54495-7	138
S54498-1	269
S54500-7	1010
S54506-6	221
S54508-2	747
S54509-0	476
S54511-2	879
S54512-0	746
S54513-9	1050
S54514-7	1048
S54515-5	749
S54516-3	384
S54519-8	1049
S54520-1	1010
S54523-6	723
S54524-4	1010
S54525-2	722
S54527-9	1020
S54529-5	324
S54533-3	78
S54534-1	77
S54538-4	610
S54541-4	438
S54544-9	858
S54548-1	365
S54551-1	1011
S54557-0	421
S54562-7	964
S54564-3	711
S54568-6	1077
S54569-4	140
S54570-8	386
S54571-6	735
S54572-4	255
S54573-2	736
S54575-9	742
S54576-7	742
S54577-5	742
S54578-3	742
S54579-1	742
S54581-3	830
S54582-1	834

Product	Page
S54584-8	837
S54585-6	741
S54586-4	747
S54587-2	284
S54588-0	275
S54589-9	835
S54591-0	158
S54592-9	879
S54593-7	746
S54594-5	746
S54595-3	148
S54598-8	664
S54599-6	735
S54600-3	737
S54601-1	854
S54603-8	1128
S54605-4	741
S54606-2	1049
S54607-0	749
S54608-9	738
S54609-7	139
S54610-0	741
S54612-7	10
S54613-5	831
S54614-3	232
S54615-1	232
S54618-6	275
S54619-4	229
S54621-6	758
S54623-2	232
S54625-9	967
S54626-7	966
S54627-5	363
S54629-1	363
S54630-5	1078
S54631-3	698
S54635-6	903
S54636-4	906
S54638-0	823
S54639-9	903
S54645-3	938
S54646-1	229
S54648-8	994
S54656-9	994
S54657-7	129
S54659-3	443
S54670-4	10
S54673-9	49
S54676-3	49
S54679-8	9
S54680-1	9
S54684-4	298
S54685-2	310
S54688-7	117
S54689-5	183
S54692-5	26
S54693-3	810
S54694-1	26
S54696-8	302
S54697-6	346
S54698-4	255
S54699-2	199
S54702-6	165
S54703-4	718
S54704-2	38
S54705-0	339
S54707-7	1117
S54708-5	163
S54709-3	73
S54710-7	883
S54711-5	890
S54712-3	869
S54713-1	25
S54715-8	26
S54716-6	287
S54718-2	48
S54719-0	27
S54720-4	25
S54721-2	83
S54722-0	764
S54723-9	1129
S54724-7	1132
S54725-5	1129
S54726-3	171
S54729-8	35
S54730-1	229
S54732-8	27
S54737-9	296
S54740-9	746
S54741-7	715
S54742-5	738
S54744-1	738
S54746-8	718
S54747-6	715
S54748-4	750
S54749-2	738
S54750-6	734
S54751-4	715
S54752-2	139
S54753-0	736
S54754-9	715
S54757-3	475
S54758-1	746
S54760-3	746
S54761-1	743
S54766-2	302
S54767-0	434
S54768-9	306
S54769-7	302
S54770-0	183
S54773-5	978
S54774-3	721
S54777-8	318
S54778-6	127
S54779-4	127
S54780-8	136
S54781-6	157
S54783-2	979
S54784-0	396
S54785-9	469
S54786-7	45
S54787-5	269
S54788-3	111
S54789-1	127
S54790-5	697
S54791-3	503
S54792-1	126
S54798-0	303
S54799-9	306
S54800-6	306
S54802-2	232
S54811-1	387
S54813-8	302
S54815-4	502
S54817-0	1049
S54818-9	1068
S54819-7	1142
S54820-0	1048
S54821-9	1141
S54822-7	1141
S54823-5	1049
S54824-3	498
S54827-8	1142
S54828-6	1142
S54832-4	476
S54833-2	105
S54834-0	1054
S54835-9	182
S54837-5	300
S54838-3	302
S54839-1	232
S54842-1	232
S54844-8	306
S54846-4	954
S54848-0	200
S54849-9	737
S54850-2	200
S54851-0	717
S54852-9	853
S54853-7	737
S54854-5	718
S54855-3	737
S54856-1	943
S54859-6	850
S54861-8	190
S54862-6	721
S54863-4	717
S54864-2	1034
S54865-0	737
S54866-9	476
S54868-5	502
S54872-3	358
S54873-1	751
S54876-6	306
S54877-4	306
S54878-2	232
S54879-0	306
S54880-4	306
S54881-2	306
S54883-9	302
S54884-7	277
S54885-5	306
S54886-3	1079
S54892-8	503
S54893-6	246
S54894-4	219
S54895-2	280
S54896-0	64
S54898-7	275
S54899-5	264
S54903-7	279
S54905-3	292
S54912-6	1043
S54915-0	498
S54916-9	498
S54917-7	1043
S54918-5	1048
S54920-7	1049
S54928-2	739
S54929-0	498
S54931-2	162
S54936-3	919
S54940-1	1048
S54942-8	1094
S54943-6	720
S54945-2	762
S54946-0	755
S54948-7	384
S54949-5	139
S54951-7	1041
S54956-8	1009
S54957-6	735
S54958-4	335
S54959-2	741
S54961-4	732
S54965-7	973
S54966-5	836
S54970-3	149
S54974-6	830
S54975-4	1003
S54976-2	1042
S54978-9	296
S54979-7	153
S54981-9	41
S54982-7	274
S54983-5	270
S54984-3	1098
S54985-1	153
S54987-8	102
S54989-4	834
S54994-0	832
S54999-1	1133
S55000-0	1133
S55001-9	158
S55002-7	831
S55003-5	830
S55004-3	363
S55007-8	363
S55009-4	716
S55011-6	769
S55017-5	769
S55018-3	769
S55022-1	768
S55024-8	769
S55025-6	769
S55026-4	210
S55027-2	111
S55028-0	112
S55031-0	111
S55032-9	502
S55034-5	500
S55036-1	503
S55038-8	498
S55039-6	499
S55043-4	503
S55044-2	499
S55045-0	462
S55046-9	462
S55047-7	462
S55048-5	462
S55049-3	462
S55050-7	463
S55051-5	462
S55052-3	463
S55053-1	1014
S55055-8	1008
S55058-2	1008
S55059-0	388
S55061-2	387
S55062-0	384
S55063-9	426
S55069-8	794
S55070-1	144
S55075-2	182
S55079-5	365
S55082-5	319
S55083-3	368
S55091-4	752
S55093-0	946
S55094-9	30
S55096-5	945
S55098-1	105
S55100-7	16
S55101-5	836
S55102-3	276
S55109-0	354
S55112-0	353
S55113-9	757
S55117-1	285
S55119-8	15
S55120-1	976
S55122-8	717
S55124-4	183
S55127-9	149
S55128-7	18
S55129-5	28
S55131-7	18
S55136-8	709
S55137-6	704
S55138-4	704
S55139-2	694
S55140-6	698
S55141-4	704
S55142-2	698
S55143-0	698
S55144-9	700
S55145-7	705
S55146-5	698
S55147-3	704
S55148-1	695
S55150-3	723
S55151-1	853
S55155-4	858
S55157-0	720
S55158-9	722
S55159-7	722
S55160-0	723
S55161-9	125
S55162-7	722
S55163-5	722
S55164-3	723
S55171-6	719
S55173-2	155
S55174-0	1006
S55175-9	716
S55176-7	720
S55177-5	146
S55182-1	857
S55184-8	1013
S55185-6	199
S55186-4	104
S55188-0	856
S55189-9	857
S55193-7	843
S55195-3	899
S55196-1	42
S55200-3	859
S55203-8	857
S55205-4	859
S55206-2	169
S55207-0	1129
S55209-7	853
S55210-0	1129
S55214-3	17
S55215-1	343
S55217-8	855
S55218-6	858
S55219-4	897
S55234-8	578
S55251-8	755
S55267-4	206
S55270-4	162
S55306-9	169
S55307-7	398
S55311-5	259
S55317-4	19
S55318-2	34
S55319-0	19
S55320-4	41
S55322-0	19
S55324-7	23
S55325-5	59
S55330-1	28
S55332-8	385
S55333-6	275
S55334-4	275
S55335-2	275
S55336-0	122
S55337-9	140
S55338-7	146
S55339-5	717
S55341-7	738
S55347-6	2
S55352-2	269
S55354-9	820
S55356-5	323
S55357-3	209
S55360-3	357
S55361-1	357
S55363-8	88
S55364-6	355
S55365-4	443
S55366-2	778
S55367-0	141
S55368-9	949
S55370-0	1008
S55371-9	379
S55373-5	157
S55374-3	753
S55375-1	822
S55377-8	815
S55378-6	738
S55379-4	951
S55380-8	747
S55381-6	748
S55383-2	748
S55385-9	718
S55387-5	750
S55388-3	716
S55389-1	466
S55390-5	1137
S55391-3	350
S55398-0	275
S55400-6	275
S55401-4	275
S55402-2	275
S55403-0	275
S55404-9	267
S55405-7	275
S55406-5	267
S55407-3	266
S55408-1	269
S55410-3	267
S55413-8	275
S55416-2	169
S55418-9	365
S55423-5	727
S55424-3	387
S55425-1	789
S55428-6	29
S55429-4	787
S55431-6	35
S55437-5	812
S55438-3	438
S55439-1	788
S55441-3	130
S55446-4	800
S55447-2	802
S55448-0	155
S55449-9	366
S55454-5	444
S55455-3	363
S55459-6	788
S55463-4	830
S55465-0	741
S55467-7	831
S55468-5	475
S55469-3	1054
S55470-7	744
S55471-5	1072
S55472-3	720
S55473-1	762
S55475-8	749
S55476-6	739
S55477-4	718
S55478-2	794
S55479-0	794
S55480-4	295
S55481-2	292
S55484-7	275
S55485-5	264
S55487-1	266
S55490-1	944
S55493-6	366
S55497-9	483
S55498-7	483
S55499-5	874
S55500-2	477
S55501-0	467
S55503-7	558
S55504-5	480
S55505-3	480
S55506-1	479
S55508-8	481
S55509-6	489
S55511-8	477
S55512-6	466
S55513-4	469
S55514-2	489
S55515-0	483
S55516-9	489
S55518-5	477
S55519-3	677
S55520-7	677
S55521-5	877
S55522-3	467
S55523-1	446
S55525-8	467
S55526-6	446
S55527-4	476
S55528-2	483
S55529-0	483
S55531-2	473
S55532-0	483
S55533-9	476
S55534-7	483
S55535-5	439
S55536-3	441
S55537-1	471
S55539-8	473
S55540-1	677
S55542-8	471
S55544-4	471
S55545-2	473
S55546-0	490
S55547-9	483
S55548-7	471
S55549-5	470
S55550-9	483
S55551-7	677
S55552-5	483
S55553-3	483
S55554-1	471
S55556-8	483

Product	Page
S55558-4	467
S55559-2	483
S55560-6	483
S55561-4	490
S55562-2	471
S55565-7	275
S55566-5	267
S55570-3	275
S55573-8	300
S55574-6	266
S55576-2	275
S55577-0	275
S55578-9	28
S55581-9	908
S55582-7	144
S55583-5	366
S55588-6	842
S55595-9	835
S55598-3	835
S55601-7	677
S55602-5	591
S55604-1	15
S55606-8	51
S55608-4	229
S55613-0	69
S55615-7	141
S55623-8	293
S55631-9	229
S55635-1	10
S55638-6	160
S55639-4	81
S55640-8	4
S55646-7	738
S55650-5	17
S55657-2	365
S55666-1	1140
S55671-8	153
S55673-4	152
S55674-2	905
S55677-7	210
S55679-3	837
S55680-7	55
S55683-1	144
S55685-8	147
S55686-6	101
S55687-4	109
S55690-4	93
S55691-2	30
S55692-0	105
S55699-8	1083
S55701-3	1083
S55704-8	979
S55705-6	979
S55706-4	979
S55707-2	979
S55708-0	979
S55709-9	978
S55710-2	979
S55711-0	979
S55713-7	727
S55714-5	979
S55715-3	138
S55718-8	501
S55730-7	444
S55747-1	286
S55749-8	82
S55753-6	86
S55755-2	4
S55757-9	356
S55760-9	765
S55761-7	293
S55763-3	907
S55764-1	170
S55767-6	483
S55768-4	889
S55769-2	728
S55770-6	833
S55774-9	559
S55775-7	717
S55776-5	719
S55777-3	749
S55778-1	834
S55781-1	499
S55812-5	33
S55816-8	273
S55822-2	850
S55823-0	432
S55825-7	433
S55826-5	305
S55827-3	846
S55828-1	272
S55830-3	834
S55831-1	65
S55837-0	290
S55838-9	441
S55842-7	253
S55845-1	440
S55849-4	693
S55852-4	794
S55857-5	233
S55862-1	328
S55868-0	814
S55871-0	229
S55872-9	999
S55873-7	763
S55875-3	288
S55876-1	218
S55882-6	906
S55883-4	1083
S55889-3	218
S55892-3	826
S55895-8	813
S55897-4	724
S55898-2	336
S55902-4	441
S55904-0	86
S55907-5	762
S55911-3	296
S55920-2	989
S55922-9	4
S55923-7	169
S55928-8	86
S55929-6	933
S55936-9	387
S55938-5	49
S55941-5	28
S55942-3	387
S55945-8	83
S55946-6	274
S55949-0	1034
S55952-0	933
S55959-8	301
S55963-6	304
S55966-0	433
S55970-9	376
S55972-5	304
S55976-8	302
S55978-4	477
S55979-2	301
S55982-2	943
S55984-9	436
S55985-7	943
S55988-1	943
S55993-8	942
S55996-2	942
S55998-9	802
S55999-7	942
S56000-6	828
S56001-4	349
S56002-2	302
S56003-0	379
S56005-7	305
S56008-1	305
S56013-8	296
S56014-6	360
S56016-2	293
S56019-7	297
S56025-1	106
S56027-8	708
S56033-2	88
S56037-5	960
S56040-5	335
S56042-1	285
S56051-0	1138
S56052-9	932
S56061-8	350
S56062-6	1087
S56063-4	79
S56064-2	272
S56067-7	263
S56070-7	305
S56072-3	14
S56073-1	272
S56075-8	160
S56077-4	272
S56080-4	290
S56081-2	441
S56090-1	178
S56092-8	4
S56103-7	861
S56104-5	296
S56106-1	350
S56108-8	58
S56111-8	250
S56112-6	691
S56115-0	296
S56118-5	355
S56120-7	378
S56122-3	437
S56123-1	389
S56125-8	999
S56126-6	105
S56127-4	238
S56130-4	73
S56132-0	62
S56134-7	75
S56135-5	226
S56136-3	961
S56140-1	365
S56144-4	296
S56146-0	376
S56148-7	376
S56152-5	964
S56159-2	99
S56164-9	63
S56168-1	341
S56180-0	77
S56184-3	90
S56189-4	50
S56191-6	275
S56193-2	83
S56194-0	384
S56195-9	347
S56196-7	243
S56197-5	188
S56198-3	286
S56203-3	318
S56208-4	344
S56210-6	344
S56224-6	290
S56226-2	360
S56228-9	181
S56229-7	293
S56231-9	821
S56233-5	462
S56235-1	789
S56237-8	1083
S56239-4	422
S56240-8	1009
S56241-6	549
S56242-4	202
S56249-1	10
S56250-5	10
S56252-1	202
S56254-8	268
S56255-6	218
S56256-4	164
S56257-2	19
S56259-9	144
S56263-7	822
S56264-5	799
S56265-3	829
S56266-1	422
S56268-8	789
S56269-6	265
S56273-4	340
S56278-5	433
S56279-3	733
S56281-5	818
S56282-3	155
S56283-1	387
S56285-8	4
S56286-6	217
S56288-2	365
S56289-0	222
S56290-4	49
S56291-2	50
S56297-1	801
S56299-8	497
S56300-5	102
S56301-3	4
S56304-8	10
S56307-2	88
S56318-8	144
S56322-6	29
S56324-2	63
S56328-5	30
S56329-3	28
S56331-5	987
S56332-3	4
S56336-6	355
S56337-4	180
S56339-0	181
S56341-2	381
S56342-0	181
S56343-9	192
S56346-3	787
S56347-1	298
S56350-1	30
S56358-7	362
S56360-9	790
S56361-7	59
S56364-1	378
S56367-6	959
S56368-4	179
S56369-2	787
S56370-6	828
S56371-4	876
S56372-2	440
S56374-9	983
S56383-8	373
S56384-6	82
S56386-2	289
S56388-9	76
S56390-0	986
S56394-3	83
S56395-1	376
S56400-1	350
S56402-8	789
S56403-6	963
S56409-5	46
S56412-5	477
S56414-1	169
S56417-6	28
S56418-4	71
S56421-4	93
S56426-5	357
S56430-3	197
S56431-1	478
S56438-9	221
S56439-7	181
S56442-7	30
S56443-5	92
S56444-3	1138
S56448-6	389
S56450-8	141
S56451-6	4
S56452-4	4
S56459-1	706
S56462-1	290
S56469-9	160
S56473-7	229
S56474-5	385
S56475-3	856
S56476-1	305
S56479-6	932
S56481-8	857
S56483-4	384
S56484-2	305
S56485-0	857
S56486-9	2
S56488-5	857
S56489-3	443
S56491-5	192
S56496-6	972
S56500-8	346
S56506-7	85
S56512-1	1099
S56516-4	1107
S56518-0	1115
S56525-3	1096
S56526-1	1110
S56531-8	1096
S56532-6	1096
S56535-0	1107
S56538-5	1112
S56542-3	1110
S56543-1	1110
S56546-6	1049
S56557-1	339
S56563-6	217
S56565-2	973
S56567-9	52
S56570-9	69
S56571-7	17
S56577-6	870
S56579-2	350
S56580-6	76
S56581-4	727
S56582-2	36
S56583-0	375
S56585-7	1052
S56587-3	328
S56588-1	1001
S56591-1	972
S56593-8	461
S56596-2	46
S56597-0	142
S56598-9	192
S56600-4	150
S56601-2	69
S56603-9	52
S56607-1	15
S56610-1	34
S56614-4	138
S56615-2	28
S56616-0	769
S56617-9	861
S56620-9	1013
S56621-7	420
S56623-3	291
S56624-1	836
S56626-8	355
S56629-2	218
S56631-4	196
S56633-0	52
S56635-7	351
S56637-3	379
S56638-1	807
S56648-9	30
S56652-7	1110
S56654-3	1124
S56657-8	19
S56658-6	852
S56659-4	1008
S56660-8	43
S56662-4	6
S56663-2	364
S56665-9	154
S56680-2	864
S56681-0	918
S56685-3	153
S56686-1	153
S56691-8	16
S56694-2	153
S56695-0	50
S56696-9	21
S56699-3	152
S56700-0	19
S56701-9	152
S56703-5	152
S56704-3	153
S56707-8	152
S56711-6	325
S56712-4	285
S56713-2	334
S56714-0	285
S56716-7	285
S56717-5	282
S56718-3	285
S56719-1	285
S56721-3	282
S56724-8	285
S56729-9	171
S56733-7	26
S56741-8	171
S56742-6	855
S56744-2	61
S56745-0	62
S56748-5	791
S56750-7	676
S56753-1	1014
S56755-8	882
S56757-4	997
S56758-2	436
S56759-0	127
S56762-0	1088
S56765-5	1089
S56773-6	975
S56780-9	882
S56786-8	411
S56788-4	1017
S56793-0	156
S56794-9	210
S56796-5	88
S56800-7	40
S56805-8	330
S56807-4	720
S56811-2	89
S56812-0	19
S56813-9	10
S56816-3	25
S56817-1	161
S56819-8	19
S56820-1	1049
S56825-2	17
S56831-7	707
S56834-1	155
S56837-6	802
S56838-4	298
S56841-4	106
S56842-2	169
S56843-0	754
S56846-5	442
S56850-3	95
S56851-1	720
S56855-4	318
S56857-0	229
S56864-3	841
S56868-6	823
S56869-4	805
S56871-6	812
S56879-1	906
S56881-3	473
S56882-1	802
S56885-6	1051
S56886-4	473
S56887-2	795
S56888-0	414
S56890-2	795
S56891-0	354
S56892-9	819
S56896-1	111
S56898-8	907
S56899-6	1009
S56900-3	957
S56904-6	356
S56905-4	220
S56906-2	263
S56908-9	272
S56909-7	259
S56910-0	259
S56911-9	272
S56912-7	259
S56913-5	272
S56914-3	272
S56917-8	272
S56918-6	266
S56919-4	302
S56921-6	301
S56922-4	302
S56923-2	433
S56925-9	229
S56926-7	267

Product	Page
S56927-5	229
S56928-3	229
S56929-1	267
S56930-5	301
S56931-3	321
S56934-8	321
S56936-4	365
S56937-2	258
S56938-0	321
S56939-9	258
S56940-2	336
S56942-9	339
S56943-7	321
S56944-5	498
S56951-8	820
S56961-5	1055
S56963-1	812
S56965-8	814
S56970-4	821
S56973-9	816
S56979-8	819
S56983-6	812
S56984-4	907
S56985-2	428
S56991-7	906
S56992-5	501
S56993-3	138
S56996-8	904
S57004-4	701
S57005-2	145
S57009-5	133
S57017-6	661
S57021-4	795
S57022-2	55
S57023-0	266
S57028-1	26
S57030-3	4
S57031-1	86
S57033-8	286
S57035-4	4
S57037-0	697
S57038-9	142
S57043-5	1085
S57044-3	884
S57045-1	933
S57047-8	26
S57048-6	1127
S57050-8	26
S57051-6	358
S57053-2	104
S57054-0	365
S57055-9	27
S57056-7	26
S57057-5	26
S57058-3	73
S57059-1	24
S57060-5	25
S57061-3	26
S57062-1	699
S57065-6	26
S57066-4	1100
S57067-2	25
S57068-0	26
S57069-9	954
S57070-2	25
S57071-0	80
S57072-9	26
S57074-5	27
S57075-3	482
S57076-1	895
S57081-8	362
S57082-6	722
S57085-0	916
S57087-7	63
S57088-5	145
S57089-3	31
S57091-5	74
S57095-8	162
S57097-4	1
S57098-2	98
S57101-6	220
S57102-4	351
S57103-2	789
S57106-7	368
S57111-3	206
S57115-6	722
S57116-4	1129
S57120-2	478
S57122-9	33
S57125-3	231
S57128-8	276
S57129-6	659
S57134-2	2
S57135-0	76
S57137-7	37
S57138-5	131
S57142-3	188
S57147-4	203
S57148-2	188
S57150-4	5
S57157-1	204
S57169-5	6
S57177-6	1121
S57178-4	327
S57179-2	252
S57182-2	4
S57193-8	317
S57194-6	993
S57195-4	765
S57197-0	316
S57198-9	413
S57199-7	807
S57200-4	789
S57201-2	414
S57202-0	233
S57203-9	788
S57204-7	823
S57205-5	811
S57206-3	468
S57207-1	811
S57209-8	788
S57219-5	1012
S57222-5	832
S57231-4	25
S57232-2	1132
S57236-5	229
S57237-3	300
S57238-1	299
S57240-3	229
S57241-1	299
S57243-8	229
S57244-6	386
S57250-0	1137
S57251-9	1137
S57252-7	1137
S57253-5	1129
S57255-1	1137
S57258-6	1137
S57260-8	277
S57261-6	1137
S57262-4	1137
S57263-2	300
S57264-0	300
S57267-5	300
S57268-3	300
S57274-8	9
S57276-4	696
S57280-2	318
S57283-7	1015
S57284-5	50
S57286-1	74
S57288-8	8
S57289-6	352
S57300-0	49
S57301-9	28
S57302-7	61
S57304-3	904
S57305-1	69
S57308-6	967
S57309-4	286
S57310-8	351
S57312-4	34
S57314-0	69
S57316-7	147
S57317-5	74
S57318-3	79
S57321-3	792
S57324-8	375
S57326-4	34
S57327-2	221
S57330-2	2
S57331-0	10
S57332-9	88
S57333-7	268
S57335-3	217
S57338-8	218
S57339-6	218
S57344-2	269
S57345-0	268
S57347-7	323
S57349-3	417
S57350-7	417
S57352-3	419
S57353-1	419
S57355-8	417
S57357-4	418
S57359-0	418
S57360-4	227
S57361-2	418
S57364-7	269
S57366-3	414
S57367-1	176
S57369-8	321
S57370-1	321
S57372-8	419
S57373-6	417
S57378-7	57
S57380-9	300
S57381-7	299
S57382-5	314
S57383-3	314
S57386-8	229
S57389-2	32
S57392-2	16
S57407-4	392
S57409-0	64
S57412-0	379
S57427-9	330
S57431-7	269
S57436-8	789
S57441-4	128
S57443-0	414
S57444-9	370
S57445-7	269
S57447-3	370
S57451-1	266
S57454-6	268
S57455-4	268
S57456-2	323
S57457-0	340
S57458-9	268
S57459-7	179
S57460-0	323
S57461-9	269
S57462-7	179
S57463-5	370
S57467-8	321
S57469-4	179
S57470-8	266
S57471-6	340
S57472-4	267
S57473-2	323
S57474-0	419
S57476-7	227
S57477-5	218
S57478-3	218
S57479-1	418
S57480-5	418
S57481-3	227
S57482-1	417
S57485-6	341
S57486-4	419
S57488-0	340
S57489-9	179
S57490-2	217
S57492-9	341
S57495-3	417
S57496-1	268
S57498-8	323
S57499-6	269
S57500-3	268
S57501-1	370
S57505-4	269
S57506-2	370
S57507-0	370
S57508-9	268
S57509-7	269
S57511-9	268
S57512-7	227
S57514-3	419
S57515-1	417
S57517-8	417
S57518-6	419
S57519-4	417
S57523-2	266
S57525-9	218
S57526-7	419
S57527-5	228
S57528-3	419
S57529-1	227
S57530-5	418
S57531-3	217
S57532-1	218
S57534-8	419
S57535-6	227
S57536-4	324
S57537-2	179
S57538-0	323
S57539-9	269
S57540-2	370
S57542-9	268
S57543-7	323
S57545-3	323
S57546-1	267
S57548-8	323
S57549-6	269
S57551-8	370
S57552-6	179
S57555-0	268
S57557-7	417
S57558-5	418
S57559-3	227
S57560-7	419
S57561-5	419
S57562-3	418
S57563-1	217
S57565-8	217
S57566-6	419
S57567-4	217
S57569-0	418
S57570-4	323
S57573-9	227
S57574-7	340
S57577-1	268
S57579-8	323
S57580-1	269
S57584-4	323
S57585-2	321
S57586-0	179
S57589-5	370
S57590-9	370
S57591-7	138
S57592-5	792
S57594-1	725
S57596-8	136
S57597-6	657
S57598-4	355
S57601-8	300
S57603-4	300
S57605-0	597
S57607-7	300
S57608-5	1084
S57609-3	1084
S57610-7	1084
S57611-5	299
S57617-4	419
S57621-2	60
S57622-0	208
S57624-7	104
S57626-3	834
S57627-1	1131
S57629-8	36
S57633-6	1097
S57634-4	659
S57636-0	351
S57638-7	417
S57639-5	439
S57640-9	428
S57641-7	228
S57644-1	228
S57647-6	580
S57648-4	597
S57650-6	228
S57652-2	302
S57653-0	438
S57654-9	440
S57655-7	398
S57656-5	398
S57658-1	559
S57661-1	963
S57666-2	150
S57669-7	56
S57670-0	417
S57671-9	439
S57672-7	426
S57673-5	426
S57674-3	557
S57675-1	559
S57678-6	439
S57679-4	439
S57680-8	440
S57681-6	302
S57684-0	63
S57687-5	789
S57688-3	1016
S57689-1	127
S57690-5	709
S57691-3	783
S57696-4	783
S57697-2	284
S57698-0	772
S57703-0	293
S57707-3	212
S57714-6	305
S57717-0	301
S57719-7	257
S57720-0	397
S57721-9	264
S57723-5	228
S57724-3	301
S57725-1	209
S57733-2	186
S57735-9	10
S57745-6	207
S57751-0	805
S57755-3	1075
S57758-8	156
S57759-6	733
S57761-8	891
S57765-0	426
S57766-9	813
S57767-7	266
S57770-7	821
S57771-5	814
S57778-2	127
S57790-1	217
S57792-8	814
S57796-0	1051
S57798-7	878
S57802-9	86
S57803-7	10
S57806-1	207
S57808-8	17
S57809-6	100
S57811-8	200
S57812-6	12
S57814-2	24
S57815-0	79
S57816-9	217
S57817-7	10
S57818-5	144
S57823-1	341
S57826-6	874
S57831-2	15
S57833-9	274
S57835-5	100
S57842-8	329
S57844-4	727
S57846-0	912
S57848-7	820
S57850-9	815
S57856-8	818
S57857-6	1055
S57858-4	814
S57859-2	1055
S57862-2	821
S57863-0	813
S57868-1	805
S57870-3	1056
S57875-4	820
S57880-0	821
S57881-9	1055
S57884-3	1079
S57887-8	1055
S57893-2	820
S57896-7	820
S57898-3	81
S57900-9	221
S57901-7	368
S57902-5	227
S57904-1	9
S57908-4	789
S57911-4	118
S57913-0	103
S57914-9	98
S57920-3	356
S57927-0	1002
S57928-9	205
S57929-7	161
S57930-0	962
S57932-7	156
S57934-3	153
S57939-4	656
S57941-6	663
S57942-4	973
S57943-2	123
S57944-0	592
S57948-3	997
S57949-1	349
S57952-1	69
S57955-6	566
S57956-4	1094
S57962-9	103
S57963-7	811
S57964-5	164
S57966-1	76
S57968-8	592
S57973-4	585
S57975-0	604
S57976-9	914
S57977-7	237
S57979-3	1015
S57982-3	372
S57983-1	953
S57985-8	279
S57987-4	883
S57988-2	392
S57990-4	81
S57993-9	792
S57997-1	202
S58000-7	174
S58002-3	855
S58003-1	967
S58005-8	228
S58009-0	196
S58010-4	1060
S58013-9	380
S58014-7	1005
S58016-3	439
S58017-1	438
S58024-4	368
S58026-0	802
S58027-9	590
S58029-5	673
S58030-9	1013
S58031-7	680
S58033-3	680
S58037-6	976
S58038-4	319
S58040-6	585
S58044-9	938
S58051-1	64
S58062-7	363
S58063-5	164

Product	Page
S58065-1	249
S58067-8	1139
S58068-6	36
S58069-4	41
S58073-2	81
S58074-0	803
S58076-7	957
S58077-5	462
S58079-1	163
S58081-3	416
S58087-2	148
S58088-0	163
S58091-0	863
S58094-5	866
S58096-1	66
S58098-8	748
S58099-6	953
S58104-6	855
S58110-0	186
S58111-9	966
S58112-7	961
S58113-5	815
S58117-8	934
S58126-7	984
S58127-5	961
S58130-5	1000
S58132-1	1011
S58136-4	995
S58138-0	820
S58141-0	995
S58143-7	70
S58144-5	1006
S58145-3	199
S58148-8	984
S58152-6	961
S58155-0	802
S58156-9	984
S58157-7	333
S58159-3	978
S58160-7	946
S58161-5	774
S58162-3	1055
S58167-4	822
S58174-7	821
S58175-5	805
S58176-3	1055
S58177-1	571
S58179-8	443
S58185-2	718
S58186-0	802
S58187-9	822
S58188-7	430
S58191-7	956
S58197-6	802
S58199-2	788
S58201-8	443
S58202-6	71
S58206-9	977
S58208-5	293
S58212-3	470
S58213-1	354
S58218-2	1054
S58219-0	468
S58220-4	996
S58223-9	1031
S58225-5	1041
S58229-8	957
S58234-4	805
S58237-9	820
S58239-5	212
S58242-5	777
S58243-3	369
S58248-4	829
S58251-4	413
S58252-2	149
S58253-0	775
S58254-9	258
S58255-7	485
S58256-5	789
S58261-1	341
S58263-8	341
S58264-6	773
S58268-9	804
S58270-0	802
S58271-9	802
S58272-7	193
S58273-5	1051
S58275-1	145
S58278-6	814
S58279-4	875
S58283-2	710
S58286-7	117
S58289-1	802
S58298-0	221
S58300-6	910
S58302-2	144
S58303-0	800
S58306-5	908
S58313-8	845
S58314-6	810
S58315-4	462
S58318-9	811
S58319-7	908
S58325-1	411
S58327-8	1006
S58332-4	1044
S58334-0	870
S58339-1	192
S58342-1	814
S58348-0	468
S58349-9	820
S58351-0	821
S58359-6	820
S58363-4	997
S58364-2	866
S58365-0	696
S58366-9	118
S58372-3	821
S58373-1	1059
S58378-2	1073
S58380-4	1035
S58382-0	266
S58383-9	266
S58384-7	1035
S58385-5	1035
S58387-1	1029
S58389-8	1029
S58390-1	1038
S58392-8	1038
S58393-6	856
S58394-4	1020
S58395-2	1020
S58396-0	1035
S58397-9	1033
S58398-7	1033
S58399-5	1033
S58400-2	1031
S58401-0	1031
S58402-9	857
S58404-5	1035
S58405-3	1031
S58406-1	228
S58408-8	228
S58411-8	267
S58412-6	222
S58415-0	305
S58416-9	301
S58417-7	300
S58418-5	305
S58419-3	1050
S58420-7	1078
S58421-5	1078
S58422-3	1078
S58423-1	1078
S58425-8	1078
S58426-6	1078
S58427-4	1077
S58428-2	1078
S58431-2	1021
S58432-0	1021
S58433-9	1021
S58434-7	314
S58435-5	311
S58436-3	1078
S58437-1	1078
S58439-8	1078
S58440-1	1024
S58442-8	1056
S58443-6	1056
S58444-4	1043
S58446-0	1077
S58447-9	1049
S58448-7	300
S58449-5	433
S58450-9	313
S58452-5	307
S58453-3	1049
S58454-1	1078
S58456-8	1078
S58458-4	1078
S58459-2	1078
S58462-2	1096
S58465-7	1112
S58467-3	1110
S58468-1	1110
S58470-3	1110
S58485-1	1096
S58489-4	1099
S58496-7	1111
S58498-3	1096
S58500-9	1099
S58501-7	1099
S58504-1	1096
S58507-6	1096
S58508-4	1112
S58512-2	1113
S58513-0	1112
S58514-9	1113
S58521-1	1106
S58530-0	1097
S58538-6	1097
S58539-4	1098
S58540-8	1098
S58541-6	1098
S58544-0	1112
S58551-3	1096
S58558-0	1106
S58560-2	1110
S58562-9	1096
S58564-5	1097
S58565-3	1106
S58566-1	1115
S58577-7	1100
S58580-7	1123
S58581-5	1097
S58582-3	1098
S58583-1	1098
S58585-8	1113
S58587-4	1097
S58591-2	1096
S58597-1	1108
S58600-5	1114
S58604-8	1112
S58610-2	1096
S58611-0	603
S58613-7	565
S58615-3	585
S58616-1	678
S58618-8	128
S58619-6	575
S58622-6	130
S58623-4	586
S58624-2	569
S58625-0	565
S58626-9	575
S58627-7	590
S58628-5	682
S58629-3	682
S58631-5	682
S58635-8	682
S58636-6	680
S58638-2	603
S58639-0	687
S58640-4	682
S58641-2	680
S58642-0	670
S58644-7	680
S58645-5	1056
S58646-3	1073
S58647-1	1073
S58649-8	1073
S58650-1	1073
S58652-8	1073
S58653-6	1037
S58654-4	1073
S58657-9	1056
S58658-7	1031
S58660-9	301
S58662-5	1142
S58669-2	1018
S58673-0	870
S58677-3	290
S58678-1	468
S58681-1	56
S58683-8	468
S58684-6	178
S58685-4	760
S58688-9	431
S58689-7	468
S58691-9	277
S58692-7	798
S58693-5	775
S58694-3	465
S58698-6	464
S58700-1	468
S58703-6	12
S58706-0	468
S58707-9	817
S58708-7	414
S58709-5	155
S58710-9	414
S58712-5	822
S58714-1	768
S58716-8	876
S58717-6	822
S58718-4	221
S58719-2	1127
S58721-4	817
S58728-1	547
S58730-3	811
S58739-7	811
S58741-9	821
S58750-8	901
S58752-4	812
S58753-2	814
S58754-0	498
S58755-9	811
S58756-7	802
S58762-1	822
S58766-4	1084
S58769-9	903
S58770-2	944
S58772-9	811
S58773-7	811
S58774-5	812
S58775-3	465
S58778-8	1045
S58782-6	919
S58784-2	802
S58787-7	557
S58790-7	916
S58792-3	485
S58793-1	811
S58795-8	811
S58796-6	811
S58797-4	811
S58798-2	230
S58799-0	811
S58800-8	811
S58803-2	19
S58806-7	745
S58807-5	155
S58808-3	191
S58810-5	140
S58811-3	803
S58815-6	878
S58817-2	908
S58818-0	155
S58819-9	821
S58820-2	442
S58822-9	802
S58823-7	368
S58825-3	804
S58826-1	802
S58828-8	813
S58831-8	215
S58832-6	913
S58833-4	424
S58834-2	785
S58835-0	29
S58838-5	17
S58839-3	88
S58841-5	203
S58843-1	792
S58845-8	795
S58846-6	147
S58847-4	89
S58849-0	10
S58850-4	790
S58851-2	795
S58852-0	794
S58853-9	799
S58856-3	790
S58857-1	789
S58859-8	173
S58860-1	279
S58862-8	279
S58864-4	173
S58865-2	188
S58871-7	329
S58883-0	796
S58885-7	794
S58895-4	195
S58897-0	195
S58901-2	184
S58902-0	228
S58903-9	1045
S58907-1	1065
S58909-8	1035
S58910-1	384
S58912-8	1031
S58913-6	1031
S58914-4	1022
S58915-2	1035
S58917-9	1045
S58918-7	303
S58920-9	272
S58921-7	1034
S58922-5	1066
S58923-3	1073
S58926-8	384
S58927-6	1066
S58928-4	266
S58929-2	299
S58931-4	266
S58932-2	384
S58934-9	300
S58935-7	305
S58936-5	1066
S58937-3	299
S58938-1	272
S58941-1	1029
S58944-6	305
S58945-4	272
S58946-2	266
S58947-0	266
S58948-9	272
S58949-7	228
S58951-9	1066
S58952-7	1065
S58953-5	1065
S58954-3	272
S58957-8	272
S58958-6	1066
S58959-4	1062
S58960-8	303
S58961-6	372
S58962-4	299
S58963-2	272
S58964-0	272
S58965-9	272
S58966-7	1065
S58967-5	263
S58968-3	305
S58969-1	228
S58970-5	1065
S58972-1	299
S58974-8	302
S58975-6	1056
S58976-4	1049
S58977-2	304
S58978-0	1059
S58980-2	272
S58981-0	1022
S58983-7	301
S58986-1	1073
S58988-8	1073
S58991-8	1049
S58992-6	933
S58995-0	572
S58996-9	680
S58997-7	603
S58998-5	566
S58999-3	680
S59000-2	682
S59001-0	687
S59003-7	572
S59004-5	586
S59005-3	687
S59006-1	603
S59008-8	687
S59009-6	592
S59011-8	680
S59012-6	683
S59013-4	603
S59014-2	687
S59015-0	580
S59017-7	683
S59018-5	680
S59019-3	591
S59020-7	591
S59021-5	591
S59022-3	588
S59023-1	603
S59025-8	682
S59026-6	591
S59027-4	603
S59028-2	584
S59029-0	687
S59030-4	589
S59031-2	589
S59032-0	602
S59033-9	603
S59034-7	592
S59035-5	129
S59037-1	580
S59039-8	592
S59040-1	579
S59042-8	566
S59043-6	116
S59044-4	602
S59045-2	569
S59046-0	116
S59047-9	599
S59048-7	594
S59049-5	591
S59050-9	687
S59051-7	603
S59052-5	603
S59053-3	596
S59054-1	116
S59056-8	587
S59057-6	592
S59058-4	687
S59059-2	603
S59060-6	481
S59061-4	470
S59062-2	469
S59063-0	481
S59064-9	1035
S59066-5	1063
S59067-3	1076
S59068-1	466
S59070-3	477
S59073-8	478
S59074-6	480
S59075-4	466
S59076-2	480
S59078-9	381
S59082-7	1035
S59083-5	1035
S59084-3	937
S59085-1	135
S59088-6	417

Product	Page
S59089-4	419
S59091-6	789
S59107-6	411
S59108-4	14
S59112-2	1085
S59114-9	486
S59115-7	1084
S59116-5	486
S59117-3	84
S59118-1	79
S59124-6	585
S59127-0	591
S59128-9	590
S59129-7	585
S59130-0	590
S59131-9	591
S59132-7	582
S59134-3	589
S59135-1	585
S59137-8	585
S59138-6	572
S59139-4	582
S59140-8	584
S59141-6	590
S59142-4	585
S59143-2	578
S59144-0	679
S59145-9	585
S59147-5	586
S59148-3	589
S59149-1	572
S59151-3	585
S59152-1	589
S59157-2	1039
S59162-9	444
S59163-7	585
S59165-3	969
S59171-8	363
S59173-4	589
S59175-0	115
S59176-9	1018
S59179-3	969
S59182-3	547
S59185-8	578
S59186-6	461
S59188-2	946
S59189-0	414
S59190-4	378
S59191-2	213
S59192-0	443
S59193-9	414
S59194-7	461
S59195-5	228
S59196-3	822
S59197-1	479
S59199-8	478
S59200-5	466
S59202-1	1067
S59204-8	481
S59205-6	481
S59206-4	481
S59207-2	479
S59208-0	479
S59209-9	479
S59210-2	1065
S59211-0	479
S59212-9	479
S59213-7	466
S59214-5	478
S59216-1	479
S59218-8	478
S59221-8	382
S59222-6	182
S59223-4	878
S59224-2	795
S59225-0	736
S59227-7	373
S59229-3	107
S59231-5	1044
S59232-3	265
S59233-1	275
S59235-8	813
S59236-6	813
S59237-4	777
S59240-4	188
S59241-2	794
S59242-0	812
S59245-5	872
S59246-3	443
S59249-8	793
S59252-8	845
S59253-6	270
S59255-2	792
S59256-0	796
S59257-9	776
S59258-7	802
S59259-5	851
S59260-9	497
S59261-7	797
S59262-5	820
S59263-3	355
S59264-1	795
S59267-6	271
S59268-4	384
S59269-2	749
S59270-6	847
S59271-4	137
S59272-2	788
S59273-0	776
S59274-9	212
S59275-7	804
S59277-3	389
S59278-1	157
S59287-0	340
S59288-9	813
S59290-0	100
S59291-9	220
S59292-7	1044
S59293-5	788
S59295-1	792
S59298-6	792
S59299-4	374
S59300-1	801
S59302-8	861
S59303-6	433
S59304-4	305
S59305-2	432
S59306-0	918
S59307-9	983
S59310-9	444
S59311-7	918
S59312-5	444
S59316-8	15
S59318-4	36
S59323-0	847
S59325-7	288
S59328-1	197
S59331-1	857
S59334-6	854
S59340-0	441
S59351-6	286
S59354-0	148
S59356-7	945
S59358-3	197
S59365-6	946
S59368-0	847
S59369-9	961
S59371-0	953
S59372-9	373
S59373-7	964
S59376-1	932
S59379-6	932
S59381-8	961
S59387-7	1031
S59388-5	1031
S59389-3	1032
S59390-7	1022
S59391-5	270
S59393-1	86
S59396-6	86
S59397-4	86
S59400-8	87
S59401-6	86
S59402-4	992
S59403-2	575
S59404-0	747
S59405-9	675
S59406-7	589
S59407-5	682
S59408-3	670
S59409-1	964
S59410-5	590
S59411-3	685
S59414-8	673
S59415-6	991
S59416-4	960
S59417-2	605
S59418-0	964
S59419-9	682
S59420-2	575
S59422-9	584
S59423-7	581
S59424-5	580
S59425-3	678
S59426-1	571
S59428-8	580
S59429-6	673
S59431-8	571
S59432-6	677
S59433-4	604
S59434-2	466
S59435-0	1065
S59436-9	479
S59438-5	1046
S59439-3	450
S59440-7	1066
S59442-3	1065
S59445-8	470
S59446-6	1065
S59448-2	479
S59449-0	465
S59451-2	283
S59452-0	286
S59454-7	283
S59455-5	286
S59456-3	285
S59457-1	283
S59459-8	283
S59460-1	285
S59462-8	257
S59463-6	257
S59465-2	1047
S59466-0	258
S59471-7	1043
S59474-1	267
S59476-8	1041
S59477-6	1046
S59478-4	1057
S59479-2	15
S59480-6	87
S59481-4	86
S59482-2	568
S59483-0	591
S59485-7	86
S59486-5	86
S59488-1	71
S59496-2	192
S59497-0	192
S59498-9	194
S59499-7	196
S59502-0	220
S59506-3	340
S59509-8	270
S59510-1	94
S59514-4	836
S59515-2	812
S59516-0	878
S59517-9	71
S59521-7	279
S59523-3	105
S59524-1	93
S59526-8	102
S59528-4	794
S59529-2	93
S59530-6	111
S59531-4	436
S59532-2	78
S59533-0	93
S59535-7	681
S59536-5	670
S59537-3	588
S59538-1	587
S59540-3	682
S59541-1	991
S59543-8	584
S59544-6	572
S59545-4	991
S59546-2	483
S59547-0	991
S59548-9	964
S59549-7	568
S59551-9	683
S59552-7	992
S59553-5	483
S59555-1	960
S59557-8	483
S59559-4	960
S59560-8	360
S59561-6	745
S59562-4	589
S59564-0	599
S59565-9	590
S59566-7	588
S59567-5	589
S59568-3	587
S59569-1	590
S59570-5	991
S59571-3	283
S59572-1	283
S59575-6	286
S59576-4	308
S59579-9	28
S59580-2	688
S59581-0	688
S59582-9	688
S59584-5	286
S59585-3	1012
S59586-1	1050
S59591-8	188
S59594-2	782
S59595-0	6
S59598-5	65
S59602-7	1073
S59603-5	1073
S59604-3	1005
S59607-8	15
S59608-6	65
S59610-8	15
S59616-7	219
S59617-5	1086
S59620-5	1080
S59621-3	388
S59622-1	310
S59627-2	869
S59629-9	192
S59634-5	217
S59639-6	213
S59641-8	20
S59643-4	1011
S59644-2	1011
S59645-0	1045
S59646-9	1022
S59647-7	266
S59649-3	309
S59650-7	385
S59651-5	194
S59655-8	967
S59657-4	780
S59658-2	975
S59659-0	1003
S59660-4	956
S59661-2	974
S59662-0	1003
S59664-7	779
S59670-1	781
S59672-8	780
S59676-0	1003
S59681-7	1003
S59682-5	956
S59683-3	785
S59684-1	781
S59686-8	780
S59687-6	781
S59688-4	293
S59690-6	754
S59691-4	974
S59692-2	974
S59694-9	1005
S59695-7	1005
S59696-5	95
S59698-1	6
S59700-7	65
S59701-5	6
S59707-4	103
S59708-2	67
S59709-0	65
S59712-0	77
S59713-9	67
S59715-5	960
S59716-3	77
S59724-4	310
S59728-7	683
S59729-5	804
S59731-7	945
S59732-5	202
S59736-8	84
S59737-6	67
S59738-4	65
S59740-6	72
S59741-4	685
S59742-2	593
S59744-9	555
S59745-7	991
S59746-5	605
S59747-3	555
S59748-1	985
S59750-3	555
S59751-1	565
S59753-8	939
S59754-6	554
S59755-4	559
S59756-2	574
S59757-0	597
S59758-9	683
S59759-7	600
S59760-0	669
S59761-9	554
S59762-7	559
S59763-5	670
S59764-3	565
S59765-1	564
S59767-8	939
S59768-6	991
S59769-4	610
S59770-8	589
S59771-6	991
S59772-4	593
S59773-2	564
S59774-0	749
S59776-7	853
S59777-5	307
S59781-3	308
S59784-8	309
S59785-6	816
S59787-2	793
S59789-9	792
S59791-0	450
S59793-7	795
S59794-5	56
S59796-1	754
S59799-6	329
S59800-3	985
S59801-1	363
S59803-8	413
S59804-6	800
S59806-2	737
S59807-0	801
S59808-9	443
S59809-7	1031
S59810-0	797
S59817-8	369
S59819-4	855
S59820-8	797
S59823-2	815
S59825-9	271
S59832-1	821
S59834-8	790
S59835-6	879
S59836-4	804
S59837-2	822
S59838-0	803
S59842-9	817
S59844-5	789
S59846-1	1000
S59848-8	828
S59849-6	802
S59854-2	812
S59856-9	821
S59860-7	821
S59863-1	216
S59867-4	902
S59868-2	502
S59870-4	797
S59876-3	797
S59880-1	461
S59882-8	442
S59883-6	31
S59885-2	468
S59886-0	1091
S59888-7	148
S59889-5	89
S59893-3	480
S59894-1	472
S59896-8	480
S59897-6	465
S59898-4	479
S59899-2	480
S59901-8	479
S59902-6	480
S59903-4	480
S59905-0	466
S59906-9	370
S59907-7	480
S59909-3	476
S59910-7	480
S59912-3	510
S59913-1	510
S59915-8	564
S59916-6	747
S59917-4	564
S59918-2	509
S59919-0	564
S59920-4	564
S59921-2	510
S59922-0	747
S59923-9	510
S59925-5	747
S59926-3	510
S59927-1	564
S59930-1	510
S59932-8	511
S59933-6	510
S59934-4	510
S59936-0	1124
S59937-9	671
S59941-7	824
S59942-5	824
S59944-1	671
S59946-8	824
S59948-4	90
S59950-6	1014
S59951-4	266
S59953-0	1072
S59954-9	341
S59955-7	1064
S59956-5	1066
S59958-1	1065
S59960-3	1066
S59961-1	1066
S59963-8	1066
S59964-6	1065
S59965-4	1058
S59971-9	189
S59972-7	1087
S59975-1	957
S59977-8	1076
S59978-6	1021
S59979-4	191
S59982-4	1112
S59986-7	95
S59988-3	761
S59989-1	1064
S59991-3	957
S59992-1	385

Product	Page
S59995-6	301
S59996-4	305
S59997-2	1073
S59998-0	971
S60005-9	24
S60008-3	209
S60009-1	209
S60017-2	82
S60023-7	15
S60028-8	62
S60029-6	56
S60032-6	115
S60033-4	309
S60034-2	383
S60035-0	248
S60036-9	384
S60037-7	422
S60038-5	1008
S60040-7	1043
S60042-3	86
S60045-8	971
S60047-4	736
S60049-0	321
S60052-0	242
S60053-9	468
S60057-1	860
S60060-1	466
S60063-6	975
S60064-4	918
S60067-9	336
S60070-9	169
S60071-7	78
S60074-1	369
S60076-8	71
S60079-2	1131
S60080-6	27
S60082-2	42
S60083-0	510
S60084-9	1009
S60086-5	1009
S60087-3	610
S60088-1	71
S60094-6	692
S60096-2	721
S60097-0	775
S60099-7	392
S60102-0	1084
S60104-7	118
S60106-3	719
S60107-1	366
S60110-1	658
S60112-8	118
S60113-6	659
S60114-4	87
S60116-0	225
S60117-9	1035
S60118-7	226
S60119-5	1048
S60120-9	1068
S60121-7	223
S60122-5	225
S60123-3	1068
S60124-1	1049
S60127-6	1048
S60128-4	1036
S60129-2	385
S60130-6	1064
S60131-4	226
S60145-4	442
S60146-2	442
S60147-0	201
S60149-7	105
S60151-9	597
S60152-7	589
S60153-5	590
S60154-3	553
S60155-1	591
S60157-8	559
S60158-6	655
S60159-4	558
S60160-8	590
S60161-6	590
S60162-4	688
S60163-2	949
S60164-0	589
S60165-9	590
S60166-7	559
S60167-5	559
S60168-3	590
S60169-1	590
S60170-5	559
S60171-3	688
S60172-1	569
S60174-8	553
S60176-4	509
S60177-2	562
S60178-0	577
S60179-9	685
S60180-2	591
S60181-0	680
S60182-9	682
S60183-7	552
S60184-5	679
S60185-3	605
S60186-1	569
S60188-8	569
S60189-6	685
S60191-8	600
S60192-6	646
S60193-4	511
S60194-2	601
S60195-0	755
S60196-9	688
S60197-7	747
S60198-5	578
S60199-3	590
S60202-7	749
S60203-5	597
S60204-3	500
S60205-1	682
S60207-8	801
S60210-8	1043
S60211-6	1047
S60213-2	469
S60214-0	294
S60215-9	1064
S60216-7	1045
S60217-5	1067
S60218-3	1067
S60219-1	1067
S60220-5	1064
S60221-3	1046
S60222-1	1067
S60225-6	1047
S60226-4	1064
S60228-0	1045
S60229-9	1066
S60230-2	1046
S60231-0	1047
S60232-9	480
S60233-7	480
S60234-5	1047
S60236-1	466
S60238-8	1055
S60241-8	480
S60242-6	479
S60243-4	1047
S60244-2	469
S60245-0	480
S60247-7	470
S60248-5	480
S60249-3	480
S60250-7	466
S60251-5	480
S60256-6	790
S60257-4	75
S60260-4	834
S60261-2	1098
S60262-0	26
S60264-7	153
S60265-5	657
S60267-1	1085
S60270-1	153
S60274-4	88
S60276-0	1139
S60277-9	949
S60278-7	35
S60283-3	657
S60284-1	378
S60290-6	1011
S60293-0	15
S60296-5	697
S60303-1	346
S60305-8	18
S60307-4	55
S60308-2	248
S60312-0	10
S60315-5	971
S60317-1	975
S60320-1	379
S60322-8	1
S60324-4	74
S60325-2	69
S60327-9	39
S60330-9	689
S60331-7	881
S60333-3	296
S60336-8	48
S60338-4	601
S60339-2	573
S60340-6	129
S60341-4	600
S60342-2	591
S60343-0	1012
S60344-9	585
S60345-7	129
S60346-5	569
S60347-3	599
S60348-1	571
S60350-3	600
S60351-1	685
S60353-8	594
S60354-6	600
S60355-4	599
S60356-2	129
S60357-0	597
S60358-9	599
S60359-7	130
S60360-0	600
S60361-9	985
S60362-7	655
S60363-5	991
S60364-3	652
S60365-1	480
S60368-6	71
S60369-4	61
S60374-0	10
S60382-1	83
S60386-4	171
S60387-2	314
S60399-6	461
S60400-3	363
S60403-8	417
S60407-0	157
S60408-9	375
S60409-7	221
S60410-0	1047
S60411-9	1065
S60412-7	1047
S60413-5	1047
S60414-3	1067
S60415-1	1047
S60417-8	1047
S60418-6	1046
S60420-8	1047
S60422-4	1066
S60423-2	1025
S60425-9	481
S60426-7	481
S60427-5	481
S60428-3	481
S60429-1	481
S60431-3	1073
S60432-1	1067
S60434-8	1075
S60435-6	1064
S60436-4	1064
S60438-0	1066
S60440-2	468
S60441-0	1064
S60442-9	481
S60443-7	481
S60444-5	481
S60445-3	267
S60446-1	127
S60448-8	559
S60449-6	699
S60451-8	579
S60452-6	610
S60453-4	992
S60454-2	129
S60455-0	573
S60456-9	129
S60457-7	992
S60458-5	600
S60459-3	646
S60460-7	685
S60461-5	130
S60462-3	597
S60463-1	580
S60465-8	599
S60466-6	570
S60467-4	559
S60468-2	665
S60469-0	665
S60470-4	665
S60471-2	654
S60472-0	557
S60473-9	665
S60474-7	649
S60475-5	555
S60476-3	646
S60477-1	1068
S60479-8	1063
S60480-1	263
S60483-6	264
S60484-4	264
S60485-2	1067
S60489-5	368
S60490-9	1036
S60492-5	442
S60496-8	374
S60497-6	755
S60498-4	1036
S60499-2	1036
S60501-8	156
S60502-6	858
S60503-4	711
S60504-2	813
S60505-0	796
S60506-9	796
S60507-7	721
S60508-5	796
S60509-3	804
S60510-7	218
S60511-5	797
S60512-3	359
S60513-1	797
S60515-8	176
S60516-6	802
S60517-4	776
S60518-2	351
S60519-0	958
S60525-5	788
S60526-3	794
S60527-1	951
S60530-1	910
S60533-6	702
S60534-4	737
S60537-9	63
S60538-7	845
S60539-5	1084
S60540-9	343
S60541-7	796
S60542-5	803
S60543-3	1056
S60544-1	946
S60546-8	841
S60547-6	812
S60549-2	443
S60550-6	841
S60551-4	780
S60554-9	803
S60555-7	644
S60558-1	217
S60560-3	793
S60561-1	701
S60566-2	843
S60567-0	1084
S60569-7	443
S60570-0	795
S60572-7	273
S60573-5	811
S60574-3	377
S60575-1	215
S60577-8	821
S60578-6	814
S60579-4	801
S60580-8	442
S60583-2	447
S60584-0	757
S60587-5	745
S60589-1	442
S60591-3	845
S60592-1	462
S60595-6	990
S60596-4	810
S60597-2	443
S60598-0	779
S60599-9	1083
S60600-6	982
S60601-4	878
S60603-0	811
S60605-7	368
S60606-5	983
S60607-3	975
S60608-1	221
S60610-3	878
S60614-6	971
S60615-4	234
S60619-7	987
S60620-0	598
S60621-9	646
S60622-7	600
S60623-5	686
S60624-3	1011
S60625-1	577
S60627-8	587
S60628-6	578
S60629-4	573
S60630-8	589
S60631-6	598
S60632-4	588
S60633-2	574
S60634-0	601
S60635-9	571
S60636-7	854
S60637-5	711
S60638-3	1011
S60639-1	991
S60640-5	678
S60642-1	685
S60644-8	685
S60645-6	130
S60646-4	578
S60647-2	600
S60648-0	854
S60649-9	578
S60650-2	600
S60651-0	605
S60652-9	587
S60653-7	681
S60654-5	601
S60655-3	572
S60656-1	679
S60658-8	605
S60659-6	591
S60661-8	578
S60662-6	951
S60663-4	582
S60664-2	584
S60665-0	705
S60666-9	582
S60667-7	600
S60668-5	680
S60669-3	601
S60670-7	580
S60671-5	581
S60672-3	598
S60673-1	588
S60675-8	854
S60676-6	601
S60677-4	678
S60678-2	597
S60679-0	601
S60680-4	599
S60681-2	589
S60683-9	685
S60684-7	1011
S60685-5	686
S60686-3	854
S60687-1	677
S60689-8	582
S60690-1	598
S60692-8	685
S60693-6	581
S60694-4	681
S60695-2	950
S60696-0	854
S60697-9	578
S60698-7	992
S60699-5	678
S60700-2	680
S60701-0	960
S60702-9	116
S60703-7	1012
S60704-5	992
S60705-3	992
S60706-1	1012
S60708-8	580
S60709-6	1012
S60711-8	580
S60712-6	571
S60713-4	582
S60715-0	790
S60716-9	800
S60718-5	1010
S60719-3	569
S60720-7	992
S60721-5	568
S60722-3	1011
S60723-1	600
S60725-8	569
S60726-6	569
S60727-4	603
S60728-2	657
S60729-0	568
S60730-4	577
S60731-2	679
S60732-0	854
S60733-9	1011
S60734-7	945
S60735-5	678
S60736-3	128
S60737-1	854
S60739-8	703
S60740-1	573
S60742-8	588
S60743-6	670
S60744-4	568
S60745-2	568
S60746-0	128
S60748-7	989
S60749-5	599
S60750-9	605
S60751-7	603
S60752-5	687
S60753-3	1011
S60754-1	128
S60756-8	670
S60757-6	649
S60758-4	568
S60759-2	579
S60760-6	678
S60761-4	1010
S60762-2	672
S60763-0	950
S60764-9	992
S60765-7	568
S60766-5	826
S60767-3	826
S60768-1	687
S60770-3	569
S60771-1	606

Product	Page
S60773-8	1012
S60774-6	670
S60775-4	687
S60776-2	687
S60777-0	603
S60778-9	568
S60779-7	569
S60780-0	567
S60781-9	601
S60782-7	568
S60783-5	569
S60784-3	1012
S60785-1	579
S60787-8	674
S60788-6	127
S60789-4	552
S60790-8	572
S60791-6	991
S60792-4	672
S60793-2	950
S60794-0	599
S60795-9	950
S60796-7	646
S60797-5	548
S60798-3	584
S60799-1	951
S60800-9	562
S60801-7	965
S60802-5	670
S60807-6	1017
S60813-0	690
S60814-9	832
S60815-7	202
S60823-8	1034
S60824-6	787
S60826-2	181
S60827-0	115
S60828-9	202
S60831-9	351
S60832-7	202
S60833-5	202
S60834-3	199
S60839-4	202
S60840-8	182
S60841-6	182
S60842-4	188
S60843-2	181
S60844-0	182
S60845-9	181
S60848-3	181
S60863-7	277
S60864-5	116
S60869-6	156
S60874-2	111
S60875-0	370
S60877-7	997
S60879-3	703
S60882-3	202
S60888-2	689
S60902-1	23
S60904-8	23
S60922-6	466
S60923-4	1049
S60924-2	1077
S60925-0	1064
S60926-9	1049
S60928-5	1045
S60929-3	1064
S60930-7	1049
S60931-5	1067
S60933-1	479
S60935-8	1013
S60936-6	654
S60937-4	548
S60938-2	705
S60939-0	938
S60940-4	846
S60941-2	686
S60942-0	950
S60943-9	960
S60944-7	1013
S60945-5	548
S60946-3	574
S60947-1	960
S60949-8	951
S60950-1	1009
S60953-6	678
S60954-4	573
S60955-2	938
S60956-0	1012
S60958-7	986
S60959-5	654
S60960-9	601
S60961-7	570
S60962-5	1013
S60964-1	654
S60966-8	671
S60967-6	674
S60968-4	548
S60969-2	548
S60970-6	939
S60971-4	1012
S60972-2	674
S60973-0	672
S60974-9	672
S60975-7	601
S60976-5	1010
S60977-3	670
S60978-1	1012
S60980-3	654
S60981-1	557
S60983-8	556
S60984-6	654
S60985-4	1009
S60987-0	670
S60988-9	548
S60989-7	986
S60990-0	960
S60991-9	556
S60992-7	557
S60993-5	939
S60994-3	674
S60995-1	560
S60997-8	568
S60998-6	1013
S60999-4	557
S61000-3	851
S61001-1	602
S61003-8	564
S61004-6	558
S61005-4	600
S61006-2	686
S61007-0	854
S61008-9	674
S61009-7	557
S61010-0	992
S61011-9	558
S61012-7	584
S61013-5	565
S61014-3	599
S61015-1	938
S61017-8	960
S61018-6	601
S61019-4	685
S61020-8	1013
S61021-6	679
S61022-4	556
S61023-2	548
S61024-0	558
S61025-9	654
S61026-7	686
S61027-5	598
S61028-3	599
S61029-1	580
S61030-5	570
S61031-3	569
S61032-1	558
S61034-8	600
S61035-6	580
S61036-4	711
S61037-2	599
S61039-9	678
S61040-2	579
S61041-0	851
S61043-7	674
S61044-5	568
S61045-3	703
S61046-1	569
S61048-8	557
S61049-6	680
S61051-8	558
S61052-6	584
S61053-4	553
S61054-2	557
S61055-0	558
S61056-9	600
S61057-7	846
S61058-5	601
S61059-3	569
S61060-7	569
S61061-5	598
S61062-3	574
S61063-1	554
S61065-8	686
S61066-6	674
S61067-4	1013
S61068-2	582
S61069-0	556
S61070-4	554
S61071-2	601
S61072-0	597
S61073-9	585
S61074-7	601
S61075-5	670
S61076-3	554
S61077-1	579
S61079-8	130
S61082-8	590
S61083-6	548
S61084-4	556
S61085-2	557
S61086-0	599
S61087-9	1010
S61088-7	568
S61089-5	558
S61090-9	601
S61091-7	939
S61092-5	651
S61093-3	846
S61094-1	580
S61096-8	598
S61097-6	950
S61098-4	674
S61099-2	600
S61101-8	577
S61102-6	597
S61103-4	583
S61104-2	706
S61105-0	991
S61106-9	703
S61107-7	582
S61108-5	584
S61109-3	580
S61110-7	674
S61111-5	570
S61112-3	558
S61113-1	582
S61115-8	565
S61117-4	686
S61118-2	950
S61119-0	601
S61120-4	1010
S61121-2	584
S61122-0	674
S61123-9	964
S61124-7	583
S61125-5	584
S61126-3	554
S61127-1	1010
S61129-8	566
S61130-1	547
S61132-8	558
S61133-6	600
S61134-4	565
S61135-2	711
S61136-0	553
S61137-9	510
S61138-7	1013
S61139-5	713
S61140-9	510
S61141-7	593
S61142-5	510
S61143-3	583
S61144-1	583
S61146-8	558
S61147-6	711
S61148-4	597
S61149-2	595
S61150-6	559
S61151-4	583
S61152-2	593
S61153-0	693
S61154-9	558
S61155-7	553
S61156-5	654
S61157-3	1013
S61158-1	582
S61160-3	597
S61161-1	583
S61163-8	981
S61164-6	559
S61165-4	1013
S61166-2	510
S61167-0	579
S61168-9	583
S61169-7	843
S61170-0	601
S61172-7	705
S61173-5	851
S61174-3	1008
S61175-1	597
S61177-8	583
S61178-6	851
S61179-4	594
S61180-8	578
S61181-6	570
S61182-4	508
S61183-2	1009
S61184-0	711
S61185-9	965
S61186-7	583
S61187-5	610
S61188-3	1013
S61189-1	510
S61190-5	510
S61191-3	705
S61192-1	598
S61194-8	573
S61195-6	683
S61196-4	573
S61197-2	510
S61198-0	950
S61199-9	594
S61200-6	605
S61201-4	593
S61202-2	605
S61203-0	604
S61204-9	578
S61205-7	573
S61206-5	509
S61207-3	594
S61208-1	584
S61210-3	610
S61211-1	479
S61213-8	481
S61214-6	467
S61216-2	481
S61217-0	481
S61218-9	370
S61219-7	481
S61220-0	479
S61221-9	482
S61222-7	480
S61223-5	480
S61224-3	480
S61225-1	467
S61227-8	465
S61228-6	481
S61229-4	467
S61230-8	481
S61236-7	240
S61238-3	25
S61244-8	10
S61245-6	651
S61246-4	1009
S61247-2	1009
S61250-2	940
S61251-0	1054
S61255-3	130
S61262-6	60
S61263-4	599
S61264-2	685
S61265-0	685
S61266-9	573
S61267-7	679
S61268-5	682
S61269-3	579
S61270-7	682
S61271-5	950
S61272-3	605
S61273-1	573
S61275-8	583
S61276-6	605
S61277-4	960
S61278-2	991
S61279-0	1013
S61280-4	583
S61281-2	992
S61282-0	582
S61283-9	685
S61284-7	685
S61285-5	558
S61286-3	587
S61287-1	583
S61289-8	687
S61290-1	846
S61292-8	568
S61293-6	706
S61294-4	705
S61295-2	557
S61296-0	597
S61297-9	583
S61298-7	1013
S61299-5	598
S61300-2	578
S61301-0	680
S61302-9	599
S61303-7	960
S61304-5	562
S61305-3	597
S61306-1	1013
S61308-8	559
S61309-6	590
S61311-8	600
S61312-6	559
S61313-4	711
S61314-2	570
S61315-0	580
S61316-9	598
S61317-7	590
S61318-5	590
S61319-3	585
S61320-7	591
S61321-5	589
S61322-3	686
S61323-1	599
S61325-8	711
S61326-6	590
S61327-4	597
S61328-2	679
S61329-0	591
S61330-4	583
S61331-2	1013
S61332-0	597
S61333-9	572
S61334-7	591
S61335-5	938
S61336-3	591
S61337-1	570
S61339-8	597
S61340-1	599
S61342-8	598
S61343-6	568
S61344-4	597
S61345-2	679
S61346-0	991
S61347-9	705
S61348-7	589
S61349-5	655
S61350-9	711
S61351-7	597
S61352-5	597
S61353-3	573
S61354-1	597
S61356-8	580
S61357-6	950
S61358-4	582
S61359-2	851
S61360-6	558
S61361-4	598
S61362-2	583
S61363-0	582
S61364-9	675
S61365-7	1051
S61366-5	1051
S61368-1	1051
S61370-3	311
S61371-1	305
S61373-8	311
S61374-6	276
S61375-4	277
S61376-2	301
S61377-0	226
S61378-9	311
S61379-7	1068
S61380-0	303
S61382-7	301
S61383-5	301
S61384-3	441
S61387-8	128
S61388-6	87
S61390-8	107
S61391-6	279
S61393-2	52
S61394-0	155
S61395-9	470
S61396-7	470
S61401-7	190
S61403-3	357
S61404-1	209
S61406-8	754
S61409-2	210
S61410-6	284
S61413-0	782
S61414-9	357
S61415-7	357
S61416-5	357
S61417-3	358
S61418-1	358
S61420-3	211
S61424-6	211
S61426-2	48
S61428-9	754
S61432-7	358
S61437-8	357
S61438-6	421
S61439-4	786
S61440-8	727
S61441-6	772
S61442-4	754
S61443-2	783
S61444-0	785
S61445-9	754
S61446-7	782
S61447-5	153
S61448-3	956
S61449-1	724
S61451-3	63
S61452-1	727
S61454-8	781
S61455-6	727
S61456-4	727
S61457-2	138
S61458-0	781
S61459-9	727
S61462-9	727
S61463-7	210
S61464-5	727
S61466-1	725
S61468-8	725
S61471-8	424
S61474-2	780
S61475-0	824
S61476-9	723

Product	Page
S61477-7	942
S61479-3	214
S61480-7	754
S61481-5	779
S61482-3	781
S61487-4	1005
S61488-2	329
S61489-0	313
S61490-4	302
S61492-0	314
S61494-7	851
S61495-5	301
S61497-1	301
S61499-8	303
S61500-5	702
S61502-1	426
S61504-8	422
S61506-4	424
S61507-2	247
S61511-0	147
S61513-7	147
S61518-8	747
S61519-6	734
S61521-8	747
S61522-6	748
S61523-4	748
S61524-2	740
S61525-0	735
S61526-9	741
S61527-7	748
S61528-5	745
S61529-3	96
S61530-7	122
S61531-5	721
S61535-8	815
S61536-6	52
S61537-4	203
S61540-4	819
S61541-2	488
S61542-0	351
S61543-9	854
S61544-7	3
S61545-5	19
S61546-3	96
S61547-1	467
S61550-1	478
S61553-6	479
S61554-4	479
S61555-2	479
S61556-0	1073
S61557-9	1077
S61558-7	1049
S61559-5	852
S61560-9	478
S61561-7	1073
S61563-3	466
S61564-1	357
S61566-8	479
S61567-6	479
S61568-4	479
S61569-2	1047
S61570-6	477
S61572-2	465
S61573-0	1047
S61574-9	479
S61575-7	479
S61576-5	1064
S61577-3	1065
S61578-1	1065
S61580-3	1065
S61581-1	1041
S61583-8	1064
S61584-6	1042
S61585-4	1042
S61586-2	1041
S61587-0	1034
S61588-9	1065
S61589-7	1066
S61590-0	1065
S61591-9	1065
S61592-7	1066
S61593-5	1066
S61594-3	672
S61595-1	950
S61597-8	584
S61598-6	685
S61599-4	599
S61600-1	594
S61602-8	598
S61603-6	851
S61604-4	1012
S61605-2	594
S61606-0	600
S61607-9	899
S61608-7	1079
S61609-5	1079
S61610-9	1077
S61611-7	925
S61612-5	992
S61613-3	593
S61614-1	679
S61616-8	239
S61617-6	208
S61622-2	781
S61626-5	721
S61627-3	306
S61630-3	713
S61631-1	556
S61634-6	740
S61635-4	739
S61636-2	377
S61637-0	858
S61638-9	303
S61640-0	1052
S61641-9	722
S61643-5	103
S61644-3	226
S61645-1	301
S61649-4	831
S61650-8	832
S61652-4	377
S61654-0	1063
S61655-9	379
S61657-5	1017
S61658-3	955
S61660-5	111
S61667-2	157
S61670-2	830
S61671-0	830
S61672-9	361
S61674-5	326
S61676-1	868
S61678-8	723
S61684-2	1137
S61686-9	375
S61691-5	41
S61696-6	250
S61697-4	868
S61703-2	477
S61704-0	44
S61705-9	1129
S61709-1	1131
S61710-5	1140
S61711-3	1131
S61714-8	157
S61716-4	157
S61717-2	42
S61719-9	833
S61722-9	331
S61723-7	301
S61725-3	949
S61726-1	937
S61728-8	852
S61731-8	54
S61732-6	56
S61733-4	423
S61736-9	310
S61737-7	725
S61738-5	971
S61739-3	788
S61740-7	138
S61741-5	473
S61742-3	1137
S61746-6	269
S61748-2	270
S61749-0	709
S61750-4	314
S61751-2	314
S61752-0	303
S61753-9	311
S61754-7	232
S61755-5	314
S61756-3	313
S61757-1	311
S61759-8	691
S61760-1	358
S61762-8	207
S61763-6	942
S61764-4	788
S61765-2	134
S61766-0	986
S61768-7	76
S61771-7	207
S61772-5	372
S61773-3	132
S61778-4	772
S61779-2	815
S61784-9	904
S61785-7	951
S61787-3	816
S61790-3	469
S61791-1	366
S61794-6	1056
S61796-2	156
S61798-9	788
S61799-7	266
S61801-2	978
S61802-0	1044
S61803-9	227
S61804-7	294
S61809-8	1091
S61810-1	266
S61813-6	779
S61814-4	953
S61819-5	1061
S61822-5	811
S61823-3	472
S61826-8	366
S61828-4	1044
S61831-4	1044
S61832-2	411
S61833-0	772
S61837-3	366
S61838-1	234
S61840-3	980
S61841-1	141
S61844-6	419
S61845-4	816
S61847-0	155
S61849-7	366
S61853-5	366
S61854-3	381
S61857-8	816
S61859-4	798
S61860-8	358
S61861-6	85
S61862-4	473
S61865-9	1061
S61866-7	433
S61868-3	375
S61869-1	1044
S61870-5	775
S61874-8	416
S61875-6	1041
S61876-4	825
S61877-2	1017
S61879-9	27
S61880-2	832
S61881-0	44
S61882-9	832
S61883-7	832
S61884-5	122
S61885-3	27
S61886-1	282
S61889-6	42
S61892-6	909
S61894-2	157
S61897-7	1095
S61898-5	503
S61902-7	158
S61905-1	868
S61909-4	1010
S61911-6	956
S61912-4	781
S61913-2	94
S61915-9	1047
S61918-3	1066
S61919-1	958
S61920-5	958
S61921-3	998
S61922-1	1079
S61925-6	1043
S61926-4	1078
S61927-2	1066
S61928-0	1045
S61929-9	998
S61930-2	1041
S61931-0	1079
S61932-9	958
S61933-7	1034
S61935-3	958
S61936-1	1051
S61938-8	958
S61939-6	958
S61941-8	1041
S61942-6	1034
S61943-4	1079
S61944-2	1079
S61945-0	1042
S61947-7	222
S61949-3	232
S61950-7	285
S61951-5	276
S61952-3	1056
S61953-1	933
S61955-8	856
S61956-6	232
S61957-4	311
S61959-0	312
S61960-4	283
S61961-2	305
S61962-0	327
S61964-7	345
S61965-5	294
S61966-3	360
S61969-8	331
S61970-1	868
S61972-8	217
S61973-6	797
S61974-4	293
S61975-2	386
S61976-0	797
S61977-9	345
S61978-7	374
S61979-5	242
S61980-9	1057
S61981-7	242
S61984-1	1038
S61987-6	283
S61988-4	285
S61990-6	711
S61993-0	60
S61994-9	33
S61995-7	34
S61996-5	34
S61997-3	34
S61998-1	34
S62003-3	771
S62004-1	771
S62006-8	772
S62007-6	726
S62008-4	783
S62009-2	832
S62010-6	771
S62011-4	213
S62014-9	285
S62015-7	421
S62017-3	357
S62018-1	754
S62020-3	357
S62021-1	771
S62023-8	63
S62024-6	754
S62028-9	358
S62030-0	1017
S62031-9	211
S62032-7	25
S62033-5	308
S62039-4	106
S62041-6	362
S62042-4	95
S62043-2	389
S62044-0	147
S62045-9	365
S62046-7	361
S62055-6	895
S62058-0	462
S62059-9	332
S62063-7	758
S62064-5	331
S62073-4	154
S62074-2	322
S62075-0	429
S62078-5	258
S62079-3	258
S62080-7	258
S62082-3	827
S62086-6	1031
S62087-4	1073
S62089-0	1073
S62090-4	353
S62091-2	1073
S62092-0	1073
S62093-9	1051
S62095-5	1051
S62096-3	1064
S62100-5	958
S62101-3	958
S62104-8	1046
S62105-6	1048
S62106-4	1065
S62107-2	1068
S62108-0	232
S62111-0	1067
S62112-9	1068
S62116-1	263
S62118-8	232
S62121-8	272
S62123-4	373
S62124-2	303
S62125-0	1068
S62127-7	1049
S62130-7	381
S62135-8	106
S62138-2	38
S62139-0	32
S62142-0	38
S62144-7	38
S62145-5	35
S62149-8	838
S62155-2	35
S62158-7	587
S62159-5	582
S62160-9	603
S62161-7	553
S62162-5	587
S62163-3	588
S62164-1	565
S62166-8	564
S62167-6	558
S62168-4	587
S62169-2	564
S62170-6	988
S62171-4	558
S62172-2	564
S62173-0	587
S62174-9	673
S62175-7	588
S62177-3	572
S62178-1	555
S62180-3	554
S62181-1	699
S62183-8	678
S62184-6	669
S62185-4	653
S62186-2	305
S62187-0	273
S62188-9	273
S62189-7	304
S62192-7	305
S62193-5	304
S62194-3	305
S62195-1	36
S62198-6	366
S62199-4	366
S62202-8	219
S62203-6	366
S62204-4	942
S62206-0	358
S62207-9	308
S62208-7	1054
S62209-5	1055
S62210-9	366
S62213-3	701
S62214-1	254
S62216-8	772
S62219-2	780
S62220-6	363
S62221-4	500
S62222-2	777
S62223-0	772
S62225-7	821
S62226-5	755
S62230-3	103
S62231-1	279
S62233-8	789
S62236-2	142
S62237-0	94
S62240-0	817
S62243-5	31
S62244-3	476
S62245-1	51
S62250-8	376
S62251-6	680
S62254-0	92
S62257-5	81
S62258-3	1076
S62261-3	430
S62264-8	814
S62271-0	794
S62272-9	803
S62275-3	1061
S62278-8	794
S62281-8	345
S62282-6	803
S62283-4	801
S62284-2	442
S62285-0	1142
S62286-9	753
S62287-7	141
S62288-5	1083
S62289-3	1057
S62293-1	916
S62295-8	213
S62296-6	1087
S62299-0	263
S62300-8	817
S62302-4	793
S62304-0	878
S62305-9	795
S62306-7	35
S62307-5	316
S62308-3	800
S62309-1	955
S62311-3	438
S62312-1	938
S62314-8	958
S62315-6	938
S62316-4	302
S62317-2	228
S62318-0	1063
S62319-9	1063
S62320-2	342
S62321-0	274
S62322-9	274
S62323-7	1068
S62324-5	938
S62325-3	957
S62326-1	417
S62328-8	439
S62329-6	439
S62331-8	1062
S62332-6	1065
S62333-4	417

Product	Page	Product	Page	Product	Page	Product	Page	Product	Page	Product	Page	Product	Page
S62334-2	428	S62429-2	129	S62521-3	703	S62622-8	1046	S62777-1	802	S62918-9	266	S63053-5	354
S62335-0	438	S62431-4	129	S62522-1	566	S62623-6	1047	S62779-8	1087	S62921-9	464	S63055-1	353
S62339-3	958	S62432-2	130	S62524-8	678	S62624-4	1068	S62783-6	576	S62924-3	797	S63058-6	559
S62341-5	958	S62433-0	130	S62525-6	566	S62625-2	232	S62784-4	239	S62925-1	21	S63059-4	678
S62342-3	958	S62434-9	649	S62526-4	563	S62626-0	305	S62785-2	914	S62927-8	486	S63060-8	442
S62343-1	472	S62435-7	571	S62527-2	685	S62627-9	1020	S62791-7	502	S62929-4	797	S63061-6	259
S62345-8	958	S62436-5	571	S62528-0	566	S62630-9	320	S62792-5	797	S62930-8	1067	S63063-2	956
S62346-6	958	S62437-3	582	S62529-9	686	S62632-5	1039	S62794-1	788	S62931-6	265	S63065-9	600
S62347-4	952	S62438-1	685	S62530-2	748	S62641-4	833	S62796-8	240	S62932-4	486	S63066-7	957
S62348-2	952	S62440-3	583	S62531-0	566	S62642-2	158	S62806-9	746	S62933-2	272	S63070-5	1014
S62349-0	952	S62441-1	580	S62533-7	1020	S62644-9	19	S62811-5	586	S62934-0	371	S63075-6	588
S62350-4	958	S62444-6	434	S62534-5	261	S62645-7	19	S62812-3	601	S62935-9	355	S63076-4	736
S62351-2	998	S62445-4	369	S62535-3	263	S62646-5	218	S62813-1	669	S62936-7	485	S63078-0	744
S62352-0	997	S62446-2	876	S62536-1	1020	S62647-3	218	S62815-8	605	S62937-5	801	S63079-9	807
S62353-9	952	S62447-0	1074	S62538-8	1065	S62651-1	56	S62816-6	573	S62938-3	486	S63081-0	817
S62354-7	958	S62448-9	1056	S62539-6	272	S62653-8	56	S62817-4	686	S62939-1	485	S63086-1	1034
S62355-5	557	S62449-7	272	S62541-8	440	S62656-2	218	S62818-2	683	S62940-5	443	S63092-6	323
S62356-3	593	S62450-0	438	S62542-6	1063	S62657-0	264	S62819-0	953	S62941-3	464	S63098-5	363
S62357-1	593	S62453-5	1062	S62543-4	305	S62659-7	268	S62820-4	587	S62942-1	990	S63099-3	956
S62359-8	590	S62455-1	1075	S62544-2	266	S62660-0	265	S62821-2	584	S62945-6	485	S63101-9	938
S62360-1	557	S62457-8	1055	S62545-0	229	S62663-5	53	S62822-0	953	S62946-4	373	S63105-1	222
S62363-6	600	S62458-6	439	S62546-9	1045	S62667-8	830	S62823-9	711	S62947-2	358	S63107-8	19
S62364-4	559	S62460-8	372	S62547-7	439	S62669-4	832	S62824-7	713	S62949-9	1066	S63108-6	253
S62365-2	580	S62462-4	1068	S62549-3	1017	S62671-6	151	S62825-5	589	S62951-0	876	S63111-6	672
S62366-0	585	S62463-2	1058	S62561-2	1064	S62675-9	1034	S62826-3	713	S62954-5	468	S63112-4	985
S62367-9	559	S62464-0	1055	S62562-0	1064	S62676-7	832	S62827-1	669	S62956-1	442	S63113-2	670
S62368-7	592	S62465-9	1067	S62563-9	1064	S62677-5	832	S62829-8	603	S62963-4	443	S63114-0	380
S62369-5	598	S62466-7	1034	S62564-7	1064	S62679-1	217	S62830-1	603	S62964-2	369	S63115-9	220
S62370-9	580	S62467-5	1068	S62565-5	1064	S62680-5	217	S62832-8	582	S62965-0	1064	S63116-7	983
S62371-7	580	S62468-3	936	S62566-3	481	S62681-3	122	S62833-6	681	S62967-7	340	S63117-5	559
S62372-5	558	S62469-1	1020	S62567-1	1066	S62682-1	947	S62835-2	1058	S62968-5	797	S63119-1	803
S62373-3	559	S62470-5	1067	S62569-8	1066	S62685-6	1013	S62840-9	1058	S62969-3	339	S63122-1	62
S62374-1	597	S62471-3	305	S62570-1	480	S62686-4	170	S62841-7	913	S62971-5	442	S63124-8	294
S62376-8	580	S62472-1	1068	S62572-8	1064	S62687-2	804	S62843-3	1067	S62972-3	812	S63125-6	271
S62377-6	593	S62474-8	1066	S62573-6	1065	S62689-9	384	S62847-6	112	S62973-1	1058	S63126-4	816
S62378-4	573	S62475-6	1036	S62574-4	1064	S62690-2	213	S62848-4	1047	S62976-6	1066	S63134-5	1052
S62380-6	948	S62476-4	1056	S62575-2	1064	S62691-0	615	S62849-2	917	S62977-4	810	S63139-6	485
S62382-2	948	S62477-2	1068	S62576-0	1046	S62692-9	1058	S62851-4	363	S62979-0	485	S63144-2	818
S62383-0	477	S62478-0	571	S62577-9	1046	S62694-5	1042	S62853-0	792	S62981-2	600	S63145-0	264
S62385-7	948	S62479-9	116	S62578-7	1046	S62696-1	1042	S62854-9	1067	S62995-2	588	S63148-5	436
S62386-5	997	S62480-2	116	S62579-5	1077	S62699-6	1044	S62857-3	786	S62996-0	748	S63149-3	559
S62387-3	477	S62481-0	556	S62581-7	1066	S62700-3	954	S62858-1	786	S63000-4	95	S63150-7	1089
S62388-1	1079	S62482-9	581	S62582-5	481	S62701-1	1039	S62860-3	252	S63001-2	434	S63153-1	79
S62390-3	948	S62483-7	582	S62583-3	1064	S62706-2	203	S62861-1	963	S63002-0	892	S63157-4	478
S62391-1	948	S62484-5	992	S62584-1	1046	S62709-7	1067	S62863-8	804	S63005-5	739	S63158-2	1018
S62393-8	479	S62485-3	571	S62586-8	1044	S62710-0	1046	S62865-4	747	S63007-1	814	S63163-9	649
S62394-6	1079	S62486-1	670	S62587-6	1045	S62711-9	938	S62866-2	818	S63010-1	703	S63165-5	688
S62395-4	997	S62488-8	116	S62588-4	1044	S62715-1	1058	S62867-0	144	S63012-8	650	S63167-1	429
S62396-2	478	S62491-8	581	S62589-2	1079	S62721-6	1040	S62868-9	755	S63013-6	681	S63175-2	588
S62397-0	948	S62492-6	988	S62590-6	1039	S62724-0	1043	S62869-7	789	S63014-4	374	S63176-0	128
S62398-9	1050	S62493-4	113	S62592-2	1047	S62725-9	1058	S62875-1	340	S63015-2	652	S63178-7	587
S62399-7	1020	S62494-2	582	S62593-0	1047	S62729-1	233	S62877-8	818	S63016-0	186	S63179-5	359
S62400-4	342	S62495-0	116	S62594-9	482	S62730-5	793	S62878-6	363	S63017-9	815	S63180-9	1041
S62402-0	305	S62496-9	1011	S62595-7	482	S62732-1	792	S62881-6	803	S63019-5	797	S63181-7	1045
S62403-9	433	S62497-7	115	S62598-1	1042	S62734-8	174	S62882-4	1067	S63021-7	372	S63183-3	1064
S62404-7	1074	S62498-5	653	S62601-5	1042	S62737-2	35	S62884-0	1067	S63023-3	369	S63184-1	1034
S62405-5	342	S62499-3	560	S62602-3	1050	S62738-0	22	S62888-3	677	S63026-8	186	S63186-8	1046
S62406-3	372	S62500-0	560	S62603-1	1064	S62740-2	1058	S62889-1	176	S63027-6	829	S63187-6	1063
S62407-1	1062	S62501-9	702	S62606-6	1066	S62741-0	68	S62894-8	326	S63030-6	339	S63188-4	1044
S62409-8	811	S62507-8	566	S62607-4	1031	S62743-7	1042	S62896-4	1058	S63031-4	559	S63189-2	1045
S62412-8	432	S62508-6	669	S62608-2	419	S62744-5	1058	S62898-0	906	S63033-0	42	S63191-4	1045
S62413-6	797	S62509-4	1012	S62609-0	415	S62746-1	1039	S62899-9	810	S63034-9	983	S63193-0	1045
S62417-9	587	S62510-8	951	S62610-4	914	S62748-8	1039	S62900-6	269	S63035-7	369	S63194-9	1048
S62418-7	671	S62512-4	670	S62611-2	229	S62755-0	1058	S62901-4	254	S63036-5	746	S63195-7	1048
S62419-5	678	S62513-2	681	S62612-0	1031	S62763-1	267	S62903-0	1067	S63037-3	140	S63196-5	1048
S62421-7	571	S62514-0	604	S62613-9	1034	S62765-8	1058	S62906-5	369	S63038-1	73	S63197-3	960
S62422-5	588	S62515-9	685	S62614-7	1036	S62768-2	1042	S62907-3	228	S63040-3	819	S63198-1	683
S62423-3	683	S62516-7	748	S62615-5	1033	S62769-0	102	S62908-1	369	S63041-1	811	S63200-7	129
S62424-1	679	S62517-5	653	S62616-3	1049	S62770-4	1040	S62910-3	797	S63045-4	372	S63201-5	750
S62426-8	655	S62518-3	566	S62617-1	1036	S62771-2	1040	S62913-8	798	S63046-2	818	S63202-3	711
S62427-6	653	S62519-1	603	S62619-8	1047	S62774-7	758	S62915-4	293	S63048-9	443	S63203-1	711
S62428-4	129	S62520-5	584	S62620-1	1047	S62775-5	776	S62916-2	485	S63050-0	442	S63206-6	572

Product	Page
S63207-4	669
S63208-2	964
S63209-0	991
S63210-4	747
S63211-2	711
S63212-0	129
S63214-7	129
S63215-5	116
S63216-3	686
S63217-1	711
S63219-8	964
S63220-1	711
S63222-8	686
S63223-6	129
S63224-4	711
S63225-2	582
S63226-0	8
S63228-7	772
S63230-9	468
S63231-7	464
S63232-5	803
S63233-3	1062
S63234-1	1047
S63236-8	156
S63237-6	938
S63238-4	1055
S63245-7	376
S63247-3	262
S63248-1	803
S63250-3	812
S63251-1	861
S63254-6	812
S63255-4	827
S63262-7	297
S63263-5	1085
S63264-3	959
S63265-1	377
S63268-6	227
S63273-2	812
S63274-0	276
S63275-9	803
S63276-7	1057
S63277-5	276
S63281-3	270
S63282-1	803
S63284-8	795
S63288-0	989
S63289-9	158
S63291-0	110
S63299-6	722
S63300-3	378
S63301-1	736
S63303-8	803
S63304-6	741
S63306-2	429
S63308-9	698
S63309-7	916
S63311-9	190
S63312-7	985
S63313-5	156
S63314-3	229
S63315-1	243
S63318-6	811
S63320-8	155
S63322-4	801
S63324-0	136
S63327-5	933
S63330-5	793
S63332-1	789
S63334-8	316
S63336-4	110
S63342-9	768
S63343-7	845
S63352-6	264
S63353-4	56
S63354-2	279
S63357-7	268
S63359-3	228
S63360-7	719
S63361-5	152
S63365-8	212
S63367-4	258
S63369-0	783
S63370-4	293
S63371-2	831
S63374-7	830
S63375-5	950
S63376-3	672
S63377-1	1009
S63379-8	605
S63380-1	687
S63382-8	553
S63383-6	604
S63384-4	570
S63385-2	699
S63386-0	1011
S63387-9	670
S63388-7	687
S63389-5	558
S63390-9	678
S63391-7	686
S63392-5	554
S63393-3	699
S63394-1	603
S63396-8	851
S63397-6	851
S63398-4	563
S63399-2	565
S63401-8	713
S63402-6	699
S63403-4	702
S63404-2	582
S63405-0	572
S63406-9	565
S63407-7	949
S63408-5	555
S63409-3	646
S63410-7	557
S63411-5	595
S63412-3	588
S63413-1	992
S63415-8	646
S63421-2	424
S63422-0	1039
S63423-9	1039
S63425-5	1028
S63426-3	1040
S63427-1	1020
S63429-8	1040
S63430-1	1039
S63432-8	1039
S63433-6	1039
S63434-4	669
S63435-2	681
S63436-0	565
S63437-9	588
S63438-7	671
S63439-5	604
S63440-9	604
S63441-7	1012
S63442-5	672
S63443-3	685
S63444-1	558
S63446-8	566
S63447-6	1012
S63448-4	593
S63449-2	1010
S63450-6	688
S63451-4	604
S63452-2	682
S63453-0	557
S63454-9	598
S63455-7	646
S63456-5	588
S63457-3	564
S63458-1	590
S63460-3	702
S63461-1	1010
S63463-8	565
S63464-6	604
S63465-4	564
S63466-2	586
S63467-0	565
S63468-9	571
S63469-7	1012
S63470-0	570
S63471-9	669
S63472-7	653
S63473-5	854
S63474-3	669
S63475-1	653
S63477-8	572
S63478-6	605
S63479-4	652
S63480-8	592
S63481-6	711
S63482-4	711
S63483-2	711
S63484-0	712
S63485-9	712
S63486-7	592
S63487-5	605
S63488-3	685
S63489-1	572
S63490-5	584
S63491-3	592
S63492-1	712
S63494-8	592
S63495-6	950
S63497-2	376
S63500-6	1058
S63501-4	354
S63502-2	348
S63503-0	284
S63505-7	217
S63506-5	555
S63507-3	1064
S63508-1	1046
S63510-3	1047
S63511-1	1066
S63513-8	1034
S63514-6	311
S63515-4	1067
S63516-2	1041
S63517-0	1064
S63518-9	1041
S63519-7	1064
S63520-0	1067
S63521-9	1066
S63522-7	1065
S63523-5	1066
S63524-3	1067
S63525-1	1045
S63527-8	1065
S63528-6	1067
S63530-8	1051
S63532-4	1065
S63533-2	1046
S63534-0	1046
S63535-9	1045
S63536-7	1046
S63537-5	958
S63540-5	958
S63541-3	1075
S63542-1	1075
S63544-8	944
S63545-6	1067
S63546-4	958
S63547-2	1067
S63548-0	1064
S63549-9	1066
S63550-2	1064
S63552-9	1067
S63553-7	1065
S63554-5	1067
S63555-3	1065
S63556-1	1066
S63558-8	958
S63559-6	1048
S63561-8	1045
S63562-6	1046
S63563-4	1043
S63564-2	1067
S63565-0	1042
S63566-9	1047
S63567-7	1047
S63568-5	1045
S63569-3	1046
S63570-7	1041
S63571-5	1041
S63572-3	1046
S63573-1	1042
S63575-8	240
S63576-6	753
S63579-0	440
S63580-4	439
S63582-0	1047
S63583-9	1048
S63584-7	1046
S63585-5	1067
S63587-1	1067
S63590-1	1046
S63592-8	959
S63593-6	1045
S63595-2	438
S63597-9	226
S63598-7	1067
S63599-5	1079
S63601-0	1041
S63602-9	1063
S63603-7	1068
S63604-5	1067
S63608-8	392
S63609-6	1061
S63611-8	1068
S63612-6	1039
S63614-2	424
S63615-0	938
S63616-9	1040
S63618-5	1039
S63619-3	852
S63620-7	852
S63621-5	1043
S63622-3	675
S63623-1	685
S63625-8	669
S63626-6	673
S63628-2	950
S63629-0	675
S63630-4	582
S63631-2	570
S63632-0	852
S63633-9	129
S63634-7	851
S63635-5	713
S63636-3	580
S63637-1	591
S63639-8	847
S63640-1	992
S63642-8	653
S63643-6	846
S63644-4	684
S63645-2	584
S63646-0	652
S63647-9	557
S63648-7	654
S63649-5	597
S63650-9	598
S63651-7	849
S63652-5	593
S63653-3	593
S63654-1	594
S63656-8	713
S63657-6	598
S63658-4	584
S63659-2	598
S63660-6	689
S63661-4	960
S63662-2	605
S63663-0	1012
S63664-9	712
S63665-7	597
S63666-5	555
S63668-1	755
S63670-3	817
S63673-8	1075
S63674-6	995
S63675-4	790
S63676-2	604
S63679-7	992
S63680-0	585
S63681-9	1088
S63685-1	791
S63688-6	433
S63691-6	756
S63693-2	557
S63694-0	480
S63698-3	960
S63699-1	589
S63730-0	220
S63734-3	217
S63743-2	221
S63748-3	227
S63749-1	378
S63758-0	221
S63762-9	106
S63763-7	15
S63764-5	1056
S63766-1	112
S63768-8	811
S63769-6	845
S63771-8	706
S63772-6	913
S63774-2	789
S63776-9	1040
S63777-7	857
S63778-5	1035
S63779-3	1033
S63780-7	1028
S63781-5	1034
S63782-3	1034
S63783-1	1028
S63785-8	263
S63786-6	272
S63787-4	1040
S63788-2	1033
S63789-0	831
S63790-4	330
S63792-0	421
S63793-9	209
S63794-7	212
S63795-5	782
S63796-3	556
S63797-1	556
S63799-8	679
S63800-5	675
S63801-3	572
S63802-1	556
S63804-8	960
S63805-6	651
S63806-4	553
S63807-2	649
S63809-9	573
S63810-2	683
S63811-0	598
S63812-9	597
S63813-7	588
S63814-5	583
S63815-3	650
S63816-1	554
S63818-8	677
S63819-6	556
S63821-8	699
S63822-6	598
S63823-4	682
S63824-2	698
S63825-0	849
S63826-9	850
S63827-7	991
S63828-5	675
S63829-3	600
S63830-7	555
S63831-5	600
S63832-3	571
S63833-1	671
S63835-8	671
S63836-6	849
S63837-4	584
S63838-2	673
S63839-0	584
S63840-4	1011
S63841-2	848
S63842-0	604
S63843-9	713
S63844-7	673
S63845-5	951
S63846-3	600
S63847-1	581
S63849-8	950
S63850-1	584
S63852-8	681
S63853-6	653
S63854-4	650
S63855-2	681
S63856-0	1028
S63859-5	847
S63860-9	847
S63862-5	962
S63863-3	377
S63866-8	957
S63867-6	794
S63868-4	794
S63869-2	255
S63870-6	1054
S63871-4	940
S63872-2	374
S63874-9	876
S63876-5	798
S63877-3	1062
S63878-1	811
S63880-3	489
S63881-1	1062
S63886-2	798
S63887-0	209
S63888-9	1033
S63889-7	817
S63890-0	1051
S63892-7	255
S63895-1	91
S63897-8	821
S63898-6	171
S63899-4	786
S63900-1	1000
S63902-8	318
S63903-6	683
S63904-4	739
S63905-2	852
S63906-0	811
S63910-9	776
S63911-7	145
S63912-5	706
S63913-3	225
S63914-1	815
S63916-8	293
S63917-6	341
S63919-2	748
S63920-6	669
S63923-0	570
S63924-9	584
S63925-7	570
S63926-5	586
S63927-3	1012
S63928-1	986
S63930-3	604
S63931-1	712
S63933-8	669
S63934-6	712
S63935-4	1009
S63936-2	986
S63937-0	688
S63938-9	988
S63940-0	606
S63941-9	1009
S63942-7	991
S63943-5	712
S63944-3	757
S63945-1	669
S63947-8	653
S63948-6	673
S63949-4	1010
S63950-8	1012
S63951-6	849
S63952-4	653
S63953-2	1010
S63954-0	665
S63955-9	852
S63956-7	951
S63957-5	1034
S63958-3	1051
S63959-1	1036
S63960-5	956
S63961-3	957
S63962-1	1061
S63964-8	957

Product Page

S63965-6......1055
S63966-4........937
S63967-2........957
S63968-0......1063
S63969-9......1034
S63970-2......1051
S63972-9......1038
S63973-7......1037
S63974-5......1051
S63975-3......1044
S63976-1........300
S63978-8........589
S63979-6........845
S63983-4........265
S63984-2........938
S63986-9........243
S63987-7........240
S63989-3........237
S63990-7........977
S63995-8........892
S63996-6........294
S63997-4........334
S63998-2........892
S64001-8........380
S64002-6........231
S64004-2........330
S64005-0........350
S64006-9......1014
S64007-7........298
S64008-5........986
S64011-5........349
S64012-3........975
S64013-1........264
S64015-8........231
S64016-6........379
S64017-4........298
S64018-2........985
S64019-0........368
S64021-2........587
S64022-0........852
S64023-9........587
S64024-7........572
S64025-5........572
S64026-3........590
S64027-1........686
S64029-8........950
S64030-1........582
S64032-8........572
S64033-6........587
S64034-4........572
S64035-2........588
S64036-0........852
S64037-9........580
S64038-7........590
S64039-5........911
S64040-9......1012
S64041-7........568
S64042-5......1012
S64043-3......1012
S64044-1........680
S64046-8........683
S64047-6........654
S64048-4........556
S64049-2........605
S64051-4........654
S64052-2........655
S64053-0........684
S64054-9......1012
S64055-7........601
S64056-5........110
S64057-3........892
S64060-3......1073
S64064-6......1058
S64065-4......1055
S64066-2......1051
S64067-0......1058
S64068-9......1055
S64069-7......1055
S64070-0......1055
S64071-9......1055
S64072-7......1055
S64075-1......1058
S64077-8......1055
S64078-6......1055
S64079-4........481
S64080-8........481
S64081-6......1055
S64082-4......1043
S64083-2......1043
S64084-0......1043
S64086-7......1043
S64087-5......1048
S64088-3......1046
S64089-1......1045
S64090-5......1047
S64091-3......1055
S64092-1......1055
S64094-8......1055
S64095-6......1055
S64096-4......1043
S64097-2......1043
S64098-0......1045
S64099-9......1048
S64100-6......1048
S64101-4......1045
S64102-2......1047
S64103-0......1045
S64104-9......1045
S64106-5......1055
S64107-3......1048
S64108-1......1047
S64110-3......1045
S64111-1......1041
S64113-8......1040
S64114-6........470
S64116-2..........89
S64117-0..........50
S64119-7..........60
S64120-0..........89
S64122-7..........16
S64123-5........368
S64124-3........200
S64125-1........388
S64127-8........379
S64128-6........353
S64130-8........982
S64134-0........203
S64138-3..........33
S64140-5..........69
S64141-3........982
S64145-6..........89
S64146-4..........16
S64149-9........239
S64150-2..........28
S64152-9..........67
S64158-8........355
S64162-6........172
S64165-0........156
S64172-3..........89
S64173-1........892
S64175-8..........16
S64176-6........967
S64177-4........967
S64178-2........588
S64179-0........300
S64180-4........936
S64181-2........299
S64182-0......1038
S64184-7........319
S64185-5......1034
S64186-3......1035
S64187-1........933
S64190-1........802
S64192-8........377
S64196-0........447
S64199-5........816
S64200-2........696
S64201-0........706
S64202-9........772
S64204-5........772
S64208-8........876
S64209-6........398
S64213-4........786
S64215-0........775
S64216-9........468
S64217-7........775
S64219-3........294
S64220-7........485
S64221-5........156
S64223-1........442
S64225-8........776
S64226-6........816
S64227-4........447
S64229-0........693
S64230-4........709
S64231-2........485
S64232-0........701
S64233-9........812
S64234-7........485
S64236-3........776
S64237-1........875
S64239-8........775
S64240-1........130
S64242-8........216
S64243-6........130
S64245-2........698
S64246-0........778
S64247-9........761
S64248-7........786
S64249-5........216
S64252-5........818
S64254-1........801
S64256-8........214
S64257-6........159
S64258-4..........45
S64259-2......1041
S64260-6........270
S64262-2........257
S64265-7........328
S64266-5........215
S64267-3......1056
S64268-1........242
S64270-3........754
S64275-4........140
S64278-9........500
S64279-7........847
S64280-0........829
S64281-9........111
S64282-7........322
S64283-5......1043
S64289-4........228
S64290-8........291
S64291-6........193
S64295-9........497
S64296-7........496
S64297-5......1028
S64298-3......1028
S64300-9........350
S64301-7........354
S64302-5........332
S64303-3........298
S64310-6........298
S64311-4........371
S64320-3......1058
S64321-1......1058
S64323-8......1058
S64324-6......1028
S64325-4......1058
S64326-2......1033
S64327-0......1055
S64328-9......1061
S64329-7......1044
S64330-0......1058
S64331-9......1057
S64332-7......1058
S64335-1........843
S64338-6........702
S64339-4........811
S64340-8........356
S64341-6........360
S64345-9............6
S64347-5........170
S64348-3..........14
S64349-1........967
S64350-5........354
S64354-8........322
S64355-6........962
S64356-4......1043
S64357-2......1028
S64358-0......1058
S64359-9........361
S64360-2......1058
S64361-0......1058
S64362-9......1043
S64363-7......1057
S64364-5......1059
S64365-3......1041
S64368-8........362
S64369-6........350
S64371-8........305
S64372-6........303
S64373-4........303
S64374-2........303
S64375-0........303
S64376-9........305
S64377-7........303
S64378-5........305
S64379-3........301
S64380-7........585
S64381-5........682
S64382-3........574
S64383-1........673
S64385-8........586
S64386-6........600
S64387-4........682
S64388-2........680
S64389-0........673
S64390-4........685
S64392-0........852
S64393-9........688
S64394-7........688
S64395-5........988
S64396-3........682
S64397-1........679
S64399-8........601
S64400-5........688
S64401-3........593
S64402-1........593
S64404-8........945
S64405-6........573
S64406-4........583
S64407-2........649
S64408-0........594
S64409-9........681
S64410-2........682
S64411-0........565
S64412-9........665
S64413-7........573
S64414-5........650
S64415-3........579
S64416-1........599
S64418-8........603
S64419-6........556
S64421-8........566
S64422-6........686
S64423-4........683
S64424-2........683
S64425-0........683
S64426-9........945
S64427-7........675
S64428-5........589
S64429-3........748
S64430-7........679
S64431-5........589
S64432-3........603
S64433-1........950
S64435-8........683
S64436-6........913
S64437-4........590
S64438-2........582
S64439-0........603
S64440-4........556
S64441-2........688
S64442-0........583
S64443-9........850
S64444-7........600
S64445-5........589
S64446-3........586
S64447-1........684
S64449-8........686
S64450-1........650
S64452-8......1009
S64453-6........603
S64454-4........950
S64455-2........600
S64456-0........600
S64457-9........693
S64458-7........917
S64459-5........767
S64460-9........767
S64461-7........916
S64462-5........604
S64463-3........558
S64464-1........573
S64466-8........590
S64467-6........650
S64468-4........605
S64470-6........650
S64471-4........711
S64472-2........563
S64473-0........965
S64474-9........566
S64475-7........555
S64476-5........603
S64477-3........757
S64478-1........601
S64480-3........849
S64481-1........605
S64483-8........992
S64485-4......1139
S64489-7........997
S64491-9........776
S64493-5........817
S64494-3........876
S64498-6........671
S64499-4........605
S64500-1........992
S64502-8........992
S64503-6........670
S64504-4........992
S64505-2........592
S64506-0........588
S64507-9........587
S64508-7........588
S64509-5........598
S64510-9........575
S64511-7........712
S64512-5........712
S64513-3........553
S64514-1........575
S64516-8........592
S64517-6........592
S64518-4........590
S64519-2........757
S64520-6........592
S64522-2......1012
S64523-0........570
S64524-9........992
S64525-7........556
S64526-5......1013
S64527-3........554
S64528-1........579
S64530-3........575
S64531-1........575
S64533-8........686
S64534-6........991
S64535-4........991
S64536-2........714
S64537-0........112
S64540-0..........72
S64544-3........876
S64545-1........768
S64547-8........768
S64555-9......1057
S64556-7........303
S64557-5......1127
S64564-8........861
S64565-6........303
S64567-2........421
S64568-0........756
S64575-3......1075
S64578-8........159
S64579-6........116
S64581-8..........65
S64582-6........146
S64586-9........721
S64588-5........245
S64589-3........266
S64590-7........827
S64591-5........745
S64592-3......1032
S64593-1........704
S64595-8........471
S64596-6........719
S64597-4......1030
S64600-8......1075
S64601-6........943
S64602-4......1035
S64603-2........706
S64604-0........719
S64605-9........475
S64606-7........467
S64607-5........706
S64608-3........719
S64609-1........706
S64610-5........253
S64615-6........153
S64616-4........154
S64618-0........421
S64619-9........211
S64620-2........279
S64621-0........186
S64622-9......1052
S64623-7........181
S64624-5........265
S64626-1........174
S64631-8........279
S64632-6........216
S64633-4........288
S64634-2........265
S64635-0........196
S64636-9........375
S64637-7........178
S64638-5........178
S64640-7........199
S64641-5......1052
S64642-3........983
S64646-6........992
S64647-4........588
S64648-2........712
S64649-0........580
S64650-4........600
S64651-2........991
S64652-0........750
S64653-9........580
S64654-7........509
S64655-5........510
S64656-3........587
S64657-1........595
S64659-8........587
S64660-1........575
S64662-8........580
S64663-6........597
S64664-4........573
S64665-2........992
S64666-0........582
S64667-9........665
S64668-7........593
S64669-5........992
S64670-9........598
S64671-7........593
S64672-5........597
S64673-3........682
S64674-1........581
S64676-8........685
S64677-6........591
S64678-4........565
S64679-2........575
S64680-6........598
S64681-4........598
S64682-2........598
S64683-0........595
S64684-9........594
S64685-7........712
S64686-5........712
S64688-1........596
S64690-3........586
S64691-1........570
S64693-8........566
S64694-6........581
S64695-4........688
S64696-2........688
S64697-0........577
S64698-9........568

Product Page

S64699-7........572
S64700-4........992
S64701-2........960
S64702-0........673
S64703-9........673
S64704-7........594
S64705-5........578
S64706-3........685
S64707-1........688
S64709-8........673
S64710-1........653
S64712-8........654
S64713-6........129
S64714-4........595
S64715-2........570
S64716-0........688
S64717-9........592
S64718-7........673
S64720-9........549
S64721-7........557
S64722-5........129
S64723-3........129
S64724-1........593
S64726-8........549
S64727-6........688
S64728-4........599
S64729-2........372
S64730-6........275
S64731-4........823
S64733-0........464
S64734-9........464
S64736-5........763
S64737-3........755
S64741-1........414
S64743-8........472
S64744-6........722
S64745-4........943
S64746-2........139
S64747-0........818
S64748-9........140
S64749-7........139
S64750-0........499
S64751-9........297
S64752-7........330
S64753-5........919
S64754-3........919
S64757-8........354
S64758-6........765
S64759-4........472
S64760-8........806
S64761-6........984
S64762-4........354
S64763-2........764
S64764-0........396
S64765-9........816
S64767-5........993
S64768-3........764
S64769-1........366
S64770-5........147
S64771-3..........40
S64772-1........156
S64774-8..........40
S64775-6..........40
S64776-4..........40
S64777-2........251
S64778-0........251
S64782-9..........29
S64786-1......1010
S64789-6........472
S64791-8........104
S64797-7........276
S64800-0........263

Product Page

S64802-7......1052
S64803-5........178
S64804-3......1052
S64805-1........188
S64807-8......1052
S64808-6........370
S64809-4........186
S64810-8........838
S64812-4........279
S64815-9......1052
S64816-7........178
S64817-5........271
S64818-3........335
S64821-3........257
S64824-8......1083
S64825-6........681
S64827-2........181
S64828-0........181
S64830-2......1008
S64831-0........265
S64832-9..........82
S64835-3......1013
S64838-8........212
S64841-8..........45
S64843-4......1079
S64846-9........286
S64848-5........173
S64850-7............4
S64853-1..........42
S64859-0........220
S64862-0........291
S64863-9........380
S64864-7........311
S64866-3........995
S64867-1..........19
S64870-1........325
S64873-6........233
S64874-4........989
S64875-2........792
S64877-9........792
S64878-7..........80
S64879-5........875
S64880-9........300
S64881-7........912
S64883-3........768
S64884-1........768
S64886-8........768
S64887-6......1028
S64889-2......1033
S64891-4........716
S64892-2........236
S64894-9........151
S64895-7........209
S64896-5........151
S64897-3........151
S64898-1..........24
S64901-5........209
S64903-1........209
S64905-8........151
S64906-6..........24
S64907-4........751
S64908-2........952
S64909-0........234
S64910-4........215
S64912-0........841
S64913-9........139
S64914-7........719
S64915-5........751
S64917-1........952
S64919-8........719
S64922-8......1080
S64923-6........741

Product Page

S64926-0........728
S64927-9........282
S64928-7........259
S64930-9........138
S64931-7........749
S64932-5........447
S64933-3......1020
S64934-1........259
S64936-8........758
S64937-6........717
S64938-4........901
S64939-2........252
S64940-6........148
S64941-4........743
S64945-7......1045
S64946-5......1046
S64947-3........654
S64948-1........562
S64950-3........686
S64951-1........598
S64953-8........598
S64954-6........605
S64955-4........650
S64956-2........565
S64957-0........565
S64958-9........681
S64959-7........509
S64960-0........683
S64961-9........991
S64962-7........712
S64963-5........552
S64967-8........693
S64968-6........565
S64969-4........553
S64970-8........859
S64972-4........214
S64974-0..........55
S64975-9........237
S64977-5........296
S64978-3........682
S64980-5........587
S64981-3........572
S64982-1........672
S64986-4........758
S64987-2........764
S64988-0........764
S64989-9..........36
S64990-2........686
S64991-0........605
S64992-9........683
S64993-7........592
S64994-5........595
S64995-3........595
S64996-1........950
S64998-8........851
S64999-6........573
S65000-5........767
S65001-3........595
S65002-1........683
S65004-8........595
S65005-6........686
S65006-4........653
S65007-2......1012
S65008-0........655
S65009-9........683
S65010-2........689
S65011-0......1012
S65012-9........611
S65013-7........672
S65014-5........681
S65015-3........682
S65016-1........925

Product Page

S65018-8........683
S65019-6........553
S65021-8........115
S65022-6........566
S65023-4........665
S65024-2........601
S65025-0........555
S65026-9........846
S65027-7........553
S65028-5........593
S65029-3........673
S65030-7........594
S65031-5........689
S65032-3........689
S65033-1........650
S65035-8........605
S65036-6......1012
S65037-4......1012
S65038-2........991
S65039-0........566
S65040-4........666
S65041-2........555
S65042-0........650
S65043-9........672
S65044-7........595
S65045-5........665
S65046-3........573
S65047-1........573
S65049-8........594
S65050-1........712
S65052-8........992
S65053-6........587
S65054-4........672
S65055-2........672
S65056-0........672
S65057-9........669
S65058-7........672
S65059-5........671
S65060-9........652
S65061-7........652
S65062-5........598
S65063-3........681
S65066-8........689
S65067-6........683
S65068-4........652
S65069-2........652
S65070-6..........75
S65073-0........764
S65074-9........556
S65075-7........686
S65077-3........991
S65078-1........464
S65080-3........464
S65083-8........946
S65084-6........946
S65085-4........464
S65086-2........597
S65087-0........938
S65088-9........684
S65089-7........566
S65090-0........946
S65091-9........946
S65092-7........946
S65093-5........945
S65094-3........946
S65095-1........925
S65097-8........946
S65098-6........592
S65099-4........585
S65100-1........605
S65102-8........249
S65103-6........244

Product Page

S65104-4........946
S65105-2........925
S65106-0........244
S65107-9........249
S65108-7........945
S65109-5........249
S65110-9........566
S65111-7........565
S65112-5........555
S65113-3........594
S65114-1........328
S65116-8..........84
S65120-6........320
S65121-4........138
S65122-2........712
S65123-0........572
S65124-9........673
S65125-7........464
S65126-5..........39
S65127-3........249
S65128-1........946
S65130-3........925
S65131-1........946
S65133-8........249
S65134-6........681
S65135-4........598
S65136-2........570
S65138-9........257
S65140-0........142
S65142-7........761
S65143-5........988
S65144-3........309
S65145-1........431
S65147-8........588
S65148-6........572
S65150-8........149
S65152-4........243
S65153-2........953
S65154-0........157
S65156-7........764
S65157-5......1011
S65158-3........849
S65159-1........587
S65160-5........583
S65162-1........412
S65167-2........335
S65169-9........365
S65173-7........209
S65174-5........858
S65187-7........899
S65188-5........733
S65191-5........320
S65193-1........859
S65195-8........111
S65199-0........764
S65200-8........764
S65203-2..........41
S65207-5........829
S65208-3........787
S65209-1........187
S65210-5........672
S65211-3........652
S65212-1........672
S65214-8........652
S65215-6........599
S65216-4........595
S65217-2........566
S65218-0........685
S65219-9........565
S65220-2........594
S65221-0........566
S65222-9........605

Product Page

S65223-7........575
S65224-5........556
S65225-3........852
S65226-1........574
S65228-8........593
S65229-6........593
S65231-8........688
S65232-6........688
S65233-4........593
S65234-2........950
S65235-0........587
S65236-9........849
S65237-7........583
S65238-5........570
S65239-3........587
S65240-7........573
S65241-5......1059
S65242-3......1050
S65243-1......1050
S65246-6........415
S65247-4......1075
S65248-2......1062
S65249-0......1075
S65250-4......1075
S65251-2......1075
S65253-9........817
S65255-5........818
S65256-3........160
S65257-1........925
S65262-8........445
S65263-6........431
S65264-4........117
S65265-2........382
S65266-0........566
S65268-7........594
S65269-5........965
S65270-9........587
S65271-7........605
S65272-5........682
S65273-3........595
S65274-1........580
S65276-8........679
S65277-6........595
S65278-4........609
S65279-2........566
S65280-6........672
S65281-4........672
S65282-2........688
S65286-5........372
S65288-1........413
S65294-6........859
S65295-4........899
S65296-2........795
S65297-0........122
S65299-7........355
S65301-2........358
S65302-0........859
S65304-7........673
S65305-5........652
S65306-3........652
S65307-1........672
S65309-8........220
S65310-1........217
S65312-8..........78
S65313-6......1063
S65314-4......1028
S65315-2......1063
S65316-0......1034
S65317-9........300
S65318-7........916
S65319-5......1063
S65320-9........842

Product Page

S65321-7........375
S65324-1........271
S65327-6........831
S65328-4........713
S65329-2........583
S65330-6........713
S65331-4........594
S65332-2........693
S65333-0........713
S65334-9........738
S65335-7........699
S65336-5........556
S65337-3........564
S65338-1........556
S65340-3........713
S65343-8........609
S65344-6........598
S65345-4........611
S65346-2........713
S65347-0........699
S65348-9........611
S65349-7........595
S65350-0........294
S65351-9........713
S65352-7........696
S65353-5........601
S65354-3........653
S65355-1........714
S65357-8........714
S65358-6........186
S65359-4........653
S65360-8........576
S65361-6........655
S65362-4........587
S65363-2........611
S65365-9........699
S65366-7........590
S65367-5........610
S65368-3........555
S65369-1......1141
S65370-5........683
S65371-3........683
S65372-1........713
S65374-8........688
S65375-6......1142
S65376-4........580
S65377-2........593
S65379-9........688
S65380-2........273
S65381-0........593
S65382-9........558
S65383-7........594
S65384-5........214
S65385-3........186
S65386-1........573
S65388-8........804
S65392-6........554
S65394-2........591
S65395-0........581
S65398-5........714
S65400-0........558
S65402-7........171
S65403-5........847
S65408-6........688
S65409-4......1011
S65410-8........654
S65411-6........559
S65412-4........831
S65415-9........816
S65416-7........703
S65417-5........553
S65418-3........704

Product Page

S65419-1......559
S65420-5......350
S65421-3......714
S65422-1......581
S65424-8......698
S65425-6......699
S65426-4......989
S65427-2......595
S65428-0......681
S65430-2......599
S65431-0......575
S65432-9......713
S65433-7......593
S65435-3......604
S65436-1......767
S65438-8......713
S65442-6......672
S65444-2......580
S65446-9......594
S65448-5......713
S65451-5......595
S65453-1......339
S65455-8......678
S65457-4......666
S65459-0......688
S65460-4......1009
S65461-2......601
S65462-0......415
S65463-9......713
S65465-5......826
S65467-1......417
S65472-8......364
S65475-2......983
S65477-9......737
S65478-7......70
S65479-5......773
S65480-9......588
S65481-7......575
S65482-5......320
S65483-3......949
S65484-1......1063
S65486-8......559
S65487-6......14
S65488-4......765
S65489-2......1039
S65490-6......948
S65491-4......1039
S65492-2......959
S65494-9......556
S65495-7......28
S65496-5......945
S65497-3......174
S65498-1......713
S65502-3......983
S65503-1......749
S65507-4......547
S65508-2......702
S65510-4......228
S65512-0......227
S65513-9......790
S65515-5......412
S65516-3......716
S65517-1......813
S65523-6......420
S65526-0......412
S65527-9......786
S65528-7......260
S65532-5......913
S65533-3......794
S65534-1......691
S65539-2......705
S65540-6......1050
S65541-4......792
S65542-2......218
S65545-7......825
S65547-3......1092
S65548-1......372
S65554-6......127
S65556-2......824
S65558-9......826
S65560-0......1052
S65561-9......991
S65563-5......825
S65564-3......576
S65567-8......826
S65572-4......1043
S65575-9......778
S65576-7......803
S65578-3......576
S65579-1......878
S65580-5......210
S65581-3......738
S65582-1......1059
S65584-8......941
S65585-6......1061
S65586-4......776
S65588-0......95
S65589-9......892
S65590-2......755
S65591-0......277
S65593-7......342
S65594-5......804
S65595-3......804
S65596-1......803
S65598-8......576
S65599-6......804
S65600-3......423
S65601-1......817
S65603-8......76
S65604-6......822
S65605-4......903
S65606-2......364
S65607-0......293
S65608-9......1033
S65611-9......1052
S65612-7......340
S65613-5......829
S65615-1......1053
S65617-8......85
S65620-8......276
S65623-2......375
S65625-9......337
S65626-7......801
S65627-5......215
S65628-3......804
S65629-1......804
S65630-5......801
S65631-3......803
S65632-1......779
S65634-8......293
S65635-6......803
S65637-2......804
S65638-0......960
S65639-9......10
S65640-2......832
S65641-0......602
S65643-7......556
S65648-8......691
S65649-6......297
S65651-8......371
S65655-0......382
S65659-3......389
S65662-3......479
S65663-1......387
S65665-8......346
S65666-6......965
S65668-2......344
S65670-4......690
S65672-0......821
S65674-7......793
S65676-3......440
S65677-1......342
S65680-1......985
S65683-6......689
S65685-2......727
S65686-0......768
S65687-9......311
S65688-7......690
S65689-5......691
S65690-9......690
S65691-7......738
S65692-5......691
S65694-1......940
S65697-6......746
S65699-2......714
S65702-6......706
S65706-9......358
S65707-7......398
S65709-3......461
S65710-7......691
S65711-5......982
S65716-6......704
S65717-4......726
S65718-2......284
S65719-0......245
S65723-9......245
S65725-5......953
S65726-3......967
S65729-8......291
S65730-1......909
S65735-2......704
S65738-7......345
S65740-9......286
S65741-7......222
S65742-5......798
S65743-3......296
S65744-1......312
S65746-8......767
S65747-6......219
S65748-4......420
S65750-6......1038
S65752-2......1011
S65754-9......945
S65755-7......826
S65757-3......712
S65760-3......714
S65763-8......788
S65764-6......384
S65765-4......713
S65766-2......726
S65768-9......602
S65769-7......58
S65770-0......587
S65772-7......587
S65773-5......24
S65774-3......575
S65775-1......852
S65777-8......986
S65778-6......593
S65779-4......714
S65780-8......916
S65781-6......596
S65782-4......276
S65783-2......673
S65784-0......594
S65785-9......592
S65786-7......588
S65787-5......647
S65789-1......380
S65791-3......906
S65794-8......953
S65795-6......145
S65797-2......765
S65800-6......754
S65803-0......470
S65804-9......475
S65806-5......753
S65807-3......388
S65808-1......1009
S65814-6......475
S65815-4......902
S65816-2......139
S65817-0......388
S65818-9......139
S65819-7......753
S65820-0......349
S65821-9......856
S65822-7......765
S65824-3......235
S65825-1......388
S65827-8......140
S65828-6......852
S65830-8......244
S65831-6......1011
S65832-4......1000
S65833-2......819
S65835-9......734
S65836-7......1001
S65837-5......871
S65838-3......38
S65840-5......83
S65842-1......1000
S65846-4......789
S65847-2......319
S65848-0......713
S65849-9......565
S65851-0......950
S65852-9......753
S65853-7......425
S65855-3......373
S65856-1......787
S65858-8......788
S65861-8......807
S65865-0......35
S65866-9......4
S65872-3......723
S65873-1......345
S65877-4......1127
S65878-2......61
S65883-9......215
S65886-3......341
S65887-1......67
S65892-8......191
S65894-4......158
S65895-2......61
S65896-0......8
S65898-7......104
S65905-3......238
S65906-1......238
S65909-6......192
S65914-2......250
S65925-8......1013
S65930-4......1052
S65932-0......1084
S65934-7......82
S65936-3......204
S65940-1......480
S65942-8......108
S65947-9......833
S65948-7......721
S65949-5......297
S65950-9......335
S65952-5......807
S65953-3......922
S65954-1......1136
S65958-4......316
S65960-6......596
S65961-4......588
S65962-2......595
S65963-0......471
S65964-9......566
S65965-7......764
S65966-5......563
S65968-1......537
S65970-3......542
S65971-1......541
S65973-8......818
S65974-6......539
S65975-4......566
S65976-2......373
S65977-0......776
S65978-9......366
S65979-7......530
S65980-0......373
S65981-9......964
S65982-7......978
S65983-5......596
S65984-3......366
S65985-1......760
S65987-8......245
S65988-6......593
S65989-4......594
S65990-8......558
S65992-4......565
S65993-2......913
S65994-0......913
S65995-9......941
S65996-7......566
S65997-5......373
S65998-3......563
S66000-0......541
S66001-9......566
S66002-7......563
S66004-3......992
S66005-1......543
S66007-8......525
S66008-6......215
S66009-4......595
S66010-8......511
S66011-6......588
S66012-4......793
S66013-2......748
S66014-0......573
S66015-9......566
S66017-5......592
S66018-3......508
S66019-1......583
S66020-5......229
S66021-3......763
S66022-1......592
S66024-8......915
S66025-6......317
S66026-4......703
S66027-2......953
S66028-0......945
S66030-2......572
S66031-0......572
S66032-9......565
S66034-5......943
S66035-3......540
S66036-1......544
S66038-8......177
S66039-6......748
S66041-8......510
S66042-6......747
S66043-4......510
S66044-2......582
S66045-0......596
S66046-9......350
S66047-7......557
S66049-3......130
S66051-5......596
S66053-1......558
S66055-8......762
S66056-6......594
S66058-2......714
S66060-4......130
S66062-0......565
S66063-9......765
S66064-7......583
S66065-5......671
S66066-3......566
S66067-1......853
S66069-8......129
S66070-1......427
S66072-8......578
S66073-6......752
S66074-4......573
S66075-2......328
S66076-0......583
S66078-7......713
S66079-5......755
S66080-9......594
S66081-7......807
S66082-5......595
S66083-3......360
S66084-1......596
S66086-8......762
S66087-6......556
S66089-2......520
S66090-6......753
S66091-4......852
S66093-0......556
S66094-9......761
S66095-7......766
S66097-3......986
S66098-1......148
S66100-7......766
S66101-5......950
S66102-3......737
S66105-8......714
S66106-6......182
S66107-4......376
S66108-2......602
S66109-0......761
S66110-4......364
S66111-2......602
S66112-0......1008
S66113-9......766
S66114-7......1037
S66115-5......712
S66116-3......148
S66117-1......93
S66119-8......584
S66120-1......148
S66123-6......714
S66124-4......15
S66126-0......148
S66127-9......598
S66128-7......148
S66129-5......365
S66130-9......595
S66132-5......60
S66133-3......574
S66136-8......78
S66137-6......713
S66138-4......78
S66139-2......766
S66140-6......602
S66141-4......766
S66142-2......148
S66143-0......672
S66144-9......149
S66145-7......148
S66146-5......574
S66148-1......828
S66150-3......599
S66151-1......328
S66153-8......852
S66154-6......195
S66155-4......205
S66156-2......594
S66157-0......436
S66158-9......799
S66159-7......742
S66161-9......188
S66162-7......594
S66165-1......699
S66169-4......602
S66170-8......773
S66171-6......343
S66172-4......671
S66173-2......946
S66174-0......649
S66175-9......573
S66177-5......369
S66178-3......985
S66179-1......701
S66180-5......998
S66181-3......140
S66182-1......247
S66185-6......181
S66186-4......320
S66187-2......689
S66188-0......959
S66189-9......705
S66191-0......726
S66194-5......726
S66199-6......700
S66200-3......1008
S66203-8......294
S66204-6......974
S66209-7......467
S66211-9......384
S66212-7......869
S66213-5......693
S66217-8......726
S66219-4......973
S66221-6......329
S66222-4......372
S66223-2......816
S66224-0......756
S66227-5......690
S66228-3......383
S66230-5......191
S66232-1......67
S66234-8......236
S66236-4......298
S66237-2......854
S66238-0......100
S66239-9......309
S66240-2......188
S66241-0......365
S66243-7......375

Product	Page
S66244-5	984
S66246-1	800
S66248-8	331
S66251-8	657
S66256-9	293
S66258-5	721
S66259-3	309
S66260-7	55
S66262-3	383
S66263-1	101
S66265-8	98
S66266-6	1010
S66269-0	171
S66270-4	757
S66271-2	798
S66273-9	207
S66274-7	301
S66275-5	340
S66277-1	384
S66284-4	345
S66285-2	296
S66287-9	345
S66289-5	170
S66290-9	346
S66292-5	388
S66293-3	1016
S66294-1	122
S66296-8	383
S66297-6	188
S66301-8	383
S66302-6	280
S66306-9	336
S66307-7	388
S66311-5	344
S66315-8	96
S66316-6	292
S66317-4	360
S66318-2	85
S66319-0	1011
S66320-4	182
S66321-2	1048
S66324-7	362
S66325-5	220
S66326-3	362
S66330-1	163
S66333-6	188
S66335-2	170
S66337-9	8
S66338-7	184
S66339-5	210
S66340-9	362
S66341-7	367
S66342-5	83
S66343-3	24
S66344-1	595
S66346-8	594
S66347-6	521
S66348-4	602
S66349-2	714
S66350-6	595
S66351-4	699
S66352-2	602
S66353-0	652
S66354-9	595
S66355-7	602
S66356-5	595
S66357-3	950
S66358-1	713
S66360-3	602
S66361-1	714
S66363-8	852
S66364-6	583
S66365-4	713
S66366-2	712
S66367-0	712
S66368-9	599
S66369-7	593
S66370-0	602
S66371-9	584
S66372-7	573
S66373-5	602
S66374-3	712
S66375-1	602
S66377-8	596
S66378-6	584
S66379-4	582
S66380-8	574
S66381-6	602
S66382-4	583
S66383-2	950
S66384-0	601
S66386-7	951
S66387-5	579
S66388-3	599
S66389-1	583
S66390-5	713
S66391-3	986
S66392-1	986
S66394-8	986
S66396-4	601
S66397-2	867
S66398-0	596
S66400-6	556
S66401-4	733
S66402-2	596
S66403-0	861
S66404-9	596
S66406-5	556
S66407-3	434
S66411-1	544
S66413-8	441
S66414-6	578
S66415-4	148
S66416-2	595
S66418-9	596
S66419-7	789
S66420-0	506
S66421-9	377
S66422-7	686
S66423-5	705
S66425-1	600
S66427-8	21
S66428-6	588
S66430-8	578
S66431-6	1141
S66432-4	594
S66433-2	1141
S66434-0	506
S66439-1	1141
S66444-8	578
S66446-4	552
S66448-0	594
S66450-2	584
S66451-0	256
S66452-9	593
S66453-7	1092
S66454-5	596
S66455-3	441
S66456-1	506
S66459-6	599
S66462-6	684
S66463-4	556
S66464-2	634
S66465-0	362
S66468-5	683
S66470-7	510
S66471-5	599
S66473-1	595
S66475-8	596
S66476-6	858
S66477-4	706
S66478-2	934
S66481-2	184
S66482-0	697
S66485-5	385
S66486-3	466
S66487-1	721
S66490-1	96
S66492-8	67
S66493-6	76
S66496-0	596
S66497-9	538
S66498-7	504
S66499-5	544
S66500-2	616
S66501-0	619
S66503-7	541
S66504-5	538
S66505-3	540
S66506-1	540
S66508-8	538
S66509-6	650
S66511-8	653
S66512-6	685
S66513-4	699
S66515-0	849
S66516-9	686
S66517-7	699
S66518-5	635
S66519-3	684
S66520-7	650
S66521-5	684
S66522-3	552
S66523-1	557
S66525-8	678
S66526-6	686
S66527-4	650
S66528-2	714
S66529-0	557
S66530-4	607
S66531-2	653
S66532-0	607
S66533-9	653
S66534-7	604
S66535-5	699
S66536-3	650
S66537-1	684
S66539-8	650
S66540-1	607
S66542-8	611
S66543-6	650
S66544-4	556
S66545-2	601
S66546-0	554
S66547-9	130
S66548-7	554
S66549-5	714
S66550-9	712
S66551-7	635
S66552-5	528
S66553-3	592
S66554-1	849
S66556-8	713
S66557-6	685
S66558-4	713
S66559-2	684
S66560-6	595
S66561-4	684
S66562-2	540
S66563-0	681
S66564-9	539
S66565-7	546
S66566-5	540
S66567-3	556
S66568-1	538
S66570-3	636
S66571-1	683
S66573-8	714
S66574-6	849
S66575-4	554
S66576-2	1008
S66577-0	528
S66578-9	543
S66579-7	573
S66580-0	699
S66581-9	553
S66582-7	556
S66583-5	650
S66584-3	938
S66585-1	543
S66587-8	593
S66588-6	635
S66589-4	538
S66590-8	601
S66591-6	650
S66592-4	553
S66593-2	615
S66594-0	573
S66595-9	1008
S66596-7	593
S66597-5	598
S66598-3	595
S66599-1	852
S66600-9	573
S66601-7	599
S66602-5	852
S66603-3	714
S66604-1	594
S66606-8	593
S66607-6	578
S66608-4	572
S66609-2	669
S66610-6	519
S66611-4	635
S66612-2	506
S66613-0	635
S66614-9	843
S66615-7	552
S66616-5	583
S66617-3	634
S66618-1	950
S66620-3	596
S66621-1	596
S66623-8	684
S66624-6	599
S66625-4	925
S66626-2	950
S66627-0	684
S66628-9	684
S66629-7	601
S66630-0	679
S66631-9	684
S66632-7	1012
S66633-5	543
S66634-3	678
S66635-1	664
S66637-8	605
S66639-4	665
S66640-8	665
S66641-6	684
S66642-4	664
S66643-2	650
S66644-0	684
S66645-9	672
S66646-7	665
S66647-5	672
S66648-3	767
S66649-1	684
S66650-5	849
S66651-3	665
S66652-1	672
S66654-8	849
S66655-6	665
S66656-4	650
S66657-2	546
S66658-0	604
S66659-9	634
S66660-2	682
S66661-0	634
S66662-9	649
S66663-7	650
S66664-5	650
S66665-3	650
S66666-1	650
S66668-8	649
S66669-6	672
S66671-8	684
S66672-6	646
S66673-4	684
S66674-2	607
S66675-0	671
S66676-9	672
S66677-7	684
S66678-5	652
S66679-3	616
S66681-5	653
S66682-3	672
S66683-1	671
S66685-8	647
S66686-6	616
S66687-4	684
S66688-2	616
S66689-0	678
S66690-4	604
S66692-0	601
S66693-9	953
S66694-7	1008
S66695-5	116
S66696-3	546
S66697-1	113
S66699-8	606
S66700-5	943
S66701-3	665
S66702-1	953
S66704-8	635
S66705-6	678
S66706-4	683
S66707-2	616
S66708-0	566
S66709-9	637
S66710-2	616
S66711-0	620
S66712-9	686
S66713-7	621
S66714-5	1097
S66715-3	633
S66716-1	665
S66719-6	684
S66721-8	849
S66722-6	116
S66723-4	849
S66724-2	953
S66726-9	679
S66727-7	518
S66728-5	847
S66729-3	582
S66730-7	530
S66731-5	621
S66732-3	611
S66733-1	518
S66735-8	684
S66736-6	604
S66737-4	518
S66738-2	678
S66739-0	579
S66740-4	129
S66741-2	684
S66742-0	543
S66743-9	1012
S66744-7	546
S66745-5	541
S66746-3	679
S66747-1	684
S66749-8	684
S66750-1	1012
S66752-8	616
S66753-6	556
S66755-2	767
S66756-0	540
S66757-9	592
S66758-7	1012
S66759-5	849
S66760-9	1012
S66761-7	636
S66762-5	635
S66763-3	679
S66764-1	678
S66766-8	665
S66767-6	1012
S66768-4	1012
S66769-2	634
S66770-6	635
S66771-4	616
S66772-2	679
S66773-0	684
S66774-9	1008
S66775-7	683
S66776-5	683
S66777-3	632
S66778-1	620
S66780-3	567
S66781-1	650
S66783-8	651
S66784-6	686
S66785-4	951
S66786-2	686
S66787-0	567
S66788-9	1012
S66789-7	686
S66790-0	849
S66791-9	1011
S66793-5	767
S66794-3	610
S66795-1	650
S66798-6	536
S66799-4	680
S66800-1	635
S66802-8	849
S66803-6	682
S66805-2	320
S66806-0	687
S66807-9	568
S66809-5	681
S66810-9	610
S66811-7	672
S66812-5	967
S66813-3	1012
S66816-8	353
S66818-4	383
S66819-2	759
S66820-6	1015
S66822-2	799
S66824-9	1005
S66827-3	934
S66828-1	330
S66830-3	1005
S66833-8	1005
S66839-7	931
S66840-0	1005
S66841-9	1005
S66842-7	1006
S66843-5	492
S66844-3	797
S66845-1	583
S66847-8	102
S66849-4	1140
S66852-4	343
S66853-2	289
S66857-5	596
S66858-3	41
S66861-3	171
S66864-8	94
S66865-6	769
S66868-0	966
S66872-9	253
S66874-5	353
S66875-3	6
S66878-8	242
S66881-8	531
S66882-6	557
S66883-4	576
S66885-0	497
S66886-9	531
S66887-7	531
S66888-5	531
S66889-3	288
S66890-7	155
S66891-5	699
S66892-3	945
S66893-1	705
S66895-8	850
S66896-6	744
S66898-2	413
S66899-0	703
S66900-8	530
S66901-6	610
S66904-0	557
S66905-9	567
S66906-7	567
S66907-5	572
S66908-3	571
S66910-5	554
S66911-3	568
S66912-1	554
S66914-8	94
S66915-6	552
S66916-4	568
S66917-2	531
S66918-0	703
S66919-9	563
S66920-2	437
S66921-0	531
S66922-9	576
S66923-7	554

Product	Page
S66924-5	568
S66928-8	219
S66929-6	219
S66931-8	6
S66932-6	567
S66933-4	576
S66934-2	576
S66935-0	576
S66936-9	568
S66937-7	219
S66938-5	576
S66939-3	567
S66940-7	566
S66941-5	571
S66942-3	576
S66945-8	703
S66946-6	505
S66947-4	633
S66948-2	568
S66949-0	428
S66950-4	571
S66951-2	567
S66952-0	576
S66953-9	576
S66954-7	576
S66955-5	976
S66956-3	568
S66957-1	531
S66959-8	531
S66960-1	567
S66962-8	185
S66963-6	576
S66964-4	576
S66966-0	703
S66967-9	568
S66968-7	703
S66969-5	24
S66970-9	531
S66971-7	576
S66972-5	557
S66973-3	554
S66974-1	557
S66976-8	95
S66977-6	557
S66978-4	505
S66979-2	576
S66980-6	576
S66981-4	531
S66982-2	31
S66983-0	568
S66984-9	531
S66985-7	576
S66988-1	1023
S66990-3	1029
S66991-1	1023
S66993-8	1063
S66994-6	1065
S66995-4	322
S66996-2	498
S66997-0	500
S66998-9	499
S66999-7	500
S67000-6	299
S67001-4	300
S67002-2	300
S67003-0	379
S67004-9	970
S67006-5	581
S67007-3	531
S67008-1	557
S67010-3	531
S67011-1	531
S67013-8	596
S67014-6	702
S67015-4	552
S67016-2	531
S67017-0	539
S67018-9	545
S67019-7	536
S67020-0	563
S67021-9	563
S67022-7	564
S67023-5	564
S67024-3	564
S67025-1	557
S67027-8	545
S67028-6	702
S67033-2	823
S67034-0	67
S67038-3	539
S67039-1	732
S67040-5	534
S67041-3	591
S67042-1	602
S67044-8	588
S67045-6	531
S67046-4	511
S67047-2	541
S67049-9	564
S67050-2	850
S67051-0	564
S67052-9	576
S67053-7	563
S67055-3	564
S67056-1	510
S67058-8	525
S67059-6	694
S67061-8	596
S67062-6	928
S67063-4	322
S67064-2	973
S67065-0	1017
S67066-9	328
S67068-5	507
S67069-3	586
S67070-7	702
S67071-5	564
S67072-3	596
S67073-1	558
S67075-8	534
S67076-6	562
S67077-4	702
S67078-2	559
S67079-0	545
S67080-4	562
S67081-2	562
S67082-0	563
S67083-9	596
S67084-7	555
S67085-5	702
S67089-8	999
S67092-8	283
S67093-6	695
S67094-4	34
S67096-0	1011
S67097-9	724
S67098-7	928
S67099-5	704
S67100-2	874
S67101-0	297
S67102-9	253
S67103-7	706
S67104-5	768
S67106-1	364
S67108-8	356
S67109-6	93
S67111-8	177
S67112-6	381
S67113-4	52
S67114-2	19
S67115-0	373
S67116-9	20
S67117-7	13
S67118-5	355
S67119-3	346
S67121-5	717
S67122-3	1014
S67125-8	324
S67127-4	937
S67128-2	988
S67129-0	101
S67131-2	495
S67132-0	385
S67133-9	1008
S67135-5	1009
S67139-8	358
S67140-1	51
S67145-2	953
S67146-0	726
S67148-7	1001
S67149-5	911
S67150-9	291
S67151-7	568
S67152-5	568
S67153-3	64
S67154-1	341
S67157-6	174
S67158-4	92
S67160-6	1011
S67162-2	329
S67163-0	208
S67164-9	344
S67165-7	721
S67166-5	567
S67167-3	774
S67168-1	1068
S67171-1	181
S67173-8	763
S67174-6	402
S67175-4	811
S67176-2	822
S67177-0	443
S67181-9	411
S67182-7	792
S67183-5	792
S67184-3	411
S67185-1	978
S67187-8	275
S67188-6	817
S67189-4	905
S67190-8	803
S67192-4	803
S67194-0	339
S67196-7	799
S67197-5	797
S67199-1	378
S67200-9	799
S67202-5	379
S67204-1	885
S67207-6	90
S67209-2	1000
S67211-4	298
S67214-9	364
S67217-3	717
S67218-1	328
S67224-6	871
S67226-2	944
S67227-0	205
S67231-9	1052
S67232-7	1032
S67233-5	1052
S67234-3	273
S67235-1	128
S67237-8	103
S67238-6	128
S67239-4	287
S67241-6	297
S67242-4	960
S67243-2	462
S67245-9	58
S67246-7	970
S67247-5	344
S67248-3	331
S67249-1	710
S67250-5	707
S67251-3	359
S67252-1	293
S67254-8	299
S67256-4	156
S67258-0	156
S67261-0	265
S67263-7	20
S67264-5	1008
S67265-3	359
S67266-1	102
S67268-8	824
S67269-6	824
S67271-8	779
S67272-6	328
S67273-4	20
S67276-9	355
S67279-3	726
S67280-7	163
S67282-3	82
S67285-8	327
S67287-4	95
S67288-2	234
S67290-4	327
S67295-5	1045
S67301-3	1046
S67307-2	1021
S67309-9	1024
S67312-9	1046
S67324-2	1046
S67325-0	1046
S67326-9	1046
S67329-3	1024
S67330-7	577
S67331-5	507
S67335-8	628
S67336-6	545
S67337-4	532
S67339-0	853
S67340-4	551
S67342-0	694
S67343-9	833
S67344-7	833
S67345-5	834
S67352-8	511
S67354-4	474
S67355-2	695
S67356-0	706
S67357-9	1043
S67361-7	492
S67363-3	641
S67367-6	596
S67368-4	875
S67370-6	492
S67371-4	687
S67373-0	529
S67375-7	364
S67376-5	376
S67377-3	976
S67380-3	1040
S67383-8	492
S67385-4	548
S67386-2	93
S67389-7	94
S67390-0	1030
S67391-9	612
S67393-5	391
S67395-1	834
S67397-8	396
S67399-4	834
S67403-6	164
S67404-4	871
S67406-0	730
S67407-9	650
S67408-7	638
S67413-3	851
S67414-1	534
S67416-8	403
S67417-6	575
S67418-4	534
S67420-6	553
S67421-4	891
S67422-2	844
S67423-0	834
S67424-9	504
S67426-5	636
S67427-3	401
S67428-1	366
S67430-3	123
S67431-1	695
S67433-8	367
S67434-6	739
S67435-4	554
S67436-2	245
S67437-0	534
S67438-9	577
S67439-7	1063
S67441-9	1060
S67442-7	722
S67443-5	925
S67444-3	521
S67448-6	403
S67449-4	735
S67451-6	835
S67452-4	734
S67453-2	735
S67455-9	517
S67456-7	822
S67457-5	806
S67458-3	825
S67459-1	527
S67460-5	912
S67461-3	551
S67462-1	674
S67464-8	428
S67465-6	1006
S67467-2	1005
S67468-0	1006
S67469-9	491
S67471-0	418
S67472-9	492
S67473-7	1040
S67476-1	536
S67478-8	409
S67481-8	565
S67483-4	129
S67484-2	521
S67485-0	433
S67486-9	577
S67487-7	737
S67488-5	16
S67490-7	558
S67491-5	669
S67492-3	735
S67493-1	540
S67495-8	545
S67497-4	1027
S67498-2	1081
S67499-0	526
S67500-8	532
S67501-6	796
S67502-4	531
S67503-2	733
S67504-0	732
S67505-9	638
S67506-7	744
S67507-5	826
S67508-3	405
S67510-5	548
S67511-3	740
S67514-8	475
S67515-6	465
S67516-4	476
S67517-2	354
S67519-9	497
S67520-2	963
S67521-0	515
S67522-9	624
S67524-5	546
S67525-3	278
S67528-8	534
S67529-6	546
S67535-0	536
S67536-9	395
S67537-7	475
S67538-5	639
S67539-3	946
S67540-7	689
S67547-4	736
S67548-2	730
S67549-0	912
S67550-4	110
S67551-2	396
S67552-0	396
S67555-5	95
S67557-1	631
S67559-8	640
S67560-1	728
S67562-8	695
S67566-0	649
S67567-9	495
S67568-7	730
S67569-5	745
S67570-9	745
S67571-7	385
S67574-1	732
S67576-8	736
S67577-6	732
S67578-4	450
S67579-2	399
S67580-6	1092
S67581-4	489
S67582-2	1081
S67583-0	744
S67584-9	401
S67585-7	751
S67586-5	484
S67587-3	603
S67590-3	805
S67591-1	751
S67593-8	519
S67594-6	409
S67595-4	389
S67596-2	825
S67597-0	577
S67598-9	489
S67599-7	577
S67600-4	410
S67601-2	127
S67603-9	874
S67604-7	416
S67605-5	399
S67606-3	448
S67607-1	401
S67609-8	282
S67610-1	399
S67612-8	399
S67613-6	448
S67614-4	806
S67615-2	490
S67617-9	806
S67618-7	552
S67619-5	452
S67621-7	452
S67622-5	401
S67624-1	1080
S67626-8	190
S67627-6	537
S67628-4	1081
S67629-2	417
S67630-6	63
S67631-4	401
S67632-2	677
S67634-9	521
S67635-7	527
S67636-5	1041
S67637-3	406
S67641-1	726
S67643-8	807
S67644-6	560
S67645-4	1012
S67646-2	743
S67647-0	1023
S67648-9	488
S67649-7	308
S67650-0	489
S67651-9	416
S67652-7	416
S67653-5	998
S67654-3	743
S67655-1	489
S67657-8	1063
S67658-6	96
S67659-4	431
S67660-8	376
S67661-6	826
S67663-2	596
S67664-0	551
S67665-9	545
S67668-3	140
S67669-1	373
S67670-5	526
S67671-3	853
S67672-1	1063
S67674-8	1025
S67675-6	823
S67676-4	830
S67677-2	825
S67678-0	431
S67681-0	972

Product	Page
S67682-9	657
S67683-7	521
S67686-1	140
S67692-6	216
S67693-4	216
S67695-0	456
S67696-9	852
S67697-7	190
S67698-5	286
S67700-0	769
S67701-9	768
S67702-7	769
S67703-5	188
S67705-1	190
S67708-6	216
S67709-4	216
S67710-8	78
S67713-2	147
S67714-0	745
S67715-9	216
S67717-5	82
S67719-1	769
S67721-3	82
S67725-6	768
S67726-4	85
S67727-2	216
S67728-0	76
S67729-9	190
S67730-2	59
S67731-0	81
S67733-7	20
S67735-3	78
S67736-1	85
S67741-8	111
S67742-6	502
S67743-4	576
S67744-2	234
S67745-0	768
S67746-9	768
S67747-7	190
S67748-5	216
S67749-3	1018
S67750-7	203
S67752-3	190
S67756-6	33
S67757-4	241
S67760-4	973
S67761-2	476
S67762-0	140
S67764-7	532
S67765-5	571
S67767-1	534
S67769-8	763
S67772-8	384
S67773-6	1024
S67775-2	472
S67781-7	332
S67782-5	938
S67784-1	499
S67786-8	821
S67789-2	248
S67790-6	248
S67792-2	361
S67793-0	710
S67794-9	201
S67796-5	215
S67797-3	215
S67798-1	290
S67802-3	220
S67803-1	265
S67805-8	258
S67807-4	261
S67809-0	260
S67810-4	224
S67811-2	260
S67815-5	265
S67816-3	463
S67817-1	828
S67822-8	267
S67824-4	270
S67826-0	263
S67828-7	389
S67831-7	246
S67832-5	97
S67833-3	420
S67836-8	356
S67837-6	1062
S67842-2	98
S67843-0	243
S67844-9	447
S67847-3	708
S67848-1	388
S67850-3	1052
S67851-1	990
S67853-8	1024
S67854-6	1052
S67855-4	1038
S67856-2	984
S67857-0	263
S67858-9	1052
S67859-7	1052
S67860-0	393
S67861-9	392
S67862-7	282
S67863-5	1050
S67864-3	1027
S67865-1	1027
S67867-8	990
S67868-6	286
S67869-4	285
S67870-8	1050
S67871-6	1027
S67872-4	839
S67873-2	66
S67874-0	1000
S67876-7	411
S67877-5	676
S67878-3	604
S67879-1	708
S67880-5	676
S67881-3	667
S67882-1	813
S67884-8	674
S67885-6	667
S67886-4	676
S67887-2	888
S67888-0	676
S67889-9	676
S67890-2	377
S67891-0	104
S67892-9	51
S67893-7	350
S67895-3	821
S67896-1	828
S67898-8	791
S67899-6	88
S67903-8	50
S67904-6	858
S67905-4	721
S67906-2	676
S67907-0	814
S67908-9	126
S67909-7	645
S67910-0	365
S67911-9	412
S67912-7	677
S67915-1	763
S67917-8	9
S67918-6	236
S67921-6	990
S67924-0	343
S67927-5	983
S67928-3	1053
S67930-5	290
S67931-3	373
S67934-8	990
S67936-4	351
S67937-2	128
S67938-0	353
S67939-9	353
S67940-2	985
S67943-7	766
S67945-3	1142
S67946-1	954
S67948-8	354
S67951-8	798
S67953-4	321
S67954-2	277
S67957-7	146
S67959-3	186
S67960-7	816
S67974-7	771
S67976-3	178
S67983-6	355
S67986-0	186
S67994-1	772
S67998-4	415
S68002-8	75
S68003-6	501
S68005-2	277
S68008-7	17
S68009-5	201
S68010-9	110
S68011-7	853
S68018-4	179
S68020-6	293
S68021-4	379
S68023-0	304
S68025-7	186
S68026-5	79
S68027-3	157
S68030-3	178
S68031-1	12
S68035-4	12
S68036-2	368
S68037-0	155
S68039-7	301
S68042-7	788
S68043-5	142
S68045-1	1
S68048-6	325
S68050-8	68
S68051-6	791
S68052-4	10
S68055-9	378
S68059-1	325
S68065-6	368
S68066-4	246
S68067-2	356
S68068-0	205
S68070-2	28
S68071-0	28
S68072-9	115
S68073-7	574
S68074-5	101
S68100-8	57
S68101-6	488
S68102-4	117
S68103-2	299
S68104-0	382
S68105-9	477
S68108-3	1016
S68109-1	299
S68110-5	287
S68112-1	157
S68114-8	479
S68116-4	386
S68117-2	690
S68119-9	450
S68120-2	456
S68121-0	307
S68123-7	304
S68124-5	299
S68125-3	301
S68126-1	303
S68128-8	462
S68131-8	210
S68132-6	1010
S68136-9	374
S68138-5	1009
S68140-7	222
S68142-3	429
S68145-8	176
S68146-6	706
S68148-2	479
S68150-4	462
S68153-9	292
S68155-5	292
S68156-3	461
S68157-1	1017
S68162-8	695
S68163-6	1025
S68164-4	706
S68166-0	387
S68169-5	436
S68171-7	136
S68173-3	45
S68174-1	222
S68176-8	447
S68177-6	695
S68178-4	146
S68181-4	931
S68184-9	250
S68188-1	1001
S68191-1	931
S68193-8	249
S68195-4	357
S68196-2	1017
S68199-7	358
S68200-4	968
S68205-5	287
S68213-6	182
S68216-0	710
S68219-5	695
S68223-3	447
S68224-1	249
S68227-6	503
S68230-6	384
S68231-4	711
S68236-5	714
S68238-1	432
S68243-8	309
S68244-6	468
S68245-4	383
S68246-2	446
S68247-0	383
S68248-9	287
S68249-7	384
S68251-9	304
S68257-8	106
S68258-6	1016
S68259-4	284
S68262-4	286
S68264-0	775
S68265-9	697
S68266-7	258
S68269-1	285
S68270-5	927
S68271-3	692
S68272-1	287
S68274-8	136
S68275-6	1009
S68276-4	1015
S68278-0	299
S68279-9	692
S68280-2	1014
S68281-0	264
S68282-9	909
S68284-5	869
S68285-3	700
S68297-7	304
S68302-7	698
S68303-5	289
S68304-3	92
S68305-1	867
S68307-8	304
S68309-4	1019
S68311-6	765
S68312-4	310
S68313-2	289
S68315-9	692
S68316-7	1041
S68317-5	182
S68318-3	1017
S68319-1	292
S68321-3	177
S68324-8	1044
S68330-2	242
S68331-0	985
S68332-9	239
S68334-5	147
S68335-3	157
S68339-6	42
S68341-8	31
S68343-4	70
S68344-2	1087
S68345-0	307
S68346-9	779
S68347-7	639
S68348-5	925
S68349-3	574
S68350-7	728
S68353-1	676
S68355-8	608
S68360-4	937
S68362-0	999
S68364-7	384
S68370-1	461
S68372-8	972
S68376-0	414
S68377-9	571
S68379-5	421
S68380-9	462
S68381-7	384
S68382-5	351
S68387-6	1016
S68388-4	137
S68390-6	928
S68392-2	110
S68396-5	204
S68401-5	805
S68403-1	401
S68405-8	544
S68406-6	574
S68407-4	928
S68411-2	521
S68414-7	728
S68415-5	544
S68416-3	727
S68419-8	581
S68423-6	959
S68424-4	712
S68426-0	528
S68430-9	562
S68432-5	807
S68438-4	959
S68440-6	673
S68444-9	728
S68447-3	42
S68448-1	1023
S68450-3	1023
S68451-1	1031
S68453-8	1023
S68454-6	1023
S68455-4	1029
S68456-2	1023
S68457-0	1023
S68458-9	1035
S68459-7	1034
S68460-0	224
S68461-9	1029
S68464-3	106
S68470-8	728
S68471-6	808
S68472-4	734
S68473-2	518
S68478-3	806
S68480-5	1036
S68481-3	1023
S68482-1	1041
S68484-8	261
S68485-6	1076
S68486-4	1050
S68487-2	1024
S68488-0	1050
S68489-9	1038
S68490-2	266
S68491-0	1038
S68492-9	1032
S68494-5	1050
S68499-6	715
S68500-3	532
S68501-1	729
S68506-2	747
S68513-5	1038
S68514-3	1032
S68515-1	1027
S68517-8	1022
S68518-6	299
S68519-4	299
S68520-8	1037
S68521-6	1027
S68522-4	273
S68528-3	528
S68532-1	629
S68536-4	693
S68537-2	917
S68540-2	1029
S68541-0	813
S68548-8	574
S68552-6	398
S68556-9	964
S68557-7	1041
S68560-7	371
S68561-5	969
S68565-8	142
S68566-6	968
S68568-2	439
S68569-0	1028
S68570-4	99
S68571-2	16
S68574-7	804
S68575-5	163
S68577-1	132
S68579-8	1063
S68580-1	376
S68582-8	35
S68586-0	865
S68587-9	1000
S68590-9	607
S68591-7	729
S68592-5	225
S68593-3	450
S68594-1	1036
S68596-8	518
S68597-6	1009
S68598-4	869
S68599-2	471
S68601-8	692
S68603-4	295
S68604-2	371
S68607-7	414
S68609-3	535
S68610-7	463
S68611-5	448
S68612-3	737
S68613-1	677
S68615-8	649
S68616-6	929
S68617-4	616
S68618-2	614
S68619-0	532
S68620-4	626
S68621-2	640
S68622-0	1029
S68623-9	1027
S68624-7	1050
S68625-5	1050
S68626-3	981
S68627-1	1032
S68629-8	1029
S68630-1	1023
S68632-8	984
S68633-6	735
S68634-4	748
S68635-2	729
S68636-0	532
S68637-9	640
S68638-7	401
S68639-5	730
S68641-7	565
S68642-5	463
S68643-3	616
S68644-1	578
S68646-8	528
S68647-6	608
S68648-4	1022
S68649-2	1032
S68650-6	1037
S68651-4	1077
S68652-2	1062
S68653-0	1025
S68654-9	1062
S68655-7	1049

Product	Page
S68656-5	730
S68657-3	752
S68658-1	807
S68660-3	550
S68661-1	1037
S68663-8	1032
S68664-6	1023
S68665-4	1032
S68666-2	1025
S68667-0	1063
S68668-9	1066
S68669-7	1027
S68670-0	532
S68671-9	964
S68672-7	508
S68673-5	733
S68675-1	622
S68677-8	965
S68680-8	621
S68681-6	621
S68682-4	615
S68683-2	636
S68684-0	610
S68685-9	638
S68687-5	929
S68688-3	640
S68690-5	608
S68691-3	622
S68692-1	667
S68694-8	532
S68695-6	607
S68696-4	491
S68697-2	733
S68698-0	928
S68701-4	929
S68706-5	608
S68708-1	806
S68713-8	524
S68715-4	557
S68716-2	402
S68718-9	544
S68721-9	1088
S68725-1	1024
S68727-8	686
S68728-6	631
S68730-8	939
S68732-4	578
S68735-9	608
S68736-7	529
S68738-3	376
S68739-1	621
S68741-3	1025
S68742-1	674
S68744-8	504
S68746-4	632
S68748-0	523
S68750-2	638
S68751-0	525
S68752-9	810
S68753-7	543
S68754-5	543
S68756-1	563
S68758-8	517
S68769-3	641
S68770-7	735
S68773-1	622
S68776-6	929
S68778-2	628
S68779-0	522
S68780-4	554
S68783-9	638
S68784-7	391
S68787-1	410
S68789-8	554
S68792-8	729
S68794-4	619
S68795-2	520
S68797-9	630
S68798-7	513
S68802-9	740
S68804-5	641
S68805-3	624
S68808-8	695
S68811-8	640
S68812-6	964
S68813-4	718
S68815-0	539
S68819-3	527
S68820-7	534
S68821-5	511
S68823-1	806
S68825-8	464
S68826-6	936
S68827-4	730
S68828-2	394
S68830-4	638
S68831-2	402
S68832-0	929
S68837-1	925
S68840-1	965
S68842-8	741
S68843-6	492
S68844-4	505
S68845-2	647
S68846-0	579
S68847-9	517
S68850-9	516
S68851-7	815
S68852-5	608
S68856-8	505
S68857-6	649
S68858-4	790
S68859-2	445
S68860-6	632
S68861-4	772
S68863-0	520
S68864-9	518
S68867-3	494
S68868-1	608
S68870-3	505
S68871-1	560
S68874-6	1036
S68875-4	505
S68876-2	912
S68877-0	607
S68878-9	607
S68879-7	172
S68881-9	174
S68882-7	968
S68883-5	726
S68885-1	708
S68887-8	44
S68888-6	776
S68889-4	724
S68890-8	723
S68891-6	819
S68893-2	1057
S68894-0	226
S68895-9	1057
S68896-7	1021
S68897-5	368
S68898-3	419
S68899-1	1030
S68900-9	966
S68901-7	1024
S68902-5	1021
S68903-3	227
S68904-1	1057
S68906-8	1043
S68907-6	1030
S68908-4	264
S68909-2	1036
S68910-6	1030
S68911-4	1030
S68912-2	1020
S68913-0	1021
S68914-9	1021
S68915-7	1030
S68916-5	1033
S68917-3	1031
S68918-1	1057
S68920-3	1029
S68921-1	936
S68923-8	1032
S68924-6	366
S68925-4	1024
S68926-2	1021
S68927-0	1030
S68928-9	155
S68929-7	1030
S68934-3	391
S68935-1	628
S68938-6	625
S68939-4	625
S68940-8	619
S68941-6	614
S68942-4	391
S68943-2	410
S68944-0	640
S68945-9	619
S68949-1	628
S68951-3	546
S68954-8	739
S68955-6	529
S68958-0	626
S68960-2	551
S68961-0	391
S68963-7	521
S68965-3	621
S68968-8	504
S68969-6	730
S68971-8	495
S68974-2	619
S68978-5	513
S68979-3	394
S68982-3	563
S68986-6	630
S68987-4	945
S68988-2	638
S68989-0	639
S68990-4	619
S68991-2	619
S68992-0	619
S68994-7	626
S68996-3	547
S68997-1	619
S68999-8	492
S69000-7	549
S69001-5	296
S69006-6	967
S69007-4	180
S69009-0	742
S69012-0	810
S69013-9	286
S69014-7	1052
S69015-5	1028
S69016-3	393
S69017-1	439
S69020-1	574
S69022-8	740
S69023-6	619
S69024-4	730
S69025-2	733
S69026-0	744
S69027-9	629
S69028-7	733
S69030-9	558
S69031-7	613
S69032-5	939
S69033-3	937
S69034-1	731
S69036-8	638
S69037-6	622
S69038-4	619
S69041-4	730
S69042-2	965
S69044-9	518
S69048-1	740
S69050-3	744
S69051-1	608
S69053-8	639
S69055-4	512
S69056-2	808
S69058-9	682
S69059-7	395
S69060-0	731
S69062-7	529
S69063-5	607
S69064-3	731
S69065-1	633
S69067-8	620
S69068-6	636
S69069-4	638
S69070-8	621
S69071-6	637
S69072-4	521
S69073-2	635
S69074-0	625
S69076-7	639
S69077-5	621
S69078-3	638
S69079-1	638
S69081-3	731
S69084-8	622
S69085-6	742
S69086-4	611
S69088-0	749
S69089-9	846
S69090-2	517
S69091-0	504
S69092-9	752
S69093-7	1040
S69094-5	646
S69095-3	551
S69096-1	636
S69098-8	728
S69100-3	1030
S69101-1	1028
S69103-8	1021
S69104-6	1029
S69105-4	223
S69106-2	1030
S69107-0	1021
S69108-9	1030
S69109-7	1030
S69110-0	1021
S69111-9	223
S69112-7	226
S69113-5	570
S69115-1	735
S69117-8	740
S69118-6	625
S69119-4	505
S69120-8	826
S69121-6	639
S69123-2	613
S69124-0	614
S69125-9	927
S69126-7	1072
S69130-5	629
S69135-6	633
S69136-4	634
S69137-2	614
S69139-9	619
S69140-2	732
S69142-9	745
S69143-7	731
S69144-5	629
S69145-3	619
S69146-1	901
S69148-8	513
S69149-6	735
S69151-8	551
S69152-6	985
S69153-4	494
S69154-2	1030
S69156-9	733
S69157-7	141
S69158-5	527
S69159-3	1048
S69160-7	507
S69161-5	544
S69163-1	716
S69165-8	653
S69166-6	617
S69169-0	845
S69170-4	391
S69171-2	614
S69172-0	939
S69173-9	628
S69174-7	517
S69175-5	547
S69177-1	638
S69179-8	745
S69180-1	618
S69182-8	607
S69184-4	676
S69185-2	733
S69186-0	574
S69187-9	928
S69188-7	846
S69194-1	1040
S69198-4	1037
S69199-2	581
S69201-8	733
S69203-4	833
S69204-2	574
S69205-0	875
S69206-9	606
S69207-7	835
S69211-5	1062
S69215-8	1030
S69217-4	730
S69219-0	620
S69221-2	1072
S69222-0	514
S69226-3	514
S69227-1	939
S69230-1	865
S69233-6	382
S69234-4	382
S69235-2	874
S69236-0	843
S69237-9	875
S69239-5	550
S69240-9	851
S69241-7	928
S69243-3	533
S69244-1	738
S69246-8	709
S69247-6	127
S69249-2	512
S69250-6	930
S69251-4	756
S69252-2	448
S69253-0	551
S69254-9	711
S69255-7	528
S69260-3	1074
S69263-8	743
S69264-6	529
S69265-4	529
S69266-2	570
S69270-0	223
S69271-9	227
S69272-7	1057
S69273-5	934
S69274-3	226
S69277-8	1033
S69278-6	1029
S69279-4	222
S69280-8	463
S69281-6	465
S69282-4	398
S69283-2	395
S69284-0	395
S69285-9	470
S69286-7	474
S69287-5	461
S69288-3	467
S69289-1	446
S69290-5	454
S69291-3	455
S69292-1	834
S69294-8	275
S69295-6	449
S69296-4	1032
S69297-2	1030
S69298-0	1053
S69299-9	788
S69301-4	338
S69302-2	978
S69303-0	337
S69304-9	671
S69305-7	926
S69306-5	644
S69307-3	652
S69310-3	561
S69313-8	241
S69315-4	70
S69316-2	236
S69317-0	695
S69318-9	619
S69320-0	504
S69321-9	752
S69322-7	639
S69323-5	619
S69324-3	513
S69325-1	581
S69327-8	626
S69328-6	929
S69329-4	534
S69332-4	733
S69333-2	528
S69334-0	560
S69335-9	927
S69336-7	715
S69337-5	639
S69338-3	513
S69339-1	625
S69340-5	507
S69341-3	930
S69344-8	926
S69345-6	723
S69346-4	614
S69348-0	617
S69349-9	756
S69351-0	614
S69353-7	641
S69354-5	626
S69355-3	614
S69356-1	528
S69358-8	563
S69359-6	636
S69361-8	744
S69362-6	1061
S69363-4	728
S69364-2	635
S69365-0	528
S69366-9	619
S69368-5	613
S69369-3	621
S69371-5	730
S69372-3	527
S69373-1	729
S69375-8	613
S69376-6	669
S69377-4	949
S69378-2	620
S69379-0	515
S69380-4	545
S69382-0	1068
S69383-9	1048
S69385-5	964
S69387-1	540
S69389-8	930
S69390-1	834
S69392-8	721
S69394-4	878
S69395-2	640
S69396-0	516
S69397-9	744
S69398-7	617
S69399-5	626
S69400-2	561
S69401-0	515
S69402-9	373
S69403-7	612
S69406-1	735
S69408-8	504
S69409-6	687
S69412-6	518
S69414-2	513
S69415-0	629
S69416-9	606
S69417-7	638
S69418-5	637
S69419-3	618
S69420-7	728
S69421-5	639
S69422-3	529
S69425-8	501
S69427-4	382
S69428-2	613

Product Page

S69429-0 581
S69430-4 731
S69432-0 619
S69433-9 1026
S69434-7 504
S69435-5 528
S69436-3 618
S69437-1 1040
S69439-8 634
S69440-1 612
S69442-8 734
S69443-6 386
S69444-4 747
S69445-2 614
S69446-0 607
S69447-9 707
S69448-7 639
S69449-5 736
S69450-9 545
S69451-7 965
S69452-5 1026
S69453-3 639
S69457-6 126
S69459-2 864
S69461-4 831
S69462-2 790
S69463-0 156
S69464-9 129
S69465-7 695
S69466-5 574
S69470-3 730
S69471-1 606
S69473-8 449
S69474-6 637
S69475-4 448
S69476-2 621
S69477-0 973
S69478-9 608
S69479-7 409
S69481-9 761
S69482-7 994
S69484-3 493
S69485-1 515
S69487-8 576
S69488-6 825
S69489-4 570
S69490-8 624
S69491-6 574
S69492-4 846
S69493-2 844
S69494-0 447
S69495-9 457
S69496-7 928
S69497-5 470
S69498-3 626
S69499-1 515
S69500-9 649
S69502-5 519
S69503-3 465
S69504-1 516
S69506-8 740
S69507-6 377
S69509-2 459
S69511-4 409
S69513-0 287
S69514-9 276
S69516-5 518
S69517-3 512
S69518-1 409
S69523-8 525
S69524-6 265
S69525-4 397
S69526-2 398
S69527-0 398
S69528-9 263
S69529-7 397
S69530-0 396
S69531-9 1022
S69532-7 474
S69533-5 1022
S69534-3 259
S69535-1 222
S69538-6 396
S69539-4 465
S69540-8 980
S69541-6 931
S69542-4 107
S69544-0 454
S69545-9 274
S69546-7 1077
S69547-5 620
S69548-3 574
S69549-1 741
S69550-5 1026
S69551-3 926
S69552-1 368
S69554-8 543
S69555-6 368
S69556-4 350
S69557-2 695
S69558-0 1032
S69560-2 1022
S69561-0 228
S69562-9 438
S69563-7 630
S69564-5 623
S69566-1 504
S69568-8 710
S69569-6 659
S69572-6 394
S69573-4 401
S69574-2 424
S69576-9 492
S69577-7 629
S69578-5 639
S69579-3 619
S69580-7 1092
S69582-3 929
S69583-1 525
S69585-8 377
S69586-6 553
S69587-4 493
S69591-2 514
S69592-0 515
S69593-9 575
S69594-7 623
S69595-5 667
S69599-8 449
S69601-3 376
S69602-1 377
S69604-8 530
S69605-6 452
S69609-9 427
S69611-0 560
S69612-9 428
S69613-7 550
S69618-8 624
S69619-6 617
S69621-8 667
S69624-2 222
S69625-0 194
S69626-9 222
S69627-7 222
S69629-3 926
S69630-7 926
S69631-5 926
S69632-3 210
S69633-1 235
S69635-8 913
S69636-6 335
S69637-4 773
S69638-2 260
S69640-4 193
S69641-2 225
S69642-0 263
S69643-9 225
S69644-7 225
S69645-5 254
S69646-3 1003
S69649-8 1003
S69652-8 446
S69653-6 367
S69654-4 223
S69655-2 931
S69656-0 225
S69657-9 1057
S69658-7 1040
S69659-5 936
S69661-7 473
S69662-5 378
S69664-1 827
S69666-8 243
S69668-4 241
S69670-6 925
S69672-2 981
S69673-0 1072
S69674-9 288
S69677-3 375
S69678-1 288
S69681-1 289
S69684-6 348
S69685-4 347
S69686-2 180
S69688-9 997
S69689-7 749
S69690-0 735
S69691-9 940
S69699-4 319
S69700-1 342
S69704-4 176
S69707-9 489
S69708-7 716
S69710-9 270
S69711-7 990
S69712-5 176
S69714-1 325
S69716-8 919
S69717-6 282
S69721-4 332
S69722-2 177
S69724-9 277
S69734-6 798
S69740-0 190
S69742-7 860
S69743-5 782
S69744-3 278
S69745-1 725
S69747-8 1053
S69749-4 457
S69750-8 243
S69754-0 225
S69755-9 22
S69757-5 278
S69759-1 172
S69760-5 177
S69764-8 468
S69765-6 360
S69766-4 177
S69767-2 367
S69768-0 367
S69769-9 176
S69770-2 169
S69771-0 170
S69774-5 236
S69776-1 767
S69781-8 176
S69782-6 800
S69783-4 153
S69784-2 21
S69785-0 2
S69787-7 27
S69788-5 211
S69789-3 47
S69790-7 23
S69792-3 17
S69793-1 211
S69795-8 210
S69796-6 225
S69797-4 64
S69799-0 211
S69800-8 211
S69801-6 398
S69803-2 64
S69808-3 1024
S69809-1 931
S69810-5 1050
S69811-3 1076
S69812-1 959
S69816-4 844
S69822-9 22
S69824-5 25
S69826-1 22
S69832-6 219
S69833-4 850
S69837-7 222
S69838-5 1049
S69839-3 193
S69840-7 263
S69841-5 194
S69842-3 781
S69843-1 925
S69845-8 243
S69846-6 163
S69847-4 410
S69848-2 571
S69849-0 408
S69850-4 427
S69851-2 126
S69852-0 1080
S69854-7 850
S69856-3 825
S69857-1 434
S69859-8 852
S69860-1 530
S69862-8 465
S69865-2 470
S69866-0 404
S69867-9 545
S69868-7 668
S69869-5 1034
S69870-9 1023
S69871-7 1029
S69872-5 1022
S69873-3 1029
S69874-1 1030
S69876-8 1030
S69877-6 1043
S69878-4 1060
S69879-2 1057
S69880-6 1030
S69881-4 1023
S69882-2 1022
S69883-0 1028
S69884-9 1022
S69885-7 1022
S69886-5 1028
S69887-3 1030
S69888-1 1032
S69890-3 506
S69891-1 823
S69897-0 551
S69898-9 448
S69901-2 644
S69902-0 818
S69903-9 509
S69905-5 532
S69906-3 529
S69909-8 378
S69910-1 625
S69913-6 451
S69915-2 626
S69916-0 813
S69918-7 560
S69919-5 560
S69920-9 625
S69923-3 320
S69924-1 198
S69926-8 173
S69927-6 23
S69928-4 292
S69929-2 716
S69930-6 286
S69931-4 932
S69935-7 625
S69936-5 623
S69940-3 539
S69946-2 563
S69947-0 641
S69948-9 982
S69949-7 1060
S69951-9 396
S69952-7 981
S69957-8 427
S69958-6 426
S69959-4 525
S69960-8 401
S69962-4 432
S69964-0 465
S69969-1 451
S69972-1 567
S69974-8 575
S69975-6 407
S69977-2 626
S69980-2 377
S69981-0 625
S69982-9 623
S69985-3 844
S69986-1 625
S69992-6 850
S69993-4 624
S69994-2 625
S69996-9 621
S69997-7 669
S69998-5 637
S69999-3 634
S70000-2 377
S70001-0 612
S70002-9 619
S70005-3 528
S70008-8 539
S70009-6 116
S70011-8 611
S70012-6 485
S70016-9 451
S70020-7 410
S70021-5 617
S70022-3 528
S70023-1 846
S70025-8 640
S70026-6 536
S70029-0 407
S70030-4 404
S70031-2 636
S70032-0 392
S70034-7 405
S70035-5 928
S70040-1 629
S70043-6 526
S70047-9 451
S70048-7 530
S70050-9 845
S70054-1 624
S70057-6 544
S70058-4 843
S70063-0 574
S70064-9 616
S70065-7 611
S70066-5 618
S70068-1 846
S70070-3 547
S70071-1 574
S70073-8 394
S70076-2 668
S70077-0 404
S70079-7 623
S70081-9 524
S70082-7 399
S70084-3 383
S70085-1 453
S70087-8 453
S70088-6 263
S70089-4 453
S70090-8 717
S70091-6 926
S70092-4 282
S70093-2 427
S70094-0 223
S70095-9 195
S70096-7 1024
S70097-5 105
S70098-3 435
S70099-1 483
S70100-9 474
S70101-7 416
S70102-5 418
S70103-3 474
S70104-1 461
S70106-8 470
S70107-6 474
S70108-4 454
S70109-2 406
S70110-6 1044
S70111-4 963
S70112-2 470
S70113-0 461
S70114-9 435
S70115-7 222
S70116-5 1050
S70117-3 1044
S70118-1 1040
S70120-3 466
S70121-1 1041
S70123-8 469
S70124-6 1029
S70125-4 1024
S70126-2 939
S70127-0 1031
S70128-9 1030
S70129-7 446
S70130-0 474
S70131-9 469
S70132-7 109
S70133-5 222
S70134-3 263
S70138-6 222
S70139-4 857
S70140-8 194
S70142-4 223
S70143-2 934
S70144-0 926
S70145-9 934
S70147-5 213
S70148-3 376
S70149-1 187
S70150-5 3
S70151-3 766
S70155-6 692
S70156-4 693
S70157-2 66
S70159-9 722
S70161-0 642
S70179-3 634
S70180-7 377
S70183-1 624
S70187-4 575
S70188-2 626
S70190-4 398
S70192-0 618
S70197-1 505
S70202-1 592
S70205-6 567
S70207-2 526
S70211-0 636
S70221-8 927
S70222-6 451
S70223-4 457
S70224-2 577
S70230-7 636
S70233-1 451
S70237-4 623
S70240-4 611
S70241-2 680
S70242-0 376
S70246-3 667
S70247-1 547
S70249-8 1014
S70250-1 18
S70253-6 831
S70254-4 723
S70256-0 978
S70257-9 397
S70258-7 395
S70260-9 447
S70261-7 408
S70263-3 695
S70264-1 611
S70266-8 536
S70267-6 537
S70268-4 537
S70269-2 581
S70271-4 109
S70273-0 189
S70275-7 440
S70276-5 1043

Product	Page	Product	Page	Product	Page	Product	Page	Product	Page	Product	Page	Product	Page
S70277-3	122	S70361-3	435	S70452-0	109	S70577-2	90	S70720-1	859	S70840-2	506	S70944-1	460
S70278-1	401	S70362-1	407	S70459-8	1019	S70578-0	219	S70722-8	797	S70841-0	760	S70946-8	633
S70280-3	435	S70364-8	418	S70460-1	1019	S70581-0	46	S70723-6	260	S70845-3	459	S70947-6	636
S70281-1	406	S70365-6	418	S70462-8	1016	S70586-1	297	S70724-4	827	S70849-6	541	S70948-4	559
S70283-8	1028	S70366-4	435	S70463-6	1019	S70589-6	28	S70725-2	985	S70853-4	457	S70949-2	586
S70284-6	436	S70367-2	225	S70469-5	1018	S70592-6	76	S70726-0	340	S70856-9	457	S70950-6	478
S70285-4	465	S70368-0	435	S70471-7	1016	S70594-2	80	S70728-7	457	S70857-7	459	S70951-4	762
S70286-2	1054	S70369-9	406	S70476-8	1016	S70596-9	85	S70731-7	912	S70859-3	530	S70952-2	696
S70287-0	406	S70370-2	435	S70477-6	962	S70598-5	76	S70733-3	566	S70860-7	530	S70953-0	760
S70288-9	1037	S70372-9	318	S70478-4	963	S70600-0	2	S70734-1	632	S70861-5	526	S70954-9	760
S70289-7	435	S70373-7	406	S70480-6	436	S70601-9	82	S70736-8	581	S70862-3	541	S70955-7	687
S70290-0	427	S70374-5	407	S70481-4	317	S70602-7	74	S70741-4	511	S70863-1	843	S70956-5	981
S70291-9	435	S70375-3	407	S70482-2	315	S70604-3	6	S70743-0	530	S70865-8	537	S70957-3	631
S70292-7	720	S70376-1	407	S70483-0	317	S70605-1	186	S70745-7	596	S70866-6	524	S70960-3	632
S70293-5	722	S70378-8	407	S70484-9	318	S70608-6	186	S70747-3	541	S70867-4	536	S70961-1	762
S70294-3	449	S70379-6	503	S70485-7	315	S70609-4	81	S70750-3	541	S70868-2	536	S70963-8	527
S70295-1	397	S70381-8	121	S70486-5	454	S70612-4	5	S70753-8	459	S70869-0	541	S70964-6	760
S70297-8	1024	S70382-6	435	S70487-3	316	S70614-0	800	S70754-6	457	S70870-4	541	S70965-4	760
S70298-6	1054	S70383-4	434	S70488-1	963	S70616-7	170	S70756-2	460	S70871-2	541	S70966-2	459
S70299-4	434	S70384-2	435	S70490-3	454	S70618-3	77	S70757-0	467	S70872-0	544	S70967-0	527
S70300-1	1025	S70385-0	407	S70491-1	725	S70619-1	81	S70758-9	464	S70873-9	541	S70968-9	648
S70302-8	406	S70386-9	407	S70493-8	658	S70621-3	194	S70760-0	480	S70874-7	560	S70969-7	627
S70303-6	963	S70387-7	407	S70494-6	660	S70624-8	84	S70762-7	542	S70876-3	540	S70970-0	539
S70304-4	1062	S70388-5	407	S70496-2	819	S70627-2	376	S70763-5	596	S70880-1	545	S70971-9	448
S70305-2	483	S70389-3	315	S70497-0	718	S70629-9	678	S70764-3	585	S70882-8	545	S70973-5	94
S70306-0	1032	S70390-7	317	S70498-9	725	S70633-7	819	S70767-8	464	S70883-6	546	S70974-3	375
S70307-9	406	S70391-5	315	S70499-7	139	S70635-3	314	S70768-6	467	S70884-4	545	S70977-8	186
S70308-7	435	S70392-3	317	S70500-4	146	S70636-1	711	S70769-4	481	S70885-2	540	S70978-6	196
S70309-5	426	S70393-1	317	S70504-7	348	S70638-8	148	S70770-8	541	S70886-0	129	S70980-8	619
S70310-9	1032	S70395-8	318	S70506-3	660	S70639-6	571	S70771-6	472	S70887-9	945	S70981-6	760
S70311-7	1022	S70396-6	317	S70510-1	656	S70641-8	17	S70773-2	473	S70888-7	458	S70983-2	627
S70312-5	397	S70397-4	474	S70512-8	656	S70646-9	278	S70774-0	459	S70889-5	556	S70985-9	640
S70313-3	426	S70398-2	445	S70513-6	104	S70647-7	691	S70775-9	469	S70890-9	467	S70987-5	635
S70314-1	263	S70399-0	454	S70514-4	659	S70648-5	727	S70776-7	459	S70891-7	535	S70988-3	627
S70316-8	383	S70400-8	455	S70515-2	660	S70651-5	192	S70777-5	993	S70892-5	533	S70989-1	641
S70317-6	1039	S70401-6	416	S70516-0	656	S70652-3	792	S70779-1	480	S70894-1	616	S70990-5	681
S70318-4	435	S70402-4	435	S70517-9	660	S70653-1	1093	S70781-3	479	S70896-8	450	S70991-3	648
S70319-2	223	S70403-2	954	S70519-5	117	S70660-4	373	S70782-1	459	S70897-6	458	S70992-1	642
S70320-6	722	S70404-0	717	S70520-9	656	S70661-2	190	S70784-8	458	S70898-4	459	S70994-8	671
S70321-4	1032	S70405-9	1028	S70521-7	659	S70662-0	255	S70787-2	473	S70899-2	596	S70995-6	760
S70322-2	1028	S70409-1	1032	S70522-5	656	S70665-5	156	S70788-0	459	S70901-8	637	S70997-2	631
S70323-0	463	S70411-3	286	S70523-3	660	S70669-8	221	S70790-2	765	S70903-4	616	S70998-0	632
S70324-9	336	S70412-1	230	S70524-1	660	S70673-6	1139	S70792-9	478	S70904-2	480	S70999-9	622
S70325-7	382	S70414-8	377	S70526-8	119	S70674-4	366	S70795-3	762	S70906-9	464	S71000-8	759
S70327-3	194	S70415-6	793	S70527-6	19	S70675-2	380	S70796-1	762	S70908-5	542	S71001-6	767
S70328-1	1028	S70416-4	101	S70530-6	205	S70676-0	170	S70799-6	459	S70909-3	511	S71003-2	641
S70330-3	448	S70418-0	48	S70532-2	204	S70677-9	990	S70800-3	459	S70910-7	542	S71004-0	627
S70334-6	871	S70419-9	1070	S70533-0	193	S70678-7	789	S70807-0	758	S70911-5	540	S71005-9	641
S70336-2	723	S70421-0	454	S70534-9	191	S70679-5	226	S70808-9	527	S70912-3	604	S71006-7	760
S70337-0	295	S70422-9	454	S70537-3	217	S70681-7	990	S70810-0	448	S70913-1	759	S71008-3	370
S70338-9	855	S70423-7	318	S70538-1	794	S70682-5	475	S70811-9	459	S70915-8	526	S71009-1	204
S70339-7	413	S70424-5	416	S70540-3	32	S70683-3	434	S70812-7	459	S70918-2	606	S71010-5	844
S70340-0	413	S70426-1	318	S70543-8	19	S70684-1	302	S70813-5	458	S70919-0	604	S71011-3	308
S70341-9	259	S70428-8	450	S70544-6	19	S70686-8	310	S70814-3	457	S70920-4	479	S71014-8	1092
S70342-7	359	S70429-6	454	S70546-2	144	S70687-6	378	S70815-1	571	S70921-2	478	S71017-2	411
S70343-5	52	S70431-8	954	S70547-0	55	S70688-4	476	S70817-8	564	S70922-0	457	S71021-0	828
S70344-3	97	S70432-6	315	S70549-7	296	S70690-6	62	S70818-6	473	S70925-5	525	S71024-5	771
S70345-1	57	S70433-4	317	S70550-0	50	S70692-2	721	S70820-8	595	S70926-3	982	S71028-8	810
S70347-8	994	S70434-2	318	S70552-7	324	S70693-0	194	S70821-6	521	S70927-1	982	S71029-6	1077
S70348-6	503	S70435-0	465	S70553-5	304	S70694-9	337	S70822-4	539	S70929-8	679	S71031-8	1072
S70349-4	121	S70436-9	315	S70555-1	62	S70695-7	380	S70823-2	458	S70932-8	527	S71033-4	567
S70350-8	121	S70437-7	315	S70557-8	63	S70697-3	304	S70824-0	521	S70933-6	628	S71034-2	458
S70351-6	121	S70438-5	954	S70558-6	75	S70698-1	304	S70825-9	542	S70934-4	635	S71035-0	575
S70352-4	435	S70439-3	315	S70562-4	80	S70701-5	314	S70826-7	450	S70935-2	642	S71036-9	533
S70353-2	122	S70440-7	406	S70563-2	76	S70705-8	91	S70827-5	459	S70936-0	627	S71038-5	714
S70354-0	121	S70441-5	318	S70564-0	386	S70706-6	76	S70828-3	458	S70937-9	640	S71040-7	539
S70355-9	449	S70442-3	426	S70565-9	85	S70710-4	209	S70831-3	452	S70938-7	642	S71041-5	554
S70356-7	318	S70443-1	406	S70567-5	353	S70711-2	266	S70832-1	457	S70939-5	616	S71042-3	538
S70357-5	317	S70445-8	316	S70572-1	79	S70712-0	771	S70835-6	458	S70940-9	509	S71043-1	760
S70358-3	315	S70446-6	1109	S70574-8	77	S70715-5	861	S70836-4	459	S70941-7	853	S71045-8	825
S70359-1	318	S70450-4	109	S70575-6	219	S70716-3	693	S70837-2	459	S70942-5	844	S71046-6	544
S70360-5	435	S70451-2	68	S70576-4	367	S70717-1	714	S70838-0	760	S70943-3	634	S71047-4	601

Product	Page
S71048-2	533
S71049-0	575
S71050-4	543
S71051-2	527
S71052-0	555
S71053-9	533
S71054-7	533
S71055-5	760
S71056-3	509
S71057-1	569
S71059-8	633
S71060-1	997
S71062-8	1005
S71066-0	159
S71067-9	661
S71069-5	527
S71070-9	689
S71071-7	689
S71072-5	749
S71074-1	965
S71076-8	622
S71077-6	982
S71078-4	982
S71079-2	671
S71081-4	761
S71082-2	760
S71083-0	760
S71084-9	994
S71085-7	569
S71086-5	569
S71087-3	520
S71088-1	535
S71090-3	670
S71093-8	764
S71094-6	1142
S71095-4	964
S71096-2	606
S71097-0	578
S71098-9	529
S71099-7	850
S71100-4	129
S71101-2	32
S71104-7	765
S71105-5	672
S71106-3	850
S71109-8	534
S71110-1	653
S71112-8	519
S71113-6	648
S71115-2	511
S71116-0	928
S71117-9	641
S71118-7	539
S71120-9	641
S71121-7	1008
S71122-5	585
S71123-3	616
S71124-1	534
S71126-8	696
S71127-6	545
S71129-2	534
S71130-6	760
S71132-2	129
S71133-0	982
S71134-9	448
S71136-5	479
S71138-1	767
S71140-3	530
S71141-1	982
S71143-8	536
S71144-6	530
S71145-4	572
S71146-2	546
S71147-0	478
S71148-9	539
S71149-7	555
S71150-0	511
S71151-9	458
S71152-7	760
S71155-1	543
S71157-8	986
S71158-6	569
S71160-8	278
S71161-6	803
S71163-2	574
S71164-0	712
S71165-9	604
S71166-7	604
S71167-5	605
S71168-3	604
S71169-1	533
S71170-5	627
S71171-3	627
S71172-1	593
S71174-8	930
S71175-6	928
S71176-4	610
S71177-2	608
S71179-9	639
S71180-2	636
S71181-0	603
S71182-9	545
S71183-7	545
S71184-5	543
S71185-3	545
S71186-1	633
S71188-8	636
S71189-6	546
S71191-8	530
S71192-6	540
S71193-4	540
S71194-2	538
S71195-0	546
S71196-9	540
S71197-7	546
S71198-5	540
S71199-3	541
S71200-0	542
S71201-9	542
S71202-7	544
S71203-5	524
S71204-3	539
S71205-1	539
S71207-8	540
S71208-6	540
S71209-4	539
S71210-8	537
S71211-6	537
S71212-4	537
S71213-2	541
S71214-0	543
S71215-9	543
S71216-7	538
S71217-5	538
S71218-3	526
S71219-1	535
S71220-5	535
S71221-3	1008
S71222-1	760
S71224-8	507
S71225-6	506
S71226-4	759
S71227-2	759
S71228-0	759
S71229-9	759
S71230-2	759
S71231-0	759
S71232-9	506
S71233-7	460
S71234-5	760
S71235-3	450
S71236-1	458
S71238-8	526
S71239-6	519
S71241-8	521
S71242-6	541
S71243-4	542
S71244-2	542
S71245-0	542
S71246-9	541
S71247-7	539
S71248-5	537
S71249-3	541
S71250-7	541
S71251-5	519
S71252-3	526
S71253-1	536
S71255-8	534
S71258-2	465
S71259-0	762
S71261-2	543
S71262-0	508
S71263-9	542
S71264-7	541
S71265-5	542
S71266-3	542
S71267-1	568
S71269-8	648
S71270-1	673
S71274-4	982
S71275-2	581
S71277-9	542
S71278-7	464
S71279-5	595
S71280-9	684
S71281-7	981
S71282-5	543
S71283-3	543
S71284-1	459
S71286-8	541
S71287-6	542
S71288-4	542
S71289-2	542
S71290-6	542
S71291-4	596
S71292-2	586
S71293-0	555
S71294-9	578
S71295-7	525
S71296-5	583
S71297-3	542
S71298-1	596
S71301-5	542
S71302-3	550
S71303-1	603
S71305-8	679
S71307-4	542
S71308-2	589
S71309-0	988
S71310-4	178
S71311-2	204
S71312-0	8
S71313-9	8
S71314-7	8
S71316-3	165
S71322-8	313
S71324-4	77
S71327-9	207
S71330-9	435
S71331-7	670
S71334-1	149
S71337-6	88
S71338-4	167
S71340-6	107
S71341-4	75
S71343-0	144
S71344-9	648
S71345-7	709
S71346-5	1008
S71347-3	817
S71348-1	567
S71350-3	688
S71351-1	634
S71353-8	633
S71354-6	627
S71356-2	622
S71357-0	679
S71358-9	673
S71359-7	767
S71360-0	620
S71361-9	684
S71362-7	610
S71363-5	674
S71364-3	760
S71365-1	673
S71367-8	673
S71368-6	629
S71372-4	367
S71373-2	530
S71374-0	572
S71375-9	128
S71376-7	521
S71377-5	541
S71378-3	530
S71379-1	541
S71380-5	568
S71381-3	506
S71382-1	594
S71384-8	524
S71385-6	539
S71386-4	506
S71387-2	541
S71388-0	538
S71389-9	930
S71390-2	759
S71391-0	550
S71392-9	575
S71393-7	562
S71394-5	826
S71395-3	953
S71396-1	457
S71398-8	464
S71399-6	457
S71401-1	460
S71403-8	460
S71404-6	459
S71405-4	458
S71406-2	409
S71407-0	460
S71408-9	457
S71409-7	457
S71415-1	532
S71417-8	520
S71418-6	926
S71419-4	936
S71422-4	1043
S71424-0	627
S71425-9	543
S71426-7	760
S71427-5	629
S71428-3	767
S71429-1	604
S71431-3	510
S71432-1	538
S71434-8	6
S71440-2	176
S71441-0	287
S71442-9	309
S71443-7	54
S71444-5	186
S71445-3	179
S71446-1	334
S71449-6	380
S71452-6	791
S71453-4	794
S71454-2	794
S71455-0	793
S71457-7	793
S71459-3	789
S71461-5	680
S71462-3	481
S71465-8	460
S71466-6	458
S71467-4	482
S71470-4	460
S71471-2	605
S71472-0	929
S71474-7	678
S71475-5	631
S71476-3	767
S71477-1	767
S71479-8	760
S71480-1	610
S71482-8	586
S71483-6	560
S71484-4	586
S71485-2	591
S71486-0	530
S71487-9	530
S71488-7	515
S71489-5	930
S71490-9	545
S71491-7	541
S71492-5	541
S71493-3	537
S71494-1	536
S71496-8	538
S71497-6	537
S71498-4	521
S71499-2	545
S71502-6	528
S71503-4	535
S71505-0	511
S71506-9	506
S71507-7	506
S71508-5	762
S71509-3	759
S71511-5	458
S71515-8	912
S71516-6	913
S71517-4	585
S71518-2	605
S71520-4	405
S71521-2	911
S71522-0	543
S71523-9	994
S71524-7	449
S71527-1	550
S71529-8	460
S71532-8	447
S71533-6	994
S71534-4	826
S71535-2	964
S71536-0	759
S71538-7	526
S71539-5	463
S71540-9	541
S71541-7	410
S71542-5	457
S71543-3	480
S71544-1	611
S71546-8	767
S71548-4	930
S71549-2	989
S71551-4	807
S71552-2	969
S71554-9	585
S71555-7	524
S71556-5	506
S71557-3	514
S71558-1	534
S71560-3	964
S71561-1	504
S71565-4	562
S71567-0	911
S71568-9	577
S71569-7	551
S71570-0	512
S71571-9	513
S71572-7	567
S71573-5	522
S71574-3	506
S71575-1	1087
S71577-8	570
S71579-4	638
S71580-8	647
S71581-6	992
S71582-4	687
S71583-2	673
S71584-0	639
S71585-9	676
S71586-7	473
S71587-5	416
S71588-3	763
S71589-1	910
S71590-5	437
S71591-3	477
S71592-1	490
S71594-8	447
S71595-6	428
S71596-4	390
S71597-2	407
S71598-0	427
S71599-9	415
S71600-6	449
S71601-4	448
S71602-2	466
S71603-0	819
S71604-9	808
S71605-7	825
S71606-5	1041
S71608-1	1041
S71610-3	497
S71611-1	1027
S71613-8	497
S71614-6	501
S71615-4	1028
S71619-7	495
S71620-0	494
S71621-9	500
S71622-7	498
S71623-5	493
S71624-3	502
S71627-8	1027
S71628-6	1040
S71630-8	1029
S71631-6	1025
S71633-2	1082
S71634-0	977
S71635-9	455
S71636-7	447
S71637-5	455
S71638-3	451
S71639-1	455
S71640-5	449
S71641-3	452
S71642-1	450
S71644-8	406
S71645-6	676
S71646-4	488
S71648-0	721
S71649-9	761
S71651-0	519
S71652-9	453
S71653-7	689
S71654-5	760
S71655-3	511
S71656-1	815
S71658-8	631
S71659-6	575
S71661-8	930
S71662-6	530
S71663-4	458
S71664-2	457
S71665-0	586
S71666-9	535
S71667-7	590
S71668-5	760
S71669-3	586
S71670-7	586
S71671-5	543
S71672-3	539
S71673-1	628
S71676-6	677
S71677-4	748
S71678-2	456
S71679-0	575
S71680-4	448
S71681-2	542
S71682-0	507
S71683-9	464
S71684-7	982
S71685-5	458
S71686-3	542
S71690-1	458
S71692-8	459
S71693-6	544
S71694-4	586
S71695-2	510
S71696-0	536
S71697-9	596
S71698-7	631
S71699-5	758
S71700-2	637
S71701-0	994
S71702-9	633
S71703-7	611
S71704-5	611
S71705-3	689
S71706-1	925
S71708-8	634
S71709-6	757
S71711-8	635
S71712-6	605

Product	Page
S71713-4	629
S71714-2	762
S71716-9	629
S71717-7	684
S71718-5	641
S71720-7	608
S71721-5	764
S71722-3	608
S71723-1	607
S71725-8	760
S71726-6	982
S71727-4	634
S71728-2	115
S71729-0	680
S71730-4	454
S71731-2	426
S71732-0	1021
S71733-9	469
S71734-7	435
S71735-5	1021
S71736-3	1037
S71737-1	315
S71739-8	1033
S71740-1	1022
S71742-8	474
S71743-6	418
S71744-4	210
S71745-2	627
S71746-0	1008
S71747-9	313
S71748-7	555
S71749-5	542
S71751-7	526
S71752-5	761
S71753-3	526
S71754-1	536
S71756-8	950
S71757-6	530
S71758-4	764
S71759-2	565
S71760-6	610
S71761-4	610
S71762-2	767
S71763-0	982
S71764-9	633
S71765-7	641
S71766-5	608
S71767-3	982
S71768-1	982
S71770-3	760
S71771-1	981
S71773-8	929
S71774-6	843
S71775-4	604
S71776-2	459
S71777-0	479
S71778-9	460
S71779-7	445
S71780-0	458
S71781-9	457
S71782-7	527
S71783-5	525
S71784-3	825
S71785-1	544
S71787-8	534
S71788-6	515
S71789-4	546
S71790-8	527
S71791-6	969
S71792-4	534
S71794-0	871
S71795-9	1088
S71796-7	546
S71797-5	592
S71798-3	602
S71803-3	390
S71804-1	390
S71806-8	391
S71807-6	391
S71808-4	390
S71809-2	445
S71810-6	390
S71811-4	772
S71812-2	392
S71813-0	824
S71814-9	404
S71815-7	461
S71816-5	410
S71817-3	419
S71818-1	413
S71820-3	410
S71821-1	615
S71823-8	627
S71824-6	667
S71825-4	696
S71826-2	695
S71827-0	810
S71828-9	810
S71829-7	809
S71830-0	807
S71831-9	806
S71832-7	454
S71833-5	806
S71834-3	532
S71835-1	823
S71837-8	377
S71838-6	236
S71843-2	744
S71844-0	941
S71845-9	401
S71848-3	622
S71849-1	965
S71850-5	619
S71851-3	402
S71852-1	536
S71854-8	641
S71855-6	528
S71858-0	667
S71859-9	941
S71860-2	630
S71861-0	631
S71862-9	744
S71863-7	939
S71864-5	1082
S71865-3	604
S71866-1	414
S71868-8	626
S71869-6	1082
S71871-8	666
S71872-6	531
S71874-2	936
S71876-9	538
S71877-7	515
S71878-5	536
S71879-3	737
S71880-7	514
S71881-5	611
S71882-3	930
S71883-1	930
S71885-8	548
S71886-6	448
S71887-4	532
S71889-0	622
S71891-2	631
S71892-0	548
S71893-9	634
S71894-7	912
S71895-5	912
S71897-1	620
S71899-8	653
S71900-5	435
S71901-3	426
S71902-1	471
S71904-8	474
S71905-6	474
S71906-4	273
S71907-2	474
S71908-0	474
S71909-9	962
S71910-2	470
S71911-0	122
S71912-9	433
S71913-7	1024
S71914-5	294
S71915-3	1024
S71916-1	854
S71918-8	877
S71919-6	723
S71921-8	397
S71922-6	398
S71923-4	962
S71924-2	453
S71925-0	454
S71926-9	397
S71927-7	405
S71928-5	406
S71929-3	418
S71930-7	426
S71931-5	406
S71932-3	435
S71933-1	963
S71935-8	418
S71936-6	434
S71937-4	434
S71938-2	715
S71939-0	715
S71940-4	715
S71941-2	715
S71942-0	406
S71943-9	723
S71944-7	715
S71945-5	1062
S71946-3	1062
S71947-1	1025
S71949-8	1032
S71950-1	1032
S71952-8	963
S71953-6	474
S71954-4	453
S71955-2	474
S71956-0	474
S71957-9	318
S71958-7	383
S71959-5	318
S71960-9	317
S71961-7	317
S71962-5	316
S71963-3	315
S71964-1	1030
S71966-8	432
S71967-6	1038
S71968-4	1026
S71969-2	1074
S71970-6	1037
S71971-4	1039
S71972-2	1023
S71973-0	1026
S71974-9	1031
S71975-7	1038
S71976-5	1038
S71977-3	926
S71978-1	1031
S71980-3	1027
S71981-1	1027
S71983-8	398
S71984-6	434
S71985-4	434
S71986-2	434
S71987-0	406
S71988-9	432
S71989-7	418
S71990-0	1059
S71991-9	1075
S71992-7	1050
S71993-5	1050
S71994-3	1074
S71995-1	1037
S71997-8	1038
S71998-6	1075
S71999-4	1075
S72000-3	1031
S72001-1	1074
S72003-8	1022
S72004-6	1037
S72005-4	1023
S72006-2	1021
S72007-0	431
S72008-9	949
S72009-7	435
S72010-0	406
S72011-9	398
S72012-7	1031
S72013-5	434
S72014-3	418
S72015-1	418
S72017-8	434
S72018-6	963
S72019-4	406
S72020-8	406
S72021-6	1020
S72022-4	406
S72023-2	406
S72024-0	397
S72025-9	398
S72026-7	395
S72027-5	963
S72028-3	103
S72029-1	855
S72031-3	930
S72032-1	825
S72034-8	534
S72035-6	530
S72036-4	759
S72037-2	390
S72038-0	844
S72041-0	177
S72042-9	860
S72043-7	866
S72044-5	979
S72045-3	708
S72049-6	971
S72051-8	1062
S72052-6	1075
S72053-4	708
S72054-2	395
S72055-0	406
S72056-9	433
S72057-7	270
S72058-5	706
S72059-3	1035
S72060-7	274
S72061-5	294
S72062-3	273
S72063-1	122
S72065-8	963
S72066-6	963
S72067-4	435
S72068-2	406
S72069-0	945
S72070-4	418
S72071-2	963
S72072-0	877
S72073-9	944
S72074-7	949
S72075-5	962
S72076-3	626
S72077-1	522
S72079-8	525
S72080-1	532
S72082-8	408
S72083-6	728
S72084-4	818
S72085-2	865
S72086-0	984
S72087-9	483
S72088-7	825
S72089-5	1030
S72091-7	727
S72092-5	736
S72093-3	735
S72096-8	523
S72097-6	395
S72098-4	471
S72101-8	464
S72103-4	651
S72104-2	632
S72105-0	416
S72106-9	653
S72107-7	655
S72108-5	655
S72109-3	484
S72110-7	616
S72111-5	473
S72112-3	130
S72113-1	545
S72115-8	518
S72116-6	668
S72118-2	578
S72119-0	451
S72120-4	483
S72121-2	449
S72123-9	806
S72124-7	533
S72125-5	472
S72126-3	618
S72127-1	475
S72129-8	636
S72130-1	639
S72132-8	449
S72133-6	911
S72134-4	911
S72135-2	912
S72136-0	912
S72137-9	912
S72138-7	687
S72140-9	560
S72141-7	450
S72142-5	544
S72143-3	530
S72144-1	621
S72146-8	452
S72147-6	454
S72148-4	470
S72149-2	965
S72150-6	599
S72152-2	952
S72153-0	930
S72154-9	964
S72155-7	524
S72156-5	543
S72157-3	925
S72158-1	634
S72160-3	807
S72161-1	627
S72163-8	129
S72164-6	440
S72165-4	543
S72166-2	424
S72167-0	879
S72168-9	640
S72169-7	401
S72170-0	681
S72171-9	819
S72172-7	805
S72173-5	504
S72175-1	408
S72177-8	396
S72178-6	398
S72179-4	877
S72180-8	1026
S72182-4	270
S72183-2	771
S72184-0	199
S72185-9	212
S72188-3	375
S72189-1	62
S72190-5	205
S72192-1	371
S72194-8	210
S72195-6	55
S72197-2	786
S72198-0	1142
S72202-2	6
S72203-0	210
S72206-5	864
S72207-3	187
S72208-1	4
S72210-3	187
S72211-1	211
S72213-8	97
S72214-6	5
S72215-4	207
S72216-2	180
S72217-0	187
S72218-9	1016
S72219-7	85
S72220-0	85
S72221-9	212
S72223-5	169
S72224-3	211
S72225-1	283
S72227-8	456
S72230-8	66
S72231-6	993
S72232-4	993
S72233-2	1089
S72235-9	342
S72236-7	180
S72238-3	319
S72239-1	193
S72240-5	318
S72242-1	42
S72246-4	331
S72247-2	808
S72250-2	253
S72251-0	796
S72252-9	69
S72253-7	231
S72254-5	342
S72258-8	730
S72259-6	743
S72261-8	742
S72265-0	655
S72271-5	815
S72272-3	866
S72273-1	560
S72276-6	523
S72278-2	729
S72279-0	648
S72280-4	575
S72281-2	504
S72282-0	505
S72283-9	547
S72285-5	668
S72286-3	640
S72287-1	647
S72289-8	633
S72290-1	445
S72293-6	716
S72294-4	551
S72295-2	576
S72296-0	729
S72297-9	522
S72298-7	560
S72299-5	540
S72300-2	448
S72302-9	1024
S72303-7	623
S72304-5	454
S72306-1	981
S72309-6	711
S72311-8	651
S72312-6	687
S72313-4	527
S72314-2	929
S72315-0	512
S72316-9	491
S72317-7	491
S72320-7	614
S72321-5	623
S72322-3	865
S72326-6	447
S72327-4	634
S72328-2	530
S72330-4	548
S72331-2	636
S72332-0	984
S72333-9	735
S72334-7	523
S72337-1	624
S72339-8	636
S72340-1	532
S72342-8	529
S72345-2	843
S72346-0	1026
S72349-5	560
S72352-5	394
S72354-1	574
S72357-6	553
S72358-4	925
S72360-6	928
S72361-4	937
S72362-2	850
S72364-9	553

Product Page

S72365-7........846
S72366-5........851
S72367-3........553
S72368-1........553
S72370-3........641
S72371-1........640
S72373-8........715
S72374-6........521
S72375-4........732
S72376-2........522
S72377-0........623
S72378-9........512
S72379-7........512
S72380-0........617
S72381-9........623
S72382-7........623
S72383-5........616
S72385-1........494
S72387-8........622
S72388-6........621
S72389-4........621
S72390-8........635
S72391-6........621
S72392-4........636
S72394-0........522
S72395-9........525
S72397-5........641
S72398-3........634
S72399-1........639
S72400-9........637
S72401-7........637
S72402-5........639
S72403-3........628
S72404-1........867
S72406-8........737
S72408-4........927
S72409-2........964
S72413-0........532
S72416-5........751
S72417-3........732
S72418-1........513
S72421-1........401
S72429-7........900
S72430-0......1023
S72431-9........571
S72432-7........638
S72434-3........400
S72437-8........637
S72439-4........402
S72440-8........680
S72441-6........450
S72442-4........638
S72444-0........733
S72445-9........634
S72446-7........400
S72447-5........624
S72449-1........536
S72450-5........667
S72451-3........929
S72452-1........640
S72454-8........640
S72455-6........400
S72456-4........459
S72457-2........457
S72458-0........666
S72459-9........640
S72461-0........653
S72465-3........623
S72466-1........634
S72468-8........469
S72469-6........468
S72471-8........459

Product Page

S72472-6........447
S72473-4........687
S72475-0........128
S72476-9........400
S72477-7........402
S72480-7........452
S72481-5........616
S72482-3........679
S72483-1........400
S72485-8........632
S72486-6........457
S72487-4........401
S72488-2........524
S72489-0........565
S72490-4........400
S72491-2........414
S72492-0........456
S72493-9........536
S72496-3........728
S72499-8........550
S72502-1........577
S72504-8........633
S72505-6........607
S72506-4........395
S72507-2........394
S72508-0........729
S72509-9........394
S72511-0........130
S72515-3........845
S72516-1........623
S72518-8........668
S72519-6........620
S72521-8........927
S72522-6........494
S72524-2........927
S72525-0........456
S72526-9........420
S72527-7........745
S72528-5........529
S72529-3........618
S72530-7........515
S72531-5........687
S72532-3........637
S72535-8........518
S72536-6........599
S72537-4........619
S72538-2........927
S72541-2........616
S72542-0........551
S72543-9........696
S72544-7........618
S72545-5........415
S72547-1........492
S72549-8........530
S72550-1........631
S72552-8........652
S72553-6........635
S72555-2........941
S72557-9........561
S72558-7......1030
S72559-5........545
S72560-9........940
S72561-7........492
S72566-8........606
S72568-4........563
S72587-0........910
S72589-7........509
S72590-0........532
S72591-9........532
S72593-5......1060
S72595-1......1070
S72598-6........400

Product Page

S72599-4........400
S72600-1........564
S72602-8........851
S72603-6........622
S72604-4........929
S72607-9........622
S72608-7........395
S72609-5........516
S72611-7........520
S72613-3........542
S72614-1........636
S72618-4........408
S72619-2......1029
S72624-9........939
S72630-3........930
S72631-1........545
S72634-6........500
S72636-2........528
S72642-7........408
S72643-5........450
S72648-6........636
S72649-4........621
S72650-8........699
S72652-4........408
S72654-0........633
S72655-9........516
S72657-5........518
S72658-3........517
S72659-1........439
S72660-5........527
S72661-3........514
S72665-6........584
S72666-4........532
S72667-2........561
S72668-0........964
S72669-9........564
S72670-2........638
S72671-0........641
S72673-7........681
S72674-5........806
S72675-3........505
S72679-6........612
S72682-6........408
S72685-0........631
S72687-7........694
S72690-7........628
S72692-3........633
S72695-8........638
S72697-4........636
S72698-2........450
S72700-8......1128
S72704-0........881
S72705-9........834
S72706-7........833
S72707-5........834
S72709-1..........43
S72711-3..........44
S72712-1......1127
S72714-8......1129
S72715-6......1128
S72717-2......1127
S72720-2......1128
S72723-7......1128
S72725-3........881
S72726-1........881
S72729-6........834
S72733-4......1126
S72734-2......1128
S72735-0......1127
S72736-9......1140
S72738-5......1140
S72740-7......1140

Product Page

S72741-5......1127
S72742-3......1140
S72743-1......1135
S72745-8........881
S72753-9........622
S72756-3........514
S72757-1........514
S72759-8........619
S72760-1........969
S72762-8........544
S72763-6........614
S72764-4........512
S72765-2........626
S72766-0........637
S72767-9........550
S72768-7........728
S72769-5........751
S72770-9........548
S72771-7........560
S72772-5........814
S72774-1........550
S72777-6........613
S72778-4........648
S72779-2........527
S72780-6........453
S72781-4........553
S72782-2........456
S72784-9........626
S72785-7........654
S72786-5........408
S72787-3........575
S72791-1........639
S72793-8........553
S72794-6........646
S72795-4........646
S72796-2........640
S72797-0........612
S72798-9........871
S72801-2........646
S72802-0........844
S72803-9........749
S72804-7........782
S72805-5........773
S72809-8........622
S72810-1........625
S72812-8........577
S72813-6........651
S72815-2........536
S72816-0........625
S72818-7........612
S72819-5........506
S72820-9........615
S72821-7........615
S72822-5........504
S72823-3........535
S72824-1........517
S72826-8........549
S72827-6........978
S72828-4........625
S72829-2........892
S72831-4........562
S72836-5........561
S72838-1........626
S72840-3........519
S72843-8........638
S72845-4........453
S72848-9........682
S72849-7........561
S72850-0......1080
S72855-1........529
S72858-6........401
S72861-6........597

Product Page

S72868-3........455
S72871-3........496
S72875-6........876
S72876-4........874
S72884-5........869
S72885-3........869
S72891-8........565
S72892-6........519
S72895-0........390
S72897-7..........44
S72899-3........978
S72901-9........959
S72903-5......1011
S72904-3........899
S72907-8........715
S72909-4........448
S72910-8........416
S72911-6........453
S72912-4........401
S72913-2........452
S72914-0........477
S72915-9........465
S72916-7........460
S72917-5........395
S72918-3........401
S72919-1........407
S72920-5........415
S72921-3........405
S72922-1........404
S72924-8........489
S72925-6........416
S72927-2........877
S72928-0........513
S72929-9........567
S72930-2........520
S72931-0........451
S72933-7........912
S72934-5........493
S72936-1........507
S72939-6........888
S72941-8........377
S72942-6........576
S72943-4........630
S72944-2........484
S72946-9........826
S72948-5........578
S72951-5........542
S72952-3........510
S72956-6........492
S72957-4........627
S72958-2........517
S72961-2........627
S72962-0........612
S72964-7........514
S72966-3........518
S72967-1........666
S72970-1........744
S72972-8........930
S72973-6........477
S72974-4........637
S72975-2........911
S72976-0........451
S72977-9........981
S72978-7........484
S72979-5........911
S72980-9........936
S72981-7........527
S72982-5........432
S72984-1........648
S72986-8........644
S72988-4........731
S72991-4........808

Product Page

S72992-2........926
S72993-0........911
S72994-9........911
S72995-7........667
S72996-5........493
S72998-1........681
S73000-9......1054
S73001-7......1053
S73002-5......1040
S73003-3......1038
S73004-1......1074
S73006-8......1031
S73007-6........706
S73008-4........706
S73010-6......1035
S73011-4........273
S73012-2........273
S73013-0........850
S73014-9........854
S73015-7........420
S73017-3........107
S73018-1........397
S73020-3........397
S73021-1........396
S73023-8........398
S73024-6........395
S73025-4......1027
S73026-2......1023
S73027-0......1026
S73028-9......1025
S73029-7........429
S73030-0........436
S73031-9........431
S73032-7........963
S73033-5........406
S73034-3........407
S73035-1........435
S73037-8........435
S73038-6........435
S73039-4........407
S73040-8........435
S73041-6........407
S73042-4........426
S73044-0........418
S73045-9........399
S73046-7......1027
S73047-5......1023
S73048-3........424
S73049-1........110
S73050-5........423
S73051-3........110
S73054-8........471
S73055-6........471
S73056-4........543
S73057-2........536
S73058-0........434
S73059-9........406
S73060-2........429
S73061-0........431
S73062-9........434
S73063-7........122
S73064-5........854
S73065-3........122
S73066-1........963
S73068-8........575
S73071-8........406
S73072-6........122
S73073-4........122
S73074-2........471
S73075-0........963
S73076-9......1022
S73077-7......1027

Product Page

S73078-5......1027
S73079-3......1027
S73080-7......1024
S73081-5......1024
S73082-3......1026
S73083-1......1022
S73085-8........474
S73086-6........901
S73088-2........465
S73089-0........731
S73090-4........681
S73091-2........930
S73092-0........525
S73093-9........670
S73094-7........657
S73095-5........496
S73096-3......1025
S73099-8........496
S73100-5........501
S73101-3........494
S73102-1........495
S73104-8........492
S73105-6........814
S73106-4........815
S73107-2........647
S73110-2......1009
S73121-8......1018
S73126-9........287
S73129-3........166
S73132-3........424
S73133-1........427
S73135-8........432
S73136-6........422
S73137-4........335
S73142-0........327
S73146-3........174
S73147-1........365
S73149-8......1088
S73150-1........999
S73152-8........718
S73153-6........988
S73154-4........348
S73156-0........729
S73157-9........981
S73158-7........933
S73159-5........514
S73160-9........538
S73161-7........629
S73163-3........530
S73164-1........501
S73167-6........627
S73170-6........612
S73173-0......1060
S73175-7......1082
S73177-3........626
S73178-1........612
S73180-3......1069
S73181-1........654
S73183-8........611
S73184-6........500
S73185-4......1070
S73186-2........632
S73187-0........548
S73188-9........455
S73189-7........567
S73190-0........666
S73192-7........630
S73193-5........515
S73195-1........548
S73197-8........627
S73199-4........527
S73202-8........134

Product Page

S73205-2........171
S73206-0........965
S73208-7......1006
S73209-5......1000
S73211-7......1008
S73213-3........458
S73214-1........581
S73216-8........871
S73218-4........384
S73220-6........772
S73221-4......1011
S73223-0........716
S73225-7........919
S73226-5........723
S73227-3........986
S73231-1........453
S73233-8........755
S73235-4........290
S73236-2........968
S73237-0........294
S73238-9........407
S73240-0........871
S73241-9..........51
S73242-7........244
S73243-5........865
S73244-3........728
S73245-1........757
S73247-8........197
S73249-4........253
S73250-8........710
S73251-6........297
S73252-4........859
S73253-2........557
S73257-5........709
S73258-3........723
S73262-1........711
S73265-6........735
S73266-4........463
S73268-0........724
S73269-9........321
S73270-2........477
S73271-0........799
S73273-7........297
S73274-5........205
S73275-3........871
S73276-1......1011
S73278-8........208
S73281-8........310
S73282-6........708
S73284-2........689
S73286-9........757
S73291-5........694
S73292-3........254
S73299-0........695
S73300-8........907
S73302-4........726
S73303-2........339
S73304-0........834
S73305-9........477
S73307-5........710
S73309-1........966
S73314-8........864
S73316-4........287
S73317-2........695
S73318-0........709
S73321-0........319
S73322-9........937
S73323-7........300
S73324-5........251
S73328-8........775
S73331-8........957
S73332-6........698
S73333-4........811
S73334-2........931
S73335-0........388
S73336-9........780
S73337-7........387
S73338-5........236
S73342-3........432
S73343-1........690
S73345-8........198
S73346-6........255
S73348-2........205
S73349-0........254
S73350-4........710
S73351-2........309
S73352-0......1033
S73353-9........931
S73354-7........327
S73355-5........689
S73356-3........405
S73357-1......1024
S73360-1........734
S73362-8........938
S73363-6........298
S73366-0........218
S73368-7........473
S73369-5........726
S73370-9........931
S73371-7........286
S73376-8........244
S73377-6........519
S73380-6........382
S73382-2........310
S73383-0........871
S73385-7........237
S73387-3........107
S73388-1........749
S73391-1........408
S73393-8........278
S73394-6........753
S73397-0........387
S73409-8........828
S73416-0........697
S73424-1........725
S73426-8........215
S73427-6........246
S73428-4........871
S73430-6........956
S73431-4........385
S73435-7........872
S73436-5........327
S73438-1........871
S73440-3........353
S73441-1........332
S73443-8........276
S73444-6........778
S73446-2........309
S73447-0........344
S73448-9..........91
S73449-7........198
S73450-0........844
S73451-9......1018
S73452-7........208
S73453-5........207
S73457-8........353
S73458-6..........91
S73459-4........782
S73460-8........931
S73462-4........692
S73463-2........246
S73465-9........778
S73466-7........707
S73468-3..........93
S73470-5........713
S73471-3........384
S73472-1........693
S73474-8........254
S73475-6........737
S73476-4..........61
S73478-0........309
S73481-0........222
S73482-9........288
S73484-5........390
S73485-3........480
S73486-1........282
S73488-8........736
S73491-8........218
S73492-6........446
S73493-4........535
S73494-2........753
S73495-0........468
S73496-9........295
S73498-5........463
S73499-3........862
S73500-0........363
S73501-9........304
S73502-7........705
S73505-1........955
S73507-8........811
S73508-6........945
S73509-4........999
S73510-8........704
S73511-6......1071
S73512-4......1008
S73513-2........928
S73514-0........310
S73515-9........778
S73517-5........337
S73520-5........864
S73521-3........812
S73522-1......1054
S73525-6........689
S73526-4........244
S73527-2........201
S73528-0........197
S73531-0........690
S73532-9......1009
S73533-7........461
S73538-8........765
S73541-8........565
S73542-6........672
S73543-4........552
S73544-2........575
S73545-0........510
S73546-9........759
S73548-5........506
S73549-3........529
S73550-7........511
S73551-5........759
S73553-1........530
S73555-8........575
S73557-4........917
S73558-2........917
S73559-0........248
S73561-2........460
S73562-0........536
S73563-9........538
S73564-7........448
S73565-5........542
S73566-3........450
S73567-1........767
S73569-8........472
S73572-8........459
S73573-6........459
S73574-4........450
S73575-2........541
S73576-0........460
S73577-9........545
S73578-7........627
S73579-5........581
S73580-9........458
S73581-7........459
S73582-5........474
S73583-3........447
S73584-1........467
S73586-8........867
S73588-4........984
S73591-4........448
S73592-2........672
S73594-9........450
S73595-7........727
S73596-5........471
S73597-3........464
S73598-1........450
S73600-7........527
S73601-5........496
S73602-3........496
S73603-1........686
S73605-8........465
S73607-4........631
S73608-2........636
S73609-0........620
S73610-4........626
S73611-2........501
S73612-0........686
S73613-9........501
S73614-7........415
S73615-5........729
S73616-3........413
S73617-1........469
S73619-8........399
S73620-1........399
S73622-8........975
S73623-6........805
S73624-4........337
S73625-2........392
S73626-0........404
S73627-9........871
S73630-9........927
S73631-7........626
S73633-3........537
S73634-1........611
S73636-8........514
S73637-6........524
S73638-4........808
S73639-2........871
S73640-6........627
S73642-2........544
S73643-0........531
S73644-9........530
S73645-7........517
S73646-5........537
S73647-3........537
S73648-1........537
S73650-3........537
S73651-1........908
S73653-8........453
S73655-4........731
S73656-2........496
S73657-0........927
S73658-9........404
S73659-7........825
S73660-0........951
S73661-9......1070
S73663-5........676
S73664-3........496
S73665-1........693
S73667-8........617
S73668-6........524
S73669-4........522
S73670-8........499
S73671-6........822
S73672-4........626
S73673-2........506
S73674-0........911
S73675-9........640
S73676-7........530
S73677-5........530
S73678-3........631
S73679-1........531
S73680-5........525
S73681-3........495
S73682-1........633
S73684-8........633
S73685-6........615
S73686-4........526
S73687-2........526
S73688-0........492
S73689-9........516
S73690-2........666
S73691-0........500
S73692-9........538
S73693-7........537
S73694-5........519
S73695-3........515
S73696-1........495
S73698-8........759
S73699-6........910
S73700-3........899
S73701-1......1061
S73703-8........826
S73704-6........611
S73705-4........527
S73706-2........548
S73708-9........495
S73709-7........414
S73710-0........456
S73711-9........456
S73712-7........616
S73714-3........414
S73715-1........501
S73717-8........606
S73718-6........531
S73719-4........507
S73720-8........523
S73721-6........618
S73722-4........614
S73723-2........493
S73724-0........996
S73725-9........901
S73726-7..........88
S73727-5........619
S73730-5........607
S73731-3........507
S73732-1........498
S73734-8........615
S73737-2......1027
S73741-0........733
S73742-9........638
S73745-3........485
S73748-8........547
S73752-6........528
S73753-4........929
S73757-7........141
S73758-5........939
S73759-3........639
S73762-3........580
S73763-1........678
S73766-6........527
S73767-4........507
S73771-2........459
S73773-9........647
S73774-7........630
S73776-3........617
S73779-8........546
S73784-4........394
S73787-9........930
S73788-7........844
S73789-5........844
S73790-9........641
S73792-5........633
S73794-1........627
S73796-8........964
S73797-6........460
S73798-4........846
S73799-2......1041
S73801-8......1042
S73802-6........497
S73805-0........354
S73806-9........670
S73807-7........670
S73808-5........670
S73810-7........492
S73811-5........494
S73812-3........484
S73813-1........677
S73815-8........410
S73818-2........495
S73819-0........633
S73821-2........929
S73823-9........696
S73824-7........941
S73825-5........495
S73826-3........927
S73827-1........929
S73829-8........730
S73832-8........844
S73833-6........847
S73834-4........845
S73835-2........654
S73836-0........729
S73837-9........730
S73838-7........846
S73840-9........729
S73841-7........513
S73842-5........739
S73843-3........729
S73844-1........637
S73846-8........844
S73847-6........929
S73849-2........844
S73850-6........639
S73851-4........845
S73852-2........695
S73853-0........497
S73854-9........417
S73855-7........405
S73857-3........448
S73858-1........445
S73860-3........428
S73861-1........453
S73863-8........494
S73864-6........847
S73865-4........845
S73866-2........502
S73872-7........502
S73873-5........497
S73874-3........498
S73875-1........641
S73877-8........626
S73880-8........976
S73881-6........847
S73882-4........844
S73883-2........617
S73885-9........407
S73887-5........410
S73888-3........405
S73890-5........675
S73891-3........447
S73892-1........671
S73894-8........569
S73895-6......1080
S73896-4......1088
S73897-2........681
S73898-0........445
S73899-9........428
S73900-6........669
S73901-4........490
S73902-2........612
S73903-0........638
S73904-9........929
S73905-7........614
S73906-5........639
S73907-3........455
S73908-1........525
S73911-1........623
S73914-6........636
S73915-4........399
S73916-2........502
S73918-9........980
S73919-7........528
S73920-0........964
S73921-9........453
S73922-7........198
S73923-5........528
S73924-3........842
S73925-1........842
S73927-8........842
S73928-6........842
S73929-4............9
S73931-6..........91
S73932-4..........86
S73935-9........818
S73936-7......1044
S73939-1........819
S73941-3..........62
S73944-8........297
S73945-6........290
S73946-4........386
S73949-9..........97
S73950-2........260
S73952-9........365
S73953-7..........20
S73954-5........967
S73955-3........307
S73956-1........476
S73958-8......1038
S73959-6......1076
S73962-6........283
S73963-4......1033
S73964-2........222
S73965-0........222
S73966-9........215
S73967-7........237
S73968-5........223
S73969-3........225
S73970-7........290
S73971-5........875
S73972-3........225
S73973-1........290
S73975-8........290
S73977-4........238
S73978-2........309

Product	Page
S73979-0	316
S73980-4	343
S73981-2	318
S73982-0	850
S73983-9	315
S73984-7	454
S73985-5	454
S73986-3	454
S73987-1	302
S73989-8	454
S73990-1	451
S73992-8	473
S73993-6	282
S73994-4	398
S73995-2	273
S73996-0	299
S73997-9	304
S73998-7	839
S73999-5	716
S74001-2	709
S74002-0	700
S74003-9	769
S74007-1	803
S74009-8	307
S74014-4	940
S74016-0	885
S74017-9	886
S74018-7	190
S74021-7	52
S74026-8	68
S74027-6	380
S74028-4	343
S74029-2	890
S74030-6	353
S74032-2	752
S74034-9	1078
S74035-7	235
S74036-5	271
S74038-1	842
S74040-3	710
S74044-6	15
S74045-4	1094
S74046-2	290
S74050-0	708
S74061-6	726
S74066-7	724
S74067-5	724
S74078-0	726
S74104-3	732
S74107-8	732
S74112-4	695
S74117-5	695
S74118-3	740
S74130-2	737
S74131-0	737
S74132-9	740
S74152-3	24
S74156-6	17
S74158-2	261
S74163-9	66
S74164-7	144
S74165-5	751
S74166-3	723
S74167-1	298
S74169-8	1128
S74170-1	428
S74174-4	836
S74176-0	1128
S74177-9	886
S74178-7	1137
S74179-5	1127
S74180-9	1137
S74181-7	1137
S74183-3	1128
S74184-1	1140
S74186-8	1126
S74187-6	1140
S74188-4	1126
S74189-2	1129
S74192-2	830
S74193-0	152
S74194-9	216
S74195-7	1128
S74196-5	1128
S74197-3	830
S74198-1	834
S74200-7	32
S74201-5	32
S74202-3	93
S74203-1	63
S74205-8	755
S74206-6	195
S74207-4	392
S74208-2	723
S74209-0	198
S74213-9	463
S74215-5	978
S74217-1	351
S74222-8	885
S74223-6	960
S74226-0	95
S74227-9	440
S74229-5	87
S74230-9	144
S74231-7	42
S74232-5	751
S74236-8	719
S74237-6	1128
S74238-4	834
S74241-4	386
S74242-2	298
S74243-0	1128
S74244-9	294
S74245-7	968
S74246-5	294
S74248-1	715
S74250-3	433
S74251-1	432
S74253-8	294
S74254-6	432
S74255-4	331
S74256-2	452
S74260-0	332
S74261-9	427
S74263-5	92
S74264-3	236
S74265-1	723
S74267-8	1010
S74268-6	111
S74269-4	57
S74270-8	796
S74271-6	1128
S74272-4	1126
S74274-0	829
S74275-9	717
S74276-7	1141
S74277-5	465
S74278-3	155
S74279-1	887
S74280-5	63
S74282-1	432
S74284-8	264
S74286-4	294
S74287-2	751
S74288-0	830
S74290-2	265
S74292-9	810
S74293-7	810
S74294-5	825
S74295-3	810
S74298-8	526
S74299-6	518
S74300-3	738
S74301-1	679
S74303-8	679
S74304-6	622
S74305-4	637
S74306-2	621
S74307-0	668
S74308-9	615
S74309-7	649
S74310-0	641
S74311-9	679
S74312-7	635
S74313-5	647
S74314-3	670
S74315-1	629
S74318-6	515
S74319-4	584
S74320-8	519
S74321-6	517
S74323-2	730
S74324-0	529
S74325-9	878
S74326-7	571
S74328-3	571
S74329-1	549
S74330-5	540
S74331-3	582
S74332-1	527
S74334-8	543
S74335-6	743
S74336-4	512
S74337-2	602
S74338-0	589
S74339-9	572
S74340-2	523
S74341-0	737
S74342-9	603
S74345-3	909
S74346-1	904
S74351-8	454
S74352-6	447
S74353-4	452
S74354-2	451
S74355-0	969
S74356-9	497
S74357-7	502
S74358-5	492
S74359-3	492
S74360-7	501
S74361-5	494
S74363-1	1048
S74365-8	814
S74366-6	807
S74367-4	806
S74368-2	814
S74370-4	813
S74371-2	807
S74372-0	846
S74373-9	409
S74375-5	416
S74377-1	415
S74379-8	410
S74380-1	437
S74382-8	438
S74383-6	410
S74384-4	404
S74385-2	416
S74386-0	466
S74387-9	764
S74388-7	455
S74389-5	462
S74390-9	264
S74391-7	269
S74392-5	531
S74393-3	854
S74394-1	407
S74396-8	434
S74397-6	433
S74398-4	485
S74401-8	724
S74404-2	1075
S74409-3	728
S74411-5	140
S74416-6	911
S74418-2	928
S74420-4	758
S74421-2	767
S74423-9	763
S74424-7	763
S74425-5	763
S74426-3	758
S74427-1	759
S74429-8	759
S74430-1	759
S74432-8	759
S74433-6	763
S74434-4	759
S74435-2	759
S74436-0	759
S74437-9	758
S74438-7	753
S74439-5	763
S74440-9	763
S74441-7	759
S74442-5	759
S74443-3	759
S74446-8	753
S74448-4	608
S74453-0	885
S74456-5	354
S74457-3	985
S74458-1	998
S74460-3	1001
S74464-6	999
S74466-2	999
S74468-9	58
S74469-7	405
S74471-9	289
S74473-5	942
S74474-3	226
S74479-4	241
S74480-8	288
S74482-4	999
S74484-0	409
S74485-9	239
S74486-7	372
S74487-5	276
S74488-3	239
S74489-1	850
S74490-5	968
S74491-3	271
S74492-1	968
S74494-8	177
S74495-6	351
S74496-4	818
S74497-2	759
S74500-6	477
S74505-7	200
S74506-5	1001
S74508-1	330
S74510-3	276
S74511-1	405
S74513-8	294
S74515-4	764
S74516-2	308
S74517-0	1000
S74518-9	313
S74519-7	384
S74520-0	1126
S74527-8	978
S74528-6	94
S74529-4	87
S74530-8	673
S74533-2	771
S74534-0	886
S74535-9	380
S74536-7	302
S74537-5	1128
S74538-3	89
S74539-1	1128
S74540-5	1127
S74541-3	1140
S74542-1	65
S74544-8	710
S74548-0	886
S74549-9	697
S74551-0	1126
S74553-7	831
S74554-5	831
S74556-1	39
S74562-6	891
S74565-0	881
S74566-9	884
S74569-3	224
S74570-7	752
S74571-5	755
S74573-1	751
S74575-8	30
S74576-6	1128
S74577-4	1127
S74581-2	490
S74585-5	486
S74586-3	487
S74587-1	1014
S74592-8	233
S74593-6	312
S74594-4	202
S74596-0	838
S74597-9	767
S74599-5	210
S74600-2	836
S74601-0	989
S74606-1	1092
S74608-8	475
S74612-6	354
S74613-4	624
S74614-2	451
S74615-0	554
S74616-9	393
S74618-5	696
S74620-7	211
S74623-1	212
S74625-8	1127
S74626-6	367
S74627-4	435
S74628-2	407
S74629-0	407
S74630-4	432
S74631-2	397
S74632-0	435
S74633-9	406
S74634-7	407
S74635-5	409
S74636-3	1027
S74637-1	261
S74639-8	716
S74640-1	714
S74642-8	716
S74643-6	716
S74644-4	717
S74645-2	715
S74646-0	716
S74647-9	716
S74648-7	717
S74649-5	454
S74650-9	407
S74651-7	407
S74652-5	455
S74653-3	435
S74654-1	877
S74656-8	393
S74657-6	393
S74658-4	391
S74659-2	391
S74660-6	273
S74661-4	423
S74662-2	294
S74663-0	294
S74664-9	261
S74665-7	342
S74666-5	259
S74667-3	264
S74668-1	266
S74670-3	263
S74671-1	261
S74673-8	261
S74674-6	260
S74675-4	224
S74676-2	373
S74677-0	228
S74678-9	1022
S74679-7	762
S74680-0	759
S74681-9	764
S74682-7	764
S74683-5	764
S74684-3	856
S74685-1	764
S74687-8	766
S74688-6	890
S74689-4	916
S74690-8	424
S74691-6	432
S74692-4	718
S74694-0	875
S74695-9	622
S74698-3	494
S74699-1	427
S74700-9	716
S74701-7	458
S74702-5	527
S74703-3	603
S74704-1	553
S74706-8	1040
S74707-6	573
S74708-4	740
S74710-6	807
S74711-4	448
S74712-2	451
S74714-9	555
S74715-7	1092
S74717-3	739
S74718-1	515
S74720-3	561
S74723-8	419
S74725-4	666
S74726-2	1032
S74727-0	520
S74728-9	547
S74729-7	437
S74732-7	561
S74733-5	522
S74737-8	503
S74738-6	639
S74740-8	730
S74741-6	730
S74742-4	731
S74743-2	729
S74744-0	731
S74745-9	975
S74748-3	492
S74751-3	612
S74752-1	448
S74754-8	454
S74755-6	453
S74757-2	492
S74759-9	494
S74760-2	499
S74761-0	494
S74763-7	501
S74764-5	454
S74766-1	508
S74771-8	1053
S74772-6	528
S74777-7	504
S74778-5	560
S74781-5	522
S74783-1	423
S74785-8	502
S74786-6	1052
S74787-4	1029
S74789-0	449
S74790-4	468
S74791-2	455
S74792-0	877
S74793-9	463
S74794-7	469
S74795-5	721
S74796-3	475
S74797-1	455
S74799-8	463
S74801-3	1006
S74802-1	353
S74804-8	296
S74806-4	221
S74807-2	375
S74808-0	1024
S74809-9	212
S74810-2	210
S74812-9	12
S74813-7	678
S74814-5	3
S74818-8	205
S74819-6	199
S74822-6	12
S74826-9	256
S74827-7	199
S74829-3	771
S74830-7	289
S74837-4	1026
S74838-2	931
S74839-0	721

Product	Page
S74840-4	313
S74841-2	243
S74842-0	931
S74845-5	885
S74847-1	763
S74850-1	367
S74853-6	690
S74855-2	993
S74856-0	531
S74857-9	845
S74860-9	716
S74861-7	469
S74864-1	766
S74868-4	382
S74869-2	978
S74870-6	653
S74871-4	999
S74872-2	980
S74873-0	978
S74874-9	980
S74876-5	980
S74880-3	955
S74883-8	245
S74884-6	709
S74886-2	488
S74887-0	374
S74888-9	208
S74889-7	222
S74890-0	486
S74891-9	314
S74892-7	313
S74893-5	312
S74895-1	280
S74899-4	312
S74900-1	312
S74903-6	385
S74905-2	231
S74907-9	833
S74911-7	314
S74912-5	382
S74914-1	197
S74916-8	231
S74917-6	280
S74919-2	1077
S74921-4	1082
S74922-2	691
S74924-9	491
S74926-5	1049
S74927-3	884
S74928-1	231
S74930-3	791
S74933-8	125
S74935-4	404
S74936-2	107
S74938-9	313
S74940-0	233
S74943-5	310
S74945-1	464
S74947-8	199
S74948-6	764
S74949-4	627
S74950-8	868
S74951-6	265
S74952-4	231
S74954-0	271
S74955-9	89
S74956-7	463
S74957-5	891
S74958-3	822
S74961-3	464
S74964-8	870
S74965-6	492
S74966-4	348
S74967-2	923
S74968-0	644
S74969-9	464
S74970-2	322
S74971-0	260
S74972-9	504
S74973-7	383
S74974-5	348
S74975-3	348
S74976-1	201
S74979-6	383
S74983-4	491
S74985-0	491
S74986-9	604
S74987-7	372
S74988-5	835
S74990-7	718
S74991-5	488
S74992-3	311
S74995-8	403
S74996-6	101
S74997-4	397
S74998-2	397
S75001-8	1027
S75002-6	1044
S75004-2	899
S75007-7	1033
S75008-5	694
S75009-3	456
S75010-7	808
S75011-5	476
S75013-1	492
S75015-8	1015
S75016-6	981
S75017-4	637
S75018-2	623
S75019-0	399
S75020-4	451
S75021-2	431
S75022-0	405
S75023-9	949
S75024-7	949
S75025-5	453
S75026-3	454
S75029-8	586
S75030-1	941
S75032-8	528
S75033-6	558
S75034-4	730
S75035-2	521
S75036-0	571
S75037-9	733
S75038-7	525
S75039-5	539
S75040-9	561
S75041-7	517
S75042-5	941
S75044-1	936
S75046-8	490
S75047-6	484
S75048-4	485
S75049-2	405
S75050-6	470
S75051-4	456
S75052-2	1081
S75053-0	437
S75054-9	438
S75055-7	427
S75056-5	470
S75057-3	806
S75058-1	810
S75060-3	806
S75061-1	813
S75063-8	808
S75064-6	814
S75065-4	825
S75066-2	496
S75067-0	493
S75068-9	493
S75070-0	1069
S75071-9	494
S75072-7	494
S75073-5	498
S75074-3	498
S75075-1	501
S75077-8	498
S75078-6	501
S75079-4	494
S75080-8	497
S75081-6	499
S75082-4	493
S75083-2	497
S75084-0	497
S75085-9	501
S75086-7	497
S75087-5	495
S75088-3	502
S75089-1	496
S75090-5	491
S75091-3	496
S75092-1	491
S75095-6	495
S75096-4	497
S75098-0	499
S75099-9	751
S75100-6	676
S75103-0	404
S75104-9	632
S75105-7	522
S75108-1	868
S75114-6	448
S75115-4	449
S75116-2	705
S75118-9	467
S75122-7	866
S75123-5	465
S75124-3	452
S75127-8	337
S75129-4	613
S75130-8	667
S75131-6	523
S75135-9	624
S75136-7	460
S75137-5	447
S75138-3	472
S75139-1	468
S75140-5	432
S75142-1	455
S75144-8	454
S75145-6	454
S75147-2	400
S75149-9	447
S75150-2	409
S75151-0	484
S75152-9	832
S75153-7	262
S75154-5	313
S75155-3	174
S75159-6	692
S75161-8	691
S75162-6	484
S75163-4	491
S75164-2	343
S75166-9	1082
S75167-7	484
S75168-5	312
S75170-7	475
S75173-1	405
S75175-8	217
S75177-4	220
S75181-2	210
S75182-0	223
S75183-9	210
S75184-7	210
S75185-5	409
S75186-3	409
S75187-1	198
S75189-8	463
S75190-1	427
S75192-8	323
S75193-6	323
S75195-2	542
S75196-0	262
S75198-7	403
S75199-5	260
S75200-2	438
S75203-7	1115
S75205-3	198
S75206-1	409
S75208-8	383
S75209-6	262
S75211-8	271
S75212-6	409
S75215-0	280
S75216-9	257
S75217-7	438
S75218-5	198
S75219-3	257
S75220-7	265
S75221-5	257
S75222-3	463
S75223-1	135
S75225-8	408
S75226-6	667
S75227-4	511
S75228-2	626
S75229-0	400
S75230-4	623
S75231-2	520
S75232-0	544
S75233-9	449
S75234-7	519
S75235-5	517
S75236-3	552
S75237-1	522
S75239-8	976
S75240-1	400
S75242-8	636
S75243-6	521
S75244-4	516
S75245-2	520
S75246-0	515
S75247-9	516
S75248-7	516
S75249-5	734
S75251-7	516
S75252-5	549
S75253-3	549
S75254-1	549
S75256-8	516
S75257-6	634
S75258-4	635
S75259-2	616
S75260-6	682
S75261-4	640
S75262-2	487
S75266-5	486
S75267-3	396
S75270-3	560
S75271-1	555
S75273-8	826
S75274-6	440
S75275-4	421
S75276-2	439
S75277-0	440
S75278-9	438
S75279-7	453
S75280-0	632
S75281-9	584
S75282-7	687
S75283-5	651
S75284-3	602
S75287-8	405
S75288-6	516
S75289-4	582
S75290-8	633
S75291-6	453
S75292-4	439
S75293-2	440
S75294-0	440
S75295-9	405
S75296-7	570
S75297-5	572
S75298-3	504
S75299-1	484
S75300-9	484
S75302-5	270
S75303-3	262
S75306-8	667
S75307-6	440
S75308-4	409
S75310-6	311
S75312-2	226
S75313-0	996
S75315-7	488
S75316-5	488
S75318-1	277
S75320-3	843
S75321-1	126
S75323-8	84
S75324-6	729
S75325-4	111
S75326-2	21
S75327-0	1022
S75328-9	1025
S75329-7	538
S75331-9	75
S75332-7	400
S75333-5	400
S75334-3	400
S75335-1	638
S75337-8	400
S75338-6	400
S75339-4	538
S75340-8	540
S75341-6	507
S75342-4	394
S75343-2	401
S75344-0	401
S75345-9	506
S75346-7	505
S75347-5	401
S75348-3	400
S75349-1	544
S75350-5	542
S75351-3	544
S75352-1	395
S75354-8	400
S75355-6	401
S75356-4	401
S75357-2	511
S75358-0	511
S75359-9	511
S75360-2	511
S75361-0	511
S75362-9	400
S75363-7	538
S75364-5	538
S75365-3	540
S75366-1	542
S75368-8	540
S75369-6	511
S75371-8	400
S75372-6	545
S75373-4	545
S75374-2	546
S75375-0	401
S75376-9	400
S75377-7	591
S75378-5	591
S75379-3	438
S75380-7	631
S75381-5	401
S75382-3	591
S75383-1	592
S75385-8	591
S75386-6	589
S75387-4	591
S75388-2	591
S75389-0	592
S75390-4	592
S75391-2	438
S75392-0	439
S75393-9	439
S75394-7	490
S75396-3	722
S75397-1	381
S75400-5	278
S75401-3	489
S75402-1	313
S75405-6	709
S75406-4	366
S75407-2	312
S75409-9	222
S75410-2	487
S75412-9	343
S75413-7	690
S75414-5	314
S75416-1	829
S75418-8	832
S75419-6	312
S75422-6	208
S75423-4	1087
S75424-2	175
S75425-0	271
S75426-9	441
S75428-5	372
S75429-3	946
S75431-5	174
S75432-3	231
S75433-1	490
S75435-8	314
S75436-6	271
S75437-4	312
S75438-2	314
S75439-0	344
S75440-4	1059
S75441-2	339
S75442-0	226
S75443-9	374
S75446-3	697
S75449-8	404
S75450-1	716
S75452-8	244
S75453-6	778
S75454-4	707
S75455-2	694
S75456-0	36
S75457-9	928
S75459-5	863
S75461-7	1001
S75464-1	293
S75466-8	92
S75468-4	325
S75469-2	874
S75470-6	325
S75471-4	711
S75473-0	874
S75474-9	314
S75475-7	409
S75477-3	312
S75478-1	314
S75480-3	490
S75484-6	280
S75485-4	491
S75486-2	793
S75487-0	403
S75488-9	231
S75490-0	490
S75491-9	231
S75492-7	488
S75493-5	485
S75495-1	485
S75498-6	723
S75499-4	491
S75500-1	381
S75502-8	484
S75503-6	832
S75504-4	491
S75505-2	1079
S75506-0	271
S75507-9	379
S75508-7	1052
S75509-5	280
S75511-7	1000
S75513-3	488
S75514-1	490
S75516-8	487
S75517-6	394
S75518-4	226
S75519-2	230
S75521-4	268
S75522-2	160
S75523-0	222
S75524-9	312
S75526-5	484
S75527-3	487
S75528-1	490
S75530-3	262
S75531-1	233
S75533-8	313
S75534-6	380
S75536-2	260
S75538-9	491
S75540-0	385
S75542-7	174
S75543-5	166
S75545-1	280
S75547-8	837
S75548-6	484
S75549-4	487

Product	Page	Product	Page	Product	Page	Product	Page	Product	Page	Product	Page	Product	Page
S75551-6	312	S75660-1	898	S75802-7	344	S75922-8	965	S76028-5	750	S76112-5	734	S76192-3	432
S75552-4	315	S75663-6	768	S75803-5	980	S75923-6	208	S76029-3	858	S76113-3	325	S76193-1	5
S75553-2	231	S75664-4	355	S75804-3	942	S75924-4	881	S76030-7	901	S76114-1	437	S76195-8	885
S75554-0	360	S75665-2	325	S75805-1	356	S75925-2	693	S76031-5	1009	S76116-8	943	S76196-6	139
S75555-9	222	S75667-9	734	S75808-6	373	S75926-0	345	S76032-3	369	S76117-6	692	S76197-4	331
S75557-5	230	S75668-7	135	S75809-4	275	S75927-9	694	S76033-1	1000	S76118-4	252	S76198-2	1024
S75558-3	315	S75669-5	923	S75810-8	327	S75928-7	713	S76035-8	875	S76119-2	244	S76200-8	477
S75559-1	491	S75670-9	348	S75811-6	1001	S75929-5	289	S76036-6	870	S76120-6	245	S76201-6	715
S75560-5	233	S75671-7	422	S75812-4	1081	S75930-9	236	S76037-4	691	S76121-4	692	S76202-4	819
S75561-3	485	S75672-5	123	S75814-0	491	S75931-7	245	S76038-2	692	S76122-2	320	S76203-2	330
S75562-1	691	S75673-3	87	S75815-9	488	S75932-5	248	S76039-0	751	S76124-9	735	S76205-9	747
S75564-8	208	S75674-1	817	S75816-7	227	S75934-1	434	S76040-4	249	S76125-7	1056	S76206-7	926
S75565-6	222	S75676-8	974	S75817-5	170	S75936-8	199	S76041-2	919	S76126-5	320	S76207-5	801
S75566-4	485	S75677-6	974	S75819-1	314	S75937-6	254	S76042-0	710	S76127-3	710	S76208-3	149
S75567-2	1060	S75679-2	882	S75820-5	315	S75938-4	697	S76043-9	691	S76128-1	698	S76209-1	1025
S75568-0	160	S75680-6	838	S75821-3	313	S75939-2	930	S76045-5	320	S76130-3	324	S76210-5	134
S75569-9	897	S75682-2	854	S75822-1	381	S75940-6	999	S76046-3	701	S76131-1	353	S76211-3	38
S75570-2	1002	S75683-0	849	S75824-8	312	S75941-4	362	S76047-1	709	S76133-8	981	S76212-1	869
S75571-0	330	S75684-9	87	S75825-6	280	S75942-2	1005	S76049-8	1036	S76134-6	726	S76214-8	728
S75572-9	898	S75685-7	52	S75826-4	228	S75943-0	385	S76050-1	722	S76135-4	244	S76215-6	468
S75573-7	1013	S75686-5	52	S75827-2	892	S75945-7	294	S76052-8	353	S76136-2	286	S76217-2	1027
S75574-5	90	S75687-3	334	S75829-9	723	S75946-5	477	S76053-6	864	S76137-0	255	S76218-0	968
S75575-3	808	S75690-3	54	S75830-2	280	S75947-3	461	S76054-4	246	S76138-9	696	S76219-9	745
S75579-6	288	S75694-6	181	S75831-0	669	S75948-1	1008	S76055-2	897	S76139-7	283	S76220-2	1014
S75582-6	898	S75695-4	257	S75832-9	244	S75950-3	1037	S76056-0	689	S76140-0	735	S76222-9	926
S75583-4	897	S75697-0	407	S75838-8	244	S75953-8	384	S76057-9	838	S76141-9	694	S76223-7	942
S75585-0	898	S75699-7	690	S75841-8	885	S75954-6	952	S76058-7	889	S76142-7	708	S76224-5	701
S75586-9	897	S75702-0	796	S75842-6	690	S75955-4	772	S76059-5	693	S76143-5	931	S76225-3	872
S75587-7	190	S75703-9	895	S75843-4	996	S75956-2	309	S76060-9	253	S76144-3	996	S76228-8	253
S75588-5	334	S75704-7	484	S75845-0	868	S75957-0	772	S76061-7	93	S76145-1	705	S76231-8	931
S75589-3	283	S75705-5	488	S75848-5	728	S75958-9	842	S76062-5	930	S76147-8	955	S76232-6	871
S75590-7	898	S75716-0	982	S75849-3	692	S75959-7	842	S76063-3	706	S76148-6	996	S76233-4	291
S75591-5	1013	S75718-7	981	S75850-7	961	S75960-0	843	S76066-8	934	S76149-4	466	S76235-0	807
S75592-3	288	S75721-7	289	S75852-3	252	S75962-7	839	S76067-6	694	S76150-8	713	S76236-9	917
S75593-1	116	S75726-8	309	S75853-1	353	S75965-1	839	S76069-2	705	S76151-6	727	S76237-7	195
S75595-8	898	S75729-2	983	S75855-8	1016	S75968-6	839	S76070-6	701	S76152-4	341	S76238-5	874
S75597-4	270	S75733-0	98	S75858-2	251	S75969-4	147	S76071-4	690	S76153-2	322	S76239-3	31
S75598-2	324	S75736-5	1105	S75859-0	251	S75970-8	187	S76072-2	696	S76154-0	234	S76243-1	877
S75599-0	897	S75737-3	133	S75860-4	966	S75971-6	195	S76073-0	709	S76155-9	724	S76245-8	1012
S75600-8	313	S75738-1	133	S75863-9	693	S75972-4	201	S76074-9	691	S76156-7	141	S76247-4	897
S75601-6	226	S75740-3	98	S75867-1	339	S75973-2	192	S76075-7	859	S76157-5	885	S76248-2	897
S75607-5	312	S75743-8	66	S75869-8	413	S75974-0	838	S76077-3	354	S76158-3	1026	S76249-0	897
S75608-3	311	S75744-6	41	S75870-1	746	S75978-3	839	S76078-1	766	S76159-1	966	S76251-2	877
S75610-5	488	S75746-2	12	S75873-6	248	S75979-1	91	S76080-3	702	S76160-5	701	S76252-0	414
S75612-1	280	S75747-0	56	S75874-4	252	S75984-8	243	S76081-1	864	S76161-3	707	S76253-9	335
S75615-6	208	S75748-9	37	S75875-2	319	S75986-4	897	S76083-8	246	S76162-1	690	S76254-7	270
S75616-4	488	S75750-0	49	S75877-9	253	S75987-2	897	S76084-6	931	S76164-8	692	S76255-5	191
S75617-2	233	S75752-7	250	S75878-7	295	S75991-0	898	S76085-4	689	S76165-6	382	S76256-3	858
S75619-9	105	S75755-1	177	S75879-5	283	S75992-9	885	S76086-2	705	S76166-4	726	S76257-1	890
S75621-0	134	S75758-6	103	S75880-9	449	S76001-3	707	S76087-0	716	S76167-2	254	S76259-8	856
S75622-9	131	S75759-4	116	S75884-1	163	S76002-1	741	S76088-9	696	S76168-0	697	S76260-1	334
S75624-5	110	S75760-8	211	S75887-6	252	S76004-8	940	S76089-7	368	S76169-9	891	S76263-6	898
S75625-3	110	S75762-4	144	S75888-4	248	S76005-6	837	S76090-0	745	S76170-2	695	S76264-4	898
S75626-1	134	S75763-2	113	S75889-2	865	S76007-2	695	S76091-9	888	S76171-0	244	S76265-2	897
S75628-8	97	S75764-0	874	S75890-6	699	S76008-0	697	S76092-7	318	S76172-9	712	S76266-0	897
S75629-6	136	S75767-5	221	S75891-4	752	S76009-9	252	S76093-5	959	S76173-7	432	S76267-9	897
S75633-4	45	S75769-1	316	S75892-2	881	S76010-2	245	S76094-3	383	S76174-5	199	S76268-7	898
S75634-2	825	S75770-5	99	S75893-0	389	S76011-0	701	S76095-1	1001	S76175-3	698	S76269-5	51
S75636-9	705	S75772-1	849	S75894-9	221	S76012-9	707	S76097-8	251	S76176-1	295	S76270-9	2
S75639-3	530	S75779-9	12	S75896-5	961	S76013-7	902	S76098-6	461	S76178-8	707	S76271-7	384
S75640-7	529	S75780-2	66	S75901-5	488	S76014-5	243	S76099-4	232	S76179-6	692	S76272-5	1072
S75641-5	175	S75784-5	99	S75902-3	779	S76015-3	864	S76100-1	450	S76181-8	331	S76274-1	982
S75643-1	374	S75785-3	51	S75903-1	244	S76016-1	707	S76102-8	951	S76182-6	1011	S76276-8	842
S75645-8	374	S75789-6	66	S75907-4	246	S76018-8	284	S76103-6	966	S76183-4	40	S76277-6	431
S75646-6	307	S75791-8	331	S75908-2	1008	S76019-6	934	S76104-4	864	S76184-2	703	S76278-4	51
S75651-2	364	S75793-4	66	S75909-0	209	S76021-8	690	S76105-2	728	S76185-0	330	S76279-2	66
S75652-0	898	S75794-2	31	S75911-2	726	S76022-6	709	S76106-0	463	S76186-9	256	S76280-6	66
S75653-9	364	S75795-0	66	S75912-0	1015	S76023-4	468	S76107-9	977	S76187-7	320	S76281-4	66
S75654-7	875	S75796-9	98	S75914-7	294	S76024-2	866	S76108-7	692	S76188-5	749	S76282-2	98
S75655-5	388	S75797-7	962	S75917-1	246	S76025-0	733	S76109-5	690	S76189-3	469	S76283-0	135
S75656-3	334	S75799-3	66	S75919-8	244	S76026-9	881	S76110-9	725	S76190-7	185	S76284-9	753
S75659-8	334	S75800-0	897	S75920-1	359	S76027-7	277	S76111-7	452	S76191-5	478	S76285-7	31

Product	Page
S76287-3	1109
S76288-1	310
S76290-3	4
S76291-1	49
S76293-8	135
S76294-6	985
S76295-4	1021
S76297-0	1104
S76300-4	693
S76301-2	204
S76302-0	321
S76303-9	871
S76304-7	230
S76305-5	373
S76310-1	324
S76313-6	634
S76314-4	465
S76315-2	246
S76316-0	368
S76317-9	677
S76318-7	634
S76319-5	665
S76320-9	246
S76321-7	819
S76322-5	289
S76323-3	291
S76324-1	707
S76326-8	562
S76327-6	382
S76329-2	716
S76330-6	1012
S76334-9	689
S76335-7	336
S76336-5	757
S76337-3	307
S76340-3	762
S76341-1	701
S76344-6	366
S76345-4	95
S76346-2	277
S76347-0	699
S76349-7	327
S76350-0	874
S76353-5	314
S76354-3	732
S76355-1	709
S76357-8	360
S76358-6	860
S76359-4	1018
S76360-8	245
S76361-6	157
S76362-4	141
S76365-9	267
S76366-7	883
S76368-3	199
S76369-1	781
S76371-3	1026
S76375-6	287
S76376-4	244
S76378-0	942
S76379-9	904
S76380-2	904
S76381-0	874
S76382-9	689
S76383-7	577
S76384-5	703
S76385-3	717
S76386-1	781
S76388-8	578
S76389-6	245
S76391-8	503
S76393-4	315
S76394-2	689
S76395-0	396
S76396-9	991
S76397-7	373
S76398-5	244
S76400-0	870
S76401-9	854
S76402-7	473
S76403-5	701
S76404-3	1050
S76405-1	474
S76407-8	469
S76408-6	465
S76409-4	121
S76410-8	719
S76411-6	543
S76412-4	546
S76413-2	838
S76414-0	285
S76415-9	1036
S76417-5	701
S76418-3	122
S76419-1	563
S76420-5	484
S76422-1	1079
S76424-8	273
S76425-6	236
S76426-4	237
S76427-2	843
S76428-0	122
S76429-9	404
S76430-2	122
S76431-0	404
S76432-9	454
S76435-3	318
S76436-1	316
S76438-8	427
S76439-6	466
S76441-8	409
S76442-6	436
S76443-4	44
S76445-0	244
S76447-7	79
S76449-3	4
S76450-7	4
S76451-5	15
S76452-3	99
S76455-8	338
S76460-4	66
S76461-2	85
S76462-0	1109
S76463-9	56
S76464-7	158
S76465-5	344
S76466-3	989
S76467-1	1018
S76469-8	706
S76470-1	358
S76472-8	48
S76473-6	707
S76474-4	358
S76475-2	294
S76476-0	78
S76477-9	1015
S76478-7	197
S76480-9	961
S76481-7	32
S76482-5	708
S76483-3	1003
S76484-1	332
S76486-8	367
S76487-6	357
S76488-4	28
S76489-2	791
S76491-4	438
S76492-2	115
S76493-0	867
S76494-9	982
S76496-5	468
S76497-3	916
S76500-7	1141
S76501-5	69
S76502-3	96
S76503-1	818
S76505-8	356
S76506-6	21
S76509-0	801
S76510-4	1093
S76511-2	697
S76512-0	646
S76513-9	816
S76514-7	817
S76515-5	830
S76516-3	949
S76517-1	664
S76519-8	888
S76520-1	1083
S76523-6	657
S76524-4	659
S76527-9	888
S76530-9	123
S76531-7	1132
S76532-5	660
S76533-3	652
S76534-1	708
S76536-8	659
S76537-6	655
S76538-4	817
S76539-2	73
S76540-6	1
S76541-4	73
S76542-2	977
S76543-0	87
S76544-9	124
S76545-7	660
S76546-5	645
S76547-3	708
S76550-3	660
S76551-1	2
S76553-8	660
S76555-4	909
S76556-2	660
S76558-9	792
S76559-7	124
S76560-0	68
S76561-9	364
S76562-7	95
S76563-5	106
S76567-8	959
S76568-6	959
S76569-4	43
S76570-8	43
S76571-6	43
S76572-4	262
S76573-2	970
S76574-0	44
S76575-9	667
S76576-7	434
S76577-5	661
S76578-3	884
S76579-1	89
S76581-3	124
S76582-1	656
S76584-8	348
S76585-6	412
S76586-4	429
S76587-2	815
S76588-0	436
S76589-9	429
S76590-2	816
S76591-0	971
S76592-9	80
S76593-7	969
S76594-5	969
S76595-3	124
S76596-1	677
S76598-8	1140
S76599-6	118
S76600-3	386
S76601-1	1003
S76604-6	321
S76605-4	321
S76608-9	956
S76609-7	247
S76611-9	325
S76613-5	244
S76615-1	694
S76620-8	957
S76621-6	244
S76622-4	301
S76623-2	385
S76624-0	97
S76625-9	244
S76626-7	247
S76629-1	374
S76630-5	244
S76631-3	446
S76632-1	471
S76634-8	1008
S76636-4	965
S76637-2	880
S76638-0	463
S76640-2	865
S76641-0	331
S76642-9	252
S76644-5	312
S76645-3	221
S76648-8	703
S76649-6	246
S76651-8	260
S76654-2	337
S76656-9	697
S76658-5	325
S76659-3	865
S76660-7	199
S76661-5	689
S76662-3	179
S76663-1	299
S76665-8	262
S76666-6	461
S76667-4	288
S76669-0	689
S76670-4	1010
S76671-2	782
S76672-0	701
S76674-7	155
S76675-5	866
S76677-1	934
S76680-1	629
S76684-4	41
S76685-2	1072
S76686-0	200
S76687-9	985
S76690-9	870
S76691-7	691
S76692-5	184
S76693-3	710
S76696-8	864
S76698-4	1009
S76699-2	968
S76701-8	39
S76702-6	850
S76706-9	944
S76707-7	816
S76708-5	691
S76709-3	1029
S76710-7	1048
S76711-5	1040
S76712-3	1008
S76713-1	842
S76715-8	38
S76718-2	773
S76719-0	275
S76720-4	690
S76721-2	955
S76722-0	578
S76724-7	696
S76725-5	954
S76726-3	865
S76727-1	697
S76729-8	955
S76730-1	774
S76733-6	282
S76734-4	387
S76735-2	198
S76736-0	185
S76738-7	723
S76740-9	703
S76741-7	221
S76744-1	241
S76746-8	285
S76747-6	747
S76749-2	999
S76750-6	1018
S76753-0	198
S76755-7	290
S76756-5	711
S76760-3	247
S76763-8	359
S76764-6	692
S76765-4	385
S76767-0	356
S76768-9	842
S76769-7	842
S76770-0	842
S76773-5	358
S76774-3	135
S76775-1	690
S76777-8	697
S76778-6	864
S76780-8	157
S76782-4	410
S76783-2	180
S76784-0	211
S76785-9	472
S76786-7	211
S76787-5	169
S76788-3	211
S76789-1	210
S76790-5	154
S76791-3	864
S76794-8	403
S76795-6	987
S76796-4	375
S76797-2	270
S76798-0	97
S76799-9	211
S76800-6	850
S76801-4	518
S76802-2	166
S76803-0	1013
S76804-9	525
S76805-7	899
S76806-5	437
S76807-3	578
S76808-1	517
S76810-3	1027
S76813-8	620
S76814-6	757
S76815-4	929
S76816-2	581
S76817-0	451
S76818-9	543
S76819-7	609
S76820-0	654
S76821-9	909
S76822-7	865
S76823-5	673
S76825-1	986
S76827-8	460
S76828-6	987
S76829-4	890
S76830-8	579
S76831-6	551
S76832-4	537
S76833-2	732
S76834-0	809
S76835-9	1015
S76836-7	455
S76837-5	539
S76838-3	908
S76839-1	715
S76840-5	853
S76841-3	537
S76842-1	465
S76844-8	736
S76845-6	1063
S76846-4	576
S76847-2	551
S76848-0	694
S76849-9	506
S76850-2	402
S76851-0	698
S76852-9	740
S76853-7	586
S76854-5	461
S76855-3	927
S76856-1	809
S76858-8	578
S76859-6	609
S76861-8	577
S76862-6	519
S76863-4	406
S76864-2	583
S76865-0	397
S76866-9	521
S76867-7	926
S76868-5	526
S76869-3	451
S76870-7	977
S76871-5	984
S76872-3	550
S76875-8	695
S76876-6	608
S76877-4	537
S76878-2	899
S76879-0	560
S76880-4	537
S76881-2	731
S76882-0	962
S76883-9	899
S76884-7	646
S76885-5	930
S76886-3	1026
S76887-1	727
S76889-8	606
S76890-1	497
S76892-8	498
S76893-6	152
S76894-4	702
S76895-2	495
S76896-0	452
S76897-9	606
S76898-7	607
S76899-5	616
S76900-2	497
S76901-0	509
S76902-9	615
S76903-7	592
S76904-5	586
S76905-3	925
S76906-1	688
S76908-8	581
S76909-6	502
S76911-8	630
S76912-6	452
S76913-4	877
S76914-2	806
S76915-0	1027
S76916-9	646
S76917-7	569
S76918-5	620
S76919-3	535
S76920-7	689
S76921-5	642
S76922-3	976
S76923-1	440
S76925-8	586
S76926-6	901
S76927-4	504
S76929-0	1070
S76930-4	496
S76931-2	531
S76932-0	726
S76933-9	592
S76934-7	438
S76935-5	753
S76936-3	555
S76937-1	535
S76939-8	506
S76940-1	655
S76942-8	612
S76943-6	501
S76944-4	612
S76945-2	671
S76946-0	560
S76947-9	618
S76948-7	500
S76949-5	496
S76950-9	527
S76951-7	740
S76952-5	635
S76953-3	540
S76954-1	400
S76956-8	845
S76957-6	620
S76958-4	474
S76959-2	107
S76960-6	670
S76961-4	773
S76962-2	774
S76963-0	786

Product Page

S76964-9.....453
S76965-7.....538
S76966-5.....602
S76967-3.....852
S76968-1.....521
S76970-3.....734
S76971-1.....452
S76973-8.....633
S76974-6.....633
S76975-4.....730
S76976-2.....408
S76977-0.....729
S76978-9.....624
S76979-7.....399
S76980-0.....533
S76981-9.....964
S76982-7.....926
S76983-5.....401
S76984-3.....544
S76985-1.....676
S76987-8.....864
S76988-6.....926
S76989-4.....899
S76990-8.....871
S76991-6.....751
S76992-4.....586
S76993-2.....529
S76994-0.....561
S76995-9.....968
S76996-7.....901
S76997-5.....732
S76998-3.....633
S76999-1.....611
S77000-0.....464
S77001-9.....464
S77003-5.....171
S77004-3.....983
S77007-8.....981
S77009-4.....982
S77013-2.....322
S77014-0.....271
S77015-9.....689
S77017-5.....197
S77018-3.....335
S77019-1.....82
S77021-3.....985
S77022-1.....984
S77025-6.....983
S77026-4.....288
S77027-2.....467
S77028-0.....276
S77029-9.....962
S77030-2.....175
S77031-0.....995
S77032-9.....379
S77033-7.....322
S77034-5.....986
S77035-3.....166
S77036-1.....215
S77038-8.....260
S77039-6.....262
S77041-8.....403
S77042-6.....377
S77043-4.....263
S77044-2.....403
S77045-0.....259
S77046-9.....424
S77047-7.....1003
S77048-5.....32
S77049-3.....259
S77050-7.....259
S77051-5.....383
S77052-3.....54
S77053-1.....97
S77055-8.....99
S77056-6.....134
S77057-4.....93
S77059-0.....66
S77060-4.....994
S77061-2.....288
S77062-0.....897
S77063-9.....999
S77064-7.....340
S77065-5.....1014
S77066-3.....490
S77067-1.....490
S77069-8.....95
S77074-4.....1094
S77076-0.....983
S77077-9.....234
S77078-7.....999
S77079-5.....981
S77080-9.....452
S77081-7.....472
S77083-3.....313
S77084-1.....231
S77086-8.....313
S77087-6.....417
S77088-4.....367
S77089-2.....1016
S77090-6.....1016
S77091-4.....864
S77092-2.....199
S77093-0.....160
S77094-9.....1025
S77095-7.....960
S77096-5.....270
S77097-3.....3
S77098-1.....771
S77100-7.....490
S77101-5.....487
S77102-3.....488
S77103-1.....360
S77106-6.....715
S77107-4.....26
S77108-2.....666
S77109-0.....357
S77110-4.....829
S77111-2.....764
S77112-0.....785
S77113-9.....466
S77114-7.....12
S77115-5.....360
S77116-3.....999
S77117-1.....381
S77119-8.....212
S77120-1.....211
S77122-8.....270
S77123-6.....771
S77124-4.....771
S77126-0.....1109
S77127-9.....137
S77128-7.....135
S77129-5.....135
S77130-9.....40
S77131-7.....1109
S77132-5.....239
S77133-3.....40
S77134-1.....1049
S77136-8.....1068
S77137-6.....1072
S77138-4.....1026
S77139-2.....1068
S77140-6.....1072
S77141-4.....1037
S77142-2.....1026
S77143-0.....1072
S77144-9.....1072
S77145-7.....428
S77146-5.....1021
S77147-3.....1070
S77148-1.....724
S77150-3.....492
S77151-1.....1072
S77153-8.....1023
S77154-6.....1023
S77155-4.....1023
S77156-2.....1023
S77157-0.....1023
S77158-9.....1023
S77159-7.....1072
S77160-0.....1026
S77162-7.....724
S77163-5.....1026
S77164-3.....1021
S77165-1.....234
S77167-8.....910
S77168-6.....1040
S77169-4.....1040
S77170-8.....1071
S77172-4.....138
S77173-2.....1025
S77175-9.....1072
S77176-7.....304
S77177-5.....1023
S77178-3.....1049
S77179-1.....1039
S77180-5.....1056
S77181-3.....138
S77182-1.....1023
S77184-8.....1039
S77185-6.....1049
S77186-4.....1072
S77187-2.....1070
S77188-0.....1049
S77189-9.....1021
S77190-2.....1024
S77191-0.....1026
S77192-9.....1049
S77193-7.....1048
S77194-5.....494
S77195-3.....1026
S77196-1.....1026
S77198-8.....1022
S77199-6.....106
S77203-8.....782
S77204-6.....927
S77205-4.....528
S77206-2.....899
S77207-0.....404
S77208-9.....528
S77209-7.....756
S77210-0.....865
S77211-9.....646
S77212-7.....394
S77213-5.....607
S77214-3.....494
S77215-1.....528
S77217-8.....809
S77218-6.....455
S77219-4.....695
S77220-8.....532
S77221-6.....844
S77222-4.....529
S77223-2.....900
S77224-0.....733
S77225-9.....875
S77226-7.....729
S77227-5.....772
S77228-3.....775
S77229-1.....518
S77230-5.....618
S77231-3.....553
S77232-1.....875
S77234-8.....911
S77235-6.....731
S77236-4.....941
S77237-2.....640
S77238-0.....1029
S77240-2.....563
S77241-0.....651
S77242-9.....475
S77243-7.....446
S77244-5.....751
S77245-3.....510
S77246-1.....640
S77248-8.....586
S77249-6.....400
S77251-8.....439
S77252-6.....751
S77253-4.....529
S77254-2.....753
S77255-0.....581
S77256-9.....757
S77257-7.....752
S77258-5.....569
S77259-3.....911
S77260-7.....844
S77261-5.....977
S77262-3.....755
S77263-1.....535
S77265-8.....736
S77266-6.....1028
S77267-4.....534
S77268-2.....926
S77269-0.....641
S77270-4.....617
S77271-2.....926
S77272-0.....930
S77273-9.....674
S77274-7.....277
S77275-5.....901
S77276-3.....409
S77277-1.....928
S77279-8.....505
S77280-1.....634
S77282-8.....607
S77283-6.....639
S77284-4.....926
S77285-2.....1071
S77286-0.....631
S77287-9.....825
S77288-7.....561
S77289-5.....678
S77290-9.....581
S77291-7.....581
S77292-5.....654
S77293-3.....535
S77294-1.....554
S77296-8.....695
S77297-6.....209
S77298-4.....534
S77299-2.....534
S77301-8.....535
S77302-6.....581
S77303-4.....535
S77304-2.....505
S77305-0.....747
S77306-9.....504
S77307-7.....1029
S77308-5.....505
S77309-3.....524
S77310-7.....478
S77311-5.....591
S77312-3.....581
S77313-1.....581
S77315-8.....544
S77316-6.....581
S77317-4.....521
S77318-2.....502
S77319-0.....566
S77320-4.....183
S77321-2.....460
S77322-0.....554
S77323-9.....555
S77324-7.....620
S77325-5.....546
S77326-3.....602
S77327-1.....536
S77329-8.....559
S77330-1.....506
S77332-8.....537
S77333-6.....446
S77334-4.....541
S77335-2.....520
S77336-0.....621
S77337-9.....728
S77338-7.....865
S77339-5.....638
S77340-9.....453
S77341-7.....740
S77342-5.....19
S77343-3.....1020
S77344-1.....569
S77346-8.....833
S77347-6.....567
S77348-4.....584
S77349-2.....654
S77350-6.....678
S77351-4.....528
S77352-2.....617
S77353-0.....318
S77354-9.....925
S77355-7.....987
S77356-5.....561
S77357-3.....810
S77358-1.....1009
S77360-3.....612
S77361-1.....730
S77363-8.....651
S77364-6.....617
S77365-4.....408
S77366-2.....807
S77367-0.....408
S77368-9.....530
S77369-7.....445
S77370-0.....569
S77371-9.....459
S77372-7.....698
S77373-5.....538
S77374-3.....527
S77375-1.....810
S77377-8.....650
S77378-6.....537
S77379-4.....577
S77380-8.....623
S77381-6.....646
S77382-4.....677
S77383-2.....681
S77384-0.....668
S77385-9.....623
S77386-7.....577
S77387-5.....549
S77388-3.....635
S77389-1.....641
S77390-5.....548
S77391-3.....382
S77392-1.....808
S77394-8.....383
S77395-6.....640
S77396-4.....737
S77397-2.....576
S77398-0.....732
S77399-9.....382
S77400-6.....484
S77401-4.....280
S77402-2.....313
S77403-0.....225
S77405-7.....280
S77406-5.....312
S77408-1.....313
S77410-3.....712
S77411-1.....459
S77413-8.....605
S77414-6.....830
S77415-4.....830
S77417-0.....170
S77418-9.....404
S77419-7.....999
S77420-0.....231
S77421-9.....172
S77422-7.....311
S77423-5.....231
S77424-3.....360
S77425-1.....1073
S77427-8.....484
S77429-4.....208
S77432-4.....983
S77434-0.....1085
S77435-9.....434
S77436-7.....358
S77437-5.....688
S77438-3.....929
S77439-1.....622
S77440-5.....614
S77441-3.....384
S77442-1.....96
S77444-8.....138
S77445-6.....230
S77447-2.....39
S77448-0.....133
S77449-9.....98
S77450-2.....361
S77451-0.....278
S77452-9.....823
S77453-7.....160
S77454-5.....172
S77455-3.....430
S77456-1.....570
S77458-8.....432
S77459-6.....488
S77461-8.....431
S77462-6.....107
S77463-4.....490
S77464-2.....428
S77465-0.....428
S77466-9.....404
S77467-7.....424
S77468-5.....425
S77469-3.....424
S77470-7.....425
S77471-5.....425
S77472-3.....110
S77473-1.....587
S77475-8.....397
S77476-6.....571
S77478-2.....461
S77479-0.....482
S77480-4.....482
S77481-2.....489
S77482-0.....489
S77484-7.....441
S77486-3.....457
S77487-1.....431
S77489-8.....243
S77490-1.....475
S77492-8.....431
S77493-6.....431
S77494-4.....410
S77495-2.....441
S77496-0.....208
S77497-9.....205
S77498-7.....204
S77499-5.....2
S77500-2.....579
S77501-0.....472
S77502-9.....445
S77503-7.....107
S77504-5.....205
S77506-1.....441
S77508-8.....425
S77509-6.....408
S77511-8.....472
S77512-6.....472
S77513-4.....475
S77514-2.....472
S77516-9.....163
S77517-7.....471
S77519-3.....572
S77521-5.....441
S77522-3.....441
S77523-1.....441
S77525-8.....410
S77526-6.....193
S77527-4.....980
S77528-2.....252
S77529-0.....587
S77530-4.....488
S77531-2.....488
S77532-0.....102
S77533-9.....441
S77534-7.....482
S77535-5.....472
S77536-3.....475
S77537-1.....219
S77539-8.....425
S77540-1.....585
S77542-8.....154
S77544-4.....1142
S77545-2.....828
S77546-0.....152
S77547-9.....154
S77551-7.....774
S77553-3.....827
S77554-1.....784
S77558-4.....786
S77560-6.....786
S77561-4.....784
S77563-0.....782
S77564-9.....772
S77565-7.....155
S77566-5.....785
S77567-3.....211
S77570-3.....785

Product	Page
S77571-1	784
S77573-8	154
S77574-6	806
S77575-4	801
S77576-2	784
S77577-0	785
S77578-9	801
S77579-7	157
S77581-9	1100
S77582-7	170
S77584-3	796
S77585-1	796
S77588-6	1079
S77589-4	258
S77591-6	255
S77593-2	792
S77595-9	24
S77596-7	1100
S77597-5	821
S77598-3	135
S77599-1	135
S77600-9	886
S77601-7	664
S77602-5	416
S77603-3	655
S77604-1	662
S77606-8	708
S77607-6	701
S77608-4	125
S77610-6	119
S77611-4	660
S77612-2	664
S77613-0	664
S77614-9	120
S77615-7	121
S77616-5	664
S77617-3	422
S77618-1	422
S77620-3	477
S77621-1	412
S77623-8	412
S77624-6	429
S77625-4	124
S77626-2	412
S77627-0	120
S77628-9	120
S77629-7	356
S77630-0	643
S77631-9	889
S77632-7	889
S77633-5	924
S77634-3	416
S77637-8	697
S77638-6	644
S77639-4	412
S77640-8	860
S77641-6	429
S77642-4	671
S77643-2	890
S77645-9	412
S77647-5	121
S77648-3	119
S77649-1	117
S77650-5	120
S77651-3	660
S77652-1	660
S77655-6	973
S77656-4	121
S77657-2	411
S77658-0	660
S77659-9	655
S77660-2	662
S77661-0	708
S77662-9	883
S77663-7	883
S77664-5	658
S77665-3	643
S77666-1	645
S77668-8	643
S77669-6	644
S77671-8	659
S77672-6	422
S77673-4	660
S77674-2	660
S77675-0	660
S77676-9	969
S77677-7	656
S77678-5	642
S77680-7	701
S77681-5	657
S77682-3	697
S77683-1	868
S77685-8	135
S77686-6	974
S77687-4	119
S77688-2	118
S77689-0	121
S77690-4	643
S77691-2	410
S77692-0	348
S77693-9	660
S77694-7	642
S77695-5	489
S77696-3	658
S77697-1	656
S77699-8	664
S77700-5	655
S77701-3	664
S77702-1	656
S77704-8	666
S77705-6	658
S77706-4	348
S77707-2	886
S77708-0	700
S77709-9	645
S77710-2	658
S77711-0	664
S77712-9	656
S77713-7	658
S77714-5	131
S77715-3	411
S77719-6	890
S77721-8	890
S77722-6	486
S77723-4	922
S77724-2	121
S77725-0	121
S77726-9	124
S77727-7	123
S77729-3	660
S77730-7	118
S77733-1	422
S77735-8	422
S77736-6	422
S77737-4	884
S77739-0	666
S77741-2	700
S77742-0	696
S77743-9	121
S77745-5	890
S77746-3	657
S77749-8	117
S77750-1	924
S77752-8	428
S77753-6	124
S77754-4	348
S77755-2	429
S77756-0	659
S77757-9	642
S77758-7	656
S77759-5	642
S77760-9	656
S77761-7	656
S77762-5	644
S77763-3	662
S77764-1	671
S77766-8	704
S77767-6	697
S77768-4	123
S77769-2	889
S77770-6	124
S77771-4	263
S77772-2	700
S77773-0	922
S77774-9	101
S77775-7	920
S77776-5	120
S77777-3	120
S77778-1	120
S77780-3	121
S77781-1	422
S77783-8	124
S77784-6	118
S77785-4	642
S77786-2	664
S77787-0	660
S77789-7	656
S77790-0	655
S77791-9	645
S77792-7	658
S77794-3	643
S77795-1	866
S77797-8	883
S77798-6	662
S77799-4	662
S77802-8	169
S77803-6	778
S77804-4	773
S77805-2	773
S77806-0	152
S77807-9	157
S77808-7	156
S77809-5	1019
S77810-9	135
S77812-5	135
S77813-3	135
S77814-1	135
S77816-8	1019
S77817-6	1013
S77821-4	409
S77823-0	785
S77824-9	783
S77825-7	330
S77826-5	1017
S77827-3	768
S77828-1	1019
S77830-3	785
S77831-1	781
S77833-8	1017
S77834-6	360
S77835-4	774
S77836-2	781
S77837-0	771
S77838-9	783
S77842-7	783
S77844-3	799
S77845-1	154
S77847-8	1019
S77848-6	800
S77849-4	774
S77850-8	801
S77851-6	786
S77852-4	780
S77854-0	800
S77855-9	252
S77856-7	213
S77857-5	783
S77858-3	212
S77861-3	253
S77862-1	278
S77866-4	809
S77867-2	1070
S77868-0	729
S77869-9	872
S77870-2	523
S77871-0	473
S77872-9	409
S77874-5	478
S77875-3	450
S77878-8	800
S77879-6	156
S77882-6	156
S77886-9	854
S77888-5	784
S77889-3	784
S77890-7	154
S77896-6	783
S77897-4	784
S77898-2	784
S77900-8	783
S77901-6	783
S77902-4	785
S77904-0	784
S77905-9	154
S77906-7	784
S77907-5	800
S77908-3	827
S77909-1	783
S77910-5	785
S77912-1	478
S77915-6	231
S77916-4	16
S77917-2	320
S77918-0	970
S77920-2	959
S77923-7	858
S77925-3	877
S77926-1	912
S77928-8	399
S77929-6	668
S77931-8	1070
S77932-6	928
S77933-4	846
S77934-2	959
S77935-0	880
S77936-9	881
S77937-7	523
S77938-5	513
S77939-3	518
S77940-7	635
S77941-5	630
S77942-3	559
S77943-1	555
S77945-8	607
S77946-6	453
S77947-4	629
S77948-2	826
S77949-0	751
S77950-4	504
S77951-2	845
S77952-0	473
S77953-9	531
S77954-7	451
S77955-5	901
S77956-3	520
S77957-1	449
S77959-8	607
S77960-1	540
S77962-8	452
S77963-6	559
S77964-4	497
S77965-2	845
S77966-0	843
S77967-9	877
S77968-7	496
S77969-5	504
S77970-9	968
S77971-7	911
S77972-5	982
S77973-3	622
S77974-1	549
S77976-8	549
S77977-6	511
S77978-4	518
S77979-2	549
S77980-6	525
S77981-4	737
S77982-2	587
S77983-0	455
S77984-9	437
S77985-7	437
S77986-5	394
S77987-3	482
S77988-1	463
S77990-3	463
S77991-1	622
S77993-8	521
S77994-6	665
S77995-4	833
S77996-2	493
S77997-0	493
S77998-9	911
S77999-7	855
S78000-6	891
S78001-4	497
S78002-2	670
S78003-0	589
S78004-9	592
S78005-7	602
S78006-5	596
S78007-3	435
S78008-1	511
S78010-3	540
S78011-1	545
S78013-8	540
S78014-6	539
S78015-4	539
S78016-2	537
S78017-0	535
S78018-9	1056
S78019-7	806
S78020-0	513
S78021-9	558
S78022-7	544
S78023-5	899
S78024-3	544
S78025-1	516
S78027-8	1029
S78028-6	889
S78029-4	451
S78030-8	846
S78031-6	521
S78032-4	520
S78033-2	738
S78034-0	413
S78035-9	716
S78036-7	544
S78037-5	540
S78038-3	525
S78039-1	448
S78040-5	460
S78041-3	764
S78042-1	427
S78044-8	865
S78045-6	414
S78046-4	751
S78047-2	752
S78048-0	752
S78049-9	552
S78050-2	495
S78051-0	455
S78052-9	547
S78053-7	437
S78054-5	416
S78055-3	419
S78056-1	620
S78058-8	525
S78059-6	610
S78061-8	549
S78062-6	408
S78063-4	546
S78064-2	634
S78065-0	1081
S78066-9	408
S78067-7	475
S78068-5	521
S78069-3	522
S78070-7	525
S78071-5	546
S78072-3	518
S78073-1	577
S78075-8	520
S78076-6	449
S78077-4	640
S78078-2	536
S78079-0	735
S78080-4	535
S78081-2	473
S78082-0	561
S78083-9	509
S78084-7	508
S78085-5	395
S78086-3	515
S78087-1	537
S78089-8	733
S78090-1	725
S78092-8	532
S78093-6	847
S78094-4	731
S78095-2	538
S78096-0	536
S78097-9	1072
S78098-7	850
S78099-5	698
S78100-2	733
S78101-0	525
S78102-9	695
S78103-7	730
S78104-5	737
S78105-3	209
S78108-8	724
S78109-6	724
S78111-8	752
S78112-6	620
S78113-4	625
S78114-2	818
S78115-0	147
S78116-9	1038
S78117-7	610
S78118-5	606
S78119-3	126
S78120-7	657
S78121-5	662
S78122-3	924
S78123-1	486
S78125-8	804
S78126-6	645
S78127-4	886
S78128-2	644
S78129-0	969
S78130-4	262
S78131-2	370
S78132-0	645
S78133-9	118
S78134-7	659
S78135-5	923
S78136-3	412
S78137-1	658
S78139-8	889
S78140-1	656
S78142-8	866
S78143-6	645
S78145-2	660
S78146-0	667
S78147-9	883
S78148-7	422
S78149-5	884
S78150-9	659
S78151-7	659
S78152-5	412
S78153-3	924
S78154-1	121
S78156-8	118
S78157-6	663
S78158-4	404
S78159-2	644
S78160-6	717
S78161-4	656
S78162-2	660
S78163-0	664
S78165-7	884
S78166-5	429
S78167-3	349
S78168-1	121
S78170-3	121
S78171-1	120
S78173-8	412
S78175-4	130
S78176-2	353
S78177-0	349
S78178-9	124
S78179-7	921
S78180-0	119
S78181-9	886
S78182-7	398
S78183-5	644
S78184-3	645
S78185-1	645
S78187-8	644
S78188-6	656
S78189-4	890
S78190-8	412
S78191-6	664
S78192-4	660

Product	Page
S80728-1	784
S80730-3	211
S80731-1	795
S80733-8	274
S80734-6	819
S80736-2	782
S80737-0	829
S80738-9	247
S80739-7	780
S80740-0	800
S80741-9	859
S80742-7	1006
S80743-5	380
S80744-3	1006
S80745-1	212
S80747-8	1017
S80748-6	1017
S80749-4	784
S80750-8	256
S80751-6	783
S80753-2	811
S80754-0	795
S80755-9	801
S80756-7	776
S80757-5	783
S80758-3	776
S80760-5	784
S80762-1	828
S80764-8	801
S80765-6	799
S80766-4	783
S80767-2	781
S80768-0	623
S80769-9	620
S80770-2	629
S80771-0	456
S80772-9	809
S80773-7	1081
S80774-5	405
S80775-3	416
S80776-1	416
S80778-8	405
S80779-6	631
S80781-8	731
S80782-6	516
S80783-4	550
S80784-2	515
S80785-0	682
S80786-9	640
S80787-7	614
S80788-5	518
S80789-3	612
S80790-7	615
S80791-5	446
S80792-3	961
S80793-1	678
S80795-8	557
S80796-6	545
S80797-4	1056
S80798-2	520
S80799-0	520
S80803-2	838
S80804-0	838
S80805-9	842
S80806-7	842
S80807-5	42
S80808-3	910
S80809-1	91
S80810-5	1094
S80811-3	1094
S80812-1	703
S80817-2	703
S80818-0	477
S80821-0	1000
S80822-9	1000
S80825-3	286
S80826-1	282
S80828-8	285
S80831-8	331
S80832-6	336
S80833-4	336
S80834-2	503
S80835-0	853
S80837-7	1095
S80838-5	698
S80839-3	1073
S80843-1	471
S80847-4	763
S80848-2	387
S80851-2	755
S80852-0	987
S80854-7	944
S80855-5	414
S80859-8	150
S80860-1	239
S80862-8	577
S80863-6	257
S80864-4	709
S80865-2	325
S80866-0	1084
S80867-9	335
S80868-7	334
S80869-5	334
S80870-9	334
S80871-7	858
S80872-5	176
S80873-3	214
S80874-1	379
S80876-8	335
S80878-4	90
S80879-2	190
S80883-0	178
S80884-9	553
S80885-7	290
S80886-5	367
S80887-3	307
S80888-1	290
S80891-1	338
S80893-8	291
S80894-6	50
S80895-4	152
S80896-2	795
S80897-0	1017
S80898-9	1018
S80899-7	796
S80900-4	1019
S80901-2	782
S80902-0	785
S80904-7	774
S80905-5	256
S80906-3	801
S80909-8	774
S80910-1	785
S80912-8	1019
S80913-6	1019
S80914-4	1019
S80915-2	784
S80916-0	784
S80917-9	264
S80918-7	325
S80919-5	331
S80920-9	377
S80921-7	359
S80922-5	386
S80923-3	753
S80924-1	332
S80926-8	1092
S80927-6	423
S80928-4	307
S80929-2	195
S80931-4	436
S80935-7	195
S80936-5	319
S80937-3	319
S80938-1	319
S80940-3	195
S80941-1	319
S80943-8	990
S80944-6	766
S80945-4	378
S80946-2	290
S80947-0	378
S80948-9	378
S80949-7	378
S80950-0	378
S80951-9	378
S80952-7	378
S80954-3	243
S80957-8	1032
S80958-6	17
S80959-4	1126
S80960-8	186
S80961-6	214
S80962-4	17
S80963-2	214
S80964-0	175
S80965-9	172
S80966-7	185
S80967-5	189
S80969-1	214
S80970-5	1126
S80971-3	187
S80972-1	207
S80974-8	280
S80975-6	200
S80976-4	214
S80977-2	200
S80979-9	1128
S80980-2	169
S80981-0	213
S80982-9	212
S80983-7	212
S80984-5	212
S80985-3	187
S80986-1	212
S80988-8	211
S80989-6	211
S80991-8	280
S80992-6	280
S80995-0	200
S80997-7	175
S80998-5	213
S81000-2	983
S81002-9	990
S81003-7	896
S81005-3	895
S81006-1	896
S81012-6	894
S81013-4	1125
S81015-0	894
S81017-7	1122
S81018-5	1107
S81020-7	894
S81021-5	894
S81022-3	894
S81026-6	994
S81027-4	988
S81028-2	986
S81029-0	986
S81031-2	986
S81032-0	477
S81035-5	836
S81036-3	1120
S81037-1	1120
S81039-8	989
S81040-1	343
S81042-8	995
S81045-2	1138
S81046-0	993
S81047-9	204
S81048-7	226
S81049-5	438
S81050-9	848
S81051-7	848
S81052-5	998
S81053-3	845
S81054-1	848
S81056-8	853
S81057-6	190
S81058-4	84
S81059-2	853
S81060-6	231
S81061-4	232
S81062-2	29
S81063-0	172
S81064-9	44
S81065-7	330
S81066-5	748
S81067-3	245
S81068-1	581
S81070-3	322
S81071-1	985
S81074-6	994
S81075-4	784
S81079-7	955
S81080-0	937
S81081-9	939
S81082-7	1005
S81083-5	939
S81084-3	853
S81085-1	1138
S81087-8	71
S81088-6	75
S81089-4	1139
S81091-6	1125
S81092-4	895
S81093-2	896
S81094-0	894
S81095-9	894
S81096-7	897
S81097-5	893
S81098-3	787
S81099-1	896
S81100-9	896
S81101-7	790
S81102-5	896
S81103-3	856
S81104-1	952
S81106-8	952
S81107-6	952
S81110-6	277
S81111-4	895
S81113-0	689
S81114-9	689
S81115-7	692
S81116-5	690
S81118-1	1110
S81120-3	1114
S81123-8	1001
S81125-4	894
S81126-2	489
S81127-0	189
S81131-9	1124
S81132-7	848
S81134-3	848
S81135-1	848
S81137-8	848
S81138-6	848
S81139-4	845
S81140-8	847
S81141-6	845
S81143-2	386
S81144-0	386
S81145-9	386
S81146-7	386
S81147-5	853
S81148-3	853
S81149-1	853
S81151-3	252
S81152-1	253
S81154-8	1139
S81155-6	896
S81156-4	1135
S81157-2	1135
S81158-0	1133
S81159-9	212
S81160-2	210
S81161-0	988
S81162-9	895
S81163-7	892
S81164-5	893
S81165-3	896
S81166-1	991
S81168-8	1008
S81169-6	845
S81171-8	845
S81172-6	845
S81173-4	896
S81174-2	1108
S81175-0	1108
S81176-9	1111
S81177-7	1120
S81178-5	892
S81179-3	892
S81180-7	29
S81181-5	29
S81182-3	142
S81183-1	110
S81185-8	232
S81186-6	129
S81187-4	1105
S81188-2	249
S81189-0	249
S81190-4	1001
S81191-2	994
S81192-0	988
S81193-9	914
S81194-7	771
S81195-5	142
S81196-3	314
S81197-1	186
S81199-8	29
S81200-5	895
S81201-3	895
S81202-1	895
S81204-8	894
S81205-6	893
S81206-4	893
S81207-2	894
S81208-0	895
S81209-9	896
S81210-2	291
S81211-0	55
S81212-9	1111
S81213-7	1101
S81214-5	1110
S81215-3	1101
S81216-1	1110
S81218-8	1110
S81219-6	1110
S81221-8	1100
S81222-6	1110
S81223-4	337
S81224-2	1110
S81225-0	962
S81226-9	967
S81227-7	858
S81228-5	857
S81229-3	858
S81230-7	852
S81233-1	896
S81235-8	894
S81236-6	896
S81237-4	1138
S81238-2	1107
S81239-0	995
S81240-4	773
S81241-2	773
S81242-0	780
S81244-7	780
S81245-5	842
S81246-3	68
S81247-1	776
S81249-8	68
S81250-1	157
S81252-8	1107
S81253-6	432
S81254-4	794
S81255-2	302
S81256-0	960
S81258-7	785
S81259-5	786
S81260-9	786
S81261-7	786
S81262-5	2
S81264-1	1016
S81266-8	158
S81268-4	290
S81270-6	857
S81271-4	62
S81272-2	1141
S81273-0	1141
S81274-9	1143
S81275-7	1141
S81276-5	1141
S81277-3	1143
S81278-1	1142
S81280-3	1141
S81281-1	1141
S81283-8	1141
S81284-6	1141
S81285-4	983
S81286-2	1137
S81287-0	1136
S81288-9	1136
S81289-7	1136
S81290-0	1139
S81293-5	1119
S81294-3	1107
S81295-1	1107
S81297-8	1122
S81298-6	1119
S81299-4	1115
S81300-1	1122
S81302-8	1122
S81303-6	1109
S81304-4	1100
S81306-0	1128
S81307-9	93
S81308-7	91
S81309-5	1134
S81310-9	1135
S81311-7	1134
S81312-5	1135
S81313-3	1135
S81314-1	1136
S81316-8	298
S81317-6	1138
S81318-4	1139
S81319-2	1133
S81320-6	1133
S81322-2	1122
S81323-0	1123
S81325-7	1105
S81326-5	1105
S81327-3	1109
S81328-1	917
S81330-3	997
S81331-1	1138
S81333-8	853
S81334-6	848
S81335-4	482
S81336-2	586
S81337-0	996
S81338-9	1120
S81340-0	786
S81341-9	1106
S81342-7	846
S81343-5	1141
S81344-3	1111
S81345-1	141
S81347-8	426
S81349-4	413
S81351-6	297
S81352-4	490
S81353-2	892
S81354-0	489
S81355-9	428
S81356-7	1108
S81358-3	1080
S81364-8	1084
S81365-6	1084
S81366-4	1107
S81367-2	1105
S81369-9	1086
S81373-7	151
S81374-5	848
S81375-3	853
S81376-1	1140
S81379-6	236
S81381-8	297
S81382-6	177
S81383-4	185
S81385-0	370
S81386-9	207
S81387-7	894
S81391-5	893
S81393-1	260
S81395-8	70
S81396-6	215
S81397-4	262
S81399-0	437
S81401-6	895
S81402-4	1113

Product Page

S81403-2......1134
S81404-0......1136
S81405-9......854
S81407-5......1134
S81408-3......1134
S81409-1......1134
S81410-5......1134
S81411-3......1134
S81412-1......1134
S81414-8......1138
S81416-4......847
S81417-2......847
S81419-9......846
S81420-2......850
S81422-9......847
S81423-7......1096
S81425-3......1103
S81426-1......1117
S81429-6......170
S81431-8......1121
S81432-6......1110
S81434-2......852
S81435-0......1071
S81436-9......848
S81440-7......854
S81441-5......490
S81442-3......490
S81445-8......1070
S81446-6......1085
S81447-4......1080
S81448-2......1085
S81449-0......1139
S81450-4......848
S81451-2......1137
S81452-0......848
S81453-9......848
S81454-7......996
S81455-5......413
S81456-3......28
S81457-1......199
S81459-8......845
S81462-8......848
S81463-6......847
S81464-4......271
S81465-2......1122
S81466-0......1106
S81468-7......1111
S81470-9......1123
S81471-7......1097
S81472-5......987
S81473-3......994
S81474-1......271
S81476-8......164
S81477-6......198
S81478-4......849
S81479-2......853
S81480-6......853
S81481-4......848
S81482-2......345
S81483-0......1092
S81484-9......70
S81485-7......265
S81486-5......893
S81487-3......893
S81490-3......893
S81491-1......893
S81493-8......893
S81494-6......893
S81495-4......339
S81496-2......1136
S81497-0......1134
S81498-9......1134
S81499-7......1136
S81500-4......1134
S81501-2......1130
S81502-0......1130
S81503-9......1130
S81504-7......1131
S81505-5......1131
S81506-3......1130
S81507-1......1130
S81509-8......1130
S81510-1......1130
S81512-8......1130
S81513-6......1132
S81514-4......1136
S81515-2......1136
S81516-0......1136
S81517-9......1135
S81518-7......1133
S81519-5......1136
S81520-9......1134
S81521-7......1136
S81522-5......1134
S81523-3......1133
S81524-1......1133
S81526-8......417
S81527-6......894
S81529-2......322
S81530-6......1134
S81532-2......1117
S81533-0......1136
S81534-9......274
S81535-7......778
S81536-5......898
S81537-3......111
S81538-1......144
S81540-3......910
S81541-1......709
S81543-8......708
S81544-6......894
S81545-4......827
S81546-2......138
S81547-0......136
S81548-9......136
S81549-7......136
S81551-9......1096
S81554-3......1096
S81555-1......1100
S81557-8......1104
S81559-4......1100
S81560-8......1104
S81561-6......1104
S81562-4......1096
S81563-2......5
S81564-0......5
S81565-9......5
S81566-7......5
S81567-5......4
S81569-1......5
S81571-3......1097
S81572-1......1097
S81574-8......1097
S81575-6......1097
S81576-4......1104
S81579-9......1139
S81580-2......1139
S81581-0......1138
S81582-9......262
S81583-7......261
S81584-5......216
S81585-3......1135
S81588-8......1135
S81589-6......896
S81591-8......895
S81592-6......116
S81593-4......35
S81594-2......262
S81595-0......261
S81596-9......262
S81597-7......238
S81599-3......897
S81601-9......1132
S81604-3......1122
S81608-6......1130
S81609-4......232
S81610-8......1136
S81611-6......1136
S81612-4......1134
S81613-2......1135
S81614-0......1136
S81615-9......1135
S81616-7......1135
S81618-3......709
S81619-1......1136
S81620-5......1136
S81621-3......1134
S81622-1......1134
S81624-8......1104
S81626-4......1142
S81627-2......1115
S81628-0......1140
S81629-9......1139
S81631-0......898
S81632-9......898
S81633-7......898
S81634-5......897
S81635-3......1101
S81636-1......1101
S81639-6......1101
S81642-6......1102
S81644-2......895
S81645-0......936
S81647-7......239
S81650-7......997
S81651-5......265
S81653-1......1122
S81655-8......768
S81656-6......239
S81658-2......1122
S81660-4......1119
S81661-2......1098
S81662-0......1106
S81663-9......1122
S81665-5......1122
S81666-3......1143
S81667-1......1141
S81669-8......1142
S81670-1......1143
S81672-8......1128
S81673-6......1143
S81674-4......1142
S81676-0......1142
S81677-9......1143
S81678-7......1141
S81680-9......1141
S81681-7......1139
S81684-1......1141
S81686-8......987
S81687-6......821
S81688-4......847
S81689-2......329
S81705-8......994
S81708-2......934
S81710-4......996
S81711-2......985
S81714-7......995
S81715-5......961
S81716-3......967
S81717-1......967
S81719-8......942
S81720-1......966
S81725-2......292
S81728-7......283
S81736-8......892
S81739-2......1120
S81740-6......1120
S81741-4......1120
S81742-2......1120
S81743-0......1120
S81744-9......1121
S81746-5......1116
S81747-3......1108
S81754-6......692
S81756-2......692
S81758-9......691
S81759-7......691
S81760-0......691
S81761-9......985
S81762-7......1128
S81764-3......1129
S81765-1......1129
S81767-8......1129
S81768-6......1130
S81771-6......1130
S81772-4......1132
S81773-2......726
S81774-0......1093
S81775-9......334
S81776-7......1132
S81777-5......1130
S81778-3......1132
S81779-1......233
S81780-5......785
S81781-3......503
S81784-8......64
S81786-4......289
S81787-2......988
S81789-9......986
S81792-9......988
S81793-7......1107
S81800-3......898
S81801-1......425
S81803-8......898
S81804-6......898
S81805-4......898
S81806-2......899
S81807-0......899
S81808-9......414
S81809-7......177
S81810-0......898
S81811-9......334
S81814-3......934
S81815-1......960
S81819-4......848
S81820-8......1129
S81821-6......1128
S81822-4......1128
S81823-2......1134
S81824-0......1134
S81825-9......1136
S81826-7......1134
S81827-5......1136
S81828-3......1134
S81829-1......1134
S81830-5......1134
S81831-3......1128
S81832-1......1128
S81834-8......1129
S81835-6......1134
S81836-4......293
S81837-2......225
S81838-0......204
S81839-9......909
S81840-2......1025
S81842-9......1128
S81845-3......1129
S81846-1......1135
S81848-8......1134
S81849-6......1134
S81851-8......1135
S81852-6......1136
S81853-4......1135
S81854-2......1140
S81855-0......1135
S81856-9......1129
S81857-7......1133
S81858-5......1133
S81859-3......1133
S81860-7......1135
S81861-5......1134
S81862-3......892
S81863-1......1134
S81865-8......1133
S81866-6......1135
S81867-4......1128
S81868-2......1127
S81869-0......1127
S81870-4......1128
S81871-2......1127
S81872-0......1106
S81873-9......1106
S81876-3......333
S81877-1......330
S81880-1......15
S81886-0......217
S81887-9......217
S81888-7......217
S81889-5......100
S81890-9......100
S81891-7......100
S81892-5......100
S81893-3......62
S81896-8......82
S81897-6......84
S81899-2......1008
S81901-8......91
S81903-4......673
S81904-2......379
S81905-0......961
S81906-9......961
S81909-3......990
S81910-7......1093
S81912-3......878
S81913-1......876
S81915-8......987
S81917-4......995
S81918-2......998
S81919-0......998
S81920-4......998
S81922-0......251
S81923-9......251
S81924-7......343
S81930-1......273
S81932-8......1132
S81933-6......1130
S81934-4......1132
S81935-2......1132
S81936-0......1133
S81937-9......1133
S81938-7......1130
S81939-5......1133
S81940-9......1130
S81941-7......1129
S81942-5......1133
S81943-3......1133
S81944-1......1130
S81946-8......1130
S81948-4......196
S81949-2......353
S81950-6......1083
S81951-4......1083
S81952-2......1080
S81953-0......986
S81954-9......185
S81955-7......1107
S81957-3......1116
S81958-1......163
S81960-3......186
S81966-2......894
S81967-0......794
S81968-9......1135
S81969-7......1136
S81970-0......1136
S81971-9......1134
S81972-7......1133
S81973-5......1136
S81974-3......1134
S81975-1......1134
S81977-8......1134
S81978-6......1100
S81979-4......1105
S81980-8......1115
S81982-4......1107
S81983-2......1108
S81985-9......1118
S81986-7......1118
S81988-3......1013
S81989-1......232
S81990-5......586
S81991-3......1136
S81992-1......1136
S81995-6......243
S81996-4......62
S81997-2......1128
S81998-0......1128
S81999-9......1128
S82000-8......1139
S82003-2......1138
S82004-0......853
S82005-9......1138
S82006-7......727
S82008-3......342
S82010-5......976
S82011-3......718
S82012-1......785
S82015-6......796
S82016-4......226
S82018-0......1131
S82019-9......845
S82020-2......846
S82021-0......366
S82022-9......809
S82023-7......991
S82024-5......998
S82025-3......998
S82026-1......308
S82028-8......343
S82031-8......892
S82032-6......892
S82033-4......892
S82034-2......1007
S82036-9......990
S82037-7......986
S82039-3......749
S82040-7......988
S82042-3......230
S82045-8......341
S82047-4......1001
S82052-0......185
S82054-7......248
S82059-8......311
S82060-1......185
S82062-8......185
S82065-2......185
S82066-0......274
S82068-7......24
S82069-5......274
S82070-9......278
S82071-7......219
S82072-5......216
S82073-3......271
S82076-8......229
S82078-4......954
S82079-2......839
S82080-6......993
S82081-4......989
S82082-2......990
S82084-9......989
S82095-4......771
S82096-2......989
S82097-0......989
S82099-7......892
S82106-3......128
S82107-1......128
S82110-1......128
S82112-8......128
S82113-6......142
S82116-0......314
S82118-7......1139
S82121-7......357
S82122-5......357
S82123-3......357
S82124-1......357
S82129-2......433
S82130-6......433
S82132-2......427
S82134-9......437
S82136-5......419
S82138-1......489
S82140-3......489
S82141-1......828
S82143-8......1136
S82150-0......710
S82152-7......503
S82154-3......499
S82155-1......503
S82157-8......499
S82158-6......503
S82159-4......1119
S82160-8......1105
S82162-4......1105
S82163-2......1114
S82164-0......1119
S82165-9......1115
S82166-7......1105
S82167-5......1105
S82168-3......1122
S82169-1......1114
S82170-5......1111
S82171-3......1115
S82172-1......1117
S82174-8......1117
S82175-6......1105

Product	Page	Product	Page	Product	Page	Product	Page	Product	Page	Product	Page	Product	Page
S82176-4	1104	S82280-9	346	S82381-3	858	S82471-2	1131	S82557-3	1110	S82642-1	307	S82728-2	1102
S82177-2	1114	S82281-7	1125	S82382-1	172	S82472-0	1133	S82558-1	1123	S82644-8	307	S82729-0	1104
S82178-0	1125	S82282-5	224	S82384-8	1100	S82473-9	1133	S82560-3	1101	S82645-6	224	S82730-4	1104
S82179-9	1092	S82283-3	224	S82385-6	1102	S82474-7	1133	S82561-1	1102	S82646-4	1114	S82731-2	1101
S82180-2	1125	S82284-1	225	S82386-4	1102	S82475-5	1131	S82563-8	1102	S82648-0	1133	S82732-0	1115
S82182-9	1125	S82286-8	225	S82387-2	1102	S82476-3	1132	S82564-6	1103	S82649-9	64	S82734-7	1113
S82183-7	1125	S82289-2	1116	S82388-0	1103	S82477-1	1130	S82565-4	1103	S82650-2	761	S82735-5	1105
S82184-5	987	S82290-6	1113	S82389-9	961	S82479-8	1130	S82566-2	1103	S82651-0	330	S82736-3	1114
S82185-3	987	S82292-2	1108	S82390-2	932	S82480-1	1130	S82567-0	1130	S82652-9	761	S82737-1	1114
S82186-1	987	S82294-9	1108	S82391-0	932	S82482-8	1130	S82568-9	1103	S82653-7	163	S82739-8	1112
S82191-8	987	S82296-5	1108	S82392-9	1121	S82483-6	1130	S82569-7	1104	S82654-5	368	S82742-8	1103
S82192-6	987	S82298-1	1108	S82394-5	1107	S82484-4	327	S82570-0	1130	S82655-3	51	S82744-4	1103
S82194-2	925	S82300-7	1100	S82395-3	1118	S82485-2	1083	S82571-9	1103	S82656-1	375	S82745-2	1112
S82196-9	941	S82301-5	1097	S82396-1	1116	S82487-9	325	S82574-3	197	S82658-8	287	S82746-0	1103
S82198-5	772	S82302-3	1118	S82398-8	1111	S82488-7	325	S82575-1	1110	S82659-6	1101	S82747-9	1105
S82199-3	36	S82303-1	1118	S82399-6	1114	S82489-5	83	S82577-8	1114	S82661-8	1103	S82748-7	1115
S82200-0	749	S82305-8	1119	S82400-3	342	S82490-9	400	S82578-6	367	S82662-6	1103	S82749-5	1104
S82201-9	38	S82307-4	1121	S82401-1	280	S82491-7	400	S82579-4	370	S82665-0	1125	S82750-9	1130
S82202-7	36	S82308-2	1121	S82407-0	251	S82492-5	859	S82580-8	198	S82666-9	1103	S82751-7	1116
S82203-5	1106	S82309-0	1103	S82408-9	252	S82493-3	859	S82581-6	377	S82667-7	1103	S82753-3	1122
S82204-3	1108	S82310-4	1121	S82411-9	994	S82494-1	859	S82582-4	367	S82668-5	1112	S82754-1	1104
S82205-1	1107	S82311-2	1118	S82412-7	771	S82496-8	170	S82583-2	1082	S82669-3	1112	S82756-8	1106
S82207-8	852	S82312-0	1118	S82413-5	771	S82497-6	1099	S82584-0	336	S82670-7	1119	S82757-6	1121
S82208-6	847	S82313-9	1110	S82414-3	778	S82498-4	1108	S82585-9	337	S82671-5	1113	S82758-4	1117
S82209-4	1062	S82314-7	1102	S82415-1	778	S82499-2	1105	S82586-7	177	S82672-3	1120	S82759-2	1120
S82210-8	201	S82315-5	1102	S82419-4	296	S82501-8	1132	S82587-5	190	S82673-1	1113	S82760-6	1120
S82212-4	1091	S82316-3	1101	S82420-8	1098	S82503-4	995	S82588-3	1072	S82675-8	1121	S82761-4	1099
S82214-0	863	S82319-8	235	S82421-6	1123	S82504-2	995	S82589-1	1120	S82676-6	1113	S82762-2	1121
S82215-9	344	S82322-8	163	S82422-4	1097	S82505-0	983	S82591-3	351	S82677-4	1120	S82763-0	1122
S82216-7	1113	S82323-6	163	S82423-2	1097	S82507-7	1099	S82592-1	351	S82678-2	1106	S82764-9	1110
S82217-5	1113	S82324-4	1135	S82424-0	1113	S82508-5	1111	S82594-8	962	S82679-0	1106	S82765-7	1102
S82218-3	1114	S82325-2	1135	S82425-9	1117	S82509-3	774	S82595-6	848	S82680-4	1120	S82766-5	1122
S82219-1	1114	S82326-0	1136	S82426-7	1107	S82510-7	1118	S82596-4	351	S82681-2	1113	S82767-3	183
S82220-5	1114	S82327-9	1136	S82427-5	1129	S82511-5	1119	S82597-2	70	S82682-0	1121	S82768-1	1106
S82221-3	1116	S82328-7	325	S82428-3	1102	S82512-3	1119	S82598-0	70	S82683-9	1121	S82770-3	857
S82222-1	1114	S82329-5	1085	S82429-1	1102	S82513-1	182	S82599-9	70	S82684-7	1114	S82771-1	1120
S82224-8	1117	S82330-9	252	S82430-5	1101	S82515-8	1106	S82600-6	848	S82686-3	1120	S82773-8	1119
S82225-6	1109	S82331-7	771	S82431-3	1102	S82516-6	1106	S82602-2	849	S82687-1	1122	S82774-6	1106
S82226-4	1115	S82334-1	898	S82432-1	1102	S82517-4	338	S82603-0	853	S82689-8	1120	S82775-4	1115
S82227-2	1132	S82337-6	899	S82434-8	1102	S82518-2	1120	S82604-9	24	S82690-1	1121	S82776-2	1115
S82228-0	988	S82338-4	899	S82435-6	1102	S82519-0	1121	S82605-7	151	S82692-8	1106	S82777-0	1116
S82230-2	337	S82342-2	898	S82436-4	1101	S82521-2	1104	S82606-5	336	S82693-6	1119	S82778-9	1132
S82231-0	843	S82344-9	897	S82437-2	1101	S82522-0	232	S82607-3	77	S82694-4	1119	S82779-7	1116
S82232-9	942	S82345-7	1113	S82438-0	1101	S82523-9	1111	S82608-1	1011	S82695-2	1121	S82780-0	1117
S82233-7	41	S82346-5	1109	S82439-9	1103	S82524-7	1116	S82610-3	371	S82696-0	1120	S82781-9	1099
S82234-5	1137	S82347-3	1100	S82440-2	1099	S82525-5	438	S82611-1	157	S82697-9	69	S82782-7	1121
S82238-8	36	S82348-1	1105	S82441-0	1100	S82526-3	1119	S82613-8	793	S82698-7	69	S82783-5	1121
S82241-8	1115	S82350-3	1114	S82442-9	1100	S82527-1	1118	S82615-4	795	S82699-5	69	S82784-3	1099
S82242-6	1115	S82351-1	1115	S82443-7	1101	S82529-8	128	S82616-2	216	S82701-0	69	S82785-1	1105
S82243-4	1116	S82353-8	1117	S82444-5	897	S82530-1	240	S82617-0	823	S82702-9	70	S82787-8	1104
S82245-0	1116	S82354-6	1105	S82445-3	895	S82532-8	241	S82618-9	823	S82703-7	70	S82788-6	339
S82246-9	1117	S82355-4	1115	S82446-1	914	S82533-6	242	S82619-7	1027	S82704-5	848	S82789-4	1125
S82247-7	1118	S82356-2	1118	S82448-8	987	S82534-4	828	S82620-0	932	S82705-3	848	S82790-8	1124
S82252-3	1098	S82358-9	1109	S82449-6	994	S82535-2	373	S82621-9	1089	S82706-1	106	S82791-6	1125
S82253-1	913	S82359-7	1111	S82451-8	987	S82536-0	1115	S82622-7	1142	S82708-8	478	S82792-4	165
S82255-8	894	S82360-0	1097	S82453-4	987	S82539-5	849	S82623-5	236	S82709-6	106	S82793-2	912
S82256-6	895	S82361-9	1115	S82454-2	1099	S82540-9	1100	S82624-3	1138	S82711-8	106	S82794-0	795
S82257-4	894	S82362-7	1112	S82455-0	1118	S82541-7	1100	S82625-1	332	S82712-6	292	S82796-7	260
S82258-2	210	S82363-5	1113	S82456-9	1132	S82542-5	1100	S82627-8	333	S82713-4	106	S82797-5	219
S82260-4	912	S82364-3	1115	S82457-7	1109	S82543-3	1100	S82628-6	6	S82714-2	106	S82798-3	773
S82261-2	743	S82365-1	1116	S82458-5	1113	S82544-1	1100	S82629-4	5	S82715-0	110	S82800-9	892
S82262-0	986	S82367-8	1102	S82459-3	1117	S82546-8	1101	S82630-8	5	S82716-9	1103	S82802-5	891
S82265-5	138	S82368-6	198	S82460-7	1098	S82547-6	1103	S82631-6	1106	S82717-7	1125	S82804-1	891
S82266-3	795	S82369-4	198	S82461-5	174	S82548-4	1101	S82632-4	1098	S82718-5	1108	S82806-8	892
S82267-1	726	S82370-8	1101	S82462-3	1112	S82549-2	1101	S82633-2	174	S82719-3	1107	S82807-6	112
S82269-8	785	S82371-6	1099	S82463-1	1112	S82550-6	1104	S82634-0	1098	S82720-7	207	S82808-4	233
S82272-8	688	S82374-0	185	S82465-8	1115	S82551-4	1104	S82635-9	1096	S82721-5	1108	S82809-2	857
S82273-6	500	S82375-9	898	S82466-6	1116	S82552-2	1100	S82636-7	198	S82722-3	1108	S82810-6	212
S82275-2	990	S82376-7	1101	S82467-4	1121	S82553-0	1104	S82637-5	307	S82723-1	1108	S82811-4	859
S82276-0	895	S82378-3	1102	S82468-2	1117	S82554-9	1104	S82638-3	1098	S82725-8	1116	S82812-2	162
S82277-9	342	S82379-1	1102	S82469-0	1132	S82555-7	1104	S82639-1	966	S82726-6	1108	S82813-0	218
S82279-5	214	S82380-5	1102	S82470-4	1132	S82556-5	1100	S82641-3	482	S82727-4	1121	S82815-7	94

Product	Page	Product	Page	Product	Page	Product	Page	Product	Page	Product	Page	Product	Page
S82816-5	233	S82905-6	561	S82992-7	1105	S83083-6	988	S83169-7	173	S83249-9	1116	S83339-8	327
S82817-3	817	S82906-4	270	S82993-5	1114	S83084-4	171	S83171-9	961	S83250-2	1122	S83340-1	176
S82818-1	828	S82907-2	779	S82994-3	1099	S83085-2	177	S83172-7	306	S83251-0	1100	S83342-8	376
S82820-3	94	S82908-0	776	S82995-1	1117	S83086-0	243	S83173-5	306	S83252-9	1120	S83343-6	895
S82821-1	94	S82909-9	230	S82997-8	1104	S83087-9	210	S83174-3	1102	S83253-7	1122	S83344-4	185
S82823-8	94	S82910-2	1088	S82998-6	1117	S83088-7	827	S83175-1	307	S83254-5	1122	S83346-0	1112
S82824-6	29	S82911-0	839	S82999-4	1117	S83089-5	987	S83177-8	855	S83255-3	1106	S83347-9	1113
S82825-4	967	S82912-9	838	S83000-3	1118	S83093-3	987	S83178-6	855	S83256-1	1115	S83348-7	1108
S82826-2	967	S82913-7	337	S83001-1	1104	S83094-1	29	S83179-4	855	S83259-6	1096	S83349-5	1122
S82827-0	170	S82914-5	1100	S83003-8	1119	S83096-8	210	S83180-8	1103	S83261-8	1122	S83350-9	1114
S82828-9	307	S82915-3	1101	S83004-6	1119	S83097-6	29	S83181-6	1101	S83262-6	1102	S83351-7	1115
S82830-0	157	S82916-1	1101	S83005-4	967	S83098-4	243	S83182-4	163	S83263-4	1103	S83352-5	1108
S82831-9	69	S82918-8	1101	S83006-2	307	S83101-8	141	S83183-2	1117	S83264-2	1118	S83353-3	1119
S82832-7	74	S82919-6	1106	S83007-0	1102	S83102-6	1109	S83184-0	1119	S83267-7	235	S83354-1	1119
S82833-5	795	S82921-8	413	S83008-9	1015	S83103-4	1113	S83185-9	1099	S83268-5	786	S83356-8	1112
S82834-3	209	S82922-6	313	S83009-7	307	S83104-2	1113	S83186-7	1118	S83269-3	1124	S83357-6	1096
S82835-1	835	S82925-0	778	S83010-0	1015	S83105-0	1114	S83187-5	162	S83270-7	896	S83358-4	1112
S82837-8	209	S82927-7	1121	S83011-9	838	S83106-9	1105	S83188-3	1127	S83271-5	894	S83359-2	1123
S82838-6	793	S82928-5	1109	S83012-7	307	S83107-7	165	S83189-1	673	S83272-3	796	S83360-6	1112
S82840-8	49	S82929-3	1121	S83013-5	307	S83108-5	1116	S83190-5	256	S83273-1	897	S83361-4	1112
S82841-6	1105	S82930-7	1102	S83014-3	172	S83109-3	783	S83191-3	1127	S83275-8	985	S83362-2	162
S82843-2	49	S82931-5	1104	S83015-1	198	S83110-7	780	S83192-1	1118	S83277-4	1132	S83363-0	1113
S82844-0	1107	S82932-3	1102	S83018-6	198	S83111-5	309	S83194-8	1117	S83278-2	796	S83364-9	1113
S82845-9	1108	S82933-1	1111	S83020-8	325	S83112-3	775	S83195-6	172	S83279-0	796	S83365-7	1114
S82846-7	1105	S82935-8	1111	S83021-6	1112	S83113-1	783	S83196-4	163	S83280-4	209	S83366-5	1114
S82847-5	1112	S82936-6	1120	S83023-2	1100	S83115-8	209	S83197-2	165	S83281-2	212	S83367-3	204
S82848-3	1111	S82937-4	1111	S83024-0	1103	S83117-4	341	S83198-0	163	S83283-9	212	S83368-1	162
S82849-1	891	S82938-2	1115	S83025-9	1112	S83118-2	781	S83199-9	166	S83284-7	800	S83370-3	1122
S82850-5	1122	S82939-0	1116	S83026-7	1119	S83119-0	1099	S83200-6	166	S83285-5	213	S83371-1	1111
S82851-3	201	S82940-4	1139	S83027-5	1119	S83120-4	833	S83201-4	313	S83286-3	990	S83373-8	1121
S82852-1	1111	S82942-0	1122	S83028-3	1104	S83121-2	1130	S83202-2	336	S83287-1	916	S83374-6	1117
S82854-8	1111	S82943-9	1116	S83029-1	1112	S83122-0	1103	S83204-9	244	S83289-8	994	S83375-4	1120
S82855-6	1111	S82944-7	1116	S83030-5	1113	S83123-9	1103	S83205-7	345	S83290-1	911	S83376-2	1119
S82856-4	1122	S82945-5	1121	S83031-3	1113	S83124-7	1103	S83206-5	1129	S83293-6	251	S83377-0	204
S82857-2	1111	S82946-3	1117	S83032-1	1106	S83125-5	1134	S83207-3	1097	S83294-4	912	S83378-9	224
S82858-0	1096	S82947-1	482	S83034-8	1113	S83126-3	1134	S83208-1	1104	S83295-2	912	S83379-7	1123
S82859-9	1116	S82949-8	1129	S83035-6	1113	S83127-1	1101	S83210-3	1130	S83297-9	710	S83380-0	1119
S82860-2	1118	S82950-1	1129	S83037-2	1099	S83129-8	1101	S83211-1	1103	S83298-7	251	S83382-7	162
S82861-0	1118	S82952-8	1132	S83038-0	1119	S83130-1	1102	S83213-8	1132	S83299-5	1117	S83383-5	186
S82862-9	1118	S82953-6	1132	S83039-9	1119	S83132-8	1101	S83214-6	1132	S83300-2	1118	S83384-3	162
S82863-7	1119	S82954-4	1132	S83040-2	1104	S83133-6	997	S83215-4	1132	S83301-0	1118	S83385-1	162
S82864-5	1119	S82955-2	1132	S83041-0	1109	S83134-4	994	S83216-2	1132	S83302-9	1099	S83387-8	162
S82865-3	289	S82956-0	1131	S83042-9	1105	S83135-2	252	S83217-0	1119	S83303-7	1108	S83388-6	162
S82866-1	1100	S82957-9	1111	S83043-7	1114	S83136-0	252	S83218-9	1096	S83304-5	1108	S83389-4	333
S82868-8	1096	S82958-7	171	S83044-5	1103	S83137-9	1118	S83219-7	1109	S83305-3	1116	S83390-8	1107
S82869-6	1097	S82959-5	173	S83045-3	1121	S83138-7	1118	S83220-0	1112	S83306-1	1108	S83391-6	233
S82871-8	310	S82960-9	165	S83046-1	1106	S83139-5	1106	S83221-9	1113	S83308-8	1108	S83392-4	1133
S82872-6	1113	S82961-7	1137	S83048-8	1106	S83140-9	1112	S83222-7	1098	S83309-6	1099	S83393-2	1107
S82873-4	1109	S82962-5	578	S83049-6	1119	S83141-7	1122	S83223-5	1098	S83312-6	373	S83394-0	1123
S82874-2	1114	S82963-3	1118	S83051-8	1106	S83142-5	1123	S83224-3	1098	S83313-4	373	S83395-9	987
S82875-0	1109	S82964-1	1104	S83053-4	1121	S83143-3	1120	S83225-1	1116	S83314-2	337	S83396-7	985
S82876-9	1114	S82967-6	295	S83054-2	1122	S83144-1	1123	S83227-8	1116	S83315-0	1123	S83397-5	988
S82877-7	1109	S82968-4	278	S83055-0	1121	S83146-8	1123	S83228-6	1117	S83316-9	198	S83398-3	1117
S82878-5	1106	S82970-6	14	S83056-9	1096	S83147-6	1098	S83229-4	1119	S83317-7	198	S83399-1	1117
S82879-3	1110	S82971-4	164	S83057-7	1096	S83148-4	1114	S83230-8	1119	S83318-5	1123	S83400-9	932
S82880-7	1114	S82972-2	183	S83058-5	1096	S83149-2	1115	S83231-6	1105	S83319-3	1120	S83401-7	293
S82882-3	1108	S82973-0	183	S83060-7	1097	S83150-6	1098	S83232-4	346	S83320-7	200	S83402-5	35
S82883-1	1108	S82974-9	242	S83061-5	1097	S83151-4	1098	S83233-2	182	S83321-5	69	S83403-3	987
S82885-8	1116	S82975-7	389	S83062-3	1097	S83152-2	1116	S83234-0	1123	S83322-3	350	S83404-1	994
S82886-6	1117	S82976-5	855	S83063-1	1112	S83153-0	1132	S83235-9	1109	S83323-1	351	S83406-8	238
S82887-4	1117	S82977-3	1123	S83066-6	1113	S83154-9	1099	S83236-7	1109	S83325-8	70	S83407-6	238
S82888-2	791	S82978-1	1129	S83067-4	1114	S83155-7	1118	S83237-5	185	S83326-6	70	S83408-4	244
S82889-0	224	S82980-3	961	S83069-0	1123	S83156-5	1124	S83238-3	1103	S83327-4	106	S83409-2	226
S82892-0	270	S82981-1	932	S83070-4	1097	S83157-3	1124	S83239-1	1103	S83328-2	106	S83410-6	237
S82893-9	778	S82983-8	198	S83071-2	1114	S83158-1	1114	S83240-5	1121	S83329-0	154	S83411-4	238
S82894-7	779	S82984-6	1104	S83072-0	1115	S83160-3	1105	S83241-3	1121	S83331-2	106	S83412-2	237
S82896-3	230	S82985-4	1107	S83073-9	1133	S83161-1	1116	S83242-1	1099	S83332-0	106	S83413-0	35
S82897-1	776	S82986-2	1107	S83074-7	1102	S83163-8	857	S83244-8	1104	S83333-9	775	S83414-9	237
S82899-8	230	S82987-0	1112	S83077-1	172	S83164-6	1111	S83245-6	1115	S83334-7	775	S83416-5	953
S82900-5	308	S82988-9	1103	S83079-8	172	S83165-4	896	S83246-4	1103	S83335-5	299	S83417-3	293
S82901-3	794	S82989-7	1105	S83080-1	165	S83166-2	931	S83247-2	1115	S83336-3	300	S83418-1	291
S82904-8	421	S82991-9	1114	S83082-8	993	S83167-0	925	S83248-0	1115	S83337-1	328	S83420-3	996

Product	Page
S83423-8	1101
S83424-6	1123
S83425-4	1113
S83426-2	1104
S83427-0	1122
S83428-9	1132
S83429-7	1116
S83430-0	1117
S83431-9	1098
S83432-7	1117
S83433-5	1099
S83434-3	1117
S83435-1	1117
S83437-8	1102
S83438-6	1102
S83439-4	1123
S83440-8	1123
S83441-6	1121
S83442-4	1099
S83443-2	1118
S83444-0	1120
S83445-9	1123
S83446-7	1099
S83447-5	1121
S83448-3	1106
S83449-1	1106
S83450-5	1120
S83451-3	1118
S83452-1	1123
S83454-8	1118
S83455-6	1099
S83456-4	1104
S83457-2	1123
S83458-0	1120
S83459-9	1119
S83460-2	191
S83461-0	988
S83463-7	280
S83464-5	993
S83465-3	989
S83466-1	986
S83468-8	571
S83469-6	988
S83471-8	991
S83472-6	1043
S83473-4	311
S83475-0	296
S83476-9	297
S83477-7	22
S83478-5	986
S83479-3	771
S83480-7	1080
S83481-5	248
S83482-3	253
S83483-1	211
S83485-8	987
S83486-6	998
S83487-4	995
S83488-2	1027
S83489-0	891
S83490-4	891
S83491-2	891
S83492-0	224
S83493-9	1133
S83494-7	1132
S83495-5	1131
S83496-3	1099
S83499-8	1102
S83500-5	166
S83502-1	1101
S83504-8	1101
S83505-6	1122
S83507-2	1123
S83508-0	1117
S83509-9	1122
S83510-2	1118
S83511-0	1099
S83512-9	1099
S83513-7	1137
S83514-5	1137
S83515-3	1137
S83516-1	1137
S83518-8	1133
S83519-6	1133
S83521-8	1129
S83522-6	1133
S83523-4	1133
S83524-2	1129
S83525-0	192
S83526-9	321
S83527-7	325
S83528-5	937
S83529-3	985
S83530-7	984
S83531-5	189
S83532-3	234
S83533-1	183
S83535-8	899
S83536-6	897
S83537-4	838
S83538-2	1123
S83539-0	1101
S83540-4	170
S83541-2	1101
S83542-0	1102
S83543-9	1101
S83544-7	322
S83545-5	322
S83546-3	893
S83547-1	896
S83549-8	787
S83550-1	204
S83552-8	914
S83553-6	265
S83554-4	1098
S83555-2	1108
S83556-0	1117
S83557-9	898
S83559-5	891
S83560-9	1098
S83562-5	1111
S83563-3	1106
S83564-1	1109
S83566-8	1110
S83567-6	28
S83568-4	158
S83569-2	1127
S83570-6	204
S83571-4	204
S83572-2	786
S83573-0	237
S83574-9	336
S83575-7	779
S83578-1	779
S83580-3	780
S83581-1	773
S83584-6	775
S83586-2	1110
S83587-0	995
S83588-9	995
S83591-9	830
S83592-7	756
S83593-5	892
S83594-3	26
S83595-1	211
S83597-8	218
S83598-6	211
S83599-4	230
S83600-1	292
S83602-8	850
S83603-6	295
S83604-4	226
S83605-2	183
S83606-0	251
S83607-9	771
S83608-7	251
S83609-5	251
S83610-9	251
S83611-7	771
S83612-5	778
S83613-3	252
S83614-1	253
S83616-8	221
S83617-6	1000
S83618-4	999
S83619-2	1000
S83622-2	961
S83623-0	961
S83624-9	178
S83625-7	260
S83626-5	773
S83627-3	224
S83628-1	230
S83630-3	270
S83631-1	270
S83633-8	277
S83634-6	270
S83635-4	779
S83636-2	956
S83637-0	982
S83638-9	841
S83639-7	841
S83640-0	839
S83641-9	998
S83642-7	999
S83643-5	1109
S83644-3	843
S83645-1	843
S83647-8	1013
S83648-6	47
S83649-4	855
S83650-8	848
S83651-6	381
S83652-4	342
S83654-0	849
S83655-9	842
S83657-5	242
S83659-1	852
S83660-5	850
S83661-3	776
S83662-1	776
S83673-7	895
S83674-5	897
S83675-3	893
S83676-1	897
S83678-8	162
S83679-6	893
S83681-8	1114
S83682-6	70
S83683-4	847
S83684-2	849
S83685-0	165
S83686-9	172
S83687-7	180
S83688-5	180
S83690-7	328
S83691-5	166
S83693-1	1130
S83695-8	1134
S83696-6	1097
S83697-4	1111
S83699-0	1110
S83700-8	1110
S83701-6	1111
S83702-4	1122
S83703-2	1123
S83704-0	1118
S83705-9	1135
S83706-7	1140
S83707-5	1140
S83708-3	1140
S83709-1	1137
S83710-5	304
S83711-3	1096
S83712-1	187
S83714-8	1123
S83715-6	1123
S83716-4	1109
S83717-2	1100
S83718-0	1105
S83719-9	1107
S83720-2	1120
S83721-0	1110
S83722-9	1123
S83723-7	1123
S83724-5	1123
S83725-3	350
S83726-1	768
S83728-8	746
S83729-6	1011
S83731-8	204
S83732-6	204
S83733-4	173
S83734-2	136
S83735-0	794
S83737-7	218
S83738-5	247
S83739-3	221
S83740-7	895
S83741-5	218
S83742-3	237
S83743-1	895
S83745-8	217
S83746-6	894
S83747-4	673
S83748-2	895
S83749-0	27
S83750-4	220
S83751-2	893
S83752-0	893
S83753-9	893
S83755-5	893
S83756-3	179
S83759-8	893
S83760-1	893
S83762-8	893
S83763-6	810
S83764-4	894
S83765-2	895
S83766-0	896
S83767-9	894
S83768-7	894
S83769-5	986
S83770-9	780
S83771-7	994
S83772-5	785
S83773-3	574
S83774-1	1122
S83776-8	1123
S83777-6	1120
S83778-4	1098
S83779-2	1099
S83780-6	177
S83781-4	1134
S83782-2	1134
S83783-0	1134
S83784-9	1072
S83785-7	798
S83786-5	1113
S83787-3	1111
S83788-1	1115
S83790-3	1115
S83791-1	1111
S83793-8	1116
S83794-6	1116
S83795-4	1120
S83796-2	1111
S83797-0	1116
S83798-9	1116
S83799-7	1108
S83800-4	1098
S83801-2	1098
S83802-0	571
S83803-9	1111
S83804-7	1108
S83805-5	1111
S83806-3	1108
S83807-1	1126
S83809-8	1126
S83810-1	264
S83812-8	1127
S83813-6	895
S83816-0	914
S83817-9	987
S83818-7	987
S83819-5	1139
S83820-9	893
S83821-7	1133
S83822-5	1135
S83823-3	1104
S83824-1	338
S83827-6	341
S83828-4	340
S83829-2	1105
S83830-6	90
S83831-4	21
S83832-2	10
S83836-5	692
S83837-3	693
S83840-3	692
S83841-1	841
S83843-8	197
S83844-6	215
S83845-4	891
S83846-2	891
S83847-0	1139
S83848-9	849
S83849-7	848
S83850-0	848
S83851-9	894
S83852-7	893
S83853-5	893
S83854-3	893
S83858-6	909
S83859-4	213
S83860-8	1114
S83861-6	1126
S83862-4	989
S83863-2	983
S83864-0	988
S83866-7	981
S83867-5	981
S83869-1	410
S83870-5	1139
S83871-3	1135
S83872-1	894
S83874-8	893
S83875-6	364
S83876-4	1130
S83877-2	1135
S83878-0	1107
S83879-9	1136
S83880-2	172
S83881-0	198
S83882-9	1115
S83883-7	1103
S83884-5	1100
S83885-3	1121
S83886-1	1121
S83888-8	74
S83889-6	293
S83891-8	298
S83892-6	292
S83893-4	298
S83894-2	289
S83895-0	380
S83896-9	297
S83897-7	291
S83899-3	29
S83900-0	1
S83901-9	10
S83902-7	346
S83903-5	896
S83904-3	854
S83905-1	1139
S83907-8	1134
S83908-6	1131
S83909-4	1130
S83910-8	1132
S83911-6	1132
S83912-4	746
S83913-2	894
S83914-0	910
S83915-9	916
S83916-7	848
S83918-3	187
S83919-1	346
S83920-5	207
S83921-3	226
S83922-1	260
S83924-8	191
S83925-6	896
S83926-4	912
S83927-2	894
S83929-9	1142
S83930-2	1137
S83931-0	1138
S83932-9	297
S83933-7	1137
S83934-5	376
S83935-3	724
S83936-1	276
S83938-8	173
S83939-6	847
S83941-8	853
S83942-6	322
S83943-4	336
S83944-2	853
S83945-0	846
S83946-9	1139
S83947-7	1107
S83948-5	1135
S83949-3	1135
S83950-7	69
S83952-3	253
S83953-1	1125
S83955-8	50
S83956-6	51
S83957-4	50
S83958-2	50
S83959-0	51
S83960-4	981
S83961-2	50
S83962-0	952
S83963-9	1096
S83964-7	1024
S83965-5	689
S83967-1	1077
S83969-8	1096
S83970-1	1097
S83972-8	1119
S83974-4	848
S83975-2	993
S83976-0	847
S83977-9	1110
S83978-7	714
S83979-5	671
S83980-9	1141
S83981-7	896
S83982-5	991
S83983-3	896
S83984-1	1112
S83986-8	197
S83987-6	1136
S83988-4	1132
S83989-2	1109
S83991-4	200
S83992-2	989
S83993-0	304
S83994-9	1006
S83995-7	297
S83996-5	1100
S83997-3	777
S83998-1	345
S84000-9	849
S84001-7	1141
S84002-5	289
S84003-3	848
S84007-6	574
S84008-4	98
S84010-6	987
S84011-4	339
S84012-2	845
S84013-0	954
S84016-5	190
S84018-1	177
S84020-3	822
S84021-1	823
S84023-8	987
S84024-6	822
S84026-2	1120
S84027-0	1130
S84028-9	842
S84029-7	780
S84030-0	1102
S84031-9	1103
S84032-7	773
S84033-5	780
S84034-3	914
S84035-1	1130
S84037-8	994
S84038-6	1116
S84039-4	1130
S84041-6	1118

Product	Page
S84042-4	1118
S84043-2	1113
S84044-0	1138
S84045-9	848
S84047-5	896
S84048-3	184
S84049-1	323
S84050-5	384
S84052-1	894
S84054-8	385
S84056-4	424
S84057-2	1138
S84058-0	842
S84059-9	988
S84060-2	988
S84063-7	252
S84065-3	304
S84068-8	896
S84069-6	896
S84071-8	983
S84072-6	158
S84073-4	917
S84074-2	894
S84076-9	990
S84077-7	986
S84078-5	981
S84081-5	936
S84082-3	289
S84083-1	831
S84086-6	297
S84087-4	293
S84088-2	74
S84089-0	350
S84090-4	26
S84091-2	26
S84092-0	152
S84093-9	24
S84094-7	26
S84095-5	1105
S84096-3	25
S84097-1	154
S84099-8	26
S84100-5	998
S84101-3	994
S84102-1	489
S84104-8	327
S84105-6	484
S84106-4	995
S84107-2	232
S84108-0	791
S84109-9	337
S84110-2	827
S84111-0	425
S84112-9	786
S84114-5	1063
S84115-3	326
S84116-1	898
S84118-8	307
S84119-6	250
S84121-8	1135
S84122-6	895
S84123-4	1100
S84124-2	185
S84125-0	185
S84126-9	193
S84127-7	896
S84128-5	896
S84129-3	914
S84130-7	997
S84131-5	830
S84133-1	199
S84135-8	199
S84136-6	876
S84137-4	217
S84138-2	810
S84139-0	205
S84140-4	111
S84142-0	719
S84143-9	36
S84200-1	985
S84202-8	709
S84203-6	379
S84204-4	378
S84205-2	380
S84206-0	203
S84207-9	297
S84209-5	166
S84211-7	164
S84212-5	174
S84213-3	166
S84214-1	165
S84216-8	183
S84217-6	380
S84219-2	300
S84221-4	199
S84223-0	199
S84224-9	360
S84226-5	436
S84227-3	758
S84228-1	331
S84230-3	199
S84231-1	207
S84233-8	290
S84234-6	297
S84235-4	285
S84236-2	846
S84237-0	163
S84238-9	166
S84239-7	318
S84240-0	287
S84241-9	162
S84242-7	167
S84243-5	318
S84244-3	318
S84245-1	316
S84247-8	333
S84248-6	333
S84250-8	273
S84251-6	274
S84253-2	337
S84254-0	280
S84255-9	310
S84256-7	168
S84257-5	280
S84258-3	832
S84259-1	738
S84260-5	375
S84262-1	1128
S84264-8	171
S84265-6	231
S84266-4	20
S84267-2	209
S84269-9	380
S84270-2	190
S84271-0	975
S84272-9	177
S84273-7	112
S84274-5	261
S84275-3	193
S84278-8	698
S84279-6	186
S84281-8	271
S84282-6	978
S84284-2	339
S84285-0	185
S84286-9	367
S84287-7	187
S84288-5	193
S84289-3	262
S84290-7	187
S84291-5	384
S84292-3	167
S84293-1	184
S84295-8	174
S84296-6	187
S84298-2	764
S84300-8	949
S84302-4	322
S84303-2	697
S84304-0	184
S84305-9	973
S84306-7	972
S84307-5	382
S84314-8	377
S84315-6	332
S84316-4	766
S84317-2	320
S84318-0	758
S84320-2	377
S84323-7	437
S84324-5	378
S84325-3	179
S84326-1	1092
S84329-6	307
S84332-6	414
S84333-4	307
S84335-0	194
S84336-9	752
S84339-3	11
S84340-7	95
S84341-5	265
S84342-3	766
S84343-1	757
S84345-8	321
S84346-6	321
S84347-4	50
S84348-2	55
S84349-0	53
S84350-4	52
S84352-0	50
S84353-9	1110
S84354-7	1109
S84356-3	53
S84363-6	859
S84365-2	858
S84366-0	975
S84367-9	969
S84368-7	67
S84370-9	1005
S84371-7	976
S84372-5	917
S84373-3	908
S84374-1	908
S84376-8	13
S84377-6	184
S84378-4	321
S84379-2	336
S84380-6	260
S84382-2	857
S84384-9	971
S84385-7	976
S84387-3	54
S84388-1	54
S84390-3	858
S84393-8	321
S84394-6	322
S84396-2	260
S84397-0	177
S84398-9	54
S84399-7	367
S84400-4	395
S84401-2	112
S84402-0	875
S84403-9	408
S84404-7	394
S84405-5	409
S84406-3	641
S84407-1	838
S84409-8	876
S84410-1	437
S84412-8	681
S84413-6	844
S84414-4	844
S84415-2	844
S84416-0	426
S84417-9	437
S84418-7	694
S84419-5	543
S84420-9	965
S84421-7	625
S84422-5	930
S84423-3	626
S84424-1	625
S84426-8	625
S84427-6	625
S84428-4	525
S84429-2	524
S84430-6	526
S84431-4	524
S84432-2	513
S84433-0	543
S84434-9	543
S84435-7	513
S84436-5	524
S84437-3	524
S84438-1	588
S84440-3	534
S84441-1	732
S84443-8	611
S84444-6	524
S84445-4	611
S84446-2	877
S84447-0	928
S84448-9	807
S84449-7	808
S84450-0	744
S84451-9	695
S84452-7	730
S84453-5	506
S84454-3	504
S84455-1	451
S84457-8	451
S84458-6	731
S84459-4	731
S84460-8	524
S84461-6	625
S84462-4	618
S84463-2	627
S84464-0	626
S84465-9	627
S84466-7	627
S84467-5	626
S84468-3	526
S84469-1	458
S84470-5	635
S84471-3	635
S84472-1	617
S84474-8	518
S84475-6	528
S84476-4	520
S84477-2	611
S84478-0	517
S84479-9	521
S84480-2	529
S84481-0	532
S84482-9	526
S84483-7	517
S84484-5	517
S84485-3	517
S84486-1	518
S84488-8	976
S84489-6	809
S84491-8	732
S84492-6	527
S84493-4	532
S84494-2	846
S84495-0	748
S84496-9	696
S84497-7	696
S84498-5	695
S84499-3	678
S84500-0	699
S84501-9	699
S84502-7	653
S84503-5	705
S84504-3	705
S84505-1	455
S84507-8	608
S84508-6	608
S84509-4	615
S84510-8	1026
S84511-6	625
S84512-4	622
S84513-2	625
S84514-0	744
S84515-9	626
S84516-7	626
S84517-5	625
S84518-3	626
S84519-1	625
S84520-5	615
S84521-3	638
S84522-1	626
S84524-8	509
S84525-6	681
S84526-4	641
S84527-2	635
S84528-0	615
S84529-9	523
S84530-2	525
S84531-0	517
S84532-9	516
S84533-7	544
S84534-5	520
S84535-3	515
S84536-1	965
S84539-6	744
S84541-8	513
S84542-6	513
S84543-4	712
S84544-2	712
S84545-0	712
S84546-9	625
S84547-7	729
S84548-5	524
S84549-3	611
S84550-7	611
S84551-5	734
S84552-3	735
S84553-1	523
S84555-8	732
S84556-6	737
S84557-4	735
S84558-2	636
S84559-0	638
S84560-4	524
S84561-2	547
S84562-0	547
S84563-9	528
S84564-7	546
S84565-5	688
S84566-3	668
S84567-1	651
S84569-8	577
S84570-1	535
S84572-8	629
S84573-6	629
S84574-4	629
S84575-2	629
S84576-0	546
S84577-9	744
S84578-7	743
S84579-5	736
S84580-9	654
S84581-7	736
S84582-5	563
S84583-3	639
S84584-1	637
S84586-8	641
S84587-6	677
S84588-4	676
S84589-2	528
S84590-6	589
S84591-4	549
S84592-2	567
S84593-0	567
S84594-9	567
S84595-7	976
S84596-5	734
S84597-3	675
S84598-1	737
S84600-7	66
S84601-5	87
S84602-3	62
S84603-1	331
S84605-8	332
S84606-6	358
S84607-4	692
S84608-2	323
S84609-0	331
S84610-4	692
S84611-2	332
S84612-0	436
S84613-9	165
S84614-7	295
S84615-5	471
S84616-3	471
S84620-1	359
S84622-8	716
S84623-6	255
S84624-4	331
S84625-2	294
S84626-0	294
S84627-9	722
S84628-7	66
S84629-5	304
S84630-9	92
S84634-1	717
S84636-8	1011
S84637-6	177
S84638-4	177
S84639-2	53
S84640-6	193
S84641-4	193
S84642-2	331
S84644-9	379
S84645-7	316
S84646-5	193
S84647-3	53
S84648-1	162
S84650-3	54
S84651-1	54
S84653-8	178
S84654-6	378
S84655-4	414
S84656-2	178
S84657-0	209
S84658-9	304
S84659-7	207
S84660-0	178
S84661-9	197
S84662-7	304
S84663-5	377
S84664-3	265
S84665-1	975
S84667-8	975
S84668-6	289
S84669-4	224
S84670-8	195
S84671-6	1010
S84672-4	278
S84673-2	378
S84674-0	177
S84675-9	194
S84676-7	178
S84677-5	165
S84678-3	168
S84679-1	846
S84680-5	177
S84681-3	176
S84682-1	179
S84684-8	54
S84685-6	316
S84686-4	193
S84688-0	314
S84689-9	1086
S84690-2	388
S84692-9	378
S84693-7	380
S84694-5	378
S84695-3	380
S84696-1	179
S84698-8	171
S84699-6	798
S84700-3	178
S84701-1	866
S84703-8	279
S84704-6	281
S84705-4	194
S84706-2	177
S84707-0	340
S84708-9	604
S84709-7	638
S84710-0	708
S84711-9	335
S84712-7	857
S84713-5	280
S84714-3	386
S84715-1	383
S84717-8	971
S84718-6	308
S84719-4	177
S84720-8	191
S84722-4	94

Product	Page
S84723-2	92
S84724-0	234
S84725-9	234
S84726-7	1080
S84727-5	572
S84728-3	535
S84729-1	577
S84730-5	569
S84731-3	573
S84732-1	562
S84734-8	601
S84735-6	686
S84736-4	685
S84737-2	651
S84738-0	676
S84739-9	676
S84740-2	632
S84741-0	659
S84742-9	1088
S84743-7	642
S84744-5	685
S84745-3	685
S84746-1	626
S84748-8	612
S84749-6	679
S84751-8	737
S84752-6	512
S84753-4	641
S84754-2	736
S84755-0	732
S84756-9	736
S84757-7	736
S84758-5	728
S84759-3	740
S84760-7	744
S84761-5	740
S84762-3	744
S84763-1	740
S84765-8	524
S84766-6	928
S84767-4	929
S84768-2	728
S84769-0	740
S84770-4	737
S84771-2	529
S84772-0	925
S84773-9	509
S84774-7	505
S84775-5	529
S84776-3	518
S84777-1	518
S84779-8	509
S84780-1	507
S84782-8	509
S84783-6	507
S84784-4	529
S84785-2	509
S84786-0	508
S84787-9	529
S84788-7	516
S84789-5	513
S84790-9	547
S84791-7	547
S84792-5	547
S84793-3	949
S84794-1	523
S84796-8	523
S84797-6	562
S84798-4	564
S84799-2	564
S84801-8	1081
S84802-6	612
S84803-4	676
S84805-0	688
S84806-9	563
S84807-7	620
S84808-5	1081
S84809-3	676
S84810-7	563
S84811-5	676
S84812-3	676
S84813-1	637
S84815-8	634
S84816-6	847
S84817-4	524
S84818-2	622
S84819-0	644
S84820-4	629
S84821-2	614
S84822-0	614
S84823-9	631
S84824-7	614
S84825-5	614
S84826-3	669
S84827-1	652
S84829-8	563
S84830-1	669
S84832-8	516
S84833-6	549
S84834-4	623
S84835-2	649
S84836-0	648
S84837-9	550
S84838-7	549
S84839-5	562
S84840-9	547
S84841-7	647
S84842-5	647
S84843-3	639
S84844-1	648
S84846-8	562
S84847-6	666
S84848-4	563
S84849-2	669
S84850-6	669
S84851-4	669
S84852-2	550
S84853-0	648
S84854-9	669
S84855-7	648
S84856-5	632
S84857-3	562
S84858-1	648
S84860-3	669
S84861-1	614
S84863-8	550
S84864-6	646
S84865-4	646
S84866-2	515
S84867-0	515
S84868-9	546
S84869-7	546
S84870-0	1126
S84871-9	550
S84872-7	648
S84873-5	516
S84874-3	514
S84875-1	647
S84877-8	550
S84878-6	550
S84879-4	648
S84880-8	647
S84881-6	46
S84882-4	46
S84883-2	46
S84884-0	45
S84885-9	73
S84886-7	45
S84887-5	45
S84888-3	45
S84889-1	46
S84890-5	235
S84891-3	295
S84892-1	1008
S84894-8	234
S84895-6	784
S84896-4	75
S84898-0	134
S84899-9	502
S84900-6	1009
S84901-4	46
S84902-2	45
S84903-0	112
S84904-9	112
S84905-7	1006
S84906-5	45
S84907-3	258
S84910-3	988
S84911-1	47
S84914-6	100
S84915-4	235
S84916-2	112
S84917-0	492
S84918-9	136
S84919-7	999
S84920-0	112
S84921-9	112
S84922-7	112
S84923-5	501
S84924-3	503
S84925-1	46
S84927-8	945
S84928-6	945
S84931-6	765
S84932-4	111
S84934-0	258
S84935-9	258
S84937-5	1054
S84938-3	1054
S84939-1	909
S84940-5	46
S84941-3	45
S84942-1	46
S84945-6	45
S84946-4	158
S84947-2	109
S84948-0	288
S84949-9	386
S84950-2	760
S84951-0	304
S84952-9	388
S84953-7	300
S84954-5	752
S84955-3	210
S84956-1	330
S84958-8	330
S84959-6	210
S84961-8	104
S84962-6	313
S84963-4	313
S84964-2	386
S84965-0	344
S84967-7	344
S84968-5	344
S84969-3	201
S84970-7	344
S84971-5	429
S84972-3	377
S84973-1	470
S84975-8	374
S84976-6	465
S84977-4	203
S84978-2	895
S84979-0	169
S84981-2	315
S84982-0	1142
S84983-9	215
S84984-7	169
S84985-5	213
S84986-3	367
S84987-1	291
S84989-8	866
S84990-1	241
S84992-8	410
S84993-6	339
S84994-4	247
S84995-2	291
S84996-0	300
S84997-9	251
S84998-7	382
S84999-5	432
S85000-4	323
S85001-2	452
S85002-0	1139
S85003-9	270
S85004-7	322
S85005-5	322
S85006-3	793
S85009-8	418
S85010-1	279
S85012-8	887
S85013-6	469
S85014-4	456
S85015-2	180
S85016-0	334
S85017-9	355
S85018-7	332
S85019-5	282
S85020-9	282
S85021-7	1011
S85022-5	998
S85023-3	184
S85024-1	184
S85026-8	502
S85027-6	289
S85028-4	179
S85029-2	177
S85030-6	286
S85031-4	185
S85033-0	366
S85034-9	292
S85035-7	318
S85036-5	377
S85037-3	194
S85038-1	184
S85040-3	51
S85041-1	52
S85043-8	317
S85044-6	179
S85045-4	180
S85046-2	179
S85047-0	648
S85048-9	752
S85049-7	648
S85050-0	450
S85051-9	930
S85052-7	522
S85054-3	621
S85055-1	630
S85057-8	530
S85058-6	630
S85059-4	513
S85060-8	515
S85061-6	514
S85062-4	532
S85063-2	631
S85064-0	615
S85065-9	512
S85066-7	621
S85067-5	637
S85068-3	638
S85069-1	631
S85070-5	620
S85071-3	607
S85072-1	621
S85074-8	621
S85075-6	620
S85076-4	620
S85077-2	617
S85078-0	617
S85079-9	513
S85080-2	620
S85081-0	617
S85082-9	621
S85083-7	620
S85084-5	612
S85085-3	617
S85086-1	618
S85088-8	623
S85089-6	622
S85091-8	630
S85092-6	619
S85093-4	512
S85094-2	618
S85095-0	612
S85096-9	614
S85097-7	631
S85098-5	622
S85099-3	618
S85100-0	843
S85101-9	631
S85102-7	646
S85103-5	647
S85105-1	647
S85107-8	646
S85108-6	630
S85109-4	637
S85110-8	648
S85111-6	648
S85112-4	546
S85113-2	646
S85114-0	648
S85115-9	648
S85116-7	647
S85117-5	936
S85118-3	549
S85119-1	647
S85120-5	549
S85121-3	615
S85122-1	812
S85124-8	546
S85125-6	515
S85126-4	621
S85127-2	642
S85128-0	641
S85129-9	648
S85130-2	611
S85131-0	634
S85132-9	631
S85133-7	632
S85134-5	743
S85135-3	844
S85136-1	728
S85138-8	535
S85139-6	514
S85141-8	642
S85142-6	514
S85143-4	513
S85144-2	531
S85145-0	514
S85146-9	513
S85147-7	641
S85148-5	531
S85149-3	1126
S85150-7	513
S85151-5	875
S85153-1	728
S85155-8	523
S85156-6	651
S85157-4	646
S85158-2	668
S85159-0	668
S85160-4	665
S85161-2	624
S85162-0	649
S85163-9	988
S85164-7	668
S85165-5	624
S85166-3	624
S85167-1	624
S85169-8	666
S85170-1	624
S85172-8	666
S85173-6	526
S85174-4	624
S85175-2	399
S85176-0	624
S85177-9	624
S85178-7	624
S85179-5	624
S85180-9	624
S85181-7	649
S85182-5	551
S85183-3	665
S85184-1	648
S85186-8	665
S85187-6	517
S85188-4	665
S85189-2	651
S85190-6	625
S85191-4	847
S85192-2	155
S85193-0	215
S85194-9	18
S85195-7	77
S85196-5	1137
S85197-3	800
S85198-1	73
S85200-7	172
S85201-5	282
S85202-3	708
S85203-1	923
S85205-8	1080
S85206-6	376
S85207-4	178
S85209-0	162
S85210-4	171
S85211-2	187
S85213-9	331
S85215-5	437
S85217-1	206
S85219-8	782
S85222-8	722
S85223-6	98
S85224-4	98
S85225-2	65
S85226-0	81
S85228-7	257
S85232-5	208
S85234-1	362
S85236-8	337
S85238-4	343
S85239-2	87
S85240-6	287
S85241-4	898
S85242-2	6
S85243-0	66
S85244-9	89
S85245-7	178
S85246-5	376
S85247-3	477
S85250-3	330
S85251-1	274
S85253-8	7
S85254-6	175
S85255-4	336
S85256-2	174
S85258-9	286
S85259-7	211
S85260-0	360
S85261-9	328
S85262-7	206
S85263-5	206
S85264-3	361
S85265-1	172
S85267-8	173
S85268-6	162
S85269-4	162
S85270-8	800
S85271-6	799
S85272-4	799
S85273-2	799
S85274-0	799
S85275-9	798
S85276-7	362
S85278-3	357
S85280-5	155
S85281-3	376
S85282-1	375
S85284-8	345
S85285-6	148
S85286-4	171
S85287-2	503
S85288-0	361
S85289-9	162
S85290-2	361
S85291-0	1132
S85292-9	366
S85293-7	259
S85294-5	100
S85295-3	334
S85296-1	213
S85299-6	1071
S85300-3	1006
S85301-1	354
S85304-6	84
S85305-4	723
S85306-2	186
S85307-0	180
S85308-9	195
S85309-7	174
S85310-0	376
S85311-9	241
S85312-7	206

Product	Page
S85313-5	206
S85314-3	164
S85315-1	243
S85317-8	206
S85318-6	172
S85319-4	361
S85320-8	799
S85321-6	798
S85322-4	213
S85324-0	668
S85325-9	547
S85326-7	944
S85327-5	562
S85328-3	669
S85329-1	668
S85330-5	606
S85331-3	561
S85332-1	562
S85334-8	547
S85335-6	561
S85336-4	562
S85337-2	562
S85338-0	876
S85339-9	508
S85340-2	628
S85341-0	635
S85342-9	615
S85343-7	613
S85344-5	628
S85345-3	850
S85346-1	623
S85348-8	701
S85349-6	636
S85351-8	615
S85352-6	561
S85353-4	623
S85354-2	615
S85355-0	614
S85356-9	648
S85357-7	647
S85358-5	647
S85359-3	615
S85360-7	667
S85361-5	665
S85362-3	636
S85363-1	615
S85365-8	549
S85366-6	613
S85367-4	562
S85368-2	649
S85369-0	615
S85370-4	563
S85371-2	622
S85372-0	666
S85373-9	622
S85374-7	615
S85375-5	615
S85376-3	516
S85377-1	515
S85379-8	550
S85380-1	647
S85382-8	563
S85383-6	647
S85385-2	612
S85386-0	614
S85387-9	612
S85388-7	514
S85389-5	514
S85390-9	613
S85391-7	613
S85392-5	613
S85393-3	612
S85394-1	612
S85396-8	613
S85397-6	514
S85398-4	613
S85399-2	514
S85401-8	515
S85402-6	613
S85403-4	870
S85404-2	843
S85405-0	612
S85406-9	694
S85407-7	613
S85408-5	613
S85409-3	694
S85410-7	613
S85411-5	613
S85412-3	614
S85413-1	613
S85415-8	694
S85416-6	698
S85417-4	514
S85418-2	613
S85419-0	514
S85420-4	514
S85421-2	512
S85422-0	514
S85423-9	614
S85424-7	613
S85425-5	516
S85426-3	613
S85427-1	613
S85429-8	694
S85430-1	512
S85432-8	518
S85433-6	549
S85434-4	546
S85435-2	548
S85436-0	627
S85437-9	619
S85438-7	620
S85439-5	611
S85440-9	620
S85441-7	620
S85442-5	646
S85443-3	617
S85444-1	614
S85446-8	620
S85447-6	647
S85448-4	647
S85449-2	1068
S85450-6	1041
S85451-4	1036
S85452-2	1069
S85453-0	396
S85454-9	516
S85455-7	577
S85456-5	554
S85457-3	554
S85458-1	549
S85460-3	402
S85461-1	397
S85463-8	486
S85464-6	414
S85465-4	261
S85466-2	277
S85467-0	567
S85468-9	448
S85469-7	449
S85470-0	735
S85471-9	604
S85472-7	1040
S85473-5	484
S85474-3	484
S85475-1	735
S85477-8	482
S85478-6	488
S85479-4	486
S85480-8	1088
S85481-6	391
S85482-4	391
S85483-2	399
S85484-0	447
S85485-9	434
S85486-7	462
S85487-5	453
S85488-3	455
S85489-1	463
S85490-5	452
S85491-3	470
S85492-1	1091
S85494-8	391
S85495-6	1081
S85496-4	1082
S85497-2	666
S85498-0	452
S85499-9	448
S85500-6	448
S85501-4	434
S85502-2	447
S85503-0	417
S85504-9	428
S85505-7	675
S85506-5	412
S85507-3	675
S85508-1	951
S85510-3	576
S85511-1	960
S85513-8	675
S85514-6	569
S85515-4	748
S85516-2	556
S85517-0	550
S85518-9	938
S85519-7	623
S85520-0	629
S85521-9	526
S85523-5	666
S85524-3	654
S85525-1	654
S85527-8	675
S85528-6	666
S85529-4	616
S85530-8	613
S85531-6	620
S85532-4	628
S85533-2	926
S85534-0	469
S85535-9	668
S85536-7	624
S85537-5	615
S85538-3	649
S85539-1	668
S85540-5	823
S85541-3	812
S85542-1	653
S85544-8	654
S85545-6	618
S85546-4	651
S85547-2	651
S85548-0	647
S85549-9	647
S85550-2	654
S85551-0	632
S85552-9	668
S85553-7	627
S85554-5	620
S85555-3	653
S85556-1	715
S85558-8	405
S85559-6	405
S85561-8	416
S85562-6	427
S85563-4	416
S85564-2	416
S85565-0	433
S85566-9	455
S85567-7	433
S85568-5	445
S85569-3	455
S85570-7	449
S85571-5	405
S85572-3	405
S85573-1	610
S85575-8	454
S85576-6	650
S85577-4	397
S85578-2	421
S85579-0	402
S85580-4	678
S85581-2	680
S85582-0	433
S85583-9	635
S85584-7	619
S85585-5	652
S85586-3	426
S85587-1	655
S85589-8	653
S85590-1	687
S85592-8	532
S85593-6	745
S85594-4	493
S85595-2	498
S85596-0	492
S85597-9	631
S85598-7	677
S85599-5	752
S85601-0	534
S85602-9	533
S85603-7	585
S85604-5	534
S85605-3	533
S85606-1	533
S85608-8	533
S85609-6	533
S85611-8	533
S85612-6	533
S85613-4	533
S85614-2	534
S85615-0	534
S85616-9	533
S85617-7	585
S85619-3	809
S85620-7	809
S85621-5	630
S85622-3	585
S85623-1	630
S85625-8	825
S85626-6	825
S85627-4	808
S85628-2	809
S85629-0	579
S85630-4	579
S85631-2	533
S85632-0	533
S85633-9	533
S85634-7	533
S85635-5	630
S85636-3	630
S85637-1	677
S85639-8	630
S85640-1	534
S85642-8	533
S85643-6	534
S85644-4	533
S85645-2	533
S85646-0	533
S85647-9	543
S85648-7	533
S85649-5	579
S85650-9	579
S85651-7	533
S85652-5	758
S85653-3	752
S85654-1	630
S85657-6	610
S85658-4	610
S85659-2	114
S85660-6	610
S85661-4	113
S85662-2	608
S85663-0	113
S85664-9	113
S85665-7	610
S85666-5	114
S85667-3	84
S85668-1	606
S85670-3	113
S85671-1	113
S85673-8	113
S85674-6	113
S85675-4	608
S85676-2	606
S85677-0	113
S85678-9	610
S85679-7	114
S85680-0	113
S85681-9	113
S85682-7	113
S85683-5	114
S85684-3	114
S85685-1	113
S85687-8	608
S85688-6	609
S85689-4	113
S85690-8	112
S85691-6	606
S85692-4	606
S85693-2	609
S85694-0	606
S85695-9	113
S85696-7	606
S85697-5	113
S85698-3	113
S85699-1	608
S85700-9	114
S85701-7	112
S85702-5	113
S85703-3	114
S85704-1	113
S85706-8	113
S85707-6	113
S85708-4	606
S85709-2	609
S85710-6	113
S85711-4	610
S85712-2	609
S85713-0	608
S85714-9	606
S85715-7	608
S85716-5	114
S85717-3	114
S85718-1	114
S85720-3	114
S85721-1	114
S85723-8	114
S85724-6	609
S85725-4	610
S85726-2	114
S85727-0	114
S85728-9	114
S85729-7	113
S85730-0	114
S85732-7	609
S85733-5	114
S85734-3	114
S85735-1	114
S85737-8	114
S85738-6	114
S85739-4	113
S85740-8	114
S85741-6	610
S85742-4	610
S85743-2	114
S85744-0	113
S85745-9	113
S85746-7	113
S85747-5	113
S85748-3	114
S85749-1	114
S85750-5	674
S85751-3	674
S85754-8	628
S85755-6	674
S85756-4	674
S85757-2	824
S85758-0	824
S85759-9	824
S85760-2	824
S85761-0	824
S85762-9	808
S85763-7	808
S85764-5	674
S85765-3	674
S85766-1	808
S85768-8	808
S85769-6	808
S85771-8	808
S85772-6	808
S85773-4	808
S85774-2	808
S85775-0	628
S85776-9	808
S85778-5	628
S85779-3	628
S85782-3	628
S85783-1	628
S85785-8	627
S85786-6	628
S85787-4	628
S85788-2	628
S85789-0	628
S85790-4	805
S85791-2	805
S85792-0	805
S85793-9	805
S85795-5	609
S85796-3	609
S85797-1	609
S85799-8	609
S85801-3	808
S85802-1	808
S85804-8	808
S85805-6	808
S85806-4	628
S85807-2	752
S85810-2	752
S85811-0	752
S85812-9	631
S85814-5	752
S85815-3	752
S85816-1	752
S85819-6	752
S85821-8	628
S85822-6	757
S85823-4	757
S85824-2	609
S85825-0	609
S85826-9	609
S85827-7	609
S85828-5	609
S85829-3	609
S85830-7	609
S85831-5	609
S85832-3	609
S85833-1	609
S85835-8	609
S85836-6	609
S85837-4	259
S85838-2	259
S85839-0	259
S85840-4	259
S85841-2	259
S85842-0	396
S85843-9	396
S85844-7	396
S85845-5	396
S85846-3	396
S85847-1	396
S85849-8	396
S85850-1	396
S85852-8	396
S85853-6	397
S85854-4	397
S85855-2	397
S85856-0	396
S85857-9	397
S85858-7	751
S85859-5	751
S85860-9	751
S85861-7	776
S85862-5	777
S85863-3	777
S85864-1	777
S85866-8	777
S85867-6	777
S85869-2	777
S85870-6	777
S85871-4	777
S85872-2	777
S85873-0	818
S85874-9	818
S85875-7	818
S85876-5	818
S85878-1	805
S85880-3	777
S85881-1	778
S85883-8	778
S85884-6	778
S85885-4	777
S85886-2	778
S85887-0	778
S85888-9	778

Product	Page
S85889-7	777
S85890-0	778
S85891-9	778
S85892-7	777
S85893-5	777
S85894-3	777
S85895-1	777
S85897-8	777
S85898-6	777
S85899-4	777
S85900-1	777
S85902-8	259
S85903-6	259
S85904-4	259
S85905-2	259
S85906-0	777
S85907-9	777
S85908-7	714
S85909-5	714
S85910-9	689
S85911-7	974
S85912-5	159
S85913-3	84
S85914-1	1138
S85916-8	970
S85917-6	141
S85918-4	161
S85919-2	833
S85920-6	14
S85921-4	714
S85923-0	71
S85925-7	252
S85926-5	717
S85927-3	115
S85928-1	714
S85930-3	1138
S85931-1	235
S85934-6	134
S85935-4	235
S85936-2	131
S85937-0	834
S85938-9	95
S85940-0	142
S85941-9	142
S85942-7	71
S85943-5	31
S85948-6	835
S85950-8	835
S85951-6	204
S85952-4	422
S85953-2	102
S85954-0	768
S85955-9	135
S85956-7	132
S85957-5	127
S85958-3	30
S85959-1	661
S85961-3	227
S85962-1	679
S85964-8	424
S85965-6	146
S85966-4	376
S85968-0	145
S85969-9	745
S85970-2	136
S85972-9	31
S85973-7	834
S85974-5	235
S85975-3	745
S85976-1	698
S85978-8	836
S85979-6	971
S85981-8	172
S85982-6	204
S85983-4	207
S85984-2	52
S85985-0	308
S85986-9	177
S85987-7	318
S85988-5	968
S85989-3	696
S85991-5	45
S85992-3	134
S85995-8	258
S85996-6	125
S85997-4	125
S85998-2	293
S85999-0	464
S86001-8	96
S86002-6	62
S86003-4	289
S86004-2	231
S86005-0	323
S86006-9	358
S86007-7	213
S86008-5	341
S86009-3	176
S86010-7	97
S86011-5	341
S86012-3	354
S86013-1	235
S86015-8	463
S86016-6	830
S86017-4	374
S86018-2	367
S86019-0	74
S86020-4	200
S86021-2	199
S86022-0	371
S86023-9	371
S86024-7	77
S86025-5	834
S86026-3	834
S86027-1	161
S86029-8	833
S86030-1	84
S86032-8	371
S86033-6	160
S86034-4	379
S86035-2	183
S86036-0	161
S86037-9	84
S86038-7	86
S86039-5	22
S86040-9	21
S86041-7	60
S86042-5	920
S86043-3	59
S86044-1	921
S86046-8	1005
S86047-6	700
S86048-4	697
S86049-2	922
S86051-4	131
S86052-2	420
S86053-0	1039
S86054-9	1105
S86055-7	96
S86056-5	53
S86057-3	336
S86058-1	131
S86060-3	553
S86061-1	379
S86063-8	374
S86064-6	211
S86066-2	366
S86067-0	1016
S86068-9	237
S86069-7	291
S86071-9	45
S86072-7	85
S86073-5	125
S86075-1	257
S86077-8	369
S86078-6	355
S86079-4	325
S86080-8	692
S86081-6	367
S86082-4	960
S86083-2	24
S86084-0	92
S86085-9	771
S86086-7	322
S86087-5	1008
S86088-3	322
S86092-1	383
S86094-8	191
S86095-6	849
S86096-4	53
S86098-0	75
S86100-6	262
S86101-4	262
S86102-2	845
S86103-0	20
S86104-9	95
S86105-7	262
S86106-5	82
S86107-3	96
S86108-1	16
S86110-3	67
S86111-1	57
S86113-8	1006
S86114-6	446
S86116-2	1017
S86117-0	1017
S86118-9	44
S86119-7	293
S86121-9	855
S86122-7	718
S86123-5	264
S86124-3	257
S86125-1	159
S86127-8	441
S86128-6	836
S86129-4	160
S86130-8	834
S86131-6	719
S86132-4	719
S86133-2	447
S86135-9	284
S86136-7	339
S86137-5	355
S86138-3	371
S86139-1	831
S86140-5	96
S86141-3	849
S86142-1	842
S86144-8	849
S86145-6	283
S86146-4	1019
S86147-2	378
S86148-0	246
S86149-9	78
S86150-2	369
S86151-0	377
S86152-9	254
S86153-7	1017
S86154-5	363
S86155-3	369
S86156-1	329
S86158-8	464
S86161-8	982
S86162-6	800
S86163-4	355
S86164-2	191
S86165-0	1017
S86167-7	158
S86168-5	128
S86169-3	159
S86170-7	159
S86171-5	159
S86172-3	159
S86173-1	835
S86175-8	159
S86176-6	158
S86177-4	159
S86178-2	158
S86179-0	835
S86180-4	159
S86181-2	159
S86182-0	159
S86183-9	159
S86184-7	107
S86185-5	129
S86186-3	107
S86187-1	159
S86189-8	159
S86190-1	158
S86192-8	159
S86193-6	159
S86194-4	836
S86195-2	158
S86198-7	1040
S86199-5	697
S86201-0	1053
S86202-9	1078
S86203-7	254
S86204-5	254
S86205-3	256
S86206-1	254
S86213-4	652
S86214-2	501
S86215-0	478
S86216-9	856
S86217-7	770
S86218-5	234
S86219-3	646
S86220-7	234
S86221-5	234
S86222-3	674
S86223-1	1073
S86225-8	993
S86226-6	388
S86227-4	1124
S86228-2	309
S86229-0	795
S86230-4	795
S86231-2	879
S86232-0	805
S86233-9	805
S86234-7	805
S86235-5	756
S86236-3	878
S86237-1	725
S86239-8	331
S86240-1	878
S86242-8	724
S86243-6	878
S86244-4	879
S86245-2	771
S86246-0	776
S86248-7	826
S86249-5	878
S86250-9	795
S86251-7	795
S86252-5	142
S86253-3	878
S86254-1	878
S86256-8	795
S86257-6	142
S86258-4	733
S86259-2	728
S86260-6	878
S86261-4	726
S86262-2	730
S86263-0	878
S86264-9	878
S86265-7	878
S86266-5	741
S86267-3	570
S86268-1	741
S86270-3	828
S86271-1	741
S86273-8	745
S86274-6	726
S86275-4	741
S86276-2	749
S86277-0	809
S86278-9	809
S86279-7	809
S86280-0	809
S86281-9	809
S86282-7	822
S86283-5	820
S86285-1	826
S86287-8	826
S86288-6	826
S86289-4	879
S86290-8	808
S86291-6	879
S86292-4	827
S86293-2	826
S86294-0	826
S86295-9	879
S86296-7	824
S86297-5	827
S86298-3	979
S86299-1	826
S86300-9	826
S86301-7	741
S86302-5	819
S86303-3	826
S86304-1	814
S86306-8	726
S86307-6	827
S86309-2	820
S86310-6	879
S86311-4	878
S86312-2	815
S86313-0	156
S86314-9	29
S86315-7	822
S86316-5	258
S86317-3	727
S86318-1	775
S86320-3	153
S86321-1	785
S86323-8	771
S86324-6	783
S86326-2	795
S86327-0	173
S86328-9	782
S86329-7	153
S86330-0	774
S86331-9	778
S86334-3	784
S86335-1	786
S86337-8	798
S86338-6	747
S86339-4	747
S86340-8	745
S86341-6	1011
S86342-4	134
S86345-9	760
S86346-7	967
S86347-5	381
S86348-3	289
S86349-1	14
S86350-5	173
S86351-3	78
S86354-8	164
S86355-6	1010
S86356-4	211
S86357-2	1071
S86358-0	318
S86359-9	173
S86360-2	289
S86361-0	173
S86362-9	172
S86363-7	168
S86364-5	318
S86365-3	821
S86366-1	188
S86368-8	171
S86369-6	1036
S86372-6	784
S86374-2	12
S86375-0	97
S86376-9	77
S86378-5	173
S86379-3	780
S86380-7	462
S86381-5	175
S86382-3	337
S86383-1	375
S86385-8	173
S86386-6	758
S86387-4	753
S86389-0	799
S86391-2	822
S86392-0	812
S86393-9	671
S86395-5	165
S86396-3	1071
S86397-1	819
S86399-8	291
S86401-3	1137
S86402-1	267
S86404-8	277
S86405-6	337
S86406-4	173
S86407-2	173
S86409-9	206
S86410-2	162
S86411-0	200
S86412-9	173
S86413-7	162
S86414-5	436
S86415-3	180
S86416-1	332
S86418-8	799
S86419-6	463
S86421-8	178
S86422-6	469
S86423-4	201
S86424-2	388
S86425-0	1060
S86426-9	357
S86427-7	332
S86428-5	6
S86429-3	1008
S86430-7	794
S86431-5	41
S86432-3	351
S86433-1	102
S86435-8	22
S86436-6	102
S86437-4	643
S86438-2	800
S86439-0	273
S86440-4	273
S86441-2	175
S86442-0	782
S86443-9	644
S86444-7	644
S86445-5	289
S86446-3	285
S86447-1	298
S86449-8	173
S86450-1	171
S86452-8	798
S86453-6	1091
S86454-4	369
S86455-2	206
S86456-0	187
S86458-7	212
S86460-9	799
S86461-7	799
S86462-5	62
S86463-3	62
S86466-8	134
S86467-6	21
S86469-2	126
S86470-6	134
S86471-4	23
S86472-2	278
S86473-0	1138
S86474-9	923
S86476-5	923
S86477-3	66
S86478-1	6
S86481-1	17
S86483-8	18
S86485-4	65
S86486-2	211
S86487-0	14
S86488-9	14
S86489-7	18
S86490-0	1107
S86491-9	98
S86493-5	186
S86494-3	382
S86495-1	17
S86497-8	52
S86498-6	87
S86499-4	720
S86500-1	211
S86502-8	6
S86503-6	1108
S86504-4	40
S86505-2	1110
S86506-0	174
S86507-9	377
S86508-7	1109

Product Page

S86509-5 329
S86510-9 164
S86511-7 164
S86512-5 273
S86513-3 798
S86514-1 206
S86516-8 1055
S86517-6 1092
S86518-4 1036
S86519-2 796
S86520-6 1137
S86521-4 821
S86522-2 786
S86523-0 172
S86525-7 212
S86526-5 197
S86528-1 782
S86530-3 164
S86531-1 1071
S86533-8 1071
S86534-6 343
S86535-4 371
S86536-2 180
S86537-0 381
S86538-9 375
S86539-7 361
S86540-0 289
S86541-9 212
S86542-7 173
S86543-5 174
S86545-1 799
S86549-4 286
S86551-6 173
S86552-4 173
S86553-2 205
S86554-0 374
S86555-9 183
S86557-5 205
S86558-3 314
S86559-1 241
S86561-3 812
S86562-1 173
S86565-6 286
S86566-4 173
S86567-2 173
S86568-0 326
S86570-2 345
S86571-0 822
S86572-9 1071
S86573-7 329
S86575-3 341
S86576-1 1138
S86578-8 164
S86579-6 173
S86581-8 286
S86582-6 361
S86583-4 361
S86584-2 361
S86585-0 172
S86586-9 503
S86587-7 205
S86588-5 310
S86589-3 1019
S86592-3 382
S86593-1 361
S86595-8 361
S86596-6 1138
S86597-4 1056
S86598-2 371
S86599-0 1138
S86600-8 21
S86601-6 4
S86602-4 25
S86603-2 21
S86604-0 18
S86605-9 21
S86606-7 42
S86607-5 150
S86609-1 52
S86611-3 5
S86612-1 25
S86614-8 24
S86615-6 37
S86616-4 1096
S86619-9 81
S86620-2 81
S86621-0 18
S86624-5 61
S86626-1 357
S86628-8 186
S86629-6 210
S86631-8 1104
S86632-6 1100
S86633-4 1108
S86635-0 37
S86636-9 22
S86638-5 1107
S86639-3 357
S86640-7 169
S86641-5 1100
S86642-3 1110
S86643-1 2
S86645-8 94
S86646-6 77
S86647-4 37
S86648-2 161
S86649-0 1111
S86650-4 73
S86651-2 107
S86652-0 1097
S86653-9 186
S86654-7 55
S86655-5 189
S86656-3 357
S86657-1 169
S86659-8 3
S86660-1 800
S86662-8 18
S86663-6 56
S86664-4 77
S86665-2 22
S86666-0 38
S86667-9 6
S86668-7 6
S86669-5 284
S86670-9 211
S86671-7 1106
S86672-5 1072
S86673-3 18
S86674-1 27
S86676-8 32
S86677-6 56
S86678-4 49
S86679-2 1111
S86680-6 24
S86681-4 28
S86682-2 66
S86683-0 186
S86685-7 32
S86686-5 18
S86687-3 20
S86690-3 49
S86691-1 38
S86693-8 1129
S86694-6 74
S86695-4 1137
S86696-2 201
S86697-0 77
S86698-9 52
S86699-7 1137
S86700-4 800
S86701-2 800
S86702-0 1132
S86703-9 388
S86704-7 835
S86705-5 186
S86706-3 373
S86707-1 799
S86709-8 1132
S86710-1 1138
S86712-8 1130
S86713-6 722
S86714-4 1130
S86715-2 1129
S86717-9 206
S86719-5 1143
S86720-9 637
S86721-7 927
S86722-5 928
S86723-3 678
S86724-1 644
S86726-8 624
S86727-6 624
S86728-4 525
S86729-2 523
S86730-6 523
S86731-4 525
S86732-2 524
S86733-0 512
S86734-9 807
S86735-7 928
S86736-5 809
S86737-3 631
S86738-1 694
S86740-3 694
S86741-1 625
S86743-8 504
S86744-6 731
S86745-4 638
S86746-2 625
S86747-0 624
S86748-9 622
S86749-7 633
S86750-0 635
S86751-9 524
S86752-7 517
S86753-5 518
S86754-3 517
S86755-1 524
S86757-8 525
S86758-6 521
S86759-4 666
S86760-8 513
S86761-6 514
S86762-4 826
S86763-2 814
S86764-0 523
S86765-9 696
S86766-7 629
S86767-5 506
S86768-3 629
S86769-1 506
S86770-5 629
S86771-3 524
S86772-1 524
S86774-8 514
S86775-6 628
S86776-4 611
S86777-2 516
S86778-0 871
S86779-9 617
S86780-2 617
S86781-0 523
S86782-9 526
S86783-7 512
S86784-5 517
S86785-3 517
S86786-1 521
S86788-8 523
S86789-6 666
S86791-8 522
S86792-6 628
S86793-4 706
S86794-2 628
S86795-0 849
S86796-9 630
S86797-7 614
S86798-5 630
S86799-3 622
S86800-0 523
S86801-9 524
S86802-7 631
S86803-5 639
S86804-3 511
S86805-1 651
S86807-8 623
S86808-6 612
S86809-4 560
S86810-8 527
S86811-6 684
S86812-4 526
S86813-2 675
S86814-0 575
S86815-9 640
S86816-7 629
S86817-5 628
S86818-3 629
S86819-1 513
S86820-5 529
S86821-3 674
S86822-1 624
S86824-8 675
S86825-6 694
S86826-4 635
S86827-2 639
S86828-0 639
S86829-9 632
S86830-2 515
S86831-0 571
S86832-9 570
S86833-7 675
S86834-5 567
S86835-3 632
S86836-1 632
S86838-8 731
S86839-6 728
S86841-8 666
S86842-6 514
S86843-4 514
S86844-2 514
S86845-0 529
S86846-9 616
S86847-7 616
S86848-5 630
S86849-3 529
S86850-7 524
S86851-5 548
S86852-3 639
S86853-1 650
S86855-8 651
S86856-6 675
S86857-4 639
S86858-2 548
S86859-0 675
S86860-4 666
S86861-2 548
S86862-0 731
S86863-9 629
S86864-7 704
S86865-5 671
S86866-3 675
S86867-1 632
S86869-8 624
S86870-1 523
S86872-8 525
S86873-6 567
S86874-4 574
S86875-2 547
S86876-0 536
S86877-9 570
S86878-7 621
S86879-5 529
S86880-9 621
S86881-7 621
S86882-5 626
S86883-3 491
S86884-1 496
S86886-8 632
S86887-6 523
S86888-4 567
S86889-2 549
S86890-6 621
S86891-4 621
S86892-2 492
S86893-0 655
S86894-9 495
S86895-7 545
S86896-5 541
S86897-3 561
S86898-1 112
S86900-7 871
S86901-5 493
S86902-3 512
S86903-1 527
S86905-8 666
S86906-6 500
S86907-4 112
S86908-2 516
S86909-0 497
S86910-4 493
S86911-2 493
S86912-0 551
S86913-9 666
S86914-7 666
S86915-5 496
S86916-3 497
S86917-1 497
S86919-8 577
S86920-1 502
S86922-8 501
S86923-6 497
S86924-4 670
S86925-2 497
S86926-0 496
S86927-9 548
S86928-7 628
S86929-5 670
S86930-9 57
S86931-7 1053
S86932-5 1006
S86934-1 869
S86936-8 744
S86937-6 750
S86938-4 84
S86939-2 410
S86940-6 327
S86941-4 170
S86942-2 170
S86943-0 170
S86944-9 764
S86945-7 726
S86946-5 58
S86947-3 745
S86948-1 241
S86950-3 64
S86953-8 995
S86955-4 984
S86956-2 984
S86957-0 743
S86959-7 169
S86960-0 169
S86961-9 59
S86962-7 255
S86963-5 771
S86964-3 257
S86965-1 105
S86968-6 944
S86969-4 461
S86971-6 252
S86972-4 43
S86973-2 43
S86974-0 43
S86975-9 43
S86976-7 129
S86977-5 965
S86978-3 993
S86980-5 984
S86981-3 984
S86982-1 984
S86984-8 744
S86985-6 444
S86988-0 43
S86989-9 170
S86990-2 252
S86991-0 64
S86992-9 71
S86993-7 461
S86994-5 1069
S86995-3 810
S86996-1 821
S86998-8 786
S87000-5 824
S87001-3 444
S87002-1 743
S87004-8 862
S87006-4 725
S87007-2 255
S87008-0 43
S87009-9 911
S87010-2 990
S87011-0 174
S87012-9 174
S87013-7 271
S87014-5 917
S87015-3 996
S87016-1 820
S87018-8 820
S87019-6 818
S87021-8 829
S87022-6 1083
S87023-4 823
S87024-2 820
S87025-0 821
S87026-9 766
S87027-7 961
S87028-5 1090
S87029-3 862
S87030-7 726
S87031-5 924
S87032-3 966
S87033-1 815
S87035-8 819
S87036-6 744
S87037-4 1002
S87038-2 1003
S87040-4 956
S87041-2 863
S87042-0 994
S87043-9 282
S87044-7 239
S87046-3 1090
S87047-1 64
S87049-8 766
S87050-1 1003
S87054-4 821
S87056-0 820
S87057-9 813
S87058-7 820
S87059-5 994
S87060-9 990
S87061-7 988
S87062-5 990
S87066-8 990
S87069-2 740
S87070-6 64
S87073-0 1090
S87074-9 1089
S87075-7 251
S87077-3 575
S87078-1 575
S87080-3 990
S87081-1 1088
S87083-8 251
S87084-6 1054
S87085-4 471
S87086-2 656
S87087-0 1090
S87088-9 1091
S87089-7 821
S87091-9 1019
S87092-7 1019
S87094-3 983
S87100-1 173
S87103-6 220
S87104-4 61
S87105-2 104
S87106-0 565
S87109-5 812
S87110-9 565
S87111-7 832
S87112-5 832
S87113-3 832
S87116-8 832
S87119-2 832
S87120-6 832
S87121-4 832
S87122-2 220
S87124-9 220
S87127-3 1006
S87128-1 586
S87130-3 1006
S87131-1 944
S87133-8 1088
S87140-0 931

Product	Page
S87141-9	949
S87142-7	30
S87143-5	30
S87144-3	142
S87145-1	16
S87147-8	38
S87148-6	125
S87149-4	876
S87150-8	112
S87151-6	28
S87152-4	900
S87153-2	52
S87154-0	54
S87155-9	149
S87156-7	725
S87157-5	804
S87158-3	252
S87159-1	253
S87160-5	59
S87161-3	888
S87162-1	888
S87165-6	1015
S87166-4	125
S87167-2	467
S87168-0	71
S87169-9	42
S87171-0	698
S87172-9	441
S87173-7	754
S87174-5	821
S87175-3	818
S87176-1	915
S87178-8	915
S87182-6	71
S87183-4	59
S87184-2	915
S87186-9	954
S87187-7	913
S87188-5	718
S87190-7	913
S87191-5	954
S87193-1	473
S87195-8	42
S87196-6	954
S87197-4	71
S87198-2	67
S87200-8	59
S87201-6	915
S87202-4	915
S87203-2	954
S87204-0	954
S87205-9	954
S87206-7	59
S87207-5	71
S87208-3	1139
S87210-5	469
S87211-3	915
S87214-8	915
S87216-4	913
S87217-2	39
S87218-0	915
S87219-9	820
S87220-2	879
S87222-9	820
S87223-7	915
S87224-5	502
S87226-1	461
S87231-8	65
S87232-6	247
S87233-4	820
S87234-2	872
S87235-0	714
S87237-7	698
S87238-5	725
S87239-3	71
S87240-7	1088
S87241-5	822
S87242-3	55
S87243-1	59
S87245-8	96
S87246-6	882
S87248-2	51
S87250-4	1071
S87251-2	64
S87252-0	341
S87253-9	341
S87255-5	63
S87257-1	64
S87259-8	820
S87260-1	820
S87263-6	820
S87264-4	820
S87265-2	1071
S87266-0	948
S87268-7	916
S87269-5	820
S87270-9	1061
S87271-7	64
S87272-5	1060
S87273-3	64
S87274-1	957
S87277-6	798
S87278-4	1055
S87279-2	64
S87280-6	64
S87281-4	302
S87282-2	915
S87283-0	255
S87284-9	55
S87287-3	55
S87288-1	64
S87290-3	39
S87291-1	137
S87293-8	1000
S87294-6	389
S87295-4	953
S87296-2	915
S87297-0	408
S87298-9	909
S87299-7	1000
S87300-4	1000
S87301-2	1001
S87302-0	998
S87303-9	1055
S87304-7	139
S87305-5	915
S87306-3	988
S87309-8	451
S87310-1	985
S87312-8	725
S87313-6	987
S87315-2	1053
S87316-0	704
S87317-9	1061
S87318-7	245
S87320-9	1034
S87321-7	499
S87322-5	881
S87324-1	225
S87326-8	1084
S87327-6	312
S87328-4	312
S87329-2	312
S87330-6	990
S87332-2	255
S87333-0	255
S87334-9	769
S87337-3	1085
S87340-3	1002
S87341-1	255
S87343-8	61
S87344-6	61
S87346-2	254
S87347-0	256
S87348-9	255
S87349-7	769
S87350-0	1060
S87351-9	302
S87352-7	39
S87353-5	39
S87357-8	972
S87358-6	39
S87360-8	139
S87362-4	952
S87363-2	958
S87364-0	32
S87365-9	1018
S87367-5	251
S87368-3	255
S87369-1	312
S87370-5	302
S87371-3	236
S87374-8	913
S87375-6	302
S87377-2	223
S87378-0	285
S87379-9	1062
S87380-2	249
S87381-0	225
S87382-9	42
S87383-7	305
S87384-5	368
S87385-3	52
S87388-8	39
S87389-6	957
S87392-6	935
S87393-4	962
S87395-0	935
S87396-9	54
S87397-7	54
S87399-3	69
S87400-0	54
S87403-5	966
S87404-3	966
S87405-1	308
S87408-6	345
S87410-8	957
S87411-6	957
S87412-4	295
S87413-2	966
S87414-0	961
S87415-9	962
S87416-7	213
S87417-5	293
S87418-3	963
S87420-5	145
S87424-8	958
S87425-6	211
S87426-4	978
S87427-2	59
S87428-0	146
S87429-9	57
S87430-2	57
S87431-0	59
S87432-9	59
S87433-7	59
S87434-5	1084
S87435-3	1033
S87436-1	725
S87438-8	57
S87439-6	934
S87441-8	1085
S87442-6	725
S87443-4	1057
S87444-2	725
S87445-0	235
S87446-9	725
S87447-7	1062
S87449-3	725
S87450-7	725
S87451-5	249
S87453-1	61
S87455-8	61
S87457-4	61
S87458-2	61
S87459-0	61
S87460-4	64
S87461-2	64
S87462-0	61
S87463-9	869
S87465-5	765
S87466-3	869
S87472-8	64
S87473-6	54
S87474-4	1088
S87475-2	1088
S87476-0	957
S87479-5	213
S87481-7	64
S87484-1	64
S87486-8	1061
S87487-6	64
S87488-4	988
S87489-2	742
S87491-4	1044
S87492-2	64
S87494-9	966
S87496-5	1053
S87497-3	1061
S87498-1	92
S87500-7	740
S87501-5	769
S87502-3	1057
S87503-1	915
S87505-8	744
S87506-6	744
S87508-2	1091
S87509-0	681
S87510-4	1057
S87511-2	1061
S87512-0	64
S87513-9	738
S87514-7	938
S87515-5	915
S87516-3	769
S87517-1	767
S87519-8	765
S87520-1	92
S87522-8	967
S87523-6	763
S87524-4	356
S87525-2	829
S87527-9	740
S87528-7	733
S87529-5	732
S87530-9	538
S87531-7	534
S87532-5	534
S87533-3	733
S87534-1	732
S87536-8	965
S87537-6	769
S87538-4	769
S87539-2	769
S87540-6	961
S87541-4	146
S87542-2	427
S87544-9	1089
S87545-7	88
S87546-5	65
S87547-3	574
S87550-3	821
S87551-1	746
S87553-8	372
S87554-6	72
S87555-4	140
S87556-2	1137
S87557-0	718
S87561-9	370
S87563-5	1139
S87567-8	1099
S87568-6	1138
S87569-4	6
S87570-8	2
S87571-6	162
S87572-4	174
S87573-2	796
S87574-0	89
S87576-7	18
S87578-3	388
S87579-1	1137
S87580-5	41
S87581-3	90
S87582-1	210
S87584-8	716
S87585-6	1130
S87586-4	186
S87587-2	1096
S87588-0	55
S87589-9	1137
S87591-0	16
S87592-9	5
S87593-7	1125
S87595-3	769
S87596-1	210
S87598-8	210
S87599-6	25
S87600-3	17
S87603-8	1097
S87605-4	22
S87606-2	160
S87607-0	160
S87608-9	6
S87609-7	27
S87610-0	27
S87611-9	498
S87612-7	497
S87613-5	502
S87614-3	498
S87615-1	112
S87617-8	127
S87618-6	961
S87619-4	64
S87620-8	944
S87621-6	931
S87622-4	254
S87623-2	332
S87624-0	169
S87625-9	206
S87626-7	207
S87627-5	362
S87628-3	169
S87629-1	310
S87630-5	310
S87631-3	275
S87632-1	962
S87634-8	278
S87635-6	371
S87636-4	267
S87637-2	820
S87638-0	1074
S87639-9	955
S87640-2	207
S87641-0	308
S87642-9	604
S87643-7	643
S87644-5	94
S87645-3	437
S87646-1	204
S87649-6	282
S87651-8	797
S87652-6	546
S87653-4	1002
S87656-9	12
S87657-7	61
S87658-5	8
S87659-3	13
S87660-7	329
S87661-5	237
S87663-1	14
S87665-8	56
S87666-6	14
S87667-4	14
S87668-2	388
S87669-0	960
S87670-4	824
S87671-2	298
S87672-0	308
S87674-7	800
S87675-5	309
S87676-3	309
S87677-1	309
S87679-8	766
S87680-1	326
S87683-6	200
S87684-4	201
S87685-2	1020
S87687-9	362
S87688-7	371
S87689-5	709
S87690-9	291
S87691-7	331
S87692-5	378
S87693-3	371
S87694-1	371
S87696-8	162
S87697-6	989
S87698-4	359
S87699-2	690
S87701-8	201
S87702-6	380
S87703-4	263
S87704-2	931
S87705-0	372
S87706-9	356
S87707-7	356
S87708-5	372
S87709-3	372
S87710-7	993
S87711-5	993
S87712-3	10
S87713-1	14
S87715-8	342
S87716-6	761
S87717-4	375
S87718-2	361
S87719-0	993
S87720-4	201
S87721-2	330
S87722-0	85
S87723-9	797
S87724-7	940
S87725-5	477
S87726-3	8
S87727-1	84
S87729-8	22
S87730-1	85
S87732-8	949
S87735-2	961
S87736-0	1006
S87737-9	121
S87739-5	362
S87740-9	362
S87741-7	207
S87742-5	389
S87743-3	371
S87744-1	242
S87747-6	297
S87748-4	371
S87749-2	372
S87750-6	237
S87751-4	362
S87752-2	62
S87753-0	16
S87754-9	83
S87755-7	5
S87756-5	216
S87757-3	79
S87758-1	17
S87760-3	408
S87761-1	345
S87763-8	10
S87764-6	1018
S87765-4	365
S87766-2	282
S87767-0	433
S87768-9	164
S87769-7	164
S87770-0	366
S87771-9	364
S87772-7	18
S87773-5	37
S87774-3	38
S87775-1	371
S87777-8	478
S87779-4	37
S87780-8	18
S87781-6	456
S87782-4	324
S87783-2	38
S87784-0	251
S87785-9	278
S87786-7	181
S87787-5	692
S87788-3	169
S87790-5	79
S87792-1	79
S87794-8	40
S87795-6	37
S87796-4	38
S87797-2	375
S87798-0	365
S87799-9	278
S87800-6	450

Product	Page
S87801-4	940
S87802-2	402
S87803-0	402
S87804-9	407
S87805-7	418
S87806-5	1022
S87807-3	432
S87808-1	420
S87810-3	1022
S87811-1	402
S87813-8	1022
S87814-6	1054
S87815-4	315
S87816-2	420
S87817-0	420
S87818-9	420
S87819-7	420
S87820-0	105
S87821-9	390
S87822-7	395
S87823-5	396
S87824-3	715
S87825-1	718
S87827-8	714
S87828-6	720
S87829-4	715
S87830-8	718
S87831-6	720
S87832-4	264
S87833-2	720
S87834-0	720
S87835-9	402
S87836-7	720
S87837-5	759
S87838-3	1054
S87839-1	275
S87840-5	717
S87841-3	719
S87842-1	420
S87844-8	390
S87845-6	391
S87846-4	759
S87847-2	715
S87848-0	940
S87849-9	856
S87850-2	715
S87851-0	764
S87852-9	762
S87853-7	762
S87854-5	720
S87855-3	761
S87856-1	724
S87858-8	764
S87859-6	727
S87861-8	723
S87862-6	764
S87863-4	725
S87864-2	275
S87865-0	224
S87866-9	232
S87867-7	1029
S87868-5	1022
S87869-3	282
S87870-7	285
S87871-5	285
S87872-3	1033
S87873-1	841
S87875-8	404
S87876-6	299
S87877-4	299
S87878-2	857
S87879-0	404
S87880-4	857
S87881-2	1054
S87882-0	1040
S87883-9	283
S87884-7	432
S87885-5	283
S87886-3	283
S87887-1	285
S87889-8	763
S87890-1	285
S87892-8	1040
S87893-6	1053
S87894-4	1009
S87895-2	264
S87896-0	1024
S87897-9	1032
S87898-7	764
S87899-5	764
S87900-2	866
S87901-0	1020
S87902-9	1032
S87903-7	1008
S87904-5	1008
S87905-3	1009
S87906-1	1024
S87908-8	724
S87909-6	408
S87911-8	1020
S87912-6	1021
S87913-4	404
S87914-2	1025
S87915-0	724
S87916-9	1020
S87917-7	1029
S87918-5	1030
S87919-3	724
S87920-7	724
S87921-5	724
S87922-3	724
S87923-1	727
S87925-8	1022
S87926-6	32
S87927-4	693
S87928-2	1022
S87929-0	49
S87930-4	50
S87931-2	279
S87932-0	55
S87935-5	767
S87936-3	759
S87937-1	756
S87939-8	756
S87940-1	752
S87942-8	758
S87943-6	751
S87944-4	759
S87945-2	327
S87946-0	371
S87947-9	371
S87948-7	213
S87949-5	329
S87950-9	753
S87951-7	940
S87952-5	967
S87953-3	344
S87954-1	880
S87957-6	201
S87958-4	114
S87959-2	112
S87960-6	309
S87961-4	1060
S87962-2	324
S87963-0	114
S87964-9	983
S87965-7	1028
S87966-5	241
S87968-1	358
S87970-3	237
S87971-1	289
S87973-8	185
S87974-6	243
S87975-4	881
S87976-2	871
S87977-0	1107
S87978-9	1105
S87979-7	372
S87980-0	1107
S87982-7	885
S87983-5	241
S87984-3	885
S87985-1	725
S87987-8	765
S87988-6	2
S87990-8	252
S87991-6	192
S87992-4	114
S87993-2	41
S87994-0	298
S87995-9	115
S87996-7	107
S87997-5	69
S87999-1	810
S88000-0	180
S88001-9	831
S88002-7	308
S88003-5	308
S88004-3	230
S88005-1	40
S88007-8	365
S88008-6	38
S88009-4	720
S88010-8	165
S88011-6	478
S88012-4	365
S88013-2	278
S88014-0	44
S88015-9	180
S88016-7	474
S88018-3	690
S88019-1	345
S88020-5	707
S88021-3	37
S88022-1	41
S88024-8	191
S88025-6	230
S88026-4	172
S88027-2	40
S88028-0	41
S88029-9	37
S88030-2	1018
S88031-0	5
S88032-9	1018
S88034-5	37
S88035-3	1027
S88036-1	472
S88038-8	177
S88039-6	527
S88041-8	774
S88042-6	41
S88043-4	40
S88044-2	1139
S88045-0	40
S88046-9	457
S88047-7	365
S88048-5	345
S88049-3	80
S88050-7	37
S88052-3	1082
S88053-1	78
S88055-8	321
S88056-6	207
S88057-4	236
S88058-2	207
S88059-0	364
S88060-4	1050
S88061-2	365
S88062-0	365
S88063-9	856
S88064-7	85
S88065-5	1040
S88066-3	1040
S88067-1	364
S88069-8	457
S88070-1	364
S88072-8	367
S88073-6	85
S88074-4	17
S88075-2	277
S88076-0	422
S88077-9	1139
S88078-7	216
S88079-5	335
S88080-9	22
S88082-5	110
S88083-3	345
S88084-1	81
S88086-8	61
S88087-6	80
S88088-4	102
S88089-2	107
S88090-6	1138
S88091-4	1137
S88092-2	106
S88093-0	80
S88094-9	101
S88095-7	60
S88096-5	131
S88097-3	15
S88098-1	831
S88100-7	380
S88101-5	1050
S88102-3	217
S88103-1	191
S88105-8	40
S88106-6	78
S88107-4	206
S88108-2	287
S88109-0	235
S88110-4	321
S88111-2	286
S88112-0	1139
S88113-9	81
S88114-7	1139
S88115-5	1138
S88116-3	11
S88117-1	80
S88119-8	1138
S88120-1	5
S88122-8	87
S88123-6	1137
S88124-4	1015
S88125-2	5
S88126-0	60
S88127-9	144
S88128-7	81
S88129-5	16
S88130-9	1015
S88131-7	62
S88132-5	83
S88133-3	62
S88134-1	1018
S88136-8	255
S88137-6	1139
S88138-4	1039
S88139-2	164
S88140-6	376
S88141-4	271
S88142-2	1139
S88143-0	336
S88144-9	1017
S88145-7	452
S88147-3	79
S88148-1	365
S88150-3	354
S88151-1	44
S88153-8	371
S88154-6	967
S88155-4	15
S88156-2	41
S88157-0	169
S88158-9	164
S88160-0	18
S88161-9	19
S88162-7	79
S88163-5	365
S88164-3	1015
S88165-1	1018
S88167-8	20
S88168-6	79
S88169-4	18
S88170-8	60
S88171-6	1138
S88172-4	78
S88173-2	478
S88174-0	453
S88175-9	1109
S88176-7	890
S88177-5	78
S88178-3	1138
S88179-1	169
S88180-5	1137
S88181-3	17
S88182-1	1138
S88184-8	171
S88185-6	1138
S88186-4	262
S88187-2	1137
S88188-0	63
S88189-9	1136
S88190-2	180
S88191-0	1138
S88192-9	4
S88193-7	325
S88194-5	5
S88195-3	11
S88196-1	353
S88198-8	365
S88199-6	2
S88200-3	1138
S88201-1	206
S88203-8	1139
S88204-6	236
S88205-4	77
S88206-2	720
S88207-0	720
S88208-9	365
S88209-7	287
S88210-0	284
S88211-9	79
S88212-7	1018
S88213-5	78
S88214-3	29
S88215-1	364
S88217-8	44
S88218-6	75
S88219-4	1073
S88220-8	1071
S88221-6	178
S88222-4	278
S88223-2	380
S88224-0	363
S88225-9	375
S88226-7	165
S88227-5	278
S88228-3	179
S88229-1	1039
S88230-5	324
S88231-3	1040
S88232-1	831
S88240-2	759
S88245-3	922
S88251-8	923
S88255-0	76
S88256-9	175
S88259-3	895
S88260-7	52
S88262-3	242
S88263-1	1060
S88265-8	1077
S88266-6	254
S88267-4	165
S88268-2	165
S88270-4	298
S88271-2	254
S88272-0	386
S88273-9	386
S88274-7	334
S88275-5	386
S88276-3	256
S88277-1	388
S88279-8	345
S88280-1	66
S88282-8	1090
S88283-6	1090
S88284-4	1086
S88285-2	1086
S88286-0	1086
S88287-9	1087
S88288-7	1088
S88289-5	1081
S88290-9	649
S88291-7	1107
S88292-5	115
S88293-3	881
S88294-1	694
S88296-8	969
S88298-4	885
S88299-2	183
S88301-8	290
S88302-6	184
S88303-4	961
S88304-2	456
S88305-0	221
S88306-9	36
S88307-7	252
S88308-5	21
S88309-3	990
S88310-7	386
S88311-5	346
S88312-3	230
S88313-1	194
S88315-8	342
S88316-6	291
S88317-4	338
S88318-2	338
S88319-0	755
S88320-4	199
S88321-2	883
S88322-0	148
S88323-9	216
S88324-7	148
S88325-5	187
S88326-3	436
S88327-1	307
S88329-8	301
S88330-1	872
S88332-8	358
S88333-6	335
S88334-4	755
S88336-0	291
S88337-9	338
S88340-9	95
S88341-7	864
S88342-5	197
S88343-3	874
S88344-1	874
S88346-8	62
S88347-6	152
S88348-4	191
S88349-2	204
S88350-6	379
S88351-4	204
S88352-2	357
S88353-0	308
S88354-9	329
S88355-7	285
S88357-3	203
S88358-1	188
S88360-3	783
S88361-1	323
S88363-8	15
S88364-6	176
S88365-4	379
S88366-2	1105
S88368-9	477
S88369-7	693
S88370-0	697
S88371-9	12
S88372-7	1096
S88373-5	173
S88374-3	80
S88375-1	73
S88377-8	159
S88378-6	56
S88379-4	80
S88380-8	907
S88381-6	866
S88382-4	867
S88383-2	867
S88384-0	868
S88385-9	144
S88386-7	354
S88387-5	94
S88389-1	1107
S88390-5	380
S88392-1	1107
S88395-6	187
S88397-2	1130
S88399-9	369
S88401-4	78
S88402-2	441
S88403-0	367

Product Page

S88404-9........106
S88405-7........74
S88406-5........85
S88407-3........83
S88408-1........100
S88410-3........33
S88411-1........80
S88413-8........61
S88414-6........33
S88415-4........389
S88416-2........330
S88417-0........83
S88418-9........106
S88421-9........107
S88422-7........357
S88423-5........363
S88424-3........362
S88425-1........2
S88427-8........1105
S88428-6........363
S88429-4........107
S88430-8........79
S88432-4........354
S88433-2........93
S88434-0........239
S88435-9........1112
S88436-7........202
S88437-5........59
S88438-3........79
S88439-1........144
S88440-5........79
S88441-3........79
S88442-1........77
S88444-8........1015
S88445-6........293
S88446-4........81
S88447-2........81
S88448-0........1019
S88450-2........91
S88451-0........93
S88452-9........21
S88453-7........12
S88454-5........60
S88455-3........478
S88456-1........1099
S88458-8........61
S88459-6........73
S88461-8........13
S88462-6........94
S88463-4........18
S88464-2........61
S88465-0........1017
S88466-9........966
S88467-7........94
S88468-5........92
S88469-3........90
S88470-7........85
S88471-5........106
S88472-3........32
S88473-1........63
S88475-8........83
S88476-6........90
S88477-4........54
S88478-2........148
S88479-0........65
S88480-4........364
S88481-2........482
S88482-0........358
S88483-9........1098
S88484-7........1111
S88485-5........291
S88486-3........80
S88487-1........867
S88489-8........43
S88490-1........1018
S88492-8........90
S88493-6........147
S88494-4........910
S88495-2........1015
S88496-0........55
S88497-9........1009
S88498-7........937
S88499-5........866
S88500-2........154
S88501-0........867
S88502-9........867
S88504-5........864
S88505-3........5
S88506-1........63
S88508-8........867
S88509-6........45
S88513-4........284
S88514-2........180
S88515-0........179
S88516-9........832
S88517-7........307
S88518-5........283
S88519-3........27
S88520-7........22
S88521-5........720
S88522-3........59
S88523-1........56
S88525-8........52
S88526-6........284
S88527-4........12
S88528-2........21
S88529-0........6
S88531-2........870
S88533-9........914
S88534-7........11
S88535-5........61
S88536-3........55
S88539-8........386
S88540-1........283
S88542-8........284
S88543-6........180
S88544-4........360
S88545-2........283
S88546-0........830
S88550-9........355
S88551-7........831
S88552-5........1132
S88553-3........282
S88554-1........283
S88556-8........12
S88557-6........53
S88559-2........127
S88561-4........355
S88562-2........149
S88563-0........180
S88565-7........284
S88566-5........149
S88567-3........284
S88570-3........187
S88571-1........283
S88574-6........194
S88575-4........190
S88576-2........263
S88577-0........322
S88578-9........193
S88579-7........769
S88580-0........185
S88581-9........193
S88582-7........185
S88583-5........380
S88585-1........837
S88587-8........890
S88588-6........874
S88589-4........208
S88590-8........969
S88591-6........971
S88592-4........975
S88593-2........972
S88594-0........198
S88595-9........1084
S88596-7........388
S88597-5........38
S88598-3........48
S88599-1........1020
S88600-9........64
S88601-7........73
S88602-5........45
S88603-3........482
S88604-1........1076
S88606-8........1020
S88607-6........1020
S88609-2........1068
S88610-6........1068
S88611-4........1025
S88612-2........234
S88613-0........1073
S88614-9........234
S88615-7........234
S88616-5........324
S88617-3........277
S88618-1........258
S88620-3........235
S88621-1........765
S88623-8........285
S88624-6........502
S88625-4........234
S88626-2........235
S88627-0........235
S88628-9........235
S88629-7........235
S88630-0........148
S88631-9........148
S88632-7........758
S88633-5........1019
S88634-3........150
S88635-1........756
S88637-8........150
S88638-6........236
S88639-4........236
S88640-8........855
S88641-6........1077
S88642-4........850
S88643-2........756
S88644-0........751
S88645-9........751
S88647-5........756
S88648-3........287
S88649-1........751
S88650-5........755
S88651-3........751
S88652-1........165
S88654-8........757
S88655-6........756
S88656-4........751
S88657-2........753
S88658-0........757
S88659-9........756
S88660-2........753
S88661-0........751
S88662-9........757
S88663-7........756
S88664-5........751
S88665-3........753
S88666-1........757
S88668-8........757
S88669-6........751
S88671-8........163
S88672-6........757
S88673-4........751
S88674-2........297
S88675-0........1077
S88676-9........297
S88678-5........289
S88679-3........261
S88680-7........216
S88681-5........1002
S88682-3........373
S88683-1........12
S88685-8........356
S88686-6........405
S88687-4........1051
S88689-0........285
S88690-4........282
S88691-2........311
S88692-0........987
S88693-9........329
S88694-7........989
S88695-5........983
S88696-3........288
S88697-1........968
S88699-8........784
S88700-5........327
S88701-3........308
S88702-1........340
S88704-8........235
S88705-6........1004
S88706-4........1004
S88707-2........753
S88708-0........308
S88709-9........209
S88710-2........230
S88711-0........179
S88712-9........432
S88713-7........150
S88714-5........757
S88715-3........757
S88716-1........757
S88718-8........756
S88719-6........844
S88721-8........855
S88722-6........856
S88723-4........846
S88724-2........751
S88725-0........759
S88726-9........756
S88727-7........758
S88728-5........751
S88729-3........757
S88730-7........854
S88731-5........759
S88732-3........762
S88733-1........756
S88735-8........756
S88736-6........756
S88738-2........1128
S88739-0........43
S88740-4........1128
S88742-0........344
S88743-9........888
S88744-7........1140
S88745-5........1139
S88746-3........1140
S88747-1........151
S88749-8........1095
S88750-1........1095
S88752-8........1095
S88753-6........1095
S88754-4........44
S88755-2........1041
S88756-0........1140
S88758-7........225
S88759-5........1057
S88760-9........40
S88761-7........769
S88762-5........436
S88763-3........1140
S88764-1........1139
S88766-8........23
S88767-6........1095
S88768-4........1094
S88769-2........1043
S88770-6........1095
S88771-4........294
S88773-0........1139
S88774-9........1139
S88775-7........1139
S88776-5........770
S88777-3........953
S88778-1........1080
S88780-3........255
S88781-1........958
S88783-8........941
S88784-6........742
S88788-9........185
S88789-7........89
S88790-0........67
S88791-9........209
S88792-7........769
S88797-8........371
S88798-6........726
S88804-4........84
S88806-0........358
S88807-9........703
S88808-7........63
S88809-5........967
S88810-9........335
S88811-7........378
S88812-5........56
S88813-3........193
S88816-8........887
S88817-6........858
S88818-4........253
S88819-2........889
S88820-6........226
S88821-4........341
S88822-2........855
S88823-0........104
S88824-9........330
S88825-7........296
S88826-5........296
S88827-3........328
S88828-1........328
S88830-3........291
S88831-1........338
S88833-8........213
S88834-6........319
S88835-4........296
S88836-2........343
S88837-0........328
S88838-9........353
S88839-7........473
S88840-0........1141
S88841-9........296
S88842-7........873
S88843-5........382
S88844-3........374
S88845-1........432
S88847-8........313
S88848-6........867
S88849-4........131
S88850-8........171
S88851-6........141
S88852-4........379
S88853-2........369
S88854-0........411
S88855-9........289
S88856-7........327
S88857-5........741
S88858-3........898
S88859-1........382
S88860-5........335
S88861-3........333
S88862-1........782
S88864-8........549
S88865-6........243
S88866-4........477
S88867-2........384
S88868-0........368
S88869-9........71
S88870-2........11
S88871-0........337
S88872-9........338
S88873-7........234
S88874-5........235
S88875-3........134
S88878-8........290
S88879-6........877
S88881-8........380
S88882-6........213
S88883-4........178
S88884-2........213
S88885-0........60
S88887-7........383
S88888-5........763
S88889-3........777
S88890-7........365
S88891-5........362
S88892-3........174
S88893-1........763
S88895-8........213
S88896-6........888
S88897-4........1142
S88898-2........361
S88899-0........938
S88900-8........717
S88901-6........344
S88902-4........344
S88903-2........379
S88904-0........260
S88905-9........857
S88906-7........67
S88907-5........703
S88908-3........396
S88909-1........405
S88910-5........441
S88911-3........827
S88914-8........308
S88915-6........213
S88916-4........365
S88917-2........313
S88918-0........365
S88919-9........213
S88920-2........742
S88921-0........213
S88923-7........214
S88924-5........389
S88925-3........319
S88926-1........114
S88928-8........452
S88929-6........451
S88931-8........445
S88932-6........446
S88933-4........452
S88934-2........452
S88935-0........452
S88936-9........451
S88937-7........452
S88938-5........452
S88939-3........544
S88940-7........399
S88941-5........544
S88942-3........542
S88943-1........469
S88945-8........729
S88946-6........639
S88947-4........542
S88948-2........781
S88949-0........773
S88950-4........773
S88951-2........729
S88952-0........399
S88953-9........399
S88954-7........399
S88955-5........564
S88956-3........674
S88957-1........544
S88959-8........112
S88960-1........607
S88962-8........609
S88963-6........524
S88964-4........537
S88965-2........512
S88966-0........511
S88967-9........511
S88968-7........511
S88969-5........537
S88970-9........539
S88971-7........564
S88972-5........563
S88974-1........564
S88976-8........564
S88977-6........588
S88978-4........591
S88979-2........595
S88980-6........512
S88981-4........512
S88982-2........564
S88983-0........537
S88984-9........539
S88985-7........540
S88986-5........555
S88987-3........560
S88988-1........559
S88990-3........555
S88991-1........555
S88993-8........555
S88994-6........552
S88995-4........579
S88996-2........602
S88997-0........537
S88998-9........544
S88999-7........539
S89000-6........586
S89001-4........586
S89002-2........589
S89003-0........579
S89004-9........579
S89005-7........602
S89006-5........440

Product	Page
S89007-3	439
S89008-1	440
S89010-3	461
S89011-1	457
S89013-8	458
S89014-6	460
S89015-4	445
S89016-2	446
S89017-0	458
S89018-9	460
S89019-7	536
S89020-0	519
S89021-9	454
S89022-7	474
S89023-5	654
S89024-3	655
S89025-1	648
S89027-8	396
S89028-6	540
S89029-4	439
S89030-8	440
S89031-6	809
S89032-4	399
S89033-2	628
S89034-0	617
S89035-9	538
S89036-7	809
S89037-5	394
S89038-3	631
S89039-1	399
S89040-5	399
S89041-3	399
S89042-1	633
S89044-8	400
S89045-6	674
S89046-4	677
S89047-2	687
S89048-0	632
S89049-9	404
S89050-2	404
S89051-0	682
S89052-9	430
S89053-7	401
S89054-5	807
S89055-3	458
S89056-1	458
S89058-8	458
S89059-6	458
S89061-8	456
S89062-6	456
S89063-4	519
S89064-2	558
S89065-0	828
S89066-9	551
S89067-7	449
S89068-5	531
S89069-3	419
S89070-7	466
S89071-5	545
S89072-3	545
S89073-1	560
S89075-8	455
S89076-6	456
S89077-4	475
S89078-2	475
S89079-0	520
S89080-4	519
S89081-2	567
S89082-0	400
S89083-9	400
S89084-7	567
S89085-5	400
S89086-3	419
S89087-1	653
S89089-8	552
S89090-1	508
S89092-8	525
S89093-6	1138
S89094-4	95
S89095-2	18
S89097-9	1142
S89098-7	213
S89099-5	3
S89100-2	38
S89102-9	1017
S89103-7	398
S89104-5	615
S89105-3	978
S89106-1	978
S89108-8	541
S89109-6	595
S89111-8	603
S89112-6	538
S89113-4	590
S89114-2	599
S89115-0	510
S89116-9	540
S89117-7	580
S89118-5	512
S89119-3	568
S89120-7	712
S89121-5	712
S89122-3	538
S89123-1	805
S89125-8	525
S89127-4	627
S89128-2	526
S89129-0	539
S89130-4	427
S89131-2	437
S89132-0	404
S89133-9	426
S89134-7	408
S89135-5	437
S89136-3	437
S89137-1	437
S89139-8	437
S89140-1	437
S89142-8	408
S89143-6	408
S89144-4	408
S89145-2	408
S89146-0	408
S89147-9	515
S89148-7	400
S89149-5	400
S89150-9	399
S89151-7	519
S89152-5	699
S89153-3	699
S89154-1	698
S89156-8	551
S89157-6	394
S89158-4	394
S89159-2	552
S89160-6	519
S89161-4	517
S89162-2	518
S89163-0	399
S89164-9	550
S89165-7	552
S89166-5	440
S89167-3	478
S89168-1	550
S89170-3	515
S89171-1	516
S89173-8	550
S89174-6	516
S89175-4	456
S89176-2	679
S89177-0	680
S89178-9	653
S89179-7	666
S89180-0	681
S89181-9	682
S89182-7	653
S89183-5	633
S89184-3	729
S89185-1	618
S89187-8	654
S89188-6	655
S89189-4	551
S89190-8	551
S89191-6	654
S89192-4	647
S89193-2	651
S89194-0	655
S89195-9	655
S89196-7	651
S89197-5	637
S89198-3	655
S89200-9	685
S89201-7	668
S89202-5	632
S89203-3	668
S89204-1	687
S89206-8	616
S89207-6	634
S89208-4	623
S89209-2	847
S89210-6	631
S89211-4	846
S89212-2	637
S89213-0	847
S89214-9	678
S89215-7	844
S89216-5	846
S89217-3	562
S89218-1	467
S89220-3	463
S89221-1	450
S89223-8	136
S89224-6	430
S89225-4	430
S89226-2	632
S89227-0	632
S89228-9	680
S89229-7	681
S89230-0	620
S89231-9	930
S89232-7	394
S89233-5	394
S89234-3	695
S89235-1	695
S89237-8	399
S89238-6	421
S89239-4	450
S89240-8	456
S89241-6	398
S89242-4	103
S89243-2	395
S89244-0	482
S89245-9	453
S89246-7	478
S89247-5	482
S89248-3	448
S89249-1	482
S89250-5	234
S89251-3	457
S89252-1	431
S89254-8	446
S89255-6	457
S89256-4	140
S89257-2	140
S89258-0	140
S89259-9	730
S89260-2	222
S89261-0	140
S89262-9	722
S89263-7	455
S89264-5	455
S89265-3	722
S89266-1	222
S89268-8	549
S89269-6	446
S89271-8	715
S89272-6	717
S89273-4	450
S89274-2	549
S89275-0	521
S89276-9	743
S89277-7	441
S89278-5	130
S89279-3	128
S89280-7	470
S89281-5	535
S89282-3	465
S89283-1	448
S89285-8	453
S89286-6	440
S89287-4	536
S89288-2	561
S89289-0	546
S89290-4	415
S89291-2	582
S89292-0	602
S89293-9	548
S89294-7	475
S89295-5	456
S89296-3	475
S89297-1	399
S89299-8	399
S89300-5	399
S89301-3	399
S89304-8	219
S89305-6	550
S89306-4	458
S89308-0	959
S89309-9	39
S89310-2	213
S89311-0	1142
S89312-9	943
S89313-7	1015
S89314-5	1013
S89315-3	721
S89316-1	633
S89318-8	351
S89319-6	392
S89321-8	189
S89322-6	362
S89323-4	752
S89324-2	1003
S89325-0	248
S89326-9	244
S89327-7	696
S89328-5	248
S89329-3	893
S89330-7	67
S89331-5	6
S89332-3	391
S89333-1	752
S89335-8	346
S89336-6	709
S89337-4	570
S89338-2	364
S89340-4	709
S89341-2	752
S89342-0	292
S89343-9	237
S89344-7	244
S89345-5	778
S89346-3	245
S89347-1	721
S89349-8	1003
S89350-1	302
S89352-8	244
S89353-6	245
S89354-4	1018
S89355-2	292
S89356-0	1003
S89357-9	292
S89358-7	1003
S89359-5	422
S89360-9	295
S89361-7	292
S89363-3	245
S89364-1	237
S89367-6	5
S89368-4	62
S89369-2	1015
S89370-6	102
S89371-4	83
S89372-2	56
S89373-0	862
S89374-9	862
S89375-7	109
S89376-5	859
S89377-3	1011
S89378-1	377
S89380-3	384
S89381-1	940
S89383-8	278
S89384-6	174
S89385-4	179
S89386-2	178
S89387-0	233
S89388-9	763
S89389-7	763
S89390-0	100
S89391-9	6
S89392-7	6
S89393-5	5
S89394-3	173
S89395-1	6
S89397-8	100
S89398-6	60
S89399-4	62
S89400-1	5
S89402-8	101
S89403-6	15
S89404-4	15
S89405-2	82
S89406-0	53
S89407-9	101
S89408-7	62
S89409-5	82
S89410-9	92
S89411-7	83
S89412-5	83
S89413-3	84
S89414-1	16
S89416-8	92
S89417-6	5
S89418-4	53
S89419-2	60
S89420-6	92
S89421-4	860
S89422-2	83
S89423-0	1015
S89424-9	1016
S89425-7	92
S89426-5	96
S89427-3	171
S89428-1	171
S89430-3	946
S89431-1	109
S89433-8	301
S89434-6	859
S89435-4	304
S89436-2	296
S89437-0	478
S89438-9	173
S89439-7	173
S89440-0	198
S89441-9	234
S89442-7	174
S89443-5	174
S89444-3	144
S89445-1	62
S89447-8	96
S89448-6	96
S89449-4	150
S89450-8	291
S89451-6	398
S89452-4	62
S89453-2	6
S89454-0	967
S89455-9	105
S89456-7	964
S89457-5	758
S89458-3	171
S89459-1	374
S89461-3	478
S89462-1	324
S89464-8	761
S89465-6	389
S89466-4	342
S89467-2	1082
S89468-0	357
S89469-9	312
S89471-0	411
S89472-9	870
S89473-7	914
S89474-5	251
S89475-3	343
S89478-8	326
S89479-6	1085
S89481-8	889
S89482-6	890
S89485-0	890
S89486-9	889
S89487-7	677
S89488-5	882
S89489-3	369
S89491-5	842
S89492-3	842
S89496-6	842
S89497-4	885
S89498-2	1057
S89499-0	64
S89501-6	1091
S89502-4	1011
S89503-2	125
S89504-0	67
S89505-9	72
S89506-7	885
S89507-5	72
S89508-3	72
S89509-1	65
S89510-5	101
S89511-3	101
S89512-1	108
S89514-8	885
S89516-4	388
S89518-0	331
S89521-0	475
S89522-9	388
S89523-7	359
S89524-5	360
S89526-1	98
S89529-6	881
S89531-8	374
S89532-6	888
S89535-0	883
S89536-9	831
S89538-5	838
S89539-3	70
S89540-7	65
S89541-5	57
S89542-3	112
S89543-1	91
S89545-8	1059
S89546-6	1059
S89547-4	889
S89549-0	885
S89554-7	79
S89556-3	70
S89557-1	147
S89559-8	277
S89560-1	140
S89562-8	1137
S89563-6	85
S89564-4	124
S89565-2	60
S89566-0	1059
S89568-7	150
S89569-5	85
S89570-9	1056
S89571-7	891
S89572-5	882
S89573-3	574
S89576-8	888
S89577-6	1092
S89579-2	882
S89580-6	889
S89581-4	888
S89584-9	93
S89585-7	842
S89586-5	842
S89587-3	388
S89588-1	388
S89590-3	330
S89591-1	888
S89594-6	883
S89595-4	691
S89596-2	691
S89597-0	691
S89598-9	838
S89599-7	888
S89602-0	881
S89604-7	388
S89605-5	752
S89607-1	889
S89609-8	889

Product	Page
S89610-1	882
S89612-8	359
S89613-6	388
S89614-4	888
S89615-2	830
S89616-0	474
S89617-9	27
S89618-7	6
S89619-5	13
S89621-7	282
S89622-5	207
S89623-3	761
S89624-1	235
S89626-8	1054
S89627-6	366
S89628-4	255
S89629-2	312
S89630-6	173
S89631-4	1027
S89632-2	389
S89633-0	891
S89634-9	708
S89635-7	1077
S89636-5	312
S89637-3	253
S89638-1	274
S89640-3	242
S89641-1	261
S89643-8	224
S89644-6	482
S89645-4	998
S89646-2	312
S89647-0	312
S89648-9	471
S89649-7	831
S89650-0	1076
S89651-9	203
S89652-7	1076
S89653-5	184
S89654-3	203
S89655-1	200
S89657-8	51
S89658-6	245
S89661-6	284
S89662-4	716
S89663-2	284
S89664-0	147
S89665-9	57
S89666-7	57
S89667-5	57
S89668-3	793
S89669-1	238
S89672-1	294
S89674-8	55
S89675-6	283
S89676-4	283
S89677-2	283
S89678-0	284
S89679-9	284
S89680-2	282
S89681-0	282
S89683-7	282
S89685-3	284
S89686-1	944
S89689-6	88
S89691-8	284
S89692-6	134
S89693-4	88
S89694-2	58
S89695-0	52
S89696-9	37
S89697-7	137
S89698-5	93
S89699-3	58
S89700-0	57
S89702-7	975
S89703-5	133
S89704-3	281
S89705-1	245
S89708-6	137
S89709-4	58
S89710-8	237
S89711-6	59
S89712-4	57
S89713-2	240
S89714-0	284
S89715-9	282
S89716-7	284
S89717-5	284
S89718-3	284
S89719-1	238
S89720-5	364
S89721-3	242
S89724-8	282
S89725-6	276
S89726-4	364
S89728-0	238
S89730-2	942
S89731-0	91
S89732-9	961
S89733-7	984
S89734-5	137
S89735-3	57
S89738-8	57
S89739-6	1050
S89743-4	137
S89745-0	57
S89747-7	975
S89748-5	60
S89749-3	55
S89751-5	133
S89752-3	136
S89753-1	133
S89756-6	137
S89757-4	710
S89758-2	136
S89759-0	133
S89760-4	137
S89763-9	57
S89765-5	137
S89766-3	137
S89767-1	136
S89770-1	137
S89772-8	140
S89773-6	443
S89775-2	137
S89776-0	57
S89777-9	238
S89778-7	443
S89779-5	707
S89780-9	562
S89781-7	402
S89782-5	912
S89783-3	402
S89784-1	619
S89786-8	505
S89787-6	1017
S89788-4	911
S89789-2	871
S89790-6	875
S89791-4	767
S89792-2	502
S89793-0	731
S89794-9	513
S89795-7	580
S89797-3	835
S89798-1	870
S89800-7	767
S89801-5	665
S89802-3	879
S89803-1	767
S89805-8	607
S89806-6	762
S89807-4	760
S89808-2	760
S89809-0	760
S89810-4	760
S89811-2	760
S89812-0	764
S89814-7	727
S89815-5	931
S89816-3	536
S89817-1	865
S89819-8	767
S89820-1	183
S89822-8	770
S89823-6	760
S89824-4	769
S89825-2	769
S89826-0	769
S89827-9	524
S89830-9	746
S89831-7	615
S89832-5	766
S89834-1	618
S89836-8	873
S89837-6	611
S89838-4	732
S89839-2	746
S89840-6	745
S89841-4	872
S89842-2	872
S89843-0	905
S89846-5	976
S89847-3	735
S89848-1	925
S89850-3	727
S89851-1	727
S89853-8	931
S89854-6	872
S89855-4	931
S89856-2	727
S89857-0	736
S89858-9	732
S89859-7	760
S89860-0	727
S89861-9	729
S89862-7	729
S89863-5	745
S89864-3	872
S89865-1	745
S89867-8	873
S89868-6	738
S89869-4	767
S89870-8	381
S89871-6	381
S89872-4	760
S89873-2	138
S89874-0	139
S89875-9	139
S89876-7	640
S89877-5	381
S89878-3	382
S89879-1	610
S89880-5	677
S89881-3	116
S89882-1	381
S89885-6	729
S89886-4	677
S89887-2	381
S89888-0	381
S89889-9	381
S89890-2	381
S89891-0	382
S89892-9	826
S89893-7	114
S89894-5	307
S89896-1	302
S89899-6	307
S89900-3	631
S89901-1	21
S89903-8	707
S89904-6	879
S89905-4	631
S89906-2	271
S89907-0	159
S89910-0	814
S89911-9	924
S89912-7	924
S89913-5	67
S89914-3	924
S89915-1	1003
S89917-8	689
S89918-6	167
S89919-4	265
S89920-8	257
S89921-6	670
S89922-4	706
S89923-2	382
S89924-0	322
S89925-9	726
S89926-7	322
S89927-5	1139
S89928-3	877
S89929-1	877
S89930-5	452
S89931-3	462
S89932-1	965
S89934-8	460
S89935-6	460
S89936-4	460
S89937-2	450
S89938-0	468
S89939-9	520
S89940-2	472
S89941-0	582
S89942-9	130
S89943-7	460
S89944-5	548
S89945-3	129
S89946-1	585
S89948-8	547
S89949-6	130
S89951-8	623
S89952-6	539
S89953-4	731
S89954-2	731
S89955-0	531
S89956-9	632
S89957-7	632
S89958-5	112
S89959-3	728
S89960-7	680
S89961-5	617
S89962-3	632
S89963-1	116
S89965-8	667
S89966-6	612
S89967-4	679
S89968-2	680
S89969-0	631
S89970-4	497
S89971-2	731
S89972-0	1026
S89973-9	731
S89974-7	737
S89975-5	111
S89976-3	274
S89977-1	1053
S89979-8	632
S89980-1	732
S89982-8	536
S89983-6	585
S89985-2	600
S89986-0	747
S89987-9	748
S89988-7	699
S89989-5	586
S89990-9	748
S89991-7	740
S89992-5	493
S89993-3	474
S89994-1	621
S89996-8	462
S89997-6	551
S89998-4	551
S89999-2	612
S90000-1	406
S90002-8	498
S90003-6	632
S90004-4	497
S90005-2	551
S90006-0	489
S90007-9	748
S90008-7	694
S90009-5	570
S90010-9	142
S90011-7	126
S90012-5	590
S90013-3	243
S90014-1	569
S90016-8	404
S90017-6	404
S90018-4	538
S90019-2	537
S90020-6	519
S90021-4	426
S90022-2	539
S90023-0	521
S90024-9	421
S90025-7	544
S90026-5	538
S90027-3	559
S90028-1	435
S90030-3	428
S90031-1	555
S90034-6	535
S90035-4	558
S90036-2	551
S90037-0	136
S90038-9	126
S90039-7	405
S90040-0	539
S90041-9	126
S90042-7	408
S90043-5	395
S90044-3	535
S90045-1	466
S90047-8	397
S90048-6	408
S90049-4	437
S90050-8	437
S90051-6	586
S90052-4	584
S90053-2	438
S90054-0	472
S90055-9	749
S90056-7	748
S90058-3	457
S90059-1	729
S90060-5	535
S90061-3	737
S90062-1	733
S90064-8	562
S90065-6	747
S90066-4	779
S90067-2	453
S90068-0	649
S90069-9	449
S90070-2	535
S90071-0	739
S90072-9	153
S90073-7	152
S90074-5	1059
S90075-3	1033
S90076-1	1027
S90078-8	369
S90079-6	773
S90081-8	1058
S90082-6	774
S90083-4	774
S90084-2	775
S90085-0	234
S90086-9	878
S90087-7	438
S90088-5	409
S90089-3	480
S90090-7	481
S90091-5	419
S90092-3	446
S90093-1	394
S90095-8	393
S90096-6	450
S90097-4	806
S90098-2	446
S90099-0	703
S90100-8	493
S90101-6	1074
S90102-4	449
S90103-2	814
S90104-0	700
S90105-9	523
S90106-7	809
S90107-5	807
S90108-3	807
S90109-1	810
S90110-5	806
S90111-3	807
S90112-1	806
S90114-8	810
S90115-6	478
S90116-4	806
S90117-2	496
S90118-0	829
S90119-9	424
S90120-2	261
S90121-0	808
S90122-9	805
S90123-7	825
S90124-5	825
S90125-3	808
S90126-1	456
S90128-8	914
S90129-6	810
S90131-8	809
S90132-6	695
S90133-4	809
S90134-2	460
S90135-0	806
S90136-9	806
S90137-7	806
S90138-5	453
S90139-3	456
S90140-7	491
S90141-5	450
S90142-3	494
S90143-1	810
S90145-8	366
S90146-6	372
S90147-4	495
S90148-2	112
S90149-0	498
S90150-4	501
S90151-2	850
S90152-0	579
S90153-9	844
S90154-7	851
S90155-5	474
S90156-3	474
S90157-1	466
S90159-8	851
S90160-1	579
S90162-8	474
S90163-6	533
S90164-4	533
S90165-2	674
S90166-0	579
S90167-9	1023
S90168-7	1024
S90169-5	1027
S90170-9	1026
S90171-7	1026
S90172-5	1056
S90173-3	1076
S90174-1	1027
S90176-8	1027
S90177-6	1074
S90178-4	1074
S90179-2	1074
S90180-6	635
S90181-4	463
S90182-2	474
S90183-0	467
S90184-9	423
S90185-7	805
S90186-5	639
S90187-3	567
S90188-1	705
S90190-3	877
S90191-1	616
S90193-8	638
S90194-6	871
S90195-4	640
S90196-2	640
S90197-0	843
S90198-9	758
S90199-7	505
S90200-4	607
S90201-2	608
S90202-0	509
S90203-9	563
S90204-7	529
S90205-5	512
S90206-3	517

Product	Page
S90207-1	523
S90209-8	512
S90210-1	522
S90212-8	523
S90213-6	512
S90214-4	508
S90215-2	508
S90216-0	507
S90217-9	507
S90218-7	1068
S90219-5	944
S90220-9	1024
S90221-7	1024
S90222-5	1033
S90223-3	1040
S90224-1	1053
S90226-8	1053
S90227-6	1031
S90228-4	851
S90229-2	127
S90230-6	1053
S90231-4	547
S90232-2	550
S90233-0	561
S90234-9	562
S90235-7	548
S90236-5	563
S90237-3	512
S90238-1	523
S90240-3	522
S90241-1	522
S90243-8	513
S90244-6	517
S90245-4	648
S90246-2	613
S90247-0	647
S90248-9	616
S90249-7	650
S90250-0	115
S90251-9	557
S90252-7	550
S90253-5	574
S90254-3	574
S90255-1	606
S90257-8	606
S90258-6	607
S90259-4	606
S90260-8	843
S90261-6	874
S90262-4	438
S90263-2	440
S90264-0	409
S90265-9	409
S90266-7	409
S90267-5	410
S90268-3	409
S90269-1	410
S90270-5	410
S90271-3	397
S90272-1	405
S90274-8	419
S90275-6	403
S90276-4	433
S90277-2	421
S90278-0	825
S90279-9	413
S90280-2	427
S90281-0	674
S90282-9	618
S90283-7	652
S90284-5	457
S90285-3	1025
S90286-1	493
S90288-8	1025
S90289-6	493
S90291-8	936
S90292-6	512
S90293-4	516
S90294-2	546
S90295-0	937
S90296-9	517
S90297-7	551
S90298-5	1053
S90299-3	547
S90300-0	561
S90301-9	549
S90302-7	523
S90303-5	522
S90304-3	734
S90305-1	730
S90307-8	727
S90308-6	729
S90309-4	740
S90310-8	928
S90311-6	522
S90312-4	522
S90313-2	529
S90314-0	522
S90315-9	928
S90316-7	522
S90317-5	522
S90318-3	517
S90319-1	512
S90320-5	930
S90321-3	638
S90322-1	607
S90324-8	502
S90325-6	502
S90326-4	641
S90327-2	112
S90328-0	495
S90329-9	496
S90330-2	492
S90331-0	499
S90332-9	495
S90333-7	452
S90334-5	492
S90335-3	491
S90336-1	492
S90338-8	491
S90339-6	492
S90341-8	493
S90342-6	493
S90343-4	493
S90344-2	491
S90345-0	491
S90346-9	493
S90347-7	493
S90348-5	493
S90349-3	491
S90350-7	492
S90351-5	493
S90352-3	1057
S90353-1	936
S90355-8	878
S90356-6	552
S90357-4	669
S90358-2	494
S90359-0	538
S90360-4	879
S90361-2	641
S90362-0	521
S90363-9	550
S90364-7	845
S90365-5	516
S90366-3	730
S90367-1	550
S90369-8	515
S90370-1	516
S90372-8	550
S90373-6	648
S90374-4	753
S90375-2	499
S90376-0	498
S90377-9	498
S90378-7	498
S90379-5	495
S90380-9	495
S90381-7	495
S90382-5	496
S90383-3	1068
S90384-1	499
S90386-8	491
S90387-6	496
S90388-4	1059
S90389-2	493
S90390-6	498
S90391-4	1068
S90392-2	496
S90393-0	499
S90394-9	495
S90395-7	493
S90396-5	501
S90397-3	501
S90398-1	501
S90400-7	496
S90401-5	501
S90402-3	499
S90403-1	491
S90405-8	500
S90406-6	494
S90407-4	1071
S90408-2	502
S90409-0	495
S90410-4	1028
S90411-2	495
S90412-0	494
S90413-9	1031
S90414-7	938
S90415-5	532
S90416-3	532
S90417-1	567
S90419-8	539
S90420-1	505
S90422-8	878
S90423-6	732
S90424-4	844
S90425-2	522
S90426-0	520
S90427-9	526
S90428-7	505
S90429-5	539
S90430-9	533
S90431-7	507
S90432-5	519
S90433-3	552
S90434-1	520
S90436-8	553
S90437-6	939
S90438-4	939
S90439-2	519
S90440-6	508
S90441-4	878
S90442-2	508
S90443-0	925
S90444-9	511
S90445-7	509
S90446-5	505
S90447-3	507
S90448-1	509
S90450-3	509
S90451-1	696
S90453-8	696
S90454-6	390
S90455-4	391
S90456-2	968
S90457-0	446
S90458-9	880
S90459-7	459
S90460-0	492
S90461-9	494
S90462-7	494
S90463-5	460
S90464-3	880
S90465-1	460
S90467-8	875
S90468-6	449
S90469-4	492
S90470-8	494
S90471-6	808
S90472-4	398
S90473-2	396
S90474-0	494
S90475-9	494
S90476-7	396
S90477-5	395
S90478-3	398
S90479-1	491
S90480-5	491
S90481-3	520
S90482-1	548
S90484-8	531
S90485-6	965
S90486-4	504
S90487-2	140
S90488-0	139
S90489-9	730
S90490-2	728
S90491-0	445
S90492-9	879
S90493-7	851
S90494-5	395
S90495-3	390
S90496-1	399
S90498-8	437
S90499-6	395
S90500-3	408
S90501-1	401
S90503-8	390
S90504-6	391
S90505-4	390
S90506-2	391
S90507-0	390
S90508-9	390
S90509-7	391
S90510-0	391
S90511-9	390
S90512-7	693
S90513-5	698
S90514-3	936
S90515-1	925
S90517-8	951
S90518-6	391
S90519-4	390
S90520-8	391
S90521-6	396
S90522-4	409
S90523-2	410
S90524-0	399
S90525-9	395
S90526-7	805
S90527-5	806
S90528-3	810
S90529-1	807
S90530-5	852
S90531-3	805
S90532-1	851
S90534-8	484
S90535-6	538
S90536-4	824
S90537-2	879
S90538-0	715
S90539-9	767
S90540-2	964
S90541-0	824
S90542-9	715
S90543-7	715
S90544-5	718
S90545-3	534
S90546-1	407
S90548-8	543
S90549-6	875
S90551-8	551
S90552-6	520
S90553-4	447
S90554-2	567
S90555-0	446
S90556-9	775
S90557-7	451
S90558-5	813
S90559-3	826
S90560-7	676
S90561-5	546
S90562-3	649
S90563-1	493
S90565-8	615
S90566-6	628
S90567-4	624
S90568-2	529
S90569-0	694
S90570-4	569
S90571-2	649
S90572-0	941
S90573-9	560
S90574-7	702
S90575-5	805
S90576-3	505
S90577-1	508
S90579-8	474
S90580-1	526
S90582-8	637
S90583-6	617
S90584-4	731
S90585-2	616
S90586-0	616
S90587-9	615
S90588-7	811
S90589-5	806
S90590-9	818
S90591-7	810
S90592-5	829
S90593-3	816
S90594-1	814
S90596-8	806
S90597-6	523
S90598-4	928
S90599-2	808
S90601-8	4
S90602-6	93
S90603-4	62
S90604-2	27
S90605-0	77
S90606-9	768
S90607-7	942
S90608-5	370
S90609-3	809
S90612-3	309
S90613-1	375
S90615-8	242
S90616-6	255
S90617-4	345
S90618-2	953
S90619-0	308
S90620-4	41
S90622-0	17
S90623-9	947
S90624-7	757
S90625-5	473
S90626-3	87
S90627-1	363
S90630-1	1003
S90632-8	361
S90633-6	173
S90634-4	363
S90638-7	332
S90639-5	13
S90640-9	251
S90641-7	361
S90644-1	754
S90646-8	370
S90647-6	60
S90648-4	869
S90650-6	162
S90651-4	247
S90652-2	238
S90653-0	256
S90657-3	87
S90661-1	168
S90663-8	166
S90664-6	841
S90665-4	1011
S90666-2	256
S90667-0	240
S90668-9	55
S90669-7	965
S90670-0	798
S90671-9	411
S90672-7	841
S90674-3	499
S90675-1	691
S90677-8	756
S90678-6	71
S90680-8	32
S90681-6	93
S90682-4	262
S90683-2	1028
S90684-0	338
S90685-9	319
S90686-7	95
S90687-5	1011
S90688-3	933
S90689-1	797
S90691-3	401
S90692-1	798
S90694-8	355
S90696-4	209
S90697-2	214
S90698-0	168
S90699-9	242
S90700-6	182
S90702-2	212
S90703-0	186
S90704-9	1008
S90705-7	1143
S90708-1	179
S90710-3	213
S90711-1	242
S90713-8	750
S90714-6	1008
S90715-4	1014
S90716-2	214
S90717-0	179
S90718-9	386
S90720-0	262
S90721-9	776
S90724-3	295
S90725-1	294
S90729-4	385
S90730-8	1013
S90731-6	234
S90732-4	346
S90733-2	175
S90734-0	441
S90735-9	227
S90736-7	312
S90738-3	234
S90739-1	798
S90740-5	233
S90741-3	233
S90742-1	376
S90744-8	227
S90745-6	311
S90746-4	233
S90747-2	233
S90748-0	233
S90749-9	379
S90753-7	876
S90754-5	910
S90755-3	312
S90759-6	373
S90761-8	227
S90762-6	251
S90763-4	876
S90764-2	376
S90765-0	229
S90766-9	754
S90767-7	276
S90769-3	226
S90770-7	225
S90772-3	224
S90773-1	277
S90775-8	233
S90776-6	233
S90777-4	233
S90778-2	224
S90779-0	225
S90783-9	276
S90784-7	274
S90785-5	856
S90786-3	778
S90787-1	439
S90792-8	876
S90793-6	856
S90794-4	288
S90795-2	693
S90796-0	693
S90797-9	274
S90798-7	276
S90799-5	916
S90800-2	397
S90802-9	469
S90803-7	414
S90804-5	263
S90805-3	264

Product	Page
S90806-1	857
S90808-8	796
S90809-6	910
S90811-8	858
S90812-6	733
S90813-4	450
S90814-2	914
S90815-0	759
S90816-9	759
S90817-7	337
S90818-5	322
S90819-3	879
S90820-7	856
S90821-5	1011
S90822-3	976
S90823-1	424
S90825-8	280
S90826-6	276
S90827-4	276
S90828-2	856
S90830-4	856
S90831-2	340
S90832-0	220
S90833-9	276
S90834-7	276
S90835-5	276
S90836-3	274
S90837-1	295
S90839-8	1013
S90840-1	253
S90842-8	781
S90843-6	432
S90845-2	857
S90847-9	321
S90848-7	468
S90849-5	425
S90850-9	447
S90851-7	856
S90852-5	415
S90853-3	425
S90854-1	876
S90856-8	874
S90858-4	260
S90859-2	874
S90860-6	758
S90861-4	763
S90862-2	266
S90863-0	216
S90864-9	910
S90865-7	914
S90866-5	856
S90867-3	796
S90868-1	856
S90870-3	690
S90871-1	425
S90873-8	147
S90874-6	229
S90875-4	225
S90876-2	798
S90877-0	718
S90878-9	426
S90879-7	425
S90880-0	415
S90881-9	219
S90882-7	433
S90883-5	425
S90884-3	426
S90885-1	415
S90887-8	718
S90888-6	469
S90889-4	414
S90890-8	793
S90891-6	205
S90892-4	415
S90893-2	415
S90894-0	783
S90895-9	396
S90896-7	718
S90897-5	233
S90898-3	876
S90899-1	165
S90900-9	425
S90901-7	781
S90902-5	415
S90903-3	1003
S90904-1	998
S90906-8	726
S90907-6	997
S90908-4	997
S90910-6	997
S90912-2	1004
S90913-0	1002
S90915-7	1005
S90918-1	1005
S90920-3	912
S90923-8	1007
S90924-6	997
S90925-4	998
S90926-2	997
S90927-0	997
S90928-9	998
S90929-7	793
S90930-0	360
S90931-9	998
S90932-7	968
S90934-3	972
S90935-1	976
S90937-8	976
S90939-4	977
S90940-8	726
S90941-6	1005
S90943-2	1093
S90945-9	944
S90947-5	968
S90948-3	968
S90949-1	941
S90950-5	957
S90951-3	989
S90956-4	989
S90957-2	955
S90958-0	955
S90961-0	955
S90963-7	955
S90964-5	417
S90966-1	1002
S90968-8	955
S90969-6	1002
S90971-8	1002
S90973-4	931
S90974-2	965
S90975-0	908
S90976-9	908
S90978-5	939
S90979-3	428
S90980-7	729
S90981-5	1086
S90982-3	957
S90983-1	942
S90986-6	445
S90987-4	904
S90988-2	875
S90989-0	723
S90990-4	526
S90991-2	518
S90992-0	526
S90993-9	516
S90994-7	819
S90995-5	723
S90996-3	476
S90997-1	851
S90999-8	877
S91000-7	1008
S91001-5	458
S91002-3	446
S91003-1	465
S91005-8	468
S91006-6	470
S91007-4	963
S91008-2	469
S91009-0	963
S91010-4	470
S91011-2	476
S91012-0	851
S91013-9	458
S91014-7	445
S91015-5	470
S91016-3	470
S91017-1	963
S91019-8	470
S91020-1	634
S91022-8	677
S91023-6	637
S91024-4	952
S91025-2	532
S91026-0	824
S91027-9	965
S91028-7	634
S91029-5	668
S91030-9	992
S91031-7	532
S91032-5	637
S91033-3	860
S91034-1	735
S91036-8	936
S91037-6	912
S91038-4	912
S91039-2	704
S91040-6	465
S91041-4	647
S91042-2	992
S91043-0	977
S91044-9	860
S91045-7	936
S91046-5	936
S91047-3	912
S91048-1	912
S91050-3	903
S91051-1	901
S91053-8	911
S91054-6	869
S91055-4	912
S91056-2	869
S91057-0	860
S91058-9	903
S91059-7	917
S91060-0	974
S91061-9	739
S91062-7	903
S91063-5	941
S91064-3	913
S91065-1	914
S91067-8	860
S91068-6	869
S91069-4	903
S91070-8	941
S91071-6	932
S91072-4	900
S91073-2	899
S91074-0	876
S91075-9	871
S91076-7	941
S91077-5	941
S91078-3	647
S91079-1	651
S91080-5	623
S91081-3	667
S91082-1	668
S91085-6	522
S91086-4	969
S91087-2	930
S91088-0	860
S91089-9	732
S91090-2	930
S91091-0	901
S91092-9	927
S91093-7	911
S91094-5	911
S91095-3	911
S91096-1	860
S91098-8	901
S91099-6	930
S91100-3	930
S91101-1	926
S91103-8	695
S91104-6	927
S91105-4	899
S91106-2	875
S91107-0	637
S91108-9	613
S91109-7	617
S91110-0	636
S91111-9	636
S91112-7	629
S91113-5	629
S91114-3	636
S91115-1	856
S91117-8	977
S91118-6	945
S91119-4	868
S91120-8	916
S91121-6	913
S91122-4	916
S91123-2	962
S91124-0	961
S91125-9	962
S91126-7	933
S91127-5	859
S91128-3	977
S91129-1	927
S91130-5	618
S91131-3	927
S91132-1	927
S91134-8	927
S91135-6	927
S91136-4	890
S91137-2	669
S91138-0	943
S91139-9	949
S91140-2	900
S91141-0	949
S91142-9	943
S91143-7	975
S91144-5	630
S91145-3	647
S91146-1	900
S91148-8	900
S91149-6	906
S91151-8	926
S91152-6	911
S91153-4	926
S91154-2	926
S91155-0	927
S91156-9	614
S91157-7	614
S91158-5	927
S91159-3	613
S91160-7	613
S91161-5	927
S91162-3	930
S91163-1	927
S91165-8	860
S91166-6	927
S91167-4	929
S91168-2	929
S91169-0	899
S91170-4	929
S91171-2	929
S91172-0	620
S91173-9	694
S91174-7	968
S91175-5	926
S91176-3	901
S91177-1	927
S91179-8	875
S91180-1	612
S91182-8	619
S91183-6	901
S91185-2	911
S91186-0	869
S91187-9	899
S91188-7	543
S91189-5	564
S91190-9	561
S91191-7	927
S91192-5	911
S91193-3	929
S91194-1	929
S91196-8	865
S91197-6	881
S91198-4	901
S91199-2	899
S91201-8	928
S91202-6	929
S91203-4	626
S91204-2	901
S91205-0	910
S91206-9	925
S91207-7	910
S91208-5	925
S91209-3	899
S91210-7	925
S91211-5	981
S91212-3	977
S91213-1	945
S91215-8	950
S91216-6	868
S91217-4	585
S91218-2	871
S91219-0	877
S91220-4	647
S91221-2	651
S91222-0	949
S91223-9	900
S91224-7	508
S91225-5	563
S91226-3	905
S91227-1	916
S91229-8	649
S91230-1	899
S91232-8	969
S91233-6	977
S91234-4	860
S91235-2	875
S91236-0	584
S91237-9	567
S91238-7	585
S91239-5	976
S91240-9	950
S91241-7	945
S91242-5	900
S91243-3	933
S91244-1	900
S91246-8	969
S91247-6	965
S91248-4	927
S91249-2	911
S91250-6	899
S91251-4	899
S91252-2	977
S91253-0	925
S91254-9	925
S91255-7	929
S91256-5	576
S91257-3	127
S91258-1	519
S91260-3	519
S91261-1	519
S91263-8	751
S91264-6	729
S91265-4	519
S91266-2	908
S91267-0	908
S91268-9	504
S91269-7	453
S91270-0	413
S91271-9	405
S91272-7	451
S91273-5	395
S91274-3	446
S91275-1	406
S91277-8	427
S91278-6	824
S91279-4	824
S91280-8	517
S91281-6	519
S91282-4	844
S91283-2	520
S91284-0	981
S91285-9	520
S91286-7	1052
S91287-5	1030
S91288-3	1070
S91290-5	503
S91291-3	1058
S91292-1	928
S91294-8	494
S91295-6	908
S91296-4	535
S91297-2	505
S91298-0	505
S91299-9	505
S91300-6	751
S91301-4	728
S91302-2	505
S91303-0	606
S91304-9	562
S91305-7	739
S91306-5	754
S91307-3	908
S91308-1	523
S91310-3	732
S91311-1	532
S91313-8	752
S91314-6	908
S91315-4	908
S91316-2	536
S91317-0	981
S91318-9	745
S91319-7	578
S91320-0	909
S91321-9	989
S91322-7	509
S91323-5	751
S91324-3	728
S91325-1	908
S91327-8	843
S91328-6	728
S91329-4	751
S91330-8	908
S91331-6	908
S91332-4	510
S91333-2	991
S91334-0	993
S91335-9	992
S91336-7	993
S91338-3	985
S91339-1	128
S91340-5	852
S91341-3	989
S91342-1	126
S91344-8	535
S91345-6	739
S91346-4	1026
S91347-2	747
S91348-0	908
S91349-9	909
S91350-2	850
S91352-9	563
S91353-7	945
S91354-5	976
S91355-3	454
S91356-1	944
S91358-8	944
S91359-6	877
S91361-8	453
S91362-6	851
S91363-4	448
S91364-2	453
S91365-0	449
S91367-7	928
S91368-5	870
S91369-3	929
S91370-7	455
S91371-5	451
S91372-3	455
S91373-1	732
S91375-8	945
S91376-6	732
S91377-4	729
S91378-2	747
S91379-0	733
S91380-4	390
S91381-2	758
S91383-9	391
S91384-7	442
S91385-5	130
S91386-3	390
S91387-1	415
S91389-8	401
S91390-1	416
S91392-8	404
S91393-6	438
S91394-4	810
S91395-2	417

Product	Page
S91396-0	846
S91397-9	435
S91398-7	407
S91399-5	825
S91400-2	427
S91401-0	981
S91402-9	996
S91403-7	996
S91404-5	981
S91405-3	981
S91406-1	356
S91408-8	195
S91409-6	943
S91411-8	760
S91413-4	766
S91414-2	224
S91416-9	944
S91418-5	15
S91419-3	758
S91420-7	278
S91422-3	1024
S91423-1	291
S91425-8	946
S91426-6	938
S91427-4	947
S91428-2	765
S91429-0	941
S91430-4	958
S91431-2	67
S91432-0	378
S91433-9	199
S91435-5	343
S91437-1	946
S91439-8	275
S91440-1	944
S91443-6	933
S91444-4	946
S91445-2	173
S91446-0	758
S91447-9	64
S91448-7	438
S91451-7	310
S91453-3	833
S91454-1	338
S91456-8	343
S91458-4	279
S91459-2	257
S91460-6	205
S91461-4	203
S91462-2	162
S91463-0	67
S91464-9	349
S91465-7	868
S91466-5	374
S91467-3	294
S91468-1	270
S91470-3	952
S91471-1	866
S91473-8	707
S91474-6	944
S91475-4	144
S91476-2	828
S91477-0	993
S91478-9	349
S91479-7	263
S91480-0	76
S91481-9	349
S91483-5	867
S91484-3	92
S91485-1	326
S91487-8	797
S91488-6	574
S91489-4	13
S91490-8	53
S91491-6	232
S91492-4	308
S91493-2	798
S91495-9	32
S91496-7	165
S91497-5	933
S91499-1	269
S91501-7	57
S91502-5	757
S91503-3	385
S91504-1	255
S91506-8	359
S91507-6	831
S91508-4	1137
S91509-2	83
S91510-6	827
S91511-4	308
S91514-9	79
S91515-7	865
S91516-5	867
S91517-3	947
S91518-1	867
S91520-3	1056
S91521-1	865
S91523-8	916
S91524-6	933
S91525-4	865
S91526-2	948
S91527-0	44
S91529-7	867
S91530-0	946
S91531-9	865
S91532-7	1053
S91533-5	865
S91534-3	1069
S91535-1	869
S91537-8	137
S91539-4	865
S91540-8	986
S91541-6	865
S91542-4	915
S91543-2	885
S91544-0	907
S91545-9	1058
S91546-7	1069
S91547-5	356
S91549-1	133
S91550-5	408
S91551-3	351
S91554-8	342
S91555-6	339
S91556-4	342
S91557-2	946
S91558-0	340
S91559-9	338
S91560-2	433
S91561-0	338
S91562-9	345
S91563-7	337
S91564-5	833
S91565-3	836
S91566-1	833
S91568-8	834
S91569-6	835
S91571-8	837
S91572-6	835
S91573-4	835
S91574-2	835
S91575-0	833
S91576-9	833
S91577-7	836
S91578-5	837
S91579-3	833
S91580-7	835
S91581-5	836
S91582-3	833
S91583-1	837
S91585-8	834
S91586-6	834
S91587-4	833
S91588-2	342
S91589-0	350
S91590-4	835
S91591-2	95
S91594-7	837
S91595-5	836
S91596-3	837
S91597-1	833
S91600-5	762
S91601-3	837
S91602-1	313
S91604-8	837
S91605-6	94
S91606-4	133
S91607-2	844
S91608-0	875
S91609-9	449
S91610-2	451
S91611-0	451
S91612-9	506
S91613-7	544
S91614-5	504
S91615-3	525
S91616-1	520
S91618-8	520
S91619-6	544
S91621-8	520
S91622-6	982
S91623-4	545
S91624-2	640
S91625-0	627
S91626-9	535
S91627-7	514
S91628-5	535
S91629-3	731
S91630-7	526
S91631-5	998
S91632-3	855
S91633-1	855
S91635-8	532
S91636-6	855
S91637-4	533
S91638-2	627
S91639-0	996
S91640-4	855
S91642-0	855
S91643-9	554
S91644-7	505
S91645-5	855
S91646-3	505
S91647-1	526
S91649-8	981
S91650-1	550
S91652-8	1017
S91653-6	396
S91654-4	403
S91655-2	403
S91656-0	409
S91657-9	419
S91658-7	707
S91659-5	391
S91660-9	391
S91661-7	390
S91662-5	392
S91663-3	392
S91664-1	390
S91666-8	390
S91667-6	390
S91668-4	390
S91669-2	390
S91670-6	758
S91671-4	392
S91672-2	392
S91673-0	433
S91674-9	404
S91675-7	410
S91676-5	394
S91677-3	874
S91678-1	394
S91680-3	855
S91681-1	394
S91683-8	394
S91684-6	806
S91685-4	823
S91686-2	810
S91687-0	813
S91688-9	805
S91689-7	816
S91690-0	534
S91691-9	977
S91692-7	977
S91693-5	859
S91694-3	865
S91695-1	844
S91697-8	851
S91698-6	843
S91699-4	850
S91700-1	851
S91702-8	851
S91703-6	585
S91704-4	561
S91705-2	560
S91706-0	744
S91707-9	735
S91708-7	744
S91709-5	740
S91710-9	515
S91711-7	519
S91712-5	609
S91713-3	637
S91714-1	949
S91716-8	548
S91717-6	548
S91718-4	992
S91719-2	637
S91720-6	635
S91721-4	548
S91722-2	548
S91723-0	760
S91724-9	992
S91725-7	640
S91726-5	636
S91727-3	513
S91728-1	512
S91730-3	546
S91731-1	513
S91733-8	637
S91734-6	728
S91735-4	984
S91736-2	561
S91737-0	729
S91738-9	882
S91739-7	178
S91740-0	144
S91741-9	74
S91742-7	16
S91743-5	86
S91745-1	16
S91747-8	225
S91748-6	361
S91749-4	166
S91750-8	278
S91751-6	46
S91752-4	357
S91753-2	765
S91754-0	765
S91755-9	277
S91756-7	765
S91757-5	355
S91758-3	355
S91759-1	148
S91760-5	798
S91761-3	278
S91762-1	765
S91764-8	150
S91765-6	278
S91767-2	45
S91768-0	46
S91769-9	45
S91770-2	83
S91771-0	771
S91772-9	147
S91773-7	359
S91775-3	780
S91776-1	278
S91778-8	213
S91779-6	278
S91781-8	780
S91782-6	763
S91783-4	784
S91784-2	776
S91785-0	778
S91786-9	780
S91787-7	372
S91788-5	784
S91789-3	780
S91790-7	270
S91791-5	784
S91792-3	73
S91793-1	779
S91795-8	145
S91796-6	780
S91797-4	776
S91798-2	782
S91799-0	780
S91801-6	46
S91802-4	46
S91803-2	47
S91804-0	45
S91805-9	376
S91806-7	2
S91807-5	1
S91808-3	46
S91809-1	1026
S91810-5	45
S91811-3	87
S91812-1	197
S91814-8	278
S91815-6	2
S91816-4	45
S91817-2	3
S91818-0	45
S91819-9	24
S91820-2	213
S91821-0	277
S91822-9	258
S91823-7	46
S91824-5	257
S91825-3	28
S91826-1	277
S91828-8	172
S91829-6	770
S91831-8	370
S91832-6	258
S91834-2	256
S91835-0	959
S91836-9	212
S91837-7	213
S91838-5	363
S91839-3	213
S91840-7	83
S91841-5	62
S91843-1	73
S91845-8	73
S91846-6	343
S91847-4	75
S91848-2	73
S91849-0	46
S91850-4	84
S91851-2	45
S91852-0	77
S91853-9	73
S91854-7	73
S91855-5	765
S91856-3	761
S91857-1	46
S91860-1	765
S91862-8	77
S91863-6	131
S91865-2	46
S91866-0	78
S91867-9	758
S91868-7	366
S91869-5	370
S91870-9	313
S91871-7	313
S91872-5	782
S91874-1	313
S91876-8	370
S91877-6	258
S91878-4	312
S91879-2	312
S91880-6	73
S91882-2	75
S91883-0	73
S91884-9	312
S91885-7	699
S91886-5	73
S91887-3	952
S91888-1	150
S91890-3	258
S91891-1	46
S91893-8	367
S91894-6	46
S91895-4	978
S91896-2	170
S91897-0	257
S91898-9	212
S91899-7	268
S91900-4	257
S91901-2	258
S91902-0	277
S91903-9	263
S91904-7	263
S91905-5	276
S91906-3	212
S91907-1	265
S91909-8	46
S91910-1	46
S91912-8	45
S91913-6	73
S91914-4	257
S91915-2	370
S91916-0	370
S91917-9	46
S91918-7	50
S91919-5	48
S91920-9	769
S91921-7	287
S91922-5	373
S91923-3	74
S91924-1	295
S91926-8	73
S91927-6	73
S91928-4	48
S91929-2	780
S91930-6	26
S91931-4	209
S91932-2	754
S91933-0	270
S91934-9	779
S91935-7	270
S91936-5	780
S91937-3	763
S91938-1	756
S91940-3	780
S91941-1	869
S91944-6	262
S91945-4	193
S91946-2	870
S91947-0	338
S91948-9	1010
S91951-9	183
S91952-7	358
S91953-5	774
S91954-3	1058
S91957-8	41
S91960-8	586
S91961-6	578
S91962-4	578
S91963-2	228
S91964-0	292
S91965-9	183
S91966-7	1089
S91967-5	308
S91969-1	966
S91971-3	1057
S91972-1	1041
S91974-8	200
S91975-6	1089
S91976-4	233
S91978-0	194
S91979-9	1092
S91980-2	935
S91981-0	668
S91983-7	230
S91984-5	316
S91985-3	316
S91986-1	966
S91989-6	180
S91991-8	179
S91993-4	268
S91994-2	228
S91996-9	278
S91997-7	230
S91998-5	193
S91999-3	262
S92000-2	139
S92001-0	139
S92002-9	876

Product Page

S92003-7......1090
S92004-5......857
S92005-3......788
S92006-1......956
S92008-8......324
S92011-8......979
S92012-6......268
S92013-4......935
S92014-2......906
S92015-0......336
S92017-7......938
S92018-5......904
S92019-3......935
S92022-3......906
S92023-1......691
S92025-8......779
S92026-6......978
S92027-4......978
S92029-0......990
S92030-4......774
S92031-2......91
S92032-0......263
S92033-9......919
S92034-7......178
S92035-5......935
S92036-3......919
S92037-1......1004
S92039-8......1085
S92040-1......1085
S92042-8......384
S92043-6......756
S92044-4......330
S92045-2......781
S92046-0......1057
S92047-9......748
S92048-7......359
S92049-5......935
S92050-9......1016
S92051-7......940
S92052-5......935
S92053-3......980
S92054-1......980
S92056-8......931
S92057-6......935
S92058-4......1057
S92059-2......1002
S92060-6......935
S92061-4......249
S92062-2......180
S92065-7......931
S92066-5......931
S92067-3......966
S92068-1......936
S92070-3......934
S92071-1......1002
S92073-8......934
S92074-6......936
S92075-4......935
S92076-2......937
S92077-0......936
S92078-9......278
S92079-7......937
S92080-0......370
S92081-9......957
S92082-7......196
S92083-5......940
S92084-3......960
S92085-1......1054
S92087-8......373
S92088-6......935
S92089-4......966
S92090-8......935

Product Page

S92091-6......1088
S92093-2......1005
S92094-0......325
S92096-7......935
S92097-5......753
S92098-3......995
S92099-1......1057
S92100-9......934
S92101-7......926
S92102-5......935
S92103-3......926
S92104-1......869
S92107-6......1093
S92108-4......278
S92109-2......105
S92110-6......230
S92111-4......260
S92112-2......1085
S92113-0......791
S92114-9......793
S92115-7......1092
S92116-5......774
S92117-3......194
S92118-1......1089
S92120-3......1089
S92121-1......278
S92123-8......1057
S92124-6......1002
S92125-4......262
S92126-2......278
S92127-0......317
S92128-9......169
S92129-7......934
S92130-0......305
S92131-9......218
S92132-7......78
S92133-5......700
S92134-3......196
S92135-1......935
S92137-8......392
S92138-6......456
S92139-4......393
S92140-8......450
S92141-6......456
S92142-4......474
S92143-2......395
S92144-0......299
S92145-9......464
S92146-7......391
S92147-5......299
S92148-3......464
S92149-1......392
S92150-5......398
S92151-3......456
S92152-1......304
S92154-8......456
S92155-6......391
S92156-4......304
S92157-2......453
S92158-0......299
S92159-9......454
S92160-2......390
S92161-0......299
S92162-9......299
S92163-7......402
S92164-5......393
S92165-3......299
S92166-1......433
S92168-8......393
S92169-6......455
S92171-8......455
S92172-6......455

Product Page

S92173-4......334
S92174-2......1090
S92175-0......402
S92176-9......402
S92177-7......402
S92178-5......467
S92179-3......450
S92180-7......406
S92181-5......396
S92182-3......420
S92183-1......402
S92185-8......936
S92186-6......230
S92187-4......934
S92188-2......934
S92189-0......264
S92190-4......934
S92191-2......936
S92192-0......271
S92193-9......957
S92195-5......230
S92196-3......936
S92197-1......303
S92199-8......442
S92200-5......136
S92201-3......1014
S92202-1......133
S92204-8......942
S92205-6......32
S92206-4......337
S92207-2......260
S92208-0......758
S92209-9......877
S92210-2......921
S92211-0......219
S92213-7......913
S92214-5......258
S92215-3......1009
S92216-1......263
S92218-8......935
S92219-6......977
S92221-8......1
S92222-6......760
S92223-4......259
S92224-2......761
S92225-0......237
S92226-9......954
S92227-7......314
S92228-5......759
S92229-3......455
S92230-7......454
S92231-5......869
S92232-3......1085
S92233-1......74
S92235-8......518
S92236-6......521
S92237-4......1084
S92238-2......1003
S92239-0......1083
S92240-4......1085
S92241-2......1084
S92242-0......1082
S92243-9......880
S92244-7......279
S92249-8......620
S92253-6......880
S92254-4......754
S92255-2......405
S92256-0......1032
S92257-9......739
S92258-7......344
S92259-5......935

Product Page

S92261-7......935
S92263-3......758
S92264-1......474
S92267-6......57
S92268-4......374
S92269-2......31
S92270-6......1059
S92271-4......1021
S92272-2......921
S92273-0......324
S92274-9......966
S92275-7......317
S92276-5......317
S92277-3......960
S92278-1......375
S92280-3......1028
S92281-1......1031
S92283-8......407
S92284-6......1086
S92285-4......710
S92286-2......4
S92288-9......704
S92289-7......795
S92290-0......718
S92291-9......309
S92292-7......1021
S92293-5......488
S92295-1......1086
S92297-8......711
S92298-6......407
S92299-4......940
S92300-1......941
S92302-8......709
S92303-6......1080
S92304-4......1030
S92305-2......1083
S92306-0......1022
S92307-9......1083
S92308-7......1080
S92309-5......1080
S92310-9......754
S92311-7......338
S92312-5......215
S92313-3......182
S92314-1......1088
S92316-8......1083
S92317-6......1084
S92318-4......315
S92319-2......936
S92320-6......1093
S92321-4......1089
S92322-2......436
S92323-0......710
S92324-9......1083
S92327-3......1081
S92328-1......522
S92330-3......1080
S92331-1......1083
S92333-8......1091
S92334-6......310
S92335-4......935
S92336-2......758
S92337-0......1092
S92338-9......317
S92339-7......1083
S92340-0......433
S92341-9......935
S92343-5......879
S92344-3......306
S92345-1......880
S92348-6......277
S92349-4......869

Product Page

S92352-4......879
S92353-2......879
S92354-0......937
S92355-9......355
S92356-7......935
S92359-1......880
S92360-5......1085
S92361-3......880
S92362-1......345
S92364-8......1093
S92366-4......1002
S92367-2......934
S92368-0......1080
S92369-9......935
S92370-2......1086
S92371-0......230
S92372-9......1083
S92373-7......166
S92374-5......193
S92375-3......935
S92376-1......757
S92378-8......1094
S92379-6......937
S92381-8......706
S92384-2......977
S92385-0......935
S92386-9......150
S92387-7......150
S92388-5......1010
S92389-3......488
S92390-7......935
S92391-5......248
S92392-3......291
S92393-1......415
S92395-8......488
S92397-4......1082
S92398-2......935
S92399-0......280
S92400-8......701
S92401-6......904
S92402-4......807
S92403-2......423
S92404-0......715
S92405-9......739
S92406-7......513
S92407-5......518
S92408-3......508
S92409-1......506
S92410-5......530
S92411-3......531
S92412-1......993
S92414-8......809
S92415-6......878
S92416-4......506
S92417-2......549
S92418-0......736
S92419-9......528
S92420-2......751
S92421-0......581
S92422-9......523
S92423-7......930
S92424-5......519
S92425-3......547
S92426-1......733
S92429-6......732
S92431-8......984
S92432-6......648
S92433-4......614
S92434-2......614
S92435-0......652
S92436-9......637
S92437-7......484

Product Page

S92438-5......616
S92439-3......667
S92440-7......617
S92441-5......153
S92443-1......1048
S92445-8......1025
S92446-6......1063
S92447-4......1025
S92448-2......1048
S92449-0......1025
S92450-4......1060
S92451-2......579
S92452-0......929
S92453-9......898
S92454-7......420
S92455-5......1028
S92456-3......173
S92457-1......410
S92459-8......448
S92460-1......805
S92462-8......385
S92463-6......385
S92464-4......821
S92465-2......1071
S92466-0......1036
S92467-9......1072
S92468-7......855
S92469-5......806
S92470-9......463
S92471-7......473
S92472-5......1060
S92473-3......1024
S92474-1......1053
S92476-8......1053
S92477-6......1069
S92478-4......1048
S92479-2......1028
S92480-6......1032
S92481-4......1031
S92482-2......1040
S92483-0......1043
S92484-9......1044
S92485-7......1033
S92486-5......1031
S92487-3......1025
S92488-1......1063
S92490-3......1025
S92491-1......1079
S92493-8......1034
S92494-6......1025
S92495-4......1031
S92496-2......1034
S92497-0......1028
S92498-9......1034
S92499-7......1045
S92500-4......1041
S92501-2......1024
S92502-0......1041
S92503-9......1045
S92504-7......1052
S92505-5......1041
S92506-3......1023
S92507-1......1022
S92509-8......1025
S92510-1......1025
S92512-8......1060
S92513-6......1069
S92514-4......1036
S92515-2......1036
S92516-0......1045
S92517-9......1063
S92518-7......547

Product Page

S92519-5......499
S92520-9......1071
S92521-7......1042
S92522-5......1069
S92523-3......933
S92524-1......1060
S92526-8......1070
S92527-6......1036
S92528-4......1048
S92529-2......1060
S92530-6......1052
S92531-4......1048
S92532-2......1069
S92533-0......1069
S92534-9......1036
S92535-7......1036
S92536-5......498
S92537-3......1032
S92538-1......1051
S92540-3......1051
S92541-1......1045
S92543-8......1063
S92544-6......1034
S92545-4......1026
S92546-2......1070
S92547-0......1032
S92548-9......1071
S92549-7......1068
S92550-0......1026
S92551-9......1048
S92552-7......1036
S92553-5......1069
S92554-3......1028
S92555-1......1070
S92557-8......1028
S92558-6......630
S92559-4......900
S92560-8......585
S92562-4......880
S92563-2......374
S92564-0......1027
S92565-9......1073
S92566-7......766
S92567-5......250
S92569-1......913
S92570-5......385
S92571-3......880
S92572-1......880
S92574-8......880
S92575-6......262
S92576-4......839
S92577-2......917
S92578-0......880
S92579-9......94
S92580-2......418
S92581-0......436
S92582-9......341
S92583-7......262
S92584-5......328
S92585-3......730
S92586-1......855
S92588-8......859
S92589-6......880
S92591-8......1076
S92592-6......757
S92595-0......982
S92596-9......904
S92598-5......456
S92599-3......189
S92601-9......879
S92602-7......436
S92603-5......880

Product	Page
S92604-3	880
S92605-1	442
S92607-8	434
S92608-6	88
S92609-4	95
S92610-8	94
S92611-6	328
S92612-4	327
S92613-2	292
S92614-0	357
S92615-9	177
S92616-7	1012
S92617-5	360
S92618-3	331
S92619-1	297
S92620-5	293
S92621-3	304
S92622-1	304
S92624-8	500
S92625-6	410
S92626-4	308
S92627-2	379
S92628-0	369
S92629-9	367
S92630-2	374
S92631-0	374
S92632-9	374
S92633-7	369
S92634-5	368
S92635-3	374
S92636-1	374
S92638-8	374
S92639-6	499
S92641-8	458
S92642-6	15
S92643-4	375
S92644-2	374
S92645-0	374
S92646-9	374
S92647-7	259
S92648-5	248
S92649-3	271
S92650-7	31
S92651-5	252
S92652-3	278
S92653-1	864
S92655-8	338
S92656-6	344
S92657-4	206
S92658-2	337
S92660-4	218
S92661-2	471
S92662-0	342
S92664-7	305
S92665-5	187
S92666-3	494
S92667-1	676
S92669-8	179
S92670-1	776
S92672-8	197
S92673-6	194
S92674-4	194
S92675-2	230
S92676-0	284
S92677-9	15
S92678-7	279
S92679-5	1002
S92680-9	165
S92681-7	172
S92683-3	934
S92684-1	934
S92686-8	936
S92687-6	935
S92688-4	935
S92689-2	935
S92690-6	311
S92691-4	196
S92692-2	196
S92693-0	1079
S92694-9	766
S92695-7	310
S92696-5	314
S92697-3	343
S92698-1	316
S92700-7	484
S92701-5	1088
S92702-3	1093
S92703-1	1088
S92705-8	1089
S92706-6	1088
S92707-4	1089
S92708-2	1089
S92709-0	1089
S92710-4	1091
S92711-2	1085
S92712-0	1086
S92713-9	1086
S92714-7	1076
S92715-5	373
S92717-1	98
S92719-8	859
S92720-1	857
S92722-8	848
S92723-6	854
S92724-4	872
S92725-2	872
S92726-0	878
S92729-5	862
S92730-9	1005
S92731-7	942
S92732-5	941
S92733-3	927
S92734-1	974
S92736-8	970
S92737-6	703
S92738-4	701
S92739-2	703
S92740-6	708
S92741-4	697
S92742-2	124
S92743-0	968
S92745-7	124
S92746-5	723
S92747-3	365
S92750-3	692
S92753-8	704
S92754-6	765
S92755-4	917
S92756-2	869
S92757-0	904
S92758-9	908
S92759-7	903
S92760-0	903
S92762-7	902
S92763-5	902
S92764-3	901
S92765-1	831
S92768-6	900
S92769-4	988
S92770-8	981
S92771-6	981
S92772-4	981
S92773-2	981
S92774-0	691
S92775-9	382
S92776-7	997
S92777-5	998
S92778-3	1002
S92779-1	999
S92780-5	998
S92781-3	998
S92784-8	998
S92785-6	1000
S92786-4	1003
S92787-2	1002
S92788-0	1003
S92789-9	1003
S92790-2	978
S92791-0	923
S92792-9	923
S92793-7	868
S92794-5	923
S92798-8	183
S92799-6	358
S92801-1	385
S92809-7	907
S92811-9	1076
S92814-3	857
S92815-1	838
S92818-6	1092
S92819-4	867
S92820-8	184
S92821-6	872
S92822-4	872
S92823-2	657
S92824-0	872
S92825-9	872
S92826-7	949
S92828-3	963
S92829-1	963
S92830-5	962
S92831-3	931
S92832-1	1005
S92834-8	375
S92835-6	248
S92836-4	971
S92837-2	969
S92838-0	132
S92839-9	888
S92840-2	195
S92841-0	290
S92842-9	745
S92843-7	921
S92844-5	1003
S92845-3	977
S92849-6	271
S92855-0	1022
S92856-9	909
S92858-5	909
S92863-1	310
S92865-8	311
S92867-4	66
S92868-2	85
S92869-0	862
S92870-4	72
S92871-2	72
S92872-0	356
S92873-9	149
S92874-7	833
S92875-5	415
S92876-3	3
S92877-1	3
S92879-8	355
S92880-1	3
S92882-8	291
S92883-6	350
S92884-4	293
S92885-2	353
S92886-0	906
S92887-9	360
S92888-7	379
S92889-5	83
S92890-9	74
S92891-7	74
S92892-5	739
S92893-3	791
S92894-1	773
S92896-8	372
S92897-6	220
S92898-4	1052
S92899-2	248
S92901-8	254
S92902-6	215
S92903-4	53
S92904-2	261
S92905-0	223
S92906-9	225
S92907-7	377
S92908-5	256
S92909-3	768
S92910-7	335
S92911-5	208
S92912-3	54
S92913-1	340
S92915-8	378
S92916-6	55
S92917-4	55
S92918-2	1082
S92919-0	1089
S92920-4	1089
S92921-2	1087
S92922-0	1084
S92923-9	1085
S92924-7	1086
S92925-5	1086
S92926-3	1081
S92927-1	1086
S92929-8	1086
S92930-1	1081
S92932-8	1082
S92933-6	1088
S92934-4	1081
S92935-2	1092
S92936-0	1080
S92937-9	1086
S92938-7	1090
S92940-9	1091
S92941-7	1090
S92942-5	1082
S92943-3	1092
S92944-1	1086
S92946-8	1090
S92947-6	1090
S92948-4	1086
S92949-2	1083
S92950-6	1091
S92951-4	1091
S92952-2	1084
S92953-0	752
S92954-9	752
S92955-7	758
S92956-5	753
S92957-3	757
S92958-1	296
S92960-3	292
S92961-1	765
S92963-8	766
S92964-6	365
S92966-2	766
S92967-0	765
S92969-7	760
S92970-0	766
S92971-9	765
S92972-7	365
S92973-5	766
S92974-3	763
S92975-1	764
S92977-8	383
S92978-6	307
S92979-4	766
S92980-8	762
S92981-6	759
S92982-4	300
S92983-2	304
S92984-0	780
S92985-9	755
S92986-7	292
S92987-5	313
S92990-5	917
S92991-3	907
S92992-1	187
S92994-8	201
S92995-6	201
S92996-4	58
S92997-2	367
S92998-0	418
S92999-9	362
S93000-8	1092
S93001-6	1087
S93003-2	1082
S93004-0	1086
S93005-9	1091
S93006-7	1082
S93007-5	1091
S93008-3	1087
S93009-1	1080
S93011-3	1087
S93014-8	1087
S93015-6	1091
S93016-4	1082
S93017-2	1087
S93018-0	1092
S93019-9	1080
S93020-2	1091
S93022-9	1090
S93023-7	1082
S93024-5	1090
S93025-3	1091
S93026-1	1091
S93028-8	1082
S93029-6	1091
S93031-8	1091
S93032-6	1080
S93033-4	1080
S93034-2	1092
S93035-0	1082
S93036-9	1082
S93037-7	1091
S93038-5	1091
S93039-3	1080
S93040-7	1091
S93041-5	1082
S93042-3	1091
S93043-1	1091
S93045-8	1087
S93048-2	1089
S93049-0	1086
S93050-4	1089
S93052-0	1082
S93053-9	1091
S93054-7	1090
S93056-3	1091
S93063-6	1089
S93064-4	1089
S93065-2	1081
S93066-0	1082
S93067-9	1080
S93068-7	1089
S93069-5	1089
S93070-9	1089
S93071-7	1089
S93072-5	1080
S93073-3	1090
S93074-1	1081
S93076-8	1089
S93077-6	1082
S93078-4	1081
S93079-2	1081
S93080-6	1081
S93081-4	1081
S93082-2	1081
S93083-0	1081
S93084-9	1081
S93085-7	1081
S93086-5	1081
S93087-3	1081
S93088-1	1081
S93090-3	1089
S93091-1	1089
S93093-8	1081
S93094-6	1089
S93095-4	1080
S93096-2	819
S93097-0	259
S93098-9	223
S93101-2	225
S93102-0	225
S93103-9	427
S93104-7	788
S93105-5	781
S93106-3	315
S93107-1	280
S93109-8	407
S93110-1	432
S93112-8	406
S93113-6	406
S93114-4	299
S93115-2	300
S93116-0	299
S93117-9	299
S93118-7	299
S93119-5	300
S93120-9	299
S93121-7	299
S93122-5	299
S93123-3	844
S93124-1	844
S93126-8	265
S93127-6	277
S93128-4	261
S93129-2	277
S93130-6	277
S93131-4	760
S93132-2	761
S93133-0	759
S93134-9	762
S93135-7	758
S93136-5	759
S93137-3	858
S93138-1	413
S93140-3	418
S93141-1	377
S93143-8	377
S93144-6	377
S93145-4	366
S93146-2	344
S93147-0	835
S93148-9	905
S93150-0	902
S93151-9	905
S93152-7	765
S93153-5	765
S93154-3	1025
S93155-1	762
S93157-8	1080
S93158-6	1086
S93159-4	1011
S93160-8	29
S93161-6	29
S93162-4	29
S93164-0	29
S93165-9	5
S93166-7	24
S93167-5	364
S93168-3	335
S93169-1	1009
S93170-5	734
S93171-3	975
S93172-1	131
S93174-8	971
S93175-6	940
S93176-4	868
S93177-2	868
S93178-0	923
S93179-9	369
S93180-2	1056
S93181-0	854
S93182-9	721
S93184-5	936
S93185-3	936
S93186-1	935
S93188-8	934
S93191-8	966
S93192-6	410
S93193-4	338
S93194-2	201
S93195-0	1087
S93196-9	1008
S93198-5	12
S93199-3	692
S93200-0	1127
S93201-9	1126
S93202-7	1126
S93203-5	1126
S93204-3	1126
S93205-1	1127
S93207-8	1127
S93208-6	1126
S93209-4	1127
S93210-8	1128
S93211-6	1127
S93212-4	1127
S93213-2	1126
S93214-0	1126
S93215-9	1127
S93216-7	1127
S93217-5	298
S93219-1	305
S93220-5	302
S93221-3	301
S93222-1	303
S93225-6	295
S93226-4	297
S93227-2	303

Product	Page
S93228-0	309
S93229-9	308
S93230-2	154
S93231-0	154
S93232-9	774
S93233-7	152
S93236-1	155
S93239-6	774
S93241-8	152
S93245-0	152
S93246-9	153
S93248-5	154
S93249-3	154
S93250-7	782
S93252-3	156
S93255-8	872
S93256-6	942
S93257-4	1080
S93258-2	941
S93261-2	968
S93262-0	134
S93263-9	134
S93264-7	969
S93266-3	969
S93267-1	133
S93269-8	135
S93270-1	886
S93272-8	723
S93275-2	903
S93276-0	904
S93277-9	988
S93278-7	994
S93279-5	989
S93280-9	983
S93281-7	987
S93282-5	988
S93284-1	1000
S93286-8	1000
S93287-6	1006
S93288-4	998
S93289-2	996
S93290-6	1006
S93296-5	993
S93298-1	989
S93300-7	1001
S93301-5	994
S93302-3	995
S93303-1	995
S93305-8	1007
S93306-6	995
S93307-4	1000
S93308-2	1086
S93309-0	1082
S93310-4	1088
S93311-2	1086
S93312-0	308
S93313-9	1126
S93314-7	792
S93315-5	468
S93316-3	643
S93317-1	1001
S93319-8	801
S93320-1	174
S93322-8	311
S93326-0	69
S93327-9	94
S93328-7	667
S93329-5	477
S93330-9	358
S93331-7	865
S93332-5	666
S93333-3	655
S93334-1	219
S93336-8	294
S93337-6	284
S93338-4	43
S93339-2	243
S93340-6	50
S93341-4	1016
S93342-2	285
S93343-0	309
S93344-9	200
S93345-7	798
S93346-5	773
S93347-3	772
S93348-1	323
S93350-3	336
S93351-1	763
S93353-8	760
S93354-6	380
S93355-4	345
S93356-2	355
S93357-0	317
S93358-9	295
S93359-7	306
S93360-0	736
S93361-9	364
S93362-7	494
S93363-5	1009
S93364-3	424
S93365-1	503
S93367-8	409
S93368-6	925
S93369-4	930
S93370-8	961
S93371-6	478
S93372-4	313
S93373-2	374
S93374-0	941
S93376-7	728
S93377-5	404
S93378-3	288
S93380-5	402
S93381-3	1069
S93382-1	927
S93384-8	420
S93385-6	420
S93386-4	288
S93387-2	954
S93388-0	996
S93389-9	169
S93390-2	421
S93391-0	288
S93392-9	640
S93394-5	193
S93395-3	296
S93396-1	1088
S93398-8	1127
S93399-6	1126
S93400-3	1126
S93401-1	1126
S93403-8	1126
S93404-6	1126
S93405-4	1127
S93406-2	1128
S93407-0	1126
S93408-9	1126
S93409-7	1126
S93410-0	1126
S93411-9	1126
S93412-7	1127
S93413-5	1126
S93415-1	1127
S93417-8	1127
S93418-6	1126
S93419-4	1126
S93420-8	1127
S93421-6	1055
S93422-4	732
S93423-2	732
S93424-0	265
S93425-9	753
S93426-7	752
S93427-5	753
S93428-3	755
S93429-1	880
S93430-5	909
S93431-3	1060
S93432-1	870
S93434-8	870
S93435-6	870
S93436-4	870
S93437-2	1072
S93438-0	1068
S93439-9	1029
S93440-2	1026
S93441-0	1026
S93442-9	1049
S93443-7	1072
S93444-5	1069
S93445-3	1072
S93446-1	1072
S93448-8	1029
S93449-6	1048
S93451-8	1025
S93452-6	1072
S93453-4	1072
S93454-2	1054
S93455-0	1071
S93456-9	869
S93457-7	870
S93459-3	942
S93461-5	868
S93462-3	863
S93463-1	861
S93465-8	862
S93466-6	860
S93467-4	862
S93468-2	860
S93469-0	861
S93470-4	860
S93471-2	349
S93472-0	385
S93473-9	364
S93474-7	364
S93475-5	737
S93476-3	750
S93477-1	727
S93479-8	875
S93480-1	981
S93482-8	763
S93483-6	758
S93485-2	901
S93486-0	903
S93487-9	905
S93488-7	905
S93490-9	905
S93491-7	765
S93492-5	763
S93494-1	124
S93496-8	125
S93497-6	734
S93498-4	149
S93499-2	473
S93501-8	1072
S93502-6	1058
S93503-4	996
S93504-2	995
S93505-0	996
S93506-9	996
S93507-7	996
S93508-5	996
S93509-3	996
S93510-7	995
S93511-5	985
S93512-3	247
S93513-1	755
S93515-8	709
S93517-4	858
S93518-2	720
S93519-0	757
S93520-4	753
S93521-2	292
S93522-0	295
S93523-9	292
S93524-7	943
S93525-5	857
S93526-3	856
S93527-1	857
S93529-8	856
S93530-1	859
S93532-8	857
S93533-6	965
S93534-4	858
S93535-2	856
S93536-0	856
S93537-9	1139
S93538-7	372
S93539-5	343
S93540-9	343
S93541-7	343
S93542-5	343
S93543-3	63
S93544-1	156
S93546-8	1016
S93547-6	1016
S93548-4	53
S93549-2	723
S93550-6	343
S93551-4	234
S93552-2	187
S93553-0	191
S93554-9	277
S93555-7	276
S93556-5	239
S93557-3	229
S93558-1	39
S93560-3	499
S93561-1	368
S93563-8	724
S93564-6	768
S93565-4	207
S93566-2	333
S93567-0	262
S93568-9	953
S93569-7	276
S93570-0	267
S93571-9	230
S93572-7	264
S93573-5	274
S93574-3	264
S93575-1	273
S93577-8	274
S93578-6	274
S93579-4	293
S93580-8	168
S93581-6	165
S93582-4	167
S93583-2	346
S93584-0	319
S93586-7	302
S93587-5	302
S93588-3	306
S93589-1	302
S93590-5	988
S93591-3	403
S93592-1	395
S93594-8	401
S93595-6	402
S93596-4	403
S93597-2	402
S93598-0	403
S93599-9	401
S93600-6	231
S93601-4	166
S93602-2	166
S93603-0	166
S93604-9	166
S93605-7	167
S93606-5	167
S93607-3	167
S93608-1	167
S93610-3	167
S93611-1	167
S93613-8	381
S93614-6	166
S93615-4	166
S93616-2	245
S93618-9	253
S93619-7	771
S93620-0	247
S93627-8	701
S93628-6	763
S93630-8	763
S93636-7	694
S93640-5	909
S93641-3	72
S93642-1	870
S93645-6	403
S93646-4	403
S93647-2	403
S93648-0	403
S93649-9	403
S93650-2	403
S93651-0	403
S93652-9	403
S93653-7	403
S93654-5	403
S93655-3	403
S93656-1	754
S93658-8	1001
S93659-6	949
S93661-8	859
S93662-6	90
S93665-0	901
S93668-5	975
S93669-3	980
S93670-7	997
S93671-5	856
S93672-3	490
S93673-1	414
S93675-8	421
S93676-6	402
S93677-4	402
S93678-2	421
S93679-0	421
S93680-4	402
S93681-2	421
S93682-0	414
S93683-9	421
S93684-7	402
S93687-1	66
S93689-8	111
S93690-1	410
S93692-8	906
S93693-6	906
S93694-4	466
S93695-2	476
S93696-0	476
S93697-9	476
S93698-7	472
S93699-5	466
S93700-2	59
S93702-9	150
S93703-7	200
S93704-5	177
S93705-3	419
S93706-1	778
S93708-8	854
S93711-8	968
S93712-6	1028
S93714-2	980
S93715-0	979
S93716-9	980
S93717-7	980
S93718-5	979
S93719-3	294
S93720-7	337
S93721-5	442
S93722-3	421
S93723-1	366
S93725-8	995
S93726-6	195
S93727-4	471
S93728-2	471
S93729-0	471
S93730-4	199
S93731-2	199
S93732-0	209
S93733-9	209
S93734-7	346
S93735-5	323
S93736-3	283
S93737-1	10
S93739-8	42
S93740-1	748
S93742-8	414
S93743-6	101
S93744-4	917
S93745-2	989
S93746-0	766
S93747-9	1003
S93748-7	766
S93750-9	250
S93751-7	996
S93752-5	275
S93753-3	901
S93754-1	905
S93756-8	905
S93757-6	905
S93759-2	974
S93760-6	975
S93761-4	928
S93764-9	765
S93767-3	907
S93768-1	904
S93770-3	903
S93771-1	763
S93773-8	1019
S93774-6	218
S93775-4	333
S93777-0	859
S93778-9	858
S93779-7	855
S93780-0	857
S93781-9	485
S93782-7	340
S93783-5	341
S93785-1	1019
S93787-8	298
S93788-6	948
S93789-4	247
S93790-8	995
S93791-6	979
S93792-4	994
S93793-2	757
S93794-0	148
S93795-9	1017
S93796-7	1017
S93797-5	1016
S93798-3	690
S93799-1	215
S93801-7	163
S93802-5	42
S93803-3	371
S93804-1	356
S93806-8	88
S93807-6	67
S93808-4	6
S93809-2	32
S93811-4	107
S93812-2	110
S93813-0	105
S93815-7	127
S93816-5	140
S93817-3	127
S93818-1	140
S93820-3	127
S93821-1	326
S93823-8	342
S93824-6	343
S93825-4	336
S93827-0	391
S93828-9	259
S93830-0	145
S93831-9	155
S93833-5	1041
S93834-3	211
S93835-1	976
S93837-8	1037
S93840-8	983
S93843-2	1088
S93844-0	1087
S93845-9	1087
S93846-7	1080
S93847-5	1053
S93850-5	1024
S93852-1	781
S93854-8	835
S93855-6	153
S93856-4	775
S93857-2	773
S93858-0	775
S93859-9	778
S93860-2	773
S93861-0	778
S93862-9	154
S93863-7	153
S93864-5	154
S93865-3	153
S93866-1	799
S93868-8	154
S93869-6	157
S93871-8	154

Product Page

S93872-6........153
S93873-4........152
S93874-2........152
S93876-9........775
S93877-7........777
S93878-5........152
S93879-3........153
S93880-7........152
S93881-5........152
S93882-3........831
S93883-1........30
S93885-8........976
S93886-6........968
S93887-4........968
S93888-2........968
S93889-0........968
S93890-4........975
S93891-2........975
S93892-0........968
S93893-9........968
S93894-7........1003
S93896-3........979
S93897-1........979
S93900-5........798
S93902-1........16
S93904-8........371
S93905-6........101
S93906-4........367
S93907-2........370
S93908-0........370
S93909-9........200
S93910-2........798
S93911-0........183
S93912-9........16
S93913-7........798
S93914-5........183
S93915-3........371
S93916-1........200
S93918-8........205
S93919-6........370
S93921-8........205
S93922-6........184
S93923-4........74
S93924-2........370
S93925-0........367
S93926-9........371
S93927-7........371
S93928-5........371
S93929-3........372
S93930-7........16
S93931-5........356
S93932-3........797
S93933-1........798
S93935-8........371
S93936-6........372
S93937-4........371
S93938-2........371
S93939-0........862
S93940-4........862
S93941-2........862
S93942-0........862
S93943-9........862
S93944-7........862
S93945-5........862
S93946-3........864
S93947-1........863
S93949-8........863
S93950-1........861
S93952-8........900
S93953-6........900
S93954-4........909
S93955-2........912
S93956-0........914
S93957-9........722
S93958-7........695
S93959-5........939
S93960-9........973
S93961-7........902
S93962-5........904
S93963-3........174
S93964-1........188
S93966-8........97
S93967-6........24
S93968-4........348
S93969-2........202
S93970-6........411
S93971-4........202
S93972-2........370
S93973-0........367
S93974-9........355
S93975-7........356
S93976-5........370
S93977-3........380
S93978-1........56
S93980-3........295
S93981-1........107
S93983-8........75
S93984-6........791
S93985-4........1074
S93986-2........865
S93987-0........865
S93988-9........864
S93989-7........865
S93990-0........867
S93992-7........865
S93993-5........372
S93994-3........887
S93995-1........886
S93997-8........252
S93998-6........253
S93999-4........882
S94000-3........915
S94001-1........909
S94003-8........879
S94004-6........246
S94005-4........935
S94006-2........690
S94007-0........690
S94008-9........690
S94009-7........763
S94010-0........763
S94011-9........914
S94012-7........915
S94013-5........913
S94014-3........914
S94015-1........744
S94017-8........903
S94018-6........942
S94019-4........954
S94020-8........1006
S94021-6........864
S94022-4........865
S94023-2........865
S94024-0........473
S94025-9........216
S94026-7........1112
S94027-5........864
S94028-3........862
S94029-1........863
S94030-5........863
S94031-3........863
S94032-1........863
S94034-8........861
S94035-6........861
S94036-4........863
S94037-2........862
S94038-0........860
S94039-9........863
S94040-2........863
S94041-0........864
S94042-9........863
S94043-7........863
S94044-5........861
S94045-3........863
S94046-1........863
S94048-8........864
S94049-6........864
S94051-8........861
S94052-6........860
S94053-4........863
S94054-2........863
S94055-0........863
S94056-9........863
S94057-7........862
S94058-5........1129
S94059-3........329
S94060-7........333
S94061-5........333
S94062-3........332
S94063-1........329
S94065-8........329
S94066-6........329
S94067-4........214
S94068-2........211
S94069-0........211
S94070-4........210
S94071-2........214
S94072-0........904
S94073-9........903
S94074-7........901
S94075-5........901
S94076-3........903
S94077-1........904
S94079-8........906
S94080-1........902
S94082-8........107
S94083-6........106
S94085-2........6
S94086-0........310
S94087-9........310
S94088-7........1034
S94090-9........271
S94091-7........967
S94092-5........775
S94094-1........153
S94096-8........367
S94097-6........367
S94098-4........92
S94099-2........1008
S94101-8........839
S94102-6........838
S94103-4........839
S94104-2........838
S94105-0........839
S94106-9........839
S94107-7........840
S94108-5........839
S94109-3........900
S94110-7........900
S94111-5........900
S94112-3........900
S94113-1........900
S94115-8........861
S94116-6........866
S94117-4........141
S94118-2........838
S94119-0........859
S94120-4........356
S94121-2........184
S94122-0........356
S94123-9........184
S94124-7........184
S94125-5........184
S94126-3........184
S94127-1........184
S94129-8........184
S94130-1........184
S94132-8........184
S94133-6........280
S94134-4........488
S94135-2........315
S94136-0........319
S94137-9........315
S94138-7........313
S94139-5........312
S94140-9........313
S94141-7........311
S94142-5........311
S94143-3........311
S94144-1........310
S94146-8........725
S94147-6........725
S94148-4........996
S94149-2........989
S94150-6........78
S94151-4........777
S94152-2........420
S94153-0........415
S94154-9........416
S94155-7........298
S94156-5........321
S94157-3........92
S94158-1........75
S94160-3........1140
S94161-1........293
S94163-8........324
S94164-6........131
S94165-4........96
S94166-2........58
S94167-0........1125
S94168-9........1125
S94169-7........1125
S94171-9........1124
S94172-7........1125
S94174-3........1125
S94175-1........1125
S94177-8........1124
S94179-4........403
S94180-8........403
S94181-6........421
S94182-4........402
S94183-2........403
S94184-0........260
S94185-9........178
S94186-7........260
S94187-5........260
S94188-3........266
S94189-1........265
S94190-5........268
S94191-3........267
S94192-1........91
S94194-8........266
S94195-6........263
S94196-4........224
S94197-2........260
S94198-0........473
S94199-9........323
S94200-6........1080
S94201-4........1081
S94202-2........1086
S94203-0........12
S94204-9........52
S94205-7........54
S94206-5........56
S94207-3........172
S94208-1........363
S94210-3........364
S94213-8........287
S94214-6........363
S94215-4........363
S94216-2........56
S94217-0........92
S94218-9........65
S94219-7........1016
S94220-0........181
S94221-9........380
S94222-7........179
S94223-5........939
S94224-3........954
S94225-1........941
S94227-8........945
S94228-6........942
S94229-4........133
S94230-8........40
S94231-6........68
S94232-4........581
S94234-0........862
S94235-9........863
S94236-7........862
S94237-5........864
S94238-3........342
S94242-1........386
S94244-8........858
S94245-6........386
S94246-4........889
S94247-2........858
S94248-0........709
S94249-9........708
S94250-2........887
S94251-0........843
S94252-9........889
S94256-1........917
S94261-8........903
S94262-6........901
S94263-4........901
S94264-2........903
S94265-0........907
S94266-9........907
S94267-7........907
S94268-5........903
S94269-3........907
S94270-7........364
S94271-5........900
S94272-3........997
S94273-1........781
S94275-8........783
S94276-6........774
S94277-4........774
S94278-2........466
S94279-0........6
S94280-4........72
S94281-2........109
S94282-0........67
S94283-9........102
S94284-7........436
S94285-5........424
S94286-3........427
S94287-1........81
S94289-8........360
S94290-1........818
S94292-8........1068
S94293-6........282
S94294-4........355
S94295-2........491
S94296-0........92
S94297-9........1011
S94298-7........243
S94299-5........58
S94300-2........171
S94301-0........171
S94302-9........171
S94303-7........171
S94304-5........171
S94305-3........171
S94306-1........171
S94308-8........171
S94309-6........170
S94311-8........171
S94312-6........171
S94313-4........170
S94314-2........281
S94315-0........261
S94316-9........80
S94318-5........82
S94319-3........83
S94320-7........1015
S94321-5........1015
S94322-3........764
S94323-1........179
S94325-8........200
S94326-6........373
S94327-4........781
S94330-4........842
S94331-2........843
S94333-9........843
S94335-5........843
S94336-3........843
S94337-1........324
S94339-8........842
S94342-8........838
S94343-6........843
S94344-4........839
S94345-2........839
S94346-0........996
S94347-9........700
S94348-7........700
S94349-5........383
S94350-9........384
S94351-7........1139
S94352-5........382
S94353-3........382
S94354-1........570
S94356-8........700
S94357-6........700
S94358-4........53
S94359-2........171
S94360-6........171
S94361-4........219
S94362-2........441
S94363-0........998
S94364-9........998
S94365-7........999
S94366-5........999
S94367-3........290
S94368-1........261
S94370-3........190
S94371-1........191
S94373-8........192
S94374-6........346
S94375-4........965
S94376-2........172
S94377-0........344
S94378-9........324
S94379-7........274
S94380-0........167
S94381-9........167
S94382-7........167
S94383-5........84
S94384-3........917
S94385-1........337
S94387-8........379
S94388-6........858
S94389-4........169
S94390-8........461
S94391-6........180
S94392-4........753
S94393-2........176
S94394-0........261
S94395-9........310
S94396-7........892
S94397-5........221
S94398-3........210
S94399-1........922
S94400-9........1095
S94401-7........1094
S94402-5........1094
S94403-3........1094
S94404-1........1094
S94406-8........1084
S94407-6........1084
S94408-4........1082
S94409-2........1084
S94410-6........1084
S94411-4........1084
S94412-2........1088
S94413-0........1082
S94414-9........1083
S94415-7........1084
S94416-5........1086
S94417-3........1060
S94418-1........1060
S94420-3........1060
S94421-1........1062
S94423-8........1094
S94424-6........1060
S94425-4........1043
S94426-2........1094
S94427-0........1094
S94428-9........1094
S94429-7........1076
S94430-0........1052
S94431-9........1095
S94433-5........1093
S94434-3........1094
S94435-1........1095
S94437-8........1095
S94439-4........1095
S94440-8........1095
S94441-6........1095
S94442-4........1094
S94443-2........1094
S94444-0........1093
S94445-9........1093
S94446-7........1093
S94447-5........1093
S94448-3........1093
S94449-1........1093
S94450-5........1093
S94451-3........1093
S94452-1........1093
S94454-8........1093
S94455-6........1093
S94456-4........1094
S94457-2........1093

Product	Page
S94459-9	863
S94461-0	862
S94462-9	862
S94463-7	863
S94464-5	862
S94465-3	864
S94466-1	862
S94468-8	363
S94469-6	363
S94472-6	274
S94473-4	827
S94474-2	357
S94475-0	1092
S94476-9	308
S94477-7	195
S94478-5	165
S94480-7	287
S94481-5	417
S94482-3	1087
S94483-1	1086
S94485-8	1084
S94486-6	1085
S94487-4	1087
S94488-2	1087
S94489-0	1094
S94490-4	996
S94491-2	996
S94492-0	386
S94493-9	158
S94494-7	346
S94495-5	315
S94496-3	315
S94497-1	759
S94499-8	810
S94500-5	205
S94501-3	765
S94502-1	344
S94504-8	494
S94505-6	239
S94506-4	909
S94507-2	909
S94508-0	886
S94509-9	858
S94510-2	521
S94511-0	824
S94512-9	258
S94513-7	1003
S94515-3	919
S94516-1	926
S94518-8	926
S94519-6	925
S94521-8	942
S94522-6	945
S94523-4	963
S94524-2	964
S94526-9	926
S94527-7	926
S94528-5	864
S94529-3	863
S94530-7	863
S94531-5	860
S94532-3	860
S94535-8	1070
S94536-6	864
S94537-4	862
S94538-2	860
S94539-0	861
S94540-4	863
S94542-0	864
S94543-9	709
S94544-7	862
S94546-3	900
S94547-1	900
S94549-8	900
S94550-1	900
S94552-8	861
S94553-6	861
S94554-4	861
S94555-2	869
S94556-0	162
S94557-9	345
S94558-7	173
S94559-5	380
S94560-9	409
S94561-7	165
S94562-5	306
S94563-3	292
S94564-1	297
S94566-8	773
S94567-6	438
S94568-4	446
S94569-2	446
S94570-6	363
S94571-4	346
S94572-2	143
S94573-0	343
S94574-9	1075
S94575-7	184
S94576-5	236
S94577-3	855
S94578-1	363
S94580-3	954
S94581-1	334
S94583-8	173
S94584-6	279
S94585-4	723
S94586-2	928
S94587-0	977
S94588-9	381
S94589-7	781
S94590-0	772
S94591-9	307
S94592-7	219
S94593-5	169
S94594-3	333
S94595-1	441
S94597-8	18
S94598-6	721
S94599-4	1075
S94600-1	472
S94602-8	330
S94603-6	425
S94604-4	402
S94605-2	402
S94606-0	403
S94607-9	402
S94608-7	421
S94609-5	403
S94610-9	763
S94611-7	467
S94612-5	306
S94613-3	306
S94614-1	773
S94616-8	775
S94617-6	775
S94618-4	288
S94619-2	288
S94620-6	289
S94621-4	292
S94622-2	291
S94623-0	309
S94624-9	311
S94625-7	311
S94626-5	847
S94627-3	847
S94628-1	847
S94630-3	108
S94631-1	876
S94633-8	38
S94634-6	866
S94635-4	873
S94636-2	873
S94637-0	869
S94638-9	869
S94639-7	346
S94640-0	940
S94641-9	994
S94642-7	1000
S94643-5	252
S94644-3	790
S94645-1	914
S94647-8	916
S94648-6	999
S94650-8	849
S94651-6	165
S94652-4	1019
S94653-2	341
S94654-0	1075
S94655-9	196
S94656-7	31
S94657-5	346
S94658-3	878
S94659-1	942
S94660-5	995
S94661-3	771
S94662-1	156
S94664-8	75
S94665-6	80
S94666-4	74
S94667-2	50
S94668-0	80
S94669-9	198
S94670-2	207
S94671-0	487
S94672-9	765
S94673-7	292
S94674-5	350
S94675-3	153
S94676-1	571
S94678-8	878
S94679-6	103
S94681-8	130
S94682-6	130
S94683-4	103
S94684-2	116
S94685-0	106
S94686-9	12
S94687-7	219
S94688-5	42
S94689-3	145
S94690-7	102
S94691-5	103
S94695-8	230
S94696-6	262
S94697-4	271
S94698-2	216
S94699-0	259
S94700-8	763
S94701-6	758
S94702-4	761
S94703-2	763
S94704-0	764
S94705-9	759
S94706-7	759
S94707-5	761
S94708-3	764
S94709-1	763
S94710-5	698
S94711-3	694
S94712-1	701
S94714-8	369
S94715-6	306
S94716-4	367
S94717-2	294
S94718-0	476
S94719-9	758
S94720-2	353
S94721-0	378
S94722-9	77
S94723-7	805
S94724-5	384
S94725-3	384
S94726-1	831
S94728-8	200
S94729-6	282
S94731-8	299
S94732-6	353
S94733-4	274
S94734-2	273
S94735-0	275
S94736-9	183
S94737-7	195
S94738-5	665
S94740-7	855
S94741-5	246
S94742-3	253
S94743-1	249
S94745-8	245
S94746-6	249
S94747-4	1005
S94748-2	933
S94749-0	300
S94750-4	264
S94751-2	264
S94752-0	22
S94753-9	184
S94754-7	1075
S94755-5	289
S94756-3	127
S94757-1	831
S94759-8	836
S94760-1	170
S94762-8	190
S94763-6	256
S94764-4	280
S94765-2	263
S94766-0	67
S94767-9	172
S94768-7	360
S94769-5	330
S94770-9	194
S94771-7	359
S94772-5	189
S94773-3	832
S94774-1	287
S94776-8	195
S94777-6	336
S94778-4	367
S94779-2	322
S94780-6	367
S94781-4	338
S94782-2	367
S94783-0	91
S94784-9	182
S94785-7	184
S94786-5	327
S94787-3	321
S94788-1	758
S94790-3	298
S94791-1	324
S94793-8	196
S94794-6	768
S94795-4	768
S94796-2	298
S94797-0	188
S94798-9	164
S94799-7	182
S94800-4	67
S94801-2	135
S94802-0	112
S94803-9	101
S94804-7	112
S94805-5	141
S94806-3	36
S94807-1	103
S94809-8	101
S94810-1	101
S94812-8	82
S94813-6	82
S94814-4	109
S94815-2	84
S94816-0	71
S94817-9	64
S94818-7	72
S94819-5	12
S94820-9	101
S94821-7	101
S94822-5	105
S94823-3	110
S94824-1	110
S94826-8	109
S94827-6	109
S94828-4	110
S94829-2	109
S94830-6	29
S94831-4	29
S94832-2	16
S94833-0	29
S94834-9	72
S94835-7	105
S94836-5	78
S94837-3	105
S94838-1	142
S94840-3	274
S94841-1	274
S94843-8	274
S94844-6	116
S94845-4	116
S94846-2	116
S94847-0	116
S94848-9	116
S94849-7	116
S94850-0	130
S94851-9	130
S94852-7	130
S94853-5	130
S94854-3	130
S94855-1	142
S94857-8	856
S94858-6	154
S94859-4	152
S94860-8	154
S94861-6	153
S94863-2	153
S94864-0	724
S94865-9	877
S94866-7	875
S94867-5	876
S94868-3	876
S94869-1	877
S94870-5	875
S94871-3	876
S94872-1	1048
S94874-8	1069
S94875-6	861
S94876-4	870
S94877-2	879
S94878-0	870
S94879-9	705
S94880-2	932
S94881-0	133
S94882-9	131
S94883-7	134
S94884-5	248
S94885-3	909
S94886-1	914
S94888-8	916
S94889-6	913
S94891-8	914
S94892-6	905
S94893-4	905
S94894-2	905
S94895-0	904
S94897-7	902
S94898-5	1085
S94899-3	906
S94900-0	989
S94901-9	96
S94902-7	500
S94903-5	700
S94904-3	248
S94905-1	326
S94907-8	411
S94908-6	410
S94909-4	411
S94910-8	291
S94911-6	291
S94912-4	288
S94913-2	377
S94914-0	378
S94915-9	377
S94916-7	434
S94917-5	731
S94918-3	211
S94919-1	496
S94920-5	767
S94921-3	343
S94922-1	307
S94924-8	1087
S94925-6	307
S94926-4	136
S94927-2	194
S94928-0	137
S94929-9	1017
S94930-2	306
S94931-0	78
S94932-9	12
S94933-7	20
S94934-5	1143
S94935-3	1143
S94936-1	1142
S94938-8	1143
S94939-6	1141
S94941-8	1141
S94942-6	1143
S94943-4	1141
S94944-2	1142
S94945-0	1142
S94946-9	1142
S94947-7	1142
S94948-5	1142
S94949-3	1143
S94950-7	136
S94951-5	692
S94952-3	58
S94953-1	989
S94955-8	981
S94956-6	985
S94957-4	991
S94958-2	250
S94959-0	237
S94960-4	177
S94961-2	164
S94962-0	750
S94963-9	750
S94964-7	739
S94965-5	739
S94966-3	58
S94967-1	1010
S94969-8	705
S94970-1	705
S94972-8	135
S94973-6	702
S94974-4	805
S94975-2	442
S94976-0	821
S94978-7	835
S94979-5	105
S94981-7	72
S94982-5	43
S94983-3	63
S94984-1	18
S94986-8	50
S94987-6	87
S94988-4	354
S94989-2	105
S94990-6	353
S94991-4	715
S94992-2	77
S94993-0	189
S94994-9	237
S94995-7	198
S94996-5	321
S94997-3	445
S94998-1	51
S95000-9	1017
S95002-5	1016
S95003-3	63
S95004-1	62
S95006-8	146
S95007-6	95
S95009-2	63
S95010-6	63
S95011-4	140
S95012-2	158
S95015-7	85
S95016-5	86
S95017-3	19
S95018-1	822
S95020-3	19
S95021-1	738
S95023-8	102
S95024-6	148
S95025-4	105
S95026-2	86
S95027-0	85
S95028-9	739
S95029-7	834
S95031-9	86
S95032-7	664
S95033-5	739
S95034-3	431
S95035-1	63
S95037-8	831

Product Page

S95038-6........831
S95039-4........830
S95040-8..........82
S95041-6..........32
S95042-4..........69
S95043-2..........58
S95044-0........350
S95046-7........109
S95047-5........716
S95048-3..........69
S95049-1..........92
S95050-5..........67
S95051-3..........77
S95055-6..........53
S95056-4..........51
S95061-0..........67
S95063-7..........66
S95064-5..........85
S95066-1..........91
S95068-8........754
S95069-6........165
S95071-8........285
S95072-6........282
S95073-4........205
S95074-2........165
S95075-0........178
S95077-7......1142
S95078-5......1142
S95080-7......1141
S95081-5......1142
S95082-3........840
S95083-1........840
S95085-8........125
S95086-6........842
S95088-2........769
S95089-0........841
S95091-2........125
S95092-0........840
S95093-9........839
S95094-7........840
S95095-5........125
S95096-3........841
S95097-1........841
S95099-8........841
S95100-5........150
S95101-3........841
S95104-8........841
S95105-6........839
S95106-4........840
S95107-2........839
S95108-0........840
S95109-9........840
S95110-2........839
S95112-9........838
S95113-7........841
S95116-1........840
S95119-6........840
S95123-4........839
S95124-2........839
S95125-0........839
S95126-9........840
S95127-7........839
S95128-5........838
S95129-3........841
S95130-7........838
S95131-5........840
S95132-3........841
S95133-1..........89
S95135-8........714
S95136-6........720
S95137-4........693
S95138-2........841

S95139-0........841
S95140-4........841
S95141-2........839
S95142-0........839
S95144-7........841
S95145-5........841
S95146-3........840
S95147-1........841
S95149-8........149
S95152-8........841
S95153-6........841
S95154-4........133
S95155-2........722
S95156-0........842
S95159-5........838
S95161-7........840
S95162-5........840
S95164-1........700
S95166-8........840
S95168-4........840
S95169-2........841
S95172-2........840
S95174-9........840
S95175-7........840
S95176-5........840
S95177-3........838
S95181-1........839
S95185-4........839
S95186-2........840
S95190-0........838
S95194-3........838
S95197-8........838
S95198-6........840
S95199-4........842
S95200-1........839
S95202-8........840
S95203-6........705
S95205-2........838
S95207-9........841
S95208-7........840
S95210-9........716
S95212-5........839
S95213-3........840
S95214-1........840
S95217-6........840
S95218-4........840
S95222-2........840
S95223-0........839
S95224-9........838
S95225-7........840
S95230-3........838
S95234-6..........98
S95236-2..........95
S95238-9..........97
S95239-7..........96
S95240-0........845
S95241-9..........89
S95242-7..........95
S95244-3........849
S95247-8..........89
S95248-6..........96
S95250-8........845
S95252-4........386
S95253-2........857
S95256-7........134
S95257-5........711
S95259-1........868
S95260-5........870
S95261-3........871
S95262-1........871
S95264-8........871
S95265-6........871

S95269-9........134
S95270-2........385
S95271-0........874
S95272-9........874
S95273-7........874
S95274-5........874
S95275-3..........68
S95279-6........576
S95282-6..........68
S95283-4........705
S95285-0........705
S95286-9........769
S95288-5........873
S95289-3........873
S95290-7........873
S95293-1........873
S95298-2........874
S95299-0........873
S95300-8........873
S95301-6........873
S95302-4........873
S95303-2........873
S95305-9........879
S95306-7........870
S95307-5........870
S95308-3......1138
S95309-1........870
S95310-5........810
S95311-3........338
S95312-1......1070
S95314-8........141
S95316-4........322
S95317-2........769
S95319-9..........52
S95323-7..........69
S95324-5......1094
S95325-3......1009
S95326-1........267
S95328-8........326
S95329-6........434
S95332-6........392
S95333-4........799
S95334-2........361
S95335-0........269
S95336-9........738
S95337-7......1071
S95338-5........726
S95339-3........376
S95340-7........365
S95341-5........755
S95342-3........364
S95343-1........329
S95345-8........502
S95346-6........724
S95348-2........369
S95349-0........339
S95350-4..........65
S95352-0........257
S95354-7........201
S95355-5..........66
S95356-3..........65
S95357-1..........64
S95359-8..........65
S95360-1......1070
S95362-8......1070
S95363-6..........14
S95364-4........397
S95366-0........335
S95367-9........768
S95368-7........420
S95370-9........762
S95371-7........329

S95372-5........376
S95373-3........176
S95374-1........420
S95376-8........373
S95377-6........646
S95379-2........329
S95380-6........327
S95381-4........197
S95382-2......1009
S95384-9..........65
S95385-7..........53
S95386-5........364
S95387-3........761
S95388-1........340
S95390-3........340
S95391-1........334
S95393-8........373
S95394-6........752
S95396-2........149
S95397-0........149
S95398-9..........95
S95400-4..........83
S95402-0......1009
S95403-9........369
S95404-7........378
S95405-5........369
S95406-3........369
S95409-8..........91
S95410-1........103
S95413-6......1088
S95414-4........369
S95415-2........328
S95416-0........378
S95417-9........364
S95419-5........414
S95420-9........467
S95421-7........799
S95424-1........268
S95426-8........116
S95427-6......1017
S95428-4........349
S95429-2........326
S95431-4........234
S95433-0........363
S95434-9........341
S95435-7........341
S95436-5........381
S95437-3........499
S95441-1........199
S95443-8........443
S95444-6........373
S95445-4........353
S95447-0......1016
S95449-7........654
S95450-0........304
S95451-9..........78
S95452-7........413
S95453-5......1094
S95455-1........186
S95458-6........477
S95459-4........688
S95461-6........354
S95462-4........339
S95463-2......1005
S95464-0........934
S95465-9........942
S95466-7......1005
S95467-5......1006
S95468-3........134
S95470-5........131
S95471-3........925
S95472-1........134

S95474-8........134
S95475-6........131
S95476-4........125
S95477-2........966
S95478-0........961
S95482-9........936
S95483-7........286
S95485-3........966
S95486-1........959
S95488-8........888
S95489-6......1006
S95491-8......1006
S95493-4......1006
S95494-2......1006
S95496-9........149
S95498-5........358
S95499-3..........91
S95501-9..........58
S95503-5........153
S95504-3..........58
S95505-1..........58
S95507-8..........58
S95508-6..........42
S95510-8........776
S95511-6........776
S95512-4..........89
S95514-0........710
S95516-7........146
S95517-5........245
S95518-3..........53
S95519-1........102
S95520-5........102
S95521-3..........77
S95522-1........307
S95524-8........291
S95525-6........291
S95526-4........334
S95528-0........371
S95529-9........338
S95530-2........288
S95535-3......1013
S95536-1......1010
S95541-8........702
S95543-4........693
S95545-0........700
S95546-9........882
S95548-5........883
S95549-3........883
S95550-7........690
S95552-3........881
S95553-1........881
S95555-8........363
S95556-6........887
S95557-4........886
S95558-2........882
S95559-0........691
S95560-4........882
S95561-2........882
S95562-0........882
S95565-5........887
S95566-3........883
S95567-1........689
S95569-8......1040
S95570-1........889
S95572-8......1046
S95573-6........882
S95574-4........888
S95582-5........746
S95583-3........721
S95584-1........691
S95586-8........721
S95587-6........716

S95588-4........761
S95591-4........382
S95592-2........383
S95593-0........383
S95594-9........382
S95596-5........846
S95598-1........276
S95600-7........195
S95601-5........282
S95602-3..........65
S95603-1..........51
S95605-8..........95
S95606-6..........64
S95608-2........191
S95609-0..........54
S95610-4..........54
S95612-0..........53
S95614-7..........56
S95615-5..........56
S95616-3........316
S95619-8..........92
S95622-8..........54
S95623-6..........53
S95624-4..........57
S95625-2........389
S95626-0........188
S95627-9..........52
S95628-7........298
S95629-5..........92
S95630-9..........52
S95632-5..........54
S95633-3..........92
S95634-1..........14
S95636-8........144
S95637-6........145
S95638-4........144
S95640-6..........84
S95641-4........196
S95644-9........204
S95645-7........195
S95646-5........388
S95647-3........234
S95648-1........371
S95650-3........654
S95651-1........165
S95654-6......1048
S95655-4........187
S95656-2........156
S95657-0......1079
S95658-9......1017
S95659-7........989
S95660-0........135
S95661-9........993
S95662-7........986
S95663-5........986
S95664-3........988
S95665-1........986
S95667-8........991
S95668-6........996
S95669-4........997
S95670-8........109
S95673-2........245
S95675-9........696
S95676-7........697
S95677-5........382
S95678-3........385
S95689-9........699
S95691-0........886
S95693-7..........56
S95695-3........317
S95696-1........315
S95698-8........310

S95699-6........815
S95700-3........316
S95701-1........296
S95703-8..........54
S95704-6..........54
S95705-4..........51
S95706-2..........51
S95708-9..........51
S95709-7........388
S95711-9......1070
S95712-7..........65
S95713-5..........65
S95714-3..........59
S95715-1..........54
S95717-8........355
S95719-4..........78
S95720-8..........53
S95721-6..........67
S95726-7........890
S95728-3........708
S95729-1........708
S95730-5........699
S95731-3........706
S95735-6........699
S95737-2........699
S95739-9........247
S95741-0........706
S95745-3........995
S95748-8........996
S95753-4........308
S95754-2........309
S95755-0......1013
S95756-9........309
S95774-7..........78
S95775-5..........53
S95776-3..........13
S95777-1..........51
S95779-8..........51
S95780-1..........53
S95782-8..........20
S95783-6............7
S95784-4..........15
S95785-2........116
S95786-0..........55
S95787-9..........55
S95788-7..........55
S95789-5......1142
S95790-9......1013
S95791-7......1069
S95792-5..........55
S95794-1........799
S95796-8........799
S95797-6......1073
S95798-4......1079
S95799-2........186
S95801-8........134
S95802-6........126
S95806-9........758
S95807-7........134
S95808-5......1012
S95809-3........134
S95811-5..........96
S95812-3..........99
S95813-1..........99
S95815-8..........65
S95816-6........388
S95817-4........287
S95818-2........757
S95822-0..........98
S95824-7..........59
S95825-5..........41
S95826-3..........44

Product Page

S95837-9.........89
S95839-5........742
S95840-9........319
S95842-5........286
S95843-3......1070
S95844-1......1069
S95848-4..........65
S95850-6........654
S95852-2........487
S95854-9........691
S95856-5......1011
S95858-1........498
S95863-8........643
S95865-4........371
S95869-7........986
S97197-9........315
S97199-5........316
S97200-2........895
S97203-7........793
S97204-5........316
S97206-1........315
S97208-8........316
S97213-4........319
S97214-2........317
S97216-9........378
S97218-5........320
S97223-1........315
S97229-0........317
S97230-4........319
S97231-2........320
S97236-3........319
S97242-8........319
S97243-6......1001
S97244-4........320
S97248-7........316
S97256-8........320
S97258-4........316
S97260-6........316
S97261-4........316
S97262-2........316
S97265-7........316
S97266-5........315
S97268-1........316
S97271-1........316
S97300-9........205
S97301-7........287
S97302-5........316
S97303-3........318
S97304-1........166
S97306-8........257
S97309-2........175
S97310-6........195
S97312-2........194
S97313-0........209
S97315-7........226
S97316-5........298
S97317-3........208
S97318-1........346
S97326-2........346
S97328-9........291
S97329-7........287
S97330-0........208
S97332-7........172
S97333-5........318
S97334-3........376
S97335-1........334
S97337-8........318
S97339-4........187
S97340-8........376
S97343-2........200
S97345-9........375
S97348-3........318

Product Page

S97349-1........298
S97351-3........319
S97354-8........175
S97355-6........175
S97360-2........298
S97363-7........317
S97364-5........317
S97365-3........175
S97366-1........175
S97368-8........172
S97369-6........175
S97371-8........175
S97374-2........175
S97376-9........175
S97377-7........167
S97378-5........167
S97381-5........167
S97382-3........188
S97386-6........188
S97388-2........758
S97389-0........768
S97390-4........381
S97391-2........346
S97392-0........346
S97393-9........381
S97394-7........175
S97395-5........172
S97396-3........175
S97397-1........175
S97400-5........175
S97401-3........175
S97402-1........167
S97405-6........167
S97406-4........315
S97408-0........189
S97410-2........189
S97411-0........197
S97412-9........309
S97413-7........167
S97414-5........309
S97415-3........200
S97418-8........768
S97422-6........381
S97423-4........346
S97426-9........378
S97431-5........503
S97432-3........208
S97433-1........315
S97435-8........189
S97436-6........189
S97438-2........381
S97439-0........363
S97440-4........376
S97443-9........319
S97444-7........280
S97446-3........188
S97449-8......1001
S97452-8........166
S97453-6........196
S97454-4........196
S97455-2........196
S97457-9........231
S97459-5........231
S97460-9........231
S97461-7........231
S97463-3........231
S97464-1........309
S97466-8........346
S97467-6........346
S97469-2........346
S97471-4........346
S97472-2........174

Product Page

S97473-0........489
S97474-9........166
S97475-7........188
S97477-3........188
S97478-1........309
S97485-4........376
S97486-2........376
S97488-9........490
S97489-7........490
S97490-0........490
S97491-9........490
S97492-7........490
S97493-5........490
S97494-3........167
S97495-1........167
S97498-6........361
S97499-4........174
S97500-1......1095
S97502-8........196
S97503-6........196
S97505-2........195
S97506-0........231
S97508-7........231
S97509-5........324
S97512-5........208
S97514-1........167
S97518-4........208
S97519-2........188
S97520-6........489
S97522-2........334
S97533-8........196
S97534-6........324
S97535-4........324
S97536-2........166
S97539-7........208
S97541-9........298
S97543-5........298
S97544-3........381
S97547-8........381
S97548-6........166
S97549-4........208
S97550-8........175
S97551-6........167
S97555-9........167
S97556-7........167
S97558-3........167
S97564-8........167

Chemical Structures Made Easy.

Aldrich Drawing Tools

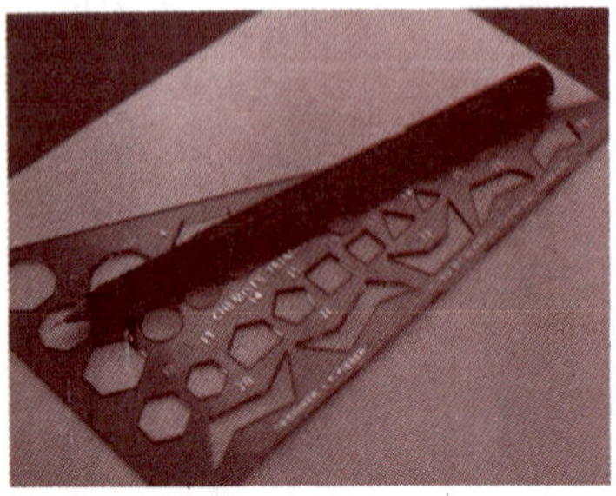

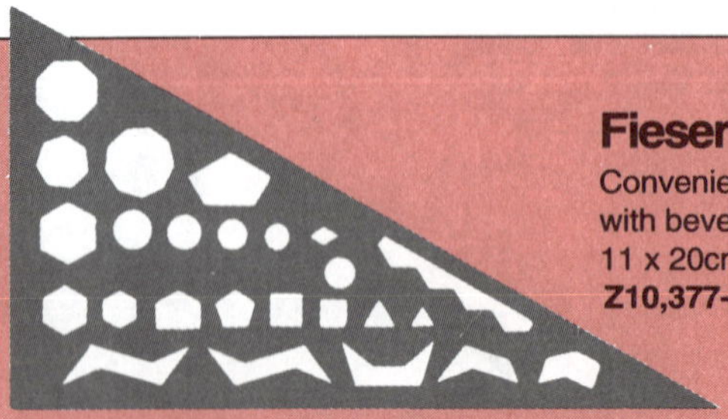

Fieser chemist's triangle

Convenient template made of sturdy plastic with bevelled edges to prevent ink smearing 11 x 20cm.
Z10,377-2

Rapidoliner technical pen

Professional quality drawing pen features a slender 0.35-mm, SS tip that is perfect for use with chemistry templates on tracing paper, drawing paper, and lineboards. Black ink is waterproof and smudgeproof when dry. Uses a prefilled, disposable, capillary ink cartridge.
Z21,852-9

Set of four technical pens
With SS tip (sizes: 0.25.0.35, 0.5, 0.7mm). Includes compass adapter, and pen holder.
Z22,048-5

Replacement drawing-ink cartridge for use with 0.35-mm tip Rapidoliner technical pen listed above. Black ink.
Z21,818-9

For more drawing tools see the 1992-1993 Aldrich Catalog Handbook of Fine Chemicals pages 1528-1535

Chemistry templates

Chemical apparatus template
Z13,519-4

Chemical process flowchart template for food, petroleum, and plastics
Z13,515-1

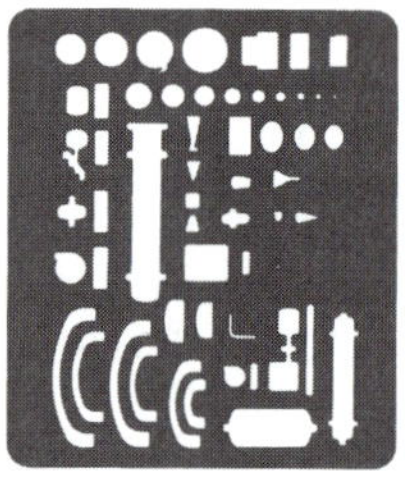

Chemical-process template
Z13,520-8

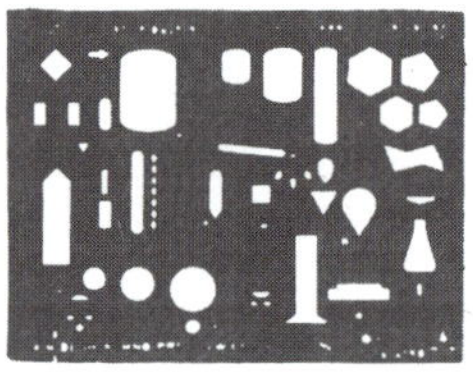

CHEM•ART Chemical apparatus and symbol template
Z15,672-8

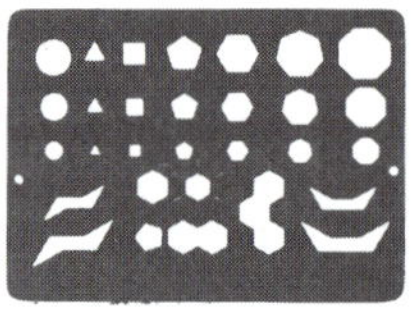

Organic-chemistry ring template
Z13,517-8

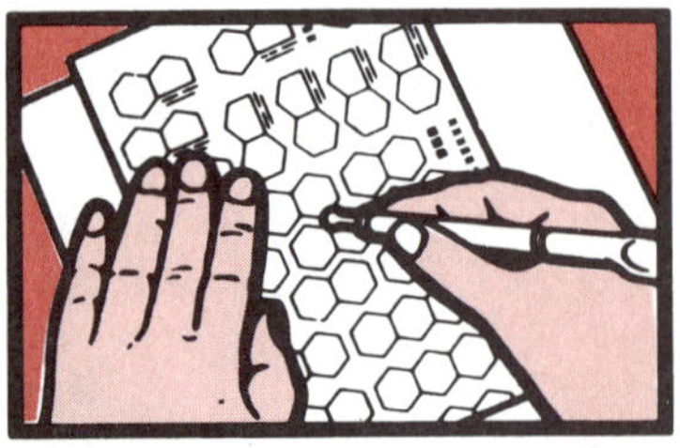

Chemical-structure transfers

You can "draw" chemical structures in minutes with dry transfers. The broad, opaque lines of the dry-transfer structural diagrams and lettering make them ideal for technical papers, monographs, theses, slides and overhead projections. The quality and precision of the lines surpass those made with drafting pen or template.

Aldrich dry-transfer symbols are printed in special ink on the underside of transparent plastic sheets. "Drawing" chemical structures is accomplished by simply burnishing (transferring) the symbols from the plastic sheet onto the paper. The symbols fit together perfectly, so no trimming is needed.

See pages 1528-1531 in the 1992-1993 Aldrich Catalog Handbook of Fine Chemicals for more information and descriptions of all dry transfers available

Benzene rings, Kekulé

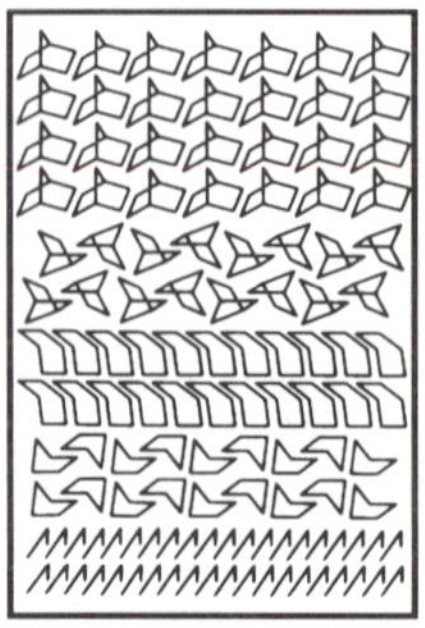

Bicyclic compounds

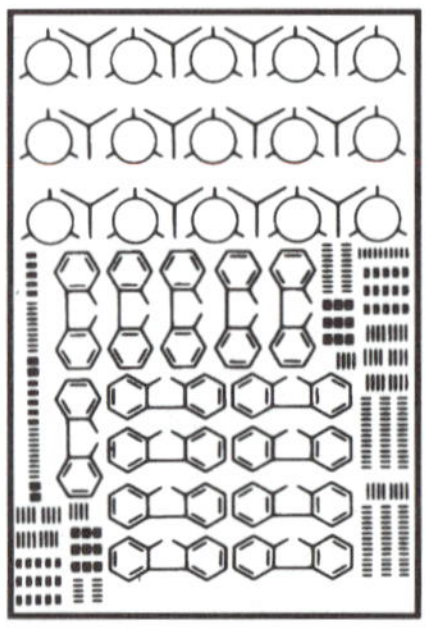

Carbazoles, Newman projection

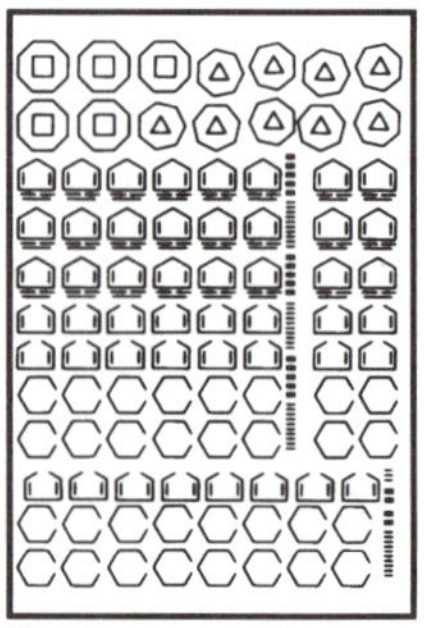

Cyclic components